GUIDE TO MICROFORMS IN PRINT

GUIDE TO MICROFORMS IN PRINT

AUTHOR TITLE

Incorporating International Microforms in Print

VOL. 1

2008

A–K

K·G·SAUR

Edited by:
Irene Izod

Bibliographic information published by the Deutsche Nationalbibliothek
The Deutsche Nationalbibliothek lists this publication in the Deutsche Nationalbibliografie;
detailed bibliographic data are available in the Internet at http://dnb.d-nb.de

Library of Congress Catalog Card Number: 61-7082

© 2008 by K. G. Saur Verlag, München
An Imprint of Walter de Gruyter GmbH & Co. KG
All Rights strictly reserved. No part of this publication may
be reproduced, stored in a retrieval system, or transmitted
in any form or by any means, electronic, mechanical,
photocopying, recording, or otherwise, without permission
in writing from the publisher.

The publisher does not assume and hereby disclaims any
liability to any party for any loss or damage caused by
errors or omissions in Guide to Microforms in Print,
whether such errors or omissions result from negligence,
accident or any other cause.

Computer-controlled data preparation
and automatic data processing
by bsix information exchange, Braunschweig, Germany

Cover art by William Pownall

Printed in the USA

ISSN 0164-0747
ISBN 978-3-598-11780-0 (2 Volumes)

Table of Contents

Vol. 1

Foreword	vi
Country-of-Publication Codes	viii
Currency Symbols	viii
Key to Abbreviated Series	ix
Abbreviations	x
Survey of Classes	xi
Survey of Subjects	xiv
Publishers and Distributors	3
Index to Publishers and Distributors	11
Author-Title List A–K	15

Vol. 2

Author-Title List L–Z	1425

Foreword

Guide to Microforms in Print is a title main entry listing, with cross-references from all authors and editors to titles. In addition, cross-references from variant authors and titles are provided where needed.

The **Author-Title Guide** provides access to international microform publications. Publications listed include books, journals, newspapers, government publications, archival material, collections and other projects currently available from microform publishers throughout the world provided they are for sale on a regular and current basis.

Alphabetical Arrangement of Entries

Main entries and cross-references are interfiled to form a single list arranged alphabetically by word according to Anglo-American practice. Personal name entries (name references) precede other kinds of entries beginning with the same word or group of words.

Entries beginning with initial letters file before words beginning with the same letter. Acronyms file as multi-letter words. The following initial articles do not file:

a, an, das, dat, dem, den, der, det, die, dit, een, eene, ein, eine, einem, einen, einer, eines, eit, el, ett, gli, het, il, la, las, le, les, lo, los, o, os, the, um, uma, un, una, unas, une, uno, unos.

U.S., Mc and St. are filed as though spelled out when they occur at the beginning of an entry.

Numbers file before letters of the alphabet; the user looking for a title that begins with the date 1981, for example, should search under "Nineteen Eighty-one" in the N alphabet as well as under "1981" in the numerical arrangement preceding the A alphabet.

Roman numerals are being changed successively to arabic numerals. Search under both the arabic numeral as well as roman numerals as though they were letters.

Authors

As many as two personal authors or editors may be listed for a given title. If more than two authors or editors are responsible for a given work, only the name of the first is provided followed by et al.

Form and Content of Entries

The main form of entry is as follows:

Title: Subtitle [Paralleltitle] / Author(s), Editor(s) - place, publisher and date of publication of the original and of the microform. [Series]. Collation information. Type of microform code. Price. [ISBN] - [ISSN]. Extra distributor or co-publisher information. Further title information. [Order no]. Publisher code. Subject classes.

A typical entry may look like this:

> **Baltic Index** = Baltikum inex / ed by
> Bildarchiv Foto Marburg – Deutsches
> Dokumentationszentrum fuer Kunstgeschichte
> Philipps- Universitaet Marburg – [mf ed
> 2005-06] – (= ser Marburger index 4) –
> 45mf (1:24) in 4 installments – 9 – silver
> €3520.00 – ISBN-10: 3-598-35596-3 –
> ISBN-13: 978-3-598-35596-7 – (sold only as
> set) – gw Saur [700]

or:

> **Iskusstvo farfora** / Ivanov, Dmitrii Dmitrievich –
> Moskva: b Gos izd-vo 1924 [mf ed 2006] –
> (= ser Russkoe dekorativnoe iskusstvo 6) –
> 1r – 1 – (filmed with: oil and the germs of
> war / by scott nearing [c1923]) – mf#5763p –
> us Wisconsin U Libr [730]

The numbers given in square brackets following the publisher's name designate the classes based on a modified Dewey Decimal Classification System, in which materials of related interest may be found in the **GMIP Subject Guide**.

Prices

Where no price is given, the user should contact the publisher directly.

The prices listed are inclusive, unless they are followed by a coded symbol as follows:

Price Code:		
	f	per fiche; e.g. $1.25f
	f	per reel; e.g. $1.05f
	y	per year; e.g. $4.00y

Type of Microform

The numerical code designation for the type of microform in which a title is available now precedes the price. The coding for the type of microform is as follows:

1	reel microfilm; 35mm [See also no. 13 & 14]
2	micro-opaque cards; 75 x 125mm (3 x 5")
3	micro-opaque cards; 6 x 9"
5	reel microfilm; 16mm [See also no. 6]
6	reel microfilm; 16mm (cartridge or cassette)
7	microfiche; 75 x 125mm (3 x 5")
8	microfiche; 9 x 12cm
9	microfiche; 105 x 148mm (4 x 6")
11	text-fiche; case bound volume with full size front and end matter, including index; text on 4 x 6" pocket-held microfiche
13	reel microfilm; 35mm (cartridge or cassette)
14	reel microfilm; 35mm in color
15	color microfiche; 105 x 148 mm (4 x 6")
16	other; description necessary
17	COM fiche

Ordering information

Publishers are identified in the main entry by an abbreviation preceded by a two character country-of-publication code in lower-case letters. This code and the abbreviation are arranged alphabetically in the section **Publishers and Distributors** where full ordering information can be found. In addition, an alphabetical **Index to Publishers and Distributors** is provided.

Please note that books listed in **GMIP** cannot be ordered through K. G. Saur.

Country-of-Publication Codes

at	Australia		mf	Mauritius
au	Austria		mx	Mexico
be	Belgium		my	Malaysia
bl	Brazil		ne	Netherlands
cc	China, People's Republic; Hong Kong		nz	New Zealand
ch	China, Republic		ph	Philippines
cn	Canada		si	Singapore
dk	Denmark		sa	South Africa
fi	Finland		sp	Spain
fr	France		sw	Sweden
gw	Germany, Federal Republic		sz	Switzerland
ie	Ireland		uk	United Kingdom
ja	Japan		us	United States
ko	Korea, Republic			

Currency Symbols

A$	Australian dollar		Kr	Swedish krona
Can$	Canadian dollar		NZ$	New Zealand dollar
DKr	Danish krone		RMBY	Renminbi yuan (People's Republic of China)
€	Euro		Y	yen (Japan)
Fmk	markka (Finland)		$	United States dollar
HK$	Hong Kong dollar		£	English pound

Key to Abbreviated Series

19th c art & architecture	19th century books on art & architecture
19th c british colonization	19th century books on british colonization
19th c economics	19th century general collection on economics
19th c evolution & creation	19th century books on evolution & creation
19th c ireland	19th century books on ireland
19th c publishing...	19th century books on publishing, the booktrade & diffusion of knowledge coll
Aasms 1968	Aas microfiche series (1968-) [aasms]
American indian periodicals...	American indian periodicals from the princeton university library
Asn 1-3	Anti-soviet newspapers 1918-1922. pt 1, 2 and 3 [asn 1-3]
Bgphma	Beitraege zur geschichte der philosophie und theologie des mittelalters (bgphma)
Biblical crit - us & gb	Biblical criticism in the united states and great britain
Bibliographies du cours...1947-66	Les bibliographies du cours de bibliotheconomie de l'universitelaval, 1947-1966
Blodgett coll	Blodgett collection of spanish civil war pamphlets
Blvs	Bibliothek des literarischen vereins in stuttgart
Books on religion...1543/44-c1800	Early printed books on religion from colonial spanish america, 1543/44-c1800
Cccm	Corpus christianorum continuatio mediaevalis (cccm)
Ccm	Chinese christian monographs coll
Ccsl	Corpus christianorum series latina (ccsl)
Coll...in the bikol language	Collection of philippine literature in the bikol language
Coll...in the bisaya language	Collection of philippine literature in the bisaya language
Coll...in the tagalog language	Collection of philippine literature in the tagalog language
Dz	Deutsche zeitschriften des 18. und 19. jahrhunderts
Fbc	Fuerstliche bibliothek corvey
German-jewish periodicals...1768-1945	German-jewish periodicals from the leo baeck institute in new york, 1768-1945
Hq	Historische quellen zur frauenbewegung und geschlechterproblematik (hq)
ILL	Instrumenta lexicologica latina
Latin american & caribbean...1821-1982	Latin american and caribbean official statistical serials, 1821-1982
Mees	Memoirs of the egypt exploration society (mees)
O & t journals	Ottoman & turkish journals & popular press
P-k&k period	Pre-kangzhan & kangzhan period preservation project
Ps 19	Periodicos seculo 19
Samp: indian books	Samp early 20th-century indian books project
Tugal	Texte und untersuchungen zur geschichte der altchristlichen literatur (tugal)
Yale coll	Yale day missions collection: selections from asia & the pacific

Abbreviations

abr	abridged	ltd ed	limited edition
ac	audiocassette(s)	mf	microfiche(s)
add	addition(s), addendum(s)	mf#	microform, order, shelf number
aka	also known as	mins	minutes
alt ed	alternative edition	misc	miscellaneous
amalg	amalgamation	ms, mss	manuscript, manuscripts
ann	annotation(s), annotated	mthly	monthly
app	appendix, appendices		
apply	apply to publisher for price	n	number(s)
approx	approximately	natl	national
aut	author	no, nos	number, numbers
		ns	new series
bibl	bibliography(ies), bibliographical		
bimthly	bimonthly	o/p	out-of-print
biogr	biographical, biography(ies)	o/s	out-of-stock
biwkly	biweekly	orig	original
bk(s)	books	os	old series
c, ca	circa	p, pgs, pp	pages
coll	collection(s)	pb	softback
comp(s)	compiler(s)	p/g	printed guide
cont	continues	pref	preface
cont by	continued by	pseud	pseudonym
corp	corporation	pt(s)	parts
cttee	committee	pub, publ	published, publisher, publishing
dept	department	r	reel(s)
d/g	digital guide	rb	looseleaf; ringbound
dist	distributor	ref	references
doc(s)	document(s)	repr	reprint
		rev	revised
ea	each		
ed(s)	editor(s); edition(s)	sb	softback
enl	enlarged	St.	Saint
		sect	section
f	microfiche(s)	semiwkly	semiweekly
facs	facsimile	ser	series
fasc	fascicule	subs	subscription; subscribers
fr	from	suppl	supplement
freq	frequency		
		tr, trans	translator, translated, translation
hb	hardback	transc	transcription(s)
ill	illustrated, illustration(s), illustrator(s)	v	volume(s)
impr	imprint	vol(s)	volume(s)
in prep	in preparation		
incl	includes, including	wkly	weekly
incorp	incorporates, incorporated		
ind	index(es), indexed	y, yr	year
inst	institute	yrly	yearly
int	introduction		
ISBN	International Standard Book Number	//	ceased publication
iss	issue(s)	*	incomplete
		[§]	see **us CRL** under *Publishers and Distributors*
lbd	leather		
lea	leaf(ves)		

Survey of Classes

Divisions marked with (*) have been modified. See Guidelines.

*GENERALITIES

000	Generalities
010	General Bibliography
011	Bibliography of Philosophy and Psychology
012	Bibliography of Religion
013	Bibliography of Social Sciences
014	Bibliography of Language and Literature Studies
015	Bibliography of Science
016	Bibliography of Technology
017	Bibliography of the Arts
018	Bibliography of Belles-Lettres
019	Bibliography of Geography and History
020	Library and Information Science
025	Archives / Brittle Books
030	Encyclopedic Works
040	Language Dictionaries
050	Generalities Dictionaries
051	Philosophy and Related Disciplines Dictionaries
052	Religion Dictionaries
053	Social Science Dictionaries
054	Linguistics and Literature Dictionaries
055	Science Dictionaries
056	Technology (Applied Science) Dictionaries
057	Dictionaries of the Arts
059	Geography and History Dictionaries
060	General Organizations/Museology
070	Journalism/Publishing/Newspapers
071	Newspapers – United States and Canada
072	Newspapers – British Isles
073	Periodicals
074	Newspapers/Periodicals – Western Europe [excl Scandinavia]
077	Newspapers/Periodicals – Eastern Europe
078	Newspapers/Periodicals – Scandinavia
079	Newspapers/Periodicals – Other Geographical areas
080	General Collections
090	Manuscripts, Book Rarities ansd Ephemera

PHILOSOPHY AND RELATED DISCIPLINES

100	Philosophy and Related Disciplines
110	Metaphysics (Speculative Philosophy)
120	Epistemology/Causation/Humankind
130	Paranormal Phenomena and Arts
140	Specific Philosophical Viewpoints
150	Psychology
160	Logic
170	Ethics (Moral Philosophy)
180	Ancient, Medieval, Eastern Philosophy
190	Modern Western Philosophy

*RELIGION

200	Religion
210	Religious Beliefs and Attitudes
220	Bible
221	Old Testament
225	New Testament
226	Gospels and Acts
227	Epistles
230	Comparative Religion
240	Christianity/Early Church and Eastern Churches
241	Catholic/Roman Catholic Church
242	Protestant Denominations/Anglican Church
243	Orthodox/Other Christian Denominations/Sects
250	Classical (Greek and Roman) Religions
260	Islam/Koran
270	Judaism
280	Indic Religions/Buddhism
290	Miscellaneous Religions

SOCIAL SCIENCES

300	Social Sciences
301	Sociology
302	Social Interaction
303	Social Processes
304	Relation of Natural Factors to Social Processes
305	Social Stratification/Cultural Ethnology/Ethnography
306	Culture and Institutions
307	Communities
310	Statistics
314	European Statistics
315	Asian Statistics
316	African Statistics
317	North American Statistics
318	South American Statistics

SOCIAL SCIENCES cont'd

319	Statistics of Other Parts of the World
320	Political Science
321	Current Affairs
322	Civil/Political Rights
323	Constitutional History and Government
324	Official Government Documents
325	Political Process/Political Parties
327	International Relations
330	Economics
331	Labor Economics
332	Financial Economics
333	Land Economics
334	Cooperatives
335	Socialism and Related Systems
336	Public Finance
337	International Economics
338	Production Economics
339	Macroeconomics and Related Topics
340	Law
341	International Law/World Organizations
342	Constitutional Law
343	Tax, Trade, Industrial Law
344	Labour, Public Safety Law
345	Criminal Law
346	Private, Commercial Law
347	Civil Procedure and Courts
348	Statutes, Regulations
350	Public Administration
355	Military Science/Military History/Military Naval Organization and Administration
360	Social Problems and Services/Associations
362	Specific Social Problems and Services
364	Criminology
365	Penal and Related Instituions
366	Associations, Clubs and Societies
368	Insurance
370	Education
373	Secondary Education
374	Adult Education
376	Education of Women
377	Schools and Religion
378	Higher Education
380	Commerce/Communications/Transportation
382	International Foreign Trade
390	Customs/Etiquette/Folklore

*LANGUAGE/LINGUISTICS/LITERATURE

400	General Language Studies
410	General Literature Studies
420	English Language and Literature
430	Germanic Languages and Literatures
440	Romance Languages and Literatures
450	Latin and Greek Languages and Literatures
460	Balto-Slavic Languages and Literatures
470	Afro-Asiatic Languages and Literatures
480	Sino-Tibetan and Japanese Languages and Literatures
490	Other Languages and Literatures

PURE SCIENCES

500	Pure Sciences
510	Mathematics
520	Astronomy and Allied Sciences
530	Physics
540	Chemistry and Allied Sciences
550	Sciences of the Earth and Other Worlds
560	Paleontology/Paleozoology
570	Life Sciences
572	Human Races
573	Physical Anthropology
574	Biology
575	Organic Evolution/Genetics
576	Microbes
577	General Nature of Life
578	Microscopy in Biology
579	Collection and Preservation of Biological Specimens
580	Botanical Sciences
590	Zoological Sciences

TECHNOLOGY (APPLIED SCIENCES)

600	Technology (Applied Sciences)
610	Medicine/Nursing
611	Human Anatomy/Cytology/Tissue Biology
612	Human Physiology
613	General and Personal Hygiene
614	Public Health and Related Topics
615	Pharmacology/Therapeutics
616	Diseases
617	Surgery
618	Gynecology/Obstetrics/Pediatrics/Geriatrics
619	Experimental Medicine
620	Engineering and Allied Operations
621	Applied Physics
622	Mining and Related Operations
623	Military and Nautical Engineering
624	Civil Engineering
625	Railroads/Roads/Highways
627	Hydraulic Engineering

TECHNOLOGY cont'd

628	Sanitary and Municipal Engineering
629	Other Branches of Engineering
630	Agriculture/Veterinary Science
631	Plant Culture
634	Forestry/Fruits
635	Horticulture/Vegetables
636	Animal Husbandry
639	Hunting, Gamekeeping, Fishing
640	Home Economics and Family Living
650	Management and Auxilliary Services
660	Chemical and Related Technologies
670	Manufacturers
680	Manufacture of Products for Specific Uses
690	Buildings

THE ARTS

700	The Arts
710	Civic and Landscape Art
720	Architecture
730	Plastic Arts/Sculpture
740	Drawing/Decorative and Minor Arts
750	Painting and Paintings
760	Graphic Arts/Printmaking and Prints
770	Photography and Photographs
780	Music
790	Recreational and Performing Art

*Belles Lettres

800	General Collections of Belles Lettres
810	Poetry
820	Drama
830	Fiction
840	Literary Essays
850	Speeches
860	Letters
870	Satire and Humor
880	Miscellaneous Writings
890	Unassigned Belles-Lettres

*GEOGRAPHY/HISTORY AND AREA STUDIES

900	Geography and History
910	General Geography/Travel
912	Atlases, Maps, Charts, Plans
914	Geography/Travel in Europe
915	Geography/Travel in Asia
916	Geography/Travel in Africa
917	Geography/Travel in North America
918	Geography/Travel in South and Central America
919	Geography/Travel Other Parts
920	General Biography
929	Genealogy/Historical Auxilliary Science
930	Ancient History/Archeology
931	Middle Ages
933	French Revolution to First World War
934	Contemporary History
939	Judaica
940	European History/Local History
941	British/Scottish/Irish History/Local History
943	Central European/German History/Local History
944	French/Monacan History/Local History
945	Italian History/Local History
946	Spanish/Portuguese History/Local History
947	Eastern European/Russian History/Local History
948	Northern European/Scandinavian History/Local History
949	History of Other Parts of Europe/Local History
950	Asian History/Local History
951	China and Chinese History/Local History
954	South Asian and Indian History/Local History
956	Middle Eastern History/Local History
959	South East Asian History/Local History
960	African History
970	History/Local History of the Americas
971	General/Local History of Canada
972	General/Local History of Mexico, Middle and South America
975	General History of the United States
976	United States: Slavery/Civil War Period
977	United States: War with Spain to Present
978	United States: Local History and Geography
980	History of Oceania
990	History of Other Areas and Worlds

Survey of Subjects

Accounting	650
Acting	790
Adult education	374
Advertising	650
African History	960
Afro-Asiatic Languages and Literatures	470
Agriculture	630
Albanian Language and Literature	460
Algebra	510
Analytical Chemistry	540
Anatomy, Human	611
Ancient History	930
Ancient Philosophy	180
Ancient World, History of	930
Anglican Church	242
Animal Husbandry	636
Anthropology	
general	301
physical	573
social	306
Anthropometry	573
Anthroposophy	290
Applied Physics	621
Archeology	930
Architecture	720
Archives	025
Area Planning (civic Art)	710
Arithmetic	510
Arts	700
dictionaries	057
Asian History	950
Associations	366
Astronomy and Allied Sciences	520
Atheism	210
Atlases	912
Australian History and Geography	980
Ballet	790
Balto-Slavic Languages and Literatures	460
Banks and Banking	332
Belles-Lettres,	800–890
collections	800
unassigned	890
Beverage Technology	660
Bible	220
Bibliography	
arts	017
Belles-Lettres	018
general	010
geography and history	019
language and literature studies	014
philosophy and psychology	011
religion	012
science	015
social sciences	013
technology	016
Bioclimatology	574
Biography	920
Biological Specimens	579
Biophysics	574
Biostatistics	574
Blacksmithing	680
Blood and Circulation	612
Book Rarities	090
Botanical Sciences	580
Bridge Engineering	624
British History	941
Brittle Books	025
Buddhism	280
Building and Housing Cooperatives	334
Building Materials	690
Buildings (Engineering)	690
Business Enterprises	338
Calendar(s)	520
Canadian History	971
Carpentry	690
Carving and Carvings	730
Catholic Church	241
Causation (Philosophy)	120
Celestial Mechanics	520
Celestial Navigation	520
Celtic Languages and Literatures	490
Central American History	972
Central European History	943
Ceramic and Allied Manufacturing Technologies	660
Ceramic Arts	730
Charts	912
Chemical Technologies	660
Chemistry and Allied Sciences	540
Child-rearing	640
Childbirth	618
China and Chinese History	951
Christianity	240
Chromolithography	760
Chronology	520
Cinematic Arts	790
Civic Art	710
Civil Engineering	624
Civil Procedure and Courts	347
Civil Rights	322
Clothing Manufacture	670
Clubs	366
Collections, General	080
Collective Bargaining	331
Commerce	380
Commercial Law	346

Communication	000
Communications	380
Communities	307
Comparative Literature	410
Comparative Religion	230
Computers	000
hardware	621
software	000
Confucianism	290
Constitutional History and Government	323
Constitutional Law	342
Consumer Cooperatives	334
Contemporary History	934
Cooperatives, Economic	334
Costume	390
Courts of Law	340
Craniology	573
Credit	332
Criminal Law	345
Criminology	364
Crystallography	540
Cultural Ethnology	305
Cultural History	900
Culture and Institutions	306
Current Affairs	321
Customs (Anthropology)	390
Cytology	611
Data Processing	000
Death Customs	390
Decorative Arts	740
Dentistry	617
Dictionaries	
language	040
subject	050–059
Dictionary Catalogs	010
Diplomatics	370
Diseases	616
Domestic Trade	380
Drama	820
Drawing	740
Early Church	240
Earth Sciences	550
Eastern Churches	240
Eastern European History	947
Eastern Philosophy	180
Economics	330–339
Education	370
Education of Women	376
Elastomers	670
Encyclopedic Works	030
Engineering and Allied Operations	620
Engineering, Other Branches	629
English Language and Literature	420
Epistemology	120
Epistles	227
Equestrian Sports	790
Essays	840

Etching	760
Ethics	170
Ethnography	305
Etiquette	390
European History	
eastern	947
general	940
northern	948
Evolution, Organic	575
Excavation Techniques, Engineering	622
Explosives Manufacturing	660
Fiction	830
Financial Economics	332
Finno-Ugaric Languages	490
Fishing	639
recreational	790
Fluid Mechanics	530
Folk Literature	390
Folklore	390
Foreign Trade	380
Forestry and Fruits	634
Fossils	560
French History	944
French Revolution to First World War	933
Fruits	634
Fur Processing	670
Furniture/Arts	740
Gamekeeping	639
Games of Chance	790
Genealogy	929
General Collections	080
General Customs	390
Generalities	000
dictionaries	050
Genetics	575
Geography	910
dictionaries	059
Geology	550
Geometry	510
Geriatrics	618
German History	943
Germanic Languages and Literatures	430
Gospels and Acts	226
Government Documents, Official	324
Graphic Arts	760
Greek Language and Literature	450
Ground Transportation	380
Gynecology	618
Heraldry	929
Higher Educaion	378
Highways (Engineering)	625
Hinduism	280
History	900–990
dictionaries	059
Holography	770
Home Economics	640

xv

Horticulture	635
Housekeeping	640
Human Beings (Philosophy)	120
Human Races	
ethnic groups	305
life sciences	572
Human Resources	331
Hunting	639
Hunting, Recreational	790
Hydraulic Engineering	627
Hydraulic-power Technology	621
Hydrology	550
Hygiene	613
Incunabula	090
Indian History	951
Indian Sub-Continent, Languages and Literatures of	490
Industrial Law	343
Information Sciences	020
Inorganic Chemistry	540
Institutions (Sociology)	306
Insurance	368
Interior Decoration	740
International Economics	337
International Foreign Trade	382
International Law	341
International Relations	327
Investment	332
Irish History	941
Islam	260
Jainism	280
Japanese Language and Literature	480
Jewish Theology	270
Journalism	070
Judaica	939
Judaism	270
Koran	260
Labor Economics	331
Labour Law	344
Land Economics	333
Landscape Art	710
Latin Language and Literature	450
Law	340
Leather Processing	670
Letters (Literary)	860
Lettish Language	460
Library Science	020
Life Sciences	570–579
Life, General Nature of	577
Linguistics	400–490
dictionaries	054
general	400
Liquid Mechanics	530

Literature	
studies of	400–490
dictionaries	054
Lithography	760
Lithuanian Language	460
Loan Institutions	332
Local Geography	910
Local History	910
Logic	160
Lumber Processing	670
Machine Engineering	621
Macroeconomics	339
Magnetism	530
Management Services	650
Manufactures	670
Manuscripts	090
Maps	912
Marriage and Family, Sociology of	306
Marxian Systems	335
Mathematics	510
Medical Sciences	610–619
Medicine, Experimental	619
Medieval Philosophy	180
Metallurgy	660
Metaphysics	110
Meteorology	550
Metrology and Standardization	380
Mexican History	972
Mezzotinting	760
Microbes	576
Microeconomics	338
Microscopy in Biology	578
Middle Ages	931
Middle American History	972
Middle Eastern History	956
Military Art and Science	355
Military Engineering	623
Military Naval Organization and Administration	355
Mineralogy	540
Mining (Engineering)	622
Miscellaneous Writings (Belles-Lettres)	880
Modern Western Philosophy	190
Monacan History	944
Money	332
Monotheism	210
Moral Philosophy	170
Movies	790
Municipal Engineering	628
Museology	060
Music	780
Musical Instruments	780
Natural Resources	333
Nautical Engineering	623
Naval Science	359
New Testament	225

New Zealand History and Geography	980
Newspapers	
British Isles	072
Eastern Europe	077
Other Geographical areas	079
Scandinavia	078
studies of	070
United States/Canada	071
Western Europe [excl Scandinavia]	074
North American History	970
Northern European History	948
Numismatics	730, 929
Nursing	610
Obstetrics	618
Occult Sciences	130
Oceania, History and Geography	980
Office Services	650
Official Government Documents	324
Old Testament	221
Operative Surgery	617
Ophthalmology	617
Orchards	630
Organic Chemistry	540
Organizations, General	060
Oriental Philosophy	180
Orthodox/Other Christian Denominations/Sects	243
Painting and Paintings	750
Paleobotany	560
Paleontology	560
Paleozoology	560
Paranormal Phenomena and Arts	130
Pathology	574
Pediatrics	618
Penal and Related Institutions	365
Performing Arts	790
Periodicals	
British Isles	072
Eastern Europe	077
General	073
Other Geographical areas	079
Scandinavia	078
studies of	070
United States/Canada	071
Western Europe [excl Scandinavia]	074
Petrology	550
Pharmacology	615
Philosophy	100–190
dictionaries	051
modern western	190
Photography and Photographs	770
Phrenology	130
Physics	530
applied	621
Physiology, Human	612
Plans	912
Plant Culture	631
Plastic Arts	730
Platonic Philosophy	180
Poetry	810
Political Process/Parties	325
Political Rights	322
Political Science	320
Polytheism	210
Portuguese History	946
Postal Communication	380
Pre-Socratic Philosophy	180
Printing	680
Printmaking	760
Prints	760
Private Law	346
Production Cooperatives	334
Production Economics	338
Protestant Denominations	242
Psychology	150
Public Administration	350
Public Finance	336
Public Health	614
Public Performances	790
Public Relations	650
Public Saftey Law	344
Publishing	070
Pulp and Paper Technology	670
Railroads	
transportation	380
engineering	625
Recreation	790
Regulations	348
Religion	200–290
dictionaries	052
Germanic	290
Greek	250
Roman	250
Road Engineering	625
Roman Catholic Church	241
Romance Languages and Literatures	440
Romanian History	949
Roofing	690
Russian History	947
Sanitary Engineering	628
Satire and Humour	870
Scandinavian History	948
Science	
dictionaries	055
pure	500
Scientology	290
Schools and Religion	377
Scottish History	941
Sculpture	730
Secondary Education	373
Secretarial skills	650
Shintoism	290
Shooting, Recreational	790
Sigillography	730
Sikhism	280

Sino-Tibetan Languages and Literatures	480	Talmudic Literature	270
Slavery	320	Tax Law	343
Slavic Languages and Literatures	460	Taxes and Taxation	336
Slide Preparation (Biology)	578	Taxidermy	579
Social Change	303	Teaching Theory	370
Social Interaction	302	Technology (Applied Science)	600
Social Problems	360	dictionaries	056
Social Processes	303	Telecommunication	380
relation of natural factors to	304	Teleology	120
Social Sciences	300	Textiles	
dictionaries	053	arts	740
Social Services	360	manufacturing	670
Social Stratification	305	Theater	790
Social Welfare	360	Theosophy	290
Socialism and Related Systems	335	Therapeutics	615
Societies	366	Tissue Biology	611
Sociology	301	Toxicology	615
Sociometry	302	Trade Law	343
Solid Mechanics	530	Trade Unions	331
Somatology	306	Transportation	380
South American History and Geography	972	Travel – General	910
South Asian History	951	African Travel	916
South East Asian History	959	Asian Travel	915
Spanish History	946	European Travel	914
Specific Social Problems	362	North American Travel	917
Specific Social Services	362	Other Parts	919
Speeches	850	South and Central American Travel	918
Sports	790		
Stage Presentations	790	Unassigned Belles-Lettres	890
State, Theory of	321	United States	
Statistics	310	general history	975
African Statistics	316	local history and geography	978
Asian Statistics	315	slavery and Civil War period	976
European Statistics	314	war with Spain to present	977
North American Statistics	317	Universities and Colleges	370
South American Statistics	318	Urban Sociology	307
Statistics of Other Parts	319		
Statutes	348	Vegetables	635
Substance Abuse		Veterinary Science	630
medical	616		
social	360	Women, Sociological Study of	305
Subsurface Resources	333	World Organizations	341
Surgery	617		
Swiss History	949	Yoga	613
		Zoological Sciences	590

Publishers and Distributors

at Australian
Australian Institute of Genealogical Studies Inc, Unit 1, 41 Railway Rd, Blackburn, Victoria 3130, PO Box 339, Blackburn, Victoria 3130, Australia / T: +61 3 9877 3789, E-mail: info@aigs.org.au, Internet: www.aigs.org.au

at Cairns
Cairns and District Family History Society Inc, c/o Mrs Marilyn Prime, PO Box 5069, Cairns, Queensland 4870, Australia / E-mail: info@cdfhs.org, Internet: www.cdfhs.org

at Cape
Cape Banks Family History Society Inc, c/o Publications Officer, PO Box 67 Maroubra, NSW 2035, Australia / E-mail: ericjmcl@iinet.net.au, Internet: http://www.capebanks.org.au/

at Hunter
Hunter Publications, 58A Gipps St, Collingwood, Victoria 3066, POB 404, Abbotsford, VIC 3067, Australia / T: +61 3 94175361, Fax: +61 3 94197154, E-mail: sales@hunter-pubs.com.au

at Northern
Genealogical Society of the Northern Territory Inc, POB 37212, Winnellie, NT 0821, Australia / T: +61 8 89817363, Fax: +61 8 89817383, E-mail: gsntinc@bigpond.net.au, Internet: http://members.iinet.net.au/~genient/ – ISBN 978-0-949124

at Pacific Mss
Pacific Manuscripts Bureau (Pambu), Australian National University, c/o Rm 4201, Coombs Bldg 9, Fellows Rd, Research School of Pacific and Asian Studies, Australian National Univ, Canberra, ACT 0200, Australia / T: +61 2 61252521, Fax: +61 2 61250198, E-mail: pambu@coombs.anu.edu.au, Internet: rspas.anu.edu.au/pambu/
Orders
ProQuest Information and Learning, 300 N Zeeb Rd, Ann Arbor, MI 48106-1346, POB 1346, Ann Arbor, MI 48106-1346, USA / T: +1 734 7614700, +1 800 5213042, Fax: +1 734 9739145, E-mail: info@il.proquest.com, Internet: www.umi.com; http://www.proquest.com/products_umi/catalogs/ – ISBN 978-0-8357
Information
Unrestricted titles available for purchase; Pacific Islands, New Zealand and Australia silver A$70 per reel, vesicular A$65 per reel; Rest of the world silver US$70 per reel, vesicular US$65 per reel

at Pascoe
W & F Pascoe Pty Ltd, 7 Hayes St, Balgowlah, NSW 2093, Australia / T: +61 2 99491133, Fax: +61 2 99494389, E-mail: pascoe@pascoe.com.au, Internet: www.pascoe.com.au – ISBN 978-0-9593336

at ProQuest
ProQuest Information and Learning, POB 181, Drummoyne NSW 1470, Australia / T: +61 2 99116660, Fax: +61 2 99116652, E-mail: julie.stevens@anz.proquest.com

at UNSW Lib
University of New South Wales Library, UNSW, Sydney, NSW 2052, Australia / T: +61 2 93852650, Fax: +61 2 93851260, Internet: http://info.library.unsw.edu.au/Welcome.html – ISBN 978-0-909283

au Morawa
Mohr Morawa, Buchvertrieb GmbH, Sulzengasse 2, 1230 Wien, Postf 260, 1101 Wien, Austria / T: +43 1 680140, Fax: +43 1 68014140, E-mail: momo@mohr-morawa.co.at; bestellung@mohrmorawa.at, Internet: www.morawa.com

be Brepols
Brepols Publishers, Begijnhof 67, 2300 Turnhout, Belgium / T: +32 14 448020, Fax: +32 14 428919, E-mail: info@brepols.com; orders@brepols.be, Internet: www.brepols.net – ISBN 978-2-503; 978-2-85006; 978-90-5622; 978-90-72100

bl Biblioteca
Fundação Biblioteca Nacional, Avenida Rio Branco, 219 – 1° andar, Rio de Janeiro, RJ 20040-008, Brazil / T: +55 21 22209367, Fax: +55 21 22204173, E-mail: adriana@bn.br, Internet: www.bn.br – ISBN 978-85-7017

cc Misc Inst
Miscellaneous Institutions, City various, China, People's Republic
Information
Lists titles of unknown/unidentified microform publishers, or publishers whose addresses have not been traced

cc Taylor
Ian Taylor Associates, Rm 1505 Zhejiang Bldg, Block 3, AnZhenXiLi, ChaoYang District, Beijing 100029, China, People's Republic / T: +86 10 64455987, E-mail: hrthome@gmail.com, Internet: www.iantaylorassociates.com

ch Transmission
Transmission Books & Microinfo Co, Ltd, 7F, No 315, Sec 3, Ho Ping E Rd, Taipei, China, Republic / T: +886 2 27361058, Fax: +886 2 27363001, E-mail: info@tts.tbmc.com.tw, Internet: www.tbmc.com.tw – ISBN 957-30801

cn Bibl Nat
Bibliothèque Nationale du Québec, Service de Microphotographie, 2275 rue Holt, Montréal, H2G 3H1, Canada / T: +1 514 8731100, +1 800 3639028, Fax: +1 514 8739312, E-mail: collectionspeciale@banq.qc.ca; reproduction@banq.qc.ca, Internet: www.banq.qc.ca
Information
16mm Can$20r; 35mm Can$60r; mf Can$10 per title for monographs Can$10 per annum for publ in a series. The bibliotheque Nationale du Quebec (BNQ) is merged with the Grande Bibliotheque du Quebec

cn Canadiana
Canadiana.org, 395 Wellington St, Rm 468, Ottawa, ON K1A 0N4, POB 2428, Stn D, Ottawa, ON K1P 5W5, Canada / T: +1 613 2352628, Fax: +1 613 2352952, E-mail: info@canadiana.org, Internet: http://www.canadiana.org/eco.php?doc=cihm – ISBN 978-0-665; 978-0-659

cn Commonwealth Imaging
Commonwealth Imaging, A Division of West Canadian Industries Group, 901 – 10th Ave S.W., Calgary, T2R 0B5, Canada / T: +1 403 2452555, Fax: +1 403 2285712, E-mail: nvehrs@westcanadian.com, Internet: www.commonwealthimaging.com
Information
Price for current newspapers on microfilm Can$99 per reel, for archived newspapers on microfilm, special colls and CLA publ Can$110 per reel, in both cases for reel count of up to 50

cn Fitzhenry
Fitzhenry & Whiteside, 195 Allstate Parkway, Markham, ON L3R 4T8, Canada / T: +1 9054779700, +1 800 3879776, Fax: +1 9054779179, +1 800 2609777, E-mail: godwit@fitzhenry.ca; ndoucet@fitzhenry.ca, Internet: www.fitzhenry.ca

cn Library and Archives
The Library and Archives of Canada/Bibliothèque et Archives Canada, Canadian Theses Service, 395 Wellington St, Ottawa, ON K1A 0N4, Canada / T: +1 613 9959481, +1 877 8969481; +1 819 9536221, Fax: +1 613 9431112; +1 819 9946904, E-mail: mel.simoneau@lac-bac.gc.ca, Internet: www.collectionscanada.ca – ISBN 978-0-612; 978-0-315

cn Library Assoc
Canadian Library Association, 328 Frank St, Ottawa, ON K2P 0X8, Canada / T: +1 613 2329625, Fax: +1 613 5639895, E-mail: orders@cla.ca, emorton@cla.ca, pwilson@cla.ca, Internet: www.cla.ca – ISBN 978-0-88802
Orders
Microfilm: Commonwealth Imaging, A Division of West Canadian Industries Group, 901 – 10th Ave S.W., Calgary, T2R 0B5, Canada / T: +1 403 2452555, Fax: +1 403 2285712, E-mail: nvehrs@westcanadian.com, Internet: www.commonwealthimaging.com
Information
Can$92.00 per reel

cn McLaren
McLaren Micropublishing Ltd, POB 972, Station F, Toronto, ON M4Y 2N9, Canada / T: +1 416 9604801, Fax: +1 416 9643745, E-mail: mmicro@interlog.com, Internet: http://www.interlog.com/~mmicro/order.htm

cn Scholarly Bk
Scholarly Book Services Inc, 127 Portland St, 3rd fl, Toronto, ON M5V 2N4, Canada / T: +1 800 8479736, Fax: +1 800 2209895, E-mail: customerservice@sbookscan.com; orders@sbookscan.com, Internet: www.sbookscan.com
Information
Canadian residents add 7% GST

cn UBC Preservation
University of British Columbia Library, Preservation Microfilming Office, 2206 East Mall, Vancouver, BC V6T 1Z3, Canada / T: +1 604 8225951, Fax: +1 604 8224789, E-mail: norman.amor@ubc.ca, Internet: www.library.ubc.ca
Orders
UNIpresses, 34 Armstrong Ave, Georgetown, ON L7G 4R9, Canada / T: +1 9058739781; +1 8778648477 – toll free, Fax: +1 9058736170; +1 8778644272 – toll free, E-mail: orders@gtwcanada.com

cn UNIpresses
UNIpresses, 34 Armstrong Ave, Georgetown, ON L7G 4R9, Canada / T: +1 9058739781; +1 8778648477 – toll free, Fax: +1 9058736170; +1 8778644272 – toll free, E-mail: orders@gtwcanada.com

fi Helsinki
Helsinki University Library, The National Library of Finland, POB 15 (Unioninkatu 36), 00014 Helsinki University, Finland / T: +358 9 19123196, Fax: +358 9 19122719; +358 15 151228 (microfilm enquiries), E-mail: kk-miko@helsinki.fi, Internet: www.lib.helsinki.fi/english/
Information
microfilm orders to: Kansalliskirjasto, Mikrokuvaus- ja konservointilaitos, Saimaankatu 6, 50100 Mikkeli

fr ACRPP
ACRPP, Association pour la Conservation et la Reproduction photographique de la Presse, Le Parc aux Vignes, 11 allée des Sarments, 77183 Croissy-Beaubourg, BP 21, 77113 Marne-La-Vallée Cedex 2, France / T: +33 1 60176810, +33 1 60177213, Fax: +33 1 60176805, E-mail: contact@acrpp.fr, Internet: www.acrpp.fr

fr Atelier National
Atelier National de Reproduction des Thèses, Université de Lille III, 9, rue Auguste Angellier, 59046 Lille Cedex, France / T: +33 3 20308673, Fax: +33 3 20542195, E-mail: anrt@univ-lille3.fr, Internet: http://we225.lerelaisinternet.com/ – ISBN 978-02-284
Information
€15.00 per title up to the 11th fiche. After this add €1.00 per fiche

fr Bibl Nationale
Bibliothèque nationale de France, Richelieu & Francois-Mitterrand Buildings, Service reproduction, Quai Francois Mauriac, 75706 Paris cedex 13, France / T: +33 1 53795959, Fax: +33 1 53794260, E-mail: reproduction@bnf.fr, Internet: www.bnf.fr – ISBN 978-2-7177
Orders
ProQuest Information and Learning, 300 N Zeeb Rd, Ann Arbor, MI 48106-1346, POB 1346, Ann Arbor, MI 48106-1346, USA / T: +1 734 7614700, +1 800 5213042, Fax: +1 734 9739145, E-mail: info@il.proquest.com, Internet: www.umi.com; http://www.proquest.com/products_umi/catalogs/ – ISBN 978-0-8357

fr CRDP
Centre Régional de Documentation Pédagogique de Franche-Comté, CRDP Besançon, 5 rue des Fusilés de la Résistance, 25003 Besançon Cédex, France / T: +33 5 81250250, +33 5 81250263 (Tainturier), Fax: +33 5 81250255, E-mail: crdp@ac-besancon.fr; crdp.sg@ac-besancon.fr25; laurent.tainturier@ac-besancon.fr, Internet: crdp.ac-besancon.fr

fr Journal Officiel
Journal Officiel, Editeur des Journaux Officiels, 26 rue Desaix, 75727 Paris Cedex 15, France / T: +33 1 40587500, +33 1 40587600, Fax: +33 1 45791784, E-mail: info@journal-officiel.gouv.fr, bdarthois@hotmail.com, Internet: www.journal-officiel.gouv.fr
Orders
N America: ProQuest Information and Learning, 300 N Zeeb Rd, Ann Arbor, MI 48106-1346, POB 1346, Ann Arbor, MI 48106-1346, USA / T: +1 734 7614700, +1 800 5213042, Fax: +1 734 9739145, E-mail: info@il.proquest.com, Internet: www.umi.com; http://www.proquest.com/products_umi/catalogs/ – ISBN 978-0-8357

gw Bundesarchiv
Bundesarchiv Koblenz, Potsdamer Str 1, 56075 Koblenz, Germany / T: +49 261 5050; +49 261 505263, Fax: +49 261 505226, E-mail: koblenz@barch.bund.de; t.koops@barch.bund.de, Internet: www.bundesarchiv.de – ISBN 978-3-89192
Information
Prices given in the entries are for the printed guide

gw Fischer
Harald Fischer Verlag GmbH, Theaterplatz 31, 91054 Erlangen, Postf 1565, 91005 Erlangen, Germany / T: +49 9131 205620, Fax: +49 9131 206028, E-mail: info@haraldfischerverlag.de, Internet: www.haraldfischerverlag.de – ISBN 978-3-89131

gw Frankfurter
Frankfurter Taschenbuch Verlag GmbH Verlag der Deutschen Hochschulschriften, Großer Hirschgraben 15, 60311 Frankfurt a M, Germany / T: +49 69 408940, Fax: +49 69 40894194, E-mail: info@haensel-hohenhausen.de; huentelmann@haensel-hohenhausen.de; vertrieb@fouque-verlag.de, Internet: www.haensel-hohenhausen.de – ISBN 978-3-89349; 978-3-8267
Information
Fouqué Literaturverlag and Medien- und Verlagsgruppe Dr. Hänsel-Hohenhausen are imprints of Frankfurter Taschenbuchverlag GmbH, a part of Frankfurter Verlagsgruppe AG August von Goethe

gw Herold
Firma Herold, Raiffeisenallee 10, 82041 Oberhaching, Germany / T: +49 89 61387112, Fax: +49 89 61387120, E-mail: herold@herold-va.de

gw Lengenfelder
Edition Helga Lengenfelder, Schönstr 51, 81543 München, Germany / T: +49 89 663845, E-mail: Lengenfelder.Edition@t-online.de, Internet: www.geist.de/lengenfelder/verlag-D.html – ISBN 978-3-89219

gw Mikrofilm
Mikrofilmarchiv der deutschsprachigen Presse eV, Königswall 18, 44122 Dortmund, Germany / T: +49 231 5023249, +49 231 5023216, +49 231 5026564 (Hr Pankratz), Fax: +49 231 5023218, E-mail: marlt@stadtdo.de, MFA@stadtdo.de, Internet: http://www.zeitungsforschung.de/
Orders
N America: ProQuest Information and Learning, 300 N Zeeb Rd, Ann Arbor, MI 48106-1346, POB 1346, Ann Arbor, MI 48106-1346, USA / T: +1 734 7614700, +1 800 5213042, Fax: +1 734 9739145, E-mail: info@il.proquest.com, Internet: www.umi.com; http://www.proquest.com/products_umi/catalogs/ – ISBN 978-0-8357
Information
Please note that only titles with master/original negatives are listed; these originals are obtainable either at source or from the various institutions which are affiliated members of Mikrofilmarchiv der deutschsprachigen Presse e.V. Diazo film costs €35 per reel, silver film 20% more; working copies are priced at €30 per reel

gw Mikropress
Mikropress GmbH, Ollenhauerstr 1, 53113 Bonn, Germany / T: +49 228 623261, Fax: +49 228 628868, E-mail: info@mikropress.de, Internet: www.mikropress.de
Orders
Mikrofilmarchiv der deutschsprachigen Presse eV, Königswall 18, 44122 Dortmund, Germany / T: +49 231 5023249, +49 231 5023216, +49 231 5026564 (Hr Pankratz), Fax: +49 231 5023218, E-mail: marlt@stadtdo.de, MFA@stadtdo.de, Internet: http://www.zeitungsforschung.de/

gw Misc Inst
Miscellaneous Institutions, see gw Mikrofilm, Germany
Orders
Mikrofilmarchiv der deutschsprachigen Presse eV, Königswall 18, 44122 Dortmund, Germany / T: +49 231 5023249, +49 231 5023216, +49 231 5026564 (Hr Pankratz), Fax: +49 231 5023218, E-mail: marlt@stadtdo.de, MFA@stadtdo.de, Internet: http://www.zeitungsforschung.de/
Information
All institutes, libraries whose masters are available for filming through Mikrofilmarchiv. Inquire at Mikrofilmarchiv for details

gw Olms
Georg Olms Verlag AG, Hagentorwall 7, 31134 Hildesheim, Germany / T: +49 5121 15010, +49 5121 150139, Fax: +49 5121 150150, +49 5121 32007, E-mail: info@olms.de; new.media@olms.de, Internet: www.olms.de – ISBN 978-3-487; 978-3-615; 978-3-296
Orders
VVA – Vereinigte Verlagsauslieferung, An der Autobahn, 33310 Gütersloh, Postf 7777, 33310 Gütersloh, Germany / T: +49 5241 803896, +49 5241 802894, Fax: +49 5241 809595, E-mail: horst.raemsch@bertelsmann.de, Internet: www.vva-online.net
USA: Lubrecht & Cramer, Ltd, International Books by Mail, 78 Front St Ste 76, Port Jervis, NY 12771, POB 3110, Port Jervis, NY 12771, USA / T: +1 800 9209334; +1 845 8565990, E-mail: books@lubrechtcramer.com, Internet: www.lubrechtcramer.com

gw ProQuest
ProQuest Information and Learning, Grüner Weg 8, 61169 Friedberg, Germany / T: +49 6031 87473, Fax: +49 6031 87469, E-mail: claudia.spengemann@proquest.co.uk

gw Rhenus
Rhenus Medien Logistik GmbH & Co. KG, Justus-von-Liebig-Str. 1, 86899 Landsberg/Lech, Germany / T: +49 8191 97000214 (books); +49 8191 97000881 (journals), Fax: +49 8191 97000594 (books); +49 8191 97000103 (journals), E-mail: degruyter@de.rhenus.com, Internet: www.rhenus.com

gw Saur
K G Saur Verlag, Ein Imprint der Walter de Gruyter GmbH & Co. KG (An Imprint of Walter de Gruyter GmbH & Co. KG), Mies-van-der-Rohe-Str 1, 80807 München, Germany / T: +49 89 769020, Fax: +49 89 76902150, +49 89 76902250, E-mail: info@degruyter.com, Internet: www.saur.de – ISBN 978-3-7940; 978-3-598
Orders
Rhenus Medien Logistik GmbH & Co. KG, Justus-von-Liebig-Str. 1, 86899 Landsberg/Lech, Germany / T: +49 8191 97000214 (books); +49 8191 97000881 (journals), Fax: +49 8191 97000594 (books); +49 8191 97000103 (journals), E-mail: degruyter@de.rhenus.com, Internet: www.rhenus.com
USA, Canada & Mexico: Walter de Gruyter Inc, POB 960, Herndon, VA 20172-0960, USA / T: +1 703 6611589, +1 800 2088144, Fax: +1 703 6611501, E-mail: degruytermail@presswarehouse.com, Internet: www.degruyter.com

gw VVA
VVA – Vereinigte Verlagsauslieferung, An der Autobahn, 33310 Gütersloh, Postf 7777, 33310 Gütersloh, Germany / T: +49 5241 803896, +49 5241 802894, Fax: +49 5241 809595, E-mail: horst.raemsch@bertelsmann.de, Internet: www.vva-online.net

ie National
The National Library of Ireland, Kildare St, Dublin 2, Ireland / T: +353 1 6030213, Fax: +353 1 6030288, E-mail: Newsplan@nli.ie; info@nli.ie, Internet: www.nli.ie; http://www.nli.ie/en/newspapers-on-microfilm.aspx
Information
new filming €150 per reel; duplication of an existing microfilm €70 per reel

ja Far Eastern
Far Eastern Booksellers, 12 Kanda Jimbocho 2-chome, Chiyoda-ku, Tokyo, POB 72, Tokyo, Japan / T: +81 3 32657532, Fax: +81 3 32654656, E-mail: info@kyokuto-bk.co.jp, Internet: www.kyokuto-bk.co.jp

ja Kinokuniya
Kinokuniya Bookstore Co Ltd, Journal Dept, 38-1 Sakuragaoka 5-chome Setagaya-ku, Tokyo 156, PO Box 55, Chitose, Tokyo 156, Japan / T: +81 3 34390161, Fax: +81 3 34390839, E-mail: psale@kinokuniya.co.jp, Internet: http://bookweb.kinokuniya.co.jp – ISBN 978-4-314

ja Maruzen
Maruzen Company Ltd, Import and Export Dept, c/o Dai-3 Maruzen Bldg, 2-16-1 Nihonbashi, Chuo-ku, Tokyo 103, POB 5050, Tokyo International 100-3191, Japan / T: +81 3 32789223, Fax: +81 3 32742270, E-mail: kawamura@maruzen.co.jp, Internet: http://www.maruzen.co.jp/home-eng/ – ISBN 978-4-621; 978-4-89580; 978-4-8395

ja UPS
UPS United Publishers Services Ltd, A member of Times Publishing Group, Singapore, 1-32-5, Higashi-Shinagawa, Shinagawa-ku, Tokyo 140-0002, Japan / T: +81 3 54797251, Fax: +81 3 54797307, E-mail: general@ups.co.jp, info@ups.co.jp, Internet: www.ups.co.jp

ja Yushodo
Yushodo Co Ltd, 29 San-ei-cho, Shinjuku-ku, Tokyo 160-0008, Japan / T: +81 3 33571411, Fax: +81 3 33515855, E-mail: ysdhp@yushodo.co.jp, Internet: www.yushodo.co.jp – ISBN 978-4-8419
Orders
N America: ProQuest Information and Learning, 300 N Zeeb Rd, Ann Arbor, MI 48106-1346, POB 1346, Ann Arbor, MI 48106-1346, USA / T: +1 734 7614700, +1 800 5213042, Fax: +1 734 9739145, E-mail: info@il.proquest.com, Internet: www.umi.com; http://www.proquest.com/products_umi/catalogs/ – ISBN 978-0-8357

ko Information
Information & Culture Korea, 473-19 Seokyo-dong, Mapo-ku, Seoul 121-842, Korea, Republic / T: +82 2 3141 4791 / 4793, Fax: +82 2 3141 7733, E-mail: ickseoul@kornet.net

ko ProQuest
ProQuest Information and Learning, Rm 1402, Doosan Bearstel 1319-11, Seocho 2-dong, Seocho-Gu, Seoul 137-072, Korea, Republic / T: +82 2 588 6045, Fax: +82 2 588 6046, E-mail: junghyun.kim@asia.proquest.com, eunkyung.kim@asia.proquest.com, Internet: www.proquest.jp

mf Russick
Katri Russick, Rue la Chaux, Mahebourg, Mauritius / T: +230 6311512; +230 2533527 (mobile), E-mail: mrussick@intnet.mu

mx Thomson
Thomson Learning Iberoamerica, Seneca No 53, Colonia Polanco, Mexico, D F C.P. 11560, Mexico / T: +52 55 15006000, Fax: +52 55 52812656, E-mail: rosa.viveros@thomson.com

my Gale
Thomson Asia Pte Ltd, Branch Office, No 3, Jalan PJS 7/19, Bandar Sunway, 46150 Petaling Jaya, Selangor Darul Ehasan, Malaysia / T: +60 3 56368351/8352, Fax: +60 3 56368302, E-mail: simon.tay@thomsonlearning.com.sg, Internet: www.gale.com/world/

my ProQuest
ProQuest Information and Learning, B909 (Block B), Phileo Damansara 1 No 9, Jalan 16/11, Jalan Damansara, 46350 Petaling Jaya, Selangor D.E, Malaysia / T: +60 3 79542880, Fax: +60 3 79583446, E-mail: kelvin.low@asia.proquest.com; richard.hollingsworth@asia.proquest.com

ne Brill
Brill Academic Publishers, Plantijnstraat 2, 2321 JC Leiden, Postbus 9000, 2300 PA Leiden, Netherlands / T: +31 71 5353500, Fax: +31 71 5317532, E-mail: cs@brill.nl; Akkermans@brill.nl, Internet: www.brill.nl – ISBN 978-90-04; 978-90-247
Orders
USA: Brill Academic Publishers Inc, 112 Water Street, Ste 400, Boston MA 02109, USA / T: +1 617 2632323, Fax: +1 617 2632324, E-mail: cs@brillusa.com

ne IDC
IDC Publishers bv, Hogewoerd 151, 2311 HK Leiden, POB 11205, 2301 EE Leiden, Netherlands / T: +31 71 5142700, Fax: +31 71 5131721, E-mail: info@idc.nl; mpijl@idc.nl, Internet: www.idc.nl
Information
IDC Publishers acquired by Brill Academic Publishers in 2005. For a distributor in your area/country, contact IDC Publishers. You can also find a list of distributors on the website (How to contact us)

ne MMF Publ
MMF Publications, Heereweg 331a, 2160 AG Lisse, POB 287, 2160 AG Lisse, Netherlands / T: +31 252 432100, Fax: +31 252 432101, E-mail: mmf@microformat.nl, Internet: www.mmfpublications.nl

ne Moran
Moran Micropublications, Singel 357, 1012 WK Amsterdam, Netherlands / T: +31 20 5286139, Fax: +31 20 6239358, E-mail: info@moranmicropublications.nl, Internet: www.moranmicropublications.nl
Information
Publisher distributes the titles of MMF, which are marked with the code MMP-A, B etc.

ne Slangenburg
Microlibrary Slangenburg Abbey, Abdijlaan 1, Slangenburg, 7004 JL Doetinchem, Netherlands / T: +31 315 298268, Fax: +31 315 298798, E-mail: info@willibrords-abbey.nl, Internet: www.willibrords-abbey.nl
Orders
N America: ProQuest Information and Learning, 300 N Zeeb Rd, Ann Arbor, MI 48106-1346, POB 1346, Ann Arbor, MI 48106-1346, USA / T: +1 734 7614700, +1 800 5213042, Fax: +1 734 9739145, E-mail: info@il.proquest.com, Internet: www.umi.com; http://www.proquest.com/products_umi/catalogs/ – ISBN 978-0-8357

nz BAB
BAB Microfilming, 6 Kathryn Av, Mt Roskill, Auckland 1004, New Zealand / T: +64 9 6259778, Fax: +64 9 6259379, Internet: http://www.micrographics.co.nz/BAB.Microfilming.htm – ISBN 978-0-908797; 978-0-908989

nz Nat Libr
National Library of New Zealand, Cnr Molesworth & Aitken St, Wellington, POB 1467, Wellington 6001, New Zealand / T: +64 4 4743151, Fax: +64 4 4743035, E-mail: David.Adams@natlib.govt.nz, information@natlib.govt.nz, Internet: www.natlib.govt.nz
Information
Inquire for prices

ph Sagun
I J Sagun Enterprises, 2 Topaz Road, Greenheights Village, Tavtay, Rizal 1901, Philippines / T: +63 2 6797266, Fax: +63 2 6588466, +63 2 6566801

sa Misc Inst
Miscellaneous Institutions, City various, South Africa
Information
Lists titles of unknown/unidentified microform publishers whose addresses have not been traced; also listed are those microform titles whose active status is difficult to establish e.g. if they are still purchasable

sa National
National Library of South Africa, Queen Victoria St 5, Cape Town 8001, POB 496, Cape Town 8000, South Africa / T: +27 21 4246320, Fax: +27 21 4233359, E-mail: Herschel.Miller@nlsa.ac.za; info@nlsa.ac.za, Internet: www.nlsa.ac.za; http://natlib1.sabinet.co.za/ – ISBN 978-0-86968
Orders
N America: ProQuest Information and Learning, 300 N Zeeb Rd, Ann Arbor, MI 48106-1346, POB 1346, Ann Arbor, MI 48106-1346, USA / T: +1 734 7614700, +1 800 5213042, Fax: +1 734 9739145, E-mail: info@il.proquest.com, Internet: www.umi.com; http://www.proquest.com/products_umi/catalogs/ – ISBN 978-0-8357
Information
Merged with South African Library to become National Library of South Africa

sa State Libr
National Library of South Africa – State Library, Staatsbiblioteek, Vermeulen St 239, Pretoria, POB 397, Pretoria 0001, South Africa / T: +27 12 3861661, Fax: +27 12 3255984, E-mail: infodesk@nlsa.ac.za; Herschel.Miller@nlsa.ac.za, Internet: www.nlsa.ac.za; http://natlib1.sabinet.co.za – ISBN 978-0-7989; 978-0-7961

sa Stellenbosch
University of Stellenbosch, Private Bag X1, Matieland 7602, South Africa / T: +27 21 808-9111, Internet: http://www.sun.ac.za/index.asp

sa U of Johannesburg
University of Johannesburg, Kingsway, Auckland Park 2092, POB 524, Auckland Park 2006, South Africa / T: +27 11 5592637, Fax: +27 11 5592191, E-mail: webmaster@uj.ac.za, Internet: http://www.uj.ac.za/

sa Unisa
University of South Africa, 340 Prellerstreet, Muckleneuk Hill, Pretoria, PO Box 392, Unisa, 0003, South Africa / T: +27 12 4292309, Fax: +27 12 4292323, E-mail: munnij@unisa.ac.za, Internet: www.unisa.ac.za – ISBN 978-0-86981; 978-1-86888
Information
Vista is a part of the publisher since january 2004. Unisa publications can be found on the webpage of the National Library of South Africa

si Info Pubs
Information Publications/Pte Ltd, 41 Kallang Pudding, Unit 04-03, Golden Wheel Bldg, Singapore 1334, Singapore / T: +65 67415166, Fax: +65 67429356

si Publishers
Publishers Marketing Services Pte Ltd, 10-C Jalan Ampas, #07-01 Ho Seng Lee Flatted Warehouse, Singapore 329513, Singapore / T: +65 62565166, Fax: +65 62530008, E-mail: raymondlim@pms.com.sg; info@pms.com.sg, Internet: www.pms.com.sg

sp Bibl Santa Ana
Biblioteca "IX Marqués de la Encomienda", Calle Ortega Munoz s/n, 32, 06200 Almendralejo (Badajoz), Spain / T: +34 924661689, +34 924661178, E-mail: colegio@csantana.com, Internet: www.csantana.com/biblioteca.htm
Information
Price per fiche: $2.00 Price per roll: $100.00

sp Boletín
Boletín Oficial del Estado, Av de Manoteras 54, 28050 Madrid, Spain / T: +34 913841624; +34 902365303, Fax: +34 913841555, Internet: www.boe.es

sp Chadwyck
Chadwyck-Healey España S.L., Juan Bravo 18, 2°C, Madrid, 28006, Spain / T: +34 915755597, Fax: +34 915759885, E-mail: mascorda@chadwyck.es

sp Cultura
Ministerio de Cultura, Servicio de Reproducción de Documentos de Archivos Estatales, Calle Serrano, 115, 3a planta, 28006 Madrid, Spain / T: +34 915628011, +34 915628458, Fax: +34 914116669, E-mail: cristina.uson@cult.mec.es, Internet: http://www.mcu.es/archivos/index.html

sw Gothenburg University
Gothenburg University Library, Göteborgs Universitetsbibliotek, Renströmsgt 4, 40530 Göteborg, Box 222, 40530 Göteborg, Sweden / T: +46 31 7731733, Fax: +46 31 163797, Internet: www.ub.gu.se – ISBN 978-91-7346; 978-91-85206

sw Kungliga
Kungl. biblioteket, National Library of Sweden, Humlegården, Stockholm, POB 5039, SE-10241 Stockholm, Sweden / T: +46 8 4634000, Fax: +46 8 4634004, E-mail: svedag@kb.se; kungl.biblioteket@kb.se, Internet: www.kb.se

sz UN Office
Sales Unit United Nations Office at Geneva, Palais des Nations, 1211 Genève 10, Switzerland / T: +41 22 917872, Fax: +41 22 9170027, E-mail: unpubli@unog.ch

uk Aberdeen UL
Aberdeen University Library, Meston Walk, Old Aberdeen AB24 3UE, United Kingdom / T: +44 1224 273600, Fax: +44 1224 273956, E-mail: library@abdn.ac.uk, Internet: http://www.abdn.ac.uk/library/index.php

uk Academic
Academic Microforms Ltd, Kirkhill House, Wick, Caithness, KW1 4DB, United Kingdom / T: +44 207 7353011, E-mail: mjgunn@academicmicro.com, Internet: www.academicmicroforms.com
Orders
ProQuest Information and Learning, 300 N Zeeb Rd, Ann Arbor, MI 48106-1346, POB 1346, Ann Arbor, MI 48106-1346, USA / T: +1 734 7614700, +1 800 5213042, Fax: +1 734 9739145, E-mail: info@il.proquest.com, Internet: www.umi.com; http://www.proquest.com/products_umi/catalogs/ – ISBN 978-0-8357
Information
Proquest licensed to act as distributor for certain titles. micropublished materials are offered for sale as 35mm silver duplicate microfilms and also as electronic media (cd-rom/dvd).

uk BedsFHS
Bedfordshire Family History Society, PO Box 214, Bedford MK42 9RX, United Kingdom / E-mail: bfhs@bfhs.org.uk; microfichesales@bfhs.org.uk, Internet: http://mysite.verizon.net/brianmpayne/publications.html

uk Bell
Bell & Howell Information and Learning, The Old Hospital, Ardingly Road, Cuckfield, West Sussex RH17 5JR, United Kingdom / T: +44 1444 445000, Fax: +44 1444 445050, E-mail: umi@umi.uk.com

uk BerksFHS
Berkshire Family History Society Research Centre, Yeomanry House, 131 Castle Hill, Reading RG1 7TJ, United Kingdom / T: + 44 118-950-9553, E-mail: ResearchCentre@BerksFHS.org.uk, Internet: http://www.berksfhs.org.uk/

uk Bowker
Bowker Ltd, Windsor Crt, East Grinstead House, East Grinstead, West Sussex RH19 1XA, United Kingdom / T: +44 1342 326972, +44 1342 336185, Fax: +44 1342 335612, E-mail: customer.services@bowker.co.uk, Internet: www.bowker.co.uk – ISBN 978-0-86291

uk British Libr
British Library National Bibliographic Service, Boston Spa, Wetherby, West Yorkshire LS23 7BQ, United Kingdom / T: +44 1937 546610, Fax: +44 1937 546586, E-mail: arthur.cunningham@bl.uk; bd-info@bl.uk, Internet: http://www.bl.uk/services/bibliographic/records.html – ISBN 978-0-7123; 978-0-900220
Orders
ProQuest Information and Learning, 300 N Zeeb Rd, Ann Arbor, MI 48106-1346, POB 1346, Ann Arbor, MI 48106-1346, USA / T: +1 734 7614700, +1 800 5213042, Fax: +1 734 9739145, E-mail: info@il.proquest.com, Internet: www.umi.com; http://www.proquest.com/products_umi/catalogs/ – ISBN 978-0-8357

uk British Libr Newspaper
British Library Newspaper Library, c/o Microfilm Sales, Colindale Av, London NW9 5HE, United Kingdom / T: +44 20 74127353, Fax: +44 20 74127379, E-mail: newspaper@bl.uk; reproductions@bl.uk, Internet: www.bl.uk/collections/micro.html; http://www.bl.uk/catalogues/listings.html
Orders
Mikrofilmarchiv der deutschsprachigen Presse eV, Königswall 18, 44122 Dortmund, Germany / T: +49 231 5023249, +49 231 5023216, +49 231 5026564 (Hr Pankratz), Fax: +49 231 5023218, E-mail: marlt@stadtdo.de, MFA@stadtdo.de, Internet: http://www.zeitungsforschung.de/

Information
Please note that not all the microfilm recorded may be available for sale from the British Library. Check with staff in the BL Reproductions to establish microfilm for purchase. Enquire for prices

uk British Library
The British Library, St Pancras, 96 Euston Rd, London NW1 2DB, United Kingdom / T: +44 1937 546060, E-mail: social-sciences@bl.uk; customer-services@bl.uk, Internet: http://www.bl.uk/contact/howto.html

uk CambsFHS
Cambridgeshire Family History Society, c/o Carol Noble, 22 St Margarets Rd, Girton, Cambridge CB3 0LT, United Kingdom / E-mail: bookstall@cfhs.org.uk, Internet: http://www.cfhs.org.uk/bookstall.html

uk Cardiganshire
Cardiganshire Family History Society, c/o Publications, PO Box 37, Aberystwyth SY23 2WL, United Kingdom / E-mail: sales@cgnfhs.org.uk, Internet: http://www.cgnfhs.org.uk/

uk Chadwyck
Chadwyck-Healey Ltd, A part of ProQuest Information and Learning, The Quorum, Barnwell Rd, Cambridge CB5 8SW, United Kingdom / T: +44 1223 215512, Fax: +44 1223 215513, E-mail: mail@chadwyck.co.uk; nic.sinclair@chadwyck.co.uk, Internet: www.proquest.co.uk – ISBN 978-0-85964
Orders
Australia & NZ: ProQuest Information and Learning, POB 181, Drummoyne NSW 1470, Australia / T: +61 2 99116660, Fax: +61 2 99116652, E-mail: julie.stevens@anz.proquest.com
Germany, Austria & Switzerland: ProQuest Information and Learning, Grüner Weg 8, 61169 Friedberg, Germany / T: +49 6031 87473, Fax: +49 6031 87469, E-mail: claudia.spengemann@proquest.co.uk
Korea, Taiwan & Philippines: ProQuest Information and Learning, Rm 1402, Doosan Bearstel 1319-11, Seocho 2-dong, Seocho-Gu, Seoul 137-072, Korea, Republic / T: +82 2 588 6045, Fax: +82 2 588 6046, E-mail: junghyun.kim@asia.proquest.com, eunkyung.kim@asia.proquest.com, Internet: www.proquest.jp
SE Asia & Asia Pacific: ProQuest Information and Learning, B909 (Block B), Phileo Damansara 1 No 9, Jalan 16/11, Jalan Damansara, 46350 Petaling Jaya, Selangor D.E, Malaysia / T: +60 3 79542880, Fax: +60 3 79583446, E-mail: kelvin.low@asia.proquest.com; richard.hollingsworth@asia.proquest.com
Spain: Chadwyck-Healey España S.L., Juan Bravo 18, 2°C, Madrid, 28006, Spain / T: +34 915755597, Fax: +34 915759885, E-mail: mascorda@chadwyck.es
USA: ProQuest Information and Learning, 300 N Zeeb Rd, Ann Arbor, MI 48106-1346, POB 1346, Ann Arbor, MI 48106-1346, USA / T: +1 734 7614700, +1 800 5213042, Fax: +1 734 9739145, E-mail: info@il.proquest.com, Internet: www.umi.com; http://www.proquest.com/products_umi/catalogs/ – ISBN 978-0-8357
Information
Became part of ProQuest Information and Learning in 1999. Inquire for prices

uk CheshireFHS
Family History Society of Cheshire (FHSC), c/o Sidney Barber, 9 Buckfast Close, Penketh, Warrington WA5 2NS, United Kingdom / E-mail: fiche.sales@fhsc.org.uk; info@fhsc.org.uk, Internet: http://www.fhsc.org.uk/

uk Eurospan
Eurospan Group, c/o c/o Turpin Distribution, 3 Henrietta St, Covent Garden, London WC2E 8LU, United Kingdom / T: +44 20 72400856, Fax: +44 20 73790609, E-mail: info@eurospangroup.com, Internet: http://www.eurospanonline.com/eurospan/index.asp

uk Glamorgan FHS
Glamorgan Family History Society Microfiche Publications, 'Millbrook', 2 Heol-y-Nant, Llandow, Cowbridge, Vale of Glamorgan CF71 7PE, United Kingdom / Internet: http://www.rootsweb.com/~wlsglfhs/fiche4.htm

uk Info Pubs
Information Publications International Ltd, White Swan House, Godstone, Surrey RH9 8LW, United Kingdom / T: +44 1883 744123, Fax: +44 1883 744024 – ISBN 978-0-902741

uk Lanarkshire
Lanarkshire Family History Society, Unit 26A, Motherwell Business Centre, Coursington Road, Motherwell, Lanarkshire ML1 1PW, United Kingdom / E-mail: society@lanarkshirefhs.org.uk, Internet: http://www.lanarkshirefhs.org.uk/

uk Matthew
Adam Matthew Publications, Pelham House, London Rd, Marlborough, Wilts SN8 2AA, United Kingdom / T: +44 1672 511921, Fax: +44 1672 511663, E-mail: info@ampltd.co.uk; kate@ampltd.co.uk; david@ampltd.co.uk, Internet: www.adam-matthew-publications.co.uk; www.ampltd.co.uk

uk Microform Academic
Microform Academic Publishers, A Division of Microform (Wakefield) Ltd, Main St, E Ardsley, Wakefield, Yorkshire WF3 2AP, United Kingdom / T: +44 1924 825700, Fax: +44 1924 871005, E-mail: mmortimer@microform.co.uk; MAP@microform.co.uk, Internet: www.microform.co.uk – ISBN 978-1-85117
Orders
Japan: Far Eastern Booksellers, 12 Kanda Jimbocho 2-chome, Chiyoda-ku, Tokyo, POB 72, Tokyo, Japan / T: +81 3 32657532, Fax: +81 3 32654656, E-mail: info@kyokuto-bk.co.jp, Internet: www.kyokuto-bk.co.jp
N-America: PraXess Associates, 71 Cassilis Ave, Bronxville, NY 10708, USA / T: +1 914 7937842, Fax: +1 914 7937842, E-mail: praxess@optonline.net, Internet: www.e-praxess.net
World, excl Japan and SE Asia: ProQuest Information and Learning, 300 N Zeeb Rd, Ann Arbor, MI 48106-1346, POB 1346, Ann Arbor, MI 48106-1346, USA / T: +1 734 7614700, +1 800 5213042, Fax: +1 734 9739145, E-mail: info@il.proquest.com, Internet: www.umi.com; http://www.proquest.com/products_umi/catalogs/ – ISBN 978-0-8357
Information
Microform Academic Publishers is a division of Microform Imaging Limited

uk Mindata
Mindata Ltd, Bathwick Hill, Bath BA2 6LA, United Kingdom / T: +44 1225 466447, Fax: +44 1225 482841, E-mail: fmoore@mindata.co.uk; info@mindata.co.uk, Internet: www.mindata.co.uk
Orders
IDC Publishers Inc, 3265 Johnson Av, Riverdale, NY 10463, USA / T: +1 718 4321400, +1 800 7577441, Fax: +1 718 4320020, E-mail: idc-us@mindspring.com
N America: ProQuest Information and Learning, 300 N Zeeb Rd, Ann Arbor, MI 48106-1346, POB 1346, Ann Arbor, MI 48106-1346, USA / T: +1 734 7614700, +1 800 5213042, Fax: +1 734 9739145, E-mail: info@il.proquest.com, Internet: www.umi.com; http://www.proquest.com/products_umi/catalogs/ – ISBN 978-0-8357
Information
Mindata Ltd acquired the titles and publishing assets of Ormonde Publishing Ltd in 1990

uk MLA
Museums Libraries Archives North West, c/o The Brew House, Wilderspool Park, Greenall's Ave, Warrington, Lancashire WA4 6HL, United Kingdom / T: +44 1925 625050, Fax: +44 1925 243453, E-mail: info@mlanorthwest.org.uk, Internet: http://www.mlanorthwest.org.uk/newsplan/

uk National
The National Archives, Ruskin Av, Kew, Richmond Surrey TW9 4DU, United Kingdom / T: +44 20 88763444, E-mail: http://www.nationalarchives.gov.uk/contact/; collectioncare@nationalarchives.gov.uk, Internet: www.nationalarchives.gov.uk; www.pro.gov.uk
Information
Offers a microfiche-on-demand service. The National Archives comprises the Public Record Office (PRO) and the Historical Manuscripts Commission (HMC)

uk NBN
NBN International, Estover Rd, Plymouth, Devon PL6 7PZ, United Kingdom / T: +44 1752 202301, Fax: +44 1752 202333, E-mail: orders@nbninternationl.com, Internet: www.nbninternational.com

uk Newsplan
NEWSPLAN 2000 Project, c/o British Library Newspaper Library, Colindale Avenue, London NW9 5HE, United Kingdom / T: +44 20 7412 7371, Fax: +44 20 7412 7373, Internet: http://www.bl.uk/collections/nplan.html
Information
NEWSPLAN is a co-operative programme between public libraries, British Library, National Library of Scotland, National Library of Wales, National Library of Ireland for the microfilming of local newspapers. It is based on 10 regions, some of wh provide local information about their respective NEWSPLAN projects via the Internet: East Midlands, Ireland, London & SE Region, NW Region, Northern Region, Scotland, South West, Wales/Cymru, West Midlands, Yorkshire & Humberside

uk NRAS
National Register of Archives for Scotland, c/o H.M. General Register House, 2 Princes St, Edinburgh EH1 3YY, United Kingdom / T: +44 131 535 1405, Fax: +44 131 535 1430, E-mail: enquiries@nas.gov.uk, Internet: http://www.nas.gov.uk/default.asp

uk Oxbow
Oxbow Books, 10 Hythe Bridge St, Oxford, OX1 2EW, United Kingdom / T: +44 1865 241249, Fax: +44 1865 794449, E-mail: oxbow@oxbowbooks.com, Internet: http://www.oxbowbooks.com/home.cfm/Location/Oxbow

uk Plymbridge
Plymbridge Distributors Ltd, c/o Plymbridge House, Estover Rd, Plymouth PL6 7PZ, United Kingdom / T: +44 1752 202301, Fax: +44 1752 202331, E-mail: orders@plymbridge.com

uk Scotland NatLib
National Library of Scotland, George IV Bridge, Edinburgh EH1 1EW, United Kingdom / T: +44 131 6233700, Fax: +44 131 6233701, E-mail: enquiries@nls.uk; r.jackson@nls.uk, Internet: http://www.nls.uk/catalogues/index.html – ISBN 978-0-902220
Information
Library offers digital copies of microfilms

uk Tay Valley
Tay Valley Family History Society, c/o The Research Centre, 179-181 Princes St, Dundee DD4 6DQ, United Kingdom / T: +44 1382 461845, Fax: +44 1382 461845, E-mail: tvfhs@tayvalleyfhs.org.uk, Internet: http://www.tayvalleyfhs.org.uk/

uk Thomson Gale/PSM
Thomson Gale/PSM, High Holborn House, 50/51 Bedford Row, London WC1R 4LR, United Kingdom / T: +44 20 70672663, Fax: +44 20 70672600, E-mail: sarah.brannan@thomsonlearning.co.uk, Internet: www.gale.com

uk Turnaround
Turnaround Publisher Services Ltd, Unit 3, Olympia Trading Estate, Coburg Rd, Wood Green, London N22 6TZ, United Kingdom / T: +44 20 88293000, Fax: +44 20 8815088, E-mail: orders@turnaround-uk.com, Internet: http://www.turnaround-psl.com/

uk Wales NatLib
The National Library of Wales, Llyfrgell Genedlaethol Cymru, Aberystwyth, Ceredigion, Wales SY23 3BU, United Kingdom / T: +44 1970 632800, Fax: +44 1970 615709, E-mail: cat@llgc.org.uk, Internet: http://www.llgc.org.uk/index.php?id=2

uk World
World Microfilms Publications Ltd, c/o Microworld House, POB 35488, St John's Wood, London NW8 6WD, United Kingdom / T: +44 207 5864499, Fax: +44 207 7221068, E-mail: microworld@ndirect.co.uk, Internet: http://www.microworld.uk.com/ – ISBN 978-1-85035; 978-1-86013

us ABHS
American Baptist Historical Society, POB 851, Valley Forge, PA 19482-0851, USA / T: +1 610 7682269; +1 610 7682374, Fax: +1 610 7682266, Internet: www.abc-usa.org/abhs/ – ISBN 978-0-910056
Orders
Scholarly Resources Inc, Thomson Gale, 104 Greenhill Ave, Wilmington, DE 19805-1897, USA / T: +1 302 6547713, +1 888 7727817, Fax: +1 302 6543871, E-mail: sales@scholarly.com, Internet: www.scholarly.com

us ACS
American Chemical Society (ACS), 1155 16th St NW, Washington, DC 20036, USA / T: +1 202 8724376, +1 800 2275558, Fax: +1 202 8726325, E-mail: m_neville@acs.org; help@acs.org, Internet: www.chemistry.org – ISBN 978-0-8412

us AGU
American Geophysical Union, 2000 Florida Ave NW, Washington, DC 20009-1277, USA / T: +1 202 4626900, +1 800 9662481, Fax: +1 202 3280566, E-mail: service@agu.org, Internet: www.agu.org – ISBN 978-0-87590

us AIA
American Institute of Architects, 1735 New York Av, NW Washington, DC 20006, USA / T: +1 800 AIA3837; +1 202 6267300, Fax: +1 800 6267547; +1 202 6267547, E-mail: infocentral@aia.org, Internet: www.aia.org – ISBN 978-1-57165; 978-1-879304; 978-0-913962
Orders
ProQuest Information and Learning, 300 N Zeeb Rd, Ann Arbor, MI 48106-1346, POB 1346, Ann Arbor, MI 48106-1346, USA / T: +1 734 7614700, +1 800 5213042, Fax: +1 734 9739145, E-mail: info@il.proquest.com, Internet: www.umi.com; http://www.proquest.com/products_umi/catalogs/ – ISBN 978-0-8357

us AIP
American Institute of Physics, Ste 1NO1, 2 Huntington Quadrangle, Melville, NY 11747-4502, USA / T: +1 516 5762444; +1 800 3446902, Fax: +1 516 3499704, E-mail: aipinfo@aip.org; dberger@aip.org, Internet: www.aip.org – ISBN 978-0-88318

us Aircraft Tech
Aircraft Technical Publishers, 101 South Hill Dr, Brisbane, CA 94005-1251, USA / T: +1 415 3309500, +1 800 2274610, Fax: +1 415 4681596, E-mail: info@atp.com, Internet: www.atp.com

us AJPC
American Jewish Periodical Center, Klau Library, c/o Hebrew Union College – Jewish Institute of Religion, 3101 Clifton Av, Cincinnati, OH 45220-2488, USA / T: +1 513 2211875, Fax: +1 513 2210321, Internet: http://www.huc.edu/libraries/; http://www.huc.edu/libraries/CN/

us Alper
Jerry Alper Inc, 271 Main St, Eastchester, NY 10707, POB 218, Eastchester, NY 10707, USA / T: +1 914 7932100, Fax: +1 914 7937811, E-mail: jalper@alperbooks.com

us ALPL
Abraham Lincoln Presidential Library, A Division of the Illinois Historic Preservation Agency, 112 N 6th St, Springfield, IL 62701-1507, USA / T: +1 217 7857942, +1 217 5247214, Fax: +1 217 7856250, E-mail: Cheryl_Schnirring@ihpa.state.il.us, Internet: www.state.il.us/hpa/lib – ISBN 978-0-912226
Information
Originally known as the Illinois State Historical Library.

us Amistad
Amistad Research Center, c/o Tilton Hall, Tulane University, 6823 St Charles Ave, New Orleans, LA 70118-5665, USA / T: +1 504 8655535, Fax: +1 504 8655580, E-mail: hdodson@tulane.edu, Internet: www.amistadresearchcenter.org
Orders
Scholarly Resources Inc, Thomson Gale, 104 Greenhill Ave, Wilmington, DE 19805-1897, USA / T: +1 302 6547713, +1 888 7727817, Fax: +1 302 6543871, E-mail: sales@scholarly.com, Internet: www.scholarly.com

us AMS Press
AMS Press, Inc, c/o Brooklyn Navy Yard, Bldg. 292, Ste 417, 63 Flushing Ave, Unit 221, Brooklyn, NY 11205, USA / Fax: +1 212 995-5413, E-mail: queries@amspressinc.com; editorial@amspressinc.com, Internet: www.amspressinc.com – ISBN 978-0-404
Information
Microform publishing program temporarily discontinued due to water damage. Please inquire as to the availability of microforms

us Archive
Archive Publishing, 1462 W 1970 N, Provo, UT 84604, USA / T: +1 801 8180881, E-mail: info@archivepublishing.com; query@archivepublishing.com, Internet: www.archivepublishing.com
Information
$5.00 per fiche. Customers with standing orders receive a 20% discount

us ATBI
Alvina Treut Burrows Institute, Inc, POB 49, Manhasset, NY 11030, USA / T: +1 516 8698457

us ATLA
American Theological Library Association, ATLA Preservation Programs, 250 S Wacker Dr, Ste 1600, Chicago, IL 60606-5889, USA / T: +1 312 4545100 (outside N America); 888 665-ATLA, Fax: +1 312 4545505, E-mail: atla@atla.com; rkracke@atla.com; sales@atla.com, Internet: www.atla.com – ISBN 978-0-8370; 978-0-7905; 978-0-524
Orders
microfilm: Scholarly Resources Inc, Thomson Gale, 104 Greenhill Ave, Wilmington, DE 19805-1897, USA / T: +1 302 6547713, +1 888 7727817, Fax: +1 302 6543871, E-mail: sales@scholarly.com, Internet: www.scholarly.com

us ATLA

Information
Inquire for prices. Offers on-demand filming service. Monographs on microfilm and microfiche are distributed by ATLA. SR (Scholarly Resources) distributes collections and serials on microfilm

us Balch
The Balch Institute for Ethnic Studies of the Historical Society of Pennsylvania, 1300 Locust St, Philadelphia, PA 19107, USA / T: +1 215 7326200, Fax: +1 215 7322680, E-mail: library@hsp.org, Internet: www.hsp.org – ISBN 978-0-944190
Orders
Scholarly Resources Inc, Thomson Gale, 104 Greenhill Ave, Wilmington, DE 19805-1897, USA / T: +1 302 6547713, +1 888 7727817, Fax: +1 302 6543871, E-mail: sales@scholarly.com, Internet: www.scholarly.com
Information
Merged with Historical Society of Pennsylvania 2002.

us Barker
Eugene C Barker Texas History Center, University of Texas, Sid Richardson Hall, Unit 2, Twenty-fifth and Red River, Austin, Texas 78712, USA / T: +1 512 4715961

us Bell
ProQuest Business Solutions Inc, 3900 Kinross Lakes Parkway, Richfield OH 44286, USA / T: +1 330 6591600, Fax: +1 330 6591601, E-mail: info@pbs.proquest.com, Internet: www.pbs.proquest.com
Information
Includes titles from Micro-Photo (Wooster, Ohio) and Micro-Photo (Cleveland, Ohio). Formerly: Bell & Howell Publications Systems

us Brill
Brill Academic Publishers Inc, 112 Water Street, Ste 400, Boston MA 02109, USA / T: +1 617 2632323, Fax: +1 617 2632324, E-mail: cs@brillusa.com

us Brook
Brookhaven Press, A Division of NMT Corporation, 2004 Kramer St, La Crosse, WI 54603, POB 2287, La Crosse, WI 54602-2287, USA / T: +1 608 7810850, Fax: +1 608 7813883, E-mail: brookhaven@normicro.com, carol.berteotti@nmt.com, Internet: www.brookhavenpress.com/micro.htm

us Brookings
Brookings Institution Press, 1775 Massachusetts Ave, NW, Washington, DC 20036, USA / T: +1 202 7976258, Fax: +1 202 7976004, E-mail: bibooks@brook.edu

us Buffalo
Buffalo and Erie County Historical Society, 25 Nottingham Court, Buffalo, NY 14216-3199, USA / T: +1 716 8739644 ext 306, Fax: +1 716 8738754, Internet: www.bechs.org

us Career
Career Guidance Foundation, College Catalog Library, 8090 Engineer Road, San Diego, CA 92111, USA / T: +1 858 5608051, 800 8542670, Fax: +1 858 2788960, E-mail: sales@cgf.org, Internet: www.collegesource.org – ISBN 978-0-89262

us Cassidy
Cassidy & Associates Inc, c/o Mr Thomas Cassidy, 288 Cliffside Dr, Torrington, CT 06790, USA / T: +1 860 4823030, Fax: +1 860 4827588, E-mail: chinacas@prodigy.net

us Chemical
Chemical Abstracts Service (CAS), A division of American Chemical Society, 2540 Olentangy River Rd, Columbus, OH 43210, POB 3012, Columbus, OH 43210-0012, USA / T: +1 614 4473600, 800 8486538, Fax: +1 614 4473713, E-mail: help@cas.org, Internet: www.cas.org

us Chicago U Pr
University of Chicago Press, 1427 East 60th St, Chicago, IL 60637, USA / T: +1 773 7027700, +1 773 7027748, 800 6212736, Fax: +1 773 7029756, E-mail: marketing@press.uchicago.edu, Internet: www.press.uchicago.edu – ISBN 978-0-226
Orders
Chicago Distribution Center, 11030 S Langley Ave, Chicago, IL 60628, USA / T: +1 773 5681550, +1 800 6212736, Fax: +1 800 6218476, +1 773 7027212, E-mail: custserv@press.uchicago.edu, Internet: www.chicagodistributioncenter.org
Europe: John Wiley & Sons Ltd, Distribution Centre, 1 Oldlands Way, Bognor Regis Sussex PO22 9SA, United Kingdom / T: +44 1243 843294, +44 1243 779777, Fax: +44 1243 843296, +44 1243 820250, E-mail: cs-books@wiley.co.uk, Internet: www.wileyeurope.com

us Chinese Res
Center for Chinese Research Materials, 10415 Willow Crest Ctr, Vienna VA 22182-1852, USA / T: +1 703 7152688, Fax: +1 703 7157913

us CIS
Congressional Information Service, Inc, 7500 Old Georgetown Rd, Ste 1300, Bethesda, MD 20814-3389, USA / T: +1 301 6541550, +1 800 6388380, Fax: +1 301 6573203, +1 301 6544033, E-mail: academicinfo@lexisnexis.com; academicinternational@lexisnexis.com, Internet: http://www.lexisnexis.com/academic/3cis/cisMnu.asp – ISBN 978-0-912380; 978-0-88692; 978-0-89093; 978-1-55655
Orders
Japan: Maruzen Company Ltd, Import and Export Dept, c/o Dai-3 Maruzen Bldg, 2-16-1 Nihonbashi, Chuo-ku, Tokyo 103, POB 5050, Tokyo International 100-3191, Japan / T: +81 3 32789223, Fax: +81 3 32742270, E-mail: kawamura@maruzen.co.jp, Internet: http://www.maruzen.co.jp/home-eng/ – ISBN 978-4-621; 978-4-89580; 978-4-8395

us Colorado Hist
Colorado Historical Society, 1300 Broadway, Denver, CO 80203-2137, USA / T: +1 303 8662305, E-mail: Rebecca.Lintz@chs.state.co.us, Publications@chs.state.co.us, Internet: www.coloradohistory.org – ISBN 978-0-942576

us Commission
General Commission on Archives & History, United Methodist Church, 36 Madison Ave, Madison, NJ 07940, POB 127, Madison, NJ 07940, USA / T: +1 973 4083189, Fax: +1 973 4083909, E-mail: research@gcah.org, Internet: www.gcah.org – ISBN 978-1-880927

us Cornell
Cornell University Library, Department Preservation & Collection Maintenance, Ithaca, NY 14853, USA / T: +1 607 2555291, Fax: +1 607 2547493, E-mail: beb1@cornell.edu, Internet: www.library.cornell.edu/preservation

us Crest
Crest Microfilm, Cedar Rapids, Iowa, USA / T: +1 800 3660077

us CRL
Center for Research Libraries, 6050 South Kenwood Ave, Chicago, IL 60637-2804, USA / T: +1 733 9554545, Fax: +1 733 9554339, E-mail: simon@crl.edu, Internet: http://www.crl.edu – ISBN 978-0-932486
Information
A consortium of North American universities, colleges & independent research libraries. Filming of rarely held research materials, and foreign & ethnic newspapers is done by special projects which are supported by subscribing institutions, administered & coordinated by CRL, and in this bibliography also listed under this umbrella institution

us CUP
CUP Services, 750 Cascadilla St, Ithaca, NY 14851, POB 6525, Ithaca, NY 14851, USA / T: +1 607 2772211, +1 800 6662211, Fax: +1 800 688 2877; +1 607 2776292, E-mail: orderbook@cupserv.org, Internet: http://www.cupserv.org/cps_customer.html

us Current
The Current Digest, 3857 N High St, Columbus, OH 43214, USA / T: +1 614 2924234, Fax: +1 614 2676310, E-mail: periodicals@eastview.com, Internet: www.currentdigest.org – ISBN 978-0-913601

us Dartmouth
Dartmouth College Library, Baker Library, Hanover, NH 03755-3525, USA / T: +1 603 6462235, Fax: +1 603 6463702, Internet: www.dartmouth.edu
Orders
ProQuest Information and Learning, 300 N Zeeb Rd, Ann Arbor, MI 48106-1346, POB 1346, Ann Arbor, MI 48106-1346, USA / T: +1 734 7614700, +1 800 5213042, Fax: +1 734 9739145, E-mail: info@il.proquest.com, Internet: www.umi.com; http://www.proquest.com/products_umi/catalogs/ – ISBN 978-0-8357

us David Brown
David Brown Book Co, POB 511, Oakville, CT 06779, USA / T: +1 860 9459329, Fax: +1 860 9459468, E-mail: david.brown.bk.co@snet.net, Internet: http://www.oxbowbooks.com/home.cfm/Location/DBBC

us East View
East View Publications, 3020 Harbor Lane N, Minneapolis, MN 55447-5137, USA / T: +1 763 5500961, +1 800 4771005, Fax: +1 763 5592931, E-mail: info@eastview.com; micro@eastview.com, Internet: www.eastview.com – ISBN 978-1-879944

us Eastman
Eastman Kodak Company, 343 State Street, Rochester, NY 14650-1206, USA / T: +1 716 7243041, +1 716 7244000, Fax: +1 716 7241985, Internet: www.kodak.com

us Ei
Engineering Information Inc, Ei, c/o Stevens Institute Campus, 1 Castle Point Terrace, Hoboken, NJ 07030-5996, USA / T: +1 201 2168500, +1 800 2211044 (Canada & USA), Fax: +1 201 2168532, E-mail: ei@ei.org; customer.support@ei.org, Internet: www.ei.org – ISBN 978-0-911820; 978-0-87394
Information
A subsidiary of Elsevier Science

us EMC
EMC International Inc, 3622 West Liberty, Ann Arbor, MI 48103, USA / T: +1 313 7696065, Fax: +1 313 7694880

us ETS
Educational Testing Service, Rosedale Rd, Princeton, NJ 08541-0001, USA / T: +1 609 9219000, Fax: +1 609 7345410, E-mail: pstanley@ets.org, Internet: www.ets.org – ISBN 978-0-88685

us Facts
Facts on File, Inc, A Subsidiary of Infobase Holdings, Inc, 132 W 31st St, 17th fl, New York, NY 10001, USA / T: +1 212 9678800, +1 800 3228755, Fax: +1 212 9679196, +1 800 6783633, E-mail: CustServ@factsonfile.com, Internet: www.factsonfile.com – ISBN 978-0-8160; 978-0-87196
Orders
Japan: UPS United Publishers Services Ltd, A member of Times Publishing Group, Singapore, 1-32-5, Higashi-Shinagawa, Shinagawa-ku, Tokyo 140-0002, Japan / T: +81 3 54797251, Fax: +81 3 54797307, E-mail: general@ups.co.jp, info@ups.co.jp, Internet: www.ups.co.jp
Philippines: I J Sagun Enterprises, 2 Topaz Road, Greenheights Village, Tavtay, Rizal 1901, Philippines / T: +63 2 6797266, Fax: +63 2 6588466, +63 2 6566801
PRC & HK: Ian Taylor Associates, Rm 1505 Zhejiang Bldg, Block 3, AnZhenXiLi, ChaoYang District, Beijing 100029, China, People's Republic / T: +86 10 64455987, E-mail: hrthome@gmail.com, Internet: www.iantaylorassociates.com
S'pore & M'sia: Publishers Marketing Services Pte Ltd, 10-C Jalan Ampas, #07-01 Ho Seng Lee Flatted Warehouse, Singapore 329513, Singapore / T: +65 62565166, Fax: +65 62530008, E-mail: raymondlim@pms.com.sg; info@pms.com.sg, Internet: www.pms.com.sg
UK, Ireland & Europe: Eurospan Group, c/o c/o Turpin Distribution, 3 Henrietta St, Covent Garden, London WC2E 8LU, United Kingdom / T: +44 20 72400856, Fax: +44 20 73790609, E-mail: info@eurospangroup.com, Internet: http://www.eurospanonline.com/eurospan/index.asp

us Fairchild Micro
Fairchild Microfilm (Publications), Inc, 7 W 34 St, New York, NY 10001, USA / T: +1 212 6303880, +1 800 9324724, Fax: +1 212 6303868, E-mail: olga.kontzias@fairchildpub.com, Internet: www.fairchildbooks.com – ISBN 978-0-87005

us FFIEC
Federal Financial Institutions Examination Council, Washington, DC, USA / Fax: 888 8820982, E-mail: ffiec-suggest@frb.gov, Internet: www.ffiec.gov

us Gov Printing Dist
United States Government Printing Office, c/o Superintendent of Documents, Washington, DC 20402-9325, USA / T: +1 202 7833238, Fax: +1 202 2750019

us GPO
United States Government Printing Office (USGPO), 732 N Capitol St NW, Washington, DC 20401, USA / T: +1 202 5121800, Fax: +1 202 5122250, E-mail: gpoinfo@gpo.gov, Internet: http://www.gpoaccess.gov/cgp/index.html – ISBN 978-0-16
Orders
United States Government Printing Office, c/o Superintendent of Documents, Washington, DC 20402-9325, USA / T: +1 202 7833238, Fax: +1 202 2750019
Information
A complete sales catalog is available in 48x microfiche

us Harvard
Harvard College Library, Preservation & Imaging Department, Harvard Yard, Widener Library, Cambridge, MA 02138, USA / T: +1 617 4953995, Fax: +1 617 4950403, E-mail: white6@fas.harvard.edu, Internet: http://preserve.harvard.edu/; http://hcl.harvard.edu/widener/services/access/photocopying.html
Information
Harvard University Library is a member of CRL (Center for Research Libraries)

us Harvard Law
Harvard Law School Library, Microform Proj, Rm 221, level 2 N, Langdell Hall, Cambridge, MA 02138, USA / T: +1 617 4954722, +1 617 4962127, E-mail: kauppi@law.harvard.edu, Internet: http://www.law.harvard.edu/library/collections/microforms/index.php

us Haworth
The Haworth Press, Inc, 10 Alice St, Binghamton, NY 13904-1580, USA / T: +1 607 7225857, +1 800 4296784, Fax: +1 607 7710012, +1 800 8950582, E-mail: getinfo@haworthpress.com, Internet: www.haworthpress.com
Orders
UK, ireland, France, Italy, Spain, greece, Belgium, Netherlands, Luxembourg, Portugal: Turnaround Publisher Services Ltd, Unit 3, Olympia Trading Estate, Coburg Rd, Wood Green, London N22 6TZ, United Kingdom / T: +44 20 88293000, Fax: +44 20 8815088, E-mail: orders@turnaround-uk.com, Internet: http://www.turnaround-psl.com/
Information
Multi-price scheme. Enquire

us Hein
William S Hein & Co Inc, Hein Bldg, 1285 Main St, Buffalo, NY 14209-1987, USA / T: +1 716 8822600, +1 800 8287571, Fax: +1 716 8838100, E-mail: mail@wshein.com; new_books@wshein.com; bjablonski@wshein.com, Internet: www.wshein.com – ISBN 978-0-8377; 978-0-87452; 978-0-89941; 978-0-930342; 978-1-57588
Information
Includes Rothman microform titles

us HRAF
Human Relations Area Files Press, Affiliate of Yale University, 755 Prospect St, New Haven, CT 06511-1225, USA / T: +1 203 7649401, Fax: +1 203 7649404, E-mail: hrafmem@minerva.cis.yale.edu – ISBN 978-0-87536

us IASWR
The Institute for Advanced Studies of World Religions (IASWR), 2020 Rt 301, Carmel, NY 10512, USA / T: +1 845 2251445, Fax: +1 845 2251485, E-mail: iaswr@aol.com, Internet: www.iaswr.org – ISBN 978-0-915078

us IDC
IDC Publishers Inc, 3265 Johnson Av, Riverdale, NY 10463, USA / T: +1 718 4321400, +1 800 7577441, Fax: +1 718 4320020, E-mail: idc-us@mindspring.com

us IHRC
Immigration History Research Center, University of Minnesota, c/o Elmer L Andersen Libray – Ste 311, 222-21st Ave South, Minneapolis, MN 55455, USA / T: +1 612 6256581, +1 612 6254800, Fax: +1 612 6260018, E-mail: myron001@tc.umn.edu, ihrc@umn.edu, Internet: www.ihrc.umn.edu
Information
$35 per reel; $30 per reel for 2r or more

us IHS
Indiana Historical Society, Preservation Imaging Department, 450 W Ohio St, Indianapolis, IN 46202-3299, USA / T: +1 317 2321882, +1 800 4471830, +1 317 2339557, Fax: +1 317 2333109, +1 317 2330857, E-mail: tmason@indianahistory.org; CBennett@indianahistory.org, Internet: www.indianahistory.org – ISBN 978-0-87195

us IL Archives
Illinois State Archives, Margaret Cross Norton Bldg, Capitol Complex, Springfield, IL 62756, USA / T: +1 217 7821083, +1 217 7824682, Fax: +1 217 5243930, E-mail: BMBailey@ILSOS.NET, Internet: http://www.cyberdriveillinois.com/departments/archives/archives.html
Orders
The Society of American Archivists, 600 South Federal St, Suite 504, Chicago, IL 60605, USA / T: +1 312 9220140, Fax: +1 312 3471452, Internet: www.archivists.org – ISBN 978-0-931828

us IMR
IMR Limited, 1591 S 19th St, Harrisburg, PA 17104, POB 1777, Harrisburg, PA 17105, USA / T: +1 717 9851000, +1 800 4462826, +1 717 9090126, Fax: +1 717 9095900, E-mail: dflinchbaugh@imrlimited.com, Internet: www.imrlimited.com

us Indiana Preservation
Indiana University Libraries, E. Lingle Craig Preservation Laboratory, 851 N Range Rd, Bloomington, IN 47408, USA / T: +1 812 8556281, Fax: +1 812 8564948, E-mail: hufford@indiana.edu, Internet: http://www.indiana.edu/~libpres/; http://www.iucat-alt.iu.edu/authenticate.cgi?status=start

us IRC
Effective Information Resource Corporation, 5132 Bolsa Ave 107, Huntington Beach, CA 92649, USA / T: +1 714 3734544, +1 800 8592800, Fax: +1 714 3730177, E-mail: info@scanorfilm.com, Internet: www.scanorfilm.com
Information
Formerly Vision Press Inc

us IRE
International Research & Evaluation, Information & Technology Transfer Database, 21098 IRE Control Center, Eagan, MN 55121-0098, USA / T: +1 612 8889635, Fax: +1 612 8889124 – ISBN 978-0-930318

us Kansas
Kansas State Historical Society, c/o Historic Preservation Office, 6425 SW 6th Av, Topeka, KS 66615-1099, USA / T: +1 785 2728681, Fax: +1 785 2728682, E-mail: jchinn@kshs.org, Internet: www.kshs.org – ISBN 978-0-87726

us Kinesology
Kinesology Publications, International Institute for Sport and Human Performance, 1243 University of Oregon, Eugene, OR 97403-1243, USA / T: +1 541 3464117, Fax: +1 541 3460935, E-mail: hheiny@uoregon.edu, Internet: kinpubs.uoregon.edu/
Information
Please apply to publisher for information on titles prior to 1991

us L of C Photodup
Library of Congress, Photoduplication Service, 101 Independence Av at First St SE, Washington, DC 20540-4574, USA / T: +1 202 7075000, +1 202 7075640, Fax: +1 202 7075844, +1 202 7071771, E-mail: dmcn@loc.gov, photoduplication@loc.gov, Internet: www.loc.gov/preserv/pds – ISBN 978-0-8444

us Law
The Lawbook Exchange, Ltd, 33 Terminal Ave, Clark, NJ 07066-1321, USA / T: +1 732 3821800, Fax: +1 732 3821887, E-mail: law@lawbookexchange.com, Internet: www.lawbookexchange.com – ISBN 978-0-9630106; 978-1-58477; 978-1-886363
Orders
William S Hein & Co Inc, Hein Bldg, 1285 Main St, Buffalo, NY 14209-1987, USA / T: +1 716 8822600, +1 800 8287571, Fax: +1 716 8838100, E-mail: mail@wshein.com; new_books@wshein.com; bjablonski@wshein.com, Internet: www.wshein.com – ISBN 978-0-8377; 978-0-87452; 978-0-89941; 978-0-930342; 978-1-57588
Information
Member: Antiquarian Booksellers Association of America and International League of Antiquarian Booksellers

us Library Micro
Library Microfilms, A division of BMI Imaging Systems, 749 W Stadium Lane, Sacramento, CA 95834, USA / T: +1 916 9246666, Fax: +1 916 9280277, E-mail: info@librarymicrofilms.com, Internet: www.librarymicrofilms.com
Orders
N America: ProQuest Information and Learning, 300 N Zeeb Rd, Ann Arbor, MI 48106-1346, POB 1346, Ann Arbor, MI 48106-1346, USA / T: +1 734 7614700, +1 800 5213042, Fax: +1 734 9739145, E-mail: info@il.proquest.com, Internet: www.umi.com; http://www.proquest.com/products_umi/catalogs/ – ISBN 978-0-8357

us Lippincott
Lippincott Williams & Wilkins, A Wolters Kluwer Company, 530 Walnut St, Philadelphia, PA 19106-3621, USA / T: +1 215 5218300, Fax: +1 215 5218902, E-mail: orders@lww.com, Internet: www.lww.com – ISBN 978-0-397

us LLMC
Law Library Microform Consortium, University of Hawaii – Windward Campus, Kaneohe, POB 1599, Kaneohe, HI 96744, USA / T: +1 808 2352200, +1 800 2354446, Fax: +1 808 2351755, E-mail: llmconsort@aol.com, Internet: www.llmc.com
Information
$4.50 per mf (1:42) $1.50 per mf (1:24)

us Lubrecht
Lubrecht & Cramer, Ltd, International Books by Mail, 78 Front St Ste 76, Port Jervis, NY 12771, POB 3110, Port Jervis, NY 12771, USA / T: +1 800 9209334; +1 845 8565990, E-mail: books@lubrechtcramer.com, Internet: www.lubrechtcramer.com

us MA Hist
Massachusetts Historical Society, 1154 Boylston St, Boston, MA 02215, USA / T: +1 617 6460502, Fax: +1 617 8590074, E-mail: bjohnson@masshist.org; blawson@masshist.org, Internet: http://www.masshist.org/in_print/microfilm.cfm – ISBN 978-0-934909
Orders
ProQuest Information and Learning, 300 N Zeeb Rd, Ann Arbor, MI 48106-1346, POB 1346, Ann Arbor, MI 48106-1346, USA / T: +1 734 7614700, +1 800 5213042, Fax: +1 734 9739145, E-mail: info@il.proquest.com, Internet: www.umi.com; http://www.proquest.com/products_umi/catalogs/ – ISBN 978-0-8357

us MD Archives
Maryland State Archives, Hall of Records, 350 Rowe Blvd, Annapolis, MD 21401, USA / T: +1 410 9743915, +1 410 2606402, Fax: +1 410 9743895; +1 410 9742525, E-mail: archives@mdsa.net, Internet: www.msa.md.gov

us MEDOC
Middle East Documentation Center (MEDOC), The University of Chicago, 5828 S University Ave, 201 Pick Hall, Chicago, IL 60637, USA / T: +1 773 7028425, Fax: +1 773 7530569, E-mail: mideast-library@uchicago.edu, msaleh@midway.uchicago.edu, Internet: www.lib.uchicago.edu/e/su/mideast/CatIntro.html
Information
Checks are payable in US$ to the University of Chicago. The date of publication refers to the Hijri calendar unless: in square brackets (Gregorian), followed by an M (Mali), by a K (Kameriye) or by an S (Semsi calendar)

us MicroColour
MicroColour Inc, POB 243, Ridgewood, NJ 07451, USA / T: +1 201 4453450, Fax: +1 201 4452924, E-mail: arah@aol.com, Internet: www.microcolour.com
Information
Issue numbers are sold in sets of 5 colour mf (with some exceptions) at $35.95

us Microfilm
Microfilm Service Bureau, 124 West 25th Street, Kearney, Nebraska, USA

us Microfilm Corp
Microfilm Corporation of Pennsylvania, 2013 Noble St, Pittsburgh, PA 15218, USA / T: +1 412 3519380

us Minn Hist
Minnesota Historical Society Press, Division of Library and Archives, 345 Kellogg Blvd W, St Paul, MN 55102-1906, USA / T: +1 651 2962264, +1 800 6477827, Fax: +1 651 2971345, E-mail: sally.rubinstein@mnhs.org; alicia.cordes@mnhs.org, Internet: www.mnhs.org/market/mhspress – ISBN 978-0-87351
Orders
Chicago Distribution Center, 11030 S Langley Ave, Chicago, IL 60628, USA / T: +1 773 5681550, +1 800 6212736, Fax: +1 800 6218476, +1 773 7027212, E-mail: custserv@press.uchicago.edu, Internet: www.chicagodistributioncenter.org

us Misc Inst
Miscellaneous Institutions, City various, USA
Information
Lists titles of unknown/unidentified microform publishers whose addresses have not been traced; also listed are those microform titles whose active status is difficult to establish e.g. if they are still purchasable

us Mitchell Int
Mitchell International, 9889 Willow Creek Rd, San Diego, POB 26260, San Diego, CA 92196-0260, USA / T: +1 619 5786550, +1 800 8547030, Internet: www.mitchell.com
Information
A part of Hellmann & Friedman LLC

us Moody
Moody's Investors Service, Inc, 99 Church St, New York, NY 10007, USA / T: +1 212 5530300, Internet: www.moodys.com

us MRTS
MRTS, Medieval & Renaissance Texts & Studies, State Univ of NY at Binghamton, Binghamton, NY 13902-6000, USA / T: +1 607 7776758, +1 800 6662211, Fax: +1 607 7772408; +1 800 6882877, E-mail: mrts@asu.edu; roy.rukkila@asu.edu; orderbook@cupserv.org, Internet: http://www.asu.edu/clas/acmrs/publications/mrts/aboutmrts.html – ISBN 978-0-86698
Orders
Europe: NBN International, Estover Rd, Plymouth, Devon PL6 7PZ, United Kingdom / T: +44 1752 202301, Fax: +44 1752 202333, E-mail: orders@nbninternationl.com, Internet: www.nbninternational.com
N America: CUP Services, 750 Cascadilla St, Ithaca, NY 14851, POB 6525, Ithaca, NY 14851, USA / T: +1 607 2772211, +1 800 6662211, Fax: +1 800 688 2877; +1 607 2776292, E-mail: orderbook@cupserv.org, Internet: http://www.cupserv.org/cps_customer.html
Information
Each issue [or vol] contains 8-10 mss, with an average total of 900-1000 folios [1800-2000p] on 40-50 fiches. Each issue incl a full scholarly guide to the mss

us Nat Archives
National Archives Trust Fund, National Archives and Records Administration, 8601 Adelphi Rd, College Park, MD 20740-6001, USA / T: +1 301 7136800, +1 866 2726272, Fax: +1 202 5016175, E-mail: inquire@nara.gov, Internet: www.archives.gov – ISBN 978-0-911333
Orders
Scholarly Resources Inc, Thomson Gale, 104 Greenhill Ave, Wilmington, DE 19805-1897, USA / T: +1 302 6547713, +1 888 7727817, Fax: +1 302 6543871, E-mail: sales@scholarly.com, Internet: www.scholarly.com
Information
$34.00 per roll for U.S. customers; outside U.S. $39.00 per reel (includes shipping)

us National Clearing
National Clearinghouse on Marital and Date Rape / Women's History Library, Women's History Research Center, 2325 Oak St, Berkeley, CA 94708-1697, USA / T: +1 510 5241582, Internet: www.ncmdr.org – ISBN 978-0-912374
Information
Titles distributed by Scholarly Resources

us National Women's
National Women's History Project, 3343 Industrial Dr, Ste 4, Santa Rosa, CA 95403, USA / T: +1 707 6362888, Fax: +1 707 6362909, E-mail: nwhp@aol.com, Internet: www.nwhp.org – ISBN 978-0-938625
Orders
Scholarly Resources Inc, Thomson Gale, 104 Greenhill Ave, Wilmington, DE 19805-1897, USA / T: +1 302 6547713, +1 888 7727817, Fax: +1 302 6543871, E-mail: sales@scholarly.com, Internet: www.scholarly.com

us NE Hist
Nebraska State Historical Society, 1500 R St, Lincoln, NE 68501, POB 82554, Lincoln, NE 68501-2554, USA / T: +1 402 4714751, Fax: +1 402 4718922, E-mail: lanshs@nebraskahistory.org, Internet: www.nebraskahistory.org

us North Dakota
State Archives and Historical Research Library, State Historical Society of North Dakota, 612 E Boulevard Ave, Bismarck, ND 58505-0830, USA / T: +1 701 3282091, Fax: +1 701 3282650, E-mail: archives@state.nd.us, wbailey@state.nd.us, Internet: www.state.nd.us/hist/sal.htm

us Northeast
Northeast Document Conservation Center, 100 Brickstone Sq, Andover, MA 01810-1494, USA / T: +1 978 4701010, Fax: +1 978 4756021, E-mail: nedcc@nedcc.org; vellis@nedcc.org, Internet: www.nedcc.org – ISBN 978-0-9634685

us Notre Dame
University Libraries of Notre Dame, 221 Hesburgh Library, Notre Dame, IN 46556, USA / T: +1 574 6316258, Fax: +1 574 6316772, E-mail: archives.1@nd.edu, Internet: www.library.nd.edu

us NTIS
National Technical Information Service (NTIS), c/o US Department of Commerce, 5285 Port Royal Rd, Springfield, VA 22161, USA / T: +1 703 4874650, +1 703 6056585, E-mail: info@ntis.gov, Internet: www.ntis.gov

us NY Public
The New York Public Library, Rare Books and Manuscripts Div, Center for the Humanities, 5th Av & 42nd St, Rm 315 R, New York, NY 10018-2788, USA / T: +1 212 9300814, Fax: +1 212 2213423, E-mail: copyservices@nypl.org, Internet: www.nypl.org – ISBN 978-0-87104
Information
The New York Public Library is a member of CRL (Center for Research Libraries)

us Ohio Hist
Ohio Historical Society, c/o Ohio Historical Ctr, Microfilm Dept, 1982 Velma Av, Columbus, OH 43211-2497, USA / T: +1 614 2972364, Fax: +1 614 2972367, E-mail: rtooker@ohiohistory.org, Internet: www.ohiohistory.org – ISBN 978-0-87758
Information
Newspaper listings $31.00 per roll. Govt records unlisted at publisher's request. Enquire there for details

us OIEAHC
Omohundro Institute of Early American History and Culture, c/o Swem Library, 1 Landrum Drive, Williamsburg, VA 23185, POB 8781, Williamsburg, VA 23187-8781, USA / T: +1 757 2211110, +1 757 2211114, Fax: +1 757 2211047, E-mail: pvhigg@wm.edu, fjteut@wm.edu, Internet: www.wm.edu/oieahc/ – ISBN 978-0-910776

us OmniSys
OmniSys Corporation, see us UMI Proquest, USA – ISBN 978-1-884719

us Oregon Hist
Oregon Historical Society Library, 1200 SW Park Av, Portland, OR 97205, USA / T: +1 503 2221741, Fax: +1 503 2212035, E-mail: orhist@ohs.org, Internet: www.ohs.org
Information
$55.00 per reel

us Oregon Lib
University of Oregon Library System, c/o Oregon Newspaper Project, 1299 University of Oregon, Eugene, OR 97403-1299, USA / T: +1 541 3460708, +1 541 3461864, Fax: +1 541 3463485, E-mail: onp@darkwing.uoregon.edu; nhelmer@darkwing.uoregon.edu, Internet: www.libweb.uoregon.edu
Information
All microfilm is 35mm; price is $58.00 per reel

us Oriental
The Oriental Institute, University of Chicago, 1155 East 58th St, Chicago, IL 60637, USA / T: +1 773 7029522, Fax: +1 773 7029853, E-mail: oi-publications@uchicago.edu, Internet: http://oi.uchicago.edu/contact/ – ISBN 978-0-918986; 978-1-885923
Orders
David Brown Book Co, POB 511, Oakville, CT 06779, USA / T: +1 860 9459329, Fax: +1 860 9459468, E-mail: david.brown.bk.co@snet.net, Internet: http://www.oxbowbooks.com/home.cfm/Location/DBBC
UK: Oxbow Books, 10 Hythe Bridge St, Oxford, OX1 2EW, United Kingdom / T: +44 1865 241249, Fax: +44 1865 794449, E-mail: oxbow@oxbowbooks.com, Internet: http://www.oxbowbooks.com/home.cfm/Location/Oxbow

us Penn Academy
Pennsylvania Academy of the Fine Arts, 118 N Broad St, Philadelphia, PA 19102, USA / T: +1 215 9727642, Fax: +1 215 9725564, E-mail: archives@pafa.org, Internet: www.pafa.org – ISBN 978-0-943836

us Penn Hist
Pennsylvania Historical and Museum Commission, Division of Publications and Sales, Third & North Sts, Harrisburg, PA 17108, POB 11466, Harrisburg, PA 17108-1466, USA / T: +1 717 783-2618, +1 800 7477790, Fax: +1 717 7878312, E-mail: RA-PHMC-Webmaster@state.pa.us, Internet: www.phmc.state.pa.us – ISBN 978-0-911124; 978-0-89271
Orders
Scholarly Resources Inc, Thomson Gale, 104 Greenhill Ave, Wilmington, DE 19805-1897, USA / T: +1 302 6547713, +1 888 7727817, Fax: +1 302 6543871, E-mail: sales@scholarly.com, Internet: www.scholarly.com

us Perceptual
Perceptual and Motor Skills, POB 9229, Missoula, MT 59807, USA / T: +1 406 7281710

us Praxess
PraXess Associates, 71 Cassilis Ave, Bronxville, NY 10708, USA / T: +1 914 7937842, Fax: +1 914 7937842, E-mail: praxess@optonline.net, Internet: www.e-praxess.net

us Presbyterian
Presbyterian Historical Society, 425 Lombard St, Philadelphia, PA 19147-1516, USA / T: +1 215 9283891, Fax: +1 215 6270509, E-mail: ntaylor@history.pcusa.org, Internet: www.history.pcusa.org – ISBN 978-0-912686
Orders
Scholarly Resources Inc, Thomson Gale, 104 Greenhill Ave, Wilmington, DE 19805-1897, USA / T: +1 302 6547713, +1 888 7727817, Fax: +1 302 6543871, E-mail: sales@scholarly.com, Internet: www.scholarly.com

us Preston Publ
Preston Publications, Inc, A Division of Preston Industries, Inc, 7800 Merrimac Av, Niles, POB 48312, Niles, IL 60714-0312, USA / T: +1 708 9650566 – ISBN 978-0-912474

us Primary
Primary Source Microfilm, 12 Lunar Dr, Woodbridge, CT 06525-2398, USA / T: +1 203 3972600, +1 800 4440799, Fax: +1 203 3978296, E-mail: Michelle.Strauch@thomson.com; dan.haverkamp@thomson.com, Internet: www.galegroup.com/psm/ – ISBN 978-0-89235

Orders
Japan: Yushodo Co Ltd, 29 San-ei-cho, Shinjuku-ku, Tokyo 160-0008, Japan / T: +81 3 33571411, Fax: +81 3 33515855, E-mail: ysdhp@yushodo.co.jp, Internet: www.yushodo.co.jp – ISBN 978-4-8419
UK and Rest of Europe: Thomson Gale/PSM, High Holborn House, 50/51 Bedford Row, London WC1R 4LR, United Kingdom / T: +44 20 70672663, Fax: +44 20 70672600, E-mail: sarah.brannan@thomsonlearning.co.uk, Internet: www.gale.com
US & Canada: Thomson Gale World Headquarters, 12 Lunar Drive, Woodbridge CT 06525-2398, USA / T: +1 800 4440799, Fax: +1 203 3973893, E-mail: sales@gale.com; fenn.quigley@gale.com, Internet: www.gale.com
Information
An imprint of Thomson Gale; contact publisher for pricing

us Princeton U Pr
Princeton University Press, 41 William St, Princeton, NJ 08540-5237, USA / T: +1 609 2584900, Fax: +1 609 2586305, E-mail: webmaster@pupress.princeton.edu; orders@cpfs.pupress.princeton.edu; whl@pupress.princeton.edu, Internet: www.pup.princeton.edu – ISBN 978-0-691
Orders
UK: University Presses of California, Columbia, and Princeton Ltd, 1 Oldlands Way, Bognor Regis, West Sussex PO22 9SA, United Kingdom / T: +44 1243 842165, Fax: +44 1243 842167, E-mail: Lois@upccp.demon.co.uk

us Ross
Norman Ross Publishing Inc, see us UMI Proquest, USA – ISBN 978-0-8287; 978-0-88354

us Roth
Roth Publishing Inc, 27500 Drake Rd, Farmington Hills, MI 48331-3535, POB 220406, Great Neck, NY 11022, USA / T: +1 516 4663675, +1 800 899ROTH, Fax: +1 516 8297746, E-mail: comments@rothpoem.com, orders@rothpoem.com – ISBN 978-0-8486; 978-0-89609

us SAA
The Society of American Archivists, 600 South Federal St, Suite 504, Chicago, IL 60605, USA / T: +1 312 9220140, Fax: +1 312 3471452, Internet: www.archivists.org – ISBN 978-0-931828

us SAE
Society of Automotive Engineers, Inc, SAE, 400 Commonwealth Dr, Warrendale, PA 15096-0001, USA / T: +1 412 7764841 – ISBN 978-1-56091

us Scholarly Res
Scholarly Resources Inc, Thomson Gale, 104 Greenhill Ave, Wilmington, DE 19805-1897, USA / T: +1 302 6547713, +1 888 7727817, Fax: +1 302 6543871, E-mail: sales@scholarly.com, Internet: www.scholarly.com
Orders
Japan: Far Eastern Booksellers, 12 Kanda Jimbocho 2-chome, Chiyoda-ku, Tokyo, POB 72, Tokyo, Japan / T: +81 3 32657532, Fax: +81 3 32654656, E-mail: info@kyokuto-bk.co.jp, Internet: www.kyokuto-bk.co.jp
Information
Scholarly Resources acquired the Michael Glazier microfilm collection in 1991. Acquired by Gale Thomson, 2004

us Scholars Facs
Scholars' Facsimiles & Reprints, Subs of Academic Resources Corp, 410 Lenawee Dr., Ann Arbor MI 48104, USA / T: +1 734 7410344, Fax: +1 734 7410344, E-mail: maxinmin@umich.edu, Internet: http://www.scholarsbooklist.com/Home_Page.html – ISBN 978-0-8201

us SD Archives
South Dakota State Archives, Subsidiary of South Dakota State Historical Society, 900 Governors Drive, Pierre, SD 57501-2217, USA / T: +1 605 7733804, Fax: +1 605 7736041

us SIAM
Society of Industrial and Applied Mathematics (SIAM), 3600 University City Science Center, Philadelphia, PA 19104-2688, USA / T: +1 215 3829800, Fax: +1 215 3867999, E-mail: siam@siam.org; grossman@siam.org, Internet: www.siam.org – ISBN 978-0-89871

us Sibley
Sibley Music Library Microform Service, c/o Eastman School of Music, 27 Gibbs St, Rochester, NY 14604-2596, USA / T: +1 716 2741305, E-mail: jmahoney@esm.rochester.edu, Internet: http://www.esm.rochester.edu/sibley/?page=illiad

us SME
Society of Manufacturing Engineers (SME), One SME Drive, Dearborn, MI 48121, POB 930, Dearborn, MI 48121-0930, USA / T: +1 313 2711500, Fax: +1 313 2712861, E-mail: kingbob@sme.org, Internet: www.sme.org – ISBN 978-0-87263

us South C Archives
South Carolina Department of Archives and History Center, 8301 Parklane Rd, Columbia, SC 29223, USA / T: +1 803 8966100, Fax: +1 803 8966198, Internet: www.state.sc.us/scdah/
Orders
Scholarly Resources Inc, Thomson Gale, 104 Greenhill Ave, Wilmington, DE 19805-1897, USA / T: +1 302 6547713, +1 888 7727817, Fax: +1 302 6543871, E-mail: sales@scholarly.com, Internet: www.scholarly.com

us South Carolina Historical
South Carolina Historical Society, Fireproof Bldg, 100 Meeting St, Charleston, SC 29401-2299, USA / T: +1 843 7233225, Fax: +1 843 7238584, E-mail: info@schistory.org, Internet: www.schistory.org
Information
Microform titles which require special permission or have restricted access are not listed

us South Caroliniana
South Caroliniana Library, University of South Carolina, 910 Sumter st, Columbia, SC 29208, USA / T: +1 803 7775183, Fax: +1 803 7775747, E-mail: fulmerh@gwm.sc.edu, Internet: www.sc.edu/library/socar/index.html

us Southern Baptist
Southern Baptist Historical Library and Archives, 901 Commerce St, Suite 400, Nashville, TN 37203-3630, USA / T: +1 615 2440344, Fax: +1 615 7824821, E-mail: bsumners@edge.net – ISBN 978-0-939804

us Stanford
Stanford University Libraries, Cecil H Green Libr, Stanford, CA 94305-6004, USA / T: +1 650 7239108, Fax: +1 650 7256874, Internet: www-sul.stanford.edu

us Striker
Paul S Striker, Aesthetic Plastic Surgery, 660 Park Av, New York, NY 10021, USA / T: +1 212 7444265, Fax: +1 212 8615800, E-mail: pss@ix.netcom.com

us Thomson Gale
Thomson Gale World Headquarters, 12 Lunar Drive, Woodbridge CT 06525-2398, USA / T: +1 800 4440799, Fax: +1 203 3973893, E-mail: sales@gale.com; fenn.quigley@gale.com, Internet: www.gale.com

us Thomson Gale
Thomson Gale World Headquarters, 27500 Drake Rd, Farmington Hills, MI 48331-3535, POB 9187, Farmington Hills, MI 48333-9187, USA / T: +1 248 6994253, Fax: +1 248 6998064, E-mail: gale.customerservice@thomson.com, Internet: www.gale.com – ISBN 978-0-8103; 978-0-7876
Orders
Asia: Thomson Asia Pte Ltd, Branch Office, No 3, Jalan PJS 7/19, Bandar Sunway, 46150 Petaling Jaya, Selangor Darul Ehasan, Malaysia / T: +60 3 56368351/8352, Fax: +60 3 56368302, E-mail: simon.tay@thomsonlearning.com.sg, Internet: www.gale.com/world/
Australia & NZ: Katri Russick, Rue la Chaux, Mahebourg, Mauritius / T: +230 6311512; +230 2533527 (mobile), E-mail: mrussick@intnet.mu
Europe & Middle East: Thomson Gale/PSM, High Holborn House, 50/51 Bedford Row, London WC1R 4LR, United Kingdom / T: +44 20 70672663, Fax: +44 20 70672600, E-mail: sarah.brannan@thomsonlearning.co.uk, Internet: www.gale.com
Latin America: Thomson Learning Iberoamerica, Seneca No 53, Colonia Polanco, Mexico, D F C.P. 11560, Mexico / T: +52 55 15006000, Fax: +52 55 52812656, E-mail: rosa.viveros@thomson.com
Information
St James Press, The Taft Group, U X L, Visible Ink Press, Primary Source Microfilm, Graham & Whiteside, Macmillan Reference USA, Oceano Grupo Editorial, Primary Source Media etc. are imprints of the publisher.

us Trans-Media
Trans-Media / The Oceana Group, 40 Cedar St, Dobbs Ferry, NY 10522, USA / T: +1 914 6931100, Fax: +1 914 6930402, E-mail: marketing@oceanalaw.com, Internet: www.oceanalaw.com – ISBN 978-0-913338; 978-0-379

us TX Culture
The University of Texas, Institute of Texan Cultures at San Antonio, 801 South Bowie St, San Antonio, TX 78205-3296, USA / T: +1 210 4582234, +1 800 7767651, Fax: +1 210 4582205, E-mail: JFavor@utsa.edu, Internet: www.lib.utsa.edu

us UF Libraries
University of Florida Libraries, Preservation Department, POB 117007, Gainesville FL 32611-7007, USA / T: +1 352 3926962, Fax: +1 352 3926597, E-mail: cathy@mail.uflib.ufl.edu; neldas@mail.uflib.ufl.edu, Internet: www.uflib.ufl.edu
Information
All microfilm is 35mm: price is $50 per reel

us UMI ProQuest
ProQuest Information and Learning, 300 N Zeeb Rd, Ann Arbor, MI 48106-1346, POB 1346, Ann Arbor, MI 48106-1346, USA / T: +1 734 7614700, +1 800 5213042, Fax: +1 734 9739145, E-mail: info@il.proquest.com, Internet: www.umi.com; http://www.proquest.com/products_umi/catalogs/ – ISBN 978-0-8357
Orders
Europe, Africa, the Mid East & Australasia: Information Publications International Ltd, White Swan House, Godstone, Surrey RH9 8LW, United Kingdom / T: +44 1883 744123, Fax: +44 1883 744024 – ISBN 978-0-902741
Germany: Mikropress GmbH, Ollenhauerstr 1, 53113 Bonn, Germany / T: +49 228 623261, Fax: +49 228 628868, E-mail: info@mikropress.de, Internet: www.mikropress.de
Latin America, Caribbean & Bermuda: EMC International Inc, 3622 West Liberty, Ann Arbor, MI 48103, USA / T: +1 313 7696065, Fax: +1 313 7694880
SE Asia & Far East: Information Publications/Pte Ltd, 41 Kallang Pudding, Unit 04-03, Golden Wheel Bldg, Singapore 1334, Singapore / T: +65 67415166, Fax: +65 67429356
UK: Bell & Howell Information and Learning, The Old Hospital, Ardingly Road, Cuckfield, West Sussex RH17 5JR, United Kingdom / T: +44 1444 445000, Fax: +44 1444 445050, E-mail: umi@umi.uk.com
Information
Inquire for prices; National Archive Publishing Company (NAPC) is the exclusive provider of UMI© Periodicals in Microform

us UN Publ
United Nations Publications, 2 United Nations Plaza, DC2 853, New York, NY 10017, USA / T: +1 212 9638302, +1 800 2539646, Fax: +1 212 3719454, E-mail: publications@un.org, Internet: http://www.un.org/english/ – ISBN 978-92-1
Orders
Europe, Africa & Middle E: Sales Unit United Nations Office at Geneva, Palais des Nations, 1211 Genève 10, Switzerland / T: +41 22 917872, Fax: +41 22 9170027, E-mail: unpubli@unog.ch

us Univ Ill Libr
University of Illinois at Urbana-Champaign, Library Science and Information Science, 1325 S Oak St, Champaign, IL 61820, USA / T: +1 217 3330950, Fax: +1 217 2448082, E-mail: uipress@uiuc.edu, Internet: www.library.uiuc.edu – ISBN 978-0-252
Information
Orders to: P.O. Box 4856, Baltimore, MD 21211, USA

us Univ Music
University Music Editions, POB 192, Fort George Station, New York, NY 10040, USA / T: +1 212 5695393, 5695340, +1 800 4482805, Fax: +1 212 5691269, E-mail: ume@universitymusicedition.com, Internet: www.universitymusicedition.com
Orders
N America: ProQuest Information and Learning, 300 N Zeeb Rd, Ann Arbor, MI 48106-1346, POB 1346, Ann Arbor, MI 48106-1346, USA / T: +1 734 7614700, +1 800 5213042, Fax: +1 734 9739145, E-mail: info@il.proquest.com, Internet: www.umi.com; http://www.proquest.com/products_umi/catalogs/ – ISBN 978-0-8357

us Univelt
Univelt Inc, POB 28130, San Diego, CA 92198-0130, USA / T: +1 760 7464005, Fax: +1 760 7463139, E-mail: ROBERTHJACOBS@compuserve.com, sales@univelt.com, Internet: http://www.univelt.com/home.html – ISBN 978-0-912183; 978-0-87703

us UPA
University Publications of America UPA, Imprint of LexisNexis Academic & Library Solutions (CIS), 7500 Old Georgetown Rd, Ste 1300, Bethesda, MD 20814-3389, USA / T: +1 301 6573200, +1 800 6926300, Fax: +1 301 6573203, Internet: http://www.lexisnexis.com/academic/upa_cis/default.asp?t=82
Information
Greenwood Press's existing microform collections were transferred to University Publications of America, which now owns and distributes them

us US Gen Account
The United States General Accounting Office, 441 G St, NW, Washington, DC 20548, USA / T: +1 202 5124800, E-mail: webmaster@gao.gov, Internet: www.gao.gov

us UW Libraries
University of Washington Libraries, c/o MSCUA, Allen Library, Box 352900, Seattle, WA 98195-2900, USA / T: +1 206 5431929, Fax: +1 206 6858049, Internet: http://catalog.lib.washington.edu

us Virginia U Pr
The University of Virginia Press, POB 400318 University Station, Charlottesville, VA 22904-4318, USA / T: +1 800 8313406; +1 434 9246070, Fax: +1 877 2886400, E-mail: njm8j@virginia.edu, Internet: www.upress.virginia.edu/ – ISBN 978-0-8139
Orders
UK, Europe & Middle E: Eurospan Group, c/o c/o Turpin Distribution, 3 Henrietta St, Covent Garden, London WC2E 8LU, United Kingdom / T: +44 20 72400856, Fax: +44 20 73790609, E-mail: info@eurospangroup.com, Internet: http://www.eurospanonline.com/eurospan/index.asp

us Western Res
The Western Reserve Historical Society, Library, 10825 East Blvd, University Circle, Cleveland, OH 44106, USA / T: +1 216 7215722, Fax: +1 216 7215702, E-mail: mike@wrhs.org, Internet: www.wrhs.org – ISBN 978-0-911704
Orders
Scholarly Resources Inc, Thomson Gale, 104 Greenhill Ave, Wilmington, DE 19805-1897, USA / T: +1 302 6547713, +1 888 7727817, Fax: +1 302 6543871, E-mail: sales@scholarly.com, Internet: www.scholarly.com
Information
$115.00 per roll of microfilm

us WHS
Wisconsin Historical Society Library and Archives, 816 State St, Madison, WI 53706-1482, USA / T: +1 608 2646460, Fax: +1 608 2646486, E-mail: archref@whs.wisc.edu; lbbessler@whs.wisc.edu; jdcooper@whs.wisc.edu, Internet: www.wisconsinhistory.org/microfilm – ISBN 978-0-87020
Information
Address orders to the Library Acquisitions Section. $80.00 per reel ($95.00 for silver halide) plus shipping, handling etc. ProQuest has licence to distribute some specific titles

us Wisconsin U Libr
University of Wisconsin Library, c/o B106d Memorial Library, 728 State St, Madison, WI 53706, USA / T: +1 608 2620897, Fax: +1 608 2622754, E-mail: arolich@library.wisc.edu, Internet: http://madcat.library.wisc.edu/index.html

Index to Publishers and Distributors

Aberdeen University Library, Old Aberdeen *see* uk Aberdeen UL
Abraham Lincoln Presidential Library, Springfield *see* us ALPL
Academic Microforms Ltd, Caithness *see* uk Academic
ACRPP, Marne-La-Vallée *see* fr ACRPP
Adam Matthew Publications, Marlborough *see* uk Matthew
Aircraft Technical Publishers, Brisbane *see* us Aircraft Tech
Alper, Eastchester *see* us Alper
Alvina Treut Burrows Institute, Inc, Manhasset *see* us ATBI
American Baptist Historical Society, Valley Forge *see* us ABHS
American Chemical Society (ACS), Washington *see* us ACS
American Geophysical Union, Washington *see* us AGU
American Institute of Architects, NW Washington *see* us AIA
American Institute of Physics, Melville *see* us AIP
American Jewish Periodical Center, Cincinnati *see* us AJPC
American Theological Library Association, Chicago *see* us ATLA
Amistad Research Center, New Orleans *see* us Amistad
AMS Press, Inc, Brooklyn *see* us AMS Press
Archive Publishing, Provo *see* us Archive
Atelier National de Reproduction des Thèses, Lille *see* fr Atelier National
Australian Institute of Genealogical Studies Inc, Blackburn *see* at Australian
BAB Microfilming, Mt Roskill, Auckland *see* nz BAB
Barker Texas History Center, Austin *see* us Barker
Bedfordshire Family History Society, Bedford *see* uk BedsFHS
Bell & Howell Information and Learning, Cuckfield *see* uk Bell
Berkshire Family History Society, Reading *see* uk BerksFHS
Biblioteca "IX Marqués de la Encomienda", Almendralejo (Badajoz) *see* sp Bibl Santa Ana
Biblioteca Nacional, Rio de Janeiro *see* bl Biblioteca
Bibliothèque nationale de France, Paris *see* fr Bibl Nationale
Bibliothèque Nationale du Québec, Montréal *see* cn Bibl Nat
Boletín Oficial del Estado, Madrid *see* sp Boletín
Bowker Ltd, East Grinstead *see* uk Bowker
Brepols Publishers, Turnhout *see* be Brepols
Brill, Boston *see* us Brill
Brill Academic Publishers, Leiden *see* ne Brill
British Library National Bibliographic Service, Wetherby *see* uk British Libr
British Library Newspaper Library, London *see* uk British Libr Newspaper
Brookhaven Press, La Crosse *see* us Brook
Brookings Institution Press, Washington, DC *see* us Brookings
Buffalo and Erie County Historical Society, Buffalo *see* us Buffalo
Bundesarchiv Koblenz, Koblenz *see* gw Bundesarchiv
Cairns and District Family History Society Inc, Cairns *see* at Cairns
Cambridgeshire Family History Society, Girton *see* uk CambsFHS
Canadian Library Association, Ottawa *see* cn Library Assoc
Canadiana.org, Ottawa *see* cn Canadiana
Cape Banks Family History Society Inc, Maroubra *see* at Cape
Cardiganshire Family History Society, Aberystwyth *see* uk Cardiganshire
Career Guidance Foundation, San Diego *see* us Career
Cassidy & Associates Inc, Torrington *see* us Cassidy
Center for Chinese Research Materials, Vienna *see* us Chinese Res
Center for Research Libraries, Chicago *see* us CRL
Centre Régional de Documentation Pédagogique de Franche-Comté, Besançon Cédex *see* fr CRDP
Chadwyck-Healey España S.L., Madrid *see* sp Chadwyck
Chadwyck-Healey Ltd, Cambridge *see* uk Chadwyck
Chemical Abstracts Service (CAS), Columbus *see* us Chemical
Colorado Historical Society, Denver *see* us Colorado Hist
Commonwealth Imaging, Calgary *see* cn Commonwealth Imaging
Congressional Information Service, Inc, Bethesda *see* us CIS
Cornell University Library, Ithaca *see* us Cornell
Crest Microfilm, Cedar Rapids *see* us Crest
CUP Services, Ithaca *see* us CUP
The Current Digest, Columbus *see* us Current
Dartmouth College Library, Hanover *see* us Dartmouth
David Brown Book Co, Oakville *see* us David Brown
East View Publications, Minneapolis *see* us East View
Eastman Kodak Company, Rochester *see* us Eastman
Educational Testing Service, Princeton *see* us ETS
Effective Information Resource Corporation, Huntington Beach *see* us IRC
EMC International Inc, Ann Arbor *see* us EMC
Engineering Information Inc, Hoboken *see* us Ei
Eurospan Group, London *see* uk Eurospan
Facts on File, Inc, New York *see* us Facts
Fairchild Microfilm (Publications), Inc, New York *see* us Fairchild Micro
Family History Society of Cheshire (FHSC), Penketh *see* uk CheshireFHS
Far Eastern Booksellers, Tokyo *see* ja Far Eastern
Federal Financial Institutions Examination Council, Washington *see* us FFIEC
Firma Herold, Oberhaching *see* gw Herold
Fischer, Erlangen *see* gw Fischer
Fitzhenry & Whiteside, Markham *see* cn Fitzhenry
Frankfurter Taschenbuch Verlag GmbH, Frankfurt a M *see* gw Frankfurter
Genealogical Society of the Northern Territory Inc, Winnellie *see* at Northern
General Commission on Archives & History, Madison *see* us Commission
Glamorgan Family History Society, Llandow, Cowbridge *see* uk Glamorgan FHS
Gothenburg University Library, Göteborg *see* sw Gothenburg University
Harvard College Library, Cambridge *see* us Harvard
Harvard Law School Library, Cambridge *see* us Harvard Law
Hein & Co Inc, Buffalo *see* us Hein
Helsinki University Library, Helsinki University *see* fi Helsinki
Human Relations Area Files Press, New Haven *see* us HRAF
Hunter Publications, Abbotsford *see* at Hunter
I J Sagun Enterprises, Rizal *see* ph Sagun
IDC Publishers bv, Leiden *see* ne IDC
IDC Publishers Inc, Riverdale *see* us IDC
Illinois State Archives, Springfield *see* us IL Archives
Immigration History Research Center, Minneapolis *see* us IHRC
IMR Limited, Harrisburg *see* us IMR
Indiana Historical Society, Indianapolis *see* us IHS
Indiana University Libraries, Bloomington *see* us Indiana Preservation
Information & Culture Korea, Seoul *see* ko Information
Information Publications International Ltd, Surrey *see* uk Info Pubs
Information Publications/Pte Ltd, Singapore *see* si Info Pubs
International Research & Evaluation, Eagan *see* us IRE
Journal Officiel, Editeur des Journaux Officiels, Paris *see* fr Journal Officiel
Kansas State Historical Society, Topeka *see* us Kansas
Kinesology Publications, Eugene *see* us Kinesology
Kinokuniya Bookstore Co Ltd, Chitose *see* ja Kinokuniya
Kungl. biblioteket, Stockholm *see* sw Kungliga
Lanarkshire Family History Society, Motherwell *see* uk Lanarkshire
Law Library Microform Consortium, Kaneohe *see* us LLMC
Lengenfelder, München *see* gw Lengenfelder
Library Microfilms, Sacramento *see* us Library Micro
Library of Congress, Washington *see* us L of C Photodup
Lippincott Williams & Wilkins, Philadelphia *see* us Lippincott
Lubrecht & Cramer, Ltd, Port Jervis *see* us Lubrecht
McLaren Micropublishing Ltd, Toronto *see* cn McLaren
Maruzen Company Ltd, Tokyo *see* ja Maruzen
Maryland State Archives, Annapolis *see* us MD Archives
Massachusetts Historical Society, Boston *see* us MA Hist
MicroColour Inc, Ridgewood *see* us MicroColour
Microfilm Corporation of Pennsylvania, Pittsburgh *see* us Microfilm Corp
Microfilm Service Bureau, Kearney *see* us Microfilm
Microform Academic Publishers, Wakefield *see* uk Microform Academic
Microlibrary Slangenburg Abbey, Doetinchem *see* ne Slangenburg
Middle East Documentation Center (MEDOC), Chicago *see* us MEDOC
Mikrofilmarchiv der deutschsprachigen Presse eV, Dortmund *see* gw Mikrofilm
Mikropress GmbH, Bonn *see* gw Mikropress
Mindata Ltd, Bath *see* uk Mindata
Ministerio de Cultura, Madrid *see* sp Cultura
Minnesota Historical Society Press, St Paul *see* us Minn Hist
Miscellaneous Institutions, City various *see* cc Misc Inst
Miscellaneous Institutions, City various *see* us Misc Inst
Miscellaneous Institutions, City various *see* sa Misc Inst
Miscellaneous Institutions, see gw Mikrofilm *see* gw Misc Inst
Mitchell International, San Diego *see* us Mitchell Int
MMF Publications, Lisse *see* ne MMF Publ
Mohr Morawa, Wien *see* au Morawa
Moody's Investors Service, Inc, New York *see* us Moody
Moran Micropublications, Amsterdam *see* ne Moran
MRTS, Binghamton *see* us MRTS
Museums Libraries Archives, Warrington *see* uk MLA
National Archives Trust Fund, College Park *see* us Nat Archives
National Clearinghouse on Marital and Date Rape / Women's History Library, Berkeley *see* us National Clearing
The National Library of Ireland, Dublin *see* ie National
National Library of New Zealand, Wellington *see* nz Nat Libr
National Library of Scotland, Edinburgh *see* uk Scotland NatLib
National Library of South Africa, Cape Town *see* sa National
National Library of South Africa – State Library, Pretoria *see* sa State Libr
National Register of Archives for Scotland, Edinburgh *see* uk NRAS
National Technical Information Service (NTIS), Springfield *see* us NTIS
National Women's History Project, Santa Rosa *see* us National Women's
NBN International, Plymouth, Devon *see* uk NBN
Nebraska State Historical Society, Lincoln *see* us NE Hist
NEWSPLAN 2000 Project, London *see* uk Newsplan
Northeast Document Conservation, Andover *see* us Northeast
Ohio Historical Society, Columbus *see* us Ohio Hist
Olms, Hildesheim *see* gw Olms
OmniSys Corporation, see us UMI Proquest *see* us OmniSys
Omohundro Institute of Early American History and Culture, Williamsburg *see* us OIEAHC
Oregon Historical Society Library, Portland *see* us Oregon Hist
The Oriental Institute, Chicago *see* us Oriental
Oxbow Books, Oxford *see* uk Oxbow
Pacific Manuscripts Bureau (Pambu), Canberra *see* at Pacific Mss
Pascoe, Balgowlah *see* at Pascoe
Pennsylvania Academy of the Fine Arts, Philadelphia *see* us Penn Academy
Pennsylvania Historical and Museum Commission, Harrisburg *see* us Penn Hist
Perceptual and Motor Skills, Missoula *see* us Perceptual
Plymbridge Distributors Ltd, Plymouth *see* uk Plymbridge
PraXess Associates, Bronxville *see* us Praxess
Presbyterian Historical Society, Philadelphia *see* us Presbyterian
Preston Publications, Inc, Niles *see* us Preston Publ
Primary Source Microfilm, Woodbridge *see* us Primary
Princeton University Press, Princeton *see* us Princeton U Pr
ProQuest Business Solutions Inc, Richfield *see* us Bell
ProQuest Information and Learning, Ann Arbor *see* us UMI ProQuest
ProQuest Information and Learning, Drummoyne NSW *see* at ProQuest
ProQuest Information and Learning, Friedberg *see* gw ProQuest
ProQuest Information and Learning, Petaling Jaya, Selangor D.E *see* my ProQuest
ProQuest Information and Learning, Seoul *see* ko ProQuest
Publishers Marketing Services Pte Ltd, Singapore *see* si Publishers
Rhenus Medien Logistik GmbH & Co. KG, Landsberg/Lech *see* gw Rhenus
Ross, see us UMI Proquest *see* us Ross
Roth Publishing Inc, Great Neck *see* us Roth
Russick, Mahebourg *see* mf Russick
Saur, München *see* gw Saur
Scholarly Book Services Inc, Toronto *see* cn Scholarly Bk
Scholarly Resources Inc, Wilmington *see* us Scholarly Res
Scholars' Facsimiles & Reprints, Ann Arbor *see* us Scholars Facs
Sibley Music Library Microform Service, Rochester *see* us Sibley
Society of Automotive Engineers, Inc, Warrendale *see* us SAE
Society of Industrial and Applied Mathematics (SIAM), Philadelphia *see* us SIAM
Society of Manufacturing Engineers (SME), Dearborn *see* us SME
South Carolina Department of Archives and History Center, Columbia *see* us South C Archives
South Carolina Historical Society, Charleston *see* us South Carolina Historical
South Caroliniana Library, Columbia *see* us South Caroliniana
South Dakota State Archives, Pierre *see* us SD Archives
Southern Baptist Historical Library and Archives, Nashville *see* us Southern Baptist
Stanford University Libraries, Stanford *see* us Stanford
State Archives and Historical Research Library, Bismarck *see* us North Dakota
Striker, New York *see* us Striker
Tay Valley Family History Society, Dundee *see* uk Tay Valley
Taylor Associates, Beijing *see* cc Taylor
The Balch Institute for Ethnic Studies of the Historical Society of Pennsylvania, Philadelphia *see* us Balch
The British Library, London *see* uk British Library
The Haworth Press, Inc, Binghamton *see* us Haworth
The Institute for Advanced Studies of World Religions (IASWR), Carmel *see* us IASWR
The Lawbook Exchange, Clark *see* us Law
The Library and Archives of Canada/Bibliothèque et Archives Canada, Ottawa *see* cn Library and Archives
The National Archives, Richmond *see* uk National
The National Library of Wales, Aberystwyth *see* uk Wales NatLib
The New York Public Library, New York *see* us NY Public
The Society of American Archivists, Chicago *see* us SAA
The United States General Accounting Office, Washington *see* us US Gen Account
The University of Texas, Institute of Texan Cultures at San Antonio, San Antonio *see* us TX Culture
The University of Virginia Press, Charlottesville *see* us Virginia U Pr
The Western Reserve Historical Society, Cleveland *see* us Western Res
Thomson Asia Pte Ltd, Petaling Jaya *see* my Gale
Thomson Gale, Farmington Hills *see* us Thomson Gale
Thomson Gale, Woodbridge *see* us Thomson Gale
Thomson Gale/PSM, London *see* uk Thomson Gale/PSM
Thomson Learning Iberoamerica, Colonia Polanco *see* mx Thomson
Trans-Media / The Oceana Group, Dobbs Ferry *see* us Trans-Media
Transmission Books & Microinfo Co, Ltd, Taipei *see* ch Transmission
Turnaround Publisher Services Ltd, London *see* uk Turnaround
UNIpresses, Georgetown *see* cn UNIpresses
United Nations Office, Genève 10 *see* sz UN Office
United Nations Publications, New York *see* us UN Publ
United States Government Printing Office, Washington *see* us Gov Printing Dist
United States Government Printing Office (USGPO), Washington *see* us GPO
Univelt Inc, San Diego *see* us Univelt
University Libraries of Notre Dame, Notre Dame *see* us Notre Dame
University Music Editions, New York *see* us Univ Music
University of British Columbia Library, Vancouver *see* cn UBC Preservation
University of Chicago Press, Chicago *see* us Chicago U Pr
University of Florida Libraries, Gainesville *see* us UF Libraries
University of Illinois at Urbana-Champaign, Champaign *see* us Univ Ill Libr
University of Johannesburg, Auckland Park *see* sa U of Johannesburg
University of New South Wales Library, Sydney *see* at UNSW Lib
University of Oregon Library System, Eugene *see* us Oregon Lib
University of South Africa, Unisa *see* sa Unisa
University of Stellenbosch, Matieland *see* sa Stellenbosch
University of Washington Libraries, Seattle *see* us UW Libraries
University of Wisconsin Library, Madison *see* us Wisconsin U Libr
University Publications of America UPA, Bethesda *see* us UPA
UPS United Publishers Services Ltd, Tokyo *see* ja UPS
VVA – Vereinigte Verlagsauslieferung, Gütersloh *see* gw VVA
Wisconsin Historical Society, Madison *see* us WHS
World Microfilms Publications Ltd, London *see* uk World
Yushodo Co Ltd, Tokyo *see* ja Yushodo

Author-Title List
A–K

Title main entries with references
from authors, editors

1, 2 and 3 john, jude, and revelation : a popular commentary upon a critical basis, especially designed for pastors and sunday schools / Eaches, Owen Philips – Philadelphia: American Baptist Publ Soc 1910 [mf ed 1989] – (= ser Clark's peoples commentary) – 1mf – 9 – 0-7905-1814-7 – (incl ind) – mf#1987-1814 – us ATLA [225]

El 1 congreso nacional de brujologia en san sebastian / Gutierrez Macias, Valeriano – Badajoz: Dip Provincial, 1973 – 1 – sp Bibl Santa Ana [946]

1 consejo sindical comarcal. delegacion comarcal de azuaga / Delegacion Provincial de Sindicatos – Azuaga: Tip Domenech, 1959 – sp Bibl Santa Ana [946]

1 [first] samuel 1-7, 1 : text- und quellenkritisch untersucht / Holtz, Kurt – Leipzig: W Drugulin, 1904 [mf ed 1985] – 1mf – 9 – 0-8370-3630-5 – (in german) – mf#1985-1630 – us ATLA [221]

1. mose 14 [erste mose vierzehn] : eine historisch-kritische untersuchung / Meinhold, Johannes – Giessen: A Toepelmann 1911 [mf ed 1989] – (= ser Beihefte zur zeitschrift fuer die alttestamentliche wissenschaft 22) – 1mf – 9 – 0-7905-1303-X – (incl bibl ref) – mf#1987-1303 – us ATLA [221]

1. on the nomination of agents formerly appointed to act in england for the colonies in north america. 2. a brief statement of the dispute between sir c metcalfe and the house of assembly of the province of canada / Falconer, Thomas – London?: Reynell and Weight, 1844 – 1mf – 9 – mf#35091 – cn Canadiana [971]

1 reunion hispano portuguesa de hematologia / Asociacion Espanola de Hematologia y Hemoterapia, 12 Reunion – Badajoz: Elvas, Doncel, I G, 1969 – sp Bibl Santa Ana [060]

1er quintetto pour deux violons, deux altos et basse / Widerkehr, Jacques – Paris: Imbault [179-?] [mf ed 1989] – 5pt on 1r – 1 – mf#pres. film 51 – us Sibley [780]

1-i vserossiiskii sezd delegatov soiuza 17-go oktiabria 8-12-go fevr 1906 g, g moskva – 1906 – 44p on 1mf – 9 – (perepechatano iz gazety slovo) – mf#RPP-182 – ne IDC [325]

1re [-12me] feuille [d'allemandes] / Dubois – A Paris, Chez l'auteur [c1770] – (incl instructions, diagrams and tunes for the dances) – mf#*ZBD-*MGO pv 23 – Located: NYPL – us Misc Inst [790]

1re these : sur la figure des planetes / Cailliatte, Charles – Paris: Gauthier-Villars 1947 [mf ed 2006] – 1r – 1 – (incl: 2e these: propositions donnees par la faculte; incl bibl ref) – mf#film mas 37490 – us Harvard [520]

1re these: etude sur les eclipses des satellites de jupiter / Obrecht, Albert – Paris: Gauthier-Villars 1884 [mf ed 2007] – 1r – 1 – (with: 2e these: propositions donnees par la faculte) – mf#film mas 37494 – us Harvard [520]

1re these: sur l'acceleration seculaire du mouvement de la lune / Puiseux, Pierre Henri – Paris: Gauthier-Villars 1879 [mf ed 2007] – 1r – 1 – (with: 2e these: propositions donees par la faculte; incl bibl ref) – mf#film mas 37494 – us Harvard [520]

1re these: sur une inegalite lunaire a longue periode due a l'action perturbatrice de mars / Gogou, Constantin – Paris: Gauthier-Villars 1882 [mf ed 2007] – 1r – 1 – (with: 2e these: propositions donees par la faculte; incl bibl ref) – mf#film mas 37494 – us Harvard [520]

1st bayona y la politica de napoleon en america... : y 2nd historia de la primera republica de venezuela... / Parra Perez, C – Madrid: Razon y Fe, 1940 – 1 – sp Bibl Santa Ana [972]

1st campeonato de pesca en la modalidad de c grupo iberduero. embalse de alcantara en aguas del rio tajo. finca la carrascosa. dia 21 de junio de 1975. reglamento y programa de actos / Campeonato de Pesca 1, 1975 – Caceres: Imp. Rodriguez, 1975 – 1 – sp Bibl Santa Ana [946]

The 1st canadian division in the battles of 1918 / Craig, J D [comp] – London: Barrs, 1919 [mf ed 1997] – 1mf – 9 – 0-665-85890-6 – mf#85890 – cn Canadiana [355]

1st Cavalry Division Association *see*
– Saber
– Saber news

1st concurso hipico nacional 26 al 28 agosto... – Caceres: Tip Extremadura, 1972 – 1 – sp Bibl Santa Ana [946]

1st congreso regional agrario de extremadura. discursos, desarrollo, ponencias, y conclusiones. caceres-badajoz noviembre de 1949 – Caceres: Tip El Noticiero, s a – sp Bibl Santa Ana [360]

1st congreso sindical agrario de extremadura 2. ponencia reginen de precios y mercados en la agricultura / Corral Acedo, Francisco et al – Caceres: Tip El Noticiero s a, 1949 – sp Bibl Santa Ana [630]

1st congreso sindical agrario de extremadura 3 ponencia. plan nacional de transformacion en el campo / Sierra, Francisco et al – Caceres: Tip El Noticiero s a, 1949 – sp Bibl Santa Ana [630]

1st congreso sindical agrario de extremadura. ponencia 4 servicios economicos de las hermandades / Sanz Catalan, Jose & Masa Campos, Antonio – Caceres: Tip El Noticieros s a, 1949 – sp Bibl Santa Ana [630]

1st congreso sindical agrario de extremadura. ponencia 5 mutualidades agricolas / Diaz Montilla, Rafael et al – Caceres: Tip El Noticieros s a, 1949 – sp Bibl Santa Ana [630]

1st congreso sindical agrario de extremadura. ponencia 6. mecanizacion e industrializacion del campo / Muriel Jimenez, Vicente et al – Caceres: Tip El Noticiero s a, 1949 – sp Bibl Santa Ana [630]

1st congreso sindical agrario de extremadura. ponencia 7. situacion de la, produccion ganadera / Moreno de Arteaga, Antonio & Blazquez Izquierdo, Jose – Caceres: Tip El Noticiero, s a, 1949 – sp Bibl Santa Ana [630]

1st congreso sindical agrario de extremadura. ponencia 7, soluciones urgentes a los problemas fundamentales del campo extremenos / Sana Catalan, Jose et al – Caceres: Tip El Noticiero, s a (1949) – sp Bibl Santa Ana [630]

1st congreso sindical agrario de extremadura, ponencia i estructura y fines del sindicalismo agrario / Donoso Cortes y Donoso Cortes, E et al – Caceres: Tip El Noticiero, s a 1949 – sp Bibl Santa Ana [946]

1st exposicion colectiva del grupo artistico el canchal, 1976 – Caceres: Tip Extremadura, 1976 – 1 – sp Bibl Santa Ana [700]

1st regiment, ovla, records, ms 2894 – 1861-62 – 1r – 1 – us Western Res [976]

1st regiment, ovla, records, ms 3965 – 1861 – 1r – 1 – us Western Res [976]

1st, Sei-Karang, 1962 Berita Research Institute of the SPA *see* Konperensi ahli2 perkebunan

1st vuelta ciclista a caceres 1970. organiza club polideportivo union. colabroa federacion extremena de ciclismo. caceres, feria de 1970 / Vuelta Ciclista – Plasencia: Graf. Sandoval, 1970 – 1 – sp Bibl Santa Ana [946]

1x2 las quinielas al alcance de todos / Correyero, M – Don Benito: Tip Trejo, 1969 – sp Bibl Santa Ana [946]

2 concurso provincial de formacion profesional, mayo 1948 / Delegacion Provincial del Frente de Juventudas, Seccion de Centros de Trabajo – Caceres: Tip El Noticiero, s a – sp Bibl Santa Ana [946]

2 congreso sindical agrario de extremadura : discursos, ponencias, conclusiones / Delegacion Provincial de Sindicatos – Badajoz: Imp. Bartolome Garcia, 1951 – sp Bibl Santa Ana [946]

2 juegos deportivos / Caceres. Delegacion Provincial de Organizaciones del Movimiento – Caceres: Imp San Guiro, 1965 – 1 – sp Bibl Santa Ana [790]

2. sonate, op 21, piano solo / Szymanowski, Karol – Wien: Universal ed c1912 [mf ed 1991] – 1r – 1 – mf#pres. film 106 – us Sibley [780]

2. timothy 2:15 : a course of study in personal work / Pope, Howard Walter – Boston: United Society of Christian Endeavor, c1910 [mf ed 1985] – 1mf – 9 – 0-8370-2611-3 – mf#1985-0611 – us ATLA [240]

2 [tweede] petrus en judas : textuitgave met inleidende studien en textueelen commentaar / Zwaan, Johannes de – Leiden: S C van Doesburgh, 1909 [mf ed 1993] – 1mf – 9 – 0-524-06585-3 – mf#1992-0928 – us ATLA [220]

2d coming – v1 n1-v2 n14 [1969 oct 6-1971 may 18/25] – 1 – mf#1051381 – us WHS [071]

2e conference au sommet : organisation des etats riverains du senegal – [Conakry]: Impr nationale "Partice Lumumba", 1970 – us CRL [960]

2eme quintetti pour deux violons, deux altos, une basse, deux cors ad-libitum [no 2 e minor] / Winter, Peter von – Paris: Naderman [180-?] [mf ed 19–] – 5pt on 1r – 1 – mf#pres. film 177 – us Sibley [780]

2nd, 3rd, 4th and 5th interim reports of the civil war workers' committee / Great Britain. Ministry of Reconstruction Civil War Workers' Committee – London: HMSO, 1918 [mf ed 19–] – 27p – mf#Z-BTZE pv350 n5 – us Harvard [331]

2nd congreso de geografia e historia hispano-americanas / ed by Bayle, Constantino – Madrid: Razon y Fe, 1921 – 1 – sp Bibl Santa Ana [972]

2nd festivales de espana. plasencia, 13 al 16 de junio 1970 – Plasencia: Imp La Victoria, 1970 – 1 – sp Bibl Santa Ana [390]

2nd gala del deporte comarcal / Delegacion de Deportes – Coriz: Gr. Planta, 1980 – 1 – sp Bibl Santa Ana [946]

2nd jornadas de promocion del deporte laboral 1976 / Obra Sindical de Educacion y Descanso – Caceres: Tip Extremadura, 1976 – 1 – sp Bibl Santa Ana [790]

2nd juegos didactico-deportivos nacionales de las escuelas normales : fase de sector. caceres 1 al 4 de abril 1971 – Caceres: Tip. Extremadura, 1971 – 1 – sp Bibl Santa Ana [370]

2tes trio fuer piano, violine und violoncello, op 20 / Bargiel, Woldemar – Leipzig: F E C Leuckart; New York: Scharfenberg & Luis [186-?] [mf ed 1992] – 1r – 1 – mf#pres. film 123 – us Sibley [780]

3 anos de lucha resumen historico de la federacion sindical de plantas electricas, gas y agua : criticas y orientaciones / Bretau, Francisco – Habana, Cuba: Imp Virtudes 97 1937 [mf ed 1991] – 1r – 1 – us UF Libraries [331]

3 dagar – Vaexjoe, 1993-94 – 9 – sw Kungliga [078]

3 dagar *see* Kronobergaren

3 discursos / Primo de Rivera, Jose Antonio – n.p, 1936? – (= ser Blodgett coll) – 9 – mf#fiche w1502 – us Harvard [946]

3 discursos americanos / Union Interamericana del Caribe – Habana, Cuba 1943 [mf ed 1991] – 1r – 1 – us UF Libraries [972]

3 feria regional del campo extremeno / Zafra. Ayuntamiento – Zafra: Comision de Ferias y Fiestas del Ayuntamiento, 1968 – sp Bibl Santa Ana [946]

3 libros en 1 / Velasquez, Atilio – Bogota, Colombia: Kelly 1963 [mf ed 1993] – 1r – 1 – (aka: la historia de las contradicciones) – us UF Libraries [972]

3 obras de teatro nuevo : chinfonia burguesa – Managua: [Academia Nicaraguerse de la Lengua] 1957 [i.e. 1958, c1957] [mf ed 1992] – (= ser Coleccion lengua. teatro 4) – 1r – 1 – (aka: tres obras de teatro nuevo) – us UF Libraries [820]

3 quatuors pour deux violons, alto & violoncelle, oeuvre 7 / Hoffmann, Heinrich Anton – Bonn: Simrock [179-?] [mf ed 1988] – 4pt on 1r – 1 – mf#pres. film 47 – us Sibley [780]

3 raid hipico internacional : badajoz-caceres-badajoz [14 y 15 de marzo 1970] / Sociedad Hipica Lebrera de Badajoz – Badajoz: Imp Comercial M Cordon, 1970 – sp Bibl Santa Ana [946]

3 sonates pour le clavecin ou pianoforte avec flute et basse / Clementi, Muzio – Offenbach s/M: Jean Andre [c1785] [mf ed 1990] – (= ser Journal de musique pour les dames) – 3pt on 1r – 1 – mf#pres. film 86 – us Sibley [780]

3 sonatines faciles : pour le clavecin avec violon [ad libitum] / Vanhal, Johann Baptist – Vienne: Artaria et Comp [182-?] [mf ed 1989] – 1r – 1 – mf#pres. film 51 – us Sibley [780]

3d support command log – v11 n2-9 [1981 feb-sep] – 1 – mf#625729 – us WHS [071]

3d Ward Chamber of Commerce [Chicago IL] *see* Central south sider

3-ia gosudarstvennaia duma : sessiia 1-ia -[2-ia]: otchet fraktsii narodnoi svobody – 1909 – 228p 4mf – 9 – mf#RPP-107 – ne IDC [325]

3me quatuor concertant pour piano, violon, alto et violoncelle / Jadin, Louis-Emmanuel – Paris: Erard [180-?] [mf ed 1989] – 4pt on 1r – 1 – mf#pres. film 48 – us Sibley [780]

3-p pacer – 1981 jan-1989 feb – 1 – mf#1496695 – us WHS [071]

3rd exposicion filatelica cacerna / CNS Vicesecretaria Provincial de O Sindicales de E D – Caceres: Tip El Noticiero, 1955 – 1 – sp Bibl Santa Ana [946]

El 4 centenario de la universidad de santo domingo (1538-1938) discurso pronunciado en el acto académico celebrado el 28 de octubre de 1938 / Ortega Frier, Julio – Ciudad Trujillo, Dominican Republic: Univ de Santo Domingo [1946] [mf ed 1991] – 1r – 1 – us UF Libraries [378]

El 4 centenario del descubrimiento de california / ed by Bayle, Constantino – Madrid: Razon y Fe, 1932 – 1 – sp Bibl Santa Ana [978]

4. comision no 1 : ganaderia / Consejo Economico Sindical Provincial – Badajoz: Imp Inca, 1965 – sp Bibl Santa Ana [630]

4. comision no 1 a : agricultura / Consejo Economico Sindical Provincial – Badajoz: Imp Inca, 1965 – sp Bibl Santa Ana [630]

4. comision no 1 b : repoblacion forestal / Consejo Economico Sindical Provincial – Badajoz: Imp Inca, 1965 – sp Bibl Santa Ana [630]

4. comision no 7 : ensenanza y formacion profesional / Consejo Economico Sindical Provincial – Badajoz: Graficas Jimenez, 1965 – sp Bibl Santa Ana [330]

4 consejo economico sindical / Laguna Sanz, Eduardo – 1st prov. Comision (ganaderia). Badajoz: Imp Inca, 1966 – 1 – sp Bibl Santa Ana [330]

4 consejo economico sindical 11th prov. comision : sanidad y asistencia social / Lopez Santamaria, Luis – Badajoz: Imp Inca, dic 1965 – 1 – sp Bibl Santa Ana [330]

4 consejo economico sindical 14th prov. comision. trabajo / Masa Godoy, Jose – Badajoz: Impenta Inca, 1966 – 1 – sp Bibl Santa Ana [330]

4 corners samaj – 1985 feb – 1 – mf#1477066 – us WHS [071]

4 de outubro – Maceio, AL. 4 out 1886; 4 out 1888 – (= ser Ps 19) – mf#PR-SOR 2433(1) – bl Biblioteca [079]

4 discursos de pilar primo de rivera – [Barcelona: Talleres Graficos de A Nuez] 1939 – 9 – mf#w1121 – us Harvard [080]

4 festivales de espana. plasencia 23-29 junio 1972 / Festivales de Espana N. Plasencia, 1972 – Plasencia: Imp La Victoria, 1972 – 1 – sp Bibl Santa Ana [390]

4 [i. e. cuatro] cuentos / Otero, Jose Manuel – Habana, Cuba: [Ediciones Belic] 1965 [mf ed 1991] – (= ser Cuadernos giron 7) – 1r – 1 – us UF Libraries [972]

Les 4 [i.e. quatre] candidats a la presidence – Paris, [1848?] – us CRL [320]

4. internationale – Marseille (F), 1941 jul-oct, 1942 jan-nov, 1943 jun-1 aug – 1 – gw Misc Inst [074]

4 marcha penitencial nocturna al santuario del palancar (caceres) – Madrid: Graf Calleja, 1967 – 1 – sp Bibl Santa Ana [240]

4 pleno : reglamento / Consejo Economico Sindical – Badajoz: Imprenta Inca, 1963 – sp Bibl Santa Ana [330]

4 pleno del consejo economico sindical. conclusiones definitivas / Organizacion Sindical de Badajoz – Badajoz: Imp Inca, 1966 – sp Bibl Santa Ana [330]

4 raid hipico internacional-badajoz-caceres-badajoz (18-19 de marzo 1971) / Sociedad Hipica Lebrera de Badajoz – Badajoz: Imp Comercial C Oudrid, 1971 – sp Bibl Santa Ana [946]

4 stimmiges choralbuch : [238 choral-melodien mit beziffertem bass. handschrift, die nach dem schriftcharakter und der orthographie zu schliessen aus der mitte des 18. jahrhdts...] / Bach, Johann Sebastian – [175-] – 8mf – 9 – mf#fiche864 – us Sibley [780]

4-H Club, Willing Workers, Reno County, KS *see* Records

4-h news – 1980 nov-1986 fall – 1 – mf#1278859 – us WHS [071]

4th asamblea plenaria del consejo economico sindical : ponencias y conclusiones / Consejo Economico Sincial – Caceres: Tip el Noticiero, 1956 – 1 – sp Bibl Santa Ana [330]

El 4th centenario de la fundacion de lima / ed by Bayle, Constantino – Madrid: Razon y Fe, 1935 – 1 – sp Bibl Santa Ana [972]

4th centenario de san pedro de alcantara – Madrid: arch ibero americano, 1961 – 1 – sp Bibl Santa Ana [241]

4th dimension – v1 n7-8 [1972 may 4-jun 19] – 1 – mf#1051385 – us WHS [071]

4th juegos nacionales de la educacion general basica / Delegacion Provincial de la Juventudes – Caceres: Sergio Dorado, 1973 – 1 – sp Bibl Santa Ana [370]

4th trans comd news – v5 n1-v6 n6 [1970 jan 31-1971 sep] – 1 – mf#1051386 – us WHS [071]

5 concurso de embellecimiento de pueblos – Badajoz: Imp Provincial, 1973 – sp Bibl Santa Ana [946]

5 cuentos de sangre / Gonzalez, Jose Luis – San Juan, Puerto Rico: Impr Venezuela 1945 [mf ed 1991] – 1r – 1 – (aka: cinco cuentos de sangre; with foreword by: francisco matos paoli) – us UF Libraries [972]

5 de septiembre – Cienfuegos, Cuba. aug-nov 1982; feb-jul 1983; feb 17 1984-1990 – 9r – 1 – us L of C Photodup [079]

5 [i.e. cinco] poetas hispanoamericanos en espana / Laredo, Alonso – Madrid: Ediciones Cultura Hispanica 1953 [mf ed 1992] – (= ser Encina y el mar 14) – 1r – 1 – (aka: cinco poetas hispanoamericanos en espana) – us UF Libraries [440]

5th exposicion regional filatelica cacerena / Vicesecretaria provincial de obras sindicales – Caceres: Tip Extremadura, 1959 – 1 – sp Bibl Santa Ana [338]

6 cornet pieces : with an introduction for the diapasons, and a fugue / Burney, Charles – London: I Walsh [1751] [mf ed 19–] – 1mf – 9 – mf#fiche 756 – us Sibley [780]

0 6 de marco *see* O seis de marco

6 duo, a deux violons. oeuvre 8 / Campioni, Carlo Antonio – Paris: Academie royale de musique [c1770] [mf ed 1988] – 2pt on 1r – 1 – mf#pres. film 44 – us Sibley [780]

6 festival iberico de musica – Badajoz: dip provincial y ayuntamiento de badajoz e institucion cultural pedro de valencia, 1978 – sp Bibl Santa Ana [780]

6 [i.e. six] sonates a une flute traversiere, un haubois ou violon & basse continue : dediees a monsieur abraham dreyer...& a madame son epouse par jean chretien schickardt. 20me ouvrage livre second / Schickhard, Johann Christian – Amsterdam: M C le Cene [1720?] [mf ed 19–] – 1r – 1 – mf#film 130 – us Sibley [780]
6 menuettes pour le piano forte / Zapf, Johann Nepomuk – Vienne: J Eder [180-?] [mf ed 1989] – 1r – 1 – mf#pres. film 51 – us Sibley [780]
6 poesias y 5 cuentos premiados / Federacion Provincial de Escritores de la Habana – [La Habana: Sociedad Colombista Panamericana 1956] [mf ed 1992] – 1r – 1 – us UF Libraries [350]
6 sonate a tre : due violini o flauti e basso, opera quinta / Locatelli, Pietro Antonio – Paris: Le Clerc [1737] [mf ed 19–] – 3pt on 1mf – 9 – mf#fiche 607, 919 – us Sibley [780]
La 6e conference economique nationale 1977 – Conakry: Impr nationale "Partice Lumumba", [1978] – us CRL [330]
6th bienal extremena de pintura – Merida: Bimilenario de Merida, Tip. Vadillo, 1976 – 1 – sp Bibl Santa Ana [946]
6th regiment : ov cavalry scrapbook, 1861-1865 – 1r – 1 – mf#B32880 – us Ohio Hist [976]
7 congreso, seccion de parasitología, la habana, cuba, enero 18 al 23, 1938 / Pan American Medical Association Congress (7th : 1938 : Havana, Cuba) – Habana, Cuba? 1938? [mf ed 1993] – 1r – 1 – (incl bibl; in spanish or english) – us UF Libraries [616]
7 days of mkt *see* Sem'dnei mkt
7 de octubre : una nueva era en el campo / Spain. Ministerio de Agricultura – Madrid, 1936 – (= ser Blodgett coll) – 9 – mf#fiche w1161 – us Harvard [946]
7 sonetos de ausencia / Arce y Valladares, Manuel Jose – [n.p.] 1945 [mf ed 1992] – 1r – 1 – (aka: siete sonetos de ausencia) – us UF Libraries [972]
7 tage *see* Die sieben tage
7 trii per violino due e cetra / Canaletti, Giovanni Battista – London: Longman, Lukey & Co [177-] [mf ed 1988] – 1r – 1 – mf#pres. film 37 – us Sibley [780]
[7e] concerto, pour la flute : [no 7, op 22, g major] / Hoffmeister, Franz Anton – Paris: Sieber [c180-?] [mf ed 1988] – 9pt on 1r – 1 – mf#pres. film 47 – us Sibley [780]
7th air force news – v6 n11-v9 n3 [1970 mar 18-1973 jan 27] – 1 – mf#1051388 – us WHS [071]
8 buecher gegen celsus, 2. bd 1. teil (bdk52 1.reihe) / Origenes (Origen) – (= ser Bibliothek der kirchenvaeter. 1. reihe (bdk 1.reihe)) – €17.00 – ne Slangenburg [240]
8 buecher gegen celsus, 3. bd 2. teil (bdk53 1.reihe) / Origenes (Origen) – (= ser Bibliothek der kirchenvaeter. 1. reihe (bdk 1.reihe)) – €15.00 – ne Slangenburg [240]
8 memories betreffende javaanse hoven, 1773-1803 – 6mf – 8 – mf#SD-102 mf 17-22 – ne IDC [959]
8 short, melodious pieces for pianoforte duet, op 13 / Kiel, Friedrich – London: Augener & Co [189-?] [mf ed 19–] – 2v on 1mf – 9 – mf#fiche 161 – us Sibley [780]
8 sidor – Stockholm, 1992- – 9 – sw Kungliga [078]
8 variations sur l'air beglueckt durch dich : pour le piano forte composees / Zapf, Johann Nepomuk – Vienne: J Eder [180-?] [mf ed 1989] – 1r – 1 – mf#pres. film 51 – us Sibley [780]
8 vsesoiuznoe soveshchanie "izuchenie i osvoenie flory i rastitelnosti vysokogorii" : tezisy dokladov / ed by Gorchakovskii, P L – Sverdlovsk: Akademiia nauk SSSR, Uralskii nauch. tsentr, v1. 1982 – us CRL [580]
8emes congres nationaux jrda-cntg – Conakry: Impr nationale "Partice Lumumba", 1975 – us CRL [320]
Das 9. buch innerhalb der pharsalia des lucan und die frage der vollendung des epos / Voegler, Gudrun – Frankfurt a.M., 1967 – 1mf – 9 – 3-89349-771-4 – gw Frankfurter [430]
'09 express – 1978 jul-1984 dec – 1 – mf#1477858 – us WHS [071]
9 to 5, Minnesota Working Women *see* Memo
9eme congres national du pdg (9th: 1972 conkary) / Parti democratique de Guinee Congres nationale – Conakry: Impr Nationale "Patrice Lumumba", 1973 – us CRL [960]
The 9th annual examination of the classes of the institute for colored youth will take place friday, may 3 1861 : at the institute building, 716 and 718 lombard street / Home for Aged and Infirm Colored Persons (Philadelphia PA) – [s.l: s.n.] 1861 [mf ed 2006] – 1r [complete] – 1 – 0-524-10538-3 – (reel also incl: a commencement address from 1864, and a commencement program from 1866) – mf#2006-s023 – us ATLA [362]
9th regiment, ov cavalry / Gatch, Asbury P – 1r – 1 – mf#B34922 – us Ohio Hist [976]
9to5 news – 1980 oct/nov-1984 nov/dec – 1 – mf#1100102 – us WHS [071]
10 20 news – v2 n9-v9 n6 [1974 nov-1981 jun] – 1 – mf#675893 – us WHS [071]
10 anos de ensenanza laboral en trujiullo / Trujillo. Centro de Ensenanza Media y Profesional Garcia de Paredes – Trujillo (Caceres): Imp. Moderna, 1960 – 1 – sp Bibl Santa Ana [370]
10 anos de labor, 1930-1940 / Costa Rica Patronato Nacional de la Infancia – San Jose, Costa Rica 1941? [mf ed 1993] – 1r – 1 – us UF Libraries [362]
10 let raboty arteli "metallist" / Vlasov, P A – Serpukhov, 1928 – 47p 1mf – 9 – mf#COR-417 – ne IDC [335]
10 sonatas for 2 german flutes or 2 violins : with a thorough bass, opera settima... / Fesch, Willem de – London: B Cooke [1732?] [mf ed 19–] – 1r – 1 – mf#pres. film 91 – us Sibley [780]
10eme congres du pdg : parti democratique de guinee – Conakry: Imprimerie Nationale "Patrice Lumumba", 1973 – us CRL [960]
10-k reports / Leasco Information Products – Bethesda, MD: Disclosure Inc, 196 - – 9 – (consists of alphabetical file of annual reports filed with the securities and exchange commission by corporations listed on the new york stock exchange and the american stock exchange) – us Misc Inst [332]
10th certamen de experiencias teatrales para la juventud / Delegacion Provincial de la Juventudes – Caceres: Imp. M. Sergio Dorado, 1973 – 1 – sp Bibl Santa Ana [946]
10th feria regional del campo extremeno, octubre 1975 / Zafra. Ayuntamiento – Zafra: ind tip extremenas, 1975 – 1 – sp Bibl Santa Ana [240]
10th Tactical Fighter Wing *see* Spartan spirit
11th campeonato de espana de la clase de...y copa nacional juvenil 1974 : instrucciones de regata / Club Nautico Lago Gabriel y Galan. Plasencia – Plasencia: Graf Sandoval, 1974 – 1 – sp Bibl Santa Ana [790]
11th comision : sanidad y asistencia social / Consejo Economico Sindical Provincial – Badajoz: Tmorenta Inca, 1965 – sp Bibl Santa Ana [330]
11th dynasty temple at deir el-bahari : pt 1 / Naville, E – London, 1907 – (= ser Mees 28 – 7mf – 8 – €16.00 – ne Slangenburg [720]
11th dynasty temple at deir el-bahari : pt 2 / Naville, E – London, 1910 – (= ser Mees 30 – 4mf – 8 – €11.00 – ne Slangenburg [720]
11th dynasty temple at deir el-bahari : pt 3 / Naville, E & Hall, H R – London, 1913 – (= ser Mees 32 – 5mf – 8 – €12.00 – ne Slangenburg [720]
11th season, 1882-83 : philharmonic society, toronto – [Toronto?: s.n. 1882?] [mf ed 1987] – 1mf – 9 – 0-665-18706-8 – mf#18706 – cn Canadiana [780]
Les 12 chansons du saguenay – Montreal: le Repertoire regional canadien, [1939] [mf ed 1991] – 1mf – 9 – mf#SEM105P1475 – cn Bibl Nat [780]
Les 12 coups de mes nuits / Daunais, Jean – Montreal: Editions Heritage, 1979 [mf ed 1993] – (= ser Coll a lire en vacances) – 2mf – 9 – mf#SEM105P1795 – cn Bibl Nat [890]
12 [i.e. doce] muertes famosas / Cuellar Vizcaino, Manuel – Habana, Cuba 195- [mf ed 1991] – 1r – 1 – us UF Libraries [972]
12 sibyllarum icones elegantissimi...crispiano passaeo zelando delineati... / Passe, C de – n.p, 1601 – 1mf – 9 – mf#O-30 – ne IDC [090]
12 sinfonie a quattro. due violine, alto, organo e violoncello, opera seconda, libro primo-[secondo] / Alberti, Giuseppe Matteo – London: printed for I Walsh [1730?] [mf ed 1988] – 1r – 1 – mf#pres. film 35 – us Sibley [780]
[12 sonatas in four parts (triosonaten) fuer 2 violinen, violoncello und basso continuo] / Pepusch, John Christopher – [171?] [mf ed 19–] – 3mf / 1r – 9,1 – mf#fiche 1032 / pres. film 61 – us Sibley [780]
12 sonates pour le violoncelle [et basse continue] / Azais, Pierre Hyacinthe – Sorese, Revel: chez l'auteur; Paris: Bignon [1780] [mf ed 198-] – 1r – 1 – mf#pres. film 36 – us Sibley [780]
12 spanish american poets : an anthology / ed by Hays, Hoffman Reynolds – New Haven CT: Yale UP; London: H Milford, OUP 1943 [mf ed 1992] – 1r – 1 – (in spanish & english on opposite pp) – us UF Libraries [810]
12e congres international de geologie : tenu a toronto, 6-14 aout 1913 / Choquette, Charles Philippe – [Canada?: s.n, 1913?] – 1mf – 9 – 0-665-65213-5 – mf#65213 – cn Canadiana [550]
12th street rag – v1 n1-7 [1963 mar 7-may 14] – 1 – mf#630982 – us WHS [071]
Das 12-uhr-blatt *see* Neue berliner zeitung
13 de julho – Aracaju, SE. 20 dez 1932 – (= ser Ps 19) – mf#PR-SOR 02913 – bl Biblioteca [079]
13 (i e trece) anos de violencia : asesinos intelectuales de gaitan, dictaduras – militarismo – alternacion / Cuellar Vargas, Enrique – Bogota, Colombia: Ediciones "Cultura Social Colombiana" 1960 [mf ed 1993] – 1r – 1 – us UF Libraries [972]
El 13 [i.e. trece] de junio en 33 numeros de ya / Canal Ramirez, Gonzalo – Bogota, Colombia: Antares 1954 [mf ed 1993] – 1r – 1 – us UF Libraries [355]
13 [i.e. trece] novelas cortas / Cuchi Coll, Isabel – Barcelona: Ediciones Rumbos 1965 [mf ed 1992] – 1r – 1 – us UF Libraries [440]
Die 13 punkte, fuer welche spanien kaempft – Barcelona, 1938. Fiche W845. (Blodgett Collection of Spanish Civil War Pamphlets) – 9 – us Harvard [946]
13 vsesoiuznoe chugaevskoe soveshchanie po khimii kompleksnykh soedinenii, moskva, 12-15 iiunia 1978 g : tezisy dokladov / Akademiia nauk SSSR, Otdelenie fiziko-khimii i tekhnologii neorganicheskikh materialov – Moskva: Nauka, 1978 – us CRL [540]
The 13th battalion of hamilton / Champion, Thomas Edward – S.l: s.n, 1897 – 1mf – 9 – mf#14666 – cn Canadiana [355]
13th census, 1910. connecticut / United States Census (1910). Population. Connecticut – [Washington, DC]: National Archives; [Bountiful, UT: AGLL, 1985?] – (= ser National archives microfilm publications) – 18r – 1 – (r127-129: Fairfield. r130: Fairfield, Hartford. r131-133: Hartford. r134: Hartford, Litchfield. r135: Litchfield, Middlesex. r136: Middlesex, New Haven. r137-140: New Haven. r141: New Haven, New London. r142: New London. r143: New London, Tolland, Windham. r144: Windham) – us Nat Archives [317]
13th census, 1910. new jersey / United States Census (1910). Population. New Jersey – [Washington, DC]: National Archives; [Bountiful, Utah: AGLL [distributor, 1985?] – (= ser National archives microfilm publications) – 46r – 1 – (r867: Atlantic. r868-869: Bergen. r870: Bergen, Cape May. r871: Burlington. r872: Burlington, Camden. r873-874: Camden. r875: Cumberland. r876-884: Essex. r885: Gloucester, Hudson. r886-894: Hudson. r895: Hunterdon, Mercer. r896-897: Mercer. r898-899: Middlesex. r900: Middlesex, Monmouth. r901: Monmouth. r902: Morris. r903: Morris, Ocean. r904-906: Passaic. r907: Passaic, Somerset. r908: Somerset, Salem. r909: Sussex, Union. r910: Union. r911: Union, Warren. r912: Warren) – us Nat Archives [317]
13th gala del deporte provincial. 1977. homenaje a la excma. diputacion provincial / Delegacion Provincial de Educacion Fisica y Deportes – Caceres: Tip. Extremadura, 1977 – 1 – sp Bibl Santa Ana [370]
14 de diciembre de 1951 : tercer aniversario de la revolucion salvadorena / El Salvador. Ministerio del Interior – San Salvador, El Salvador: Impr Nacional 1952 [mf ed 1992] – 1r – 1 – us UF Libraries [972]
Los 14 el auxiliar del quinielista por caceres / Martin y Bueso, Tomas – Imprenta Sanguino. 1953 – sp Bibl Santa Ana [946]
14 jamaican short stories – 1st ed. Kingston, Jamaica: Pioneer Press [1950] [mf ed 1991] – 1r – 1 – us UF Libraries [830]
Le 14 juillet : action populaire, 3 actes / Rolland, Romain – Paris: editions des cahiers 1902 [mf ed 1994] – (= ser Cahiers de la quinzaine 3/11) – 1r – 1 – us UF Libraries [820]
[1-4] sonatas of three parts for two violins, a violoncello and thorough bass for the harpsichord made out of geminianis solos / Barsanti, Francesco – London: B Cooke [173-] [mf ed 1990] – 3pt on 1r – 1 – mf#pres. film 75 – us Sibley [780]
The 14th annual commencement exercises of the institute for colored youth will take place 5th and 6th days, eleventh mo 1st and 2d (thursday and friday, november 1st and 2d) / Home for Aged and Infirm Colored Persons (Philadelphia PA) – [s.l: s.n.] 1866 [mf ed 2006] – 1r [complete] – 1 – 0-524-10540-5 – (reel also incl: "the 9th annual examination.... (1861), and a commencement address from 1864) – mf#2006-s025 – us ATLA [370]
14th census of population, 1920. new york / United States Census (1910). Population. New York – [Washington, DC]: Bureau of the Census, Micro-film Laboratory; Bountiful, UT: AGLL, 1992? – (= ser National archives microfilm publications) – 200r – 1 – (incl census fr new york city and other counties in new york state) – us Nat Archives [317]
15 buecher ueber die dreieinigkeit, 11. bd (bdk13 2.reihe) : buch 1-7 / Augustinus (Augustine, Saint, Bishop of Hippo) – (= ser Bibliothek der kirchenvaeter. 2. reihe (bdk 2.reihe)) – €14.00 – ne Slangenburg [241]
15 buecher ueber die dreieinigkeit, 12. bd (bdk14 2.reihe) : buch 8-15 / Augustinus (Augustine, Saint, Bishop of Hippo) – (= ser Bibliothek der kirchenvaeter. 2. reihe (bdk 2.reihe)) – €15.00 – ne Slangenburg [241]
O 15 de novembro : orgam republicano – Sao Paulo, SP. 30 jul, ago, 08 nov 1895 – (= ser Ps 19) – mf#P18,01,37 – bl Biblioteca [325]
15 sezd vkp(b) i kooperatsiia / Vladilenkina, E – 1928 – 24p on 1mf – 9 – mf#COR-148 – ne IDC [335]
15 to 18 : report of the central advisory council for education (england) (crowther report), 1959-1960 – (= ser Parliamentary and non-parliamentary reports on education) – 2v on 9mf – 9 – £63 / $126 – (aka: the crowther report) – mf#f17009 – uk Microform Academic [370]
15 uhr aktuell – Berlin, 1998 13 okt-2000 18 feb [gaps] [mf ed 2004] – 5r – 1 – gw Mikrofilm [074]
15-i doklad v komissiiu imperatorskogo moskovskogo obshchestva selskogo khoziaistva po voprosu o khutorakh i sovrenmennykh usloviiakh krestianskogo khoziaistva / Stolypin, D A – 1884 – 10p 1mf – 9 – mf#COR-112 – ne IDC [335]
15th campeonato nacional federal de canaricultura y pajaros exoticos e indigenas y 7th concurso exposicion de la u c e... / Federacion Ornitologica Espanola – Caceres: Tip La Minerva, 1962 – 1 – sp Bibl Santa Ana [590]
15th certamen nacional del ahorro / Caja Postal de Ahorros – Badajoz, SL. 1947 – 1 – sp Bibl Santa Ana [946]
16th gala del deporte provincial / Delegacion Provincial Consejo Superior de Deportes – Caceres – 1 – sp Bibl Santa Ana [946]
16th Guam Legislature *see* A reassessment of guam's political relationship with the united states
16th national convention discussion bulletin – 1956 nov 1-1957 jan 15 – 1 – mf#3177742 – us WHS [071]
17 districto : orgao politico, noticioso e commercial – Diamantina, MG: Typ do 17 Districto, 12 jul-dez 1885, jan-maio,13 out 1886 – (= ser Ps 19) – mf#P31,03,13 – bl Biblioteca [321]
17 opuscules / Valdes, Juan de; ed by Betts, John Thomas – London: Truebner, 1882 [mf ed 1990] – 1mf – 9 – 0-7905-6384-3 – (english trans fr spanish and italian) – mf#1988-2384 – us ATLA [220]
17th century correspondence of the stuart monarchs : from the victoria & albert museum – 1r – 1 – mf#96629 – uk Microform Academic [025]
18 de julio : dos anos de guerra – Bilbao, 1939 – (= ser Blodgett coll) – 9 – mf#fiche w837 – us Harvard [946]
18th Century Society [New Alexandria PA]. French and Indian War Associators for Re-Enactments *see* F and i war
19 bahman danishju I *see* Bisu-yi azadi
19 bahman danishju i *see* Azarakhsh
19 bahman danishju'i – London. shumarah-'i 1-7. bahman 1354-farvardin 1356 [jan/feb 1976-mar/apr 1977] – 1r – 1 – $53.00 – (r also incl: azaraksh, bisu-yi azadi; sitiz) – us MEDOC [079]
19 bahman danishju-i *see* Sitiz
O 19 de dezembro *see* – O dezenove de dezembro
19 de outubro : orgam estudantil – Fortaleza, CE: [s.n.] 30 jul-20 set 1891 – (= ser Ps 19) – mf#P18B,03,37 – bl Biblioteca [321]
19 pueblo news – 1976 sep-oct – 1 – mf#2302335 – us WHS [071]
Le 19e siecle : journal politique quot. – Paris: Edmond About, 10 nov 1871-juin 1921 – (= ser Dix Neuvieme Siecle) – 1 – fr ACRPP [320]
Le 19e siecle tableaux des premieres annees : bonaparte et pie 7; le concordat de 1801 / Gosselin, Auguste – Quebec: impr de L J Demers et frere...1901 [mf ed 1985] – (= ser Le dix-neuvieme siecle) – 1mf – 9 – mf#SEM105P467 – cn Bibl Nat [241]
19th century paintings – (= ser Christie's pictorial archive new york) – 131mf – 9 – $1135.00 – 0-907006-77-9 – (almost 8000 reproductions) – uk Mindata [750]
20 ans apres, le club des 21 en 1879 : courte biograhie de chacun de ses menbres / Baillairge, Charles P Florent – s.l: s.n, 1899? – 1mf – 9 – mf#00964 – cn Canadiana [920]
20 century british history – Oxford, England 1990- – 1,5,9 – ISSN: 0955-2359 – mf#18524 – us UMI ProQuest [941]
20 cuentos de manuel del cabral / Cabral, Manuel del – Buenos Aires: [s.n] 1951 [mf ed 1992] – 1r – 1 – (aka: veinte cuentos de manuel del cabral) – us UF Libraries [440]
20 de julio : capitulos sobre la revolucion de 1810 / Posada, Eduardo – Bogota, Colombia: Impr de Arboleda & Valencia 1914 [mf ed 1993] – (= ser Biblioteca de historia nacional (academia colombiana de historia) 13) – 1r – 1 – us UF Libraries [972]
20 (i.e. viente) rabulas en flux ensayo de picaresca / Herrera, Flavio – Montevideo: [s.n.] 1946 (Montevideo: Impr Artigas) [mf ed 1991] – 1r – 1 – us UF Libraries [830]

20 let raboty nazarevskoi arteli / Vlasov, P A – 1929 – 70p 1mf – 9 – mf#COR-416 – ne IDC [335]

20 oktobar – Belgrade, Yugoslavia. feb 1945-mar 1952 – 3r – 1 – us L of C Photodup [949]

Le 20 siecle – Hauvre, France. 12 feb, 27 mar-8 may, 8 sep, 20 nov, 31 dec 1915; 1916-18 mar 1919 (very imperfect) – 6r – 1 – (very imperfect from 5 nov 1916 to 12 may 1918; publ in paris, 25 feb 1919 onward in brussels) – uk British Libr Newspaper [074]

20-e gody : Stanovlenie i razvitie novoi ekonomicheskoi politiki / Gorinov, M M & Tsakunov, S V – M, 1991 – 1mf – 9 – mf#REF-27 – ne IDC [332]

21 anos de estadisticas dominicanas 1936-1956 / Dominican Republic. Direccion General de Estadistica y Censos – (= ser Latin american & caribbean...1821-1982) – 6mf – 9 – uk Chadwyck [318]

21 greatest treasures : ancient slavic manuscripts from the moscow state university library – 1200s-1500s [mf ed Norman Ross Publ] – 21 titles on 183mf – 9 – (individual titles listed separately) – us UMI ProQuest [460]

21 [i.e. veintiun] anos de poesia colombiana, 1942-1963 / ed by Echeverri Mejia, Oscar & Bonilla Naar, Alfonso – Bogota: Editorial "Stella" [1964] [mf ed 1993] – 1r – 1 – us UF Libraries [810]

21 vsesoiuznoe soveshchanie po fizike nizkikh temperatur : tezisy dokladov. akademiia nauk sssr, nauchnyi sovet po probleme "fizika nizkikh temperatur" [and] akademiia nauk ussr, fiziko-tekhnicheskii institut nizkikh temperatur – Kharkov: Nauchnyi sovet, 1980 – us CRL [530]

21st century afro review – 1995 spring; 1996; 1997 fall – 1 – mf#3430289 – us WHS [071]

21st century in space [aasms58] – 1990 – (= ser Aasms 1968) – 1paper on 1mf – 9 – $10.00 – 0-87703-316-1 – (suppl to v70, advances) – us Univelt [629]

21st loan exhibition of paintings in the art gallery, phillips square : beginning february 20th, 1899 / Art Association of Montreal – Montreal?: s.n, 1899? (Montreal: D Bentley) – 1mf – 9 – mf#64650 – cn Canadiana [700]

21st ward renaissance – 1987 dec – 1 – mf#5004771 – us WHS [071]

22 de janeiro de 1903 – Manoas-AM, 22 jan 1903 – (= ser Ps 19) – mf#PR-SPR 0897(1) – bl Biblioteca [079]

22 nien ch'uan kuo yun tung ta hui tsung pao kao / Ch'uan kuo yun tung ta hui (1933: Nanking, China) – Shang-hai: Chung-hua shu chu, min kuo 23 nien [1934] – (= ser P-k&k period) – us CRL [790]

O 23 de julho *see* O vinte tres de julho

24 emblemata dat zijn zinnebeelden / Drijfhout, A E – Bussum: CAJ van Dishoeck NV, 1932 – 1mf – 9 – mf#O-593 – ne IDC [090]

Le 24 fevrier : journal de la republique democratique et des reformes sociales = Vingt-quatre fevrier – [Paris]: Lange Levy [mar 1850] – 1r – 1 – us CRL [074]

24 menuettos for two violins and a bass / Schwindl, Friedrich – [London]: A Hummell [177-?] [mf ed 1989] – 3pt on 1r – 1 – mf#pres. film 51 – us Sibley [780]

24 neue deutsche erzaehler / ed by Kesten, Hermann – 2.aufl. Berlin: G Kiepenheuer, 1929 [mf ed 1993] – 421p – 1 – (incl bibl ref) – mf#8364 – us Wisconsin U Libr [830]

24 ore – Milan, Italy 15 feb 1950-22 april 1962 (daily) [imperfect] – 55r – 1 – uk British Libr Newspaper [072]

25 anos a traves del estado de antioquia / Gomez Barrientos, Estanislao – Medellin (Colombia), 1927 – 1 – sp Bibl Santa Ana [946]

25 chorale, mit achterly general baessen / Kittel, Johann Christian – 1791 [mf ed 19–] – 1r – 1 – mf#pres. film 31 – us Sibley [780]

25 de marco : orgao do partido conservador – Curitiba, PR. 25 mar-17 nov 1876 – (= ser Ps 19) – mf#PR-SOR 0586(1) – bl Biblioteca [325]

25 fables des animaux : vray miroir exemplaire... / Perret, E – Anvers: Christophle Plantin, 1578 – 1mf – 9 – mf#O-1935 – ne IDC [090]

25 [i. e. veinte y cinco] anos de historia colombiana, 1853 a 1878, del centralismo a la federacion / Perez Aguirre, Antonio – Bogota, Colombia: Editorial Sucre 1959 [mf ed 1993] – 1r – 1 – (incl bibl) – us UF Libraries [972]

25 jahre gewerkverein christlicher bergarbeiter / Imbusch, H – Essen, 1919 – 1 – gw Mikropress [331]

25 sermon...san juan de la mata / Solano de Figueroa y Altamirano, Juan – 1670 – 9 – sp Bibl Santa Ana [241]

25th Infantry Division Association *see* Tropic lightning flashes

25th sermon en la festividad del glorioso patriarca san juan de la mata, fundador de la orden de la santisima trinidad... / Solano de Figueroa y Altamirano, Juan – Madrid: Joseph Fernandez de Buendia, 1670 – 1 – sp Bibl Santa Ana [241]

26 juegos escolares nacionales. fase facional de baloncesto. categorias : juvenil e infantil – Caceres: Imp M Sergio Dorado, 1974 – 1 – sp Bibl Santa Ana [946]

26 plus – New York. 1973-73 – 1 – ISSN: 0091-410X – mf#8565.01 – us UMI ProQuest [380]

26th concurso hipico nacional 1969 – Caceres: Imp Moderna, 1963 – 1 – sp Bibl Santa Ana [946]

El 26th congreso internacional de americanistas / ed by Bayle, Constantino – Madrid: Razon y Fe, 1935 – 1 – sp Bibl Santa Ana [970]

El 27 / Pena El 27 – Caceres: Imp. Fernandez, 1971 – 1 – sp Bibl Santa Ana [946]

27 de noviembre de 1871 / Valdes Dominguez, Fermin – 6 ed. Habana, Cuba: Impr de Rambla y Bouza 1909 [mf ed 1991] – 1r – 1 – (1st ed [1873] has title: los voluntarios de la habana) – us UF Libraries [972]

El 27 junio 1975 / Pena El 27 – Plasencia: imp. iersa, 1975 – 1 – sp Bibl Santa Ana [946]

27eme anniversaire du partie democratique de guinee – Conakry: Impr nationale "Partice Lumumba", 1974 – us CRL [325]

28 discours chrestiens, touchant l'estat du monde et de l'eglise de dieu / Goulart, S – [Geneve], Stoer, 1591 – 4mf – 9 – mf#PFA-164 – ne IDC [240]

28 news – v36 n10-v30 [i.e. 40] n7 [1966 nov-1971 sep] – 1 – mf#633516 – us WHS [071]

29 lets go – v5 n70 [1945 mar 15] – 1 – mf#632129 – us WHS [071]

Der 29 psalm ausgelegt / Bugenhagen, J – [Wittemberge], 1542 – 1mf – 9 – mf#TH-1 mf 184 – ne IDC [242]

29er – 1964 feb-1982 oct – 1 – mf#630106 – us WHS [071]

30 caricaturas de la guerra / Spain. Ministerio de Propaganda – Valencia, 1937 – (= ser Blodgett coll) – 9 – mf#fiche w1189 – us Harvard [946]

El 30 de febrero (vida de un hombre interino) / Laguerre, Enrique A – San Juan de Puerto Rico: Biblioteca de autores puertorriquenos 1943 [mf ed 1992] – 1r – 1 – us UF Libraries [830]

30 [i.e. treinta] poemas / Cortes, Alfonso – Managua, Nicaragua: El Hilo Azul 1952 [mf ed 1992] – (= ser Coleccion poesia de america) – 1r – 1 – us UF Libraries [810]

30 mois de ma vie, quinze mois avant et quinze mois apres mon voyage au congo : ou ma justification des infamies debitees contre moisuivie de details nouveaux et curieux sur les moeurs et les usages des habitans du bresil et de buenos-ayres, et d'une description de la colonie patagonia / Douville, J B – Paris 1833 [mf ed Hildesheim 1995-98] – 1v on 3mf – 9 – €90.00 – 3-487-26845-0 – gw Olms [910]

30 ovi history – Brinkerhoff – 1r – 1 – mf#B33402 – us Ohio Hist [976]

30 yuniyu – Khartoum, Sudan, may 20-jun 19; jul 14-dec 15 1990; jan 1-may 11; may 25-aug 10 1991 – 1r – 1 – us CRL [960]

[31 cantatas, for one voice with a thorough bass] / Scarlatti, Alessandro – [1725?] [mf ed 19–] – 1r – 1 – mf#film 980 – us Sibley [780]

O 31 de agosto / Biblioteca Publica do Para – Belem, PA. 31 ago 1889 – (= ser Ps 19) – bl Biblioteca [079]

[31]p metabolic responses to activity of nonspecifically trained muscle tissue : in elite endurance athletes and in healthy, sedentary subjects as observed by (31)p magnetic spectroscopy / Brown, Richard L & Klug, Gary A – 1992 – 3mf – 9 – $18.00 – us Kinesology [612]

31-phosphorous, nuclear magnetic resonance spectroscopy studies of exercising human muscle / Marsh, Gregory D & Taylor, A W – 1992 – 2mf – 9 – $12.00 – us Kinesology [612]

32b – v44 n2-v45 n7 [1976 apr/may-1977 aug/sep] – 1 – mf#607437 – us WHS [071]

32b-32j – 1977 oct-1984 – 1 – mf#607432 – us WHS [071]

32d aadcom news – v25 n2, 4-4, 7-9 [1984 feb, apr-may, jul-sep]; v27 n6 [1986 jun] – 1 – mf#1345496 – us WHS [071]

33 – Newark. 1963-1976 (1) 1971-1976 (5) 1976-1976 (9) – (cont by: 33 metal producing) – ISSN: 0040-6155 – mf#5081 – us UMI ProQuest [660]

33 metal producing – Cleveland. 1977- – 1,5,9 – (cont: 33) – ISSN: 0149-1210 – mf#5081.01 – us UMI ProQuest [660]

35 nhac mrok a ja nann ne – Ran Kun: Prann kra re nhan a sam lvhan u ci thana 1982 [mf ed 1990] – 1 – (in burmese & english; title on added t.p.: 35th anniversary martyrs' day) – mf#mf-10289 seam reel 137/4 [§] – us CRL [959]

35 trabajos periodisticos : publicaciones de la secretaria de educacion, direccion de cultura / Lara, Justo de – Habana, Cuba: [Cultural, s.a.] 1935 [mf ed 1991] – (= ser Grandes periodistas cubanos 1) – 1r – 1 – us UF Libraries [972]

Les 35 votes principaux de l'assemblee nationale constituante – Paris, 1849 – us CRL [323]

38 [i.e. trente-huit] jours de voyage le general antoine simon president de la republique, dans le departement du sud / Laforest, Antoine & Osson, Rocher – Port-Au-Prince, Haiti: Impr de l'Abeille 1910 [mf ed 1991] – 1r – 1 – us UF Libraries [918]

The 39 articles of our established church – 1571 : the original latin, collated with early editions / ed by Budd, Henry – London: Edward Lumley, [18–?] – 1mf – 9 – 0-8370-8802-X – (in english and latin) – mf#1986-2802 – us ATLA [240]

39 men for one woman : an episode of the colonization of canada / Chevalier, Henri-Emile – New York: J Bradburn, 1862 [mf ed 1991 – 4mf – 9 – 0-665-90615-3 – mf#90615 – cn Canadiana [830]

40 acres and a mule – 1967 nov-1972 sep; 1979 dec; 1980 mar, jun, aug/sep-oct/nov – 1 – mf#545143 – us WHS [071]

40 saraser tam sutr / Ravivans Govid – Bhnam Ben: Pannagar Rasmi Kambuja 2506 [1963], [cover 1965] [mf ed 1990] – 1r with other items – 1 – (in khmer; title on cover in french: 40 dictees khmeres: pour toutes les classes) – mf#mf-10289 seam reel 100/06 [§] – us CRL [480]

'42 rebellion : an authentic review of the great upheaval of 1942 / Sahaya, Govinda – Delhi: Rajkamal Publ, 1947 – (= ser Samp: indian books) – us CRL [954]

50 [fuenfzig] thesen zu den kirchlichen fragen der gegenwart als positive gegenantwort auf die paepstliche einladung zum concil – Neuwied: J H Heuser, 1869 [mf ed 1986] – 1mf – 9 – 0-8370-8938-7 – (in german) – mf#1986-2938 – us ATLA [241]

Die 50 homilien des makarios / Doerries, H – Berlin, 1964 – 7mf – 8 – €15.00 – ne Slangenburg [243]

50 let potrebitelskoi kooperatsii v rossii : istoricheskii ocherk i sovremennoe sostoianie / Kheisin, M L – 1915 – 55p 1mf – 9 – mf#COR-135 – ne IDC [335]

50 t and t teamster telecast – v19 n10-v23 n9 [1978 dec-1982 nov/dec] – 1 – mf#965194 – us WHS [071]

52 questions on the nationalization of canadian railways / Fabius – Toronto: J M Dent, c1918 [mf ed 1996] – 2mf – 9 – 0-665-80239-0 – mf#80239 – cn Canadiana [380]

55 jours de gestion de cajuste bijou secretaire d'etat des finances & du commerce, du 5 aout au 30 septembre, 1903 / Haiti Departement des finances – Port-Au-Prince, Haiti: E Chenet 1904 [mf ed 1991] – 1r – 1 – us UF Libraries [336]

57. morale sinne-beelden, aen sijne hoogheydt, den doorluchtigen ende hoogh-gheboren vorst Fredrick Hendrick prince van Orangien... / [Barbonius, J] – t'Amsterdam: Theunis Jacobsz., 1641 – 2mf – 9 – mf#O-3029 – ne IDC [090]

64 bezondere zinne-beelden... – Amsterdam: Wed: Jacobus van Egmont, [c1780] – 1mf – 9 – mf#O-3022 – ne IDC [090]

El "65" [i.e. sesenta y cinco] en revista : datos historicos, relatos y anecdotas, tipos y cuentos del regimiento / Padron, Antonio E – New York NY: Las Americas Pub Co 1961 [mf ed 1993] – 1r – 1 – us UF Libraries [355]

65 news – 1958 jan-1971 jun; 1971 jul-1982 sep; 1983 jun-aug – 1 – mf#408468 – us WHS [071]

67 special reports and bibliographies / Iron and Steel Institute – 370mf – 7 – (full title list available on request) – mf#540 – uk Microform Academic [670]

70 disputationes theologicae adversus pontificios / Hommius, F – Lugduni Batavorum, 1614 – 6mf – 9 – mf#PBA-193 – ne IDC [240]

70 horas tragicas : los fusilamiento de junio en la argentina – Montevideo, (1957) – 1 – us CRL [972]

72 dnia pervogo russkogo parlamenta / Tsitron, A – 1906 – 165p 2mf – 9 – mf#RPP-48 – ne IDC [325]

73 amateur radio – Peterborough NH 1987-89 – 1,5,9 – (cont: 73 for radio amateurs international ed; cont by: 73 amateur radio today) – ISSN: 0889-5309 – mf#2393.06 – us UMI ProQuest [380]

73 amateur radio – Peterborough NH 1975-77 – 1,5,9 – (cont: 73 magazine for radio amateurs; cont by: 73 magazine for radio amateurs) – mf#2393.06 – us UMI ProQuest [380]

73 amateur radio today – Peterborough NH 1991- – 1,5,9 – (cont: 73 amateur radio) – ISSN: 1052-2522 – mf#2393.06 – us UMI ProQuest [380]

73 amateur radio's technical journal : international edition – Peterborough NH 1983-84 – 1,5,9 – (cont: 73 magazine for radio amateurs; cont by: 73 for radio amateurs international ed) – ISSN: 0745-080X – mf#2393.06 – us UMI ProQuest [380]

73 magazine for radio amateurs – Peterborough NH 1979-1981+ – 1 – (cont: 73 amateur radio; cont by: 73 amateur radio's technical journal [international ed]) – mf#2393.06 – us UMI ProQuest [380]

73 magazine for radio amateurs – Peterborough NH 1960-73 – 1,5,9 – (cont by: 73 amateur radio; former title(s): amateur radio 73 [oct 1960-dec 1971]) – ISSN: 0098-9010 – mf#2393.06 – us UMI ProQuest [380]

75 favorite square dance calls : [for dancers and teachers] / McVicar, Wes – Niagara Falls, NY: G V Thompson [c1949] – mf#*ZBD-*MGO pv 24 – Located: NYPL – us Misc Inst [790]

76 bicentennial courier – v1 n1-7 [1971 jan-sep] – 1 – mf#407161 – us WHS [071]

76 chester county bicentennial courier – v6 n1-8 [1976 apr-nov] – 1 – mf#407163 – us WHS [071]

76 pennsylvania bicentennial courier – v1 n9-v5 n3 [1971 dec-1975 3rd quarter] – 1 – mf#407162 – us WHS [071]

80 micro – Peterborough NH 1982-88 – 1,5,9 – (cont: 80 microcomputing) – ISSN: 0744-7868 – mf#12201.01 – us UMI ProQuest [000]

80 microcomputing – Peterborough NH 1980-82 – 1,5,9 – (cont by: 80 micro) – ISSN: 0199-6789 – mf#12201.01 – us UMI ProQuest [000]

80's theoretical journal of the communist workers party, usa – v1 n1-v2 n3 [1980 may-1981 dec] – 1 – mf#626195 – us WHS [071]

Le 89 du clerge / Cayla, Jean-Mamert – Paris: E Dentu 1861 [mf ed 1995] – 1r – 1 – mf#c2440 – us Harvard [241]

90 nian xin wen zi liao jian ji mu lu *see* Xianggang bao zhang jian bao

95th edition – 1971 dec 13-1980 oct 10 – 1 – mf#671860 – us WHS [071]

99 news – v19 n1-6 [1993 jan/feb-nov/dec]; v20 n1-4 [1994 jan/feb-jul/aug] – 1 – mf#1701657 – us WHS [071]

100 ganu baky samramn / Jym Bau – Bhnam Ben: Panagar P'ut Nan, 2504 [1961] [mf ed 1990] – [ill] 1r with other items – 1 – (in khmer) – mf#mf-10289 seam reel 102/05 [§] – us CRL [480]

100 jahre luechower heimatzeitung *see* Elbe-jeetzel-zeitung

100 jahre wallraf-richartz museum 1861-1961 / Wallraf-Richartz-Museum. Cologne – 1961 – 1mf – 9 – uk Chadwyck [700]

100 melodias folcloricas : documentario musical nordestino / Araujo, Alceu Maynard – Sao Paulo, Brazil: Ricordi Brasileira S A 1957, c1958 – 1r – 1 – (aka: cem melodias folcloricas) – us UF Libraries [780]

100 silberne blechbestecke : erlebte heitere geschichten / Wiese, Ernst August – Leipzig: Amthorsche Verlagsbuchhandlung, 1942 [mf ed 1992] – 63p (ill) – 1 – (ill by rudolf haupt) – mf#7770 – us Wisconsin U Libr [870]

101 things for adult bible classes to do / Moninger, Herbert – Cincinnati, O[hio]: Standard Pub Co 1911 [mf ed 1993] – (= ser Christian church (disciples of christ) coll) – 1mf [ill] – 9 – 0-524-06486-5 – mf#1991-2586 – us ATLA [240]

102 monitor – Washington DC 1975-1981 (1,5,9) – ISSN: 0090-3574 – mf#7919 – us UMI ProQuest [324]

103 dnia vtoroi dumy / Tsitron, A – 1907 – 188p 3mf – 9 – mf#RPP-49 – ne IDC [325]

105 forerunner – 1979 jan 11-1981 mar 12 – 1 – mf#1743162 – us WHS [071]

115 versetten und cadenzen fuer die orgel in den gewoehnlichen 8 kirchen tonarten...ch 1 / Eberlin, Johann Ernst – Muenchen: Falter & Sohn [180-] [mf ed 19–] – 1mf – 9 – mf#fiche 449 – us Sibley [780]

125 ovi opdyke tigers : regimental history, 1895 – 1r – 1 – mf#B33080 – us Ohio Hist [355]

The 125th anniversary report of the american sunday school union / American Sunday School Union – 1942 – 1 – $58.72 – mf#0894 – us Southern Baptist [242]

127th tactical fighter wing – 1985 sep-1988 jan – 1 – mf#1840721 – us WHS [071]

Der 128 psalm vom glueck, segen, gedeien der eheleut / Corvinus, A – [Hildensheim], 1543 – 3mf – 9 – mf#TH-1 mf 354-356 – ne IDC [242]

135 Medical Regiment [Organization] *see*
– Bull sheet
– Chaplains' bulletin
– Christmas bulletin

136th feria de santiago para toda clase de ganados 1971 / Casatejada. Ayuntamiento – Caceres: Imp La Minerva, 1971 – 1 – sp Bibl Santa Ana [390]

139 fotografias del movimiento nacional en sevilla – Sevilla, 193? – (= ser Blodgett coll) – 9 – mf#fiche w800 – us Harvard [946]

141st feria de santiago 1976 – Caceres: Imp. La Minerva, 1976 – 1 – sp Bibl Santa Ana [390]

142nd feria de santiago para toda clase de ganados / Casatejada. Ayuntamiento – Caceres: Imp. La Minerva, 1977 – sp Bibl Santa Ana [390]

144th feria de santiaago. julio de 1979 / Casatejada. Ayuntamiento – Caceres: Tip La Minerva, 1979 – 1 – sp Bibl Santa Ana [390]

165 update – 1981 jun-1983 mar – 1 – mf#711907 – us WHS [071]

171 – 1976 mar-1980 jun – 1 – mf#630277 – us WHS [071]

172p *see* Von der musica und den meistersaengern

174 news – n111-130 [1978 apr-1983 jul] – 1 – mf#1745872 – us WHS [071]

178th thunderer – v23 n6-12 [1981 jun-dec]; v24 n1-5, 7-12 [1982 jan-may, jul-dec]; v25 n1-2, 4-8, 10-11 [1983 jan-feb, apr-aug, oct-nov]; v26 n8, 11-12 [1984 aug, nov-dec]; v27 n1-3, 5-6, 8, 10-11 [1985 jan-mar, may-jun, aug, oct-nov]; v18 n? – 1 – mf#1520636 – us WHS [071]

180 [i.e. ciento ochenta] dias en el frente / Arango Uribe, Arturo – Manizales, Colombia: Tip Cervantes [1933] [mf ed 1993] – 1r – 1 – us UF Libraries [972]

191st Fighter Interceptor Group *see* Newsletter

199 news – v17 n2-v24 n5 [1982 mar-1988 dec]; v24 n6-9 [1989 sep-dec]; v25 n1-9 [1990 jan-dec]; 1991 feb-jun; 1992 feb-may – 1 – mf#1051405 – us WHS [071]

200e anniversaire de la decouverte du mississipi sic par jolliet et le p marquette : soiree litteraire et musicale a l'universite laval le 17 juin 1873 – Quebec: L H Huot, 1873 – 1mf – 9 – mf#25220 – cn Canadiana [917]

211 park st : a newsletter of the afro-american cultural center at yale – 1985 apr – 1 – mf#4882459 – us WHS [071]

233 register – 1975 apr-1982 aug – 1 – mf#691523 – us WHS [071]

255 stranits maiakovskogo. knj 1 / Mayakovsky, Vladimir – Moskva: Gos. izd-vo, 1923 [mf ed 2002]+ – 1r – 1 – (varying form of title: dvesti piat'desiat piat' stranits maiakovskogo. filmed with: listy sada morii / [e i rerikh], (paris 1924) & other titles) – mf#5214 – us Wisconsin U Libr [080]

309 review – 1982 aug-sep; 1983 mar, may-jun, oct-dec; 1984 jan-may – 1 – mf#1222521 – us WHS [071]

311 log sheet – 1954 jan-1958 oct – 1 – mf#3162504 – us WHS [071]

314-o-gram : official publication of atomic energy lodge 314, iamaw – 1971 feb; 1972 oct-1986 mar 14 – 1 – mf#1223120 – us WHS [071]

325 news – 1979 jun-1983 dec; 1984 dec; 1985 jun, sep, dec; 1986 may, dec; 1987 mar v1 n1-v18 n4, 5-v28 n3 [1944 jun-1961 dec; 1964 may-1976 sep]; v29 n1 [1977 nov] – 1 – mf#958145 – us WHS [071]

328 news digest – v18 n1; v19 n3 [1977 fall; 1978 winter] – 1 – mf#657700 – us WHS [071]

328 news digest – v24 n1-4; ns: v1 n1 [1982 spring-winter; 1983 spring] – 1 – mf#657724 – us WHS [071]

328a news digest – v20 n2; v22 n1-v23 n1 [1979 fall; 1980 fall-1981 fall] – 1 – mf#657715 – us WHS [071]

338 news – 1968 jan-1984 sep/oct – 1 – mf#412274 – us WHS [071]

340 leader – 1962 apr-1974 dec; 1974 dec-1979 mar/apr – 1 – mf#641282 – us WHS [071]

349th globe – v3 n5-7, 11 [1981 jun-aug, nov]; v4 n2-10 [1982 feb-oct] – 1 – mf#1065183 – us WHS [071]

408 news – 1983 feb-1994 oct/dec – 1 – mf#1051408 – us WHS [071]

444 forum – 1980 jul/aug-1992 jul/sep – 1 – mf#1051409 – us WHS [071]

454 news – v29 n10-v34 n4;v35 n5-v39 n1; [1976 dec-1981 apr;1982 may-1986 spring] – 1 – mf#920065 – us WHS [071]

494 relay – n117-147 [1953 dec-1961 dec]; n147-148 [1962 may-1962 sep] – 1 – mf#1051410 – us WHS [071]

525 geistliche lieder und psalmen : welche in den christlichen gemeinen und versammlungen auch bey ausstheilung der hochwuerdigen sacrament gesungen werden moegen – Nuernberg: Dieterich 1599 – (= ser Hqab. literatur des 16. jahrh.) – 10mf – 9 – €90.00 – (with ind) – mf#1599a – gw Fischer [780]

535 leader – 1979 may/jun-1981 jul/aug – 1 – mf#900933 – us WHS [071]

588's voice – 1986-90 – 1 – mf#2744915 – us WHS [071]

600 dias con fidel tres misiones en la habana / Lequerica Velez, Fulgencio – 1 ed. Bogota, Colombia: Ediciones Mito [1961?] [mf ed 1993] – 1r – 1 – (aka: seiscientos dias con fidel) – us UF Libraries [327]

689 news – v9 n12-v21 n3 [1982 dec-1995 mar] – 1 – mf#1051415 – us WHS [071]

700 years of hollidays / Holliday, Omar – 1939 – 1 – $10.00 – mf#0741 – us Southern Baptist [390]

751 aero mechanic – v40 n5-v41 n11 [1981 may 13-1985 dec] – 1 – mf#999984 – us WHS [629]

807 teamster – v25 n1-43 [1979 jan-1984 jul] – 1 – mf#717351 – us WHS [071]

816 express – v9 n1-v22 n1 [1966 jan-1977 mar] – 1 – mf#568762 – us WHS [071]

830 reporter – v19 n10-v22 n1 [1979 jan-1983 jan] – 1 – mf#1313279 – us WHS [071]

870 news : a publication of united food and commercial workers union, local 870 – v19 n3, 5-7, 10-12 [1988 mar, may-jul, oct-dec]; v20 n1-4; 6-12 [1989 jan-apr, jun-dec]; v21 n1-7 [1990 jan-jul] – 1 – mf#2699801 – us WHS [071]

880 news and views – 1975 jan-1978 dec; 1979 sep, dec; 1980 mar-1983 sep – 1 – mf#685818 – us WHS [071]

888 leader – v13 n1-v20 n1 [1975 apr-1982 mar] – 1 – mf#670061 – us WHS [071]

899 good land = tierra buena – v1 n1-4 [1976 sep-dec] – 1 – mf#676442 – us WHS [071]

900 = Neuf cents – Rome, Florence. n1-5. 1926-27 – 1 – fr ACRPP [073]

9:02 times – 1978 jul-aug; 1978 dec; 1979 may-jun; 1979 sep; 1979 dec – 1 – mf#708542 – us WHS [071]

917 *see* Geschichte und beschreibung von newfoundland und der kueste labrador

992's news and views – v1-3; ? [1979 feb-jul; oct-1980 jul] – 1 – mf#653744 – us WHS [071]

992's news and views – v4 n2-v5 n3; v1-4 [1974 may-1975 oct; 1976 jun/jul; 1977 jan-aug; 1978 dec] – 1 – mf#679630 – us WHS [071]

1000 jours de resistance nationale : le drame d'une guerre imposee a un peuple pacifique – Phnom Penh: Khmer Republic [mf ed 1989] – 1r with other items – 1 – (special iss of french ed of khmer republic sep 1973) – mf#mf-10289 seam reel 021/01 [§] – us CRL [959]

1000 tage westfront : die erlebnisse eines einfachen soldaten / Wallenborn, Franz – Leipzig: Hesse & Becker, 1929 [mf ed 1993] – 280p – 1 – mf#7830 – us Wisconsin U Libr [830]

1001 home ideas – Palm Coast FL 1986-1991 (1,5,9) – ISSN: 0278-0844 – mf#14396,02 – us UMI ProQuest [640]

1033 news and views – v21 n10-v24 n9 [1977 feb-1985 sep] – 1 – mf#920069 – us WHS [071]

1045 news – v35 n1-v44 n1 [1984 feb-1993 feb]; v6 n1-v34 n6 [1955 feb-1983 dec] – 1 – mf#920271 – us WHS [071]

1092 review – 1981 jan-apr, jun, oct-dec; 1982 feb, may-jun, aug-sep, nov – 1 – mf#122266 – us WHS [071]

1100's gate way – 1987 jun-1988 jul – 1 – mf#2699854 – us WHS [071]

1190 bulletin / United Steelworkers of America – 1979 jul-nov; 1980-aug, dec; 1981 mar-may, aug; 1982 mar – 1 – mf#649015 – us WHS [331]

1212 reporter : voice of the tri-state steelworkers crucible-midland – v1 n1-v3 n3 [1976 jul-1979 mar] – 1 – mf#524843 – us WHS [071]

1245 news – v14 n7-v20 n2 [1974 apr-1981 oct] – 1 – mf#615885 – us WHS [071]

1262 banner / United Food and Commercial Workers International Union – 1975-1979 oct, 1979 nov-1989 jan – 2r – 1 – (cont by: local 1262 banner) – mf#666103 – us WHS [331]

1288 grapevine / Retail Clerks Union – v1 n1 [1975 dec]; v2 n10 [1976 oct] – 1 – mf#679111 – us WHS [331]

1293 tuerk-rus muharebesi hakayikindan hulasa-i vukuat-i harbiye / Pasa, Sueleyman – [Istanbul], 1324 [1906] OFC1115 – (= ser Ottoman histories and historical sources) – 2mf – 9 – $40.00 – us MEDOC [956]

1297 [1880] sene-i hicriyesine mahsus salname-i kameri / Tevfik, Ebuezziya – Istanbul: Mihran Matbaasi – (= ser Ottoman histories and historical sources) – 5mf – 9 – $75.00 – us MEDOC [956]

1324 balkan harbinde sark ordusu kumandant abdullah pasa hatirati – Istanbul: Erkan-i Harbiye Mektebi Matbaasi, 1336 [1924] – (= ser Ottoman histories and historical sources) – 4mf – 9 – $60.00 – us MEDOC [956]

1335 senesi tuerk sanayi sergisi katalogu – Istanbul: Tuerk Duenyasi Matbaasi, 1335 [1919] – 1mf – 9 – $25.00 – us MEDOC [380]

1360 newsline / Local 1360, Retail Clerks Union (Camden NJ) – 1979 mar-jun, 1980 jan-nov – 1r – 1 – (cont by: newsline (united food and commercial workers union local 1360 (marlton, nj)) – mf#1337436 – us WHS [380]

1397 rank and file – 1980 feb-dec; 1981 sep – 1 – mf#626854 – us WHS [071]

1428 message – 1956 aug-1977 dec – 1 – mf#555646 – us WHS [071]

1444 monitor – 1981 jan-aug; 1981 jul-1992 may/jun; 1986 mar/apr-1992 may/jun – 1 – mf#676608 – us WHS [071]

1557 labor journal – v13 n6-v18 n1 [1975 jun-1981 oct] – 1 – mf#622096 – us WHS [331]

1582 [pontificale romanum ad omnes pontificias caeremonias : quibus nunc utitur sacro sancta r. e. accommodatum] gregorio 13. pont. max. pontificale romanvm ad omnes pontificias caeremonias... – rev enl ed, Venetiis: Iuntas 1582 – (= ser Hqab. literatur des 16. jahrh.) – 6mf – 9 – €70.00 – mf#1582b – gw Fischer [780]

1600 college union voice – v1 n1-v13 n2-3 [1966 jun-1976 sep/oct] – 1 – mf#642447 – us WHS [378]

1650 newsletter – 1980 mar 7-1984 dec 7; 1984 jan 4-1988 jan 7 – 1 – mf#961388 – us WHS [071]

1776 gazette – v1 n1-52 [1976] – 1 – mf#187472 – us WHS [071]

1793 beitrag zur geheimen geschichte der franzoesischen revolution : mit besonderer beruecksichtigung danton's und challier's; zugleich als berichtigung der in den werken von thiers und mignet enthaltenen schilderungen / Funck, Friedrich – Mannheim 1843 [mf ed Hildesheim 1995-98] – 1v on 3mf – 9 – €90.00 – 3-487-26272-X – gw Olms [944]

1804 [i. e. dix-huit cent quatre] economiquement / Roy, Fernand Alix – Port-Au-Prince, Haiti: Impr "Les Presses libres" 1960 [mf ed 1992] – 1r – 1 – us UF Libraries [330]

1818 [i. e. mil ochocientos diez y ochio] : guerra de independencia / Vergara y Velasco, Francisco Javier – Bogota, Colombia: Editorial Kelly 1960 [mf ed 1993] – (= ser Academia colombiana de historia biblioteca eduardo santos 23) – 1r – 1 – us UF Libraries [972]

1828 nouveaux memoires secrets : pour servir a l'histoire de notre temps / Musset-Pathay, Victor Donatien de – Paris 1829 [mf ed Hildesheim 1995-98] – 1v on 3mf – 9 – €90.00 – 3-487-26140-5 – gw Olms [880]

1830 federal population census for indiana : index / Alig, Tobey – 1981 – 9 – $12.00 – us IHS [317]

1837 : my connection with it / Brown, Thomas Storrow – Quebec: R Renault, 1898 – 1mf – 9 – (originally publ in the New Dominion Monthly, v4 n1) – mf#00318 – cn Canadiana [971]

1840 : muerte de santander / Academia Colombiana de Historia – Bogota, Colombia: Editorial Cromos 1940 [mf ed 1993] – 1r – 1 – us UF Libraries [972]

1840 federal population census, indiana : index / Genealogy Division, Indiana State Library – 1975 – 9 – $12.00 – us IHS [317]

1841 census surname index : the county of angus including the burgh of dundee / Binnie, William [comp] – Dundee: Tay Valley Family History Society c1999 – 4mf – 9 – uk Tay Valley [929]

1841 et 1941 : ou, aujourd'hui et dans cent ans. revue fantastique en deux actes, a grand spectacle / Cogniard, Theodore & Muret, Theodore Cesar – Paris: Beck [1841?] [mf ed 1994] – (= ser Repertoire dramatique des auteurs contemporains 182) – 1r – 1 – us UF Libraries [820]

1841 street index for metropolitan london middlesex/surrey – [mf ed 1996] – 9 – at Australian [941]

1842 [mil huit cent quarante-deux] au cap tremblement de terre / Delorme, Demesvar; ed by Lambert, Jean M – Cap-Haitien, Haiti: Impr du Progres 1942 [mf ed 1991] – 1r – 1 – us UF Libraries [550]

1851 / Calendario de Extremadura – Madrid: Est.Tip.de los Sres. Martinez y Minuesa, 1850 – 1 – (1849, 1852, 1833, 1828, 1845, 1857) – sp Bibl Santa Ana [946]

1851 census enumerators' returns / Anderson, Michael [comp] – [mf ed Chadwyck-Healey] – 145mf – 9 – (with p/g) – uk Chadwyck [314]

1851 census index for the county of lanarkshire [carmunock, east kilbride and libberton] / Lanarkshire Family History Society – [Lanark?]: Lanarkshire Family History Society c1998 – 1mf – 9 – uk Lanarkshire [929]

1851 census index for the county of lanarkshire [carstairs, crawford, culter, dalziel] / Lanarkshire Family History Society – [Lanark?]: Lanarkshire Family History Society c1998 – 1mf – 9 – uk Lanarkshire [929]

1851 census index for the county of lanarkshire [shotts] / Lanarkshire Family History Society – [Lanark?]: Lanarkshire Family History Society c1998 – 1mf – 9 – uk Lanarkshire [929]

1851 census of berkshire, vol 2, hungerford : index and transcript – Reading: Berkshire Family History Society c2000 – 5mf – 9 – (incl parishes: avington, chilton foliat, east garston, east shefford, hungerford, inkpen, kintbury, lambourn, shalbourne, west shefford, west woodhay) – uk BerksFHS [314]

1851 census of berkshire, vol 11, easthampstead : index and transcript – Reading: Berkshire Family History Society c2002 – 4mf + notes – 9 – (title fr notes; incl binfield, bracknell, easthampstead, sandhurst, sandhurst military college, warfield, winkfield) – uk BerksFHS [314]

1851 census of berkshire, vol 1:2, newbury : index and transcript – Reading: Berkshire Family History Society c2000 – 4mf – 9 – (title fr accompanying bklet of notes; incl parishes: crookham heath, enborne, hampstead marshall, newbury town, newbury workhouse, newtown (hants), sandleford) – uk BerksFHS [314]

1851 census of berkshire, vol 1:3, speen : index and transcript – Reading: Berkshire Family History Society c2002 – 4mf + notes – 9 – (title fr notes; incl bagnor, benham buslett, boxford, chieveley, cold blow, curridge, donnington, downend, easton, furzehill, hoe benham, leckhampstead, nalderhill, oare, ownham, shaw, snelsmore, speen, stockcross, welford, westbrook, weston, wickham, winterbourne) – uk BerksFHS [314]

1851 census of berkshire, vol 5:1, wantage : index and transcript. – Reading: Berkshire Family History Society c2000 – 4mf – 9 – (incl parishes: childrey, denchworth, east challow, east hanney, goosey, grove, letcombe regis, letcombe bassett, north denchworth, sparsholt, wantage, wantage union workhouse, west challow, west hanney, westcot) – uk BerksFHS [314]

1851 census of berkshire, vol 8:1, reading st mary : index and transcript – Reading: Berkshire Family History Society c2001 – 4mf + notes – 9 – (incl reading st mary, southcote) – uk BerksFHS [314]

1851 census of berkshire, vol 8:2, reading st lawrence & reading st giles : index and transcript – Reading: Berkshire Family History Society c2001 – 6mf + notes – 9 – (incl reading st mary, reading st lawrence, whitley) – uk BerksFHS [314]

1860 census index / Franklin Co, OH – 2r – 1 – mf#B31130-31131 – us Ohio Hist [317]

1860 india census index – Indianapolis: Indiana Historical Society, 1981 – 159mf – 9 – $5.00mf – us IHS [315]

1869 times – v1 n1-v8 n3 [1975 nov-1983 may] – 1 – mf#916665 – us WHS [071]

1870 ballot for priority of choices in selection of country sections *see* Northern territory land orders/adelaide and london registers 1870 ballot

1883 voters list of the municipality of the township of east zorra – [Embro, Ont?: s.n, 1883?] [mf ed 1995] – 1mf – 9 – 0-665-88922-4 – mf#88922 – cn Canadiana [325]

1884 voters' list, municipality of the township of west zorra – [Embro, Ont?: s.n, 1884?] [mf ed 1995] – 1mf – 9 – 0-665-88889-9 – mf#88889 – cn Canadiana [325]

1885 clinton voters' list : john callander, clerk – [Clinton, Ont?: s.n, 1885? [mf ed 1995] – 1mf – 9 – 0-665-88915-1 – mf#88915 – cn Canadiana [325]

1887 departures from port darwin *see* Port darwin shipping, 1887, 1898, 1899 and 1900

1887 inward shipping for the northern territory *see* Port darwin shipping, 1887, 1898, 1899 and 1900

1889 : die erste erhebung der bergarbeiter. zur erinnerung an den grossen bergarbeiterstreik vor 20 jahren / Bredenbeck, Anton – Dortmund 1905 – 1 – gw Mikropress [331]

1889 shipping to port darwin *see* Port darwin shipping, 1887, 1898, 1899 and 1900

1894 settler's guide, province of quebec / Flynn, Edmund James [comp] – S.l: s.n, 1894? – 2mf – 9 – mf#01455 – cn Canadiana [917]

1896 guide du colon : province de Quebec... / Flynn, Edmund James [comp] – S.l: s.n, 1896? – 2mf – 9 – mf#59150 – cn Canadiana [630]

1898 inward shipping for the northern territory *see* Port darwin shipping, 1887, 1898, 1899 and 1900

1899 inward/outward shipping for the northern territory *see* Port darwin shipping, 1887, 1898, 1899 and 1900

1900 census schedules / U.S. Bureau of the Census – (= ser 1900 federal population census) – 1854r – 1 – (arr by state or territory and thereunder by county. names of large cities may appear. not in strict alphabetical order. printed catalog available) – mf#T623 – us Nat Archives [317]

1900 [millenovecento]-1901 [millenovecentuno] – New York NY, 1901* – 1r – 1 – (italian periodical) – us IHRC [073]

1900 shipping to port darwin *see* Port darwin shipping, 1887, 1898, 1899 and 1900

1910 census schedules / U.S. Bureau of the Census – (= ser 1910 federal population census) – 1784r – 1 – (arr by state or territory and thereunder by county. names of large cities may appear. not in strict alphabetical order. printed catalog available) – mf#T624 – us Nat Archives [317]

1910 soundex/miracode / U.S. Bureau of the Census – (= ser 1910 federal population census) – 4642r – 1 – (alabama 140r t1259 (soundex). arkansas 139r tl260 (miracode). california 272r t1261 (miracode). florida 84r t1262 (miracode). georgia 174r t1263 (soundex). illinois 491r t1264 (miracode). kansas 145r t1265 (miracode). kentucky 194r t1266 (miracode). louisiana 132r t1267 (soundex). michigan 253r t1268 (miracode). mississippi 118r t1269 (soundex). missouri 285r t1270 (miracode). north carolina 178r t1271 (miracode). ohio 418r t1272 (miracode). oklahoma 143r t1273 (miracode). pennsylvania & philadelphia county 688r t1274 (miracode). south carolina 93r t1275 (soundex). tennessee 142r t1276 (soundex). texas 262r t1277 (soundex). virginia 183r t1278 (miracode). west virginia 108r t1279 (miracode)) – us Nat Archives [317]

1915 index to the northern territory times and gazette – [mf ed 1993] – 1mf – 9 – A$5.50 – 0-949124-66-4 – at Northern [079]

1916 and 1917 index to the northern territory times and gazette – [mf ed 1996] – 2mf – 9 – A$11.00 – 0-949124-69-9 – at Northern [079]

1918 and 1919 index to the northern territory times and gazette – [mf ed 1996] – 1mf – 9 – A$5.50 – 0-949124-72-9 – at Northern [079]

1920 census schedules / U.S. Bureau of the Census – (= ser 1920 federal population census) – 2076r – 1 – (arr by state or territory, and thereunder by county, and finally by enumeration district. the states are arr alphabetically; however, alaska, guam and american samoa, hawaii, military and naval schedules, the panama canal, puerto rico, and the virgin islands (taken in 1917) are listed last. there was no separate indian schedule for 1920) – mf#T625 – us Nat Archives [317]

1920 index to the northern territory times and gazette – [mf ed 1996] – 1mf – 9 – A$8.80 – 0-949124-85-0 – at Northern [079]

1921 index to the northern territory times and gazette – [mf ed 1996] – 1mf – 9 – A$5.50 – 0-949124-92-3 – at Northern [079]

1924 : a magazine of the arts – Woodstock, New York. n1-4. jul-dec 1924 – 1 – us NY Public [700]

1924 supplement to rosbrook on new york corporations (2d ed) and bender's corporation manual / Rosbrook, Alden Ivan – Albany: Bender 1924 – 1 – mf#LL-1471 – us L of C Photodup [346]

1926 senesi ziraat istatistikleri – Istanbul, 1926 – 1mf – 9 – $25.00 – (tuerkiye cumhuriyeti ziraat vekaleti) – us MEDOC [350]

"The 1927 excavations at beisan" / Rowe, Alan – University of Pennsylvania Museum Journal: Jun 1928 – 9 – $10.00 – us IRC [932]

1933 : a year magazine – Philadelphia. June c 1933-Dec 1933 Apr 1934 – 1 – us NY Public [073]

1937 (1 jan-dec) / economische opstellen uit de inheemsche pers – Batavia, [1938] – 3mf – 8 – mf#SE-1283 – ne IDC [959]

1945 : ein jahr in dichtung und bericht / Rauschning, Hans; ed by Rauschning, Hans – Frankfurt/M: Fischer, c1965 – (= ser Fischer buecherei 663) – 262p – 1 – (incl bibl ref) – mf#8156 – us Wisconsin U Libr [430]

1956 population census, 8th april 1956 : report / Taylor, D H – Maseru: Basutoland Govt, 1958 – us CRL [316]

1960 agricultural census, basutoland / Morojele, C M H – Maseru: Statistics Dep. pt1-6. 1962-65 – us CRL [630]

1960 census of british virgin islands / Jamaica Dept of Statistics – Kingston, Jamaica [1961?] [mf ed 1993] – 1r – 1 – (aka: west indies population census) – us UF Libraries [350]

1966 population census report : kingdom of lesotho / Lesotho. Bureau of Statistics – Maseru: the Bureau, 1966? – us CRL [316]

1967 general elections manifestoes – New Delhi: Educational Resources Ctr, 1967 – us CRL [325]

1970 colorado comprehensive outdoor recreation plan / Colorado. Dept of Natural Resources – [Denver]: Div of Game, Fish & Parks, [1970?] – us CRL [790]

1970 decennial census guide and microfiche / U.S. Bureau of the Census – 3857mf – 9 – $6380.00 silver $5425.00 diazo – (coll incl printed aid. subsets incl: 1970 census of housing (incl block reports and maps): $3950 silver, $3355 diazo. census of population: $1670 silver, $1420 diazo. census of population and housing: $1470 silver $1250 diazo) – us CIS [317]

1976 colorado comprehensive outdoor recreation plan / Colorado. Div of Parks & Outdoor Recreation – [Denver: Comprehensive Planning], 1975 – us CRL [790]

The 1979 federal district court time study / Flanders, Steven – Washington: FJC, Oct 1980 – 2mf – 9 – $3.00 – mf#LLMC 95-825 – us LLMC [347]

1980 decennial census guide and microfiche / U.S. Bureau of the Census – 6828mf – 9 – $12,910.00 silver $10,975.00 diazo – (Basic coll excl block reports and maps 3767mf $6,295 silver $5,350 diazo. Subsets include: 1980 Census of Housing: $1,755 silver, $1,490 diazo. Census of Population: $2,330 silver, $1,980 diazo. Census of Population and Housing (excludes Block Reports and Maps): $2,915 silver $2,480 diazo. Block Reports and Maps: $6,615 silver $5,625 diazo. Guide avaliable free with mf, $195 if purchased separately) – us CIS [317]

The 1981 bankruptcy court time study – Washington: FJC, Nov 1982 – 2mf – 9 – $3.00 – mf#LLMC 95-832 – us LLMC [346]

1988 presidential election polls : the gallup/conus reports / Gallup Organization – 1990 – 5r – 1 – $650.00 – (With printed guide) – mf#S3210 – us Scholarly Res [977]

1990 decennial census guide and microfiche / U.S. Bureau of the Census – 4216mf – 9 – $9,995.00 silver $8,495.00 diazo – (subsets include: 1990 census of housing: $2,295 silver $1,950 diazo. census of population: $2,865 silver $2,435 diazo. census of population and housing: $8,995 silver $7,645 diazo. census tract and block numbering area maps only: $2,970 silver $2,525 diazo. related commerce dept, gao, & congressional background reports (guides not included) $580 silver $495 diazo. guide free with mf, $345 if purchased separately) – us CIS [317]

The 2001 presidential awardees for excellence in mathematics and science teaching : views from the blackboard: hearing...house of representatives, 107th congress, 2nd session, march 20 2002 / United States. Congress. House. Committee on Science. – Washington: US GPO 2002 [mf ed 2002] – 1mf – 9 – (incl bibl ref) – us GPO [370]

The 2002 winter olympics in salt lake city, utah : cooperation between federal, state, local and private agencies to address public safety concerns: hearing...united states senate, 107th congress, 1st session, may 31 2001, salt lake city, utah / United States. Congress. Senate. Committee on the Judiciary – Washington: US GPO 2002 [mf ed 2002] – 1mf – 9 – 0-16-067083-7 – us GPO [350]

The 2003 budget : a review of the hhs health care priorities: hearing...house of representatives, 107th congress, 2nd session, march 13 2002 / United States. Congress. House. Committee on Energy and Commerce. Subcommittee on Health – Washington: US GPO 2002 [mf ed 2002] – 1mf – 9 – 0-16-068657-1 – us GPO [350]

20/20 vision – 1984 jul-1987 oct – 1 – mf#1534226 – us WHS [071]

2108 news – 1984 jul-1992 dec – 1 – mf#352088 – us WHS [071]

2227 news – v10 n6-v17 n5 [1975 jun-1983 dec] – 1 – mf#713040 – us WHS [071]

3657 newsletter : voice of local 3657, uswa – afl-cio/clc – v1 n1-2 [1976 jan/feb-mar/apr] – 1 – mf#673281 – us WHS [071]

3657 pointer : voice of local 3657, uswa – afl-cio/clc – v1 n3-5 [1976 may/jun-oct/dec] – 1 – mf#673283 – us WHS [071]

4,000 dollars reward! : a proclamation = 4,000 piastres de recompense!: proclamation – [Bas-Canada]: John Charlton Fisher et William Kemble, [1837] (mf ed 1988) – 1mf – 9 – mf#SEM105P922 – cn Bibl Nat [971]

$4,000 reward : ...by his excellency the right honourable archibald, earl of gosford...a proclamation, whereas, by information upon oath, it appears that Louis Joseph Papineau...is charged with the crime of high treason... – Quebec?: s.n, 1837? [mf ed 1986] – 1mf – 9 – 0-665-55271-8 – mf#55271 – cn Canadiana [971]

4080 : the hip hop monthly for the greater bay area – 1993 dec-1995 dec; 1996 feb-1997 aug – 1 – mf#2901318 – us WHS [780]

5296 bay news – v1 n1-v3 n4 [1975 oct-1977 jun] – 1 – mf#673259 – us WHS [071]

5512 dispatch – iss n1-n9 [1979 mar 14-dec 5], v2 n1-v6 n9 [1980; jan-1984 sep] – 1 – mf#922991 – us WHS [071]

5698 beacon – v1 n1-v5 n6 [1980 feb-1985 dec] – 1 – mf#1124939 – us WHS [071]

11574 and more – v4 n2,7-v8 n4 1980 feb/mar, jul/aug-1983 oct] – 1 – mf#1125097 – us WHS [071]

11593 – v1 n2-v5 [special iss] [1976 sep-1980 fall] – 1 – mf#3162544 – us WHS [071]

19806 chronicle – 1951 apr-1961 dec; 1962 jan-1974 feb – 1 – mf#692707 – us WHS [071]

1504-1904 : festschrift zum gedaechtnis philipps des grossmuetigen, landgrafen von hessen, geboren am 13. november 1504 / ed by Verein fuer Hessische Geschichte und Landeskunde – Kassel: G Dufayel 1904 [mf ed 1996] – (= ser Zeitschrift des verein fuer hessische geschichte und landeskunde 38 [nf v28]) – 1r – 1 – (incl bibl ref) – mf#film mas 26638 – us Harvard [943]

1734-1884 : sesqui-centennial...silver springs presbyterian church / Ferguson, Thomas James – 1885 – 1 – $50.00 – us Presbyterian [242]

1868-1918 : address by the hon wm c edwards to the rockland employees on the 23rd august, 1919 / Edwards, William Cameron – [Ottawa?: s.n, 1919] – 1mf – 9 – 0-665-78159-8 – (in english and french) – mf#78159 – cn Canadiana [331]

1876-1896, die ersten zwanzig jahre der bayreuther buehnenfestspiele / Chamberlain, Houston Stewart – Bayreuth: L Ellwanger vorm Th Burger 1896 – 1mf – 9 – mf#wa-14 – ne IDC [780]

1877-1878 osmanli-rus seferinde osmanli kumandanlari – Istanbul: Matbaa-yi Ebueziyya, 1329 [1912] – (= ser Ottoman histories and historical sources) – 1mf – 9 – $25.00 – us MEDOC [956]

The 1919-20 breasted expedition to the near east / Marcanti, Ruth – 1977 – 2mf – 9 – $25.00f – 0-226-69473-9 – (30p accompanying text) – us Chicago U Pr [915]

2000-2001 edexpress : direct loan – [Washington DC]: SFA University, US Dept of Education, Office of Educ Research & Improvement... [c2000] [mf ed 2001] – 6mf – 9 – us GPO [378]

2000-2001 edexpress : Pell training. – [Washington DC]: SFA University, US Dept of Education, Office of Educ Research & Improvement...[c2000] [mf ed 2001] – 4mf – 9 – us GPO [378]

A abelho da china – [Macao]: Na Typographia do Governo, sep 12 1822-mar 20 1823 – 1r – 1 – us CRL [951]

A aurore macaense – Macao: [Typographia Armenia, jan 14 1843-feb 3 1844 – 1r – 1 – us CRL [079]

A Beckett, Arthur William *see* The comic guide to the royal academy, for 1864

A Beckett, Gilbert Abbott [Joint pseud: Gemini] *see* The comic guide to the royal academy, for 1864

A bogaerts historische reizen door d'oostersche deelen van asia : zynde eene historische beschryving dier koninkryken en landschappen... – t'Amsterdam: By Nicolaas ten Hoorn, boekverkoper 1711 – (= ser [Travel descriptions from south africa, 1711-1938]) – 7mf – 9 – mf#zah-36 – ne IDC [910]

A c williams papers *see* Williams, a c, papers, ms 593

A caza de testamentos. una pieza mayor / Bayle, Constantino – Madrid: Razon y Fe, 1925 – 1 – sp Bibl Santa Ana [972]

A celebrar en caceres del 5 al 11 de octubre de 1969 / Caceres. 34th Asamblea de la Federacion Espanola de Centros de Iniciativas y Turismo – Plasencia: Imp. Sanguino, 1969 – 1 – sp Bibl Santa Ana [338]

A cim kon kri : a novel / Mra Cakra, Cha ra kri – Ran kun: Tan Mon Kri ca up tuik 1959 [mf ed 1990] – 1 – (in burmese) – mf#mf-10289 seam reel 180/3 [§] – us CRL [830]

A d bacup and district advertiser – [NW England] Bacup Lib 9 dec 1967-12 feb 1970 – 1 – uk MLA; uk Newsplan [072]

A d united church of christ – New York NY 1972-83 – 1,5,9 – ISSN: 0190-9207 – mf#9291 – us UMI ProQuest [240]

A d united presbyterian – New York NY 1972-83 – 1,5,9 – ISSN: 0191-2275 – mf#9290 – us UMI ProQuest [242]

A deux / Filion, Laetitia – 2ieme mille. [Levis]: le Quotidien ltee, [1937] [mf ed 1994] – 2mf – 9 – mf#SEM105P2207 – cn Bibl Nat [071]

A dhi pa ti lam : short stories / Aon Kyo Cann – Ran kun: U Kyo Nnvan 1975 [mf ed 1990] – 1 – (in burmese) – mf#mf-10289 seam reel 152/4 [§] – us CRL [830]

A diable-vent : legendaire du bas-saint-laurent et de la vallee de la matapedia / Gauthier-Chasse, Helene – Montreal: Quinze, cop 1981 [mf ed 1994] – 2mf – 9 – (with ind and glos) – mf#SEM105P2203 – cn Bibl Nat [390]

A e f : organe hebdomadaire de defense et de propagande de l'afrique equatoriale francaise – Brazzaville, Congo Republic 12 jun 1943-12 aug 1945 – 4r – 1 – (wanting: n60,67,74,82,88-91,94,129) – uk British Libr Newspaper [079]

A exposicao do venerando corpo do apostolo das indias, s francisco xavier, em 1878 : noticia historica / Albuquerque, Viriato Antonio Caetano Bras de – Nova-Goa: Typographia da "Cruz", 1879 [mf ed 1995] – (= ser Yale coll) – 100p (ill) – 1 – 0-524-09884-0 – (in portuguese) – mf#1995-0884 – us ATLA [241]

A f craig & co ltd engineers and ironfounders, paisley – [Edinburgh: NRAS [1984?] – 1mf – 9 – mf#n2551 – uk NRAS [620]

A floyer-acland, esq, london – [Edinburgh: NRAS 1987 – 1mf – 9 – mf#n2892 – uk NRAS [920]

A fuego lento / Bobadilla, Emilio – Barcelona: Impr de Henrich 1903 [mf ed 1990] – (= ser Biblioteca de novelistas del siglo 20) – 1r – 1 – us UF Libraries [440]

A i welders ltd, welding machine manufacturers, inverness – [Edinburgh: NRAS [1984?] – 1mf – 9 – mf#n2566 – uk NRAS [600]

A impresna – Rio de Janeiro, RJ: Typ Central, 27 set 1879 – (= ser Ps 19) – mf#P29,03B,05 – bl Biblioteca [079]

a ioanne busaeo s j...redacta *see* Opera pia et spiritualia...

A j b jardine-paterson esq – [Edinburgh: NRAS 1989 – 1mf – 9 – mf#n3054 – uk NRAS [340]

A & j robertson (granite) ltd, aberdeen – [Edinburgh: NRAS 1989 – 1mf – 9 – mf#n3036 – uk NRAS [622]

A ka ta khvan, mre pran ta van / Wun, U – Ran kun: Re khyamcan Ca pe 1974 [mf ed 1993] – on pt of 1r – 1 – mf#11052 r647 n8 – us Cornell [959]

A kay da mi lha ke si / Man Su Rin – Ran Kun: Su rassati ca pe tuik 1973 [mf ed 1990] – 1 – (in burmese) – mf#mf-10289 seam reel 161/3 [§] – us CRL [959]

A khre kham rhe hon su te sa na panna / Kyo Van, Manussa – Ran kun: Ca pe bi man 1977 [mf ed 1990] – (= ser Prann su lak cvai ca can) – 1 – (in burmese) – mf#mf-10289 seam reel 135/2 [§] – us CRL [930]

A khyac lvan su-le phru phru / Aon Chve, Mon, Takkasuil – Ran kun: Mui Ve Pe Tuik 1973 [mf ed 1990] – 1 – mf#mf-10289 seam reel 152/5 [§] – us CRL [959]

A khyac nhan cit panna / Tak Tui – Ran kun: San Bhava pum nhip tuik 1965 [mf ed 1990] – 1 – (in burmese) – mf#mf-10289 seam reel 168/2 [§] – us CRL [150]

A kirkman, esq, east calder – [Edinburgh: NRAS 1987 – 1mf – 9 – mf#n2951 – uk NRAS [929]

A kirkpatrick & sons wholesale butchers, glasgow – [Edinburgh: NRAS 1990 – 1mf – 9 – mf#n3177 – uk NRAS [380]

A kkhara sankhya sac a thak tan mya a phui / Bha Rup, U – Ran kun: Bhasa pran ca pe a san 1955 [mf ed 1990] – 1 – (in burmese) – mf#mf-10289 seam reel 182/2 [§] – us CRL [510]

A la brunante : contes et recits / Faucher de Saint-Maurice – Montreal: Duvernay, et Dansereau, 1874 – 4mf – 9 – mf#03056 – cn Canadiana [971]

A la conquete de la liberte : les constitutions du canada / DeCelles, Alfred Duclos – Paris: Comite " France-Amerique", 1914 [mf ed 1985] – 1mf – 9 – mf#SEM105P518 – cn Bibl Nat [323]

A la decouverte de shakespeare / Lefranc, Abel – Paris: Michel [1945-50] [mf ed 1984] – 2v on 1r [ill] – 1 – (with bibl footnotes; filmed with goethe schiller uber das theater / goethe, johan w) – mf#1316 – us Wisconsin U Libr [420]

A la diabla : versos / Villoch, Federico – Habana: A Miranda y c1893 (mf ed 19–) – (= ser Biblioteca de "el figaro") – mf#Z-639 – us NY Public [810]

A la feuille de rose : maison turque / Maupassant, Guy de – [Montreal]: Ed d'Orphee, 2000 [mf ed 2002] – 2mf – 9 – mf#SEM105P3435 – cn Bibl Nat [790]

A la futura madre. barcelona, 1930 / Roca Puig, P – Madrid: Razon y Fe, 1930 – 1 – sp Bibl Santa Ana [946]

A la legion cantos de amor y de dolor de espana por... / Gordo Moreno, Angel – Alcala de Henares: Imp. Talleres Penitenciarios Alcala, 1941 – sp Bibl Santa Ana [780]

A la memoire de charles gill / Doucet, Louis-Joseph – Quebec: [s.n.], 1920 – 1mf – 9 – 0-659-90904-9 – mf#990904 – cn Canadiana [920]

A la memoire de l'honorable charles seraphin rodier : avocat, ex-maire de montreal, membre du conseil legislatif de la province de quebec... – S.l: s.n, 1874? – 1mf – 9 – mf#05671 – cn Canadiana [920]

A la memoire de soeur de la nativite de la congregation de notre-dame : decede a la villa-maria, jeudi, 23 decembre, 1875 – Montreal?: s.n, 1876 – 2mf – 9 – (incl english text) – mf#02381 – cn Canadiana [241]

A la memoire du rev messire joseph f l duhamel : secretaire du diocese d'ottawa – Ottawa?: s.n, 1881 [mf ed 1979] – 2mf – 9 – 0-665-99998-4 – (incl english text) – mf#00003 – cn Canadiana [920]

A la recherche du grand-axe : contribution aux etudes transsahariennes, avec quarante-deux photographies et cinq cartes / Gradis, Gaston – Paris: Plon.Nourrit, [1924] – 1 – us CRL [960]

A la rencontre du christ / Dragon, Antonio – Val Racine, Chicoutimi: Maison de retraites fermees, [1964?] ([Barcelona]: Tip Cat Casals]) [mf ed 1992] – 3mf – 9 – mf#SEM105P1624 – cn Bibl Nat [240]

A la sombra de fouche : pequeno proceso de las izquierdas en colombia / Diaz, Antolin – Bogota, Colombia: Editorial ABC 1937 [mf ed 1993] – 1r – 1 – us UF Libraries [320]

A la sombra de los olivos (breviario intimo) poesias / Lamarche, Juan Bautista – Ciudad Trujillo, Dominican Republic: R D Escuela Salesiana de "Artes Graficas" 1953 [mf ed 1992] – 1r – 1 – us UF Libraries [440]

A la tres-excellente majeste de la reine : nous soussignes, eveques, vicaires generaux, cures et autres membres du clerge catholique du diocese de quebec... – S.l: s.n, 1838? – 1mf – 9 – mf#43258 – cn Canadiana [971]

A la tres-reverende sur sainte-ursule, superieure generale de la congregation notre-dame : amour, respect, reconnaissance a ma mere: hommage et salut a sainte fortunata – S.l: s.n, 1865? – 1mf – 9 – mf#56451 – cn Canadiana [810]

A la veillee : contes et recits / Faucher de Saint-Maurice – Montreal: Cadieux & Derome, 1878 – 2mf – 9 – mf#27140 – cn Canadiana [971]

A la veillee : contes et recits / Faucher de Saint-Maurice – Quebec: C Darveau, 1877 – 3mf – 9 – mf#03057 – cn Canadiana [971]

A la veillee : contes et recits / Faucher de Saint-Maurice – [Quebec?: s.n.], 1880 – 3mf – 9 – 0-665-90975-6 – mf#90975 – cn Canadiana [971]

A la veillee : contes et recits / Faucher de Saint-Maurice – [Quebec?: s.n.], 1882 – 3mf – 9 – 0-665-90982-9 – mf#90982 – cn Canadiana [971]

A la veillee : contes et recits / Faucher de Saint-Maurice – [Quebec?: s.n.], 1883 – 3mf – 9 – 0-665-90974-8 – mf#90974 – cn Canadiana [971]

A la virgen de los chinatos, la santisima virgen de la luz. abril 1970 – Plasencia: Imp. La Victoria, 1970 – 1 – sp Bibl Santa Ana [946]

A l'abbaye de montmajour : les origines de la reforme de saint-maur / Benoit, F – Marseille, 1927 – (= ser Memoires de l'institut historique se provence 4) – €3.00 – ne Slangenburg [241]

A l'approche du soir du monde : piece en trois actes et un prologue / Reignier, Fabien – Paris: O Lieutier [1946], c1945 [mf ed 1994] – 1r – 1 – us UF Libraries [820]

A l'armee : discours du general le flo, prononce a l'assemblee nationale le 11 mai 1849 – [Paris], 1849 – us CRL [355]

A l'honorable conseil de l'instruction publique, etc – S.l: s.n, 1876? – 1mf – 9 – mf#57827 – cn Canadiana [350]

A l'ile du diable enquete d'un reporter aux iles du salut et a cayenne / Hess, Jean – Paris: Nilsson – P Lamm [1898] [mf ed 1991] – 1r – 1 – us UF Libraries [918]

A l'ombre d'angkor : notes et impressions sur les temples inconnus de l'ancien cambodge / Groslier, George – Paris: A Challamel [mf ed 1989] – 1r with other items – 1 – mf#mf-10289 seam reel 007/02 [§] – us CRL [720]

A los intelectuales de espana : discurso pronunciado...partido comunista de espana... / Hernandez, Jesus – Barcelona: El Partido 1937 [mf ed 1977] – (= ser Blodgett coll) – 1mf – 9 – mf#w940 – us Harvard [946]

A los lectores antiguos y nuevos de "razon y fe" / Bayle, Constantino – Madrid: Razon y Fe, 1926 – 1 – sp Bibl Santa Ana [946]

A lum kon kra rai la : a novel / Lha, Lu thu U – Mantale: Lu thu Sa tan Ca tuik [195-?] [mf ed 1990] – 1 – (in burmese) – mf#mf-10289 seam reel 191/2 [§] – us CRL [830]

A lup sa ma sa muin / San Nnvan – Ran kun: Ba ma Rhe chon Pum nhip Tuik [195-?] [mf ed 1990] – 1 – (in burmese) – mf#mf-10289 seam reel 186/1 [§] – us CRL [331]

A m a archives of industrial health – Chicago IL 1950-60 – 1 – ISSN: 0567-3933 – mf#407 – us UMI ProQuest [360]

A m e church review – Nashville TN 1968- – 1,5,9 – ISSN: 0360-3725 – mf#3102 – us UMI ProQuest [240]

A m h gladstone, esq – [Edinburgh: NRAS 1989 – 1mf – 9 – mf#n3051 – uk NRAS [920]

A m kir konkoly-alapitvanyu astrophysikai observatorium kisebb kiadvanyai / Konkoly-Alapitvanyu Budapest-Svabhegyi m. kir. Astrofizikai Obszervatorium – Budapest: [s.n.] 1 (1901)-10 (1906) (irreg) [mf ed 2006] – 10v on 1r – 1 – (cont by: kleinere veroeffentlichungen des o-gyallaer astrophysikalischen observatoriums stiftung von konkolys; in hungarian & german; publ varies) – mf#film mas 37492 – us Harvard [520]

A m mackay : pioneer missionary of the church missionary society in uganda / ed by Harrison, J W [Mrs] – New York: A C Armstrong, 1890 [mf ed 1986] – 2mf – 9 – 0-8370-6270-5 – mf#1986-0270 – us ATLA [920]

A m von thuemmel's saemmtliche werke – Leipzig: G J Goeschen, 1853-54 [mf ed 1993] – 8v in 4 – 1 – mf#8706 – us Wisconsin U Libr [802]

A Ma, Lu thu Do *see* Rvhe don ton chon pa mya

A ma, Lu thu Do *see* Lu thu khyac sa myha lu thu u lha

A macdonald esq, waternish, skye – [Edinburgh: NRAS 1993 – 1mf – 9 – mf#n3371 – uk NRAS [912]

A maria santisima que bajo la advocacion de nuestra senora de la piedad se venera en almendralejo – Madrid: Imp. Asilo Huerfanos, 1902 – 1 – sp Bibl Santa Ana [946]

A mason esq, dundee – [Edinburgh: NRAS 1988 – 1mf – 9 – mf#n2960 – uk NRAS [920]

A me pro so pum : [short stories] / So Jan, U – Ran Kun: Ran On Ca pe 1966A [mf ed 1990] – 1 – (in burmese) – mf#mf-10289 seam reel 193/7 [§] – us CRL [830]

A messieurs les electeurs de la division de rougemont / Dessaulles, L A – [S.l: s.n, 1858?] – 1mf – 9 – 0-665-34766-9 – mf#34766 – cn Canadiana [325]

A mi tierra! suite musicana para gran orquestra / Perez Casas, Bartolome – [191-?] [mf ed 19–] – 1r – 1 – mf#pres. film 34 – us Sibley [780]

A mme honore mercier, fils (ma fille jeanne) a l'occasion de son mariage, 21 avril 1903 / Frechette, Louis – [Montreal?: s.n, 1903?] – 1mf – 9 – 0-665-97296-2 – mf#97296 – cn Canadiana [810]

A myui sa a ja nann khon chan kri sakhan mra / Khyac Mon, Sa khan – Ran Kun: Bo sa to Ca pe: Mi sa Ca pe Phyan khyi re 1979 [mf ed 1990] – (= ser Mi sa cu ca can 1) – 1 – (in burmese; incl bibl ref) – mf#mf-10289 seam reel 138/6 [§] – us CRL [959]

A myui sa ca pe chu ra ka bya mya [1967] / Myui Mran Chve – Ran Kun: Ca pe bi man a phvai 1972 [mf ed 1990] – 1 – (in burmese) – mf#mf-10289 seam reel 166/4 [§] – us CRL [959]

A myui sa mi mya kuiy kak kram khuin re panna / Van Me – Ran Kun: Ca pe bhi man 1954 [mf ed 1990] – 1 – (in burmese) – mf#mf-10289 seam reel 168/7 [§] – us CRL [613]

A myui sa thok lham re a phvai upade – Ran kun: Prann thon cu Chuirhay lac sammata Mran ma nuin nam to 1975 [mf ed 1995] – on pt of 1r – 1 – mf#11052 r2002 n12 – us Cornell [959]

A n forbes esq, stoke-on-trent – [Edinburgh: NRAS 1987 – 1mf – 9 – mf#n2870 – uk NRAS [941]

A na gat sasana re / Janakabhivamsa, Ashin – Mantale: Lvan Pum nhip tuik 1975 [mf ed 1994] – on pt of 1r – 1 – mf#11052 r1735 n4 – us Cornell [959]

'A nila i ra dovot' : methodist mission magazine – 1909-1973 – 5r – 1 – (in the kuonua or tinota tuno language; first appeared in feb 1909; some iss for 1911, 1928-31, 1933-34, 1936, 1938-49, 1961 & 1962 were not available for microfilming) – mf#pmb647 – at Pacific Mss [240]

A nnatra bhava rhu khan mya / Tak Tui – Rangoon: Nu Yin Press & Pub House 1963 [mf ed 1990] – 1 – (in burmese) – mf#mf-10289 seam reel 193/4 [§] – us CRL [959]

A nos bienfaiteurs *see* La mission du kiang-nan

A nos fideles abonnes : nous vous adressons sous le meme pli que contient cette lettre votre compte d'abonnement au bulletin des recherches historiques... – S.l: s.n, 1895? – 1mf – 9 – mf#53477 – cn Canadiana [971]

A nu panna a thve thve prassana / So Jan, U – Ran Kun: A Sac Ca pe rip mruim 1967 [mf ed 1990] – 1 – (in burmese) – mf#mf-10289 seam reel 171/6 [§] – us CRL [790]

A o akanya ka ga kopafatso? / South Africa. Department of Health [Departement van Gesondheid] – Pretoria: Dept of Health [1977?] [mf ed Pretoria, RSA: State Library [199-]] – 4p [ill] on 1r with other items – 5 – (in tswana) – mf#op 06684 r24 – sa State Libr [360]

A o nagana ka ngopofatso? = Do you consider sterilization? / South Africa. Department of Health [Departement van Gesondheid] – Pretoria: Dept of Health [1977?] [mf ed Pretoria, RSA: State Library [199-]] – 4p [ill] on 1r with other items – 5 – (in northern sotho; also available in afrikaans, english, sotho, tsonga, tswana, venda & zulu) – mf#op 06685 r24 – us CRL [360]

A O Smith Corporation *see* Impact

A orillas del filosofar / Fernandez Spencer, Antonio – Ciudad Trujillo, Dominican Republic: Arquero 1960 [mf ed 1992] – (= ser Coleccion arquero 6) – 1r – 1 – us UF Libraries [100]

A orillas del orinoco y a orillas del tamesis / Bayle, Constantino – Madrid: Razon y Fe, 1930 – 1 – sp Bibl Santa Ana [946]

A orillas del sueno : novela / Fabbiani Ruiz, Jose – Merida, Venezuela: [Tall Graf Universitarios 1959] [mf ed 1993] – (= ser Publicaciones del departamento de extension cultural de la universidad de los andes 68) – 1r – 1 – us UF Libraries [378]

A Philip Randolph Senior Center [New York NY] *see* Challenge

A phru kui mann su a nak chui sa nann : [short stories] / Man Kyo – Ran kun: Pu gam Ca up 1980 [mf ed 1990] – 1 – (in burmese) – mf#mf-10289 seam reel 179/3 [§] – us CRL [830]

A pie y descalzo de trinidad a cuba, 1870-1871 (recuerdos de campana) / Roa, Ramon M – Miami [FL]: Club del Libro Latinoamericano 1977 [mf ed 1991] – 1r – 1 – (incl ind) – us UF Libraries [972]

A pleines voiles / Universite Laval (Quebec). Ecole superieure des pecheries. Service social economique – Sainte-Anne-de-la-Pocatiere: le Service. v1 n1 janv 1945-v24 n11/12 nov/dec 1967 (mthly) [mf ed 1991] – 1 – (cont by: pleines voiles) – mf#SEM35P366 – cn Bibl Nat [300]

A. por el excmo. sr. conde de montijo... sra. dona mariana enriquez / Garcia Jalon, Miguel – 1741 – 9 – sp Bibl Santa Ana [946]

A pra thai ka ne van ni nhan a khra vatthu tui mya : [short stories] / Su Sin – Ran Kun: Ca pe Bi man 1981 [mf ed 1990] – (= ser Prann su lak cvai ca can) – 1 – (in burmese) – mf#mf-10289 seam reel 171/3 [§] – us CRL [830]

A propos de la campagne 'anti-superstitieuse' / Roumain, Jacques – Port-Au-Prince, Haiti: Impr de l'Etat 1944? [mf ed 1991] – 1r – 1 – us UF Libraries [230]

A propos de la legislation sur l'inventaire / Alexandre, Pierre C – Port-Au-Prince, Haiti: Au Service de ladministration haitienne 1951 [i. e. 1952] [mf ed 1993] – 1r – 1 – us UF Libraries [343]

A propos de la separation des eglises et de l'etat *see* Disestablishment in france

A propos de l'ete du serpent / Francoeur, Lucien – Talence, [France]: Editions du Castor astral, 1980 [mf ed 1993] – 2mf – 9 – mf#SEM105P1807 – cn Bibl Nat [440]

A propos d'education : lettres a m l'abbe baillarge du college de joliette / Frechette, Louis – ed rev augm. Montreal: Desaulniers, 1893 [mf ed 1971] – 1r – 5 – mf#SEM16P43 – cn Bibl Nat [370]

A propos d'education : lettres a m l'abbe baillarge du college de joliette / Frechette, Louis – Montreal?: Impr Desaulniers, 1893 – 2mf – 9 – mf#59166 – cn Canadiana [370]

A propos des eglises baptistes d'haiti : extrait de l'haitien, des 1er et 7eme octobre 1927 / Paultre, Hector – Port-Au-Prince, Haiti: A Heraux [1927?] [mf ed 1991] – 1r – 1 – us UF Libraries [242]

A propos du centenaire / Bobo, Rosalvo – Cap-Haitien, Haiti: Impr La Conscience 1903 [mf ed 1991] – (= ser Bibliotheque libertaire) – 1r – 1 – us UF Libraries [972]

A proposito de los condes de la gomera "fue antano regular la sucesion de este titulo" / Darias y Padron, Dacio V – Madrid, s.i., 1956 – 1 – sp Bibl Santa Ana [946]

A proposito de un documento. noticias sobre los sanchez de badajoz / Rodriguez Amaya, Esteban – Badajoz: Imp. Provincial, 1945 – 1 – sp Bibl Santa Ana [946]

A quoi bon la societe aberdeen / Aberdeen and Temair, Ishbel Gordon, Marchioness of – Montreal: J Lovell, 1897 – 1mf – 9 – mf#04000 – cn Canadiana [366]

A r e journal – Virginia Beach VA 1972-80 – 1,5,9 – mf#7485 – us UMI ProQuest [130]

A r gray ltd, provision wholesales and retailers, aberdeen – [Edinburgh: NRAS [1984?] – 1mf – 9 – mf#n2585 – uk NRAS [338]

A & r miller & company, iron founders, edinburgh – [Edinburgh: NRAS 1993 – 1mf – 9 – mf#n3401 – uk NRAS [660]

A r penck : geisteshaltung einer zeichensprache / Puettmann, Natalie – (mf ed 1997) – 4mf – 9 – €56.00 – 3-8267-2391-0 – mf#DHS 2391 – gw Frankfurter [740]

A rak kron a pay makya phui / Lha Thvan Phru, U – Ran kun: Capay Ca pe 1984 [mf ed 1990] – 1 – (in burmese) – mf#mf-10289 seam reel 148/7 [§] – us CRL [615]

A rhan janakabhivamsa e dasama kyam ca anagat sasana re / Janakabhivamsa, Ashin – Mantale: Lvan Pum nhip tuik 1975 [mf ed 1995] – on pt of 1r – 1 – mf#11052 r1960 n4 – us Cornell [959]

A rhan janakabhivamsa e nok chumchay la mrat buddha / Janakabhivamsa, Ashin – Mantale Mrui': Mon 'On Khan 1976 [mf ed 1995] – on pt of 1r – 1 – mf#11052 r2002 n7 – us Cornell [280]

A ritschl's philosophische und theologische ansichten / Fluegel, Otto – 3. aufl. Langensalza: H Beyer & Soehne, 1895 [mf ed 1990] – 1mf – 9 – 0-7905-7582-5 – (incl bibl ref) – mf#1989-0807 – us ATLA [242]

A ruin capay : [a novel] / On Lan – Ran kun: Mran ma padesa Ca pe tuik 1960 [mf ed 1990] – 1 – (in burmese) – mf#mf-10289 seam reel 184/4 [§] – us CRL [830]

A s falardeau et a e aubry / Casgrain, Henri-Raymond – Montreal: Librairie Beauchemin, ltee, 1912 [mf ed 1986] – 2mf – 9 – mf#SEM105P715 – cn Bibl Nat [920]

A sa grandeur monseigneur charles larocque : eveque de st hyacinthe / Dessaulles, L A – s.l: s.n, 1868? – 1mf – 9 – mf#02160 – cn Canadiana [240]

A sa majeste victoria 1ere, reine d'angleterre et imperatrice des indes / Frechette, Louis – S.l: s.n, 1897? – 1mf – 9 – mf#28303 – cn Canadiana [810]

A sac mran ba ma samuin / Than Tun – Mantale: Mra Kan Sa Cap 1975 [mf ed 1994] – on pt of 1r – 1 – mf#11052 r1715 n4 – us Cornell [959]

A sai mha phrat ca re may / Mran Ni – Mran: Tan Mon 1969 [mf ed 1990] – 1 – (in burmese) – mf#mf-10289 seam reel 161/6 [§] – us CRL [959]

A salamanca diak / Espronceda, Jose de – Maqyar Helikon, 1961 – 1 – sp Bibl Santa Ana [946]

A. shattucks book / Shattuck, Abel – Manuscript, ca. 1801. Solos for treble instrument, some with second parts. Includes: "Hail Columbia," "Irish Washerwoman," and "Yankee Doodle." MUSIC 1975 – 1 – us L of C Photodup [780]

A solas con mi alma / Ramos Aparicio, Juan – Caceres: Imprenta Moderna, 1955 – sp Bibl Santa Ana [780]

A son eminence le cardinal simeoni, prefet de la s c de la propagande / Lafleche, Louis Francois – [Trois-Rivieres, Quebec: s.n, 1887?] [mf ed 1994] – 1mf – 9 – 0-665-94570-1 – mf#94570 – cn Canadiana [241]

A son excellence dom henri smeulders : commissaire apostolique a quebec / Tardivel, Jules Paul – Quebec?: s.n, 1883? – 1mf – 9 – mf#24639 – cn Canadiana [241]

A son excellence dom henri smeulders : commissaire apostolique au canada / Livernois, Victor – Quebec?: s.n, 1883? – 1mf – 9 – mf#09131 – cn Canadiana [241]

A son excellence dom henri smeulders : commissaire apostolique au canada – S.l: s.n, 1883? – 1mf – 9 – mf#00865 – cn Canadiana [241]

A son excellence dom henri smeulders : commissaire apostolique au canada – S.l: s.n, 1884? – 1mf – 9 – mf#08073 – cn Canadiana [241]

A son excellence dom henri smeulders : commissaire apostolique au canada: excellence, les soussignes, mus par le seul desir d'etre a la ste eglise et a leur partie... – [S.l: s.n, 188-?] – 1mf – 9 – 0-665-91937-9 – mf#91937 – cn Canadiana [241]

A son excellence mgr le commissaire apostolique au canada / Lafleche, Louis Francois – [Trois-Rivieres, Quebec: s.n, 1883?] [mf ed 1994] – 1mf – 9 – 0-665-94571-X – mf#94571 – cn Canadiana [241]

A son excellence monseigneur donat sbarretti : archeveque d'ephese, delegue apostolique au canada, et aux reverendissimes peres du premier concile plenier du canada – [Quebec]: [s.n.], [1909] (mf ed 1986) – 1mf – 9 – mf#SEM105P690 – cn Bibl Nat [241]

A st-fabien, en s'amusant, cuisinons / Belanger, Claudine – St-Fabien: AFEAS de Saint-Fabien, 1978 [mf ed 1998] – 2mf – 9 – mf#SEM105P3024 – cn Bibl Nat [640]

L'a t a c : journal de l'amicale tonkinoise des anciens combattants et mutiles de la grande guerre – Hanoi: IDEO. v9 n182-v10 n207 (1937-feb 1938); v10 n210-212 [apr-may 1 1938] [mf ed Hanoi, Vietnam: National Library of Vietnam 1997] – 1r – 1 – (master neg held by crl) – mf#mf-11795 seam – us CRL [355]

A t b metallurgie (acta technica belgica) – Brussels, Belgium 1973-90 – 1,5,9 – ISSN: 0001-2696 – mf#8040 – us UMI ProQuest [660]

A t metcalf, geologist – [Edinburgh: NRAS 1990 – 1mf – 9 – mf#n3207 – uk NRAS [550]

A table! : recette[s] d'antan et d'aujourd'hui / Morin, Pierre – Quebec: Pleine page enr, 1996 [mf ed 1999] – 2mf – 9 – mf#SEM105P3067 – cn Bibl Nat [640]
A thu a thvak tui ca pa cuik pyui re / Lvan U – Ran kun: Ca pe Biman 1984 [mf ed 1990] – (= ser Ca pe biman thut prann su lak cvai ca can) – 1 – (in burmese; with bibl) – mf#mf-10289 seam reel 176/1 [§] – us CRL [630]
A thve thve sippam san pra nann / A Vin, U – Ran kun: Bhasa pran ca pe a san 1953 [mf ed 1990] – (= ser Mula tan chara lak cvai ca can; a mhat 3) – 1 – (in burmese) – mf#mf-10289 seam reel 179/5 [§] – us CRL [370]
A tous les electeurs du bas canada : les efforts qui vous avez faits a la derniere election n'ont pas ete inutiles... / Bdard, Pierre-Stanislas – [S.l: s.n, 18–?] [mf ed 1984] – 1mf – 9 – 0-665-48183-7 – mf#48183 – cn Canadiana [325]
A tous les electeurs du bas canada – [Quebec: Charles Le Francois, 1810?] [mf ed 1986] – 1mf – 9 – 0-665-57521-1 – mf#57521 – cn Canadiana [325]
A travers la litterature canadienne-française : premiere serie / Leopold, frere – Montreal: Freres des ecoles chretiennes, 1928 [mf ed 1991] – 3mf – 9 – (pref by fr. marie-victorin) – mf#SEM105P1412 – cn Bibl Nat [440]
A travers la vie / Michel, Louise – Paris: Fayard, 1894 – (= ser Les femmes [coll]) – 2mf – 9 – mf#10684 – fr Bibl Nationale [640]
A travers l'amerique : nouvelles et recits / Biart, Lucien – Paris: A Hennuyer, 1876? – 5mf – 9 – mf#03582 – cn Canadiana [917]
A travers l'amerique : nouvelles et recits / Biart, Lucien – Paris: Bibliotheque du Magasin des demoiselles, 1876? – 5mf – 9 – mf#28684 – cn Canadiana [917]
A travers le fouta-diallon et le bambouc (soudan occidental) / Noirot, Ernest – Paris, 1885? – 1 – us CRL [960]
A travers le fouta-diallon et le bambouc (soudan occidental) / Noirot, Ernst – Paris: E Flammarion, [1885?] – us CRL [960]
A travers les deserts de la tartarie et les neiges du thibet : curieuses aventures d'une caravane / Huc, Evariste Regis – Lille: Maison du Bon Livre, [c1900] [mf ed 1995] – (= ser Yale coll) – 501p (ill) – 1 – 0-524-10091-8 – (1st ed publ in 1850 under title: souvenirs d'un voyage dans la tartarie, le thibet et la chine; joseph gabet was huc's companion on the jouney to tibet) – mf#1995-1091 – us ATLA [241]
A travers les grandes terres a ble du nord ouest canadien / Bouthillier-Chavigny, Charles, vicomte de – S.l: s.n, 1893? – 1mf – 9 – mf#02182 – cn Canadiana [630]
A travers les registres : notes recueillies / Tanguay, Cyprien – Montreal: Librairie St-Josephe: Cadieux & Derome, 1886 [mf ed 1978] – 3mf – 9 – mf#SEM105P5 – cn Bibl Nat [350]
A travers l'histoire des ursulines de quebec / Roy, Pierre-Georges – Levis: [s.n.], 1939 [mf ed 1988 – 3mf – 9 – mf#SEM105P929 – cn Bibl Nat [241]
A un gran partido, una gran organizacion : discurso / Checa, Pedro – [Barcelona]: Partido comunista de Espana 1937 – 9 – mf#w794 – us Harvard [335]
A. v. badajoz la tienda asilo. poema – 1889 – 9 – sp Bibl Santa Ana [810]
A V Williams Jackson *see* History of indi
A varias tintas / Gutierrez Cunado, Antolin – Caceres: Ed. Extremadura, 1929 – 1 – sp Bibl Santa Ana [946]
A Vin, U *see* A thve thve sippam san pra nann
A vrai dire : le journal de la communaute etudiante de l'ithq – [Montreal]: [s.n.] v1 n1 (dec 1984)-v1 n3 (avril/mai 1985 (irreg) [mf ed 1988] – 1mf – 9 – (cont: hotel. cont by: mille-feuille) – mf#SEM105P1085 – cn Bibl Nat [640]
A w schlegel's lectures on german literature : from gottsched to goethe / ed by Fiedler, Hermann Georg & Toynbee, George – Oxford: B Blackwell, 1944 [mf ed 1993] – 96p (ill) – 1 – (given at the university of bonn & taken down by george toynbee in 1833; together with toynbee's 'continuation to heine'; with int, notes & portrait; incl bibl ref & ind) – mf#8232 – us Wisconsin U Libr [430]
A w schlegels vorlesungen ueber schoene litteratur und kunst / ed by Minor, J – Heilbronn: Henninger, 1884 [mf ed 1993] – (= ser Deutsche litteraturdenkmale des 18. und 19. jahrhunderts 17-19) – 3v – 1 – (int by j minor) – mf#8676 reel 2 – us Wisconsin U Libr [410]
A w sloan esq, bonnyrigg (jacobite journal of capt john maclean, 1745-46) – [Edinburgh: NRAS 1990 – 1mf – 9 – mf#n3234 – uk NRAS [920]
A warwai ure iesu karisito : translation of gospel stories into the blanche bay dialect, and rev walker's reflections on the work of the missionary, written on his return from new britain, pmb / Walker, Francis Trafford – 1927-30 – 1r – 1 – mf#pmb1264 – at Pacific Mss [226]
A100 *see* Natal star / journal of commerce / agriculture / a100
A-356 site and the florida archaic / Clausen, Carl Jon – 1964 [mf ed 1992] – 1r – 1 – us UF Libraries [978]
Aa, Abraham Jacob von der *see* Beschrijving der nederlandsche bezittingen in oost-indie
Aaas science books / American Association for the Advancement of Science – Washington. 1965-75 – 1,5 – (cont by: science books and films) – ISSN: 0036-8253 – mf#5926.01 – us UMI ProQuest [500]
Aabe energy news – 1990 annual review-1991 spring – 1 – mf#4775539 – us WHS [071]
Aace international transactions – Morgantown WV 1997- – 1,5,9 – (cont: transactions of aace international) – ISSN: 1528-7106 – mf#10810.02 – us UMI ProQuest [620]
Aachener allgemeine zeitung – Aachen, Germany 17 jan 1916-30 jun 1919 (imperfect) – 1 – uk British Libr Newspaper [074]
Aachener anzeiger 1848 – Aachen DE, 1848 9 apr-1849 – 1r – 1 – (title varies: 9 sep 1851: echo der gegenwart; filmed by other misc inst: 1881 1 jan-1 jul, 1850-1935 [179r]) – gw Misc Inst [074]
Aachener anzeiger 1871 – Aachen DE, 1885, 1887-1939 30 sep, 1940-43 – 1 – (title varies: 1 aug 1878: aachener anzeiger-politisches tageblatt) – gw Misc Inst [074]
Aachener anzeiger-politisches tageblatt *see* Aachener anzeiger 1871
Aachener fremdenblatt – Aachen DE, 1848-1849 1 jan – 1r – 1 – gw Misc Inst [074]
Aachener nachrichten – Aachen DE, 1963-1969 10 apr, 1970-1976 19 feb – 67r – 1 – (filmed by mikropress: 1946 16 jul-1949 (gaps) [2r] order#6563; filmed by misc inst: 1945 24 jan-1962; 1968- [ca 8r/yr]; 1945-47 [gaps]) – gw Mikrofilm; gw Mikropress; gw Misc Inst [074]
Aachener volkszeitung – Aachen DE, 1958-1961 10 jan, 1965 24 jun – 1 – (title varies: 1996 6 mar: aachener zeitung; filmed by bnl: 946 20 apr-1952 26 nov [17r]; filmed by other misc inst: 1946 22 feb-1957, 1961-78; 1972- [ca 8r/yr]) – gw Misc Inst [074]
Aachener volkszeitung *see* Heinsberger volkszeitung 1882
Aachener zeitung *see* Journal de la roer
Aacn clinical issues in critical care nursing – Baltimore MD 1992-94 – 1,5,9 – (cont by: aacn clinical issues) – ISSN: 1046-7467 – mf#19707.01 – us UMI ProQuest [610]
Aadcom news – v28 n1-2, 4-6 [1987 feb/mar-mar/apr, jul-sep]; v28 n8 [1988 jan] – 1 – mf#1345497 – us WHS [071]
Aaec notebook – n4-31 [1967 jan-1972 jul] – 1 – mf#1051424 – us WHS [071]
Aafli news – 1977 aug/sep-1984 jul – 1 – mf#809698 – us WHS [071]
Aagri news – v1 n3-4 [1996 winter, spring] – 1 – mf#3621803 – us WHS [071]
Aahs newsletter – 1967 mar-1982 2nd quarter – 1 – mf#614797 – us WHS [071]
Aajanann khon chon kri didut u ba khyui athuppatti / Sankha – Yan Kun: Khat Mahi Cape 1976 [mf ed 1990] – 1 – (in burmese) – mf#mf-10289 seam reel 143/1 [§] – us CRL [920]
Aakjar, Svend *see*
– Maal og vagt
– Mont
Aal, Johannes *see* Tragoedie johannis des taeufers
Aalborg stiftsbog – 1958-92 [complete] – inquire – mf#ATLA S0768 – us ATLA [073]
Aalborg stiftstidende – Aalborg, Denmark 3 jan-23 jun 1945 – 3r – 1 – uk British Libr Newspaper [078]
Aalc reporter – 1983-1996 mar – 1 – mf#1109815 – us WHS [071]
Aalders, G *see* The problem of the book of jonah
Aalders, Willem Jan *see* Schleiermacher's reden ueber die religion als proeve van apologie
Aalener volkszeitung *see* Schwaebische zeitung [main edition]
Aall, Anathon *see*
– Geschichte der logosidee in der christlichen litteratur
– Geschichte der logosidee in der griechischen philosophie
Aall, Anathon et al *see* Philosophische abhandlungen
Aalst, G van *see* Alexius afanasevic dmitrievskij
Aalten en bredevoort in vervlogen tijden / Rots, B D – Aalten, 1950 – €7.00 – ne Slangenburg [240]
Aan de rand de dibese / Denolf, Prosper – Brussels, 1954 – 1 – us CRL [960]
Aana journal / American Association of Nurse Anesthetists – Park Ridge IL 1974+ – 1,5,9 – (cont: journal of the american association of nurse anesthetists) – ISSN: 0094-6354 – mf#2083,01 – us UMI ProQuest [610]
Aang angagin, aang angaginas – 1982 sep-1988 aug – 1 – mf#1051497 – us WHS [071]
Aangeboden door de "nillmij" / Nieuw-Semarang "De Heuvelstad" – Weltevreden, 1917 (ill) – 1mf – 8 – mf#SE-1449 – ne IDC [959]
Aanmerkingen op otto van veens zinnebeelden der goddelijke liefde / Bruin, Cornelis de – t'Amsterdam: Theodorus Dankerts, 1726 – 2mf – 9 – mf#O-564 – ne IDC [090]
Aanteekeningen eener reis door de binnenlanden van zuid-afrika : van port-elizabeth naar de kaapstad, gedaan in 1823 / Theunissen, J B N – Oostende: Th Vermeirsch Drukker en Boekhandelaar 1824 [mf ed Cape Town: South African Lib 1993] – (= ser [Travel descriptions from south africa, 1711-1938]) – 3mf – 9 – mf#mf.959 – ne IDC [916]
Aanteekeningen ter toelichting van den strijd over de praedestinatie in het gereformeerd protestantisme : pt 2: heinrich bullinger / Gooszen, M A – N.S. 20, 1909. p 393-454 – (= ser Geloof en vrijheid) – 1mf – 9 – mf#PBU-446 – ne IDC [242]
Aantekeningen van fvha de stuers gehouden bij het overbrengen van diponegoro van magelang naar batavia, 1830 – 1mf – 8 – mf#SD-102 mf 46 – ne IDC [959]
Aaohn journal : official journal of the american association of occupational health nurses / American Association of Occupational Health Nurses – Thorofare NJ 1986+ – 1,5,9 – (cont: occupational health nursing) – ISSN: 0891-0162 – mf#11478.02 – us UMI ProQuest [610]
Aaps news – 1983 jan/feb-1994 dec – 1 – mf#1051426 – us WHS [071]
Aar mit gebrochener schwinge : clemens brentano, annette von droste-huelshoff / Schneider, Reinhold – 2. aufl. Heidelberg: F H Kerle 1948 [mf ed 1993] – 1r – 1 – (incl bibl ref. filmed with: der erzieherische gehalt in j j breitingers "critischer dichtkunst" / vorgelegt von jakob braeker) – mf#8525 – us Wisconsin U Libr [430]
AarcTimes / American Association for Respiratory Care – Irving TX 1987+ – 1,5,9 – (cont: aartimes) – ISSN: 0893-8520 – mf#14134,01 – us UMI ProQuest [616]
Aardahl, Anya *see* The influence of height on body image, self-confidence, and performance of female basketball players
Aarhus amtstidende – Aarhus, Denmark 2 jan-18 jun 1945 (imperfect) – 2r – 1 – uk British Libr Newspaper [078]
Aaron, Elizabeth A *see* The oxygen cost of exercise hypernea
Aarp modern maturity – Washington DC 2002+ – 1,5,9 – (cont: modern maturity [library ed]) – ISSN: 1538-5981 – mf#21260,01 – us UMI ProQuest [618]
Aarsskrift for danske kvinders missionsfond / Danske kvinders missionsfond – Cedar Falls IA: Dannevirke's trykkeri 1919-36 [annual] [mf ed 2005] – 18v on 1r – 1 – (publ by dannevirke trykkeri 1919-24; corporate body on 1921 iss called: danske kvinders missionsforening; some pgs damaged) – mf#2005c-s046 – us ATLA [242]
L'aart de batir des maisons de campagne... / Briseux, C E – Paris, 1743 – 2v on 27mf – 9 – mf#O-1161 – ne IDC [720]
Aartimes : an official publication of the american association for respiratory therapy / American Association for Respiratory Therapy – Irving TX 1983-86 – 1,5,9 – (cont by: aarctimes) – ISSN: 0195-1777 – mf#14134.01 – us UMI ProQuest [616]
Aas/aiaa astrodynamics conference, 1975 [aasms26] – 1976 – (= ser Aasms 1968) – 59papers on 36mf – 9 – $40.00 – 0-87703-142-8 – (suppl to v33, advances) – us Univelt [629]
Aas/aiaa astrodynamics conference, 1977 [aasms27] – (= ser Aasms 1968) – 73papers on 31mf – 9 – $50.00 – 0-87703-241-6 – us Univelt [629]
Aas/aiaa astrodynamics conference, 1979 [aasms32] – 1979 – (= ser Aasms 1968) – 27papers on 13mf – 9 – $20.00 – 0-87703-139-8 – (suppl to v40, advances) – us Univelt [629]
Aas/aiaa astrodynamics conference [aasms37] – 1981 – (= ser Aasms 1968) – 41papers on 21mf – 9 – $40.00 – 0-87703-163-0 – (suppl to v46, advances) – us Univelt [629]
Aas/aiaa astrodynamics specialist conference, 1973 [aasms21] – (= ser Aasms 1968) – 44papers on 28mf – 9 – $35.00 – 0-87703-238-6 – us Univelt [629]
Aatcc review – Research Triangle Park NC 2001+ – 1,5,9 – (cont: textile chemist & colorist & american dyestuff reporter) – ISSN: 1532-8813 – mf#31387 – us UMI ProQuest [670]
Aauw bulletin/ – v44 n1-v47 n3 [1980 sep-1984 dec] – 1 – mf#914897 – us WHS [071]
Aauw journal / American Association of University Women – Washington DC 1906-78 – 1,5 – (cont by: graduate woman) – ISSN: 0001-0278 – mf#904.02 – us UMI ProQuest [376]
Ab dafuer, nach afrika : eine bubengeschichte aus vergangenen tagen / Weller, Anton Friedrich Tuedel – Muenchen: F Eher [194-?] [mf ed 1991] – (= ser Soldaten-kameraden! 24) – 1r – 1 – (filmed with: josef weinheber / franz koch) – mf#2957p – us Wisconsin U Libr [830]
Ab initio molecular orbital theoretical studies of the molecular complexes of boron trifluoride and oxygen electron donor ligands / Modau, Eric – Pretoria: Vista University 2001 [mf ed 2001] – 4mf – 9 – (incl bibl ref) – mf#mfm15177 – sa Unisa [540]
Ab urbe condita libri 1.2.21.22 : adiunctae sunt partes e selectae ex libris 3.4.5.6.7.26.29 / Livy; ed by Zingerle, Anton – Wien: Tempsky, 1906 [mf ed 1987] – 1 – (unter mitwirkung von a scheindler, fuer den schulgebrauch. filmed with: king arthur and the table round / newell, w w) – mf#2081 – us Wisconsin U Libr [930]
Aba bank marketing – Washington DC 2001+ – 1,5,9 – ISSN: 1539-7890 – mf#10211.02 – us UMI ProQuest [332]
Aba bankers news – Washington DC 2000+ – 1,5,9 – ISSN: 1530-1125 – mf#20216.01 – us UMI ProQuest [332]
ABA Commission of Inquiry *see* Notes of evidence
Aba Commission of Inquiry *see* Notes of evidence taken by the commission of inquiry appointed to inquire into the disturbance in the calabar and owerri provinces, december, 1929
Aba criminal justice section *see* American criminal law review
Aba journal – Chicago IL 1984+ – 1,5,9 – (cont: american bar association journal) – ISSN: 0747-0088 – mf#1897.01 – us UMI ProQuest [340]
Aba journal – v1-87. 1915-2001 – 9 – $1942.00 set – (title varies: v1-69 1915-83 as american bar association journal) – ISSN: 0747-0088 – mf#100361 – us Hein [340]
Aba trust & investments – Washington DC 2001+ – 1,5,9 – mf#26480.01 – us UMI ProQuest [332]
Aba trust letter – Washington DC 2001+ – 1,5,9 – mf#20284.01 – us UMI ProQuest [332]
Abab, Addis *see* Report on census of population, 10-11 sep, 1961
Ababio, Ernest Peprah *see* Effects of change in inputs in policy-making for the south african public service
Les ababua (congo belge) / Halkin, Joseph – Bruxelles: A DeWit [etc], 1911 – us CRL [960]
Abacus – Oxford, England 1965+ – 1,5,9 – ISSN: 0001-3072 – mf#9848 – us UMI ProQuest [650]
Abacus, Inc et al *see* Kerista
Abacus of the altitude and azimuth of the pole star / Deville, Edouard – Ottawa: J Hope 1906 – 1mf – 9 – 0-665-88185-1 – mf#88185 – cn Canadiana [520]
Abad, L V De *see* Impuestos especiales del empresito
Abad Mendez, Ramon Antonio *see* Florilegio de sonetos
Abad muslimin – Djakarta: Jajasan Kebudajaan Muslimin Indonesia 1965- (wkly) [mf ed Washington, DC]: LOC, Photoduplication Service] – 3r [oct 4 1965-dec 1 1966] – 1 – (iss also have islamic calendar designations; iss by: jajasan kebudajaan muslimin indonesia oct 4-dec 1 1966) – mf#mf-9732 – us CRL [079]
Abad y Perfecto, Joaquin *see* An macalolodoc na buhay ni florentina sa cahadean nin alemania
Abad Yllana, Manuel *see* Carta pastoral del illustrissimo senor d don manuel abad yllana
Abadia, Ignacio *see* Resumen sacado del inventario general...de los arneses de la real academia
Abadie, A *see* Itineraire topographique et historique des hautes-pyrenees principalement des etablissemens thermaux de cauterets, saint-sauveur...
Abadie, Charles A *see* Nouveaux riches
Abadilla, A G *see* Maikling katha ng 20 pangunahing awtor
Abadilla, A G et al *see* Ang maikling kathang tagalog
Abaelardus, Petrus *see* Theologia
Abaelardus, Petrus (Abelard, Peter) *see* Epitome theologiae christianae
O abaeteense – Abaete, PA: Typ do Abaeteense, 15 ago 1884 – (= ser Ps 19) – bl Biblioteca [079]
Abafi, Lajos [Ludwig] *see* Geschichte der freimaurerei in oesterreich-ungarn

Abagabe b'ankole – Kampala: Eagle Press, 1955 – 9 – (filmed with: rowe, john a "selected articles on the bataka" (bukalasa, 1922-23). kagwa, apolo "the clans of the baganda (mengo, 1949). gray, j m "mutesa of buganda" (1934)) – us CRL [960]
Abakama ba bunyoro kitara : translation (p. 33-162) / Rowe, J A – Kampala, Uganda, 1964 – (filmed with: daudi chawa 2: kabaka of buganda – why sir apolo kaggwa..resigned [kampala, 1964]) – us CRL [960]
Abalo, J L *see* Gran crisis y la necesidad de una confederacion pa...
Abalobi *see*
- Sekuthedlwe
- Sengikhasa
- Sengingabala
- Sengithethuthu

Abalone Alliance *see* It's about times
La abanderada de 1868 candelaria figueredo (hija de perucho) : autobiografia / Figueredo, Candelaria – Havana: Comision Patriotica "Pro Himna Nacional" a la Mujer Cubana 1929 [mf ed 1991] – 1r – 1 – us UF Libraries [972]
Abandoned! : a narrative of the fearful adventure that befell a young man in the service of the hudson bay co / Andrews, Morgan – S.l: s.n, 1899? – 1mf – 9 – 0-665-15099-7 – mf#15099 – cn Canadiana [830]
Abanindranath tagore : his early work / ed by Chakravorty, Ramendranath – Calcutta: Art Section, Indian Museum: Distributors, Visva Bharati, 1951 – (= ser Samp: indian books) – us CRL [490]
Abantu batho – Johannesburg. 1930-31 [mf ed Cape Town: SA library 1985] – 1r – 1 – (absorbed: moromica, 1912 and: umlomo wa bantu, 1913) – mf#MS00452 – sa National [079]
Abascal, Jesus *see* Soroche y otros cuentos
Abasolo, Juan Antonio *see* Fray juan antonio abasolo de la regular observancia de nuestro seraphico padre san francisco...
Abastecimiento de aguas *see* Aforos en la sierra de montanchez. ano 1818
Abastumnis astropizikuri observatoria *see* Biulleten' [Abastumnis astropizikuri observatoria]
Abate, Giuseppe *see* La casa natale di s. francesco...secolo 13. roma (1966)
L'abatis / Savard, Felix-Antoine – Montreal: Fides, cop. 1943 ([Montreal]: Imprimerie Saint-Joseph) [mf ed 1992] – 9 – mf#SEM105P1600 – cn Bibl Nat [971]
Abau-Aly al-Husayn ibn Abdallah ibn Sainaa *see* A compendium on the soul
Abba hillel silver papers, 1909-1989, [bulk 1914-1963] / Silver, Abba Hillel – [mf ed 1994] – 8 ser 236r – 1 – (ser1: general correspondence 1914-69 [r1-99]; ser2: harold p manson file 1932-49 [r100-109]; ser3: personal correspondence 1914-64 [r110-144]; ser4: sermons 1915-63 [r145-170]; ser5: writings 1909-63 [r171-187]; ser6: speaking engagements 1917-63 [r188-210]; ser7: personal miscellaneous 1908-89 [r211-225]; ser8: scrapbooks 1909-64 [r226-235]. guide sold separately: d3497.g $40) – us Western Res [939]
Abba Tecle Mariam Semhary Selam *see* De ss sacramentis secundum ritum aethiopicum
Abbadie, A d' *see*
- Catalogue raisonne de manuscrits ethiopiens
- Geographie de l'ethiopie. ce que j'ai entendu, faisant suite...ce que j'ai vu

Abbas, Ahmed *see* La loi municipale du 5 avril 1884 et l'algerie
Abbas al-cAzzawi *see* Ta'rikh al-ciraq baina 'l-ihtilalain
Abbas, Khwaja Ahmad *see*
- An indian looks at america
- Invitation to immortality
- Rice and other stories
- Tomorrow is ours!

Abbatessa, Giovanni Battista. *see* Ghirlanda di varii fiori
L'abbaye de lobbes : depuis les origines jusqu'en 1200 / Warichez, J – Louvain, 1909 – €28.00 – ne Slangenburg [241]
L'abbaye de rossano : contribution a l'histoire de la vaticane / Batiffol, Pierre – Paris: Alphonse Picard, 1891 [mf ed 1992] – viii/xl/182p on 1mf – 9 – 0-524-03454-0 – (in french. incl bibl ref) – mf#1990-0997 – us ATLA [241]
L'abbaye exempte de cluny et le saint-siege (afm22) / Letonnelier, G – 1923 – (= ser Archives de la france monastique (afm)) – €11.00 – ne Slangenburg [241]
Abbayes et prieures de l'ancienne france (afm10) : tom 3: provinces ecclesiastiques d'auch et de bordeaux / Beaunier, dom; ed by Besse, J M – (ed Besse) 1910 – (= ser Archives de la france monastique (afm)) – 6mf – 8 – €18.00 – ne Slangenburg [241]
Abbayes et prieures de l'ancienne france (afm12) : tom 4: provinces ecclesiastiques d'alby, de narbonne et de toulouse / Beaunier, dom – (ed Besse) 1911 – (= ser Archives de la france monastique (afm)) – €17.00 – ne Slangenburg [241]
Abbayes et prieures de l'ancienne france (afm14) : tom 5: province ecclesiastique de bourges / Beaunier, dom – (ed Besse) 1912 – (= ser Archives de la france monastique (afm)) – €15.00 – ne Slangenburg [241]
Abbayes et prieures de l'ancienne france (afm15) : tom 6: province ecclesiastique de sens / Beaunier, dom – (ed Besse) 1913 – (= ser Archives de la france monastique (afm)) – €11.00 – ne Slangenburg [241]
Abbayes et prieures de l'ancienne france (afm17) : tom 7: province ecclesiastique de rouen / Beaunier, dom – (ed Besse). 1914 – (= ser Archives de la france monastique (afm)) – €14.00 – ne Slangenburg [241]
Abbayes et prieures de l'ancienne france (afm19) : tom 7: province ecclesiastique de tours / Beaunier, dom – (ed Besse) 1920 – (= ser Archives de la france monastique (afm)) – €17.00 – ne Slangenburg [241]
Abbayes et prieures de l'ancienne france (afm36) : tome 9: province ecclesiastique de vienne / Besse, J et al – 1932 – (= ser Archives de la france monastique (afm)) – €14.00 – ne Slangenburg [241]
Abbayes et prieures de l'ancienne france (afm37) : tom 10: province ecclesiastique de lyon, pt 1 / Beyssac, J – 1933 – (= ser Archives de la france monastique (afm)) – €11.00 – ne Slangenburg [241]
Abbayes et prieures de l'ancienne france (afm45) : tom 12: province ecclesiastique de lyon, pt 3 / Laurent, J & Claudon, F – 1941 – (= ser Archives de la france monastique (afm)) – €25.00 – ne Slangenburg [241]
L'abbe bourassa / DeCelles, Alfred Duclos – Ottawa: J Hope, 1905 – 1mf – 9 – 0-665-72869-7 – mf#72869 – cn Canadiana [241]
L'abbe de broglie : sa vie – ses oeuvres / Largent, Augustin – 2e ed. Paris: Bloud & Barral 1900 [mf ed 1986] – 1mf – 9 – 0-8370-7473-8 – (in french; incl bibl ref) – mf#1986-1473 – us ATLA [241]
The abbe de lamennais and the liberal catholic movement in france / Gibson, William – London; New York: Longmans, Green, 1896 – 1mf – 9 – 0-7905-9382-3 – mf#1989-2607 – us ATLA [241]
L'abbe eusebe renaudot : essai sur sa vie et son oeuvre liturgique / Villien, A – Paris 1904 – 4mf – 8 – €11.00 – ne Slangenburg [241]
L'abbe gabriel richard : cure de detroit: conference donnee a l'universite laval / Dionne, Narcisse Eutrope – Quebec: Impr de S A Demers, 1902 [mf ed 1985] – 1mf – 9 – mf#SEM105P503 – cn Bibl Nat [241]
L'abbe h-r casgrain / Laflamme, Joseph Clovis Kemner – [Quebec]: [s.n.], [1905?] (mf ed 1986) – 1mf – 9 – mf#SEM105P736 – cn Bibl Nat [241]
L'abbe joseph aubry / Chandonnet, Thomas Aime – Montreal?: s.n, 1875 – 2mf – 9 – mf#00554 – cn Canadiana [241]
L'abbe j-p lapauze : ven de l ang, o de bordeaux: cure de bonzac et galgon, archipretre de fronsac, au diocese de bordeaux 1750-1792 / Brun, Guillaume, abbe – Bordeaux: Feret & Fils 1903 – 2mf – 9 – mf#vrl-166 – ne IDC [366]
Un abbe part en guerre contre un sulpicien / Lanctot, Gustave – Montreal: G Ducharme, 1943 [mf ed 1993] – 1mf – 9 – mf#SEM105P1966 – cn Bibl Nat [971]
Abbeloos, Jean Baptiste *see* De vita et scriptis sancti jacobi, batnarum sarugi in mesopotamia episcopi
Abbetmeyer, Charles et al *see* Four hundred years
Abbey, Charles J *see*
- The english church in the eighteenth century
- Religious thought in old english verse

Abbey, Charles John *see* The english church and its bishops, 1700-1800
Abbey, Richard *see*
- Diuturnity
- Diuturnity, or, the comparative age of the world

Abbildungen und beschreibungen neuer und seltener thiere und pflanzen in syrien und im westlichen taurus gesammelt / Kotschy, C G T – Stuttgart, 1843 – 6mf – 9 – mf#8393 – ne IDC [590]
Abbondio-Kuenzle, Christine *see* Chrut und uchrut im seelegaertli
Abbonis de bello parisiaco libri 3 (mgh7:1.bd) – 1871 – (= ser Monumenta germaniae historica 7: scriptores rerum germanicarum in usum scholarum (mgh7)) – €5.00 – ne Slangenburg [240]
Abbot, Charles Greeley *see* The earth and the stars
Abbot engineer – Camp Abbot OR, 1943-44 [wkly] [mf ed 1972] – 1r – 1 – us Oregon Lib [355]
Abbot, Ezra *see*
- The authorship of the fourth gospel
- Critical greek and english concordance of the new testament
- Dr. william smith's dictionary of the bible
- The literature of the doctrine of a future life

Abbot, Ezra et al *see*
- The fourth gospel
- Notes on scriveners' "plain introduction to the criticism of the new testament"

Abbot, Francis Ellingwood *see*
- Impeachment of christianity
- Is romanism real christianity?
- Scientific theism
- The way out of agnosticism

Abbot, G *see* The reasons which doctour hill hath brought
Abbotsford sumas and matsqui news – British Columbia, CN. 1923- – 4r/y – 1 – Can$93.00r – cn Commonwealth Imaging [071]
Abbotsford tribune – 1923 jan 11-1925 dec 31; 1928, 1930-32; 1933-37; 1938-43 sep 30; 1943 oct 7-1948; 1949-52; 1953-58; 1959-63 jan 31 – 1 – mf#911667 – us WHS [071]
Abbott, Albert Holden *see* Thoughts on philosophy
Abbott, Anstice *see* Indian idylls
Abbott, Austin *see*
- Abbott's new york digest
- A brief for the trial of criminal cases
- A brief on the modes of proving the facts most frequently in issue.

Abbott, Benjamin Vaughan *see*
- Abbott's new york digest
- Abbott's reports of cases in admiralty for the southern district of new york
- Abbott's reports of u.s. circuit and district court decisions
- Judge and jury
- The patent laws of all nations
- A treatise upon the united states courts.

Abbott, Caroline Luxburg *see* Hin und zurueck
Abbott, Charles *see*
- A treatise of the law relative to merchant ships and seamen.
- A treatise of the law relative to merchant ships and seamen

Abbott, Edith *see*
- Historical aspects of the immigration problem
- Immigration
- The wages of unskilled labor in the united states, 1850-1900

Abbott, Edwin Abbott *see*
- The anglican career of cardinal newman
- Apologia
- Bible lessons
- Cambridge sermons
- Clue
- Contrast
- The corrections of mark adopted by matthew and luke
- The fourfold gospel
- From letter to spirit
- Illusion in religion
- Indices to diatessarica
- Johannine vocabulary
- Kernel and the husk
- Light on the gospel from an ancient poet
- The message of the son of man
- Miscellanea evangelica
- Notes on new testament criticism
- Onesimus
- Paradosis
- Philomythus
- St thomas of canterbury
- The son of man
- Through nature to christ

Abbott, Ernest Hamlin *see* Religious life in america
Abbott, Frank F *see* A history and description of roman political institutions
Abbott, George Frederick *see*
- The holy war in tripoli
- Israel in europe
- Macedonian folklore

Abbott, Jacob *see*
- The corner-stone
- History of queen elizabeth

Abbott, Je'Anna Lanza *see*
- Journal of convention and event tourism
- Journal of convention and exhibition management

Abbott, John *see*
- The keys of power
- Sind

Abbott, John Stevens Cabot *see*
- The history of christianity
- History of marie antoinette

Abbott, Lyman *see*
- America in the making
- Christianity and social problems
- Christ's secret of happiness
- The epistle of paul the apostle to the romans
- The evolution of christianity
- The gospel according to luke
- The great companion
- Henry ward beecher
- An illustrated commentary of the gospel according to matthew
- An illustrated commentary on the acts of the apostles
- An illustrated commentary on the gospel according to st john
- An illustrated commentary on the gospels according to mark and luke
- Impressions of a careless traveler
- Jesus of nazareth
- Laicus
- The life and literature of the ancient hebrews
- Old testament shadows of new testament truths
- The other room
- The personality of god
- Problems of life
- Reminiscences
- The rights of man
- The roman catholic question
- Seeking after god
- The spirit of democracy
- Study in human nature
- The supernatural
- The theology of an evolutionist

Abbott, Lyman et al *see*
- Henry ward beecher as his friends saw him
- The new puritanism
- The problem of human destiny
- The prophets of the christian faith

Abbott, Mary *see* Collected field reports on the phonology of basari
Abbott, Ouida Davis *see*
- General properties of some tropical and sub-tropical fruits of florida
- Nutritional anemia and its prevention
- Utilization and storage of florida grapes

Abbott, S J *see* Convent jubilee memorial
Abbott, Thomas Kingsmill *see*
- The codex rescriptus dublinensis of st matthew's gospel
- A critical and exegetical commentary on the epistles to the ephesians and to the colossians
- "Do this in remembrance of me," should it be, "offer this"?
- The english bible and our duty with regard to it
- Essays chiefly on the original texts of the old and new testaments

Abbott, W *see*
- Covert from the tempest
- Felgate archer

Abbottempo – Montreal, Canada 1963-65 – 1,5,9 – mf#1908 – us UMI ProQuest [610]
Abbott's appeals decisions : unreported / New York. (State) – v1-4. 1852-69 (all publ) – 28mf – 9 – $42.00 – (a pre-nrs title) – mf#LLMC 80-053 – us LLMC [340]
Abbott's new cases / New York. (State) – v1-32. 1876-94 (all publ) – 213mf – 9 – $319.00 – mf#LLMC 78-094 – us LLMC [340]
Abbott's new york digest / Abbott, Benjamin Vaughan & Abbott, Austin – New York: Voohris. v1-8. 1813-69 (all publ) – 67mf – 9 – $100.00 – mf#LLMC 79-518 – us LLMC [348]
Abbott's practice reports / New York. (State) – os: v1-19. 1854-65; ns: v1-16. 1865-75 (all publ) – 227mf – 9 – $340.00 – mf#LLMC 78-090 – us LLMC [340]
Abbott's reports of cases in admiralty for the southern district of new york / Abbott, Benjamin Vaughan – Boston: Little-Brown. 1v. 1857 (all publ) – (= ser Early federal nominative reports) – 7mf – 9 – $10.50 – mf#LLMC 81-427 – us LLMC [324]
Abbott's reports of u.s. circuit and district court decisions / Abbott, Benjamin Vaughan – New York: Diossy, 1863-1871 – (= ser Early federal nominative reports) – 14mf – 9 – $21.00 – mf#LLMC 81-428 – us LLMC [347]
Abbott's weekly and illustrated news – Chicago. Ill. v. 1, no. 2-15; 6, no. 8-14. 1933-1934 – 1 – us NY Public [071]
Abbrege de la doctrine evangelique et papistique / Bullinger, Heinrich – [Geneve], Iean Crespin, 1558 – 1mf – 9 – mf#PBU-167 – ne IDC [240]
Abbrege des methodes de traicter des controverses / Veron, F – [Paris], 1636 – 9mf – 9 – mf#CA-152 – ne IDC [240]
Abbs, J *see* Twenty-two years' missionary experience in travancore
L'abc – Port-au Prince: Impr Vve J Chenet, 1re annee, n2-28. 27 mars-31 juli 1897 – us CRL [072]
Abc – Madrid: Prensa Espanola, 1956-feb 1993 – 1 – us CRL [074]
Abc – Madrid, Spain. 1942-55 – 84r – 1 – us L of C Photodup [074]
Abc, americans before columbus – v1 n1 [1963 oct]; v2 n4 [1964 dec] – 1 – mf#1109393 – us WHS [071]
The abc catechism *see*
- Chin pu wen ta [ccm265]

Abc del alcoholismo / Curado Garcia, Blas – Barcelona: Offset industrial, S.A. 1972 – 1 – sp Bibl Santa Ana [362]
Abc des petits canadiens / Maxine – Montreal: Editions Albert Levesque, 1933 [mf ed 1994] – 1mf – 9 – (ill by j-arthur lemay) – mf#SEM105P2150 – cn Bibl Nat [810]
Abc des wissens fuer die denkenden / Douai, Adolf – Leipzig: Genossenschaftsbuchdruckerei, 1875 [mf ed 19–] – (= ser Sammlung social-politischer Schriften, Bd 26) – mf#*ZT-SFC pv 35 n23 – us NY Public [300]
Abc diario musico : critical notices of the principal musicians of the period, resident in england – Bath: printed for aut...1780 [mf ed 19–] – 1mf – 9 – mf#fiche38 – us Sibley [780]

L'abc du hatha-yoga pour enfants de 6 a 12 ans / Favre, Norette & Bastien, Gilles – [Montreal]: le Cercle du livre de France, [1958?] [mf ed 1998] – 1mf – 9 – (pref by f gilles bastien) – mf#SEM105P2810 – cn Bibl Nat [613]

Abc [lisbon] – Lisbon, Portugal 15 Jul 1920-24 Sep 1931 – 12r – 1 – (with undated specimen iss) – uk British Libr Newspaper [074]

Abc [madrid] – Madrid, Spain 20 aug-31 oct 1914; 17 jul 1915-10 aug 1919; 2,6,8 sep 1936; 3 jan, 30 sep-26 oct 1937; 18 jul, 11,15,17,18,22-25,29,30 aug 1939; 1 aug 1940-28 feb 1993 – 1 – (wanting: dec 1946, jan, feb, dec 1951, jan 1952, 1-15 dec 1955) – uk British Libr Newspaper [074]

Abc (miami fl) : [...organo del interamerican bureau anti-comunista labor] – Miami FL: Publicidad CACUGA 1971- (wkly) [mf ed 1999] – (= ser Cuban heritage collection) – 1r – 1 – us UF Libraries [335]

Abc news – 1958-62; 1958-69 nov; 1963-65 – 1 – mf#1051431 – us WHS [071]

Abc news – [Eugene OR: Active Bethel Citizens, (mthly) [mf 1974-83] [mf ed [1984?]] – 3r – 1 – (cont by: active bethel citizens (eugene, or) newsletter) – us Oregon Lib [071]

Abc news transcripts – New York NY 1 jan 1970-29 dec 1995 – 9 – uk British Libr Newspaper [071]

The abc of colonization : in a series of letters...no 1: addressed to the gentlemen forming the committee of the family colonization loan society... / Chisholm, Caroline – London, 1850 – (= ser 19th c books on british colonization) – 1mf – 9 – mf#1.1.531 – uk Chadwyck [330]

The abc of options and arbitrage / Nelson, Samuel Armstrong – New York: S A Nelson, 1904 (mf ed 19–) – (= ser The Wall Street library) – 87p – mf#ZT-520 – us NY Public [332]

The abc of the government of the united states / Morse, Perley – NY: Perley Morse & Co, 1916 – 1mf – 9 – $1.50 – mf#LLMC 96-057 – us LLMC [323]

The abc of the irish land question / Castletown, Bernard Edward Barnaby Fitzpatrick, 2nd Baron – London, 1881 – (= ser 19th c ireland) – 1mf – 9 – mf#1.1.2195 – uk Chadwyck [333]

The abc of the knights of the ku klux klan / Simmons, W J – Atlanta: The Klan, c1917 – us CRL [366]

The abc of wall street / ed by S.A.N. – NY: Doubleday, Page & Co, 1916 – 2mf – 9 – $3.00 – mf#LLMC 96-061 – us LLMC [332]

The abc (or three hundred character) catechism : being a statement of the fundamentals of christian doctrine in simple style and with the use of only 303 separate characters... / Price, Philip Francis – 2nd ed. Shanghai: Chinese Tract Society, 1917 [mf ed 1995] – (= ser Yale coll; Catechism series 1) – 26p (ill) – 1 – 0-524-09945-6 – (in chinese) – mf#1995-0945 – us ATLA [240]

The abc railway and steamboat travellers' guide – Montreal: J T Robinson, [1880-188-? or 189-?] – 9 – mf#P04042 – cn Canadiana [380]

Abc teacher – v1 n1-v8 n1 [1972 dec-1980 sep/oct] – 1 – mf#647071 – us WHS [071]

Abc wisconsin merit shop talk – 1988 feb-1993 dec – 1 – mf#2481806 – us WHS [071]

Abdala – v1 n4 [1971 jul] – 1 – mf#1584696 – us WHS [071]

Abd-al-Karim Ibn-Akibat Ibn-Muhammad Bulaki al-Kasmiri *see* Voyage de l'inde a la mekke

Abdallae beidavaei historia sinensis, persice e gemino manuscripto edita, latine... / Mueller, A – Jenae: prostat apud Johannem Bielkivm, [1689] – 2mf – 9 – mf#HT-689 – ne IDC [910]

Abdallah, A *see* L'ordinamento liturgico die gabriele 5, 88e patriarca copto

Abdallah, Yohanna B *see* Wayao'we

Abderrahman ben Abdallah ben Imran ben Amir Es-sa'di *see* Documents arabes relatifs a l'histoire du soudan

Abdias / Stifter, Adalbert – Frankfurt am Main: H Cobet 1947 [mf ed 1991] – 1r – 1 – (filmed with: der dichter der geharnschten venus / von albert koster) – mf#2902p – us Wisconsin U Libr [830]

Abdo, Ada *see* Mateo y las sirenas

Abdominal imaging – Dordrecht, Netherlands 1993+ – 1,5,9 – ISSN: 0942-8925 – mf#19607 – us UMI ProQuest [616]

The abduction of mary ann smith, by the roman catholics : and her imprisonment in a nunnery, for becoming a protestant / Mattison, Hiram – Jersey City, NJ: publ by aut [1868] [mf ed 1984] – 2mf – 9 – 0-8370-0776-3 – mf#1984-4141 – us ATLA [240]

Abduelhamid'in kaygulari / Sami, Ebuesuereyya – Istanbul: Kader Matbaasi, 1330 [1914] – (= ser Ottoman histories and historical sources) – 2mf – 9 – $40.00 – us MEDOC [956]

Abduelkadir, Mehmed *see* Muekemmel ve mufassal tuerkiye cumhuriyeti atlasi

Abduh, Muhammad *see* Al-curwa al-wuthqa

Abdul baha in egypt / Sohrab, Mirza Ahmad – London: Rider [1929] [mf ed 1985] – 1r – 1 – (filmed with: the astrology of personality / rudhyar, d) – mf#1342 – us Wisconsin U Libr [290]

Abdul baha on divine philosophy / Abdul-Baha; ed by Fraser-Chamberlain, Isabel – Boston, MA: Tudor Press, 1916 [mf ed 1991] – 1mf – 9 – 0-524-00813-2 – mf#1990-2059 – us ATLA [290]

Abdul baha on divine philosophy / Chamberlain, Isabel Fraser – Boston: Tudor Press c1918 [mf ed 1987] – 1r – 1 – (at the suggestion of abdul baha, these notes on divine philosophy, together with a short introductory history, have been comp & publ by isabel fraser chamberlain. filmed with: germanische mythologie / mogk, e & other titles) – mf#1891 – us Wisconsin U Libr [290]

Abdul-Baha *see*
- Abdul baha on divine philosophy
- Some answered questions
- Ten days in the light of acca
- Traveller's narrative written to illustrate the episode of the bab

Abdullah abdurahman family papers, 1906-1962 – Chicago: Uni of Chicago, Photodup Dept, [1977] – us CRL [920]

'Abdullah al-Rumi, Esrefoglu *see* The divan project

'Abdu'l-Latif, Sayyid *see* The influence of english literature on urdu literature

Abdur Rahman, A F M *see* Institutes of mussalman law

Abecedario pittorico... / Orlandi, P A – Bologna, 1704 – 8mf – 9 – mf#O-382 – ne IDC [700]

Abeel, David *see* Journal of a residence in china and the neighbouring countries from 1830 to 1833

L'abeille – [Quebec: Petit Seminaire de Quebec. v1 n1 27 juil 1848-v14 n38 23 juin 1881] – 9 – (with ind) – mf#P04944 – cn Canadiana [370]

L'abeille *see* La minerve litteraire

L'abeille francoise ou nouveau recueil, de morceaux brillans, des auteurs francois les plus celebres : ouvreage utile a ceux qui etudient la langue francoise, et amusant pour ceux qui la connoissent... / Nancrede, Joseph – A Boston: de l'impr de Belknap et Young...1792 – 4mf – 9 – (int in french and english) – mf#41882 – cn Canadiana [440]

L'abeille medicale – Montreal: T Berthiaume. v1 n1 janv 1879-v4 n1 janv 1882 – 9 – mf#P05196 – cn Canadiana [610]

L'abeille paroissiale – Montreal: Granger Greres. n1 15 avril 1895-n14 mai 1896 – 9 – mf#P04024 – cn Canadiana [440]

L'abeille pour les enfans ou lecons francaises : a l'usage des ecoles – Montreal: H Ramsay, 18–? – 2mf – 9 – (in english and french; incl bibl ref) – mf#46072 – cn Canadiana [440]

La abeja : periodico politico y de agricultura, artes, industria, commercio, instruccion y beneficiencia – Rio de Janeiro, RJ. 04-11 jun 1868 – (= ser Ps 19) – mf#DIPER – bl Biblioteca [079]

La abeja espanola – Anos 1812-1813 (12-IX/31-VIII) – 49mf – 9 – sp Cultura [946]

Abeken, Bernhard Rudolf *see*
- Goethe in den jahren 1771 bis 1775
- Goethe in meinem leben

Abel : en plattdeutsch stueckschen merrn ut de marsch un merrn ut't leben / Trede, Paul – 2. opl. Garding: H Luehr & Dircks [1896] [mf ed 1991] – 1r – 1 – (filmed with: der junge tieck und seine marchenkomodien / von kathe brodnitz) – mf#2913p – us Wisconsin U Libr [390]

Abel, Carl *see*
- Koptische untersuchungen
- Linguistic essays
- Narrative of a journey in the interior of china
- Ueber den gegensinn der urworte

Abel, Charles William *see* Savage life in new guinea

Abel, Felix-Marie *see*
- Bethleem
- Une croisiere autour de la mer morte
- Geographie de la palestine, vols 1-2
- Jerusalem nouvelle. fascicule 1 et 2, aelia capitolina, le saint-sepulcre et le mont des oliviers
- Les livres des maccabees

Abel, Karl Friedrich *see*
- Six sonatas, for a violin, a violoncello, & base [sic] with a thorough base [sic] for the harpsichord, opera 9
- Six symphonies a deux violins, deux hautbois, deux cors de chasse, alto viola e basse...

Abel larkin family papers, 1790-1895 / Larkin, Abel – [mf ed 1986] – 4r – 1 – (correspondence, legal & financial papers, land deeds and agreements...of abel & stillman larkin, early residents of southeast ohio. abel was also a justice of the peace; stillman, a local historian) – mf#ms924 – us Western Res [978]

Abel, Ludwig *see* Keilschrifttexte

Abelard : sa lutte avec saint bernard / Vacandard, Elphege – Paris: A. Roger et F. Chernoviz, 1881 – 1r – 1 – 0-8370-0608-2 – (incl bibl ref) – mf#1984-T098 – us ATLA [240]

Abelard : sa vie, sa philosophie & sa theologie / Remusat, Charles de – nouv ed. Paris: Didier, 1855 [mf ed 1990] – 2v on 3mf – 9 – 0-7905-7074-2 – (incl bibl ref) – mf#1988-3074 – us ATLA [240]

Abelard and the origin and early history of universities / Compayre, Gabriel – New York: Charles Scribner, 1893 [mf ed 1989] – (= ser The great educators) – 1mf – 9 – 0-7905-4261-7 – (incl bibl ref) – mf#1988-0261 – us ATLA [378]

Abelard, Peter *see* Lettres completes d'abelard et heloise

A abelha : periodico da sociedade pharmaceutica brasileira – Rio de Janeiro, RJ: Typ de Paula Brito, jul 1862-jan 1864 – (= ser Ps 19) – mf#P17,01,63 – bl Biblioteca [615]

A abelha : semanario scientifico, industrial e litterario – Rio de Janeiro, RJ: Empresa Nacional do Diario, 12 jan-30 jun 1856 – (= ser Ps 19) – 1,5,6 – mf#P01,04,30 – bl Biblioteca [079]

Abelha do itaculumy – Ouro Preto, MG: Officina Patricia de Barboza e C, jan 1824-jul 1825 – (= ser Ps 19) – mf#P0,03,19-20 – bl Biblioteca [320]

Abelha pernambucana – Pernambuco: Typ Fidedigna, 24 abr 1829-31 ago 1830 – (= ser Ps 19) – mf#P19,03,09 – bl Biblioteca [320]

Abella Rodriguez, Arturo *see* Florero de llorente

Abella, V M de *see* Vade-mecum filipino, o manual de la conversacion familiar espanol-tagalog

Abellard, Alexandre Charles *see* David

Abelly, L *see*
- Traitte des heresies, contenant les causes des heresies, les moeurs et artifices...
- La vie du venerable serviteur de dieu vincent de paul

Abelman, Jonathan *see* Sefer zikhron yehonatan

Abelous, Louis David et al *see* Les catacombes de rome – souvenirs de rome

Abelson, J *see* The immanence of god in rabbinical literature

Aben, Tersur Akuma *see* Doctrine of divine immutability as god's constancy

Abenaki indian legends, grammar and place names / Masta, Henry Lorne – Victoriaville: l'impr de Victoriaville, 1932 [mf ed 1992] – 2mf – 9 – (foreword by a irving hallawell) – mf#SEM105P1640 – cn Bibl Nat [390]

Les abenaquis : habitat et migrations (17e et 18e siecles) / Sevigny, Paul-Andre – Montreal: Editions Bellarmin, 1976 [mf ed 1995] – 3mf – 9 – mf#SEM105P2389 – cn Bibl Nat [305]

Der abend – Berlin, DE. 1928-32; 1946-49; 1955-70 – 64r – 1 – us L of C Photodup [074]

Das abend blatt – New York etc. v1. no. 1-v9. no. 2337. Oct 15 1894-Apr 12 1902 – 1 – us NY Public [071]

Das abend blatt – New York. N.Y. The evening paper. 1894-1902 – 1 – us AJPC [071]

Der abend / duesseldorfer stadtanzeiger – Duesseldorf DE, apr-may 1933 [gaps] – 1r – 1 – gw Misc Inst [074]

Der abend ohne gefolge : eine pratergeschichte / Brehm, Bruno – Stuttgart: Deutsche Volksbuecher 1943 [mf ed 1989] – (= ser Wiesbadener volksbuecher 292) – 1r – 1 – (filmed with other titles) – mf#7066 – us Wisconsin U Libr [830]

Abend post – Milwaukee WI. 1892 sep 22-dec; 1893 jan-jun; 1893 jul-dec; 1894 jan-jun; 1894 jul-dec; 1895 jan-jun; 1895 jul-dec; 1896 jan-apr; 1896 may-aug; 1896 sep-dec; 1897 jan-apr – 11r – 1 – (cont by: germania [milwaukee, wi: 1873: daily]; germania und abend=post) – mf#1166840 – us WHS [071]

Der abend / westdeutsche abendzeitung fuer rhein und ruhr – Duesseldorf DE, apr-may 1933 [gaps] – 1r – 1 – gw Misc Inst [074]

Abendblatt – Duesseldorf DE, sep 30 1880-feb 1 1881 – 1r – 1 – gw Misc Inst [074]

Abendblatt *see* Telegramm-zeitung

Die abendburg : chronika eines goldsuchers in zwoelf abenteuern / Wille, Bruno – Jena: E Diederichs 1909 [mf ed 1996] – 1r – 1 – (filmed with: buch der gedichte / anton wildgans) – mf#4064p – us Wisconsin U Libr [830]

Abendecho *see* Hamburg-altonaer volksblatt

Abendglocken : gedichte / Giorg, Kara – Chicago: Koelling & Klappenbach [19–] [mf ed 1989] – 1r – 1 – (filmed with: fliegt der blaufuss? / otto brues) – mf#7092 – us Wisconsin U Libr [810]

Die abendlaendische messe : vom fuenften bis zum achten jahrhundert / Probst, Ferdinand – Muenster i. W.: Aschendorff, 1896 – 2mf – 9 – 0-7905-5960-9 – (incl bibl ref) – mf#1988-1960 – us ATLA [240]

Abendlaendische palaestinapilger des ersten jahrtausends und ihre berichte / Baumstark, A – Koeln, 1906 – 1mf – 9 – mf#H-2968 – ne IDC [915]

Abendlaendische palaestinapilger des ersten jahrtausends und ihre berichte : eine kulturgeschichtliche skizze / Baumstark, Anton – Koeln: JP Bachem, 1906 [mf ed 1989] – 1mf – 9 – 0-7905-4131-9 – (incl bibl ref) – mf#1988-0131 – us ATLA [240]

Die abendlaendische spekulation des zwoelften jahrhunderts in ihrem verhaeltnis zur aristotelischen und juedisch-arabischen philosophie / Schneider, A – 1915 – (= ser Bgphma 17/4) – €5.00 – ne Slangenburg [100]

Abendland : unabhaengige deutsche europaeische stimmen fuer christliche gesellschaftserneuerung – Prag (CZ), 1938 feb – 1 – gw Misc Inst [240]

Das abendland : central organ fuer alle zeitgemaessen interessen des judenthums – Prague, Bruenn: Daniel Ehrmann. v1-6. 1864-69 – (= ser German-jewish periodicals...1768-1945, pt 2) – 1r – 1 – $125.00 – mf#B1 – us UPA [270]

Das abendland – Prag (CZ), Bruenn, 1864-68 – 1r – 1 – gw Misc Inst [077]

Abendlied / Schubert, Franz – [1817?] [mf ed 19–] – 1mf – 9 – (words by johann mayrhofer; filmed with: die blumensprache) – mf#fiche 29 – us Sibley [780]

Das abendmahl im neuen testament / Seeberg, Reinhold – 2. durchges aufl. Berlin: Edwin Runge, 1907 – 1mf – 9 – 0-7905-0514-2 – mf#1987-0514 – us ATLA [225]

Das abendmahl in neuen testament / Eichhorn, Albert – Leipzig: J C B Mohr, 1898 – (= ser Hefte zur "christlichen welt") – 1mf – 9 – 0-524-05578-5 – mf#1992-0438 – us ATLA [225]

Abendpost : ausgabe sued-hannover – Goettingen DE, 1947 6 feb-1949 3 mar – 1r – 1 – gw Mikrofilm [074]

Abendpost – Hannover DE, 1947 6 feb-1948 – 1 – gw Misc Inst [074]

Abendpost – chicago ed. Chicago, IL: Glogauer & Co – 1 – (iss for 1904, 1906-07, 1939 filmed consecutively with: sonntagpost 1904, 1906-07, 1922-38, 1939) – us CRL [071]

Abendpost – Weimar DE, 1947 – 1r – 1 – (missing: n244-247; filmed by misc inst: 1947 jul 7-dec [1r]) – mf#6521 – gw Mikropress [074]

Abendpost – Chicago IL (USA), 1939, 1962 26 jun-24 sep, 1963 2 jan-1 apr, 1964 24 jun-22 sep, 1965 7 apr-1966 23 mar, 1966 13 jun-23 jul, 1966 19 oct-1967, 1968 3 apr-1970 30 sep, 1971 5 jan-30 sep, 1972-84, 1986-1987 18 dec, 1989 6 jan-22 nov – 1 – (title varies: 1951?: abendpost und milwaukee deutsche zeitung) – gw Misc Inst [071]

Abendpost frankfurt, nachtausgabe *see* Frankfurter neue presse / nachtausgabe

Abendpost nachtausgabe *see* Frankfurter neue presse / nachtausgabe

Abendpost & sonntagpost – Omaha NE 1889-1979 (1) – ISSN: 0896-3762 – mf#10213 – us UMI ProQuest [071]

Abendpost und milwaukee deutsche zeitung – 1968 jun 25-dec 19; 1968 dec 20-69 jun 11; 1969 dec 8-70 jun 8; 1969 jun 12-dec 5; 1970 jun 9-dec 31; 1971 aug 18-72 may 24; 1971 jan 5-aug 7; 1972 may 25-73 feb 28, mar 2-sep 28, oct-74 mar, apr-sep, oct-dec; 1975 jan-dec; 1976 jan-aug-77 feb; 1977 mar-aug; 1977 sep-78 feb; 1978 mar-aug 1978 jul-dec; 1979 jan-jun, jul-dec; 1980 jan-jun, jul-dec; 1981 jan-may, jun-dec; 1982 jan-apr, may-dec; 1983 jan-dec 9 – 1 – mf#855972 – us WHS [071]

Abendpost und milwaukee deutsche zeitung – 1984 aug 1-dec 28; 1985; 1986-87 jun, jul 3-1988 dec; 1989; 1990 – 1 – mf#1166321 – us WHS [071]

Abendpost und milwaukee deutsche zeitung *see*
- Abendpost
- Sonntagspost

Abendpost-nachtausgabe – Frankfurt/M DE, 1949 28 jan-1966 30 apr, 1976-1988 12 dec – [ca 4r/yr, until 1966 41r] – 1 – (until 30 apr 1966: abendpost; also ausg s; filmed by bnl: 1948 1 oct-1952 25 nov [16r], lacks: 1949) – gw Misc Inst [074]

Abend-zeitung – Dresden DE, 1805 – 1 – (filmed by other misc inst: 1847 jul-dez [1r]) – gw Misc Inst [074]

Die abendzeitung – Muenchen DE, 1948 6 may-1950 – 3mf=5df – 9 – (until n41 1948: die tageszeitung; title varies: 1 jul 1950: abendzeitung. filmed by misc inst: 1951- [ca 8r/yr]) – gw Mikrofilm; gw Misc Inst [074]

Abenteuer eines blaustruempfchens / Heyse, Paul – Stuttgart: C Krabbe [1897] [mf ed 1990] – 1r – 1 – (ill by carl zopf. filmed with: der wollmarkt / von h clauren) – mf#2725p – us Wisconsin U Libr [830]

Abenteuer ernster leute / ed by Scheer, Maximilian – Berlin: Aufbau-Verlag, 1961 [mf ed 1993] – 247p – 1 – mf#8365 – us Wisconsin U Libr [830]

Abenteuer in japan : roman / Brod, Max – Amsterdam: A de Lange 1938 [mf ed 1989] – 1r – 1 – (filmed with: ein gelegenheitsgedicht von brockes / friedrich gundolf) – mf#7089 – us Wisconsin U Libr [830]

Abenteuer und schwaenke : alten meistern nacherzaehlt / Baumbach, Rudolf – Stuttgart: J G Cotta, 1904 [mf ed 1989] – 170p – 1 – mf#6983 – us Wisconsin U Libr [880]

Eine abenteuerliche reise : roman = La folle aventure / Lichtenberger, Andre – Leipzig: Philipp Reclam jun., [19–?] – 1r – 1 – us Wisconsin U Libr [830]

Der abenteuerliche simplicissimus / Grimmelshausen, Hans Jakob Christoph von; ed by Kelletat, Alfred – vollstaend ausg. Muenchen: Winkler, 1956 [mf ed 1993] – 681p/22pl (ill) – 1 – (nach den ersten drucken des "simplicissimus teutsch" und der "continuatio" von 1669, ill fr first complete works ed of 1683-84) – mf#8451 – us Wisconsin U Libr [830]

Der abenteuerliche simplicissimus und andere schriften / Grimmelshausen, Hans Jakob Christoph von; ed by Keller, Adelbert von – Stuttgart: Litterarischer Verein, 1854-62 [mf ed 1993] – (= ser Blvs 33-34, 65-66) – 4v – 1 – mf#8470 reels 7, 13 – us Wisconsin U Libr [430]

Abentheuerliche, doch wahrhafte schicksale zu wasser und zu lande : von ihm selbst treu und einfach erzaehlt und herausgegeben / Staehelin, Johann J – St Gallen 1811 [mf ed Hildesheim 1995-98] – 1v on 2mf – 9 – €60.00 – 3-487-27611-9 – gw Olms [910]

Aberaman, glamorgan, parish church of st margaret : baptisms 1882-1930 – 6mf – 9 – £7.50 – uk Glamorgan FHS [929]

Aberavon, glamorgan, parish church of st mary : baptisms 1696-1900, burials 1691-1900, marriages 1696-1837 – 2mf – 9 – £2.50 – uk Glamorgan FHS [929]

Aberavon, st mary, monumental inscriptions – 2mf – 9 – £2.50 – uk Glamorgan FHS [929]

Aberbargoed, (mon), caersalem baptist, monumental inscriptions – 1mf – 9 – £1.25 – uk Glamorgan FHS [929]

Aberdare, calvaria welsh baptist, monumental inscriptions : and aberaman saron congregational; and cwmbach abernant-y-groes unitarian – 1mf – 9 – £1.25 – uk Glamorgan FHS [242]

Aberdare chapels, glamorgan : baptisms 1788-1951, burials 1809-1836 – 1mf – 9 – £1.25 – uk Glamorgan FHS [929]

Aberdare ebenezer, nebo, salem : baptisms 1790-1851 & membership roll; & hen-dy-cwrdd baptisms 1788-1915; & tabernacle – with members baptisms 1915-1965, burials 1900-1901, marriages 1915-1917 – 1mf – 9 – £1.25 – uk Glamorgan FHS [929]

Aberdare & gelligaer – (= ser 1861 census returns [glamorgan]) – 10mf – 9 – £12.50 – uk Glamorgan FHS [314]

Aberdare & gelligaer s r d s – (= ser 1851 census returns [glamorgan]) – 6mf – 9 – £7.50 – uk Glamorgan FHS [314]

Aberdare, glamorgan, parish church of st elvan : baptisms, 1863-1889, marriages 1855-1918 – [Glamorgan]: GFHS [mf ed c1999] – 1mf – 9 – £2.50 – uk Glamorgan FHS [929]

Aberdare, glamorgan, parish church of st fagans : baptisms 1854-1911, burials 1855-1942, marriages 1854-1925 – 2mf – 9 – £2.50 – uk Glamorgan FHS [929]

Aberdare, glamorgan, parish church of st john baptist : baptisms 1734-1900, burials 1734-1924, marriages 1717-1925 – 6mf – 9 – £7.50 – uk Glamorgan FHS [929]

Aberdare leader – [Wales] LLGC 7 jun 1902-dec 1950 [mf ed 2004] – 47r – 1 – uk Newsplan [072]

Aberdare, st john the baptist, monumental inscriptions – 2mf – 9 – £2.50 – uk Glamorgan FHS [929]

Aberdare, trecynon, ebenezer independent, monumental inscriptions : and trecynon hen-dy-cwrdd unitarian – 1mf – 9 – £1.25 – uk Glamorgan FHS [929]

Aberdare, trecynon, st fagan, monumental inscriptions – 1mf – 9 – £1.25 – uk Glamorgan FHS [929]

Aberdeen advertiser – [Scotland] Aberdeen: W Jolly & Sons 26 oct 1917-15 mar 1918 (wkly) [mf ed 2003] – 1r – 1 – uk Newsplan [072]

Aberdeen and northern friendly society – [Edinburgh: NRAS 1985 – 1mf – 9 – mf#n2488 – uk NRAS [366]

Aberdeen and Temair, Ishbel Gordon, Marchioness of *see*
- A quoi bon la societe aberdeen
- "A dieu"
- Lady aberdeen's address
- The national council of women of canada
- President's closing address at the fifth annual meeting of the national council of women of canada
- What is the use of the victorian order of nurses for canada?
- Where dwells "our lady of the sunshine?"

Aberdeen and Temair, Ishbel Gordon, marchioness of *see*
- Address by lady aberdeen
- Canada's late premier

Aberdeen Area Genealogical Society [SD] *see* Tree climber

Aberdeen Association *see*
- The mission of the old magazine
- Report of the aberdeen association 1898

Aberdeen Auxiliary Bible Society *see* Fifth report of the aberdeen auxiliary bible society

Aberdeen, banff and kincardine people's journal – [Scotland] Aberdeenshire, Dundee: J Leng 4 may 1861-8 sep 1877, jan 1909-dec 1950 (wkly) [mf ed 2003] – 116r – 1 – (missing: 1864; cont by: people's journal for aberdeen, banff, kincardine, and northern counties [1877]; aberdeen people's journal (aberdeen, banff and kincardine) [jan 1909-dec 1913]; people's journal aberdeen, banff and kincardine [jan 1914-dec 1950]) – uk Newsplan [072]

Aberdeen, banff, and kincardine people's journal – Dundee: J Leng -1877 (wkly) [mf ed 2003] – 10r – 1 – (cont by: people's journal for aberdeen, banff, kincardine, and northern counties) – uk Newsplan [072]

Aberdeen catholic herald – [Scotland] Aberdeen: Scottish Observer & Herald Ltd 1 aug 1936-6 jan 1939 (wkly) [mf ed 2004] – 5r – 1 – (absorbed by: scottish catholic herald) – uk Newsplan [241]

Aberdeen chamber of commerce – [Edinburgh: NRAS 1991 – 1mf – 9 – mf#n3255 – uk NRAS [380]

Aberdeen chronicle – [Scotland] Aberdeen: printed by Alex, Aberdein & Co 9 oct 1806-dec 1832 (wkly) [mf ed 2003] – 1351v on 13r – 1 – (cont by: aberdeen herald, and general advertiser for the counties of aberdeen, banff, & kincardine) – uk Newsplan [072]

Aberdeen citizen – [Scotland] Aberdeen: Aberdeen Citizen (Aberdeen) Publ Society 24 may 1926-24 aug 1928 (daily ex sun) [mf ed 2003] – 8r – 1 – uk Newsplan [072]

Aberdeen constitutional and general advertiser for the counties of aberdeen, banff, and kincardine – [Scotland] Aberdeen: J Davidson 16 sep 1837-26 jul 1844 (wkly) [mf ed 2003] – 7v on 2r – 1 – (cont: aberdeen observer) – uk Newsplan [072]

Aberdeen daily free press – Aberdeen: printed & publ by Alexander Marr 1872-74 (wkly) [mf ed 1995-96] – 675v on 3r – 1 – (cont: aberdeen free press, and north of scotland advertiser; cont by: daily free press (aberdeen, scotland)) – uk Scotland NatLib [072]

Aberdeen daily journal – Aberdeen: Aberdeen & North of Scotland Newspaper & Print Co Ltd 1901-22 (daily ex sun) – 6691v – 1 – (cont: aberdeen journal, and daily advertiser for the north of scotland; merged with: aberdeen free press to form: aberdeen press and journal) – uk Scotland NatLib [072]

The aberdeen doctors : a notable group of scottish theologians of the first episcopal period, 1610-1638, and the bearing of their teaching on some questions of the present time / Macmillan, Donald – London: Hodder & Stoughton, 1909 – 1mf – 9 – 0-7905-5113-6 – mf#1988-1113 – us ATLA [240]

Aberdeen evening express – [Scotland] Aberdeen: Aberdeen & North of Scotland Newspaper & Print Co (Ltd) 1879-1899, jan 1895-dec 1950 (daily ex sun) [mf ed 2004] – 267r – 1 – (absorbed: northern evening news; cont by: evening express (aberdeen, scotland) [apr 1899-dec 1950]) – uk Newsplan [072]

Aberdeen free press – Aberdeen: A Marr 1901-22 (daily ex sun) [mf ed 2001] – 1 – (cont: daily free press (aberdeen, scotland); merged with: aberdeen daily journal to become: aberdeen press and journal) – uk Scotland NatLib [072]

Aberdeen free press, and north of scotland advertiser – Aberdeen: printed & publ by Alexander Marr 1869-72 (wkly) [mf ed 1995] – 777v – 1 – (cont: aberdeen free press, peterhead, fraserburgh, and buchan news, and north of scotland advertiser; cont by: aberdeen daily free press) – uk Scotland NatLib [072]

Aberdeen free press and north of scotland review : a general advertiser for aberdeen and the northern counties – Aberdeen: printed & publ by Arthur King 1853-55 (wkly) [mf ed 1995] – 104v on 3r – 1 – (cont: north of scotland gazette; cont by: aberdeen free press, peterhead, fraserburgh, and buchan news, and north of scotland advertiser) – uk Scotland NatLib [072]

Aberdeen free press, peterhead, fraserburgh, & buchan news and north of scotland advertiser – Aberdeen: A King 1855-69 (wkly) [mf ed 1995] – 744v – 1 – (cont: aberdeen free press and north of scotland review; cont by: aberdeen free press, and north of scotland advertiser) – uk Scotland NatLib [072]

Aberdeen, G *see* An inquiry into the principles of beauty in grecian architecture

Aberdeen, George Hamilton Gordon, comte de *see* Copy of a despatch, and its enclosures

Aberdeen guardian and northern counties chronicle – [Scotland] Aberdeen: printed & publ by J Duffas 11 sep 1869-12 mar 1870 (wkly) [mf ed 2003] – 27v on 1r – 1 – uk Newsplan [072]

Aberdeen herald – Aberdeen, Scotland 20 aug 1988-22 apr 1989 – 1 – (discontinued; replaced by: aberdeen post) – uk British Libr Newspaper [072]

The aberdeen herald – Aberdeen SA, jul 20 1889-nov 28 1891 (wkly) [mf ed Cape Town: SA library 1986] – 2r – 1 – mf#MS00429 – sa National [079]

Aberdeen herald and general advertiser for the counties of aberdeen, banff and kincardine – Aberdeen, Scotland 1 sep 1832-11 nov 1876 – 1 – (amalg with: aberdeen weekly free press & subsequently publ as: herald & weekly free press) – uk British Libr Newspaper [072]

Aberdeen herald and post – Aberdeen, Scotland 26 aug 1989-19 jun 2002 – 1 – (discontinued; repl by: aberdeen citizen) – uk British Libr Newspaper [072]

Aberdeen journal [1750] – Aberdeen: J Chambers 2 jan 1750-29 aug 1768 (wkly) – 973v – 1 – (cont: aberdeen's journal; absorbed: aberdeen intelligencer (1757); cont by: aberdeen journal & north-british magazine) – uk Scotland NatLib [072]

Aberdeen labour elector : a weekly record of the labour movement – London: P C Standing 1893 (wkly) [mf ed 1988] – 35v on 1r – 1 – (cont by: aberdeen standard) – uk Aberdeen UL [325]

Aberdeen market co – [Edinburgh: NRAS 1990 – 1mf – 9 – mf#n3222 – uk NRAS [338]

Aberdeen medico-chirurgical society – [Edinburgh: NRAS 1988 – 1mf – 9 – mf#n2969 – uk NRAS [617]

Aberdeen new shaver – [Scotland] Aberdeen: printed by G Leith & Co jul 1837-1 jul 1840 (mthly) [mf ed 2003] – 25v on 1r – 1 – uk Newsplan [072]

Aberdeen observer : a commercial and political journal – [Scotland] Aberdeen: J Davidson & Co 27 may 1829-11 mar 1936 (wkly) [mf ed 2003] – 3r – 1 – (ceased with iss sep 8 1837; cont by: aberdeen constitutional and general advertiser for the counties of aberdeen, banff, and kincardine) – uk Newsplan [072]

Aberdeen pirate : being a weekly miscellany, intended to profit and amuse the public – [Scotland] Aberdeen: printed by R Edward & Co 15 sep 1833-29 aug 1833, 31 oct 1833, 28 nov 1833, 20 feb 1834 (mthly) [mf ed 2003] – 1r – 1 – uk Newsplan [072]

Aberdeen post – Aberdeen, Scotland 29 apr-19 aug 1989 – 1 – (discontinued; repl by: aberdeen herald & post) – uk British Libr Newspaper [072]

Aberdeen press and journal – Aberdeen: Aberdeen Newspapers Ltd 1922-39 (daily ex sun) [mf ed 2003] – 73r – 1 – (formed by union of: aberdeen daily journal and: aberdeen free press; cont by: press and journal; imprint varies) – uk Newsplan [072]

Aberdeen press and journal – [Scotland] Aberdeen : Aberdeen Newspapers Ltd 1 dec 1922-dec 1934 (daily ex sun) [mf ed 2003] – 73r – 1 – (formed by union of: aberdeen daily journal and: aberdeen free press; cont by: press and journal; imprint varies) – uk Newsplan [072]

Aberdeen review and north of scotland advertiser – [Scotland] Aberdeen City 1844 [mf ed 2003] – 1r – 1 – uk Newsplan [072]

Aberdeen shaver : a new monthly miscellany – [Scotland] Aberdeen: printed by J Anderson & Co 19 sep 1833-apr 1839 (mthly) [mf ed 2003] – 64v on 1r – 1 – uk Newsplan [072]

Aberdeen standard – [Scotland] Aberdeen: printed...by J Mackinnon 7 sep 1893-3 feb 1894 (wkly) [mf ed 2003] – 1r – 1 – (cont by: aberdeen labour elector) – uk Newsplan [072]

Aberdeen trades council, 1876-1951 – (= ser Labour party in britain, origins and development at local level. series 1) – 5r – 1 – (int by doris m hatvany) – mf#97284 – uk Microform Academic [331]

Aberdeen trawlowners and traders engineering company ltd – [Edinburgh: NRAS 1987 – 1mf – 9 – mf#n2783 – uk NRAS [338]

Aberdeen university club, london – [Edinburgh: NRAS 1985 – 1mf – 9 – mf#n2682 – uk NRAS [378]

Aberdeen weekly free press – [Scotland] Aberdeen: A Marr 18 may 1872-11 nov 1876 (wkly) [mf ed 2003] – 5r – 1 – (began in 1853; cont by: weekly free press [jan 1875-nov 1876]) – uk Newsplan [072]

Aberdeen weekly news and advertiser for the north of scotland – [Scotland] Aberdeen: J Smith aug 1869-dec 1878, 1886 (wkly) [mf ed 2003] – 8r – 1 – (cont by: aberdeen weekly news [1878]; weekly news for aberdeen and the north [1886]) – uk Newsplan [072]

Aberdeen-angus journal – St Joseph MO 1919-79 – 1,5,9 – ISSN: 0001-3161 – mf#177 – us UMI ProQuest [636]

Aberdeen's concrete construction – Addison IL 1991-98 – 1,5,9 – (cont: concrete construction; cont by: concrete construction) – ISSN: 1051-5526 – mf#6545.02 – us UMI ProQuest [690]

Aberdonian – [Scotland] Aberdeen: The Aberdonian 14 oct 1937-24 feb 1938 (wkly) [mf ed 2003] – 20v on 1r – 1 – uk Newsplan [072]

Abergavenny chronicle – [Wales] Monmouth 16 dec 1871-dec 1950 [mf ed 2003] – 70r – 1 – (missing: 1882, 1897, 1890) – uk Newsplan [072]

Abergavenny echo and brynmawr district advertiser – [Wales] LLGC 23 apr-dec 1897 [mf ed 2004] – 1r – 1 – uk Newsplan [072]

Abergele and pensarn visitor – Clwyd, Wales 17 jul 1869-7 aug 1886, 20 nov 1886-24 oct 1936 – 1 – (cont by: abergele visitor [31 oct 1936-14 sep 1979]) – uk British Libr Newspaper [072]

Abergele gazette & journal – [Wales] LLGC 22 mar 1873-7 feb 1874 [mf ed 2004] – 1r – 1 – uk Newsplan [072]

Der aberglaube des mittelalters : ein beitrag zur culturgeschichte / Schindler, Heinrich Bruno – Breslau [Wroclaw]: Wilh Gottl Korn, 1858 [mf ed 1992] – 1mf – 9 – 0-524-02610-6 – (incl bibl ref) – mf#1990-3060 – us ATLA [130]

Aberglaube oder volksweisheit? : der wahre sinn der bauernbraeuche / Fischer, Hanns – Leipzig : Milde [1939] [mf ed Bloomington IN: Indiana Uni Lib, Preservation Dept 1984] – 292p on 1r [ill] – 1 – us Indiana Preservation [390]

Aberglaube, sage und maerchen bei grimmelshausen / Amersbach, Karl – Baden-Baden: E Koelblin, 1891-1893 [mf ed 1990] – 2pts in 1v – 1 – (incl bibl ref) – mf#7423 – us Wisconsin U Libr [390]

Aberglaube und volksmedizin im lande der bibel / Canaan, Taufik – Hamburg: L Friederichsen, 1914 [mf ed 1991] – (= ser Abhandlungen des hamburgischen kolonialinstituts 20; Voelkerkunde, kulturgeschichte und sprachen 12) – 1mf – 9 – 0-524-01598-8 – mf#1990-2537 – us ATLA [610]

Aberi *see* Spanish-basque political periodicals

Aberkenfig, (tondu), glamorgan parish church of st john : baptisms 1868-1915, burials 1868-1933 – [Glamorgan]: GFHS [mf ed 1999] – 1mf – 9 – £1.25 – uk Glamorgan FHS [929]

Aberkenfig, (tondu), st john, monumental inscriptions – 1mf – 9 – £1.25 – uk Glamorgan FHS [929]

Aberlin, Joachim *see* Bibel oder heilige geschrifft

Abernethy, Alonzo *see* A history of iowa baptist schools

Aberpergwm, st cadoc & blaengwrach, st mary, glamorgan, parish churches : baptisms 1849-1921, 1895-1925; burials 1849-1957, 1895-1988 – [Glamorgan]: GFHS [mf ed [2001?]] – 1mf – 9 – £1.25 – uk Glamorgan FHS [929]

Abersychan and blaenavon critic – [Wales] LLGC 15 mar-20 nov 1879 [mf ed 2004] – 1r – 1 – (cont by: weekly argus) – uk Newsplan [072]

Abert, Friedrich Philipp *see* Das wesen des christentums nach thomas von aquin

Abert, Hermann *see* Goethe und die musik

Aberystwyth chronicle and illustrated times – [Wales] LLGC 9 jun-22 dec 1855 [mf ed 2002] – 1r – 1 – uk Newsplan [072]

Aberystwyth times, cardiganshire chronicle and merionethshire news – [Wales] LLGC 2 oct 1868-25 jun 1870 [mf ed 2002] – 1r – 1 – uk Newsplan [072]

Abessinien : eine landeskunde nach reisen und studien in den jahren 1907-1913 / Rein, G Kurt – Berlin: D Reimer (E Vohsen), 1918-20 – 1 – us CRL [960]

Abetti, Giorgio *see*
- History of astronomy
- Le stelle e i pianeti

Abeyesooriya, Samson *see*
- De fonseka family of kalutara
- Life of lady catherine de soysa
- Who's who of ceylon

Die abfassung des galaterbriefs vor dem apostelkonzil : grundlegende untersuchungen zur geschichte des urchristentums und des lebens pauli / Weber, Valentin – Ravensburg: Hermann Kitz, 1900 – 1mf – 9 – 0-8370-9662-6 – (incl bibl ref and index of biblical citations) – mf#1986-3662 – us ATLA [227]

Die abfassung des philipperbriefes in ephesus : mit einer anlage ueber roem. 16, 3-20 als epheserbrief / Feine, Paul – Guetersloh: C Bertelsmann, 1916 – (= ser Beitraege zur foerderung christlicher theologie) – 1mf – 9 – 0-524-04454-6 – mf#1992-0123 – us ATLA [227]

Die abfassungszeit der schriften tertullians / Noeldechen, Ernst – Leipzig, 1888 – (= ser Tugal 1-5/2a) – 3mf – 9 – €7.00 – ne Slangenburg [240]

Die abfassungszeit des galaterbriefes : ein beitrag zur neutestamentlichen einleitung und zeitgeschichte / Steinmann, Alphons – Muenster i.W: Aschendorff, 1906 – 1mf – 9 – 0-8370-6779-0 – (incl indes) – mf#1986-0779 – us ATLA [227]

Abfassungszeit und abschluss des psalters zur pruefung der frage nach makkabaeerpsalmen : historisch-kritisch untersucht / Ehrt, Carl – Leipzig: Johann Ambrosius Barth, 1869 [mf ed 1985] – 1mf – 9 – 0-8370-3037-4 – (incl ind) – mf#1985-1037 – us ATLA [221]

Abfertigung des ubiqvistischen predigers d philippi nicolai zu hamburg / Pierius, U – Bremen, 1603 – 1mf – 9 – mf#TH-1 mf 1275 – ne IDC [242]

Abgasuntersuchungen an einem pflanzenoelbetriebenen vorkammer-dieselmotor / Fischer, Tim – (mf ed 1995) – 2mf – 9 – €40.00 – 3-8267-2127-6 – mf#DHS 2127 – gw Frankfurter [628]

Abhandlung ueber den bau der thatwoerter im koptischen / Ewald, Heinrich – Goettingen: Dieterich, 1861 [mf ed 1989] – (= ser Sprachwissenschaftliche abhandlungen 1) – 1mf – 9 – 0-8370-1192-2 – (incl bibl ref) – mf#1987-6022 – us ATLA [470]

Abhandlung ueber des aethiopischen buches henokh : entstehung sinn und zusammensetzung / Ewald, Heinrich – Goettingen: Dieterich, 1854 [mf ed 1984] – 1mf – 9 – 0-8370-0403-9 – mf#1984-0072 – us ATLA [221]

Abhandlung ueber die frage : ob die musik bey dem gottesdienste der christen zu dulden, oder nicht... / Albrecht, Johann Lorenz – Berlin: F W Birnstiel 1764 [mf ed 19–] – 1mf – 9 – mf#fiche971 – us Sibley [780]

Abhandlung ueber die kriegskunst der tuerken...desgleichen derjenigen voelker...als griechen, armenier, araber, kurden... / Hayne, J C G – Berlin, 1783. 2v – 6mf – 9 – mf#AR-1654 – ne IDC [956]

Abhandlung ueber die rechenkunst oder practische arithmetik zum gebrauch fuer schulen / ed by Benner, Enos – 3. verb. u verm. Aufl. Sumnytaun (USA) 1853 (mf ed 1994) – 2mf – 9 – €31.00 – 3-8267-3029-1 – mf#DHS-AR 3029 – gw Frankfurter [510]

Eine abhandlung von den musicalischen intervallen und geschlechten / Scheibe, Johann Adolph – Hamburg: Auf Kosten des Verfassers 1739 [mf ed 19–] – 3mf – 9 – mf#fiche 898 – us Sibley [780]

Abhandlung von der fuge : nach den grundsaetzen und exempeln der besten deutschen und auslaendischen meister entworfen von friedrich wilhelm marpurg / Marpurg, Friedrich Wilhelm – Berlin: A Haude & J C Spener 1753-54 [mf ed 19–] – 2v on 12mf – 9 – mf#fiche 422 – us Sibley [780]

Abhandlungen / Deutscher Seefischereiverein – v1-13 1897-1922 – 1 – us CRL [639]

Abhandlungen : gelesen in der koeniglichen akademie der wissenschaften / Schleiermacher, Friedrich [Ernst Daniel]; ed by Braun, Otto – Leipzig: F Meiner [1911?] [mf ed 1991] – 1mf – 9 – 0-524-00389-0 – (incl bibl ref) – mf#1989-3089 – us ATLA [170]

Abhandlungen / Naturwissenschaftlicher Verein zu Bremen – Bremen: G A v Halem. Bd1-32[1866/1868-1951] [mf ed 1979] – 9r – 1 – (publ suspended 1943-48; publ varies: im selbstverlag des naturwissenschaftlichen vereins zu bremen 1976-) – mf#film mas c575 – us Harvard [580]

Abhandlungen der churfuerstlich-baierischen akademie der wissenschaften muenchen : historische abhandlungen – 1(1736)-10(1776) – 9 – €223.00 – (neue philosophische abhandlungen 1(1778)-4(1785) €136) – ne Slangenburg [500]

Abhandlungen der koenigl akademie der wissenschaften zu berlin aus dem jahre 1884 – €44.00 – ne Slangenburg [500]

Abhandlungen der koeniglichen akademie der wissenschaften in berlin,1804-1907 : physikalische, mathematische, philosophische, historisch-philologische klasse – Berlin, 1815-1907 – 2462mf – 8 – mf#H-641 – ne IDC [500]

Abhandlungen der mathematisch-physischen classe der koeniglich saechsischen gesellschaft der wissenschaften – Leipzig: Weidmannsche Buchhandlung, [1849- v15 1890] – 1r – 1 – us CRL [500]

Abhandlungen des staatswirthschaftlichen instituts zu marburg / ed by Jung-Stilling, Johann Heinrich & Robert, Carl Wilhelm – Offenbach 1790 – (= ser Dz) – 1v on 2mf – 9 – €60.00 – mf#k/n2759 – gw Olms [330]

Abhandlungen und beobachtungen durch die oekonomische gesellschaft zu bern gesammelt – Bern 1762-7 – (= ser Dz) – 3-14jg on 52mf – 9 – €520.00 – (jg 1 & 2 under title: der schweitzerischen gesellschaft in bern s) – mf#k/n2874 – gw Olms [330]

Abhandlungen zu goethes leben und werken / Duentzer, Heinrich – Leipzig: E Wartig, 1885 [mf ed 1991] – 2v – 1 – mf#7544 – us Wisconsin U Libr [430]

Abhandlungen zur geschichte der mathematischen wissenschaften – v1-30. 1877-1913 – 1,5 – $216.00 – (in german) – mf#0001 – us Brook [510]

Abhandlungen zur mittleren und neueren geschichte – v1-83. 1907-39 – 9 – $360.00 – (in german) – mf#0002 – us Brook [900]

Abhandlungen zur orientalischen und biblischen literatur / Ewald, Georg Heinrich August – Goettingen: Dieterich, 1832 [mf ed 1986] – 1mf – 9 – 0-8370-9059-8 – mf#1986-3059 – us ATLA [470]

Abhayacharana Dasa [comp] *see* The indian ryot, land tax, permanent settlement, and the famine

Abhedananda, Swami *see*
- Divine heritage of man
- How to be a yogi
- India and her people

Abhedananda, swami *see*
- Great saviors of the world
- The ideal of education
- The path of realization
- Science of psychic phenomena
- Spiritual unfoldment

Abhidhammatthasangaha : traite de metaphysique bouddhique / Vimalappanna, Brah Gru – Phnom-Penh: Impr nouvelle A Portail 1927 [mf ed 1990] – 1r with other items – 1 – (title & text in khmer; added t.p. in french) – mf#mf-10289 seam reel 114/8 [§] – us CRL [280]

Abhijan – Bishnupur. v1-32 n40. jun 29 1941-mar 1973 – 7r – 1 – us CRL [954]

The abhijnanasakuntala of kalidasa : with the commentary styled arthadyotanika' of ra'ghavabhatta / ed by Kale, Moreshvar Ramchandra – 2nd rev ed. Bombay: Oriental Publ, 1902 – (= ser Yale coll; S k press sanskrit series 1) – 1 – 0-524-09182-X – (in sanskrit. english trans, crit and explanatory notes by ed) – mf#1995-0182 – us ATLA [490]

Abib – New York. 1915 – 1 – us AJPC [073]

Abich, H *see* Aus kaukasischen laendern

Abicht, Friedrich K *see* Der kreis wetzlar

Abicht, Joh Heinrich *see* Philosophisches journal

Abicht, Joh Heinrich et al *see* Neues philosophisches magazin

Abiding faith for a nation in crisis *see* Ta shih tai ti tsung chiao hsin yang (ccm340)

Abiding knowledge of christian truth *see* Chen tao ch'ang shih [ccc202]

The abiding value of the old testament / Robinson, George Livingston – New York: Young Men's Christian Association Press, 1911 – 1mf – 9 – 0-7905-0262-3 – mf#1987-0262 – us ATLA [221]

Abilene. Kansas *see* Ordinance book

Ability grouping in harare secondary schools : its effect on instruction, learning and social stratification / Chisaka, Bornface Chenjerai – Uni of South Africa 2000 [mf ed Johannesburg 2000] – 5mf – 9 – (incl bibl ref) – mf#mfm15076 – sa Unisa [373]

The ability of sarcoplasmic reticulum to regulate intracellular calcium following a fatiguing bout of exercise / Favero, Terence G & Klug, Gary A – 1990 – 1mf – 9 – $4.00 – us Kinesology [612]

The ability of undergraduate physical education majors to verbally identify and visually discriminate critical elements of select sport skills / Edkins, C E – 1991 – 2mf – 9 – $8.00 – us Kinesology [150]

Abingdon and reading herald – Abingdon, England 27 aug 1870-6 oct 1883 – 1 – (wanting: jan-may 1871, 1874, 1903; cont: reading & abingdon herald [14 may-20 aug 1870]; cont by: abingdon herald [13 oct 1883-30 apr 1910]) – uk British Libr Newspaper [072]

Abingdon express, berks, bucks, oxon, hants and wilts observer – Abingdon, England 5 feb 1887-30 jun 1888 – 1r – 1 – (discontinued) – uk British Libr Newspaper [072]

Abingdon free press and didcot news – Abingdon, England 6 jun 1902-24 mar 1916 [mf 1912] – 1 – (missing: 1911) – uk British Libr Newspaper [072]

Abingdon herald – Abingdon, England 4 apr 1974- [mf 1986-] – 1 – (cont: north berks herald [6 jan 1961-28 mar 1974]) – uk British Libr Newspaper [072]

Abingdon herald – Abingdon, England 13 oct 1883-30 apr 1910 – 1 – (wanting: jan-may 1871, 1874, 1903; cont: abingdon & reading herald [27 aug 1870-6 oct 1883]; cont by: north berks herald [7 may 1910-8 sep 1939]) – uk British Libr Newspaper [072]

Abingdon virginian – 1870 oct 28, dec 2; 1871 jan 20-27, mar 17, apr 14-28, jul 28-aug 18, oct 6; 1880 dec 10 – 1 – mf#881888 – us WHS [071]

Abington 1691-1849 – Oxford, MA (mf ed 1994) – (= ser Massachusetts vital record transcripts to 1850) – 27mf – 9 – 0-87623-196-2 – (mf 1t: deaths 1721-71. mf 1t-37: births 1691-1759. mf 3t: marriages & intents 1712-61. mf 4t-5t: births 1759-88. mf 5t: deaths 1772-96. mf 6t: births 1727-96. mf 6t-7t: intentions 1761-1800. mf 7t-8t: births 1777-99. mf 8t: deaths. mf 8t-9t: marriages 1760-99. mf 9t-11t: births 1800-21. mf 11t: out-of-town marriages 1715-99. mf 11t-12t: intentions 1801-21. mf 12t-13t: marriages 1794-1820; deaths 1800-21. mf 14t-20t: births 1821-49; deaths 1821-44; intentions 1821-49; marriages 1821-44. mf 25t: vitals 1837-49. mf 25t-26t: marriages 1844-49. mf 26t-27t: deaths 1844-49) – us Archive [978]

Abington 1691-1892 – Oxford, MA (mf ed 1992) – (= ser Massachusetts vital records) – 79mf – 9 – 0-87623-159-8 – (mf 1-5: town & vital 1691-1777. mf 6-9: town records 1712-66. mf 9-10: vital records 1691-1777. mf 11-15: town & vital 1727-1820. mf 16-22: town & vital 1715-1856. mf 23-25: birth & death 1821-51. mf 26-32: town records 1766-1821. mf 33-38: town records 1821-52. mf 39-43: town records 1822-1918. mf 44-49: marriages 1821-60. mf 50-51: birth index 1850-61. mf 52-54: births 1851-61. mf 55-56: birth index 1862-91. mf 57-60: births 1862-91. mf 61-62: marr index 1850-61. mf 63-64: marriages 1852-62. mf 65-66: marr index 1862-91. mf 67-71: marriages 1863-91. mf 72: death index 1850-61. mf 73-74: deaths 1851-61. mf 75-76: death index 1862-91. mf 77-79: deaths 1862-91) – us Archive [978]

Abington Abbey *see* Chronicon monasterii de abington (rs2)

Abington, great 1599-1950 – (= ser Cambridgeshire parish register transcript) – 4mf – 9 – £5.00 – uk CambsFHS [929]

Abington, little 1599-1950 – (= ser Cambridgeshire parish register transcript) – 4mf – 9 – £5.00 – uk CambsFHS [929]

Abington pigotts 1599-1950 – (= ser Cambridgeshire parish register transcript) – 3mf – 9 – £3.75 – uk CambsFHS [929]

Abismo...y discurso...virgen maria / Guerrero, Alonso – 1686 – 9 – sp Bibl Santa Ana [240]

Abissinia : giornale di un viaggio / Vigoni, P – Milano, 1881 – 4mf – 9 – mf#NE-20215 – ne IDC [916]

L'abissinia settentrionale e le strade che vi conducono da massaua / Cecchi, A – Milano, 1888 – 1mf – 9 – mf#NE-20197 – ne IDC [916]

Abitibi magazine *see* Abitibi-dimanche

Abitibi-dimanche : l'hebdo-magazine du nord-ouest quebecois – Val-d'Or: [s.n.]. v1 n1 30 juil 1972-v2 n18 2 dec 1973 (wkly) [mf ed 1973] – 2r – 1 – (repl by: abitibi magazine, section faisant partie integrante de l'echo d'amos, l'echo d'abitibi-ouest, l'echo abitibien l'echo d'amos, l'echo d'abitibi-ouest, l'echo de malartic et l'echo de la baie james, a partir du 12 dec 1973) – mf#SEM35P11 – cn Bibl Nat [073]

Abito y armadura espiritual...con privilegio imperial. 1544 / Cabranes, Diego de – 9 – sp Bibl Santa Ana [240]

Der abituriententag : die geschichte einer jugendschuld / Werfel, Franz – Berlin: P Zsolnay 1928 [mf ed 1992] – 1r – 1 – (filmed with: bonzen und rebellen / tudel weller) – mf#7936 – us Wisconsin U Libr [830]

Die ablaesse, ihr wesen und gebrauch / Beringer, fr – Paderborn. v1-2. 1921-1922 – €40.00 – ne Slangenburg [240]

Ablaing van Giessenburg, Willem Jan, baron d' *see* De duitsche orde

Ablanedo, Juan Bautista *see* Cuestion de cuba

Der ablass : seine geschichte und bedeutung in der heilsoekonomie / Groene, Valentin – Regensburg: G.J. Manz, 1863 – 1mf – 9 – 0-7905-5401-1 – (incl bibl ref) – mf#1988-1401 – us ATLA [240]

Der ablassstreit / Dieckhoff, August Wilhelm – Gotha: F A Perthes, 1886 – 1mf – 9 – 0-7905-5871-8 – (incl bibl ref) – mf#1988-1871 – us ATLA [240]

The able minister / Evans, Hugh – A sermon preached in Broadmead before the Britol Education Society. 1773 – 1 – 5.00 – us Southern Baptist [242]

Abm review : official organ – Sydney: D S Ford. v1-64. 1910-74 [qrterly] [mf ed 2003] – 64v on 8r – 1 – mf#2003-s042 – us ATLA [242]

Abner creek baptist church – New York. 1972-1974 (1) 1972-1972 (5) (9) – 2r – 1 – $102.06 – mf#6656 – us Southern Baptist [242]

Abo underrattelser – Turku, Finland 13 oct 1917-28 jan 1920, 1-31 dec 1944 – 1 – uk British Libr Newspaper [078]

Aboda zara : der mischnatraktat "goetzendienst" / ed by Strack, Hermann Leberecht – Berlin: H Reuther, 1888 [mf ed 1985] – (= ser Schriften des institutum judaicum in berlin 5) – 1mf – 9 – 0-8370-2083-2 – (incl bibl ref, ind of hebrew words & names) – mf#1985-0083 – us ATLA [939]

La abolicion de la esolavitud en el ordern economico / Labra y Cadrana, Rafael Maria de – Madrid: J. Noguera, 1873 – 1 – us Wisconsin U Libr [305]

Abolicion oficial del laicismo en las escuelas / Bayle, Constantino – Madrid: Razon y Fe, 1937 – 1 – sp Bibl Santa Ana [370]

Abolicionista : orgao litterario e noticioso dos typographos da "regeneracao" – Desterro, SC. 05 out-dez 1884; 01 mar 1885 – (= ser Ps 19) – mf#UFSC/BPESC – bl Biblioteca [079]

O abolicionista – Bahia: Typ do Correio da Bahia, 30 abr"jul 1871; 01 mar-15 abr 1872 – (= ser Ps 19) – mf#P18B,02,17 – bl Biblioteca [079]

O abolicionista : orgao da caixa emancipadora maranhense marques rodrigues – Recife, PE: Typ Universal, 20 jul-10 ago 1883 – (= ser Ps 19) – bl Biblioteca [079]

O abolicionista : orgao da sociedade brasileira contra a escravidao – Rio de Janeiro, RJ: Typ da Gazeta de Noticias, 01 nov 1880-01 dez 1881 – (= ser Ps 19) – mf#P10,02,27 – bl Biblioteca [079]

O abolicionista – Teresina, P: Typ do Telephone, 08 out-19 dez 1884 – (= ser Ps 19) – mf#P11,03,01 – bl Biblioteca [079]

Abolicionista do amazonas – Manaus, AM. 04 maio-10 jul 1884 – (= ser Ps 19) – 1,5,6 – bl Biblioteca [079]

Abolition and emancipation – [mf ed Marlborough 1996] – 6pt – 1 – (pt 1: papers of thomas clarkson, william lloyd garrison, zachary macaulay, harriet martineau, harriet beecher stowe & william wilberforce fr the huntington library [10r] £850; pts2,3: slavery coll fr the merseyside maritime museum, liverpool [24r,20r] £2100, £1800; pt4: the granville sharp papers fr gloucestershire record office [30r] £2700; pt5: papers of thomas clarkson fr the british library, london [5r] £450; pt6: papers of william wilberforce, william smith, iveson brookes, francis corbin & related records fr the rare books, mss & special coll library, duke university [17r] £1550; with d/g) – uk Matthew [322]

Abolition intelligencer and missionary magazine – 1822 may-nov; 1823 feb-mar – 1 – mf#801801 – us WHS [071]

Abolition intelligencer and missionary magazine / ed by Crow, John Finley – La Salle IL may 7 1822-apr 1823 – 1 – mf#4406 – us UMI ProQuest [071]

The abolition of slavery : the right of the government under the war power / Adams, John Quincy et al; ed by Garrison, William Lloyd – Boston: RF Wallcutt, 1862 – 1mf – 9 – 0-524-01940-1 – mf#1990-0529 – us ATLA [976]

The abolition of the roman jurisdiction / Creighton, Mandell – London: SPCK 1896 [mf ed 1992] – (= ser Church historical society (series) 7) – 1mf – 9 – 0-524-05497-5 – mf#1990-1492 – us ATLA [941]

Abolitionist – 1970 mar-1972 feb – 1 – mf#1051535 – us WHS [071]

Abolitionist : or record of the new england anti-slavery society – La Salle IL 1833 – 1 – mf#3912 – us UMI ProQuest [976]

Der abolitionist : organ fuer die bestrebungen der internationalen foederation zur bekaempfung der staatlich reglementierten prostitution / ed by Scheven, Katharina – 1902-33 [mf ed 1997] – (= ser Hq 28) – 32v on 35mf – 9 – €210.00 – 3-89131-140-0 – (later subtitles: organ des deutschen verbandes zur foerderung der sittlichkeit; organ des bundes fuer frauen- und jugendschutz) – gw Fischer [322]

L'abolitionniste francais – Paris. 1844-50 – 1 – fr ACRPP [073]

Los aborigenes de costa rica / Gagini, Carlos – San Jose de Costa Rica: Impr Trejos hermanos 1917 [mf ed 1993] – 1r – 1 – us UF Libraries [306]

Los aborigenes del pais de cuyo / Cabrera, Pablo – Cordoba (Argentina), 1929; Madrid: Razon y Fe, 1931 – 1 – sp Bibl Santa Ana [305]

Aboriginal american authors and their productions : especially those in the native languages / Brinton, Daniel Garrison – Philadelphia: D Brinton 1883 – 1mf – 9 – 0-665-00278-5 – (incl ind) – mf#00278 – cn Canadiana [490]

Aboriginal Multi-media Society of Alberta *see* Windspeaker

The aboriginal tribes of hyderabad / Furer-Haimendorf, Christoph von – London; New York: Macmillan & Co, 1943-1948 – (= ser Samp: indian books) – us CRL [307]

The aboriginal tribes of hyderabad... / Fuerer-Haimendorf, Christoph von – London: Macmillan & Co., 1943 – 1 – us Wisconsin U Libr [305]

The aborigines of australia : being an account of the institution for their education at poonindie, south australia / Hale, Matthew Blagden – London, [1889] – (= ser 19th c books on british colonization) – 2mf – 9 – mf#1.1.6168 – uk Chadwyck [370]

Aborigines of jamaica / Cundall, Frank – Kingston, Jamaica: Institute of Jamaica 1934 [mf ed 1992] – (= ser Jamaican pamphlets 1) – 1r – 1 – (repr fr: handbook of jamaica for 1933-34) – us UF Libraries [306]
Aborigines of jamaica / Sherlock, Philip Manderson – Kingston, Jamaica: Institute of Jamaica 1939 [mf ed 1991] – 1r – 1 – us UF Libraries [306]
The aborigines of northern formosa : a paper read before the north china branch of the royal asiatic society, shanghai, 18th june, 1874 / Taintor, Edward C – Shanghai, 1874 – (= ser 19th c books on china) – 1mf – 9 – mf#7.1.60 – uk Chadwyck [305]
The aborigines of the highlands of central india / Mazumdar, Bijay Chandra – Calcutta: Published by the University of Calcutta, 1927 – (= ser Samp: indian books) – us CRL [307]
Aborigines' protection society : publications about native tribes around the world – 1837-1909 – 8r – 1 – (incl in coll are the annual reports fr 1839-1908; the aborigines' friend fr 1847-1909 (originally known as the colonial intelligencer, or aborigines' friend and aborigines' friend, and the colonial intelligencer); and a vol containing 32 of the society's pamphlets from 1896-1908) – Dist. us UMI ProQuest – uk Academic [305]
The aborigines' protection society – 1997 – ca 3r – 1 – ca £220.00 – 1-897955-59-6 – uk Academic [366]
Aborigines Protection Society Committee. London *see* On the british colonization of new zealand
Aborigines Protection Society, London *see*
– The bechunanas, the cape colony, and the transvaal
– Canada west and the hudson's-bay company
– England and her colonies considered in relation to the aborigines
– The new-zealand government and the maori war of 1863-64
– Report on the indians of upper canada
The aborigines – "so called" – and their future / Ghurye, Govind Sadashiv – Poona: Gokhale Institute of Politics and Economics, 1943 – (= ser Samp: indian books) – us CRL [305]
Abou naddara – Paris. 1878-90, 1898-99 – 1 – fr ACRPP [073]
Aboussouan, B *see* Le probleme politique syrien
About an old new england church : an address on "the good old days" published as a souvenir of the 150th anniversary of the congregational church of sharon, connecticut / Lee, Gerald Stanley – Sharon, CT: WW Knight, 1891 [mf ed 1993] – 1mf – 9 – 0-524-08576-5 – mf#1993-3161 – us ATLA [242]
About arts and crafts = L'art et l'artisanat – 1977 winter-1982 v5 n3 [1982] – 1 – mf#1051537 – us WHS [071]
About, Edmond *see*
– Le fellah souvenirs d'egypte
– Greece and the greeks of the present day
– Handbook of social economy
– Mariage de paris
About etching / Haden, Francis Seymour – 4th ed. [London] 1879 – (= ser 19th c art & architecture) – 2mf – 9 – mf#4.2.135 – uk Chadwyck [760]
About face – 1970 jul 4; v1 n6 [1970 sep 12]; v2 n1 [1971 jan 15] – 1 – mf#720552 – us WHS [071]
About face – n1-2, 4-5 [1969 mar-jul 4]; n1-5 [1969 mar-jul] – 1 – mf#720791 – us WHS [071]
About face! – n1 [1971 may]; v2 n1-v3 n5 [1972 feb-1973 jun];; n1 [1971 may]; v2 n3 [1972 feb]; v3 n2 [1973 mar]; n1; v2 n1-v3 n5 [1971 may; 1972 feb-1973 jun; 1975 jun] – 1 – mf#720786 – us WHS [071]
About hebrew manuscripts / Adler, Elkan Nathan – London, New York: Oxford UP, 1905 [mf ed 1988] – 1mf – 9 – 0-7905-0301-8 – (incl ind) – mf#1987-0301 – us ATLA [470]
About persia and its people : a description of their manners, customs, and home life... / Knanishu, Joseph – Rock Island IL: Lutheran Augustana Book Concern 1899 [mf ed 1992] – 1mf – 9 – 0-524-04771-5 – mf#1991-2157 – us ATLA [306]
About the house – London, England 1962-92 – 1,5,9 – ISSN: 0001-3242 – mf#8594 – us UMI ProQuest [780]
About the house – London, Friends of Covent Garden. v1- nov 1962- (qrtly) – 1 – mf#*ZAN-*MD13 – Located: NYPL – us Misc Inst [790]
About the jews since bible times : from the babylonian exile till the english exodus / Magnus, Katie, Lady – London: C Kegan Paul, 1881 [mf ed 1988] – 1mf – 9 – 0-7905-0099-X – (incl bibl ref) – mf#1987-0099 – us ATLA [939]
About town – v4 n25 [1980 dec 18]; [1981 feb 12-1984 dec 13] – 1 – mf#1047646 – us WHS [071]
Abou-Zeid, A M *see* A study of some age-set systems of north and east africa
Above ground – v1 n1-7 [1969 aug-1970 mar]; [final iss 1970 may] – 1 – mf#720561 – us WHS [071]
Abraham : recent discoveries and hebrew origins / Woolley, Leonard – Farber & Farber, 1936 – 9 – $12.00 – us IRC [221]
Abraham : studien ueber die anfaenge des hebraeischen volkes / Dornstetter, Paul – Freiburg i B, St Louis MO: Herder, 1902 [mf ed 1989] – (= ser Biblische studien 7/1-3) – 1mf – 9 – 0-7905-1933-X – mf#1987-1933 – us ATLA [221]
Abraham : the typical life of faith / Breed, David Riddle – Chicago: F H Revell, 1886 [mf ed 1989] – 1mf – 9 – 0-7905-2887-8 – mf#1987-2887 – us ATLA [221]
Abraham a Sancta Clara *see*
– Auf, auf ihr christen
– Etwas fuer alle
– Grillen und pillen aus abraham a sancta clara
– Judas der erz-schelm fuer ehrliche leut
– Neue predigten
– Neun neue predigten
Abraham a sancta clara (1644-1709) : zur zweihundertsten wiederkehr seines todestages: eine auswahl aus seinen werken mit einer einleitung versehen / Keller, Gottfried [comp] – Bern: G Grunau 1909 [mf ed 1988] – 1r – 1 – (filmed with: ueber die schriftstellerische thaetigkeit thomas abbt's / dr geisler) – mf#6934 n9 – us Wisconsin U Libr [240]
Abraham a Santa Clara *see* Werke, in auslese
Abraham als babylonier, joseph als aegypter : der weltgeschichtliche hintergrund der biblischen vaetergeschichten auf grund der keilinschriften / Winckler, Hugo – Leipzig: J C Hinrichs, 1903 [mf ed 1990] – 1mf – 9 – 0-7905-3500-9 – mf#1987-3500 – us ATLA [221]
Abraham and his age / Tomkins, Henry George – London, New York: Eyre & Spottiswoode, 1897 [mf ed 1988] – (= ser Bible student's library 6) – 1mf – 9 – 0-7905-0402-2 – (incl bibl ref & ind) – mf#1987-0402 – us ATLA [221]
Abraham and the patriarchal age / Duff, Archibald – London: J M Dent; Philadelphia: J B Lippincott [190-?] [mf ed 1989] – (= ser The temple series of bible handbooks) – 1mf – 9 – 0-7905-0648-3 – (incl bibl ref & ind) – mf#1987-0648 – us ATLA [221]
Abraham, Fritz et al *see* Jahresberichte der geschichtswissenschaft
Abraham, his life and times / Deane, William John – New York: Fleming H Revell, [189-?] [mf ed 1986] – 1mf – 9 – 0-8370-9932-3 – (incl bibl ref) – mf#1986-3932 – us ATLA [221]
Abraham, isaak und jakob / Lotz, Wilhelm – Berlin: Edwin Runge, 1910 [mf ed 1989] – (= ser Biblische zeit- und streitfragen 5.ser/10) – 1mf – 9 – 0-7905-2727-8 – mf#1987-2727 – us ATLA [221]
Abraham, joseph, and moses in egypt : being a course of lectures / Kellogg, Alfred Hosea – New York: Anson D F Randolph; London: Truebner, 1887 [mf ed 1989] – 1mf – 9 – 0-7905-1072-3 – (incl bibl ref) – mf#1987-1072 – us ATLA [221]
Abraham Joshua Heschel *see* Ohev yisra'el...
Abraham, Karl *see* Dreams and myths
Abraham lincoln papers – (mf ed 1959) – (= ser Presidential papers microfilm) – 97r – 1 – (with guide) – Dist. us Scholarly Res – us L of C Photodup [975]
Abraham lincoln quarterly – Springfield IL 1948-52 – 1 – mf#189 – us UMI ProQuest [976]
Abraham, Robert *see* To the editor of the carlisle journal
Abraham, Roy Clive *see*
– A dictionary of the tiv language
– The grammar of tiv
– Hausa literature, and the hausa sound system
– Language of the hausa people
– The principles of idoma
– The principles of tiv
Abraham schellenberg papers / Schellenberg, Theodore R – 1875-1921 – 1 – us Kansas [920]
Abraham the faithful / Royer, Galen Brown – Elgin IL: Brethren Pub House 1907 [mf ed 1992] – 1mf – 9 – 0-524-03858-9 – mf#1990-4905 – us ATLA [221]
Abraham, the friend of god : a study from old testament history / Dykes, James Oswald – London: James Nisbet, 1877 [mf ed 1989] – 1mf – 9 – 0-7905-0987-3 – (incl bibl ref) – mf#1987-0987 – us ATLA [221]
Abraham und seine zeit / Doeller, Johannes – 1. & 2. aufl. Muenster i W: Aschendorff, 1909 [mf ed 1993] – (= ser Biblische zeitfragen 2.folge/1) – 1mf – 9 – 0-524-06123-8 – mf#1992-0790 – us ATLA [221]
Abrahams, Glenis Gail *see* Dionysiac as a determining factor in nietzsche's philosophy of superman
Abrahams, Israel *see*
– Chapters on jewish literature
– Jewish life in the middle ages
– Judaism
– A short history of jewish literature
Abrahams, Peter *see* Return to goli
Abraham...Sancta Clara *see*
– Besonders meublirt- und gezierte todten-capelle
– [Centifolium stultorum] hundert weniger eine thorheit in eben so vielen kupfern vorgestellt...
– Etwas fuer alle
– Heilsames gemisch gemasch
– Huy! und pfuy! der welt
– Ein karn voller narren
– Mercurialis oder wintergruen
– Mercurialis oder winter-gruen
– Mercurialis of winter-groen
– Neu-eroeffnete welt-galeria
– Nuttelyk mengelmoes, bestaande uyt alderhande zeldzame en wonderlyke geschiedenissen
– Stella ex jacob orta maria
– De welvoorziene wynkelder
– Wohl angefuellter wein-keller
– Zeedelyke geschiedenisse of aangenaam winter-groen
[Abraham...Sancta Clara] *see*
– Centi-folium stultorum in quarto
– Mala gallina, malum ovum
Abrahamson, Laurentius Gustav *see* Jubel-album
Abram, Annie *see*
– English life and manners in the later middle ages
– Social england in the fifteenth century
Abramov, A V et al *see* Nastol'nyi entsiklopedicheskii slovar'
[Abramovich, N I] *see* Padenie dinastii
Abramovicha, D I *see* Paterik kievo-pecherskago monastyria
Abrams, Charles *see* Charles abrams
Abramski, Shmuel *see*
– Mesilah ba-'aravah
– Parashah be-toldot ha-negev
Abranches, Dunshee De *see* Revolta da armada e a revolucao rio grandense
Abrantes, Laure J d' *see* Memoires de madame la duchesse d'abrantes
Abrasive engineering – New York NY 1955-74 – 1,5 – ISSN: 0001-3277 – mf#1651 – us UMI ProQuest [620]
Abrasives / DeSapio, Vincent – Washington DC: Office of Industries, US International Trade Commission [1995] [mf ed 1995] – 1mf – 9 – (incl bibl ref) – us GPO [338]
Abravanel, Isaac et al *see* Mikraot ketanot 613 mitsvot ha-torah
Abraxas – 1974 mar 1-oct – 1 – mf#1109431 – us WHS [071]
Abraxas : studien zur religionsgeschichte des spaetern altertums / Dieterich, Albrecht – Leipzig: B G Teubner, 1891 [mf ed 1989] – 1mf – 9 – 0-7905-1590-3 – (incl bibl ref and ind) – mf#1987-1590 – us ATLA [450]
Abray, William Ewart *see* Critical evaluation of the ontario department of education concert plan
Abrege chronologique : ou histoire des decouvertes faites par les europeens dans les differentes parties du monde; extrait des relations les plus exactes et des voyageurs les plus veridiques / Barrow, John – Paris 1766 [mf ed Hildesheim 1995-98] – 12v on 36mf – 9 – €360.00 – 3-487-29928-3 – gw Olms [910]
Abrege complet de l'histoire sainte : a l'usage des ecoles / Gosselin, David – Quebec: J A Langlais, 1886 – 1mf – 9 – mf#53656 – cn Canadiana [220]
Abrege complet de l'histoire sainte : a l'usage des ecoles, deuxieme cours / Gosselin, David – Quebec: J A Langlais, 1887 – 2mf – 9 – mf#53657 – cn Canadiana [220]
Abrege complet de l'histoire sainte : a l'usage des ecoles, troisieme cours / Gosselin, David – Quebec: J A Langlais, 1887 – 2mf – 9 – mf#53658 – cn Canadiana [220]
Abrege d'agriculture : vol 1: regne mineral: le sol / Champoux, Gerard – Montreal, Quebec: Librairie J-A Parent, 1942 [mf ed 1995] – 2mf – 9 – (pref by jean-charles magnan) – mf#SEM105P2519 – cn Bibl Nat [630]
Abrege d'arithmetique decimale : contenant les operations du calcul, des quatre premieres regles...a l'usage des ecoles chretiennes – Trois-Rivieres [Quebec]: G Stobbs, 1836 [mf ed 1984] – 1mf – 9 – 0-665-43129-5 – mf#43129 – cn Canadiana [510]
Abrege de geographie commerciale et historique : contenant un precis d'astronomie selon le systeme de copernic... – 2e rev augm ed. Paris: Chez l'Auteur, et chez Jh Moronval [et 3 autres], 1833 [mf ed 1983] – 3mf – 9 – 0-665-39878-6 – mf#39878 – cn Canadiana [520]
Abrege de la geographie de l'ile d'haiti : contenant des notions topographiques sur les autres antilles / Fortunat, Dantes – 3. rev corr augm ed. Paris, France: Librairie Classique G Guerin et Cie [1889] [mf ed 1990] – 1r – 1 – us UF Libraries [918]
Abrege de la geographie du canada : a l'usage des ecoles de cette province – [Montreal?: s.n.] 1881 [mf ed 1985] – 1mf – 9 – 0-665-37438-0 – mf#37438 – cn Canadiana [917]
Abrege de la geographie du canada : a l'usage du college de st pierre, a chambly – Montreal?: s.n, 1831 (Montreal: L Duvernay) – 1mf – 9 – mf#32257 – cn Canadiana [917]
Abrege de la grammaire francaise / Lafrance, Charles Joseph Levesque – Quebec: C Darveau, 1865 [mf ed 1984] – 2mf – 9 – 0-665-45449-X – mf#45449 – cn Canadiana [440]
Abrege de la grammaire selon l'academie / Bonneau, B – 26e rev corr ed. Quebec: Departement de l'instruction publique, 1879 [mf ed 2001] – 2mf – 9 – mf#SEM105P3292 – cn Bibl Nat [440]
Abrege de la vie de bernadette, soeur marie bernard / Raymond, Henri, abbe – Montreal: J Rolland, 1879 – 1mf – 9 – mf#04001 – cn Canadiana [241]
Abrege de la vie des plus fameux peintres... / [Dezallier d'Argenville, A J] – Paris, 1745-1752. 3v – 30mf – 9 – mf#0-225 – ne IDC [700]
Abrege de la vie des saints : chacune suivie de trois reflexions – Saint-Philippe: impr Ecclesiastique, 1825 [mf ed 1974] – 1r – 5 – mf#SEM16P63 – cn Bibl Nat [240]
Abrege de la vie du bienheureux jean de britto de la compagnie de jesus – 3e ed. [Montreal: s.n.] 1854 [mf ed 1983] – 1mf – 9 – 0-665-43249-6 – mf#43249 – cn Canadiana [920]
Abrege de l'exposition de la doctrine chretienne : cours elementaire / Reticius, frere – Montreal: les Freres des ecoles chretiennes, 1942 [mf ed 1986] – 6mf – 9 – mf#SEM105P712 – cn Bibl Nat [241]
Abrege de l'histoire d'haiti : redige pour l'enseignement de l'histoire d'haiti dans les ecoles / Robin, Enelus – 6.ed. Port-Au-Prince, Haiti: A A Heraux 1902 [mf ed 1993] – 1r – 1 – us UF Libraries [972]
Abrege de l'histoire ecclesiastique : contenant les evenemens considerables de chaque siecle: avec des reflexions / Racine, Bonaventure – nouv ed. Cologne [Allemagne]: Aux depens de la Compagnie. 13v. 1752.54 [mf ed 1984] – 13v on 1mf – 9 – mf#47831 – cn Canadiana [240]
Abrege de l'histoire generale des voyages : contenant ce qu'il y a de plus remarquable, de plus utile and de mieux avere dans les pays ou les voyageurs ont penetre... / Harpe, Jean F de la – Paris 1780 [mf ed Hildesheim 1995-98] – 23v+1 atlas on 81mf – 9 – €810.00 – 3-487-29940-2 – gw Olms [910]
Abrege de l'histoire generale des voyages : contenant ce qu'il y a de plus remarquable, de plus utile et de mieux avere dans les pays ou les voyageurs ont penetre... / Harpe, Jean-Francois de la – Paris: Hotel de Thou... 1780-1802 [mf ed 1985] – 32v on 1mf – 9 – 0-665-48817-3 – mf#48817 – cn Canadiana [910]
Abrege de l'histoire generale des voyages : contenant ce qu'il ya de plus remarquable, de plus utile et de mieux avere dans pays... / Harpe, Jean-Francois de la – nouv corr ed. Paris: Chez etienne Ledoux, libraire 1820 [mf ed 1984] – 24v on 1mf – 9 – 0-665-45341-8 – mf#45341 – cn Canadiana [910]
Abrege de l'histoire sainte, de l'histoire de france et de l'histoire du canada : a l'usage des commencants – [Quebec?: s.n.] 1864 [mf ed 1984] – 2mf – 9 – 0-665-46069-4 – mf#46069 – cn Canadiana [221]
Abrege de l'histoire sainte, de l'histoire de france, et de l'histoire du canada : a l'usage des commencants – [mf ed 1999] – 9 – mf#SEM105P – cn Bibl Nat [944]
Abrege de theologie sociale d'apres les grands auteurs / Hourcade, Laurent – Paris: Vicet Amat, 1909 [mf ed 1986] – vii/615p on 2mf – 9 – 0-8370-7158-5 – (in french. incl bibl ref) – mf#1986-1158 – us ATLA [241]
Abrege des memoires pour servir a l'histoire du jacobinisme / Barruel, Augustin – Londres 1800 [mf ed Hildesheim 1995-98] – 1v on 3mf – 9 – €90.00 – 3-487-26293-2 – gw Olms [933]
Abrege des principaux episodes de la revolution du 15 janvier 1908 / Dorneval, E – Port-Au-Prince, Haiti: A A Heraux [1908?] [mf ed 1991] – 1r – 1 – us UF Libraries [972]
Abrege d'histoire d'haiti / Dorsinville, Luc – Port-Au-Prince, Haiti: Impr de l'Etat 1960-61 [mf ed 1993] – 2v on 1r – 1 – us UF Libraries [370]
Abrege du catechisme de perseverance : ou, expose historique, dogmatique, moral et liturgique de la religion depuis l'origine du monde jusqu'a nos jours / Gaume, Jean – Montreal: Cadieux & Derome, [1885?] – 6mf – 9 – 0-665-91887-9 – mf#91887 – cn Canadiana [241]
Abrege du code penal irlandais : suivi de quelques actes publics du gouvernement britannique, a l'egard de la religion catholique en canada – [St-Charles-sur-Richelieu, Quebec?]: J P Laboureur, 1835 [mf ed 1984] – 1mf – 9 – 0-665-43131-7 – mf#43131 – cn Canadiana [345]
Abrege du petit catechisme de quebec pour les petits enfants – S.l: s.n, 18–? – 1mf – 9 – mf#52752 – cn Canadiana [241]

Abrege du veritable christianisme theorique et pratique : ou recueil de maximes chretiennes tant de foi, que de piete et de conduite spirituelle / Labadie, Jean de – Amsterdam, 1670 – 7mf – 9 – mf#PPE-185 – ne IDC [240]

Abrege historique de la devotion au sacre-coeur – Quebec: Impr l'Action sociale limitee, 1921 [mf ed 1998] – 3mf – 9 – mf#SEM105P2989 – cn Bibl Nat [241]

Abrege historique des principaux traits de la vie de confucius : celebre philosophe chinois / Amiot, Joseph Marie – Paris: Chez L'auteur et chez M Ponce, [178-] [mf ed 1995] – (= ser Yale coll) – [50]p/24pl – 1 – 0-524-09716-X – (in french) – mf#1995-0716 – us ATLA [290]

Abrege methodique des principes heraldiques : ou du veritable art du blason / Menestrier, C F – Lyons: Amaulry, 1686 – 3mf – 9 – mf#O-03 – ne IDC [090]

Abreje istoua, daiti, 1492-1945 / Deroche, F Louis – Port-Au-Prince, Haiti: Comite protestant d'alphabetisation et de litterature [194-?] [mf ed 1991] – (= ser Konesans se riches) – 1r – 1 – us UF Libraries [972]

Abreu, Francisco *see* Verdad manifiesta

Abreu Gomez, Ermilo *see*
- Escritores de costa rica
- Ruiz de alarcon, bibliografia critica
- Sala de retratos

Abreu, Joao Capistrano De *see*
- Caminhos antigos e povoamento do brasil
- Capitulos de historia colonial, 1500-1800

Abreu, Sylvio Froes *see*
- Distrito federal e seus recursos naturais
- Riqueza mineral do brazil

Abricht, J *see* Divine emblems

An abridged account of the state of religion in china and cochinchina – London: Keating, Brown, and Keating, 1809-11 [mf ed 1995] – (= ser Yale coll) – 2v in 1 – 1 – 0-524-09607-4 – (with int pref, ann and reflections by trans) – mf#1995-0607 – us ATLA [951]

The abridged cambodian chronicle : [a thai version of cambodian history] = Chronique abregee du cambodge / Osborne, Milton E et al – Paris [1968?] [mf ed 1989] – 1r with other items – 1 – (french trans of phongsawadan khamen yang yo]: p[197]-203) – mf#mf-10289 seam reel 015/19 [§] – us CRL [959]

Abridged grammars of the languages of the cuneiform inscriptions : containing 1: a sumero-akkadian grammar, 2: an assyro-babylonian grammar, 3: a vannic grammar, 4: a medic grammar, 5: an old persian grammar / Bertin, George – London: Truebner, 1888 [mf ed 1986] – (= ser Truebner's coll of simplified grammars of the principal asiatic and european languages 17) – 1mf – 9 – 0-8370-8563-2 – mf#1986-2563 – us ATLA [470]

Abridged history of canada / Withrow, William Henry – Toronto: W Briggs, Montreal: C W Coates, [1887?] [mf ed 1982] – 3mf – 9 – (also: an outline history of canadian literature by g mercer adam) – mf#34656 – cn Canadiana [971]

Abridged index medicus – Washington DC 1973-97 – 1,5,9 – ISSN: 0001-3331 – mf#6285 – us UMI ProQuest [610]

Abridged scientific publications from the kodak research laboratories / Eastman Kodak Co. Research Laboratories – v1-41. 1913-64 – 1 – $270.00 – mf#0190 – us Brook [530]

An abridgement of military law / Winthrop, W – Washington, DC: W H Morrison, 1887 – 5mf – 9 – $7.50 – (this is the 1st of eight eds of the abridgement) – mf#LLMC 96-094 – us LLMC [355]

An abridgement of murray's english grammar and exercise : with questions, adapted to the use of schools and academies; also an appendix containing rules and observations for writing with perspicuity and accuracy / Murray, Lindley – Montreal: Armour & Ramsay, 1847 [mf ed 1984] – 2mf – 9 – 0-665-38144-1 – mf#38144 – cn Canadiana [420]

An abridgment of christian doctrine – [Montreal?: s.n.] 1836 [mf ed 1984] – 1mf – 9 – 0-665-44138-X – mf#44138 – cn Canadiana [241]

An abridgment of christian doctrine : published for the use of the diocese of quebec – Montreal: J Brown, 1812 [mf ed 1971] – 1r – 5 – mf#SEM16P28 – cn Bibl Nat [241]

An abridgment of the compiled laws of the state of michigan, 1897. / Michigan. Laws, Statutes, etc – Lansing, MI: Smith, 1899. 977p. L.C. copy imperfect: p. 975-977 wanting. LL-951 – 1 – us L of C Photodup [348]

Abridgment of the debates of congress from 1789-1856 / U.S. Congress – v1-16 – 9 – $722.00 – mf#0604 – us Brook [324]

Abril Amores, Eduardo *see*
- Adentro
- Aguila acecha
- Bajo la garra
- Surcos de redencion

Abril, Manuel *see* Felipe trigo. exposicion y glosa de su vida, su filosofia, su moral, su arte, su estilo

Abril, P *see* Apuntamientos...de como se deben reformar...las doctrinas y la manera de...

Abril, Xavier *see* Antologia de la poesia moderna hispanoamericana

Abriss der deutschen dichtung : nebst verschiedenen anhaengen / Roehl, Hans – 5. aufl. Leipzig: B G Teubner, 1929 [mf ed 1993] – vii/166p – 1 – (incl ind) – mf#8282 – us Wisconsin U Libr [430]

Abriss der deutschen literaturgeschichte in tabellen / Schmitt, Fritz – 3rd rev ed. Frankfurt am Main: Athenaeum-Verlag 1965 [mf ed 1993] – (= ser Athenaeum paperback) – 1r – 1 – (incl bibl ref & ind; filmed with: poesie der demokratie / klaus r scherpe) – mf#3177p – us Wisconsin U Libr [430]

Abriss der einleitung zum alten testament in tabellenform : an stelle der dritten ausgabe von hertwig's einleitungstabellen / ed by Kleinert, Paul – Berlin: G W F Mueller, 1878 [mf ed 1985] – 1mf – 9 – 0-8370-3917-7 – (incl bibl & ind) – mf#1985-1917 – us ATLA [221]

Abriss der geographie, statistik und geschichte des preussischen staatesein lehr- und lesebuch fuer schule und haus / Uvermann, M – Leipzig [1845] [mf ed Hildesheim: 1995-98] – (= ser Fbc) – 170p on 2mf – 9 – €60.00 – 3-487-29580-6 – gw Olms [943]

Abriss der gesamten kirchengeschichte / Herzog, Johann Jakob – 2. verm verb aufl. Erlangen: E Besold, 1890-92 [mf ed 1990] – 2v on 4mf – 9 – 0-7905-4974-3 – (incl bibl ref) – mf#1988-0974 – us ATLA [240]

Abriss der geschichte der evangelisch-lutherischen synode von ohio : u a staaten in einfacher darstellung: von ihren ersten anfaengen bis zum jahre 1846 / Spielmann, Christian [comp] – Columbus OH: Ohio Synodal-Druckerei 1880 [mf ed 1991] – 1mf – 9 – 0-524-01094-3 – mf#1990-4059 – us ATLA [242]

Abriss der geschichte des alttestamentlichen schrifttums *see* Outline of the history of the literature of the old testament

Abriss der geschichte israels und judas / lieder der hudhailiten, arabisch und deutsch / Wellhausen, Julius – Berlin: Georg Reimer, 1884 [mf ed 1986] – 1mf – 9 – 0-8370-7753-2 – (in german & arabic) – mf#1986-1753 – us ATLA [939]

Abriss der geschichte wiener mechitharisten-congregation und ihrer wirksamkeit : aus anlass des 50-jaehrigen jubilaeums der grundsteinlegung zu ihrem neuen kloster... – Wien: Mechitharisten-Buchdruckerei, 1887 [mf ed 1986] – 1mf – 9 – 0-8370-8060-6 – mf#1986-2060 – us ATLA [241]

Abriss der vergleichenden religionswissenschaft / Achelis, Thomas – 2. umgearb aufl. Leipzig: G J Goeschen 1908 [mf ed 1992] – (= ser Sammlung goeschen) – 1mf – 9 – 0-524-02419-7 – (incl bibl ref) – mf#1990-3003 – us ATLA [230]

Abriss des biblischen aramaeisch : grammatik, nach handschriften berichtigte texte, woerterbuch / Strack, Hermann Leberecht – Leipzig: JC Hinrichs, 1896 [mf ed 1986] – 1mf – 9 – 0-8370-7193-3 – mf#1986-1193 – us ATLA [470]

Abriss einer geschichte der deutschen arbeiterliteratur / Stieg, Gerald & Witte, Bernd – 1. aufl. Stuttgart: E Klett, c1973 [mf ed 1993] – (= ser Literaturwissenschaft-gesellschaftswissenschaft) – 201p – 1 – (incl bibl ref and ind) – mf#8292 – us Wisconsin U Libr [430]

Abriss einer geschichte der evangelischen kirche auf dem europaeischen festlande im 19. jahrhundert / Zahn, Adolf – Stuttgart: JB Metzler, 1886 [mf ed 1986] – 1mf – 9 – 0-8370-8878-X – (incl bibl and ind) – mf#1986-2878 – us ATLA [242]

Abriss einer geschichte der protestantischen missionen : von der reformation bis auf die gegenwart: ein beitrag zur neueren kirchengeschichte / Warneck, Gustav – 7. aufl. Berlin: Martin Warneck, 1901 [mf ed 1986] – 1mf – 9 – 0-8370-6627-1 – (incl bibl ref & ind) – mf#1986-0627 – us ATLA [242]

Abriss einer geschichte der protestantischen missionen *see* Outlines of a history of protestant missions

Ein abriss eines christlich-politischen printzens in 101. sinn-bildern... / Saavedra Faxardo, Didaco de – Amsterdam: Bey Johann Janssonio, dem Jungern, 1655 – 11mf – 9 – mf#O-1883 – ne IDC [090]

Abroad for the bible society / Ritson, John Holland – London: Robert Culley, [1910?] [mf ed 1991] – 1mf – 9 – 0-524-01403-5 – mf#1990-0402 – us ATLA [240]

Abschatz, Johann Erasmus Assmann, Freiherr von *see* Anemons und adonis blumen

Abschid der stette zuerich bern vnnd sant gallen, von wegen der widertaeuffer auszgangen – N p, 1527 – 1mf – 9 – mf#ME-7 – ne IDC [242]

Abschied : einer deutschen tragoedie erster teil, 1900-1914 / Becher, Johannes Robert – Berlin: Aufbau-Verlag, 1949 [mf ed 1989] – 421p – 1 – mf#7040 – us Wisconsin U Libr [820]

Abschied vom paradies : ein roman unter kindern / Thiess, Frank – Stuttgart: I Engelhorn 1927 [mf ed 1991] – 1r – 1 – (filmed with: gotti und gotteli / rudolf von tavel) – mf#2910p – us Wisconsin U Libr [830]

Abschied von mariampol : roman / Brandt, Rolf – Berlin: Scherl, c1936 [mf ed 1989] – 250p – 1 – mf#7063 – us Wisconsin U Libr [830]

Abschied von mariampol *see*
- Liebe auf oesel
- Mutter maria
- Wir deutsch-amerikaner

Das abschiedskonzert : erzaehlung / Czibulka, Alfons von – Stuttgart: J G Cotta, 1944 [mf ed 1989] – 257p – 1 – mf#7161 – us Wisconsin U Libr [830]

The absence of precision in the formularies of the church of england scriptural : and suitable to a state of probation: in eight sermons / Bode, John Ernest – Oxford: J Wright, 1855 [mf ed 1990] – (= ser Bampton lectures 1855) – 1mf – 9 – 0-7905-3640-4 – mf#1989-0133 – us ATLA [242]

Absente reo / Dougall, Lily – London: Macmillan, 1910 [mf ed 1990] – 1mf – 9 – 0-7905-7293-1 – mf#1989-0518 – us ATLA [240]

Abshire, absher notes – v1 n1-v4 n2 [1984 may/jun-1987 jul/aug] – 1 – mf#1533924 – us WHS [071]

Abside – San Luis Potosi, Mexico 1975-78 – 1,5,9 – ISSN: 0001-3382 – mf#451 – us UMI ProQuest [079]

Absolute oder relative wahrheit der heiligen schrift? : dogmatische-kritische untersuchung einer neuen theorie / Egger, Franz – Brixen: A Weger, 1909 [mf ed 1985] – 1mf – 9 – 0-8370-3031-5 – (incl bibl ref) – mf#1985-1031 – us ATLA [220]

Absolute religion : a view of the absolute religion, based on philosophical principles and the doctrines of the bible / Upham, Thomas Cogswell – New York: G P Putnam, 1873 [mf ed 1985] – 1mf – 9 – 0-8370-5650-0 – mf#1985-3650 – us ATLA [230]

Die absolute religion : oder, die vollendete offenbarung gottes in der religion der menschheit / Noack, Ludwig – Darmstadt: Carl Wilhelm Leske, 1846 [mf ed 1989] – 2mf – 9 – (incl bibl) – mf#1985-1491 – us ATLA [240]

Das absolute und die vergeistigung der einzelnen indogermanischen religionen / Asmus, Paul – Halle: CEM Pfeffer, 1877 [mf ed 1991] – (= ser Indogermanische religion in den hauptpunkten ihrer entwickelung 2) – 1mf – 9 – 0-524-01248-2 – mf#1990-2284 – us ATLA [200]

Die absolutheit des christentums und die religionsgeschichte : vortrag gehalten auf der versammlung der freunde der christlichen welt zu muehlacker am 3. oktober 1901 / Troeltsch, Ernst – 2., durchgesehene Aufl. Tuebingen: Mohr, 1912 – 1mf – 9 – 0-7905-3985-3 – mf#1989-0478 – us ATLA [240]

Absolvo te : roman / Viebig, Clara – Berlin: E Fleischel 1907 [mf ed 1989] – 1r – 1 – (filmed with: matthias claudius / urban roedl) – mf#2481p – us Wisconsin U Libr [830]

Absolvta de christi...sacramentis tractatio / Bullinger, Heinrich – London, Stephan Myerdman, 1551 – 3mf – 9 – mf#PBU-158 – ne IDC [240]

Absonderliche charaktere bei wilhelm raabe / Jansen, Werner – Greifswald, 1914 (mf ed 1994) – 1mf – 9 – €24.00 – 3-8267-3085-2 – mf#DHS-AR 3085 – gw Frankfurter [430]

The absorbent mind / Montessori, Maria – Madras, India: Theosophical Pub House, 1949 – (= ser Samp: indian books) – us CRL [150]

The abstainer – Halifax, NS: J Barnes, [1856-1874] – 9 – mf#P04951 – cn Canadiana [073]

The abstract and concrete in education : the word, the image, the reality / Baillairge, Charles P Florent – S.l: s.n, 1897? – 1mf – 9 – mf#17067 – cn Canadiana [370]

Abstract and constructivist art : subject collections – (= ser Art exhibition catalogues on microfiche) – 87 catalogues on 106mf – 9 – £670.00 – (individual titles not listed separately) – uk Chadwyck [700]

Abstract of a course of ten lectures on municipal administration in montreal : delivered in connection with the educational work of the young men's christian association of montreal, 1895-6 / Ames, Herbert Brown et al – S.l: s.n, 1896? – 1mf – 9 – mf#51739 – cn Canadiana [350]

Abstract of a historical sketch of canadian institutions for the insane / Burgess, Thomas Joseph Workman – S.l: s.n, 1899 – 1mf – 9 – mf#05984 – cn Canadiana [366]

Abstract of four lectures on buddhist literature in china / Beal, Samuel – London: Truebner, 1882 [mf ed 1991] – 1mf – 9 – 0-524-00817-5 – (incl bibl ref) – mf#1990-2063 – us ATLA [280]

Abstract of proceedings for the academic year... – 1941,1942,1949,1950 [complete] – inquire – 1 – mf#ATLA 1944-S524 – us ATLA [370]

Abstract of proceedings for the year... – 1891/92-1940/41 [complete] – inquire – 1 – mf#ATLA 1994-S523 – us ATLA [073]

Abstract of statistical returns in judicial matters for... – [Quebec: s.n., ca 1861] [mf ed 1992] – 9 – (cont by: extraits des rapports statistiques judiciaires pour...) – mf#SEM105P1686 – cn Bibl Nat [317]

Abstract of statistics 1961-1970/1972 / Belize. Central Planning Unit – (= ser Latin american & caribbean...1821-1982) – 11mf – 9 – (1964 not publ. 1961, 1963, 1967 not available) – uk Chadwyck [318]

An abstract of statistics of the leeward islands, windward islands and barbados / West Indies University. Institute of Social and Economic Research. (Eastern Caribbean) – (= ser Latin american & caribbean...1821-1982) – 4mf – 9 – uk Chadwyck [318]

Abstract of systematic theology / Boyce, James Petigru & Kerfoot, Franklin Howard – Philadelphia: American Baptist Pub Soc, c1899 [mf ed 1991] – 2mf – 9 – 0-7905-9143-X – (1st printed 1882. incl bibl ref) – mf#1989-2368 – us ATLA [242]

An abstract of the annual reports and correspondence of the society for promoting christian knowledge : from the commencement of its connexion with the east india missions, a d 1709, to the present day – London: The Board of the SPCK, 1814 [mf ed 1995] – (= ser Yale coll) – xvi/730p – 1 – 0-524-09065-3 – mf#1995-0065 – us ATLA [240]

An abstract of the arguments on the catholic question / MacKenna, Theobald – London, 1805 – (= ser 19th c ireland) – 1mf – 9 – mf#1.1.2189 – uk Chadwyck [241]

Abstract of the census of the population : and other statistical returns of prince edward island: taken in the year 1861 – [s.l: s.n, 1861?] [mf ed 1984] – 2mf – 9 – 0-665-46121-6 – mf#46121 – cn Canadiana [317]

Abstract of the census of the population and other statistical returns of prince edward island : taken in the year 1871, under the act 33d victoria, cap 6 – [Charlottetown, PEI?: s.n.] 1871 [mf ed 1983] – 2mf – 9 – 0-665-43336-0 – mf#43336 – cn Canadiana [317]

Abstract of the collection and business laws of colorado territory / Morrison, Robert Stewart – Denver: Rocky Mountain News, 1872. 53,2p. LL-932 – 1 – us L of C Photodup [346]

An abstract of the douay catechism – Quebec: printed by Wm Brown, 1778 [mf ed 1983] – 1mf – 9 – 0-665-38463-7 – mf#38463 – cn Canadiana [241]

Abstract of the former articles of faith confessed by the original baptist church – (Free Will Baptist 1812) – 1 – $5.00 – us Southern Baptist [242]

An abstract of the history of the old and new testaments : divided into three parts / Challoner, Richard – 6th London ed 1st ed [Montreal?: s.n.] 1828 [mf ed 1983] – 3mf – 9 – 0-665-44850-3 – mf#44850 – cn Canadiana [220]

An abstract of the loix de police : or, public regulations for the establishment of peace and good order, that were of force in the province of quebec – London: printed Charles Eyre & William Strahan, 1772 [mf ed 1984] – 1mf – 9 – mf#SEM105P370 – cn Bibl Nat [360]

Abstract of the militia act at present in force : and of the duties thereby imposed on the officers and militiamen – Quebec: printed by P E Desbarats...1821 [mf ed 1994] – 1mf – 9 – mf#SEM105P1902 – cn Bibl Nat [355]

An abstract of the most material parts of an act... : intituled an act for making, repairing and altering the highways and bridges within this province... / Bas-Canada – Montreal: printed at the Office of the Morning Courier, 1839 [mf ed 1982] – 1mf – 9 – mf#SEM105P145 – cn Bibl Nat [348]

Abstract of the proceedings of the church society of the archdeaconry of new brunswick / Church Society of the Archdeaconry of New Brunswick – [Saint John NB?: s.n.] 1837 [mf ed 1983] – 1mf – 9 – 0-665-43833-8 – mf#43833 – cn Canadiana [240]

An abstract of the several royal edicts and declarations, and provincial regulations and ordinances : that were in force in the province of quebec in the time of the french government and the commissions of the several gouvernours-general and intendants of the said province during the same period / Cugnet, Francois Joseph – London: printed by Charles Eyre & William Strahan, 1772 – 1mf – 9 – mf#SEM105P371 – cn Bibl Nat [348]

An abstract of the statutory law of corporations as respects their formation, officers, meetings, liability of members, etc / Collins, Fred K – Cleveland, Schenck, 1881. 79 p. LL-494 – 1 – us L of C Photodup [340]

An abstract of those parts of the custom of the viscounty and provostships of paris : which were received and practifed in the province of quebec, in the time of the french government – London: printed by Charles Eyre & William Strahan, 1772 [mf ed 1984] – 1mf – 9 – mf#SEM105P368 – cn Bibl Nat [348]

Abstracts and reviews (1965-1999) – weekly briefings (1968-1999) / Royal Institute of Chartered Surveyors – 14r – 1 – £660.00 – (incl printed guide) – mf#RCI – uk World [520]

Abstracts for social workers – Washington DC 1965-77 – 1,5,9 – (cont by: social work research and abstracts) – ISSN: 0001-3412 – mf#5852.01 – us UMI ProQuest [360]

Abstracts of book reviews in current legal periodicals – repr 1987. v1-12 (1974-87) – 9 – $174.00 set – 0-89941-584-9 – (none publ aug 1977 through jun 1979) – mf#110541 – us Hein [340]

Abstracts of declassified documents – Washington DC 1947-48 – 1 – mf#8570 – us UMI ProQuest [020]

Abstracts of english studies – Oxford, England 1958-91 – 1,5,9 – ISSN: 0001-3560 – mf#1603 – us UMI ProQuest [420]

Abstracts of folklore studies – Columbus OH 1963-75 – 1,5 – ISSN: 0001-3587 – mf#5897 – us UMI ProQuest [390]

Abstracts of health care management studies – Chicago IL 1979-87 – 1,5,9 – (cont: abstracts of hospital management studies) – ISSN: 0194-4908 – mf#10526,01 – us UMI ProQuest [362]

Abstracts of hospital management studies – Chicago IL 1965-78 – 1,5,9 – (cont by: abstracts of health care management studies) – ISSN: 0001-3595 – mf#10526.01 – us UMI ProQuest [362]

Abstracts of jamaica wills, 1625-1792 : from the british library, add. ms. 34184 – (= ser British records relating to america in microform) – 1r – 1 – (int by richard s dunn) – mf#96661 – uk Microform Academic [972]

Abstracts of magnetical observations made at the magnetical observatory, toronto, canada west : during the years 1856 to 1862, inclusive and during parts of the years 1853, 1854 and 1855 – [Toronto?: s.n.] 1863 [mf ed 1984] – 2mf – 9 – 0-665-46127-5 – mf#46127 – cn Canadiana [530]

Abstracts of north american geology – Washington DC 1966-71 – 1,5 – ISSN: 0001-3625 – mf#3292 – us UMI ProQuest [550]

Abstracts of oregon donation land claims, 1852-1903 – (= ser Records of the bureau of land management) – 6r – 1 – (with printed guide) – mf#M145 – us Nat Archives [333]

Abstracts of papers presented to the american mathematical society / American Mathematical Society – Providence RI 1980+ – 1,5,9 – ISSN: 0192-5857 – mf#12051 – us UMI ProQuest [510]

Abstracts of photographic science and engineering literature – Springfield VA 1962-72 – 1 – ISSN: 0001-3633 – mf#6700 – us UMI ProQuest [770]

Abstracts of service records of naval officers ('records of officers'), 1798-1893 / U.S. Navy. Bureau of Naval Personnel – (= ser Records relating to service in the united states navy and united states marine) – 19r – 1 – mf#M330 – us Nat Archives [355]

Abstracts of service records of naval officers (records of officers), 1829-1924 – (= ser Records of the bureau of naval personnel) – 18r – 1 – (with printed guide) – mf#M1328 – us Nat Archives [355]

Abstracts of statistics 1956-1969 / Barbados. Statistical Service – (= ser Latin american & caribbean...1821-1982) – 12mf – 9 – uk Chadwyck [318]

Abstracts of the financial reports / Massachusetts. Comptroller's Division – 1922-77. 56 fiches. (Harvard Law School Library Collection.) – 9 – us Harvard Law [324]

Abstracts of washington donation land claims, 1855-1902 – (= ser Records of the bureau of land management) – 1r – 1 – (with printed guide) – mf#M203 – us Nat Archives [333]

Abstracts of working papers in economics – Cambridge, England 1991-95 – 1,5,9 – ISSN: 0951-0079 – mf#17114 – us UMI ProQuest [330]

Abstracts of world medicine – London, England 1947-71 – 1,5 – mf#1365 – us UMI ProQuest [610]

Abstracts on crime and juvenile delinquency – 1968-1985 – 9 – $2700.00 set – 0-89941-411-7 – (with cum ind. price incl hb user ind) – mf#400600 – us Hein [345]

Abstracts on hygiene – Wallingford, England 1976-80 – 1,5,9 – (cont by: abstracts on hygiene and communicable diseases) – ISSN: 0001-3692 – mf#11151.01 – us UMI ProQuest [613]

Abstracts on hygiene and communicable diseases – London, England 1981-89 – 1,5,9 – (cont: abstracts on hygiene) – ISSN: 0260-5511 – mf#11151.01 – us UMI ProQuest [616]

Abstrakte arbeit und abstraktwerden der kunst / Fiebig, Wilfried – Berlin, 1974 (mf ed 1994) – 4mf – 9 – €45.00 – 3-89349-998-9 – mf#DHS-AR 998 – gw Frankfurter [700]

Abt bischof waldo (tab10-11) : begruender des goldenen zeitalters der reichenau / Munding, E – 1924 – (= ser Texte und arbeiten. beuron (tab). beitraege zur ergruendung des aelteren lateinischen und christlichen schrifttums und gottesdienstes) – €7.00 – ne Slangenburg [931]

Abteilung neuere deutsche Literaturgeschichte *see* Neue deutsche forschungen

Abtics : abstract and book title index card service – London, England 1960-71 – 5,9 – ISSN: 0001-3404 – mf#40000 – us UMI ProQuest [020]

Der abtruennige : roman / Brachvogel, Carry – Berlin: Vita Deutsches Verlagshaus, 1907, c1905 [mf ed 1989] – 347p – 1 – mf#7061 – us Wisconsin U Libr [830]

Abu al-Alar al-Maarri *see* The quatrains of abu'l-ala

Abu 'ali Al-hasan Ibn Al-haytham *see* Die psychologie alhazens

Abu Amran Musa *see* Die bibelexegese moses maimunis

Abu Salih *see* The churches and monasteries of egypt and some neighboring countries

Abu Salih, the Armenian *see* The churches and monasteries of egypt and some neighbouring countries

Abu Talib Han *see* Voyages du prince persan mirza aboul taleb khan en asie, en afrique, en europe

Abu telfan : oder, die heimkehr vom mondgebirge: roman / Raabe, Wilhelm Karl – Berlin: Aufbau-Verlag 1961 [mf ed 1995] – 1r – 1 – (filmed with: gertrud von loden / c quandt) – mf#3706p – us Wisconsin U Libr [830]

Abu Ya-'qub Ishaq Ibn Sulayman Al-isra'ili *see* Die philosophische lehren des isaak ven salomon israeli

Abudacnus, Jos *see* Historia jacobitarum

Abudarham, David Ben *see* Sefer abudarham

Abulfedae *see* Annales muslemici arabice et latine

The abundant life / Jones, Rufus Matthew – London: Headley, [1908?] – 1mf – 9 – 0-7905-7847-6 – mf#1989-1072 – us ATLA [240]

Abundant life magazine – Tulsa OK 1975-79 – 1,5,9 – mf#10336 – us UMI ProQuest [240]

Abundant living – 1982 oct/nov – 1 – mf#4026936 – us WHS [071]

Abusch, Alexander *see* Literatur und wirklichkeit

Abu-Tammam Haabib Ibn-Aus ata-Tai *see* Kitab asa al-hamasa maa sarh...abi-zakarija...at-tibrizi wa-arbaa faharis

Die abwege, oder irrungen und versuchungen : gutwilliger und frommer menschen, aus beystimmung des gottseeligen alterthums angemercket / Arnold, Gottfried – Frankfurt: T Fritsch, 1708. Chicago: Dep of Photodup, U of Chicago Lib, 1971 (1r); Evanston: American Theol Lib Assoc, 1984 (1r) – 1 – 0-8370-0396-2 – mf#1984-B258 – us ATLA [240]

Die abwehr – Warnsdorf [Varnsdorf CZ], 1926 1 may-1929 21 apr – 7r – 1 – (filmed by misc inst: 1926 1 may-1938 [34r]) – gw Mikrofilm; gw Misc Inst [077]

Aby-astorpstidningen – AEngelholm, 1898-1901 – 9 – sw Kungliga [078]

Abydos / Petrie, W M – London, 1902-1904. 3v – 12mf – 9 – mf#NE-20351 – ne IDC [956]

Abydos : pt 1 / Flinders Petrie, W M – London, 1902 – (= ser Mees 22) – 9mf – 8 – €18.00 – ne Slangenburg [930]

Abydos : pt 2 / Flinders Petrie, W M – London, 1903 – (= ser Mees 24) – 8mf – 8 – €17.00 – ne Slangenburg [930]

Abydos : pt 3 / Ayrton, Edward R et al – London, 1904 – (= ser Mees 25) – 9mf – 8 – €18.00 – ne Slangenburg [930]

Abyssinia : through the lion-land to the court of the lion of judah / Vivian, H – London, 1901 – 5mf – 9 – mf#NE-20216 – ne IDC [916]

Abyssinia and its people : or, life in the land of prester john / Hotten, J C – London, 1868 – 5mf – 9 – mf#NE-20202 – ne IDC [916]

Abyssinia on the eve / Farago, L – London, New York, 1935 – 4mf – 9 – mf#NE-20220 – ne IDC [956]

Abyssinian – 1964 feb 9-sep – 1 – mf#4714226 – us WHS [071]

The abyssinian at home / Walker, C H – London, 1933 – 3mf – 9 – mf#NE-20244 – ne IDC [956]

The abyssinian church / Dowling, Theodore Edward – London: Cope & Fenwick, [1909?] – 1mf – 9 – 0-7905-4407-5 – (incl bibl ref) – mf#1988-0407 – us ATLA [240]

Abyssinian news – v1 n10-12 [1990 apr-jun]; v2 n1-3 [1990 jul-nov] – 1 – mf#1702387 – us WHS [071]

Abyssinian news letter – 1982 winter; 1983 fall-84 spring – 1 – mf#4881422 – us WHS [071]

Abyssinian papyrus – 1994 feb – 1 – mf#4839070 – us WHS [071]

Abyssinian scrapbooks : comprised of clippings and photographs relating to his diplomatic mission to ethiopia in 1903-1904 / Skinner, Robert Peet – Annapolis, MD: US Naval Academy Library, 1963 – 9 – us CRL [960]

L'abyssinie et les italiens / Castonnet des Fosses, Henri – Paris: P Tequi, 1897 – 1 – us CRL [960]

Abz [aktuelle bilder-zeitung] – Duesseldorf DE, 1959 3 jan-28 mar – 1r – 1 – (title varies: 1952 n48: abz illustrierte; filmed by misc inst: 1948-58 [10r]) – gw Mikrofilm; gw Misc Inst [074]

Abz [arbeit in bild und zeit] – Berlin DE, 1933 n1-19 – 1 – gw Misc Inst [331]

Abz illustrierte *see* Abz [aktuelle bilder-zeitung]

AC Australia. New South Wales *see* Statistical blue books 1822-1894

AC Australia. Tasmania *see* Statistical blue books 1822-1847

Ac ukazatele zhurnalenykh statei po ekonomicheskim voprosam za desiatiletie 1904-1913 gg. 1 – Kiev, 1915 – 7mf – 9 – mf#R-7033 – ne IDC [077]

A-c views – 1947 nov 6-1964 dec – 1 – mf#1051435 – us WHS [071]

A-c viewsletter – 1964 jan 31-1976 dec 3 – 1 – mf#1109396 – us WHS [071]

Aca news – 1938 aug 20-1947 dec; 1948-66; 1967-1971 apr – 1 – mf#1109397 – us WHS [071]

The acacia / Wilkins, Harriet Annie – [Hamilton, Ont?: s.n.] 1860 [mf ed 1982] – 2mf – 9 – 0-665-36139-4 – (incl ind) – mf#36139 – cn Canadiana [810]

Academe / American Association of University Professors – Washington DC 1979+ – 1,5,9 – (cont: aaup bulletin) – ISSN: 0190-2946 – mf#974,01 – us UMI ProQuest [378]

Academe – Washington DC 1967-78 – 1,5 – ISSN: 0001-3749 – mf#6517 – us UMI ProQuest [378]

Academia – v1 n1-3, 5, 7-9 [1895 dec 1-1896 jan 3, feb 1, apr 1, jun 1, oct 1-nov 1] – 1 – mf#1775716 – us WHS [071]

Academia Altorfina *see*
- Emblemata anniversaria academiae altorfinae studiorum iuventutis exercitandorum causa proposita et variorum orationibus exposita
- Emblemata anniversaria academiae noribergensis, quae est altorffii
- Epitome emblematum panegyricorum academiae altorfinae

Academia Colombiana De Historia *see* Homenaje al profesor paul rivet

Academia Colombiana de Historia *see* 1840

Academia de Bellas Artes de San Fernando *see* Homenaje a eugenio hermoso

Academia De Ciencias Medicas, Fisicas Y Naturales *see* Estado actual 1939

Academia De La Historia De Cuba *see* Homenaje a los academicos de honor

Academia de la Historia Madrid *see* Archivo documental espanol

Academia de la Historia. Madrid *see*
- Boletin
- Memorial historico espana9ol
- Memorial historico espanol

La Academia de Musica y Banda Municipal de Caceres *see* Reglamento

Academia popular : seminario de instruccao e recreio para o povo – Pernambuco: Typ de M F de Faria. 03 maio-15 jun 1863 – (= ser Ps 19) – mf#P17,02,160 – bl Biblioteca [370]

Academia Portuhuesa da Historia *see* Guia de bibliografia historica portuguesa

Academia que se celebro en badajoz en casa de don manuel meneses – 1684 – 9 – sp Bibl Santa Ana [946]

Academia Socialista Preparatoria...Escuelas Populares de Guerra *see* Circular de la agrupacion socialista de valencia y regulamento de la academia

L'academia todesca della architectura, scultura e pittura : oder deutsche akademie der edlen bau- bild- und mahleren kuenste... / Sandrart, J von – Nuernberg, Frankfurt, 1675-79 – 2v on 35mf – 9 – mf#0-424 – ne IDC [700]

Academiae caesareo-leopoldinae naturae curiosorum ephemerides *see* Miscellanea curiosa sive ephemeridum medico-physicarum germanicarum academiae caesareo-leopoldinae naturae curiosorum

Academic and conduct records of cadets, 1881-1908 *see* United states naval academy registers of delinquencies, 1846-1850 and 1853-1882, and academic and conduct records of cadets, 1881-1908

Academic annals of painting, sculpture, and architecture, published by authority of the royal academy of arts, 1805-1806, 1807, 1808-1809 / Hoare, Prince – London, 1809 – (= ser 19th c art & architecture) – 1mf – 9 – mf#4.1.125 – uk Chadwyck [700]

Academic correspondence, 1803... : with the academies of vienna and st petersburg / Hoare, Prince – London 1804 – (= ser 19th c art & architecture) – 1mf – 9 – mf#4.2.1643 – uk Chadwyck [700]

Academic emergency medicine – Oxford, England 1994+ – 1,5,9 – ISSN: 1069-6563 – mf#21605 – us UMI ProQuest [610]

Academic medicine – 1989+ – 1,5,9 – (cont: journal of medical education) – ISSN: 1040-2446 – mf#171,01 – us UMI ProQuest [610]

Academic monthly = Hsueh shu yueh pao – v3 n1-10. feb-nov 1899* – (= ser Chinese christian coll 55) – 1r – 1 – mf#ATLA S0296K – us ATLA [370]

Academic performance, attendance, and schedule rigor of extracurricular participants and nonparticipants / Patranella, Kenneth W – 1987 – 2mf – 9 – $8.00 – us Kinesology [370]

Academic psychiatry – Arlington VA 1989+ – 1,5,9 – (cont: journal of psychiatric education) – ISSN: 1042-9670 – mf#11182,01 – us UMI ProQuest [616]

Academic questions (aq) – Piscataway NJ 1987+ – 1,5,9 – ISSN: 0895-4852 – mf#16937 – us UMI ProQuest [378]

Academic therapy – Austin TX 1965-90 – 1,5,9 – (cont by: intervention in school and clinic) – ISSN: 0001-396X – mf#6344.01 – us UMI ProQuest [370]

The academical year at king's college : begins with michaelmas term, september 1st, and ends with trinity term, july 1st dating, therefore, from september 1, 1845... – [S.l: s.n, 1847?] [mf ed 1984] – 1mf – 9 – 0-665-46798-2 – mf#46798 – cn Canadiana [378]

Academician – La Salle IL 1818-20 – 1 – mf#3533 – us UMI ProQuest [378]

O academico : jornal juridico, litterario e noticioso – Sao Paulo: Typ do Ypiranga, 07 jun-19 nov 1868 – (= ser Ps 19) – mf#P46,06,35 – bl Biblioteca [073]

O academico : orgam dos estudantes da fac de sciencias jur e sociaes de manaus – Manaus, AM. 28 set-dez 1926; jan-jun, set-out 1927; fev, set 1928 – (= ser Ps 19) – mf#P11B,06,33 – bl Biblioteca [370]

O academico : periodico scientifico, litterario e especialmente medico – Rio de Janeiro, RJ: Typ Fluminense, jul 1855-set 1856 – (= ser Ps 19) – mf#P01B,05,10 – bl Biblioteca [079]

L'academie canadienne-francaise – Montreal: [s.n.], 1955 [mf ed 1986] – 1mf – 9 – mf#SEM105P691 – cn Bibl Nat [366]

Academie des inscriptions et belles lettres, Paris *see* Recueil des historiens des gaules et de la france

Academie des inscriptions et belles-lettres de paris : repertoire d'epigraphie semitique – Paris, 1900-1929. v1-7 – 32mf – 9 – (missing: 1906/1907 v2; 1919 v4) – mf#NE-20002 – ne IDC [956]

Academie des jeux historiques : contenant les jeux de l'histoire de france... – Paris: Le Gras, 1718 – 5mf – 9 – mf#0-59 – ne IDC [090]

Academie des Sciences, Belles-lettres et arts. Besancon *see* Proces-verbaux et memoires

Academie des Sciences d'Outre-Mer *see* Comptes rendus

Academie des Sciences Morales et Politiques *see* Le suffrage des femmes

Academie des Sciences. Russie *see* Comptes rendus

Academie du sacre-coeur, grand'mere : album-souvenir des noces d'argent, 1902-1927 – [Grand'Mere: J P Emile Dessureault...1927?] [mf ed 1998] – 2mf – 9 – mf#SEM105P2985 – cn Bibl Nat [241]

L'Academie Imperiale des Sciences. St. Petersbourg *see* Melanges asiatiques, tires de l'academie imperiale des sciences de st. petersbourg

Academie royale des arts du Canada *see* Catalogue 1900

Academie Royale des Sciences *see* Descriptions des arts et metiers

Academie royale des sciences d'outre-mer. Classe des sciences morales et politiques *see* Memoires in-80

Academische festrede am hundertjaehrigen geburtstage friedrich schleiermachers, dem 21. november 1868 : an der christian-albrechts-universitaet / Thomsen, Nicolaus – Kiel: C F Mohr 1868 [mf ed 1991] – 1mf – 9 – 0-524-00172-3 – mf#1989-2872 – us ATLA [190]

Academus : periodico politico, scientifico e litterario – Recife, PE: Typ da Provincia, 15 maio-15 jun 1876 – (= ser Ps 19) – bl Biblioteca [079]

Academy : lutherans in profession – v1-40. 1943-86 [complete] – 4r – 1 – (cont: lutheran scholar) – ISSN: 0024-7502 – mf#ATLA S0732 – us ATLA [242]

The academy – Pictou, NS: [s.n, 1884?- 18– or 19–] – 9 – mf#P04866 – cn Canadiana [378]

Academy and literature – La Salle IL 1869-1916 – 1 – mf#4181 – us UMI ProQuest [420]

The academy annual / Halifax Academy – [Halifax, NS?]: Pub by the students, [1896?-19–] – 9 – mf#A02625 – cn Canadiana [378]

Academy architecture and architectural review – London, England 1889-1931 – 1 – mf#5896 – us UMI ProQuest [720]

The academy for princes / Norlie, Olaf Morgan – Minneapolis: Augsburg Pub House, 1917 – 1mf – 9 – 0-524-07582-4 – (incl bibl ref) – mf#1991-3202 – us ATLA [377]

The academy gossip / Mount Allison Wesleyan Academy – Sackville, N.B.: Chignecto Post, [1871-18– or 19–] – 9 – ISSN: 1190-7339 – mf#P04540 – cn Canadiana [378]

Academy notes / Blackburn, Henry – London 1875-1900 – (= ser 19th c art & architecture) – 42mf – 9 – mf#4.2.204 – uk Chadwyck [700]

Academy of General Dentistry *see* Journal – academy of general dentistry

Academy of management executive – Ithaca NY 1987-89 – 1,5,9 – (cont by: executive) – ISSN: 0896-3789 – mf#16369.03 – us UMI ProQuest [650]

Academy of management executive – Ithaca NY 1994+ – 1,5,9 – (cont: executive) – ISSN: 1079-5545 – mf#16369.03 – us UMI ProQuest [650]

Academy of management journal – Ithaca NY 1958+ – 1,5,9 – ISSN: 0001-4273 – mf#5393 – us UMI ProQuest [650]

Academy of management review – Ithaca NY 1976+ – 1,5,9 – ISSN: 0363-7425 – mf#10726 – us UMI ProQuest [650]

Academy of marketing science. journal – Newbury Park CA 1973+ – 1,5,9 – ISSN: 0092-0703 – mf#10335 – us UMI ProQuest [650]

Academy of National Sciences Philadelphia. Entomological Section *see* Entomological news und proceedings

Academy of parish clergy : journal – v1-5. apr 1971-aug 1975* – 1r – 1 – (superseded by: sharing the practice) – mf#ATLA T0002 – us ATLA [240]

Academy of Political Science *see* Political science quarterly

Academy of political science. proceedings of the academy of political science – New York NY 1910-1992 – 1,5,9 – ISSN: 0065-0684 – mf#831 – us UMI ProQuest [320]

Academy of religion and psychical research : journal – v2-12. 1979-89 [complete] – 3r – 1 – mf#ATLA S0841 – us ATLA [073]

Academy of Sciences Library, St Petersburg *see* Nep rare editions

Academy of Sciences of the USSR *see* Mathematical notes of the academy of sciences of the ussr

Academy of Sciences of the USSR. Division of Chemical Sciences *see* Bulletin of the academy of sciences of the ussr...

Academy proceedings in earth and planetary sciences / Indian Academy of Sciences – Bangalore, India 1999-2004 – 1,5,9 – (cont: proceedings. earth and planetary sciences / indian academy of sciences) – mf#12438.02 – us UMI ProQuest [550]

Academy reflections – 1978 mar/apr-jun/jul; 1979 jan/mar – 1 – mf#5132415 – us WHS [071]

Academy triforium – v1 n1-v8 n1 [1973 feb-1980 feb] – 1 – mf#135412 – us WHS [071]

Academy triforium newsletter of the wisconsin academy of sciences, arts and letters – 1973 feb-1980 feb; 1888-1891 [transactions v8] – 1 – mf#671693 – us WHS [071]

Acadia athenaeum – Wolfville [NS]: Students of Acadia College, [1874-1961] [mf ed v1 n1 nov 1874-v26 n8 jun 1900] – 9 – mf#P05016 – cn Canadiana [378]

Acadia College. Associated Alumni. Executive Committee *see*
– The sixteenth annual report of the executive committee
– The twenty-second annual report of the executive committee...and, addresses in memoriam, relating to the life and labors of the late Rev J M Cramp

Acadia College. Halifax *see* Jubilee...and memorial exercises

Acadia Provident Association *see* Prospectus

The acadian exile and sea shell essays / Clark, Jeremiah Simpson – [Charlottetown, PEI?: s.n.] 1902 – 1mf – 9 – 0-659-91196-5 – mf#9-91196 – cn Canadiana [810]

Acadian Genealogical and Historical Association *see* Etoile d'acadie

Acadian recorder – Halifax, Canada 6 jul 1839, 9 aug 1851, 16 jan, 24 mar 1913-30 jun 1922 – 26r – 1 – uk British Libr Newspaper [071]

Acadian recorder – Halifax, NS. 1817-69 – 9r – 1 – ISSN: 1181-3466 – cn Library Assoc [071]

The acadian scientist – Wolfville, NS: Acadian Science Club, [1883?-1884] – 9 – (cont by: canadian science monthly) – mf#P04007 – cn Canadiana [500]

The acadians of louisiana and their dialect / Fortier, Alcee – S.l: s.n, 1891? – 1mf – 9 – (incl bibl ref) – mf#51861 – cn Canadiana [305]

Acadie and the acadians / Roth, David Luther – Philadelphia: Lutheran Pub Soc 1890 [mf ed 1993] – 1mf – 9 – 0-524-07591-3 – mf#1991-3211 – us ATLA [242]

L'acadie nouvelle – Caraquet, New Brunswick, CN. 1984- – 12r/y – 1 – Can$1065.00 – cn Commonwealth Imaging [071]

L'acadien – Moncton, NB. 1913-26 – 6r – 1 – ISSN: 1486-7532 – cn Library Assoc [070]

Les acadiens a moncton : un siecle et demi de presence francaise au coude / Brun, Regis – Moncton: R Brun, 1999 [mf ed 2001] – 9 – cn Bibl Nat [971]

Acariden aus egypten und dem sudan / Traegardh, I – Upsala, 1901 – 3mf – 8 – mf#1559 – ne IDC [960]

The acarn-hog economy of the oak wood lands... / Parsons James, J – New York: American geographical Society, 1962 – 1 – sp Bibl Santa Ana [338]

Acbes newsletter – 1984 may-jun; 1986 apr, jun, aug, oct – 1 – mf#4798765 – us WHS [071]

Accademia della Crusca Firenze *see* Vocabolario degli accademici della crusca

Accademia Gelati *see* Rime de gli academici gelati di bologna

Accademia nazionale Luigi Cherubini di Musica. Lettere e Arti figurative *see* Atti

Accademia Occulti *see* Rime de gli academici occulti con le loro imprese et discorsi

Accademie e biblioteche d'italia – Rome, Italy 1927+ – 1,5,9 – ISSN: 0001-4451 – mf#2147 – us UMI ProQuest [020]

Accao colonial – Porto: Accao colonial, jun 30 1930-jan 1934 – us CRL [074]

Accao nacional – Lourenco Marques: [Accao Nacional de Mocambique, apr 5-aug 6, sep 11-24 1926 (reel 15); oct 1-8, nov 5-dec 3 1926 (reel 18) – us CRL [074]

The acceleration phase in the baseball pitching sequence / Simeone, Mark – 1997 – 1mf – 9 – $4.00 – mf#PE 3774 – us Kinesology [612]

Accelerator – v1 n1-to date [1974 jun-to date] – 1 – mf#639074 – us WHS [071]

Accelerators, spectrometers, detectors and associated equipment *see* Nuclear instruments and methods in physics research, sect a

Accent – Needham MA 1976-96 – 1,5,9 – ISSN: 0192-7507 – mf#11778 – us UMI ProQuest [730]

Accent on living – Bloomington IL 1956-2001 – 1,5,9 – ISSN: 0001-4508 – mf#7406 – us UMI ProQuest [071]

Accent/l a – 1987 may-1990 nov – 1 – mf#2690981 – us WHS [071]

Acceptable service (what it really is) : illustrated and enforced from holy writ by sarah r geldard, 1875; honble f dillon, 1880; rev geo wright, 1809 / Geldard, Sarah R – Fergus, Ont: s.n, 1880 – 1mf – 9 – mf#33728 – cn Canadiana [240]

Acceptance of cessions of certain samoan islands : report of the senate committee on territorial and insular possessions / American Samoa. US Congress – 70th Congress 1st sess. Senate report no 984. n.p. 3 May 1928 – 1mf – 9 – $1.50 – mf#LLMC 82-100C Title 33 – us LLMC [327]

Acceptances and order for commissions in the records of the department of state, 1789-1828 / U.S. Dept of State – (= ser General records of the department of state) – 2r – 1 – mf#T645 – us Nat Archives [324]

Access – Washington DC 1975-85 – 1,5,9 – ISSN: 0149-9262 – mf#11715 – us UMI ProQuest [380]

Access and engagement : program design and instructional approaches for immigrant students in secondary school / Walqui, Aida – McHenry IL: Delta Systems Co Inc; [Washington DC]: CAL, ERIC, c2000 [mf ed 2000] – 3mf – 9 – (incl bibl ref) – us GPO [373]

Access center newsletter – [1978 feb 17-1978 may 15] – 1 – mf#678865 – us WHS [071]

Access control & security systems – Shawnee Mission KS 2001+ – 1,5,9 – mf#7706,05 – us UMI ProQuest [690]

Access news – [1978 aug-1980 summer] – 1 – mf#678860 – us WHS [071]

Access to an open polar sea : in connection with the search after sir john franklin and his companions / Kane, Elisha Kent – New York: Baker, Godwin, 1853 [mf ed 1984] – 1mf – 9 – 0-665-45220-9 – mf#45220 – cn Canadiana [919]

Access to energy – 1984 sep-1993 aug – 1 – mf#305833 – us WHS [071]

An accidence, or gamut, of painting in oil / Ibbetson, Julius Caesar – 2nd ed. London 1828 – (= ser 19th c art & architecture) – 1mf – 9 – mf#4.2.1223 – uk Chadwyck [750]

Accident analysis and prevention – Oxford, England 1969+ – 1,5,9 – ISSN: 0001-4575 – mf#49000 – us UMI ProQuest [360]

El accidente de trabajo en la historia y en la realidad espanola / Rodriguez Bautista, Ambrosio – Caceres: Tip. El Noticiero, 1950 – sp Bibl Santa Ana [331]

Les accidents du travail : memoire soumis le 14 novembre 1961 a l'honorable rene hamel, ministre du travail... / Confederation des syndicats nationaux. Quebec – [Montreal?]: la Confederation des syndicats nationaux...[1961?] (mf ed 1995) – 1mf – 9 – mf#SEM105P2488 – cn Bibl Nat [362]

Accidents of an antiquary's life / Hogarth, David George – London: Macmillan, 1910 [mf ed 1989] – 1mf – 9 – 0-7905-3262-X – mf#1987-3262 – us ATLA [915]

Accioly, Breno *see* Cogumelos

Accioly, Hildebrando Pompeo Pinto *see*
– Limites do brasil
– Reconhecimento do brasil pelos estados unidos da a...

Accion – Asuncion, Paraguay: [s.n, [ano 1: n1-3. epoca: ano 25: n134 (abr 1969-nov 1992)] – 4r – 1 – us CRL [073]

Accion chilena – Santiago. v. 1-7. no. 1. Jan. 1934-Jan. 1938 – 1 – us NY Public [073]

Accion continental – Miami FL: Jesus Retureta 1973- [1973 mar 27-oct 30] (mthly) [mf ed 1999] – (= ser Cuban heritage collection) – 1r – 1 – (in spanish & english) – us UF Libraries [071]

Accion cubana : [dios, patria y familia] – Luxemburgo [i.e. Luxembourg]: Rocaner (1960 oct 20-1962 may 26) (biwkly) [mf ed 1999] – (= ser Cuban heritage collection) – 1r – 1 – (began in nov 1959) – us UF Libraries [320]

Accion cuentos / Rodas, Hector Ovidio – Guatemala: Editorial del Ministerio de Educacion Publica "Jose de Pineda Ibarra" 1959 [mf ed 1991] – (= ser Coleccion contemporaneos 49) – 1r – 1 – us UF Libraries [972]

Accion Cultural Popular (Colombia) (Radio Program) *see* Revolucion violenta?

Accion educativa del gobierno federal del... / Mexico. Secretaria de Educacion Publica – Mexico: La Secretaria, [1952/54-1954/55] – us CRL [370]

Accion (miami, fl) : [organo oficial del movimiento nacionalista cristiano] – Miami FL: Movimiento Nacionalista Cristiano, Seccion Cubana de Joven America 1964 may-1970 apr (mthly) [mf ed 1999] – (= ser Cuban heritage collection) – 1r – 1 – (began in dec 1963) – us UF Libraries [320]

Accion Regional Extremena *see* Arex

Accolti, B *see*
– ...De bello a christianis contra barbaros gesto pro christi sepvlchro et ivdaea recvperandis
– La gverra fatta da christiani contra barbari per la ricvperatio...

Accolti Gil Vitale, Nicola *see* La giovinezza di hamann

Accompaniment to gregorian chant in medieval harmony / Farrell, Gerard J – U of Rochester 1951 [mf ed 19–] – 2mf – 9 – mf#fiche 1165 – us Sibley [780]

Accomplishments of the clean air act, as amended by the clean air act amendments of 1990 : hearing...house of representatives, 107th congress, 2nd session, may 1 2002 / United States. Congress. House. Committee on Energy and Commerce. Subcommittee on Energy and Air Quality – Washington: US GPO 2002 [mf ed 2002] – 2mf – 9 – 0-16-068796-9 – (incl bibl ref) – us GPO [344]

L'accomplissement des propheties / [Jurieu, P] – Rotterdam, 1686 2v – 4mf – 9 – mf#PRS-154 – ne IDC [240]

Accomplissement des propheties... / Du Moulin, P – Sedan, 1624 – 5mf – 9 – mf#CA-128 – ne IDC [240]

The accompt rekenynge and confession of the faith... / Zwingli, H – Geneua [recte: Emden], 1555 – 2mf – 9 – mf#PBU-529 – ne IDC [240]

L'accord americano-haitien du 7 aout 1933 : et le pouvoir legislatif d'haiti precede d'une lettre de la ligue internationale des femmes / Dehoux, Lorrain – Port-Au-Prince, Haiti: Impr haitienne 1933 [mf ed 1991] – 1r – 1 – us UF Libraries [327]

L'accord passe et conclvd tovchant la matiere des sacremens... / Bullinger, Heinrich – Geneve, Iehan Crespin, 1551 – 1mf – 9 – mf#PBU-263 – ne IDC [240]

An account and history of the oregon territory : together with a journal of an emigrating party across the western prairies of america and to the mouth of the columbia river / Wilkes, George – 2nd ed. London: W Lott, 1846 [mf ed 1983] – 2mf – 9 – 0-665-44922-4 – mf#44922 – cn Canadiana [978]

Account book / Elfe, Thomas – [mf ed Spartanburg SC: Reprint Co, 1981] – 1v on 12mf – 9 – mf#51-049 – us South Carolina Historical [380]

Account book / Ewing, Clymer and Co. Westport, Mo – 1839-40 – 1 – us Kansas [025]

Account book / Grant, George – 1873-77 – 1 – us Kansas [920]

Account book / Grinstead, William – 1850-1951 – 1 – us Kansas [920]

Account book / Grinter, Moses – 1855-82 – 1 – us Kansas [920]

Account book / Roberts, Isaac N – 1857-78 – 1 – us Kansas [978]

Account book, 1861 / South Carolina. Militia. Brigade, 4th – [mf ed Spartanburg SC: Reprint Co [1981?]] – 1mf – 9 – mf#51-519 – us South Carolina Historical [976]

Account book of his expenses abroad, 1712-1718 and 1714-1718 – (= ser Holkham library, the house, park and art colls 733,734) – 1r – 1 – mf#95972 – uk Microform Academic [640]

Account books / Krall, William – 1900-1925, Farm account books of William Krall, Atchison County, KS – 1 – us Kansas [380]

Account books and miscellaneous papers / Harrouff, George – 1874-1944 – 1 – (diary. 1872-1944. 1) – us Kansas [920]

Account books of his expenses, 1707-1718 – (= ser Holkham manuscripts 732b) – 1r – 1 – mf#774 – uk Microform Academic [090]

Account, guide and form book for administrators and executors in the state of ohio / Gale, John T – Columbus, O.: Ruggles-Gale Co., 1895. 72,51p. LL-18 – 1 – us L of C Photodup [348]

An account, historical, political, and statistical, of the united provinces of rio de la plata : with an appendix, concerning the usurpation of monte video by the portuguese and brazilian governments / Nunez, Ignacio B – London 1825 [mf ed Hildesheim 1995-98] – 1v on 3mf – 9 – €90.00 – 3-487-26854-X – (trans fr spanish) – gw Olms [972]

Account of a tour in normandy : undertaken chiefly for the purpose of investigating the architectural antiquities of the duchy, with observations on its history, on the country, and on its inhabitants / Turner, Dawson – London 1820 [mf ed Hildesheim 1995-98] – 2v on 5mf – 9 – €100.00 – 3-487-29719-1 – gw Olms [914]

An account of a voyage for the discovery of a north-west passage by hudson's streights... : performed in the years 1746 and 1747, in the ship california / [Drage] – London, 1748-1749. 2v – 14mf – 9 – mf#N-194 – ne IDC [919]

Account of a voyage of discovery to the north-east of siberia, the frozen ocean, and the north-east sea / Sarycev, Gavriil A – London 1806-07 [mf ed Hildesheim 1995-98] – 2v on 2mf – 9 – €60.00 – 3-487-26442-0 – gw Olms [910]

Account of a voyage of discovery to the north-east of siberia, the frozen ocean, and the north-east sea / Sarytschew, G A – London, 1806-1807. 2v – 4mf – 9 – mf#N-381 – ne IDC [915]

An account of a voyage to establish a colony at port philip in bass's strait : on the south coast of new south wales, in his majesty's ship calcutta, in the years 1802-3-4 / Tuckey, James H – London 1805 [mf ed Hildesheim 1995-98] – 1v on 2mf – 9 – €60.00 – 3-487-26799-3 – gw Olms [919]

An account of a voyage to india, china etc in his majesty's ship caroline : performed in the years 1803-4-5; interspersed with descriptive sketches and cursory remarks / Johnson, James – London 1806 [erschienen] 1807 [mf ed Hildesheim 1995-98] – 1v on 1mf – 9 – €40.00 – 3-487-26443-9 – gw Olms [915]

An account of a voyage to new south wales : to which is prefixed a detail of his life, trials, speeches etc / Barrington, George – London 1810 [mf ed Hildesheim 1995-98] – 1v on 3mf – 9 – €90.00 – 3-487-26806-X – gw Olms [919]

Account of a voyage to the western coast of africa : performed by his majesty's sloop favourite, in the year 1805; being a journal of the events which happened to that vessel, from the time of her leaving england till her capture by the french, and the return of the author / Spilsbury, Francis B – London 1807 [mf ed Hildesheim 1995-98] – 1v on 1mf – 9 – €40.00 – 3-487-26437-4 – gw Olms [916]

Account of a west indian sanatorium and a guide to barbados / Moxly, Joseph Henry Sutton – London: S Low, Marston, Searle, & Rivington 1886 [mf ed 1990] – 1r – 1 – us UF Libraries [918]

An account of an embassy to the court of the teshoo lama, in tibet : containing a narrative of a journey through bootan, and part of tibet... / Turner, S – London, 1800 – 10mf – 9 – mf#H-6153 – ne IDC [915]

An account of an embassy to the court of the teshoo lama, in tibet : containing a narrative of a journey through bootan, and part of tibet / Turner, Samuel – London 1800 [mf ed Hildesheim 1995-98] – 1v on 6mf – 9 – €120.00 – 3-487-27217-2 – gw Olms [915]

An account of an embassy to the kingdom of ava : in the year 1795 / Symes, Michael – Edinburgh 1827 [mf ed Hildesheim 1995-98] – 2v on 4mf – 9 – €120.00 – 3-487-27457-4 – gw Olms [959]

Account of an expedition from pittsburgh to the rocky mountains : performed in the years 1819, 1820 / James, E – London, 1823. 3v – 12mf – 9 – mf#H-6158 – ne IDC [917]

Account of an expedition from pittsburgh to the rocky mountains : preformed in the years 1819, 1820; by order of the hon j c calhoun, secretary of war, under the command of major s h long, of the u s top. engineers / James, Edwin – London 1823 [mf ed Hildesheim 1995-98] – 3v on 9mf – 9 – €180.00 – 3-487-27136-2 – gw Olms [917]

Account of an insurrection of the negro slaves in the colony of demarara : which broke out on the 18th of august, 1823 / Bryant, Joshua – Demarara, 1824 – (= ser 19th c books on british colonization) – 2mf – 9 – mf#1.1.7612 – uk Chadwyck [972]

An account of assam / Wade, John Peter; ed by Sharma, Benudhar – Assam: R Sharma, 1927 – (= ser Samp: indian books) – us CRL [954]

Account of bermuda, past and present / Ogilvy, John – Hamilton, Bermuda: S Nelmes 1883 [mf ed 1990] – 1r – 1 – us UF Libraries [972]

An account of jamaica : and its inhabitants / Stewart, John – London 1808 [mf ed Hildesheim 1995-98] – 1v on 2mf – 9 – €60.00 – 3-487-26927-9 – gw Olms [972]

Account of marriages solemnized by tolaver robertson since oct 1842 – South Carolina. 34p – 1 – $5.00 – us Southern Baptist [242]

Account of my travels : through the united states of america and in great britain and other parts of europe in the years 1801-1802-1803 and 1804... / Brisbane, William – 1807 [mf ed Charleston SC, 1981] – 1v on 7mf – 9 – mf#51-016A – us South Carolina Historical [910]

An account of six years residence in hudson's-bay : from 1733 to 1736, and 1744 to 1747: containing a variety of facts, observations and discoveries... / Robson, Joseph – London: printed for T Jeffreys...1759 [mf ed 1984] – 3mf – 9 – 0-665-38701-6 – mf#38701 – cn Canadiana [917]

Account of some recent discoveries in hieroglyphical literature and egyptian antiquities : including the author's original alphabet, as extended by mr champollion, with a translation of five unpublished greek and egyptian manuscripts / Young, Thomas – London: Murray 1823 – (= ser Whsb) – 2mf – 9 – €30.00 – mf#Hu 367 – gw Fischer [930]

An account of south-west barbary : containing what is most remarkable in the territories of the king of fez and morocco / Ockley, Simon – London: J Bowyer: H Clements, 1713 [mf ed 1986] – 3mf – 9 – (in arabic and english) – mf#SEM105P554 – cn Bibl Nat [916]

An account of the abiponesan equestrian people of paraguay / Dobrizhoffer, Martin – London 1822 [mf ed Hildesheim 1995-98] – 3v on 9mf – 9 – €180.00 – 3-487-26843-4 – gw Olms [972]

Account of the abolition of female infanticide in guzerat : with considerations on the question of promoting the gospel in india / Cormack, John – London, 1815 – (= ser 19th c books on british colonization) – 5mf – 9 – mf#1.1.8338 – uk Chadwyck [306]

An account of the american baptist mission to the burman empire : in a series of letters, addressed to a gentleman in london / Judson, Ann H – London 1823 [mf ed Hildesheim 1995-98] – 1v on 2mf – 9 – €60.00 – 3-487-27455-8 – gw Olms [242]

An account of the american baptist mission to the burman empire : in a series of letters, addressed to a gentleman in london / Judson, Ann Hasseltine – London: Printed for J Butterworth & son, 1823 [mf ed 1995] – (= ser Yale coll) – xv/334p – 1 – 0-524-09165-X – mf#1995-0165 – us ATLA [242]

An account of the american baptist mission to the burman empire... / Judson, A H – London, 1823 – 4mf – 9 – mf#SE-20144 – ne IDC [915]

Account of the american church mission in shanghai and the lower yangtse valley / Huntington, M C & Barbour, A G – New York: Domestic and Foreign Missionary Society, Church Missions House, 1900 [mf ed 1995] – (= ser Yale coll) – 55p (ill) – 1 – 0-524-10250-3 – mf#1996-1250 – us ATLA [242]

Account of the american church mission in shanghai and the lower yangtse valley / Huntington, M C & Barbour, A G – New York: E & J B Young, 1898 [mf ed 1995] – (= ser Yale coll) – 26p – 1 – 0-524-10169-8 – mf#1995-1169 – us ATLA [240]

Account of the ancient flemish school of painting : translated from his description of the netherlands... / Guicciardini, L – London, 1795 – 1mf – 9 – mf#O-989 – ne IDC [750]

An account of the arctic regions : with a history and description of the northern whale-fishery / Scoresby, William – Edinburgh 1820 [mf ed Hildesheim 1995-98] – 2v on 8mf – 9 – €160.00 – 3-487-27071-4 – gw Olms [990]

An account of the battle of chateauguay : being a lecture delivered at ormstown, march 8th, 1889 / Lighthall, William Douw – Montreal: W Drysdale & Co, 1889 [mf ed 1981] – 1mf – 9 – 0-665-13997-7 – (with some local and personal notes by w patterson) – mf#13997 – cn Canadiana [355]

An account of the canadian protest : against the introduction into canada of musical examinations by outside musical examining bodies / ed by Canadian Protesting Committee – Toronto: The Committee, 1899 [mf ed 1980] – 1mf – 9 – 0-665-00443-5 – mf#00443 – cn Canadiana [780]

An account of the captivity of elizabeth hanson, late of kachecky in new-england : who, with four of her children, and servant-maid, was taken captive by the indians, and carried into canada... / Hanson, Elizabeth – new ed. London: printed & sold by James Phillips...1782 [mf ed 1984] – 1mf – 9 – 0-665-44949-6 – mf#44949 – cn Canadiana [978]

An account of the churches in rhode-island : presented...28th annual meeting of the rhode-island baptist state convention, providence... / Jackson, Henry – Providence: GH Whitney, 1854 [mf ed 1993] – (= ser Baptist coll) – 1mf – 9 – 0-524-06472-5 – mf#1990-5246 – us ATLA [242]

An account of the colony of van diemen's land principally designed for the use of emigrants / Curr, Edward – London 1824 [mf ed Hildesheim 1995-98] – 1v on 2mf – 9 – €60.00 – 3-487-26794-2 – gw Olms [980]

An account of the commencement, and present state of the capuchin mission in tibet : and two other neighbouring kingdoms in the year 1741 / Penna di Billi, F O della – London, 1745-1747. v4 – 1mf – 9 – mf#A-271 – ne IDC [915]

An account of the conquest of guatemala in 1524 / Alvarado, Pedro de; ed by Mackie, Sedley J – New York: The Cortes Society, 1924 – 1 – sp Bibl Santa Ana [972]

Account of the county of cumberland / Denton, John – 17th c – (= ser Holkham library manuscript books 760) – 1r – 1 – (with ind) – mf#97107 – uk Microform Academic [914]

An account of the cultivation and manufacture of tea in china : derived from personal observation during an official residence in that country from 1804 to 1826 / Ball, Samuel – London: printed for Longman, Brown, Green, & Longmans, 1848 – (= ser 19th c books on china) – 5mf – 9 – mf#7.1.6 – uk Chadwyck [630]

An account of the danes and norwegians in england, scotland, and ireland / Worsaae, Jens Jakob Asmussen – London: J Murray 1852 [mf ed 1991] – 1r [ill] – 1 – (with numerous wood-cuts. filmed with: die sending der frau in der deutschen geschichte / kohler, r) – mf#1272 – us Wisconsin U Libr [941]

An account of the different existing systems of sanskrit grammar : being the vishwanath narayan mandlik gold medal prize-essay for 1909 / Belvalkar, Shripad Krishna – Bombay: University of Bombay, 1915 – (= ser Samp: indian books) – us CRL [490]

An account of the discoveries of the portuguese in the interior of angola and mozambique : from original manuscripts: to which is added a note by the author, on a geographical error of mungo park in his last journal into the interior of africa / Bowditch, Thomas Edward – London: J Booth, 1824 – 1 – us CRL [916]

An account of the discoveries of the portuguese in the interior of angola and mozambique : to which is added a note by the author, on a geographical error of mungo park, in his last journal into the interior of africa / Bowdich, Thomas E – London 1824 [mf ed Hildesheim 1995-98] – 1v on 2mf – 9 – €60.00 – 3-487-27231-8 – gw Olms [960]

An account of the empire of china / Navarette, D F – London, 1704. v1 – 10mf – 9 – mf#HT-672 – ne IDC [915]

An account of the empire of china... / Escalante, B de – London, 1745. v2 – 2mf – 9 – mf#HT-676 – ne IDC [915]

An account of the english colony in new south wales : from its first settlement in jan 1788 to aug 1801... / Collins, David – London, 1804 – (= ser 19th c books on british colonization) – 7mf – 9 – mf#1.1.2436 – uk Chadwyck [980]

An account of the english colony in new south wales : with remarks on the dispositions, customs, manners & c of the native inhabitants of that country. to which are added, some particulars of new zealand... / Collins, D – London, 1798-1802. 2v – 12mf – 9 – mf#H-6165 – ne IDC [917]

An account of the excavations at tell atchana / Woolley, Leonard – Oxford, 1955 – 9 – $18.00 – us IRC [930]

An account of the expenditure of the office and establishment of lord gosford, as governor general and commissioner in canada, for one year – [London, England: s.n., 1838] (mf ed 1991) – 1mf – 9 – mf#SEM105P1391 – cn Bibl Nat [336]

An account of the facts which appeared on the late enquiry into the loss of minorca : from authentic papers / The Monitor – London: Printed for J Scott...1757 – 2mf – 9 – (the monitor was a wkly newspaper, publ in london, england) – mf#20235 – cn Canadiana [946]

An account of the gold coast of africa : from the royal commonwealth society library / Meredith, H – London, 1812 – 18mf – 7 – mf#2989 – uk Microform Academic [330]

An account of the gold coast of africa : with brief history of the african company / Meredith, Henry – London 1812 [mf ed Hildesheim 1995-98] – 1v on 2mf – 9 – €60.00 – 3-487-27285-7 – gw Olms [960]

An account of the history and manufacture of...terra cotta / Blashfield, John Marriott – London 1855 – (= ser 19th c art & architecture) – 2mf – 9 – mf#4.2.266 – uk Chadwyck [730]

An account of the island of ceylon : containing its history, geography, natural history...to which is added, the journal of an embassy to the court of candy / Percival, R – London, 1803 – 8mf – 9 – mf#Z-287 – ne IDC [915]

An account of the island of jersey : containing a compendium of its ecclesiastical, civil, and military history / Plees, W – Southampton 1817 [mf ed Hildesheim 1995-98] – 1v on 3mf – 9 – €90.00 – 3-487-27940-1 – gw Olms [941]

An account of the island of newfoundland : with the nature of its trade and method of carrying on the fishery, with reasons for the great decrease of that most valuable branch of trade / Williams, Griffith – S.l: Printed for Capt Thomas Cole and sold by W Owen...1765 – 1mf – 9 – mf#18730 – cn Canadiana [639]

An account of the journey of the peres boures, fontenay, gerbillon, le comte, and vesdelou : from the port of ning po to peking / Halde, J B du – London, 1741. 1v – 2mf – 9 – mf#HT-510 – ne IDC [915]

An account of the kingdom of caubul : and its dependencies in persia, tartary, and india comprising a view of the afghaun nation, and a history of the dooraunee monarchy / Elphinstone, Mountstuart – London 1815 [mf ed Hildesheim 1995-98] – 1v on 8mf – 9 – €160.00 – 3-487-27260-1 – gw Olms [956]

An account of the kingdom of caubul : and its dependencies, in persia, tartary, and india; comprising a view of the afghaun nation and a history of the dooraunee monarchy / Elphinstone, Mountstuart – new rev ed. London: Richard Bentley, 1842 [mf ed 1995] – (= ser Yale coll) – 2v (ill) – 1 – 0-524-09441-1 – mf#1995-0441 – us ATLA [915]

An account of the kingdom of nepal : and of the territories annexed to this dominion by the house of gorkha / Buchanan, Francis – Edinburgh 1819 [mf ed Hildesheim 1995-98] – 1v on 4mf – 9 – €120.00 – 3-487-27256-3 – gw Olms [954]

An account of the kingdom of nepaul : being the substance of observations made during a mission to that country, in the year 1793 / Kirkpatrick, William – London 1811 [mf ed Hildesheim 1995-98] – 1v on 5mf – 9 – €100.00 – 3-487-27254-7 – gw Olms [954]

An account of the last battle of panipat and of the events leading to it / Kashiraj – London, New York: Oxford University Press, 1926 – (= ser Samp: indian books) – us CRL [954]

Account of the life and writings of james bruce : author of travels to discover the source of the nile, in the years 1768-1773 / Murray, Alexander – Edinburgh: A Constable & Co [etc], 1808 – us CRL [916]

An account of the life and writings of s irenaeus, bishop of lyons and martyr : intended to illustrate the doctrine, discipline, practices...of the gnostic heretics, during the 2nd century / Beaven, James – London: Rivington, 1841 [mf ed 1990] – 1mf – 9 – 0-7905-4843-7 – mf#1988-0843 – us ATLA [240]

An account of the life, character etc of the rev samuel parris, of salem village : and of his connection with the witchcraft delusion of 1692 / Fowler, Samuel Page – Salem: W Ives & G W Pease, Printers, 1857 [mf ed 1991] – (= ser Congregational coll) – 1mf – 9 – 0-524-01330-6 – mf#1990-4079 – us ATLA [242]

An account of the life of dr william augustus carleton : apostle to the west / Hayatt, Alice Nelson – 1905-80 – 1 – $5.00 – us Southern Baptist [242]

An account of the lives and works of the most eminent spanish painters, sculptors and architects... / Palomino, [A A] – London, 1739 – 2mf – 9 – mf#O-1060 – ne IDC [700]

An account of the loss of the wesleyan missionaries : messrs white, hillier, truscott, oke, and jones...in the maria mail boat, off the island of antigua, in the west indies, feb 28 1826 / Jones [Mrs] – New-York: G Lane & PP Sandford, 1841 [mf ed 1991] – (= ser Methodist coll) – 1mf – 9 – 0-524-01937-1 – mf#1990-4161 – us ATLA [242]

An account of the manners and customs of the modern egyptian : switten in egypt during the years 1833, 34, and 35, partly from notes made during a former visit to that country in the years 1825, 26, 27, and 28 / Lane, Edward W – London 1836-37 [mf ed Hildesheim 1995-98] – 2v on 6mf – 9 – €120.00 – 3-487-27367-5 – gw Olms [960]

Account of the musical performances in westminster abbey and the pantheon, may 26th, 27th, 29th : and june the 3d and 5th, 1784. in commemoration of handel / Burney, Charles – London: printed for the benefit of the Musical Fund & sold by T Payne 1785 [mf ed 19–] – 6mf – 9 – mf#fiche 380 – us Sibley [780]

An account of the native africans in the neighbourhood of sierra leone... / Winterbottom, T – London, 1803. 2v – 14mf – 9 – mf#A-357 – ne IDC [916]

An account of the natives of the tonga islands, in the south pacific ocean : with an original grammar and vocabulary of their language / Mariner, William – London 1817 [mf ed Hildesheim 1995-98] – 2v on 7mf – 9 – €140.00 – 3-487-26770-5 – gw Olms [919]

An account of the pelew islands / Keate, George – Basil [ie Paris] 1789 [mf ed Hildesheim 1995-98] – 1v on 3mf – 9 – €90.00 – 3-487-27444-2 – gw Olms [919]

An account of the pelew islands : situated in the western parts of the pacific ocean / Wilson, H – London, 1789 – 5mf – 9 – mf#H-6171 – ne IDC [919]

An account of the principalities of wallachia and moldavia : with various political observations relating to them / Wilkinson, William – London 1820 [mf ed Hildesheim 1995-98] – (= ser Fbc) – 2mf – 9 – €60.00 – 3-487-29087-1 – gw Olms [947]

An account of the printed text of the greek new testament : with remarks on its revision upon critical principles / Tregelles, Samuel Prideaux – London: Samuel Bagster, 1854 [mf ed 1988] – 1mf – 9 – 0-7905-0404-9 – (incl bibl ref & ind) – mf#1987-0404 – us ATLA [225]

Account of the proceedings and doings of the government commissioners : against the unfortunate settlers upon the indian lands in the townships of tuscarora and oneida, in the years of our lord 1846 and 1847 / Cheshire, F J – [Hamilton, Ont? s.n.] 1847 [mf ed 1983] – 1mf – 9 – 0-665-44280-7 – mf#44280 – cn Canadiana [307]

Account of the proceedings of h m s enterprise from behring strait to cambridge bay / Collinson, R – London, 1855. v25 – 1mf – 9 – mf#N-167 – ne IDC [910]

An account of the remarkable musical talents of several members of the wesley family / Winters, William – London: F Davis, 1874 [mf ed 1990] – 1mf – 9 – 0-7905-7201-X – (incl bibl ref) – mf#1988-3201 – us ATLA [780]

An account of the rise and progress of mahometanism : with the life of mahomet and a vindication of him and his religion from the calumnies of the christians / Stubbe, Henry; ed by Shairani, Mahmud Khan – London: Luzac, 1911 [mf ed 1991] – 1mf – 9 – 0-524-01931-2 – (int & app by ed) – mf#1990-2744 – us ATLA [260]

Account of the russian discoveries between asia and america : to which are added, the conquest of siberia, and the history of the transactions and commerce between russia and china / Coxe, William – London, 1787 – 10mf – 9 – mf#N-177 – ne IDC [915]

Account of the russian discoveries between asia and america microform : to which are added, the conquest of siberia, and the history of the transactions and commerce between russia and china / Coxe, William – 4th enl ed. London: Cadell & Davies 1803 [mf ed 1980] – 1r [ill] – 1 – (filmed with: hunting and hunted in the belgian congo / cooper, r d) – mf#8686 – us Wisconsin U Libr [910]

An account of the transactions of his majesty's mission to the court of persia : in the years 1807-11 / Brydges, Harford J – London 1834 [mf ed Hildesheim 1995-98] – 2v on 6mf – 9 – €120.00 – 3-487-27584-8 – gw Olms [956]

An account of the united states of america : derived from actual observation, during a residence of four years in that republic: including original communications / Holmes, Isaac – London [1823] [mf ed Hildesheim 1995-98] – 1v on 3mf – 9 – €90.00 – 3-487-27167-2 – gw Olms [975]

An account of the voyages by the order of his present majesty for making discoveries in the southern hemisphere... / Hawkesworth, J – London, 1773. 3v – 18mf – 9 – mf#H-6109 – ne IDC [910]

An account of the voyages undertaken by the order of his present majesty for making discoveries in the southern hemisphere : and successively performed by commodore byron, captain carteret, captain wallis... / Hawkesworth, John – London 1785 [mf ed Hildesheim 1995-98] – 4v on 12mf – 9 – €120.00 – 3-487-26661-X – gw Olms [910]

An account of the work of the north india mission of the presbyterian church of america for the year 1906-1907 / ed by Forman, Henry – Ajmer: Scottish Mission Industries, [1907?] [mf ed 1993] – (= ser Presbyterian coll) – 1mf – 9 – 0-524-07239-6 – mf#1991-2980 – us ATLA [242]

An account of timbuctoo : from the royal commonwealth society library / Shabeeny, H A S – 1820 – 16mf – 7 – mf#3032 – uk Microform Academic [916]

An account of timbuctoo and housa : territories in the interior of africa... / Jackson, J G – London, 1820 – 10mf – 9 – mf#A-143 – ne IDC [916]

An account of timbuctoo and housa : territories in the interior of africa / Jackson, James G – London 1820 [mf ed Hildesheim 1995-98] – 1v on 4mf – 9 – €120.00 – 3-487-27301-2 – (with crit and expl notes) – gw Olms [960]

An account of travels into the interior of southern africa, in the years 1797 and 1798... / Barrow, J – London, 1801 – 5mf – 9 – mf#HT-6 – ne IDC [916]

An account of tunis : of its government, manners, customs, and antiquities; especially of its productions, manufactures, and commerce / MacGill, Thomas – London 1816 [mf ed Hildesheim 1995-98] – 1v on 2mf – 9 – €60.00 – 3-487-27334-9 – gw Olms [960]

An account of various silver and copper medals : presented to the north american indians by the sovereigns of england, france, and spain, from 1600 to 1800 and especially of five such medals of george 1 of great britain / Hayden, Horace Edwin – Wilkes-Barre, PA: s.n, 1886 – 1mf – 9 – mf#51811 – cn Canadiana [730]

An account showing the amount of collections and disbursements of the new england conference,for the year ending june 12 1823 / Methodist Episcopal Church – Providence: Miller 1823 – 1r – 1 – $35.00 – mf#um-15 – us Commission [242]

Accountability and ethics in the south african public administration / Sindane, Abakholwa Moses – Pretoria: Vista University 2002 [mf ed 2002] – 4mf – 9 – (incl bibl ref) – mf#mfm15209 – sa Unisa [350]

Accountancy – London, England 1928+ – 1,5,9 – ISSN: 0001-4664 – mf#2162 – us UMI ProQuest [650]

Accountancy, 1958-84 : the journal of the institute of chartered accountants – 35r – 1 – mf#95984 – uk Microform Academic [650]

Accountancy age – London, England 1975-89 – 1,5,9 – ISSN: 0001-4672 – mf#8880 – us UMI ProQuest [650]

Accountancy ireland – Dublin, Ireland 1975+ – 1,5,9 – ISSN: 0001-4699 – mf#8595 – us UMI ProQuest [650]

Accountant – London, England 1975+ – 1,5,9 – ISSN: 0001-4710 – mf#10491 – us UMI ProQuest [650]

Accountants digest – Boca Raton FL 1973-88 – ISSN: 0001-4737 – mf#10202 – us UMI ProQuest [650]

Accountant's magazine – Edinburgh, Scotland 1976-93 – 1,5,9 – (cont by: ca magazine) – ISSN: 0001-4761 – mf#11283.01 – us UMI ProQuest [650]

Accountant's magazine : edinburgh - London. v1-87. 1897-1984 – 1 – $3000.00 – us Alper [330]

Accountants review *see* Commercial accountant, the.../accountants review, 1947-77

Accounting / Japan Society of Accounting – v1 n1-v55 n2. 1917-44 – 324 iss on 54r – 1 – Y400,000 – (with 228p guide. in japanese) – ja Yushodo [650]

Accounting and business research – London, England 1970+ – 1,5,9 – ISSN: 0001-4788 – mf#9976 – us UMI ProQuest [650]

Accounting and finance – Oxford, England 1986+ – 1,5,9 – ISSN: 0810-5391 – mf#15002,02 – us UMI ProQuest [650]

Accounting and the public interest – Sarasota FL 2000+ – 1,5,9 – ISSN: 1530-9320 – mf#32310 – us UMI ProQuest [650]

Accounting department management report – New York NY 2002+ – 1,5,9 – ISSN: 1541-111X – mf#22803,01 – us UMI ProQuest [650]

Accounting forum – New York NY 1974-86 – 1,5,9 – ISSN: 0001-4818 – mf#10279 – us UMI ProQuest [650]

Accounting historians journal – University MS 1991+ – 1,5,9 – ISSN: 0148-4184 – mf#18978,01 – us UMI ProQuest [650]

Accounting horizons – Sarasota FL 1987+ – 1,5,9 – ISSN: 0888-7993 – mf#16393 – us UMI ProQuest [650]

Accounting, management and information technologies – New York NY 1991-2000 – 1,5,9 – ISSN: 0959-8022 – mf#49616.01 – us UMI ProQuest [650]

Accounting, organizations and society – Oxford, England 1976+ – 1,5,9 – ISSN: 0361-3682 – mf#49090 – us UMI ProQuest [650]

Accounting research – London, 1948-98+ – 21r – 1 – £970.00 – uk World [650]

Accounting review – Sarasota FL 1926+ – 1,5,9 – ISSN: 0001-4826 – mf#8788 – us UMI ProQuest [650]

Accounting series releases / U.S. Securities and Exchange Commission – n1-195. 6 jan 1937-18 jan 1973 (all publ) – (= ser The sec release series preceding the sec docket) – 6mf – 9 – $9.00 – mf#LLMC 84-359 – us LLMC [346]

Accounting technology – New York NY 1994+ – 1,5,9 – (cont: computers in accounting) – ISSN: 1068-6452 – mf#16340,01 – us UMI ProQuest [000]

Accounting today – New York NY 1991+ – 1,5,9 – ISSN: 1044-5714 – mf#18748 – us UMI ProQuest [650]

Accounts / Pickens, Francis Wilkinson – [mf ed Spartanburg SC: Reprint Co [1981?]] – 1mf – 9 – mf#51-122 – us South Carolina Historical [332]

Accounts and claims : settled by the second auditor of the treasury department relating to the arsenal at harper's ferry, 1817-1851 / U.S. Treasury Dept – (= ser Records of the accounting officers of the department of the treasury) – 62r – 1 – mf#M1678 – us Nat Archives [336]

Accounts and expenses of the households of henry 6, the 3rd earl of stafford, the 3rd duke of buckingham, edward duke of buckingham, william malvern and francis devereux in the 15th, 16th and 17th centuries – (= ser Archives of the marquess of bath, longleat house, warminster, wiltshire) – 1r – 1 – mf#96778 – uk Microform Academic [640]

Accounts and expenses of the households of the earl of warwick, (1420-1), the duke of richmond (1527-8), the duke of buckingham (1506-7) and edward seymour (1538-41) in the 15th and 16th centuries – (= ser Archives of the marquess of bath, longleat house, warminster, wiltshire) – 3r – 1 – mf#96779 – uk Microform Academic [640]

Accounts and papers relating to the building kedleston hall, derbyshire, c1758-70 : from the archives of viscount scarsdale, kedleston hall – 2r – 1 – (int by lethe harris) – mf#96845 – uk Microform Academic [640]

Accounts audited of claims growing out of the revolution in south carolina / South Carolina. Dept of Archives and History – 165r – 1 – $75.00r – Out-of-state orders: us Scholarly Res – us South C Archives [975]

Accounts of all monies paid and payable by the canada company : under the existing contracts for the sale to them of part of the crown reserves and other lands in upper canada – [London, England: s.n, 1831] (mf ed 1996) – 1mf – 9 – mf#SEM105P2740 – cn Bibl Nat [971]

Accounts of british trade in america / Blathwayt, William – 1682-1714 – 1 – us L of C Photodup [975]

Accounts of chemical research – [Easton, PA]: ACS. v1(1968)-v22(1989) [mthly] – 1,5,6,9 – mf#0001-4842 – us ACS [540]

Accounts of religious revivals in many parts of the united states from 1815 to 1818 / Bradley, Joshua [comp] – Albany: GJ Loomis, 1819 [mf ed 1992] – 1mf – 9 – 0-524-03811-2 – mf#1990-1127 – us ATLA [240]

Accounts of the kitchen gardens, woods and plantations, water boats etc, 1743-1759 *see* Catalogue of the manuscripts and some early printed books in the library at holkham

Accounts of the net revenue and expenditure of the province of lower canada : for the years 1825, 1826, and 1827 – London: Printed by William Clowes, 1828 [mf ed 1984] – 1mf – 9 – mf#SEM105P397 – cn Bibl Nat [336]

Accounts of the works of art bought in rome by matthew brettingham 1747 – (= ser Holkham library, the house, park and art colls 744) – 1r – 1 – mf#771 – uk Microform Academic [700]

Accounts relating to the foreign trade and navigation of india / India. Dept. of Commercial Intelligence and Statistics – Delhi. 1869 70-1908. (Scattered issues wanting) – 1 – us L of C Photodup [954]

Accounts relating to trade and navigation 1847/48-1964 – [mf ed Chadwyck-Healey] – (= ser British government publications...1801-1977) – 125r – 1 – uk Chadwyck [337]

Accoustical phenomena as they relate to the performance and manufacture of the modern valve trumpet / Grocock, Robert – U of Rochester 1950 [mf ed 19–] – 1r – 1 – mf#film 892 – us Sibley [780]

Accra. Gold Coast Public Relations Dept *see* Achievement in the gold coast

Accrington express and weekly courier of coming events – [NW England] Accrington Lib 31 jul 1903-23 dec 1904 – 1 – uk MLA; uk Newsplan [072]

Accrington free press – [NW England] Accrington Lib 1858-oct 1860 – 1 – uk MLA; uk Newsplan [072]

Accrington gazette and north east lancashire observer – [NW England] Accrington Lib 15 jan 1881-22 feb 1890 – 1 – (title change: accrington division gazette [mar 1890-19 feb 1916]; accrington gazette [26 feb 1916-24 feb 1927]) – uk MLA; uk Newsplan [072]

Accrington guardian – [NW England] Accrington Lib 12 jan 1861-11 apr 1863 – 1 – uk MLA; uk Newsplan [072]

Accrington guardian and church and oswaldtwistle times – [NW England] Accrington Lib oct 1879-dec 1880 – 1 – uk MLA; uk Newsplan [072]

Accrington herald – [NW England] Accrington Lib 22, 29 jan 1870 – 1 – uk MLA; uk Newsplan [072]

Accrington observer – [NW England] Accrington Lib 1887-9 apr 1892 – 1 – (title change: accrington observer & times [16 apr 1892-]) – uk MLA; uk Newsplan [072]

Accrington reporter – [NW England] Accrington Lib 2 may-26 dec 1868 – 1 – uk MLA; uk Newsplan [072]

Accrington star – [NW England] Accrington Lib 14 nov 1894-24 jan 1895 – 1 – uk MLA; uk Newsplan [072]

Accrington times – [NW England] Accrington Lib 29 dec 1866-1891 – 1 – uk MLA; uk Newsplan [072]

Accrington weekly advertiser – [North West] Accrington Lib 24 may 1889-31 may 1890 [incomplete] – 1 – (title change: accrington advertiser [7 jun 1890-8 feb 1896]; weekly advertiser [15 feb-dec 1896]; accrington advertiser [4-11 jan 1898]; [22 mar 1898-2 nov 1899]; northern morning news & accrington advertiser [14 jan-11 mar 1898]; accrington advertiser & northern morning news [24 nov 1899]; advertiser & northern morning news [28 nov 1899-19 may 1905]; accrington advertiser & northern morning news [26 may-dec 1905, 1911, 1914, 1915]) – uk MLA; uk Newsplan [072]

Acculturation antecedents and outcomes associated with international and domestic student-athlete adjustment to college / Ridinger, Lynn L – 1998 – 3mf – 9 – $12.00 – mf#PSY 2085 – us Kinesology [305]

Accumulated oxygen deficit among highly conditioned female rowers during a 2,000 meter race simulation / Pripstein, Laura – 1997 – 1mf – 9 – $4.00 – mf#PH 1604 – us Kinesology [612]

Accuracy – v1 n1-4-v3 n1-4 [1978-80] – 1 – mf#642471 – us WHS [071]

Accuracy of a treadmill scoring system for prediction of coronary artery disease in female subjects / Sheehan, Laurieanne & Sanborn, Charlotte F – 1991 – 1mf – 9 – $4.00 – us Kinesology [612]

The accuracy of heart rate as an indicator of metabolic rate while performing step aerobics / Hartman, Greta B – University of North Carolina at Chapel Hill, 1995 – 1mf – 9 – $4.00 – mf#PH1463 – us Kinesology [612]

Accuracy of individual scores expressed in percentile ranks : classical test theory calculations / Rogosa, David Roth – Los Angeles CA: Center for the Study of Evaluation...Uni of California, Los Angeles; [Washington DC]: US Dept of Education, Office of Educ Research & Improvement...[2000] [mf ed 2000] – 1mf – 9 – us GPO [370]

Accuracy of perceived heaviness and perceived joint placement in normal and injured shoulder joints / Spoerl, J J – 1991 – 1mf – 9 – $4.00 – us Kinesology [612]

The accuracy of various indirect determinations of body composition : comparison with a multicomponent criterion model / Wegner, Michael S – Oregon State University, 1995 – 2mf – 9 – $8.00 – mf#PE 3677 – us Kinesology [617]

Accuracy of year-1, year-2 comparisons using individual percentile rank scores : classical test theory calculations / Rogosa, David Roth – Los Angeles CA: Center for the Study of Evaluation...Uni of California, Los Angeles; [Washington DC]: US Dept of Education, Office of Educ Research & Improvement...[2000] [mf ed 2000] – 1mf – 9 – us GPO [370]

L'accusateur public / ed by Esquiros, Alphonse – Paris: Impr de Lacour [jun 11/14-21/25 1848] (semiwkly) – (= ser French revolution of 1848. newspapers) – 1r – 1 – us CRL [074]

L'accusateur revolutionnaire : journal des ouvriers / ed by Douhet-Rathail – Paris: A Rene [apr 2 1848] (wkly) – (= ser French revolution of 1848. newspapers) – 1r – 1 – us CRL [074]

Accusations against bulgaria : official documents presented to the peace conference / Paris Peace Conference (1919-1920). Bulgarian Delegation – [S.l: s.n, 1919] [mf ed 2000] – 1r – 1 – mf#29094 – us Harvard [933]

Ace (miami, fl) – Miami FL: Agrupacion de Caimitenses Exiliados 1985- (1985 mar-1989 jan) (irreg) [mf ed 1999] – (= ser Cuban heritage collection) – 1r – 1 – us UF Libraries [071]

Ace news / Licking Co. Heath – aug 1977-jun 1984 [wkly] – 3r – 1 – mf#B29491-29493 – us Ohio Hist [071]

Acebal, Sergio *see* Historia de un homre insignificante

Acedo, Federico *see*
- Correspondiente en caceres de la r.a. de la historia
- De los nombres atribuidos a trujillo
- Fallecimiento en caceres
- Guia de trujillo

Acemel, Isidoro *see* Partida de bautismo del p. andres de guadalupe

Acena Duran, Ramon *see* Itinerario

Acephale – Paris. n1.24 juin 1936; no. double. 21 janv 1937; n 3-4. juil 1937; no. 5. juin 1939; n.s., n1. 1938 – 1 – fr ACRPP [073]

Acerbi, G *see* Travels through sweden, finland, and lapland, to the north cape, in the years 1798 and 1799

Acerbi, Giuseppe *see* Voyage au cap-nord, par la suede, la finlande et la laponie

Acerca de lo que necesita villaclara para ser ciudad modelo sugestiones a un comite / Vidaurreta, Antonio Julio – Santa Clara, Cuba: Ediciones Culturales "Publicidad" 1943 [mf ed 1991] – 1r – 1 – (aka: lo que necesita villaclara para ser ciudad modelo; first publ in la publicidad, santa clara, may 31 1943) – us UF Libraries [972]

Acerrellos, R S *see* Die freimaurerei in ihrem zusammenhang

L'acetylene – Paris, France 1 jun 1901-20 jun 1903 – 1r – 1 – (cont by: revue generale de l'acetylene [12 jul 1903-25 dec 1908]) – uk British Libr Newspaper [074]

Aceuchal *see*
- Fiestas patronales de nuestra senora la santisima virgen de la soledad
- Museo taurino de mahizflor. catalogo guia 1950

Acevedo, Alfonso de *see*
- Commentarii juris civilis.
- Commentariorum continuatio ad leges regias.
- Commentariorum iuris civilis
- Commentariorum iuris civilis, tomus quintus
- Commentariorum iuris civilis, tomus secundus
- Commentariorum iuris civilis, tomus sextus
- Commentarium juris civilis...salmantical, didacuscusio, 1591
- Consilia...perfecta...per johannem de acevedo
- Opera doctoris...in his paniae regias constitutiones
- Tractatus de curia pisana..

Acevedo, Alonso de *see* Creacion del mundo

Acevedo, E O *see* Changes in cognitive appraisals and metabolic indices of physical exertion during at two-hour run

Acevedo Latorre, Eduardo *see* Colaboradores de santander en la organizacion de l...

Acevedo Y Laborde, Rene *see* Menores e incapacitados

The ach index of nutritional status / Franzen, Raymond H & Palmer, G T – 1934 – 1mf – 9 – $3.00 – us Kinesology [612]

Acha action / American College Health Association – Rockville MD 1977-81 – 1,5,9 – ISSN: 0002-7952 – mf#11548 – us UMI ProQuest [366]

Die achaemenideninschriften zweiter art / ed by Weissbach, Franz Heinrich – Leipzig: J C Hinrichs, 1890 – 1mf – 9 – 0-8370-7749-4 – (text in german and elamite; commentary in german) – mf#1986-1749 – us ATLA [470]

Achaintre, Nicolas L *see* Histoire genealogique et chronologique de la maison royale de bourbon

Achard, Claude Francois *see* Dictionare de la provence et du comte-venaissin

Achard, Marcel *see* Malborough s'en va-t-en guerre

Achard, Micheline *see* La petite souris grise suivi de, nicolas va a la chasse

Achard, Paul *see*
- Celestine

Acharya, Ananda *see* Cakrasakha

Acharya, Prasanna Kumar *see*
- A dictionary of hindu architecture
- Elements of hindu culture and sanskrit civilization
- Glories of india on indian culture and civilization

Ach-chiheb – Constantine. 1925-27; 1929-aout 1939 – 1 – fr ACRPP [073]

Acheen, and the ports on the north and east coasts of sumatra / Anderson, J – London, 1840 – 3mf – 9 – mf#SE-20210 – ne IDC [915]

Achelis, E Chr *see* Lehrbuch der praktischen theologie

Achelis, Ernst Christian *see*
- Die bergpredigt nach matthaeus und lucas
- Der dekalog als katechetisches lehrstuck
- Die entstehungszeit von luther's geistlichen liedern
- Lehrbuch der praktischen theologie
- Zur symbolfrage

Achelis, Hans *see*
- Acta ss nerei et achillei
- Die aeltesten quellen des orientalischen kirchenrechtes
- Die canones hippolyti
- Das christentum in den ersten drei jahrhunderten
- Exegetische und homiletische schriften
- Hippolytsstudien
- Hippolytstudien
- Die martyrologien
- Das symbol des fisches und die fischdenkmaeler der roemischen katakomben
- Die syrische subintroductae
- Virgines subintroductae

Achelis, Johannes *see* Der religionsgeschlichtliche gehalt der psalmen

Achelis, Thomas *see*
- Abriss der vergleichenden religionswissenschaft
- Die entwicklung der ehe
- Grundzuege der lyrik goethes
- Ueber mythologie und cultus von hawaii

Acher- und buehler bote *see* Acher-bote

Acher-bote – Buehl, Achern, Karlsruhe DE, 1988- – 9r/yr – 1 – (title varies: 13 jun 1896: buehler bote (im wechsel mit acher-bote), 2 apr 1899: acher- und buehler bote, 1 jan 1936: mittelbadischer bote, 29 oct 1949: acher- und buehler bote. regional ed of badische neueste nachrichten, karlsruhe) – gw Misc Inst [074]

Acherland : e psalm / Wueest, Josef – Luzern: E Haag 1928 [mf ed 1991] – 1r – 1 – (poems in swiss-german. filmed with: volk, ich breche deine kohle! / otto wohlgemuth) – mf#2964p – us Wisconsin U Libr [810]

Achermann, Franz Heinrich *see* Daemonentaenzer der urzeit

Acherner zeitung – Offenburg DE, 1951 3 jan-1964 17 apr, 1965 20 feb-1979 30 sep – 55r – 1 – gw Misc Inst [074]

Acher-rench-zeitung – Oberkirch DE, 1983 1 jun- – ca 10r/yr – 1 – gw Misc Inst [074]

Achery, Luc d' *see*
- Acta sanctorum o s b
- Veterum aliquot scriptorum

Acheson, George *see* Biological study of the tap water in the school of practical science, toronto

L'acheteuse : piece en trois actes / Passeur, Steve – Paris, France: Librairie Gallimard 1930 [mf ed 1994] – 1r – 1 – us UF Libraries [820]

Achievement – Sevenoaks, England 1975+ – ISSN: 0001-4907 – mf#1399 – us UMI ProQuest [337]

Achievement in the gold coast : aspects of development in a british west african territory / Accra. Gold Coast Public Relations Dept – Accra, 1951 – us CRL [960]

Achievement motivation among anglo-american and hawaiian physical-activity participants : individual differences and social contextual factors / Hayashi, Carl T – 1994 – 3mf – $12.00 – us Kinesology [150]

Achieving high quality reading and writing in an urban middle school : the case of gail slatko / Manning, Tanya – Albany NY: National Research Center on English Learning & Achievement; [Washington DC]: US Dept of Education, Office of Educ Research & Improvement...[2000] [mf ed 2000] – 1mf – 9 – us GPO [373]

Achill missionary herald and western witness – Achill Island, Ireland 31 jul 1837-8 jun 1869 – 1 – uk British Libr Newspaper [072]

Achille : [dramma per musica in due atti di gamerra. musica] / Paer, Ferdinando – [n.d. 180-?] [mf ed 19–] – 2v on 1r – 1 – mf#pres. film 19 – us Sibley [780]

Achille et polixene : tragedie dont le prologue et les quatre derniers actes ont este mis en musique par p. collasse...et le premier acte par feu mre j b lully... / Collasse, Pascal – Paris: C Ballard 1687 [mf ed 19–] – 1r – 1 – (prologue & last 4 acts by collasse; 1st act by j b lully) – mf#film 1975 – us Sibley [780]

Achilleis *see* Bucolica...

Achilles, Alexander et al *see* Protokolle der kommission fuer die zweite lesung des entwurfs des buergerlichen gesetzbuchs, im auftrage des reich-justizamts

Achilles, Paula *see* Brasil de oeste

Achilli, Giacinto *see* Dealings with the inquisition

Achim von arnim / Seidel, Ina – Stuttgart: J G Cotta, c1944 [mf ed 1988] – (= ser Die dichter der deutschen) – 95p – 1 – mf#6956 – us Wisconsin U Libr [430]

Achim von arnim und bettina brentano / ed by Steig, Reinhold – Stuttgart: J G Cotta, 1913 [mf ed 1993] – (= ser Achim von arnim und die ihm nahe standen 2) – 1 – (incl bibl ref & ind) – mf#8467 – us Wisconsin U Libr [430]

Achim von arnim und clemens brentano / ed by Steig, Reinhold – Stuttgart: J G Cotta, 1894 [mf ed 1993] – (= ser Achim von arnim und die ihm nahe standen 1) – 1 – (incl bibl ref & ind) – mf#8467 – us Wisconsin U Libr [920]

Achim von arnim und die ihm nahe standen / ed by Steig, Reinhold & Grimm, Herman Friedrich – Stuttgart: J G Cotta, 1894-1913 [mf ed 1993] – 3v on 1r – 1 – (incl bibl ref & ind. filmed with: achim von arnim und clemens brentano & other titles) – mf#3417p – us Wisconsin U Libr [920]

Achim von arnim und jacob und wilhelm grimm / ed by Steig, Reinhold – Stuttgart: J G Cotta, 1904 [mf ed 1993] – (= ser Achim von arnim und die ihm nahe standen 3) – 1 – (incl bibl ref & ind) – mf#8467 – us Wisconsin U Libr [430]

Achim von arnims werke / Steig, Reinhold [comp] – Leipzig: Insel-Verlag, [1911] [mf ed 1993] – 3v on 1r – 1 – (incl bibl ref) – mf#8197 – us Wisconsin U Libr [802]

Achimer kreisblatt *see* Neues wochenblatt fuer die amtsbezirke achim und thedinghausen 2 jan 1878

Achimowin – 1974 feb-1987 spring – 1 – mf#515364 – us WHS [071]

Achinskii krai : Prosvet-osvedom, bespartiin gaz / ed by Smirnov, M S – Achinsk [Enis gub]: Komis po rasprostraneniiu osvedom lit sredi naseleniia 1919 [1919 6 iiulia-] – (= ser Asn 1-3) – n1-33 [1919] [gaps] item 16, on reel n6 – 1 – (suppl: telegrammy [asn-1.394]) – mf#asn-1.016 – ne IDC [077]

Achintre, Auguste *see*
- Apropos [sic]
- Cantate
- Cantate, la confederation
- Portraits et dossiers parlementaires du premier parlement de quebec

Achiote de la comarca cuentos / Perez Cadalso, Eliseo – Guatemala: Editorial del Ministerio de Educacion Publica "Jose de Pineda Ibarra" 1959 [mf ed 1992] – (= ser Coleccion contemporaneos 50) – 1r – 1 – us UF Libraries [830]

Achirnja satoe pembalesan / Siem, Hwat San – Soerabaia: Tan's Drukkery, 1934 [mf ed 1998] – (= ser Penghidoepan 110) – 1r – 1 – (coll as pt of the colloquial malay collection. filmed with: poetri satrija dewi, atawa, resia madjapait / h s t) – mf#10001 – us Wisconsin U Libr [830]

Achleitner, Arthur *see*
- Angela
- Berggeschichten
- Der finanzer
- Gruene brueche
- Halali!
- Im gruenen tann
- In den bergen, da lauert der wildschuetz
- In treue fest

Achleitner, Richard *see* Die unentwegten

Achper healthy lifestyles journal / Australian Council for Health, Physical Education and Recreation – Hindmarsh, Australia 1994+ – 1,5,9 – (cont: achper national journal) – ISSN: 1321-0394 – mf#10250,03 – us UMI ProQuest [613]

Achper national journal – Hindmarsh, Australia 1984-93 – 1,5,9 – (cont: australian journal for health, physical education & recreation: ajhper. cont by: achper healthy lifestyles journal) – ISSN: 0813-2283 – mf#10250.03 – us UMI ProQuest [613]

Die acht gesichter am biwasee : japanische liebesgeschichten / Dauthendey, Max – Muenchen: A Langen, G Mueller, c1911 [mf ed 1989] – 184p – 1 – mf#7170 – us Wisconsin U Libr [830]

Acht lieder / Goethe, Johann Wolfgang von – Wetzlar: Rathgeber 1857 [mf ed 1990] – (= ser Bibliothek der deutschen literatur) – 1r – 1 – (with ann by theodor bergk. filmed with: die lyrischen meisterstuecke von johann wolfgang von goethe & other titles) – mf#7318 – us Wisconsin U Libr [810]

Der acht und sechzigste psalm : ein denkmal exegetischer noth und kunst zu ehren unsrer ganzen zunft / Reuss, Eduard – Jena: Friedrich Mauke, 1851. Chicago: Dep of Photodup, U of Chicago Lib, 1978 (1r); Evanston: American Theol Lib Assoc, 1984 (1r) – 1 – 0-8370-1131-0 – (incl bibl ref) – mf#1984-T117 – us ATLA [220]

[Acht variationen ueber ein franzoesisches lied] : in e fuer klavier zu 4 haenden, op 10 / Schubert, Franz – [1818 sep] [mf ed 1988] – 1r – 1 – (purchased fr otto liepmannssohn [13 sep 1930]; publ in: franz schubert: neue ausgabe saemtliche werke 7/1, 1) – mf#pres. film 7 – us Sibley [780]

Acht-en-dertig konstige zinnebeelden *see* Met dichtkundige uitleggingen verrykt

Achterfeld, Johann Heinrich *see* Christkatholische dogmatik

Achtsiedel : roman / Bauer, Josef Martin – Berlin: Propylaeen-Verlag, c1937 [mf ed 1989] – (= ser Soldatenbuecherei 4) – 315p – 1 – mf#6982 – us Wisconsin U Libr [830]

Acht-uhr-abendblatt *see* National-zeitung 1848

Der achtundsechzigste psalm : mit besonderer ruecksicht auf seine alten uebersetzer und neueren ausleger / Grill, Julius – Tuebingen: H Laupp, 1883 – 1mf – 9 – 0-8370-3397-7 – (incl bibl ref) – mf#1985-1397 – us ATLA [220]

Ein achtundvierziger : erlebtes und gedachtes / Wagner, Philipp – Brooklyn, NY: J Wagner, 1882 – 1r – 1 – us Wisconsin U Libr [920]

Achtzehn monate in spanien / Mohr, Wilhelm – Koeln 1876 [mf ed Hildesheim 1995-98] – 2v on 5mf – 9 – €100.00 – 3-487-29853-8 – gw Olms [914]

Achtzehnhundertneun : die politische lyrik des kriegsjahres / ed by Arnold, Robert Franz & Wagner, Karl – Wien: Literarischer Verein, 1909 [mf ed 1993] – (= ser Schriften des literarischen vereins in wien 11) – xxvii/482/16p – 1 – (incl bibl ref) – mf#8308 reel 3 – us Wisconsin U Libr [810]

Achyuta Menon, Chelnat *see* Kali-worship in kerala

Aci materials journal – Farmington Hills MI 1987+ – 1,5,9 – ISSN: 0889-325X – mf#16080 – us UMI ProQuest [690]

Aci ran kham ca = Report of the vernacular and vocational education reorganization committee, 1936 / Burma. Tuin Ran Bhasa Panna nhan Asak Mve Vam Kron Panna Re Pru Pran Ci Mam Mhu Komiti – Rangoon: Supt Govt Print & Stationery, Burma 1940 [mf ed 1990] – 1 – (in burmese) – mf#mf-10289 seam reel 145/8 [§] – us CRL [370]

Aci ran kham ca nhan ovada cariya chara to mya i ovada katha = [Report of the national education in buddhist monasteries enquiry committee] / Burma. Bhun To Kri Kron Mya Tvan Tuin Ran Bhasa Panna San Kra Mhu Cum Cam Re Aphvai – Rangoon: Supt Govt Print & Stationery, Burma 1948 [mf ed 1990] – 1 – (in burmese) – mf#mf-10289 seam reel 157/6 [§] – us CRL [280]

Aci structural journal – Farmington Hills MI 1987+ – 1,5,9 – ISSN: 0889-3241 – mf#16079 – us UMI ProQuest [690]

Acid deposition and the environment : the international annual "grey literature" environmental reference collection – 410mf – 9 – $1,804.00 coll – (pt 1: basic set to 1988 347mf c39-13201. pt2: 1989 update 63mf c39-13202. printed guide available) – mf#C39-13200 – us Primary [360]

Acik soez – Kastamonu: Kastamonu Matbaasi, 1919-28. Sahib-i Imtiyaz: Ahmed Hamdi n32 (1 subat 1336 [1920],141-142,1406,2271,2297 (15 temmuz 1928) – (= ser O & t journals) – 1mf – 9 – $25.00 – us MEDOC [956]

Acimak / Guentekin, Resat Nuri – Istanbul: Aksam Matbaasi, 1928 – (= ser Ottoman literature, writers and the arts) – 2mf – 9 – $40.00 – us MEDOC [470]

Acis et galatee : pastorale heroique / Lully, Jean Baptiste – Paris: Christophe Ballard 1686 [mf ed 19–] – 1r – 1 – (libretto by jean galbert de campistron) – mf#film 1217 – us Sibley [780]

Acivilicao catolica e os erros modernos / Donoso Cortes, Juan Francisco – Petropolis: Editora Vozes limitada, 1960 – 1 – sp Bibl Santa Ana [241]

Acka, Sohuily Felix *see* Droit et science dans la pensee de hans kelsen (contribution a la theorie pure du droit)

Acker, Doris M *see* Bibliography of recorded music for dance

Acker, L van *see*
- Opera omnia (cccm 52)
- Opera omnia (cccm-pb 52)

Acker und gartenbau zeitung : nebst landwirth, deutscher farmer – 1905-06; 1907-09; 1910-11 feb 4, 1916-17 jun 2 – 1 – mf#3072636 – us WHS [635]

Ackerman, George Everett *see* Man, a revelation of god

Der ackermann aus boehmen – 2. aufl. Leipzig: S Hirzel, 1954 [mf ed 1993] – (= ser Altdeutsche quellen heft 1) – xxiii/68p – 1 – (incl bibl ref) – mf#8398 – us Wisconsin U Libr [430]

Ackermann, Johannes *see* Tolstoi und das neue testament

Ackermann, Manie *see* Coping with tension

Ackermann, Rudolph *see* Ackermann's new drawing book

Der ackermann und der tod : [ein streit- und trostgespraech vom tode aus dem jahre 1400] / Tepl, Johannes von – Berlin: H Kuepper [1939] [mf ed 1996] – (= ser Quellen der geistigen ueberlieferung) – 1r – 1 – (german trans of middle high german text. filmed with: filmed with: die elsaessischen sagen / th maurer [ed]) – mf#4196p – us Wisconsin U Libr [820]

Ackermann's new drawing book – London 1809 – (= ser 19th c art & architecture) – 1mf – 9 – mf#4.2.772 – uk Chadwyck [740]

Ackermann's 'repository of arts' : and other periodicals / National Art Library – (= ser The art periodicals coll at the v and a museum, 1750-1920, pt 1) – 45r (19 col) – 1 – £3000.00 – (incl: ackermann's 'repository of arts' 1909-28 [14r] £1150; artistic japan 1888-91 [2r] £180; the beau monde 1806-08 – magazine of the fine arts 1905-06 [2r] £180; nature and art 1866-87 [1r] £100; american art review 1880-81 [1r] £65; annales du musee et de l'ecole moderne des beaux arts 1801-22 [5r] £270; annals of the fine arts 1817-20 [2r] £100; arnold's magazine of the fine arts 1831-44 [3r] £150; the artist's repository 1787-90 [1r] £65; les beaux arts 1843-45 [1r] £65; le cabinet de l'amateur et de l'antiquaire 1842-46 [2r] £100; the century guild hobby horse 1884-93 [2r] £100; memorie per le belle arte 1785-88 [1r] £100; le musee artistique et litteraire 1879-81 [2r] £100; the portfolio 1870-93 [5r] £270; somerset house gazette 1824 – le studio 1833 – l'art dans le deux mondes 1890-91 [1r] £65; with p/g) – mf#VAR – uk World [700]

Ackermann's repository of arts, literature, commerce, manufactures, fashions and politics – 3 Series in 40v. 1809-1829 – 1 – (1st series, 14v. 2nd series, 14v. 3rd series, 12v) – us AMS Press [700]

Ackerman's juvenile forget-me-not – 1830-32 – (= ser English gift books and literary annuals, 1823-1857) – 10mf – 9 – uk Chadwyck [800]

Ackland, Thomas Suter *see* The story of creation as told by theology and by science

The acknowledged doctrines of the church of rome : being an exposition of roman catholic doctrines as set forth by esteemed doctors of the said church, and confirmed by repeated publication, with the sanction of bishops and ministers of her communion – London: Charles Gilpin, 1850 – 2mf – 9 – 0-8370-8325-7 – mf#1986-2325 – us ATLA [241]

The acknowledgment of deeds, containing all the statutes, territorial and state, of illinois. / Hunt, John Eddy – 1st ed. Chicago, 1896. 206p. LL-611 – 1 – us L of C Photodup [348]

Ackworth, New Hampshire.First Baptist Church *see* Records

Acl forum – 1920 dec-1921 sep – 1 – mf#492305 – us WHS [071]

Acland, Henry Wentworth, Baronet *see* The oxford museum

Acland, Hugh Dyke *see* Glorious recovery by the vaudois of their valleys, from the original with a compendious history of that people, previous and subsequent to that event, by hugh dyke acland

Acls newsletter – New York NY 1949-97 – 1,5,9 – ISSN: 1041-5963 – mf#9638 – us UMI ProQuest [000]

Aclyou in action / American Civil Liberties Union – 1986 feb, oct, dec; 1987 mar, aug; 1988 jan, may, nov – 1r – 1 – (cont: san diego aclu news; cont by: aclu in action) – mf#1611619 – us WHS [322]

Acm computing surveys – New York NY 1969+ – 1,5,9 – (cont: computing surveys) – ISSN: 0360-0300 – mf#12685,01 – us UMI ProQuest [000]

Acm transactions on computer systems – New York NY 1986+ (1,5,9) – ISSN: 0734-2071 – mf#16265 – us UMI ProQuest [000]

Acm transactions on database systems – New York NY 1978+ – 1,5,9 – ISSN: 0362-5915 – mf#12686 – us UMI ProQuest [000]

Acm transactions on graphics – New York NY 1986+ – 1,5,9 – ISSN: 0730-0301 – mf#16266 – us UMI ProQuest [000]

Acm transactions on information systems – New York NY 1989+ – 1,5,9 – (cont: acm transactions on office information systems) – ISSN: 1046-8188 – mf#16267,01 – us UMI ProQuest [000]

Acm transactions on mathematical software – New York NY 1978+ – 1,5,9 – ISSN: 0098-3500 – mf#12687 – us UMI ProQuest [000]

Acm transactions on office information systems – New York NY 1986-88 – 1,5,9 – (cont by: acm transactions on information systems) – ISSN: 0734-2047 – mf#16267.01 – us UMI ProQuest [000]

Acm transactions on programming languages and systems – New York NY 1986+ – 1,5,9 – ISSN: 0164-0925 – mf#16268 – us UMI ProQuest [000]

Aco : [a novel] / Tan Tay – Ran kun: U Kyaw of Shumawa Co 1961 [mf ed 1990] – (= ser Rhu ma va ca up 144) – 1 – (in burmese) – mf#mf-10289 seam reel 189/4 [§] – us CRL [830]

Acompanando a francisca sanchez resumen de una vida junto a ruben dario / Conde, Carmen – 1.ed. Nicaragua: [Editorial Union] 1964 [mf ed 1992] – 1r – 1 – us UF Libraries [972]

Aconteceu : especial / Centro Ecumenico de Documentacao e Informacao. Rio de Janeiro – Rio de Janeiro: CEDI, n10 12-16 apr 1982, apr 1983-1986 – (suppl to: aconteceu. filmed together n202-289) – us CRL [073]

Aconteceu – Rio de Janeiro: CEDI, [n202-581(oct 1982-1991)] (biwkly) – 10r – 1 – us CRL [073]

Acords del ple extraordinari del comite nacional de la union general de trabajadores : valencia, 27-30 d'octubre del 1937 / Union General de Trabajadores de Espana. Comite Nacional – Barcelona: Edicions UGT 1937 – 9 – mf#w701 – us Harvard [331]

Acorn – v8 n1-v12 n2 [1974 feb-1978 may] – 1 – mf#400903 – us WHS [071]

Acorn news – v7 n5-v21 n12 [1979 may-1994 dec] – 1 – mf#1110767 – us WHS [071]

Acornley, John Holmes *see*
- The colored lady evangelist
- A history of the primitive methodist church in the united states of america

Acorns : a publication of the oak lawn historical society – v5 n2-v8 n1 [1981 dec-1988 jun] – 1 – mf#1051636 – us WHS [978]

Acosta, Agustin *see*
- Ala
- Islas desoladas

Acosta, Aurelio *see* Sobreviviente del glorioso liberalismo colombiano...

Acosta, C *see*
- Tractado de las drogas, y medicinas de las indias orientales...
- Tratado de las drogas y medicinas de las indias orientales con sus plantas...

Acosta Hoyos, Luis Eduardo *see* Sesenta anos de la universidad

Acosta, J *see* De natura novi orbis libri duo et de promulgatione evangeli apud...

Acosta, Joaquin *see* Descubrimiento y colonizacion de la nueva granada

Acosta Leon, Raul D *see*
- Glorioso pasado historico de camaguey, 1868-1878 y...
- Revolucion en camaguey

Acosta, Oscar *see*
- Poesia
- Rafael heliodoro valle
- Tiempo detenido

Acosta Rubio, Raoul *see*
- Amor libre
- Ensayo biografico batista

Acosta Saignes, Miguel *see* Estudios de etnologia antigua de venezuela

Acosta Y Albear, Francisco De *see*
- Compendio historico
- Memoria sobre el estado actual de cuba

Acosta y Alguilar, Juan Antonio de *see* El lic d juan antonio de acosta y aguilar

Acotaciones para la historia de un libro / Reyes Monroy, Jose Luis – 1.ed. [Guatemala]: Editorial del Ministerio de Educacion Publica "Jose de Pineda Ibarra" 1960 [mf ed 1993] – 1r – 1 – (incl text of el puntero apuntado con apuntes breves) – us UF Libraries [070]

Acoustical physics – v1- 1955- – 1,5,6 – us AIP [530]

An acoustical study of brou vowels / Miller, John Daniel – [New York? 1967?] [mf ed 1989] – [ill] 1r with other items – 1 – mf#mf-10289 seam reel 023/13 [§] – us CRL [480]

Acquainted with grief / Shaw, George – [4th ed] USA: Caxton Press, c1906 [mf ed 1992] – (= ser Christian & missionary alliance coll) – 1mf – 9 – 0-524-02139-2 – mf#1990-4205 – us ATLA [220]

Acquire : the magazine for collectors – 1973-76 – 1,5,9 – (cont by: collector editions quarterly) – mf#8740.02 – us UMI ProQuest [790]

The acquirements and principal obligations and duties of the parish priest : being a course of lectures. delivered at the university of cambridge... / Blunt, John James – 4th ed. London: J Murray, 1861 – 1mf – 9 – 0-524-05367-7 – (incl bibl ref) – mf#1991-2273 – us ATLA [240]

Acquiring knowledge in initial teacher education : reading, writing, practice and the pgce course / Squirrell, Gillian – [mf ed Wakefield: Microform Ltd 1990] – (= ser Library & information research report) – 2mf – 9 – 0-7123-3216-2 – uk Microform Academic [370]

Acquisition of cuba : ...delivered in the house of representatives, feb 10 1859 / Taylor, Miles – [s.l.]: T McGill, printer [1859?] [mf ed 1990] – 1r – 1 – us UF Libraries [972]

The acquisitions librarian / ed by Katz, Bill – v1- 1989- – 1, 9 ($135.00 in US $189.00 outside hardcopy subsc) – us Haworth [020]

Acquoy, J G R *see* Het klooster te windesheim en zijn invloed

Acquoy, Johannes Gerhardus Rijk *see*
- Handleiding tot de kerkgeschiedvorsching en kerkgeschiedschrijving
- Middeleeuwsche geestelijke liederen en leisen

Acramavasika parva – Calcutta: Bharata Press, 1895 [mf ed 1993] – (= ser The mahabharata of krishna-dwaipayana vyasa) – 1mf – 9 – 0-524-08005-4 – (trans chiefly by kesari mohan ganguli) – mf#1991-0227 – us ATLA [490]

O acre : orgam dos interesses geraes – Sena Madureira, AC. 05-30 de jul 1916 – (= ser Ps 19) – mf#P25,01,17 – bl Biblioteca [079]

O acre : orgao dos interesses acreanos – Xapuri, AC: Impresso nas Officinas do Boletim Official, 24 jun-01 out 1907; 16 mar-01 jun 1913 – (= ser Ps 19) – mf#P25,01,16 – bl Biblioteca [079]

An acre of green grass : a review of modern bengali literature / Bose, Buddhadeva – Bombay: Orient Longmans, 1948 – (= ser Samp: indian books) – us CRL [490]

Acreano : orgao do partido autonomista acreano – Empreza, AC: Impresso nas Officinas do Acreano, 15 nov 1907-26 jun 1912 – (= ser Ps 19) – 1,5,6 – mf#P25,01,23 – bl Biblioteca [321]

Acremant, Albert *see* Gertrude et mon ceur

Acres of diamonds / Conwell, Russell Herman – New York: Harper, c1915 [mf ed 1993] – 1mf – 9 – 0-524-08271-5 – (life achievements by robert shackleton. with autobiogr note) – mf#1993-3026 – us ATLA [920]

Acrobatic enchainements and hints on presentation : the works of judy cholerton / Association of American Dancing – 3rd ed. [Derby, England, n.d.] – 1 – mf#*ZBD-*MGO pv 26 – Located: NYPL – us Misc Inst [790]

Across africa / Cameron, VL – London, 1877. 2v – 17mf – 9 – mf#A-168 – ne IDC [916]

Across america and asia : notes of a five years' journey around the world and of residence in arizona, japan and china / Pumpelly, R – New York: Leypoldt & Holt, 1870 – 6mf – 9 – mf#HT-116 – ne IDC [910]

Across central america / Boddam Whetham, John Whetham – London: Hurst & Blackett 1877 [mf ed 1990] – 1r – 1 – us UF Libraries [918]

Across central america / Boddam-Whetham, John Whetham – London: Hurst and Blackett, 1877 [mf ed 1987] – xii/353p – 1 – mf#8371 – us Wisconsin U Libr [918]

Across china on foot : life in the interior and the reform movement / Dingle, Edwin John – Bristol: J W Arrowsmith; London: Simpkin, Marshall, Hamilton, Kent, [1911] [mf ed 1995] – (= ser Yale coll) – xvi/445p (ill) – 1 – 0-524-09538-8 – mf#1995-0538 – us ATLA [915]

Across india at the dawn of the 20th century / Guinness, L E – London, 1898 – 3mf – 9 – mf#HTM-74 – ne IDC [915]

Across newfoundland with the governor : a visit to our mining region and; this newfoundland of ours, being a series of papers on the natural resources and future prospects of the colony / Harvey, Moses – St. John's Nfld?: s.n, 1879 – 2mf – 9 – mf#06784 – cn Canadiana [622]

Across the board – New York NY 1984+ – 1,5,9 – ISSN: 0147-1554 – mf#14582.02 – us UMI ProQuest [650]

Across the continent via the canadian pacific railway : a lecture delivered...23rd march, 1887 / Beaugrand, Honore – Montreal?: s.n, 1887? – 1mf – 9 – mf#02978 – cn Canadiana [917]

Across the desert : a life of moses / Campbell, Samuel Miner – Philadelphia: Presbyterian Board of Pub, c1873 [mf ed 1986] – 1mf – 9 – 0-8370-9534-4 – mf#1986-3534 – us ATLA [221]

Across the north pole to america / Gromov, Mikhail – Moscow: Foreign Languages Pub House, 1939 [mf ed 2000] – 1r – 1 – mf#29077 – us Harvard [919]

Across the subarctics of canada : a journey of 3,200 miles by canoe and snowshoe through the barren lands / Tyrrell, J W – London: T Fisher Unwin, [1893] – 6mf – 9 – mf#N-542 – ne IDC [917]

Across the vatna jokull : or, scenes in iceland; being a description of hitherto unknown regions / Watts, William Lord – London: Longmans & Co 1876 [mf ed 1987] – 1r – 1 – (filmed with: dara shukoh / qanungo, k) – mf#1823 – us Wisconsin U Libr [914]

Across yunnan : a journey of surprises; including an account of the remarkable french railway line now completed to yunnan-fu / Little, Archibald John – London: Sampson Low, Marston, 1910 [mf ed 1995] – (= ser Yale coll) – 164p (ill) – 1 – 0-524-09223-0 – mf#1995-0223 – us ATLA [915]

Acsm bulletin – Gaithersburg MD 1989+ – 1 – ISSN: 0747-9417 – mf#12485,01 – us UMI ProQuest [900]

Acsus newsletter – 1977 aug-1982 feb – 1 – mf#622718 – us WHS [071]

Act – v1 n1-v2 n2 [1970 jan-sep] – 1 – mf#720556 – us WHS [071]

An act amalgamating the port dover and lake huron, the stratford and huron and the georgian bay and wellington railway companies as the grand trunk, manitoulin, georgian bay and lake erie railway company – [Toronto?: s.n, 1881] [mf ed 1991] – 1mf – 9 – 0-665-90550-5 – mf#90550 – cn Canadiana [380]

An act authorising the establishment of mutual insurance companies in the several districts of upper canada : together with the resolutions and by-laws adopted by the stockholders and directors of the mutual fire insurance company of the district of niagara, established at st catharines – [s.l: s.n.] 1836 [mf ed 1983] – 1mf – 9 – 0-665-42544-9 – mf#42544 – cn Canadiana [368]

An act authorizing the formation of corporations for manufacturing, mining – New York, Banks, 1875. 94 p. LL-478 – 1 – us L of C Photodup [343]

The act authorizing the formation of corporations for manufacturing, mining, mechanical, chemical. / New York. (State). Laws, Statutes, etc – New York: Baker, Voorhis, 1884. 84p. LL-1685 – 1 – us L of C Photodup [343]

Act books of the archbishops of canterbury : 1663-1914 – 11r – 1 – £495.00 – mf#ACB – uk World [241]

An act concerning bankrupts and the administration of their effects / Canada (Province) – Kingston: printed by S Derbishire and G Desbarats, 1843 (mf ed 1983) – 1mf – 9 – (with ind) – mf#SEM105P200 – cn Bibl Nat [324]

The act concerning corporations in the state of new jersey, approved april 7, 1875 / Corbin, William Horace – Jersey City: Linn, 1889. 108p. LL-227 – 1 – us L of C Photodup [348]

An act concerning corporations (revision of 1896) : taking effect july 4, 1896, in the state of new jersey / Corbin, William Horace – 10th ed. Newark: Soney & Sage, 1897. 109p. LL-839 – 1 – us L of C Photodup [346]

An act for appointing commissioners to inquire into the losses occasioned by the late destructive fires in this province – Fredericton [NB]: G K Lucrin, 1826 [mf ed 1984] – 1mf – 9 – 0-665-46117-8 – mf#46117 – cn Canadiana [346]

An act for granting certain powers to the british american land company – [London?]: Haslan & Bischoff [1834?] [mf ed 1984] – 1mf – 9 – 0-665-46116-X – mf#46116 – cn Canadiana [346]

An act for limiting the time of service in the army : passed 21st june 1847 / Canada (Province) – Quebec: printed by J N Duquet, 1865 [mf ed 1983] – 4mf – 9 – mf#SEM105P318 – cn Bibl Nat [355]

An act for making a rail-road from lake champlain to the river st lawrence – Montreal: printed by Andrew H Armour & Co, 1835 [mf ed 1994] – 1mf – 9 – mf#SEM105P1979 – cn Bibl Nat [380]

An act for the abolition of feudal rights and duties in lower canada : 18 vict cap 3 / Canada. Laws, Statutes, etc – Quebec: printed by Stewart Derbishire & George Desbarats, 1854 [mf ed 1983] – 1mf – 9 – mf#SEM105P265 – cn Bibl Nat [348]

An act for the better establishment and maintenance of public schools in upper-canada : and for repealing the present school act, 12th victoria, cap 83 = [Acte pour mieux etablir et maintenir les ecoles publiques dans le haut-canada et revoquer l'acte des ecoles actuelles / Canada (Province) – Montreal: printed by Stewart Derbishire & George Desbarats, 1849 [mf ed 1996] – 1mf – 9 – mf#SEM105P1854 – cn Bibl Nat [370]

An act for the construction of water works in the city of hamilton – [Canada?: s.n, 1856?] [mf ed 1992]] – 1mf – 9 – 0-665-94676-7 – mf#94676 – cn Canadiana [343]

An act further to amend the judicature acts of lower canada = Acte pour amender les actes de judicature du bas-Canada / Canada (Province) – [S.l: s.n, 1858?] (mf ed 1995) – 1mf – 9 – mf#SEM105P2035 – cn Bibl Nat [348]

Act of 1800 bankruptcy case files of the u.s. district court of maryland, 1800-1803 / U.S. Circuit and District Courts – (= ser Records of district courts of the united states) – 2r – 1 – (with printed guide) – mf#M1031 – us Nat Archives [346]

Act of 1800 bankruptcy records of the us district court for the eastern district of pennsylvania, 1800-1806 / U.S. Circuit and District Courts – (= ser Records of district courts of the united states) – 24r – 1 – (with printed guide) – mf#M993 – us Nat Archives [346]

Act of 1800 bankruptcy records of the u.s. district court for the southern district of new york, 1800-1809 / U.S. District Court – (= ser Records of district courts of the united states) – 11r – 1 – (with printed guide) – mf#M933 – us Nat Archives [346]

Act of 1935 and amendments, 1939-67 *see* Us social security administration. act of 1935 and amendments, 1939-67

The act of baptism in the history of the christian church / Burrage, Henry S – Philadelphia: American Baptist Publication Society, c1879 – 1mf – 9 – 0-7905-4444-X – mf#1988-0444 – us ATLA [242]

Act of incorporation and by-laws : (for distribution amongst the members) / Bank of Montreal. Pension Fund Society – Montreal?: Gazette, 1885 – 1mf – 9 – mf#10386 – cn Canadiana [332]

Act of incorporation, bye-laws, rules and regulations / British American Friendly Society of Canada – [Montreal?: s.n.] 1856 [mf ed 1987] – 1mf – 9 – 0-665-63122-7 – mf#63122 – cn Canadiana [336]

Act of incorporation, by-laws, and list of members...established, 1864 / Ottawa Natural History Society – [Ottawa?: s.n.] 1866 [mf ed 1984] – 1mf – 9 – 0-665-23332-9 – mf#23332 – cn Canadiana [500]

Act of incorporation, by-laws, and list of shareholders... / Victoria Skating Club (Montreal, Quebec) – [Montreal?: s.n.] 1862 [mf ed 1984] – 1mf – 9 – 0-665-46288-3 – mf#46288 – cn Canadiana [790]

Act of incorporation, declaration, constitution, rules, canons and by-laws of the synod of the diocese of niagara : with standing resolutions, etc... / Church of England. Diocese of Niagara – [Hamilton, Ont?: s.n.] [mf ed 1981] – 4mf – 9 – mf#08840 – cn Canadiana [242]

The act of uniformity : a measure of liberation / Hancock, Thomas – London: SPCK 1898 [mf ed 1993] – (= ser Church historical society (series) 48) – 1mf – 9 – 0-524-05503-3 – mf#1990-1498 – us ATLA [348]

An act respecting the militia : 27 vict, cap 2 – Quebec: G Desbarats & M Cameron, 1863 [mf ed 1984] – 1mf – 9 – 0-665-46103-8 – mf#46103 – cn Canadiana [343]

An act respecting the militia, extracted from consolidated statutes of canada : proclaimed and published under the authority of the act 22 vict cap 29 ad 1859 / Canada (Province) – Quebec: printed by Stewart Derbishire & George Desbarats, 1861 [mf ed 1983] – 1mf – 9 – mf#SEM105P181 – cn Bibl Nat [348]

An act respecting the preservation of the public health : 22 victoriae, cap 28 – [s.l.]: printed for the Bureau of Agriculture and Statistics, 1866 [mf ed 1984] – 1mf – 9 – 0-665-46104-6 – mf#46104 – cn Canadiana [344]

An act respecting the preservation of the public health : 22 victoriae, cap 28 [i.e. 38] / Canada (Province) – [Ottawa?]: printed for the Bureau of Agriculture and Statistics, 1866 [mf ed 1983] – 1mf – 9 – mf#SEM105P198 – cn Bibl Nat [614]

An act respecting the sale and management of the public lands : 23 vict, cap 2 – Quebec: S Derbishire & G Desbarats, 1860 [mf ed 1984] – 1mf – 9 – 0-665-45120-2 – mf#45120 – cn Canadiana [343]

An act to abolish imprisonment for debt and for the punishment of fraudulent debtors in lower canada and for other purposes / Canada (Province) – Montreal: printed by Stewart Derbishire & George Desbarats, 1849 [1993] – 1mf – 9 – mf#SEM105P1977 – cn Bibl Nat [345]

An act to abolish imprisonment for debt, and for the punishment of fraudulent debtors, in lower canada and for other purposes : 12 victoriae, cap 42 – Montreal: S Derbishire & G Desbarats, 1849 [mf ed 1984] – 1mf – 9 – 0-665-46115-1 – mf#46115 – cn Canadiana [346]

An act to amend : and reduce into one act, the militia laws of this province – Toronto: R Stanton, 1838 [mf ed 1984] – 1mf – 9 – 0-665-46113-5 – mf#46113 – cn Canadiana [343]

An act to amend an act intituled "an act for the construction of water works in the city of hamilton" – Hamilton: Times Steam Press, [1860?] [mf ed 1992]] – 1mf – 9 – 0-665-94682-1 – mf#94682 – cn Canadiana [343]

An act to amend the acts relating to the grand trunk railway company of canada / Canada (Province) – Quebec: printed by Stewart Derbishire & George Desbarats, 1855 [mf ed 1993] – 1mf – 9 – mf#SEM105P1761 – cn Bibl Nat [380]

An act to amend the acts relating to the grand trunk railway company of canada / Canada (Province) – [S.l: s.n, 1858?] (mf ed 1993) – 1mf – 9 – mf#SEM105P1767 – cn Bibl Nat [380]

An act to amend the laws in force respecting the sale of intoxicating liquors and the issue of licenses therefor [sic] : and otherwise for repression of abuses resulting from such sale, 27 & 28 vict, cap 18 – Quebec: G Desbarats & M Cameron, 1864 [mf ed 1984] – 1mf – 9 – 0-665-46112-7 – mf#46112 – cn Canadiana [344]

Act to approve the compact of free association with palau : 99th congress 2nd session / U.S. Congress – P.L. 99-658, nov 14, 1986 – (= ser Republic Of Belau (Palau) – Status Negotiations With The U.S.) – 1mf – 9 – $1.50 – mf#LLMC 82-100G, Title 32 – us LLMC [327]

An act to authorize the grand trunk railway company of canada to construct a bridge over the river st clair at sarnia / Canada (Province) – Toronto: S Derbishire & G Desbarats, [1858 ?] (mf ed 1993) – 1mf – 9 – mf#SEM105P1766 – cn Bibl Nat [380]

An act to authorize the granting of charters of incorporation to manufacturing, mining, and other companies, and amendments : 27-28 victoria, cap 23; 29 victoria, cap 20 – Ottawa: M Cameron, 1866 [mf ed 1984] – 1mf – 9 – 0-665-46110-0 – mf#46110 – cn Canadiana [346]

An act to consolidate and amend the several acts relating to the niagara and detroit rivers railway co : both before and since the amalgamation of the companies forming that company, 22 victoriae, cap 90 – Toronto: S Derbishire & G Desbarats, 1859 [mf ed 1984] – 1mf – 9 – 0-665-46111-9 – mf#46111 – cn Canadiana [343]

An act to define seigniorial rights in lower canada : and to facilitate the redemption thereof = [Acte seigneurial de 1853] / Canada (Province) – [S.l: s.n., 185-?] [mf ed 1995] – 1mf – 9 – mf#SEM105P2797 – cn Bibl Nat [343]

An act to enable the members of the united church of england and ireland in canada to meet in synod : together with the canons, passed by the synod of the diocese of toronto... – [Toronto?: s.n.], 1857 [mf ed 1984] – 1mf – 9 – 0-665-46107-0 – mf#46107 – cn Canadiana [242]

An act to establish a uniform system of bankruptcy throughout the united states / Blumenstiel, Alexander – New York, Blumenstiel 1880 75 p. LL-1457 – 1 – us L of C Photodup [346]

Act to explain and amend the laws relating to lands holden in free and common soccage in the province of lower canada *see* Bill intituled an act to explain and amend the laws relating to lands holden in free and common soccage in the province of lower canada

An act to extend the charter of the bank of upper canada, and to increase the capital stock thereof – [s.l: s.n, 1842?] [mf ed 1984] – 1mf – 9 – 0-665-46106-2 – mf#46106 – cn Canadiana [332]

An act to grant additional aid to the grand trunk railway company of canada / Canada (Province) – Toronto: S Derbishire & G Desbarats, [1856?] (mf ed 1993) – 1mf – 9 – mf#SEM105P1765 – cn Bibl Nat [380]

An act to incorporate a company for the construction of a ship canal : to connect the waters of lake champlain and the river saint lawrence, 12th victoriae, cap 180 – Montreal: S Derbishire & G Desbarats, 1849 [mf ed 1984] – 1mf – 9 – 0-665-46283-2 – mf#46283 – cn Canadiana [380]

An act to incorporate the drummond and arthabaska counties railway company : 23 vict cap 111 – Quebec: S Derbishire & G Desbarats, 1860 [mf ed 1984] – 9 – 0-665-46123-2 – mf#46123 – cn Canadiana [380]

An act to incorporate the montreal mining company : 10 and 11 vict, cap 68 – [Montreal?: s.n.] 1850 [mf ed 1984] – 1mf – 9 – 0-665-46105-4 – mf#46105 – cn Canadiana [346]

Act to incorporate the quebec fire-assurance company : to which are added, by-laws, rules and regulations of the said company... / Bas-Canada – 2nd ed. Quebec: printed by P E Desbarats, 1827 [mf ed 1994] – 1mf – 9 – mf#SEM105P2211 – cn Bibl Nat [368]

An act to make a new and more convenient subdivision of the province into counties : for the purpose of effecting a more equal representation thereof in the assembly than heretofore = Acte pour faire une division nouvelle et plus commode de la province en comtes, afin d'avoir une representation [sic] dans l'assemblee plus egale que ci-devant – [s.l]: [s.n.], [1829?] (mf ed 1985) – 1mf – 9 – (in french and english) – mf#SEM105P529 – cn Bibl Nat [348]

An act to make more ample provision for the incorporation of the town of three rivers – Toronto: S Derbishire & G Desbarats, 1857 [mf ed 1984] – 1mf – 9 – 0-665-46281-6 – mf#46281 – cn Canadiana [350]

An act to make temporary provision for the government of lower canada = Acte pour etablir des dispositions temporaires pour le gouvernement du bas-Canada / Grande-Bretagne – Montreal: A H Armour & H Ramsay, 1838 [mf ed 1993] – 1mf – 9 – (in french and english) – mf#SEM105P1908 – cn Bibl Nat [323]

An act to provide for the better organization of agricultural societies in lower canada – Quebec: S Derbishire & G Desbarats, 1852 [mf ed 1984] – 1mf – 9 – 0-665-43132-5 – mf#43132 – cn Canadiana [630]

An act to provide for the organization and regulation of certain business corporations / McMaster, Robert Bach – New York, Baker, Voorhis, 1875. 33, xvii p. LL-706 – 1 – (new york: baker, voorhis, 1884 50p II-668. new york: baker, voorhis, 1887 94p II-684. new york: baker, voorhis, 1890 II-919) – us L of C Photodup [348]

An act to regulate the inspection and measurement of timber, masts, spars, deals, staves and other articles of a like nature in the ports of quebec and montreal : and for other purposes relative to the same / Canada (Province) – Kingston: R Stanton...1841 [mf ed 1983] – 1mf – 9 – mf#SEM105P174 – cn Bibl Nat [348]

An act to renew the charter of the bank of montreal : and to increase its capital stock – Montreal: Lovell & Gibson, 1842 [mf ed 1984] – 1mf – 9 – 0-665-46125-9 – mf#46125 – cn Canadiana [332]

An act to repeal, alter, and amend, the laws now in force for the regulation of the several macadamized roads within this province – [Toronto?: s.n, 1840?] [mf ed 1984] – 1mf – 9 – 0-665-46286-7 – mf#46286 – cn Canadiana [344]

Acta academiae scientiarum imperialis petropolitanae – Petropoli, 1777-1782. v 1-6 – 133mf – 9 – mf#R-5816 – ne IDC [077]

Acta amazonica – Manaus Amazonas, Brazil 1971+ – 1,5,9 – ISSN: 0044-5967 – mf#8161 – us UMI ProQuest [574]

Acta anatomica – Basel, Switzerland 1966-74 – 1,5,9 – (cont by: cells tissues organs: in vivo, in vitro) – ISSN: 0001-5180 – mf#2043.01 – us UMI ProQuest [574]

Acta apostolicae sedis – 19(1909)-60(1968) – 920mf – 9 – €1753.00 – ne Slangenburg [226]

Acta apostolicae sedis – Romae, 1909-1942. v1-34 – 590mf – 8 – mf#H-194 – ne IDC [240]

Acta apostolicae sedis – Rome, Italy 1909+ – 1,5,9 – ISSN: 0001-5199 – mf#3202 – us UMI ProQuest [241]

Acta apostolorum : graece et latine edidit actus apostolorum extra canonem receptum addidit / ed by Hilgenfeld, A – Berolini, 1899 – 6mf – 8 – €14.00 – ne Slangenburg [250]

Acta apostolorum : sive, lucae ad theophilum liber alter: editio philologica apparatu critico, commentario perpetuo, indice verborum illustrata / Blass, Friedrich Wilhelm – Goettingen: Vandenhoeck & Ruprecht, 1895 [mf ed 1986] – 1mf – 9 – 0-8370-9529-8 – (comm in latin & greek; text in greek. incl ind) – mf#1986-3529 – us ATLA [226]

Acta apostolorum : sive, lucae ad theophilum liber alter: secundam formam quae videtur romanum / ed by Blass, Fridericus – Lipsiae [Leipzig]: B G Teubner, 1896 [mf ed 1985] – 1mf – 9 – 0-8370-2366-1 – mf#1985-0366 – us ATLA [226]

Acta apostolorum apocrypha : post constantinum tischendorf / ed by Lipsius, Richard Adelbert & Bonnet, Max – Lipsiae: Apud Hermannum Mendelssohn, 1891 [mf ed 1989] – 2v in 3 on 3mf – 9 – 0-8370-1197-3 – (in greek & latin. incl bibl ref) – mf#1987-6027 – us ATLA [226]

Acta apostolorum graece et latine : secundum antiquissimos testes / Hilgenfeld, Adolf – Berolini: Sumptibus Georgii Reimeri, 1899 [mf ed 1986] – 1mf – 9 – 0-8370-9546-8 – (incl bibl ref & ind) – mf#1986-3546 – us ATLA [226]

Acta apostolorvm annotationum lucae lossij, in novum testamentum iesu christi nazaraeni, duodecim capitum actorum apostolicorum / Lossius, L – Francoforti, 1558. v3 – 7mf – 9 – mf#TH-1 mf 865-871 – ne IDC [242]

Acta archelai und das diatessaron tatians *see* Die altercatio simonis iudaei et theophili christiani

Acta astronautica – Oxford, England 1974+ – 1,5,9 – (cont: astronautica acta) – ISSN: 0094-5765 – mf#49001 – us UMI ProQuest [520]

Acta audiologica y foniatrica hispanoamericana – Mexico City, Mexico 1972 – 1,5 – ISSN: 0515-2747 – mf#7771 – us UMI ProQuest [617]

Acta biologica venezuelica – Caracas, Venezuela 1975-78 – 1,5,9 – ISSN: 0001-5326 – mf#7261 – us UMI ProQuest [574]

Acta borussica ecclesiastica, civilia, literaria – Koenigsberg (Kaliningrad RUS), 1730-32 – 1 – gw Misc Inst [077]

Acta borussica neue folge : 1. reihe: die protokolle des preussischen staatsministeriums 1817-1934/38 / ed by Kocka, Juergen et al – 12v on 1161mf – 9 – diazo €3980.00 silver €4600.00 – (v1: 1817-29 [mf ed 2001]; v2: 1830-40 [mf ed 2004]; v3: 1840-48 [mf ed 2000]; v4: 1848-58 [mf ed 2003]; v5: 1858-66 [mf ed 2001]; v6: 1867-78 in prep; v7:1879-90 [mf ed 1999]; v8: 1890-1900; v9: 1900-09 in prep; v10: 1909-18 [mf ed 1999]; v11: 1918-25 [mf ed 2000]; v12: 1925-1934/38 in prep) – gw Olms [943]

Acta botanica neerlandica – Oxford, England 1987-90 – 1,5,9 – ISSN: 0044-5983 – mf#16732 – us UMI ProQuest [580]

Acta capitularia congregationis de septem fontibus – 1847-1891 – 4mf – 8 – €11.00 – ne Slangenburg [241]

Acta capituli generalis romae : in conventu s. mariae super mineruam celebrati. in festo sanctissime, pentecostes 9. iunij anno 1612 / Papiensis, Seraphino Sicco – Mexici: Ex oficina Ihoannis [sic] Ruyz 1613 – (= ser Books on religion...1543/44-c1800: ordenes, etc: dominicos) – 1mf – 9 – mf#crl-186 – ne IDC [241]

Acta capituli generalis ulyssiponae / Dominic, Saint – Mexici: Apud Bachalauru[m] Iannem de Alcacar, anno 1619 – (= ser Books on religion...1543/44-c1800: ordenes, etc: dominicos) – 1mf – 9 – mf#crl-188 – ne IDC [241]

Acta capituli provincialis celebrati in hoc imperiali s p n dominici mexicano conventu : die quinta decima mensis maij, anni ab incarnatione domini millesimi septingentisimi septeguagesimi tertij / Dominicans. Provincia de Santiago de Mexico – [Mexico City]: Ex typis D Joseph Jauregi [1774?] – (= ser Books on religion...1543/44-c1800: ordenes, etc: dominicos) – 1mf – 9 – mf#crl-195 – ne IDC [241]

Acta capituli windeshemensis / ed by Woude, Utig S v. d. – 's-Gravenhage, 1953 – (= ser Kerkhistorische studien 4) – 5mf – 8 – €12.00 – ne Slangenburg [241]

Acta chirurgica *see* European journal of surgery

Acta colloquij montis belligartensis / [Andreae d A, J] – Tvbingae, 1587 – 7mf – 9 – mf#TH-1 mf 30-36 – ne IDC [242]

Acta conciliorum : et epistolae decretales ac constitutiones summorum pontificum / Harduinus, Johannes – Parisiis. v1-11. 1714-1715 – €865.00 – ne Slangenburg [227]

Acta conciliorum oecumenicorum / ed by Schwartz, Eduard – Berolini. tom 1-4. 1927 ss – 4v – €334.00 – ne Slangenburg [240]

Acta cytologica – St Louis MO 1957+ – 1,5,9 – ISSN: 0001-5547 – mf#1570 – us UMI ProQuest [574]

Acta da assemblea dos membros da communidade portugueza de bombaim : jurisdiccionados do exmo e rmo sr arcebispo de cranganor, primeiro bispo de damao, reunida no dia 18 de novembro de 1888 [microform] / Assemblea dos membros da communidade portugueza (1888: Bombay) – Bombaim: Typographia do "Anglo-lusitano," 1888 [mf ed 1995] – (= ser Yale coll) – 47p – 1 – 0-524-10048-9 – (in portuguese) – mf#1995-1048 – us ATLA [241]

Acta de 22 de octubre de 1918 / Comision de Monumentos – Madrid: Fortanet, 1919. B.R.A.H. lxxiv/pp. 268-272 – 1 – sp Bibl Santa Ana [946]

Acta de constitucion y reglamento de "el alba" sociedad obrera de socorros mutuos de aldea del ca no – (Caceres): Tip. La Minerva Cacerena de Serafin Ronda s.a., 1915? – 1 – sp Bibl Santa Ana [350]

El acta de la separacion dominicana y el acta de independencia de los estados unidos de america / Rodriguez Demorizi, Emilio – Ciudad Trujillo, Dominican Republic: Impr "La Opinion" 1943 [mf ed 1991] – 1r – 1 – (with bibl footnotes) – us UF Libraries [972]

Acta de la sesion de 1 de octubre de 1920 / Comision de Monumentos – Madrid: Ed. Reus, 1921. B.R.A.H. 78. pp. 88-90. Tambien: 12 de noviembre de 1920 – 1 – sp Bibl Santa Ana [946]

Acta de la sesion de 9 de diciembre de 1918 / Subcomision de monumentos – Madrid: Fortanet, 1919. B.R.A.H. 74. pp. 287-289 – 1 – sp Bibl Santa Ana [946]

Acta de la sesion de 22 de marzo de 1920 / Comision de Monumentos – Madrid: Fortanet, 1920. B.R.A.H. 77. pp. 86-92 y 383 – 1 – sp Bibl Santa Ana [946]

Acta de la sesion del 14 de marzo de 1920 / Comision de Monumentos – Madrid: Fortanet, 1920. B.R.A.H. 77. pp. 365-379 – 1 – sp Bibl Santa Ana [946]

Acta der provinciale en particuliere synoden : gehouden in de noordelijke nederlanden, 1572-1620 / ed by Reitsma, A S & Veen, S D Van – Groningen. v1-8. 1892-1899 – 65mf – 8 – €124.00 – ne Slangenburg [242]

Acta des colloquij : zwischen den wuertenbergischen theologen vnd d ioanne pistorio, zu baden gehalten / Heerbrand, J – Tuebingen, 1590 – 5mf – 9 – mf#TH-1 mf 600-604 – ne IDC [242]

Acta diabetologica latina – Milan, Italy 1977-91 – 1,5,9 – ISSN: 0001-5563 – mf#11544.01 – us UMI ProQuest [616]

Acta electronica – Paris. 1956-61 – 1 – fr ACRPP [073]

Acta eruditorum / ed by Raabe, Paul – Leipzig 1682-1782 [mf ed Hildesheim 1981] – 117v on 850mf – 9 – diazo €3400.00 silver €4980.00 – (cont by: nova acta eruditorum, leipzig 1732-82; afterword by ed) – gw Olms [500]

Acta et decreta / Catholic Church. Province of Calcutta (India). Concilium Provinciale (1st: 1894) – Calcuttae: Cath Orphan 1905 [mf ed 1992] – 1mf – 9 – 0-524-03543-1 – mf#1990-4738 – us ATLA [241]

Acta et decreta conciliorum : coll lacensus 7: concilium vaticanum 1 – Freiburg Brg, 1890 – €61.00 – ne Slangenburg [241]

Acta et decreta primae provincialis synodi tokiensis a.d. 1895 : cum mutationibus a s cong de propaganda fide inductis – Hongkong: Typis Societatis Missionum ad Exteros, 1896 [mf ed 1995] – (= ser Yale coll) – 56p – 1 – 0-524-10019-5 – (in latin) – mf#1995-1019 – us ATLA [241]

Acta et decreta sacrosancti et oecumenici concilii vaticani : die 8 dec 1869 a ss d n pio p 9 inchoati – Friburgi Brisgoviae [Freiburg i.B.]: Herder, 1871 [mf ed 1986] – 1mf – 9 – 0-8370-7031-7 – mf#1986-1031 – us ATLA [241]

Acta et scripta theologorvm vvirtembergensivm : et patriarchae constantinopolitani d hieremiae... / [Andreae, A J] – Vvitebergae, 1584 – 5mf – 9 – mf#TH-1 mf37-41 – ne IDC [243]

Acta et scripta theologorvm vvirtembergensivm, et patriarchae constantinopolitani d hieremiae / [Andreae d A, J] – Vvitebergae, 1584 – 5mf – 9 – mf#TH-1 mf 37-41 – ne IDC [242]

Acta et verba / Devot, Justin – Paris: F Pichon 1893 [mf ed 1990] – 1r – 1 – us UF Libraries [340]

Acta final / Symposium para evaluacion y defensa de los recursos naturales de la republica dominicana : (1st : 1959 : santo domingo, dominican republic) – Ciudad Trujillo, Dominican Republic: Libreria Dominicana 1959 [mf ed 1993] – 1r – 1 – us UF Libraries [333]

Acta forestalia fennica : arbeiten der forstwissenschaftlichen gesellschaft in finnland / Suomen Metsaetieteellinen Seura et al – Helsingforsiae: Die Gesellschaft 1913-99 (irreg) [1-35(1913-29) [mf ed 2005] – 11r [ill] – 1 – (lacks 28-30, 32-33; cont: communicationes instituti forestalis fenniae, issn 0358-9609; absorbed by: silva fennica [helsinki, finland: 1967]; in english, finnish, french, german & swedish; with ind) – ISSN: 0001-5636 – mf#film mas c6034 – us Harvard [634]

Acta ginecologica – v1-20. 1950-69 – 1 – us AMS Press [616]

Acta haematologica – Basel, Switzerland 1966+ – 1,5,9 – ISSN: 0001-5792 – mf#2045 – us UMI ProQuest [616]

Acta handlungen : legation vnd schriffte: so durch den herrn philipsen landgraven zu hessen etc / Corvinus, A – Wittemberg, 1536 – 2mf – 9 – mf#TH-1 mf 349-350 – ne IDC [242]

Acta hepato-gastroenterologica – Stuttgart, Germany 1975-78 – 1,5,9 – (cont by: hepato-gastroenterology) – ISSN: 0300-970X – mf#10151.01 – us UMI ProQuest [616]

Acta histochemica et cytochemica – Kyoto, Japan 1972+ – 1,5,9 – ISSN: 0044-5991 – mf#7937 – us UMI ProQuest [574]

Acta historico-ecclesiastica nostri temporis : oder gesammlete nachrichten und urkunden zu der kirchengeschichte unsrer zeit / ed by Schneider, Christian Wilhelm – Weimar 1774-90 – (= ser Dz. abt theologie) – 13v[=t1-100] on 88mf – 9 – €880.00 – mf#k/n2119 – gw Olms [240]

Acta in consistorio secreto habito a sanctissimo domino nostro pio divina providentia papa sexto feria 6 : decembris 1778... / Pius 6, Pope – Mexici: Impr Typis Haered Lic D Josephi de Jauregui 1780 – (= ser Books on religion...1543/44-c1800: papas [cartas apostolicas, etc]) – 1mf – 9 – mf#crl-373 – ne IDC [241]

Acta informatica – Dordrecht, Netherlands 1971+ – 1,5,9 – ISSN: 0001-5903 – mf#13128 – us UMI ProQuest [000]

Acta leprologica – nos 1-37 1960-69 – 1 – us AMS Press [616]

Acta litteraria antiqua – Leipzig DE, 1715-16 – 1 – gw Misc Inst [400]

Acta martyrum : opera ac studio – Ratisbonae [Regensburg]: G Josephi Manz, 1859 [mf ed 1986] – 2mf – 9 – 0-8370-6938-6 – (incl ind) – mf#1986-0938 – us ATLA [240]

Acta martyrum et sanctorum (syriace) / ed by Bedjan, Paul – Paris. v7. 1890-1897 – 9 – €99.00 – ne Slangenburg [241]

Acta martyrum selecta : und andere urkunden aus der verfolgungszeit der christlichen kirche = Ausgewaehlte maertyreracten / Gebhardt, Oscar von – Berlin: A Duncker, 1902 [mf ed 1990] – 1mf – 9 – 0-7905-5211-6 – (in greek, latin & german; incl bibl ref) – mf#1988-1211 – us ATLA [240]

Acta materialia – Oxford, England 1996+ – 1,5,9 – (cont: acta metallurgica et materialia) – ISSN: 1359-6454 – mf#49002,02 – us UMI ProQuest [660]

Acta mechanica – Dordrecht, Netherlands 1991+ – 1 – ISSN: 0001-5970 – mf#13256 – us UMI ProQuest [621]

Acta metallurgica – Oxford, England 1953-89 – 1,5,9 – (cont by: acta metallurgica et materialia) – ISSN: 0001-6160 – mf#49002.02 – us UMI ProQuest [660]

Acta metallurgica et materialia – Oxford, England 1990-95 – 1,5,9 – (cont: acta metallurgica; cont by: acta materialia) – ISSN: 0956-7151 – mf#49002.02 – us UMI ProQuest [660]

Acta meteorologica sinica = Ch'i hsiang hsueh pao – Beijing, China 2001+ – 1,5,9 – ISSN: 0894-0525 – mf#22003 – us UMI ProQuest [550]

Acta mythologica apostolorum / Lewis, Agnes Smith – London: C J Clay, 1904 [mf ed 1991] – (= ser Horae semiticae 3) – 1mf – 9 – 0-8370-1998-2 – (in arabic) – mf#1987-6385 – us ATLA [226]

Acta neurochirurgica – Dordrecht, Netherlands 1985+ – 1,5,9 – ISSN: 0001-6268 – mf#13257 – us UMI ProQuest [617]

Acta neurologica belgica – Brussels, Belgium 1975-80 – 1,5,9 – mf#10527 – us UMI ProQuest [616]

Acta neuropathologica – Dordrecht, Netherlands 1981+ – 1,5,9 – ISSN: 0001-6322 – mf#13100 – us UMI ProQuest [616]

Acta oceanologica sinica = Hai yang hsueh pao – Oxford, England 1985-93 – 1,5,9 – ISSN: 0253-505X – mf#49587 – us UMI ProQuest [550]

Acta ofte handelinghen des nationalen synodi... : ghehouden...tot dordrecht, anno 1618 ende 1619 / [Heinsius, D] – Dordrecht, 1621. 3v – 21mf – 9 – mf#PBA-101 – ne IDC [240]

Acta oncologica – Abingdon, Oxfordshire 1979+ – 1,5,9 – (cont: acta radiologica oncology) – ISSN: 0284-186X – mf#6042,04 – us UMI ProQuest [616]

Acta Orientalia *see* Ediderunt societatas orientales batava...

Acta osnabrugensia : oder beytraege zu den rechten und geschichten von westfalen. insonderheit vom hochstifte osnabrueck / ed by Lodtmann, Justus Friedrich August – Osnabrueck 1778-82 – (= ser Dz. historisch-geographische abt) – 2pt on 5mf – 9 – €100.00 – mf#k/n1095 – gw Olms [323]

Acta paediatrica – Abingdon, Oxfordshire 1992+ – 1,5,9 – (cont: acta paediatrica scandinavica) – ISSN: 0803-5253 – mf#2174,01 – us UMI ProQuest [618]

Acta paediatrica scandinavica – Abingdon, Oxfordshire 1921-91 – 1,5,9 – (cont by: acta paediatrica) – ISSN: 0001-656X – mf#2174.01 – us UMI ProQuest [618]

Acta pathologica japonica – Oxford, England 1972-87 – 5,9 – ISSN: 0001-6632 – mf#7164.01 – us UMI ProQuest [610]

Acta pauli : aus der heidelberger koptischen papyrushandschrift nr 1 / ed by Schmidt, Carl – 2. erw ausg. Leipzig: JC Hinrichs, 1905 [mf ed 1991] – 1v on 1mf – 9 – 0-8370-1949-4 – (text in german & coptic; discussion in german) – mf#1987-6336 – us ATLA [090]

Acta pauli. tafelband : aus der heidelberger koptischen papyrushandschrift nr. 1 / ed by Schmidt, Carl – Leipzig: JC Hinrichs, 1904 – (= ser Veroeffentlichungen aus der Heidelberger Papyrus-Sammlung) – 1r – 1 – 0-7905-8323-2 – mf#1987-B001 – us ATLA [090]

Acta pharmaceutica nordica – Oxford, England 1989-92 – 1,5,9 – ISSN: 1100-1801 – mf#17721 – us UMI ProQuest [615]

Acta pharmaceutica suecica – Stockholm, Sweden 1977-88 – 1,5,9 – mf#11578 – us UMI ProQuest [615]

Acta physica polonica – v1-48. 1932-73 – 9 – $1080.00 – mf#0008 – us Brook [530]

Acta physico-medica academiae caesareae leopoldino-carolinae naturae curiosorum *see* Miscellanea curiosa sive ephemeridum medico-physicarum germanicarum academiae caesareo-leopoldinae naturae curiosorum

Acta physiologica et pharmacologica latinoamericana – Buenos Aires, Argentina 1989-90 – 1 – (cont: acta physiologica latinoamericana; cont by: acta physiologica, pharmacologica et therapeutica latinoamericana) – ISSN: 0326-6656 – mf#8789.02 – us UMI ProQuest [612]

Acta physiologica latinoamericana – Buenos Aires, Argentina 1950-83 – 5,9 – (cont by: acta physiologica et pharmacologica latinoamericana) – ISSN: 0001-6764 – mf#8789.02 – us UMI ProQuest [612]

Acta physiologica, pharmacologica et therapeutica latinoamericana – Buenos Aires, Argentina 1991+ – 1 – (cont: acta physiologica et pharmacologica latinoamericana) – mf#8789,02 – us UMI ProQuest [612]

Acta physiologica scandinavica – Oxford, England 1954+ – 1,5,9 – ISSN: 0001-6772 – mf#6924.01 – us UMI ProQuest [612]

Acta physiologica scandinavica – Stockholm, Sweden. 1940-61 – 8r – 1 – sw Kungliga [612]

Acta pii pp 10 : modernismi errores reprobantis – Oeniponte [Innsbruck]: Feliciani Rauch, 1907 [mf ed 1985] – 1mf – 9 – 0-8370-2447-1 – mf#1985-0447 – us ATLA [241]

Acta polytechnica scandinavica : applied physics series ph – Helsinki, Finland 1976+ – 1,5,9 – (cont: acta polytechnica scandinavica: physics including nucleonics series) – ISSN: 0355-2721 – mf#7064,01 – us UMI ProQuest [530]

Acta polytechnica scandinavica : physics including nucleonics series – Helsinki, Finland 1958-75 – 1,5,9 – (cont by: acta polytechnica scandinavica: applied physics series ph) – ISSN: 0001-6888 – mf#7064.01 – us UMI ProQuest [530]

Acta polytechnica scandinavica: chemical technology and metallurgy series – Helsinki, Finland 1986+ – 1,5,9 – (cont: acta polytechnica scandinavica: chemistry including metallurgy series) – ISSN: 0781-2698 – mf#7059.02 – us UMI ProQuest [540]

Acta polytechnica scandinavica: chemistry including metallurgy series – Helsinki, Finland 1958-82 – 1,5,9 – (cont by: acta polytechnica scandinavica: chemical technology & metallurgy series) – ISSN: 0001-6853 – mf#7059.02 – us UMI ProQuest [540]

Acta polytechnica scandinavica: civil engineering and building construction series – Helsinki, Finland 1958+ – 1,5,9 – ISSN: 0355-2705 – mf#7060 – us UMI ProQuest [690]

Acta polytechnica scandinavica: electrical engineering series – Helsinki, Finland 1958+ – 1,5,9 – ISSN: 0001-6845 – mf#7061 – us UMI ProQuest [621]

Acta polytechnica scandinavica: mathematics and computer science series – Helsinki, Finland 1975-93 – 1,5,9 – (cont: acta polytechnica scandinavica: mathematics & computing machinery series; cont by: acta polytechnica scandinavica: mathematics & computing in engineering series) – ISSN: 0355-2713 – mf#7062.02 – us UMI ProQuest [510]

Acta polytechnica scandinavica: mathematics and computing in engineering series – Helsinki, Finland 1994+ – 1,5,9 – (cont: acta polytechnica scandinavica: mathematics & computer science series) – ISSN: 1237-2404 – mf#7062,02 – us UMI ProQuest [510]

Acta polytechnica scandinavica: mathematics and computing machinery series – Helsinki, Finland 1958-71 – 1 – (cont by: acta polytechnica scandinavica: mathematics & computer science series) – ISSN: 0001-6861 – mf#7062.02 – us UMI ProQuest [510]

Acta polytechnica scandinavica: mechanical engineering series – Helsinki, Finland 1958+ – 1,5,9 – ISSN: 0001-687X – mf#7063 – us UMI ProQuest [621]

Acta psychologica – Oxford, England 1935+ – 1,5,9 – ISSN: 0001-6918 – mf#42005 – us UMI ProQuest [150]

Acta radiologica – Abingdon, Oxfordshire 1987+ – 1,5,9 – (cont: acta radiologica: diagnosis) – ISSN: 0284-1851 – mf#2542,01 – us UMI ProQuest [616]

Acta radiologica: diagnosis – Abingdon, Oxfordshire 1921-86 – 1,5,9 – (cont by: acta radiologica) – ISSN: 0567-8056 – mf#2542.01 – us UMI ProQuest [616]

Acta radiologica: oncology – Abingdon, Oxfordshire 1984-86 – 1,5,9 – (cont: acta radiologica: oncology, radiation, physics, biology; cont by: acta radiologica. oncology) – ISSN: 0349-652X – mf#6042.04 – us UMI ProQuest [615]

Acta radiologica: oncology, radiation, physics, biology – Abingdon, Oxfordhsire 1978 – 1,5,9 – (cont: acta radiologica: therapy, physics, biology; cont by: acta radiologica oncology, radiation therapy, physics & biology) – ISSN: 0348-5196 – mf#6042,04 – us UMI ProQuest [500]

Acta radiologica: oncology, radiation therapy, physics and biology – Abingdon, Oxfordshire 1980-83 – 1,5,9 – (cont: acta radiologica oncology, radiation therapy, physics & biology; cont by: acta oncologica) – ISSN: 0349-652X – mf#6042,04 – us UMI ProQuest [616]

Acta radiologica: therapy, physics, biology – Abingdon, Oxfordshire 1970-76 – 1,5,9 – (cont by: acta radiologica: oncology, radiation, physics, biology) – ISSN: 0567-8064 – mf#6042.04 – us UMI ProQuest [530]

Acta sanctae sedis – 1(1865)-41(1908/09) – 386mf – 9 – €736.00 – ne Slangenburg [240]

Acta sanctae sedis : ephemerides romanae a ssmo d n pio pp x authenticae et offícales apostolicae – Killen TX 1865-1908 – 1 – mf#3156 – us UMI ProQuest [240]

Acta sanctorum / ed by Bollandus, J & Henschenius, G – ed novissima. Paris. v1-69. 1863-1940 – €3206.00 set – (individual titles also listed) – ne Slangenburg [240]

Acta sanctorum – aprilis – (v1 1-10 1866 €52. v2 11-21 1866 €54. v3 22-30 1866 €56) – ne Slangenburg [240]

Acta sanctorum – augustus – (v1 1-4 1867 €50. v2 5-12 1867 €40. v3 13-19 1867 €42. v4 20-24 1867 €46. v5 25-26 1868 €54. v6 27-31 1868 €46) – ne Slangenburg [240]

Acta sanctorum belgii selecta / ed by Ghesquiere, J – Bruxellis. v1-6. 1783-1794 – 123mf – 8 – €235.00 – ne Slangenburg [240]

Acta sanctorum – februarius – (v1 1-6 1863 €52. v2 7-16 1864 €48. v3 17-28 1864 €42) – ne Slangenburg [240]

Acta sanctorum – januarius – (v1 1-11 1863 €44. v2 12-21 1863 €42. v3 22-31 1863 €42) – ne Slangenburg [240]

Acta sanctorum – julius – (v1 1-3 1867 €37. v2 4-9 1867 €44. v3 10-14 1867 €44. v4 15-19 1868 €42. v5 20-25 1868 €43. v6 25-28 1868 €40. v7 29-31 1868 €46) – ne Slangenburg [240]

Acta sanctorum – junius – (v1 1-6 1867 €60. v2 7-11 1867 €32. v3 12-15 1867 €32. v4 16-19 1867 €50. v5 20-21 1867 €54. v6 22-24 1867 martyrologium usuardi monachi €46. v7 25-30 1867 €54) – ne Slangenburg [240]

Acta sanctorum – maius – (v1 1-5 1866 €48. v2 5-11 1866 €44. v3 12-16 1866 €42. v4 17-19 1866 €32. v5 20-24 1866 €37. v6 24-28 1866 €52. v7 29-31 1866 €52. propyleum ad septem tomos, maii 1866 €38) – ne Slangenburg [240]

Acta sanctorum – martius – (v1 1-8 1865 €50. v2 9-18 1865 €52. v3 19-31 1866 €52) – ne Slangenburg [240]

Acta sanctorum martyrum orientalium et occidentalium / Assemanius, S E – Romae. v1-2. 1748 – 2v on 36mf – 9 – €69.00 – ne Slangenburg [240]

Acta sanctorum – november – (v1 1-3 1887 €52. v2/1 4 1894 €46. v2/2 4 1931 €37. v3 5-8 1910 €50. v4 9-10 1920 €38. propyleum ad acta sanctorum novembris 1902 €31. propyleum ad acta sanctorum decembris 1940 €35) – ne Slangenburg [240]

Acta sanctorum o s b / ed by Achery, Luc d' & Mabillon, Jean – Lut. Parisiorum. v1-9. 1668-1701 – €662.00 – ne Slangenburg [240]

Acta sanctorum o s b / ed by d'Achery, L & Mabillon, J – Lut Parisiorum. v1-9. 1668-1701 – 9v on 347mf – 8 – €662.00 – ne Slangenburg [241]

Acta sanctorum o s b / Mabillon, Jean – Venetiis. v1-9. 1733 – 9v on 314mf – 8 – €599.00 – ne Slangenburg [241]

Acta sanctorum – october – (v1 1-2 1866 €42. v2 3-4 1866 €56. v3 5-7 1868 €54. v4 8-9 1866 €57. v5 10-11 1866 €46. v6 12-14 1906 €37. v7/1 15 1869 €44. v7/2 16 1869 €21. v8 17-20 1866 €57. v9 21-22 1869 €50. v10 23-24 1869 €50. kalendarii, oct v11, 25 1870 €57. v12 26-29 1884 €52. v13 29-31 1883 €52. vol complectens auctoria octobris, 1877 €14. index hagiologicus actorum sanctorum decem priorum mensium 1875 €29) – ne Slangenburg [240]

Acta sanctorum ordinis st benedicti / ed by Mabillon, Jean & Ruinart, Thierry – Seculum 1-6. (500-1100). Paris, 1668-1701 [mf ed Hildesheim 1996] – 9v on 96mf – 9 – diazo €410.00 silver €498.00 – gw Olms [931]

Acta sanctorum – september – (v1 1-3 1868 €42. v2 4-6 1869 €46. v3 7-11 1868 €50. v4 12-14 1868 €44. v5 15-18 1866 €56. v6 19-24 1867 €52. v7 25-28 1867 €50. v8 29-30 1867 €46) – ne Slangenburg [240]

Acta scientiarum mathematicarum – Budapest, Hungary 1976+ – 1,5,9 – ISSN: 0001-6969 – mf#9178 – us UMI ProQuest [510]

Acta seminarii philologici erlangensis – Erlangae. v1-5. 1878-1891 – 42mf – 8 – mf#H-358 – ne IDC [400]

Acta Societatis Ophthalmologicae Japonicae *see* Nippon ganka gakkai zasshi

Acta societatis pro fauna et flora fennica – Helsingforsiae, 1875-1934. v1-57 – 331mf – 9 – mf#8553 – ne IDC [590]

Acta sociologica – Abingdon, Oxfordshire 1983+ – 1,5,9 – ISSN: 0001-6993 – mf#13020 – us UMI ProQuest [301]

Acta ss d n pii pp 9 : ex quibus excerptus est syllabus editus die 8 decembris 1864 – Romae: Typis Rev Camerae Apostolicae, 1865 [mf ed 1986] – 1mf – 9 – 0-8370-8463-6 – mf#1986-2463 – us ATLA [241]

Acta ss nerei et achillei / Achelis, Hans – Leipzig, 1893 – (= ser Tugal 1-11/2)) – 2mf – 9 – €5.00 – ne Slangenburg [240]

Acta ss nerei et achillei : text und untersuchung / Achelis, Hans – Leipzig: J C Hinrichs, 1893 [mf ed 1989] – (= ser Tugal 11/2) – 1mf – 9 – 0-7905-1800-7 – (discussion in german, text in greek & latin. incl bibl ref & ind) – mf#1987-1800 – us ATLA [240]

Acta Synodi Nationalis... *see* Dordrechti habitae anno 1618 et 1619...

Acta synodi nationalis dordracense 1618-1619 – Leiden, 1620 – 40mf – 8 – €76.00 – ne Slangenburg [240]

Acta synodi tridentinae : cum antidoto / Calvin, J – [Geneva: Jean Girard], 1547 – 4mf – 9 – mf#CL-27 – ne IDC [240]

Acta tropica – Oxford, England 1990+ – 1,5,9 – ISSN: 0001-706X – mf#42470 – us UMI ProQuest [616]

Acta van de nederl synode der zestiende eeuw / ed by Rutgers, F – Utrecht, 1889 – (= ser Marnix vereeniging 2/3) – 12mf – 8 – €27.00 – ne Slangenburg [242]

Acta victoriana – v116 n1 – 1 – mf#5265748 – us WHS [071]

Acta vitaminologica et enzymologica – Milan, Italy 1947-85 – 1,5,9 – ISSN: 0300-8924 – mf#8742 – us UMI ProQuest [574]

Acta zoologica et pathologica antverpiensia – Antwerp, Belgium 1974-80 – 1,5,9 – (backfiles: v55-57 1953-73*, v76-82 1981-feb 1992*// (1). former title(s): societe royale de zoologie d'anvers. bulletins de la societe royale de zoologie d'anvers (juil 1953-mars 1966)) – ISSN: 0001-7280 – mf#7519 – us UMI ProQuest [636]

Actas / Mexico. Convencion Nacional Bautista – v1-2. 1903-20, 1921-27 – 1 – $27.93 – us Southern Baptist [242]

Actas capitulares (1479-1510) – Cordoba – 1r – 5,6 – sp Cultura [946]

Actas capitulares (anno 1518) – Avila – 1r – 5,6 – sp Cultura [946]

Actas capitulares (anno 1520) – Avila – 1r – 5,6 – sp Cultura [946]

Actas capitulares (anno 1522-1542) – Avila – 1r – 5,6 – sp Cultura [946]

Actas capitulares (anno 1541-1599) – Avila – 1r – 5,6 – sp Cultura [946]

Actas capitulares (anno 1543-1556) – Caceres – 1r – 5,6 – sp Cultura [946]

Actas capitulares sobre el gran capitan (anno 1498-1512) – Cordoba – 1r – 5,6 – sp Cultura [946]

Actas das sessoes da sociedade de geographia de lisboa – Lisboa: Imprensa Nacional. v1-13. 1876/81-1893 – 6r – 1 – us CRL [946]

Actas de la comision de armamento y defensa de la provincia de extremadura – 1835 – 9 – sp Bibl Santa Ana [355]

Actas de sessao / Angola. Conselho Legislativo – 1935-36, 1956-Nov. 11, 1963 – 1 – us NY Public [324]

Actas del cabildo de caracas : tomo 1: caracas, 1943 / Bayle, Constantino – Madrid: Razon y Fe, 1946 – 1 – sp Bibl Santa Ana [946]

Actas del congreso : decimocuarto congreso internacional de las entidades fiscalizadoras superiores, washington, dc, oct de 1992 – Washington DC: General Accounting Office de los Estados Unidos [1992?] [mf ed 1997] – 3mf – 9 – us GPO [336]

Actas y trabajos del segundo congresso medico nacional, habana, febrero 24-28 de 1911 / Congreso medico nacional cubano (2nd : 1911 : Havana, Cuba) – Habana, Cuba: La Universal 1911 [mf ed 1990] – 1r – 1 – us UF Libraries [610]

Acte d'amendement des municipalites et des chemins du bas-canada, de 1856 : avec sommaire et index = The lower Canada municipal and road amendment act of 1856 / Canada. Laws, Statutes, etc – Toronto: Impr par S Derbishire et G Desbarats, 1856 [mf ed 1983] – 1mf – 9 – (with summary and ind) – mf#SEM105P185 – cn Bibl Nat [348]

L'acte d'amendement seigneurial de 1859 : 22 victoriae, cap 48 = The seigniorial amendment act of 1859 – Toronto: imprime par Stewart Derbishire & George Desbarats, 1859 [mf ed 1983] – 1mf – 9 – mf#SEM105P274 – cn Bibl Nat [348]

L'acte des bois / Rutledge, Jean Jacques – [Montreal?: s.n, 1845?] [mf ed 1991] – 1mf – 9 – 0-665-90448-7 – (in dble clms) – mf#90448 – cn Canadiana [343]

Acte des municipalites et des chemins de 1855... : les actes de la representation parlementaire...et les actes seigneuriaux... = Lower Canada municipal and road act 1855...the parliamentary representation acts... / Canada (Province) – Quebec: Impr par Stewart Derbishire & George Desbarats, 1855 [mf ed 1983] – 3mf – 9 – mf#SEM105P192 – cn Bibl Nat [348]

Acte d'incorporation et constitution et reglements du grand conseil de l'association catholique de bienfaisance mutuelle du canada et de ses succursales, revisee en août 1896 / Association catholique de bienfaisance mutuelle du Canada. Grand conseil – Levis: imprime par Mercier & cie, [1896?] [mf ed 1994] – 9 – mf#SEM105P – cn Bibl Nat [366]

L'acte municipal du bas canada de 1860 : 23 vict cap 61 = The lower canada municipal act of 1860 / Canada (Province) – Quebec: impr par Stewart Derbishire et George Desbarats... 1860 – 2mf – 9 – mf#SEM105P195 – cn Bibl Nat [348]

Acte pour abroger certaines lois y mentionnees pour mieux pourvoir a la defense de cette province et pour en regler la milice / Canada (Province)– Montreal: Stewart Derbishire & George Desbarats, 1846 [mf ed 1993] – 1mf – 9 – mf#SEM105P1784 – cn Bibl Nat [323]

Acte pour abroger certains actes y mentionnes : et pour amender, refondre et resumer en un seul acte les diverses dispositions des statuts maintenant en vigueur pour regler les elections des membres qui representent le peuple de cette province a l'assembl / Canada (Province) – Quebec: Stewart Derbishire & George Desbarats, 1851 [mf ed 1992] – 2mf – 9 – (with ind) – mf#SEM105P1459 – cn Bibl Nat [325]

Acte pour abroger certains actes y mentionnes et etablir de meilleures dispositions relativement a l'admission des arpenteurs et a l'arpentage des terres en cette province / Canada (Province) – Quebec: impr Augustin Cote, 1855 – 1mf – 9 – mf#SEM105P1789 – cn Bibl Nat [323]

Acte pour amender et consolider les dispositions de l'ordonnance pour incorporer la cite et ville de montreal : et d'une certaine ordonnance et de certains actes amendant cette ordonnance, et pour investir de certains autres pouvoirs la corporation de la dite cite de montreal / Canada (Province) – Toronto: impr par Stewart Derbishire & George Desbarats, 1851 [mf ed 1995] – 1mf – 9 – mf#SEM105P1558 – cn Bibl Nat [350]

Acte pour amender et refondre les differents actes concernant le notariat / Canada (Province) – Quebec: impr par Charles-Francois Langlois, [1875] (mf ed 1990) – 1mf – 9 – mf#SEM105P1279 – cn Bibl Nat [348]

Acte pour amender l'acte municipal refondu du bas-canada / Canada (Province) – Quebec: impr par Stewart Derbishire & George Desbarats, 1861 [mf ed 1990] – 1mf – 9 – mf#SEM105P1278 – cn Bibl Nat [348]

Acte pour amender les actes de judicature du bas-canada : 20 victoriae, cap 44 / Canada (Province) – Toronto: impr par Stewart Derbishire & George Desbarats, 1857 [1995] – 1mf – 9 – mf#SEM105P2028 – cn Bibl Nat [347]

Acte pour amender les lois en force : concernant la vente des liqueurs enivrantes et l'octroi de licences a cet effet et pour reprimer autrement les abus resultant de ce commerce / Canada (Province) – Quebec: impr par George Desbarats & Malcom Cameron, 1864 [mf ed 1983] – 1mf – 9 – mf#SEM105P182 – cn Bibl Nat [348]

Acte pour amender les lois relatives a la milice de cette province, et les rendre parmanentes : 22 victoriae, cap 18 / Canada (Province) – Toronto: impr par Stewart Derbishire & George Desbarats, 1859 [mf ed 1983] – 1mf – 9 – mf#SEM105P178 – cn Bibl Nat [971]

Acte pour augmenter la representation du peuple de cette province en parlement : 16 vict cap 152 = An act to enlarge the representation of the people of this province in parliament / Canada (Province) – Quebec: imprime par Stewart Derbishire & George Desbarats...1854 [mf ed 1999] – 1mf – 9 – mf#SEM105P3151 – cn Bibl Nat [325]

Acte pour etablir des dispositions temporaires pour le gouvernement du bas-Canada *see* Anno primo victoriae reginae, magnae britanniae et hiberniae

Acte pour faire de plus amples dispositions pour l'incorporation de la ville des trois-rivieres : 20 victoria, cap 129 / Canada (Province) – Toronto: impr par Stewart Derbishire & George Desbarats, 1857 [mf ed 1983] – 1mf – 9 – mf#SEM105P184 – cn Bibl Nat [348]

Acte pour incorporer la compagnie d'assurance de quebec contre le feu – Quebec: Impr par ordre de la dite assemblee, par A Cote, 1862 – 1mf – 9 – mf#48152 – cn Canadiana [368]

Acte pour la decision sommaire des petites causes / Bas-Canada – Quebec: impr par MM Frechette & cie, 1834 [mf ed 1995] – 1mf – 9 – mf#SEM105P1855 – cn Bibl Nat [348]

Acte pour l'abolition des droits et devoirs feodeaux dans le bas-canada : 18 vic cap 3 = An act for the abolition of feudal rights and duties in lower Canada / Canada (Province) – Quebec: impr par Stewart Derbishire & George Desbarats, 1854 [mf ed 1983] – 1mf – 9 – mf#SEM105P264 – cn Bibl Nat [348]

Acte pour pourvoir a la decision sommaire des petites causes, dans le bas-canada : cap 19, 7 victoria, 1843 / Canada (Province) – Kingston: impr par S Derbishire & G Desbarats, 1844 [mf ed 1983] – 1mf – 9 – mf#SEM105P189 – cn Bibl Nat [348]

Acte pour pourvoir plus amplement a l'incorporation de la ville de st hyacinthe : et pour etendre ses limites: 16 victoria, cap 236 / Canada – Quebec: impr par Stewart Derbishire & George Desbarats, 1853 [mf ed 1992] – 1mf – 9 – mf#SEM105P1559 – cn Bibl Nat [348]

Acte pour regler la milice de cette province et pour abroger les actes maintenant en force a cette fin : 18 vict cap 77 / Canada (Province) – Quebec: impr par Stewart Derbishire & George Desbarats...1855 [mf ed 1982] – 1mf – 9 – mf#SEM105P180 – cn Bibl Nat [348]

Die acten des karpus, des papylus und der agathonike / Harnack, Adolf von – Leipzig, 1888 – (= ser Tugal 1-3/4b) – 1mf – 9 – €3.00 – ne Slangenburg [240]

Acten, urkunden und nachrichten zur neuesten kirchengeschichte / ed by Schneider, Christian Wilhelm – Weimar 1789-93 – (= ser Dz. abt theologie) – v1-3 on 18mf – 9 – €180.00 – mf#k/n2225 – gw Olms [240]

Acten van de classicale en synodale vergaderingen der verschillenden gemeenten in het land van cleef, sticht van keulen en aken 1571-1589 (de werken..2,2) / ed by Janssen, H Q & Toorenbergen, J J van – Utrecht, 1882 – (= ser De werken der marnix-vereeniging) – 8 – 3mf – ne Slangenburg [242]

Acten van de colloquia der nederlansche gemeenten in engeland 1575- (de werke...2/1) / ed by Toorenbergen, J J van – Utrecht, 1872 – (= ser De werken der marnix-vereeniging) – ne Slangenburg [242]

Actensammlung zur geschichte der zuercher reformation in den jahren 1519-1533 / Egli, E – Zuerich, 1879 – 10mf – 9 – mf#ZWI-28 – ne IDC [240]

Actes : premier colloque de bande dessinee de montreal / Colloque de bande dessinee de Montreal – [Montreal]: Analogon, 1986 [mf ed 2003] – 3mf – 9 – mf#SEM105P7240 – cn Bibl Nat [323]

Actes and monuments of matters most speciall and memorable, happenyng in the church : with an universall history of the same... / Foxe, J – London, 1583. 2v – 39mf – 9 – mf#PW-13 – ne IDC [240]

Les actes apocryphes de l'apotre andre : les actes d'andre et de mathias, de pierre et d'andre et les textes apparentes / Flamion, Joseph – Louvain: Bureaux du Recueil, 1911 [mf ed 1993] – (= ser Recueil de travaux [universite catholique de louvain (1835-1969)] 33) – 1mf – 9 – 0-524-05669-2 – (incl bibl ref) – mf#1992-0519 – us ATLA [920]

Actes authentiques des eglises reformees : de france, germanie, grande bretaigne, pologne, hongrie, pais bas. etc. touchant la paix et charite fraternelle / Blondel, D – Amsterdam, 1655 – 2mf – 9 – mf#PRS-116 – ne IDC [242]

Actes concernant l'education et les ecoles dans le bas-canada : etant les chapitres 15, 16 et 17 des statuts refondus pour le bas-canada proclames et publies en vertu de l'acte 23 vic cap 56, ad 1860 / Canada (Province) – Quebec: impr par Stewart Derbishire & George Desbarats, 1861 [mf ed 1983] – 1mf – 9 – mf#SEM105P183 – cn Bibl Nat [370]

Les actes de la journee imperiale, tenue en la cite de regespourg, aultrement dicte ratispone... : desquelz l'inventoire sera recite en la paige suyvante / [Calvin, J] – [Geneva: Jean Girard], 1541 – 5mf – 9 – mf#CL-19 – ne IDC [240]

Actes de la societe du chemin de fer ottoman – 1889-1913 – 38mf – 9 – $625.00 – (in french and ottoman) – us MEDOC [338]

Les actes de paul et ses lettres apocryphes : introduction, textes, traduction et commentaire / Vouaux, Leon – Paris: Letouzey & Ane, 1913 [mf ed 1990] – (= ser Les apocryphes du nouveau testament) – 1mf – 9 – 0-8370-1782-3 – (text in greek & french; comm in french) – mf#1987-6170 – us ATLA [226]

Actes de philippe 1er, dit le noble : comte et marquis de namur (1196-1212) / Walraet, M – Brussel, 1949 – 8mf – 8 – €17.00 – ne Slangenburg [920]

Les actes de sa saintete le pape jean 23 octobre 1958-janvier 1962 : bibliographie analytique / Saint-Philippe-Andre, soeur – 1963 [mf ed 1979] – (= ser Bibliographies du cours...1947-66) – 3mf – 9 – (with ind; pref by monsieur le chanoine achille couture) – mf#SEM105P4 – cn Bibl Nat [241]

Actes d'education elementaire : et pour l'etablissement d'ecoles normales suivis des circulaires y relatives nos 9, 12 et 15 et des instructions et tableaux du surintendant de l'education pour le bas-canada / Canada (Province) – Quebec: Stewart Derbishire & George Desbarats, 1852 [mf ed 1992] – 1mf – 9 – mf#SEM105P1521 – cn Bibl Nat [370]

Actes d'education elementaire : et les circulaires y relatives nos 9 et 12 du surintendant de l'education pour le bas-canada = Statutes relating to elementary education with the circulars nos 9 and 12 of the superintendent of education for lower Canada / Canada (Province) – Montreal: impr par Stewart Derbishire & Georges Desbarats, 1849 [mf ed 1984] – 2mf – 9 – mf#SEM105P323 – cn Bibl Nat [370]

Les actes des apotres – Paris [1789]-1791 [mf ed Hildesheim 1995-98] – 11v on 63mf – 9 – €630.00 – 3-487-26297-5 – gw Olms [226]

Les actes des apotres – Pelletiers, etc. Paris (VII-XI). 1790-91 – 1 – fr ACRPP [240]

Les actes des apotres : traduction nouvelle avec introduction et notes / Loisy, A – Paris, 1925 – 6mf – 8 – €14.00 – ne Slangenburg [226]

Actes des comtes de namur de la premiere race (946-1196) / Rousseau, F – Bruxelles, 1936 – €19.00 – ne Slangenburg [949]

Actes des etats generaux des anciens pays-bas : tom 1 (1424-1477) / Cuvelier, J – Bruxelles, 1948 – €23.00 – ne Slangenburg [240]

Les actes des martyrs de l'eglise copte : etude critique / Amelineau, Emile – Paris: E Leroux, 1890 [mf ed 1990] – 1mf – 9 – 0-7905-5801-7 – mf#1988-1801 – us ATLA [243]

Les actes des martyrs de l'egypte : tires des manuscrits coptes de la bibliotheque vaticane et du musee borgia – Paris: E Leroux, 1886 [mf ed 1990] – 1mf – 9 – 0-7905-6068-2 – (no more publ) – mf#1988-2068 – us ATLA [240]

Actes des municipalites du bas-canada / Canada (Province) – Toronto: impr par Stewart Derbishire & George Desbarats, 1851 [mf ed 1983] – 2mf – 9 – mf#SEM105P170 – cn Bibl Nat [348]

Actes des princes-eveques de liege : hugues de pierrepont 1200-1229 / Poncelet, E – Bruxelles, 1941 – €27.00 – ne Slangenburg [240]

Actes du 4e congres international d'histoire des religions : tenua leide du 9e-13e sep 1912 – Leide: EJ Brill, 1913 [mf ed 1991] – 1mf – 9 – 0-524-01505-8 – (in french, german & english) – mf#1990-2481 – us ATLA [200]

Actes du colloque sur la participation des communautes culturelles au devenir du quebec – [Quebec]: Maison internationale de Quebec, 1991 [mf ed 1999] – 3mf – 9 – mf#SEM105P3116 – cn Bibl Nat [305]

Les actes du concile de trente : avec le remede contre la poison / Calvin, J – [Geneva: Jean Girard], 1548 – 4mf – 9 – mf#CL-51 – ne IDC [240]

Actes et deliberations du premier congres catholique canadien francais tenu a quebec les 25, 26, et 27 juin 1880 / Congres catholique canadien francais (1er : 1880 : Quebec) – Montreal: E Senecal, 1880 – 5mf – 9 – mf#00126 – cn Canadiana [241]

Les actes et ordonnances revises du bas-canada / Canada (Province) – Montreal: impr par S Derbishire et G Desbarats, 1845 [mf ed 1998] – 8mf – 9 – mf#SEM105P1992 – cn Bibl Nat [348]

Actes pour promouvoir l'education dans le bas-canada / Canada (Province) – Toronto: impr par Stewart Derbishire & George Desbarats, 1857 [mf ed 1983] – 1mf – 9 – mf#SEM105P202 – cn Bibl Nat [370]

Actes relatifs aux chemins a barrieres et ponts dans et pres quebec / Canada (Province) – Quebec: impr par Stewart Derbishire & George Desbarats, 1853 [mf ed 1990] – 1mf – 9 – mf#SEM105P1277 – cn Bibl Nat [348]

Actes relatifs aux pouvoirs : aux devoirs et a la protection des juges de paix dans le bas-canada avec un index analytique complet = Acts relating to the powers, duties and protection of justices of the peace in lower Canada... / Canada (Province) – Quebec: impr par Stewart Derbishire & George Desbarats, 1853 [mf ed 1983] – 2mf – 9 – (with ind) – mf#SEM105P175 – cn Bibl Nat [348]

Actes relatifs aux pouvoirs : aux devoirs et a la protection des juges de paix dans le bas canada, avec un index analytique complet / Canada (Province) – Toronto: impr par S Derbishire & G Desbarats, 1858 [mf ed 1983] – 2mf – 9 – mf#SEM105P177 – cn Bibl Nat [348]

Les actes seigneuriaux savoir : l'acte seigneurial de 1854...l'acte d'amendement seigneurial de 1855...l'acte d'amendement seigneurial de 1856...avec un index copieux = The seigniorial acts: viz. the seigniorial act of 1854...the seigniorial amendment act of 1855... – Toronto: impr par S Derbishire & G Desbarats, 1856 [mf ed 1983] – 1mf – 9 – mf#SEM105P267 – cn Bibl Nat [348]

Acti del congresso internazionale di scienze storiche – (Roma, 1-9 apr 1903) v3 Roma, 1906 – 13mf – 9 – €3.00 – ne Slangenburg [500]

Acting on a vision : agency conversion at kfi, millinocket, maine / Walker, Pam – Syracuse NY: Center on Human Policy, Syracuse Uni; [Washington DC]: US Dept of Education, Office of Educ Research & Improvement...[2000] [mf ed 2001] – 1mf – 9 – us GPO [362]

Acting the dance : an application of the stanislavski acting method / Plumlee, Linda K – 1989 – 106p 2mf – 9 – $8.00 – us Kinesology [790]

L'action – Manchester NH: Franco-American Publ Corp [ca 1950]- (wkly) – 1 – (ceased in 197-?) – mf#SEM35P313 – cn Bibl Nat [071]

L'action – [Paris]: Impr Balitout, apr 6,8-9 1871 – (Filmed as part of: Commune de Paris newspapers. Newspapers on these reels are filmed chronologically, not alphabetically) – us CRL [074]

L'action : journal des etudiants(es) en sciences comptables de l'uqam – Montreal: Groupe Action. v1 n1 (12 sep 1983)- [biwkly] [mf ed 1988] – 9 – mf#SEM105P990 – cn Bibl Nat [500]

L'action – Paris. 29 mars 1903-1924 – 1 – fr ACRPP [073]

L'action – Port-au-Prince, Haiti: Impr du Petit Impartial, 1 annee n2-106. 1juin 1929-5 juin 1930 – us CRL [079]

L'action – Port-au-Prince, Haiti: L'Action, 2eme annee n101-5eme annee n363. 9 sep 1948-25 aout 1952 – us CRL [079]

Action – 1980 jan-1993 dec – 1 – mf#3316389 – us WHS [071]

Action – 1982 winter-1986 autumn – 1 – mf#1363045 – us WHS [071]

Action – 1984 jan 11-1994 dec 23 – 1 – mf#916671 – us WHS [071]

Action – 1984 nov/dec-1988 sep/oct – 1 – mf#1051665 – us WHS [071]

Action – Leopoldville 1964(nov 28)-1965(jan 9) – 27mf – 9 – (cont as: le populaire [leopoldville 1965(aug 21)-1966(jan 15)]. cont as: afrique populaire [leopoldville 1966(feb 5)-1966(nov 19)]) – mf#A-666 – ne IDC [079]

Action – [New York: Action], feb 26 1973-feb 25 1974 – 1r – 1 – us CRL [071]

Action – New York, NY. 1969-83 – 1 – us AJPC [071]

Action – Paris. 9 sept 1944-9 mai 1952 – 1 – fr ACRPP [073]

Action – v1 n1-v2 n20 [1968 jul 12-1971 jun/jul] – 1 – mf#1095451 – us WHS [071]

Action – v1 n2-v3 n2 [1945 jan-1947 mar] – 1 – mf#971629 – us WHS [071]

Action – v13 n1 [1973 apr]; v14 n2 [1975 dec]; v15-16 [1976-77]; v17 n1,3-5 [1978 apr,oct-dec]; v18 n1-9,11 [1979 jan-oct, dec]; v22 n7 [1984 nov/dec]; v23 n1-3 [1985 jan/feb-aug]; v24 n1-6 [1986 jan/feb-nov/dec]; v25 n3 [1987 – 1 – mf#632465 – us WHS [071]

Action – v4 n1-to date [1971 jan 18-to date]; 1980 jan-1993 may – 1 – mf#630439 – us WHS [071]

Action – West Hollywood CA 1966-75 – 1,5 – ISSN: 0001-7361 – mf#10525 – us UMI ProQuest [790]

Action *see* Fascist and anti-fascist newspapers

Action alert – n2-9 [1974 jul 18-1975 oct 15]; [1976 dec 10-1977 jun 24] – 1 – mf#615857 – us WHS [071]

Action and reaction – [New York: Arabic-English Newspaper Inc]. apr 4 1977-apr 1 1983 – 1r – 1 – us CRL [071]

L'action antimilitariste – Marseille. Mens. sept 1904-janv 1905 – 1 – fr ACRPP [320]

Action at pft – v3 n4-28 [1977 oct 5-1978 aug 30, 1978 nov 8-dec 5]; v4-23; [1980 feb-1981 jun]; [1981 jul-1982 feb 2] – 1 – mf#905958 – us WHS [071]

Action bulletin – [1967 apr 10]; n4-15 [1967 aug 4-1969 feb]; ns: n1-3; [1969 may 26-aug 1] – 1 – mf#942257 – us WHS [071]

Action bulletin – v1 n1-v9 n4 [1973 dec-1981]; v9 n5 [1982 jan]-v22 n3 [1993 dec] – 1 – mf#624422 – us WHS [071]

L'action chretienne des etudiants russes : messager orthodoxe – Paris, 1965-1972 – 28mf – 9 – mf#R-10540 – ne IDC [243]

Action concertee cablodistribution : rapport final / Fonds FCAC – [Sainte-Foy]: le Fonds, 1981 [mf ed 2003] – 1mf – 9 – mf#SEM105P3570 – cn Bibl Nat [971]

L'action d'art – Paris. no.1-18. Fevr-dec 1913.Mq. no. 7,12 – 1 – fr ACRPP [700]

L'action de marieville – Waterloo. 1re annee n1 (24 jan 1951) (bimthly) [mf ed 1973] – 1r – 1 – (ceased publ in 1951?) – mf#SEM35P12 – cn Bibl Nat [971]

L'action directe – Paris. n4-31. fevr-sept 1908 – 1 – fr ACRPP [325]

L' action du batiment *see* Moniteur de l'entreprise et de l'industrie

L'action francaise – Paris, Lyon. Quot. 21 mars 1908-24 aout 1944 – 1 – fr ACRPP [074]

L'action francaise – Paris, may 13 1938-jun 6 1940; mar 26-dec 17 1941 – 9r – 1 – us CRL [074]

Action in teacher education – Reston VA 1978+ – 1,5,9 – ISSN: 0162-6620 – mf#11661 – us UMI ProQuest [370]

L'action indochinoise – no. 1. Saigon. 23 aout 1928 – 1 – fr ACRPP [079]

L'action liberale – Montreal: Societe de publication L'Action liberale. 1re annee n1 (11 oct 1919) [wkly] [mf ed 1992] – 6mf – 9 – (ceased in 192-?) – mf#SEM105P1572 – cn Bibl Nat [071]

Action linkage networker – n1-35 [1986 oct-1989 nov] – 1 – mf#1616796 – us WHS [071]

L'action nationale – Port-au-Prince. oct 30 1931-jun 19 1934 (incomplete) – 1 – us NY Public [079]

Action news – v4 n45-v12 n22 [1974 aug 23-1982 oct 29]; v13 n2-4; [1983 feb 11-mar 11] – 1 – mf#618623 – us WHS [071]

Action, oct-dec 1931, feb 1936-jun 1940 – (= ser Fascism and reactions to fascism in britain 1918-1989) – 5r – 1 – mf#97591 – uk Microform Academic [072]

Action of the commission of assembly in professor smith's case : explained and vindicated / Adam, John – 3d ed. Glasgow: David Bryce, 1881. Princeton: Speer Lib, and Dep of Photodup, U of Chicago Lib, 1978 (1r); Evanston: American Theol Lib Assoc, 1984 (1r) – (= ser Case of william robertson smith in the free church of scotland) – 1 – 0-8370-0629-5 – mf#1984-6274 – us ATLA [242]

Action of the free church commission ultra vires : a reply to the "action of the commission explained and vindicated by the rev j adam, d.d." / Blackie, Walter Graham – Glascow: Blackie, 1881. Princeton: Speer Lib, and Dep of Photodup, U of Chicago Lib, 1978 (1r); Evanston: American Theol Lib Assoc, 1984 (1r) – (= ser Case of william robertson smith in the free church of scotland) – 1 – 0-8370-0632-5 – mf#1984-6289 – us ATLA [242]

L'action regionaliste – Paris. fevr 1902; 1903-juin juil 1914; fevr-sept oct 1920; 1933-mars 1940; 1953-61; janv mars-juil sept 1968 – 1 – fr ACRPP [073]

Action report – v3 n2-v5 n11 [1975 feb-1977 nov] – 1 – mf#667598 – us WHS [071]

Action sociale de la femme et le livre francais – v1-39 10 Apr 1902-Nov 1939. Jan 1903 wanting – 1 – $109.00 – us L of C Photodup [305]

L'action syndicale – Lens. 17 janv 1904-2 oct 1910. – 1 – (suite de: reveil syndical) – fr ACRPP [320]

L'action syndicale *see* Le reveil syndical

Actiongram – 1980 dec 18 [cambodian crisis campaign] – 1 – mf#665547 – us WHS [071]

Actions and reactions / Kipling, Rudyard – Toronto: Macmillan, 1909 [mf ed 1995] – 9 – 0-665-77250-5 – mf#77250 – cn Canadiana [880]

The actis and deidis of schir william wallace / Henry the Minstrel – 1570 – 9 – $25.00 – us Scholars Facs [810]

La actitud internacional: ceguera para el verdadero peligro – Madrid, 1936? Fiche W 702. (Blodgett Collection of Spanish Civil War Pamphlets) – 9 – us Harvard [946]

Active pacifist : a newsletter of citizens [sic] actionfor lasting security – 1987 oct – 1 – mf#2297761 – us WHS [071]

Actividades de educacion fundamental en la provincia de badajoz / Ministerio de Educacion Nacional. Junta Nacional contra el analfabetismo – Madrid: Sucesores de Rivadeneyra S.A. – sp Bibl Santa Ana [370]

Actividades dos missionarios... / Castelo-Branco, Fernando A – Madrid: Archivo Ibero Americano, 1960 – 1 – sp Bibl Santa Ana [240]

Activist – Buffalo NY 1969-71 – 1 – mf#7786 – us UMI ProQuest [320]

Activist : newspaper of buffalo youth against war and fascism – 1969 sep-1971 jun – 1 – mf#1109533 – us WHS [320]

Activist – Oberlin OH 1960-75 – 1,5 – ISSN: 0001-7590 – mf#2328 – us UMI ProQuest [073]

L'activitea1 de l'assemblea1e parlementaire europea1ene / European Parliamentary Assembly – Paris, dec 1959 jan 1960-aug sep 1961 – 1 – us NY Public [341]

Activites en geometrie : symetrie: document de travail – [Quebec: Ministere de l'education...1979] [mf ed 1996] – 1mf – 9 – mf#SEM105P2530 – cn Bibl Nat [510]

Les activites et les publications du centre canadien des cercles lacordaire : et sainte-jeanne d'arc depuis sa fondation en decembre 1939 / Lacroix, Denise – 1953 [mf ed 1979] – (= ser Bibliographies du cours...1947-66) – 1mf – 9 – (pref by ubald villeneuve) – mf#SEM105P4 – cn Bibl Nat [012]

Activites ludiques, sensorielles et naturalistes : aux cycles 2 et 3 – 1999 – 44mf+36p+annexes – 9 – €14.48 – mf#250B0139 – fr CRDP [333]

Activities, adaptation and aging : the journal of activities management / ed by Couture, Linea – ISSN: 0192-4788 – National association of activity professionals members – us Haworth [618]

Activity location cards of the fleet post office, san francisco, california, 1940-1945 / U.S. Post Office – (= ser Records of naval districts and shore establishments) – 2r – 1 – mf#T1015 – us Nat Archives [380]

Activity of business under the national policy / Liberal-Conservative Party – [Toronto?: s.n, 1887?] [mf ed 1993] – 1mf – 9 – 0-665-91474-1 – (in dble clms. original iss in ser: facts for the people n29) – mf#91474 – cn Canadiana [325]

Acto academico de homenaje a justo sierra : en ocasion de la entrega de la medalla conmemorativa del centenario del ilustre educador enviada...(9 de diciembre de 1949) / Universidad de Santo Domingo – Ciudad Trujillo, Dominican Republic: Pol Hnos [1950] [mf ed 1991] – 1r – 1 – us UF Libraries [378]

Acto academico para rendir tributo a la memoria de s.s. el papa paulo iiio : fundador de la universidad en ocasion del 400o aniversario de su muerte : 10 de noviembre, 1949 / Universidad de Santo Domingo – Ciudad Trujillo, Dominican Republic: Editora Montalva 1949 [mf ed 1991] – 1r – 1 – us UF Libraries [241]

Acton 1734-1849 – Oxford, MA (mf ed 1994) – (= ser Massachusetts vital record transcripts to 1850) – 7mf – 9 – 0-87623-197-0 – (mf 1t-3t: births & deaths 1735-1816. mf 3t-4t: intentions 1734-1813. mf 4t: marriages 1739-1812. mf 4t-5t: births 1791-1851. mf 5t: deaths 1801-44; out-of-town marriages 1738-99. mf 5t-6t: intentions 1813-50. mf 6t-7t: marriages 1812-49. mf 7t: births & deaths 1844-49) – us Archive [978]

Acton 1735-1910 – Provo UT (mf ed 2003) – (= ser Massachusetts vital records) – 19v on 94mf – 9 – 0-87623-431-7 – (mf1-11: town records 1735-97. mf12-21: town records 1798-1818. mf22-33: town records 1819-39. mf34-45: town records 1838-62. mf46-47: rebellion record 1861-65. mf48-51: militia 1810-23. mf52-58: vital records 1735-1844. mf58: death index 1800-44. mf59: marriage intentions 1737-63. mf60-61: marriages 1738-1844. mf62-65: vital records 1734-1812. mf66-69: vital records 1740-1850. mf70-73: birth index 1845-1931. mf74-76: marriage index 1844-1941. mf77-78: death index 1844-1941. mf81-82: vital records 1844-57. mf83-88: vital records 1858-91. mf89-90: births 1892-1910. mf91-92: marriages 1892-1915. mf93-94: deaths 1892-1913) – us Archive [978]

Acton, Eliza *see*
- The english bread-book for domestic use
- Poems
- The voice of the north

Acton free press – Acton, ON. 1875-95 – 7r – 1 – ISSN: 0834-5767 – cn Library Assoc [071]

Acton gazette – Ealing, London 21 oct 1938-9 sep 1988 [mf 8 sep 1939-dec 1950, 1977-88] – 1 – (incorp with: gazette (ealing borough ed); during the printers' strike of 1959 replaced by "middlesex county press" cont as: the acton gazette, shepherds bush gazette, hammersmith post; cont: acton gazette & express, etc [5 apr 1918-14 oct 1938]) – uk British Libr Newspaper [072]

Acton gazette and general district advertiser – [London & SE] Ealing: 1 apr 1871-dec 1873, 1874-6 mar 1880 – 1 – (cont as: acton, chiswick & turnham green gazette [13 mar 1880-1 oct 1892]; acton & chiswick gazette [8 oct 1892-13 jul 1900]; acton gazette [20 jul 1900-29 mar 1918]; chiswick gazette [5 apr 1918-14 oct 1938]; acton gazette [21 oct 1938-1 sep 1939]; acton gazette & express (incorp in ealing gazette 16 sep 1988-1 sep 1989) [8 sep 1939-9 sep 1988]) – uk Newsplan [072]

Acton, Harriet *see* Poems

Acton, Henry *see* Rational and spiritual religion the one thing needful

Acton, John Emerich Edward Dalberg Acton, Baron *see*
- Lectures on modern history
- Lord acton and his circle

Acton (of Aldenham), John Emerich Edward Dalberg, 1st Baron *see*
- Historical essays and studies
- The history of freedom

[Acton-] acton rooster – CA. nov 1900-dec 1913 – 1r – 1 – $60.00 – mf#C02000 – us Library Micro [071]

Acton, Rose *see* Poems

Acton, st mary – [South Cheshire FHS] – (= ser Cheshire monumental inscriptions) – 5mf – 9 – £5.50 – mf#410 – uk CheshireFHS [929]

Actor *see* Eighteenth century journals

Actors and managers of the english and american stage : from the bram stoker collection, the shakespeare centre library, stratford-upon-avon and the john forster collection at the national art library / Stoker, Bram – 2-series coll – 40r – 1 – (ser 1: the papers of henry irving and ellen terry 30r c35-12610; ser 2: the papers of david garrick – correspondence from the john forster coll at the national art library, victoria and albert museum, london 10r c35-12611; with p/g for both series) – mf#C35-12600 – us Primary [790]

Actors and managers of the english and american stage *see* Actors and managers of the english and american stage

Actors' Equity Association *see* Equity news

Actors' equity association : the magazine and executive committee minutes of the actors' equity association – 1913-81 [mf ed Chadwyck-Healey] – 73r – 1 – (actors' equity association magazine 1915-81 [255mf]. actors' equity association council minutes, 1913-70 and executive commitee minutes, 1918-24 [73mf]) – uk Chadwyck [790]

Actors' equity association magazine : equity – equity news – dec 1915-spr 1973 (Equity) aug 1973-dec 1981 (Equity News) – 255mf – 9 – uk Chadwyck [790]

Actorvm apostolicorvm liber doctis simus et utilissimus scholijs illustratus / Sarcerius, E – Basileae, 1540 – 12mf – 9 – mf#TH-1 mf 1291-1302 – ne IDC [242]

Actos oficiales del gobierno provisorio de los estados unidos de colombia : recopilados conforme a lo dispuesto por el decreto de 7 de abril de 1862 – Bogota, Colombia: Echeverria Hermanos 1862 [mf ed 1993] – 1r – 1 – us UF Libraries [323]

Actouka, Marcelino *see* Land tenure and power / a basis for community development planning on ponape

Actrascope – 1978 jan-nov/dec – 1 – mf#496400 – us WHS [071]

The acts and epistles aprokos [complete] : vetkovskoe sobranie [vetka collection] – late 1400s-early 1500s – 10mf – 9 – (russian version) – us UMI ProQuest [090]

Acts and ordinances of the governor and council of new south wales : and acts of parliament enacted for, and applied to, the colony, with notes and index / Callaghan, Thomas – Sydney: W J Row. 2v + app. 1844 – 17mf – 9 – $25.00 – mf#LLMC 96-002 – us LLMC [323]

Acts and pastoral epistles : timothy, titus, and philemon – London: J M Dent; Philadelphia: J B Lippincott 1902 [mf ed 1989] – (= ser The temple bible) – 1mf – 9 – 0-7905-1856-2 – mf#1987-1856 – us ATLA [226]

Acts of english martyrs : hitherto unpublished / Pollen, John Hungerford – London: Burns & Oates, 1891 [mf ed 1990] – 1mf – 9 – 0-7905-5556-5 – (pref by john morris) – mf#1988-1556 – us ATLA [240]

Acts of parliament / Scotland. Parliament – 1124-1707 – 4r – 1 – $200.00 – us Trans-Media [324]

Acts of parliament and bench table orders of the inner temple / England. Inns of Court – London, 1926 – 2mf – 9 – $3.00 – mf#LLMC 84-269 – us LLMC [324]

The acts of saint mary magdalene considered in a series of discourses : as illustrating certain important points of doctrine / Stretton, Henry – London: Joseph Masters, 1848 [mf ed 1989] – 1mf – 9 – 0-7905-2331-0 – mf#1987-2331 – us ATLA [242]

Acts of the anti-slavery apostles / Pillsbury, Parker – Concord, NH: Clague, Wegman, Schlicht, 1883 [mf ed 1992] – 2mf – 9 – 0-524-04967-X – mf#1990-1370 – us ATLA [976]

Acts of the apostles = Apostelgeschichte / Harnack, Adolf von – London: Williams & Norgate; New York: GP Putnam 1909 – (= ser Beitraege zur einleitung in das neue testament; Crown theological library) – 1mf – 9 – 0-8370-3481-7 – (incl bibl ref; in english) – mf#1985-1481 – us ATLA [226]

Acts of the apostles : the teaching of the holy scriptures / Young, Emanuel Sprankel – Elgin IL: Bible Student Co 1915 [mf ed 1992] – 1mf – 9 – 0-524-03945-3 – mf#1990-4939 – us ATLA [226]

Acts of the apostles : with notes, critical, explanatory, and practical / Cowles, Henry – New York: D Appleton, 1883 [mf ed 1985] – 1mf – 9 – 0-8370-2751-9 – mf#1985-0751 – us ATLA [226]

The acts of the apostles / Alexander, Joseph Addison – 3rd ed. New York: Scribner, 1866, c1857 – 3mf – 9 – 0-7905-0600-9 – mf#1987-0600 – us ATLA [226]

The acts of the apostles / Andrews, Herbert Tom – New York: Fleming H Revell; London: A Melrose, [1908?] – 1mf – 9 – 0-7905-1320-X – (incl ind) – mf#1987-1320 – us ATLA [226]

The acts of the apostles : or, the history of the church in the apostolic age = Apostelgeschichte / Baumgarten, Michael – Edinburgh: T & T Clark 1854 [mf ed 1989] – (= ser Clark's foreign theological library. new series 2-4) – 3v on 3mf – 9 – 0-7905-1571-7 – (trans fr german by a j w morrison; v3 trans by theodor mayer) – mf#1987-1571 – us ATLA [226]

The acts of the apostles : an exegetical and doctrinal commentary = Apostelgeschichten / Lechler, Gotthard Victor – New York: Charles Scribner, c1866 – (= ser Theologisch-homiletisches bibelwerk) – 2mf – 9 – 0-8370-6751-0 – (in english) – mf#1986-0751 – us ATLA [226]

The acts of the apostles : being the greek text – London: Macmillan, 1911 – 1mf – 9 – 0-524-06786-4 – mf#1992-0949 – us ATLA [226]

The acts of the apostles : a course of sermons / Maurice, Frederick Denison – London, New York: Macmillan, 1894 – 1mf – 9 – 0-7905-1235-1 – mf#1987-1235 – us ATLA [226]

The acts of the apostles : an exposition / Gaebelein, Arno Clemens – New York City: Publication Office "Our Hope", [1912?] – 1mf – 9 – 0-7905-1387-0 – mf#1987-1387 – us ATLA [226]

The acts of the apostles : an exposition / Rackham, Richard Belward – London: Methuen, 1901 – (= ser Oxford commentaries) – 2mf – 9 – 0-8370-1367-4 – mf#1987-6056 – us ATLA [226]

The acts of the apostles : an exposition for english readers on the basis of professor hackett's commentary on the original text / Green, Samuel Gosnell – London: J Heaton. 2v. 1862 – 2mf – 9 – 0-8370-9868-8 – (incl ind) – mf#1986-3868 – us ATLA [226]

The acts of the apostles – London: Joseph Masters, 1856 – 2mf – 9 – 0-8370-1832-3 – mf#1987-6220 – us ATLA [226]

The acts of the apostles : a popular commentary upon a critical basis, especially designed for pastors and sunday schools / Clark, George Whitefield – new rev ed. Philadelphia: American Baptist Publ Soc, 1896 – 1mf – 9 – 0-524-05663-3 – mf#1992-0513 – us ATLA [226]

The acts of the apostles / Ripley, Henry Jones – stereotyped ed. Boston: Gould and Lincoln, 1869 – 1mf – 9 – 0-524-07114-4 – mf#1992-1030 – us ATLA [226]

The acts of the apostles – London: Printed for the British and Foreign Bible Society, 1890 – 1mf – 9 – (trans into the teni (or slave) language by w c bompass) – mf#16178 – cn Canadiana [290]

The acts of the apostles : with commentary / Plumptre, Edward Hayes – 3rd ed. London: Cassell, Petter, Galpin, [1879?] – (= ser New testament commentary for schools) – 2mf – 9 – 0-524-05288-3 – mf#1992-0389 – us ATLA [226]

The acts of the apostles : with introduction and notes / Page, Thomas Ethelbert – London: Macmillan, 1895 – 1mf – 9 – 0-8370-4656-4 – (incl bibl ref, glossary, index) – mf#1985-2656 – us ATLA [226]

The acts of the apostles : with introduction, notes, and maps / Lindsay, Thomas Martin – Edinburgh: T & T Clark. 2v. [1884-85?] – 2mf – 9 – 0-7905-1131-2 – (incl indes) – mf#1987-1131 – us ATLA [226]

The acts of the apostles : with maps, introduction and notes / Lumby, Joseph Rawson – stereotyped ed. Cambridge: University Press; New York: Macmillan [distributor], 1897 – 1mf – 9 – 0-8370-6753-7 – (incl bibl ref and indexes) – mf#1986-0753 – us ATLA [226]

The acts of the apostles : with notes critical and practical – London: George Bell, 1898 – 2mf – 9 – 0-524-04785-5 – mf#1992-0205 – us ATLA [226]

The acts of the apostles, the epistles and the revelation of st john the divine : a comparison of the text as it is given in the protestant and roman catholic bible versions in the english language, in use in america / ed by Firth, Frank Jones – New York: Fleming H Revell, c1912 – 2mf – 9 – 0-8370-1977-X – mf#1987-6364 – us ATLA [220]

Acts of the chief superintendent explained and vindicated – Toronto?: s.n, 1868 (Toronto: Hunter, Rose) – 9 – mf#23584 – cn Canadiana [370]

Acts of the church, 1531-1885 : the church of england her own reformer, as testified by the records of her convocations... / Joyce, James Wayland – London: J Whitaker, 1886 [mf ed 1990] – 1mf – 9 – 0-7905-4764-3 – (incl bibl ref) – mf#1988-0764 – us ATLA [242]

Acts of the general assembly of the commonwealth of kentucky / Kentucky. Laws, Statutes, etc – Frankfort. On film: 1900-72. LL-060 – 1 – us L of C Photodup [348]

The acts of the holy spirit : being an examination of the active mission and ministry of the spirit of god, the divine paraclete, as set forth in the acts of the apostles / Pierson, Arthur Tappan – New York: Fleming H Revell, 1898, c1895 [mf ed 1989] – 1mf – 9 – 0-7905-1782-5 – mf#1987-1782 – us ATLA [226]

Acts of the italian parliament, 1848-1870 – Microcard Editions – 1320mf (20:1) – 9 – $4600.00 – us UPA [323]

The acts of the parliaments of scotland, 1124-1707 – Edinburgh, 1814-44 – 12v on 5r – 1 – $700.00 – 0-89093-030-9 – us UPA [348]

Acts of the privy council of england : colonial series, 1613-1783 / Grant, W L & Munro, James – London, 1908-12 – 6v on 2r – 1 – $285.00 – 0-89093-031-7 – us UPA [323]

Acts of the privy council of england, colonial series / Great Britain. Privy Council – v1-6. 1613-1783 – 9 – $126.00 – mf#0250 – us Brook [324]

Acts of the privy council of england, new series / Great Britain. Privy Council – v1-43. 1542-1628 – 1 – $564.00 – mf#0251 – us Brook [941]

Acts relating to car trusts, as in force down to august 1, 1893 / Rawle, Francis – Philadelphia, 1893. 111 p. LL-1067 – 1 – us L of C Photodup [340]

The acts relating to common schools and also separate schools in ontario / ed by Hodgins, J George – Toronto: Hunter, Rose, 1870 – 2mf – 9 – mf#06798 – cn Canadiana [370]

Acts relating to the grand trunk railway and for the prevention of accidents on railways / Canada (Province) – Toronto: printed by Stewart Derbishire & George Desbarats, 1857 [mf ed 1995] – 1mf – 9 – mf#SEM105P1762 – cn Bibl Nat [380]

Acts relating to the powers, duties and protection of justices of the peace in lower canada : with a full synoptical index = Actes relatifs aux pouvoirs, aux devoirs et a la protection des juges de paix dans le bas-Canada / Canada (Province) – Quebec: printed by S Derbishire & G Desbarats, 1853 [mf ed 1983] – 2mf – 9 – mf#SEM105P176 – cn Bibl Nat [348]

The acts to regulate commerce : indexed and digested / Hamlin, Charles S – Boston: Little Brown, 1907 – 5mf – 9 – $7.50 – mf#LLMC 80-525 – us LLMC [346]

Actu – Marseilles, France 2,18 aug, 4,18 oct -27 dec 1942; 10-31 jan, 14 feb-14 mar, 9 may, 3,10 oct, 12 dec 1943; 7,21 may 1944 – 1 – uk British Libr Newspaper [074]

Actuacion de la junta de senoras de la cruz roja de badajoz durante la campana de africa de 1921 y 1922 / Badajoz, Cruz Roja – Badajoz: La Libertad, 1923 – 1 – sp Bibl Santa Ana [946]

Actuacion policial de unnextremeno en el siglo 17 / Munoz de San Pedro, Miguel – D. Luis de Tapia y Paredes. Badajoz: Imp. Diput. Provincial. Sep. Rev. Est. Extremenos – 1 – sp Bibl Santa Ana [320]

Actual panorama economico agricola de el salvador : visto desde el mirador de la direccion de estudios economicos y estadistica / Choussy, Felix – El Salvador: C A Ministerio de agricultura y ganaderia [1952] [mf ed 1992] – 1r – 1 – us UF Libraries [380]

Actualidad colonial – Lisbon: Editorial Cosmos, n1-3. jan-mar/apr 1935 – us CRL [074]

Actualidad linguistica de francisco sanchez de las brozas / Salinero, Fernando G – Badajoz: Imprenta Diputacion Provincial, 1973 – sp Bibl Santa Ana [410]

Actualidad liturgica – Mexico: Obra Nacional de la Buena Prensa, n20-109. 1978-92 – 6r – 1 – us CRL [079]

Actualidad (miami, fl) – Miami FL: Publicidad Lopez 1970- (1970 sep 27-1985 may 27) (gaps) [semimthly] [mf ed 1999] – 2r – 1 – (numerical designation ceases with: ano 2 n35 (dic 31 1971)) – us UF Libraries [071]

Actualidad pastoral – Buenos Aires: Actualidad Pastoral. v7 n68-v25 n195 [feb 1974-92] – 5r – 1 – mf#mf-9982 lamp – us CRL [079]

La actualidad y la nueva presidenciah : paginas de politica oriental, tres fragmentos / Sienra Carranza, Jose Manuel – Montevideo: Tip Gimenez, 1911 (mf ed 2001) – 1r – 1 – (1st and 2nd vols originally publ: buenos aires: otero & cia, impresores, 1910; alt title: republica oriental del uruguay, and: actualidad y la proxima presidencia) – mf#Z-9436 – us NY Public [972]

Actualidad y la proxima presidencia *see* La actualidad y la nueva presidenciah

A actualidade : jornal politico, litterario e noticioso – Rio de Janeiro, RJ: Typ de Paula Brito, 22 jan 1859-28 abr 1864 – (= ser Ps 19) – mf#P18A,06,01 – bl Biblioteca [079]

A actualidade : orgao imparcial – Valenca, RJ, 16 ago 1900 – (= ser Ps 19) – 1,5,6 – bl Biblioteca [079]

A actualidade – Sao Joao del Rei, MG, 23 jun 1898 – (= ser Ps 19) – bl Biblioteca [079]

Actualidade – Natal, RN: Typ Liberal, 15 out 1884 – (= ser Ps 19) – bl Biblioteca [079]

Actualidade : orgao do partido liberal – Vitoria, ES: Typ da Actualidade, 02 fev, jun-29 dez 1878 – (= ser Ps 19) – mf#P11B,05,18 – bl Biblioteca [325]

Actualidades (miami beach, fl) – Miami Beach FL: J A Salazar (1982 aug-1996 nov 3) (irreg) [mf ed 1999] – (= ser Cuban heritage collection) – 1r – 1 – us UF Libraries [071]

Actualite – Toronto, Canada 1989+ – 1,5,9 – ISSN: 0383-8714 – mf#17814 – us UMI ProQuest [073]

L'actualite de paul e magloire / Piquion, Rene – [Port-au-Prince, Haiti] H Deschamps [1950?] [mf ed 1991] – 1r – 1 – us UF Libraries [920]

L'actualite juridique – Paris. nov 1945-49. A partir de 1950, divise en deux parties – 5 – (droit administratif. 1950-92. 5. propriete immobiliere. 1950-92. 5) – fr ACRPP [340]

L'actualite litteraire, artistique, scientifique – Paris. n1-26. 2 juin-24 nov 1861 – 1 – fr ACRPP [073]

L'actualitee – Port-au-Prince: Ambiard, 1re annee n1-2eme annee n90. 26 mai 1906-22 fevr 1908 – us CRL [079]

Actualites africaines – Leopoldville: Impr de l'Avenir, jun 4,18-jul 9 1960 – us CRL [079]

Actualites de kivu – Bukavu, mar 17-may 26, jun 9-aug 11 1962 – us CRL [079]

Actualites port-au-princiennes : (un hommage rendu aux professionnels, aux commercants, aux industriels et aux grands planteurs) / Rosemond, Ludovic – Port-Au-Prince, Haiti 1944 [mf ed 1991] – 1r – 1 – us UF Libraries [650]

Actuarial review – Arlington VA 1991-94 – 1 – ISSN: 1046-5081 – mf#12639 – us UMI ProQuest [360]

Act-up : the aids coalition to unleash power / New York Public Library – ca 80r – 1 – (coll traces history of the gay rights movement and the transformation of the movt during the aids crisis. consists of memoranda, correspondence, large amounts of ephemera, the minutes of meetings and video tape of meetings and demonstrations of the new york chapter of act-up) – us Primary [305]

Actus apostolorum : secundum editionem sancti hieronymi – Oxonii: E Typographeo Clarendoniano, 1905 [mf ed 1990] – (= ser Novum testamentum domini nostri iesu christi latine 3/1) – 3mf – 9 – 0-8370-1849-8 – mf#1987-6236 – us ATLA [226]

Actwu labor unity : the official publication of the amalgamated clothing and textile workers union, afl-cio – 1976 sep-1982 mar/apr; 1978 jul-1982 mar/apr – 1 – mf#1399706 – us WHS [071]

Acuarelas / Candray, Jose Eulalio – San Salvador, El Salvador: Sintesis [1956] [mf ed 1991] – 1r – 1 – us UF Libraries [972]

Acueducto y luz electrica de santiago de los caballeros replica a don agustin acevedo / Lithgow, A W – Puerto Plata RD: Tip El Porvenir de Viuda Castellanos e Hijos 1907 [mf ed 1990] – 1r – 1 – us UF Libraries [350]

Acuerdos del extinguido cabildo de buenos aires / Mallie, Augusto S – Madrid: Razon y Fe, 1926 – 1 – sp Bibl Santa Ana [972]

Acuerdos del extinguido cabildo de buenos aires... : sevie 3, tomo 7 (1782-85) / Mallie, Augusto S – Buenos Aires, 1930; Madrid: Razon y Fe, 1931 – 1 – sp Bibl Santa Ana [350]

Acuerdos del extinguido cabildo de buenos aires. serie 2, tomo 4 (1719 a 1722). buenos aires, 1927 / Mallie, Augusto S; ed by Bayle, Constantino – Madrid: Razon y Fe, 1928 – 9 – sp Bibl Santa Ana [972]

Acuerdos del extinguido cabildo de buenos aires. serie 2, tomo 6 1729-1733; serie 3, tomo 5 1774-76; serie 4, tomo 5 1812-1813. buenos aires, 1918-1928 / Mallie, Augusto S – Madrid: Razon y Fe, 1929 – 1 – sp Bibl Santa Ana [946]

Acuerdos del extinguido cabildo de buenos aires. serie 2, tomo 9 y serie 3, tomo 9 / Bayle, Constantino – Buenos Aires, 1931; Madrid: Razon y Fe, 1933 – 1 – sp Bibl Santa Ana [972]

Acuna, Angelina *see*
- Fiesta de luciernagas
- Madre america

Acuna, Cristobal de *see*
- Descubrimiento del amazonas
- Nuevo descubrimiento del gran rio de las amazonas

Acuna de Figueroa, Francisco Esteban *see* Obras completas

Acus – annual reports / Administrative Conference of the US (ACUS) – 1969 thru 1993 – 35mf – 9 – $52.00 – (lacking: 1981) – mf#LLMC 94-335 – us LLMC [340]

Acus recommendations and reports / U.S. Administrative Conference of the United States – Washington: GPO. v1-4. 8 Jan 1968-31 Dec 1977 – 191mf – 9 – $286.00 – (with composite index in v4 to all publ works prior to dec 1977, including the reports of 2 temporary conferences of 1953 + 1961. followed by annual reports and recommendations in 1 or 2 vols, from 1978-1992. lacking: 1990. updates planned) – mf#LLMC 82-207 – us LLMC [324]

Acusado a la inquisicion / Esquivel, Antonio – Madrid: Graf. Calleja, 1969 – 1 – sp Bibl Santa Ana [240]

Acushnet 1860-1900 – Oxford, MA (mf ed 1992) – (= ser Massachusetts vital records) – 5mf – 9 – 0-87623-137-7 – (mf 1: births 1860-85. mf 2: births 1886-1900. mf 2: marriages 1860-67. mf 3: marriages 1867-99. mf 4: marriages 1899-1900. mf 4: deaths 1860-85. mf 5: deaths 1886-1900) – us Archive [978]

Acute cardiovascular responses of cardiac patients to dynamic variable resistance exercise of varying intensity / Stralow, C R – 1991 – 1mf – 9 – $4.00 – us Kinesology [612]

The acute effect of a six-hour fast on exercise performance / Maffucci, Dawn – 1998 – 1mf – 9 – $4.00 – mf#PH 1621 – us Kinesology [612]

The acute effect of heading in soccer on postural stability and cognitive functioning / Miller, Amy E – 1997 – 1mf – 9 – $4.00 – mf#PE 3813 – us Kinesology [612]

Acute effect of incremental exercise on leptin in normal humans / Torjman, Marc – 1997 – 3mf – 9 – $12.00 – mf#PH 1586 – us Kinesology [612]

The acute effects of aerobic versus resistance exercise on mood enhancement / Rosenfeld, Stacey M – 1998 – 2mf – 9 – $8.00 – mf#PSY 2032 – us Kinesology [790]

The acute effects of conservative surgery plus radiotherapy on the functional capacity and psychological well being of women with early stage breast cancer / Reid, Dana Claire – 1997 – 2mf – 9 – $8.00 – mf#HE 602 – us Kinesology [616]

The acute effects of moderate itensity circuit weight training on lipid-lipoprotein profiles / Lee, Y – 1991 – 2mf – 9 – $8.00 – us Kinesology [612]

Acute effects of strength training on cardiorespiratory parameters during subsequent aerobic exercise / Wallis, Jason D – Oregon State University, 1995 – 1mf – 9 – mf#PH 1514 – us Kinesology [612]

The acute physiological responses to walking with and without power poles in patients with cardiac disease / Walter, Patrick – University of Wisconsin-La Crosse, 1995 – 1mf – 9 – mf#PH 1515 – us Kinesology [612]

Acute poliomyelitis (infantile paralysis) – Pretoria: Dept of Health 1948 [mf ed Pretoria, RSA: State Library [199-]] – 11p [ill] on 1r with other items – 5 – mf#op 00872 r25 – us CRL [618]

Acvaghosha's discourse on the awakening of faith in the mahaayaana / Asvaghosa – Chicago: Open Court, 1900 [mf ed 1993] – 2mf – 9 – 0-524-07931-5 – (english trans fr chinese by teitaro suzuki) – mf#1991-0181 – us ATLA [280]

Acwa scrapbooks and press releases, 1910-1961 *see* Records of the amalgamated clothing workers of america

Acwamedha parva – Calcutta: Bharata Press, 1894 [mf ed 1993] – (= ser The mahabharata of krishna-dwaipayana vyasa) – 1mf – 9 – 0-524-08006-2 – (trans chiefly by kesari mohan ganguli) – mf#1991-0228 – us ATLA [490]

Aczi *see* The divan project

Ad acta colloquii montisbelgardensis tubingae edita ia / Beza, Theodor de – Geneve, Le Preux, 1587-1588 – 6mf – 9 – mf#PFA-113 – ne IDC [240]

Ad astra = To the stars – Washington DC 1989+ – 1,5,9 – (cont: space world) – ISSN: 1041-102X – mf#17079 – us UMI ProQuest [629]

Ad bartholomaei latomi rhetoris calumnias... / Dathenus, P – Francofurti, 1560 – 7mf – 9 – mf#PBA-161 – ne IDC [240]

Ad bellarmini disputationes responsio / Daneau, Lambert – Geneve, Le Preux, 1596 – 17mf – 9 – mf#PFA-136 – ne IDC [240]

Ad bilibaldum pykraimervm, de eucharistia, ioannis husschin, responsio posterior / Oecolampadius, J – Basileae, And Cratander, 1527 – 2mf – 9 – mf#PBU-375 – ne IDC [240]

...Ad caesarem oratio pro christiana repv... / Coptius, F – Romae, 1523 – 1mf – 9 – mf#H-8230 – ne IDC [956]

Ad christianos principes de svscepto pro christiana rep contra turcas bello communiter conficiendo... / Ginius, L – Senis, 1572 – 1mf – 9 – mf#H-8328 – ne IDC [956]

Ad clerum : advices to a young preacher / Parker, Joseph – London: Hodder & Stoughton, 1873 [mf ed 1984] – 4mf – 9 – 0-8370-0810-7 – mf#1984-4170 – us ATLA [240]

Ad & d : automotive design and development – Torrance CA 1977-80 – 1,5,9 – (cont: autoproducts) – ISSN: 0164-4904 – mf#7934,01 – us UMI ProQuest [629]

Ad d ioan zuiccium...epistola : accessit...antilogia, ad...gasparis schuenckfeldij argumenta / Vadian, J – Tigvri, [Christoph] Frosch[auer, 1540] – 3mf – 9 – mf#PBU-403 – ne IDC [240]

Ad danielis hofmanni demonstrationes ad oculum... / Beza, Theodor de – Geneve, Vignon, 1586 – 2mf – 9 – mf#PFA-109 – ne IDC [240]

Ad forum – New York NY 1982-85 – 1,5,9 – (cont by: adweek national marketing ed) – ISSN: 0274-6328 – mf#13435 – us UMI ProQuest [650]

Ad futuram rei memoriam / Clement 12, Pope – [Mexico City: s.n.] julij 1732 – (= ser Books on religion...1543/44-c1800: papas [cartas apostolicas, etc]) – 1mf – 9 – mf#crl-365 – ne IDC [241]

Ad gilberti genebrardi accusationem / Beza, Theodor de – Geneve, Vignon, 1585 – 2mf – 9 – mf#PFA-110 – ne IDC [240]

Ad harnacks wesen des christentums fuer die christliche gemeinde / Walther, Wilhelm – Leipzig: A Deichert (Georg Boehme), 1901 [mf ed 1985] – 1mf – 9 – 0-8370-5701-9 – mf#1985-3701 – us ATLA [240]

Ad historiam cassinensis accessiones / Gattula, Erasmus – Venetiis. v1-3. 1734 – €76.00 – ne Slangenburg [241]

Ad Hoc Committee on Indian Education et al *see* Early american newsletter

Ad illustrissimos germaniae principes et optimates liberarum at imperialium ciuitatum oratio...de restituenda pace in germanico imperio caeterisque politijs... / Bibliander, T – Basileae, Ioannes Oporinus, [1553] – 1mf – 9 – mf#PBU-483 – ne IDC [240]

Ad ioannis cochlei de canonicae scriptvrae... authoritate libellum...responsio / Bullinger, Heinrich – Tigvri, [Christoph] Froschover, 1544 – 2mf – 9 – mf#PBU-146 – ne IDC [240]

Ad libros commetariorum d. joannis oecolampadii...praefatio / Bullinger, Heinrich – [Genevae], 1558 – 1mf – 9 – mf#PBU-698 – ne IDC [240]

Ad magnificos...ministros...in polonia...praefatio / Bullinger, Heinrich – [Tigvri, 1568] – 1mf – 9 – mf#PBU-689 – ne IDC [240]

Ad majorem dei gloriam! : die vorgeschichte des ausstandes von 1910/11 in ponape / Fritz, Georg – Leipzig: Dieterich'sche Verlagsbuchhandlung, 1912 [mf ed 1995] – (= ser Yale coll) – 107p – 1 – 0-524-09135-8 – (in german) – mf#1995-0135 – us ATLA [980]

Ad monachos dehortationes / Trithemius, Ioan – Romae, 1898 – 10mf – 8 – €35.00 – ne Slangenburg [241]

Ad nominis christiani socios consultatio qu nam ratione turcarum dira potentia reppelli possit ac debeat...populo christiano... / Bibliander, T – Basileae, [Nicolaus Brylinger, 1542] – 2mf – 9 – mf#PBU-479 – ne IDC [240]

Ad nurse – Franksville WI 1986-89 – 1,5,9 – (cont by: advancing clinical care) – ISSN: 0887-2198 – mf#16482.01 – us UMI ProQuest [610]

Ad omniu ordinum reip : christianae principes uiros populumq christianum, relatio fidelis theodori bibliandri... / Bibliander, T – Basileae, [Ioan. Oporinus], 1545 – 3mf – 9 – mf#PBU-575 – ne IDC [240]

Ad physiologum : eiusdem in die festo palmarum sermo... / Epiphanius – Romae, Antwerpiae: Plantin, 1587-88 – 4mf – 9 – mf#O-841 – ne IDC [090]

Ad reverendissimum in christo patrem et d.d. antonium galeaz : de bentivolis sedis apostolicae prothonotarium b.m. johannis spadarii in musica humilimi... / Spataro, Giovanni – 1491 – 1v on 2mf – 9 – mf#fiche 196 – us Sibley [780]

Ad sacrum convivium... / Lasso, Rudolph de – 1617 – (= ser Mssa) – 1mf – 9 – €20.00 – mfchl 469 – gw Fischer [780]

Ad septem accvsationis capita...responsio / Bullinger, Heinrich – Tigvri, Christoph Froschover, 1575 – 2mf – 9 – mf#PBU-254 – ne IDC [240]

Ad suam historiam aethiopicam antehac editam commentarius. : in quo multa breviter dicta fulsius narrantur... / Ludolfi, Iobi (alias Leutholf dicti) – Francofurti ad Moenum: Sumptibus Johannis David Zunneri: Typis: Martini Jacqueti, 1691 – 9 – (filmed with: his relatio nova de hodierno habessiniae statu...) – us CRL [960]

Ad testamentvm d ioannis brentii...responsio / Bullinger, Heinrich – Tigvri, Chri[stoph] Froschover, 1571 – 1mf – 9 – mf#PBU-243 – ne IDC [240]

Ad tractationem de ministrorum evangelii gradibus ab saravia editam responsio / Beza, Theodor de – [Geneve], Le Preux, 1592 – 3mf – 9 – mf#PFA-118 – ne IDC [240]

Ad virum nobilem de cultu confucii philosophi et progenitorum apud sinas / [Dez, J] – n.p, 1700 – 1mf – 9 – mf#HTM-226 – ne IDC [910]

Ada beeson farmer : a missionary heroine of kuang si, south china / Farmer, Wilmoth Alexander – Atlanta, GA: Foote & Davies, 1912 [mf ed 1992] – (= ser Christian & missionary alliance coll) – 1mf – 9 – 0-524-02389-1 – mf#1990-4291 – us ATLA [920]

Ada field notes / Field, M J – Chicago: [Cooperative Africana Microfilming Project, CRL], 1973 (mf ed) – us CRL [960]

Ada legislative newsletter – ns: [v1], n1-v8 n8 [1972 aug 1-1980 jun 15] – 1 – mf#579497 – us WHS [071]

Ada world – 1967 feb-1976 oct; 1977 jan-1985 apr/may – 1 – mf#1004517 – us WHS [071]

Adabi – Alexandria: Ahmad Zaki Abu Shadi. v1 n1-v2 n3. jan 1936-mar 1937 [complete] – (= ser Arabic journals and popular press) – 1r – 1 – $300.00 – us MEDOC [956]

Adair, Patrick *see* True narrative of the rise and progress of the presbyterian church in ireland (1623-1670)

Adair, Samuel Lyle and Florella (Brown) Family *see* Collection

Adalbert gyrowetz (1763-1850) : kapellmeister der k.k. hoftheater in wien. mit einem katalog der buehnenwerke / Fischer-Wildhagen, Rita – (mf ed 1999) – 4mf – 9 – €56.00 – 3-8267-2623-5 – mf#DHS 2623 – gw Frankfurter [780]

Adalbert stifter : sein leben in selbstzeugnissen, briefen und berichten – Berlin: Verlag des Druckhauses Tempelhof, 1947, c1946 – 445/[1]p/[4]lea/[16]pl (ill) – 1 – (incl bibl ref) – mf#8886 – us Wisconsin U Libr [920]

Adalbert stifter und wien / Eltz-Hoffmann, Lieselotte von – [Wien]: Wiener Verlag 1946 [mf ed 1995] – 1r – 1 – (with 14 reproductions of old paintings. filmed with: abdias / adalbert stifter) – mf#3748p – us Wisconsin U Libr [920]

Adalbert stifters "witiko" / Conrath, Annemarie – Wuerzburg: K Triltsch 1942 [mf ed 1995] – 1r – 1 – (incl bibl ref. filmed with: abdias / adalbert stifter) – mf#3748p – us Wisconsin U Libr [430]

Adalbert von weislingen : schauspiel in fuenf aufzuegen: erster teil der zweiteiligen theaterbearbeitung des goetz von berlichingen von 1819 / Goethe, Johann Wolfgang von; ed by Kilian, Eugen – Leipzig: Klinkhardt & Biermann, 1919 [mf ed 1990] – 1r – 1 – (filmed with: das volkslied und sein einfluss auf goethe's lyrik / j suter) – mf#7320 – us Wisconsin U Libr [820]

Adalbertus samaritanus (mgh quellen..: 3.bd) : praecepta dictaminum – 1961 – (= ser Monumenta germaniae historica. quellen zur geistesgeschichte des mittelalters (mgh quellen...)) – €5.00 – ne Slangenburg [931]

Adallis, Diogenes *see* Greek study

Adam : recit dramatique / Lewisohn, Ludwig – Paris, France: Editions Excelsior 1933 [mf ed 1994] – 1r – 1 – us UF Libraries [440]

Adam, Adela Marion *see*
– The religious teachers of greece
– The vitality of platonism

Adam, Adolf *see* Der postillon von lonjumeau

Adam And Charles Black (Firm) *see* Black's tourist guide to derbyshire

Adam and eve : history or myth? / Townsend, Luther Tracy – Boston: Chapple Publ, 1904 [mf ed 1985] – 1mf – 9 – 0-8370-5558-X – mf#1985-3558 – us ATLA [221]

Adam and the adamite : or, the harmony of scripture and ethnology / M'Causland, Dominick – London, 1864 – (= ser 19th c evolution & creation) – 4mf – 9 – mf#1.1.11600 – uk Chadwyck [573]

Adam and the adamite : or, the harmony of scripture and ethnology / M'Causland, Dominick – London: Richard Bentley, 1872 [mf ed 1985] – 1mf – 9 – 0-8370-4215-1 – mf#1985-2215 – us ATLA [221]

Adam Anglicus *see* Opera

Adam bede / Eliot, George – Toronto: G N Morang, 1902 [mf ed 1996] – 6mf – 9 – 0-665-79752-4 – mf#79752 – cn Canadiana [830]

Adam, Ch Et Al *see* Descartes, par ch adam, e brehier, l brunschvicg

Adam, Charles *see* Etudes sur les principaux philosophes

Adam, David Stow *see* Cardinal elements of the christian faith

Adam de Saint Victor *see* The liturgical poetry of adam of st victor

Adam drawings in the victoria and albert museum – 1r – 1 – $135.00 – 1-900853-45-0 – uk Mindata [720]

Adam, George Jefferys *see* Behind the scenes at the front

Adam, Graeme Mercer *see*
– An algonquin maiden
– Canada, historical and descriptive
– The canadian north-west
– Catalogue of the books in the library of the law society of upper canada
– Handbook of commercial union
– Illustrated quebec
– The librarian's [key]
– Muskoka illustrated
– Prominent men of canada
– Reform in the education office
– Toronto, old and new

Adam, Graeme Mercer [comp] *see* A history of upper canada college

Adam, Heribert *see* Sudafrika

Adam, J *see*
– Evangelische kirchengeschichte der elsaessischen territorien bis zur franzoesischen revolution
– Evangelische kirchengeschichte der stadt strassburg bis zur franzoesischen revolution

Adam, James *see*
– The religious teachers of greece
– The vitality of platonism

Adam, John *see*
– Action of the commission of assembly in professor smith's case
– An exposition of the epistle of james

Adam, John Douglas *see* Paul in everyday life

Adam, Louis *see* Sonate pour le fortepiano, op 8, no 3

Adam, Lucien *see* En quoi la langue esquimaude differe-t-elle grammaticalement des autres langues de l'amerique du nord?

Adam melchior : vitae germanorum theologorum. / Bullinger, Heinrich – Heidelberg, Johann Georg Geyder, 1620 – 1mf0mf – 9 – mf#PBU-415 – ne IDC [240]

Adam, Melchior *see* Vitae germanorum, 1615-1620

Adam of Evesham *see* Magna vita s hugonis, episcopi lincolniensis (rs37)

Adam, Paul *see* Le taureau de mithra

Adam Scotus (The Premonstratensian) *see* Opera

Adam smith – (= ser The Goldsmiths'-Kress Library of Economic Literature) – 58r – 1 – us Primary [330]

Adam und christus : roem 5, 12-21: eine exegetische monographie / Dietzsch, August – Bonn: Adolph Marcus, 1871 [mf ed 1985] – 1mf – 9 – 0-8370-3563-5 – (incl bibl ref) – mf#1985-1563 – us ATLA [220]

Adam und eva : ein biblisches lehrstueck ueber werden und wesen der ersten menschen / Goettsberger, Johann – 1. & 2. aufl. Muenster in Westf: Aschendorff, 1910 [mf ed 1989] – (= ser Biblische zeitfragen 3/11) – 1mf – 9 – 0-7905-3136-4 – (incl bibl ref) – mf#1987-3136 – us ATLA [230]

Adam und quain : im lichte der vergleichenden mythenforschung / Boeklen, Ernst – Leipzig: J C Hinrichs, 1907 [mf ed 1985] – (= ser Mythologische bibliothek 1/2-3) – 1mf – 9 – 0-8370-2400-5 – mf#1985-0400 – us ATLA [220]

Adam von Bremen *see* Adam's von bremen hamburgische kirchengeschichte

Adamantia : the truth about the south african diamond fields: or, a vindication of the right of the orange free state to that territory... / Lindley, Augustus F – London, 1873 – (= ser 19th c british colonization) – 5mf – 9 – mf#1.1.5873 – uk Chadwyck [343]

Adamastor : poems / Campbell, Roy – London: Faber & Faber ltd 1930 – (= ser [Travel descriptions from south africa, 1711-1938]) – 2mf – 9 – mf#zah-29 – ne IDC [810]

Adamaua : bericht ueber die expedition des deutschen kamerun-komitees in den jahren 1893/94 / Passarge, Siegfried – Berlin: D Reimer, 1895 – 1 – us CRL [960]

Adami, A *see* Discurso medico sobre el verdadero metodo de curar las viruelas...

Adami, Giuseppe *see* Balilla

Adami, John George *see* Medical contributions to the study of evolution

Adamic, Louis *see* House in antigua

Adamnan, Saint *see*
– Life of saint columba or columbkille
– Prophecies, miracles and visions of st. columba

Adams 1878-1894 – Oxford, MA (mf 2nd ed 2000) – (= ser Massachusetts vital records) – 143mf – 9 – 0-87623-411-2 – (mf 38-41: vital records 1763-1847. mf 42-45: births 1844-53. mf 45: marriages 1844-54. mf 45-47: deaths 1844-53. mf 48-54: births 1854-78. mf 52, 55-60: marriages 1854-78. mf 61-65: deaths 1853-78. mf 66-69: birth index 1763-1878. mf 69-72: marr index 1781-1878. mf 73-76: death index 1781-1860. mf 1-3: death index 1878-1907. mf 9-13: deaths 1878-1907. mf 14-15: marriage index 1878-1900. mf 20-24: marriages 1878-1900. mf 25-27: birth index 1878-1897. mf 33-37: births 1878-1897. mf 77-78: proprietors 1765-73. mf 79-89: town meetings 1778-1832. mf 90-104: town & militia 1832-54. mf 105-120: town & militia 1855-65. mf 120-130: town & militia 1865-72. mf 131-143: town & militia 1872-1877) – us Archive [978]

Adams advertiser – 1914 jan 22-1915 apr 15; 1915 apr 22-1916 oct 5; 1916 oct 12-1918 mar 16; 1918 feb 23; 1918 mar 23-1919 aug 2; 1919 aug 9-1920 dec 25; 1922 jan 7-dec 30 – 1 – mf#912526 – us WHS [071]

Adams, Archibald G *see* T'an tao lu (ccm1)

Adams, Arthur Prince *see* Bible harmony

Adams, Arvil V *see* Expediting settlement of employee grievances in the federal sector

Adams, Brooks *see*
– The emancipation of massachusetts
– The law of civilization and decay

Adams, Charles Christopher *see* Ecological survey of isle royale, lake superior

Adams, Charles Coffin *see* The bible

Adams, Charles Francis *see*
– Antinomianism in the colony of massachusetts bay, 1636-1638
– Massachusetts
– Papers of charles francis adams 2, 1861-1933

Adams, Charlotte *see* William woodland

Adams, Charlotte Hannah *see* The mind of the messiah

Adams, Clayton (Mrs) *see* The church universal

Adams co. independent – Littlestown, PA., 1860-1942 – 13 – $25.00r – us IMR [071]

Adams Co. Manchester *see*
– Gazette
– Signal

Adams Co. Peebles *see*
– Adams county news
– Press

Adams Co. West Union *see*
– Adams county democrat
– Adams county news
– Courier of liberty
– Democratic union
– Intelligencer series
– People's defender
– Village register

Adams, Coker *see* Principles of the purchas case

Adams county atlas, 1880 : by caldwell – 1r – 1 – mf#B30575 – us Ohio Hist [978]

Adams county [city directory] : listing – 1898/1899 – 1 – mf#3394030 – us WHS [978]

Adams county democrat / Adams Co. West Union – jan 1847-oct 1851, (1852-mar 1860) [wkly] – 3r – 1 – mf#B5575-5577 – us Ohio Hist [071]

Adams county democrat – Hastings, NE: Richard Thompson. 42v. v1 n1. jul 10 1880-v42 n50. apr 27 1923 (wkly) [mf ed 1883-92,1895-1923 (gaps) filmed [1969?-77?]] – 12r – 1 – (cont by: hastings democrat) – us NE Hist [071]

Adams County Historical Society (NE) *see* Historical news

Adams county news / Adams Co. Peebles – v1 n2. oct 1890-dec 1891 [wkly] – 1r – 1 – mf#B29326 – us Ohio Hist [071]

Adams county news / Adams Co. West Union – may-dec 1928, sep 1929-dec 1946 (8r); jan 1947-dec 1983 (15r); apr 1903-dec 1920 scattered (4r) [wkly] – 1 – mf#B10125-10132; B13484-13498; B11689-11692 – us Ohio Hist [071]

Adams county press – 1865 jun 30-1868 may 31; 1868 jun 1-1872 dec 31; 1873 jan 4-1875 nov27; 1875 dec 4-1877 jun 23; 1877 jun 30-1878 dec 14; 1878 dec 21-1880 jun 26; 1880 jul 3-1881 dec 31; 1882 jan 7-1883 jun 2; 1883 jun 9-1884 sep 27; 1884 oct 4-1886 mar 6; 1886 mar 13-1887 sep 10; 1887 sep 17-1889 mar 9; 1889 mar 16-1890 aug 30; 1890 sep 6-1892 mar 12; 1892 mar 19-1893 sep 9; 1893 sep 16-1895 mar 16; 1895 mar 23-1896 sep 5; 1896 sep 12-1898 feb 26; 1898 mar 5-1899 aug 26; 1899 sep 2-01 mar 2; 1901 mar 9-1902 sep 6; 1902 sep 13-1904 apr 9; 1904 apr 16-1905 oct 21; 1905 oct 28-1906 sep 1; 1906 sep 8-1908 feb 1; 1908 feb 8-1909 aug 21; 1909 aug 28-1911 mar 11; 1911 mar 18-1912 oct 12; 1912 oct 19-1914 jun 27; 1914 jul 4-1915 dec 25; 1916 jan 1-1917 jul 14; 1917 jul 21-1918 feb 16 – 1 – mf#929899 – us WHS [071]

Adams county record : [official paper of adams county] – Hettinger, Adams County, ND: Record Printing Co. apr 23 1907 (wkly) [mf ed jun 6 1907-dec 25 2000 with gaps] – 1 – (missing: 1907: apr 25-may 30 1989; jan 24 publ as: adams county record and hettinger headlight, apr 1-aug 12 1909 (usually on a secondary masthead); absorbed: hettinger headlight and hettinger tribune) – mf#00773-14416+ – us North Dakota [071]

Adams county reporter – 1904 oct 28 – 1 – mf#1043719 – us WHS [071]

Adams county times – Adams, Friendship WI. [1929 jul 12/1930 jul 11]-[2005 sep/dec] [few gaps] – 100r – 1 – (cont: adams times [adams, wi]) – mf#912534 – us WHS [071]

Adams county times : [official paper of adams county 1908] – Reeder, Adams Co, ND: Herbert Lewis & R A Lucas, jan 31 1908; -v2 n23 apr 9 1909 (wkly) – 1 – (missing: 1908 dec 4; 1909 feb 12; cont by: reeder times) – mf#03720 – us North Dakota [071]

Adams county tribune – Bucyrus, Adams County, ND: Wm A Stager. v1 n1 jul 23 1908-v1 n39 apr 15 1909 (wkly) [mf ed jul 23 1908-apr 15 1909] – 1 – (missing: 1908 nov 26, dec 10) – mf#06264++ – us North Dakota [071]

The adams county voice – Kenesaw, NE : Mr & Mrs Phil E Douglas. v44 n1. dec 15 1938- (wkly) [mf ed dec15 1938-aug 20 1942 (gaps)] – 3r – 1 – (cont: kenesaw progress) – us NE Hist [071]

Adams, David M *see* A brief history of claar congregation

Adams, David M et al *see* Temperance addresses

Adams, Deborah *see* The relative effectiveness of three instructional strategies on the learning of an overarm throw for force

Adams, Elenor B *see* Don diego quijada...

Adams, Ellinor Davenport *see*
– Colonel russell's baby
– A girl of to-day
– Miss secretary ethel

Adams, Elmer E et al *see* Papers

Adams, Emma Hildreth *see* John of wycliffe, the morning star of the reformation

Adams, Ephraim *see* The iowa band

Adams family newsletter – v7 n1-v12 n1 [1985 spring-1990 spring] – 1 – mf#1802473 – us WHS [071]

Adams, Faried *see*
– Crown vs. adams and 29 others
– Treason trial evidence

Adams, Florence Chenoweth *see* Structural form and analysis of ernest bloch's schelomo

Adams, Francis Alexandre *see* Who rules america?

Adams, Francis William Lauderdale *see* Mass of christ

Adams, Frank Dawson *see*
– The artesian and other deep wells on the island of montreal
– Description of a series of thin sections of typical rocks
– An experimental investigation into the flow of marble
– Geology of a portion of the laurentian area to the north of montreal
– An investigation into the elastic constants of rocks
– Laurentian area to the north and west of st jerome
– Mcgill and science
– Memoir of sir j william dawson
– The monteregian hills
– Nodular granite from pine lake, ontario
– Notes on the iron ore deposits of bilbao, northern spain
– Notes on the lithological character of some of the rocks
– Notes on the ore-deposit of the treadwell mine, alaska / On the microscopical character of the ore of the treadwell mine, alaska
– Obituary, sir john william dawson
– On a new alkali hornblende and a titaniferous andradite
– On some canadian rocks containing scapolite
– On some granites from british columbia
– On the amount of internal friction developed in rocks during deformation
– On the geology of the st clair tunnel
– On the igneous origin of certain ore deposits
– On the need of a topographical survey of the dominion of canada / on a new nepheline rock from the province of ontario, canada
– On the origin and relations of the grenville and hastings series in the canadian laurentian
– Our mineral resources
– Report on the geology of a portion of the laurentian area lying to the north of the island of montreal
– Ueber das norian oder ober-laurentian von canada

Adams, George Burton *see* Civilization during the middle ages

Adams, George F et al *see* History of baptist churches in maryland

Adams, Hannah *see* A dictionary of all religions and religious denominations

Adams, Henry *see*
– Are christ and belial united? are the church and the world agreed?
– The bible versus infidelity
– The cause of the degradation of man
– "Close communion"
– The demon alcohol, the great man-slayer
– A drunkard's experience at home and abroad
– Esther
– The fourth anniversary of the spring garden road home of the first baptist church, halifax, ns, lord's day, april 12, 1891
– The henry adams papers, 1843-1938
– "Inconsistency"
– Infant baptism
– Pedo-baptist bulwarks of the baptists' position
– A sermon on cards, dancing, theatres and carnivals
– A sermon on lotteries
– Tahiti
– A true picture of the effects of intemperance

Adams, Henry Austin *see* Orations of henry austin adams

Adams, Henry Cadwallader *see* History of the jews
Adams, Henry G *see* Original poems
Adams, Herbert B *see* Johns hopkins university studies in historical and political science
Adams, Herbert Baxter *see* The church and popular education
Adams, J *see*
– The country from cape palmas to river congo
– Remarks on the country extending from cape palmas to the river congo
Adams, J A *see* History of platte river
Adams, J N *see*
– Bibliography of eighteenth-century legal literature
– A bibliography of nineteenth-century legal literature
Adams, James Edward *see* The missionary pastor
Adams, Jenny L *see* Maximal power output on the bicycle ergometer
Adams, John *see*
– Israel's ideal
– The man among the myrtles
– Novanglus, and massachusettensis
– Remarks on the country extending from cape palmas to the river congo
– Sermons in accents
– Sermons in syntax
– Sketches taken during ten voyages to africa, between the years 1786 and 1800
– Works of john adams
Adams, John Coleman *see*
– Christian types of heroism
– The doctrine of equity
– Hosea ballou and the gospel renaissance of the nineteenth century
– Universalism and the universalist church
Adams, John Greenleaf *see*
– Lectures on universalism to inquirers after christian truth
– Memoir of rev. john moore
– Memoir of thomas whittemore, d.d
– The sabbath school melodist
– Talks about the bible to the young folks
– The universalism of the lord's prayer
– The universalist church
Adams, John Quincy *see*
– Baptists, the only thorough religious reformers
– Baptists, thorough religious reformers
– The birth of mormonism
– Correspondence between john quincy adams, esquire, president of the united states, and several citizens of massachusettes
– A history of auburn theological seminary, 1818-1918
– Letters and opinions of the masonic institution
– South sea memories
Adams, John Quincy et al *see* The abolition of slavery
Adams, Joseph *see* Ten thousand miles through canada
Adams, Karen A *see* Measurement of student-athletes' perceptions of their intercollegiate head coaches on conceptual, human, and technical managerial skills
Adams, M *see* British attitude to german colonial development, 1880-1885
Adams, Mark J *see* The perception of high school athletes and coaches in regard to individual and team efficacy in basketball
Adams, Maurice Bingham *see*
– Artists' homes
– Examples of old english houses and furnitur
Adams, Myron *see* Creation of the bible
Adams, Nehemiah *see*
– Church pastorals
– Discussion of the scripturalness of future endless punishment
– God is love
The adams papers, 1639-1889 – [mf ed 1954-59] – 608r – 1 – (contributions of the many adams family members in the political, social, and economic spheres of public life are documented in more than 300,000 mss pp. pt1 contains diaries of john, john quincy, and charles francis adams. pt2 foll with letterbooks of these three statesmen. pt3 is organized by generation, & thereafter by individual. pt4 contains letters received by the family & other loose papers, arranged chronologically fr 1639-1889) – us MA Hist [975]
Adams, R *see*
– The narrative of...a sailor
– Young gentlemen and lady's explanatory monitor
Adams, R J *see* The lord's supper in baptist churches
Adam's rib / Herschberger, Ruth – New York: Harper & Row [1970, c1948] [mf ed 1994] – 1r – 1 – us UF Libraries [170]
Adams, Richard Newbold *see* Encuesta sobre la cultura de los ladinos en guatemala
Adams, Robert *see* The narrative of robert adams
Adams, Robert Chamblet *see*
– History of the united states in rhyme
– Illustrated story of the union in rhyme
– Pioneer pith
– Travels in faith from tradition to reason
Adams, Samuel Houston *see* Comparing tort liability knowledge of future teacher coaches and current practicing teacher coaches
Adams, Sarah Fuller *see* Vivia perpetua
Adams, Seymour Webster *see* Address before the society of religious inquiry of granville college
Adams soehne : roman / Wilbrandt, Adolf von – 2. aufl. Berlin: W Hertz 1890 [mf ed 1991] – 1r – 1 – (filmed with: jedermann / ernst wiechert) – mf#3031p – us Wisconsin U Libr [830]
Adams times – 1923 jan-jun 2; 1923 jun 9-1924 oct 25; 1924 nov 1; 1925 jan 3-1926 may 22; 1926 may 29-1927 oct 28; 1927 nov 4-1929 feb 22; 1929 mar 1-jul 5 – 1 – mf#4757202 – us WHS [071]
Adam's von bremen hamburgische kirchengeschichte = Gesta hammaburgensis ecclesiae pontificum / Adam von Bremen; ed by Wattenbach, Wilhelm – 2. aufl. Leipzig: Dyk, 1888 [mf ed 1992] – (= ser Die geschichtschreiber der deutschen vorzeit 2/44) – 1mf – 9 – 0-524-02909-1 – (in german fr latin) – mf#1990-0725 – us ATLA [240]
Adams, W A *see* Romeo und julia auf dem dorfe
Adams, W B *see* A moral and political sketch of the united states of north america
Adam's weekly courant – Chester, England 17 jan-23 may 1739, 10 mar 1747-29 nov 1748, 4 mar 1766-4 aug 1767, 19 jul 1768, 30 may, 3 oct 1769-16 may 1775, 12 mar 1776-29 dec 1778, 4 jan 1780-27 dec 1785, 17 jul 1792 – 1 – (cont by: chester courant & anglo-welsh gazette [4 jan 1825-20 sep 1831]) – uk British Libr Newspaper [072]
Adam's weekly courant with news both foreign and domestick – [NW England] Chester, Manchester ALS 14 nov-4 aug 1778* – 1 – (title change: chester courant & anglo-welsh gazette [19 nov 1793-1806, 1809-24, 1847-20 sep 1831]; chester courant & advertiser for north wales [12 dec 1860-25 dec 1861, 14 jan-16 dec 1863; 15 jan-17 dec 1974*, 1880-95, 1898-7 sep 1982]; cheshire observer & general advertiser for cheshire & north wales [may 1854-jun 1863]; cheshire observer & chester, birkenhead, crewe & north wales times [4 jul-19 dec 1863, 4 jul 1863-sep 1847, 1869, jan-feb 1870]; cheshire observer [1900-14, 1916-36, 1975-77, jan- 6 jul 1979]; 13 jul 1979-1986]) – uk MLA; uk Newsplan [072]
Adams weekly globe – Adams, Gage County, NE: E W Varner. -v86 n29. feb 26 1981 [mf ed jul 12 1950-feb 26 1981 (gaps)] – 12r – 1 – (suspended oct 28 1943-may 2 1946; iss for apr 10 1975-dec 18 1980 incl regional weekend magazine suppl; cont: adams globe) – us NE Hist [071]
Adams, William *see*
– Conversations of jesus christ with representative men
– A discourse on the life and services of professor moses stuart
– East meets west
– Flowers of modern voyages and travels
Adams, William Henry *see* Souvenir story of the georgian bay
Adams, William Henry Davenport *see*
– Egypt past and present
– Recent polar voyages
– St paul
– Witch, warlock, and magician
Adams, William O *see* Musick book
Adams-Acton, Marion (Hamilton) *see* Golden days
Adamson, Robert *see*
– The development of modern philosophy
– Fichte
– On the philosophy of kant
– Pure logic and other minor works
Adamson, Robert M *see* The christian doctrine of the lord's supper
Adamson, Thomas *see* The spirit of power
Adamson, William *see* Gospel of evolution
Adamus, Franz *see* Familie Wawroch
Adamus, John *see* Zastaw w prawie litewskiem 15 i 16 wieku
Adamus, M *see* Vitae germanorum theologorum, qvi superiori seculo ecclesiam christi voce scriptisque propagarunt et propugnarunt. congestae et ad annum usque 1618 deductae
Adana – (= ser Vilayet salnames) – 9 – (1287 [1870] def'a 1 2mf $225; 1312 [1894] 4mf $150) – us MEDOC [956]
Adana ticaret rehberi = Guide commercial d'adana / Oguz [Arik], Remzi – Istanbul: Cihan Biraderler Matbaasi, 1340 [1924] – 5mf – 9 – $75.00 – us MEDOC [380]
Adanson, Michel *see*
– Familles des plantes
– Histoire naturelle du senegal
Adaptation francaise du message de son excellence monsieur le marechal lon nol, president de la republique khmere : a propos de la constitution du nouveau gouvernement [fait a phnom-penh, le 15 octobre 1972] / Lon Nol – [Phnom-Penh: s.n. 1972?] [mf ed 1989] – 1r with other items – 1 – (In Khmer & French) – mf#mf-10289 seam reel 026/13 [§] – us CRL [342]
Adaptation of the electoral law of 1890 to cuba and porto rico : royal decree of 1897 – Washington: GPO, 1899 – 1mf – 9 – $1.50 – mf#LLMC 92-307 – us LLMC [972]
Adapted physical education specialists' perceptions and role in the consultation process / Lytle, Rebecca K – 1999 – 2mf – 9 – $8.00 – mf#PE 3972 – us Kinesology [790]
Adapting the offense efficiency rating in basketball to accommodate the 3-point goal / Stauffer, Bryan E – 1998 – 2mf – 9 – $8.00 – mf#PE 3891 – us Kinesology [790]
Adapting to dynamically changing balance threats : differentiating young, healthy older adults and unstable older adults / Lin, Sang-I – 1997 – 2mf – 9 – $8.00 – mf#PSY 2033 – us Kinesology [612]
Adaptive beamforming applied to hydroacoustic communications / Stiller, Christoph – (mf ed 1994) – 1mf – 9 – €30.00 – 3-8267-2001-6 – mf#DHS 2001 – gw Frankfurter [621]
Adar, Zvi *see* Mishnat ha-rambam
Adarkar, Bhalchandra Pundlik *see*
– If war comes
– The indian fiscal policy
– The indian monetary policy
Adarkar, Bhaskar Namdeo *see* The indian tariff policy
Adat istiadat daerah kalimantan tengah – [Palangka Raya]: Tim adat Istiadat P3KD Kalimantan Tengah 1977/1978tc1979 [mf ed [1982?] – on pt of 1r – 1 – mf#12418 n2 – us Cornell [350]
'Adat Tsadikim *see* Sefer 'adat tsadikim
Adcock, Adam Kennedy *see* The glorious gospel
Adcock, Arthur St John *see* Gods of modern grub street
Addams, Jane et al *see* Women at the hague
Addenda et emendanda ad f ehrle historia bibliothecae romanorum pontificum, tomus 1 / Pelzer, A – Romae, 1947 – €18.00 – ne Slangenburg [241]
Addenda of the remainder of the furniture...of ralph bernal / Christie, Manson and Woods, Ltd, London – [London?] 1855] – (= ser 19th c art & architecture) – 1mf – 9 – mf#4.2.389 – uk Chadwyck [740]
Adderley, James et al *see* Oxford house papers. second series
Adderley, Joseph C *see* American tung tree
Addicion al parecer del r.p. fr...acerca de una eleccion que dio en doce de diciembre de 1639 por el mismo padre a peticion del padre fr. sebastian de moratilla vicario de la santa casa de guadalupe y professo della / Virgen, Juan de la – S.l., s.i., s.a. 1640 – 1 – sp Bibl Santa Ana [240]
Addiction – Abingdon, Oxfordshire 1993+ – 1,5,9 – (cont: british journal of addiction) – ISSN: 0965-2140 – mf#13420,03 – us UMI ProQuest [362]
Addiction & recovery – 1990-93 – 1,5,9 – (cont: alcoholism and addiction and recovery life; cont by: behavioral health management) – ISSN: 1052-4614 – mf#16861.05 – us UMI ProQuest [362]
Addictive behaviors – Oxford, England 1975+ – 1,5,9 – ISSN: 0306-4603 – mf#49004 – us UMI ProQuest [150]
Addington, John Gellibrand Hubbard *see* Census of religions
Addin-Hon *see* Torias
Addis and geller collections from the school of oriental and african studies, london *see* China through western eyes
Addis, William Edward *see*
– Anglican misrepresentations
– Anglicanism and the fathers
– The book of job and the book of ruth
– The deuteronomical writers and the priestly documents
– Hebrew religion to the establishment of judaism under ezra
– The oldest book of hebrew history
Addis zaman – Addis Ababa, Ethiopia. jun 7 1941-10 sep 1971 – 1 – us CRL [960]
Addison, Charles Greenstreet *see*
– Addison on torts
– A treatise on the law of contracts
Addison, Charles Morris *see* A book of offices and prayers for priest and people
Addison, Daniel Dulany *see*
– The clergy in american life and letters
– Episcopalians
Addison, Joseph *see* Mr davin on "fanning in church"
Addison, Julia *see*
– Crow's nest farm
– Effie vernon
– Evelyn lascelles
– The molyneux family
Addison, Maine. Indian River Baptist Sewing Circle *see* Records
Addison, Maine. Second Baptist Church *see* Records
Addison on torts / Addison, Charles Greenstreet – 6th ed., Toronto, Carswell, 1890. 935 p. LL-897 – 1 – us L of C Photodup [340]
Addison's reports / Pennsylvania. Superior Court – 1v. 1791-99 – (= ser Pre-nrs nominative reports) – 2mf – 9 – $9.00 – mf#LLMC 84-192 – us LLMC [340]
Additamenta ad synopsim theologiae pro anno 1908 / Tanquerey, Adolphe – Romae: Desclee & Socii, 1908 [mf ed 1986] – 1mf – 9 – 0-8370-8391-5 – (incl bibl ref) – mf#1986-2391 – us ATLA [241]
Additional answer to the libel : with some account of the evidence that parts of the pentateuchal law are later than the time of moses / Smith, William Robertson – [2d ed] Edinburgh: David Douglas, 1878. Princeton: Speer Library, and Dep of Photodup, U of Chicago Lib, 1978 (1r); Evanston: American Theol Lib Assoc, 1984 (1r) – 1 – 0-8370-0640-6 – (incl bibl ref) – mf#1984-6270 – us ATLA [220]
Additional by-laws, rules, regulations and ordinances of the common-council of the city of montreal – [Montreal?: s.n, 1833?] [mf ed 1984] – 1mf – 9 – 0-665-46496-7 – mf#46496 – cn Canadiana [350]
Additional notes on the birds of haiti and the dominican republic / Wetmore, Alexander – Washington DC 1934 [mf ed 1991] – 1r – 1 – us UF Libraries [590]
Additional notes on the geology and palaeontology of ottawa and vicinity / Ami, Henry Marc – Ottawa?: s.n, 1886 – 1mf – 9 – mf#56361 – cn Canadiana [550]
Additional observations on penal jurisprudence, and the reformation of criminals : containing remarks on prison discipline, in reply to an article in the edinburgh review, and on the punishment of criminals by solitary confinement... / Roscoe, William – London, 1823 – (= ser 19th c evolution & creation) – 3mf – 9 – mf#1.1.6 – uk Chadwyck [365]
Additional poems / Dewart, Edward Hartley – S.l: s.n, 1892? – 1mf – 9 – mf#05744 – cn Canadiana [810]
Additiones ad dictionarium japonicum / Collado, Diego – [s.l.]: [s.n.], [s.a.] – (= ser Whsb) – 3mf – 9 – €40.00 – mf#Hu 327a – gw Fischer [480]
Additiones ad varias resolutiones. / Ayllon Laynez, Juan – 1653 – 9 – sp Bibl Santa Ana [946]
Additions and corrections to the book of genesis / Driver, Samuel Rolles – 1909 – 9 – $10.00 – us IRC [221]
Additions et corrections a la faune coleopterologique de la province de quebec, 1879 / Provancher, Leon – Quebec: C Darveau, 1879 [mf ed 1974] – 1r – 5 – mf#SEM16P196 – cn Bibl Nat [590]
Addon, Esther *see* The forest grange
Ad-dourra al-faakhira = La perle precieuse de ghazaalai / Ghazzali – Leipzig: G Kreysing, 1877 [mf ed 1986] – 1mf – 9 – 0-8370-7698-6 – (in french & arabic; no more publ; incl bibl ref) – mf#1986-1698 – us ATLA [260]
Address / Casgrain, Thomas Chase – [Vancouver?: s.n, 1915?] – 1mf – 9 – 0-665-73749-1 – mf#73749 – cn Canadiana [933]
Address : delivered...jan 21 1901 / Bilgrami, Syed Husain – Cawnpore: CC Mission Press, 1901 [mf ed 1991] – 1mf – 9 – 0-524-01414-0 – mf#1990-2409 – us ATLA [260]
Address / Stebbins, Rufus Phineas – [s.l: s.n.] 1857 [mf ed 1993] – 1mf – 9 – 0-524-08532-3 – mf#1993-1062 – us ATLA [230]
An address : the place of baptists in protestant christendom / Bliff, G Ripley – 1 – $5.31 – us Southern Baptist [242]
An address : pronounced on the first tuesday of march 1831 / Thacher, Peter Oxenbridge – Boston: Hilliard, Gray, Little & Wilkins, 1831 [mf ed 1993] – 1mf – 9 – 0-524-08653-2 – (incl biogr footnotes) – mf#1993-2113 – us ATLA [230]
Address at the convocation of the university of toronto : june 10th, 1890 / Blake, Edward – [Toronto?: s.n.], 1890 [mf ed 1980] – 1mf – 9 – 0-665-02174-7 – mf#02174 – cn Canadiana [370]
Address at the dedication of the new library building of gammon theological seminary, atlanta, georgia / Payne, Charles Henry – [Georgia?: s.n. 1889?] [mf ed 2006] – 1r [complete] – 1 – (reel also incl: catalogue of the gammon theological seminary, a dedication, & another address) – mf#2006-s007 – us ATLA [080]
An address before the commercial exchange of des moines, iowa, thursday, december 20, 1894 / Brown, William Carlos – [S.l: s.n, 1894?] [mf ed 1980] – 1mf – 9 – 0-665-00970-4 – mf#00970 – cn Canadiana [380]
Address before the grafton and coos bar association / Chandler, William Eaton – Concord: Republican, 1888. 38p. LL-10 – 1 – us L of C Photodup [340]

Address before the historical society of pennsylvania, 28th january, 1848 : on the occasion of opening the hall in the athenæum / Reed, William Bradford – Philadelphia: C Sherman, Printer, 1848 (mf ed 19–) – (= ser Pennsylvania history on microfilm) – 51p – mf#ZH-IAG pv241 n14 – us NY Public [978]

Address before the imperial institute of great britain on the 10th of march, 1898 / Bouthillier-Chavigny, Charles, vicomte de – Montreal: [s.n.], 1898 [mf ed 1980] – 1mf – 9 – 0-665-02529-7 – mf#02529 – cn Canadiana [080]

Address before the society of religious inquiry of granville college : granville, july 7th, 1850 / Adams, Seymour Webster – Cleveland: Smead & Cowles, 1850 [mf ed 1993] – 1mf – 9 – 0-524-08246-4 – mf#1993-3001 – us ATLA [240]

[Address, business and telephone directories of poland] : donated by jewish genealogy society to slavic and baltic division of the new york public library – [New York: Slavic & Baltic Div, NYPL, 1996] – 15r – 1 – (alt title: ksiega adresowa przemyslu galicyjskiego. ksiega adresowa stol. miasta lwowa. ksiega adresowa krol. stol. miasta lwowa. ksiega adresowa przemyslu fabrycznego w krolestwie polskiem. ksiega adresowa przemyslu, handlu i finansow. ksiega adresowa polski (wraz z w.m. gdanskiem) dla handlu, przemyslu, rzemiosl, i rolnictwa. polski przemysl i handel. skorowidz przemyslowo-handlowy krolestwa galicyi. urzedowy spis abonentow panstwowej sieci telefonicznej okregu krakowskiej i katowickiej dyrekcyj poczt i telegrafow oraz sieci zaglebia dabrowskiego polskiej akcyjnej spolki telefonicznej i abonentow miast niem. bytomia, gliwic i zabrza. ksiega adresowa handlu, przemyslu, rolnictwa i wolnych zawodow wojewodztwa stanislawowskiego i tarnopolskiego. adressen-buch der handel- und gewerbetreibenden sowie der actien-gessellschaften der osterreichisch-ungarischen monarchie) – mf#Slav. Reserve 96-7788 – us NY Public [914]

[Address, business and telephone directories of poland donated by jewish genealogy society to slavic and baltic division of the new york public library] – [New York : Slavic & Baltic Div, NYPL, 1996] – 15r – 1 – (ksiega adresowa przemyslu galicyjskiego, 1901 (iu microform master n80-0687/1, 1r); ksiega adresowa stol. miasta lwowa, 1897 (iu microform master n80-0541/1, 1r, microfilm by biblioteka jagiellonska, 1979); ksiega adresowa król. stol. miasta lwowa, 1916 (microfilm by lc, 1r); ksiega adresowa przemyslu fabrycznego w królestwie polskiem, 1906 (iu microform master n89-0683/1, 1rl); ksiega adresowa przemyslu, handlu i finansow, 1922 (1r, microfilm made by stanford u); ksiega adresowa polski (wraz z w.m. gdanskiem) dla handlu, przemyslu, rzemiosl, i rolnictwa, 1926-1927 (3r, microfilm made by stanford u), 1929 (iu microform master n81-0228/1, 2r, microfilm made by biblioteka narodowa, warsaw, 1982); urzedowy spis abonentów panstwowej sieci telefonicznej okregu krakowskiej i katowickiej dyrekcyj poczt i telegrafów oraz sieci zaglebia dabrowskiego polskiej akcyjnej spólki telefonicznej i abonentów miast niem. bytomia, gliwic i zabrza, 1930 (microfilm by lc, 1r); ksiega adresowa handlu, przemyslu, rolnictwa i wolnych zawodów województwa stanislawowskiego i tarnopolskiego, 1931 (iu microform master n80-0682/1, 1r, microfilm made by biblioteka narodowa, warsaw, 1982); polski przemysl i handel, 1930 (iu microform master n80-0600/1, 1r); skorowidz przemyslowo-handlowym królestwa galicyi, 1933 (iu microform master n80-0691, 1r); adressen-buch der handel- und gewerbetreibenden sowie der actien-gesellschaften der österreichisch-ungarischen monarchie / zusammengestellt und herausgegeben von leopold kastner, wien : a hoelder, 1877 (british library shelfmark pp 2440c, mic.c.1259, 1r, microfilm made by the british library, 1986)) – mf#Slav Reserve 96-7788 – Located: NYPL – us Misc Inst [939]

Address by ex-ald e a macdonald, of toronto, ontario : delivered in faneuil hall, boston mass...on friday, september 23rd, 1892 – [Toronto?: s.n.], 1892 [mf ed 1984] – 1mf – 9 – 0-665-09330-6 – mf#09330 – cn Canadiana [971]

Address by lady aberdeen : president of the aberdeen association, at a public meeting, ottawa, 1898 / Aberdeen and Temair, Ishbel Gordon, marchioness of – [S.l: s.n, 1898?] [mf ed 1979] – 1mf – 9 – 0-665-00765-5 – mf#00765 – cn Canadiana [366]

Address by owen d young : given at the testimonial dinner tendered him by the business men of new york at the waldorf astoria hotel, december 11th, 1924 – [S.l: s.n, 1924?] (mf ed 19–) – us Harvard [933]

Address by rev john fraser, of kincardine : in the debate on instrumental music at the synod in montreal, june 12th, 1868 – [Kincardine, Ont?: s.n, 1868?] – 1mf – 9 – 0-665-88959-3 – mf#88959 – cn Canadiana [780]

Address by the provincial council to the people of ontario : dealing mainly with separate schools / Equal Rights Association for the Province of Ontario – Toronto: The Association, [1889?] [mf ed 1980] – 1mf – 9 – 0-665-02918-7 – mf#02918 – cn Canadiana [370]

An address delivered... : in the city of st john, dominion of canada, 4th july, 1883 / Peyster, John Watts de – New York?: C H Ludwig, 1883 – 1mf – 9 – mf#02641 – cn Canadiana [971]

Address delivered april 18, 1937 / Cannon, Walter Bradford – New York: Medical Bureau... 1937 [mf ed 1977] – (= ser Blodgett coll) – 9 – mf#w772 – us Harvard [946]

An address delivered at the opening of queens sic college, 1853 / George, James – S.l: s.n, 1853? – 1mf – 9 – mf#43322 – cn Canadiana [378]

Address delivered at the opening of the session of 1862-63 / Langton, John – [s.l: s.n, 1862?] [mf ed 1984] – 1mf – 9 – 0-665-45330-2 – mf#45330 – cn Canadiana [080]

An address delivered at the opening of the training school for nurses : at the general public hospital in st john, on october 4th, 1888 / Bayard, William – St John, NB?: s.n, 1888? – 1mf – 9 – mf#05979 – cn Canadiana [610]

Address delivered before the alumni association of the university of the state of missouri / Elkins, Stephen Benton – New York: Press of Styles & Cash, 1885 (mf ed 19–) – 36p – (alt title: industrial question in the united states) – mf#ZT-TB+ pv104 n13 – us Harvard [331]

An address delivered before the canadian club, ottawa, december, 1907 *see* The future of canada / a perplexed imperialist / the canadian flag etc

An address delivered before the senior class : in divinity college, cambridge...15 july, 1838 / Emerson, Ralph Waldo – Boston: James Munroe, 1838 [mf ed 1993] – (= ser Unitarian/universalist coll) – 1mf – 9 – 0-524-07871-8 – mf#1991-3416 – us ATLA [243]

An address delivered before the university of nashville, 1839 / Howell, Robert Boyte C – 28p – 1 – $5.00 – us Southern Baptist [242]

Address delivered by rev g h atkinson, dd, before the chamber of commerce of the state of new-york / Atkinson, George Henry; ed by upon the possession, settlement, climate and resources of oregon and the northwest coast, including some remarks upon alaska, december 3d, 1868 – New York?: s.n, 1868 (New York: J W Amerman) – 1mf – 9 – mf#14066 – cn Canadiana [917]

Address delivered by the rev h burges : at the opening of the second winter session of the three rivers literary association, on the 2d november 1842 – [Trois-Rivieres, Quebec?: s.n.] 1842 [mf ed 1983] – 1mf – 9 – 0-665-43091-4 – mf#43091 – cn Canadiana [080]

Address delivered by william morton grinnell : at the trust conference in cooper union, february 23d, 1900 – New York: Evening Post job print house, [1900] (mf ed 19–) – 8p – mf#ZT-TN pv34 n2 – us Harvard [080]

Address delivered in boston music hall, wednesday evening, january 31, 1894 / Blake, Edward – Boston: Municipal Council of the Irish National Federation of Boston and Vicinity, 1894? [mf ed 1980] – 1mf – 9 – 0-665-02296-4 – mf#02296 – cn Canadiana [941]

An address delivered in chancellors hall, state education building, albany, ny / Cardozo, Benjamin Nathan – Albany: Evory, 1925. 11p. LL-2265 – 1 – us L of C Photodup [340]

Address delivered in convocation hall, queen's college, kingston, april 28th, 1885 / Fleming, Sandford – Ottawa: Citizen, 1885 [mf ed 1980] – 1mf – 9 – 0-665-03122-X – mf#03122 – cn Canadiana [378]

Address delivered in st mary's church, st john's, nf : 22nd may, 1898, to lodge dudley, soe / Botwood, Edward – [S.l: s.n.], 1898 [mf ed 1980] – 1mf – 9 – 0-665-00908-9 – mf#00908 – cn Canadiana [360]

An address delivered in the chapel of the general theological seminary of the protestant episcopal church in the united states on friday, nov 13th 1852 : the bishop of montreal's address at the general theological seminary, 1852 / Fulford, Francis – New York: Church Depository, 1852 [mf ed 1983] – 1mf – 9 – 0-665-44526-1 – (incl bibl ref) – mf#44526 – cn Canadiana [242]

An address delivered in the first church, salem : at the funeral services of charles w upham, june 18, 1875 / Ellis, George Edward – Salem, MA: publ by the family, 1875 [mf ed 1980] – 1mf – 9 – 0-665-05235-9 – (and the sermon preached on the succeeding sabbath by james t hewes) – mf#05235 – cn Canadiana [242]

An address delivered in the first parish, beverly, october 2, 1867, on the two-hundredth anniversary of its formation / Thayer, Christopher Toppan – Boston: Nichols and Noyes, 1868. Chicago: Dep of Photodup, U of Chicago Lib, 1972 (1r); Evanston: American Theol Lib Assoc, 1984 (1r) – 1 – 0-8370-0290-7 – mf#1984-B316 – us ATLA [240]

Address, delivered may 8, 1878, at the annual commencement of the cincinnati law school / Drake, Charles Daniel – Washington City, McGill, 1878. 18 p. LL-410 – 1 – us L of C Photodup [340]

An address delivered on saturday, the 16th march, 1878, in old st andrew's church, toronto : on the occasion of the formal withdrawal of the congregation therefrom and the final closing of that edifice as their place of worship... / Barclay, John – Toronto?: s.n, 1878 – 1mf – 9 – mf#07071 – cn Canadiana [240]

An address delivered on the 5th april, 1855 : before the senatus and students of queen's college on conferring the degree of doctor of medicine / George, James – Kingston, Ont?: Daily News, 1855 – 1mf – 9 – mf#22545 – cn Canadiana [610]

Address delivered on the 30th day of october, ad 1883 : on the occasion of the opening of a law school in connection with dalhousie college, halifax, nova scotia / Archibald, Adams George – [Halifax, N.S.?: s.n, 1883?] [mf ed 1980] – 1mf – 9 – 0-665-02473-8 – mf#02473 – cn Canadiana [378]

An address, delivered to the inhabitants of the county of stanstead at a public meeting of that county : held at the north meeting-house in stanstead on thursday, 24th of apr 1834 / Child, Marcus – [s.l: s.n, 1834?] [mf ed 1984] – 1mf – 9 – 0-665-44252-1 – mf#44252 – cn Canadiana [080]

An address from the charleston association... : calling for the organization of the state baptist convention of south carolina / Baptist Association. Charleston, South Carolina – 46p – 1 – $5.00 – us Southern Baptist [242]

Address from the committee of synod : to the office-bearers and members of the presbyterian church of canada, on the subject of the commemoration of the westminster assembly / Presbyterian Church of Canada...with the Church of Scotland. Synod – [Kingston, Ont?: s.n.] 1843 [mf ed 1985 – 1mf – 9 – 0-665-26751-7 – mf#26751 – cn Canadiana [242]

Address from the working men's association *see* Political tracts and pamphlets... 19th c

Address given by a kelly evans : at a meeting...toronto, june 7th, 1905 to form an association for the better protection of the game and fish of the country... – [Toronto?: s.n, 1905?] [mf ed 1996] – 1mf – 9 – 0-665-80831-3 – mf#80831 – cn Canadiana [639]

An address historic and reminiscent...mt nebo presbyterian church / McCollough, Andrew W – 1905 – 9 – $50.00 – us Presbyterian [242]

Address in behalf of the china mission / Boone, William Jones – New York: printed by W Osborn, 1837 – (= ser 19th c books on china) – 1mf – 9 – mf#7.1.20 – uk Chadwyck [951]

An address, intended when written : to have been delivered before the district conference of the baltimore district / Jennings, Samuel K – Baltimore: Sands 1828 – 1r – 1 – $35.00 – mf#um-15 – us Commission [242]

Address of cuba to the united states – New York: Comes, Lawrence 1873 [mf ed 1990] – 1r – 1 – us UF Libraries [327]

Address of gov william gilpin of colorado territory : before the santa fé historical society [i.e. the historical society of new mexico], jan 20th 1863 / Gilpin, William – [Santa Fé] The New Mexican Print, 1863 (mf ed 19–) – mf#*ZH-IAG pv 147 n8 – us NY Public [972]

Address of hon. daniel h. chamberlain to the graduating class at the commencement exercises of columbia college law school. / Chamberlain, Daniel Henry – New York: Evening Post Job Printing Office, 1886. 19p. LL-425 – 1 – us L of C Photodup [340]

Address of james bicheno francis, president of the american society of civil engineers : at the thirteenth annual convention of the society, at montreal, june 15, 1881 / Francis, James Bicheno – Lowell, Mass.?: Stone, Bacheller & Livingston, 1881 – 1mf – 9 – mf#57118 – cn Canadiana [620]

Address of president lluis companys to the parliament of catalunya, march 1st, 1938 – [Barcelona?: s.n. 1938?] – 9 – (trans fr catalan) – mf#w815 – us Harvard [350]

Address of the british american association to the electors of the province of new brunswick – [Saint John, NB?: s.n, 1865?] [mf ed 1987] – 1mf – 9 – 0-665-61740-2 – mf#61740 – cn Canadiana [978]

Address of the canadian campbells to the marquess of lorne – Ottawa?: s.n, 1882 – 1mf – 9 – mf#25067 – cn Canadiana [080]

Address of the hamilton branch of the british american league : with the by-laws for the guidance of the association / British American League Hamilton Branch – [Hamilton, Ont?: s.n.], 1849 [mf ed 1987] – 1mf – 9 – 0-665-63123-5 – mf#63123 – cn Canadiana [366]

Address of the lord bishop of niagara : and other papers contributed on the occasion of the 40th anniversary of the synod of the diocese of niagara – [Hamilton, Ont?: Spectator Print], 1915 – 1mf – 9 – 0-665-73995-8 – mf#73995 – cn Canadiana [080]

Address of the members of the philadelphia anti-slavery society to their fellow citizens / Philadelphia Anti-Slavery Society – Philadelphia: The Society, Board of Managers, 1835 (mf ed 1976) – 1r – 1 – mf#Sc Micro R-2401 – us NY Public [976]

Address of the president, hon. william wirt henry, delivered at the ninth annual meeting, held at the hot springs of virginia, august 3, 4 and 5, 1897 / Henry, William Wirt – Richmond: Goode Printing Co, 1897 – 1 – mf#LL-1197 – us L of C Photodup [340]

Address of the president of the chicago and north western railway company : to the stock and bondholders at the annual meeting... – 1860 – 1 – mf#536452 – us WHS [338]

Address of the retiring president of "the association of medical superintendents of american institutions for the insane" / Clark, Daniel – S.l: s.n, 1892? – 1mf – 9 – mf#01628 – cn Canadiana [616]

Address of the rev abbe j c k laflamme...vice-president of the royal society of canada : delivered at a public meeting of the society held at queen's hall, montreal, wednesday, 27th may, 1891 – Toronto: Rowsell & Hutchison; Montreal: E Picken, 1891 – 1mf – 9 – mf#08301 – cn Canadiana [370]

Address of the rev atticus g haygood of the methodist episcopal church, south : at the 4th annual opening of the gammon school of theology, atlanta, ga, oct 27 1886 / Haygood, Atticus Greene – 1r – 1 – (filmed with: circular of the gammon school of theology) – mf#2006-s006 – us ATLA [370]

An address on build up canada / Fleming, Sandford – [Toronto?: s.n, 1904?] [mf ed 1995] – 1mf – 9 – 0-665-74266-5 – mf#74266 – cn Canadiana [380]

An address on congregationalism : as affected by the declarations of the advisory council of feb 1876 / Storrs, Richard Salter – New York: AS Barnes [1876?] [mf ed 1991] – (= ser Congregational coll) – 1mf – 9 – 0-524-00857-4 – mf#1990-4017 – us ATLA [242]

An address on imperial federation, at cambridge, jun 4 1885 / Young, Frederick – London 1885 – (= ser 19th c british colonization) – 1mf – 9 – mf#1.1.4701 – uk Chadwyck [080]

An address on supposed miracles : delivered monday, sep 20 1875, before the new york ministers' meeting of the m e church / Buckley, James Monroe – NY: Hurd & Houghton, 1875 [mf ed 1985] – 1mf – 9 – 0-8370-2506-0 – mf#1985-0506 – us ATLA [240]

An address on the anglo-saxon coronation forms : and on the word protestant in the coronation oath: at st mary redcliffe church, sun, june 22 1902 / Browne, George Forrest – London: SPCK 1902 [mf ed 1993] – 1mf – 9 – 0-524-05488-6 – (incl ind) – mf#1990-1483 – us ATLA [941]

An address on the formation of rifle associations for defensive purposes : delivered in the town hall, guleph, on wednesday evening, the 15th of aug 1866 / Howitt, Dr – [Guelph, Ont?: s.n, 1866?] [mf ed 1984] – 1mf – 9 – 0-665-45114-8 – mf#45114 – cn Canadiana [971]

Address on the government control of corporations and combinations of capital / Washburn, Charles Grenfill – [S.l: s.n., 1911?] (mf ed 19–) – 22p – mf#ZT-TN pv54 n7 – us Harvard [332]

An address on the humanities and mathematics : delivered...sep 19 1856 / Shepardson, Daniel – [Cincinnati]: Cincinnati Teachers' Assoc, 1856 [mf ed 1993] – 1mf – 9 – 0-524-08529-3 – mf#1993-1059 – us ATLA [370]

Address on the life and character of william cranch, delivered january 10th, 1907, by alexander b. hagner at the request of the bar association of the district of columbia / Hagner, Alexander Burton – Washington, D.C., 1913. 36 p. LL-469 – 1 – us L of C Photodup [340]

An address on the missionary character / Smith, Eli – Boston: Printed by Perkins & Marvin, 1840. Chicago: Dep of Photodup, U of Chicago Lib, 1970 (1r); Evanston: American Theol Lib Assoc, 1984 (1r) – 1 – 0-8370-0480-2 – mf#1984-B121 – us ATLA [240]

An address on the necessity of a liberal education : delivered in barton, on friday evening, oct 30 1857 / Juvenis – [Hamilton, Ont?: s.n.] 1857 [mf ed 1994] – 1mf – 9 – 0-665-94630-9 – mf#94630 – cn Canadiana [370]
An address on the occasion of the laying of the corner-stone : december 21st, '88 (founder's day) / Crogman, William Henry – Atlanta: Clark UP 1889 [mf ed 2006] – 1r [complete] – 1 – (reel also incl: catalogue of the gammon theological seminary, an additional address, & a dedication) – mf#2006-s007 – us ATLA [080]
An address on water in relation to disease / Bayard, William – [St John?: s.n, 1901?] – 1mf – 9 – 0-659-92178-2 – mf#9-92178 – cn Canadiana [350]
An address on woman's work in the church : before the presbytery of new albany / Heckman, George C – Madison, IN: Courier Steam Print House, 1875 [mf ed 1984] – (= ser Women & the church in america 81) – 1mf – 9 – 0-8370-1224-4 – mf#1984-2081 – us ATLA [305]
Address presented to mr u e archambault : on the eve of his departure for europe by the citizens of montreal, 27th november 1883 – S.l: s.n, 1883? – 1mf – 9 – mf#61456 – cn Canadiana [370]
An address to all the colored citizens of the united states / Meachum, John B – 1846 – 9 – $50.00 – us Presbyterian [240]
An address to parliament on the duties of great britain to india : in respect of the education of the natives, and their official employment / Cameron, Charles Hay – London, 1853 – (= ser 19th c books on british colonization) – 3mf – 9 – mf#1.1.6369 – uk Chadwyck [323]
An address to seamen : delivered at portland, 28 oct 1821 / Payson, Edward – 1 – $5.00 – us Southern Baptist [242]
An address to students of law in the united states (circular) / Hoffman, David – Baltimore: Toy, 1824. 15p. LL-383 – 1 – us L of C Photodup [340]
Address [to the british association for the advancement of science] / Dawson, John William – [S.l: s.n, 1886?] [mf ed 1980] – 1mf – 9 – 0-665-02239-5 – (incl bibl ref and ind) – mf#02239 – cn Canadiana [500]
An address to the citizens of bath : in reference to a speech delivered at the guildhall, on the 29th of june / Cockburn, William Sarsfield Rossiter – Bath?: s.n, 1837 – 1mf – 9 – mf#21581 – cn Canadiana [080]
An address to the committee of the county of york : on the state of public affairs / Hartley, David – London: Printed for J Stockdale... 1781 – 1mf – 9 – mf#20620 – cn Canadiana [350]
Address to the geographical section of the british association / Lefroy, John Henry – [London?: s.n, 1884?] [mf ed 1980] – 1mf – 9 – 0-665-08641-5 – mf#08641 – cn Canadiana [910]
Address to the graduating class, 1911, of the unitrinian school of personal harmonizing : founded by mary perry king at moonshine, twilight park, in the catskills / Carman, Bliss – [New York?: s.n.], 1911 – 1mf – 9 – 0-665-77778-7 – mf#77778 – cn Canadiana [080]
An address to the inhabitants of new brunswick, nova scotia, in north america : occasioned by the mission of two ministers, john james, and charles william milton / Bradford, John – London: Printed for and sold by Hughes & Walsh...1788 – 1mf – 9 – mf#20697 – cn Canadiana [240]
An address to the members of the methodist episcopal church : by a meeting of methodists, held in pittsburgh – Pittsburgh: Andrews 1827 – 1r – 1 – $35.00 – mf#um-15 – us Commission [242]
An address to the people of british america : upon subjects relating to the progress of the people and the improvement of the country / McDonald, A – [s.l.]: A McDonald, 1853 [mf ed 1984] – 1mf – 9 – 0-665-46139-9 – mf#46139 – cn Canadiana [971]
Address to the presbyteries of the presbyterian church in the united states of america / Craven, Elijah Richardson – [New York: s.n, 1884?] [mf ed 1992] – (= ser Presbyterian coll) – 1mf – 9 – 0-524-05538-6 – mf#1990-5142 – us ATLA [242]
Address to the public : containing a review of the charges exhibited against lord viscount melville, which led to the resolutions of the house of commons, on the 8th april, 1805 – London: J Hatchard, J Asperne, R Bickerstaff, 1805 – (= ser 19th c general coll on politics) – 1mf – 9 – mf#1.1.2 – uk Chadwyck [323]
Address to the public concerning political opinions, and plans lately adopted to promote religion in scotland / Haldane, Robert – 1800 – 1 – $5.00 – us Southern Baptist [242]
An address to the public introducing a letter to the rev mr pollard : with reference to his recent attacks upon the universalists and unitarians of london, c w / Gunn, Marcus – [s.l: s.n, 1853] [mf ed 1983] – 1mf – 9 – 0-665-44937-2 – mf#44937 – cn Canadiana [243]
Address to the section of anthropology of the british association / Tylor, Edward Burnett – [London?: s.n, 1884?] [mf ed 1982] – 1mf – 9 – 0-665-28615-5 – mf#28615 – cn Canadiana [301]
Address to the sessions and congregations : under the inspection of the united presbyterian synod in canada, relative to the temporal support of the ministers of the gospel / United Presbyterian Church in Canada. Synod – [Cayuga ON?: s.n.] 1854 [mf ed 1987] – 1mf – 9 – 0-665-53089-7 – mf#53089 – cn Canadiana [242]
An address to the suffolk north association of congregational ministers / Lesley, J Peter – Boston: Wm Crosby & HP Nichols, 1849 [mf ed 1991] – 1mf – 9 – 0-7905-9304-1 – mf#1989-2529 – us ATLA [242]
An address to the unemployed workmen of yorkshire and lancashire : on the present distress, and on machinery / Baines, Edward – London: James Ridgway & Effingham Wilson; Leeds: Edward Baines, 1826 – (= ser 19th c economics) – 1mf – 9 – mf#1.1.287 – uk Chadwyck [331]
An address to those who have been baptized in infancy : and who have not yet joined themselves to the church by partaking of the sacramental supper / George, James – Toronto: printed and publ by H Scobie, 1841 – 1mf – 9 – mf#42643 – cn Canadiana [242]
An address to william tudor, esq author of letters on the eastern states : intended to prove the calumny and slander of his remarks on the olive branch / Carey, Mathew – Philadelphia: M Carey & Son...1821 – 1mf – 9 – mf#43394 – cn Canadiana [320]
Address upon the progress of medical science : read before the new brunswick medical society / Bayard, William – St John, NB: s.n, 1871 – 1mf – 9 – mf#05978 – cn Canadiana [610]
An address upon the progress of medicine, surgery and hygiene, during the last 100 years : delivered by request of the st john mechanics' institute, on feb 4th, 1884 / Bayard, William – St John, NB?: s.n, 1884? – 1mf – 9 – mf#05977 – cn Canadiana [610]
An address upon the use and abuse of alcoholic drinks / Bayard, William – [St John, NB?: s.n, 1882?] [mf ed 1980] – 1mf – 9 – 0-665-02996-9 – mf#02996 – cn Canadiana [360]
Address-buch fuer die koeniglich-preussischen fuerstenthuemer ansbach und bayreuth... *see* Die amtskalender der fraenkischen fuerstentuemer ansbach und bayreuth [1737-1801]
Address...dickinson college / Nisbet, Charles – June, 1798 – 1 – $50.00 – us Presbyterian [240]
Addresse de l'association d'annexion de montreal au peuple du canada / Association d'annexion de Montreal – [Montreal?: s.n.], 1849 [mf ed 1985] – 1mf – 9 – 0-665-48264-7 – mf#48264 – cn Canadiana [971]
Addresse de l'honorable louis joseph papineau aux electeurs de la cite de montreal – S.l: s.n, 1851? – 1mf – 9 – mf#43735 – cn Canadiana [325]
Addresses / Arthur, William; ed by Strickland, William Peter – New-York: Carlton & Phillips, 1856 [mf ed 1990] – 1mf – 9 – 0-7905-3989-6 – mf#1989-0482 – us ATLA [242]
Addresses / Brooks, Phillips – Boston: C E Brown c1893 – 1 – (int by julius h ward. filmed with: armorial insignia of the princes of wales / rother, g c) – mf#2062 – us Wisconsin U Libr [080]
Addresses : delivered by the right worshipful dis't deputy grand master alfio de grassi and very worshipful brother rev j d gibson, st john's lodge, columbus...on the occasion of the dedication of the new masonic hall, aurora – Toronto: S E Horne for the Rising Sun Lodge, n129, GRC, 1866 – 1mf – 9 – mf#63529 – cn Canadiana [080]
Addresses / Drummond, Henry – Toronto: T Eaton, c1891 – 2mf – 9 – (brief sketch of aut by w j dawson) – mf#06071 – cn Canadiana [080]
Addresses / Drummond, Henry – Philadelphia: Henry Altemus [1891?] [mf ed 1991] – 1mf – 9 – 0-7905-9265-7 – mf#1989-2490 – us ATLA [242]
Addresses / McNeill, John – New York: Fleming H Revell, [1890?] [mf ed 1986] – 1mf – 9 – 0-8370-7087-2 – mf#1986-1087 – us ATLA [240]
Addresses and correspondence, 1857-1908 / Blyden, Edward Wilmot – 1 – us CRL [920]
Addresses and discourses : historical and religious: with a paper on bishop berkeley / Beardsley, Eben Edwards – Cambridge: Riverside Press, 1892 [mf ed 1992] – (= ser Anglican/episcopal coll) – 1mf – 9 – 0-524-03607-1 – mf#1990-4767 – us ATLA [242]
Addresses and lectures... / New Orleans Baptist Theological Seminary – 1953-70 – 1 – $43.82 – (faculty addresses, layne lectures, carver-barnes lectures, evangelism lectures, missionary days, tharp lectures, founders day addresses) – us Southern Baptist [242]
Addresses and sermons for preachers *see* Chiao hui li shih [ccm145]
Addresses at the annual meeting of the new west education commission : ...oct 14 1890 in the first congregational church, chicago / Gunsaulus, Frank Wakeley et al – Chicago: [s.n.] 1890 [mf ed 1992] – 1mf – 9 – 0-524-03230-0 – mf#1990-0858 – us ATLA [230]
Addresses at the celebration of the 250th anniversary of the westminster assembly : by the general assembly of the presbyterian church in the usa / Jackson, Sheldon et al; ed by Roberts, William Henry – Philadelphia: Presbyterian Board of Publ & Sabbath-School Work, 1898 [mf ed 1990] – 1mf – 9 – 0-7905-6546-3 – mf#1988-2546 – us ATLA [242]
Addresses before the members of the bar, of worcester county, massachusetts: by joseph willard, october 2, 1829; emory washburn, february 7, 1856; dwight foster, october 3, 1878 / Worcester County. Massachusetts Bar – Worchester, Hamilton, 1879. 250 p. LL-1565 – 1 – us L of C Photodup [340]
Addresses before the new york state conference of religion – 1903-15 [complete] – 1r – 1 – mf#ATLA S0903 – us ATLA [200]
Addresses delivered at agincourt, april 2nd and may 7th, 1878 : an able exposition of the cause of the hard times: the banks, loan companies and importing merchants chiefly to blame / Bradford, Robert – Toronto: Morton & McLean, [1878?] [mf ed 1979] – 1mf – 9 – 0-665-00232-7 – mf#00232 – cn Canadiana [332]
Addresses delivered at richmond, vermont, june 28, 1895 : in memory of austin hazen / Green, Salmon et al – Middletown, CT: Pelton & King, 1895 [mf ed 1993] – (= ser Congregational coll) – 1mf – 9 – 0-524-06089-4 – mf#1991-2402 – us ATLA [242]
Addresses delivered at the 40th anniversary of the boards of home missions, foreign missions and church extension of the general synod of the evangelical lutheran church at harrisburg... – Philadelphia: Lutheran Publ Society, c1909 [mf ed 1986] – 1mf – 9 – 0-8370-6152-0 – mf#1986-0152 – us ATLA [242]
Addresses delivered at the centennial celebration of the general assembly of the presbyterian church : ...may 24th 1888 – Philadelphia: publ...for the 100th General Assembly by MacCalla, c1888 [mf ed 1992] – (= ser Presbyterian coll) – 1mf – 9 – 0-524-02693-9 – mf#1990-4400 – us ATLA [242]
Addresses delivered at the inauguration of rev j w nevin : ...mercersburg, pa, may 20th 1840 / Helffenstein, Jacob et al – Chambersburg, PA: printed at the Office of Publication of the German Reformed Church, 1840 [mf ed 1993] – 1mf – 9 – 0-524-08764-4 – mf#1993-3269 – us ATLA [240]
Addresses delivered at the observance of the 100th anniversary of the establishment of the harvard divinity school : cambridge, massachusetts, oct 5 1917 – Cambridge: Harvard University, 1917 [mf ed 1993] – (= ser Unitarian/universalist coll) – 1mf – 9 – 0-524-07751-7 – mf#1991-3319 – us ATLA [240]
Addresses delivered at the world's congress and general missionary conventions of the church of christ : ...chicago in sep 1893 – Chicago: SJ Clarke, 1893 [mf ed 1991] – (= ser Christian church (disciples of christ) coll) – 1mf – 9 – 0-524-01219-9 – mf#1990-4077 – us ATLA [240]
Addresses, essays, lectures / Broadus, John Albert – 1851-95. v1-2. 844p – 1 – $37.98 – us Southern Baptist [242]
Addresses, inaugurals and charges : delivered... sep 1st and nov 24th 1858 / Kurtz, Benjamin et al – Baltimore: T Newton Kurtz 1859 [mf ed 1992] – 1mf – 9 – 0-524-04772-3 – mf#1991-2158 – us ATLA [242]
Addresses of henry russell pritchard : with biographical sketch / Pritchard, Henry Russell & Tyler, B B – Cincinnati, OH: Standard Pub Co, c1898 [mf ed 1992] – (= ser Christian church (disciples of christ) coll) – 1mf – 9 – 0-524-02491-X – mf#1990-4350 – us ATLA [240]
Addresses of rev drs park, post, and bacon : at the anniversary...may, 1854 – New York: publ for the american congregational union [by] clark, austin & smith, 1854 [mf ed 1993] – (= ser Congregational coll) – 1mf – 9 – 0-524-06963-8 – mf#1990-5327 – us ATLA [242]
Addresses of rev drs sturtevant and stearns at the anniversary of the american congregational union, may, 1855 – Andover: Warren F Draper, 1855 [mf ed 1993] – 1mf – 9 – 0-524-08593-5 – mf#1993-3178 – us ATLA [242]
Addresses of the lord bishop of ontario, visitor, and rev canon bedford jones, lld, warden, at the inaugural conversazione, november 16, 1876 – Ottawa: Citizen Print & Pub Co, 1877 [mf ed 1987] – 1mf – 9 – 0-665-28435-7 – mf#28435 – cn Canadiana [242]
Addresses on foreign missions / Storrs, Richard Salter – Boston: American Board of Commissioners for Foreign Missions, 1900 [mf ed 1986] – 1mf – 9 – 0-8370-6399-X – mf#1986-0399 – us ATLA [240]
Addresses on historical and literary subjects : in continuation of 'studies in european history' = Akademische vortraege / Doellinger, Johann Joseph Ignaz von – London: J Murray, 1894 [mf ed 1990] – 1mf – 9 – 0-7905-4730-9 – (english trans by margaret warre) – mf#1988-0730 – us ATLA [940]
Addresses on the acts of the apostles / Benson, Edward White – London, New York: Macmillan, 1901 [mf ed 1989] – 2mf – 9 – 0-7905-1024-3 – (incl bibl ref & ind) – mf#1987-1024 – us ATLA [226]
Addresses on the art of pleading, delivered to the glasgow legal and speculative society on september 29, 1860 / Moncreiff, James Moncreiff – London: Griffin, 1860. 23p. LL-2239 – 1 – us L of C Photodup [340]
Addresses on the gospel of st john : delivered...oct 21 1903 and may 11 1904 – Providence, RI: ...St John Conference Cttee, 1906 [mf ed 1989] – 2mf – 9 – 0-7905-2077-X – (incl ind) – mf#1987-2077 – us ATLA [226]
Addresses on the occasion of the gathering in the gordon memorial college, khartoum on the 20th february 1945 : to celebrate the inaugration of the college in its new form – [Khartoum, 1945] – us CRL [080]
Addresses, petitions etc : from the kings and chiefs of sudan (africa,) and the inhabitants of sierra leone, to his late majesty, king william the fourth, and his excellency, h d campbell – London, 1838 – (= ser 19th c books on british colonization) – 1mf – 9 – mf#1.1.3779 – uk Chadwyck [960]
Addresses presented to his excellency major general sir john colborne, kcb, lieut governor of upper canada : on the occasion of his leaving the province – Toronto: R Stanton, 1836 – 1mf – 9 – 0-665-53942-8 – mf#53942 – cn Canadiana [971]
Addresses to the dispersed of judah / Livermore, Harriet – Philadelphia: printed by L R Bailey, 1849 [mf ed 1984] – (= ser Women & the church in america 169) – 1mf – 9 – 0-8370-1422-0 – mf#1984-2169 – us ATLA [270]
Addresses to the people of ireland : on the degradation and misery of their country / Ensor, George – Dublin, 1823 – (= ser 19th c ireland) – 1mf – 9 – mf#1.1.1560 – uk Chadwyck [941]
Addressing, analyzing, and challenging social issues and problems in the coaching profession : a survey of ncaa division 2 women basketball coaches / Berg, Theresa A – 1997 – 1mf – 9 – $4.00 – mf#PE 3899 – us Kinesology [790]
Addressing school violence : practical strategies and interventions / Juhnke, Gerald A – Greensboro NC: Caps Publ, ERIC Counseling & Student Services Clearinghouse c2000 [mf ed 2000] – 1mf – 9 – us GPO [362]
Addrich im moos : historischer roman / Zschokke, Heinrich – 1. aufl. Berlin: Buchverlag Der Morgen 1966 [mf ed 1995] – 1r – 1 – (filmed with: der zerbrochene krug / heinrich zschokke & other titles) – mf#3766p – us Wisconsin U Libr [830]
Addy, Sidney Oldall *see* Church and manor
Ade, antología del tabaco / [Compania colombiana de tabaco, Bogota] – [Bogota, Colombia 1944] [mf ed 1993] – 1r – 1 – us UF Libraries [810]
Ade, George *see*
– Fables in slang
– Slim princess
Adeb-i sedat / Pasa, Ahmet Cevdet – Istanbul: Karabet ve Kasbar Matbaasi, 1303 [1886] – (= ser Ottoman literature, writers and the arts) – 1mf – 9 – $25.00 – us MEDOC [470]
O adejo litterario : jornal de instrucao e recreio – Para: Typ Commercial de A J R Guimaraes, 27 dez 1857 – (= ser Ps 19) – bl Biblioteca [440]

Adel und untergang : [poems] / Weinheber, Josef – 5. aufl. Wien: A Luser c1934 [mf ed 1991] – 1r – 1 – (filmed with: wassermann: sein kampf um wahrheit / walter goldstein) – mf#3037p – us Wisconsin U Libr [810]

Adelaide advertiser – Adelaide, jan 1897-dec 1899 – 9r – 1 – diazo A$594.00 silver A$643.50 – at Pascoe [079]

Adelaide free press – Adelaide SA, 1905-11 – 1 – (merged with: the enterprise to become: adelaide free press and bedford enterprise; in afrikaans & english) – mf#MS00416 – sa National [079]

Adelaide law review – Adelaide, Australia 1973+ – 1,5,9 – ISSN: 0065-1915 – mf#9476 – us UMI ProQuest [340]

Adelaide list – list of the numbers and names of the holders of preliminary land orders in the northern territory *see* Northern territory land orders/adelaide and london registers 1870 ballot

Adelaide observer – Adelaide, Australia 6 jan 1844-26 dec 1846; 24 apr 1847; 6 may, 15 jul 1848; 4 aug 1849; 16 mar 1850; 8 mar 1851-18 sep 1852; 1 oct 1853; 2 sep 1854; 22 sep 1855; 10 jan, 28 mar, 25 apr, 5 sep 1857; 2 feb 1861-31 dec 1870; 3 jan 1880-31 dec 1904 – 1 – (1851,52 imperfect; cont by: observer [7 jan 1905-24 jun 1922, 4 jan-28 aug 1930]) – uk British Libr Newspaper [079]

Adelaide Opinion *see* Adelaide times

Adelaide opinion – Adelaide. 3 may-29 nov 1882 (wkly) [mf ed Cape Town: SA library 1985] – 1r – 1 – (filmed with: adelaide phoenix and adelaide times) – mf#MS00374 – sa National [079]

Adelaide opinion *see* Adelaide phoenix

Adelaide phoenix – Adelaide. jan 13-mar 17 1883 [mf ed Cape Town: SA library 1985] – 1r – 1 – (filmed with: adelaide opinion and adelaide times) – mf#MS00374 – sa National [079]

Adelaide phoenix *see*
- Adelaide opinion
- Adelaide times

Adelaide recorder – Adelaide. 1885-87. Cape Town: SA Library – sa National [079]

The adelaide recorder – Adelaide, SA. 1 jul 1885-8 nov 1887 – 1r – 1 – sa National [079]

Adelaide register allotment *see* N[orthern] t[erritory] adelaide register for land orders

Adelaide review – Adelaide – 4r – at Pascoe [079]

Adelaide standard – Adelaide. jan 2 1878-feb 22 1882 (wkly) [mf ed Cape Town: SA library 1985] – 2r – 1 – mf#MS00375 – sa National [079]

Adelaide times – Adelaide, Australia 2 oct 1848-31 dec 1856, 19 jun-30 dec 1857 (imperfect) – 1 – uk British Libr Newspaper [079]

Adelaide times – Adelaide. 14 jul-29 dec 1883 (wkly) [mf ed Cape Town: SA library 1985] – 1r – 1 – (filmed with: adelaide opinion and adelaide phoenix) – mf#MS00374 – sa National [079]

El adelantado de la florida pedro menendez de aviles / Camin, Alfonso – Mexico: Revista Norte 1944 [mf ed 1991] – 1r – 1 – us UF Libraries [978]

El adelantado hernando de soto : breve noticias, nuevos documentos para su biografia y relacion de los que le acompanaron a la florida / Solar y Taboada, Antonio & Rigula y Ochotorena, Jose de – Badajoz: Ediciones Arqueros, 1929 – 1 – sp Bibl Santa Ana [946]

Adelante – Villafranca de los Barros 1920 – 5 – sp Bibl Santa Ana [074]

Adelante libre – Orlando FL: Amado F Hernandez 1974 jul-1977 may (mthly) [mf ed 1999] – (= ser Cuban heritage collection) – 1r – 1 – (began in 1972; missing: 1974 aug-nov; 1975 mar-jun, sep-nov; 1976 jan-feb, apr-may, oct-dec; 1977 feb, apr) – us UF Libraries [071]

Adelante raza – 1973 jul-1976 jun – 1 – mf#345243 – us WHS [071]

Adelante (tampa, fl) : official organ of the movimiento democratico de liberacion cubana – Tampa FL: Movimiento Democratico de Liberacion Cubana sep 1967 (mthly) [mf ed 1999] – (= ser Cuban heritage collection) – 1r – 1 – (began in 1967) – us UF Libraries [320]

Adelard of Bath *see* Des adelard von bath traktat de eodem et diverso

Adelardo lopez de ayala / Blanco Garcia, Francisco – Madrid: Saenz de Jubera, 1909 – sp Bibl Santa Ana [440]

Adelbert chamisso's werke / Chamisso, Adelbert von; ed by Hitzig, Julius Eduard – 2. aufl. Leipzig: Weidmann 1842 [mf ed 1993] – 6v in 3 on 1r [ill] – 1 – (incl bibl ref) – mf#8531 – us Wisconsin U Libr [430]

Adelbert de chamisso de boncourt / Brun, Xavier – Lyon: A Waltener 1896 [mf ed 1989] – 1r – 1 – (incl bibl. filmed with: prosit neujahr! / l angely) – mf#7151 – us Wisconsin U Libr [430]

Adelbert von chamisso's werke – 2. aufl. Leipzig: Weidmann 1842 [mf ed 1993] – 6v on 1r – 1 – mf#8531 – us Wisconsin U Libr [800]

Adelbertus, frere *see* Geographie du cours elementaire

Adele et dorsan : opera en deux actes / Dalayrac, Nicolas & Marsollier – 2.ed. paris: chez vente, libraire...an 11 [1802 or 1803] [mf ed 1994] – 1r – 1 – us UF Libraries [790]

Adelong argus – Adelong, jan 1899-dec 1905 – 2r – 1 – A$132.00 vesicular A$143.00 silver – at Pascoe [079]

Adelong mining journal – Adelong, oct 1858-sep 1860 – 1r – A$44.18 vesicular A$49.68 silver – at Pascoe [079]

Adelong & tumut express – Adelong, apr 1900-dec 1954 – 12r – 1 – A$858.00 vesicular A$924.00 silver – at Pascoe [079]

Adelpha *see* Cycling for old and young

Adelphi – uk Chadwyck [073]

Adelphi Organization *see* Stelle group letter

Adelphi papers – Abingdon, Oxfordshire 1963+ – 1,5,9 – ISSN: 0567-932X – mf#3175 – us UMI ProQuest [327]

Adelung, F von *see*
- Kritisch-literarische uebersicht der reisenden in russland bis 1700
- Siegmund freiherr von herberstein

[Adelung, F von] *see* Ueber die aelteren auslaendischen karten von russland bis 1700

Adelung, Johann Christoph *see* Magazin fuer die deutsche sprache

Aden chronicle – Aden, Yemen, 7 jan-11 feb, 7 apr-19 may, 3 nov 1960-20 dec 1962; 9 jan 1964-14 april 1966 (imperfect) – 10r – 1 – uk British Libr Newspaper [079]

Adenauer kreis- und wochenblatt *see* Kautionsfreies kreis-wochenblatt fuer den kreis adenau und umgegend

Adeney, John Howard *see* The jews of eastern europe

Adeney, Walter Frederic *see*
- A century's progress in religious life and thought
- The christian conception of god
- The construction of the bible
- Ezra, nehemiah, and esther
- The greek and eastern churches
- The hebrew utopia
- How to read the bible
- St luke
- The song of solomon and the lamentations of jeremiah
- The theology of the new testament
- Thessalonians and galatians
- The virgin birth and the divinity of christ
- Women of the new testament

Adentro : bien adentro del alma cubana / Abril Amores, Eduardo – Manzanillo, Cuba: Editorial El Arte 1945 [mf ed 1993] – 1r – 1 – us UF Libraries [972]

Ader, Jean J *see*
- Napoleon devant ses contemporains
- Paris revolutionnaire
- Resume de l'histoire du bearn, de la gascogne superieure et des basques

Adern in marmor : gedichte / Berninger, Gertrud – Wien: W Andermann, 1944 [mf ed 1989] – 62p – 1 – mf#7013 – us Wisconsin U Libr [810]

Aderson, Susan McMurray *see* Journal of elder abuse and neglect

Ad'ge psalie – Nal'chik, USSR. Apr 1991-1992 – 2r – 1 – us L of C Photodup [077]

Adger, John Bailey *see* The collected writings of james henley thornwell, d.d., ll. d

Adhelmi opera (mgh1:15.bd) / ed by Ehwald, R – 1919 – (= ser Monumenta germaniae historica 1: scriptores – auctores antiquissimi) – €38.00 – ne Slangenburg [240]

Adherent – Seattle WA 1978-83 – 1,5,9 – ISSN: 0360-9588 – mf#11912 – us UMI ProQuest [331]

Adhesives age – New York NY 1958-97 – 1,5,9 – ISSN: 0001-821X – mf#2494 – us UMI ProQuest [660]

Adhin, Jan Handsdew *see* Development planning in surinam in historical pers...

Adhipati lam / On Kyo Cann – Ran kun: Phran khyi re Re khyamcan Ca pe 1975 [mf ed 1993] – on pt of 1r – 1 – mf#11052 r584 n6 – us Cornell [959]

Adhortatio ad omnes...verbi dei ministros / Bullinger, Heinrich – Tigvri, Christoph Froschover, 1572 – 1mf – 9 – mf#PBU-246 – ne IDC [240]

Adhuc stat! : la franc-maconnerie en dix demandes et reponses: pour l'edification du peuple et de ses amis / Henne am Rhyn, Otto – St-Gall: Scheitlin & Zollikofer 1865 – 1mf – 9 – mf#vrl-147 – ne IDC [366]

Adi Koro, Raden Pandji *see* Geschiedenis van kalilah en daminah

Adi parva – Calcutta: Bharata Press, 1884 [mf ed 1993] – (= ser The mahabharata of krishna-dwaipayana vyasa) – 2mf – 9 – 0-524-08007-0 – (trans chiefly by kesari mohan ganguli) – mf#1991-0229 – us ATLA [490]

Adicion a la relacion descriptiva de los mapas planos...archivo general de indias / Torre Revello, Jose; ed by Bayle, Constantino – Madrid: Razon y Fe, 1928 – 9 – sp Bibl Santa Ana [910]

Adicion a un folleto / Torre Isunza de Hita, Pedro – Cabra: Imp. Manuel Cordon, 1928 – sp Bibl Santa Ana [946]

Adicion al discurso...provando que no se deve sangrar en el sarampion... / Saavedra, J – Granada, 1626 – 1mf – 9 – sp Cultura [616]

Adicion del inventario del museo de la comision provincial de monumentos historicos y artisticos de badajoz / Solar y Taboada, Antonio – Badajoz: Minerva Extremena, 1919 – 1 – sp Bibl Santa Ana [060]

Adieu paris : journal d'une evacuee canadienne, 10 mai-17 juin 1940 / Routier, Simone – Ottawa: Ed du droit, 10 mai-17 juin 1940 [mf ed 1975] – 1r – 5 – mf#SEM16P250 – cn Bibl Nat [920]

Adieux d'adolphe monod a ses amis et a l'eglise *see* Adolphe monod's farewell to his friends and to the church

Les adieux lamentables du general cavaignac au peuple fracais – Paris [1849] – (= ser Pamphlets and periodicals relating to the French Revolution of 1848) – 1r – 1 – us CRL [944]

Adimari, A *see* Esequie dell'ill mo & ecc mo principe don francesco medici celebrate dal ser mo don cosimo 2, gran duca di toscana 4

Adinah – Shumarah-'i. 1-30,32. azar 1364-isfand 1367 [nov 1985-feb 1989] – 1r – 1 – $53.00 – us MEDOC [079]

Adindiin = Light – Cortez CO: Navajo Christian Reading 1964-69 [mf ed 2007] – (= ser Religious periodical literature of the hispanic and indigenous peoples of the americas, 1850-1985) – 1r – 1 – (in english & navajo) – mf#2007i-s018 – us ATLA [240]

Adinegoro, D [comp] *see* Kamoes bahasa indonesia-nippon dan nippon indonesia

El adios a ruben dario / Teja Zabre, Alfonso – Mexico: D F 1941 [ed 1991] – (= ser Cuadernos de letras 1) – 1r – 1 – us UF Libraries [920]

Adiramled / Hamilton Co. Wyoming – dec 1900-nov 1903 [mthly] – 1r – 1 – (some non-ohio) – mf#B29886 – us Ohio Hist [071]

Adisesa *see* The paramarthasara of adi sesa

Aditi; indisk-orientalsk ballet i to akter (anden akt i to afdelinger). musiken af fr. rung. dekorationerne af w. guellich. kostumerne tegnede af pietro krohn. opfort forste gang i marts 1880 / Hansen, Emil – Kobenhavn: J H Schubothes Boghandel [1880?] – 1 – mf#*ZBD-*MGTZ pv 2-Res – Located: NYPL – us Misc Inst [790]

Das adjektiv bei ulrich von lichtenstein / Lucas, Wilhelm – Greifswald, 1914 (mf ed 1995) – 2mf – 9 – €31.00 – 3-8267-3111-5 – mf#DHS-AR 3111 – gw Frankfurter [430]

The adjudged cases on insanity as a defense to crime. / Lawson, John Davison – St. Louis, Thomas, 1884. 953 p. LL-332 – 1 – us L of C Photodup [345]

Adjumenta oratoris sacri : seu, divisiones, sententiae et documenta de iis christianae vitae veritatibus... / Schouppe, Francois Xavier – Bruxellis: H Goemaere, 1867 [mf ed 1986] – 2mf – 9 – 0-8370-7504-1 – mf#1986-1504 – us ATLA [241]

Adjustment of labor-management disputes in california / California. State Conciliation Service – 1948-69. 11 fiches. (Harvard Law School Library Collection – 9 – us Harvard Law [331]

Der adjutant : eine erzaehlung / Frentz, Hans – Leipzig: Ruethig 1941 [mf ed 1989] – 1r – 1 – (filmed with: ein glaubensbekenntniss / ferdinand freilingrath) – mf#7269 – us Wisconsin U Libr [830]

Adjutant General of the State of the State of Nebraska *see* Roster of nebraska volunteers from 1816 to 1869

Adkins, Frank *see* Disciples and baptists

Adl on the frontline – New York NY 1991+ – 1,5,9 – (cont: adl bulletin) – ISSN: 1061-5202 – mf#19564 – us UMI ProQuest [320]

Adlem, Willem Louis Johan *see* Administratiewe aspekte van dorpstigting

Der adler – Berlin DE, 1939-44 [gaps] – 1 – gw Misc Inst [074]

Der adler – Vienna, Austria jul 7-25 1933 [mf ed Norman Ross] – 1r – 1 – mf#nrp-1937 – us UMI ProQuest [074]

Adler, A *see* Die reform des judentums

Adler, C *see* Oriental studies published in commemoration of the fortieth anniversary of paul haupt as director of the john hopkins university

Adler, Carl *see* Mundartlich heiteres

Adler, Elkan Nathan *see*
- About hebrew manuscripts
- Auto de fe and jew

Adler, Emma *see* Goethe und frau v stein

Adler, Felix *see* An ethical philosophy of life

Adler, Friedrich *see* J'accuse

Adler, Guido et al *see*
- Denkmaeler der tonkunst in oesterreich
- Ludwig van beethoven

Adler, Hans *see* Kampf dem tode

Adler, Hermann *see* Father's barmitzvah exhortation

Adler, Howard *see* Journal of human resources in hospitality and tourism

Adler im sueden – Muenchen DE, 1942 – 1r – 1 – gw Misc Inst [074]

Adler, John Hans *see* Finanzas publicas y el desarrollo economico de gua...

Adler, K *see* Apologia m casparis aqvilae

Adler, Lazurus *see* Vortrage zur forderung der humanitat

Adler, Mortimer Jerome *see* How to think about war and peace

Adler, Ottilie *see* Friedrich und caroline perthes

Adler ueber land und meer : [frontzeitung der luftflotte 2.] – Germany 1/2 may 1940 – 1 – uk British Libr Newspaper [355]

Der adlerflug im romischen konserkrationszeremoniell / Geyer, Ursula – Bonn, 1967 – 1 – gw Mikropress [930]

Adler-Rudel, Salomon *see* Juedische arbeits- und wanderfuersorge

Adlershofer tageblatt – Berlin DE, 1925 jan-mar – 1r – 1 – gw Misc Inst [074]

Adloff, Virginia McLean *see* Biographical file of african leaders

Adlung, Jacob *see* Anleitung zu der musikalischen gelahrtheit

Adlung, Jakob *see*
- Anleitung zur musikalischen gelahrtheit
- Musica mechanica organoedi..

Administering the federal judicial circuits : a survey of chief judges' approaches and procedures / Wheeler, Russell R & Nihan, Charles W – Washington: FJC, Aug 1982 – 1mf – 9 – $1.50 – mf#LLMC 95-348 – us LLMC [340]

Administracion de espana en el reinado de los reyes catolicos / Montero de Espinosa, Luis – Madrid, 1858 – 1 – sp Bibl Santa Ana [350]

La administracion de sacramentos en toledo despues del cambio de rito / Garcia, Alf I – 1958 – 2mf – 8 – €5.00 – ne Slangenburg [240]

Administracion...reyes catolicos / Montero de Espinosa, Luis – 1858 – 9 – sp Bibl Santa Ana [946]

Administrasi negara *see* Lembaga administrasi negara

Administratiewe aspekte van dorpstigting : met besondere verwysing na die departement plaaslike bestuur van die transvaalse provinsiale administrasie / Adlem, Willem Louis Johan – U of South Africa 1979 [mf ed Pretoria: Universiteit van Suid-Afrika 1980] – 5mf – 9 – (summary in afrikaans & english; incl bibl) – sa Unisa [350]

Administration and policy in mental health – Dordrecht, Netherlands 1989+ – 1,5,9 – (cont: administration in mental health) – ISSN: 0894-587X – mf#11427,01 – us UMI ProQuest [362]

L'administration chapleau – Montreal: s.n, 1881 – 1mf – 9 – mf#11943 – cn Canadiana [325]

L'administration chapleau – Montreal: s.n, 1881 – 1mf – 9 – mf#00014 – cn Canadiana [325]

Administration in mental health – Dordrecht, Netherlands 1972-89 – 1,5,9 – (cont by: administration and policy in mental health) – ISSN: 0090-1180 – mf#11427.01 – us UMI ProQuest [362]

Administration in social work : the quarterly journal of human services management / ed by Ginsberg, Leon – mf#0364-3107 – us Haworth [650]

Administration of a revolution / Goodsell, Charles T – Cambridge MA: Harvard UP 1965 [mf ed 1992] – (= ser Harvard political studies) – 1r – 1 – (incl bibl ref & notes) – us UF Libraries [972]

The administration of cecil john rhodes as prime minister of the cape colony, 1890-1896 / Jenkins, Stanley John – [Cape Town], 1951 – us CRL [960]

Administration of crown lands department under the mowat government : eleven years of efficient and economical government – [S.l: s.n.], 1883 [mf ed 1980] – 1mf – 9 – 0-665-06460-8 – mf#06460 – cn Canadiana [380]

The administration of estates / Sheard, Terence – Rev. Toronto: Life Underwriters Association of Canada. 1942. 133p. LL-2322 – 1 – us L of C Photodup [340]

Administration of health and physical education in colleges / Hughes, William Leonard – New York: A S Barnes, 1935 (mf ed 1996) – 1r – 1 – mf#*Z-7783 – us NY Public [378]

Administration of justice during the muslim rule in india : with a history of the origin of the islamic legal institutions / Husain, Wahed – [Calcutta]: University of Calcutta, 1934 – (= ser Samp: indian books) – us CRL [260]

Administration of justice in a large appellate court : the ninth circuit innovations project / Cecil, Joe S – Washington: FJC, 1985 – 2mf – 9 – $3.00 – mf#LLMC 95-320 – us LLMC [347]

The administration of lieut-governor simcoe viewed in his official correspondence / Cruikshank, Ernest Alexander – s.l: s.n, 1891? – 1mf – 9 – mf#03631 – cn Canadiana [971]

The administration of mysore under sir mark cubbon, 1834-1861 / Venkatasubba Sastri, Kasi Nageswara – London: George Allen & Unwin, 1932 – (= ser Samp: indian books) – us CRL [350]

The administration of sir james craig : a chapter in canadian history / Cruikshank, Ernest Alexander – Ottawa: printed for the Royal Society of Canada, 1909 – 1mf – 9 – 0-665-74459-5 – (incl some text in french) – mf#74459 – cn Canadiana [971]

The administration of the sultanate of dehli / Qureshi, Ishtiaq Husain – Lahore: Sh Muhammad Ashraf, 1944 – (= ser Samp: indian books) – us CRL [954]

Administration papers / Uganda Institute of Public Administration – 1 – us CRL [960]

Administration report : census of india – Peshwar: North-West Frontier Prov Govt Press, 1911-31 – 1 – us CRL [315]

Administration & society – Newbury Park CA 1976+ – 1,5,9 – (cont: journal of comparative administration) – ISSN: 0095-3997 – mf#12641,01 – us UMI ProQuest [303]

The administration's proposed legislation on creating a department of homeland security : hearing before the committee on agriculture, house of representatives, 107th congress, 2nd session, june 26 2002 / United States. Congress. House. Committee on Agriculture – Washington: US GPO 2002 [mf ed 2002] – 2mf – 9 – 0-16-068672-5 – (incl bibl ref) – us GPO [343]

Administrative and procedural aspects of the federal reserve board/department of the treasury proposed rule concerning competition in the real estate brokerage and management markets : hearing...house of representatives, 107th congress, 2nd session, may 16 2002 / United States. Congress. House. Committee on the Judiciary. Subcommittee on Commercial and Administrative Law – Washington: US GPO 2002 [mf ed 2002] – 3mf – 9 – 0-16-068803-5 – (incl bibl ref) – us GPO [342]

Administrative archives, 1833-1969 / Catholic Archdiocese of Papeete – 60r – 1 – mf#pmb1080 – at Pacific Mss [241]

Administrative Conference of the US (ACUS) *see*
- Acus – annual reports
- Administrative procedure sourcebook
- Colloquium on regulatory design in theory and practice
- Directory of administrative hearing facilities
- Drafting federal grant statutes
- Executive control of rulemaking
- Expediting settlement of employee grievances in the federal sector
- Federal administrative law judge hearings, statistical reports
- Federal user fees
- Government in the sunshine act
- A guide to federal agency rulemaking
- Judicial review under the clean air act and federal water pollution control act
- Legislative veto of agency rules after u.s. v. chadha
- La mediacion
- Mediation
- Negotiated rulemaking sourcebook
- The ombundsman
- Regulatory agency chairmen and the regulatory process

Administrative Conference of the US (ACUS). Subcomm on Administrative Practice and Procedure od the Senate Judiciary Committee *see* Temporary administrative conference of the us

Administrative decisions under employer sanctions : and unfair immigration-related employment practices laws of the u s – v1-7 n1-999. mar 1988-jun 1998 – 102mf – 9 – $153.00 – (vols added as they become available) – mf#llmc99-009 – us LLMC [344]

The administrative economy of the fine arts in england / Edwards, Edward – London 1840 – (= ser 19th c art & architecture) – 4mf – 9 – mf#4.2.779 – uk Chadwyck [700]

The administrative histories of us civilian agencies – 2 colls – 68r – 1 – (world war 2 56r c39-27341. the korean war 12r c39-27342. each comes with detailed guide arranged by author, title, subject, with added entries) – mf#C39-27340 – us Primary [975]

Administrative history, general headquarters, united states army forces, pacific / U.S. Army. Far East Command – 6 Apr 1945-31 Dec 1946. 25v – 1 – 69.00 – us L of C Photodup [977]

Administrative law bulletin *see* Administrative law review (aba)

Administrative law judges decisions / U.S. Dept of Labor – v1-7 n5. 1987-sep/oct 1993 [all publ] – 123mf – 9 – $184.00 – mf#llmc 95-002 – us LLMC [342]

Administrative law review (aba) – v1-51. 1949-1999 – 9 – $791.00 set – (title varies: v1-12 1949-1960 as administrative law bulletin) – ISSN: 0001-8368 – mf#100041 – us Hein [340]

Administrative management – New York NY 1940-82 – 1,5,9 – ISSN: 0001-8376 – mf#700.02 – us UMI ProQuest [650]

Administrative management – New York NY 1986-88 – 1,5,9 – ISSN: 0884-5905 – mf#700,02 – us UMI ProQuest [650]

Administrative problems of british india / Chailley-Bert, Joseph – London: Macmillan and Co, 1910 – (= ser Samp: indian books) – (trans by william meyer) – us CRL [954]

Administrative problems of instrumental teaching / Livermore, John Sisson – U of Rochester 1946 [mf ed 19–] – 2mf – 9 – mf#fiche 1206 – us Sibley [780]

Administrative procedure sourcebook : statutes and related materials / Administrative Conference of the US (ACUS) – 1st ed 1985. Washington: GPO, 1985 (all publ) – 11mf – 9 – $16.50 – mf#LLMC 94-335A – us LLMC [348]

Administrative procedure sourcebook : statutes and related materials / Administrative Conference of the US (ACUS) – 2nd ed 1992. Washington: GPO, 1992 (all publ) – 11mf – 9 – $16.50 – mf#LLMC 94-335B – us LLMC [348]

Administrative report of the census of cochin – Ernakulam: Printed at the Cochin Govt Press, pt3. 1901 – (= ser Census of india) – us CRL [315]

Administrative report of the census operations in the bombay presidency – Bombay: Govt Central Press, 1901 – 1 – us CRL [324]

Administrative rules and regulations of the government of guam, 1975 / Bohn, John A – Agana: Guam Law Revision Commission. 2v. 1975-82 – 39mf – 9 – $58.00 – (with revisions for 1981-82) – mf#LLMC 82-100B Title 5 – us LLMC [324]

Administrative science quarterly – New York NY 1956+ – 1,5,9 – ISSN: 0001-8392 – mf#5139 – us UMI ProQuest [350]

Administrative structure in large district courts : a report to the conference of metropolitan district chief judges / Dubois, Philip – Washington: FJC, Dec 1981 – 1mf – 9 – $1.50 – mf#LLMC 95-802 – us LLMC [347]

Administrative structure of athletic departments and the impact of title ix / Derouin, Barbara – 1981 – 9 – us Kinesology [790]

Administrative system of the marathas : from original sources / Sen, Surendra Nath – Calcutta: University of Calcutta, 1923 – (= ser Samp: indian books) – us CRL [350]

Administrator's digest – Baltimore MD 1965-82 – 1,5,9 – (cont by: library administrator's digest) – ISSN: 0001-8422 – mf#2738.01 – us UMI ProQuest [020]

Administrator's notebook – Chicagom IL 1952-95 – 1,5,9 – ISSN: 0001-8430 – mf#1842 – us UMI ProQuest [370]

Administrators swap shop – Washington DC 1971-78 – 1,5,9 – ISSN: 0567-9559 – mf#9907 – us UMI ProQuest [370]

Der admiral : drei novellen / Moeller, Eberhard Wolfgang – Muenchen: A Langen, G Mueller 1937 [mf ed 1990] – (= ser Die kleine buecherei 85) – 1r – 1 – (filmed with: deutschlands traum, kampf und sieg / hans minckwitz.) – mf#2837p – us Wisconsin U Libr [830]

Admiral nimitz command summary : running estimate and survey, 1941-1945 / U.S. Navy. Historical Center. Operational Archives Branch – 1985 – 3r – 1 – $390.00 – mf#S1162 – us Scholarly Res [355]

Admiral of the fleet : sir provo w p wallis, ccb, etc: a memoir / Brighton, John George – London: Hutchinson, 1892 [mf ed 1979] – 4mf – 9 – 0-665-00272-6 – mf#00272 – cn Canadiana [355]

Admiral of the fleet, sir geoffrey phipps hornby gcb : a biography / Egerton, Fred, mrs – Edinburgh: W Blackwood, 1896 [mf ed 1980] – 5mf – 9 – 0-665-02891-1 – (incl ind) – mf#02891 – cn Canadiana [920]

Admiralty case files of the u.s. district court for the eastern district of pennsylvania, 1789-1840 / U.S. Circuit and District Courts – (= ser Records of district courts of the united states) – 18r – 1 – (with printed guide) – mf#M988 – us Nat Archives [347]

Admiralty case files of the u.s. district court for the eastern district of virginia, 1801-1861 / U.S. District Court – (= ser Records of district courts of the united states) – 18r – 1 – (with printed guide) – mf#M1300 – us Nat Archives [347]

Admiralty case files of the u.s. district court for the northern district of california, 1850-1900 – (= ser Records of district courts of the united states) – 401r – 1 – (with printed guide) – mf#M1249 – us Nat Archives [347]

Admiralty case files of the us district court for the southern district of new york, 1790-1842 / U.S. District Court – (= ser Records of district courts of the united states) – 62r – 1 – (with printed guide) – mf#M919 – us Nat Archives [347]

Admiralty final record books and minutes for the us district court, district of south carolina, 1790-1857 / U.S. District Court – (= ser Records of district courts of the united states) – 4r – 1 – (with printed guide) – mf#M1182 – us Nat Archives [347]

Admiralty final record books of the u.s. district court for the southern district of florida (key west), 1828-1911 – (= ser Records of district courts of the united states) – 19r – 1 – (with printed guide) – mf#M1360 – us Nat Archives [347]

Admiralty final record books, us district court, eastern district of north carolina, 1858-1907 / U.S. District Court – (= ser Records of district courts of the united states) – 1r – 1 – (with printed guide) – mf#M1429 – us Nat Archives [347]

Admiralty jurisdiction: report / New Zealand. Special Law Reform Committee on Admiralty Jurisdiction – Wellington 1972. 65p. LL-4199 – 1 – us L of C Photodup [355]

Admiranda orbis christiani : tom 1-2: post editionem venetam secunda in germania / Bagatta, J B – Aug Vindelicorum & Dilingae, 1700 – €94.00 – ne Slangenburg [240]

Admiranda romanarum antiquitatum veteris sculpturae vestigia...notis i p bellorii illustrata / Bartoli, P – Romae, 1693 – 5mf – 9 – mf#O-1088 – ne IDC [730]

The admissibility and evaluation of scientific evidence in court / Faurie, Annari – Uni of South Africa 2000 [mf ed Pretoria: UNISA 2000] – 3mf – 9 – (incl bibl ref) – mf#mfm14720 – sa Unisa [347]

The admissibility of evidence of "system" in criminal cases; a reading delivered in the hall of the honourable society of the middle temple by the autumn reader, sir fred e. pritchard, on thursday, 21 november, 1957 / Pritchard, Fred Eills – London: Flint, 1958?. 34p. LL-2261 – 1 – us L of C Photodup [345]

Admitted alien crew lists of vessels arriving at san francisco, 1896-1921 – (= ser Immigrant and passenger arrivals) – 8r – 1 – mf#M1436 – us Nat Archives [975]

Admonitio de praecipvis capitibvs controversiarvm de coena domini / Eitzen, P von – [Hamburg], 1561 – 1mf – 9 – mf#TH-1 mf 403 – ne IDC [242]

Admonitio paterna pauli 3 romani pontificis ad invictiss : caesarem carolum 5 qua eum castigat, quod se lutheranis praebuerit nimis facilem: deinde quod tum in cogenda synodo, tum in definiendis fidei controversiis aliquid potestatis sibi sumpserit / [Calvin, J] – [Basle: Robert Winter], 1545 – 1mf – 9 – mf#CL-48 – ne IDC [241]

An admonition to the parliament / Field, J – n.p., 1572 – 1mf – 9 – (missing: title pg) – mf#PW-12 – ne IDC [240]

The admonitions of an egyptian sage / Gardiner, A H – Leipzig, 1909 – 4mf – 9 – mf#NE-20045 – ne IDC [470]

Adobes in san mateo county / Bowman, J N – San Mateo Co, CA. – 1r – 1 – $50.00 – mf#B40259 – us Library Micro [978]

Adolescence – San Diego CA 1966+ – 1,5,9 – ISSN: 0001-8449 – mf#2544 – us UMI ProQuest [640]

The adolescence of an airline / McGregor, Gordon R – Montreal: [Air Canada], 1970 [i.e. 1980] [mf ed 1995] – 4mf – 9 – mf#SEM105P2353 – cn Bibl Nat [380]

La adolescencia como evasion y retorno / Arevalo, Juan Jose – [2.ed] Guatemala: Tip nacional 1945 [mf ed 1993] – 1r – 1 – us UF Libraries [373]

Adolescens academicus sub institutione salomonis / Musart, Ch – Duaci: Typis Balthasaris Bellari, 1633 – 8mf – 9 – mf#O-3247 – ne IDC [090]

Adolescent and pediatric gynecology – Oxford, England 1988-95 – 1,5,9 – ISSN: 0932-8610 – mf#16973.01 – us UMI ProQuest [618]

Adolescent perceptions of mentoring : a phenomenological approach in recreation / Hayes, Jennifer M – 1999 – 1mf – 9 – $4.00 – mf#RC 535 – us Kinesology [306]

Adolescent psychiatry – Hillsdale NJ 1979+ – 1,5,9 – ISSN: 0065-2008 – mf#12195 – us UMI ProQuest [616]

Adolescente y nubes. poemas, 1947-1954 / Fernandez Mejia, Abel – Ciudad Trujillo, Dominican Republic: Editores Pol 1958 [mf ed 1990] – 1r – 1 – us UF Libraries [810]

Adolescents in unitarian churches / Harrington, Donald – Chicago, 1938. Chicago: Dep of Photodup, U of Chicago Lib, 1971 (1r); Evanston: American Theol Lib Assoc, 1984 (1r) – 1 – 0-8370-0327-X – mf#1984-B172 – us ATLA [243]

Adolf bruell's populaerwissenschaftliche monatsblaetter [...] – Frankfurt/M DE, 1881-1908 – 3r – 1 – (missing: 1901, 1907) – gw Misc Inst [500]

Adolf bruell's populaerwissenschaftliche monatsblaetter zur belehrung ueber das judenthum fuer gebildete aller confessionen – Frankfurt: Adolf Bruell. v1-28. 1881-1908 – (= ser German-jewish periodicals...1768-1945, pt 1) – 3r – 1 – $325.00 – (lacking: v21 (1901, except n10) + v27 (1907)) – mf#B333 – us UPA [270]

Adolf diesterweg : lichtstrahlen aus seinen schriften – Leipzig: F A Brockhaus, 1875 [mf ed 1989] – vi/231p – 1 – (biogr int by eduard langenberg) – mf#7177 – us Wisconsin U Libr [430]

Adolf friedrich graf von schack : ein poetisches charakterbild / Manssen, W J – Stuttgart: J B Metzler 1888 [mf ed 1991] – 1r – 1 – (trans fr dutch. filmed with: heiterer guckkasten / bruno wolfgang) – mf#2865p – us Wisconsin U Libr [430]

Adolf friedrich graf von schack als uebersetzer / Walter, Erich – Leipzig: M Hesse, 1907 [mf ed 1992] – (= ser Breslauer beitraege zur literaturgeschichte 10) – 1 – mf#8014 reel 1 – us Wisconsin U Libr [430]

Adolf glassbrenner : ein beitrag zur geschichte des 'jungen deutschland' und der berliner lokaldichtung / Rodenhauser, Robert – Nikolassee: M Harrwitz 1912 [mf ed 1990] – 1r [ill] – 1 – (incl ind & bibl ref. filmed with: stilprobleme in gessners kunst und dichtung / rufolf strasser) – mf#7309 – us Wisconsin U Libr [430]

Adolf reichwein (1898-1944) : leben und werk des politischen paedagogen im widerstand gegen das ns-regime unter besonderer beruecksichtigung seiner auseinandersetzung mit kultur, politik, wirtschafts- und sozialproblemen ostasiens / Wittig, Horst E – (mf ed 1993) – 2mf – 9 – €40.00 – 3-89349-762-5 – mf#DHS 762 – gw Frankfurter [370]

Adolf wilbrandt als dramatiker / Scharrer, Eduard – Muenchen: Hans Sachs-Verlag 1912 [mf ed 1991] – 1r – 1 – (incl bibl ref. filmed with: wielands romane / f bobertag) – mf#2960p – us Wisconsin U Libr [430]

Adolph germer papers – (= ser Cio and industrial unionism in america; Research colls in labor studies) – 9 – $2890.00 – 1-55655-026-X – us UPA [331]

Adolph, Karl *see*
- Am ersten mai
- Daughters of vienna

Adolphe et clara : ou, les deux prisonniers: comedie en un acte et en prose, melee d'ariettes / Marsollier – Paris: Au magasin de pieces de theatre 1803 [mf ed 1994] – 1r – 1 – us UF Libraries [820]

Adolphe monod's farewell to his friends and to the church = Adieux d'adolphe monod a ses amis et a l'eglise – New York: Robert Carter, 1858 [mf ed 1993] – 1mf – 9 – 0-524-08521-8 – (in english) – mf#1993-1051 – us ATLA [242]

Adolphus and ellis' reports : reports of cases argued and determined in the court of king's bench... / Adolphus, John L & Ellis, Thomas F – v1-12. 1834-41. London: Saunders and Benning, 1835-42 (all publ) – 133mf – 9 – $199.00 – (this series became "queen's bench reports" with the accession of victoria in 1838) – mf#LLMC 84-741 – us LLMC [324]

Adolphus and ellis' reports, new series / Adolphus, John L & Ellis, Thomas F – v1-18. 1841-52. London: Saunders & Benning, 1843-56 (all publ) – 215mf – 9 – $322.00 – mf#LLMC 84-742 – us LLMC [324]

Adolphus, John L *see*
- Adolphus and ellis' reports
- Adolphus and ellis' reports, new series
- Barnewell and adolphus' reports

Adonde van los cefalomos? / Arango, Angel – Habana, Cuba: Ediciones R 1964 [mf ed 1992] – 1r – 1 – us UF Libraries [830]

Adoniram judson : a biography / Judson, Edward – Philadelphia: American Baptist Publ Society, c1894 [mf ed 1986] – 1mf – 9 – 0-8370-6671-9 – (incl list of adoniram judson's publ & ind) – mf#1986-0671 – us ATLA [242]

Adoniram judson : his life and labours / Judson, E – London, 1883 – 7mf – 9 – mf#HTM-93 – ne IDC [910]

Adoniram judson gordon : a biography with letters and illustrative extracts drawn from unpublished or uncollected sermons and addresses / Gordon, Ernest B – 2nd ed. New York: F H Revell, c1896 [mf ed 1990] – 1mf – 9 – 0-7905-5221-3 – mf#1988-1221 – us ATLA [920]

Adonis und esmun : eine untersuchung zur geschichte des glaubens an auferstehungsgoetter und an heilgoetter / Baudissin, W W – Leipzig, 1911 – €31.00 – ne Slangenburg [250]

Adonis und esmun : eine untersuchung zur geschichte des glaubens an auferstehungsgoetter und an heilgoetter / Baudissin, Wolf Wilhelm, Graf von – Leipzig: J C Hinrichs, 1911 [mf ed 1989] – 2mf – 9 – 0-7905-0547-9 – (incl ind) – mf#1987-0547 – us ATLA [230]

Adopted child : a muscial drama in two acts / Attwood, Thomas – London: printed by M Clementi & Co [1806?] [mf ed 1988] – 1r – 1 – mf#pres. film 44 – us Sibley [780]

Adoption and legitimation of children / Joyce, Joseph Asbury – Oakland Oakland Tribune Publishing Co., 1890. 73 p. LL-1202 – 1 – us L of C Photodup [340]

Adoption counseling as a model for churches seeking a staff person / Lowry, Robert Louis – Princeton, New Jersey, 1976. Chicago: Dep of Photodup, U of Chicago Lib, 1976 (1r); Evanston: American Theol Lib Assoc, 1984 (1r) – 1 – 0-8370-1289-9 – mf#1984-T011 – us ATLA [240]

Die adoption im altbabylonischen recht / David, M – Leipzig, 1927 – (= ser Leipziger rechtwissenschaftliche Studien) – 2mf – 9 – (leipziger rechtwissenschaftliche studien. v23) – mf#NE-419 – ne IDC [930]

Adoption quarterly : innovations in community and clinical practice, theory, and research / ed by Finley, Gordon E & Wrobel, Gretchen Miller – ISSN: 1092-6755 – us Haworth [360]

Adorers of the Blood of Christ. Wichita, KS see Annals

The adornment of the spiritual marriage; the sparkling stone; the book of supreme truth = Selections. 1916 / Ruusbroec, Jan van; ed by Underhill, Evelyn – London: JM Dent; New York: EP Dutton, 1916 – 1mf – 9 – 0-7905-9862-0 – mf#1989-1587 – us ATLA [240]

Adquisiciones y donaciones : [exposicion organizada por los amigos del museo de bellas artes de caracas y el museo de bellas artes de caracas] / Museo de Bellas Artes (Venezuela) – [s.l]: MBA [1963?] [mf ed 1993] – 1r – 1 – us UF Libraries [700]

Adr and settlement in the federal district courts : a sourcebook for judges and lawyers: (a joint project of the fjc and the cpr institute for dispute resolution) / Plapinger, Elizabeth et al – 1996 – 4mf – 9 – $6.00 – mf#llmc99-028 – us LLMC [347]

Adrario, A see Per la vittoria dell' armata christiana...

Adrastea / ed by Herder, Johann Gottfried – Leipzig 1801-03 – (= ser Dz. abteilung literatur) – 6v[=12st] on 17mf – 9 – €170.00 – mf#k/n4635 – gw Olms [073]

Adrem – 1990-92 [complete] – 1r – 1 – mf#ATLA S0929 – us ATLA [073]

Adresboeken voor den nederlandschen boekhandel, [1828]1848-1896 see Nineteenth-century directories for the dutch book trade

Adressbuch fuer den deutschen buchhandel und verwandte geschaefts-zweige, 1839-1900 see Nineteenth-century directories for the dutch book trade

Adresse a mm les electeurs du comte de lotbiniere : suivie de divers documents / Amyot, Gail (Guillaume) – [S.l: s.n, 187-?] [mf ed 1979] – 1mf – 9 – 0-665-00886-4 – mf#00886 – cn Canadiana [320]

Adresse aux electeurs municipaux de la ville de longueuil – [S.l: s.n, 18–] [mf ed 1979] – 1mf – 9 – 0-665-00777-9 – mf#00777 – cn Canadiana [325]

Adresse de bienvenue par m baillairge a la section de montreal des architectes du canada : lors de l'assemblee annuelle de la societe tenue au chateau frontenac...quebec, le 2 octobre, 1895 / Baillairge, Charles P Florent – Quebec?: s.n, 1895? [mf ed 1979] – 1mf – 9 – 0-665-00883-X – mf#00883 – cn Canadiana [720]

Adresse de l'association du commerce libre, au peuple du canada – S.l: s.n, 1846? – 1mf – 9 – mf#49032 – cn Canadiana [380]

Adresse de l'hon juge wurtele aux petits jures, le 2 octobre 1897 : et allocution au defendeur lors de la sentence, le 14 octobre 1897, dans le proces pour libelle de la reine vs w a grenier – Montreal: C Theoret, 1897? – 1mf – 9 – mf#26285 – cn Canadiana [347]

Adresse de m c beausoleil : candidat national, aux electeurs du comte berthier / Beausoleil, Cleophas – [Quebec (Province)]: s.n, [1886?] [mf ed 1980] – 1mf – 9 – 0-665-03528-4 – mf#03528 – cn Canadiana [971]

Adresse de son exc le president de la republique a l'occasion de la fete nationale de l'agriculture et du travail, le 1er mai, 1942 / Lescot, Elie – Port-Au-Prince, Haiti: Impr de l'etat [1942] [mf ed 1991] – 1r – 1 – us UF Libraries [630]

Adresse des associes de la temperance de longueuil au rev pere chiniquy – Montreal: Bureau des melanges religieux, 1848 [mf ed 1984] – 1mf – 9 – 0-665-44852-X – mf#44852 – cn Canadiana [170]

Die adresse des epheserbriefs des paulus / Harnack, Adolf von – [Berlin?]: Verlag der koeniglichen Akademie der Wissenschaften in Commission bei Georg Reimer, 1910 – 1mf – 9 – 0-8370-9628-6 – (incl bibl ref) – mf#1986-3628 – us ATLA [227]

Adresse particulierement aux membres canadiens elus pour le prochain parlement provincial / Estimauville, Robert Anne d', chevalier de Beaumochel – [s.l: s.n, 1827?] [mf ed 1984] – 1mf – 9 – 0-665-44273-4 – mf#44273 – cn Canadiana [342]

Adret, Solomon Ben Abraham see Hidushe ha-rashba

Adrian, Heinrich see Der saelden hort

Adrian, Johann V see Skizzen aus england

Adrian, Michigan. First Baptist Church see Records

Adriana – Quebec: Impr franciscaine missionnaire, 1927 [mf ed 1998] – 3mf – 9 – (trans fr italian) – mf#SEM105P2880 – cn Bibl Nat [241]

Adriani, J H see
– Guido fridolin verbeek [i e verbeck]
– Het land der morgenkalmte

Adrianov, A V see Sibirskaia zhizn'

Adriatische rosemund / Zesen, Philipp von; ed by Jellinek, M Hermann – Halle: M Niemeyer, 1899 [mf ed 1993] – (= ser Neudrucke deutscher literaturwerke des 16. und 17. jahrhunderts 160-163) – I/270p – 1 – (incl bibl ref) – mf#8413 reel 7 – us Wisconsin U Libr [830]

Adrzejewski, B W see Somali modes of thought and communication

Yr adsain – [Wales] Denbighshire ion 1910-9 ion 1945 [mf ed 2004] – 35r – 1 – uk Newsplan [072]

Adso Dervensis see De ortu et tempore antichristi. opera hagiographica

Aduersus cuiusdam sacramentarii falsam criminationem, ivsta defensio / Westphal, J aus Hamburg – Francoforti, 1555 – 2mf – 9 – mf#TH-1 mf 1474-1475 – ne IDC [242]

Adult see Radical leader, 1888

Adult basic education – Lanham MD 1991+ – 1,5,9 – (cont: adult literacy and basic education) – ISSN: 1052-231X – mf#18782 – us UMI ProQuest [374]

The adult bible class : its organization and work / Pearce, William Cliff – Philadelphia: Westminster Press, 1910 – 1mf – 9 – 0-524-04740-5 – mf#1991-2145 – us ATLA [220]

Adult bible class quarterly – jan 1915-61 – 1 – $305.69 – us Southern Baptist [220]

Adult education – Newbury Park CA 1950-89 – 1 – (cont by: adult education quarterly) – ISSN: 0001-8481 – mf#268.01 – us UMI ProQuest [374]

Adult education – Leicester, England 1926-89 – 1,5,9 – (cont by: adults learning) – ISSN: 0001-849X – mf#10228 – us UMI ProQuest [374]

Adult education: a plan for development, report of the committee of inquiry, 1973 – 4mf – 9 – mf#87030 – uk Microform Academic [324]

An adult education program for orissa, india / Osgood, William Cyril – Corvallis, Or: Oregon State College, 1950 – (= ser Samp: indian books) – us CRL [374]

Adult education quarterly – Newbury Park CA 1984+ – 1,5,9 – (cont: adult education) – ISSN: 0741-7136 – mf#268,01 – us UMI ProQuest [374]

Adult foster care journal – Lawrence KS 1987-88 – 1,5,9 – (cont by: adult residential care journal) – ISSN: 8756-6559 – mf#16145.01 – us UMI ProQuest [362]

Adult leader – Nashville TN 1968-80 – 1,5,9 – mf#3277 – us UMI ProQuest [240]

Adult leadership – Lanham MD 1952-77 – 1,5,9 – ISSN: 0001-8554 – mf#920 – us UMI ProQuest [374]

Adult learning – Lanham MD 1989+ – 1,5,9 – (cont: lifelong learning) – ISSN: 1045-1595 – mf#17359 – us UMI ProQuest [374]

Adult learning – Toronto, Canada 1936-39 – 1 – ISSN: 0701-3507 – mf#1177 – us UMI ProQuest [374]

Adult license revocations – 1980 aug-1982; 1983-84; 1985-88 – 1 – mf#550532 – us WHS [071]

Adult literacy and parenting outcomes of a rural, home-based program / Meehan, Merrill L [et al] – Charleston VA: AEL Inc; [Washington DC]: US Dept of Education, Office of Educ Research & Improvement...[2000] [mf ed 2000] – 1mf – 9 – us GPO [374]

Adult literacy & basic education – Orlando FL 1977-90 – 1,5,9 – (cont by: adult basic education) – ISSN: 0147-8354 – mf#11338 – us UMI ProQuest [374]

Adult literacy training in the border/kei region of the eastern cape / Moodly, Adele Leah – Uni of South Africa 2000 [mf ed Johannesburg 2000] – 6mf – 9 – (incl bibl ref) – mf#mfm14823 – sa Unisa [374]

Adult residential care journal – Lawrence KS 1989-96 – 1,5,9 – (cont: adult foster care journal) – ISSN: 0899-1995 – mf#16145,01 – us UMI ProQuest [362]

Adult services – Chicago IL 1961-72 – 1,5 – mf#3168 – us UMI ProQuest [020]

Adult teacher – Nashville TN 1947-68 – 1 – mf#2108 – us UMI ProQuest [374]

El adulterio : (estudio doctrinal, practico y de jurisprudencia, sobre la ley de 6 de febrero de 1930...del codigo penal vigente en cuba) / Azcarate, Carlos – Habana, Cuba: Impr "El Siglo 20" 1932 [mf ed 1993] – (= ser Biblioteca del colegio de abogados de la habana 6) – 1r – 1 – us UF Libraries [345]

Adults learning – Leicester,England 1989+ – 1,5,9 – (cont: adult education) – ISSN: 0955-2308 – mf#17416 – us UMI ProQuest [374]

Adults with learning disabilities and their perspectives of physical activity and recreation / Youngblood, Joseph O – 1999 – 2mf – 9 – $8.00 – mf#RC 527 – us Kinesology [790]

Advance – 1885 mar 19-1887 apr 21; 1885 mar 26-1886 apr 1 – 1 – mf#891596 – us WHS [071]

Advance – 1896 jul 8-1902 aug 30 – 1 – mf#3185222 – us WHS [071]

Advance – 1898 feb 4-1899 jun 30; 1899 jul 7-1901 jan 25; 1901 feb-1902 dec – 1 – mf#954654 – us WHS [071]

Advance – 1898 sep 17 – 1 – mf#851147 – us WHS [071]

Advance – 1915 nov 19-1921 may 20; 1921 jun 3-1925 may 22; 1925 jun 5-1929 may 30; 1929 jun 13-1932 feb 17 – 1 – mf#1166099 – us WHS [071]

Advance – 1917 mar 9-1920 oct 22; 1920 oct 29-1923 dec; 1924-38; 1939-1941 dec 15; 1942-1944 dec 15; 1945-1947 dec 15; 1948-64; 1965-1967 dec 15; 1968 jan 1-1970 dec 26; 1968 jan 1-1970 dec 26; 1971-73; 1974-1976 jun; 1968 jan 1-1970 dec 26 – 1 – mf#3185216 – us WHS [071]

Advance – 1976 nov18-1977 oct; 1977 nov-1978 aug 10; 1978 aug 17-1978 dec; 1979-1996 aug – 1 – mf#1003594 – us WHS [071]

Advance – 1980 mar 20-1980 dec; 1981; 1982 jan-1982 oct 14 – 1 – mf#1003597 – us WHS [071]

Advance : canada's forward-looking youth paper – Toronto: Advance Press Committee. v1-3 n6; nov 1 1961-jul 1963// – 1r – 1 – Can$110.00 – cn McLaren [071]

Advance – Cape Town, SA: Competent Publ & Printing, [1952-nov 6-oct 21 1954] – 1 – us CRL [079]

Advance – Chicago, IL. 1934-49 [complete] – 5r – 1 – mf#ATLA S0868 – us ATLA [073]

Advance : official organ of the amalgamated clothing workers of america / Amalgamated Clothing Workers of America – New York NY 30 dec 1921-16 nov 1928 (imperfect) – 3r – 1 – uk British Libr Newspaper [680]

Advance – Port Louis, Mauritius 6 jan 1958-29 feb 1968 (imperfect) – 1 – uk British Libr Newspaper [079]

Advance – St Louis MO 1973-75 – 1,5,9 – ISSN: 0001-8570 – mf#8532 – us UMI ProQuest [240]

Advance see
– Guardian
– The messenger

The advance – Niagara-on-the-Lake Ontario, CN. jan 1919-dec 1988 – 33r – 1 – cn Commonwealth Imaging [071]

The advance – St Louis, MO. v2-21. 1934- 54 [complete] – 1r – 1 – ISSN: 0001-8570 – mf#ATLA S0532 – us ATLA [073]

The advance – Wilmington, DE: P H Murray. v2 n47. sep 22 1900 (mf ed 1947) – 1r – 1 – us L of C Photodup [071]

The advance – Zurich, Ontario, CN. jan-dec 1986 – 1r – 1 – cn Commonwealth Imaging [071]

The advance see Prace

Advance and labor leaf – 1885-1889 – 1 – mf#3185223 – us WHS [071]

The advance and retreat of the roman catholic priests, at carlow – Dublin, 1825 – (= ser 19th c ireland) – 1mf – 9 – mf#1.1.2261 – uk Chadwyck [241]

Advance argus – Greenville, PA. -w 1898-1923. 9 rolls – 13 – $25.00r – us IMR [071]

Advance australia – Melbourne 21 apr 1914-15 dec 1919* – 2r – 1 – (wanting: may, jun 1914; feb, aug, sep 1915; apr, dec 1916; oct 1917; jan 1918; sep 1919) – uk British Libr Newspaper [079]

Advance in the antilles : the new era in cuba and porto rico / Grose, Howard Benjamin – New York: Presbyterian Home Missions 1910 [mf ed 1986] – (= ser Forward mission study courses) – 1mf – 9 – 0-8370-6058-3 – (incl ind) – mf#1986-0058 – us ATLA [240]

The advance of science in the last half century see Half-century of science

Advance reporter / Williams Co. Stryker – v1 n1. jan 1978-dec 1992 [wkly] – 6r – 1 – mf#B32535-32540 – us Ohio Hist [071]

Advanced australia : a short account of australia on the eve of federation / Galloway, William Johnson – [London], 1899 – (= ser 19th c books on british colonization) – 3mf – 9 – mf#1.1.7318 – uk Chadwyck [980]

An advanced catechism of catholic faith and practice : based upon the third plenary council catechism / O'Brien, Thomas John – Chicago IL: John B Oink c1913 [mf ed 1992] – 1mf – 9 – 0-524-05384-7 – (incl bibl ref) – mf#1991-2290 – us ATLA [241]

Advanced course in yogi philosophy and oriental occultism / Ramacharaka, yogi – Chicago, IL: Yogi Pub Soc, 1909 [mf ed 1991] – 1mf – 9 – 0-524-00727-6 – mf#1990-2055 – us ATLA [180]

Advanced drug delivery reviews – Oxford, England 1996+ – 1,5,9 – ISSN: 0169-409X – mf#42548 – us UMI ProQuest [362]

Advanced energy projects fy... : research summaries / United States. Dept of Energy. Division of Advanced Energy Projects – Washington, DC: US Dept of Energy...Office of Energy Research [mf ed 1987] – annual – 9 – us GPO [333]

Advanced engineering informatics – Oxford, England 2002+ – 1,5,9 – ISSN: 1474-0346 – mf#42697,02 – us UMI ProQuest [621]

Advanced functional materials – Hoboken NJ 2001+ – 1,5,9 – (cont: advanced materials for optics and electronics) – ISSN: 1616-301X – mf#19064,01 – us UMI ProQuest [621]

Advanced management journal – Corpus Christi 1936-84 – 1,5,9 – (cont by: sam advanced management journal) – ISSN: 0362-1863 – mf#1959.01 – us UMI ProQuest [650]

Advanced manufacturing engineering [ame] – Oxford, England 1988-91 – 1,5,9 – ISSN: 0951-5232 – mf#17210 – us UMI ProQuest [620]

Advanced materials abstracts – Oxford, England 1989-91 – 1,5,9 – mf#49579 – us UMI ProQuest [620]

Advanced materials for optics and electronics – Hoboken NJ 1992-93 – 1,5,9 – ISSN: 1057-9257 – mf#19064.01 – us UMI ProQuest [621]

Advanced materials & processes – Novelty OH 1985+ – 1,5,9 – ISSN: 0882-7958 – mf#14991 – us UMI ProQuest [620]

Advanced nuclear research: hearing. / U.S. Congress. House. Committee on Science and Astronautics. Subcommittee on Aeronautics and Space Technology – Washington, Govt. Print. Off., 1974. 75 p. LL-2362 – 1 – us L of C Photodup [340]

Advanced technology libraries – Millwood NY 1972+ – 1,5,9 – ISSN: 0044-636X – mf#7806 – us UMI ProQuest [020]

Advanced tennis / Bowers, Chester – New York NY: Macmillan Co 1940 [mf ed 1994] – 1r – 1 – us UF Libraries [790]

Advancement of science – London, England 1939-71 – 1,5,9 – ISSN: 0001-866X – mf#1278 – us UMI ProQuest [600]

Advance-press – 1896 nov25-1897;1898-1900 apr 26 – 1 – mf#968102 – us WHS [071]

Advances in behaviour research and therapy – Oxford, England 1977-94 – 1,5,9 – ISSN: 0146-6402 – mf#49539 – us UMI ProQuest [150]

Advances in colloid and interface science – Oxford, England 1967+ – 1,5,9 – ISSN: 0001-8686 – mf#42006 – us UMI ProQuest [540]

Advances in contraception – Dordrecht, Netherlands 1999 – 1,5,9 – ISSN: 0267-4874 – mf#16761 – us UMI ProQuest [613]

Advances in enzyme regulation – Oxford, England 1963+ – 1 – ISSN: 0065-2571 – mf#49006 – us UMI ProQuest [612]

Advances in free radical biology and medicine – Oxford, England 1985-86 – 1,5,9 – ISSN: 8755-9668 – mf#49479 – us UMI ProQuest [574]

Advances in neuroimmunology – Oxford, England 1991-93 – 1,5,9 – ISSN: 0960-5428 – mf#49620 – us UMI ProQuest [616]

Advances in physics – Abingdon, Oxfordshire 1992+ – 1,5,9 – ISSN: 0001-8732 – mf#17347 – us UMI ProQuest [530]

Advances in physiology education – Bethesda MD 1998+ – 1,5,9 – ISSN: 1043-4046 – mf#17272 – us UMI ProQuest [612]

Advances in plastics technology – Hoboken NJ 1981 – 1,5,9 – (cont by: advances in polymer technology) – ISSN: 0272-9504 – mf#13076.01 – us UMI ProQuest [660]

Advances in polymer technology – Hoboken NJ 1982+ – 1,5,9 – (cont: advances in plastics technology) – ISSN: 0730-6679 – mf#13076,01 – us UMI ProQuest [660]

Advances in skin & wound care – Ambler PA 2000+ – 1,5,9 – ISSN: 1527-7941 – mf#26676,02 – us UMI ProQuest [610]

Advances in space research – Oxford, England 1981+ – 1,5,9 – ISSN: 0273-1177 – mf#49540 – us UMI ProQuest [629]

Advances in thanatology – Brooklyn NY 1977+ – 1,5,9 – (cont: journal of thanatology) – ISSN: 0196-1934 – mf#7826,01 – us UMI ProQuest [610]

Advances in the biosciences – Oxford, England 1967-92 – 1,5,9 – ISSN: 0065-3446 – mf#49007 – us UMI ProQuest [574]

Advances in tunnelling technology and subsurface use = Developpement des travaux en souterrain – Oxford, England 1981-84 – 1,5,9 – ISSN: 0275-5416 – mf#49385 – us UMI ProQuest [624]

Advancing clinical care – Franksville WI 1990-91 – 1,5,9 – (cont: ad nurse) – ISSN: 1042-9565 – mf#16482,01 – us UMI ProQuest [610]

Advantages of imperial federation : a lecture delivered at a public meeting held in toronto on january 30th, 1891... / Grant, George Monro – [S.l: s.n.], 1891 [mf ed 1980] – 1mf – 9 – 0-665-05115-8 – mf#05115 – cn Canadiana [320]

The advantages of life assurance to the working classes : being a lecture delivered to the mechanics' institute and library association of quebec / Cook, John – Montreal?: Armour & Ramsay, 1848 – 1mf – 9 – mf#33465 – cn Canadiana [368]

Advent and ascension : or, how jesus came and how he left us / Faunce, Daniel Worcester – New York: Eaton & Mains [c1903] [mf ed 1984] – 3mf – 9 – 0-8370-0925-1 – (incl bibl ref) – mf#1984-4247 – us ATLA [220]

Advent christian missions – v58 n7-v61 n11 [1976 jul/aug-1979 dec] – 1 – mf#1018627 – us WHS [243]

Advent christian news – 1977 jul 15, sep 25-1983 – 1 – mf#629814 – us WHS [243]

Advent christian witness – 1981 may-1985 dec; v26 n3-v27 n12 [1978 mar-1979 dec] – 1 – mf#1218262 – us WHS [243]

Advent christian witness to the world – 1981 may-1983 may – 1 – mf#1218262 – us WHS [243]

The advent hope in st paul's epistles / Robinson, Joseph Armitage – London, New York: Longmans, Green, 1911 [mf ed 1990] – 1mf – 9 – 0-7905-3468-1 – mf#1987-3468 – us ATLA [227]

Adventist heritage – Loma Linda CA 1974-96 – 1,5,9 – ISSN: 0360-389X – mf#9335 – us UMI ProQuest [242]

Adventist heritage – v5-6 [1978 summer-1979 winter] – 1 – mf#202883 – us WHS [243]

Die adventsperikopen : exegetisch-homiletisch erklaert / Keppler, Paul Wilhelm von – Freiburg im Bresgau; St Louis MO: Herder 1899 [mf ed 1989] – (= ser Biblische studien 4/1) – 1mf – 9 – 0-7905-2722-7 – mf#1987-2722 – us ATLA [225]

Adventure for god / Brent, Charles Henry – New York: Longmans, Green 1905 [mf ed 1990] – (= ser The bishop paddock lectures 1904) – 1mf – 9 – 0-7905-3549-1 – mf#1989-0042 – us ATLA [242]

Adventure in costa rica / Lundberg, Donald E – [1st ed] Tallahassee FL: Dixie Publ c1960 [mf ed 1993] – 1r – 1 – us UF Libraries [972]

The adventurer – v1-2. 1752-54 – 1 – us AMS Press [420]

Adventurers of bermuda : a history of the island from its discovery until the dissolution of the somers island company in 1684 / Wilkinson, Henry Campbell – 2nd ed. London, New York: OUP 1958 [mf ed 1992] – 1r – 1 – us UF Libraries [918]

Adventures amidst the equatorial forests and rivers of south america also in the west indies and the wilds of florida : to which is added "jamaica revisited" / Stuart, Villiers – London, England: J Murray 1891 [mf ed 1990] – 1r – 1 – us UF Libraries [918]

Adventures among books / Lang, Andrew – London, New York: Longmans, Green, 1905 [mf ed 1991] – 1mf – 9 – 0-7905-7892-1 – mf#1989-1117 – us ATLA [420]

The adventures and sufferings of john r jewitt, only survivor of the ship boston, during a captivity of nearly three years among the savages of nootka sound : with an account of the manners, mode of living, and religious opinions of the natives... / Jewitt, John – Edinburgh 1824 [mf ed Hildesheim 1995-98] – 1v on 2mf – 9 – €60.00 – 3-487-27123-0 – gw Olms [920]

Adventures in faith / Ober, Charles Kellogg – New York: Association Press, 1915 [mf ed 1991] – 1mf – 9 – 0-7905-9828-0 – mf#1989-1553 – us ATLA [200]

Adventures in mashonaland, by two hospital nurses / Blennerhassett, R & Sleeman, L – London, 1893 – 4mf – 9 – mf#HTM-17 – ne IDC [916]

Adventures in new guinea / Chalmers, James – [London]: Religious Tract Society, 1889 [mf ed 1986] – (= ser The rts library – illustrated) – 1mf – 9 – 0-8370-8327-3 – mf#1986-2327 – us ATLA [980]

Adventures in nyasaland. / Fotheringham, L Monteith – London. 1891 – 1 – us CRL [960]

Adventures in patagonia : a missionary's exploring trip / Coan, Titus – New York: Dodd, Mead, 1880 [mf ed 1993] – 1mf – 9 – 0-524-08223-5 – (int by henry m field) – mf#1993-1008 – us ATLA [918]

Adventures in samoa / Bassett, Henry Lawrence – 1890-93 – 1r – 1 – mf#pmb doc45 – at Pacific Mss [980]

Adventures in siam in the 17th century / Hutchinson, E W – v18. 1940 – 1r – 1 – mf#97074 – uk Microform Academic [915]

Adventures in the wilds of north america / Lanman, Charles; ed by Weld, Charles Richard – London: Longman, Brown, Green & Longmans, 1854 [mf ed 1984] – 4mf – 9 – 0-665-45221-7 – mf#45221 – cn Canadiana [917]

Adventures in tibet : including the diary of miss annie r taylor's remarkable journey from tau-chau to ta-chien-lu through the heart of the "forbidden land" / Carey, William – Chicago: Student Missionary Campaign Library, [1901] [mf ed 1995] – (= ser Yale coll) – 285p (ill) – 1 – 0-524-09225-7 – mf#1995-0225 – us ATLA [915]

The adventures of a bric-a-brac hunter / Hall, Herbert Byng – London: Tinsley Bros, 1868 – (= ser 19th c art & architecture) – 3mf – 9 – mf#4.1.98 – uk Chadwyck [740]

The adventures of a protestant in search of a religion / Iota – New York, Montreal: D & J Sadlier & Co, 1879 [mf ed 1990] – 4mf – 9 – mf#SEM105P1235 – cn Bibl Nat [242]

The adventures of a protestant in search of a religion / Waller, John Francis – New York: D & J Sadlier, 1879 [mf ed 1986] – 1mf – 9 – 0-8370-6956-4 – mf#1986-0956 – us ATLA [242]

The adventures of a serf's wife among the mines of siberia / Agar, Mrs – London: T Cautley Newby, 1866 – (= ser 19th c women writers) – 4mf – 9 – mf#5.1.22 – uk Chadwyck [830]

Adventures of an attorney in search of practice / Warren, Samuel – Chicago: James Cockcroft & Co, 1872 – 5mf – 9 – $7.50 – mf#LLMC 95-191 – us LLMC [340]

Adventures of british seamen in the southern ocean : displaying the striking contrasts which the human character exhibits in an uncivilized state / ed by Murray, Hugh – Edinburgh 1827 [mf ed Hildesheim 1995-98] – 1v on 3mf – 9 – €90.00 – 3-487-26774-8 – gw Olms [910]

The adventures of don juan de ulloa, in a voyage to calicut : soon after the discovery of india, by vasco de gama – London [1825] [mf ed Hildesheim 1995-98] – 1v on 4mf – 9 – €120.00 – 3-487-26950-3 – gw Olms [915]

The adventures of hatim tai : a romance / At-Tai – London 1830 [mf ed Hildesheim 1995-98] – 1v on 3mf – 9 – €90.00 – 3-487-27941-X – gw Olms [910]

The adventures of miss kang *see* Kang hsiao chieh (ccm327)

The adventures of robert drury, during fifteen years captivity on the island of madagascar : containing a description of that island...; to which is added, a vocabulary of the madagascar language; written by himself, and now carefully revised and corrected from the original copy / Defoe, Daniel – Hull 1807 [mf ed Hildesheim 1995-98] – 1v on 3mf – 9 – €90.00 – 3-487-27238-5 – gw Olms [920]

Adventures of robinson crusoe / Defoe, Daniel – Chicago, IL. 1930? – 1r – 1 – us UF Libraries [830]

Adventures of robinson crusoe / Defoe, Daniel – London, England. 1862 – 1r – 1 – us UF Libraries [830]

Adventures of robinson crusoe / Defoe, Daniel – London, England. 1864 – 1r – 1 – us UF Libraries [830]

Adventures of robinson crusoe / Defoe, Daniel – London, England. 1880? – 1r – 1 – us UF Libraries [830]

Adventures of robinson crusoe / Defoe, Daniel – London, England. 1884 – 1r – 1 – us UF Libraries [830]

Adventures of robinson crusoe / Defoe, Daniel – London, England. Between 1882 And 1890 – 1r – 1 – us UF Libraries [830]

Adventures of robinson crusoe / Defoe, Daniel – Springfield, MA. 1927? – 1r – 1 – us UF Libraries [830]

Adventures of robinson crusoe – London, England. 1821 – 1r – 1 – us UF Libraries [830]

Adventures of robinson crusoe – London, England. 1826? – 1r – 1 – us UF Libraries [830]

Adventures of robinson crusoe / Mcgovern, mary Harriet – Racine, WI. 1917 – 1r – 1 – us UF Libraries [420]

Adventures of robinson crusoe – New Haven, CT. 1825 – 1r – 1 – us UF Libraries [830]

Adventures of robinson crusoe of york, mariner / Defoe, Daniel – London, England. 1894 – 1r – 1 – us UF Libraries [830]

Adventures of the christian soul : being chapters in the psychology of religion / Saunders, Kenneth James – Cambridge: University Press, 1916 [mf ed 1991] – 1mf – 9 – 0-524-01918-5 – (pref by w r inge) – mf#1990-2731 – us ATLA [150]

The adventures of the gooroo noodle : a tale in the tamil language / Beschi, Costantino Giuseppe – Allahabad: Panini Office, 1915 – (= ser Samp: indian books) – (trans by benjamin babington) – us CRL [390]

Adventures of the gooroo paramartan : a tale in the tamul langugage: accompanied by a translation and vocabulary together with an analysis of the first story / Beschi, Costanzo Giuseppe – London: Richardson 1822 – (= ser Whsb) – 3mf – 9 – €40.00 – (trans by benjamin guy babington) – mf#Hu 273 – gw Fischer [490]

Adventures on the mosquito shore / Squier, Ephraim George – New York: Worthington Co 1891 [mf ed 1991] – 1r – 1 – (1st ed publ 1855, under title: waikna; or, adventures on the mosquito shore. by samuel a bard [pseud]) – us UF Libraries [918]

Adventus regni : being sermons chiefly on the parables of the kingdom / Lilley, Alfred Leslie – London: Francis Griffiths, 1907 [mf ed 1991] – 1mf – 9 – 0-7905-8828-5 – mf#1989-2053 – us ATLA [242]

Adversaria critica sacra : with a short explanatory introduction / Scrivener, Frederick Henry Ambrose – Cambridge: University Press; New York: Macmillan [dist] 1893 [mf ed 1985] – 1mf – 9 – 0-8370-5198-3 – mf#1985-3198 – us ATLA [225]

The adversary : his person, power, and purpose / Matson, William A – New York: WB Ketcham, c1891 – 1mf – 9 – 0-7905-8517-0 – mf#1989-1742 – us ATLA [210]

Adverse drug reaction bulletin – Baltimore MD 1973+ – 1,5,9 – ISSN: 0044-6394 – mf#8330 – us UMI ProQuest [615]

Adverse drug reactions and acute poisoning reviews – Auckland, New Zealand 1982-90 – 1,5,9 – (cont by: adverse drug reactions & toxicological reviews) – ISSN: 0260-647X – mf#14048.02 – us UMI ProQuest [615]

Adverse drug reactions and toxicological reviews – Auckland, New Zealand 1991+ – 1,5,9 – (cont: adverse drug reactions & acute poisoning reviews) – ISSN: 0964-198X – mf#14048.02 – us UMI ProQuest [615]

Adversus elipandum : formae tplila 19 / Liebanensis, Beatus & Oxomensis, Eterius – 1984 – (= ser ILL – ser a; Cccm 59)) – 6mf+54p – 9 – €30.00 – 2-503-60592-3 – be Brepols [400]

Adversus omnes haereses libri 14 / Castro, Alf A – Antverpiae, 1556 – 41mf – 8 – €79.00 – ne Slangenburg [240]

Adversus omnia catabaptistarum prava dogmata / Bullinger, Heinrich – Tigvri, Christoph Froschover, 1535 – 5mf – 9 – mf#PBU-547 – ne IDC [240]

Adversvs anabaptistas libri 6 / Bullinger, Heinrich – Tigvri, Christoph Froschouer, 1560 – 7mf – 9 – mf#PBU-212 – ne IDC [242]

Advertencias : para los confessores de los naturales / Baptista, Juan – en Mexico: En el Conuento de Sanctiago Tlatilulco, Por M Ocharte 1600-01 – (= ser Books on religion...1543/44-c1800: evangelizacion) – 2v on 4mf – 9 – mf#crl-339 – ne IDC [241]

Advertencias a la istoria (sic) de merida / Gomez Bravo, Juan – 1638 – 9 – sp Bibl Santa Ana [946]

Advertencias y obligaciones...rejon / Trejo, Luis de – 1639 – 9 – sp Bibl Santa Ana [946]

Advertentie blad *see* De zuid-afrikaan

Advertisement : or syllabus of means for rectifying, settling, and consummating all our foreign and all our domestic interests, with those of the world at large, on a permanent basis / Edwards, George – [London: 1813?] – (= ser 19th c general coll on politics) – 1mf – 9 – mf#1.1.18 – uk Chadwyck [330]

Advertiser – 1925 sep 11-27 jun 9; 1927 jun 16-29 may 23; 1929 may 30-30 dec 31; 1931 jan 1-1931 jun 25 – 1 – mf#944125 – us WHS [071]

Advertiser – 1990 may 4-dec 26; 1991 jan-dec; 1992 jan-dec; 1993 jan-dec – 1 – mf#3137164 – us WHS [071]

Advertiser – Ashfield, jan 1899-dec 1908 – 4r – A$338.32 vesicular A$360.32 silver – at Pascoe [079]

Advertiser : for the districts of kirkintilloch, kilsyth, & campsie – [Scotland] East Dunbartonshire, [Kirkintilloch]: Kirkintilloch, Kilsyth, & Campsie Printing Office 2 & 9 aug 1862 (wkly) [mf ed 2004] – 1r – 1 – (cont by: budget for the districts of kirkintilloch, kilsyth, & campsie) – uk Newsplan [072]

Advertiser / Lucas Co. Manhattan – (sep 1838-mar 1841) [irreg] – 1r – 1 – mf#B34530 – us Ohio Hist [071]

Advertiser / Medina Co. Lodi – apr 1977-nov 1987// [wkly] – 10r – 1 – mf#B29576-29585 – us Ohio Hist [071]

Advertiser / Medina Co. Lodi – v1 n1. (sep 1955-aug 1978) [wkly] – 12r – 1 – mf#B33132-33143 – us Ohio Hist [071]

Advertiser / Richland Co. Plymouth – v1 n1. nov 1853-sep 1855 (1r); 1914-2/1920,3-7/1924,7/1926-3/1988 (27r) [wkly] – 1r – 1 – mf#B30077; B32227-32253 – us Ohio Hist [071]

Advertiser / Ross Co. Chillicothe – (aug 1840-jan 1853) spotty (1r); jul 1882-jun 1886, 1888-sep 1895 (6r) [wkly] – 1 – mf#B1226; B10615-10620 – us Ohio Hist [071]

Advertiser – Valley City, ND: E P Getchell, -aug 10-24, 1934 (wkly) – 1 – mf#11453 – us North Dakota [071]

Advertiser *see* Daily coast mail

The advertiser – City of Cebu: Advertiser Press, jan 9 1942-mar 7 1942 – us CRL [079]

The advertiser – New York. N.Y. 1921, 1922 – 1 – us AJPC [071]

Advertiser, and news of the week – Portadown, Ireland 10 feb-15 sep 1887, 16 jan 1890-10 jun 1893 (imperfect) [mf 1890-96] – 1 – (cont by: portadown recorder, & weekly advertiser [17 jun 1893-29 jan 1904]) – uk British Libr Newspaper [072]

Advertiser and ohio phoenix / Hamilton Co. Cincinnati – jan 1835-apr 1837 (poor quality) [semiwkly] – 2r – 1 – mf#B1242-1243 – us Ohio Hist [071]

Advertiser (ashton edition) – [NW England] Ashton Lib 26 apr, 17 may-dec 1979; 1980-83 held at advertiser (stockport office) – 1 – (missing: 14 feb 1980-22 may 1980; title change: advertiser (tameside ed) [1984]) – uk MLA; uk Newsplan [072]

Advertiser [barnet] – Barnet, England 25 apr 1996-23 jun 1999 [1986-2001] – 1 – (cont: barnet advertiser [3 apr 1986-18 apr 1996]; cont by: advertiser (barnet) 25 apr 1996-23 jun 1999) – uk British Libr Newspaper [072]

Advertiser (bonnyrigg, scotland) – Bonnyrigg: Scottish Country Press 4th nov 1982- (wkly) [mf ed 6 jan 1994-] – 1 – (formed by union of: dalkeith advertiser and: south midlothian advertiser (1938)) – uk Scotland NatLib [072]

Advertiser for blairgowrie, rattray, coupar-angus, alyth, kirriemuir, strathmore and stormont – Blairgowrie: D Christie 1859-61 (wkly) [mf ed 2003] – 1r – 1 – (cont: blairgowrie advertiser and coupar-angus and alyth journal; cont by: blairgowrie advertiser and strathmore and stormont news) – uk Newsplan; uk Scotland NatLib [072]

Advertiser for shettleston, tollcross and surrounding districts : weekly journal of literature and local interest – [Scotland] Glasgow: D Barr 31 mar 1896 [mf ed 2003] – 1v on 1r – 1 – uk Newsplan [072]

Advertiser, for the counties of louth, meath, dublin, monaghan and cavan – Drogheda, Ireland 18 mar 1896-30 sep 1908 – 1 – (cont by: drogheda advertiser [7 oct 1908-14 dec 1929]) – uk British Libr Newspaper [072]

Advertiser (oldham ed) – [NW England] Oldham apr 1982-1984 – 1 – uk MLA; uk Newsplan [072]

Advertiser (prestwich ed) – [NW England] Prestwich nov 1982-1984, 1-14 jan 1992-dec 1996 – 1 – uk MLA; uk Newsplan [072]

Advertiser [queens park etc] – London, England 10 oct 1889-29 jul 1920 – 1 – (cont by: queen's park advertiser & west london star [5 aug 1920-20 may 1938]) – uk British Libr Newspaper [072]

Advertiser series / Butler Co. Hamilton – nov 1821-oct 1827 [wkly] – 2r – 1 – mf#B25610-25611 – us Ohio Hist [071]

Advertiser series / Cuyahoga Co. Berea – v1 n1. 6/1868-80,1882-12/1908 (all damaged) [wkly] – 11r – 1 – mf#B34759-34769 – us Ohio Hist [071]

Advertiser & social news – [London & SE] Wandsworth 16 nov 1901-2 may 1903 [mf ed 2003] – 2r – 1 – uk Newsplan [072]

Advertiser weekly news – Barking, England 15 jul 1981-10 feb 1982 – 1 – (cont by: barking & dagenham weekly news [17 feb-29 apr 1982]) – uk British Libr Newspaper [072]

Advertiser [west manchester ed] – [NW England] Manchester ALS 1983-84 – 1 – uk MLA; uk Newsplan [072]

Advertiser-news – Sutton, NE: E P Burnett. -new v8 n36. feb 20 1903; v24 n37-38. feb 27-mar 6 1903 (wkly) [mf ed 1895-1903 (gaps) filmed [1974?]] – 4r – 1 – (cont by: sutton news. issues for sep 6 1895-feb 20 1903 also called old v10-old v16) – us NE Hist [071]

Advertiser's weekly – London, England 1950-56 – 1 – ISSN: 0001-8880 – mf#476 – us UMI ProQuest [650]

Advertising : the social and economic problem / French, George – New York: The Ronald Press Co 1915 [mf ed 1985] – 1r – 1 – (filmed with: egypt, cyprus and asiatic-turkey, j l) – mf#6845 – us Wisconsin U Libr [650]

Advertising age [electronic media edition] – Detroit MI 1982 – 1,5,9 – (cont by: electronic media) – ISSN: 0744-6675 – mf#13840.02 – us UMI ProQuest [380]

Advertising age [midwest region edition] – Detroit MI 1930+ – 1,5,9 – ISSN: 0001-8899 – mf#347 – us UMI ProQuest [650]

Advertising age's business marketing – Detroit MI 1994-99 – 1,5,9 – (cont: business marketing; cont by: b to b) – ISSN: 1087-948X – mf#348.03 – us UMI ProQuest [650]

Advertising and sales promotion – Detroit MI 1953-73 – 1,5 – (cont by: promotion) – ISSN: 0001-8937 – mf#1174.01 – us UMI ProQuest [650]

Advertising research foundation. annual conference proceedings – New York NY 1975-77 – 1,5,9 – ISSN: 0568-0352 – mf#10340 – us UMI ProQuest [650]

Advertising world – London, England dec 1901-jul/aug 1940 [mf oct, dec 1901] – 1 – (incl registration iss dated oct 1901; fr mar 1941-oct 1943 iss as suppl to: world's press news) – uk British Libr Newspaper [650]

Advertissement a tous bons et loyaux subiectz du roy, ecclesiastiques, nobles, et du tiers estat : pour n'estre surprins et circonuenuz par les propositions colores... – Paris: Pour I Dallier 1567 [mf ed 1980] – 1r – 1 – (filmed with: advertissement a tous bons et loyaux subiectz du roy [paris: pour i dallier 1567]) – mf#106 – us Wisconsin U Libr [944]

Advertissement contre l'astrologie, qu'on appelle judiciaire : et autres curiositez qui regnent aujourd'huy au monde / Calvin, J – Geneve: Jean Girard, 1549 – 1mf – 9 – mf#CL-76 – ne IDC [240]

Advertissement sur la censure qu'ont faicte les bestes de sorbonne, touchant les livres qu'ilz appellent heretiques / [Calvin, J] – [Geneva: Jean Girard], 1544 – 1mf – 9 – mf#CL-23 – ne IDC [240]

Advertissement sur le faict du concile de trente : faict pan mil cing sens soixante quatre / [Mesnil, Jean Baptiste du] – impr nourellement. s.l: s.n. 1567 [mf ed 1980] – 1r – 1 – (filmed with: advertissement a tous bons et loyaux subiectz du roy [paris: pour i dallier 1567]) – mf#106 – us Wisconsin U Libr [900]

Advertissement tresutile du grand proffit qui reviendroit a la chrestiente s'il se faisoit inventoire de tous les corps sainct... / Calvin, J – Geneva: Jean Girard, 1543 – 2mf – 9 – mf#CL-73 – ne IDC [240]

Advice literature in america : pt 1: the schlesinger collection of etiquette and advice books from the arthur and elizabeth schlesinger library on the history of women in america, radcliffe institute for advanced study, harvard university – 1747-1900 – 15r – 1 – £1400.00 – (with d/g) – uk Matthew [390]

Advice to proprietors : on the care of valuable pictures painted in oil – London 1835 – (= ser 19th c art & architecture) – 1mf – 9 – mf#4.2.1203 – uk Chadwyck [750]

Advice to students having in view the christian ministry addressed to them at the academy at britsol / Evans, C – 1770 – 1 – $5.00 – us Southern Baptist [242]

Advice to young men, and (incidentally) to young women, in the middle and higher ranks of life : in a series of letters, addressed to a youth, a bachelor, a lover, a husband, a citizen or a subject / Cobbett, William – New York: J Doyle, 1833 [mf ed 1984] – 3mf – 9 – 0-665-44102-9 – mf#44102 – cn Canadiana [390]

Adviento y sermones varios / Duran de Montijo, Juan – 1722. 5v – 9 – sp Bibl Santa Ana [240]

Advis de la magnifiqve et triomphante entree du seigneur marc antoine colonne... – Paris, 1572 – 1mf – 9 – mf#H-8189 – ne IDC [956]

Advis et devis de la sovrce de lidolatrie et tyrannie papale : par qvelle practiqve et finesse les papes sont en si haut degre montez... / Bonivard, Francois – Geneve: Iules Guillaume Fick, 1856 [mf ed 1986] – 1mf – 9 – 0-8370-8005-3 – (in middle french) – mf#1986-2005 – us ATLA [241]

Advis et devis des lengues : suivis de lamartigenee, cest a dire de la source de peche / Bonivard, Francois – Geneve: Jules-Guillaume Fick, 1865 [mf ed 1986] – 1mf – 9 – 0-8370-8564-0 – mf#1986-2564 – us ATLA [490]

Advis novvellement venvs de messine... – Paris, 1565 – 1mf – 9 – mf#H-8164 – ne IDC [956]

Adviser : or, vermont evangelical magazine – La Salle 1809-1815 – 1 – mf#3534 – us UMI ProQuest [240]

The adviser : a book for young people – Toronto: Ontario Temperance & Prohibitory League, 1873 [mf ed 1987] – 2mf – 9 – 0-665-08071-9 – mf#08071 – cn Canadiana [360]

Advisor – 1968 sep 19-1969 aug 8 – 1 – mf#554657 – us WHS [071]

Advisor : journal of the american family foundation / American Family Foundation – v1 n1-v6 n3 [1979 aug-1984 apr/may] – 1 – mf#978399 – us WHS [071]

Advisor – v8 n1-v16 n2 [1978 jan/feb-1986 mar/apr] – 1 – mf#894001 – us WHS [071]

The advisory commission of the council of national defense, 1916-1918... *see* Minutes of the meetings of the council of national defense, 1916-1921 / the advisory commission of the council of national defense, 1916-1918...

Advisory committee notes to the federal rules of evidence that may require clarification / Capra, Daniel J – 1998 – 1mf – 9 – $1.50 – mf#llmc99-035 – us LLMC [347]

Advisory Council for the Northern Sudan. Khartoum *see* Proceedings, 1st-2nd sessions 1944, 4th-8th sessions nov 1944-1948

Advisory Council on Historic Preservation *see* Trusteeship termination

Advisory councils / Batchelor, Henry – Glasgow, Scotland. 1875 – 1r – 1 – us UF Libraries [240]

Advocate – 1889 sep – 1 – mf#3185233 – us WHS [071]

Advocate – 1898 sep 29-dec 31 – 1 – mf#947507 – us WHS [071]

Advocate – 1904 jun 9 – 1 – mf#5259194 – us WHS [071]

Advocate – 1938 nov4-1939 mar 18; jun 10, nov-1940 apr – 1 – mf#5468405 – us WHS [071]

Advocate – 1977 jul-1982 oct; 1982 nov-1988 nov/dec – 1 – mf#621570 – us WHS [071]

Advocate – Portland OR: E D Cannaday, [wkly] – 1 – (absorbed: mt scott herald (1914-23)) – us Oregon Lib [071]

Advocate – Queenston and Toronto, Canada. 1824-34 – (= ser Colonial advocate) – 2r – 1 – (also: colonial advocate) – cn Library Assoc [071]

Advocate / Ashland Co. Loudonville – (6/1873-98,6/09-5/1911) many gaps [wkly] – 2r – 1 – mf#B10370-10371 – us Ohio Hist [071]

Advocate / Ashland Co. Loudonville – 3/1873-3/75,3/87-3/89,3/90-9/1920 [wkly] – 17r – 1 – (a democrat newspaper) – mf#B2552-2568 – us Ohio Hist [071]

Advocate – Burnie, 1923-feb 1997 – at Pascoe [079]

Advocate – c1899 feb 25-02 jun 28; 1902 jul 5-1905 sep 30; 1905 oct 7-1908 dec 31; 1909 jan 7-1912 mar 21; 1912 mar 28-oct 3 – 1 – mf#1131074 – us WHS [071]

Advocate – Los Angeles CA. n211- [biwkly], mar 1977- [yrly] – 1 – (cont: los angeles advocate) – ISSN: 0001-8996 – mf#OCLC 7927377 – us Wisconsin U Libr [071]

Advocate / Crawford Co. Crestline – (sep 1869-1979) many gaps (44r); jan 1980-dec 1982 (2r); jan 6, 1988-dec 25 1991 (3r) [wkly] – 44r – 1 – (also b10990) – mf#B10827-10869; B12371-12372; B31649-31651 – us Ohio Hist [071]

Advocate / Crawford Co. Crestline – jan 1983-dec 1987 [wkly] – 4r – 1 – mf#B34770-34773 – us Ohio Hist [071]

Advocate / Darke Co. Greenville – dec 1966-dec 1968 [daily] – 11r – 1 – mf#B7731-7741 – us Ohio Hist [071]

Advocate – Southampton, England 19 jun 1897-28 mar 1925 – 1 – (discontinued; cont: shirley & freemantle advocate, & parish of millbrook news [20 jan 1894-12 jun 1897]) – uk British Libr Newspaper [072]

Advocate / Fulton Co. Archbold – jan 1900-dec 1907 (poor quality) [wkly] – 3r – 1 – mf#B559-561 – us Ohio Hist [071]

Advocate : a journal for military defense counsel – v1-16 n2. 1969-84 (all publ) – (= ser Military law coll) – 9 – (cont by: the army lawyer) – mf#LLMC 84-237 – us LLMC [355]

Advocate / Licking Co. Newark – feb 7-mar 31, 1927 (gap filler) [daily] – 1r – 1 – mf#B10372 – us Ohio Hist [071]

Advocate / Licking Co. Newark – jan 1822-apr 1824 [wkly] – 1r – 1 – mf#B209 – us Ohio Hist [071]

Advocate / Licking Co. Newark – jan-dec 1909 (gap filler) [daily] – 3r – 1 – mf#B204-206 – us Ohio Hist [071]

Advocate / Licking Co. Newark – may 1850-aug 1854; (may 1866-sep 1871) [wkly] – 3r – 1 – mf#B3421; B12748-12749 – us Ohio Hist [071]

Advocate – Los Angeles CA 1999+ – 1,5,9 – ISSN: 0001-8996 – mf#18623.01 – us UMI ProQuest [071]

Advocate / Madison Co. Plain City – sep 1897-oct 1908, nov 1909-1934 [wkly] – 16r – 1 – mf#B7102-7117 – us Ohio Hist [071]

Advocate / Madison Co. Plain City – v1 n1. apr 1952-dec 1970; jan 1985-dec 1993 [wkly] – 15r – 1 – mf#B33780-33794 – us Ohio Hist [071]

Advocate / Madison County. Plain City – jan 1971-dec 1984 – 11r – 1 – mf#B8026-8036 – us Ohio Hist [071]

Advocate / Marion Co. Marion – (oct 1973-oct 1979) scattered [irreg] – 1r – 1 – mf#B10525 – us Ohio Hist [331]

Advocate / Perry Co. Somerset – may 1867-feb 1869 [wkly] – 1r – 1 – mf#B5531 – us Ohio Hist [071]

Advocate – v1 n1-v2 n2 [1979 feb-1980 may] – 1 – mf#639941 – us WHS [071]

Advocate – v1-2 n11 [1901 jun-1903 apr]; 1902 may – 1 – mf#966634 – us WHS [071]

Advocate – v6 n3-v9 n46 [1894 jan 17-1987 nov17] – 1 – mf#1037418 – us WHS [071]

Advocate – Cleveland, OH, may 15 1915-jun 1 1918 – 2r – 1 – (weekly african-american republican newspaper) – mf#(M) 34 C9.3 085 – us Western Res [071]

Advocate : a weekly law journal – St Paul, Minneapolis, Chicago. v1-2. 1888-90 [all publ] – 1 – $50.00 set – mf#408760 – us Hein [340]

Advocate *see* Mt scott herald

The advocate – Bassett, NE: W F Bowser (wkly) [mf ed may 17 1895-dec 20 1895 (gaps) filmed 1979]] – 1r – 1 – (cont: newport advocate) – us NE Hist [071]

The advocate – Fergus, Ont: J Coram, [1885-1889] – 9 – mf#P06138 – cn Canadiana [071]

The advocate – Minneapolis. v1-2. 1888-90 (all publ) – 8mf – 9 – $12.00 – mf#LLMC 84-395 – us LLMC [073]

The advocate : a novel / Heavysege, Charles – Montreal: R Worthington, 1865 – 2mf – 9 – mf#48293 – cn Canadiana [830]

The advocate – Toronto: L P Kribs, [1894-189-or 19–] – 9 – mf#P04183 – cn Canadiana [073]

The advocate – Idaho State Bar. v1-29. 1957-86 – 66mf – 9 – $99.00 – (updated regularly. missing: v16 no 11, v24 nos 11-12, v26 nos 1-6) – mf#LLMC 84-396 – us LLMC [340]

Advocate and labor news – 1972 mar 9-aug 10 – 1 – mf#3069370 – us WHS [071]

Advocate and press – New Bloomfield, PA., 1889-1932 – 13 – $25.00r – us IMR [071]

Advocate and valley vista *see* [Alhambra-] post-advocate

Advocate courier *see* Courier

Advocate daily bulletin – 1861 apr 24-sep 10 – 1 – mf#947640 – us WHS [071]

Advocate for the dead / Brand, Joel – London, England. 1958, C1956 – 1r – 1 – us UF Libraries [880]

Advocate for the testimony of god : as it is written in the books of nature and revelation – La Salle IL 1835-39 – 1 – mf#3711 – us UMI ProQuest [210]

Advocate (idaho state bar journal) – v1-44. 1957-2001 – 9 – $484.00 set – ISSN: 0515-4987 – mf#400730 – us Hein [340]

Advocate of peace – jun 1837-dec 1932 – (= ser The library of world peace studies) – 260mf – 9 – $1705.00 – us UPA [320]

Advocate of peace and christian patriot – La Salle IL 1828-29 – 1 – mf#4407 – us UMI ProQuest [190]

Advocate of science : a popular scientific journal – Ann Arbor MI 1833-34 – 1 – mf#3913 – us UMI ProQuest [500]

Advocate of science and annals of natural history – Ann Arbor MI 1834-35 – 1 – mf#3914 – us UMI ProQuest [500]

Advocate: or irish industrial journal – Dublin, Ireland 28 oct 1848-28 jan 1860 – 11r – 1 – uk British Libr Newspaper [072]

Advocate (portland, or) *see* Mt scott herald

Advocate series / Licking Co. Newark – jan 1874-dec 1911 [wkly, semiwkly, wkly] – 19r – 1 – mf#B11153-11171 – us Ohio Hist [071]

Advocate (suffolk) – v1-27. 1968-1997 – 5,6,9 – $240.00 set – (v1-16 1968-85 in reel $72. v17-27 1986-97 in mf $114) – ISSN: 0568-0425 – mf#100051 – us Hein [340]

Advocate (vancouver) – v1-59. 1943-2001 – 9 – $836.00 set – ISSN: 0044-6416 – mf#115361 – us Hein [340]

Advocate-news – Bridgetown, Barbados. 1976 10 Year Edition – 1r – 1 – us UF Libraries [071]

Advocate-tribune – Bloomington, NE: Clyde & Helen Shade. v76 n46. jan 24 1957-v79 n30. oct 1 1959 (wkly) [mf ed jan 24 1957-oct 1 1959 (gaps) filmed in 1979]] – 1r – 1 – (cont: bloomington advocate-tribune) – us NE Hist [071]

Advocate-tribune – Bloomington, NE: Crane & Co. 12v. v53 n53. dec 7 1933-v64 n48. aug 16 1945 (wkly) – 4r – 1 – (lacks: oct 22 1936. cont: bloomington advocate. cont by: bloomington advocate-tribune) – us NE Hist [071]

The advocate-tribune – Bloomington, NE: H M Crane. v41 n5. oct 13 1922-v49 n21. jan 29 1931 (wkly) [mf ed oct 13 1922] – 5r – 1 – (lacks oct 3 1929. formed by union of: bloomington advocate and franklin county tribune. cont by: bloomington advocate) – us NE Hist [071]

Advogado rui barbosa / Nogueira, Rubem – Rio De janeiro, Brazil. 1949 – 1r – 1 – us UF Libraries [972]

Adwaitism and the religions of the east : a small treatise on the principles of adwaitism... / Khedkar, Raghunath Vithal – Kolhapur: Shri Venkateshwar Press, 1913 [mf ed 1991] – 1mf – 9 – 0-524-01565-1 – mf#1990-2519 – us ATLA [280]

Adweek – New York NY 2003+ – 1,5,9 – ISSN: 1549-9553 – mf#34420 – us UMI ProQuest [650]

Adweek [eastern ed] – New York NY 1985+ – 1,5,9 – ISSN: 0199-2864 – mf#15417,01 – us UMI ProQuest [650]

Adweek magazines' technology marketing – New York NY 2001+ – 1,5,9 – (cont: mc technology marketing intelligence) – ISSN: 1536-2272 – mf#18869,04 – us UMI ProQuest [650]

Adweek [midwest ed] – New York NY 1991+ – 1,5,9 – ISSN: 0276-6612 – mf#18872,01 – us UMI ProQuest [650]

Adweek [national marketing ed] – New York NY 1985 – 1,5,9 – (cont by: adweek's marketing week: national marketing ed) – ISSN: 0888-3718 – mf#16149.02 – us UMI ProQuest [650]

Adweek [southeast ed] – New York NY 1991+ – 1,5,9 – ISSN: 8756-6389 – mf#18871,01 – us UMI ProQuest [650]

Adweek [southwest ed] – New York NY 1992-96 – 1,5,9 – ISSN: 0746-892X – mf#18873,01 – us UMI ProQuest [650]

Adweek [western ed] – New York NY 1995-96 – 1,5,9 – mf#18806 – us UMI ProQuest [650]

Adweek's marketing week [national marketing ed] : national marketing ed – New York NY 1987-1991 – 1,5,9 – (cont: adweek: national marketing ed; cont by: brandweek) – ISSN: 0892-8274 – mf#16149.02 – us UMI ProQuest [650]

Ady, Julia Mary *see* The art annual for 1894 sir edward burne-jones, bart

Ady, Julia Mary (Cartwright) *see*
– Jules bastien-lepage
– Mantegna and francia

Adye, Stephen P *see* A treatise on courts-martial

Ae Kru, U *see* Britisyha mran ma khet ca krann tuik mya a khre a ne, 1826-1947

Ae Kruin, Do *see* Ca mrin vuin

Ae Nuin, U *see* Chay nhac ra si mran ma rui ra ra si pvai to mya

Ae Son, U *see* Lak tve a sum khya sak se kham upade

Aea advocate / Arizona Education Association – Phoenix AZ 1974+ – 1 – (cont: arizona educator advocate) – ISSN: 0194-8849 – mf#11260,01 – us UMI ProQuest [370]

L'aechitecture francoise : or recueil des plans, elevations, coupes et profils des eglises, palais...de france... / Mariette, J; ed by Hautecoeur, M L – Paris, 1727 – 3v on 36mf – 9 – (repr of original) – mf#O-360 – ne IDC [720]

Der aechte schwarzwaelder – Freiburg Br DE, 1832 2 may-19832 29 dec – 1r – 1 – gw Misc Inst [074]

Die aechtheit der pastoralbriefe : mit besonderer ruecksicht auf den neuesten angriff von herrn dr. baur / Baumgarten, Michael – Berlin: Ludwig Dehmigke, 1837 – 1mf – 9 – 0-7905-0965-2 – (incl bibl ref) – mf#1987-0965 – us ATLA [227]

Aedc journal / American Economic Development Council – Washington DC 1981 – 1,5,9 – (cont: aidc journal) – ISSN: 0279-6430 – mf#2313,01 – us UMI ProQuest [650]

Aedes walpolianae : or, a description of the collection of pictures at houghton hall in norfolk... / Walpole, H – London, 1752 – 4mf – 9 – mf#O-1173 – ne IDC [700]

Aeds journal – Eugene OR 1967-1986 – 1,5,9 – (cont by: journal of research on computing in education) – ISSN: 0001-1037 – mf#10379.02 – us UMI ProQuest [370]

Aeds monitor – Eugene OR 1962-85 – 1,5,9 – (cont by: monitor) – ISSN: 0001-1045 – mf#10380.01 – us UMI ProQuest [370]

Aef nouvelle – Brazzaville. aout 1948-nov 1949 – 1 – fr ACRPP [079]

Aegean breeze – 1981 apr 30-1983 nov3; 1984 feb 16-1986 apr 25; 1986 may-1989 jun – 1 – mf#947923 – us WHS [071]

Aegidii gutbirii lexicon syriacum : omnes novi testamenti syriaci dictiones et particulas complectens / Gutbier, Aegidius; ed by Henderson, Ebenezer – Londini [London]: Sumptibus Samuelis Bagster, 1836 [mf ed 1993] – 1mf – 9 – 0-524-07121-7 – mf#1992-1037 – us ATLA [225]

Aegis – n22-1942 [1978 sep/oct-1987] – 1 – mf#35699 – us WHS [071]

Aegis, and western courier – Castlebar, Ireland 23 jun 1841-26 nov 1842 – 1 – uk British Libr Newspaper [072]

The aegis and western courier – Castlebar. Ireland. -w. 23 Jun 1841-26 Nov 1842. (43 ft) – 1 – uk British Libr Newspaper [072]

Aegypten / Winterer, H – Guben, 1915 – 2mf – 9 – mf#ILM-2129 – ne IDC [956]

Aegypten, oder sitten, gebraeuche, trachten und denkmaeler der aegypter / Breton de LaMartiniere, Jean – Pesth 1817 [mf ed Hildesheim 1995-98] – 4v on 5mf – 9 – €100.00 – 3-487-27363-2 – gw Olms [960]

Aegypten und aegyptisches leben im altertum *see* Life in ancient egypt

Aegypten und die bibel : die urgeschichte israels im licht der aegyptischen mythologie / Voelter, Daniel – 2. neubearb aufl. Leiden: E J Brill, 1904 [mf ed 1986] – 1mf – 9 – 0-8370-9276-0 – (incl bibl ref) – mf#1986-3276 – us ATLA [221]

Aegypten-index : bilddokumentation zur kunst in aegypten = Egyptian index. pictorial documentation on art in egypt / ed by Bildarchiv Foto Marburg – Deutsches Dokumentationszentrum fuer Kunstgeschichte Philipps-Universitaet Marburg – [mf ed 1997] – 111mf (1:24) – 9 – silver €2500.00 – ISBN-10: 3-598-33634-9 – ISBN-13: 978-3-598-33634-8 – gw Saur [700]

Aegypten's stellung in der religions- und culturgeschichte / Nippold, Friedrich – Berlin: C G Luederitz 1869 [mf ed 1991] – (= ser Sammlung gemeinverstaendlicher wissenschaftlicher vortraege 4/82) – 1mf – 9 – 0-524-01803-0 – mf#1990-2651 – us ATLA [290]

Aegyptiaca : oder beschreibung des zustandes des alten und neuen aegypten nach eigenen, in den jahren 1801 und 1802 angestellten beobachtungen / Hamilton, William R – Weimar 1814 – 2mf – 9 – €16.00 – gw Olms [930]

Aegyptiaca oder beschreibung des zustandes des alten und neuen aegypten : nach eigenen, in den jahren 1801 und 1802 angestellten beobachtungen / Hamilton, William – Weimar 1814 [mf ed Hildesheim 1995-98] – 1v on 2mf – 9 – €60.00 – ISBN-10: 3-487-26532-X – ISBN-13: 978-3-487-26532-2 – gw Olms [930]

Aegyptisch-aramaeische inschriften / Lauth, Franz Joseph – [s.l: s.n.] 1878 [mf ed 1986] – 1mf – 9 – 0-8370-7164-X – mf#1986-1164 – us ATLA [470]

Aegyptische abendmahlsliturgien des ersten jahrtausends / Schermann, Theodor – Paderborn, 1912 – €12.00 – ne Slangenburg [240]

Aegyptische chrestomathie / Erman, A – Berlin, 1904 – 3mf – 9 – mf#NE-20385 – ne IDC [930]

Aegyptische goldschmiedearbeiten / Schaefer, H – Berlin, 1910 – 6mf – 9 – mf#NE-20404 – ne IDC [930]

Aegyptische grabsteine und denksteine aus athen und konstantinopel / Poertner, B – Strassburg, 1908 – 1mf – 9 – mf#NE-391 – ne IDC [956]

Aegyptische grabsteine und denksteine aus sueddeutschen sammlungen / Spiegelberg, W & Poertner, B – Strassbourg, 1902 – 6mf – 9 – mf#NE-390 – ne IDC [956]

Aegyptische grammatik / Erman, A – Berlin, 1928 – 4mf – 9 – mf#NE-20379 – ne IDC [470]

Eine aegyptische koenigstochter : historischer roman / Ebers, Georg – Stuttgart: Deutsche Verlags-Anstalt, 1893 [mf ed 1993] – (= ser Georg ebers gesammelte werke 1-2) – 2v – 1 – mf#8554 reel 1 – us Wisconsin U Libr [830]

Aegyptische kunstgeschichte von den aeltesten zeiten bis auf die eroberung durch die araber / Bissing, F W von – Berlin-Charlottenburg, 1934. 3v – 12mf – 9 – mf#NE-20419 – ne IDC [930]

Aegyptische lesestuecke... / Sethe, K – Leipzig, 1928 – 2mf – 9 – mf#NE-486 – ne IDC [956]

Aegyptische nachrichten – Kairo (ET), 1912 3 jan- 31 dec – 1 – (tw a in franzoesischer sprache) – gw Misc Inst [077]

Die aegyptische religion / Erman, A – Berlin, 1905 – 3mf – 9 – mf#NE-20392 – ne IDC [290]

Die aegyptische religion *see* Handbook of egyptian religion

Die aegyptische sammlung des museum-meermanno-westreenianum im haag / Spiegelberg, W – Strassburg, 1896 – 1mf – 9 – mf#NE-392 – ne IDC [956]

Das aegyptische todtenbuch der 18 bis 20 dynastie / Naville, E – Berlin, 1886 – 35mf – 9 – mf#NE-20023 – ne IDC [956]

Aegyptische urkunden aus den koeniglichen museen zu berlin... – 137mf – 8 – mf#H-106 – ne IDC [956]

Das aegyptische verbum in altaegyptichen, neuaegyptischen und koptischen / Sethe, K – Leipzig, 1899-1902. 3v – 20mf – 9 – mf#NE-495 – ne IDC [470]

Aegyptisches glossar / Erman, A – Berlin, 1904 – 2mf – 9 – mf#NE-20383 – ne IDC [956]

Die aegyptologie : abriss der entzifferungen und forschungen auf dem gebiete der aegyptischen schrift, sprache und alterthumskunde / Brugsch, Heinrich Karl – Leipzig: W Friedrich, 1891 – 2mf – 9 – 0-524-02199-6 – (incl bibl ref) – mf#1990-2873 – us ATLA [930]

Die aegyptologie und die buecher mosis / Scholz, Anton – Wuerzburg: L Woerl, 1878 – 1mf – 9 – 0-7905-3411-8 – (incl bibl ref) – mf#1987-3411 – us ATLA [221]

Aegyptologische randglossen zum alten testament / Spiegelberg, W – Strassburg, Schlesier und Schweikhardt, 1904 – 1mf – 9 – mf#NE-482 – ne IDC [956]

Aegyptus : rivista italiana di egittologia e di papirologia – 1(1920)-20(1940) – 152mf – 9 – €287.00 – mf#470 – ne Slangenburg [930]

Aelfrida : drama in fuenf aufzuegen / Loewenberg, Jakob – Hamburg: M Glogau 1919 [mf ed 1990] – 1r – 1 – (filmed with: unrast / gerhard lorenz) – mf#2830p – us Wisconsin U Libr [820]

Aelia et mysis : ou, l'atellane. ballet-pantomime en deux actes. musique de henri potier. decorations de mm. cambon, thierry et desplechin. represente pour la premiere fois, a paris, sur le theâtre de l'academie imperiale de musique, le mercredi 21 septembre 1853 / Mazilier, Joseph – Paris: V Jonas, libraire-éditeur de l'Opéra, 1853 – 1 – mf#*ZBD-*MGTZ pv 2-Res – us NY Public [790]

Aelianus, Claudius *see* Opera

Aelred of Rievaulx, Saint *see* Lives of s ninian and s kentigern

Der aeltere prophetismus : bis auf die heldengestalten von elia und elisa / Koenig, Eduard – Berlin: Edwin Runge, 1905 – 1mf – 9 – 0-7905-0502-9 – mf#1987-0502 – us ATLA [221]

Der aeltere vedanta : geschichte, kritik und lehre / Walleser, Max – Heidelberg: C Winter, 1910 – 1mf – 9 – 0-524-02621-1 – (incl bibl ref) – mf#1990-3071 – us ATLA [240]

Die aeltere wormser briefsammlung (mgh epistolae 2:3.bd) – 1949 – (= ser Monumenta germaniae historica epistolae 2. die briefe der deutschen kaiserzeit (mgh epistolae 2)) – €7.00 – ne Slangenburg [241]

Die aelteren juedischen feste : mit einer kritik der gesetzgebung des pentateuch / George, Johann Friedrich Leopold – Berlin: E H Schroeder, 1835 – 1mf – 9 – 0-7905-0028-0 – (incl bibl ref) – mf#1987-0028 – us ATLA [270]

Aelterer deutscher 'macer' / ortolf von baierland: 'arzneibuch' / 'herbar' des bernhard von breidenbach / faerber- und maler-rezepte (cima13) : die oberrheinische medizinische sammelhandschrift des kodex berleburg. farbmikrofiche-edition der handschrift berleburg, fuerstl sayn-wittgensteinsche bibliothek, cod rt 2/6 – (mf ed 1991) – (= ser Codices illuminati medii aevi (cima) 13) – 105p on 7 color mf – 15 – €370.00 – 3-89219-013-5 – (int & descriptions by werner dressendoerfer et al) – gw Lengenfelder [615]

Die aelteste agende des bistums muenster / Stapper, R – Muenster, 1906 – €7.00 – ne Slangenburg [931]

Der aelteste englische marienhymnus "on god ureisun of ure lefdi" / Marufke, Willy – Leipzig: Quelle & Meyer, 1907 [mf ed 1992] – (= ser Breslauer beitraege zur literaturgeschichte. neue folge 3) – 74p – 1 – (written in 1st pt of 13th c in the berkshire or wiltshire dialect. generally ascribed to edmund rich, archbishop of canterbury. incl bibl ref) – mf#8014 reel 2 – us Wisconsin U Libr [420]

Das aelteste evangelium : kritische untersuchung der zusammensetzung, des wechselseitigen verhaeltnisses, des geschichtlichen werths und des ursprungs der evangelien nach matthaeus und marcus = Oudste evangelie / Scholten, Johannes Henricus – Elberfeld: R L Friderichs, 1869 – 1mf – 9 – 0-8370-5134-7 – (incl bibl ref. in german) – mf#1985-3134 – us ATLA [220]

Das aelteste germanische christentum : oder, der sogen. "arianismus" der germanen. vortrag / Schubert, Hans von – Tuebingen: Mohr, 1909 – 1mf – 9 – 0-7905-6880-2 – (incl bibl ref) – mf#1988-2880 – us ATLA [240]

Das aelteste liturgiebuch der lateinischen kirche (tab26-28) / Dold, Alban – 1936 – (= ser Texte und arbeiten. beuron (tab). beitraege zur ergruendung des aelteren lateinischen und christlichen schrifttums und gottesdienstes) – €12.00 – ne Slangenburg [240]

Der aelteste sohn : roman / Bethusy-Huc, Valeska von Reiswitz-Kaderzin, Graefin von [pseud: Moritz von Reichenbach] – Stuttgart: Deutsche Verlags-Anstalt, 1889 [mf ed 1993] – (= ser Deutsche romanbibliothek. salon-ausgabe 2. jahrg/5) – 232p – 1 – mf#8512 – us Wisconsin U Libr [830]

Die aelteste terminologie der juedischen schriftauslegung : ein woerterbuch der bibelexegetischen kunstsprache der tannaiten / Bacher, Wilhelm – Leipzig: J C Hinrichs, 1899 – 1mf – 9 – 0-8370-2134-0 – (in hebrew incl bibl ref) – mf#1985-0134 – us ATLA [220]

Die aeltesten apologeten : texte mit kurzen einleitungen / ed by Goodspeed, Edgar Johnson – Goettingen: Vandenhoeck & Ruprecht, 1914 [mf ed 1990] – 1mf – 9 – 0-7905-5219-1 – (in greek & latin. int in german) – mf#1988-1219 – us ATLA [240]

Die aeltesten biographien des heiligen norbert / Rosenmund, R – Berlin, 1874 – 2mf – 8 – €5.00 – ne Slangenburg [241]

Die aeltesten quellen des orientalischen kirchenrechtes / Achelis, Hans – Leipzig, 1891 – (= ser Tugal 1-6/4) – 5mf – 9 – €12.00 – ne Slangenburg [240]

Die aeltesten roemischen sacramentarien und ordines / Probst, Ferdinand – Muenster i.W.: Aschendorff, 1892 – 1mf – 9 – 0-8370-7185-2 – mf#1986-1185 – us ATLA [240]

Die aeltesten tanzlehrbuecher / Michel, Artur – [Bruenn: fuer den Verfasser als Manuskript gedruckt, R M Rohrer, 1938?] – 1 – mf#*ZBD-*MGO pv 9 – Located: NYPL – us Misc Inst [790]

Aenchbacher, L E *see* The relationship between physical activity and self-rated depression in free-living women aged 60 years and older

Aeneas sylvius piccolomineus, qui postea pius 2 p m, de viris illustribus / Pius 2, Pope – Stuttgardiae: Sumtibus Societatis literariae stuttgardiensis, 1842 [mf ed 1993] – (= ser Blvs 1/3) – 1 – (incl bibl ref) – mf#8470 reel 1 – us Wisconsin U Libr [945]

Aeneas sylvius piccolomini als papst pius 2 : sein leben und einfluss auf die literarische cultur deutschlands / ed by Weiss, Anton – Graz: Ulr Moser (J Meyerhoff), 1897 [mf ed 1986] – 1mf – 9 – 0-8370-7275-1 – (incl bibl ref) – mf#1986-1275 – us ATLA [241]

Aeneas Sylvius Piccolomini [Pius 2, Pope] *see*
– Commentaria, libri 13
– Opuscula 30...

Aeneid *see* Works / aeneid

Aenetius, Theophilus *see* Peri ethon pragmateia

Aenwysinge der misverstanden van g melder / Ruse, H – Amsterdam, 1670 – 1mf – 9 – mf#OA-165 – ne IDC [720]

Aepinus, J *see* D ioannes aepini in psalmum 16 commentariu

[Aepinus, J] *see* Bekentniss vnnd erklerung auffs interim durch der erbarn stedte

Aequationes mathematicae – Dordrecht, Netherlands 1989+ – 1 – ISSN: 0001-9054 – mf#13939 – us UMI ProQuest [510]

Aereboe *see* Nung yeh cheng ts'e

Aereoplastes theosophicus... / Oraeus, H – Francofurti: Apud Jacobum de Zetter, 1620 [1621] – 4mf – 9 – mf#O-704 – ne IDC [090]

Aerial odyssey / Powell, E Alexander – New York, USA. 1936 – 1r – 1 – us UF Libraries [972]

Aero mechanic – 1942 may 14-1944; 1945-72 – 1 – mf#999980 – us WHS [629]

Aero 'n' photos – v1 n1-v2 n2 [1979 may-1981] – 1 – mf#676049 – us WHS [629]

Aero philatelist annals – v21 n2-v25 n2 [1980 jan-1982 jan] – 1 – mf#656626 – us WHS [760]

Aerobic and anaerobic performance measures in active and inactive young and middle-aged males / Ecker, KR – 1990 – 2mf – 9 – $8.00 – us Kinesology [612]

Aerobic certification / Jefferis, Shelly J – California State University, Northridge, 1995 – 1mf – 9 – mf#PE 3655 – us Kinesology [370]

Aerobic exercise related to functional aerobic capacity, repetitive/interfering behavior, and platelet serotonin concentration of individuals with autism / Schmidt, Gordon J – 1989 – 221p 3mf – 9 – $12.00 – us Kinesology [617]

Aerobic fitness testing and feeling states among 9 to 11 year old students / Bonfiglio – 2000 – 71p on 1mf – 9 – $5.00 – mf#HE 677 – us Kinesology [150]

Aerobic responses to 12 weeks of exerstriding or walking training in sedentary adult women / Larkin, James M & Butts, Nancy Kay – 1992 – 1mf – $4.00 – us Kinesology [612]

Aerobic responses to 12 weeks of training on various modes of home exercise equipment in sedentary adults / Angelini, Marc A – University of Wisconsin-La Crosse, 1995 – 1mf – 9 – $4.00 – mf#PH1451 – us Kinesology [613]

Aerodynamics of the curve-ball : an investigation of the effects of angular velocity on baseball trajectories / Alaways, LeRoy W – 1998 – 2mf – 9 – $8.00 – mf#PE 4035 – us Kinesology [790]

Aero-Graphic Corporation *see* Florida from the air

Aerolites and religion / Harvey, Arthur – [S.l: s.n, 1896?] – 1mf – 9 – 0-665-91401-6 – mf#91401 – cn Canadiana [210]

Aeronaut : a periodical paper, by an association of gentlemen – La Salle IL 1816-22 – 1 – mf#3535 – us UMI ProQuest [073]

Aeronautica brasileira / Barros, Domingos – Rio De Janeiro, Brazil. 1940 – 1r – 1 – us UF Libraries [972]

Aeronautical and miscellaneous notebooks, c1799-1826 / Cayley, George; ed by Hodgson, J E – 1933 – (= ser Newcomen society extra publication 3) – 4mf – 7 – mf#86575 – uk Microform Academic [629]

Aeronautical journal – London, England 1897+ – 1,5,9 – ISSN: 0001-9240 – mf#1281 – us UMI ProQuest [629]

Aeronautical quarterly – London, England 1949-83 – 1,5,9 – ISSN: 0001-9259 – mf#1280 – us UMI ProQuest [629]

Aeronautical Society of America *see* Bulletin

Aeronautics – London. v. 1-45. Aug 1932-Mar 1962 – 1 – 90.00 – us L of C Photodup [629]

Aeronautics, 1939-60 – 26r – 1 – mf#517 – uk Microform Academic [629]

Aeronautics and space reports to the congress, 1958-84 – 8r – 1 – $1260.00 – 0-89093-985-3 – (with p/g) – us UPA [629]

Aeronautique canadienne – Montreal: Aeronautique canadienne, [ca 1938] (wkly) [mf ed 2001] – 9 – (ceased in 194-?) – mf#SEM105P3374 – cn Bibl Nat [629]

Aeroplane – London, England 1958-68 – 1 – mf#1276 – us UMI ProQuest [629]

Aeroplane monthly – Cheam, England 1974-91 – 1,5,9 – mf#8448 – us UMI ProQuest [629]

Aerosol age – Parsippany NJ 1956-91 – 1,5,9 – (cont by: spray technology & marketing) – ISSN: 0001-9291 – mf#2100 – us UMI ProQuest [680]

Aerosol science and technology – Abingdon, Oxfordshire 1982-96 – 1,5,9 – ISSN: 0278-6826 – mf#42397 – us UMI ProQuest [660]

Aerospace america – Reston VA 1984+ – 1,5,9 – (cont: astronautics and aeronautics) – ISSN: 0740-722X – mf#1605,01 – us UMI ProQuest [629]

Aerospace century 21 [aasms54] – 1987 – (= ser Aasms 1968) – 11 papers on 6mf – 9 – $25.00 – 0-87703-278-5 – (suppl to v64, advances) – us Univelt [629]

Aerospace engineering [reston] – Reston VA 1942-63 – 1 – mf#5069 – us UMI ProQuest [629]

Aerospace engineering [warrendale] – Warrendale PA 1986-2001* – 1,5,9 – (cont: sae in aerospace engineering; backfiles: v3-5 mar 1983-85*) – ISSN: 0736-2536 – mf#16289,01 – us UMI ProQuest [629]

Aerospace facts – Brigham City UT 1965-79 – 1,5,9 – ISSN: 0001-9356 – mf#7938 – us UMI ProQuest [629]

Aerospace historian – Andrews Air Force Base MD 1954-88 – 1,5,9 – (cont by: air power history) – ISSN: 0001-9364 – mf#1519.01 – us UMI ProQuest [629]

Aerospace international [bonn] – Bonn, Germany 1972-81 – 1,5,9 – ISSN: 0001-9372 – mf#6701,01 – us UMI ProQuest [629]

Aerospace international [london] – London, England 1999+ – 1,5,9 – (cont: aerospace) – mf#11492,01 – us UMI ProQuest [629]

Aerospace [london] – London, England 1977-96 – 1,5,9 – ISSN: 0305-0831 – mf#11492.01 – us UMI ProQuest [629]

Aerospace management [new york] – New York NY 1958-64 – 1 – ISSN: 0568-0670 – mf#1109 – us UMI ProQuest [629]

Aerospace management [philadelphia] – Philadelphia PA 1966-71 – 5 – ISSN: 0001-9399 – mf#6337 – us UMI ProQuest [650]

Aerospace material specifications (ams) / Society of Automotive Engineers – 9 – $2,125.00 – (monthly update service is available for $1725. (ams) + update service are available as a set for $3075) – us SAE [629]

Aerospace medicine and biology – Washington DC 1974-79 – 1,5,9 – ISSN: 0001-9410 – mf#3139 – us UMI ProQuest [574]

Aerospace medicine and biology: an annotated bibliography – Washington DC 1952-63 – 1 – ISSN: 0001-9410 – mf#6286 – us UMI ProQuest [610]

Aerospace power journal – Washington DC 2000+ – 1,5,9 – (cont: airpower journal) – ISSN: 1535-4245 – mf#16177,02 – us UMI ProQuest [629]

Aerospace research and development – 9 – $40.00 – us Univelt [629]

Aerospace safety – Washington DC 1950-80 – 1,5,9 – (cont by: flying safety) – ISSN: 0001-9429 – mf#6287.01 – us UMI ProQuest [629]

Aerospace standards / Society of Automotive Engineers – 9 – $1,575.00 – (monthly update service is available for $800. when purchased together the price for both is $1895) – us SAE [629]

Aerospace [washington dc] – Washington DC 1963-87 – 1,5,9 – ISSN: 0001-9321 – mf#10065 – us UMI ProQuest [629]

Aerosvet – Belgrade, Yugoslavia 12 jan 1958-15 dec 1962 – 2r – 1 – uk British Libr Newspaper [077]

Aertnys, Josef *see* Supplementum ad tractatum de 7 decalogi praecepto secundum jus civile gallicum

Die aerzte – Hamburg, 1785-86 – 2r – 1 – (cont by: die deutsche gesundheits-zeitung, 1786) – gw Misc Inst [610]

AFFIRMATIONS

Aerzte in ost- und westpreussen : leben und leistung seit dem 18. jahrhundert / ed by Scholz, Harry & Schroeder, Paul – Wuerzburg: Holzner Verlag, 1970 – (= ser Ostdeutsche beitraege aus dem goettinger arbeitskreis 48) – x/330p (ill) – 1 – (incl bibl ref and ind) – mf#8098 reel 8 – us Wisconsin U Libr [610]
Aerztliche missionen / Christlieb, Theodor – Guetersloh: C Bertelsmann, 1889 [mf ed 1990] – 1mf – 9 – 0-7905-5454-2 – (incl bibl ref) – mf#1988-1454 – us ATLA [610]
Aerztliche praxis – Munich, Germany 1977-79 – 1 – ISSN: 0001-9534 – mf#9843 – us UMI ProQuest [610]
Aerztliches standesrecht : eine darstellung fuer klinik und praxis / Ratzel, Rudolf – Frankfurt/Main: Kommentator Verlag, 1990 (mf ed 1996) – 2mf – 9 – €31.00 – 3-8267-9685-3 – mf#DHS 9685 – gw Frankfurter [340]
Aerztliches vereinsblatt fuer deutschland – Potsdam DE, 1872-82 – 2r – 1 – gw Misc Inst [610]
Aeschylus *see*
– Perses
– Prometheus bound
Aescoly, Aaron Zeev *see* Recueil de textes falachas
Aesculapian register – La Salle 1824 – 1 – mf#4408 – us UMI ProQuest [610]
Aesopi et aviani fabulae / physiologus (cima48) : farbmikrofiche-edition der handschrift hamburg, staats- und universitaetsbibliothek, cod 47 in scrinio – (mf ed 2003) – (= ser Codices illuminati medii aevi (cima) 48) – 90p on 3 color mf – 15 – €240.00 – 3-89219-048-8 – (description & ind by helga lengenfelder) – gw Lengenfelder [390]
The aesthetic and miscellaneous works of frederick von schlegel : comprising letters on christian art, an essay on gothic architecture... / Schlegel, Friedrich von – London: HG Bohn, 1849 [mf ed 1990] – (= ser Bohn's standard library) – 2mf – 9 – 0-7905-7610-4 – (english trans by e j millington) – mf#1989-0835 – us ATLA [802]
Aesthetic as science of expression and general linguistic = Estetica come scienza dell'espressione e linguistica generale / Croce, Benedetto – London, New York: Macmillan, 1909 [mf ed 1990] – 1mf – 9 – 0-7905-3776-1 – (english trans fr italian by douglas ainslie) – mf#1989-0269 – us ATLA [110]
Aesthetic papers – 9 – us Scholars Facs [700]
Aesthetic papers – 1849 – 1 – us AMS Press [110]
Aesthetic plastic surgery – Dordrecht, Netherlands 1981-96 – 1,5,9 – ISSN: 0364-216X – mf#13129 – us UMI ProQuest [617]
Aesthetic standards in old time dancing in southwest virginia : african-american and european-american threads / Spalding, Susan E & Dixon-Gottschild, Brenda – 1993 – 4mf – 9 – $16.00 – us Kinesology [790]
Aesthetic surgery journal – Oxford, England 1998+ – 1,5,9 – ISSN: 1090-820X – mf#22105,02 – us UMI ProQuest [617]
Aesthetica. nociones de la belleza y de las artes / Gomez Bravo, Vicente – Madrid: Editorial Razon y Fe, 1934 – 1 – sp Bibl Santa Ana [700]
Aesthetics in bridge design / Young, Clarence Richard – [S.l: s.n, 1911?] [mf ed 1991] – 1mf – 9 – 0-665-99504-0 – mf#99504 – cn Canadiana [624]
Aesthetik des jahrmarkts : utopie in der gegenwart? / Kletzka, Renate – (mf ed 1995) – 1mf – 9 – €30.00 – 3-8267-2268-X – mf#DHS 2268 – gw Frankfurter [700]
Aesthetische feldzuege : dem jungen deutschland gewidmet / Wienbarg, Ludolf; ed by Dietze, Walter – Hamburg: Hoffmann & Campe 1834 [mf ed 1992] – 1r – 1 – (incl bibl ref. filmed with: studien zur erbetheorie und erbeaneignung / claus traeger) – mf#3256p – us Wisconsin U Libr [430]
Das aesthetische programm in goethes schriften zur meteorologie / Kaufmann, Dorothee – (mf ed 1997) – 1mf – 9 – €30.00 – 3-8267-2411-9 – mf#DHS 2411 – gw Frankfurter [700]
Aesthetische rundschau – Vienna, Austria jan 1866-may 1867 [mf ed Norman Ross] – 1r – 1 – mf#nrp-1923 – us UMI ProQuest [700]
Aesthetische wahrnehmung als urspruengliche erkenntnis : eine kunstphilosophische studie zum werk von james turrell / Schuermann, Eva – (mf ed 1998) – 2mf – 9 – €40.00 – 3-8267-2576-X – mf#DHS 2576 – gw Frankfurter [110]
Die aesthetischen konzeptionen von john keats und g.w.f. hegels vorlesungen ueber die aesthetik : zur dialektik des aesthetischen / Scholze, Werner – Frankfurt a.M., 1975 – 2mf – 9 – 3-89349-671-8 – gw Frankfurter [110]
Aeternitatis prodromus mortis nuntius... / Drexelius, H – Coloniae, Antverpia, 1633-45. 3v – 6mf – 9 – mf#O-1560 – ne IDC [090]
Aethiopie, empire des negres blancs / Liano, A – Paris, [1929] – 4mf – 9 – mf#NE-20224 – ne IDC [960]
Die aethiopische bibeluebersetzung : ihre herkunft, art, geschichte und ihr wert fuer die alt- und neutestamentliche wissenschaft / Heider, August – Leipzig: Eduard Pfeiffer, 1902 – 1mf – 9 – 0-8370-3547-3 – mf#1985-1547 – us ATLA [220]
Die aethiopische uebersetzung des zacharias, erstes heft : text zum ersten male herausgegeben, prolegomena, commentar / Kramer, Friedrich Oswald – Leipzig: Doerffling & Franke, 1898 – 1mf – 9 – 0-8370-3991-6 – (incl bibl ref) – mf#1985-1991 – us ATLA [221]
Aethiops *see* Bulletin ge'ez
Aethiopum servus : a study in christian altruism / Petre, Maude Dominica – London: Osgood, McIlvaine, 1896 [mf ed 1990] – 1mf – 9 – 0-7905-6821-7 – mf#1988-2821 – us ATLA [241]
Aetiopathogenetische vorstellung des lumbago-ischialgie-syndroms in der deutschen medizinischen wochenschrift von 1900 bis 1991 : entstehung eines paradigmas / Lutz, Gabriele K – (mf ed 1995) – 3mf – 9 – €49.00 – 3-8267-2223-X – mf#DHS 2223 – gw Frankfurter [616]
The aetna – Montreal: W H Orr, [1868-1911] – 9 – mf#P04962 – cn Canadiana [360]
Aetude sur la naturalisation en algerie / Rouard de Card, E – Paris, 1881 – 1mf – 9 – mf#ILM-3148 – ne IDC [960]
Aetudes critiques sur divers textes des 10 et 11 siecles / Lair, J – Paris, 1899. 2v – 15mf – 9 – mf#H-2986 – ne IDC [956]
Aetudes juridiques du probleme de l'egypte / Saleh Hussein, I – Paris, 1931 – 4mf – 9 – mf#ILM-1932 – ne IDC [956]
Af argentiner erd / Alpersohn, Marcos – Buenos Ayres, Argentina. 1931 – 1r – 1 – us UF Libraries [939]
Af di vegn zu der nayer shul – 1924-1928 – 1 – us NY Public [073]
Af en endnu levendes papirer / Kierkegaard, Soeren – Kobenhavn: CA Reitzel, 1838 [mf ed 1990] – 1mf – 9 – 0-7905-3786-9 – mf#1989-0279 – us ATLA [190]
Af literarishe temes / Dunets, Kh – Minsk, Belarus. 1934 – 1r – 1 – us UF Libraries [939]
Af of l auto worker / United Automobile Workers of America International Union – 1939-55 – (= ser Labor union periodicals, pt 1: the metal trades) – 2r – 1 – $430.00 – 1-55655-225-4 – us UPA [331]
Afa organizer – 1940 may – 1 – mf#3910318 – us WHS [071]
Afa-bundeszeitung *see* Mitteilungsblatt der arbeitsgemeinschaft freier angestelltenverbaende
Afak / ed by Kamil, A – Istanbul: Mahmut Bey Matbaasi, Matbaa-i Osmaniye, 1882-83. n1-7 (20 zilhicce 1299-23 cemazielevvel 1300 [1882-83]) – (= ser O & t journals) – 3mf – 9 – $55.00 – us MEDOC [956]
Afanador, Gonzalo *see* Derechos y garantias del procesado
Afanasev, A N *see* Russkie satiricheskie zhurnaly
Afanas'ev, IA A *see* Semirechenskie oblastnye vedomosti
Afanas'ev, N I *see* Sovremenniki
Afar / American Friends of the Angolan Revolution – [New York]: AFAR, n2-3. apr-jun 1970 – us CRL [071]
Die afar sprache / Reinisch, L – Wien, 1885-1887. 3v – 1mf – 9 – (missing: 1886-1887 v1-2) – mf#NE-20256 – ne IDC [956]
Afbeelding der menschelyke bezigheden : bestaande in hondert onderscheiden printverbeeldingen... / Luyken, Jan & Luyken, Caspar – Amsterdam: Reinier en Josua Ottens, n.d. – 2mf – 9 – mf#O-3115 – ne IDC [090]
Afbeelding van 't stadt huys van amsterdam... / Campen, J van – Amsterdam, 1661 – 3mf – 9 – mf#OA-86 – ne IDC [720]
Afbeeldingen van minne : emblemata amatoria, emblemes d'amour / [Heinsius, D] – Leyden: I Marcusz, 1613 – 2mf – 9 – (titlepg missing) – mf#O-3079 – ne IDC [090]
Af-beeldinghe van d'eerste eeuwe der societyt iesu : voor ooghen ghestelt door de duyts-nederlantsche provincie der selver societeyt – t'Antwerpen: Inde Plantijnsche Druckerije, 1640 – 13mf – 9 – mf#O-646 – ne IDC [090]
Afbeelsels der voornaemste gebouwen uyt alle die philips vingboons geordineert heeft / Vingboons, P – Amstelerdam, 1648 – 4mf – 9 – mf#OA-290 – ne IDC [720]
Afectos espirituales de la venerable madre y obser / Castillo Y Guevara, Francisca Josefa De – Bogota, Colombia. v1-2. 1956 – 1r – 1 – us UF Libraries [972]
A-feng / Ling, Hsi – Shang-hai: Chung-hua, Min kuo 20 [1931] – (= ser P-k&k period) – us CRL [480]
Afevork, G J *see* Guide du voyageur en abyssinie
Affable savages / Huxley, Francis – New York, USA. 1957 – 1r – 1 – us UF Libraries [972]
L'affaire clerfeyt. / Clerfeyt, Joseph Maximilian Louis – Bruxelles: Vanderauwera, 1873 – 354p – 1 – mf#LL-4030 – us L of C Photodup [340]
L'affaire de la baie des chaleurs devant la commission royale : appreciation de la preuve – Quebec?: s.n, 1891 – 1mf – 9 – mf#02578 – cn Canadiana [380]
Affaire de la consolidation... / Haiti. Departement De La Justice – Port-Au-Prince, Haiti. 1906 – 1r – 1 – us UF Libraries [972]
Affaire de w a grenier, proprietaire du journal "la libre parole", accuse de libelle par l'honorable j israel tarte, ministre des travaux publics : plaidoyer de mtre h c st-pierre, c r pour la poursuite... – Montreal: C Theoret, 1897? – 2mf – 9 – mf#13125 – cn Canadiana [347]
Affaire des fourrures – [S.l: s.n, 18–?] [mf ed 1980] – 1mf – 9 – 0-665-02579-3 – mf#02579 – cn Canadiana [380]
Affaire des tanneries : seance du 25 novembre 1875 – [S.l: s.n, 1875?] [mf ed 1980] – 1mf – 9 – 0-665-02582-3 – mf#02582 – cn Canadiana [971]
Affaire d'honneur / Hibbert, Fernand – Port-Au-Prince, Haiti. 1916 – 1r – 1 – us UF Libraries [972]
Affaire d'or / Mongrolle – Montmartre, France. 1854 – 1r – 1 – us UF Libraries [440]
L'affaire dreyfus : ligue francaise pour la defense des droits de l'homme et du citoyen – Paris, 1906 – 2v – 1 – (1908 3v LL-4034) – mf#LL-4035 – us L of C Photodup [345]
Affaire du dock de bizoton / Terlonge, Windsor – Port-Au-Prince, Haiti. 1909 – 1r – 1 – us UF Libraries [972]
Affaire guibord : dame brown, appelante vs la fabrique de montreal, intimee / Brown, Henriette – [Montreal?: s.n.], 1871 [mf ed 1986] – 1mf – 9 – 0-665-10538-X – (incl english text) – mf#10538 – cn Canadiana [340]
Affaire guibord : discours de f x a trudel, ecr, prononce les 28 et 29 mars et le 1er avril 1870 / Trudel, Francois-Xavier-Anselme – [Montreal?: s.n.], 1870 [mf ed 1981] – 1mf – 9 – 0-665-24874-1 – mf#24874 – cn Canadiana [340]
Affaire guibord : jugement de l'hon juge johnson / Johnson, Francis – [Montreal? : s.n. 1875?] [mf ed – 1mf – 9 – 0-665-93858-6 – mf#93858 – cn Canadiana [340]
Affaire guibord : jugement des lords du comite judiciaire du conseil prive sur l'appel de dame henriette brown vs les cure et marguilliers de l'oeuvre et fabrique de notre-dame de montreal, au canada, prononce le 21 nov 1874 / Grande-Bretagne. Privy Council – Londres?: s.n, 1874? [mf ed 1981] – 1mf – 9 – 0-665-04380-5 – mf#04380 – cn Canadiana [340]
Affaire guibord : question de refus de sepulture / Brown, Henriette – Montreal: des presses a vapeur de La Minerve, 1870 [mf ed 1991] – 2mf – 9 – (int by oscar dunn) – mf#SEM105P1449 – cn Bibl Nat [917]
Affaire guibord : question de refus de sepulture... – Montreal: Presses a vapeur de la Minerve, 1870 [mf ed 1979] – 3mf – 9 – 0-665-00043-X – (incl english text) – mf#00043 – cn Canadiana [340]
Affaire guibord : question de refus de sepulture: rapport de la cause avec le texte du jugement de son honneur le juge mondelet – [Montreal?: s.n.], 1870 [mf ed 1983] – 3mf – 9 – 0-665-29053-5 – mf#29053 – cn Canadiana [340]
Affaire haitiano-dominicaine – Port-Au-Prince, Haiti. 193-? – 1r – 1 – us UF Libraries [972]
Affaire luders / Menos, Solon – Port-Au-Prince, Haiti. 1898 – 1r – 1 – us UF Libraries [972]
Affaire maunder / Cauvin, Leger – Port-Au-Prince, Haiti. 1887 – 1r – 1 – us UF Libraries [972]
Affaire riel : discours de l'honorable e j flynn, solicitor-general: prononce devant l'assemblee legislative le 29 avril 1886 – Quebec?: s.n, 1886? – 1mf – 9 – mf#30194 – cn Canadiana [971]
Affaire shortis : presidence de l'hon juge mathieu; plaidoyer de mtre h c saint-pierre, cr pour la defence de valentine shortis accuse de meurtre / Saint-Pierre, Henri Cesaire – Montreal?: C O Beauchemin, 1896 – 6mf – 9 – mf#13660 – cn Canadiana [345]
Affaire-pelletier : la reine vs. prudent pelletier, proces pour meurtre, novembre 1853 – Quebec?: A Cote, 1853 – 2mf – 9 – mf#22415 – cn Canadiana [345]
Les affaires : hebdomadaire d'information financiere, industrielle et commerciale – Montreal, Canada 12 jun 1972, 28 oct 1974-5 jan 1996 – 1 – uk British Libr Newspaper [071]
Affaires communales / Lamy, Amilcar F – Port-Au-Prince, Haiti. 1950 – 1r – 1 – us UF Libraries [972]
Affaires de la plata : extrait de la correspondance de m eugene guillemot, pendant sa mission dans l'amerique du sud – [Paris?: s.n.] 1849 [mf ed 1984] – 1mf – 9 – 0-665-44938-0 – mf#44938 – cn Canadiana [327]
Affaires d'haiti (1883-1884) / Janvier, Louis Joseph – Paris, France. 1885 – 1r – 1 – us UF Libraries [972]
Affaires entre la ville et la fabrique de longueuil concernant les taxes sur l'eglise – Montreal: [s.n.], 1888 [mf ed 1980] – 1mf – 9 – 0-665-04006-7 – mf#04006 – cn Canadiana [971]
Affairs – v1 n1-v3 n1 [1982 mar-1984] ns: v1 n1 [1985 may] – 1 – mf#1032727 – us WHS [071]
The affairs of a tribe : a study in tribal dynamics / Majumdar, Dhirendra Nath – Lucknow: Published for the Ethnographic and Folk Culture Society, UP by Universal Publishers, 1950 – (= ser Samp: indian books) – us CRL [307]
The affairs of the canadas : in a series of letters / Ryerson, Egerton – London?: s.n, 1837 (London: J King) – 1mf – 9 – mf#40621 – cn Canadiana [971]
Affaitati, C *see*
– L'ortolain villa e l'accurato giardiniere in citta...
– L'ortolano in villa e l'accurato giardiniere in citta...
L'affame – n1-6. Marseille. mai-aout 1884 – 1 – fr ACRPP [073]
Affani crudeli : a favorite song, sung by madame mara, at the king's theater, hay market, in the serious opera gli giochi d'agrigento [a pasticcio] = Giochi d'agrigento / Federici, Vincenzo – London: T Skillern [1793?] [mf ed 1988] – 1r – 1 – mf#pres. film 40 – us Sibley [780]
Affarsvarlden – Stockholm, Sweden 1977-79 – 1,5,9 – mf#10230 – us UMI ProQuest [332]
Affectionate invitation to the holy communion – London, England. 18– – 1r – 1 – us UF Libraries [240]
Affective and cognitive performance due to exercise training : an examination of individual different variables / Lochbaum, Marc R – 1998 – 219p on 3mf – 9 – $15.00 – mf#PSY 2162 – us Kinesology [613]
Affectos divinos con emblemas sagradas por el p. po de salas de la compania de jesus... / Hugo, H – Valladolid: por Greo. de Bedoya, 1658 – 7mf – 9 – mf#O-1891 – ne IDC [090]
Affectus amantis christum iesum : seu, exercitium amoris erga dominum iesum pro tota hebdomada / Chastelain, Pierre – Paris: Denis Bechet, 1648 [mf ed 1995] – 6mf – 9 – mf#SEM105P956 – cn Bibl Nat [241]
Die affen des grossen friedrich : oder, eine geschichte von handel und fahne / Bruees, Otto – Berlin: G Grote, 1944 [mf ed 1989] – (= ser Grote'sche sammlung von werken zeitgenoessischer schriftsteller 243) – 227p – 1 – mf#7090 – us Wisconsin U Libr [830]
Die affenschande : deutsche satiren von sebastian brant bis bertolt brecht / ed by Berger, Karl Heinz – 2. aufl. Berlin: Eulenspiegel, 1969 – 499p (ill) – 1 – (incl bibl ref. ill by renate totzke-israel) – us Wisconsin U Libr [870]
Affiches alsaciennes = Elsaessischer anzeiger – Colmar / Elsass (F), 1881-87, 1889-1897 27 mar [gaps] – fr ACRPP [074]
Affiches, annonces, avis divers de toulouse et du haut-languedoc – Toulouse. 1785-93 – 1 – (devenu: journal universel et affiches de toulouse et du languedoc; devenu: journal et affiches du department de haute-garonne) – fr ACRPP [073]
Affiches, annonces et avis divers de jever *see* Jeverische woechentliche anzeigen und nachrichten
Affiches, annonces et avis divers de lubek *see* Luebeckische anzeigen
Affiches de strasbourg – Strasbourg (Strasbourg F), 1789-92, 1799 23 sep-1801 23 aug – 1 – (with gaps; title varies: later: petites affiches de strasbourg, 23 sep 1800: feuille decadaire du bas-rhin [bilingual]) – fr ACRPP; gw Misc Inst [074]
Affiches du bas-rhin. niederrheinische anzeigen – Strassburg (Strasbourg F), 1797 23 sep-1798 17 mar, 1799 1 feb-21 feb, 1808 2 apr-29 jun [gaps] – 1 – fr ACRPP [074]
Affilia – Newbury Park CA 1988+ – 1,5,9 – ISSN: 0886-1099 – mf#17050 – us UMI ProQuest [360]
Affiliate – 1993 jan-1998 may – 1 – mf#3240662 – us WHS [071]
Les affinites electives de goethe : essai de commentaire critique / Francois-Poncet, Andre – Paris: F Alcan, 1910 – 1r – 1 – (incl bibl ref) – us Wisconsin U Libr [430]
Affirmations – v1-9. 1974-82 [complete] – 1r – 1 – (supersedes: sisterhood) – mf#ATLA S0343B – us ATLA [240]

The affirmative intellect : an account of the origin and mission of the american spirit / Ferguson, Charles – New York: Funk & Wagnalls, 1901 – 1mf – 9 – 0-7905-8649-5 – mf#1989-1874 – us ATLA [301]
Affirmez vous! : petit guide d'entraînement aux aptitudes sociales / Alberti, Robert Edward & Emmons, Michael L – Saint-Hyacinthe: Edisem, 1978 [mf ed 1996] – 2mf – 9 – (trans by wilfrid pilon. original title: your perfect right [san luis obispo, us: impact 1974]) – mf#SEM105P2644 – cn Bibl Nat [150]
Afflicted saviour rising from the depths in the garment of praise / Ferguson, Archibald – Alyth, Scotland. 1869 – 1r – 1 – us UF Libraries [240]
The afflictions of the righteous : as discussed in the book of job and in the light of the gospel / Macleod, William B – London, New York: Hodder & Stoughton, [1913] – 1mf – 9 – 0-8370-6272-1 – mf#1986-0272 – us ATLA [220]
Affolter, Ludwig *see* Ueber die vernehmung der parteien als zeugen im civilprozess.
Affonse Celso de Assis Figueiredo *see* El emperador d pedro 2 y el instituto historico (5)
Affonso arinos / Lima, Alceu Amoroso – Rio De Janeiro, Brazil. 1922 – 1r – 1 – us UF Libraries [972]
Affonso Celso, Affonse Celso De Assis Figueiredo *see* Oito annos de parlamento
Affonso Celso, Affonso Celso De Assis Figueiredo *see* Visconde de ouro preto
Affonso Celso, Affonso Celso de Assis Figueiredo, conde de *see* Porque me ufano do meu paiz
L'affranchi – [Paris]: Impr Schiller, apr 2-4, 6-7, 9-10, 12-13, 15, 17, 19-20, 23-25 1871 – 2r – 1 – (Filmed as pt of: Commune de Paris newspapers. Newspapers on these reels are filmed chronologically, not alphabetically) – us CRL [074]
Affray at brownsville, august 13 and 14 1906, court-martial, macklin : proceedings of a general court-martial convened at headquarters, department of texas, april 15 1907, in the case of captain edgar a macklin – Washington: GPO, 1907 – 3mf – 9 – $4.50 – mf#LLMC 97-006 – us LLMC [347]
Affreuse tentative de corruption / Hilbey, Constant – Paris: Bureau du Journal de Sans-Culottes [1849] – 1r – 1 – us CRL [360]
Afge collective bargaining report – 1980 oct 31-1985 jan 31 – 1 – mf#1054275 – us WHS [331]
Afge local 1336 newsletter – 1994 jul-dec – 1 – mf#4358155 – us WHS [071]
Afgezet naar recht en waarheid : een antwoord op de brochure van h wierenga... / Fortuin, K W – Grand Rapids, MI: Wm B Eerdmans, 1925 [mf ed 1993] – (= ser Reformed church coll) – 1mf – 9 – 0-524-06179-3 – (in dutch) – mf#1991-2435 – us ATLA [242]
Afghan mellat – Kabul, Afghanistan 1970-72 [mf ed Norman Ross] – 1 – mf#nrp-575 – us UMI ProQuest [079]
Afghan serials : decades of coverage from afghan newspapers and periodicals – 1931-1988 [mf ed Norman Ross Publ] – 4 titles on 79r – 1 – (individual titles listed separately) – us UMI ProQuest [079]
Afghanistan : the making of us policy, 1973-1990 – [mf ed Chadwcyk-Healey] – (= ser National security archive, washington dc: the making of us policy) – 15,000+ p on 424mf – 9 – (with 2v p/g & ind) – uk Chadwyck [327]
Afghanistan *see* Rasmi jarida
Afghanistan and south africa / Frere, Bartle – Pretoria, South Africa. 1969 – 1r – 1 – us UF Libraries [960]
Afghanistan. Ministry of Commerce *see* Exports of merchandise from afghanistan 1959/60-1973/74
Afgoden-dienst der jesuiten in china : waar over fy nog heden beschuldigt worden aan het hof van romen / Mauritius, Joannes – Amsterdam: Jacobus Borstius, 1711 [mf ed 1995] – (= ser Yale coll) – 626p (ill) – 1 – 0-524-09052-1 – (in dutch) – mf#1995-0052 – us ATLA [241]
Afgoderye der oost-indische heydenen / Baldaeus, Philippus; ed by Jong, Albert Johannes de – 'S-Gravenhage: M Nijhoff, 1917 [mf ed 1992] – 1mf – 9 – 0-524-03112-6 – (in dutch) – mf#1990-3165 – us ATLA [280]
Afhandlinger og foredrag om menigheden / Sverdrup, Georg; ed by Helland, Andreas – Minneapolis MI: Frikirkens Boghandels Forlag 1910 [mf ed 1993] – 1mf – 9 – 0-524-06322-2 – mf#1991-2495 – us ATLA [242]
Afike yehudah / Edil, Yehudah Leyb – Lvov, Ukraine. 1912 – 1r – 1 – us UF Libraries [939]
Afirmacoes nacionalistas / Melo, Mario – Rio De janeiro, Brazil. 1942 – 1r – 1 – us UF Libraries [972]
Afisha tim = Poster "tim" – Moscow. n1-4. 1926-27 – 3mf – 9 – us UMI ProQuest [790]
Afiyet – Istanbul: Luesyen Matbaasi, 1913-14 – (= ser O & t journals) – 3mf – 9 – $150.00 – (added title: dilber kontes. sahib-i imtiyaz ve mueduerue: sisak ferid, muharriri: avanzade m suleyman n1-62 (24 tesrinievvel 1329-24 kanunisani 1330 [1913-14])) – us MEDOC [956]
Afkar : unique literary, cultural and family journal – Karachi: Maktab-i-Afkar – 8r – 1 – mf#1116 – us Wisconsin U Libr [490]
Afl auto worke – 1939 sep 13-1950; 1951-1956 feb – 1 – mf#1109401 – us WHS [625]
Afl auto worker – 1956 mar-jun – 1 – mf#1096046 – us WHS [625]
Afl cannery reporter – v1 n1-v8 n10 [1945 nov9-1954 may] – 1 – mf#1109404 – us WHS [660]
Afl cannery reporter, 1949-1954 / united dairy farmer, 1941, 1944-1945 / California State Council of Cannery Unions, Teamsters & United Mine Workers – (= ser Labor union periodicals, pt 3: food and agricultural industries) – 1r – 1 – $210.00 – 1-55655-613-6 – us UPA [660]
Afl news-reporter – 1951 dec 5-1954; 1955 – 1 – mf#969941 – us WHS [071]
Afl rank-and-file federationist – New York. v1-2. 1934-35 – (= ser Radical periodicals in the united states, 1881-1960. series 2) – 1r – 1 – $115.00 – us UPA [320]
Afl weekly news service – v41 n1-1948 [n2073-2120; 1951 jan 2-nov 27] – 1 – mf#859478 – us WHS [071]
Aflak – Tehran. sal-i 1, shumarah-'i 8-14. 6 murdad-8 shahrivar 1304 [28 jul-30 aug 1925] – 1r – 1 – $53.00 – (r incl: khalq, nahid, and sitarah-'i subh; cont by: khalq) – us MEDOC [079]
Aflak *see*
– Nahid
– Sitarah-'i subh
AFL-CIO *see*
– Education update
– F and b topics
AFL-CIO [American Federation of Labor and Congress of Industrial Organizations] *see*
– Collective bargaining report
– Community
– Coordinated collective bargaining quarterly [cbq]
– Correio operario norteamericano
– In public service
– Insight
– Iud coordinated bargaining quarterly
– Iud spotlight on health and safety
– King county labor news
– Labeletter
– Labor and investments
– Labor chronicle
– Labor law reform
– Labor looks at the...congress
– Labor news conference
– Labor world
– Maritime in the nation's press
– Memo from cope and afl-cio legislative alert!
– Metaletter
– Milwaukee labor press
– News
– Newsletter
– Nordamerikanische arbeiter nachrichten
– Noticiero obrero norteamericano
– Notiziario sindacale del norte america
– Nouvelles des syndicats nord-americains
– Policy resolutions adopted by the constitutional convention
– Political memo from cope
– Record of proceedings of the milwaukee county labor council, afl-cio
– Scanner
– Scanner and king county labor news
– South carolina labor news
– Spotlight
– Statistical and tactical information report
– Viewpoints mscl
AFL-CIO [American Federation of Labor and Congress of Industrial Organizations]. Library *see* American federation of labor and congress of industrial organizations
AFL-CIO [American Federation of Labor-Congress of Industrial Organizations] *see*
– Builders
– Digest
AFL-CIO [American Federation of Labor-Congress of Industrial Organizations] Community File [San Francisco CA] *see* Dateline
Afl-cio american federationist / American Federation of Labor – Silver Springs MD 1977-82// – 1,5,9 – (cont: american federationist) – ISSN: 0149-2489 – mf#2232.01 – us UMI ProQuest [331]
Afl-cio education news and views – 1956 jan-1958 dec; 1959-1961 apr – 1 – mf#1109402 – us WHS [370]
Afl-cio free trade union news – v30 n1-v39 n9 [1975 jan-1984 sep] – 1 – mf#781714 – us WHS [331]
Afl-cio freigewerkschaftliche nachrichten – jan 1977-dez 1979 – 1 – mf#348747 – us WHS [331]
AFL-CIO Human Resources Development Institute *see* Hrdi advisory
Afl-cio international affairs bulletin – 1956 aug-1957 oct – 1 – mf#1051447 – us WHS [327]
Afl-cio legislative alert – 1984 feb 20-1994 oct 17 – 1 – mf#991850 – us WHS [071]
Afl-cio news – 1955 dec 10-1956; 1957 jan 5-1958 dec 27; 1959-66; 1967 jan 7-1968 dec 21; 1969 jan 4-1970 dec 26; 1971 jan 9-1972 dec 23; 1973 jan 6-1975 jun 28; 1975 jul 5-1977 dec 24; 1978 jan 7-1979 sep 29; 1979 oct-1980 dec; 1981-84; 1985 jan 5-1987 jun; 1987 jul 4-1990 dec; 1991-93; 1994-1996 oct – 1 – mf#775321 – us WHS [071]
Afl-cio news – 1956 jan-jan 7 – 1 – mf#2467655 – us WHS [071]
Afl-cio news – Silver Springs MD 1955-96 – 1 – (cont by: america @ work) – ISSN: 0001-1185 – mf#3199 – us UMI ProQuest [331]
Afl-cio news – v1 n6, 12, 16 [1956 jan 14, feb 25, mar 24] – 1 – mf#1051450 – us WHS [071]
Afl-cio noticiario do sindicalismo livre – 1978 aug-1979 dec; 1980 aug-1984 jul/aug – 1 – mf#638925 – us WHS [071]
Afl-cio noticiero del movimiento sindical libre – 1978 mar-1979 dec; 1980 apr-1984 jul/aug – 1 – mf#638928 – us WHS [071]
Afl-cio notiziario del movimento sindacale libero – v33 n9-v34 n12 [1978 sep-1979 dec] – 1 – mf#638930 – us WHS [071]
Afl-cio nouvelles des syndicats libres – v33 n1-v34 n12 [1978 jan-1979 dec] – 1 – mf#638923 – us WHS [071]
Afl-cio press – 1960 jan-1961 jun – 1 – mf#3215189 – us WHS [071]
Afl-cio proceedings – Silver Springs MD 1955-95 – 1,5,9 – ISSN: 0569-4515 – mf#3284 – us UMI ProQuest [331]
Afl-cio publications – Silver Springs MD 1956-94 – 1,5,9 – ISSN: 0569-4523 – mf#9494 – us UMI ProQuest [331]
Afonskie listki – nos1-350. 1899-1906 (complete) – (= ser Corpus of russian orthodox periodicals) – 1r – 1 – mf#ATLA S0193A – us ATLA [243]
Afonskii paterik ili zhizneopisaniia sviatykh, na sviatoi afonskoi gore prosiiavshchikh / Azariia, Monakh – 1876 – 2v 10mf – 9 – mf#R-18258 – ne IDC [243]
Afonso, Jose Nuno De Sousa *see* Tabelas para a conversao de coordenadas geograficas no elipsoide
Aforismos de luz y caballero / Garcia Barcena, Rafael – Habana, Cuba. 1945 – 1r – 1 – us UF Libraries [972]
Aforismos de luz y caballero / Luz Y Caballaero, Jose De La – Habana, Cuba. 1960 – 1r – 1 – us UF Libraries [972]
Aforismos y apuntaciones / Luz Y Caballero, Jose De La – Habana, Cuba. 1945 – 1r – 1 – us UF Libraries [972]
Aforos en la sierra de montanchez. ano 1818 / Abastecimiento de aguas – Caceres: Tip. El Noticiero, 1818 – 1 – sp Bibl Santa Ana [946]
Aforos practicados en las cuencas de los rios guadiana – 1881 – 9 – sp Bibl Santa Ana [910]
Afp exchange – Bethesda MD 1999+ – 1 – (cont: tma journal) – ISSN: 1528-4077 – mf#15782,02 – us UMI ProQuest [332]
Afram communique – 1985 jul 3 [repr 1997]; 1987 mar, apr 1-2, dec 14; 1988 jan, jan; 28, mar 3, apr 14; 1999 mar 6 [repr] – 1 – mf#4851164 – us WHS [071]
Afram newsletter – 1975 autumn-1984; 1985-95 – 1 – mf#3398378 – us WHS [071]
Aframerican report – v1 n3 [undated] – 1 – mf#3170482 – us WHS [071]
Afranio peixoto / Ribeiro, Leonidio – Rio De Janeiro, Brazil. 1950 – 1r – 1 – us UF Libraries [972]
Africa : antropologia della stirpe camitica (specie eurafricana) / Sergi, Giuseppe – Torino: F Bocca, 1897 – 9 – us CRL [306]
Africa : being an accurate description of the regions of aegypt, barbary, lybia, and billedulgerid, the land of negroes, guinee, aethiopia, and the abyssines... / Ogilvy, J – London, 1670 – 61mf – 9 – mf#A-126 – ne IDC [916]
Africa / Conder, Josiah – London 1829 [mf ed Hildesheim 1995-98] – 3v on 8mf – 9 – €160.00 – 3-487-27395-0 – gw Olms [960]
Africa : containing a description of the manners and customs, with some historical particulars of the moors of the zahara, and of the negro nations between the rivers senegal and gambia / ed by Shoberl, Frederick – London [1821] [mf ed Hildesheim 1995-98] – 4v on 8mf – 9 – €160.00 – 3-487-27387-X – gw Olms [960]
Africa : drame en cinq actes, en vers / Deschamps, Edouard – Paris: E Dentu, 1893 – 9 – us CRL [820]
Africa – Lisboa: Cultura Nacional Editora [may 14 1932-apr 22 1933;jan 25 1934] (wkly) – 1r – 1 – us CRL [079]
Africa / Molnar, Thomas Steven – New York, USA. 1965 – 1r – 1 – us UF Libraries [960]
Africa : organe independent le plus africain de defense des interets des peuples africaines – Paris: T-G Konyate, [dec 1 1935;jun 1936-aug/sep 1938] (mthly) – 1r – 1 – us CRL [321]
Africa : orgao oficial do movimento nacionalista african – Lisboa, Portugal: Industrias Graficas, [ser.8 v21 n882 (1931)] – (= ser Coll of lusophone african newspapers and serials) – 1r – 1 – us CRL [960]
Africa : revista quinzenal de cultura e propaganda colonial – Lourenco Marques: A Alves da Silva, Manuel Santana [v1 n1-2(feb-mar 1936)] (mthly) – (= ser Coll of lusophone african newspapers and serials) – 1r – 1 – us CRL [079]
Africa / Suggate, Leonard Sydney – London, England. 1929 – 1r – 1 – us UF Libraries [960]
Africa : through its own music – Transvaal, South Africa. 19–? – 1r – 1 – us UF Libraries [780]
Africa : to-day and to-morrow / African Academy of Arts and Research. New York – New York. April 1945 – 1 – us NY Public [960]
Africa – Washington DC 1931-69 – 1 – ISSN: 0065-3802 – mf#6661 – us UMI ProQuest [200]
Africa / Woddis, Jack – London, England. 1961 – 1r – 1 – us UF Libraries [960]
Africa, 1928-81 : the journal of the international african institute – 19r 74mf – 1,9 – mf#364 – uk Microform Academic [960]
Africa, 1941-1961 / U.S. Office of Strategic Services & U.S. State Dept – (= ser Oss/state department intelligence and research reports 13) – 11r – 1 – $1690.00 – (with p/g) – us UPA [327]
Africa, 1946-1976 / U.S. Central Intelligence Agency – (= ser Cia research reports) – 3r – 1 – $480.00 – 0-89093-423-1 – (with p/g) – us UPA [327]
Africa a traves del pensamiento espanol / Flores Morales, Angel – Madrid, Spain. 1949 – 1r – 1 – us UF Libraries [960]
Africa advancing / Davis, Jackson – New York, USA. 1945 – 1r – 1 – us UF Libraries [960]
Africa and asia : portuguese manuscript maps in the rhodes library at groote schuur – Pretoria: National Film Board 1979 – 16mf – 9 – 0-7989-0077-6 – (int in portuguese & english entitled: ancient portuguese maps in the collection at groote schuur / by a teixeira da mota; text in english & portuguese; maps chiefly by joao teixeira albernas ii, c1680; originally iss in portuguese [1977]) – mf#mf.979 – sa State Libr [520]
Africa and middle east : background reports. world communist movement 7.3.1958-17.5.1972 / Radio Free Europe – 5mf – 9 – mf#R-17087 – ne IDC [956]
Africa and the american negro : addresses and proceedings of the congress on africa held under auspices of the stewart missionary foundation for africa of gammon theological seminary..cotton states and international exposition – dec 13-15 1895 – 9 – us CRL [960]
Africa and the brussels geographical conference / Banning, Emile Theodore – London: Low, Marston, Searle & Rivington, 1877 – 9 – us CRL [916]
Africa and world peace / Padmore, George – London: Martin Secker and Warburg, 1937 (mf ed 1977) – 1r – 1 – mf#ZZ-15285 – us NY Public [327]
Africa as i have known it / Maugham, Reginald Charles Fulke – New York, USA. 1969 – 1r – 1 – us UF Libraries [960]
Africa company town : the social history of a wartime planning experiment / Ericksen, Ephraim Gordon – Dubuque, Iowa: W C Brown Book Co [1964] – 1r – 1 – us CRL [307]
Africa contra el colonialismo / Bayo, Armando – Habana, Cuba. 1962 – 1r – 1 – us UF Libraries [960]
Africa currents – Harare, Zimbabwe 1975-81 – 1,5,9 – ISSN: 0306-8412 – mf#11520 – us UMI ProQuest [327]
Africa debates – 1967 apr – 1 – mf#5266237 – us WHS [071]
Africa described : in its ancient and present state, including accounts from bruce, ledyard, lucas...and others, down to the recent discoveries by major denham, dr oudney, and captain clapp / Hofland, Barbara – London 1828 [mf ed Hildesheim 1995-98] – 1v on 2mf – 9 – €60.00 – 3-487-27394-2 – gw Olms [960]
Africa digest – Harare Zimbabwe 1952-74 – 1,5,9 – ISSN: 0001-9798 – mf#6962 – us UMI ProQuest [960]
Africa do sul sob el-rei d. manuel, 1469-1521 / Welch, Sidney R – Lourenco marques, Mozambique. 1950 – 1r – 1 – us UF Libraries [960]
Africa [edinburgh] – Edinburgh, Scotland 1985+ – 1,5,9 – ISSN: 0001-9720 – mf#15393 – us UMI ProQuest [305]

Africa en el pensamiento de donoso cortes / Donoso Cortes, Juan Francisco – Madrid, 1955 – 1 – sp Bibl Santa Ana [920]

Africa et bucolica / Petrarch [Francesco Petrarca] – 16th c – (= ser Holkham library manuscript books 429) – 1r – 1 – mf#96930 – uk Microform Academic [450]

Africa, facts and forecasts / Maisel, Albert Q – New York, USA. 1943 – 1r – 1 – us UF Libraries [960]

Africa from early times to 1800 / Mcewan, Peter J M – London, England. 1968 – 1r – 1 – us UF Libraries [960]

Africa insight – Pretoria, South Africa 1979+ – 1,5,9 – ISSN: 0256-2804 – mf#12021,01 – us UMI ProQuest [321]

Africa institute. bulletin : english edition – Pretoria, South Africa 1963+ – 1,5,9 – ISSN: 0001-981X – mf#7042 – us UMI ProQuest [301]

Africa Institute Of South Africa *see* Botswana

Africa letter – Calcutta, India 1971-72 – 1,5 – ISSN: 0044-6491 – mf#7869 – us UMI ProQuest [079]

Africa of albert schweitzer / Joy, Charles Rhind – London, England. 1958 – 1r – 1 – us UF Libraries [960]

Africa portugueza = Afrique portugaise – Lisboa: Caetano de Magalhas, [nov 4-25,1877] (wkly) – (= ser Coll of lusophone african newspapers and serials) – 1r – 1 – us CRL [079]

Africa quarterly – New Delhi, India 1961+ – 1,5,9 – ISSN: 0001-9828 – mf#6961 – us UMI ProQuest [079]

Africa rediviva : or, the occupation of africa by christian missionaries of europe and north america / Cust, Robert Needham – London, 1891 – (= ser 19th c books on british colonization) – 2mf – 9 – mf#1.1.996 – uk Chadwyck [240]

Africa Research Group *see* Armed struggle in southern africa

Africa seen by american negroes – Paris, France. 1958 – 1r – 1 – us UF Libraries [960]

Africa south / De Blij, Harm J – Evanston, IL. 1962 – 1r – 1 – us UF Libraries [960]

Africa south of the sahara / Grove, Alfred Thomas – Oxford, England. 1967 – 1r – 1 – us UF Libraries [960]

Africa south of the sahara / Kingsnorth, G W – Cambridge, England. 1962 – 1r – 1 – us UF Libraries [960]

Africa speaks : a collection of original verse with an introduction on "poetry in africa" – 2nd ed. Accra: Guinea Press, [1960?] – us CRL [810]

Africa theological journal – Arusha, Tanzania 1985-93 – 1,5,9 – ISSN: 0253-9322 – mf#15121 – us UMI ProQuest [240]

Africa through western eyes – [mf ed Marlborough 1996] – (= ser 1) – 4pt – 1 – (pts1,2: original manuscripts from the royal commonwealth society library at cambridge university library [9r,6r] £850, £570; pt3: papers of cameron, cruikshank, livingstone, moffatt, park & stanley from the national library of scotland [10r] £950 [mf ed summer 2003]; pt4: papers of sir john kirk (1832-1922) from the national library of scotland [10r] £925 [mf ed winter 2003/4]; pt5: papers of frederick william hugh migeod & mary ward from the royal commonwealth society library at cambridge university library [13r] £1250; with d/g) – uk Matthew [960]

Africa today – Denver CO 2006+ – 1,5,9 – (cont: africa today) – ISSN: 0001-9887 – mf#5714.01 – us UMI ProQuest [321]

Africa today – Denver CO 2006+ – (cont by: africa today) – ISSN: 0001-9887 – mf#5714.01 – us UMI ProQuest [321]

Africa waiting : or, the problem of africa's evangelization / Thornton, Douglas Montagu – New York: Student Volunteer Movt for Foreign Missions, 1899 [mf ed 1986] – 1mf – 9 – 0-8370-6530-5 – (incl app) – mf#1986-0530 – us ATLA [240]

Africa [washington] – Washington DC 1971-86 – 1,5,9 – ISSN: 0044-6475 – mf#6109 – us UMI ProQuest [300]

African : a journal of african affairs / Universal Ethiopian Students' Association – New York. v1-6 n5. 1937-48 [all publ] – (= ser Black journals, series 2) – 1r – 1 – $200.00 – us UPA [960]

O african : jornal publicada em beneficio da colonia portugueza em africa – Lisboa: Narciso Feyo [dec 1884] – (= ser Coll of lusophone african newspapers and serials) – 1r – 1 – us CRL [079]

O african – S Thome: A A Mendes [mar 14 1909-jun 23 1910] (wkly) – (= ser Coll of lusophone african newspapers and serials) – 1r – 1 – us CRL [079]

African abstracts, 1950-1972 – 6r – 1 – (quarterly review publ by the international african inst of ethnographic, social and linguistic studies) – mf#575 – uk Microform Academic [960]

African Academy of Arts and Research. New York *see* Africa

African advance / Rhodesia Mission, East Africa – [Old Umtali, Rhodesia]: Rhodesia Mission Conference 1916-19 [v1-3 (1916-19)] [mf ed 2004] – 3v on 1r – 1 – (lacks v1 n6; cont: rhodesia missionary advocate) – mf#2004c-s005 – us ATLA [242]

African advertiser *see* Natal chronicle / african advertiser

African affairs – Oxford, England 1984+ – 1,5,9 – ISSN: 0001-9909 – mf#14338,02 – us UMI ProQuest [321]

African affairs, 1901-2000 : the journal of the royal african society – 37r 219mf – 1,9 – mf#2683 – uk Microform Academic [960]

African agenda : a voice of afro-american opinion – v1 n3-v6 n2 [1972 mar-1977 may] – 1 – mf#185362 – us WHS [071]

African American Catholic Pastoral Center [Oakland CA] *see* Center news

African american chronicle – 1991 aug-1993 nov; 1997 winter – 1 – mf#4712681 – us WHS [071]

African american forum : newsletter, the office of afro-american affairs, indiana university/ bloomington – 1994 fall-1996 spring – 1 – mf#3430321 – us WHS [071]

African american genealogy group newsletter – 1997 [fall]; 1998 spring, summer, winter – 1 – mf#4862333 – us WHS [929]

African american journal – 1989 nov 22-1990 nov 30 – 1 – mf#4339723 – us WHS [071]

African american networker – 1996 apr – 1 – mf#4026926 – us WHS [071]

African American Parents Coalition for Quality Education *see* Can american parents coalition for quality education

The african american press collection – 1830-84 titles – 1 – (coll can be purchased in its entirety or as individual units) – us UMI ProQuest [305]

African american regional – 1993 spring; 1994 apr; 1995 jan-may/jun – 1 – mf#2848835 – us WHS [071]

African american review – Terre Haute IN 1992+ – 1,5, 9 – (cont: black american literature forum) – ISSN: 1062-4783 – mf#5044,02 – us UMI ProQuest [420]

African american voices in the academy : newsletter / Michigan State University – v3 n1-2 [1992 fall-1993 spr] – 1r – 1 – mf#3944230 – us WHS [378]

African american women on tour : aawot conference – 2000 – 1 – mf#4996250 – us WHS [305]

African americans on wheels – 1996 summer-2000 sep – 1 – mf#3746588 – us WHS [305]

African and colonial journals *see*
- African review of mining, finance and commerce, 1892-1904
- African times and orient review, 1912-1914, 1917-1918
- Colonial enterprise
- Colonial gazette, 1838-1847

African archives of the united society for the propagation of the gospel, index / United Society for the Propagation of the Gospel. Archives – 1r – 1 – mf#96878 – uk Microform Academic [220]

African arts = Arts d'afrique – Los Angeles CA 1971+ – 1,5,9 – ISSN: 0001-9933 – mf#6069 – us UMI ProQuest [700]

The african as suckling and as adult / Ritchie, J F – (= ser Institute for social research, university of zambia. papers 9) – 2mf – 9 – mf#363/7 – uk Microform Academic [960]

African awakening / Davidson, Basil – London, England. 1955 – 1r – 1 – us UF Libraries [960]

The african awakening / Davidson, Basil – London: Cape, 1955. 262p. illus – 1 – us Wisconsin U Libr [960]

African background / Gelfand, Michael – Cape Town, South Africa. 1965 – 1r – 1 – us UF Libraries [960]

African betrayal / Darlington, Charles F – New York, USA. 1968 – 1r – 1 – us UF Libraries [960]

African biographical archive (afba) = Archives biographiques africaines (afba) / Herrero Mediavilla, Victor [comp] – [mf ed 1994-97] – 457mf (1:24) in 12 installments – diazo €10,060.00 (silver €11,080 ISBN: 978-3-598-33100-8) – ISBN-10: 3-598-33101-0 – ISBN-13: 978-3-598-33101-5 – (with printed ind) – gw Saur [960]

African books newsletter – Calcutta, India 1966-74 – 1 – ISSN: 0001-9941 – mf#2308 – us UMI ProQuest [020]

African business – London, England 1978+ – 1,5,9 – ISSN: 0141-3929 – mf#12162 – us UMI ProQuest [337]

African business series – [Legon]: Economic Res Div, Uni College of Ghana, n1-2. 1959-60 – (filmed with other titles) – us CRL [338]

African census reports – [Washington, DC] Library of Congress Photoduplication Service, [19–?] – 2r – 1 – us CRL [316]

African challenge / Jones, Arthur Creech – London, England. 1952 – 1r – 1 – us UF Libraries [960]

African chronicle – Durban: [s.n.] jun 1908-nov 1921; apr 1928-jul 1930 (wkly) – 12r – 1 – (contributions in english and tamil) – mf#MS00370 – sa National [079]

African churches of tanzania / Ranger, Terence O – Nairobi, Kenya. 1969 – 1r – 1 – us UF Libraries [960]

African Civilization Society *see*
- Annual reports, 1859-1867
- Constitution of the african civilization society

African colonizer – London, England 22 feb 1840-27 mar 1841 – 1 – uk British Libr Newspaper [072]

African colonizer *see* African times and orient review, 1912-1914, 1917-1918

African commentary – Florence MA 1989-91 – 1,5,9 – ISSN: 1045-2303 – mf#18033 – us UMI ProQuest [305]

African commentary : a journal of people of african descent – v1 iss 2 [1989 nov]; v2 iss 1/2-6, 7 [1990 jan/feb-jun, aug] – 1 – mf#1538252 – us WHS [305]

African communist – Johannesburg, South Africa 1973+ – 1,5,9 – ISSN: 0001-9976 – mf#8332 – us UMI ProQuest [335]

African connection newspaper – 1991 dec 27/jan 5-1999 jun 10/25; 1999 jun 10-2000 jun 5; 2000 jul 3/17-dec 20/jan 10 – 1 – mf#3058520 – us WHS [071]

African contrasts / Shepherd, Robert Henry Wishart – Cape Town, South Africa. 1947 – 1r – 1 – us UF Libraries [960]

African conversation-piece / Leith-Ross, Sylvia – London: Hutchison, [1944] – 9 – us CRL [960]

African crucible / Gelfand, Michael – Cape Town, South Africa. 1963 – 1r – 1 – us UF Libraries [960]

African cultural summaries / Murdock, George Peter – New Haven, CT. 1958 – 1r – 1 – us UF Libraries [960]

African culture and its relation to the training of african nurses – [Pretoria: South African Nursing Assoc, 1952?] – 9 – us CRL [960]

African dances of the witwatersrand gold mines / Tracey, Hugh – Johannesburg, South Africa. 1952 – 1r – 1 – us UF Libraries [960]

African development – London, England 1966-76 – 1,5,9 – (cont by: new african development) – ISSN: 0001-9984 – mf#9142.02 – us UMI ProQuest [337]

African development and education in southern rhodesia / Parker, Franklin – Columbus, OH. 1960 – 1r – 1 – us UF Libraries [960]

The african drum – Johannesburg SA, 1951-85 – 33r – 1 – (sale subject to copyright restrictions. title varies: drum) – sa National [079]

African drums / Puleston, Fred – New York, USA. 1930 – 1r – 1 – us UF Libraries [960]

African eagle – Harare, Zimbabwe 6 jan 1959-25 jan 1962 – 6r – 1 – (incorp with: bwalo) – uk British Libr Newspaper [079]

African education / Horrell, Muriel – Johannesburg, South Africa. 1963 – 1r – 1 – us UF Libraries [370]

African educator – 1972 dec – 1 – mf#4990574 – us WHS [071]

African eldorado / Boxer, Charles Ralph – Salisbury, Zimbabwe. 1966 – 1r – 1 – us UF Libraries [960]

African elephant / Dittebrandt, Hazel – Johannesburg, South Africa. 1970 – 1r – 1 – us UF Libraries [960]

African explains apartheid / Ngubane, Jordan K – New York, USA. 1963 – 1r – 1 – us UF Libraries [960]

African expositor – Raleigh, nc. oct 1880, jan 1883, jan 1884, jan 1888 – 1 – us ABHS [960]

African expression – 1982 dec – 1 – mf#5132558 – us WHS [071]

African factory worker / Pietermaritzburg. University Of Natal. Dept. Of Economics – Cape Town, South Africa. 1950 – 1r – 1 – us UF Libraries [960]

African family life – Johannesburg, South Africa. 1968 – 1r – 1 – us UF Libraries [960]

African fest journal – 1992 jul – 1 – mf#2576815 – us WHS [071]

African figurines / Cory, Hans – London, England. 1956 – 1r – 1 – us UF Libraries [960]

African folk-lore / Bleek, Wilhelm Heinrich Immanuel – s.l, s.l, v118–? – 1r – 1 – us UF Libraries [390]

African game trails / Roosevelt, Theodore – New York, USA. 1910 – 1r – 1 – us UF Libraries [790]

African genesis / Frobenius, Leo – New York, USA. 1937 – 1r – 1 – us UF Libraries [960]

African giant / Cloete, Stuart – Boston, MA. 1955 – 1r – 1 – us UF Libraries [960]

African giant / Cloete, Stuart – London, England. 1957 – 1r – 1 – us UF Libraries [960]

African hands / Scutt, Joan – London, England. 1961 – 1r – 1 – us UF Libraries [960]

African herald – 1992 sep-dec; 1993 jan-1994 dec; 1995 jan-1996 dec; 1997 jan-1998 dec; 1998 feb; 2000 jan-dec – 1 – mf#2569933 – us WHS [071]

The african herald *see* The sierra leone gazette, 1808-10 and 1817-27

African heritage studies association newsletter – 1977 feb – 1 – mf#4990551 – us WHS [071]

The african historian – Ife, Nigeria: University of Ife, Historical Society [v2 n1-v3 n1 (may 1966-may 1969)] – 1r – 1 – us CRL [960]

African ideas of god / Smith, Edwin William – London, England. 1950 – 1r – 1 – us UF Libraries [960]

African image / Mphahlele, Ezekiel – London, England. 1962 – 1r – 1 – us UF Libraries [960]

The african in canada / the maroons of jamaica and nova scotia / Hamilton, James Cleland – S.l: s.n, 189-? – 1mf – 9 – mf#05348 – cn Canadiana [305]

African intelligencer – La Salle IL 1820 – 1 – mf#3536 – us UMI ProQuest [079]

African interlude / Holleman, J F – Cape Town, South Africa. 1958 – 1r – 1 – us UF Libraries [960]

African journal – 5 jul 1849-25 sep 1851 (wkly) [mf ed Cape Town: SA library 1978] – 1r – 1 – (microform ed incl sam sly's african journal fr jun 1 1843 onwards) – sa National [079]

African journal *see* Sam sly's african journal

African journal, 1853-1856 / Livingstone, David – Berkeley, CA. v1-2. 1963 – 1r – 1 – us UF Libraries [960]

African journal of ecology – Oxford, England 1979+ – 1,5,9 – ISSN: 0141-6707 – mf#15501,01 – us UMI ProQuest [574]

African journal of medicine and medical sciences – Ibadan, Nigeria 1980-90 – 1,5,9 – ISSN: 0309-3913 – mf#15502,01 – us UMI ProQuest [610]

African journal of tropical hydrobiology and fisheries – Jinja, Uganda 1971-72 – 1 – ISSN: 0002-0036 – mf#8328 – us UMI ProQuest [574]

African languages in school / Conference On The Teaching Of African Languages In Schools – Salisbury, Zimbabwe. 1964 – 1r – 1 – us UF Libraries [370]

African letter – n1606-09, 1612-13, 1615-16, 1618-19, 1621-22 [1991 apr, 16/31-jun 1/15, jul 16/31-aug 1/15, sep 1/15-16/30, oct 16/31-nov1/15, dec 1/15-16/31], n1703-04, 1706-07, 1709-11, 1713, 1716 [1992 feb 1/15-16/29, mar 16/31-apr 1/15, may 1/1 – 1r – 1 – mf#2171864 – us WHS [071]

African letters in the archives of the united society for the propagation of the gospel, calendar, 1837-1896 / United Society for the Propagation of the Gospel. Archives – 1r – 1 – mf#96041 – uk Microform Academic [220]

African literature and the universities / Moore, Gerald – Ibadan, Nigeria. 1965 – 1r – 1 – us UF Libraries [470]

African Literature Association *see*
- Conference papers
- Papers of the...conference of the african literature association...

African literature in rhodesia / National Creative Writers Conference (1964: Ranche House College) – Gwelo, Zimbabwe. 1966 – 1r – 1 – us UF Libraries [470]

African local government in british east and central africa... / Howman, H Roger G – Pretoria, South Africa. 1963 – 1r – 1 – us UF Libraries [350]

African mail – Lusaka, Zambia 8 mar 1960-20 feb 1962 – 1 – (cont by: central african mail [27 feb 1962-25 may, 1/3 jun, 18 oct 1963]) – uk British Libr Newspaper [079]

The african mail – v. 1-16. 4 Oct 1903-5 Jan 1917. v. 1, no. 1-77; v. 4, no. 183 wanting – 1 – us L of C Photodup [960]

African market – 1993 jan, mar-apr, jul – 1 – mf#4878584 – us WHS [071]

African memoranda : from the royal commonwealth society library / Beaver, P – 1805 – 12mf – 9 – mf#2980 – uk Microform Academic [960]

African memoranda : relative to an attempt to establish a british settlement on the island of bulama on the western coast of africa, in the year 1792... / Beaver, Philip – London 1805 [mf ed Hildesheim 1995-98] – 1v on 6mf – 9 – €120.00 – 3-487-26740-3 – gw Olms [916]

African Methodist Episcopal Church *see*
- Christian recorder
- Journal of religious education of the african methodist episcopal church
- Journal of the...session (after organization) of the new jersey annual conference of the african methodist episcopal church
- Journal of the...session of the alabama annual conference of the african methodist episcopal church
- Minutes of the...session of the illinois annual conference of the african methodist episcopal church

- Minutes of the...session of the tennessee annual conference of the african methodist episcopal church
- Missionary magazine
- Missionary magazine of the women's missionary society of the african methodist episcopal church
- Official journal of the...session of the new jersey annual conference of the african methodist episcopal church
- Proceedings of the...annual session of the illinois conference of the african methodist episcopal church
- Trumpet
- Voice of missions
- Year book
- Year book of negro churches

African Methodist Episcopal Church. Baltimore Conference *see*
- Journal of proceedings of the...session of the baltimore annual conference of the african methodist episcopal church in the united states of america
- Minutes of the baltimore annual conference of the african methodist episcopal church
- Minutes of the...annual conference of the african methodist episcopal church for the baltimore district
- Minutes of the...session of the baltimore annual conference of the african methodist episcopal church
- Minutes of the...session of the baltimore annual conference of the african methodist episcopal church in the united states of america
- Proceedings of the annual conference of the african methodist episcopal church
- Session of the baltimore annual conference of the african methodist episcopal church in the united states of america

African Methodist Episcopal Church. Board of Education *see*
- Annual report of the board of education of the african m e church
- Report of the board of education of the african m e church

African Methodist Episcopal Church. Central Alabama Conference *see* Minutes of the...annual session of the central alabama conference of the african m e church

African Methodist Episcopal Church. Colorado Conference *see*
- Minutes of the...annual session of the colorado conference of the african methodist episcopal church
- Minutes of the...session of the colorado annual conference of the african m e church

African Methodist Episcopal Church. Division of Christian Education *see* Journal of religious education of the african methodist episcopal church

African Methodist Episcopal Church. Finance Dept *see* Quadrennial report of the department of finance to the...session (centennial) of the general conference, african methodist episcopal church

African Methodist Episcopal Church. Financial Board *see* Quadrennial report of the financial board to the...session of the general conference of the african methodist episcopal church

African Methodist Episcopal Church. General Conference *see*
- General conference of the african m e church
- Journal of the...quadrennial session of the general conference of the african m e church
- Journal of the...session and the...quadrennial session of the general conference of the african methodist episcopal church in the united states
- Minutes of the general and annual conferences of the african methodist episcopal church
- Official minutes of the...session of the general conference of the african methodist episcopal church
- Proceedings of the...general conference of the african m e church
- The...session, and the...quadrennial session of the general conference of the african methodist episcopal church
- The...quadrennial session of the general conference of the african methodist episcopal church

African Methodist Episcopal Church. General Conference (7th : 1844 : Pittsburgh, PA) *see* Seventh general conference

African Methodist Episcopal Church in America *see* Star of zion

African Methodist Episcopal Church. Indiana Conference *see*
- Journal of proceedings of the...annual conference of the african methodist episcopal church for the district of indiana
- Journal of proceedings of the...session of the indiana annual conference of the african methodist episcopal church
- Minutes of the...session of the indiana annual conference of the african methodist episcopal church
- Official minutes of the...annual session of the indiana conference of the african methodist episcopal church
- Proceedings of the...session of the indiana conference of the african methodist episcopal church

African Methodist Episcopal Church. Iowa Conference *see* Minutes of the...annual session of the iowa annual conference of the african methodist episcopal church

African Methodist Episcopal Church. Louisiana Conference *see*
- Minutes of the...annual conference of the african methodist episcopal church for the louisiana district
- Minutes of the...session of the louisiana annual conference of the african methodist episcopal church

African Methodist Episcopal Church. Michigan Conference *see*
- Journal of the...annual session of the michigan conference of the african methodist episcopal church
- Minutes of the...annual session of the michigan conference of the african methodist episcopal church

African Methodist Episcopal Church. Missouri Conference *see*
- Journal of proceedings of the...annual conference of the african m e church for the district of missouri
- Session of the missouri annual conference of the african m e church

African Methodist Episcopal Church. North Ohio Conference. Woman's Mite Missionary Society *see* Proceedings of the...annual session of the north ohio conference branch woman's mite missionary society of the african methodist episcopal church

African Methodist Episcopal Church. Ohio Conference *see*
- Journal of proceedings of the...session of the ohio annual conference of the african methodist episcopal church
- Journal of the...annual session of the ohio conference of the african methodist episcopal church
- Minutes of the ohio annual conference of the african methodist episcopal church
- Minutes of the...session of the ohio annual conference of the african m e church

African Methodist Episcopal Church. Ohio Conference. Woman's Mite Missionary Society *see* Proceedings of the...annual convention of the ohio conference branch, women's mite missionary society

African Methodist Episcopal Church. Richmond District Conference *see* Minutes of the richmond, virginia, district conference and sunday school convention of the ame church

African Methodist Episcopal Church. Seventh Episcopal District. School of Religion Pastoral Clinic *see*
- Program of the school of religion pastoral clinic and congress of youth of the 7th episcopal district of the african methodist episcopal church
- Program of the...annual school of religion-pastoral clinic, congress of youth and...annual laymen's league of the african methodist episcopal church, seventh episcopal district#

African Methodist Episcopal Church. South Ohio Conference *see*
- Minutes of the...annual session of the south ohio conference of the african methodist episcopal church
- Minutes of the...annual session of the south ohio conference of the african methodist episcopal church, third episcopal district
- Proceedings of the...annual session of the south ohio conference of the african methodist episcopal church

African Methodist Episcopal Church. South Ohio Conference. Women's Mite Missionary Society *see*
- Official report of the...annual session
- Proceedings of the south ohio conference branch of the african methodist episcopal church
- Proceedings of the...annual convention of the south ohio conference branch of the african methodist episcopal church

African Methodist Episcopal Church. Virginia Conference *see*
- Journal
- Minutes of the virginia annual conference of the african methodist episcopal church in the united states of america
- Minutes of the...annual session of the virginia conference of the african methodist episcopal church, united states of america
- Minutes of the...session of the virginia annual conference of the african methodist episcopal church
- Proceedings of the virginia conference,...annual session, african m e church

African Methodist Episcopal Church. West Tennessee Conference *see*
- Journal of proceedings of the...annual session of the west tennessee conference of the african methodist episcopal church
- Minutes of the...annual session of the west tennessee conference of the african methodist episcopal church
- Proceedings of the...annual session of the west tennessee conference of the african methodist episcopal church

African Methodist Episcopal Church. Western North Carolina Conference *see* The...session of the western n c annual conference of the african methodist episcopal church

African Methodist Episcopal Church. Women's Home and Foreign Missionary Society *see* Programme of the state convention of the woman's home and foreign missionary society

African Methodist Episcopal Church. Women's Parent Mite Missionary Society *see* Minutes and reports of the...quadrennial convention of the african methodist episcopal church

African Methodist Episcopal Zion Church *see*
- Minutes of the north carolina annual conference of the african methodist episcopal zion church in america
- Minutes of the...annual session of the central north carolina conference of the african methodist episcopal zion church
- Minutes of the...session of the virginia annual conference of the african methodist episcopal zion church
- Missionary seer
- Star of zion
- Year book

African Methodist Episcopal Zion Church. Cape Fear Conference *see* Minutes of the...annual session of the cape fear annual conference of the a m e zion church

African Methodist Episcopal Zion Church. Church Extension Dept *see*
- Annual report of...corresponding secretary-treasurer
- Semi-annual report of..., corresponding secretary-treasurer

African Methodist Episcopal Zion Church. General Conference *see*
- Official journal of the daily proceedings of the...quadrennial session, of the general conference of the african methodist episcopal zion church
- Official journal,...quadrennial session, general conference

African Methodist Episcopal Zion Church. New Jersey Conference *see* Minutes of the...annual session of the new jersey conference of the african methodist episcopal zion church

African Methodist Episcopal Zion Church. New York Conference *see*
- Minutes of the annual conferences of the wesleyan methodist episcopal zion church in america
- Minutes of the several annual conferences of the african methodist episcopal church in america
- Minutes of the yearly conferences of the african methodist episcopal zion church in america
- Minutes taken at the yearly conference of the african methodist episcopal church in america

African Methodist Episcopal Zion Church. North Carolina Conference *see* Directory of the...session of the north carolina annual conference of the african methodist episcopal zion church

African Methodist Episcopal Zion Church. Philadelphia and Baltimore Conference *see* Minutes of the...session of the philadelphia and baltimore annual conference of the african, methodist, episcopal, zion church

African Methodist Episcopal Zion Church. Philadelphia Conference *see* Minutes of the philad'a district annual conference of the african m e zion church in america

African military accounts : war office route books, military reports and information precis for british africa – 2pt – 9 – (pt1: 1869-1912 from the ministry of defence library, whitehall, london 136mf [87075]. pt2: 1906-1933 from the royal artillery institution library, london 70mf [87287]. int by d c dorward) – uk Microform Academic [355]

An african millionaire : episodes in the life of the illustrious colonel clay / Allen, Grant – New York: E Arnold, 1897 [mf ed 1984] – 4mf – 9 – 0-665-32086-8 – mf#32086 – cn Canadiana [920]

African mine labour at the messina copper mines : the struggle for economic and social survival: 1905-1960 / Malunga, Wilson Felix – Pretoria: Vista University 2002 [mf ed 2002] – 7mf – 9 – (incl bibl) – mf#mfm15159 – sa Unisa [622]

African missionary heroes and heroines / Kumm, Hermann Karl Wilhelm – New York: Macmillan, 1917 [mf ed 1993] – (= ser Christian church (disciples of christ) coll) – 1mf – 9 – 0-524-06430-X – (incl bibl ref) – mf#1991-2552 – us ATLA [240]

African missions : impressions of the south, east, and centre of the dark continent / O'Rorke, Benjamin Garniss – London: SPCK; New York: ES Gorham, 1912 [mf ed 1990] – 1mf – 9 – 0-7905-7018-1 – (pref by j taylor smith) – mf#1988-3018 – us ATLA [240]

African missions, education and the road to independence : the sum in nigeria, the cameroons, chad, sudan and other african territories – 5pt – 1 – (pt1: mss papers fr centre for the study of christianity in the non-western world, new college, university of edinburgh, 1898-1960 [12r] £1100; pt2: lightbearer, 1905-91 [12r] £1100; pt3: newsletters, 1940-89, publ & annual reports 1908-79 [28r] £2650; pt4: lantern slides, slides & photos [7r] £675; pt5: publ in hausa [11r] £1050; with d/g) – uk Matthew [960]

The african missions of the white fathers – Quebec: [White Fathers, 1909?-194-?] – 9 – mf#P04287 – cn Canadiana [241]

African morning post – Accra, Ghana 4 jun 1935, 1 oct 1937-30 sep 1938, 4 jan-28 feb 1939, 2 oct 1950-27 mar 1953 – 5r – 1 – uk British Libr Newspaper [079]

The african national times – Accra. Ghana. 14, 25, 18 Aug; 11, 22, 25 Sep; 2, 16 Oct 1948 – 1 – us NY Public [079]

African nationalism / Sithole, Ndabaningi – Cape Town, South Africa. 1959 – 1r – 1 – us UF Libraries [960]

African Nationalist Pioneer Movement *see* Extra

African news digest – v8 n6 [1997 jun] – 1 – mf#1056488 – us WHS [321]

African news weekly – 1993 jan 22-dec 3; 1994 jan 21-dec 30; 1995 jan 20-jun 30; 1995 jul 7-dec 22; 1996 jan 15/21-jun 24/30; 1996 jul-nov17 – 1 – mf#2658591 – us WHS [071]

African notebook / Schweitzer, Albert – Bloomington, IL. 1958, C1939 – 1r – 1 – us UF Libraries [960]

African notes – Ibadan: University of Ibadan, Institute of African Studies. v1-5 n3. oct 1963-jan 1970 (varies) – 1 – us CRL [960]

African observer : a monthly journal illustrative of the general character, and moral and political effects of negro slavery – Philadelphia. n1-12. 1827-28 [all publ] – (= ser Black journals, series 1) – 4mf – 9 – $55.00 – us UPA [976]

African official statistical serials, 1867-1982 : general statistical compendia – economic, financial, social and demographic statistics – issued by the governments of nearly every african country... – [mf ed Chadwyck-Healey] – 2085mf – 9 – (individual titles also listed and may be purchased separately) – uk Chadwyck [316]

African opposition in south africa / Feit, Edward – Palo Alto, CA. 1967 – 1r – 1 – us UF Libraries [960]

The african orthodox churchman – [s.l.]: African Orthodox Church Publ, 1929- [bimthly, mthly] [mf ed 2004] – 1r – 1 – (mf: v1 n3-v6 x'mas no [mar 1929-dec 1938]; [n.s.] v1 [dec 1939-dec 1948] lacks v1 n7,11-12, v2 n1,6-12, v3-5, 1940-44. began in 1929? dec 1939 called christmas no & carries no vol designation; dec 1945-dec 1948 called [n.s.] v1. official organ of: province of south africa 1929-may 1930; province of south & central africa dec 1938; province of south & central africa, & rhodesia dec 1939-dec 1948) – mf#2004-s014 – us ATLA [243]

African pamphlets : microfilm record [being inventory of nigerian pamphlets on film in camp] / University of Ibadan. Library – Ibadan: University of Ibadan Library [1963-66] – 1r – 1 – us CRL [960]

African People's Socialist Party et al *see* Burning spear

African political ephemera, 1958-1966 / Bartlett, Robert E – [S.l, s.n, 19–?] – 1r – 1 – us CRL [960]

African political systems / Fortes, Meyer – London, England. 1950 – 1r – 1 – us UF Libraries [960]

African political sytems / Fortes, Meyer – London, England. 1958 – 1r – 1 – us UF Libraries [960]

African portraits / Cloete, Stuart – London, England. 1946 – 1r – 1 – us UF Libraries [960]

The african preacher : an authentic narrative / White, William Spottswood – Philadelphia: Presbyterian Board of Pub, c1849 [mf ed 1993] – (= ser Presbyterian coll) – 2mf – 9 – 0-524-07925-0 – mf#1991-3470 – us ATLA [242]

African quest newspaper – 1997 jul-1998 dec; 1999 – 1 – mf#5014526 – us WHS [071]

African repository – La Salle 1825-92 – 1 – mf#4570 – us UMI ProQuest [976]

African researches : from the royal commonwealth society library. proceedings of the association for promoting the discovery of the interior parts of africa / Park & Hornemann, Friedrich K – v2. 1792-1802 – 13mf – 7 – mf#2979 – uk Microform Academic [916]

African review *see* Business and financial papers, 1780-1939

African review of mining, finance and commerce, 1892-1904 – [mf ed Marlborough 1996] – (= ser Business and financial papers series) – 23r – 1 – £2100.00 – uk Matthew [073]

African revolution / Cameron, James – New York, USA. 1961 – 1r – 1 – us UF Libraries [960]

African revolution *see* Revolution africa, latin america, asia

African sectional committee of the manchester chamber of commerce and industry : minutes, 1892-1926 – 6r – 1 – (with int & ind) – mf#97530 – uk Microform Academic [380]
African shopper – 1996 may – 1 – mf#3641480 – us WHS [071]
The african sketch-book / Reade, Winwood – London: Smith, Elder & Co, 1873 – 1 – us CRL [960]
The african slave trade / Buston, Thomas Fowell – London, 1839 – 1 – us CRL [305]
African societies in southern africa – New York, USA. 1969 – 1r – 1 – us UF Libraries [305]
African spectrum – 2000 jan-dec – 1 – mf#4819653 – us WHS [071]
African spirit – 1989 sep-1991 apr – 1 – mf#4841684 – us WHS [071]
African studies association annual meeting papers – Atlanta, GA: African Studies Association, [31st (1988)] (annual) – 2r – 1 – us CRL [960]
African studies review – New Brunswick NJ 1989+ – 1,5,9 – ISSN: 0002-0206 – mf#18291,01 – us UMI ProQuest [960]
African sun times – [1996 may 1/15, jul 16/31, oct 16/30]; 1997 jun 20-dec 16; 1998; 1999 jan 11/17-oct 4/10, dec; 2000 jan 13/19-jun 29/jul 5; jul 6/12-dec 14/20; 2001 jan 11/17-jun 21/27 – 1 – mf#3573885 – us WHS [071]
African switzerland = Basutoland of today / Rosenthal, Eric – Cape Town, South Africa. 1948 – 1r – 1 – us UF Libraries [960]
African teachers' journal – n1 (1952)-n37 (1961) [mf ed Pretoria: State Library 1974] – 30mf – 9 – (title changed to: teachers' journal fr n38 (1961)-n51 (1965)) – ISSN: 0303-1233 – mf#MFS00267 – sa National [370]
African telegraph and gold coast mirror – London, England 14 nov 1914-25 feb 1915, dec 1918-dec 1919 [mf dec 1918] – 1 – (not publ between 25 feb 1915 & dec 1918) – uk British Libr Newspaper [072]
African theology : a bibliography / ed by Parratt, John – [Zomba : Dept of Religious Studies, Chancellor College, University of Malawi, 1983] (mf ed 198-) – (= ser Sources for the Study of Religion in Malawi, No 8) – 1r – 1 – mf#Sc Micro R-4102 n9 – us NY Public [290]
African tightrope / Alexander, Henry Templer – New York, USA. 1966, 1965 – 1r – 1 – us UF Libraries [960]
African times – 1994 jul 1/15-1996 dec 15/31; 1997 jan 15/31-1998 dec 15/31; 1999 jan 1/15-dec 15/31; 2000 jan 1/15-dec 15/31 – 1 – mf#3058633 – us WHS [071]
African times – New York. v1-3. n16. dec. 1948-may 1951 – 1 – us NY Public [960]
African times and orient review : a monthly (weekly) journal devoted to the interests of the coloured races of the world – London, England jul 1912-nov/dec 1913, 24 mar-18 aug 1914, mid-jan 1917-mid-oct 1918 – 1 – (not publ between nov/dec 1913 & 24 mar 1914 or between 18 aug 1914 & mid-jan 1917) – uk British Libr Newspaper [305]
African times and orient review, 1912-1914, 1917-1918 : and the african colonizer, 1840-1841 – [mf ed Marlborough 1996] – (= ser Business and financial papers series) – 2r – 1 – £185.00 – uk Matthew [960]
African torch – iss 3-5 [1995 oct-dec] – 1 – mf#5009250 – us WHS [071]
African town crier – 1994 oct 28-97 apr/may – 1 – mf#3421684 – us WHS [071]
African trade review – Madrid. Dec 1967-Oct 1975; Oct 1977-Oct 1978 – 1 – 92.00 – us L of C Photodup [380]
African traders in kumasi / Garlick, Peter C – [Legon, Ghana?] Economic Research Division, University College of Ghana, 1959 – 1r – 1 – us CRL [380]
African trading : or, the trials of william narh ocansey, of addah, west coast of africa, river volta / Ocansey, John Emanuel – Liverpool, [England]: J Looney, 1881 (mf ed 1978) – 1r – 1 – mf#ZZ-15996 – us NY Public [380]
African tragedy / Dhlomo, Rolfes Robert Reginald – Lovedale, South Africa. 1928? – 1r – 1 – us UF Libraries [960]
African voice in southern rhodesia, 1898-1930 / Ranger, Terence O – Evanston, IL. 1970 – 1r – us UF Libraries [960]
African voice in southern rhodesia, 1898-1930 / Ranger, Terence O – London, England. 1970 – 1r – 1 – us UF Libraries [960]
African voices – n1-3 [1993 apr-oct/nov]; n5, 7-8 [1993 feb/mar, jul/aug-dec]; n9-10, 12 [1995 mar-summer, oct/nov]; n13 [1996 winter] – 1 – mf#2864494 – us WHS [071]
African wastes reclaimed : illustrated in the story of the lovedale mission / Young, Robert – London: JM Dent, 1902 [mf ed 1992] – (= ser Presbyterian coll) – 1mf – 9 – 0-524-04390-6 – (incl bibl ref) – mf#1991-2094 – us ATLA [240]
African wildlife – Johannesburg, South Africa 1977+ – 1,5,9 – ISSN: 0002-0273 – mf#11426 – us UMI ProQuest [639]
African women / Simons, Harold – Evanston, IL. 1968 – 1r – 1 – us UF Libraries [305]
African world – 1988 nov – 1 – mf#4851897 – us WHS [071]
African world – v10-112. 1905-30 – 1 – (some missing issues) – us L of C Photodup [960]
African world – v1 n1-v4 n10 [1971 aug 24-1975 jun] – 1 – mf#810701 – us WHS [071]
African youth – Binghamton, NY: AYMLU. [v1 n4-9/10 (jun/jul-dec 1976)] (mthly) – 1r – 1 – us CRL [079]
Africana : revista mensal ilustrada – Lourenco Marques: Antonio Sebastiao de Vasconcelos, [n1-9(apr-dec1933)] (mthly) – (= ser Coll of lusophone african newspapers and serials) – 1r – 1 – us CRL [321]
Africana archives : in microfilm at northwestern university library / Finnegan, Gregory Allan [comp] – Evanston, IL: Melville J Herskovits Library of African Studies, Northwestern University Library, 1982 (mf ed 1982) – 1r – 1 – mf#ZZ-22925 – us NY Public [960]
Africana libraries newsletter – Boston: African Studies Library, Boston University, [n1-16 (jul 1975-mar1978)] (irreg) – 1r – 1 – us CRL [020]
Africana libraries newsletter – Urbana, IL: African Studies Program, University of Illinois, [n34-48 (jul 1983-dec1986)] (irreg) – 1r – 1 – us CRL [020]
Africana notes and news = Africana aantekeninge – Johannesburg, South Africa 1976-79 – 1,5,9 – ISSN: 0002-032X – mf#10378 – us UMI ProQuest [960]
Africana repository / Kennedy, Reginald Frank – Cape Town, South Africa. 1965 – 1r – 1 – us UF Libraries [960]
African-american archaeology : newsletter of the african-american archaeology network – spring 1990-93 – 1 – mf#3640909 – us WHS [305]
African-american baptist annual reports, 1865-1990s – 104r – 1 – $130.00r – (guide also sold separately $25. state-by-state breakdown: alabama 13r (r1-13); arkansas 3r (r14-16); california/washington 1r (r17); district of columbia/maryland/ pennsylvania 1r (r18); florida 7r (r19-25); georgia 16r (r26-41); illinois 1r (r42); indiana/iowa 1r (r43); kansas 1r (r44); kentucky 2r (r45-46); louisiana 3r (r47-49); mississippi 5r (r50-54); missouri 2r (r55-56); new england 2r (r57-58); new jersey/new york 1r (r59); north carolina 8r (r60-67); ohio 4r (r68-71); south carolina 7r (r72-78); tennessee/oklahoma 3r (r79-81); texas 8r (r82-89); virginia 12r (r90-101); west virginia 2r (r102-103); pre-national bodies/western regional bodies 1r (r104)) – mf#D3441 – us ABHS [242]
African-american bookselling : a specialty bookseller – 1991 spring – 1 – mf#4798871 – us WHS [070]
African-american business and consumer magazine – 1995 jun – 1 – mf#3419016 – us WHS [338]
African-American Council on Africa *see* Newsletter
African-American Institute of Islamic Research *see* Network
African-american journal – 1993 apr 1-1994 dec 15 – 1 – mf#2844041 – us WHS [071]
African-american reader newspaper – v1 n3 [1992 sep]; v1 n10-1912 [1993 apr-jun]; v2 n1-4; [1993 jul-oct] – 1 – mf#2690982 – us WHS [071]
African-american traveler – v1 n6 [1988?] – 1 – mf#2690985 – us WHS [071]
African-american voice – 1993 sep 29-dec 29; 1994 jan 5/11-dec 28/jan 3 1995; 1995 jan 4-dec 27; 1996 jan 3-dec 25; 1997 jan 1/7-dec 30; 1998 jan 7, 14, 21, mar 4, apr 1, 8, 15, dec 9, 16, 23, 30; 1999 jan 13-dec 29 – 1 – mf#2837302 – us WHS [071]
African-americanist – 1992-96 – 1 – mf#3099538 – us WHS [071]
Africania de la musica folklorica de cuba / Ortiz, Fernando – Habana, Cuba. 1950 – 1r – 1 – us UF Libraries [780]
The africanist – Maseru, Lesotho: Dept of Publicity & Information, Pan Africanist Congress, [jan1965-mar/apr 1968] (mthly) – 1r – 1 – us CRL [320]
O africano – Lourenco Marques: Gremio Africano, [mar 1909-jan 3 1920] – 5r – 1 – (sample issue: dec 25 1908) – us CRL [960]
Africanos no brasil / Nina Rodrigues, Raymundo – Sao Paulo, Brazil. 1935 – 1r – 1 – us UF Libraries [960]
Africans in the new world, 1493-1834 : from the john carter brown library at brown university – [mf ed Marlborough 2003] – 30r – 1 – £2800.00 – (with d/g) – uk Matthew [976]
African's religion / Gelfand, Michael – Cape Town, South Africa. 1966 – 1r – 1 – us UF Libraries [200]
Africanus, L *see* The history and description of africa...
African-usa [magazine] – 1992 nov-1995 may – 1 – mf#3059067 – us WHS [071]
Africa's luminary – v1-3 mar 15 1839-dec 17 1841 [complete] – 1r – 1 – mf#ATLA B0124 – us ATLA [073]
Africa's red harvest / Lessing, Pieter – New York, USA. 1962 – 1r – 1 – us UF Libraries [960]
The africo-american presbyterian vols 46-59 (1925-1938) partial – Atlanta, GA – 4r – 1 – $200.00 – us Presbyterian [242]
Afrika : der dunkle erdtheil im lichte unserer zeit / Schweiger-Lerchenfeld, Amand, Freiherr von – Wien: A Hartleben 1886 [mf ed 1984] – 1r [ill] – 1 – (filmed with: greek life and thought / larue van hook [1923] & other titles) – mf#11115 – us Wisconsin U Libr [960]
Afrika delili / Muhsin, Mehmet – Kahire: Al-Felah Ceridesi Matbaasi, 1312 [1895] – (= ser Ottoman histories and historical sources) – 15mf – 9 – $250.00 – us MEDOC [956]
Afrika en deszelfs bewoners, volgens de nieuwste ontdekkingen : een werk ter bevordering der kennis van landen en volken en van derzelver voortbrengsels en handel / Kampen, Nicolaas Godfried – Haarlem: Francois Bohn 1828 – (= ser [Travel descriptions from south africa, 1711-1938]) – 3v on 14mf – 9 – mf#zah-24 – ne IDC [916]
Afrika kwetu = African weekly – Zanzibar, Tanzania 8 jan 1959-1 feb 1962 – 1r – 1 – uk British Libr Newspaper [079]
Afrika mix extra : radio afrika report – 1992 aug/oct – 1 – mf#4882191 – us WHS [071]
Afrika must unite – v1 n4 [1972 mar/apr]; v2 n13-14 [1973 nidra-kukadzi] – 1 – mf#2847420 – us WHS [071]
Afrika nachrichten – Leipzig DE, 1938-39 – 1r – 1 – gw Misc Inst [960]
Afrika segodnia – Moscow, Russia. 1962 – 1r – 1 – us UF Libraries [960]
Afrika unbound – 1992 sep 17-oct 29; 1993 apr-dec; 1994 mar, oct-nov; 1995 feb-dec/1996 jan – 1 – mf#3521204 – us WHS [071]
Afrikaander Bond en Boerenvereeniging, South Africa *see* Official documents of the africander bond and farmers' association
Die afrikaans van kharkams / Links, Thomas Hodson – U of the Western Cape 1983 [mf ed S.l: s.n. 1983] – 4mf – 9 – (incl bibl) – sa Misc Inst [470]
De afrikaansche boerenvriend – Colesberg SA, jul 1 1882-jun 30 1883 [wkly] [mf ed Cape Town: SA library 1986] – 1r – 1 – (absorbed by: colesberg advertiser) – mf#ms00431 – sa National [079]
De afrikaansche voorstander – Port Elizabeth SA, may 3 1849-mar 21 1850? (wkly) [mf ed Cape Town: SA library 1982] – 1r – 1 – mf#ms00432 – sa National [079]
Afrikaanse bevolking van belgisch-kongo en van ruanda-urundi / Kerken, Georges Van Der – Gent, Belgium. 1952 – 1r – 1 – us UF Libraries [960]
Die afrika-literatur in der zeit von 1500 bis 1750 / Paulitschke, Phillip Viktor – Vienna. 1882 – 1r – 1 – mf#388 – uk Microform Academic [960]
Afrikan : an international magazine for the progress and unity of people of afrikan descent – 1996/1997 2nd iss – 1 – mf#3641585 – us WHS [071]
Afrikan Information Bureau and al *see* Mazungumzo
Afrikan Students for Afrikan Liberation *see* Black rap
Die afrikaner – 1 jan 1905-29 dec 1931 [mf ed Cape Town: SA library [1905]] – 33r – 1 – sa National [079]
Die afrikaner – Cradock: [s.n.], 1933- (Cradock: Gedruk deur die here White & Boughton) – 1 – (issues for 1950-61 filmed with midland news and karoo framer 1926-27, 1929-61, & with middellandsche afrikaander 1928) – us CRL [079]
Die afrikaner *see* Middellandsche afrikaander, de
Afrikaner and african nationalism / Munger, Edwon S – London, England. 1967 – 1r – 1 – us UF Libraries [960]
Afrikaner bond / Davenport, T R H – Cape Town, South Africa. 1966 – 1r – 1 – us UF Libraries [960]
Afrikaner en sy geskiedenis / Jaarsveld, Floris Albertus van – Kaapstad, South Africa. 1959 – 1r – 1 – us UF Libraries [960]
Afrikaner's interpretation of south african history / Jaarsveld, Floris Albertus van – Cape Town, South Africa. 1964 – 1r – 1 – us UF Libraries [960]
Afrikanische trauerspiele : cleopatra, sophonisbe / Lohenstein, Daniel Casper von; ed by Just, Klaus Guenther – Stuttgart: Hiersemann 1957 [mf ed 1993] – (= ser Blvs 294) – 58r – 1 – (incl bibl ref. filmed with: tuerkische trauerspiel / daniel caspar von lohenstein) – mf#3420p – us Wisconsin U Libr [820]
Afrikanische verkehrssprachen / Heine, Bernd – Koln, Germany. 1968 – 1r – 1 – us UF Libraries [470]
Afrikans for Communications *see* Uhuru sasa
L'afrique : ou histoire, moeurs, usages et coutumes des africains / Geoffroy de Villeneuve, Rene – Paris 1814-21 [mf ed Hildesheim 1995-98] – 7v on 14mf – 9 – €140.00 – 3-487-27396-9 – gw Olms [916]
Afrique – 1991 jul-1992 dec; 1993-96; 1997-1999 jun – 1 – mf#2521501 – us WHS [071]
Afrique – no. 1-274; MQ no. 94, 107, 182, 249. Alger. avr. 1924-61 – 1 – fr ACRPP [960]
Afrique = Perspectives politiques / Paul, Edouard C – Port-Au-Prince, Haiti. 1963 – 1r – 1 – us UF Libraries [960]
L'afrique centrale francaise : recit du voyage de la mission, appendice par pellegrin, et al / Chevalier, Auguste – Paris: A Challamel, 1907 [mf ed 1987] – 776p (ill) – 1 – mf#7852 – us Wisconsin U Libr [916]
L'afrique chretienne / Leclercq, Henri – 2e ed. Paris: Victor Lecoffre, 1904 [mf ed 1986] – (= ser Bibliotheque de l'enseignement de l'histoire ecclesiastique) – 2v on 3mf – 9 – 0-8370-7643-9 – (in french. incl bibl ref) – mf#1986-1643 – us ATLA [240]
L'afrique contemporaine – Paris. avr mai 1962-nov dec 1979 – 1 – fr ACRPP [073]
L'afrique de marmol... / Marmol [y Carvajal] – Paris, 1667 – 3v on 34mf – 9 – mf#A-123 – ne IDC [916]
Afrique d'espression francaise et madagascar – Paris, France. 1966 – 1r – 1 – us UF Libraries [960]
L'afrique du sud / Aubert, Georges – Paris: Flammarion, [1898] – 1 – us CRL [960]
Afrique equatoriale, orientale et australe / Maurette, Fernand – Paris, France. 1938 – 1r – 1 – us UF Libraries [960]
L'afrique et le monde : hebdomadaire independant d'interet general – Bruxelles: Wellens-Pay [sep 7 1950-nov 24 1960] (wkly) – (= ser Congolese newspaper coll) – 5r – 1 – us CRL [079]
Afrique exploree et civilisee – Geneve-Paris-Bruxelles. juil 1879-juil 1894 – 1 – fr ACRPP [960]
Afrique exploree et civilisee – Killen TX 1880-94 – 1 – mf#3326 – us UMI ProQuest [960]
L'afrique fantome / Leiris, Michel – [Paris]: Gallimard, [1934] – 1 – us CRL [960]
L'afrique francaise : bulletin mensuel du comite de l'afrique francaise et du comite du maroc – v1-50, 61-65. 1891-1940, 1952-56 – 1 – (missing: v50 n1, 4-12) – us L of C Photodup [960]
L'afrique francaise pour tous / Cros, Louis – Paris: Albin Michel, 1928 [mf ed 1976] – 1r – 1 – mf#ZZ-14577 – us NY Public [330]
L'afrique francaise pour tous comment aller & que faire en afrique francaise? : quatorze colonies a mettre en valeur mauritanie, senegal, cote d'ivoire, dahomey, soudan, niger, tchad, congo, guinee, togo, cameroun... / Cros, Louis – Paris: Albin Michel [19-] – (= ser [Travel descriptions from south africa, 1711-1938]) – 7mf – 9 – mf#zah-76 – ne IDC [916]
L'afrique noire occidentale : esquisse des cadres geographiques / Gautier, Emile Felix – Paris: E Larose, 1935 – 1 – us CRL [960]
Afrique, nous t'ignorons! / Matip, Benjamin – Paris, R Lacoste [1956] – 1r – 1 – us CRL [960]
Afrique nouvelle – Dakar. n1288. 1960-15 juin 1972 – 1 – fr ACRPP [073]
L'afrique occidentale : algerie, mzab, tildikelt / Soleillet, Paul – Avignon: F Sequin aine, 1887 – 1 – us CRL [960]
L'afrique occidentale francaise / Francois, Georges Alphonse Florent Octave – Paris: E Larose, 1907 – 1 – us CRL [960]
L'afrique occidentale francaise : par l'atlantique ou par le sahara? / Tuaillon, Georges – Paris: Charles-Lavauzelle, 1936 – 1 – us CRL [960]
L'afrique occidentale francaise dans la litterature francaise (depuis 1870) / Lebel, Roland – Paris: E Larose, 1925 – 1 – us CRL [300]
Afrique orientale : abyssinie / Raffray, A – Paris, 1880 – 6mf – 9 – mf#NE-20288 – ne IDC [916]
L'afrique pittoresque : le continent africain et les iles. lectures choisies – 2e ed. Paris: C Delagrave, 1890 – 1 – us CRL [960]
Afrique, terre qui meurt : la degradation des sols africains sous l'influence de la colonisation / Harroy, Jean-Paul – Bruxelles: M Hayez, 1944 – 1 – us CRL [960]
Afriscope – Lagos, Nigeria 1974-82 – 1,5,9 – ISSN: 0044-667X – mf#1361 – us UMI ProQuest [338]
Afro American Genealogical and Historical Society *see* Newsletter
Afro news – 1995 aug-1998 dec – 1 – mf#3603674 – us WHS [071]
Afro scholar newsletter – 1984 fall-1985 spring/summer; 1987 spring, fall; 1991 fall – 1 – mf#1081268 – us WHS [071]
Afro times – 1987 jul 18-1988 jun; 1988 jul-1989 jun; 1989 jul-1990 jun-dec; 1991-95; 1996 jan 6-dec 28; 1997; 1998 jan-dec; 1999 jan 2-jun 26; 1999 jul 3-dec 25/31; 2000 jan-dec; 2001 jan 6/12-jun 30/jul 6 – 1 – mf#1345283 – us WHS [071]

Afro-am – 1982 apr/jun-jul/sep; 1983 jan/mar, jul/sep; 1986 jan/mar-apr/jun, dec; 1987 jan/mar – 1 – mf#3199631 – us WHS [071]

Afro-america – 1966 sep/oct; 1967 jul/aug – 1 – mf#1051719 – us WHS [071]

Afroamerica / Franco, Jose L – Havana, Cuba. 1961 – 1r – 1 – us UF Libraries [972]

Afroamerica – Mexico: Instituto International de Estudios Afroamericanos. [v1-2 n3 1945-jan 1946] (biannual) – 1r – 1 – us CRL [978]

Les afro-americains – Dakar: IFAN, 1952 [1953] – 1 – us CRL [960]

Afro=american advance – Minneapolis, St Paul. v1 n16- may 27 1899- (mf ed 1947) – (= ser Negro newspapers on microfilm) – 1r – 1 – (formed by: twin=city american, and: colored citizen) – us L of C Photodup [071]

Afro-american art history newsletter – 1987 spring-fall – 1 – mf#3055961 – us WHS [740]

The afro-american citizen – Charleston, SC: Citizen Pub Co. v1 n38. jan 17 1900 (wkly) [mf ed 1947] – (= ser Negro Newspapers on Microfilm) – 1r – 1 – us L of C Photodup [071]

Afro-American Clubwoman's Project *see* Collected records

Afro-american consumer – 1970 apr-may, sep – 1 – mf#4862453 – us WHS [071]

Afro-american folksongs : a study in racial and national music / Krehbiel, Henry Edward – New York: G Schirmer, c1914 [mf ed 1990] – 1mf – 9 – 0-7905-5243-4 – (incl bibl ref) – mf#1988-1243 – us ATLA [780]

Afro-american gazette – 1991-1993 dec 20; 1994 jan-1995 aug – 1 – mf#2699173 – us WHS [071]

Afro-American Historical and Genealogical Society *see*
- Newsletter
- Nj-aahgs newsletter

Afro-American Historical Association of the Niagara Frontier *see* Historically speaking

The afro-american history series / ed by Whiteman, Maxwell – 1978 – 58mf – 9 – $261.00 – mf#S1842 – Dist. us Scholarly Res – us L of C Photodup [305]

Afro-american journal – v1 n1-v5 n5 [1973 feb-1977 1st qrt] – 1 – mf#294293 – us WHS [071]

Afro-american Museum of Detroit *see* Gallery

Afro-american studies department newsletter – v1 n2-v4 n1 [1961 apr 28-1978 mar] – 1 – mf#928663 – us WHS [305]

Afro-american times – 1987 mar 14-jul 11 – 1 – mf#1269028 – us WHS [071]

Afro-americana – New York NY 1969-70 – 1 – ISSN: 0002-0583 – mf#10111 – us UMI ProQuest [305]

The afro-asian journalist – Djakarta, 1964-1965 – 15mf – 9 – (missing: 1965 v2(1)) – mf#SE-521 – ne IDC [950]

Afro/carib news – 1995 jan-jul – 1 – mf#3421694 – us WHS [071]

Afro-hawaii news – 1987 jun-1989 may; v2 n12-v5 n7 [1989 feb 1/28-1991 dec 1/31]; v2 n2-4, 6-8 [1988 jul-aug 15/30, sep 15/30-oct 15/31] – 1 – mf#1533692 – us WHS [071]

Afro-hispanic review – 1991+ – 1,5,9 – ISSN: 0278-8969 – mf#19148 – us UMI ProQuest [305]

Afro-independent – St Paul, MN: v1 n16. sep 22 1888 [mf ed 1947] – (= ser Negro Newspapers on Microfilm) – 1r – 1 – us L of C Photodup [071]

Afro-world briefs – 1985 jan, jun, dec; 1986 jun, dec – 1 – mf#2881549 – us WHS [071]

Afsa news – 1976 sep-1984 sep – 1 – mf#893963 – us WHS [071]

Afsc reporter – v2 n2-v5 n3 [1971 mar-1978 jan/feb/mar] – 1 – mf#403989 – us WHS [071]

Afschriften uit het oudarchief van modjokerto 1819-1850 met inventaris – 20mf – 8 – mf#SD-102 mf 23-43 – ne IDC [959]

Afschriften van eenige brieven en telegrammen gewisseld tisschen... – Pretoria, South Africa. 1967 – 1r – 1 – us UF Libraries [960]

Afschriften van portugese archiefstukken voornamelijk betreffende vestigingen in de molukken en op malakka, 1515-1532 – 5mf – 8 – mf#SD-102 mf 3-7 – ne IDC [959]

Afschriften van portugese brieven over de oudste missieposten in oost-indie, 1650 – 2mf – 8 – mf#SD-102 mf 8(1-2)-8(2-2) – ne IDC [959]

Afscme – 1978 feb-1984 jun – 1 – mf#916442 – us WHS [071]

Afscme 93 news – 1986 feb-oct/nov, 1986 convention rprt – 1 – mf#2607651 – us WHS [071]

Afscme 1695 union news – v14 n1-v17 n4 [1980 feb-1983 jun] – 1 – mf#637180 – us WHS [071]

Afscme bulletin – 1967 may 16 – 1 – mf#637172 – us WHS [071]

Afscme council 66 news – v5 n2-v5 n5 [1986 mar/apr-sep/oct] – 1 – mf#1289153 – us WHS [071]

Afscme local 1363 news – v1 n1-v3 n2 [1974 sep-1978 oct/nov] – 1 – mf#492665 – us WHS [071]

Afscme local 2059 : [newsletter] – v3 n5-v5 n6 [1980 may-1983 nov/dec] – 1 – mf#1043319 – us WHS [071]

Afscme ohio council 8 news – 1978 jun-1979 summer – 1 – mf#624679 – us WHS [071]

Afscme ohio people council 8 – v2 n3-to date [1977 aug-nov] – 1 – mf#62468 – us WHS [071]

Afscme reports : official publication of the american federation of state, county, and municipal employees-wisconsin councils 24, 40, and 48 – v1 n1-v3 n6 [1984 aug-1988 dec]; v4 n7-v9 n19 [1989 jan-1994 dec] – 1 – mf#1051454 – us WHS [350]

Afscme steward – v1 n1-v1 n4 [1979 oct/nov-1980 apr/may] – 1 – mf#637045 – us WHS [071]

Afsluttende uvidenskabelig efterskrift til de philosophiske smuler : mimisk-pathetisk-dialektisk sammenskrift, existentielt indlaeg / Climacus, Johannes; ed by Kierkegaard, Soeren – Kobenhavn: CA Reitzel, 1846 [mf ed 1990] – 2mf – 9 – 0-7905-7411-X – mf#1989-0636 – us ATLA [240]

Aft action – n1-1942 [1977 sep 2-1984 jul 21] – 1 – mf#916668 – us WHS [071]

Aft in the news – 1977 may/jun-1979 summer – 1 – mf#611740 – us WHS [071]

Aft news – 1978 jan-jul 4 – 1 – mf#633791 – us WHS [071]

Aft news clips – 1980 summer-1981; 1982-84 fall; 1985 winter/spring – 1 – mf#611744 – us WHS [071]

Aftenposten – Oslo: [s.n., jul 1938-] – 1 – us CRL [079]

After coronado, spanish exploration northeast of new mexico, 1696-1727 : documents from the archives of spain, mexico and new mexico...norman, 1935 / Thomas, Alfred B – Madrid: Razon y Fe, 1936 – 1 – sp Bibl Santa Ana [917]

After coronado. spanish exploration northeast of new mexico, 1696-1727... / Bernab y Thomas, Alfred – Madrid: Missionalia Hispanica, 1945 – 1 – sp Bibl Santa Ana [917]

After dark – North Hollywood CA 1960-83 – 1,5,9 – ISSN: 0002-0702 – mf#3228 – us UMI ProQuest [790]

After death : an examination of the testimony of primitive times respecting the state of the faithful dead, and their relationship to the living / Luckock, Herbert Mortimer – 5th ed. New York: T Whittaker, 1886 [mf ed 1991] – 1mf – 9 – 0-7905-8507-3 – mf#1989-1732 – us ATLA [240]

After death : is there a postmortem probation? / Randles, Marshall – London: Charles Kelly, 1904 [mf ed 1991] – 1mf – 9 – 0-7905-9595-8 – mf#1989-1320 – us ATLA [240]

After death – what? : or, hell and salvation. considered in the light of science and philosophy / Platt, William Henry – 2nd rev enl ed. San Francisco: A Roman, 1878 [mf ed 1991] – 1mf – 9 – 0-7905-8874-9 – mf#1989-2099 – us ATLA [210]

After death, what? : a scholarly exposition of a vitally interesting question that has deeply agitated thinking men and women from time immemorial / Peters, Madison Clinton – New York: Christian Herald, c1908 [mf ed 1993] – (= ser Baptist coll) – 1mf – 9 – 0-524-07636-7 – mf#1991-3243 – us ATLA [110]

After england–we / Maloney, Arnold Hamilton – Boston, MA. 1949 – 1r – 1 – us UF Libraries [972]

After fifty years : or, an historical sketch of the guntur mission of the evangelical lutheran church of the general synod in the usa / Wolf, Luther Benaiah – Philadelphia: Lutheran Pub Soc, c1896 [mf ed 1986] – 1mf – 9 – 0-8370-6542-9 – (incl ind) – mf#1986-0542 – us ATLA [242]

After fifty years : or, letters of a grandfather: on occasion of the jubilee of the free church of scotland in 1893 / Blaikie, William Garden – London, New York: T Nelson, 1893 [mf ed 1989] – 1mf – 9 – 0-7905-4142-4 – mf#1988-0142 – us ATLA [242]

After leaving mr. mackenzie / Rhys, Jean – New York, USA. 1931 – 1r – 1 – us UF Libraries [420]

After mother india / Field, Harry Hubert – New York: Harcourt, Brace & Co [c1929] [mf ed 1984] – 1r [ill] – 1 – (filmed with: the king of court poets / gardner, e g) – mf#1303 – us Wisconsin U Libr [306]

After pentecost, what? : a discussion of the doctrine of the holy spirit in its relation to modern christological thought / Campbell, James Mann – New York: Fleming H Revell, 1897 [mf ed 1986] – 1mf – 9 – 0-8370-9767-3 – (incl ind) – mf#1986-3767 – us ATLA [240]

After school journal – 1990 spring – 1 – mf#4841719 – us WHS [071]

After seventeen years : a picture album, with a little tale / ed by Hagin, Fred Eugene – Japan: [s.n.], 1917 [mf ed 1995] – (= ser Yale coll) – 32p (ill) – 1 – 0-524-10083-7 – mf#1995-1083 – us ATLA [920]

After the arusha declaration / Nyerere, Julius Kambarage – Dar Es Salaam, Tanzania. 1967 – 1r – 1 – us UF Libraries [960]

After twenty-five years : a plea and a plan for the help of poor children in the juniata valley / Emmert, David – [s.l: s.n, 1905?] [mf ed 1992] – 1mf – 9 – 0-524-02885-0 – mf#1990-4476 – us ATLA [242]

After whitsitt what? / Mitchell, S C – 1 – $5.00 – us Southern Baptist [242]

After work : home reading for the family circle – La Salle 1874-87 – 1 – mf#4744 – us UMI ProQuest [640]

After you, columbus / Mielche, Hakon – London, England. 1950 – 1r – 1 – us UF Libraries [972]

After-action report, third us army : 1 august 1944-9 may 1945 / U.S. Army – 3r – 1 – $390.00 – mf#S1651 – Center for Military History – us Scholarly Res [355]

After-death communications / Bazett, L Margery – New York: H Holt and Co, 1920 [mf ed 1986] – (= ser Psychic series) – 119p – 1 – (int by j arthur hill) – mf#1676 – us Wisconsin U Libr [130]

Afterimage – Rochester NY 1972+ – 1,5,9 – ISSN: 0300-7472 – mf#10712 – us UMI ProQuest [770]

The aftermath : based on original records, 1818-1826 / Choksey, Rustom Dinshaw – Bombay: New Book Co, 1950 – (= ser Samp: indian books) – us CRL [320]

The after-math of a revolution : being the inaugural address as president of the united empire loyalists association, delivered november 12th, 1896 / Ryerson, George Sterling – Toronto: W Briggs, 1896 – 1mf – 9 – mf#12907 – cn Canadiana [975]

Afternoons in the college chapel : short addresses to young men on personal religion / Peabody, Francis Greenwood – Boston: Houghton, Mifflin, c1898 [mf ed 1991] – 1mf – 9 – 0-7905-9568-0 – mf#1989-1293 – us ATLA [243]

Afterschool alert : poll report: a report of findings from the 1999 mott foundation – [Washington DC: Afterschool Alliance]: [Washington DC]: US Dept of Education, Office of Educ Research & Improvement...[2000] [mf ed 2000] – 1mf – 9 – us GPO [362]

An afterword *see* The master of the isles / an afterword / a robin song / the tragedy of willow / the faithless lover / the faithful love

Aftistes et repertoire des scenes de saint-domingu / Fouchard, Jean – Port-Au-Prince, Haiti. 1955 – 1r – 1 – us UF Libraries [972]

Aftonbladet – Goteborg, Sweden. 1908-30 – sw Kungliga [078]

Aftonbladet – Stockholm: Stockholms-Tidningen tr., jul 1938-aug 1939; dec 1943-jun 1945; 1948-jun 1950 – 1 – us CRL [079]

Aftonbladet – Stockholm, Sweden. 1830- – 1 – sw Kungliga [078]

Aftonbladet newsbills – Stockholm, 1935-78 – 9 – sw Kungliga [078]

Aftonposten – Goteborg, Sweden. 1951-56 – 40r – 1 – sw Kungliga [078]

Aftontidningen – Stockholm, 1889-90 – 9 – sw Kungliga [078]

Aftontidningen – Stockholm, Sweden. 1909-20 – 48r – 1 – sw Kungliga [078]

Aftontidningen – Stockholm, Sweden. 1942-56 – 135r – 1 – sw Kungliga [078]

Aftontidningen newsbills – Stockholm, 1942-56 – 7r – 1 – sw Kungliga [078]

Aftontidningen semiweekly edition – Stockholm, 1914-20 – 9 – sw Kungliga [078]

Aftra, sag, seg san diego take – n1-5 [1978 nov-1980 jan] – 1 – mf#644057 – us WHS [071]

Aftra san diego newsletter – n1-n65 [1971 may-1978 may] – 1 – mf#644052 – us WHS [071]

Afva evaluations / American Film & Video Association – Fort Atkinson WI 1988-92 – 1,5,9 – (cont: efla evaluations) – mf#9916,01 – us UMI ProQuest [790]

Afyon-karahisarda nur – Afyon, 1924-28. Sahibi: Ahmed Sami [Onur], Mueduerue: Tahir Hayreddin. n3-52 (gaps). 1 haziran 1340 [1924]-15 haziran 1928 – (= ser O & t journals) – 6mf – 9 – $125.00 – us MEDOC [956]

Afzal Iqbal *see* My life, a fragment

Ag chem & commercial fertilizer – Parsippany NJ 1946-73 – 1,5 – ISSN: 0092-0037 – mf#60 – us UMI ProQuest [630]

Ag consultant – Willoughby OH 1986-99 – 1,5,9 – (cont: ag consultant and fieldman) – ISSN: 0894-7155 – mf#1415,04 – us UMI ProQuest [630]

Ag consultant and fieldman – Willoughby OH 1980-86 – 1,5,9 – (cont by: ag consultant; cont: agri-fieldman and consultant) – ISSN: 0199-6460 – mf#1415.04 – us UMI ProQuest [630]

Aga monthly – Washington DC 1984-88 – 1,5,9 – (cont: american gas association monthly; cont by: american gas) – ISSN: 0885-2413 – mf#10410.02 – us UMI ProQuest [550]

Aga, Silahdar Findiklili Mehmet *see* Silahdar tarihi

Die agada der babylonischen amoraeer / Bacher, Wilhelm – Frankfurt, 1913 – €10.00 – ne Slangenburg [240]

Die agada der babylonischen amoraeer : ein beitrag zur geschichte der agada und zur einleitung in den babylonischen talmud / Bacher, Wilhelm – 2., durch Ergaenzungen und Berichtigungen verm Aufl. Frankfurt aM: J Kauffmann, 1913 – 1mf – 9 – 0-8370-2135-9 – mf#1985-0135 – us ATLA [270]

Die agada der palaestinensischen amoraeer / Bacher, Wilhelm – Strassburg. v1-3. 1892-1899 – €67.00 – ne Slangenburg [240]

Die agada der tannaiten / Bacher, Wilhelm – 8 – (v1: von hillel bis akiba, strassburg 1884 8mf €16. v2: von akiba's tod bis zum abschluss der mischna, strassburg 1890 10mf €19) – ne Slangenburg [270]

Agadat bereshit – Krakow, Poland. 1902 – 1r – 1 – us UF Libraries [939]

Agadat ma'amarim / Krochmal, Abraham – Lemberg, Ukraine. 1885 – 1r – 1 – us UF Libraries [939]

Agafonov, V K *see* Izdanie gruppy sotsialistov-revoliutsionerov

Against the current – v1 n1-3:1 [1980 fall-1985 winter]; ns: v1 n1-v2 n5 [1986 jan-1987 dec] – 1 – mf#632003 – us WHS [071]

Against the world / Brown, Douglas – Garden City, USA. 1968 – 1r – 1 – us UF Libraries [960]

Against these three / Cloete, Stuart – Garden City, USA. 1945 – 1r – 1 – us UF Libraries [960]

The agamasastra of gaudapada / Gaudapada Acarya; ed by Bhattacharya, Vidhushekhara – Calcutta: University of Calcutta, 1943 – (= ser Samp: indian books) – (trans and ann by ed) – us CRL [280]

The agape and the eucharist in the early church : studies in the history of the christian love-feasts / Keating, John Fitzstephen – London: Methuen, 1901 [mf ed 1989] – 1mf – 9 – 0-7905-1173-8 – (incl bibl ref & ind) – mf#1987-1173 – us ATLA [240]

Agape, dans le n t / Spicq, C – Paris, 1958 – 3v on 20mf – 8 – €38.00 – ne Slangenburg [225]

L'agape dans l'eglise primitive / Ermoni, Vincent – Paris: Bloud, 1906 [mf ed 1992] – (= ser Science et religion) – 62p on 1mf – 9 – 0-524-03462-1 – (in french. incl bibl ref) – mf#1990-1005 – us ATLA [210]

Agapov, D V *see* Alfavitnyi katalog russkikh knig po matematike, vyshedshikh v rossii s nachala knigopechataniia do poslednego vremeni

Agar, J *see* American orator's own book

Agar, Mrs *see* The adventures of a serf's wife among the mines of siberia

The agaria / Elwin, Verrier – [Bombay]: Humphrey Milford: Oxford University Press, 1942 – (= ser Samp: indian books) – (foreword by sarat chandra roy) – us CRL [307]

Agar-O'connell, R M *see*
- Iintsomi

Agarwal, S N *see* The two worlds

Agarwal, Shriman Narayan *see* The medium of instruction

Agarwala, Amar Narain *see*
- A critique of the industrialists' plan
- Gandhism
- Health insurance in india
- Insurance finance
- Social insurance planning in india
- Some economic issues of transition and planning in india
- The ukcc and india

Agassiz, british columbia : the home of the dominion experimental farm – [Agassiz, BC?]: Agassiz Board of Trade, [19–?] – 1mf – 9 – 0-665-98344-1 – mf#98344 – cn Canadiana [917]

Agassiz, Lewis *see* A journey to switzerland

Agassiz, Louis *see*
- Bibliographia zoologiae et geologiae
- An essay on classification
- Lake superior
- Monographie des poissons fossiles du vieux gres rouge, ou systeme d,vonien (old red sandstone) des eles britanniques et de russie
- Viagem ao brasil, 1865-1866

Agathiae Myrinaei *see* Historiarum libri 5 (cshb1)

Agathiae Scholastici *see* De imperio et rebus gestis iustiniani (cbh4)

Agawam 1855-1892 – Oxford, MA (mf ed 1987) – (= ser Massachusetts vital records) – 17mf – 9 – 0-87623-011-7 – (mf 1-8: births, marriages, deaths 1855-92. mf 9-11: index to births 1855-92. mf 12-14: index to marriages 1855-92. mf 15-17: index to deaths 1855-92) – us Archive [978]

Agazzari, Agostino *see*
– Il primo libro de madrigaletti a tre voci
– Sacrae cantiones...liber quartus
– Sacrae laudes de jesu...liber secundus
– Sacrarum cantionum. liber tertius
– Sacrarum cantionum...liber 2
– Sacrarum cantionum...liber primus

Agba eyiogbe : a santeria guide / Lopes Valdes, Rafael L – s.l, s.l, 1960 – 1r – 1 – us UF Libraries [972]

Agc news and comment – 1988 aug 22-1994 dec – 1 – mf#1655370 – us WHS [071]

Agc reporter – [1973 jan/feb/mar-1975 oct/nov/dec] – 1 – mf#342976 – us WHS [071]

Agder bispedomme : arbok – v1-33. 1951-83 [complete] – 3r – 1 – mf#ATLA S0510 – us ATLA [073]

Age – Melbourne, 1854-1977 – 730r – 1 – at Pascoe [079]

Age – summer 1984-87 – 1 – mf#1051792 – us WHS [071]

Age – v1 n1-1926 [1879 jan 4-jun 28] – 1 – mf#361936 – us WHS [071]

The age – 1854- mthly updates – 1 – us Primary [070]

The age – Melbourne, 1859- – 24r per y – 1 – us UMI ProQuest [071]

The age – Melbourne, Australia. -d. Jan 1941-Oct 1974. 581 reels – 1 – uk British Libr Newspaper [072]

The age – York, PA., 1890-1900 – 13 – $25.00r – us IMR [071]

Age and ageing – Oxford, England 1985+ – 1,5,9 – ISSN: 0002-0729 – mf#15304 – us UMI ProQuest [618]

The age and the church : being a study of the age, and of the adaptation of the church to its needs / Stuckenberg, John Henry Wilbrandt – Hartford, CT: Student Publ, c1893 [mf ed 1990] – 1mf – 9 – 0-7905-6633-8 – mf#1988-2633 – us ATLA [240]

The age and the gospel : four sermons preached...hulsean lecture, 1864 / Moore, Daniel – London: Rivingtons, 1865 [mf ed 1989] – 1mf – 9 – 0-7905-1251-3 – (incl bibl ref) – mf#1987-1251 – us ATLA [240]

The age and the ministry : a sermon. delivered to the students of horton college... / Webb, James – Leeds: John Heaton, 1851 – 1mf – 9 – 0-524-07924-2 – mf#1991-3469 – us ATLA [240]

Age differences in performance of a coincident anticipation task: application of a modified information processing model / Williams, Kathleen – 1982 – 9 – $12.00 – us Kinesology [150]

Age [edinburgh, scotland] – [Scotland] Edinburgh: W Bryson 2 jan 1858-31 mar 1860 (wkly) [mf ed 2004] – 3r – 1 – uk Newsplan [072]

Age [london] : the leading weekly conservative journal – London, England 15 may 1825-1 oct 1843 (wkly) [mf 1828-30] – 8r – 1 – (amalg with: the argus & subsequently publ as: the age & argus) – uk British Libr Newspaper [072]

Age [melbourne] – Melbourne, Australia 17 oct 1854- – 1 – uk British Libr Newspaper [079]

Age newsletter – v1 n1 [1979 fall]; v1 n2-v1 n3 [1980/81 winter-1981 spring] – 1 – mf#668966 – us WHS [071]

The age of charlemagne (charles the great) / Wells, Charles Luke – New York: Christian Literature Co, 1898 [mf ed 1990] – (= ser Ten epochs of church history 4) – 2mf – 9 – 0-7905-6273-1 – mf#1988-2273 – us ATLA [931]

The age of charlemagne (charles the great) / Wells, Charles Luke – New York: The Christian Literature Co., 1898. xix,472p.(Ten Epochs of Church History IV). Bibliography – 1 – us Wisconsin U Libr [900]

The age of disfigurement / Evans, Richardson – London: Remington, 1893 (mf ed 19–) – 112p – mf#ZM-3-MAR pv61 n8 – us Harvard [650]

The age of erasmus : lectures / Allen, P S – Oxford: Clarendon Press; New York: Oxford University Press, 1914 – 1mf – 9 – 0-7905-4308-7 – mf#1988-0308 – us ATLA [190]

The age of faith / Bradford, Amory Howe – Boston: Houghton, Mifflin, 1900 – 1mf – 9 – 0-8370-4869-9 – (incl bibl ref and index) – mf#1985-2869 – us ATLA [240]

The age of hildebrand / Vincent, Marvin Richardson – New York: Christian Literature, 1896 – (= ser Ten Epochs Of Church History) – 2mf – 9 – 0-7905-6142-5 – (incl bibl ref) – mf#1988-2142 – us ATLA [240]

The age of hus / Workman, Herbert Brook – London: Charles H Kelly, 1902 – (= ser Books for bible students) – 1mf – 9 – 0-524-00665-2 – (incl bibl ref and ind) – mf#1990-0165 – us ATLA [240]

Age of innocence / Wharton, Edith – New York, USA. 1920 – 1r – 1 – us UF Libraries [830]

Age of introduction and current frequency of participation in league and casual bowlers / Duray, Nicholas A – 1982 – 9 – $4.00 – us Kinesology [790]

Age of jackson / Schlesinger, Arthur Meier – Boston, MA. 1945 – 1r – 1 – us UF Libraries [975]

The age of revolution : being an outline of the history of the church from 1648 to 1815 / Hutton, William Holden – London: Rivingtons, 1908 [mf ed 1990] – (= ser The church universal 7) – 1mf – 9 – 0-7905-5342-2 – (incl bibl ref) – mf#1988-1342 – us ATLA [240]

The age of romanticism / Rosenwald, Henry M – New York: F Ungar Pub Co, c1959 [mf ed 1993] – (= ser Living german literature 3) – 189p – 1 – (incl bibl ref and ind) – mf#8186 – us Wisconsin U Libr [430]

The age of schism : being an outline of the history of the church from a d 1304 to a d 1503 / Bruce, Herbert – London: Rivingtons, 1907 [mf ed 1992] – (= ser The church universal 5) – 1mf – 9 – 0-524-02700-5 – (incl bibl ref) – mf#1990-0681 – us ATLA [931]

The age of the crusades / Ludlow, James Meeker – New York: Christian Literature Co, 1896 [mf ed 1992] – (= ser Ten epochs of church history 6) – 1mf – 9 – 0-524-02703-X – (incl bibl ref) – mf#1990-0684 – us ATLA [931]

The age of the fathers : being chapters in the history of the church during the fourth and fifth centuries / Bright, William – London, New York: Longmans, Green, 1903 – 3mf – 9 – 0-7905-4492-X – mf#1988-0492 – us ATLA [240]

The age of the great western schism / Locke, Clinton – New York: Christian Literature Co., 1896 – (= ser Ten Epochs Of Church History) – 1mf – 9 – 0-7905-5175-6 – mf#1988-1175 – us ATLA [240]

The age of the imperial guptas / Banerji, Rakhal Das – [Varanasi]: Benares Hindu University, 1933 – (= ser Samp: indian books) – us CRL [954]

The age of the maccabees : with special reference to the religious literature of the period / Streane, Annesley William – London, New York: Eyre and Spottiswoode, 1898 – (= ser Bible student's library) – 1mf – 9 – 0-8370-9988-9 – (incl bibl ref and ind) – mf#1986-3988 – us ATLA [221]

Age of the nandas and mauryas / ed by Sastri, K A Nilakanta – Banaras: Publ for the Bharatiya Itihas Parishad by Motilal Banarsidas, 1952 – (= ser Samp: indian books) – us CRL [930]

The age of the renascence : an outline sketch of the history of the papacy from the return from avignon to the sack of rome (1377-1527) / Dyke, Paul van – New York: Christian Literature, 1897 [mf ed 1990] – (= ser Ten epochs of church history 7) – 1mf – 9 – 0-7905-6026-7 – mf#1988-2026 – us ATLA [931]

The age of unreason : being a reply to thomas paine, robert ingersoll, felix adler, o b frothingham, and other american rationalists / Brann, Henry Athanasius – 2nd ed. New York: Martin B Brown, 1881, c1880 [mf ed 1991] – 1mf – 9 – 0-7905-9145-6 – mf#1989-2370 – us ATLA [140]

The age of wyclif / Workman, Herbert Brook – London: Charles H Kelly, 1901 – (= ser Books for bible students) – 1mf – 9 – 0-524-00666-0 – (incl bibl ref) – mf#1990-0166 – us ATLA [240]

Agecv express : le communique officiel des etudiants – [Valleyfield]: [s.n.], [ca 1975]- (irreg) [mf ed 1988] – 9 – (ceased in 197-?) – mf#SEM105P1002 – cn Bibl Nat [073]

Aged christian's hope / Goode, William – London, England. 1813 – 1r – 1 – us UF Libraries [240]

Aged disciple – London, England. 18– – 1r – 1 – us UF Libraries [240]

Aged minister's encouragement to his younger brethren / Wilson, Daniel – London, England. 1821 – 1r – 1 – us UF Libraries [242]

Aged rector's valedictory address to his parishioners... / Wilson, Harry Bristow – London, England. 1853 – 1r – 1 – us UF Libraries [240]

Aged widow – London, England. 18– – 1r – 1 – us UF Libraries [240]

Ageefep : journal des etudiants et des etudiantes de la fep / Association generale des etudiants de la Faculte de l'education permanente de l'Universite de Montreal – v1 n1 sep 1985-v1 n2 nov 1985 (mthly) [mf ed 1989] – 1mf – 9 – (cont by: revue de l'ageefep) – mf#SEM105P1164 – cn Bibl Nat [378]

Ageing and society – Cambridge, England 1987+ – 1,5,9 – ISSN: 0144-686X – mf#16514 – us UMI ProQuest [300]

Ageing research reviews – Oxford, England 2002+ – 1,5,9 – ISSN: 1568-1637 – mf#42889 – us UMI ProQuest [618]

Agence de presse libre du Quebec *see*
– Bulletin de l'agence de presse libre du quebec
– Bulletin populaire

Agence economique and financiere [agefi] – Paris, France 23 may 1942-12 may 1944 (very imperfect) – 1r – 1 – uk British Libr Newspaper [330]

Agence economique des territoires francais sous mandat – n.s., no. 1-26. 1926-28. devenu: Togo-Cameroun. 1929-janv 1937. devenu: Cameroun. Paris. avr-dec 1937 – 1 – (devenu: togo-cameroun. 1929-janv 1937; devenu: cameroun. paris. avr-dec 1937) – fr ACRPP [073]

Agencia Cooperativa De Exportacion De Azucar (Havana) *see* Estatutos de la agencia cooperativa de exportacion

The agency and mercantile, shipping, agricultural, advertising and general register – Freetown. Sierra Leone. -w. 24 Oct-21 Nov 1884, 31 Dec 1886-15 Apr 1887. (8 ft) – 1 – uk British Libr Newspaper [072]

The agency contract and the power of attorney applied to business in indonesia / Pratomo, R – Djakarta, 1972. LL-10032 – 1 – us L of C Photodup [346]

The agency of a.b. steinberger in the samoan islands : message from the president of the united states to the congress / Grant, President – 44th Congress 2nd sess. House Exec Doc no 44 24 Feb 1877. n.p,n.d. – 2mf – 9 – $3.00 – mf#LLMC 82-100C Title 42 – us LLMC [327]

Agency of human means in the propagation of the gospel / Russell, Michael – Edinburgh, Scotland. 1828 – 1r – 1 – us UF Libraries [240]

Agency sales – Lake Forest CA 1973+ – 1,5,9 – ISSN: 0749-2332 – mf#9623 – us UMI ProQuest [650]

Agend buechlein fuer die pfarrherren auff dem land / Dietrich, Veit – Nuernberg 1569 – (= ser Hqab. literatur des 16. jahrh.) – 3mf – 9 – €40.00 – mf#1569b – gw Fischer [780]

Agend buechlein fuer die pfarrherren auff dem land / Dietrich, Veit – Nuernberg: VomBerg & Newber 1553 – (= ser Hqab. literatur des 16. jahrh.) – 3mf – 9 – €40.00 – mf#1553a – gw Fischer [780]

Agenda – 1973 jun 7-1974 dec; 1975 jan-1976 jun; 1976 jul-1978 oct 12 – 1 – mf#913108 – us WHS [071]

Agenda – 1973 mar 22-29; 1973 apr 19; 73 apr 25; 1974 may 2-1975 apr 17; 1975 apr 24-1976 oct 28; 1976 nov4-1978 apr 27; 1978 may 4-1978 dec 28; 1979 jan 4-oct 25 – 1 – mf#999740 – us WHS [071]

Agenda – 1992 mar-nov; 1993 jan/feb-1994 aug/sep; 1995 may/jun – 1 – mf#2662098 – us WHS [071]

Agenda : d. i. kirchenordnung...fuer die diener der kyrchen in hertzog heinrichen zu sachsen v. g. h. fuerstenthumb gestellet – Jena: Richtzenhan 1580 – (= ser Hqab. literatur des 16. jahrh.) – 2mf – 9 – €30.00 – mf#1580a – gw Fischer [240]

Agenda : d i kyrchenordnung wie sich die pfarrherrn vnd seelsorger in jren ampten vn diensten halten sollen; fur die diener der kyrchen in hertzog heinrichen zu sachssen v g h fuerstenthumb gestellet – Leipzig: Wolrab 1540 – (= ser Hqab. literatur des 16. jahrh.) – 2mf – 9 – €30.00 – mf#1540a – gw Fischer [240]

Agenda : dat is ordninge der hilligen kerckenempter unde ceremonien wo sich de parrherren, seelsorgere unde kerckendenere in erem ampte holden schoelen – Stettin: Kellner [15]91 – (= ser Hqab. literatur des 16. jahrh.) – 10mf – 9 – €90.00 – mf#1591a – gw Fischer [780]

Agenda : das ist kirchenordnung, wie sich die pfarrherrn vnd seelsorger in jren ampten vnd diensten halten sollen, fuer die diener der kirchen in hertzog heinrichen zu sachsen v g h fuerstenthumb gestellet – Leipzig: Rhambaw 1564 – (= ser Hqab. literatur des 16. jahrh.) – 2mf – 9 – €30.00 – mf#1564a – gw Fischer [240]

Agenda – London, England 1959+ – 1,5,9 – ISSN: 0002-0796 – mf#6612 – us UMI ProQuest [400]

Agenda and diaries / Duncan, Irma – 3v. 1921-24 (mf ed 1990) – 1r – 1 – mf#*ZBD-507 – us NY Public [790]

Agenda book of the south african indian congress conference – [Johannesburg: The Congress]. v20 1952; v22 1956 – 1 – us CRL [960]

Agenda book of the...provincial conference / Natal Indian Congress – 1940-41, 1944, 1947-48, 1951-55, 1957 – 1 – us CRL [960]

Agenda coloniensis ecclesiae – 1614 – 9mf – 8 – €18.00 – ne Slangenburg [240]

Agenda communis : die aelteste agende in der dioezese ermland und im deutschordenstaate preussen / Kolberg, A – Braunsberg, 1903 – 3mf – 8 – €7.00 – ne Slangenburg [931]

Agenda czeska to gest spis o ceremoniich a poradcych cyrkewniich : kterak se slowem bossiim a swatostmi krystowymi lidu w kralowstwii czeskem prawdu ewanglium swateho magiicymu a milugiicymu posluhowati ma – Lipsste 1581 – (= ser Hqab. literatur des 16. jahrh.) – 3mf – 9 – €40.00 – mf#1581a – gw Fischer [780]

Agenda das ist : kirchenordnung, wie es im fuerstenthumb hessen mit verkuendigung goettliches worts, reichung der heiligen sacramenten und andern christlichen handlungen und ceremonien gehalten werden soll – Marpurgk: Colbius 1574 – (= ser Hqab. literatur des 16. jahrh.) – 6mf – 9 – €70.00 – (p16 & 92 missing) – mf#1574a – gw Fischer [240]

Agenda dat is ordninge der hilligen kerckenempter unde ceremonien wo sich de parrherren, seelsorgere unde kerckendenere in erem ampte holden schoelen – [Wittenberg]: [Schwertel] 1569 – (= ser Hqab. literatur des 16. jahrh.) – 10mf – 9 – €90.00 – mf#1569a – gw Fischer [780]

Agenda der dioezese naumburg von 1502 / ed by Schoenfelder, A – Paderborn, 1904 – 1mf – €3.00 – ne Slangenburg [241]

Agenda der dioezese schwerin von 1521 / ed by Schoenfelder, A – Paderborn, 1906 – €7.00 – ne Slangenburg [241]

Agenda du forestier pour... / Societe forestiere de Franche-Comte & Belfort – Besancon: P Jacquin (annual) [2. annee (1901)] [mf ed 2005] – 1r [ill] – 1 – (cont by: carnet-agenda du forestier pour...; began with 1re ed [1900]?) – mf#film mas 36970 – us Harvard [634]

Agenda for the commonwealth : a government plan for the northern marianas / Mantel, Howard N & Calvert, Stephen R – New York: Inst of Public Admin, dec 1977 – 7mf – 9 – $10.50 – mf#llmc82-100j, title 20 – us LLMC [350]

Agenda scholastica / ed by Haehn, Johann Friedrich – Berlin 1749-52 – (= ser Dz) – 10st on 6mf – 9 – €120.00 – mf#k/n604 – gw Olms [370]

Agenda seu obsequiale, simul ac benedictionale, iuxta ritum et normam ecclesiae et episcopatus constantiensis – [Dilingae]: Mayer 1570 – (= ser Hqab. literatur des 16. jahrh.) – 4mf – 9 – €50.00 – mf#1570a – gw Fischer [780]

Agenda Sindical *see* Badajoz-71

Agend-buechlein fuer die pfarrherrn auff dem land / Dietrich, Veit – Franckfurdt am Mayn: Guelfferich 1546 – (= ser Hqab. literatur des 16. jahrh.) – 3mf – 9 – €40.00 – mf#1546a – gw Fischer [240]

Agende der allgemeinen, evangelisch-lutherischen synode von ohio und andern staaten – Columbus OH: Schulze & Gassmann 1870 [mf ed 1993] – 1mf – 9 – 0-524-07238-8 – mf#1991-2979 – us ATLA [242]

Agende fuer christliche gemeinden des lutherischen bekenntnisses / Loehe, Wilhelm – 2. verm aufl. Noerdlingen: C H Beck 1853-59 [mf ed 1993] – 2v on 5mf – 9 – 0-524-07096-2 – mf#1991-2919 – us ATLA [242]

Agenet : a national resource for social workers in the field of aging – Washington DC: National Assoc of Social Workers [1993] [mf ed 1994] – 1mf – 9 – us GPO [360]

Agenskalna baptist church, riga : fifty years of the agenskalna baptist church of riga = Rigas agenskalna baptistu draudses 50 gadi – Riga. 152p. 1934 – 1 of 6 items on reel – 1 – mf#6346 n3 – us Southern Baptist [242]

Der agent : roman / Lindau, Paul – Breslau: S Schottlaender; New York: G E Stechert 1899 [mf ed 1995] – 1r – 1 – (filmed with: lichtenberg / paul requadt) – mf#3691p – us Wisconsin U Libr [830]

El agente de matrimonios / Lopez de Ayala, Adelardo – Madrid: Imprenta de Jose Rodriguez, 1862 – sp Bibl Santa Ana [780]

Agents and actions – Dordrecht, Netherlands 1969-94 – 1,5,9 – (cont by: inflammation research) – ISSN: 0065-4299 – mf#5141.01 – us UMI ProQuest [615]

The agents' companion – London, Ont: Companion Pub Co, [1874-18–?] – 9 – mf#P04342 – cn Canadiana [070]

Agenty moskovskogo strakhovogo ot ognia obshchestva – M, 1872 – 1mf – 9 – mf#REF-415 – ne IDC [332]

Agenty russkogo dela *see* Vivos voco!

Agenutemagen / Union of New Brunswick Indians – 1972 nov – 1r – 1 – mf#1051729 – us WHS [305]

Age-related changes in the release point, velocity and acceleration in girls' overarm throwing performance / Yan, Jin H & Payne, V Gregory – 1992 – 1mf – 9 – $4.00 – us Kinesology [612]

The ages before moses : a series of lectures on the book of genesis / Gibson, John Monro – New York: Anson D F Randolph, c1879 [mf ed 1985] – 1mf – 9 – 0-8370-3264-4 – mf#1985-1264 – us ATLA [221]

Ages of christendom before the reformation / Stoughton, John – London: Jackson & Walford, 1857 [mf ed 1990] – (= ser The congregational lecture 1855) – 2mf – 9 – 0-7905-5977-3 – (incl bibl ref) – mf#1988-1977 – us ATLA [240]

AgExporter – Washington DC 1989+ – 1,5,9 – (cont: foreign agriculture) – ISSN: 1047-4781 – mf#16928 – us UMI ProQuest [630]

The age/york press – York, PA. -d 1897-1898 – 13 – $25.00r – us IMR [071]

Aggarawala, Om Prakash *see* Fundamental rights and constitutional remedies

L'agglomeration dakaroise; quelques aspects sociologiques et demographiques / Institut Francais d'Afrique Noire – Saint Louis, Senegal, 1954 – 1 – us CRL [316]

Aggregates of polls, property and taxes / Massachusetts. Dept. of Corporations and Taxation – 1861-1940. 94 fiches. (Harvard Law School Library Collection.) – 9 – us Harvard Law [324]

Aggression and its relationship with performance : perspectives of professional hockey players / Lauer, Larry L – 1998 – 2mf – 9 – $8.00 – mf#PSY 2026 – us Kinesology [150]

Aggression and violent behavior – Oxford, England 1996+ – 1,5,9 – ISSN: 1359-1789 – mf#49635 – us UMI ProQuest [150]

Aggressive irreligion / Eoule, Rowland Edmund Prothero – Oxford, England. 1886 – 1r – 1 – us UF Libraries [240]

Aggrey we africa / Macartney, William M – Pietermaritzburg, South Africa. 19– – 1r – 1 – us UF Libraries [960]

Aghnides, Nicolas Prodromou *see* Mohammedan theories of finance

Agia, Miguel *see* Servidumbres personales de indios

Agikuyu folk tales / Njururi, Ngumbu – London, England. 1966 – 1r – 1 – us UF Libraries [390]

Agila *see* Agila p'alante

Agila p'alante / Agila – Trujullo: Imp. Gexme, 1981 – 1 – sp Bibl Santa Ana [946]

Aging – Washington DC 1951-96 – 1,5,9 – ISSN: 0002-0966 – mf#2119 – us UMI ProQuest [618]

Aging and body composition : a 12 year longitudinal study of middle-aged and elderly women / Williams, Bryce C – 1996 – 1mf – 9 – $4.00 – mf#PH 1565 – us Kinesology [618]

Aging and work – Washington DC 1978-84 – 1,5,9 – ISSN: 0161-2514 – mf#11717 – us UMI ProQuest [618]

Agippa von Nettesheim, Heinrich Cornelius *see* Three books of accult philosophy

Agir pour l'insertion : initiatives d'insertion par l'economique au quebec / Bordeleau, Daniele & Valadou, Christian – Montreal: Institut de formation en developpement economique communautaire, 1995 [mf ed 2002] – 2mf – 9 – mf#SEM105P3485 – cn Bibl Nat [338]

L'agitateur – I, no. 1-12. Marseille. mars-mai 1892; II, no. 1-6. janv-fevr 1893; n.s., no. 1-2. fevr-mars 1897 – 1 – fr ACRPP [073]

Agitation in ireland : from a landlord's point of view / Staples, Robert, Jr – London, 1880 – (= ser 19th c ireland) – 1mf – 9 – mf#1.1.1902 – uk Chadwyck [333]

Agitator – 1884 may 10 – 1 – mf#3177704 – us WHS [071]

Agitator – 1901-09; 1910-1913 feb 8 – 1 – mf#949811 – us WHS [071]

Agitator – Cleveland, OH, feb 1858-apr 1 1860 – 1r – 1 – (monthly, later weekly free love/spiritualist newspaper) – us Western Res [071]

Agitator – Russian Federation 1961-67 – 1 – ISSN: 0320-7161 – mf#16200 – us UMI ProQuest [077]

Der agitator – Berlin DE, 1870 1 apr-1876 29 sep – 5r – 1 – (title varies: 2 jul 1871: neuer social-demokrat) – mf#177 – gw Mikropress [329]

Agitator (coos bay, or) – Coos Bay OR: F B Cameron, [wkly] – 1 – (cont by: sunday morning bee) – us Oregon Lib [071]

Agla – Montijo 1935 – 5 – sp Bibl Santa Ana [074]

Aglaia – New York. 1925+ (1) 1968+ (5) 1975+ (9) – 37mf – 9 – mf#1448 – ne IDC [077]

Aglaia of melos / Willson, Beckles – [Canada?: s.n, 1914?] – 1mf – 9 – 0-665-66047-2 – mf#66047 – cn Canadiana [810]

Aglaja : [a long poem] / Puchner, Rudolph – Milwaukee WI: C N Caspar 1887 [mf ed 1992] – 1r – 1 – (filmed with: heiterer guckkasten / bruno wolfgang) – mf#2865p – us Wisconsin U Libr [810]

Aglio, Augustine *see* Architectural ornaments

[Aglionby, W] *see* Painting illustrated in three diallogues...

Aglipay y Labayan, Gregorio *see* Biblia filipina

Agmazine – 1947 sep; 1948 jun; 1949 oct; 1950 feb-1961 nov; 1961 dec-1982 oct – 1 – mf#642651 – us WHS [071]

L'agneau de dieu : entretiens sur quelques textes des livres de saint jean / Blanc, Joseph – Rome: Institut biblique pontifical, 1913 [mf ed 1993] – xx/262p on 1mf – 9 – 0-524-05713-3 – (in french. incl bibl ref) – mf#1992-0556 – us ATLA [225]

Agnelli, J *see* Galleria di pitture dell'...tommaso ruffo, vescovo di palestrina, e di ferrara...

Agnes : trauerspiel / Braunfels, Ludwig; ed by Mahr, August C – Frankfurt/M: Freier Deutscher Hochstift; [Palo Alto, CA]: Stanford Universitaet, 1928 [mf ed 1989] – 1 – mf#7089p – us Wisconsin U Libr [820]

Agnes bernauer : ein deutsches trauerspiel in fuenf akten / Hebbel, Friedrich; ed by Evans, Marshall Blakemore – Boston: D C Heath, c1912 [mf ed 1995] – (= ser Heath's modern language series) – xxxiii/163p/1pl – 1 – (in german. int and notes by ed in english) – mf#8764 – us Wisconsin U Libr [820]

Agnes de chaillot : comedie / Biancolelli, Pierre-Francois – 2nd ed. Paris: chez Francois Flahaut, 1723 [mf ed 1991] – 1mf – 9 – mf#SEM105P1362 – cn Bibl Nat [870]

Agnes harcourt or "for his sake" : a canadian story illustrative of the power of a child's life – Montreal: Montreal Women's Printing Office, 1879 [mf ed 1980] – 1mf – 9 – 0-665-04007-5 – (repr fr: woman's work in canada) – mf#04007 – cn Canadiana [920]

Agnesian – 1966 jan/jul-1984 may – 1 – mf#894013 – us WHS [071]

Agnetheler wochenblatt – Agnetheln (Agnita) RO, 18 mar 1922-29 nov 1924 – 1r – 1 – (cont by: agnetheler zeitung) – Dist. gw Mikrofilm – gw Misc Inst [077]

Agnew, David Carnegie A *see*
- Englishmen introduced to the free church of scotland
- Protestant exiles from france..
- The theology of consolation

Agnew, Emily C *see*
- The merchant prince and his heir
- Saint mary and her times

Agnew, Mary *see* The pestilence that walketh in darkness

Agni : and other poems and translations / Paratiyar – Madras: Bharati Prachur Alayam, 1937 – (= ser Samp: indian books) – us CRL [810]

Agni bawana *see* Pakem kedjaksaan tinggi

L'agnistoma : description complete de la forme normale du sacrifice de soma dans le culte vedique / Caland, Willem & Henry, Victor – Paris: E Leroux, 1906-07 [mf ed 1992] – 2v (ill) on 2mf – 9 – 0-524-04325-6 – (in french. incl bibl ref) – mf#1990-3309 – us ATLA [280]

Agno, Jehan D' *see* Gendarme par telephone

Agno, Lydia Navarro *see* Kung paano namumuhay at gumagawa ang mga tao

The agnostic gospel : a review of huxley on the bible: with related essays / Parker, Henry Webster – New York: John B Alden, 1896, c1895 [mf ed 1985] – 1mf – 9 – 0-8370-4667-X – mf#1985-2667 – us ATLA [210]

Agnostic journal (and secular review), 1889-1907 – (= ser Periodicals connected with owenite socialism and its successors in secularist, freethought and allied movements, 1834-1916) – 18r – 1 – mf#97174 – uk Microform Academic [210]

Agnostic looks at life : challenges of a militant pen / Haldeman-Julius, Emanuel – Girard KS: Haldeman-Julius Co c1926 [mf ed 1986] – (= ser Big blue book 25) – 1r – 1 – (filmed with: the wisdom of life: being the first part of arthur schopenhauer's aphorismen zur lebensweisheit / trans, with pref by t bailey saunders [194-]) – mf#10758 – us Wisconsin U Libr [210]

Agnosticism : a doctrine of despair: a baccalaureate sermon, june 27, 1880 / Porter, Noah – New York: American Tract Society, [1880?] [mf ed 1985] – 1mf – 9 – 0-8370-5117-7 – mf#1985-3117 – us ATLA [210]

Agnosticism / Flint, Robert – New York: C Scribner, 1903 [mf ed 1991] – (= ser Croall lectures 1887-88) – 2mf – 9 – 0-7905-9376-9 – (incl bibl ref) – mf#1989-2601 – us ATLA [210]

Agnosticism : sermons preached in st peter's, cranley gardens, 1883-4 / Momerie, Alfred Williams – 4th rev ed. Edinburgh: William Blackwood, 1891 [mf ed 1985] – 1mf – 9 – 0-8370-4462-6 – mf#1985-2462 – us ATLA [221]

Agnosticism and theism in the nineteenth century : an historical study of religious thought. six lectures / Armstong, Richard Acland – London: P Green, 1905 [mf ed 1990] – 207p – 1 – (int by philip h wicksteed) – mf#7478 – us Wisconsin U Libr [210]

Agnosticism of hume and huxley : with a notice of the scottish school / McCosh, James – New York: Scribner 1884 [mf ed 1991] – 1mf – 9 – 0-7905-9802-7 – mf#1989-1527 – us ATLA [210]

Agnosticism writ plain / Gould, F J – London, England. 18– – 1r – 1 – us UF Libraries [240]

An agnostic's apology : and other essays / Stephen, Leslie – London, 1893 – (= ser 19th c evolution & creation) – 4mf – 9 – mf#1.1.11594 – uk Chadwyck [140]

An agnostic's apology : and other essays / Stephen, Leslie – New York: G P Putnam; London: Smith, Elder, 1893 [mf ed 1985] – 1mf – 9 – 0-8370-5398-6 – mf#1985-3398 – us ATLA [210]

An agnostic's progress / Palmer, William Scott – London, New York: Longmans, Green, 1906 [mf ed 1990] – 1mf – 9 – 0-7905-9561-3 – mf#1989-1286 – us ATLA [210]

Agnostos theos : untersuchungen zur formengeschichte religioeser rede / Norden, E – Leipzig-Berlin, 1913 – 8mf – 8 – €17.00 – ne Slangenburg [210]

Agnostos theos : untersuchungen zur formengeschichte religioeser rede / Norden, Eduard – Leipzig: BG Teubner, 1913 [mf ed 1991] – 4mf – 9 – 0-7905-8346-1 – (in german & greek. incl bibl ref) – mf#1987-6445 – us ATLA [225]

Agonia antillana / Araquistain, Luis – Madrid, Spain. 1928 – 1r – 1 – us UF Libraries [972]

La agonia del principe de la paz. discurso... 1923-1924 / Ossorio, Angel – Madrid: Est. Tip. Anonima Mefar, 1923 – 1 – sp Bibl Santa Ana [972]

Agonistes – Prague, CS. Dec 1953-Jan 1954 – 1r – 1 – us L of C Photodup [077]

Agora – Luanda, Angola. 1998 nov 07- 1999 may 01 – 1r – 1 – (1998 nov 21, dec - 1999 jan 02, jan 16-feb 27, apr 17) – us UF Libraries [071]

Agorastes, Phil *see* Amphioxus and ascidian

Agostini, Enzo *see* La france et le canada

Agostini, Jules *see* Au cambodge

Agostini, Lodovico *see* Canones, et echo sex vocibus

Agostini, Victor *see*
- Bibijaguas
- Dos viajes
- Hombres y cuentos

Agostino da Montefeltro, padre *see* Die wahrheit

Agosto Mendez, J M *see* Libro del centenario del congreso de angostura

Agra akhbar – Agra, India. 1947-53 – 1r – 1 – us L of C Photodup [079]

Agramer tagblatt – Zagreb, Croatia 15 feb 1909-30 apr 1912; 1-31 dec 1913; 2 jan-28 feb, 27 jun-27 jul 1914; 21 feb-2 apr 1919 – 1 – (most iss are incomplete) – uk British Libr Newspaper [077]

Agramer tagblatt – Zagreb, Yugoslavia. 1-7 Jun 1918 – 1r – 1 – us L of C Photodup [949]

Agramonte, Elpidio *see* Ritmo recondito, poemas

Agramonte Y Pichardo, Roberto Daniel *see*
- Biografia del dictador garcia moreno
- Biologia contra la democracia
- Filosofo y la comprension internacional
- Grandes momentos de la filosofia en cuba

Agrapha : aussercanonische evangelienfragmente / Resch, A – Leipzig, 1899 – (= ser Tugal 1-5/4a) – 8mf – 9 – €17.00 – ne Slangenburg [225]

Agrapha : aussercanonische evangelienfragmente / Resch, Alfred [comp] – Leipzig: JC Hinrichs, 1889 [mf ed 1990] – (= ser Tugal 5/4) – 1mf – 9 – 0-7905-3402-9 – (in german, latin & greek. with app: das evangelienfragment von fajjum by adolf von harnack) – mf#1987-3402 – us ATLA [240]

Agrapha : aussercanonische paralleltexte zu den evangelien / Resch, A – Leipzig, 1893-97 – (= ser Tugal 1-10/1.2.3.4.5) – 32p – 9 – €61.00 – ne Slangenburg [225]

Agrapha : aussercanonische schriftfragmente / Resch, A – Leipzig, 1906 – (= ser Tugal 2-30/3.4) – 7mf – 9 – €15.00 – ne Slangenburg [225]

Agrapha : aussercanonische schriftfragmente / ed by Resch, Alfred – Leipzig: JC Hinrichs, 1906 [mf ed 1989] – (= ser Tugal 30/3-4) – 1mf – 9 – 0-7905-1676-4 – (in german, greek & latin. incl bibl ref & ind) – mf#1987-1676 – us ATLA [221]

Agrapha, neue oxyrhynchuslogia / ed by Klostermann, Erich – Bonn: A Marcus & E Weber 1904 [mf ed 1992] – (= ser Kleine texte fuer theologische vorlesungen und uebungen 11/3) – 1mf – 9 – 0-524-04754-5 – (text in greek & latin, notes in german) – mf#1992-0196 – us ATLA [220]

Agrarian periodicals in the united states, 1920-1960 – Greenwood Press – 41 titles on 25r – 1 – $3905.00 – us UPA [630]

The agrarian reform – (Spain. Embajada. United States). Washington, DC. n.d. Fiche W 1176. (Blodgett Collection of Spanish Civil War Pamphlets) – 9 – us Harvard [946]

Agrarnaia politika tsarskogo pravitel'stva i krest'ianskii pozemel'nyi bank / Baturinskii, D A – M, 1925 – 3mf – 9 – mf#REF-263 – ne IDC [332]

Agrarnaia problema v sviazi s krestianskim dvizheniem / Peshekhonov, A V – 1906 – 136p 2mf – 9 – mf#RPP-201 – ne IDC [325]

Agrarnaia programma partii narodnoi svobody i ee posleduiushchaia razrabotka / Chernenkov, N N – 1907 – 79p 1mf – 9 – mf#RPP-132 – ne IDC [325]

Agrarnye programmy rossiiskikh politicheskikh partii v 1917 g / Morokhovets, E A – 1929 – 168p 2mf – 9 – mf#RPP-60 – ne IDC [325]

Agrarnyi vopros / Liubimov, L I – Kharkov, 1918 – 77p 1mf – 9 – mf#RPP-200 – ne IDC [325]

Agrarnyi vopros : protokoly zasedanii agrarnoi komissii, 11-13 fevr 1907 g s dokl i prilozh – Spb, 1907 – 458p 6mf – 9 – mf#RPP-110 – ne IDC [325]

Agrarnyi vopros i kooperatsiia : doklad, chitannyi v v krainskim...20 dek 1905 g / Krainskii, V V – 1906 – 47p 1mf – 9 – mf#COR-52 – ne IDC [335]

Agrarnyi vopros i kooperatsiia / Oganovskii, N P – 1917 – 62p 1mf – 9 – mf#COR-80 – ne IDC [335]

Agrarnyi vopros i sotsial-demokratiia / Ratner, M B – 1908 – 251p 3mf – 9 – mf#RPP-153 – ne IDC [325]

Agrarnyi vopros i sovremennyi moment : izvlechenie iz lektsii, prochitannoi 30 aprelia v universitete shaniavskogo v moskve / Chernov, V – n.d. – 23p 1mf – 9 – mf#RPP-254 – ne IDC [325]

Agrarnyi vopros o zemle i zemelnykh poriadkakh / Kondratev, N D – 1917 – 64p 1mf – 9 – mf#COR-199 – ne IDC [335]

Agrarnyi vopros v rossii i ego reshenie v programmakh razlichnykh partii / Rozhkov, N – 1906 – 41p 1mf – 9 – mf#RPP-39 – ne IDC [325]

Agrawala, Vasudeva Sharana *see* India as known to panini

Agraz / Cordero, Carmen – Camaguey, Cuba. 1946 – 1r – 1 – us UF Libraries [972]

Agreement between : the government of quebec and the societe d'energie de la baie james...and: the grand council of the crees (of quebec) and the james bay crees and the northern quebec inuit association... / Convention de la Baie James et du Nord quebecois (1975) – [Montreal: Conseil excecutif: Negociations Indiens Inuit de la Baie James, 1974] [mf ed 1985] – 11mf – 9 – mf#SEM105P448 – cn Bibl Nat [971]

Agreement between his majesty's government and the french government : respecting the boundary line between syria and palestine from the mediterranean to el hamme – London, 1923 – 1mf – 9 – mf#J-28-69 – ne IDC [956]

Agreement between palestine and syria and the lebanon amending the agreement of february 2, 1926, regarding frontier questions, nov 3 1938 – London, 1939 – 1mf – 9 – mf#J-28-103 – ne IDC [956]

Agreement between palestine and syria and the lebanon to facilitate good neighbourly relations in connection with frontier questions signed at jerusalem, feb 2 1926 – London, 1927 – 1mf – 9 – mf#J-28-192 – ne IDC [956]

Agreement between: the government of quebec and the societe d'energie de la baie james and the societe de developpement de la baie james and the commission hydroelectrique de quebec (hydro-quebec) and... / Quebec (Province). Conseil executif et al – [Montreal: Conseil executif: Negociations indiens Inuit de la Baie James, 1974] (mf ed 1984) – 11mf – 9 – mf#SEM105P448 – cn Bibl Nat [333]

The agreement between union seminary and the general assembly / Prentiss, George Lewis – New York: A.D.F. Randolph, c1892 [mf ed 1990] – 1mf – 9 – 0-7905-5794-0 – mf#1988-1794 – us ATLA [242]

The agreement recently concluded between his britannic majesty and his highness the amir of transjordan – Jerusalem, 20 feb 1928 – 1mf – 9 – mf#J-28-174 – ne IDC [956]

Agreements and subject files of the office of synthetic rubber, 1941-1953 / U.S. Reconstruction Finance Corporation – 41r – 1 – mf$_g$49 – us Nat Archives [338]

L'agrement des concerts de la rue feydeau – Paris: Laurens 1795 [mf ed 19–] – 1mf – 9 – mf#fiche57 – us Sibley [780]

Les agrements et le rythme : leur representation graphique dans la musique vocale francaise du 18e siecle...etude theorique et pratique contenant de nombreux textes et exemples des maitres francais... / Arger, Jane – Paris: Rouart, Lerolle [1921] [mf ed 19–] – 1r – 1 – mf#film 1351 – us Sibley [780]

L'agression vietcong et nord-vietnamienne contre la republique khmere : nouveaux documents – Phnom-Penh: Ministere de l'information 1971 [mf ed 1989] – 1r with other items – 1 – mf#mf-10289 seam reel 016/23 [§] – us CRL [959]

Les agressions sud-vietnamiennes des 7 et 8 mai 1964 contre le territoire cambodgien / Cambodia. Departement de l'information – [Phnom Penh 1964] [mf ed 1989] – [ill] 1r with other items – 1 – mf#mf-10289 SEA M reel 014/08 [§] – us CRL [327]

Agri marketing – Ann Arbor MI 1989+ – 1,5,9 – ISSN: 0002-1180 – mf#15342 – us UMI ProQuest [650]

Agri Research, Inc *see* Economic and technical feasibility of increased ma

Agribusiness – Hoboken NY 1985+ – 1,5,9 – ISSN: 0742-4477 – mf#14802 – us UMI ProQuest [630]

Agricola, F et al *see* Oratio de bello adversvs tvrcam, ad ferdinandum vngariae and bohemiae regem, and principes germaniae

Agricola, Martin *see*
– Musica choralis deudsch
– Musica figuralis deudsch
– Musica instrumentalis deudsch
– Musica instrumentalis deudschynn welcher begriffen ist wie mannoch dem gesange auff mancherley pfeiffen lernen sol auch wie auff die orgel harfen lauten geigen vnd allerley instrument vnd seytenspiel noch der rechtgegruendten tabelthur sey abzusetzen
– Rudimenta musices...
– Von den proporcionibus

Agricola's sprichwoerter : ihr hochdeutscher ursprung und ihr einfluss auf die deutschen und niederlaendischen sammler... / Latendorf, Friedrich – Schwerin: Baerensprung 1862 [mf ed Bloomington IN: Indiana Uni Lib, Preservation Dept 1984] – 1r – 1 – us Indiana Preservation [390]

L'agriculteur – Montreal: De Montigny, [1857-1862?] – 9 – mf#P04665 – cn Canadiana [630]

L'agriculteur – Saint-Boniface, Man: A Gauvin, [1889-1891] – 9 – mf#P04262 – cn Canadiana [630]

Agriculteur *see* Revue agricole, manufacturiere, commerciale et de colonisation

L'agriculteur canadien – Montreal: H A Chapet, [1886-189- ou 19–] – 9 – mf#P04081 – cn Canadiana [630]

El agricultor – Aldeanueva del Camino 1908 – 5 – sp Bibl Santa Ana [630]

O agricultor – Juiz de Fora, MG. set 1897 – (= ser Ps 19) – bl Biblioteca [630]

O agricultor brazileiro : jornal do fazendeiro – Rio de Janeiro, RJ: Typ de Nicolau Lobo Vianna & Filhos, nov 1853-out 1854 – (= ser Ps 19) – mf#P01B,05,09 – bl Biblioteca [630]

O agricultor progressista – Rio de Janeiro, RJ: Typ Nacional, 21 jul-11 dez 1881 – (= ser Ps 19) – mf#P05,04,05 – bl Biblioteca [630]

Agricultura – Madrid, Spain 1975-89 – 1,5,9 – ISSN: 0002-1334 – mf#8519 – us UMI ProQuest [630]

Agricultura metodica...naturaleza / Zepeda y Vivero, Juan A – 1791 – 9 – sp Bibl Santa Ana [630]

Agricultura subdesenvolvida – Petropolis, Brazil. 1969 – 1r – 1 – us UF Libraries [630]

Agricultural administration – Oxford, England 1974-86 – 1,5,9 – (cont by: agricultural administration & extension) – ISSN: 0309-586X – mf#42007.01 – us UMI ProQuest [630]

Agricultural administration and extension – Oxford, England 1987-88 – 1,5,9 – (cont: agricultural administration) – ISSN: 0269-7475 – mf#42007,01 – us UMI ProQuest [630]

Agricultural age *see* Oregon farmer-stockman

Agricultural and forest meteorology – Oxford, England 1964+ – 1,5,9 – ISSN: 0168-1923 – mf#42239 – us UMI ProQuest [550]

Agricultural and Industrial Exhibition (1st : 1879 : Toronto, Ont) *see* The authorized catalogue of the first annual exhibition of the agricultural and industrial exhibition association of toronto

Agricultural and Industrial Exhibition [1871: Quebec, Quebec] *see* Prize list for the agricultural and industrial exhibition

Agricultural and Industrial Exhibition (1878: Truro, Nova Scotia) *see* General regulations and prize list...

Agricultural and Industrial Exhibition (1880: Kentville, Nova Scotia) *see* General regulations and prize list...

Agricultural and Rural Workers *see* Rural worker

The agricultural and social state of ireland in 1858 / Miller, Thomas – Dublin, 1858 – (= ser 19th c ireland) – 1mf – 9 – mf#1.1.5921 – uk Chadwyck [630]

Agricultural college organization in land-grant institutions / Fleming, Samuel Todd – Gainesville, FL. 1932 – 1r – 1 – us UF Libraries [630]

An agricultural conciliation commissioner's guide / Rood, John Romain – Detroit: Detroit Law Book Co. 1934. 66p. LL-1104 – 1 – us L of C Photodup [340]

Agricultural data / Goebel, Rubye K – s.l, s.l, 1936 – 1r – 1 – us UF Libraries [978]

Agricultural development in tanzania / Seminar For Agricultural Officers On Africultural Development... – London, England. 1965 – 1r – 1 – us UF Libraries [960]

Agricultural economics – Washington DC 1986+ – 1,5,9 – ISSN: 0169-5150 – mf#42498 – us UMI ProQuest [630]

Agricultural economics research – Washington DC 1975-86 – 1,5,9 – (cont by: journal of agricultural economics research) – ISSN: 0002-1423 – mf#1846.01 – us UMI ProQuest [630]

Agricultural education : a lecture delivered... charlottetown, pe island, on thursday evening, january 17, 1884 / Ferguson, Donald – [S.l: s.n, 1884?] [mf ed 1980] – 1mf – 9 – 0-665-03087-8 – mf#03087 – cn Canadiana [630]

Agricultural education magazine – Alexandria VA 1929+ – 1,5,9 – ISSN: 0732-4677 – mf#10338 – us UMI ProQuest [630]

Agricultural engineer – Bedford, England 1973-95 – 1,5,9 – (cont by: landwards) – ISSN: 0308-5732 – mf#8678.01 – us UMI ProQuest [630]

Agricultural engineering – St Joseph MI 1920-94 – 1,5,9 – ISSN: 0002-1458 – mf#767 – us UMI ProQuest [630]

Agricultural experiment station correspondence, 1906-1936 / University Of Florida Archives. Public Records Coll – Gainesville, FL. Series 89, 9.1b-9.5b. 1906-36 – 5r – 1 – us UF Libraries [630]

Agricultural experiments / Depass, Jas. P – Lake City, FL. 1891 – 1r – 1 – us UF Libraries [630]

Agricultural finance review – Washington DC 1938+ – 1,5,9 – ISSN: 0002-1466 – mf#5755 – us UMI ProQuest [630]

Agricultural gazette / British Solomon Islands Protectorate – v1-3. 1933-36 – 1r – 1 – mf#pmb doc460 – at Pacific Mss [630]

Agricultural history – Berkeley CA 1927+ – 1,5,9 – ISSN: 0002-1482 – mf#901 – us UMI ProQuest [630]

Agricultural history review – Berkshire, England 1953+ – 1,5,9 – ISSN: 0002-1490 – mf#10687 – us UMI ProQuest [630]

Agricultural history series – Washington DC 1941-43 – 1 – mf#5756 – us UMI ProQuest [630]

Agricultural improvement by the education of those who are engaged in it as a profession : addressed, very respectfully, to the farmers of canada / Evans, William – [Montreal?: s.n.], 1837 [mf ed 1983] – 2mf – 9 – 0-665-44464-8 – mf#44464 – cn Canadiana [630]

Agricultural journal and transactions of the lower canada agricultural society – Montreal: Lovell & Gibson, [1847?-1853] – 9 – (cont: the canadian agricultural journal) – mf#P04880 – cn Canadiana [630]

Agricultural marketing – Washington DC 1956-71 – 1 – ISSN: 0002-1547 – mf#3191 – us UMI ProQuest [630]

Agricultural museum – La Salle IL 1810-12 – 1 – mf#3537 – us UMI ProQuest [630]

Agricultural opportunities in charlotte county, florida – Punta Gorda, FL. 1937 – 1r – 1 – us UF Libraries [630]

Agricultural outlook – Washington DC 1978+ – 1,5,9 – ISSN: 0099-1066 – mf#11789 – us UMI ProQuest [630]

Agricultural outlook digest – Washington DC 1974-75 – 1,5,9 – mf#7426 – us UMI ProQuest [630]

The agricultural paper of canada : office of the farmer's advocate and home magazine, london, ont...i beg to inform you that your subscription expires... – S.l: s.n, 18–? – 1mf – 9 – mf#53224 – cn Canadiana [630]

The agricultural question : a letter to his excellency samuel rowe...and how to double the revenue and trade, and improve the sanitary condition of the west africa settlements / Lardner, Henry Harold – London, 1880 – (= ser 19th c british colonization) – 1mf – 9 – (with biogr sketch) – mf#1.1.4971 – uk Chadwyck [630]

Agricultural reporter *see* Natal commercial advertiser / agricultural reporter

Agricultural research – Washington DC 1953+ – 1,5,9 – ISSN: 0002-161X – mf#1711 – us UMI ProQuest [630]

Agricultural science review – Washington DC 1963-73 – 1,5,9 – ISSN: 0002-1652 – mf#2412 – us UMI ProQuest [630]

Agricultural situation – Washington DC 1922-79 – 1,5,9 – ISSN: 0002-1660 – mf#372 – us UMI ProQuest [630]

Agricultural statistics : united kingdom 1867-1975 – [mf ed Chadwyck-Healey] – (= ser British government publications...1801-1977) – 9r – 1 – uk Chadwyck [630]

Agricultural statistics / U.S. Dept of Agriculture – 1936-72 – 1 – $474.00 – mf#0614 – us Brook [630]

Agricultural statistics – Washington DC 1936+ – 1,5,9 – ISSN: 0082-9714 – mf#5791 – us UMI ProQuest [630]

Agricultural statistics by plot to plot enumeration in bengal, 1944-45 / Bengal Govt. Dept of Agriculture, Forest and Fisheries – Alipore: Bengal Govt Press, 1946-[194-?]. pts1-3 – 1 – us CRL [954]

Agricultural statistics of florida / Florida. Dept Of Agriculture – Tallahassee, FL. 1938 – 1r – 1 – us UF Libraries [630]

Agricultural survey, 1949-50 : report / Basutoland. Dept of Agriculture – Maseru: Basutoland Govt, 1952 – 1r – 1 – (various titles) – us CRL [630]

An agricultural survey of southern rhodesia / Rhodesia and Nyasaland. Ministry of Agriculture – Salisbury, Southern Rhodesia, Govt Printer [1959?] – 1r – 1 – us CRL [630]

Agricultural surveys / West Indies – Port-Of-Spain, Trinidad And Tobago. 1959-1960 – 1r – 1 – us UF Libraries [071]

Agricultural surveys / West Indies (Federation) Ministry of Natural Resources and Agriculture – Port-of-Spain 1959-60 [mf ed 1961] – 1 – (filmed with: st christopher, nevis & anguilla. dept of agriculture. report, 1960 [port-of-spain [1960?]] and: engledow, frank leonard: report on agriculture...port-of-spain [1945]) – us UF Libraries [630]

Agricultural systems – Oxford, England 1976+ – 1,5,9 – ISSN: 0308-521X – mf#42008 – us UMI ProQuest [630]

Agricultural unionist *see* Sharecropper's voice, 1935-1937 / southern farm leader, 1936 / stfu news, 1938-1939 / tenant farmer, 1941-1942 / farm worker, 1943-1944 / farm labor news, 1946-1951 / the union farmer, 1952-1953 / agricultural unionist, 1952-1954

Agricultural wastes – London, England 1979-86 – 1,5,9 – (cont by: biological wastes) – ISSN: 0141-4607 – mf#42009.01 – us UMI ProQuest [630]

Agricultural water management – Oxford, England 1976+ – 1,5,9 – ISSN: 0378-3774 – mf#42010 – us UMI ProQuest [630]

The agriculturalist and canadian journal *see* The canada farmer

L'agriculture : lettre aux fideles / Emard, Joseph-Medard – Valleyfield [Quebec]: Bureaux de la chancellerie, 1915 [mf ed 1994] – 1mf – 9 – 0-665-73217-1 – mf#73217 – cn Canadiana [630]

Agriculture – (= ser The Goldsmiths'-Kress Library of Economic Literature) – 382r – 1 – us Primary [630]

Agriculture : farms, livestock, and crops : pinell / Hunter, C M – s.l, s.l, 1936 – 1r – 1 – us UF Libraries [630]

Agriculture – Norwich, England 1949-72 – 1,5 – ISSN: 0002-1695 – mf#622 – us UMI ProQuest [630]

Agriculture *see* Natal star / journal of commerce / agriculture / a100

Agriculture and environment – Oxford, England 1974-82 – 1,5 – ISSN: 0304-1131 – mf#42105 – us UMI ProQuest [333]

Agriculture and farming, 1610-1900 – [mf ed Marlborough 1993] – 9 – (pt1: manuals & textbooks a-d [127mf] £850; pt2: f-y [134mf] £875; with d/g) – uk Matthew [630]

Agriculture and politics in meru-arusha / University College of Dar es Salaam. History Dept. – [Dar es Salaam] Uni of Dar es Salaam, Photographic Unit, [19–?] – 1 – us CRL [960]

L'agriculture au point de vue de l'emigration et de l'immigration / Barnard, Edouard Andre – Montreal: des presses a vapeur de La Minerve, [1872] [mf ed 1980] – 1mf – 9 – 0-665-00909-7 – mf#00909 – cn Canadiana [630]

Agriculture bulletin – Edmonton, Canada 1973-74 – 1 – (cont by: agriculture & forestry bulletin) – ISSN: 0568-9074 – mf#7501.01 – us UMI ProQuest [630]

L'agriculture dans la province de quebec : comment l'ameliorer, conferences / Barnard, Edouard-Andre – Saint-Hyacinthe, Quebec?: s.n, 1896 – 1mf – 9 – mf#00910 – cn Canadiana [630]

Agriculture decisions / U.S. Dept of Agriculture – v1-58. 1942-99 – 1135mf – 9 – $1702.00 – (updates planned) – mf#llmc 78-200 – us LLMC [340]

L'agriculture des regions froides de quebec / Chapais, Jean Charles – Chicoutimi [Quebec: s.n.], 1914 – 1mf – 9 – 0-665-75970-3 – mf#75970 – cn Canadiana [630]

Agriculture, ecosystems and environment – Oxford, England 1983+ – 1,5,9 – ISSN: 0167-8809 – mf#42107 – us UMI ProQuest [333]

Agriculture & equipment international – Horne, England 1993+ – 1,5,9 – (cont: agriculture international) – mf#16061,01 – us UMI ProQuest [630]

Agriculture et elevage : au congo belge et dans les colonies tropicales et subtropicales – Bruxelles: impr G Bothy, [1.annee:[n1]-14.annee n5 (19 fevr 1927-mai 1940)] (mthly) – 4r – 1 – us CRL [630]

Agriculture: experiments with fertilizers – Lake City, FL. 1888 – 1r – 1 – us UF Libraries [630]

The agriculture gazette *see* Natal mercantile advertiser / the agriculture gazette

Agriculture in public schools : an address delivered...ontario teachers' association at their thirtieth annual convention held at niagara-on-the lake, august, 1890 / Bryant, John Ebenezer – Toronto: Warwick, 1891 [mf ed 1979] – 1mf – 9 – 0-665-00303-X – mf#00303 – cn Canadiana [630]

Agriculture in volusia county, florida – Deland, FL. 1936 – 1r – 1 – us UF Libraries [630]

Agriculture international – Horne, England 1985-91 – 1,5,9 – (cont by: agriculture & equipment international) – ISSN: 0269-2457 – mf#16061.01 – us UMI ProQuest [630]

Agriculture section circular tan / Commercial Advisory Foundation in Indonesia – Djakarta, 1970-1972(44) – 21mf – 9 – (missing: 1972(34)) – mf#SE-1383 – ne IDC [959]

The agriculturist and canadian journal – Toronto: Pub...by Brewer, McPhail, 1848 – 9 – mf#P04332 – cn Canadiana [630]

Agri-fieldman – Willoughby OH 1973-76 – 1,5,9 – (cont: farm technology & agri-fieldman; cont by: agri-fieldman & consultant) – ISSN: 0276-394X – mf#1415.04 – us UMI ProQuest [630]

Agri-fieldman and consultant – Willoughby OH 1976-79 – 1,5,9 – (cont: agri-fieldman; cont by: ag consultant and fieldman) – ISSN: 0190-2423 – mf#1415.04 – us UMI ProQuest [630]

Agrikov, P A *see* Izvestiia vremennogo komiteta samarskoi torgovo-promyshlennoi palaty

L'agrippe : journal etudiant du cegep d'alma – Alma: l'Agrippe PIP, [ca 1980]-v3 n1 aout 1982 [mf ed 1988] – 9 – (cont: l'ardoise (alma, quebec)) – mf#SEM105P971 – cn Bibl Nat [073]

Agri-view – 1984 jan-dec; 1985 jan-dec; 1986 jan-dec; 1987 jan-dec; 1988 jan-may 20 – 1 – mf#715837 – us WHS [630]

Agri-view – Marshfield WI. 1984 feb 24-jun 29, 1984 jul-dec, 1985 jan-dec, 1986 jan-dec, 1987 jan-dec, 1988 jan-may 20 – 9r – 1 – (cont: agri-view [marshfield, wi: eastern ed]) – mf#715837 – us WHS [630]

Agri-view – Marshfield WI. 1979 oct 5-dec 31, 1980 jan-dec, 1981 jan-dec, 1982 jan-dec, 1983 jan-jun, 1983 jul-oct 20 – 10r – 1 – (cont: agri-view [marshfield, wi: northern ed]; cont by: agri-view [marshfield, wi: northwest ed: 1983]; agri-view [marshfield, wi: north central ed]) – mf#590998 – us WHS [630]

Agri-view – Marshfield WI. 1983 dec 31-1984 dec, 1985 jan-dec, 1986 jan-dec, 1987 jan-dec, 1988 jan-may 20 – 9r – 1 – (cont: agri-view [marshfield, wi: northwest ed]) – mf#692895 – us WHS [630]

Agri-view – Marshfield WI. 1983 oct 28-1984 jun, 1984 jul-dec, 1985 jan-dec, 1986 jan-dec, 1987 jan-dec; 1988 jan-may 20 – 9r – 1 – (cont: agri-view [marshfield, wi: northwest ed]) – mf#692892 – us WHS [630]

Agri-view – Marshfield WI. 1983 oct 27-dec, 1984 jan-dec, 1985 jan-dec, 1986 jan-dec, 1987 jan-dec, 1988 jan-dec, 1989 jan-dec, 1990 jan-dec, 1991 jan-dec, 1992 jan-jun – 18r – 1 – (cont: agri-view [marshfield, wi: southern ed]) – mf#692888 – us WHS [630]

Agri-view – Marshfield WI. 1984 feb 23-jun, 1984 jul-dec, 1985 jan-dec, 1986 jan-dec, 1987 jan-dec, 1988 jan-may 19 – 19r – 1 – (cont: agri-view [marshfield, wi: southern ed]) – mf#715835 – us WHS [630]

Agri-view – Marshfield WI. 1981 nov 5-1982 apr, 1982 may-aug, 1982 sep-dec, 1983 jan-jun, 1983 jul 7-oct 20 – 5r – 1 – (cont by: agri-view [marshfield, wi: southeast ed]; agri-view [marshfield, wi: northeast ed]) – mf#689657 – us WHS [630]

Agri-view – Marshfield WI. 1981 feb-dec, 1982 jan-dec, 1983 jan-oct – 7r – 1 – (cont by: agri-view [marshfield, wi: southwest ed]; agri-view [marshfield, wi: south central ed]) – mf#674512 – us WHS [630]

Agri-view – Marshfield WI. 1978 oct 20-dec 29, 1979 jan-sep 28 – 3r – 1 – (cont: dairyland agri-view; cont by: agri-view (marshfield, wi: northwest ed]) – mf#689623 – us WHS [630]

O agrocultor : semanario independente e noticioso – Rio do Sul, SC. 28 jul 1928 – (= ser Ps 19) – mf#UFSC/BPESC – bl Biblioteca [079]

Agro-ecosystems – Oxford, England 1975-82 – 1,5 – ISSN: 0304-3746 – mf#42106 – us UMI ProQuest [574]

Agroforestry systems – Dordrecht, Netherlands 1989+ – 1,5,9 – ISSN: 0167-4366 – mf#16762 – us UMI ProQuest [634]

Agro-kooperativnyi kruzhok / ed by Orleanskii, V L & Krivosheina, P I – 1929 – 2mf – 9 – mf#COR-242 – ne IDC [335]

Agromyzidae of florida / Spencer, Kenneth A – Gainesville, FL. 1973 – 1r – 1 – us UF Libraries [574]

Agronomie – Lausanne, Switzerland 1991-97 – 1,5,9 – ISSN: 0249-5627 – mf#42462 – us UMI ProQuest [630]

Agrupacion Trabajadores Latinoamericanos Sindicalistas *see* Atlas (buenos aires, argentina)

Agua ausente / Pou, Angel Neovildo – n.p, n.p,1959 – 1r – 1 – us UF Libraries [972]

Agua de coco / Herrera Velado, Francisco – San Salvador, El Salvador. 1955 – 1r – 1 – us UF Libraries [972]

Agua de juventa / Coelho Netto, Henrique – Porto, Portugal. 1925 – 1r – 1 – us UF Libraries [972]

Agua en el silencio, poesia / Morales Santos, Francisco – Antigua, Guatemala. 1961 – 1r – 1 – us UF Libraries [972]

Agua, fuerza y luz / Rivas, Pedro – Tegucigalpa, Mexico. 1945 – 1r – 1 – us UF Libraries [972]

Agua suelta / Palma, Marigloria – San Juan, Puerto Rico. 1942 – 1r – 1 – us UF Libraries [972]

Aguacatec texts (phrases and sentences) : mechanically recorded and transcribed / Andrade, Manuel Jose – Chicago: University of Chicago Library, 1976 – (= ser Microfilm coll of manuscripts on cultural anthropology) – 371p – us Chicago U Pr [490]

Aguacatec vocabulary : with grammatical notes / McArthur, H S – Chicago: University of Chicago Library, 1976 – (= ser Microfilm coll of manuscripts on cultural anthropology) – 511p – us Chicago U Pr [490]

Aguado, Pedro De *see*
- Historia de venezuela
- Recopilacion historical

Aguafuerte / Sierra Berdecia, Fernando – San Juan, Puerto Rico. 1963 – 1r – 1 – us UF Libraries [972]

Aguas azoadas.. / Bejarano y Sanchez, Eloy – 1888 – 9 – sp Bibl Santa Ana [946]

Aguas bicarbonatadas calcicas de alange / Berben, Abdon – Madrid: Imp. Leonardo Minon e hijos, 1895 – sp Bibl Santa Ana [331]

Aguas Monreal, Mariano *see*
- Programa de cuadros de historia natural
- Tratado elemental de historia natural

Aguas turbias / Dobles, Fabian – San Jose, Costa Rica. 1943 – 1r – 1 – us UF Libraries [972]

Aguascalientes. Mexico (State) *see* Periodico oficial

Aguayo, Alfredo Miguel *see*
- Geografia de cuba para uso de las escuelas
- Guia didactica de la escuela nueva
- Tratado de psicologia pedagogica
- Tres grandes educadores cubanos
- Universidad y sus problemas

Agudah news reporter – New York, NY. 1974-78 – 1 – us AJPC [071]

Agudat agadot / Horowitz, Chaim Meir – Berlin, Germany. 1881 – 1r – 1 – us UF Libraries [939]

Agudelo Ramirez, Luis Eduardo *see* Guerrilleros intelectuales

Aguero, Cristobal de *see* Alaarij tobiticha, nicolla-aba yobibennizaa, ni yabaaqui copabijtoo, nipecaa yyeeni ouijchirijni loquiraani

Aguero De Costales, Corina *see* Mi ofrenda

Aguero, Luis *see* De aqui para alla

Aguero Vives, Eduardo *see*
- Julia soler
- Prosa, teatro, verso

Aguero Y Estrada, Francisco *see* Biografia de joaquin de aguero

Aguessy, C *see* Contribution a l'etude de l'histoire de l'ancien royaume de porto-nova

Agui-agui nin mapaladon na mag-agom na si padre juan, bta maria asin papa silvestre sa reinong astrucia – Nueva Caceres: Libreria Mariana 1910 [mf ed Bloomington IN: Indiana Uni Lib, Preservation Dept 1984] – (= ser Coll...in the bikol language) – 1r – 1 – us Indiana Preservation [490]

Aguiar, Armando De *see* Portugueses do brasil

Aguiar De Mariani, Maruja *see* Poesias para ninos

Aguiar, Pinto De *see*
- Bancos no brasil colonial
- Brasil

Aguiar, T *see* Apologia pro consilio medicinali in diminute visiones adversus duas epitolas...

Aguila acecha / Abril Amores, Eduardo – Santiago, Cuba. 1921 – 1r – 1 – us UF Libraries [972]

El aguila extremena – v1. 1899 – 9 – sp Bibl Santa Ana [800]

Aguila, Gilberto R *see* Flechazos

Aguilar, Carlos H *see* Religion y magia entre los indios de costa rica de origen sureno

Aguilar Derpich, Juan *see*
- Al cantio de un gallo
- De alquilan cuartos amueblados

Aguilar, Faustino *see*
- Anf lihim ng isang pulo
- Ang lihim ng isang pulo

Aguilar, Francisco de *see*
- Historia de la nueva espana
- Relacion de la conquista de la nueva espana

Aguilar Gallegos, Manuel *see*
- Oracion funebre...por los fallecidos...de la guerra civil
- La romeria espanola al vaticano en el ano de 1876

Aguilar, Grace *see*
- The days of bruce
- Home influence
- Home scenes and heart studies
- The mother's recompense
- The vale of cedars
- The women of israel

Aguilar Gutierrez, Antonio *see* Panorama de la legislacion civil de mexico

Aguilar, Leopoldo *see* Contratos civiles

Aguilar Machado, Alejandro *see*
- Miscelanea
- Opiniones y discursos

Aguilar, Manuel R *see* Inquientudes profanas

Aguilar, Octavio *see* Juez olaverri y juan canastuj

Aguilera Camacho, Alberto *see* Derecho agrario colombiano

Aguilera De Leon, Carlos *see* Libro-centenario, 1835-1935, conmemorativo del ani

Aguilera, Fito *see* Rosca, s a

Aguilera, Miguel *see*
- Ensenanza de la historia en colombia
- Marco fidel suarez

Aguilera Patino, Luisita V *see* Secreto de antatura

Aguilera Y Aguilera, Jose Ernesto *see* Ondas

Aguilera Y Cespedes De Ferrer, Gertrudis *see* Alimentos y nutricion en graficas y cantos popular

Aguinaldo lirico de la poesia puertorriquena / Rosa-Nieves, Cesareo – San Juan, Puerto Rico. v1-31957 – 1r – 1 – us UF Libraries [972]

Aguinaldo puertorriqueno de 1843 – San Juan, Puerto Rico. 1946 – 1r – 1 – us UF Libraries [972]

Aguinaldos / Gonzalez Ricardo, Rogelio – Habana, Cuba. 1959 – 1r – 1 – us UF Libraries [972]

Aguirre Acha, Jose *see* Los andes al amazonas

Aguirre, Agustin *see* Derecho hipotecario

Aguirre, Jose M *see* Honduras

Aguirre, Mirta *see* Presencia interior

Aguirre, Pedro Antonio de *see*
- Sentencia apostolica definitiva de la precedencia en todos actos publicos y privados
- Transito gloriosissimo de n sra la santissima virgen maria

Aguirre-Sacasa, Roberto *see*
- "Caddie woodlawn", adapted by greg gunning from the novel by carol ryrie brink
- "Ramona quimby", adapted by len jenkin from the ramona books by beverly cleary

Aguja / Morales, Raul – San Jose, Costa Rica. 1955 – 1r – 1 – us UF Libraries [972]

Agular, M Maria Esperanza *see* Estudio bibliografico de don manuel eduardo de gorostiza

Agundez, Antonio *see* Formularios para los fiscales municipales y comarcales

Agundez Fernandez, Antonio *see*
- Biografia de caceres
- Notas para la historia de la ciudad de badajoz a fines del siglo 18
- Segualazion orizzontali in spague

Agundez Ferrandez, Antonio *see* Sintesis biografica de caceres

Agursky, Samuel *see* Kamf kegn "bund"

Agus, Giuseppe *see*
- Allemands danced at the kings-theatre in the hay-market by mr slingsby and sig-ra radicate
- Six notturnos for two violins and a violoncello obbligato
- Six sonatas

Agus, Joseph *see* Epistola beati pauli apostoli ad romanos

Agustin agualongo y su tiempo / Ortiz, Sergio Elias – Bogota, Colombia. 1958 – 1r – 1 – us UF Libraries [972]

Agustin de iturbide : emperador de mejico... / Mestas, Alberto de – Madrid: Razon y Fe, 1939 – 1 – sp Bibl Santa Ana [972]

Agustin, M *see* Libro de los secretos de agricultura, casa de campo pastoril

Agva news – 1961 jan-1965; 1966 spring-1966 summer – 1 – mf#1051457 – us WHS [071]

Agyei, Samuel Kwasi *see* A guide to records relating to ghana in repositories in the u.k. excluding the public record office

Ah chi sa : the favorite trio in the comic opera il barbier di seviglia / Paisiello, Giovanni – London: printed for Rt Birchall...[179-?] [mf ed 1989] – 1r – 1 – (in italian) – mf#pres. film 50 – us Sibley [780]

Ah! enfin! / Clairville, M – Paris, France. 1848? – 1r – 1 – us UF Libraries [440]

Ah non lasciarmi no bell' idol mio : sung by madam mara in didone abandonata / Mortellari, Michele – London: Longman & Broderip [1786] [mf ed 19–] – 1r – 1 – (in italian; libretto by metastasio) – mf#pres. film 73 – us Sibley [780]

Ah! parlate, che forse tacendo *see* Le sacrifice d'abraham

Aha newsletter / American Historical Association – Washington DC 1962-81 – 1,5,9 – (cont by: aha perspectives: newsletter of the american historical association incl eib notices) – ISSN: 0001-138X – mf#8092.02 – us UMI ProQuest [975]

Aha perspectives : newsletter of the american historical association including eib notices / American Historical Association – Washington DC 1983 – 1,5,9 – (cont: aha newsletter; cont by: perspectives: newsletter of the american historical association) – ISSN: 0745-0516 – mf#8092.02 – us UMI ProQuest [975]

Ahagon, Marques de Laurencin *see* Series de los mas importantes documentos del archivo y biblioteca del excmo. sr. duque de medinaceli, elegidos por su encargo y publicados a sus expensas...

Ahali – Edirne, Sofya: Olyadsamaryaviza Matbaasi, 1919-21. Sermuharriri: Mehmed Behcet. n69. 9 nisan 1920 – (= ser O & t journals) – 1mf – 9 – $25.00 – us MEDOC [956]

Ahali – Samsun: Ahali Matbaasi. Sahib-i Imtiyaz: Ismail Cenani n133. 14 eyluel 1338 [1922], 176/222, 187/223. 9 mart 1926 – (= ser O & t journals) – 1mf – 9 – $25.00 – us MEDOC [956]

Ahangar – Tehran. sal-'i 1, shumarah-'i 1-16. 27 farvardin 1358-16 murdad 1358 [16 apr 1979-7 aug 1979] – 1r – 1 – $53.00 – us MEDOC [079]

Aharit yerushalayim / Rothberg, Marcus – Berdichev, Ukraine. 1901 – 1r – 1 – us UF Libraries [939]

Ahasver in rom : eine dichtung in sechs gesaengen / Hamerling, Robert – 23. aufl. Hamburg: Verlagsanstalt und Druckerei (vormals J F Richter), 1892 [mf ed 1993] – 279p – 1 – mf#8670 – us Wisconsin U Libr [810]

Ahasverus : der ewige jude / Zirus, Werner – Berlin: W de Gruyter & Co 1930 [mf ed 1993] – 1r – 1 – (incl bibl ref & ind. filmed with: stoff- und motivgeschichte der deutschen literatur / ed by paul merker und gerhard luedtke) – mf#3000p – us Wisconsin U Libr [430]

Ahasverus, der ewige jude / Zirus, Werner – Berlin: W de Gruyter 1930 [mf ed Bloomington IN: Indiana Uni Lib, Preservation Dept 1984] – 1r – 1 – us Indiana Preservation [390]

Ahauser kreiszeitung [mf ed 2004] – Ahaus 1933 1 jan-31 mar – 1r – 1 – (filmed by misc inst: 1952-57 [gaps]) – gw Mikrofilm; gw Misc Inst [074]

Ahavat ha-kadmonim – Yerushalayim: [s.n.] 649 [1888 or 1889] [mf ed 1987] – 1r – 1 – mf#1953 – us Wisconsin U Libr [270]

Ahavat tsiyon / Mapu, Abraham – Tel-Aviv, Israel. 193-? – 1r – 1 – us UF Libraries [939]

Ahavat tsiyon / Mapu, Abraham – Warsaw, Poland. 1889 – 1r – 1 – us UF Libraries [939]

Ahavat tsiyon / Mapu, Abraham – Warsaw, Poland. 1903 – 1r – 1 – us UF Libraries [939]

Ahchanaulla, Khanabahadura *see* History of the muslim world

'Ahd Al-Baqi, Muhammad Fu'ad *see* Mujam al-mufahras li-alfaz al-qur'an al-karim

Ahead of the herd – 1983 apr 1-1985 dec 9; 1986 jan 15-1987 aug 20 – 1 – mf#1094307 – us WHS [071]

Ahearn, Jr, Frederick L *see* Journal of religion and spirituality in social work

Ahenk – Izmir: Ahenk Yurdu Matbaasi, 1895-1928. Sahib-i Imtiyaz: Mehmed Necati, Ali Nazmi. n1257 (21 eyluel 1316 [1900] – (= ser O & t journals) – 1mf – 9 – $25.00 – us MEDOC [956]

Ahepan – 1978 jan/mar-1982 summer/fall; 1983 fall; 1984 spring-fall – 1 – mf#571867 – us WHS [071]

Ahern, John *see*
- Mon premier livre
- Pedagogic organization of schools from the regulations of the catholic committee
- Principles of book-keeping

Ahern, Michael Joseph *see* Notes pour servir a l'histoire de la medecine dans le bas-canada...commencement du 19e siecle

Die ahhijava-urkunden / Sommer, F – Muenchen, 1932 – (= ser Abh bayerischen akademie der wissenschaften 6) – 11mf – 9 – (abh bayerischen akademie der wissenschaften. philosophisch-historische abt n.s. v6) – mf#NE-434 – ne IDC [956]

Ahiasaf – Warsaw. 1-13. 1893-1923 – 1 – us NY Public [073]

The ahiman rezon : or book of the constitution of the right, worshipful grand lodge of free and accepted masons of pennsylvania – Philadelphia, 1902 – 3mf – 9 – $4.50 – mf#LLMC 92-186 – us LLMC [366]

Ahkam al-aradi da'ibis'al-murr – 5mf – 9 – mf#NE-1583 – ne IDC [956]

Ahkam-i sal – sene 1276 [1859] – (= ser Ottoman histories and historical sources) – 1mf – 9 – $25.00 – us MEDOC [956]

The ahkimal awawtom – 1927-31 – (= ser American indian periodicals... 1) – 3mf – 9 – $95.00 – us UPA [305]

Ahl, Frances Norene *see* Two thousand miles up the amazon

Ahlander, Julie D *see* Audience enjoyment of dance performance improvisation as affected by improvisational structures and audience education

Ahlborn, Luise Jaeger *see* Anonym

Ahlefeld, Charlotte von *see* Tagebuch auf einer reise durch einen theil von baiern, tyrol und oestreich

Ahlener volkszeitung – Ahlen DE, 1988- – ca 6r/yr – 1 – (bezirksausgabe von westfaelischer tageszeitung / westfaelische nachrichten, muenster; title varies: jul 1940: westfaelische tageszeitung, 1950: westfaelische nachrichten, 2 jan 1998: ahlener zeitung) – gw Misc Inst [074]

Ahler, Stanley A et al *see* The archeology of the white buffalo robe site

Ahlers, Rudolf *see*
- Thomas torsten
- Das weite land

Ahmad, Bashiruddin Mahmud *see* The islamic mode of worship

Ahmad, Ghulam *see*
- The teachings of islam
- Teachings of islam

Ahmad ibn hanbal and the mihna : a biography of the imam including an account of the mohammadan inquisition called the mihna / Patton, Walter Melville – Leiden: E. J. Brill, 1897. Chicago: Dep of Photodup, U of Chicago Lib, 1971 (1r); Evanston: American Theol Lib Assoc, 1984 (1r) – 1 – 0-8370-0542-6 – (incl bibl ref and index) – mf#1984-B285 – us ATLA [260]

Ahmad ibn Muhammad ibn Kathir, Al Farghani *see* Mvhamedis alfragani arabis chronoligca et astronomica elementa...

Ahmad, Jamil-ud-Din *see* Some recent speeches and writings of mr jinnah

Ahmad Khan, Sayyid *see* Syed ahmed bahadoor, c.s.i., on dr. hunter's "our indian mussulmans, are they bound in conscience to rebel against the queen?"

Ahmad, Z A [comp] *see*
- National language for india
- Philosophy of socialism

Ahmadiya movement / Walter, Howard Arnold – Calcutta, India. 1918 – 1r – 1 – us UF Libraries [320]

The ahmadiya movement / Walter, H A – Calcutta: Association Press, 1918 – (= ser The Religious Life of India) – 1mf – 9 – 0-524-05887-3 – (incl bibl ref) – mf#1991-0015 – us ATLA [280]

Ahmad-ul-Umri, Turkoman *see* The lady of the lotus

Ahmann, Chester F *see* Nutritional study of the white school children in five representative counties of florida

Ahmed, Mustafa Ali b. (Geliboeluelue) *see* Kuenh uel-ahbar

Ahmed, Umaru *see* Introduction to classical hausa and the major dialects

Ahmednagar und golconda : ein beitrag zur eroerterung der missionsprobleme des weltkrieges / Oepke, Albrecht – Leipzig: Doerffling & Franke, 1918 [mf ed 1995] – (= ser Yale coll) – viii/160p – 1 – 0-524-09288-5 – (in german) – mf#1995-0288 – us ATLA [954]

Ahmet, Servili Hafiz *see* Ahter-i kebir

Ahn : the voice of the african-american community in paradise – v5 n8-v10 n2 [1992 jan 1/31-97 win] – 1r – 1 – (cont: afro-hawaii news) – mf#2341663 – us WHS [305]

Ahnapee record – 1897 mar 18-sep 16; [1873 jun 13-1875]; 1876-90; 1891 jan 1-1894 jan 18; 1894 jan 25-1897 mar 11 – 1 – mf#912556 – us WHS [071]

Ahnas el medineh : (heracleopolis magna) / Naville, E – London, 1894 – (= ser Mees 11) – 7mf – 8 – €16.00 – (filmed with: j j tylor and l griffith: the tomb of paheri at el kab) – ne Slangenburg [930]

Ahnas el medineh : the tomb of pakeri at el kab / Naville, E et al – London, 1904 – (= ser Egypt exploration fund mem) – 4mf – 9 – mf#NE-387 – ne IDC [956]

Die ahnen : roman / Freytag, Gustav – Leipzig: S Hirzel, 1884 [mf ed 1990] – 6v – 1 – mf#7077 – us Wisconsin U Libr [830]

Die ahnen : roman / Freytag, Gustav – Leipzig: S Hirzel. 3v. 1875 – 1r – 1 – us Wisconsin U Libr [830]

Ahnenbuechlein / Finckh, Ludwig – Stuttgart: Strecker & Schroeder [1921?] [mf ed 1990] – 1r – 1 – (filmed with: double, double, toil and trouble / lion feuchtwanger) – mf#7237 – us Wisconsin U Libr [430]

Der ahnenkultus und die urreligion israels / Grueneisen, Carl – Halle a S: Max Niemeyer, 1900 [mf ed 1985] – 1mf – 9 – 0-8370-3415-9 – (incl ind) – mf#1985-1415 – us ATLA [939]

Der ahnenring / Finckh, Ludwig – Goerlitz: C A Starke 1943 [mf ed 1989] – 1r [ill] – 1 – (filmed with: double, double, toil and trouble / lion feuchtwanger) – mf#7237 – us Wisconsin U Libr [890]

Die ahnfrau : trauerspiel in fuenf aufzuegen / Grillparzer, Franz – 3. Aufl. Stuttgart, Berlin: J G Cotta, 1902 – 1r – 1 – us Wisconsin U Libr [200]

Ahnung und aussage / Heiseler, Bernt von – Muenchen: Verlag Koesel-Pustet, 1939 [mf ed 1993] – 251p – 1 – mf#8232 – us Wisconsin U Libr [430]

Ahogado / Sanchez Borbon, Guillermo – Panama, Panama. 1957 – 1r – 1 – us UF Libraries [972]

Ahora : cuba primero – Miami, FL. 1970 sep 15-1971 jan 16 – 1r – 1 – us UF Libraries [071]

Ahora : un periodico para hoy – Hialeah, FL. 1982 jan 20-feb 03 – 1r – 1 – us UF Libraries [071]

Ahora : la voz del pueblo – Miami, FL. 1975 oct 24-1976 dec 31 – 1r – 1 – us UF Libraries [071]

Ahora! – v1 n2 [1970 oct 13]; v4 n10, 11, 12, 13, 15 [1973 jul 10, 23, aug, oct/nov, nov/dec]; v5 n2, 4, 6 [1974 feb/mar, apr/may, jul/aug) – 1 – mf#1267160 – us WHS [071]

El ahorro y la politica social / Allue Salvador, Miguel – Caceres: tip editorial extremadura, 1956 – 1 – sp Bibl Santa Ana [320]

Ahot yehudah / Polishts, Yehudah – Jerusalem, Israel. 1931 – 1r – 1 – us UF Libraries [939]

Ahrendt, Christine. see Analysis of the second quartet of bela bartok

Ahrenhoerster, Greg see Take me out to the ballgame

Ahrweiler kreisblatt – Bad Neuenahr-Ahrweiler DE, 1861-1866 16 dec – 1 – gw Misc Inst [074]

Ahter-i kebir / Ahmet, Servili Hafiz – Istanbul: Matbaa-i Ahmed Ihsan, 1321 [1903] – (= ser Ottoman literature, writers and the arts) – 18mf – 9 – $290.00 – us MEDOC [470]

Ahuizote / Castillo R & Cesar, A – Guatemala, ? – 1r – 1 – us UF Libraries [972]

Ahuma, S R B Attoh see Memoirs of west african celebrities

Ahumada, P see Question en la qual se intenta averiguar como y de que venas y de que parte...se deba sangrar

Ahzab – Tehran, 1980- . sal-i 1, shumarah-'i 1-18,20-31. 9 day 1359-13 murdad 1360 [30 dec 1980-2 aug 1981] – 1r – 1 – $53.00 – us MEDOC [079]

Ai / Ai, Wu – Kuei-lin: Ta ti t'u shu kung ssu, 1943 – (= ser P-k&k period) – us CRL [480]

Ai, Ch'ing see
- Fan fa-hsi-ssu
- Hsiang t'ai yang
- Hsien kei hsiang ts'un ti shih
- Hsueh li tsuan
- Huo pa
- K'uang yeh
- Li ming ti t'ung chih
- Pei fang
- Shih lun
- Ta yen ho

Ai ch'ing san pu ch'u / Pa, Chin – Shang-hai: Liang yu t'u shu kung ssu, [Min kuo 26 [1937]] – (= ser P-k&k period) – us CRL [830]

Ai ch'ing ti san pu ch'u / Pa, Chin – Shang-hai: K'ai ming shu tien, Min kuo 30 [1941] – (= ser P-k&k period) – us CRL [830]

Ai ch'ing ti san pu ch'u / Pa, Chin – Shang-hai: Liang yu t'u shu yin shua kung ssu, Min kuo 25 [1936] – (= ser P-k&k period) – us CRL [830]

Ai jen ju chi (ccm286) = Love others as yourself / Surdam, T Janet – Hong Kong, 1958 [mf ed 198?] – (= ser Ccm 286) – 1 – mf#1984-b500 – us ATLA [230]

Ai kuo shih ko / Su-min pien – Shang-hai: Shen chou kuo kuang she, 1933 – (= ser P-k&k period) – us CRL [810]

Ai magazine – Menlo Park 1985+ – 1,5,9 – ISSN: 0738-4602 – mf#15622 – us UMI ProQuest [000]

Ai mei hsiao cha / Hsu, Chih-mo – Shang-hai: Liang yu t'u shu yin shua kung ssu, 1945 – (= ser P-k&k period) – us CRL [480]

Ai meng ying / Ch'en, Ch'uan – Ch'ung-ch'ing: Tsai ch'uang ch'u pan she, Min kuo 33 [1944] – (= ser P-k&k period) – us CRL [810]

Ai para ton iordanen laurai kalamoonos kai agiou gerasimou : kai oi bioi tou agiou gerasimou kai kyriakou tou anachooretou / ed by Koikylides, K M – Jerusalem, 1902 – 3mf – 8 – €7.00 – ne Slangenburg [243]

Ai & society – Dordrecht, Netherlands 1987+ – 1,5,9 – ISSN: 0951-5666 – mf#16971 – us UMI ProQuest [000]

Ai, Ssu-ch'i see
- Chih shih ti ying yung: tu shu wen ta ti erh chi
- Hsin che hsueh lun chi

Ai te sheng li (ccc148) = Victory of love / Ho, David Chi-hsing – Hong Kong, 1954 [mf ed 198?] – (= ser Ccm 148) – 1 – mf#1984-b500 – us ATLA [230]

Ai ti tsou ch'u / Tan-ch'un – [China]: Huang mo she, 1933 – (= ser P-k&k period) – us CRL [810]

Ai tukutuku vakalotu – 1897-1903 – 1r – 1 – mf#pmb doc199 – at Pacific Mss [980]

Ai tukutuku vakalotu – 1908-17 – 1r – 1 – mf#pmb doc200 – at Pacific Mss [980]

Ai tukutuku vakalotu – 1917-25 – 1r – 1 – mf#pmb doc201 – at Pacific Mss [980]

Ai tukutuku vakalotu – 1926-30 – 1r – 1 – mf#pmb doc202 – at Pacific Mss [980]

Ai tukutuku vakalotu – jun-dec 1935 – 1r – 1 – mf#pmb doc204 – at Pacific Mss [980]

Ai tukutuku vakalotu – mar 1937-mar 1960 – 1r – 1 – mf#pmb doc205 – at Pacific Mss [980]

Ai tukutuku vakalotu – sep 1930-may 1935 – 1r – 1 – mf#pmb doc203 – at Pacific Mss [980]

Ai, Wu see
- Ai
- Feng jao ti yuan yeh ti 1 pu, ch'un t'ien
- Meng ya
- Nan kuo chih yeh
- T'ao huang
- Tuan lien
- Wen hsueh shou ts'e
- Yao yuan ti hou fang
- Yeh ching

Ai yu tz'u = With love and irony / Lin, Yutang – [Shang-hai?]: Wen yu ch'u pan she, Min kuo 30 [1941] – (= ser P-k&k period) – us CRL [305]

Aiaa bulletin / American Institute of Aeronautics and Astronautics – Reston VA 1964-75 – 1,5,9 – ISSN: 0001-1444 – mf#5073 – us UMI ProQuest [629]

Aiamie tipadjimoin masinaigan ka ojitogobanen kaiat ka niinaisi mekate8ikonaieigobanen kanactageng / Mathevet, Jean-Claude – Moniang [i.e. Montreal]: O ki magabikicoton John Lovell, ate mekate8ikonaieikamikong, kanactageng, 1859 [mf ed 1984] – 4mf – 9 – 0-665-46360-X – mf#46360 – cn Canadiana [221]

Aich newsletter – 1975-1985 – 1 – mf#618714 – us WHS [071]

Aicher, Georg see Das alte testament in der mischna

[Aicher, O] see Theatrum funebre

Aichinger, Gregor see
- Sacrae cantiones, 4, 5, 6, 8 & 10 vocum
- Sacrae dei laudes...

Aicpa tax section newsletter – Jersey City NJ 2002+ – 1,5,9 – mf#32341 – us UMI ProQuest [650]

The aic's as interlocutors for black theology in south africa / Molobi, Masilo Sonnyboy – Uni of South Africa 2000 [mf ed Johannesburg 2000] – 2mf – 9 – mf#mfm14916 – sa Unisa [240]

Aid and abet newsletter – n1-3 [1984 jan-sep]; n4-5 [1985 may-dec]; n6 [1986: jan]; n7-8 [1987 may-oct]; n9-11 [1988 jan-sep] – 1 – mf#1051744 – us WHS [071]

Aid for children – [Madrid]: Ministerio de trabajo y asistensia social [193-] – (= ser Blodgett coll) – 1v – 9 – mf#w704 – us Harvard [362]

Aid for old people – [Madrid]: Ministerio de Trabajo y Asistencia social [193-] – (= ser Blodgett coll) – 1v – 9 – mf#w705 – us Harvard [362]

An aid to national defence / Cameron, Donald Roderick – Toronto: C B Robinson, 1890? – 1mf – 9 – mf#10336 – cn Canadiana [355]

Aida pollution insurance bulletin – Chicago IL 1980 – 1,5,9 – mf#12544 – us UMI ProQuest [368]

Aidc journal / American Industrial Development Council – Washington DC 1966-80 – 1,5,9 – (cont by: aedc journal) – ISSN: 0001-155X – mf#2313.01 – us UMI ProQuest [338]

Aidit, D N see Problems of the indonesian revolution

Aids care – Abingdon, Oxfordshire 1994+ – 1,5,9 – ISSN: 0954-0121 – mf#20901 – us UMI ProQuest [616]

Aids education and prevention – New York NY 1989+ – 1,5,9 – ISSN: 0899-9546 – mf#17419 – us UMI ProQuest [616]

Aids interfaith : new york newsletter – 1994 apr, jul-oct, dec; 1995 jan-jun – 1 – mf#5306448 – us WHS [071]

Aids research archives : observations on social and political change 1980-90 – New York, 1994 – 9 – $675.00 – (documents the history of aids growth and treatment as seen through the eyes of alternative media) – us Alper [301]

Aids to faith : a series of theological essays / Mansel, Henry Longueville et al; ed by Thomson, William – New York: D Appleton, 1862 [mf ed 1984] – 1r – 1 – 0-8370-0298-2 – (incl bibl ref. also available in mf) – mf#1984-B317 – us ATLA [210]

Aids to reflection in the formation of a manly character : on the several grounds of prudence, morality and religion / Coleridge, Samuel Taylor – new rev ed. Liverpool: E Howell, 1873 [mf ed 1990] – 1mf – 9 – 0-7905-7330-X – (with ind and trans of greek & latin quotations by thomas fenby. 1st printed 1825) – mf#1989-0555 – us ATLA [230]

Aids to scripture study / Gardiner, Frederic – Boston: Houghton, Mifflin, 1890 [mf ed 1985] – 1mf – 9 – 0-8370-3231-8 – (incl ind) – mf#1985-1231 – us ATLA [220]

Aids to the devout study of criticism : pt 1: the david-narratives pt 2: the book of psalms / Cheyne, Thomas Kelly – New York: Thomas Whittaker, 1892 [mf ed 1985] – 1mf – 9 – 0-8370-2640-7 – mf#1985-0640 – us ATLA [221]

Aids to the study and use of law books: a selected list, classified and annotated, of publications relating to law literature, law study and legal ethics / Hicks, Frederick Charles – New York: Baker, Voorhis, 1913. 129p. L.C. copy imperfect: supplement wanting. LL-220 – 1 – us L of C Photodup [340]

Aids to the study of acts see Shih t'u hsing chuan chih yen chiu (ccm236)

Aids to the study of dante / Dinsmore, Charles Allen – Boston: Houghton Mifflin, 1903 [mf ed 1989] – 2mf – 9 – 0-7905-4558-6 – (incl bibl ref) – mf#1988-0558 – us ATLA [440]

Aids to the study of german theology / Matheson, George – 2nd ed. Edinburgh: T & T Clark, 1876 [mf ed 1986] – 1mf – 9 – 0-8370-8768-6 – (incl ind) – mf#1986-2768 – us ATLA [240]

Aiello, Kimberly A see Differences in physiological and mechanical properties of stair climbing on three different apparatus

Ai-fan-ssu-tun hu sheng (ccm90) : reports from the 2nd assembly of the world council of churches, august 15-31 1954 = Evanston speaks / Chen, Chu – Hong Kong, 1956 [mf ed 198?] – (= ser Ccm 90) – 1 – mf#1984-b500 – us ATLA [240]

Aifld briefs – 1988 sep 26; 1989 sep 28, nov22, nov27 [action bulletin]; 1990 jan 24, feb 21, mar 9, apr 12, may 30, jul 13, sep 20; 1991 feb 6, oct 18 [special report], nov1 [special report], nov4 – 1 – mf#1683719 – us WHS [071]

Aifld report – 1964 jan-1976 sep; 1977 jan/feb-1991 3rd qtr – 1 – mf#203540 – us WHS [071]

Aigentliche beschreibung der raisz : so er vor diser zeit gegen auffgang inn die morgenlaender, fuernemlich syrian, iudaeam, arabiam, mesopotamiam, babyloniam, assyriam, armeniam... / Rauwolff, L – [Franckfurt am Mayn], 1582 – 6mf – 9 – mf#AR-1973 – ne IDC [915]

L'aigle republicaine – Paris: A Rene [n1-2(1848)] (wkly) – (= ser French revolution of 1848. newspapers) – 1r – 1 – (le gerant: guillemain) – us CRL [074]

l'aiglon (lac etchemin, quebec) – Beauceville-Est: [s.n.], [ca 1938?]-juil 1973// – 1 – cn Bibl Nat [071]

Aignan, Etienne see Extraits des memoires relatifs a l'histoire de france

L'aiguillon – Mont-Joli: C-B Beaudet. v1 n1 juin 1943-v3 n5 oct 1945 (mthly) [mf ed 1990] – 1r – 1 – (suspended: mai-sep 1945) – mf#SEM 35P342 – cn Bibl Nat [073]

Aiguino da Brescia, Illuminato see
- La illuminata de tutti i tuoni di canto fermo
- Il tesoro illuminato di tutti i tuoni di canto figurato

Aiha journal [american industrial hygiene association journal] – Fairfax VA 2003+ – 1,5,9 – mf#10121,02 – us UMI ProQuest [360]

Aihaj – Fairfax VA 2000+ – 1,5,9 – (cont: american industrial hygiene association journal) – ISSN: 1529-8663 – mf#10121.02 – us UMI ProQuest [360]

Aiken, Catharine see Methods of mind-training

Aiken, Charles Augustus see The proverbs of solomon

Aiken, Charles Francis see The dhamma of gotama the buddha and the gospel of jesus the christ

Aiken, Charles Francis et al see India and buddhism

Aiken first baptist church – Aiken, SC. 1814-jun 10 1889 – 1 – $29.61 – us Southern Baptist [242]

Aiken, Janet Rankin see English, present and past

Aiken, John see Miscellaneous papers

Aiken, Joseph Daniel see Typescript of travel diary

Aikenhead and Crombie (Firm) see Aikenhead and crombie's handy book of builders' hardware

Aikenhead and crombie's handy book of builders' hardware : in which will be found illustrated and described a portion of the leading goods which we always keep in stock – [Toronto?: s.n, 18–?] [mf ed 1986] – 1mf – 9 – 0-665-50680-5 – (incl ind) – mf#50680 – cn Canadiana [690]

Aikenhead Hardware Co see Catalogue and price list of armstrong patent tool holders for turning, planing and boring metals

Aikens, Asa see Practical forms, with notes and references explanatory of the law governing the cases to which they are applicable.

Aiken's reports / Vermont. Supreme Court – v1-2. 1825-1828 (all publ) – (= ser Vermont Supreme Court Reports) – 10mf – 9 – $15.00 – (a pre-nrs title) – mf#LLMC 90-311 – us LLMC [347]

Aikin, Edmund see
- Designs for villas and other rural buildings
- An essay on the doric order of architecture
- Plans, elevation, section

Aikin, John see England described

Aikin, Lucy see
- Epistles on women
- Robinson crusoe in words of one syllable

Aikin's heads of chemistry see Miscellaneous papers

Aikman, Duncan see All-american front

Aikman, James see An historical account of covenanting in scotland

Ai-k'o-hua-shih see T'ieh lu ching chi yuan li

Les ailes – n1-984, inc. Paris. 1921-6 juin 1940 – 1 – fr ACRPP [073]

Les ailes qui montent : hommage du nouvel an 1919 / Tremblay, Jules – [Ottawa?: s.n.], 1918 – 1mf – 0-665-77159-2 – mf#77159 – cn Canadiana [810]

Ai-lun-k'ai ti li hun lun / Yun, Jang pien – Shang-hai: Pei hsin shu chu, 1929 – (= ser P-k&k period) – us CRL [306]

Aim : the bulletin of the american independent movement / American Independent Movement – 1966 may-1969 jul 15 – 1 – mf#492053 – us WHS [071]

The aim and scope of philosophy of religion : three cambridge lectures / Tennant, Frederick Robert – London: SPCK, 1913 – 1mf – 9 – 0-7905-9708-X – mf#1989-1433 – us ATLA [200]

Aim, racial harmony and peace – 1977 nov-1986 winter; 1986 spring-1990 winter; 1991 spring-1994 winter – 1 – mf#811523 – us WHS [071]

L'aimable faubourien – Paris: [s.n, [may 1849] (mthly) – (= ser Periodicals relating to the french revolution of 1848) – 1r – 1 – us CRL [944]

Aimes, Hubert Hillary Suffern see A history of slavery in cuba, 1511 to 1868

Aims and aids for girls and young women on the various duties of life : including physical, intellectual, and moral development... / Weaver, George Sumner – New York: Fowler and Wells, 1856 [mf ed 1987] – 1r – 1 – mf#7175 – us Wisconsin U Libr [640]

The aims of a theological seminary : an address. delivered before the alumni association of the theological seminary, new brunswick, n.j... / Hartranft, Chester David – New York: Board of Publication of the Reformed Church in America, 1878 – 1mf – 9 – 0-524-08378-9 – mf#1993-3078 – us ATLA [240]

Ain christlicher sendprieff an frauw anna / Bugenhagen, J – [Augsburg, 1525] – 1mf – 9 – mf#TH-1 mf 158 – ne IDC [242]

Ain lobliche ordnug der fuerstlichen stat wittenberg... see Die wittenberger und leisniger kastenordnung, 1522, 1523

Ain schoen neuw gaistlich lobgsang : im thon. es fleugt ain voegele leyse; ain anders lied, kumpt her zu mir spricht gotes sun, im thon. was woell wir aber heben an, das best das mir gelernet han – Augspurg [c1550] – (= ser Hqab. literatur des 16. jahrh.) – 1mf – 9 – €20.00 – mf#1550a – gw Fischer [780]

Der ain vn[d] neintzichst psalm troestlich in der gemain zu der zeyt der pestilentz zu singen – [Augsburg?] [vor 1537] – (= ser Hqab. literatur des 16. jahrh.) – 1mf – 9 – €20.00 – mf#1537 – gw Fischer [780]

Ainslee's – New York NY 1898-1926 – 1 – mf#6127 – us UMI ProQuest [400]

Ainslie, Peter see
- Among the gospels and the acts
- Christ or napoleon – which?
- The message of the disciples for the union of the church
- Towards christian unity

Ainslie, Rosalynde see Collaborators

Ainslie, Whitelaw see An historical sketch of the introduction of christianity into india

Ainsworth, Barbara E see
- Factors associated with the recall of physical activity
- The relationship between social physique anxiety and physical activity

Ainsworth, Barbare E see Use of the exercise benefits/barriers scale in north carolina department of correction employees

Ainsworth, Henry see Annotations on the pentateuch

Ainsworth herald – Ainsworth, NE: Geo A Miles. v11 n42. jun 7 1900-dec 17 1903// (wkly) – 1r – 1 – (cont: ainsworth home rule; absorbed by: ainsworth star-journal (1893)) – us Bell [071]

Ainsworth home rule – Ainsworth, NE: Geo A Miles, 1897-v11 n41. may 31 1900 (wkly) – 1r – 1 – (cont: home rule (ainsworth ne); cont by: ainsworth herald) – us Bell [071]

Ainsworth, John Dawson see John ainsworth, pioneer kenya administrator, 1864-1946

The ainsworth journal – Ainsworth, NE: Leroy Hall. v1 n1. jul 3 1884-may 7 1891// (wkly) – 2r – 1 – (merged with: ainsworth star to form: star-journal (ainsworth ne). v2 n3-v3 n45 also called whole no 54-149) – us Bell [071]

Ainsworth, Percy Clough see The silences of jesus – st paul's hymn to love

Ainsworth star – Ainsworth, NE: T J Smith, aug 1886-v5 n40. may 7 1891 (wkly) – 3r – 1 – (merged with: ainsworth journal to form: star-journal (ainsworth ne)) – us Bell [071]

Ainsworth star see The ainsworth journal

Ainsworth star-journal – Ainsworth, NE: Robert H and Gwyneth L Tyler. v74 n2. jul 12 1956- (wkly) – 31r – 1 – (cont: ainsworth star-journal and brown county democrat; iss for v95 n50- accompanied by suppl sandhill advertiser) – us NE Hist [071]

The ainsworth star-journal – Ainsworth, NE: T J Smith. 57v. v7 n41. may 25 1893-v63 n43. may 31 1945 (wkly) – 32r – 1 – (cont: star-journal (ainsworth ne); absorbed: ainsworth herald and: long pine journal (1899); merged with : brown county democrat (ainsworth ne) to form: ainsworth star-journal and brown county democrat; iss for oct 29 1936 incorrectly dated nov 29 1936; suppls accompany some iss) – us Bell [071]
Ainsworth star-journal and brown county democrat – Ainsworth, NE: H B Tyler. v63 n49. jun 7 1945-v74 n1. jul 5 1956 (wkly) – 9r – 1 – (formed by the union of: ainsworth star-journal and: brown county democrat (ainsworth ne); cont by: ainsworth star-journal; cont numbering of: ainsworth star-journal) – us Bell [071]
Ainsworth trading post – 1983 oct-1987 oct – 1 – mf#1476944 – us WHS [071]
Ainsworth, William Francis *see* Travels in the track of the ten thousand greeks
Ainsworth, William Harrison *see* The good old times
Ainsworth's magazine – La Salle IL 1842-54 – 1 – mf#4197 – us UMI ProQuest [420]
Ain't i a woman? – Iowa City IA 1970-73 – 1 – ISSN: 0044-6939 – mf#7939 – us UMI ProQuest [305]
The ainu and their folk-lore / Batchelor, John – London: Religious Tract Society, 1901 – 2mf – 9 – 0-524-04857-6 – mf#1990-3419 – us ATLA [390]
The ainu of japan : the religion, superstitions, and general history of the hairy aborigines of japan / Batchelor, J – London, 1892 – 4mf – 9 – mf#HTM-10 – ne IDC [915]
L'aioli – Avignon, France 7 jan 1891-27 dec 1899 – 2r – 1 – (discontinued) – uk British Libr Newspaper [072]
L'aioli – Avignon, France 7 jan 1891-27 dec 1899 – 2r – 1 – (discontinued) – uk British Libr Newspaper [074]
Aiollo, G et al *see* Biulleteni diktatury tsentrokaspiia i prezidiuma vremennogo ispolnit k-ta
Aion [daily ed] – Athens, Greece 2 jul 1880-15 jun 1885 (imperfect) – 6r – 1 – (in greek) – uk British Libr Newspaper [074]
Aion-aionios : an excursus on the greek word rendered everlasting, eternal, etc, in the holy bible / Hanson, John Wesley – Chicago: Jansen, McClurg, 1880 [mf ed 1986] – 1mf – 9 – 0-8370-9240-X – (incl bibl ref & ind) – mf#1986-3240 – us ATLA [450]
Aipla quarterly journal – v1-28. 1973-2000 – 9 – $432.00 set – (title varies: v1-11 (1973-83) apla quarterly journal) – ISSN: 0883-6078 – mf#100641 – us Hein [340]
Air allemand varie pour le violon avec accompagnement de piano / Fontaine, Antoine Nicolas Marie – [18-?] [mf ed 19–] – 2pt on 1r – 1 – mf#pres. film 33 – us Sibley [780]
Air and space – Washington DC 1978-83 – 1,5,9 – mf#11718 – us UMI ProQuest [629]
Air and waste – Pittsburgh PA 1993-94 – 1,5,9 – (cont: journal of the air & waste management association; cont by: journal of the air & waste management association) – ISSN: 1073-161X – mf#6210.04 – us UMI ProQuest [333]
Air and water pollution – Oxford, England 1958-66 – 1,5,9 – ISSN: 0568-3408 – mf#49008 – us UMI ProQuest [333]
Air avec 24 variations pour l'etude de la flute, op 1 / Jusdorf, J C – 2e ed. Offenbach sur la Mein: Jean Andre [1809] [mf ed 1989] – 1r – 1 – mf#pres. film 50 – us Sibley [780]
Air cargo – Oak Brook IL 1959-71 – 1,5,9 – ISSN: 0568-3432 – mf#1755 – us UMI ProQuest [380]
Air cargo magazine – Newark NJ 1977-82 – 1,5,9 – (cont: cargo airlift; cont by: air cargo world) – ISSN: 0148-7469 – mf#242.02 – us UMI ProQuest [380]
Air cargo world – Newark NJ 1983+ – 1,5,9 – (cont: air cargo magazine) – ISSN: 0745-5100 – mf#242.02 – us UMI ProQuest [380]
Air castle don : or, from dreamland to hardpan / Ashley, Barnas Freeman – Chicago: Laird & Lee, 1896? – 4mf – 9 – mf#27223 – cn Canadiana [830]
Air conditioning, heating & refrigeration news – Troy MI 1926+ – 1,5,9 – ISSN: 0002-2276 – mf#758 – us UMI ProQuest [621]
Air cooled news – Cazenovia NY 1953+ – 1,5,9 – ISSN: 0515-8095 – mf#11384 – us UMI ProQuest [621]
Air de nina *see* Ouverture, romance et choeur de l'opera de nina
Air des deux jumeaux de bergame... : varie pour le violon et...avec un accompagnement de pianoforte...oeuvre 31, no 3 / Desaugiers, Marc Antoine – Paris: Erard [179-?] [mf ed 19–] – 1r – 1 – mf#film 2541 – us Sibley [780]
Air engineering – Troy MI 1959-69 – 1 – ISSN: 0568-3459 – mf#1956 – us UMI ProQuest [629]
Air favori varie : pour le piano forte / Latour, T – 2e ed. Offenbach s/M: Jean Andre [1805] [mf ed 1992] – 1r – 1 – mf#pres. film 113 – us Sibley [780]
Air force – v.1-46. Sept 21, 1918-Dec 1963 – 1 – 989.00 – us L of C Photodup [355]
Air force civil engineer – Washington DC 1972-83 – 1,5,9 – ISSN: 0002-2357 – mf#6345.01 – us UMI ProQuest [629]
Air force comptroller – Washington DC 1967+ – 1,5,9 – ISSN: 0002-2365 – mf#6650 – us UMI ProQuest [629]
Air force engineering & services quarterly – Washington DC 1984-86 – 1,5,9 – ISSN: 0883-0193 – mf#6345,01 – us UMI ProQuest [629]
Air Force Institute of Technology (US) *see* Educator
Air force law review – Washington DC 1974+ – 1,5,9 – (cont: jag law review) – ISSN: 0094-8381 – mf#5750,01 – us UMI ProQuest [355]
Air force law review / U.S. Army. Air Force – v1-29. mar 1958-88 – (= ser Military law coll) – 9 – (v1-6 n5 titled: u.s. air force jag bulletin; v6 no 6-v15 n2 titled: jag law review; updates planned) – mf#LLMC 80-527 – us LLMC [355]
Air force magazine – Arlington VA 1926+ – 1,5,9 – ISSN: 0730-6784 – mf#1541 – us UMI ProQuest [355]
Air force recruiter – v27 n5-v29 n9 [1981 may-1983 sep]; v30 n9-v34, n12 [1984 sep-1988 dec] – 1 – mf#1703396 – us WHS [355]
Air force research review – Washington DC 1962-71 – 1 – ISSN: 0029-6902 – mf#6288 – us UMI ProQuest [355]
Air Force Sergeants Association *see*
– Lobby ledger
– State legislative bulletin
Air force times – 1942-67. Some wanting – 1 – us L of C Photodup [629]
Air force times – Springfield VA 1965+ – 1,5,9 – ISSN: 0002-2403 – mf#1760 – us UMI ProQuest [355]
Air law review – New York University. v1-12. 1930-1941 (complete) – 61r – 1 – $91.00 – mf#LLMC 95-111 – us LLMC [341]
Air law review – v1-12. 1930-41 – 1 – $120.00 – mf#0010 – us Brook [341]
Air laws and treaties of the world / U.S. Library of Congress – Washington: GPO. 3v. 1965 (all publ) – 48mf – 9 – $72.00 – mf#LLMC 81-408 – us LLMC [341]
Air line pilot – 1932 apr 5-1949 may; 1949 jun-1955; 1956-60 – 1 – mf#1051751 – us WHS [380]
Air line pilot – Herndon VA 1932+ – 1,5,9 – ISSN: 0002-242X – mf#2766 – us UMI ProQuest [380]
Air Line Pilots' Association *see* Council 12 banner
Air Line Stewards and Stewardesses Association, International *see* Service aloft
Air Ministry Aeronautical Research Committee *see* Reports and memoranda of the air ministry aeronautical research committee
Air navigation / Weems, Philip Van Horn – New York, USA. 1938 – 1r – 1 – us UF Libraries [380]
Air of mozart : with variations for a violin and violoncello / Cohen, E G W – London [n.p. c1800] [mf ed 1989] – 2pt on 1r – 1 – mf#pres. film 45 – us Sibley [780]
Air pollution : all known articles relating to air pollutants and air pollution controls – 139r – 1 – $6950.00 – (18,454 articles. accession cards, aut cards, subject cards available) – mf#B70065 – us Library Micro [360]
Air pollution control association. journal of the air pollution control association – Pittsburgh PA 1951-86 – 1,5,9 – (cont by: japca) – ISSN: 0002-2470 – mf#6210.04 – us UMI ProQuest [366]
Air pollution titles – University Park PA 1967-91 – 1,5 – ISSN: 0002-2497 – mf#7667 – us UMI ProQuest [333]
Air power – The Air Force quarterly. v. 1-7. 1953-60 – 1 – 60.00 – us L of C Photodup [629]
Air power historian – v. 1-10. 1954-63 – 1 – us L of C Photodup [629]
Air power history – Andrews AF Base MD 1989+ – 1,5,9 – (cont: aerospace historian) – ISSN: 1044-016X – mf#1519,01 – us UMI ProQuest [629]
Air progress – Chatsworth CA 1972-81 – 1,5,9 – ISSN: 0002-2500 – mf#6583.02 – us UMI ProQuest [629]
Air pulse – 1986 sep-1987 feb; 1987 mar 6-jun; 1987 jul-dec; 1988 jan-dec; 1989 jan-sep; 1989 oct-1990 jan; 1981 may-dec; 1982 jan 8-sep 24; 1982 oct-1983 jun; 1983 jul-1984 oct; 1984 nov-1985 apr; 1985 may-nov; 1986 jan-aug – 1 – mf#550920 – us WHS [071]
Air quality data report – 1973-1980 ? – 1 – mf#367146 – us WHS [360]
Air rescue information letter – 1951-56 – 1 – us L of C Photodup [629]
Air reservist – Washington DC 1972-86 – 1,5,9 – (cont by: citizen airman: the official magazine of the air national guard & air force reserve) – ISSN: 0002-2535 – mf#7421.01 – us UMI ProQuest [355]
Air scoop – 1982 mar-1983 jun; 1983 jul-1984; 1985 jan-jun; 1985 jul-1986 feb; 1986 mar-1987 mar; 1987 apr-oct; 1987 nov-1988 jun; 1988 dec-1989 apr; 1988 jul-nov; 1989 may-oct; 1989 nov-1990 feb – 1 – mf#705209 – us WHS [071]
Air scoop – 1983 apr 1-1988 jun – 1 – mf#1520884 – us WHS [071]
Air service and air corps news letter, 1918-1935 / U.S. Army – 5r – 1 – $650.00 – mf#S1652 – us Scholarly Res [355]
Air & space power journal – Washington DC 2002+ – 1,5,9 – ISSN: 1555-385X – mf#16177.02 – us UMI ProQuest [629]
Air transport interchange – v1 n1-v6 n9 [1979 aug-1984 nov/dec] – 1 – mf#965146 – us WHS [380]
Air transport world – Cleveland OH 1975+ – 1,5,9 – ISSN: 0002-2543 – mf#10339 – us UMI ProQuest [380]
Air university dispatch – v. 1-17. 1947-63 – 1 – us L of C Photodup [629]
Air university quarterly – v. 1-13. 1947-62 – 1 – us L of C Photodup [629]
Air university review / U.S. Army. Air Force – v1-38 n 2. 1947-mar 1987 (all publ) – (= ser Military law coll) – 372mf – 9 – $558.00 – mf#LLMC 80-528 – us LLMC [073]
Air university review [united states ed] – Maxwell AFB AL 1947-87 – 1,5,9 – (cont by: airpower journal) – ISSN: 0002-2594 – mf#1772 – us UMI ProQuest [629]
Air varie pour violon et violoncelle / Barni, Camille – Paris: Hanry [c1806] [mf ed 1988] – 2pt on 1r – 1 – mf#pres. film 44 – us Sibley [780]
Air weather service observer – v1-10 Nov 1954-63. Missing 1962 – 1 – $20.00 – us L of C Photodup [550]
Airay, Henry *see* Lectures upon the whole epistle of st paul to the philippians
Airconditioning & refrigeration business – Cleveland OH 1944-80 – 1,5,9 – (cont by: contracting business) – ISSN: 0002-2640 – mf#797.01 – us UMI ProQuest [690]
Aircraft : a journal published in the interest of aviation – London. v. 1, no. 1-v. 3, no. 33. Aug. 1916-Dec. 25, 1918. (incomplete) – 1 – us NY Public [380]
Aircraft cannibalization : an expensive appetite?: hearing...house of representatives, 107th congress, 1st session, may 22 2001 / United States. Congress. House. Committee on Government Reform. Subcommittee on National Security, Veterans Affairs, and International Relations. – Washington: US GPO 2002 [mf ed 2002] – 2mf – 9 – 0-16-068855-8 – (incl bibl ref) – us GPO [355]
Aircraft engineering – v. 1-35. 1929-63 – 1 – 589.00 – us L of C Photodup [629]
Aircraft engineering and aerospace technology – Bradford, England 2001+ – 1,5,9 – mf#27885,01 – us UMI ProQuest [629]
Aircraft shop stewards national council journal *see* Journals of the labour movement in trade and industry
Aircrafter – v1 n1-v5 n4 [1978 aug-1984 dec]; v5 n5 [1985 dec] – 1 – mf#1131886 – us WHS [629]
Aircrafter – v18 n2-v21 n3 [1974 apr-1978 may] – 1 – mf#496401 – us WHS [629]
Aircraftsman – v12 n13-17, 19-20 [1981 jul 2-aug 27, sep 24-oct 8]; v14 n13-14, 17, 19-20, 22-1925 [1983 jun 30-jul 14, aug 25, sep 22-oct 6, nov3-dec 15]; v15 n1, 5, 7-[1984 jan 12, mar 8, apr 12-sep 20]; v16 n20-21, 25 [1985 oct 10-14, dec 5] – 1 – mf#1295822 – us WHS [629]
Airdie and coatbridge telegraph *see* Weekly telegraph
Airdrie advertiser and linlithgowshire standard – Airdrie, Scotland 19 may 1883-18 jan 1902 – 1 – (cont: advertiser for airdrie, coatbridge, bathgate & wishaw [4 jul 1868-12 may 1883]; cont by: airdrie & coatbridge advertiser [25 jan 1902-]) – uk British Libr Newspaper [072]
Airdrie and coatbridge advertiser – Airdrie: Baird & Hamilton 1902- (wkly) [mf ed 6 jan 1995-] – 1 – (cont: airdrie advertiser, and linlithgowshire standard; absorbed: coatbridge express) – ISSN: 1353-4327 – uk Scotland NatLib [072]
Airdrie and coatbridge advertiser – Airdrie, Scotland 25 jan 1902- [mf 1986-] – 1 – (cont: airdrie advertiser, & linlithgowshire standard [19 may 1883-18 jan 1902]) – uk British Libr Newspaper [072]
Airdrie journal – [Scotland] North Lanarkshire, Airdrie: printed & publ in the Luminary Office nov 1850 & feb 1851 [mf ed 2004] – 1r – 1 – uk Newsplan [072]
Airdrie literary album : or, weekly repository of original and select literature / ed by Bell, G M – [Scotland] North Lanarkshire, Airdrie: sold by G H Stark 6 dec 1828-16 may 1829 (wkly) [mf ed 2003] – 24v on 1r – 1 – uk Newsplan [410]
Aire, el agua y el arbol / Garron De Doryan, Victoria – San Jose, Costa Rica. 1962 – 1r – 1 – us UF Libraries [972]
Aires monteros / Saenz Morales, Ramon – Managua, Nicaragua. 1947 – 1r – 1 – us UF Libraries [972]
Airfinance journal – London, England 1991+ – 1,5,9 – ISSN: 0143-2257 – mf#19527 – us UMI ProQuest [332]
Airi, Raghunath *see* Concept of sarasvati (in vedic literature)
Airlift – 1985 nov-1987 may – 1 – mf#1542228 – us WHS [071]
Airlift dispatch – 1983 jan 7-1984 aug; sep-1985 jun; jul-1986 sep; oct-1987 sep; oct-1988 sep; oct-1989 – 1 – mf#1278382 – us WHS [071]
Airlifter – 1981 apr 30-1982 jun; jul-1983 jul; 1983 jul 7-1984; 1985; 1986 jan-1987 may; jun-1988 jan; feb-oct; nov-1989 jun; jul-1990 feb – 1 – mf#627588 – us WHS [071]
Airlifter – v3 n12 [1981 jun 1] – 1 – mf#612372 – us WHS [071]
Airline executive – Shawnee Mission 1983-89 – 1,5,9 – (cont by: airline executive international) – ISSN: 0278-6702 – mf#14110.01 – us UMI ProQuest [380]
Airline executive international – Shawnee Mission KS 1990-91 – 1,5,9 – (cont: airline executive) – ISSN: 1051-631X – mf#14110,01 – us UMI ProQuest [380]
Airman – v. 1-7. Aug 1957-63 – 1 – 85.00 – us L of C Photodup [355]
Airman – Washington DC 1957+ – 1,5,9 – ISSN: 0002-2756 – mf#7243 – us UMI ProQuest [629]
Airport area : culver city, el segundo, gardena, redondo beach, santa monica, etc. – 1975- – (= ser California telephone directory coll) – 33r – 1 – $1650.00 – mf#P00001 – us Library Micro [917]
Airpower journal – Washington DC 1987-98 – 1,5,9 – (cont: air university review / [united states ed]) – ISSN: 0897-0823 – mf#16177.02 – us UMI ProQuest [629]
Air-raid over barcelona : [poem] / Richardson, Stanley – [s.l: s.n. 193-] – 9 – mf#w1138 – us Harvard [810]
Airs a quatre parties : avec la basse-continue, et quelques-uns a trois en forme de motets a la fin du livre, sur la paraphrase de quelques pseaumes et cantiques de messire anthoine godeau... / Du Mont, Henry – Paris: par Robert Ballard...1663. 8v [mf ed 1993] – 2mf – 9 – mf#SEM105P2021 – cn Bibl Nat [780]
Airs detaches de melide : ou, le navigateur: comedie mise en musique / Philidor, Francois Danican – A Paris: aux adresses ordinaires; A Bruxelles: Chez J L De Boubers... [mf ed 1989] – 1r – 1 – (in french) – mf#pres. film 50 – us Sibley [780]
Air-scoop – 1988 may-jul, oct-dec; 1989 mar, may-jun – 1 – mf#2230327 – us WHS [071]
Airscoop – 1959 oct-1974; 1975-1988 sep – 1 – mf#1330691 – us WHS [071]
Airscoop – 1981 apr 30-1983 oct 27; nov-1988 jun; jul-1989 dec 22 – 1 – mf#1046601 – us WHS [071]
Airscoop – 1981 may-1989 dec – 1 – mf#1702448 – us WHS [071]
Airtides – 1981 may-dec; 1982 jan-sep; oct-1983 jun; 1983 jul-1984 dec 14; 1985 jan-1987 apr 24; 1987 may-1988 apr; may-1989 may 19 – 1 – mf#648343 – us WHS [071]
Airwaves – 1979 aug 15; 1980 jan 15 – 1 – mf#4862652 – us WHS [071]
Airwinger – v26 n12, 20 [1983 mar 24, may 19]; v27 n26-1931, 34, 37, 40-41 [1984 jul 12-aug 16, sep 6, 27, oct 18-1925]; v28 n3, 8, 11, 13-14, 45 [1985 jan 24, feb 28, mar 21, apr 11-18, nov27]; v29 n30, 33, 35, 41-42, 45, 47 [1986 aug 7, 28, sep 11 – 1 – mf#1567842 – us WHS [071]
'Airy fairy lilian' / Hungerford, Margaret Wolfe – London: Smith, Elder & Co. 3v. 1879 – (= ser 19th c women writers) – 12mf – 9 – mf#5.1.109 – uk Chadwyck [830]
Airy, George Biddell *see*
– Gravitation
– Notes on the earlier hebrew scriptures
Airy, George Biddell et al *see* Astronomical and magnetical and meteorological observations made at the royal observatory, greenwich, in the year...
Ais newsletter – v10 n4, 12 [1976 apr, dec]; v11 n6 [1977 jun]; v12 n5; [1979 may]; v14 n3-8, 11-1912 [1980 mar-aug, nov-dec]; v15 n1-1912 [1981]; v16 n1-6 [1982 jan-jun] – 1 – mf#618628 – us WHS [071]
Aise steel technology / Association of Iron and Steel Engineers – Pittsburgh PA 1999+ – 1,5,9 – (cont: iron & steel engineer) – ISSN: 1528-5855 – mf#1473,01 – us UMI ProQuest [660]

Aisse, Charlotte E see Lettres de mademoiselle aisse a madame c...

The aitareya brahmanam of the rigveda : containing the earliest speculations of the brahmans on the meaning of the sacrificial prayers, and on the origin, performance and sense of the rites of the vedic religion / ed by Haug, Martin – Bombay: Govt Central Book Depot, 1863 – 2mf – 9 – 0-524-06371-0 – mf#1990-3538 – us ATLA [280]

Aitchison, David see Scottish presbyterianism not presbyterian

Aitken, George Atherton see The life and works of john arbuthnot...

Aitken, James see The book of job

Aitken, Roger see Brotherly-kindness and unity essential to the christian character

Aitken, William Benford see The dominion of canada

Aitken, William Hay Macdowall Hunter see
– Around the cross
– Derby mission, november, 1873

Aitkin, George see Solomons tok tok

Aixala Casellas, Jose see Luces de otono

Aix-en-provence (france) – Bibliotheque Mejanes. Vue du Chateau de Saint Thome [1709?] – (filmed with: pinto, m r relacas) – us CRL [944]

Aiyangar, S Krishnaswami see History of the nayaks of madura

Aiyappan, A see The manley collection of stone age tools

Aiyar, P S see Conflict of races in south africa

Ai-yin-hsi see Ti erh tz'u shih chieh ta chan chung ti ching chi wen t'i

Aizman, A la see Osnovy i praktika garantiinogo strakhovaniia

Aizpura, Aizpuru see
– Idealismos de verdad y de belleza

Aja – Banaras, India. Mar 1947-1959; 1961-94 – 185r – 1 – us L of C Photodup [079]

Ajisafe, Ajayi Kolawole see
– History of abeokuta
– The laws and customs of the yoruba people

Ajmer : historical and descriptive / Sarda, Har Bilas, Diwan Bahadur – Ajmer: Fine Art Print Press, 1941 – (= ser Samp: indian books) – us CRL [954]

Ajot : the american journal of occupational therapy – Bethesda MD 1978-79 – 1,5,9 – (cont: american journal of occupational therapy; cont by: american journal of occupational therapy) – ISSN: 0161-326X – mf#10224.02 – us UMI ProQuest [615]

Ajr [american journal of roentgenology] – Leesburg VA 1976+ – 1,5,9 – (cont: american journal of roentgenology, radium therapy, & nuclear medicine) – ISSN: 0361-803X – mf#1574,01 – us UMI ProQuest [616]

Akademi Angkatan Bersendjata Republik Indonesia see Madjalah

Akademi angkatan udara / Putera angkasa – Jogjakarta, 1964-1965 – 2mf – 9 – (missing: 1964, v1(1-3); 1964(6-end)) – mf#SE-774 – ne IDC [959]

Akademi popular see Jajasan akademi populer

Akademicheskie izvestiia – New York. 1960-1973 (1) 1971-1973 (5) – 45mf – 9 – mf#1676 – ne IDC [077]

Die akademie der grazien : eine wochenschrift zur unterhaltung des schoenen geschlechts / ed by Schuetz, Christian Gottfried – Halle 1774-80 – (= ser Dz) – 5v on 15mf – 9 – €150.00 – mf#k/n6477 – gw Olms [640]

Akademie der schoenen redekuenste / ed by Buerger, Gottfried August – Berlin, Goettingen 1797-98 – (= ser Dz. abt literatur) – 2v on 4mf – 9 – €120.00 – mf#k/n4576 – gw Olms [400]

Akademie fuer deutsches recht (bestand r 61) bd 9 / ed by Werhan, Walter – 2. aufl 1976 – 110p – (p/g o/p) – gw Bundesarchiv [342]

Akademiia Dukhovnaia. Leningrad see Opisanie knig grazhdanskoi pechati

Akademiia nauk : izvestiia postoiannoi tsentralenoi seismicheskoi komissii – Pg., L., 1902-1924. v1-7(3) – 160mf – 9 – (missing: v6(2); v7(1)) – mf#R-2323 – ne IDC [077]

Akademiia Nauk. Otdelyeniye Russakavo Yazyka i Slovesnosti. St Petersburg see Sbornik

Akademiia nauk Soiuza SSR see Zhurnal obshchei biologii

Akademiia Nauk. SSSR see
– Botanicheskii zhurnal
– Dagestansik filial, makhach-kala
– Sbornik po russkomu yazyku i slovesnosti

Akademiia nauk SSSR. Astronomicheskii institut see Astronomicheskii tsirkuliar

Akademiia nauk SSSR et al see Mineralogiia zolota

Akademiia nauk SSSR, Geologicheskii institut see K pervoi mezhdunarodnoi palinologicheskoi konferentsii, takson, ssha

Akademiia Nauk Sssr Institut Istorii see Nueva historia de los paises coloniales y dependie

Akademiia nauk SSSR. Institut mezhdunarodnogo rabochego dvizheniia see Problemy massovogo rabochego i obshchedemokraticheskogo dvizheniia v italii

Akademiia nauk SSSR. Institut morfologii zhivotnykh im A N Severtsova see Problemy pochvennoi zoologii

Akademiia Nauk. SSSR Institut Russkovo Yazyka see Dialektologicheski sektor

Akademiia nauk SSSR, Institut vysokomolekuliarnykh soedinenii i Biblioteka Akademii nauk SSSR see Bibliograficheskii ukazatel rabot nauchnykh sotrudnikov instituta vysokomolekuliarnykh soedinenii an sssr

Akademiia nauk SSSR. Ministerstvo geologii SSSR, Institut mineralogii, geokhimii i kristallokhimii redkikh elementov (IMGRE) see Redkie elementy

Akademiia nauk SSSR. Nauchnyi sovet po elektronnoi mikroskopii... et al see Materialy 9 vsesoiuznoi konferentsii po elektronnoi mikroskopii, 29 oktiabria-2 noiabria 1973 g., tbilisi

Akademiia nauk SSSR, Nauchnyi sovet po probleme "Biokhimiia zhivotnykh i cheloveka"... see Sovremennye problemy biokhimii dykhaniia i klinika

Akademiia nauk SSSR, Ordena Lenina Institut geokhimii i analiticheskoi khimii im see Mezhdunarodnyi geokhimicheskii kongress (1st: 1971: moscow, rsfsr)

Akademiia nauk SSSR, Otdelenie fiziko-khimii i tekhnologii neorganicheskikh materialov see 13 vsesoiuznoe chugaevskoe soveshchanie po khimii kompleksnykh soedinenii, moskva, 12-15 iiunia 1978 g

Akademiia nauk SSSR, Otdelenie ordena Lenina Instituta khimicheskoi fiziki see Gorenie i vzryv

Akademiia nauk SSSR, Sibirskoe otdelenie see Chislennye metody mekhaniki sploshnoi sredy

Akademiia nauk SSSR, Sibirskoe otdelenie AN SSSR, Institut neorganicheskoi khimii see Vsesoiuznaia shkola "primenenie matematicheskikh metodov dlia opisaniia i izucheniia khimicheskikh ravnovesii," g novosibirsk, 9-13 fevralia 1976 g: tezisy dokladov

Akademiia nauk SSSR. Sibirskoe otdelenie, Gosudarstvennaia publichnaia nauchno-tekhnicheskaia biblioteka see Iz istorii knigi, bibliotechnogo dela i bibliografii v sibiri

Akademiia nauk SSSR. Sibirskoe otdelenie, Institut istorii, filologii i filosofii see Materialy po arkheologii sibiri i dalnego vostoka

Akademiia nauk SSSR. Sibirskoe otdelenie, Institut teplofiziki see Teplo- i massoperenos v absorbtsionnykh apparatakh

Akademiia nauk SSSR, Sibirskoe otdelenie, Ordena Trudovogo Krasnogo Znameni Institut kataliza see Katalizatory, soderzhashchie nanesennye kompleksy

Akademiia nauk SSSR, Vychislitelnyi tsentr see Obrabotka simvolnoi informatsii

Akademiia Nauk. URSR. Kiev. see Zvidomlennya

Akademiia Nauk. URSR. Kiev. Instytut Elektrozvaryuvannya see Trudy po avtomaticheskoi svarke pod flyusom

Akademiia Nauk Ursr, Kiev Instytut Ievreis'koi Proletars'koi see Bibliologisher zamlbukh

Akademiia Nauk. URSR. Kiev. Istorychna Sektziya see Komissiya istorychnoi pisennosty ukrayins'ki narodni dumy

Akademiia Navuk Belaruskai Ssr see Rabonim in dinst fun finants-kapital

Akademische festrede zu grillparzers hundertstem geburtstage : gehalten in der aula des carolinums von august sauer / Sauer, August – Prag: J G Calve 1891 [mf ed 1990] – 1r – 1 – (filmed with: franz grillparzer / adalbert faulhammer) – mf#2689p – us Wisconsin U Libr [430]

Akademische predigten / Holtzmann, Heinrich Julius – Leipzig: FA Brockhaus, 1873 [mf ed 1993] – 1mf – 9 – 0-524-06841-0 – mf#1992-0983 – us ATLA [242]

Akademische vortraege see
– Addresses on historical and literary subjects
– Studies in european history

Akademischer beobachter – Muenchen DE, 1929 – 1 – gw Misc Inst [378]

Akademiska dzive – n1-1916 [1958-1974]; n17-32 [1975-90] – 1 – mf#681827 – us WHS [071]

Akademiya nauk SSSR. Doklady see
– Doklady. biochemistry
– Doklady. biological sciences
– Doklady. biophysics
– Doklady. botanical sciences
– Doklady. chemical technology
– Doklady. chemistry
– Doklady. mathematics
– Doklady. physical chemistry

Akali – Jullundur, India. 1951-56; Apr-Jun 1971 – 5r – 1 – us L of C Photodup [079]

Akamaba in british east africa / Lindblom, Gerhard – Uppsala, Sweden. 1920 – 1r – 1 – us UF Libraries [960]

Akamba stories / Mbiti, John S – Oxford, England. 1966 – 1r – 1 – us UF Libraries [960]

Akan religion and the christian faith / Williamson, S G – Accra, 1965 – 4mf – 8 – €11.00 – ne Slangenburg [230]

Akan (twi-fante) language collection / Warren, Denise M – Chicago: Uni of Chicago, Photodup Dept, [19–?] – 1 – us CRL [490]

Akana – v1 n1-v2 n13 [1982 dec 6-nov8 1983] – 1 – mf#658977 – us WHS [071]

Akaroa mail – 1877-1939; apr-dec 1973; 1974-87 – 80r – 1 – mf#70.3 – nz Nat Libr [079]

Akashvani – New Delhi. v.7-26. May 1942-Oct 1961 – 1 – us L of C Photodup [410]

Akatanshi takalisha / Chimolula, A – Cape Town, South Africa. 1957 – 1r – 1 – us UF Libraries [960]

Akbar / Binyon, Laurence, 1869-1943 – [Edinburgh]: Peter Davies Ltd, 1932 – (= ser Samp: indian books) – us CRL [920]

Akbar Ali, Sheikh see Iqbal, his poetry and message

Akbar and the rise of the mughal empire / Malleson, George Bruce – Oxford: Clarendon Press, 1908 [mf ed 1995] – (= ser Yale coll; Rulers of india 2) – 204p – 1 – 0-524-09995-2 – mf#1995-0995 – us ATLA [954]

Akbar the great mogul 1542-1605 / Smith, Vincent Arthur – Oxford: Clarendon Press, 1917 [mf ed 1995] – (= ser Yale coll) – xv/504p (ill) – 1 – 0-524-09388-1 – mf#1995-0388 – us ATLA [954]

Akbar, the great mogul, 1542-1605 / Smith, Vincent Arthur – Oxford: Clarendon Press, 1926 – (= ser Samp: indian books) – us CRL [920]

Aked, Charles Frederic see Changing creeds and social struggles

Aken, Adolf Friedrich see Die grundzuege der lehre von tempus und modus im griechischen

Akerman, John Yonge see Remains of pagan saxondom

Akerman, William see "The bible and the square"

Akers, Peter see Introduction to biblical chronology

Akhbar – Alger. janv-sept 1897, 1902-05, 1909-fevr 1934 – 1 – fr ACRPP [073]

Akhbar al-ahad = Sunday news – Bi-Khartum: al-Tajammu al-Masihi al-Sudani, [sep 5 1985-jun 1 1986] (wkly) – 1r – 1 – us CRL [079]

Akhbar al-'alam al-islami – Mecca, Saudi Arabia. 1977-1992 – 15r – 1 – us L of C Photodup [079]

Akhbar al-sabah – al-Khartum: Mu'assasat Dar Akhbar al-Sabah lil-Sihafah, [nov 12 1988-feb 5 1989] (wkly) – 1r – 1 – us CRL [079]

Akhbar al-suq = Sudanese business – [Khartoum?]: Akhbar al-Suq, [may 13-jun 24 1989] (wkly) – 1r – 1 – us CRL [338]

Akhbar-i iran – Tehran, 197?- . dawrah-'i jadid, sal-i 6, shumarah-'i 10-38. 26 day 1358-22 mihr 1360 [26 jan 1980-14 oct 1981] – 1r – 1 – $53.00 – us MEDOC [079]

Akhbar-i jabhah-'i milli-i iran – [Tehran]: Kumisyun-i Intisharat va tablighat-i Jabhah-'i Milli-i Iran. shumarah-'i 1-48. 14 bahman 1340-17 urdibihisht 1342 [3 feb 1942-7 may 1963] – 1r – 1 – $53.00 – us MEDOC [079]

Akhinson, I see Praktishe pedologye

Akhir al-anba' – [Khartoum]: Dar al-Sahafah, [apr 19 1986-dec 10 1988] (daily) – 1r – 1 – us CRL [079]

The akhmaim fragment of the apocryphal gospel of st peter / ed by Swete, Henry Barclay – London, New York: Macmillan, 1893 – 1mf – 9 – 0-7905-8303-8 – (incl bibl ref) – mf#1987-6408 – us ATLA [226]

Akhmanov, A S see Nadzor za kreditnymi uchrezhdeniiami

Akhsanya shel torah / Hebrew Gymnasiumn (Berlin, Germany) – Berlin, Germany. 1921 – 1r – 1 – us UF Libraries [939]

Akhtar – Istanbul. sal-i 2, adad 61-sal-i 3, adad 17 [11 jan-4 apr 1877]; sal-i 5, shumarah-i 1-50 [25 dec 1878-10 dec 1879]; sal-i 6, shumarah-i 1-49 [17 dec 1879-29 oct 1880]; sal-i 7, shumarah-i 1-48 [9 dec 1880-17 nov 1881]; sal-i 8, shumarah-i 1-51 [24 nov 1881-8 nov 1882] – 2r – 1 – $155.00 – us MEDOC [079]

Akiba : ein palastinensischer gelehrter aus dem zweiten nachchristli / Funk, Samuel – M-Theresiopel, Yugoslavia. 1896 – 1r – 1 – us UF Libraries [956]

Akibat Ibn-Muhammad Bulaki al-Kasmiri, Abd-al-Karim Ibn see Voyage de l'inde a la mekke

Akif bey / Kemal, Namik – Istanbul, 1290 [1873] – (= ser Ottoman literature, writers and the arts) – 2mf – 9 – $40.00 – us MEDOC [470]

Akif, Mehmed see Sebil uer-resad

Akif, Munse'at-i see
– The divan project

The akikuyu : their customs, traditions and folklore / Cagnolo, C – Nyeri, Kenya: Printed by Akikuyu in the Mission Printing School, 1933 – 1 – us CRL [306]

Akili newsletter – v1 n1-v2 n4 [1993 may-1995 mar] – 1 – mf#2864504 – us WHS [071]

Akimoto, S see Keloearga dan roemah tangga nippon

Akin – Guemuelcine [Komotini] GR. n17,29,31,49,53,79. 31 mayis 1957-17 ekim 1958 – (= ser O & t journals) – 1mf – 9 – $25.00 – us MEDOC [956]

Akindele, Adolphe see Contribution a l'etude de l'histoire de l'ancien royaume de porto-nova

Akins, Ann S see Dancing in dixie's land

Akinyotu, Adetunji see A bibliography on development planning in nigeria, 1955-1968

Akkadische goetterepitheta / Tallqvist, K – (= ser Studia Orientalia) – 6mf – 9 – (studia orientalia, helsingforsiae 1938 v7) – mf#NE-403 – ne IDC [956]

Die akkadische sprache : vortrag gehalten auf dem fuenften internationalen orientalisten congresse zu berlin / Haupt, Paul – Berlin: A Asher, 1883 – 1mf – 9 – 0-8370-7636-6 – mf#1986-1636 – us ATLA [470]

Der akkadische wettergott in mesopotamien / Schlobies, H – Leipzig, 1925 – (= ser Mitteilungen der altorientalischen Gesellschaft) – 1mf – 9 – (mitteilungen der altorientalischen gesellschaft v1 pt3) – mf#NE-20101 – ne IDC [930]

Aklat ng pagluluto : hinan-go sa lalong bantog at dakilang aklat ng pagluluto sa gawing europa at sa pilipinas... / Ignacio, Rosendo – Maynila: J Martinez 1919 – [ill] – 1 – (on reel with : sevilla, jose n: ag akalat ng tagalog) – mf#1707 reel 2 n4 – us Wisconsin U Libr [640]

Aklida de-rahame / Rotski, Aizik Leb – Vilna, Lithuania. 1913 – 1r – 1 – us UF Libraries [939]

Akron business and economic review – Akron OH 1970-91 – 1,5,9 – ISSN: 0044-7048 – mf#8214 – us UMI ProQuest [338]

Akron city directory, 1903 – 1r – 1 – mf#B30662 – us Ohio Hist [978]

Akron law review – v1-34. 1967-2001 – 5,6,9 – $655.00 set – (v1-18 1969-85 reel $253.00. v19-34 mf $402.00) – ISSN: 0002-371X – mf#101581 – us Hein [340]

Akron, OH see Selections (1825-1928)

Akron reporter see Miscellaneous newspapers of washington county

Akron tax journal – University of Akron. v1-15. 1983-2000 – 9 – $170.00 set – ISSN: 1044-4130 – mf#109311 – us Hein [336]

Akron telephone books (1929-1989) – 26r – 1 – mf#B31342-31367 – us Ohio Hist [978]

Akropolis : griechische tageszeitung in deutschland – Bonn DE, 1978 1 sep-1989 1 aug – 31r – 1 – gw Misc Inst [074]

Die akroterfiguren des tempels der athener auf delos / Wester, Ursula – Heidelberg, 1969 – 2mf – 9 – 3-89349-675-0 – gw Frankfurter [730]

Aksakov, Sergei Timofeevich see Izbrannye sochineniia

Aksakova, K S et al see Moskovskii sbornik

Aksam guenesi – Istanbul: Maarifet Matbaasi, 1926 – (= ser Ottoman Literature, Writers and the Arts) – 6mf – 9 – $90.00 – us MEDOC [470]

Aksara : a forgotten chapter in the history of indian philosophy / Modi, Prataprai Mohanlal – [Baroda?: sn], 1932 – (= ser Samp: indian books) – us CRL [180]

Aksaranuson chabap mahathirarat ramluk – [Krung Thep Maha Nakhon]: Khana aksonsat, Chulalongkon Mahawitthayalai 2496 [1953] [mf ed 1993] – on pt of 1r – 1 – mf#11052 r1114 n2 – us Cornell [490]

Aksariyat – West Germany: Sazman-i Fida'yan-i Khalq-i Iran Dar Kharij Az Kishvar. shumarah-i 70-148. 8 murdad 1364-11 isfand 1365 [9 aug 1985-2 mar 1987] – 1r – 1 – $53.00 – (missing: n71-76, 109, 115, 118-119, 147) – us MEDOC [079]

Akselrod, P see Rossiiskaia sotsial-demokraticheskaia rabochaia partiia

Akselrod, P et al see "iskra" za dva goda

Akselrod, P V see Borba sotsialisticheskikh i burzhuaznykh tendentsii v russkom revoliutsionnom dvizhenii

Aksi Press see Ekspres

Aks-i sada. anadolu sesleri – Samsun. n1023. 25 kanunisani 1336 [1920] – (= ser O & t journals) – 1mf – 9 – $25.00 – us MEDOC [956]

Akta grodzkie i ziemskie z czasow rzeczypospolitej polskiej z archiwum tak zwanego bernardynskiego we lwowie – v. 1-25. 1868-1935 – 1 – us L of C Photodup [943]

Die akte adalbert stifter / ed by Mueller, Joachim – Weimar: Kommissionsvertrieb durch den Volksverlag Weimar, [1961] – 1 – us Wisconsin U Libr [430]

Die akte arno holz / ed by Klein, Alfred – Weimar: Kommissionsvertrieb: Aufbau-Verlag, [1965] – 1 – us Wisconsin U Libr [430]

Die akte detlev von liliencron / ed by Kirsten, Wulf – Weimar: Kommissionsvertrieb: Aufbau-Verlag, [1968] – 1 – us Wisconsin U Libr [430]

Die akte eduard moerike / ed by Reuter, Hans-Heinrich – Weimar: Kommissionsvertrieb durch den Volksverlag Weimar, [1962] – 1 – us Wisconsin U Libr [430]

Die akte johannes schlaf / ed by Baete, Ludwig – Weimar: Kommissionsvertrieb: Aufbau-Verlag, [1966] [mf ed 1993] – (= ser Veroeffentlichungen aus dem archiv der deutschen schillerstiftung 10) – 50p – 1 – mf#8343 – us Wisconsin U Libr [430]

Die akte louise von francois / ed by Motekat, Helmut – Weimar: Kommissionsvertrieb: Aufbau-Verlag, [1963] – 1 – us Wisconsin U Libr [430]

Die akte ludwig feuerbach / ed by Dobbek, Wilhelm – Weimar: Kommissionsvertrieb durch den Volksverlag Weimar, [1962] – 1 – us Wisconsin U Libr [430]

Die akte max kretzer / ed by Tschoertner, Heinz Dieter – Weimar: Kommissionsvertrieb: Aufbau-Verlag [1969] [mf ed 1993] – (= ser Veroeffentlichungen aus dem archiv der deutschen schillerstiftung 14) – 2r – 1 – mf#3338p – us Wisconsin U Libr [430]

Die akte otto ludwig / ed by Mueller, Joachim – Weimar: Kommissionsvertrieb: Aufbau-Verlag, [1965] – 1 – us Wisconsin U Libr [430]

Die akte paul zech / ed by Mueller, Joachim – Weimar: Kommissionsvertrieb: Aufbau-Verlag, [1966] – 1 – us Wisconsin U Libr [430]

Die akte theodor daeubler / ed by Mueller, Joachim – Weimar: Kommissionsvertrieb: Aufbau-Verlag, [1967] – 1 – us Wisconsin U Libr [430]

Die akte wilhelm raabe / ed by Richter, Helmut – Weimar: Kommissionsvertrieb durch den Volksverlag Weimar [1963] [mf ed 1993] – (= ser Veroeffentlichungen aus dem archiv der deutschen schillerstiftung 4) – 2r – 1 – mf#3338p – us Wisconsin U Libr [430]

Die akten aus dem buero erich honecker [partei und staat in der ddr pt 2] = Records from the office of erich honecker / ed by Stiftung Archiv der Parteien und Massenorganisationen der DDR im Bundesarchiv – [mf ed 2004] – (= ser Partei und staat in der ddr – akten aus der stiftung archiv der parteien und massenorganisationen der ddr im bundesarchiv 2; Socialist power in the gdr – records from the foundation archive of parties and mass organisations in the federal archives) – 1216mf (1:24) – 9 – silver €3980.00 – ISBN-10: 3-598-35533-5 – ISBN-13: 978-3-598-35533-2 – (incl guidebook) – gw Saur [934]

Die akten aus dem buero walter ulbricht [partei und staat in der ddr pt 1] = Records from the office of walter ulbricht / ed by Stiftung Archiv der Parteien und Massenorganisationen der DDR im Bundesarchiv – [mf ed 2004] – (= ser Partei und staat in der ddr – akten aus der stiftung archiv der parteien und massenorganisationen der ddr im bundesarchiv 1; Socialist power in the gdr – records from the foundation archive of parties and mass organisations in the federal archives) – 1006mf (1:24) – 9 – silver €3580.00 – ISBN-10: 3-598-35531-9 – ISBN-13: 978-3-598-35531-8 – (incl guidebook) – gw Saur [934]

Die akten der edessenischen bekenner curjas, samonas und abibos / Gebhardt, Oscar von – Leipzig, 1911 – (= ser Tugal 3-37/2) – 5mf – 9 – €12.00 – ne Slangenburg [240]

Die akten der edessenischen bekenner gurjas, samonas und abibos : aus dem nachlass von oscar von gebhardt / ed by Dobschuetz, Ernst von – Leipzig: J C Hinrichs, 1911 [mf ed 1989] – (= ser Tugal 37/2) – 1mf – 9 – 0-7905-1700-0 – (in german, greek & latin. incl bibl ref & ind) – mf#1987-1700 – us ATLA [240]

Akten der parteikanzlei der nsdap : rekonstruktion eines verlorengegangenen bestandes = Files of the national socialist party chancellery: a reconstruction of lost records / Institut fuer Zeitgeschichte Muenchen – [mf ed 1983-92] – 491mf (1:48) + suppl vols – 9 – silver €5000.00 – ISBN-10: 3-598-30260-6 – ISBN-13: 978-3-598-30260-2 – gw Saur [325]

Akten der prinzipalkommission des immerwaehrenden reichstages zu regensburg 1663-1806 / Haus-, Hof- und Staatsarchiv, Wien. Reichskanzlei – [mf ed 1990-93] – 5118mf (1:24) + suppl vol – 9 – silver €12,600.00 – ISBN-10: 3-598-33080-4 – ISBN-13: 978-3-598-33080-3 – (int by hans booms) – gw Saur [323]

Die akten des vogelsangs : erzaehlung / Raabe, Wilhelm Karl – Leipzig: P Reclam, 1944 – 1r – 1 – us Wisconsin U Libr [830]

Die akten des vogelsangs / Raabe, Wilhelm Karl – Berlin: Otto Janke, 1896 – 1r – 1 – us Wisconsin U Libr [830]

Die akten des vogelsangs / Raabe, Wilhelm Karl – 3. Aufl. Berlin: Otto Janke, 1904 (mf ed 1990) – 1r – 1 – (filmed with: campagne in frankreich) – us Wisconsin U Libr [830]

Die akten ferdinand freiligrath und georg herwegh / ed by Kaiser, Bruno – Weimar: Kommissionsvertrieb durch den Volksverlag Weimar, [1963] – 1 – us Wisconsin U Libr [430]

Die akten gustav falke und max dauthendey / ed by Mueller, Joachim – Weimar: Kommissionsvertrieb: Aufbau-Verlag, [1907] – 1 – us Wisconsin U Libr [430]

Akten ueber die krankheit von heinrich heines vater – Kiel: Wissenschaftliche Gesellschaft fuer Literatur und Theater 1928 [mf ed 1990] – (= ser Mitteilungen der wissenschaftlichen gesellschaft fuer literatur und theater 6/1) – 1r – 1 – (incl bibl ref. filmed with: briefe von heinrich heine an heinrich laube / ed by eugen wolff) – mf#2707p – us Wisconsin U Libr [920]

Aktenstuecke die altkatholische bewegung betreffend : mit einem grundriss der geschichte derselben / Friedberg, Emil – Tuebingen: H Laupp, 1876 [mf ed 1986] – 2mf – 9 – 0-8370-8983-2 – (incl bibl) – mf#1986-2983 – us ATLA [241]

Aktenstuecke zum concil : das infallibilitaetsschema und die minoritaetsgutachten / Vatican Council. 1st – Stuttgart: J G Cotta, 1870 [mf ed 1986] – 1mf – 9 – 0-8370-8479-2 – (in german & latin) – mf#1986-2479 – us ATLA [241]

Die aktenstuecke zum frieden von s germano 1230 (mgh epistolae 4:4.bd) – 1926 – (= ser Monumenta germaniae historica epistolae 4. epistolae selectae (mgh epistolae 4)) – €5.00 – ne Slangenburg [241]

Aktepe, M Muenir *see* Tarih-i lutfi

Aktion – Porto Alegre (BR), 1933-1937 10 oct – 1 – gw Misc Inst [079]

Die aktion – Berlin DE, 1914-23, 1925-28 – 1 – (filmed by other misc inst: 1939-43 [gaps]) – gw Misc Inst [074]

Die aktion – Paris (F), 1933 4 may-21 dec – 1 – fr ACRPP [074]

Aktionen, bekenntnisse, perspektiven : berichte und dokumente vom kampf um die freiheit des literarischen schaffens in der weimarer republik / ed by Deutschen Akademie der Kuenste zu Berlin. Sektion Dichtkunst und Sprachpflege. Abt Geschichte der sozialistischen Literatur – Berlin, Weimar: Aufbau-Verlag, c1966 [mf ed 1993] – 674p/[42pl] (ill) – 1 – (comp, int and ann by friedrich albrecht et al. incl bibl ref and ind) – mf#8264 – us Wisconsin U Libr [430]

Aktionsarten mit graduierender semantik in der russischen sprache der gegenwart / Suchy, Anke – 1995 – 2mf – 9 – 3-8267-2130-6 – mf#DHS 2130 – gw Frankfurter [460]

Das aktionsbuch / ed by Pfemfert, Franz – Berlin-Wilmersdorf: Verlag der Wochenschrift Die Aktion 1917 [mf ed 1993] – 1r – 1 – (filmed with: das elternhaus / herbert roch [comp]) – mf#3392p – us Wisconsin U Libr [840]

Aktiv : die wirtschaftszeitung, die jeder versteht – Koeln DE 1972-2002 – 29mf=58df – 1 – (various eds filmed) – gw Mikrofilm [074]

Aktives zuhoeren und behalten : eine empirische untersuchung / Haensel, Anette – (mf ed 1992) – 2mf – 9 – €49.00 – 3-89349-482-0 – mf#DHS 482 – gw Frankfurter [150]

Aktivist *see* Der gwk-aktivist

Aktivitaet und aktivierbarkeit von polyphenoloxidasen in embryogenen und nicht-embryogenen suspensionskulturen von euphorbia pulcherrima willd. ex. klotzsch / Grotkass, Carolin – (mf ed 1997) – 2mf – 9 – €40.00 – 3-8267-2402-X – mf#DHS 2402 – gw Frankfurter [574]

Aktivitas kabupaten sukohardjo tahun dinas guna pedoman kerdja tahun dinas – Sukohardjo, 1968 – 1mf – 9 – mf#SE-1942 – ne IDC [950]

Aktsioner – Moscow, Russia 1860 [mf ed Norman Ross] – 1 – mf#nrp-1057 – us UMI ProQuest [077]

Aktsionernye kommercheskie banki v rossii / Levin, I I – Pg, 1917 – 6mf – 9 – mf#REF-308 – ne IDC [332]

Aktsionernye kommercheskie banki v rossii v 1886 godu : statisticheskii etiud / Rafalovich, L – Spb, 1887 – 1mf – 9 – mf#REF-277 – ne IDC [332]

Aktsionernye kompanii v rossii / Shepelev, L E – L, 1973 – 7mf – 9 – mf#REF-165 – ne IDC [332]

Aktsionernye zemel'nye banki i krest'ianskii bank v 1905-1916 gg / Raiskii, Iu L – Kursk, 1973. v25 – 1mf – 9 – mf#REF-503 – ne IDC [332]

Aktsionernye zemel'nye banki kak odno iz zven'ev sviazi mezhdu finansovym kapitalom i pomeshchich'im zemlevladeniem v rossii / Raiskii, Iu L – Kursk, 1975. v43 – 1mf – 9 – mf#REF-504 – ne IDC [332]

Aktsionernye zemel'nye banki v rossii vo vtoroi polovine 19-nachale 20 veka : aftoreferat dissertatsii na soiskanie uchenoi stepeni doktora istoricheskikh nauk / Raiskii, Iu L – 1983 – 1mf – 9 – mf#REF-502 – ne IDC [332]

Der aktuar salzmann : goethe's freund und tischgenosse in strassburg: eine lebens-skizze, nebst briefen von goethe, lenz, l. wagner, michaelis, hufeland u.a. / Stoeber, August – Frankfurt/M: Th Voelcker 1855 [mf ed 1990] – 1r – 1 – (incl bibl ref. filmed with: goethe: vier reden / albert schweitzer) – mf#2676p – us Wisconsin U Libr [920]

Aktuell – Magdeburg DE, 1967-1990 apr [gaps] – 4r – 1 – (spezialbaukombinat) – gw Misc Inst [620]

Aktuell informiert – Dessau DE, 1976 may – 1r – 1 – (notes: magnetbandfabrik) – gw Misc Inst [621]

Aktuelle chirurgie – Stuttgart, Germany 1976 – 1 – (cont by: viszeralchirurgie; former title(s): actuelle chirurgie (maerz 1966-nov 1973)) – ISSN: 0001-785X – mf#10152.01 – us UMI ProQuest [617]

Aktuelle dermatologie – Stuttgart, Germany feb 1975+ – 1,5,9 – ISSN: 0340-2541 – mf#10149 – us UMI ProQuest [616]

Aktuelle gerontologie – Stuttgart, Germany 1975-76 – 1,5,9 – (cont: actuelle gerontologie) – ISSN: 0300-5704 – mf#10153,01 – us UMI ProQuest [618]

Die aktuelle hallesche umschau – Halle S DE, 1962 5 oct-1965 7 dec, 1966 5 apr-1969 26 nov – 1r – 1 – (later: die aktuelle wochenzeitung) – gw Misc Inst [074]

Aktuelle neurologie – Stuttgart, Germany 1976+ – 1,5,9 – ISSN: 0302-4350 – mf#10154 – us UMI ProQuest [616]

Der aktuelle pressedienst *see* Das wichtigste der woche

Aktuelle traumatologie – Stuttgart, Germany 1975+ – 1,5,9 – (former title(s): actuelle traumatologie (feb 1971-73)) – ISSN: 0044-6173 – mf#10155 – us UMI ProQuest [617]

Aktuelle urologie – Stuttgart, Germany 1975+ – 1,5,9 – (former title(s): actuelle urologie. (1970-74)) – ISSN: 0001-7868 – mf#10156 – us UMI ProQuest [616]

Die aktuelle wochenzeitung *see* Die aktuelle hallesche umschau

Aktueller bildschirm – Stassfurt DE, 1968-1971 [gaps], 1975-1989 oct [gaps], 1989 dec-1990 8 may – 3r – 1 – (fernsehgeraetewerk) – gw Misc Inst [621]

Aktueller fernsehdienst [afd] – Hamburg DE, 1955 15 sep-1978 mar – 13r – 1 – (publ in frankfurt/main since mar 1958, in bad homburg vor der hoehe since 3 aug 1966, in muenchen since 1 jun 1971, in berlin since 11 jul 1977) – gw Mikrofilm [790]

Akty / Russia. Kavkazskaya Arkheograficheskaya Kommissiya. Tiflis – v. 1-12. 181904 – 1 – us NY Public [324]

Akty istoricheskie, sobrannye i izdannye arkheograficheskoi komissiei – Spb., 1841-1872. 17v+suppl – 289mf – 9 – mf#1172 – ne IDC [077]

Akty iuridicheskie, ili sobranie form starinnogo deloproizvodstva – London. 1944-1993 (1) 1971-1993 (5) 1976-1993 (9) – 22mf – 9 – mf#1332 – ne IDC [077]

Akty, izdavaemye vilenskoi arkheograficheskoi komissiei dlia razbora i izdaniia drevnikh aktov – Vilna, 1865-1915. 39 v – 832mf – 9 – mf#1147 – ne IDC [077]

Akty moskovskogo gosudarstva, izdannye imperatorskoi akademiei nauk / ed by Popov, N A – Spb., 1890-1901. 3 v – 63mf – 9 – mf#1069 – ne IDC [077]

Akty, otnosiashchiesia k istorii iuzhnoi i zapadnoi rossii, sobrannye i izdannye arkheograficheskoi komissiei – El Paso. 1905+ (1) 1905+ (5) 1905+ (9) – 236mf – 9 – mf#1068 – ne IDC [077]

Akty, otnosyashchiesya do yuridicheskago byta drevnei rossii / Russia. Arkheograficheskaia Kommissiia – St Petersburg. 1857-1884 – 1 – us NY Public [930]

Akty, sobrannye v bibliotekakh i arkhivakh russkoi imperatorskoi arkheograficheskoi ekspeditsiei imperatorskoi akademii nauk – Spb., 1836-1838. 4 v – 73mf – 9 – mf#R-112 – ne IDC [077]

Akty yuridicheskiye / Russia. Arkheograficheskaia Kommissiia – St Petersburg. 1838 and Index, 1840 – 1 – us NY Public [930]

Akusoka lingenasici / Mpofu, I N – Cape Town, South Africa. 1958 – 1r – 1 – us UF Libraries [960]

Akustische zeitschrift – Killen TX 1936-44 – 1 – mf#185 – us UMI ProQuest [621]

Akwamu, 1650-1750 : a study of the rise and fall of a west african empire / Wilks, Ivor – 1958 – 1r – 1 – us CRL [959]

Akwesasne Library Cultural Center *see* Ka ri wen ha wi

Akwesasne notes – 1969 feb-1976 early winter; 1977 mid winter-1984 spring; 1984 spring-1990 – 1 – mf#29505 – us WHS [071]

Akwesasne notes – Akwesasne NY 1969-96 – 1,5,9 – ISSN: 0002-3949 – mf#7156 – us UMI ProQuest [305]

Akzent und diphthongierung / Schmitt, Alfred – Heidelberg, Germany. 1931 – 1r – 1 – us UF Libraries [960]

Akzeptanzprobleme beim einsatz innovativer informationstechnologie im buero- und verwaltungsbereich analysiert am beispiel eines industriebetriebes / Wagner, Albert – Erlangen-Nuernberg, 1983 (mf ed 1994) – 3mf – 9 – €38.00 – 3-89349-902-4 – mf#DHS-AR 902 – gw Frankfurter [650]

Al cantio de un gallo / Aguilar Derpich, Juan – Habana, Cuba. 1962 – 1r – 1 – us UF Libraries [972]

Al exc[ellentissi]mo senor don christoval protocarrero, guzman, luna, enriquez de almanfa, pacheco, acuna, funes de villalpando... / Catholic Church Diocese of Santiago [Chile]. Synod [1763] – [Concepcion: s.n. 1645] – (= ser Books on religion...1543/44-c1800: concilios y sinodos) – 3mf – 9 – mf#crl-406 – ne IDC [241]

Al kalam – Bangalore, India. 1946-Jun 1953; 1957-60; 1962; Jun-Dec 1963 – 12r – 1 – us L of C Photodup [079]

Al kalam – Mangalore, India. 1961 – 1r – 1 – us L of C Photodup [079]

Al' kods (sviatoi gorod) / russko-palestinskii golos : palestinskii golos, zateriannyi v rossii – Moscow, Russia. n34(7/20/92), n36(8/19/92), n38(10/10/92), n40(4)-42(6)(okt 1992-noiab 1992), n1/2[7/8](ianv 1993)-n2[69](iiun 1996) – 1 – mf#mf-12248 (reel 1-2) – us CRL [077]

Al lado del guadiana / Vaca Morales, Francisco – Badajoz: Arqueros, 1943 – 1 – sp Bibl Santa Ana [946]

Al margen del plan peynado / Garcia Godoy, Federico – La Vega, Dominican Republic. 1922 – 1r – 1 – us UF Libraries [972]

Al Marrekoshi, Abdo-'L-Wahid *see* The history of the almohades

Al padre miestro fray ioseph barrasa de la orden de nuestra senora de la merced : y su pocurador gerenal de los prounicia de lima, en los reynos del piru... / Barrasa, Joseph – [Lima?: s.n. 1665?] – (= ser Books on religion...1543/44-c1800: colegios religiosos) – 1mf – 9 – mf#crl-358 – ne IDC [241]

Al pairo, y otros cuentos / Montero Madrigal, Jorge – San Jose, Costa Rica. 1971 – 1r – 1 – us UF Libraries [972]

Al pais / Congreso De Municipios Dominicanos – Santiago, Dominican Republic. 1944 – 1r – 1 – us UF Libraries [972]

Al pasar ante las horas / Sanchez Arjona, Vicente – Sevilla: Graficas T., Tomo 1. 1955. Tomo 2-5 – 1 – (pensamientos vulgares) – sp Bibl Santa Ana [946]

Al pueblo de cuba – New York, USA. 1898 – 1r – 1 – us UF Libraries [972]

Al rey nro sor por la provincia de la compania de jesus de la nueva espana en satisfacion de un libro de el visitador obispo d iuan de palafox y mendoca : publicado en nombre de el dean y cabildo de su iglesia catedral de la puebla de los angeles / Rojas Villandrando, Agustin de – [Mexico City: s.n. 1647] – (= ser Books on religion...1543/44-c1800: jesuitas) – 3mf – 9 – mf#crl-217; crl-461 – ne IDC [241]

Al serenissimo d. cosimo 2 : quarto gran duca di toscana, il di 25 maggio 1621 – Venetia: Appresso il Ciotti, 1621 – 1mf – 9 – mf#0-1569 – ne IDC [090]

Al sindicato de la comunidad de labradores (proyecto de cooperativa vitico-alcoholera en almendralejo) / Luengo, Juan – Badajoz: Uceda Hermanos, 1905 – 1 – sp Bibl Santa Ana [946]

Al son de mi mejorana / Gonzalez Bazan, Carlos R – Panama, Panama. 1953 – 1r – 1 – us UF Libraries [972]

Al zurriagazo zurribanda / Gallardo, Bartolome Jose – 1822 – 9 – sp Bibl Santa Ana [946]

Ala / Acosta, Agustin – Habana, Cuba. 1958 – 1r – 1 – us UF Libraries [972]

Ala kulturas biroja biletens – n1-n14 [1959-66] – 1 – mf#664180 – us WHS [071]

Ala santisima virgen de la luz 1978 / Asociacion Benefica Ntra. Sra. de la Luz. Malpartida de Plasencia – Imp. Sanchez Rodrigo, 1978 – sp Bibl Santa Ana [240]

Ala santisima virgen del rosario. actos y festejos en su honor, 1946 / Madronera. Ayuntamiento – Caceres: Tip. El Noticiero – sp Bibl Santa Ana [240]

Alaarij tobiticha, nicolla-aba yobibennizaa, ni yabaaqui copabijtoo, nipecaa yyeeni ouijchirijni loquiraani / Aguero, Cristobal de – [Mexico City: s.n. 1666] – (= ser Books on religion...1543/44-c1800: doctrina cristiana, obras de devocion) – 6mf – 9 – mf#crl-51 – ne IDC [241]

Alabama – (= ser General education board: the early southern program) – 13r – 1 – $1690.00 – us Scholarly Res [370]

Alabama : code of alabama – Charlottesville: Michie Company, 1975-Aug 1996+1999 update – 9 – $2,021.00 set – mf#401720 – us Hein [348]

Alabama : periodico noticioso, critico e alusivo – Pernambuco: Typ Liberal, 27 jun 1863 – (= ser Ps 19) – mf#P17,02,159 – bl Biblioteca [079]

Alabama : session laws of american states and territories – 1818-2001 – 9 – $2432.00 set – mf#402480 – us Hein [348]
Alabama *see*
– Biennial reports (a-g), 1882-1936
– Reports pre-nrs
– State reports, post-nrs
Alabama Academy of Science *see* Journal of the alabama academy of science
Alabama appellate courts / Alabama. Supreme Court – Montgomery, Ellis, 1968. 24 p. LL-2280 – 1 – (2nd ed. montgomery, 1970 34p ll-2276) – us L of C Photodup [347]
Alabama Archaeological Society *see* Stones and bones newsletter
Alabama attorney general reports and opinions – 1882-2002 – 6,9 – $1200.00 set – (1882-1980 v178 on reel $560. v179-266 1980-2002 mf $640) – mf#408100 – us Hein [340]
Alabama baptist – Birmingham. 1946-48; 1952-55; 1962-76 – 1 – us L of C Photodup [242]
Alabama baptist – feb 4 1843-dec 1846, 1854-1963 – 1 – $2,455.63 – us Southern Baptist [242]
Alabama baptist – v6 n24, 31, 39 [1848 aug 24, sep 22, nov 17] – 1 – mf#671438 – us WHS [242]
Alabama bar bulletin – 1939 [all publ] – 9 – $15.00 set – mf#100081 – us Hein [340]
Alabama christian advocate *see* Methodist christian advocate
Alabama Committee for Equal Justice *see* Release
Alabama Dental Association *see* Journal of the alabama dental association
Alabama family history – v4 n1-4 – 1 – mf#697678 – us WHS [929]
Alabama family history and genealogy news – 1981-1982 – 1 – mf#1120061 – us WHS [929]
Alabama historical chronicle – 1975 mar 3-1977 feb 14 – 1 – mf#1051794 – us WHS [978]
Alabama intelligencer and state rights expositor – 1833 mar 2, 1835 jul 18-1925, aug 15-oct 17, nov 7-14, dec 5 – 1 – mf#853608 – us WHS [071]
Alabama Labor Council, AFL-CIO *see* Legislative newsletter
Alabama law journal – Montgomery. v1-4. 1882-85 [all publ] – (= ser Historical legal periodical series) – 1 – $60.00 set – mf#408770 – us Hein [340]
The alabama law journal – Montgomery, AL. v1-4. 1882-85 (complete) – 16mf – 9 – $24.00 – (cont by a modern series in 1925 but for copyright reasons is not offered at present by llmc) – mf#LLMC 82-900 – us LLMC [340]
Alabama law review – v1-50. 1948-1999 – 9 – $787.00 set – ISSN: 0002-4279 – mf#100111 – us Hein [340]
Alabama lawyer – v1-62. 1940-2001 – 9 – $870.00 set – ISSN: 0002-4287 – mf#100121 – us Hein [340]
Alabama librarian – Montgomery AL 1949+ – 1,5,9 – ISSN: 0002-4295 – mf#2277 – us UMI ProQuest [020]
Alabama news digest – 1938 nov 3-1939 apr; 1939 may-1940; 1941-42; 1943-1944 jun; 1944 jan 29-1945 may 24; 1946 dec 5-1947 nov5; 1947 nov12-50 oct 13; 1950 oct 20-1956 jan 13 – 1 – mf#870846 – us WHS [071]
Alabama review – Tuscaloosa AL 1989+ – 1,5,9 – ISSN: 0002-4341 – mf#17983 – us UMI ProQuest [978]
Alabama State Bar Association *see* Reports
Alabama. State Bar Association *see* Proceedings
Alabama state bar association reports – nos 1-35. 1880-1912 – 84mf – 9 – $126.00 – (regular updates) – mf#LLMC 84-397 – us LLMC [340]
Alabama State College *see* Tones and overtones
Alabama state journal – 1875 apr 2, 16, 30-may 21, jun 4-11, 25-jul 9 – 1 – mf#870902 – us WHS [071]
Alabama State Teachers Association *see* Journal
An alabama student : and other biographical essays / Osler, William – Toronto: Oxford University Press Canadian Branch, 1908 – 4mf – 9 – 0-665-86135-4 – mf#86135 – cn Canadiana [610]
Alabama. Supreme Court *see* Alabama appellate courts
Alabama supreme court reports – Ann Arbor MI 1820-86 – 1,5,9 – mf#1743 – us UMI ProQuest [347]
Alabama supreme court reports – v1-214. 1840-1926 – 1845mf – 9 – $2767.00 – (pre-nrs: v1-80 1840-87 [696mf] $1044.00; updates planned as they fall out of copyright) – mf#llmc82-980 – us LLMC [347]
Alabama watchman – v1 n1 [1820 aug 8] – 1 – mf#853657 – us WHS [071]
Alabanza – Santo Domingo: impr Amigo del Hogar. [n34-131 feb1980-1998] (6 times/yr) – 3r – 1 – us CRL [240]
Alabanza a la memoria / Lara Cintron, Rafael – Ciudad Trujillo, Dominican Republic. 1958 – 1r – 1 – us UF Libraries [972]
Alabanza de mexico / Olivares, Armando – Guanajuato, Mexico. 1962 – 1r – 1 – us UF Libraries [972]
Alabanza en la torre de ciales / Corretjer, Juan Antonio – San Juan, Puerto Rico. 1965 – 1r – 1 – us UF Libraries [972]
Alabanzas, conversaciones, 1951-1955 / Fernandez Retamar, Roberto – Mexico City? Mexico. 1955 – 1r – 1 – us UF Libraries [972]
Alabaster box – London, England. 18– – 1r – 1 – us UF Libraries [240]
Alabaster, Chaloner *see* The truine powers
Alabaster, Henry *see*
– The wheel of the law
Alabastros / Camin, Alfonso – Mejico, Mexico. 1919 – 1r – 1 – us UF Libraries [090]
Alachua County (FL) Board Of Public Instruction *see* Financial facts concerning alachua county public s...
Alachua County (FL) Chamber Of Commerce *see* Great bowl of alachua
Alachua county news – s.l, s.l, 193-? – 1r – 1 – us UF Libraries [978]
Alachua gazette – Alachua, FL. 1891 mar 19,26;Apr 23;may 21;Jul 02;aug 06;nov 05 – 1r – 1 – us UF Libraries [071]
Alachua, the garden county of floridia, its resources and advantage / Ashby, John W – Gainesville, FL. 1888 – 1r – 1 – us UF Libraries [630]
Aladino : ossia, il talismano. ballo magico fantastico in cinque quadri d'invenzione e composizione del coreografo antonio monticini da rappresentarsi nell' i. e r. teatro de' sigg. accademici immobili, il carnevale 1850-51 / Monticini, Antonio – Firenze: Tip Galletti, 1851? – 1 – mf#*ZBD-*MGTZ pv 2-Res – Located: NYPL – us Misc Inst [790]
Al-adwa – al-Khartum: [s.n.] sep 9 1986-jun 6 1987 – 2r – 1 – us CRL [079]
al-Afghani, Jamal ad-Din *see* Al-curwa al-wuthqa
Alafia : a magazine of the black arts – 1971 winter – 1 – mf#1214436 – us WHS [071]
Alagoas (Brazil) Governor *see* Relatorios dos presidentes, 1a republica, 1890-1930
Alagoas (Brazil) President *see* Relatorios dos presidentes, epoca do imperio
Alah ku te ku ta / Nanda, Mon – Yan Kun: Mo Kvan Cape 1976 [mf ed 1990] – 1 – (in burmese) – mf#mf-10289 seam reel 169/2 [§] – us CRL [959]
Al-ahali – Baghdad, Iraq. 1960-mar 28 1961 – 1 – us CRL [079]
Alahambra – 1909-34; 1981; 1992-94; 1993- – (= ser California telephone directory coll) – 31r – 1 – $1550.00 – mf#P00002 – us Library Micro [917]
Alahou – Helu 1-1910 [1979 nov-1980 dec] – 1 – mf#1061247 – us WHS [071]
Al-ahrar – Cairo, Egypt. Nov 14 1977-July 30 1979; Aug 4 1980-1991 – 7r – 1 – (scattered issues lacking) – us L of C Photodup [079]
Al-aihhah *see* Al-azhar
Alain-Fournier *see* Verbes 'dire' en grac ancien
Alaior baptisms – Minorca, Spain. v 1-131667-1806 – 5r – 1 – us UF Libraries [946]
Alaior deaths – Minorca, Spain. V1-91670-1816 – 4r – 1 – us UF Libraries [946]
Alaior maifests – Minorca, Spain. 1762-3,1768 – 1r – 1 – us UF Libraries [946]
Alaior marriages – Minorca, Spain. v1-51585-1824 – 2r – 1 – us UF Libraries [946]
Al-alam-al-ahmar – Paris. nos1-4. mai-aout 1926 – 1 – fr ACRPP [073]
Alamanni, C *see* Summa...d thomae aquinatis...in ordinem cursus philosophici accommodata
Alameda *see* Oakland/alameda
[Alameda-] alameda argus – CA. dec 6 1877-dec 1885; 1887-88; 1890-jun 1910; 1911-12 – 33r – 1 – $1980.00 – mf#B02003 – us Library Micro [071]
[Alameda-] alameda county gazette – CA. aug 1856-feb 1875 – 2r – 1 – $120.00 – mf#B02004 – us Library Micro [071]
[Alameda-] alameda encinal – CA. sep 21 1869; nov 1 1869; sep 16 1869-dec 31 1874 – 27r – 1 – $1620.00 – mf#B02005 – us Library Micro [071]
[Alameda-] alameda journal – CA. jun 7 1935-apr 23 1937; apr 1987- – 10r – 1 – $600.00 (subs $90/y) – mf#B05025 – us Library Micro [073]
[Alameda-] alameda times star – CA. 1909-dec 1956, an 1959- – 566r – 1 – $33,960.00 (subs $480/y) – (aka: evening times star and daily argus) – mf#BC02007 – us Library Micro [071]
[Alameda county-] alameda city directories including berkeley and oakland – CA. 1869-1901 – 26r – 1 – $1560.00 – mf#D001 – us Library Micro [917]
[Alameda-] community services calendar / alameda housing news – CA. 1948-1955 – 1r – 1 – $60.00 – mf#B06001 – us Library Micro [071]
[Alameda county-] alameda, contra costa, monterey, san benito, san mateo, santa clara and santa cruz counties – CA. 1879 – 1r – 1 – $50.00 – mf#D002 – us Library Micro [978]
The alameda corridor project : its successes and challenges: hearing...house of representatives, 107th congress, 1st session, april 16 2001 / United States. Congress. House. Committee on Government Reform. Subcommittee on Government Efficiency, Financial Management and Intergovernmental Relations – Washington: US GPO 2002 [mf ed 2002] – 2mf – 9 – 0-16-068388-2 – us GPO [380]
Alameda county – 1901; 1903-05; 1929-34 – (= ser California telephone directory coll) – 13r – 1 – $650.00 – mf#P00003 – us Library Micro [917]
[Alameda county-] dalton's san francisco, oakland, alameda business directories – CA. 1880; 1887-1888 – 4r – 1 – $200.00 – mf#D003 – us Library Micro [978]
[Alameda county-] hayward and san leandro city directories – CA. 1925-1926; 1931-1934; 1938; 1940; 1946-1948 – 6r – 1 – $300.00 – mf#D004 – us Library Micro [917]
[Alameda-] daily star – CA. 1908 – 1r – 1 – $60.00 – mf#B06002 – us Library Micro [071]
[Alameda-] daily times – CA. 1908 – 1r – 1 – $60.00 – mf#B06003 – us Library Micro [071]
Alameda housing news *see* [Alameda-] community services calendar / alameda housing news
[Alameda-] island journal – CA. may 1982-jan 1 1984 – 1r – 1 – $60.00 – mf#B06004 – us Library Micro [073]
[Alameda-] sun – CA. nov 1965-dec 1967 (wkly) – 1r – 1 – $60.00 – mf#B02006 – us Library Micro [071]
[Alameda-] the carrier : u s naval air station – CA. jul 1948-sep 1979 – 5r – 1 – $300.00 – mf#B06000 – us Library Micro [071]
Alami, Solomon *see* Mishpete ha-shem
Alamilla, Guillermo *see* Mi viaje a europa
Alamillo Salgado, Ildefonso *see* El brocense
O alamire – Bragança, SP: Typ Alamire, 25 ago, nov 1880; 26 abr 1881 – (= ser Ps 19) – mf#P18,01,65 – bl Biblioteca [079]
[Alamitos bay-] local enterprise – CA. jul 1964-jun 1965 – 1r – 1 – $60.00 – mf#H04000 – us Library Micro [071]
[Alamo-] tri-valley news – CA. sep 1973-jun 1980 (daily) – 80r – 1 – $4800.00 – (cont by: livermore-the news) – mf#B02005 – us Library Micro [071]
Alamosa county miscellaneous newspapers – Alamosa, CO (mf ed 1991) – 1r – 1 – (alamosa courier (jan-dec 1907); alamosa empire (apr 1909-dec 1910); alamosa news (aug 13 1942-dec 9 1943); colorado independent (jan 6 1883); the expositer (sep 27 1883-dec 27 1883); garrison tribune (nov 24 1892)) – mf#MF Z99 Al11 – us Colorado Hist [071]
Alamosa courier *see* Alamosa county miscellaneous newspapers
Alamosa empire *see* Alamosa county miscellaneous newspapers
Alamosa news *see* Alamosa county miscellaneous newspapers
Alan, Miriam *see* Wednesday's child
Al-anba – Jerusalem 1968 oct-1985 jan – (= ser Palestine newspapers) – 50r – 1 – (ceased in 1985) – mf#j-93-1/3 – ne IDC [079]
Al-anba' = Alanba – al-Khartum: [s.n, [1989- . [jan 21-apr 19 1989] (wkly) – 1r – 1 – us CRL [079]
Aland, Kurt *see*
– Kleine schriften 1
– Kleine schriften 2
– Studia patristica
Al-andalus : revista de las escuelas de estudios arabes... – Madrid, Granada, 1933-1955. v1-20 – 203mf – 9 – mf#NE-101 – ne IDC [956]
Aland-Cross-Danielou *see* Studia evangelica
Alandskii kongress : vneshniaia politika rossii v kontse severnoi voiny / Feigina, S A – 1959 – 6mf – 9 – mf#R-10524 – ne IDC [947]
Alange / Puerto Reyna, Juan Antonio – Sevilla: Eulogio de las Heras, 1913 – 1 – sp Bibl Santa Ana [946]
Alange / Puerto Reyna, Juan Antonio – Sevilla: Eulogio de las Heras, 1914 – 1 – sp Bibl Santa Ana [946]
Alange. noticias historicas acerca de esta villa y de sus famosos banos / Puerto Reyna, Juan Antonio – Sevilla: Prida, 3rd ed. 1925 – 1 – sp Bibl Santa Ana [946]
Alanka : [poems] / Tara, Da gun – Ran kun: Khyui Phru Ca pe, Mon Tan San 1962 [mf ed 1990] – (= ser Kabya pon khyup) – 1 – (in burmese) – mf#mf-10289 seam reel 201/2 [§] – us CRL [810]
Al-anwar *see* Arab newspapers
Alapont, Giuseppe *see* The lives of the blessed leonard of port maurice and of the blessed nicholas fattore
Alarcon De Folgar, Romelia *see*
– Plataforma de cristal
– Poemas de la vida simple
Alarcon, Pedro Antonio de *see* Una visita al monasterio de yuste
Alarm – 1975 apr-1976 sep – 1 – mf#647873 – us WHS [071]
Alarm – 1980 jan-1983 dec – 1 – mf#1477122 – us WHS [071]
Alarm – Chicago IL 22 aug 1885, 16 jun-7 jul 1888 – 1 – uk British Libr Newspaper [071]
Alarm : kampfblatt gegen volksbetrug – Berlin DE, 1931 8 jan-24 dec, 1932 7 jan-29 dec, 1933 5 jan-16 feb – 1 – gw Misc Inst [074]
Alarm : mitteilungsblatt der liga fuer menschenrechte – Porto Alegre (BR), 1937 12 feb-1 may – 1r – 1 – gw Misc Inst [322]
Alarm – v1-3 n3 [1884 oct 4-1886 apr 24]; ns: v1 n1-47 [1887 nov5-1889 feb 2] – 1 – mf#1109592 – us WHS [071]
Alarm : working people's international association – ser1: v1-3 n3 1884-86 [all publ]. ser2: v1-2 n1-47 1887-89 [all publ] – (= ser Radical periodicals in the united states, 1881-1960. series 2) – 1r – 1 – $115.00 – us UPA [335]
Alarm *see* Die westfront
Alarm in zion / Whitefield, George – London, England. 1803 – 1r – 1 – us UF Libraries [240]
Alarm signal – Silver Spring MD 1980-83 – 1,5,9 – (cont: signal) – ISSN: 0199-6835 – mf#11019,01 – us UMI ProQuest [360]
Alarm to the unconverted / Alliene, Joseph – Belfast, Northern Ireland. 1816 – 1r – 1 – us UF Libraries [240]
Alarm ueber tage : roman / Broeker, Heinz – Breslau: Gauverlag Niederschlesien c1940 [mf ed 1995] – 1r – 1 – (filmed with: der tod des vergil / hermann broch) – mf#3808p – us Wisconsin U Libr [830]
Les alarmes de l'episcopat justifiees par les faits : lettre a un cardinal par mgr. l'evaeque d'orleans / Dupanloup, Felix – Paris: Charles Douniol, 1868 – 1mf – 9 – 0-8370-7539-4 – (incl bibl ref) – mf#1986-1539 – us ATLA [370]
Alarming cry – 1954 jan-autumn; 1962?-1981 spring – 1 – mf#342980 – us WHS [071]
Alarms in regard to popery / Campbell, George – London, England. 1840 – 1r – 1 – us UF Libraries [240]
Alas, Leopoldo *see*
– Obras completas
– Publicidad y los bienes muebles
Las alas rotas / Costa Duran, Maria – Barcelona, 1933; Madrid: Razon y Fe, 1933 – 1 – sp Bibl Santa Ana [999]
Alas vestis – 1972 fall-1986 apr – 1 – mf#1329789 – us WHS [071]
Alash *see* Sary arka
al-Ashqar, Riyad *see*
– Mizan al-quwa al-askariyah bayna al-duwal al-arabiyah wa-israil fi al-thamaninat
– Muahadah al-misriyah al-israiliyah wa-abaduha al-istiratijiyah wa-al-askariyah
Al-'asifah : sawt harakat al-tahrir al-watani al-filastini – [S.l.]: Harakat al-Tahrir al-Watani al-Filastini, [al-'Adad 8-128 (Ukt.1967-14/2/1969)] (irreg) – 1r – 1 – us CRL [079]
Alaska – Anchorage AK 1992+ – 1,5,9 – ISSN: 0002-4562 – mf#18624,01 – us UMI ProQuest [978]
Alaska : its southern coast and the sitkan archipelago / Scidmore, Eliza Ruhamah – Boston: D Lothrop & Co [c1885] [mf ed 1987] – 1r [ill] – 1 – mf#8624 – us Wisconsin U Libr [917]
Alaska : session laws of american states and territories – 1913-2001 – 9 – $1061.00 set – mf#402490 – us Hein [348]
Alaska *see*
– Reports and opinions
– State reports, pre-nrs
Alaska and missions on the north pacific coast / Jackson, Sheldon – New York: Dodd, Mead, c1880 [mf ed 1986] – 1mf – 9 – 0-8370-6580-1 – mf#1986-0580 – us ATLA [240]
Alaska and the gold fields of nome, port clarence, golovin bay, kougarok, the klondike, and other districts / North American Transportation and Trading Co – [S.l: s.n, 1900?] [mf ed 1981] – 2mf – 9 – mf#15727 – cn Canadiana [622]
Alaska and the gold fields of the yukon : great northern railway: the klondike cook inlet [an]d other mining regions – Chicago: Poole Bros, [1898?] [mf ed 1982] – 1mf – 9 – mf#17934 – cn Canadiana [622]
Alaska and the gold fields of the yukon, koyukuk, tanana, klondike and their tributaries / North American Transportation and Trading Co – [S.l: s.n, 1898?] [mf ed 1983] – 2mf – 9 – mf#15728 – cn Canadiana [622]

Alaska and the klondike : a journey to the new eldorado, with hints to the traveller... / Heilprin, Angelo – London: C A Pearson, 1899 [mf ed 1980] – 5mf – 9 – 0-665-05536-6 – mf#05536 – cn Canadiana [917]
Alaska and the klondike : the new gold fields and how to reach them / Wells, Harry Laurenz – [Portland, Or?: s.n.], 1897 [mf ed 1980] – 2mf – 9 – mf#16437 – cn Canadiana [917]
Alaska and the klondike gold fields : containing a full account of the discovery of gold, enormous deposits of the precious metal, routes traversed by miners, how to find gold, camp life at klondike / Harris, A C – Chicago: Monroe Book Co, [1897?] [mf ed 1981] – 7mf – 9 – mf#15188 – cn Canadiana [622]
Alaska and the klondike gold fields : containing a full account of the discovery of gold, enormous deposits of the precious metal, routes traversed by miners, how to find gold, camp life at klondike / Harris, A C – [Chicago?: s.n, 1897?] [mf ed 1980] – 7mf – 9 – mf#15187 – cn Canadiana [622]
Alaska attorney general reports and opinions – 1917-1995 – 6,9 – $1240.00 set – (1917-1968 on reel $245. 1979-1995 on mf $995. 1969-78 not available) – mf#408110 – us Hein [340]
Alaska baptist messenger – 1946-48, 1952-91 – 1 – $190.08 – us Southern Baptist [242]
Alaska bar rag – v1-25. 1978-2001 – 9 – $400.00 set – mf#401700 – us Hein [340]
Alaska business and industry – Anchorage AK 1969-84 – 1,5,9 – (cont: alaska industry) – mf#7464,01 – us UMI ProQuest [338]
Alaska business monthly – Anchorage AK 1985+ – 1,5,9 – ISSN: 8756-4092 – mf#14900 – us UMI ProQuest [650]
Alaska Cannery Workers' Association *see* International examiner
Alaska Commercial Co *see* To the klondike gold fields and other points of interest in alaska
Alaska conveyance news – 1984 apr-1986 sep – 1 – mf#921677 – us WHS [071]
Alaska daily empire – 1926 feb 8-dec 7 – 1 – mf#790811 – us WHS [071]
'Alaska file' of the office of the secretary of the treasury, 1868-1903 / U.S. Fish and Wildlife Service – (= ser Records of the fish and wildlife service) – 25r – 1 – (with printed guide) – mf#M720 – us Nat Archives [639]
Alaska file of the revenue cutter service, 1867-1914 / U.S. Coast Guard – (= ser Records of the united states coast guard) – 20r – 1 – (with printed guide) – mf#M641 – us Nat Archives [360]
Alaska file of the special agents division of the department of the treasury, 1867-1903 / U.S. Dept of the Treasury – 16r – 1 – (with p/g) – mf#m802 – us Nat Archives [336]
The alaska friend – Douglas AK: J E Connett, 1893- [mthly] mf v1 n1-6 1893-may 1894 filmed 2003] – 1r – 1 – mf#2003-s057 – us ATLA [071]
Alaska. Governor's Office *see* Chronological files of the alaskan governor, 1884-1913
Alaska herald – San Francisco. Calif. v. 1, no. 1-v. 8, no. 196. Mar 1868-Mar 1876 – 1 – us NY Public [071]
Alaska highway news – Fort St John, British Columbia, CN. mar 1944-mar 1977 – 28r – 1 – cn Commonwealth Imaging [071]
Alaska history news – 1970 dec-1986 feb – 1 – mf#642834 – us WHS [978]
Alaska journal of commerce – Anchorage AK 1999+ – 1,5,9 – ISSN: 1537-4963 – mf#17827,03 – us UMI ProQuest [380]
Alaska law journal – v1-9. 1963-71 [all publ] – 9 – $80.00 set – mf#114141 – us Hein [340]
Alaska law review – v1-18. 1984-2001 – $295.00 set – (v1-10 1984-93 roll $147v; v11-18 1994-2001 fiche $148v; supersedes: ucla, alaska law review) – ISSN: 0883-0568 – mf#109541 – us Hein [340]
Alaska law review (ucla) *see* Alaska law review
Alaska medicine – Anchorage AK 1959+ – 1,5,9 – ISSN: 0002-4538 – mf#2460 – us UMI ProQuest [610]
Alaska miner – 1899 jun 3 – 1 – mf#853298 – us WHS [622]
Alaska Native Brotherhood *see* Too proud to serve
Alaska native claims appeals board decisions and orders / U.S. Dept of the Interior – sept 1975-sun 1982 [all publ] – 35mf – 9 – $52.00 – mf#llmc 82-201 – us LLMC [343]
Alaska native times : official publication of the non-resident alaska natives in the 13th region – v5 n1-v9 n4 [1979 jan-81 oct/nov] – 1 – mf#641632 – us WHS [071]
Alaska. Office of the Secretary *see* Correspondence of the secretary of alaska, 1900-1913
Alaska pine leader – n1-4 [1949 mar 14-may 9] – 1 – mf#681558 – us WHS [071]
Alaska Public Employees Association *see*
– News 'n views
– Reporter
Alaska reports – v1-4. 1884-1914 – 12mf – 9 – $54.00 – mf#LLMC 86-101 – us LLMC [340]
Alaska review – Anchorage AK 1963-72 – 1 – ISSN: 0002-4554 – mf#10377 – us UMI ProQuest [978]
Alaska spotlight – 1952 jul 28-1968 sep 21-nov 20 – 1 – mf#870120 – us WHS [071]
Alaska statutes annotated – Superseded vols. 1980- – 9 – enquire for prices – mf#401730 – us Hein [348]
Alaska. Supreme Court *see* Alaska supreme court reports
Alaska supreme court reports / Alaska. Supreme Court – v1-6. 1884-1923 – 18mf – 9 – $81.00 – (no pre-nrs vols. add vols planned) – mf#LLMC 86-101 – us LLMC [347]
Alaska. (Territory). Attorney General's Office *see* Report
Alaska, the eldorado of the midnight sun : marvels of the yukon; the klondike discovery; fortunes made in a day;... / Hall, Edward Hagaman – New York: Republic Press, 1897 [mf ed 1980] – 1mf – 9 – (incl ind) – mf#14011 – cn Canadiana [917]
The alaska treaty / Miller, David Hunter – (= ser General records of the department of state) – 1r – 1 – mf#T1024 – us Nat Archives [975]
Alaska Women's Resource Center *see* Sourceline
The alaskan boundary line / Glass, David – S.l: s.n, 1899? – 1mf – 9 – mf#15021 – cn Canadiana [917]
Alaskan philatelist – 1960 jan 15-1979 mar – 1 – mf#497197 – us WHS [760]
Alaskan russian church archives – 1730-1930 – 1 – 8095.00 – us L of C Photodup [240]
The alaskan russian church archives : records of the russian orthodox greek catholic church of north america–diocese of alaska – Washington: Manuscript Division, Library of Congress, 1984- – 1r – 1 – us L of C Photodup [241]
Alaska-Yukon-Klondike Gold Syndicate *see* Klondike gold miners of the alaska-yukon-klondike gold syndicate
Al-aswar / ed by Ibrahim, Hanna et al – Akka al-Qadimah, Israel: Maktab al-Aswar, 1988- . n1-14. spr 1988-93 – (= ser Arabic journals and popular press) – 2r – 1 – $345.00 – us MEDOC [073]
Al-atma al-sihyuniyah fi miyah al-urdun wa-al-litani / al-Din al-Khayru, Izz – [al-Qahirah]: Jamiat al-duwal al-arabiyah, al-munazzamah al-arabiyah lil-tarbiyah wa-al-thaqafah wa-al-Ulum, Mahad al-buhuth wa-al-dirasat al-arabiyah, 1977. – us CRL [079]
Alaways, LeRoy W *see* Aerodynamics of the curve-ball
Al-ayyam : al-khartum: sharikat al-ayy am lil-sihafah al-mahdurah, [mar 27 1987-jun 29 1989] – 9r – us CRL [079]
Al-Ayyubi, Ilyas *see* Ta'rikh misr fi 'ahd al-khidiw isma'il basha
Al-azhar – Cairo: Hasan Rifqi & Ibrahim Mustafa, 1890-93. yr 3 n3-yr 4 n12. n.d. 1890?-93? – (= ser Arabic journals and popular press) – 1r – 1 – $300.00 – (first iss marks title change fr al-sihhah (incl in r) with no break in numbering) – us MEDOC [079]
Al-azhar *see* Al-sihhah
L'alba = Corriere della sera – Newport, RI: Russo Pub Co [jan 11 1936-aug 14 1937] (wkly) – 1r – 1 – (weekly independent newspaper consolidated with il corriere della sera (the evening courier)) – us CRL [071]
L'alba – Newport, RI. 1913-1935 (1) – mf#66224 – us UMI ProQuest [071]
Alba C, Manuel Maria *see* Intruduccion al estuido de las lenguas indigenas d...
Alba compartida / Menendez Alberdi, Adolfo – Habana, Cuba. 1964 – 1r – 1 – us UF Libraries [972]
Alba, Jose *see* Perdigon y perrunilla
L'alba sociale – Ybor City-Tampa FL, 1901* – 1r – 1 – (italian periodical) – us IHRC [073]
Alba [stirling, scotland] : paipear-naidheachd etc – [Scotland] Stirling: Scott, Learmonth & Allan 3 d'en fhaoilteach 1920-an dudlachd 1921 (wkly, mthly) [mf ed 2004] – 30v on 1r – 1 – uk Newsplan [072]
Al-badil – al-khartum: al-badil, [sep 9 1985-feb 11 1989] – 1r – us CRL [079]
Al-Badisi, "Abd al-Haqq ibn Ismail" *see* El'maqsad (vies des saintes du rif)
Al-balad – al-khartum: Dar al-balad lil-tibaah wa-al-nashr, feb 11-may 1 1989 – 1r – us CRL [072]
Albaladejo, Mariano *see* Alta mar
Al-balagh al-usbu'i – Cairo, 1926-30. v1 n1-v4 n174. 21 jumada al-ula 1345-20 safar 1349 [26 nov 1926-16 jul 1930] [all publ] – (= ser Arabic journals and popular press) – 3r – 1 – $395.00 – us MEDOC [073]
Alban stolz in seiner entwicklung als schriftsteller / Hulshof, Franz – Graz: Waechter-Verlag 1931 [mf ed 1995] – 1r – 1 – (incl bibl ref & ind. filmed with: ut'n knick / julius stinde) – mf#3752p – us Wisconsin U Libr [430]
Albanesi, Effie Adelaide Maria *see*
– The fault of one
– The kingdom of a heart
– "Margery daw"
– "My pretty jane 1"
– The woman who came between
Albanesische grammatik *see* Vergleichungstafeln der europaeischen stamm-sprachen und sued-, west-asiatischer
Albania – Bruxelles. mars 1897-1907, 1909 – 1 – fr ACRPP [949]
Albania *see* Revue albanaise
Albanian Orthodox Diocese of America *see* Drita e vertete
L'albanie libre – Milan, Italy mar 1947-30 may 1959 – 1 – (in french, english, italian & albanian) – uk British Libr Newspaper [074]
Albany agenda – 1978 oct 19-1980 dec; 1980-91 – 1 – mf#913112 – us WHS [071]
[Albany-] albany community news – CA. dec 1976-may 1983 – 1r – 1 – $60.00 – mf#B06005 – us Library Micro [071]
[Albany-] albany times – CA. oct 1935-oct 1938; oct 1944-dec 1948; 1979 – 18r – 1 – $1080.00 – mf#B06006 – us Library Micro [071]
Albany bouquet and literary spectator – La Salle IL 1835 – 1 – mf#3716 – us UMI ProQuest [420]
Albany centinel – 1799 dec 27-1801 jan 2 – 1 – mf#780594 – us WHS [071]
Albany citizen – Albany OR: G D Arnold, [wkly] [mf ed 1962] – 1r – 1 – (began and ceased in 1910?) – us Oregon Lib [071]
Albany Club *see* Members and shareholders 1891-92
The albany club, incorporated 1882 – Toronto: E H Harcourt, [1906?] – 1mf – 9 – mf#86512 – cn Canadiana [366]
The albany club, incorporated 1882 – Toronto?: s.n, 1889? – 1mf – 9 – mf#04008 – cn Canadiana [366]
The albany club, incorporated 1882 – Toronto?: s.n, 1899? – 1mf – 9 – mf#06454 – cn Canadiana [366]
Albany daily democrat – Albany OR: Stites & Nutting, 1888- [daily] [mf ed 1965-66] – 12r – 1 – (related to wkly eds: state rights democrat (albany, or), 1888-1900 and: albany democrat (albany, or: 1900), 1900-12 and: albany weekly democrat, 1912-13 and: semi-weekly democrat (albany, or), 1921-1922. semi-wkly ed: semi-weekly democrat (albany, or), 1913-21. cont: daily evening albany democrat; cont by: albany democrat (albany, or: 1922)) – us Oregon Lib [071]
Albany daily democrat *see*
– Albany democrat (albany, or: 1900)
– Albany weekly democrat
– Semi-weekly democrat
Albany democrat (albany, or: 1900) – Albany OR: F P Nutting, 1900-12 [wkly] [mf ed 1964-66] – 4r – 1 – (related to: albany daily democrat; cont: state rights democrat (albany, or); cont by: albany weekly democrat) – us Oregon Lib [071]
Albany democrat (albany, or: 1922) – Albany OR: W L Jackson & R R Cronise, -1925 [daily] [mf ed 1966] – 6r – 1 – (related to: semi-weekly democrat (albany, or). cont: albany daily democrat. merged with: albany evening herald to form: albany evening herald and the albany democrat) – us Oregon Lib [071]
Albany democrat-herald *see*
– Mid-valley sunday
– Weekly democrat (albany, or)
Albany democrat-herald (albany, or) – Albany OR: W L Jackson & R R Cronise, 1925- [daily ex sun] – 1 – (related to: weekly democrat (albany, or), mid-valley sunday; cont: albany evening herald and the albany democrat) – us Oregon Lib [071]
Albany evening democrat – Albany OR: Brown & Stewart, -1876 [daily ex sun] – 1r – 1 – (related to wkly ed: state rights democrat (1865-1900); cont by: daily albany democrat (1876-18-?)) – us Oregon Lib [071]
Albany evening democrat *see* State rights democrat (albany, or)
Albany evening herald – Albany OR: C C Page, -1925 [daily ex sun] [mf ed 1967] – 10r – 1 – (cont: morning daily herald; merged with: albany democrat (1922) and: albany evening herald and the albany democrat) – us Oregon Lib [071]
Albany evening herald and the albany democrat – Albany OR: W L Jackson & R R Cronise, 1925 [daily] [mf ed 1963] – 1r – 1 – (related to: semi-weekly democrat (1913-26); merger of: albany evening herald (-1925) and: albany democrat (-1925); cont by: albany democrat-herald (1925-)) – us Oregon Lib [071]
Albany first baptist church : church minutes – London. 1840-1843 (1) – 1 – $114.57 – (lacks: jul 1860-jun 1866, nov 1866-may 1870, jan 1927-feb 1933) – mf#5239 – us Southern Baptist [242]
Albany first baptist church. albany, georgia – jul 1848-sep 1987 – 1 – $299.16 – (lacking: 1861-99, 1901-16, 1937-42. history: 1836-1906. incl: ladies aid soc records, 1895-1901) – us Southern Baptist [242]
Albany gazette – 1799 dec 30-1800 dec 15 – 1 – mf#780595 – us WHS [071]
Albany. Georgia. First Baptist Church *see* The way
Albany herald – 1925 jan 1-dec 17; 1925 dec 24-1928 oct 11; oct 18-1931 jun 18; 1931 jun 25-1934 jan 12; jan 18-oct 4; oct 11-1936 apr 23; apr 30-1937 sep 23; sep 30-1939 feb 23; mar 2-1940 oct 24; oct 31-1942 dec 31; 1943-64; 1965 jan 6-1967 mar 8; mar 15-1969 oct 30; nov6-1972 jul 27; aug 3-1973 dec 27; 1974 jan 3-1976 jun 24; jul 1-1977 dec 29; 1978-94; 1995 jan-nov16 – 1 – mf#914191 – us WHS [071]
Albany herald *see* Disseminator
Albany herald (albany, or) – Albany OR: A S Pottinger et al [wkly] – 1 – (began in 1879? merged with: disseminator (harrisburg, or) to form: weekly herald=disseminator (-1904)) – us Oregon Lib [071]
Albany inquirer – Albany OR: Haley & Stinson, 1862-63 [wkly] – 1r – 1 – (began in 1862; ceased in 1863; cont: oregon democrat (1859-62); cont by: oregon democrat (1863-64)) – us Oregon Lib [071]
Albany journal – 1879 feb 12-1882 feb 11; feb 18-1885 may 23; may 30-1888 may 5; may 12-1889 jul 20;1890 oct 18-1891 oct 31 – 1 – mf#913609 – us WHS [071]
Albany journal – 1895 nov 7-1896 feb 27 – 1 – mf#913602 – us WHS [071]
Albany journal – Albany OR: Albany Printing & Pub Co, 1864- [wkly] [mf ed 1964] – 1r – 1 – us Oregon Lib [071]
Albany law environmental outlook *see* Albany law environmental outlook journal
Albany law environmental outlook journal – v1-5. 1995-2000 – 9 – $74.00 – ISSN: 1085-3634 – mf#116481 – us Hein [344]
Albany law journal – La Salle IL 1870-1908 – 1 – mf#4624 – us UMI ProQuest [340]
Albany law journal – v1-70. 1870-1908 [all publ] – (= ser Historical legal periodical series) – 5 – $770.00 set – mf#408780 – us Hein [340]
The albany law journal : a weekly record – v1-70. 1870-1909 – 140mf – 9 – $630.00 – (v1-62 weekly. v63-70 monthly. titles added regularly) – mf#LLMC 82-901 – us LLMC [340]
Albany law review – Albany NY 1980+ – 1,5,9 – ISSN: 0002-4678 – mf#12612,02 – us UMI ProQuest [340]
Albany law review – Albany Law School: v1-19. 1931/32-1955 – 52mf – 9 – $78.00 – (no additions poss for copyright reasons) – mf#LLMC 95-110 – us LLMC [340]
Albany law review – v1-64. 1931-2001 – 5,6,9 – $1030.00 set – (v1-49 1931-85 $613.00 reel or mf. v50-61 1985-98 $417.00 mf) – ISSN: 0002-4678 – mf#100131 – us Hein [340]
Albany news – 1979-83 – 1r – 1 – mf#12.16 – nz Nat Libr [079]
[Albany-] news – CA. apr 1986-may 1988 – 1r – 1 – $60.00 – mf#B06008 – us Library Micro [071]
[Albany-] news review – CA. mar 1964-may 1965 – 1r – 1 – $60.00 – mf#B06009 – us Library Micro [071]
Albany. Presbytery (Pres. Church in the USA) *see* Records, 1790-1797
Albany register – 1799 dec 31-1800 dec 30 – 1 – mf#780596 – us WHS [071]
Albany register – Albany OR: C van Cleve, 1868- [wkly] – 1 – (cont: albany weekly register (-1868)) – us Oregon Lib [071]
Albany review – Killen TX 1903-08 – 1 – mf#2901 – us UMI ProQuest [420]
Albany southwest georgian – 1993 mar 4/6-dec 23/25; 1994 jan 6/8-dec 29/31; 1995 jan 5-dec 28; 1996 jan 4-dec 26; 1997 jan 2/4-97 dec 31/1998 jan 7 – 1 – mf#2682184 – us WHS [071]
Albany. Synod (Pres. Church in the USA) *see* Minutes
[Albany-] the enterprise – CA. aug 1982-dec 1983 – 6r – 1 – $360.00 – mf#B02006 – us Library Micro [071]
[Albany-] the journal – CA. apr 1988-1995 – 9r – 1 – $540.00 – mf#B06007 – us Library Micro [073]
[Albany-] times journal – CA. oct 1979-may 1984 – 5r – 1 – $300.00 – (see berkeley) – mf#B06010 – us Library Micro [073]
Albany vindicator – 1885 jan 8-1887 apr 7; apr 14-1890 jun 12; jun 19-1893 jul 6; jul 13-1894; 1895-96; 1897-1898 aug 11; aug 18-1901 sep 12; sep 19-1904 oct 13; oct 20-1907 oct 10; oct 17-1910 nov10; nov17-1914 jan 8; jan 15-1916 dec 28; 1917 jan 4-1919 dec 25; 1920 jan 1-1922 dec 7; dec 14-1924 dec 25 – 1 – mf#914185 – us WHS [071]

Albany weekly democrat – Albany OR: Wm H Hornbrook, 1912-13 [wkly] [mf ed 1964] – 1r – 1 – (related to: albany daily democrat; cont: albany democrat (1900-12); cont by: semi-weekly democrat (1913-26)) – us Oregon Lib [071]

Albany weekly herald – Albany OR: W A Shewman, Jr, 1909- [wkly] [mf ed 1967] – 1r – 1 – (cont: weekly herald (1904-09)) – us Oregon Lib [071]

Albany weekly journal – v1 n1-1926 [1865 oct 12-1866 may 3] – 1 – mf#913598 – us WHS [071]

Albarellos, Juan *see* Efemerides burgalescas

Al-barid al-gaza'iri – no. 1-4. Alger. aout-sept 1913 – 1 – fr ACRPP [079]

Albarran, Ramon *see* Los torpedos en la guerra maritima

Albaspinaeus, Gabr *see* Observationum libri duo

Al-bassir – Paris. 1881-82 – 1 – fr ACRPP [079]

Al-bayan – Cairo: 'Abd al-Rahman al Barquqi, 1911-21. v1 n1-v9 n5. 24 aug 1911-nov 1921 – (= ser Arabic journals and popular press) – 2r – 1 – $750.00 – (missing: v5-6) – us MEDOC [956]

Albayrak – Erzurum, 1913-21. Mes'ul Mueduerue: Suelayman Necati. n49. 4 kanuniewel 1335 [1919] – (= ser O & t journals) – 1mf – 9 – $25.00 – us MEDOC [956]

Albbote und rundschau *see* Intelligenz-blatt fuer die oberaemter ehingen und muensingen

Albee, Ernest *see*
– The beginnings of english utilitarianism
– A history of english utilitarianism

Al-beirak-al-ahmar / Parti communiste S F IC – Paris. n1 sep 1926 – 1 – fr ACRPP [335]

Albemarle : a monthly review – La Salle IL 1892 – 1 – mf#4198 – us UMI ProQuest [420]

Albemarle, George Thomas Keppel, Earl *see* Memoirs of the marquis of rockingham and his contemporaries

Albemarle, William Coutts Keppel, Earl of *see* British columbia and vancouver's island

Albenino, Nicolas de *see*
– Verdadera relacion de lo sussedido en los reynos...del peru...introduccion de jose toribio medina. paris 1930
– Verdadera relacion delo sussedido enlos reynos e provincias del peru desde la yda deellos del virey blasco nunes vela...y muerte de goncalo picarro. sevilla 1549

Albeniz, Isaac *see* Pavana-capricho, op 12

Alber, E *see*
– Das ehbuechlin
– Ivdicivm erasmi alberi, de spongia erasmi roterod

[Alber, E] *see*
– Ein dialogus oder gespraech etlicher personen vom interim
– Die grosse wolthat so unser herre gott durch d martinum luther der welt erzeiget

Albergati Capacelli, Pirro *see* Cantate morali a voce sola del conte pirro albergati. opera terza

Albermontius, F *see* Symmetria iuridico-austriaca continens viva themidis & avstriae oscvla

Alberni valley times – Port Alberni, British Columbia, CN. nov 1967- – 4r/y – 1 – Can$93.00r – cn Commonwealth Imaging [071]

L'albero di diana : dramma giocoso in 2 atti / Martin y Soler, Vicente – [1787?] [mf ed 19–] – 1r – 1 – mf#pres. film 20 – us Sibley [780]

Alberry, A J *see* Avicenna on theology

Albers, B *see* Consuetudines monasticae

Albers, Br *see*
– S ambrosii mediolanensis episcopi. de obitu satyri fratris laudatio funebris
– S pachomii abbatis tabennensis regulae monasticae. s orsiesii doctrina de institutione monachorum

Albers, Emanuel *see* Die quellenberichte in josua 1-12

Albers, Jan *see*
– Bau und test eines wirbelstrom-septums fuer delta
– Untersuchungen elektrochemisch erzeugter adsorbate auf pt und pd-einkristallen in einer uhv-anlage

Albers, Robert H *see* Journal of ministry in addiction and recovery

Albert *see* Asschepoester, groot toover-ballet in drie bedrijven

Albert, Albrecht *see* Kid

Albert anderssons affarsblad – Uddevalla, Sweden. 1899-1902 – 1r – 1 – sw Kungliga [078]

Albert bitzius : lebensbild eines republikaners, nach seinem handschriftlichen nachlasse / Gotthelf, Jeremias [Albert Bitzius]; ed by Balmer, Hans – Bern: Nydegger & Baumgart, 1888 [mf ed 1993] – viii/259p (ill) – 1 – mf#8518 – us Wisconsin U Libr [920]

Albert college times – Belleville [Ont]: The College, [1889-18– or 19–] [mf ed v1 n1 mar 1 1889] – 9 – mf#P04001 – cn Canadiana [378]

Albert duerer : his life and works / Thausing, Moritz – London 1882 – (= ser 19th c art & architecture) – 10mf – 9 – mf#4.1.267 – uk Chadwyck [700]

Albert ehrhardt und die erforschung der griechisch-byzantinischen hagiographie / Winkelmann, F – Berlin, 1971 – (= ser Tugal 5-111) – 2mf – 9 – €5.00 – ne Slangenburg [240]

Albert eichhorn und die religionsgeschichtliche schule / Gressmann, Hugo – Goettingen: Vandenhoeck & Ruprecht, 1914 [mf ed 1989] – 1mf – 9 – 0-7905-2966-1 – mf#1987-2966 – us ATLA [200]

Albert, Eugen d' *see* Scirocco

Albert family newsletter – 1986-july/aug 1988 – 1 – mf#1322431 – us WHS [929]

Albert, Felix Richard *see*
– Die bluetezeit der deutschen predigt im mittelalter, 1100-1400
– Die geschichte der predigt in deutschland bis auf karl den grossen, 600-814
– Seit wann giebt es eine predigt in deutscher sprache?

Albert gazette – Burgersdorp, South Africa. 6 jan 1894-24 dec 1898 – 3r – 1, 16 diazo available at reduced price – sa National [072]

Albert james myer, founder of the army signal corps: a biographical study / Scheips, Paul Joseph – 1965 – 1 – us Kansas [920]

Albert john luthuli and the south african race conflict / Callan, Edward – Kalamazoo, MI. 1962 – 1r – 1 – us UF Libraries [960]

Albert john luthuli and the south african race conflict / Callan, Edward – Kalamazoo, MI. 1965 – 1r – 1 – us UF Libraries [960]

Albert knapp als dichter und schriftsteller : mit einem anhang unveroeffentlicher jugendgedichte / Knapp, Martin – Tuebingen: Mohr 1912 [mf ed 1992] – 1mf – 9 – 0-524-04820-7 – (incl bibl ref) – mf#1992-2049 – us ATLA [430]

The albert levitt papers – (= ser American legal manuscripts from the harvard law school library) – 42r – 1 – $7405.00 – 0-89093-807-5 – (guide only $175) – us UPA [341]

Albert, Maria *see* Bibliographie de mere isabelle sormany

Albert, Mary *see*
– Brooke finchley's daughter
– The diamond shoe buckles
– The luckiest man in the world

The albert memorial / Dafforne, James – London [1878?] – (= ser 19th c art & architecture) – 3mf – 9 – mf#4.2.739 – uk Chadwyck [720]

Albert moore : his life and works / Baldry, Alfred Lys – London 1894 – (= ser 19th c art & architecture) – 3mf – 9 – mf#4.2.190 – uk Chadwyck [920]

The albert record – Burgersdorp SA, 7 mar-22 aug 1885 – 1r – 1 – sa National [079]

Albert tessier et son oeuvre – [mf ed Montreal, 1974] – 1r – 1 – (coll: depouillement: tavi (cinema et photos), loisirs (photographies) (conference – 1947), cinema tessier, 1936-1958) – mf#SEM35P95 – cn Bibl Nat [790]

Albert times *see* The burghersdorp gazette

The albert times – Burgersdorp SA, 4 jan 1855-31 dec 1859 – 4r – 1 – mf#MS00264 – sa National [079]

Albert von beham und regesten papst innocenz 4 / ed by Hoefler, Constantin – Stuttgart: Literarischer Verein, 1847 [mf ed 1993] – (= ser Blvs 16/2) – xxiv/223p – 1 – (incl bibl ref) – mf#8470 reel 4 – us Wisconsin U Libr [943]

Albert von sachsen : sein lebensgang und sein kommentar zur nikomachischen ethik des aristoteles / Heidingsfelder, G – 1927 – (= ser Bgphma 22/3-4) – €7.00 – ne Slangenburg [170]

Albert walter gilhrist / Staid, Mary Evangelista – s.l, s.l, 1950 – 1r – 1 – us UF Libraries [090]

Alberta farmer... *see* Weekly herald

Alberta Genealogical Society *see* Relatively speaking

Alberta historical review – Calgary AB 1953-74 – 1,5,9 – (cont by: alberta history) – ISSN: 0002-4783 – mf#7401.01 – us UMI ProQuest [971]

Alberta history – Calgary AB 1975+ – 1,5,9 – (cont: alberta historical review) – ISSN: 0316-1552 – mf#7401,01 – us UMI ProQuest [971]

Alberta labour – v3 n1-v4 n5 [1978 jan-1979 nov/dec] – 1 – mf#470894 – us WHS [331]

Alberta law reports – 1st series: v1-26. 1908-33 (all publ) – 181mf – 9 – $271.00 – (cont by 2nd series which is not offered by llmc) – mf#LLMC 81-018 – us LLMC [340]

Alberta law review – Edmonton AB 1955+ – 1,5,9 – ISSN: 0002-4821 – mf#2422 – us UMI ProQuest [340]

Alberta. Legislative Assembly *see* Journals

Alberta Native Communications Society *see* Native people

Alberta nonpartisan – Alberta, CN. oct 1917-jul 1919 – 1r – 1 – cn Commonwealth Imaging [071]

Alberta RN – Edmonton AB 2001+ – 1,5,9 – ISSN: 1481-9988 – mf#26735,01 – us UMI ProQuest [610]

Alberta scrapbook hansard – Alberta, CN. 1906-64 – 20r (complete coll) – 1 – (newspaper articles concerning the activities of the alberta legislative assembly) – cn Commonwealth Imaging [324]

Alberta social credit chronicle – Alberta, CN. jul 1934-jan 1936 – 1r – 1 – cn Commonwealth Imaging [071]

"Alberta" Society for the Preservation of Indian Identity *see*
– Nations' ensign
– Nations' ensign/ pow wow west
– Native ensign

Alberta statutes, session laws and revisions – 1906 Consolidated Statutes-1945 Parliament 2nd sess. 1906-45 – 331mf – 9 – $496.00 – (updates planned) – mf#LLMC 90-110 – us LLMC [348]

Alberta Teachers' Association *see* Ata magazine

Alberta teachers' association. ata news – Edmonton AB 1974+ – 1 – ISSN: 0001-267X – mf#10093 – us UMI ProQuest [370]

Alberta tribune – Calgary, Canada 1 aug 1896-16 dec 1899 (imperfect) – 3r – 1 – uk British Libr Newspaper [071]

Alberta-Marie, soeur *see* Bibliographie de monsieur le cure l boisseau

Albertan – Calgary, Alberta, CN. jan 1897-dec 1980 – 497r – 1 – cn Commonwealth Imaging [071]

Albertan *see* Morning albertan

Albertazzi Avendano, Jose *see* Palabras al viento (reconstrucciones de discursos)

Alberti, Conrad *see*
– Die alten und die jungen
– Bettina von arnim
– Gustav freytag
– Ludwig boerne
– Die rose von hildesheim

Alberti de bezanis abbatis s laurentii cremonensis chronica (mgh7:3.bd) – 1908 – (= ser Monumenta germaniae historica 7: scriptores rerum germanicarum in usum scholarum (mgh7)) – €7.00 – ne Slangenburg [240]

Alberti, Eduard *see* Lexikon der schleswig-holstein-lauenburgischen und eutinischen schriftsteller von 1829-1882

Alberti, Fra Leandro *see* Descrittione di tutta italia

Alberti, Giuseppe Matteo *see* 12 sinfonie a quattro. due violine, alto, organo e violoncello, opera seconda, libro primo-[secondo]

Alberti, I *see* Geist des musikalischen kunstmagazins

Alberti, Innocenzo *see* Il terzo libro de madrigali a quattro voci

Alberti, L B *see*
– La architectura
– L'architettura di leon batista alberti tradotta in lingua fiorentina da c bartoli
– De pictura praestantissima, et numquam satis laudata arte libri tres...
– De re aedificatoria...
– De re aedificatoria dece
– Los diez libros de architectura
– I dieci libri de l'architettura
– Nahere und ausgereitetere nachrichten von denen einen theil...

[Alberti, L B] *see*
– L'Architecture et art de bien bastir du...
– Leone battista alberti's kleinere kunsttheoretische schriften...
– Leonis baptiste alberti de re aedificatoria incipit...

Alberti, Lodewijk *see* De kaffers aan de zuid kust van afrika

Alberti Magni (Albertus Magnus, Saint (Albert the Great)) *see* Opera omnia

Alberti, Robert Edward *see* Affirmez vous!

Albertina / Bompiani, Valentino – Paris, France. 1948 – 1r – 1 – us UF Libraries [440]

Albertini, Gioacchino *see* La mia sposa

Albertinus, A *see*
– Hiren schleifer
– Hirnschleiffer

Albertinus, Edm *see* De eucharistiae sive coenae dominicae sacramento libri tres

Alberts des grossen verhaeltnis zu plato / Gaul, L – 1913 – (= ser Bgphma 12/1) – €7.00 – ne Slangenburg [180]

Alberts, Juergen *see* Arbeiteroeffentlichkeit und literatur

Alberts, Wilhelm *see*
– Gustav frenssen

Albertura dos portos / Pinho, Wanderley – Salvador, Brazil. 1961 – 1r – 1 – us UF Libraries [972]

Albertus de Ferrariis *see* Opuscula de horis canonicis. de defectibus occurentibus in missa

Albertus magnus : beitraege zu seiner wuerdigung / Hertling, G von – 1914 – (= ser Bgphma 14/5-6) – €11.00 – ne Slangenburg [100]

Albertus magnus : beitraege zu seiner wuerdigung / Hertling, Georg, Graf von – 2. aufl. Muenster i. W: Aschendorff, 1914 [mf ed 1990] – (= ser Beitraege zur geschichte der philosophie des mittelalters) – 1mf – 9 – (incl bibl ref) – mf#1988-1534 – us ATLA [180]

Albertus magnus ((bgphma16 : de animalibus libri 26 / Stadler, H – 1920 – (= ser Bgphma) – €27.00 – (nach der koelner urschrift, zweiter band: buch 13-26 und die indices enthaltend) – ne Slangenburg [100]

Albertus magnus, de animalibus libri 26 : nach der koelner urschrift. erster band: buch 1-12 enthaltend / Stadler, H – 1916 – (= ser Bgphma 15) – €31.00 – ne Slangenburg [100]

Albertus, Magnus, Saint *see* Heinrich mynsinger von den falken, pferden und hunden

Albertus Magnus, Saint (Albert the Great) *see* Opera omnia

Albertus-Universitaet zu Koenigsberg i. Pr. Sternwarte *see* Veroeffentlichungen der universitaets-sternwarte koenigsberg pr

Albertype Company, New York *see* St augustine

Alberuni's india : an account of the religion, philosophy, literature, geography...of india about a d 1030 / Biruni, Muhammad ibn Ahmad – London: K Paul, Trench, Truebner, 1910 [mf ed 1992] – (= ser Truebner's oriental series) – 2v on 3mf – 9 – 0-524-03476-1 – (incl bibl ref. english ed with notes & ind by edward c sachau) – mf#1990-3218 – us ATLA [390]

Albes, Edward *see* Rio de janeiro, the fair capital of brazil

Die albigenser : freie dichtungen / Lenau, Nicolaus – Stuttgart: J G Cotta, 1842 [mf ed 1995] – vi/253p – 1 – mf#8799 – us Wisconsin U Libr [810]

Albina weekly courier – Albina OR: W N Carter, -1894 [wkly] – 1 – us Oregon Lib [071]

Albinana, Jose Maria *see* Confinado en las hurdes (una victima de la inquisicion republicana)

Albinana Sanz, J *see* Los crimines del caciquismo. la tragedia de el pobo. defensa del medico...

Albion – Boone NC 1983+ – 1,5,9 – ISSN: 0095-1390 – mf#13915 – us UMI ProQuest [940]

Albion – [NW England] Liverpool 1827-oct 1871 – 1 – (title change: liverpool weekly albion [nov 1871-dec 1872, jan 1874-mar 1887]) – uk MLA; uk Newsplan [072]

The albion – New York. June 22 1822-Dec 28 1833; Jan 14 1834-Sept 29 1855; Jan 5 1856-Dec 26 1863; Mar 28 1868-Dec 18 1869; Jan 7 1871-Dec 30 1871; and Oct 25 1873-Jan 30 1875. Not collated – 1 – us NY Public [071]

Albion and erin : a voice from the english side of the irish question...by geo ambrose mcneill of new brunswick...jan 1886 / McNeill, George Ambrose – Toledo, Ohio?: s.n, 1886? – 1mf – 9 – mf#49102 – cn Canadiana [941]

Albion argus – Albion, NE: C C Barnes. -n14. apr 1 1948 (wkly) – 22r – 1 – (cont: boone county argus. absorbed: boone county blade. absorbed by: albion weekly news. vol numbering dropped with jan 23 1919) – us Bell [071]

Albion daily critic – Albion, NE: Critic Publ Co. v1 n1. jan 28 1896-mar 1896// (daily) – 1r – 1 – us Bell [071]

Albion daily news – Albion, NE: A W Ladd, apr 22 1898 (daily) – 1r – 1 – us Bell [071]

Albion Mines (Nova Scotia) *see* Special rules for the conduct and guidance of the persons acting in the management and of all persons employed in or about the albion mines

Albion news – Albion, NE: Jack Lough. v70 n1. oct 21 1948)- (wkly) – 33r – 1 – (cont: albion weekly news; absorbed: boone companion; iss for v70 n12 jan 6 1949-accompanied by suppl: nowadays; companion of: boone companion oct 30 1958-may 14 1963) – us Bell [071]

Albion semi-weekly news – Albion, NE: A W Ladd (semiwkly) [mf ed v8 n41. apr 13 1887-may 1887 (gaps)] – 1r – 1 – (cont: boone county news. cont by: albion weekly news) – us NE Hist [071]

The albion w tourgee papers : from the personal papers collections – [mf ed UMI] – 60r – 1 – us UMI ProQuest [975]

Albion weekly news – Albion, NE: A W Ladd. -v69 n53. oct 14 1948 (wkly) – 40r – 1 – (cont: albion semi-weekly news; absorbed: petersburg index (petersburg ne) jan 28 1943, cedar rapids leader-outlook apr 1 1943 and: albion argus apr 1 1948; cont by: albion news; vol numbering irregular for sep 22 1921-jan 1 1925; iss for aug 21 1924 incorrectly dated aug 28 1924; special historical iss publ oct 26 1939) – us Bell [071]

Albion's voice – v1 n4 [1970 jul]; 1970 nov – 1 – mf#1582941 – us WHS [071]

Albis, Victor H *see* Elitros

Albo, Jamy M *see* Cerebral blood flow responses to a cognitive challenge in an older population

Albo, Joseph *see* Sefer 'ikarim

Albo-albo : po konferencji moskiewskiej / Mackiewicz, Stanislaw – Londyn: Nakl. autora, 1943 (mf ed 19–) – (alt title: po konferencji moskiewskiej) – mf#ZQ-153 – us NY Public [943]

Alboise Du Pujol, Jules Edward *see*
– Idiote

Al'bom uchastnikov vserossiiskoi promyshlennoi i khudozhestvennoi vystavki v nizhnem novgorode, 1896 – Spb, 1896 – 15mf – 9 – mf#REF-469 – ne IDC [332]

La alborada – Manila: Sr Rafael Corpus [v1 n6 (nov 9 1901)] (wkly) – 1r – 1 – us CRL [079]

Alboreda, A M *see* Historia de montserrat

Albores.ensayos / Gutierrez, Miguel – 1881 – 9 – sp Bibl Santa Ana [840]

Albornoz, Alvaro de *see*
- En los caminos de la libertad
- El fascismo y las armas y las letras espanolas

Albornoz, Orlando *see* Maestro y la educacion en la sociedad venezolana

Alboroto y motin de mexico del 8 de junio de 1692. mexico, 1932 / Siguenza y Gongora, Carlos – Madrid: Razon y Fe, 1935 – 1 – sp Bibl Santa Ana [972]

Albrecht, Carl *see* Die wissenschaftlich geordnete weltansicht

Albrecht, Chr *see* Schleiermacher's liturgik

Albrecht durers schriftlicher nachlass – Berlin, Germany. 1910 – 1r – 1 – us UF Libraries [720]

Albrecht, Erwin *see* Wegerecht

Albrecht, Friedrich *see* Deutsche schriftsteller in der entscheidung

Albrecht, Guenter *see* Deutsche schwaenke

Albrecht, Guenter et al *see* Lexikon deutschsprachiger schriftsteller

Albrecht hallers tagebuecher seiner reisen nach deutschland, holland und england : 1723-1727 / ed by Hirzel, Ludwig – Leipzig: S Hirzel 1883 [mf ed 1990] – 1r – 1 – (filmed with: albrecht von haller / stephen d'irsay) – mf#2696p – us Wisconsin U Libr [914]

Albrecht, Hellmuth F G *see* La epica juglaresca alemana del siglo 12

Albrecht, Helmut *see* Entwurf und erprobung eines konzepts fuer die ltg in der lehrerbildung an paedagogischen hochschulen

Albrecht, Hermann *see* Der praezeptoratsvikari

Albrecht, J W *see* Tractatus physicus de effectibus musices in corpus animatum...

Albrecht, Johann Lorenz *see*
- Abhandlung ueber die frage
- Gruendliche einleitung in die anfangslehren der tonkunst
- Musica mechanica organoedi..

Albrecht, Joseph *see* Conrads von weinsberg, des reichs-erbkaemmerers, einnahmen- und ausgaben-register von 1437 und 1438

Albrecht, Julius *see* Ausgewaehlte kapitel zu einer hans-sachs-grammatik

Albrecht, Karl *see* Shaar ha-shir

Albrecht, Luitgard *see* Der magische idealismus in novalis' maerchentheorie und maerchendichtung

Albrecht, Otto *see*
- Die evangelische gemeinde miltenberg und ihr erster prediger
- Luthers katechismen

Albrecht, Paul *see* Leszing's plagiate

Albrecht ritschl and his school / Mackintosh, Robert – London: Chapman & Hall, 1915 [mf ed 1991] – (= ser The great christian theologies) – 1mf – 9 – 0-524-00060-3 – (incl bibl ref) – mf#1989-2760 – us ATLA [242]

Albrecht ritschl und seine schueler : im verhaeltnis zur theologie, zur philosophie und zur froemmigkeit unsrer zeit / Wendland, Johannes – Berlin: Georg Reimer, 1899 [mf ed 1991] – 1mf – 9 – 0-7905-8970-2 – mf#1989-2195 – us ATLA [242]

Albrecht ritschls anschauung von evangelischem glauben und leben : ein vortrag...am 19. dezember 1899 in der basler aula / Vischer, Eberhard – Tuebingen: J C B Mohr 1900 [mf ed 1992] – (= ser Sammlung gemeinverstaendlicher vortraege und schriften aus dem gebiet der theologie und religionsgeschichte 18) – 1mf – 9 – 0-524-05523-8 – mf#1990-1518 – us ATLA [242]

Albrecht ritschls leben / Ritschl, Otto – Freiburg i B: JCB Mohr, 1892-96 [mf ed 1991] – 2v on 3mf – 9 – 0-524-00595-8 – (incl bibl ref) – mf#1990-0095 – us ATLA [242]

Albrecht von halberstadt und ovid im mittelalter / Bartsch, Karl – Quedlinburg, Leipzig: G Basse, 1861 [mf ed 1993] – (= ser Bibliothek der gesammten deutschen national-literatur von der aeltesten bis auf die neuere zeit sect1/38) – cclx/501p – 1 – (incl bibl ref and ind) – mf#8438 reel 8 – us Wisconsin U Libr [410]

Albrecht von haller : a physician – not without honor / Reed, Charles Bert – [Chicago]: Chicago Literary Club, 1915 [mf ed 1993] – (= ser Club papers (chicago literary club)) – 56p (ill) – 1 – mf#8669 – us Wisconsin U Libr [610]

Albrecht von haller : eine studie zur geistesgeschichte der aufklaerung / Irsay, Stephen d' – Leipzig: G Thieme 1930 [mf ed 1991] – (= ser Arbeiten des instituts fuer geschichte der medizin an der universitaet leipzig 1) – 1r [ill] – 1 – (incl ind. filmed with: haller als philosoph / heinrich ernst jenny & other titles) – mf#2696p – us Wisconsin U Libr [612]

Albrecht von haller und seine bedeutung fuer die deutsche cultur : vortrag, gehalten in der literarischen gesellschaft zu danzig / Lissauer, Abraham – Berlin: Luederitz, 1873 [mf ed 1993] – 39p – 1 – mf#8034 – us Wisconsin U Libr [430]

Albrecht von hallers sprache in ihrer entwicklung dargestellt / Kaeslin, Hans – [s.l: s.n.] 1892 (Brugg: Buchdruckerei "Effingerhof") [mf ed 1990] – 1r – 1 – (incl bibl ref. filmed with: albrecht von haller / stephen d'irsay) – mf#2696p – us Wisconsin U Libr [430]

Albrechts von scharfenberg juengerer titurel / ed by Wolf, Werner – Berlin: Akademie-Verlag, 1955- [mf ed 1993] – (= ser Deutsche texte des mittelalters 45, 55, 61, 73, 77, 79) – (ill) – 1 – (in middle high german; comm & notes in german; incl bibl ref) – mf#8623 reel 12 – us Wisconsin U Libr [430]

Albrechtsberger, Johann Georg *see*
- Johann georg albrechtsbergers... gruendliche anweisung zur composition
- Kurzgefasste methode den generalbass zu erlernen
- Quatuor pour le clavecin, ou fortepiano, deux violons et basse
- Quintuor pour trois violons, alto et basse
- Sechs fugen fuer pianoforte oder orgel
- Sei quartetti con fughe per diversi stromenti
- Six fugues pour le pianoforte
- Six quatuors en fugues, a deux violons, taille & basse, oeuvre second
- Six quatuors pour deux violons, alto et violoncelle, oeuvre 21
- Six trios concertans pour violon, viola et violoncelle, oeuvre 9eme
- Trois sextuors pour deux violons, deux altos, violoncelle et basse, op 13, no 1 & no 2

Albree, George *see* Things of the kingdom

Albright, Cindy W *see* Comparison of two methods of training special olympics volunteers to teach and coach bowling

Albright, M Catharine *see* Letters from india

Albright, Philip H *see* Original solo concertos for the double bass.

Albright, W F *see*
- Archaeology and the religion of israel
- The archaeology of palestine
- The archaeology of palestine and the bible

Der albtalbote *see* Mittelbadischer courier

Albuerme Brea, P E *see*
- Ignis

L'album – Paris. 1821-mars 1823; nov 1828-aout 1829 – 1 – fr ACRPP [073]

Album aus paris / Lewald, August – Hamburg 1832 [mf ed Hildesheim: 1995-98] – (= ser Fbc) – 2v on 4mf – 9 – €120.00 – 3-487-29635-7 – gw Olms [914]

Album conmemorativo del quincuagesimo aniversario / Havana. Colegio De Belen – Habana, Cuba. 1904 – 1r – 1 – us UF Libraries [972]

L'album de la famille girouard / Girouard, Desire – [S.l: s.n, 1906?] – 1mf – 9 – 0-665-72366-0 – mf#72366 – cn Canadiana [929]

Album de la grande guerre / Comite de recrutement canadien-francais (Montreal, Quebec) – Montreal, [entre 1915 et 1917] (mf ed 1986) – 1mf – 9 – mf#SEM105P731 – cn Bibl Nat [355]

Album de la minerve : journal de la famille – Montreal: [s.n.] v1 n1 1 janv 1872-v3 n28 9 juil 1874 (mthly) [mf ed 1976] – 1r – 1 – mf#SEM35P136 – cn Bibl Nat [073]

Album de la paz y el trabajo / Paz, Ireneo – Mexico: Impr Litografia y Encuadernacion de I Paz 1911? [mf ed 1977] – 1r [ill] – 1 – us Misc Inst [972]

Album de la revue canadienne – Montreal: La Revue, 1899 [mf ed 1980] – 11mf – 9 – 0-665-06574-4 – mf#06574 – cn Canadiana [860]

Album de oro de puerto rico / Monteagudo, Antonio M – Habana, Cuba. 1939 – 1r – 1 – us UF Libraries [972]

Album de terre-sainte / Frederic, de Ghyvelde, pere – [Quebec (Province)]: [s.n.] 2v [1905?] (mf ed 1985) – 2mf – 9 – mf#SEM105P539 – cn Bibl Nat [915]

Album de vistas de costa rica / Zamora, Fernando – San Jose, Costa Rica. 1909 – 1r – 1 – us UF Libraries [972]

Album del bardo. coleccion de articulos...de varios autores – 9 – sp Bibl Santa Ana [800]

Album del estado mayor del cuartel general... – Habana, Cuba. 1912 – 1r – 1 – us UF Libraries [972]

Album del sesquicentenario / Ortega Ricaurte, Daniel – Bogota, Colombia. 1960? – 1r – 1 – us UF Libraries [972]

Album der basler missionsgesellschaft : achtzig ansichten von der goldkueste (westafrika) nach originalaufnahmen des missionars / Ramseyer, Friedrich August – Neuenburg: Attinger, 1895. Chicago: Dep of Photodup, U of Chicago Lib, 1971 (1r); Evanston: American Theol Lib Assoc, 1984 (1r) – 1 – 0-8370-0478-0 – mf#1984-B220 – us ATLA [240]

Album des familles – Ottawa: [Bureaux de "l'Album des familles"] [mf ed 5e annee n2 1er fevr 1880-6e annee n6 1er juin 1881; 7e annee n1 1er janv 1882-9e annee n6 1er juin 1884] – 9 – mf#P04009 – cn Canadiana [440]

Album des legendes – Paris. Mens. 1894-95 – 1 – (devenu: le livre des legendes) – fr ACRPP [073]

Album des peres du concile oecumenique du vatican commence le 8 decembre 1869 / Desmarais, Louis Elie – [Montreal]: publie par L E Desmarais, photographe, [ca 1876] [mf ed 1980] – 1mf – 9 – 0-665-02394-4 – mf#02394 – cn Canadiana [241]

Album des thueringerwaldes zum geleit und zur erinnerung / Schwerdt, Heinrich – Leipzig [1859] [mf ed Hildesheim: 1995-98] – (= ser Fbc) – viii/324p on 2mf – 9 – €60.00 – 3-487-29511-3 – gw Olms [914]

Album do domingo – Porto Alegre, RS: Typ Album do Domingo, 07 abr 1878-09 mar 1879 – (= ser Ps 19) – bl Biblioteca [079]

Album historico de la primera asamblea filipina fotografias reproducidas de la revista filipina / Tuohy, Anthony R [comp] – Manila: [s.n.] 1908 [mf ed 1984] – 1r [ill] – 1 – (spanish suppl of the far eastern review) – mf#6475 – us Wisconsin U Libr [323]

Album historique publie a l'occasion des fetes du cinquantenaire de la paroisse de sainte-agathe-des-monts, 1861-1911 / Grignon, Edmond – [Montreal?: s.n.], 1912 – 3mf – 9 – 0-665-74041-7 – mf#74041 – cn Canadiana [917]

L'album industriel – Montreal: T Berthiaume. 1ere annee n1 8 dec 1894-1ere annee n26 1er juin 1895 – 9 – mf#P04013 – cn Canadiana [670]

Album literario espanol – 1846 – 9 – sp Bibl Santa Ana [800]

Album litteraire de la revue canadienne : lectures du soir – nouv ser. Montreal: Bureaux de la Revue canadienne. 3e annee 1re livraison janv 1848-3e annee 12e livraison dec 1848 (mthly) [mf ed 1978] – 1r – 5 – (cont: album litteraire et musical de la revue canadienne; cont by: album litteraire et musical de la minerve) – mf#SEM16P306 – cn Bibl Nat [410]

Album litteraire et musical de la revue canadienne : bibliotheque des familles ou recueil choisi de romans... – Montreal: [s.n.] 1 ere annee, 1 ere livr janv 1846-2 ieme annee, 12 ieme livr dec 1847 (mthly) [mf ed 1978] – 1r – 5 – mf#SEM16P306 – cn Bibl Nat [073]

Album litterario : periodico instructivo e recreativo – Rio de Janeiro, RJ: Typ de Pinheiro & C, 15 ago 1860-01 abr 1861 – (= ser Ps 19) – mf#P17,01,81 – bl Biblioteca [440]

The album (london) – jul 1822-jan 1824 – (= ser 19th c british periodicals) – r23 – 1 – us Primary [073]

L'album musical – Montreal: A Filiatreault, [1882-1884] – 9 – mf#P04149 – cn Canadiana [780]

The album of language : illustrated by the lord's prayer in one hundred languages... / Naphegyi, Gabor – Philadelphia: J B Lippincott, 1869 – 1mf – 9 – 0-524-08088-7 – mf#1992-1148 – us ATLA [400]

Album of new songs by american composers – New York: Breitkopf & Hartel [c1898] [mf ed 1991] – 1r – 1 – mf#pres. film 96 – us Sibley [780]

Album of niagara falls – Portland, ME: Chisholm, 18-?] [mf ed 1979] – 1mf – 9 – 0-665-00022-7 – mf#00022 – cn Canadiana [917]

An album of the attorneys of rhode island : with a portrait and brief record of the life of each / Bowler, Ernest Constant – Bethel, Me., News Publishing Co., 1904. 208 p. LL-722 – 1 – us L of C Photodup [340]

Album of the table rock, niagara falls, c w and sketches of the falls, etc / ed by Menzies, George – [Niagara, Ont?: s.n.], 1846 – 2mf – 9 – 0-665-93887-X – mf#93887 – cn Canadiana [917]

Album paleographicum 17 provinciarum / Dekker, C et al – 1992 – 1mf+448p+300 facs – 9 – €67.70 – 90-72100-45-X – (Single titles in palaeography, manuscript studies and book history) – be Brepols [940]

Album (poesias) / Rosado H de Sotomayor, Jose – Plasencia: Tip. Jose Hotiveros, 1934 – 1 – sp Bibl Santa Ana [810]

Album simbolico : homenaje de los poetas... / Ateneo Dominicano – Ciudad Trujillo, Dominican Republic. 1957 – 1r – 1 – us UF Libraries [972]

Album, Simon Hirsch *see* Sefer divre emet

L'album souvenir des noces d'argent de la societe saint-jean-baptiste du college saint-joseph, Memramcook, NB : histoire, morceaux, poesies, portraits, gravures, biographies, discours, rapports, lettres, statistiques, statuts et reglements, convention, etc – S.l: s.n, 1894? – 5mf – 9 – (incl english text; incl ind) – mf#13776 – cn Canadiana [366]

L'album souvenir des noces d'argent de la societe saint-jean-baptiste du college saint-joseph, Memramcook, NB : histoire, morceaux, poesies, portraits, gravures, biographies, discours, rapports, lettres, statistiques, statuts et reglements, convention etc – Montreal?: [s.n.], 1894? – 5mf – 9 – (with english text; incl ind) – mf#02395 – cn Canadiana [366]

Album verses : and other poems / Davin, Nicholas Flood – [Ottawa?: s.n.], 1882 [mf ed 1980] – 1mf – 9 – 0-665-02590-4 – mf#02590 – cn Canadiana [810]

Album-souvenir : publiees a l'occasion de l'exposition de la societe d'agriculture de gentilly, tenue a gentilly, le 16 aout 1928 – [Quebec (Province): s.n, 1928? [mf ed 1997] – 2mf – 9 – mf#SEM105P2822 – cn Bibl Nat [971]

Album-souvenir a l'occasion des fetes de l'annee centenaire de notre fidele alma mater : le cher couvent de saint-cesaire, les 29, 30 juin et 1er juillet 1957 – [Saint-Cesaire?: s.n, 1957?] (mf ed 2001) – 2mf – 9 – (incl english text) – mf#SEM105P3403 – cn Bibl Nat [241]

Album-souvenir du 3e centenaire du quebec, 1608-1908 / Dion, Albert, abbe – [Quebec]: [s.n.], 1908 [mf ed 1985) – 1mf – 9 – mf#SEM105P479 – cn Bibl Nat [971]

Album-souvenir du 3e centenaire du quebec, 1608-1908 / Dion, Albert, abbe – [Quebec]: [s.n.] [1912?] (mf ed 1985) – 1mf – 9 – mf#SEM105P480 – cn Bibl Nat [971]

Album-souvenir offert par le departement... / Haiti. Departement Des Travaux Publics – s.l, s.l, 1960 – 1r – 1 – us UF Libraries [972]

Album-souvenir publie a l'occasion de la consecration de la cathedrale de saint-jerome : presidee par mgr charles valois, eveque de saint-jerome le 13 mai 1978 – [Saint-Jerome?: s.n, 1978] (mf ed 1994) – 2mf – 9 – mf#SEM105P2247 – cn Bibl Nat [241]

Albuquerque Coelho, Duarte De *see* Memorias diarias da guerra do brasil, 1630-1638

Albuquerque Felner, Alfredo De *see* Angola

The albuquerque indian – 1905-06 – (= ser American indian periodicals... 1) – 9 – $95.00 – us UPA [305]

[Albuquerque-] rayas – NM. 1978-1979 – 1r – 1 – $60.00 – mf#R04985 – us Library Micro [071]

Albuquerque, Viriato Antonio Caetano Bras de *see*
- A exposicao do venerando corpo do apostolo das indias, s francisco xavier, em 1878
- Exposicao do venerando corpo do glorioso apostolo das indias, s francisco xavier, em 1890

Alburas / Liendo, Arturo – Habana, Cuba. 1935 – 1r – 1 – us UF Libraries [972]

Alburquerque. Ayuntamiento *see*
- Ferias y fiestas en honor de la santisima virgen de carrion, patrona de alburquerque
- Ferias y fiestas...septiembre, 1960 en honor de la santisima virgen de carrion
- Festejos en honor de la santisima virgen de carrion, 1977

Alburquerque. Hermandad de Ntra. Sra. de Carrion *see* Reglamento de la...patrona de alburquerque

Albury banner – jan 3 1896-may 25 1950 – 2r – A$133.23 vesicular A$144.23 silver – at Pascoe [079]

Albury border post and wodonga advertiser – Albury, Australia 7 jun 1889-3 oct 1902 – 26r – 1 – uk British Libr Newspaper [079]

Albury herald – Albury, jan 1899-apr 1900 – 1r – A$75.50 vesicular A$81.00 silver – at Pascoe [079]

Albyn [i.e. Andrew Shiels] *see*
- Dupes and demagogues
- An eye to the ermine
- John walker's courtship
- Letter to eliza
- My mother
- The preface
- Retribution
- Rusticating in reality
- The sabbath in dartmouth
- The water lily
- The witch of the westcot

Alc newsletter – 1943 mar 22-1945 jul 3 – 1 – mf#1109411 – us WHS [071]

Las alcabolas de alburquerque o los celebres baldios / Duarte Insua, Lino – Badajoz: Dip. Provincial, 1946 – 1 – (sep de la rev de estudios extremenos) – sp Bibl Santa Ana [340]

Alcaide, Jose *see* Victor rojas

Alcais, Abel *see* Figures et recits de carthage chretienne

Alcala, Galiano, Dionisio *see* Cuba en 1858

Alcala, Manuel *see* Cesar cortes

El alcalde de zalamea / Calderon de la Barca, Pedro – 1849 – 9 – sp Bibl Santa Ana [820]

El alcalde de zalamea / Calderon de la Barca, Pedro – Madrid: Rivadeneyra, 1849 – 1 – sp Bibl Santa Ana [946]

Alcance misional de la liturgia del canaculo o el problema de la adaptacion / Morillo Trivino, Santiago – Granada: Imp. F. Roman, 1945 – 1 – sp Bibl Santa Ana [240]

Alcancia del artesano / Feijoo, Samuel – Santa Clara, Cuba. 1958 – 1r – 1 – us UF Libraries [972]

Alcaniz, Florentino *see*
- Las cruzadas del corazon de jesus
- Los cruzados del corazon de jesus. avisos practicos para su fundacion y organizacion

Alcantara – Caceres: I C Brocense -2006 – 5 – sp Bibl Santa Ana [074]

Alcantara. Ayuntamiento *see*
- Estatutos del patronato de viviendas sociales de alcantara
- Tradicionales ferias y fiestas de primavera, 1978
- Tradicionales ferias y fiestas de primavera... 1979

Alcantara (Caceres) / Junta Provincial de Turismo – Vitoria: Tip. Fournier, s.a. – 1 – (fotos gudiol) – sp Bibl Santa Ana [338]

Alcantara Machado, Jose De *see*
- Brasilio machado
- Vida e morte do bandeirante

Alcantara, San Pedro de *see* Tratado de la oracion y meditacion

Alcantara. Spain *see* Ordenanzas municipales

Alcanzar Anguita, Eufrasio *see* Tecnica y peritacion caligrafica

Alcaraz Segura, Lorenzo *see* Los ninos

Alcarotti, G F *see* Del viaggio di terra santa

Alcatel telecommunications review [english ed] – Paris, France 1996+ – 1,5,9 – (cont: electrical communication [english ed]) – ISSN: 1267-7167 – mf#4,01 – us UMI ProQuest [380]

La alcazaba almohade de badajoz / Torres Balbas, Leopoldo – Madrid. C.S.I.C. Granada. Al-Anadalus, 8, vol 6, fasc. 1. 1941 – 1 – sp Bibl Santa Ana [946]

Alcazaba of merida : early muslim architecture / Hernandez, Felix – Oxford: Clasendon, 1940 – 1 – sp Bibl Santa Ana [720]

Alcazar Alenda, Jose Maria *see*
- Carta pastoral
- Carta pastoral del...con ocasion del 4th centenario de la muerte de hernando cortes y del homenaje de espana a nuestra senora de guadalupe

El alcazar de toledo / Sanchez Arjona, Vicente – Sevilla: Imprenta Zambrano, s.a. – 1 – sp Bibl Santa Ana [946]

Alcazar Molina, Cayetano *see* Los virreinatos en el siglo 18. madrid, 1945

Alceste : tragedie en musique / Lully, Jean Baptiste – 2e ed, Paris: J B C Ballard 1716 [mf ed 19–] – 1r – 1 – (libretto by philippe quinault) – mf#film 508 – us Sibley [780]

Alcester chronicle – [West Midlands] Worcestershire jan 1923-dec 1950 [mf ed 2004] – 30r – 1 – uk Newsplan [072]

Alcester chronicle, etc – Alcester, England 2 apr 1864- [mf 1986-] – 1 – (wanting: jan-may, nov, dec 1912; 30 jul-23 dec 1986; wanting: apr-dec 1966 [see redditch indicator & alcester chronicle for iss publ during this period]) – uk British Libr Newspaper [072]

Alcester gazette – Alcester, England 29 apr-30 dec 1864 – 1 – uk British Libr Newspaper [072]

Alcestis : a poetry quarterly – New York. v. 1 no. 1-4 Oct 1934-July 1935 – 1 – us NY Public [420]

The alchemy of happiness = Kimiya-yi saadat / Al-Ghazzali – New York: E P Dutton 1910 [mf ed 1991] – (= ser Wisdom of the east series (new york, ny)) – 1mf – 9 – 0-524-00837-X – (trans fr hindustani by claud field) – mf#1990-2083 – us ATLA [260]

The alchemy of thought / Jacks, Lawrence Pearsall – London: Williams and Norgate, 1910 – 1mf – 9 – 0-7905-9231-2 – mf#1989-2456 – us ATLA [100]

Alcheringa : ethnopoetics – Boston MA 1976-80 – 1,5,9 – ISSN: 0044-7218 – mf#10362 – us UMI ProQuest [400]

Alciato, Andrea *see*
- Andreae alciati emblemata cum commentariis claudii minois...
- Andreae alciati emblematum libellus
- Clarissimi viri d. andreae alciati emblematum libellus, vigilanter recognitus...
- Clarissimi viri d. andreae alciati emblematum libri duo/ [and] in d andreae alciati emblemata succincta commentariola...
- Diverse imprese accomodate a diverse moralit...
- Los emblemas de alciato
- Emblemata
- Emblemata andreae alciati iurisconsulti clarissimi
- Emblemata andreae alciati...imaginibusque... illustrata
- Emblemata d a alciati denuo ab ipso autore recognita...
- Emblemata v c andreae alciati mediolanensis iurisconsulti
- Emblemata...cum imaginibus plerisque restitutis ad mentem auctoris
- Emblematum liber
- Les emblemes...mis en rime francoyse
- Liber emblematum d. andreae alciati
- Livret des emblemes de maistre andre alciat mis en rime francoyse et presente a monseigneur ladmiral de france
- Omnia andreae alciati v c emblemata
- Viri clarissimi d andreae alciati iusrisconsultiss

Alcide al brivo : Festa theatrale per le felicissime nozze delle...l'arciduca giuseppe d'austria e la principessa isabella di borbone... / Hasse, Johann Adolf; ed by f – Lipsia: B C Breitkopf e figlio 1763 [mf ed 19–] – 1r – 1 – mf#film 704 – us Sibley [780]

Alcimi ecdicii aviti viennensis episcopi opera quae supersunt (mgh1:6/2) / ed by Reiper, R – 1883 – (= ser Monumenta germaniae historica 1: scriptores – auctores antiquissimi) – €23.00 – ne Slangenburg [240]

Alcina : ein heroisch-allegorisches ballett von der erfindung und ausfuehrung des herrn joseph trafieri. aufgefuehrt in den k. k. hoftheatern 1798 / Trafieri, Giuseppe – Wien: Bey Matthias Andreas Schmidt [1798?] – 1 – (ballet scenario, in german and italian) – mf#*ZBD-*MGTZ pv 5-Res – Located: NYPL – us Misc Inst [790]

Alcindor, Fernand *see* Contribution du nord-ouest a l'independance nation

Alciphron *see* Alciphronis rhetoris epistularum libri 4

Alciphronis rhetoris epistularum libri 4 / Alciphron – Lipsiae, Germany. 1905 – 1r – 1 – us UF Libraries [090]

Alcmeonidas / Sancho, Alfredo – San Salvador, El Salvador. 1961 – 1r – 1 – us UF Libraries [972]

Alcobendes, Severiano *see* Las misiones franciscanas en china

Alcock, Deborah *see*
- The roman students or on the wings of the morning
- The romance of protestantism
- The seven churches of asia

Alcock, George Augustus *see* Key to the hebrew psalter

Alcock, John Congreve *see* Observations concerning the nature and origin of the meetings of the twelve judges for the consideration of cases reserved from the circuits

Alcock, Rutherford *see*
- Art and art industries in japan
- Catalogue of works of industry and art

Alcoforado, Pedro Guedes *see* Tupi na geografia fluminense

Alcohol : and other drug abuse news memo – v9 n11 [1976 nov]-v16 [1983] – 1 – mf#1066081 – us WHS [360]

Alcohol / Martinez Sobral, Enrique – Guatemala, Guatemala. 1962 – 1r – 1 – us UF Libraries [972]

Alcohol – Oxford, England 1986+ – 1,5,9 – ISSN: 0741-8329 – mf#49536 – us UMI ProQuest [362]

Alcohol against the bible : and the bible against alcohol / Shrewsbury, William J – London, England. 1841 – 1r – 1 – us UF Libraries [220]

Alcohol and alcoholism : international journal of the medical council on alcoholism – Oxford, England 1983+ – 1,5,9 – (cont: british journal on alcohol & alcoholism; cont by: alcohol & alcoholism: international journal of the medical council on alcoholism) – ISSN: 0735-0414 – mf#25177 – us UMI ProQuest [616]

Alcohol and drug research – Oxford, England 1980-87 – 1,5,9 – ISSN: 0883-1386 – mf#49334 – us UMI ProQuest [362]

Alcohol as a medicine / Higginbottom, J – London, England. 18– – 1r – 1 – us UF Libraries [010]

Alcohol as a medicine / Watkins, Thomas C – [Hamilton, Ont?: s.n, 188-?] [mf ed 1994] – 1mf – 9 – 0-665-94624-4 – (original iss in ser: prohibition series. incl bibl ref) – mf#94624 – cn Canadiana [615]

Alcohol health and research world – Washington DC 1973-98 – 1,5,9 – (cont by: alcohol research & health) – ISSN: 0090-838X – mf#12110.01 – us UMI ProQuest [362]

Alcohol in health and disease / Bucke, Richard Maurice – London [Ont]: J Bryce, 1880 [mf ed 1979] – 1mf – 9 – 0-665-00322-6 – mf#00322 – cn Canadiana [360]

Alcohol in the sanctuary / Tinling, J F B – London, England. 1889? – 1r – 1 – us UF Libraries [010]

Alcohol Information and Referral Center *see* Vida, luces y sombras

Alcohol research & health – Washington DC 1999+ – 1,5,9 – (cont: alcohol health & research world) – ISSN: 1535-7414 – mf#12110,01 – us UMI ProQuest [362]

The alcoholic beverage laws of the district of columbia, rev. to january 1, 1948 / District of Columbia. Laws, Statutes, etc – Washington, Division of Printing and Publications, Govt. of the District of Columbia, 1948 58 p. LL-381 – 1 – us L of C Photodup [348]

Alcoholism : clinical and experimental research – v5-20. 1981-96 – 16r – 1,5,6,9 – $95.00r – us Lippincott [360]

Alcoholism – Zagreb, Croatia 1973+ – 1,5,9 – ISSN: 0002-502X – mf#7237 – us UMI ProQuest [616]

Alcoholism & addiction – 1988 – 1,5,9 – (cont: alcoholism & addiction magazine; cont by: alcoholism & addiction & recovery life) – ISSN: 0899-8043 – mf#16861.05 – us UMI ProQuest [362]

Alcoholism & addiction magazine – 1987 – 1,5,9 – (cont by: alcoholism & addiction) – ISSN: 0884-1403 – mf#16861.05 – us UMI ProQuest [362]

Alcoholism & addiction & recovery life – 1989 – 1,5,9 – (cont: alcoholism & addiction; cont by: addiction & recovery) – ISSN: 1053-3923 – mf#16861.05 – us UMI ProQuest [362]

Alcoholism and drug dependence / South Africa. Department of Health [Departement van Gesondheid] – [Pretoria: Dept of Health 1975?] [mf ed Pretoria, RSA: State Library [199-]] – 12p [ill] on 1r with other items – 5 – mf#op 06440 r23 – us CRL [362]

Alcoholism treatment quarterly : the practitioner's quarterly for individual, group, and family therapy / ed by McGovern, Thomas F – mf#0734-7324 – us Haworth [360]

Alcollarin. Ayuntamiento *see* Fiestas en honor de los emigrantes. alcollarin, 7 y 8 agosto 1971

Alconetar / Sanchez Loro, Domingo – Caceres: Garcia Floriano, 1947 – 1 – sp Bibl Santa Ana [946]

Alcool, alcoolisme, milieu de travail : recherche effectuee en 1978 aupres de diverses industries de la province de quebec / Gauthier, Yves & Dorman, Alain – [Quebec]: Sobriete du Canada, 1978 [mf ed 1996] – 2mf – 9 – mf#SEM105P2655 – cn Bibl Nat [362]

Alcool et alcoolisme (causeries sur l'intemperance) / Rousseau, Edmond – 2nd ed. Quebec: la Cie de publication "Le Soleil", 1906 [mf ed 1992] – 3mf – 9 – (compositions inedites de ludger larose; lettre-preface de louis-nazaire begin; lettre-preface du delphis brochu) – mf#SEM105P1670 – cn Bibl Nat [360]

Alcorani seu legis mahometi et evangelistarum cocordiae liber, in quo de calamitatibus orbi christiano imminentibus tractatur / Postel, G – Parisiis, 1543 – 2mf – 9 – mf#H-8269 – ne IDC [956]

Alcott, Amos Bronson *see* Essays on education together with the town reports for 1859-1861

Alcott, Louisa May *see*
- Little women

Alcott, William Andrus *see*
- The beloved physician
- Letters to a sister

Alcts newsletter – Chicago IL 1990-98 – 1,5,9 – (cont: rtsd newsletter) – ISSN: 1047-949X – mf#17556 – us UMI ProQuest [020]

Alcuesca R. Ayuntamiento *see* Grandes fiestas en honor de la virgen del rosario, 1977

Alcuescar *see* Comision de monumentos, antiguedades romanas

Alcuescar. Ayuntamiento *see*
- Fiestas en honor de la santisima virgen del rosario, 1971
- Grandes fiestas en honor de la santisima virgen del rosario. 1974

Alcuin and the rise of the christian schools / West, Andrew Fleming – New York: Scribner, 1892 [mf ed 1990] – (= ser The great educators) – 1mf – 9 – 0-7905-6797-0 – (incl bibl ref) – mf#1988-2797 – us ATLA [377]

Alcuin club collection – v1-40. 1899-1958 – 7r – 1 – £399 / $798 – (r1: v1-9; r2: v10-12; r3: v13-15; r4: v16-21; r5: v22-30; r6: v31-36; r7: v37-40; mf vols also listed individually) – mf#96818 – uk Microform Academic [242]

Alcuin of york : lectures / Browne, George Forrest – London: SPCK; New York: ES Gorham, 1908 [mf ed 1989] – 1mf – 9 – 0-7905-4190-4 – mf#1988-0190 – us ATLA [241]

Alcuin und sein jahrhundert : ein beitrag zur christlich-theologischen literargeschichte / Werner, Karl – Paderborn: Ferdinand Schoeningh, 1876 [mf ed 1986] – 1mf – 9 – 0-8370-7037-6 – (incl bibl ref & ind) – mf#1986-1037 – us ATLA [241]

Alcune lettere latine del suddetto padre toccanti l'istesse materie / [Grueber, J] – Firenze, 1697 – 1mf – 9 – mf#HT-561 – ne IDC [910]

Al-curwa al-wuthqa / ed by al-Afghani, Jamal ad-Din & Abduh, Muhammad – Rajab, 1346 – 6mf – 9 – mf#NE-20321 – ne IDC [956]

Alcvne lettere delle cose del giappone – Roma, 1584 – 2mf – 9 – mf#H-8364 – ne IDC [956]

L'alcyon – Rio de Janeiro, RJ: Imprimerie de Cremiere, 20 mar 1841 – (= ser Ps 19) – mf#P14,04,24 n01 – bl Biblioteca [079]

Al-da'ayah : offical journal of the general organization for the prohibition of alcoholic beverages – Cairo: Muhammad 'Abd-al-mun'im Ibrahim al-Muhami. yr 1 n1-6. 15 jul-15 dec 1944 – (= ser Arabic journals and popular press) – 1r – 1 – $200.00 – us MEDOC [956]

Aldabadas / Juez Nieto, Antonio – Badajoz: Tip. Clasica, 1944 – sp Bibl Santa Ana [946]

Al-dajaj – Cairo: Ahmad Zaki Abu Shadi. v1 n1-v2 n12. jan 1932-dec 1933 [complete] – (= ser Arabic journals and popular press) – 1r – 1 – $300.00 – us MEDOC [956]

al-Dajjani, Ahmad Sidqi *see* Masirat al-shab al-filastini wa-afaq al-sira al-arabi al-israili fi al-thamaninat

Aldama, J A *see* Morillo, santiago. las iglesias cristianas de oriente. texto de teologia oriental. granada, 1946

Aldama, Miguel De *see* Facts about cuba

Aldana, Abelardo *see* Chile and the chilians

Aldana, Cosme de *see* Ottavas y canciones espirituales

Aldana, Francisco de *see* Epistolario poetico completo

Alday, Francois *see* Trois quatuors pour deux violons, alto et violoncelle, oeuvre 8 [no 1]

Alday y Aspee, Manuel de *see*
- Oracion
- Oracion que el ill[ustrissi]mo senor d d manuel de alday y aspee, del consejo de s m obispo de santiago de chile

Aldea de Trujillo. Ayuntamiento *see* Ferias y fiestas en honor de san isidro labrador. mayo de 1951

Aldeacentenera *see* Sociedad de cazadores de aldeacentenera. reglamento

Aldeanueva del Camino. Ayuntamiento *see* Ordenanzas municipales

Aldeburgh corporation letter books, 1625-63 – 1r – 1 – mf#65864 – uk Microform Academic [941]

Aldef *see* Europeo en el tropico

Alden, George J *see* Florida

Den alderheijlichsten naem voor een nieu-jaer-gift geschoncken... / Poirters, Adrianus – Antwerpen: By de weduwe ende erfgenahmen van Ian Cnobbart, 1647 – 3mf – 9 – mf#O-405 – ne IDC [090]

Alderley and wilmslow advertiser and general repertory of news from chelford to cheadle – [North West] Accrington, Wilmslow Lib 7 aug 1874-23 sep 1876 – 1 – (title change: alderley & wilmslow advertiser & knutsford gazette & general repertory of news from sandbach to cheadle [mobberley, styal, lindow, handforth, gatley, morley, holmes chapel & chelford]; alderley & wilmslow advertiser & east & mid-cheshire gazette [16 dec 1876-30 jun 1883]; alderley & wilmslow advertiser & knutsford gazette...chelford & goostrey [30 sep-9 dec 1876]; alderley & wilmslow advertiser & guardian, east & mid-cheshire gazette [jul 1883-13 mar 1942]; alderley & wilmslow & sandbach advertiser [20 mar 1942-26 nov 1943]; alderley & wilmslow & knutsford advertiser [3 dec 1943-jul 1981]; alderley & wilmslow advertiser [aug 1981-1987]) – uk MLA; uk Newsplan [072]

Alderley and wilmslow advertiser and general repertory of news from chelford to cheadle – [NW England] Altrincham aug 1981-1985 – 1 – (title change: wilmslow express advertiser) – uk MLA; uk Newsplan [072]

Alderley edge cemetery – [Macclesfield Ferrets] – (= ser Cheshire monumental inscriptions) – 4mf – 9 – £5.00 – mf#396 – uk CheshireFHS [929]

Alderley, st mary : burials 1600-1942 – [North Cheshire FHS] – (= ser Cheshire church registers) – 7mf – 9 – £8.00 – mf#256 – uk CheshireFHS [929]

Alderley, st mary – [North Cheshire FHS] – (= ser Cheshire monumental inscriptions) – 2mf – 9 – £2.75 – mf#116 – uk CheshireFHS [929]

Aldershot and district town crier – Aldershot, England ns: sep-dec 1932, jun 1933-jul 1939 – 1 – uk British Libr Newspaper [072]

Aldershot camp gazette – Aldershot, England 24 jul 1880-26 mar 1881 – 1 – (incorp with: hants & surrey times) – uk British Libr Newspaper [072]

Aldershot courier – Aldershot, England 7 jan-28 oct 1987, 6 jan 1988- – 1 – (wanting: feb, mar, jul 1987) – uk British Libr Newspaper [072]

Aldershot military gazette – Aldershot, England 6 aug 1859-23 jun 1860 – 1 – (cont by: sheldrake's aldershot military gazette [30 jun 1860-26 sep 1913]) – uk British Libr Newspaper [355]

Aldershot news – Aldershot, England 23 jun 1894- [mf 1910, 1977-] – 1 – (wanting: 1897; fr 21 oct 1969-25 feb 1972 tues & fri edns entitled: aldershot midweek news & aldershot weekend news respectively; fr 25 jul 1972-2 apr 1985 tue ed entitled: aldershot news & mail; fr 16 apr 1985 tue ed entitled: aldershot mail) – uk British Libr Newspaper [072]

Aldershot review and farnborough observer – Aldershot, England 2 jun 1898-13 jul 1899 [mf 1899] – 1 – uk British Libr Newspaper [072]
Alderson, Edward H *see* Barnewell and alderson's reports
Aldford, st john the baptist – (= ser Cheshire monumental inscriptions) – 1 – 9 – £2.50 – mf#1a – uk CheshireFHS [929]
Al-difac – Jaffa, Jerusalem, 1934-1966 – 57r – 1 – (missing: 1934(jan-mar); 1935(dec); 1939(jan-aug); 1948(may)-1949(feb.); 1952(sep)-1955(jan)) – mf#J-93-2 – ne IDC [956]
Al-diffah al-gharbiyah – Chicago: The West Bank Publ, 1993 – 1 – (filmed by the university of chicago library photodup laboratory for the middle eastern microfilm project at the center for research libraries) – us CRL [071]
Al-dimuqrati – [Khartoum?]: Dar al-Arabi lil-Tabaah wa-al-Nashr, nov 28 1985; apr 28-dec 17 1987 – 1r – us CRL [956]
al-Din al-Khayru, Izz *see* Al-atma al-sihyuniyah fi miyah al-urdun wa-al-litani
Aldine : the art journal of america – La Salle IL 1870-79 – 1 – mf#5201 – us UMI ProQuest [700]
Aldinger, P *see* Die neubesetzung der deutschen bistuemer unter papst innocenz 4. 1243-1254
Al-Djami'ah *see* Institut agama islam negeri al-djami'ah
Al-djazair – Alger. n1-19, dec 1904-avr 1905. mq. n3-4, 12 – 1 – fr ACRPP [073]
Aldousari, Badi *see* The history and philosophy of sport in islam
Aldre biskotsel i sverige och danmar / Sandklef, Albert – Goteborg: Elanders Boktryckeri 1937 [mf ed Bloomington IN: Indiana Uni Lib, Preservation Dept 1984] – 1r – 1 – us Indiana Preservation [390]
Aldred, Guy A *see*
– Commune
– Council
– News from spain
Aldred, Guy Alfred *see* Regeneracion!
Aldredge, Robert Croom *see* Weather observers and observations at charleston, south carolina, 1670-1871
Aldrete y Soto, L *see*
– Crisol de la verdad ilustrado con divinas y humanas letras, padres y doctores...
– Defensa de la astrologia y conjeturas
– Discurso del cometa del ano 1680
– Luz de medicina y respuesta a las objeciones puestas a la universal
– La verdad acrisolada en las letras divinas y humanas...respondiendo al auto...
Aldrich, Annie Charlotte Catharine *see*
– Daisy beresford
– The future marquis
– A maid called barbara
Aldrich, Bertha *see* Florida sea shells
Aldrich, Edgar *see* Trusts and monopolies
Aldrich family – 1982 winter-1984 fall; 1987 spring-1989 fall – 1 – mf#1209838 – us WHS [929]
Aldrich, Jeremiah Knight *see* A critical examination of the question in regard to the time of our saviour's crucifixion
Aldrich, Nelson W *see* Papers
Aldrich, Thomas Bailey *see* Poems of thomas bailey aldrich
Ale ksovim / Ettinger, Solomon – Wilno, Lithuania. 1925 – 1r – 1 – us UF Libraries [939]
Aleander und luther auf dem reichstage zu worms : ein beitrag zur reformationsgeschichte / Hausrath, Adolf – Berlin: G Grote, 1897 [mf ed 1990] – 1mf – 9 – 0-7905-6106-9 – mf#1988-2106 – us ATLA [242]
Aleandro, Girolamo *see* Die depeschen des nuntius aleander vom wormser reichstage 1521
Aleantara. Ayuntamiento *see* Ferias y fiestas de primavera 1972
Alegacion al derecho...duque de medinaceli / Santos Cuenda, Juan y Consortes – 1877 – 9 – sp Bibl Santa Ana [946]
Alegacion en derecho de don juan de arguello sobre su prision – 1736? Incompleto – 9 – sp Bibl Santa Ana [946]
Alegacion en derecho...virgen del puerto... plasencia / Garcia Mora, Jose – 1892 – 9 – sp Bibl Santa Ana [240]
Alegacion...dehesa de la serena...pleito...marques de perales / Torres, Manuel de – 1796 – 9 – sp Bibl Santa Ana [946]
Alegaciones en favor del clero, estado eclesiastico i secular, espanoles e indios del obispado de la puebla de los angeles : sobre las doctrinas que en execucion del s concilio de trento, cedulas... – [Puebla: s.n. 1648?] – (= ser Books on religion...1543/44-c1800: obispos: obispos de puebla) – 7mf – 9 – mf#crl-397 – ne IDC [241]
Alegacoes da camara municipal de lourenco marques na accao... / Seica, Serafim Gomes De – Lourenco Marques, Mozambique. 1917 – 1r – 1 – us UF Libraries [960]
Alegato contra una grave falta de insensibilidad historica / Spain. Servicio Espanol de Informacion – [Madrid: El Servicio 1936] [mf ed 1977] – (= ser Blodgett coll) – 9 – mf#w708 – us Harvard [327]
Alegato de buena prueba presentado / Cabrera de la Rocha, Juan – 1841 – 9 – sp Bibl Santa Ana [946]
Alegato persentado (sic) a nombre de... / Luna Y Parra, Jose – Habana, Cuba. 1875 – 1r – 1 – us UF Libraries [972]
Alegato presentado a nombre de la.... / Compania De Caminos De Hierro De La Habana – Habana, Cuba. 1875 – 1r – 1 – us UF Libraries [972]
Le alegrezze fatte in venetia per la miracolosa vittoria ottenuta dalla santissima liga... 1571 – [1571] – 1mf – 9 – mf#H-8179 – ne IDC [956]
Alegria, Claribel *see* Tres cuentos
Alegria de andar / Zamacois, Eduardo – Madrid, Spain. 1930 – 1r – 1 – us UF Libraries [972]
Alegria de proteo / Buesa, Jose Angel – Habana, Cuba. 1948 – 1r – 1 – us UF Libraries [972]
Alegria, Jose S *see*
– Cartas a florinda
– Cincuenta anos de literatura puertorriquena
– Retablos de la aldea
– Rosas y flechas
Alejandro-De Leon, Daniel *see* Comparison of skinfold measurements under normally hydrated and dehydrated conditions in females ages to 54
Alekanndr kuprin / Boraisha, Menahem – New York, USA. 1919 – 1r – 1 – us UF Libraries [939]
Aleksander Debski : zycie i dzialalnosc, 1857-1935 / Barlicki, Norbert – Warszawa: Stowarzyszenie B Wiezniow Politycznych, 1937 (mf ed 19–) – (= ser Harvard Slavic humanities preservation microfilm project) – xiv/293p – (incl ind) – mf#ZQ-188 – us NY Public [920]
Aleksandr Mikhailovich *see* Religion of love
Aleksandrov *see* Oborona strany
Aleksandrov, A *see* Polnyi angliisko-russkii slovar'
Aleksandrov, M *see* Sbornik zakonodatelnykh materialov, instruktsii i raziasnenii po kooperatsii invalidov
Aleksandrovskii, Iu V *see* Polozhenie o gorodskikh obshchestvennykh bankakh
Alekseev, A M *see* Optimalnoe perspektivnoe planirovanie v otrasliakh promyshlennogo proizvodstva
Alekseev, A S *see* Manifest 17 oktiabria 1905 g i politicheskoe dvizhenie, ego vyzvavshee
Alekseev, M P *see* Ocherki istorii ispano-russkikh literaturnykh otnoshenii
Aleksii *see*
– Kitaiskaia biblioteka i uchenye trudy chlenov imp rossiiskoi dukhovnoi i diplomaticheskoi missii v g pekine...
– O vozrozhdenii kreshcheniem"
Alele, Joseph *see* Solomon islands diaries
Alem – Istanbul: Bekir Efendi Matbaasi, 1908-09. Sahib-i Imtiyaz ve Mueduerue: Yakovalizade Arif. n1-12. 29 kanunisani 1324-21 mayis 1325 [1908-09] – (= ser O & t journals) – 5mf – 9 – $75.00 – us MEDOC [956]
O alem parayba : folha dedicada aos interesses sociaes – Alem Paraiba, MG: Typ do Alem Parayba, 12 maio 1881; 24 jan 1886 – (= ser Ps 19) – mf#P11B,03,88 – bl Biblioteca [079]
Aleman Bolanos, Gustavo *see* Centenario de la guerra nacional de nicaragua cont...
Aleman Y Martin, Ricardo M *see* Sociedades mercantiles en el derecho vigente
Alemania y el mundo ibero-americano / Ibero-Amerikanisches Institut – Berlin, Germany. 1939 – 1r – 1 – us UF Libraries [972]
Der alemanne – Freiburg Br DE, 1931 11 nov-1945 20 apr [gaps] – 1 – (regional ed available) – gw Misc Inst [074]
Alemannia zeitschrift fuer sprache, litteratur und volkskunde des Elsasses und Oberrheins / ed by Birlinger, Anton – Bonn 1875-92 – 64mf – 9 – diazo €198.00 silver €238.00 silver – gw Olms [430]
Alemannische heimat – Freiburg Br DE, 1934 21 jan-1940 27/28 jan – 1 – gw Misc Inst [074]
Alemany Bolufer, Jose *see* Estudio elemental de gramatica historica de la lengua castellana
Alemar, Luis E *see* Santa domingo, ciudad trujillo
Alembert, J le Rond d' *see* Encyclopedie
Alembert, Jean Le Rond d' *see*
– Elemens de musique theorique et pratique
– Systematische einleitung in die musikalische setzkunst
Alem-i nisvan – Bakhchisarai, Ukraine 1906-07 [mf ed Norman Ross] – 1r – 1 – mf#nrp-2220 – us UMI ProQuest [077]
Alencar Araripe, Tristao De *see* Historia da provincia do ceara
Alencar, Jose Martiniano de *see*
– Iracema
– Minas de prata
– Paginas avulsas
– Senhora
– Sonhos de ouro
– Til
– Tronco do ipe
Aleppo *see* The travels of macarius, patriarch of antioch
Aler, Jan *see* Im spiegel der form
Alero / Maderal, Luis – Habana, Cuba. 1957 – 1r – 1 – us UF Libraries [972]
Alert : amherst edition – Amherst, NE: Epley & Krewson. 1v. may 1897-v1 n34. dec 15 1897 (wkly) – 1r – 1 – (cont by: miller gazette) – us Bell [071]
Alert – Maryborough, apr 1902-apr 1919 – 1r – A$33.40 vesicular A$38.90 silver – at Pascoe [079]
Alert : miller edition – Amherst, NE: [Epley & Krewson] 1v. may 1897-v1 n34. dec 17 1897 (wkly) – 1r – 1 – (cont by: miller gazette) – us Bell [071]
Alert! – 1984-v3 n3 [1985 jul/aug] – 1 – mf#1440997 – us WHS [071]
Alert! magazine – Rocky Hill IL 1992+ – 1,5,9 – mf#19855 – us UMI ProQuest [650]
Alerta – 1981 feb-1982 jul – 1 – mf#656583 – us WHS [071]
Alerta – Miami, FL. 1970 feb 19-1978 feb 24 – 9r – 1 – (gaps) – us UF Libraries [071]
Alerta – Guatemala [s.n.] ano1- n1- 1963- (daily ex mon) oct 1 1972- (wkly) [mf ed 197?-1982] – 16r – 1 – mf#2573 – us Wisconsin U Libr [079]
Alerta – Havana, Cuba. 9 apr-3 sep 1945 [gaps] – 1r – 1 – uk British Libr Newspaper [079]
Alerta – Rio Piedras, PR: American Federation Govt Employees, Local n2408 -1982 (San Juan, PR) (mthly) [mf ed 1991] – 3v on 1r – 1 – (text in spanish) – mf#P83-1510 n83-448 – us Wisconsin U Libr [331]
L'alerte : l'hebdomadaire de la renovation francaise – Nice, France 24 sep 1940-6 dec 1941, 3 oct 1942-2 oct 1943 (imperfect) – 2r – 1 – uk British Libr Newspaper [072]
L'alerte – no. 29-47. Lyon. avr 1936-37 – 1 – fr ACRPP [073]
Alerte : journal hebdomadaire independant – Coquilhatville: L Ilufa. [dec 9 1961-apr 20 1963] (wkly) – 1r – 1 – us CRL [079]
Ales, A D *see* Etude sur les origines de la penitence chretienne
Ales, Adhemar d' *see*
– L'edit de calliste
– La theologie de tertullien
Alessandri, Felice *see* Six sonatas for two violins and a thorough bass for the harpsichord
Alessandro nell'indie : vergleichende studien zu melodiebildung und affekt in der opernarie an der schwelle zur klassik / Holzbauer, Martin & Kraehe, Ignaz – (mf ed 2000) – 4mf – 9 – €56.00 – 3-8267-2737-1 – mf#DHS 2737 – gw Frankfurter [780]
Aletas de tiburon / Serpa, Enrique – Habana, Cuba. 1963 – 1r – 1 – us UF Libraries [972]
Aletazos dominicanos / Manon, Dario A – Mexico City? Mexico. 1936 – 1r – 1 – us UF Libraries [972]
Alethian critic : or error exposed – Ann Arbor MI 1804-06 – 1 – mf#5205 – us UMI ProQuest [240]
Alewijn, A] *see* Boertige en ernstige minnezangen
Alewyn, A. *see* Vermeerderde zede en harpgezangen
Alex wiley's newsletter – 1950 dec 14-1962 dec 25 – 1 – mf#1051818 – us WHS [071]
Alexander : drama / Baumann, Hans – Jena: E Diederichs, c1941 [mf ed 1989] – 157p – 1 – mf#6983 – us Wisconsin U Libr [820]
Alexander / Etzenbach, Ulrich von; ed by Toischer, Wendelin – Stuttgart: Litterarischer Verein, 1888 (Tuebingen: H Laupp) [mf ed 1993] – (= ser Blvs 183) – xxii/870p – 1 – (middle high german text. int in german) – mf#8470 reel 38 – us Wisconsin U Libr [810]
Alexander : ein hoefischer versroman des 13. jahrhunderts / Ems, Rudolf von; ed by Junk, Victor – Leipzig: K W Hiersemann, 1928-29 [mf ed 1993] – (= ser Blvs 272, 274) – 2v – 1 – mf#8470 reels 54-55 – us Wisconsin U Libr [830]
Alexander, Alfonso *see* Sandino
Alexander, Archibald *see*
– Biographical sketches of the founder and principal alumni of the log college
– A brief compend of bible truth
– A history of colonization on the western coast of africa
– Theories of the will in the history of philosophy
– A theory of conduct
Alexander, Archibald Browning Drysdale *see*
– The canon of the old and new testaments ascertained
– Christianity and ethics
– The ethics of st. paul
– Evidences of the authenticity, inspiration, and canonical authority of the holy scriptures
– A history of the israelitish nation
– Outlines of moral science
– Practical sermons
– Practical truths
– Some problems of philosophy
– Theories of the will in the history of philosophy
– Thoughts on religious experience
– Universalism false and unscriptural
Alexander campbell : leader of the great reformation of the 19th century / Grafton, Thomas William – St Louis: Christian Publ, 1897 [mf ed 1990] – 1mf – 9 – 0-7905-5835-1 – (int by herbert l willett) – mf#1988-1835 – us ATLA [242]
Alexander campbell and christian liberty : a centennial volume on his controlling ideas, enforced by his own words / Egbert, James – centennial ed, 1809-1909. St Louis: Christian Publ Co, 1909 [mf ed 1990] – 1mf – 9 – 0-7905-7931-6 – (incl bibl ref) – mf#1989-1156 – us ATLA [240]
Alexander campbell and the general convention : a history of the rise of organization among the disciples of christ / Moore, Allen Rice – St Louis, MO: Christian Board of Publ, c1914 [mf ed 1992] – (= ser Christian church (disciples of christ) coll) – 1mf – 9 – 0-524-02263-1 – mf#1990-4270 – us ATLA [242]
Alexander campbell as a preacher : a study / McLean, Archibald – New York: Fleming H Revell, c1908 [mf ed 1993] – (= ser Christian church (disciples of christ) coll) – 1mf – 9 – 0-524-06270-6 – mf#1991-2461 – us ATLA [240]
Alexander campbell's theology : its sources and historical setting / Garrison, Winfred Ernest – St Louis: Christian Publ Co, 1900 [mf ed 1990] – 1mf – 9 – 0-7905-4964-6 – mf#1988-0964 – us ATLA [242]
Alexander campbell's tour in scotland : how he is remembered by those who saw him then / Chalmers, Thomas – Louisville, KY: Guide Printing, 1892 [mf ed 1993] – (= ser Christian church (disciples of christ) coll) – 2mf – 9 – 0-524-07859-9 – mf#1991-3404 – us ATLA [242]
Alexander, Charles Beatty *see* Notes on the new york law of life insurance
Alexander crummell collection : from the holdings of the schomburg center for research in black culture, manuscripts, archives and rare books division: the new york public library, astor, lenox and tilden foundations – 1995 – 10r – 1 – $850.00 – (guide which covers all coll under "civil rights advocates" sold separately for $20 d3305.g3) – mf#D3305P11 – Dist. us Scholarly Res – us L of C Photodup [240]
Alexander, Disney *see* Church of christ and sunday school extension
Alexander, F J *see* In the hours of meditation
Alexander, Gilchrist G *see*
– Lao-tsze, the great thinker
– The temple of the nineties
Alexander, Gross *see*
– The epistles to the colossians and to the ephesians
– Ritual of the methodist episcopal church, south
– The son of man
Alexander, Gross et al *see* A history of the methodist church, south, the united presbyterian church, the cumberland presbyterian church, and the presbyterian church, south, in the united states
Alexander harper family papers *see* Family papers ms 3231
Alexander, Henry Carrington *see* The life of joseph addison alexander, d.d
Alexander, Henry Templer *see* African tightrope
Alexander Hepple archive, 1940-1963 – Chicago, IL: University of Chicago, Photoduplication Dept, 1968 – 1r – 1 – us CRL [025]
Alexander heriot mackonochie : a memoir / Towle, Eleanor A; ed by Russell, Edward Francis – 2nd ed. New York: E & J B Young, 1890 [mf ed 1990] – 1mf – 9 – 0-7905-6129-8 – mf#1988-2129 – us ATLA [920]
Alexander, Horace Gundry *see*
– India since cripps
– The indian ferment
– New citizens of india
Alexander, J A *see* The psalms translated and explained
Alexander, James E *see*
– Transatlantic sketches
– Travels to the seat of war in the east, through russia and crimea, in 1829
Alexander, James McKinney *see* The islands of the pacific
Alexander, James Russell *see* Route planning algorithm for a network with dynamic topology
Alexander, James W *see* Forty years' familiar letters of james w alexander
Alexander, James Waddel *see*
– Consolation
– Faith
– Forty years' familiar letters of james w. alexander, d.d
– A geography of the bible
– The life of archibald alexander
– The missionary offering
Alexander, James Waddel et al *see* The new york pulpit in the revival of 1858
Alexander, Jeffery L *see* Validity of a single-stage submaximal treadmill walking test for predicting vo 2 max in college students
Alexander, John *see* Reasons for becoming a baptist

Alexander, Joseph Addison *see*
- The acts of the apostles
- The earlier prophecies of isaiah
- Essays on the primitive church offices
- A geography of the bible
- The gospel according to mark
- The gospel according to matthew
- Isaiah
- The later prophecies of isaiah
- Notes on new testament literature and ecclesiastical history
- The psalms

Alexander, Levy *see* Alexander's hebrew ritual...

Alexander, Lindsay *see* Jesus, the source of spiritual blessing to men

Alexander mackenzie's reise nach dem noerdlichen eismeere : vom 3. jun bis 12. sep 1798 / Mackenzie, Alexander – Weimar 1802 [mf ed Hildesheim 1995-98] – 1v on 1mf – 9 – €40.00 – 3-487-26601-6 – gw Olms [919]

Alexander, Magnus Washington [comp] *see* Cost of health supervision in industry

Alexander minorita (mgh quellen..:1.bd) : expositio in apocalypsim – 1955 – (= ser Monumenta germaniae historica. quellen zur geistesgeschichte des mittelalters (mgh quellen...)) – €21.00 – ne Slangenburg [931]

Alexander Murray / Bell, Robert – [S.l: s.n, 1892?] [mf ed 1979] – 1mf – 9 – 0-665-00100-2 – (repr fr: canadian record of science) – mf#00100 – cn Canadiana [500]

Alexander, P Y *see*
- Darwin and darwinism pure and mixed
- More loose links in the darwinian armour

Alexander, Padinjarethalakal Cherian *see*
- Buddhism in kerala
- The dutch in malabar

Alexander pierre tureaud papers – 1909-72 – ca 58r – 1 – ca $7540.00 – (from the coll of the amistad research center. guide also sold separately $40 s3518.g) – mf#S3518 – us Scholarly Res [305]

Alexander, Robert Jackson *see* Venezuelan democratic revolution

Alexander, Robert L *see* The architecture of russell warren

Alexander, Samuel Davies *see* The presbytery of new york, 1738 to 1888

Alexander, Sarah *see* A voice from the wilderness

Alexander, Shana *see* Mkazi wokamba nkhani

Alexander, Sidney Arthur *see* The christianity of st paul

Alexander, Thomas P *see* Diary

Alexander turnbull library biographies index – [Wellington, NZ]: National Library of New Zealand, 1995 – 271mf+1bk – 9 – (with guide; completed over last 75yrs, ind contains c200,000 cards, giving references to about 150,000 individuals from the new zealand and the pacific) – nz Nat Libr [980]

Alexander turnbull library catalogue / National Library of New Zealand – [Wellington, NZ]: The Library, oct 1984- – 9 – mf#0112-3467 – nz Nat Libr [017]

Alexander und gilgamos / Meissner, Bruno – Leipzig: Eduard Pfeiffer, [19-?] [mf ed 1986] – 1mf – 9 – 0-8370-7088-0 – (incl bibl ref) – mf#1986-1088 – us ATLA [230]

Alexander varian jr letters *see* Varian, alexander, jr. letters, ms 3141

Alexander viets griswold allen, 1841-1908 / Slattery, Charles Lewis – New York: Longmans, Green, 1911 [mf ed 1991] – 1mf – 9 – 0-524-01015-3 – mf#1990-0292 – us ATLA [242]

Alexander, William *see*
- Costumes et vues de la chine
- The divinity of our lord
- The epistles of st john
- Invitation to sinners to escape from coming wrath...
- The leading ideas of the gospels
- The life insurance company
- Primary convictions
- St paul at athens
- Verbum crucis
- Vues de la chine et de la tartarie
- The witness of the psalms to christ and christianity

Alexander, William Addison *see* A digest of the acts and proceedings of the general assembly of the presbyterian church in the united states

Alexander, William Lindsay *see*
- Anglo-catholicism not apostolical
- Christ and christianity
- Cyclopaedia of biblical literature
- Good man
- Logical analysis of the epistle of paul to the romans
- Look to the end
- A system of biblical theology
- Zechariah

Alexander, William Menzies *see* Demonic possession in the new testament

Alexander-Armstrong, J *see* Songs of the new world

Alexander's hebrew ritual... / Alexander, Levy – London, England. 1819 – 1r – 1 – us UF Libraries [240]

Alexander's magazine – Boston. v1-7. 1905-09 [all publ] – (= ser Black journals, series 1) – 29mf – 9 – $280.00 – us UPA [305]

Alexander's magazine – Killen TX 1905-09 – 1 – mf#3341 – us UMI ProQuest [976]

Alexandra herald – 1902-31; 1933-39 – 36r – 1 – mf#83.6 – nz Nat Libr [079]

Alexandre, Charles *see* Dictionnaire francais-grec...

Alexandre, J B H *see* Patrie et les conspirations

Alexandre, Pierre *see*
- Europe et jupiter, concert francois a deux voix
- Langues et langage en afrique noire

Alexandre, Pierre C *see* A propos de la legislation sur l'inventaire

Alexandre vinet : histoire de sa vie et de ses ouvrages / Rambert, Eugene – 4e ed. Lausanne: G Bridel, 1912 [mf ed 1991] – 2mf – 9 – 0-7905-9448-X – (pref & notes by ph bridel. 1st printed 1875. incl bibl ref) – mf#1989-2673 – us ATLA [242]

Alexandri neckam de naturis rerum, libri duo (rs34) : with neckam's poem de laudibus divinae sapientiae / ed by Wright, T – 1863 – (= ser The rolls series (rs)) – €21.00 – ne Slangenburg [931]

The alexandria argus – Alexandria, NE: R B Enslow, 1894-v79 n52. dec 28 1972 (wkly) – 26r – 1 – us Bell [071]

Alexandria Black History Resource Center *see* Hayti

Alexandria herald – 1818 dec 4 – 1 – mf#881689 – us WHS [071]

Alexandria herald – Alexandria, NE: F E Matson. v9 n11. jul 15 1892 (wkly) – 1r – 1 – (cont: thayer county herald) – us NE Hist [071]

The alexandria news – Alexandria, NE: Babcock & Abbott (wkly) – 1r – 1 – us Bell [071]

Alexandrian and carthaginian theology contrasted : the hulsean lectures, 1892-93 / Heard, John Bickford – Edinburgh: T & T Clark; New York: Scribner [dist], 1893 [mf ed 1990] – 1mf – 9 – 0-7905-4810-0 – mf#1988-0810 – us ATLA [240]

De alexandrijnsche vertaling van het dodekapropheton / Schuurmans Stekhoven, J Z – Leiden: E J Brill, 1887 [mf ed 1986] – viii/137p on 1mf – 9 – 0-8370-9270-1 – (incl bibl ref and ind) – mf#1986-3270 – us ATLA [221]

Die alexandrinische uebersetzung des buches hosea, heft 1 : ein beitrag zu den septuaginta-studien und der auslegung des propheten hosea / Treitel, Leopold – Karlsruhe: A Bielefeld, 1887 – 1mf – 9 – 0-8370-5568-7 – mf#1985-3568 – us ATLA [221]

Die alexandrinische uebersetzung des buches jesaias : eine rectorsrede / Scholz, Anton – Wuerzburg: Leo Woerl, 1880 – 1mf – 9 – 0-8370-5137-1 – (incl bibl ref) – mf#1985-3137 – us ATLA [221]

Alexiadis libri 15 (cshb38) / Annae Commenae – Bonnae, 1839 – (= ser Corpus scriptorum historiae byzantinae (cshb)) – €18.00 – (graeca ad codd. fidem nunc primum recensuit, novam interpretationem latinam subiecit, c dugangii commentarios suasque annotationes addidit lud schopenus) – ne Slangenburg [243]

Alexiadis libri 15 (cshb49) / Annae Commenae – Bonnae. v2. 1878 – (= ser Corpus scriptorum historiae byzantinae (cshb)) – €29.00 – ne Slangenburg [243]

Alexis : ou, l'erreur d'un bon pere / Dalayrac, Nicolas – Paris, France. 1802 – 1r – 1 – us UF Libraries [440]

Alexis de Barbezieux, pere *see*
- Histoire de la province ecclesiastique d'ottawa et de la colonisation dans la vallee de l'ottawa, vol 1
- Histoire de la province ecclesiastique d'ottawa et de la colonisation dans la vallee de l'ottawa, vol 2
- Histoire de la province ecclesiastique d'ottawa et de la colonisation dans la vallee de l'ottawa, vols 1 and 2
- Sermon du r p alexis
- Sermon sur le socialisme
- Un voyage a la guadeloupe

Alexis, Jacques Stephen *see* Romancero de las estrellas

Alexis, M G *see*
- Le congo belge illustre
- La traite des negres et la croisade africaine, choix raisonne de documents relatifs a la question de lesclavage africain et comprenant la lettre encyclique de leon 13 sur lesclavage

Alexis, pere *see*
- Une ame sacerdotale, le chanoine michel
- Histoire de la province ecclesiastique d'ottawa
- Un voyage a la guadeloupe

Alexis, R P *see* L'etat religieux et politique de la france contemporaine

Alexis, Stephen *see*
- Black liberator
- Introduction a l'instruction economique morale et...
- Negre masque

Alexis, Willibald *see*
- Erinnerungen
- Herbstreise durch scandinavien
- Wanderungen im sueden
- Der werwolf

Alexius afanasevic dmitrievskij : biografische gegevens en zijn liturgische leer vooral over het liturgisch typikon / Aalst, G van – Tilburg, 1956 – €5.00 – ne Slangenburg [920]

Aley, Peter *see* Eduard moerikes kuenstlerisches selbstverstaendnis. im spiegel seiner gedichte "die elemente", "goettliche reminiszenz" und "neue liebe"

Alfabetic order table / Cutter, Charles Ammi – Boston, MA. 1887 – 1r – 1 – us UF Libraries [090]

Alfabeto christiano : which teaches the true way to acquire the light of the holy spirit / Valdes, Juan de – London: Bosworth & Harrison, 1861 [mf ed 1993] – 1mf – 9 – 0-524-08658-3 – (english by benjamin b wiffen) – mf#1993-2118 – us ATLA [240]

Alfabeto de la lengua primitiva de espana : y explicacion de sus mas antiquos monumentos de incripciones y medallas / Erro y Aspiroz, Juan B – Madrid: Imprenta de repulles 1806 – (= ser Whsb) – 4mf – 9 – €50.00 – mf#Hu 459 – gw Fischer [440]

Al-fajr – Jerusalem, 1972-1993 – 66r – 1 – mf#J-93-3 – ne IDC [956]

Al-Falah *see* Dewan pimpinan pusat djam'ijatul muslimin indonesia

Alfalfa herald – Overton, NE: J W Dunaway. -v4 n6. jul 22 1904 (wkly) [mf ed 1902-04 (gaps)] – 3r – 1 – (cont by: overton herald) – us NE Hist [071]

Alfar moruno de badajoz / Melida, Jose Ramon & Fita, Fidel – Madrid: Fortanet, 1912. B.R.A.H. 60. pp. 161-162 – 1 – sp Bibl Santa Ana [946]

Al-farabi (alpharabius), des arabischen philosophen leben und schriften / Steinschneider, M – Spb, 1869 – (= ser Memoires de l'Academie Imperiale des Sciences de St Petersbourg) – 10mf – 8 – (memoires de l'academie imperiale des sciences de st petersbourg, s7 v13) – mf#H-131 – ne IDC [956]

Alfaric, Prosper *see*
- Les ecritures manicheennes
- L'evolution intellectuelle de saint augustin

Alfaro De Jimenez, Isabel *see* Indias y espanolas

Alfaro, Ricardo J *see*
- Commentary on pan american problems
- Costa rica y panama
- Diccionario de anglicismos
- Panorama internacional de america

Alfasi, Yitshak *see*
- Rabi mi-kotsk
- Rishonim le-tsiyon

Al-fatah – Alexandria: Hind Nawfal, 1892-94. yr 1 pts1-12. 1 jumada I 1310-9 ramadan 1311 [20 nov/tishrin 2 1892-16 mar/adhar 1894] [complete] – (= ser Arabic journals and popular press) – 1r – 1 – $300.00 – us MEDOC [956]

Al-fatat – New York, dec 12 1917-nov 15 1918; may 24-31 jun 21-jul 5 1919 – 1r – 1 – us CRL [071]

Al-fatawa al-'alamgiriyah. futawa alemgiri : a collection of opinions and precepts of mohammedan law – Calcutta, Education Press. 6v. 1828-35 – (= ser Fatawa Al-'Alamgiriyah) – 1 – mf#LL-12025 – us L of C Photodup [340]

Al-fateh revolution in ten years – Libya? Libya. 1979 – 1r – 1 – us UF Libraries [090]

Alfavitnyi katalog russkikh knig po matematike, vyshedshikh v rossii s nachala knigopechataniia do poslednego vremeni / Agapov, D V – Orenburg, 1908 – 2mf – 8 – mf#R-7049 – ne IDC [947]

Alfavitnyi spisok / Russia. Tsentral'nyi Komitet Tsenzury Inostrannoi – 1866-69, 1882-92, 1894, 1896-99 – 1 – $69.00 – us L of C Photodup [324]

Alfavitnyi spisok periodicheskikh izdanii rossiiskoi imperii – Spb., 1899 – 2mf – 9 – mf#R-5858 – ne IDC [077]

Al-fawa'id al-sihhiyah – Cairo: Dr Shalhub, 1891-93. yr 1 pt1-yr 2 pt 6. 29 rabi' II 1309-6 dhu al-Qa'dah 1310 [dec 1891-jun 1893] – (= ser Arabic journals and popular press) – 1r – 1 – $200.00 – (ceased publ, resumed in 1902) – us MEDOC [956]

Alfelder zeitung – Alfeld DE, 1987- – 5r/yr – 1 – gw Misc Inst [074]

Alferez real / Palacios, Eustaquio – Bogota, Colombia. 1954 – 1r – 1 – us UF Libraries [972]

Alfero, Giovanni Angelo *see* La prima parte del 'faust' di wolfgang goethe

al-Fil, Muhammad Rashid *see* Al-takhtit al-zirai li-mintaqat al-wafrah

O alfinete : orgao dos interesses da parachia do espirito santo – Rio de Janeiro, RJ: Typ Particular do Espectador, 17 mar-15 abr 1883 – (= ser Ps 19) – mf#P05,04,08 – bl Biblioteca [241]

Alfiyya (quintessence de la grammaire arabe) / Ibn Malik – 1833 – (= ser Royal asiatic society oriental translation fund. old series) – 1r – 1 – mf#400 – uk Microform Academic [470]

Alfonso 2 y lugartenencia infante pedro (anno 1289-1295) – Barcelona – 1r – 5,6 – sp Cultura [946]

Alfonso 4 *see* Majoricarum (anno 1416-1458)

Alfonso 10 el Sabio, Rey de Castilla *see*
- Cantigas de santa maria
- Codigo de las siete partidas del rey d. alfonso el sabio glosadas por el lic. gregorio lopez de tovar
- Quinta partida, com. de gregorio lopez
- El sabio
- Setena partida, com. de gregorio lopez
- Sexta partida, com. de gregorio lopez

Alfonso de Orozco, Beato *see*
- Commentaria quaedam in cantica canticorum
- Declamationes quadragesimales.

Alfonso, Domingo *see*
- Poemas del hombre comun
- Sueno en el papel

Alfonso, Francisco *see* Disputationes in...aristotelis de anima

Alfonso, Manuel F *see* Cuba before the world

Alfonso, Paco *see* Yari-yari, mama olua

Alfonsov, I V *see* Ukazatel k "izvestiiam obshchestva arkheologii, istorii i etnografii pri imperatorskom kazanskom universitete" za 1878-1905 gody

Alfonsus vargas toletanus : und seine theologische einleitungs- lehre / Kuerzinger, J – 1930 – (= ser Bgphma 22/5-6) – €12.00 – ne Slangenburg [100]

Alford 1759-1850 – Oxford MA (mf ed 1994) – (= ser Massachusetts vital record transcripts to 1850) – 2mf – 9 – 0-87623-198-9 – (mf 1t: intentions 1808, 1810, 1820-49; marriages 1803, 1820, 1841-44; births & deaths 1759-1825; deaths 1842-43; births 1839-42. mf 2t: births 1843-50; marriages 1843-50; deaths 1843-50) – us Archive [978]

Alford 1759-1906 – Oxford MA (mf ed 1988) – (= ser Massachusetts vital records) – 12mf – 9 – 0-87623-077-X – (mf 1-4: town records 1773-1835. mf 5-6: town & vital records 1759-1844. mf 7: births 1843-74. mf 8: marriages 1843-75. mf 8: deaths 1843-74. mf 9: deaths 1875-1981. mf 10: deaths 1981-86. mf 11: marriages 1875-1987. mf 12: births 1875-1906) – us Archive [978]

Alford, Bradley Hurt *see* Old testament history and literature

Alford, C *see* Island of tobago, the west indies

Alford, C R *see* Pope's late bull

Alford, Charles Richard *see* Church of rome

Alford, Elizabeth Mary *see*
- The fair maid of taunton
- The romance of coombehurst
- Stanhurst

Alford, Fanny *see* Life, journals and letters of henry alford, d.d., late dean of canterbury

Alford, Gilbert K *see* Chapman chatter

Alford, Henry *see*
- Audi alteram partem
- The book of genesis, and part of the book of exodus
- The consistency of the divine conduct in revealing the doctrines of redemption
- Essays and addresses chiefly on church subjects
- The greek testament
- Homilies on the former part of the acts of the apostles
- How to study the new testament
- Life, journals and letters of henry alford, d.d., late dean of canterbury
- Life of duty
- Meditations in advent

Alford, Loyal Adolphus *see* The mystic numbers of the word

Alford, M W *see* The scriptural doctrine of the trinity

Alford, mablethorpe and sutton-on-sea standard – Boston, England 30 jun 1989-26 feb 1993 [mf 1986-93] – 1 – (discontinued. repl by local edns: alford standard & mabelthorpe & sutton-on-sea standard; cont: lincolnshire standard [12 may-16 jun 1989]) – uk British Libr Newspaper [072]

Alford, Marian Margaret Cust, viscountess *see* Needlework as art

Alfred – Sydney 1835 – 1r – 1 – A$27.50 vesicular A$33.00 silver – at Pascoe [079]

The alfred advocate – Portland, ME: Libby & Smith, nov 4 1915-sep 28 1916 – 1r – 1 – us CRL [071]

The alfred and westminster evening gazette – London. -d. 12 May 1810-31 Dec 1811. (3 reels) – 1 – uk British Libr Newspaper [072]

Alfred Booker (Firm) *see*
- Books and pamphlets
- Catalogue of miscellaneous books by auction

Alfred cheney johnston studio portraits / U.S. Library of Congress. Prints and Photographs Division – Movie stars, theatrical performers of the 1920's and 1930's; studies for advertisements. 245 images. 1 reel. P&P8782 – 1 – us L of C Photodup [770]

Alfred doeblin : im Buch, zu haus, auf der strasse – 1.-3. aufl. Berlin: S Fischer, 1928 [mf ed 1989] – 177p/4pl – 1 – mf#7180 – us Wisconsin U Libr [430]

Alfred hitchcock's mystery magazine – Norwalk CT 1978+ – 1,5,9 – ISSN: 0002-5224 – mf#11961 – us UMI ProQuest [820]

Alfred Laliberte et son oeuvre – [mf ed 1977] – 1r – 1 – mf#SEM35P149 – cn Bibl Nat [730]

Alfred meissner – franz hedrich : geschichte ihres literarischen verhaeltnisses auf grundlage der briefe, die alfred meissner seit dem jahre 1854 bis zu seinem tode 1885 an franz hedrich geschrieben / Hedrich, Franz – Berlin: Otto Janke 1890 [mf ed 1995] – 1r – 1 – mf#3696p – us Wisconsin U Libr [860]

Alfred pellan et son oeuvre – [mf ed 1973] – 4r – 1 – mf#SEM35P1 – cn Bibl Nat [760]

Alfred stevens / Armstrong, Walter – [Paris] 1881 – (= ser 19th c art & architecture) – 1mf – 9 – mf#4.2.588 – uk Chadwyck [750]

Alfred stevens and his work / Stannus, Hugh Hutton – London 1891 – (= ser 19th c art & architecture) – 8mf – 9 – mf#4.2.587 – uk Chadwyck [750]

Alfred the great : a sketch and seven studies / Draper, Warwick Herbert – London: E Stock 1901 [mf ed 1987] – 1r – 1 – (pref by right rev j percival. filmed with: ancestral stories and traditions of great families...of english history / timbs, j & other titles) – mf#1869 – us Wisconsin U Libr [941]

Alfredo, Benjamim *see* The informal sector and its taxation system in mozambique

Alfredo do valle cabral / Rodrigues, Jose Honorio – Rio De janeiro, Brazil. 1954 – 1r – 1 – us UF Libraries [972]

Alfredo victoria, chacal de jacagua / Lugo, Pompilio – Ciudad Trujillo, Dominican Republic. 1946 – 1r – 1 – us UF Libraries [972]

Alfreton journal and east derbyshire advertiser – [East Midlands] Derbyshire 21 jan 1870]-dec 1906 [mf ed 2004] – 11r – 1 – (missing: 1871, 1896-98) – uk Newsplan [072]

Alfriend, Mary Bethell *see* San luis of apalache

Alfwar och skamt – 1841-43 – (aka: sundsvalls tidning) – sw Kungliga [078]

Algae : united states exploring expedition. during the years 1838-1842 under the command of charles wilkes / Bailey, J W & Harvey, W H – New York. 1866-1906 (1) – 2mf – 9 – mf#5466 – ne IDC [910]

Al-garidah – Cairo. Sept 1 1909-Aug 30 1913 – 1 – us NY Public [079]

Algarotti, F *see* An essay on painting...

Al-gaza'ir : revue algerienne d'education sociale – Al Djazaier, Alger. n1-2. oct-nov 1908 – (= ser Gazair) – 1 – fr ACRPP [073]

Algazi, Solomon Nissim *see* Sefer lehem setarim

Algebra and logic – Dordrecht, Netherlands 1968+ – 1,5,9 – ISSN: 0002-5232 – mf#10875 – us UMI ProQuest [510]

Algebraische strukturen in einfachen warteschlangen-netzen / Knaup, Werner – (mf ed 1994) – 3mf – 9 – €49.00 – 3-89349-878-8 – mf#DHS 878 – gw Frankfurter [510]

Algemeen dagblad van nederlandsch indie – 1873-86 – 323mf – 9 – €1060.00 – mf#m180 – ne MMF Publ [079]

Algemeen geillustreerd weekblad / Het Leven – Amsterdam. v1-35. 1906-1940 – 651mf – 9 – mf#H-2034 – ne IDC [700]

Algemeen handelsblad *see* Nieuwe rotterdamsche courant, 1845-1970

Algemeen handelsblad, 1828-1970 – 9807mf – 9 – €28,000.00 set silver (€25,580 set diazo) – (merged with: nieuwe rotterdamsche courant in 1970 to form nrc handelsblad; individual yrly vols) – mf#M400 – ne MMF Publ [074]

Algemeen overzicht van de staatkundige gesteldheid van nederl indie / Politiek verslag 1852, 2 – 12mf – 8 – mf#SD-100 mf 14-25 – ne IDC [959]

Algemeen overzicht van de staatkundige gesteldheid van nederl indie, 1839-1840 – 7mf – 8 – mf#SD-100 mf 1-7 – ne IDC [959]

Algemeen wijsgeerig, geschiedkundig en biographisch woordenboek voor vrijmetselaren – Amsterdam: C L Brinkman [1844?] – 3v on 12mf – 9 – mf#vrl-4 – ne IDC [366]

D'algemeene bouwkunde : volgens d'antyke en hedendaagse manier... / Goeree, W – Amsterdam, 1681 – 3mf – 9 – mf#OA-84 – ne IDC [720]

Algemeene serie / Mededeelingen van het Algemeen Proefstation der AVROS (Sumatra Planters Association) – Batavia, 1948 – 1mf – 9 – mf#SE-844 – ne IDC [959]

Algemene konst- en letter-bode, voor meer-en min-geoefenden : behelzende berigten uit de geleerde waereld, van alle landen... – Haarlem 1788-1860 – 72v on 633mf – 9 – (main cultural magazine for the netherlands 1788-1860, bringing info on all events of significance in dutch science) – mf#8604/1 – ne IDC [500]

A'alger a tombouctou, des rives de la loire aux rives du niger / More, Rene le – Paris, 1913 – 1 – us CRL [960]

Alger, Horatio *see* Helping himself

Alger republicain – Algiers 2 oct 1943-28 jul 1945 (imperfect) – 2r – 1 – uk British Libr Newspaper [079]

Alger sous la domination francaiseson etat present et son avenir / Pichon, Louis A – Paris 1833 [mf ed Hildesheim 1995-98] – 1v on 4mf – 9 – €120.00 – 3-487-25965-6 – gw Olms [960]

Alger tableau du royaume : de la ville d'alger et de ses environs; etat de son commerce, de ses forces de terre et de mer; description des moeurs et des usages du pays; precedes d'une introduction historique sur les differentes / Renaudot – Paris 1830 [mf ed Hildesheim 1995-98] – 1v on 2mf – 9 – €60.00 – 3-487-27349-7 – gw Olms [960]

Alger, William Rounseville *see*
– A critical history of the doctrine of a future life
– The school of life

Alger, William Rounseville et al *see* [Unitarian practical theology]

Algeria *see*
– Al-jaridah al-rasmiyah
– Journal official de la republique algerienne
– Journal officiel
– Moniteur algerien

Algeria. Service Central de Statistique *see* Annuaire statistique de l'algerie 1926-1964

Algeria. Service de la Statistique Generale *see*
– Statistique generale de l'algerie
– Statistique generale de l'algerie 1867-1925

Algeria. Service de la Statistique Generale *see* Annuaire statistique de l'algerie

L'algerie ancienne et moderne : depuis les temps les plus recules jusqu'a nos jours; comprenant le bombardement de tanger, la prise de mogador, la bataille d'isly et le glorieux combat de djemma-gazouat / Galibert, Leon – Paris 1846 [mf ed Hildesheim 1995-98] – 1v on 8mf – 9 – €160.00 – 3-487-27344-6 – gw Olms [960]

L'algerie d'abord – no. 1-2. Alger. juil 1955 – 1 – fr ACRPP [073]

Algerie ouvriere – Alger. 1930-aout 1939 – 1 – fr ACRPP [073]

Algerie-actualite – Algiers, Algeria [mf 1965-66] – 1 – uk British Libr Newspaper [079]

L'algerien en france – no. 1-64. Paris. juil 1950-55 – 1 – fr ACRPP [944]

Algerie-nouvelle – Alger. juil 1946-sep 1955. B.N. Jo. 86525 – 1 – (suite de: la lutte sociale) – fr ACRPP [079]

Algernon champion – Algernon, NE: B Watkins, 1886 (wkly) – 1r – 1 – (cont by: mason city advocate) – us NE Hist [071]

Algernon graham earl of kingsbury / Law, Elizabeth Susan, Baroness Colchester – [London]: printed by Spottiswoode & Co, 1872 – (= ser 19th c women writers) – 4mf – 9 – mf#5.1.28 – uk Chadwyck [830]

Alger-republicain – Alger-republicain. oct 1938-oct 1939; fevr-avr 1940; oct 1943-sept 1955; oct 1962-64 – 1 – fr ACRPP [079]

Al-ghadd – [Khartoum?]: al-Ghadd, apr 19, may 17, jun 7, 1989 – 1r – 1 – us CRL [956]

Al-ghazalah – Cairo: Jufani Zananiri, 1896-98. yr 1 n1-24. 2 jun 1896-1 jun 1897 – (= ser Arabic journals and popular press) – 1r – 1 – $200.00 – us MEDOC [956]

Al-Ghaziri, Bernard Ghobaeira *see* Rome et l'eglise syrienne-maronite d'antioche (517-1531)

Al-Ghazzali *see* The alchemy of happiness

Algiers mission band journal – 1910-19 [mf ed 2001] – (= ser Christianity's encounter with world religions, 1850-1950) – 1r – 1 – (filmed with: story of... 1916-19 [2001-s018]) – mf#2001-s017 – us ATLA [240]

Algo / Suarez Estrada, Jesus Manuel – Santa Clara, Cuba. 1955 – 1r – 1 – us UF Libraries [972]

Algo mas sobre las bulas alejandrinas / Bayle, Constantino – Madrid: Razon y Fe, 1946 – 1 – sp Bibl Santa Ana [946]

Algo pasa en la calle / Quiroga, Elena – Barcelona, Spain. 1960 – 1r – 1 – us UF Libraries [090]

Algo sobre la republica dominicana / Lopez, Nicolas F – Quito, Ecuador. 1948 – 1r – 1 – us UF Libraries [972]

Algo sobre los discipulos y seguidores de zurbaran (1) / Torres Martin, Ramon – Badajoz: Imprenta Diput. Provincial, 1964 – sp Bibl Santa Ana [946]

Algo sobre los discipulos y seguidores de zurbaran (2) / Torres Martin, Ramon – Badajoz: Imprenta Diputacion Provincial, 1965 – sp Bibl Santa Ana [946]

Algol – Brooklyn NY 1973-78 – 1,5,9 – (cont by: starship) – ISSN: 0002-5364 – mf#9671.01 – us UMI ProQuest [400]

The algoma district : and that part of the nipissing district north of the mattawan river, lake nipissing and french river, their resources, agricultural and mining capabilities – [Toronto?: s.n.], 1884 [mf ed 1985] – 2mf – 9 – 0-665-53948-7 – mf#53948 – cn Canadiana [333]

Algoma herald – 1913 nov 20-1914 oct 29; nov5-1918 jan 31 – 1 – mf#958910 – us WHS [071]

The Algoma Land and Colonization Co *see* Algoma! the new Ontario!! the new northwest!!! happy homes and fertile farms! land for the landless! homes for the homeless!

Algoma missionary news – Sault Ste Marie, Ont: [Algoma Missionary Press, 1883?-1956] – 9 – mf#P04377 – cn Canadiana [242]

Algoma missionary news and shingwauk journal – Sault Ste Marie, OT [apr 1 1877-jun 1880] (mthly) – 1r – 1 – Can$110.00 – (publ by the shingwauk home, this paper was devoted to the "civilization, education and christian training of indian children...") – cn McLaren [240]

Algoma press – v1 n1-v4 n51 [1897 oct 6-1901 sep 19] – 1 – mf#914667 – us WHS [071]

Algoma quarterly – [Sault Ste. Marie, Ont: s.n, 1874-1877] [mf ed sep 1 1874-mar 1 1876] – 9 – mf#P04349 – cn Canadiana [242]

Algoma record – 1897 sep 23-1900 jan 26; feb 2-1901 jun 30; jul 5-1902 oct 31; nov 7-1904 apr 22; apr 29-1905 dec 15; dec 22-1907 aug 9; aug 16-1909 mar 26; apr 2-1910 aug 19; aug 26-1912 mar 15; mar 22-1913 oct 24; oct 31-1915 apr 16; apr 23-1916 dec 1; dec 8-1918 feb 1 – 1 – mf#1001408 – us WHS [071]

Algoma record-herald – 1918 feb 1-jul 28; jul 5-1919 dec 26; 1920 jan 2-1921 jun 17; jun 24-1922 nov 3; nov 10-1924 feb 15; feb 22-1925 jul 17; jul 24-1927 jan 14; jan 21-1928 jul 20; jul 27-1930 jan 10; jan 17-1931 aug 14; aug 21-1933 mar 24; mar 31-1934 nov9; nov16-1936 jun 5; jun 12-1937 dec 31; 1938 jan 7-1939 may 5; may 12-dec 29; 1940-43; 1946-1948 may; jun-1950; 1951-62; 1963 jan-1964 jun; jul 1965 dec; 1966 jan 6-sep 15; sep 22-1967 may 25; jun 1-1968 feb 15; feb 22-oct 24; oct 31-1969 jul 17; jul 24-1970 apr 2; apr 9-dec 10; dec 17-1971 aug 12; aug 19-1972 apr 5; apr 12-oct 25; nov-1973 apr; may-oct; nov-1974 apr; 1974 may-nov; dec-1975 jun; jul-1976 jan; feb-aug; sep-1977 mar; apr-nov; dec-1978 may; 1978 jun-dec; 1979 jan-dec; 1980 jan-dec; 1981 jan-dec; 1982 jan-dec; 1983 jan-dec; 1984 jan-dec; 1985 jan-dec; 1986 jan-dec; 1987 jan-dec; 1988 jan-dec; 1989 jan-dec; 1990 jan-dec; 1991 jan-dec; 1992 jan-dec; 1993 jan-dec; 1994 jan-dec; 1995 jan-dec; 1996 jan-dec; 1997 jan-dec; 1998 jan-dec; 1999 jan-dec; 2000 jan-dec – 1 – mf#1001404 – us WHS [071]

Algoma review – Sault Ste Marie, Ont: J E Dudley, [1882?-18– or 19–] [mf ed v1 n3 jul 1 1882] – 9 – mf#P04341 – cn Canadiana [320]

Algoma! the new Ontario!! the new northwest!!! happy homes and fertile farms! land for the landless! homes for the homeless! : algoma farmers testify / The Algoma Land and Colonization Co – 1st ed. [Sault Ste Marie, Ont?: s.n, 1892?] [mf ed 1980] – 1mf – 9 – mf#07082 – cn Canadiana [630]

Algoma unionist – 1977 jul/aug-1982 mar/apr – 1 – mf#626817 – us WHS [071]

Algoma west : its mines, scenery and industrial resources / Roland, Walpole – [Toronto?: s.n.], 1887 [mf ed 1981] – 3mf – 9 – (pt 1: topographical and historical notes: nipigon lake and river to thunder bay; pt 2: history, location and development of our new mines; pt 3: geology of algoma west) – mf#12850 – cn Canadiana [622]

Algonquin indian tales / Young, Egerton Ryerson – New York; Toronto: F H Revell, [1903?] – 4mf – 9 – 0-665-86452-3 – mf#86452 – cn Canadiana [390]

An algonquin maiden : a romance of the early days of upper canada / Adam, Graeme Mercer – London: S Low, Marston, Searle & Rivington, 1887 [mf ed 1982] – 3mf – 9 – (incl: a selection fr the list of books publ by sampson low, marston, searle & rivington) – mf#36079 – cn Canadiana [830]

An algonquin maiden : a romance of the early days of upper canada / Adam, Graeme Mercer – Montreal: J Lovell, Toronto: Williamson, 1887, c1886 [mf ed 1979] – 3mf – 9 – 0-665-00008-1 – mf#00008 – cn Canadiana [830]

Algora y Pontes, Loreto M *see* Naciones elementales de aritmetica

Algumas reflexoes em resposta a reaccao ultramontana em portugal : ou a concordata de 21 de fevereiro por alexandre herculano / Lavradio, Antonio de Almedia Portugal Soares, marques de – Lisboa: Typ de Mathias Jose Marques de Silva, 1859 [mf ed 1995] – (= ser Yale coll) – 86p – 1 – 0-524-10092-6 – (in portuguese) – mf#1995-1092 – us ATLA [241]

Algunas actas capitulares de la provincia de san gabriel al principio del siglo 17 (anos 1601-1608) / Barrado Manzano, Arcangel – Madrid: Archivo Ibero-Americano, 1960 – 1 – sp Bibl Santa Ana [946]

Algunas ideas sobre el engrandecimiento de caceres / Castel, Joaquin – 1898 – 9 – sp Bibl Santa Ana [000]

Algunas paginas del expediente...construccion de un cementerio / Fregenal de la Sierra. Spain – 1882 – 9 – sp Bibl Santa Ana [324]

Algunas reformas en la isla de cuba / Saco, Jose Antonio – London, England. 1865 – 1r – 1 – us UF Libraries [972]

Algunos aspectos de la obra administrativa del pre / Morel, Emilio A – Santo Domingo, Dominican Republic. 1932 – 1r – 1 – us UF Libraries [972]

Algunos aspectos juridicos de la controversia... / Carrillo, Alfonso – Guatemala, Guatemala. 1948 – 1r – 1 – us UF Libraries [972]

Algunos datos sobre la tragedia de euzkadi – [Madrid: s.n. 1937] [mf ed 1980] – (= ser Blodgett coll) – 1mf – 9 – mf#w709 – us Harvard [946]

Algunos ensayos / Melendez Munoz, Miguel – San Juan, Puerto Rico. 1958 – 1r – 1 – us UF Libraries [972]

Algunos juicios de escritos guatemaltecos... / Biblioteca Nacional De Guatemala – Guatemala, Guatemala. 1959 – 1r – 1 – us UF Libraries [972]

Algunos puntos de historia acerca de la historia del coloniaje en el ecuador / ed by Bayle, Constantino – Madrid: Razon y Fe, 1927 – 1 – sp Bibl Santa Ana [972]

Algunos rasgos del hombre extremeno / Caba, Pedro – Badajoz: Dip. Provincial, 1966 – sp Bibl Santa Ana [946]

Algunos rasgos sobre como debiera organizarse la lucha antituberculosas en el nuevo estado espanol nacional sindicalista / Merino Hompanera, Jose – Caceres: Tip. La Minerva Cacerena, 1938 – sp Bibl Santa Ana [946]

Algunos versos : con un retrato de su autor por j. moreno villa / Diez Canedo, Enrique – Madrid: s.i., 1924 – 9 – sp Bibl Santa Ana [999]

Alguns numeros acerca do desenvolvimento da colonia de angola... – Lisboa, Portugal. 1936 – 1r – 1 – us UF Libraries [960]

Al-hadah al-siyasi – [al-Khartum, al-Sudan]: Hizb al-Bath al-Arabi al-Ishtirak, aug 15 1985-jun 29 1989] – 14r – 1 – us CRL [079]

Al-hadarah – Constantinople: 'Abd al-Hamid al-Zahrawi and Shakir al-Hanbali. n1-145. 4 rabi' al-Thani 1328-14 safar 1331 [14 apr 1910-24 jan 1913] – (= ser Arabic journals and popular press) – 1r – 1 – $350.00 – (missing: n1 p1-2; xerographic copy of top of n1, p1 inserted; incomplete n1 p3-4, bottom of pg lost) – us MEDOC [956]

Alhajadito / Asturias, Miguel Angel – Buenos Aires, Argentina. 1961 – 1r – 1 – us UF Libraries [972]

Alhambra – Belawan, nos 1-2 – 1mf – 9 – (missing: no 1) – mf#SE-3500 – ne IDC [950]

[Alhambra-] post-advocate – CA. oct 8 1898-sep 1902; oct 1903-1915; may 12 1916-may 3 1918 – 396r – 1 – $23,760.00 (subs $50/y) – (aka: advocate and valley vista, daily alhambra advocate, alhambra progress) – mf#H03133 – us Library Micro [071]

Al-haqa'iq – Damascus: al-Sayyid 'Abd al-Qadir al-Iskandarani. v1 pts 1-12. 1 sha'ban 1328-1 rajab 1329 [7 Ab 1910-28 haziran 1911] [complete] – (= ser Arabic journals and popular press) – 1r – 1 – $200.00 – us MEDOC [956]

Al-haqiqah : sawt munamzamat al-masihiyin al-dimuqratiyin – Bayrut: Munazamat al-Masihiyin al-Dimuqratiyin, feb 1980-nov 1981 – 1r – 1 – us CRL [956]

Al-haqq – Cairo: Yusuf Manqaryus, 1894-1910. yr 1 n1-50. 21 barmuda 1610-29 barmahat 1611 [coptic era][28 apr 1894-6 apr 1895] – (= ser Arabic journals and popular press) – 1r – 1 – $200.00 – us MEDOC [956]

Al-Harizi, Judah Ben Solomon *see* Tahkemoni

Al-hatif – al-Najaf: [s.n] [v1-19 may 3 1935-apr 1 1954] (bimthly) – 11r – 1 – us CRL [079]

Al-hayat *see* Arab newspapers

Al-hidayah – Cairo, Istanbul: 'Abd al-'aziz Jawish, 1910-14. yr 1 pt 1-yr 4 pt 11. muharram 1328-dhu al-Qadah 1331 [feb 1910-oct 1913] – (= ser Arabic journals and popular press) – 1r – 1 – $750.00 – (missing: yr 3 pts6-12; yr 4 pt9 not publ) – us MEDOC [956]

Al-hikma – Judeo-arabe. n1-32. Constantine, juil 1922-mai 1923. mq n1-5, 7-13, 19 – (= ser Hikma) – 1 – fr ACRPP [073]

Al-hikmah – Cairo: 'Abd al-'Aziz Nazmi, 1904-21. yr 1 pts1-12. 1 rabi' 2 1322-rabi' 2 1323 [15 jun 1904-jun 1905] – (= ser Arabic journals and popular press) – 1r – 1 – $250.00 – us MEDOC [956]

Al-hikmah – Cairo: Matba'at al-Amanah, 1937-40. yr 1 n1-yr 3 n10. 1 tut 1654-1 ba'una 1656 [coptic era][11 sep 1937-8 jun 1940] – (= ser Arabic journals and popular press) – 1r – 1 – $250.00 – us MEDOC [956]

Al-hilf al-atlasi wa-al-sharq al-awsat / Buhayri, Marwan – Bayrut: Muassasat al-Dirasat al-Filastiniyah, 1982 – 1r – 1 – us CRL [079]

Ali, Abdullah Yusuf *see*
– A cultural history of india during the british period
– Life and labour of the people of india
– The making of india
– Medieval india

Ali, Ahmed *see* Twilight in delhi

Ali akbar dihhuda (1879-1956) : leben, werk und wirkung zwischen politik und wissenschaft / Moghaddam, Abdollah Golijani – 1993 – 3mf – 9 – €50.11 – 3-89349-806-0 – mf#DHS 806 – gw Frankfurter [920]

Ali, Ameer *see* The personal law of the mahommedans

Ali baba : against the forty thieves – Istanbul: Sabahattin Ali. n1-4. 1947 [all publ] – (= ser O & t journals) – 1mf – 9 – $25.00 – us MEDOC [073]

Ali bey's el abassi reisen in afrika und asien in den jahren 1803 bis 1807 / Badia y Leblich, Domingo – Weimar 1816 [mf ed Hildesheim 1995-98] – 2v on 6mf – 9 – €120.00 – 3-487-26521-4 – (trans fr french) – gw Olms [910]

Ali iktisad meclisi raporlari : birinci ictima devresi: 1 mart 1928-19 mart 1928 – [Istanbul]: Tuerk Ocaklari Merkez Heyeti Matbaasi, 1928 – 1mf – 9 – $25.00 – us MEDOC [350]

Ali, Khalid Ismail *see* Studien ueber homonyme wurzeln im arabischen mit besonderer beruecksichtigung des mucvgam maqayis al-luga von ahmad ibn faris (ge 395/1005)

Ali, Mohamed *see*
– My life, a fragment
– Select writings and speeches of maulana mohamed ali

Ali, Muhammad *see*
– Muhammad, the prophet
– The religion of islam

Ali, Mustafa *see* Heft meclis

Ali, Mustafa bin Ahmet *see* Eser-i eslaftan heft meclis

Ali, Syed Ameer *see*
– Islam
– The life and teachings of mohammed

Ali, Syud Amir *see* Woman in islam

Ali the lion : ali of tebeleni, pasha of jannina, 1741-1822 / Plomer, William – London: J Cape [1936] [mf ed 1986] – 1r [ill] – 1 – (filmed with: 400 years of freethought / putnam, samuel p) – mf#1807 – us Wisconsin U Libr [954]

Ali-aba business law course materials journal / American Law Institute et al – Philadelphia PA 2000+ – 1,5,9 – ISSN: 1536-4445 – mf#12548,01 – us UMI ProQuest [346]

Ali-aba business law course materials journal – v1-25. 1976-2001 – 9 – $650.00 set – (title varies: v1-24 n3 1976-2000 as ali-aba course materials journal) – ISSN: 0145-6342 – mf#100151 – us Hein [340]

Ali-aba course materials journal / American Law Institute et al – Philadelphia PA 1980+ – 1,5,9 – ISSN: 0145-6342 – mf#12548.01 – us UMI ProQuest [340]

Ali-aba course materials journal *see* Ali-aba business law course materials journal

ali-aba estate planning course materials – v1-7. 1995-2001 – 9 – $135.00 – ISSN: 1086-8206 – mf#116291 – us Hein [340]

Los aliados – Malaga, Spain. 15 May-30 Oct 1915.-w. 9 ft – 1 – uk British Libr Newspaper [072]

Alias Pozas, Isabel Margarita *see* El valle del jerte, sus realidades y esperanzas

Ali-baba : ou les quarante voleurs extermines par une esclave / Galland, Antoine – 5 ed. Montreal: la Librairie Beauchemin limitee, [1955?] – 1mf – 9 – (ill by Vernier) – cn Bibl Nat [470]

Alibert, Francois Paul *see* Cyclope

Alice – 1968 may 18-1970 apr; v1 n1-v8 n1 [1968 may 16-1970 apr] – 1 – mf#1109604 – us WHS [071]

Alice / Lytton, Edward Bulwer Lytton, Baron – Boston, MA. 189- – 1r – 1 – us UF Libraries [090]

Alice drive baptist church – Sumter Co, SC. 1956-60 – 1 – $20.16 – us Southern Baptist [242]

Alice in wonderland / Carroll, Lewis – New York, USA. 1946 – 1r – 1 – us UF Libraries [830]

Alice Rouse Donaldson Educational Self Help Center *see* Newsletter

Alice times – Alice SA, 4 mar 1874-[1930] – 26r – 1 – mf#MS00289 – sa National [079]

The alice times – Alice, SA. 7 mar 1874-24 dec 1930 – 42r – 1 – sa National [079]

Alice warner : a novel / Allen, John (Mrs) – London: F V White & Co, successors to Samuel Tinsley & Co. 2v. 1881 – (= ser 19th c women writers) – 6mf – 9 – mf#5.1.118 – uk Chadwyck [830]

Alice's adventures in wonderland / Carroll, Lewis – Philadelphia, PA. 1895 – 1r – 1 – us UF Libraries [830]

Alichmore trumpet – [Scotland] Perthshire, Alichmore: printed by A M'Ginty 29 aug 1891-13 may 1893 [mf ed 2004] – 1r – 1 – (preceded by an iss called n183,457 & dated aug 29 1891; cont by: crieff trumpet [4 mar-may 1893]) – uk Newsplan [072]

Alicia larde de venturino... – Barcelona, Spain. 1924 – 1r – 1 – us UF Libraries [972]

Al-ictisam – Cairo, 1977-1981 – 18mf – 9 – mf#NE-20323 – ne IDC [956]

Alien critic – Portland OR 1973-74 – 1 – (cont by: science fiction review) – mf#7469.01 – us UMI ProQuest [420]

The alien in american law : a monthly survey of legislation, literature and decisions / ed by Jadeson, Sam – v1-5 apr 1954-dec 1958 [all publ] – 19mf – 9 – $28.50 – mf#llmc95-243 – us LLMC [342]

Alien labor program in guam : hearing before the special study subcommittee of the house judiciary committee / Guam. US Congress – 93rd Congress 1st sess. 9 Aug 1973. Washington: GPO, 1973 – 3mf – 9 – $4.50 – mf#LLMC 82-100B Title 20 – us LLMC [324]

The alien transvaal : a moral review / Russell, Annie – London [1885] – (= ser 19th c british colonization) – 2mf – 9 – mf#1.1.7393 – uk Chadwyck [960]

Alienado no direito civil brasileiro / Nina Rodrigues, Raymundo – Sao Paulo, Brazil. 1939 – 1r – 1 – us UF Libraries [972]

L'alienation mentale devant la justice criminelle / Gaultier, D Z – [Sorel]: impr au Sorelois, 1883 [mf ed 1992] – 1mf – 9 – mf#SEM105P1619 – cn Bibl Nat [360]

The aliened american – Cleveland, OH. v1 n1. apr 9 1853- [mf ed 1947] – (= ser Negro Newspapers on Microfilm) – 1r – 1 – us L of C Photodup [071]

Les alienes devant la loi : etude medico-legale / Villeneuve, George – Montreal: E Senecal, 1900 – 2mf – 9 – mf#36602 – cn Canadiana [344]

Alienes Urosa, Julian *see* Problemas de la economia de la paz

Alienist and neurologist – St Louis, MO: EV E Carreras, Steam Printer, Publ & Binder. [v18 (1897); v22 (1901)] (qrtly) – 1 – us CRL [616]

Aliens index (births) 1888-1922 (deaths) 1875-1922 / overlanders (or drovers) arriving in NT 1879-1883 / mining permits 1896-1911 – [mf ed 1986] – 1mf – 9 – A$5.50 – 0-949124-17-6 – (filmed with: mining permits 1896-1911 [no. date, lessee, location, amt sq area of permit, mnt no. (minister of the northern territory approval number)]) – at Northern [980]

Aliens naturalized in new zealand 1843-1916 : names, etc of alien friends who have been naturalized in new zealand – Wellington, 1918 – 2mf – NZ$9.00 – 0-908797-08-7 – (an alphabetical listing of full names, occupation, residential location and date of naturalization. plus copy of the acts and ordinances publ in the statutes of new zealnd, 1844-70 in chronological order) – mf#NZNB N1583 – nz BAB [980]

Aliens or americans? / Grose, Howard Benjamin – New York: Young People's Missionary Movt c1906 [mf ed 1990] – (= ser Forward mission study courses) – 1mf – 9 – 0-7905-5230-2 – (incl bibl ref) – mf#1988-1230 – us ATLA [320]

Aliesch, Peter *see* Studien zu thomas hardy's prosastil

Alifba – Paris: Kitab-i Alifba, 1982-83. dawrah-'i jadid jild-i 1-4. zimistan 1361-payiz 1362 [winter 1982-fall 1983] – 1r – 1 – $53.00 – us MEDOC [079]

Alig, Tobey *see* 1830 federal population census for indiana

Al-ikhwan al-muslimun – Cairo: Salih [Mustafa] 'Ashmawi, 1942-? v1 n1-10,13-14,16-24, v3 n50,66,79,81, v4 n82-85,87-88,90-91,93-95, v6 n218. 17 sha'ban 1361-20 dhu al-Hijjah 1367 [29 aug 1942-23 oct 1948] – (= ser Arabic journals and popular press) – 1r – 1 – $600.00 – (cont: al-nadhir; r also incl: jaridat al-ikhwan al-muslimin and al-nadhir) – us MEDOC [079]

Al-ikhwan al-muslimun *see* Jaridat al-ikhwan al-muslimin

Al-ilm wa-al-tiknulujiya fi al-sira al-arabi-al-israili / Zahlan, Antwan – Bayrut: Muassasat al-Dirasat al-Filastiniyah, 1981 – 1r – 1 – us CRL [956]

Alilot 'al rof'im yehudiyim / Muntner, Sussmann – Jerusalem, Israel. 1953 – 1r – 1 – us UF Libraries [939]

Alimacani : original indian name for fort george island – s.l, s.l, 193-? – 1r – 1 – us UF Libraries [978]

Alimentacion en los tropicos / Castro, Josue De – Mexico City? Mexico. 1946 – 1r – 1 – us UF Libraries [972]

La alimentacion racional del ganado / Diaz Montilla, Rafael – Badajoz: Graficas Iberia, 1944 – sp Bibl Santa Ana [946]

Alimentacion y nutricion en colombia / Bejarano, Jorge – Bogota, Colombia. 1950 – 1r – 1 – us UF Libraries [972]

Alimentary pharmacology and therapeutics – Oxford, England 1987+ – 1,5,9 – ISSN: 0269-2813 – mf#15612 – us UMI ProQuest [615]

Alimentos y nutricion en graficas y cantos popular / Aguilera Y Cespedes De Ferrer, Gertrudis – Habana, Cuba. 1944 – 1r – 1 – us UF Libraries [972]

Alimonda *see* Il dogma dell' immacolata

Alin, Folke *see* Studier oefver schleiermachers uppfattning af det evangeliska skapelsebegreppet

Aline, reine de golconde / Vial, Jean-Baptiste-Charles – Paris, France. 1815 – 1r – 1 – us UF Libraries [440]

Alinea : le journal etudiant...pour ouvrir des horizons – [Joliette]: [s.n.], [ca 1987]- (irreg) [mf ed 1988] – 9 – (cont: bloc (joliette, quebec)) – mf#SEM105P964 – cn Bibl Nat [071]

Alingsas nyheter – Alingsas, 1918-20 – 2r – 1 – sw Kungliga [078]

Alingsas tidning – Alingsas, Sweden. 1888-1978 – 156r – 1 – sw Kungliga [078]

Alingsas tidning – Alingsas, Sweden. 1888- – 1 – (elfsborgs lans tidning, 1980-82, gota alvdalsnyheterna, 1979-82, nya lerums tidning, 1980-82, lerums tidning, 1982) – sw Kungliga [078]

Alingsas weckoblad – Alingsas, Sweden. 1865-88 – 11r – 1 – sw Kungliga [078]

Aliotta, Antonio *see* The idealistic reaction against science

Aliquo, David *see* Comparison of the association of cervical spinal canal stenosis and intervertebral foraming canal stenosis and transient upper extremity parasthesias

Aliran islam – Bandung, 1948-1951(24) – 12mf – 9 – (missing: 1948, v1(1); 1949, v2(5-8, 11-15); 1950, v4(17-23)) – mf#SE-326 – ne IDC [950]

Al-irfan / Majallah Ilmiyah Adabiyah Akhlaqiyah Ijtima iyah. Saydun, Lebanon; ed by Al-Zayn, Ahmad 'Arif – Saydun, LE. v1-77. 1909-93 – (= ser Arabic journals and popular press) – 9 – $50.00ea – (v1 1909 8mf. v2 1910 9mf. v3 1911 13mf. v4 1911-12 6mf. v5 1913-14 7mf. v6 1920-21 8mf. v7 1921-22 9mf. v8 1922-23 12mf. v9 1923-24 14mf. v10 1924-25 16mf. v11 1925-26 18mf. v12 1926-27 10mf. v13-16 1927-28 9mf per v. v17-22 1929-31 10mf per v. v23 1932-33 11mf. v24-25 1933-35 15mf per v. v26 1935-36 12mf. v27 1936-37 13mf. v28 1937-38 17mf. v29 1938-39 14mf. v30 1940-41 7mf. v31 1942-45 9mf. v32 1945-46 17mf. v33 1946-47 19mf. v34-35 1947-48 12mf per v. v36 1949 17mf. v37-38 1950-51 18mf per v. v39 1951-52 19mf. v40-41 1952-54 18mf per v. v42 1954-55 19mf. v43-44 1955-57 17mf per v. v45-50 1957-63 15mf per v. v51 1963-64 17mf. v52-54 1964-67 16mf per v. v55 1967-68 17mf. v56 1968-69 18mf. v57 1969-70 22mf. v58 1970-71 18mf. v59 1971 16mf. v60 1972 24mf. v61 1973 21mf. v62 1974 19mf. v63 1975 21mf. v64 1976 3mf. v65 1977 18mf. v66 1978 12mf. v67 1979 15mf. v68 1980 13mf. v69 1981 10mf. v70 1982 14mf. v71 1983 16mf. v72 1984 13mf. v73-74 1985-86 11mf per v. v75 1987 7mf. v76 1992 19mf. v77 1993 8mf. missing: v64 n1, 4-10; v66 n6-7) – us MEDOC [073]

Alisan *see* Avrupa bizi nasil taniyor

Al-ishtira kiyah – [Khartoum?]: al-Hizb al-Ishtiraki al-Islami, jan 28 1988-jun 3 1989 – 1r – 1 – us CRL [956]

Al'islaah = The reform – New York NY, 1949-58, 1962-77 – 5r – 1 – (arabic newspaper) – us IHRC [071]

Al-islah [mombasa, kenya] : [the only muslim paper published in kenya] – Mombasa: Al-Islah [1932-33] [mf ed [Nairobi]: Kenya National Archives Photographic Service 1970] – feb 29 1932-oct 30 1933 on 1r with other items – 1 – (in swahili & arabic) – mf#mf-1550 camp r43 – us CRL [079]

Al-islam – Cairo: Ahmad 'Ali al-Shadhili al-Azhari, 1894-1913. yr 10 n1-yr 11 n12. dhu al-Qa'dah 1329?-dhu al-Hijjah 1331 [oct 1911?-dec 1913] – (= ser Arabic journals and popular press) – 1r – 1 – $275.00 – us MEDOC [956]

Alison, Archibald *see*
– Discourse
– Discourse, preached in the episcopal chapel
– Discourse preached in the episcopal chapel
– Remarks on the administration of criminal justice in scotland

Alison, Francis *see*
– Miscellaneous manuscripts, biographical data, and documents
– Sermons

Al-ittihad – al-Khartum: al-Sharikah al-Arabiyah li-Khadamat al-Ilam wa-al Malumat, oct 11 1987-jul 18 1988 – 3r – 1 – us CRL [956]

Al-ittihad – Haifa, 1944-1999 – 99r – 1 – mf#J-93-4 – ne IDC [956]

Al-ittihad – Plainfield IN 1972-76 – 1,5,9 – mf#7074 – us UMI ProQuest [305]

Al-ittihad = L'union – Paris. n2-3 – (= ser Ittihad) – 1 – (titre francais: l' union sept-oct 1880; titre et texte en caracteres arabes) – fr ACRPP [073]

Al-ittihad al-isra'ili – Cairo: Jam'iyat al-Ittihad al-Isra'ili lil-Qurra'in, 1924-29. yr 1 n1-yr 4 n25. 16 nisan 5684-12 nisa 5688 [jewish era][20 apr 1924-3 apr 1928] – (= ser Arabic journals and popular press) – 1r – 1 – $450.00 – us MEDOC [956]

Al-ittihadi – Khartoum. dec 21 1985-jun 15 1989 – us CRL [079]

Alivardi and his times / Datta, Kalikinkar – [Calcutta]: University of Calcutta, 1939 – (= ser Samp: indian books) – us CRL [954]

Alive – St Louis MO 1969-80 – 1,5,9 – ISSN: 0002-5461 – mf#7526 – us UMI ProQuest [240]

Alivio de los sedientos... / Micon, F – Barcelona, 1576 – 9 – sp Cultura [610]

Alivizatos, Amilkas S *see* Die kirchliche gesetzgebung des kaisers justinian 1

Aliwal north observer *see* The aliwal north standard / aliwal north standard

The aliwal north standard / aliwal north standard – Aliwal North SA, 1870-74 – 1 – mf#MS00433 – sa National [079]

Alix, Juan Antonio *see* Decimas

'Aliyah Ha-Sheniyah / Kalai, David – Tel-Aviv, Israel. 1946 – 1r – 1 – us UF Libraries [939]

'Aliyah Veha-Hatsalah Bi-Shenot Ha-Sho'ah / Dobkin, Eliahu – Jerusalem, Israel. 1945 – 1r – 1 – us UF Libraries [939]

'Aliyat Tuvyah / Kohn, Tobias – Warsaw, Poland. 1885 – 1r – 1 – us UF Libraries [720]

Aliye divan-i harb-i oerfiyesinde tedkik olunan mesele-i siyasiye hakkinda izahat – Istanbul: Tanin Matbaasi, 1332 [1916] – (= ser Ottoman histories and historical sources) – 3mf – 9 – $60.00 – us MEDOC [956]

'Aliyoth Eliyahu / Levin, Joshua Herschel – Vilna, Lithuania. 1892 – 1r – 1 – us UF Libraries [939]

al-Jabarti *see* Merveilles biographiques et historiques

Al-Jadid : a monthly digest of culture and the arts in the arab world – Los Angeles, CA: International Desktop Publ [v1 n1-5 jun-nov 1993] (mthly) – 1r – 1 – us CRL [700]

Al-jadid : shahriyah lil-adab wa-al-'ulum wa-al-funun – Haifa, Israel: Hanna Naqqarah. mujallad 1, 'adad 1-mujallad 40 'adad 9. 1953-91// – (= ser Arabic journals and popular press) – 11r – 1 – $950.00 – (missing: mujallad 12 'adad 11-12; mujallad 30 'ada 3-4 never publ; the publ skipped v32 and 33 without skipping any time. publ ceased aft mujallad 40 'adad 9) – us MEDOC [073]

Aljadid : a record of arab culture and arts = Jadid magazine – Los Angeles, CA: Nagam Cultural Project, 1995- . [v1 n1-v4 n23 nov 1995-spring 1998] (qrtly) – 1r – 1 – us CRL [700]

Al-jamahiriyah – Tripoli, Libya. Sept 1980-Jan 1986; 1987 – 8r – 1 – us L of C Photodup [079]

Al-jami'ah al-arabiyyah – Jerusalem, 1927-1935 – 15r – 1 – mf#J-93-9 – ne IDC [956]

Al-jaridah – [Khartoum]: Dar al-Jaridah. sep 14 1987-aug 17 1988 – 3r – 1 – us CRL [079]

Al-jaridah al-rasmiyah – 1970-79 – (= ser Jaridah Al-Rasmiyah) – 12r – 1 – (1980-) – us L of C Photodup [324]

Al-jaridah al-rasmiyah / Algeria – 1970-77. 6 reels – 1 – us L of C Photodup [324]

Al-jaridah al-rasmiyah / Jordan – 1970-78 – 1 – (1979-. ca $50y) – us L of C Photodup [324]

Al-jaridah al-rasmiyah / Lebanon – 1940-45 – 1 – (1971-74 12r. 1975-) – us L of C Photodup [324]

Al-jaridah al-rasmiyah / Libya – 1970-78 – 1 – (1979-. ca $50y) – us L of C Photodup [324]

Al-jaridah al-rasmiyah / Morocco – 1970-79 – 29r – 1 – $580.00; outside North America add $1.25r – (1980-. ca $65y) – us L of C Photodup [324]

Al-jaridah al-rasmiyah / Oman – Apr 1973-Dec 1979 – 4r – 1 – $75.00; outside North America add $1.25r – (1980-. ca $20y) – us L of C Photodup [324]

Al-jaridah al-rasmiyah / Qatar – 1970-79 – 1 – 69.00 – (1980-. ca $20y) – us L of C Photodup [324]

Al-jaridah al-rasmiyah / Syria – 1970-79 – 139r – 1 – $2780.00; outside North America add $1.25r – (1980-. ca $250y) – us L of C Photodup [324]

Al-jaridah al-rasmiyah / United Arab Emirates – Dec. 1971-1977 – 2r – 1 – $45.00; outside North America add $1.25r – (1978-. ca $20y) – us L of C Photodup [324]

Al-jaridah al-rasmiyah / Yemen – 1971-77 – 2r – 1 – $45.00; outside North America add $1.25r – (1978-. ca $20y) – us L of C Photodup [324]

Al-jaridah al-rasmiyah li-hukumat dubayy wa-tawabi'iha – 1974-77 – 1r – 1 – (1978-) – us L of C Photodup [324]

Al-jaridah al-rasmiyah lil-jumhuriyah al-suriyah – Journal officiel de la republique syrienne – Damascus, Syria, 1925, 1929, 1931, 1933, 1940-48, 1950-51, 1959-64 – (= ser Arabic research materials) – 82r – 1 – $5140.00 – (several iss missing. in french and arabic) – us MEDOC [956]

Al-jaridah al-rasmiyah lil-mamlakah al-urduniyah al-hashimiyah – apr-dec 1972 – 1r – 1 – (cont: al-jaridah al-rasmiyah li-imarat sharq al-urdun) – mf#LL-02156 – us L of C Photodup [079]

Al-jaridah al-rasmiyah. (official gazette) / Egypt – Mar 1958-79 – 32r – 1 – $736.00; outside North America add $1.25r – (1980- ca $30y) – us L of C Photodup [324]

Al-jawa'ib – Istanbul: Ahmad Faris Shidyaq, 1861-18?. n3-1162 (5 Dhu al-Hijjah 1277-19 Muharram 1301 [14 Jun 1861-20 Nov 1883]) – 3mf – 9 – $400.00 – (missing iss: 5, 6, 11, 16, 33, 37, 40, 66, 68, 74, 76, 78-81, 98, 117, 118, 188, 190, 225, 245, 276, 289, 336, 375, 381, 906, 1160; n1009 never publ. a74.1-3; amp 215-217) – us MEDOC [956]

Aljibe / Gerena Bras, Gaspar – San Juan, Puerto Rico. 1959 – 1r – 1 – us UF Libraries [972]

Al-jins al-latif – Cairo: Malikah Sa'd. yr 1 n2-yr 5 n10. aug 1908-apr 1913 – (= ser Arabic journals and popular press) – 1r – 1 – $200.00 – (missing: yr 1 n6) – us MEDOC [956]

Al-jughrafiya bayna al-ilm al-tatbiqi wa-al-wazifah al-ijtimaiyah / al-Sharnubi, Muhammad Abd al-Rahman – [al-Kuwayt]: Qism al-Jughrafiya bi-Jamiat al-Kuwayt wa-al-Jamiyah al-Jughrafiyah al-Kuwaytiyah, [1981] – us CRL [956]

Al-karmah – Cairo: Habib Jirjis al-Shammas, 1904-14. v1 n1-v12 n9. 1 tut 1621-24 ba'una 1642 [coptic era][11 sep 1904-1 jul 1926] – (= ser Arabic journals and popular press) – 3r – 1 – $850.00 – (missing: v7. ceased publ, resumed 1923-31) – us MEDOC [956]

Alkartu *see* Spanish-basque political periodicals

Al-katib – Jerusalem, 1979-93. mujallad 1 'adad 1-mujallad 14 'adad 150. nov 1979-feb 1993 – (= ser Arabic journals and popular press) – 7r – 1 – $371.00 – us MEDOC [073]

Al-kawkab al-gaza-iri *see* Kawkab ifriqiya

Alker, Emmerich *see* Die chronologie der buecher der koenige und paralipomenon

Alker, Ernst *see*
– Die deutsche literatur im 19. jahrhundert, 1832-1914
– Franz grillprarzer
– Geschichte der deutschen literatur

Alkestis : Schauspiel in vier aufzuegen / Wesendonk, Mathilde – Oldenburg: Schulzesche Hof-Buchdruckerei 1898 – 1mf – 9 – (ed fr the greek by mathilde wesendonck) – mf#mw-7 – ne IDC [820]

Alkestis d'apres euripide / Rivollet, Georges – Paris, France. 1901 – 1r – 1 – us UF Libraries [440]

Al-khartum – [Khartoum]: Dar al-Khartum lil-Sihafah, sep 12 1988-jun 29 1989 – 5r – 1 – us CRL [079]

Al-khasais al-jimruflujiyah li-nahr al-sahl al-faydi : maa dirasah an al-nil fi misr al-wusta / Jad, Taha Muhammad – [al-Kuwayt]: Qism al-Jughrafiya bi-Jamiat al-Kuwayt wa-al-Jamiyah al-Jughrafiyah al-Kuwaytiyah, [1981] – us CRL [956]

al-Khayr, Yahya Muhammad Shaykh Abu *see* Zahf al-rimal bi-mintaqat al-ahsa

Al-khulafa ar-rashidun : or, the four rightly-guided khalifas / Sell, Edward – 2nd ed. London: Christian Literature Society for India, 1913 [mf ed 1992] – (= ser The islam series) – 1mf – 9 – 0-524-02543-6 – (incl bibl ref. 1st printed 1909) – mf#1990-3038 – us ATLA [260]

Alkibiades : drama in fuenf akten / Bauernfeld, Eduard von – Dresden: L Ehlermann, 1889 [mf ed 1993] – 49p – 1 – mf#8509 – us Wisconsin U Libr [820]

Al-kindi : genannt "der philosoph der araber": ein vorbild seiner zeit und seines volkes / Fluegel, Gustav – Leipzig: In Commission bei F A Brockhaus, 1857 [mf ed 1986] – 1mf – 9 – 0-8370-7696-X – (text in german; bibl in german & arabic) – mf#1986-1696 – us ATLA [180]

al-Kindi, Abd al-Masih *see* The apology of al kindy

Alkoholdebatt – Stockholm, Sweden 1975-77 – 1 – ISSN: 0002-550X – mf#2746 – us UMI ProQuest [362]

Al-kuds – Jerusalem, 1968-1999 – 137r – 1 – mf#J-93-5 – ne IDC [956]

Al-kuschairis darstellung des sufitums : mit uebersetzungs-beilage und indices / Hartmann, Richard – Berlin: Mayer & Mueller, 1914 [mf ed 1991] – (= ser Tuerkische bibliothek 18) – 1mf – 9 – 0-524-02019-1 – mf#1990-2794 – us ATLA [260]

Al-kuwayt al-yawm / Kuwait – 1970-79 – 1 – 782.00 – (1980-. ca $160y) – us L of C Photodup [324]

All about african violets / Free, Montague – Garden City, USA. 1951 – 1r – 1 – us UF Libraries [580]

All about issues – 1982 jun-1986 mar; apr-1989 dec – 1 – mf#1238821 – us WHS [321]

All about jesus / Dickson, Alexander – New York: Robert Carter, 1878, c1875 [mf ed 1985] – 1mf – 9 – 0-8370-2906-6 – mf#1985-0906 – us ATLA [240]

All about Lok Tilak – Madras: BG Paul & Co, [1922] – (= ser Samp: indian books) – us CRL [920]

All about victoria, british columbia / Emberson, Alfred – [Victoria BC]: Victoria Print & Pub, 1916 [mf ed 1997] – 2mf – 9 – 0-665-84601-0 – mf#84601 – cn Canadiana [917]

All africa is standing up – v2 n3-5, 7-8 [1978 apr-jun, sep-oct/nov]; flyer, 1978 may 20 – 1 – mf#2576780 – us WHS [321]

All american – 1974 aug-1982 dec; 1983-87 – 1 – mf#696773 – us WHS [071]

All american university one act plays – Franklin, OH. 1931 – 1r – 1 – us UF Libraries [820]

All around the house : or, how to make homes happy / Beecher, Henry Ward (Mrs) – Toronto: J Robertson, 1881 – (= ser Robertson's cheap series) – 2mf – 9 – mf#03541 – cn Canadiana [640]

All chicago city news – Cook. 1981 feb 12-1984 dec 24, 1985 jan 31-1986 jul 29 – 2r – 1 – mf#961485 – us WHS [071]

All City Employees Association of Los Angeles et al *see* Pueblo

All Cooperating Assembly *see*
– Main stream
– Scoop

All examination questions used for twelve years, in the regular courses in columbian university / Howe, Frank Clifford – Washington, D.C. 1889. 93p. L.C. copy imperfect: p. 1-2 wanting. LL-1074 – 1 – us L of C Photodup [340]

All Florida / Florida. Bureau Of Immigration – Tallahassee, FL. 1926 – 1r – 1 – us UF Libraries [978]

All florida magazine – Ocala, FL. 1965-1968 – 6r – 1 – (gaps) – us UF Libraries [071]

All glory to the blood of jesus : devotion to the precious blood, followed by a choice selection of prayers and exercises in its honor – Montreal: [s.n.] 1887 [mf ed 1984] – 1mf – 9 – 0-665-46400-2 – mf#46400 – cn Canadiana [241]

All hallows' in the west – [Yale, BC?: All Hallows' Canadian School, 1899?-1901] [mf ed v1 n2 michaelmas-tide, 1899-v3 n1 ascension-tide, 1901; v3 n3 christmas-tide, 1901] – 9 – ISSN: 1190-7320 – mf#P04504 – cn Canadiana [242]

All hands : the bureau of naval personnel informatoin bulletin – Jun 1945-63 – 1 – $320.00 – us L of C Photodup [355]

All hands – Washington DC 1952+ – 1,5,9 – ISSN: 0002-5577 – mf#6829 – us UMI ProQuest [355]

All hands abandon ship – 1970 jun-1972 oct/dec – 1 – mf#964659 – us WHS [071]

All in one : all useful science and profitable arts in one book of jehovah aelohim / Bampfield, Francis – 1677 – 1 – $11.20 – us Southern Baptist [242]

The all india ayurvedic medical council bill, 1965 / Sharma, Anant Tripath – [S.l: s.n, 1965?] – 1r – 1 – us CRL [615]

All Indian Pueblo Council *see*
– New mexico's 19 pueblo news
– Pueblo news

All ireland review – Dublin. v1-7. n3. jan 6 1900-jan 1907 – 1 – us NY Public [073]

All ireland review – Dublin, Ireland 6 jan 1900-dec 1906 – 1 – (fr 6 jan 1900 to 14 sep 1901 publ at kilkenny) – uk British Libr Newspaper [072]

All news – v1 n2-v3 n5 [1984 oct 19-1987 mar 20]; 1987 may – 1 – mf#1044976 – us WHS [321]

All of grace : an earnest word with those who are seeking salvation by the lord jesus christ / Spurgeon, Charles Haddon – Chicago: Bible Institute Colportage Association, [18–?] [mf ed 1986] – 1mf – 9 – 0-8370-9906-4 – mf#1986-3906 – us ATLA [240]

All, or none – London, England. 18– – 1r – 1 – us UF Libraries [240]

All or nothing / Beresford, John Davys – Indianapolis, IN. 1928 – 1r – 1 – us UF Libraries [090]

All outdoors – New York. v1-9 n5. fall 1913-feb 1922 [all publ?] – (= ser Sports periodicals, 1822-1922) – 4r – 1 – $760.00 – us UPA [790]

"All right!" – London, England: publ for the Baptist Tract Society by Elliot Stock 18– – 1r – 1 – us UF Libraries [240]

All round the world : adventures in europe, asia, africa and america / Gillmore, Parker – London: Chapman and Hall, 1871 – 3mf – 9 – (ill by Sidney P Hall) – mf#32858 – cn Canadiana [910]

All saints' sermons, 1905-1907 / Inge, William Ralph – London: Macmillan, 1907 [mf ed 1990] – 1mf – 9 – 0-7905-7345-8 – mf#1989-0570 – us ATLA [242]

All select – iss n1-11. fall 1943-fall 1946 – 15 – mf#001MV-002MV; 042MV – us MicroColour [740]

All she wrote – 1981 nov/dec – 1 – mf#622099 – us WHS [071]

All the articles of the darwin faith / Morris, Francis Orpen – London, 1875 – (= ser 19th c evolution & creation) – 1mf – 9 – mf#1.1.11620 – uk Chadwyck [575]

All the best in bermuda, the bahamas, puerto rico / Clark, Sydney – New York, USA. 1965 – 1r – 1 – us UF Libraries [972]

All the best in central america / Clark, Sydney – New York, USA. 1946 – 1r – 1 – us UF Libraries [972]

All the best in central america / Clark, Sydney – New York, USA. 1952 – 1r – 1 – us UF Libraries [972]

All the best in central america / Clark, Sydney – New York, USA. 1961 – 1r – 1 – us UF Libraries [972]

All the best in cuba... / Clark, Sydney – New York, USA. 1946 – 1r – 1 – us UF Libraries [972]

All the best in south america / Clark, Sydney – New York, USA. 1957 – 1r – 1 – us UF Libraries [972]

All the best in south america west coast / Clark, Sydney – New York, USA. 1947 – 1r – 1 – us UF Libraries [972]

All the best in the caribbean / Clark, Sydney – New York, USA. 1948 – 1r – 1 – us UF Libraries [972]

All the way : bulletin of forsyth county defense league / Forsyth County Defense League – 1987 jul 9-1992 mar – 1 – mf#1893583 – us WHS [366]

All the way to abenab / Haythornthwaite, Frank – London, England. 1956 – 1r – 1 – us UF Libraries [960]

All the year round – A weekly journal conducted by Charles Dickens. v1-76. 1859-95 – 1 – us AMS Press [420]

All the year round – La Salle IL 1859-95 – 1 – mf#5202 – us UMI ProQuest [420]

All the year round in japan / Ballard, Susan – Westminster: Society for the Propogation of the Gospel in Foreign Parts, 1913 [mf ed 1995] – (= ser Yale coll) – 55p (ill) – 1 – 0-524-09559-0 – mf#1995-0559 – us ATLA [950]

"All things are possible to him that believeth" – London, England: Baptist Tract Society 18– – 1r – 1 – us UF Libraries [240]

All through the Gandhian era / Iyengar, A S – Bombay: Hind Kitabs, 1950 – (= ser Samp: indian books) – us CRL [954]

All winners – iss n1-21. sum 1941-win 1946-47 – 15 – (n20 not publ) – mf#003MV-006MV – us MicroColour [740]

Alla en caracas / Vallenilla Lanz, Laureano – Caracas, Venezuela. 1954 – 1r – 1 – us UF Libraries [972]

Alla sacra real maesta di federigo augusto... signor cardinale annibale albani... : ragguaglio delle solenni esequie fatte celebrare in roma nella basilica di s clemente – Roma, 1733 – 2mf – 9 – mf#0-1129 – ne IDC [700]

Alla tedesca : a favorite air. arranged as a rondo [for piano] / Dussek, Johann Ladislaus – London: printed by Lavenu & Mitchel [1806?] [mf ed 1989] – 1r – 1 – mf#pres. film 45 – us Sibley [780]

Alla terra dei galla : narrazione della spedizione bianchi in africa nel 1879-1880 / Bianchi, G – Milano, 1884 – 10mf – 9 – mf#NE-20182 – ne IDC [916]

Allaback, Nicole J *see* A comparison of physiologic responses when exercising on five exercise modalities at a self-selected exercise intensity

All-African News Service *see*
– News file

Allah, Hajji Sayyid Farraj *see* Uqyanus

Allahabad, Oudh *see* Democracy not suited to india

Allain, Ernest *see* L'eglise et l'enseignement populaire

Allaire, Jean-Baptiste-Arthur *see* Nos saints patrons

Allais, Alphonse *see* Silverie

Allam, Paul F W *see* System 5

All-american front / Aikman, Duncan – New York, USA. 1940 – 1r – 1 – us UF Libraries [972]

Allan, Alexander *see* Power of the civil magistrate in matters of religion

Allan, Alexander M *see* Before the mast and behind the pulpit

Allan, Charles Wilfrid *see* Our entry into hunan

Allan, Diane E *see* Gender differences in sport centrality

Allan, George William *see* Notes on the ornithology of the seasons

Allan, John et al *see* The cambridge shorter history of india

Allan morrison papers, 1940-1968 : from the holdings of the schomburg center for research in black culture, manuscripts, archives and rare books division: the new york public library, astor, lenox and tilden foundations – 1995 – 3r – 1 – $255.00 – (guide which covers all coll under "literature and the arts" sold separately for $20 d3305.g6) – mf#D3305P23 – Dist. us Scholarly Res – us L of C Photodup [070]

Allan, William et al *see* Land holding and land usage among the plateau tonga of mazabuka district

Allard, Alberic *see* Histoire de la justice criminelle

Allard, Gaston *see* Bref expose historique des recherches en industrie laitiere faites dans la province de quebec

Allard, Paul *see*
– L'art paien sous les empereurs chretiens
– Les dernieres persecutions du troisieme siecle (gallus, valerien, aurelien)
– Les esclaves chretiens
– Esclaves, serfs et mainmortables
– Ten lectures on the martyrs

L'allarme – Sommerville MA, 1916* – 1r – 1 – (italian periodical) – us IHRC [073]

Allart de Meritens, Hortense *see* La femme et la democratie de notre temps

Allason, Thomas *see* Picturesque views of the antiquities of pola

Al-lata'if – Cairo: Shahin Makaryus, 1886-96. v1 n1-v9 n12 ([11 sha'ban] 1303-10 rajab 1314 (15 may) 1886-15 dec 1896]) [complete] – (= ser Arabic journals and popular press) – 2r – 1 – $875.00 – (no iss publ 15 apr 1893-15 jan 1895) – us MEDOC [956]

Allatius, L *see*
– Breviarium historicum
– De ecclesiae occidentalis et orientalis perpetua consensione
– Historia
– Symmicta sive opuscula graeca et latina

Al-layali – Alger. 5nos. 1936-37 – (= ser Layali) – 1 – fr ACRPP [073]

Allberry, C R C *see* Manichean manuscripts in the chester beatty collection

Alldeutsche blaetter *see* Mitteilungen des allgemeinen deutschen verbandes

Alldeutscher verband flugschriften – Munich. v1-25. 1897-1906 – 1 – us NY Public [073]

Alldridge, Thomas Joshua *see* The sherbro and its hinterland

Alle de gedichten van hieronymus sweerts – Amsterdam: Cornelis Sweerts, 1697 – 10mf – 9 – mf#0-3175 – ne IDC [090]

Alle de wercken : so ouden als nieuwe, van de heer jacob cats, ridder, oudt raedpensionaris van hollandt, etc – t'Amsterdam: lan lacobsz Schipper, 1655 – 18mf – 9 – mf#0-1539 – ne IDC [090]

Alle den volcke – v28-34. 1934-40 [complete] – 1r – 1 – ISSN: 0002-5666 – mf#ATLA S0545 – us ATLA [073]

Alle kirchen gesaeng vnd gebeet des gantzen jars : von der hailigen christenlichen kirchen angenommen vnd bissher in loeblichem brauch erhalten / alles verteutscht vnnd laengest durch m. – Augspurg: Ulhart 1563 – (= ser Hqab. literatur des 16. jahrh.) – 13mf – 9 – €105.00 – ([1.] vom introit der mess biss auff die complent, darneben die benedeyung der liecht, der palm, des feuers, des osterstocks, der tauff, und der kreuetter [1563]; [2.] von dem ersten sonntage nach der heiligen dryfaeltigkeit, biss auf das advent [1563]) – mf#1563a – gw Fischer [780]

Alle kirchen gesang vnd gebeth des gantzen iars : von der heyligen christlichen kirchen angenommen...nu wider vbersehen...vnd zirlicher verdeutscht, auch ynn vielen stuecken gemehrt. durch m christophorum flurheym von kytzingen – Leyptzigk: Thanner – (= ser Hqab. literatur des 16. jahrh.) – 13mf – 9 – €105.00 – ([1] vom introit der mess bis auff die complent. darneben die benedeyung der liecht, der palm, des fewers, des osterstocks, der tawff und der krewter 1529; [2] das ander teyl der kirchengeseng: von dem ersten sontage nach der heyligen dreifaltigkeit bis auff das advent 1529) – mf#1529 – gw Fischer [780]

Alle kirchengesaeng und gebett des gantzen jars : von der heiligen christenlichen kirchen angenommen unnd bissher inn loeblichem brauch erhalten / alles verteutscht und laengest durch m. christophorum flurhaim von kitzinge[n] gemehret – Dilingen: Mayer – (= ser Hqab. literatur des 16. jahrh.) – 3mf – 9 – €105.00 – (contents: [1.] vom introit der mess biss auff die complend [1571]; 2. von dem erste[n] sontage nach der heiligen dreyfaltigkeit biss auff das advent [1571]) – mf#1571a – gw Fischer [780]

Alle schriften und buecher / Thomas a Kempis (Thomas Hemerken) – Coellen, 1713 – €46.00 – (trans by adamum jacobs) – ne Slangenburg [241]

Alle wasser boehmens fliessen nach deutschland / Bodenreuth, Friedrich – Berlin: H von Hugo, 1937 [mf ed 1992] – 347p – 1 – mf#7461 – us Wisconsin U Libr [830]

Alle wipfel rauschen heimat : roman / Zenker, Wolfgang – Leipzig: O Janke 1943 [mf ed 1992] – 1r [ill] – 1 – (filmed with: die bruecke / heinrich zerkaulen & other titles) – mf#3067p – us Wisconsin U Libr [830]
Die allegemeinen grundsaetze des obligationenrechts in dem entwurfe eines buergerlichen gesetzbuches fuer das deutsche reich / Seuffert, Lothar, Ritter von – Berlin: J Guttentag, 1889 – (= ser Civil law 3 coll; Beitraege zur erlaeuterung und beurtheilung des entwurfes eines buergerlichen gesetzbuches fuer das deutsche reich) – 1mf – 9 – (incl bibl ref) – mf#LLMC 96-606 – us LLMC [346]
Alleghenian : newsletter of the western pennsylvania african american historical and genealogical society / Western Pennsylvania African American Historical and Genealogical Society – 1993 fall, 1994 winter, fall, spring/summer – 1 – mf#2955204 – us WHS [929]
The alleghenian – Pittsburgh, PA. -w 1893-1896 – 13 – $25.00r – us IMR [071]
Allegheny County Industrial Union *see* Industrialist
Allegheny County Socialist Party *see* Iron city socialist
Allegheny Observatory *see* Publications of the allegheny observatory of the university of pittsburgh
Allegiance to the church / Dodsworth, W – London, England. 1841 – 1r – 1 – us UF Libraries [240]
The allegorical drama of calderon : an introduction to the autos sacramentales / Parker, Alexander Augustine – Oxford: Dolphin Book, 1943 – 1 – us Wisconsin U Libr [440]
Die allegorie in ihrer exegetischen anwendung bei maimonides / Goldberger, Philipp – Wien: A Fanto, 1894 – 1mf – 9 – 0-8370-3331-4 – mf#1985-1331 – us ATLA [221]
Die allegorie in kunst, wissenschaft und kirche / Bornemann, Wilhelm – Freiburg i. B.: J C B Mohr, 1899 – 1mf – 9 – 0-7905-5810-6 – (incl bibl ref) – mf#1988-1810 – us ATLA [100]
Allegorische personen zum gebrauche der bildenden kuenstler : mit kupfern von bernhard rode / Ramler, K W – Berlin: Akademische Kunst-und Buchhandlung, 1788 – 2mf – 9 – mf#O-1262 – ne IDC [700]
Allegro – New York. 1943-1976 – 1 – us L of C Photodup [780]
Allegro – v24 n1-v28 n2 [1949 nov-1953 dec]; v74 n8-1980 [1974 sep-1980]; 1981 jan-1985 dec; 1986 jan-1989 jun; jul/aug-1991 dec; 1992-94 – 1 – mf#573416 – us WHS [071]
Allegro qumran collection on microfiche / ed by Brooke, George J – Manchester Museum 1955-62 [mf ed Leiden: E J Brill/IDC, may 1996] – 30mf – 9 – €759 / $1025 – ISBN-10: 90-04-10558-1 – ISBN-13: 978-90-04-10558-4 – ne Brill [270]
Allehanda – Lindesberg, Sweden 1882-85 – 2r – 1 – sw Kungliga [078]
Allehanda *see* Malmo allehanda
Allehanda karlskoga-degerfors *see* Nerikes allehanda
Die all-einheit : grundlinien der welt- und lebensanschauung im geiste goethes und spinozas / Kronenberg, Moritz – Stuttgart: Strecker, 1924 – 1r – 1 – us Wisconsin U Libr [100]
Alleluia : a hymnal for use in schools, in the home... / ed by Sheppard, Franklin L – Philadelphia: Westminster Press, 1915 [mf ed 1993] – (= ser Presbyterian coll) – 4mf – 9 – 0-524-06665-5 – mf#1991-2720 – us ATLA [242]
Allemagne – Paris. 1949-avr 1967 – 1 – fr ACRPP [073]
Alleman, George Mervin *see* A critique of some philosophical aspects of the mysticism of jacob boehme
Alleman, Herbert Christian *see*
– The bible
– The book and the message
Allemands danced at the kings-theatre in the hay-market by mr slingsby and sig-ra radicate : to which is added mr slingsby's hornpipe. set for the german flute, violin or harpsichord / Agus, Giuseppe – London: Welcker [1767?] [mf ed 198-] – 1r – 1 – mf#pres. film 36 – us Sibley [780]
Allemannische gedichte / Hebel, Johann Peter; ed by Heilig, Otto – Heidelberg: C Winter 1902 [mf ed 1990] – 1r – 1 – (german text & phonetic transcr on opposite pp; filmed with: friedrich hebbel und die gegenwart / wilhelm tideman) – mf#2706p – us Wisconsin U Libr [810]
Allemannische lieder : nebst worterklaerung und einer allemannischen grammatik / Hoffmann von Fallersleben, August Heinrich – 5. verb verm ausg. Mannheim: F Bassermann 1843 [mf ed 1991] – (= ser Bibliothek der deutschen literatur) – 1r – 1 – (filmed with: der grosse baum / herbert von hoerner) – mf#2727p – us Wisconsin U Libr [810]
Allemeier, Meredith Frances *see* Ciau athletes' use and intentions to use performance enhancing drugs
Allen, Abel Leighton *see* The message of new thought
Allen, Alexander Viets Griswold *see*
– Christian institutions
– The continuity of christian thought
– Freedom in the church
– Jonathan edwards
– Religious progress
Allen, Alexander Viets Griswold et al *see* Jonathan edwards
Allen and Co, W T *see* [W t allen and company's volume of designs
Allen, Andrew James Campbell *see* The church catechism
Allen, B F *see* History of san mateo county
Allen, Charles Bruce *see* Cottage building
Allen, Charles Edwin *see* Rev jacob bailey
Allen, Charles H *see* "Chinese" gordon, r e, c b
Allen, Charles William *see* The land prospector's manual and field-book
Allen Co. Bluffton *see*
– Linking ring
– News
Allen Co. Delphos *see* Kleeblatt
Allen Co. Lima *see*
– Allen county republican-gazette series
– Bulletin – strike paper
– Early newspapers
– Reporter series
– Times democrat
– Times-democrat series
Allen Co. Spencerville *see*
– Journal
– Journal news series
– Journal-news
Allen County. Kansas. School District 23 *see* Records
Allen county lines – 1979 mar-1988 jun – 1 – mf#1336334 – us WHS [071]
Allen county reporter – v34 n1-v41 n2/3 [1978-85] – 1 – mf#1099930 – us WHS [071]
Allen county republican-gazette series / Allen Co. Lima – (1889-01,03-07,09-1915) [wkly] – 18r – 1 – mf#B10327-10344 – us Ohio Hist [071]
Allen, David M *see*
– The comparison of resting metabolic rate in trained vs. untrained females
– A kinetic and kinematic comparison of the grab start and track start in swimming
Allen diary 1835-1837 *see* Travel diary
Allen, Donna *see* Coverage of the spiritual dimension of health in personal health textbooks in higher education
Allen, Edgar Leonard *see* Christianity and society
Allen, Edith Hedden *see* Home missions in action
Allen, Ethan *see*
– Clergy in maryland of the protestant episcopal church since the independence of 1783
– Reason, the only oracle of man
Allen, Frank Gibbs *see* The old-path pulpit
Allen, Fred C *see* Handbook of the new york state reformatory at elmira
Allen, Gardner Weld *see* Our navy and the west indian pirates
Allen, George *see* The andover fuss, or, dr. woods versus dr. dana, on the imputation of heresy against professor park respecting the doctrine of original sin
Allen gewalten zum trotz : lebenskaempfe, niederlagen, arbeitssiege / Carle, Erwin – 20. aufl. Stuttgart: R Lutz 1940, c1922 [mf ed 1989] – (= ser Lutz' memoiren-bibliothek 6/4) – 1r – 1 – mf#7143 – us Wisconsin U Libr [880]
Allen, Grant *see*
– An african millionaire
– An army doctor's romance
– At market value
– The attis of caius valerius catallus
– The beckoning hand
– Biographies of working men
– Blood royal
– The british barbarians
– Charles darwin
– Cities of belgium
– Colin clout's calendar
– The colour-sense
– The desire of the eyes
– The duchess of powysland
– The duchess of powysland, vol 1
– The duchess of powysland, vol 2
– The duchess of powysland, vol 3
– Dumaresq's daughter
– The european tour
– The evolution of the idea of god
– The evolutionist at large
– Falling in love
– Flashlights on nature
– Florence
– Flowers and their pedigrees
– For maimie's sake
– Force and energy
– The great taboo
– The hand of god
– Hilda wade
– In all shades
– In memoriam
– In nature's workshop
– Incidental bishop
– Ivan greet's masterpiece, etc
– Kalee's shrine
– Linnet
– The lower slopes
– Magdalen tower
– Michael's crag
– The miscellaneous and posthumous works of henry thomas buckle, vol 1
– The miscellaneous and posthumous works of henry thomas buckle, vol 2
– Miss cayley's adventures
– The natural history of selborne
– Paris
– Physiological aesthetics
– Plant life
– Recalled to life
– The return of aphrodite
– The scallywag
– Science in arcady
– A splendid sin
– The story of the plants
– Strange stories
– The tents of shem
– A terrible inheritance
– This mortal coil
– Tidal thames
– Tom, unlimited
– Twelve tales
– Under sealed orders
– Venice
– Vignettes from nature
– What's bred in the bone
– The white man's foot
– The woman who did
Allen, Grant [Cecil Power] *see*
– Babylon
– Babylon, vol 1
– Babylon, vol 2
– Babylon, vol 3
Allen, Grant [Olive Pratt Rayner] *see*
– Rosalba
– The type-writer girl
Allen, H N *see* Things korean
Allen, Hamilton Ford *see* The infinitive in polybius compared with the infinitive in biblical greek
Allen, Heidi *see* Stages of motif writing development in third grade children
Allen, Henry Justin *see* Venezuela
Allen, Herbert J *see* Early chinese history
Allen, I M *see* The us baptist annual register
Allen, J A *see* Bulletin of the nuttall ornithological club
Allen, J S *see* Based on byzantinische zeitschrift
Allen, Jacob D *see*
– The musings of uncle jake
– Poems
Allen, James Stewart *see* The negroes in a soviet america
Allen, James T *see* John g paton
Allen, Joe W *see* Neodesha scrapbook
Allen, Joel Asaph *see* History of north american pinnipeds
Allen, John *see*
– Reply to dr lingard's vindication
– State churches and the kingdom of christ
Allen, John (Mrs) *see* Alice warner
Allen, John Slater *see* From apollyonville to the holy city
Allen, Joseph Antisell *see*
– Daydreams by a butterfly
– Orangism, catholicism, and sir francis hincks
– The religion of the pope and primitive christianity
– A reply to the speech of the hon edward blake against the orange incorporation bill
– The true and romantic love-story of colonel and mrs hutchinson
Allen, Joseph H *see* Greek reader
Allen, Joseph Henry *see*
– Antichrist
– Christian history in its three great periods
– Hebrew men and times
– An historical sketch of the unitarian movement since the reformation
– A history of the unitarians and the universalists in the united states
– Our liberal movement in theology
– Outline of christian history, a.d. 50-1880
– Sequel to "our liberal movement"
Allen, Kristen L *see* Differences in intrinsic risk factors for injured and non-injured athletes
Allen, L *see*
– Architecture
– Boom in orlando 1923-1936
– Dairies
– Groveland, lake county, florida
– History of lake county
– History of orange county
– Lake county, florida
– Leesburg, lake county, florida
– Orlando zoos
– Points of interest in lake county
– Small communities in lake county, florida
Allen, Leslie Henri [comp] *see* Bryan and darrow at dayton
Allen, Marie-B *see* Elie goulet de la societe des ecrivains canadiens
Allen, Mary *see* On the cards or the return of the princess
Allen, Mary Moncrief Simons *see* Travel diary
Allen, Melissa S *see* The roles of popular entertainment dance during the great depression
Allen memorial art museum. bulletin – Oberlin OH 1944+ – 1,5,9 – ISSN: 0002-5739 – mf#6722 – us UMI ProQuest [060]
Allen memorial baptist church – Grover, NC: Kings Mountain Assoc, 1947-oct 1963 – 1 – $13.86 – us Southern Baptist [242]
Allen, Nathan *see* Lecture
Allen news – Allen, Dixon County, NE: News Pub Co, -v53 n26. nov 18 1948 (wkly) [mf ed jan 21913-may 6 1948 (gaps)] – 10r – 1 – (absorbed by: wakefield republican. suspended publ during world war ii) – us NE Hist [071]
Allen newsletter – v1 n1-v3 n4 [1975 oct-1978 jan]; v3 n5 [1982 nov]; v3 n6-7, 9-10 [1983 feb-mar, may-jun] – 1 – mf#379632 – us WHS [071]
Allen, Otis *see* Memoir of otis allen.
Allen, P S *see*
– The age of erasmus
– Opus epistolarum
Allen, Percy Stafford *see* Transactions of the third international congress for the history of religions
Allen, Philip Schuyler *see*
– In longfellows pantoffeln
– Wilhelm mueller and the german volkslied
Allen, Phoebe *see*
– The boys of priors dean
– Like to a double cherry
– Minon
– Old iniquity
– Thanksgiving tabernacle
Allen, Richard *see* Brief vindications of an essay to prove singing of psalms, etc
Allen, Ross *see* Fishes of silver springs, florida
Allen, Samuel E S *see* Explorations among the watershed rockies of canada
Allen, Stephen Merrill *see*
– The life of rev john allen
– Religion and science
Allen, Sue *see*
– Victorian bookbindings
Allen, Thomas *see* History and antiquities of london, westminster, southwark and parts adjacent
Allen, Thomas Coley *see* Problem of city government
Allen, Thomas George *see* Horus in the pyramid texts
Allen, Thomas Gilchrist *see* Psychic research and gospel miracles
Allen, W *see*
– A narrative of the expedition...to the river niger, in 1841
– Studies in african land usage in northern rhodesia
Allen, W C *see* A brief unpublished history of the baptists of south carolina, 1683-1937
Allen, Wilkes *see* Apollo
Allen, William *see* Brief remarks upon the carnal and spiritual state of man
Allen, William Francis *see* Essays and monographs
Allen, William K *see* Lactate threshold in masters athletes as compared to young athletes
Allen, William Osborne Bird *see* Two hundred years
Allen, William Stannard *see* Phonetics in ancient india
Allen, Willoughby Charles *see*
– A critical and exegetical commentary on the gospel according to s matthew
– Introduction to the books of the new testament
Allensteiner kreisblatt – Allenstein (Olsztyn) PL, 4 jan 1851-55, 1857-59*, 1863-69, 2 jul-29 sep 1912, 1 jan-31 mar 1914, 1 jul 1920-30 jun 1921* – 12r – 1 – (cont by: allensteiner zeitung, 1881) – Dist. gw Mikrofilm – gw Misc Inst [077]
Allensteiner volksblatt – Olsztyn, Poland 6 jan 1918-6 aug 1919 (imperfect) – 3r – 1 – uk British Libr Newspaper [077]
Allensworth, Allen *see* Papers
Allenton baptist church *see* Eureka central baptist church. eureka, missouri
The allentown democrat – Allentown, PA. -d 1879-1918 – 13 – $25.00r – us IMR [071]
Allentown teacher – 1978 oct-1981 aug – 1 – mf#633792 – us WHS [071]
Allenwood, Pennsylvania.White Deer Baptist Church *see* Minutes and members
Aller praktik grossmutter / Fischart, Johann; ed by Braune, Wilhelm – Halle a/S: M Niemeyer 1876 [mf ed 1993] – (= ser Neudrucke deutscher literaturwerke des 16. und 17. jahrhunderts 2) – 11r – 1 – (filmed with: neudrucke deutscher literaturwerke des 16. und 17. jahrhunderts) – mf#3387p – us Wisconsin U Libr [430]
Allergnaedigst privilegierte anzeigen – Vienna, Ghelen, Austria jan 1771-jun 1776 [mf ed Norman Ross] – 2r – 1 – mf#nrp-2033 – us UMI ProQuest [074]
Der allergnaedigste privilegierte saechsische postillon *see* Der privilegirte churfuerstlich saechsische postillion
Allergy information association. newsletter – Toronto ON 1973-74 – 1 – ISSN: 0705-0984 – mf#8711 – us UMI ProQuest [616]

Allerhand humore : kleinbaeuerliches, grossstaedtisches und gefabeltes / Anzengruber, Ludwig – Leipzig: Breitkopf und Haertel, 1883 [mf ed 1988] – 204p – 1 – mf#6940 n12 – us Wisconsin U Libr [880]

Allerhand slag lued : vertelln / Fehrs, Johann Hinrich – Braunschweig: G Westermann [19–?] [mf ed 1989] – 1r – 1 – (filmed with: neun essays / von karl federn) – mf#7233 – us Wisconsin U Libr [880]

Allerhand ungezogenheiten / Blumenthal, Oscar – 4. aufl. Leipzig: E J Guenther, 1876 [mf ed 1989] – 1 – mf#7036 – us Wisconsin U Libr [880]

Allerhoechst privilegirte schleswig-holsteinische anzeigen – Glueckstadt: J W Augustin. nf: v29-43(1865-79) (annual) [mf ed 1978] – 8r – 1 – (began in 1837) – mf#film mas c392 – us Harvard [943]

Allerlei gereimtes / Fontane, Theodor; ed by Rost, Wolfgang – Dresden: C Reissner, 1932 [mf ed 1989] – xvi/247p – 1 – mf#7248 – us Wisconsin U Libr [810]

Allerneueste europaeische welt- und staats-geschichte – Erfurt DE, 1744 1 jan-21 sep – 1r – 1 – gw Misc Inst [940]

Allerneueste mannigfaltigkeiten : eine gemeinnuetzige wochenschrift / ed by Otto, Johann Friedrich Wilhelm – Berlin 1782-84 – (= ser Dz) – 3 jge on 15mf – 9 – €150.00 – mf#k/n5605 – gw Olms [073]

Allers : illustrerad familjejournal – Kobenhavn, Denmark; Helsingborg, Sweden. 1879-1955 – 1 – sw Kungliga [078]

De allerverborgenste geheimen van de zeven hoofdgraaden der vrijmetselaarij ontdekt : of, het waare roozenkruis: uit het engelsch vertaald, waarbij gevoegd is de geschiedenis der noachiten, uit het hoogduitsch in het fransch vertaald... – [s.l: s.n] 1805 – 2mf – 9 – mf#vrl-3 – ne IDC [366]

Aller-zeitung – Gifhorn DE, 1852 4 aug-1863, 1865-66, 1868-77, 1879-95, 1897-1945 6 apr – 78r – 1 – (filmed by other misc inst: 1987- [7r/yr]) – gw Misc Inst [074]

Alles um goethe : kleine aufsaetze und reden / Wahl, Hans; ed by Wahl, Dora – Weimar: G Kiepenheuer, 1956 [mf ed 1993] – (= ser Gustav kiepenheuer buecherei) – 192p/4pl (ill) – 1 – (incl bibl ref) – mf#8652 – us Wisconsin U Libr [430]

Alles um liebe : goethes briefe aus der ersten haelfte seines lebens / ed by Hartung, Ernst – Ebenhausen bei Muenchen: W Langewiesche-Brandt, 1913 [mf ed 1993] – (= ser Buecher der rose 2) – 1 – mf#8607 – us Wisconsin U Libr [430]

Alles um liebe : goethes briefe aus der ersten haelfte seines lebens / ed by Hartung, Ernst – Duesseldorf: W Langewiesche-Brandt 1907 [mf ed 1990] – (= ser Die buecher der rose 2) – 1r [ill] – 1 – (filmed with: correspondence of fraulein gunderode and bettine von arnim) – mf#2781p – us Wisconsin U Libr [860]

Alles unsinn : deutsche ulk- und scherzdichtung von ehedem bis momentan / ed by Seydel, Heinz – Berlin: Eulenspiegel-Verlag, 1969 [mf ed 1993] – 317p – 1 – (incl ind) – mf#8361 – us Wisconsin U Libr [870]

Alletz, Pons A *see* Ceremonial du sacre des rois de france

The all-father : sermons preached in a village church / Newnham, Philip Hankinson – 2d ed. London; New York: Longmans, Green, 1891. Beltsville, Md: NCR Corp, 1978 (3mf); Evanston: American Theol Lib Assoc, 1984 (3mf) – 9 – 0-8370-1063-2 – mf#1984-4409 – us ATLA [240]

Allfeld, J B *see* Tristan und isolde von richard wagner

Allgaeu sturm *see*
– Leutkircher wochenblatt
– Schwaebische zeitung [main edition]

Der allgaeuer : kempter tagblatt – Kempten DE, 1946 20 sep-1948 27 nov [many gaps] – 1r – 1 – (filmed by misc inst: 1945 13 dec-1968 30 sep [74r]; title varies: 1 oct 1968: allgaeuer zeitung / ke [regional ed of: augsburger allgemeine], 1 sep 1981: allgaeuer zeitung [regional ed of: augsburger allgemeine]) – gw Mikrofilm; gw Misc Inst [074]

Allgaeuer anzeigeblatt – Immenstadt (Allgaeu) DE, 1983 1 jun- – ca 8r/yr – 1 – (bezirksausgabe von allgaeuer zeitung, kempten) – gw Misc Inst [074]

Allgaeuer bote *see*
– Leutkircher wochenblatt
– Schwaebische zeitung [main edition]

Allgaeuer volksfreund *see* Leutkircher wochenblatt

Allgaeuer zeitung : kaufbeurer tagblatt – Kaufbeuren DE, 1983 1 jun- – ca 9r/yr – 1 – (main ed in kempten) – gw Misc Inst [074]

Allgaeuer zeitung / ke *see* Der allgaeuer

Allgeier, A *see* Der palimpsestpsalter im cod sangallensis 91 (tab21-24)

Allgeier, Arthur *see* Uber doppelberichte in der genesis

Allgemeine annalen der gewerbskunde – Leipzig DE, 1803-04 – 1 – gw Mikrofilm [943]

Allgemeine arbeiter-zeitung – Frankfurt/M DE, 1848 18 may-10 jun – 1r – 1 – gw Misc Inst [331]

Allgemeine arbeiterzeitung – Budapest (H) 1870 – 1r – 1 – gw Misc Inst [331]

Allgemeine auswanderungs-zeitung – Rudolstadt DE, 1846 29 sep-1856 22 dec – 9r – 1 – (with suppls) – gw Misc Inst [074]

Allgemeine automobil-zeitung und officielle mittheilungen des oesterreichischen automobil-club – Vienna, Austria 7 jan 1900-26 dec 1909 – 19r – 1 – (wanting: 1903-05) – uk British Libr Newspaper [629]

Allgemeine badzeitung – Baden-Baden DE, 1849 16 feb-23 jun – 1r – 1 – (title varies: apr 2 1849: mittelrheinische zeitung) – gw Misc Inst [790]

Allgemeine bauzeitung / ed by Foerster, C F L – Wien, 1836-1918. v1-83. ind 1836-1885; 1874-1894 – 727mf – 9 – mf#O-1738 – ne IDC [720]

Allgemeine berg- und huettenmaennische zeitung, etc – Quedlinburg, Germany 4 jan 1859-24 dec 1863 – 2 1/2r – 1 – uk British Libr Newspaper [622]

Allgemeine bibliothek der biblischen litteratur / ed by Eichhorn, Johann Gottfried – Leipzig 1787-1801 – (= ser Dz. abt theologie) – 10v [zu je 6st] on 71mf – 9 – €710.00 – mf#k/n2218 – gw Olms [220]

Allgemeine bibliothek fuer das schul- und erziehungswesen in deutschland / ed by Boeckh, Christian Gottfried – Noerdlingen 1773-86 – (= ser Dz) – 11v on 44mf – 9 – €440.00 – mf#k/n620 – gw Olms [370]

Allgemeine casseler vereins-zeitung – Kassel DE, 1913 4 jan-29 mar – 1r – 1 – gw Misc Inst [074]

Das allgemeine concil und seine bedeutung fuer unsere zeit / Ketteler, Wilhelm Emmanuel, Freiherr von – 4.auf. Mainz: Franz Kirchheim, 1869 [mf ed 1986] – 1mf – 9 – 0-8370-8177-7 – mf#1986-2177 – us ATLA [241]

Das allgemeine concil von vatican : zwei hirtenschreiben / Rauscher, Joseph Othmar – Wien: Wilhelm Braumueller, 1870 [mf ed 1986] – 1mf – 9 – 0-8370-8372-9 – mf#1986-2372 – us ATLA [241]

Allgemeine correspondenz *see* Deutsche korrespondenz

Allgemeine deutsche arbeiter-zeitung *see* Arbeiter-zeitung

Allgemeine deutsche bibliothek : literarische zeitschrift – Berlin DE, Stettin (Szczecin PL), 1766-94 – 913mf – 9 – mf#6494 – gw Mikropress [430]

Allgemeine deutsche bibliothek / ed by Nicolai, Friedrich – Berlin, Stettin [1770ff: Kiel] 1765-96 – (= ser Dz) – 141v + 86 suppl vol on 559mf – 9 – €3354.00 – mf#k/n248 – gw Olms [020]

Allgemeine deutsche bibliothek / ed by Nicolai, Friedrich – Berlin, Stettin, Kiel, 1765-96 [mf ed Hildesheim 1992-98] – 141v+86v suppl on 559mf – 9 – €3354.00 diazo – gw Olms [430]

Allgemeine deutsche bibliothek [adb] / ed by Nicolai, Friedrich – Berlin/Stettin 1765-96 [mf ed 1993] – 770mf – 9 – €4000 diazo €4800 silver – 3-89131-108-7 – (filmed with: neue allgemeine deutsche bibliothek [berlin/ stettin sp kiel 1793-1806], intelligenzblaetter [1793-1800], with app: gustav parthey: "die mitarbeiter an friedrich nicolai's allgemeiner deutscher bibliothek") – gw Fischer [020]

Allgemeine deutsche biografie – Leipzig, 1875-1912. 56v – 499mf – 9 – mf#H-3026 – ne IDC [700]

Allgemeine deutsche lehrerzeitung *see* Berliner paedagogische zeitung

Allgemeine deutsche musik-zeitung / ed by Reinsdorf, Otto et al – [mf ed 1988] – 670mf (1:24) – 9 – silver €3150.00 – ISBN-10: 3-598-32530-4 – ISBN-13: 978-3-598-32530-4 – gw Saur [780]

Das allgemeine deutsche pfennig-magazin – Danzig (Gdansk PL), 1834 jan-jun – 1r – 1 – gw Misc Inst [730]

Allgemeine deutsche polytechnische zeitung – Berlin, Germany 4 jan 1873-26 dec 1874 – 1 – (cont by: engineering d a polytechnische zeitung [2 jan- 25 dec 1875, 4 jan 1879-30 dec 1882, 5 jan-27 dec 1884]) – uk British Libr Newspaper [378]

Allgemeine deutsche real-encyclopaedie fuer die gebildeten staende [conversations-lexikon] (ael1/ 35.13) – 7th ed. Leipzig 1827 [mf ed 1997] – (= ser Das brockhaus conversations-lexikon 1796-1898 (ael1/35)) – 12v on 75mf – 9 – €570.00 – 3-89131-263-6 – gw Fischer [030]

Allgemeine deutsche real-encyclopaedie fuer die gebildeten staende [conversations-lexikon] (ael1/ 35.15) – 7th ed. Leipzig 1830 [mf ed 1997] – (= ser Das brockhaus conversations-lexikon 1796-1898 (ael1/35)) – 12v on 99mf – 9 – €570.00 – 3-89131-277-6 – gw Fischer [030]

Allgemeine deutsche real-encyclopaedie fuer die gebildeten staende [conversations-lexikon] (ael1/ 35.18) – 10th ed. Leipzig 1851-55 [mf ed 1997] – (= ser Das brockhaus conversations-lexikon 1796-1898 (ael1/35)) – 15v on 85mf – 9 – €640.00 – 3-89131-267-9 – gw Fischer [030]

Allgemeine deutsche real-encyclopaedie fuer die gebildeten staende [conversations-lexikon] (ael1/35.8) – 5th ed. Leipzig 1820 [mf ed 1997] – (= ser Das brockhaus conversations-lexikon 1796-1898 (ael1/35)) – 10v on 61mf – 9 – €590.00 – 3-89131-258-X – gw Fischer [030]

Allgemeine deutsche real-encyclopaedie fuer die gebildeten staende [conversations-lexikon] (ael1/35.11) – 6th ed. Leipzig 1824 [mf ed 1997] – (= ser Das brockhaus conversations-lexikon 1796-1898 (ael1/35)) – 10v on 60mf – 9 – €590.00 – 3-89131-261-X – gw Fischer [030]

Allgemeine deutsche real-encyclopaedie fuer die gebildeten staende [conversations-lexikon] (ael1/35.12) : supplementband fuer die besitzer der fuenften und frueheren auflagen – Leipzig 1824 [mf ed 1997] – (= ser Das brockhaus conversations-lexikon 1796-1898 (ael1/35)) – 7mf – 9 – €100.00 – 3-89131-262-8 – gw Fischer [030]

Allgemeine deutsche real-encyclopaedie fuer die gebildeten staende [conversations-lexikon] (ael1/35.16) – 8th ed. Leipzig 1833-37, 1839 [mf ed 1997] – (= ser Das brockhaus conversations-lexikon 1796-1898 (ael1/35)) – 12v+ind on 74mf – 9 – €610.00 – 3-89131-265-2 – gw Fischer [030]

Allgemeine deutsche real-encyclopaedie fuer die gebildeten staende [conversations-lexikon] (ael1/35.17) – 9th ed. Leipzig 1843-48 [mf ed 1997] – (= ser Das brockhaus conversations-lexikon 1796-1898 (ael1/35)) – 15v on 75mf – 9 – €610.00 – 3-89131-266-0 – gw Fischer [030]

Allgemeine deutsche real-encyclopaedie fuer die gebildeten staende [conversations-lexikon] (ael1/35.19) – 11th ed. Leipzig 1864-68 [mf ed 1997] – (= ser Das brockhaus conversations-lexikon 1796-1898 (ael1/35)) – 90mf – 9 – €820.00 – 3-89131-268-7 – gw Fischer [030]

Allgemeine deutsche real-encyclopaedie...(ael1/ 35.14) : supplementband fuer die besitzer der sechsten und frueheren auflagen und der neuen folgen – Leipzig 1829 [mf ed 1997] – (= ser Das brockhaus conversations-lexikon 1796-1898 (ael1/35)) – 7mf – 9 – €100.00 – 3-89131-264-4 – gw Fischer [030]

Allgemeine deutsche schulzeitung – Berlin DE, 1889-90 – 1r – 1 – gw Misc Inst [370]

Allgemeine deutsche zeitung fuer rumaenien *see* Neuer weg

Allgemeine einleitung, 1. bd (bdk17 1.reihe) : exameron / Ambrosius – (= ser Bibliothek der kirchenvaeter. 1. reihe (bdk 1.reihe)) – €15.00 – ne Slangenburg [240]

Allgemeine einleitung in die schriften des neuen testaments *see* The gospel records

Allgemeine encyclopaedie der wissenschaften und kuenste / Ersch, Johann S – [mf ed 1996] – 310mf (1:24) – 9 – silver €4590.00 – ISBN-10: 3-598-33511-3 – ISBN-13: 978-3-598-33511-2 – (sect 1: v1-99 leipzig 1818-82; sect 2: v1-43 leipzig 1827-89; sect 3: v1-25 leipzig 1830-50) – gw Saur [030]

Allgemeine encyclopaedie der wissenschaften und kuenste (ael1/33) / Ersch, Johann Samuel & Gruber, Johann Gottfried – Leipzig 1818-89 [mf ed 1995] – (= ser Archiv der europaeischen lexikographie, abt 1: enzyklopaedien) – 167v on 462mf – 9 – €3840.00 – 3-89131-214-8 – gw Fischer [030]

Allgemeine familien-zeitung – Stuttgart DE, 1869-72 – 1r – 1 – gw Misc Inst [640]

Allgemeine fleischer zeitung – Berlin, Germany 1 nov 1916 nov-31 dec 1918 (imperfect) – 5r – 1 – uk British Libr Newspaper [636]

Allgemeine forstzeitschrift – vol. 1-12. 1946-57. (Scattered issues lacking) – 1 – us L of C Photodup [634]

Allgemeine frauenzeitung – Vienna, Stuttgart, Leipzig jan-dec 1871 – 1r – nrp-2037 – us UMI ProQuest [074]

Allgemeine geographie der insel rhodos / Heffter, Moritz Wilhelm – Brandenburg: Wiesike [Druck] 1827 – (= ser Whsb) – 1mf – 9 – €20.00 – mf#Hu 471 – gw Fischer [914]

Allgemeine geschichte der literatur des mittelalters im abendlande / Ebert, Adolf – Leipzig: F C W Vogel 1874-87 [mf ed 1979] – 3v on 1r – 1 – (incl bibl ref & ind; 1. bd: geschichte der christlich-lateinischen litteratur von ihren anfaengen bis zum zeitalter karis des grossen [1874]; 2. bd: die lateinische literatur vom zeitalter karis des grossen bis zum tode karis des kahlen [1880]; 3. bd: die nationalen literaturen von ihren anfaengen und die lateinische literatur vom tode karls des kahlen bis zum beginne des elften jahrhunderts [1887]) – mf#film mas 9227 – us Harvard [410]

Allgemeine geschichte der morgenlaendischen sprachen und litteratur worinnen von sprache und litteratur der armener... / Wahl, S F G – Leipzig, 1784 – 8mf – 9 – mf#AR-1578 – ne IDC [956]

Allgemeine geschichte der musik / Forkel, Johann Nikolaus – Leipzig: Im Schwickertschen verlage 1788-1801 [mf ed 19–] – 2v on 2mf – 9 – mf#fiche 460 – us Sibley [780]

Allgemeine geschichte der philosophie / Deussen, Paul – Leipzig. v1/1-3. 1920 (v11/1 7mf v1/2 7mf v1/3 13mf) – 8 – €52.00 set – ne Slangenburg [100]

Allgemeine geschichte des priesterthums / Lippert, J – Berlin. bd1-2. 1883-1884 – €44.00 – ne Slangenburg [241]

Allgemeine geschichte des priesterthums / Lippert, Julius – Berlin: Theodor Hofmann, 1883-84 [mf ed 1989] – 2v on 3mf – 9 – 0-7905-2980-7 – (incl bibl ref) – mf#1987-2980 – us ATLA [200]

Allgemeine geschichte des zeitungswesens / Salomon, Ludwig – 1907 – 1 – gw Mikropress [943]

Allgemeine gewerbe-zeitung – Berlin DE, 1874 2 oct-1875, 1877-79 – 1r – 1 – gw Mikrofilm [074]

Allgemeine handlungszeitung – Nuernberg DE, 1818-29 – 8r – 1 – gw Misc Inst [380]

Allgemeine handwerker- und gewerbe-zeitung *see* Allgemeine handwerkerzeitung

Allgemeine handwerkerzeitung – Muenchen DE, 1887-1919 – 1 – (title varies: 1917: allgemeine handwerker- und gewerbe-zeitung) – gw Misc Inst [640]

Allgemeine historische bibliothek / ed by Gatterer, Johann Christoph – Halle 1767-71 – (= ser Dz. historisch-geographische abt) – 16pt on 35mf – 9 – €350.00 – mf#k/n1041 – gw Olms [943]

Allgemeine illustrierte zeitung – Leipzig DE, 1865-69 – 1 – gw Misc Inst [074]

Allgemeine illustrirte judenzeitung, carmel – Budapest (H), 1860-1861 21 jun – 1r – 1 – gw Misc Inst [939]

Allgemeine industriezeitung *see* Generalanzeiger fuer fabrikbedarf 1912

Allgemeine juedische wochenzeitung – Duesseldorf DE, 1976-97 – 21r – 1 – (1998 subsc) – mf#1005 – gw Mikropress [939]

Allgemeine juedische wochenzeitung – Duesseldorf, Bonn DE, 1946 15 apr-1982 – 1 – (title varies: juedisches gemeindeblatt fuer die nordrhein-provinz und westfalen, juedisches gemeindeblatt fuer die britische zone, juedisches gemeindeblatt – allgemeine zeitung der juden in deutschland, allgemeine wochenzeitung der juden in deutschland, allgemeine unabhaengige juedische wochenzeitung, allgemeine juedische wochenzeitung. suppl also available: allgemeine juedische illustrierte, sep 1950-may 1968, 1990-; fr 9 feb 1973 publ in bonn) – gw Misc Inst [939]

Allgemeine juedische wochenzeitung *see* Allgemeine juedische wochenzeitung

Allgemeine juristische bibliothek / ed by Malblanc, Julius Friederich & Siebenkees, Johann Christian – Nuernberg 1781-86 [mf ed 1992-98] – (= ser Dz) – 6v[=12st] on 17mf – 9 – €170.00 – mf#k/n2563 – gw Olms [340]

Allgemeine kino-boerse – Leipzig DE, 1919 n5-1922 18 sep [gaps] – 1r – 1 – gw Mikrofilm [790]

Die allgemeine kirchenordnung, fruehchristliche liturgien und kirchliche ueberlieferung / ed by Schermann, Theodor – Paderborn: F Schoeningh, 1914-16 [mf ed 1991] – (= ser Studien zur geschichte und kultur des altertums 3) – 3v on 2mf – 9 – 0-524-01666-6 – (incl bibl ref. discussion in german, text in greek & latin) – mf#1990-0487 – us ATLA [240]

Allgemeine kirchen-zeitung – 3(1824)-11(1832) – 287mf – 9 – €547.00 – ne Slangenburg [240]

Allgemeine kirchliche zeitschrift – 1(1860)-9(1868) – 107mf – 9 – €202.00 – ne Slangenburg [240]

Allgemeine kritische geschichte der religionen / Meiners, Christoph – Hannover: Helwing, 1806-07 [mf ed 1993] – 2v on 4mf – 9 – 0-524-06689-2 – mf#1990-3550 – us ATLA [230]

Allgemeine kunst-chronik – Wien (A), Muenchen DE, 1888, 1890-91, 1894-95 – 5r – 1 – gw Misc Inst [700]

Allgemeine laender- und voelkerkunde : ein lehr- und hausbuch fuer alle staende; nebst einem abriss der physikalischen erdbeschreibung / Berghaus, Heinrich K – Stuttgart 1837-44 [mf ed Hildesheim 1995-98] – 6v on 32mf – 9 – €320.00 – 3-487-29954-2 – gw Olms [910]

Das allgemeine landrecht fuer die preussischen staaten : in seiner jetzigen gestalt – 3., verb Aufl. Berlin: C Heymann, 1896 – (= ser Civil law 3 coll) – 6mf – 9 – (incl ind) – mf#LLMC 96-522 – us LLMC [348]

Die allgemeine lehre, die taufe, die firmung und die eucharistie / Oswald, Johann Heinrich – 2. verb aufl. Muenster: Aschendorff, 1864 [mf ed 1992] – (= ser Lehre von den heiligen sakramenten der katholischen kirche 1; Roman catholic coll) – 2mf – 9 – 0-524-04556-9 – mf#1991-2120 – us ATLA [241]

Allgemeine linguistische alphabet *see* Standard alphabet

Allgemeine literatur-zeitung / ed by Bertuch, Friedrich Johann Justin et al – Halle 1804-49 – (= ser Dz) – jg 1-45 on 611mf – 9 – €3666.00 – mf#k/n478 – gw Olms [400]

Allgemeine literatur-zeitung / ed by Schuetz, Christian Gottfried & Ersch, Johann Samuel – Halle 1804-49 [mf ed 1996] – 770mf – 9 – €4350 diazo €5220 silver – 3-89131-109-5 – (mit ergaenzungsblaettern, intelligenzblaettern, registern & kupferstichen) – gw Fischer [430]

Allgemeine literatur-zeitung / ed by Schuetz, Christian Gottlieb et al – Jena/Leipzig 1785-1803 – (= ser Dz) – 526mf – 9 – €3156.00 – mf#k/n388 – gw Olms [400]

Allgemeine litterarische rundschau *see* National-zeitung 1848

Allgemeine musikalische zeitung – Leipzig. v. 1-8. Oct. 1798-Sept. 1806 – 1 – us NY Public [780]

Allgemeine musikalische zeitung / ed by Rochlitz, G Fink et al – Leipzig. v1-50 + index. 1798-1848 – 11 – $660.00 – us Univ Music [780]

Allgemeine musikalische zeitung – v. 1-50. 1798-1848. n.s. 1-3. 1863-65. s.3. 1-17. 1866-82. Index, 1798-1848 – 1 – 544.00 – us L of C Photodup [780]

Allgemeine musikalische zeitung mit besonderer rucksicht – Auf den Osterreichischen Kaiserstaat. Vienna. 1817-1824 – 1 – us NY Public [780]

Allgemeine musikalische zeitung, mit besonderer rucksicht auf den oesterreichischen kaiserstaat – Wien. 1817-21. 1 reel – 1 – us L of C Photodup [780]

Allgemeine musik-zeitung – Berlin, Leipzig etc. v. 8-20, 26-70, no. 6. 1881-Mar 19 1943 – 1 – us NY Public [780]

Allgemeine nachrichten fuer pommerellen – Briesen (Wabrzezno PL), 1928 16 oct-1930 28 jun, 1931 3 jan-14 nov – 1 – (cont: briesener zeitung) – gw Misc Inst [077]

Allgemeine naturgeschichte und theorie des himmels *see* Kant's cosmogony

Allgemeine oder-zeitung – Breslau (Wroclaw PL), 1846 2 apr-1847 30 jun, 1848-50 (morgenblatt) – 11r – 1 – (title change: neue oder-zeitung, mar 27 1849; filmed by other misc inst: 1846-1847 jun, 1848-55 [21r]) – gw Misc Inst [077]

Allgemeine photographen zeitung – Muenchen DE, 1896-97 [single pgs] – 1r – 1 – gw Mikrofilm [770]

Allgemeine politische nachrichten – Essen DE, 1883 16 may-1944 31 aug – 192r – 1 – (title varies: 1 jan 1860: essener zeitung, 15 may 1883: rheinisch-westfaelische zeitung; filmed by other misc inst: 1871 apr-dec) – gw Misc Inst [320]

Allgemeine politische zeitung fuer die provinz preussen – Danzig (Gdansk PL), 1838, 1841, 1843, 1845 – 1 – gw Misc Inst [077]

Allgemeine preussische staats-zeitung – Berlin DE, 1819-1945 14 apr – 553r – 1 – (filmed by misc inst: 1840 [2r], 1848 [2r], 1935-38 [gaps], 1940 n85, 1941 n148; title varies: 1 jul 1843: allgemeine preussische zeitung, may 1848: preussischer staats-anzeiger, 1 jul 1851: koeniglich-preussischer staatsanzeiger, jul 1871: deutscher reichsanzeiger und preussischer staatsanzeiger; with suppls: zentralhandelsregister 1940-44 278 (gaps) [also: handelsregister, 5r]) – gw Mikropress; gw Misc Inst [074]

Allgemeine preussische zeitung *see*
- Allgemeine preussische staats-zeitung
- Neueste berliner morgenzeitung

Allgemeine realencyclopaedie (ael1/16) : oder conversations-lexikon fuer das katholische deutschland / ed by Binder, Wilhelm – Regensburg (Manz) 1846-50 (mf ed 1993] – (= ser Archiv der europaeischen lexikographie, abt 1: enzyklopaedien) – 10v+2 suppl vol on 124mf – 9 – €710.00 – 3-89131-104-4 – (int by otmar seeman) – gw Fischer [030]

Allgemeine reise-encyklopaedien auszugen aus den groesseren bisher erschienenen reisewerken : zur unterhaltenden belehrung in der laender-, voelker- und naturkunde; ein buch fuer gebildete leser, fuer lehrende und lernende in allen staenden – Leipzig 1810-11 [mf ed Hildesheim 1995-98] – 6v on 17mf – 9 – €170.00 – 3-487-26461-7 – gw Olms [910]

Allgemeine religionsgeschichte / Orelli, Conrad von – [2. aufl] Bonn: A Marcus & E Weber, 1911-13 [mf ed 1992] – 2v on 3mf – 9 – 0-524-04165-2 – (incl bibl ref) – mf#1990-3295 – us ATLA [200]

Der allgemeine rheinische anzeiger – Karlsruhe DE, 1837 16 dec-1838 26 sep – 1r – 1 – gw Misc Inst [074]

Allgemeine schlosser- und maschinenbauer zeitung – Luebeck DE, 1919-29 – 7r – 1 – mf#9708 – gw Mikropress [620]

Allgemeine schutzhuettenzeitung fuer die ostalpen – Gaishorn (A), Wien (A), 1929/30-1943 – 2r – 1 – (publ in vienna since 1940) – gw Misc Inst [790]

Allgemeine slawische zeitung – Vienna, Austria jan-dec 1848 [mf ed Norman Ross] – 1r – 1 – mf#nrp-1924 – us UMI ProQuest [074]

Die allgemeine sonntagszeitung – Duesseldorf, Wuerzburg DE, 1956-88 – 1 – gw Misc Inst [074]

Allgemeine sport-zeitung – Vienna, Austria jul 1880-sep 1927 [mf ed Norman Ross] – 41r – 1 – mf#nrp-1925 – us UMI ProQuest [790]

Allgemeine staats-korrespondenz als zeitgemaesse reihefolge der zeitschrift der rheinische bund – Aschaffenburg 1814-15 – (= ser Dz. historisch-politische abt) – 3v on 9mf – 9 – €180.00 – mf#k/n1891 – gw Olms [320]

Der allgemeine teil des deutschen buergerlichen rechts / Tuhr, Andreas von – Leipzig: Duncker & Humblot. 2v in 3. 1910-18 – (= ser Civil law 3 coll) – 21mf – 9 – (contents: 1. bd. allgemeine lehren und personenrecht; 2. bd. die rechtserheblichen tatsachen, insbesondere das rechtsgeschaeft. incl bibl ref & ind) – mf#LLMC 96-541 – us LLMC [346]

Allgemeine theater-chronik : organ fuer das gesamtineresse der deutschen buehnen und ihrer mitglieder – Leipzig DE, 1852 1 oct-1873 24 mar – 5r – 1 – mf#12507 – gw Mikropress [790]

Allgemeine theorie der schoenen kuenste / Sulzer, Johann Georg – Leipzig. 4v. 1792-1794 – 56mf – 9 – mf#0-438 – ne IDC [700]

Allgemeine theorie der schoenen kuenste / Sulzer, Johann Georg – Leipzig. v1-5. 1720-79 – 9 – $120.00 – mf#0574 – us Brook [780]

Allgemeine unabhaengige juedische wochenzeitung *see* Allgemeine juedische wochenzeitung

Allgemeine unterhaltende reise-bibliothek : oder sammlung der besten und neuesten reisebeschreibungen / Fischer, Christian A – Berlin 1806-09 [mf ed Hildesheim 1995-98] – 4v on 13mf – 9 – €130.00 – 3-487-29929-1 – gw Olms [910]

Allgemeine vereins-zeitung – Kassel DE, 1907 23 mar-7 nov – 1r – 1 – gw Misc Inst [360]

Die allgemeine vergleichende religionswissenschaft : im akademischen studium unserer zeit: eine akademische antrittsrede / Hardy, Edmund – Freiburg i B: Herder, 1887 [mf ed 1991] – 1mf – 9 – 0-524-00885-X – mf#1990-2108 – us ATLA [230]

Allgemeine volkszeitung – Vienna, Austria aug 1868-nov 1873 [mf ed Norman Ross] – 7r – 1 – (liberal democratic) – mf#nrp-1926 – us UMI ProQuest [074]

Allgemeine volkszeitung, arbeiterblatt – Vienna, Austria jul-dec 1868 [mf ed Norman Ross] – 1r – 1 – (liberal democratic) – mf#nrp-1928 – us UMI ProQuest [074]

Allgemeine weltgeschichte : mit besonderer beruecksichtigung des geistes und culturlebens der voelker und mit benutzung der neueren geschichtlichen forschungen / Weber, Georg – Leipzig: Engelmann 1857-80 [mf ed 1979] – 15v on 7r – 1 – mf#film mas c468 – us Harvard [900]

Allgemeine wiener musik-zeitung / ed by Schmidt, August – Vienna, 1841-48 – 11 – $185.00 – (a categorical ind incl for each vol yr) – us Univ Music [780]

Allgemeine wochen-chronik – Bremen DE, 1854 6 aug-1855 7 jan – 1r – 1 – gw Misc Inst [074]

Allgemeine wochenzeitung der juden in deutschland *see* Allgemeine juedische wochenzeitung

Allgemeine zeitschrift von deutschen fuer deutsche / ed by Schelling, Friedrich Wilhelm Joseph – Nuernberg 1813 – (= ser Dz. abt literatur) – 1v[=4iss] on 4mf – 9 – €120.00 – mf#k/n4734 – gw Olms [074]

Allgemeine zeitung – Augsburg, Munich, Germany. aug 3 1847-dec 31 1855; jan 1877-dec 31 1922 [daily] – 1 – mf#ZY 73-1 – us NY Public [074]

Allgemeine zeitung – Berlin DE, 1945 8 aug-11 nov – 1r – 1 – gw Misc Inst [074]

Allgemeine zeitung – Windhuk (Windhoek NAM), 6 jan 1997-2000, 20 dec 2000-30 jun 2003 – 16r – 1 – (filmed by misc inst: 1972-79; 1982-96; 2001) – gw Mikrofilm; gw Misc Inst [079]

Allgemeine zeitung : kreisblatt fuer den kreis coesfeld – Coesfeld DE, 1981- – 7r/yr – 1 – gw Mikrofilm [074]

Allgemeine zeitung – Mainz, Germany. Nov 1966-1967; May 1973-1980 – 98r – 1 – us L of C Photodup [074]

Allgemeine zeitung – Munich, Germany 9 sep 1798-1 mar 1925 [mf 1798-1847, 1850-1929] – 1 – (originally publ at tuebingen; fr 13 dec 1831-29 sep 1882 publ at augsburg; cont: neueste weltkunde [jan- 8 sep 1798]; cont by: a z am morgen: allgemeine zeitung [3 mar 1925-11 aug 1926]) – uk British Libr Newspaper [074]

Allgemeine zeitung *see*
- Der kriegsbote
- Lahrer wochenblatt 1796
- Leipziger dorfanzeiger
- Taeglicher anzeiger

Allgemeine zeitung der lueneburger heide *see* Nachrichten fuer uelzen und die umgegend

Allgemeine zeitung des judenthums – Leipzig, Berlin DE, 1846-50 – 1 – (later: allgemeine zeitung des judentums, fr 1891 publ in berlin [1837-1922 n9]; cont by: c-v-zeitung; with suppl: literaisches und homiletisches beiblatt 1838-1839 jun) – gw Misc Inst [939]

Allgemeine zeitung des judenthums – Leipzig. 1837-Apr 28 1922. Incomplete – 1 – us NY Public [074]

Allgemeine zeitung (mannheim) *see* Badische abend-zeitung

Allgemeinen deutschen zeitung, bukarest *see* Karpaten-rundschau

Allgemeinen Konferenz der Deutschen Sittlichkeitsvereine *see* Moderne realistische litteratur im lichte der ethik und aesthethik

Die allgemeinen lehren des buergerlichen rechts des deutschen reichs und preussens / Dernburg, Heinrich – 2., unveraend Aufl. Halle (Saale): Waisenhaus, 1902 – (= ser Civil law 3 coll; Archiv fuer theorie und praxis des allegemeinen deutschen handels- und wechselrecht) – 6mf – 9 – (incl bibl ref and index) – mf#LLMC 96-580 – us LLMC [346]

Allgemeiner anzeiger – Meisenheim DE, 1992-95 – 36r – 1 – (bezirksausgabe von rhein-zeitung, koblenz) – gw Misc Inst [074]

Allgemeiner anzeiger – Halver DE, 1960 30 jun-1972 3 dec – 47r – 1 – (filmed by misc inst: 1949 29 oct-1957, 1992- [6r/yr]; until 1957 20r]; 1932-1933 30 jun, 1958-1960 30 aug) – gw Mikrofilm; gw Misc Inst [074]

Allgemeiner anzeiger – Rees DE, 1931-33 – 4r – 1 – (filmed with suppl) – gw Misc Inst [074]

Allgemeiner anzeiger – Dresden DE, 1881-91, 1893-1901, 1903-22 – 106r – 1 – (title varies: 2 aug 1887: loebtauer anzeiger, 22 sep 1904: dresdner westendzeitung, 1905: elbtal-abendpost) – gw Misc Inst [074]

Allgemeiner anzeiger – Koeln DE, 1849 1 apr-30 dec – 1r – 1 – (title varies: 26 sep 1850: allgemeiner anzeiger fuer rheinland-westphalen; filmed by other misc inst: 1855-67 [gaps]) – gw Misc Inst [074]

Allgemeiner anzeiger der deutschen *see* Der anzeiger

Allgemeiner anzeiger fuer allendorf, bad sooden und umgegend – Bad Sooden-Allendorf DE, 1911 3 jan-1912 31 aug, 1912 5 sep-1920 31 mar – 11r – 1 – (title varies: 28 sep 1912: tageblatt und allgemeiner anzeiger fuer allendorf, bad sooden, das werratal und umgegend) – gw Misc Inst [074]

Allgemeiner anzeiger fuer den kreis dannenberg-luechow *see* Zeitung fuer das wendland

Allgemeiner anzeiger fuer die amtsgerichtsbezirke hessisch-lichtenau, grossalmerode, spangenberg und umgegend – Hessisch-Lichtenau DE, 1897 2 oct-1916 9 nov [gaps] – 9r – 1 – (title varies: 3 jul 1909: allgemeiner anzeiger; incl suppl: illustrierter familien-freund 1898 20 feb-1902 28 dec [gaps]) – gw Misc Inst [943]

Allgemeiner anzeiger fuer die kreise wolmirstedt und neuhaldensleben – Wolmirstedt DE, 1870-75, 1878-79 – 1 – gw Misc Inst [074]

Allgemeiner anzeiger fuer rheinland-westphalen *see* Allgemeiner anzeiger

Allgemeiner anzeiger und zeitung an der aller und boehme *see* Walsroder wochenblatt

Allgemeiner Arbeiter-bund von Nord-Amerika *see* Sociale republik

Allgemeiner arbeiter-kalender – Budapest (H), 1877, 1887, 1889, 1891-93 – 1r – 1 – gw Misc Inst [331]

Allgemeiner bonner anzeiger fuer industrie, handel und gewerbe – Bonn DE, 1859 24 dec 24, 1860 1 jan-28 jun [gaps] – 1 – gw Misc Inst [380]

Allgemeiner deutscher Arbeiter-Verein *see* Protokoll der generalversammlung

Allgemeiner deutscher arbeiterverein – Coburg DE, 1865-73 – 1 – gw Misc Inst [331]

Allgemeiner deutscher arbeiterverein protokolle – Frankfurt/M, Berlin, Hannover DE – (minutes: generalversammlung: frankfurt-main 30 nov 1865-1 dec 1866; 1868; generalversammlung: berlin 19-25 may 1871; vorstandssitzung: hannover 3 mar 1872; generalversammlung: hannover 26 may-5 jun 1874; generalversammlung: berlin 18-24 may 1873) – mf#4946 – gw Mikropress [331]

Allgemeiner deutscher Gewerkschaftsbund *see*
- Jahrbuch
- Korrespondenzblatt..

Allgemeiner deutscher Gewerkschaftsbund. Ortsausschuss Berlin *see* Geschaeftsbericht

Allgemeiner deutscher Gewerkschaftsbund. Ortsausschuss Halle *see* Bericht...ortsausschuss halle a.s., sowie des arbeitersekretariats halle a.s., fuer das jahr...

Allgemeiner deutscher Gewerkschaftsbund. Ortsausschuss Muenchen *see* Jahrbuch..

Allgemeiner deutscher literaturkalender – 1879-82 [mf ed 1991] – 380mf – 9 – €990 diazo €1188 silver – 3-89131-041-2 – (filmed with: deutscher literaturkalender 1883-1902, kuerschners deutscher literaturkalender 1903-1917, sold singly: yrs1-13 1879-91 €30y, yrs14-39 1892-1917 €40y) – gw Fischer [430]

Allgemeiner Eisenbahnerverband *see* Protokoll des delegiertentages

Allgemeiner Evangelisch-Protestantischer Missionsverein *see* Jahresbericht der ostasien-mission

Allgemeiner frauenkalender : handbuch fuer frauenbestrebungen, frauenvereine, lehranstalten, berufs-, fortbildungs- und gewerbeschulen / Morgenstern, Lina – v1-2. 1885-86 [mf ed 1995] – (= ser Hq 17) – 7mf – 9 – €120.00 – 3-89131-129-X – (filmed with: die frauenbestrebungen unserer zeit: allgemeiner frauenkalender. culturhistorisches, biographisches und statistisches jahrbuch [v3 1887]) – gw Fischer [305]

Allgemeiner Freier Angestelltenbund *see*
- Die angestellten-bewegung
- Niederschrift vom gewerkschaftskongress
- Protokoll vom afa-gewerkschaftkongress

Allgemeiner Heimarbeiterschutz-Kongress. 1st, Berlin. 1904 *see* Protokoll der verhandlungen..

Allgemeiner kreisanzeiger – Wesel DE, 1855 jul-dec, 1865, 1868, 1870 – 1 – (later: kreisanzeiger, 1869: weseler zeitung; incl suppl: rheinischer bote 1919 20 apr-1929 sep [gaps], 1930 jan-29 mar) – gw Misc Inst [074]

Allgemeiner litterarischer anzeiger 1796-1801 / litterarische blaetter 1802-1806 / neuer litterarischer anzeiger 1806-1808 / allgemeines register 1811 – Leipzig / Nuernberg / Muenchen sp Tuebingen / Berlin-Stettin [mf ed 1992] – 97mf – 9 – €360 diazo €432 silver – 3-89131-055-2 – gw Fischer [430]

Allgemeiner oberschlesischer anzeiger – Breslau (WrocLaw PL), Ratibor, 1818, 1830-31, 1832 jul-dez, 1836-38 – 1 – gw Misc Inst [077]

Allgemeiner oeconomischer : oder landwirts-kalender auf das jahr 1770[-71] / ed by Sprenger, Balthasar – Stuttgart 1770-71 – (= ser Dz) – 3mf – 9 – €90.00 – mf#k/n2902 – gw Olms [630]

Allgemeiner Richard Wagner-Verein *see*
- Mitgliederverzeichnis des allgemeinen richard wagner-vereins fuer das jahr 1891
- Ueber die naeheren und ferneren aufgaben des allgemeinen richard wagner-vereins und ueber einen festspiel-cycklus...im jahre 1884

Allgemeiner Schutzkongress fuer alle in der Schiffahrt und im Schiffbau Beschaeftigten *see* Protokoll der verhandlungen

Allgemeiner wohnungs- und immobilien-anzeiger fuer duesseldorf und umgebung – Duesseldorf DE, 1908-1914 1 aug – 4r – 1 – gw Misc Inst [333]

Allgemeiner zittauer anzeiger – Zittau DE, 1864 23 jan-1868 28 jun – 1 – (title varies: 1 jan 1868: zittauer zeitung) – gw Misc Inst [074]

Allgemeines anzeigeblatt fuer doernigheim, hochstadt und umgebung [...] – (Maintal-) Doernigheim DE, 1929 3 jan-1931 jun, 1932 apr-1934 – 7r – 1 – gw Misc Inst [074]

Allgemeines archiv fuer die laender- und voelkerkunde / ed by Hirsching, Friedrich Karl Gottlob – Leipzig 1790-91 – (= ser Dz. historisch-geographische abt) – 2v on 6mf – 9 – €120.00 – mf#k/n1225 – gw Olms [305]

Allgemeines berliner intelligenzblatt – Berlin DE, 1860, 1861 apr-jul – 4r – 1 – gw Misc Inst [074]

Allgemeines bucher-lexicon / Heinsius, Wilhelm – 19v. 1812-94 – 1 – us L of C Photodup [430]

Allgemeines buecher-lexikon / Heinsius, Wilhelm – 19v. in 26 pts. 1812-94 – 1,9 – us AMS Press [430]

Allgemeines buergerliches gesetzbuch fuer gesammten deutschen erblaender der oesterreichischen monarchie – Wien: K K Hof- und Staatsdruckerey. 3v in 1. 1811 – (= ser Civil law 3 coll) – 10mf – 9 – (incl bibl ref and index) – mf#LLMC 96-617 – us LLMC [348]

Allgemeines chronikon fuer handlung, kuenste, fabriken und manufakturen /.../ – Ronneburg DE, 1797 – 1r – 1 – gw Misc Inst [670]

Allgemeines converstions-taschenlexikon (ael1/31) : oder real-encyklopaedie der fuer die gebildeten staende notwendigen kenntnisse und wissenschaften – Quedlinburg/Leipzig 1828-33 [mf ed 1995] – (= ser Archiv der europaeischen lexikographie, abt 1: enzyklopaedien) – 65v on 67mf – 9 – €1020.00 – 3-89131-212-1 – (int by otmar seemann) – gw Fischer [030]

Allgemeines deutsches lieder-lexikon / Bernhard, W – repr Leipzig. 4v. 1844-46 – 11 – $75.00 set – (coll of all the german lieder and folk-songs in alphabetical order) – us Univ Music [780]

Allgemeines deutsches sach-woerterbuch aller menschlichen kenntnisse und fertigkeiten (ael1/22) : mit den erklaerungen der aus andern sprachen entlehnten ausdruecke und der weniger bekannten kunstwörter in verbindung mit mehreren gelehrten / ed by Liechtenstern, Joseph, freiherr von – Meissen 1824-34 [mf ed 1994] – (= ser Archiv der europaeischen lexikographie, abt 1: enzyklopaedien) – 10v+1 suppl vol on 43mf – 9 – €650.00 – 3-89131-172-9 – gw Fischer [030]

Allgemeines deutsches volksblatt [...] – Darmstadt DE, 1795-apr 1796 – 1 – gw Misc Inst [074]

Allgemeines deutsches volks-conversations-lexikon und fremdwoerterbuch (AEL1/23) : ein unentbehrliches handbuch fuer jedermann – Hamburg 1844-49 [mf ed 1994] – (= ser Archiv der europaeischen lexikographie, abt 1: enzyklopaedien) – 8v on 73mf – 9 – €390.00 – 3-89131-173-7 – gw Fischer [030]

Allgemeines europaeisches journal / ed by Frantzky, Fr Jos Th – Bruenn 1795-98 – (= ser Dz. historisch-geographische abt) – 4 jg [=12v] on 94mf – 9 – €940.00 – mf#k/n1281 – gw Olms [074]

Allgemeines friedberger wochenblatt fuer stadt- und landleute – Friedberg, Hessen DE, 1809 2 oct, 1811-1835 1 aug – 1 – (title varies: 1819: gemeinnuetziges wochenblatt fuer friedberg und die gegend, 1828: wochenblatt fuer friedberg und die gegend, 13 sep 1834: friedberger wochenblatt zu den wetterauer anzeigen) – gw Mikrofilm [074]

Allgemeines gelehrten-lexicon / Joecher, Christian Gottlieb – Leipzig, Delmenhorst und Bremen 1750-1897 [mf ed Hildesheim 1984] – (= ser Die schriftsteller- und gelehrtenlexika des 17., 18., und 19. jahrhunderts) – 4v on 134mf – 9 – diazo €498.00 – gw Olms [430]

Allgemeines, helvetisches, eydgenoessisches oder schweizerisches lexicon (ael1/8) / Leu, Hans Jakob – Zuerich 1747-95 [mf ed 1992] – (= ser Archiv der europaeischen lexikographie, abt 1: enzyklopaedien) – 171mf – 9 – €710.00 – 3-89131-077-3 – (filmed with: suppl zum allgemeinen helvetisch-eidgenoessischen lexicon herrn buergermeisters leu herausgegeben von hans jacob holzhalb [zug 1786-95] 6v; int by otmar seemann) – gw Fischer [030]

Allgemeines historisches lexicon in welchem das leben und die thaten derer patriarchen, propheten, apostel...vorgestellet werden (ael1/37) – 3rd rev ed. Leipzig 1730-32 [mf ed 1997] – (= ser Archiv der europaeischen lexikographie, abt 1: enzyklopaedien) – 4v on 51mf – 9 – €300.00 – 3-89131-276-8 – gw Fischer [030]

Allgemeines historisches lexikon (ael1/44.9) / ed by Buddeus, Johann Franz – Leipzig 1709, suppl vol 1714 [mf ed 1998] – (= ser Le grand dictionnaire historique [deutsche ausgaben]) – 4pt+suppl vol on 32mf – 9 – €260.00 – 3-89131-328-4 – gw Fischer [900]

Allgemeines historisches lexikon (ael1/44.10) / ed by Buddeus, Johann Franz – Leipzig 1709, suppl vol 1740 [mf ed 1998] – (= ser Le grand dictionnaire historique [deutsche ausgaben]) – 4v+suppl vol on 60mf – 9 – €490.00 – 3-89131-329-2 – gw Fischer [900]

Allgemeines historisches magazin / ed by Boysen, Friedrich Eberhard – Halle 1767-70 – (= ser Dz. historisch-geographische abt) – 6st on 13mf – 9 – €130.00 – mf#k/n1043 – gw Olms [900]

Allgemeines historisch-statistisch-geographisches handlungs- post- und zeitungs-lexicon [...] – Erfurt DE, 1804 [a-e]; 1805 [f-i]; 1806 [k-l]; 1810 [m] – 3r – 1 – gw Misc Inst [900]

Allgemeines intelligenz- oder wochenblatt fuer das land breisgau und die ortenau *see* Freyburger zeitung

Allgemeines intelligenz- oder wochenblatt fuer saemtliche hochfuerstliche lande – Karlsruhe DE, 1775 4 may-1777, 1779/80, 1787-1814, 1816-1825 30 jun, 1826 jul-dec, 1831-55 – 1 – (title varies: 1803: provinzialblatt der badischen markgrafschaft, 1808: grossherzoglich badisches mittelrheinisches provinzial-blatt, 1831: grossherzoglich badisches anzeige-blatt fuer kinzig, murg- und pfinzkreis, 1832: grossherzoglich badisches anzeige-blatt fuer den mittel-rheinkreis; with suppl: 1810-14, 1816-18, 1820-23, 1825-26) – gw Misc Inst [074]

Allgemeines juedisches familienblatt – Leipzig DE, 1926-1933 n14 – 3r – 1 – gw Misc Inst [939]

Allgemeines landrecht fuer die preussischen staaten : in verbindung mit den ergaenzenden verordnungen / ed by Mannkopff, A J – Berlin: A Nauck. 2v in 7bks+index vol. 1837-38 (51mf); Berlin: A Nauck.+suppl vol (8mf) – (= ser Civil law 3 coll) – 59mf – 9 – (incl bibl ref) – mf#LLMC 96-565 – us LLMC [348]

Allgemeines landrecht fuer die preussischen staaten – Neue Aufl. Berlin: G C Nauck. v1-5. 1832 – (= ser Civil law 3 coll) – 32mf – 9 – (vol 5, register, lacks an edition statement and is dated 1828) – mf#LLMC 96-554 – us LLMC [348]

Allgemeines lexicon der kuenste und wissenschaften (ael1/42) / Jablonski, Johann Theodor – Leipzig 1721, Koenigsberg/Leipzig 1748, Koenigsberg/Leipzig 1767 [mf ed 1998] – (= ser Archiv der europaeischen lexikographie, abt 1: enzyklopaedien) – 36mf – 9 – €310.00 set – 3-89131-341-1 – ((ael1/42.1): 1st ed, 1721 [9mf] isbn: 3-89131-316-0 €110; (ael1/42.2): 2nd ed, 1748 [12mf] isbn: 3-89131-317-9 €140; (ael1/42.3)3rd ed, 1767 [15mf] isbn: 3-89131-318-7 €160) – gw Fischer [030]

Allgemeines litteraturarchiv fuer geschichte, geographie und statistik / ed by Canzler, Friedrich Gottlob – Leipzig 1791, 1793-95 – (= ser Dz) – 12mf – 9 – €120.00 – mf#k/n2762 – gw Olms [900]

Allgemeines magazin fuer die buergerliche baukunst / ed by Huth, Gottfried – Weimar 1789-96 – (= ser Dz) – 2v on 11mf – 9 – €110.00 – mf#k/n4092 – gw Olms [720]

Allgemeines register *see* Allgemeiner litterarischer anzeiger 1796-1801 / litterarische blaetter 1802-1806 / neuer litterarischer anzeiger 1806-1808 / allgemeines register 1811

Allgemeines register ueber die goettingischen gelehrten anzeigen von 1753 bis 1782 / ed by Ekkard, Friedrich – Goettingen 1784-85 [mf ed Hildesheim 1980] – 25mf – 9 – €98.00 diazo €118.00 silver – gw Olms [430]

Allgemeines repertorium fuer die theologische literatur und kirchliche statistik – 1(1833)-80(1853) – 405mf – 9 – €772.00 – ne Slangenburg [240]

Allgemeines repertorium fuer empirische psychologie und verwandte wissenschaften / ed by Mauchart, Imm Dav – Nuernberg 1792-1801 – (= ser Dz. abt philosophie) – 6v on 15mf – 9 – €150.00 – mf#k/n569 – gw Olms [150]

Allgemeines theater-lexikon (ael1/49) : oder encyclopaedie alles wissenswerthen fuer buehnenkuenstler, dilettanten und theaterfreunde – 1846 [mf ed 2002] – (= ser Archiv der europaeischen lexikographie, abt 1: enzyklopaedien) – 29mf – 9 – €190.00 – 3-89131-385-3 – gw Fischer [790]

Allgemeines ueber die hebraeische dichtung und ueber das psalmenbuch / Ewald, Heinrich – new ed. Goettingen: Vandenhoeck & Ruprecht, 1866 [mf ed 1984] – 9mf – 9 – 0-8370-1112-4 – (incl bibl ref) – mf#1984-4476 – us ATLA [470]

Allgemeines volksblatt – Koeln DE, 1845 jan-sep – 1 – gw Misc Inst [074]

Allgemeines, vollstaendiges neuhebraeisch-deutsches woerterbuch : mit inbegriff aller in den talmudischen schriften und in der neuen literatur ueberhaupt vorkommenden fremdwoerter / Schulbaum, Moses – Lemberg: Michael Wolf, 1880 [mf ed 1986] – 1mf – 9 – 0-8370-7506-8 – (in hebrew with occasional phrases in greek or latin) – mf#1986-1506 – us ATLA [470]

Allgemeinwissenschaftliche und literarische zeitschriften des 17. und 18. jahrhunderts – [mf ed 1977-81] – 1632mf – 9 – diazo €6400.00 silver €7200.00 – (individual titles also listed separately) – gw Olms [410]

Allianca – Sao Carlos, SP. 26 jan 1878 – (= ser Ps 19) – bl Biblioteca [079]

L'alliance – Paris. 1 2 fevr-5 juil 1846 – 1 – fr ACRPP [944]

Alliance – Lincoln, NE: Alliance Pub Co. 1v. v1 n1. jun 12-v1 n25. dec 7 1889 (wkly) [mf ed lacks jul 19 filmed 1962?] – 1r – 1 – (cont by: farmers' alliance) – us NE Hist [071]

The alliance – Denver, CO: Colorado Alliance of Business, jul 1981-spr 1983 (mf ed 1993) – 1r – 1 – mf#MF Al51a – us Colorado Hist [071]

Alliance boomerang – Crawford, NE: S I Meseraull (wkly) [mf ed v2 n41. jun 1 1892] – 1r – 1 – (cont by: crawford gazette) – us NE Hist [071]

Alliance daily times-herald – Alliance, NE: Gene Kemper. 23v. v63 n13. aug 1 1950-v85 n146. nov 18 1972 (daily ex sun) [mf ed aug1 1950-nov 18 1972 (gaps)] – 86r – 1 – (cont: alliance times and herald; cont by: alliance times-herald) – us NE Hist [071]

L'alliance democratique : puis bulletin interieur du parti – Paris. 1934-39; 1947-avr 1955 – 1 – fr ACRPP [335]

Alliance for Catholic Tradition *see* War is now!

Alliance for Life *see*
- National newsletter
- National pro-life news
- Pro-life news/canada

Alliance herald – Alliance, NE: T J O'Keefe. 21v. v9 n8. feb 21 1902-v29 n62. jun 30 1922 (semiwkly) – 6r – 1 – (merged with: alliance semi-weekly times to form: alliance semi-weekly times and alliance herald) – us Bell [071]

The alliance herald – Alliance, NE: D S Dusenbery. v17 n52. apr 20 1894 (wkly) [filmed 1973] – 1r – 1 – (cont: nuckolls county herald. cont by: nuckolls county herald (nelson, ne 1894)) – us NE Hist [071]

The alliance herald – Alliance, NE: T J O'Keefe. 21v. v9 n8. feb 21 1902-v29 n62. jun 30 1922 (semiwkly) – 9r – 1 – (merged with: alliance semi-weekly times to form: alliance semi-weekly times and alliance herald; cont: hemingford herald; absorbed: pioneer grip) – us Bell [071]

L'alliance israelite universelle : paix et droit – Paris, 1921-janv mars 1940 – 1 – fr ACRPP [939]

Alliance leader / National Alliance of Postal Employees [US] – 1961 mar/apr-1982 sep/oct – 1r – 1 – (cont: new york alliance leader) – mf#1534166 – us WHS [380]

Alliance life – Colorado Springs CO 1987+ – 1,5,9 – (cont: alliance witness) – ISSN: 1040-6794 – mf#8790,01 – us UMI ProQuest [240]

Alliance life – Nyack NY: Christian & Missionary Alliance. v122- 1987- [mthly] [mf ed 2003-] – (= ser Christian and missionary alliance) – 8r – 1 – mf1034 – us ATLA [240]

L'alliance nationale – Montreal: La Societe de secours mutuels l'"Alliance nationale" 1895-[1919] – 9 – mf#P04163 – cn Canadiana [360]

Alliance news – London, England 4 jan 1862-28 dec 1905 – 1 – (publ in manchester fr 1854-1919; cont: alliance weekly news [28 jul 1855-28 dec 1861]; cont by: alliance news & temperance reformer [4 jan 1906-may/jun 1991]) – uk British Libr Newspaper [072]

The alliance news – Alliance, NE: Alliance Printing Co (wkly) [mf ed feb 8 1917] – 1r – 1 – (cont by: antioch news (antioch, ne)) – us NE Hist [071]

Alliance of Poles of America *see* Records, ms p.p.

Alliance of the Reformed Churches Holding the Presbyterian System *see* Selections for the service of praise

Alliance of the rockies – v10 n34-v13 n36 [1902 dec 20-jan 28; 1905] – 1 – mf#1051843 – us WHS [071]

Alliance of transylvania saxons, series 1-3 : records – Cleveland OH, 1902-81 – 37r – 1 – (wkly publ is the volksblatt. records incl convention minutes. in german (1r); related coll: ser 2 1957-81 [1r] minutes of grand officers' meetings. minutes of 1970-80 restricted. in english; ser 3 1923-81 [35r] application files nos 42-23958. restricted. in english) – us IHRC [366]

Alliance republicaine democratique / Parti republicain democratique et social – Paris. 21 fevr 1902-janv 1921 – 1 – fr ACRPP [320]

Alliance sem-weekly times – Alliance, NE: H J Ellis. v14 whole n908. jun 2 1903-jun 30 1922 (semiwkly) [mf ed with gaps] – 28r – 1 – (cont: alliance times; absorbed: pioneer grip; merged with: alliance herald to form: alliance semi-weekly times and alliance herald) – us NE Hist [071]

Alliance sun – Lyons, NE: Goodell & Carter, apr 1891-v4 n36. aug 12 1892 (wkly) [mf ed with gaps filmed 1988] – 1r – 1 – (cont: logan valley sun. cont by: logan valley sun (1892)) – us NE Hist [071]

Alliance times and herald – Alliance, NE: Ben J Sallows, may 1 1923-jul 28 1950 (semiwkly) [mf ed with gaps] – 44r – 1 – (cont: alliance semi-weekly times and the alliance herald; cont by: alliance daily times-herald) – us NE Hist [071]

Alliance to End Repression *see* Progress report

The alliance tribune – O'Neill, NE: C S Evans & Son, 1890 (wkly) [mf ed -1892 (gaps) filmed 1973] – 1r – 1 – (issues for aug 7 1891-oct 21 1892 called also whole n59-123) – us NE Hist [071]

The alliance weekly : a journal of christian life and missions – New York: A B Simpson. v37-92. 1911-57 [wkly] [mf ed 1995-2003] – (= ser Christian and missionary alliance) – 56v on 27r – 1 – (lacks: ind for v50-53. incl unpubl cumulative ind for 1925-57) – mf#1995-s300 – us ATLA [240]

Alliance witness – Colorado Springs CO 1973-86 – 1,5,9 – (cont by: alliance life) – ISSN: 0745-3256 – mf#8790.01 – us UMI ProQuest [240]

The alliance witness – New York NY: [Christian & Missionary Alliance] v93-122. 1958-87 [biwkly] [mf ed 2003] – (= ser Christian and missionary alliance) – 30v on 14r – 1 – (with unpubl annual ind at beginning of ea vol except v122) – mf1033 – us ATLA [240]

Alliance work in western china and tibet / Christie, William – rev ed. New York: Christian and Missionary Alliance, 1913 [mf ed 1992] – (= ser Christian & missionary alliance coll) – 1mf – 9 – 0-524-03699-3 – mf#1990-4804 – us ATLA [240]

Alliance-independent – Lincoln, NE: [Alliance Pub Co] 2v. v4 n3. jun 30 1892-v5 n38. mar 8 1894 (wkly) [mf ed with gaps filmed 1962?] – 1r – 1 – (cont: farmers' alliance and nebraska independent. cont by: wealth makers of the world) – us NE Hist [071]

Allibaco, W A *see* The philosophic and scientific ultimatum

Allibone, Samuel Austin *see*
- An alphabetical index to the new testament
- The union bible companion

Allied Council of Senior Citizens of Wisconsin *see* Wisconsin senior citizen

Allied high command papers, 1943-45 : from material collected by the historian, david irving – 8r – 1 – mf#97276 – uk Microform Academic [941]

Allied industrial worker – 1956 dec-1964 dec; 1965-93 – 1 – mf#1051844 – us WHS [331]

Allied Printing Trades Council *see*
- Trades unionist

Allied Printing Trades Council et al *see* Labor herald

Allied printing trades journal – 1903 may, oct – 1 – mf#4967017 – us WHS [680]

Allied propaganda in world war 2 : the complete record of the political warfare executive (fo898) from the public record office / ed by Taylor, Philip M – [mf ed 2003] – ca 175r – 1 – us Primary [150]

Allied propaganda of the first world war *see* First world war: a documentary record, series 1

Alliene, Joseph *see* Alarm to the unconverted

L'Allier, Jean-Paul *see* Notes pour l'allocution de m jean-paul l'allier

Allier, Raoul *see* Les troubles de chine et les missions chretiennes

Allies, Mary Helen *see* Pius the seventh, 1800-1823

Allies, Thomas William *see*
- The holy see and the wandering of the nations
- Leaves from st. john chrysostom
- A life's decision
- Royal suprmacy viewed in reference to the two spiritual...
- St peter, his name and his office

Alligator times – 1977 sep-1979 may; 1981 feb-1983 jun – 1 – mf#703601 – us WHS [071]

Alligator times – Hollywood, FL. V1 N3-1974 Jul-1983 jun – 1r – 1 – (scattered issues only filmed; filmed with later title, seminole tribune) – us UF Libraries [071]

Allighan, Garry *see*
- Curtain-up on south africa
- Verwoerd

Allihn, F H Th *see* Zeitschrift fuer exacte philosophie im sinne des neuern philosophischen realismus

Allihn, Friedrich Heinrich Theodor *see* Der verderbliche einfluss der hegelschen philosophie

Allin, Thomas *see*
- The augustinian revolution in theology
- Exposition of the principles of church-government adopted by the me...
- Immortality of the soul
- The question of questions

Allin, Thomas et al *see* The jubilee of the methodist new connexion

Allinson, William J *see* Memorials of rebecca jones

Alliott, James Bingham (Mrs) *see*
- The dowager lady tremaine
- "Thou shalt not surely die"
- A woman with a history in her face

Allis, O T *see* The five books of moses

Allis-Chalmers Corporation *see* Dialog

Allis-chalmers engineering review – Milwaukee WI 1936-76 – 1,5,9 – ISSN: 0002-6123 – mf#678 – us UMI ProQuest [620]

Allison, Charles Wm Benjamin *see* Wooster college debating society

Allison, L C *see* Discovery of tripoli

Allison, Leon McDill *see* The doctrine of scripture in the theology of john calvin and francis turretin

Allison, Leonard A *see* The rev oliver arnold, first rector of sussex, n b

Allison, Leslie K *see* Relationships between postural control system impairments and disabilities

Allison Peers, E *see* The church in spain

Allison, R V *see* Stimulation of plant response on the raw peat soils of the florida everglades

Allison, Ruth *see* Analysis of a poulenc trio [for bassoon, oboe, and piano]

Allison, Samuel Buell *see* Teacher's robinson crusoe

Allison, Walter Leslie *see* The sadhs

Allison, William Henry *see*
- Baptist councils in america
- Inventory of unpublished material for american religious history in protestant church archives and other repositories

Den allmaenna religionshistorien och den kyrkliga teologien : intraedesfoerelaesning / Soederblom, Nathan – Uppsala: W. Schultz, [1901?] – 1mf – 9 – 0-7905-6319-3 – mf#1988-2319 – us ATLA [240]

Allmanna journalen – Stockholm, Sweden. 1813 – 1r – 1 – sw Kungliga [078]

Allmers, Hermann *see*
- Dichtungen
- Fromm und frei
- Marschenbuch

Allmers, Robert *see* Kampf um thurant

Allnatt, Charles F B *see* Which is the true church?

Allnatt, Elizabeth *see*
- Autumn gatherings
- Sebie dorr

Allo, E B *see*
- St paul

Allo police – Montreal: Societe de publication Merlin limitee. v1 n1 28 fevr 1953- (wkly) [mf ed 1988-] – 1 – mf#SEM35P306 – cn Bibl Nat [360]

Alloa advertiser – Alloa, Scotland 18 may 1850-8 sep 1972* – 1 – (cont by: alloa advertiser-journal [13 sep 1972-27 aug 1976]) – uk British Libr Newspaper [072]

Alloa advertiser journal – Alloa, Scotland 13 sep 1972-27 aug 1976 – 1 – (cont: alloa advertiser [18 may 1850-8 sep 1972]; cont by: alloa & hillfoots advertiser journal [1 sep 1976-23 mar 1988]) – uk British Libr Newspaper [072]

Alloa and hillfoots advertiser – Alloa: Alloa Printing & Publ Co 25 mar 1988- (twice wkly, wkly) [mf ed 1 jul 1994-] – 1 – (cont: alloa & hillfoots advertiser journal [1 sep 1976-23 mar 1988] publ 25th mar 1988-27th mar 1996 on wed & fri; wed ed has title: alloa & hillfoots mid-week advertiser, iss for 1st aug 1990-27th mar 1996 have title: midweek advertiser) – ISSN: 0962-7596 – uk Scotland NatLib [072]

Alloa and hillfoots advertiser journal – Alloa, Scotland 1 sep 1976-23 mar 1988 – 1 – (cont: alloa advertiser-journal [13 sep 1972-27 aug 1976]; cont by: alloa & hillfoots advertiser [25 mar 1988-]) – uk British Libr Newspaper [072]

Alloa and hillfoots' searchlight – [Scotland] Tillicoultry, Clackmannan: printed...by W M Bett oct 1925-jun 1926 (wkly) [mf ed 2003] – 9v on 1r – 1 – (cont by: searchlight (tillicoultry, scotland)) – uk Newsplan [072]

Alloa and hillfoots wee county news – Alloa: Wee County Publ Ltd 1998- (wkly) [mf ed 2001-] – 1 – (cont: wee county news) – ISSN: 1359-3390 – uk Scotland NatLib [072]

Alloa journal – Alloa, Scotland 19 feb 1916-7 sep 1972 – 1 – (incorp with: alloa advertiser & subsequently publ as: alloa advertiser-journal; cont: alloa journal & clackmannanshire advertiser [26 feb 1859-12 feb 1916]) – uk British Libr Newspaper [072]

Alloa journal and clackmannanshire advertiser – Alloa, Scotland 26 feb 1859-12 feb 1916 – 1 – (cont: clackmannanshire advertiser, & alloa journal of news, etc [1 feb 1851-19 feb 1859]; cont by: alloa journal [19 feb 1916-7 sep 1972]) – uk British Libr Newspaper [072]

Allo!...allo! ici la creche : plaidoyers et nouvelles / Germain, Victorin – 5e mille. Quebec: chez l'auteur, 1940 [mf ed 1990] – 3mf – 9 – mf#SEM105P1297 – cn Bibl Nat [360]

Allocucoes do presidente da academia de medicina / Couto, Miguel – Rio De Janeiro, Brazil. 1923 – 1r – 1 – us UF Libraries [972]

Allocution de monsieur richard beaulieu : sous-ministre adjoint, ministere des affaires municipales du quebec, devant l'association quebecoise des techniques de l'eau, a montreal, le lundi, 13 fevrier 1967, a midi trente – [Quebec: Ministere des affaires municipales, 1967] (mf ed 1995) – 1mf – 9 – mf#SEM105P2497 – cn Bibl Nat [360]

Allocution prononcee a l'ouverture du congres de l'enseignement secondaire tenu a quebec, juin 1914 / Emard, Joseph-Medard – Valleyfield [Quebec: s.n.] 1914 [mf ed 1994] – 1mf – 9 – 0-665-73183-3 – mf#73183 – cn Canadiana [377]

Allocutions to the clergy and pastorals of the late right rev dr moriarty, bishop of kerry / Moriarty, David – Dublin: Browne & Nolan; London: Burns & Oates, 1884 [mf ed 1986] – 1mf – 9 – 0-8370-7006-6 – mf#1986-1006 – us ATLA [241]

Allodi, L *see* In regulam sancti benedicti commentarium nunc primum editum

Allom, Elizabeth Anne *see*
- Death scenes and other poems
- Sea-side pleasures

Allom, T *see* China

Allom, Thomas *see* Constantinople and the scenery of the seven churches of asia minor

Allometric scaling of bench press strength by body mass and lean body mass in college-age men / Parker, Robert G – 156p on 2mf – 9 – $10.00 – mf#PE 4146 – us Kinesology [612]

Allometric scaling of grip strength by body mass and lean body mass in college-age men and women / Lee, Siu Y – Springfield College, 1995 – 2mf – 9 – $8.00 – mf#PE3606 – us Kinesology [612]

Allon, Henry *see*
- Christ, the book, and the church
- Church of the future
- Congregationalism

Allonville, Armand F d' *see* Memoires tires des papiers d'un homme d'etat

Allor, Karin M *see* Perceived competence and attraction to physical activity in a diverse population of fifth graders

Allostock, unitarian chapel 1690-1823 – [North Cheshire FHS] – (= ser Cheshire church registers) – 1mf – 9 – £3.00 – mf#344 – uk CheshireFHS [929]

Allotria : [poems] / Boetticher, Georg – Leipzig: P Reclam, [1908?] [mf ed 1989] – 92p – 1 – mf#7055 – us Wisconsin U Libr [810]

Alloway, Mary Wilson *see*
- Crossed swords
- Famous firesides of french canada

Alloys and their industrial applications / Law, Edward F – London, England. 1909 – 1r – 1 – us UF Libraries [660]

All-Peoples Congress *see* Bulletin of the all-peoples congress

All-Peoples Congress et al *see* Pam-apc newsletter

The all-round route guide : the hudson river, trenton falls, niagara, toronto, the thousand islands and the river st lawrence, ottawa, montreal, quebec... – Montreal?: Montreal Print & Pub Co, 1869 – 2mf – 9 – mf#26300 – cn Canadiana [917]

All's well – Fayetteville arkansas. v1, n9-v. 13, n5. aug 1921-dec 1935 – 1 – us NY Public [073]

Allsopp, Henry *see* An introduction to english industrial history

Allston, Joseph Blyth [comp] *see* Life and times of james l petigru

Allston, Washington *see* Washington allston papers, 1800-1843

Alltagsleben in china : bilder aus dem chinesischen volksleben: nach schilderungen von e j dukes und a fielde; frei nach dem englischen von luise ohler / Dukes, Edwin Joshua & Fielde, Adele Marion – Basel: Missionsbuchhandlung, 1892 [mf ed 1995] – (= ser Yale coll) – 229p (ill) – 1 – 0-524-09196-X – (in german) – mf#1995-0196 – us ATLA [951]

Alltagsleben in london : ein skizzenbuch / Rodenberg, Julius – Berlin 1860 [mf ed Hildesheim 1995-98] – (= ser Fbc) – 2mf – 9 – €60.00 – 3-487-27954-1 – gw Olms [914]

Allton-alton association newsletter – v1 n1-v7 n2 [1974 mar-1980 nov] – 1 – mf#665386 – us WHS [366]

Allue Salvador, Miguel *see* El ahorro y la politica social

All-union union of evangelical christians-baptists : congresses – Moscow, 1966, 1979 – 1r – 1 – $7.12 – us Southern Baptist [420]

Allured's cosmetics and toiletries : c & t – Carol Stream IL 1999+ – 1,5,9 – ISSN: 1530-1338 – mf#2538,02 – us UMI ProQuest [660]

Allwardt, Henry August *see* Die jetzige lehre der synode von missouri von der ewigen wahl gottes

Allwood, Philip *see* Brief remarks on "the declaration of the catholic bishops...

Ally – n1-41 [1968 feb-1972 aug] – 1 – mf#964678 – us WHS [071]

The ally – A newspaper for servicemen. no. 1-15. 1968-69 – 1 – us AMS Press [071]

Allyn, Avery *see* Ritual and illustrations of freemasonry, and the orange and odd fellows' societies [a]

Allyn, Jack *see*
- Jonathan and his continent
- Jonathan et son continent

Allyn k. ford collection of historical manuscripts – Over 1500 letters, cards and documents spanning five centuries (1472-1970) and several continents. 5 reels, including filmed inventory – 1 – $150.00; $30.00r – us Minn Hist [900]

Allyn, Rose *see* Fairy tales

Alm, R von der *see* Die urheile heidnischer und juedischer schriftsteller der vier ersten christlichen jahrhunderte

Alma / Domenech De Calvo, Carmen – Habana, Cuba. 1960 – 1r – 1 – us UF Libraries [972]

Alma blaetter – 1888 nov-dec; 1889-1892 sep 22; 1895-1905; 1906-1910 jun 9 – 1 – mf#915752 – us WHS [071]

Alma center herald – 1898 feb 2-1899 sep 20 – 1 – mf#914654 – us WHS [071]

Alma center news – 1906 jul 6-1907; 1908-29; 1930-1933 aug 17 – 1 – mf#914657 – us WHS [071]

El alma cristiana de cortes. en el centenario de su muerte / Bayle, Constantino – Madrid: Razon y Fe, 1948 – 1 – sp Bibl Santa Ana [920]

Alma cubana a traves de sus poetas / Garcia Kohly, Mario – Madrid, Spain. 1928 – 1r – 1 – us UF Libraries [440]

The alma daily herald – Alma, NE: J M Hiatt & J D Hurd. v1 n1. aug 9 1881-sep 1881 (daily)// – 1r – 1 – us NE Hist [071]

Alma dominicana / Garcia Godoy, Federico – Santo Domingo, Dominican Republic. 1911 – 1r – 1 – us UF Libraries [972]

Alma emerita – Merida 1908 – 5 – sp Bibl Santa Ana [074]

Alma en los labios / Trigo, Felipe – Madrid: Renacumiento, 7th ed 1920 – sp Bibl Santa Ana [946]

Alma extremena – Caceres 1905-06 – 2 iss – 5 – sp Bibl Santa Ana [074]

Alma guajira / Salinas Y Lopez, Marcelo – Habana, Cuba. 1942 – 1r – 1 – us UF Libraries [972]

The alma herald – Alma, Harlan County, NE: Hiatt & Hurd. 1880- (wkly) [mf ed aug 18 1881-feb 9 1882 (gaps)] – 2r – 1 – us NE Hist [071]

Alma John Workshops Association *see* Newsletter

Alma journal – 1863 jun 25-1864 dec 1 – 1 – mf#1001363 – us WHS [071]

Alma journal and beef slough advocate – 1868 jun 4-aug 13 – 1 – mf#1001357 – us WHS [071]

Alma llanera / Gonzalez Herrera, Edelmira – San Jose, Costa Rica. 1946 – 1r – 1 – us UF Libraries [972]

Alma mining record *see* Miscellaneous newspapers of park county

Alma nova : quinzenario academico – S Vicente: Sociedade de Tip e Publicidade, [apr 27, dec 7 1933] (bimthly) – (= ser Coll of lusophone african newspapers and serials) – 1r – 1 – us CRL [370]

O alma nova : orgao do 'almas novas' grupo educativo, dramatico e recreativo – Lisboa: Almas Novas [apr 9 1939] – (= ser Coll of lusophone african newspapers and serials) – 1r – 1 – us CRL [790]

The alma record – [Alma], NE: Arthur Kimberling. v23 n14. feb 13 1914-v34 n30. jun 26 1925 (wkly) – 4r – 1 – (cont: shaffer's alma record. absorbed by: harlan county journal. vol numbering irregular dec 9 1921-aug 18 1922) – us Bell [071]

Alma weekly express – 1868 oct 23-1873 may 29; 1869 jul 29-aug 26; sep 2-1870 dec 15; 1873 jun 5-1876 may 4; 1874 jan 15-1875 dec 30; 1876 jan 6-1879 may 22; may 29 – 1 – mf#986072 – us WHS [071]

Alma weekly record – Alma, NE: Furse Bros, -1907// (wkly) – 2r – 1 – (cont: weekly record (alma ne); cont by: shaffer's alma record; publ as: alma record oct 27-dec 22 1899) – us Bell [071]

The alma weekly record – Alma, NE: Furse Bros. [mf ed apr 291898-oct 7 1904 (gaps)] – 1r – 1 – (cont: weekly record (alma, ne). cont by: shaffer's alma record. publ as: alma record, oct 27 1899-dec 22 1899) – us NE Hist [071]

Alma y paisaje / Villaronga, Luis – San Juan, Puerto Rico. 1954 – 1r – 1 – us UF Libraries [972]

Alma y tierra, problemas cubanos / Fernandez Vega, Wifredo – Habana, Cuba. 1928 – 1r – 1 – us UF Libraries [972]

Al-mabahith – Tripoli, SY: Jirji & Samu'il Yanni. yr 1 n11-18. 24 rabi' 2-15 sha'ban 1327 [15 ayyar/mar-1 aylul/sep 1909] – (= ser Arabic journals and popular press) – 1r – 1 – $200.00 – us MEDOC [956]

Almada, Lourenco Vaz De Almada *see* Notas sobre a viagem de sua alteza real

The almafilian – St Thomas, Ont: Alma College, [188-?-189- or 19–] – 9 – ISSN: 1190-6251 – mf#P04113 – cn Canadiana [378]

Al-magrib – Alger. n1-32. avr-juil 1903 – (= ser Magrib) – 1 – (mq no. 20-24) – fr ACRPP [073]

Al-magrib al-arabi – Alger – (= ser Magrib al-arabi) – 1 – (ed. francaise. sept 1947-mai 1949. ed. arabe. juin 1947-mai 1949, mars-mai 1956) – fr ACRPP [073]

Al-magrib el-maghrib – n1-38. Alger, 1930-31 – (= ser Magrib El-Maghrib) – 1 – fr ACRPP [073]

Almagro Basch, Martin *see*
- Excavaciones de ruinas de epoca visigoda en la aldea de san pedro de merida
- Guia de merida
- Merida. guide de la ville et de ses monuments

Almagro, Martin *see* Origen y formacion del pueblo hispano

Al-majallah al-misriyah – Cairo: Khalil Mitran, 1900-02. yr 1 n1-yr 3 n18. i jun/haziran 1900-1 jun 1909 – (= ser Arabic journals and popular press) – 1r – 1 – $775.00 – (ceased publ, resumed 1909-?) – us MEDOC [956]

Almanac panorama – Los Angeles, CA. 1980-86 – 1 – us AJPC [071]

Almanaccando : bilder aus italien / Hevesi, Ludwig – Stuttgart 1888 [mf ed Hildesheim 1995-98] – (= ser Fbc) – 3mf – 9 – €90.00 – 3-487-29259-9 – gw Olms [914]

Almanach : oder uebersicht der neuesten fortschritte in den spekulativen und positiven wissenschaften – Erfurt DE, 1802-07 – 6r – 1 – gw Misc Inst [500]

Almanach administrativo, historico e mercantil da provincia – Manaus, AM: Typ do Amazonas, 1884 – (= ser Ps 19) – 1,5,6 – bl Biblioteca [350]

Almanach agricole des cultivateurs pour l'annee... / Compagnie d'assurance agricole du Canada – Montreal?: La Compagnie, 18—18– ou 19– – 9 – mf#A01274 – cn Canadiana [630]

Almanach contenant une liste alphabetique des cites, villes, villages, paroisses et cantons de la province de quebec... = Almanac containing an alphabetical list of the cities, towns, villages, parishes and townships of the province of quebec... – Levis, Quebec: Mercier, 18—19– – 9 – (in french and english) – mf#A01275 – cn Canadiana [971]

Almanach das familias – Bahia: Lith typ de J G Tourinho, 1877 – (= ser Ps 19) – 1,5,6 – bl Biblioteca [640]

Almanach de gotha : annuaire genealogique, diplomatique et statistique – 1767-1863 – 9 – $3036.00 – (in french. 1864-1943 $1188 [0017]) – mf#0016 – us Brook [929]

Almanach de la guadeloupe et dependances – Basse-Terre. 1832-36; 1838-41; 1843-50 – 1 – fr ACRPP [972]

Almanach de la litterature, du theatre et des beaux-arts – Paris. 1853-69 – 1 – fr ACRPP [410]

Almanach de la montagne – Paris, 1849 – (= ser Pamphlets and periodicals relating to the French Revolution of 1848) – 1r – 1 – us CRL [944]

Almanach de la question sociale et de la libre pensee – Paris. 1891-1900; 1902-03 – 1 – fr ACRPP [073]

L'almanach de la semaine agricole pour... – Montreal: Duvernay, 1870-1871?//? – 9 – mf#A00182 – cn Canadiana [630]

Almanach der buecherstube – Muenchen: H Stobbe, 1918- [mf ed 1993] – (ill) – 1 – mf#8359 – us Wisconsin U Libr [430]

Almanach des dames, pour l'annee 1807 / Plamondon, Louis – Quebec: Nouvelle-Imprimerie, [1806?] (mf ed 1974) – 1r – 5 – mf#SEM16P193 – cn Bibl Nat [030]

Almanach des femmes – publ by Jeanne Deroin; London: J Watson, 1854 – (= ser Les femmes [coll]) – 2mf – 9 – (in english & french) – mf#8650 – fr Bibl Nationale [640]

Almanach des femmes = Women's almanac for 1853 – 2eme annee. publ. by Jeanne Deroin, London: J Watson, 1853 – (= ser Les femmes [coll]) – 3mf – 9 – (in english & french) – mf#8649 – fr Bibl Nationale [305]

Almanach des societes saint-jean-baptiste du canada et des etats-unis pour l'annee 1884 : cinquantieme anniversaire de la fondation de la societe: premiere annee – Montreal: J B Rolland, 1884? – 2mf – 9 – mf#54380 – cn Canadiana [030]

Almanach des traditions populaires – Paris. 3v. 1882-1884 – 7mf – 8 – mf#H-1374 – ne IDC [400]

Almanach du barreau : livre de references contenant le nom et l'adresse des juges, sherifs, protonotaires, avocats, notaires, huissiers...de la province de quebec... – Montreal: E Senecal & Fils, 1887- (irreg) [mf ed 1988] – 2mf – 9 – (only 1887 filmed) – mf#SEM105P912 – cn Bibl Nat [030]

L'almanach du monde qui chante contenant tous les derniers succes de la chanson : paroles et musique – Montreal: le Passe-temps, [190-]- [mf ed 2001] – 2mf – 9 – (ceased 1929?) – mf#SEM105P398 – cn Bibl Nat [780]

L'almanach du peuple : compilation de faits et chiffres a l'usage des electeurs du canada, supplement a la gazette de montreal, hommage des editeurs – Montreal: La Gazette, 1892? – 1mf – 9 – mf#42982 – cn Canadiana [971]

Almanach du peuple de beauchemin et payette pour l'an... – Montreal: Beauchemin et Payette, 1855 [mf ed 1857 filmed 1988] – 1mf – 9 – (ceased 186-?; cont by: l'almanach du peuple) – mf#SEM105P895 – cn Bibl Nat [030]

Almanach fuer aerzte und nichtaerzte / ed by Gruner, Christian Gottfried – Jena 1782-97 – (= ser Dz) – 31mf – 9 – €310.00 – mf#k/n3595 – gw Olms [610]

Almanach fuer das jahr [...] : almanach der psychoanalyse / Bibliotheque de la Societe Psychanalytique de Paris – Wien (A), 1926-38 – 1 – fr ACRPP [616]

Almanach fuer freunde der schauspielkunst auf das jahr... – Berlin: Gedruckt bei J Sittenfeld 1837-91 (annual) [mf ed 1979] – 17v on 11r – 1 – (began with 1836; ceased with 17. jahrg [1853]; cont by: deutscher buehnen-almanach [berlin, germany: 1854]) – mf#film mas c487 – us Harvard [790]

Almanach fuer freunde der schauspielkunst jahrgang (1)-(6), jahrgang 7-10 = German theatre almanach and yearbooks / ed by Wolff, Ludwig – [mf ed 1988] – 463mf (1:24) – 9 – diazo €1896.00 (silver €2400 isbn: 978-3-598-32323-2) – ISBN-10: 3-598-32324-7 – ISBN-13: 978-3-598-32324-9 – (cont as: wolff's almanach fuer freunde der schauspielkunst: jg 11, 12-17; deutscher buehnen-almanach: jg 18-57, berlin 1837-1893) – gw Saur [790]

Almanach fuer freunde der theologischen lektuere ueberhaupt und der gelehrten vaterlandsgeschichte insonderheit / ed by Waldau, Georg Ernst – Nuernberg 1781-83 – (= ser Dz. abt theologie) – 4mf – 9 – €120.00 – mf#k/n2145 – gw Olms [943]

Almanach fuer prediger, die lesen, forschen und denken / ed by Horrer, Georg Adam et al – Weissenfels, Leipzig 1785-93 – (= ser Dz. abt theologie) – 8 jg on 16mf – 9 – €160.00 – mf#k/n2201 – gw Olms [240]

Almanach general du commerce de la guadeloupe – Basse-Terre. 1843 – 1 – fr ACRPP [972]

Almanach historique et chronologique des spectacles – devenu: Nouveau calendrier historique des theatres de l'Opera et des comedies francaise et italienne et des foires. devenu: Les Spectacles de Paris, ou suite du calendrier historique et chronologique des theatres. devenu: Almanach des spectacles de Paris. Paris. 1752-1815 (interruptions entre 1794 et l'an VIII puis entre l'an IX et 1815) – 1 – fr ACRPP [790]

L'almanach judiciaire, agricole et municipal de la province de quebec pour annee 1877 – [Quebec?: s.n.], 1876 [mf ed 1984] – 1mf – 9 – 0-665-43052-3 – mf#43052 – cn Canadiana [030]

L'almanach judiciaire, agricole et municipal de la province de quebec pour annee bissextile 1876 – [Quebec?: s.n.], 1875 [mf ed 1984] – 1mf – 9 – 0-665-43051-5 – mf#43051 – cn Canadiana [030]

L'almanach judiciaire, agricole, scolaire, municipal et commercial de la province de quebec pour 1874 – [Quebec?: s.n.], 1873 [mf ed 1984] – 1mf – 9 – 0-665-43049-3 – mf#43049 – cn Canadiana [030]

L'almanach judiciaire, agricole, scolaire, municipal et commercial de la province de quebec pour 1875 – [Quebec?: s.n.], 1874 [mf ed 1984] – 1mf – 9 – 0-665-43050-7 – mf#43050 – cn Canadiana [030]

Almanach judiciaire de la province de Quebec : contenant les noms des juges de la puissance du canada, les protonotaires, sherifs... / Audette, Louis Arthur & Dunn, Thomas William Shea [comp] – Levis [Quebec]: Mercier, 1888 [mf ed 1979] – 1mf – 9 – 0-665-00062-6 – (incl ind) – mf#00062 – cn Canadiana [340]

Almanach judiciaire et commercial pour l'annee 1871 / Belanger, Jules [comp] – [Quebec?: s.n.], 1871 [mf ed 1984] – 1mf – 9 – 0-665-43054-X – mf#43054 – cn Canadiana [346]

Almanach litterario alagoano das senhoras – Jaragua, AL. mar 1888; jan 1889 – (= ser Ps 19) – 1,5,6 – bl Biblioteca [079]

Almanach national : annuaire de la republique francaise / France – 1695-1769 – 9 – $1122.00 – mf#0212 – us Brook [944]

Almanach oder uebersicht der neuesten fortschritte in den spekulativen und positiven wissenschaften – Erfurt DE, 1802-07 – 6r – 1 – gw Misc Inst [500]

Almanach turc : ou tableau de l'empire ottomanou l'on trouve tout ce qui concerne la religion, la milice, le gouvernement civil des turcs, et les grandes charges et dignites de l'empire, les differentes intrigues du serail / Porte, Joseph de la – Paris [1760] [mf ed Hildesheim 1995-98] – 3mf – 9 – €90.00 – 3-487-29132-0 – gw Olms [931]

Almanach-journal de l'ecole et du couvent – Joliette, Quebec?: s.n, 1887?-19–? – 9 – mf#A01273 – cn Canadiana [030]

Almanachs et annuaires de la ville de quebec de 1780 a 1900 / Carrier, Nicole – 1964 [mf ed 1979] – (= ser Bibliographies du cours...1947-66) – 2mf – 9 – (with ind) – mf#SEM105P4 – cn Bibl Nat [917]

The almanack of the fine arts for the year 1850 / Buss, Robert William – London 1850 – (= ser 19th c art & architecture) – 3mf – 9 – mf#4.2.455 – uk Chadwyck [700]

The almanack of the fine arts for the year 1852 / Buss, Robert William – London 1852 – (= ser 19th c art & architecture) – 3mf – 9 – mf#4.2.456 – uk Chadwyck [700]

Almanacks, 1855-1901 *see* Papers relating to the rochdale equitable pioneers

Almanak / Huria Kristen Batak Protestant – Medan, 1965-1972 – 16mf – 9 – mf#SE-150=1 – ne IDC [950]

Almanak – Canada. jan 1895-dec 1954 – 6r – 1 – (in icelandic) – cn Commonwealth Imaging [071]

O almanak – Ceara: Typ da Aurora Cearense, 25 ago 1867 – (= ser Ps 19) – mf#P17,01,34 – bl Biblioteca [079]

Almanak administrativo, mercantil e industrial da corte e da capital da provincia do Rio de Janeiro com os municipios de campos e de santos = Almanak laemmert – Rio d Janeiro: E & H Laemmert, 1872 (annual) – 2r – 1 – us CRL [030]

Almanak administrativo, mercantil e industrial da corte e da capital da provincia do Rio de Janeiro inclusive alguns municipios da provincia, e a cidade de santos = Almanak laemmert – Rio de Janeiro: E & H Laemmert, 1873-1874 (annual) – 4r – 1 – us CRL [030]

Almanak administrativo, mercantil e industrial da corte e provincia do Rio de Janeiro = Almanak laemmert – Rio de Janeiro: E & H Laemmert, 1848-1871 (annual) – 37r – 1 – us CRL [350]

Almanak administrativo, mercantil e industrial da corte e da capital da provincia do Rio de Janeiro inclusive a cidade de santos, da provincia de s paulo = Almanak laemmert – Rio de Janeiro: E & H Laemmert, 1875-82 (annual) – 16r – 1 – us CRL [350]

Almanak administrativo, mercantil e industrial do imperio do Brazil = Almanak laemmert – Rio de Janeiro: H Laemmert. [v40-46 1883-1889] (annual) – 17r – 1 – us CRL [350]

Almanak administrativo, mercantil e industrial do rio de janeiro – Rio de Janeiro: E H Laemmert. v1-4. 1844-47 – 1 – us CRL [030]

Almanak Angkatan Perang *see* Usaha pegawai nasional indonesia

Almanak "asia-raya" : tahoen ke-1 – Djakarta, Asia-Raya, Bagian Penerbitan (2603) – (= ser Indonesian imprints 1942-1945) – 3mf – 9 – mf#SE-2002 mf192-194 – ne IDC [959]

Almanak "asia-raya" : tahoen ke-2 – Djakarta, Djawa Sjinboen Sja (2604) – 280p 3mf – 9 – mf#SE-2002 mf195-197 – ne IDC [959]

Almanak dai toa (asia timoer raja) disoesoen oleh hassan noel 'arifin – Medan, Toko Boekoe "Antara" (2602) – 36p 1mf – 9 – mf#SE-2002 mf198 – ne IDC [959]

O almanak de goa para o anno bissexto de 1840 : com varias noticias historicas, ecclesiasticas, civis, politicas, e outras nocoens uteis a todo a genero de pessoas / Peres, Caetano Joao – Bombaim: Typographia Portugueza do Pregoeiro, [1839?] [mf ed 1995] – (= ser Yale coll) – vi/ix/362p (ill) – 1 – 0-524-10268-6 – (in portuguese) – mf#1996-1268 – us ATLA [241]

Almanak de goyaz – Goias: Typ Perseveranca, 1887 – (= ser Ps 19) – bl Biblioteca [079]

Almanak djawatan pendidikan kedjuruan – Djakarta, 1960 – 4mf – 9 – mf#SE-625 – ne IDC [959]

Almanak Indonesia *see* Pustaka djaja

Almanak indonesia – Djakarta, 1968. v1-2 – 32mf – 9 – mf#SE-1306 – ne IDC [959]

Almanak Kristen *see* Pustaka kristen

Almanak lembaga-lembaga negara dan kepartaian – Djakarta, 1961 – 9mf – 9 – mf#SE-244 – ne IDC [959]

Almanak municipal de barbacena – Barbacena, MG, 1897 – (= ser Ps 19) – bl Biblioteca [350]

Almanak organisasi KONI Pusat / Komite Olahraga Nasional Indonesia – Djakarta, 1969 – 6mf – 9 – mf#SE-1752 – ne IDC [959]

Almanak pegawai negeri – Djakarta, 1954-1956 – 16mf – 9 – mf#SE-204 – ne IDC [959]

Almanak pemerintah daerah propinsi sumatera utara – Medan, 1969 – 16mf – 9 – mf#SE-1850 – ne IDC [959]

Almanak perdagangan indonesia = The commercial year book of indonesia / Yin-ni shang yeh nien chien – Djakarta, 1955 – 13mf – 9 – mf#SE-2713 – ne IDC [959]

Almanak soeara asia – Soerabaja, Soeara Asia, 2604 – 216p 3mf – 9 – mf#SE-2002 mf205-207 – ne IDC [959]

Almanak tani – Weltevreden, Djakarta, 1925-1958 – 36mf – 9 – (missing: 1932-54) – mf#SE-607 – ne IDC [959]

Almanak "tjerdas" – Medan, 1950 – 3mf – 9 – mf#SE-609 – ne IDC [950]

Almanak umum nasional / Endang – Djakarta, 1954-1960 – 55mf – 9 – mf#SE-611 – ne IDC [959]

Almanak van het leidsche studentencorps : 119e jaarg. 1933 – Leiden, 1932 – 5mf – 8 – mf#SE-1439 – ne IDC [949]

Almanak veteran ri markas daerah legiun veteran ri – Medan, 1966 – 4mf – 9 – mf#SE-1309 – ne IDC [950]

Almanak wanita *see* Balapan

Al'manakh inostrannoi proletarskoi literatury / ed by Vygodskii, David Isaakovich – Leningrad: Izd-vo "Krasnaia gazeta", 1929 [mf ed 2002] – (= ser Literaturnaia studiia "reztsa") – 1r – 1 – (Filmed with: k biografii adama mizkevicha v 1821-1829 godakh / fedor verzhbovskii [teodor wierzbowski], (1898)) – mf#5239 – us Wisconsin U Libr [800]

Almanakh tsum 20 yorikn yubileum... – Buenos Aires, Argentina. 1942? – 1r – 1 – us UF Libraries [939]

Almanaque del maestro / Pimentel y Donaire, Miguel – 1892 – 9 – sp Bibl Santa Ana [030]

Almanaque filipino i guia de forasteros para el ano de... – Manila: Impr de D Jose Maria Dayot por Tomas Oliva, [1835?-] – 1r – 1 – us CRL [959]

al-Manis, Walid Abd Allah *see* Tafsir al-shari lil-tamaddun

Almanzar, Armando *see* Pulso de la ciudad

Al-manzum – Cairo: Ahmad Najib Qanawi, 1892-93. yr 1 pts1-24. 25 rabi' 2 1310-22 rabi 2 1311 [15 nov 1892-1 nov 1893] (complete) – (= ser Arabic journals and popular press) – 1r – 1 – $200.00 – us MEDOC [956]

Al-mar'ah al-'arabiyah – Damascus: Ittihad al-'Amm al-Nisa fi al-Qutr al-'Arabi al-Suri. n98-394. 5 kanun al-Thani 1977-tishrin al-Awwal 1998 – (= ser Arabic journals and popular press) – 9r – 1 – $1500.00 – (missing: n120-142,318-319,322,324,329) – us MEDOC [956]

Almaraz. Ayuntamiento *see* Fiestas de san roque en almaraz 1980

Almas rebeldes : drama en cuatro actos / Ramos, Jose Antonio – Barcelona: A Lopez, 1906 (mf ed 19–) – (= ser Teatro antiguo y moderno) – mf#Z-712 – us NY Public [820]

Al-masirah – [Khartoum]: al-Masirah, jul 18 1990-feb 21 1994 – 3r – 1 – us CRL [960]

Al-mawakib – Nazareth, 1984- . mujallad 1-11. jan 1984-dec 1994 – (= ser Arabic journals and popular press) – 3r – 1 – $265.00 – (missing: mujallad 9 n1-4, 9-12) – us MEDOC [073]

Al-mawqif – al-Khartum: al-Mawqif, n2-7. sep 28-nov 23 1988 – 1r – 1 – us CRL [079]

Al-mawqudhah – Cairo: Muhammad Tawfiq al-Azhari, 1905. v1 n1-3. 20 apr-20 jun 1905 – (= ser Arabic journals and popular press) – 1r – 1 – $775.00 – (cont & cont by: humarat munyat; in 1905 title changes to: al-mawqudhah and then back to humarat munyati; r also incl: humarat munyati) – us MEDOC [956]

Al-mawqudhah *see* Humarat munyati

Al-maydan : journal social, economique et politique algerian – Constantine. n1-28. juil 1937-mars 1938. mq n1,10,13,16,17 – (= ser Maydan) – 1 – fr ACRPP [073]

Almeida, A Tavares De *see* Oeste paulista

Almeida, Aluisio De *see* Revolucao liberal de 1842

Almeida, Antonio De *see*
- Bushmen and other non-bantu peoples of angola
- Subsidio para o estudo da colonizacao dos dembos

Almeida, Antonio Ramos De *see* Para a compreensao da cultura no brasil

Almeida, Ferrand Pimentel d' *see* O sentimento da natureza no fausto de goethe

Almeida, Guilherme de *see*
- Do sentimento nacionalista na poesia brasileira
- Homens e factos de uma revolucao

Almeida, Joao De *see* Sul d'angola

Almeida, Jose Americo de *see*
- Ano do nego
- A parahyba e seus problemas
- Vice-reinado de d luiz d'almeida portugal

Almeida, Lourival Nobre De *see* Comunidade luso-brasileira

Almeida, M de *see* Historia geral de ethiopia a alta ou abassin

Almeida, Ruy *see* Poesia e os cantadores do nordeste

Almena broadcaster – 1937 apr 22-38 oct 27; 1938 nov 3-1940 may 30; jun 6-1941 jul 3 – 1 – mf#915741 – us WHS [071]

Almenak "Waspada" *see* Jajasan penerbit pesat

Almendralejo *see*
- Feria de agosto, 1927
- Feria de las mercedes en almendralejo
- Feria de las mercedes en almendralejo de 1913. concurso de ganaderia
- Ferias y fiestas. agosto de 1945
- Padron general del ano 1897
- Revista de ferias, 1943. festividad de nuestra senora de la piedad

Almendralejo asociacion de adoradores de jesus sacramentado y practicas religiosas / Canciones – Almendralejo: Luciano Carballar, 1901 – 1 – sp Bibl Santa Ana [240]

Almendralejo. El Obrero Extremeno *see* Reglamento de la sociedad cooperativa y de socorros mutuos. el obrero extremeno. almendralejo

Almendralejo. ferias y fiestas 1942. nuestra senora de la piedad – 1 – sp Bibl Santa Ana [946]

Almendralejo, Pedro de *see* Escudo de las indulgencias de la religion...de s francisco

Almendralejo. Spain *see*
- Estatutos de la cofradia del santisimo sacramento
- Reglamento de guardas de la comunidad de labradores de almendralejo

Almendrallucas, Bernardo *see* Doctrinas sociales, superadas por...

Almendros, Herminio *see* Oros viejos

Almenrausch und edelweiss : erzaehlung aus dem bairischen hochgebirge / Schmid, Herman – 2.aufl. Leipzig: Keil [18–?] – (= ser Gesammelte schriften. volks- und familien-ausgabe 20) – 1 – (bound with: die gasselbuben and das muencher kindeln) – mf#film mas c438 – us Harvard [880]

Almeras, Henri d' *see* La femme amoureuse dans la vie et dans la litterature

Al-midan – Khartum: Dar al-Ayam lil-Tibaah wa-al-Nashr, jul 15 1985-jun 1989 – 12r – us CRL [916]

Almidon / Cuadra, Manolo – Managua, Nicaragua. 1945 – 1r – 1 – us UF Libraries [972]

El almirante de castilla hasta las capitulaciones de santa fe : sevilla, 1944 / Perez Embrid, Florentino – Madrid: Razon y Fe, 1946 – 1 – sp Bibl Santa Ana [946]

Almirante saldanha e a revolta da Armada / Souza E Silva, Augusto Carlos De – Rio De Janeiro, Brazil. 1936 – 1r – 1 – us UF Libraries [972]

Al-misbah – Oran. n1-34. juin 1904-fevr 1905 – (= ser Misbah) – 1 – fr ACRPP [073]

Al-mithaq – Jerusalem, 1980-1986 – 12r – 1 – mf#J-93-10 – ne IDC [956]

Al-mi'yar : garida adabiya intiqadiya fukahiya – Alger. n1-9. dec 1932-avr 1933 – (= ser Miyar) – 1 – fr ACRPP [073]

Almkvist, H *see* Die bischari-sprache tu-bedawie in nordost-afrika beschreibend und vergleichend dargestellt

Al-moayad – Cairo. nov 19 1907-apr 1914 – 1 – us NY Public [073]

Almoharin. Sociedad Mutua de Criadores de Ganado Vacuno *see* Reglamento de la sociedad mutua de criadores de ganado vacuno

Almoina, Jose *see* Biblioteca erasmista de diego mendez

Almon bennett's platform bee house : with full instructions, patented, may 17 1858 / Bennett, Almon – [Hamilton, Ont?: s.n, 1858?] [mf ed 1994] – 1mf – 9 – 0-665-94612-0 – mf#94612 – cn Canadiana [630]

Almon, John *see*
- The remembrancer, or impartial repository of public events from 1775-1784, together with "prior documents," 1764-1775
- A review of the reign of george 2

Almond press – 1924 nov 21-27; 1928-1931 oct 30 – 1 – mf#916086 – us WHS [071]

Almoner : a periodical religious publication – La Salle IL 1814-15 – 1 – mf#3538 – us UMI ProQuest [240]

Al-montada : christian news bulletin – n1-118. apr 1967-85 – 3r – 1 – (lacks some pp) – mf#atla s0175 – us ATLA [240]

The almost christian discovered : or, the false professor tried and cast / Mead, Matthew – New York: Lewis Colby, 1850 – 1mf – 9 – 0-8370-7169-0 – mf#1986-1169 – us ATLA [240]

Almost forgotten, never told / Green, Lawrence George – Cape Town, South Africa. 1965 – 1r – 1 – us UF Libraries [960]

Almost protestant, and the almost romanist / Cumming, J – London, England. 1852? – 1r – 1 – us UF Libraries [240]

Almp newsletter – 1981 fall-1982 winter – 1 – mf#711897 – us WHS [071]

Almsgiving / Baugh, Folliott – London, England. 1842 – 1r – 1 – us UF Libraries [240]

Almsgiving and the offertory – London, England. 1852 – 1r – 1 – us UF Libraries [240]

Al-Mubassir *see* Le mobacher

Al-mubassir – Alger, sept 1847-70, 1912-26 – (= ser Mubassir) – 1 – (arabic text. bilingual: 1859-1861) – fr ACRPP [073]

Al-Muhailani, Abdul-Rahman S *see* The influence of physical conditioning and deconditioning upon cardiac structure of males and females

Al-muhit – Cairo: 'Awad Wasif, 1902-14. yr 1 n1-yr 12 n10. i jan 1903-dec 1914 [complete] – (= ser Arabic journals and popular press) – 2r – 1 – $1,200.00 – (sample iss with ind 1 nov 1902) – us MEDOC [956]

Al-munadil – al-Khartum: Hizb al-Bath al-Arabi al-Ishtiraki, Munazzamat al-Sudan, sep 28 1985-apr 9 1989 – 2r – us CRL [079]

Al-munazer – Sao Paulo, SP. 03 jan 1900 – (= ser Ps 19) – mf#P18,01,97 – bl Biblioteca [079]

Al-muntahab min ta'ri-h halab / Ibn-al-Adim, Umar Ibn-Ahmad; ed by Freytag, Georg Wilhelm Friedrich – Lutetiae Parisiorum: Typographia Regia 1819 – (= ser Whsb) – 4mf – 9 – €50.00 – (ann by ed) – mf#Hu 479 – gw Fischer [410]

Al-muqattam al-usbu'i – Cairo: Ya'qub Sarruf, 1889-. yr 1 iss1-45. 27 jumada 2 1306-11 jumada [27 feb/shubat 1889-3 jan/kanun 2 1890] – (= ser Arabic journals and popular press) – 1r – 1 – $350.00 – us MEDOC [956]
Al-muqtabas – Damascus: Muhammad Kurd 'Ali, 1907-17. [daily] n1-1464 (feb 1907-apr 15 1914) – (= ser Arabic journals and periodicals) – 8r – 1 – $500.00 – (cont by & cont: al-ummah; cont by: al-qabas; reels also contain al-ummah and al-qabas) – us MEDOC [079]
Al-muqtabas – Damascus, 1906-1911. v1-6 – 52mf – 9 – mf#NE-20327 – ne IDC [956]
Al-muqtabas – Damascus: Muhammad Kurd 'Ali, 1906-17 [mthly] – (= ser Arabic journals and periodicals) – 114r – 1 – $1700.00 – (v1-9 [1906-17] on mf) – us MEDOC [079]
Al-musaadah al-amirikiyah li-israil : al-ribat al-hayawi / Situfar, Tumas R – [Beirut?]: Muassasat al-Dirasat al-Filastiniyah, 1983 – 1r – us CRL [950]
Al-muslimun – Bangil, jun/jul, 1963-1964(1) nos 1-6 – 4mf – 9 – mf#SE-389 – ne IDC [950]
Al-mustaqbal – Paris. n1-151. mars 1916-19 – (= ser Mustaqbal) – 1 – fr ACRPP [073]
Al-mustathmir – tusdiruha al-Hayah al-Ammah lil-Istithmar al-Khartum: al-Hayah, n1-25. jul 1 1992-mar 3 1994 – 1r – us CRL [073]
Al-mutamar – al-Khartum: [s.n., dec 3 1991-may 12 1992 – 1r – us CRL [073]
al-Mutawwa, Subhi *see* Al-takhtit al-zirai li-mintaqat al-wafrah
Al-muwatin – Baghdad, [jun 6-aug 1962] – us CRL [079]
Al-muwazzaf – Cairo: Amin Khayrat al-Ghandur (the Union of Egyptian Government Workers), 1936-39. yr 1 v1 n1-yr 3 v3 n9. jan 1936-sep 1938 – (= ser Arabic journals and popular press) – 2r – 1 – $700.00 – us MEDOC [956]
Al-nadhir – Cairo: Salih Mustafa 'Ashmawi, 1938-39? v1 n1-10,35, v2 n4-5,19-20,27,33,39,41. 30 rabi' I 1357-23 shawwal 1358 [30 may 1938-6 dec 1939] – (= ser Arabic journals and popular press) – 1r – 1 – $600.00 – (cont: jaridat al-ikhwan al-muslimin; cont by: al-ikhwan al-muslimun; r incl both) – us MEDOC [956]
Al-nadhir *see* Jaridat al-ikhwan al-muslimin
Al-naft al-arabi wa-al-qadiyah al-filastiniyah / Qarm, Jurj – Bayrut: Mu'assasat al-Dirasat al-Filastiniyah, 1979 – 1r – us CRL [956]
Al-naft al-arabi wa-qadiyat filastin fi al-thamaninat / Saigh, Yusuf Abd Allah – Bayrut: Muassasat al-Dirasat al-Filastiniyah, 1986 – 2r – us CRL [956]
Al-nahar – Khartoum, Sudan dec 17 1987-mar 14 1989 – 3r – us CRL [079]
Al-nahar – Jerusalem, 1986-1995 – 30r – 1 – (missing: 1986(26); 1987(289): 1988(apr 6, 12, 29-30; jun 24, 29); 1989-1990; 1991(1403, 1443, 1459, 1576, 1619, 1679, 1683)) – mf#J-93-11 – ne IDC [956]
Al-nahdah – [al-Khartum, Sudan]: Dar al-Thaqafah il-Nashr wa-al-lan al-Mahdudah, nov 14-dec 10 1988 – 1r – us CRL [079]
Al-nahla – Beyrouth. n1. mai 1870 – (= ser Nahla) – 1 – fr ACRPP [073]
Al-nashrah al-ammah – Aden, Yemen, jan 3 1983-may 15 1984 – 6r – us CRL [079]
Al-nashrah al-yawmiyah lil-anba – Aden, Yeman, nov 1-28 1981; feb 10-17, apr 11, 15-20, 29, may 2-12, jul 20, 24-27, aug 18-24, dec 23, 26, 29 1982 – 1r – us CRL [079]
Al-nibras – Beirut. v1-2. 1909-1910 – 10mf – 9 – mf#NE-20334 – ne IDC [956]
Al-nida – [Omdurman, Sudan]: Dar al-Sham lil Tibaah wa-al-Nashr, sep 29 1986-feb 16 1988 – 1r – us CRL [079]
Alnutts irish land schedule *see* Irish land schedule
Al-nuzhah – Assiut, Alexandria: Jurji al-Khayyat, 1866. yr 1 n3-19. 7 barmahat 1602 [coptic era]/9 jumada 2 1303/15 mar 1886-7 hatur 1603/19 safar 1304/15 nov 1886 – (= ser Arabic journals and popular press) – 1r – 1 – $175.00 – us MEDOC [956]
Alnwick advertiser – Alnwick, England 21 apr 1983-30 apr 1992 [mf 1986-92] – 1 – (incorp with: northumberland gazette; cont: alnwick & morpeth advertiser [10 mar-7 apr 1983]) – uk British Libr Newspaper [072]
Alnwick and county gazette – [NE England] Northumberland jan 1886-dec 1950 [mf ed 2004] – 63r – 1 – (missing: 1896-97, 1909; cont by: alnwick & county gazette & guardian [jan 1924-dec 1942]; northumberland & alnwick gazette [jan 1943-dec 1947]; northumberland gazette [jan 1948-dec 1950]) – uk Newsplan [072]
Alnwick and county gazette – Alnwick, England 5 jan 1884-16 jun 1923 – 1 – (wanting: 1896, 1897; cont: alnwick & county gazette & general record [10 nov-29 dec 1883]; cont by: alnwick & county gazette & guardian [23 jun 1923-24 dec 1942]) – uk British Libr Newspaper [072]
Alnwick and county gazette and guardian – Alnwick, England 23 jun 1923-24 dec 1942) – 1 – (cont: alnwick & county gazette [5 jan 1884-16 jun 1923]; cont by: northumberland & alnwick gazette [1 jan 1943-4 apr 1947]) – uk British Libr Newspaper [072]
Alnwick guardian and county advertiser – Alnwick, England 6 feb 1886-16 jun 1923 [mf 1910, 1911] – 1 – (incorp with: alnwick & county gazette; wanting: 1896, 1897) – uk British Libr Newspaper [072]
Alnwick mercury – Alnwick, England [ns] 1 jun 1854-1 dec 1864, 31 dec 1864-29 dec 1883 – 1 – (incorp with: alnwick & county gazette) – uk British Libr Newspaper [072]
ALOC *see* The story of a dark plot
Alocucion : que el santisimo senor nuestro pio por la divina providencia papa 6. tuvo en el consistorio secreto: el dia 13. de noviembre de 1775... / Catholic Church. Pope [1775-1799: Pius 6] – Mexico: impr D Jose de Juaregui [1776] – (= ser Books on religion...1543/44-c1800: ordenes, etc: dominicos) – 1mf – 9 – mf#crl-430 – ne IDC [241]
Alocucion a los actores argentinos. manuscrito / Garcia Lorca, Federico – 1mf – 9 – sp Cultura [850]
ALOG : army logistician – Washington DC 1985-86 – 1,5,9 – (cont: army logistician; cont by: army logistician) – mf#5713.02 – us UMI ProQuest [355]
Aloha breeze – Hillsboro OR: Argus Enterprises] -1983 [wkly] – 7r – 1 – us Oregon Lib [071]
Aloha news – Aloha OR: S M Brown, 1927-51 [wkly] – 6r – 1 – (merged with: tigard sentinel and: beaverton enterprise and: multnomah press to form: valley news (1951-62)) – us Oregon Lib [071]
Aloha news *see*
– Beaverton enterprise
– Tigard sentinel
Aloha times – Beaverton OR: Valley Pub Inc, 1974- [semiwkly] [mf ed 1978-79] – 10r – 1 – us Oregon Lib [071]
Alois Blumauer's saemmtliche werke : und handschriftlicher nachlass – Wien: M Stern, 1884 [mf ed 1989] – 4v in 2 (ill) – 1 – (erste, vollstaendige gesammt-ausgabe mit vorwort, einleitung und anmerkungen...) – mf#7035 – us Wisconsin U Libr [802]
Aloja, Ada D' *see* Informe sobre la investigacion antropologico-demog...
Alone in the wide, wide world : a musically illustrated service / Andrews, J R – Toronto: W Briggs, 1891 [mf ed 1980] – 1mf – 9 – 0-665-02409-6 – mf#02409 – cn Canadiana [780]
Alone in the wilderness / Knowles, Joseph – Toronto: Copp, Clark, c1913 [mf ed 1996] – 4mf – 9 – 0-665-76899-0 – (ill by aut) – mf#76899 – cn Canadiana [790]
Along the florida reef / Holder, Charles Frederick – New York, USA. 1892 – 1r – 1 – us UF Libraries [574]
Along the lines at the front : a general survey of baptist home and foreign missions / Bainbridge, William Folwell – Philadelphia: American Baptist Pub Soc, c1882 [mf ed 1992] – (= ser Baptist coll) – 1mf – 9 – 0-524-05071-6 – mf#1991-2195 – us ATLA [242]
Along the north arm – n1 [1949 apr 25] – 1 – mf#681682 – us WHS [071]
Along the towpath – 1960 mar-1961 may – 1 – mf#681377 – us WHS [071]
Along the way – 1981 jun-1986 sep – 1 – mf#1279119 – us WHS [071]
Alonso, Amado *see*
– Gramatica castellana
Alonso Chacon, Joseph *see* Tradiciones y memorias historiales de don gonzalo de stuniga
Alonso Cortes, Narciso *see* Espronceda. ilustraciones biograficas y criticas
Alonso de la Avecilla, Pablo *see* Canciones guerreras
Alonso de la Vera Cruz, fray *see*
– Recognitio, summularum reuerendi patris illdephonsi a vera cruce augustiniani artium ac sacrae theologiae doctoris apud indorum indytam mexicum primarij in academia theologiae moderatoris
– Speculum coniugiorum
Alonso de Llerena *see* Oracion funebre...5 de julio de 1736
Alonso fernandez de barrantes. su testamento (1390) apuntes genealogicos de su casa / Ciadoncha, Marques de – Madrid: Tip. Arch., 1931. B.R.A.H. 99, pp. 225-267 – sp Bibl Santa Ana [920]
Alonso Getino, G *see* Incendio de conventos en espana y supresion de colegios y misiones espanolas en ultramar
Alonso golfin. leyenda / Hurtado de Mendoza, Publio – 1894 – 9 – sp Bibl Santa Ana [830]
Alonso, Isidoro *see*
– Iglesia en peru y bolivia
– Iglesia en venezuela y ecuador
Alonso, Julio *see*
– Vias ferreas. asiento y conservacion...
– Vias ferreas. asiento y conservacion. atlas
Alonso, Longinos *see* Deberes y facultades de los alcaldes de barrio
Alonso, M *see* Exposicao sobre los livros de beato dionisio areopagita
Alonso, Maria Rosa *see* Residente en venezuela
Alonso perez de guzman / Justiniano Arribas, Juan – 1896 – 9 – sp Bibl Santa Ana [920]
Alonso Pujol, Guillermo *see* Parlamento
Alonso Quintero, Elfidio *see* Europeo en el caribe
Alonso qvijano el bveno / Motta Salas, Julian – Bogota, Colombia. 1930 – 1r – 1 – us UF Libraries [972]
Alonso y de los Ruizes de Fontecha, J *see*
– Diez privilegios para mujeres prenadas...con un diccionario medico
– Disputationes medicae de anginorum ...
Alor – Badajoz 1950-58 – 5 – sp Bibl Santa Ana [074]
Alos, J *see* Pharmaco-medica dissertatio de viperiis trochiscis
Aloysio Maria a Carpo *see* Caeremoniale iuxta ritum romanum
Alpayim shanah ve-shanah / Navon, Aryeh – Merhavyah, Israel. 1949 – 1r – 1 – us UF Libraries [939]
Der alpbote *see* Intelligenz-blatt fuer die oberaemter ehingen und muensingen
Alpca newsletter – 1975 feb-1983 jun; 1983 aug-1987 dec – 1 – mf#1362462 – us WHS [071]
Die alpen / Haller, Albrecht von; ed by Betteridge, Harold T – Berlin: Akademie-Verlag, 1959 – 1r – 1 – us Wisconsin U Libr [430]
Die alpen in natur- und lebensbildern / Berlepsch, Hermann A – Leipzig 1861 [mf ed Hildesheim 1995-98] – (= ser Fbc) – viii/441p on 3mf [ill] – 9 – €90.00 – 3-487-29394-3 – gw Olms [914]
Alpengegenden niederoesterreichs und obersteyermarks im bereiche der eisenbahn von wien bis muerzzuschlag / Weidmann, Franz C – Wien 1862 [mf ed Hildesheim: 1995-98] – (= ser Fbc) – 269p on 2mf – 9 – €60.00 – 3-487-29435-4 – gw Olms [380]
Alpenklaenge und lawinendonner : [poems] / Waelti, C – 2. ausg. Thun und Aarau: J J Christen 1844 [mf ed 1991] – 1r – 1 – (half-title: freie lieder aus der schweiz. filmed with: richard wagner / hans von wolzogen) – mf#2974p – us Wisconsin U Libr [810]
The alpenstock : or, sketches of swiss scenery and manners 1825-26 / Latrobe, Charles – London 1829 [mf ed Hildesheim 1995-98] – 3mf – 9 – €90.00 – 3-487-29332-3 – gw Olms [914]
Alpenwanderungen : fahrten auf hohe und hoechste alpenspitzen / Grube, August W – Oberhausen [u.a.] 1873 [mf ed Hildesheim 1995-98] – (= ser Fbc) – 2v on 4mf – 9 – €120.00 – 3-487-29351-X – gw Olms [914]
Alpenzeitung – Bozen (I), 1938, 1939 2 may-31 dec, 1940 25 apr 1941 [gaps] – 7r – 1 – gw Misc Inst [074]
Alper, Rebekah *see* Pirpure mahapekhah
Alpers, Paul *see* Karl goedeke, sein leben und sein werk
Alpers, Wilhelm *see* Die heldenbraut
Alperschn, Marcos *see* Dreisig yor in argentine
Alpersohn, Marcos *see*
– Af argentiner erd
– Galuth
– Sheloshim shenoth ha-hithyashvuth
Alpert, Benjamin M *see*
– Outline of new york criminal law
– Outline of the law of private corporations.
Alpes-libres – Houtes-Alpes, France 1944 – 1 – (in french) – us UMI ProQuest [934]
Alpha – Miami, FL. 1979 feb-Apr – 1r – 1 – us UF Libraries [071]
O alpha : hebdomadario litterario, scientifico, noticioso e industrial – Rio Claro, SP: Typ Rio Clarense, 06 jan, 10 fev 1878 – (= ser Ps 19) – mf#P18,01,67 – bl Biblioteca [079]
Alpha gram – v1 n1-1913 [1984 feb 16-may 9] – 1 – mf#1477263 – us WHS [071]
Alpha news – 1953 jun; 1968 spring – 1 – mf#5286997 – us WHS [071]
Alpha paper : a publication of the wynne family and kinsmen association – 1978 sep/oct-1979 nov/dec; 1981-1984 apr – 1 – mf#637200 – us WHS [929]
Alpha Phi Alpha Fraternity *see* Sphinx
Alpha spirit – 1989 spring-1990 summer – 1 – mf#4712872 – us WHS [071]
The alphabet : an account of the origin and development of letters / Taylor, Isaac – London: Kegan Paul, Trench. 2v. 1883 – 2mf – 9 – 0-8370-9117-9 – (incl bibl ref and index) – mf#1986-3117 – us ATLA [400]
The alphabet – Brockville [Ont]: McMullen & Co [185-?] [mf ed 1993] – 1mf – 9 – 0-665-91345-1 – mf#91345 – cn Canadiana [420]
Alphabet of fascist economics : a critique of the bombay plan of economic development of india / Parikha, Govardhana – Calcutta: Renaissance Publishers, [1944] – (= ser Samp: indian books) – us CRL [339]
Alphabetical card manifests of alien arrivals at alexandria bay, cape vincent, champlain... : new york, july 1929-april 1956 / U.S. Immigration and Naturalization Service – (= ser Records of the immigration and naturalization service) – 3r – 1 – mf#m1481 – us Nat Archives [975]
Alphabetical card name indexes to the compiled service records of volunteer soldiers who served in union organizations not raised by states or territories : excepting the veterans reserve corps and the u.s. colored troops – (= ser Records of the adjutant general's office, 1780's-1917) – 36r – 1 – (with printed guide) – mf#M1290 – us Nat Archives [355]
An alphabetical catalogue of plates / Boydell, John & Boydell, Josiah – London 1803 – (= ser 19th c art & architecture) – 1mf – 9 – mf#4.2.589 – uk Chadwyck [760]
Alphabetical catalogue of the library of parliament : being an index to the classified catalogues printed in 1857 and 1858, and to the books since added to the library, up to 1st march, 1862 = Catalogue alphabetique de la bibliotheque du parlement: comprenant l'index des catalogues methodiques publies en 18 / Canada (Province). Parlement. Bibliotheque – Quebec: Hunter, Rose & cie, 1862 [mf ed 1994] – 4mf – 9 – mf#SEM105P2181 – cn Bibl Nat [020]
Alphabetical catalogue of the library of the hon the legislative council of canada : authors and subjects / Canada (Province). Parlement. Conseil legislatif. Bibliotheque – Montreal: printed by James Starkes & Co, 1845 [mf ed 1983] – 3mf – 9 – mf#SEM105P156 – cn Bibl Nat [020]
Alphabetical index of the births, marriages and deaths recorded in providence – v3 [1879-35] – 1 – mf#4861275 – us WHS [929]
An alphabetical index of the code of civil procedure of lower canada / Coutlee, Louis William – Montreal: Dawson Bros; Quebec: M L Cremazie, 1870 [mf ed 1980] – 1mf – 9 – mf#SEM105P47 – cn Bibl Nat [348]
Alphabetical index of words occurring in the aitareya braahmanam / ed by Josi, Visvanatha Balkrishna – 1st ed. Bombay: Govt Central Book Depot [dist] 1916 [mf ed 1992] – 1mf – 9 – 0-524-02534-7 – (in sanskrit) – mf#1990-3029 – us ATLA [490]
Alphabetical index to canadian border entries through small ports in vermont, 1895-1924 – (= ser Records of the immigration and naturalization service) – 6r – 5 – mf#m1462 – us Nat Archives [975]
Alphabetical index to declarations of intention of the us district court for the southern district of new york, 1917-1950 / U.S. District Court – (= ser Records of district courts of the united states) – 111r – 1 – mf#M1675 – us Nat Archives [347]
Alphabetical index to petitions for naturalization of the us district court for the southern district of new york, 1824-1941 / U.S. District Court – (= ser Records of district courts of the united states) – 102r – mf#M1676 – us Nat Archives [347]
Alphabetical index to petitions for naturalization of the us district court for the western district of new york, 1906-1966 / U.S. District Court – (= ser Records of district courts of the united states) – 20r – 1 – mf#M1677 – us Nat Archives [347]
An alphabetical index to the laws of canada : being a ready reference to the statutes / Glackemeyer, Edouard Claude – Toronto?: Lovell & Gibson, 1859 – 1mf – 9 – mf#10816 – cn Canadiana [348]
An alphabetical index to the new testament : common version / Allibone, Samuel Austin – Philadelphia: American Sunday-School Union, c1868 [mf ed 1985] – 1mf – 9 – 0-8370-2079-4 – mf#1985-0079 – us ATLA [225]
An alphabetical list of engravings declared at the office of the printsellers' association, london : ...since its establishment in 1847 to the end of 1885 / Friend, George William [comp] – London: printed for the Printsellers' Assoc [1847-90] – (= ser 19th c art & architecture) – 2v on 7mf – 9 – mf#4.1.95 – uk Chadwyck [760]
An alphabetical list of the feasts and holidays of the hindus and muhammadans / Imperial Record Dept. India – Calcutta: Superintendent of Govt Printing, 1914 – us CRL [230]
Alphabetical listing of employer subject to wisconsin's unemployment compensation law – 1959-71 – 1 – mf#697272 – us WHS [071]
Alphabetical listing of nurserymen, nursery stock dealers, turf nurseries – 1973-80 – 1 – mf#525840 – us WHS [635]
Alphabetical manifest cards of alien and citizen arrivals at fort fairfield, maine, ca 1909-april 1953 / U.S. Immigration and Naturalization Service – (= ser Records of the immigration and naturalization service) – 1r – 1 – mf#m2064 – us Nat Archives [975]

Alphabetical manifest cards of alien arrivals at [...] / U.S. Immigration and Naturalization Service – (= ser Records of the immigration and naturalization service) – 1 – (calais, maine c1906-52 [5r] m2042; jackman, maine c1909-53 [3r] m2046; van buren, maine c1906-52 [1r] m2065. vanceboro, maine c1906-dec 24 1952 [13r] m2071) – us Nat Archives [975]

Alphabetical record : engineers and superintendents, etc, and the principal public works on which they have reported or been employed: canada, 1779 to 1891 / Baillairge, George Frederick – [S.l: s.n, 1891?] [mf ed 1980] – 1mf – 9 – 0-665-02219-0 – mf#02219 – cn Canadiana [620]

Alphabetical series of defense documents presented for evidence and rejected by the international military tribunal for the far east, 1945-1947 / World War 2. Defense Section – (= ser Records of allied operational and occupation headquarters, world war 2) – 3r – 1 – mf#M1694 – us Nat Archives [355]

Alphabetischer katalog der staatsbibliothek zu berlin preussischer kulturbesitz / Deutsche Staatsbibliothek Berlin. Musiksammlung – [Berlin, Hildesheim mf ed 1990] – (= ser Die europaeische musik) – 467mf [1:42] – 9 – diazo €2840.00 silver €3000.00 – gw Olms [780]

Alphabetischer musikalienkatalog der pfaelzischen landesbibliothek speyer – [mf ed 1991] – 55mf (1:42) + suppl – 9 – diazo €1260.00 – ISBN-10: 3-598-33195-9 – ISBN-13: 978-3-598-33195-4 – (incl suppl) – gw Saur [780]

Alphabetischer zentralkatalog der zuercherischen bibliotheken – [mf ed Hildesheim 1990] – 1357mf (1:42) – 9 – diazo €4900.00 – gw Olms [020]

Alphabetisches verzeichniss der sich in j schmidt's mondcharte befindlichen objecte : zusammengestellt nach der 'kurzen erlaeuterung zu schmidt's mondcharte' / Hildesheimer, L – [Odessa?: s.n.] 1885 (Odessa: Druck von A Schultze) [mf ed 2007] – 1r – 1 – mf#film mas 37813 – us Harvard [520]

Alphabetisch-statistisch-topographische uebersicht aller doerfer, flecken, staedte und andern orte der koenigl preuss provinz schlesien : mit einschluss der ganzen jetzt zur provinz gehoerenden markgrafthums ober-lausitz, und der grafschaft glatz / Knie, Johann – Breslau 1830 [mf ed Hildesheim: 1995-98] – (= ser Fbc) – xxiii/1079p on 12mf – 9 – €120.00 – 3-487-29583-0 – gw Olms [914]

Alphabetum divini amoris / Gerson, Jean de (Jean Charlier) – Lovanii, c1483 – €5.00 – ne Slangenburg [242]

Alphabetum tibetanum missionum apostolicorum commodo editum / Giorgi, A A – Romae: typis sacrae congregationis de propaganda fide, 1762 – 10mf – 9 – mf#HT-641 – ne IDC [915]

Alphabetvm arabicvm – Romae, 1592 – 1mf – 9 – mf#H-8221 – ne IDC [470]

Alphandery, Paul *see* Les idees morales chez les heterodoxes latins au debut du 13e siecle

Alphonse desjardins : pionnier de la coopération d'epargne et de credit en amerique: volume-souvenir du cinquantieme anniversaire de la caisse populaire de levis / Vaillancourt, Cyrille – Levis: Editions le Quotidien, 1950 – 1r – 5 – (pref by chanoine philibert grondin) – mf#SEM16P215 – cn Bibl Nat [332]

Alphonse, pere *see* Sainte catherine de sienne

Alphonsus, Joao *see* Rola-moca

Alphorn : illustrirtes schweizer familienblatt – Luzern (CH), 1889 n2-52 – 1 – (aka: das alphorn) – gw Misc Inst [640]

Das alphorn *see* Alphorn

Le alpi *see* Miscellaneous newspapers of las animas county, reel 1

Alpic newsletter – v5 n7 [1979 nov/1980 jan]; v7 n1-3; [1981 jan-spring/summer]; 1982-1989 mar – 1 – mf#656576 – us WHS [071]

Alpina : eine schrift, der genauern kenntniss der alpen gewiedmet / ed by Salis-Marschlins, Carl U von – Winterthur 1806-09 [mf ed Hildesheim 1995-98] – (= ser Fbc) – 4v on 13mf – 9 – €130.00 – 3-487-29392-7 – gw Olms [914]

Alpine post – St Moritz (CH) 29 oct 1892-30 mar 1901 – 9r – 1 – (in english & german; cont as: engadin express & alpine post wh is entered under samaden) – uk British Libr Newspaper [074]

Alport, Cuthbert James Mccall Alport *see* Sudden assignment

The alps, switzerland, and the north of italy / Williams, Charles – London: J. Cassell, 1854. viii,633p. map. illus – 1 – us Wisconsin U Libr [949]

Al-qabas – [al-Khartum, al-Sudan]: al-Ikhwan al Muslimun, apr 25 1988-jun 28 1989 – 2r – 1 – us CRL [079]

Al-qabas – Damascus: Muhammad Kurd 'Ali, 1907-17. [daily] n1-1464 (feb 1907-apr 15 1914) – (= ser Arabic journals and periodicals) – 8r – 1 – $500.00 – (cont & cont by: al-muqtabas; reels also contain al-ummah and al-qabas) – us MEDOC [079]

Al-qabas *see* Al-ummah

Al-qadiyah al-filastiniyah fi al-istiratijiyah al-amirikiyah : al-mushkilat wa-al-khiyarat / Pranger, Robert J – [Beirut]: Muassasat al-Dirasat al-Filastiniyah, 1983 – 1r – 1 – us CRL [956]

[Al-qanun fi al-tibb] : [a system of medicine, and other works] / Husain ibn Abdallah – Rome. 3pts. 1593 – 19mf – 9 – (typographia medicea, rome 1593 3pts) – mf#H-8435 – ne IDC [956]

Al-qiblah – [al-Khartum, Sudan]: Hayat Ihyaal-Nashat al-Islami, nov 21 1987-feb 9 1989 – 1r – 1 – us CRL [073]

Al-qutr al-misri – Cairo: Ahmad Hilmi, 1908-09. yr 1 n1-10. 22 rabi' I-27 jumada 1326 [24 apr-26 jun 1908] – (= ser Arabic journals and popular press) – 1r – 1 – $200.00 – us MEDOC [956]

Al-rai – [Khartoum]: Dar al-Rai, mar 10 1988-may 1989 – 2r – 1 – us CRL [079]

Al-raiat-al-hamra : organe du parti communiste (sfic) – Paris: [s.n.], jul 1930 – (filmed with: les continents and 11 other titles) – us CRL [320]

Al-ra'id al-rasmi. / Tunisia – 1970-77 – 12r – 1 – $240.00; outside North America add $1.25r – (1978-. ca $40y) – us L of C Photodup [324]

Al-raida [al-ra'idah] – Beirut: Institute for Women's Studies in the Arab World, Beirut University College/Lebanese American University, 1976- . v1 n1-v15 n81. may 1976-spr 1998 – (= ser Arabic journals and popular press) – 1r – 1 – $450.00 – (in english) – us MEDOC [956]

Alraune : die geschichte eines lebenden wesens / Ewers, Hanns Heinz – special ed. Berlin: Sieben Staebe-Verlags- und Druckerei-gesellschaft 1928 [mf ed 1985] – 1r – 1 – (filmed with: t'ai p'ing chun kuang-hsi.../ chien, yu-wen) – mf#6684 – us Wisconsin U Libr [830]

Alraunenmaeren / List, Guido – Linz: Oesterreichische Verlagsanstalt [1903] [mf ed Bloomington IN: Indiana Uni Lib, Preservation Dept 1984] – 1r – 1 – us Indiana Preservation [390]

Al-rawi – Cairo: Butrus Hanna, 1893- . yr 1 pt 1-yr 2 pt 2. 15 jan 1893-15 mar 1894/8 barmahat 1610 [coptic era]/8 ramadan 1311 – (= ser Arabic journals and popular press) – 1r – 1 – $200.00 – us MEDOC [956]

Al-ra'y al-akhar – Fort Worth, TX: Lonestar, Inc, 1994- [n1-v3 n12(sep 20 1994-oct 1997)] (mthly) – 1r – 1 – us CRL [073]

Al-rayah – Khartoum, jun 29 1985-jun 29 1989 – 13r – 1 – us CRL [079]

Already on the left – v1 n1-v2 n1 [1970 aug-1971 apr]; v1 n1-v2 n1 [1970 aug-1971 apr] – 1 – mf#720767 – us WHS [071]

Alrededor del problema unionista de centro-america / Mendieta, Salvador – Barcelona, Spain. v1-2. 1934 – 1r – 1 – us UF Libraries [972]

Alresford : essays for the times / Newnham, William Orde – London: Longmans, Green, 1891 [mf ed 1985] – 1mf – 9 – 0-8370-3928-2 – mf#1985-1928 – us ATLA [240]

Al-riyadh – al-Riyadh, Saudi Arabia: Muassasat al-Yamamah al-Sahafiyah, [1972-74]; [1977-78]; [1979-] – us CRL [079]

Als eskimo unter den eskimos / Klutschak, H W – Wien, Pest, Leipzig, 1881 – 6mf – 9 – mf#N-283 – ne IDC [919]

Als ich jung noch war : neue geschichten aus der waldheimat / Rosegger, Peter – Leipzig: L Staackmann 1895 [mf ed 1995] – (= ser Bibliothek der deutschen literatur) – 1r – 1 – (filmed with: bergpredigten / p k schwaechen) – mf#8854 – us Wisconsin U Libr [920]

Als ik eens nederlander was... / Soerjaningrat, R M Soewardi; ed by Het Inlandsch Comite tot herdenking van Neerlands honderdjarige vrijheid – Bandoeng, 1913 – 1mf – 8 – mf#SE-1428 – ne IDC [949]

Als landrat in ostpreussen : ragnit-allenstein / Pauly, Walter – Wuerzburg: Holzner-Verlag 1957 [mf ed 1992] – (= ser Ostdeutsche beitraege aus dem goettinger arbeitskreis 8) – 10r – 1 – (filmed with: ostdeutsche beitraege aus dem goettinger arbeitskreis) – mf#3180p – us Wisconsin U Libr [350]

Als oesterreich zerfiel : 1848 / Bartsch, Rudolf Hans – Wien: C W Stern, 1905 [mf ed 1995] – 337p – 1 – mf#8971 – us Wisconsin U Libr [830]

Als schriftsteller leben : gespraeche mit peter handke, franz xaver kroetz, gerhard zwerenz, walter jens, peter ruehmkorf, guenter grass / Arnold, Heinz Ludwig – 1. ausg. Reinbek bei Hamburg: Rowohlt, 1979 [mf ed 1993] – (= ser Neue buch 18) – 154p – 1 – mf#7849 – us Wisconsin U Libr [430]

Als seekadett nach fernost : ein buch fuer jungen, das von kriegsschiffen, seefahrt und ausland erzaehlt / Fuchs, Hans – 5. aufl. Stuttgart: Loewes Verlag F Carl 1942 [mf ed 1990] – 1r [ill] – 1 – (ill by heinz schubel. filmed with: liebeskampfe / hermann friedrichs) – mf#7279 – us Wisconsin U Libr [880]

Alsa forum *see* Legal studies forum

al-Sabbah, Amal Yusuf al-Adhabi *see* Al-tadadat al-sukkaniyah al-hadithah

L'alsace – Muelhausen / Elsass (Mulhouse F), 1980-83 – 1 – (bilingual. filmed by misc inst: 1983-85) – fr ACRPP; gw Misc Inst [074]

L'alsace – Colmar (F), 1869-1870 n38 – 1r – 1 – gw Misc Inst [074]

L'alsace : nouvelle description historique et topographique des deux departemens du rhin / Aufschlager, Johann F – Strasbourg 1826-28 [mf ed Hildesheim 1995-98] – 8mf – 9 – €160.00 – 3-487-29707-8 – gw Olms [914]

L'alsace et la lorraine comment elles redeviendront francaises / Barthelemy, Hippolyte – Paris 1887 [mf ed Hildesheim 1995-98] – 1v on 1mf – 9 – €40.00 – 3-487-25948-6 – gw Olms [914]

L'alsace libere : bulletin departemental d'informations du bas-rhin (organe alsacien de la resistance – organe de la resistance) – Strasbourg, France 29 nov, 3 dec 1944, 12 may-10/11 jun, 4 jul-31 dec 1945 (imperfect) – 1r – 1 – uk British Libr Newspaper [934]

L'alsace lundi – Mülhausen / Elsass (Mulhouse F), 1972-74, 1978-85 – 1 – gw Misc Inst [074]

L'alsace-lorrain – Paris (F), 1880-1902, 1903 [gaps] – 6r – 1 – gw Misc Inst [074]

L'alsacien – Strasbourg. 1848-avr 1849 – 1 – fr ACRPP [073]

L'alsacien / elsaessische volks- und handelszeitung – Colmar / Elsass (F), 1871-72 – 1 – (title varies: 1872 n267: elsaessische volkszeitung und colmarer anzeiger) – gw Misc Inst [074]

Al-sadaqah – Cairo: s.n.: aug 6-21, oct 30-nov 6, dec 11 1952; jan 1, mar 19, apr 16-may 14, may 28-jun 4, jun 16, jul 9 1953; sep-dec 1962 – 1r – mf#MF-11304 MEMP – us CRL [079]

Al-sa'ih – New York – 9mf – 9 – mf#NE-20332 – ne IDC [956]

Al-samir al-saghir – Cairo: Jam'iyat al-Ta'lif al-Ilmiyah, 1897-1900. yr 1 n1-34. 21 oct 1897/12 baba 1614 [coptic era]/24 jumada 1 1315-21 sep 1898/12 tut 1615/4 jumada 1 1316 – (= ser Arabic journals and popular press) – 1r – 1 – $175.00 – us MEDOC [956]

Al-sayyad – Sharbin, UA: Muhammad Ahmad Ghayth al-Sharbini. yr 1 n1-12. 22 safar-20 dhu al-Qa'dah 1344. [10 sep 1925-1 jun 1926] [complete] – (= ser Arabic journals and popular press) – 1r – 1 – $200.00 – us MEDOC [956]

Alsdorf, Ludwig *see* Deutsch-indische geistesbeziehungen

Al-sha'b – Cairo, Egypt. May 1 1979-1991 – 8r – 1 – us L of C Photodup [079]

Al-shabibah – al-Khartum: Ittihad al-Shabab al-Sudani. aug 30 1986-jun 24 1989 – 1r – us CRL [079]

Al-shacb – Jerusalem, 1972-1992 – 57r – 1 – mf#J-93-6 – ne IDC [956]

Al-shammashah – Umm Durman: al-Shammashah, apr 14, 1986-jun 29, 1989 – 2r – us CRL [079]

al-Sharnubi, Muhammad Abd al-Rahman *see* Al-jughrafiya bayna al-ilm al-tatbiqi wa-al-wazifah al-ijtimaiyah

Al-sharq – al-Khartum: Dar al-Sudani lil-Tibaah wa-al-Nashr, 1988: may 9, 31, jun 14-27, jul 11, aug 1; 1989: feb 13-27 – 1r – us CRL [079]

Al-sharq / Majallah shahriyah ta ni bi-Shu"un al-Adab wa-al-Fann wa-al-Fikr – Jerusalem, Tasdur 'an Sahifat al-Anba'. al-sanah 2, 'adad al-awwal-al-sanah 23, al-'adad al-rabi. jun 1971-nov/dec 1993 – (= ser Arabic journals and popular press) – 4r – 1 – $700.00 – us MEDOC [073]

Alshekh / Mishnah – Warsaw, Poland. 1872 – 1r – 1 – us UF Libraries [939]

Al-shifa' – Cairo: Shibli al-Shumayyil, 1886-91. yr 1 n1-yr 4 n12. 15 shubat/feb 1886-1 jan/kanun 2 1891 [11 jumada 1 1303-20 jumada 1 1308] – (= ser Arabic journals and popular press) – 1r – 1 – $425.00 – us MEDOC [956]

Al-shihab – Cairo: Hasan al-Banna (Jama'at al-Ikhwan al-Muslimin) 1947-49. yr 1 n1-5. 1 muharram-1 jumada 1 1367 [14 nov 1947-11 mar 1948] – (= ser Arabic journals and popular press) – 1r – 1 – $200.00 – us MEDOC [956]

Al-shita' – Cairo: Salim al-'Anhuri, 1906- . yr 1 pts 1-4. 1 jan/kanun 2 1906-n.d. [1907?] [complete] – (= ser Arabic journals and popular press) – 1r – 1 – $200.00 – us MEDOC [956]

Al-sihafah – Khartoum: al-Sihafah. 1971-jun 1973; 1974; jul 1975-apr 1978; jul 1978-aug 1979; nov 1979-aug 3 1986 – us CRL [079]

Al-sihhah – Cairo: Hasan Rifqi, Ibrahim Mustafa, 1887-90. yr 2 n1-yr 3 n2. n.d.- 1888?-90? – (= ser Arabic journals and popular press) – 1r – 1 – $300.00 – (with yr 3 n3, title changes to: al-azhar with no break in numbering; r also incl: al-azhar) – us MEDOC [079]

Al-sin wa-al-qadiyah al-filastiniyah, 1976-81 / Musallam, Sami – Bayrut, Lubnan: Muassasat al-Dirasat al-Filastiniyah, 1982 – 1r – us CRL [999]

Alsinet, J *see*
- Nuevas utilidades de la quina
- Nuevo metodo para curar flatos, hipocondrias, vapores y ataques hystericos...

Alsino : novela / Prado, Pedro – 2nd ed. Santiago, Chile: Nascimento 1928 [mf ed 1985] – 1r – 1 – (filmed with: dorothy south / eggleston, g c) – mf#6713 – us Wisconsin U Libr [830]

Al-siyasah – al-Khartum: Farah lil-Tibaah wa-al-Nashr. jun 30 1986-jun 29 1989 – 18r – us CRL [079]

Al-siyasah al-maiyah li-israil / Dayfis, Uri et al – Bayrut, Lubnan: Muassasat al-Dirasat al-Filastiniyah, 1980 – 1r – us CRL [950]

Al-siyasi – Cairo, Egypt. May 15 1977-1979; July 27 1980-Nov 1987; Jan 1988-1991 – 15r – 1 – us L of C Photodup [079]

Al'skii, M *see* Nashi finansy za vremia grazhdanskoi voiny i nepa

Alsleben, A *see* Johann fischarts geschichtklitterung (gargantua)

Also sprach zarathustra 120-130 / Nietzsche, Friedrich Wilhelm – Leipzig, Germany. no date – 1r – 1 – us UF Libraries [190]

Alsted, J H *see*
- Definitiones theologicae secundum ordinem locorum communium traditae
- Distinctiones per universum theologiam sumtae ex canone sacrarum literarum
- Metaphysica
- 'Rakouws catechismus met sijn Onder-soeck
- Scientiarum omnium encyclopaediae
- Theatrum scholasticum
- Theologia catechetica
- Theologia didactica
- Theologia naturalis...
- Theologia naturalis exhibens augustissimam naturae scholam
- Theologia prophetica exhibens...
- Theologiaa polemica
- Theologiae casuum

Alston, Leonard *see*
- Education and citizenship in india
- Modern constitutions in outline
- Stoic and christian in the second century
- The white man's work in asia and africa

Alston, R C *see*
- Nineteenth century books on linguistics collection
- Nineteenth century books on publishing, the booktrade and the diffusion of knowledge collection
- Nineteenth century women writers collection

Al-sudan al-hadith – Khartoum, Sudan. Aug 1989-1990 – 4r – 1 – us L of C Photodup [079]

Al-sudan al-yawm – [Sudan: s.n.] jan 13-mar 16 1989 – 1r – us CRL [960]

Al-sudani – al-Khartum: Dar al-Sudani, jun 3 1985-jun 29 1989 – 13r – us CRL [079]

Al-sufur – Cairo: 'Abd al-Hamid Hamdi, 1915-? v2 n52-v4 n201. 24 rajab 1334-15 sha'ban 1337 [26 may 1916-15 may 1919] – (= ser Arabic journals and popular press) – 1r – 1 – $775.00 – (missing: n65,78,135,148,195) – us MEDOC [956]

Al-sumud – Beirut, Lebanon: Jabhat al-quwa al-filastiniyah al-rafidah li-al-hulul al-istislamiyah. jun 19 1969-oct 21 1977 – 1r – us CRL [079]

Al-suwar al-mutaharrikah – Cairo: Muhammad Tawfiq, 1923-25. v1 n1-v3 n73. 31 may 1923-4 jun 1925 [complete] – (= ser Arabic journals and popular press) – 1r – 1 – $700.00 – us MEDOC [956]

Alt, Albrecht *see*
- Die griechischen inschriften der palaestina tertia westlich der "araba"'
- Israel und aegypten

Alt, Albrecht et al *see* Alttestamentliche studien

Alt, Axel *see* Der tod fuhr im zug

Der alt gloub / Bullinger, Heinrich – [Zuerich, Christoffel Froschouer], 1539 – 2mf – 9 – mf#PBU-132 – ne IDC [240]

Der alt gloub... / Bullinger, Heinrich – Tigvri, Froschouer, 1544 – 2mf – 9 – mf#PBU-134 – ne IDC [240]

Alt, Heinrich *see* Das kirchenjahr des christlichen morgen- und abendlandes

Alt, Karl *see* Goethes faust

Alt, Karl Hermann *see*
- Goethe und seine zeit
- Studien zur entstehungsgeschichte von goethes dichtung und wahrheit

Alt und neu schreibkalender – Stettin (Szczecin PL), 1646, 1685, 1694, 1697 – 1 – (aka: auch: alter und newer schreib-calender. since 1639 publ in stettin & rostock) – gw Misc Inst [077]

[Alta-] alta advocate – CA. 1897-1906; jan 1907-sep 1907; feb 1932-1937; feb 1963-jan 19 – 15r – 1 – $900.00 – (comic sect only: jun 1932-jan 1938 1r $50) – mf#C03727 – us Library Micro [071]

Alta extremadura : carnestolendas / Gutierrez Macias, Valeriano – Badajoz: Imp. Dip. Provincial, 1968 – sp Bibl Santa Ana [946]

Alta frequenza – Milano, Italy 1976-82 – 1,5,9 – ISSN: 0002-6557 – mf#2021 – us UMI ProQuest [621]

Alta mar / Albaladejo, Mariano – Habana, Cuba. 1951 – 1r – 1 – us UF Libraries [972]

Al-tadadat al-sukkaniyah al-hadithah : dirasah tatbiqiyah ala duwal al-khalij al-arabi / al-Sabbah, Amal Yusuf al-Adhabi – al-Kuwayt: Qism al-Jughrafiya bi-Jamiat al-Kuwayt wa-al-Jamiyah al-Jughrafiyah al-Kuwaytiyah, 1984 – us CRL [956]

[Altadena-] altadena : the weekly – CA. nov 1929-jan 1930; mar 1930-jun 1930; nov 1930-apr 1935, jul 1 – 43r – 1 – $2580.00 – (aka: the altadena press; altadena chronicle; pasadena: the weekly) – mf#H03134 – us Library Micro [071]

Altaegyptische tempelinschriften in den jahren 1863-1865 an ort und stelle gesammelt / Duemichen, J – Leipzig, 1867 – 9mf – 9 – mf#NE-369 – ne IDC [933]

Die altaegyptischen goetter und goettersagen / Strauss und Torney, Victor von – Heidelberg: C Winter, 1889 – 2mf – 9 – 0-524-04537-2 – mf#1990-3371 – us ATLA [290]

Altai / ed by Ornatskii, P V – Biisk [Alt gub]: [s.n.] / 1918 [1918 9 iiunia-] – (= ser Asn 1-3) – n1-129 [1918] [gaps] item 3, on reel n2 – 1 – (suppl: izvestiia po biiskomu i karakorum-altaiskomu uezdam [asn-1.176]) – mf#asn-1.003 – ne IDC [077]

Altaiskaia mysl' : Obshchestv-lit sots-demokrat gaz / ed by Raiunets, K L – Barnaul [Alt gub]: T-vo Alt izd-vo 1919 [1919 9 marta-1919 [?]] – (= ser Asn 1-3) – n1-127 [1919] [gaps] item 4, on reel n3 – 1 – (cont: novyi altaiskii luch) – mf#asn-1 004 – ne IDC [077]

Altaiskaia pravda – Barnaul, Russia 1973-88 [mf ed Norman Ross] – 1r – 1 – mf#nrp-259 – us UMI ProQuest [077]

Altaiskaia torgovo-promyshlennaia gazeta – Barnaul, Russia 1911-13 [mf ed Norman Ross] – 1 – mf#nrp-260 – us UMI ProQuest [077]

Altaiskie gubernskie izvestiia : Organ Alt gub komissariata / ed by Vinokurov, N M – Barnaul [Alt gub]: [s n] 1919 [1918 28 iiulia-1919 9 fevr] – (= ser Asn 1-3) – n1-43 [1918] n1-9 [1919] [gaps] item 5, on reel n3 – 1 – (cont by: altaiskii vestnik) – mf#asn-1 005 – ne IDC [077]

Altaiskii krai : organ obshchestv -polit i koop mysli / ed by Norin, G A – Biisk [Alt gub]: Biis kredit soiuz i Alt gornyi soiuz kooperativov 1919 [1919 20 fevr-] – (= ser Asn 1-3) – n1-102 [1919] [gaps] item 7, on reel n4 – 1 – (cont: dumy) – mf#asn-1 007 – ne IDC [077]

Altaiskii krestianin – Barnaul, 1912-1918 – 96mf – 9 – (missing: 1916(32); 1917(7-12, 17, 24-25, 27, 44, 46-48)) – mf#COR-545 – ne IDC [077]

Altaiskii luch : obshchestv -lit, polit, sotsial-demokrat gaz / ed by Kodor, I I – Barnaul [Alt gub]: Kom RSDRP 1918 [1918 14 [1] fevr-25 sent – (= ser Asn 1-3) – n10-116 [1918] [gaps] item 8, on reel n4 – 1 – mf#asn-1 008 – ne IDC [077]

Altaiskii ogonek : Izd neperiod: Odnodn gaz v "Den' Narodnogo Obrazovaniia" v Barnaule / ed by Shubkin, N F – Barnaul [Alt gub]: Kul't-prosvet soiuz Alt kraia 1919 – (= ser Asn 1-3) – 1v item 9, on reel n4 – 1 – mf#asn-1 009 – ne IDC [077]

Altaiskii vestnik : ofits gub organ / ed by Kondratenko, V S & Kurskii, S M – Barnaul [Alt gub]: [s n] 1919 [1919 11 fevr-] – (= ser Asn 1-3) – n1-147 [1919] [gaps] item 6, on reel n3,4 – 1 – (cont: altaiskie gubernskie izvestiia) – mf#asn-1 006 – ne IDC [077]

Al-takhtit al-zirai li-mintaqat al-wafrah / al-Fil, Muhammad Rashid & al-Mutawwa, Subhi – [al-Khalidiyah, al-Kuwayt]: Qism al-Jughrafiya bi-Jamiat al-Kuwayt wa-al-Jamiyah al-Jughrafiyah al-Kuwaytiyah, [1983] – us CRL [900]

Altamira, Rafael *see* Tecnica de investigacion...

Altamirano, Carlos Luis *see* Funeral de un sueno

Altamirano, Ignacio Manuel *see* Paisajes y leyendas

Altaner, Bruno *see*
– Dietrich von bern in der neueren literatur
– Kleine schriften

Al-taqaddum – [Khartoum: s.n., mar 22-29, apr 12, 1988(– 1r – us CRL [073]

The altar : a service book for sunday schools / Bartholomew, John Glass – new and enl ed. Boston: Universalist Pub House, [1865?] – 1mf – 9 – 0-524-03315-3 – mf#1990-4675 – us ATLA [240]

Altar lights : their history and meaning – London, England. 18- – 1r – 1 – us UF Libraries [240]

Die altarabische mondreligion und die mosaische ueberlieferung / Nielsen, Ditlef – Strassburg: KJ Truebner, 1904 – 1mf – 9 – 0-524-04532-1 – (incl bibl ref) – mf#1990-3366 – us ATLA [290]

Altarbuch...altkatholiken – Bonn, 1959 – 6mf – 8 – €14.00 – ne Slangenburg [241]

Al-tariq – Cairo: Muhammad Siraj al-Din, Husayn Muhammad Ghannam, 1929-31. yr 1 n1-yr 3 n9. 1 ramadan 1347-1 jumada 1 1350 [feb 1929-sep 1931] – (= ser Arabic journals and popular press) – 1r – 1 – $500.00 – us MEDOC [956]

Altars prohibited by the church of england / Goode, William – London, England. 1844 – 1r – 1 – us UF Libraries [241]

Das altarwerk zu lauenstein und die anfaenge des barock in sachsen / Carus, Victor A – Stuttgart, 1912 (mf ed 1993) – 1mf – 9 – €24.00 – 3-89349-268-2 – mf#DHS-AR 125 – gw Frankfurter [720]

Alt-asiatische gottes- und weltideen in ihren wirkungen auf das gemeinleben der menschen : fuenf oeffentliche vortraege / Bluntschli, Johann Caspar – Noerdlingen: CH Beck, 1866 [mf ed 1990] – 1mf – 9 – 0-7905-7271-0 – mf#1989-0496 – us ATLA [200]

Al-tatbiq al-handasi lil-kharait al-jiyumurfulujiyah / Farhan, Yahya Isa – [al-Kuwayt]: Qism al-Jughrafiya bi-Jamiat al-Kuwayt wa-al-Jamiyah al-Jughrafiyah al-Kuwaytiyah, [1980] – us CRL [999]

Al-tawhid – Cairo, 1973-1981 – 54mf – 9 – (gaps) – mf#NE-20324 – ne IDC [956]

Al-tawzi al-jughrafi lil-sukkan fi al-yaman / Sadi, Abbas Fadil – [al-Kuwayt]: Qism al-Jughrafiya bi-Jamiat al-Kuwayt wa-al-Jamiyah al-Jughrafiyah al-Kuwaytiyah, [1983] – us CRL [915]

Altbabylonische keilschrifttexte : zum gebrauche bei vorlesungen / ed by Winckler, Hugo – Leipzig: Eduard Pfeiffer, 1892 [mf ed 1986] – 1mf – 9 – 0-8370-8635-3 – (in akkadian) – mf#1986-2635 – us ATLA [470]

Altbabylonische rechtsurkunden aus der zeit der hammurabi-dynastie / Daiches, Samuel – Leipzig: J C Hinrichs 1903 [mf ed 1986] – (= ser Leiziger semitischen studien 1/2) – 1mf – 9 – 0-8370-7688-9 – (text in german & akkadian. comm in german. incl bibl ref) – mf#1986-1688 – us ATLA [930]

Ein altbabylonischer felderplan : nach mittheilungen von f.v. scheil / ed by Eisenlohr, August – Leipzig: J C Hinrichs, 1896 – 1mf – 9 – 0-8370-8570-5 – (in german and sumerian. incl bibl ref) – mf#1986-2570 – us ATLA [470]

Altbayerische sagen / Hofmiller, Josef [comp] – Altoetting: Verlag A Coppenrath, 1949 – (= ser Buecher der heimat 4) – 1 – mf#8300 – us Wisconsin U Libr [390]

Der altchinesische monotheismus : vortrag. gehalten in evang. verein zu berlin... / Strauss und Torney, Victor von – Heidelberg: C Winter, 1885 – (= ser Sammlung von Vortraegen fuer das deutsche Volk) – 1mf – 9 – 0-524-03304-8 – mf#1990-3189 – us ATLA [290]

Die altchinesische reichsreligion : vom standpunkte der vergleichenden religionsgeschichte / Happel, Julius – Leipzig: O Schulze, 1882 – 1mf – 9 – 0-524-01366-7 – (incl bibl ref) – mf#1990-2378 – us ATLA [230]

Die altchristliche grabeskunst : ein versuch der einheitlichen auslegung / Styger, Paul – Muenchen, 1927 (mf ed 1993) – 2mf – 9 – €31.00 – 3-89349-314-X – mf#DHS-AR !/= – gw Frankfurter [730]

Die altchristliche literatur und ihre erforschung seit 1880 : allgemeine uebersicht und erster literaturbericht (1880-1884) / Ehrhard, Albert – Freiburg im Breisgau; St. Louis, Mo.: Herder, 1894 – (= ser Strassburger Theologische Studien) – 1mf – 9 – 0-7905-5872-6 – (incl bibl ref) – mf#1988-1872 – us ATLA [240]

Die altchristliche literatur und ihre erforschung seit 1880. allgemeine uebersicht und erster literaturbericht (1880-1884) / Ehrhard, A – Strassburg. StrThS I, 4-5. 1894 – €12.00 – ne Slangenburg [240]

Die altchristliche literatur und ihre erforschung von 1884-1900 : erste abteilung. die vornicaenische literatur / Ehrhard, A – Freiburg i. Br. StrThS. suppl v1. 1900 – €21.00 – ne Slangenburg [240]

Die altchristliche litteratur und ihre erforschung von 1884-1900 / Ehrhard, Albert – Freiburg im Breisgau; St. Louis, Mo.: Herder, 1900 – (= ser Strassburger Theologische Studien) – 2mf – 9 – 0-7905-5873-4 – (incl bibl ref) – mf#1988-1873 – us ATLA [240]

Altchristliche liturgische stuecke aus der kirche aegyptens : nebst einem dogmatischen brief des bischofs serapion von thmuis / Wobbermin, Georg – Leipzig: J C Hinrichs 1899 [mf ed 1989] – (= ser Tugal 17/3b) – 1mf – 9 – 0-7905-1848-1 – (together with: zur ueberlieferung des philostorgios by ludwig jeep, discussion in german & greek, texts in greek; incl bibl ref) – mf#1987-1848 – us ATLA; ne Slangenburg [240]

Altchristliche malerei und altkirchliche literatur : eine untersuchung ueber den biblischen cyklus der gemaelde in den roemischen katakomben / Hennecke, Edgar – Leipzig: Veit, 1896 [mf ed 1990] – 1mf – 9 – 0-7905-6295-2 – (in german, greek & latin. incl bibl ref) – mf#1988-2295 – us ATLA [700]

Altchristliche sagen ueber das leben jesu und der apostel : mit einem anhang – juedische sagen ueber das leben jesu / Couard, Ludwig – Guetersloh: Bertelsmann, 1909 [mf ed 1985] – 1mf – 9 – 0-8370-2749-7 – mf#1985-0749 – us ATLA [225]

Altchristliche staedte und landschaften 1, konstantinopel (324-450) / Schultze, V – Leipzig, 1913 – €14.00 – ne Slangenburg [240]

Altchristliche und moderne gedanken ueber frauenberuf : drei aufsaetze / Mausbach, Joseph – 4.-7. verb aufl. M Gladbach: Volksvereins Verlag, 1910 [mf ed 1990] – (= ser Apologetische tagesfragen 6) – 1mf – 9 – 0-7905-6763-6 – (incl bibl ref. 1st-3rd ed publ 1906) – mf#1988-2763 – us ATLA [305]

Die altdeutsche genesis : nach der wiener handschrift = Genesis / ed by Dollmayr, Viktor – Halle/S: M Niemeyer Verlag, 1932 [mf ed 1993] – (= ser Altdeutsche textbibliothek 31) – xi/183p/[1]pl – 1 – (middle high german. int in german. incl bibl ref) – mf#8193 reel 3 – us Wisconsin U Libr [221]

Altdeutsche maerchen, sagen und legenden : treu nacherzaehlt und fuer jung und alt / ed by Bechstein, Reinhold – 2. verm aufl. Leipzig: O A Schulz, 1877 [mf ed 1989] – 248p/6pl (ill) – 1 – mf#7002 – us Wisconsin U Libr [390]

Altdeutsche novellen – Berlin: Erich Reiss, c1912 [mf ed 1993] – 2v – 1 – (trans by leo greiner) – mf#8380 – us Wisconsin U Libr [430]

Der altdeutsche physiologus : die millstaetter reimfassung und die wiener prosa (nebst dem lateinischen text und dem althochdeutschen physiologus) / ed by Maurer, Friedrich – Tuebingen: Niemeyer, 1967, c1966 [mf ed 1993] – (= ser Altdeutsche textbibliothek n67) – x/95p – 1 – (incl bibl ref) – mf#8193 reel 6 – us Wisconsin U Libr [430]

Altdeutsche quellen : heft 1 (1937)-heft 4 (1957) / ed by Pretzel, Ulrich – Leipzig: S Hirzel, 1937-57 – 4v (ill) – 1 – mf#8377 – us Wisconsin U Libr [430]

Altdeutsche textbibliothek – Halle a.S: M Niemeyer, 1882- [mf ed 1993] – n1- – 1 – mf#8193 – us Wisconsin U Libr [800]

Die altdeutschen bruchstuecke des tractats des bischof isidorus von sevilla de fide catholica contra judaeos : nach der pariser und wiener handschrift mit abhandlung und glossar / ed by Weinhold, Karl – Paderborn: F Schoeningh 1874 – (= ser Bibliothek der aeltesten deutschen litteratur-denkmaeler v6) – 1 – (text & trans on opposite pp) – mf#8437 reel 2 – us Wisconsin U Libr [430]

Altdeutsches lesebuch in neudeutscher sprache / Simrock, Karl – Stuttgart: J G Cotta 1854 [mf ed 1993] – (= ser Bibliothek der deutschen literatur) – 1r – 1 – (with an overview of literary history by trans karl simrock. filmed with: altdeutsche novellen / trans fr middle high german by leo greiner) – mf#3370p – us Wisconsin U Libr [430]

Der alte anfang und die urspruengliche form von cyprians schrift ad donatum / Goetz, K G – Leipzig, 1899 – (= ser Tugal 2-19/1c) – 1mf – 9 – €36.00 – ne Slangenburg [240]

Der alte anfang und die ursprungliche form von cyprian's schrift ad donatum *see* Die todestage der apostel paulus und petrus

Alte bekannte aus dem new yorker deutschen viertel : [anecdotes] / Stuerenburg, E – New York: E Steiger [1886] [mf ed 1991] – (= ser Steiger's deutsche bibliothek 4) – 1r – 1 – (filmed with: totenhorn-sudwand / karl hans strobl) – mf#2907p – us Wisconsin U Libr [880]

Alte einblattdrucke / ed by Clemen, Otto – Bonn: A Marcus & E Weber, 1911 [mf ed 1992] – (= ser Kleine texte fuer vorlesungen und uebungen 86) – 1mf – 9 – 0-524-05310-3 – (in german & latin. incl bibl ref) – mf#1990-1428 – us ATLA [090]

Der alte glaube und die wahrheit des christentums / Schmidt, Wilhelm – Berlin: Wiegandt & Grieben, 1891 [mf ed 1990] – 1mf – 9 – 0-7905-7611-2 – mf#1989-0836 – us ATLA [240]

Die alte heidelberger liederhandschrift / ed by Pfeiffer, Franz – Stuttgart: Litterarischer Verein, 1844 [mf ed 1993] – (= ser Blvs 9/3) – xii/295p/1pl – 1 – mf#8470 reel 2 – us Wisconsin U Libr [780]

Die alte heimat : erzaehlung / Doerfler, Peter – Berlin-Schildow: E Sicker, 1944 – 1r – 1 – us Wisconsin U Libr [830]

Alte hoch- und niederdeutsche volkslieder : mit abhandlung und anmerkungen / ed by Uhland, Ludwig – 3. aufl. Stuttgart: J G Cotta [1892?] [mf ed 1991] – (= ser Cotta'sche bibliothek) – 4v on 1r – 1 – mf#2921p – us Wisconsin U Libr [780]

Die alte kirche *see* History of the christian church, a d 1-600

Alte meister des orgelspiels / Straube, J – Leipzig 1929 – €7.00 – ne Slangenburg [780]

Alte nester : zwei buecher lebensgeschichten [a novel] / Raabe, Wilhelm Karl – 2. aufl. Berlin: O Janke 1897 [mf ed 1995] – 1r – 1 – (filmed with: gertrud von loden / c quandt) – mf#3706p – us Wisconsin U Libr [830]

Alte schule : drei novellen, bertram vogelweid, ohne liebe / Ebner-Eschenbach, Marie von – Leipzig: H Fikentscher, H Schmidt & H Guenther [1928] [mf ed 1993] – 2r – 1 – (filmed with: [saemtliche werke] / [ebner-eschenbach]) – mf#8570 reel 2 – us Wisconsin U Libr [830]

Alte schule : erzaehlungen / Ebner-Eschenbach, Marie von – Berlin: Gebrueder Paetel 1897 [mf ed 1993] – 1r – 1 – (filmed with: erzaehlungen / marie von ebner-eschenbach) – mf#8572 – us Wisconsin U Libr [880]

Das alte testament : seine entstehung und seine geschichte / Thomsen, Peter – Leipzig: B G Teubner 1918 [mf ed 1992] – (= ser Aus natur und geisteswelt 669) – 1mf – 9 – 0-524-05638-2 – (incl bibl ref) – mf#1992-0493 – us ATLA [221]

Das alte testament bei johannes : ein beitrag zur erklaerung und beurtheilung der johanneischen schriften / Franke, A H – Goettingen: Vandenhoeck & Ruprecht, 1885 [mf ed 1989] – 1mf – 9 – 0-7905-3131-3 – mf#1987-3131 – us ATLA [225]

Das alte testament im christlichen religionsunterrichte / Meltzer, Hermann – Gotha: E J Thienemann, 1899 [mf ed 1986] – 1mf – 9 – 0-8370-7721-4 – (incl bibl ref & ind) – mf#1986-1721 – us ATLA [221]

Das alte testament im evangelischen religionsunterricht / Boehm, Friedrich – Berlin: R Gaertners, 1895 [mf ed 1986] – (= ser Wissenschaftliche beilage zum jahresbericht der friedrichs-werderschen oberrealschule zu berlin easter 1895) – 1mf – 9 – 0-8370-7847-4 – mf#1986-1847 – us ATLA [221]

Das alte testament im evangelischen religionsunterricht / Floering, Friedrich – Giessen: J Ricker, 1895 [mf ed 1986] – (= ser Vortraege der theologischen konferenz zu giessen 9) – 1mf – 9 – 0-8370-7861-X – (incl bibl ref) – mf#1986-1861 – us ATLA [221]

Das alte testament im lichte der neuesten assyrisch-babylonischen endeckungen / Mueller, Alois – Frankfurt a. M: A Foesser, 1896 [mf ed 1989] – (= ser Frankfurter zeitmaesse brochueren 17/4) – 1mf – 9 – 0-7905-1535-0 – (incl bibl ref) – mf#1987-1535 – us ATLA [221]

Das alte testament im lichte des alten orients : handbuch zur biblisch-orientalischen altertumskunde / Jeremias, Alfred – Leipzig: J C Hinrichs, 1904 [mf ed 1993] – 1mf – 9 – 0-524-05732-X – (incl bibl ref) – mf#1992-0575 – us ATLA [221]

Das alte testament im lichte des alten orients *see* The old testament in the light of the ancient east

Das alte testament im neuen testament : ueber die citate des alten testaments im neuen testament: und ueber den opfer- und priesterbegriff im alten und neuen testamente / Tholuck, August – 5. verb aufl. Gotha: Friedrich Andreas Perthes, 1861 [mf ed 1986] – 1mf – 9 – 0-8370-7431-2 – mf#1986-1431 – us ATLA [225]

Das alte testament im neuen testament : ueber die citate des alten testaments im neuen testament und ueber den opfer- und priesterbegriff im alten und neuen testamente / Tholuck, August – 6. verm aufl. Gotha, F A Perthes 1868 [mf ed 1980] – 1r – 1 – (zwei beilagen zu dem kommentar ueber den brief an die hebraeer) – mf#mflm2011 – us Harvard [220]

Das alte testament in der johanneischen apokalypse / Schlatter, Adolf von – Guetersloh: C Bertelsmann, 1912 [mf ed 1989] – (= ser Beitraege zur foerderung christlicher theologie 16/6) – 1mf – 9 – 0-7905-3218-2 – mf#1987-3218 – us ATLA [225]

Das alte testament in der mischna / Aicher, Georg – Freiburg i B; St Louis, MO: Herder, 1906 [mf ed 1989] – (= ser Biblische studien 11/4) – 1mf – 9 – 0-7905-2160-1 – (incl ind) – mf#1987-2160 – us ATLA [221]

Das alte testament in predigten und bibelstunden : das erste buch mose seinem heimgang / Schlosser, Gustav; ed by Scriba, Otto – Bielefeld: Velhagen & Klasing 1892 [mf ed 1985] – 1mf – 9 – 0-8370-5099-5 – mf#1985-3099 – us ATLA [221]

Das alte testament und der christliche glaube : ein wort zur verstaendigung / Wilke, Fritz – Leipzig: Dieterich, 1911 [mf ed 1989] – 1mf – 9 – 0-7905-2338-8 – (incl bibl ref) – mf#1987-2338 – us ATLA [225]

Das alte testament und die kritik : oder, die hauptprobleme der alttestamentlichen forschung / Gasser, Johann Conrad – Stuttgart: D Gundert, 1906 [mf ed 1990] – 1mf – 9 – 0-7905-3373-1 – (incl bibl ref) – mf#1987-3373 – us ATLA [221]

Das alte testament und die naechstenliebe / Nikel, Johannes – 1.+2. aufl. Muenster i W: Aschendorff, 1913 [mf ed 1989] – (= ser Biblische zeitfragen 6/11-12) – 1mf – 9 – 0-7905-2856-8 – (incl bibl ref) – mf#1987-2856 – us ATLA [221]

Der alte und angenommene schottische ritus und friedrich der grosse / Begemann, Wilhelm – Berlin: E S Mittler 1913 – 2mf – 9 – mf#vrl-36 – ne IDC [366]

Das alte und das neue buergerliche recht deutschlands : mit einschluss des handelsrechts historisch und dogmatisch dargestellt / Engelmann, Arthur – Berlin: J J Heines, 1899 – (= ser Civil law 3 coll) – 9mf – 9 – (incl bibl ref and index) – mf#LLMC 96-552 – us LLMC [346]

Das alte und das neue china / Voskamp, Carl John – Berlin: Berliner evang Missionsgesellschaft, 1914 [mf ed 1995] – (= ser Yale coll) – 124p – 1 – 0-524-09736-4 – (in german) – mf#1995-0736 – us ATLA [951]

Der alte und der neue glaube : ein bekenntniss als antwort auf david friedrich strauss / Weis, Ludwig – Berlin: F Henschel, 1873 [mf ed 1986] – 1mf – 9 – 0-8370-7436-3 – mf#1986-1436 – us ATLA [240]

Der alte und der neue glaube : ein bekenntniss von david friedrich strauss / Huber, Johannes – Noerdlingen: CH Beck, 1873 [mf ed 1991] – 1mf – 9 – 0-7905-8663-0 – mf#1989-1888 – us ATLA [140]

Der alte und der neue glaube / Strauss, David Friedrich – 7.aufl. Bonn: Emil Strauss, 1874 [mf ed 1985] – 1mf – 9 – 0-8370-5446-X – mf#1985-3446 – us ATLA [140]

Alte und moderne kunst – Vienna, 1956- [mf ed Chadwyck-Healey, 1956-75] – (= ser Art periodicals on microform) – 6r – 1 – uk Chadwyck [720]

Alte und neue angriffe auf das alte testament : ein rueckblick und ausblick / Nikel, Johannes – Muenster i W: Aschendorff, 1908 [mf ed 1989] – (= ser Biblische zeitfragen 1/1) – 1mf – 9 – 0-7905-2857-6 – mf#1987-2857 – us ATLA [220]

Alte und neue aramaeische papyri / Staerk, Willy – Bonn: A Marcus & E Weber, 1912 [mf ed 1986] – (= ser Kleine texte fuer vorlesungen und uebungen 94) – 1mf – 9 – 0-8370-7342-1 – mf#1986-1342 – us ATLA [470]

Alte und neue gedichte / Huch, Ricarda Octavia – Leipzig: Insel-Verlag [1920?] [mf ed 1990] – 1r – 1 – (filmed with: einer baut einen dom / carl maria holzapfel) – mf#2733p – us Wisconsin U Libr [810]

Alte und neue geschichten aus baiern / Schmid, Herman – 2.aufl. Leipzig: Keil [18–?] – (= ser Gesammelte schriften. volks- und familien-ausgabe 8,14-16,18,26-27,35) – 8v – 1 – (v7 bound with: die tuerken in muenchen; v8 also numbered nf3; bound with: muetze und krone v1) – mf#film mas c438 – us Harvard [880]

Alte und neue kinder-lieder und reime / Wesendonk, Mathilde [comp] – Berlin: Zimmermann 1890 – 3mf – 9 – (with 15 pictures and "inititalen" by a gude-scholz) – mf#mw-8 – ne IDC [780]

Alte und neue quellen zur geschichte des taufsymbols und der glaubensregel / Caspari, Carl Paul – Christiana, 1879 – 6mf – 8 – €14.00 – ne Slangenburg [240]

Alte und neue quellen zur geschichte des taufsymbols und der glaubensregel / Caspari, Carl Paul – Christiania [Oslo]: Mallingsche Buchdruckerei, 1879 [mf ed 1985] – 1mf – 9 – 0-8370-2605-9 – (incl additions & corr) – mf#1985-0605 – us ATLA [240]

Alte und neue zeit in tsimo : der kreisstadt vom hinterlande in tsingtau / Lutschewitz, W – Berlin: Berliner ev Missionsgesellschaft, 1910 [mf ed 1995] – (= ser Yale coll) – 163p (ill) – 1 – 0-524-09162-5 – (in german) – mf#1995-0162 – us ATLA [951]

Der alte weg zum alten gott : gedanken und betrachtungen ueber wichtige fragen des christlichen glaubens / Bruckner, Albert – Schkenditz: W Schaefer, 1903 [mf ed 1985] – 1mf – 9 – 0-8370-2490-0 – (pref by otto kirn) – mf#1985-0490 – us ATLA [240]

Das alte westasien *see* The history of babylonia and assyria

Altekar, A S *see* Bibliography of indian coins

Altekar, Anant Sadashiv *see*
- The position of women in hindu civilisation
- Sources of hindu dharma
- State and government in ancient india
- The vakataka-gupta age

Die alten lateinischen thomasakten / Zelzer, K – 1977 – (= ser Tugal 5-122) – 3mf – 9 – €7.00 – ne Slangenburg [226]

Die alten petrusakten im zusammenhang der apokryphen apostellitteratur : nebst einem neuentdeckten fragment / Schmidt, Carl – Leipzig: J C Hinrichs, 1903 – (= ser Tugal) – 1mf – 9 – 0-7905-1734-5 – (incl ind) – mf#1987-1734 – us ATLA [470]

Die alten petrusakten im zusammenhang mit der apokryphen apostellitteratur / Schmidt, Carl – Leipzig, 1903 – (= ser Tugal 2-24/1) – 3mf – 9 – €7.00 – ne Slangenburg [240]

Die alten streitfragen gegenueber dem entwurfe eines buergerlichen gesetzbuches fuer das deutsche recht / Meischeider, Emil – Berlin, Leipzig: J Guttentag, 1889 – (= ser Civil law 3 coll; Beitraege zur erlaeuterung und beurtheilung des entwurfes eines buergerlichen gesetzbuches fuer das deutsche reich) – 2mf – 9 – mf#LLMC 96-603 – us LLMC [346]

Die alten und die jungen : dramatisches genrebild in einem akt / Lorm, Hieronymus – Berlin: L Kolbe, 1862 – 1 – us Wisconsin U Libr [820]

Die alten und die jungen : sozialer roman / Alberti, Conrad – Leipzig: W Friedrich, [190-] [mf ed 1992] – 2v in 1 – 1 – mf#7792 – us Wisconsin U Libr [830]

Altenberg, Peter *see*
- Mein lebensabend
- Was der tag mir zutraegt
- Wie ich es sehe

Altenburg, Clarence E *see* Modern conquistador in south america

Der altenglische regius-psalter : eine interlineaversion in hs royal 2 b 5 des brit mus / ed by Roeder, F – Halle, 1904 – 6mf – 8 – €14.00 – ne Slangenburg [240]

Alter / Dineson, Jacob – Varshe, Poland. 19– – 1r – 1 – us UF Libraries [939]

Das alter der babylonischen astronomie / Jeremias, Alfred – 2. erw aufl. Leipzig: J C Hinrichs, 1909 [mf ed 1989] – (= ser Im kampfe um den alten orient 3) – 1mf – 9 – 0-7905-2117-2 – (incl bibl ref & ind) – mf#1987-2117 – us ATLA [520]

Das alter des menschengeschlechts : nach der heiligen schrift, der profangeschichte und der vorgeschichte / Schanz, Paul – Freiburg i B, St Louis, MO: Herder, 1896 [mf ed 1989] – (= ser Biblischen studien 1/2) – 1mf – 9 – 0-7905-2692-1 – (incl bibl ref) – mf#1987-2692 – us ATLA [221]

Alter, Franz Carl *see* Ueber die tagalische sprache

Alter, Isaac Meir *see* Hidushe ha-rim 'al shalosh bavot

Alter katalog der musikdrucke / Oesterreichische Nationalbibliothek Wien. Musiksammlung – [mf ed Hildesheim 1985] – (= ser Die europaeische musik) – 249mf – 9 – diazo €2100.00 silver €2380.00 – gw Olms [780]

Alter und herkunft des achikar-romans und sein verhaeltnis zu aesop *see* Beitraege zur erklaerung und kritik des buches tobit

Altera pars selctissimarum cantionum / Lasso, Orlando di – 1587 – (= ser Mssa) – 9mf – 9 – €105.00 – mfchl 282 – gw Fischer [780]

Alterations in 72 kilodalton stress protein levels following eccentrically biased exercise / Sim, James D & Noble, Earl G – 1992 – 2mf – $8.00 – us Kinesology [612]

The alterations in the ordinal of 1662 : why were they made? / Firminger, Walter Kelly – London: SPCK 1898 [mf ed 1992] – (= ser Church historical society (series) 31) – 1mf – 9 – 0-524-05501-7 – (incl bibl ref) – mf#1990-1496 – us ATLA [242]

Die altercatio simonis iudaei et theophili christiani : nebst untersuchungen ueber die antijuedische polemik in der alten kirche / Harnack, Adolf von – Leipzig: JC Hinrichs, 1883 [mf ed 1989] – (= ser Tugal 1/3) – 1mf – 9 – 0-7905-1708-6 – (filmed with: die acta archelai und das diatessaron tatians by adolf von harnack and: der arethascodex paris gr 41...by oscar von gebhardt. in german, greek & latin. incl bibl ref) – mf#1987-1708 – us ATLA [226]

Die altercatio simonis judaei et theophili christiani und die acta archeali und das diatesseron tatians / Harnack, Adolf von – Leipzig, 1883 – (= ser Tugal 1-1/3) – 3mf – 9 – €7.00 – ne Slangenburg [240]

The altered self : an exploration of the processes of self-identity reconstruction by people who acquire a brain injury / Hutchinson, Susan L – 1996 – 3mf – 9 – $12.00 – mf#PSY 2030 – us Kinesology [150]

Altern in der arbeitsgesellschaft : ueber die soziale konstruktion des hoeheren lebensalters / Huf, Stefan – 1995 – 1mf – 9 – 3-8267-2129-2 – mf#DHS 2129 – gw Frankfurter [360]

Alternating current engineering practically treated / Raymond, Edward Brackett – New York, USA. 1907 – 1r – 1 – us UF Libraries [621]

Alternative : an american spectator – v4 n1-v8 n10 [1970 nov-1975 sep] – 1 – mf#1051862 – us WHS [071]

Alternative / Committee for Non-violent Revolution – 1-3 2+5. 1948-51 [all publ] – (= ser Radical periodicals in the united states, 1881-1960. series 1) – 1mf – 9 – $85.00 – us UPA [303]

Alternative – v1 n2 [1969 dec]; v1 n3-4 [1970 jan-feb] – 1 – mf#1582953 – us WHS [071]

Alternative – v1-v2 n5 [1976 feb 10-1977 mar 15] – 1 – mf#372302 – us WHS [071]

Alternative [1972] – Arlington VA 1972-74 – 1 – (cont by: alternative: an american spectator) – ISSN: 0044-7382 – mf#7782.02 – us UMI ProQuest [073]

Alternative: an american spectator – Arlington VA 1975-77 – 1 – (cont: alternative; cont by: american spectator) – mf#7782.02 – us UMI ProQuest [073]

Alternative Concepts, Inc *see* Los angeles weekly news

Alternative dispute resolution in a bankruptcy court : the mediation program in the southern district of california / Hartwell, Steven & Bermant, Gordon – Washington: GPO, 1988 – 2mf – 9 – $3.00 – mf#LLMC 95-339 – us LLMC [346]

Alternative futures – Troy NY 1978-81 – 1,5,9 – ISSN: 0162-9786 – mf#12052 – us UMI ProQuest [320]

Alternative higher education – Dordrecht, Netherlands 1976-84 – 1,5,9 – (cont by: innovative higher education) – ISSN: 0361-6851 – mf#11172.01 – us UMI ProQuest [378]

Alternative investment news – London, England 2001+ – 1,5,9 – ISSN: 1544-7596 – mf#32363 – us UMI ProQuest [332]

Alternative law journal – v1-26. 1974-2 – 9 – $405.00 set – (title varies: v1-16 1974-91 as legal services bulletin) – ISSN: 0817-3516 – mf#401241 – us Hein [340]

Alternative lifestyles – Dordrecht, Netherlands 1982-85 – 1,5,9 – (cont by: lifestyles) – ISSN: 0161-570X – mf#14128.02 – us UMI ProQuest [640]

Alternative media – v10 n1-v16 n1 [1978 spring-[1986 winter] – 1 – mf#1265744 – us WHS [071]

Alternative Media Information Center *see* Mediactive

Alternative Press Committee *see* Phoenix

Alternative press revue – 1973 mar-dec – 1 – mf#772621 – us WHS [071]

Alternative sources of energy – Tulsa OK 1971-87 – 1,5,9 – (cont by: independent power) – ISSN: 0146-1001 – mf#9535.02 – us UMI ProQuest [333]

Alternative structures for bankruptcy appeals / McKenna, J A & Wiggins, E – 2000 – 2mf – 9 – $1.50 – mf#llmc99-047 – us LLMC [346]

Alternatives – 1992 winter; 1995 jan-feb, apr-jun/jul, sep/oct-nov/dec; 1996 feb/mar, apr, may/jun – 1 – mf#3192256 – us WHS [071]

Alternatives : a journal of world policy – v1-26. 1974-2001 – 5,6,9 – $665.00 set – (v1-10 1974-85 in reel $160. v11-26 1985-2001 in mf $505) – ISSN: 0304-3754 – mf#100821 – us Hein [320]

Alternatives – n3, 5-7 [1971 jun 18, aug 3-sep 18] – 1 – mf#1582957 – us WHS [071]

Alternatives and solutions – 1993 fall; 1994 fall/winter; 1995 summer/fall – 1 – mf#2844044 – us WHS [071]

Alternatives [boulder] – Boulder CO 1979+ – 1,5,9 – ISSN: 0304-3754 – mf#12219 – us UMI ProQuest [320]

Alternatives journal – Waterloo ON 1996+ – 1,5,9 – (cont: alternatives) – ISSN: 1205-7398 – mf#7598,01 – us UMI ProQuest [333]

Alternatives journal – n1-10, 12-24, 27-32 [1971 aug 23-1972 may 1; Sep 1/15-1973 mar; 1/15; apr 16/30-aug 23] – 1 – mf#812136 – us WHS [071]

Alternatives news magazine – n1-3 [1971-1972 winter] – 1 – mf#715426 – us WHS [071]

Alternatives newsletter – prelim iss; v1 n1-v2 n1 [1971] – 1 – mf#453133 – us WHS [071]

Alternatives [oberlin] – Oberlin OH 1976-77 – 1,5,9 – mf#12353 – us UMI ProQuest [320]

The alternatives of faith and unbelief / Stanford, Charles – 2nd ed. [London]: Religious Tract Society, 1888 [mf ed 1985] – 1mf – 9 – 0-8370-5366-8 – (incl bibl ref) – mf#1985-3366 – us ATLA [210]

Alternatives [waterloo] – Waterloo ON 1971-95 – 1,5,9 – (cont by: alternatives journal) – ISSN: 0002-6638 – mf#7598.01 – us UMI ProQuest [333]

Alternativet – Stockholm, 1988-92 – 9 – (title changes to: miljomagasinet in 1992) – sw Kungliga [078]

Alternativet *see* Miljoemagasinet

Die alterthumer des volkes israel *see* The antiquities of israel

Der altertuemelnde stil in den ersten drei baenden von gustav freytags 'ahnen' / Posern, Armin – Greifswald: H Adler Inh E Panzig 1913 [mf ed 1989] – 1r – 1 – (filmed with: raetsel um herta / hermann freyberg) – mf#7273 – us Wisconsin U Libr [430]

Altes herz geht auf die reise : roman / Fallada, Hans – Berlin: Rowohlt, c1936 [mf ed 1989] – 1r – 1 – (filmed with: ein kleiner deutscher / ernst dittmer) – mf#7178 – us Wisconsin U Libr [830]

Altes und neues aus den herzogthümern bremen und verden / ed by Partje, Johann Heinrich – Stade 1769-81 – (= ser Dz. historisch-geographische abt) – v1-12 on 36mf – 9 – €360.00 – mf#k/n1051 – gw Olms [943]

Altes und neues aus spanien / Minutoli, Julius von – Berlin 1854 [mf ed Hildesheim 1995-98] – 2v on 4mf – 9 – €120.00 – 3-487-29856-2 – gw Olms [914]

Altes und neues in deutscher bibel : oder, vergleichung der bibelverdeutschung d m luthers mit ihrer berichtigung durch d j j v meyer / Stier, Rudolf – Basel: Felix Schneider, 1828 [mf ed 1986] – 1mf – 9 – 0-8370-9987-0 – mf#1986-3987 – us ATLA [220]

Altes und neues pommerland – Stargard (Stargard Szczecinski PL), 1721-22 – 1r – 1 – gw Misc Inst [077]

Altevogt, Heinrich *see* Labor improbus

Altfriesisches woerterbuch / Wiarda, Tileman Dothias – Aurich: Winter 1786 – (= ser Whsb) – 6mf – 9 – €70.00 – mf#Hu 159 – gw Fischer [430]

Altfriesisches worterbuch / Richthofen, Karl Otto Johannes Theresius – Gottingen, Germany. 1840 – 1r – 1 – us UF Libraries [430]

Altgermanische religionsgeschichte / Meyer, Richard Moritz – Leipzig: Quelle & Meyer, 1910 [mf ed 1992] – 2mf – 9 – 0-524-04344-2 – (incl bibl ref) – mf#1990-3328 – us ATLA [290]

Altgermanische religionsgeschichte. erster band / Helm, Karl – Heidelberg: C Winter, 1913 [mf ed 1991] – (= ser Germanische bibliothek 1/5/2; Religionswissenschaftliche bibliothek 5) – 1mf – 9 – 0-524-01369-1 – (incl bibl ref) – mf#1990-2381 – us ATLA [290]

Althaus, Friedrich *see* The roman journals of ferdinand gregorovius 1852-1874

Althaus, Paul *see*
- Die heilsbedeutung der taufe im neuen testamente
- Die prinzipien der deutschen reformierten dogmatik

Al-thawrah al-shabiyah – al-Khartum: Harakat al-lijan al-thawriyah fi al-Sudan, oct 9 1985-jun 29 1989 – 2r – us CRL [079]

Alt-heidelberg : schauspiel in 5 aufzuegen / Meyer-Foerster, Wilhelm – Berlin: A Scherl c1902 [mf ed 1990] – (= ser Woche 2) – 1r [ill] – 1 – (filmed with: der anti-necker j h mercks und der minister fr k v moser / richard loebell) – mf#2834p – us Wisconsin U Libr [820]

Altheim, Franz *see* Die krise der alten welt im 3. jahrhundert n. zw. und ihre ursachen

Althens, Margaret Magdalen *see* Christian character exemplified

Die althochdeutsche benediktinerregel des cod sang 916 / Benedict, Saint, Abbot of Monte Cassino; ed by Daab, Ursula – Tuebingen: M Niemeyer, 1959 [mf ed 1993] – (= ser Altdeutsche textbibliothek n50) – 304p – 1 – (latin and old high german in opposite columns. int in german) – mf#8193 reel 5 – us Wisconsin U Libr [430]

Der althochdeutsche isidor : nach der pariser handschrift und den monseer fragmenten / Isidore of Seville, Saint; ed by Eggers, Hans – Tuebingen: Max Niemeyer, 1964 [mf ed 1993] – (= ser Altdeutsche textbibliothek n63) – xix/77p – 1 – (parallel latin and old high german text. int in german. incl bibl ref) – mf#8193 reel 6 – us Wisconsin U Libr [430]

Althochdeutsche lesestuecke / Wackernagel, Wilhelm – Basel: H Richter 1875 [mf ed 1993] – 1r – 1 – (incl notes on vocabulary. filmed with: tristan und isolde / gottfried von strassburg & other titles) – mf#8504 – us Wisconsin U Libr [890]

Die althochdeutschen poetischen denkmaeler / Groseclose, J Sidney – Stuttgart: J B Metzler, 1976 [mf ed 1993] – (= ser Sammlung metzler. abt d, literaturgeschichte 140) – xiii/111p – 1 – (incl bibl and ind) – mf#8166 – us Wisconsin U Libr [430]

Alticchiero par made jwcdr / Rosenberg-Orsini, J – Padoue, 1787 – 3mf – 9 – mf#GDI-23 – ne IDC [700]

Das altindische neu- und vollmondsopfer in seiner einfachsten form : mit benutzung handschriftlicher quellen dargestellt / Hillebrandt, Alfred – Jena: Gustav Fischer, 1879 – 1mf – 9 – 0-524-07140-3 – mf#1991-0070 – us ATLA [280]

Alting, H *see* Theologia historica

Alting, J *see* Opera omnia theologica

La altisima / Trigo, Felipe – Madrid: Renacimiento, 8th ed 1907 – sp Bibl Santa Ana [946]

Altisraelitische kultstaetten / Gall, August, Freiherr von – Giessen: J Ricker 1898 [mf ed 1985] – (= ser Beihefte zur zeitschrift fuer die alttestamentliche wissenschaft 3) – 1mf – 9 – 0-8370-3224-5 – mf#1985-1224 – us ATLA [221]

Altisraelitische ueberlieferung in inschriftlicher beleuchtung *see* The ancient hebrew tradition as illustrated by the monuments

Altissiodorensis, Guillermus *see* Summa aurea in 4 libros sentent

Altitalische inschriften / Jacobson, Hermann – Bonn: A Marcus & E Weber 1910 [mf ed 1992] – (= ser Kleine texte fuer theologische und philologische vorlesungen und uebungen 57) – 1mf – 9 – 0-524-05460-6 – mf#1990-3486 – us ATLA [400]

Altjuedische gleichnisse und die gleichnisse jesu / Fiebig, Paul – Tuebingen: J C B Mohr (Paul Siebeck), 1904 [mf ed 1989] – 1mf – 9 – 0-7905-0883-4 – (incl bibl ref) – mf#1987-0883 – us ATLA [225]

Altjuedische liturgische gebete / ed by Staerk, Willy – Bonn: A Marcus & E Weber 1910 [mf ed 1992] – (= ser Kleine texte fuer theologische und philologische vorlesungen und uebungen 58) – 1mf – 9 – 0-524-04704-9 – (text in hebrew, notes in german. incl bibl ref. int by ed) – mf#1990-3413 – us ATLA [270]

Die altkanaanaeischen fremdworte und eigennamen im aegyptischen / Burchardt, M – Leipzig, 1909-1910. 2pts – 4mf – 9 – mf#NE-20034 – ne IDC [470]

Der altkatholicismus / Buehler, Christian – Leiden: EJ Brill, 1880 – 1mf – 9 – 0-8370-8484-9 – mf#1986-2484 – us ATLA [241]

Der altkatholicismus : eine denk- und schutzschrift an das evangelische deutschland / Beyschlag, Willibald – 2. Aufl. Halle a S: In Commission bei Eugen Strien, 1883 – 1mf – 9 – 0-8370-8403-2 – (incl bibl ref) – mf#1986-2403 – us ATLA [241]

Der altkatholicismus : geschichte seiner entwicklung, inneren gestaltung und rechtlichen stellung in deutschland / Schulte, Johann Friedrich von – Giessen: E. Roth, 1887 – 2mf – 9 – 0-7905-8073-X – (incl bibl ref) – mf#1988-6054 – us ATLA [241]

Der altkatholicismus : eine geschichtliche studie / Foerster, Theodor – Gotha: Friedrich Andreas Perthes, 1879 – 1mf – 9 – 0-8370-8423-7 – (incl bibl ref) – mf#1986-2423 – us ATLA [241]

Die altkatholische kirche des erzbisthums utrecht : geschichtliche parallele zur altkatholischen gemeindebildung in deutschland / Nippold, Friedrich – Heidelberg: Fr Bassermann, 1872 – 1mf – 9 – 0-7905-6249-9 – (incl bibl ref) – mf#1988-2249 – us ATLA [241]

Die Alt-Katholische Kirche in Deutschland *see* Die alt-katholische kirche in deutschland

Die alt-katholische kirche in deutschland : kirchliches jahrbuch / Die Alt-Katholische Kirche in Deutschland – 1974-85 [complete] – 1r – 1 – mf#ATLA S0649 – us ATLA [241]

Alt-katholische kirchenzeitung – Berlin, GW. v19-25. 1966-74, 1982-89 [complete] – 4r – 1 – (cont: der alt-katholik) – mf#ATLA S0517 – us ATLA [241]

Altkircher kreisblatt – Altkirch, DE. 1880-5 aug 1914 – 1 – fr ACRPP [944]

Die altkirchliche christologie / Lobstein, Paul – Leipzig: Fr. Wilh. Grunow, 1896 – 1r – 1 – 0-8370-0556-6 – mf#1984-6061 – us ATLA [240]

Altkirchliche christologie und der evangelische heilsglaube *see* Collected works

Altkreta : kunst und handwerk in griechenland / Bossert, Helmuth Theodor – Berlin, Germany. 1923 – 1r – 1 – us UF Libraries [720]

Altlateinische inschriften / Diehl, Ernst – Bonn: A Marcus & E Weber 1909 [mf ed 1992] – (= ser Kleine texte fuer theologische und philologische vorlesungen und uebungen 38-40) – 1mf – 9 – 0-524-04510-0 – (texts in latin & greek, notes in german & latin. incl bibl ref) – mf#1990-3344 – us ATLA [450]

Die altlateinischen biblischen cantica (tab29-30) / Schneider, H – 1938 – (= ser Texte und arbeiten. beuron (tab). beitraege zur ergruendung des aelteren lateinischen und christlichen schrifttums und gottesdienstes) – €11.00 – ne Slangenburg [220]

Die altlateinischen texte des proverbien-buches (tab32-33) / Schildenberger, J – 1941 – (= ser Texte und arbeiten. beuron (tab). beitraege zur ergruendung des aelteren lateinischen und christlichen schrifttums und gottesdienstes) – €11.00 – ne Slangenburg [221]

Altmaennersommer : drei geschichten um ein thema / Schaefer, Wilhelm – Muenchen: A Langen, G Mueller 1942 [mf ed 1996] – 1r – 1 – (filmed with: jenseits der augen / emil sandt) – mf#9262 – us Wisconsin U Libr [830]

Der altmaerker – Stendal DE, 1926 jul-aug, 1927 jul-aug – 2r – 1 – gw Misc Inst [074]

Altmaerker volksfreund – Stendal DE, 1919 7 apr-1923 10 nov – 8r – 1 – gw Mikrofilm; gw Misc Inst [074]

Altmaerkisch niedersaechsische rundschau – Wittingen DE, 1928 4 aug-1931 30 apr, 1932 9 aug-31 dec – 8r – 1 – (title varies: 2 may 1929: niedersaechsisch-altmaerkische rundschau, 9 aug 1932: rundschau) – gw Misc Inst [074]

Altmaerkische volkszeitung – Salzwedel DE, 1962 8 mar-1965 22 dec – 1r – 1 – gw Misc Inst [074]

Altmaerkisches intelligenz- und leseblatt – Stendal DE, 1885-1900 – 1 – gw Misc Inst [074]

Altmann, J G *see* Versuch einer historischen und physischen beschreibung der helvetischen eisbergen

Altmann, Otto *see* Tegen den stroom

Altmann-Gottheiner, Elisabeth et al *see* Jahrbuch der frauenbewegung

Die altmark im dreissigjaehrigen kriege / Zahn, Wilhelm – Halle a. S: Verein fuer Reformationsgeschichte 1904 [mf ed 1990] – (= ser Schriften des vereins fuer reformationsgeschichte 21/80) – 1mf – 9 – 0-7905-5319-8 – mf#1988-1319 – us ATLA [943]

Altmark stimme – Stendal DE, 1962 7 feb-1967 29 mar – 1r – 1 – gw Misc Inst [074]

Altmark-zeitung – Gardelegen, Kloetze DE, 1992- – 8r/yr – 1 – (main ed in salzwedel) – gw Misc Inst [074]

Altmuehl-bote : kelheimer zeitung – Kelheim DE, 1952 jul-1972 14 jun – 239r – 1 – (bezirksausgabe von mittelbayerische zeitung, regensburg; since 15 jun 1972: mittelbayerische zeitung) – gw Misc Inst [074]

Altneuland : monatschrift fuer die wirtschaftliche erschliessung palaestinas – Berlin, Germany 1905-07 [mf ed Norman Ross] – 1r – 1 – (with: jahresbericht der gesellschaft zur foerderung der wissenschaft des judenthums, berlin, 1905-13) – mf#nrp-293 – us UMI ProQuest [339]

Alt-neuoettinger anzeiger – Altoetting, Burghausen/Salzach DE, 1978 1 sep- – ca 9r/yr – 1 – (bezirksausgabe von passauer neue presse, passau) – gw Misc Inst [074]

Altnordische grammatik / Wimmer, Ludvig Frands Adalbert – Halle, Germany. 1871 – 1r – 1 – us UF Libraries [430]

Altnordisches handbuch / Brenner, Oskar – Leipzig, Germany. 1882 – 1r – 1 – us UF Libraries [430]

'Alto Esta E Alto Mora' / Neves, Guilherme Santos – Vitoria, Brazil. 1954 – 1r – 1 – us UF Libraries [972]

O alto jurua : orgam do municipio – Cruzeiro do Sul, AC. 12 ago-30 dez 1913 – (= ser Ps 19) – mf#P25,01,24 – bl Biblioteca [350]

O alto rio doce – Alto Rio Doce, MG. 11 dez 1894 – (= ser Ps 19) – bl Biblioteca [079]

Alto sentir / Ulloa Zamora, Alfonso – San Jose, Costa Rica. 1953 – 1r – 1 – us UF Libraries [972]

Altolaguirre, Angel de *see*
- Coleccion de las memorias o relaciones que escribieron los virreyes del peru acerca del estado en que dejaban las cosas generales del reino, tomo 2
- Don pedro de alvarado, conquistador del reino de guatemala. madrid, 1927
- Hernando cortes (estudio de un caracter) por el teniente general marques de polariega
- Prueba historica de la inocencia de d. hernando cortes en la muerte de su esposa, de juan palacios. informe
- Los restos de hernando cortes (de luis... obregon)

Alton industrial-williamson county / Baptist Associations. Illinois – 1971-78 – 3r – 1 – $107.80 – (37 associations, alphabetically arr) – us Southern Baptist [242]

Alton, Johann *see*
- Anseis von karthago
- Li romans de claris et laris
- Le roman de marques de rome

Alton telegraph and democratic review – 1842 apr 29 – 1 – mf#976042 – us WHS [071]

Altonaer buerger-zeitung – Hamburg DE, 1929-1941 31 may – 1 – (title varies: 9.8.1924: altonaer neueste nachrichten, 20 jun 1925: altonaer nachrichten, 1 apr 1938: hamburger neueste zeitung / altonaer nachrichten; filmed by other misc inst: 1924 9 aug-1938 2 jan) – gw Misc Inst [074]

Altonaer mercur *see* Staats- und gelehrte zeitung des koeniglichen daenischen unpartheyischen correspondenten

Altonaer nachrichten 1850 – Hamburg DE, 1917 sep-dec – 1r – 1 – (title varies: 1 jan 1856: nordischer courier und altonaer nachrichten, after 1863: altonaer nachrichten) – gw Misc Inst [074]

Altonaer neueste nachrichten *see* Altonaer buerger-zeitung

Altonaer privilegirte adress-comtoir-nachrichten *see* Koeniglich privilegirte altonaer adress-comtoir-nachrichten

Altonaer tageblatt *see* Schleswig-holsteinische zeitung

Altonaer tageblatt 1908? – Hamburg DE, 1930-43 – 34r – 1 – gw Misc Inst [074]

Altonaische relation – (Hamburg-) Altona DE, 1673-74 [single iss], 1683 aug-1684 jun, 1685-1686 nov, 1687-88, 1689 [single iss], 1694 apr-jun, 1695 jan-oct, 1696 – 1 – (with gaps) – gw Misc Inst [074]

Altonaischer mercurius *see* Staats- und gelehrte zeitung des koeniglichen daenischen unpartheyischen correspondenten

Altoona times – Altoona, PA. -d. 1884-1919 – 13 – $25.00r – us IMR [071]

Altoona tribune – 1941 aug 7-43 jan 28; 1943 feb 4-1944; aug 31; sep 7-1945 oct 4 – 1 – mf#916274 – us WHS [071]

Altoona tribune – Altoona, PA. -d. 1889-1957 – 13 – $25.00r – us IMR [071]

Altorientalische forschungen / Winckler, Hugo – Leipzig: Eduard Pfeiffer, 1893-1905 [mf ed 1989] – 3v on 4mf – 9 – 0-7905-3058-9 – (incl bibl ref & ind) – mf#1987-3058 – us ATLA [470]

Altorientalische texte und bilder zum alten testamente / ed by Gressmann, Hugo et al – Tuebingen: J C B Mohr (Paul Siebeck), 1909, c1905 [mf ed 1989] – 2v on 2mf – 9 – 0-7905-0995-4 – (incl bibl) – mf#1987-0995 – us ATLA [221]

Altorientalischer und israelitischer monotheismus : ein wort zur revision der entwicklungsgeschichtlichen auffasung der israelitischen religionsgeschichte / Baentsch, Bruno – Tuebingen: J C B Mohr, 1906 [mf ed 1985] – 1mf – 9 – 0-8370-2149-9 – mf#1985-0149 – us ATLA [210]

[Los altos-] herald american – CA. 1955-1956 – 6r – 1 – $360.00 – mf#H04023 – us Library Micro [071]

[Los altos-] local enterprise – CA. 1959-1967 – 16r – 1 – $960.00 – mf#H04024 – us Library Micro [071]

[Los altos-] town crier – CA. 1973-76 [wkly] – 11r – 1 – $550.00 – mf#B02366 – us Library Micro [071]

Altosmanischen anonymen chroniken / ed by Giese, Friedrich – Breslau, [1922] – (= ser Ottoman histories and historical sources) – 14mf – 9 – $230.00 – us MEDOC [956]

Altoviti, Giovanni *see* Essequie della sacra cattolica e real maest...

Die altpersischen keilinschriften : in umschrift und uebersetzung / ed by Weissbach, Franz Heinrich & Bang, Willy – Leipzig: J C Hinrichs, 1908 – 1mf – 9 – 0-8370-7750-8 – (incl bibl ref. text in german and old persian; commentary in german. issued in parts) – mf#1986-1750 – us ATLA [470]

Die altpersischen keilinschriften / Spiegel, Friedrich – 2. verm Aufl. Leipzig: Wilhelm Engelmann, 1881 – 1mf – 9 – 0-8370-7668-4 – (text in german and old persian; commentary in german) – mf#1986-1668 – us ATLA [470]

Altpreussische monatsschrift – Koenigsberg 1864-1923 [mf ed 1991] – 439mf – 9 – €2310.00 – 3-89131-039-0 – gw Fischer [943]

Altpreussische volkszeitung *see* Intelligenzblatt fuer litthauen

Alt-ratingen – Ratingen DE, 1925 apr-1930 n8 – 1r – 1 – gw Misc Inst [074]

Altrichter, Gertrud *see* Mirko, der knecht

Altrincham and bowden guardian – [NW England] Altrincham, Sale Lib jan 1874-dec 1903 – 1 – (title change: altrincham, bowden & hale guardian [1904-14 may 1948 (1917, 1918 & 1937 incomplete)]; [21 may 1948-1969]; altrincham guardian [1970-82]; altrincham & sale guardian [nov 1982-1990]) – uk MLA; uk Newsplan [072]

Altrincham county express – [NW England] Altrincham, Sale Lib 9 apr 1959-jun 1963 – 1 – uk MLA; uk Newsplan [072]

Altrincham division chronicle and cheshire county news – [NW England] Ashton, Sale Lib 24 jun 1887-19 dec 1890 [incomplete] – 1 – uk MLA; uk Newsplan [072]

Altrincham rd – [North Cheshire FHS] – (= ser 1891 census surname/location indexes [cheshire]) – 2mf – 9 – £4.50 – mf#179 – uk CheshireFHS [929]

Altrincham rd, 1881 surname/location index – (= ser 1881 census transcripts/indexes [cheshire]) – 2mf – 9 – £4.00 – mf#56 – uk CheshireFHS [929]

Altrincham rd (ho 107/2162-63) – [North Cheshire FHS] – (= ser 1851 census surname/location indexes [cheshire] [cheshire]) – 2mf – 9 – £3.00 – mf#174 – uk CheshireFHS [929]

Altrincham rd (rg 11/3502-3503) – (= ser 1881 census transcripts/indexes [cheshire]) – 1mf – 9 – £2.50 – mf#52 – uk CheshireFHS [929]

Altrincham rd (rg 11/3504) – (= ser 1881 census transcripts/indexes [cheshire]) – 1mf – 9 – £2.50 – mf#53 – uk CheshireFHS [929]

Altrincham rd (rg 11/3505, pt) – (= ser 1881 census transcripts/indexes [cheshire]) – 1mf – 9 – £2.50 – mf#54 – uk CheshireFHS [929]

Altrincham rd (rg 11/3506-3507, pt) – (= ser 1881 census transcripts/indexes [cheshire]) – 1mf – 9 – £2.50 – mf#55 – uk CheshireFHS [929]

Altrincham, st george – (= ser Cheshire monumental inscriptions) – 1mf – 9 – £2.50 – mf#415 – uk CheshireFHS [929]

Das altrussische heiligenbild. die ikone / Hackel, A A – Noviomagi, 1936 – €11.00 – ne Slangenburg [243]

Alts in eyn lebn / Bailin, Israel Ber – New York, USA. 1970 – 1r – 1 – us UF Libraries [939]

Altschul, Jakob *see* Der geist des hohen liedes

Die altsemitischen inschriften von sendschirli in den koeniglichen museen zu berlin : text in hebraeischer umschrift, uebersetzung, commentar, grammatischer abriss und vocabular / Mueller, David Heinrich – Wien: Alfred Hoelder, 1893 – 1mf – 9 – 0-8370-7315-4 – mf#1986-1315 – us ATLA [930]

Altsheler, Joseph Alexander *see* A soldier of manhattan

Die altsyrische evangelienuebersetzung und tatians diatessaron, besonders in ihrem gegenseitigen verhaeltnis untersucht... / Hjelt, Arthur – Leipzig: Deichert, 1901 – 1r – 1 – 0-8370-0353-9 – mf#1984-B413 – us ATLA [220]

Die altsyrischen evangelien in ihrem verhaeltnis zu tatians diatessaron / Vogels, Heinrich Joseph – Freiburg i B, St Louis MO: Herder, 1911 – (= ser Biblische studien) – 1mf – 9 – 0-7905-2994-7 – (incl bibl ref) – mf#1987-2994 – us ATLA [220]

Alttestamentliche kritik und christenglaube : ein wort zum frieden / Koenig, Eduard – Bonn: Eduard Weber (Julius Flittner), 1893 [mf ed 1985] – 1mf – 9 – 0-8370-3962-2 – (incl bibl ref) – mf#1985-1962 – us ATLA [221]

Die alttestamentliche offenbarung / Koeberle, Justus – 2nd rev ed. Wismar i M: Hans Bartholdi, 1908 [mf ed 1985] – 1mf – 9 – 0-8370-3950-9 – (earlier ed iss under title: zum kampfe um das alte testament) – mf#1985-1950 – us ATLA [221]

Die alttestamentliche opfercultus... *see* Sacrificial worship of the old testament

Der alttestamentliche prophetismus : drei studien / Sellin, Ernst – Leipzig: A Deichert, 1912 – 1mf – 9 – 0-7905-2134-2 – (incl bibl ref) – mf#1987-2134 – us ATLA [221]

Alttestamentliche religions-geschichte / Loehr, Max – Leipzig: G J Goeschen 1911 [mf ed 1989] – (= ser Sammlung goeschen) – 1mf – 9 – 0-7905-1426-5 – (incl ind) – mf#1987-1426 – us ATLA [221]

Die alttestamentliche schaetzung des gottesnamens und ihre religionsgeschichtliche grundlage / Giesebrecht, Friedrich – Koenigsberg: Thomas & Oppermann, 1901 – 1mf – 9 – 0-8370-3268-7 – mf#1985-1268 – us ATLA [210]

Die alttestamentliche spruchdichtung : rede / Baudissin, Wolf Wilhelm, Graf von – Leipzig: S Hirzel, 1893 – 1mf – 9 – 0-7905-0548-7 – mf#1987-0548 – us ATLA [221]

Alttestamentliche studien / Gumpach, Johannes von – Heidelberg: J C B Mohr, 1852 [mf ed 1989] – 1mf – 9 – 0-7905-1665-9 – (in german & hebrew. incl bibl ref) – mf#1987-1665 – us ATLA [221]

Alttestamentliche studien : rudolf kittel zum 60. geburtstag / Alt, Albrecht et al – Leipzig: J C Hinrichs, 1913 [mf ed 1990] – (= ser Beitraege zur wissenschaft vom alten testament 13) – 1mf – 9 – 0-7905-3300-6 – (incl bibl ref & ind. in german, hebrew & greek) – mf#1987-3300 – us ATLA [221]

Alttestamentliche theologie / Riehm, Eduard; ed by Pahncke, K – Halle: Eugen Strien, 1889 [mf ed 1985] – 1mf – 9 – 0-8370-4898-2 – (incl bibl ref) – mf#1985-2898 – us ATLA [221]

Alttestamentliche theologie *see* Old testament theology

Der alttestamentliche unterbau des reiches gottes / Boehmer, Julius – Leipzig: JC Hinrichs, 1902 – 1mf – 9 – 0-8370-2396-3 – (contains ind of biblical citations) – mf#1985-0396 – us ATLA [221]

Alttestamentliche untersuchungen, erstes heft / Riedel, Wilhelm – Leipzig: A Deichert (Georg Boehme), 1902 [mf ed 1985] – 1mf – 9 – 0-8370-4895-8 – (incl bibl ref) – mf#1985-2895 – us ATLA [221]

Die alttestamentliche weisagung von der vollendung des gottesreiches *see* The old testament prophecy of the consummation of god's kingdom

Die alttestamentliche wissenschaft in ihren wichtigsten ergebnissen : mit beruecksichtigung des religionsunterrichts / Kittel, Rudolf – Leipzig: Quelle & Meyer, 1910 – 1mf – 9 – 0-7905-1126-6 – (incl ind) – mf#1987-1126 – us ATLA [221]

Die alttestamentliche wissenschaft und die religionsgeschichte : rede zum antritt des rektorates der koeniglichen friedrich-wilhelms-universitaet in berlin / Baudissin, Wolf Wilhelm – Berlin: Gustav Schade, 1912 – 1mf – 9 – 0-7905-1921-6 – mf#1987-1921 – us ATLA [221]

Die alttestamentlichen citate bei paulus : textkritisch und biblisch-theologisch gewuerdigt / Vollmer, Hans – Freiburg i.B: J C B Mohr (Paul Siebeck), 1895 – 1mf – 9 – 0-8370-6440-6 – (incl bibl ref and index) – mf#1986-0440 – us ATLA [221]

Die alttestamentlichen lektionen der griechischen kirche / Rahlfs, Alfred – Berlin: Weidmann, 1915 – (= ser Mitteilungen des septuaginta-unternehmens) – 1mf – 9 – 0-8370-1778-5 – mf#1987-6166 – us ATLA [221]

Alttestamentlichen untersuchungen, 1. buch / Bachmann, Johannes – Berlin: S Calvary, 1894 [mf ed 1985] – 1mf – 9 – 0-8370-2137-5 – (incl app) – mf#1985-0137 – us ATLA [221]
Altteutsche schauspiele / ed by Mone, Franz Joseph – Quedlinburg, Leipzig: G Basse, 1841 [mf ed 1993] – (= ser Bibliothek der gesammten deutschen national-literatur von der aeltesten bis auf die neuere zeit sect1/21) – 217p – 1 – mf#8438 reel 5 – us Wisconsin U Libr [820]
Altturkestanische volksweisheit / Brockelmann, Carl – Berlin, 1920 – 1mf – 8 – mf#U-352 – ne IDC [956]
Altube, Gregorio de *see* El excmo sr d xavier maria de munibe, conde de penaflorida
Altun yurt – Adana: Yeni Adana Matbaasi, 1923-? Mueduer-i Mes'ul: Agah Tugrul. n1-6. 15 mayis-11 tesrinievvel 1339 [1923] – (= ser O & t journals) – 3mf – 9 – $75.00 – us MEDOC [956]
Alturas de america / Llorens Torres, Luis – Rio Piedras, Puerto Rico. 1954 – 1r – 1 – us UF Libraries [972]
[Alturas-] modoc county republican – CA. apr 13 1906-jul 23 1915 – 3r – 1 – $180.00 – mf#BC02012 – us Library Micro [071]
[Alturas-] modoc county times – CA. aug 2 1928-may 2 1929 (incomplete); may 16 1929-dec 27 – 6r – 1 – $360.00 – mf#B02008 – us Library Micro [071]
[Alturas-] new era – CA. feb 1901-aug 1925 – 10r – 1 – $600.00 – mf#BC02009 – us Library Micro [071]
[Alturas-] plaindealer – CA. apr 27 1906-dec 1940 – 11r – 1 – $660.00 – mf#B02010 – us Library Micro [071]
[Alturas-] plaindealer and modoc county – CA. sep 19 1913-52 – 23r – 1 – $1380.00 – mf#C02011 – us Library Micro [071]
[Alturas-] the modoc county record – CA. feb 11 1937-aug 1987; jan 1988- (wkly) – 44r – 1 – $2640.00 (subs $90/y) – mf#BC02007 – us Library Micro [071]
[Alturas-] tulelake reporter – CA. 1935-63; 1965- – 33r + – 1 – $1980.00 (subs $50/y) – mf#B02013 – us Library Micro [071]
Altus Air Force Base (OK) *see* Galaxy
Altwegg, Wilhelm *see* Johann peter hebel
Altweibersommer : aus einem zeitlosen tagebuch; die prinzessin von banalien; meine kinderjahre; meine erinnerungen an grillparzer; am ende / Ebner-Eschenbach, Marie von – Leipzig: H Fikentscher, H Schmidt & H Guenther [1928] [mf ed 1993] – 2r – 1 – (filmed with: [saemtliche werke] / [ebner-eschenbach]) – mf#8570 reel 2 – us Wisconsin U Libr [880]
Altweimarische liebes- und ehegeschichten / Boehlau, Helene – Stuttgart: J Engelhorn 1897 [mf ed 1989] – 1r – 1 – (filmed with: schriften / johann jakob bodmer) – mf#7042 – us Wisconsin U Libr [830]
Alt-weimars abend : briefe und aufzeichnungen aus dem nachlasse der graefinnen egloffstein / ed by Egloffstein, Hermann, Freiherr von – Muenchen: Beck, 1923 [mf ed 1990] – vi/624p/[7pl] – 1 – (incl ind) – mf#7076 – us Wisconsin U Libr [860]
Alt-wien in geschichten und sagen fuer die reifere jugend / Bermann, Moriz – Wien [u.a.] 1865 [mf ed Hildesheim: 1995-98] – (= ser Fbc) – 199p on 2mf – 9 – €60.00 – 3-487-29461-3 – gw Olms [390]
Alu Like Information Office et al *see* Native hawaiian
Aluin, Juan *see* Vida de la venerable sierra de dios maria de s francisco, llamada comunmente la rozas...
Alum creek lake cemetery relocations, 1973 – 1r – 1 – mf#B25947 – us Ohio Hist [978]
Alumbaugh allies – v1 n1-4 [1978 mar-dec] – 1 – mf#429500 – us WHS [071]
Los alumbrados espanoles de los siglos 16 y 17 / Lorca, B – Madrid: Razon y Fe, 1934 – 1 – sp Bibl Santa Ana [946]
L'aluminium, l'acetylene, l'or & l'argent – Paris, France aug/oct 1898-15 oct 1900 – 1 – (cont: l'aluminium, l'or et l'argent [jan 1895-jul 1898]; cont by: aluminium et l'acetylene [1 nov-15 dec 1900]) – uk British Libr Newspaper [540]
Aluminium workers news digest *see* Cio news
Aluminium Workers of America *see* Cio news
Aluminium world – New York NY oct 1894-dec 1902 – 1 – (cont by: metal industry [jan 1903-dec 1904]) – uk British Libr Newspaper [660]
Aluminum Goods Manufacturing Co *see* Mixing bowl
Aluminum workers news digest – 1943 jul-1944 may – 1 – mf#1051872 – us WHS [071]
Al-ummah – Damascus: Muhammad Kurd 'Ali, 1907-17. [daily] n1-1464 (feb 1907-apr 15 1914) – (= ser Arabic journals and periodicals) – 8r – 1 – $500.00 – (cont & cont by: al-muqtabas; reels also contain al-ummah and al-qabas) – us MEDOC [079]
Al-ummah = Nation – Mogadishu [Somalia]: M I Amin [nov 13 1967-apr 26 1969] (mthly) – 1r – 1 – us CRL [960]
Al-ummah *see* Al-qabas
Alumnae Association of the Baptist WMU Training School. Louisville, Kentucky *see* Annual bulletin
Alumni Association of Hunter College *see* Wistarians
Alumni Association of Malcolm-King College *see* Maak
Alumni Association of the University of Wisconsin *see* View
Alumni news – 1977 feb; 1980 jul-oct; 1981 jan-apr; 1983 jan-apr, oct; 1984 jan-apr, oct; 1985 jan-jul; 1986 apr-fall; 1987 jul; 1988 jan-jul; 1989 jul-oct; 1990 jan/apr; 1992 spring/summer-fall/winter; 1993 fall/winter; 1994 sep – 1 – mf#2691184 – us WHS [071]
Alumni news – v2 n1-v14 n2 [1971 feb-1983 may] – 1 – mf#641428 – us WHS [071]
Alumni newsletter – v1 n1-v1 n4 [1969 oct-1970 sep] – 1 – mf#641434 – us WHS [071]
Alumni souvenir : illustrating buildings and faculties of the university of toronto and affiliated colleges / Aylsworth, M B [comp] – Arts and Divinity ed. [Toronto?]: M B Aylsworth, 1892 [mf ed 1980] – 1mf – 9 – 0-665-02446-0 – mf#02446 – cn Canadiana [378]
Al-'urwa al-wutqa = Le lien indissoluble – Paris. n1-18. mars-oct 1884 – (= ser Urwa al wutqa) – 1 – (reedition de 1958) – fr ACRPP [073]
Al-usbu – al-Khartum: Dar al-usbu ul-tibaah wa-al-nashr, apr 1986-jun 1989 – 15r – us CRL [079]
Aluwihare, Bernard Herbert *see* Pamplets
Alva, Bartolome de *see* Confessionario mayor y menor en la lengua mexicana
Alva citizen and hillfoots free press – [Scotland] Alloa: Steadman & Co 14 may 1895-8 jun 1897 (wkly) [mf ed 2003] – 109v on 1r – 1 – uk Newsplan [072]
Alva, florida : lee county / Lamme, Corinne W – s.l, s.l, 1936 – 1r – 1 – us UF Libraries [978]
Alva, Joachim *see* Men and supermen of hindustan
Alva y Viamont, D *see* El perfecto capitan, instruido en la disciplina militar y nueva ciencia de artilleria
Alvanley, William Arden, Baron *see* The state of ireland considered
Alvar garcia de santa maria / Cantera y Burgos, Francisco – Madrid. 1951 – 1 – us CRL [946]
Alvar saints : their lives and teachings / Bharati, Shuddhananda – Ramachandrapuram, Trichy Dist: Anbu Nilayam, 1942 – (= ser Samp: indian books) – us CRL [280]
Alvarado De Ricord, Elsie *see* Estilo y densidad en la poesia de ricardo j bermu...
Alvarado Garaicoa, Toedoro *see*
– Vasco nunez de balboa. adelantado de la costa del mar del sur
– Vida, pasion y muerte de vasco nunez de balboa (descubridor del oceano pacifico) y francisco de orellana (descubridor del rio amazonas)
Alvarado Garcia, Ernesto *see*
– Historia de centro-america
– Odisea de leoncio prado en honduras
Alvarado, Huberto *see*
– Exploracion de guatemala
– Sombras de sal
Alvarado, Lisandro *see*
– Datos etnograficos de venezuela
– Glosarios del bajo espanola en venezuela
– Historia de la revolucion federal en venezuela
Alvarado, Manuel *see* Discurso que...pronuncio el 23 de abril.
Alvarado, Pedro de *see* An account of the conquest of guatemala in 1524
Alvarado Pinetta, Rony Stanley *see* Transformacion agraria en guatemala
Alvarado Quiros, Alejandro *see*
– Discursos pronunciados en las recepciones
– Nuestra tierra prometida
– Ya se oyen los claros clarines
Alvarado, Rafael *see* Cuestion de belice (conferencia)
Los alvarados en el – nuevo mundo / Solar y Taboada, Antonio & Rigula y Ochotorena, Jose de – Madrid: Tip. Rev. Arch, Bib. y Museos, 1934 – 1 – sp Bibl Santa Ana [946]
Alvarenga, Oneyda *see* Cateretes do sul de minas gerais
Alvares, Didac *see* De auxiliis divinae gratiae
Alvarez, Alfred *see* Shaping spirit
Alvarez Amandi, Justo *see* La catedral de oviedo
Alvarez, Antonio *see* Noneto
Alvarez, Arturo *see*
– Ataide, antonio de
– Un curioso manuscrito sobre el convento de san onofre de la lapa (badajoz) (su biblioteca y sacristia en el siglo 16)
– Guadalupe, arte, devocion y...
– Guadalupe en la america andina. madrid, 1969
– Las municipalidades hispano-portuguesas
– Los pilares de la hispanidad se forjaron en guadalupe
– Tradicion conceptista en la provincia betica
Alvarez Baragano, Jose *see*
– Amor original
– Para el 26 de julio
– Poemas escogidos
– Poesia
Alvarez Bravo, Armando *see* Azoro
Alvarez, C *see*
– Cheo alvanez
– Guerra y marina, epoca de carlos 1 de espana...
– Valladolid. archivo general de simancas. secretaria de guerra (s. 18)...hojas de servicios de america
Alvarez de la Rivera, Senen *see* Biblioteca historico-genealogica asturiana...
Alvarez de Sotomayor, Agustin *see* Memoria sobre la cria caballar
Alvarez del Vayo, Juan *see* Poesias
Alvarez del Vayo, Julio *see*
– Deux discours
– L'espagne accuse
– Speech delivered by his excellence don julio alvarez del vayo, minister of foreign affairs
Alvarez del vayo's answer to the british charge d'affaires – [s.l: s.n. 1937] [mf ed 1980] – (= ser Blodgett coll) – 1mf – 9 – mf#w713 – us Harvard [946]
Alvarez Elizondo, Pedro *see* Presidente arevalo y el retorno a bolivar
Alvarez, F *see* Comercio y comerciantes, y sus proyecciones
Alvarez, Francois *see* Regimiento contra la peste
Alvarez Garzon, Juan *see* Clavijos
Alvarez Guerra, Andres *see*
– Credito nacional o sea hacienda publica
– Descripcion y diseno del trillo
– Invento ceres o sea metodo de proceder... propio por diez anos
– Tercer cuaderno de los inventos ceres
Alvarez Guerra, Juan *see* Correcciones al trillo inventado por don juan alvarez guerra, executadas por don juan francisco gutierrez
Alvarez, Jose Ma. Leyendas *see* Barcelona, 1933
Alvarez Joven, Arturo *see* La gitana extremena y otros poemas
Alvarez Lejarza, Emilio *see* Ensayo biografico del procer jose leon sandoval
Alvarez Lencero, H *see* Canciones en carne viva
Alvarez Lencero, Luis *see*
– Hombre. grabados de francisco mateos
– El surco de la sangre
Alvarez Madariaga, Luz *see* Contratos y cuasicontratos mineros en las legislaciones sudamericanas
Alvarez Magana, Manuel *see* Antologia poetica
Alvarez Medina, Felipe *see* Poesias
Alvarez Nazario, Manuel *see*
– Arcaismo vulgar en el espanol de puerto rico
– Elemento afronegroide en el espanol de puerto rico
Alvarez Pedroso, Antonio *see* Miguel de aldama
Alvarez Pedroso, Armando *see* Nueva revision de algunos de los...
Alvarez Puga, Miguel *see* Ancla para tu voz
Alvarez, Ramon *see* Geografia de venezuela
Alvarez, Ricardo *see* Psiquiatria en venezuela desde la epoca precolombi
Alvarez Rubiano, Pablo *see* Pedrerias davila. contribucion al estudio de la figura del "gran justador", gobernador de castilla del oro y nicaragua. madrid, 1934
Alvarez Saenz de Buruaga, Jose *see*
– Anfiteatro
– Datos para el estudio de las antiguedades de merida
– El escudo de merida y su origen romano
– La fundacion de merida
– Localizacion de la reliquia de la cabeza de santa eulalia
– Merida en el siglo 27 (continuacion de la "historia de la ciudad de merida" de moreno de vargas)
– Merida y los viajeros
– Miscelanea emeritense del s. 16
– Nuevas aportaciones al estudio de la necropolis oriental de merida
– El palacio del duque de la roca, de merida
– Las ruinas de emerita y de italica a traves de nebrija y rodrigo caro
Alvarez, Santiago *see* El pueblo de galicia, contra el fascismo
Alvarez, Segis *see*
– La juventud y los campesinos
– Nuestra organizacion y nuestros cuadros
Alvarez Silva, Ramon *see*
– Methode pour l'enseignement de l'espagnol en haiti
– Volutas
Alvarez Soler, Margot *see* Poemas del amor mas puro, y otros poemas
Alvarez Suarez, Augustin Enrique *see* Educacion moral
Alvarez Tabio, Fernando *see* Teoria general de la constitucion cubana
Alvarez Y Alvarez De La Cadena, Luis *see* Mexico
Alvarez y Saenz de Buruaga, Jose *see* Un nucleo de neterramientos romanos en la campina de merida
Alvaro cordobes, opera (siecle 10) – Cordoba – 1r – 5,6 – sp Cultura [946]
Alvarus Pelagius (Alvaro Pelayo) *see* De planctu ecclesiae
Alvensleben, Maximilian, Baron von *see* With maximilian in mexico
Alverdes, Paul *see*
– Dank und dienst
– Die flucht
– Reinhold
– Vergeblicher fischzug
– Die verwandelten
– Das winterlager
Alverez, Felix *see* "Perfiles sacerdotales". barcelona, edit. hernando. 1959...
Alverstone, Viscount *see* Recollections of bench and bar
Alves, Albano *see* Dicionario portugues-chisena e chisena-portugues
Alves, Aluizio *see*
– Angicos
– Sem odio e sem medo
Alves, Castro *see*
– Espumas flutuantes
– Poesias completas
Alves, Mario *see* Nobrega e a civilizacao brasileira
Alves, P A *see* Biblia ia ana
Alves, Raul *see* Canastra
Alves, William *see* Lectures on the epistle of paul the apostle to the ephesians, chapter 1
O alvicareiro : periodico critico, commercial, noticioso e moral – Natal, RN. 05 nov 1880 – (= ser Ps 19) – bl Biblioteca [073]
Alvin ailey revelations : the newsletter of the friends of alvin ailey – 1986 spring; 1988 fall; 1989 fall; 1990 summer-fall; 1991 spring; 1992 spring; 1993 spring, winter – 1 – mf#4862536 – us WHS [071]
The alvina treut burrows research collection on composition / Svobodny, Dolly – 1900-86 – 1500 titles on 300mf – 9 – (printed card indexes included) – us ATBI [370]
Alviola, Uldarico *see* Felicitas
Alvira alias orea / Henderson, J Duff – Toronto: Hunter, Rose, 1899 [mf ed 1980] – 4mf – 9 – 0-665-05539-0 – mf#05539 – cn Canadiana [830]
Alvo advance – Alvo, NE: Interstate Pub Co (wkly) – 9r – 1 – us Bell [071]
A alvorada – Fortaleza, CE: Typ Universal, 3 jun 1894 – 1,5,6 – mf#P18B,03,01 – bl Biblioteca [079]
A alvorada : jornal semanario, orgao do parti do republicano portuguese – S Pedro do Sul: T P Nova Estabelecimento, [jan 31-mar 2, mar 13-16, mar 30-apr 6 1913] (wkly) – (= ser Coll of lusophone african newspapers and serials) – 1r – 1 – us CRL [079]
A alvorada : orgao democratico – Piracicaba, SP: Typ de Joaquim Espiridiao de Almeida Proenca, 16 jun-28 nov 1880 – (= ser Ps 19) – 1 – mf#P18,01,68 – bl Biblioteca [079]
A alvorada : periodico litterario e noticioso – Taarauaca, AC: Officinas d'O Municipio, 14 jul,out 1913; jan 1914; mar-maio, ago-out, dez 1915; abr, jul-ago, out 1916; fev-mar 1917; 28 fev 1919 – (= ser Ps 19) – mf#P25,01,25 – bl Biblioteca [440]
Alvorada : orgam do collegio conceicao – Sao Joao del Rei, MG: Typ da Gazeta Mineira, 28 mar 1886 – (= ser Ps 19) – mf#P17,02,88 – bl Biblioteca [440]
Alvorada : orgao da democracia, litterario e recreativo – Sao Joao da Barra, RJ: Typ da Alvorada, 12 fev 1878 – (= ser Ps 19) – mf#P05,04,10 – bl Biblioteca [079]
Alvorada – Rio de Janeiro, RJ: Typ de Serafim Jose Alves, 20 jul 1879 – (= ser Ps 19) – mf#P05,04,09 – bl Biblioteca [440]
O alvorada : semanario republicano-democratico – Inhambano: Jose Flores [dec 1 1912-nov 20 1913] (wkly) – (= ser Coll of lusophone african newspapers and serials) – 1r – 1 – us CRL [079]
Alvord, Emery Delmont *see* Development of native agriculture and land tenure in southern...
Alvsborgs nyheter – Alingsas, 1909-31 – 20r – 1 – sw Kungliga [078]
Alvsborgs nyheter – Vanersborg, Sweden. 1982-92 – 1 – sw Kungliga [078]
Alvsborgsposten – Uddevalle, Sweden. 1978-95 – 1 – sw Kungliga [078]
Al-wahah – al-Khartum: al-Wahah, jul, oct-nov 1986; jan 1987 – 1r – us CRL [950]
Alwan – Khartum: Husayn Khujuli [1984 oct-1989 jun] (daily) – 13r – 1 – us CRL [079]
Alwan salamat – [Khartoum] Alwan Salamat, [mar24-may 8 1988] (wkly) – 1r – 1 – us CRL [079]
Alwaqai aliraqiya: official gazette of the republic of iraq / Iraq – 1970-78. Formerly: Iraq weekly gazette – 1 – $92.00 – (1979-. ca 20.00y) – us L of C Photodup [324]
Al-waqa'i al-iraqiyah / Iraq – 1970-79 – 1 – $230.00 – (1980-. ca $30.00y) – us L of C Photodup [324]
Al-waqai al-misriya : egytian bulletin / Egypt – 1887-91, 1901, 1907-17 – (= ser Waqai Al-Misriyah) – 81r – 1 – (1919-26, 1928-49, 1961-66, 1970-79 63r. 1980-.) – us L of C Photodup [324]

Alward, Silas *see* The jubilee year
Al-watan – al-Khartum: Sayyid ahmad khalifah, apr 2 1988-jun 29 1989 – 4r – us CRL [074]
Al-watani al-ittihadi – Umm durman: al-hizb al-watani al-ittihadi, oct 23 1985-may 10 1987 – us CRL [079]
Al-watwany – Moroni, Comoros. sep 6 1985-oct 11 1991 – 4r – 1 – (cont: l'echo des comores) – us L of C Photodup [079]
Aly, Wolfgang *see* Geschichte der griechischen literatur
Alyeska reports – v1-3 n3 [1975 jul-1977 oct] – 1 – mf#305845 – us WHS [071]
Alyth gazette and merchants' advertiser – [Scotland] Alyth: A Lunan jan 1925-dec 1950 (wkly) [mf ed 2003] – 12r – 1 – (merged with: alyth guardian and advertiser to form: alyth gazette and guardian [jan 1927-dec 1950]) – uk Newsplan [072]
Alyth guardian and district advertiser – [Scotland] Alyth: T M'Murray 7 nov 1884-may 1925 (wkly) [mf ed 2003] – 23r – 1 – (cont by: alyth guardian and advertiser [jan 1887-may 1925]) – uk Newsplan [072]
Al-zahf al-akhdar – Tripoli, Libya. 1980-1984; 1987 – 6r – 1 – us L of C Photodup [079]
Al-zahra' – Cairo, 1924-1928. v1-5 – 39mf – 9 – mf#NE-20329 – ne IDC [956]
Al-Zayn, Ahmad 'Arif *see* Al-irfan
Alzheimer's care quarterly – Baltimore MD 2000+ – 1,5,9 – ISSN: 1525-3279 – mf#32162 – us UMI ProQuest [616]
Al-zilal – Khartoum, Sudan. jun 17, 1993 – 1r – us CRL [999]
Alzog, Johannes *see* Manual of universal church history
Am anderen morgen : roman / Bauer, Josef Martin – Muenchen: R Piper, c1949 [mf ed 1995] – 377p – 1 – mf#8972 – us Wisconsin U Libr [830]
Am da malshk ga na damsh st john. ligi = The gospel according to st john – London: SPCK, 1889 [mf ed 1980] – 1mf – 9 – 0-665-00125-8 – (trans into zimshian) – mf#00125 – cn Canadiana [225]
Am da malshk ga na damsh st luke. ligi = The gospel according to st luke – London: SPCK, [1887] [mf ed 1980] – 1mf – 9 – (text in tsimshian) – mf#14251 – cn Canadiana [225]
Am da malshk ga na damsh st mark. ligi = The gospel according to st mark – London: SPCK, [1887] [mf ed 1981] – 1mf – 9 – (text in tsimshian) – mf#14253 – cn Canadiana [225]
Am da malshk ga na damsh st matthew. ligi = The gospel according to st matthew – London: SPCK, [1885] [mf ed 1980] – 1mf – 9 – (text in tsimshian) – mf#14256 – cn Canadiana [225]
Am dreilaendereck – Zittau DE, 1961, 18 aug-1967, 28 mar – 3r – 1 – (filmed by other misc inst: 1962 24 jan-1966 1 nov [1r]; title varies: 6 jun 1962: dreilaendereck, publ in dresden) – gw Misc Inst [074]
Am engineering data base in order by county and state / U.S. National Technical Information Service – Monthly – 9 – us NTIS [000]
Am engineering data base in order by state / U.S. National Technical Information Service – Monthly.Secondarily, in order by city – 9 – us NTIS [000]
Am ersten mai : eine tragikomoedie der arbeit aus friedenstagen / Adolph, Karl – Leipzig: Neuer Akademischer Verlag, 1919 [mf ed 1995] – 41p – 1 – mf#8918 – us Wisconsin U Libr [820]
Am euphrat und tigris : reisenotizen aus dem winter 1897-1898 / Sachau, E – Leipzig, 1900 – 2mf – 9 – mf#AR-1964 – ne IDC [915]
Am fenster : jugenderinnerungen / Federer, Heinrich – Berlin: G Grote 1927 [mf ed 1989] – (= ser Grote'sche sammlung von werken zeitgenoessischer schriftsteller 170) – 1r – 1 – (filmed with: gustav falke / friedrich castelle) – mf#7229 – us Wisconsin U Libr [880]
Am goldenen steig : und andere erzaehlungen aus dem bayer- und boehmerwald / Schmidt, Maximilian – 2. aufl. Reutlingen: Ensslin & Laiblin [1899?] [mf ed 1995] – 1r – 1 – (filmed with: sueden und norden / hermann schmid) – mf#3738p – us Wisconsin U Libr [880]
The am ha-aretz : the ancient hebrew parliament / Sulzberger, Mayer – Philadelphia: J H Greenstone, 1909 – 1mf – 9 – 0-8370-5465-6 – (incl indes) – mf#1985-3465 – us ATLA [270]
Am heiligen quell – Muenchen DE, 1932/33-1939/40 – 1 – (title varies: 1933/34 n5: am heiligen quell deutscher kraft; until 1931 suppl to: ludendorffs volkswarte) – gw Misc Inst [074]
Am i a christian? / Geissler, Mortiz – London, England. 18– – 1r – 1 – us UF Libraries [240]
Am i a christian or am i not? – s.l, England?. 18– – 1r – 1 – us UF Libraries [240]
Am i going to heaven? – London, England. 18– – 1r – 1 – us UF Libraries [240]
Am i jew or gentile? : read and see / Davies, Thomas Alfred – New York: E H Coffin, c1889 [mf ed 1985] – 1mf – 9 – 0-8370-2842-6 – (incl add) – mf#1985-0842 – us ATLA [221]
Am kachelofen / Bloesch, Hans – Bern: Gute Schriften, 1945 [mf ed 1993] – 90p – 1 – (biog aft by rudolf hunziker) – mf#8520 – us Wisconsin U Libr [890]
Am kamin : novellen und erzaehlungen / Schmid, Herman – 2.aufl. Leipzig: Keil [18–?] – (= ser Gesammelte schriften. volks- und familien-ausgabe 2) – 1 – (bound with: tannengruen and erzstufen) – mf#film mas c438 – us Harvard [800]
Am khyi bhvay ra i kambha / Tvan Aon, U – Ran kun: Bhasa pran ca pe a san: Burma Translation Soc 1951 [mf ed 1990] – 1 – (in burmese) – mf#mf-10289 seam reel 176/8 [§] – us CRL [550]
Am leben entlang : gedichte und balladen / Baum, Kurt – Amerika-Ausg. Chicago: Gutenberg Pub Co, 1933 [mf ed 1989] – 52p (ill) – 1 – (ill by willy knapp) – mf#6982 – us Wisconsin U Libr [810]
Am rheinesstrand *see* Benrather tageblatt
'Am Yisra'el Ba-Tefutsot / Rozenberg, A – Yerushalayim, Israel. 1944 – 1r – 1 – us UF Libraries [939]
Ama, Do *see*
– Chara kri sakhan kuiyto mhuin
– Rvhe don ton chon pa mya
Ama martire ase uganda / Streicher, Henri – s.l, s.l, 1926? – 1r – 1 – us UF Libraries [960]
Amabile, Luigi *see* Il santo officio della inquisizione in napoli
Amabilis Dominguez, Manuel *see* Arquitectura precolombina en mexico
Amaculo ase lovedale : lovedale music / Bokwe, John Knox – 5th ed. Lovedale, South Africa : [Lovedale Press], 1922 (mf ed 1993) – 1r – 1 – (xosa or english words) – mf#Sc Micro R-7073 – us NY Public [240]
Amade *see* Voyage en espagne
Amadeo, Francisco L *see* Luciernagas
Amadeo Gely, Teresa *see* Biografia de lucas amadeo antomarchi en relacion...
Amadis : erstes buch / ed by Keller, Adelbert von – Stuttgart: Literarischer Verein, 1857 [mf ed 1993] – (= ser Blvs 40) – 482p – 1 – (incl bibl ref and ind) – mf#8470 reel 9 – us Wisconsin U Libr [830]
Amado Blanco, Luis *see* Dona velorio
Amado, Gilberto *see* Grao de areia e estudos brasileiros
Amado, Jorge *see*
– Cavaleiro da esperanca
– De como o mulato porciuncula descarregou seu defunto
– Sao jorge dos ilheus
– Violent land
Amado, Manuel *see*
– Compendio historicos de las vidas de los santos...del orden de predicadores
– Dios y espana
– La monarquia y la religion triunfantes de los sofismas
– Los siete dias de la pasion o lecciones practicas de virtud
Amador : orgao do clube terpsychore – Rio de Janeiro, RJ. 14 ago 1886 – (= ser Ps 19) – bl Biblioteca [079]
Amador *see* El dorado/amador
O amador : periodico litterario do club dramatico goncalves leite – Rio de Janeiro, RJ. 08 set 1888 – (= ser Ps 19) – mf#DIPER – bl Biblioteca [790]
Amador, Armando *see* Origen, auge y crisis de una dictadura
[Amador county-] amador, el dorado, placer and sacramento counties – CA. 1884-1885 – 1r – 1 – $50.00 – mf#D005 – us Library Micro [978]
Amador, Jorge *see* Presencia en lejania
Amador prospector – v1 n1-4 [1982] – 1 – mf#656230 – us WHS [071]
Amador/el dorado – 1928-38, 1992- – (= ser California telephone directory coll) – 13r – 1 – $650.00 – mf#P00004 – us Library Micro [917]
Yr amaethydd – [Wales] Isle of Anglesey ion 1845-hydref 1846 [mf ed 2003] – 1r – 1 – uk Newsplan [072]
Amagram – 1982 jan, apr-1984 – 1 – mf#1109618 – us WHS [071]
Amalarii episcopi opera / Hanssens, J M – Roma, 1948 – 8 – €73.00 – (liturgica omnia 1948 studi e testi 138, tom1 10mf; studi e testi 139 tom2 15mf; studi e testi 140 tom3 13mf) – ne Slangenburg [241]
Amal'ezulu / Vilakazi, B Wallet – Johannesburg, South Africa. 1960 – 1r – 1 – us UF Libraries [960]
Amalgamated Association of Iron and Steel Workers *see* Proceedings of the...annual convention of the national lodge, a a of i and s w
Amalgamated Association of Iron, Steel and Tin Workers *see* Amalgamated journal
Amalgamated Association of Iron, Steel, and Tin Workers of North America *see* Journal of proceedings of the...annual convention of the international lodge, a a of i s and t w of north america
Amalgamated Association of Street and Electric Railway Employees of America *see*
– International president's report to the...convention of the amalgamated associationof street and electric railway employes of america
– International president's report to the...convention of the amalgamated associationof street railway employees of america
– Wages of motormen, conductors and bus operators
– Wages of the motormen and conductors
Amalgamated Association of Street and Electric Railway Employees of America et al *see* Motorman, conductor and motor coach operator
Amalgamated Association of Street, Electric Railway and Motor Coach Employees of America *see* Union leader
Amalgamated Association of Street, Electrical Railway and Motor Coach Employees of America et al *see* In transit
Amalgamated Association of Street Railway Employees of America *see* Year book of the amalgamated association of street railway employees of america
Amalgamated Clothing and Textile Workers Union *see*
– Labor times
– Labor unity
– Social justice
Amalgamated Clothing Workers of America *see*
– Advance
– Voice
Amalgamated Food and Allied Workers Union *see* Union reporter
Amalgamated Food Workers et al *see* Free voice of the amalgamated food workers
Amalgamated Food Workers of America *see* Free voice of the amalgamated food workers
Amalgamated Glass Workers' International Association of America *see* Glassworker
Amalgamated Jewelry, Diamond and Watchcase Workers Union *see* Jewelry workers bulletin
Amalgamated journal – 1899 oct 6-1900 dec 27; 1901-08; 1909-1910 jun; jul-1912 jun; jul-1913; 1914-22; 1923-1924 jun; jul-1925; 1926-1927 jun;jul-1928; 1929-1930 jun; jul-1931; 1932-1933 jun; jul-1934; 1935-1936 jun; jul-1937; 1938-1939 jun; 1939 jul-1940; 1941-1942 aug – 1 – mf#783029 – us WHS [071]
Amalgamated journal / Amalgamated Association of Iron, Steel and Tin Workers – 1988-1942 – (= ser Labor union periodicals, pt 1: the metal trades) – 29r – 1 – $6060.00 – 1-55655-226-2 – us UPA [331]
Amalgamated Lithographers of America *see*
– Lithographer's bulletin, 1901-1904 / official publication of the lithographer's international protective and beneficial association, 1910-1913 / the lithographer's journal, 1918-1955
– Lithographers journal
Amalgamated Meat Cutters and Butcher Workmen of North America *see*
– Bulletin of the amalgamated meat...
– Butcher workman
– Butchers' 532 review
– Butchers' union local n120
– Chit 'n chatter
– Dist local 340 reporter
– Journal
– Local 525 news
– Local 538 newsletter
– Local 593's allied progress report
– Local p-6 news
– Meat of the matter
– Meatcutters local 81 newsletter
– Newsletter
– Official journal
– Official journal of the amalgamated meatcutters and butcher workmen
– P-3 pacer
– Packinghouse worker
– Seventy-three
Amalgamated Meat Cutters and Butcher Workmen of North America et al *see*
– District record
– District union 427 voice
Amalgamated Meat Cutters and Butcher Workmen of North America. Local 304 [Sioux Falls, SD] et al *see* News and views, afl-cio, local no 304
Amalgamated Meat Cutters and Retail Food Store Employees Union *see*
– Local 342 forefront
– Local 342 news
Amalgamated Meat Cutters, Butcher Workmen and Affiliated Crafts of North America *see* Voice of local 1
Amalgamated news – v11 n5-v25 n2 [1962 feb-1981 may] – 1 – mf#345248 – us WHS [071]
Amalgamated sheet metal workers' journal – v1 n19 [1896 dec 10], v2 n3-v6 [1897 may 10-1901 nov 15] – 1 – mf#1405073 – us WHS [071]
Amalgamated Society of Carpenters and Joiners *see* Minutes of the proceedings of the advisory council of the north american continent
Amalgamated Textile Workers of America *see* New textile worker
Amalgamated Transit Union *see* Union topics
Amalgamated Woodworkers' International Union of America et al *see* International wood worker
Amalgamation of the general and particular baptist in england / Baptist Missionary Society. Archives. London – 225p. 1889-91 – 1 – $7.87 – us Southern Baptist [242]
The amalgamation of the two branches of the legal profession. / Saunders, Cornelius Thomas – London: Butterworths, 1870. 32p. LL-2245 – 1 – us L of C Photodup [340]
Amalgamation of unions and proposed modifications in the poor-law (ireland) / Chichester, Charles Raleigh – Dublin, London: James Duffy & Sons, 1879 – (= ser 19th c economics) – 1mf – 9 – mf#1.1.464 – uk Chadwyck [344]
Amalgamated Marine Workers' Union. London, England *see* Marine worker
Amalie fuerstin von gallitzin / Brentano, Maria Rafaela – 3. Aufl. Freiburg, 1920 (mf ed 1993) – 2mf – 9 – €24.00 – 3-89349-207-0 – mf#DHS-AR 96 – gw Frankfurter [920]
Amaliens erholungsstunden – 1790-92 [mf ed 1999] – (= ser Hq 42) – 3v on 24mf – 9 – €230.00 – 3-89131-359-4 – gw Fischer [305]
Amals tidning – Amal, Sweden. 1874-85 – 4r – 1 – sw Kungliga [078]
Amals weckoblad – Amal, Sweden. 1846-47, 1856-74 – 4r – 1 – sw Kungliga [078]
Amalsposten – Amal, Sweden. 1883-97 – 10r – 1 – sw Kungliga [078]
Amalthea : zeitschrift fuer wissenschaft und geschmack / ed by Erhard, Christian Daniel – Leipzig 1788-89 – (= ser Dz) – v1(=st1-3); v2[=st1.2] on 5mf – 9 – €100.00 – mf#k/n5779 – gw Olms [000]
Amana : the community of true inspiration / Shambaugh, Bertha Maud Horack – Iowa City, IA: State Historical Society of Iowa, c1908 [mf ed 1990] – 1mf – 9 – 0-7905-6566-8 – (incl bibl ref) – mf#1988-2566 – us ATLA [242]
Amana society bulletin – v56 n9-v58 n31 [1987 apr 30-1989 sep 31 – 1 – mf#1612159 – us WHS [071]
Amand, D *see* Fatalisme et liberte dans l'antiquite grecque
Amandebele kamzilikazi / Sithole, Ndabaningi – Cape Town, South Africa. 1956 – 1r – 1 – us UF Libraries [960]
Amanecer : reflexion cristiana en la nueva nicaragua – Managua, Nicaragua: Centro Ecumenico Antonio Valdivieso [n1-84 (mayo 1981-dic 1994)] (qrtly) – 2r – 1 – us CRL [240]
Amann, Emile *see* Le protevangile de jacques et ses remaniements latins
Amann, Paul *see* Goethe
Amano, Haruko [comp] *see* Ohraimono bunrui shusei 2
Amans en poste : ou, la magicienne supposee / Caigniez, Louis-Charles – Paris, France. 1804 – 1r – 1 – us UF Libraries [440]
Amans, Georges *see* Pour seduire les femmes
Amant de coeur / Verneuil, Louis – Paris, France. 1921 – 1r – 1 – us UF Libraries [440]
Amant malheureux / Arnould, Auguste Jean Francois – Paris, France. 1844 – 1r – 1 – us UF Libraries [440]
Amants puerils / Crommelynck, Fernand – Paris, France. 1921 – 1r – 1 – us UF Libraries [440]
Amar : verbo intransitivo / Andrade, Mario De – Sao Paulo, Brazil. 1944 – 1r – 1 – us UF Libraries [972]
Amar, Jules *see* The human motor
Amar y Arguedas, J *see*
– Instruccion curativa de las calenturas conocidas...como tabardillo
– Instruccion curativa de las viruelas
– Instruccion curativa y preservativa de los dolores de costado y pulmones
Amaral, Aracy A *see*
– Artes plasticas na semana de 22
– Blaise cendrars no brasil e os modernistas
Amaral, Azevedo *see*
– Brasil na crise actual
– Estado autoritario e a realidade nacional
Amaral, Braz Do *see*
– Historia da bahia do imperio a republica
– Historia da independencia na bahia
Amaral, Ilidio Do *see* Ensaio de um estudo geografico da rede urbana de angola
Amaral, Leonidas Do *see* Prodromos da campanha presidencial
Amaral, Louis *see* Outro brazil
Amaral, Tancredo do *see*
– O estado de sao paulo..
– A historia de sao paulo ensinada pela biographia dos saus vultos mais notaveis
Amarante La Tarde, Antonia De *see* Monumentos principais do distrito federal

Amaranth – Saint John, NB: R Shives, [1841?-1843] [mf ed v1 n1 jan 1841-v3 n12 dec 1843] – 9 – (incl ind) – mf#P04643 – cn Canadiana [410]
Amaranth [1847] : a semi-monthly publication devoted to polite literature, science, poetry, and amusement – La Salle IL 1847 – 1 – mf#3757 – us UMI ProQuest [073]
Amaranth Energies *see*
– Razzberry
– Razzberry radicle
Amaranth, or masonic garland – La Salle IL 1828 – 1 – mf#3915 – us UMI ProQuest [366]
Amaranth, or token of remembrance – La Salle IL 1847-55 – 1 – mf#3824 – us UMI ProQuest [073]
Amaranthes *see* Nutzbares, galantes und curioeses frauenzimmer-lexicon...von amaranthes
Amarga, Naranja *see* The settling of bertie merian
Amari-i kishvar – 1358 [1979-80] sal-i 1 – (= ser Vilayet salnames) – 22mf – 9 – $350.00 – us MEDOC [956]
Amarillo Genealogical Society *see* Reflector
Die amarna-zeit *see* The tell el amarna period
Die amarnazeit : palaestina und aegypten in der zeit israelitischer wanderung und siedelung / Miketta, Karl – 1. & 2. aufl. Muenster i W: Aschendorff 1908 [mf ed 1992] – (= ser Biblische zeitfragen 1/10) – 1mf – 9 – 0-524-05583-1 – (incl bibl ref) – mf#1992-0443 – us ATLA [930]
Amaroc news – Coblenz, germany. v1, n1-v. 2, n255. apr. 21, 1919-dec. 31, 1920 – 1 – us NY Public [073]
Amaron, Calvin Elijah *see* Le retour de l'emigre
Amartiri a ku uganda / Streicher, Henri – Lilongwe, Malawi. 1951? – 1r – 1 – us UF Libraries [960]
Amas news – 1985 mar, jun, sep; 1986 feb, oct; 1987 mar, oct – 1 – mf#4852700 – us WHS [071]
Amasiah the son of zichri / Babington, W P – Manningtree, England. 1860 – 1r – 1 – us UF Libraries [240]
Amasya tarihi / Huseyin, Husameddin – Istanbul: Hikmet Matba'asi and Istikbal Matba'asi. 4v. 1912-35 – (= ser Ottoman Histories and Historical Sources) – 20mf – 9 – $320.00 – (history of amasya from pre-ottoman times. last half of v4 in latin script) – us MEDOC [956]
Amasya'da emel – Amasya, 1921-23. Sahib-i Imtiyaz: Mehmed Sirri. n24. 13 temmuz 1338 [1922],76,80,81,85,86. 4 tesrinievvel 1339 [1923] – (= ser O & t journals) – 1mf – 9 – $25.00 – us MEDOC [956]
Amat, Joan Carles *see* Guitarra espanola, y vandola, en dos maneras de guitarra, castellana y valenciana de cinco ordenes
Amat, Juan Carlos *see* Quatre cents aforismes cathalans...
Amateur acting / Angus, J Keith – Toronto: Musson Book Co, [1880?] [mf ed 1993] – 2mf – 9 – 0-665-91432-6 – mf#91432 – cn Canadiana [790]
The amateur athlete – New York. v1-4 n2. mar 1896-sep 1897 [all publ] – (= ser Sports periodicals, 1822-1922) – 1r – 1 – $315.00 – us UPA [790]
Amateur athletic union of the united states. info aau – Lake Buena Vista FL 1973-92 – 1,5,9 – (cont: aau news) – ISSN: 0279-9863 – mf#8556,01 – us UMI ProQuest [790]
Amateur cine world – Hemel Hempstead, England 1950-66 – 1,5,9 – mf#525 – us UMI ProQuest [790]
Amateur contest to-night! $5.00 prize : london opera house, london, canada...the old time favorite drama "east lynne"... – S.l: s.n, 18–? – 1mf – 9 – mf#51911 – cn Canadiana [790]
The amateur gentleman : a romance / Farnol, Jeffery – Toronto: W Briggs [1911?] [mf ed 1999] – 5mf – 9 – 0-659-90260-5 – mf#9-90260 – cn Canadiana [830]
Amateur master file / U.S. National Technical Information Service – Twice a year. Data pertaining to amateur licensees and listed in call sign sequence – 9 – us NTIS [000]
Amateur master file supplement / U.S. National Technical Information Service – Monthly – 9 – us NTIS [000]
Amateur photographer – London, England 1985-90 – 1,5,9 – ISSN: 0002-6840 – mf#11258 – us UMI ProQuest [770]
The amateur photographer – London: [Hazell, Watson & Viney, 1884-1908] v1-47 (n1-1231) oct 10 1884-may 5 1908 (wkly) – 25r – 1 – us CRL [770]
The amateur photographer and photographic news – London: Hazell, Watson & Viney. v47-67 n1232-1758. may 12 1908-jun 10 1918 – 1 – us CRL [760]
Amateur work : a monthly magazine of the useful arts and sciences – Killen TX 1901-07 – 1 – mf#3212 – us UMI ProQuest [073]
The amateura – [Upper Dorchester, NB]: U D L & M Society, [1888?-1889?] – 9 – ISSN: 1190-6618 – mf#P04537 – cn Canadiana [420]
The amateur's assistant : or a series of instructions in sketching / Clark, John – London 1826 – (= ser 19th c art & architecture) – 2mf – 9 – mf#4.2.396 – uk Chadwyck [740]
Amator Patriae *see* An appeal to capitalists
Amauta – Lima. n1-32. sept. 1926-aug. sept. 1930 – 1 – us NY Public [073]
Amavo / Jolobe, James – Johannesburg, South Africa. 1947 – 1r – 1 – us UF Libraries [960]
Ama-xosa [the] : life and customs / Soga, John Henderson – Lovedale: Lovedale Press 1931 – (= ser [Travel descriptions from south africa, 1711-1938]) – 6mf – 9 – mf#zah-11 – ne IDC [916]
Amaya Delgado, Manuel *see* Tratado de las asfixias o muertes asfixiantes
Amaya Roldan, Martin *see* Historia de chita
The amazing argentine : a new land of enterprise / Fraser, John Foster – London, Toronto: Cassell, 1914 – 5mf – 9 – 0-665-76867-2 – mf#76867 – cn Canadiana [972]
Amazing grace – v1 n7-v2 n10 [1970 apr-1971 dec] – 1 – mf#1106515 – us WHS [071]
Amazing science fiction – New York. v20-49. feb 1946-nov 1975 – (= ser Science fiction periodicals, 1926-1978. series 2) – 27r – 1 – $3905.00 – us UPA [830]
Amazing stories – Chicago. v1-19. apr 1926-dec 1945 – (= ser Science fiction periodicals, 1926-1978. series 1) – 22r – 1 – $2750.00 – us UPA [830]
Amazing stories annual – Chicago, 1927 [all publ] – (= ser Science fiction periodicals, 1926-1978. series 1) – 1r – 1 – $95.00 – us UPA [830]
Amazing stories quarterly – Chicago. v1-7. winter 1928-fall 1934 [all publ] – (= ser Science fiction periodicals, 1926-1978. series 1) – 2r – 1 – $325.00 – us UPA [830]
Amazon – 1973-1979 nov; dec-1983 jan; 1984 feb/mar – 1 – mf#186311 – us WHS [071]
Amazon / Haskins, Caryl Parker – Garden City, USA. 1943 – 1r – 1 – us UF Libraries [972]
Amazon : river of promise / Malkus, Alida – New York, USA. 1970 – 1r – 1 – us UF Libraries [972]
The amazon – v1, n1-. Milwaukee: Amazon Collective, May 1972 – 1 – us Wisconsin U Libr [073]
Amazon quarterly – Oakland CA 1972-75 – 1,5 – mf#7940 – us UMI ProQuest [305]
Amazon throne / Harding, Bertita Leonarz – Indianapolis, IN. 1941 – 1r – 1 – us UF Libraries [972]
Amazona de canas : novela / Sanchez Gomez, Gregorio – Cali, Colombia. 1958 – 1r – 1 – us UF Libraries [972]
Amazonas / Lopes Goncalves, Augusto Cezar – New York, USA. 1904 – 1r – 1 – us UF Libraries [972]
Amazonas : sua historia / Jobim, Anisio – Sao Paulo, Brazil. 1957 – 1r – 1 – us UF Libraries [972]
Amazonas (Brazil) Governor *see* Relatorios dos presidentes, 1a republica, 1891-1930
Amazonas (Brazil) President *see* Relatorios dos presidentes, epoca do imperio, 1852-1889
Amazonas commercial : publicacao diaria do commercio, artes industria – Manaos, AM. 10 mar 1895; jul 1897; fev, 26 maio 1900 – (= ser Ps 19) – 1,5,6 – bl Biblioteca [380]
Amazonfloden / Christmas, Walter – Kobenhavn, Denmark. 1892 – 1r – 1 – us UF Libraries [972]
Amazonia – Belem, PA. 09 mar-26 maio 1888 – (= ser Ps 19) – mf#DIPER – bl Biblioteca [079]
Amazonia. a terra e o homem / Lima, Araujo – 2a. ed. Sao Paulo. 1937 – 1 – us CRL [972]
Amazonia, aspectos economicos / Versissimo De Mattos, Jose – Rio De Janeiro, Brazil. 1892 – 1r – 1 – us UF Libraries [972]
Amazonia brasileira / Biblioteca Nacional (Brazil) – Rio De Janeiro, Brazil. 1969 – 1r – 1 – us UF Libraries [972]
Amazonia colombiana / Salemanca T, Demetrio – Bogota, Colombia. 1916 – 1r – 1 – us UF Libraries [972]
Amazonia cyclopica / Hurley, Jorge – Rio De Janeiro, Brazil. 1931 – 1r – 1 – us UF Libraries [972]
Amazonia [jul 1884-mar 1885] *see* Correio da manha
Amazonia, maranhao, nordeste / Macedo, Duarte Ribeiro De – Belo Horizonte, Brazil. 1970 – 1r – 1 – us UF Libraries [972]
Amazonia, paraiso e inferno / Silva, Renato Ignacio Da – Sao Paulo, Brazil. 1970 – 1r – 1 – us UF Libraries [972]
Amazonia que eu vi : obidos-tumucumaque / Cruls, Gastao – Sao Paulo, Brazil. 1945 – 1r – 1 – us UF Libraries [972]
Amazonia que os portugueses revelaram / Reis, Arthur Cezar Ferreira – Rio De Janeiro, Brazil. 1957 – 1r – 1 – us UF Libraries [972]
Amazulu / Jenkinson, Thomas B – London, England. 1884 – 1r – 1 – us UF Libraries [960]
Amazulu / Jenkinson, Thomas B – Pretoria, South Africa. 1968 – 1r – 1 – us UF Libraries [960]
Ambar / Salles Diaz, margarita – Habana, Cuba. 1960 – 1r – 1 – us UF Libraries [972]
Ambassade au thibet et au boutan : contenant des details tres-curieux sur les moeurs, la religion, les productions et le commerce du thibet, du boutan et des etats voisins; et une notice sur les evenemens qui s'y sont passes jusqu'en 1793 / Turner, Samuel – Paris 1800 [mf ed Hildesheim 1995-98] – 3v on 6mf – 9 – €120.00 – 3-487-27625-9 – gw Olms [951]
Ambassade de charles-ambroise messabarba : patriarche d'alexandrie, vers l'empereur kang-hi / Viani, S – Paris, 1749-1761. v20 – 2mf – 9 – mf#HT-678 – ne IDC [910]
L'ambassade de d garcias de silva figveroa en perse... – Paris, 1667 – 6mf – 9 – mf#ILM-1216 – ne IDC [956]
Ambassade du mareschal de bassompierre en espagne l'an 1621 / Bassompierre, Francois de – Cologne 1668 [mf ed Hildesheim 1995-98] – 1v on 2mf – 9 – €60.00 – 3-487-25830-7 – gw Olms [946]
Ambassade du mareschal de bassompierre en suisse l'an 1625 / Bassompierre, Francois de – Cologne 1668 [mf ed Hildesheim 1995-98] – 2v on 8mf – 9 – €160.00 – gw Olms [949]
Ambassadeur / Scribe, Eugene – Paris, France. 1828 – 1r – 1 – us UF Libraries [440]
Ambassador for intermediates – mar 1932-41 – 1 – $138.46 – us Southern Baptist [242]
Ambassador guggenheim and the cuban revolt / Cuban Information Bureau, Washington, DC – Washington, DC. 1931 – 1r – 1 – us UF Libraries [972]
Ambassador leader – Brotherhood Commission publ, SBC, apr 1959-67 – 1 – $52.85 – us Southern Baptist [242]
Ambassador life – 1946-67 – 1 – $277.48 – us Southern Baptist [242]
Ambassador review – 1976 jun – 1 – mf#1701311 – us WHS [071]
Ambassadorial and secret service reports on revolutionary and napoleonic france, 1781-1786 : pro class f027, france, general correspondence – v1-20 – 17r – 1 – mf#C39-19600 – us Primary [940]
Ambassadorial relazione, 1565 *see* Extracts from regole brievi della volgare grammatica
Ambedkar, Bhimrao Ramji *see*
– Pakistan or partition of india
– Ranade, gandhi and jinnah
– The untouchables
– What congress and gandhi have done to the untouchables
– Who were the shudras?
Ambedkar refuted : what congress and gandhi have done to the untouchables / Rajagopalachari, Chakravarti – Bombay: Hind Kitabs, 1946 – (= ser Samp: indian books) – us CRL [305]
Amberger volksblatt – Amberg/Oberpf DE, 1952 jul- – 1 – (bezirksausgabe von mittelbayerische zeitung, regensburg) – gw Misc Inst [074]
Amberley, John Russell, Viscount *see* An analysis of religious belief
Ambidexterity and mental culture / Macnaughton-Jones, Henry – New York: Rebman Co [n.d.] [mf ed 1987] – 1r [ill] – 1 – (filmed with: viga-glum's saga) – mf#1888 – us Wisconsin U Libr [150]
Ambiente axiologico de la teoria pura del derecho / Carrillo, Rafael – Bogota. Univ. Nacional de Colombia. 1947. 92p. LL-4068 – 1 – us L of C Photodup [340]
Ambiente penal de la violencia / Umana Luna, Eduardo – Bogota, Colombia. 1962 – 1r – 1 – us UF Libraries [972]
Ambio – Lawrence KS 1972-91 – 1,5,9 – ISSN: 0044-7447 – mf#49282 – us UMI ProQuest [333]
Ambit – London, England 1972+ – 1,5,9 – ISSN: 0002-6972 – mf#7641 – us UMI ProQuest [400]
Amblard, Arturo *see* Notas coloniales
Ambler, Charles Henry *see* Sectionalism in virginia from 1776-1861
Ambler, Pennsylvania. Mount Pleasant Baptist Church *see* Records
Ambon beroept zich op recht en trouw / ed by Bureau Zuid-Molukken – Den Haag, 1950 nos 4, 8 – 1mf – 9 – mf#SE-1289 – ne IDC [959]
Ambon en de a r-partij : de vrijheidsstrijd van de republiek der zuid-molukken / Gerbrandy, P S – Kampen, 1956 – 1mf – 8 – mf#SE-1419 – ne IDC [959]
Ambon en de ar-partij de vrijheidsstrijd van de republiek der zuid-molukken / Gerbrandy, P S – Kampen, 1956 – 1mf – 9 – mf#SE-1419 – ne IDC [950]
Ambon nu! : [vier reportages] / Kloosterhuis, H – Wageningen, 1968 – 1mf – 8 – mf#SE-1451 – ne IDC [959]
The ambon question : facts and appeal / Leimena, J – n.p, 1950 – 1mf – 8 – mf#SE-1290 – ne IDC [959]
Amboyna's struggle against the lies of djocja / Lokollo, P W – The Hague, 1950 – 1mf – 9 – mf#SE-1594 – ne IDC [959]
Das ambraser liederbuch vom jahre 1582 / ed by Bergmann, Joseph – Stuttgart: Literarischer Verein, 1845 [mf ed 1993] – xiv/400p – 1 – (incl ind) – mf#8470 reel 3 – us Wisconsin U Libr [780]
Ambri, M [comp] *see* Dongeng-dongeng sasakala, kenging ngempelkeun moh
Ambrogi, Arturo *see*
– Jeton
– Marginales de la vida
– Muestrario
– Paginas escogidas
Ambrogio, Amelli D *see* S leone magno e l'oriente
Ambroise, Fernand *see* General magloire ambroise a-t-il ete tue ou s'est-...
Ambrose : archbishop of milan / Telford, John – London, England. 18– – 1r – 1 – us UF Libraries [241]
Ambrose, Saint *see* Epistola ad quintum fratrem...
Ambrose, Saint, Bishop of Milan *see* Selections
Ambrosia – 1980 oct – 1 – mf#2847550 – us WHS [071]
The ambrosian liturgy : the ordinary and canon of the mass according to the rite of the church of milan – London: Cope and Fenwick, 1909 – (= ser Christian Liturgies) – 1mf – 9 – 0-524-03010-3 – mf#1990-4532 – us ATLA [240]
Ambrosini, Gaspare *see* Marx, mazzini e l'internazionale socialista
Ambrosio de Montanchez *see* Miscelanea sagrada de varios discursos panegiricos
Ambrosio-films – Berlin DE, 1910 p93-1911 – 1 – gw Mikrofilm [790]
Ambrosius *see* Allgemeine einleitung, 1. bd (bdk17 1.reihe)
Ambrosius (Ambrose of Milan, Saint) *see* Pflichtenlehre und ausgewaehlte kleinere schriften, 3. bd (bdk32 1.reihe)
Ambrosius (Ambrose, Saint) *see* Lukaskommentar, 2. bd (bdk21 1.reihe)
Ambrosius blaurer / Pressel, T – Elberfeld, R L Friderichs, 1861 – 2mf – 9 – mf#PBU-458 – ne IDC [240]
Ambrosius, Johanna *see* Gedichte
Ambrosius, T *see* Introductio in chaldaicam linguae, syriacae atque armenica et dece alias linguas
Ambrosius von mailand als kirchenpolitiker / Campenhausen, H von – Berlin, 1929 – €14.00 – ne Slangenburg [241]
Het ambt bij calvijn / Goumaz, Louis – Franeker: T. Wever, 1964 – 1r – 1 – 0-8370-1592-8 – mf#1984-T023 – us ATLA [242]
Ambts-brieven : 1802-1842 / Falck, Anton Reinhard – 's Gravenhage [i.e. Hague]: W P van Stockum 1878 [mf ed 1986] – 1r – 1 – (filmed with: religion and culture / schleiter, f) – mf#6688 – us Wisconsin U Libr [327]
Ambushes and surprises : being a description of some of the most famous instances of the leading into ambush and the surprise of armies, from the time of hannibal to the period of the indian mutiny / Malleson, George Bruce – London: W H Allen 1885 [mf ed 1986] – 1r [ill] – 1 – (filmed with: la religion de l'empereur julien / farney, r) – mf#1725 – us Wisconsin U Libr [355]
Ambuyamuderere / Mutswairo, Solomon M – London, England. 1967 – 1r – 1 – us UF Libraries [960]
Amc journal – Washington DC 1915-94 – 1,5,9 – ISSN: 0891-6209 – mf#10300 – us UMI ProQuest [622]
Amc news – 1987 oct-dec; 1988 jan-feb, apr-jul, sep-oct, dec; 1989 jan, mar-apr, jun-oct, dec]; 1990 mar-apr, jul-dec; 1991 jan-jun, aug, oct-dec; 1992 mar – 1 – mf#2539456 – us WHS [071]
Amcabey – Istanbul. cilt 1 sayi 1-cilt 3 sayi 69. 5 kanunievvel 1942-25 mart 1944 [5 dec 1942-25 mar 1944] – (= ser O & t journals) – 10mf – 9 – $165.00 – us MEDOC [956]
Amccom quarterly – v2 n2-3 [1984 apr-jul]; v4 n2 [1988 jul]; v5 n1-3; [1989 jan-jul]; v6 n1-4 [1990 jan-oct]; v7 n1, 3-4 [1991 jan, jul-oct]; v8 n1-3 [1992 jan-jul] – 1 – mf#1051467 – us WHS [071]
Amddiffynydd y gweithiwr – Merthyr Tydfil, Wales 8 aug 1874-13 nov 1875 – 1 – (Discontinued) – uk British Libr Newspaper [072]
Amddiffynydd y gweithiwr – [Wales] LLGC 8 awst 1874-13 tach 1875 [mf ed 2004] – 1r – 1 – uk Newsplan [072]
A.m.d.g. : la vida en un colegio de jesuistas [a novel] / Perez de Ayala, Ramon – [Santiago, Chile]: Empresa Letras [1936?] [mf ed 1994] – (= ser Les grandes escritores 59) – 1r – 1 – us UF Libraries [830]
L'ame amante de son dieu / Hugo, Hermannus & Vaenius, Othon – Paris: Libraires Associes, 1790 – 3mf – 9 – mf#O-58 – ne IDC [090]

L'ame amante de son dieu... / Hugo, Hermannus & Vaenius, Othon – Utrecht: Herm & Joh Besseling, 1750 – 4mf – 9 – mf#O-3092 – ne IDC [090]

[L'ame amante de son dieu...] / Hugo, Hermannus & Vaenius, Othon – Cologne: Chez Jean de la Pierre, 1717 – 4mf – 9 – mf#O-311 – ne IDC [090]

Ame christian recorder – 1963 jan 1-1964 sep 22; 1964 sep 29-1966 dec; 1968 jul 30-dec 17; 1967 jan 3-1968 jul 23; 1968 dec 24-1970 dec 29; 1971 jan-1972 jun; 1972 jul-1974 may; 1974 jun-dec 30; 1975 jan-1976 dec; 1977 jan-1979 aug; 1979 sep-1984 nov 19; 1988 jan 11-1989 dec 25; 1990 jan 8-1991 dec 23; 1992 jan 6-1993 dec 20 – 1 – mf#1050181 – us WHS [071]

L'ame du pygmee d'afrique / Trilles, H – Paris: Editions du Cerf, 1945 – 1 – us CRL [290]

L'ame d'un bon roi : ou choix d'anecdotes et de pensees de henri 4... / Costard, Jean – Londres [i.e. Paris 1775 [mf ed Hildesheim 1995-98] – 1v on 1mf – 9 – €40.00 – ISBN-10: 3-487-26110-3 – ISBN-13: 978-3-487-26110-2 – gw Olms [944]

L'ame d'un peuple africain : les bambara, leur vie psychique, ethique, sociale, religieuse / Henry, Joseph – Muenster i W: Aschendorff, 1910 [mf ed 1991] – (= ser Bibliotheque-anthropos 1,2) – v/238p/24pl (ill) on 1mf – 9 – 0-524-01553-8 – (in french) – mf#1990-2507 – us ATLA [390]

L'ame d'un peuple africain, les bamabara, leur vie psychique, ethique, sociale, religieuse / Henry, Jos – Munster: Verlag der Aschendorffschen Buchhandlung, 1910 – 1 – us CRL [306]

Ame en folie / Curel, Francois De – Paris, France. 1920 – 1r – 1 – us UF Libraries [440]

L'ame est immortelle *see* Evidence for a future life

L'ame humaine : existence et nature / Coconnier, Marie Thomas – Paris: Perrin, 1890 [mf ed 1991] – vii/495p on 2mf – 9 – 0-7905-8773-4 – (in french) – mf#1989-1998 – us ATLA [110]

Ame qui meurt / Papillon, Pierre – Port-Au-Prince, Haiti. 1950 – 1r – 1 – us UF Libraries [972]

Une ame sacerdotale, le chanoine michel / Alexis, pere – Ottawa: l'Echo de S Francois, [1912] (mf ed 1986) – 1mf – 9 – (pref by l c raymond) – mf#SEM105P687 – cn Bibl Nat [241]

Amedra : roman de moeurs negres du congo belge / Delhaise, Arnould M L – Bruxelles: Renaissance d'Occident, 1926 – 1 – us CRL [390]

El amel : organe des travailleurs nord-africains – Paris: Bideau [apr1932] (mthly) – 1r – 1 – us CRL [074]

Amelia island and fort clinch / Shepherd, Rose – s.l, s.l, 1939 – 1r – 1 – us UF Libraries [978]

Amelia island early history – s.l, s.l, 193-? – 1r – 1 – us UF Libraries [978]

Amelia smith / Taunay, Alfredo D'escragnolle Taunay – Sao Paulo, Brazil. 1930 – 1r – 1 – us UF Libraries [972]

Amelineau, Emile *see*
- Les actes des martyrs de l'eglise copte
- Essai sur le gnosticisme egyptien
- Essai sur l'evolution historique et philosophique des idees morales dans l'egypte ancienne
- La geographie de l'egypte a l'epoque copte
- Histoire de la sepulture et des funerailles dans l'ancienne egypte
- Histoire des monasteres de la basse-egypte
- Les idees sur dieu dans l'ancienne egypte
- Les moines egyptiens
- Monuments pour servir a l'histoire de l'egypte chretienne
- Monuments pour servir a l'histoire de l'egypte chretienne au 4e siecle
- La morale egyptienne quinze siecles avant notre ere
- Prolegomenes a l'etude de la religion egyptienne
- Le tombeau d'osiris

Ameller, C *see* Elementos de geometria y fisica experimental...

Amelli, A M *see* Miniature sacre e profane dell' an1023, illustranti l'enciclopedia medioevale di rabamauro

Amelotte, D *see* La vie du pere charles de condren

Amelung, Heinz *see* Briefwechsel zwischen clemens brentano und sophie mereau

Amelung, Heinz [comp] *see* Goethe als persoenlichkeit

Amelunxen, C P *see* Geschiedenis van curacao

Les amen de monsabre : lecture faite au cercle ville-marie de montreal / Beaubien, Charles Philippe – Montreal: E Senecal, 1892 – 1mf – 9 – mf#03514 – cn Canadiana [240]

"Amen" in old testament liturgical texts : a study of its meaning and later development as a plea for ecumenical understanding / Flor, Elmer Nicodemo – Uni of South Africa 2000 [mf ed Johannesburg 2000] – 4mf – 9 – mf#mfm14910 – sa Unisa [221]

Amen sugerencias liturgicas / Aradillas Agudo, Antonio – Madrid: Studium, 1966 – sp Bibl Santa Ana [240]

Amendements... : passes 11 jul [sic] 1861, sanctionnes 12 sep 1861 / Commissaires du havre de Montreal – [Montreal?: s.n.] 1861 [mf ed 1983] – 1mf – 9 – 0-665-44018-9 – mf#44018 – cn Canadiana [343]

The amending of the federal constitution / Orfield, Lester Bernhardt – Ann Arbor: University of Michigan, 1942. 242p. LL-1136 – 1 – us L of C Photodup [342]

Amending the microenterprise for self-reliance act of 2000 and the foreign assistance act of 1961, increasing assistance for the poorest people in developing countries... : markup before the committee on international relations, house of representatives, 107th congress, 2nd session, on h.r. 4073 and h.r. 3969, april 2002 / United States. Congress. House. Committee on International Relations – Washington: US GPO 2002 [mf ed 2002] – 2mf – 9 – us GPO [327]

Amendment to the organic act of guam : hearing before the subcommittee on territorial and insular affairs... / Guam. US Congress – 92nd Congress 2nd sess. 14 Sep 1972. Washington: GPO, 1972 – 1mf – 9 – $1.50 – mf#LLMC 82-100B Title 16 – us LLMC [342]

Amendments to articles 2 and 36, uniform code of military justice : hearings before the military personnel subcommittee of the committee on armed services – House of Rep. 96th Congress, 1st session, June 11-12 1979. Washington: GPO, 1979 – 2mf – 9 – $3.00 – mf#LLMC 96-079 – us LLMC [348]

Amendments to the constitution of north carolina, 1776-1974 / Sanders, John L – Chapel Hill: Institute of Government, University of North Carolina at Chapel Hill, 1975. LL-2395 – 1 – us L of C Photodup [342]

Amenidades...de la vera alta y baxa / Azedo de la Berrueza, Gabriel – 1891 – 9 – sp Bibl Santa Ana [946]

The amens of christ / Bowen, George – Boston, MA: McDonald & Gill, c1886 – 1mf – 9 – 0-524-04791-X – mf#1992-0211 – us ATLA [220]

Amentet : an account of the gods, amulets and scarabs of the ancient egyptians / Knight, Alfred Ernest – London: Longmans, Green, 1915 [mf ed 1992] – 1mf – 9 – 0-524-02088-4 – mf#1990-2852 – us ATLA [290]

Amenumey, Divine Edem Kobla *see* The ewe people and the coming of european rule, 1850-1914

The amenyah archives on ada history / Amenyah, Jacob Dosoo – Chicago, IL: [Cooperative Africana Microfilming Project, Center for Research Libraries] 1973 – 1 – us CRL [960]

Amenyah, Jacob Dosoo *see* The amenyah archives on ada history

Amer, Carlos *see* Cuba y la opinion publica

L'amer du chene : ou, avenir de l'europe d'apres le passe et le present – [Paris]: b Impr de J Frey [jun 1848] – (= ser French revolution of 1848. newspapers) – 1r – 1 – us CRL [074]

Amerasia – v1-11 n2,7. 1937-47 [all publ] – (= ser Radical periodicals in the united states, 1881-1960. series 1) – 60mf – 9 – $560.00 – us UPA [303]

Amerasia journal – Los Angeles CA 1973+ – 1,5,9 – ISSN: 0044-7471 – mf#7751 – us UMI ProQuest [305]

America – 1888 oct 11, nov1-dec 27; 1889 jan 3-1931, feb 14, 28, mar 14; 1889 apr 4, may 9, jun 13-20, jul 11-sep 26; 1889 oct 3-1890 mar 27; 1890 apr 3-sep 25 – 1 – mf#1092224 – us WHS [071]

America / Cuyahoga Co. Cleveland – mar-apr 1918, jun 1918-oct 1922 [daily] – 6r – 1 – (inrumanian) – mf#B7096-7101 – us Ohio Hist [071]

America – Detroit, MI. 1906-66 – 1 – us CRL [071]

America : jornal noticioso, litterario e scientifico – 01 agio 1870-20 mar 1871 – (= ser Ps 19) – bl Biblioteca [073]

America : life and labor / American National Women's Trade Union League – 1911-21 – (= ser Women's periodical and manuscripts coll) – 4r – 1 – £200.00 – mf#ALL – uk World [331]

America – New York NY 1909+ – 1,5,9 – ISSN: 0002-7049 – mf#320 – us UMI ProQuest [240]

America : or a general survey of the political situation of the several powers of the western continent with conjectures on their future prospects / Everett, Alexander H – London 1828 [mf ed Hildesheim 1995-98] – 1v on 3mf – 9 – €90.00 – 3-487-27183-4 – gw Olms [975]

America : organ al romanilor den statele unite si in special al bisericilor gr-orientale / ed by Peter Lucaci – Cleveland, OH: Moise Balea, [sep 28 1906-1966] (mthly) – 75r – 1 – us CRL [071]

America : patria de cain / Pereda, Diego De – Habana, Cuba. 1933 – 1r – us UF Libraries [972]

America : a sketch of the political, social, and religious character of the united states of north america, in two lectures / Schaff, Philip – New York: Scribner, 1855 [mf ed 1991] – 1mf – 9 – 0-524-00784-5 – (in english) – mf#1990-0216 – us ATLA [975]

America : su geografia, su historia / Fernandez Pesquero, J – Madrid: Razon y Fe, 1930 – 1 – sp Bibl Santa Ana [972]

America *see*
- Coleccion de documentos ineditos, relativos al descubrimiento, conquista y organizacion de las antiguas posesiones espanolas de america y oceania
- Coleccion de libros y documentos referentes a la historia de america

La america – 1-15. 1910-25 – 1 – us NY Public [073]

La america – Madrid. Spain. -w. 8 Mar 1857-13 Mar 1875. (12 reels) – 1 – uk British Libr Newspaper [074]

La america – New York, NY. 1910-25 – 1 – us AJPC [071]

America 1883 – the american visitor 1884 – the american eagle 1885-86 – american humorist and storyteller 1888 – (= ser American newspapers and periodicals published in the uk and europe 18th-20th centuries) – 1r – 1 – £55.00 – uk World [071]

America, 1935-1946 : the photographs of the farm security administration and the office of war information in the prints and photographs division of the library of congress... / U.S. Library of Congress – [mf ed Chadwyck-Healey, 1981] – 1574mf – 9 – (arrangement by region: northeastern states 434mf. midwestern states 254mf. northwestern states 153mf. southern states 301mf. southwestern states 205mf. farwestern states 159mf. foll sects not available separately but are incl in the complete coll: general usa 1558 photos, canada & alaska 146 photos, virgin islands & puerto rico 2200 photos. with p/g & ind) – uk Chadwyck [975]

America – a new march / Gram, Hans – Printed from movable type as a supplement to Vol. III, 1791, of "The Massachusetts Magazine." Scored on three staves, two treble and one bass, instrumentation unspecified. MUSIC 123, Item 3 – 1 – us L of C Photodup [780]

America abroad – 1891-1907 – (= ser American newspapers and periodicals published in the uk and europe 18th-20th centuries) – 2r – 1 – £95.00 – uk World [072]

America achter theil : in welchem erstlich beschrieben wirt das...koenigreich guiana...item, eine kurtze beschreibung der vmbligenden landtschafften... – Gedruckt zu Franckfurt am Mayn: Durch Matthaeum Becker, 1599 [mf ed 1994] – (= ser America 6) – 3mf – 9 – 0-665-94744-5 – mf#94744 – cn Canadiana [972]

America and brittania : peace. a new march / Taylor, Raynor – Composed by R. Taylor (and so arranged as to harmonize perfectly with Washingtons march played both together). Philadelphia: G. Willig n.d.. MUSIC 3082, Item 11 – 1 – us L of C Photodup [780]

America and europe – New York, USA. 1896 – 1r – 1 – us UF Libraries [025]

America and her problems = Les etats-unis d'amerique / Estournelles de Constant, Paul Henri Benjamin, baron d' – New York: Macmillan, 1915 [mf ed 1990] – 2mf – 9 – 0-7905-5987-0 – (in english) – mf#1988-1987 – us ATLA [306]

America and her resources : or a view of the agricultural, commercial, manufacturing, financial, political, literary, moral and religious capacity and character of the american people / Bristed, John – London 1818 [mf ed Hildesheim 1995-98] – 1v on 3mf – 9 – €90.00 – 3-487-27173-7 – gw Olms [975]

America and the americans / Baxter, William Edward – London; New York: G Routledge, 1855 [mf ed 1986] – 4mf – 9 – 0-665-47918-2 – (incl bibl ref) – mf#47918 – cn Canadiana [917]

America and the americans : a narrative of a tour in the united states and canada; with chapters on american home life / Craib, Alexander – Paisley [Scotland]: A Gardner, 1892 [mf ed 1980] – 4mf – 9 – 0-665-00273-4 – mf#00273 – cn Canadiana [917]

America and the americans : the theatres, the streets, the cars, the newspapers... / Offenbach, Jacques – London: W Reeves, [1877?] [mf ed 1982] – 1mf – 9 – mf#32912 – cn Canadiana [917]

America, and the americans / Boardman, James – London 1833 [mf ed Hildesheim 1995-98] – 1v on 3mf – 9 – €90.00 – 3-487-27182-6 – gw Olms [975]

America and the asiatic world / Mathews, Shailer – New York: Church Peace Union, [1916?] [mf ed 1991] – (= ser The church and international peace 9) – 1mf – 9 – 0-7905-9336-X – mf#1989-2561 – us ATLA [327]

America and the british colonies : an abstract of all the most useful information relative to the united states of america, and the british colonies of canada, the cape of good hope, new south wales, and van diemen's island / Kingdom, William – London 1820 [mf ed Hildesheim 1995-98] – 1v on 3mf – 9 – €90.00 – 3-487-26685-7 – gw Olms [327]

America, britain and the war of independence : the papers and correspondence of sir jeffrey, 1st baron amherst (1717-97) from the amherst mss in the kent archives office – 16r – 1 – (with printed guide) – mf#C39-16300 – us Primary [975]

America del sud / Bryce, James Bryce, Viscount – New York, USA. 1914 – 1r – 1 – us UF Libraries [972]

America e o libertador / Bolivar, Simon – Caracas, Venezuela. 1953 – 1r – 1 – us UF Libraries [972]

America en fin de siglo / Serrano De Wilson, Emilia, Baronesa – Barcelona, Spain. 1897 – 1r – 1 – us UF Libraries [972]

America en paris – Paris. n1-33. 1891-mai 1892 – 1 – fr ACRPP [073]

America en tiempo de felipe 2 segun el cosmografo cronista juan lopez de velasco. el territorio espanol de ifni / Beltran y Rozpide, Ricardo; ed by Bayle, Constantino – Madrid: Razon y Fe, 1928 – 9 – sp Bibl Santa Ana [970]

America for all – 1932 aug 6-nov 5; 1932 sep 3, 17, oct 3, 8, 29, nov 5 – 1 – mf#3910342 – us WHS [071]

America herold / lincoln freie presse – Omaha NE (USA), 1972-1982 9 apr – 1 – gw Misc Inst [071]

America illustrada – Sao Paulo, SP: Typ Brasil de Carlos Gerke & Cia, fev 1898; jul-ago 1899 – (= ser Ps 19) – mf#P18,01,69 – bl Biblioteca [079]

America in the east : a glance at our history, prospects, problems, and duties in the pacific ocean / Griffis, William Elliot – New York: A S Barnes, 1899 [mf ed 1990] – 1mf – 9 – 0-7905-4799-6 – mf#1988-0799 – us ATLA [327]

America in the making / Abbott, Lyman – New Haven: Yale UP; London: Oxford UP, 1911 [mf ed 1989] – (= ser Yale lectures on the responsibilities of citizenship) – 1mf – 9 – 0-7905-4300-1 – mf#1988-0300 – us ATLA [320]

America indigena – v1-31. 1941-71 – 1 – us AMS Press [306]

La america indigena, tomo 1 : el hombre americano. los pueblos de america. barcelona, 1935 / Pericot y Garcia, Luis – Madrid: Razon y Fe, 1936 – 1 – sp Bibl Santa Ana [970]

America latina y su enrique jose varona / Entralgo, Elias Jose – Habana, Cuba. 1951 – 1r – 1 – us UF Libraries [972]

America latine : males de origem / Bomfim, Manoel Jose Do – Rio De Janeiro, Brazil. 1903 – 1r – 1 – us UF Libraries [972]

America libre – Miami, FL. 1967 jun 01-1975 nov 24 – 7r – 1 – (gaps) – us UF Libraries [071]

America of jose marti / Marti, Jose – New York, USA. 1953 – 1r – 1 – us UF Libraries [972]

America of to-morrow = Amerique de demain / Klein, Felix – Chicago: A C McClurg, 1911 [mf ed 1990] – 1mf – 9 – 0-7905-4988-3 – (english by e h wilkins. int note by charles r henderson) – mf#1988-0988 – us ATLA [917]

America or rome : christ or the pope / Brandt, John Lincoln – Toledo, OH: Loyal Pub Co, 1895 [mf ed 1992] – 2mf – 9 – 0-524-03757-4 – (incl bibl ref. int by w j h traynor & j g white) – mf#1990-1104 – us ATLA [241]

America or rome, which? / Christian, John Tyler – Louisville, KY: Baptist Book Concern, 1895 [mf ed 1993] – (= ser Baptist coll) – 1mf – 9 – 0-524-07674-X – mf#1991-3259 – us ATLA [241]

America picturesque and descriptive, vol 1 / Cook, Joel – Philadelphia: H T Coates. 3v. 1900 [mf ed 1980] – 7mf – 9 – 0-665-05241-3 – mf#05241 – cn Canadiana [917]

America picturesque and descriptive, vol 2 / Cook, Joel – Philadelphia: H T Coates. 3v. 1900 [mf ed 1980] – 7mf – 0-665-05242-1 – mf#05242 – cn Canadiana [917]

America picturesque and descriptive, vol 3 / Cook, Joel – Philadelphia: H T Coates. 3v. 1900 [mf ed 1980] – 8mf – 9 – 0-665-05243-X – (incl ind) – mf#05243 – cn Canadiana [917]

America picturesque and descriptive, vols 1-3 / Cook, Joel – Philadelphia: H T Coates. 3v.1900 – 1mf – 9 – 0-665-05240-5 – mf#05240 – cn Canadiana [917]

America through a st andrean's spectacles / Sloan, A D – [St Andrews]: St Andrews Citizen, 1909 [mf ed 1992] – 1mf – 9 – 0-524-05446-0 – mf#1990-1478 – us ATLA [917]

America today – 1977 mar-1979 feb – 1 – mf#498360 – us WHS [071]

America @ work – Silver Spring MD 1996+ – 1,5,9 – (cont: afl-cio afl-cio news) – ISSN: 1091-594X – mf#25994 – us UMI ProQuest [331]

America y hostos / Comision Pro Celebracion Del Centenario... – Habana, Cuba. 1939 – 1r – 1 – us UF Libraries [972]

America y la 'hilea amazonica' / Bustamente Yepez, Marco A – Guayaquil, Ecuador. 1948 – 1r – 1 – us UF Libraries [972]

America y otras paginas / Pagan, Bolivar – San Juan, Puerto Rico. 1922 – 1r – us UF Libraries [972]

Americae das fuenffte buch : vol schoener vnerhoerter historien – [Frankfurt am Main: s.n, 1595] [mf ed 1994] – (= ser America 5) – 9 – 0-665-94741-0 – mf#94741 – cn Canadiana [972]

America-herold – 1924 dec 4-1925 aug 20; 1925 aug 27-1926 may 20; 1926 may 27-27 feb 24; 1927 mar 5-dec 29 – 1 – mf#1131011 – us WHS [071]

America-herold – 1924 dec 4-1925 jul 30; 1925 aug 6-1926 apr 29; 1926 may 6-dec 30; 1927 jan 6-dec 29 – 1 – mf#1131019 – us WHS [071]

America-herold – 1926 dec 23 [v54 n37] – 1 – mf#1131010 – us WHS [071]

America-herold – 1967 sep 27-1969 jul 23; jul 30-1971 may 26; jun-1972 dec; 1973 jan-1974 jun; jul-1975 dec; 1976 jan-1977jun; jul-1978 dec; 1979 jan 3-mar 30 – 1 – mf#663738 – us WHS [071]

America-herold – Omaha, NE: Tribune Pub Co. 90. jahrg n14. 10 jun 1964-101v iss 51. 30 mar 1979 (wkly) [mf ed 1975-79 filmed in 1979-80] – 12v on 3r – 1 – (in german; cont: america-herold, lincoln freie presse, und heimatbote; some irregularities in numbering; sunday ed: sonntagspost (winona, mn), 10 juni 1964-24 mai 1970; merged with: sonntagspost (winona, mn), to form: america-herold und sonntagspost) – us NE Hist [071]

America-herold – Winona WI (USA), 1924 2 dec-1926, 1929-35, 1937-1939 29 nov [gaps] – 7r – 1 – (title varies: 1929: america-herold und lincoln freie presse) – gw Misc Inst [071]

America-herold und lincoln freie presse *see* America-herold

America-herold und sonntagspost – 1979 apr 6-dec 28; 1980-81; 1982 jan-may 28 – 1 – mf#663735 – us WHS [071]

America-herold und sonntagspost – Omaha, NE: Tribune Pub Co. 5v. v101 iss52. 6 apr 1979-v105 iss7. 28 mai 1982 (wkly) [mf ed 1980-82] – 3r – 1 – (in german. formed by the union of: america-herold (omaha ne), and sonntagspost (winona mn). merged with: buffalo volksfreund, and: california freie prese, and: cincinnati kurier, and: deutsche wochen schrift, and: milwaukee-herold (wkly), and: volkszeitung-tribuene, and: welt-post und der staats-anzeiger, to form: amerika woche) – us NE Hist [071]

America-latina – London, England 15 feb 1915-apr 1920 – 1 – (fr 1 sep 1916 onward set contains both london & paris edns) – uk British Libr Newspaper [072]

American – 1816 apr 26 – 1 – mf#858790 – us WHS [071]

American – 1845 jan 11 – 1 – mf#960393 – us WHS [071]

American – Cleveland, OH: F J & J F Svoboda, [1918-sep 17 1936]; 1937-jun 3 1939 – 1 – us CRL [071]

American – Antler, Bottineau Co, ND: A J Drake. v1 n1 may 27 1905-v1 n8 jul 15 1905 (wkly) – 1 – (cont by: antler american) – mf#03941 – us North Dakota [071]

American / Greene Co. Yellow Spring – jun 1953-apr 1954 [wkly] – 1r – 1 – mf#B4363 – us Ohio Hist [320]

American / Hamilton Co. Cincinnati – (feb 1830-may 1832) [wkly] – 1r – 1 – mf#B11014 – us Ohio Hist [071]

American – Manila, Philippines: Chofre & Co 1898 – 4v on 2r – 1 – us L of C Photodup [079]

American – v1 n25 [1898 nov 12] – 1 – mf#3230702 – us WHS [071]

The american – jan 10, 1893-oct 6, 1894 – 1 – us NY Public [073]

The american – 1832-1836. Incomplete – 1 – (name changed from new york american for the country) – us NY Public [073]

American: a national journal – La Salle IL 1880-1900 – 1 – mf#5204 – us UMI ProQuest [975]

American Academy in Rome *see* Papers and monographs

American Academy of Arts and Sciences. Boston *see* Memoirs of the american academy of arts and sciences

American academy of arts and sciences. bulletin – Cambridge MA 1977+ – 1,5,9 – ISSN: 0002-712X – mf#11447 – us UMI ProQuest [500]

American academy of arts and sciences memoirs – Microcard Editions. v1-4; ns: v1-3 – 142mf (20:1) – 9 – $720.00 – us UPA [060]

American Academy of Audiology *see* Journal of the american academy of audiology

American academy of business, cambridge *see* Journal of american academy of business, cambridge

American academy of dental science. transactions – Killen TX 1889-1900 – 1 – mf#5207 – us UMI ProQuest [617]

American academy of dermatology. journal – Oxford, England 1979+ – 1,5,9 – ISSN: 0190-9622 – mf#11875 – us UMI ProQuest [616]

American Academy of Medicine. Conference on Prevention of Infant Mortality. 1909, New Haven, CT *see* Papers and discussions

American Academy of Nurse Practitioners *see* Journal of the american academy of nurse practitioners

American Academy of Ophthalmology and Otolaryngology *see* Transactions american academy of ophthalmology and otolaryngology

American academy of orthopaedic surgeons bulletin – Rosemount IL 1966+ – 1,5,9 – ISSN: 1049-9741 – mf#2091 – us UMI ProQuest [617]

American Academy of Otolaryngology *see* Otolaryngology – head and neck surgery

American Academy of Physician Assistants *see* Jaapa

American Academy of Political and Social Science *see*
- Annals of the american academy of political and social science
- The initiative, referendum and recall.

American academy of psychiatry and the law. bulletin – Bloomfield CT 1972-96 – 1,5,9 – (cont by: journal of the american academy of psychiatry & the law) – ISSN: 0091-634X – mf#12354.01 – us UMI ProQuest [344]

American academy of psychoanalysis. journal – New York NY 1973+ – 1,5,9 – ISSN: 0090-3604 – mf#11046.01 – us UMI ProQuest [150]

American Academy of Religion *see* Newsletter of the afro-american religious history group of the american academy of religion

American academy of religion. journal – Oxford, England 1933+ – 1,5,9 – ISSN: 0002-7189 – mf#1480 – us UMI ProQuest [200]

American advance – v1 n3-v3 n26 [1911 apr 15-1913 jul 5] – 1 – mf#926184 – us WHS [071]

The american advocate of peace – v1-2. 1834-36 – (= ser The library of world peace studies) – 7mf – 9 – $105.00 – us UPA [320]

American aeronaut – 1974 nov6-1980; 1981-1989 jun – 1 – mf#554658 – us WHS [629]

American agent and broker – Maryland Heights MO 1980+ – 1,5,9 – ISSN: 0002-7200 – mf#11869,01 – us UMI ProQuest [368]

American agriculturist – Carol Stream IL 1976+ – 1,5,9 – (cont: american agriculturist, rural new yorker) – ISSN: 0161-8237 – mf#1985,02 – us UMI ProQuest [630]

American agriculturist and the rural new yorker – Carol Stream IL 1842-1975 – 1,5,9 – (cont by: american agriculturist, rural new yorker) – ISSN: 0002-7219 – mf#1985.02 – us UMI ProQuest [630]

The american agriculturist law book, a compendium of every day law, for farmers, mechanics, business men, manufacturers, etc.. / Corey, Henry Bascom – New York: American Agriculturist, 1885. 404,xxp. LL-1550 – 1 – us L of C Photodup [340]

American agriculturist, rural new yorker – Carol Stream IL 1975-76 – 1 – (cont: american agriculturist & the rural new yorker; cont by: american agriculturist) – ISSN: 0002-7219 – mf#1985.02 – us UMI ProQuest [630]

The american aid program in cambodia : a decade of co-operation, 1951-1961 / United States. Agency for International Development – Phnom Penh [1961?] [mf ed 1989] – [ill] 1r with other items – 1 – mf#mf-10289 SEA M reel 027/02 [§] – us CRL [337]

American aircraft modeler – Reno NV 1973-75 – 1,5 – ISSN: 0002-7227 – mf#7557 – us UMI ProQuest [790]

American almanac and repository of useful knowledge – La Salle IL 1830-61 – 1 – mf#3823 – us UMI ProQuest [030]

American almanac and treasury of facts, statistical, financial, and political – Killen TX 1878-89 – 1 – mf#2561 – us UMI ProQuest [317]

American Almanac Collection [Library of Congress] *see* Metropolitan catholic almanac and laity's directory for the year of our lord...

American alpine journal – Golden CO 1929+ – 1,5,9 – ISSN: 0065-6925 – mf#2379 – us UMI ProQuest [790]

The american amateur photographer – Brunswick, ME: [s.n. v1-18. 1889-1906] – (= ser The American Amateur Photographer And Camera And Dark-Room) – 6r – 1 – (issues for 1905-06 filmed with: american amateur photographer and camera & dark-room, jan-jun 1907) – us CRL [770]

The american amateur photographer and camera and dark-room – New York: American Photographic Pub Co, 1907 [v19 n1-6 (1907)] (mthly) – 1r – 1 – us CRL [770]

American ancestry : giving the name and descent in the male line – 12v. 1887-99 – 1 – us L of C Photodup [920]

The american and colonial gazette – 1888 – (= ser American newspapers and periodicals published in the uk and europe 18th-20th centuries) – 1r – 1 – £55.00 – uk World [072]

American and english annotated cases – 1901-18 – 31r – 1 – $1,150.00 – us Trans-Media [340]

American and english annotated cases – Northport, NY: Thompson Co. v1-21. 1906-11 (all publ) – 287mf – 9 – $430.00 – (title merged in 1912 with the american state reports, to become the american annotated cases) – mf#LLMC 84-695 – us LLMC [348]

American and english corporation cases – 48v. 1884-95 – 9 – $1248.00 – mf#0039 – us Brook [346]

American and english corporation cases – Northport, NY/Charlotteville: Thompson, Michie Co. 1st series: v1-48 + digest for vl-40. 1833-94. New series: v1-19. 1896-1904 (all publ) – 567mf – 9 – $850.00 – mf#LLMC 80-432 – us LLMC [346]

American and english decisions in equity – Philadelphia: M Murphy Co. v1-10 + index for v1-5. 1895-1904 (all publ) – 92mf – 9 – $138.00 – mf#LLMC 84-696 – us LLMC [342]

The american and english encyclopaedia of law – v. 1-31. 1887-96 – 1 – us L of C Photodup [340]

American and english encyclopedia of law and practice – 1st ed. Northport, NY: Ed Thompson. v1-31. 1877-96 (all publ) – 373mf – 9 – $559.00 – mf#LLMC 82-507 – us LLMC [340]

American and english encyclopedia of law and practice – 2nd ed. Northport, NY: Ed Thompson. v1-32 + suppl v1-5. 1896-1905 (all publ) – 508mf – 9 – $762.00 – mf#LLMC 82-508 – us LLMC [340]

American and english patent cases – Washington: C R Brodix. v1-20. 1662-1890 (all publ) – 143mf – 9 – $214.00 – (v1-3 cover english court reports and v4-20 cover reports of the us supreme court) – mf#LLMC 81-420 – us LLMC [346]

American and english railroad cases – Northport, NY/Charlottesville: Thompson/Michie 1st series: v1-61 + 4 index/digest vols; 1881-95 + new series: v1-68 + 6 index vols; 1894-1913 (all publ) – (= ser Railroad Reports) – 1205mf – 9 – $1807.00 – (new series: v24-68 1902-13 entitled: railroad reports) – mf#LLMC 80-433 – us LLMC [343]

American and Foreign Christian Union *see* The story of the madiai

American and foreign christian union. annual report – New York, 1850-60 [mf ed 2001] – (= ser Christianity's encounter with world religions, 1850-1950) – 1r – 1 – (merger of 3 societies) – mf#2001-s141 – us ATLA [240]

American and panamanian general claims arbitration / United States – Washington, DC. 1934 – 1r – us UF Libraries [972]

American animal hospital association. journal – Denver CO 1965+ – 1,5,9 – ISSN: 0587-2871 – mf#2765 – us UMI ProQuest [636]

American annals of education – La Salle IL 1826-39 – 1 – mf#3918 – us UMI ProQuest [370]

American annals of the deaf – Washington DC 1847+ – 1,5,9 – ISSN: 0002-726X – mf#2236 – us UMI ProQuest [616]

American annotated cases – New York, San Francisco: Thompson Co, Bancroft-Whitney. v1912A-1916B (19 bks). 1912-16 (all publ) – 285mf – 9 – $427.00 – (title changes to: annotated cases – american and english in mid-1916) – mf#LLMC 84-695B – us LLMC [340]

American annual register – Ann Arbor MI 1825-33 – 1 – mf#2772 – us UMI ProQuest [978]

American Anthropological Association *see* Memoir

American anthropologist – Berkeley CA 1888+ – 1,5,9 – ISSN: 0002-7294 – mf#1829 – us UMI ProQuest [301]

American anthropologist – v31 [1929] – 1 – mf#146545 – us WHS [071]

American antiquarian and oriental journal – 1894 jan-nov (v16] – 1 – mf#4327781 – us WHS [071]

American antiquarian and oriental journal – La Salle IL 1878-1914 – 1 – mf#3886 – us UMI ProQuest [073]

American Antiquarian Society *see*
- Archaeologia americana
- Proceedings of the american antiquarian society

American antiquarian society. almanac : aas newsletter – Worcester MA 1998+ – 1,5,9 – (cont: american antiquarian society. news-letter of the american antiquarian society) – ISSN: 1098-7878 – mf#10337,01 – us UMI ProQuest [000]

American antiquarian society. news-letter of the american antiquarian society – Worcester MA 1968-91 – 1 – ISSN: 0569-2229 – mf#10337.01 – us UMI ProQuest [000]

American antiquarian society. proceedings – Worcester MA 1880+ – 1,5,9 – ISSN: 0044-751X – mf#186 – us UMI ProQuest [366]

American antiquities and discoveries in the west : being an exhibition of the evidence that an ancient population of partially civilized nations differing entirely from those of the present indians peopled america many centuries before its discovery by columbus / Priest, Josiah [comp] – 2nd rev ed. [Albany, NY?: s.n.], 1833 [mf ed 1985] – 5mf – 9 – 0-665-49463-7 – mf#49463 – cn Canadiana [930]

American antiquities and discoveries in the west : being an exhibition of the evidence that an ancient population of partially civilized nations differing entirely from those of the present indians peopled america many centuries before its discovery by columbus... / Priest, Josiah [comp] – 5th ed. [Albany, NY?: s.n.], 1838 [mf ed 1983] – 5mf – 9 – mf#39382 – cn Canadiana [930]

American antiquities and discoveries in the west : being an exhibition of the evidence that an ancient population of partially civilized nations differing entirely from those of the present indians peopled america many centuries before its discovery by columbus... / Priest, Josiah [comp] – [Albany, NY?: s.n.], 1833 [mf ed 1983] – 5mf – 9 – mf#42311 – cn Canadiana [930]

American antiquity – Washington DC 1935+ – 1,5,9 – ISSN: 0002-7316 – mf#11734 – us UMI ProQuest [930]

American anti-slavery reporter / American Anti-Slavery Society – New York. n1-8. 1834 [all publ] – (= ser Black journals, series 1) – 2mf – 9 – $45.00 – us UPA [976]

American anti-slavery reporter – La Salle IL 1834 – 1 – mf#4147 – us UMI ProQuest [976]

American Anti-Slavery Society *see*
- American anti-slavery reporter
- Anti-slavery examiner
- Anti-slavery record
- Anti-slavery tracts

American anti-slavery society. annual report – New York. n1-28. 1834-60 [all publ] – (= ser Black journals, series 2) – 1r – 1 – $200.00 – (n8-21 never publ) – us UPA [976]

American apollo – 1801 dec 2 – 1 – mf#871383 – us WHS [071]

American apollo – La Salle IL 1792 – 1,5,9 – mf#3500 – us UMI ProQuest [520]

American appeal – v1-8 n2,48. 1920-27 [all publ] – (= ser Radical periodicals in the united states, 1881-1960. series 1) – 1r – 1 – $200.00 – us UPA [303]

American Appraisal Co *see* Clipboard

American archaeologist – Killen TX 1897-99 – 1 – mf#2560 – us UMI ProQuest [975]

American architect and architecture – La Salle IL 1876-1938 – 1 – mf#4636 – us UMI ProQuest [720]

American architect and building news – Boston, New York. v1-118. 1876-1920 – 2257mf – 9 – mf#0-1206 – ne IDC [720]

American architectural books : based on the henry-russell hitchcock bibliography of the same title and "a list of architectural books available in america before the revolution" by helen park – Woodbridge CT: Research Publ 1976 – 128r – 1 – us CRL [720]

American archives of rehabilitation therapy – Paramus NJ 1953-87 – 1,5,9 – ISSN: 0002-7324 – mf#2155 – us UMI ProQuest [617]

American archivist – Chicago IL 1987+ – 1,5,9 – ISSN: 0360-9081 – mf#16424 – us UMI ProQuest [025]

The american archivist / Society of American Archivists – v1-49. 1938-86 – 1 – $280.00 – us SAA [025]

American art – Chicago IL 1991+ – 1,5,9 – (cont: smithsonian studies in american art) – ISSN: 1073-9300 – mf#17039,01 – us UMI ProQuest [700]

American art in the barbizon mood / National Collection of Fine Arts. Smithsonian Institution; ed by Birmingham, Peter – 1976 – 2 color mf – 15 – $35.00f – 0-226-69413-5 – us Chicago U Pr [760]

American art journal – New York NY 1969+ – 1,5,9 – ISSN: 0002-7359 – mf#6721 – us UMI ProQuest [700]

American art pottery – 1978 jul; 1979 jan-1984 jul – 1 – mf#821823 – us WHS [071]

American art review – Killen TX 1879-81 – 1 – mf#3354 – us UMI ProQuest [700]

American art review *see* Ackermann's 'repository of arts'

American art union bulletin – 1847-53 [mf ed Chadwyck-Healey] – (= ser Rare 19th century american art journals) – 15mf – 9 – uk Chadwyck [700]

American artisan : the warm heating and sheet metal journal – Cleveland OH 1900-70 – 1 – mf#280 – us UMI ProQuest [690]

American artist – New York NY 1937+ – 1,5,9 – ISSN: 0002-7375 – mf#1425 – us UMI ProQuest [700]
American Assembly *see* United states and africa
American Association For Health, Physical Education, And... *see* Selected volleyball articles
American Association for Higher Education *see* College and university bulletin
American association for higher education. aahe bulletin – Washington DC 1979+ – 1,5,9 – (cont: college & university bulletin) – ISSN: 0162-7910 – mf#6980,01 – us UMI ProQuest [378]
American Association for Respiratory Care *see* AarcTimes
American Association for Respiratory Therapy *see* Aartimes
American Association for the Advancement of Science *see*
– Aaas science books
– Proceedings
American association for the advancement of science. bulletin – Washington DC 1942-74 – 1 – mf#8728 – us UMI ProQuest [500]
American association for the education of the severely/profoundly handicapped. aaesph review – Baltimore MD 1975-80 – 1,5,9 – ISSN: 0147-4375 – mf#12077.03 – us UMI ProQuest [370]
American association of bovine practitioners. conference / american association of bovine practitioners conference: [proceedings] – Rome GA 2002+ – 1,5,9 – mf#30698.03 – us UMI ProQuest [636]
American association of cereal chemists. journal – St Paul MN 1915-23 – 1 – ISSN: 0095-9847 – mf#5132 – us UMI ProQuest [660]
American Association of Colleges of Nursing *see* Journal of professional nursing
American association of colleges of pharmacy teachers' seminar. proceedings – Alexandria VA 1950-63 – 1 – mf#1657 – us UMI ProQuest [378]
American Association of Cost Engineers *see* Cost engineering
American association of cost engineers. aace bulletin – Morgantown WV 1958-79 – 1,5,9 – ISSN: 0001-0049 – mf#8768.01 – us UMI ProQuest [620]
American association of cost engineers. transactions – Morgantown WV 1975+ – 1,5,9 – (cont by: transactions of aace international) – ISSN: 1074-7397 – mf#10810.02 – us UMI ProQuest [620]
American association of equine practitioners proceedings – Lexington KY 1956-96 – 1,5,9 – ISSN: 0065-7182 – mf#5967 – us UMI ProQuest [636]
American Association of General Passenger and Ticket Agents *see* Official guide of the railways and steam navigation lines of the united states, porto rico, canada, mexico and cuba
American Association of Law Libraries *see* Law library package plan
American Association of Nurse Anesthetists *see*
– Aana journal
– Journal of the american association of nurse anesthetists
American Association of Occupational Health Nurses *see* Aaohn journal
American Association of Oral and Maxillofacial Surgeons *see* Journal of oral and maxillofacial surgery
American association of petroleum geologists. aapg bulletin – Tulsa OK 1917+ – 1,5,9 – ISSN: 0149-1423 – mf#903 – us UMI ProQuest [550]
American association of teacher educators in agriculture. journal – College Station TX 1978-88 – 1,5,9 – (cont by: journal of agricultural education) – ISSN: 0002-7480 – mf#11422.01 – us UMI ProQuest [630]
American association of teachers of slavic and east european languages. aatseel's newsletter – Berkeley CA 1973-74 – 1 – ISSN: 0001-0251 – mf#8715 – us UMI ProQuest [460]
American Association of University Professors *see* Academe
American association of university professors. aaup bulletin – Washington DC 1915-78 – 1,5,9 – (cont by: academe) – ISSN: 0001-026X – mf#974.01 – us UMI ProQuest [378]
American Association of University Women *see*
– Aauw journal
– Outlook [1989]
American association of university women madison branch bulletin – 1977 sep-1980 may – 1 – mf#914887 – us WHS [071]
American Association on Indian Affairs *see* News-letter
American Association on Indian Affairs et al *see* Indian affairs
American Astronautical Society *see*
– Complete aas microfiche series collection
– Out-of-print aas books on microfiche
The American Astronomical Society *see* Bulletin
American Astronomical Society *see* The astrophysical journal
American atheist – 1978-88 – 1 – mf#515940 – us WHS [210]
American athenaeum : a repository of belles lettres, science and the arts – La Salle IL 1825-26 – 1 – mf#4410 – us UMI ProQuest [410]
American athlete and cycle trade review – Philadelphia. v1-15 n13. mar 1887?-mar 1895 (freq varies) [all publ] – (= ser Sports periodicals, 1822-1922) – 6r – 1 – $1075.00 – us UPA [790]
American autobiographies : autobiographies cited in louis kaplan's bibliography of american autobiographies – 9 – $990.00 per series – (ser1a: 1676-1825 135 titles [0024]. ser1b: 1828-41 135 titles [0018]. ser1c: 1842-50 133 titles [0019]. ser2: 1851-1900 101 titles [0020]. ser3: 1851-1900 101 titles [0021]. ser4: 1851-1900 101 titles [0022]. ser5: 1851-1900 95 titles [0023]) – us Brook [920]
American Automobile Association *see*
– Northwestern tour book
– South central tour book
– Southwestern tour book
– Tour book
– Wisconsin motor news
American baha'i – 1970 jan-apr; oct-1974 feb; 1975 jul; 1976 may-jun, aug-1977 feb – 1 – mf#361933 – us WHS [290]
American baker – Minnetonka MN 1950-60 – 1 – mf#375 – us UMI ProQuest [660]
American balance – 1837 aug 19-1839 feb 21 – 1 – mf#1238893 – us WHS [071]
American Bankers Association *see* Banking
American bankers association. aba banking journal – New York NY 1980+ – 1,5,9 – (cont: banking) – ISSN: 0194-5947 – mf#6992,01 – us UMI ProQuest [332]
American bankruptcy law journal – Lexington SC 1972+ – 1,5,9 – ISSN: 0027-9048 – mf#6484 – us UMI ProQuest [346]
American bankruptcy law journal – v1-74. 1926-2000 – 9 – $885.00 set – (titles varies: v 1-39 (1926-65) as journal of the national association of referees in bankruptcy; v 40-44 (1966-70) as journal of the national conference of referees in bankruptcy) – ISSN: 0027-9048 – mf#101501 – us Hein [346]
American bankruptcy reports, annotated – Albany: M Bender. v1-49 + digest nos 1 + 2. 1899-1923 (all publ) – 560mf – 9 – $840.00 – (cont: national bankruptcy news and reports. title followed by a new series v1-8 87mf $130 llmc 82-405) – mf#LLMC 82-404 – us LLMC [346]
American baptist – 1970 apr-1973 may; jun-1974 dec; 1975 jan-1976 dec; 1977 jan-1979 dec; 1980-87 – 1 – mf#707223 – us WHS [071]
American baptist – Valley Forge PA 1910-92 – 1,5,9 – (cont by: american baptists in mission) – ISSN: 0002-757X – mf#1979.01 – us UMI ProQuest [242]
American baptist – New York. v7-28. 1850-72 [complete] – 9r – 1 – (cont: crusader and mission) – ISSN: 0002-757X – mf#ATLA R0103 – us ATLA [242]
American Baptist Churches in the USA *see* Forum
American baptist flag – Missouri. 1875-1937 – 1 – $1,436.19 – us Southern Baptist [242]
American Baptist Foreign Mission Societies *see* Missionary correspondence
The American Baptist Historical Society *see* Chronicle
American Baptist Home Mission Societies *see* Annual reports and directories
The American Baptist Missionary Union *see* Missionary jubilee
American Baptist Missionary Union *see* Annual report of the board of managers of the...
American Baptist Missionary Union in Japan. Conference of the Missionaries *see*
– Minutes of the conference of missionaries of the american baptist missionary union in japan
– Minutes of the...annual conference of missionaries of the american baptist missionary union in japan
The american baptist preaching of the seventeenth and eighteenth centuries : an address. delivered in boston before the american baptist historical society... / Bailey, Silas – Philadelphia: Press of the Society, 1858 – 1mf – 9 – 0-524-08252-9 – mf#1993-3007 – us ATLA [242]
American Baptist Publication Society *see* The baptist harp
American baptist quarterly – Valley Forge PA 1985+ – 1,5,9 – (cont: foundations) – ISSN: 0745-3698 – mf#15659 – us UMI ProQuest [242]
American baptist register for 1852 / Burrows, J Lansing – 1 – $19.11 – us Southern Baptist [242]
American Baptist Theological Seminary *see* Bulletins
American Baptist Theological Seminary. Nashville, Tennessee *see* Catalogs
American baptist yearbook – 1841-1940 – 1 – $646.46 – us Southern Baptist [242]
American baptists in mission – Valley Forge PA 1992+ – 1,5,9 – (cont: american baptist) – mf#1979,01 – us UMI ProQuest [242]
American Bar Association *see*
– Canons of professional ethics
– Report of the special committee, appointed to consider and report whether the present delay and uncertainty in judicial administration can be lessened.
American bar association annual reports – v1-85. 1878-1960 – 759mf – 9 – $1138.00 – (no additions poss for copyright reasons) – mf#LLMC 81-400 – us LLMC [340]
American bar association archive collection – Inception-1999 – 9 – $42,850.00 set – (inception-1985 $27,995.00 set. 1986-99 price varies per yr) – mf#402270 – us Hein [340]
American Bar Association. Committee on Canons of Professional Ethics *see* Report
American Bar Association. Committee on Unauthorized Practice of the Law *see* Compendium.
American bar association journal – Chicago IL 1915-83 – 1,5,9 – (cont by: aba journal) – ISSN: 0002-7596 – mf#1897.01 – us UMI ProQuest [340]
American bar association journal – v1-12. 1915-26 – 101mf – 9 – $151.00 – (updates planned) – mf#LLMC 90-316 – us LLMC [340]
American bar association journal *see* Aba journal
American bar association. reports of the annual meetings – v1-50. 1878-1925 – 1 – $918.00 – mf#0025 – us Brook [340]
American bar association. section of antitrust law proceedings *see* Antitrust law journal
American bar association. section of corporation, banking and business law proceedings – 1939-50 – 1 – $60.00 set – mf#100211 – us Hein [346]
American bar association. section of international and comparative law bulletin – v1-10. 1957-66 – 1 – $60.00 set – mf#100251 – us Hein [341]
American bar association. section of international and comparative law proceedings – 1942-65 – 9 – $150.00 set – mf#100261 – us Hein [341]
American bar association: section of real property, probate and trust law *see* Real property probate and trust journal (aba)
American bar association section of taxation *see* Tax lawyer (aba)
American bar association visit to england, scotland and ireland, 1924 : memorial volume – New York: ABA Committee on Publications, 1926 – 6mf – 9 – $9.00 – mf#LLMC 92-127 – us LLMC [340]
American bar foundation research journal – Chicago IL 1982-87 – 1,5,9 – (cont by: law & social inquiry) – ISSN: 0361-9486 – mf#13519.01 – us UMI ProQuest [340]
American bar foundation research journal *see* Law and social inquiry
American bar foundation research reporter – Chicago IL 1979-83 – 1,5,9 – mf#12104,01 – us UMI ProQuest [340]
American bar news – Chicago IL 1975-76 – 1,5,9 – ISSN: 0002-760X – mf#10723 – us UMI ProQuest [340]
American bee journal – Hamilton IL 1861+ – 1,5,9 – ISSN: 0002-7626 – mf#317 – us UMI ProQuest [630]
American beef producer – Englewood CO 1919-72 – 1,5 – ISSN: 0002-7634 – mf#1110 – us UMI ProQuest [636]
American behavioral scientist [abs] – Newbury Park CA 1957+ – 1,5,9 – ISSN: 0002-7642 – mf#1598 – us UMI ProQuest [300]
American benedictine review – Richardton ND 1950+ – 1,5,9 – ISSN: 0002-7650 – mf#2224 – us UMI ProQuest [241]
American bible society : annual report – 1817-1988 – 1 – (lacks some pp) – mf#atla s0670 – us ATLA [220]
The american bible society and the baptists : or, the question discussed, shall the whole word of god be given to the heathen? / Wyckoff, William Henry – 2nd ed. New York: John R Bigelow, 1842 [mf ed 1989] – 1mf – 9 – 0-7905-2578-X – mf#1987-2578 – us ATLA [242]
American bibliopolist – Killen TX 1869-77 – 1 – mf#3355 – us UMI ProQuest [070]
American biographical archive (aba). supplement : to series 1 and 2 = Amerikanisches biographisches archiv. supplement zu folge 1 und 2 / Baillie, Laureen [comp] – [mf ed 2001-02] – 119mf (1:24) in 2 installments – 9 – diazo €2030.00 (silver €2460 isbn: 978-3-598-33800-7) – ISBN-10: 3-598-33797-3 – ISBN-13: 978-3-598-33797-0 – (with printed ind) – gw Saur [975]
American biographical archive (aba1) = Amerikanisches biographisches archiv (aba1) / Gibbs, Nanette [comp] – [mf ed 1986-91] – 1842mf (1:24) – 9 – diazo €10,060.00 (silver €11,080 isbn: 978-3-598-30951-9) – ISBN-10: 3-598-30950-3 – ISBN-13: 978-3-598-30950-2 – (with printed ind) – gw Saur [975]
American biographical archive. series 2 (aba2) = Amerikanisches biographisches archiv. neue folge (aba) / Baillie, Laureen [comp] – [mf ed 1993-96] – 734mf (1:24) in 12 installments – 9 – diazo €10,060.00 (silver €11,080 isbn: 978-3-598-33534-1) – ISBN-10: 3-598-33520-2 – ISBN-13: 978-3-598-33520-4 – (with printed ind) – gw Saur [920]
American biographical archive to 2 (aba3) = Amerikanisches biographisches archiv bis 2 (aba) / Baillie, Laureen [comp] – [mf ed 2002-05] – 551mf (1:24) in 12 installments – 9 – diazo €10,060.00 (silver €11,080.00 isbn: 978-3-598-34811-2) – ISBN-10: 3-598-34810-X – ISBN-13: 978-3-598-34810-5 – (with printed ind) – gw Saur [920]
American biology teacher – Reston VA 1938+ – 1,5,9 – ISSN: 0002-7685 – mf#1535 – us UMI ProQuest [370]
American Birkebeiner Ski Foundation *see* Birch scroll
American black male – 1989 oct-nov; 1990 jan-feb; 1991 nov – 1 – mf#4851571 – us WHS [071]
American blue book 1905-06 *see* Anglo-american and continental courier 1903 – the american blue book 1905-06
The american board and american slavery : speech of theodore tilton, in plymouth church, brooklyn, january 25, 1860 / Tilton, Theodore – 3rd ed. New-York: John A Gray, 1860 – 1mf – 9 – 0-524-08691-5 – mf#1993-3216 – us ATLA [976]
American board in china, 1830-1950 : review and appraisal / Goodsell, Fred Field – Boston: United Church Board for World Ministries, 1969 – 1r – 1 – 0-8370-0583-3 – mf#1984-B325 – us ATLA [240]
The american board missions in the near east from the annual report of... / American Board of Commissioners for Foreign Missions – [Boston?: s.n] [annual] [mf 1921-26 filmed 2003] – 1r – 1 – (began in 1921? filmed with earlier titles. iss for 1921 and 1923-25 fr annual report of the american board of commissioners for foreign missions) – mf#2003-s049 – us ATLA [240]
The american board missions in turkey and the balkans from the annual report of... / American Board of Commissioners for Foreign Missions – [Boston?: s.n] 1915-18 [annual] [mf ed 2003] – 4v on 1r – 1 – (filmed with earlier and later titles: The american board missions in turkey from the annual report of...; and: The american board missions in the near east from the annual report of...) – mf#2003-s – us ATLA [240]
The american board missions in turkey from the annual report of... / American Board of Commissioners for Foreign Missions – [Boston?: s.n] -194 [annual] [mf 1912-14 filmed 2003] – 1r – 1 – (filmed with later titles: The american board missions in turkey and the balkans from the annual report of...; and: The american board missions in the near east from the annual report of...) – mf#2003-s047 – us ATLA [240]
American Board of Commissioners for Foreign Missions *see*
– The american board missions in the near east from the annual report of...
– The american board missions in turkey and the balkans from the annual report of...
– The american board missions in turkey from the annual report of...
– General report of the deputation sent by the american board to china in 1907
American board of commissioners for foreign missions. annual reports – v1-171. 1810-1982 – 9 – $1747.00 – mf#0026 – us Brook [327]
American Board of Commissioners for Foreign Missions. Foochow Mission *see* Report of the jubilee year of the foochow mission of the a b c f m, 1896
American Board of Commissioners for Foreign Missions. North China Mission *see* An eventful year in north china
American board of commissioners for foreign missions. yearbook – Boston, 1917-57 [mf ed 2] – (= ser Christianity's encounter with world religions, 1850-1950) – 5r – 1 – (with: woman's board of missions 1917-26) – mf#2-s183-186 – us ATLA [240]
American book collector – 1950 sep-1976 jul/aug – 1 – mf#146586 – us WHS [071]
American book publishing record – New York NY 1960+ – 1,5,9 – ISSN: 0002-7707 – mf#3437 – us UMI ProQuest [070]
American Breweriana Association *see* Journal of the american breweriana association
American Bridge Association *see* Mashariki
American buddhist – 1957 apr 1-1974 mar – 1 – mf#400510 – us WHS [071]
American builder – New York NY 1905-69 – 1 – mf#288 – us UMI ProQuest [070]
American bulletin : the white man's viewpoint – New York. v. 1-2, no. 30. 28 Mar 1935-3 Nov 1936 – 1 – us NY Public [073]

American Bureau of Industrial Research *see* Manuscript collections on the early american labor movement, 1862-1908

American business – New York NY 1933-60 – 1 – mf#447 – us UMI ProQuest [338]

American business communication association. abca bulletin – 1973-84 – 1,5,9 – (cont by: bulletin of the association for business communication) – ISSN: 0001-0383 – mf#8090.02 – us UMI ProQuest [650]

American Business Consultants, Inc *see* Counterattack

American business law : with legal forms / Sullivan, John J – New York/London: D Appleton, 1909 – 5mf – 9 – $7.50 – mf#LLMC 92-170 – us LLMC [346]

American business law journal – Oxford, England 1963+ – 1,5,9 – ISSN: 0002-7766 – mf#2428 – us UMI ProQuest [346]

American canals : bulletin of the american canal society – 1972 mar-1982 nov – 1 – mf#653715 – us WHS [071]

American capsule news – n310-484 [1962 feb 10-1965 jul 10] – 1 – mf#1051904 – us WHS [071]

The american caravan: a yearbook of american literature – New York. 1927-1936 – 1 – us NY Public [800]

American cardiology – Lausanne, Switzerland 1958+ – 1,5,9 – ISSN: 0002-9149 – mf#1973 – us UMI ProQuest [616]

American Carpatho-Russian Orthodox Greek Catholic Diocese in USA *see* Church messenger

American Carpatho-Russian Youth *see* Kalendar

American carpatho-russian youth annual – Ligonier, Pittsburgh, PA: American Carpatho-Russian Youth. [1954, 1956] – 1 – us CRL [305]

American cartographer – Gaithersburg MD 1974-89 – 1,5,9 – (cont by: cartography & geographic information systems) – ISSN: 0094-1689 – mf#12484.02 – us UMI ProQuest [520]

American cases on contract. / Huffcut, Ernest Wilson – 2d ed. Albany, N.Y.: Banks, 1900. 898p. LL-644 – 1 – us L of C Photodup [346]

The american catalogue – 21v. 1880-1911 – 1,9 – us AMS Press [010]

The american catalogue of books / Kelly, James – 2v. 1866-71 – 1,9 – us AMS Press [010]

American catholic historical researches – La Salle IL 1884-1912 – 1 – mf#5210 – us UMI ProQuest [929]

American Catholic Philosophical Association *see* Proceedings of the american catholic philosophical association

American catholic philosophical quarterly – Bronx NY 1990+ – 1,5,9 – (cont: new scholasticism) – ISSN: 1051-3558 – mf#430,01 – us UMI ProQuest [100]

American catholic quarterly review – La Salle IL 1876-1924 – 1 – mf#5211 – us UMI ProQuest [241]

The american catholic quarterly review, and "the faith of our forefathers" : the case as it stands / Stearns, Edward Josiah – New York: Thomas Whittaker, 1880 – 1mf – 9 – 0-8370-8068-1 – mf#1986-2068 – us ATLA [241]

American cause – 1975 mar-1981 nov/dec – 1 – mf#625724 – us WHS [071]

American (central point, or) – Central Point OR: E C Galt, -1936 [wkly] – 1 – (cont by: central point american (central point, or)) – us Oregon Lib [071]

American ceramic society bulletin – 1933+ – 1,5,9 – ISSN: 0002-7812 – mf#1543 – us UMI ProQuest [660]

American ceramic society. journal – Washington DC 1918+ – 1,5,9 – ISSN: 0002-7820 – mf#1542 – us UMI ProQuest [660]

American challenger – v4 n3-v10 n3 [1968 mar-1974 apr] – 1 – mf#1051906 – us WHS [071]

The american chamber of commerce journal/ the chamber – 35v. illus. -m. Ceased publ. with v35, no.10 in Oct 1959? Suspended publ. 194?-Dec 1945 – 1 – us Wisconsin U Libr [380]

American Chamber of Commerce. Liverpool *see* Minute books of the american chamber of commerce in liverpool, 1801-1908

American Chamber of Commerce (Liverpool, England) *see* Minutes

American Chemical Society. Division of Fuel Chemistry *see* Preprints of papers

American Chemical Society. Division of Petroleum Chemistry *see* Preprints of papers

American chemical society. journal – Washington DC 1879+ – 1,5,9 – (killen tx [1879-1906] #5801) – ISSN: 0002-7863 – mf#50013 – us UMI ProQuest [540]

American child / child labor bulletin / National Child Labor Committee – v1-37 1919-55 and: child labor bulletin v1-7 1912-19 – (= ser Social welfare periodicals) – 86mf – 9 – $500.00 – (forerunner: child labor bulletin) – us UPA [331]

American childhood – Stamford CT 1916-58 – 1 – ISSN: 0731-1559 – mf#1116 – us UMI ProQuest [370]

American Chiropractic Association *see* Journal of the american chiropractic association

American chiropractic association. aca journal of chiropractic – Arlington VA 1975-81 – 1,5,9 – (cont by: journal of chiropractic) – ISSN: 0044-7609 – mf#10579.02 – us UMI ProQuest [617]

American choices : a report of the institute for independent education, inc – 1985 dec; 1986 jun, oct; 1987 aug; 1988 dec; 1989 dec; 1990 oct; 1991 jun, oct – 1 – mf#5132064 – us WHS [071]

American Choral Directors Association *see* The choral journal

American christian expositor – v1. 1831-32 – 1 – $50.00 – us Presbyterian [978]

The american christian record : containing the history, confession of faith, and statistics of each religious denomination in the united states and europe, a list of all clergymen with their post office address, etc., etc., etc – New York: WRC Clark & Meeker, 1860 – 2mf – 9 – 0-524-07548-4 – mf#1991-3168 – us ATLA [240]

American christian rulers : or, religion and men of government... / Giddings, Edward Jonathan – New York: Bromfield, c1890 [mf ed 1990] – 2mf – 9 – 0-7905-8007-1 – mf#1988-8007 – us ATLA [975]

The american church dictionary and cyclopedia / Miller, William James – 2nd ed. New York: Thomas Whittaker, c1901 – 1mf – 9 – 0-8370-9160-8 – (incl ind) – mf#1986-3160 – us ATLA [052]

American church institute for negroes. annual report / report negro education in wartime – 1906-42 [mf ed 2] – (= ser Christianity's encounter with world religions, 1850-1950) – 3r – 1 – mf#2-s086-088 – us ATLA [242]

American church monthly – Killen TX 1857-58 – 1 – mf#2773 – us UMI ProQuest [240]

The american church monthly – v1-42. 1917-37 (complete) – Inquire – mf#ATLA 1994-S535 – us ATLA [240]

American church news – 1968 jan-1975 jul; 1975 n7/8-1977 jan – 1 – mf#167918 – us WHS [071]

American church news – Berkeley CA 1955-76 – 1,5,9 – (cont by: new oxford review) – ISSN: 0002-791X – mf#2223.01 – us UMI ProQuest [240]

American Church Union *see* New oxford review

American churchman – v6 n8 [1888 may]; v11 n3-1910 [1893 apr-dec]; v18 n4-v19 n9 [1900 jan-1901 jun] – 1 – mf#868112 – us WHS [071]

American cinematographer – Hollywood CA 1921+ – 1,5,9 – ISSN: 0002-7928 – mf#2115 – us UMI ProQuest [790]

American citizen – Topeka, KS: v1 n1. feb 23 1888-1909? [mf ed 1947] – (= ser Negro Newspapers on Microfilm) – 1r – 1 – us L of C Photodup [071]

The american citizen – Des Moines IA, 1923-72 – 25r – 1 – (italian newspaper) – us IHRC [071]

The american citizen – New York. N.Y. 1912-14 – 1 – us AJPC [071]

The american citizen : official organ of the "order sons of italy in america" – Omaha, NE: American Citizen Pub Co, 1923-dec 1985// [mf ed 1938-85 (gaps)] – 11r – 1 – (earlier issues chiefly in italian with some english; later issues chiefly in english with some italian) – us NE Hist [071]

American city – Shawnee Mission KS 1909-75 – 1,5,9 – (cont by: american city & county) – ISSN: 0002-7936 – mf#12.01 – us UMI ProQuest [710]

American city & county – Shawnee Mission KS 1975+ – 1,5,9 – (cont: american city) – ISSN: 0149-337X – mf#12,01 – us UMI ProQuest [350]

American civil church law / Zollmann, Carl – New York: Columbia University, 1917 [mf ed 1993] – (= ser Studies in history, economics and public law 78; Roman catholic coll) – 2mf – 9 – 0-524-07601-4 – (incl bibl ref) – mf#1991-3221 – us ATLA [346]

American Civil Defense Association et al *see* Journal of civil defense

American civil law journal – New York. v1. 1873 – 1 – $45.00 – (all publ) – mf#100381 – us Hein [347]

American Civil Liberties Union *see*
- Aclyou in action
- Bulletin of the american civil...
- Civil liberties
- Civil liberties quarterly
- News release
- Open forum
- Weekly bulletin
- Weekly news bulletin

American civil liberties union archives (aclu) – 1912-50 [1995] – 293r – 1 – $38,090.00 set – (1912-27 (r1-50); 1927-33 (r51-100); 1933-37 (r101-150); 1937-41 (r151-200); 1941-48 (r201-250); 1948-50 (r251-293). with guide d3306.g ($75 if purchased without entire coll)) – mf#D3306 – princeton university, the new york public library, and the aclu – us Scholarly Res [322]

American civil liberties union archives (aclu), series 2 : project files – 1950-90 [mf ed 2002] – 50r – 1 – (guide must be purchased separately $65) – mf3521.p02 – us Scholarly Res [322]

American civil liberties union archives (aclu), series 4 : legal case files, 1933-1990 [bulk dates 1960-1984] – 1950-90 [mf ed 2002] – 618r – 1 – (guide must be purchased separately $75 s3521.g) – mf3521.p04 – princeton university and the aclu – us Scholarly Res [322]

American civil liberties union archives (aclu), series 4 : subject files – 1950-90 [mf ed 2002] – 358r – 1 – $46,540.00 set – (divided into 5 categories: freedom of belief, expression, & association: loyalty & security 1939-81 bulk 1947-68 [38r]; academic freedom 1947-85 bulk 1947-73 [20r]; church & state 1947-86 [18r]; access to government information 1951-77 [1r]; right to license 1954-72 bulk 1954-1961 [1r]; labor & business 1937-78 [9r]; censorship 1939-89 bulk 1947-73 [28r]; military rights 1946-83 [8r]; assembly & public protest 1949-84 [3r]; deprogramming 1975-77 [2r]; freedom of movement 1942-78 bulk 1947-64 [15r]; environment & civil liberties 1970-78 [1r]; mass communications 1945-88 bulk 1948-68 [30r]; miscellaneous 1950-79 [15r]. due process of law: military justice 1947-73 bulk 1960-72 [4r]; government due process 1947-73 bulk 1957-69 [22r]; government legislation 1938-64 [22r]; police practices 1950-82 [9r]; court proceedings 1949-76 bulk 1958-69 [17r]: right to privacy 1939-88 [4r]; wiretapping & surveillance 1942-80 [2r]; prisoner's rights 1955-85 [3r]; japanese-american internment 1942-55 [1r]; children's rights 1953-87 [2r]. miscellaneous 1955-72 [3r]; mental health issues 1941-78 [7r]. equality before the law: women's rights 1953-84 [5r]; civil rights 1943-79 bulk 1958-70 [17r]; poverty & civil liberties 1960-79 [2r]; miscellaneous 1956-75 [1r]; native americans 1947-77 [6r]; voting rights 1941-75 [4r]; lesbian & gay rights 1953-87 [1r]. international civil liberties 1942-82 [34r]. miscellaneous 1921-80 [4r]. guide must be purchased separately $65 s3521.g) – mf3521.p03 – princeton university & the aclu – us Scholarly Res [322]

American coin-op – Detroit MI 1974+ – 1,5,9 – (cont: coin-op) – ISSN: 0092-2811 – mf#1675,01 – us UMI ProQuest [660]

American collector – 1975 nov-1977 nov; dec-1979; 1980-1982 mar; apr-1983 oct – 1 – mf#342977 – us WHS [071]

American college / Sharpless, Isaac – Garden City, NY: Doubleday, Page, 1915 [mf ed 1990] – 1mf – 9 – 0-7905-6435-1 – (incl bibl ref) – mf#1988-2435 – us ATLA [378]

American college athletics / Savage, Howard J – 1929 – 8mf – 9 – $24.00 – us Kinesology [790]

American College Health Association *see*
- Acha action
- Journal of american college health
- Journal of the american college health association

American College of Chest Physicians *see* Bulletin of the american college of chest physicians

American College of Nutrition *see* Journal of the american college of nutrition

American College of Surgeons *see* Journal of the american college of surgeons

American college of surgeons. bulletin – Chicago IL 1979+ – 1,5,9 – ISSN: 0002-8045 – mf#12391 – us UMI ProQuest [617]

American college of surgeons. surgical forum – Chicago IL 1979+ – 1,5,9 – ISSN: 0071-8041 – mf#12442 – us UMI ProQuest [617]

American Colonization Society *see* Records

The american colonization society – 323r – 1 – $11,305.00 – (with guide) – Dist. us Scholarly Res – us L of C Photodup [976]

American colonization society. annual report – Washington DC. 1st-91/93rd. 1818-1908/1910 [all publ] – (= ser Black journals, series 2) – 3r – 1 – $585.00 – us UPA [975]

American commercial beacon and norfolk and portsmouth daily advertiser – 1818 jun 17, dec 12 – 1 – mf#887679 – us WHS [071]

American Committee for Cape Verde, Inc *see* Tchuba newsletter

American Committee for Solidarity with the Vietnamese People *see* Solidarity!

American Committee for the Fourth International Workers League [US] *see* Bulletin of international socialism

American committee for the protection of the foreign born bulletin *see* Civil liberties publications

American Committee On Dependent Territories *see* Informe elevado al consejo de la...

American Committee on East-West Accord *see* East-west outlook

American Committee on the History of the Second World War *see* Newsletter

The american commonwealth / Bryce, James Bryce, Viscount – 2nd rev ed. London: MacMillan & Co. 2v. 1891 – 16mf – 9 – $24.00 – mf#LLMC 95-063 – us LLMC [323]

The american commonwealth / Bryce, James Bryce, Viscount – 3rd ed. New York: The Macmillan Co, 1908 – 18mf – 9 – $27.00 – mf#LLMC 92-168 – us LLMC [323]

American Communications Association *see*
- Local 40 news
- Proceedings of the...national convention

American Communist Workers Movement [Marxist-Leninist] *see*
- People's america daily news
- Workers' advocate

American communities and co-operative colonies / Hinds, William Alfred – 2nd rev Chicago: CH Kerr, 1908 [mf ed 1990] – 1mf – 9 – 0-7905-7002-5 – (incl bibl ref) – mf#1988-3002 – us ATLA [978]

American Concrete Institute *see* Journal of the american concrete institute

American conference of academic deans. proceedings – Washington DC 1945-78 – 1,5,9 – ISSN: 0065-7905 – mf#5754 – us UMI ProQuest [378]

American Conference of Therapeutic Selfhelp/ Selfhealth/Social Action Club *see* Constructive action newsletter

The american conflict : an address spoken... 22nd december, 1864 / Cordner, John – Montreal?: J Lovell, 1865 – 1mf – 9 – mf#33339 – cn Canadiana [976]

American congregational association library. bulletin of the congregational library – Boston MA 1991+ – 1,5,9 – ISSN: 0010-5821 – mf#15161,01 – us UMI ProQuest [242]

American congregational year-book for the year 1857 – New York: Calkins & Stiles, 1857 [mf ed 1993] – (= ser Congregational coll) – 1mf – 9 – 0-524-06746-5 – mf#1990-5272 – us ATLA [242]

American congregationalism in the 19th century and entering the 20th / Willey, Samuel Hopkins – San Francisco: George Spaulding, [1902?] [mf ed 1991] – 1mf – 9 – 0-524-01639-9 – mf#1990-4103 – us ATLA [242]

American Congress of Churches (1885: Hartford, CT) *see* Proceedings of the hartford meeting, 1885

American Congress of Liberal Religious Societies *see* New unity

American congress on surveying and mapping. bulletin – Gaithersburg MD 1941-44 – 1,5,9 – (cont by: surveying & mapping) – mf#12487.03 – us UMI ProQuest [624]

The american constitutional system : an introduction to the study of the american state / Willoughby, Westel Woodbury – New York: The Century Co, 1919 – 4mf – 9 – $6.00 – mf#LLMC 95-080 – us LLMC [323]

American constitutions : constituting the constitutions of each state in the union of the united states / Hough, Franklin B – Albany: Weed-Parsons. 2v. 1871 – 6mf – 9 – $9.00 – mf#LLMC 84-257 – us LLMC [323]

An american continental commercial union or alliance / Douglas, Stephen Arnold; ed by Cutts, J Madison – Washington: s.n, 1889 – 1mf – 9 – (pref by ed) – mf#42815 – cn Canadiana [380]

The american contractor – Chicago: B. Edwards [v16,19-20,22-39 (1895; 1898-99; 1901-10)] (wkly) – 39r – 1 – us CRL [690]

The american conveyancer; containing a large variety of legal forms and instruments, adapted to popular wants and professional use throughout the united states. / Curtis, George Ticknor – New ed. Boston, Little, Brown, 1847. 283 p. LL-1089 – 1 – us L of C Photodup [340]

American corporation cases – Chicago: E B Myers. 10v + digest. 1872-88 – 1 – $180.00 – mf#0028 – us Brook [346]

American corporation cases – Chicago, IL: E B Myers Co. v1-10 + digest. 1868-87 (all publ) – 88mf – 9 – $132.00 – mf#LLMC 80-431 – us LLMC [346]

American correctional association : proceedings of annual congresses – 1874-1964 – 379mf – 9 – $568.00 – (indexes for 1870, 1874, 1876 and 1884-1934. no meetings held in 1871-73, 1875 or 1877-82. name of organization varies: 1871-1907 – the national prison association. 1908-54 – the american prison association. lacking: congress of 1870. report 1918. also missing p63-64, 137-138 1897. p323-338 1920) – mf#LLMC 84-267 – us LLMC [340]

American corrective therapy journal – San Diego CA 1947-87 – 1,5,9 – (cont by: clinical kinesiology) – ISSN: 0002-8088 – mf#3189.01 – us UMI ProQuest [615]

American correspondance in the palmerston papers, 1835-41 and 1846-50 : from the british library, add. mss. 48495 and 48575 – (= ser British records relating to america in microform) – 1r – 1 – (int by ged martin) – mf#96663 – uk Microform Academic [975]
American correspondence of james bryce, 1871-1922 – Bodleian Library [Bodley Ms] – (= ser BRRAM series) – 7r – 1 – £469 / $938 – (with int by d s porter) – mf#r96098 – uk Microform Academic [920]
American correspondence of the royal society of arts, 1755-1840 – Royal Society of Arts – (= ser BRRAM series) – 2r – 1 – £134 / $268 – (int by d g c allan) – mf#r04808 – uk Microform Academic [700]
American Council Against Nazi Propaganda *see* Hour
American Council of Christian Churches *see* Christian accent
American Council of Learned Societies. (Committee on Far Eastern Studies) *see* The bulletin of far eastern bibliography
American council on consumer interests. newsletter – Columbia MO 1973-95 – 1,5,9 – (cont by: consumer news & reviews) – ISSN: 0010-9975 – mf#9141.01 – us UMI ProQuest [380]
American Council On Education Comittee On Religion And Education *see* Relation of religion to public education
American council on industrial arts teacher education. yearbook – Ypsilanti MI 1952-86 – 1,5,9 – (cont by: yearbook council on technology teacher education (us)) – ISSN: 0084-6333 – mf#10486.01 – us UMI ProQuest [378]
American courier – 1939 apr 13 – 1 – mf#1165558 – us WHS [071]
American craft – New York NY 1980+ – 1,5,9 – (cont: craft horizons with craft world) – ISSN: 0194-8008 – mf#773,01 – us UMI ProQuest [740]
American crafts council. acc outlook – New York NY 1975-76 – 1 – ISSN: 0002-810X – mf#9966 – us UMI ProQuest [740]
American criminal law quaterly *see* American criminal law review
American criminal law review – v1-38. 1962-2 – 1,5,6 – $801.00 – (v 1-31 (1962-94) $540r. v32-35 (1994-98) $97mf; title varies: v 1-9 (1962-70) as american criminal law quaterly, publ jointly with aba criminal justice section up to jan 1 1986) – ISSN: 0164-0364 – mf#102301 – us Hein [345]
American criminal law review – Washington DC 1962+ – 1,5,9 – ISSN: 0164-0364 – mf#6716 – us UMI ProQuest [345]
American criminal reports – Callaghan & Co. v1-15. 1878-1909 – 126mf – 9 – $189.00 – (with ind) – mf#LLMC 84-263 – us LLMC [345]
American critic and general review – La Salle IL 1820 – 1 – mf#4411 – us UMI ProQuest [420]
American cultural history, 1607-1829 / Knapp, Samuel Lorenzo – 9 – us Scholars Facs [420]
American dairy review – Troy MI 1939-81 – 1,5,9 – ISSN: 0002-8169 – mf#2471 – us UMI ProQuest [630]
American decisions – San Francisco: Bancroft-Whitney. v1-100. 1760-1869 (all publ) – (= ser Trinity series 1) – 938mf – 9 – $1407.00 – mf#LLMC 78-032 – us LLMC [340]
American decorative art – (= ser Christie's pictorial archive new york) – 45mf – 9 – $400.00 – 0-907006-62-0 – (over 2600 illustrations) – uk Mindata [740]
American defender – v2 n1-2 [1987 jan-feb] – 1 – mf#1497179 – us WHS [071]
American defense – n1-1942 [1982 mar-1987 feb] – 1 – mf#1533291 – us WHS [071]
American Defense Preparedness Association *see*
– Common defense
– Technical bulletin
American democracy vs the spanish hierarchy / North American Committee to Aid Spanish democracy – [New York 1937?] – 9 – mf#w716 – us Harvard [230]
American democrat – Carlisle, PA. -w 1851-62 – 3 rolls – 13 – $25.00r – us IMR [071]
American demographics – Overland Park KS 1979+ – 1,5,9 – ISSN: 0163-4089 – mf#13396 – us UMI ProQuest [304]
American Dental Association *see* Journal of oral surgery
American Dental Association. Division of Educational Measurements *see* Annual report: dental auxiliary education
American dental association. journal – Chicago IL 1913+ – 1,5,9 – ISSN: 0002-8177 – mf#602 – us UMI ProQuest [617]
American dental association news – Chicago IL 1970+ – 1,5,9 – ISSN: 0895-2930 – mf#7636 – us UMI ProQuest [617]
American Dental Hygienists' Association *see* Journal of the american dental hygienists' association
American Deserters Committee *see*
– Minutes of the center steering committee meeting
– Newsletter
– Paper grenade newsletter of the american deserters
– Report
– Second front
American dietetic association. journal – Oxford, England 1925+ – 1,5,9 – ISSN: 0002-8223 – mf#11400 – us UMI ProQuest [613]
American digest system : century edition – St Paul: West Publ Co. v1-50. 1658-1896 (all publ) – 753mf – 9 – $1129.00 – mf#LLMC 79-406 – us LLMC [348]
American digest system – 1658-1966 – 9 – $7425.00 set – (decennial eds also available) – mf#402170 – us Hein [348]
American digest system : first decennial edition – St Paul: West Publ Co. v1-25. 1897-1906 (all publ) – 548mf – 9 – $822.00 – mf#LLMC 84-388 – us LLMC [348]
The american directory and who's who in europe – 1922-25 – (= ser American newspapers and periodicals published in the uk and europe 18th-20th centuries) – 1r – 1 – £55.00 – uk World [975]
American documents in the murraythwaite collection – Edinburgh (Scotland): Scottish Record Office [mf ed 1987] – 1r – 1 – mf#45-349 – us South Carolina Historical [380]
The american draught player : or, the theory and practice of the scientific game of checkers: simplified and illustrated with practical diagrams containing upwards of seventeen hundred games and positions / Spayth, Henry – 5th rev corr ed. New York: Dick & Fitzgerald, 1869, c1860 – us CRL [790]
American druggist – New York NY 1871-1999 – 1,5,9 – ISSN: 0190-5279 – mf#3123 – us UMI ProQuest [615]
American drycleaner – Detroit MI 1992+ – 1 – ISSN: 0002-8258 – mf#9893 – us UMI ProQuest [660]
American dyestuff reporter – New York NY 1917-99 – 1,5,9 – ISSN: 0002-8266 – mf#781 – us UMI ProQuest [660]
American eagle – 1824 nov 22 – 1 – mf#873077 – us WHS [071]
American eagle – Estero, FL. 1965 jun-1981 – 7r – us UF Libraries [071]
American eagle – La Salle IL 1847 – 1 – mf#4412 – us UMI ProQuest [071]
American eagle – Westfield, NY. v2 n23. may 7 1833 – 1r – 1 – (single iss of this anti-masonic newspaper; other titles: eagle; chautauque phenix) – mf#09 C4.2 007 – us Western Res [071]
The american eagle : official organ of the knights of pythias and order of calanthe of missouri – St Louis: R A Hudlin, 1894-1907? [mf ed 1947] – (= ser Negro Newspapers on Microfilm) – 1r – 1 – us L of C Photodup [071]
American eagle 1885-86 *see* America 1883 – the american visitor 1884 – the american eagle 1885-86 – american humorist and storyteller 1888
American eaglet of the north carolina american party – 1973 jan-1977 feb; 1978 apr-1980 may – 1 – mf#342978 – us WHS [071]
American ecclesiastical law : the law of religious societies, church government and creeds, disturbing religious meetings, and the law of burial grounds in the united states / Tyler, Ransom Hebbard – Albany: W Gould, 1866, c1865 [mf ed 1990] – 2mf – 9 – 0-7905-8163-9 – mf#1988-6110 – us ATLA [346]
American ecclesiastical review – Washington DC 1889-1975 – 1,5,9 – ISSN: 0002-8274 – mf#1929 – us UMI ProQuest [240]
The american ecclesiastical year-book : containing, 1. the present religious statistics of the world. 2. a brief religious history of all denominations in all countries during the past year / Schem, Alexander Jacob – New-York: H Dayton, 1860 – 1mf – 9 – 0-7905-8149-3 – mf#1988-6096 – us ATLA [240]
American eclectic – La Salle IL 1841-42 – 1 – mf#3919 – us UMI ProQuest [370]
American economic association quarterly – La Salle IL 1886-1910 – 1 – mf#5212 – us UMI ProQuest [330]
American Economic Development Council *see* Aedc journal
American economic review – Nashville TN 1911+ – 1,5,9 – ISSN: 0002-8282 – mf#533 – us UMI ProQuest [330]
American economist – New York NY 1957+ – 1,5,9 – ISSN: 0569-4345 – mf#5337 – us UMI ProQuest [330]
American editorial review – (West Newton, Mass.). n1 (May 1989)-n90 (29 Apr 1994) – 45ft – (missing: n22; n67) – us AJPC [270]
American education – Washington DC 1964-85 – 1,5,9 – ISSN: 0002-8304 – mf#1696 – us UMI ProQuest [370]
American education society. quarterly journal – La Salle IL 1827-46 – 1 – mf#4415 – us UMI ProQuest [370]
American educational research journal – Washington DC 1964+ – 1,5,9 – ISSN: 0002-8312 – mf#1801 – us UMI ProQuest [370]
American educationist and western school journal – Killen TX 1852 – 1 – mf#3480 – us UMI ProQuest [370]
American educator – Washington DC 1977+ – 1,5,9 – ISSN: 0148-432X – mf#11615 – us UMI ProQuest [370]
American electrical cases, annotated – Albany: M Bender: v1-9. 1873-1908 (all publ) – 94mf – 9 – $141.00 – (v7 contains an index/ digest for v1-7) – mf#LLMC 82-400 – us LLMC [340]
American emperor / Brown, Rose (Johnston) – New York, NY. 1945 – 1r – us UF Libraries [972]
American enterprise – Washington DC 1990+ – 1,5,9 – ISSN: 1047-3572 – mf#17755 – us UMI ProQuest [071]
American entomologist – Lanham MD 1990+ – 1,5,9 – (cont: bulletin of the entomological society of america) – ISSN: 1046-2821 – mf#9056,01 – us UMI ProQuest [590]
The american episcopal church in china / Richmond, Annette B – New York: Domestic and Foreign missionary Society of the Protestant Episcopal Church in the USA, 1907 [mf ed 1995] – (= ser Yale coll) – xi/170p – 1 – 0-524-09258-3 – mf#1995-0258 – us ATLA [242]
American ethnic – v6 n1/5, n6/12; v7 n1/3 [1978 spring, fall/winter; 1979 spring) – 1 – mf#635815 – us WHS [071]
American ethnologist – Berkeley CA 1974+ – 1,5,9 – ISSN: 0094-0496 – mf#11754 – us UMI ProQuest [301]
American evangelical lutheran mission files 1875-1919 / General Synod of the Evangelical Lutheran Church in the United States. Board of Foreign Missions – [mf ed 2004] – 3r – 1 – (records divided into 4 sects; sect 1 consists of printed pamphlets & bklets about the general synod work in india, arranged alphabetically; sect 2 contains general files; sect 3 & 4 consist of printed report bklets for the american evangelical lutheran mission, also incl missionary rosters & statistics; arranged in chronological order) – mf#xa0093r – us ATLA [242]
The american evangelists, d.l. moody and ira d. sankey : in great britain and ireland / ed by Hall, John & Stuart, George Hay – New York: Dodd & Mead, c1875 – 2mf – 9 – 0-524-08326-6 – mf#1993-1021 – us ATLA [240]
American examiner – New York. N.Y. 1958-68 – 1 – us AJPC [071]
The american exile in canada *see* Amex
American expositor – La Salle IL 1850 – 1 – mf#3920 – us UMI ProQuest [975]
American expressionistic drama : containing analyses of three outstanding american plays... / Rama Murthy, V – 1st ed. Delhi: Doaba House [1970] [mf ed 1986] – 1r – 1 – (filmed with: chapters in the early history of the church of wells / church, c m) – mf#7141 – us Wisconsin U Libr [420]
American Ex-Prisoners of War, inc *see* Ex-pow bulletin
American fabian – 1895 feb-1900 jan – 1 – mf#763925 – us WHS [071]
American fabian – v1-5 n2,11. 1895-1900 [all publ] – (= ser Radical periodicals in the united states, 1881-1960. series 1) – 10mf – 9 – $105.00 – us UPA [335]
American Face Brick Association *see* Brickwork in italy
American Family Foundation *see*
– Advisor
– Cult observer
American family physician [1961] – Leawood KS 1961-69 – 5 – mf#5873 – us UMI ProQuest [610]
American family physician [1970] – Leawood KS 1970+ – 1,5,9 – (cont: american family physician/gp) – ISSN: 0002-838X – mf#5888,01 – us UMI ProQuest [610]
American Family Records Association *see* Family records today
American farmer – 4th ser: v10 [1854 jul-1855 jun] – 1 – mf#1478796 – us WHS [630]
American farmer : devoted to agriculture, horticulture – La Salle IL 1819-97 – 1 – mf#4413 – us UMI ProQuest [630]
American farmers' magazine – La Salle IL 1848-59 – 1 – mf#3921 – us UMI ProQuest [630]
The american farmer's pictorial cyclopedia of live stock : embracing horses, cattle, swine, sheep and poultry...: being also a complete stock doctor / Periam, Jonathan & Baker, Austin Hart – Toronto: Best Bros, 1888 [mf ed 1994] – 14mf – 9 – 0-665-94688-0 – mf#94688 – cn Canadiana [636]
The american federal system / Smellie, Kingsley B – London: Williams & Norgate, 1928 – 2mf – 9 – $3.00 – mf#LLMC 95-068 – us LLMC [323]
American federal tax reports – v1-4. 1880-1924; New York: Prentice-Hall 1924- – 52mf – 9 – $78.00 – (updates planned as add vols fall out of copyright) – mf#llmc90-330 – us LLMC [342]
American Federation of Arts *see* Magazine of art
American federation of arts catalogue – New York, 1929-1975 – (= ser Art exhibition catalogues on microfiche) – 57 catalogues on 76mf – 9 – £560.00 – (individual titles not listed separately) – uk Chadwyck [700]
American Federation of Full Fashioned Hosiery Workers Locked Out by the Allen A Co *see* Kenosha hosiery worker
American Federation of Government Employees *see*
– Courier
– Employee's advocate
– Government standard
– Informer
– Local 1867 united together
– Local 2324 news
– News and views
– Newsletter
– Oxford blues
– Spirit of 1760/ social security local 1760, afge, new york, ny
– Union response
– Union window
American federation of government employees newsletter / Franklin Co. Columbus – 1963-1977 [irreg] – 1r – 1 – mf#B10621 – us Ohio Hist [331]
American Federation of Grain Millers *see* Grain millers news
American Federation of Hoisiery Workers *see* Hosiery worker
American Federation of Jewish Fighters, Camp Inmates, and Nazi Victims *see* Martyrdom and resistance
American Federation of Labor *see*
– Afl-cio american federationist
– Cio news
– Federated railwayman
– Labor information
– Labor union
– List of organizations affiliated with the american federation of labor
– Official news
– Official report of proceedings of the...annual convention of the american federation of labor
– Official...union label directory
– Pro [1890]ceedings of the...annual convention of the pennsylvania state branch of the american federation of labor
– Proceedings of the new york state branch of the american federation of labor
– Report of proceedings of the...annual convention of the american federation of labor
– Southern labor review
– Trades and labor news
– Union label catalogue-directory
– Weekly news letter
– Weekly news service
American Federation of Labor and Congress of Industrial Organizations *see*
– Economic trends and outlook
– Labor's economic review
American federation of labor and congress of industrial organizations : pamphlets / AFL-CIO [American Federation of Labor and Congress of Industrial Organizations]. Library – 1889-1955 – 19r – 1 – $3195.00 – (afl pamphlets, 1889-1955 12r $2205. cio pamphlets, 1935-55 7r $1330) – us UPA [331]
American federation of labor annual meetings and reports – 1st-74th conventions. 1881-1955 (all publ) – 9 – mf#LLMC 84-345 – us LLMC [331]
American Federation of Labor et al *see* Labor journal
American Federation of Labor. Missouri State Federation of Labor *see* Labor herald
American federation of labor. proceedings – Washington DC 1881-1955 – 1 – mf#5952 – us UMI ProQuest [331]
American federation of labor records – (= ser Research colls in labor studies) – 9 – $22,095.00 coll – 0-89093-895-4 – (pt1: strikes & agreements file, 1898-1953 55r isbn 0-89093-895-4 $8725. pt2: president's office files ser a: william green papers, 1934-52 38r isbn 0-89093-896-2 $6050. minutes of the executive council of the american federation of labor pt1: 1893-1924 (with vote books, 1892-1924) 22r isbn 1-55655-377-3 $3925. pt2: 1925-55 (with vote books, 1925-54) 19r isbn 1-55655-378-1 $3395. with p/g) – us UPA [331]
American Federation of Labor Unions of Monroe County et al *see* Labor news
American Federation of Musicians *see*
– Hi-notes
– International musician
– Musical news
– Score
American Federation of Musicians et al *see* Overture
American Federation of Musicians of the United States and Canada *see* International musician

American Federation of Radio Artists *see* Stand by!

American Federation of State, County and Municipal Employees *see*
- New jersey public employee beacon
- Public employee press
- Wisconsin county and city employees union news

American Federation of State, County, and Municipal Employees *see*
- City hospital worker
- Fiscal and staff news
- Michigan afscme

American Federation of State, County and Municipal Employees [AFSCME] *see* Council 66 news

American Federation of Teachers *see*
- F y i
- Inside your schools
- Newsletter
- On campus
- Public service reporter
- Semi-monthly bulletin
- United faculty newsletter
- University guardian
- Utd today
- Wtg news

American Federation of Technical Engineers *see* Engineers' outlook

American Federation of Television and Radio Artists *see*
- On the air
- Open mike
- San diego aftra-sag newsletter

American federationist – Silver Spring MD 1894-1975 – 1,5,9 – (cont by: afl-cio american federationist) – ISSN: 0002-8428 – mf#2232.01 – us UMI ProQuest [331]

American fencing – Colorado Springs CO 1949+ – 1,5,9 – ISSN: 0002-8436 – mf#7179 – us UMI ProQuest [790]

American fiction, 1774-1910 – 1849r in 37 units (complete coll) – 1 – (coll offers material tracking the evolution of american literature. coll chronologically produced in 5v. cumulative aut ind as well as title listings are available. v1: 1774-1850, units 1-4 184r. v2: 1851-75, units 5-12 405r. v3: 1876-1900, units 13-28 808r. v5: 1901-05, units 29-33 267r. v5: 1906-10, units 34-37 185r) – mf#C35-28530 – us Primary [420]

American fiction, 1911-1920 : publications from the william s charvat collection – [mf ed 2003] – 502r in 10 units – 1 – (unit 1: [anonymous]-dorothy donnell calhoun 50r. unit 2: harvey reeves calkins-louis dodge 50r. unit 3: anna mooney doling-maccown greenlee 50r. unit 4: jackson gregory-aunt jemimy [pseud] 50r. unit 5: c a (charles augustus) jenkins-elwin lorraine 50r. unit 6: g w (george william) lose-clarence edward mulford 50r. unit 7: mulier-nina wilcox putnam 50r. unit 8: kate milner rabb-b m bower 50r. unit 9: b m bower-lucille van slyke 50r. unit 10: virginia terhune van de water-x q zuss 52r) – us Primary [420]

The american field – New York, Chicago: American Field Publ Co. v16-20 jul 1980-83. v25-28 1886-87. v33-34 1890. v39-42 1893-94. v53-56 1900-01 – us CRL [073]

American fights and fighters : stories of the first five wars of the united states from the war of the revolution to the war of 1812 / Brady, Cyrus Townsend – New York: McClure, Phillips, 1900 [mf ed 1980] – 5mf – 9 – 0-665-03715-5 – (incl ind) – mf#03715 – cn Canadiana [975]

American film – New York NY 1975-92 – 1,5,9 – ISSN: 0361-4751 – mf#11838 – us UMI ProQuest [790]

American Film & Video Association *see* Afva evaluations

American fisheries society. transactions – Bethesda MD 1872-1945 – 1 – ISSN: 0002-8487 – mf#2258 – us UMI ProQuest [639]

American flag – 1847 aug 28 – 1 – mf#857221 – us WHS [071]

American flag – Matamoros, Mexico. v1 n54. nov 28, 1846 – 1 – (biwkly newspaper publ in mexico by americans during the mexican war) – mf#(M) 80 M2.1 001 – us Western Res [071]

American Flint Glass Workers' Union *see* Quarterly report of national secretary of the american flint glass workers union

American Flyer Collectors Club *see* Collector

American Folklore Society *see*
- Journal of american folk-lore
- Memoirs of the american folk-lore society

American folksong texts / Gordon, Robert Winslow – 1 – us L of C Photodup [780]

American foreign language teacher – Detroit MI 1970-71 – 1 – ISSN: 0044-5665 – mf#6936 – us UMI ProQuest [410]

American foreign policy / Division ofCarnegie Endowment For International Peace Monograph – Washington, DC. 1920 – 1r – us UF Libraries [327]

American foreign policy / U.S. Dept of State. Historical Office – 1950-67 – 1 – $360.00 – mf#0621 – us Brook [327]

American foreign policy and treaty index microfiche library – 1993- – 9 – apply for price – (access to us government documents on world affairs) – us CIS [327]

American foreign policy series : basic documents / U S Dept of State – 5bks. 1941-49, 1950-55, 1977-80 [all publ] – 697mf – 9 – $1046.00 – (current documents: 1956-67, 1981-84; none publ 1968-80. 697mf $1046 llmc 80-909; updates available) – mf#LLMC 80-909 – us LLMC [327]

American forensic association. journal – River Falls WI 1964-88 – 1,5,9 – (cont by: argumentation & advocacy) – ISSN: 0002-8533 – mf#6485.01 – us UMI ProQuest [614]

American forestry association. proceedings – Washington DC 1882-97 – 1 – mf#7189 – us UMI ProQuest [634]

American forests – Washington DC 1895+ – 1,5,9 – ISSN: 0002-8541 – mf#6554 – us UMI ProQuest [634]

American Foundry Society *see* Transactions of the american foundry society

American Foundrymen's Society *see* Transactions of the american foundrymen's society

American freedman – Killen TX 1866-69 – 1 – mf#3081 – us UMI ProQuest [976]

American freedmen's bulletin – Killen TX 1864-66 – 1 – mf#3346 – us UMI ProQuest [976]

American Freedom from Hunger Foundation *see* Catalyst

American freeman – n1741-1878 [1929 apr 13-1931 nov 29]; n1882,1889-95, 1901, 1904-05 [1931 dec 26, 1932 feb 13-mar 26, may 7, may 28-jun 4]; n2074 [1945 jul] – 1 – mf#964093 – us WHS [071]

American Freeman Association *see* Republic

American Friends of Spanish Democracy *see* The persecution of protestants in fascist spain

American Friends of the Angolan Revolution *see* Afar

American Friends of the Chinese People *see* China today

American Friends Service Committee *see*
- Equity newsletter
- Final draft/only for life
- Peacework
- Quaker service
- Quaker service bulletin
- Relay

American fruit grower – Willoughby OH 1897+ – 1,5,9 – ISSN: 0002-8568 – mf#319 – us UMI ProQuest [634]

American G I Forum *see* Forumeer

American Game Protective Association *see* Clip sheet

American gardener – Alexandria VA 1996+ – 1,5,9 – (cont: american horticulturist) – ISSN: 1087-9978 – mf#7008,01 – us UMI ProQuest [635]

American gas – Washington DC 1989-2001 – 1,5,9 – (cont: aga monthly) – ISSN: 1043-0652 – mf#10410,02 – us UMI ProQuest [550]

American gas association monthly – Washington DC 1919-84 – (cont by: aga monthly) – ISSN: 0002-8584 – mf#10410.02 – us UMI ProQuest [550]

American gazette 1768-70 – the american magazine 1851-52 – (= ser American newspapers and periodicals published in the uk and europe 18th-20th centuries) – 1r – 1 – £55.00 – uk World [072]

American genealogist – Demorest GA 1922+ – 1,5,9 – ISSN: 0002-8592 – mf#104 – us UMI ProQuest [929]

American Geographical Society of New York *see* Bulletin'of the american geographical...

American geographical society of new york. bulletin – New York NY 1852-1915 – 1 – ISSN: 0190-5929 – mf#5214 – us UMI ProQuest [917]

American Geophysical Union *see* Eos

American girl – New York NY 1917-79 – 1,5,9 – ISSN: 0002-8630 – mf#9807 – us UMI ProQuest [370]

An american girl in london / Duncan, Sara Jeannette – Toronto: Williamson, 1891 [mf ed 1980] – 4mf – 9 – 0-665-05290-1 – (ill by f h townsend) – mf#05290 – cn Canadiana [914]

American glass review – Clifton NJ 1972-99 – 1,5,9 – ISSN: 0002-8649 – mf#8380 – us UMI ProQuest [740]

American glass worker – Oittsburgh, PA: Barrows & Osborne, may 15 1885 (sample iss). v1 n1-26 jun 5 1885-jun 21 1886 – us CRL [670]

American gleanor and virginia magazine – La Salle IL 1807 – 1 – mf#3539 – us UMI ProQuest [240]

American Gold Star Mothers, Inc *see* Gold star mother

American government and politics / Beard, Charles A – New York: The Macmillan Co, 1911 – 8mf – 9 – $12.00 – mf#LLMC 92-167 – us LLMC [323]

American grange bulletin – 1906 apr/may-oct – 1 – mf#1424866 – us WHS [071]

American grange bulletin and scientific farmer – 1896 may 28-1903 dec 24; 1904 jan 14-1906 jan – 1 – mf#1424866 – us WHS [071]

American greek testaments : a critical bibliography of the greek new testament as published in america / Hall, Isaac Hollister – Philadelphia: Pickwick, 1883 [mf ed 1989] – 1mf – 9 – 0-7905-1147-9 – (incl ind) – mf#1987-1147 – us ATLA [225]

American guardian – 1931 apr 3-1936 dec 25; 1937 jan 1-1942 jan 1 – 1 – mf#780600 – us WHS [071]

The american guide – Little Rock, AR: American Guide Pub Co, 1889 (wkly) [mf ed 1947] – (= ser Negro Newspapers on Microfilm) – 1r – 1 – us L of C Photodup [071]

American guide series – Microcard Editions – (= ser Research colls in travel and exploration) – 419mf (20:1-24:1) – 9 – $1625.00 – us UPA [917]

American guide series / U.S. Federal Writers' Project – 468 titles available as a set. 1936-47 – (= ser Federal Writers' Project) – 1,9 – us AMS Press [800]

American gymnasia and athletic record : a monthly journal of rational, physical training – Boston. v1-4 n8. sep 1904-apr 1908 [all publ] – (= ser Sports periodicals, 1822-1922) – 1r – 1 – $210.00 – (subtitle varies) – us UPA [790]

American Hamilton *see* Hamilton conveyor

American harmony : containing a variety of airs suitable for divine worship, on thanksgivings, ordinations, christmas, fasts, funerals and other occasions / Holden, Oliver – Together with a number of psalm tunes, in three and four parts. Boston: Thomas and Andrews, 1792. MUSIC 123, Item 1 – 1 – us L of C Photodup [780]

American health – New York NY 1986-93 – 1,5,9 – ISSN: 0730-7004 – mf#15197.02 – us UMI ProQuest [613]

American health care association. journal – Washington DC 1983-86 – 1,5,9 – (cont by: provider) – ISSN: 0360-4969 – mf#14122.01 – us UMI ProQuest [366]

American Heart Association, Inc *see* Arteriosclerosis

American heart journal – Oxford, England 1925+ – 1,5,9 – ISSN: 0002-8703 – mf#1884 – us UMI ProQuest [616]

American hebrew – New York NY 21 nov 1879-14 may 1880, 16 nov 1884-6 nov 1885 – 1 – (cont by: american hebrew and jewish messenger: a weekly journal for the jewish home [19 may 1905-21 apr 1922]; american hebrew [28 apr 1922-21 apr 1950]) – uk British Libr Newspaper [270]

American hebrew almanac – Philadelphia. 1881 – 1 – us AJPC [030]

American hebrew and jewish messenger : a weekly journal for the jewish home – New York NY 19 may 1905-21 apr 1922 – 1 – (cont: american hebrew [21 nov 1879-14 may 1880, 16 nov 1884-6 nov 1885]; cont by: american hebrew [28 apr 1922-21 apr 1950]) – uk British Libr Newspaper [270]

The american herald – 1873-75 – (= ser American newspapers and periodicals published in the uk and europe 18th-20th centuries) – 4r – 1 – £55.00 – uk World [072]

American Heritage *see* Rfk, his life and death

American heritage – New York NY 1949+ – 1,5,9 – ISSN: 0002-8738 – mf#887 – us UMI ProQuest [975]

American heritage of invention & technology – New York NY 1985+ – 1,5,9 – ISSN: 8756-7296 – mf#15347 – us UMI ProQuest [600]

American heroes on mission fields : brief missionary biographies / ed by Haydn, Hiram Collins – New York: American Tract Society, c1890 [mf ed 1986] – 1mf – 9 – 0-8370-6239-X – mf#1986-0239 – us ATLA [240]

American hero-myths : a study in the native religions of the western continent / Brinton, Daniel Garrison – Philadelphia: HC Watts, 1882 [mf ed 1991] – 1mf – 9 – 0-524-00697-0 – (incl bibl ref) – mf#1990-2025 – us ATLA [290]

American hero-myths : a study in the native religions of the western continent / Brinton, Daniel Garrison – Philadelphia: H Watts, 1882 [mf ed 1980] – 3mf – 9 – 0-665-02440-1 – mf#02440 – cn Canadiana [290]

American Historical Association *see*
- Aha newsletter
- Aha perspectives
- Captured german documents filmed at berlin, 1960
- Directory of affiliated societies
- Employment information bulletin
- Perspectives
- Proceedings of the american historical association

American historical association. annual report – Washington DC 1884+ – 1,5,9 – ISSN: 0065-8561 – mf#1654 – us UMI ProQuest [975]

American historical magazine – La Salle IL 1836 – 1 – mf#3715 – us UMI ProQuest [975]

American historical periodicals before "american historical review" 1741-1895 – 1r – 1 – mf#B25884 – us Ohio Hist [978]

American historical register and monthly gazette of the historic, military and patriotic-hereditary society of the united states of america – La Salle IL 1894-97 – 1,5,9 – mf#3880 – us UMI ProQuest [366]

American historical review – Washington DC 1895+ – 1,5,9 – ISSN: 0002-8762 – mf#413 – us UMI ProQuest [975]

American Historical Society of Germans from Russia *see* Work paper

American history – 1994+ – 1,5,9 – (cont: american history illustrated) – ISSN: 1076-8866 – mf#12015.01 – us UMI ProQuest [975]

American history and culture : report files of the us national park service – 1930+ [mf ed Chadwyck-Healey] – over 8000mf – 9 – (also available by region: north atlantic; mid-atlantic; national capital; southeast; midwest; southwest; rocky mountain; western; pacific north west and alaska. with guide: cultutal resources management bibliography [7116mf]) – uk Chadwyck [975]

American history illustrated – 1966-94 – 1,5,9 – (cont by: american history) – ISSN: 0002-8770 – mf#12015.01 – us UMI ProQuest [975]

American home – New York NY 1928-78 – 1,5,9 – ISSN: 0002-8789 – mf#886 – us UMI ProQuest [640]

American home economics association. ahea action – Alexandria VA 1974-93 – 1,5,9 – (cont by: action) – ISSN: 0194-7176 – mf#10748.01 – us UMI ProQuest [640]

American Home Missionary Society *see* Home missionary

The american home missionary society papers – 1816-1894 [mf ed 2002] – 385r – 1 – $49,920.00 set – (coll divided into 5 series: ser 1: incoming correspondence 1816-1893 [277r]: new england 34r, middle atlantic 54r, southern & border 21r, old northwest 135r, plains & rockies 50r, pacific 11r, canada & foreign 2r. ser 2: outgoing correspondence 1826-94 [93r]. ser 3: administrative material 1821-93 [1r]. ser 4: annual reports 1826-1936 [4r]. ser 5: the home missionary 1828-1909 [9r]. with guide [$40 when purchased separately]) – mf#d3621 – us Scholarly Res [242]

American home news – 1918-19 – (= ser American newspapers and periodicals published in the uk and europe 18th-20th centuries) – 1r – 1 – £55.00 – uk World [072]

American horologist and jeweler – Denver CO 1936-78 – 1,5,9 – (cont by: watch & clock review) – ISSN: 0002-8797 – mf#6705.03 – us UMI ProQuest [730]

American horticulturist – Alexandria VA 1924-95 – 1,5,9 – (cont by: american gardener) – ISSN: 0096-4417 – mf#7008.01 – us UMI ProQuest [635]

The american house traveller's guide for river st lawrence and the cities of montreal, quebec and ottawa – Montreal?: D Rose, 1872 – 1mf – 9 – mf#33891 – cn Canadiana [917]

American Humane Association *see* National humane review

American humane magazine – Englewood CO 1976-78 – 1,5,9 – ISSN: 0149-5224 – mf#1510,01 – us UMI ProQuest [636]

American humorist and storyteller 1888 *see* America 1883 – the american visitor 1884 – the american eagle 1885-86 – american humorist and storyteller 1888

The american hungarian review – St Louis MO 1973 – 1,5 – ISSN: 0002-8835 – mf#7489 – us UMI ProQuest [500]

American hunter – Fairfax VA 1988+ – 1,5,9 – ISSN: 0092-1068 – mf#16725 – us UMI ProQuest [790]

American imago – Baltimore MD 1939+ – 1,5,9 – ISSN: 0065-860X – mf#1837 – us UMI ProQuest [150]

American immigrant autobiographies, part 1 : manuscript autobiographies from the immigration history research center, university of minnesota – 7r – 1 – $1260.00 – 1-55655-052-9 – (with p/g) – us UPA [304]

American import export bulletin – Philadelphia PA 1934-80 – 1,5,9 – (cont by: american import export management) – ISSN: 0002-886X – mf#311.08 – us UMI ProQuest [380]

American import export management – Philadelphia PA 1982-84 – 1,5,9 – (cont: american import export bulletin; cont by: american import-export management's global trade executive) – ISSN: 0279-4470 – mf#311.08 – us UMI ProQuest [380]

The american in paris / Janin, Jules – London 1843 [mf ed Hildesheim 1995-98] – 2mf – 9 – €60.00 – 3-487-29684-5 – gw Olms [914]

The american in paris during the summer : being a companion to the "winter in paris" / Janin, Jules – London 1844 [mf ed Hildesheim 1995-98] – 2mf – 9 – €60.00 – 3-487-29685-3 – gw Olms [914]

American in the east : a glance at our history, prospects, problems, and duties in the pacific ocean / Griffis, William Elliot – New York: A.S. Barnes, 1899 – 1mf – us ATLA [978]
American Independent Movement see Aim
The american indian / Association on American Indian Affairs – 1943-58/9 – (= ser American indian periodicals... 2) – 14mf – 9 – $125.00 – us UPA [305]
The american indian – New York. no. 1-8. Nov. 1927-May 1932 – 1 – us NY Public [970]
The american indian / Society of Oklahoma Indians – 1926-31 – (= ser American indian periodicals... 1) – 12mf – 9 – $115.00 – us UPA [305]
The american indian : what and whence / Campbell, John – [Toronto?: Ontario Pub Co?, 1894?] [mf ed 1981] – 1mf – 9 – mf#16173 – cn Canadiana [305]
American Indian and Alaska Native Periodicals Research Clearinghouse et al see Native press research journal
American Indian Archeological Institute see Artifacts
American Indian Center see
– Voice of the american indian community
– Warrior
American Indian Center (Honolulu HI) see Honolulu drum
American indian center news – v1 n1 [1978 mar] – 1 – mf#660730 – us WHS [071]
American indian center newsletter – v1 n2 [i.e. 1]-v2 n6 [1978 may-1979 jun] – 1 – mf#660736 – us WHS [071]
American Indian Center of Omaha see Honga
American indian constitutions, laws and treaties – v1-21 – 1 – $108.00 – mf#0031 – us Brook [342]
American indian cultural group newsletter / San Quentin State Prison – 1968-72 – (= ser American indian periodicals... 2) – 4mf – 9 – $95.00 – us UPA [305]
American indian culture and research journal – Los Angeles CA 1974+ – 1,5,9 – ISSN: 0161-6463 – mf#11579 – us UMI ProQuest [305]
American Indian Defense Association see American indian life
American Indian Ethnohistoric Conference see Newsletter
American Indian Fund et al see Indian affairs
American Indian Historical Society see
– Wassaja
– Wassaja. the indian historian
American indian index – 1953-68 – (= ser American indian periodicals... 1) – 17mf – 9 – $125.00 – us UPA [305]
American indian journal – Institute for the Development of Indian Law. v1-9. 1975-87 (all publ) – 9 – $102.00 set – (none publ: 1983-85) – ISSN: 0145-7993 – mf#110521 – us Hein [340]
American indian journal – Oklahoma City OK 1977-82 – 1,5,9 – ISSN: 0145-7993 – mf#11412 – us UMI ProQuest [340]
American indian law newsletter – 1968-78 – (= ser American indian periodicals... 1) – 29mf – 9 – $200.00 – us UPA [340]
American indian law review – University of Oklahoma. v1-24. 1973-2000 – 5,6,9 – $330.00 – (v1-11 1973-83 $99r. v12-24 1985-2000 $231mf) – ISSN: 0094-002X – mf#100391 – us Hein [340]
American indian libraries newsletter – Chicago IL 1985+ – 1,5,9 – ISSN: 0193-8207 – mf#12516 – us UMI ProQuest [020]
American indian life / American Indian Defense Association – 1925-36 – (= ser American indian periodicals... 2) – 6mf – 9 – $95.00 – us UPA [305]
American indian magazine – Killen TX 1913-20 – 1 – mf#7043 – us UMI ProQuest [305]
American indian magazine : quarterly journal of the society of american indians – 1913-20 – (= ser American indian periodicals... 1) – 25mf – 9 – $175.00 – us UPA [305]
American indian mission association. proceedings of the semi-annual meeting... – Louisville KY: Monsarrat's Steam Power Press 1845 [mf ed 2007] – (= ser Religious periodical literature of the hispanic and indigenous peoples of the americas, 1850-1985) – 1r – 1 – mf#2007i-s008 – us ATLA [366]
American indian news – 1968-73 – (= ser American indian periodicals... 1) – 3mf – 9 – $95.00 – us UPA [305]
American indian news – v1 n22 [1976 sep 10] – 1 – mf#626770 – us WHS [071]
The american indian on the new trail : the red man of the united states and the christian gospel / Moffett, Thomas Clinton – New York: Missionary Education Movt of the US & Canada, 1914 [mf ed 1990] – 1mf – 9 – 0-7905-5539-5 – (incl bibl ref) – mf#1988-1539 – us ATLA [305]
American indian periodicals from the princeton university library, 1839-1982 : pt 1 – Clearwater Publishing Co – 95 titles on 2286mf+2r – 9,1 – $12,245.00 – (with p/g. titles also listed individually) – us UPA [305]
American indian periodicals from the princeton university library, 1839-1982 : pt 2 – Clearwater Publishing Co – 34 titles on 401mf+1r – 9,1 – $2450.00 – (with p/g. titles also listed individually) – us UPA [305]
American indian periodicals from the state historical society of wisconsin : 1884-1981 – Clearwater Publ Co – 41 titles on 13r – 1 – $1445.00 – (with p/g) – us UPA [305]
American Indian Press Association see News service
American indian press association : 1 and 2 – (= ser American indian periodicals... 1) – 9 – $95.00 ea – us UPA [070]
American Indian Program [Phelps-Stokes Fund] see Roundup
American indian quarterly – Lincoln NE 1982+ – 1,5,9 – ISSN: 0095-182X – mf#14185 – us UMI ProQuest [390]
American Indian Student Association [Northridge CA] see Newsletter
American Indian Studies Dept. CSU San Francisco see Newspaper and newsletter collection "akwesasne news thru treaty counsel news"
American Indian Travel Commission et al see Indian travel newsletter
American Indian Women's Service League et al see Northwest indian news
American indian workshop newsletter – 1980 apr-1987 mar – 1 – mf#1288406 – us WHS [071]
The american indiana – Blairsville, PA., 1826 – 13 – $25.00r – us IMR [071]
American indonesian chamber of commerce inc information bulletin – New York, [195?]1955-1971 nos 1-986 – 93mf – 9 – (several iss missing) – mf#SE-703 – ne IDC [959]
American Industrial Development Council see Aidc journal
American industrial evolution from the frontier to the factory : its social and political effects / Ebert, Justus – New York: New York Labor News Co, 1907 (mf ed 19–) – 88p – mf#ZT-SFC pv72 n4; ZT-SFC pv88 n5; ZT-SFC pv88 n2 – us Harvard [331]
American industrial hygiene association. journal – Fairfax VA 1940-99 – 1,5,9 – (cont by: aihaj) – ISSN: 0002-8894 – mf#10121.02 – us UMI ProQuest [366]
American industry – Great Neck NY 1972+ – 1 – ISSN: 0002-8908 – mf#7145 – us UMI ProQuest [338]
American institute for marxist studies. aims newsletter – San Jose CA 1973-85 – 1,5,9 – ISSN: 0001-1622 – mf#6580 – us UMI ProQuest [335]
American Institute of Accountants see Bulletin of the american institute of accountants
American Institute of Aeronautics and Astronautics see
– Aiaa bulletin
– American institute of aeronautics and astronautics. aiaa student journal
American institute of aeronautics and astronautics. aiaa journal – Reston VA 1963+ – 1,5,9 – ISSN: 0001-1452 – mf#1600 – us UMI ProQuest [629]
American institute of aeronautics and astronautics. aiaa student journal / American Institute of Aeronautics and Astronautics – Reston VA 1963+ – 1,5,9 – ISSN: 0001-1460 – mf#5072 – us UMI ProQuest [629]
American Institute of Architects see
– Journal of the american institute of architects
– Papers from the american institute of architects library and archives
American institute of architects. aia journal – New York NY 1944-83 – 1,5,9 – (cont by: architecture : the aia journal) – ISSN: 0001-1479 – mf#114.01 – us UMI ProQuest [720]
The american institute of architects (aia) library and archives microform collections – 4 colls – 1,5 – $14,200.00 coll – (indvidual colls also listed separately) – us UMI ProQuest [720]
American institute of biological sciences. biological sciences curriculum study newsletter – Colorado Springs CO 1972-77 – 1,5,9 – ISSN: 0005-3295 – mf#7084 – us UMI ProQuest [574]
American institute of chemical engineers. aiche journal – New York NY 1955+ – 1,5,9 – ISSN: 0001-1541 – mf#12576 – us UMI ProQuest [660]
American Institute of Electrical Engineers see Transactions
American institute of electrical engineers. transactions – Piscataway NJ 1884-1964 – 1,5 – mf#14107 – us UMI ProQuest [621]
American Institute of Hypnosis see Journal of the american institute of hypnosis
American Institute of Industrial Engineers see Technical papers, institute conference and convention
American institute of industrial engineers. aiie transactions – Abingdon, Oxfordshire 1969-81 – 1,5,9 – (cont by: iie transactions) – ISSN: 0569-5554 – mf#3180.01 – us UMI ProQuest [670]
American institute of industrial engineers. technical papers, institute conference and convention – Norcross GA 1954-73 – 1,5 – ISSN: 0163-5573 – mf#3181.01 – us UMI ProQuest [670]
American institute of instruction. annual meeting – Killen TX 1830-1908 – 1 – mf#2774 – us UMI ProQuest [060]
American institute of mining and metallurgical engineers. transactions – New York NY 1940-47 – 1 – mf#588 – us UMI ProQuest [620]
American Institute of Mining Engineers see Programme, montreal, february 21st to 25th, 1893
American institute of mining, metallurgical, and petroleum engineers. transactions – New York NY 1958-69 – 1 – ISSN: 0096-4778 – mf#1663 – us UMI ProQuest [622]
American Institute of Physics see Computers in physics
American institute of planners. journal – Chicago IL 1965-78 – 1,5,9 – (cont by: journal of the american planning association) – ISSN: 0002-8991 – mf#2001.02 – us UMI ProQuest [710]
American Institution of Hemoeopathy see Pharmacopeia
The american intercollegiate football spectacle, 1869-1917 / Lewis, Guy M – 1964 – 4mf – 9 – $16.00 – mf#PE 4051 – us Kinesology [790]
American investor – New York NY 1972-74 – 1,5 – ISSN: 0002-9025 – mf#6656 – us UMI ProQuest [332]
American iris society. bulletin – Tulsa OK 1977-80 – 1,5,9 – ISSN: 0747-4172 – mf#10623 – us UMI ProQuest [635]
The american irish : and their influence on irish politics / Bagenal, Philip Henry Dudley – London, 1882 – (= ser 19th c ireland) – 4mf – 9 – mf#1.1.7375 – uk Chadwyck [941]
American Iron and Steel Association see Bulletin
American Iron and Steel Institute see
– Steel
– Steel facts
American issue – 1919-20; 1919 apr 12-1921 mar 26; 1921 jan-1923 dec; 1922 jan 7-1924 may 10; may 17-1926 dec 25; 1927 jan 8-1929 may 25; jun 8-1931 dec 19; 1929 oct-1938 jun; 1932 jan 2-1933 apr 1 – 1 – mf#1109675 – us WHS [071]
The american jefferson – Brookville, PA., 1826 – 13 – $25.00r – us IMR [071]
American jesuits : new york 1934 / Walsh, James – Madrid: Razon y Fe, 1935 – 1 – sp Bibl Santa Ana [241]
The american jew as patriot, soldier and citizen / Wolf, Simon; ed by Levy, Louis Edward – Philadelphia: Levytype; New York: Brentano's, 1895 [mf ed 1990] – 2mf – 9 – 0-7905-3625-0 – (incl bibl ref) – mf#1989-0118 – us ATLA [305]
The american jewess – New York. N.Y. 1895-99 – 1 – us AJPC [071]
American jewish archives – Cincinnati OH 1980-96 – 1,5,9 – ISSN: 0002-905X – mf#12676.01 – us UMI ProQuest [939]
The american jewish chronicle – New York. N.Y. 1916-18 – 1 – us AJPC [071]
American Jewish Committee see Shvartser bukh
American Jewish Congress see Congress monthly
American jewish congress-news – New York, NY. 1964-86 – 1 – us AJPC [071]
American jewish joint distribution committee – New York, NY. 1953-84 – 1 – us AJPC [071]
American jewish ledger – Newark, N.J. – 1 – (v8 n7 (sep 1952)) – us AJPC [270]
American jewish news – New York. N.Y. 1918-19 – 1 – us AJPC [071]
American jewish outlook – Pittsburgh. Pa. 1958-62 – 1 – us AJPC [071]
American jewish pulpit : a collection of sermons... – Cincinnati, OH. 1881 – 1r – us UF Libraries [939]
American jewish review – Atlanta. Ga. 1913-15 – 1 – us AJPC [071]
American jewish review – Buffalo. N.Y. 1917-25 – 1 – us AJPC [071]
American jewish times – Howell, N.J. v3, no. 1 (Sept. 1990); v3, no. 7 (Mar. 1991); v4, no. 8 (Apr. 1992) – us AJPC [071]
American jewish world – Minneapolis St. Paul. Minn. 1958-67 – 1 – us AJPC [071]
The american jews' annual – Cincinnati. 1884 85-97 – 1 – us AJPC [939]
American journal of agricultural economics – Oxford, England 1919+ – 1,5,9 – ISSN: 0002-9092 – mf#1936 – us UMI ProQuest [630]
American journal of archaeology – Boston MA 1885+ – 1,5,9 – ISSN: 0002-9114 – mf#422 – us UMI ProQuest [930]
American journal of archaeology and of the history of the fine arts – Cambridge, New York, London, 1885-1896, v1-11; 1897-1920, v1-24; ind 1897-1906, v1-10 – 420mf – 9 – mf#0-1207 – ne IDC [933]
American journal of art therapy – Montpelier VT 1961+ – 1,5,9 – ISSN: 0007-4764 – mf#1820 – us UMI ProQuest [700]
American journal of audiology – Rockville MD 1994+ – 1,5,9 – ISSN: 1059-0889 – mf#20817 – us UMI ProQuest [617]
American journal of chinese medicine [1973] – San Francisco CA 1973-77 – 1,5,9 – (cont by: comparative medicine east & west) – ISSN: 0090-2942 – mf#10055.02 – us UMI ProQuest [610]
American journal of chinese medicine [1979] – San Francisco CA 1979+ – 1,5,9 – (cont: comparative medicine east & west) – ISSN: 0192-415X – mf#10055.02 – us UMI ProQuest [610]
American journal of clinical biofeedback – Bern, Switzerland 1978-84 – 1,5,9 – (cont by: clinical biofeedback & health) – ISSN: 0190-4019 – mf#11712.01 – us UMI ProQuest [610]
American journal of clinical hypnosis – Bloomingdale IL 1958+ – 1,5,9 – ISSN: 0002-9157 – mf#2293 – us UMI ProQuest [615]
American journal of clinical nutrition – Bethesda MD 1952+ – 1,5,9 – ISSN: 0002-9165 – mf#1963 – us UMI ProQuest [613]
American journal of clinical oncology : cancer clinical trials – Baltimore MD 1993+ – 1,5,9 – ISSN: 0277-3732 – mf#18696,01 – us UMI ProQuest [616]
American journal of clinical pathology – Baltimore MD 1931+ – 1,5,9 – ISSN: 0002-9173 – mf#94 – us UMI ProQuest [616]
American journal of community psychology – Dordrecht, Netherlands 1989+ – 1,5,9 – ISSN: 0091-0562 – mf#12399 – us UMI ProQuest [150]
American journal of comparative law – Ann Arbor MI 1952+ – 1,5,9 – ISSN: 0002-919X – mf#6486 – us UMI ProQuest [340]
American journal of criminal justice [ajcj] – Norfolk VA 1985+ – 1,5,9 – (cont: southern journal of criminal justice) – ISSN: 1066-2316 – mf#13350,01 – us UMI ProQuest [345]
American journal of criminal law – Austin TX 1972+ – 1,5,9 – ISSN: 0092-2315 – mf#10070 – us UMI ProQuest [345]
American journal of criminal law – University of Texas. v1-27. 1972-2000 – 5,6,9 – $508.00 – (v1-12 1972-84 $176r. v13-27 1985-2000 $332mf) – ISSN: 0092-2315 – mf#100781 – us Hein [345]
American journal of critical care – Alisa Veijo CA 1997+ – 1,5,9 – ISSN: 1062-3264 – mf#24237 – us UMI ProQuest [610]
American journal of dance therapy – Dordrecht, Netherlands 1991+ – 1,5,9 – ISSN: 0146-3721 – mf#17540 – us UMI ProQuest [790]
American journal of digestive diseases – Dordrecht, Netherlands 1934-78 – 1,5 – (cont by: digestive diseases & sciences) – ISSN: 0002-9211 – mf#53.01 – us UMI ProQuest [616]
American journal of diseases of children – Chicago IL 1911-93 – 1,5,9 – (cont by: archives of pediatrics & adolescent medicine) – ISSN: 0002-922X – mf#408.01 – us UMI ProQuest [618]
American journal of drug and alcohol abuse – Monticello NY 1981+ – 1,5,9 – ISSN: 0095-2990 – mf#12923 – us UMI ProQuest [362]
American journal of economics and sociology – Oxford, England 1941+ – 1,5,9 – ISSN: 0002-9246 – mf#1084 – us UMI ProQuest [338]
American journal of education [1847] – Killen TX 1847 – 1 – mf#4611 – us UMI ProQuest [370]
American journal of education [1855] – La Salle IL 1855-82 – 1 – mf#3873 – us UMI ProQuest [370]
American journal of education [1869] – Killen TX 1869-1904 – 1 – mf#4813 – us UMI ProQuest [370]
American journal of education [1979] – Chicago IL 1980+ – 1,5,9 – (cont: school review) – ISSN: 0195-6744 – mf#137,01 – us UMI ProQuest [370]
American journal of emergency medicine [ajem] – Philadelphia PA 1983+ – 1,5,9 – ISSN: 0735-6757 – mf#14729 – us UMI ProQuest [610]
American journal of enology and viticulture – Davis CA 1989+ – 1,5,9 – ISSN: 0002-9254 – mf#13557,02 – us UMI ProQuest [660]
American journal of epidemiology – Oxford, England 1985+ – 1,5,9 – ISSN: 0002-9262 – mf#770 – us UMI ProQuest [614]
American journal of evaluation – Oxford, England 1998+ – 1,5,9 – (cont: evaluation practice) – ISSN: 1098-2140 – mf#17052,02 – us UMI ProQuest [300]
American journal of family law – v1-9. 1987-1995 – 9 – $250.00 set – ISSN: 0891-6330 – mf#111831 – us Hein [346]
American journal of family therapy – Abingdon, Oxfordshire 1979+ – 1,5,9 – (cont: international journal of family counseling) – ISSN: 0192-6187 – mf#8240,02 – us UMI ProQuest [362]

American journal of forensic psychiatry – v1-22. 1978-2001 – 5,6,9 – $520.00 set – (v1-5 1978-84 in reel $75. v6-22 1985-2001 in mf $445) – ISSN: 0163-1942 – mf#101551 – us Hein [614]

American journal of forensic psychology – v1-19. 1983-2001 – 5,6,9 – $495.00 set – (v1-2 1983-84 in reel $49. v3-19 1985-2001 in mf $446) – ISSN: 0733-1290 – mf#109091 – us Hein [614]

American journal of gastroenterology – Oxford, England 1949+ – 1,5,9 – ISSN: 0002-9270 – mf#5 – us UMI ProQuest [616]

American journal of gastroenterology – v77-91. 1982-96 – 15r – 1,5,6,9 – $110.00r – us Lippincott [616]

American journal of geriatric psychiatry – Arlington VA 1993+ – 1,5,9 – ISSN: 1064-7481 – mf#20183 – us UMI ProQuest [618]

American journal of health behavior – Star City WV 1996+ – 1,5,9 – (cont: health values) – ISSN: 1087-3244 – mf#12296,01 – us UMI ProQuest [613]

American journal of health education – Reston VA 2001+ – 1,5,9 – (cont: journal of health education) – mf#7254,03 – us UMI ProQuest [613]

American journal of health studies – College Station TX 2001+ – 1,5,9 – ISSN: 1090-0500 – mf#19365,02 – us UMI ProQuest [360]

American journal of health-system pharmacy [ajhp] – Bethesda MD 1995+ – 1,5,9 – (cont: american journal of hospital pharmacy) – ISSN: 1079-2082 – mf#1923,01 – us UMI ProQuest [362]

American journal of homeopathia – La Salle IL 1835 – 1 – mf#3922 – us UMI ProQuest [615]

American journal of homeopathy [1838] – La Salle IL 1838-39 – 1 – mf#3924 – us UMI ProQuest [615]

American journal of homoeopathy [1846] – Killen TX 1846-54 – 1 – mf#3923 – us UMI ProQuest [615]

American journal of hospital pharmacy – Bethesda MD 1943-94 – 1,5,9 – (cont by: american journal of health-system pharmacy [ajhp]) – ISSN: 0002-9289 – mf#1923.01 – us UMI ProQuest [615]

American journal of human genetics – Chicago IL 1949+ – 1,5,9 – ISSN: 0002-9297 – mf#2740 – us UMI ProQuest [573]

American journal of hypertension – Oxford, England 1993+ – 1,5,9 – ISSN: 0895-7061 – mf#42751 – us UMI ProQuest [616]

American journal of infection control – Oxford, England 1980+ – 1,5,9 – ISSN: 0196-6553 – mf#12232,03 – us UMI ProQuest [614]

American journal of international law – v1-95. 1907-2001 – 9 – $3035.00 set – ISSN: 0002-9300 – mf#100421 – us Hein [341]

American journal of international law – Washington DC 1907+ – 1,5,9 – ISSN: 0002-9300 – mf#1763 – us UMI ProQuest [341]

American journal of intravenous therapy & clinical nutrition – New York NY 1983-84 – 1,5,9 – (cont by: intravenous therapy news) – ISSN: 0195-0282 – mf#14121.04 – us UMI ProQuest [615]

American journal of jurisprudence – Notre Dame IN 1956-82 – 1,5,9 – ISSN: 0065-8995 – mf#2722 – us UMI ProQuest [340]

American journal of jurisprudence – v1-46 (incl 20 yr index). 1956-2 – 5,6,9 – $614.00 set – (v1-29 1956-84 in reel $259; v30-46 1985-2 in mf $355; incl 20-yr ind; cont as: natural law forum) – ISSN: 0065-8995 – mf#100791 – us Hein [340]

American journal of kidney diseases – Philadelphia PA 1995+ – 1,5,9 – ISSN: 0272-6386 – mf#21079 – us UMI ProQuest [616]

American journal of knee surgery – Thorofare NJ 1992-94 – 1 – ISSN: 0899-7403 – mf#17271.01 – us UMI ProQuest [617]

American journal of law and medicine – Boston MA 1979+ – 1,5,9 – ISSN: 0098-8588 – mf#12087 – us UMI ProQuest [346]

American journal of law and medicine – v1-27 (1975-2001) – 5,6,9 – $734.00 set – (v1-10 1975-85 in reel $176. v11-27 1985-2001 in mf $558) – ISSN: 0098-8588 – mf#100801 – us Hein [344]

American journal of legal history – v1-43. 1957-99 – 1,5,6 – $525.00 – (v1-36 1957-92 in reel $345. v37-43 1993-99 in mf $180) – ISSN: 0002-9319 – mf#100431 – us Hein [340]

American journal of maternal child nursing *see* Mcn

American journal of mathematics – Baltimore MD 1878+ – 1,5,9 – ISSN: 0002-9327 – mf#1913 – us UMI ProQuest [510]

American journal of medical quality – v1-11. 1986-96 – 11r – 1,5,6,9 – $65.00r – us Lippincott [610]

American journal of medical technology – Bethesda MD 1934-83 – 1,5,9 – ISSN: 0148-8759 – mf#2759 – us UMI ProQuest [619]

American journal of medicine – Oxford, England 1946+ – 1,5,9 – ISSN: 0002-9343 – mf#1964 – us UMI ProQuest [610]

American journal of mental deficiency – Washington DC 1962-87 – 1,5,9 – (cont by: american journal on mental retardation [ajmr]) – ISSN: 0002-9351 – mf#1558.01 – us UMI ProQuest [616]

American journal of music and musical visitor – La Salle IL 1840-46 – 1 – mf#4148 – us UMI ProQuest [780]

American journal of neurology and psychiatry – Killen TX 1882-85 – 1 – mf#3210 – us UMI ProQuest [616]

American journal of numismatics – Killen TX 1866-1907 – 1 – mf#4859 – us UMI ProQuest [730]

American journal of nursing – Baltimore MD 1900+ – 1,5,9 – ISSN: 0002-936X – mf#835 – us UMI ProQuest [610]

American journal of obstetrics and gynecology – Oxford, England 1920+ – 1,5,9 – ISSN: 0002-9378 – mf#1881 – us UMI ProQuest [618]

American journal of occupational therapy [1947] – Bethesda MD 1947-77 – 1,5,9 – (cont by: ajot the american journal of occupational therapy) – ISSN: 0002-9386 – mf#10224.02 – us UMI ProQuest [615]

American journal of occupational therapy [1980] – Bethesda MD 1980+ – 1,5,9 – (cont: ajot: the american journal of occupational therapy) – ISSN: 0272-9490 – mf#10224,02 – us UMI ProQuest [615]

American journal of ophthalmology – Oxford, England 1884+ – 1,5,9 – ISSN: 0002-9394 – mf#99 – us UMI ProQuest [617]

American journal of optometry and archives of american academy of optometry – Baltimore MD 1924-73 – 1,5,9 – (cont by: american journal of optometry & physiological optics) – ISSN: 0002-9408 – mf#100.02 – us UMI ProQuest [617]

American journal of optometry and physiological optics – Baltimore MD 1974-75 – 1,5 – (cont: american journal of optometry and archives of american academy of optometry) – ISSN: 0093-7002 – mf#100,01 – us UMI ProQuest [617]

American journal of orthodontics – Oxford, England 1915-86 – 1,5,9 – (cont by: american journal of orthodontics & dentofacial orthopedics) – ISSN: 0002-9416 – mf#1880.01 – us UMI ProQuest [617]

American journal of orthodontics and dentofacial orthopedics – Oxford, England 1986+ – 1,5,9 – (cont: american journal of orthodontics) – ISSN: 0889-5406 – mf#1880,01 – us UMI ProQuest [617]

American journal of orthopsychiatry – Washington DC 1930+ – 1,5,9 – ISSN: 0002-9432 – mf#5839 – us UMI ProQuest [616]

American journal of otolaryngology – Philadelphia PA 1979+ – 1,5,9 – ISSN: 0196-0709 – mf#12044 – us UMI ProQuest [617]

American journal of otology – Baltimore MD 1979-2000 – 1,5,9 – ISSN: 0192-9763 – mf#12998.01 – us UMI ProQuest [617]

American journal of pastoral counseling / ed by Dayringer, Richard – (cont: journal of pastoral psychotherapy; cont: journal of religion in psychotherapy) – mf#1094-6098 – us Haworth [360]

American journal of pathology – Bethesda MD 1925+ – 1,5,9 – ISSN: 0002-9440 – mf#8777 – us UMI ProQuest [614]

American journal of perinatology – New York NY 1983+ – 1,5,9 – ISSN: 0735-1631 – mf#12997 – us UMI ProQuest [618]

American journal of pharmaceutical education – Alexandria VA 1937+ – 1,5,9 – ISSN: 0002-9459 – mf#1524 – us UMI ProQuest [615]

American journal of pharmacy and the sciences supporting public health – Philadelphia PA 1952-78 – 1,5,9 – ISSN: 0002-9467 – mf#1.02 – us UMI ProQuest [615]

American journal of philately – Killen TX 1868-1906 – 1 – mf#4610 – us UMI ProQuest [760]

American journal of philology – Baltimore MD 1880+ – 1,5,9 – ISSN: 0002-9475 – mf#2097 – us UMI ProQuest [400]

American journal of philology – Baltimore. v1-67. 1880-1946 – 598mf – 8 – mf#700c – ne IDC [400]

American journal of photography – Killen TX 1882-1900 – 1 – mf#4634 – us UMI ProQuest [770]

American journal of physical medicine and rehabilitation – v1-75. 1922-96 – 1,5,6,9 – $90.00r – us Lippincott [617]

American journal of physics – v1-. 1933- – 1,5,6 – us AIP [530]

American journal of physiology [1898] – Bethesda MD 1898+ – 1,5,9 – ISSN: 0002-9513 – mf#776 – us UMI ProQuest [612]

American journal of physiology: cell physiology – Bethesda MD 1977+ – 1,5,9 – ISSN: 0363-6143 – mf#11160 – us UMI ProQuest [612]

American journal of physiology: endocrinology and metabolism – Bethesda MD 1982+ – 1,5,9 – ISSN: 0193-1849 – mf#11945 – us UMI ProQuest [612]

American journal of physiology: endocrinology, metabolism and gastrointestinal physiology – Bethesda MD 1977-78 – 1,5,9 – ISSN: 0363-6100 – mf#11156 – us UMI ProQuest [612]

American journal of physiology: gastrointestinal and liver physiology – Bethesda MD 1989+ – 1 – ISSN: 0193-1857 – mf#11946 – us UMI ProQuest [612]

American journal of physiology: heart and circulatory physiology – Bethesda MD 1977+ – 1,5,9 – ISSN: 0363-6135 – mf#11157 – us UMI ProQuest [612]

American journal of physiology: lung cellular and molecular physiology – Bethesda MD 1989+ – 1,5,9 – ISSN: 1040-0605 – mf#17273 – us UMI ProQuest [612]

American journal of physiology: regulatory, integrative and comparative physiology – Bethesda MD 1977+ – 1,5,9 – ISSN: 0363-6119 – mf#11158 – us UMI ProQuest [612]

American journal of physiology: renal, fluid and electrolyte physiology – Bethesda MD 1977-97 – 1,5,9 – (cont by: american journal of physiology: renal physiology) – ISSN: 0363-6127 – mf#11159.01 – us UMI ProQuest [612]

American journal of physiology: renal physiology – Bethesda MD 1998+ – 1 – (cont: american journal of physiology: renal, fluid & electrolyte physiology) – mf#11159,01 – us UMI ProQuest [612]

American journal of police : an interdisciplinary journal of theory and research – Bradford, England 1985-96 – 1,5,9 – ISSN: 0735-8547 – mf#15308 – us UMI ProQuest [360]

American journal of political science – Oxford, England 1973+ – 1,5,9 – (cont: midwest journal of political science) – ISSN: 0092-5853 – mf#2539,01 – us UMI ProQuest [320]

American journal of potato research – Orono ME 1998+ – 1 – (cont: american potato journal) – ISSN: 1099-209X – mf#231,01 – us UMI ProQuest [630]

American journal of preventive medicine [1988] – Oxford, England 1988-97 – 1,5,9 – (cont by: american journal of preventive medicine [1998]) – ISSN: 0749-3797 – mf#17033 – us UMI ProQuest [613]

American journal of preventive medicine [1998] – Oxford, England 1998+ – 1,5,9 – (cont: american journal of preventive medicine [1988]) – ISSN: 0749-3797 – mf#42796 – us UMI ProQuest [613]

American journal of proctology – New York NY 1950-77 – 1,5,9 – (cont by: american journal of proctology, gastroenterology & colon & rectal surgery) – ISSN: 0002-9521 – mf#1996.02 – us UMI ProQuest [617]

American journal of proctology, gastroenterology & colon & rectal surgery – New York NY 1978-84 – 1,5,9 – (cont: american journal of proctology; cont by: gastroenterology & endoscopy news) – ISSN: 0162-6566 – mf#1996.02 – us UMI ProQuest [617]

American journal of progressive therapeutics – La Salle IL 1897-1906 – 1 – mf#5729 – us UMI ProQuest [615]

American journal of psychiatry – Arlington VA 1844+ – 1,5,9 – ISSN: 0002-953X – mf#746 – us UMI ProQuest [616]

American journal of psychoanalysis – Dordrecht, Netherlands 1941+ – 1,5,9 – ISSN: 0002-9548 – mf#2394 – us UMI ProQuest [616]

American journal of psychology – Champaign IL 1887+ – 1,5,9 – ISSN: 0002-9556 – mf#121 – us UMI ProQuest [150]

American journal of psychotherapy – Bronx NY 1989+ – 1,5,9 – ISSN: 0002-9564 – mf#18267 – us UMI ProQuest [615]

American journal of public health – Washington DC 1911+ – 1,5,9 – ISSN: 0090-0036 – mf#8 – us UMI ProQuest [360]

American journal of public hygiene – Killen TX 1891-1910 – 1 – ISSN: 0272-2313 – mf#4782 – us UMI ProQuest [614]

The american journal of religious psychology and education – v1-4. 1904-11 [complete] – 1r – 1 – mf#ATLA 1994-S518 – us ATLA [377]

American journal of respiratory and critical care medicine – New York NY 1994+ – 1,5,9 – (cont: american review of respiratory disease) – ISSN: 1073-449X – mf#102,01 – us UMI ProQuest [610]

American journal of roentgenology, radium therapy, and nuclear medicine – Leesburg VA 1906-75 – 1,5,9 – (cont by: ajr [american journal of roentgenology]) – ISSN: 0002-9580 – mf#1574 – us UMI ProQuest [616]

American journal of science – New Haven CT 1818-1962 – 1 – ISSN: 0002-9599 – mf#517 – us UMI ProQuest [500]

The american journal of science and arts – New Haven, 1818/1819-1845, v1-50; 1846-1870, v1-50; 1871-1895, v1-50; 1896-1920, v1-5; 1921-1940, v1-38 – 1378mf – 9 – mf#8624 – ne IDC [720]

American journal of semitic languages and literatures – Killen TX 1884-1910 – 1 – mf#5655 – us UMI ProQuest [470]

The american journal of semitic languages and literatures – v1-58. mar 1884-oct 1941 [complete] – 9r – 1 – (supersedes: hebraica v1-11) – mf#ATLA S0003 – us ATLA [470]

American journal of small business – Oxford, England 1976-89 – 1,5,9 – (cont by: entrepreneurship theory & practice: et&p) – ISSN: 0363-9428 – mf#13012.01 – us UMI ProQuest [650]

American journal of social psychiatry – Levittown PA 1984-87 – 1,5,9 – ISSN: 0277-8173 – mf#14094 – us UMI ProQuest [616]

American journal of sociology – Chicago IL 1895+ – 1,5,9 – ISSN: 0002-9602 – mf#492 – us UMI ProQuest [301]

American journal of speech-language pathology – Rockville MD 1994+ – 1,5,9 – ISSN: 1058-0360 – mf#20818 – us UMI ProQuest [616]

American journal of sports medicine – Newbury Park CA 1986+ – 1,5,9 – ISSN: 0363-5465 – mf#16280,01 – us UMI ProQuest [617]

American journal of surgery – Lausanne, Switzerland 1898+ – 1,5,9 – ISSN: 0002-9610 – mf#1965 – us UMI ProQuest [617]

American journal of surgical pathology – Baltimore MD 1993+ – 1,5,9 – ISSN: 0147-5185 – mf#18692 – us UMI ProQuest [617]

American journal of tax policy – University of Alabama. v1-15. 1982-98 – 9 – $250.00 set – (ceased with v15/1) – ISSN: 0739-7569 – mf#101641 – us Hein [343]

American journal of the medical sciences – Baltimore MD 1828+ – 1,5,9 – ISSN: 0002-9629 – mf#1754 – us UMI ProQuest [610]

American journal of theology – Killen TX 1897-1920 – 1 – mf#3308 – us UMI ProQuest [200]

The american journal of theology – Chicago, 1(1897)-24(1920) – 326mf – 9 – €621.00 – ne Slangenburg [200]

American journal of theology & philosophy – Highlands NC 1985+ – 1,5,9 – ISSN: 0194-3448 – mf#15392 – us UMI ProQuest [200]

American journal of trial advocacy – Birmingham AL 1979+ – 1,5,9 – ISSN: 0160-0281 – mf#12035 – us UMI ProQuest [347]

American journal of veterinary research – Schaumburg IL 1940+ – 1,5,9 – ISSN: 0002-9645 – mf#160 – us UMI ProQuest [636]

American journal on mental retardation [ajmr] – Washington DC 1987+ – 1,5,9 – (cont: american journal of mental deficiency) – ISSN: 0895-8017 – mf#1558,01 – us UMI ProQuest [616]

American journalism review [ajr] – College Park MD 1994+ – 1,5,9 – (cont: washington journalism review [wjr]) – ISSN: 1067-8654 – mf#12946,03 – us UMI ProQuest [070]

American journals of george townsend fox, 1831-68 – South Shields Public Library – (= ser BRRAM series) – 1r – 1 – £67 / $134 – (with int by bernard crick) – mf#r02186 – uk Microform Academic [880]

American jubilee – Killen TX 1854-55 – 1 – mf#3082 – us UMI ProQuest [976]

American jubilee – New York. n1-12. 1854-55 [all publ] – (= ser Black journals, series 1) – 2mf – 9 – $45.00 – us UPA [305]

American judaism – New York. N.Y. 1951-65 – 1 – us AJPC [939]

American judaism – New York NY 1951-67 – 1 – mf#1537 – us UMI ProQuest [270]

American jurisprudence – 1st series compl. collection; 2nd series superseded vol only – 9 – enquire for prices – (filming in process) – mf#402471 – us Hein [340]

American jurisprudence. / Andrews, Charles – New Haven, Conn., Hoggson & Robinson, 1898. 30 p. LL-426 – 1 – us L of C Photodup [340]

American jurist and law magazine – Boston. v1-28. 1829-1843 – (= ser Historical legal periodical series) – 1 – $275.00 – (all publ) – mf#408800 – us Hein [340]

American jurist and law magazine – La Salle IL 1829-43 – 1 – mf#4155 – us UMI ProQuest [340]

The american jurist and law magazine – v1-28. 1829-43 (all publ) – 162mf – 9 – $243.00 – mf#LLMC 95-885 – us LLMC [340]

American Karakoram Expedition (1st: 1938) *see* Fve miles high

American Kennel Club *see* Stud book register

American kennel club stud book register – New York NY 1878+ – 1,5,9 – ISSN: 0162-2013 – mf#8212 – us UMI ProQuest [636]

The american kitchen magazine – Boston, MA: Home Science Publ Co. v3 n6-v18 n6. sep 1895-mar 1903 – 1 – us CRL [640]

American labor – 1979 nov-1992 mar – 1 – mf#2540075 – us WHS [071]

American labor banner – 1919 nov 2-32 aug 6; v2 n5 [1930 nov 29] – 1 – mf#2696840 – us WHS [071]
American Labor Conference on International Affairs *see* Modern review
American Labor Education Service *see* Annual report...
American labor legislation review – v1-32 no 4. 1911-41 (all publ) – 193mf – 9 – $289.00 – (missing: v1-3, v5-16) – mf#LLMC 84-346 – us LLMC [344]
American labor legislation review – New York. v1-32. 1911-42 – 9 – $372.00 – mf#0032 – us Brook [331]
American Labor Union *see* Voice of labor
American Labor Union Educational Society *see* Voice of labor
American labor union journal – v1 n8-n11; v1 n13-v2 n41 [1902 nov 27-dec 18; 1903 jan 1-1904 dec] – 1 – mf#1218215 – us WHS [071]
American laborer : devoted to the cause of protection to home industry – La Salle IL 1842-43 – 1 – mf#4354 – us UMI ProQuest [331]
American ladies' magazine – La Salle IL 1828-36 – 1 – mf#3925 – us UMI ProQuest [640]
American laundry digest – Detroit MI 1974-96 – 1,5,9 – ISSN: 0002-9718 – mf#9894 – us UMI ProQuest [660]
American law / Andrews, James deWitt – 2d ed. Chicago, Callaghan,1908. 2 v. LL-1409 – 1 – us L of C Photodup [348]
American law / Hilliard, Francis – New York: Peloubet, 1881. 2v. LL-747 – 1 – us L of C Photodup [340]
American law / Hilliard, Francis – New York: Ward & Peloubet, 1877-78. 2v. LL-1684 – 1 – us L of C Photodup [340]
American law and procedure – Chicago: LaSalle Ext University. v1-14. 1910 (all publ) – 70mf – 9 – $106.00 – (v1-12 ed by james p hall. v13-14 by james b andrews) – mf#LLMC 84-381 – us LLMC [340]
American Law Association *see* Catalogue.
American Law Institute *see* Code of criminal procedure: preliminary draft
American Law Institute et al *see*
– Ali-aba business law course materials journal
– Ali-aba course materials journal
American law institute microfiche archives *see* American law institute microfiche publications
American law institute microfiche publications – 9 – $26,813.00 set – 0-89941-412-5 – (backfile thru update 1999 n3; title varies: backfile thru 1993 n2 as american law institute microfiche archives; price incl guide to the american law institute publ) – mf#400450 – us Hein [340]
American Law Institute. Philadelphia *see*
– The institute
– A study of the business of the federal courts.
American law institute proceedings – 1st-76th. 1923-99 – 9 – $1875.00 set – mf#401081 – us Hein [340]
The american law institute, proceedings – v1-4. 1922/23-1926 – 14mf – 9 – $21.00 – (lacking: v1; updates planned as add vols fall out of copyright) – mf#llmc97-580 – us LLMC [340]
American law institute. proceedings at the annual meeting – v1-21. 1923-44 – 9 – $420.00 – mf#0033 – us Brook [340]
American law institute-american bar association. committee on continuing professional education review – Philadelphia PA 1974+ – 1,5,9 – ISSN: 0044-7560 – mf#8592 – us UMI ProQuest [340]
American law journal – v1-11. 1842-52 (all publ) – 9 – (title varies: v1-7 titled " pennsylvania law journal". v8-11 titled "american law journal, new series") – mf#LLMC 82-9031 – us LLMC [340]
The american law journal : and miscellaneous repertory – v1-6. 1808-17 [all publ] – 35mf – 9 – $52.00 – (series ed by john e hall, but publ varied) – mf#llmc95-881 – us LLMC [340]
The american law journal – Columbus, OH. v1-2. 1884-85 (all publ) – 10mf – 9 – $15.00 – mf#LLMC 84-398 – us LLMC [340]
American law journal [1808] – La Salle IL 1808-10 – 1 – mf#4543 – us UMI ProQuest [340]
American law journal [1842] – La Salle IL 1842-52 – 1 – mf#4141 – us UMI ProQuest [340]
The american law journal and miscellaneous reporter – v1-3. 1808-10 (all publ) – 9 – mf#LLMC 84-3991 – us LLMC [340]
American law journal, new series *see* American law journal
American law journal (philadelphia) – v1-4. 1848-52 – (= ser Historical legal periodical series) – 1 – $85.00 – mf#100481 – us Hein [340]
The american law list – v1-15. 1898-1910 (all publ) – 119mf – 9 – $178.00 – (missing: v1. v11 n3. v13 no 2. v15) – mf#LLMC 84-400 – us LLMC [340]
American law magazine – La Salle IL 1843-46 – 1 – mf#4149 – us UMI ProQuest [340]
American law magazine – Philadelphia. v1-6. 1843-46 (all publ) – (= ser Historical legal periodical series) – 1 – $85.00 – mf#100501 – us Hein [340]
The american law magazine – Chicago, IL. v1-2 n2. 1882-83 (all publ) – (= ser Central law journal) – 6mf – 9 – $9.00 – (merged into: the central law journal llmc 82-912) – mf#LLMC 84-401 – us LLMC [340]
The american law magazine – Philadelphia. v1-6. 1843-46 – 9 – mf#LLMC 82-905 – us LLMC [340]
The american law of real property / Hilliard, Francis – 3d ed. New York: Banks, Gould, 1855. 2v. LL-1699 – 1 – us L of C Photodup [346]
The american law of real property. 3d ed / Tiedeman, Christopher Gustavus – St. Louis, Thomas, 1906. 1017 p. LL-1026 – 1 – us L of C Photodup [346]
The american law primer, for public and private schools, families, and the unprofessional generally / Leahy, Daniel F – San Francisco, Domestic Publishing Co., 1889. 181 p. LL-208 – 1 – us L of C Photodup [346]
The american law record – Cincinnati, OH. v1-15. 1872-8 (all publ) – 45mf – 9 – $202.00 – mf#LLMC 82-906 – us LLMC [340]
American law register and review *see* University of pennsylvania law review
The american law register and review – v1-9. ns: v1-51 + digest. 1852-1912 – 177mf – 9 – $796.00 – (missing: ns v50. cont: the american law journal llmc 82-903. cont by: the university of pennsylvania law review) – mf#LLMC 82-970 – us LLMC [340]
The american law review – v1-63 n3. 1866-1929 (all publ) – 203mf – 9 – $913.00 – (title merged into: the united states law review publ by us review corporation ny. not offered by llmc for copyright reasons) – mf#LLMC 82-907 – us LLMC [340]
American law school review – v1-10. 1902-47 – 1 – $240.00 – mf#0034 – us Brook [340]
The american law school review : an intercollegiate law journal – St Paul: West Publ Co. 1902-47 (all publ) – 97mf – 9 – $145.00 – (lacking: v7) – mf#LLMC 84-402 – us LLMC [340]
American law times reports – Washington/New York: American Law Times Assoc/Cox/Hurd & Houghton. 1st series: v1-6. 1868-74. ns: v1-4. 1874-77 (all publ) – 69mf – 9 – $103.00 – (v1-2 of os entitled: the law times (u.s.) courts reports; cont by: the law and enquity reporter after v6) – mf#LLMC 81-412 – us LLMC [347]
The american lawyer – New York. v1-16. 1896-1908 – 106mf – 9 – $159.00 – mf#LLMC 82-908 – us LLMC [340]
American lawyer [1893] – La Salle IL 1893-1908 – 1 – mf#5219 – us UMI ProQuest [340]
American lawyer [1979] – New York NY 1979-91 – 1,5,9 – ISSN: 0162-3397 – mf#12142 – us UMI ProQuest [340]
The american lawyer and businessman's formbook... / Beadle, Delos White – New York: Ensign, Bridgman & Fanning, 1856 – 4mf – 9 – $6.00 – mf#LLMC 91-501 – us LLMC [346]
American leader – 1935 oct 25, dec 6-1936 may 15 – 1 – mf#1269588 – us WHS [071]
American leading cases : being select decisions of american courts, in several departments of law; with especial reference to mercantile law / Hare, John Innes Clark – 4th ed. Philadelphia: Johnson, 1857. 2v. LL-882 – 1 – us L of C Photodup [347]
American leading cases : being select decisions of the american courts in several departments of law with special reference to mercantile law / Clarke, J I & Wandle, H B – 5th enl ed. Philadelphia: T & T W Johnson. v1-2. 1871 – 20mf – 9 – $30.00 – mf#LLMC 95-115 – us LLMC [347]
American leather chemists association. journal – Lubbock TX 1951+ – 1,5,9 – ISSN: 0002-9726 – mf#761 – us UMI ProQuest [660]
American Legion *see*
– Colorado legion magazine
– Colorado service star
– Hoosier legionaire
– Iowa legionaire
– Milwaukee county legionnaire
– Milwaukee legion news
– National news of the american legion auxiliary
– National speakers' information service
– Nebraska legion news
– Pelicanaire
– Schlitz milwaukee post no 411
– Speakers' information service
– Tennessee legionnaire
– Torch
– Transmitter
– West virginia legionnaire
– Your four forty niner
American legion – Indianapolis IN 1926+ – 1,5,9 – ISSN: 0886-1234 – mf#6556 – us UMI ProQuest [366]
American legion councillor / Lucas Co. Toledo – v1 n1. apr 1921-dec 1945 [wkly] – 9r – 1 – mf#B33931-33939 – us Ohio Hist [071]
American legion. massachusetts dept official post – MA Dept, 1922 – 1 – mf#2689188 – us WHS [071]
American legion news – reel 1: jun 16 1931-oct 6 1934 – 1 – (some iss in mss form; oct 25 1932-sep 10 1934 filmed with: speakers' information) – mf#945772 – us WHS [071]
American legion press / Lucas Co. Toledo – jan 1946-dec 1965,jun 1966-nov 1986 [wkly, biwkly] – 13r – 1 – mf#B33960-33972 – us Ohio Hist [071]
American legion weekly – Indianapolis IN 1919-26 – 1 – mf#8558 – us UMI ProQuest [366]
American Legislators' Association. Council of State Governments *see* Book of the states
American letters of sir horace plunkett, 1883-1932 – Oxford: Plunkett Foundation for Co-Operative Studies – (= ser BRRAM series) – 2r – 1 – £134 / $268 – (int by bernard crick) – mf#r02826 – uk Microform Academic [630]
American letters of the unitarian college, manchester, 1751-1907 / Unitarian College, Manchester – (= ser BRRAM series) – 1r – 1 – £47 / $94 – (int by b w clapp) – mf#r95665 – uk Microform Academic [243]
American liberal – v1 n1-23 [1932 dec 1-1933 nov 1] – 1 – mf#1051963 – us WHS [071]
American libertarian – v1 n1-v4 n4 [1986 jul-1989 oct] – 1 – mf#1581734 – us WHS [071]
American libraries – Chicago IL 1970+ – 1,5,9 – ISSN: 0002-9769 – mf#5731 – us UMI ProQuest [020]
American libraries book procurement center accessions list indonesia / US Library of Congress – Djakarta, 1964-1972. v1-7 – 75mf – 9 – (missing: 1972 v7(8-9, 12)) – mf#SE-1973 – ne IDC [959]
American Library Association *see*
– Computer output microfilm (com) hardware and software: the state of the art
– Micrographics education for librarians
– Some current reprographic concerns related to interlibrary loan
– A survey of telefacsimile use in libraries in the united states
American library association. ala bulletin – Chicago IL 1907-69 – 1 – ISSN: 0364-4006 – mf#951 – us UMI ProQuest [020]
American library association. health and rehabilitative library services division. hrlsd journal – Chicago IL 1976-78 – 1,5,9 – (cont: health & rehabilitative library services division journal) – ISSN: 0196-7371 – mf#11123,02 – us UMI ProQuest [020]
American library association. health and rehabilitative library services division journal – Chicago IL 1976 – 1,5,9 – (cont: health & rehabilitative library services; cont by: hrlsd journal) – ISSN: 0196-738X – mf#11123.02 – us UMI ProQuest [020]
American library association. library administration division newsletter – Chicago IL 1975-78 – 1,5,9 – (cont by: lama newsletter) – ISSN: 0093-7972 – mf#10533.01 – us UMI ProQuest [020]
A(merican) l(ibrary) a(ssociation) portrait index : index to portraits contained in printed books and periodicals / ed by Lane, W C & Browne, N E – Washington, DC, 1906 – 29mf – 9 – mf#O-120 – ne IDC [700]
American library association. reference and adult services division update – Chicago IL 1980-96 – 1,5,9 – (cont by: reference & user services association rusa update) – ISSN: 0198-8344 – mf#12523.01 – us UMI ProQuest [020]
American library association. resources and technical services division newsletter – Chicago IL 1979-89 – 1,5,9 – (cont by: alcts newsletter) – ISSN: 0360-5906 – mf#12524 – us UMI ProQuest [020]
American literary annuals and gift books, 1825-1865 : based on the bibliography of the same, by ralph thompson – 58r – 1 – (coll contains 469 titles representing literature and art from the pre-civil war period) – mf#C35-14300 – us Primary [880]
American literary gazette and publishers' circular – La Salle IL 1855-72 – 1 – mf#3758 – us UMI ProQuest [070]
American literary history – Oxford, England 1989+ – 1,5,9 – ISSN: 0896-7148 – mf#17034 – us UMI ProQuest [400]
American literary magazine – La Salle IL 1847-49 – 1 – mf#3759 – us UMI ProQuest [420]
American literary manuscripts, 1650-1850 : from the huntington library san marino, california – 6r – 1 – (covers major and minor writers. incl printed guide) – mf#C35-22300 – us Primary [420]
American literature : a journal of literary history, criticism and bibliography – Durham NC 1929+ – 1,5,9 – ISSN: 0002-9831 – mf#1022 – us UMI ProQuest [420]
American literature abstracts – San Jose CA 1967-72 – 1,5,9 – ISSN: 0002-984X – mf#3443 – us UMI ProQuest [400]
American Lithuanian Roman Catholic Women's Alliance et al *see* Moteru dirva
American lumberman – 1899 jan-dec; 1900 jan-dec; 1901 jan-dec; 1902 jan-dec; 1903 jan-dec; 1904 jan-dec; 1905 jan-dec; 1906 jan-apr 7-jun 9 – 1 – mf#1386568 – us WHS [071]
American lumberman and building products merchandiser – 1899 jan-dec; 1900 jan-dec; 1901 jan-dec; 1902 jan-dec; 1903 jan-dec; 1904 jan-dec; 1905 jan-dec; 1906 apr 7-jun9 – 1 – mf#1051967 – us WHS [071]
American lung association bulletin – New York NY 1973-83 – 1,5,9 – (cont: national tuberculosis and respiratory disease association bulletin) – ISSN: 0092-5659 – mf#1974,01 – us UMI ProQuest [616]
American lutheran – v6-49. 1923-66 [complete] – 9r – 1 – (superseded by: lutheran forum) – ISSN: 0024-7456 – mf#ATLA S0520 – us ATLA [242]
American lutheran biographies : or, historical notices of over three hundred and fifty leading men of the american lutheran church: from its establishment to the year 1890 / Jensson, Jens Christian – Milwaukee WI: J C Jensson c1890 [mf ed 1992] – 3mf – 9 – 0-524-03013-8 – (incl bibl ref) – mf#1990-4535 – us ATLA [242]
American Lutheran Church *see*
– American lutheran church
– Right now
American lutheran church : almanac yearbook / American Lutheran Church – 1931-87 [complete] – 11r – 1 – mf#ATLA S0668 – us ATLA [242]
American lutheran church. eastern district : reports and actions – v1-26. 1961-86 – 4r – 1 – (lacks some pp) – mf#atla s0395 – us ATLA [242]
American lutheran mission work in china / Syrdal, Rolf Arthur – Madison, NJ, 1942. Chicago: Dep of Photodup, U of Chicago Lib, 1978 (1r); Evanston: American Theol Lib Assoc, 1984 (1r) – 1 – 0-8370-0709-7 – mf#1984-T106 – us ATLA [242]
American lutheran survey : a review of world progress and problems – v1-20 n9. oct 1914-jun 1928 – 15r – 1 – (lack some pp) – mf#atla s0092 – us ATLA [242]
American machinist [1877] – Cleveland OH 1877-1985 – 1,5,9 – (cont by: american machinist & automated manufacturing [am]) – ISSN: 0002-9858 – mf#30.02 – us UMI ProQuest [621]
American machinist [1964] – Cleveland OH 1964+ – 1,5,9 – (cont: american machinist & automated manufacturing [am]) – ISSN: 1041-7958 – mf#30,02 – us UMI ProQuest [621]
American machinist & automated manufacturing [am] – Cleveland OH 1986-87 – 1,5,9 – (cont: american machinist; cont by: american machinist) – ISSN: 0886-0335 – mf#30.02 – us UMI ProQuest [621]
American madura mission : jubilee volume, 1834-1884 – Madras: American Madura Mission, 1886 [mf ed 1995] – (= ser Yale coll) – 83p/viii (ill) – 1 – 0-524-09112-9 – (incl ind) – mf#1995-0112 – us ATLA [954]
American magazine [1741] : or a monthly view of the political state of the british colonies – La Salle IL 1741 – 1 – mf#3503 – us UMI ProQuest [071]
American magazine [1787] : containing a miscellaneous collection of original and other valuable essays in prose and verse – La Salle IL 1787-88 – 1 – mf#3502 – us UMI ProQuest [080]
American magazine [1815] : a monthly miscellany devoted to literature, science, history, biography and the arts – La Salle IL 1815-16 – 1 – mf#3540 – us UMI ProQuest [071]
American magazine [1841] : and repository of useful literature devoted to science, literature, and arts, and embellished with numerous engravings – La Salle IL 1841-42 – 1 – mf#3926 – us UMI ProQuest [073]
American magazine [1876] – La Salle IL 1876-1906 – 1 – mf#5221 – us UMI ProQuest [073]
American magazine [1884] – Killen TX 1884-88 – 1 – mf#2867 – us UMI ProQuest [071]
American magazine 1851-52 *see* American gazette 1768-70 – the american magazine 1851-52
American magazine and historical chronicle – Killen TX 1743-46 – 1 – mf#3501 – us UMI ProQuest [071]
American magazine and monthly chronicle for the british colonies – La Salle IL 1757-58 – 1 – mf#3505 – us UMI ProQuest [073]

American magazine of civics – La Salle IL 1892-97 – 1 – mf#5222 – us UMI ProQuest [350]

American magazine of useful and entertaining knowledge – La Salle IL 1834-37 – 1 – mf#3714 – us UMI ProQuest [390]

American magazine of wonders : and marvellous chronicle – La Salle IL 1809 – 1 – mf#3541 – us UMI ProQuest [073]

American magazine; or general repository – La Salle IL 1769 – 1 – mf#3504 – us UMI ProQuest [100]

American Marathi Mission *see* Memorial papers of the american marathi mission, 1813-1881

American marine engineer – 1950 apr-1962 jan; feb-1967; 1968-83 – 1 – mf#783031 – us WHS [071]

American maritime cases : published under the auspices of the maritime law association of the united states – Baltimore, 1923- [1923 pt1-1925 pt2] – 54mf – 9 – $81.00 – (updates planned as add vols fall out of copyright) – mf#llmc97-586 – us LLMC [341]

American maritime officer – 6 1 1971-1975 aug; 6 2 1975 oct-1979; 6 3 1980-1984 [misc iss]; 6 4 1985 jan-jul – 1 – mf#1289556 – us WHS [071]

American marketing association. conference proceedings – Chicago IL 1954+ – 1,5,9 – mf#5966 – us UMI ProQuest [650]

American masonic register and ladies' and gentlemen's magazine – La Salle IL 1820-23 – 1 – mf#3542 – us UMI ProQuest [366]

American masonic register and literary companion : being a periodical devoted to masonry, arts and science, biography – La Salle IL 1839-47 – 1 – mf#4620 – us UMI ProQuest [071]

American mass line : newspaper of the american communist workers movement [marxist-leninist] – v1 n1-8 [1970 jun 15-aug 31]; v1 n9 [1971 jan 27] – 1 – mf#1107173 – us WHS [335]

American material from the tredegar park muniments, 1719-1825 – Aberystwyth: National Library of Wales – (= ser BRRAM series) – 7r – 1 – £469 / $938 – (with p/g; int by clare taylor) – mf#r97073 – uk Microform Academic [920]

American material in the archives of the united society for the propagation of the gospel / United Society for the Propagation of the Gospel. Archives – (= ser British records relating to america in microform) – 1 – (int by isobel pridmore; ser a: letter books, v1-26 1702-37 8r [95869]; ser b: letter books, v1-25 1701-86 14r [95870]; ser c: copy letter books, v1-15 18th century 5r [95891]; ind to ser a, b and c 1r [96112]) – uk Microform Academic [220]

American material in the liverpool papers : from the papers of charles jenkinson, 1st earl of liverpool (1727-1808) and robert jenkinson, 2nd earl of liverpool (1770-1828) – British Library add. mss. 38190-489 – (= ser BRRAM series) – 3r – 1 – £201 / $402 – (int by geoffrey seed) – mf#r04987 – uk Microform Academic [025]

American mathematical monthly – Washington DC 1894+ – 1,5,9 – ISSN: 0002-9890 – mf#10928 – us UMI ProQuest [510]

American Mathematical Society *see* Abstracts of papers presented to the american mathematical society

American mathematical society. bulletin – Providence RI 1894-1978 – 1,5,9 – ISSN: 0002-9904 – mf#13410 – us UMI ProQuest [510]

American mathematical society. bulletin [new series] – Providence RI 1979+ – 1,5,9 – ISSN: 0273-0979 – mf#13411 – us UMI ProQuest [510]

American mathematical society. notices – Providence RI 1954+ – 1,5,9 – ISSN: 0002-9920 – mf#1679 – us UMI ProQuest [510]

American mathematical society. proceedings – Providence RI 1950+ – 1,5,9 – ISSN: 0002-9939 – mf#13412 – us UMI ProQuest [510]

American mathematical society. transactions – Providence RI 1900+ – 1,5,9 – ISSN: 0002-9947 – mf#3258 – us UMI ProQuest [510]

American mechanics' magazine : containing useful original matter, on subjects connected with manufactures, the arts and sciences – La Salle IL 1825-26 – 1 – mf#3708 – us UMI ProQuest [621]

American medical and philosophical register : or, annals of medicine, natural history, agriculture, and the arts – La Salle IL 1810-14 – 1 – mf#4414 – us UMI ProQuest [610]

American Medical Association *see*
- Archives of ophthalmology
- Citation
- Jama

American medical association. ama archives of neurology and psychiatry – Killen TX 1919-59 – 1 – ISSN: 0096-6886 – mf#69 – us UMI ProQuest [616]

American medical digest – Killen TX 1882-88 – 1 – mf#5224 – us UMI ProQuest [610]

American Medical Informatics Association *see* Journal of the american medical informatics association (jamia)

American medical intelligencer : a concentrated record of medical science and literature – La Salle IL 1837-42 – 1 – mf#3927 – us UMI ProQuest [610]

American medical monthly – Killen TX 1854-62 – 1 – mf#3152 – us UMI ProQuest [610]

American medical news – Chicago IL 1958+ – 1,5,9 – ISSN: 0001-1843 – mf#1162 – us UMI ProQuest [610]

American medical periodicals, 1797-1900 : from the national library of medicine and other major institutions – 750r in 15 units of 50r ea – 1 – $78,750.00 coll $5,250.00 per unit – (coll offers more than 1,200 19th-century medical periodicals from the holdings of the national library of medicine and other major institutions. guide available) – mf#C39-29310 – us Primary [610]

American medical recorder – La Salle IL 1818-29 – 1 – mf#4595 – us UMI ProQuest [610]

American medical review and journal of original and selected papers in medicine and surgery – La Salle IL 1824-26 – 1 – mf#3704 – us UMI ProQuest [610]

American Medical Technologists *see*
- Amt events
- Journal of the american medical technologists

American medical times – Killen TX 1860-64 – 1 – mf#3211 – us UMI ProQuest [610]

American medical women's association. journal – Alexandria VA 1946+ – 1,5,9 – ISSN: 0098-8421 – mf#2523 – us UMI ProQuest [610]

American mental health counselors association. amhca journal – Alexandria VA 1979-87 – 1,5,9 – (cont by: journal of mental health counseling) – ISSN: 0193-1830 – mf#11726.01 – us UMI ProQuest [362]

American merino – 1882 jan-1883 dec – 1 – mf#1051970 – us WHS [071]

American messenger – 6 1 v53 n8 [1895 aug] – 1 – mf#203558 – us WHS [071]

American meteorological journal : a monthly review of meteorology and allied branches of study – Killen TX 1884-96 – 1 – mf#3357 – us UMI ProQuest [550]

American methodism / Hurst, John Fletcher – New York: Eaton & Mains, 1903 [mf ed 1992] – (= ser Methodist coll) – 3v on 4mf – 9 – 0-524-03714-0 – mf#1990-4819 – us ATLA [242]

American methodism : its divisions and unification / Neely, Thomas Benjamin – New York: Fleming H Revell, c1915 [mf ed 1992] – (= ser Methodist coll) – 1mf – 9 – 0-524-04060-5 – (incl bibl ref) – mf#1990-4968 – us ATLA [242]

American methodism / Scudder, Moses Lewis – Hartford, CT: Scranton, 1867 [mf ed 1992] – (= ser Methodist coll) – 2mf – 9 – 0-524-02844-3 – mf#1990-4465 – us ATLA [242]

American methodist ladies' centenary association : its connectional character / Smart, James S – [Cincinnati: Poe & Hitchcock, 1866] [mf ed 1981] – (= ser Women & the church in america 203) – 1mf – 9 – 0-8370-1633-9 – mf#1984-2203 – us ATLA [242]

American metropolitan magazine – La Salle IL 1849 – 1 – mf#4355 – us UMI ProQuest [073]

American midland naturalist – Notre Dame IN 1909+ – 1,5,9 – ISSN: 0003-0031 – mf#964 – us UMI ProQuest [500]

The American Midland Naturalist *see* Devoted to natural history, primarily that of the prairie states

American military newspapers on microfilm, pt 1 – 162 newspapers on 742r – 1 – $55,000.00 $90.00r – State Historical Society of Wisconsin – us UMI ProQuest [071]

American millenarian and prophetic review – La Salle IL 1843-44 – 1 – mf#5223 – us UMI ProQuest [240]

American mineralogical journal – La Salle IL 1810-14 – 1 – mf#4544 – us UMI ProQuest [550]

American mining code / Copp, Henry Norris – Washington, DC, 1882. 226p. LL-66 – 1 – (9th ed. washington, dc, 1896 214p) – mf#ll-65 – us L of C Photodup [348]

American Ministries International *see* Native times

The american mission in egypt, 1854 to 1896 / Watson, Andrew – Pittsburgh: United Presbyterian Board of Publ, 1898, c1897 – 2mf – 9 – 0-8370-7351-0 – (incl ind) – mf#1986-1351 – us ATLA [240]

The american mission in the sandwich islands : a vindication and an appeal, in relation to the proceedings of the reformed catholic mission at honolulu / Ellis, William – London: Jackson, Walford, and Hodder, 1866 – 1mf – 9 – 0-524-04070-2 – mf#1991-2015 – us ATLA [241]

American missionary – 1860-1927 – (= ser American indian periodicals... 1) – 327mf – 9 – $1980.00 – us UPA [240]

An american missionary : a record of the work of rev william h judge / Judge, Charles Joseph – 4th ed. Maryknoll: Catholic Foreign Mission Soc c1907 [mf ed 1992] – 1mf – 9 – 0-524-05084-8 – mf#1991-2208 – us ATLA [240]

The american missionary – New York: American Missionary Association, 1846-1934 [mthly ex aug] [mf ed New Orleans [1975?] – 15r – 1 – (absorbed: congregational work and: pilgrim missionary in apr 1909. iss with: congregationalist (boston, ma 1921), jan 1929-feb 1930, and with: congregationalist and herald of gospel liberty, mar 1930-mar 1934 as a mthly magazine. merged with: missionary herald (boston, ma) to form: missionary herald at home and abroad (boston, ma 1934). vols for apr 1924-dec 1928, official organ of the congregational missionary societies for the home field. publ by the american missionary association and, apr 1909-28, by the congregational home missionary society and other congregational missionary societies. guide may be purchased separately $25) – mf#d3623 – us Amistad [243]

The american missionary – 1846-1934 – 15r – 1 – $1950.00 set – (guide may be purchased separately $25) – mf#d3623 – us Scholarly Res [240]

The american missionary – v1-63 n.s. 1-25. 1857-mar 1933* – 19r – 1 – (lacks some iss) – mf#ATLA S0227 – us ATLA [240]

American Missionary Association *see*
- Annual report of the american missionary association
- Manuscripts

The american missionary association archives – 1839-82 [mf ed 2002] – 261r – 1 – $33,930.00 set – (guide may be purchased separately $40) – mf#d3622 – us Amistad [240]

American mizrachi women – New York, NY. 1974-86/87. Continued by: The Amit Woman – 1 – us AJPC [071]

American model printer *see* Printer's circular and stationers' and publishers' gazette, 1866-1888 / american model printer, 1879-1882 / craftsman, 1884-1888

American monitor : or the republican magazine – La Salle IL 1785 – 1 – mf#3507 – us UMI ProQuest [073]

American monthly magazine [1824] – La Salle IL 1824 – 1 – mf#3543 – us UMI ProQuest [071]

American monthly magazine [1829] – La Salle IL 1829-31 – 1 – mf#3867 – us UMI ProQuest [071]

American monthly magazine [1833] – La Salle IL 1833-38 – 1 – mf#3928 – us UMI ProQuest [073]

American monthly magazine and critical review – La Salle IL 1817-19 – 1 – mf#3544 – us UMI ProQuest [500]

American monthly review [1795] – La Salle IL 1795 – 1 – mf#3506 – us UMI ProQuest [420]

American monthly review [1832] – La Salle IL 1832-33 – 1 – mf#3713 – us UMI ProQuest [420]

American moral and sentimental magazine – La Salle IL 1797-98 – 1 – mf#3508 – us UMI ProQuest [420]

American motorsport international : [newsletter] – 1987 oct-1988 aug/sep – 1 – mf#3362701 – us WHS [071]

American museum : and repository of arts and science, as connected with domestic manufactures and national industry – La Salle IL 1822-23 – 1 – mf#4546 – us UMI ProQuest [060]

The american museum – v1-12 1787-92 – 1 – us AMS Press [975]

American museum in britain, manuscripts from the... 1650-1903 : from the american museum, claverton manor, bath – (= ser British records relating to america in microform) – 2r – 1 – (with int by g m candler) – mf#97121 – uk Microform Academic [970]

American museum novitates – New York NY 1981+ – 1,5,9 – ISSN: 0003-0082 – mf#12600 – us UMI ProQuest [060]

American museum of literature and the arts – La Salle IL 1838-39 – 1 – mf#4356 – us UMI ProQuest [060]

American Museum of Natural History *see*
- Anthropological papers of the american museum of natural history
- Bulletin of the american museum of natural history

American museum, or, universal magazine – La Salle IL 1787-92 – 1 – mf#3509 – us UMI ProQuest [060]

American music – Champaign IL 1983+ – 1,5,9 – ISSN: 0734-4392 – mf#14267 – us UMI ProQuest [780]

American music : a volume of tunes, transcribed in quadrate notes – [Virginia, c1750-70] [mf ed 19–] – 6mf / 1r [1989] – 9,1 – mf#fiche 1158 / pres. film 62 – us Sibley [780]

American music teacher – Cincinnati OH 1951+ – 1,5,9 – ISSN: 0003-0112 – mf#2160 – us UMI ProQuest [780]

American musical journal – La Salle 1834-35 – 1 – mf#3712 – us UMI ProQuest [780]

American musical magazine [1786] – La Salle IL 1786-87 – 1,5,9 – mf#3510 – us UMI ProQuest [780]

American musical magazine [1801] – La Salle IL 1801 – 1 – mf#4545 – us UMI ProQuest [780]

American Musicological Society (International Congress : 1939) *see* Papers read at the international congress of musicology

American Mutuality Foundation *see* Full life magazine

American myths and legends / Skinner, Charles Montgomery – Philadelphia, PA. v1-2. 1903 – 1r – us UF Libraries [390]

American Name Society *see* Names

American national red cross. annual report – Washington, 1910-76 – 1 – $144.00 – mf#0035 – us Brook [366]

American National Standards Institute *see* Magazine of standards

American national standards institute. ansi reporter – New York NY 1974-93 – 1,5,9 – ISSN: 0038-9676 – mf#9173 – us UMI ProQuest [071]

American National Women's Trade Union League *see* America

American natural history, 1769-1865 – 264r – 1 – $27,720.00 coll – (fr bibl of american natural history 1769-1865 by max meisel. guide accompanies coll) – mf#C39-29330 – us Primary [500]

American naturalist – Chicago IL 1872+ – 1,5,9 – ISSN: 0003-0147 – mf#3 – us UMI ProQuest [500]

American negligence cases / ed by Hamilton, T F – New York: Remick & Schilling. v1-17. 1789-1897 (all publ) – (= ser American negligence reports. current series) – 179mf – 9 – $268.00 – (a complete coll of all reported negligence cases decided in the federal courts of the us, the courts of last resort of all the states and territories from the earliest times...topically arranged with notes of english cases and annotations. publ as a retrospective companion set to: american negligence reports, current series) – mf#LLMC 84-697 – us LLMC [347]

American negligence digest : from 1897-1907 / Eagle, Walter – New York: Remick & Schilling. 1v (all publ) – 18mf – 9 – $27.00 – (a digest of all the negligence cases contained in the series of american negligence reports, v1-20 inclusive) – mf#LLMC 84-699B – us LLMC [348]

American negligence digest : from the earliest times to 1902 / Hook, Alfred J – Remick & Schilling. 1v. 1902 – 7mf – 9 – $10.50 – mf#LLMC 84-699A – us LLMC [340]

American negligence reports, current series / ed by Gardner, John M – New York: Remick & Schilling. v1-21. 1897-1910 (all publ) – 190mf – 9 – $285.00 – (the "current series" references the fact that "american negligence cases" was being publ simultaneously to provide retrospective coverage in this subject area; title cont by: negligence compensation cases, annotated; individual titles under this series also listed separately) – mf#LLMC 84-698 – us LLMC [348]

American negro – v1 n6 [1956 apr] – 1 – mf#5131918 – us WHS [071]

The american negro – Springfield, MO, 1890 [mf ed 1947] – (= ser Negro Newspapers on Microfilm) – 1r – 1 – us L of C Photodup [071]

American Negro Academy. Washington, DC *see* Occasional papers

American negro historical society collection – 1790-1905 [mf ed Wilmington 1998] – 12r – 1 – $1560.00 set – (guide may be purchased separately $25) – mf3362 – historical society of pennsylvania – us Scholarly Res [976]

American Negro Labor Congress *see* The liberator

American Nephrology Nurses' Association *see* Anna journal

American neptune – Salem MA 1941+ – 1,5,9 – ISSN: 0003-0155 – mf#1028 – us UMI ProQuest [071]

The american news – 1876-77 – (= ser American newspapers and periodicals published in the uk and europe 18th-20th centuries) – 1r – 1 – £55.00 – uk World [072]

American news analyst – v4 n10-v6 n8 [1974 oct-1976 oct] – 1 – mf#345255 – us WHS [071]

American news report – Djakarta, 1961-1965 – 218mf – 9 – mf#SE-1311 – ne IDC [950]

American newspaper directory – 14 [1882]; 24 [1892]; 27 [1895] – 1 – mf#1727552 – us WHS [030]

American Newspaper Guild *see*
- Guild reporter
- Wages and conditions in american newspaper guild

American Newspaper Publishers Association. Bureau of Advertising *see* Community advertising

American newspapers and periodicals published in the uk and europe : 18th-20th centuries – 40r – 1 – £1850.00 set – (individual titles listed separately) – mf#ANP – uk World [072]

American nineteenth century art : subject collections – (= ser Art exhibition catalogues on microfiche) – 42 catalogues on 60mf – 9 – £440.00 – (individual titles not listed separately) – uk Chadwyck [700]

American nonconformist – 1893 jan 19-1895 jun 13; jun 16; 1895 jan 20-1896 apr 2 – 1 – mf#857232 – us WHS [071]

American nonconformist and industrial liberator – 1892 jun 16 – 1 – mf#857229 – us WHS [071]

American nonconformist and kansas industrial liberator – 1886 nov 11, 1890 jul 10 – 1 – mf#857227 – us WHS [071]

American notary – Tallahassee FL 1966-80 – 1,5,9 – ISSN: 0044-7773 – mf#7794 – us UMI ProQuest [340]

American notes for general circulation / Dickens, Charles – New York: Harper, 1842 [mf ed 1983] – 2mf – 9 – 0-665-44211-4 – mf#44211 – cn Canadiana [917]

American notes for general circulation / Dickens, Charles – Paris 1842 [mf ed Hildesheim 1995-98] – 1v on 2mf – 9 – €60.00 – 3-487-27222-9 – gw Olms [880]

American nurse – Silver Spring MD 1972+ – 1,5,9 – ISSN: 0098-1486 – mf#13588,01 – us UMI ProQuest [610]

American nurseryman – Chicago IL 1904+ – 1,5,9 – ISSN: 0003-0198 – mf#2107 – us UMI ProQuest [635]

American Nurses' Foundation *see* Nursing research report

American observer – New York NY 1931-72 – 1,5,9 – ISSN: 0003-0201 – mf#220 – us UMI ProQuest [378]

The american occupation in germany, 1918-1923 – 2r – 1 – $260.00 – (with printed guide) – mf#S1679 – us Scholarly Res [943]

American Oil Chemists' Society *see* Journal of the american oil chemists' society

American opinion – Appleton WI 1958-85 – 1,5,9 – ISSN: 0003-0236 – mf#1907 – us UMI ProQuest [321]

The american orator / Munn, Lewis C – Boston: Tappan & Whitmore, 1853 – 5mf – 9 – $7.50 – mf#LLMC 91-505 – us LLMC [340]

American orator's own book / Agar, J – Auburn, NY. 1853 – 1r – us UF Libraries [025]

American organist [1918] – New York NY 1918-70 – 1 – ISSN: 0003-0260 – mf#680 – us UMI ProQuest [780]

American organist [1967] – New York NY 1967+ – 1,5,9 – ISSN: 0164-3150 – mf#13408 – us UMI ProQuest [780]

American Oriental Society *see* Journal of the american oriental society

American orthoptic journal – Madison WI 1951+ – 1,5,9 – ISSN: 0065-955X – mf#1744 – us UMI ProQuest [617]

American Osteopathic Association *see* Jaoa – the journal of the american osteopathic association

American paint and coatings journal – Richmond VA 1991-97 – 1 – (cont by: apcj faxnews) – ISSN: 0098-5430 – mf#14786.02 – us UMI ProQuest [660]

American painters : with eighty-three examples of their work engraved on wood / Sheldon, George William – London [1884] – (= ser 19th c art & architecture) – 4mf – 9 – mf#4.2.1410 – uk Chadwyck [700]

American paintings – (= ser Christie's pictorial archive new york) – 124mf – 9 – $1080.00 – 0-907006-67-1 – (almost 7500 pictures, drawings & sculpture incl latin american works) – uk Mindata [750]

American paper industry – Glenview IL 1919-85 – 1,5,9 – (cont by: paper industry) – ISSN: 0003-0333 – mf#6604.02 – us UMI ProQuest [670]

American papers of ralph carr, 1741-1778 – Newcastle upon Tyne: Northumberland Record Office – (= ser BRRAM series) – 1r – 1 – £67 / $134 – (with p/g; int by w e minchinton) – mf#r96783 – uk Microform Academic [337]

American papers of sir charles richard vaughan, 1774-1849 – Oxford: Codrington Library, All Souls' College – (= ser BRRAM series) – 12r – 1 – £804 / $1608 – (with p/g & finding list; int by richard simmons) – mf#r97600 – uk Microform Academic [327]

American papers of w s lindsay, 1861-66 : from the private collection of p a r lindsay – (= ser BRRAM series) – 1r – 1 – £67 / $134 – mf#r97440 – uk Microform Academic [920]

American Park and Outdoor Art Association *see* Report

American patriot – vl n1-4 [1976/1977 winter-1977 fall] – 1 – mf#290520 – us WHS [071]

American pen – New York NY 1970-74 – 1,5 – ISSN: 0003-0376 – mf#6278 – us UMI ProQuest [400]

American people's journal of science, literature and art – La Salle IL 1850 – 1 – mf#4357 – us UMI ProQuest [071]

American Pharmaceutical Association *see*
- Journal of the american pharmaceutical association
- Journal of the american pharmaceutical association : apha

American pharmacy – Washington DC 1978-95 – 1,5,9 – (cont: journal of the american pharmaceutical association; cont by: journal of the american pharmaceutical association [apha]) – ISSN: 0160-3450 – mf#18.03 – us UMI ProQuest [615]

American pheasant and waterfowl society bulletin – Newark DE 1973 – 1 – mf#7011 – us UMI ProQuest [639]

American Philatelic Research Library *see* Philatelic literature review

American Philatelic Society et al *see* Pmcc bulletin

American philosophical association. proceedings and addresses – Newark DE 1927+ – 1,5,9 – ISSN: 0065-972X – mf#2528 – us UMI ProQuest [100]

American Philosophical Society *see* Transactions of the american philosophical society

American Philosophical Society. Committee of History, Moral Sciences, and General Literature *see* Transactions of the historical and literary committee of the american philosophical society held at philadelphia for promoting useful knowledge

American philosophical society. proceedings – Philadelphia PA 1838+ – 1,5,9 – ISSN: 0003-049X – mf#830 – us UMI ProQuest [100]

American philosophy : the early schools / Riley, Woodbridge – New York: Dodd, Mead, 1907 [mf ed 1991] – 2mf – 9 – 0-7905-8723-8 – (incl bibl ref) – mf#1989-1948 – us ATLA [190]

The american philosophy of law. / Le Buffe, Francis Peter – 4th ed. New York, Crusader Press, 1947. 418 p. LL-304 – 1 – us L of C Photodup [340]

The american philosophy pragmatism : critically considered in relation to present-day theology / Huizinga, Arnold van Couthen Piccardt – Boston: Sherman, French, 1911 – 1mf – 9 – 0-7905-3917-9 – mf#1989-0410 – us ATLA [190]

American photo – New York NY 1990+ – 1,5,9 – (cont: american photographer) – ISSN: 1046-8986 – mf#17562 – us UMI ProQuest [770]

American photo-engraver *see* Plate makers' criterion, 1907-1909 / american photo-engraver, 1908-1955

American photographer – New York NY 1980-89 – 1,5,9 – (cont by: american photo) – ISSN: 0161-6854 – mf#12577 – us UMI ProQuest [770]

American photography – New York NY 1950-53 – 1 – ISSN: 0097-577X – mf#714 – us UMI ProQuest [770]

American Physical Education Association *see* Proceedings

American physical education review – 1-34. 1896-1929 – 368mf – 9 – $1,104.00 – us Kinesology [790]

The American Physical Society *see* Bulletin

American Physical Society *see* Bulletin

American Physiological Society *see*
- Journal of applied physiology
- Physiological reviews

American pioneer – La Salle IL 1842-43 – 1 – mf#4605 – us UMI ProQuest [975]

The american plan of government : the constitution as interpreted by accepted authorities / Bacon, Charles W – 4th rev ed. New York, London: G P Putnam's Sons, 1921 – 6mf – 9 – $9.00 – mf#LLMC 95-088 – us LLMC [323]

American planning association. journal – Chicago IL 1979+ – 1,5,9 – (cont: journal of the american institute of planners) – ISSN: 0194-4363 – mf#2001,02 – us UMI ProQuest [710]

American Podiatric Medical Association *see* Journal of the american podiatric medical association

American podiatry association. journal – Bethesda MD 1907-84 – 1,5,9 – (cont by: journal of the american podiatric medical association) – ISSN: 0003-0538 – mf#6915.01 – us UMI ProQuest [617]

American poems – 1793 – 9 – us Scholars Facs [810]

American poet – Charleston IL 1972-77 – 1,5,9 – ISSN: 0003-0546 – mf#6689 – us UMI ProQuest [420]

American poetry, 1609-1870 : based on the harris collection of american poetry and plays, brown university – 426r – 1 – (filmed alphabetically within 3 chronological segments: 1609-1820, 1821-50, 1851-70. includes printed guide) – mf#C35-222000 – us Primary [810]

American poetry review – Philadelphia PA 1972+ – 1,5,9 – ISSN: 0360-3709 – mf#6880 – us UMI ProQuest [420]

American poets and their theology / Strong, Augustus Hopkins – Philadelphia: Griffith & Rowland, 1916 [mf ed 1991] – 2mf – 9 – 0-7905-9689-X – mf#1989-1414 – us ATLA [420]

American Political Items Collectors [Organization] *see*
- Jugates
- Political bandwagon

American political report – 1977 sep 16, 1978 jan 7-1982 jul; 1982 aug 13-1987 dec 25 – 1 – mf#645288 – us WHS [071]

American political science association. division of educational affairs. dea news – Washington DC 1975-77 – 1,5,9 – mf#10931.01 – us UMI ProQuest [320]

American political science association. proceedings – Washington DC 1904+ – 1,5,9 – mf#2988 – us UMI ProQuest [320]

American political science review – Washington DC 1906+ – 1,5,9 – ISSN: 0003-0554 – mf#791 – us UMI ProQuest [320]

American politics – v1 n3-1910 [1984 jan-aug] – 1 – mf#1344661 – us WHS [071]

American politics quarterly – Newbury Park CA 1983-2000 – 1,5,9 – ISSN: 0044-7803 – mf#14004.01 – us UMI ProQuest [321]

American politics research – Newbury Park CA 2001+ – 1,5,9 – (cont: american politics quarterly) – ISSN: 1532-673X – mf#14004,01 – us UMI ProQuest [320]

American Pomological Society *see* Journal of american pomological society

American populist and journal of freedom – 1986 may 1-nov/dec – 1 – mf#1212385 – us WHS [071]

The american portion of the historical library of victor morin... : comprising american voyages and explorations / Morin, Victor – New York: American Art Association, Anderson Galleries inc, 1931 [mf ed 1987] – 1mf – 9 – mf#SEM105P758 – cn Bibl Nat [020]

The american post – Paterson. N.J. 1965-67 – 1 – us AJPC [071]

American postal worker : the official publicationof the american postal workers union – 1984 may-1987 dec – 1 – mf#1217547 – us WHS [071]

American Postal Workers Union *see*
- Breaktime
- Cincinnati fed
- Coastal line
- Harr-penn dispatch
- Hi-lites
- Hollywood lantern and quill
- Jax postal worker
- Keystone area local news and views
- Lantern and quill
- Louisville fed
- Mid-west torch
- New union paper
- News service
- Notes and quotes
- Other side
- Outlook
- Philadelphia postal worker
- Postal choragus
- Rebel
- Reporter
- Seattle apwu news
- Union call
- Watt ever

American Postal Workers Union et al *see*
- Bulletin of the american postal...
- Press on

American Postal Workers Union of Wisconsin *see* Wisconsin badger bulletin

American potato journal – Orono ME 1923-97 – 1,5,9 – (cont by: american journal of potato research) – ISSN: 0003-0589 – mf#231.01 – us UMI ProQuest [630]

American potpourri : multi-ethnic books for children and young adults: a bibliography based on the acquisitions of the educational materials center / Billings, Mary DeWitt et al – Washington: Dept of Health, Education, & Welfare, Office of Education, jan 1977 (mf ed. Bethlehem, PA : Mid-Atlantic Preservation Service, 1989) – 1mf – 9 – mf#Sc Micro F-11808 – Dist. us Gov Printing Dist – Located: NYPL – us Misc Inst [370]

American practice reports : official leading cases in all state and federal courts / ed by Ray, Charles A – Washington: Washington Law Book Co. v1-2. 1897-98 (all publ) – 18mf – 9 – $27.00 – mf#LLMC 84-700 – us LLMC [347]

American practitioner – La Salle IL 1870-85 – 1 – mf#5050 – us UMI ProQuest [610]

American Presbyterian Hospital (Xiangtan, Hunan Sheng, China) *see* Annual report for...of the american presbyterian hospital, siangtan

American presbyterian review – La Salle IL 1859-71 – 1 – mf#5225 – us UMI ProQuest [242]

American presbyteriana 1 – 1 – $50.00 – us Presbyterian [242]

American presbyteriana 2 – China Mission Resources. 1900-48 – 1 – $150.00 – us Presbyterian [242]

American presbyteriana 3 – Evangelism Resources – 1 – $50.00 – us Presbyterian [242]

American presbyterianism : its origin and early history / Briggs, Charles Augustus – New York: Charles Scribner, 1885 [mf ed 1989] – 2mf – 9 – 0-7905-4101-7 – (incl bibl ref) – mf#1988-0101 – us ATLA [242]

American presbyterians – Philadelphia PA 1985-96 – 1,5,9 – (cont: journal of presbyterian history [1980]; cont by: journal of presbyterian history [1997]) – ISSN: 0886-5159 – mf#12638,02 – us UMI ProQuest [242]

American press – Berea OH 1964-72 – 1,5,9 – ISSN: 0003-0600 – mf#5829 – us UMI ProQuest [680]

American pressman, 1890-1955 : service bureau news bulletin of the international printing pressmen and assistants' union of north america, 1937- 1952 / International Printing Pressman and Assistants' Union of North America – (= ser Labor union periodicals, pt 2: the printing trades) – 26r – 1 – $5425.00 – us UPA [680]

American primers – 17th c-1930s – 1401mf (20:1) – 9 – $9785.00 – us UPA [370]

American primitive methodist magazine – 1862 jan-1866 jan – 1 – mf#1051984 – us WHS [071]

American printer – Shawnee Mission KS 1981+ – 1,5,9 – (cont: american printer and lithographer) – ISSN: 0744-6616 – mf#814,02 – us UMI ProQuest [680]

American printer and lithographer [1978] – Shawnee Mission KS 1978-81 – 1,5,9 – (cont: inland printer, american lithographer; cont by: american printer) – ISSN: 0192-9933 – mf#814.02 – us UMI ProQuest [680]

American printer & lithographer [1955] – Shawnee Mission KS 1955-58 – 1 – mf#932 – us UMI ProQuest [680]

American prints, 1870-1950 / Baltimore Museum of Art; ed by Johnson, Robert Flynn – 1976 – 1 color mf – 15 – $30.00f – 0-226-68824-0 – (34p accompanying text) – us Chicago U Pr [760]

The american prisoner : [novel] / Phillpotts, Eden – Toronto: G N Morang, 1904 – 5mf – 9 – 0-659-90457-8 – mf#9-90457 – cn Canadiana [830]

American probate reports – New York: Baker & Voorhis. v1-8. 1875-95 (all publ) – 58mf – 9 – $87.00 – (cont by: probate reports annotated) – mf#LLMC 82-403 – us LLMC [340]

An american progressive : elizabeth glendower evans – 11r – 1 – (includes complete listing) – mf#C36-28160 – us Primary [320]

American prospect – Boston MA 1991+ – 1,5,9 – ISSN: 1049-7285 – mf#19638 – us UMI ProQuest [321]

American protection and canadian reciprocity / Haliburton, Robert Grant – [S.l: s.n, 1875?] [mf ed 1986] – 1mf – 9 – 0-665-28353-9 – mf#28353 – cn Canadiana [380]

The american protective association explained : its principles, methods and objects for the instruction of members and those who wish to become such – Brooklyn, NY: Office of the Primitive Catholic, [190-?] [mf ed 1992] – 1mf – 9 – 0-524-04126-1 – mf#1990-1196 – us ATLA [240]

American Protestant League *see*
- Protestant herald
- Protestant nation

American psychiatric association area 2 council. bulletin – New York NY 1972-94 – 1,5,9 – mf#7594 – us UMI ProQuest [366]

American Psychological Association *see* Apa monitor

American psychologist – Washington DC 1946+ – 1,5,9 – ISSN: 0003-066X – mf#1152 – us UMI ProQuest [150]

American Public Health Association *see* Health situation in florida

American public opinion – [s.l: s.n. 1939] [mf ed 1980] – (= ser Blodgett coll) – 1mf – 9 – mf#w719 – us Harvard [946]

The american public school : a genetic study of principles, practices and and present problems / Finney, Ross Lee – New York: Macmillan, 1921 – (= ser The modern teacher's series) – xvi/355p – 1 – us Wisconsin U Libr [370]

The american pulpit : sketches, biographical and descriptive, of living american preachers, and of the religious movements and distinctive ideas which they represent / Fowler, Henry – New York: J M Fairchild, 1856 – 2mf – 9 – 0-7905-4639-6 – mf#1988-0639 – us ATLA [240]

The american pulpit: sketches, biographical / Fowler, Henry T – 1856 – 1 – $50.00 – us Presbyterian [920]

American quarter horse journal – Amarillo TX 2000+ – 1,5,9 – ISSN: 1538-3490 – mf#15131,01 – us UMI ProQuest [636]

American quarterly – Baltimore MD 1949+ – 1,5,9 – ISSN: 0003-0678 – mf#1031 – us UMI ProQuest [071]

American quarterly observer – La Salle IL 1833-34 – 1 – mf#3929 – us UMI ProQuest [071]
American quarterly review – La Salle IL 1827-37 – 1 – mf#3930 – us UMI ProQuest [420]
American question : a letter, from a calm observer to a noble lord, on the subject of the late declaration relative to the orders in council – London: Printed by A J Valpy... 1812 – 1mf – 9 – mf#20969 – cn Canadiana [380]
The american radicalism collection : from the holdings of the american radicalism collection, special collections, michigan state university libraries – [mf ed 2003] – ca 112r in 4pts – 1 – (pt1: leftist politics and anti-war movements ca 46r. pt2: the religious and radical right ca 12r. pt3: race, gender, and the struggle for justice and equal rights ca 35r. pt4: twentieth-century social, economic, and environmental movements ca 19r) – us Primary [320]
American Radio Relay League *see* Qst
American Radio Telegraphists Association *see*
– Communications journal
– People's press [arta edition]
American railroad and corporation reports (or reporter) / ed by Lewis, John – Chicago: Myers & Co. v-12. 1868-96 (all publ) – 108mf – 9 – $162.00 – mf#LLMC 84-701 – us LLMC [380]
American railroad journal – v. 1-74. 1832-1900 – 1 – 980.00 – us L of C Photodup [380]
American railway reports : a collection of all reported decisions related to railways / Truman, J Henry – New York: Cockcroft. v1-21. 1872-81 (all publ) – 136mf – 9 – $204.00 – (although issued between 1872-81, title retrospectively covers all r.r. cases prior to 1881) – mf#LLMC 84-702 – us LLMC [380]
American Railway Union *see* Railway times
American railways under government operation and the financial outlook / Mitchell, Charles Edwin – [Montreal: s.n, 1919?] – 1mf – 9 – 0-665-87991-1 – mf#87991 – cn Canadiana [380]
American rationalist – St Louis MO 1956+ – 1,5,9 – ISSN: 0003-0708 – mf#5885 – us UMI ProQuest [140]
American Real Estate and Urban Economics Association *see* Journal of the american real estate and urban economics association
American record guide – Cincinnati OH 1935+ – 1,5,9 – ISSN: 0003-0716 – mf#8569 – us UMI ProQuest [780]
American recorder – St Louis MO 1960+ – 1,5,9 – ISSN: 0003-0724 – mf#1999 – us UMI ProQuest [780]
American Red Cross *see* Cross section
American red cross youth news – Washington DC 1919-75 – 1,5,9 – ISSN: 0003-0740 – mf#6662 – us UMI ProQuest [362]
American referee and cycle trade journal – 1897-99 – (= ser American newspapers and periodicals published in the uk and europe 18th-20th centuries) – 4r – 1 – £180.00 – uk World [072]
American reformed horse book / Dadd, George H – New York, NY. 1889 – 1r – us UF Libraries [636]
American register [1806] : or general repository of history, politics and science – La Salle IL 1806-10 – 1 – mf#3545 – us UMI ProQuest [071]
American register [1817] : or summary review of history, politics and literature – La Salle IL 1817 – 1 – mf#3546 – us UMI ProQuest [071]
American rehabilitation – Washington DC 1979+ – 1,5,9 – ISSN: 0362-4048 – mf#12109 – us UMI ProQuest [362]
American religion / Weiss, John – Boston: Roberts, 1871 [mf ed 1985] – 1mf – 9 – 0-8370-5781-7 – mf#1985-3781 – us ATLA [200]
American repertory of arts, science, and manufactures – La Salle IL 1840-42 – 1 – mf#4569 – us UMI ProQuest [071]
American repertory of arts, science, and useful literature – La Salle IL 1830-32 – 1 – mf#4150 – us UMI ProQuest [071]
American report – New York NY 1970-74 – 1 – ISSN: 0003-0767 – mf#6046 – us UMI ProQuest [071]
American reports – New York/San Francisco: Bancroft-Whitney. v1-60. 1870-87 (all publ) – (= ser Trinity Series, Part 2) – 679mf – 9 – $1018.00 – mf#LLMC 78-036 – us LLMC [340]
The american republic and its government : an analysis of the government of the united states with a consideration of its fundamental principles and of its relations to the states and territories / Woodburn, James A – 2nd rev. ed. New York/London: G P Putnam's Sons, 1916 – 5mf – 9 – $7.50 – mf#LLMC 95-087 – us LLMC [323]
The american republic, its constitution, tendencies and destiny / Brownson, Orestes Augustus – New York: P O'Shea, 1865 – 5mf – 9 – $7.50 – mf#LLMC 95-094 – us LLMC [323]
American republican, and baltimore daily clipper – Baltimore, Maryland. Nov 11-Dec 31 1944; 1945-1946 – 3r – 1 – us L of C Photodup [071]
American Rescue Workers *see* Rescue herald
American review, and literary journal – La Salle IL 1801-02 – 1 – mf#3547 – us UMI ProQuest [420]
American review of canadian studies – Washington DC 1979+ – 1,5,9 – ISSN: 0272-2011 – mf#12073,01 – us UMI ProQuest [327]
American review of eastern orthodoxy – v22 n3 [1976 may/jun], v25 n1-v26 n6 [1976 may/jun-1980 nov/dec] – 1 – mf#629904 – us WHS [243]
American review of history and politics : and general repository of literary and state papers – La Salle IL 1811-12 – 1 – mf#3548 – us UMI ProQuest [071]
American review of international arbitration – v1-7 (1990-96) – 9 – $172.00 set – ISSN: 1050-4109 – mf#113821 – us Hein [341]
American review of public administration [arpa] – Newbury Park CA 1988+ – 1,5,9 – ISSN: 0275-0740 – mf#15703,01 – us UMI ProQuest [350]
American review of respiratory disease – New York NY 1917-93 – 1,5,9 – (cont by: american journal of respiratory & critical care medicine) – ISSN: 0003-0805 – mf#102.01 – us UMI ProQuest [616]
American Revision Committee *see*
– Historical account of the work of the american committee of revision of the authorized english version of the bible
– The holy bible
American revolution – n1-1912 [1972-77] – 1 – mf#384129 – us WHS [975]
The american revolution – 1734mf (24:1) – 9 – $7510.00 – us UPA [975]
American Revolution Bicentennial Administration *see* Bicentennial times
American Revolution Bicentennial Authority of Oklahoma *see* Oklahoma bicentennial newsletter
American Revolution Bicentennial Commission *see* Bicentennial news
American Revolution Bicentennial Commission of Connecticut *see* Connecticut bicentennial gazette
American Revolution Bicentennial Commission of Texas *see*
– Bicentennial in texas
– Emergence '76
American Revolution Bicentennial Commission of Wisconsin *see*
– Calendar of wisconsin's bicentennial events
– Wisconsin ledger
American revolution, british pamphlets relating to the... 1764-83 – (= ser British records relating to america in microform) – 49r – 1 – (with guide. int by colin bonwick and thomas r adams) – mf#97122 – uk Microform Academic [975]
American revolution, documents relating to the... 1775-83 : from the national maritime museum, greenwich – (= ser British records relating to america in microform) – 4r – 1 – (with guide. int by roger knight) – mf#97045 – uk Microform Academic [025]
The american revolution in context : a collection of original documents from archives in britain and the usa – 4pt on 6r – 1 – £300.00 – (pt 1: debates in the parliament of great britain on the american revolution, 1765-83. pt 2: the boston gazette for the main period of the revolution, 1761-76. pt 3: diplomacy of the american revolution pt 4: documents of the revolution, by franklin, jefferson, washington etc) – mf#ARC – uk World [975]
American rifleman – Fairfax VA 1923+ – 1,5,9 – ISSN: 0003-083X – mf#10634 – us UMI ProQuest [790]
American rights in samoa : message from the president of the u.s. to the congress / Cleveland, President – 3rd Congress 1st sess. House Exec Doc No 238 2 Apr 1888. Washington: GPO, n.d. – 4mf – 9 – $6.00 – mf#LLMC 82-100C Title 7 – us LLMC [327]
American Rochdale Union et al *see* Montana equity news
American rocket society. ars journal – Reston VA 1930-62 – 1 – mf#5067 – us UMI ProQuest [629]
American rondo / Holst, Gustav – London: J & L Ballo [184?] [mf ed 1988] – 1r – 1 – mf#pres. film 47 – us Sibley [780]
American roofer and building improvement contractor – Bolinas CA 1950-55 – 1 – ISSN: 0003-0880 – mf#435 – us UMI ProQuest [690]
American ruling cases, annotated – Chicago: National Law Book Co. v1-5. 1920 (all publ) – 79mf – 9 – $119.00 – mf#LLMC 95-043 – us LLMC [340]
American russian falcon *see* Amerikanskij russkij sokol sojedinenija
American Sabbath Tract Society *see* Sabbath recorder
American salesman – Burlington IA 1987+ – 1,5,9 – ISSN: 0003-0902 – mf#15704 – us UMI ProQuest [650]
American Samoa *see*
– The agency of a.b. steinberger in the samoan islands
– American rights in samoa
– American samoa congressional hearings, 1928
– American samoa legislature, session laws
– American samoa reports, 1st series
– American samoan commission
– The american samoan commission's visit to samoa, september-october 1931
– Annual reports of the governor of american samoa to the secretary of the interior
– Code of american samoa, 1946 edition
– Code of american samoa, 1973 edition
– Revised constitution of american samoa, 1967
– Samoan affairs
– Us insular areas, application of relevant provisions of the us constitution
American samoa : a descriptive and historical profile / Tansill, William R – Washington: Lib of Congress, 1974 – 1mf – 9 – $1.50 – mf#LLMC 82-100C Title 45 – us LLMC [980]
American samoa : hearing before the subcommittee on territorial and insular affairs of the house committee on interior and insular affairs / American Samoa. US Congress – 93rd Congress 1st sess 18 Apr 1973. Washington: GPO, 1973 – 1mf – 9 – $1.50 – mf#LLMC 82-100C Title 37 – us LLMC [327]
American samoa : hope and despair, 1947-1952 / McGrew, W L – n.p., n.d – 1mf – 9 – $1.50 – mf#LLMC 82-100C Title 46 – us LLMC [980]
American samoa : report of a special sub-committee on territorial and insular affairs of the house committee on interior and insular affairs / American Samoa. US Congress – 84th Congress 1st sess. Comm print no 4 Nov 1954. Washington: GPO, 1954 – 1mf – 9 – $1.50 – mf#LLMC 82-100C Title 34 – us LLMC [327]
American samoa : working paper prepared by the secretariat for the general assembly's special committee on the situation with regard to the implementation of the declaration on the granting of independence to colonial countries and peoples, june 7 1974 – 1mf – 9 – $1.50 – mf#LLMC 82-100C Title 39 – us LLMC [324]
American samoa administrative code, 1982 : a codification of the administrative rules of american samoa / American Samoa. Government – Seattle: Book Publ Co, 1982- – 30mf – 9 – $45.00 – (with suppl thru mar 1988 and a vol of superseded materials samoa) – mf#LLMC 82-100C Title 15 – us LLMC [324]
American Samoa. Bar Association *see* The samoan pacific law journal
American samoa congressional hearings, 1928 : joint hearings before the senate committee on territorial and insular possessions and the house committee on insular affairs / American Samoa – 70th Congress 1st sess 17-21 Jan 1928. Washington: GPO, 1928 – 2mf – 9 – $3.00 – mf#LLMC 82-100C Title 32 – us LLMC [327]
American Samoa. Constitutional Convention *see* Proposed revised constitution of american samoa, 1986
American Samoa. Dept of Interior *see* The application of federal laws in american samoa, guam, the northern mariana islands, and the virgin islands
American Samoa. Executive Branch *see* Codification of the regulations and orders for the government of american samoa
American Samoa. Government *see*
– American samoa administrative code, 1982
– American samoan code annotated
– The asg report
American Samoa. Govt *see* Records of the government of american samoa, 1900-1958
American Samoa. High Court *see* American samoa reports, 2nd series
American Samoa. Interim Research Section *see* Revised code of american samoa, 1961 edition
American samoa legislature, session laws : 8th legislature 1963; 10-11th legislature 1968-1969; 13-20th legislature 1973-1988 / American Samoa – Legislative Reference Bureau, 1963-88 – 81mf – 9 – $121.50 – mf#LLMC 82-100C Title 6 – us LLMC [324]
American Samoa. Office of Samoan Information *see* Amerika samoa
American samoa reports, 1st series : 1900-1975 / American Samoa – Equity Publ Co. v1-4. 1977-78 – 36mf – 9 – $54.00 – (incl ind for v1-4 and 1982) – mf#LLMC 82-100C Title 4 – us LLMC [324]
American samoa reports, 2nd series / American Samoa. High Court – High Court of American Samoa. v1-25. 1983-94 – 43mf – 9 – $64.00 – mf#LLMC 82-100C Title 4 – us LLMC [340]
American samoa treaties : with the united kingdom, august 28 and september 2 1879; with the united kingdom, the united states and germany, september 29 1883; and the final act of the conference on the affairs of samoa (treaty of berlin) june 14 1889 – 4mf – 9 – $6.00 – mf#LLMC 82-100C Title 49 – us LLMC [324]
American Samoa. US Congress *see*
– Acceptance of cessions of certain samoan islands
– American samoa
– Current problems in american samoa
– Jurisdiction of submerged lands in american samoa, guam, and the virgin islands
– Staff study on american samoa
American Samoa. US Congress. Subcomm National Parks and Insular Affairs *see* Legislative history of the omnibus insular areas act of 1979-1980
American Samoa. US Senate *see* Us senate study mission to eastern (american) samoa
American samoan code annotated : 1981 edition / American Samoa. Government – Seattle: Book Publ Co, 1982-87? – 32mf – 9 – $48.00 – mf#LLMC 82-100C Title 13 – us LLMC [324]
American samoan commission : hearings before the commission appointed by the president in accordance with public resolution no 89, 70th congress....accepting the cession of certain islands of of the samoan group, september-october, 1931, honolulu and pago pago / American Samoa – Washington: GPO, 1931 – 6mf – 9 – $9.00 – mf#LLMC 82-100C Title 8 – us LLMC [327]
The american samoan commission's visit to samoa, september-october 1931 / Moore, Reuel S & Farrington, Joseph R – Washington: GPO, 1931 – 1mf – 9 – $1.50 – mf#LLMC 82-100C Title 29 – us LLMC [327]
American saturday courier – July 16, 1831-1835 – 1 – us CRL [073]
American scholar – Washington DC 1932+ – 1,5,9 – ISSN: 0003-0937 – mf#936 – us UMI ProQuest [071]
The american scholar / Parker, Theodore; ed by Cooke, George Willis – centenary ed. Boston: American Unitarian Association, c1907 – 6mf – 9 – 0-524-07448-8 – mf#1991-3108 – us ATLA [975]
American school board journal – Alexandria VA 1891+ – 1,5,9 – ISSN: 0003-0953 – mf#842 – us UMI ProQuest [370]
American school & university – Shawnee Mission KS 1928+ – 1,5,9 – ISSN: 0003-0945 – mf#1631 – us UMI ProQuest [370]
American schools of oriental research. annual – Boston MA 1919+ – 1,5,9 – ISSN: 0066-0035 – mf#3224 – us UMI ProQuest [950]
American schools of oriental research. bulletin – Boston MA 1919+ – 1,5,9 – ISSN: 0003-097X – mf#3129 – us UMI ProQuest [950]
American schools of oriental research. bulletin...supplemental studies – Boston MA 1974-86 – 1,5,9 – ISSN: 0145-3661 – mf#8385 – us UMI ProQuest [950]
American schools of oriental research. newsletter – Boston MA 1948+ – 1,5,9 – ISSN: 0361-6029 – mf#8384 – us UMI ProQuest [950]
American scientific affiliation. journal – Ipswich MA 1949-86 – 1,5,9 – (cont by: perspectives on science & christian faith) – ISSN: 0003-0988 – mf#2263.01 – us UMI ProQuest [230]
American scientist – Research Triangle Park NC 1913+ – 1,5,9 – ISSN: 0003-0996 – mf#788 – us UMI ProQuest [500]
American scotsman – Chicago: American Scotsman Co (mthly) [mf ed 1994] – 1r – 1 – uk Scotland NatLib [072]
American secondary education – Ashland OH 1970+ – 1,5,9 – ISSN: 0003-1003 – mf#11870 – us UMI ProQuest [373]
American securities : practical hints on the tests of stability and profit, for the guidance and warning of british investors – London: M Nephews, 1860 (mf ed 19–) – 32p – mf#ZV-TPG pv67 n13 – us Harvard [332]
American Security Council. Coalition for Peace through Strength *see* Coalition insider
American Security Council Foundation *see* Peace through strength report
The american senator : [a novel] / Trollope, Anthony – Toronto: Belford, 1877 – 5mf – 9 – mf#34027 – cn Canadiana [830]
American sentinel – 1982 sep 7-1986 sep 22; oct 6-1990 jul 30; sep 21-1992 jun 17 – 1 – mf#1507274 – us WHS [071]
American sentinel – Westminster, Maryland. 1856-1928 – 1 – us MD Archives [071]
American series / Butler Co. Hamilton – sep 1963-jan 1971 (poor inking) [wkly, biwkly] – 2r – 1 – (an african-american newspaper) – mf#B29348-29349 – us Ohio Hist [071]
American series / Licking Co. Newark – (8/1858-4/1870), 10-11/1873 (scattered, damaged) [wkly] – 1r – 1 – mf#B29561 – us Ohio Hist [071]

American Servicemen's Union *see* O d d

American Servicemen's Union. Committee for GI Rights *see* Bond

American Servicemen's Union [Fort Lewis, WA] *see* Fed up!

American sheet music collection to 1830 (am-1) – 1 – $1188.00 – mf#0036 – us Brook [780]

American shipper [1977] – Jacksonville FL 1976-90 – 1,5,9 – (cont: florida journal of commerce, american shipper; cont by: american shipper international) – ISSN: 0160-225X – mf#8118.04 – us UMI ProQuest [380]

American shipper [1992] – Jacksonville FL 1992+ – 1,5,9 – (cont: american shipper international) – ISSN: 1074-8350 – mf#8118,04 – us UMI ProQuest [380]

American siberia / Powell, J C – Chicago, IL. 1891 – 1r – us UF Libraries [978]

American sketches – London 1827 [mf ed Hildesheim 1995-98] – 1v on 3mf – 9 – €90.00 – 3-487-27181-8 – gw Olms [880]

The american slave code in theory and practice : its distinctive features shown by its statutes, judicial decisions, and illustrative facts / Goodell, William – 2nd ed. New York: American & Foreign Anti-Slavery Society, 1853 – 5mf – 9 – $7.50 – mf#LLMC 92-101 – us LLMC [348]

American slavery distinguished from the slavery of english theorists : and justified by the law of nature / Seabury, Samuel – New York: Mason Brothers, 1861. Chicago: Dep of Photodup, U of Chicago Lib, 1970 (1r); Evanston: American Theol Lib Assoc, 1984 (1r) – 1 – 0-8370-0582-5 – mf#1984-B143 – us ATLA [240]

American social and religious conditions / Stelzle, Charles – New York: F H Revell, c1912 [mf ed 1990] – 1mf – 9 – 0-7905-6085-2 – mf#1988-2085 – us ATLA [301]

American social dance technique syllabus for the rumba, samba, mambo, and tango / Holman, Curt W – 1996 – 2mf – 9 – $8.00 – mf#PE 3826 – us Kinesology [790]

American socialist – 1876 mar 30-1879 dec 25 – 1 – mf#814063 – us WHS [071]

American socialist : devoted to the enlargement and perfection of home – La Salle IL 1876-79 – 1 – mf#5226 – us UMI ProQuest [335]

American socialist – v1-4. 1914-17 [all publ] – (= ser Radical periodicals in the united states, 1881-1960. series 1) – 1r – 1 – $200.00 – us UPA [335]

American society for artificial internal organs. asaio journal – Baltimore MD 1978-85 – 1,5,9 – ISSN: 0162-1432 – mf#11848 – us UMI ProQuest [617]

American Society for Conservation Archeology *see* Report

American Society for Engineering Education *see* Proceedings of the american society for engineering education

American Society for Information Science *see* Journal of the american society for information science

American Society for Information Science and Technology *see* Bulletin of the american society for information science and technology

American society for information science. bulletin – Silver Spring MD 1974+ – 1,5,9 – ISSN: 0095-4403 – mf#11872.01 – us UMI ProQuest [020]

American Society for Mass Spectrometry *see* Journal of the american society for mass spectrometry

American society for metals. asm transactions quarterly – Novelty OH 1920-69 – 1,5 – ISSN: 0097-3912 – mf#1158 – us UMI ProQuest [660]

American society for microbiology. asm news – Washington DC 1938+ – 1,5,9 – ISSN: 0044-7897 – mf#8382 – us UMI ProQuest [576]

American society for preventive dentistry. journal – Chicago IL 1970-77 – 1,5,9 – ISSN: 0093-4518 – mf#8725 – us UMI ProQuest [617]

American society for promoting the civilization and general improvement of the indian tribes within the united states annual report – La Salle IL 1824 – 1 – mf#4052 – us UMI ProQuest [303]

American Society for Psychical Research *see* Journal of the american society for psychical research

American society for psychical research. proceedings – New York NY 1907-74 – 1,5 – ISSN: 0096-8927 – mf#5707 – us UMI ProQuest [130]

American society for quality control. annual quality congress transactions – Milwaukee WI 1947-91 – 1,5,9 – (cont by: quality congress annual quality congress) – mf#6225.04 – us UMI ProQuest [620]

American Society for Technical Aid to Spanish Democracy *see* Spain is fighting for you

American society for testing and materials. astm standardization news – Conshohocken PA 1973-84 – 1,5,9 – (cont by: standardization news [sn]) – ISSN: 0090-1210 – mf#7858.01 – us UMI ProQuest [620]

American society for testing and materials. proceedings – Conshohocken PA 1899-1981 – 1,5,9 – ISSN: 0066-0515 – mf#1948 – us UMI ProQuest [620]

American society in europe – 1891-92 – (= ser American newspapers and periodicals published in the uk and europe 18th-20th centuries) – 1r – 1 – £55.00 – uk World [072]

American Society Of African Culture *see* Southern africa in transition

American Society of Association Executives *see* Association management

American Society of Chartered Life Underwriters *see*
- Clu forum report
- Clu journal
- Journal of the american society of clu

American society of church history : papers – v1 n1-8. 1888-96; v2 n1-9. 1906-33 [complete] – 2r – 1 – mf#ATLA S0001 – us ATLA [240]

American Society of Civil Engineers *see*
- Journal of the structural division
- Newsletter
- Standard time
- Transportation engineering journal of asce

American society of civil engineers. collected journals – Reston VA 1983+ – 1,5,9 – (cont: proceedings of the american society of civil engineers) – mf#13356 – us UMI ProQuest [624]

American Society of Civil Engineers. Construction Division *see* Journal of the construction division

American Society of Civil Engineers. Energy Division *see* Journal of the energy division

American Society of Civil Engineers. Engineering Mechanics Division *see* Journal of the engineering mechanics division

American Society of Civil Engineers. Environmental Engineering Division *see* Journal of the environmental engineering division

American Society of Civil Engineers. Geotechnical Engineering Division *see* Journal of the geotechnical engineering division

American Society of Civil Engineers. Hydraulics Division *see* Journal of the hydraulics division

American Society of Civil Engineers. Power Division *see* Journal of the power division

American society of civil engineers. proceedings – Reston VA 1873-1982 – 1,5,9 – ISSN: 0097-417X – mf#833 – us UMI ProQuest [624]

American Society of Civil Engineers. Soil Mechanics and Foundations Division *see* Journal of the soil mechanics and foundations division

American Society of Civil Engineers. Surveying and Mapping Division *see* Journal of the surveying and mapping division

American society of civil engineers. transactions – Reston VA 1872+ – 1,5,9 – ISSN: 0066-0604 – mf#10096 – us UMI ProQuest [624]

American Society of Civil Engineers. Urban Planning and Development Division *see* Journal of the urban planning and development division

American Society of Civil Engineers. Water Resources Planning and Management Division *see* Journal of the water resources planning and management division

American Society of Civil Engineers. Waterway, Port, Coastal, and Ocean Division *see* Journal of the waterway, port, coastal and ocean division

American Society of Civil Engineers. Waterways, Harbors, and Coastal Engineering Division *see* Journal of the waterways, harbors and coastal engineering division

American Society of CLU & ChFC *see* Journal of the american society of clu and chfc

American society of colonial families, Boston *see* Colonial

American society of composers, authors and publishers. ascap in action – New York NY 1979-93 – 1,5,9 – ISSN: 0197-7849 – mf#12471 – us UMI ProQuest [780]

American society of composers, authors and publishers. ascap today – New York NY 1974-78 – 1,5,9 – ISSN: 0001-2424 – mf#9127 – us UMI ProQuest [780]

American Society of Echocardiography *see* Journal of the american society of echocardiography

American Society of Equity *see*
- Equity news
- Wisconsin equity news

American Society of Heating, Refrigerating and Air Conditioning Engineers *see* Ashrae journal

American society of international law proceedings – v1-95. 1907-2001 – 9 – $1300.00 set – ISSN: 0272-5037 – mf#100551 – us Hein [341]

American society of international law. proceedings at its annual meeting – Washington DC 1907-69 – 1 – ISSN: 0272-5045 – mf#7206.01 – us UMI ProQuest [341]

American society of international law. proceedings of the annual meeting – Washington DC 1982+ – 1,5,9 – ISSN: 0272-5037 – mf#7206,01 – us UMI ProQuest [341]

American society of lubrication engineers. asle transactions – Park Ridge IL 1958-87 – 1,5,9 – (cont by: tribology transactions) – ISSN: 0569-8197 – mf#6768.01 – us UMI ProQuest [621]

American society of mechanical engineers. transactions – New York NY 1880-1958 – 1 – ISSN: 0097-6822 – mf#574 – us UMI ProQuest [621]

American society of safety engineers. asse journal – Des Plaines IL 1956-74 – 1 – (cont by: professional safety) – mf#9910.01 – us UMI ProQuest [366]

American Society of Saint Caecilia *see* Caecilia

American society of sugar beet technologists. journal – Denver CO 1938-80 – 1,5,9 – ISSN: 0003-1216 – mf#2474.01 – us UMI ProQuest [660]

American society of university composers. proceedings – New York NY 1966-77 – 1,5,9 – ISSN: 0066-0701 – mf#8358 – us UMI ProQuest [780]

American sociological review – Washington DC 1936+ – 1,5,9 – ISSN: 0003-1224 – mf#1060 – us UMI ProQuest [301]

American sociologist – Piscataway NJ 1965+ – 1,5,9 – ISSN: 0003-1232 – mf#2477 – us UMI ProQuest [301]

American soldier – v1 n1, n7 [1898 sep 10, oct 22]; v1 n1, n7 [1898 sep 10, oct 22]; v2 n1-7 [1898 sep 10-oct 22] – 1 – mf#846763 – us WHS [071]

The american soldier – Manila, Philippine Islands: American Pub Co [oct 8 1898-jan 1 1899] (wkly) – 1r – 1 – us CRL [355]

American spa – Cleveland OH 2001+ – 1 – mf#26653 – us UMI ProQuest [613]

American spectator – Arlington VA 1979+ – 1,5,9 – (cont: alternative: an american spectator) – ISSN: 0148-8414 – mf#7782,02 – us UMI ProQuest [071]

American spectator – v1-4. 1932-37 [all publ] – (= ser Radical periodicals in the united states, 1881-1960. series 1) – 1r – 1 – $200.00 – us UPA [303]

The american spectator: a literary newspaper – New York. v. 1-4, no. 48. Nov. 1932-May 1937 – 1 – us NY Public [420]

American speech – Durham NC 1955+ – 1,5,9 – ISSN: 0003-1283 – mf#924 – us UMI ProQuest [400]

American speech and hearing association. asha – Rockville MD 1959-79 – 1,5,9 – (cont by: american speech language hearing association. asha) – ISSN: 0001-2475 – mf#12776.01 – us UMI ProQuest [362]

American speech language hearing association. asha – Rockville MD 1979-1999 – 1,5,9 – (cont: american speech and hearing association. asha) – ISSN: 0001-2475 – mf#12776,01 – us UMI ProQuest [362]

American spirit – Washington DC 2002+ – 1,5,9 – mf#6765,01 – us UMI ProQuest [970]

American sportswear and knitting times – Summit NJ 1997-2000 – 1,5,9 – (cont: knitting times) – mf#3380,01 – us UMI ProQuest [680]

American srbobran – Pittsburgh PA, 1906-12, 1918-40 – 28r – 1 – (serbian newspaper) – us IHRC [071]

American stage of to-day – New York, NY. 1910 – 1r – us UF Libraries [790]

American standard – 1811 nov 20 – 1 – mf#881633 – us WHS [071]

American standard – San Francisco. v. 1, no. 9, 11, 13-16, 18-24, 26; v. 2, no. 31, 35, 49, 53; v. 3, no. 3. oct. 19, 1888-jan. 4, 1890 – 1 – us NY Public [073]

American state papers – Washington DC 1789-1838 – 1 – mf#2582 – us UMI ProQuest [324]

American state papers, 1789-1838 / U.S. Congress – Washington: Gale & Seaton. v1-38. 1789-1838 – 1 – $600.00 – mf#0605 – us Brook [324]

American state reports – San Francisco: Bancroft-Whitney. v1-140. 1886-1911 (all publ) – (= ser Trinity series, pt 3) – 1585mf – 9 – $2377.00 – (individual titles under this coll also listed separately) – mf#LLMC 78-038 – us LLMC [340]

American state reports prior to national reporter system – 1106 reels – 1 – $33,500.00 – us Trans-Media [340]

American state trials : criminal cases – St Louis: Thomas Law Book Co. v1-4. 1914-15 – 40mf – 9 – $60.00 – (add vols to be filmed) – mf#LLMC 90-352 – us LLMC [345]

American statesmen : an interpretation of our history and heritage / Griggs, Edward Howard – Croton-on-Hudson, NY: Orchard Hill Press, 1927 (mf ed 19–) – mf#Z-1419 – us NY Public [975]

The american stationer : a journal devoted to the interests of the stationery and fancy goods trades = Stationer. – New York: Redman & Kenny [v[8-62] 1880-1907] (wkly) – 53r – 1 – us CRL [680]

American Statistical Association *see* Journal of the american statistical association

American statistical association business and economic statistics section. proceedings – Alexandria VA 1974-89 – 1,5,9 – ISSN: 0066-0736 – mf#10052 – us UMI ProQuest [330]

American statistical association. proceedings of the section on survey research methods – Alexandria VA 1978-80 – 1,5,9 – mf#11769 – us UMI ProQuest [310]

American statistical association. social statistics section. proceedings – Alexandria VA 1974-92 – 1,5,9 – ISSN: 0066-0752 – mf#10046 – us UMI ProQuest [317]

American statistical association. statistical computing section. proceedings – Alexandria VA 1977-80 – 1,5,9 – mf#11635 – us UMI ProQuest [317]

American statistician – Alexandria VA 1947+ – 1,5,9 – ISSN: 0003-1305 – mf#1000 – us UMI ProQuest [317]

American statistics index microfiche library – 1974- – 9 – apply for prices – (includes significant statistical publications of the u.s. federal government. both depository and non-depository publications. from early 1960's onwards) – us CIS [317]

American Street Railway Association. Convention (14th : 1895 : Montreal, Quebec) *see* Souvenir and official programme

American street railway decisions – Brooklyn, NY: American Street Railroad Assoc. v1-2. 1841-64 (all publ) – 12mf – 9 – $18.00 – (covers us and canada. title originally intended to be retrospective reprint of all street railway cases from 1841 to date but only managed coverage for the years 1841-64) – mf#LLMC 84-703 – us LLMC [380]

American studies – Lawrence KS 1960+ – 1,5,9 – ISSN: 0026-3079 – mf#9767 – us UMI ProQuest [300]

American studies international – Washington DC 1981+ – 1,5,9 – ISSN: 0883-105X – mf#13455 – us UMI ProQuest [975]

American sunbeam – 1978 mar 6-1979; 1980; 1982 jan-jun; jul-1983 jun; jul-1984 jun; jul-1986 jun 9 – 1 – mf#498182 – us WHS [071]

American Sunday School Union *see*
- The 125th anniversary report of the american sunday school union
- Annual reports, 1825-1835
- The sunday school times

American sunday-school teachers' magazine and journal of education – La Salle IL 1823-24 – 1 – mf#4699 – us UMI ProQuest [240]

The american sunday-school union and the "union principle" : in reply to a. in the episcopal recorder / Tyng, Stephen Higginson – New York: John A Gray, 1855 – 1mf – 9 – 0-524-08626-5 – mf#1993-1076 – us ATLA [240]

The american supreme court as an international tribunal / Smith, Hebert Arthur – New York: Oxford 1920. 123p. LL-1496 – 1 – us L of C Photodup [341]

American surgeon – Atlanta GA 1935+ – 1,5,9 – ISSN: 0003-1348 – mf#8866 – us UMI ProQuest [617]

The American Swedish Foundation *see* The john ericsson collection of the american swedish historical foundation

American Symphony Orchestra League *see* Inter-orchestra bulletin

American teacher – 1955-65; 1966 jan-jun – 1 – mf#360449 – us WHS [071]

American teacher – 1966 sep-1968; 1969-71; 1972-1975 jun; sep-1977; 1978-1984 nov; dec-1993 jan; 1992 dec-1996 nov; dec-1999 nov – 1 – mf#193665 – us WHS [370]

American teacher – Washington DC 1912+ – 1,5,9 – ISSN: 0003-1380 – mf#221 – us UMI ProQuest [370]

American telegraphe – 1795 sep 9-oct 14 28 – 1 – mf#845947 – us WHS [071]

American Telephone and Telegraph Co *see* Classified index of rate cases, years 1925, 1926, 1927

American telephone and telegraph co. annual reports – 1900-65 – 1 – $96.00 – mf#0038 – us Brook [380]

The american temperance cyclopaedia of history, biography, anecdote, and illustration / Wakeley, Joseph Beaumont – New York: National Temperance Soc & Pub House, 1875 [mf ed 1990] – 1mf – 9 – 0-7905-8052-7 – mf#1988-6033 – us ATLA [975]

American Temperance Union *see* Journal

An american text-book of obstetrics for practioners and students / Cameron, James Chalmers et al; ed by Norris, Richard Cooper & Dickinson, Robert Latou – Philadelphia: W B Saunders, 1895 – 13mf – 9 – (incl ind) – mf#13608 – cn Canadiana [618]

An american text-book of pathology : for the use of students and practitioners of medicine and surgery / ed by Hektoen, Ludvig & Riesman, David – Philadelphia: W B Saunders & Co, 1901 – us CRL [617]
American theatre – New York NY 1992+ – 1,5,9 – ISSN: 8750-3255 – mf#19173 – us UMI ProQuest [790]
American theatre periodicals of the nineteenth and early twentieth centuries – (= ser Theatre periodical series) – 19r – 1 – (some titles incl: the thespian mirror 1805-06, the new york clipper annual 1874-1901, the cann-leighton official theatrical guide 1896-1971, the burr mcintosh monthly guide 1896-1971, and gus hill's national theatrical directory 1914-15) – mf#C35-12111 – us Primary [790]
American Theological Library Association *see* American theological library association
American theological library association : summary of proceedings / American Theological Library Association – v1-19. 1947-65 [complete] – 2r – 1 – mf#ATLA S0121 – us ATLA [020]
American theosophist / ed by Rogers, Louis William – Albany NY: [s.n.] 1908-09 [mthly] [mf ed 2006] – v1-2 (1908-09) on 1r – 1 – (lacks: v1 n2-4,7-10 (apr-jul 1908, oct 1908-jan 1909) [covers missing]) – mf#2006c-s013 – us ATLA [290]
The american theosophist : official journal of the american theosophist society – Wheaton IL: American Theosophical Soc. 1933-96 [mthly] [mf ed 1982-2003] – 64v on 14r – 1 – (some iss in combined form) – ISSN: 0003-1402 – mf0434a – us ATLA [290]
The american theosophist – Los Angeles CA, 1913-14 (mthly) [mf ed 2003] – 3v on 2r – 1 – (some iss have title: american theosophist and theosophical messenger. suppl with some iss) – mf051 – us ATLA [290]
The american theosophist *see*
– The messenger
– The theosophical messenger
American theosophist and theosophical messenger *see* The american theosophist
American thought : from puritanism to pragmatism / Riley, Woodbridge – New York: H Holt, 1915 [mf ed 1991] – 1mf – 9 – 0-7905-9611-3 – (incl bibl ref) – mf#1989-1336 – us ATLA [190]
American Tract Society *see*
– Freedman
– Freedman's journal
American tract society. annual report – 1826-2000 [mf ed 2001] – (= ser Christianity's encounter with world religions, 1850-1950) – 12r – 1 – mf#2001-s162-165/166/167 – us ATLA [240]
The american tract society, et al., vs. lydia g. atwater, et al / Smith, Palmer Cone – Circleville, Ohio: Van Cleaf & Dresbach 1877?. 26p. LL-27 – 1 – us L of C Photodup [340]
American trade in munitions of war *see* Cargo of the "wilhelmina" / american trade in munitions of war / sinking of the "frye"
American trade review 1902 *see* Anglo-saxon 1899 – american trade review 1902 – anglo-american traveler 1902-03
American trademark cases / Cox, Rowland – Cincinnati, Clark, 1871. 782 p. LL-548 – 1 – us L of C Photodup [346]
American Train Dispatchers Association *see* Train dispatcher
American tramp and underworld slang / Irwin, Godfrey – New York, NY. 1931 – 1r – us UF Libraries [420]
American transcendental quarterly – Kingston RI 1978+ – 1,5,9 – ISSN: 0149-9017 – mf#12856,01 – us UMI ProQuest [130]
The american travel promotion act : hearing... house of representatives, 107th congress, 2nd session on h.r. 3321, may 23, 2002 / United States. Congress. House. Committee on Energy and Commerce. Subcommittee on Commerce, Trade, and Consumer Protection – Washington: US GPO 2002 [mf ed 2002] – 1mf – 9 – 0-16-068813-2 – us GPO [343]
American traveler, 1 – v49 n8-11, 13-18, 21, 24 [1988 apr 25-jun 10, jul 5-sep 12, oct 31, dec 9]; v50 n2, 14, 17-18, 26-27 [1989 jan 30, jul 3, 17-1931, dec 4-22]; v51 n3-7, 9 [1990 feb 26-jun 11, dec 10]; v52 n2-4 [1991 feb 18-apr 8] – 1 – mf#2540136 – us WHS [071]
[American traveller] *see* Letters
American tribune series / Licking Co. Newark – (6/1903-09,7/11-6/18,25-1/1927) [daily] – 46r – 1 – mf#B10272-10317 – us Ohio Hist [071]
American tribune series / Licking Co. Newark – jan 1899-apr 1911 [wkly] – 9r – 1 – mf#B10902-10910 – us Ohio Hist [071]
American tropics / Corlett, William Thomas – Cleveland, OH. 1908 – 1r – us UF Libraries [972]
American Trucking Associations *see* Transport topics
American tung oil – Valparaiso, FL. v1-4. 1935-1938 – 2r – us UF Libraries [630]
American tung oil news – Valparaiso, FL. 1934/1935 – 1r – us UF Libraries [630]
American tung tree / Adderley, Joseph C – Pensacola, FL. 1936 – 1r – us UF Libraries [630]
American turf register and sporting magazine – La Salle IL 1829-44 – 1 – mf#3931 – us UMI ProQuest [790]
American twentieth century art : subject collections – (= ser Art exhibition catalogues on microfiche) – 225 catalogues on 282mf – 9 – £1,480.00 – (individual titles not listed separately) – uk Chadwyck [700]
American uniform marriage and marriage license act / National Conference of Commissioners on Uniform State Laws – Williamsport, Pa.: Railway Printing Co. 1911. 29p. LL-950 – 1 – us L of C Photodup [348]
American union – 1863 oct 23; 1864 aug 18 – 1 – mf#851243 – us WHS [071]
American union / Jefferson Co. Steubenville – may 1850-aug 1859 [wkly] – 3r – 1 – mf#B5545-5547 – us Ohio Hist [071]
American unionist – Salem OR: W A McPherson & Wm Morgan, -1869 [wkly] – 1 – (related to: daily american unionist; absorbed: oregon statesman (oregon city, or); cont by: weekly oregon unionist) – us Oregon Lib [071]
American unionist *see*
– Daily american unionist
– Daily oregon statesman
American unionist (salem, or) *see* Oregon statesman (oregon city, or)
American Unitarian Association *see* Seventh report
American unitarian association anniversary : report, and proceedings for 1848 and 1849 / Dewey, Orville – Boston: Wm Crosby & H P Nichols, 1848-49 [mf ed 1993] – 1v on 1mf – 9 – 0-524-08676-1 – mf#1993-3201 – us ATLA [243]
American unitarian eucharistic faith / Laws, John Wallace – Chicago, 1938. Chicago: Dep of Photodup, U of Chicago Lib, 1971 (1r); Evanston: American Theol Lib Assoc, 1984 (1r) – 1 – 0-8370-0382-2 – mf#1984-B173 – us ATLA [243]
American unitarian interest in the study of non-christian religions / Hammon, John Kohlsaat – 1945 – 1r – 1 – 0-8370-1519-7 – mf#1984-B189 – us ATLA [243]
American Unity League *see* Tolerance
American universal magazine – La Salle IL 1797-98 – 1 – mf#4358 – us UMI ProQuest [073]
American University *see* Us army area handbook for brazil
American university international law review – v1-16. 1986-2 – 9 – $438.00 set – (title varies: v1-12 (1986-97) as american university journal of international law and policy) – ISSN: 0888-630X – mf#110481 – us Hein [341]
American university journal of gender and the law – 9 – (title varies: see american university journal of gender, social policies and the law) – mf#115402 – us Hein [342]
American university journal of gender and the law *see* American university journal of gender, social policy and the law
American university journal of gender, social policies and the law *see* American university journal of gender and the law
American university journal of gender, social policy and the law – v1-9. 1993-2 – 9 – $144.00 set – (title varies: v1-6 1993-98 as: american university journal of gender and the law) – ISSN: 1068-428X – mf#115401 – us Hein [342]
American university journal of international law *see* American university international law review
American university law review – v1-50. 1952-2 – 5,6,9 – $1031.00 set – (v1-34 1952-85 in reel $506; v35-50 1985-2 in mf $525; title varies: v1-5 1952-56 as intramural law review) – ISSN: 0003-1453 – mf#100561 – us Hein [340]
American university law review – Washington DC 1978+ – 1,5,9 – ISSN: 0003-1453 – mf#11913,01 – us UMI ProQuest [340]
American University (Washington, DC) Foreign Area Studies Division *see*
– Area handbook for colombia
– Area handbook for venezuela
American urban life and health, 1883-1914 : reports of the charity organization society of new york – 3r – 1 – us Primary [614]
American vegetable grower – Willoughby OH 1953+ – 1,5,9 – ISSN: 0741-9848 – mf#1186 – us UMI ProQuest [634]
American versus english methods of bridge designing – [Tokyo?: s.n, 1886?] [mf ed 1980] – 1mf – 9 – 0-665-02536-X – mf#02536 – cn Canadiana [624]
American Veterans of World War 2 and Korea *see* National amvet
American veterinary medical association. journal – Schaumburg IL 1877+ – 1,5,9 – ISSN: 0003-1488 – mf#159 – us UMI ProQuest [636]
American visitor 1884 *see* America 1883 – the american visitor 1884 – the american eagle 1885-86 – american humorist and storyteller 1888
The american visitors' news and register and colonial gazette – 1893-97 – (= ser American newspapers and periodicals published in the uk and europe 18th-20th centuries) – 3r – 1 – £140.00 – uk World [072]
American vocational journal – Alexandria VA 1926-78 – 1,5,9 – (cont by: voced) – ISSN: 0003-1496 – mf#2203.03 – us UMI ProQuest [374]
American voice – 1976-1977 jan – 1 – mf#203307 – us WHS [071]
American volunteer – Carlisle, PA. -w 1842-1905; 1814-1909. 24 rolls – 13 – $25.00r – us IMR [071]
American waldensian aid society : newsletter – 1953-89 [complete] – 2r – 1 – mf#ATLA S0404 – us ATLA [240]
The american war : with some suggestions towards effecting an honorable peace / Knight, Thomas Frederick – [Halifax, NS?: [s.n.] 1864 [mf ed 1984] – 1mf – 9 – 0-665-45327-2 – mf#45327 – cn Canadiana [976]
American watchman and delaware advertiser – 1825 apr 22, 1827 aug 28-1931 – 1 – mf#846035 – us WHS [071]
American water works association. journal – Denver CO 1914+ – 1,5,9 – ISSN: 0003-150X – mf#170 – us UMI ProQuest [333]
American way – Dallas TX 1972-73 – 1,5 – ISSN: 0003-1518 – mf#7222 – us UMI ProQuest [917]
American Way Features *see* Straight talk
American west – Tucson AZ 1964-90 – 1,5,9 – ISSN: 0003-1534 – mf#1850 – us UMI ProQuest [975]
The american west : overland journeys, 1841-1880 – (= ser Research colls on the american west) – 663mf – 9 – $6660.00 – 1-55655-711-6 – us UPA [975]
American whig review – v1-16. 1845-52 – 1 – (formerly: the american review) – us AMS Press [073]
American whig review – La Salle IL 1845-52 – 1 – mf#3932 – us UMI ProQuest [325]
American wives and english husbands : a novel / Atherton, Gertrude – Toronto: Copp, Clark, 1898 [mf ed 1980] – 4mf – 9 – 0-665-03988-3 – mf#03988 – cn Canadiana [830]
American women : from selected americana from sabin's dictionary of books relating to america from its discovery to the present time – 291mf – 9 – mf#C36-28790 – us Primary [305]
American women's magazine – 1925-37, 1961-77 – (= ser American newspapers and periodicals published in the uk and europe 18th-20th centuries) – 8r – 1 – £450.00 – (not publ 1937-60) – uk World [072]
American wood worker – 1895 may 15-1895 aug – 1 – mf#3256925 – us WHS [071]
American Workers Party *see*
– Labor action
– Labor age
American writers and compilers of sacred music / Metcalf, Frank Johnson – New York, NY. 1925 – 1r – us UF Libraries [780]
American writers of to-day / Vedder, Henry Clay – new ed. New York: Silver, Burdett, 1910 [mf ed 1990] – 1mf – 9 – 0-7905-6027-5 – mf#1988-2027 – us ATLA [420]
American youth – Warren MI 1960-74 – 1,5 – ISSN: 0003-1542 – mf#1187 – us UMI ProQuest [305]
American Youth Hostels, inc *see* Hosteling in wisconsin
American zionist – New York, NY. 1921-79 – 1 – us AJPC [071]
American zionist – New York NY 1969-85 – 1,5,9 – ISSN: 0003-1550 – mf#3327 – us UMI ProQuest [321]
American zionist – New York, NY. Nov/Dec 1979-Oct/Nov 1985. Ceased publication – 1 – us AJPC [071]
American Zionist Emergency Council *see* A report of activities, 1940-1946
American zoologist – Oxford, England 1961+ – 1,5,9 – ISSN: 0003-1569 – mf#2201.01 – us UMI ProQuest [590]
Americana : magazine of pictorial satire – New York: American Group Inc, feb 1932-nov 1933 – 1 – us CRL [870]
Americana – New York NY 1973-93 – 1,5,9 – ISSN: 0090-9114 – mf#9712 – us UMI ProQuest [975]
American-arab affairs – Washington DC 1984-91 – 1,5,9 – (cont by: middle east policy) – ISSN: 0731-6763 – mf#14796 – us UMI ProQuest [327]
American-Canadian Genealogical Society of New Hampshire et al *see* Genealogist
L'americanisme / Houtin, Albert – Paris: Emile Nourry, 1904 [mf ed 1986] – vi/497p on 2mf – 9 – 0-8370-8827-5 – (incl ind) – mf#1986-2827 – us ATLA [241]
Americanismo literario / Garcia Godoy, Federico – Madrid, Spain. 1917 – 1r – us UF Libraries [972]
Americanisms, old and new : a dictionary of words, phrases and colloquialisms peculiar to the united states, british america, the west indies, etc, etc / Farmer, John Stephen – London: privately printed by T Poulter, 1889 [mf ed 1980] – 7mf – 9 – 0-665-02949-7 – mf#02949 – cn Canadiana [420]
The americanization of carl schurz / Easum, Chester Verne – Chicago, IL: The University of Chicago Press, c1929 [mf ed 1970] – (= ser Library of american civilization 15506) – xi/374p on 1mf – 9 – us Chicago U Pr [975]
The americanization of the augustana lutheran church / Lund, Gene Jessie – [1954] Chicago: Department of Photodup, U of Chicago Lib, 1965 (1r); Evanston: American Theol Lib Assoc, 1984 (1r) – 1 – 0-8370-0428-4 – mf#1984-B024 – us ATLA [242]
American-jewish life – 1975 may 30-1977 dec 16; 1978 jan 27-1979 dec; 1980 jan 25-1982 mar 26; apr 30-1984 dec 21; 1985 jan 25-1987 jul; aug 28-1989 sep 18 – 1 – mf#345247 – us WHS [071]
O americano : jornal politico e litterario – Rio de Janeiro, RJ: Typ de R Ogier, 07 jul-25 out 1831 – (= ser Ps 19) – mf#P2,4,20 – bl Biblioteca [073]
O americano : jornal politico, litterario e noticioso – Sao Bento do Sapucai, SP: Typ do Americano, 26 nov 1876 – (= ser Ps 19) – mf#P18,01,108 – bl Biblioteca [073]
O americano : orgao do partido liberal – Cachoeira, BA: Typ do Americano, 12 maio 1872; mar 1878; jan 1882; out 1883; ago 1884; jan-fev 1885; mar 1886 – (= ser Ps 19) – mf#P11,02,04 – bl Biblioteca [325]
O americano : orgao republicano – Manaus, AM: [s.n.] 21 nov-05 dez 1889 – (= ser Ps 19) – mf#P11,01,41 – bl Biblioteca [320]
O americano : periodico official, politico e litterario – Alegrete, RS: Typ Republicana Rio Grandense, 24 set 1842-01 mar 1843 – (= ser Ps 19) – mf#P03A,04,22 n02 – bl Biblioteca [321]
O americano : periodico politico, litterario, critico, noticioso – Pernambuco, 22 ago 1867 – (= ser Ps 19) – bl Biblioteca [321]
O americano – Rio de Janeiro, RJ: Typ Brasiliense de Francisco Manoel Ferreira, 11 out 1847-15 out 1851 – (= ser Ps 19) – mf#P14,01,16-18 – bl Biblioteca [079]
O americano – Sao Joao del Rei, MG: Typ de Pimentel, 16 jan-02 maio 1840 – (= ser Ps 19) – mf#P17,02,57 – bl Biblioteca [972]
O americano : semanario politico e de litteratura – Recife, PE: Typ do Commercio, maio-dez 1870; fev-nov 1871 – (= ser Ps 19) – bl Biblioteca [079]
The americans as they are : described in a tour through the valley of the mississippi / Sealsfield, Charles – London 1828 [mf ed Hildesheim 1995-98] – 1v on 2mf – 9 – €60.00 – 3-487-27180-x – gw Olms [917]
Americans before columbus – v2 iss 1-4 [1969 dec/1970 jan-aug/dec]; v2 iss 4 [1970 aug/dec]; v3 iss 1-2 [1971 jan/jul-aug/oct]; v4 iss 1 [1972 jan/jun]; v7 iss 3, 4-6 [1976 sep, 1977 mar-sep] – 1 – mf#26497 – us WHS [071]
Americans for democratic action papers, 1932-1973 – [mf ed Chadwyck-Healey] – 142r – 1 – (coll traces the evolution of a modern political movt that supported civil rights, the un, international control of atomic energy & global democracy. with p/g ed by jack t ericson) – uk Chadwyck [977]
Americans for effective law enforcement. liability reporter – Chicago IL 1978+ – 1,5,9 – (cont: aele law enforcement legal liability reporter) – ISSN: 0271-5481 – mf#10709,01 – us UMI ProQuest [346]
Americans for haganah – New York, N.Y. – (v1 n1 (15 aug. 1947)-v1 n7 (15 dec. 1947); cont by: haganah speaks) – us AJPC [270]
Americans in panama / Scott, William Rufus – New York, NY. 1912 – 1r – us UF Libraries [972]
Americans in process : a settlement study / ed by Woods, Robert Archey – Boston: Houghton, Mifflin, 1902 [mf ed 1991] – 1mf – 9 – 0-524-00807-8 – mf#1990-0239 – us ATLA [975]
Americans United for Separation of Church and State *see* Church and state
American-Scandinavian Foundation *see*
– Scan
American-scandinavian review – New York NY 1913-74 – 1,5 – (cont by: scandinavian review) – ISSN: 0003-0910 – mf#847.01 – us UMI ProQuest [305]
Americas : cuadernos de divulgacion historica / Gallegos, Gerardo – Habana, Cuba. 1945 – 1r – us UF Libraries [972]
Americas 2001 – v1 n1-8 [1987 jun/jul-1988 oct/nov] – 1 – mf#1554020 – us WHS [071]

America's architectural history : key titles from the seventeenth and eighteenth centuries – 128r – 1 – (previous title: american architectural books; incl printed guide) – us Primary [720]

America's community banker – Washington DC 1995-99 – 1,5,9 – (cont: savings & community banker; cont by: community banker) – ISSN: 1082-7919 – mf#19538.02 – us UMI ProQuest [332]

Americas [english ed] – Washington DC 1949+ – 1,5,9 – ISSN: 0379-0940 – mf#522 – us UMI ProQuest [970]

America's future – New Rochelle NY 1959+ – 1,5,9 – ISSN: 0003-1593 – mf#2249 – us UMI ProQuest [321]

America's greatest (plus) pocket comics – iss n1-8 may 1941-sum 1943 (america's greatest); iss n1,4 (harvey) aug 1941, jan 1942 (pocket comics) – 15 – mf#001FA-002FA – us MicroColour [740]

America's menace, or, the enemy within : (an epitome): including "america, my america", the most powerfully appealing patriotic poem ever penned: a clarion call to patriotic action / Simmons, William Joseph – Atlanta, GA: Bureau of Patriotic Books, c1926 – us CRL [355]

America's middle east policy : kissinger, carter and the future / Kerr, Malcolm H – Beirut: Institute for Palestine Studies 1980 [mf ed University of Chicago Library Photoduplication Lab for the Middle Eastern Microfilm Project at CRL 1993] – 1r – 1 – us CRL [327]

America's national game : historic facts concerning the beginning, evolution, development and popularity of baseball, with personal reminiscences of its vicissitudes, its victories and its votaries / Spalding, Albert G – New York: American Sports Publ Co, 1911 – (cartoons by homer c davenport) – us CRL [790]

America's network – Cleveland OH 1994+ – 1,5,9 – (cont: telephone engineer & management) – ISSN: 1075-5292 – mf#20691 – us UMI ProQuest [380]

America's present opportunity in india / Hiwale, Anand S – Boston: Arakelyan Press, [1907] [mf ed 1995] – (= ser Yale coll) – 216p – 1 – 0-524-09039-4 – (pref by david n beach) – mf#1995-0039 – us ATLA [954]

Americas [spanish ed] – Washington DC 1976+ – 1,5,9 – ISSN: 0379-0975 – mf#11077 – us UMI ProQuest [970]

America's textile industries – Atlanta GA 2000 – 1,5,9 – (cont: america's textiles international; cont by: textile industries) – ISSN: 1528-9311 – mf#16860.05 – us UMI ProQuest [670]

America's textiles international – Atlanta GA 1989-99 – 1,5,9 – ISSN: 0890-9970 – mf#16860.05 – us UMI ProQuest [670]

America's triumph at panama / Avery, Ralph Emmett – Chicago, IL. 1913 – 1r – us UF Libraries [972]

America's wonderlands : a pictorial and descriptive history of our country's scenic marvels as delineated by by pen and camera / Buel, James William – Vancouver: J MacGregor, 1894 [mf ed 1980] – 6mf – 9 – mf#03822 – cn Canadiana [917]

America's wonderlands : a pictorial and descriptive history of our country's scenic marvels as delineated by pen and camera / Buel, James William – Philadelphia: Historical Pub Co, 1893 [mf ed 1983] – 6mf – 9 – mf#26715 – cn Canadiana [917]

Americke delnicke listy = American workingmen's news – Cleveland, OH: Press Comm of the Bohemian Socialist Org of Cleveland, 1909-53 (jan 5 1945-46; 1951-mar 27 1953) – 1 – us CRL [071]

Americke delnicke listy / Cuyahoga Co. Cleveland – jun 1918-1941, 1947-50 [wkly] – 16r – 1 – (in czech) – mf#B3919-3934 – us Ohio Hist [071]

Americke listy – New York: Universum Publ Inc. roc1 cis1. 16 list 1962-nov 19 1989// (wkly) [mf ed -1982 (gaps) filmed 1983] – 6r – 1 – (absorbed: new-yorske listy (1886). publ in new york 1962-jul 8 1966; in perth amboy nj, jul 15 1966- . roc1 consists of 7 issues; roc2 begins with issue for jan 4 1963. vol designation ceases with roc8 cis39 26 zari 1969) – us NE Hist [071]

Americki hrvat – Pittsburgh PA, 1946* – 1r – 1 – (croatian newspaper) – us IHRC [071]

Americki hrvatski glasnik = American croatian herald – Chicago IL 28 oct 1953-26 sep 1956 (imperfect) – 2r – 1 – uk British Libr Newspaper [071]

Americki hrvatski glasnik = American croatian herald – Chicago, IL: Croatian Pub Co, may 14 1947-sep 1956 (wkly) – 5r – 1 – us CRL [071]

Americo lugo / Garcia Lluberes, Alcides – Ciudad Trujillo, Dominican Republic. 1954 – 1r – us UF Libraries [972]

Americus *see* Where to emigrate and why

"Amerika" : bilder und skizzen aus dem geistigen, gesellschaftlichen und geschaftlichen leben in den vereinigten staaten – v2 n37-72 [1882 mar 20-1883 mar 10] – 1r – 1 – mf#1052012 – us WHS [071]

Amerika – Madison WI. 1899, 1899 jan 4, 1900-02, 1903-1904 jun 24, jul-1905, 1906-18, 1919-1922 jul – 12r – 1 – (cont: amerika og norden) – mf#915876 – us WHS [071]

Amerika – Chicago IL, Madison WI. 1886 jun 30-1891, 1892-95, 1896-1897 oct 13 – 3r – 1 – (cont by: amerika og norden) – mf#915799 – us WHS [071]

Amerika – Goteborg, Sweden. 1869-72 – 2r – 1 – sw Kungliga [073]

Amerika – New York. N.Y. 1909 – 1 – us AJPC [071]

Amerika : die politischen, socialen und kirchlich-religioesen zustaende der vereinigten staaten von nordamerika / Schaff, Philip – 2. verm aufl. Berlin: Wiegandt & Grieben, 1858 [mf ed 1990] – 1mf – 9 – 0-7905-6772-5 – (with suppl) – mf#1988-2772 – us ATLA [975]

Amerika : ein roman / Asch, Sholem – Berlin: W Borngraeber, 1911 [mf ed 1988] – 146p – 1 – mf#6958 – us Wisconsin U Libr [830]

Amerika – St Louis MT (USA), 1922 5 dec-1924 27 jun – 2r – 1 – gw Misc Inst [071]

Amerika : die stilbildung des neuen bauens in den vereinigten / Neutra, Richard Joseph – Wien, Austria. 1930 – 1r – us UF Libraries [720]

Amerika : wochenblatt der amerika – St Louis MO 16 sep 1874 – 1 – uk British Libr Newspaper [071]

Amerika be-sifrat yisrael / Silber, Mendel – St Louis, MO. 1928 – 1r – us UF Libraries [939]

Amerika dargestellt durch sich selbst – Leipzig [mf ed Hildesheim 1995-98] – 3v on 13mf – 9 – €130.00 – 3-487-27220-2 – gw Olms [975]

Amerika, der aufgang einer neuen welt / Keyserling, Hermann Alexander, Graf von – Stuttgart, Berlin: Deutsche Verlags-Anstalt 1931 [mf ed 1986] – 1r – 1 – (filmed with: la femme et le feminisme / jacobs, a h) – mf#1743 – us Wisconsin U Libr [975]

Amerika herold und sonntagspost – Omaha NE (USA), 1979-1982 9 apr – 1 – (cont by: amerika-woche, chicago) – gw Misc Inst [071]

Amerika: lithuanian weekly – New York NY 4 jan 1946-18 jul 1947, 3 sep 1948-13 apr 1951 – 1 – (wanting: 10 jan, 13 jun 1947, 12,19 jan, 23 mar 1951) – uk British Libr Newspaper [071]

Amerika og norden – 1897 oct 20-1898 dec 28 – 1 – mf#915826 – us WHS [071]

Amerika samoa : official government periodical / American Samoa. Office of Samoan Information – Pago Pago: Office of Samoan Information. v1-2 no 1. Jun 1973-Jul 1974 – 3mf – 9 – $4.50 – mf#LLMC 82-100C Title 17 – us LLMC [324]

Amerika teekaija = The american pilgrim – V1-6. New York, 1918-1923. Estonian. 1,186p – 1 – $47.86 – us Southern Baptist [242]

Amerika woche – 1982 jun 3-1983 jul; aug-1984 jun; jul-1985 apr; may-1986 feb; mar-dec; 1987 jan-sep; oct-1988 jun; jul-1989 apr; may-dec; 1990 jan-dec; 1991 jan-dec; 1992 jan 4-dec 26; 1993 jan-dec 25; 1995 jan-dec; [1996 jan 6-dec 28] – 1 – mf#638810 – us WHS [071]

Amerikaansche en continentale opvattingen omtrent het vraagstuk der naamlooze vennootschap : openbare les... / Schmey, Fritz Ernst – Haarlem: De Erven F Bohn NV, 1935 (mf ed 19–) – 36p – mf#ZT-TN pv125 n6 – us Harvard [338]

Amerikabladet – Orebro, Sweden. 1869-70 – 1 – sw Kungliga [078]

Amerikai magyar hirlap / Mahoning Co. Youngstown – jan 1920-mar 1942 [wkly] – 9r – 1 – (in hungarian) – mf#B4482-4490 – us Ohio Hist [071]

Amerikai magyar hirlap / Trumbull Co. Youngstown – jan 1920-mar 1942 [wkly] – 9r – 1 – (in hungarian) – mf#B4482-4490 – us Ohio Hist [071]

Amerikai magyar vilag = American hungarian world – Cleveland, [OH]: Associated Hungarian Press, oct 8 1972-apr 16 1978 (wkly) – 5r – 1 – us CRL [071]

Amerikai magyar vilag = Hungarian daily world – New York, oct 25 1937-mar 8 1938 – 1 – us CRL [071]

Amerikai magyar vilag = Hungarian daily world – New York [oct 25 1937-mar 8 1938] (daily (ex sunday and certain hols) – 1 – us CRL [071]

Der amerika-muede : amerikanisches kulturbild / Kuernberger, Ferdinand – Frankfurt a.M: Meidinger 1855 [mf ed 1992] – (= ser Bibliothek der deutschen literatur) – 1r – 1 – (filmed with: wiedergeboren / gertrud kunzemann) – mf#7558 – us Wisconsin U Libr [917]

Amerikan suomalainen kirkko / Rautanen, Viljam – Hancock MI: Suomalais-Luteerilainen Kustannusliike 1911 [mf ed 1992] – 1mf – 9 – 0-524-02492-8 – mf#1990-4351 – us ATLA [242]

Amerikan suometar : kansallista siwistystae ja kirkollista elamaeae barrastawain amerikan suomalaisten aeaenenkannattaja = Finnish genius of america dec 4 1917- – Hancock, Mich.: Finnish Lutheran Book Concern, [1955] (wkly) – 2r – 1 – us CRL [071]

Amerikana sanedesa – Bombay, India. 7 Apr 1951-26 Dec 1952 – 1r – 1 – us L of C Photodup [079]

Amerikana vartahara – Bombay, India. 7 Apr 1951-26 Dec 1952 – 1r – 1 – us L of C Photodup [079]

Der amerikaner – New York, N.Y. 1904-62. Yiddish – 1 – us L of C Photodup [071]

Amerikanisch deutsche encyclopaedie (ael1/13) / ed by Prescott, Thomas H – Columbus 1860 [mf ed 1996] – (= ser Archiv der europaeischen lexikographie, abt 1: enzyklopaedien) – 12mf – 9 – €80.00 – 3-89131-099-4 – gw Fischer [030]

Die Amerikanische Armee *see*
- Hessische post 1945
- Koelnischer kurier

Amerikanische bibliothek / ed by Ebeling, Christoph Daniel – Leipzig 1777 – (= ser Dz. historisch-geographische abt) – 4st on 4mf – 9 – €120.00 – mf#k/n1091 – gw Olms [020]

Das amerikanische duell : roman / Seeliger, Ewald Gerhard – Berlin: Ullstein 1916 [mf ed 1996] – 1r – 1 – (filmed with: am alltag vorbei / peter scher) – mf#4029p – us Wisconsin U Libr [830]

Amerikanische gedichte *see*
- Damian
- Vermischte schriften, und, amerikanische gedichte

Amerikanische kirchengeschichte : seit der unabhaengigkeitserklaerung der vereinigten staaten / Nippold, Friedrich – 3. umgearb aufl. Berlin: Wiegandt & Schotte, 1892 [mf ed 1992] – (= ser Handbuch der neuesten kirchengeschichte 4) – 1mf – 9 – 0-524-03243-2 – (incl bibl ref) – mf#1990-0871 – us ATLA [240]

Die amerikanische nordpol-expedition / Bessels, Emil – Leipzig, 1879 – 12mf – 9 – mf#N-122 – ne IDC [919]

Die amerikanische nordpol-expedition : mit zahlreichen illustrationen in holzschnitt, diagrammen und einer karte in farbendruck / Bessels, Emil – Leipzig, Engelmann, 1879 – 8mf – 9 – mf#03579 – cn Canadiana [919]

Amerikanische plattdeutsche post *see* Plattdeutsche post

Amerikanische reisebilder : mit besonderer beruecksichtigung der dermaligen religioesen und kirchlichen zustaende der vereinigten staaten / Pfleiderer, Joh Gottlob – Bonn: J. Schergens, 1882. Chicago: Dep of Photodup, U of Chicago Lib, 1973 (1r); Evanston: American Theol Lib Assoc, 1984 (1r) – 1 – 0-8370-0550-7 – mf#1984-B357 – us ATLA [240]

Amerikanische schulzeitung – Louisville, KY: Henry Knofel, 1870-75. NS: v1-2 n8 oct 1873-may 1875 – us CRL [370]

Amerikanische schweizer zeitung : organ der schweizer in amerika = American swiss gazette – New York NY: Swiss Publ Co [etc] 1868- [mf ed 1980] – 4r – 1 – (began publ in 1868; in german, french & italian) – us Balch [071]

Amerikanische turnzeitung – 1885 jan 4-1886 dec 26; 1887 jan 2-1891 feb 15; 1891 feb 22-1894 jun 3; jun 10-1897 oct 31; nov7-1901 feb 10; feb 17-1904 apr 24; may 1-1906 aug 26; sep 2-1908 jun 7; jun 14-1909 dec 26; 1910-1911 may 21; may 28-1912 sep 29; 1912 oct 6-14 mar 29; 1914 apr 5-1915 aug 29; sep 5-17 jan 28; 1917 feb 4-dec 30 – 1 – mf#891566 – us WHS [071]

Der Amerikanischen 12. Heeresgruppe fuer die deutsche Zivilbevoelkerung *see* Frankfurter presse

Die amerikanischen antitrust-gesetze : eine untersuchung ihres wesens und zweckes in der modernen amerikanischen marktwirtschaft / Jenny, Hans Heinrich – Affoltern am Albis, 1952. 182 p. LL-592 – 1 – us L of C Photodup [346]

Amerikanisches / Kist, Leopold – Mainz 1871 [mf ed Hildesheim 1995-98] – 1v on 5mf – 9 – €100.00 – 3-487-27166-4 – gw Olms [975]

Amerikanisches archiv / ed by Remer, Julius August – Braunschweig 1777-78 – (= ser Dz. historisch-geographische abt) – 3v on 7mf – 9 – €140.00 – mf#k/n1089 – gw Olms [025]

Amerikanisches magazin : oder authentische beitraege zur erdbeschreibung, staatskunde und geschichte von amerika, besonders aber der vereinten staaten / ed by Hegewisch, Dietrich Hermann et al – Hamburg 1795-97 – (= ser Dz. historisch-geographische abt) – 1v on 5mf – 9 – €100.00 – mf#k/n1290 – gw Olms [975]

Amerikanisches skizzebuechelche : eine epistel in versen / Asmus, Georg – 2. aufl. New York: Willmer & Rogers News Co: American News Co, 1874 [mf ed 1988] – 95p – 1 – (missing t.p. supplied) – mf#6968 – us Wisconsin U Libr [810]

Amerikanisches skizzebuechelche : zweite epistel in versen / Asmus, Georg – Coeln: E H Mayer, 1885 [mf ed 1988] – 127p – 1 – mf#6968 – us Wisconsin U Libr [810]

Amerikanisch-lutherische evangelien postille : predigten ueber die evangelischen pericopen des kirchenjahrs / Walther, Carl Ferdinand Wilhelm – 10. aufl. St Louis: Lutherischer Concordia, [1870?] [mf ed 1993] – 1mf – 9 – 0-524-08581-1 – mf#1993-3166 – us ATLA [226]

Amerikanisch-lutherische pastoraltheologie / Walther, Carl Ferdinand Wilhelm – 4. aufl. St Louis MO: Concordia Pub House 1897 [mf ed 1993] – 1mf – 9 – 0-524-06561-6 – mf#1991-2645 – us ATLA [242]

Amerikanisch-lutherische schul-praxis / Lindemann, Johann Christoph Wilhelm – 2. unveraend aufl. St Louis: Lutherischer Concordia-Verlag 1888 [mf ed 1992] – 1mf – 9 – 0-524-04437-6 – mf#1991-2102 – us ATLA [242]

Amerikanismus, fortschritt, reform : ihr zusammenhang, zweck, und verlauf in amerika, frankreich, england, und deutschland / Braun, Carl – Wuerzburg: Goebel & Scherer, 1904 [mf ed 1990] – 1mf – 9 – 0-7905-3759-1 – (incl bibl ref) – mf#1989-0252 – us ATLA [241]

Amerikanski slovenec – Chicago IL, 1891-1924, 1891-1946 – 34r – 1 – (slovenian newspaper) – us IHRC [071]

Amerikanski srbobran – 1943-1945 aug 17 – 1 – mf#702437 – us WHS [071]

Amerikanskie izvestiia : organ rossiikikh rabochikh organizatsii soedinennykh shtatov i kanady – New York, 1922: mar 22, may 17, jul 5-19, aug 16-30, sep 20-27, oct 11-25, nov 1-15, dec 6-27. 1923: jan 10-oct 3, oct 17, nov 21-dec 26. 1924: jan 3-feb 13, feb 27-jun 11, jul 2-aug13 – us CRL [071]

Amerikanskii pravoslavnyi viestnik = Russian orthodox american messenger – New York [etc]: Pravoslavnaia amerikanskaia missiia, sep 1 1896-aug 27 1897; 1897-1930 – us CRL [071]

Amerikanskij russkij sokol sojedinenija = American russian falcon – Homestead, PA: Greek Catholic Union of USA, 1926-36 – 1 – us CRL [073]

Amerikansko slovenske noviny – 1904 may 18 – 1 – mf#866639 – us WHS [071]

Amerikansko slovenske noviny – Pittsburgh PA, 1893-1904 – 3r – 1 – (slovak newspaper) – us IHRC [071]

Amerikansko-russkii kalendar – Filadelfiia, Izd. Obshchestva russkikh bratstv v Soedinennykh Shtatakh Sievernoi Ameriki, 1911, 1940, 1950 – 1 – us CRL [520]

Amerikaposten – Stockholm, Sweden. 1925 – 1r – 1 – sw Kungliga [078]

Amerika's gesetze / Lehmann, Ignaz – St. Louis, Witter 1857 176 p. LL-424 – 1 – us L of C Photodup [348]

Amerikas Latviesu Jaunatnes Apvieniba *see* Brivibas talcinieks

Amerikas vestnesis – 1961 mar 24, 31, apr 28, 29 – 1 – mf#1443439 – us WHS [071]

Amerikas vestnesis – Boston: Amerikas Vestnesis Inc, aug 19 1958-jul 1966 – 2r – 1 – us CRL [071]

Amerikas vestnesis : nacionals laikraksts: latvian newspaper – Boston MA: [s.n.] n1 [oct 11 1955]- [mf ed 1980] – 1r – 1 – (in latvian; set incomplete) – us Balch [071]

Amerikas zhina – New York. 1926-1933. (incomplete) – 1 – us NY Public [073]

Amerika-woche – Chicago IL (USA), 1982- – 1 – (cont: amerika herold und sonntagspost) – gw Misc Inst [071]

Amerikos lietuvis = The lithuanian of america – Worcester, MA: M Paltanavicia, dec 13 1917-apr 3 1937 – 3r – 1 – us CRL [071]

Amerindian : american indian review – Chicago IL 1972-74 – 1,5,9 – ISSN: 0003-164X – mf#7843 – us UMI ProQuest [305]

The amerindian – 1952-74 – (= ser American indian periodicals... 1) – 11mf – 9 – $105.00 – us UPA [305]

L'amerique avant christophe colomb : resume des travaux de quelques antiquaires / Dunn, Oscar – Montreal?: E Senecal, 1875 – 1mf – 9 – mf#06714 – cn Canadiana [910]

L'amerique avant les europeens / Desdevises du Dezert, Theophile – Caen [France]: F Le Blanc-Hardel, 1878 [mf ed 1980] – 1mf – 9 – 0-665-04283-3 – mf#04283 – cn Canadiana [305]

L'amerique du nord pittoresque : ouvrage redige par une reunion d'ecrivains americains / Bryant, William C – Paris: A Quantin, 1880 [mf ed 1980] – 9mf – 9 – 0-665-00305-6 – (incl ind; trans by Benedict-Henry Revoil) – mf#00305 – cn Canadiana [917]

L'amerique et les travaux americains en 1866 / Cortambert, Richard – [Paris: s.n.], 1867 – 1mf – 9 – 0-665-90628-5 – mf#90628 – cn Canadiana [917]
L'amerique latine – n1-260. Paris. 1923-oct 1924; juil 1926-27. mq no. 49, 90, 95-183, 185-191, 193-208, 234, 237 – 1 – fr ACRPP [972]
L'amerique septentrionale et meridionale : ou description de cette grande partie du monde – Paris: E Ledoux, 1835 [mf ed 1982] – 8mf – 9 – mf#33981 – cn Canadiana [917]
Ameriska domovina – Cleveland, Cuyahoga, OH: J Debevec, 1919- [biwkly] – 1r – 1 – (cont: clevelandska amerika; in slovenian & english) – us Western Res [071]
Amersbach, Karl *see* Aberglaube, sage und maerchen bei grimmelshausen
Amerta *see* Dinas purbakala
Amery echo – v1 n1-v3 n9 [1889 jun 14-1891 jul 30] – 1 – mf#916277 – us WHS [071]
Amery free press – 1906 oct 18-1909; 1910-19; 1920 jan-sep 2; 1921 sep 8-24; 1925-38; 1935 jan-sep 12; 1939-1943 sep 2; oct 7-1945; 1946-62; 1963-1964 jun; jul-1965; 1966 jan 6-dec 29; 1967 jan 5-oct 12; oct 19-1968 jul 11; jul 18-1969 apr 24; may 1-1970 feb 12; feb 19-sep 24; oct 1-1971 apr 15; apr 22-oct 28; nov 4-1972 jun 29; jul 6-dec 28; 1973 jan-dec; 1974 jan-dec; 1975 jan-dec; 1976 jan-dec; 1977 jan-dec; 1978 jan-dec; 1979 apr-dec; 1980 apr-dec; 1981 apr-dec; 1982 jan-dec; 1983 jan-dec; 1984 jan-dec; 1985 jan-dec; 1986 jan-dec; 1987 jan-dec; 1988 apr-dec; 1989 apr-dec; 1990 apr-dec; 1991 apr-dec; 1992 apr-dec; 1993 apr-dec; 1994 apr-dec; 1995 jan-mar – 1 – mf#982784 – us WHS [071]
Amery, Leopold Stennett *see*
– The framework of the future
– India and freedom
Ameryka = America – Filadelfiia, PA: Provydinie, [dec 7 1917-1918; 1920-1921; 1923-1924; 1932] (triwkly) – 7r – 1 – us CRL [071]
Ameryka echo – Toledo, OH: A A Paryski, dec 1917-oct 24 1922 (10r); jan 1 1922-dec 25 1927 (7r); mar 22 1931; may 3-24 1931 – 1 – (sunday issue only) – us CRL [071]
Ameryka-echo – 1903 – 1 – mf#2737542 – us WHS [071]
Ames, A H *see*
– Revelation of st john the divine
– The revelation of st john the divine
Ames, Charles Gordon et al *see* What do unitarians believe and teach?
Ames, Daniel T *see* Ames on forgery.
Ames, Edward Scribner *see*
– The divinity of christ
– The higher individualism
– The psychology of religious experience
Ames, Herbert Brown *see*
– Canadian political history
– "The city below the hill"
Ames, Herbert Brown et al *see* Abstract of a course of ten lectures on municipal administration in montreal
Ames, James Barr *see*
– A selection of cases in equity jurisdiction.
– A selection of cases on pleading, with references and citations
– A selection of cases on the law of torts
Ames, Jennie M *see* Genealogical records of austin bearse (or bearce) of barnstable, cape cod, massachusetts, usa, a.d. 1638 to a.d. 1933
Ames on forgery. / Ames, Daniel T – Boston, Boston Book, 1901. 293 p. LL-285 – 1 – us L of C Photodup [340]
Ames primitives : contribution a l'etude du sentiment religieux chez les paiens animistes / Burnier, Theophile – Paris: Societe des missions evangeliques, 1922 – 1 – us CRL [290]
Ames, William *see*
– Conscience with the power and cases there of
– The marrow of sacred divinity...
– The substance of christian religion
Die amesa spentas : ihr wesen und ihre urspruengliche bedeutung / Geiger, Bernhard – Wien: A Hoelder, 1916 – (= ser Sitzungsberichte der philosophisch-historischen klasse der kaiserlichen akademie der wissenschaften) – 1mf – 9 – 0-524-02079-5 – (incl bibl ref) – mf#1990-2843 – us ATLA [280]
Amesbury 1685-1849 – Oxford MA (mf ed 1994) – (= ser Massachusetts vital record transcripts to 1850) – 1v on 19mf – 9 – 0-87623-199-7 – (mf 1t: marriages 1686-89. mf 1t-2t: births 1685-1723. mf 2t-3t: deaths 1686-1737. mf 3t: marriages 1728-42. mf 3t-7t: births 1700-83. mf 5t: deaths 1729-70. mf 7t-8t: marriages 1701-62; intentions 1700-27; mf 8t: births, deaths 1718-81; intentions 1760-84. mf 8t-12t: births 1742-1847. mf 12t: marriages 1700-1821; intentions 1700-06. mf 12t-15t: intentions 1784-1849. mf 14t: deaths 1774-1846. mf 15t-16t: marriages 1820-44. mf 16t: marriages 1726-42; out-town marriages 1696-1799. mf 16t-18t: births 1843-49. mf 18t: marriages 1843-49. mf 19t: deaths 1843-49) – us Archive [978]
Amethyst – 1832-34 – (= ser English gift books and literary annuals, 1823-1857) – 13mf – 9 – uk Chadwyck [800]
Amex : the american expatriate in canada – v1. 1968-69 – 1 – (formerly: the american exile in canada) – us AMS Press [073]
Amex-canada – Washington DC 1969-77 – 1,5,9 – ISSN: 0003-1674 – mf#7730 – us UMI ProQuest [071]
Amfilokhii, Arkhimandrit *see*
– Opisanie iurevskogo evangeliia 1118-1128 g
– Opisanie voskresenskoi novoierusalimskoi biblioteki...s prilozheniem snimkov so vsekh pergamennykh rukopisei i nekotorykh pisannykh na bumage
Amfiteatrov, A *see* Zhurnal politicheskii i literaturnyi
Amfiteatrov, Aleksandr Valentinovich *see* Znakomyia muzy
Amfiteatrov, V A *see* laltinskii kur'er
Amfiteatrov-Kadashev, Vladimir Aleksandrovich *see* Zum-zum
An amharic reader / Eadie, J I – Cambridge, 1924 – 3mf – 9 – mf#NE-20249 – ne IDC [956]
Die amharische sprache / Praetorius, F – Halle, 1879 – 6mf – 9 – mf#NE-20255 – ne IDC [470]
Amherst 1747-1891 – Oxford, MA (mf ed 1987) – (= ser Massachusetts vital records) – 28mf – 9 – 0-87623-012-5 – (mf 1-5: genealogical records: b,m,d 1747-1843. mf 6-8: births, marriages, deaths 1747-1851. mf 9-11: births, marriages, deaths 1843-51. mf 12-18: births 1851-91, vol 3. mf 19-23: marriages 1851-91, vol 4. mf 24-28: deaths 1851-91, vol 5) – us Archive [978]
Amherst advertiser – 1890 sep 16-dec 23 – 1 – mf#916285 – us WHS [071]
Amherst advocate – 1893 feb 22-1895; 1898 feb 9-apr 13; 1903 jan 8-dec 31; 1904 jan 14-dec 29; 1905 jan 5-dec 21; 1906 jan 4-1907 aug 8; aug 15-1909 mar 25; apr 1-1910 oct 27; 1910 nov 3-1912 may 30; jun 6-1914 jan 29; feb 5-1915 aug 26; sep 2-1917 mar 29;apr 5-1919 apr 10; apr 17-1920 oct 14; oct 21-1922 apr 13; apr 20-1923 sep 27; oct 4-1925 jan 15; jan 22-1926 jun 17; jun 24-1927 dec 15; dec 22-1929 jun 13 [1]; jun 13 [2]-1930 dec 25; 1931 jan 1-1932 jul 21; jul 28-1934 jan 11; jan 18-1935 aug 1; aug 8-1937 mar 11; 1937 mar 18-1938 sep 29; oct 6-1939 dec 28; 1940-64; 1965 jan-1967 aug 17; aug 24-1968 feb 22 – 1 – mf#961902 – us WHS [071]
Amherst announcer (newsletter) and bulletins / Tonawanda. New York. Amherst Baptist Church – 1964-86 – 1 – $45.18 – (newsletter aug 1964-jul 1986; bulletins aug 1964-jul 1986) – us Southern Baptist [242]
Amherst College. Amherst, MA *see* Catalog
Amherst college black studies newsletter – v1 n1 [1989] – 1 – mf#5320201 – us WHS [305]
The amherst fairplay – Amherst, NE: M P McElroy. v1 n1. oct 1910- (wkly) – 1r – 1 – us NE Hist [071]
Amherst farmer – v1 n1 [1884 fair time] – 1 – mf#4753414 – us WHS [071]
Amherst, Massachusetts. First Baptist Church *see* Records
Amherst pioneer – 1884 mar 22-apr 2; 1886 dec 25 – 1 – mf#4755695 – us WHS [071]
The amherst reflector *see* The reflector
The amherst times – Amherst, NE: J H Bratton. v1 n15. aug 18 1893 (wkly) [mf ed 1996] – 1r – 1 – us NE Hist [071]
Amherst, William Joseph *see* The history of catholic emancipation and the progress of the catholic church in the british isles
Amhurst, Jeffrey, 1st Baron *see* Official papers, 1740-83
Ami – 1978 feb 1-1980 jan 1 – 1 – mf#630797 – us WHS [071]
L'ami de la charte en prison ou un mois de retraite : suivi de notes explicatives & historiques / Mangin, Victor – Nantes 1827 [mf ed Hildesheim 1995-98] – 1v on 1mf – 9 – €40.00 – ISBN-10: 3-487-26274-6 – ISBN-13: 978-3-487-26274-1 – gw Olms [365]
L'ami de la jeunesse studieuse – Hanoi: Ecole Tri Duc. v1-24 [sep 15? 1931-sep 1 1932] [mf ed Hanoi, Vietnam: National Library of Vietnam 1997] – 1r – 1 – (master neg held by crl) – mf#mf-11808 seam – us CRL [079]
L'ami de la patrie : ou journal de la liberte francaise – Paris. avr 1796-mars 1798. (fragm., 107 no. entre no. 54-725) – 1 – fr ACRPP [944]
L'ami de la religion et du roi – Trois-Rivieres [Quebec]: L Duvernay, [1820-182-?] – 9 – mf#P04068 – cn Canadiana [200]
L'ami des arts : ou justification de plusiers grands hommes... / Decroix, Jacques Joseph Marie – Amsterdam, Paris: Marchands 1776 [mf ed 19–] – 5mf – 9 – mf#fiche 1175 – us Sibley [440]
Lami des citoyens – Paris. n 5, 9-10, 20-21, 36, 57-59, 61-63. aout 1791-janv 1793 – 1 – fr ACRPP [073]
Ami des femmes / Dumas, Alexandre – Paris, France. 1895 – 1r – us UF Libraries [440]
L'ami des foyers chretiens : deutsch-franzoesische zeitung – Metz (F), 1972-78 – 1 – gw Misc Inst [074]
Ami des lois – 1810 jan 18 – 1 – mf#861281 – us WHS [071]
L'ami des louis – New Orleans, LA. 1813-1824 (1) – mf#68740 – us UMI ProQuest [071]
Ami des Monuments et des Arts Parisiens et Francais *see* Organe du comite des monuments francais
L'ami d'israel – Strasbourg: Societe des amis d'Israel de Bale, v1-84 (1836-1997) [mf ed 2005] – 131v on 9r – 1 – (publ: basel, 1995-1997; iss also by: fondation suisse eglise-judaisme, 1995-; lacks: several iss) – mf#2006c-s053 – us ATLA [270]
L'ami du chanteur : nouveau recueil de romances, melodies, chansons et chansonnettes avec musique – Montreal: E Hardy...1895 [mf ed 1980] – 3mf – 9 – mf#SEM105P58 – cn Bibl Nat [780]
L'ami du clerge – Paris: V Goupy & Jourdan. v1-78. 1878-1968 [wkly] [mf ed 2002-03] – 78v on 42r – 1 – (lacks: few iss. suspended: aug 6 1914-apr 10 1919, and sep 7 1939-oct 10 1946. incl ind to suppl: l'ami du clerge paroissial) – mf1019 – us ATLA [241]
L'ami du clerge paroissial – Langres, France: Rallet-Bideaud. v1-68. 1888-1968 [biwkly] [mf ed 2003] – 68v on 27r – 1 – (lacks: v26 n6; v37 n33; v47 n9. publ suspended: sep 1914-mar 1919; sep 1939-sep 1946. with ind) – mf#2003-s020 – us ATLA [241]
L'ami du clerge paroissial *see* L'ami du clerge
L'ami du peuple – Paris. mai 1928-oct 1937; mq 1er janv-12 mars 1934 – 1 – (ed. du soir. janv-juin 1930. 1) – fr ACRPP [944]
L'ami du peuple : journal politique quotidien – Paris, France 23-29 apr 1871 – 1 – uk British Libr Newspaper [320]
L'ami du peuple – n1-24. Paris. 20 juil-15 sept 1793 – 1 – fr ACRPP [944]
L'ami du peuple : ou le defenseur des patriotes persecutes – Paris, sep 1794-oct 1797 – 1 – fr ACRPP [944]
L'ami du peuple = Der volksfreund – Strassburg (Strasbourg F), 1972- – 1 – (deutsch-franzoesische zeitung) – gw Misc Inst [074]
L'ami du peuple en 1848 – Paris, France 27 feb-14 may 1848 – 1 – (discontinued) – uk British Libr Newspaper [074]
L'ami du peuple en 1848 : journal paraissant le jeudi et le dimanche matin – Paris: Schneider [feb 27-may 14 1848] (semiwkly) – (= ser French revolution of 1848. newspapers) – 1r – 1 – us CRL [074]
L'ami du roi – Toulouse. aout 1815-juin 1819 – 1 – fr ACRPP [073]
L'ami du roi, des francais, de l'ordre et surtout de la verite – Paris. Quot. juin 1790-aout 1792 – 1 – fr ACRPP [073]
Ami, Henry Marc *see*
– Additional notes on the geology and palaeontology of ottawa and vicinity
– Annual address of the ottawa field-naturalists' club...
– A biographical sketch of george mercer dawson...
– A brief biographical sketch of sir john william dawson
– Canada and newfoundland
– Catalogue of silurian fossils from arisaig, nova scotia
– Contribution to the palaontology sic of the post-pliocene deposits of the ottawa valley
– Esquisse geologique du canada
– Flora temiscouatensis
– Note on the occurence of bellinurus grandaevus, a new species of palaeozoic limuloid crystaceans recently described by prof t r jones and dr henry woodward, from the eo-carboniferous rocks of riversdale, nova scotia
– Notes and comments
– Notes and descriptions of some new or hitherto unrecorded species of fossils from the cambro-silurian (ordovician) rocks of the province of quebec
– Notes bearing on the devono-carboniferous problem in nova scotia and new brunswick
– Notes on, and the precise geological horizon of siphonotreta scotica, davidson
– Notes on fossils from the utica formation at point-a-pic, murray river, murray bay (que), canada
– Notes on some of the fossil organic remains in the geological formations and outliers of the ottawa palozoic basin
– Notes on the geology and palontology of the rockland quarries and vicinity, in the county of russell, ontario, canada
– On the geology of quebec and environs
– On the occurrence of scolithus in rocks of the chazy formation about ottawa, ontario
– On the sequence of strata forming the quebec group of logan and billings
– Preliminary lists of the organic remains
– Progress of geological work in canada during 1898
– Progress of geological work in canada during 1899
– Resources of the country between quebec and winnipeg along the line of the grand trunk pacific railway, with map
– Sir john william dawson
– Sketch of the life and work of the late dr alfred r c selwyn...etc,
– Synopsis of the geology of canada
– Synopsis of the geology of montreal
– The utica slate formation
– The utica terrane in canada
Ami, Henry Marc [comp] *see* List of contributions to geology, palontology, etc
Ami newsletter – 1980 feb 1-1988 jun – 1 – mf#707278 – us WHS [071]
Amiable baptist church – Glenmora, LA. 1829-1977 – 1 – $63.90 – us Southern Baptist [242]
Amiable baptist church – Rapides Parish, LA. 1829-1904 – 1 – $13.59 – us Southern Baptist [242]
Amiable, Louis *see*
– La franc-maconnerie et la magistrature en france a la vieille de la revolution
– Une loge maconnique d'avant 1789
Amiama, Manuel A *see*
– Notas sobre derecho constitucional
– Viaje, ensayode novela de la vida capitalena
Amiaud, Arthur *see*
– Les inscriptions de salmanasar 2 roi d'assyrie
– Tableau compare des ecritures babylonienne et assyrienne archaiques et modernes
Amicus *see*
– An authentic history of the prayer book, its five revisions, and the periods at which they were made
– Celine
– L'ecole de medecine et de chirurgie de montreal, faculte de medecine de l'universite-victoria
Amicus journal – New York NY 1985-2001 – 1,5,9 – ISSN: 0276-7201 – mf#15196.01 – us UMI ProQuest [333]
Amida / Berchem, M van & Strzygowski, J – Heidelberg, 1910. 2pts – 8mf – 9 – mf#AR-1868 – ne IDC [956]
"Amida buddha unsere zuflucht" : urkunden zum verstaendnis des japanischen sukhavati-buddhismus / Haas, Hans – Leipzig: Dieterich, 1910 [mf ed 1991] – (= ser Religions-urkunden der voelker 2/1) – viii/187p/11pl on1mf – 9 – 0-524-01494-9 – (incl texts sacred to the jodo-shu and the jodo shinshu) – mf#1990-2470 – us ATLA [280]
Amid-i sevda – n[1]-6. 1325 [1907] – (= ser O & t journals) – 2mf – 9 – $40.00 – us MEDOC [956]
Amidst timiskiming [sic] and kipawa pines : unexcelled for sport with canoe, rod and gun / Jones, W M – Ottawa: Mortimer, 1906 – 1mf – 9 – 0-665-97724-7 – mf#97724 – cn Canadiana [639]
Amiens, Gerard d' *see* Der roman von escanor
Amiet, R *see* The benedictonals of freising (hbs88)
Amiga world – Peterborough NH 1985-95 – 1,5,9 – ISSN: 0883-2390 – mf#14947 – us UMI ProQuest [000]
El amigo de las leyes – Ano 1812, (2-xi/1-xii) – 2mf – 9 – sp Cultura [946]
El amigo de las leyes – Ano 1814, (15-ii/3-v) – 2mf – 9 – sp Cultura [946]
Amigo del hogar – Santo Domingo: Misioneros del Sagrado Corazon de Jesus. [ano 38 n390-ano 57 n610 (enero 1983-dic 1998)] (mthly) – 7r – 1 – us CRL [241]
Amigo desconocido nos aguarda / Dominguez, Franklin – Ciudad Trujillo, Dominican Republic. 1958 – 1r – us UF Libraries [972]
O amigo do escravo : orgao abolicionista – Rio de Janeiro, RJ: Typ Camoes, 27 out 1883; 17 jan 1884 – (= ser Ps 19) – mf#P05,04,12 – bl Biblioteca [320]
O amigo do povo : jornal do povo – Rio de Janeiro, RJ: Typ Vera Cruz, 01,15,19 jul 1877 – (= ser Ps 19) – mf#P11,08,05 – bl Biblioteca [320]
O amigo do povo : jornal politico, commercial e noticioso – Rio de Janeiro, RJ. 06 fev 1873 – (= ser Ps 19) – mf#P05,04,13 – bl Biblioteca [321]
O amigo do povo : orgao do partido democratico – Anchieta, ES. 22 fev 1891 – (= ser Ps 19) – mf#P11B,05,17 – bl Biblioteca [325]
O amigo do rei da nacao – Rio de Janeiro, RJ: Typ Real, 1821 – (= ser Ps 19) – mf#P01,03,16 – bl Biblioteca [320]
O amigo dos homens – Rio de Janeiro, RJ: Typ dos Santos & Companhia, 03 mar 1844; out-05 dez 1847 – (= ser Ps 19) – mf#P3A,03,15-16 – bl Biblioteca [200]
Amigo y Bertran, L *see*
– Apologia...del agua de la vida...en que se hace examen y juicio de los papeles...
– Respuesta de andres davila...a la apologia en defensa de la medicina substancial
Amigoe – Willenstad, Curacao. v106 n73- v11 n121. 1989 apr-1998 jun – 42r – (gaps) – us UF Libraries [079]

Amigos de la romeria de san pedro 1975 / Torrejoncillo. Ayuntamiento – Caceres: Imp. La Minerva, 1975 – 1 – sp Bibl Santa Ana [946]
Amigos de la romeria de san pedro 1977 / Torrejoncillo. Ayuntamiento – Caceres: Tip. Extremadura, 1977 – 1 – sp Bibl Santa Ana [946]
Amigos de la romeria de san pedro, 1978 / Torrejoncillo. Ayuntamiento – Tip. Extremadura, 1978 – sp Bibl Santa Ana [060]
Amino acid digestibility of feedstuffs for pigs / Viljoen, Johannes – Stellenbosch: U of Stellenbosch 1998 [mf ed 1998] – 4mf – 9 – mf#mf.1304 – sa Stellenbosch [636]
Amino functionalized polymers by anionic and controlled free radical polymerization methods / Ndawuni, Mzikayise Patrick – Uni of South Africa 2000 [mf ed Johannesburg 2000] – 3mf – 9 – (incl bibl ref) – mf#mfm14995 – sa Unisa [540]
Amiot, J J M *see*
– Eloge de la ville de moukden et de ses environs
– Monument de la transmigration des tourgouthes des bords de la mer caspienne, dans l'empire de la chine
Amiot, Joseph Marie *see* Abrege historique des principaux traits de la vie de confucius
Amir, Arie *see* Report on an agricultural survey trip to uganda
Amir Khusraw Dihlavi *see* The campaigns of 'ala'u'd-din khilji
Amira, K von *see* Der stab in der germanischen rechtssymbolik
Amira, Karl von *see* Das endinger judenspiel
Amiral de grimouard au port-au-prince / Grimouard, Henri – Paris, France. 1937 – 1r – us UF Libraries [972]
Amiri baraka from black arts to black radicalism – (= ser The black power movement 1) – 9r – 1 – $1740.00 – 1-55655-834-1 – (with p/g; filmed fr personal coll of dr komozi woodard) – us UPA [934]
Amis / Dreyfus, Abraham – Paris, France. 1909 – 1r – us UF Libraries [440]
Amis comme avant / Jeanson, Henri – Paris, France. 1930 – 1r – us UF Libraries [440]
Les amis de dieu au quatorzieme siecle / Jundt, Auguste – Paris: Sandoz & Fischbacher, 1879 [mf ed 1990] – 445p on 2mf – 9 – 0-7905-5288-4 – (in french) – mf#1988-1288 – us ATLA [920]
Les amis de rabelais et de la deviniere – Tours. 1951-72 – 5 – fr ACRPP [440]
Amis du 16e siecle et de la Pleiade *see* Revue de la renaissance
The amistad : a journal of good news – Atlanta GA: Amistad Publ Co (mthly ex jul & aug) [mf ed 2004] – 1r – 1 – (began in 1924? publ 1934-mar 1937 in atlanta [ga]; occt 1937-may 1938 in worcester [ma]; new series begins in oct 1937 with v1; lacks: apr, aug 1935, mar, jul, aug, nov-dec 1936, feb 1937) – mf#2004-s035 – us ATLA [242]
Amistad funesta / Marti, Jose – Mexico City?, Mexico. 1958 – 1r – us UF Libraries [972]
Amistad research center news – v1 n1-v4, n1 [1971 aug-1975 sep] – 1 – mf#671689 – us WHS [071]
Amity standard – Amity OR: W C Depew, 1910-75 [wkly] [mf ed 1960-77] – 17r – 1 – us Oregon Lib [071]
Amj: agricultural machinery journal – Sutton, England 1976-87 – 1,5,9 – ISSN: 0002-1539 – mf#11285 – us UMI ProQuest [630]
Amleto = Hamlet / Shakespeare, William – Milano: Longanesi [1971] [mf ed 19–] – 1r – 1 – (trans into italian by eugenio montale) – us OmniSys [820]
Amman, J *see*
– Icones livianae...
– Kuenstliche und wolgerissene figuren, der fuernemhsten evangelien...
– Kuenstliche...figuren von allerlai jagt und weidwerck, allen liebhabern der maler kunst, auch goltschmieden, bildthawern...
– Kunstbuechlein...
– Wapen und stammbuch darinnen der keys. maiest. chur und fuersten, graffen, freyherrn, deren vom adel
[Amman, J] *see* Stam und wapenbuch hochs und niders standts
Amman sun – n89:39-51 [1989 sep 27-dec 20]; n90: 01-17, 19-26, 28-31; [1990 jan 3-apr 25, may 9-jun 27, jul 1-aug 22] – 1 – mf#1613522 – us WHS [071]
Ammann, Johann Josef *see* Volksschauspiele aus dem boehmerwalde
Ammenhausen, Konrad von *see* Schachzabelbuch (cima58)
Die ammen-uhr : aus des knaben wunderhorn – Leipzig: Mayer und Wigand, [1843?] – us Wisconsin U Libr [730]
Ammergauer zeitung – Oberammergau DE, 1907-09 – 2r – 1 – gw Misc Inst [074]
"Ammi-my people" : containing an elucidation of the principles of the christian religion, as taught by christ and his apostles and practiced by the people of god in all ages / Shoup, William J – West Alexandria OH: Shoup c1905 [mf ed 1992] – 2mf – 9 – 0-524-02752-8 – mf#1990-4427 – us ATLA [240]
Ammirato, S, the Elder *see* Il rota overo dell'imprese. dialogo del s. scipione ammirato
Ammisca file : the magruder mission to china / U.S. Army – v. 1-7. 11 Jul 1941-Jun 1942 – 1 – us L of C Photodup [951]
Ammo – v18 n1-v23 n12 [1977 aug-1985 jan] – 1 – mf#349618 – us WHS [071]
Ammon, Blasius *see*
– Missae cum breves tum quatuor vocum laudatissimae concinnata
– Sacrae cantiones, quas vulgo moteta vocant...
Ammon, H. *see* Imitatio crameriana sive exercitium pietatis domesticum
Ammon Hennacy House of Hospitality *see* Catholic agitator
Ammon, Hermann *see* Daemon faust
Ammon, Wolfgang *see* Libri tres odarum ecclesiasticarum, de sacris cantionibus, in ecclesiis germanicis
Ammonia toxicity in the fertilization of shade tobacco / Borda, Eugene – s.l, s.l? . 1940 – 1r – us UF Libraries [630]
Ammsa : aboriginal multi-media society of alberta – v1 n2, 10 [1983 mar 25, may 20] – 1 – mf#1083327 – us WHS [366]
Ammundsen, Valdemar *see*
– Soeren kierkegaards ungdom
– Den unge luther
Ammunition – 1943 apr-1948 dec; 1949-54; 1955-1957 jun; 1976 – 1 – mf#416774 – us WHS [071]
Amner, John *see* Sacred hymnes
Amnesty action – 1980 mar-1988 aug – 1 – mf#203273 – us WHS [327]
Amnesty international newsletter – London: Amnesty International Publ. v1- 1971- [mf ed 1978-] – 4r – 1 – (mthly 1974- , qrterly 1971-) – mf#911 – us Wisconsin U Libr [322]
Amnesty International USA *see*
– Bulletin of amnesty international usa
– Labor news
– Matchbox
Amok : novellen einer leidenschaft / Zweig, Stefan – [Stockholm]: S Fischer, 1950, c1946 [mf ed 1996] – 386p – 1 – mf#9365 – us Wisconsin U Libr [830]
Amok : a story / Zweig, Stefan – New York: Viking Press, 1931 [mf ed 1992] – 121p – 1 – mf#7801 – us Wisconsin U Libr [830]
Amola, Aisa Aisa *see* Anglo-punjabi dictionary
O amolador – Rio Grande do Sul: Typ do Amolador, 12 abr-jun 1874; abr-25 dez 1875 – (= ser Ps 19) – bl Biblioteca [079]
Amon, Johann Andreas *see*
– Quatuor pour flute, violon, alto & violoncelle obliges, oeuvre 84
– Trois quatuors concertans pour hautbois, violon, alto & violoncelle, op 92
– Trois quatuors concertans pour l'alto, violin, viola & violoncelle, oeuvre 15me
– Trois quatuors concertants pour flute, violon, alto et violoncelle. [quatuor no 2]
– Trois quatuors pour flute, violon, alto et violoncelle concertans, oeuvre 42
– Trois sonates pour le forte-piano avec accompagnement d'un violon, op 11
– Trois sonates pour le piano forte avec accompagnement de violon & violoncelle obliges. oeuvre 76
– Trois sonates pour le piano-forte avec accompagnement de flute et violoncelle obliges, oeuvre 48
– Trois trios concertans pour violon, alto et basse, op 8
Among asia's needy millions : journal of a visit to the far east / Corey, Stephen Jared – Cincinnati: Foreign Christian Missionary Society, c1915 [mf ed 1992] – (= ser Christian church (disciples of christ) coll) – 1mf – 9 – 0-524-04546-1 – mf#1991-2110 – us ATLA [240]
Among central african tribes : journal of a visit to the congo mission / Corey, Stephen Jared – 2nd ed. Cincinnati: Foreign Christian Missionary Society, c1912 [mf ed 1993] – (= ser Christian church (disciples of christ) coll) – 1mf – 9 – 0-524-06400-8 – mf#1991-2522 – us ATLA [240]
Among friends – 1984 mar-1986 oct; 1988 mar-oct; v3 n1-4 [1987 jan-apr]; v4 n1-v6 n2 – 1 – mf#1295166 – us WHS [071]
Among friends / Crothers, Samuel McChord – Boston: Houghton Mifflin, 1910 [mf ed 1993] – 1mf – 9 – 0-524-08275-8 – mf#1993-3030 – us ATLA [370]
Among hills and valleys in western china : incidents of missionary work / Davies, Hannah – London: S W Partridge & Co, 1901 [mf ed 1995] – (= ser Yale coll) – 326p (ill) – 1 – 0-524-09308-3 – (int by isabella lucy bishop) – mf#1995-0308 – us ATLA [951]
Among india's students / Wilder, Robert Parmelee – New York: Fleming H Revell, c1899 [mf ed 1986] – 1mf – 9 – 0-8370-6798-7 – mf#1986-0798 – us ATLA [240]
Among the americans and a stranger in america / Holyoake, George Jacob – Chicago: Belford, Clarke, 1881 [mf ed 1986] – 3mf – 9 – 0-665-39190-0 – mf#39190 – cn Canadiana [917]
Among the bantu nomads / Brown, John Tom – London, England. 1926 – 1r – us UF Libraries [960]
Among the boers : or, notes of a trip to south africa in search of health / Nixon, John – London 1880 – (= ser 19th c british colonization) – 4mf – 9 – mf#1.1.10046 – uk Chadwyck [960]
Among the brahmins and pariahs / Sauter, Johannes A – London: T Fisher Unwin, 1924 – (= ser Samp: indian books) – (transl from the german by bernard miall) – us CRL [305]
Among the brigands / De Mille, James – Boston: Lee & Shepard, 1875 – (= ser The young dodge club) – 4mf – 9 – 0-665-90778-8 – mf#90778 – cn Canadiana [830]
Among the burmans : a record of fifteen years of work and its fruitage / Cochrane, H P – New York, Chicago, Toronto, London, Edinburgh, 1904 – 4mf – 9 – mf#HTM-39 – ne IDC [915]
Among the burmans : a record of fifteen years of work and its fruitage / Cochrane, Henry Park – New York: F H Revell, c1904 [mf ed 1989] – 1mf – 9 – 0-7905-4258-7 – mf#1988-0258 – us ATLA [306]
Among the cannibals of new guinea : being the story of the new guinea mission of the london missionary society / McFarlane, Samuel – Philadelphia: Presbyterian Board of Publ & Sabbath-School Work, [1888?] [mf ed 1986] – 1mf – 9 – 0-8370-6280-2 – mf#1986-0280 – us ATLA [240]
Among the clouds, 1877-1917 – Mt Washington NH: H M & F Burt, 1877-1908; by R H Buckler, 1910-17 [mf ed Dartmouth College 1974. v1-39 n53. jul 20 1877-sep 14 1917] – (= ser Dartmouth college library colls: views of 18th and 19th-century life in new hampshire) – 10r – 1 – (lacks: scattered. first daily newspaper printed on summit of mt washington. not publ 1908-10. summary incl souvenir iss) – us Dartmouth [071]
Among the dark-haired race in the flowery land / Drake, S B – London, 1897 – 2mf – 9 – mf#HTM-51 – ne IDC [910]
Among the dark-haired race in the flowery land / Drake, Samuel B – London: Religious Tract Society, 1897 [mf ed 1995] – (= ser Yale coll) – 158p (ill) – 1 – 0-524-09467-5 – mf#1995-0467 – us ATLA [951]
Among the eskimos of labrador / Hutton, S K – London, 1912 – 8mf – 9 – mf#N-265 – ne IDC [917]
Among the esquimaux : or, adventures under the arctic circle / Ellis, Edward Sylvester – Philadelphia: Penn Pub Co, 1894 – 4mf – 9 – mf#14957 – cn Canadiana [830]
Among the forest trees : or, how the bushman family got their homes: being a book of facts and incidents of pioneer life in upper canada, arranged in the form of a story / Hilts, Joseph Henry – Toronto: [s.n.], 1888 [mf ed 1980] – 5mf – 9 – 0-665-05620-6 – mf#05620 – cn Canadiana [971]
Among the gospels and the acts : being notes and comments covering the life of christ in the flesh... / Ainslie, Peter – Baltimore: Temple Seminary Press 1908 [mf ed 1989] – 1mf – 9 – 0-7905-0480-4 – (incl bibl ref & ind) – mf#1987-0480 – us ATLA [226]
Among the hindus and creoles of british guyana / Bronkhurst, H V P – London, England. 1888 – 1r – us UF Libraries [972]
Among the huts in egypt : scenes from real life / Whately, Mary Louisa – 2nd ed. London: Seeley, Jackson, & Halliday 1872 [mf ed 1987] – 1r [ill] – 1 – (filmed with: the gangas of talkad / krishna rao, m v) – mf#1870 – us Wisconsin U Libr [960]
Among the idolmakers / Jacks, Lawrence Pearsall – London: Williams & Norgate, 1911 [mf ed 1990] – 1mf – 9 – 0-7905-7779-8 – mf#1989-1004 – us ATLA [290]
Among the indians of the paraguayan chaco : a story of missionary work in south america / Grubb, W Barbrooke – London, 1904 – 3mf – 9 – mf#HTM-73 – ne IDC [918]
Among the lushais / Anderson, Herbert – London: Carey Press, 1914 [mf ed 1995] – (= ser Yale coll) – viii/[43]p (ill) – 1 – 0-524-09279-6 – mf#1995-0279 – us ATLA [954]
Among the matabele : with a new chapter on the 'ma-shuna'...with portraits of lobengula and khama... / Carnegie, D – London, 1894 – 2mf – 9 – mf#HTM-31 – ne IDC [916]
Among the mongols / Gilmour, J – London, [1888] – 5mf – 9 – mf#HTM-65 – ne IDC [915]
Among the mongols / Gilmour, James – London: Religious Tract Society, [1888?] [mf ed 1995] – (= ser Yale coll) – xviii/383p (ill) – 1 – 0-524-09990-1 – mf#1995-0990 – us ATLA [951]
Among the pimas : or, the mission to the pima and maricopa indians – Albany, NY: Ladies' Union Mission School Assoc, 1893 [mf ed 1986] – 1mf – 9 – 0-8370-6612-3 – mf#1986-0612 – us ATLA [242]
Among the primitive-bakongo : a record of 30 years' close intercourse with the bakongo and other tribes of equatorial africa... / Weeks, John H – London: Seeley, Service, 1914 – 1 – us CRL [301]
Among the telugoos : illustrating mission work in india / Harpster, Mary Julia – Philadelphia: Lutheran Pub Soc 1902 [mf ed 1992] – 1mf – 9 – 0-524-04551-8 – mf#1991-2115 – us ATLA [242]
Among the theologies / Orcutt, Hiram – new ed. Boston: De Wolfe, Fiske, 1892, c1888 [mf ed 1985] – 1mf – 9 – 0-8370-3935-5 – mf#1985-1935 – us ATLA [200]
Among the tibetans / Bird, Isabella Lucy' – New York; Chicago: Fleming H Revell, [1894] [mf ed 1995] – (= ser Yale coll) – 159p (ill) – 1 – 0-524-09224-9 – (ill by edward whymper) – mf#1995-0224 – us ATLA [915]
Among the wild ngoni : being some chapters in the history of the livingstonia mission in british central africa / Elmslie, Walter Angus – New York: Fleming H Revell, 1899 [mf ed 1986] – 1mf – 9 – 0-8370-6571-2 – (incl ind) – mf#1986-0571 – us ATLA [240]
Among the wild tribes of the afghan frontier : a record of sixteen years' close intercourse with the natives of the indian marches / Pennell, T L – London, 1909 – 5mf – 9 – mf#HT-109 – ne IDC [915]
Among the wild tribes of the afghan frontier : a record of sixteen years' close intercourse with the natives of the indian marches / Pennell, Theodore Leighton – 4th ed. London: Seeley, Service, 1912 [mf ed 1990] – 1mf – 9 – 0-7905-5788-6 – mf#1988-1788 – us ATLA [307]
Among the women of the punjab... / Young, M – London, 1916 – 2mf – 9 – mf#HT-163 – ne IDC [915]
Among the women of the sahara / Pommerol, Jean – London, England. 1900 – 1r – us UF Libraries [305]
Among the zulus and amatongas / Leslie, David – New York, NY. 1969 – 1r – us UF Libraries [960]
Among unknown eskimo / Bilby, J W – London, 1923 – 7mf – 9 – mf#N-124 – ne IDC [919]
O amor – Cataguases, MG. 04 mar 1897 – (= ser Ps 19) – bl Biblioteca [079]
Amor a terra / Guimaraes, Osias – Rio de Janeiro, Brazil. 1941 – 1r – us UF Libraries [972]
El amor al libro... / Redonet y Lopez, Doriga; ed by Bayle, Constantino – Madrid: Razon y Fe, 1928 – 9 – sp Bibl Santa Ana [946]
Amor alos enemigos en el antiguo testamento / Fernandez Fernandez, Juan – Madrid: R.E.E.B., 1928 – 1 – sp Bibl Santa Ana [946]
O amor ao progresso – Rio de Janeiro, RJ: Typ Lobao, 16 ago 1878 – (= ser Ps 19) – mf#P17,01,79 – bl Biblioteca [440]
Amor como ella / Diaz Martinez, Manuel – Habana, Cuba. 1961 – 1r – us UF Libraries [972]
Amor de la patria / Forner Segarra, Juan Pablo – 1794 – 9 – sp Bibl Santa Ana [946]
Amor dei intellectualis : eine religions-philosophische studie / Wyneken, Gustav Adolf – Greifswald: J Abel 1898 [mf ed 19–] – 1r – 1 – mf#film mas 8133 – us Harvard [140]
Amor en pugna / Thabuteau, Amysan – Port-Au-Prince, Haiti. 1946 – 1r – us UF Libraries [972]
Amor en tierra y mar / Ordonez Arguello, Alberto – San Salvador, El Salvador. 1964 – 1r – us UF Libraries [972]
El amor familiar...poesias / Sainz y Gomez, Manuel – 1828 – 9 – sp Bibl Santa Ana [810]
Amor libre / Acosta Rubio, Raoul – Camaguey, Cuba. 1932 – 1r – us UF Libraries [972]
Amor original / Alvarez Baragano, Jose – Habana, Cuba. 1955 – 1r – us UF Libraries [972]
Amor perfecto / Sanchez Varona, Ramon – Habana, Cuba. 1948 – 1r – us UF Libraries [972]
Amor und psyche : eine dichtung in sechs gesaengen / Hamerling, Robert – Hamburg: J F Richter, [188-?] [mf ed 1996] – 133p – 1 – (original title ill by e a fischer-coerlin) – mf#9669 – us Wisconsin U Libr [810]
Amor y caridad / Bustamante Arellana, Carlos – San Jose, Costa Rica. 1962 – 1r – us UF Libraries [972]
Amor y flores / Sanchez Arjona, Vicente – Sevilla: Imprenta Zambrano, 1955 – 1 – sp Bibl Santa Ana [830]
Amor y martirio / Hurtado de Mendoza, Publio – 1874 – 9 – sp Bibl Santa Ana [946]
Amora, Antonio Soares *see* Romantismo, 1833-1838/1878-1881
Amora, Paulo *see* Bernardes

Amore e blasone : azione coreodrammatica in 5 parti e 7 quadri. musica del mo. cesare casiraghi. da rappresentarsi al teatro grande di brescia nella stagione di carnovale 1872-1873 / Pulini, Giovanni – [Brescia, 1873] – 1 – mf#*ZBD-*MGTZ pv 3-Res – Located: NYPL – us Misc Inst [790]

Amores Gonzalez, Meliton *see*
– Angel rodriguez (alias) er periodista...
– Mis amores

Amoretti, Giovanni Vittorio *see* Saggi critici

Amoris divini emblemata studio et aere othonis vaeni concinnata / Vaenius, O – Antverpiae: Ex officina Martini Nuti & Ioannis Meursi, 1615 – 2mf – 9 – mf#O-448 – ne IDC [090]

Amoris divini emblemata studio et aere othonis vaeni concinnata / Vaenius, O – Antverpiae: Ex officina Plantiniana Balthasaris Moreti, 1660 – 2mf – 9 – mf#O-790 – ne IDC [090]

Amoris divini et humani antipathia e varijs sacrae scripturae locis... – Paris: Guillaume le Noir, 1628 – 3mf – 9 – mf#O-1812 – ne IDC [090]

Amoris divini et humani effectus varii sacrae scripturae sanctorumq – Antverpiae: Apud Michaelem Snijders, 1626 – 1mf – 9 – mf#O-541 – ne IDC [090]

Amorosamente / Lopez Suria, Violeta – Madrid, Spain. 1960 – 1r – us UF Libraries [972]

O amor-perfeito : jornal critico jecoso e instructivo – Rio de Janeiro, RJ: Typ Classica de F A de Almeida, 07 out-09 dez 1849 – (= ser Ps 19) – mf#P03A,03,21 – bl Biblioteca [410]

Amort, E *see* Philosophia pollingana ad normam burgundicae

Amort, Eus *see* Vetus disciplina canonicorum regularium et saecularium

Amorum emblemata... – Antverpiae: Venalia apud auctorem, 1608 – 3mf – 9 – mf#O-3264 – ne IDC [090]

Amorum emblemata, figuris aeneis incisa studio othonis vaeni batavo-lugdunensis – Antverpiae: Venalia apud auctorem, 1608 – 4mf – 9 – mf#O-447 – ne IDC [090]

Amory, Thomas Coffin *see* The life of admiral sir isaac coffin, baronet

Amos : an essay in exegesis / Mitchell, Hinckley Gilbert Thomas – Boston: N J Bartlett, 1893 [mf ed 1985] – 1mf – 9 – 0-8370-4453-7 – mf#1985-2453 – us ATLA [221]

Amos : an essay in exegesis / Mitchell, Hinckley Gilbert Thomas – rev ed. Boston; New York: Houghton, Mifflin, 1900 – 1r – 1 – 0-8370-0316-4 – mf#1984-B343 – us ATLA [221]

Amos : metrisch bearbeitet / Sievers, Eduard & Guthe, Hermann – Leipzig: B G Teubner, 1907 [mf ed 1986] – (= ser Abhandlungen der philologisch-historischen classe der koenigl saechsischen gesellschaft der wissenschaften 23/3) – 1mf – 9 – 0-8370-7427-4 – mf#1986-1427 – us ATLA [221]

Amos : oversat og fortolket / Michelet, S – Kristiania [Oslo]: H Aschehoug, 1893 [mf ed 1985] – 1mf – 9 – 0-8370-4425-1 – (in norwegian) – mf#1985-2425 – us ATLA [221]

Amos, hosea, isaiah (1-39) : and micha – Oxford: Clarendon, 1909 [mf ed 1989] – (= ser The hebrew prophets for english readers 1) – 1mf – 9 – 0-7905-2523-2 – mf#1987-2523 – us ATLA [221]

Amos, Sheldon *see*
– The science of law
– A systematic view of the science of jurisprudence

Amos und hosea : ein kapitel aus der geschichte der israelitischen religion / Valeton, Josue Jean Philippe – Giessen: J Ricker, 1898 [mf ed 1985] – 1mf – 9 – 0-8370-5612-8 – (in german. incl bibl ref) – mf#1985-3612 – us ATLA [221]

Amos und hosea / Nowack, Wilhelm – Tuebingen: J C B Mohr 1908 [mf ed 1989] – (= ser Religionsgeschichtliche volksbuecher fuer die deutsche christliche gegenwart 2/9) – 1mf – 9 – 0-7905-1546-6 – (in german) – mf#1987-1546 – us ATLA [221]

Amos und hosea : zwei zeugen gegen die anwendung der evolutionstheorie auf die religion israels / Oettli, Samuel – Guetersloh: C Bertelsmann, 1901 [mf ed 1989] – (= ser Beitraege zur foerderung christlicher theologie 5/4) – 1mf – 9 – 0-7905-3208-5 – mf#1987-3208 – us ATLA [221]

Amour : quand tu nous tiens! / Coolus, Romain – Paris, France. 1932, c1922 – 1r – us UF Libraries [440]

Amour a l'anglaise / Jacquelin, Jacques Andre – Paris, France. 1816 – 1r – us UF Libraries [440]

L'amour a paris, nouveaux memoires, 1 : l'amour criminel / Goron, Marie Francois – Paris, Flammarion, 1899 – (= ser Les femmes [coll]) – 4mf – 9 – mf#9497 – fr Bibl Nationale [880]

L'amour a paris, nouveaux memoires, 2 : les industries de l'amour / Goron, Marie Francois – Paris, Flammarion, 1899 – (= ser Les femmes [coll]) – 5mf – 9 – mf#9498 – fr Bibl Nationale [880]

L'amour a paris, nouveaux memoires, 3 : les parias de l'amour / Goron, Marie Francois – Paris, Flammarion, 1899 – (= ser Les femmes [coll]) – 4mf – 9 – mf#9499 – fr Bibl Nationale [880]

L'amour a paris, nouveaux memoires, 4 : le marche aux femmes / Goron, Marie Francois – Paris: Flammarion, 1899 – (= ser Les femmes [coll]) – 4mf – 9 – mf#9500 – fr Bibl Nationale [880]

Amour dans tous les quartiers / Clairville, M – Paris, France. 1845? – 1r – us UF Libraries [440]

L'amour de la patrie : ordre, bienfaisance, bonnes moeurs, instruction, art, industrie / ed by Legrand, mme – Paris: A Rene [apr 16 1848] – (= ser French revolution of 1848. newspapers) – 1r – 1 – us CRL [074]

L'amour du coeur de jesus : ou le veritable tresor de l'ame inspire par des exemples a la jeunesse et aux familles chretiennes – Quebec?: L Brousseau, 1883 – 2mf – 9 – mf#04010 – cn Canadiana [240]

Amour du prochain / Valdagne, Pierre – Paris, France. 1900 – 1r – us UF Libraries [440]

L'amour et la guerre / Perret, Paul – Paris: Ollendorff, 1892 – (= ser Les femmes [coll]) – 4mf – 9 – mf#11814 – fr Bibl Nationale [305]

Amour et la raison / Pigault-Lebrun – Paris, France. 1805 – 1r – us UF Libraries [440]

Amour, Jean d' *see* Remarques et suggestions sur la reorganisation des tribunaux

L'amour, le mariage, la justice selon le koran / Bachir, Ali – Paris: Nilsson, 1914 – (= ser Les femmes [coll]) – 2mf – 9 – fr Bibl Nationale [306]

Un amour vrai / Conan, Laure – Montreal: Leprohon & Leprohon, [1879?] [mf ed 1984] – 1mf – 9 – 0-665-16878-0 – mf#16878 – cn Canadiana [830]

Amours congolaises / Raulin, G de [pseud] – Paris: A Michel, 1881 – 1 – us CRL [360]

Amours et aventures de casanova / Casanova di Seingalt, Giacomo G – Paris: Simon, 1890-91 – (= ser Les femmes [coll]) – 10v on 30mf – 9 – mf#8515-24 – fr Bibl Nationale [920]

Amours, Joseph-Arthur *see* Ou allons-nous?

Ampac sustainer – 1982 feb, apr, jun, aug, dec; 1983 jan-apr, jun, sep nov; 1984 jan, apr-aug; 1985 feb-mar, may-jun, aug; 1986 feb – 1 – mf#1223122 – us WHS [071]

Ampera : madjalah bulanan perusahaan daerah sumatra utara – Medan, 1967 v1(1-2) – 2mf – 9 – mf#SE-1312 – ne IDC [950]

Ampera review – Djakarta, 1964-1969 – 35mf – 9 – (missing: 1967, v4(6-12); 1968, v5(1-12)) – mf#SE-523 – ne IDC [959]

Amper-bote – Dachau DE, 1877-1944 – 1 – gw Misc Inst [074]

Ampere, Jean-Jacques *see* Promenade en amerique

Amphilochius von ikonium : in seinem verhaeltnis zu den grossen kappadoziern / Holl, Karl – Tuebingen: J C B Mohr, 1904 [mf ed 1990] – 1mf – 9 – 0-7905-5535-2 – mf#1988-1535 – us ATLA [240]

Amphioxus and ascidian : our gelatinous ancestors: how the missing links were discovered and made known / Agorastes, Phil – Toronto: [s.n,] 1878 [mf ed 1979] – 1mf – 9 – 0-665-00778-7 – mf#00778 – cn Canadiana [810]

Amphitheatrum sapientiae aeternae / Khunrath, Heinrich – [Hamburg?: s.n.] 1595 – 1 – (the 1653 ed (based on the 1609 edition ed by erasmus wolfart) has 3 additional plates, & the 1595 plates have been redone but lack the wide borders with copius notes in various languages) – mf#2140p – us Wisconsin U Libr [130]

Amphitryon : ein lustspiel nach moliere / Kleist, Heinrich von – [Wien]: Phaidon Verlag 1924 [mf ed 1995] – 1r [ill] – 1 – (with original lithographs by laszlo gabor. filmed with: kleist's hermannsschlacht : ein gedicht auf osterreich / adam muller-guttenbrunn & other titles) – mf#3650p – us Wisconsin U Libr [820]

Amphitryon : or, the two sosia's: a comedy, as it is acted at the theatre royal / Dryden, John & Purcell, Henry – London: printed for J Tonson...& M Tonson 1691 [mf ed 19–] – 2mf – 9 – mf#fiche 883 – us Sibley [780]

Amphlett, George Thomas *see* History of the standard bank of south africa ltd, 1862-1913

Amphon angelicus : a work of many compositions for one, two, three and four voices: with several accompagnements of instrumental musick... / Blow, John – London: printed by W Pearson for aut 1700 [mf ed 19–] – 1r – 1 – mf#film 41 – us Sibley [780]

Amphora – San Francisco CA 1970-72 – 1 – mf#7514 – us UMI ProQuest [810]

Amphoux, Henri *see*
– Essai sur la doctrine socinienne
– Essai sur l'histoire du protestantisme au havre et dans ses environs
– Michel de l'hospital et la liberte de conscience au 16e siecle

Ampir ke noraka / Jo, Boen Ek & Liem, Poen Kie – Batavia: Goedang Tjerita, 1948 [mf ed 1998] – (= ser Goedang tjerita 6; Ngo bie wie kiam kek 3) – 1r – 1 – (coll as pt of the colloquial malay collection; indonesian trans of chinese novel possibly entitled emei wei jianke, or the fierce sword-fighters from emei shan mountain [salmon, claudine. literature in malay by the chinese of indonesia. paris: editions de la maison des sciences de l'homme, c1981]; filmed with: lajangan biroe / im yang tjoe) – mf#10005 – us Wisconsin U Libr [830]

Ampla dimostrazione degli armoniali musicale tuoni trattato teorico-prattico de fra frances calegari min : convente di venezia... / Calegari, Francesco Antonio – [1732] [mf ed 19–] – 4mf – 9 – mf#fiche 10 – us Sibley [780]

Ample discovrs : et advis de l'estat and assiette des armees chrestiennes and turquesques – Paris, 1572 – 1mf – 9 – mf#H-8188 – ne IDC [956]

Ampliaciones...historia de merida de moreno... fernandez / Plano y Garcia, Pedro – 1894 – 9 – sp Bibl Santa Ana [946]

Amplifier – v29 n1-v38 n1 [1975 jan-1986 feb] – 1 – mf#643772 – us WHS [621]

Amplior consideratio decreti synodalis tridentini : de authentica doctrina ecclesiae dei. de latina ueteri translatione sacrorum librorum... / Bibliander, T – [Basel], 1551 – 2mf – 9 – mf#PBU-579 – ne IDC [240]

Ampo – Tokyo, Japan 1973+ – 1,5,9 – ISSN: 0003-2026 – mf#9309 – us UMI ProQuest [079]

Ampthill – (= ser Bedfordshire parish register series) – 2mf – 9 – £5.00 – uk BedsFHS [929]

Ampthill, st andrew old churchyard monumental inscriptions – Arthur Weight Matthews 1915 – (= ser Bedfordshire parish register series) – 1mf – 9 – £1.25 – uk BedsFHS [929]

Amram, David Werner *see* Leading cases in the bible

Amrit Kaur, Rajkumari *see*
– Challenge to women
– To women

Amrita bazar patrika – Calcutta, India. 1962-95 – 147r – 1 – us L of C Photodup [079]

Amrita bazar patrika = Sunday amrita bazar patrika – Calcutta: T K Biswas, [1922-jul 4 1946; jan-may 1947; feb 1948-may1951; jul -dec 1951] (daily) – 146r – 1 – us CRL [079]

Amru te pati mah prnn su pati sui – [Yan Kun: Mran ma Chui Rhy Lac Lam Cin Pati, Pati Cnn Rmu Re Bahui Komati Thana Khyup 1962?] [mf ed 1990] – 1 – (in burmese) – mf#mf-10289 seam reel 157/7 [§] – us CRL [959]

[Amrullah, A M K, hadji] *see* Islam di soematera

Amsblatt : german baptist – 142p. 1914-1945 – 1 – $5.00 – us Southern Baptist [242]

Amsden, Lionel George *see* Principles and practices of refraction

Amsdorff, N von *see*
– Antwort, glaub vnd bekentnis auff das schoene vnd liebliche interim
– Das doctor martinus kein adiaphorist gewesen ist
– Das doctor pomer vnd doctor maior mit iren adiaphoristen ergernis vnnd zurtrennung angericht vnnd den kirchen christi vnueberwintlichen schaden gethan haben
– Ein kurtzer vnterricht auff d georgen maiors antwort
– Das die propositio "gute werck sind zur seligkeit schedlich" ein rechte ware christliche propositio sey

[Amsdorff, N von] *see*
– Bekentnis vnterricht vnd vermanung der pfarrhern vnd prediger der christlichen kirchen zu magdeburgk
– Confessio et apologia pastorum

Amse, Corina *see*
– Bibliographie der arbeiten von prof dr helmut breuer und dr maria weuffen
– Koeduktion

Yr amseroedd – [Wales] 30 rhag 1882-rhag 1884 [mf ed 2003] – 1r – 1 – (cont by: yr amseroedd wythnosol [ion-27 mehefin 1885]) – uk Newsplan [072]

Yr amseroedd etc – [Wales] Isle of Anglesey 1882-85 – 5r – 1 – uk Newsplan [072]

Amsterdam. International Institute of Social History *see* Archive of rudolf rocker, 1894-1958

Amsterdam Toonkunst-Bibliotheek BT 61/580-212-H-12 : the first of a four-volume catalog of musical incipits kept in 18th-century Europe / DeWitt, Jean [Jeannette] – U of Rochester 1977 [mf ed 19–] – 4mf – 9 – (with ind; incl bibl ref) – mf#fiche 1169 – us Sibley [780]

Den amsterdamschen hermus / Weyerman, J C – Amsterdam. v1-2. 1722-1723 – €31.00 – ne Slangenburg [949]

Amsterodamum monogrammon / Plemp, C G – Amsterodami: Apud Ioannem Walschardum, 1616 – 2mf – 9 – mf#O-3146 – ne IDC [090]

Amt events / American Medical Technologists – Park Ridge IL 1989+ – 1 – ISSN: 0746-9217 – mf#13914 – us UMI ProQuest [610]

Amt und geist im kampf : studien zur geschichte des urchristentums / Luetgert, Wilhelm – Guetersloh: C Bertelsmann, 1911 [mf ed 1991] – (= ser Beitraege zur foerderung christlicher theologie 15/4-5) – 1mf – 9 – 0-7905-9316-5 – mf#1989-2541 – us ATLA [225]

Amtliche bekanntmachungen – Altena DE, 25 aug 1945-26 oct 1949 – 1 – (cont by: amtliche bekanntmachungen fuer die stadt altena, aemter luedenscheid, nachrodt und neuenrade, 31 oct 1947; amtliche bekanntmachungen fuer den kreis altena, 20 nov 1948) – Dist. gw Mikrofilm – gw Misc Inst [350]

Amtliche bekanntmachungen – Muehldorf a. Inn DE, 1946 4 jan-1990 – 188r – 1 – (title varies: 27 aug 1949: muehldorfer anzeiger) – gw Mikrofilm [350]

Amtliche bekanntmachungen – Ludwigsburg DE, 1945 apr 21-1955 – 3r – 1 – (title varies: 6 apr 1946: amtsblatt fuer die stadt und den kreis ludwigsburg) – gw Misc Inst [350]

Amtliche bekanntmachungen der stadt bad nauheim – Bad Nauheim DE, 1945 29 mar-1948 22 dec – 1r – 1 – gw Mikrofilm [350]

Amtliche bekanntmachungen der stadt friedberg – Friedberg, Hessen DE, 1945 3 apr-1949 23 jul – 1 – gw Mikrofilm [350]

Amtliche bekanntmachungen fuer den kreis grimma – Grimma DE, 1945 1 nov-1951 27 apr – 2r – 1 – gw Misc Inst [350]

Amtliche bekanntmachungen fuer den kreis hofgeismar – Hofgeismar DE, 1876-1908 – 5r – 1 – (title varies: 1807: kreisblatt) – gw Misc Inst [350]

Amtliche mecklenburgische anzeigen *see* Mecklenburgische landesnachrichten

Amtliche mitteilungen – Schwaebisch Hall DE, 1943 1 jul-1946 30 mar – 1r – 1 – gw Misc Inst [350]

Amtliche mitteilungen des saalkreises – Halle S DE, 1947 21 nov-1951 – 1 – gw Misc Inst [350]

Amtliche nachrichten / Austria. Bundesministerium fuer soziale Verwaltung – v1-21. 1945-65 – 9 – $330.00 – mf#0094 – us Brook [350]

Amtliche nachrichten / Austria. (Lower) – Vienna. 1959-1966 – 1 – us NY Public [324]

Amtliche nachrichten fuer das general-gouvernement elsass – Strassburg (Strasbourg F), 1870 n1-11 – 1 – (with suppls: strassburger blaetter 1878; strassburger handelsblatt 1872-75 [gaps]) – gw Misc Inst [074]

Amtlicher anzeiger bergisch gladbach – Bergisch Gladbach DE, 1947 30 sep-1948 5 jul – 1 – gw Misc Inst [074]

Amtlicher anzeiger fuer das land ratzeburg – Schoenberg DE, 1919 3 jun-31 dec – 1r – 1 – gw Misc Inst [074]

Amtlicher anzeiger fuer den landkreis cassel – Kassel DE, 1926-1932 feb, 1946-72 – 2r – 1 – gw Misc Inst [074]

Amtlicher anzeiger im siegkreis – Siegburg DE, 1946 9 feb-1950 21 mar – 1 – gw Misc Inst [350]

Amtliches kreisblatt des koenigsberger landkreises – Koenigsberg (Kaliningrad RUS), 1863 9 may-1865 27 dec 27, 1866 3 mar-19 dec, 1907 1 jan-19 dec, 1912, 1920-29 – 10r – 1 – (with gaps) – gw Misc Inst [350]

Amtliches kreisblatt fuer den koenigsberger kreis – Koenigsberg (Chojna PL), 1868-69, 1872 – 1 – gw Misc Inst [350]

Amtliches kreisblatt fuer den kreis geldern *see* Kreis-blatt

Amtliches kreisblatt fuer den kreis land hadeln – Otterndorf DE, 1932 1 oct-1934 30 jun – 1r – 1 – gw Misc Inst [350]

Amtliches kreisblatt fuer den kreis wolmirstedt – Wolmirstedt DE, 1875 – 1 – gw Misc Inst [074]

Amtliches kursblatt der wiener boerse – Vienna, Austria 15 nov 1948-28 oct 1968 – 24r – 1 – uk British Libr Newspaper [332]

Amtliches mitteilungsblatt – Hagen, Westf DE, 1945 23 may-1950 4 mar – 1r – 1 – gw Misc Inst [350]

Amtliches mitteilungsblatt der stadt rheydt – Moenchengladbach DE, 1948 11 jun-1949 [gaps], 1951 1 jun-1974 – 4r – 1 – (title varies: vor 1 jun1951: rheydter amtsblatt) – gw Misc Inst [350]

Amtliches nachrichtenblatt des kreises norderdithmarschen – Heide, Holst DE, 1946 12 feb-1949 27 dec – 1 – (aka: kreisblatt fuer norderdithmarschen) – gw Misc Inst [074]

Amtliches niederunger kreisblatt *see* Kreisblatt des koenigl[ichen] landraths-amt der niederung

Amtliches preussisch eylauer kreisblatt *see* Preussisch eylausches kreisblatt

Amtliches verkuendigungsblatt fuer den kreis waldshut – Waldshut DE, 1870-72 – 1 – gw Misc Inst [350]

Amtliches verkuendigungsblatt fuer die grossh amtsbezirke breisach, emmendingen, ettenheim, stadt- und landamt freiburg, kenzingen, st blasien, staufen, triberg und waldkirch – Freiburg Br DE, 1863 – 1 – gw Misc Inst [074]

Amtliches verordnungs- und anzeigenblatt fuer den kreis rotenburg in hannover – Rotenburg/ Hannover DE, 1947 3 jan-1948 – 1 – (filmed with: rotenburger anzeiger) – gw Misc Inst [350]

Amts- und intelligenzblatt fuer den oberamtsbezirk muensingen *see* Intelligenz-blatt fuer die oberaemter ehingen und muensingen

Amts- und intelligenzblatt fuer den oberamtsbezirk saulgau *see* Der oberlaender

Amts- und wochen-blatt fuer den bezirk des koeniglichen landgerichts naila *see* Nailaer wochenblatt

Amtsblatt – Schwaebisch Hall DE, 1946 6 apr-1950 28 mar – 1r – 1 – gw Misc Inst [350]

Das amtsblatt – v1-6. jan 1909-15 june 1914 – 1r – 1 – (german language newspaper produced in rabaul) – mf#PMB Doc402 – at Pacific Mss [074]

Amtsblatt der europaeischen union *see* The official journal of the european union

Amtsblatt der koeniglich oppelnschen regierung – Oppeln 1816-1940 [mf ed Hildesheim 2003] – (= ser Die preussischen regierungsamtblaetter – schlesien 3) – 851mf – 9 – diazo €1798.00 – gw Olms [943]

Amtsblatt der koeniglich preussischen regierung zu allenstein fuer das jahr.../ – Allenstein (Olsztyn PL), 1913 – 1 – (title varies: 4 jan 1919: amtsblatt der regierung zu allenstein, 3 jan 1925: amtsblatt der preussischen regierung zu alleinstein. filmed by other misc inst: 1905 1 nov-1909, 1910 feb-1944 9 dec) – gw Misc Inst [350]

Amts-blatt der koeniglich preussischen regierung zu frankfurt an der oder / amtsblatt *see* Amts-blatt der koenigl[ichen] preuss[ischen] regierung von der neumark / amtsblatt

Amtsblatt der koeniglich preussischen regierung zu marienwerder 1817-1940 *see* Amtsblatt der koeniglich westpreussischen regierung zu marienwerder 1811-1816 / amtsblatt der koeniglich preussischen regierung zu marienwerder 1817-1940

Amtsblatt der koeniglich westpreussischen regierung zu marienwerder 1811-1816 / amtsblatt der koeniglich preussischen regierung zu marienwerder 1817-1940 – Marienwerder 1817-1940 [mf ed Hildesheim 2002] – (= ser Die preussischen regierungsamtblaetter – westpreussen 2) – 642mf – 9 – diazo €1698.00 – gw Olms [943]

Amtsblatt der koeniglichen breslauischen regierung – Breslau 1811-1941 [mf ed Hildesheim 2003] – (= ser Die preussischen regierungsamtblaetter – schlesien 3) – 843mf – 9 – diazo €1738.00 – gw Olms [943]

Amtsblatt der koeniglichen brombergischen regierung – Bromberg 1815-1919 [mf ed Hildesheim 2003] – (= ser Die preussischen regierungsamtblaetter – posen 2) – 626mf – 9 – diazo €1298.00 – gw Olms [943]

Amtsblatt der koeniglichen liegnitzschen regierung – Liegnitz 1811-1940 [mf ed Hildesheim 2003] – (= ser Die preussischen regierungsamtblaetter – schlesien 2) – 767mf – 9 – diazo €1498.00 – gw Olms [943]

Amts-blatt der koeniglichen liegnitzschen regierung von schlesien – Liegnitz (Legnica PL), 1891 – 1r – 1 – gw Misc Inst [350]

Amts-blatt der koeniglichen litthauischen regierung / amtsblatt – Gumbinnen (Gussew RUS), 1811 [gaps], 1812-13, 1815 [gaps] – 4r – 1 – gw Misc Inst [947]

Amts-blatt der koeniglichen ostpreussischen regierung – Koenigsberg (Kaliningrad RUS), 1813-14, 1830 – 3r – 1 – gw Misc Inst [324]

Amts-blatt der koenigl[ichen] preuss[ischen] regierung von der neumark / amtsblatt – Koenigsberg (Chojna PL), Frankfurt/O DE, 1811 1 jun-1848, 1850-1943 – 1 – (title varies: 13 mar 1816: amts-blatt der koeniglich preussischen regierung zu frankfurt an der oder / amtsblatt, 7 dec 1918: amts-blatt der regierung zu frankfurt an der oder / amtsblatt) – gw Misc Inst [943]

Amts-blatt der koeniglichen regierung von pommern / amtsblatt – Stargard (Stargard Szczecki PL), Stettin (Szczecin PL), 1811 1 may-1942 – 1 – (title varies: 29 jan 1817: amts-blatt der koeniglichen regierung zu stettin / amtsblatt; 16 nov 1918: amts-blatt der preussischen regierung zu stettin / amtsblatt; fr 1818 publ in stettin) – gw Misc Inst [350]

Amtsblatt der koeniglichen regierung zu allenstein – 1905-44 [mf ed Hildesheim 2003] – (= ser Die preussischen regierungsamtblaetter – ostpreussen 1) – 169mf – 9 – €448.00 – gw Olms [943]

Amtsblatt der koeniglichen regierung zu danzig / Danzig. Regierungsbezirk – Danzig. 1868-1918. (Scattered issues wanting) – 1 – 646.00 – us L of C Photodup [943]

Amtsblatt der koeniglichen regierung zu danzig 1816-1920 / staatsanzeiger fuer danzig 1921-1939 / amtsblatt des reichstatthalter in danzig-westpreussen – Danzig nov 1939-dez 1941 [mf ed Hildesheim 2003] – (= ser Die preussischen regierungsamtblaetter – westpreussen 1) – 505mf – 9 – diazo €1380.00 – gw Olms [943]

Amtsblatt der koeniglichen regierung zu erfurt – Erfurt 1816-1943 [mf ed Hildesheim 2003] – (= ser Die preussischen regierungsamtblaetter – sachsen 1) – 576mf – 9 – diazo €1498.00 – gw Olms [943]

Amts-blatt der koeniglichen regierung zu erfurt / oeffentlicher anzeiger – Erfurt DE, 1817-59, 1866-1871 2 sep, 1873-80, 1882-83, 1885-91, 1893-94, 1897-1922 30 sep, 1924-1933 30 sep, 1938 7 oct-1943 – 1 – (title varies: 30 nov 1918: amts-blatt der regierung zu erfurt / oeffentlicher anzeiger, 2 jan 1932: amts-blatt der preussischen regierung zu erfurt / oeffentlicher anzeiger; with suppl: sonderbeilage 1913-1914 28 nov, 1915-16) – gw Misc Inst [943]

Amtsblatt der koeniglichen regierung zu gumbinnen – 1811-1942 [mf ed Hildesheim 2003] – (= ser Die preussischen regierungsamtblaetter – ostpreussen 2) – 579mf – 9 – diazo €1498.00 – gw Olms [943]

Amtsblatt der koeniglichen regierung zu koenigsberg – 1811-1940 [mf ed Hildesheim 2003] – (= ser Die preussischen regierungsamtblaetter – ostpreussen 3) – 926mf – 9 – diazo €2498.00 – gw Olms [943]

Amtsblatt der koeniglichen regierung zu koeslin – 1816-1943 [mf ed Hildesheim 2003] – (= ser Die preussischen regierungsamtblaetter – pommern 1) – 443mf – 9 – diazo €1148.00 – gw Olms [943]

Amtsblatt der koeniglichen regierung zu memel – 1921-39 [mf ed Hildesheim 2003] – (= ser Die preussischen regierungsamtblaetter – ostpreussen 4) – 116mf – 9 – diazo €448.00 – gw Olms [943]

Amtsblatt der koeniglichen regierung zu neumark / frankfurt o – 1810-1943 [mf ed Hildesheim 2003] – (= ser Die preussischen regierungsamtblaetter – brandenburg 1) – 619mf – 9 – diazo €1498.00 – gw Olms [943]

Amtsblatt der koeniglichen regierung zu posen 1816-1919 / amtsblatt der regierung zu posen 1940-1943 – [mf ed Hildesheim 2003] – (= ser Die preussischen regierungsamtblaetter – posen 1) – 794mf – 9 – diazo €1698.00 – gw Olms [943]

Amtsblatt der koeniglichen regierung zu stettin – Stettin 1811-1942 [mf ed Hildesheim 2003] – (= ser Die preussischen regierungsamtblaetter – pommern 2) – 580mf – 9 – diazo €1498.00 – gw Olms [943]

Amtsblatt der koeniglichen regierung zu stralsund – Stralsund 1818-1932 [mf ed Hildesheim 2003] – (= ser Die preussischen regierungsamtblaetter – pommern 3) – 353mf – 9 – diazo €1380.00 – gw Olms [943]

Amts-blatt der koeniglichen regierung zu stralsund / oeffentlicher anzeiger – Stralsund DE, 1818 5 apr-1889, 1892-94, 1897, 1900-05, 1907-08, 1910-15, 1917-1920 20 nov, 1923-30, 1932 2 jan-24 sep – 1 – (title varies: 11 jan 1919?: amts-blatt der regierung zu stralsund / oeffentlicher anzeiger; 3 jan 1920?: amts-blatt der preussischen regierung zu stralsund /oeffentlicher anzeiger. incl special suppl 1913-15, 1917) – gw Misc Inst [943]

Amtsblatt der koeniglichen reichenbacher regierung – Reichenbach 1816-20 [mf ed Hildesheim 2003] – (= ser Die preussischen regierungsamtblaetter – schlesien 1) – 14mf – 9 – diazo €44.80 – gw Olms [943]

Amtsblatt der landesverwaltung mecklenburg-vorpommern *see* Herzoglich mecklenburg-schwerinsches officielles wochenblatt

Amtsblatt der landwirtschaftskammer in thueringen *see* Der thueringer landbund

Amtsblatt der preussischen regierung zu koenigsberg / Koenigsberg. Regierungsbezirk – v. 55-126. Koenigsberg. 1865-1936. (Wanting scattered issues) – 1 – us L of C Photodup [943]

Amtsblatt der preussischen regierung zu sigmaringen – Sigmaringen DE, 1918-33 – 6r – 1 – gw Misc Inst [350]

Amtsblatt der regierung zu allenstein *see* Amtsblatt der koeniglich preussischen regierung zu allenstein fuer das jahr.../

Amtsblatt der regierung zu coeln / Cologne. Regierungsbezirk – Coeln. 1874-1943. (Scattered issues wanting) – 1 – us L of C Photodup [943]

Amts-blatt der regierung zu erfurt / oeffentlicher anzeiger *see* Amts-blatt der koeniglichen regierung zu erfurt / oeffentlicher anzeiger

Amtsblatt der regierung zu posen 1940-1943 *see* Amtsblatt der koeniglichen regierung zu posen 1816-1919 / amtsblatt der regierung zu posen 1940-1943

Amtsblatt der regierung zu schleswig / Schleswig. Regierungsbezirk – Schleswig, 1869-1943 – 1 – mf#04850 JS – us L of C Photodup [943]

Amtsblatt der regierungsstelle zu schneidemuehl – Schneidemuehl 1919-41 [mf ed Hildesheim 2003] – (= ser Die preussischen regierungsamtblaetter – grenzmark westpreussen – posen 1) – 53mf – 9 – €168.00 – gw Olms [943]

Amtsblatt der stadt muelheim an der ruhr – Muelheim DE, 1950-62 – 2r – 1 – gw Misc Inst [350]

Amtsblatt der stadt stuttgart – June 1, 1945-1968- – 1 – us NY Public [324]

Amtsblatt des regierungspraesidenten in litzmannstadt – Lodz (PL), 1941 n1-15 – 1r – 1 – gw Misc Inst [323]

Amtsblatt des reichskommissars fuer die besetzten rheinischen gebiete – Koblenz DE, 1920-1923 17 apr (n15) – 1r – 1 – mf#4889 – gw Mikropress [350]

Amtsblatt des reichstatthalter in danzig-westpreussen *see* Amtsblatt der koeniglichen regierung zu danzig 1816-1920 / staatsanzeiger fuer danzig 1921-1939 / amtsblatt des reichstatthalter in danzig-westpreussen

Amtsblatt des saarlandes / Saarland – Saarbrucken. Dec 1947-1969- – 1 – us NY Public [324]

Amts-blatt fuer das koenigliche kreisgericht und oberamt zu hechingen *see* Wochenblatt fuer das fuerstenthum hohenzollern-hechingen

Amts-blatt fuer den bezirk der koeniglichen landdrostei lueneburg *see* Oeffentliche anzeigen fuer das koenigliche-westphaelische departement der nieder-elbe

Amtsblatt fuer die stadt und den kreis ludwigsburg *see* Amtliche bekanntmachungen

Die amtskalender der fraenkischen fuerstentuemer ansbach und bayreuth [1737-1801] / ed by Kiel, Rainer-Maria – [mf ed 2000] – 193mf – 9 – €980.00 – 3-89131-360-8 – (filmed with: hoch-fuerstlich brandenburg-onoltzbachischer address- und schreib-calender (1737-1769); hochfuerstlich-brandenburgisch-culmbachischer...address- und schreib-calender (1738-1768); hochfuerstlicher brandenburg-onolzbach- und culmbachischer genealogischer calender und addresse-buch (1770-1791); address-buch fuer die koeniglich-preussischen fuerstenthuemer ansbach und bayreuth bzw addresshandbuch fuer die fraenkischen fuerstenthuemer ansbach und bayreuth (1796-1801)) – gw Fischer [943]

Die amtsschelle – Moenchengladbach DE, 1975-83 – 3r – 1 – (title varies: 1949, n45: amtliche moenchengladbacher mitteilungen; 1975: amtsblatt der stadt moenchengladbach) – gw Misc Inst [074]

Amtszeitung – Dortmund DE, 1902 3 jan-1904 30 dec, 1906 2 jan-1913 30 dec, 1915 2 jan-1920, 1922-1923 25 may, 1925 2 jan-24 dec, 1926 5 jan-dec, 1928 3 jan-dec, 1934 2 jan-29 nov, 1936 30 jan-30 jun, 1937 2 jan-1941 31 may, 1953 1 jul-31 dec, 1957 3 jul-1960 10 oct, 1961-1990 (z.t. nur lokalteil) – 1 – (title varies: 1940: amtszeitung und martener zeitung, 15 nov 1949: dortmunder nord-west-zeitung, covers: (dortmund-) luetgendortmund) – gw Misc Inst [074]

Amtszeitung und martener zeitung *see* Amtszeitung

Amuchastegui, Carlos J *see* Curso de literatura hispanoamericana

'Amude Arazim / Margaliyot, Yesha'y Asher Zelig – Jerusalem, Israel. 1932 – 1r – 1 – us UF Libraries [939]

Amulet – 1826-36 – (= ser English gift books and literary annuals, 1823-1857) – 35mf – 9 – uk Chadwyck [800]

Amulets : illustrated by the egyptian collection in university college, london / Petrie, W M – London, 1914 – 3mf – 9 – mf#NE-20366 – ne IDC [930]

L'amulette / Meyer, Conrad Ferdinand – Geneve: H Robert, 1898 [mf ed 1995] – xii/346p/1pl – 1 – (trans fr german by h s. pref by gaspard vallette) – mf#8823 – us Wisconsin U Libr [830]

Amunategui, Miguel Luis *see*
- La dictadura de o'higgins
- Vida de don andres bello

Amunategui Solar, Domingo *see*
- Formacion de la nacionalidadcchilena. santiago de chile, 1943
- Historia social de chile

Amunategui Y Solar, Domingo *see* Emancipacion de hispanoamerica

Amundsen, Edward *see* In the land of the lamas

Amundsen, R *see* Nordvestpassagen

Amur – Irkutsk, Russia 1861 [mf ed Norman Ross] – 1 – mf#nrp-552 – us UMI ProQuest [077]

Amur – Irkutsk, 1860-1862 – 77mf – 9 – (missing: 1860 (37, 40, 47); 1861(59-60, 71-72)) – mf#R-1509 – ne IDC [077]

Amurru : the home of the northern semites: a study showing that the religion and culture of israel are not of babylonian origin / Clay, Albert Tobias – Philadelphia: Sunday School Times, 1909 [mf ed 1986] – 1mf – 9 – 0-8370-9852-1 – (incl bibl ref & ind) – mf#1986-3852 – us ATLA [939]

Amurskaia gazeta – Blagoveschensk, Russia 1895 [mf ed Norman Ross] – 1 – mf#nrp-343 – us UMI ProQuest [077]

Amurskaia pravda – Blagoveschensk, Russia 1973-88 [mf ed Norman Ross] – 5r – 1 – mf#nrp-344 – us UMI ProQuest [077]

Amurskaia zhizn' : Ezhedn vnepartiin demokrat gaz / ed by Rodionov, N N – Blagoveshchensk [Amur obl]: Amur obl zems uprava 1918-19 [1918 17 okt-] – (= ser Asn 1-3) – n1-62 [1918] n3-163, [1919] [gaps] item 10, on reel n4 – 1 – mf#asn-1 010 – ne IDC [077]

Amurskii liman [primor obl] : bespart gaz / ed by Komarovskii, V A – Nikolaevsk-na-Amure: I S Kaptsan 1918-20 [1908 10 avg-] – (= ser Asn 1-3) – n1048 [1918]-n1252 [1920] [gaps] item 11, on reel n5 – 1 – mf#asn-1 011 – ne IDC [077]

Amurskoe ekho / ed by Brodovikov, A M – Blagoveshchensk [Amur obl]: [s n] 1918-19 [1915 3 fevr-] – (= ser Asn 1-3) – n839 [1918]-n1040 [1919] [gaps] item 13, on reel n5,6 – 1 – mf#asn-1 013 – ne IDC [077]

Amurskoe slovo : ezhedn vnepartiin gaz / ed by Gubanov, N F – Blagoveshchensk [Amur obl]: [s n] 1919 [1919 29 [16] iiunia-] – (= ser Asn 1-3) – n3-11 [1919] [gaps] item 12, on reel n5 – 1 – mf#asn-1 012 – ne IDC [077]

Les amusemens de la hollande : avec des remarques nouvelles et particulieres sur le genie, moeurs et caracteres de la nation – La Haye 1739-40 [mf ed Hildesheim 1995-98] – (= ser Fbc) – 4mf – 9 – €120.00 – 3-487-29655-1 – gw Olms [914]

Amusemens des bains de bade en suisse, de schintznach et de pfeffers : avec la description, et la comparaison de leurs eaux avec celles des bains de schwalbach and autres de l'empire... / Merveilleux, David F de – Londres 1739 [mf ed Hildeheim 1995-98] – (= ser Fbc) – 2mf – 9 – €60.00 – 3-487-29395-1 – gw Olms [914]

Amusement business – New York NY 1972+ – 1,5,9 – ISSN: 0003-2344 – mf#6382 – us UMI ProQuest [790]

Am-xtra – 1988 oct/nov-1989 mar/apr – 1 – mf#3362712 – us WHS [621]

Amx-tra – 1975 sep-1988 mar/apr – 1 – mf#2539451 – us WHS [071]

Amyntor, Gerhard von [pseud of: Dagobert von Gerhardt] *see* Hypochondrische plaudereien

Amyot, Gail (Guillaume) *see* Adresse a mm les electeurs du comte de lotbiniere

Amyot, Michel *see* Bibliographie analytique de l'ileaux-coudres

Amyraut, M *see*
- De secessione ab ecclesia romana deque ratione pacis inter evangelicos...
- Discours de la souverainete des rois
- La morale chrestienne
- Traite des religions contre ceux qui les estiment indifferentes
- Traite des religions contre ceux qui les extiment indifferentes

An – Kazan, Russia dec 1912-18 [mf ed Norman Ross] – 4r – 1 – mf#nrp-660 – us UMI ProQuest [077]

An den christlichen adel deutscher nation : von des christlichen standes besserung / Luther, Martin; ed by Benrath, Karl – Halle: Verein fuer Reformationsgeschichte, 1884 [mf ed 1990] – (= ser [Schriften des vereins fuer reformationsgeschichte] 4) – 1mf – 9 – 0-7905-4658-2 – (in german. int & ann by ed) – mf#1988-0658 – us ATLA [241]

An den christlichen adel deutscher nation von des christlichen standes besserung / Luther, Martin; ed by Braune, Wilhelm – Halle a/S: M Niemeyer 1877 [mf ed 1993] – (= ser Neudrucke deutscher literaturwerke des 16. und 17. jahrhunderts 4) – 11r – 1 – (int by ed. filmed with: neudrucke deutscher literaturwerke des 16. und 17. jahrhunderts) – mf#3387p – us Wisconsin U Libr [430]

An den durchlaeuchtigesten fuersten und herren, herrn philippen...von der fuersichtigkeyt gottes... / Zwingli, H – Zuerich, Christoffel Froschouer, 1531 – 3mf – 9 – mf#PBU-530 – ne IDC [240]

An den durchlaeuchtigesten...herrn allbrechten... : ein sendbrieff... / Bullinger, Heinrich – [Zuerich, Christoffel Froschouer], 1532 – 1mf – 9 – mf#PBU-112 – ne IDC [240]

An den grossmaechtigsten und durchlauchtigsten adel deutscher nation / Murner, Thomas; ed by Voss, Ernst Karl Johann Heinrich – Halle a/S: M Niemeyer 1899 [mf ed 1993] – (= ser Flugschriften aus der reformationszeit 13; Neudrucke deutscher literaturwerke des 16. und 17. jahrhunderts 153) – 11r [ill] – 1 – (incl bibl ref. filmed with: neudrucke deutscher literaturwerke des 16. und 17. jahrhunderts) – mf#3387p – us Wisconsin U Libr [430]

An der grenze : roman = Ved graendsen / Gjellerup, Karl Adolph – Leipzig: Quelle & Meyer 1919 [mf ed 1990] – 1r – 1 – (filmed with: hermann von gilm / arnulf sonntag) – mf#7302 – us Wisconsin U Libr [830]

An der pforte der zukunft : allegorische dichtung / Friedrichs, Hermann – Zuerich: Verlags-Magazin (J Schabelitz), 1889 ([mf ed 1989] – 1r – 1 – (filmed with: gustav freytag / hans zuchhold) – mf#7278 – us Wisconsin U Libr [810]

An der schwelle des zwanzigsten jahrhunderts : rueckblicke auf das letzte jahrhundert deutscher kirchengeschichte / Seeberg, Reinhold – Leipzig: A Deichert, 1901 [mf ed 1991] – 1mf – 9 – 0-524-00122-7 – (1st printed 1900) – mf#1989-2822 – us ATLA [242]

An die aufgeloes'te preussische national-versammlung *see* Bettina von arnims polenbroschuere

An die durchlauchtige [!] fuersten teutscher nation zu ougspurg versammlot ein sendtbrief...die schelckwort eggens...betreffendt / Zwingli, H – [Zuerich, Christoph Froschauer, 1530] – 1mf – 9 – mf#PBU-669 – ne IDC [240]

An die korinther 1 / Lietzmann, Hans – Tuebingen: J C B Mohr, 1907 [mf ed 1990] – (= ser Handbuch zum neuen testament 3/1) – 1mf – 9 – 0-7905-3385-5 – (incl bibl ref) – mf#1987-3385 – us ATLA [227]

An die souveraine der rheinischen konfoederation : ueber das denselben zugesprochene recht, ihren staaten eigene landesbischoefe und eine bischoefliche dioezesan-einrichtung... / Frey, Franz Andreas – Bamberg, Wuerzburg, 1813 (mf ed 1994) – 1mf – 9 – €24.00 – 3-8267-3024-0 – mf#DHS-AR 3024 – gw Frankfurter [240]

An einen juengling im felde : drei briefe / Johannsen, Christa – Leipzig: P List c1943 [mf ed 1990] – 1r – 1 – (filmed with: eddystone / wilhelm jensen) – mf#2743p – us Wisconsin U Libr [830]

An ha nhu't bao – Cantho. 20 sept 1917-16 nov 1933 – 1 – (puis an ha bao) – fr ACRPP [073]

An, Lan *see* Wang yu tsao (ccm257)

An nam tap chi – n1-48. 1er juil 1926-1er juin 1931; N.S., n1-9. 16 janv 1932-1er mars 1933 – 1 – fr ACRPP [073]

An rhein und ruhr *see* Aufwaerts

An, Todjin *see* Pedang kilat dari gunung thian san

An yun / Wang, Tu-ch'ing – Shang-hai: Kuang ming shu chu, 1931 – (= ser P-k&k period) – us CRL [840]

Ana de jesus y la herencia teresiana...roma, 1968 / Moriones, Ildefonso – Madrid: Graf. Calleja, 1968 – 1 – sp Bibl Santa Ana [946]

Anabaptism : from its rise at zwickau to its fall at muenster 1521-36 / Heath, Richard – London: Alexander & Shepheard, 1895 [mf ed 1986] – 1mf – 9 – 0-8370-8908-5 – (incl bibl) – mf#1986-2908 – us ATLA [242]

Anabaptism from its rise at zwickau to its fall at munster / Heath, Richard – 1521-36 – 1 – $7.84 – us Southern Baptist [242]

Anabaptisticum et enthusiasticum pantheon geistliches rust-hauss wider die alten quacker – Zurich, Switzerland. 1702 – 1 – $28.00 – us Southern Baptist [242]

Anabolic steroids : knowledge, attitude, and behavior in college age students / Munro, R – 1991 – 2mf – 9 – $8.00 – us Kinesology [150]

Anacalypsis : an attempt to draw aside the veil of the saitic isis, or, an inquiry into the origins of languages, nations, and religions / Higgins, Godfrey – London: Longman, Rees, Orme etc, 1836 – 1 – us CRL [290]

Anacaona / Burr-Reynaud, Frederic – Port-Au-Prince, Haiti. 1911 – 1r – us UF Libraries [972]

L'anacharsis francais : ou description historique et geographique de toute la france / Malo, Charles – Paris 1822-23 [mf ed Hildesheim 1995-98] – 4v on 14mf – 9 – €140.00 – 3-487-29782-5 – gw Olms [944]

Anacker, Heinrich *see*
- Der aufbau
- Einkehr
- Die fanfare
- Lieder aus stille und stuermen
- Die trommel
- Ueber die maas, ueber schelde und rhein!
- Ein volk – ein reich – ein fuehrer!
- Von beilen, barten und haeckchen microform
- Wir wachsen in das reich hinein

Anaconda assayer *see* Miscellaneous newspapers of teller county

Anaconda labor-socialist – v1 n7 [1903 mar 28] – 1 – mf#3177701 – us WHS [071]

Anacreon : ou, enfant cheri des dames / Dupeuty, M (Charles) – Paris, France. 1838 – 1r – us UF Libraries [440]

Anada Marga Yoga Society *see* Sadvipra

Anadolu : istatistiki, iktisadi, askeri cografya / Cemal, Mehmed – Istanbul: Matbaa-yi Askeri, 1336 [1920] – (= ser Ottoman histories and historical sources) – 4mf – 9 – $60.00 – us MEDOC [956]

Anadolu – Izmir, 1912-28. sahib ve sermuharriri: haydar ruesdu [oektem] n4092. 21 mayis 1928 – (= ser O & t journals) – 1mf – 9 – $25.00 – us MEDOC [956]

Anadolu tib mecmuasi – Ankara: Yeni Guen Matbaasi, 1921-22. Sahib-i Imtiyaz ve Mueduer-i Mes'ul: Muallim Dr Ekrem Hayri. n3-4. 15 subat 1338-10 mart 1338 [1922] – (= ser O & t journals) – 1mf – 9 – $25.00 – us MEDOC [956]

Anadolu'da kalem – Ankara, 1921. Sahib-i Imtiyaz: Hueseyin Suad; Mueduer-i Mes'ul: Kemal Salih. n1-2. 21-28 mayis 1337 [1921] – (= ser O & t journals) – 1mf – 9 – $25.00 – us MEDOC [956]

Anadolu'da ortodoksluk sadasi – Kayseri, 1922-23. Sahib-i Imtiyaz: Istimat Zihni. n11. 21 tesrinievvel 1338 [1922] – (= ser O & t journals) – 1mf – 9 – $25.00 – us MEDOC [956]

Anadolu'da peyam-i sabah – Ankara: Yeniguen Matbaasi, OEgud Matbaasi, 1920-22. Mueduer-i Mes'ul: Aka Guenduez. Numarasiz. 23 kanunievvel 1337 [1921] – (= ser O & t journals) – 1mf – 9 – $25.00 – us MEDOC [956]

Anadolu'da tanin / Serif, Ahmet – Istanbul: Tanin Matbaasi, 1325 [1909] – (= ser Ottoman histories and historical sources) – 5mf – 9 – $75.00 – us MEDOC [956]

Anadu, Edith C *see* Factors affecting risk perception about drinking water and response to public notification

Anaesthesia – Oxford, England 1960+ – 1,5,9 – ISSN: 0003-2409 – mf#1314 – us UMI ProQuest [617]

Anaesthesia and intensive care – Edgecliff, Australia 1991+ – 1,5,9 – ISSN: 0310-057X – mf#13843 – us UMI ProQuest [610]

Anaesthesist – Dordrecht, Netherlands 1981+ – 1,5,9 – ISSN: 0003-2417 – mf#13102 – us UMI ProQuest [617]

Anafor – Istanbul: Ahmediye Matbaasi, 1918-? Sahib-i Imtiyaz: Kemal Ibrahim n1. 5 kanunievvel 1334 [1918] – (= ser O & t journals) – 1mf – 9 – $25.00 – us MEDOC [956]

Anagat e kray pvan kale mya : [short stories] / Man Nvay Sin – Ran kun: Cape Biman 1987 [mf ed 1990] – (= ser Cape Biman thut prann su lak cvai ca can) – 1r with other items – 1 – (in burmese) – mf#mf-10289 seam reel 152/1 [§] – us CRL [830]

Anagrammas en aplauso y gloria de la concepcion purissima de maria senora nuestra : concebida sin la culpa original... / Mora, Juan Antonio de – en Mexico: En la imprenta...Miguel de Rivera, en la Empedradillo, ano de 1731 – (= ser Books on religion...1543/44-c1800: milagros y culto de la virgen) – 4mf – 9 – mf#crl-88 – ne IDC [241]

[Anaheim-] anaheim independent – CA. 1980 – 1r – 1 – $60.00 – mf#R04000 – us Library Micro [071]

[Anaheim-] anaheim news progress – CA. mar 7 1963-nov 20 1963 – 1r – 1 – $60.00 – mf#R04001 – us Library Micro [071]

[Anaheim-] daily news (anaheim edition) – CA. mar-oct 1959 – 2r – 1 – $120.00 – mf#R04002 – us Library Micro [071]

[Anaheim-] gazette – CA. oct 21 1876-oct 10 1877; oct 17 1879-oct 7 1882; 1944-22+ r – 1 – $1320.00 – mf#RC02015 – us Library Micro [071]

[Anaheim hills-] anaheim hills news – CA. oct 1992- – 10+ r – 1 – $600.00 (subs $250/y) – mf#R04004 – us Library Micro [071]

[Anaheim-] the bulletin – CA. mar 1966-may 1970; sep 1992- – 59r – 1 – $3540.00 (subs $500/y) – mf#RC02014 – us Library Micro [071]

[Anaheim-] the news – CA. mar-jul 1962 – 2r – 1 – $120.00 – mf#R04003 – us Library Micro [071]

Anais / Encontro De Geologos (1st: 1966: Porto Alegre, Brazil) – Porto Alegre, Brazil. 1966 – 1r – us UF Libraries [550]

Anais / Reuniao De Fitossanitaristas Do Brasil (10th:1966) – Rio de Janeiro, Brazil. 1967 – 1r – us UF Libraries [972]

Anais da provincia de s pedro / Fernandes Pinheiro, Jose Feliciano – Rio de Janeiro, Brazil. 1946 – 1r – us UF Libraries [972]

Anais do observatorio astronomico da universidade de coimbra : 1. seccao: fenomenos solares / Universidade de Coimbra. Observatorio Astronomico – Coimbra: Impr de Universidade 1929- (annual) [mf ed 2001] – t1-11(1929-39) on 4r [ill] – 1 – (imprint varies) – mf#film mas c4680 – us Harvard [520]

Anais do primeiro congresso de historia de revolucao / Congresso De Historia da Revolucao de 1894 (1st) – Curitiba, Brazil. 1944 – 1r – us UF Libraries [972]

Anais do seminario o ensino da protecao a saude materna e i / Seminario o Ensino da Protecao a Saude Materna E I – Rio de Janeiro, Brazil. 1970 – 1r – us UF Libraries [972]

Anais paulistas de medicina e cirurgia – Sao Paulo, Brazil 1932+ – 1,5,9 – ISSN: 0003-245X – mf#704 – us UMI ProQuest [617]

Anak bentara – Endeh, [1952]-1961 – 3mf – 9 – (missing: [1952]-1959/1960, v1-8(1, 3, 6-10?)-1960/1961, v9(1-4)) – mf#SE-870 – ne IDC [950]

Anaknja saorang desa / Phoa, Gin Hian – Soerabaia: Tan's Drukkerij, 1935 [mf ed 1998] – (= ser Penghidoepan 127) – 1r – 1 – (coll as pt of the colloquial malay collection. filmed with: multi-millionair / ong khing han) – mf#10002 – us Wisconsin U Libr [830]

Analecta / Sukthankar, Vishnu Sitaram – Poona: V S Sukthankar Memorial Edition Committee, 1945 – (= ser Samp: indian books) – us CRL [490]

Analecta ante-nicaena / Bunsen, Christian Karl Josias, Freiherr von – London: Longman, Brown, Green & Longmans, 1854 [mf ed 1990] – (= ser Christianity and mankind. philological section 5-7) – 3v on 4mf – 9 – 0-7905-5022-9 – mf#1988-1022 – us ATLA [240]

Analecta bollandiana – 1(1882)-60(1940) – 568mf – 9 – €926.00 – ne Slangenburg [073]

Analecta bollandiana – Killen TX 1882-1943 – 1 – ISSN: 0003-2468 – mf#1138 – us UMI ProQuest [900]

Analecta bollandiana *see* Les khazars dans la passion de s abo de tiflis

Analecta hierosolymitikes stachyologias / Papadopoulos-Kerameoos, A – St Petersburg. v1-5. 1891-1898 – 72mf – 8 – €137.00 – ne Slangenburg [240]

Analecta hymnica medii aevi / ed by Dreves, Guido Maria – Leipzig 1886-1922 – v1-55 on 265mf – 9 – €505.00 – ne Slangenburg; ne IDC [240]

Analecta juris pontificii : recueil de dissertations sur differents subjets de droit canonique, liturgie, theologie et histoire – Geneve, 1879-1888. v18-27 – 239mf – 8 – mf#H-209 – ne IDC [240]

Analecta monumentorum omnis aevi vindobonensia / Kollarius, A F – Vindobonae. v1-2. 1761-1762 – €124.00 – ne Slangenburg [240]

Analecta novissima spicilegio solesmensi parata / Pitra, J-B – Tusculi. v1-2. 1885-1888 – 9 – €75.00 – (v1 21mf. v2 18mf) – ne Slangenburg [240]

Analecta premonstratensia – 1(1925)-21(1945) – 162mf – 9 – €309.00 – ne Slangenburg [073]

Analecta reformatoria / Egli, E – Zuerich, 1899 – 2mf – 9 – mf#ZWI-60 – ne IDC [242]

Analecta reformatoria / ed by Egli, E – Zuerich, 1899-1901. 2 pts – 4mf – 9 – mf#PBU-449 – ne IDC [242]

Analecta reformatoria / Egli, Emil – Zuerich: Zuercher & Furrer, 1899-1901 [mf ed 1993] – (= ser Presbyterian coll) – 2v on 4mf – 9 – 0-524-07412-7 – (incl bibl footnotes) – mf#1991-3072 – us ATLA [242]

Analecta romana : kirchengeschichtliche forschungen in roemischen bibliotheken und archiven / Laemmer, Hugo – Schaffhausen: Verlag der Fr Hurter'schen Buchh, 1861 [mf ed 1986] – 1mf – 9 – 0-8370-8033-9 – (discussion in german; texts in latin & italian. incl bibl ref) – mf#1986-2033 – us ATLA [241]

Analecta (rotterham) – 9 nov 1822-15 feb 1823 – (= ser 19th c british periodicals) – reel 48 – 1 – (filmed with: the cheap magazine (haddington), 1813-14) – us Primary [073]

Analecta sacra et classica spicilegio solesmensi parata / Pitra, J-B – Paris. v1-8. 1876-1891 – 8 – €265.00 – (lacking: v7) – ne Slangenburg [240]

Analecta sacri ordinis fratrum praedicatorum : seu vetera ordinis monumenta recentioraque acta – 1(1893)-10(1902) – 169mf – 9 – €592.00 – ne Slangenburg [241]

Analecta zur aeltesten geschichte des christentums in rom / Harnack, Adolf von – Leipzig, 1905 – (= ser Tugal 2-28/2b) – 1mf – 9 – €3.00 – ne Slangenburg [240]

Analecta zur septuaginta, hexapla und patristik / Klostermann, Erich – Leipzig: A Deichert (Georg Boehme), 1895 – 1r – 1 – 0-8370-0155-2 – mf#1984-B037 – us ATLA [221]

Analectes : pour servir a l'histoire eccles. de la belgique. 1e section – 1(1864)-40(1914) – 9 – €675.00 – (2e sect 1(1894)-7(1905) €33) – ne Slangenburg [240]

Analectes de l'ordre de premontre – 1(1905)-10(1914) – 56mf – 9 – €106.00 – ne Slangenburg [241]

Analectic magazine – La Salle IL 1813-20 – 1 – mf#4416 – us UMI ProQuest [810]

The analects of confucius = Lun yue / Confucius – Yokohama: WE Soothill, 1910 – 11mf – 9 – 0-524-07933-1 – (in english) – mf#1991-0183 – us ATLA [180]

Analekten fuer das studium der exegetischen und systematischen theologie – Leipzig. bd 1-4. 1813-1822 – 4v on 26mf – 9 – €46.00 – ne Slangenburg [240]

Analekten zur textkritik des alten testaments : neue folge / Perles, Felix – Leipzig: G Engel, 1922 [mf ed 2002] – 1r – 1 – (in german & hebrew. with ind) – mf#b00651 – us ATLA [221]

Anales / Colombia. Senado – Bogota. 1903-13 Aug 1945 Session Extraordinary.Incomplete – 1 – us NY Public [972]

Anales / Sociedad rural Argentina. Buenos Aires – v.1-82. 1866-1948. LC lacks v.10-14, 16, 78 & scattered issues – 1 – us L of C Photodup [630]

Anales / Venezuela. Universidad Central. Caracas – 1-41. 1900-July 1956 – 1 – us L of C Photodup [025]

Anales, 1959 / Colegio Oficial de Farmaceuticos. Caceres – Caceres, 1960 – 1 – sp Bibl Santa Ana [615]

Anales biograficas – Madrid: Tip. Pasejo del Comercio 8, 1914 – 1 – sp Bibl Santa Ana [920]

Anales de antropologia – v1-6. 1940-70. 1971- – 1 – us AMS Press [301]

Anales de ingeniera – Bogota, Colombia 1949-50 – 1 – mf#582 – us UMI ProQuest [620]

Anales de la academia de ciencias medicas, fisicas – Habana, Cuba. v1-86. 1864-1947 – 24r – us UF Libraries [972]

Anales de la academia de ciencias medicas, fisicas / Torriente-Brau, Zoe De La – Habana, Cuba. v1-2. 1974 – 1r – us UF Libraries [972]

Anales de la fudacion de la habana / Rousset, Ricardo V – Habana, Cuba. 1919 – 1r – us UF Libraries [972]

Anales de la guerra de cuba / Pirala, Antonio – Madrid, Spain. v1-2. 1895-98 – 1r – us UF Libraries [972]

Anales de la literatura espanola contemporanea – Boulder CO 1981+ – 1,5,9 – ISSN: 0272-1635 – mf#12849,02 – us UMI ProQuest [440]

Anales de la sociedad cientifica argentina – Buenos Aires, Argentina. no date – 1r – us UF Libraries [079]

Anales de los xahil – Mexico City?, Mexico. 1946 – 1r – us UF Libraries [972]

Anales del museo nacional de buenos aires – Buenos Aires: Impr de Juan A Alsina, 1991; ser 3 n5 1905 – 1r – 1 – us CRL [060]

Anales del paraiso / Sanchez Arjona, Vicente – Sevilla: Imprenta Alvarez, Tomo 1-8. 1955, 1956 – 1 – sp Bibl Santa Ana [946]

Anales diplomaticos y consulares de colombia / Colombia Ministerio de Relaciones Exteriores – Bogota, Colombia. v1-8. 1900-1958 – 4r – us UF Libraries [972]

Anales eclesiasticos venezolanos / Navarro, Nicolas Eugenio – Caracas, Venezuela. 1951 – 1r – us UF Libraries [972]

Anales eclesiasticos venezolanos. caracas, 1929 / Navarro, Nicolas E – Madrid: Razon y Fe, 1930 – 1 – sp Bibl Santa Ana [240]

Anales estadisticos de la republica de guatemala 1882-1883, tomo 1-2 – Guatemala, 1883-1884 – 9mf – 9 – sp Cultura [318]

Anales minorum...tomus 32 (1671-1680) / Pandzic, Basilio – Madrid: Graf. Calleja, 1966 – 1 – (roma, 1964) – sp Bibl Santa Ana [946]

Anales religiosos de colombia – Santafe de Bogota, Colombia: [s.n.] 1883-86 [mf ed 2007] – (= ser Religious periodical literature of the hispanic and indigenous peoples of the americas, 1850-1985) – 3v on 1r – 1 – (lacks: v2 n42 [jul 15 1885], v2 n46 [sep 15 1885], v3 n54-55 [mar 15-apr 1 1886], v3 n58 [may 15 1886], v3 n65 [sep 1 1886]; damaged: v1 n17 [jul 1884] p265, v2 n28 [dec 1884] p59,61,63) – mf#2007h-s041 – us ATLA [241]

Anales...medicina...y biografico / Chinchilla, Anastasio – v3. 1848 – 9 – sp Bibl Santa Ana [920]

Analfabetismo en puerto rico / Rodriguez Bou, Ismael – San Juan, Puerto Rico. 1945 – 1r – us UF Libraries [972]

Analise do intercambio comercial, brasil-reino-uni / Confederacao Nacional da Industria Departamento E – Rio de Janeiro, Brazil. 1969 – 1r – us UF Libraries [338]

Analisi del quaresimale del padre paolo segneri / Malmusi, Giuseppe – Torino: Giacinto Marietti, 1879 [mf ed 1986] – 1mf – 9 – 0-8370-6917-3 – mf#1986-0917 – us ATLA [241]

Analisis – Buenos Aires, Argentina 16 aug, 6 dec 1965-16 feb 1970 (imperfect) – 1 – uk British Libr Newspaper [079]

Analisis de la contestacion del diputado de provincia don antonio concha y del libelo informativo...d joaquin rodriguez leal... / Ceresoles, Mauricio – Caceres: Imp. D. Lucas de Burgos, 1839 – 1 – sp Bibl Santa Ana [946]

Analisis de la poblacion protegida por el seguro s / Franky Vasquez, Pablo – Bogota, Colombia. 1964 – 1r – us UF Libraries [972]

Analiz otchetnosti promyslovykh i promyslovo-kreditnykh kooperativov / Simonovich, I A – 1930 – 104p 2mf – 9 – mf#COR-446 – ne IDC [335]

Analog science fiction & fact – Norwalk CT 1960+ – 1,5,9 – ISSN: 1059-2113 – mf#6221 – us UMI ProQuest [400]

Die analogie von natur und geist als stilprinzip in novalis' dichtung / Feng, Chih – Heidelberg: A Lippl, 1935 – 1r – 1 – (incl bibl ref) – us Wisconsin U Libr [430]

The analogies of being as embodied in and upon this orb : shewn to be the only inductive base of divine revelation... / Wood, Joseph – London:Frederick Farrah, 1867 – 1mf – 9 – 0-8370-5899-6 – mf#1985-3899 – us ATLA [210]

Analogy and the scope of its application in language / Wheeler, Benjamin Ide – [s.l]: John Wilson, 1887 [mf ed 1986] – 1mf – 9 – 0-8370-8236-6 – mf#1986-2236 – us ATLA [400]

Analogy considered as a guide to truth : and applied as an aid to faith / Buchanan, James – Edinburgh: Johnstone, Hunter, 1864 [mf ed 1984] – 7mf – 9 – 0-8370-0949-9 – (incl bibl ref and app) – mf#1984-4326 – us ATLA [230]

The analogy of existences and christianity / Wallace, Charles J – London: Hodder & Stoughton, 1892 [mf ed 1985] – 1mf – 9 – 0-8370-5688-8 – mf#1985-3688 – us ATLA [210]

The analogy of religion natural and revealed / Butler, Joseph – London:J M Dent; New York:E P Dutton, [19–] [mf ed 1985] – 1mf – 9 – 0-8370-2558-3 – (int by ronald bayne) – mf#1985-0558 – us ATLA [210]

The analogy of revelation and science established in a series of lectures : delivered...1833 / Nolan, Frederick – Oxford: J H Parker, 1833 [mf ed 1989] – (= ser Bampton lectures 1833) – 1mf – 9 – 0-7905-1493-1 – mf#1987-1493 – us ATLA [210]

Analogy which subsists between the british constitution in its... – Edinburgh, Scotland. 1840 – 1r – us UF Libraries [240]

Analyse – Zeitz DE, 1950 24 apr-1951 nov, 1952-1960 mar" 1960 may-nov, 1961-1969 nov, 1970-1985 nov, 1986-91 – 11r – 1 – (with gaps. mineraloelwerk luetzkendorf; publ in halle, saal) – gw Misc Inst [550]

Die analyse – Dessau DE, 1954 jul-1990 [gaps] – 7r – 1 – (notes: gaerungschemie) – gw Misc Inst [540]

Analyse bibliographique de reverende mere sainte-louise-de-marillac... / Neiges, Marie-des, soeur – 1962 [mf ed 1978] – (= ser Bibliographies du cours...1947-66) – 1mf – 9 – (with ind; pref by andre leveille) – mf#SEM105P4 – cn Bibl Nat [920]

Analyse chronologique relative a la concession du 25 fevrier 1661 : appelee la seigneurie de l'isle aux oeufs / Bouchette, Joseph – Quebec: L'Evenement, 1868 [mf ed 1980] – 1mf – 9 – 0-665-02533-5 – mf#02533 – cn Canadiana [971]

Analyse der fetalen herzfrequenz mit methoden der chaostheorie / Maris, Bartolomeus – (mf ed 1994) – 1mf – 9 – €30.00 – 3-8267-2053-9 – mf#DHS 2053 – gw Frankfurter [612]

Analyse der offenbarung johannis / Wellhausen, Julius – Berlin: Weidmann, 1907 [mf ed 1985] – (= ser Abhandlungen der koeniglichen gesellschaft der wissenschaft zu goettingen. philologisch-historische klasse. neue folge 9/4) – 1mf – 9 – 0-8370-5768-X – (incl bibl ref) – mf#1985-3768 – us ATLA [225]

Analyse des reponses faites par les deputes de s g mgr l'eveque de montreal en 1867 et 1868 – Lyon, France?: s.n, 1869 [mf ed 1985] – 3mf – 9 – 0-665-08701-2 – mf#08701 – cn Canadiana [241]

Analyse du kandjour : recueil des livres sacres au tibet / Csoma, Alexandre, de Koros – Paris: E Leroux, 1881 – 1 – us CRL [951]

Analyse d'un entretien sur la conservation des etablissemens du bas-canada, des lois, des usages, etc de ses habitans / Viger, Denis Benjamin – Montreal: J Lane, 1826 [mf ed 1971] – 1r – 5 – mf#SEM16P90 – cn Bibl Nat [971]

L'analyse grammaticale et l'analyse logique : aux brevets de capacite, a l'ecole normale et a l'ecole primaire intermediaire et superieur / Magnan, Charles-Joseph – Quebec: J A Langlais, [1907?] – 2mf – 9 – 0-665-73320-8 – mf#73320 – cn Canadiana [440]

Analyse logique : [petit traite sur l'analyse logique, l'architecture, la perspective et l'emploi du subjonctif] – [Ste Anne de la Pocatiere, Quebec?: s.n.] 1864 [mf ed 1983] – 1mf – 9 – 0-665-39834-4 – mf#39834 – cn Canadiana [440]

Analyse statistique des donnees : concernant le nombre de cadres superieurs et d'adjoints dans la fonction publique quebecoise / Depatie, Francine – [Quebec]: Ministere de la fonction publique, 1971 [mf ed 1999] – 2mf – 9 – mf#SEM105P3112 – cn Bibl Nat [317]

Analyse zu kardinalsymptomen im langzeitverlauf des morbus meniere : vertigo, schwerhoerigkeit, tinnitus / Kolbe, Ulrich – [mf ed 2001] – 9 – €30.00 – 3-8267-2769-X – mf#DHS2769 – gw Frankfurter [616]

Analyses of iron ores from pictou county, nova scotia, dominion of canada : red hematite, specular ore, brown hematite, spathose ore – Montreal?: Morton, Phillips & Bulmer, 1885? – 1mf – 9 – mf#67849 – cn Canadiana [660]

Analyses of the documentary evidence introduced by the prosecution before the international military tribunal for the far east, 1946-1948 / World War 2. International Prosecution Section – (= ser Records of allied operational and occupation headquarters, world war 2) – 6r – 1 – mf#M1697 – us Nat Archives [355]

Analyses of the orange / Pickell, J M – Lake City, FL. 1892 – 1r – us UF Libraries [634]

Analysing the managerial skills of entrepreneurs in the pretoria area / Radipere, Simon – Pretoria: Vista University 2003 [mf ed 2003] – 3mf – 9 – (incl bibl ref) – mf#mfm15266 – sa Unisa [650]

Analysing the understanding of the implementation of the curriculum 2005 : by grade four educators in mangaung / Dyantyi, Vuyo Cedric – Pretoria: Vista University 2003 [mf ed 2003] – 2mf – 9 – (incl bibl ref) – mf#mfm15225 – sa Unisa [370]

Analysis – Oxford, England 1933+ – 1,5,9 – ISSN: 0003-2638 – mf#11847 – us UMI ProQuest [140]

Analysis and critique of "christ the transformer of culture" in the thought of h richard niebuhr / Wittmer, Michael Eugene – Grand Rapids MI: Calvin Theological Seminary, 2000 [mf ed 2003] – 1r – 1 – $130.00 – mf#D00003 – us ATLA [230]

An analysis and critique of leonardo boff's theology and social ethics / Mattos, Luiz Roberto Franca de – Grand Rapids MI: Calvin Theological Seminary, 2001 [mf ed 2001] – 1r – 1 – $130.00 – mf#D00001 – us ATLA [230]

The analysis and decision of summary judgement motions : a monograph on rule 56 of the federal rules of civil procedure / Schwarzer, William W et al – Washington: FJC, 1991 – 2mf – 9 – $3.00 – mf#LLMC 95-382 – us LLMC [347]

An analysis and evaluation of the administrative budget statement between 1984 and 1995 for the south korean ministry of culture and sports / Kim, Sangho – 1997 – 1mf – 9 – $4.00 – mf#PE 3842 – us Kinesology [790]

An analysis and evaluation of the courses in elementary school supervision offered in colleges, normal schools and universities / Selke, Erich – Minnesota, 1933 (mf ed 1994) – 1mf – 9 – €24.00 – 3-8267-3090-9 – mf#DHS-AR 3090 – gw Frankfurter [370]

Analysis and exposition of hebrews 7 1-8 – London, England. 1841 – 1r – us UF Libraries [240]

Analysis and intervention in developmental disabilities – Oxford, England 1981-86 – 1,5,9 – ISSN: 0270-4684 – mf#49388 – us UMI ProQuest [610]

Analysis and proof texts of dr julius mueller's system of theology / Smith, Henry Boynton [comp] – New York: JM Sherwood, 1868 [mf ed 1991] – 1mf – 9 – 0-7905-9041-7 – mf#1989-2266 – us ATLA [240]

Analysis fidei catholicae : hoc est ratio methodica eam in universum fidem... / Gregorius de Valencia – Ingolstadii, 1585 – 5mf – 9 – mf#CA-75 – ne IDC [241]

Analysis logica in epistolam apostoli pauli ad romanos *see* Logical analysis of the epistle of paul to the romans

Analysis of a poulenc trio [for bassoon, oboe, and piano] / Allison, Ruth – 1947 – 2mf – 9 – mf#fiche1163 – us Sibley [780]

Analysis of a social situation in modern zululand / Gluckman, M – (= ser Institute for social research, university of zambia. papers 28) – 2mf – 1 – mf#4734 – uk Microform Academic [960]

An analysis of ancient domestic architecture : exhibiting the best existing examples / Dollman, Francis Thomas & Jobbins, John Richard – London [1861-63] – (= ser 19th c art & architecture) – 6mf – 9 – mf#4.2.1404 – uk Chadwyck [720]

Analysis of ankle inversion with 20 cm drops onto a laterally tilted force plate in braced and unbraced conditions / Slack, Robert W – 1997 – 1mf – 9 – $4.00 – mf#PE 3795 – us Kinesology [612]

An analysis of association between respiration rate and finger temperature in normals and raynaud's subjects given thermal biofeedback / Spalding, Thomas W – 1982 – 3mf – 9 – $12.00 – us Kinesology [610]

An analysis of athletic department operations at the dean smith center / Heeden, Matthew – 1999 – 1mf – 9 – $4.00 – mf#PE 3949 – us Kinesology [790]

An analysis of backgrounds of professional baseball players / Jones, Kenneth W – 1998 – 1mf – 9 – $4.00 – mf#PE 3898 – us Kinesology [790]

Analysis of bartok's sixth string quartet / Bailey, Robert Wayne – U of Rochester 1952 [mf ed 19–] – 2mf – 9 – mf#fiche337, 527 – us Sibley [780]

The analysis of beauty... / Hogarth, W – London, 1810 – 3mf – 9 – mf#O-1178 – ne IDC [700]

Analysis of beginning string ensemble methods / Arnold, Charles D – U of Rochester 1935 [mf ed 1938] – 2mf – 9 – mf#fiche972 – us Sibley [780]

Analysis of blackstone's commentaries on the laws of england / Bailey, Henry – Charleston SC: [H Bailey] 1822 [mf ed Spartanburg SC: Reprint Co, dist, 1981] – 28mf – 9 – mf#51-010 – us South Carolina Historical [348]

An analysis of bultmann's nonform-critical criteria used in evaluating authenticity in the synoptic gospels / Kwik, Robert Julius – Princeton, NJ: 1966. Chicago: U of Chicago Lib, 1975 (1r); Evanston: American Theol Lib Assoc, 1984 (1r) – 1 – 0-8370-0501-9 – mf#1984-B483 – us ATLA [226]

An analysis of butler's analogy of religion : and three sermons on human nature / Angus, Joseph – London: Religious Tract Society [1882?] [mf ed 1990] – 1mf – 9 – 0-7905-3519-X – mf#1989-0012 – us ATLA [240]

Analysis of decisions / New York. State Labor Relations Board – No1-36, 1937-72. 63 fiches. (Harvard Law School Library Collection.) – 9 – $ – us Harvard Law [330]

Analysis of documentary evidence / Supreme Commander for the Allied Powers. International Prosecution Section – Tokyo. On film: Doc. 1-3379, 3500-3517, 4001-4097; 1946-48. LL-023 – 1 – us L of C Photodup [340]

The analysis of enrollment patterns and student provile characteristics at a small rural new england university 1978-1988 / Holmes, ML – 1990 – 5mf – 9 – $20.00 – us Kinesology [378]

An analysis of exit surveys of student-athletes : at the university of north carolina at chapel hillfrom 1994 to 1999 / Saunches, Nicole – 2000 – 106p on 2mf – 9 – $10.00 – mf#PE 4154 – us Kinesology [150]

Analysis of fertilizers / Pickell, J M – Lake City, FL. 1889 – 1r – us UF Libraries [630]

An analysis of gothick architecture / Brandon, Raphael & Brandon, Joshua Arthur – London 1847 – (= ser 19th c art & architecture) – 6mf – 9 – mf#4.2.931 – uk Chadwyck [720]

Analysis of hurricane problems in coastal areas of florida – Jacksonville, FL. 1961 – 1r – us UF Libraries [550]

Analysis of in-vivo meniscal kinematic motion of the non-injured knee / Porter, Scott T – 1994 – 1mf – $4.00 – us Kinesology [612]

An analysis of johann lydwig krebs' eight chorale preludes for organ with trumpet or oboe / Pedde, Dennis R – 1981 – 1 – $5.00 – us Southern Baptist [780]

Analysis of l'ascension for organ by olivier messiaen / Fort, Robert Edwin – U of Rochester 1956 [mf ed 19–] – 3mf – 9 – mf#fiche 89 – us Sibley [780]

The analysis of moral man : an outline of the conditions of human righteousness / Stevans, C M – Chicago: Popular Pub, c1900 [mf ed 1986] – 1mf – 9 – 0-8370-7510-6 – mf#1986-1510 – us ATLA [170]

Analysis of music education in three new york state cities / Truitt, Austin H – U of Rochester 1953 [mf ed 198-?] – 5mf – 9 – (with app & bibl) – mf#fiche 905 – us Sibley [780]

An analysis of national athletic training association accredited education program facilities for the athletic trainer / Sabo, James M – 1994 – 5mf – $20.00 – us Kinesology [370]

An analysis of ncaa division 1-a football sports information director's experiences with independent internet sites / Stepp, Thomas – 2000 – 48p on 1mf – 9 – $5.00 – mf#PE 4155 – us Kinesology [302]

An analysis of new testament history : embracing the criticism and interpretation of the original text, the authenticity of its several books, a harmony chronologically arranged, and a copious historical index / Pinnock, William Henry – 12th ed. Cambridge: J Hall, 1869 [mf ed 1985] – 1mf – 9 – 0-8370-4757-9 – mf#1985-2757 – us ATLA [225]

Analysis of ornament : the characteristics of styles / Wornum, Ralph Nicholson – [8th ed]. London 1893 – (= ser 19th c art & architecture) – 2mf – 9 – mf#4.1.415 – uk Chadwyck [740]

An analysis of pain and injury experiences of intercollegiate athletes based on gender and four sport status variables / Yoder, Kelly J – 1998 – 1mf – 9 – $4.00 – mf#PSY 2018 – us Kinesology [612]

Analysis of paley's view of the evidences of christianity – Harlow, England. 1810 – 1r – us UF Libraries [240]

An analysis of referrals received by a psychiatric unit in a general hospital / Dor, Marlene – Uni of South Africa 2000 [mf ed Pretoria: UNISA 2000] – 3mf – 9 – (incl bibl ref) – mf#mfm14729 – sa Unisa [362]

An analysis of religious belief / Amberley, John Russell, Viscount – New York: D M Bennett. 1878 [mf ed 1982] – 745p – 1 – (incl ind) – mf#5774 – us Wisconsin U Libr [210]

Analysis of reports filed by consumer finance and consumer in stallment loan companies / Illinois. Consumer Credit Division – 1930-79. 14 fiches. (Harvard Law School Library Collection.) – 9 – $ – us Harvard Law [336]

An analysis of scripture history : with examination questions / Pinnock, William Henry – 18th ed. Cambridge: J Hall; London: Whittaker, 1871 [mf ed 1985] – 2mf – 9 – 0-8370-4758-7 – (incl ind) – mf#1985-2758 – us ATLA [220]

An analysis of selected attendance factors in the world league of american football / Bryan, Carlton H Jr & Billing, John E – 1992 – 1mf – 9 – $4.00 – us Kinesology [790]

Analysis of selected respiratory and cardiovascular characteristics of wind instrument performer / Middlesworth, Jane L van – U of Rochester 1978 [mf ed 19–] – 1r – (with bibl ref) – mf#film 2555 – us Sibley [780]

An analysis of student-athletes' experiences since leaving the university of north carolina / Sabo, Tim – 1999 – 1mf – 9 – $4.00 – mf#PE 3946 – us Kinesology [790]

An analysis of swimming economy as assessed by a comparison between vo2 values and arm stroke index / Sharar, Brian D – 1989 – 63p 1mf – 9 – $4.00 – us Kinesology [612]

An analysis of texts of scripture : the better to understand the true fact or spiritual meaning that the revelator intended to convey to the reader by means of the letter / Fowler, Josiah – Castalia, Erie Co, OH: [s.n.], 1881 [mf ed 1993] – 1mf – 9 – 0-524-06332-X – mf#1992-0870 – us ATLA [220]

An analysis of the backgrounds professional baseball players / Randall, Jeff – 2000 – 50 On 1mf – 9 – $5.00 – mf#PE 4163 – us Kinesology [790]

An analysis of the bernoulli lift effect as a propulsive component of swimming strokes / Ferrell, M D – 1991 – 1mf – 9 – $4.00 – us Kinesology [790]

Analysis of the book of isaiah : pt 1st. introduction / Lord, Eleazar – New York: John A Gray, 1861 [mf ed 1985] – 1mf – 9 – 0-8370-4178-3 – (no more publ) – mf#1985-2178 – us ATLA [220]

Analysis of the church music curriculum of selected protestant seminaries / Irwin, E Robert – U of Rochester 1967 [mf ed 19–] – 4mf – 9 – mf#fiche 218, 314 – us Sibley [780]

Analysis of the constitution of the commonwealth of the northern mariana islands : n.a – Saipan: n.p. dec 6 1976 – 3mf – 9 – $4.50 – mf#llmc82-100j, title 12 – us LLMC [323]

An analysis of the current judging methods used in competitive ballroom dancing as well as comparisons to competitive pairs figure skating and ice dancing / Keck, Mary L B – 1998 – 2mf – 9 – $8.00 – mf#PE 3996 – us Kinesology [790]

An analysis of the diagnostic and prescriptive expertise of level 2 and examiner downhill ski instructors / Young, Ben – 1999 – 2mf – 9 – $8.00 – mf#PE 3973 – us Kinesology [370]

Analysis of the draft compact of association : prepared for the members of the micronesian independence commission / Uludong, F T – n.p, n.d, submitted to the Congress of Micronesia on aug 25, 1972 – (= ser Micronesia: prelude to the constitutional convention) – 1mf – 9 – $1.50 – (typescript) – mf#LLMC 82-100F, Title 46 – us LLMC [323]

Analysis of the dumbarton oaks concerto for chamber orchestra by igor stravinsky / Dowdakin, James Daniel – U of Rochester 1953 [mf ed 19–] – 1r – 1 – mf#film 2551 – us Sibley [780]

An analysis of the exposition of the creed : written by the right rev father in god john pearson...late lord bishop of chester / Mill, William Hodge – Cambridge: University Press, 1874 [mf ed 1991] – 1mf – 9 – 0-7905-9035-2 – mf#1989-2260 – us ATLA [240]

An analysis of the factors that influence fan attendance at minor league baseball games / Freiling, Howard P – 1996 – 1mf – 9 – $4.00 – mf#PSY 1943 – us Kinesology [790]

Analysis of the galic language / Shaw, William – 2nd ed. Edinburgh: Jamieson 1778 – (= ser Whsb) – 2mf – 9 – €30.00 – mf#Hu 091 – gw Fischer [490]

Analysis of the gospels / Bowman, Hervey Meyer – Canada: [s.n.], c1919 – 1mf – 9 – 0-665-73617-7 – mf#73617 – cn Canadiana [226]

An analysis of the hermeneutic of the southern baptist convention sermon in selected periods of biblical controversy : 1925, 1963, 1970 / Dortch, John Douglas – 1982 – 1 – $5.92 – us Southern Baptist [242]

Analysis of the human rights and gender consequences of the new south african constitution and bill of rights : with regard to the recognition and implementation of muslim personal law (mpl) / Moosa, Najma – U of the Western Cape 1996 [mf ed S.l: s.n. 1996] – 9mf – 9 – sa Misc Inst [342]

Analysis of the instructional ecology in tutorial tennis settings / England, Kathleen M & Tannenhill, Deborah – 1993 – 4mf – $16.00 – us Kinesology [790]

Analysis of the later symphonic style of edmund rubbra / Milligan, Stuart – U of Rochester 1966 [mf ed 19–] – 5mf – 9 – (with app & bibl) – mf#fiche 49 – us Sibley [780]

An analysis of the laws of england : english, irish and french editions / Blackstone, William – (= ser The yale law library blackstone coll) – 21mf – 9 – mf#LLMC 82-800 titles 217-224 – us LLMC [343]

Analysis of the main sewer system of the inner cape metropolitan region : with the aid of a geographical information system / Sinske, Stefan – Stellenbosch: U of Stellenbosch 1998 [mf ed 1998] – 3mf – 9 – mf#mf.1281 – sa Stellenbosch [628]

Analysis of the medium / Simone, Carol M de – U of Rochester 1963 [mf ed 19–] – 2mf – 9 – mf#fiche 90 – us Sibley [780]

Analysis of the moravian chorales / Rauch, Ralph Frederick – U of Rochester 1952 [mf ed 19–] – 1mf – 9 – mf#fiche 235, 539 – us Sibley [780]

An analysis of the motivational impact of a health risk appraisal and a college health education course utilizing a lifestyle theme on selected health behaviors / Cottrell, Randall R – 1982 – 3mf – 9 – $12.00 – us Kinesology [790]

Analysis of the part writing technic in the later works of vaughan wiliams / Fisher, Charles Milton – U of Rochester 1941 [mf ed 19–] – 2mf – 9 – mf#fiche 1196 – us Sibley [780]

Analysis of the phenomena of the human mind / Mill, James – new ed. London: Longmans, Green, Reader & Dyer, 1869 [mf ed 1991] – 2v on 2mf – 9 – 0-524-00288-6 – mf#1989-2988 – us ATLA [150]

An analysis of the primary use of church sports programs in anderson, indiana / Hensley, Tammy – 1998 – 1mf – 9 – $4.00 – mf#PE 3855 – us Kinesology [790]

An analysis of the principles of equity pleading / Lube, Denis George – San Francisco, Bancroft-Whitney, 1886. 283 p. L.C. copy imperfect: p. 3-4, 11-12 wanting. LL-117 – 1 – us L of C Photodup [340]

An analysis of the processes used by athletic directors to evaluate the head coaches of men's and women's basketball teams : at national collegiate athletic association division 1, 2 and 3 colleges and universities in pennsylvania / Overton, Reginald F – 1997 – 4mf – 9 – $16.00 – mf#PE 3820 – us Kinesology [370]

Analysis of the proposed constitution of the federated states of micronesia / ed by Political Status Commission of the Marshall Islands – 1976? – (= ser Micronesian constitutional convention, 1975) – 5mf – 9 – $7.50 – (draft; unpag) – mf#LLMC 82-100F, Title 97 – us LLMC [323]

Analysis of the relationship between exercise and heart rate variability in trained and untrained individuals / Tonkins, William P – 1999 – 1mf – 9 – $4.00 – mf#PH 1675 – us Kinesology [612]

Analysis of the second quartet of bela bartok / Ahrendt, Christine. – U of Rochester 1946 [mf ed 19–] – 1r / 3mf – 1,9 – mf#film 760 / fiche241 – us Sibley [780]

An analysis of the social structure of a western town : a specimen study according to small and vincent's method / Dunn, Arthur William – Chicago: The University of Chicago Press, 1896 [mf ed 1970] – (= ser Library of american civilization 40081) – 53p on 1mf – 9 – (repr fr charities review) – us Chicago U Pr [307]

Analysis of the twelve preludes and fugues of franz reizenstein / Hennicke, Gayle Watts – U of Rochester 1972 [mf ed 19–] – 3mf – 9 – mf#fiche 553, 733 – us Sibley [780]

Analysis of training protocols for challenge course instructors / Novak, Jeremy D – 1999 – 1mf – 9 – $4.00 – mf#PE 3990 – us Kinesology [370]

An analysis of unsportsmanlike behavior and ejections : in the member high schools of the north carolina high school athletic association / Lee, Karin A – 1997 – 1mf – 9 – $4.00 – mf#PE 3760 – us Kinesology [790]

An analysis of visual reaction time, dynamic reaction activity, and depth perception of males wearing color eye shields / Wilson, J E – 1991 – 1mf – 9 – $4.00 – us Kinesology [150]

Analysis of ward's errata of the protestant bible / Ryan, Edward – Dublin, Ireland. 1808 – 1r – us UF Libraries [242]

Analysis of watson's theological institutes : designed for the use of students and examining committees / McClintock, John – New York: Eaton & Mains, [1842?] [mf ed 1991] – 1mf – 9 – 0-524-00201-0 – mf#1989-2901 – us ATLA [240]

An analysis on self-talk and self-confidence with female tennis players / Stokes, Hilary Gail – 1998 – 2mf – 9 – $8.00 – mf#PSY 2017 – us Kinesology [150]

Analyst – Cambridge, England 1956+ – 1,5,9 – ISSN: 0003-2654 – mf#1207 – us UMI ProQuest [540]

Analyst : a quarterly journal of science, literature, natural history and the fine arts – La Salle IL 1834-40 – 1 – mf#5227 – us UMI ProQuest [073]

Analyst; or, mathematical museum – La Salle IL 1808-14 – 1 – mf#3549 – us UMI ProQuest [510]

Analytic psychology / Stout, George Frederick – London: S Sonnenschein; New York: Macmillan 1896 [mf ed 1987] – (= ser Library of philosophy) – 2v on 1r – 1 – (filmed with: from boston to bareilly and back / butler, w) – mf#2088 – us Wisconsin U Libr [150]

Analytica chimica acta – Oxford, England 1947+ – 1,5,9 – ISSN: 0003-2670 – mf#42011 – us UMI ProQuest [540]

Analytical abstracts – Cambridge, England 1954+ – 1,5,9 – ISSN: 0003-2689 – mf#1209 – us UMI ProQuest [540]

An analytical and practical grammar of the english language : with an appendix on prosody, punctuation etc / Davies, Henry William – Toronto: J Campbell, 1868 [mf ed 1991] – 3mf – 9 – 0-665-90622-6 – (original iss in ser: canadian national series of school books) – mf#90622 – cn Canadiana [420]

Analytical and quantitative cytology – St Louis MO 1983-84 – 1,5,9 – (cont by: analytical & quantitative cytology & histology) – ISSN: 0190-0471 – mf#13595.01 – us UMI ProQuest [574]

Analytical and quantitative cytology and histology – St Louis MO 1985+ – 1,5,9 – (cont: analytical and quantitative cytology) – ISSN: 0884-6812 – mf#13595,01 – us UMI ProQuest [574]

Analytical chemistry – v1- 1929- – 1,5,6,9 – us ACS [540]

Analytical communications / Royal Society of Chemistry (Great Britain) – Cambridge, England 1996-99 – 1,5,9 – (cont: analytical proceedings royal society of chemistry (great britain)) – ISSN: 1359-7337 – mf#11249,04 – us UMI ProQuest [540]

An analytical concordance to the holy scriptures : or, the bible presented under distinct and classified heads or topics / ed by Eadie, John – Boston: Gould & Lincoln, 1857 [mf ed 1992] – 8mf – 9 – 0-524-02775-7 – mf#1987-6469 – us ATLA [220]

Analytical ethnology : the mixed tribes in great britain and ireland examined / Massy, Richard Tuthill – London, 1855 – (= ser 19th c ireland) – 3mf – 9 – mf#1.1.7481 – uk Chadwyck [941]

Analytical exposition of the absurdity and iniquity of the oaths / Steele, Thomas – London, England. 1829 – 1r – us UF Libraries [240]

Analytical foundations of celestial mechanics / Wintner, Aurel – Princeton NJ: Princeton UP; London: Humphrey Milford, OUP 1941 [mf ed 2005] – (= ser Princeton mathematical series 5) – 1r – 1 – (incl bibl ref & ind) – mf#film mas 36377 – us Harvard [520]

Analytical grammar of shona / Fortune, George – London, England. 1955 – 1r – us UF Libraries [470]

An analytical history of the patent policy of the department of health, education, and welfare / U.S. Dept of Health, Education and Welfare – Washington, Govt. Print. Off., 1961. 93 p. LL-2306 – 1 – us L of C Photodup [346]

Analytical index to the civil code of lower canada – Ottawa: G E Desbarats, 1867 [mf ed 1984] – 2mf – 9 – mf#SEM105P432 – cn Bibl Nat [348]

Analytical proceedings. chemical society (great britain). analytical division – Cambridge, England 1980 – 1,5,9 – (cont: proceedings of the analytical division of the chemical society; cont by: analytical proceedings royal society of chemistry (great britain)) – ISSN: 0144-557X – mf#11249.04 – us UMI ProQuest [540]

Analytical proceedings. royal society of chemistry (great britain). analytical division – Cambridge, England 1980-95 – 1,5,9 – (cont: analytical proceedings chemical society (great britain); cont by: analytical communications) – ISSN: 0144-557X – mf#11249.04 – us UMI ProQuest [540]

The analytical reader : a short method for learning to read and write chinese / Martin, William Alexander Parsons – Shanghai: Presbyterian Mission Press, 1897 [mf ed 1995] – (= ser Yale coll) – 204p – 1 – 0-524-09878-6 – mf#1995-0878 – us ATLA [480]

Analytical review : or, history of literature, domestic and foreign – La Salle IL 1788-99 – 1 – mf#4199 – us UMI ProQuest [410]

Analytical studies of sugar cane grown in florida / Lemon, J M – s.l, s.l? . 1925 – 1r – us UF Libraries [630]

Analytical study of nielsen's commotio / Burchill, James Frederick – U of Rochester 1973 [mf ed 19–] – 2mf – 9 – mf#fiche419 – us Sibley [780]

Analytical study of published clarinet sonatas by american composers / Carlucci, Joseph Barry – U of Rochester 1957, c1958 [mf ed 19–] – 6mf – 9 – mf#fiche387 – us Sibley [780]

Analytical study of selected piano works by edward macdowell / Lien, Beatrix – U of Rochester 1940 [mf ed 19–] – 1r / 2mf – 1,9 – mf#film 1016 / fiche 100 – us Sibley [780]

Analytical study of the symphony in e minor by ralph vaughan williams / Tomasick, Paul – U of Rochester 1955 [mf ed 1986] – 2mf – 9 – (with bibl) – mf#fiche 1199 – us Sibley [780]

Analytical study of the timbre of the clarinet / Gibson, Oscar Lee – U of Rochester 1938 [mf ed 19–] – 2mf – 9 – mf#fiche 1205 – us Sibley [780]

An analytical survey of zulu poetry : both traditional and modern / Kunene, Mazisi – [Durban?; 1961?] – 1r – 1 – us CRL [470]

An analytical synopsis of the criminal code and of the canada evidence act / Crankshaw, James – Montreal: C Theoret, 1899 [mf ed 1979] – 2mf – 9 – 0-665-00096-0 – mf#00096 – cn Canadiana [345]

Analytical view of the principal plans of church reform / Bloomfield, S T – London, England. 1833 – 1r – us UF Libraries [240]

An analytic-critical reflection on an integrated arts education curriculum in a multicultural south africa / Nevhutanda, Ntshengedzeni Alfred – Uni of South Africa 2000 [mf ed Johannesburg 2000] – 5mf – 9 – (incl bibl ref) – mf#mfm14865 – sa Unisa [370]

O analytico – Oeiras, PI: Typ Provincial, 09 out-09 nov 1848 – (= ser Ps 19) – mf#P17,02,135 – bl Biblioteca [079]

The analytics of a belief in a future life / Gratacap, Louis Pope – New York: J Pott, 1888 – 1mf – 9 – 0-7905-3880-6 – mf#1989-0373 – us ATLA [210]

Analyzed new york decisions and citations, 1914-1917 : covering duplicate reports / Kreidler, Charles Ray – Rochester, Williamson Law Book Co., 1916. 832 p. LL-269 – 1 – us L of C Photodup [340]

Analyzing computer applications in national collegiate athletic association's men's basketball programs / Eaton, Scott W – 1999 – 244p on 3mf – 9 – $15.00 – mf#PE 4192 – us Kinesology [000]

Analyzing the private contributions among collegiate letterwinners / Yablunosky, Matthew S – University of North Carolina at Chapel Hill, 1995 – 1mf – 9 – $4.00 – mf#PE3625 – us Kinesology [790]

Analyzing wholesale distribution costs / Millard, James William – [Washington, DC: GPO], 1927 (mf ed 19–) – 12 leaves – mf#ZT-TB+ pv484 n4 – us Harvard [650]

Anamnesis sive commemoratio / Tamayo de Salazar, Juan – v1-2. 1651. v3, 1655 – 9 – (v5 1658. v6 1659) – sp Bibl Santa Ana [946]

Anand, Mulk Raj *see*
- The barber's trade union and other stories
- The big heart
- The bride's book of beauty
- The golden breath
- The hindu view of art
- Homage to tagore
- Indian short stories
- The indian theatre
- The king-emperor's english
- Lament on the death of a master of arts
- Letters on india
- Lines written to an indian air
- Marx and engels on india
- On education
- Persian painting
- Private life of an indian prince
- Seven summers
- The story of india
- The tractor and the corn goddess
- Two leaves and a bud
- Untouchable

Ananda Acharya *see*
- Brahmadarsanam
- Saki, the comrade
- Snow-birds

Ananda bazar patrika – Calcutta, India. Apr 1944-85 – 201r – 1 – us L of C Photodup [079]

Ananda candra [rhc ra cu rakhuin vesali man] / Cam Sa On, U – Yan Kun: Cape Biman [1975] [mf ed 1990] – 1r with other items – 1 – (in burmese) – mf#mf-10289 seam reel 082/02 [§] – us CRL [959]

Ananda Metteyya *see* An outline of buddhism

Ananda ranga pillai : the 'pepys' of french india / Srinivasachari, Chidambaram S – Madras: P Varadachary & Co, 1940 – (= ser Samp: indian books) – us CRL [920]

The ananda temple at pagan / Duroiselle, Charles – Delhi: Manager of Publications, 1937 – (= ser Samp: indian books) – us CRL [720]

Anandalahari = Wave of bliss / Sankaracarya – Madras: Ganesh & Co, 1924 – (= ser Samp: indian books) – (transl with commentary by arthur avalon) – us CRL [810]

Anandan, P M *see* Kamalist turkey

Anandasramasamskrtagranthavalih *see* Srivedavyasapranitamahabharatantargata srimadbhagavadgita

Ananles del tajo lisboa / Coronado, Carolina – Lisboa 1875 – 1 – (filmed with: kyodo kenkyu shi / inostrantsev, k) – mf#2190 – us Wisconsin U Libr [946]

Anantanpilla, Pi *see* Vidya prakasika

Anantasuriya : anahc rahc ra [i.e. 800] prynn atham amaht / Magha – Yan Kun: Tao Wan Cape 1973 [mf ed 1990] – 1r with other items – 1 – (in burmese) – mf#mf-10289 seam reel 160/7 [§] – us CRL [480]

Anantha Krishna Iyer, L Krishna, Diwan Bahadur *see*
- The cochin tribes and castes
- Lectures on ethnography
- The mysore tribes and castes...

Anao – Para, 25 ago 1890 – (= ser Ps 19) – bl Biblioteca [079]

O anao : periodico noticioso, social, critico e litterario – Pernambuco, 21 jan 1863 – (= ser Ps 19) – bl Biblioteca [079]

The anaphora or great eucharistic prayer / Frere, Walter H – London, 1938 – 4mf – 8 – €11.00 – ne Slangenburg [241]

Anaphorae syriacae / Pontificii Studiorum Orientalium – Roma. v1-2. 199-1951 – €52.00 – ne Slangenburg [241]

The anaphoras of the ethiopic liturgy / Harden, J M – London, 1928 – 3mf – 8 – €7.00 – ne Slangenburg [243]

Anaplasis – Athens, Greece 15 may 1922-25 mar 1937 – 7r – 1 – (in greek) – uk British Libr Newspaper [074]

L'anarchia – New York NY, 1918* – 1r – 1 – (italian periodical) – us IHRC [073]

L'anarchico – New York NY, 1888* – 1r – 1 – (italian periodical) – us IHRC [073]

L'anarchie – Paris (F), 1905 13 apr-1914 16 jul [gaps] – 1 – fr ACRPP [320]

Anarchie im drama : kritik und darstellung der modernen dramatik / Diebold, Bernhard – Frankfurt/M: Frankfurter Verlags-Anstalt A-G 1922 [mf ed 1993] – 1r [ill] – 1 – (incl bibl ref & ind. filmed with: cyanen) – mf#3383p – us Wisconsin U Libr [820]

Der anarchist – Berlin: Verlag & Red des "Anarchist". v1-5 n7. mar 1903-oct 1907 – 1 – us CRL [320]

Der anarchist : organ zur propaganda des anarchismus und sozialismus – Leipzig. v1-3. oct 2 1909-dec 1911 – 1 – (literarisches beiblatt: v5 n5 apr 1913) – us CRL [335]

The anarchist – Glasgow, Scotland. -w. 3 May 1912-17 Jan 1913. 26 ft – 1 – uk British Libr Newspaper [072]

Anarchist Association of the Americas *see* Emancipation

Anarchist black dragon – n2-n10 [1978 sep-1982 spring] – 1 – mf#635841 – us WHS [071]

Anarchist Communist Federation [Regina, SK] *see* Flashpoint

Anarchist pamphlets, 1830-1985 : from the labadie collection, university of michigan and the library of congress – [mf ed Chadwyck-Healey] – 2601mf – 9 – (pamphlets are a vital primary source for the study of the anarchist movement) – uk Chadwyck [320]

The anarchist press in britain : the publications of freedom press, 1928-1976 – 16r – 1 – mf#C39-16400 – us Primary [070]

Anarchist-Communist Federation [ACF-NA] *see* Free passage

Anarchist-Communist Federation of North-America *see* North american anarchist

Anarchists – 1894 feb 10 – 1 – mf#3177699 – us WHS [071]

Anarcho-Feminists *see* Siren

Anarchy in worship / Begg, James – Edinburgh, Scotland. 1875 – 1r – us UF Libraries [240]

Anargharaghavam / Murari – Mumbayyam: Nirnayasagarakhyayantralayadhipatina, 1937 [mf ed 1985] – 395p – 1 – (in sanskrit) – mf#9103 – us Wisconsin U Libr [820]

Anarkhiia i anarkhisty / Ivanovich, S – 1917 – 31p 1mf – 9 – mf#RPP-81 – ne IDC [325]

Anarkhisty / Ravich-Cherkasskii, M – Kharkov, 1930 – 68p 1mf – 9 – mf#RPP-89 – ne IDC [325]

Anarkhisty, maksimalisty i makhaevtsy : anarkh techeniia v pervoi russkoi revoliutsii / Gorev, B I – 1918 – 69p 1mf – 9 – mf#RPP-79 – ne IDC [325]

Anarkhisty v rossii / Zalezhskii, V – 1930 – 80p 1mf – 9 – mf#RPP-80 – ne IDC [325]

Anarkhizm / Borovoi, A – 1918 – 169p 2mf – 9 – mf#RPP-76 – ne IDC [325]

Anarkhizm / Goldman, E – 1921 – 116p 2mf – 9 – mf#RPP-77 – ne IDC [325]

Anarkhizm i rabochii klass : anarkhizm v rossii. sotsializm i sotsial-demokratiia. kommunizm anarkhistov. ikh taktika. anarkhizm i sotsialist. rabochee dvizhenie / [Litvinova, L F] – Taganrog, 1917 – 1mf – 9 – mf#RPP-91 – ne IDC [325]

Anarkhizm i sotsializm / Plekhanov, G V – n.d. – 80p 1mf – 9 – mf#RPP-88 – ne IDC [325]

Anarkhizm v rossii : kak istoriia razreshila spor mezhdu anarkhistami i kommunistami v russkoi revoliutsii / Iaroslavskii, E – 1939 – 120p 2mf – 9 – mf#RPP-93 – ne IDC [325]

Anarkhizm v rossii : ot bakunina do makhno / Gorev, B – 1930 – 143p 2mf – 9 – mf#RPP-78 – ne IDC [325]

Anarquismo da colonia cecilia / Sousa, Newton Stadler De – Rio de Janeiro, Brazil. 1970 – 1r – us UF Libraries [972]

El anarquismo militante y la realidad espanala / Montseny, Frederica – [Barcelona]: Oficina de Propaganda [1937] – (= ser Blodgett coll) – 9 – mf#w1059 – us Harvard [946]

Anasagasti, Pedro de *see* Francisco de asis busca al hombre...

Anashim ve-sofrim / Zitron, Samuel Leib – Warszawa, Poland. 1922 – 1r – us UF Libraries [939]

Anastasii bibliothecarii *see* Chronographia [cshb39,40]

Anastasion, Georgios *see* First greek colony in america

Anastasis : or, the doctrine of the resurrection of the body / Bush, George – New York: Wiley & Putnam, 1845, c1844 [mf ed 1989] – 1mf – 9 – 0-7905-1082-0 – mf#1987-1082 – us ATLA [240]

Anastasius Bibliothecaris *see*
- Historia de vitis romanorum pontificum
- Historia ecclesiastica sive chronographia tripertita

Anastasius gruen : verschollenes und vergilbtes aus dessen leben und wirken / ed by Radics, Peter von – Leipzig: H Foltz, 1879 [mf ed 1993] – 200p – 1 – (incl bibl ref) – mf#8504 – us Wisconsin U Libr [920]

Anastasius gruen : verschollenes und vergilbtes aus dessen leben und wirken / Radics, Peter von – Leipzig: H Foltz 1879 [mf ed 1993] – 1r – 1 – (incl bibl ref. filmed with: althochdeutsche lesestuecke / von wilhelm wackernagel) – mf#8504 – us Wisconsin U Libr [430]

Anastasius gruen's gesammelte werke / ed by Frankl, Ludwig August – Berlin: G Grote 1907 [mf ed 1988] – 5v on 1r – 1 – (filmed with: aus einer ganz kleinen garnison – baal / von ferd avenarius & other titles) – mf#6970 – us Wisconsin U Libr [802]

Anastasius "wechwyser", bullingers "huysboeck" en calvins "institutie"... / Oorthuys, G – Leiden, E J Brill, 1919 – 3mf – 9 – mf#PBU-448 – ne IDC [242]

Anastassopoulou, Itheoni *see* Causalite et creation

Anatolia through the ages / Schmidt, Eric F – 1931 – 9 – $10.00 – us IRC [930]

Anatolian times – 1981 may-1985 dec 6; 1985 dec 13-1988 sep 30 – 1 – mf#1043609 – us WHS [071]

Anatolien : skizzen und reisebriefe aus kleinasien (1850-1859) / Mordtmann, A D – Hannover, 1925 – 7mf – 9 – mf#AR-1811 – ne IDC [915]

Anatolius, Bishop of Mohilew and Mstislaw *see* The greek catholic faith

Anatomia da renuncia / Carli, Gileno De – Rio de Janeiro, Brazil. 1962 – 1r – us UF Libraries [972]

Anatomia del corazon / Guerrero Y Pallares, Teodoro – Habana, Cuba. 1858 – 1r – us UF Libraries [611]

Anatomia et laboratorium veri christiani : to gest: cztwery knihy v praw, m krestanstwy...[i.e.: vier buecher von dem wahren christenthum...] / [Arndt, J] – Prague: Karlsspergka, 1617 – 9mf – 9 – mf#O-99 – ne IDC [090]

Anatomical drawings / Royal Library. Windsor Castle – (= ser Leonardo drawings) – 8 colour 1 bw mf – 15,9 – $330.00 – 0-907716-02-4 – (84 drawings, 224 details; 14 b&w images incl 7 in ultra-violet; with ind) – uk Mindata [740]

Anatomical studies of the bones and muscles, for the use of artists / Flaxman, John – London 1833 – (= ser 19th c art & architecture) – 4mf – 9 – mf#4.2.1657 – uk Chadwyck [611]

Anatomicheskii atlas dlia etudentov i vrachei / Toldt, Carl – Berlin, Germany. v1-3. 1921 – 1r – us UF Libraries [500]

Anatomie de la messe / Moulin, P du – Geneve. 2v. 1636-40 – 4mf – 9 – mf#PRS-174 – ne IDC [240]

Anatomie de la messe / Moulin, P du – Geneve, Sedan. 2v. 1636, 1639 – 3mf – 9 – mf#PRS-144 – ne IDC [240]

L'anatomie du calvinisme... / Gaultier, J – Lyon. 1621 – 9mf – 9 – mf#CA-100 – ne IDC [242]

Anatomische, histologische, histomorphologische und ethologische untersuchungen zur tiergerechtheit am beispiel des kaninchens / Drescher, Birgit – (mf ed 1997) – 3mf – 9 – €49.00 – 3-8267-2421-6 – mf#DHS 2421 – gw Frankfurter [630]

Anatomy and embryology – Dordrecht, Netherlands 1980-84 – 1,5,9 – ISSN: 0340-2061 – mf#13104,02 – us UMI ProQuest [612]

Anatomy of aparteid – London, England. 1960? – 1r – us UF Libraries [322]

Anatomy of apartheid / ed by Randall, Peter – Johannesburg: Study Project on Christianity in Apartheid Soc, 1970 – us CRL [322]

An anatomy of atheism : as demonstrated in the light of the constitution and laws of nature / Moore, Homer H – Cincinnati: Cranston & Stowe [c1890] [mf ed 1984] – 4mf – 9 – 0-8370-1061-6 – (incl ind) – mf#1984-4411 – us ATLA [210]

The anatomy of humane bodies: with figures drawn after the life...and curiously engraven.. / Cowper, William – Oxford: Printed at the Theater, for Sam. Smith and Benj. Walford, London, 1698. 1v. illus. With: Les Oeuvres de Jean Baptiste van Helmont. Lyons, 1671 – 1 – us Wisconsin U Libr [611]

The anatomy of negation / Saltus, Edgar – New York: Brentano, [1886?] – 1mf – 9 – 0-8370-6364-7 – mf#1986-0364 – us ATLA [210]

The anatomy of pattern / Day, Lewis Foreman – London 1887 – (= ser 19th c art & architecture) – 2mf – 9 – mf#4.2.117 – uk Chadwyck [740]

The anatomy of scepticism : an examination into the causes of the progress which scepticism is making in england / Girdlestone, Robert Baker – London: W Hunt; Oxford: Slatter & Rose, [1863?] [mf ed 1990] – 1mf – 9 – 0-7905-3377-4 – (incl bibl ref) – mf#1987-3377 – us ATLA [210]

Anatomy of south africa / Hudson, William – Cape Town, South Africa. 1966 – 1r – us UF Libraries [960]

The anatomy of south african misery / Kiewiet, C W de – London, New York: Oxford UP, 1956 – us CRL [321]

Anatomy of the dicotyledons / Metcalfe, Charles Russell – Oxford, England. v1-2. 1950 – 1r – us UF Libraries [580]

The anatomy of the osmundaceae / Faull, Joseph Horace – [Toronto?: s.n, 1902?] [mf ed 1998] – 1mf – 9 – 0-665-99441-9 – (repr fr: botanical gazette, v32) – mf#99441 – cn Canadiana [580]

Anatomy of woody plants / Jeffrey, Edward Charles – Chicago, IL. 1917 – 1r – us UF Libraries [580]

Anaya, Jose Lucas *see* Doctrina christiana, y platicas doctrinales

Anbind- oder fangbriefe / Spangenberg, Wolfhart; ed by Behrend, Fritz – Tuebingen: Gedruckt fuer den Litterarischen Verein in Stuttgart 1914 [mf ed 1993] – (= ser Blvs) – 58r – 1 – (in verse; incl bibl ref & ind. filmed with: der laubacher barlaam / ed by adolf perdisch) – mf#3420p – us Wisconsin U Libr [800]

Anbind- oder fangbriefe / Spangenberg, Wolfhart; ed by Behrend, Fritz – Tuebingen: Gedruckt fuer den Litterarischen Verein in Stuttgart, 1914 [mf ed 1993] – (= ser Blvs 262) – xvi/249p – 1 – mf#8470 reel 53 – us Wisconsin U Libr [860]

Anbruch : oesterreichische zeitschrift fuer musik – v. 1-19. 1919-37 – 1 – 75.00 – us L of C Photodup [780]

Der anbruch : flugblaetter aus der zeit – Wien (A), Berlin DE, 1917-22 – 1 – (fr 1919 publ in berlin) – gw Misc Inst [074]

Anburey, Thomas *see*
- Journal d'un voyage fait dans l'interieur de l'amerique septentrionale
- Voyages dans les parties interieures de l'amerique

Ancash: tradiciones y cuentos / Mendoza, Mauro G – Lima, Peru. 1958 – 1r – us UF Libraries [972]

Ancelet-Hustache, J *see*
- Mechtilde de magdebourg
- La vie mystique d'un monastere de dominicaines au moyen age d'apres la chronique de toess

Ancelle, J *see* Les explorations au senegal et dans les contrees voisines depuis l'antiquite jusqu'a nos jours

Ancelot, Francois *see*
- Dieu vous benisse!
- Gabrielle
- Louis 9
- Lucienne, ou, dix heures du soir
- Mancini, ou, la famille mazarin
- Six mois en russie

Ancelot, Marguerite-Louise-Virginia *see*
- Marguerite
- Pere marcel

Ancessi, Victor *see* Job et l'egypte

Ancestors : yours and mine – v1 nl-v8 n4 [1975 feb-1982 nov] – 1 – mf#657276 – us WHS [929]

Ancestors unlimited edition – 1979 mar-1983; 1984 mar-1989 – 1 – mf#474635 – us WHS [929]

Ancestor-worship and japanese law / Hozumi, Nobushige – 2nd rev ed. Tokyo: Maruzen Kabushiki-Kaisha, 1912 [mf ed 1991] – 1mf – 9 – 0-524-01179-6 – mf#1990-2255 – us ATLA [340]

Ancestor-worship and japanese law / Hozumi, Nobushige – 6th rev ed. Tokyo: The Hokuseido Press 1940 [mf ed 1987] – 1r – 1 – (ed rev by shigeto hozumi; with bibl footnotes. filmed with: etude sur catulle / couat, a h) – mf#1965p – us Wisconsin U Libr [290]

Ancestral charts of george addison throop, deborah goldsmith : many historically interesting letters from the old travelling bag / Smith, Olive Cole; ed by Throop, James Addison – East St Louis: Throop & Son, 1934 – 1r – 1 – mf#Throop family – us Western Res [920]

Ancestral circle – 1978 nov-1982 spring – 1 – mf#646761 – us WHS [929]

Ancestral news – v1 n1-v9 n4 [1976 jan-1984 fall] – 1 – mf#845219 – us WHS [929]

Ancestral pursuit – v1 n1 [1987 aug] – 1 – mf#1609418 – us WHS [929]

Ancestral stories and traditions of great families illustrative of english history / Timbs, John – London: Griffith & Farran 1869 [mf ed 1987] – 1r – 1 – (filmed with: alfred the great / draper, w h) – mf#1869 – us Wisconsin U Libr [929]

Ancestral worship : a revised edition of an essay...shanghai, may 10-24 1877 / Yates, Matthew Tyson – Shanghai: American Presbyterian Mission Press, 1878 [mf ed 1992] – 1mf – 9 – 0-524-02947-4 – mf#1990-3159 – us ATLA [290]

Ancestree House *see* Midwest ancestree quarterly

Ancestry of general sir william fenwick williams of kars : and incidentally a maternal line of the present marquis of donegal; including geneological sketches of the historic annapolis royal families of winniett, dyson, williams and walker... / Savary, Alfred William – [Exeter?: s.n.], 1911 – 1mf – 9 – 0-665-76028-0 – (repr fr "the genealogist") – mf#76028 – cn Canadiana [929]

L'ancetre americain du droit compare : la doctrine du juge story / Lambert, Edouard – Paris: Recueil Sirey, 1947 – 350p – 1 – mf#LL-4076 – us L of C Photodup [340]

Anchieta na capitania de sao vicente / Machado, Antonio De Alcantara – Rio de Janeiro, Brazil. 1929 – 1r – us UF Libraries [972]

Anchor of the soul – Kelso, Scotland. 18– – 1r – us UF Libraries [240]

[Anchorage-] alaska herald – AL. mar 1 1868-apr 19 1872 (wkly) – 3r – 1 – $180.00 – (incl the u.s. constitution in russian) – mf#B01000 – us Library Micro [071]

Anchorage gazette – 1992 dec-1993 jan/feb – 1 – mf#2667832 – us WHS [071]

The anchorite and other stories / Kincaid, Charles Augustus – Bombay: Oxford University Press, 1922 – (= ser Samp: indian books) – us CRL [830]

Anchorline – 1990 spring – 1 – mf#1789366 – us WHS [071]

El anciano *see* Miscellaneous newspapers of las animas county, reel 2

Anciaux, Leon *see*
- Le lingala vehiculaire
- La participation des belges a l'oeuvre coloniale des hollandais aux indes orientales

L'ancien et le nouveau monde – Paris: Impr de Boule [apr 6 1848] – (= ser French revolution of 1848. newspapers) – 1r – 1 – us CRL [944]

L'ancien monde et le christianisme *see* The ancient world and christianity

L'ancien quebec, descriptions, nos archives, etc / Bechard, Auguste – Quebec?: [Belleau], 1890 – 2mf – 9 – mf#03535 – cn Canadiana [917]

Ancien regime in turmoil? : commerce, politics and society in france, c1682-1793: the gazette manuscrite, 1775-1793 & related sources from the john rylands university library of manchester – [mf ed Marlborough 1991] – 9r – 1 – £850.00 – (with d/g) – uk Matthew [944]

L'ancien testament dans ses rapports : avec le nouveau et la critique moderne / Meignan, Guillaume Rene – Paris: Victor Lecoffre, 1896 [mf ed 1993] – 2mf – 9 – 0-524-06576-4 – (in french) – mf#1992-0919 – us ATLA [221]

L'ancienne liturgie gallicane : son origine et sa formation en provence aux 15e et 16e siecles / Thibaut, J-B – Paris, 1929 – €7.00 – ne Slangenburg [241]

L'ancienne liturgie romaine : le rite lyonnais / Buenner, D – Paris-Lyon, 1934 – 5mf – 8 – €12.00 – ne Slangenburg [241]

Ancienne revue des revues *see* La revue des revues

L'ancienne version latine des questions sur la genese de philon d'alexandrie : 1. ed critique / Petit, F – Berlin, 1973 – (= ser Tugal 5-113) – 2mf – 9 – €5.00 – ne Slangenburg [180]

L'ancienne version latine des questions sur la genese de philon d'alexandrie : 2. commentaire / Petit, F – Berlin, 1973 – (= ser Tugal 5-114) – 4mf – 9 – €11.00 – ne Slangenburg [180]

Les anciennes cotes du lac saint-louis : avec un tableau complet des anciens et nouveaux proprietaires = The old settlements of lake st louis with a list of the old and new proprietors / Girouard, Desire – Montreal?: s.n, 1892 – 2mf – 9 – (in french and english) – mf#03459 – cn Canadiana [971]

Anciennes coutumes claustrales / Gougaud, L – Liguge, 1930 – 2mf – 8 – €5.00 – ne Slangenburg [241]

Anciennes litteratures chretiennes 2. la litterature syriaque / Duval, R – 3e ed. Paris, 1907 – €17.00 – ne Slangenburg [240]

Les anciennes liturgies 2. l'ancien sacramentaire de l'eglise 1 / Grancolas, M J – Paris, 1699 – 16mf – 8 – €31.00 – ne Slangenburg [241]

Les anciennes liturgies 3. l'ancien sacramentaire de l'eglise 2 / Grancolas, M J – Paris, 1699 – 8mf – 8 – €17.00 – ne Slangenburg [241]

Anciennes relations des indes et de la chine : de deux voyageurs mahometans, qui y allerent dans le neuvieme siecle / Renaudot, E – Paris: Jean-Baptiste Coignard, 1718 – 5mf – 9 – mf#HT-693 – ne IDC [915]

Les anciens canadiens / Aubert de Gaspe, Philippe – Quebec: Desbarats & Derbishire, 1863 [mf ed 1974] – 1r – 5 – mf#SEM16P98 – cn Bibl Nat [440]

Anciens canaux khmers dans les provinces de treang et de baphnom / Paris, Pierre – [n.p. 1931] [mf ed 1989] – 1r with other items – 1 – (extraits du bulletin de l'ecole francaise d'extreme-orient, t31 & 41) – mf#mf-10289 seam reel 017/08 [§] – us CRL [930]

Les anciens missionnaires de l'acadie : devant l'histoire / Bourgeois, Phileas Frederic – Shediac, NB: Presses du moniteur Acadien, [1910?] – 2mf – 9 – 0-665-71724-5 – (with bibl ref) – mf#71724 – cn Canadiana [241]

Les anciens missionnaires de l'acadie devant l'histoire / Bourgeois, Phileas Frederic – Shediac: des Presses du Moniteur Acadien, [1910?] (mf ed 1988) – 2mf – 9 – mf#SEM105P930 – cn Bibl Nat [241]

Les anciens postes du lac saint-louis / Girouard, Desire – Levis Quebec: P-G Roy, 1895 – 1mf – 9 – mf#03430 – cn Canadiana [971]

Anciens royaumes de la zone interlacustre meridionale / Hertefelt, Marcel D' – London, England. 1962 – 1r – us UF Libraries [960]

Ancient aboriginal trade in north america / Rau, Charles – Washington: Govt Print Off, 1873 [mf ed 1980] – (= ser Archiv fuer anthropologie (braunschweig, 1872)) – 1mf – 9 – (originally in german; repr fr: report of the smithsonian institution for 1872; incl bibl ref) – mf#09233 – cn Canadiana [380]

Ancient america : in notes on american archaeology / Baldwin, John Dennison – New York: Harper, 1871 [mf ed 1986] – 4mf – 9 – 0-665-55871-6 – (incl ind and publ list) – mf#55871 – cn Canadiana [930]

ANCIENT

Ancient america : in notes on american archaeology / Baldwin, John Dennison – New York: Harper, 1872 [mf ed 1980] – 4mf – 9 – 0-665-02482-7 – mf#02482 – cn Canadiana [930]

Ancient and mediaeval india / Manning [Mrs] [Speir, Charlotte] – London: WH Allen, 1869 [mf ed 1992] – 2v on 2mf – 9 – 0-524-04526-7 – (incl bibl ref) – mf#1990-3360 – us ATLA [954]

Ancient and medieval church history (to a d 1517) / Newman, Albert Henry – Philadelphia: American Baptist Pub Soc, 1906 [mf ed 1991] – (= ser A manual of church history) – 2mf – 9 – 0-524-01233-4 – (incl bibl) – mf#1990-0372 – us ATLA [240]

Ancient and modern furniture / Small, John William – Edinburgh [1883] – (= ser 19th c art & architecture) – 1mf – 9 – mf#4.2.1765 – uk Chadwyck [740]

The ancient and modern history of china : comprising an account of its government and laws, religion, population, revenue, productions... – London: Edward Gover, 1840 – (= ser 19th c books on china) – 2mf – 9 – mf#7.1.5 – uk Chadwyck [951]

The ancient and modern history of the maritime ports of ireland / Marmion, Anthony – [London], 1855 – (= ser 19th c ireland) – 7mf – 9 – mf#1.1.6392 – uk Chadwyck [941]

Ancient and modern scottish songs, heroic ballads, etc / Herd, David – Glasgow, Scotland. v1-2. 1869 – 1r – us UF Libraries [780]

Ancient arabia : the hanged poems, the koran in translations / Johnson, Frank Ernest et al – New York: Parke, Austin & Lipscomb c1917 [mf ed 1992] – (= ser The sacred books and early literature of the east 5) – 2mf – 9 – 0-524-05068-6 – (incl bibl ref) – mf#1991-0007 – us ATLA [810]

Ancient Arabic Order of the Nobles of the Mystic Shrine for North America *see*
– Tripoli tattler
– Zor zephyr

Ancient art and ritual / Harrison, Jane Ellen – New York: Henry Holt, c1913 [mf ed 1991] – (= ser Home university library of modern knowledge 70) – 1mf – 9 – 0-524-00888-4 – mf#1990-2111 – us ATLA [110]

The ancient art stoneware of the low countries and germany / Solon, Louis Marc Emmanuel – London 1892 – (= ser 19th c art & architecture) – 9mf – 9 – mf#4.2.1078 – uk Chadwyck [730]

Ancient assyria / Johns, Claude Hermann Walter – Cambridge: University Press; New York: G P Putnam 1912 [mf ed 1989] – (= ser The cambridge manuals of science and literature) – 1mf – 9 – 0-7905-1167-3 – (incl ind) – mf#1987-1167 – us ATLA [930]

Ancient babylonia / Johns, Claude Hermann Walter – Cambridge: University Press 1913 [mf ed 1992] – (= ser The cambridge manuals of science and literature) – 1mf – 9 – 0-524-05217-4 – mf#1992-0350 – us ATLA [930]

"Ancient babylonian tablets" / Poebel, Arno – University of Pennsylvania Museum Journal: June 1913 – 9 – $10.00 – us IRC [930]

Ancient ballads and legends of hindustan / Dutt, Toru – Allahabad: Kitabistan, 1941 – (= ser Samp: indian books) – us CRL [780]

Ancient british and irish churches / Cathcart, William – 1894 – 1 – $13.51 – us Southern Baptist [720]

The ancient british and irish churches : including the life and labors of st patrick / Cathcart, William – Philadelphia: American Baptist Publ Soc 1894 [mf ed 1989] – 1mf – 9 – 0-7905-4447-4 – (incl bibl ref) – mf#1988-0447 – us ATLA [240]

Ancient Brotherhood of Satan *see* Brimstone

Ancient buddhism in japan / Visser, Marinus Willem De – Paris, France. v1-2. 1928-1935 – 1r – us UF Libraries [280]

The ancient castles of england and wales – London 1825 [mf ed Hildesheim 1995-98] – 2v on 4mf – 9 – €120.00 – 3-487-28821-4 – (ill by william woolnoth) – gw Olms [720]

The ancient catholic church : from the accession of trajan to the fourth general council (a d 98-451) / Rainy, Robert – New York: C Scribner, 1902 [mf ed 1990] – (= ser International theological library) – 2mf – 9 – 0-7905-5313-9 – (incl bibl ref) – mf#1988-1313 – us ATLA [241]

Ancient channels of the ottawa river / Ells, Robert Wheelock – [Ottawa?: s.n, 1901?] – 1mf – 9 – 0-665-78221-7 – mf#78221 – cn Canadiana [550]

Ancient christianity and the doctrines of the oxford tracts / Taylor, Isaac – Philadelphia: Herman Hooker, 1840 [mf ed 1992] – 2mf – 9 – 0-524-03433-8 – (inlc only pt1-3 of 8pt originally publ in london, 1839-40) – mf#1990-0987 – us ATLA [240]

Ancient christianity exemplified : in the private, domestic, social, and civil life of the primitive christians and in the original institutions, offices, ordinances, and rites of the church / Coleman, Lyman – Philadelphia: Lippincott, Grambo, 1852 [mf ed 1989] – 2mf – 9 – 0-7905-0978-4 – (incl ind) – mf#1987-0978 – us ATLA [240]

Ancient christians' principle / Turford, Hugh – London, England. 1819 – 1r – us UF Libraries [240]

The ancient church : from the captivity to the coming of christ / Pond, Enoch – Boston: Massachusetts Sabbath School Soc, 1851 [mf ed 1993] – 1mf – 9 – 0-524-05687-0 – mf#1992-0537 – us ATLA [939]

The ancient church : its history, doctrine, worship, and constitution, traced for the first three hundred years / Killen, William Dool – new rev ed. New York: ADF Randolph, c1883 [mf ed 1990] – 2mf – 9 – 0-7905-5658-8 – mf#1988-1658 – us ATLA [240]

Ancient church of ireland / Gargan, Denis – Dublin, Ireland. 1864 – 1r – us UF Libraries [240]

The ancient church of shobdon, herefordshire : illustrated and described / Lewis, George Robert – London: Pelham Richardson, 1852 – (= ser 19th c art & architecture) – 2mf – 9 – mf#4.1.141 – uk Chadwyck [720]

Ancient church orders / Maclean, Arthur John – Cambridge: University Press; New York: GP Putnam, 1910 – (= ser Cambridge handbooks of liturgical study) – 1mf – 9 – 0-7905-1191-6 – (incl ind) – mf#1987-1191 – us ATLA [240]

Ancient cities : from the dawn to the daylight / Wright, William Burnet – Boston: Houghton, Mifflin, 1887 [mf ed 1993] – 1mf – 9 – 0-524-08614-1 – mf#1993-0049 – us ATLA [930]

Ancient cities and empires : their prophetic doom read in the light of history and modern research / Gillett, Ezra Hall – Philadelphia: Presbyterian Publ Cttee; New York: ADF Randolph, c1867 [mf ed 1990] – 1mf – 9 – 0-7905-4908-5 – mf#1988-0908 – us ATLA [220]

The ancient cities of the new world : being travels and explorations in mexico and central america from 1857-1882 = Les anciennes villes du nouveau monde(1885) / Charnay, Desire – London: Chapman & Hall Ltd, 1887 – (= ser 19th c art & architecture) – 6mf – 9 – mf#4.1.139 – uk Chadwyck [720]

Ancient city – St Andrews, FL. 1850-1854 – 1r – us UF Libraries [071]

The ancient city of quebec – Quebec?: Canadian Pacific Railway Co, c1894 – 1mf – 9 – mf#04011 – cn Canadiana [917]

The ancient civilization of angkor / Pym, Christopher – New York, New American Library [1968] [mf ed 1989] – [ill] 1r with other items – 1 – (with bibl) – mf#mf-10289 seam reel 022/01 [§] – us CRL [930]

Ancient collects and other prayers / Bright, William – 2nd enl ed. Oxford: J H & Jas Parker, 1862 [mf ed 1989] – 1mf – 9 – 0-7905-4153-X – mf#1988-0153 – us ATLA [240]

Ancient complaint applied to modern exigencies / Sieveright, James – Edinburgh, Scotland. 1848 – 1r – us UF Libraries [240]

The ancient coptic churches of egypt / Butler, Alfred Joshua – Oxford: Clarendon Press 1884 [mf ed 1986] – 2v on 2mf [ill] – 9 – 0-8370-7611-0 – (incl bibl ref & ind) – mf#1986-1611 – us ATLA [243]

The ancient coptic churches of egypt / Butler, Alfred Joshua – Oxford 1884 – (= ser 19th c art & architecture) – 10mf – 9 – mf#4.2.1171 – uk Chadwyck [720]

Ancient cures, charms, and usages of ireland : contributions to irish lore / Wilde, Jane Francesca (Elgee) – London, 1890 – (= ser 19th c ireland) – 3mf – 9 – mf#1.1.2596 – uk Chadwyck [640]

The ancient east / Hogarth, David George – London: Williams and Norgate, [1914?] – (= ser Home university library of modern knowledge) – 1mf – 9 – 0-524-05614-5 – (incl bibl ref) – mf#1992-0469 – us ATLA [930]

The ancient east / ed by Hutchison, J – London, David Nutt, no.1-6, 1901-03. -irr. No more published. Includes bibliographies. – 1 – us Wisconsin U Libr [956]

Ancient egypt : her testimony to the truth of the bible / Osburn, William – London: S Bagster, 1846 [mf ed 1992] – 1mf – 9 – 0-524-05237-9 – (incl bibl ref) – mf#1992-0370 – us ATLA [221]

Ancient egypt : its antiquities, religion, and history to the close of the old testament period / Trevor, George – Boston: American Tract Society, [1863?] [mf ed 1992] – 1mf – 9 – 0-524-04596-8 – mf#1992-0184 – us ATLA [220]

Ancient egypt – London, 1914-1934. v1-19 – 61mf – 9 – mf#NE-359 – ne IDC [930]

Ancient egypt and the ancient near east *see* History of glass

Ancient egypt and the east / British School of Archaeology, Egypt – London, 194-35 [mf ed 2000] – (= ser Christianity's encounter with world religions, 1850-1950) – 2r – 1 – mf#2000-s007-008 – us ATLA [930]

Ancient egypt in the light of modern discoveries / Osborn, Henry Stafford – Cincinnati: Robert Clarke, 1883 [mf ed 1989] – 1mf – 9 – 0-7905-1619-5 – (incl bibl ref & ind) – mf#1987-1619 – us ATLA [930]

Ancient Egyptian Arabic Order of the Nobles of the Mystic Shrine of North and South America *see* Pyramid

Ancient egyptian dances / Lexova, Irena; ed by Lexa, Frantisek – Praha, Czechoslovakia: Oriental institute 1935 [mf ed 1987] – 1r [ill] – 1 – (with drawings made fr reproductions of ancient egyptian originals by milada lexova; trans by k haltmar) – mf#7015 – us Wisconsin U Libr [390]

The ancient egyptian doctrine of the immortality of the soul / Wiedemann, Alfred – New York: GP Putnam; London: H Grevel, 1895 – 1mf – 9 – 0-7905-0533-9 – (incl bibl ref) – mf#1987-0533 – us ATLA [930]

The ancient empires of the east / Sayce, Archibald Henry – New York: Charles Scribner, 1884 [mf ed 1988] – 1mf – 9 – 0-7905-0282-8 – mf#1987-0282 – us ATLA [930]

L'ancient et le futur quebec : projet de son excellence lord dufferin: conference faite a la salle victoria le 19 janvier 1876 / Buies, Arthur – Quebec?: s.n, 1876 [mf ed 1981] – 1mf – 9 – mf#24123 – cn Canadiana [971]

Ancient facts and fictions concerning churches and tithes / Selborne, Roundell Palmer, Earl of – 2nd ed. London; New York: Macmillan, 1892 [mf ed 1990] – 1mf – 9 – 0-7905-6830-6 – (1st publ 1888. incl bibl ref) – mf#1988-2830 – us ATLA [333]

Ancient gaza / Petrie, William Matthew Flinders – BSA, 1933 – 9 – $10.00 – us IRC [930]

Ancient gaza: 1 / Petrie, William Matthew Flinders – BSA, 1931 – 9 – $10.00 – us IRC [930]

The ancient geography of india / Cunningham, Alexander – London: Truebner, 1871 – 2mf – 9 – 0-524-08157-3 – mf#1991-0287 – us ATLA [900]

The ancient half-timbered houses of england / Habershon, Matthew – London 1836 – (= ser 19th c art & architecture) – 3mf – 9 – mf#4.2.759 – uk Chadwyck [720]

Ancient hebrew names : notes on their significance and historic value / Jeffreys, Letitia D – London: James Nisbet, 1906 [mf ed 1985] – 1mf – 9 – 0-8370-3776-X – (incl ind. pref by archibald henry sayce) – mf#1985-1776 – us ATLA [221]

The ancient hebrew tradition as illustrated by the monuments : a protest against the modern school of old testament criticism = Altisraelitische ueberlieferung in inschriftlicher beleuchtung / Hommel, Fritz – New York: E & JB Young, 1897 [mf ed 1992] – 1mf – 9 – 0-524-04462-7 – (english by edmund mcclure & leonard crossle. incl bibl ref) – mf#1992-0131 – us ATLA [221]

The ancient hebrews : with an introductory essay concerning the world before the flood / Mills, Abraham – New York: AS Barnes, 1874 [mf ed 1992] – 1mf – 9 – 0-524-04582-8 – mf#1992-0170 – us ATLA [221]

Ancient history for colleges and high schools / Myers, Philip Van Ness – Boston: Ginn & Co 1890, c1888 – 1 – us CRL [930]

Ancient history from the monuments : the history of babylonia / Smith, George; ed by Sayce, Archibald Henry – London: SPCK; New York: E & J B Young, [1877?] [mf ed 1990] – 1mf – 9 – 0-7905-3417-7 – mf#1987-3417 – us ATLA [930]

The ancient history of china : to the end of the chou dynasty / Hirth, Friedrich – New York: Columbia University Press, 1911 [mf ed 1995] – (= ser Yale coll) – xx/383p – 1 – 0-524-09228-1 – mf#1995-0228 – us ATLA [951]

The ancient history of the near east : from the earliest times to the battle of salamis / Hall, Harry Reginald – 3rd and rev ed. London: Methuen, 1916 – 7mf – 9 – 0-7905-8305-4 – (incl bibl ref) – mf#1987-6410 – us ATLA [930]

Ancient history of universalism : from the time of the apostles to the fifth general council / Ballou, Hosea – Boston: Universalist Pub House, 1872, c1871 [mf ed 1991] – 1mf – 9 – 0-7905-8759-9 – (notes by alber st john chambre & thomas jefferson sawyer) – mf#1989-1984 – us ATLA [243]

Ancient ideals : a study of intellectual and spiritual growth from early times to the establishment of christianity / Taylor, Henry Osborn – 2nd ed. New York: Macmillan, 1913 [mf ed 1991] – 2v on 3mf – 9 – 0-524-00792-6 – (1st printed 1896) – mf#1990-0224 – us ATLA [930]

Ancient ideals / Taylor, Henry Osborn – New York, NY. v1-2. 1913 – 1r – us UF Libraries [100]

Ancient ideals in modern life / Besant, Annie Wood – Benares: Theosophical Pub Society, 1901 – (= ser Samp: indian books) – us CRL [280]

Ancient india : from the earliest times to the first century ad / Rapson, Edward James – Cambridge: University Press, 1914 – (= ser Samp: indian books) – us CRL [930]

Ancient india : history of ancient india for 1000 years in four volumes, from 900 bc to 100 ad / Shah, Tribhuvandas L – Baroda: Shashikant & Co, 1938-1941 – (= ser Samp: indian books) – us CRL [930]

Ancient india : its language and religions / Oldenberg, Hermann – 2nd ed. Chicago: Open Court Publ, 1898 [mf ed 1995] – (= ser Yale coll) – 110p – 1 – 0-524-09005-X – mf#1995-0005 – us ATLA [930]

Ancient india and indian civilization / Masson-Oursel, Paul et al – London: Kegan Paul, Trench, Truebner & Co, 1934 – (= ser Samp: indian books) – us CRL [930]

Ancient india as described by megasthenaes and arrian : being a translation of the fragments of the indika of megasthenaes collected by dr schwanbeck, and of the first part of the indika of arrian / Megasthenes & Arrian – Calcutta: Thacker, Spink, 1877 [mf ed 1992] – 1mf – 9 – 0-524-03480-X – (trans & ann by john watson mccrindle. with int & notes) – mf#1990-3222 – us ATLA [954]

Ancient india as described by megasthenaes and arrian : being a translation of the fragments of the indika of megasthenes collected by dr schwanbeck and of the first part of the indika of arrian / McCrindle, J W – Calcutta: Chuckervertty, Chatterjee & Co, 1926 – (= ser Samp: indian books) – us CRL [930]

Ancient indian chronology : illustrating some of the most important methods / Sengupta, Prabodh Chandra – Calcutta: University of Calcutta, 1947 – (= ser Samp: indian books) – us CRL [930]

Ancient indian colonies in the far east / Majumdar, Ramesh Chandra – Lahore: Punjab Sanskrit Book Depot, 1927-1937 – (= ser Samp: indian books) – us CRL [930]

Ancient indian fasts and feasts / Mukerji, Abhay Charan – Calcutta: Macmillan & Co, 1932 – (= ser Samp: indian books) – us CRL [390]

Ancient indian life / Raya, Yogesacandra – Calcutta: T R Sen: To be had of Sen Ray & Co, 1948 – (= ser Samp: indian books) – us CRL [930]

Ancient indian numismatics / Bhandarkar, Devadatta Ramakrishna – Calcutta: University of Calcutta, 1921 – (= ser Samp: indian books) – us CRL [730]

Ancient indian theater : an interpretation of bharata's second adhyaya / Mankada, Dolararaya Ram – Vallabh Vidyanagar: Charutar Prakashan, 1950 – (= ser Samp: indian books) – us CRL [790]

Ancient jaffna : being a research into the history of jaffna from very early times to the portuguese period / Rasanayagam, C – [Jaffna?: sn, 1926] – (= ser Samp: indian books) – us CRL [954]

Ancient jerusalem / Merrill, Selah – F.H. Revell, 1908 – 9 – $15.00 – us IRC [939]

Ancient jerusalem : a new investigation into the history, topography and plan of the city, environs, and temple / Thrupp, Joseph Francis – Cambridge: Macmillan, 1855 [mf ed 1993] – 1mf – 9 – 0-524-05700-1 – (incl bibl ref) – mf#1992-0550 – us ATLA [939]

Ancient kaurawa flags : with apologies to a certain "note" to rebut ignorant calumnies, and in the interests of ceylon history / De Soysa, A H T – Colombo: Ceylon Examiner Press, 1930 – 1 – us CRL [954]

Ancient khmer empire / Briggs, Lawrence Palmer – Philadelphia: American Philosophical Soc [mf ed 1989] – (= ser American philosophical society. transactions. new series 41/1) – [ill] 1r with other items – 1 – (with bibl) – mf#mf-10289 seam reel 001/03 [§] – us CRL [930]

Ancient laws and institutes of england : ancient latin version of the anglo-saxon laws – London, 1840 – (= ser Monumenta ecclesiastica anglicana) – €52.00 – ne Slangenburg [242]

Ancient laws and institutes of england / Thorpe, B – 1r – 1 – $50.00 – us Trans-Media [941]

Ancient laws and institutes of wales / Owen, Anevrin – 1r – 1 – $50.00 – us Trans-Media [941]

Ancient laws and institutes of wales – London, 1841 – €84.00 – (with english trans of welsh text) – ne Slangenburg [340]

Ancient laws of ireland – Dublin, London, 1865-1901 – 6v on 2r – 1 – $285.00 – 0-89093-027-9 – us UPA [340]

ANDERSON

The ancient life-history of the earth : a comprehensive outline of the principles and leading facts of palaeontological science / Nicholson, Henry Alleyne – [Edinburgh] 1877 – (= ser 19th c evolution & creation) – 5mf – 9 – mf#1.1.6441 – uk Chadwyck [560]
The ancient liturgy of the church of england : according to the uses of sarum york hereford and bangor and the roman liturgy / Maskell, W – 3rd ed. Oxford, 1882 – €17.00 – ne Slangenburg [242]
Ancient meeting-houses : or, memorial pictures of nonconformity in old london / Pike, Godfrey Holden – London: Passmore & Alabaster, 1870 [mf ed 1991] – 2mf – 9 – 0-524-01238-5 – mf#1990-0377 – us ATLA [240]
Ancient mexico / Peterson, Frederick A – New York, NY. 1962 – 1r – us UF Libraries [972]
Ancient models : or, hints on church-building / Anderson, Charles Henry John, 9th Baronet – [new ed] London 1841 – (= ser 19th c art & architecture) – 3mf – 9 – mf#4.1.455 – uk Chadwyck [720]
Ancient monuments and holy writ / Walsh, William Pakenham – Dublin, Ireland. 1878 – 1r – us UF Libraries [240]
The ancient monuments, temples and sculptures of india / Burgess, James – London 1897 – (= ser 19th c art & architecture) – 9mf – 9 – mf#4.2.1186 – uk Chadwyck [720]
Ancient Order of United Workmen *see* Proceedings of the session of the grand lodge of wisconsin, a o u w
Ancient Order of United Workmen. Delphos Lodge No. 129 *see* Minute books
Ancient oriental cylinder and other seals : with a description of the collection of mrs william h moore / Eisen, Gustavus A – 1940 – 9 – $6.00f – 0-226-19527-9 – us Oriental [930]
Ancient pagan and modern christian symbolism / Inman, Thomas & Newton, John – 2nd rev enl ed. New York: P Eckler, 1915 [mf ed 1992] – 1mf – 9 – 0-524-01964-9 – (essay by john newton) – mf#1990-2755 – us ATLA [700]
Ancient poems, ballads and songs of the peasantry of england / ed by Bell, Robert – London: J W Parker 1857 [mf ed Bloomington IN: Indiana Uni Lib, Preservation Dept 1984] – 252p on 1r – 1 – us Indiana Preservation [810]
Ancient practice and proposed revival of diocesan synods in england / Pound, William – London, England. 1851 – 1r – us UF Libraries [240]
Ancient records of egypt : historical documents from the earliest times to the persian conquest / ed by Breasted, James Henry – Chicago: University of Chicago Press, 1906-07 [mf ed 1990] – 5v on 5mf – 9 – 0-8370-1660-6 – (incl bibl ref) – mf#1987-6090 – us ATLA [930]
Ancient religion and modern thought / Lilly, William Samuel – London: Chapman & Hall, 1884 [mf ed 1985] – 1mf – 9 – 0-8370-4130-9 – (incl bibl ref & ind) – mf#1985-2130 – us ATLA [230]
Ancient reliques : or, delineations of monastic, castellated, and domestic architecture / Storer, James Sargant – London 1812-13 – (= ser 19th c art & architecture) – 6mf – 9 – mf#4.2.1376 – uk Chadwyck [720]
Ancient rome / Lanciani, Rodolfo – Houghton, Mifflin. 1890 – 9 – $15.00 – us IRC [930]
Ancient rome and modern america : a comparative study of morals and manners / Ferrero, Guglielmo – New York: Putnam, c1914 [mf ed 1991] – 1mf – 9 – 0-524-01110-9 – mf#1990-0324 – us ATLA [340]
Ancient sacrifice / Newman, Francis William – London, England. 1874 – 1r – us UF Libraries [290]
Ancient science : or, secrets of pyramids, walls and temples: to which is added a short review of piazzi smith's "our inheritance in the great pyramid" / MacDonald, Dugald – Montreal: Gazette Printing Co, 1901 [mf ed 1985] – 1mf – 9 – mf#SEM105P463 – cn Bibl Nat [510]
The ancient scriptures and the modern jew / Baron, David – London: Hodder and Stoughton, 1900 – 1mf – 9 – 0-7905-0665-3 – (incl bibl ref) – mf#1987-0665 – us ATLA [220]
Ancient sepulchral monuments... : over six hundred examples from various countries / Brindley, William & Weatherley, William Samuel – London 1887 – (= ser 19th c art & architecture) – 8mf – 9 – mf#4.2.1435 – uk Chadwyck [720]
Ancient sermons for modern times / Asterius of Amasea, Bishop of Amasea – New York: Pilgrim Press, c1904 [mf ed 1991] – 1mf – 9 – 0-7905-7678-3 – (english by galusha anderson & edgar johnson goodspeed) – mf#1989-0903 – us ATLA [240]
Ancient ships / Torr, Cecil – Cambridge: University Press 1894 [mf ed 1987] – 1r [ill] – 1 – (ill in plates 1 & 7 by j a burt and those in 8 by h w bennett. filmed with: german psychology of today / ribat, t) – mf#1974 – us Wisconsin U Libr [623]
Ancient slavic manuscripts from the moscow state university library – 183mf coll – 9 – $1,500.00 coll – (coll contains 21 ancient mss from the early 1200's to the early 1500's. enquire for individual listings. comes with printed guide) – us UMI ProQuest [090]
The ancient strait at nipissing / Taylor, Frank Bursley – [S.l: s.n, 1893?] [mf ed 1986] – 1mf – 9 – 0-665-58328-1 – mf#58328 – cn Canadiana [550]
Ancient symbol worship : influence of the phallic idea in the religions of antiquity / Westropp, Hodder Michael & Wake, Charles Staniland – 2nd ed. New York: JW Bouton, 1875 [mf ed 1992] – 1mf – 9 – 0-524-02623-8 – (incl bibl ref. int, add notes & app by alexander wilder) – mf#1990-3073 – us ATLA [290]
Ancient symbolism among the chinese / Edkins, Joseph – London: Treubner; Shanghai: Society for the Diffusion of Christian and General Knowledge among the Chinese, 1889 [mf ed 1995] – (= ser Yale coll) – 26p – 1 – 0-524-09004-1 – mf#1995-0004 – us ATLA [390]
Ancient syriac documents relative to the earliest establishment of christianity in edessa and the neighbouring countries : from the year after our lord's ascension to the beginning of the 4th century / ed by Cureton, William – London: Williams & Norgate, 1864 [mf ed 1989] – 1mf – 9 – 0-7905-4220-X – (text in syriac, trans in english, french & latin) – mf#1988-0220 – us ATLA [240]
The ancient syriac version of the epistles of saint ignatius to saint polycarp, the ephesians, and the romans : together with extracts from his epistles, collected from the writings of severus of antioch, timotheus of alexandria, and others / Cureton, William – London: Rivingtons, 1845 – 1r – 1 – 0-8370-0354-7 – mf#1984-B069 – us ATLA [227]
Ancient times – 1973 mar-1982 winter – 1 – mf#641442 – us WHS [071]
Ancient times : a history of the early world / Breasted, James Henry – Boston: Ginn, c1916 [mf ed 1991] – 2mf – 9 – 0-524-01880-4 – (incl bibl ref) – mf#1990-0507 – us ATLA [930]
Ancient towers and doorways / Galletly, Alexander – London 1896 – (= ser 19th c art & architecture) – 4mf – 9 – mf#4.2.1027 – uk Chadwyck [720]
Ancient, William Johnson *see* The cross
The ancient wisdom : an outline of theosophical teachings / Besant, Annie Wood – London: Theosophical Pub Soc, 1897 [mf ed 1992] – 1mf – 9 – 0-524-02712-9 – mf#1990-3115 – us ATLA [290]
The ancient wisdom : an outline of theosophical teachings / Besant, Annie Wood – London: Theosophical Pub Society, 1910 – (= ser Samp: indian books) – us CRL [180]
The ancient world and christianity = L'ancien monde et le christianisme / Pressense, Edmond de – London: Hodder & Stoughton, 1888 [mf ed 1989] – 2mf – 9 – 0-7905-1498-2 – (incl bibl ref & ind) – mf#1987-1498 – us ATLA [230]
Ancient York Masons. Palmetto Grand Lodge *see* Minutes of the annual communication
Ancient-babylonian temple records in the columbia university library / ed by Arnold, William R – New York: [s.n.], 1896 [mf ed 1986] – 1mf – 9 – 0-8370-8480-6 – (text in akkadian. int in english) – mf#1986-2480 – us ATLA [470]
Ancizar, Manuel *see* Editoriales del neo-granadino
Ancla para tu voz / Alvarez Puga, Miguel – Camaguey, Cuba. 1953 – 1r – us UF Libraries [972]
Ancona, J S *see* Hints for the valuation of ecclesiastical and other property
Ancsa – v1 n1-3 [1978 nov-1979 jan] – 1 – mf#858261 – us WHS [071]
Ancsa news – v1 n4-v5 n2 [1979 feb-1982 mar/apr] – 1 – mf#838352 – us WHS [071]
And gazelles leaping / Ghose, Sudhindra Nath – London: Michael Joseph, 1949 – (= ser Samp: indian books) – (with illustrations by arnakali e carlile) – us CRL [890]
And many believed / Ramquist, Grace Bess Chapman – Kansas City, MO. 1951 – 1r – us UF Libraries [972]
and other papers upon the chinese philosophy in japan *see* Japanese philosopher
...And so we played : memory, place and the brooklyn dodgers / Hoyte, Thor A – 1998 – 1mf – 9 – $6.00 – mf#PE 4002 – us Kinesology [790]
A and t register – [1985 feb-1986 dec 5]; 1989 jan 13, 20, feb 3; 1968 sep 27-1973 feb 23; 1973 mar 2-1975 may 2; 1975 aug 22-1977 dec 9; 1978 jan 10-1979 dec 11; 1980 jan 11-1982 apr 30, jun30, aug 31-dec 7; 1983-84; 1987 jan 16-1988 nov 18; 1989 jan 13-feb 3; 1997 nov13, dec 4 – 1 – mf#1132063 – us WHS [071]
And the years roll by / Solomon, Frank H – Cape Town, South Africa. 1953 – 1r – us UF Libraries [960]
And thou shalt teach them / Eldridge, Paul – New York, NY. 1947 – 1r – us UF Libraries [025]
Anda, Diane de *see*
– Journal of ethnic and cultural diversity in social work
– Journal of multicultural social work
Andacht- und gebetbuch (cima46) : farbmikrofiche-edition der handschrift hannover, kestner-museum, inv wm ue 22 – (mf ed 1998) – (= ser Codices illuminati medii aevi (cima) 46) – 51p on 3 color mf – 15 – €245.00 – 3-89219-046-1 – (int by hans-walter stork. description by helga lengenfelder) – gw Lengenfelder [090]
Andachten / Jordan, Wilhelm – Frankfurt/M: W Jordan, 1877 [mf ed 1995] – 237p – 1 – mf#8796 – us Wisconsin U Libr [810]
Andachtsbuch der orthodox-katholischen... : kirche des morgenlands – Berlin, 1895 – €21.00 – (in german and slavic) – ne Slangenburg [243]
Andachts-buechlein zu ehren der elf martyrer-schaaren aus der gesellschaft jesu : die pius 9. am st petrustage 1867 selig gesprochen / Weninger, Francis Xavier – Cincinnati OH: Benziger 1868 [mf ed 1986] – 1mf – 9 – 0-8370-6848-7 – mf#1986-0848 – us ATLA [241]
Andachtsbuechlein zu ehren der sechsundzwanzig [sic] japanesischen martyrer / Weninger, Francis Xavier – Cincinnati: Fredewest & Donnerberger, 1863 [mf ed 1986] – 1mf – 9 – 0-8370-6850-9 – mf#1986-0850 – us ATLA [241]
Andachtsbuechlein zu ehren der seligen vierzig martirer der gesellschaft jesu, ignatius azevedo und seiner gefaehrten / Weninger, Francis Xavier – Cincinnati: "Wahrheitsfreundes", 1855 [mf ed 1986] – 1mf – 9 – 0-8370-6849-5 – mf#1986-0849 – us ATLA [241]
Andachtsbuechlein zu ehren der zwei seligen, johannes und andreas : priester und maertyrer der gesellschaft jesu / Weninger, Francis Xavier – St Louis, MO: "Tageschronik", 1854 [mf ed 1986] – 1mf – 9 – 0-8370-6851-7 – mf#1986-0851 – us ATLA [241]
Andachtsbuechlein zu ehren des seligen peter claver : priester der gesellschaft jesu und apostel von central-amerika / Weninger, Francis Xavier – Cincinnati: "Wahrheits-Freundes", 1852 [mf ed 1986] – 1mf – 9 – 0-8370-6852-5 – mf#1986-0852 – us ATLA [241]
Andaechtige gedancken zur vermeidung des boesen und vollbringung des guten : aus dem buch weeg des ewigen lebens r p antonii sucquet soc iesugezogen – Wienn: [Gedruckt bey Leopold Voigt], 1681 – 2mf – 9 – mf#O-1921 – ne IDC [090]
Andala, Ruardus *see* Examen ethicae clar
Andancas e tropecos / Luz, Anysio Cerqueira – Rio de Janeiro, Brazil. 1969 – 1r – us UF Libraries [972]
Andanzas de hernan cortes / Valle-Arizpe, Artemio del – Mexico: Editorial Diana, 1978 – sp Bibl Santa Ana [910]
Andanzas y malandanzas / Rivas Bonilla, Alberto – San Salvador, El Salvador. 1955 – 1r – us UF Libraries [972]
Andanzas y observaciones / Rodriguez, Miguel – Ciudad Trujillo, Dominican Republic. 1944 – 1r – us UF Libraries [972]
Der ander theil teutscher lieder mit 5 stimmen / Lasso, Orlando di – 1573 – (= ser Mssa) – 2mf – 9 – €60.00 – mfchl 296 – gw Fischer [780]
Anderdon, William Henry *see*
– Contending for the faith
– Fasti apostolici
– Letter to the parishioners of st margaret's leicester
Das andere deutschland : La otra alemania = La otra alemania – Buenos Aires [RA], Montevideo [ROU], 1938 may-dec [gaps], 1939-1949 10 jan – 2r – 1 – (publ in montevideo, uruguay, dec 15 1943-apr 25 1944 due to ban) – gw Misc Inst [079]
Das andere deutschland *see* Der pazifist
Die andere zeitung – Hamburg DE, 1956 5 jan-1965 23 dec – 7r – 1 – (filmed by misc inst: 1955 12 may-29 dec, 1966 6 jan-22 dec, 1967 5 jan-1969 27 feb) – gw Mikrofilm; gw Misc Inst [074]
Andernacher buergerblatt fuer stadt und umgegend – Andernach, Neuwied DE, 1837 5 jan-26 jun, 1855-62 – 1 – (later: andernacher buergerblatt) – gw Misc Inst [350]
Anders, Artur *see* Wien
Anders ist der neue tag : gedichte / Becher, Johannes Robert; ed by Berger, Uwe – Berlin: Aufbau-Verlag, 1960 [mf ed 1995] – (= ser Taschenbuch des aufbau-verlages 87) – 193p – 1 – mf#8973 – us Wisconsin U Libr [810]
Anders, Rainer-Elk *see* Globalization and the russian far east prospects for intergration
Andersen Air Force Base [Guam] *see* Tropic topics
Andersen, Hans Christian *see*
– Fairy tales and stories
– In spain
– Mit livs eventyr
– Sammtliche marchen
Andersen, Joachim *see* Ballade et danse des sylphes, op 5
Andersen, Rasmus *see*
– Daabsminder fra herrens tjeneste i kirke og mission
– Den evangelisk-lutherske kirkes historie i amerika
Andersen, Steven J *see* The effects of acquaintance rape prevention programming on male athletes' sexual and dating attitudes
A andersen's (eines gebornen daenen) kleine fuss-reise durch einen theil von seeland / Feldborg, Andreas A – Weimar 1807 [mf ed Hildesheim 1995-98] – 1v on 1mf – 9 – €40.00 – 3-487-26553-2 – (trans fr english) – gw Olms [914]
Anderson, A H *see*
– Brief sketch of british honduras
– A narrative of the british embassy to china, in the years 1792, 1793 and 1794
Anderson, Aeneas *see* Relation de l'ambassade du lord macartney a la chine, dans les annees 1792, 1793 et 1794
Anderson, Alexander *see* Notebooks
Anderson, Alexander Caulfield *see*
– The dominion at the west
– Hand-book and map to the gold region of frazer's and thompson's rivers
– Notes on the indian tribes of british north america and the northwest coast
Anderson, Alice Jean *see* Study of tempo rubato
Anderson, Amy *see* The effect of a wilderness therapy program on youth-at-risk, as measured by locus of control and self-concept
Anderson baptist church – Texas, 1844-1911 – 1 – $53.55 – us Southern Baptist [242]
Anderson Bros. Charlottesville, VA *see* Law examinations, embracing examination papers from the year 1869 to 1894
Anderson, C W *see* Pardon case file n39-242, wichita iww case defendents
Anderson, Chandler P *see* Diaries
Anderson, Charles *see*
– New readings of old parables
– Outlines of a plan submitted to her majesty's government
– A true and impartial account of the actions fought at chippawa and lundy's lane during the last war with the united states
Anderson, Charles Henry John, 9th Baronet *see* Ancient models
Anderson, Charles Loftus Grant *see*
– Old panama and castilla del oro
Anderson, Charles Palmerston *see* Letters to laymen
Anderson, Christopher *see*
– The annals of the english bible
– An appeal by a lancashire liberal against the unjust operation of the irish land act
Anderson, Christopher. *see* Historical sketches of the ancient native irish and their descendants
Anderson county, sorth carolina : hopewell baptist church – 1457p – 1 – $65.57 – (church records 1868-90, 1963-88. wmu 1906-26, sunday school 1958-60, history 1803-1981, scrapbooks 1893-1987) – us Southern Baptist [242]
Anderson, Daphne L *see* The role of external non-rigid ankle bracing in the prevention of inversion injuries
Anderson, David R *see* The ombudsman
Anderson, Duncan *see*
– Lays of canada
– The newer districts of ontario
– Scottish folk-lore
Anderson family courier – v1 n1-v2 n4 [1985 apr-1987 jan] – 1 – mf#930370 – us WHS [071]
Anderson family papers – 1802-1905. In Kansas State Historical Society. Guide – 1 – us Kansas [920]
Anderson first baptist church – Anderson Co, SC. 1637p. 1869-1990 (incomplete) – 1 – $73.67 – us Southern Baptist [242]
Anderson first baptist church – Anderson, MO. 1852-1907, 1924-32, apr 1950-61 – 1 – $29.07 – us Southern Baptist [242]
Anderson, Florence Bennett *see* Religious cults associated with the amazons
Anderson, Frances *see* Colonial poems
Anderson, Francine M *see* Effect of exercise on bone mineral density of the forearm in premenarcheal girls
Anderson, Frank Maloy *see* Handbook for the diplomatic history of europe, asia and africa
Anderson, Gene *see* Railroad car design and pullman car data
Anderson, Herbert *see* Among the lushais
Anderson, Hugh *see* The annals of the english bible
Anderson, I M *see* Our first decade in china, 1905-1915
Anderson Imbert, Enrique *see*
– Critica interna
– Estudios sobre escritores de america
– Literatura hispanoamericana

Anderson, J *see*
- Acheen, and the ports on the north and east coasts of sumatra
- Correspondence for the introduction of cochineal insects from america...
- English intercourse with siam in the seventeenth century
- Mission to the east coast of sumatra, in m.dccc.xxiii...

Anderson, Jack *see* Mccarthy
Anderson, James *see* The improvement of agriculture
Anderson, James Drummond *see* The peoples of india
Anderson, James S. M *see* The history of the church of england in the colonies and foreign dependencies of the british empire
Anderson, James Stuart Murray *see* The history of the church of england
Anderson, John *see*
- The course of creation
- Mission to the east coast of sumatra, in 1823

Anderson, John Corbet *see* Old testament and monumental coincidences
Anderson, Laura J *see* Impact of training patterns on incidence of illness and injury
Anderson, Lewis Flint *see* The anglo-saxon scop
Anderson, Lillian S *see* Up and down the virgin islands
Anderson, Lynn S *see* The effects of a miniumum impact camping slide-tape program on wilderness visitors' awareness of minimum impact camping
Anderson, Mary *see* History of the efts summary
Anderson, Maxwell *see* Mary of scotland
Anderson, Michael [comp] *see* 1851 census enumerators' returns
Anderson, Nels *see* The hobo
Anderson, Paul Gerard *see* Investigation into the effect of race and politics on the development of south african sport (1970-1979)
Anderson, Paula *see* University of wisconsin-la crosse adult fitness/cardiac rehabilitation graduate program assessment
Anderson, Philip, d. 1857 *see* English in western india
Anderson, Rasmus Bjoern *see* Norse mythology
Anderson, Rasmus Bjorn *see* Viking tales of the north
Anderson, Robert *see*
- The bible and modern criticism
- The buddha of christendom
- Christianized rationalism and the higher criticism
- Daniel in the critics' den
- A doubter's doubts about science and religion
- Fighting the mill creeks
- The gospel and its ministry
- Pseudo-criticism
- Rudiments of tamul grammar
- The silence of god

Anderson, Robert et al *see* The church, the people, and the age
Anderson, Robert Phillips *see* The story of christian endeavor
Anderson, Robert Stuart Guthrie *see* Kirk folk
Anderson, Robert Woodruff *see* Tea and sympathy
Anderson, Rufus *see*
- Foreign missions
- History of the mission of the american board of commissioners for foreign missions to the sandwich islands
- History of the missions of the american board of commissioners for foreign missions in india
- History of the missions of the american board of commissioners for foreign missions to the oriental churches
- Observations upon the peloponnesus and greek islands

Anderson, Samuel Gilmore *see* Woman's sphere and influence
Anderson, Sara May *see* Korean folk songs
Anderson, Thomas Fenwick *see*
- Nova scotia, the land of evangeline and the tourist's paradise
- Vacation days in nova scotia

[Anderson-] valley news – CA. 1910-68 (wkly) – 24r – 1 – $1440.00 – mf#B02016 – us Library Micro [071]
Anderson, W *see* The scottish nation
Anderson, Wayne F *see* Income tax administration in the state of israel
Anderson, William *see*
- Can extreme voluntaryism be made an open question?
- Descriptive and historical catalogue...japanese and chinese
- Japanese wood engravings
- No priests
- On the conversion of heat into work
- Opening of the case
- The pictorial arts of japan
- Regeneration

Anderson, William B *see* Far north in india
Anderson, William G *see*
- The enrollment and persistence of african-american doctoral students in physical education and related disciplines
- A multi-case study of first year athletic trainers at the high school level

Anderson, William Henry *see* Luther
Anderson, William James *see*
- Architecture of the renaissance in italy
- The architecture of the renaissance in italy
- The archives of canada
- Canadian history
- Canadian history and biography
- Education and pictou academy
- "Evangeline" and "the archives of nova scotia"
- Holiwell's tourist guide to quebec
- Two chapters in the life of f m, h r h edward, duke of kent
- The valley of the chaudiere

Anderson, William James [comp] *see* The gold fields of the world
Anderson, William L *see* A history of the descendents of jacob and maria eva harshbarger of switzerland
Anderson's dictionary of law – Chicago: T H Flood & Co – 12mf – 9 – $18.00 – (judicial definitions, words, phrases, maxims, principles of law comprising a dictionary and compendium of american and english jurisprudence) – mf#LLMC 87-301 – us LLMC [340]
Andersonville baptist church – Anderson Co, SC. 306p. 1968-82 – 1 – $13.77 – us Southern Baptist [242]
Andersson, Charles J *see* Reisen in suedwest-afrika bis zum see ngami in den jahren 1850 bis 1854
Andersson, Charles John *see*
- Lake ngami
- Notes of travel in south africa
- Okavango river

Andersson, Efraim *see* Messianic popular movements in the lower congo
Andersson, L G *see* List of reptiles and batrachians collected by the swedish zoological expedition to egypt, the sudan and the sinaitic peninsula
Andersson, Otto *see* Musik och musikinstrument
Anderton, Basil *see* Report on the annual meeting of the library association, held in belfast, 1894
Los andes al amazonas / Aguirre Acha, Jose – La Paz, Bolivia. 1927 – 1r – us UF Libraries [972]
Andeutungen zu ausfluegen : von einem halben tag bis zu vier tagen mittelst der beiden von wien auslaufenden eisenbahnen / Weidmann, Franz C – Wien 1842 [mf ed Hildesheim: 1995-98] – (= ser Fbc) – 131p on 1mf – 9 – €40.00 – 3-487-29439-7 – gw Olms [380]
Andhra janatha – Hyderabad, India. 1962-Jun 1973 – 24r – 1 – (in telugu language) – us L of C Photodup [079]
Andhra jyoti – Vijayawada, India. Jul-Sept 1966 – 1r – 1 – (in telugu language) – us L of C Photodup [079]
Andhra patrika – Madras, India. 26 Apr 1944-Aug 1983; May 1991 – 128r – 1 – (telugu language) – us L of C Photodup [079]
Andhra prabha – Madras, India. 18 Aug 1950-1958 – 18r – 1 – (telugu language) – us L of C Photodup [079]
Andhra prabha – Vijayawada, India. Oct 1964-1976; May 1977-1990 – 96r – 1 – (telugu language) – us L of C Photodup [079]
Andhra Pradesh (India) *see* Business rules and secretariat instructions
Andino, Manuel *see*
- Mirando vivir
- Obra del gobierno del doctor quinonez-molina
- Vocacion de marino

Andishah – Tehran. shumarah-'i 1-5. farvardin 1358-urdibihish 1359 [mar 1979-apr 1980] – 1r – 1 – $53.00 – us MEDOC [079]
Andishah-'i azad – Tehran: Kanun-i Nivisandagan-i Iran, [1980-]. dawrah-'i jadid, sal-i 1, shumarah-'i 1-6. 30 bahman 1358-15 khurdad 1359 [19 feb-30 may 1980] – 1r – 1 – $53.00 – us MEDOC [079]
Andizhanskaia pravda – Andizhan, Uzbekistan 1974-88 [mf ed Norman Ross] – 1r – 1 – mf#nrp-186 – us UMI ProQuest [077]
Andosilla Salazar, V *see* Libro...en que se prueba...con claridad el mal...que corre por espana...ser nuevo
Andover 1647-1849 – Oxford MA (mf ed 1994) – (= ser Massachusetts vital record transcripts to 1850) – 1v on 30mf – 9 – 0-87623-200-4 – (mf 1t-2t: births 1651-1704. mf 2t: deaths 1650-1700; marriages 1647-1700. mf 3t-6t: births 1704-57. mf 6t-8t: deaths 1701-64. mf 8t-10t: marriages 1701-99. mf 10t-11t: intentions 1704-38. mf 11t-14t: births 1757-1801. mf 14t: deaths 1764-99. mf 14t-16t: intentions 1738-1800. mf 17t-19t: births 1800-44. mf 19t-21t: deaths 1800-44. mf 21t-23t: marriages 1800-43. mf 23t-25t: intentions 1800-49. mf 25t-27t: births 1843-49. mf 27t-28t: marriages 1844-49; mf 28t-29t: deaths 1844-49. mf 30t: out-of-town marriages 1674-1799; births 1766-1843; marriages 1827-1841; deaths 1839) – us Archive [978]
Andover 1647-1905 – Oxford, MA (mf ed 1998) – (= ser Massachusetts vital records) – 242mf – 9 – 0-87623-397-3 – (mf 1-9: births & index 1649-1844. mf 10-14: marriages & index 1647-1844. mf 15-19: deaths & index 1650-1844. mf 20-32: vital records: 1647-1803. mf 33-34: out-of-town marriges 1668-1799; mf 35-46: births 1649-1844. mf 40,46-47: deaths 1701-1844. mf 48-53,58: births & deaths 1800-44. mf 54-61: marriages & intentions 1701-1850. mf 62-83: land records 1667-1824. mf 84-85: town records 1660-1707. mf 86-116: town meetings 1709-1855. mf 117-165: tax assessments 1679-1851. mf 166-168: warnings out 1790-93. mf 169-171: pew deeds 1793-1827. mf 172-193: pauper records 1815-1916. mf 193-194: voters 1877-84. mf 195-201: marriage intentions 1850-98. mf 201-203: intentions index 1850-98. mf 204-208: vital records 1843-55. mf 209-216: births 1843-81. mf 217-223: marriages 1853-96. mf 224-233: deaths 1844-1914. mf 234-237: births 1882-1905. mf 238-241: intentions & index 1898-1908. mf 241-242: marriages 1897-1905, 1812-15) – us Archive [978]
Andover advertiser and north west hants gazette – Andover, England 1 jan 1858-29 dec 1871, 4 jan 1884-26 jun 1987 [mf 1858-1950, 1986-] – 1 – (cont by: andover weekend advertiser [3 jul 1987-20 apr 1990]; andover advertiser [27 apr 1990-]; (ludgershall & tidworth ed) [5 feb 1988-20 apr 1990]; (whitchurch & overton ed) [27 apr 1990-8 mar 1991]) – uk British Libr Newspaper [072]
Andover advertiser and north west hants gazette [stockbridge ed] – [SW England] Andover, Hampshire 27 apr 1990-8 mar 1991 [mf ed 2004] – 1 – uk Newsplan [072]
Andover advertiser [stockbridge ed] – Andover, England 27 apr 1990-8 mar 1991 – 1 – uk British Libr Newspaper [072]
The andover case : with an introductory historical statement, a careful summary of the arguments of the respondent professors, and the full text of the arguments of the complainants and their counsel – Boston: Stanley and Usher, 1887 – 3mf – 9 – 0-524-07347-3 – mf#1990-5384 – us ATLA [240]
The andover controversy / Andover Theological Seminary. Trustees – [S.l: s.n., 1886?] – 1mf – 9 – 0-524-03312-9 – mf#1990-4672 – us ATLA [240]
The andover fuss, or, dr. woods versus dr. dana, on the imputation of heresy against professor park respecting the doctrine of original sin / Allen, George – Boston: Tappan and Whittemore, 1853 – 1mf – 9 – 0-7905-7915-4 – mf#1989-1140 – us ATLA [240]
The andover heresy : in the matter of the complaint against egbert c smyth and others... / Smyth, Egbert Coffin – Boston: Cupples, Upham, 1887 – 1mf – 9 – 0-7905-6504-8 – mf#1988-2504 – us ATLA [240]
Andover review : a religious and theological monthly – La Salle IL 1884-93 – 1 – mf#4129 – us UMI ProQuest [200]
Andover Theological Seminary. Trustees *see* The andover controversy
Andrada, Bonifacio Jose Tamm De *see* Parlamentarismo e a evolucao brasileira
Andrade, Almir De *see*
- Contribuicao a historia administrativa do brasil
- Forca, cultura e liberdade

Andrade, Carlos Drummond De *see* Confissoes de minas
Andrade Coello, Alejandro *see* Tres poetas de la musica
Andrade, Francisco Alves De *see* Renato braga, in memoriam
Andrade, Ignacio *see* Por que triunfo la revolucion restauradora?
Andrade, J I de *see* Cartas escriptas da india e da china nos annos de 1815 a 1835...
Andrade, Jose Hermogenes De *see* Organizacao social e politica brasileira
Andrade, Lopes De *see* Introducao a sociologia das secas
Andrade, Manuel Jose *see*
- Aguacatec texts (phrases and sentences)
- Folklore de la republica dominicana

Andrade, Mario De *see*
- Amar
- Aspectos da literatura brasileira
- Liberte pour l'angola
- Movimento modernista

Andrade, Nuno Ferreira De *see* Contos e cronicas
Andrade, Olimpio De Souza *see* Historia e interpretacao de 'os sertoes'
Andrade, Raul *see* Internacional negra en colombia
Andrade, Rodrigo Melo Franco *see* Rio-branco e gastao da cunha
Andrae, Tor *see* Die person muhammeds in lehre und glauben seiner gemeinde
Andre, E *see* Species des hymenopteres d'europe and d'algerie...
Andre ernest modeste gretry : complete collection of works / ed by Gevaert, F A et al – Leipzig: Breitkopf & Haertel. v1-49. 1884-1936 – 11 – $585.00 set – us Univ Music [780]
Andre gide / Souday, Paul – Paris, France. 1927 – 1r – us UF Libraries [440]
Andre, J Lewis *see* Chests, chairs, cabinets and old english woodwork
Andre, Johann Anton *see* Die weiber von weinberg
Andre, Le Chapelain *see* Art of courtly love
Andre, Louis Edouard Tony *see*
- Les apocryphes de l'ancien testament
- L'eglise evangelique reformee de florence
- L'esclavage chez les anciens hebreux
- Le prophete agge

Andre, Louis Edward Tony *see* Etat critique du texte d'agee
Andre reboucas atraves de sua auto-biografia / Verissimo, Ignacio Jose – Rio de Janeiro, Brazil. 1939 – 1r – us UF Libraries [972]
Andre, Valere *see* Synopsis juris canonici per erotemata digesti et enucleati.
Andrea cesalpino of arezzo *see* The circulation of the blood
Andrea da Barberino *see*
- Guerino detto il meschino
- Guerino il meschino

Andrea del sarto / Guinness, H – London 1899 – (= ser 19th c art & architecture) – 3mf – 9 – mf#4.2.1506 – uk Chadwyck [750]
Andrea delfin : novelle / Heyse, Paul – Leipzig: Insel-Verlag [19–?] [mf ed 1990] – 1r – 1 – (filmed with: der wollmarkt / h clauren) – mf#2725p – us Wisconsin U Libr [830]
[Andreae, A J] *see* Acta et scripta theologorvm vvirtembergensivm
Andreae alciati emblemata cum commentariis claudii minois... / Alciato, Andrea – Patavij: Typis Pauli Frambotti, 1661 – 14mf – 9 – mf#O-117 – ne IDC [090]
Andreae alciati emblematum libellus / Alciato, Andrea – Parisiis: Excudebat Christianus Wechelus, sub scuto Basileiensi,in vico Iacobaeo, 1534 – 2mf – 9 – mf#O-107 – ne IDC [090]
Andreae d A, J *see*
- Bericht von der vbiquitet
- Ein christliche predigt, von christlicher einigkeit der theologen augspuergischer confession
- Colloquium de peccato originis inter d iacobum andreae, et m matthiam flaccivm illyricum
- Drey vnd dreissig predigen von den fuernembsten spaltungen in der christlichen religion
- Epitome colloqvii montisbelgartensis inter d iacobvm andreae, et d theodorum bezam
- Fuenff predigen
- Gruendtlicher bericht auff johann sturmij
- Methodvs concionandi
- Passional buechlein
- Sechs christlicher predig ueber den ein vnnd fuenffzigsten psalmen dauids
- Sechs christlicher predig von den spaltungen
- Vier christliche predigten vom wucher darinnen neben der summarischen ausslegung ettlicher euangelien im aduent
- Zehen predig von den sechs hauptstucken christlicher lehr

[Andreae d A, J] *see*
- Acta colloquij montis belligartensis
- Acta et scripta theologorvm vvirtembergensivm, et patriarchae constantinopolitani d hieremiae
- Gruendtlicher warhafftiger vnd bestendiger bericht

Andreae werckmeisters...cribvm mvsicvm : oder musikalisches seib, darinen einige maengel eines halb gelehrten componisten vorgestellet... / Werckmeister, Andreas – Quedlinburg, Leipzig: T P Calvisius 1700 [mf ed 198-?] – 2mf – 9 – mf#fiche 859 – us Sibley [780]
Andreas a s Victore *see* Expositio hystorica in librum regum
Andreas auf der fahrt : erzaehlung / Koll, Kilian – Muenchen: A Langen/G Mueller, c1938 [mf ed 1989] – 1 – mf#70 – us Wisconsin U Libr [880]
Andreas, F C *see* The book of the mainyo-i-khard
Andreas, Fred *see* Das vollkommene verbrechen
Andreas gryphius lateinische und deutsche jugenddichtungen : ergaenzungsband mit einer bibliographie der gryphius-drucke / Gryphius, Andreas; ed by Wentzlaff-Eggebert, Friedrich-Wilhelm – Leipzig: K W Hiersemann, 1938 [mf ed 1993] – (= ser Blvs 287) – 1 – mf#8470 reel 57 – us Wisconsin U Libr [810]
Andreas gryphius lustspiele / ed by Palm, Hermann – Stuttgart: Litterarischer Verein, 1878 (Tuebingen: H Laupp) [mf ed 1993] – (= ser Blvs 138) – 584p – 1 – (incl bibl ref) – mf#8470 reel 29 – us Wisconsin U Libr [820]
Andreas gryphius lyrische gedichte / ed by Palm, Hermann – Stuttgart: Litterarischer Verein, 1884 (Tuebingen: H Laupp) [mf ed 1993] – (= ser Blvs 171) – 610p – 1 – (incl bibl ref) – mf#8470 reel 35 – us Wisconsin U Libr [810]

Andreas gryphius trauerspiele / ed by Palm, Hermann – Stuttgart: Litterarischer Verein in Stuttgart, 1882 (Tuebingen: H Laupp) [mf ed 1993] – (= ser Blvs 162) – 814p – 1 – (early modern german text. some trans fr latin and dutch) – mf#8470 reel 33 – us Wisconsin U Libr [820]

Andreas gryphius und seine herodes-epen : ein beitrag zur charakteristik des barockstils / Gnerich, Ernst – Leipzig: M Hesse, 1906 [mf ed 1992] – (= ser Breslauer beitraege zur literaturgeschichte 2) – xvi/229p – 1 – (incl repr of gryphius' herodis furiae & rachelis lachrymae, carmine heroico, cantatae, ploratae (glogoviae [1634]) and dei vindicis impetus et herodis interitus (dantisci [1635]). incl bibl ref) – mf#8014 reel 1 – us Wisconsin U Libr [430]

Andreas hofer im liede / ed by Frankl, August – Innsbruck: Wagnerische Universitaetas-Buchhandlung, 1884 [mf ed 1993] – xii/171p/[1pl] – 1 – (incl bibl ref) – mf#8576 – us Wisconsin U Libr [430]

Andreas osang : erzaehlung / Leppa, Karl Franz – Karlsbad: Kraft 1943 [mf ed 1990] – 1r – 1 – (filmed with: hengst maestoso austria / arthur-heinz lehmann) – mf#2819p – us Wisconsin U Libr [830]

Andreas osiander : leben und ausgewaehlte schriften / Moeller, Wilhelm – Elberfeld: RL Friderichs, 1870 [mf ed 1991] – (= ser Leben und ausgewaehlte schriften der vaeter und begruender der lutherischen kirche 5) – 2mf – 9 – 0-524-00577-X – (incl bibl ref) – mf#1990-0077 – us ATLA [242]

Andreas-Salome, Lou *see*
- Im zwischenland
- Ma
- Menschenkinder
- Ruth

Andree, Fritz *see* Wirkungs- und erinnerungsstaetten des dichters hoffmann von fallersleben in wort und bild

Andree hofer : geschichtliches trauerspiel in fuenf aufzuegen / Auerbach, Berthold – Leipzig: G Wigand, 1850 [mf ed 1988] – 165p – 1 – mf#6968 – us Wisconsin U Libr [820]

Andree, Richard *see*
- Die flutsagen
- Volkskunde der juden

Andreev, A I *see* Petr velikii

Andreev, I E et al *see* Entsiklopedicheskii slovar'

Andreevskii, A F *see* Sbornik uzakonenii, pravil i pravitel'stvennykh rasporiazhenii a takzhe neobkhodimykh svedenii po predmetam vedeniia gosudarstvennoi komissii pogasheniia dolgov...

Andreevskii flag – 300 : organ derzhavnoi partii – Moscow, Russia. n1(iiul '93), n1[4]-2[5](1994), n1(sen 97), n12(dek 97), spetsvyp (11 avg 1998), n7(7-21 sen 1998) – 1 – mf#mf-12248 (reel 2) – us CRL [077]

Andreevskii, I *see* O namestnikakh, voevodakh i gubernatorakh

Andreevsky, Alexander von *see* Der weg zum gral

Andreossy, Antoine F *see* Constantinople et le bosphore de thrace

Andres bello / Lira Urquieta, Pedro – Mexico City?, Mexico. 1948 – 1r – us UF Libraries [440]

Andres, J *see*
- Confvsion de la secte de mvhamed
- Opera chiamata confvsione della setta machvmetana...

Andres, Joh Bonaventura *see* Magazin fuer prediger

Andres Marcos, Teodoro *see* Vitoria y carlos 5th en la soberania hispanoamericana. salamanca, 1937

Andres Martinez, Gregorio *see* Carta de pedro ponce de leon, obispo de plasencia, a felipe 2, sobre las reliquias y librerias de su obispado y sus actividades literarias

Andres, Stefan et al *see* Der moerderbock

Andres, Stefan Paul *see*
- Gaeste im paradies
- Der gefrorene dionysus
- Die hochzeit der feinde
- Die liebesschaukel
- Requiem fuer ein kind
- Die sintflut
- Wir sind utopia

Andres vila – Minorca, Spain. v262-v267. 1745-1777 – 2r – us UF Libraries [946]

Andresen, Carl *see* Die lehre von der wiedergeburt auf theistischer grundlage

Andresen, Karl Gustaf *see* Ueber deutsche volksetymologie

Andreu, Enrique *see* Cosas que usted debe conocer

Andreu Iglesias, Cesar *see*
- Derrotados
- Derrumbe
- Gota de tiempo

Andreve, Guillermo *see* Cuestiones legislativas

Andrew, A L *see* The samoan settlement of 1899

Andrew castagne : or, adventure of an old mariner of the brigantine swordfish wrecked in the gulf of st lawrence in 1867 – [Montreal?: s.n.], 1882 [mf ed 1979] – 1mf – 9 – 0-665-00015-4 – mf#00015 – cn Canadiana [910]

Andrew castagne : or, adventure of an old mariner of the brigantine swordfish wrecked in the gulf of st lawrence in 1867 / Thiboutot, A – [Quebec?: s.n.], 1881 [mf ed 1982] – 1mf – 9 – 0-665-17945-6 – mf#17945 – cn Canadiana [910]

Andrew dickson white papers, 1832-1918 – [mf ed ProQuest] – 149r – 1 – (with p/g) – us UMI ProQuest [327]

Andrew, Elizabeth (Wheeler) *see* The queen's daughters in india

Andrew grant's, doctor's der arzneikunde, beschreibung von brasilien – Weimar 1814 [mf ed Hildesheim 1995-98] – 1v on 2mf – 9 – €60.00 – 3-487-26533-8 – gw Olms [615]

Andrew jackson account books, 1845-1877 – Nashville, TN. 1845-77 – 2v on 1r – 1 – (containing data on the purchase, sale, birth, marriage, & death of slaves at the hermitage, 1845-77, & a receipt book of andrew 2 & sarah jackson, 1845-77) – mf#ms1880 – us Western Res [976]

Andrew jackson miscellany, ca 1793-1867 / Miller, Otto [comp] – [mf ed 1991] – 1r – 1 – mf#ms2066 – us Western Res [978]

Andrew jackson papers – 78r – 1 – $2,730.00 – Dist. us Scholarly Res – us L of C Photodup [975]

Andrew, John A *see* The papers of john a andrew, 1772-1889

Andrew johnson papers – (mf ed 1960) – (= ser Presidential papers microfilm) – 55r – 1 – (with guide) – Dist. us Scholarly Res – us L of C Photodup [975]

Andrew, Lucy Brett *see* Practical patent procedure

Andrew, Paige G *see* Journal of map and geography libraries

Andrew peterson papers / Peterson, Andrew – 1854-98 – 3r – 1 – $90.00 – (incl filmed inventory) – us Minn Hist [920]

Andrew smith and natal / Kirby, Percival Robson – Cape Town, South Africa. 1955 – 1r – us UF Libraries [960]

Andrewes, Lancelot *see*
- Devotions of bishop andrews
- Library of anglo catholic theology

Andrews, Bruce *see* Guide to swaziland

Andrews, C C *see* Brazil

Andrews, C F *see*
- Letters to a friend
- Mahatma gandhi at work
- Mahatma gandhi, his own story

Andrews, Caesar *see* Field diary

Andrews, Charles *see*
- American jurisprudence.
- A guide to the manuscript materials for the history of the united states to 1783
- A guide to the materials for american history to 1783 in the public record office, london

Andrews, Charles Freer *see*
- The challenge of the north-west frontier
- India and britain
- India and the simon report
- The indian earthquake
- Mahatma gandhi's ideas
- North india
- The opium evil in india
- The renaissance in india
- The rise and growth of the congress in india
- Sadhu sundar singh
- The true india
- Zaka ullah of delhi

Andrews, Charles McLean *see*
- Guide to the manuscript materials for the history of the united states to 1783
- Guide to the materials for american history, to 1783, in the public record office of great britain

Andrews, Charles Wesley *see* A review of recent judicial decisions in england touching the sacraments

Andrews, Elisha Benjamin *see* Gospel from two testaments

Andrews, Emerson *see* Living life

Andrews, George Arthur *see*
- Efficient religion
- What is essential?

Andrews, George William Scott *see* Penicillin and other antibiotics

Andrews, Herbert Tom *see* The acts of the apostles

Andrews, Horace *see* Manual of the laws and courts of the united states.

Andrews, J R *see* Alone in the wide, wide world

Andrews, James deWitt *see* American law

Andrews, Jean et al *see* A study of four african-american families reading to their young deaf children [1996-1997]

Andrews, John *see* Presbyterianism vs. universal[i]sm

Andrews, John D *see* Eight years in the toils

Andrews, John Nevins *see*
- History of the sabbath and first day of the week
- The three messages of revelation 14, 6-12

Andrews, Joseph *see* Journey from buenos ayres

Andrews, Marian *see* Down the village street scenes in a west country hamlet by christopher hare

Andrews, Marian [pseud Christopher Hare] *see* Broken arcs

Andrews, Matthew T *see* Comrades of the road

Andrews, Morgan *see* Abandoned!

Andrews, Robert D *see* Truth about pirates

Andrews, Samuel James *see*
- Christianity and anti-christianity in their final conflict
- God's revelations of himself to men
- The life of our lord upon the earth
- Man and the incarnation
- William watson andrews

Andrews, Steven J *see* Effects of high versus low glycemic index-rated carbohydrate foods on exercise performance and fat

Andrews, Wilbur William *see*
- Nature and self-sacrifice
- Our national sin

Andrews, William *see*
- Bygone church life in scotland
- The church treasury of history, custom, folk-lore, etc
- Curiosities of the church
- Curious church customs and cognate subjects
- Ecclesiastical curiosities
- Legal lore
- The lifeboat and other poems
- Old church life

Andrews, William Darling *see* Swimming and life-saving

Andrews, William Eusebius *see*
- The catholic school book
- A critical and historical review of fox's book of martyrs
- Critical remarks on the discussion on the indiscriminate circulatio...
- Second letter to the vicar of blackburn

Andrews, William Watson *see* The principles of soil fertility applied to the worn-out dyked lands

Andreyev, Leonid *see* Ashmodai

Andria, Giovanni Antonio Cirullo d' *see* Il quinto libro de madrigali a cinque voci

Andriani, Giuseppe *see* Socialismo e comunismo in toscana tra il 1846 e il 1849

Andrian-Werburg, Ferdinand, Freiherr von *see* Der hoehencultus

Andriessen, A. *see* Plegtige inhuldiging vn zijne doorlugtigste hoogheid, willem karel henrik friso

Andrieu, M *see*
- Immixtio et consecratio
- Melanges en l'honneur de mgr m andrieu
- Les ordines romani du h m a

Andrieux, Francois G J S *see* Reve du mari, ou, le manteau

Andrist, Ralph K *see* Steamboats on the mississippi

Andromeda – 1971 sep 8-1972 feb 2 – 1 – mf#1052066 – us WHS [071]

Andromeda strain / Crichton, Michael – New York, NY. 1970, c1969 – 1r – us UF Libraries [830]

Androuet du Cerceau, Jacques *see*
- Lecons de perspective positive
- Le premier volvme des plus excellents bastiments de france

Androutsos, Chrestos *see*
- Dokimion symbolikes ex epopseos orthodoxou
- The validity of english ordinations from an orthodox catholic point of view

Andrus, James Russell *see* Mran ma ci pva re

Andrushchenko, K K *see* Primorskaia zhizn'

Andruson, L I *see* Verba

Andy warhol's interview – New York NY 1975-77 – 1 – (cont by: interview) – ISSN: 0020-5109 – mf#10591.01 – us UMI ProQuest [700]

Los andzsheleser tegliche tsayt = The jewish times – Los Angeles, CA.1921 – 1 – us AJPC [071]

[Aneau, B] *see*
- Imagination poetique...
- Picta poesis

Anecdota ambrosiana / Muratori, Lodovico A – Mediolani/Patavii. v1-4. 1697-1713 – 4v on 23mf – 9 – €44.00 – ne Slangenburg [240]

Anecdota graeca / Muratori, Lodovico A – Patavii, 1709 – €15.00 – ne Slangenburg [240]

Anecdota maredsolana / Morin, G – Maredsoli-Oxford. v1-3. 1893-1894 – 3v on 37mf – 8 – €71.00 – ne Slangenburg [240]

Anecdota oxoniensia : semitics series – pt 1-3 – 9 – €46.00 – (pt 5-7 €134) – ne Slangenburg [270]

Anecdota oxoniensia : semitic series – pt1-12 – 9 – €180.00 – (lacking: pt4.94) – ne Slangenburg [270]

Anecdota sacra et profana ex oriente et occidente allata : sive, notitia codicum graecorum, arabicorum, syriacorum, copticorum, hebraicorum, aethiopicorum, latinorum / Tischendorf, Constantin von – Lipsiae: Sumptibus Hermanni Fries, 1861 [mf ed 1986] – 1mf – 9 – 0-8370-9319-8 – (in greek & latin. incl ind) – mf#1986-3319 – us ATLA [220]

Anecdota syriaca / Land, J P N – Lugduni Batavorum. v1-4. 1862-1875 – 54mf – 9 – €103.00 – ne Slangenburg [240]

Anecdotal life of sir john macdonald / Biggar, Emerson Bristol – Montreal: John Lovell & Son; New York: US Book Co; London: Edward Stanford, 1891 [mf ed 1979] – 4mf – 9 – 0-665-00133-9 – mf#00133 – cn Canadiana [920]

Anecdotario epico del generalismo trujillo / Suarez Vasquez, Ramon – Ciudad Trujillo, Dominican Republic. 1957 – 1r – us UF Libraries [972]

Anecdotario martiano / Quesada Y Miranda, Gonzalo De – Habana, Cuba. 1900 – 1r – us UF Libraries [972]

Anecdotas centroamericanos / Garcia, Miguel Angel – San Salvador, El Salvador. 1955 – 1r – us UF Libraries [972]

Anecdotas misionales... / Corredor Garcia, Antonio – Madrid: Arch. Ibero Americano, 1965 – 1 – sp Bibl Santa Ana [240]

Anecdote lives of wits and humourists / Timbs, John – London: R Bentley 1872 [mf ed 1986] – 2v on 1r – 1 – (filmed with: literary anecdotes and contemporary reminiscences of professor porson and others / barker, e h) – mf#1571p – us Wisconsin U Libr [420]

Anecdotes americaines : ou histoire abregee des principaux evenements arrives dans le nouveau monde... / [Dejean, M] – Paris: Chez Vincent, impr-libr...1776 [mf ed 1983] – 9mf – 9 – 0-665-44214-9 – (incl ind) – mf#44214 – cn Canadiana [917]

Anecdotes du regne de louis 16 / Nougaret, Pierre J – Paris 1776 [mf ed Hildesheim 1995-98] – 1v on 4mf – 9 – €120.00 – 3-487-26185-5 – gw Olms [944]

Anecdotes du seizieme siecle : ou intrigues de cour, politiques, et galantes; avec les portraits de charles 9, henri 3 et henri 4, rois de france et de navarre / Caumont de LaForce, Charlotte R de – Amsterdam 1741 [mf ed Hildesheim 1995-98] – 2v on 4mf – 9 – €120.00 – 3-487-26124-3 – gw Olms [944]

Anecdotes historiques, singulieres et interessantes du regne de louis 14 – Amsterdam [u a] 1770 [mf ed Hildesheim 1995-98] – 2v on 6mf – 9 – €120.00 – 3-487-26099-9 – gw Olms [944]

Anecdotes illustrative of new testament texts – New York: A C Armstrong, 1884 [mf ed 1993] – (= ser The clerical library) – 1mf – 9 – 0-524-08226-X – mf#1993-2001 – us ATLA [225]

Anecdotes of a life on the ocean : being a portion of the experiences of twenty-seven years' service in many parts of the world / Cowan, David – 3rd rev ed. [Montreal?: s.n.], 1876 [mf ed 1982] – 3mf – 9 – mf#26990 – cn Canadiana [910]

Anecdotes of aurangzib and historical essays / Sarkar, Jadunath – Calcutta: MC Sarkar & Sons, 1912 – (= ser Samp: indian books) – us CRL [954]

Anecdotes of eminent painters in spain : during the sixteenth and seventeenth centuries... / Cumberland, R – London. 2v. 1782 – 4mf – 9 – mf#0-991 – ne IDC [750]

Anecdotes of painting in england : with some account of the principal artists / Walpole, Horace, Earl of Orford – London [1849] – (= ser 19th c art & architecture) – 16mf – 9 – mf#4.2.1456 – uk Chadwyck [750]

Anecdotes of painting in england... / Walpole, H – Strawberry Hill. 4v. 1762-1771 – 13mf – 9 – mf#0-1079 – ne IDC [750]

Anecdotes of printers who have resided or been born in england / Edwards, Edward – London 1808 – (= ser 19th c art & architecture) – 4mf – 9 – mf#4.2.1083 – uk Chadwyck [760]

Anecdotes of the arts in england : or, comparative remarks on architecture, sculpture, and painting / Dallaway, James – London 1800 – (= ser 19th c art & architecture) – 6mf – 9 – mf#4.2.1703 – uk Chadwyck [700]

Anecdotes of the bombay mission for the conversion of the hindoos... / Hall, G – London, 1836 – 3mf – 9 – mf#HTM-76 – ne IDC [915]

Anecdotes of the rev j w fletcher : late vicar of madeley, shropshire – London, England. 18– – 1r – us UF Libraries [240]

Anecdotes of the wesleys : illustrative of their character and personal history / Wakeley, Joseph Beaumont – New York: Carlton & Lanahan, 1871 [mf ed 1992] – (= ser Methodist coll) – 1mf – 9 – 0-524-04943-2 – (incl bibl ref) – mf#1992-2064 – us ATLA [920]

Anecdotes of the wesleys : illustrative of their character and personal history / Wakeley, Joseph Beaumont – New York: Carlton & Lanahan; Cincinnati, Hitchcock & Walden 1869 [mf ed 1984] – 1r – 1 – (int by rev j m'clintock. filmed with: sogno d'una notte d'estate / shakespeare, william) – mf#6716 – us Wisconsin U Libr [920]

Anecdotes secretes du dix-huitieme siecle redigees : avec soin d'apres la correspondance secrete, politique et litteraire; pour fair suite aux memoires de bachaumont; ouvrage qui contient, outre une infinite de faits curieux et peu connus, un choix de vaudevilles, / ed by Nougaret, Pierre J – Paris 1808 [mf ed Hildesheim 1995-98] – 2v on 6mf – 9 – €120.00 – 3-487-25840-4 – gw Olms [440]

O anecdotista : semanario dedicado aos homens de espirito – Rio de Janeiro, RJ: Typ da Mentira, 21-28 out 1882 – (= ser Ps 19) – mf#DIPER – bl Biblioteca [079]

Aneka / Purnama – Djakarta, 1963-1967 – 19mf – 9 – (missing: 1963/1964, v1-2(1-14); 1967, v3(11, 12)) – mf#SE-954 – ne IDC [959]

Aneka olahraga – [Djakarta], 1950-1966 – 110mf – 9 – (several iss missing) – mf#SE-871 – ne IDC [959]

Aneka warta Bamunas Djaya *see* Badan musjawarah pengusaha nasional swasta dci djakarta raya

Anekdoten (zum groessten theil unbekannt) von napoleon : zur erlaeuterung seiner denk- und gemuethsart und seiner thaten / Ireland, William Henry – Leipzig 1823-28 [mf ed Hildesheim 1995-98] – 2v on 24mf – 9 – €240.00 – 3-487-26348-3 – gw Olms [944]

Anem taon dalem noraka / Tan, Boen Soan – Soerabaia: Tan's Drukkerij, 1935 [mf ed 1998] – (= ser Penghidoepan 126) – 1r – 1 – (coll as pt of the colloquial malay collection. filmed with: multi-millionair / ong khing han) – mf#10002 – us Wisconsin U Libr [830]

Anemons und adonis blumen / Abschatz, Johann Erasmus Assmann, Freiherr von; ed by Mueller, Guenther – Halle: Niemeyer, 1929 [mf ed 1993] – (= ser Neudrucke deutscher literaturwerke des 16. und 17. jahrhunderts 274-277) – xc/76p – 1 – (incl bibl ref) – mf#8413 reel 10 – us Wisconsin U Libr [830]

Anene, J C O *see* The establishment and consolidation of imperial government in southern nigeria, 1891-1904

Aner, Karl *see*
- Der aufklaerer, friedrich nicolai
- Aus den briefen des paulus nach korinth
- Goethes religiositaet

Anerio, Giovanni Francesco *see* Antiphonae, sev sacrae cantiones

Anesaki, Masaharu *see*
- Buddhist and christian gospels
- Buddhist art in its relation to buddhist ideals
- Nichiren, the buddhist prophet
- Religious history of japan

Anesthesia and analgesia – Baltimore MD 1983+ – 1,5,9 – ISSN: 0003-2999 – mf#13964,01 – us UMI ProQuest [617]

Anesthesia and analgesia – v76-83. 1993-1996 – 1,5,6,9 – $110.00r – us Lippincott [617]

Anesthesia progress – Lawrence KS 1967-95 – 1,5,9 – ISSN: 0003-3006 – mf#2516 – us UMI ProQuest [617]

Anesthesie, analgesie, reanimation – Paris, France 1935-81 – 1,5,9 – ISSN: 0003-3014 – mf#5109 – us UMI ProQuest [617]

Anesthesiology – Baltimore MD 1940+ – 1,5,9 – ISSN: 0003-3022 – mf#2315 – us UMI ProQuest [617]

Anesthesiology clinics of north america – Philadelphia PA 1987+ – 1,5,9 – ISSN: 0889-8537 – mf#13377,01 – us UMI ProQuest [617]

Anethan, Eleanora Mary (Haggard), Baronne d' *see* His chief's wife

Anexion de la republica de haiti / Hudicourt, Pierre L – Santiago, Chile. 1923? – 1r – us UF Libraries [972]

Anf lihim ng isang pulo : nobelang tagalog / Aguilar, Faustino – Manila. 1958 – 1 – us CRL [490]

Die anfaenge der beginen : ein beitrag zur geschichte der volksfroemmigkeit und des ordenswesens im hochmittelalter / Greven, Joseph – Muenster i W: Aschendorff, 1912 – (= ser Vorreformationsgeschichtliche Forschungen) – 1mf – 9 – 0-7905-7232-X – (incl bibl ref) – mf#1988-3232 – us ATLA [940]

Die anfaenge der christlichen kirche und ihrer verfassung : erster band, buch 1 bis 3 nebst einer beilage ueber die echtheit der ignatianischen briefe: ein geschichtlicher versuch / Rothe, Richard – Wittenberg: Zimmermann, 1837 – 2mf – 9 – 0-7905-0268-2 – (in german, greek, and latin. no more publ. incl bibl ref) – mf#1987-0268 – us ATLA [240]

Die anfaenge der deutschen literatur : vorkarlisches schrifttum im deutschen suedostraum / Klein, Karl Kurt – Muenchen: Verlag des Suedostdeutschen Kulturwerks, c1954 [mf ed 1992] – (= ser Veroeffentlichungen des suedostdeutschen kulturwerks. reihe b, wissenchaftliche arbeiten 3) – 142p – 1 – (incl bibl ref and ind) – mf#8166 – us Wisconsin U Libr [430]

Die anfaenge der gegenreformation in den niederlanden / Kalkoff, Paul – Halle a S: Verein fuer Reformationsgeschichte 1903 [mf ed 1990] – (= ser Schriften des vereins fuer reformationsgeschichte) – 1mf – 9 – 0-7905-5290-6 – (incl bibl ref) – mf#1988-1290 – us ATLA [949]

Die anfaenge der reformation in den preussischen landen ehemals polnischen anteils bis zum krakauer frieden, 8. april 1525 / Boetticher, Paul – Ober-Glogau [Gnogowek]: E Radek, 1894 – 1mf – 9 – 0-524-04947-5 – (incl bibl ref) – mf#1990-1350 – us ATLA [242]

Die anfaenge der tuebinger theol quartalschrift – Rottenburg, 1938 – €5.00 – ne Slangenburg [200]

Die anfaenge des christenthums : beitraege zum verstandiss des neuen testaments: ein vortrags-cyclus. gehalten im berliner unions-verein... / Holtzmann, Heinrich Julius et al – Berlin: A Haack, 1877 – 1mf – 9 – 0-524-02644-0 – mf#1990-0668 – us ATLA [240]

Die anfaenge des christenthums in der stadt rom / Schmidt, K – [Heidelberg?: s.n., 1879?] – (= ser Sammlung von Vortraegen fuer das deutsche Volk) – 1mf – 9 – 0-524-04149-0 – mf#1990-1219 – us ATLA [240]

Die anfaenge des erasmus : humanismus und "devotio moderna" / Mestwerdt, P – Leipzig, 1917 – 10mf – 8 – €19.00 – ne Slangenburg [140]

Die anfaenge des heiligenkults in der christlichen kirche / Lucius, Ernst; ed by Anrich, Gustav – Tuebingen: Mohr, 1904 – 2mf – 9 – 0-7905-5004-0 – (incl bibl ref) – mf#1988-1004 – us ATLA [240]

Die anfaenge des katholischen christentums und des islams : eine religionsgeschichtliche untersuchung / Bestmann, Hugo Johannes – Noerdlingen: CH Beck, 1884 – 1mf – 9 – 0-524-01168-0 – (incl bibl ref) – mf#1990-2244 – us ATLA [241]

Die anfaenge des nationalen jahweglaubens : ein beitrag zur israelitischen religionsgeschichte / Bewer, Julius August – Gotha: F A Perthes, [ca 1904] – 1mf – 9 – 0-8370-2323-8 – mf#1985-0323 – us ATLA [270]

Die anfaenge unserer religion / Wernle, Paul – 2., verb und verm Aufl. Tuebingen: JCB Mohr (Paul Siebeck), 1904 – 2mf – 9 – 0-8370-5792-2 – mf#1985-3792 – us ATLA [240]

Anfangsgruende der theoretischen musik / Marpurg, Friedrich Wilhelm – Leipzig: J G I Breitkopf 1757 [mf ed 19–] – 4mf – 9 – mf#fiche 423 – us Sibley [780]

Anfangs-gruende des general-basses : nach mathematischer lehr-art abgehandelt und vermittelst einer hierzu erfundenen maschine auf das deutlichste vorgetrager... / Mizler von Kolof, Lorenz Christoph – Leipzig: zu finden bey dem Verfasser [1739] [mf ed 19–] – 3mf – 9 – mf#fiche 619, 810 – us Sibley [780]

Anfangsgruende einer littauischen grammatick : in ihrem natuerlichen zusammenhange entworfen / Ruhig, Paul Friedrich – Koenigsberg: Hartung 1747 – (= ser Whsb) – 2mf – 9 – €30.00 – mf#Hu 199a – gw Fischer [460]

Anfiteatro / Alvarez Saenz de Buruaga, Jose – Merida: Imp. Rodriguez, s.a. – 1 – sp Bibl Santa Ana [820]

Anfiteatro amazonico / Morais, Raimundo De – Sao Paulo, Brazil. 1938 – 1r – us UF Libraries [972]

El anfiteatro romano de merida. memoria (1916 a 1918) / Melida, Jose Ramon – Madrid, 1919 – 1 – sp Bibl Santa Ana [946]

El anfiteatro y el circo romanos de merida : memoria / Melida, Jose Ramon – Madrid: rev arch bibl mus, 1921 – 1 – sp Bibl Santa Ana [946]

Anfitriao / Silva, Antonio Jose Da – Rio de Janeiro, Brazil. 1939 – 1r – us UF Libraries [972]

Anfora sedienta / Valle, Rafael Heliodoro – Mexico City, Mexico. 1922 – 1r – us UF Libraries [972]

Anforas, de amor y de dolor, de meditacion... / Ochao-Alcantara, Antonio – Tegucigalpa, Honduras. 1936 – 1r – us UF Libraries [972]

Anfossi, Pasquale *see* Nitteti

Ang atong kabilin – Siyudad sa Sugbu: [s.n.], nov 11 1930-dec 25 1934 – 8r – 1 – us CRL [079]

Ang bayang pilipino – Manila: Ang Bayang pilipino. ano1 n22. 28 mayo 1914 (wkly) [mf ed 1985] – (= ser Philippine labor publications 1/9) – (in tagalog with brief sections in english and spanish) – mf#6580 reel 1 n9 – us Wisconsin U Libr [079]

Ang cabuhi ni d pedro tabios cag ni manuela lilay sa imperiong roma – Mandurriao: Panayana 1912 [mf ed Bloomington IN: Indiana Uni Lib, Preservation Dept 1984] – (= ser Coll...in the bisaya language 2) – 1r – 1 – us Indiana Preservation [490]

Ang cahambalhambal na pagca guho ng Troya : catotohanang buhay nang principe paris na anac nang haring priamo at nang reyna esilva sa caharian nang troya na hinango sa tunay na historia at tinula sa lubos na catiyagaan / Ignacio, Cleto R – Maynila: P Sayo 1916 [mf ed Bloomington IN: Indiana Uni Lib, Preservation Dept 1984] – (= ser Coll...in the tagalog language 2) – 1r – 1 – us Indiana Preservation [490]

Ang camatuoran – Sugbu, Philippines: [s.n.], jan 2 1894-dec 27 1911 – us CRL [950]

Ang diwa ng mga salawikain : 60 salawikain na may tig-isang kwentong tula / Santos, Lope K – Maynila: Bookmark [1988] c1953 [mf ed 1993] – on pt of 1r – 1 – mf#11052 r556 n3 – us Cornell [490]

Ang kaibigan ng bayan – Barasoian, [Philippines: s.n., nov 1,10, dec 13 1898; jan 3, feb 4 1899 – us CRL [079]

Ang kaliwanagan – Maynila: [s.n., nov 6 1900] – us CRL [079]

[Ang kapanimalusan. libretto] / Guinto, Mariano – Cebu, I F: Falek 1913 [mf ed Bloomington IN: Indiana Uni Lib, Preservation Dept 1984] – (= ser Coll...in the bisaya language 1) – 1r – 1 – us Indiana Preservation [780]

Ang kapatid ng bayan – maynila [i.e. Manila, Philippines]: P H Poblete, jan 13 1900; mar 30 1901; jan 9 1902 – us CRL [079]

Ang laa sa bugay: dulang inawitan: duha ka acto ug tulo ka cuadro / Buyser y Aquino, Fernando – Cebu: Falek 1915 [mf ed Bloomington IN: Indiana Uni Lib, preservation Dept 1984] – (= ser Coll...in the bisaya language) – 1r – 1 – us Indiana Preservation [490]

Ang lihim ng isang pulo : nobelang tagalog (kasaysayang ukol sa mga unang panahon) / Aguilar, Faustino – 2. pagkalimbag. Maynila: [Benipayo Press], 1958, c1927 – us CRL [950]

Ang lipang kalabaw – Maynila: Lipang kalabaw, may 28 1932 [mf ed 1986] – (= ser Philippine labor publications 1/13; Rare philippine newspapers and serials 5; Philippine alternative newspapers) – 1 – mf#6581 reel 5 n12 – us Wisconsin U Libr [079]

Ang macahanoclog nga cabuhi sang haring villarba, reina d a maria, cag sang tatlo nila ca anac, d pedro, d juan, cag d a isabela, sa guin harian sa teruel – Mandurriao, Iloilo: Panayana 1913 [mf ed Bloomington IN: Indiana Uni Lib, Preservation Dept 1984] – (= ser Coll...in the bisaya language 2) – 2v on 1r – 1 – (imperfect: p33-48 v1 wanting) – us Indiana Preservation [490]

Ang macahanoclog nga vida sang duha ca mag utud nga capid nga si portivillar cag si lucibar nga bata sang haring cesar sa isa ca pastora sa guin harian sa normandia – Iloilo: La Editorial 1910 [mf ed Bloomington IN: Indiana Uni Lib, Preservation Dept 1984] – (= ser Coll...in the bisaya language 2) – 1r – 1 – us Indiana Preservation [490]

Ang maikling kathang tagalog / Abadilla, A G et al – Quezon City: Bede's Publishing House, 1967, c1954 [mf ed 1987] – viii/307p – 1 – (originally publ in 1954) – mf#6774 – us Wisconsin U Libr [830]

Ang mga salmos *see* Bible. o t psalms. bisaya. 1912

Ang palad, palad gayud : mga sugilanong binisaya / Duterte, Vicente – Cebu, I F: Falek 1912 [mf ed Bloomington IN: Indiana Uni Lib, Preservation Dept 1984] – (= ser Coll...in the bisaya language 1) – 1r – 1 – us Indiana Preservation [490]

Ang pulahan : sugilanong binisaya / Enarem – Sugbo, Sugbo, K P: Falek 1919 [mf ed Bloomington IN: Indiana Uni Lib, Preservation Dept 1984] – (= ser Coll...in the bisaya language 1) – 1r – 1 – us Indiana Preservation [490]

Ang, Siauw Tan *see* Dewi telaga warna

Ang suga – Sugbu [Philippines]: Vicente Sotto. sep 30-nov 18 1903 – 1r – us CRL [999]

Ang tibay – Manila: Ang Tibay, [1940?] [mf ed 1985] – (= ser Philippine labor publications 1/5) – 1 – (in tagalog) – mf#6580 reel 1 n5 – us Wisconsin U Libr [338]

Ang today – v16 n7-v17 n12 [1981 jul-1982 dec] – 1 – mf#646453 – us WHS [071]

Ang tuburan sang bulauan, ang cahoy nga nagarauta, cag pispis nga nagahambal – Mandurriao: Panayana 1913 [mf ed Bloomington IN: Indiana Uni Lib, Preservation Dept 1984] – (= ser Coll...in the bisaya language 2) – 1r – 1 – us Indiana Preservation [490]

Angalip : mran ma san pum kri lak cvai mu mhan / Ke-na gumi Ra An Chan – Ran kun: Sasana hi ta Ca pum nhip Tuik 1930 [mf ed 1990] – 1r with other items – 1 – (in burmese; at head of title: english and burmese spelling book) – mf#mf-10289 seam reel 155/3 [§] – us CRL [040]

Angalip mran ma cac samuin / Bhun Kyo, Nat mok – Ran kun: Pu gam 1975 [mf ed 1994] – on pt of 1r – 1 – mf#11052 r1719 n1 – us Cornell [959]

Ange dans le monde et le diable a la maison / Courcy, Frederic De – Paris, France. 1838? – 1r – us UF Libraries [440]

L'ange d'astarte : etude sur la seconde inscription d'oum-el-awamid / Berger, Philippe – [s.l: s.n, 1879?] [mf ed 1989] – 1mf – 9 – 0-7905-2461-9 – mf#1987-2461 – us ATLA [470]

Angebauer, Karl *see* Ovambo

Der angebliche evangelienkommentar des theophilius von antiochien / Harnack, Adolf von – Leipzig, 1883 – (= ser Tugal 1-1/4b) – 2mf – 9 – €5.00 – ne Slangenburg [240]

Der angebliche exzessive realismus des duns scotus / Minges, P – 1908 – (= ser Bgphma 7/1) – €5.00 – ne Slangenburg [110]

Der angebliche turmbau zu babel, die erlebnisse der familie abrahams und die beschneidung / Jedlicska, Johann – Wien: Friedrich Jasper 1903 [mf ed 1993] – 1mf [ill] – 9 – 0-524-05808-3 – mf#1992-0635 – us ATLA [221]

Angel de piedra / Stolk, Gloria – Caracas, Venezuela. 1962 – 1r – us UF Libraries [972]

Angel de sodoma / Hernandez Cata, Alfonso – Madrid, Spain. 1929 – 1r – us UF Libraries [972]

Angel policiano-silvae nutritia / Poliziano, Angelo; ed by Sanchez de las Brozas, Francisco – 1596 – 9 – sp Bibl Santa Ana [450]

Angel rodriguez (alias) er periodista... / Amores Gonzalez, Meliton – Astorga: Imp. y Lit. de Sierra, 1925 – 1 – sp Bibl Santa Ana [946]

Angel street / Hamilton, Patrick – New York, NY. 1942 – 1r – us UF Libraries [830]

Angel y las imagenes / Centeno Guell, Fernando – San Jose, Costa Rica. 1953 – 1r – us UF Libraries [972]

Angela : tiroler novelle / Achleitner, Arthur – Leipzig: Hesse & Becker, [19–] [mf ed 1988] – 96p – 1 – mf#6934 n12 – us Wisconsin U Libr [830]

Angela borgia : novelle / Meyer, Conrad Ferdinand – Leipzig: H Haessel, 1923 [mf ed 1996] – 233p/162p/95p – 1 – (incl: huttens letzte tage: eine dichtung; engelberg: eine dichtung) – mf#9721 – us Wisconsin U Libr [800]

The angela davis trial with index / Meiklejohn Civil Liberties Institute – 13 reels – 1 – $500.00 – us Trans-Media [340]

Angela de Fulginio *see* Thesaurus angelae de fulginio

Angela luisa / Torregrosa, Angela Luisa – San Juan, Puerto Rico. 1956 – 1r – us UF Libraries [920]

Angela of Foligno *see* Book of divine consolation of the blessed angela of foligno

Angelerius, Q T *see* Epidemiologia

[Los angeles-] aircraft times – CA. Sept 1941-Feb 1946 – 4r – 1 – $240.00 – mf#C02367 – us Library Micro [071]

[Los angeles-] aztlan chicano journal – CA. 1970-1974 – 1r – 1 – $60.00 – mf#R03280 – us Library Micro [305]

[Los angeles-] beirut times – CA. 1985- – 9r – 1 – $540.00 (subs $50y) – mf#H04045 – us Library Micro [071]

[Los angeles-] california afl-cio – CA. 1981-1992 – 11r – 1 – $660.00 – mf#R03282 – us Library Micro [071]

[Los angeles-] california cultivator – CA. 1892-1947 – 54r – 1 – $3240.00 – mf#B02369 – us Library Micro [071]

[Los angeles-] california eagle – CA. 1943-51 – 15r – 1 – $900.00 – mf#C02370 – us Library Micro [071]

[Los angeles-] california farmer : (southern edition) – 1949-jun 1991 – (= ser Wine & agriculture coll) – 54r – 1 – $2700.00 – mf#B02371 – us Library Micro [640]

[Los angeles-] california farmer : southern edition – CA. 1949-91 – (= ser Wine & agriculture coll) – 54r – 1 – $3240.00 – (cont: pacific rural press; see also san francisco, fresno) – mf#C02371 – us Library Micro [071]

[Los angeles-] california jewish bulletin – CA. 1933-34 (California Jewish Review) 1924-29 – 1r – 1 – $60.00 – mf#B02372 – us Library Micro [071]

[Los angeles-] california jewish press – CA. jun 1956-jan 1969 – 4r – 1 – $240.00 – mf#B02373 – us Library Micro [071]

[Los angeles-] california magyarsag – CA. jun 21 1957- – 14r – 1 – $1540.00 (subs $50/y) – (in hungarian) – mf#C02374 – us Library Micro [071]

[Los angeles-] california oil world – CA. 1909-12; 1915-23 – 8r – 1 – $480.00 – mf#C02375 – us Library Micro [071]
[Los angeles-] central avenue news – CA. Nov 1910-May 1911 – 1r – 1 – $60.00 – mf#C02376 – us Library Micro [071]
[Los angeles-] central news wave – CA. 1980- – 84r – 1 – $5040.00 (subs $300y) – mf#H04049 – us Library Micro [071]
[Los angeles-] chicano law review – CA. 1972-1975 – 1r – 1 – $60.00 – mf#R03284 – us Library Micro [071]
[Los angeles-] city news – CA. 1972-1975 – 1r – 1 – $60.00 – mf#H03285 – us Library Micro [071]
[Los angeles-] civic center news – CA. 1978-1981 – 2r – 1 – $120.00 – mf#R04055 – us Library Micro [071]
[Los angeles-] con safos – CA. 1968-69 – (= ser Chicano studies library serial) – 1r – 1 – $60.00 – mf#R02377 – us Library Micro [071]
[Los angeles county-] arcadia city directories : including monrovia, duarte, sierra madre, temple city, bradbury and east pasadena – CA. 1950-1992 (Fiche) – 238r – 1 – $595.00 – mf#D054 – us Library Micro [917]
Los Angeles County Employees Association *see* County employee
Los Angeles County Federation of Labor *see* Los angeles citizen
[Los angeles county-] hollywood directories – CA. 1906; 1928 – 2r – 1 – $100.00 – mf#D053 – us Library Micro [790]
[Los angeles county-] kern, los angeles, san bernardino, san diego, san luis obispo, santa barbara and ventura counties – CA. 1875 – 1r – 1 – $50.00 – mf#D044 – us Library Micro [978]
[Los angeles county-] pasadena city dirctories – CA. 1881-1976 – 95r – 1 – $4750.00 – mf#D051 – us Library Micro [917]
[Los angeles-] courier march field district – CA. 1934-1935 – 1r – 1 – $60.00 – mf#R04056 – us Library Micro [071]
[Los angeles-] daily commerce – CA. 1980- – 84r – 1 – $5040.00 (subs $300y) – mf#H04050 – us Library Micro [071]
[Los angeles-] daily commercial news – CA. 1976-1993 – 63r – 1 – $3780.00 – mf#R03288 – us Library Micro [071]
[Los angeles-] daily journal – CA. 1888- – 1000r – 1 – $110,000.00 (subs $800y) – mf#HC02392 – us Library Micro [073]
[Los angeles-] daily news – CA. 1870-1872 – 3r – 1 – $180.00 – mf#C03289 – us Library Micro [071]
[Los angeles-] daily record – CA. Mar 1895-1905 – 21r – 1 – $1260.00 – mf#RC02378 – us Library Micro [071]
[Los angeles-] downtown news – CA. 1972-1992 – 1r – 1 – $60.00 – mf#R04057 – us Library Micro [071]
[Los angeles-] eastside journal – CA. 1935-1977 – 7r – 1 – $420.00 – mf#R03290 – us Library Micro [073]
[Los angeles-] eastside sun – CA. 1971 – 2r – 1 – $120.00 – mf#R02379 – us Library Micro [071]
[Los angeles-] el malcriadito – CA. 1975-1976 – 1r – 1 – $60.00 – mf#R03291 – us Library Micro [071]
[Los angeles-] five cities times press recorder – CA. 1986-1988 – 18r – 1 – $1080.00 – mf#R04058 – us Library Micro [071]
[Los angeles-] happy days – CA. 1933-1934 – 1r – 1 – $60.00 – mf#R03294 – us Library Micro [071]
[Los angeles-] herald express – CA. 1945-1960 – 196r – 1 – $11,760.00 – mf#C02384 – us Library Micro [071]
[Los angeles-] il leone – CA. 1971-1983 – 1r – 1 – $60.00 – mf#C03297 – us Library Micro [071]
[Los angeles-] illustrated daily news – CA. 1926-1927 – 20r – 1 – $1200.00 – mf#R03298 – us Library Micro [071]
[Los angeles-] independent – CA. 1988- – 36r – 1 – $2160.00 (subs $300y) – mf#H04046 – us Library Micro [071]
[Los angeles-] international daily news – CA. 1983- – 132r – 1 – $7920.00 (subs $600y) – mf#H04047 – us Library Micro [071]
[Los angeles-] journal of commerce and independent review – CA. 1975 – 4r – 1 – $240.00 – mf#R03300 – us Library Micro [380]
[Los angeles-] journal of commerce review – CA. 1976-1980 – 29r – 1 – $1740.00 – mf#R03301 – us Library Micro [380]
[Los angeles-] korea times – CA. 1970- – 284r – 1 – $17,040.00 (subs $590y) – mf#H03302 – us Library Micro [071]
[Los angeles-] la opinion diaro popular independiente – CA. 1926-70 – 197r – 1 – $11,820.00 – mf#C02386 – us Library Micro [071]
[Los angeles-] la prensa – CA. 1917-22; 1967-70 – (= ser Chicano studies library serial) – 3r – 1 – $180.00 – mf#R02388 – us Library Micro [071]
[Los angeles-] latin quarter – CA. 1974-1975 – 1r – 1 – $60.00 – mf#R04060 – us Library Micro [071]
[Los angeles-] lesbian tide – CA. 1977-1980 – 1r – 1 – $60.00 – mf#R04061 – us Library Micro [071]
[Los angeles-] marin county – 1904-34; 1992- – (= ser California telephone directory coll) – 34r – 1 – $1700.00 – mf#P00061 – us Library Micro [917]
[Los angeles-] new american woman – CA. 1916-1918 – 1r – 1 – $60.00 – mf#C03313 – us Library Micro [071]
[Los angeles-] nommo – CA. 1977-82 – 2r – 1 – $120.00 – mf#R02395 – us Library Micro [071]
[Los angeles-] northwestern – 1948-71; 1975-87; 1989- – (= ser California telephone directory coll) – 100r – 1 – $5000.00 – mf#P00059 – us Library Micro [917]
[Los angeles-] oil, paint and drug reporter – CA. v183-192 – 1r – 1 – $60.00 – mf#R03315 – us Library Micro [071]
[Los angeles-] pacific citizen – CA. 1929- – 30r – 1 – $1800.00 (subs $50y) – (also publ in salt lake city and san francisco) – mf#B02398 – us Library Micro [071]
[Los angeles-] panorama – CA. n669-681. 1980-1994 – 26r – 1 – $1560.00 – mf#R04063 – us Library Micro [071]
[Los angeles-] south los angeles bulletin – CA. 1933-1945; 1948-1959; 1961-1983 – 66r – 1 – $7260.00 – mf#H04027 – us Library Micro [071]
[Los angeles-] southern california business – CA. 1939-1982 – 6r – 1 – $360.00 – mf#R03317 – us Library Micro [071]
[Los angeles-] southern california industrial news – CA. 1960-1982 – 15r – 1 – $900.00 – mf#R03318 – us Library Micro [071]
[Los angeles-] southern california teamsters – CA. 1943-1985 – 19r – 1 – $1140.00 – mf#R03319 – us Library Micro [071]
[Los angeles-] stamp collection – CA. 1847-1911 – 1r – 1 – $60.00 – mf#R04064 – us Library Micro [071]
[Los angeles-] sunday chronicle – CA. 1987 – 6r – 1 – $360.00 (subs $50y) – mf#R04065 – us Library Micro [071]
[Los angeles-] the evening news – CA. 1905-1907 – 9r – 1 – $540.00 – mf#R03292 – us Library Micro [071]
[Los angeles-] the heritage southwest jewish press – CA. 1959- – 25r – 1 – $1500.00 (subs $50y) – mf#C02385 – us Library Micro [071]
[Los angeles-] the king's business – CA. 1910-1970 – 12r – 1 – $720.00 – mf#R04059 – us Library Micro [071]
[Los angeles-] the reflex – CA. jun-sept 1935; jan, mar 1936 – 1r – 1 – $60.00 – mf#B02400 – us Library Micro [071]
[Los angeles-] the tidings – CA. 1895- – 94r – 1 – $5640.00 (subs $50y) – mf#H03323 – us Library Micro [071]
[Los angeles-] ucla daily bruin – CA: UCLA, 1945-58; Mar-May 1972 – 12r – 1 – $720.00 – mf#C02403A – us Library Micro [378]
Los Angeles Union Label Council *see* Bulletin of the los angeles...
[Los angeles-] united progressive news – CA. 1935 – 1r – 1 – $60.00 – mf#C03325 – us Library Micro [071]
[Los angeles-] upton sinclair's epic news – CA. may 1934-sep 1947 – 3r – 1 – $180.00 – (aka: end poverty paper) – mf#C02402 – us Library Micro [071]
[Los angeles-] voice 660 – CA. 1977-1984 – 2r – 1 – $120.00 – mf#R03326 – us Library Micro [071]
[Los angeles-] west end independent – CA. 1990 – 1r – 1 – $60.00 – mf#R04066 – us Library Micro [071]
Angelic wisdom concerning the divine love / Swedenborg, Emanuel – Boston, MA. 1906 – 1r – us UF Libraries [240]
Angelic wisdom concerning the divine love and the divine wisdom / Swedenborg, Emanuel – London, England. 1856 – 1r – us UF Libraries [240]
Angelica kauffmann : a biography / Gerard, Frances A – London 1892 – (= ser 19th c art & architecture) – 5mf – 9 – mf#4.2.460 – uk Chadwyck [750]
Angelica. Presbytery (Pres. Church in the USA) *see* Minutes, 1828-1856
Angelicum – Rome. v.13-20, 1936-43. Incomplete – 1 – us L of C Photodup [240]
Angelicum ac divinum opus musice / Gaffurio, Franchino – 1508 – (= ser Mssa) – 2mf – 9 – €35.00 – mfchl 52 – gw Fischer [780]
Angelicum ac divinum opus musice... / Gaffurius, Franchinus – [Colophon: Impressum Mediolani per Gotardum de pote Anno Salutis Millesimo quin getesimo octauo die sextadecima septembris ...] [mf ed 19–] – 3mf / 1r – 9, 1 – mf#fiche 723 / film 746 – us Sibley [780]
Angelin, Justin P *see* Expedition du louxor
Angeline de montbrun / Conan, Laure – [Quebec?: s.n.] 1884 [mf ed 1984] – 4mf – 9 – 0-665-00738-8 – mf#00738 – cn Canadiana [830]
Angelini Bontempi, Giovanni Andrea *see* Historia mvsica
Angelini, Carlo Antonio *see* Favourite solo for the violin and harpsicord
Angelini, Marc A *see* Aerobic responses to 12 weeks of training on various modes of home exercise equipment in sedentary adults
Angelique arnauld : abbess of port royal / Martin, Frances – 2nd ed. London: Macmillan, 1873 [mf ed 1986] – (= ser The sunday library for household reading) – 1mf – 9 – 0-8370-8696-5 – mf#1986-2696 – us ATLA [241]
Angelis, Pedro de *see*
– Basilicae veteris vaticanae descriptio...
– Coleccion de obras y documentos relativos a la historia antigua y moderna de las provincias del rio de la plata
Angell, Douglas *see* Examination of the theory and practice of church music in unitarian societies
Angell, Joseph K *see* The united states law intelligencer and review
Angell, Joseph Kinnicut *see* A treatise on the right of property in tide waters and in the soil and shores threrof
Angell, Norman *see* You and the refugee
Angelloz, Joseph-Francois *see* Goethe
The angel-messiah of buddhists, essenes, and christians / Bunsen, Ernest de – London: Longmans, Green, 1880 – 1mf – 9 – 0-7905-0560-6 – (incl bibl ref and index) – mf#1987-0560 – us ATLA [230]
Angelner landpost *see* Schleibote
Angelo da Picitono *see* Fior angelico di musica
Angelomontana : blaetter aus der geschichte von engelberg: jubilaeumsgabe fuer abt leodegar 2 / Cavelti, Sigisbert et al – Gossau St G: JG Cavelti-Hangartner, 1914 [mf ed 1992] – 2mf – 9 – 0-524-03333-1 – (incl bibl ref; incl contr by sigisbert cavelti) – mf#1990-0914 – us ATLA [241]
Angelos : archiv fuer neutestamentliche zeitgeschichte und kulturkunde – 1(1925)-4(1932) – 21mf – 9 – €40.00 – ne Slangenburg [225]
Angelov, Vasil G *see* Dobri vesti
Angels and demons according to lactantius (sca3) / Schneweis, E – Washington DC, 1944 – (= ser Studies in christian antiquity (sca)) – 4mf – 9 – €11.00 – ne Slangenburg [230]
The angels and their ministrations / Patterson, Robert Mayne – Philadelphia: Westminster Press, 1900 [mf ed 1992] – 1mf – 9 – 0-524-05414-2 – (incl bibl ref) – mf#1992-0424 – us ATLA [240]
[Angels camp-] record – CA. 1908-18 – 5r – 1 – $300.00 – mf#B02017 – us Library Micro [071]
[Angels camp-] the calaveras californian – CA. 1994- – 1r – 1 – $60.00 (subs $50/y) – mf#B02014 – us Library Micro [071]
The angels of god / Dunn, Lewis Romaine – New York: Phillips & Hunt; Cincinnati: Walden & Stowe, 1881, c1880 [mf ed 1989] – 1mf – 9 – 0-7905-3013-9 – mf#1987-3013 – us ATLA [220]
Angels of mons / Machen, Arthur – London, England. 1915 – 1r – us UF Libraries [025]
Angels of the battlefield : a history of the labors of the catholic sisterhoods in the late civil war / Barton, George – Philadelphia, PA: Catholic Art, 1897 [mf ed 1990] – 1mf – 9 – 0-7905-5623-5 – mf#1988-1623 – us ATLA [241]
Angelus *see* Primera antologia de poetas pacenses
Angelus silesius saemtliche poetische werke : und eine auswahl aus seinen streitschriften / ed by Ellinger, Georg – Berlin: Propylaeen Verlag, [1923?] [mf ed 1993] – 2v – 1 – (with biogr sketch) – mf#8456 – us Wisconsin U Libr [810]
Angelus silesius und seine mystik / Seltmann, C – Breslau: G P Aderholz, 1896 [mf ed 1993] – 208p – 1 – (incl bibl ref) – mf#7663 – us Wisconsin U Libr [430]
Angely, Louis *see*
– Neuestes komisches theater
– Prosit neujahr!
Angenehme stunden – n1-13. 1887 [complete] – (= ser Mennonite serials coll) – 1r – 1 – mf#ATLA 1993-S000 – us ATLA [073]
Anger, Alfred *see* Dichtung des rokoko
Anger, D *see*
– Les dependances de l'abbaye de saint-germain-des-pres
Anger management for substance abuse and mental health clients : participant workbook / Reilly, Patrick M – Rockville MD: US Dept of Health & Human Services...2003 [mf ed 2003] – 1mf – 9 – us GPO [360]
Anger, Rudolf *see* Vorlesungen ueber die geschichte der messianischen idee
Anger, William Henry *see*
– Be your own lawyer
– Be your own lawyer, or, secrets of the law office
– Business manual
Angerburger kreis-blatt – Angerburg (Wgorzewo PL), 1855 10 feb-1865 26 aug, 1866 10 feb-1867 21 dec, 1868 11 feb-1872 16 feb, 1882 4 oct-1885, 1888-89, 1892-1903, 1906-07 – 10r – 1 – (with gaps) – gw Misc Inst [077]
Angerburger kreiszeitung – Angerburg (Wgorzewo PL), 1939-1941 31 may [gaps] – 5r – 1 – gw Misc Inst [077]
Angermanlands nyheter harnosandsposten – Haernoesand, Stockholm, Sweden. 1951-53 – 1 – sw Kungliga [078]
Angermuender anzeiger – Angermuende DE, 1870 & 1873, 1875-1934 jun, 1934 oct-dec – 1 – (title varies: 2 jan 1858: angermuender kreisblatt, 20 jan 1870: angermuender zeitung und kreisblatt, 1 jan 1935: angermuender tageblatt) – gw Misc Inst [074]
Angermuender kreisblatt *see* Angermuender anzeiger
Angers, Auguste Real *see* Assemblee legislative de quebec
Angers. France. Cathedrale *see* Cartulaire noir de la cathedrale d'angers.
Angers. France. St. Laud (Church) *see* Cartulaire du chapitre de saint-laud d'angers.
Angers, Francois-Real *see*
– Les revelations du crime
– Les revelations du crime ou cambray et ses complices
Angers, Philippe *see* Les seigneurs et premiers censitaires de st-georges-beauce et la famille pozer
Ange's von gardane kaiserl franz gesandtschafts-sekretaers tagebuch : einer reise durch die asiatische tuerkei nach persien, und wieder zurueck nach frankreich in den jahren 1807 und 1808 – Weimar 1809 [mf ed Hildesheim 1995-98] – 1v on 1mf – 9 – €40.00 – 3-487-26545-1 – (trans fr french; with ann) – gw Olms [915]
Die angestellten-bewegung / Allgemeiner Freier Angestelltenbund – 1921-25; 1928 31. Berlin. (Serial publications of German trade unions in the Memorial Library, University of Wisconsin-Madison) – 1 – us Wisconsin U Libr [331]
Angestellten-zeitung – Teplitz (Teplice CZ), 1921-24 – 1r – 1 – gw Misc Inst [331]
Angeville, A d' *see* Essai sur la statistique de la population francaise
Angewandte dogmatik : oder, polemik und irenik / Lange, Johann Peter – Heidelberg: K Winter, 1852 [mf ed 1991] – (= ser Christliche dogmatik 3) – 1mf – 9 – 0-524-00053-0 – mf#1989-2753 – us ATLA [240]
Anggaran belandja kotapradja djakarta-raya – Djakarta, 1956 – 16mf – 9 – mf#SE-207 – ne IDC [959]
Anggaran daerah propinsi sumatera tengah – Bukittingi, 1957 – 9mf – 9 – mf#SE-106=6 – ne IDC [950]
Anggaran daerah swatantra tingkat i sumatera barat : west sumatra (province) – Padang, 1958-1959 – 15mf – 9 – mf#SE-250 – ne IDC [959]
Anggaran dasar serikat-serikat Berita-negara RI *see* Indonesia
Anggaran keuangan daerah istimewa Jogjakarta *see* Jogjakarta, indonesia (city)
Angicos / Alves, Aluizio – Rio de Janeiro, Brazil. 1940 – 1r – us UF Libraries [972]
Angiologica – Basel, Switzerland 1966-73 – 1,5,9 – (cont by: blood vessels) – ISSN: 0003-3189 – mf#2046.02 – us UMI ProQuest [611]
Angioma of the head : (from the surgical clinic of the montreal general hospital) / Armstrong, George E – S.l: s.n, 1896? – 1mf – 9 – mf#37720 – cn Canadiana [617]
Angkatan 45 / Djiwa 45 – Djakarta, 1965-1966 – 3mf – 9 – (missing: 1965/1966(1-10)) – mf#SE-1416 – ne IDC [959]
Angkatan bersendjata – Djakarta, Indonesia: Edisi Pusat, 1965-1993 – 78r – 1 – us L of C Photodup [079]
Angkatan darat / Sari attensia – Djakarta, 1960-1964 – 25mf – 9 – (missing: 1960, v1-6(1-6); v7(8-9, 11-12); 1961, v8(1-2, 4, 7-12); 1962, v9(3, 5-12); 1963, v10(3-12); 1964, v11(3-end)) – mf#SE-596 – ne IDC [959]
Angkatan Darat Madjalah Angkatan Darat Menjambut pembukaan kembali AMN *see* Indonesia
Angkatan Darat Madjalah Angkatan Darat Penerangan Angkatan Darat *see* Indonesia
Angkatan darat pusat perpustakaan madjalah : indonesia – Bandung, 1961-1971 – 24mf – 9 – (missing: 1961(1-2); 1966(15-end); 1967-1971) – mf#SE-590 – ne IDC [959]

Angkatan darat pusat sedjarah militer madjalah sedjarah militer angkatan darat : indonesia – Bandung, [195?]1960-1965 – 26mf – 9 – (missing: [195?], v1-4; 1961, v9; 1963, v14) – mf#SE-1520 – ne IDC [959]

Angkatan darat republik indonesia / Madjalah H U B – Djakarta, 1956-1959 – 17mf – 9 – (missing: 1956, v1(1-2); 1957, v2(1, 12); 1958, v3(1-11); 1959, v4(2-3, 5, 9-12)) – mf#SE-592 – ne IDC [959]

Angkatan kepolisian komando antar daerah kepolisian 1 : sumatera laporan kriminil sumatera – Medan, 1966-1967 – 2mf – 9 – mf#SE-1521 – ne IDC [950]

Angkatan perang corps pulisi militer gadjah mada : indonesia – Djakarta, 1950-1968 – 91mf – 9 – (missing: several issues) – mf#SE-587 – ne IDC [959]

Angkatan udara angkasa ass dir penerangan, departement angatan udara ri : indonesia – Djakarta, 1950-1971 – 212mf – 9 – (missing: several issues) – mf#SE-581 – ne IDC [959]

Angkor / MacDonald, Malcolm – London [England]: J Cape [1958] [mf ed 1989] – [pl] 1r with other items – 1 – (with 112 photos by loke wan tho & aut) – mf#mf-10289 seam reel 023/02 [§] – us CRL [959]

Angkor : eine reise nach den ruinen von angkor / Suter, Hugo – Berlin: D Reimer [E Vohsen] 1912 [mf ed 1989] – [pl/ill] 1r with other items – 1 – mf#mf-10289 seam reel 021/11 [§] – us CRL [915]

Angkor / Vasileiou, I – Paris: A Moranace [1971] [mf ed 1989] – [pl] 1r with other items – 1 – (trans fr greek by maximo portassi) – mf#mf-10289 seam reel 020/07 [§] – us CRL [930]

Angkor, an introduction / Coedes, George – Hong Kong, New York: OUP 1963 [mf ed 1989] – [ill] 1r with other items – 1 – (trans of: pour mieux comprendre angkor & edited by emily floyd gardiner; photos by george bliss) – mf#mf-10289 seam reel 023/04 [§] – us CRL [915]

Angkor and the khmer empire / Audric, John – London [England]: R Hale [1972] [mf ed 1989] – [ill] 1r with other items – 1 – (with bibl) – mf#mf-10289 seam reel 023/10 [§] – us CRL [830]

Angkor, fransk indo-kina / Arentz, Erik – Oslo [Norway]: Some 1931 [mf ed 1989] – [pl/ill] 1r with other items – 1 – mf#mf-10289 seam reel 022/09 [§] – us CRL [915]

Angkor: guide henri parmentier / Parmentier, Henri – Phnom-Penh: EKLIP [anciennement Albert Portail] 1960 [mf ed 1989] – 1r with other items – 1 – mf#mf-10289 seam reel 022/11 [§] – us CRL [915]

Angkorskaia imperiia : sotsialno-ekonomicheskii i gosudarstvennyi stroi kambodzhi v 9-14 vv / Sedov, Leonid – Moskva [Russia]: Nauka 1967 [mf ed 1989] – 1r with other items – 1 – mf#mf-10289 seam reel 030/03 [§] – us CRL [959]

Anglade, A *see* The dolmens of the pulney hills

Anglais au moyen age *see* English wayfaring life in the middle ages (14th century)

Une anglaise intellectuelle en france sous la restauration : miss mary clark / Smith, Marion Elmina – Paris: Champion, 1927 – (= ser Les femmes [coll]) – 2mf – 9 – mf#8124 – fr Bibl Nationale [920]

Anglaises pour rire / Sewrin, M – Paris, France. 1822 – 1r – us UF Libraries [440]

Angle orthodontist – Appleton WI 1988+ – 1,5,9 – ISSN: 0003-3219 – mf#1133 – us UMI ProQuest [617]

Anglebert, Jean Henry *see* Pieces de clavecin, composees par j henry d'anglebert...avec la maniere de les jouer

Angleria, Camillo *see* La regola del contraponto, e della musical compositione...

Angleria, Pedro Martir *see*
- De orbe novo
- De rebus oceanicis & orbe novo decada tres
- Fuentes historicas sobre colon y america
- Premiere decade du orbe novo

The angler's guide to eastern canada : showing where, when and how to fish for salmon, bass, ouananiche and trout / Chambers, Edward Thomas Davies – Quebec: "Morning Chronicle", 1898? – 2mf – 9 – mf#02148 – cn Canadiana [790]

Angles, Higini *see* Codex el musical de las huelgas

L'angleterre chretienne avant les normands / Cabrol, Fernand – 2e ed. Paris: V Lecoffre, 1909 [mf ed 1990] – (= ser Bibliotheque de l'enseignement de l'histoire ecclesiastique) – 1mf – 9 – 0-7905-6801-2 – (in french. incl bibl ref) – mf#1988-2801 – us ATLA [240]

L'angleterre ou description historique et topographique du royaume uni de la grande-bretagne : contenant: les comtes de la principaute de galles, des royaumes d'ecosse, d'irlande et d'angleterre, les iles orcades, shetland, etc / Depping, Georges – Paris 1824 [mf ed Hildesheim 1995-98] – 6v on 12mf – 9 – €120.00 – 3-487-28888-5 – gw Olms [914]

Angleton first baptist church – Angleton, TX. 1908-38 – 1 – $18.09 – us Southern Baptist [242]

Anglia rediviva : englands recovery / Sprigg, Joshua – 1647 – 9 – us Scholars Facs [941]

Anglican and episcopal history – Austin TX 1987+ – 1,5,9 – (cont: historical magazine of the protestant episcopal church) – ISSN: 0896-8039 – mf#442,01 – us UMI ProQuest [242]

Anglican baptisms 1900-1939 and confirmations 1900-1947 – [mf ed 1999] – 1mf – 9 – A$5.50 – 0-949124-86-9 – at Northern [242]

Anglican bishops versus the catholic hierarchy – London, England. 1851 – 1r – us UF Libraries [242]

The anglican career of cardinal newman / Abbott, Edwin Abbott – London; New York: Macmillan, 1892 – 3mf – 9 – 0-7905-5560-3 – (incl bibl ref) – mf#1988-1560 – us ATLA [241]

The anglican cathedral church of saint james : mount zion jerusalem / Johns, J W – London 1842 – (= ser 19th c art & architecture) – 3mf – 9 – mf#4.2.1609 – uk Chadwyck [720]

The anglican church : or, the introduction and continuity of the christian faith in the british isles / Cole, Robert Henry – New York: James Pott, 1892 – 1mf – 9 – 0-524-02565-7 – mf#1990-4377 – us ATLA [241]

Anglican church architecture / Barr, James – [2nd ed]. Oxford 1843 – (= ser 19th c art & architecture) – 3mf – 9 – mf#4.2.401 – uk Chadwyck [720]

The anglican church in south america / Every, E F – London: S.P.C.K., 1915 – 1mf – 9 – 0-7905-4418-0 – mf#1988-0418 – us ATLA [241]

The anglican church in the nineteenth century : indicating her relative position to dissent in every form, and presenting a clear and unprejudiced view of puseyism and orthodoxy = Zustaende der anglicanischen kirche / Uhden, Hermann Ferdinand – London: Hatchard, 1844 – 1mf – 9 – 0-524-04183-0 – (in english) – mf#1990-4987 – us ATLA [241]

Anglican Church of Canada *see* Canadian churchman

Anglican church registers index 1902-1953 – [mf ed 1999] – 1mf – 9 – A$8.80 – 0-949124-87-7 – (darwin christ church marriages 1902-42; darwin christ church marriages 1946-52; darwin christ church marriages 1952-53; alice springs marriages 1936-49) – at Northern [980]

Anglican claims in the light of history : a paper read before the catholic truth society of ottawa, on the 12th december, 1893, in reply to a lecture intituled "roman methods of controversy" delivered by the rev w j muckleston, on the 15th may, 1893 / Pope, Joseph – Ottawa?: s.n, 1893? – 1mf – 9 – mf#11973 – cn Canadiana [242]

The anglican communion (sect f) / duty of the church to the young (sect g) : speeches and discussions together with the papers published for the consideration of the congress / Pan-Anglican Congress 1908 – London: SPCK; New York: E S Gorham, 1908 [mf ed 1986] – 2mf – 9 – 0-8370-9096-2 – mf#1986-3096 – us ATLA [242]

The anglican episcopate and the american colonies / Cross, Arthur Lyon – New York: Longmans, Green, 1902 – (= ser Harvard Historical Studies) – 1mf – 9 – 0-7905-4219-6 – (incl bibl ref) – mf#1988-0219 – us ATLA [240]

Anglican liberalism / Handley, Hubert et al – New York: G P Putnam 1908 [mf ed 1991] – (= ser Crown theological library 24) – 1mf – 9 – 0-7905-9955-4 – mf#1989-1680 – us ATLA [242]

The anglican ministry : its nature and value in relation to the catholic priesthood / Hutton, Arthur Wollaston – London: C Kegan Paul, 1879 – 2mf – 9 – 0-7905-4762-7 – mf#1988-0762 – us ATLA [241]

Anglican misrepresentations / Addis, William Edward – London, England. 1872 – 1r – us UF Libraries [241]

Anglican orders : a speech...oct 15 1896 / Browne, George Forrest – London: SPCK 1896 [mf ed 1993] – (= ser Church historical society (series) 17) – 1mf – 9 – 0-524-05530-0 – (incl ind) – mf#1990-5134 – us ATLA [242]

The anglican reformation / Clark, William – New York: Scribner, 1900 – (= ser Ten Epochs of Church History) – 2mf – 9 – 0-524-01647-X – mf#1990-0468 – us ATLA [242]

The anglican reformation / Clark, William Robinson – New York: Christian Literature, 1897 – (= ser Ten epochs of church history 10) – 6mf – 9 – (incl ind) – mf#05259 – cn Canadiana [242]

Anglican registers of burials : darwin burials 1908-27, 1933-41; alice springs burials 1934-68 – [mf ed 1999] – 1mf – 9 – A$5.50 – 0-949124-88-5 – at Northern [242]

The anglican revival / Overton, John Henry – London: Blackie 1897 [mf ed 1990] – (= ser The victorian era series) – 1mf – 9 – 0-7905-5780-0 – mf#1988-1780 – us ATLA [242]

Anglican theological review – 12(1929-1930)-34(1952) – 141mf – 9 – €269.00 – ne Slangenburg [242]

Anglican theological review – Evanston IL 1918+ – 1,5,9 – ISSN: 0003-3286 – mf#1843 – us UMI ProQuest [242]

Anglicanism and reunion : sermon...jun 14 1908... / Henson, Hensley – London: Hugh Rees, 1908 [mf ed 1993] – 1mf – 9 – 0-524-08384-3 – mf#1993-3084 – us ATLA [242]

Anglicanism and the fathers / Addis, William Edward – London, England. 1872 – 1r – us UF Libraries [242]

Anglicanism considered in its results / Dodsworth, William – London, England. 1851 – 1r – us UF Libraries [242]

Anglican-ritualism as seen by a catholic and foreigner : a series of essays / Martin, Paulin – London: Burns & Oates, 1881 [mf ed 1992] – 1mf – 9 – 0-524-03446-X – (with app by paulin martin) – mf#1990-4706 – us ATLA [230]

Anglicanus, Clemens *see*
- Holy eucharist
- John foster, (the "essayist,") vindicated from the aspersions of mr...
- Remarks upon mr evanson's preface to his translation

Anglicanus, Julius *see* Missionary bishops

Anglicanus scotched / Dods, Marcus – Edinburgh, Scotland. 1828 – 1r – us UF Libraries [241]

L'anglicisme voila l'ennemi : causerie faite au cercle catholique de quebec le 17 decembre 1879 / Tardivel, Jules Paul – Quebec?: "Canadien", 1880 – 1mf – 9 – mf#24458 – cn Canadiana [440]

Anglicismes et canadianismes / Buies, Arthur – Quebec: C Darveau, 1888 [mf ed 1979] – 1mf – 9 – 0-665-00336-6 – mf#00336 – cn Canadiana [440]

Anglicismes et canadianismes / Buies, Arthur – Quebec: C Darveau...1888 [mf ed 1979] – 2mf – 9 – mf#SEM105P19 – cn Bibl Nat [440]

Anglim, John *see* Palau's strategic position places democracy at risk

Anglistische forschungen / ed by Hoops, J – Heidelberg. v1-61. 1901-1925 – 234mf – 8 – mf#H-173 – ne IDC [420]

Anglo african – Grahamstown SA, 1855-70 – 1 – sa National [079]

Anglo american : a journal of literature, news, politics, the drama, fine arts, etc – La Salle IL 1843-47 – 1 – mf#4555 – us UMI ProQuest [420]

Anglo Persian Oill Co *see* Oil exploration work in papua and new guinea

Anglo portuguese negotiations relating to bombay, 1660-1677 / Khan, Shafa'at Ahmad – London: Oxford University Press, [1922?] – (= ser Samp: indian books) – us CRL [327]

Anglo saxon review – La Salle IL 1899-1901 – 1 – mf#3898 – us UMI ProQuest [420]

The anglo-african – Lagos. Nigeria. -w. Jun-Dec 1863, Jul 1864-Dec 1865. (Imperfect). (37 ft) – 1 – uk British Libr Newspaper [079]

Anglo-african magazine – Ann Arbor MI 1859-60 – 1 – mf#3083 – us UMI ProQuest [079]

Anglo-american and continental courier 1903 – the american blue book 1905-06 – (= ser American newspapers and periodicals published in the uk and europe 18th-20th centuries) – 1r – 1 – £55.00 – uk World [072]

Anglo-american bible revision – New York: American Sunday School Union, 1879 [mf ed 1985] – 1mf – 9 – 0-8370-2084-0 – (incl ind) – mf#1985-0084 – us ATLA [220]

The anglo-american illustrated news – 1909-14 – (= ser American newspapers and periodicals published in the uk and europe 18th-20th centuries) – 5r – 1 – £220.00 – uk World [072]

Anglo-american magazine – La Salle IL 1843 – 1 – mf#3933 – us UMI ProQuest [420]

The anglo-american magazine – Toronto: T MacLear, 1852-[1855] – 9 – mf#P04640 – cn Canadiana [410]

The anglo-american matrimonial journal – Toronto?: s.n, 1886-18– or 19– – 9 – ISSN: 1190-6960 – mf#P04015 – cn Canadiana [306]

Anglo-american political influences on rui barbosa / Pires, Homero – Rio de Janeiro, Brazil. 1949 – 1r – us UF Libraries [972]

The anglo-american sabbath : an essay read before the national sabbath convention, saratoga, august 11, 1863 / Schaff, Philip – [New York: New York Sabbath Committee, 1863?] – (= ser Document of the New York Sabbath Committee) – 1mf – 9 – 0-524-08560-9 – mf#1993-2085 – us ATLA [240]

Anglo-american times – London, 27 Oct 1865-7 Nov 1896 – 35r – 1 – uk British Libr Newspaper [072]

Anglo-american traveler 1902-03 *see* Anglo-saxon 1899 – american trade review 1902 – anglo-american traveler 1902-03

Anglo-assamese relations, 1771-1826 : a history of the relations of assam with the east india company from 1771 to 1826, based on original english and assamese sources / Bhuyan, Suryya Kumar – Gauhati: Dept of Historical and Antiquarian Studies in Assam, 1949 – (= ser Samp: indian books) – us CRL [954]

The anglo-californian – 1896-98 – (= ser American newspapers and periodicals published in the uk and europe 18th-20th centuries) – 1r – 1 – £55.00 – uk World [072]

Anglo-canadian copyright : with special reference to the canadian act of 1889 / Clayton, Henry R – London; New York: Novello, Ewer, [1889?] [mf ed 1980] – 1mf – 9 – 0-665-03053-3 – (repr fr: the musical times) – mf#03053 – cn Canadiana [346]

Anglo-catholic theory / Price, Bonamy – London, England. 1852 – 1r – us UF Libraries [241]

Anglo-catholicism / Foster, Alfred Edye Manning – London: TC & EC Jack, [1914?] [mf ed 1992] – (= ser The people's books 73; Anglican/episcopal coll) – 1mf – 9 – 0-524-03612-8 – (incl bibl ref) – mf#1990-4772 – us ATLA [242]

Anglo-catholicism not apostolical : being an inquiry into the scriptural authority of the leading doctrines advocated in the tracts for the times... / Alexander, William Lindsay – Edinburgh: Adam & Charles Black, 1843 [mf ed 1992] – 2mf – 9 – 0-524-04946-7 – mf#1990-1349 – us ATLA [241]

Anglo-Catholicus, Presbyter *see* Puseyism, not a popish bane, but a catholic antidote

Anglo-celt – Cavan, Ireland 6 feb 1846-29 apr 1858, 10 dec 1864-20 nov 1869, 2 jul 1870-8 nov 1873, 10 jan 1885- [mf 1986-] – 1 – uk British Libr Newspaper [072]

The anglo-colorado mining and milling guild – 1898-1912 – (= ser American newspapers and periodicals published in the uk and europe 18th-20th centuries) – 4r – 1 – £180.00 – uk World [072]

The anglo-egyptian sudan / Great Britain. War Office. General Staff. Geographical Section – Rev. Dec. 1921. London, 1921 – 1 – us CRL [916]

Anglo-egyptian sudan handbook series – London: H M Stationery Off. Monographic series. No 2 – 1 – us CRL [916]

Anglo-indian poetry / Seshadri, P – Benares: Indian Bookshop, 1928 – (= ser Samp: indian books) – us CRL [410]

Anglo-indian rule historically considered : a lecture delivered at the taylor institution, apr 28 1876 / Owen, Sidney James – Oxford 1876 – (= ser 19th c british colonization) – 1mf – 9 – mf#1.1.2517 – uk Chadwyck [954]

Anglo-indian studies / Mitra, Siddha Mohana – London, New York: Longmans, Green & Co, 1913 – (= ser Samp: indian books) – us CRL [327]

Anglo-israel : or, the british nation the lost tribes of israel / Poole, William Henry – Toronto?: Bengough, 1879 – 1mf – 9 – mf#12071 – cn Canadiana [939]

Anglo-israel : or, the saxon race proved to be the lost tribes of israel: in nine lectures / Poole, William Henry – Toronto: W Briggs; Montreal: C W Coates, 1889 – 8mf – 9 – (int by william henry withrow) – mf#12072 – cn Canadiana [939]

The anglo-israel ensign – Truro, NS: J Ross, [1880-188-?] – 9 – ISSN: 1190-6758 – mf#P04299 – cn Canadiana [939]

Anglo-japanese gazette, 1902-1909 – [mf ed Marlborough 1991] – (= ser Asian journals; Business and financial papers series) – 4r – 1 – £380.00 – (r3: jul 1902-dec 1904 [v1 n1-v5 n30]; r4: jan-dec 1905 [v6 n31-v7 n42]; r5: jan 1906-dec 1907 [v8 n43-v11 n66]; r6: jan 1908-jun 1909 v12 n67-v14 n84; with d/g) – uk Matthew [380]

Anglo-jewish pamphlets from the jewish theological seminary – Clearwater Publ Co – (= ser Research colls in judaica) – 628mf (24:1) – 9 – $4185.00 – (with p/g) – us UPA [270]

Anglo-latin satirical poets and epigrammatists of the twelfth century (rs59) / ed by Wright, T – (= ser The rolls series (rs)) – (v1 1872 €17. v2 1872 €19) – ne Slangenburg [410]

Anglo-maori warder – Auckland, NZ. 1848 – 1r – 1 – mf#11.67 – nz Nat Libr [079]

Das anglonormannische erbfolgesystem / Brunner, Heinrich – Leipzig, Duncker & Humblot, 1869. 88 p. LL-105 – 1 – us L of C Photodup [346]

Anglo-panjabi dictionary of legal terms = Kanuni samketa da angrezi-panjabi kosha / Singh, Bhagat – Patiala: Mahikama Pañjabi, 1953 – (= ser Samp: indian books) – us CRL [340]

Anglo-panjabi technical terms : school subjects = Angrezi-panjabi sanketawali: sakula-mazamuna – Patiala: Mahikama Panjabi, 1953 – (= ser Samp: indian books) – us CRL [056]

Anglo-portuguese news – Lisbon, Portugal. 1937-92 – 17r – 1 – us L of C Photodup [074]
Anglo-portuguese relations in south-central africa 1890-1900 / Warhurst, Philip R – London, England. 1962 – 1r – us UF Libraries [960]
Anglo-punjabi dictionary = Aingalo panjabi dikashanari / Amola, Aisa Aisa – Ammritasara: Bha Catara Singha Jiwana Singha Pusatakam Wale, [19–] – (= ser Samp: indian books) – us CRL [040]
Anglo-russian *see* Russian journals
The anglo-russian – London. -m. Jul 1897-Jul 1914. (2 reels) – 1 – uk British Libr Newspaper [072]
Anglo-saxon – Ottawa, Canada 1 dec 1887-dec 1896, jun 1897, may 1899-jan 1900 – 1 – uk British Libr Newspaper [071]
The anglo-saxon – Ottawa: [s.n, 1887-1900?] – 9 – mf#P05013 – cn Canadiana [071]
Anglo-saxon 1899 – american trade review 1902 – anglo-american traveler 1902-03 – (= ser American newspapers and periodicals published in the uk and europe 18th-20th centuries) – 1r – 1 – £55.00 – uk World [072]
Anglo-saxon and mediaeval manuscript collection / Corpus Christi College. Cambridge – 7 sect – 155r – 1 – £8750.00 – (coll consists of mss deposited by archbishop parker on his death in 1575; incl canterbury gospels, peterborough psalter, chaucer's troilus & criseyde, anglo-saxon chronicles, dover bible & the bury bible; sect 1: theology [48r] £2750; sect 2: bible & biblical studies [17r] £1000; sect 3: law mss [7r] £430. sect 4: literature & music with greek & latin classics [18r] £1050; sect 5: history [38r] £2150; sect 6: medica' & natural sciences [11r] £650; sect. 7: secular studies [16r] £750) – uk World [090]
Anglo-saxon bibles and "the book of cerne" / ed by Doane, A N – [mf ed Tempe AZ, 2002] – (= ser ASMMF) – 1112 folios – 9 – $120.00v / $76.00v [institution] ($96v / £60v if part of subsc) – mf#mr187 – us MRTS [090]
The anglo-saxon charms / Grendon, Felix – 1909 – 1mf – 9 – 0-524-01362-4 – mf#1990-2374 – us ATLA [941]
Anglo-saxon chronicle – London, England. 1953 – 1r – us UF Libraries [941]
Anglo-saxon chronicle (rs23) : according to the several original authorities / ed by Thorpe, B – 1861 – (= ser The rolls series (rs)) – 2v – €27.00 – (trans by ed) – ne Slangenburg [941]
The anglo-saxon church : its history, revenues, and general character / Soames, Henry – 4th rev, augm, and corr ed. London: JW Parker, 1856 – 1mf – 9 – 0-7905-5966-8 – mf#1988-1966 – us ATLA [240]
Anglo-Saxon Federation of America *see*
– Bulletin of the anglo-saxon...
– Destiny
– Messenger of the covenant
Anglo-saxon gospels / ed by Liuzza, Roy M & Doane, A N – [mf ed Binghamton NY, 1995] – (= ser ASMMF) – 1100 folios – 8 – $120.00v / £76.00v [institution] ($96v / £60v if part of subsc) – 0-86698-183-7 – mf#mr144 – us MRTS [090]
The anglo-saxon scop / Anderson, Lewis Flint – [Toronto]: University Library, 1903 – (= ser University of toronto studies. philological series 1) – 1mf – 9 – 0-665-75651-8 – mf#75651 – cn Canadiana [420]
Anglo-saxon superiority / Demolins, Edmond – London, England. 1899 – 1r – us UF Libraries [941]
The anglo-saxon version of the book of psalms : commonly known as the paris psalter / Bruce, James Douglas – Baltimore: Modern Language Assoc of America, 1894 [mf ed 1985] – 1mf – 9 – 0-8370-2488-9 – mf#1985-0488 – us ATLA [221]
Anglo-Scotus *see* Opera
Anglo-scotus again / Lockhart, John – Newcastle upon Tyne, England. 1834 – 1r – us UF Libraries [240]
Anglo-sikh relations : chapters from j d cunningham's "history of the sikhs" / Cunningham, Joseph Davey; ed by Banerjee, Anil Chandra – Calcutta: A Mukherjee & Co, 1949 – (= ser Samp: indian books) – us CRL [954]
The anglo-telugu primer / Narasayya, Maddali Lakshmi – 2nd ed. Madras: Higginbotham, 1869 – 1 – us CRL [490]
Angly, Edward [comp] *see* Oh yeah?
Angola / Albuquerque Felner, Alfredo De – Coimbra, Portugal. 1933 – 1r – us UF Libraries [960]
Angola : bulletin d'information – Leopoldville: Comite executif du Front national de liberation de l'Angola, [v1, n4/5-6, 8 (oct 15/31-nov 15, dec 15/31 1963] (bimthly) – (= ser Coll of lusophone african newspapers and serials) – 1r – 1 – us CRL [079]
Angola : bulletin d'information de la representation du gouvernement revolutionnaire angolais en rau – Cairo: GRAE [n2 (mar 1964)] – 1r – 1 – us CRL [960]
Angola : coracao do imperio / Santos, Alfonso Costa Valdez Thomaz Dos – Lisboa, Portugal. 1945 – 1r – us UF Libraries [960]
Angola : cultura e revolucao. bulletin bilingue du centro de estudos angolanos / Centro de Estudos Angolanos – [Alger]: O Centro. [feb 4, oct 4, 1964; mar 8 1966] – (= ser Liberation movements in lusophone Africa) – 1r – 1 – us CRL [960]
Angola : curso de extensao universitaria, ano lectivo de 1963-1964 / Universidade Tecnica De Lisboa Instituto Superior De Ciencias – Lisboa, Portugal. 1964 – 1r – us UF Libraries [960]
Angola : essa desconhecida / Pires, Antonio – Luanda, Angola. 1964 – 1r – us UF Libraries [960]
Angola : eu quero falar contigo / Mota, Mario – Lisboa, Portugal. 1962 – 1r – us UF Libraries [960]
Angola / Ferreira Pinto, Julio – Lisboa, Portugal. 1926 – 1r – us UF Libraries [960]
Angola / Gonzaga, Norberto – Lisboa, Portugal. 1967 – 1r – us UF Libraries [960]
Angola / Jack, Homer Alexander – New York, NY. 1960 – 1r – us UF Libraries [960]
Angola : portos e transportes / Castro, Eduardo Gomes De Albuquerque – Luanda, Angola. 1968 – 1r – us UF Libraries [960]
Angola : revista mensal ilustrada – Loanda: "Angola", [n2-4 (feb-apr1923)] – (= ser Coll of lusophone african newspapers and serials) – 1r – 1 – us CRL [079]
Angola clef de l'afrique / Valahu, Magur – Paris, France. 1966 – 1r – us UF Libraries [960]
Angola. Conselho Legislativo *see* Actas de sessao
Angola do eu coracao / Falcato, Joao – Lisboa, Portugal. 1961 – 1r – us UF Libraries [960]
Angola flash – New York: The Union. [v1, n1, jul 1971] – (= ser Liberation movements in lusophone Africa) – 1r – 1 – us CRL [960]
Angola in flames / Panikkar, Kavalam Madhusudan – New York, NY. 1962 – 1r – us UF Libraries [960]
Angola in perspective / Egerton, F Clement C – London, England. 1957 – 1r – us UF Libraries [960]
Angola informations : bulletin d'information – [Alger]: Mission d'Alger du Gouvernement rbevolutionnaire de l'Angola en exil, GRAE, [n 7 jan 8 1964; n9 jun 21 1964; n12-15 nov 30 1964-apr 24 1965] – (= ser Liberation movements in lusophone Africa) – 1r – 1 – (front national de liberation de l'angola) – us CRL [960]
Angola, mozambique, guinee-bissau et le colonialisme portugais : extraits de la presse de langue portugaise d'europe et d'afrique – [Maisons-Alfort: A Hadad]. [n1-10/12 jun 1972-sep 1973] (irreg) – (= ser Liberation movements in lusophone Africa) – 1r – 1 – us CRL [960]
Angola na africa deste tempo / Rebelo, Horacio De Sa Viana – Lisboa, Portugal. 1961 – 1r – us UF Libraries [960]
Angola operaria – Kinshasa. [n2-3 1971; n1-8 1972; n1/2 1973; unnumbered 1973?] (irreg) – 1 – us CRL [321]
Angola perante uma conspiracao internacional / Diogo, Alfredo – Luanda, Angola. 1961 – 1r – us UF Libraries [960]
Angola, prospects for durable peace and economic reconstruction : hearing...house of representatives, 107th congress, 2nd session, june 13 2002 / United States. Congress. House. Committee on International Relations. Subcommittee on Africa – Washington: US GPO 2002 [mf ed 2002] – 1mf – 9 – 0-16-068787-X – us GPO [327]
Angola. Reparticao de Estatistica Geral *see*
– Anuario estatistico 1933-1973
– Anuario estatistico de angola
Angola und seine seehafen / Sendler, Gerhard – Hamburg, Germany. 1967 – 1r – us UF Libraries [960]
Angolais / Davezies, Robert – Paris, France. 1965 – 1r – us UF Libraries [960]
O angolense – Loanda: [s.n, [sep 16 1907] (wkly) – (= ser Coll of lusophone african newspapers and serials) – 1r – 1 – us CRL [079]
Angolite – 1979 mar/apr-1982 jul/aug; 1982 sep/oct-1984; 1985-1987 jun; 1987 jul/aug-1989 – 1 – mf#653710 – us WHS [071]
Angouleme, Marie-Therese Charlotte *see* Memoires particuliers
Angoulvant, Gabriel Louis *see* La pacification de la cote d'ivoire, 1908-1915
Angove, John *see* In the early days
Der angriff – Berlin DE, 1927 4 jul-1932 apr, 1932 sep-1945 21 apr – 1 – (filmed by misc inst: 1931 jan-jun [2r]; 1932 1 apr-sep [1r]) – mf#1776 – gw Mikropress; gw Misc Inst [074]
Angry men, laughing men / Brown, Wenzell – New York, NY. 1947 – 1r – us UF Libraries [972]
Angst : novelle / Zweig, Stefan – Leipzig, Germany. 1925 – 1r – us UF Libraries [830]
Die angst in den interpretationen der existenzphilosophie und der tiefenpsychologie / Streck, Friedrich Karl – Frankfurt a.M., 1978 – 2mf – 9 – 3-89349-377-8 – gw Frankfurter [120]
Anguenot, Joelle *see* Du sol a l'arbre
Anguiano, Angel *see* Tratado de cosmografia
Anguish / Ramos, Graciliano – New York, NY. 1946 – 1r – us UF Libraries [972]
Anguita Valdivia, Jose *see* Apuntes biograficos de don juan carrillo sanchez
Angular leaf spot and fruit rot of cucumbers caused by bacterium / Weber, George F – Gainesville, FL. 1929 – 1r – us UF Libraries [634]
Angulo, D *see* Historia del arte hispanoamericano. tomo 1. barcelona, 1945
Angulo Guridi, Javier *see*
– Iguaniona
Angulo-Kinsler, Rosa M *see* Exploration and control of leg movements in infants
Angus, J Keith *see* Amateur acting
Angus, Joseph *see*
– An analysis of butler's analogy of religion
– The bible hand-book
– Christ our life
– Six lectures on regeneration
Angus journal – St Joseph MO 1979+ – 1,5,9 – ISSN: 0194-9543 – mf#11969 – us UMI ProQuest [636]
Angus, Samuel *see*
– The environment of early christianity
– The sources of the first ten books of augustine's de civitate dei
Angustia / Viciedo Arteche, Ignacio – Miami, FL. 1962 – 1r – us UF Libraries [972]
Angustia y evasion de julian del casal / Portuondo, Jose Antonio – Habana, Cuba. 1937 – 1r – us UF Libraries [972]
Anh chi yeu dau : truyen dai / Dinh, Tien Luyen – [Saigon]: Nguyen-Dinh Vuong 1974 [mf ed 1992] – on pt of 1r – 1 – mf#11052 r386 n9 – us Cornell [830]
Anh lua cuoi duong ham : truyen dai / Mai Thao – [Saigon]: Anh Loc 1974 [mf ed 1992] – on pt of 1r – 1 – mf#11052 r392 n5 – us Cornell [830]
Anh sang – Hue [n1-52] – 1r – 1 – mf#mf 1935 – us CRL [079]
Anh Tho *see* Mua xuan, mau xanh
Anhaltin, C M *see* Architectura
O anhanguera – Sao Paulo, SP: Typ Americana, 18 jul-29 ago 1869 – (= ser Ps 19) – mf#P18,02,25 – bl Biblioteca [410]
Anhelos de un ciudadano / Calderon, Jose Tomas – San Salvador, El Salvador. 1951 – 1r – us UF Libraries [972]
Anhelos y esperanzas / Ferrer Hernandez, Gabriel – San Juan, Puerto Rico. 1962 – 1r – us UF Libraries [972]
O anhembi : jornal dedicado aos interesses do municipio – Tiete, SP. 23 mar 1879 – (= ser Ps 19) – bl Biblioteca [073]
Anhorn, Roland *see* Sozialstruktur und disziplinarindividuum
An-hui i nien lai chih nung ts'un chiu chi chi tiao ch'a – [China]: Kai t'ing, 1936 – (= ser P-k&k period) – us CRL [630]
An-hui jih-pao – Hofei, Anhwei. June 1, 1952-Oct 14 1962. 3 reels. Incomplete – 48.75 – 1 – us Chinese Res [079]
An-hui sheng t'ung chi nien chien – [China: An-hui sheng t'ung chi nien chien wei yuan hui, 1934] – (= ser P-k&k period) – us CRL [315]
An-hui tuan ching chi tiao ch'a tsung pao kao shu – [China: T'ieh tao pu ts'ai wu ssu tiao ch'a k'o] – (= ser P-k&k period) – us CRL [339]
Ani und anav in den psalmen / Rahlfs, Alfred – Goettingen: Dieterich, 1892 [mf ed 1985] – 1mf – 9 – 0-8370-4827-3 – mf#1985-2827 – us ATLA [221]
Anicet-Bourgeois, Auguste *see*
– Jacques coeur, l'argentier du roi
– Mademoiselle de la faille
– Nonne sanglante
– Pascal et chambord
– Perruquiere de meudon
Animadversiones...acerca de la receta del unguento de mercurio / Jimenez Guillen, F – Sevilla, 1626 – 1mf – 9 – sp Cultura [610]
Animadversions on dr haweis' impartial and succinct history / Milner, Isaac – Cambridge, England. 1800 – 1r – us UF Libraries [240]
Animadversions upon sir william hamilton's pamphlet / Cunningham, William – Edinburgh, Scotland. 1843 – 1r – us UF Libraries [240]
Animal and vegetable physiology considered with reference to natural theology / Roget, Peter Mark – London, 1834 – (= ser 19th c evolution & creation) – 14mf – 9 – mf#1.1.10851 – uk Chadwyck [110]
Animal behaviour – Oxford, England 1953+ – 1,5,9 – ISSN: 0003-3472 – mf#8455 – us UMI ProQuest [590]
Animal conservation – Oxford, England 2001+ – 1 – ISSN: 1367-9430 – mf#27996 – us UMI ProQuest [639]
Animal feed science and technology – Oxford, England 1976+ – 1,5,9 – ISSN: 0377-8401 – mf#42012 – us UMI ProQuest [636]
Animal genetics – Oxford, England 1970+ – 1,5,9 – (cont: animal blood groups and biochemical genetics) – ISSN: 0268-9146 – mf#6950,01 – us UMI ProQuest [575]
Animal health & nutrition – Mt Morris IL 1985-88 – 1,5,9 – (cont by: large animal veterinarian covering health and nutrition) – ISSN: 0896-4807 – mf#14647.04 – us UMI ProQuest [636]
Animal husbandry in the caribbean / Livestock Conference, Port-of-Spain, Trinidad – Port-of-Spain, Trinidad and Tobago. 1950 – 1r – us UF Libraries [636]
Animal intelligence : experimental studies / Thorndike, Edward Lee – New York: The Macmillan Co 1911 [mf ed 1987] – 1r [ill] – 1 – (filmed with: the power and beauty of superb womanhood / mcfadden, b) – mf#10645 – us Wisconsin U Libr [150]
Animal kingdom – Bronx NY 1975-89 – 1,5,9 – (cont by: wildlife conservation) – ISSN: 0003-3537 – mf#15198.01 – us UMI ProQuest [639]
The animal kingdom considered anatomically, physically and philosophically : the organs of generation, and the formation of the foetus in the womb, after which follow chapters on the breasts and the periosteum / Swedenborg, Emanuel – Bryn Athyn, PA: Academy of the New Church, 1928 – 1r – 1 – us CRL [612]
Animal law – v1-8. 1995-2002 – 9 – (filming in process) – mf#117001 – us Hein [342]
Animal learning and behavior – Austin TX 1973+ – 1,5,9 – ISSN: 0090-4996 – mf#7027.01 – us UMI ProQuest [150]
Animal nutrition and health – Mt Morris 1985 – 1,5,9 – (cont by: animal health & nutrition) – ISSN: 0003-3553 – mf#14647.04 – us UMI ProQuest [636]
Animal painters of england from the year 1650 / Gilbey, Walter. 1st Baronet – London 1900 – (= ser 19th c art & architecture) – 9mf – 9 – mf#4.2.1548 – uk Chadwyck [750]
Animal products their preparation, commercial uses and value / Simmonds, Peter Lund – [London: Comm of Council on Education, 1877] – (= ser 19th c economics) – 5mf – 9 – mf#1.1.347 – uk Chadwyck [680]
Animal remains from harappa / Prashad, Baini – Delhi: Manager of Publ, 1936 – (= ser Samp: indian books) – us CRL [930]
Animal reproduction science – Oxford, England 1978+ – 1,5,9 – ISSN: 0378-4320 – mf#42013 – us UMI ProQuest [636]
Animal rights / Dolan, Edward F – New York, NY. 1986 – 1r – us UF Libraries [636]
Animal symbolism in ecclesiastical architecture / Evans, E P – London, 1896 – 5mf – 9 – mf#0-1243 – ne IDC [700]
Animal symbolism in ecclesiastical architecture / Evans, Edward Payson – New York: Henry Holt, 1896 [mf ed 1989] – 1mf – 9 – 0-7905-4471-7 – (incl bibl ref) – mf#1988-0471 – us ATLA [720]
Animal tales / Borchardt, Bernard F – s.l, s.l? . 1936 – 1r – us UF Libraries [636]
Animal Trap Company of America *see* Snap judgment
Animals – Boston MA 1868+ – 1,5,9 – ISSN: 0030-6835 – mf#8935 – us UMI ProQuest [636]
Animals' agenda – Baltimore MD 1992+ – 1,5,9 – ISSN: 0892-8819 – mf#18626.01 – us UMI ProQuest [639]
Animals of canada : fishes, birds and furred animals / Buies, Arthur – Ottawa?: s.n, 1900? [mf ed 1981] – 1mf – 9 – mf#15119 – cn Canadiana [590]
Animals' rights considered in relation to social progress : with a bibliographical appendix / Salt, Henry Stephens – New York: Macmillan & Co., 1894. – 1r – 1 – us CRL [303]
Animals without backbones / Buchsbaum, Ralph Morris – Chicago, IL. 1948 – 1r – us UF Libraries [590]
An animated molecule and its nearest relatives : an essay read before the american association of medical superintendents of asylums for the insane, at washington, dc, on the 10th of may, 1878 / Clark, Daniel – [Toronto?: s.n.], 1878 [mf ed 1980] – 1mf – 9 – 0-665-00660-8 – mf#00660 – cn Canadiana [616]
Animism : the seed of religion / Clodd, Edward – London: Constable, 1918, c1906 [mf ed 1992] – (= ser Religions ancient and modern) – 1mf – 9 – 0-524-05767-2 – (incl bibl ref) – mf#1991-0010 – us ATLA [200]
Das anionentransportprotein in der erythrozytenmembran der regenbogenforelle : untersuchung zur topographie des erythroiden bande 3-proteins / Stirnberg, Barbara – (mf ed 1996) – 2mf – 9 – €40.00 – 3-8267-2373-2 – mf#DHS 2373 – gw Frankfurter [574]

Aniq Filali, Rabea *see* Deux modes satiriques

Anis – Kabul, Afghanistan 1941-89 [mf ed Norman Ross] – 1 – mf#nrp-577 – us UMI ProQuest [079]

Anis al-jalis – Alexandria: Princess Aleksandra Avierino and Labibah Hashim, 1898-1908. yr 1 pt 1-yr 6 pt 12. 31 jan 1898-31 dec 1903 – (= ser Arabic journals and popular press) – 1r – 1 – $950.00 – us MEDOC [079]

Anishinaabe giigidowin – v1 n1-v6 n1 [1976 jan-1982: summer] – 1 – mf#363839 – us WHS [071]

Anisimov, M *see* Snabzhenie derevni sredstvami proizvodstva i selskokhoziaistvennaia kooperatsiia

Anisimov, V A *see* Delo [irkutsk: 1918]

Anita : the cuban spy / Willets, Gilson – London, England. 1898 – 1r – us UF Libraries [972]

Anita : souvenirs d'un contre-guerillas / Beaugrand, Honore – S.l: s.n, 18-? – 1mf – 9 – mf#03522 – cn Canadiana [830]

Anixter, Judah Eliezer *see* Hidushe avi

Anizan, Felix *see* Le dieu au coeur qui rayonne. paris, 1928

Anjaria, J J *see* The indian rural problem

Anjaria, Jashwantrai Jayantilal *see* An essay on gandhian economics

Anjou, Lars Anton *see* The history of the reformation in sweden

Anjuman : consolidated holdings – Tabriz, Gilan, Yazd, Isfahan, [1907-11] – 2r – 1 – $250.00 – (incl ind) – us MEDOC [079]

Anjuman – Tabriz. sal-i 1, shumarah-i 38-108. 17 zu'l hijjah 1324-26 jumada al-avval 1325 [1 feb 1907-7 jul 1907] and sal-i 2, shumarah-i 8. 22 ramazan 1326 [18 oct 1908] – 1r – 1 – $110.00 – (r incl umid) – us MEDOC [079]

Anjuman *see* Umid

Ankara – (= ser Vilayet salnames) – 9 – (1307m [1891] 5mf $75; 1311 [1893] 6mf $90; 1318 [1890] 5mf $75; 1325 [1907] 7mf $110) – us MEDOC [956]

Ankara : aksam haberleri = Ankara, aksam haberleri – Ankara: Ulus Mueessesesi. [jan1 1950-dec10 1952] (daily) – 6r – 1 – us CRL [079]

Ankara : edition francaise hebdomadaire du "hakimiyeti milliye" – Ankara, Turkey 22 mar 1934-6 jan 1938, 15 sep 1938-6 jun 1940 – 4r – 1 – (wanting: n7,105,116,119,181,243,309,323) – uk British Libr Newspaper [079]

Anker geschlippt : geschichte eines marineoffiziers / Dincklage-Campe, Friedrich, Freiherr von – Leipzig: Max Hesse [18–?] [mf ed 1993] – 1r [ill] – 1 – (filmed with: gedichte / deinhardstein) – mf#8539 – us Wisconsin U Libr [830]

Anker, Johan *see* Opperman se verwerking van die nasionale geskiedenis in enkele van sy gedigte

Anketell, Cyrus P *see*
– A pronouncing english-tamil dictionary with abbreviations

Anking newsletter – Wuhu, Anhwei, China: The American Episcopal Diocese of Anking. v18-22 n1 autumn 1937-oct 1941; ns: v2, 21-22 n2 1945(?)-jun 1948 (frequency varies) – (= ser Missionary periodicals from the china mainland) – 1r – 1 – $165.00 – (title varies) – us UPA [242]

Anklag vnd ernstliches ermanen gottes... / Bullinger, Heinrich – [Zuerich, Christoph Froschauer, 1528] – 1mf – 9 – mf#PBU-102 – ne IDC [240]

Die anklage – Bad Woerishofen DE, 1953 n1-15, 1954-1957 n3 – 1 – gw Misc Inst [074]

Ankle strength and rate of force development : implications for balance control / Hall, Courtney D – 1997 – 2mf – 9 – $8.00 – mf#PH 1626 – us Kinesology [612]

Anklesaria, E T D *see* The bandahishn

The ankole agreement, 1962 – Endagaano y'Ankole eya 1962. Entebbe, Uganda Protectorate, 1962. 34 p. LL-2291 – 1 – us L of C Photodup [960]

Ankoru tosako / Delaporte, Louis – Tokyo [Japan]: Heibonsha, Showa 45 [1970] [mf ed 1989] – (= ser Toyo bunko 162) – 1r with other items – 1 – mf#mf-10289 seam reel 030/13 [§] – us CRL [720]

Ankuendigung einer neuen ausgabe der griechischen uebersezung [sic] des alten testaments / Lagarde, Paul de – Goettingen: Dieterich, 1882 [mf ed 1990] – 1mf – 9 – 0-8370-1728-9 – (in german & greek) – mf#1987-6124 – us ATLA [221]

Die anlage des menschen zur religion : vom gegenwaertigen standpunkte der voelkerkunde aus / Happel, Julius – Haarlem: De Erven F Bohn, 1877 – 1mf – 9 – 0-7905-0951-2 – (includes bibliographies) – mf#1987-0951 – us ATLA [210]

Anlagen zu den motiven des entwurfs eines familienrechts fuer das deutsche reich *see* Entwurf eines familienrechts fuer das deutsche reich

Anleitung, auf die nuetzlichste und genussvollste art die schweiz zu bereisen / Ebel, Johann G – Zuerich 1804-05 [mf ed Hildesheim 1995-98] – (= ser Fbc) – 4v on 9mf – 9 – €180.00 – 3-487-29372-2 – gw Olms [914]

Anleitung auf die nuetzlichste und genussvollste art in der schweitz zu reisen / Ebel, J G – Zuerich, 1793. 2v – 5mf – 9 – mf#HT-266 – ne IDC [914]

Anleitung englische contretaenze mit zwei wuerfeln zu componiren so viele man will : ohne etwas von der musik oder der composition zu verstehen par w a mozart = Instruction to compose without the least knowledge of music so much countrydances as one pleases, by throwing a certain number with two dice / Mozart, Wolfgang Amadeus – Bonn: N Simrock [1798] [mf ed 1989] – 1r – 1 – mf#pres. film 50 – us Sibley [780]

Anleitung zu der musikalischen gelahrtheit : worinn von der theorie und praxis der alten und neuen musik, von den musikalischen instrumenten, besonders der orgel, nachricht gegeben... / Adlung, Jacob; ed by Hiller, Johann Adam – 2. aufl, Dresden, Leipzig: In der Breitkopfischen buchhandlung 1783 [mf ed 19–] – 22mf – 9 – mf#fiche761 – us Sibley [780]

Anleitung zur erfindung der melodie und ihrer fortsetzung / Daube, Johann Friedrich – Wien: Christian Gottlob Taeubel 1797-98 [mf ed 19–] – 2v on 3mf – 9 – mf#fiche 909 – us Sibley [780]

Anleitung zur musik ueberhaupt und zur singkunst besonders mit uebungsexempeln erlaeutert... / Marpurg, Friedrich Wilhelm – Berlin: A Weaver 1763 [mf ed 19–] – 3mf – 9 – mf#Fiche 424 – us Sibley [780]

Anleitung zur musikalischen gelahrtheit : worinn von der theorie und praxis der alten und neuen musik, von den musikalischen instrumenten, besonders der orgel... / Adlung, Jakob – 2.aufl, Dresden, Leipzig: In der Breitkopfischen buchhandlung 1783 [mf ed 19–] – 22mf – 9 – mf#fiche 761 – us Sibley [780]

Anleitung zur practischen musik : vor neuangehende saenger und instrumentspieler / Petri, Johann Samuel – Lauban: J C Wirthgen 1767 [mf ed 19–] – 3mf – 9 – mf#fiche 638, 639, 826 – us Sibley [780]

Anleitung zur singcomposition / Marpurg, Friedrich Wilhelm – Berlin: G A Lange 1758 [mf ed 19–] – 4mf – 9 – (1.hauptstueck: von dem prosodischen ausdruck eines textes, oder von der mechanik der singcomposition. i. vom tonmasse der sylben und den klangfuessen ueberhaupt. ii. vom tonmasse in der deutschen sprache besonders. iii. von den klangfuessen der deutschen sprache besonders. iv. von den verschiedenen versarten der deutschen sprache. v. von der beschaffenheit musikalischer verse. vi. von der aufloesung der prose in poetische klangfuesse. vii. von dem sylbenmaasse und den klangfuessen der lateinischen sprache. viii. von dem, was man in der composition eines italiaenischen singtexts, in ansehung der sprache, zu beobachten hat) – mf#fiche 425 – us Sibley [780]

Anleitung zur singekomposition : mit oden in verschiedenen sylbenmaassen begleitet / Kirnberger, Johann Philipp – Berlin: G I Decker 1782 [mf ed 19–] – 3mf – 9 – mf#fiche 501 – us Sibley [780]

Anley, Charlotte *see* The prisoners of australia

Anmerkungen ueber die komposition der offenbarung johannis / Schmidt, Paul Wilhelm – Freiburg i.B: JCB Mohr, 1891 [mf ed 1990] – 1mf – 9 – 0-7905-3410-X – (incl bibl ref) – mf#1987-3410 – us ATLA [225]

Anmerkungen zu konrads trojanerkrieg / Bartsch, Karl – Stuttgart: Litterarischer Verein, 1877 [mf ed 1993] – (= ser Blvs 133) – xxx/489p – 1 – mf#8470 reel 28 – us Wisconsin U Libr [930]

Anmerkungen zu konrads trojanerkrieg / Bartsch, Karl – Stuttgart: Litterarischer Verein, 1877 (Tuebingen: L F Fues) [mf ed 1993] – (= ser Blvs 133) – xxx/489p – 1 – mf#8470 reel 28 – us Wisconsin U Libr [430]

Anmuthiger und nuetzlicher zeitvertreib fuer den buerger- und bauernstand [...] – Leipzig DE, 1792 – 1r – 1 – gw Misc Inst [074]

Ann arbor argus – v1 n2-1940 [1969 feb 13-1971 jun] – 1 – mf#764019 – us WHS [071]

Ann arbor review – Ann Arbor MI 1967-79 – 1,5,9 – ISSN: 0003-3731 – mf#7704 – us UMI ProQuest [400]

Ann arbor sun – Ann Arbor, MI: Rainbow People's Party, 19–1975) [mf ed 19–] – (= ser Underground newspaper coll) – 1 – (cont by: sun jul 16 1975) – us Bell [071]

Ann arbor trail by rail and lake – n1-17 [1978-1982] – 1 – mf#615768 – us WHS [917]

Ann h judson : a memorial / Wyeth, Walter Newton – Cincinnati: the author, 1888 [mf ed 1984] – (= ser Women & the church in america 139; Missionary memorials 1) – 1mf – 9 – 0-8370-1401-8 – mf#1984-2139 – us ATLA [305]

The anna elizabeth dickinson collection – 25r – 1 – $875.00 – Dist. us Scholarly Res – us L of C Photodup [976]

Anna first baptist church – Anna, IL. aug 1910-nov 1988 – 1 – $163.71 – us Southern Baptist [242]

Anna giustiniani : un dramma intimo di cavour / Codignola, Arturo – 2nd ed. [Milano?]: Garzanti, 1945 – (= ser Piccola collana storica) – mf#Z-491 – us NY Public [920]

Anna journal / American Nephrology Nurses' Association – Pitman NJ 1994-99 – 1,5,9 – (cont by: nephrology nursing journal) – ISSN: 8750-0779 – mf#21622.03 – us UMI ProQuest [610]

Anna maria islander – Anna Maria, FL. 1958-1985 – 38r – (gaps) – us UF Libraries [071]

Anna maria islander press – Anna Maria, FL. 1986 jan-1989 jun – 7r – us UF Libraries [071]

Anna maria islander press – Anna Maria, FL. 1989 jul-1999 jul – 2r – us UF Libraries [071]

Anna maria van schurman / Schotel, Gilles Dionysius Jacobus – 's Hertogenbosch: Muller, 1853 [mf ed 1990] – 1mf – 9 – 0-7905-6950-7 – (in dutch, french, latin & italian) – mf#1988-2950 – us ATLA [920]

Anna pavlova, 1882-1931 : et la danse de son temps / Opera de Paris. Bibliotheque, archives et musee – Paris, 1956. "Exposition organisee par l'Institut Choregraphique, les Amis de la Bibliotheque-Musee de la Danse et Serge Lifar" – 1 – mf#*ZBD-*MGO pv 19 – Located: NYPL – us Misc Inst [790]

Anna und greite : novelle / Fischer, Kurt W – feldpost-ausg. Berlin: Nordland Verlag c1943 [mf ed 1989] – 1r – 1 – (filmed with: der tod vor dem spiegel / edmund finke) – mf#7242 – us Wisconsin U Libr [830]

Annae Commenae *see*
– Alexiadis libri 15

Annae commenae porphyrogenitae caesarissae alexias (cbh10) : sive de rebus ab alexio imperatore vel ejus tempore gestis / ed by Possinus, P – Parisiis, 1651 – (= ser Corpus byzantinae historiae (cbh)) – €48.00 – ne Slangenburg [241]

Annaes brasilienses de medicina : jornal d'academia imperial de medicina do rio de janeiro – Rio de Janeiro, RJ: Typ de Francisco de Paula Brito, out 1849-set 1854; mar-out 1856; mar 1857-jun 1885 – (= ser Ps 19) – mf#P19A,03,32-37 – bl Biblioteca [610]

Annaes da academia philosophica – Rio de Janeiro, RJ, 1858 – (= ser Ps 19) – mf#P02,02,08 – bl Biblioteca [100]

Annaes do ensaio academico – Sao Paulo, SP: Typ Litteraria, 01 out 1862; maio-set 1863 – (= ser Ps 19) – mf#P17,02,241 – bl Biblioteca [079]

Annaes fluminense de sciencias, artes e literatura – Rio de Janeiro, RJ: Typ de Santos e Sousa, jan 1822 – (= ser Ps 19) – 1,5,6 – mf#P01,03,11 – bl Biblioteca [079]

Annaes maritimos : periodico consagrado aos interesses da marinha – Rio de Janeiro, RJ: Typ Lobo Vianna & Filhos, 14 maio-23 nov 1861 – (= ser Ps 19) – mf#93,03,09 – bl Biblioteca [380]

An-nahar arab report and memo – Beirut: An-Nahar Press Services. v5 n1-v9 n21 jan 1981-aug 9 1985 – 5r – us CRL [956]

Annalen / Vienna. Naturhistorisches Museum – v1-55 1886-1947. Scattered issues lacking. 11 reels – 1 – us L of C Photodup [580]

Die annalen asurnazirpals (884-860 v. chr.) / Ashurnasirpal 2 – Muenchen: F Straub, 1885 – 1mf – 9 – 0-8370-7841-5 – mf#1986-1841 – us ATLA [470]

Annalen der aeltern deutschen litteratur / Panzer, G W F – 3v. 1788-1805 – 1,9 – us AMS Press [430]

Annalen der baierischen litteratur / ed by Schmidt, Ignatz – Nuernberg 1781-83 – (= ser Dz) – 3v on 7mf – 9 – €140.00 – mf#k/n330 – gw Olms [430]

Annalen der braunschweig-lueneburgischen churlande / ed by Jacobi, Andreas Ludolph et al – Hannover 1787-95 – (= ser Dz. historisch-geographische abt) – 9jge[zu je 4st] on 46mf – 9 – €460.00 – mf#k/n1183 – gw Olms [943]

Annalen der brittischen geschichte / ed by Archenholtz, Johann Wilhelm von – Hamburg 1788-1800 – (= ser Dz. historisch-geographische abt) – 20v on 61mf – 9 – €610.00 – mf#k/n1211 – gw Olms [941]

Annalen der buergerlichen tugend [...] – Flensburg, Leipzig DE, 1792, 1796 – 1r – 1 – gw Misc Inst [390]

Annalen der deutschen literatur : geschichte der deutschen literatur von den anfaengen bis zur gegenwart: eine gemeinschaftsarbeit zahlreicher fachgelehrter / Burger, Heinz Otto – Stuttgart: J B Metzler, 1952 [mf ed 1993] – 882p – 1 – (incl ind) – mf#8122 – us Wisconsin U Libr [430]

Annalen der forst- und jagd-wissenschaft / ed by Gatterer, Christoph Wilhelm Jakob et al – Darmstadt: Bey C W Leske 1811- (qrtly) [v1-6(1811-21) (gaps)] [mf ed 2005] – 6v on 2r [ill] – 1 – (cont: forst-archiv zur erweiterung der forst- und jagd-wissenschaft und der forst- und jagd-literatur; cont by: allgemeine forst und jagdzeitung, issn 0002-5852; ceased wth 1822; v1-6 called also v1-4 of: annalen der societaet der forst- und jagdkunde; Publ: Marburg und Cassel: bei J C Krieger 1815-21) – mf#film mas 99999 – us Harvard [634]

Annalen der fortschritte der landwirtschaft in theorie und praxis / ed by Thaer, Albrecht – Berlin 1811-12 – (= ser Dz) – 2jge[zu je 2 bdn] on 17mf – 9 – €170.00 – mf#k/n3081 – gw Olms [630]

Annalen der geographie und statistik / ed by Zimmermann, Eberhard August Wilhelm – Braunschweig 1790-92 – (= ser Dz. historisch-geographische abt) – 3v on 13mf – 9 – €130.00 – mf#k/n1224 – gw Olms [910]

Annalen der gesammten litteratur – Erlangen 1788-89 – (= ser Dz) – 2jge on 15mf – 9 – €150.00 – mf#k/n416 – gw Olms [400]

Annalen der gesammten litteratur *see* Compendium hisitoriae literariae novissimae

Annalen der gesetzgebung und rechtsgelehrsamkeit in den kgl preussischen staaten / ed by Klein, Ernst Ferdinand – Berlin, Stettin 1788-1809 – (= ser Dz) – 26v on 69mf – 9 – €690.00 – mf#k/n2589 – gw Olms [323]

Annalen der k k sternwarte in wien / Universitaet Wien. Sternwarte; ed by Littrow, Joseph Johann – Wien: A Strauss 1821-80 (annual, irreg) [mf ed 2000] – 63v, 1821-79 on 11r [ill] – 1 – (cont by: annalen der k k universitaets-sternwarte in wien [waehring]; publ: I sommer 1848-75; j n vernay 1876-79; iss for neue folge v4-12 contain: storia celeste del r. osservatorio di palermo dal 1792 al 1813 by giuseppe piazzi, in 9v) – mf#film mas c4632 – us Harvard [520]

Annalen der k k universitaets-sternwarte in wien [waehring] / Universitaet Wien. Sternwarte; ed by Weiss, Edmund & Hepperger, Josef von – Wien: J N Vernay 1884-1927 (irreg) [mf ed 2000] – v2-25(1882-1927) [gaps?] on 6r [ill] – 1 – (cont: annalen der k k sternwarte in wien; cont by: annalen der universitaets-sternwarte wien; v1 never publ; some nos iss out of chronological sequence) – mf#film mas c4633 – us Harvard [520]

Annalen der kaiserlichen universitaets-sternwarte in strassburg / Kaiserliche Universitaets-Sternwarte in Strassburg; ed by Becker, Ernst & Bauschinger, Julius – Karlsruhe: G Braun 1896-1926 (irreg) [mf ed 2000] – v1-5 n3(1896-1926) [gaps?] on 3r [ill] – 1 – (cont by: annales d l'observatoire de strasbourg; suppls accompany some iss) – mf#film mas c4646 – us Harvard [520]

Annalen der katholischen indianer-missionen von amerika / ed by Bureau der Katholischen Indianer-Missionen – Washington DC: [Bureau of Catholic Indian Missions] 1877- [biannual] [mf ed 2006] – v1 n1 (jul 1877) [complete] on 1r – 1 – mf#2005i-s001 – us ATLA [241]

Annalen der koeniglichen sternwarte bei muenchen / K Sternwarte [Munich, Germany]; ed by Lamont, Johann von et al – Muenchen: F S Huebschmann 1848-76 (irreg) [mf ed 2000] – v1-21(1848-76)=v16-36 on 10r [ill] – 1 – (cont in pt by: neue annalen der k sternwarte in bogenhausen bei muenchen; has suppls called 1.-14. supplementband; imprint varies) – mf#film mas c4507 – us Harvard [520]

Annalen der leidenden menschheit in zwanglosen heften / ed by Hennings, August Adolph Friedrich – Altona 1795-1801 – (= ser Dz. historisch-politische abt) – 10iss on 30mf – 9 – €300.00 – mf#k/n1287 – gw Olms [900]

Annalen der literatur und kunst in den oesterreichischen staaten *see* Annalen der oesterreichischen literatur

Annalen der naturgeschichte / Link, J H F – Quebec. 1976+ (1,5,9) – 2mf – 9 – mf#11203 – ne IDC [590]

Annalen der oesterreichischen literatur – Regensburg 1802 [mf ed 1991] – (= ser Wiener literaturzeitschriften der ersten haelfte des 19. jahrhunderts) – 169mf – 9 – €510.00 – 3-89131-036-6 – (filmed with: annalen der literatur und kunst in den oesterreichischen staaten [wien 1803-05]; neue annalen der literatur des oesterreichischen kaiserthumes [wien 1807-08]; annalen der literatur und kunst in dem oesterreichischen kaiserthume [wien 1809]; annalen der literatur und kunst des in- und auslandes [wien 1810]; annalen der literatur und kunst in dem oesterreichischen kaiserthume [wien 1811-12]) – gw Fischer [430]

Annalen der physik – Heidelberg, Germany 1799-1997 – 1,5,9 – ISSN: 0003-3804 – mf#3460 – us UMI ProQuest [530]

Annalen der rechte des menschen und buergers und der voelker / ed by Schmalz, Theodor Anton Heinrich – Koenigsberg 1794 – (= ser Dz. abt philosophie) – 2iss on 2mf – 9 – €60.00 – mf#k/n573 – gw Olms [322]

Annalen der reisen, der geographie und geschichte : in original-aufsaetzen und uebersetzungen aus fremden sprachen herausgegeben – Berlin 1809 [mf ed Hildesheim 1995-98] – 1v on 3mf – 9 – €90.00 – 3-487-26462-5 – gw Olms [910]

Annalen der st joseph's congregatie van mill hill – Roosendaal: St Joseph's Missiehuis. v31 n7-v46 n4. nov 1920-jun 1935 – 1 – (iss for nov 1920-dec 1923 filmed with: annalen van het missiehuis te roosendaal en het studiehuis te tilburg, aug 1919-oct 1920. iss for 1933-may 1935 filmed with: st joseph's congregatie van mill hill, jul 1935 and: mill hill, aug 1935-nov 1936) – us CRL [240]

Annalen der sternwarte in leiden / Sterrewacht Leiden; ed by Sande Bakhuyzen, Ernst Frederik van de – Harlem: J Enschede & Soehne 1868-1915 (irreg) [mf ed 2000] – v1-9 n2(1868-1915) on 4r [ill] – 1 – (cont by: annalen van de sterrewacht te leiden; imprint varies: v2-9 [haag: m nijhoff]) – mf#film mas c4625 – us Harvard [520]

Annalen der teutschen akademien / ed by Franz, Friedr Christian et al – Leipzig 1790-91 – (= ser Dz. historisch-geographische abt) – 2st on 2mf – 9 – €60.00 – mf#k/n1223 – gw Olms [943]

Annalen der wetterauischen gesellschaft fuer die gesammte naturkunde – Bellaire. 1970-1976 (1) 1975-1976 (5) 1975-1976 (9) – 77mf – 9 – mf#8618 – ne IDC [590]

Annalen des ackerbaus / ed by Thaer, Albrecht – Berlin 1805-11 – (= ser Dz) – 6jge[zu je 2 bdn] + ind on 60mf – 9 – €600.00 – mf#k/n3056 – gw Olms [630]

Annalen des historischen vereins fuer den niederrhein – Heft 1(1855)-Heft 140(1942) – 588mf – 9 – €1121.00 – ne Slangenburg [943]

Annalen des theaters / ed by Bertram, Christian August von – Berlin 1788-97 – (= ser Dz) – 20iss on 13mf – 9 – €130.00 – mf#k/n4217 – gw Olms [790]

Annalen van de sterrewacht te leiden / Sterrewacht Leiden; ed by Sitter, Willem de – Den Haag: M Nijhoff 1913- (irreg) [mf ed 2000] – deel 10-22(1913-66) [gaps?] on 6r [ill] – 1 – (cont as an independent publ while cooperating with: publications of the astronomical institute of the university of amsterdam; recherches astronomiques de l'observatoire d'utrecht; and: publications of the astronomical laboratory at groningen; to form: bulletin of the astronomical institutes of the netherlands; cont: annalen der sternwarte in leiden; some iss publ out of chronological sequence) – mf#film mas c4626 – us Harvard [520]

Annalen van het missiehuis te rozendaal – Rozendaal: Het Missiehuis. v1-26 n8. [may 1890-1915] – 1 – (iss for may 1914-dec 1915 filmed with: annalen van het missiehuis te roosendaal en het studiehuis te tilburg, 1916-jul 1919) – us CRL [240]

Annalen vd bosscha-sterrenwacht, lembang [java] / Observatorium Bosscha – Bandoeng: Gebrs Kleijne & Co. N.v 1933- [irreg] [mf ed 2002] – 3r – 1 – (ceased with v9 pt5 in 1961; many iss publ out of chronological order; chiefly in english with some german & dutch; imprint varies) – mf#film mas c5304 – us Harvard [520]

Annales / Bogor. Indonesia. Kebun Raja – v1-51. 1876-1949 – 9 – $510.00 – mf#0109 – us Brook [959]

Annales / France. Assemblee Nationale – juin 1916-oct 1917; 9 fevr, 19 mars, 19 avr 1940 – 1 – (comptes rendus in extenso des elections des presidents de la republique 1871-1953) – fr ACRPP [323]

Annales / Institut Technique du Batiment et des Travaux Public – Paris. n205-312.1965-73 – 5 – fr ACRPP [073]

Annales / Missions de la Societe de Marie – Lyon. Puis: De l'Oceanie. 1875-1921 – 1 – fr ACRPP [240]

Annales / La Societe Jean-Jacques Rousseau – Geneve. v1-38 1905-1969/71 – 5 – fr ACRPP [190]

Annales. 1st and 2nd series *see* Toulouse, universite. faculte des sciences de toulouse. annales. 1st and 2nd series

Annales a mundi exordio usque ad obitum alexii commeni imper (cbh9) / Michaelis Glycae; ed by Labbe, Ph – Parisiis, 1660 – (= ser Corpus byzantinae historiae (cbh)) – €38.00 – ne Slangenburg [243]

Annales abbatiae : sancti petri blandiniensis / ed by Putte, F van de – Gandavi, 1842 – €19.00 – ne Slangenburg [241]

Annales academiae scientiarum fennicae. series a : v medica – Helsinki, Finland 1974-80 – 1,5,9 – ISSN: 0066-1996 – mf#9787 – us UMI ProQuest [610]

Annales aevi carolini et saxonici. chronica et gesta aevi saxonici. historiae aevi carolini et saxonici (mgh5:4.bd) – 1841 – (= ser Monumenta germaniae historica 5: scriptores in folio (mgh5)) – €44.00 – ne Slangenburg [220]

Annales aevi suevici (mgh5:16.bd) – 1859 – (= ser Monumenta germaniae historica 5: scriptores in folio (mgh5)) – €40.00 – ne Slangenburg [240]

Annales aevi suevici (mgh5:17.bd) – 1861 – (= ser Monumenta germaniae historica 5: scriptores in folio (mgh5)) – €46.00 – ne Slangenburg [240]

Annales aevi suevici (mgh5:19.bd) – 1866 – (= ser Monumenta germaniae historica 5: scriptores in folio (mgh5)) – €40.00 – ne Slangenburg [240]

Annales aevi suevici (mgh5:24.bd) : suppl tom 16 et 17 – 1879 – (= ser Monumenta germaniae historica 5: scriptores in folio (mgh5)) – €46.00 – (chronica minora saec. 12 et 13. gesta saec. 12 et 13. 1880) – ne Slangenburg [240]

Annales agronomiques – Montrouge, France 1950-51 – 1 – ISSN: 0003-3829 – mf#555 – us UMI ProQuest [630]

Annales althahenses maiores (mgh7:4.bd) – ed 2a. 1891 – (= ser Monumenta germaniae historica 7: scriptores rerum germanicarum in usum scholarum (mgh7)) – €7.00 – ne Slangenburg [240]

Annales anabaptistici, hoc est, Historia universalis de anabaptistarum origine... / Ott, J H – Basileae, apud Wehrenfelsium, 1672 – 5mf – 9 – mf#ZWI-46 – ne IDC [242]

Annales archeologiques – Paris. v1-28. 1844-1881 – 232mf – 9 – mf#0-1209 – ne IDC [930]

Annales artistiques et litteraires – Paris. mars 1888-90 – 1 – fr ACRPP [700]

Annales bertiniani (mgh7:5.bd) – 1883 – (= ser Monumenta germaniae historica 7: scriptores rerum germanicarum in usum scholarum (mgh7)) – €11.00 – ne Slangenburg [240]

Annales camaldulenses o s b / Mittarelli, J B & Costadoni, A – Venetiis. v1-9. 1750-73 – 9v on 289mf – 9 – €551.00 – ne Slangenburg [240]

Annales cambriae (ad 444-1288) (rs20) / ed by Williams, J, ab Ithel – 1860 – (= ser The rolls series (rs)) – €11.00 – ne Slangenburg [931]

Annales (cbh22) / Joannis Zonarae; ed by Cange, C du – Parisiis. v1. 1686 – (= ser Corpus byzantinae historiae (cbh)) – €56.00 – (v2 parisiis 1687 €48) – ne Slangenburg [243]

Annales chirurgiae et gynaecologiae – Helsinki, Finland 1976-80 – 1,5,9 – (cont: annales chirurgiae et gynaecologiae fenniae) – ISSN: 0355-9521 – mf#5975,01 – us UMI ProQuest [618]

Annales chirurgiae et gynaecologiae fenniae – Helsinki, Finland 1971-75 – 1,5,9 – (cont by: annales chirurgiae et gynaecologiae) – ISSN: 0003-3855 – mf#5975.01 – us UMI ProQuest [618]

Annales, chronica, genealogiae, catalogi (suppl tom1-12, pars1) (mgh5:13.bd) : gesta aevi carolini et saxonici (suppl tom 2 et 4) – 1881 – (= ser Monumenta germaniae historica 5: scriptores in folio (mgh5)) – €42.00 – ne Slangenburg [220]

Annales cisterciensium / Manrique, Angel – Lyon. v1-4. 1642-49 – v1 29mf v2 32mf v3 33mf v4 40mf – 8 – €255.00 – ne Slangenburg [241]

Les annales coloniales – Paris. 1914-juil 1917; avr 1921; 1924 – 1 – fr ACRPP [073]

Les annales coloniales – Paris. v32-40, Jan 1931-June 1940 – 1 – 92.00 – us L of C Photodup [944]

Les annales commerciales – Cap-Haitien: Impr du Progres [sep 15 1917-aug 30 1918] – 8mf – 9 – us CRL [380]

Annales criminelles canadiennes – Montreal: Societe de publ des Annales criminelles canadiennes. v1 n1 15 nov 1896-v1 n3 15 dec 1896 [mf ed 1988] – 3mf – 9 – mf#P04002 – cn Canadiana [360]

Annales (cshb27) / Michaelis Glycae; ed by Bekkeri, Imm – Bonnae, 1836 – (= ser Corpus scriptorum historiae byzantinae (cshb)) – €23.00 – ne Slangenburg [243]

Annales (cshb42,43) / Ioannis Zonarae; ed by Pinderi, Maur – Bonnae. v1-2. 1841-44 – (= ser Corpus scriptorum historiae byzantinae (cshb)) – €42.00 – ne Slangenburg [243]

Annales d'anatomie pathologique – Paris, France 1924-80 – 1,5,9 – ISSN: 0003-3871 – mf#3394 – us UMI ProQuest [574]

Annales de cardiologie et d'angeiologie – Oxford, England 2001+ – 1 – ISSN: 0003-3928 – mf#24553,04 – us UMI ProQuest [616]

Annales de chimie appliquee – puis et Revue de chimie analytique reunies. Paris. 1896-1957 – 1 – fr ACRPP [540]

Annales de chimie – science des materiaux – Paris, France 1968-81 – 1,5,9 – ISSN: 0151-9107 – mf#3420 – us UMI ProQuest [540]

Annales de chirurgie – Oxford, England 2001 – 1 – ISSN: 0003-3944 – mf#24361 – us UMI ProQuest [617]

Annales de chirurgie infantile – Paris, France 1968-77 – 1,5,9 – (cont by: chirurgie pediatrique) – ISSN: 0003-3952 – mf#3409.01 – us UMI ProQuest [617]

Annales de chirurgie plastique et esthetique – Ann Arbor MI 2001 – 1 – ISSN: 0294-1260 – mf#24363,01 – us UMI ProQuest [617]

Annales de demographie internationale – Paris, France 1877-81 – 1 – mf#5149 – us UMI ProQuest [304]

Annales de dermatologie et de syphiligraphie – Paris, France 1869-1976 – 1,5,9 – ISSN: 0003-3979 – mf#3396.01 – us UMI ProQuest [616]

Annales de dermatologie et de venereologie – Paris, France 1977-80 – 1,5,9 – ISSN: 0151-9638 – mf#3396,01 – us UMI ProQuest [616]

Annales de la confederation universelle des amis de la verite *see* La bouche de fer

Annales de la facult, des sciences de marseille – Pittsburgh. 1972-1977 (1) 1972-1977 (5) 1977-1977 (9) – 106mf – 9 – mf#7470 – ne IDC [590]

Annales de la litterature et des arts – Paris. 1820-29 (I-XXXIV) – 1 – fr ACRPP [410]

Annales de la patrie francaise – Paris. n1-125. mai 1900-juil 1905 – 1 – fr ACRPP [073]

Annales de la propagation de la foi – Lyon. v1-119. 1827-1947 – 9 – $1518.00 – (in french) – us Brook [240]

Annales de la propagation de la foi pour la province de quebec – [Montreal?: s.n.], 1877-1886 [mf ed [nouv ser], 1er n(fevr 1877)-nouv ser, 29e n(juin 1886)] (mf ed 1991) – 55mf – 9 – (cont by: annales de la propagation de la foi pour les provinces de quebec et de montreal; merger of: annales de la propagation de la foi pour le diocese de montreal, and: rapport sur les missions du diocese de quebec, du diocese de rimouski) – mf#P04988 – cn Canadiana [241]

Annales de la propagation de la foi pour la province de quebec et de montreal – Montreal: Gebhardt-Berthiaume. nouv ser: 30e n(oct 1886-nouv ser: 141e n(oct 1923] [mf ed 1991] – 223mf – 9 – (cont: annales de la propagation de la foi pour la province de quebec; cont by: annales de la propagation de la foi) – mf#P04989 – cn Canadiana [241]

Annales de la propagation de la foi pour le diocese de montreal – [Montreal: L'Oeuvre], 1874-1876 – 9 – (fait suite a: rapport de l'association de la propagation de la foi pour le diocese de montreal; fusionne avec: rapport sur les missions du diocese de quebec, du diocese de rimouski, et devient: annales de la propagation de la foi pour la province de quebec) – mf#P04255 – cn Canadiana [241]

Annales de la religion – Paris. 1795-1803 (1-18) – (= ser Memoires pour servir a l'histoire du 18e siecle.) – 1 – fr ACRPP [240]

Annales de la societe belge de medecine tropicale / Societe belge de medecine tropicale – Antwerpen, Belgium 1975-80 – 1,5,9 – ISSN: 0365-6527 – mf#7327 – us UMI ProQuest [574]

Annales de la societe historique et archeologique a maestricht – tom 1-2. (1854-1858) – 2v on 14mf – 9 – €27.00 – ne Slangenburg [930]

Annales de la societe st-jean-baptiste de quebec : deliberations du comite de regie, grandes demonstrations, receptions officielles, celebrations d'anniversaires... / Chouinard, Honore Julien Jean Baptiste [comp] – Quebec?: s.n, 1881-1903 – 4v on 1mf – 9 – mf#06191 – cn Canadiana [366]

Annales de l'assemblee nationale : compte-rendu in extenso des seances, annexes / France. Assemblee Nationale – v1-45. 12 feb 1871-8 mar 1876 – 9 – $1428.00 – mf#0217 – us Brook [324]

Annales de l'association des pretres-adorateurs – [Montreal: L'Association, 1898-1936] – 9 – (cont by: annales des pretres adorateurs; ceased 1902?) – mf#P04034 – cn Canadiana [241]

Annales de l'extreme orient et de l'afrique – v. 1-15. 1878-91 – 1 – us L of C Photodup [950]

Annales de l'institut d'etudes orientales – Paris, Alger, 1934-1962. v1-20 – (= ser Faculte des Lettres de l'Universite d'Alger) – 103mf – 8 – mf#NE-102 – ne IDC [956]

Annales de l'institut d'etudes orientales / L'Universite d'Alger – Alger. (I-XX, n.s. I). 1934-62, 1964 – 1 – fr ACRPP [378]

Annales de l'institut pasteur [1887] – Lausanne, Switzerland 1887-1972 – 1,5 – ISSN: 0020-2444 – mf#5108 – us UMI ProQuest [610]

Annales de l'institut pasteur. immunologie – Oxford, England 1980-88 – 1,5,9 – (cont by: research in immunology) – ISSN: 0769-2625 – mf#42398.01 – us UMI ProQuest [616]

Annales de l'institut pasteur. microbiologie – Oxford, England 1980-88 – 1,5,9 – (cont by: research in microbiology) – ISSN: 0769-2609 – mf#42399.01 – us UMI ProQuest [576]

Annales de l'institut pasteur. virologie – Oxford, England 1980 – 1,5 – (cont by: research in virology) – ISSN: 0769-2617 – mf#42400.01 – us UMI ProQuest [576]

Annales de l'observatoire astronomique de tokyo : appendice / Tokyo Tenmondai – Tokyo: L'Observatoire 1889-1922 (irreg) [mf ed 2001] – [1]-13(1915-23) on 1r – 1 – (cont by: annals of the tokyo astronomical observatory. appendix; title in french & japanese; text in english) – mf#film mas c5055 – us Harvard [520]

Annales de l'observatoire de besancon : astronomie et geophysique / Universite de Besancon. Observatoire – Besancon, France: Impr et lithographie Jacques et Demontrond (irreg) [mf ed 2001] – ns: t1-6(1934-64) on 1r – 1 – (cont by: universite de besancon: observatoire – annales de l'observatoire de besancon; editor: 1934-59 rene baillaud; 1964 jean delhaye; imprint varies) – mf#film mas c4699 – us Harvard [520]

Annales de l'observatoire de bordeaux / Observatoire de Bordeaux [France]; ed by Rayet, George Antoine Pons & Picart, Luc – Paris: Gauthier-Villars; Bordeaux: Ferret et fils 1885- (irreg) [mf ed 2000] – t1-17(1885-1933) on 8r [ill] – 1 – (ceased with t18 n5 [1966]) – mf#film mas c4658 – us Harvard [520]

Annales de lobservatoire de moscou – M., 1874-1884, v1-10; N.S. 1886-1890, v1-2 – 119mf – 9 – mf#R-1648 – ne IDC [077]

Annales de l'observatoire de nice / France. Bureau des longitudes – Paris: Gauthier-Villars 1887-1911 (irreg) [mf ed 2001] – t1-14 (1887-1911) on 11r – 1 – (some nos iss out of sequence; t1 iss in 1899; t13 complete in 1pt) – mf#film mas c4671 – us Harvard [550]

Annales de l'observatoire de paris. memoires – Paris: Gauthier-Villars 1868-1925 (irreg) [mf ed 1999] – 24v on 12r [ill] – 1 – (cont: annales de l'observatoire imperial de paris. memoires; v9-14 publ under the direction of urbain jean joseph le verrier; v15-20 of [e] mouchez; v21-25, of f tisserand) – mf#film mas c4271 – us Harvard [520]

Annales de l'observatoire de paris. observations / Observatoire de Paris – Paris: Gauthier-Villars. 37v. 1871-17 (annual) [mf ed 2001] – t23 (1867)-1907 on 20r – 1 – (cont: annales de l'observatoire imperial de paris. observations; vols for 1867-1868/69 called also t23-24; none publ for 1894-96; some vols publ out of sequence) – mf#film mas c4674 – us Harvard [520]

Annales de l'observatoire de strasbourg / Observatoire de Strasbourg – Paris: Gauthier-Villars 1926- (irreg) [mf ed 2000] – t1-4 fasc 3(1926-59) on 1r – 1 – (cont: annalen der kaiserlichen universitaets-sternwarte in strassburg; imprint varies) – mf#film mas c4645 – us Harvard [520]

Annales de l'observatoire de toulouse / Observatoire de Toulouse; ed by Petit, F et al – Toulouse: C Douladoure 1863- (irreg) [mf ed 2000] – t1-20(1863-1950) on 9r [ill] – 1 – (t10 [1933] incl: bulletin de l'observatoire de toulouse [n1-38[1923-28]]; t1 [1863] covers observations made: 1846-62; iss under variant names: t1 [1863]: observatoire de toulouse; t1-8 [1880-1912]: observatoire astronomique, magnetique et meteorologique de toulouse; t9-20 [1914-1950]: observatoire astronomique et meteorologique de toulouse) – mf#film mas c4670 – us Harvard [520]

Annales de l'observatoire du houga / Observatoire du Houga – Le Houga [Gers]: Observatoire 1942-61 [irreg] [mf ed 2003] – 3v on 1r – 1 – (v2 iss in 2pts) – mf#film mas c5751 – us Harvard [520]

Annales de l'observatoire imperial de paris – Paris: Mallet-Bachelier 1855-59 (annual) [mf ed 1999] – 5v on 3r [ill] – 1 – (cont by: annales de l'observatoire imperial de paris. memoires) – mf#film mas c4269 – us Harvard [520]

Annales de l'observatoire imperial de paris. memoires – Paris: Mallet-Bachelier 1861-66 (irreg) [t6-8(1861-66)] [mf ed 1999] – 3v on 2r [ill] – 1 – (cont: annales de l'observatoire imperial de paris; cont by: annales de l'observatoire de paris. memoires) – mf#film mas c4270 – us Harvard [520]

Annales de l'observatoire imperial de paris. observations / Observatoire de Paris – Paris: Mallet-Bachelier 1858-67 (irreg) [mf ed 2001] – t1-22 (1800/1829-1866) on 10r [ill] – 1 – (cont by: annales de l'observatoire de paris. observations; none publ for 1830-36; some vols publ out of sequence) – mf#film mas c4673 – us Harvard [520]

Annales de l'observatoire imperial de rio de janeiro / Imperial Observatorio do Rio de Janeiro – Rio de Janeiro: Typ & lithographie Lombaerts & Cie 1882-89 (irreg) [mf ed 2001] – t1-4(1882-89) on 3r [ill] – 1 – (ceased with t4 [1889]?; in french; t3 & 4 pt2 in french & portuguese) – mf#film mas c4689 – us Harvard [520]

Annales de l'observatoire national d'athenes / Ethnikon Asteroskopeion Athenon; ed by Eginitis, Demetrios – Athenes: Impr nationale 1896- (irreg) [mf ed 2001] – t1-12 (1896-1932) on 6r – 1 – (ea vol iss in 2 pts: pt1: memoires; pt2: observations; suspended publ 1917-25? imprint varies) – mf#film mas c4676 – us Harvard [550]

Annales de l'observatoire royal de belgique / Observatoire royal de Belgique – Bruxelles: M Hayez [1922- (irreg) [mf ed 1999] – t1-5(1922-49) on 2r [ill] – 1 – (formed by the union of: annales de l'observatoire royal de belgique. annales astronomiques, and: annales de l'observatoire royal de belgique. physique du globe; imprint varies) – mf#film mas c4314 – us Harvard [520]

Annales de l'observatoire royal de belgique. nouvelle serie. annales astronomiques / Observatoire royal de Belgique – Bruxelles: Observatoire royal de Belgique 1878-1920 [mf ed 1999] – 14v on 8r [ill] – 1 – (cont by: annales de l'observatoire royal de belgique; title varies: 1878, annales de l'observatoire royal de bruxelles. nouvelle serie. astronomie; 1879-1887, annales de l'observatoire royal de bruxelles. nouvelle serie. annales astronomiques; imprint varies) – mf#film mas c4300 – us Harvard [520]

Annales de l'observatoire royal de bruxelles – Bruxelles: M Hayez 1834-77 (irreg) [mf ed 1999] – 25v on 13r [ill] – 1 – (cont by: annales de l'observatoire royal de belgique. nouvelle serie. annales astronomiques; v1 iss in 2pt; imprint varies; t1-23 ed by adolphe quetelet; t24 by ernest quetelet) – mf#film mas c4299 – us Harvard [520]

Annales de l'ordre de ste-ursule : formant la continuation de l'histoire generale du meme institut depuis la revolution francaise jusqu'a nos jours... – [Clermont-Ferrand, France?: s.n.] 1857 [mf ed 1985] – 5v on 1mf – 9 – 0-665-48927-7 – mf#48927 – cn Canadiana [241]

Annales de marie – Lyon. 1925-juil 1940 – 1 – fr ACRPP [240]

Annales de medecine interne – Paris, France 1968-79 – 1,5,9 – ISSN: 0003-410X – mf#3391 – us UMI ProQuest [616]

Annales de mon observatoire / Observatoire prive Lucien Libert [Le Havre, France]; ed by Libert, Lucien – Le Havre: Impr Micaux [1901- (irreg) [mf ed 2001] – n1-24(1901-06) on 1r [ill] – 1 – (n6-[8] of: publications de l'observatoire prive lucien libert iss also as suppl n2-4 to: annales; n4-24 called also 2.-5. annee) – mf#film mas c4694 – us Harvard [520]

Annales de normandie – Caen. 1951-71. tb: 1951-71 – 5 – fr ACRPP [944]

Annales de notre dame du sacre-coeur – 1866-1975 – 35r – 1 – mf#pmb doc330 – at Pacific Mss [241]

Annales de paleontologie – Paris, France 1971-91 – 1,5,9 – ISSN: 0753-3969 – mf#3428 – us UMI ProQuest [560]

Annales de parasitologie humaine et comparee – Paris, France 1968-81 – 1,5,9 – ISSN: 0003-4150 – mf#3416 – us UMI ProQuest [616]

Annales de philosophie chretienne – v1-60. 1830-60 – 1 – $570.00 – (in french) – us Brook [240]

Annales de philosophie chretienne – Paris, 1(1830)-166(1913) – 9 – €2848.00 – (lacking: 99,104,110-120,144) – ne Slangenburg [240]

Annales de philosophie chretienne – Paris. oct 1905-juin 1913 – 1 – fr ACRPP [190]

Annales de philosophie chretienne : recueil periodique... – Paris, 1830-1912/1913. v1-165 – 1526mf – 8 – (missing:1861 v62; 1879-1880 v97-98; 1882 v104(p227-295),v109-112;1887-1888 v115-116;1889-1890 v118-119; 1907 v153; 1909 v158) – mf#5512 – ne IDC [100]

Annales de radio electricite – Rocquencourt Yvelines, France 1945-68 – 1 – ISSN: 0365-5008 – mf#5105 – us UMI ProQuest [621]

Annales d'endocrinologie – Paris, France 1968-81 – 1,5,9 – ISSN: 0003-4266 – mf#3415 – us UMI ProQuest [616]

Annales des mines : ou recueil de memoires sur l'exploitation des mines, et sur les sciences qui s'y rapportent – Paris: Chez Treuttel et Wurtz,. [ser 12 v13 1928] (mthly) – 1r – 1 – us CRL [622]

Annales des missions de la societe de marie : afterwards annales des missions de l'oceanie – 1877-1886 – 1r – 1 – mf#pmb doc194 – at Pacific Mss [240]

Annales des missions de la societe de marie : afterwards annales des missions de l'oceanie – 1877-1892 – 1r – 1 – mf#pmb doc195 – at Pacific Mss [240]

Annales des missions de la societe de marie : afterwards annales des missions de l'oceanie – 1886-1912 – 1r – 1 – mf#pmb doc198 – at Pacific Mss [240]

Annales des missions de la societe de marie : afterwards annales des missions de l'oceanie – 1893-1902 – 1r – 1 – mf#pmb doc196 – at Pacific Mss [240]

Annales des missions de la societe de marie : afterwards annales des missions de l'oceanie – 1903-21 – 1r – 1 – mf#pmb doc197 – at Pacific Mss [240]

Annales des missions de la societe de marie : formerly annales de la societe de marie – 1853-75 – 1r – 1 – mf#pmb doc174 – at Pacific Mss [240]

Annales des missions de la societe de marie : formerly annales de la societe de marie – jan 1925-jul 1940 – 1r – 1 – mf#pmb doc175 – at Pacific Mss [240]

Annales des missions de la societe de marie : formerly annales de la societe de marie – jan 1928-nov 1932 – 1r – 1 – mf#pmb doc176 – at Pacific Mss [240]

Annales des missions de la societe de marie : formerly annales de la societe de marie – jan 1930-nov 1933 – 1r – 1 – mf#pmb doc177 – at Pacific Mss [240]

Annales des missions de la societe de marie : formerly annales de la societe de marie – jan 1934-jul 1936 – 1r – 1 – mf#pmb doc178 – at Pacific Mss [240]

Annales des missions de la societe de marie : formerly annales de la societe de marie – jan 1937-jul 1940 – 1r – 1 – mf#pmb doc179 – at Pacific Mss [240]

Annales des missions de la societe de marie, afterwards annales des mission de l'oceanie – Lyons: Societe de Marie 1853-1921 [1887-1892] – 1r – 1 – mf#pmb doc195 – at Pacific Mss [241]

Annales des ponts et chaussees : memoires et documents relatifs a l'art des constructions... – Paris: C Dunod, [1888?] [mf ed 1980] – 2mf – 9 – 0-665-04585-9 – mf#04585 – cn Canadiana [620]

Annales des ponts et chaussees : memoires et documents relatifs a l'art des constructions et au services de l'ingenieur – Paris. Lts. 1-2. Tables generales; Series 1-11 1831-1941 – 1 – us NY Public [624]

Annales des sacres-coeurs / Congregation des Sacres-Coeurs et de l'Adoration – Paris. 1928, 1936-mai 1940 – 1 – fr ACRPP [241]

Annales des sacrescoeurs – 1894-95 – 1r – 1 – mf#pmb doc181 – at Pacific Mss [240]

Annales des sacrescoeurs – 1896-98 – 1r – 1 – mf#pmb doc182 – at Pacific Mss [240]

Annales des sacrescoeurs – 1899-1901 – 1r – 1 – mf#pmb doc183 – at Pacific Mss [240]

Annales des sacrescoeurs – 1902-04 – 1r – 1 – mf#pmb doc184 – at Pacific Mss [240]

Annales des sacrescoeurs – 1905-07 – 1r – 1 – mf#pmb doc185 – at Pacific Mss [240]

Annales des sacrescoeurs – 1908-10 – 1r – 1 – mf#pmb doc186 – at Pacific Mss [240]

Annales des sacrescoeurs – 1911-13 – 1r – 1 – mf#pmb doc187 – at Pacific Mss [240]

Annales des sacrescoeurs – 1914-22 – 1r – 1 – mf#pmb doc188 – at Pacific Mss [240]

Annales des sacrescoeurs – 1923-26 – 1r – 1 – mf#pmb doc189 – at Pacific Mss [240]

Annales des sacrescoeurs – 1927-30 – 1r – 1 – mf#pmb doc190 – at Pacific Mss [240]

Annales des sacrescoeurs – 1931-35 – 1r – 1 – mf#pmb doc191 – at Pacific Mss [240]

Annales des sacrescoeurs – 1936 – 1r – 1 – mf#pmb doc192 – at Pacific Mss [240]

Annales des sacrescoeurs – 1937-may 1940 – 1r – 1 – mf#pmb doc193 – at Pacific Mss [240]

Annales des sciences naturelles. botanique et biologie vegetale – Paris; France 1968-79 – 1,5,9 – ISSN: 0003-4320 – mf#3421 – us UMI ProQuest [580]

Annales des sciences naturelles. zoologie et biologie animale – Paris; France 1968-80 – 1,5,9 – ISSN: 0003-4339 – mf#3425 – us UMI ProQuest [590]

Annales des voyages, de la geographie et de l'histoire : ou collection des voyages nouveaux les plus estimes, traduits de toutes les langues europeennes...; accompagnee d'un bulletin ou l'on annonce toutes les decouvertes, recherches et entreprises... – Paris 1808-14 [mf ed Hildesheim 1995-98] – 24v on 72mf – 9 – €720.00 – 3-487-29895-3 – gw Olms [910]

Annales d'histoire economique et sociale – v1-10. 1929-38 – 9 – $535.00 – (in french) – mf#0040 – us Brook [300]

Annales d'histoire sociale *see* Annales d'historie economique et sociale

Annales d'historie economique et sociale – 1929-38 – 5 – (devenu: annales d'histoire sociale. 1939-45; devenu: annales economies, societes, civilisations; 1946- paris. 1929-78, 1985-93, tb: 1929-51, 1949-68) – fr ACRPP [300]

Annales d'oto-laryngologie et de chirurgie cervico-faciale – Paris, France 1968-81 – 1,5,9 – ISSN: 0003-438X – mf#3395 – us UMI ProQuest [617]

Annales du bureau des longitudes / France. Bureau des longitudes et al – Paris: Gauthier-Villars 1877- (irreg) [mf ed 2000] – t1-12(1877-1949) on 6r [ill] – 1 – (t1-6 [1877-1903] iss in assoc with l'observatoire astronomique de montsouris, t12 [1949]- with centre national de la recherche scientifique) – mf#film mas c4667 – us Harvard [520]

Annales du cabinet de lecture paroissial de montreal – Montreal: [s.n.], 1857 [mf ed 1989] – 3mf – 9 – (cont by: echo du cabinet de lecture paroissial de montreal; ceased 1858?) – mf#P04072 – cn Canadiana [971]

Annales du monastere de notre-dame de charite du bon pasteur d'angers, dit asile sainte darie a montreal 1870-1900 / Asile Sainte-Darie (Montreal, Quebec) – Montreal: [s.n, 1900?] [mf ed 1979] – 3mf – 9 – 0-665-00897-X – mf#00897 – cn Canadiana [241]

Annales du musee et de l'ecole moderne des beaux arts *see* Ackermann's 'repository of arts'

Annales du musee guimet, vols 1-17 – Paris – 8 – (v1 1880 13mf €25. v2 1881 19mf €37. v3 1881: schlagintweit, emile de: le bouddhisme au tibet 14mf €27. v4 1882 11mf €21. v5 1883: fragments extraits du kandjour, traduits du tibetain par m leon feer 19mf €37. v6 1884: la lalita vistara, traduit du sanskrit par ph e foucaux. premiere partie: traduction francaise 14mf €27. v7 1884 17mf €32. v8 1885: le yi: king ou livre des changements de la dynastie des tsheou, traduit par p l f philastre, premiere partie 16mf €31. v9 1886: m e lefebure, les hypogees royaux de thebes. premiere division: le tombeau de seti ler 11mf €21. v10 1887 21mf €40. v11 1886: j j m de groot, les fetes annuellement celebrees a emoui, ie partie 15mf €29. v12 1886: j j m de groot, les fetes annuellement celebrees a emoui, 2e partie 15mf €29. v13 1888: ch schoebel, le rayamana au point de vue religieux, philosophique et moral 8mf €17. v14 1887: m e amelineau, essai sur le gnosticisme egyptien 11mf €21. v15 1887: la siao hio ou morale de la jeunesse avec le commentaire de tchen siuen, trad du chinois par c de harlez 12mf €23. v16 1889: m e lefebure, les hypogees de thebes. seconde division. notices des hypogees. troisieme division, tombeau de ramses 4 16mf €31. v17 1889: e amelineau, histoire de saint pakhome et de ses communautes 26mf €50) – ne Slangenburg [240]

Annales du musee guimet, vols 1-32 – Paris. v1-32. 1880-1909 – 426mf – 8 – (missing: v21,22,24) – ne Slangenburg [240]

Annales du musee guimet, vols 18-32 – Paris – 8 – (v18 1891: avadana-cataka, cent legendes (bouddhiques) traduites duk sanskrit par leon feer 17mf €32. v19 1892: le lalita vistara, 2e partie, traduit du sanskrit par ph e foucaux 8mf €17. v20 1891: textes taoistes traduits des originaux chinois et commentes par c de harlez 13mf €25. v23 1893: le yi: king, 2e partie, traduit du chinois par p l f philastre €38. v25 1894: e amelineau, histoire des monasteres de la basse-egypte... vies des saints paul, etc 16mf €31. v26, 1e partie, 1894: la coree ou tchosen par m le colonel chaille-long-bey 3mf €7. v26, 3e partie, 1897: a. gayet, l'exploration des ruines d'antinoe et la decouverte d'un temple de ramses 2 4mf €11. v26, 4e partie, 1900: p lefevre-pontalis, recueil de talismans laotiens 2mf €5. v27, 1e partie, 1895: l fournerau, le siam ancien 16mf €31. v28 1896: e amelineau, histoire de la sepulture et des funerailles dans l'ancienne egypte 14mf €27. v29 1896: e amelineau, histoire de la sepulture et des funerailles dans l'ancienne egypte, v2 19mf €37. v30 1902: g legrain et e naville, l'alle nord du pylone d'amenophis 3 10mf €19. v31, 1e partie, 1907: e fonssagrives, si-ling. etude sur les tombeaux de l'ouest de la dynastie des ts'ing 7mf €15. v31, 2e partie, 1908: l founerau, le siam ancien 8mf €17. v32 1909: catalogue du musee guimet. galerie egyptienne 11mf €21) – ne Slangenburg [240]

Annales du senat et de la chambre des deputes du 8 mars 1876-28 dec 1880 / France. Assemblee Nationale – v1-51. 1876-80 – 9 – $1428.00 – mf#0214 – us Brook [944]

Annales du service des antiquites de l'egypte – Le Caire, 1900-1925. v1-25+index – 120mf – 9 – mf#H-389 – ne IDC [930]

Annales du service meteorologique de l'indochine – Hai Phong: E Bruzon 1932-34, 1940 – 1r – 1 – mf#mf-12544 seam – us CRL [550]

Les annales du t s rosaire – Cap-de-la Magdeleine, Quebec: s.n, 1892?-1918 – 9 – mf#P04043 – cn Canadiana [241]

Les annales du theatre et de la musique – 1875-1916 – 1 – $234.00 – (in french) – mf#0043 – us Brook [790]

Annales d'urologie – Oxford, England 2001 – 1,5,9 – ISSN: 0003-4401 – mf#24359 – us UMI ProQuest [616]

Annales ecclesiastici / Baronius, Caesar; ed by Theiner, A – Barri-Ducis/Parisiis, 1864-83 – 37v on 577mf – 8 – €1100.00 – ne Slangenburg [240]

Annales ecclesiastiques de 1846 a 1860 : ou, histoire resumee de l'eglise catholique pendant les dernieres annees / Chantrel, Joseph – Paris: Gaume Freres & J Duprey 1861 [mf ed 1992] – 2mf – 9 – 0-524-03492-3 – (in french) – mf#1990-4714 – us ATLA [241]

Annales economies, societes, civilisations *see* Annales d'historie economique et sociale

Annales et chronica aevi carolini (mgh5:1.bd) – 1826 – (= ser Monumenta germaniae historica 5: scriptores in folio (mgh5)) – €35.00 – ne Slangenburg [241]

Annales et chronica aevi salici (mgh5:5.bd) – 1844 – (= ser Monumenta germaniae historica 5: scriptores in folio (mgh5)) – €31.00 – ne Slangenburg [220]

Annales et chronica aevi salici (mgh5:10.bd) : vitae aevi carolini et saxonici – 1852 – (= ser Monumenta germaniae historica 5: scriptores in folio (mgh5)) – €32.00 – ne Slangenburg [240]

Annales et chronica italica aevi suevici (mgh5:31.bd) – 1903 – (= ser Monumenta germaniae historica 5: scriptores in folio (mgh5)) – €40.00 – ne Slangenburg [240]

Annales et historiae *see* Trionfi

Annales et historiae de rebus belgicis / Grotius, H – Amstelaedami, 1657 – €50.00 – ne Slangenburg [240]

Annales ferdinandei : oder kaiser ferdinands 2. leben und denkwuerdigen geschichten / Khevenhiller, Chr – Leipzig. v1-14. 1721-22 – 14v on 354mf – 8 – €675.00 – ne Slangenburg [920]

Annales forestieres et metallurgiques – [Paris: Au Bureau des Annales forestieres 1855-65] [t14-24.annee (1855-65)] [mf ed 2005] – 11v on 3r [ill] – 1 – (frequency varies 1855-juin 1865, irreg juil-dec 1865; cont: annales forestieres; tome 14, 1 called also serie nouvelle) – mf#film mas c6047 – us Harvard [634]

Annales francaises de chronometrie / ed by Baillaud, Rene et al – Besancon: Impr Millet freres [1931]- (qrtly) [mf ed 2000] – 1931-51 [gaps] on 5r [ill] – 1 – (cont: bulletin chronometrique of the observatoire national de besancon; superseded by: annales francaise de chronometrie et de micromecanique [issn 0066-2143]; numbering begins with n3-4 [3.-4. trimestre 1931]; 2nd ser called also 17.-35. annee; organ of various institutions. from 1931-41: observatoire national de besancon, institut de chronometrie de l'universite de besancon, and societe chronometrique de france; from 1942- organ of several additional institutions, among them: centre technique de l'industrie horlogere) – mf#film mas c4664 – us Harvard [520]

Annales fuldenses sive annales regni francorum orientalis (mgh7:7.bd) – 1891 – (= ser Monumenta germaniae historica 7: scriptores rerum germanicarum in usum scholarum (mgh7)) – €16.00 – ne Slangenburg [240]

Annales gauloises – Paris-Besancon. mars 1889-sept 1892 – 1 – fr ACRPP [944]

Les annales haitienne : publication. – Port-au-Prince: Impr du "Messager" [n2-3,5 (1903)] (irreg) – 1mf – 9 – us CRL [972]

Annales heremi deiparae matris monasterii in helvetia / Hartmann, Christ – Freiburg Br, 1612 – 13mf – 8 – €25.00 – ne Slangenburg [241]

Annales hildesheimenses (mgh7:8.bd) – 1878 – (= ser Monumenta germaniae historica 7: scriptores rerum germanicarum in usum scholarum (mgh7)) – €5.00 – ne Slangenburg [240]

Annales hirsaugienses / Trithemius, Ioan – Typis Monasterii S Galli. v1-2. 1690 – €168.00 – ne Slangenburg [241]

Annales historiques de la revolution de l'amerique latine : accompagnees de documents a l'appui, de l'annee 1808 jusqu'a la reconnaissance par les etats europeens de l'independance de ce vaste continent / Calvo, Carlos – Paris: A Durand; Madrid: C Bailley-Bailliere. 5v. 1864-67 [mf ed 1984] – 5v on 1mf – 9 – 0-665-48533-6 – mf#48533 – cn Canadiana [972]

Annales historiques de la revolution francaise – Paris. v18-53. 1946-81 – 1 – fr ACRPP [944]

Annales hydrographiques – 1848-50 – 1r – 1 – mf#pmb doc220 – at Pacific Mss [550]

Annales hydrographiques – 1850-52 – 1r – 1 – mf#pmb doc221 – at Pacific Mss [550]

Annales hydrographiques – 1852-53 – 1r – 1 – mf#pmb doc222 – at Pacific Mss [550]

Annales hydrographiques – 1854-56 – 1r – 1 – mf#pmb doc223 – at Pacific Mss [550]

Annales hydrographiques – 1857-59 – 1r – 1 – mf#pmb doc224-225 – at Pacific Mss [550]

Annales hydrographiques – 1860-66 – 6r – 1 – mf#pmb doc226-231 – at Pacific Mss [550]

Annales hydrographiques – 1867-68 – 1r – 1 – mf#pmb doc232 – at Pacific Mss [550]
Annales hydrographiques – 1869-71 – 1r – 1 – mf#pmb doc233 – at Pacific Mss [550]
Annales hydrographiques – 1872-78 – 4r – 1 – mf#pmb doc234-237 – at Pacific Mss [550]
Annales hydrographiques – 1879-83 – 1r – 1 – mf#pmb doc238 – at Pacific Mss [550]
Annales hydrographiques – 1884-86 – 1r – 1 – mf#pmb doc239 – at Pacific Mss [550]
Annales hydrographiques – 1887-90 – 1r – 1 – mf#pmb doc240 – at Pacific Mss [550]
Annales hydrographiques – 1891-94 – 1r – 1 – mf#pmb doc241 – at Pacific Mss [550]
Annales hydrographiques – 1895-99 – 1r – 1 – mf#pmb doc242 – at Pacific Mss [550]
Annales hydrographiques – 1900-04 – 1r – 1 – mf#pmb doc243 – at Pacific Mss [550]
Annales hydrographiques – 1905-08/10 – 1r – 1 – mf#pmb doc244 – at Pacific Mss [550]
Annales hydrographiques – 1911-16 – 1r – 1 – mf#pmb doc245 – at Pacific Mss [550]
Annales hydrographiques – 1917-1919/20 – 1r – 1 – mf#pmb doc246 – at Pacific Mss [550]
Annales hydrographiques – 1921-25/26 – 1r – 1 – mf#pmb doc247 – at Pacific Mss [550]
Annales hydrographiques – 1927/28-1931/32 – 1r – 1 – mf#pmb doc248 – at Pacific Mss [550]
Annales hydrographiques – 1933-37 – 1r – 1 – mf#pmb doc249 – at Pacific Mss [550]
Annales hydrographiques – 1938/39-1947 – 1r – 1 – mf#pmb doc250 – at Pacific Mss [550]
Annales hydrographiques – 1948-50 – 2r – 1 – mf#pmb doc251-252 – at Pacific Mss [550]
Annales hydrographiques – 1950-52 – 1r – 1 – mf#pmb doc253 – at Pacific Mss [550]
Annales hydrographiques – 1953-55 – 1r – 1 – mf#pmb doc254 – at Pacific Mss [550]
Annales hydrographiques – 1956-58 – 1r – 1 – mf#pmb doc255 – at Pacific Mss [550]
Annales hydrographiques – 1959/60-1963/64 – 1r – 1 – mf#pmb doc256 – at Pacific Mss [550]
Annales hydrographiques – 1965/66-1967/68 – 1r – 1 – mf#pmb doc257 – at Pacific Mss [550]
Annales hydrographiques – Paris. 1848-1968 – 1 – fr ACRPP [550]
Annales imperii occidentis brunsvicenses (768-1005) / Leibniz, Gottfried Wilhelm von; ed by Pertz, G H – Hannoverae. v1-3. 1843-46 – €79.00 – ne Slangenburg [240]
Annales italici aevi suevici (mgh5:18.bd) – 1863 – (= ser Monumenta germaniae historica 5: scriptores in folio (mgh5)) – €44.00 – ne Slangenburg [240]
Annales marbacenses qui dicuntur (mgh7:9.bd) : accedunt annales alsatici breviores – 1907 – (= ser Monumenta germaniae historica 7: scriptores rerum germanicarum in usum scholarum (mgh7)) – €7.00 – (cronica hohenburgensis cum continuatione et additamentis neoburgensibus) – ne Slangenburg [240]
Annales medicinae experimentalis et biologiae fenniae – Helsinki, Finland 1949-73 – 1,5,9 – (cont by: medical biology) – ISSN: 0003-4479 – mf#1694.01 – us UMI ProQuest [619]
Annales mettenses priores (mgh7:10.bd) – 1905 – (= ser Monumenta germaniae historica 7: scriptores rerum germanicarum in usum scholarum (mgh7)) – €7.00 – (accedunt addit annalium mettensium posteriorum) – ne Slangenburg [240]
Annales minores aevi saxonici (mgh5:3.bd) : chronica minora aevi saxonici. annales, chronica, historiae aevi saxonici – 1839 – (= ser Monumenta germaniae historica 5: scriptores in folio (mgh5)) – €46.00 – ne Slangenburg [220]
Annales minorum : seu trium ordinum a s francisco institutorum / Waddingo, Luca; ed by Fonseca, J M – 2nd ed, Romae: tom 1-16. 1731-1736 – 9 – €1428.00 – ne Slangenburg [240]
Annales minorum : seu trium ordinum a s francisco institutorum (continuati) – Romae, Anconae, Neapoli, Ad Claras Aquas. tom 18-25. 1740-1886 – 9 – €686.00 – ne Slangenburg [240]
Annales monasterii s albani *see* Chronia monastereii s albani 5 [rs28]
Annales monastici (rs36) / ed by Luard, H R – (= ser The rolls series (rs)) – (v1: de margam, theokesberia (tewkesbury) et burton 1864 €19. v2: de wintonia (winchester) et waverleia (waverly) 1864 €18. v3: de dunstaplia (dunstable) et bermundeseia (bermondsey) 1866 €4. v4: de osneneia (osney), chronicon thomas wykes, et de wigornia (worcester) 1869 €23. v5: ind and glos 1869 €18) – ne Slangenburg [241]
Annales muslemici arabice et latine : opera et studiis j j reiskii / Abulfedae – Hafniae. tom v1-5. 1789-1794 – 112mf – 9 – €214.00 – ne Slangenburg [260]
Annales mycologici editi in notitiam scientiae mycologicae universalis – v1-42. 1903-44 – 9 – $462.00 – mf#0044 – us Brook [580]
Annales ordinis cartusiensis : ab anno 1084 ad annum 1429 auctore d carolo le couteux o cart – Monstrolii. v1-8. 1888-91 – 8v on 77mf – 8 – €147.00 – ne Slangenburg [241]
Annales ordinis cartusiensis : tom 1. complectens ea quae ad institutionem, disciplinam et observantias ordinis spectant – Correriae, 1687 – €35.00 – ne Slangenburg [241]
Annales ordinis s benedicti / Mabillon, Jean – Lutetiae Parisorum. v1-6. 1703-1733 – 6v on 289mf – 8 – €551.00 – ne Slangenburg [241]
Annales paderbornenses / Schaten, N S J – ed altera. Monast. Westphalorum, 1774-1775 – €92.00 – ne Slangenburg [240]
Annales paediatrici – Basel, Switzerland 1966 – 1 – mf#2047 – us UMI ProQuest [618]
Annales patriotiques et litteraires de la france : et affaires politiques de l'europe, journal libre – Paris. oct 1789-94 – 1 – fr ACRPP [073]
Annales pharmaceutiques francaises – Paris, France 1968-80 – 1,5,9 – ISSN: 0003-4509 – mf#3423 – us UMI ProQuest [615]
Annales philosophiques, politiques et litteraires – La Salle IL 1807 – 1 – mf#3550 – us UMI ProQuest [190]
Les annales politiques et litteraires – Paris, France 7 jan 1900-2 dec 1903, 10 nov 1907-29 mar 1908, 7 jan 1912-26 jul 1914, 2 dec 1914, 3 jan 1915-1922 (imperfect) – 25r – 1 – uk British Libr Newspaper [074]
Les annales politiques et litteraires – Paris. -w 7 Jan 1900-27 Dec 1903; 10 Nov 1907-29 March 1908; 7 Jan 1912-26 Jul 1914; 27 Dec 1914-31 Dec 1922. Imperfect. 25 reels – 1 – uk British Libr Newspaper [074]
Annales politiques, morales et litteraires – Paris. No.1-1277. 16 dec 1815-15 juin 1819. mq no. 1089, 1091, 1201 – 1 – fr ACRPP [073]
Annales politiques, sociales, litteraires et artistiques *see* Les hommes du jour
Annales poloniae (mgh7:11.bd) – 1866 – (= ser Monumenta germaniae historica 7: scriptores rerum germanicarum in usum scholarum (mgh7)) – €7.00 – ne Slangenburg [240]
Annales regni francorum inde ab a 741 usque ad a 829 (mgh7:6.bd) : qui dicuntur annales laurissenses maiores et einhardi – 1895 – (= ser Monumenta germaniae historica 7: scriptores rerum germanicarum in usum scholarum (mgh7)) – €11.00 – ne Slangenburg [240]
Annales svltanorvm othmanidarvm : a tvrcis sva lingva scripti... / Leunclavius, J – Francofvrdi, 1596 – 6mf – 9 – mf#H-8388 – ne IDC [956]
Les annales teresiennes – Montreal: Beauchemin & Valois, [1880-19–] – 9 – mf#P04000 – cn Canadiana [241]
Annales theosophiques : receuil trimestriel de conferences et de travaux originaux – Paris: Publ theosophiques. v1-7. 1908-14 [qrterly] [mf ed 2003] – 7v on 1r – 1 – (lacks: v6 n2,4) – mf#2003-s100 – us ATLA [290]
Annales typographici ab artis inventae origine ad annum 1500 – 1793-97. vand Annales typographici ab anno 1501 ad annum 1536 – 11v – 1,9 – us AMS Press [010]
Annales typographici ab artis inventae origine ad annum 1500-1664 / Maittaire, Michael – 1719-41. vand Annalium typographicorum, supplementum. 1789. 6 v – 1,9 – us AMS Press [010]
Annales typographici colonienses : handschrift aus der ersten haelfte des 19. jahrhunderts / Buellingen, Ludwig von – Koeln [mf ed 1997] – 5v on 42mf – 9 – €370.00 – 3-89131-242-3 – (aus den bestaenden der universitaets- und stadtbibliothek koeln) – gw Fischer [090]
Annales xantenses et annales vedastini (mgh7:12.bd) – 1909 – (= ser Monumenta germaniae historica 7: scriptores rerum germanicarum in usum scholarum (mgh7)) – €5.00 – ne Slangenburg [240]
Annali del museo pitre – [Palermo]: Palumbo 1950-64 (annual) [mf ed 1984] – 1r – 1 – (cont: archivio per lo studio delle tradizioni popolari; publ: universita di palermo, istituto di storia delle tradizioni popolari, 1954/1956; banco di sicilia, fondazione per l'incremento economico, culturale e turistico della sicilia ignazio mormino, 1960-64) – us Indiana Preservation [390]
Annali dell'islam, 1-10 / ed by Caetani, L – Milano, 1905-1926. v1-10+index – 122mf – 9 – mf#NE-307 – ne IDC [260]
Annali di statistica / Italy. Istituto Centrale di Statistica – ser. 1, v. 1-10-ser. 6, v. 1-38. 1871-1936. (scattered volumes wanting) – 1 – us L of C Photodup [945]
Annali d'italia dal principio dell'era volgare sino all'anno 1749 / Muratori, Lodovico A – Milano. v1-4. 1744-1749 – 4v on 170mf – 9 – €324.00 – ne Slangenburg [945]
The annalist : a magazine of finance, commerce and economics – New York. v. 1-56. Jan 20 1913-Oct 24 1940 – 1 – us NY Public [330]
Annalium et chronicorum aevi carolini continuatio *see* Scriptores rerum sangalliensium. annalium et chronicorum aevi carolini continuatio. historiae aevi carolini (mgh5:2.bd)
Annals / Adorers of the Blood of Christ. Wichita, KS – 1901-1985 – 1 – us Kansas [240]
Annals and antiquities of rajasthan : or the central and western rajput states of india / Tod, James; ed by Crooke, William – London, New York: Oxford University Press, 1920 – (= ser Samp: indian books) – us CRL [930]
Annals and memoirs of the court of peking : (from the 16th to the 20th century) / Backhouse, Edmund & Bland, John Otway Percy – London: William Heinemann, 1914 [mf ed 1995] – (= ser Yale coll) – x/531p (ill) – 1 – 0-524-09204-4 – mf#1995-0204 – us ATLA [951]
Annals and statistics of the united presbyterian church / Mackelvie, William – Edinburgh: Oliphant, 1873 [mf ed 1991] – (= ser Presbyterian coll) – 2mf – 9 – 0-524-01739-5 – (incl bibl ref) – mf#1990-4131 – us ATLA [242]
Annals de la propagation de la foi pour les provinces de quebec et de montreal / Society for the Propagation of the Faith – 1877-1923 – (= ser American indian periodicals... 1) – 160mf – 9 – $1060.00 – us UPA [240]
The annals in brief of the st andrew's society of quebec : with the act of incorporation and the constitution and by-laws of the society / Harper, John Murdoch – Quebec: The Society, 1906 – 1mf – 9 – 0-665-75598-8 – mf#75598 – cn Canadiana [366]
Annals of a quiet neighbourhood / Macdonald, George – New York, NY. 1872 – 1r – us UF Libraries [978]
Annals of allergy – Arlington Heights IL 1943-94 – 1,5,9 – (cont by: annals of allergy, asthma, and immunology) – ISSN: 0003-4738 – mf#2442.01 – us UMI ProQuest [616]
Annals of allergy, asthma, and immunology – Arlington Heights IL 1995+ – 1,5,9 – (cont: annals of allergy) – ISSN: 1081-1206 – mf#2442,01 – us UMI ProQuest [616]
The annals of ashurbanapal (5 rawlinson pl 1-10) – Leiden: E J Brill, 1903 [mf ed 1986] – (= ser Semitic study series 2) – 1mf – 9 – 0-8370-7762-1 – (in akkadian & english. incl bibl ref) – mf#1986-1762 – us ATLA [470]
Annals of biomedical engineering – Malden MA 1992+ – 1,5,9 – ISSN: 0090-6964 – mf#21492 – us UMI ProQuest [610]
Annals of clinical biochemistry – London, England 1975+ – 1,5,9 – ISSN: 0004-5632 – mf#10690 – us UMI ProQuest [612]
Annals of clinical research – Helsinki, Finland 1971-88 – 1,5,9 – ISSN: 0003-4762 – mf#5976 – us UMI ProQuest [619]
Annals of congress, 1789-1824 / U.S. Congress – 1st to 18th Cong., 1st Sess., 42 books. All published. Followed by Congressional Debates.80-033 – 343mf – 9 – $515.00 – mf#LLMC 80-033 – us LLMC [323]
Annals of discrete mathematics – Lausanne, Switzerland 1977-81 – 1,5 – mf#42014 – us UMI ProQuest [510]
Annals of dyslexia – Baltimore MD 1982+ – 1,5,9 – (cont: bulletin of the orton society) – ISSN: 0736-9387 – mf#12837,01 – us UMI ProQuest [370]
Annals of economic and social measurement – Cambridge MA 1972-78 – 1,5,9 – ISSN: 0044-832X – mf#11430 – us UMI ProQuest [300]
Annals of emergency medicine – Oxford, England 1980+ – 1,5,9 – ISSN: 0196-0644 – mf#12584,01 – us UMI ProQuest [610]
Annals of english presbytery : from the earliest period to the present time / M'Crie, Thomas – London: James Nisbet, 1872 [mf ed 1992] – 1mf – 9 – 0-524-02405-7 – mf#1990-0608 – us ATLA [242]
Annals of epidemiology – Oxford, England 1990+ – 1,5,9 – ISSN: 1047-2797 – mf#42680 – us UMI ProQuest [614]
Annals of human biology – Abingdon, Oxfordshire 1989+ – 1 – ISSN: 0301-4460 – mf#17279 – us UMI ProQuest [574]
Annals of human genetics – Oxford, England 1989+ – 1,5,9 – ISSN: 0003-4800 – mf#16516,01 – us UMI ProQuest [575]
Annals of internal medicine – Philadelphia PA 1927+ – 1,5,9 – ISSN: 0003-4819 – mf#360 – us UMI ProQuest [616]
Annals of iowa – Iowa City IA 1975+ – 1,5,9 – ISSN: 0003-4827 – mf#10422 – us UMI ProQuest [978]
Annals of iowa – Third series. 1893-1969 – 1 – us AMS Press [978]
Annals of lloyd's register : being a sketch of the origin, constitution, and progress of lloyd's register of british and foreign shipping / Lloyd's Register of British and Foreign Shipping (Firm) – London [Wyman & sons, printers] 1884 [mf ed 1987] – 1r [ill] – 1 – (filmed with: the magic of the middle ages / rydberg, v) – mf#7144 – us Wisconsin U Libr [380]
Annals of loch ce (rs54) : a chronicle of irish affairs from ad 1014-1590 / ed by Hennessy, W M – v1-2. 1871 – (= ser The rolls series (rs)) – €23.00v – ne Slangenburg [931]
Annals of mathematical statistics – Beachwood OH 1930-72 – 1,5 – ISSN: 0003-4851 – mf#763 – us UMI ProQuest [510]
Annals of mathematics – Princeton NJ 1884+ – 1,5,9 – ISSN: 0003-486X – mf#232 – us UMI ProQuest [510]
Annals of medicine – Abingdon, Oxfordshire 1989+ – 1,5,9 – ISSN: 0785-3890 – mf#17073 – us UMI ProQuest [610]
Annals of my early life, 1806-1846 : with occasional compositions in latin and english verse / Wordsworth, Charles – London, New York: Longmans, Green, 1891 [mf ed 1991] – 1mf – 9 – 0-524-00618-0 – mf#1990-0118 – us ATLA [242]
Annals of my life, 1847-1856 / Wordsworth, Charles; ed by Hodgson, W Earl – London, New York: Longmans, Green, 1893 [mf ed 1990] – 1mf – 9 – 0-7905-8256-2 – mf#1988-8119 – us ATLA [242]
Annals of natal, 1495-1845 / Bird, John – Cape Town, South Africa. v1-2. 1965 – 1r – us UF Libraries [960]
Annals of nature : or, annual synopsis of new genera and species of animals in north america – La Salle IL 1820 – 1 – mf#3551 – us UMI ProQuest [574]
Annals of nature : or annual synopsis of new genera and species of animals, plants, etc discovered in north america / Rafinesque-Schmaltz, C S – Tokyo. 1957-1989 (1) 1972-1980 (5) 1974-1980 (9) – 1mf – 9 – mf#7651 – ne IDC [590]
Annals of neurology – Hoboken NJ 1977+ – 1,5,9 – ISSN: 0364-5134 – mf#13447 – us UMI ProQuest [616]
Annals of nuclear energy – Oxford, England 1975+ – 1,5,9 – (cont: annals of nuclear science & engineering) – ISSN: 0306-4549 – mf#49010,02 – us UMI ProQuest [621]
Annals of nuclear science and engineering – Oxford, England 1974 – 1,5,9 – (cont: journal of nuclear energy; cont by: annals of nuclear energy) – ISSN: 0302-2927 – mf#49010.02 – us UMI ProQuest [621]
Annals of occupational hygiene – Oxford, England 1959+ – 1,5,9 – ISSN: 0003-4878 – mf#49011 – us UMI ProQuest [362]
Annals of oncology – Oxford, England 1990+ – 1,5,9 – ISSN: 0923-7534 – mf#18595 – us UMI ProQuest [616]
Annals of oriental literature : london – jun 1820-feb 1821 – (= ser 19th c british periodicals) – reel 49 – 1 – (filmed with: the essex literary journal (chelmsford), jun 1838-may 1839; the evangelical penny magazine (london), 15 dec 1832) – us Primary [480]
Annals of otology, rhinology and laryngology – St Louis MO 1892+ – 1,5,9 – ISSN: 0003-4894 – mf#3184 – us UMI ProQuest [617]
Annals of pharmacotherapy – Cincinnati OH 1992+ – 1,5,9 – (cont: dicp) – ISSN: 1060-0280 – mf#6492,02 – us UMI ProQuest [615]
Annals of plastic surgery – Baltimore MD 1978+ – 1,5,9 – ISSN: 0148-7043 – mf#12497 – us UMI ProQuest [617]
Annals of probability – Beachwood OH 1973+ – 1,5,9 – ISSN: 0091-1798 – mf#6928 – us UMI ProQuest [510]
Annals of pure and applied logic – Oxford, England 1970+ – 1,5,9 – ISSN: 0168-0072 – mf#42108 – us UMI ProQuest [510]
Annals of regional science – Dordrecht, Netherlands 1989+ – 1,5,9 – ISSN: 0570-1864 – mf#17113 – us UMI ProQuest [338]
The annals of rural bengal / Hunter, William Wilson – London: Smith, Elder, 1868 [mf ed 1995] – (= ser Yale coll) – xvi/475p – 1 – 0-524-09293-1 – mf#1995-0293 – us ATLA [954]
Annals of s paul's cathedral / Milman, Henry Hart; ed by Milman, Arthur – 2nd ed. London: J Murray, 1869 [mf ed 1990] – 2mf – 9 – 0-7905-8144-2 – (1st publ 1868) – mf#1988-6091 – us ATLA [240]
Annals of saint joseph – v11 n1-v13 n12 [1899 mar-02 feb] – 1 – mf#470506 – us WHS [978]
Annals of science – Abingdon, Oxfordshire 1988+ – 1,5,9 – ISSN: 0003-3790 – mf#17314 – us UMI ProQuest [500]
Annals of southern methodism for 1856 / ed by Deems, Charles Force – Nashville, TN: Stevenson & Owen, c1857 [mf ed 1992] – (= ser Methodist coll) – 1mf – 9 – 0-524-06959-X – mf#1990-5323 – us ATLA [242]
Annals of sporting and fancy gazette : a magazine entirely appropriated to sporting subjects and fancy pursuits... – London. v1-13. jan 1822-jun 1828 [all publ] – (= ser Sports periodicals, 1822-1922) – 2r – 1 – $510.00 – us UPA [790]
Annals of st anne de beaupre – Quebec: The Directors of Levis College, [1876?-189- or 19–] – 9 – (cont by: annals of good saint anne de beaupre; suspended: mai 1877-mai 1888; ceased 1920?) – mf#P04207 – cn Canadiana [241]

Annals of st louis in its early days under the french and spanish dominations / Billon, Frederic Louis [comp] – St Louis [MO]: F Billon, 1886 [mf ed 1980] – 6mf – 9 – 0-665-00427-3 – (incl ind) – mf#00427 – cn Canadiana [978]

Annals of st louis in its territorial days, from 1804 to 1821 : being a continuation of the authors's previous work, "the annals of the french and spanish period" / Billon, Frederic Louis – St Louis [MO]: printed for aut, 1888 [mf ed 1980] – 6mf – 9 – (incl ind) – mf#06509 – cn Canadiana [978]

Annals of statistics – Beachwood OH 1973+ – 1,5,9 – ISSN: 0090-5364 – mf#6929 – us UMI ProQuest [310]

Annals of surgery – Baltimore MD 1885+ – 1,5,9 – ISSN: 0003-4932 – mf#1568 – us UMI ProQuest [617]

Annals of the american academy of political and social science / American Academy of Political and Social Science – Newbury Park CA 1890+ – 1,5,9 – ISSN: 0002-7162 – mf#757 – us UMI ProQuest [320]

Annals of the american pulpit – v1-9 (complete) – 1 – $254.17 – us Southern Baptist [242]

Annals of the astronomical observatory of harvard college – Cambridge [MA]: Metcalf & Co 1856- (irreg) [v1-120 n1(1856-1954)] [mf ed 1998] – 46r [pl/ill] – 1 – (v114 & 117 never publ; iss designated variously as pt or no; commonly known as: hco annals) – mf#film mas c3863 – us Harvard [520]

Annals of the astrophysical observatory of the smithsonian institution / Smithsonian Astrophysical Observatory – Washington: Govt Print Off 1900- (irreg) [v1-7(1900-54)] [mf ed 1998] – 3r [ill] – 1 – (v1 iss also as exhibit b of a report...of the smithsonian institution with title: appropriations expended, results reached, and present condition of the work of the astrophysical observatory, washington, 1902; and again, in a bureau ed with title: annals of the astrophysical observatory v1; suspended 1933-41) – ISSN: 1059-5600 – mf#film mas c3884 – us Harvard [520]

Annals Of The Cakchiquels *see* Memorial de tecpan atitlan

Annals of the cape observatory / Royal Observatory, Cape of Good Hope – Cape Town: D Gill. 23v. 1898-1968 (irreg) [mf ed 1999] – v1-16(1898-1950) on 1r [ill] – 1 – (absorbed in part by: royal observatory annals [issn 0080-4371] 1961; royal observatory bulletins [issn 0080-438x] 1963; some vols publ out of chronological sequence; publ varies) – mf#film mas c4327 – us Harvard [520]

Annals of the classis of bergen of the reformed dutch church : and of the churches under its care, including the civil history of the ancient township of bergen in new jersey / Taylor, Benjamin Cook – 3rd ed. New York: Board of Publ...Reformed Protestant Dutch Church, c1857 [mf ed 1992] – 2mf – 9 – 0-524-03662-4 – mf#1990-1090 – us ATLA [242]

Annals of the dearborn observatory of northwestern university / Dearborn Observatory – Evanston IL: [The Observatory?] 1915- (irreg) [mf ed 1999] – v1-7(1915-58) [gaps?] on 2r [ill] – 1 – ISSN: 1061-6853 – mf#film mas c4368 – us Harvard [520]

Annals of the delhi badshahate : being a translation of the old assamese chronicle padshah-buranji / Bhuyan, Suryya Kumar – Gauhati: Govt of Assam, Dept of Historical and Antiquarian Studies, 1947 – (= ser Samp: indian books) – us CRL [954]

Annals of the diocese of adelaide / Norris, William – London: printed for the Society for the Propagation of the Gospel, 1852 [mf ed 1993] – (= ser Anglican/episcopal coll) – 1mf – 9 – 0-524-07134-9 – mf#1990-5341 – us ATLA [242]

Annals of the diocese of new zealand – London: SPCK, 1847 [mf ed 1992] – (= ser Anglican/episcopal coll) – 1mf – 9 – 0-524-06692-2 – mf#1990-5263 – us ATLA [242]

Annals of the disruption : with extracts from the narratives of ministers who left the scottish establishment in 1843 / Brown, Thomas – Edinburgh: Macniven & Wallace, 1884 [mf ed 1990] – 2mf – 9 – 0-7905-5759-2 – mf#1988-1759 – us ATLA [242]

Annals of the early caliphate : from original sources / Muir, William – London: Smith, Elder, 1883. Chicago: Dep of Photodup, U of Chicago Lib, 1968 (1r); Evanston: American Theol Lib Assoc, 1984 (1r) – 1 – 0-8370-0463-2 – (incl bibl ref and ind) – mf#1984-B077 – us ATLA [260]

Annals of the early settlers association of cuyahoga – v1 n1,3,5 [1880, 1882, 1884] – 1 – mf#5266116 – us WHS [978]

Annals of the emperor charles 5th = Guerras de mar del emperador carlos 5 / Lopez de Gomara, Francisco; ed by Merriman, Roger Bigelow – Oxford: Clarendon 1912 [mf ed 1987] – 1r – 1 – (spanish text with english trans; int & notes by ed. filmed with: complete poetical works / rossetti, d g) – mf#1990 – us Wisconsin U Libr [946]

Annals of the emperor charles 5th / Lopez de Gomara, Francisco; ed by Merriman, Roger Bigelow – Oxford: Clarendon Press, 1912 – sp Bibl Santa Ana [940]

The annals of the english bible / Anderson, Christopher; ed by Anderson, Hugh – new rev ed. London: Jackson, Walford & Hodder, 1862 [mf ed 1989] – 2mf – 9 – 0-7905-0784-6 – (incl bibl ref & ind) – mf#1987-0784 – us ATLA [220]

Annals of the entomological society of america / Entomological society of america. annals – Lanham MD 1908+ – 1,5,9 – ISSN: 0013-8746 – mf#117 – us UMI ProQuest [590]

Annals of the evangelical association of north america : and history of the united evangelical church / Stapleton, Ammon – Harrisburg, PA: Publ House...United Evangelical Church, 1900, c1896 [mf ed 1990] – 2mf – 9 – 0-7905-8117-5 – mf#1988-6079 – us ATLA [242]

Annals of the history of computing – Washington DC 1989-91 – 1,5,9 – ISSN: 0164-1239 – mf#16975.01 – us UMI ProQuest [000]

Annals of the icrp / International Commission on Radiological Protection – Oxford, England 1977-94 – 1,5,9 – ISSN: 0146-6453 – mf#49080 – us UMI ProQuest [616]

Annals of the "low-church" party in england : down to the death of archbishop tait / Proby, William Henry Baptist – London: JT Hayes, 1888 [mf ed 1990] – 3v on 3mf – 9 – 0-7905-7020-3 – mf#1988-3020 – us ATLA [242]

Annals of the lowell observatory – Boston: Houghton, Mifflin 1898- [v1-3(1898-1905)] [mf ed 1999] – 2r [ill] – 1 – (imprint varies; v2: cambridge: university press) – mf#film mas c4146 – us Harvard [520]

Annals of the observatory of lund / Lunds universitet. Observatoriet – Lund, Sweden: The Observatory 1926-62 (irreg) [mf ed 2000] – n1-18:1(1926-62) on 3r [ill] – 1 – (suppls accompany some vols; imprint varies; some iss repr fr various scientific publ) – mf#film mas c4644 – us Harvard [520]

Annals of the propagation of the faith – Three-Rivers [Quebec]: Publ for the Institution, 1877-[19–?] – 9 – (publ: pontifical society for the propagation of the faith; ceased 188-?) – ISSN: 1191-2421 – mf#P04250 – cn Canadiana [241]

Annals of the rheumatic diseases – London, England 1939+ – 1,5,9 – ISSN: 0003-4967 – mf#1330 – us UMI ProQuest [616]

Annals of the royal college of physicians *see* Archives of the royal college of physicians, 1518-1988

Annals of the royal college of surgeons of england / Royal College of Surgeons of England – London, England 1958-76 – 1,5,9 – ISSN: 0035-8843 – mf#5324 – us UMI ProQuest [617]

Annals of the royal observatory, edinburgh / ed by Copeland, Ralph – Glasgow: printed by J Hedderwick & sons...1902-10 (quadrennial) [mf ed 1999] – 3v on 1r [ill] – 1 – (v3 [1910] has imprint: edinburgh, printed for his majesty's stationery office by neill & co) – mf#film mas c4275 – us Harvard [520]

Annals of the swedes on the delaware / Clay, Jehu Curtis – 3rd ed. Chicago: Swedish Historical Society of America, 1914 [mf ed 1992] – 1mf – 9 – 0-524-03514-8 – mf#1990-1019 – us ATLA [978]

Annals of the tokyo astronomical observatory : appendix / Tokyo Tenmondai – Tokyo: [The Observatory] [1926- (irreg) [mf ed 2001] – n14-54(1926-1937) on 1r [ill] – 1 – (cont: annales de l'observatoire astronomique de tokyo. appendice; title in english & japanese; text in english) – mf#film mas c5055 – us Harvard [520]

Annals of the tokyo astronomical observatory – Tokyo: The Observatory 1937- (irreg) [mf ed 2001] – 2nd ser: v1-3 n3 (1937-53) on 1r [ill] – 1 – (merged with: publications of the international latitude observatory of mizusawa, and: tokyo astronomical bulletin, to form: publications of the national astronomical observatory of japan; cont: annales de l'observatoire astronomique de tokyo; ceased with v22 n1 in 1988; suspended 1940-48; in english, french or german) – ISSN: 0082-4704 – mf#film mas c5054 – us Harvard [520]

Annals of the united states christian commission / Moss, Lemuel – 1868 – 1 – $50.00 – us Presbyterian [240]

Annals of the united states christian commission / Moss, Lemuel – Philadelphia: J B Lippincott, 1868 [mf ed 1990] – 2mf – 9 – 0-7905-8042-X – (incl bibl ref) – mf#1988-6023 – us ATLA [976]

The annals of the war : illustrated by a selection of historical ballads / Harper, John Murdoch – London, Toronto: Musson Book Co, [c1913?] – (= ser Studies in verse and prose) – 4mf – 9 – 0-665-74535-4 – mf#74535 – cn Canadiana [810]

Annals of thoracic surgery [1965] – Oxford, England 1965-89 – 1,5,9 – (cont by: annals of thoracic surgery) – ISSN: 0003-4975 – mf#10966 – us UMI ProQuest [617]

Annals of thoracic surgery [1989] – Oxford, England 1989+ – 1,5,9 – (cont: annals of thoracic surgery) – ISSN: 0003-4975 – mf#42585 – us UMI ProQuest [617]

Annals of tourism research – Oxford, England 1973+ – 1,5,9 – ISSN: 0160-7383 – mf#49410 – us UMI ProQuest [338]

Annals of tropical medicine and parasitology – Abingdon, Oxfordshire 1907-40 – 1 – ISSN: 0003-4983 – mf#175 – us UMI ProQuest [616]

Annals of ulster otherwise annals of senat : a chronicle of irish affairs from ad 431, to ad 1540 / ed by Hennesy, William M & MacCarthy, B – Dublin, 1887-1901 – 4v on 43mf – 8 – €82.00 – (with trans and notes) – ne Slangenburg [941]

Annals of vascular surgery – Dordrecht, Netherlands 1986+1,5,9) – 1,5,9 – ISSN: 0890-5096 – mf#18089 – us UMI ProQuest [617]

Annals of witchcraft in new england : and elsewhere in the united states, from their first settlement / Drake, Samuel Gardner – Boston: W Elliot Woodward, 1869 [mf ed 1990] – (= ser Woodward's historical series 8) – 1mf – 9 – 0-7905-5694-4 – mf#1988-1694 – us ATLA [130]

L'annam – Saigon. n63-182. mai 1926-fevr 1928 – 1 – fr ACRPP [073]

L'annam nouveau – Hanoi. 1931-avr 1942 – 1 – fr ACRPP [073]

L'annam nouveau – Hanoi: Impr Trung-Bac Tan-Van 1932, 1937, 1940-41 – 1r – 1 – mf#mf-11809 seam – us CRL [079]

Annamaig : eine dorfgeschichte aus dem baireuther land / Raithel, Hans – 3. aufl. Leipzig: C F Amelang 1920 [mf ed 1991] – 1r – 1 – (filmed with: mutter marie / heinrich mann) – mf#2844p – us Wisconsin U Libr [830]

Annambhatta *see* Tarkasamgraha-dipika on tarkasamgraha

L'annam-tonkin – Haiphong: Impr d'Extreme-Orient, jul 2-aug 30 1908 – 1r – 1 – mf#mf-4105 seam – us CRL [079]

Annan, Robert *see* Exposition and defense of the westminster assembly's confession

Annan, William *see*
- The difficulties of arminian methodism
- High church episcopacy
- Letters on psalmody

Annand, Edward *see*
- Annee de la premiere communion ou guide des enfants durant l'annee de la premiere communion
- Christian baptism

Annand, William *see*
- Confederation
- Letter from hon william annand to the electors of hants
- The speeches and public letters of the hon joseph howe

Annandale, Charles *see*
- The works of robert burns
- The works of robert burns

Annandale herald – Lockerbie: Dumfriesshire Newspapers Ltd 1971- (wkly) [mf ed 1999-] – 1 – (cont: annandale herald and record) – ISSN: 1356-8671 – uk Scotland NatLib [072]

Annandale herald and moffat news – [Scotland] Lockerbie: J Halliday 5 jan 1893-dec 1950 (wkly) [mf ed 2004] – 38r – 1 – (cont: annandale herald and southern advertiser; merged with: annandale record to form: annandale herald and record) – uk Newsplan [072]

Annandale observer – Annan: Dumfriesshire Newspapers Ltd (wkly) [mf ed 1999-] – 1 – (cont: annandale observer and advertiser) – ISSN: 1356-868X – uk Scotland NatLib [072]

Annandale record – [Scotland] Lockerbie: Cuthbertson & Laidlaw jun 1920-jun 1926 (wkly) [mf ed 2004] – 316v on 7r – 1 – (merged with: annandale herald and moffat news to form: annandale herald and record) – uk Newsplan [072]

Annapolis, Maryland *see* The maryland gazette, 1745-1839

Annapolis times – 1995 feb 20/26; 1996 mar 11/17-dec 27/1997 jan 2; 1997 jan 3/9-dec 26/1998 jan 1; 1998; 1999 jan 15/21, feb 19/25; 1999 jan 2/8-feb 19/25 – 1 – mf#3544682 – us WHS [071]

Anne, Theodore *see*
- Madrid
- Memoires, souvenirs et anecdotes sur l'interieur du palais de charles 10
- La prisonniere de blaye

L'anneau du nibelung : siegfried / Kufferath, Maurice – Bruxelles: Schott, Paris: Fischbacher, Leipzig: O Junne 1891 – (= ser Le theatre de r wagner de tannhaeuser a parsifal) – 2mf – 9 – mf#wa-58 – ne IDC [780]

Annecke, Wilhelm *see* Max dauthendey als dramatiker

Une annee : ou la france depuis le 27 juillet 1830 jusqu'au 27 juillet 1831 / Jailly, Hector de – Paris 1831 [mf ed Hildesheim 1995-98] – 1v on 2mf – 9 – €60.00 – ISBN-10: 3-487-26052-2 – ISBN-13: 978-3-487-26052-5 – gw Olms [944]

Une annee a londres / Defauconpret, Auguste – Paris 1819 [mf ed Hildesheim 1995-98] – 1v on 2mf – 9 – €60.00 – 3-487-27965-7 – gw Olms [914]

Annee, Antoine *see* Le livre noir de messieurs delavau et franchet

Annee au ministere de l'agriculture et de l'interieur / Legitime, Francois Denis – Paris, France. 1883 – 1r – us UF Libraries [630]

L'annee benedictine / Blemur, R M J de – Paris, 1667 – €143.00 – ne Slangenburg [241]

L'annee biologique : comptes rendus des travaux de biologie generale – Paris, 1921 22-1925 26 – 1 – us L of C Photodup [574]

L'annee de la premiere communion : apprets, veille, lendemain du grand jour, confirmation, perseverance, a l'usage des ecoles, des pensionnats, des academies – Montreal: Granger, [1895?] [mf ed 1986] – 5mf – 9 – 0-665-53624-0 – mf#53624 – cn Canadiana [241]

Annee de la premiere communion ou guide des enfants durant l'annee de la premiere communion : en usage dans les ecoles, pensionnats et academies / Annand, Edward – 2e augm ed. Montreal: Granger freres, editeurs, 1898 [mf ed 1980] – 6mf – 9 – 0-665-04109-8 – mf#04109 – cn Canadiana [241]

Une annee de revolution : d'apres un journal tenu a paris en 1848 / Normanby, Constantine H of – Paris 1858 [mf ed Hildesheim 1995-98] – 2v on 11mf – 9 – €110.00 – ISBN-10: 3-487-26017-4 – ISBN-13: 978-3-487-26017-4 – gw Olms [944]

L'annee litteraire – Paris, 1754-90 – 1 – fr ACRPP [410]

L'annee litteraire – Paris, 1849 – 1 – fr ACRPP [410]

L'annee litteraire et dramatique – Paris, 1859-69 – 11v on 102mf – 8 – mf#H-1375 – ne IDC [410]

Une annee memorable de la vie d'auguste de kotzebue – Berlin 1802 [mf ed Hildesheim 1995-98] – 2v on 8mf – 9 – €160.00 – 3-487-28958-X – (trans fr german) – gw Olms [880]

L'annee missionaire, 1931 / Lesourd, Paul – Madrid: Razon y Fe, 1932 – 1 – sp Bibl Santa Ana [240]

Annee musicale – Paris 1911-13 – v1-3 on 17mf – 9 – mf#kp-556/1 – ne IDC [780]

Annee philosophique – v1-24. 1890-1913 – 1 – $240.00 – mf#0046 – us Brook [100]

L'annee politique economique et cooperative *see* L'annee politique et etrangere

L'annee politique et economique *see* L'annee politique et etrangere

L'annee politique et etrangere – n1-60. 1925-juin 1940 – 1 – (devenu: l'annee politique economique et cooperative n80-93 nov dec 1947-fevr 1950 sic. devenu: l'annee politique et economique n94-230 paris mars 1950-72) – fr ACRPP [327]

L'annee religieuse de montreal pour 1864 – [Montreal?: s.n, 1864?] [mf ed 1983] – 1mf – 9 – 0-665-44884-8 – mf#44884 – cn Canadiana [241]

L'annee sociologique – Paris. 13v. (Scattered issues lacking) – 1 – 80.00 – us L of C Photodup [300]

Annee terrible / Vigoureaux, G – Jeremie, Haiti. 1909 – 1r – us UF Libraries [972]

O annel – Sao Paulo, SP. 30 out 1897 – (= ser Ps 19) – mf#P17,02,214 – bl Biblioteca [079]

Annener zeitung – Witten DE, 1950-57 – 11r – 1 – (filmed by misc inst: 1958 8 jan-1960 30 jun, 1885 26 sep-1941 30 jul, 1942-1943 31 mar, 1949 22 oct-1960 12 apr, 1960 1 jul-1961 4 feb (?)) – gw Mikrofilm; gw Misc Inst [074]

Annesley, Alexander *see* A compendium of the law of marine insurance

Annesley, G *see* Voyages and travels to india...

Annesley, Rosa *see* Voices of the wind

Annett, Edward Aldridge *see* The natural method of bible teaching for india

Annette, the metis spy : a heroine of the n w rebellion / Collins, Joseph Edmund – Toronto: Rose, 1886 [mf ed 1980] – 2mf – 9 – 0-665-02152-6 – mf#02152 – cn Canadiana [920]

Annette und levin : zur jahrhundertfeier der droste, meersburg 24. mai 1948 / Heselhaus, Clemens – Muenster-Westfalen: Aschendorff 1948 [mf ed 1996] – 1r [ill] – 1 – (filmed with: der gutsverkauf / karl domanig) – mf#4260p – us Wisconsin U Libr [430]

Annette von droste-huelshoff : eine auswahl / ed by Heselhaus, Clemens – Muenchen: C Hanser 1948 [mf ed 1989] – 1r – 1 – (aft by ed; filmed with: ottjen alldag un sien moorhex / georg droste) – mf#7186 – us Wisconsin U Libr [800]

Annette von droste-huelshoff : die entdeckung des seins in der dichtung des neunzehnten jahrhunderts / Heselhaus, Clemens – Halle (Saale): M Niemeyer 1943 [mf ed 1989] – 1r [ill] – 1 – (filmed with: der impressionismus in der lyrik der annette von droste-hulshoff / gerhard fruhbrodt) – mf#7190 – us Wisconsin U Libr [430]

Annette von droste-huelshoff : ihre dichterische entwicklung und ihr verhaeltnis zur englischen literatur / Badt-Strauss, Bertha – Leipzig: Quelle & Meyer, 1909 [mf ed 1992] – (= ser Breslauer beitraege zur literaturgeschichte. neue folge 7) – 96p – 1 – (incl bibl ref) – mf#8014 reel 2 – us Wisconsin U Libr [410]

Annette von droste-huelshoff / Ramsay, Tamara – Stuttgart: J G Cotta, c1938 [mf ed 1989] – (= ser Die dichter der deutschen) – 98p – 1 – mf#7190 – us Wisconsin U Libr [430]

Annette von droste-huelshoff als erzaehlerin : realismus und objektivitaet in der "judenbuche" / Heitmann, Felix – Münster i.W: Aschendorff 1914 [mf ed 1989] – 1r – 1 – (filmed with: der impressionismus in der lyrik der annette von droste-hulshoff / gerhard fruhbrodt) – mf#7190 – us Wisconsin U Libr [430]

Annette von droste-huelshoff als westfaelische dichterin / Scholz, Wilhelm von – Muenchen, 1897 (mf ed 1995) – 1mf – 9 – €24.00 – 3-8267-3135-2 – mf#DHS-AR 3135 – gw Frankfurter [430]

Annette von droste-huelshoff im spiegel der zeitgenoessischen kritik / Raab, Karl – Münster i. Westf: Regensberg [1933?] [mf ed 1989] – 1r – 1 – (incl ind. filmed with: der impressionismus in der lyrik der annette von droste-hulshoff / gerhard fruhbrodt) – mf#7190 – us Wisconsin U Libr [430]

Annette von droste-huelshoff im spiegel der zeitgenoessischen kritik / Raab, Karl – Muenster, 1933 (mf ed 1992) – 1mf – 9 – €24.00 – 3-89349-100-7 – mf#DHS-AR 100-7 – gw Frankfurter [430]

Annette von droste-huelshoff in der schweiz / Scheiwiller, Otmar – Einsiedeln (Schweiz): Benziger [19–] [mf ed 1989] – 1r [ill] – 1 – (filmed with: die briefe der annette von droste-hulshoff / ed by karl schulte kemminghausen) – mf#7189 – us Wisconsin U Libr [430]

Annette von droste-huelshoff in ihren beziehungen zu goethe und schiller und in der poetischen eigenart ihrer gereiften kunst / Freund, Anna – Muenchen: Kastner Callwey 1915 [mf ed 1989] – 1r – 1 – (incl bibl ref. filmed with: die briefe der annette von droste-hulshoff / ed by karl schulte kemminghausen) – mf#7189 – us Wisconsin U Libr [430]

Annette von droste-huelshoff und ihr verhaeltnis zur romantik / Lucke, Hans – Paderborn: F Schoeningh 1927 [mf ed 1989] – 1r – 1 – (filmed with: der impressionismus in der lyrik der annette von droste-hulshoff / gerhard fruhbrodt) – mf#7190 – us Wisconsin U Libr [430]

Annexation : the ideas of the late william h seward / Glen, Francis Wayland – [S.l: s.n, 1893?] [mf ed 1980] – 1mf – 9 – 0-665-04374-0 – mf#04374 – cn Canadiana [971]

Annexation : or union with the united states, is the manifest destiny of british north america / Monro, Alexander – [Saint John, NB?: s.n.], 1868 [mf ed 1985] – 1mf – 9 – 0-665-33839-2 – mf#33839 – cn Canadiana [971]

Annexation and british connection : address to brother jonathan / Lett, William Pittman – [Ottawa?: s.n.], 1889 [mf ed 1980] – 1mf – 9 – mf#09034 – cn Canadiana [971]

The annexation manifesto of 1849 / Association d'annexion de Montreal – Montreal: D English & co, 1881 [mf ed 1980] – 1mf – 9 – 0-665-02403-7 – mf#02403 – cn Canadiana [971]

Annexation of burma / Banerjee, Anil Chandra – Calcutta: A Mukherjee & Bros, 1944 – (= ser Samp: indian books) – us CRL [954]

Annexation to the united states : is it desirable? and is it possible? – [Halifax, NS?: s.n.], 1868 [mf ed 1981] – 1mf – 9 – mf#23508 – cn Canadiana [971]

Annexations to sierra leone : and their influence on british trade with west africa / Harris, John M – London, [1883] – (= ser 19th c books on british colonization) – 1mf – 9 – mf#1.1.4973 – uk Chadwyck [380]

Annexion : conference; l'union continentale / Rouilliard, Jean-Baptiste – [S.l: s.n, 1893?] [mf ed 1980] – 1mf – 9 – 0-665-03740-6 – mf#03740 – cn Canadiana [971]

L'annexion du congo a la belgique et le droit international / Brunet, Roger – Paris: Jouve et Cie, 1911 – 334p – 1 – mf#LL-12009 – us L of C Photodup [341]

Annexion du texas : nouveaux documents americains / Jollivet, Adolphe – [Paris?: A Jollivet], 1845 [mf ed 1982] – 1mf – 9 – (incl bibl ref) – mf#36686 – cn Canadiana [975]

The annie adams field papers, 1852-1912 – [mf ed 1981] – 3r – 1 – (with p/g. coll of annie adams fields' diaries and memoirs provides rare glimpse into new england's literary society) – us MA Hist [420]

Annie besant : an autobiography – 2nd ed. London: T Fisher Unwin, 1908 [mf ed 1992] – 1mf – 9 – 0-524-02069-8 – (incl bibl ref) – mf#1990-2833 – us ATLA [920]

Annihilationism not of the bible : being an examination of the principal scriptures in controversy between evangelical christians and annihilationists... / George, Nathan Dow – Boston: J P Magee, 1870 [mf ed 1984] – 4mf – 9 – 0-8370-1031-4 – (incl ind) – mf#1984-4388 – us ATLA [240]

Anniston Army Depot [AL] *see* Tracks

Anniversaire de la proclamation de la republique francaise – [Paris, 1849?] – (= ser Pamphlets and periodicals relating to the French Revolution of 1848) – 1r – 1 – us CRL [944]

Anniversary – 1829 – (= ser English gift books and literary annuals, 1823-1857) – 4mf – 9 – uk Chadwyck [800]

Anniversary number of the chignecto post and borderer – Sackville, NB: [s.n.], 1895 [mf ed 1980] – 1mf – 9 – mf#05880 – cn Canadiana [321]

Anniversary of the american peace society : proceedings at the annual meeting – [s.l: s.n.] 1856 [mf ed 1993] – 1mf – 9 – 0-524-08215-4 – mf#1993-1000 – us ATLA [230]

Anniversary sermon, 1889 : preached by rev w t herridge, bd in st andrews church, sunday evening, dec 1st / Herridge, William Thomas – Ottawa: St Andrew's Society of Ottawa, 1889 [mf ed 1980] – 1mf – 9 – 0-665-05556-0 – mf#05556 – cn Canadiana [360]

Anniversary sermon of the church pastoral-aid society / Shirley, W A – s.l, England. 1844? – 1r – us UF Libraries [240]

Anniversary sermon of the royal humane society / Valpy, R – London, England. 1802 – 1r – us UF Libraries [240]

Anniversary sermon preached in knox church, november 30th, 1890 / Farries, Francis Wallace – [S.l: s.n, 1891?] [mf ed 1980] – 1mf – 9 – 0-665-02952-7 – mf#02952 – cn Canadiana [242]

Anno 1791 fing es an : des braven schneidergesellen franz bersling abenteuerlicher kampf gegen napoleon in fuenf weltteilen – Leipzig: P Reclam, [1938?] [mf ed 1989] – 320p (ill) – 1 – (aft by konrad krause) – mf#7013 – us Wisconsin U Libr [430]

O anno novo – Joinvile, SC. 24 dez 1931; 01 jan 1933 – (= ser Ps 19) – bl Biblioteca [079]

Anno primo victoriae reginae, magnae britanniae et hiberniae : at the parliament begun and holden at westminster, on the 15th day of nov, anno domini 1837... – Quebec: J C Fisher & W Kemble, 1838 [mf ed 1985 – 1mf – 9 – 0-665-05099-2 – (in english with french trans) – mf#05099 – cn Canadiana [342]

Anno primo victoriae reginae, magnae britanniae et hiberniae : au parlement commence et tenu a westminster, le quinzieme jour de novembre, anno domini 1837... = At the parliament begun and holden at westminster, on the fifteenth day of november, anno domini 1837 / Grande-Bretagne – Quebec: impr par John Charlton Fisher et William Kemble...1838 [mf ed 1999] – 1mf – 9 – (in french and english) – mf#SEM105P3158 – cn Bibl Nat [348]

Anno regni decimo quarto, georgii 3, regis, chap 83 : [a]cte qui regle plus solidement le gouvernement de la province de quebec dans l'amerique septentrionale – [S.l: s.n, 18–?] [mf ed 1985] – 1mf – 9 – 0-665-50690-2 – mf#50690 – cn Canadiana [323]

Anno vicesimo-tertio victoriae reginae : cap 61: acte concernant les municipalites et les chemins dans le bas canada / Canada (Province) – [Quebec?: s.n, 1865?] (mf ed 1999) – 2mf – 9 – mf#SEM105P3159 – cn Bibl Nat [348]

Annoncen-blatt (general-anzeiger) fuer marburg und umgebung *see* Generalanzeiger fur marburg und umgebung 1887

Annonces, affiches et avis divers – 3mai 1752-25 fevr 1761 – 1 – (devenu: affiches, annonces et avis divers puis ou journal general de france. 4 mars 1761-84; devenu: journal general de france 1785-10 aout 1792; voir aussi: supplement au journal general de france) – fr ACRPP [073]

L'annonceur – [Quebec]: L Recio, [1880] – 9 – mf#P04845 – cn Canadiana [380]

Annonsblad foer gotland – Visby, Sweden. 1879-83 – 1r – 1 – sw Kungliga [078]

Annonsbladet – Boras, 1919-26 – 2r – 1 – sw Kungliga [078]

Annonstidning foer upland och westmanland – Enkoeping, 1885 – 1r – 1 – sw Kungliga [078]

Annotaciones in galeni interpretes quibus varii locis, in quos hactenus impegerunt... / Laguna, A de – Venecia, 1548 – 2mf – 9 – sp Cultura [610]

The annotated bible : being a household commentary upon the holy scriptures, comprehending the results of modern discovery and criticism / Blunt, John Henry – London; New York: Rivingtons, 1878-1882 – 21mf – 9 – 0-8370-1423-9 – mf#1987-6063 – us ATLA [220]

Annotated cases – american and english – New York, San Francisco: Thompson Co, Bancroft-Whitney. v1916C-1918E (13v). 1916-18 (all publ) – 185mf – 9 – $277.00 – mf#LLMC 84-695C – us LLMC [340]

Annotated constitution of india / Basu, Durga Das – Calcutta: Das Gupta & Co, 1953 – (= ser Samp: indian books) – us CRL [323]

The annotated corporation laws of all the states, generally applicable to stock corporations / Cumming, Robert Cushing – Albany, Lyon, 1899-1903. 5 v. LL-1442 – 1 – us L of C Photodup [346]

Annotated guide to women's periodicals in the us – v1 n1-2 (1982 feb-jul) [1]; v2 n1-v4 n1 (1983 apr-1985 may) [2] – 1 – mf#976474 [1]; 979534 [2] – us WHS [073]

Annotated list of commercial food fishes found in... – s.l, s.l? . 193-? – 1r – us UF Libraries [639]

The annotated proofs : from the forster collection in victoria and albert museum, london / Dickens, Charles – 3r – 1 – mf#96739 – uk Microform Academic [830]

The annotated sale catalogues of puttick and simpson : from the british library, london – 4pts. 1846-70 – 96r – 1 – (pt1: 1846-56 22r c35-10510. pt2: 1856-63 32r c35-10511. pt3: 1864-67 22r c35-10512. pt4: 1867-71 20r c35-10513) – mf#C35-10500 – us Primary [700]

Annotated sale catalogues of puttick and simpson, 1846-1871 – Pt 1: 1846-56. Pt 2: 1856-63. Pt 3: 1864-67. Pt 4: 1867-71 – (= ser Literature, Music and Art: The Annotated Sale Catalogues of Puttick and Simpson) – 1 – (previous title: literature, music and art: the annotated sale catalogues of puttick and simpson) – us Primary [780]

The annotated scottish communion office : an historical account of the scottish communion office and of the communion office of the protestant episcopal church of the usa / Dowden, John – Edinburgh: R Grant; New York: T Whittaker, 1884 [mf ed 1990] – 1mf – 9 – 0-7905-6049-6 – (incl bibl ref) – mf#1988-2049 – us ATLA [242]

Annotated statutes and rules of trial practice and appellate procedure in south dakota and north dakota / Deland, Charles Edmund – Pierre, Carter, 1896. 581 p. LL-680 – 1 – us L of C Photodup [348]

Annotated time table : with information as to all cpr routes / Canadian Pacific Railway – [Montreal?: CPR, 1890?] [mf ed 1983] – 1mf – 9 – mf#13391 – cn Canadiana [380]

Annotated time table : with information as to cpr transcontinental routes / Canadian Pacific Railway – S.l: s.n, 1892?] [mf ed 1981] – 1mf – 9 – (incl ind) – mf#14041 – cn Canadiana [380]

Annotated time table : with information as to cpr transcontinental routes / Canadian Pacific Railway – Memo ed. [Montreal?: CPR?, 1899?] [mf ed 1981] – 2mf – 9 – 0-665-25971-9 – mf#25971 – cn Canadiana [380]

Annotated time table : with information as to cpr transcontinental routes / Canadian Pacific Railway – [Montreal?]: CPR, 1898? [mf ed 1980] – 2mf – 9 – 0-665-00442-7 – mf#00442 – cn Canadiana [380]

Annotated time table : with information as to cpr transcontinental routes / Canadian Pacific Railway – [Montreal?: CPR?, 1900?] [mf ed 1981] – 2mf – 9 – 0-665-16953-1 – mf#16953 – cn Canadiana [380]

Annotated time table : with information as to cpr transcontinental routes / Canadian Pacific Railway – [Montreal?: s.n, 1893?] [mf ed 1981] – 2mf – 9 – 0-665-14572-1 – mf#14572 – cn Canadiana [380]

Annotated time table : with information as to cpr transcontinental routes / Canadian Pacific Railway – [s.l: s.n, 1893?] [mf ed 1984] – 1mf – 9 – 0-665-14713-9 – mf#14713 – cn Canadiana [380]

Annotated time table : with information as to cpr transcontinental routes / Canadian Pacific Railway – [S.l: s.n, 1896?] [mf ed 1981] – 1mf – 9 – mf#15041 – cn Canadiana [380]

Annotated time table : with information as to cpr transcontinental routes – [Memo ed]. [Montreal: s.n, 1900] [mf ed 1981] – 2mf – 9 – mf#01126 – cn Canadiana [380]

Annotation – v2 n2-v10 n2 [1974 spring-1982 dec] – 1r – 1 – mf#1259715 – us WHS [071]

Annotation of the code of the trust territory of the pacific islands (ttpi), 1952 – n.p.,n.d. – 1mf – 9 – $1.50 – mf#LLMC 82-100F Title 87 – us LLMC [348]

Annotationes in novum testamentum / Osiander, Lucas – Tuebingen, 1592 – 36mf – 8 – €69.00 – ne Slangenburg [225]

Annotationes io bvgenhagij pomerani in epistolas pauli / Bugenhagen, J – Basilae, 1525 – 4mf – 9 – mf#TH-1 mf 130-133 – ne IDC [242]

Annotationes ivsti ionae : in acta apostolorvm cvm indice / Jonas, J – Basileae, 1525 – 2mf – 9 – mf#TH-1 mf 803-804 – ne IDC [242]

Annotationes piae ac doctae in euangeliu ioannis / Oecolampadius, J – Basileae, Bebel et Cratander, 1533 – 9mf – 9 – mf#PBU-387 – ne IDC [240]

Annotationes piissimae doctissimaeque in ioseam, ioelem, amos, abdiam etc / Oecolampadius, J – Basileae, Cratander, 1535 – 7mf – 9 – mf#PBU-391 – ne IDC [240]

Annotations on some of the messianic psalms / Rosenmueller, Ern Frid Car – Edinburgh: Clark, 1841 [mf ed 1992] – (= ser The biblical cabinet 32) – 2mf – 9 – 0-524-04922-X – mf#1992-0265 – us ATLA [221]

Annotations on the acts of the apostles / Stellhorn, Frederick William – New York: Christian Literature, 1896 [mf ed 1989] – (= ser The lutheran commentary 6) – 1mf – 9 – 0-7905-2089-3 – (incl ind) – mf#1987-2089 – us ATLA [226]

Annotations on the epistles of paul to 1. corinthians 7-16, 2. corinthians and galatians / Jacobs, Henry Eyster et al – New York: Christian Literature, 1897 [mf ed 1989] – (= ser The lutheran commentary 8) – 1mf – 9 – 0-7905-3027-9 – mf#1987-3027 – us ATLA [227]

Annotations on the epistles of paul to the ephesians, philippians, colossians, thessalonians / Horn, Edward Traill & Voigt, A G – New York: Christian Literature, 1896 [mf ed 1989] – (= ser The lutheran commentary 9) – 1mf – 9 – 0-7905-3026-0 – mf#1987-3026 – us ATLA [227]

Annotations on the epistles of paul to the romans and 1. corinthians, chaps 1.-6 / Jacobs, Henry E – New York: Christian Literature, 1896 [mf ed 1989] – (= ser The lutheran commentary 7) – 1mf – 9 – 0-7905-3080-5 – mf#1987-3080 – us ATLA [227]

Annotations on the epistles to timothy, titus and the hebrews : and on philemon / Wolf, Edmund Jacob & Horn, Edward Traill – New York: Christian Literature, 1897 [mf ed 1989] – (= ser The lutheran commentary 10) – 2mf – 9 – 0-7905-3119-4 – mf#1987-3119 – us ATLA [227]

Annotations on the general epistles of james, peter, john, and jude / Weidner, Revere Franklin – New York: Christian Literature, 1897 [mf ed 1989] – (= ser The lutheran commentary 11) – 1mf – 9 – (incl bibl & ind) – mf#1987-2399 – us ATLA [227]

Annotations on the gospel according to st john / Spaeth, Adolph – New York: Christian Literature, 1896 [mf ed 1989] – (= ser The lutheran commentary 5) – 1mf – 9 – 0-7905-2080-X – mf#1987-2080 – us ATLA [226]

Annotations on the gospel according to st luke / Baugher, Henry Louis – New York: Christian Literature, 1896 [mf ed 1989] – (= ser The lutheran commentary 4) – 2mf – 9 – 0-7905-3071-6 – mf#1987-3071 – us ATLA [226]

Annotations on the gospel according to st mark / Haas, John Augustus William – New York: Christian Literature, 1895 [mf ed 1989] – (= ser The lutheran commentary 3) – 1mf – 9 – 0-7905-3023-6 – (incl bibl ref) – mf#1987-3023 – us ATLA [226]

Annotations on the gospel according to st matthew / Schaeffer, Charles Frederick – New York: Christian Literature, 1895 [mf ed 1989] – (= ser The lutheran commentary 1-2) – 2v on 2mf – 9 – 0-7905-2057-5 – mf#1987-2057 – us ATLA [226]

Annotations on the mutiny act... : with some observations on the practice of courts-martial / M'Naghten, Captain – London: Stevens & Sons, 1828 – 3mf – 9 – $4.50 – mf#LLMC 89-027 – us LLMC [347]

Annotations on the pentateuch : or the five books of moses, the psalms of david, and the song of solomon / Ainsworth, Henry – London: Blackie, 1843 [mf ed 1990] – 2v on 4mf – 9 – 0-7905-3421-5 – mf#1987-3421 – us ATLA [221]

Annotations on the revelation of st john the divine / Weidner, Revere Franklin – New York: Christian Literature, 1898 [mf ed 1989] – (= ser The lutheran commentary 12) – 1mf – 9 – 0-7905-2206-3 – (incl ind) – mf#1987-2206 – us ATLA [225]

Annotations on the sacred writings of the hindus : being an epitome of some of the most remarkable and leading tenets in the faith of that people... / Sellon, Edward – new ed. London: [s.n.], 1902 [mf ed 1992] – 1mf – 9 – 0-524-03683-7 – (incl bibl ref) – mf#1990-3261 – us ATLA [280]

Annotationvm scholasticarvm lvcae lossii lvnebvrgensis in novvm testamentvm iesv christi nazareni / Lossius, L – Franc, 1562. v5 – 7mf – 9 – mf#TH-1 mf 880-886 – ne IDC [242]
Announcement : history of the city of toronto and york county, toronto's jubilee year: a complete historical and descriptive representation of our provincial metropolis – S.l: s.n, 1883? – 1mf – 9 – mf#39549 – cn Canadiana [971]
[Announcement] / Chalif Russian Normal School of Dancing – New York [1907] – 1 – mf#*ZBD-*MGO pv 2 n6 – Located: NYPL – us Misc Inst [790]
Announcement of the russian missionary society, inc : with "facts about rev william fetler". a reply to the booklet entitled, "pastor w fetler's reply to rev e e shields" – Chicago, 1930 – 1 item of several on a reel – 1 – mf#2204-4 f – us Southern Baptist [242]
Announcements / National Indian Law Library – 1980 fall, v7 n1-v9 n1 [1981 may-1983 spr] – 1r – 1 – (cont by: narf legal review) – mf#412066 – us WHS [322]
Announcements / Native American Rights Fund – 1972-82 – (= ser American indian periodicals... 1) – 6mf – 9 – $95.00 – us UPA [305]
Annuaire / French Guiana – juin 1873-1908 – 1 – fr ACRPP [073]
Annuaire... / Cercle catholique de Quebec – Quebec: Le Cercle, 1878?-18– ou 19– – 9 – mf#A00081 – cn Canadiana [241]
Annuaire de documentation colonial comparee = Yearbook of compared colonial documentation – 1927-38 – 9 – $1502.00 – mf#0047 – us Brook [320]
Annuaire de la noblesse de france et de masions souveraince de l'europe – 1 – (v1-75 1843-1925 $600 [0050]. v76-90 1926-60 $144 [0049]) – us Brook [929]
Annuaire de la noblesse de france et d'europe – v. 50-72. 1894-1922. (v. 59 wanting) – 1 – 73.00 – us L of C Photodup [920]
L'annuaire de la rive sud de la banlieue de montreal = Lovell's south shore montreal suburban directory – Montreal: John Lovell & Son. v5 1966 [mf ed 1999] – (cont: lovell's south shore directory) – mf#SEM35P474 – cn Bibl Nat [971]
Annuaire de la russe 1904-1911 *see* Ezhegodnik rossii 1904-1911
Annuaire de la venerie francaise – Paris. 1891-94, 1897, 1901 – 1 – fr ACRPP [639]
L'annuaire de laval = Lovell's laval directory – Montreal: John Lovell & Son. v4 1968 [mf ed 1999] – 1r – 1 – (cont: lovell's ile jesus directory) – mf#SEM35P472 – cn Bibl Nat [971]
Annuaire de l'economie politique et de la statistique – v1-55. 1844-98 – 1 – $510.00 – (in french) – mf#0048 – us Brook [300]
Annuaire de legislation francaise et etrangere – Paris. Title Varies: 1870 71-1955, Annuaires de legislation etrangere. On film: v1-62, 1870-1936. LL-0283 – 1 – us L of C Photodup [350]
L'annuaire de l'ile jesus de lovell... = Lovell's ile jesus directory – Montreal: John Lovell & Son. v1 1958/1959 [mf ed 1999] – 1r – 1 – (cont by: lovell's ile jesus directory) – mf#SEM35P470 – cn Bibl Nat [030]
Annuaire de l'institut de philologie et d'histoire orientales, tom 2 – Melanges Bidez T1+2. (1934) – 20mf – 9 – €38.00 – ne Slangenburg [480]
Annuaire des archives israelites – Paris. sept 1884-sept 1925 – 1 – fr ACRPP [939]
Annuaire des cinq departements de l'ancienne normandie – 14th-40th annee. 1848-74 – 1 – $360.00 – (in french) – mf#0051 – us Brook [940]
Annuaire des eaux et forets – Paris: Bureau de la Revue des eaux et forets 1862- (annual) [annee 1862-1864, 4e-53e annee (1865-1914)] [mf ed 2005] – (= ser 1886-1890: bibliotheque du forestier) – 6r [ill] – 1 – (none publ in 1871; vol for 1872 iss in combined form as: 10e et 11e annee; none publ 1915-1922; title varies slightly) – mf#film mas c6037 – us Harvard [634]
Annuaire des journaux – 1881-82, 1884, 1886-87, 1890, 1892, 1894, 1896-98, 1900, 1906, 1910, 1912, 1914. Suppl. 1889-1913 – 1 – fr ACRPP [073]
Annuaire des postes de l'empire francaise – 1859-79 – 1 – us L of C Photodup [944]
Annuaire israelite pour la suisse *see* Juedisches jahrbuch fuer die schweiz
Annuaire oriental du commerce de l'industrie, de l'administration et de la magistrature 9me annee 1889-1890 / Cervati, Raphael C – Encres d'Imprimerie Ch. Lorilleaux & Cie, Paris; Typographie et Lithographie J Pallamary, Constantinople – 21mf – 9 – $335.00 – us MEDOC [380]
Annuaire statistique / Egypt. Maslahat al-Ihsa wa-al-Ta'dad – 1937-38 – 1 – us CRL [324]
Annuaire statistique / Quebec (Province). Bureau des statistiques – Quebec: [le Bureau] 1re annee 1914-44e ed 1961 (annual) [mf ed 1988] – 6mf – 9 – (cont by: annuaire du quebec) – mf#SEM105P893 – cn Bibl Nat [317]
Annuaire statistique 1962, 1967 / Niger. Direction de la Statistique – (= ser African official statistical serials, 1867-1982) – 5mf – 9 – uk Chadwyck [316]
Annuaire statistique 1901-1959 / Egypt. Maslahat al-Ihsa wa-al-Ta'dad – (= ser African official statistical serials, 1867-1982) – 317mf – 9 – uk Chadwyck [316]
Annuaire statistique 1958-1963, 1969 / Congo (formerly French Congo). Service National de la Statistique, des Etudes Demographiques et Economiques – (= ser African official statistical serials, 1867-1982) – 7mf – 9 – uk Chadwyck [316]
Annuaire statistique 1965-1975 / Benin. Institut nationale de l'Analyse Economique – (= ser African official statistical serials, 1867-1982) – 16mf – 9 – (1966, 1968, 1970-72, 1974 not available) – uk Chadwyck [316]
Annuaire statistique 1966-1975 / Chad. Sous Direction de la Statistique et des Etudes Economiques – (= ser African official statistical serials, 1867-1982) – 9mf – 9 – (1971, 1973 not publ. 1967, 1968 not available) – uk Chadwyck [316]
Annuaire statistique 1968-74 / Mauritania. Direction de la Statistique et des Etudes Economiques – (= ser African official statistical serials, 1867-1982) – 19mf – 9 – uk Chadwyck [316]
Annuaire statistique 1969-1975 / Burundi. Departement des Etudes et Statistiques – (= ser African official statistical serials, 1867-1982) – 9mf – 9 – us Chadwyck [316]
Annuaire statistique de la belge 1870-1962 / Belgium. Ministere des Affaires Economiques et des Classes MoyennesInstitut National de Statistique – (= ser European official statistical serials, 1841-1984) – 398mf – 9 – (1946; 1952-54; 1956-57; 1963-65 not repr) – uk Chadwyck [314]
Annuaire statistique de la cote d'ivoire 1975 / Ivory Coast. Ministere du Plan – (= ser African official statistical serials, 1867-1982) – 3mf – 9 – uk Chadwyck [316]
Annuaire statistique de la france / France. Institut National de la Statistique et des Etudes Economiques – v1-78. 1878-1973 – 9 – $1020.00 – mf#0223 – us Brook [314]
Annuaire statistique de la france 1878-1965 / France. Institut National de la Statistique et des Etudes Economiques – (= ser European official statistical serials, 1841-1984) – 499mf – 9 – uk Chadwyck [314]
Annuaire statistique de la grece 1930-1939 *see* Statistike epeteris tes hellados 1930-1939
Annuaire statistique de la guadeloupe 1949/1953-1967/1970 / France. Institut National de la Statistique et des Etudes Economiques – (= ser Latin american & caribbean...1821-1982) – 11mf – 9 – (1949-53 not available) – uk Chadwyck [318]
Annuaire statistique de la guyane 1947/1952-1961/1970 / France. Institut National de la Statistique et des Etudes Economiques – (= ser Latin american and caribbean statistical serials, 1821-1982) – 2mf – 9 – (1947-52; 1957-59; 1961-70 not available) – uk Chadwyck [318]
Annuaire statistique de la martinique 1952-1969/1972 / France. Institut National de la Statistique et des Etudes Economiques – (= ser Latin american & caribbean...1821-1982) – 13mf – 9 – uk Chadwyck [318]
Annuaire statistique de la republique centrafricaine 1962 / Central African Republic. Direction de la Statistique et de la Conjoncture – (= ser African official statistical serials, 1867-1982) – 2mf – 9 – uk Chadwyck [316]
Annuaire statistique de la republique du mali 1963-1973 / Mali (formerly French Sudan). Service de la Statistique Generale de la Comptabilite Nationale et de la Mecanographie – (= ser African official statistical serials, 1867-1982) – 25mf – 9 – (1967 not publ) – uk Chadwyck [316]
Annuaire statistique de la republique tchecoslovaque 1934-1938 / Czechoslovakia. L'Office Statistique de La Republique Tchecoslovaque – (= ser European official statistical serials, 1841-1984) – 22mf – 9 – uk Chadwyck [314]
Annuaire statistique de la suisse 1891-1965 *see* Statistisches jahrbuch der schweiz 1891-1965
Annuaire statistique de la tunisie 1940-1971 / Tunisia. Service des Statistiques – (= ser African official statistical serials, 1867-1982) – 56mf – 9 – uk Chadwyck [316]
Annuaire statistique de l'afrique equatoriale francaise 1936-1955 / French Equatorial Africa. Haut Commissariat – (= ser African official statistical serials, 1867-1982) – 8mf – 9 – uk Chadwyck [316]
Annuaire statistique de l'afrique occidentale francaise 1949-1954 / French West Africa. Direction des Services de la Statistique Generale et de la Mecanographie – (= ser African official statistical serials, 1867-1982) – 22mf – 9 – uk Chadwyck [316]
Annuaire statistique de l'algerie / Algeria. Service de Statistique Generale – 1926-37 – 1 – 69.00 – us L of C Photodup [316]
Annuaire statistique de l'algerie 1926-1964 / Algeria. Service Central de Statistique – (= ser African official statistical serials, 1867-1982) – 112mf – 9 – (1961, 1962 not publ) – uk Chadwyck [316]
Annuaire statistique de madagascar 1938-1951 / Malagasy Republic (Madagascar). Service de Statistique Generale – (= ser African official statistical serials, 1867-1982) – 2mf – 9 – uk Chadwyck [316]
Annuaire statistique du maroc 1925-1976 / Morocco. al'Maslahah al-Markaziyah lil-Ihsa'iyat – (= ser African official statistical serials, 1867-1982) – 103mf – 9 – (1930, 1931 not available) – uk Chadwyck [316]
Annuaire statistique du togo 1966-1973 / Togo. Direction de la Statistique – (= ser African official statistical serials, 1867-1982) – 15mf – 9 – uk Chadwyck [316]
Annuaire-almanach du commerce – Paris, Didot-Bottin. [1892] – 2r – 1 – us CRL [380]
Annuaire-almanach du commerce de l'industrie, de l'administration et de la magistrature 4me annee 1883 / Cervati, Raphael C – Constantinople: J Pallamary, 1883 – 17mf – 9 – $280.00 – us MEDOC [380]
Annual / Baptist Associations – 1 – (baptist associations. landmark missionary baptist associational minutes of the pacific coast: cooperative of calif., 1952-1972; northern calif., 1952-1960; costal cooperative, 1961-1972; valley, 1961-1970; central valley, 1952-1972; costal area, 1958-1961, 1968-1969, 1972; cooperative of oregon, 1960-1963, 1969, 1971, 1972 (a.w.w.), 1965-1968 (scott), 1970 (albany), 1963-1971, washington) – us Southern Baptist [242]
Annual / Baptist Associations – 1 – $105.56 – (baptist associations. landmark missionary baptist associational minutes of the pacific coast direct mission: california, 1932, 1934-1941, 1944, 1946-1971; northern california, 1940-1954, 1961-1971; central california, 1951-1958, 1960-1965; missionary baptist churches of southern calif., 1940-1941, 1943-1970) – us Southern Baptist [242]
Annual / Baptist Associations. Alabama – 1815-1958 – 1 – us Southern Baptist [242]
Annual / Baptist Associations. Arizona – 1950-72 – 1 – us Southern Baptist [242]
Annual / Baptist Associations. Arkansas – 1854-1964 – 1 – us Southern Baptist [242]
Annual / Baptist Associations. California – 1944-66 – 1 – us Southern Baptist [242]
Annual / Baptist Associations. Colorado, Southeastern, Arkansas Valley – 1953, 1957, 1959, 1960 – 1 – $5.00 – us Southern Baptist [242]
Annual / Baptist Associations. Florida – 1843-1964 – 1 – us Southern Baptist [242]
Annual / Baptist Associations. Georgia – 1880-1957 – 1 – us Southern Baptist [242]
Annual / Baptist Associations. Illinois – 1844-1990 – 1 – us Southern Baptist [242]
Annual / Baptist Associations. Illinois. Chicago Southern Baptist – 1957-65 – 1 – $69.30 – (incl associational minutes, minutes of: executive committee, missions committee, the chicago baptist banner present newsletter of the csba, 1958-1965) – us Southern Baptist [242]
Annual / Baptist Associations. Indiana, Brownstown – 1836, 1839-42, 1844-46, 1849, 1850-63, 1865-68, 1874, 1879-1905 – 1 – $23.20 – us Southern Baptist [242]
Annual / Baptist Associations. Kentucky – 1785-1958 – 1 – us Southern Baptist [242]
Annual / Baptist Associations. Kentucky. Laurel River – 1959-66 – 1 – $11.90 – us Southern Baptist [242]
Annual / Baptist Associations. Louisiana – 1842-1987 – 1 – $51.28 – us Southern Baptist [242]
Annual / Baptist Associations. Maryland – 1934-69 – 1 – us Southern Baptist [242]
Annual / Baptist Associations. Michigan. Motor Cities – Jul 1951-Aug 1957 – 1 – $5.53 – us Southern Baptist [242]
Annual / Baptist Associations. Mississippi – 1811-1952 – 1 – us Southern Baptist [242]
Annual / Baptist Associations. Missouri – 1824-1957 – 1 – us Southern Baptist [242]
Annual / Baptist Associations. Missouri. Bethel Association. United Baptist – 1816-1941 – 1 – $17.22 – us Southern Baptist [242]
Annual / Baptist Associations. Missouri. Meramec Landmark Baptist Association – 1922-82 – 1 – $60.55 – us Southern Baptist [242]
Annual / Baptist Associations. Missouri. Mineral Area – 1958-81 – 1 – $49.07 – (formerly franklin association, 1969) – us Southern Baptist [242]
Annual / Baptist Associations. New Mexico – 1871-1970 – 1 – us Southern Baptist [242]
Annual / Baptist Associations. New Mexico – 1927-48, misc reel – 1 – $20.16 – (southwestern 1942, 1944. pecos valley, 1944, 1947. rio grande 1944, 1947. plains 1944, 1948. lincoln 1906, 1910. southeastern 1941, 1944-45. central 1927. portales 1930, 1943-45. estanica valley 1939-41, 1946. organization 1939) – us Southern Baptist [242]
Annual / Baptist Associations. New Mexico. Santa Fe – 1948-70 – 1 – $20.30 – (formerly: atomic, name change 1956)) – us Southern Baptist [242]
Annual / Baptist Associations. New York and Vermont. Miscellaneous Associations – 1832-40 – 1 – $11.20 – us Southern Baptist [242]
Annual / Baptist Associations. North Carolina – 1806-1949 – 1 – us Southern Baptist [242]
Annual / Baptist Associations. North Carolina. Chowan Association – 1884-90 – 1 – $6.48 – (manuscript minutes) – us Southern Baptist [242]
Annual / Baptist Associations. North Carolina. Eastern Association – 1869-86. mss minutes – 1 – $5.68 – us Southern Baptist [242]
Annual / Baptist Associations. North Carolina. Flat River Association – 1828-59 – 1 – $7.44 – (manuscript minutes) – us Southern Baptist [242]
Annual / Baptist Associations. North Carolina. Tar River Association – 1831-90. mss minutes – 1 – $42.08 – us Southern Baptist [242]
Annual / Baptist Associations. Ohio – 1950-75 – 1 – us Southern Baptist [242]
Annual / Baptist Associations. Oklahoma – 1890-1957 – 1 – us Southern Baptist [242]
Annual / Baptist Associations. Pennsylvania. Philadelphia Baptist Association – 1707-1965 – 1 – $450.00 – us Southern Baptist [242]
Annual / Baptist Associations. Primitive. Alabama – 1848-1955 – 1 – us Southern Baptist [242]
Annual / Baptist Associations. Primitive: Arkansas, California, Arizona, Delaware, 1844-1950 – 1 – us Southern Baptist [242]
Annual / Baptist Associations. Primitive. Florida – 1874-1949 – 1 – us Southern Baptist [242]
Annual / Baptist Associations. Primitive. Georgia – 1810-1952 – 1 – us Southern Baptist [242]
Annual / Baptist Associations. Primitive. Illinois, Iowa, Indiana, Kansas – 1845-1951 – 1 – (incomplete) – us Southern Baptist [242]
Annual / Baptist Associations. Primitive. Kentucky – 1 – us Southern Baptist [242]
Annual / Baptist Associations. Primitive: Louisiana, Maine, Maryland, Michigan, Mississippi, Missouri, New York, Nebraska – 1 – us Southern Baptist [242]
Annual / Baptist Associations. Primitive. North Carolina – 1 – us Southern Baptist [242]
Annual / Baptist Associations. Primitive: Ohio, Oklahoma, Pennsylvania, South Carolina – 1 – us Southern Baptist [242]
Annual / Baptist Associations. South Carolina – 1775-1970 – 1 – us Southern Baptist [242]
Annual / Baptist Associations. South Carolina. Edgefield. Quarterly Meetings – 1877-86 – 1 – $5.00 – us Southern Baptist [242]
Annual / Baptist Associations. Tennessee – 1786-1972 – 1 – us Southern Baptist [242]
Annual / Baptist Associations. Texas – 1843-1954 – 1 – us Southern Baptist [242]
Annual / Baptist Associations. Texas, Liberty – 1883-1981 – 1 – $60.06 – us Southern Baptist [242]
Annual / Baptist Associations. Virginia – 1787-1959 – 1 – us Southern Baptist [242]
Annual / Baptist Associations. Washington. Mount Pleasant – 1904 – 1 – $5.00 – us Southern Baptist [242]
Annual / Baptist Associations. Washington. Puget Sound – 1871-88 – 1 – $15.05 – (lacking: 1874, 1875) – us Southern Baptist [242]
Annual / Baptist State Conventions. (American Baptist). Connecticut – 1824, 1829-1975 – 1 – $543.97 – us Southern Baptist [242]
Annual / Baptist State Conventions. (American Baptist). Pennsylvania – 1826-1975 – 1 – $650.09 – us Southern Baptist [242]
Annual / Baptist State Conventions. Kentucky – 1981 – 1 – $22.12 – us Southern Baptist [242]
Annual / Baptist State Conventions. Louisiana – 1958-81 – 1 – $220.92 – us Southern Baptist [242]
Annual / Baptist State Conventions. (Southern Baptist). Alabama – 1823-1958 – 1 – $645.22 – us Southern Baptist [242]
Annual / Baptist State Conventions. (Southern Baptist). Alaska – 1946-62 – 1 – $29.82 – (lacking: 1950) – us Southern Baptist [242]
Annual / Baptist State Conventions. (Southern Baptist). Arizona – 1928-72 – 1 – $134.40 – us Southern Baptist [242]
Annual / Baptist State Conventions. (Southern Baptist). Arkansas – 1848-1964 – 1 – $360.36 – us Southern Baptist [242]

Annual / Baptist State Conventions. (Southern Baptist). California – 1941-67 – 1 – $146.72 – us Southern Baptist [242]
Annual / Baptist State Conventions. (Southern Baptist). District of Columbia – 1876-1969 – 1 – $321.37 – us Southern Baptist [242]
Annual / Baptist State Conventions. (Southern Baptist). Florida – 1854-1990 – 1 – $1144.08 – us Southern Baptist [242]
Annual / Baptist State Conventions. (Southern Baptist). Georgia – 1822-1981 – 1 – $1116.43 – us Southern Baptist [242]
Annual / Baptist State Conventions. (Southern Baptist). Illinois – 1834-44, 1847-1901 – 1 – $39.83 – us Southern Baptist [242]
Annual / Baptist State Conventions. (Southern Baptist). Illinois – 1907-53 – 1 – $236.81 – us Southern Baptist [242]
Annual / Baptist State Conventions. (Southern Baptist). Kentucky – 1837-1981 – 1 – $867.02 – us Southern Baptist [242]
Annual / Baptist State Conventions. (Southern Baptist). Louisiana – 1850-1980 – 1 – $582.40 – us Southern Baptist [242]
Annual / Baptist State Conventions. (Southern Baptist). Maryland – 1836-1958 – 1 – $296.10 – us Southern Baptist [242]
Annual / Baptist State Conventions. (Southern Baptist). Mississippi – 1836-1951 – 1 – $407.40 – us Southern Baptist [242]
Annual / Baptist State Conventions. (Southern Baptist). Missouri – 1834-1952 – 1 – $611.31 – us Southern Baptist [242]
Annual / Baptist State Conventions. (Southern Baptist). New Mexico – 1914-54 – 1 – $168.28 – (with proceedings of meetings of new mexico baptist convention, 1900-1909, and the organization of the baptist general convention of new mexico, 1910) – us Southern Baptist [242]
Annual / Baptist State Conventions. (Southern Baptist). North Carolina – 1830-1955 – 1 – $596.96 – us Southern Baptist [242]
Annual / Baptist State Conventions. (Southern Baptist). Ohio – 1954-75 – 1 – $93.66 – us Southern Baptist [242]
Annual / Baptist State Conventions. (Southern Baptist). Oklahoma – 1905-52 – 1 – $316.26 – us Southern Baptist [242]
Annual / Baptist State Conventions. (Southern Baptist). South Carolina – 1821-1970 – 1 – $732.26 – us Southern Baptist [242]
Annual / Baptist State Conventions. (Southern Baptist). Tennessee – 1875-1972 – 1 – $790.23 – us Southern Baptist [242]
Annual / Baptist State Conventions. (Southern Baptist). Texas – 1848-1987 – 1 – $1018.01 – us Southern Baptist [242]
Annual / Baptist State Conventions. (Southern Baptist). Virginia – 1824-1955 – 1 – $604.66 – us Southern Baptist [242]
Annual / Baptist Woman's Missionary Union. Kentucky – 1929-65 – 1 – $66.43 – us Southern Baptist [242]
Annual / Baptist Woman's Missionary Union. Louisiana – 1899-1957 – 1 – $114.24 – us Southern Baptist [242]
Annual / Baptist Woman's Missionary Union. Mississippi – 1878-1960 – 1 – $112.35 – us Southern Baptist [242]
Annual / Baptist Woman's Missionary Union. South Carolina – 1882-1975 – 1 – $230.86 – us Southern Baptist [242]
Annual / Baptist Woman's Missionary Union. Texas. Henderson County – 1922-47 – 1 – $5.00 – us Southern Baptist [242]
Annual / German Baptist Convention – 1849-1960 – 1 – $154.00 – us Southern Baptist [242]
Annual abstract of statistics 1928-1977 – [mf ed Chadwyck-Healey] – (= ser British government publications...1801-1977) – 191mf – 9 – uk Chadwyck [314]
Annual abstract of statistics 1947-1968 / Jamaica. Dept of Statistics – (= ser Latin american & caribbean...1821-1982) – 32mf – 9 – (1964 not publ. 1947 not available) – uk Chadwyck [318]
Annual abstract of statistics 1960-1973 / Nigeria. Federal Office of Statistics – (= ser African official statistical serials, 1867-1982) – 38mf – 9 – (1962 not publ) – uk Chadwyck [316]
Annual abstract of statistics of the united kingdom, 1840-1985 : prepared by the central statistical office in london – 18r – 1 – mf#96123 – uk Microform Academic [324]
Annual address of the bishop of huron to the synod of the diocese, june 16, 1891 / Baldwin, Maurice Scollard – [London, Ont?: s.n.], 1891 – 1mf – 9 – 0-665-89227-6 – (incl bibl ref) – mf#89227 – cn Canadiana [242]
Annual address of the conference to the methodist societies... – London, England. 1841 – 1r – us UF Libraries [242]
Annual address of the conference to the methodist societies... – London, England. 1843 – 1r – us UF Libraries [242]
Annual address of the conference to the methodist societies... – London, England. 1844 – 1r – us UF Libraries [242]
Annual address of the ottawa field-naturalists' club... : delivered november 28th, 1899 / Ami, Henry Marc – S.l: s.n, 1900? – 1mf – 9 – mf#08079 – cn Canadiana [500]
Annual address of the president, mr e j hearn, barrister, etc : with a catalogue of the publications for sale by this branch / Hearn, Edward J – Toronto?: Catholic Register, 1898 – 1mf – 9 – mf#08312 – cn Canadiana [241]
Annual address of the victoria institute / Kirk, John – London, England. 1872 – 1r – us UF Libraries [240]
Annual bibliography of english language and literature – v1-33. 1920-1958 – 9 – $282.00 – mf#0052 – us Brook [420]
The annual biography and obituary for the year – v1, 1817-v21, 1837. London: Longman, Hurst, Rees, Orme, and Brown, 1817- . 21v. illus – 1 – us Wisconsin U Libr [920]
Annual bulletin / Alumnae Association of the Baptist WMU Training School. Louisville, Kentucky – 1916-mar 1965 – 1 – $60.34 – us Southern Baptist [242]
Annual catalogue / Christiansburg Industrial Institute – Cambria VA: Press of the Christiansburg Industrial Institute 1910/11-1929/30 [mf ed 2005] – 1r – 1 – (lacks: 17th-18th (1913/1914-1914/1915), 20th (1915/1916), 24th (1921-1922), 29th (1926/1927); some pgs damaged; managed by friend's freedmen's association of philadelphia 1925/1926-1929/1930; cont: christiansburg industrial institute. catalogue of the christiansburg industrial institute) – mf#2005-s105 – us ATLA [366]
Annual catalogue of kemper hall, kenosha, wisconsin / Episcopal Church – 1872-73, 1876-77, 1881, 1886, 1888, 1890, 1892, 1895-98, 1907, 1911, 1913-15, 1919, 1926, 1931, 1933-37, 1940-44 – 1r – 1 – mf#698108 – us WHS [242]
Annual catalogue of the officers, faculty and students of the university of ottawa / University of Ottawa – Ottawa?: The University, 1889-1890 – 9 – mf#A01572 – cn Canadiana [378]
Annual catalogues of british official and parliamentary publications, 1894-1909 / Great Britain. Stationery Office – 1895-1910 [mf ed Chadwyck-Healey] – (= ser Stationery office catalogues of government publications, 1894-1970 and consolidated indexes to government publications, 1936-1970) – 12mf – 9 – 0-85964-016-7 – uk Chadwyck [324]
Annual catalogues of british official and parliamentary publications, 1910-1919 / Great Britain. Stationery Office – 1911-20 [mf ed Chadwyck-Healey] – (= ser Stationery office catalogues of government publications, 1894-1970 and consolidated indexes to government publications, 1936-1970) – 8mf – 9 – 0-85964-017-5 – uk Chadwyck [324]
Annual commencement of the institute for colored youth at association hall / Home for Aged and Infirm Colored Persons (Philadelphia PA) – [Philadelphia?: Institute for Colored Youth?] [annual] [mf ed 2006] – 24th-41st (1879-96) on 1r – 1 – (lacks: v25th-33rd (1880-88); v35th-36th (1890-91)); v38th (1893)) – mf#2006-s026 – us ATLA [370]
Annual communication / Freemasons – 113th [1990 jun 28/30] – 1r – 1 – mf#5004691 – us WHS [071]
Annual conference proceedings / East African Universities Social Science Conference – Dar es Salaam, Provisional Council for the Social Sciences in East Africa. [1st 1970] – 1r – 1 – us CRL [300]
Annual convention : officers' reports, proceedings of convention / New Jersey State Federation of Labor – 55th-58th [1933-37] – 1r – 1 – (cont: official proceedings of convention...annual congress, new jersey state federation of labor) – mf#3144690 – us WHS [331]
Annual convention : [program] / National Association for the Advancement of Colored People – 36th [1980], 41st [1985], 43rd [1987], 45th [1989], 50th [1994] – 1r – 1 – mf#5004714 – us WHS [305]
Annual convention / Wisconsin building and loan league – 42d [1938] – 1r – 1 – mf#406179 – us WHS [332]
Annual convention of the new york protective associations : affiliated with district assembly 49, k of l / Knights of Labor – 1895-96 – 1r – 1 – (cont: reunion, picnic and games of the new york protective associations under the auspices of district 49 assembly, knights of labor; cont by: official journal of the new york protective associations affiliated with d.a. 49, knights of labor) – mf#3162054 – us WHS [366]
Annual Convention Of The State Farmers' Union Of Florida *see* Proceedings of the annual convention of the state farmers' union of florida
Annual conventions – 6th-7th, 11th [1928-29, 1933] – 1r – 1 – mf#2699023 – us WHS [060]
Annual discourse delivered by edwin jacob... before the fredericton atheneum, february 21, 1853 – Fredericton NB: J Simpson, 1853 – 1mf – 9 – mf#45431 – cn Canadiana [080]
Annual events : lee county / Hanson, W Stanley – s.l, s.l? . 1936 – 1r – us UF Libraries [978]
Annual florida events / Goebel, Rubye K – s.l, s.l? . 1936 – 1r – us UF Libraries [978]
Annual florida events / Leonard, Agnes Mckenna – s.l, s.l? . 1936 – 1r – us UF Libraries [978]
Annual general meeting / Bank of Montreal – [S.l.]: The Bank, [1816?-1945?] – 9 – mf#A02233 – cn Canadiana [332]
Annual guidance index / Science Research Associates – Chicago IL 1949-57 – 1 – ISSN: 0402-5202 – mf#357 – us UMI ProQuest [020]
Annual law register of the united states – La Salle IL 1821-22 – 1 – mf#4418 – us UMI ProQuest [323]
Annual lespedeza for florida pastures – Gainesville, FL. 1942 – 1r – us UF Libraries [630]
Annual medical report / Kenya. Medical Dept – Nairobi. [1920-1922] – 1 – us CRL [610]
Annual medical report – Nairobi, 1912-19; 1920-22 – 1 – us CRL [610]
Annual meeting of the korea mission of the methodist episcopal church / Methodist Episcopal Church. Korea Mission – [s.l]: ...Trilingual Press, 10th-12th (1895-96) [mf ed 2006] – 3v on 1r – 1 – (10th & 11th sess both held in 1895; cont: methodist episcopal church. korea mission. minutes of the...annual meeting of the korea mission of the m e church; cont by: methodist episcopal church. korea mission. official minutes of the...annual meeting of the korea mission of the methodist episcopal church) – mf#2006c-s041 – us ATLA [242]
Annual meeting of the korea mission of the methodist episcopal church / Methodist Episcopal Church. Korea Mission – [s.l]: ...Trilingual Press, 10th-12th (1895-96) [mf ed 2006] – 3v on 1r – 1 – (cont: methodist episcopal church. korea mission. minutes of the...annual meeting of the korea mission of the m e church; cont by: methodist episcopal church. korea mission. official minutes of the...annual meeting of the korea mission of the methodist episcopal church) – mf#2006c-s041 – us ATLA [242]
Annual meeting of the royal college of dental surgeons, ontario : address / Beers, William George – [S.l: s.n, 1890?] [mf ed 1980] – 1mf – 9 – 0-665-02976-4 – (repr fr: dominion dental journal, april, 1890) – mf#02976 – cn Canadiana [617]
Annual meeting of the society / Society of the 28th Wisconsin Volunteer Infantry – 5th-7th [1887-1889] – 1r – 1 – (cont: minutes of the...annual reunion of the society...; cont by: proceedings of...annual meeting of the society...) – mf#2821 – us WHS [355]
Annual meeting of the woman's conference of the methodist episcopal church in korea / Methodist Episcopal Church. Korea Woman's Conference – [s.l: s.n.] 1899 – 1v on 1r – 1 – (2nd report iss in: official minutes of the...annual meeting, korea mission, methodist episcopal church; cont by: methodist episcopal church. korea woman's conference. annual report of the korea woman's missionary conference of the methodist episcopal church) – mf#2006c-s035 – us ATLA [242]
Annual meeting of the woman's conference of the methodist episcopal church in korea / Methodist Episcopal Church. Korea Woman's Conference – [s.l: s.n] 1899 [mf ed 2006] – 1v on 1r – 1 – (2nd report iss in: official minutes of the...annual meeting, korea mission, methodist episcopal church; cont by: methodist episcopal church. korea woman's conference. annual report of the korea woman's missionary conference of the methodist episcopal church) – mf#2006c-s035 – us ATLA [242]
Annual minutes of the...session of the colorado texas (colored) conference of the methodist protestant church / Methodist Protestant Church (US: 1830-1939). Colorado Texas (Colored) Conference – [s.l: s.n.] -1935 [mf ed 2004] – 57th (1935) [complete] on 1r – 1 – (cont: methodist protestant church (us: 1830-1939). colorado texas district, colored. minutes of the...annual conference of the methodist protestant church, colorado-texas district; cont by: methodist protestant church (us: 1830-1939). colorado texas (colored) conference. proceedings of the...annual session of the colorado texas (colored) conference of the methodist protestant church) – mf#2004-s096 – us ATLA [242]
Annual mission meeting and meeting of missionaries,...annual session / Methodist Episcopal Church, South. Korea Conference – Yokohama, Japan: Fukuin Printing Co, 18th-20th (1914-1916) [mf ed 2006] – 1r – 1 – (cont: methodist episcopal church, south. korea conference. minutes of the...annual meeting of the korean mission of the methodist episcopal church, south; minutes of the meeting of missionaries from 1922-1924 cont in: journal of the korea annual conference, methodist episcopal church, south,...session; lacks: 20th (1916) p55-56; damaged: 19th (1915) p77) – mf#2006c-s049 – us ATLA [242]
Annual mission meeting and meeting of missionaries,...annual session / Methodist Episcopal Church, South. Korea Conference – Yokohama, Japan: Fukuin Printing Co, 18th-20th (1914-16) [mf ed 2006] – 1r – 1 – (minutes of the meeting of missionaries from 1922-1924 continued in: journal of the korea annual conference, methodist episcopal church, south,...session; cont: methodist episcopal church, south. korea conference. minutes of the...annual meeting of the korean mission of the methodist episcopal church, south; lacks: 20th (1916) p55-56) – mf#2006c-s049 – us ATLA [242]
Annual of the palestine exploration fund / The Palestine Exploration Fund – London, 1914-1915 – 5mf – 9 – mf#H-2502 – ne IDC [956]
Annual officials' bulletin / Wisconsin Interscholastic Athletic Association – 1st-23rd [1935/36-58/59, 1962/63] – 1r – 1 – (cont by: wiaa official handbook) – mf#683529 – us WHS [790]
Annual progress report (abridged) of the superintendent, muhammadan and british monuments, archaeological survey of india, northern circle... – Allahabad, United Provinces: Govt Press, 1917-18 – 1r – 1 – us CRL [930]
Annual progress report of the superintendent, archaeological survey of india, northern circle, muhammadan and british monuments... / Archaeological Survey of India. Northern Circle – Allahabad, United Provinces: Govt Press, [1920-1921] (annual) – 1r – 1 – us CRL [930]
Annual progress report of the superintendent, hindu and buddhist monuments, northern circle... – Lahore, Punjab: Supt Govt Printing, 1919-21 – 1r – 1 – us CRL [930]
Annual progress report of the superintendent...for the year ending... / Archaeological Survey of India. Northern Circle. Muhammadan and British Monuments – Allahabad, United Provinces: Govt Press, 1911-16 – 1r – 1 – us CRL [930]
Annual progress report upon state forest administration in south australia for the year... / South Australia. Woods and Forests Dept – Adelaide: E Spiller, govt printer 1882-1928 (annual) [1881/82-1927/28] [mf ed 2005] – 47v on 2r [ill] – 1 – (cont: south australia. forest board (-1882). woods and forests; cont by: south australia. woods and forests dept. annual report of the woods and forests department for the year...) – mf#film mas c6041 – us Harvard [634]
The annual record of the connectional sunday-school union of the african methodist episcopal church – [s.l]: publ by the Union 1884- [mf ed 2004] – 1st n (1884) on 1r – 1 – (damaged: 1st n (1884) p1) – mf#2004-s073 – us ATLA [242]
The annual register of the baptist denomination in north america, 1790-1794 / Asplund, John – 1 – 6.16 – us Southern Baptist [242]
Annual report / Associated Charities of Milwaukee – 1885, 1888, 1890, 1893-99, 1907-09, 1911-15, 1916 [jan, sep], 1916/17-1919/20 – 1r – 1 – (cont by: annual report, family welfare association (milwaukee wi)) – mf#5379651 – us WHS [366]
Annual report / Canada. Dept of Indian Affairs – 1880-1936 – 1 – $220.00 – us L of C Photodup [350]
Annual report / Canada Permanent Building and Savings Society – Toronto: The Society, 1856?-1874? – 9 – mf#A01700 – cn Canadiana [332]
Annual report / Canada Permanent Loan and Savings Company – Toronto?: The Company, 1875?-1900 – 9 – mf#A01701 – cn Canadiana [332]
Annual report / Central State Hospital (Petersburg VA) – Richmond: Division of Purchase & Print. 61st (1930-1931) [mf ed 2005] – 1r – 1 – mf#2005-s040 – us ATLA [362]
Annual report / Civil Rights League. Cape Town – 1951/52-1965/66 – 1 – us CRL [960]
Annual report / Clark Electric Cooperative – 1939-53 – 1 – mf#3183268 – us WHS [333]
Annual report – 1960/61-64/65 – 1 – (cont: annual report of indian education for...to office of indian affairs. wisconsin dept of public instruction; cont by: annual report of indian education in wisconsin under state contract. wisconsin dept of public instruction) – mf#609068 – us WHS [370]
Annual report / De Beers Consolidated Mines – [Kimberley, SA: De Beers Consolidated Mines?]. 1st-32nd. 1888/89-1919/20 – 1 – us CRL [622]

Annual report / Family Service of Milwaukee – 1944/45-1948/49, 1952/53-1957/58 – 1r – 1 – (cont: annual report, family welfare association (milwaukee wi)) – mf#2697712 – us WHS [362]

Annual report / Family Welfare Association (Milwaukee WI) – 1919/20-1922/23, 1924/25-1926/27, 1928/29-1929/30, 1933/34-1936/37, 1938/39-1943/44 – 1r – 1 – (cont: annual report, associated charities of milwaukee; cont by: annual report, family service of milwaukee) – mf#5379651 – us WHS [350]

Annual report / Florida Geological Survey – Tallahassee, FL. 1st-23rd/24th. 1907/08-1930-32 – 4r – us UF Libraries [500]

Annual report / Harambee Ombudsman Project – 1986/87 – 1r – 1 – mf#2442282 – us WHS [350]

Annual report / Indian Institute of Science. Bangalore – Bangalore, India: The Institute [55th-60th 1963/64-1968/69] – 2r – 1 – us CRL [500]

Annual report / Institute for Colored Youth at Cheyney – [s.l: s.n.] [annual] [mf ed 2006] – 53rd-74th (1905-12) on 1r – 1 – (lacks: (1906-07); report for 1905- called 53rd-; for 1908/1909-1912 called 71st-74th year; cont: institute for colored youth (philadelphia, pa). annual report of the board of managers of the institute for colored youth; cont by: cheyney training school for teachers. annual report of the cheyney training school for teachers (institute for colored youth)) – mf#2006-s021 – us ATLA [370]

Annual report / Lake Superior District Power Co – 1923-80 – 58v on 1r – 1 – mf#2529557 – us WHS [333]

Annual report / Middleton Baptist Church (Middleton WI) – 1960/61-1989/90 – 1r – 1 – (cont by: high point church (madison wi). annual report) – mf#1606345 – us WHS [242]

Annual report / Nantucket Maria Mitchell Association – Nantucket MA: The Association 1903- (annual) [mf ed 1999] – 2r [ill] – 1 – (report yr ends jan 31, beginning with jan 31 1947 [45th annual report] through jan 31 1955; report yr ends dec 31, beginning with dec 31 1955; iss for 1903-05 have title: report of the nantucket maria mitchell association) – mf#film mas c4159 – us Harvard [520]

Annual report / National Consumers' League – 2nd-8th [1900/01-1906/07] – 1 – (cont by: report, national consumers' league) – mf#1383469 – us WHS [380]

Annual report / New Jersey. Commission on the Urban Colored Population – [Newark: s.n.], 1943 (mf ed 19–] – 1r – 1 – (ceased publ with report for 1944?) – mf#Sc Micro R-2453 – us NY Public [305]

Annual report / Philadelphia Association for the Protection of Colored Women – [Philadelphia PA?: s.n.] 1919/1920,1922 [mf ed 2005] – 1r [ill] – 1 – (cont: report of...year's work) – mf#2005-s034 – us ATLA [362]

Annual report / Rhodesia and Nyasaland. Secretary for African Affairs – Zomba, Govt Printer [1957-1959] – 1r – 1 – us CRL [350]

Annual report / Rockefeller Foundation. China Medical Board – New York NY: Offices of the Board. 1st-12th. 1914/15-1926 [annual] [mf ed 2003] – 1r – 1 – (ceased in 1927?) – mf#2003-s059 – us ATLA [060]

Annual report / Swaziland Staff Training Institute – [S.l: s.n, [1967-1968/1970] – 1r – 1 – us CRL [350]

Annual report / Underhill Society of America – 3rd-40th [1895-1932] – 1r – 1 – (cont: annual report of the secretary, underhill society of america) – mf#1114294 – us WHS [366]

Annual report / U.S. Board of Indian Commissioners – 1-63. 1869-1932 – 1 – $69.00 – us L of C Photodup [305]

Annual report / U.S. Bureau of American Ethnolgy – v. 1-48. 1879 80-1931 – 1 – us AMS Press [306]

Annual report / Victory Mutual Life Insurance Co – 1938 dec 31, 1942 dec 31 – 1r – 1 – mf#5286980 – us WHS [368]

Annual report / Western Publishing Co – 1943-77 – 1r – 1 – mf#1114446 – us WHS [070]

Annual report... / American Labor Education Service – 1946-48, 1952, 1955-57 – 1r – 1 – mf#3397847 – us WHS [331]

Annual report... / Wisconsin Animal Diagnostic Laboratories – 1959-62 – 1r – 1 – (cont by: annual report of the wisconsin animal health laboratories) – mf#529816 – us WHS [619]

Annual report... / Wisconsin Animal Health Laboratories – 1963-64 jun – 1r – 1 – (cont: annual report of the wisconsin animal diagnostic laboratories; cont by: report of wisconsin animal health laboratories) – mf#543864 – us WHS [071]

Annual report... / Wisconsin Soldiers' Home, Milwaukee – 1864/65-1865/66 – 1r – 1 – (cont by: annual report of the northwestern branch, national home for disabled volunteer soldiers) – mf#597266 – us WHS [360]

Annual report... / Women's Trade Union League of New York – 1906/07, 1924/25, 1931/32, 1936/37-41/42, 1944-48, 1949/50-1950/51 – 1 – mf#3136006 – us WHS [331]

[Annual report] – 1939/40-1942/43; 1944/45-1955/56; 1963/64-1981 – 1 – mf#60628 – us WHS [071]

Annual report (1880) / Oxford Mission to Calcutta – Oxford: Parker. 1st. 1880 [annual] [mf ed 2002] – 1r – 1 – (filmed with later title: oxford mission to calcutta. report) – mf#2002-s099 – us ATLA [242]

Annual report (1890) / Oxford Mission to Calcutta – Oxford: W R Bowden, 1890-1968 [annual] [mf ed 2003] – 78v on 3r – 1 – mf#2002-s101 – us ATLA [242]

Annual report, 1893 – Lake City, FL. 1894 – 1r – us UF Libraries [630]

Annual report 1915 – Boston: World Peace Foundation 1915 [mf ed 1992] – (= ser World peace foundation pamphlet series 5/6/2) – 1mf – 9 – 0-524-03250-5 – mf#1990-0878 – us ATLA [327]

Annual report 1916 – Boston: World Peace Foundation 1917 [mf ed 1992] – (= ser World peace foundation pamphlet series 7/2) – 1mf – 9 – 0-524-03251-3 – mf#1990-0879 – us ATLA [327]

Annual report: advanced dental education – Chicago IL 1972-73 – 1,5,9 – ISSN: 0147-0264 – mf#8936 – us UMI ProQuest [617]

Annual report: dental auxiliary education / American Dental Association. Division of Educational Measurements – Chicago IL 1967-73 – 1 – ISSN: 0145-5370 – mf#8593 – us UMI ProQuest [617]

Annual report: dental education – Chicago IL 1973-74 – 1,5,9 – ISSN: 0147-0256 – mf#8584 – us UMI ProQuest [617]

Annual report for... / Catholic Truth Society of Ottawa – Ottawa: The Society, 1892?-189- or 19– – 9 – (issues reproduced: 1891/92, 1893-1896) – mf#A00079 – cn Canadiana [366]

Annual report for the year... / Oxford Mission to Calcutta – London: s.n.] 1969- [annual] [mf 1968 filmed 2003] – 1r – 1 – mf#2002-s102 – us ATLA [242]

Annual report for the year ended... : of the joint east african board for promoting the agricultural, commercial and industrial development of kenya, nyasaland, tanganyika, uganda and zanzibar / Joint East African Board – [London]: The Board. v15 1938 – 1r – 1 – us CRL [338]

Annual report for the year ending... / Consumers' League of Philadelphia – 6th-7th [1906-07], 9th [1909], 11th [1911] – 1r – 1 – (cont: annual report of the council, philadelphia branch consumers' league of pennsylvania, consumers' league of pennsylvania) – mf#3144615 – us WHS [380]

Annual report for...and minutes of the general council / Christian and Missionary Alliance. General Council – [s.l.] 60th-96th yr. 1946-82 [mf ed 2003] – 37v on 7r – 1 – mf#2003-s106 – us ATLA [240]

Annual report for...of the american presbyterian hospital, siangtan / American Presbyterian Hospital (Xiangtan, Hunan Sheng, China) – Shanghai: Presbyterian Mission Press [annual] [mf ed 2006] – 1913 [complete] on 1r – 1 – mf#2006c-s025 – us ATLA [360]

Annual report from...to... / Home for Aged Colored People [Chicago IL] – [Chicago IL]: s.n.] 1921- [mf ed 2005] – 28th-56th (1920/21-1948/49) on 1r – 1 – (lacks: 36th-38th,40th-41st,45th-46th]; cont: home for aged and infirm colored people. annual report of the home for aged and infirm colored people) – mf#2005-s102 – us ATLA [362]

Annual report, horticulture, cereals, stocks, etc – Lake City, FL. 1891 – 1r – us UF Libraries [630]

Annual report of american indian education in wisconsin under state contract with the federal bureau of indian – 30th [1976/77] – 1r – 1 – (cont: annual report of indian education in wisconsin under state contract, wisconsin. dept of public instruction) – mf#609072 – us WHS [370]

Annual report of bethany mission for colored children / Bethany Mission for Colored Children – [Philadelphia PA?: Bethany Mission for Colored Children] 3rd (1860) [[mf ed 2005] – 1r with other item – 1 – (cont by: bethany mission for colored people. annual report of the bethany mission for colored people) – mf#2005-s110 – us ATLA [242]

Annual report of district assembly no 30, k of l / Knights of Labor – 9th [1887] – 1r – 1 – (cont by: report of the...annual session of district assembly no 30, k of l) – mf#3185339 – us WHS [360]

Annual report of indian education for...to office of indian affairs – 1952/53-1959/60 – 1r – 1 – (cont: annual report to the office of indian affairs for..., wisconsin. dept of public instruction; cont by: annual report, wisconsin. indian education) – mf#609069 – us WHS [370]

Annual report of indian education in wisconsin under state contract – 1965/66-1974/75 – 1r – 1 – (cont: annual report, wisconsin. indian education; cont by: annual report of american indian education in wisconsin under state contract with the federal bureau of indian affairs, wisconsin. dept of public instruction) – mf#609071 – us WHS [370]

Annual report of pierre fortin, esq : stipendiary magistrate, commander of the expedition for the protection of the fisheries in the gulf of st lawrence...1864 – Quebec?: Hunter, Rose, 1865 – 1mf – 9 – mf#37753 – cn Canadiana [639]

Annual report of secretary-treasurer of the illinois state federation of labor – 4th [1905], 38th [1920] – 1r – 1 – mf#3198469 – us WHS [331]

Annual report of statistics of railways *see* Us interstate commerce commission. annual report of statistics of railways

Annual report of the... / Chicago and North Western Railway Co – 2nd-3rd [1861-62], 6th [1865], 4th [1863], 6th [1865] [duplicate copy] – 2r – 1 – (cont: address of the president of the chicago and north western railway company to the stock and bondholders at the annual meeting..., chicago and north western railway company; cont by: report of the chicago and north western railway company) – mf#1238866 – us WHS [380]

Annual report of the... / Christian and Missionary Alliance – South Nyack NY: Christian Alliance Pub Co, 1898-1936 [annual] [mf 1st-35th 1897/8-1935 filmed 2003] – 40v on 4r – 1 – mf#2003-s104 – us ATLA [240]

Annual report of the... / Methodist Episcopal Church. Pacific Japanese Mission – San Francisco CA: [s.n] [annual] [mf 1905 filmed 2003] – 1r – 1 – mf#2003-s117 – us ATLA [242]

Annual report of the american missionary association / American Missionary Association – New York: American Missionary Association, 1st [1847]- – 5r – 1 – $650.00 set – (guide may be purchased separately $15) – mf#d3624 – us Amistad [240]

Annual report of the archaeological survey, bengal circle... – Calcutta: Bengal Secretariat Press, 1905 – 1r – 1 – us CRL [930]

Annual report of the archaeological survey of india, frontier circle for... – Peshawar, N-W Frontier Province: Govt Press, 1097-21 – 1r – 1 – us CRL [930]

...Annual report of the association for the care of coloured orphans / Association for the Care of Coloured Orphans (Cheyney, PA) – Philadelphia: printed for the Association by J Rakestraw, 1st-55th (1837-91) [mf ed 2005] – 55v on 1r – 1 – (cont: association for the care of coloured orphans (cheyney, pa). report of the association for the care of coloured orphans; cont by: association for the care of coloured orphans (cheyney, pa). annual report of the shelter for colored orphans; lacks: 9th (1845) [front cover]) – mf#2005-s018 – us ATLA [362]

Annual report of the bethany mission for colored people / Bethany Mission for Colored Children – Philadelphia: E Ketterlinus' Steam Power Printing House 14th-29th (1872-1886) [mf ed 2005] – 1r with other item – 1 – (imprint varies; lacks: 15th-17th, 20th-22nd, 24th-26th, 28th; cont: bethany mission for colored children. annual report of bethany mission for colored children) – mf#2005-s111 – us ATLA [242]

Annual report of the board and superintendent of the institute for deaf, dumb and blind colored youths of the state of texas / Institute for Deaf, Dumb and Blind Colored Youths of the State of Texas – Austin: von Boeckmann-Jones 1902- (annual) [mf ed 2005] – 1r – 1 – (cont: institute for deaf, dumb and blind colored youths of the state of texas. annual report of the trustees and superintendent of the institute for deaf, dumb and blind colored youths of the state of texas; mf lacks: 16th, 18th, 20th, 22th, 24th (1903, 1905, 1907, 1909, 1911)) – mf#2005-s082 – us ATLA [362]

Annual report of the board of commissioners of state industrial school for colored youths / State Industrial School for Colored Youths (LA) – Baton Rouge LA: [State Industrial School for Colored Youths] 1949- [1st-2nd (1948/1949-1949/1950] [mf ed 2005] – 1r – 1 – mf#2005-s096 – us ATLA [365]

Annual report of the board of education of the african m e church / African Methodist Episcopal Church. Board of Education – Athens GA: Gardner Book & Job Printer [mf ed 2004] – 1889/90 on 1r – 1 – (damaged: (1889/90) [front cover]; cont: african methodist episcopal church. board of education. report of the board of education of the african m e church [2004-s058] filmed on same reel) – mf#2004-s059 – us ATLA [242]

Annual report of the board of managers of the... : issued in lieu of the regular report covering 1903: a statement of the affairs of the... / Shelter for Aged and Infirm Colored Persons of Baltimore City – Baltimore: Steam Press of W K Boyle, 1883- [annual] [mf ed 2004] – 1r – 1 – (mf: 1st-54th [1883-1936] lacks 50th, 52nd) – mf#2004-s013 – us ATLA [360]

Annual report of the board of managers of the... : with proceedings of the annual / American Baptist Missionary Union – 69th and 70th – 1r – 1 – (cont: annual report of the american baptist board of foreign missions; cont by: annual report, with the proceedings of the annual meetings) – mf#5396086 – us WHS [242]

Annual report of the board of managers of the home for aged and infirm colored persons : also, the act of incorporation and rules, with list of officers / Home for Aged and Infirm Colored Persons (Philadelphia PA) – Philadelphia: Merrihew & Son 1874- [annual] [mf ed 2005] – 10th-52nd (1874-1916) on 1r – 1 – (lacks: 30th-32nd,35th-39th,41st-43rd,45th-48th, 50th]; imprint varies; cont: home for aged and infirm colored persons (philadelphia, pa). proceedings of the...annual meeting of the home for aged and infirm colored persons, held...) – mf#2006-s017 – us ATLA [362]

Annual report of the board of managers of the institute for colored youth / Home for Aged and Infirm Colored Persons (Philadelphia PA) – Philadelphia: Ringwalt & Brown, Steam-Power Book and Job Printers 1867- [annual] [mf ed 2006] – 15th-51st (1867-1903) on 1r – 1 – (lacks: 35th,40th,46th p7-8]; damaged: v16th (1868) [last pg (unnumbered)]; cont: institute for colored youth (philadelphia, pa). objects of the institute for colored youth, with a list of the officers and students, and the annual report of the board of managers; cont by: institute for colored youth at cheyney. annual report) – mf#2006-s020 – us ATLA [362]

Annual report of the british columbia board of trade / British Columbia Board of Trade – [Victoria, BC?]: The Board, [1880?]- – ISSN: 1189-0533 – mf#A00200 – cn Canadiana [380]

Annual report of the cheyney training school for teachers (institute for colored youth) / Cheyney Training School for Teachers – [Cheyney PA?: s.n.] 1913- [annual] [mf ed 2006] – 75th-81st (1913-19) on 1r – 1 – (ceased in 1933? lacks: 76th-77th,80th (1914-15, 1918); named changed in 1934 to: state teachers college at cheyney; cont: institute for colored youth at cheyney. annual report) – mf#2006-s022 – us ATLA [370]

Annual report of the chicago hebrew mission / Chicago Hebrew Mission – [Chicago IL: Chicago Hebrew Mission], 2nd-3rd (1889-91) [mf ed 2005] – 1r – 1 – (cont: hebrew christian mission. annual report of the hebrew christian mission) – mf#2005C-s006 – us ATLA [270]

Annual report of the christiansburg industrial institute / Christiansburg Industrial Institute – [Cambria VA?]: Christiansburg Industrial Institute Press 40th-60th (1903-1923) [mf ed 2005] – 1r – 1 – (under the auspices of: friends' freedmen's association, philadelphia, pa; lacks: 56th (1919?)) – mf#2005-s106 – us ATLA [366]

Annual report of the city treasurer of the city of quebec : balance sheets, statements and other documents of the quebec corporation and water works for the civic year 1883-84 – [Quebec?: s.n.] 1884 [mf ed 1984] – 2mf – 9 – 0-665-43318-2 – mf#43318 – cn Canadiana [350]

Annual report of the collector of internal revenue to the honorable secretary of finance and justice of the government of the philippine islands / Philippines. Collector of Internal Revenue – Manila: Bureau of Printing, [2nd-46h (1906-1952/1953)] – 2r – 1 – us CRL [336]

Annual report of the colored home and hospital / Colored Home and Hospital (New York NY) – [New York: s.n.] 1880- (New York: S Angell) [mf ed 2005] – 46th-57th (1885/86-1896/97) on 1r – 1 – (lacks: 40th-44th,47th-49th,52nd-54th,56th]; cont: colored home (new york, ny) annual report of the colored home; cont by: lincoln hospital and home (new york, ny) annual report of the lincoln hospital and home) – mf#2004-s017 – us ATLA [360]

Annual report of the colored woman's league of washington, dc / Colored Woman's League of Washington, DC – [Washington DC]: F D Smith. 4th-5th [1897-98] [mf ed 2005] – 1r – 1 – (damaged: 5th (1898) p5-8) – mf#2005-s095 – us ATLA [305]

Annual report of the commission of home missions to colored people / Episcopal Church. Commission of Home Missions to Colored People – [New York?] 7th-12th. 1872-1876/77 (annual) [mf ed 2004] – 1r – 1 – (began in 1866?) – mf#2004-s060 – us ATLA [242]

Annual report of the commissioner of labor – 6th [1890] and 7th [1891, v2] – 1r – 1 – (cont: annual report of the commissioner of labor, united states. bureau of labor; cont by: annual report of the commissioner of labor (1903)) – mf#5166485 – us WHS [331]

Annual report of the common, academic and normal and model schools in nova scotia for the year ending october 31st... / Nova Scotia. Council of Public Instruction – [Halifax, NS?]: The Council, [1864?-18– or 19–] – mf#A01891 – cn Canadiana [350]

Annual report of the council of the indian institute of science, bangalore / Indian Institute of Science. Bangalore – Bangalore: The Institute, [1912]-1964. [39th-54th 1947/48-1962/63] – 3r – 1 – us CRL [500]

Annual report of the council, philadelphia branch... / Consumers' League of Pennsylvania – 1901-05 – 1r – 1 – (cont by: annual report for the year ending...) – mf#3144581 – us WHS [380]

Annual report of the director / Mount Wilson and Palomar Observatories et al – [Pasadena CA]: The Observatories [1948- (annual) [1947/1948]-1968-1969 [mf ed 1999] – 1r – 1 – (cont: mount wilson observatory. annual report of the director...issn 1057-4360; cont by: hale observatories. annual report of the director...issn 1057-4387; reports for 1956-57 through 1968-69 have title: annual report of the director...issn 1057-4379) – mf#film mas C4225 – us Harvard [520]

Annual report of the director... / Association for the Study of Negro Life and History, Inc – 1922/23-1943/44 – 1r – 1 – mf#5146478 – us WHS [305]

Annual report of the director of the astronomical observatory of harvard college / Harvard College Observatory – Cambridge MA: University Press (annual) [32nd]-104th(1877-1949)] [mf ed 1999] – 1r – 1 – (iss also in the annual reports of the president of harvard university; reports for 1846-55 were publ in v1 of the annals of the observatory; for 1856-1876 in v8; reports for 1859-1864 were publ also in the reports of the committee of the overseers appointed to visit the observatory; reports for 1888- [43d-] incl also in its miscellaneous papers 1888-; numbering begins with 34th annual report [1879]; report yr ends: 1877-93, oct 31; 1894-19 sep 30; title varies slightly) – ISSN: 0888-9880 – mf#film mas c4227 – us Harvard [520]

Annual report of the director of the mount wilson observatory / Hale, George Ellery et al – [Washington]: Carnegie Institution of Washington. 43v (1906-47) [1904/05-1946/47] [mf ed 1999] – 2r – 1 – (repr fr yearbook of the carnegie institution of washington; cont by: mount wilson and palomar observatories. annual report of the director – mount wilson and palomar observatories; report of director of the solar observatory, mount wilson, california 1904/05-1905/06; annual report of the director / mount wilson solar observatory of the carnegie institution of washington 1906/07-1912/13; annual report of the director of the mount wilson solar observatory 1913/14-1916/17; iss under earlier names of the observatory as follows: 1904/05-1905/06, 1909, solar observatory, mount wilson, california; 1907/08, 1910-1916/17, mount wilson solar observatory) – mf#film mas c4224 – us Harvard [520]

Annual report of the director to the board of trustees of experimental station for the year / Depass, Jas P – Lake City, FL. 1890 – 1r – us UF Libraries [630]

Annual report of the director to the council / Indian Institute of Science – [Bangalore: The Institute. 1st-2nd. 1908/10-1910/11 – 1 – us CRL [500]

Annual report of the executive board of the friends' association of philadelphia and its vicinity : for the relief of colored freedmen / Friends' Association of Philadelphia and Its Vicinity, for the Relief of Colored Freedmen. Executive Board – Philadelphia: Ringwalt & Brown. 3rd-8th (1866-1871) [mf ed 2005] – 1r – 1 – (lacks: 6th-7th (1869-70); cont: report of the executive board of the friends' association of philadelphia and its vicinity, for the relief of colored freedmen) – mf#2005-s060 – us ATLA [362]

Annual report of the executive committee of the institute for the training of colored ministers : at tuskaloosa, alabama, to the general assembly of the presbyterian church in the united states – Tuskaloosa AL: Institute for the Training of Colored Ministers. 14th 1891 (annual) [mf ed 2004] – 1r – 1 – (began in 1878? filmed with: journal of the...session of the alabama annual conference of the african methodist episcopal church [order051]) – mf#2004-s052 – us ATLA [242]

Annual report of the exeter diocesan board of education / Exeter Diocesan Board Of Education – Exeter?, England. 1839 – 1r – us UF Libraries [240]

Annual report of the general missionary committee and the book and tract work... : including a report...pertle springs, mo, may 26 1890 / German Baptist Brethren (US). General Church Erection and Missionary Committee – [s.l: s.n, 1890?] [mf ed 1992] – 1mf – 9 – 0-524-04174-1 – mf#1990-4978 – us ATLA [242]

Annual report of the hebrew christian mission : work among the jews in chicago / Hebrew Christian Mission – [Chicago IL: Hebrew Christian Mission] 1888 [mf ed 2005] – 1v on 1r – 1 – (cont by: chicago hebrew mission. annual report of the chicago hebrew mission) – mf#2005C-s005 – us ATLA [270]

Annual report of the home for aged and infirm colored people / Home for Aged and Infirm Colored People [Chicago IL] – [Chicago IL]: s.n.] [mf ed 2005] – 1st 24th-27th (1898; 1916/17-1919/20) on 1r – 1 – (incorp 1898; cont by: home for aged colored people. annual report from...to...; some pgs damaged) – mf#2005-s101 – us ATLA [362]

Annual report of the house of s michael and all angels for young colored cripples / House of St Michael and All Angels – Philadelphia: House of St Michael... 1887- [1st-17th (1887-1903)] [mf ed 2005] – 1r – 1 – (lacks: 3rd-5th,6th-7th,9th-15th]; cont by: house of st michael and all angels. annual report of the house of st michael and all angels for colored cripple children) – mf#2005-s097 – us ATLA [362]

Annual report of the house of st michael and all angels for colored cripple children / House of St Michael and All Angels – Philadelphia: House of St Michael... 30th-44th (1916-1930) [mf ed 2005] – 1r – 1 – (lacks: 32nd-35th,39th-43rd (1918-21, 1925-29); cont: house of st michael and all angels. annual report of the house of s michael and all angels for young colored cripples) – mf#2005-s098 – us ATLA [362]

Annual report of the indian commission to the domestic committee of the board of missions / Episcopal Church. Office of the Indian Commission – [s.l]: Indian Commission, c1872- [mf ed 2003-04] – 2nd-6th annual report on 1r – 1 – (also incl 1st-5th report of the missionary bishop of niobrara) – mf#2003-s044 – us ATLA [242]

Annual report of the industrial home for colored girls / Industrial Home for Colored Girls [Peaks VA] – Peake's Turnout VA: [s.n.] 1916-20 [annual] [mf ed 2004] – 5v on 1r – 1 – (mf: 1st-5th [1916-20]. report yr for 1915/16-1917/18 is irreg; for 1918/19-1919/20 ends mar 1) – mf#2004-s006 – us ATLA [365]

Annual report of the korea woman's conference of the methodist episcopal church / Methodist Episcopal Church. Korea Woman's Conference – Seoul: Press of the Methodist Publ House, 11th-32nd (1909-30) [mf ed 2006] – 1r – 1 – (cont: methodist episcopal church. korea woman's conference. reports read at the...annual session of the korea woman's conference of the methodist episcopal church; cont by: methodist episcopal church. woman's foreign missionary society. annual report of the members of the woman's foreign missionary society in korea; lacks: 14th (1912) p15-16; 18th (1916) p31-31; 19th (1917); 22nd (1920)) – mf#2006c-s038 – us ATLA [242]

Annual report of the korea woman's conference of the methodist episcopal church / Methodist Episcopal Church. Korea Woman's Conference – Seoul: Press of the Methodist Publ House, 11th-32nd (1909-30) [mf ed 2006] – 1r – 1 – (cont: methodist episcopal church. korea woman's conference. reports read at the...annual session of the korea woman's conference of the methodist episcopal church; cont by: methodist episcopal church. woman's foreign missionary society. annual report of the members of the woman's foreign missionary society in korea; lacks: 14th (1912) p15-16; 18th (1916) p31-31; 19th (1917); 22nd (1920); damaged: 14th (1912) p13-14) – mf#2006c-s038 – us ATLA [242]

Annual report of the korea woman's missionary conference of the methodist episcopal church / Methodist Episcopal Church. Korea Woman's Conference – Seoul: Methodist Publ House, 3rd-5th (1901-03) [mf ed 2006] – 3v on 1r – 1 – (cont: methodist episcopal church. korea woman's conference. annual meeting of the woman's conference of the methodist episcopal church in korea; cont by: methodist episcopal church. korea woman's conference. reports read at the...annual session of the korea woman's conference of the methodist episcopal church) – mf#2006c-s036 – us ATLA [242]

Annual report of the korea woman's missionary conference of the methodist episcopal church / Methodist Episcopal Church. Korea Woman's Conference – Seoul: Methodist Publishing House, 3rd-5th (1901-03) [mf ed 2006] – 3v on 1r – 1 – (cont: methodist episcopal church. korea woman's conference. annual meeting of the woman's conference of the methodist episcopal church in korea; cont by: methodist episcopal church. korea woman's conference. reports read at the...annual session of the korea woman's conference of the methodist episcopal church) – mf#2006c-s036 – us ATLA [242]

Annual report of the labour commissioner for the year... / Swaziland. Labour Dept – [Swaziland: Labour Dept [1964] – 1r – 1 – us CRL [316]

Annual report of the lincoln hospital and home / Lincoln Hospital and Home (New York NY) – New York: Knickerbocker Press 1901/02- [mf ed 2004] – 76th-83rd (1915-1922) on 1r – 1 – (lacks: v78th-80th (1917-1919); cont: colored home and hospital (new york ny) annual report of the colored home and hospital [issn 0735-097X]) – mf#2004-s018 – us ATLA [360]

Annual report of the london missionary society / The London Missionary Society – London, 1796-1939/1940. n1-145 – 480mf – 9 – (missing: 1799-1814 n5-20) – mf#H-2142 – ne IDC [956]

Annual report of the members of the woman's foreign missionary society in korea / Methodist Episcopal Church. Woman's Foreign Missionary Society – Seoul, Korea: YMCA Press, 33rd-34th (1931-32) [mf ed 2006] – 1r – 1 – (cont: methodist episcopal church. korea woman's conference. annual report of the korea woman's conference of the methodist episcopal church) – mf#2006c-s039 – us ATLA [242]

Annual report of the members of the woman's foreign missionary society in korea / Methodist Episcopal Church. Woman's Foreign Missionary Society – Seoul, Korea: YMCA Press, 33rd-34th (1931-32) [mf ed 2006] – 1r – 1 – (cont: methodist episcopal church. korea woman's conference. annual report of the korea woman's conference of the methodist episcopal church) – mf#2006c-s039 – us ATLA [242]

Annual report of the..., michigan historical collections / Bentley Historical Library – 1977/78-1979/80 – 1 – (cont: bentley library annual) – mf#808357 – us WHS [978]

Annual report of the north carolina hospitals board of control to his excellency,...governor of north carolina / North Carolina Hospitals Board of Control – [s.l: s.n.] 1944- [mf ed 2005] – 1944 [complete] on 1r – 1 – (report for 1944 submitted to: j melville broughton; vol for 1944 incl separately paginated report titled: annual report of the general business manager) – mf#2005-s140 – us ATLA [362]

Annual report of the northwestern branch... / National Home for Disabled Volunteer Soldiers – 1873/74-1884/85 – 1r – 1 – (cont: annual report of the wisconsin soldiers' home, milwaukee) – mf#592698 – us WHS [360]

Annual report of the port royal relief committee / Port Royal Relief Committee – Philadelphia: Merrihew & Thompson. 1st mar 26 1863 [annual] [mf ed 2004] – 1v on 1r – 1 – (no more publ; filmed with: the bystander [brooklyn ny: christ church cathedral] mthly [mf v1 n7 jun 1927] 2004-s009; began in 1926? ceased in 1927?) – mf#2004-s008 – us ATLA [240]

Annual report of the presbyterian mission hospital, miraj / Presbyterian Mission Hospital (Miraj, India) – Mysore: Wesleyan Mission Press 1913/1914, 1929, 1931 [mf ed 2005] – 1r [ill] – 1 – (Iss for 1932/33 a special memorial iss to Dr Wanless) – mf#2005c-S034 – us ATLA [242]

Annual report of the president of the java bank – Batavia, 1941-1951 – 18mf – 9 – mf#SE-295 – ne IDC [959]

Annual report of the provincial commissioners / Nyasaland. Native Administration – Zomba: Govt Printer, 1945-46 – 1 – us CRL [960]

Annual report of the r i w indian association / Rhode Island Women's Indian Association – [Providence RI?: s.n.] [mf ed 2006] – 2nd-6th (1886-90) [complete] on 1r – 1 – mf#2006i-s008 – us ATLA [366]

Annual report of the registrar general 1839-1920 – [mf ed Chadwyck-Healey] – (= ser British government publications...1801-1977) – 18r – 1 – uk Chadwyck [314]

Annual report of the removal of rough and detrimental fish by state and contract fishermen – 1947-78 – 1r – 1 – (cont by: annual report of the removal of rough and detrimental fish in wisconsin inland waters by state and contract fishermen) – mf#504516 – us WHS [639]

Annual report of the secretary / Underhill Society of America – 3rd-10th [1895-02] – 1r – 1 – (cont by: annual report, underhill society of america) – mf#2226889 – us WHS [360]

Annual report of the secretary for african education for the year ended... / Rhodesia, Southern. Division of African Education – [Salisbury: Govt Printer] – 1 – mf#1252 – us Wisconsin U Libr [370]

Annual report of the several departments of the city government of halifax, nova scotia : for the municipal year 1861-62 / Halifax (NS). City Council – [Halifax NS: s.n.] 1862 [mf ed 1984] – 1mf – 9 – 0-665-45059-1 – mf#45059 – cn Canadiana [350]

Annual report of the shelter for colored children / Association for the Care of Coloured Orphans (Cheyney, PA) – [Cheyney PA? Association for the Care of Coloured Orphans], 99th-109th (1935-45) [mf ed 2005] – 11v on 1r – 1 – (cont: association for the care of coloured orphans (cheyney, pa.). annual report of the shelter for colored orphans; cont by: association for the care of coloured orphans (cheyney, pa.). annual report of the shelter for colored girls; lacks: 100th (1936)) – mf#2005-s020 – us ATLA [362]

Annual report of the shelter for colored girls / Association for the Care of Coloured Orphans (Cheyney, PA) – [Cheney PA?: Association for the Care of Coloured Orphans], 110th-114th (1946-50) [mf ed 2005] – 1r – 1 – (cont: association for the care of coloured orphans (cheyney, pa.). annual report of the shelter for colored children; lacks: 112th (1948)) – mf#2005-s021 – us ATLA [362]

Annual report of the shelter for colored orphans / Association for the Care of Coloured Orphans (Cheyney, PA) – Philadelphia: WM H Pile's Sons, Printers, 56th-98th (1892-1934) [mf ed 2005] – 43v on 1r – 1 – (cont: association for the care of coloured orphans (cheyney, pa). annual report of the association for the care of coloured orphans; cont by: association for the care of coloured orphans (cheyney, pa). annual report of the shelter for colored children; lacks: 67th (1902) p13-14; 78th (1914); 97th (1933)) – mf#2005-s019 – us ATLA [362]

Annual report of the shelter for orphans of colored soldiers and friendless colored children / Shelter for Orphans of Colored Soldiers and Friendless Colored Children (Baltimore MD) – Baltimore: printed by Daugherty, Maguire & Wright. 1st-2nd. 1868-69 [mf ed 2005] – 1r – 1 – mf#2005-s026 – us ATLA [362]

Annual report of the society... / Society for the History of the Germans in Maryland – 8th/10th-13th/14th [1899/1900] – 1r – 1 – (cont by: society for the history of the germans in maryland...report) – mf#772009 – us WHS [305]

Annual report of the society for the relief of worthy aged indigent colored persons / Society for the Relief of Worthy Aged Indigent Colored Persons – New-York: M Day 1840-43 [mf ed 2004] – 1st-4th (1840-43) [complete] on 1r – 1 – (cont by: society for the support of the colored home (new york, ny). annual report of the society for the support of the colored home [issn 0735-0988]) – mf#2004-s015 – us ATLA [360]

Annual report of the society for the support of the colored home / Society for the Support of the Colored Home (New York NY) – New York: Publ for the Society by J S Taylor 1844-51 [mf ed 2004] – 5th-11th (1844-1850/51) on 1r – 1 – (damaged: v8th (1847/48) p3; cont: society for the relief of worthy aged indigent colored persons. annual report of the society for the relief of worthy aged indigent colored persons; cont by: colored home (new york, ny) annual report of the colored home) – ISSN: 0735-0988 – mf#2004-s016 – us ATLA [360]

Annual report of the spiritual and financial state of the mission – London, England. 1807 – 1r – us UF Libraries [240]

Annual report of the st james square congregation of the presbyterian church in canada, toronto... / St James Square Congregation (Toronto, ON) – Toronto?: The Congregation, 1879-1894 – 9 – (title varies slightly; issues reproduced: 1878-1893) – mf#A00854 – cn Canadiana [242]

Annual report of the state agricultural and industrial school, industry, new york / New York (State). State Agricultural and Industrial School – [Industry, NY] 86v. 1st-86th 1849-1933/34 (yrly) [mf ed 2nd-86th 1850-1933/34 filmed 1993] – 4r – 1 – (title varies: 1849-1883/84: annual report of the board of managers of the western house of refuge for juvenile delinquents (with slight variations); 1887/88-1905/06: annual report of the board of managers of the state industrial school) – us NY Public [362]

Annual report of the state historian / New York (State) – v2 – 1r – 1 – mf#2817957 – us WHS [978]

Annual report of the trustees and superintendent of the institute for deaf, dumb and blind colored youths of the state of texas / Institute for Deaf, Dumb and Blind Colored Youths of the State of Texas – Austin: B Jones, State Printers -1901 [9th-14th (1896-1901)] [mf ed 2005] – 1r – 1 – (imprint varies; cont: institute for deaf, dumb and blind colored youths of the state of texas. report of the trustees and superintendent of the institute for deaf, dumb and blind colored youths of the state of texas; cont by: institute for deaf, dumb and blind colored youths of the state of texas. annual report of the board and superintendent of the institute for the deaf, dumb and blind colored youths of the state of texas) – mf#2005-s081 – us ATLA [362]

Annual report of the trustees of the industrial school for boys at shirley for the year ending... – Boston: The School, 1910- (mf ed 1st 1909- filmed 1993) – 1r – 1 – (filmed together with other titles) – mf#*ZAN-11321 n1 – us NY Public [360]

Annual report of the virginia industrial home for colored girls / Virginia Industrial Home for Colored Girls – Peak's Turnout VA: [s.n.] 1921- [annual] [mf ed 2004] – 1r – 1 – (mf: 6th-24th [1921-39]. report yr for 1920/21-1926/27 ends mar 1; for 1927/28-1930/31, feb 28; for 1931/32, feb 29; for 1932/33-1938/39?, jun 30) – mf#2004-s007 – us ATLA [365]

Annual report of the vulcan society, inc – 1959 – 1r – 1 – mf#4765230 – us WHS [366]

Annual report of the wisconsin state board of dental examiners – v26th(1909/10]-48th(1931/32] – 1r – 1 – (cont by: directory of registered dentists and dental hygienists in wisconsin) – mf#601707 – us WHS [617]

Annual report of the woods and forests department for the year... / South Australia. Woods and Forests Dept – Adelaide: H Weir, govt printer 1929-35 (annual) [1928/29-1934/35] [mf ed 2005] – 7v on 1r [ill] – 1 – (cont: south australia. woods and forests dept. annual progress report upon state forest administration in south australia for the year...; cont by: south australia. woods and forests dept annual report on the operations of the woods and forests department for the year...) – mf#film mas 36324 – us Harvard [634]

Annual report of the...1873- / Burmah Baptist Missionary Convention – Rangoon: American Mission Press, c1873- [annual] [mf 8th-43rd 1872/73-1908 filmed 2003] – 1r – 1 – mf#2003-s094 – us ATLA [242]

Annual report of the...1892- / International Missionary Alliance – New York, 1892-1894/95 [annual] [mf ed 2003] – 4v on 1r – 1 – (incl suppl: year book of the christian alliance and the international missionary alliance, 1893) – mf#2003-s101 – us ATLA [240]

Annual report of the...1866-1871 / Burmah Baptist Missionary Convention – Rangoon: American Mission Press. 1st-6th. 1865/66-1870/71 [annual] mf ed 2003] – 6v on 1r – 1 – mf#2003-s092 – us ATLA [242]

Annual report of the...for the fiscal year ending may 31st... / Chicago and North Western Railway Co – 14th [1873], 15th-44th [1874-1903], 51st-53rd [1910-12], 55th-57th [1914-16] – 2r – 1 – (cont: report of the chicago and north western railway company, chicago and north western railway company. cont by: annual report...) – mf#153128 – us WHS [380]

Annual report of wisconsin state elections board – 1974/75-1976 jul/dec – 1r – 1 – (cont by: biennial report of wisconsin state elections board) – mf#543220 – us WHS [325]

Annual report of...corresponding secretary-treasurer / African Methodist Episcopal Zion Church. Church Extension Dept – [Birmingham AL?: s.n.] 1940/41-1944/45 [mf ed 2005] – 1r – 1 – (semi-annual reports update annual reports) – mf#2005-s023 – us ATLA [242]

Annual report on exchange restrictions / International monetary fund. annual report on exchange restrictions – Washington DC 1950-78 – 1,5,9 – (cont by: annual report on exchange arrangements and exchange restrictions) – ISSN: 0085-2163 – mf#6525 – us UMI ProQuest [332]

Annual report on native administration / Rhodesia and Nyasaland – Zomba, Govt Press. [1955-1957] – 1 – us CRL [350]

Annual report on public instruction in burma for the year... – Rangoon, Burma: Superintendent, Govt Print 1924-32 [1922/23-1930/31] (annual except every 5th yr) [mf ed New Haven CT: SEAsia Coll, Yale Uni Lib 1991] – 8v on 1r [ill] – 1 – (every 5th report iss as pt of a 5-yr cumulation called: quinquennial report on public instruction in burma for the years; filmed consecutively with: quinquennial report on public instruction in burma for the years...n7-8; filmed with: report on public instruction in burma for the year...[rangoon, burma: 1932], and: annual report on public instruction in burma for the year...[rangoon, burma: 1935], 1935/36-1939/40; cont: reports on public instruction in burma for the year...; cont by: report on public instruction in burma for the year...[rangoon, burma: 1932]) – mfeam/crl reel 3 – us CRL [350]

Annual report on public instruction in burma for the year... – Rangoon, Burma: Superintendent, Govt Print 1935/36- [mf ed New Haven CT: SEAsia Coll, Yale Uni Lib 1991] – 1r – 1 – (report for 1935-36 cumulated with: report on public instruction in burma for the year...[rangoon, burma: 1932], 1932-33-1934-35, in a 5-yr cumulation [incl report for 1936-37] called: quinquennial report on public instruction in burma for the years...1932-33 to 1936-37; iss for 1935/36-1939/40 filmed with: annual report on public instruction in burma for the year..., and: report on public instruction in burma for the year...[rangoon, burma: 1932]; iss for 1935/36-1939/40 filmed consecutively with: quinquennial report on public instruction in burma for the year..., n9; cont: annual report on public instruction in burma for the year...) – mfeam/crl reel 3 – us CRL [350]

Annual report on the operations of the woods and forests department for the year... / South Australia. Woods and Forests Dept – Adelaide: F Trigg, govt printer 1936-[1983?] (annual) [1935/36-1948/49:[lacks 1941/42]] [mf ed 2005] – 1r [ill] – 1 – (cont: south australia. woods and forests dept. annual report of the woods and forests department for the year...; report yr ends jun 30; title varies slightly) – ISSN: 0728-7801 – mf#film mas 36324 – us Harvard [634]

Annual report to members / Oconto Electric Cooperative (WW) – 6th-48th [1942-1984] – 1r – 1 – mf#964681 – us WHS [333]

Annual report to the board of health and social services / Lincoln Boys School (WI) – 1970/71-1973/74 – 1r – 1 – mf#202696 – us WHS [360]

Annual report to the general council / Christian and Missionary Alliance. General Council – New York. 50th-59th yr. 1936-45 [mf ed 2003] – 10v on 1r – 1 – mf#2003-s105 – us ATLA [240]

Annual report to the members... / Barron County Electric Cooperative – 4th-18th [1939-53] – 1r – 1 – mf#2985519 – us WHS [334]

Annual report to the office of indian affairs for... – 1947/48-1951/52 – 1r – 1 – (cont by: annual report of indian education for...to office of indian affairs, wisconsin. dept of public instruction) – mf#607655 – us WHS [305]

Annual report...for the year... / Archaeological Dept. Southern Circle. Madras – Madras: Printed by the Superintendent, Govt Press, 1909/10-1911/12 – 1r – 1 – us CRL [930]

Annual report...for the year ending october 31st... / Church Bible and Prayer Book Society – Toronto: Church of England Pub Co, 1899?-19– – 9 – mf#A00805 – cn Canadiana [240]

Annual report...held with the baptist state convention at... / Woman's Baptist Foreign Missionary Society of Wisconsin – 1st-7th [1878-1884], 15th [1892] – 1r – 1 – mf#669768 – us WHS [242]

Annual reports / Baptist North America General Conference – 1851-1971 – 1 – $766.29 – (eastern 1851-85. western 1859-82. general 1865-83. annual 1894-1917) – us Southern Baptist [242]

Annual reports / Great Northern Railway Company – 1880-1968 – 4r – 1 – $120.00; $35.50r – us Minn Hist [380]

Annual reports / India. Archaeological Survey – 1904-68 – 1 – $227.00 – us L of C Photodup [930]

Annual reports : the isthmian canal commission, 1907-1914; the governor of the panama canal, 1915-1951; the panama canal company and the government of the canal zone, 1952-1979 – Washington: GPO, 1907-79 (all publ) – 266mf – 9 – $399.00 – mf#LLMC 82-100D Title 21 – us LLMC [324]

Annual reports / North American Baptist General Convention – 23,870p. 1851-1879 – 1 – $835.45 – (eastern, 1851-85. western, 1859-82. general, 1865-93. annual, 1894-1917) – us Southern Baptist [242]

Annual reports / Northern Pacific Railway Company – 1870-1968 – 7r – 1 – $210.00; $35.00r – us Minn Hist [380]

Annual reports / U.S. Army. Air Force – Washington, DC. 1917-36 – 1 – us NY Public [355]

Annual reports *see* Puerto rico. governor. annual reports

Annual reports, 1800-2000 *see* Anti-slavery international

Annual reports, 1817-1856 / Presbyterian Church in the U.S.A. Board of Missions – 1 – $50.00 – us Presbyterian [240]

Annual reports, 1818-1824 / Philadelphia Sunday and Adult School Union – 1 – $50.00 – us Presbyterian [240]

Annual reports, 1820-1918 / Presbyterian Church in the U.S.A. Board of Education – 1 – $100.00 – us Presbyterian [240]

Annual reports, 1825-1835 / American Sunday School Union – 1 – $50.00 – us Presbyterian [240]

Annual reports, 1833-1958 / Presbyterian Church in the U.S.A. Board of Foreign Missions – 1 – $100.00 – us Presbyterian [240]

Annual reports, 1839-1923 / Presbyterian Church in the U.S.A. Board of Publication and Sabbath School Work – 1 – $50.00 – us Presbyterian [240]

Annual reports, 1855-1875 / Presbyterian Church in the U.S.A. General Assembly. Trustees. Committee on the Relief Fund for Disabled Ministers and the Widows and Orphans of Deceased Ministers – 1 – $50.00 – (also: board of relief for disabled ministers... annual reports, 1876-1912) – us Presbyterian [240]

Annual reports, 1855-1923 / Presbyterian Church in the U.S.A. Board of Church Erection – 1 – $150.00 – us Presbyterian [720]

Annual reports, 1856-1870 / Presbyterian Church in the U.S.A. Board of Church Extension (Old School) – 1 – $50.00 – us Presbyterian [240]

Annual reports, 1857-1870 / Presbyterian Church in the U.S.A. Board of Domestic Missions (Old School) – 1 – $50.00 – us Presbyterian [240]

Annual reports, 1859-1867 / African Civilization Society – 1 – $50.00 – us Presbyterian [240]

Annual reports, 1862-1870 / Presbyterian Church in the U.S.A. General Assembly. Committee on Home Missions (New School) – 1 – $50.00 – us Presbyterian [240]

Annual reports, 1866-1923 / Presbyterian Church in the U.S.A. Board of Missions for Freedmen – 1 – $150.00 – us Presbyterian [240]

Annual reports, 1871-1923 / Presbyterian Church in the U.S.A. Board of Home Missions – 1 – $200.00 – us Presbyterian [240]

Annual reports, 1871-1923 / Presbyterian Church in the U.S.A. Woman's Board of Foreign Missions – 1 – $100.00 – us Presbyterian [240]

Annual reports, 1872-1920 / Presbyterian Church in the U.S.A. Woman's Presbyterian Board of Missions of the Northwest – 1 – $150.00 – us Presbyterian [240]

Annual reports, 1872-1923 / Presbyterian Church in the U.S.A. Woman's Foreign Missionary Society – 1 – $150.00 – us Presbyterian [240]

Annual reports, 1876-1912 / Presbyterian Church in the U.S.A. Board of Relief for Disabled Ministers and the Widows and Orphans of Deceased Ministers – 1 – $50.00 – us Presbyterian [240]

Annual reports, 1882-1923 / Presbyterian Church in the U.S.A. Board of Temperance – 1 – $50.00 – us Presbyterian [240]

Annual reports, 1883-1920 / Presbyterian Church in the U.S.A. Woman's Presbyterian Board of Foreign Missions of the Southwest – 1 – $100.00 – (with presbyterian church in the u.s.a. woman's north pacific presbyterian board of missions. annual reports, 1887-1920) – us Presbyterian [240]

Annual reports, 1884-1918 / Presbyterian Church in the U.S.A. College Board – 1 – $50.00 – us Presbyterian [240]

Annual reports, 1887-1920 / Presbyterian Church in the U.S.A. Woman's North Pacific Presbyterian Board of Missions – 1 – $100.00 – (with presbyterian church in the u.s.a. woman's presbyterian board of foreign mission of the southwest. annual reports, 1883-1920) – us Presbyterian [240]

Annual reports, 1898-1927 / Presbyterian Church in the U.S.A. Woman's Board of Home Missions – 1 – $50.00 – us Presbyterian [240]

Annual reports, 1913-1927 / Presbyterian Church in the U.S.A. Board of Ministerial Relief and Sustenation – 1 – $100.00 – (also: board of pensions, annual reports 1928-58; united presbyterian ch. in the u.s.a., board of pensions, annual reports 1958-83) – us Presbyterian [240]

Annual reports, 1914-1946 / United Presbyterian Church of North America. Board of Education – 1 – $50.00 – us Presbyterian [240]

Annual reports, 1919-1923 / Presbyterian Church in the U.S.A. General Board of Education – 1 – $50.00 – us Presbyterian [240]

Annual reports, 1924-1957 / Presbyterian Church in the U.S.A. Board of National Missions – 1 – $200.00 – us Presbyterian [240]

Annual reports, 1924-1958 / Presbyterian Church in the U.S.A. Board of Christian Education – 1 – $150.00 – us Presbyterian [240]

Annual reports, 1946-1958 / United Presbyterian Church of North America. Board of Christian Education – 1 – $50.00 – us Presbyterian [240]

Annual reports, 1958-1973 / United Presbyterian Church in the U.S.A. Board of National Missions – 1 – $100.00 – us Presbyterian [240]

Annual reports (a-g) / U.S. – 1870-1970. 13 reels – 1 – $35.00r – us Trans-Media [340]

Annual reports and balance sheets, 1895-1921 *see* Women's trade union league papers

Annual reports and directories / American Baptist Home Mission Societies – 1832-1981. Single reels available – 1 – $452.60 – us ABHS [240]

Annual reports and related published papers / National Museum and Art Gallery of Papua New Guinea – 1963-77 – 1r – 1 – mf#pmb doc461 – at Pacific Mss [060]

Annual reports by common carriers to the interstate commerce commission, 1888-1914 / U.S. Interstate Commerce Commission – (= ser Records Of The Interstate Commerce Commission) – 1348r – 5 – mf#T913 – us Nat Archives [380]

Annual reports for the year ending 31st december... / St James' Square Presbyterian Church (Toronto, ON) – Toronto?: The Church, 1895-19– – 9 – (imprint varies; issues reproduced: 1894-1900) – mf#A00876 – cn Canadiana [242]

Annual reports of fleets and task forces of the us navy, 1920-1941 / U.S. Navy – (= ser General records of the department of the navy, 1798-1947) – 15r – 1 – (with printed guide) – mf#M971 – us Nat Archives [355]

Annual reports of the attorney general / U.S. Dept of Justice – 1871-1993 – 422mf – 9 – $633.00 – (incl ind/digest. no report publ for 1945. updates planned) – mf#llmc 79-410 – us LLMC [342]

Annual reports of the board of indian commissioners to the secretary of the interior – 1870-1931 – (= ser American indian periodicals... 1) – $315.00 – us UPA [305]

Annual reports of the conservative party, great britain, 1867-1982 – 10r – 1 – mf#96622 – uk Microform Academic [320]

Annual reports of the department of the navy, 1822-1866 / U.S. Navy – (= ser Records Of The Bureau Of Yards And Docks) – 8r – 1 – (with printed guide) – mf#M1099 – us Nat Archives [355]

Annual reports of the economic development administration, commerce department – 1979-83 – (= ser Patent and trademark office, commissioner of patents decisions) – 13mf – 9 – $19.50 – mf#LLMC 95-023 – us LLMC [346]

Annual reports of the executive committee of the indian rights association, inc / Indian Rights Association. Executive Committee – 1883-1934 – (= ser American indian periodicals... 1) – 45mf – 9 – $315.00 – us UPA [322]

Annual reports of the federal communications commission / U.S. Federal Communications Commission – 1935-84 – 93mf – 9 – $139.00 – (lacking: 1976) – mf#LLMC 81-222 – us LLMC [340]

Annual reports of the federal election commission – 1975-84, 1990 – 18mf – 9 – $27.00 – (lacking: 1976. add vols planned) – mf#LLMC 90-371 – us LLMC [340]

Annual reports of the federal judicial center – 1st-2nd [1968-69] & 1971, 1973-80 – 10mf – 9 – $15.00 – mf#llmc99-050 – us LLMC [340]

Annual reports of the governor of american samoa to the secretary of the interior : 1952/53-1981 / American Samoa – Office of the Governor, 1953-82 – 40mf – 9 – $60.00 – mf#LLMC 82-100C Title 5 – us LLMC [324]

Annual reports of the governor of guam, 1938-1981 / Guam. (Commonwealth) – 62mf – 9 – $93.00 – (no reports issued for 1942-50. lacking: 1978) – mf#LLMC 82-100B Title 23 – us LLMC [324]

Annual reports of the governors of guam, 1901-1941 / U.S. Navy. Office of Naval Records and Library – (= ser General records of the department of the navy, 1798-1947) – 3r – 1 – (with printed guide) – mf#M181 – us Nat Archives [324]

Annual reports of the independent labour party, 1893-1932 – 3r – 1 – mf#96971 – uk Microform Academic [320]

Annual reports of the indian female normal school and instruction society *see* Church missionary society archive, section 2

Annual reports of the interstate commerce commission / U.S. Interstate Commerce Commission – 1st-94th. 1887-1980 – 311mf – 9 – $466.00 – mf#LLMC 81-229 – us LLMC [380]

Annual reports of the labour party, britain, 1900-99 – 25r – 1 – mf#95791 – uk Microform Academic [325]

Annual reports of the librarian of congress / U.S. Library of Congress – 1866-1966 – 1 – $320.00 – us L of C Photodup [020]

Annual reports of the national association for promotion of technical education, 1888-1894 and 1897-1907 – 1r – 1 – mf#97151 – uk Microform Academic [370]

Annual reports of the national labor relations board / U.S. National Labor Relations Board – 1st-48th. 1936-83 – (= ser National labor relations board decisions) – 144mf – 9 – $216.00 – mf#LLMC 81-233 – us LLMC [331]

Annual reports of the national society, 1812-1900 – 172mf – 7 – mf#87155 – uk Microform Academic [941]

Annual reports of the primitive methodist missionary society / The Primitive Methodist Missionary Society – London, 1844-1927. n1-84. 1843-1927 – 126mf – 9 – (missing: n50-66) – mf#H-2738 – ne IDC [956]
Annual reports of the record commissioners of boston, 1876-1909 / Boston (Mass). Registry Dept – v1-39. 1876-1909 – 9 – $498.00 – mf#0113 – us Brook [978]
Annual reports of the scottish trades union congress, 1897-1979 / Scottish Trades Union Congress – 22r – 1 – mf#97148 – uk Microform Academic [331]
Annual reports of the secretary of the navy, 1821-1901 / U.S. Navy. Office of the Secretary – 1988 – 25r – 1 – $3250.00 – (incl printed guide) – mf#S3166 – U.S. Naval Historical Center – us Scholarly Res [355]
Annual reports of the secretary of the treasury on the state of the finances – 2pt – 1 – (pt1: 1790-1910 14r isbn 0-89093-006-6 $2180. pt2: 1911-74 17r isbn 0-89093-106-2 $2645) – us UPA [336]
Annual reports of the united states to the trusteeship council of the u.n. – 1947-93 – (= ser The U.N. Trusteeship Mandate) – 149mf – 9 – $223.00 – (reports on the us admin of the ttpi. reports of the us navy 1947/48-1951) – mf#LLMC 82-100F title 11 – us LLMC [324]
Annual reports of the us patent office – 1906-25; 1949-83 – (= ser Patent And Trademark Office, Commissioner Of Patents Decisions) – 313mf – 9 – $470.00 – (updates planned) – mf#LLMC 94-205 – us LLMC [346]
Annual reports of the war department, 1822-1907 / U.S. War Dept – 164r – 1 – mf#M997 – us Nat Archives [324]
Annual reports of the wesleyan education committee, 1838-1901 – 138mf – 7 – mf#87162 – uk Microform Academic [370]
Annual reports of the wesleyan methodist missionary society / The Wesleyan Methodist Missionary Society – London, 1789-1947 – 511mf – 9 – (financial suppl 1939-1947) – mf#H-2749c – ne IDC [242]
Annual reports of the world's central banks : current collection, 1984-1993 – [mf ed Chadwyck-Healey] – 9 – (also available by region & individual country) – uk Chadwyck [332]
Annual reports of the world's central banks : retrospective collection, 1946-1983 – [mf ed Chadwyck-Healey] – 9 – (available as complete coll or by region: africa 53mf. asia 24mf. australasia 9mf. europe & near east 51mf. latin america & caribbean 53mf. middle east 24 mf. north america 3mf) – uk Chadwyck [332]
Annual reports on the conferences on international arbitration / Lake Mohonk Conference on International Arbitration – Clearwater Publ Co, 1895-1916 – 56mf – 9 – 0-88354-003-7 – (ind vol available separately) – us UPA [341]
Annual report...to the national convention / Workmen's Circle (US) – 4th-7th [1904/1905-07] – 1r – 1 – mf#3147388 – us WHS [331]
Annual reunion / Third Wisconsin Veteran Infantry Association – 27th-1937th [1918-27] – 1r – 1 – (cont: proceedings of the...annual reunion of the association of the third regiment wisconsin infantry veteran volunteers held at...) – mf#2806290 – us WHS [355]
Annual reunion for...of the... / Twelfth Wisconsin Infantry Association – 1902-03 – 1r – 1 – (cont: reunion of the twelfth wisconsin infantry; cont by: story of reunion of the twelfth wisconsin infantry) – mf#3599457 – us WHS [355]
Annual reunion of huntley national association / Huntley National Association – 9th-1934th [1955-80] – 1r – 1 – mf#637641 – us WHS [366]
Annual revenue and expenditure of lower canada : from its constitution to the period of the union / Canada (Province). Parlement. Assemblee legislative – Montreal: Lovell and Gibson, 1847 [mf ed 1983] – 1mf – 9 – mf#SEM105P302 – cn Bibl Nat [336]
Annual review and history of literature – La Salle IL 1802-08 – 1 – mf#5229 – us UMI ProQuest [410]
Annual review in automatic programming – Oxford, England 1960-92 – 1,5,9 – ISSN: 0066-4138 – mf#49013.01 – us UMI ProQuest [000]
Annual review of addictions research and treatment – Oxford, England 1991-93 – 1,5,9 – ISSN: 0955-663X – mf#49622 – us UMI ProQuest [362]
Annual review of anthropology – Palo Alto CA 1974+ – 1,5,9 – ISSN: 0084-6570 – mf#9919 – us UMI ProQuest [301]
Annual review of astronomy and astrophysics – Palo Alto CA 1965+ – 1,5,9 – ISSN: 0066-4146 – mf#5899 – us UMI ProQuest [520]
Annual review of banking law – Boston University. v1-5. 1982-86 – 9 – $85.00 set – mf#110961 – us Hein [346]
Annual review of biochemistry – Palo Alto CA 1932+ – 1,5,9 – ISSN: 0066-4154 – mf#230 – us UMI ProQuest [574]
Annual review of biophysics and bioengineering – Palo Alto CA 1972-77 – 1,5,9 – ISSN: 0084-6589 – mf#8225.02 – us UMI ProQuest [574]
Annual review of cell biology – Palo Alto CA 1985-88 – 1,5,9 – ISSN: 0743-4634 – mf#14963.01 – us UMI ProQuest [574]
Annual review of chronopharmacology – Oxford, England 1984-90 – 1,5,9 – ISSN: 0743-9539 – mf#49480 – us UMI ProQuest [615]
Annual review of earth and planetary science – Palo Alto CA 1973+ – 1,5,9 – ISSN: 0084-6597 – mf#9706 – us UMI ProQuest [550]
Annual review of ecology and systematics – Palo Alto CA 1973+ – 1,5,9 – ISSN: 0066-4162 – mf#11806.01 – us UMI ProQuest [574]
Annual review of energy – Palo Alto CA 1976-90 – 1,5,9 – (cont by: annual review of energy & the environment) – ISSN: 0362-1626 – mf#11807.02 – us UMI ProQuest [333]
Annual review of energy and the environment – Palo Alto CA 1991+ – 1,5,9 – (cont: annual review of energy) – ISSN: 1056-3466 – mf#11807.02 – us UMI ProQuest [333]
Annual review of entomology – Palo Alto CA 1956+ – 1,5,9 – ISSN: 0066-4170 – mf#5094 – us UMI ProQuest [590]
Annual review of fish diseases – Oxford, England 1991-96 – 1,5,9 – ISSN: 0959-8030 – mf#49618 – us UMI ProQuest [639]
Annual review of fluid mechanics – Palo Alto CA 1969+ – 1,5,9 – ISSN: 0066-4189 – mf#9705 – us UMI ProQuest [530]
Annual review of genetics – Palo Alto CA 1967+ – 1,5,9 – ISSN: 0066-4197 – mf#7696 – us UMI ProQuest [575]
Annual review of immunology – Palo Alto CA 1983+ – 1,5,9 – ISSN: 0732-0582 – mf#13391 – us UMI ProQuest [616]
Annual review of jazz studies – Piscataway NJ 1982-88 – 1,5,9 – ISSN: 0731-0641 – mf#12885 – us UMI ProQuest [780]
Annual review of materials science – Palo Alto CA 1975+ – 1,5,9 – ISSN: 1531-7331 – mf#11808.01 – us UMI ProQuest [620]
Annual review of medicine – Palo Alto CA 1950+ – 1,5,9 – ISSN: 0066-4219 – mf#5093 – us UMI ProQuest [610]
Annual review of microbiology – Palo Alto CA 1947+ – 1,5,9 – ISSN: 0066-4227 – mf#5092 – us UMI ProQuest [576]
Annual review of neuroscience – Palo Alto CA 1978+ – 1,5,9 – ISSN: 0147-006X – mf#11809 – us UMI ProQuest [612]
Annual review of nuclear science – Palo Alto CA 1952-77 – 1,5,9 – ISSN: 0066-4243 – mf#2094.01 – us UMI ProQuest [530]
Annual review of nutrition – Palo Alto CA 1981+ – 1,5,9 – ISSN: 0199-9885 – mf#13358 – us UMI ProQuest [613]
Annual review of pharmacology – Palo Alto CA 1961-75 – 1,5,9 – (cont by: annual review of pharmacology & toxicology) – ISSN: 0066-4251 – mf#5091.01 – us UMI ProQuest [615]
Annual review of pharmacology and toxicology – Palo Alto CA 1976+ – 1,5,9 – (cont: annual review of pharmacology) – ISSN: 0362-1642 – mf#5091,01 – us UMI ProQuest [615]
Annual review of physical chemistry – Palo Alto CA 1950+ – 1,5,9 – ISSN: 0066-426X – mf#5087 – us UMI ProQuest [540]
Annual review of physiology – Palo Alto CA 1939+ – 1,5,9 – ISSN: 0066-4278 – mf#5088 – us UMI ProQuest [612]
Annual review of phytopathology – Palo Alto CA 1963+ – 1,5,9 – ISSN: 0066-4286 – mf#5900 – us UMI ProQuest [574]
Annual review of plant physiology – Palo Alto CA 1950-87 – 1,5,9 – (cont by: annual review of plant physiology & plant molecular biology) – ISSN: 0066-4294 – mf#5089.02 – us UMI ProQuest [574]
Annual review of plant physiology and plant molecular biology – Palo Alto CA 1988-2001 – 1,5,9 – (cont: annual review of plant physiology) – ISSN: 1040-2519 – mf#5089.02 – us UMI ProQuest [574]
Annual review of population law – v1- – (inquire for info) – mf#119061 – us Hein [340]
Annual review of psychology – Palo Alto CA 1950+ – 1,5,9 – ISSN: 0066-4308 – mf#5090 – us UMI ProQuest [150]
Annual review of public health – Palo Alto CA 1980+ – 1,5,9 – ISSN: 0163-7525 – mf#12725 – us UMI ProQuest [614]
Annual review of sociology – Palo Alto CA 1975+ – 1,5,9 – ISSN: 0360-0572 – mf#11810 – us UMI ProQuest [301]
Annual review of the progress of south dakota / South Dakota State Historical Society – 1909, 1911-17 – 1r – 1 – mf#1103248 – us WHS [978]
Annual school directory for... / Saint Croix County (WI) – 1st-11th [1910/19-1920/21] – 1r – 1 – (cont: list of school district clerks, st croix co, wi, list of teachers in st croix co, wi; cont by: department of education, st croix county, hammond, wi) – mf#5192576 – us WHS [370]
The annual sermon before the american sunday-school union : delivered...may 10 1857 / Eastburn, Manton – Philadelphia: American Sunday-School Union, [1857?] [mf ed 1993] – 1mf – 9 – 0-524-08289-8 – mf#1993-3044 – us ATLA [242]
Annual sermons and reports of the society for the propagation of the gospel, 1701-1845 / United Society for the Propagation of the Gospel. Archives – 7r – 1 – mf#96767 – uk Microform Academic [220]
Annual session / Big Creek Association (TN) – Martin TN: Cayces & Turner [mf ed 2005] – (= ser African american baptist serials 2005-s114-s128) – 26th (1906) [complete] on 1r – 1 – (reel incl other primitive baptist association titles) – mf#2005-s123 – us ATLA [242]
Annual session / Cypress Creek Association of Primitive Baptists – Martin TN: Cayces & Turner [mf ed 2005] – (= ser African american baptist serials 2005-s114-s128) – 35th (1906) [complete] on 1r – 1 – (aka: proceedings of the cypress creek association; reel incl other primitive baptist association titles) – mf#2005-s122 – us ATLA [242]
Annual session of the iowa state grange, p of h / Patrons of Husbandry – 14th [1883], 16th-17th [1885-86] – 1r – 1 – (cont: report of proceedings of the...annual session of the iowa state grange of the patrons of husbandry; cont by: session of the iowa state grange, patrons of husbandry) – mf#3433192 – us WHS [366]
Annual single numbers series alphabetical index, 1949-1958 : file e103-e104 / Genealogical Society of the Northern Territory – [mf ed 1992] – 2mf – 9 – A$11.00 – 0-949124-73-7 – (ind n1317-1980) – at Northern [980]
Annual single numbers series alphabetical index, 1959-1965 : file e104 – [mf ed 1992] – 4mf – 9 – $22.00 – 0-949124-82-6 – (ind n1981-2646; plus add ind for n2616, 3435, 3442 & 3491) – at Northern [980]
Annual single numbers series alphabetical index, 1965-1969 : file e104 / Genealogical Society of the Northern Territory – [mf ed 1992] – 3mf – 9 – A$16.50 – 0-949124-75-3 – (ind n2647-3267) – at Northern [980]
Annual single numbers series alphabetical index, 1969-1976 : file e104 / Genealogical Society of the Northern Territory – [mf ed 1992] – 2mf – 9 – A$11.00 – 0-949123-76-5 – (ind n3268-3681; plus add ind n2616, 3435, 3442 & 3491) – at Northern [980]
Annual single numbers series alphabetical index, 1971-1976 : file e105 / Genealogical Society of the Northern Territory – [mf ed 1992] – 2mf – 9 – A$11.00 – 0-949124-77-X – (ind n71/001-71/004, 72/001-72/091, 73/001-73/195, 1974 cyclone tracey, 75/001-75/123, 76/001-76/200; plus add annual single nos series ind n73/084 – paspalis m t file e105 (alphabetical ind); plus add annual single nos series indn75/108 and 76/094 file e105 alphabetical ind) – at Northern [980]
Annual single numbers series alphabetical index, 1977-1978 : file ntrs-f404 / Genealogical Society of the Northern Territory – [mf ed 1993] – 3mf – 9 – A$16.50 – 0-949124-78-8 – (ind n77/001-77/232, 78/001-78/217; plus add for ind n78/195 longbottom l r) – at Northern [980]
Annual single numbers series alphabetical index, 1979-1980 : file ntrs f404 / Genealogical Society of the Northern Territory – [mf ed 1993] – 2mf – 9 – A$11.00 – 0-949124-79-6 – (ind n79/001-79/233, 80/001-80/129; plus add for ind n80/109) – at Northern [980]
Annual single numbers series alphabetical index, 1981-1983 : file ntrs-f404 / Genealogical Society of the Northern Territory – [mf ed 1993] – 2mf – 9 – A$11.00 – 0-949124-80-X – (ind n81/001-81/083, 82/001-82/081, 83/001-83/098) – at Northern [980]
Annual single numbers series alphabetical index, 1984-1986 / Genealogical Society of the Northern Territory – [mf ed 1993] – 2mf – 9 – A$11.00 – 0-949124-81-8 – (ind n84/001-84/105, 85/001-85/086, 86/001-86/094) – mf#item 38 – at Northern [980]
Annual single numbers series alphabetical index, 1987-1989 / Genealogical Society of the Northern Territory – [mf ed 1993] – 2mf – 9 – A$11.00 – 0-949124-68-0 – (ind n87/001-87/102, 88/001-88/112, 89/001-89/102) – at Northern [980]
Annual single numbers series alphabetical index, 1990-1991 / Genealogical Society of the Northern Territory – [mf ed 1993] – 2mf – 9 – A$11.00 – 0-949124-70-2 – (ind n90/001-90/136, 91/001-91/149) – at Northern [980]
Annual single numbers series alphabetical index, 1992-1993 / Genealogical Society of the Northern Territory – [mf ed 1993] – 1mf – 9 – A$5.50 – 0-949124-71-0 – (ind n92/001-92/133, 93/001-93/034 [incomplete]) – at Northern [980]
Annual statement of the overseas trade of the united kingdom 1853-1975 – 1854/55-1979 – (= ser British government publications...1801-1977) – 95r – 1 – (jan-june, oct-dec 1940 + 1941-43 not publ) – uk Chadwyck [380]
Annual statement of the sea-borne trade and navigation of burma with foreign countries and indian ports / Burma. Customs Dept – Rangoon [annual] [mf ed New Haven CT: SEAsia Coll, Yale Uni Lib 1991] – 15r – 1 – (report yr ends mar 31; publ suspended 1924-1927? center has: mf-7101 seam 1890/91-1948/49; lacks: 1907/08, 1909/10-1910/11, 1912/13, 1940/41-1944/45) – us CRL [380]
Annual statement of the trade and commerce of st louis – St Louis: St Louis Merchants' Exchange etc, [1865-1923] – 1 – us CRL [380]
Annual statement respecting the canadian pacific railway / Tupper, Charles – [Ottawa?: s.n.], 1884 [mf ed 1981] – 1mf – 9 – mf#24875 – cn Canadiana [380]
Annual statistical bulletin – Maseru, Basutoland: The Bureau [1963/1964] – 1r – 1 – us CRL [316]
Annual statistical bulletin – Maseru, Lesotho: The Bureau [1965-1973] – 1r – 1 – us CRL [316]
Annual statistical bulletin / Organization of Petroleum Exporting Countries – 1966-77 – 1 – us L of C Photodup [330]
Annual statistical bulletin *see* National accounts
Annual statistical bulletin 1963-1973 / Lesotho (formerly Basutoland). Bureau of Statistics – (= ser African official statistical serials, 1867-1982) – 12mf – 9 – uk Chadwyck [316]
Annual statistical bulletin 1966-1976 / Swaziland. Central Statistical Office – (= ser African official statistical serials, 1867-1982) – 16mf – 9 – (1969 not publ) – uk Chadwyck [316]
Annual statistical digest 1935/1951-1973/1974 / Trinidad and Tobago. Central Statistical Office – (= ser Latin american & caribbean...1821-1982) – 59mf – 9 – uk Chadwyck [318]
Annual statistical digest 1963-1965 / Nigeria. Eastern Region. Statistics Division – (= ser African official statistical serials, 1867-1982) – 4mf – 9 – uk Chadwyck [316]
Annual statistical digest 1966-1974 / St Lucia. Development, Planning and Statistics Division – (= ser Latin american & caribbean...1821-1982) – 9mf – 9 – (1972-73 not available) – uk Chadwyck [318]
Annual statistical digest 1968-1976 / Sierra Leone. Central Statistics Office – (= ser African official statistical serials, 1867-1982) – 10mf – 9 – (1972-75 not publ) – uk Chadwyck [316]
Annual statistical report / New York Produce Exchange – New York. 1902-1929 – 1 – us NY Public [317]
Annual statistical survey of the electronics industry 1969, 1970, 1972-1974 – [mf ed Chadwyck-Healey] – (= ser British government publications...1801-1977) – 7mf – 9 – uk Chadwyck [338]
Annual survey of american law *see* New york university annual survey of american law
Annual survey of english law – 1928-40 – 9 – $180.00 – mf#0053 – us Brook [342]
Annual synod minutes and journals, 1854-1945 : together with miscellaneous correspondence, 1869-1899 / Methodist Church in Fiji – 4r – 1 – (restricted access) – mf#PMB1138 – at Pacific Mss [242]
Annual trade report / British Somaliland. Customs and Excise Dept – Aden, Govt Printer [1952, 1955-1957] – 1 – (british somaliland misc govt publications, 1944-1957) – us CRL [380]
The annual volunteer and service militia list of canada : 1st march 1866 / Canada (Province). Departement de la milice – Ottawa: Printed by G E Desbarats, 1866 [mf ed 1983] – 2mf – 9 – mf#SEM105P319 – cn Bibl Nat [355]
Annual wiaa handbook / Wisconsin Interscholastic Athletic Association – 1969/70-1973/74 – 1r – 1 – (cont: wiaa official handbook; cont by: official handbook of the wisconsin interscholastic athletic association) – mf#682894 – us WHS [790]
Annual wiaa yearbook / Wisconsin Interscholastic Athletic Association – v52-59 [1974/75-81/82], v60-64 [1982/83-1986/87] – 2r – 1 – (con: annual wiaa handbook) – mf#682897 – us WHS [790]
Annual year book / Wisconsin Interscholastic Athletic Association – v29-33 [1952-56], v34-39 [1957-62], v40-42(1963-65] – 3r – 1 – mf#682876 – us WHS [790]
Annuals / Arkansas Central Baptist Associations – 2302p. 1919-80 – 1 – $109.00 – (lacks: 1926, 1927, 1936) – us Southern Baptist [242]

Annuals : ascension / Baptist Associations. Louisiana – 1959-1989 – 1r – 1 – $49.52 – us Southern Baptist [242]

Annuals : district thirteen = Baptist association. texas – 1954-1963 – 1r – 1 – $16.40 – us Southern Baptist [242]

Annuals / Baptist Associations. Flint River. Alabama – 352p. 1814-66 – 1 – us Southern Baptist [242]

Annuals / Baptist Associations. Illinois. Saline – 1955-1990 – 1 reel – 1 – $83.68 – us Southern Baptist [242]

Annuals / Baptist Associations. Jefferson County. Tennessee – 744p. 1971-80 – 1 – (nashville 1977-78 264p. sweetwater 1882-90 1893 158p) – us Southern Baptist [242]

Annuals / Baptist Associations. Kentucky. Blood River – 1969-1990 – 1 reel – 1 – $57.20 – us Southern Baptist [242]

Annuals / Baptist Associations. Kentucky. Greenup – 1959-1990 – 1 reel – 1 – $80.88 – us Southern Baptist [242]

Annuals / Baptist Associations. Tennesse. Sweetwater/Eastanalle – 1830-1884 – 1 reel – 1 – $16.56 – us Southern Baptist [242]

Annuals / Baptist Associations. Tennessee. Big Hatchie – 1981-1987 – 1r – 1 – $14.64 – us Southern Baptist [242]

Annuals / Baptist Associations. Tennessee. Concord – 1866, 1900, 1957-1979 – 1 reel – 1 – $58.32 – us Southern Baptist [242]

Annuals / Baptist Associations. Utah-Idaho, Treasure Valley – 568p. 1977-88 – 1 – $22.72 – (continues: boise valley) – us Southern Baptist [242]

Annuals : district of columbia = Baptist state conventions. (southern baptist) – Oct. 1906-Mar. 1921 – 1r – 1 – $21.68 – us Southern Baptist [242]

Annuals / Baptist Woman's Missionary Union. Auxiliary to Southern Baptist Convention – 1889-1978 – 1 – $308.84 – us Southern Baptist [242]

Annuals : bayou macon / Baptist Associations. Louisiana – 1951-1988 – 1r – 1 – $70.49 – us Southern Baptist [242]

Annuals : beauregard / Baptist Associations. Louisiana – 1955-1987 – 1r – 1 – $75.92 – us Southern Baptist [242]

Annuals : big creek / Baptist Associations. Louisiana – 1955, 1957-1987 – 1r – 1 – $64.88 – us Southern Baptist [242]

Annuals / Big Hatchie. Tennessee. Baptist Associations – 378p. 1976-80 – 1 – us Southern Baptist [242]

Annuals / Black Baptist Convention – 1842-1975 – 1 – $1587.00 – ((american baptist missionary convention, 1842, 1852-54, 1857-60, 1869, 1871-72, 1877, 1879; the american national baptist convention, 1889-91, national baptist educational convention, 1892; national baptist convention, 1897-1905, 1907-12, 1914-15)) – us Southern Baptist [242]

Annuals : bossier / Baptist Associations. Louisiana – 1957-1969 – 1r – 1 – $20.32 – us Southern Baptist [242]

Annuals : caldwell / Baptist Associations. Louisiana – 1934-1991 – 1r – 1 – $84.32 – us Southern Baptist [242]

Annuals : carey / Baptist Associations. Louisiana – 1958-1989 – 1r – 1 – $86.00 – us Southern Baptist [242]

Annuals : central louisiana / Baptist Associations. Louisiana – 1956-1987 – 1r – 1 – $62.16 – us Southern Baptist [242]

Annuals : charleston / Baptist Associations. South Carolina – 1971-1986 – 1r – 1 – $45.36 – us Southern Baptist [242]

Annuals : charleston / Baptist Associations. South Carolina – 1987-1990 – 1r – 1 – $18.56 – us Southern Baptist [242]

Annuals : chester / Baptist Associations. South Carolina – 1971-1991 – 1r – 1 – $153.76 – us Southern Baptist [242]

Annuals : chesterfield / Baptist Associations. South Carolina – 1971-1990 – 1r – 1 – $85.12 – us Southern Baptist [242]

Annuals : colleton / Baptist Associations. South Carolina – 1971-1981 – 1r – 1 – $24.24 – us Southern Baptist [242]

Annuals : colleton. union of 1st and 2nd division / Baptist Associations. South Carolina – 1873-1878 – 1r – 1 – $5.00 – us Southern Baptist [242]

Annuals : dallas / Baptist Associations. Texas – 1980-1987 – 1r – 1 – $78.48 – us Southern Baptist [242]

Annuals : deer creek / Baptist Associations. Louisiana – 1922, 1957-1991 – 1r – 1 – $79.04 – us Southern Baptist [242]

Annuals : delta / Baptist Associations. Louisiana – 1958-1991 – 1r – 1 – $55.28 – us Southern Baptist [242]

Annuals : district eight / Baptist Associations. Texas – 1952, 1955-1963 – 1r – 1 – $58.32 – us Southern Baptist [242]

Annuals : district eleven / Baptist Associations. Texas – 1952, 1955-1964 – 1r – 1 – $58.40 – us Southern Baptist [242]

Annuals : district fifteen / Baptist Associations. Texas – 1957-1963 – 1r – 1 – $64.96 – us Southern Baptist [242]

Annuals : district five / Baptist Associations. Texas – 1947, 1952, 1955-1963 – 1r – 1 – $90.16 – us Southern Baptist [242]

Annuals : district four / Baptist Associations. Texas – 1952-1964 – 1r – 1 – $91.44 – us Southern Baptist [242]

Annuals : district fourteen / Baptist Associations. Texas – 1953-1963 – 1r – 1 – $86.16 – us Southern Baptist [242]

Annuals : district nine / Baptist Associations. Texas – 1955-1963 – 1r – 1 – $66.64 – us Southern Baptist [242]

Annuals : district one / Baptist Associations. Texas – 1952-1963 – 1r – 1 – $20.48 – us Southern Baptist [242]

Annuals : district seven / Baptist Associations. Texas – 1955-1962 – 1r – 1 – $24.00 – us Southern Baptist [242]

Annuals : district seventeen / Baptist Associations. Texas – 1955-1964 – 1r – 1 – $11.20 – us Southern Baptist [242]

Annuals : district six / Baptist Associations. Texas – 1951-1962 – 1r – 1 – $64.24 – us Southern Baptist [242]

Annuals : district sixteen / Baptist Associations. Texas – 1952, 1955-1963 – 1r – 1 – $79.20 – us Southern Baptist [242]

Annuals : district ten / Baptist Associations. Texas – 1952-1963 – 1r – 1 – $61.84 – us Southern Baptist [242]

Annuals : district three / Baptist Associations. Texas – 1948-1964 – 1r – 1 – $21.92 – us Southern Baptist [242]

Annuals : district two / Baptist Associations. Texas – 1956-1963 – 1r – 1 – $7.20 – us Southern Baptist [420]

Annuals : eastern louisiana / Baptist Associations. Louisiana – 1958-1990 – 1r – 1 – $56.48 – us Southern Baptist [420]

Annuals / Florida. Baptist Associations – Chipola. 2380p. 1925-79 – 1 – (formerly: jackson county -name changed 1946) – us Southern Baptist [242]

Annuals : frio river / Baptist Associations. Texas – 1963-1988 – 1r – 1 – $56.08 – us Southern Baptist [242]

Annuals : liberty / Baptist Associations. Louisiana – 1957-1973 – 1r – 1 – $31.84 – us Southern Baptist [242]

Annuals : little bethel / Baptist Associations. Kentucky – 1979-1987 – 1r – 1 – $26.08 – us Southern Baptist [242]

Annuals : louisiana / Baptist Associations. Louisiana – 1955-1987 – 1r – 1 – $51.28 – us Southern Baptist [242]

Annuals : luther rice / Baptist Associations. Louisiana – 1951-1987 – 1r – 1 – $69.84 – us Southern Baptist [242]

Annuals : madison / Baptist Associations. Louisiana – 1958-1989 – 1r – 1 – $39.92 – us Southern Baptist [242]

Annuals / Missouri Primitive Baptist Associations – 790p. Cuivre Siloam 1879, 1892-93, 1913; Nodway 1939; Salem 1939; Two River 1939 – 1 – us Southern Baptist [242]

Annuals / Montgomery. Maryland. Baptist Associations – 756p. 1969-80 – 1 – us Southern Baptist [242]

Annuals : morehouse / Baptist Associations. Louisiana – 1955-1989 – 1r – 1 – $64.40 – us Southern Baptist [242]

Annuals / National Baptist Convention – 1897-1905, 1907-12, 1914-15 – 1 – $173.95 – us Southern Baptist [242]

Annuals / National Baptist Convention of America, Unincorporated – 1916-17, 1920-30, 1938, 1950, 1952-73, 1975 – 1 – $375.20 – us Southern Baptist [242]

Annuals / National Baptist Convention. U.S.A., Inc – 1916-75 – 1 – $943.07 – us Southern Baptist [242]

Annuals : navosta river / Baptist Associations. Texas Bma – 1970-1989 – 1r – 1 – $21.28 – us Southern Baptist [420]

Annuals / New Mexico Baptist Convention – 206p – 1 – us Southern Baptist [242]

Annuals / North Carolina Negro Associations – 1 – $470.12 – us Southern Baptist [242]

Annuals / Oklahoma Baptist Associations – 40,944p – 1 – us Southern Baptist [242]

Annuals : pacific / Baptist Associations. California – 1984-1987 – 1r – 1 – $12.40 – us Southern Baptist [242]

Annuals / Southern Baptist Convention – 1845-1984 – 1 – $2456.50 – (incl ind 1845-1965) – us Southern Baptist [242]

Annuals : st tammany / Baptist Associations. Louisiana – 1955, 1958-1987 – 1r – 1 – $61.84 – us Southern Baptist [242]

Annuals / Tennessee Baptist Missionary and Educational Convention – 1974-76 – 1 – $7.35 – us Southern Baptist [242]

Annuals / Tennessee. Central. Baptist Associations – 790p. 1958-78 – 1 – (grainger county 1972-78 372p) – us Southern Baptist [242]

Annuals : vernon / Baptist Associations. Louisiana – 1955-1987 – 1r – 1 – $62.72 – us Southern Baptist [242]

Annuals : webster-claiborne / Baptist Associations. Louisiana – 1958-1991 – 1r – 1 – $63.28 – us Southern Baptist [242]

Annuario / Conservatorio di Musica Luigi Cherubini. Florence – v1-10 1898-1913/14 – 1 – $23.00 – us L of C Photodup [780]

Annuario administrativo e litterario do gabinete portuguez de leitura... – Recife, PE: Typ Universal, 1854 – (= ser Ps 19) – mf#P17,02,158 – bl Biblioteca [079]

Annuario dell'africa italiana e delle isole italiane dell'egeo / Instituto fascista dell'Africa italiana. Roma – Roma: Societa an. tipografica Castaldi. v13-14. 1938/39-1940 – 1 – us CRL [960]

Annuario delle colonie italiane – Roma: Cooperativa tipografica "Castaldi" 1926-1927 [v1-2 1926-1927] (annual) – 1r – 1 – us CRL [945]

Annuario delle colonie italiane e dei paesi vicini / Instituto coloniale fascista. Roma – Roma: Societa an. tipografica Castaldi. v3-10. 1928-35 – 1 – us CRL [960]

Annuario delle colonie italiane, isole italiane dell'egeo, paesi dell'africa – Roma: Societa an tipografica Castaldi, 1936 [v11 1936] (annual) – 1r – 1 – us CRL [945]

Annuario dell'impero italiano / Instituto coloniale fascista. Roma – Roma: Societa an. tipografica Castaldi. v12. 1937 – 1 – us CRL [960]

Annuario industrial – Rio de Janeiro, RJ: Typ Perseveranca, 1871 – (= ser Ps 19) – mf#DIPER – bl Biblioteca [338]

Annuario publicado pelo imperial observatorio do rio de janeiro para o anno de... / Imperial Observatorio do Rio de Janeiro – Rio de Janeiro: O Observatorio 1884-89 (annual) [mf ed 2001] – [1]-5 anno(1885-89) on 1r – 1 – (cont by: anuario para o ano de...publicado pelo observatorio nacional riode janeiro) – mf#film mas c5051 – us Harvard [520]

Annuario statistico italiano / Italy. Direzione Generale Della Statistica – v. 1-22. 1878-1926. (1-2 serie) – 1 – us L of C Photodup [945]

Annuario statistico italiano 1878-1965 / Italy. Istituto Centrale di Statistica – (= ser European official statistical serials, 1841-1984) – 384mf – 9 – uk Chadwyck [314]

Annuity market news – New York NY 2001+ – 1,5,9 – ISSN: 1525-2221 – mf#32345,01 – us UMI ProQuest [332]

O annunciador – Rio de Janeiro, RJ. 03-25 fev 1850 – (= ser Ps 19) – mf#P15,01,52 – bl Biblioteca [079]

O annunciador : semanario noticioso e orgam de propaganda de preparados pharmaceuticos – Aracati, CE. 05 dez 1920; 1921 – (= ser Ps 19) – mf#P18B,03,71 – bl Biblioteca [615]

O annunciador *see* Periodico dos pobres

An-nur – En Nour. Alger. n1-78. 1931-33 – 1 – fr ACRPP [073]

Annus ecclesiasticus graecorum-slavicus 1863 / Martinow, J – Bruxelles, 1963 – 14mf – 8 – €27.00 – ne Slangenburg [243]

Annus saecularis societatis iesu adumbratus ex anno temporali a gymnasio tricoronato ubiorum... – N.p.: Sumptibus Hermanni Mylii, 1640 – 5mf – 9 – mf#O-1523 – ne IDC [090]

Annus sanctus : hymns of the church for the ecclesiastical year – London, New York: Burns & Oates; New York: Catholic Publ Co [dist] 1884 [mf ed 1986] – 2mf – 9 – 0-8370-7106-2 – (incl ind & app) – mf#1986-1106 – us ATLA [241]

Annus symbolicus divisus in menses 12 : diebus singulis dans curiosas sententias ad animum recreandum... / [Redel, A C] – Augustae: Typis Antonii Nepperschmidii, [1695] – 1mf – 9 – mf#O-1815 – ne IDC [090]

Annus symbolicus, emblemmatice, et versu leonino : quemcumque statum hominum incitans ad animum pie recreandum / Redel, A C – Augustae Vindelicorum: JP Steudner, [1695] – 2mf – 9 – mf#O-1463 – ne IDC [090]

Un ano de accion sindical en la provincia / Anton Crespo, Emilio – Badajoz: Graficas Jimenez, 1964 – sp Bibl Santa Ana [946]

Un ano de accion sindical en la provincia de badajoz / Anton Crespo, Emilio – Imp. Inca, 1966 – sp Bibl Santa Ana [946]

Ano de gobierno, 1950-1951 / Colombia Ministerio De Gobierno – Bogota, Colombia. v1-2. 1951 – 1r – us UF Libraries [972]

Un ano de vida serradillana por un amante de serradilla / Sanchez Rodrigo, Agustin – Serradilla (Caceres): Imp. de el Cronista, s.a. – 9 – sp Bibl Santa Ana [910]

Ano do nego / Almeida, Jose Americo de – Rio de Janeiro, Brazil. 1968 – 1r – us UF Libraries [972]

El ano meteorologico 1879 / Fuertes Acevedo, Maximo – 1880 – 9 – sp Bibl Santa Ana [550]

El ano meteorologico 1881 / Fuertes Acevedo, Maximo – 1882 – 9 – sp Bibl Santa Ana [550]

El ano pedagogico hispanoamericano / Bayle, Constantino – Madrid: Razon y Fe, 1921 – 1 – sp Bibl Santa Ana [370]

Ano terrible del 87 / Pedreira, Antonio Salvador – Mexico City?, Mexico. 1948 – 1r – us UF Libraries [972]

The anointing of the sick in scripture and tradition : with some considerations on the numbering of the sacraments / Puller, Frederick William – 2nd rev ed. London: SPCK 1910 [mf ed 1993] – (= ser Church historical society (series) 77) – 1mf – 9 – 0-524-06358-3 – mf#1990-1541 – us ATLA [220]

Anois – Dublin, Ireland 2 sep 1984-jun 1996 [mf 1986-96] – 1 – (in gaelic) – uk British Libr Newspaper [072]

Anoka herald – Anoka, NE: E H McNeil. v1 n1. mar 6 1903- (wkly) – 3r – 1 – (absorbed: spencer tribune) – us Bell [071]

Anom na bulan na mata asin anom na bulan man na torog nin encantada reina mora caya daing pag-aram na-aquian nin principe d francisco – [Nueva Caceres]: Libreria Mariana [190-?] [mf ed Bloomington IN: Indiana Uni Lib, Preservation Dept 1984] – (= ser Coll...in the bikol language) – 1r – 1 – (aka: reina mora) – us Indiana Preservation [490]

Anonimo *see* Grandes de espana. primera serie

O anonimo – Rio de Janeiro, RJ: Typ do Diario, 04 maio-13 jul 1840 – (= ser Ps 19) – mf#P14,4,21 – bl Biblioteca [321]

Der anonimo morellia (marcanton michiel's notizia d'opere del disegno) / Frimmel, T – Wien, 1888. v1 – 2mf – 9 – mf#O-517 – ne IDC [700]

Anonym / Ahlborn, Luise Jaeger – Stuttgart: Deutsche Verlags-Anstalt, 1889 [mf ed 1993] – (= ser Deutsche romanbibliothek 2. jahrg/6) – 188p – 1 – mf#8459 – us Wisconsin U Libr [830]

Anonym : oder, die papierne welt: schauspiel in fuenf aufzuegen / Gutzkow, Karl – [S.l: s.n, 1858] [mf ed 1993] – 79p – 1 – mf#8668 – us Wisconsin U Libr [820]

Anonym quarterly – v1. 1968-1969 – 1 – us AMS Press [800]

L'anonyme – [Paris] [may 11-12 1871] – (= ser Commune de paris newspapers) – 1 – us CRL [074]

Anonymous letter / Gosse, P H – London, England. 18-- – 1r – us UF Libraries [240]

Anotaciones y documentos sobre la campana del alto / Suarez, Nicolas – Barcelona, Spain. 1928 – 1r – us UF Libraries [972]

Another brownie book / Cox, Palmer – New York: Century, c1890 – 2mf – 9 – mf#17003 – cn Canadiana [830]

Another gospel / Carr, T W – London, England. 1840 – 1r – us UF Libraries [240]

Another horrid massacre – s.l, s.l? . 193-? – 1r – us UF Libraries [978]

Another mother for peace – 1967-75 – (= ser The library of world peace studies) – 4mf – 9 – $105.00 – us UPA [320]

Anotnii *see* Razgovor "pravoslavnago i pashkovtsa o svyashchennom pisanii i predaniyakh tserkovnykh"

Anq – Washington DC 1992+ – 1,5,9 – ISSN: 0895-769X – mf#18628 – us UMI ProQuest [400]

Anquetil, Louis P *see*
- L'esprit de la ligue
- L'intrigue du cabinet
- Louis 14

Anquetil, Louis Pierre *see* Precis de l'histoire universelle

El anrah and abydos / Randall-Maciver, D & Mace, A C – London, 1903 – (= ser Mees 23) – 10mf – 8 – €19.00 – ne Slangenburg [930]

Anregungen fuer kunst, leben und wissenschaft – Leipzig DE, 1856-61 – 3r – 1 – gw Misc Inst [073]

Anregungen zur heilung des weltelends / Szilassy, Gyula, baro – Berlin: Verlag Neues Vaterland, E Berger & Co, 1921 [mf ed 1987] – (= ser Flugschriften des bundes neues vaterland. [n f] 22/23) – 24p – 1 – mf#6929 n22-23 – us Wisconsin U Libr [933]

Anrich, Gustav *see*
- Die anfaenge des heiligenkults in der christlichen kirche
- Das antike mysterienwesen in seinem einfluss auf das christentum
- Martin bucer

Ans: advances in nursing science – Baltimore MD 1978- – 1,5,9 – ISSN: 0161-9268 – mf#12727 – us UMI ProQuest [610]

Ansaldo, Matheo *see*
- Carta edificante del h augustin de valenziaga
- Copia augmentada de la carta de edificacion del v p sebastian de estrada

Ansari – Delhi, India. Oct 1942-Apr 1950 – 14r – 1 – us L of C Photodup [079]

Ansbacher morgenblatt fuer stadt und land – Ansbach DE, 1848-49 – 1 – gw Misc Inst [074]

Ansbachische monatsschrift / ed by Buettner, H E et al – Schwabach 1793-94 – (= ser Dz. historisch-geographische abt) – 3v on 12mf – 9 – €120.00 – mf#k/n1279 – gw Olms [074]

Ain anschalg wie man dem tuercke widerstand thun mag... – n.p, 1522 – 1mf – 9 – mf#H-8137 – ne IDC [956]

Die anschauung augustins ueber christi person und werk : unter beruecksichtigung ihrer verschiedenen entwicklungsstufen und ihrer dogmengeschichtlichen stellung / Scheel, Otto – Tuebingen: J.C.B. Mohr, 1901 – 2mf – 9 – 0-7905-6258-8 – (incl bibl ref) – mf#1988-2258 – us ATLA [240]

Die anschauung vom heiligen geiste bei luther : eine historisch-dogmatische untersuchung / Otto, Rudolf – Goettingen: Vandenhoeck und Ruprecht, 1898 – 1mf – 9 – 0-7905-7993-6 – (incl bibl ref) – mf#1989-1278 – us ATLA [242]

Anschauungen vom wesen deutscher kunst : im selbstzeugnis bildender kuenstler 1800-1860 / Herzog, Hildegard – Jena: E Diederich, [1937?] [mf ed 1993] – (= ser Deutsche arbeiten der universitaet koeln 15) – 74p – (incl bibl ref) – mf#8215 reel 2 – us Wisconsin U Libr [750]

Anschauungsformen in der deutschen dichtung des 18. jahrhunderts : rahmenschau und rationalismus / Langen, August – Jena: E Diederich, 1934 [mf ed 1993] – (= ser Deutsche arbeiten der universitaet koeln 6) – 131p/[3pl] (ill) – 1 – (incl bibl ref) – mf#8215 reel 1 – us Wisconsin U Libr [430]

Anschel, Leo *see* Gemeindeblatt der israelitischen religionsgemeinde dresden

Der anschluss – Vienna, Austria jan 1927-aug 1933 [mf ed Norman Ross] – 1r – 1 – mf#nrp-1938 – us UMI ProQuest [074]

Anseaume, Louis *see* Tableau parlant

Anseis von karthago / ed by Alton, Johann – Stuttgart: Litterarischer Verein, 1892 (Tuebingen: H Laupp, Jr) [mf ed 1993] – (= ser Blvs 194) – 606p – 1 – (incl bibl ref and ind) – mf#8470 reel 40 – us Wisconsin U Libr [810]

Ansell, David Abraham *see* Political generosity

Anselm and his work / Welch, Adam Cleghorn – Edinburgh: T & T Clark 1901 [mf ed 1992] – (= ser The world's epoch-makers) – 1mf – 9 – 0-524-04629-8 – (incl bibl ref) – mf#1990-1289 – us ATLA [240]

Anselm kiefer : historienmalerei nach auschwitz / Fenne, Christina – (mf ed 2000) – 3mf – 9 – €49.00 – 3-8267-2716-9 – mf#DHS 2716 – gw Frankfurter [750]

Anselm of Canterbury, Saint *see* Opera omnia

Anselm, Saint, Archbishop of Canterbury *see*
- The devotions of saint anselm, archbishop of canterbury
- Meditations and prayers to the holy trinity and our lord jesus christ

Anselm, Saint, Archbishop of Canterbury et al *see* Selections from the literature of theism

[Anselme de Sainte Marie] *see* Le palais de l'honneur

Anselmi, S, Cantuariensis Archiepiscopi *see* Opera omnia

Anselmo enterprise – Anselmo, NE: Orin B Winter. v24 n45. feb 1 1940-n45. mar 20 1947 (wkly) – 3r – 1 – (cont: enterprise (anselmo ne); merged with: merna messenger to form: enterprise-messenger (merna ne); iss for aug 6-27 1943 lack vol numbering; vol numbering ceased with nov 2 1945) – us Bell [071]

Anselmo news – Anselmo, NE: C B Whitehead, 1937 (wkly) [mf ed v1 n46. jun 30 1938] – 1r – 1 – (cont: doniphan herald) – us NE Hist [071]

Anselmo, Otacilio *see* Padre cicero, mito e realidade

Anselm's theory of the atonement / Foley, George Cadwalader – New York: Longmans, Green 1909, c1908 [mf ed 1989] – (= ser The bohlen lectures 1908) – 1mf – 9 – 0-7905-2652-2 – (incl bibl) – mf#1987-2652 – us ATLA [240]

Anselms von laon systematische sentenzen : 1. teil: texte / Bliemetzrieder, Fr – 1919 – (= ser Bgphma 18/2-3) – €11.00 – ne Slangenburg [140]

Ansgar lutheran – v1-33. 14 dec 1927-1960 – 17r – 1 – (merged with: lutheran herald to form lutheran standard; lacks some pp) – mf#atla s0288 – us ATLA [242]

Anshe shem / Buber, Solomon – Krakow, Poland. 1895 – 1r – us UF Libraries [939]

Anshelm, V *see* Die berner-chronik des valerius a

Ansichten auf der neuesten reise nach rom / Weidmann, Franz – St Gallen 1821 [mf ed Hildesheim 1995-98] – (= ser Fbc) – 1mf – 9 – €40.00 – 3-487-29227-0 – gw Olms [914]

Ansichten aus den deutschen alpen : ein lehrbuch fuer alpenreisende, ein naturgemaelde fuer alle freunde der natur / Mueller, Karl – Halle 1858 [mf ed Hildesheim 1995-98] – (= ser Fbc) – 3mf – 9 – €90.00 – 3-487-29380-3 – gw Olms [914]

Ansichten der hauptstadt des franzoesischen kayserreichs vom jahre 1806 an / Pinkerton, John – Amsterdam 1807/1808 [mf ed Hildesheim 1995-98] – 2v on 6mf – 9 – €120.00 – 3-487-29648-9 – gw Olms [914]

Ansichten ueber aesthetik und literatur : seine briefe an christian gottfried koerner (1793-1830) / Humboldt, Wilhelm, Freiherr von; ed by Jonas, Fritz – Berlin: L Schleiermacher 1880 [mf ed 1991] – 1r – 1 – (incl ind. filmed with: ricarda huch / gertrud baumer) – mf#2734p – us Wisconsin U Libr [840]

Ansichten und beobachtungen ueber religion und kirche in england / Sack, Karl Heinrich – Berlin: Realschulbuchhandlung, 1818 [mf ed 1991] – 1mf – 9 – 0-524-00598-2 – mf#1990-0098 – us ATLA [240]

Ansichten und umrisse aus den reise-mappen zweier freunde / Elsholtz, Franz von – Berlin [u.a.] 1831 [mf ed Hildesheim 1995-98] – (= ser Fbc) – 2v on 5mf – 9 – €100.00 – 3-487-27762-X – gw Olms [910]

Ansichten von England : aus dem franzoesischen / Pillet, Rene M – Jena 1816 [mf ed Hildesheim 1995-98] – (= ser Fbc) – 3mf – 9 – €90.00 – 3-487-28814-1 – gw Olms [914]

Ansichten von italien : waehrend einer reise in den jahren 1815 und 1816 / Friedlaender, Ludwig H – Leipzig 1819-20 [mf ed Hildesheim 1995-98] – (= ser Fbc) – 2v on 6mf – 9 – €120.00 – 3-487-29300-5 – gw Olms [914]

Ansichten von louisiana : nebst einem tagebuche einer, im jahre 1811, den missouri-fluss aufwaerts gemachten reise / Brackenridge, Henry M – Weimar 1818 [mf ed Hildesheim 1995-98] – 1v on 1mf – 9 – €40.00 – 3-487-26514-1 – gw Olms [917]

Ansichten von paris – Zuerich 1809 [mf ed Hildesheim 1995-98] – 2v on 5mf – 9 – €100.00 – 3-487-29664-0 – gw Olms [914]

Ansichten von paris im jahr 1809 [achtzehnhundertneun] / Uklanski, Carl T von – Berlin 1801 [mf ed Hildesheim: 1995-98] – (= ser Fbc) – 2v on 6mf – 9 – €120.00 – 3-487-29667-5 – gw Olms [914]

Ansikte av sten / Kessle, Gun – Stockholm [Sweden]: PAN/Norstedt [mf ed 1989] – [ill] 1r with other items – 1 – (with bibl) – mf#mf-10289 seam reel 023/06 [§] – us CRL [930]

An-Ski, S *see* Kol kitve s an-ski

Ansley chronicle – Ansley, NE: Tom Wright. -v19 n[24] sep 26 1902 (wkly) – 3r – 1 – (cont: chronicle (ansley ne); merged with: citizen (ansley ne) to form: chronicle-citizen) – us Bell [071]

Ansley chronicle *see* The citizen

The ansley courier – [Callaway, NE: Robert M Jensen] 1v. [v1 n1. may 20 1985]-v1 n30. jan 9 1986 (wkly) [mf ed 1986] – 1r – 1 – (issues for jun 6 and 13 1985 misdated may 27 1985 and misnumbered v1 n2) – us NE Hist [071]

Ansley herald – Ansley, NE: Thomas Wright. 67v. v25 n38. apr 28 1916-v91 n27. mar 1 1985 (wkly) – 12r – 1 – (cont: herald (ansley ne); publ in sargent ne aug 7 1980-sep 24 1981; and in broken bow ne oct 1 1981-mar 1 1985; accompanied by a mthly suppl: magazine of the grasslands jun 1982-feb 1984) – us Bell [071]

Ansley herald – Ansley, NE: Thomas Wright. 67v. v25 n38. apr 28 1916-v91 n27. mar 1 1985 (wkly) [mf ed jul 26 1956-mar 1 1985 (gaps) filmed 1969-85] – 8r – 1 – (publ in sargent, ne aug 7 1980-sep 24 1981; and in broken bow, ne oct 1 1981-mar 1 1985. accompanied by a mthly suppl: magazine of the grasslands, jun 1982-feb 1984. cont by: herald) – us NE Hist [071]

The ansley reporter – Ansley, NE: J C Hargrave. [mf ed v1 n18. apr 8 1887 filmed 1999] – 1r – 1 – us NE Hist [071]

Anson, Adelbert *see*
- A "church farm" in assiniboia, north-west, canada
- The consolidation of the church in canada
- Love for the church
- Our colonies and our church

Anson burlingame and the first chinese mission to foreign powers / Williams, Frederick Wells – New York: Charles Scribner's Sons, 1912 [mf ed 1995] – (= ser Yale coll) – x/370p (ill) – 1 – 0-524-09529-9 – mf#1995-0529 – us ATLA [951]

Anson, Wiliam Reynell *see* Ballads en termes de la ley

Anson, William *see* Principles of the english law of contract and of agency

Ansonian / Darke Co. Ansonia – (may 1926-feb 1948) very scattered [?] – 1r – 1 – mf#B29200 – us Ohio Hist [071]

Anspach, Frederick Rinehart *see* A discourse on systematic benevolence

Anspach, Lewis A *see* A history of the island of newfoundland

Der ansporn – Wittenberg DE, 1955 8 oct-1993 jul [gaps] – 5r – 1 – (stickstoffwerk piesteritz) – gw Misc Inst [600]

Ansprachen fuer christliche muettervereine / Leinz, Anton – 2. verb aufl. Freiburg i.B: Herder [1912?] [mf ed 1986] – 1mf – 9 – 0-8370-7232-8 – mf#1986-1232 – us ATLA [240]

Anstadt, Peter *see*
- Life and times of rev s s schmucker
- Luther, zinzendorf, wesley

Ansted, David Thomas *see* Stars and the earth

Anstey, H *see* Munimenta academica (rs50)

Anstey, Roger *see* King leopold's legacy

Anstey, Thomas Chisholm *see*
- Crime and government at hong kong
- Guide to the laws of england affecting roman catholics

Anstey, Vera *see*
- The economic development of india
- The trade of the indian ocean

Anstruther (Miss) *see* Sweet idolatry

Answer : official newsletter of the bay shore classroom teachers association – 1980 jan-1985 jun – 1 – mf#1363046 – us WHS [370]

Answer – v26 n1-v28 n2 [1980 jan-1982 feb; 1982 mar-apr?] [1]; n66-1993 (1970 2nd qtr-1977 1st qtr) [2] – 1 – mf#626202 [1]; 345714 [2] – us WHS [360]

Answer by her majesty's government to the memorial transmitted to s... / Graham, J R G – Edinburgh, Scotland. 1843? – 1r – us UF Libraries [240]

Answer of a good conscience / Rutherford, James – London, England. 18– – 1r – us UF Libraries [240]

Answer of the great church of constantinople to the papal encyclical on union = Batheos thlibetai / ed by Metallenos, Eustathios – [s.l: s.n.] [1896?] [mf ed 1990] – 1mf – 9 – 0-7905-4670-1 – (in greek with english trans) – mf#1988-0670 – us ATLA [241]

Answer of the rev henry esson to the charges and statements of a committee of the session of st gabriel street church, montreal : with an appendix containing correspondence, evidence in his vindication, etc – Montreal?: s.n, 1832 – 3mf – 9 – mf#50247 – cn Canadiana [242]

Answer to addresses from clergy of the diocese of exeter / Phillpotts, Henry – London, England. 1856 – 1r – us UF Libraries [240]

Answer to dr buchanan's speech in moving his overture – Glasgow, Scotland. 1874 – 1r – us UF Libraries [240]

Answer to dr kidd's appeal to the public / Grammaticus – Aberdeen, Scotland. 1830 – 1r – us UF Libraries [240]

Answer to hugh miller and theoretic geologists / Davies, Thomas Alfred – New York: Rudd & Carlton, 1860 [mf ed 1985] – 1mf – 9 – 0-8370-2843-4 – mf#1985-0843 – us ATLA [210]

Answer to mr binney's reply to "remarks" on his treatise on the habeas corpus / Wharton, George Mifflin – Philadelphia, Campbell, 1862. 8 p. LL-1646 – 1 – us L of C Photodup [340]

An answer to mr dalton's pamphlet on the irish question / Carden, Andrew – Dublin, 1865 – (= ser 19th c ireland) – 1mf – 9 – mf#1.1.1920 – uk Chadwyck [330]

Answer to no 1 of "essays and reviews" / Marshall, John George – Halifax, NS: s.n, 1862 – 1mf – 9 – mf#49650 – cn Canadiana [220]

An answer to richard allen's essay / Claridge, R – London, 1697 – 1 – $5.00 – us Southern Baptist [242]

An answer to sir thomas more's dialogue : the supper of the lord after the true meaning of john 6. and 1 cor. 11. and wm tracy's testament expounded by william tyndale, martyr, 1536 / Tyndale, William; ed by Walter, Henry – Cambridge: University Press, 1850 [mf ed 1986] – 1mf – 9 – 0-8370-7272-7 – (incl ind) – mf#1986-1272 – us ATLA [240]

An answer to some strictures in brown's sequel to campbell's history of yarmouth / Campbell, John Roy – [S.l.]: McMillan, 1889 [mf ed 1980] – 1mf – 9 – 0-665-02019-8 – mf#02019 – cn Canadiana [971]

An answer to the abbe dubois : in which the various wrong principles, misrepresentations, and contradictions, contained in his work, entitled "letters on the state of christianity in india," are pointed out... / Townley, Henry – London: Printed by R Clay, and sold by F Westley, 1824 [mf ed 1995] – (= ser Yale coll) – viii/214p – 1 – 0-524-09242-7 – mf#1995-0242 – us ATLA [240]

Answer to the amended libel / Smith, William Robertson – 2d ed. Edinburgh: David Douglas, 1879. Princeton: Speer Lib, and Dep of Photodup, U of Chicago Lib, 1978 (1r); Evanston: American Theol Lib Assoc, 1984 (1r) – (= ser Case of william robertson smith in the free church of scotland) – 1 – 0-8370-0600-7 – (incl bibl ref) – mf#1984-6288 – us ATLA [240]

Answer to the charge delivered by the lord bishop of lincoln / Eustace, John Chetwode – London, England. 1813 – 1r – us UF Libraries [240]

Answer to the dean of faculty's "letter to the lord chancellor" / Dunlop, Alexander – Edinburgh, Scotland. 1839 – 1r – us UF Libraries [240]

An answer to the difficulties in bishop colenso's book on the pentateuch / Turner, Jonathan Baldwin – London: Rivingtons, 1863 [mf ed 1984] – (= ser Biblical crit – us & gb 50) – 1mf – 9 – 0-8370-0219-2 – mf#1984-1050 – us ATLA [221]

Answer to the form of libel : now before the free church presbytery of aberdeen / Smith, William Robertson – [4th ed.] Edinburgh: David Douglas, 1878. Chicago: Dep of Photodup, U of Chicago Lib, 1978 (1r); Evanston: American Theol Lib Assoc, 1984 (1r) – (= ser Case of william robertson smith in the free church of scotland) – 1 – 0-8370-0642-2 – (incl bibl ref) – mf#1984-6281 – us ATLA [240]

Answer to the form of libel now before the free church presbytery... / Smith, W Robertson – Edinburgh, Scotland. 1878 – 1r – us UF Libraries [242]

Answer to the lord chancellor's question / Robberds, John Gooch – London, England. 1825 – 1r – us UF Libraries [240]

Answer to the protest of the free church – Edinburgh, Scotland. 1846 – 1r – us UF Libraries [240]

An answer to the question, why are you a wesleyan methodist? : to which is added, an examination of a tract entitled "tracts for the people, no 4 – methodism as held by wesley" / Peck, George – 2nd ed. New York: Carlton & Lanahan [18–] [mf ed 1984] – 3mf – 9 – 0-8370-0784-4 – mf#1984-4115 – us ATLA [242]

Answer to the speech of the dean of st paul's against subscription / Napier, Joseph – London, England. 1865? – 1r – us UF Libraries [240]

Answer to the world / Sharp, William – London, England. 1806 – 1r – us UF Libraries [240]

Answer to two letters addressed to the late right hon george canning / Shannon, Richard Q – London, England. 1828 – 1r – us UF Libraries [240]

The answere of mr richard hooker to a supplication preferred by mr walter travers to the hh lords of the privie counsell – Oxford: Joseph Barnes, 1612 – 1mf – 9 – mf#PW-15 – ne IDC [240]

An answere to a certen libel intituled : an admonition to the parliament... / Whitgift, J – London: Henrie Bynneman, 1572 – 4mf – 9 – mf#PW-58 – ne IDC [240]

Answered or unanswered / Vaughan, Louisa – Wichita: Missionary Press, [1917] [mf ed 1995] – (= ser Yale coll) – 128p (ill) – 1 – 0-524-09569-8 – mf#1995-0569 – us ATLA [951]

Answers to cuthbert's exercises in arithmetic, pts 1 and 2 : first, second, third, fourth and fifth classes / Cuthbert, W Nelson – Toronto: Copp, Clark, 1894 – 1mf – 9 – mf#13320 – cn Canadiana [510]

Answers to economic problems : a monthly commentary from the department of economics of northwood institute – v1 n1-v12 n9 [1975 jul-1986 sep] – 1 – mf#1098875 – us WHS [330]

Answers to "essays and reviews" / Marshall, John George – Halifax, NS: s.n, 1862 [mf ed 1983] – 3mf – 9 – 0-665-38223-5 – mf#38223 – cn Canadiana [210]

Answers to everyday questions / Cadman, Samuel Parkes – New York, NY. 1930 – 1r – us UF Libraries [025]

Answers to objections against the catholic religion = Reponses courtes et familieres aux objections les plus repandues contre la religion / Segur, Louis Gaston – Shermerville, IL: Society of the Divine Word [18–] [mf ed 1986] – 1mf – 9 – 0-8370-6837-1 – (in english) – mf#1986-0837 – us ATLA [241]

Answers to prayer – London, England. 18– – 1r – us UF Libraries [240]

Answers to the programmes on teaching and agriculture for elementary school, model school and academy diplomas / Langevin, Jean – 1st english ed. [Quebec?: s.n.] 1864 [mf ed 1984] – 1mf – 9 – 0-665-45450-3 – (also available in french) – mf#45450 – cn Canadiana [370]

Answers to the questions suggested by the regents of the university of the state of new york : for the examination of candidates for the degree of bachelor of laws / Seabury, Samuel – New York, 1894. 36p. LL-1384 – 1 – us L of C Photodup [340]

Answers to virginia bar examinations / Hairston, Samuel W – Richmond, Appeals Press, 1932 514 p. LL-895 – 1 – us L of C Photodup [340]

Ant sad opera theologica / Chandieu, A S – 1614 – 41mf – 8 – €79.00 – ne Slangenburg [240]

Antaisaka / Deschamps, Hubert Jules – Tanarive, Madagascar . 1936 – 1r – us UF Libraries [960]

Antalya – Antalya. Sahib ve Muharriri: Mehmed Emin. n125. 31 kanunisani 1339 [1923] – (= ser O & t journals) – 1mf – 9 – $25.00 – us MEDOC [956]

Antalya'da anadolu – Antalya, 1920-19? Sahib ve Sermuharriri: Haydar Ruestue. n378 7 mart 1338 [1922] 403-405,464,533 12 eyluel 1338 [1922] – (= ser O & t journals) – 1mf – 9 – $25.00 – us MEDOC [956]

Antapodosis. homelia paschalis. historia ottonis. relatio de legatione constantinopolitana : formae tplila 105 / Liudprandus Cremonensis – [mf ed 2002] – (= ser ILL – ser a; Cccm 156)) – 7mf+151p – 9 – €61.00 – 2-503-64562-3 – be Brepols [400]

Antapologia : sive examen atque refutatio totius apologiae remonstrantium... / Trigland, J – Amstelodami, 1664 – 9mf – 9 – mf#PBA-350 – ne IDC [240]

Antara *see*
– Ichtisar tahunan
– Inside indonesia features

Antarctic journal of the united states – Washington DC 1966-96 – 1,5,9 – ISSN: 0003-5335 – mf#6289 – us UMI ProQuest [990]

Antarctic science – Cambridge, England 1989+ – 1,5,9 – ISSN: 0954-1020 – mf#17104 – us UMI ProQuest [550]

Antarctica sun times – v4:iss 3-9 [1990 nov7-dec 21]; v4:iss 11, 13 [1991: jan 11, feb 1]; v5:iss 2-3, 6-8 [1991 nov3-10, dec 1-15]; v5:iss 11-14 [1992 jan 5-26] – 1 – mf#1831860 – us WHS [071]

Ante la conciencia de america / Zelaya, Antonio – San Jose, Costa Rica. 1938 – 1r – us UF Libraries [972]

Ante la crisis del hombre contemporaneo / Valtierra, Angel – Bogota, Colombia. v1-2. 1956 – 1r – us UF Libraries [025]

Ante la pena de muerte / Hoenigsberg, Julio – Barranquilla, Colombia. 1962 – 1r – us UF Libraries [972]

Ante los barbaros / Vargas Vila, Jose Maria – s.l, s.l? . 19– – 1r – us UF Libraries [972]

Ante todo, esposos / Aradillas Agudo, Antonio – Madrid: Editorial La Muralla, 1969 – sp Bibl Santa Ana [240]

Antecedentes del seguro social en guatemala / Garcia Laguardia, Jorge Mario – Guatemala, 1964 – 1r – us UF Libraries [972]

Antecedentes hispano-medioevales de la poesia tradicional argentina / Carrizo, Juan Alfonso – Buenos Aires: Estudios Hispanicos 1945 [mf ed Bloomington IN: Indiana Uni Lib, Preservation Dept 1984] – 1r – 1 – us Indiana Preservation [440]

Antecedentes historicos de la subversion universal / Torrente Ballester, Gonzalo – Barcelona: [Editora Nacional] 1939 – (= ser Cuadernos de orientacion politica) – 9 – mf#w1231 – us Harvard [320]

Antecedentes que debio tener a la vista el marques – New York, NY. 1875 – 1r – us UF Libraries [972]

The antediluvian history, and narrative of the flood : as set forth in the early portions of the book of genesis / Rendell, E D – 2nd rev ed. London: F Pitman, [1864?] – 1mf – 9 – 0-8370-4839-7 – (incl bibl ref and index) – mf#1985-2839 – us ATLA [220]

L'antegnata in tavolatura de ricercari d'organo / Antegnati, Costanzo – 1608 – (= ser Mssa) – 1mf – 9 – €20.00 – mfchl 167 – gw Fischer [780]

Antegnati, Costanzo *see*
– L'antegnata in tavolatura de ricercari d'organo
– L'arte organica

Der anteil des volkes an der messliturgie im frankenreiche / Nickl, Georg – Innsbruck, 1930 – 2mf – 8 – €5.00 – ne Slangenburg [240]

Antelope county eagle – Neligh, NE: Wellman & Leake. v1 n1. jan 19 1881-v1 n25. jul 13 1881 (wkly) [mf ed with gaps filmed 1958] – 1r – 1 – (cont by: neligh advocate) – us NE Hist [071]

Antelope herald – Antelope OR: E M Shutt, 1892- [wkly] [mf ed 1963] – 1r – 1 – (ceased in 1910?) – us Oregon Lib [071]

Antelope Indian Circle Archives *see* Indian archives

Antelope tribune – Neligh, NE: James R Cary. 12v. v8 n29. may 21 1887-v11 n24. mar 11 1898 (wkly) [mf ed with gaps] – 4r – 1 – (cont: neligh republican. absorbed by: yeoman. added new ser numbering with v1 n2 jun 1 1887; dropped old ser numbering foll v14 n5 nov 30 1892, continuing with new ser) – us NE Hist [071]

Antelope valley – 1992- – (= ser California telephone directory coll) – 3r – 1 – $150.00 – mf#P00005 – us Library Micro [917]

[Antelope valley-] antelope valley press – CA. jun 17 1927-1933, jan 1935-aug 1 – 228r – 1 – $13,680.00 – (aka: palmdale reporter) – mf#H03269 – us Library Micro [071]

[Antelope valley-] ledger gazette – CA. 1896-1909 (scats), 1914-15, 1917-28, may 1929-jun 1983 – 134r – 1 – $8040.00 – (aka: antelope valley press) – mf#H04001 – us Library Micro [071]

Antelope valley press *see* [Antelope valley-] ledger gazette

Ante-nicene christianity, a d 100-325 / Schaff, Philip – new rev enl ed. New York: Scribner, 1886 [mf ed 1992] – (= ser History of the christian church 2) – 3mf – 9 – 0-524-03423-0 – (incl bibl ref) – mf#1990-0977 – us ATLA [240]

Antenna – 1981 may 1-1985 dec 20; 1986 jan 10-1987 dec 18; 1988 jan 8-1989 dec 22; 1990 jan 12-1991 dec 20; 1992 jan 10-nov27; 1993 jan 8-dec 17 – 1 – mf#1048991 – us WHS [071]

Antenna – 1991 jan-1994 dec; v35 n6 [1986 jun]-1990 dec – 1 – mf#1053960 – us WHS [071]

Antennes : la revue quebecoise des communications / Quebec. (Province). Ministere des communications. Service des communications – Quebec: le Service. v1 n1 1er trimestre 1976-v6 n21 1er trimestre 1981 [mf ed 1978-82] – 1r – 5 – mf#SEM16P227 – cn Bibl Nat [380]

Anteo / Labrador Raiz, Enrique – Habana, Cuba. 1940 – 1r – us UF Libraries [972]

Ante-proyecto de estatuto organico / Universidad De La Habana – Habana, Cuba. 1934 – 1r – us UF Libraries [972]

Antes, John *see* Tre trii, per due violini and violoncello, op 3

Antevs, E V *see* Results of dr e mjobergs swedish scientific expeditions to australia 1910-13

[Anthem collection] – [16–] [mf ed 19–] – 1r – 1 – (incl anthems by: john blow, pelham humfrey, orlando gibbons, henry purcell) – mf#film 257 – us Sibley [780]

Anthems : from winchester cathedral library / Wesley, Samuel Sebastian – v1. 1853 – 1r – 1 – mf#625 – uk Microform Academic [240]

Anthes, R *see* Die felseninschriften von hatnub nach den aufnahmen georg moellers

Anthoine de Saint-Joseph, Antoine I *see* Historischer versuch ueber den handel und die schiffahrt auf dem schwarzen meere

Anthologia anthropologica / Frazer, James George – London, England. 1938 – 1r – us UF Libraries [301]

Anthologia graeca : sive poetarum graecorum lusus – Lipsiae. v1-12 + index. 1794-1814 – 1 – $162.00 – mf#0054 – us Brook [450]

Anthologia graeca carminum christianorum / Christ, W & Paranikas, M – Lipsiae, 1871 – 7mf – 8 – €15.00 – ne Slangenburg [240]

Anthologie *see* La critique internationale

Anthologie aus den werken von ernst moritz arndt : mit der bibliographie und dem portraet des verfassers – Hildburghausen, New York: Verlag des Bibliographischen Instituts, [18–?] [mf ed 1988] – (= ser National-bibliothek der deutschen classiker 17) – 187p – 1 – mf#6954 – us Wisconsin U Libr [800]

Anthologie aus den werken von johann gottlieb fichte : mit der biographie des verfassers / Fichte, Johann Gottlieb – [Hildburghausen, New York: Verlag des Bibliographischen Instituts 18–?] [mf ed 1988] – (= ser National-bibliothek der deutschen classiker 17) – 1r – 1 – (filmed with: ludwig anzengruber / sigismund friedmann) – mf#6954 – us Wisconsin U Libr [800]

Anthologie des poetes canadiens – Montreal: [s.n.] 1920 [mf ed 1999] – 4mf – 9 – 0-659-91971-0 – mf#9-91971 – cn Canadiana [810]

Anthologie du folklore haitien / Bastien, Remy – Mexico City?, Mexico. 1946 – 1r – us UF Libraries [390]

L'anthologie du folklore musical d'espagne / Garcia Matos, Manuel – Madrid: Hispavor, s.a., 1960 – 1 – sp Bibl Santa Ana [780]

Anthology of african and malagasy poetry in french / Wake, Clive – London, England. 1965 – 1r – us UF Libraries [440]

An anthology of german poetry, 1830-1880 / Bithell, Jethro – New York: Rinehart, 1947 [mf ed 1993] – cviii/211p – 1 – (int in english, poems in german. incl bibl ref) – mf#8350 – us Wisconsin U Libr [810]

An anthology of modern arabic poetry / Megally, S – Np, 1974 – 2mf – 9 – mf#NE-383 – ne IDC [956]

Anthology of modern indian poetry / ed by Goodwin, Gwendoline – London: J Murray, 1927 – (= ser Samp: indian books) – us CRL [810]

Anthology of spanish american literature / International Institute Of Ibero-American Literature – New York, NY. 1946 – 1r – us UF Libraries [440]

Anthony, Alfred Williams *see* An introduction to the life of jesus

Anthony Champagne et al [comp] *see* A directory of oral history interviews related to the federal courts

Anthony. Kansas. Grace Episcopal Church *see* Parish records

Anthony, Katharine Susan *see* Mothers who must earn

Anthony, Ryan M *see* Is fast walking an adequate aerobic training stimulus for male and female cardiac patients?

Anthony, Susan B *see* The papers of elizabeth cady stanton and susan b. anthony

Anthony van dyck : an historical study of his life and works / Cust, Lionel Henry – London 1900 – (= ser 19th c art & architecture) – 11mf – 9 – mf#4.2.1344 – uk Chadwyck [750]

Anthony wayne herald / Lucas Co. Toledo – jan 1982-oct 1991 [wkly] – 6r – 1 – mf#B34195-34200 – us Ohio Hist [071]

Anthony's photographic bulletin for... – New York: E & H T Anthony & Co. v1-32. feb 1870-1901 – 8r – 1 – us CRL [770]

Anthouard, Albert Francois Ildefonse D' *see* Expedition de madagascar en 1895

Anthracnose of the pomelo / Hume, H Harold – Lake City, FL. 1904 – 1r – us UF Libraries [634]

Anthropological institute of new york. journal – Killen TX 1871-72 – 1 – mf#3359 – us UMI ProQuest [301]

Anthropological papers / Modi, Jivanji Jamshedji – Bombay: British India Press, 1911 – (= ser Samp: indian books) – us CRL [305]

Anthropological papers of the american museum of natural history / American Museum of Natural History – New York NY 1981+ – 1,5,9 – ISSN: 0065-9452 – mf#12599 – us UMI ProQuest [301]

Anthropological papers of the american museum of natural history – v12-13; v17-18; v21-22 – 1 – mf#756769 – us WHS [060]

Anthropological, philological, geographical, historical : and other writings original / Keane, Augustus Henry – [London?], 1897 – (= ser 19th c publishing...) – 1mf – 9 – mf#3.1.50 – uk Chadwyck [070]

Anthropological quarterly – Washington DC 1928+ – 1,5,9 – ISSN: 0003-5491 – mf#3296 – us UMI ProQuest [301]

Anthropological records – Berkeley, CA. v1-25. 1937-67 – 1 – $180.00 – mf#0055 – us Brook [301]

Anthropological religion / Mueller, Friedrich Max – London, New York: Longmans, Green, 1892 [mf ed 1990] – (= ser Gifford lectures 1891) – 2mf – 9 – 0-7905-7535-3 – mf#1989-0760 – us ATLA [210]

Anthropological reports / Chinnery, E W P – 1925-30 – 1r – 1 – mf#pmb doc3 – at Pacific Mss [301]

Anthropological reports / Papua Territory – 1921-23 – 1r – 1 – mf#pmb doc303 – at Pacific Mss [301]

Anthropological review – Killen TX 1863-70 – 1 – mf#2777 – us UMI ProQuest [301]

Anthropologie – Oxford, England 1968-80 – 1,5,9 – ISSN: 0003-5521 – mf#3426 – us UMI ProQuest [301]

Anthropologie bolivienne / Chervin, Arthur – Paris: Imprimerie nationale, Librairie H Le Soudier 1907-08 [mf ed Bloomington IN: Indiana Uni Lib, Preservation Dept 1984] – 3v on 1r – 1 – us Indiana Preservation [301]

Die anthropologie der araber im zehnten jahrhundert n chr / Dieterici, Friedrich – Leipzig: JC Hinrichs, 1871 [mf ed 1992] – 1mf – 9 – 0-524-04328-0 – (in german) – mf#1990-3312 – us ATLA [180]

Die anthropologie des apostels paulus und ihre stellung innerhalb seiner heilslehre : nach den vier hauptbriefen / Luedemann, Hermann – Kiel: Universitaets-Buchhandlung, 1872 – 1mf – 9 – 0-8370-4192-9 – (incl bibl ref) – mf#1985-2192 – us ATLA [240]

Anthropology / Kroeber, A L – New York, NY. 1923 – 1r – us UF Libraries [301]

Anthropology / Marett, Robert Ranulph – New York, NY. 1912 – 1r – us UF Libraries [301]

Anthropology and education quarterly – Washington DC 1979+ – 1,5,9 – ISSN: 0161-7761 – mf#12214,02 – us UMI ProQuest [370]

Anthropology and humanism quarterly – Berkeley CA 1990-93 – 1,5,9 – ISSN: 0193-5615 – mf#18229 – us UMI ProQuest [301]

Anthropology and the classics : six lectures / Evans, Arthur; ed by Marett, Robert Ranulph – Oxford: Clarendon Press; New York: H Frowde [dist] 1908 [mf ed 1990] – 1mf – 9 – 0-7905-5822-X – (incl bibl ref) – mf#1988-1822 – us ATLA [450]

Anthropology in north america / Boas, Franz et al – New York: G E Stechert & Co 1915 [mf ed 1990] – 1r [ill] – 1 – (incl bibl ref. filmed with: inbreeding & outbreeding / east, e m) – mf#7436 – us Wisconsin U Libr [301]

Anthropology newsletter – v20 n2-v24 n9 [1979 feb-1983 dec]; v25 n1-v28 n9 [1984-87] – 1 – mf#202269 – us WHS [301]

Anthropology of florida / Hrdlicka, Ales – Deland, FL. 1922 – 1r – us UF Libraries [301]

Anthropology of the syrian christians / Iyer, Anantha Krishna et al – Ernakulam: Cochin Govt Press, 1926 – (= ser Samp: indian books) – us CRL [306]

Anthropology today – Oxford, England 1985+ – 1,5,9 – (cont: royal anthropological institute news) – ISSN: 0268-540X – mf#15021 – us UMI ProQuest [301]

Anthropometry and physical examination / Seaver, Jay W – 1909 – 5mf – 9 – $15.00 – us Kinesology [790]

Anthropomorphe auffassung des gebaeudes und seiner teile : sprachlich untersucht an quellen aus der zeit von 1525-1750 / Brzoska, Maria – Jena: E Diederich, 1931 [mf ed 1993] – (= ser Deutsche arbeiten der universitaet koeln 3) – 70p – 1 – (incl bibl ref) – mf#8215 reel 1 – us Wisconsin U Libr [430]

Anthropomorphism / Newman, Francis William – Ramsgate, England. 1870 – 1r – us UF Libraries [210]

Anthropos – Sankt Augustin, Germany 1978+ – 1,5,9 – ISSN: 0257-9774 – mf#11824 – us UMI ProQuest [400]

Anthroposophus *see* Clerical sketches

Anti darwin : or some reasons for not accepting his hypothesis / Suckling, Horatio John – Twickenham, 1884 – (= ser 19th c evolution & creation) – 3mf – 9 – mf#1.1.1586 – uk Chadwyck [575]

Antia, E E K *see* Karnamak i artakhshir papakan

Anti-achitophel / ed by Jones, Harold W – 1682 – 9 – us Scholars Facs [810]

Antiaircraft Replacement Training Center [TX] *see* Camp wallace trainer

Anti-apartheid alert – v1 n2-v6 n2 [1986 feb-1991: summer] – 1 – mf#1053969 – us WHS [320]

Anti-arminianisme : or the church of englands old antithesis to new arminianisme / Prynne, W – Ed 2. London, 1630 – 5mf – 9 – mf#PW-26 – ne IDC [241]

Anti-arminianisme : the surplice, crosse in baptisme... – n.p., 1622 – 2mf – 9 – mf#PW-27 – ne IDC [242]

Antiaxiomas morales, medicos... / Diez de Leiva, F – Madrid, 1682 – 3mf – 9 – sp Cultura [610]

Anti-Bakke Decision Coalition [US] *see* Seize the time

Anti-bellarminus contractus / Vorstius, C – Hannoviae, 1610 4v – 9mf – 9 – mf#PBA-339 – ne IDC [240]

Antibiotikakonzentration im kieferknochen : eine vergleichende uebersicht der wissenschaftlichen literatur / Sembol, Maryla – (mf ed 2000) – 2mf – 9 – €40.00 – 3-8267-2675-8 – mf#DHS 2675 – gw Frankfurter [617]

Die antibourgeoise aesthetik des jungen baudelaire : untersuchungen zum "salon de 1846" / Oehler, Dolf – Frankfurt a.M., 1975 – 2mf – 9 – 3-89349-795-1 – gw Frankfurter [440]

Anti-bread tax circular – [NW England] Manchester ALS 21 apr 1841-sep 1843 – 1 – uk MLA; uk Newsplan [072]

L'antica mvsica ridotta alla moderna prattica : con la dichiaratione, et con gli essempi de i tre generi, con le loro spetie / Vicentino, Nicola – Roma: A Barre 1555 [mf ed 19–] – 1r / 7mf – 1,9 – mf#film 696 / fiche 857 – us Sibley [780]

O anti-charlatao – Rio de Janeiro, RJ: Typ do Brasil de J J da Rocha, 27 jun-29 ago 1846 – (= ser Ps 19) – mf#P01B,05,13 – bl Biblioteca [610]

Antichrist : including the period from the arrival of paul in rome to the end of the jewish revolution / Renan, Ernest; ed by Allen, Joseph Henry – Boston: Roberts Bros, 1897 [mf ed 1989] – (= ser Beginnings of christian history 4) – 2mf – 9 – 0-7905-2685-9 – (english trans of l'antechrist by ed; incl bibl ref) – mf#1987-2685 – us ATLA [240]

Antichrist / Nietzsche, Friedrich Wilhelm – New York, NY. 1920 – 1r – us UF Libraries [210]

Antichrist : or, the spirit of sect and schism / Nevin, John Williamson – New York: JS Taylor, 1848 [mf ed 1990] – 1mf – 9 – 0-7905-9535-4 – mf#1989-1240 – us ATLA [240]

Antichrist *see* Apokalypse / ars moriendi / biblia pauperum / antichrist / fabel vom kranken loewen / kalendarium und planetenbuecher / historia david (mxt2)

Der antichrist / Preuss, Hans – Berlin: E Runge 1909 [mf ed 1989] – (= ser Biblische zeit- und streitfragen 5/4) – 1mf – 9 – 0-7905-2681-6 – (incl bibl ref) – mf#1987-2681 – us ATLA [220]

Antichrist and other sermons / Figgis, John Neville – London: Longmans, Green, 1913 [mf ed 1990] – 1mf – 9 – 0-7905-5034-2 – mf#1988-1034 – us ATLA [242]

Antichrist dethroned / Cotter, Joseph R – London, England. 1828 – 1r – us UF Libraries [240]

Der antichrist in den vorchristlichen juedischen quellen / Friedlaender, Moriz – Goettingen: Vandenhoeck & Ruprecht, 1901. Chicago: Dep of Photodup, U of Chicago Lib, 1971 (1r); Evanston: American Theol Lib Assoc, 1984 (1r) – 1 – 0-8370-0494-2 – (incl bibl ref) – mf#1984-B279 – us ATLA [240]
Der antichrist in der ueberlieferung des judentums, des neuen testaments und der alten kirche : ein beitrag zur auslegung der apocalypse / Bousset, Wilhelm – Goettingen: Vandenhoeck und Ruprecht, 1895 – 1r – 1 – 0-8370-1510-3 – (incl bibl ref and indexes) – mf#1984-B283 – us ATLA [240]
Der antichrist in der ueberlieferung des judentums, des neuen testaments und der alten kirche *see* The antichrist legend
The antichrist legend : a chapter in christian and jewish folklore = Der antichrist in der ueberlieferung des judentums, des neuen testaments und der alten kirche / Bousset, Wilhelm – London: Hutchinson, 1896 [mf ed 1989] – 1mf – 9 – 0-7905-3312-X – (incl bibl ref. english by wilhelm bousset) – mf#1987-3312 – us ATLA [230]
Anti-christian cults : an attempt to show that spiritualism, theosophy, and christian science are devoid of supernatural powers and are contrary to the christian religion / Barrington, Arthur H – Milwaukee, WI: Young Churchman, c1898 [mf ed 1991] – 1mf – 9 – 0-524-00686-5 – mf#1990-2014 – us ATLA [290]
Antichristus / Gwalther, R – Zuerich, Froschouer, 1546/1547 – 3mf – 9 – mf#PBU-291 – ne IDC [240]
Antichristvs, hoc est dispvtatio lenis et perspicva de anti-christo... / Wolf, J – Tigvri, Ioannes Vvolph, 1592 – 1mf – 9 – mf#PBU-661 – ne IDC [240]
Un antico catalogo greco de' romani pontefici inedito / Mercati, Giovanni – Roma: Tipografia vaticana, 1891 – 1mf – 9 – 0-8370-8362-1 – mf#1986-2362 – us ATLA [240]
Anticolonialismo, marxismo y portugal / Pattee, Richard – Mexico City?, Mexico. 1967 – 1r – us UF Libraries [960]
Anti-corn law circular – Manchester ALS 16 apr 1839-1841 – 1 – uk MLA; uk Newsplan [072]
Anti-corrosion methods and materials – Bradford, England 2001+ – 1,5,9 – ISSN: 0003-5599 – mf#31581 – us UMI ProQuest [660]
Anticosti Co *see* Prospectus
Anticosti en 1900 / Baillairge, Charles – [S.l: s.n, 1900?] [mf ed 1980] – 1mf – 9 – 0-665-02374-X – mf#02374 – cn Canadiana [917]
Anti-darwinism / M'cann, Jas – Glasgow, Scotland. 1869 – 1r – us UF Libraries [230]
The antidote – Montreal: [s.n., 1892-1893] – 9 – ISSN: 1190-7266 – mf#P04066 – cn Canadiana [420]
The antidote – or Protestant Guardian. Dublin. Ireland. -w. 25 Jan 1823, 24 Jan, 14 Feb, 24 Apr, 12 Jun-9 Oct 1824, 22 Jan-2 Apr 1825. (22 ft) – 1 – uk British Libr Newspaper [072]
Antidote to deism.. / Ogden, Uzal – 2v. 1795 – 1 – $50.00 – us Presbyterian [210]
The antidote to dr ryerson's scriptural rights, etc in two parts : n1 – relating to children; n2 – do. to adults: shewing the error of the positions on which his assumption is founded, that attendance at class meeting is not a proper condition of membership in the wesleyan methodist church / Wilkinson, Henry – London, ON?: s.n, 1855 (London, C W Ont: H E Newcombe) – 1mf – 9 – mf#34129 – cn Canadiana [242]
Antidote to rev h j van dyke's pro-slavery discourse : "american slavery has no foundation in the scriptures"; delivered in the m e church, mount vernon, new york, on sunday, jan 13 1861 / Boole, William H – New York: E Jones, Printers 1861 [mf ed 2005] – 1r [complete] – 1 – 0-524-10514-6 – mf#b00727 – us ATLA [221]
Antidote to the errors of universalism : or, a scriptural and common sense review of modern universalism. together with strictures on restorationism as contained in rev. j.m. austin's review of "universalism another gospel" / Winfield, Aaron Burr – Auburn, NY: Derby, Miller, 1850 [mf ed 1992] – 1mf – 9 – 0-524-06940-9 – mf#1990-3566 – us ATLA [240]
Antidote to the poison of popery : in the writings and conduct of professors nevin and schaff... / Janeway, Jacob Jones – New Brunswick, NJ: J Terhune, 1856 [mf ed 1991] – 1mf – 9 – 0-524-01227-X – mf#1990-0366 – us ATLA [241]
Antidotvm contra impivm et blasphemvm dogma matthiae flacii illyrici / Hesshusen, T – Ienae, 1572 – 5mf – 9 – mf#TH-1 mf 662-666 – ne IDC [242]
Antier, Benjamin *see* Femmes
Antietam times – v3 n4-7 [1990 aug 1-nov1]; v4 n2, 5 [1991 feb 1, may] – 1 – mf#1789391 – us WHS [071]
Anti-evolution : girardeau vs woodrow / Martin, James L – [s.l: s.n, 1888] [mf ed 1985] – 1mf – 9 – 0-8370-4298-4 – mf#1985-2298 – us ATLA [210]
Antifaschistische front = Le front antifasciste – Kopenhagen (DK), Paris (F), 1933 12 mar-14 sep – 1r – 1 – (cont: weltfront gegen imperialistischen krieg [...], paris) – gw Misc Inst [934]
Antifonario visigotico... / Brou, Louis & Vives, D Jose – Madrid: Archivo Ibero Americano, 1960 – 1 – sp Bibl Santa Ana [946]
Antifonario visigotico mozarabe de la catadral de leon / Brou y J Vives, L – MHS. Barcelona. v1. 1959 – €32.00 – ne Slangenburg [241]
The anti-foreign riots in china in 1891 : with an appendix – Shanghai: North China Herald, 1892 – 1r – 1 – 0-8370-1506-5 – mf#1984-B370 – us ATLA [951]
Anti-gallican : or standard of british loyalty, religion and liberty – La Salle IL 1804 – 1 – mf#4608 – us UMI ProQuest [240]
Antigo banner – 1919 dec 5-1921 dec 21; 1922 jan 1-1925 apr 10; 1925 apr 17-1928 aug 25; 1928 aug 31-1931 dec 31; 1932 jan 1-1937 jul 15 – 1 – mf#916522 – us WHS [071]
Antigo daily journal – 1924 apr 10-1976 may-jun [with gaps] – 1 – mf#1124532 – us WHS [071]
Antigo herald – 1919 dec-20; 1921-23 dec 21 – 1 – mf#916793 – us WHS [071]
Antigo herold – 1901 sep 3-1902; 1903-05; 1906-08; 1909-1912 may; 1912 jun-14; 1915-17; 1918-1919 nov 28 – 1 – mf#916759 – us WHS [071]
Antigo journal – 1913 nov 21-14 nov 13; 1914 nov 20-1916 nov 24; 1916 dec 1-19 aug 1; 1919 aug 8-22 mar 10; 1922 mar 17-1923 dec 28 – 1 – mf#961336 – us WHS [071]
Antigo journal, the antigo republican – 1911 sep 29-1913 mar 14; 1913 mar 21-nov 14 – 1 – mf#961334 – us WHS [071]
Antigo Milk Products Co-operative *see* Dairy topics
Antigo republican – 1889 feb 7-1890 may 15-1910 aug 25-11 sep 14 [with gaps] – 1 – mf#961328 – us WHS [071]
Antigone : d'apres sophocle / Chancerel, Leon – Paris, France. 1941 – 1r – us UF Libraries [450]
Antigone : tragoedie in 5 akten / Hasenclever, Walter – 9. aufl. Berlin: P Cassirer 1919 [mf ed 2001] – 1r – 1 – (filmed with: aus der schmiede des glücks / otto hartwich) – mf#10557 – us Wisconsin U Libr [820]
[Antigono] : dramma per musica rappresentarsi nel regio electtoral teatro all corte di dresda nel carnovale dell'anno 1744 / Hasse, Johann Adolf – [n.p. 177-?] [mf ed 19–] – 1r – 1 – mf#pres. film 214 – us Sibley [780]
Antigrapheus sive conscientia hominis coram s.s.mo maximiliano : electore bavaro illustrata / Drexelius, H – Coloniae Agrip: Apud Jodocum Ralcovium, 1655 – 2mf – 9 – mf#O-1563 – ne IDC [090]
La antigua biblioteca jesuitica de cordoba / Cabrera, Pablo – Cordoba (Argentina), 1930; Madrid: Razon y Fe, 1932 – 1 – sp Bibl Santa Ana [020]
Antigua certa de hermandad entre plasencia y talavera / Berjano, Daniel Escobar – Madrid: Tip. Fortanet, 1899 – sp Bibl Santa Ana [946]
Antigua herald and gazette – St Johns, Antigua 6 sep 1839, 10 jan-3 apr 1840, 29 jan 1847-9 dec 1848 – 1r – 1 – (several iss are mutilated) – uk British Libr Newspaper [079]
Antigua Laws, Statutes, Etc *see* Revised laws of antigua
Antigua magnet – St John's, Antigua 4 jan 1930-31 dec 1932, 1 aug 1933-30 jul 1938, 3 jan-24 nov 1939, 10-26 feb 1940 – 9r – 1 – (several iss are mutilated) – uk British Libr Newspaper [079]
Antigua news notes – St John's, Antigua 15 mar 1909-28 jan 1911 (imperfect) – 2r – 1 – (discontinued) – uk British Libr Newspaper [079]
Antigua observer – St Johns, Antigua 30 nov 1848, 9 dec 1870-27 dec 1888, 2-30 may, 4-25, 28 sep 1889-11 jun 1903 – 10 1/2r – 1 – uk British Libr Newspaper [079]
Antigua sirena / Tapia Y Rivera, Alejandro – Mexico City?, Mexico. 1959 – 1r – us UF Libraries [972]
Antigua standard – St Johns, Antigua 2 jul 1883-5 mar 1890, 5 jul-27 dec 1890, 3 jan 1891-22 mar 1902, 19 jul-27 dec 1902, 3 jan 1903-11 jul 1908 (imperfect) – 1 – (discontinued) – uk British Libr Newspaper [079]
Antigua star – St John's, Antigua 1 apr 1937-30 jul 1938, 4 jan-24 nov 1939, 10-26 feb 1940 – 3r – 1 – uk British Libr Newspaper [079]
Antigua. Statistics Division *see* Statistical yearbook 1975-1976
Antigua times – St Johns, Antigua 8 aug 1863; 5,12 mar 1864; 8 oct 1870; 10 dec 1870-27 jun 1883 – 1 – (discontinued) – uk British Libr Newspaper [079]
Antigua weekly register – St John's, Antigua 29 aug 1848; 8 may 1860; 8 jan 1861; 8 mar, 19 jul, 9 aug, 15 nov 1864; 31 jan 1865; 13 dec 1870-23 nov 1875; 11 jun 1878-26 dec 1882 – 1 – (discontinued; cont: weekly register (ns): 11 dec 1838, 14 may 1839-14 apr 1840) – uk British Libr Newspaper [079]
Antiguas culturas mexicanas / Krickeberg, Walter – Mexico City?, Mexico. 1961 – 1r – us UF Libraries [972]
Antiguas epigrafes de tanger, jerez y arcos de la frontera (merida) / Fita, Fidel – Madrid: Tip. Fortanet, 1896 – 1 – sp Bibl Santa Ana [946]
Las antiguas ferias de medina del campo / Espejo, Cristobal – Vallodolid, 1908 – 1 – us CRL [930]
Antiguas historias de los indios quiches de guatem... – Mexico City?, Mexico. 1965 – 1r – us UF Libraries [972]
Antiguedad, Alfredo R *see* Jose antonio en la carcel de madrid [del 14 de marzo al 6 de junio de 1936]
Antiguedad y limites del obispado de coria : nuevo estudio / Escobar Prieto, Eugenio – Madrid: Tip. Fortanet, 1912 – sp Bibl Santa Ana [946]
Antiguedades cacerenas / Ramon y Fernandez, Jose – Valladolid, 1944-45 – 1 – sp Bibl Santa Ana [946]
Antiguedades de espana propugnadas en las noticias de sus reyes y condes... / Berganza, F – Madrid, 1719 – 24mf – 9 – sp Cultura [930]
Antiguedades de extremadura / Viu, Jose de – 1852 – 9 – sp Bibl Santa Ana [930]
Las antiguedades de las ciudades de espana, tomo 9 / Morales, Ambrosio – Madrid: Benito Cano, 1792 – 1 – sp Bibl Santa Ana [946]
Antiguedades de merida / Forner y Segarra, Agustin F – 1893 – 9 – sp Bibl Santa Ana [930]
Antiguedades de merida recientemente remitidas al museo arqueologico nacional por la comision de monumentes de aquella ciudad / Rada Delgado, Juan de Dios de la – 1 – sp Bibl Santa Ana [930]
Antiguedades de torrecillas (alcuescar) / Sanguino y Michel, Juan – Madrid: Fortanet, 1911. B.R.A.H. 59, 1911, pp. 439-456 – sp Bibl Santa Ana [946]
Las antiguedades de...ciudades de espana / Morales, Antonio y otros – 1712 – 9 – sp Bibl Santa Ana [946]
Antiguedades extremenas : la audiencia territorial de extremadura / Duarte Insua, Lino – Badajoz Tip. La Alianza, 1935 – 1 – sp Bibl Santa Ana [946]
Antiguedades neogranadinas / Uricoechea, Ezequiel – Bogota, Colombia. 1936 – 1r – us UF Libraries [972]
Antiguedades romanas de alcuescar / Comision de Monumentos de Caceres – Madrid: Tip. de Fortanet, 1900 – sp Bibl Santa Ana [946]
Antiguedades romanas del cortijo de las virgenes, cerca de baena / Sanguino y Michel, Juan – Madrid: Fortanet, 1913. B.R.A.H. 62, pp. 483-486 – sp Bibl Santa Ana [946]
Antiguedades y santos...de alcanatara / Arias de Quintanaduenas, Jacinto – 1661 – 9 – sp Bibl Santa Ana [946]
Los antiguos diputados de cuba – Habana, Cuba. 1979 – 1r – us UF Libraries [079]
Anti-haeckel : eine replik nebst beilagen / Loofs, Friedrich – Halle a. S: Max Niemeyer, 1900 [mf ed 1985] – 1mf – 9 – 0-8370-4175-9 – (incl bibl ref) – mf#1985-2175 – us ATLA [230]
Anti-higher criticism : or, the testimony to the infallibility of the bible / Osgood, Howard et al; ed by Munhall, Leander Whitcomb – New York: Hunt & Eaton, 1894, c1893 [mf ed 1985] – 1mf – 9 – 0-8370-4540-1 – mf#1985-2540 – us ATLA [220]
La antihistoria extremena / Munoz de San Pedro, Miguel – Badajoz: Imp. de la Diputacion Provincial, 1969 – sp Bibl Santa Ana [946]
Antihumanismus in der westdeutschen literatur : situation und alternative / Reinhold, Ursula – 1. aufl. Berlin: Dietz, 1971 [mf ed 1993] – 244p – 1 – (incl bibl ref) – mf#8281 – us Wisconsin U Libr [430]
Anti-infidel – La Salle IL 1831 – 1 – mf#4701 – us UMI ProQuest [073]
Antiischaemische und haemodynamische effekte nach akuter und einwoechiger gabe von ramipril im vergleich mit plazebo und isosorbiddinitrat bei patienten mit koronarer herzkrankheit / Kroneisen, Antonia – Mainz: Gardez, 1996 (mf ed 1996) – (= ser Neue medizinische Bibliothek) – 2mf – 9 – €31.00 – 3-8267-9659-4 – mf#DHS 9659 – gw Frankfurter [615]
Anti-jacobin : or weekly examiner – La Salle IL 1797-98 – 1 – mf#5231 – us UMI ProQuest [320]
The anti-jacobin : or, weekly examiner – n1-36. 1797-98 – 1 – us AMS Press [073]
Antijacobin review and protestant advocate : or, monthly political, and literary censor – La Salle IL 1798-1821 – 1 – mf#4200 – us UMI ProQuest [242]
Anti-janus : an historico-theological criticism of the work, entitled "the pope and the council" by janus / Hergenruether, Joseph – Dublin: WB Kelly; New York: Catholic Pub Soc, 1870 [mf ed 1986] – 1mf – 9 – 0-8370-9246-9 – (english trans & int by james burton robertson. incl bibl ref & ind. german version also available isbn: 0-8370-9245-0 [mf ed 1986]) – mf#1986-3246 – us ATLA [241]
Antijovio / Jimenez De Quesada, Gonzalo – Bogota, Colombia. 1952 – 1r – us UF Libraries [972]
L'anti-juif – Paris. 11 aout 1898-5 avr 1903, 1899-1900, inc – 1 – fr ACRPP [939]
L'antijuif : organe de la ligue antisemitique de france – Paris, France 21 aug 1898-2 nov 1902 – 2 1/2r – 1 – uk British Libr Newspaper [321]
Antike fluchtafeln / Wuensch, Richard [comp] – Bonn: A Marcus & E Weber 1907 [mf ed 1993] – (= ser Kleine texte fuer theologische vorlesungen und uebungen 20) – 1mf – 9 – 0-524-06234-X – (texts in greek & latin. discussion in german) – mf#1991-0027 – us ATLA [450]
Antike heilungswunder : untersuchungen zum wunderglauben der griechen und roemer / Weinreich, Otto – Giessen: A Toepelmann, 1909 [mf ed 1992] – (= ser Religionsgeschichtliche versuche und vorarbeiten 8/1) – 1mf – 9 – 0-524-02329-8 – (incl bibl ref) – mf#1990-2952 – us ATLA [250]
Antike jesus-zeugnisse / Aufhauser, Johannes Baptist – Bonn: A Marcus & E Weber, 1913 [mf ed 1992] – (= ser Kleine texte fuer vorlesungen und uebungen 126) – 1mf – 9 – 0-524-05427-4 – (in german, latin, greek & hebrew) – mf#1990-1459 – us ATLA [240]
Antike jesus-zeugnisse (kit126) / Aufhauser, Johannes Baptist – Bonn, 1913 – 1mf – 8 – €4.00 – ne Slangenburg [240]
Das antike mysterienwesen in seinem einfluss auf das christentum / Anrich, Gustav – Goettingen: Vandenhoeck & Ruprecht, 1894 [mf ed 1989] – 1mf – 9 – 0-7905-4243-9 – (incl bibl ref) – mf#1988-0243 – us ATLA [250]
Antike und antikes lebensgefuehl im werke gerhart hauptmanns / Voigt, Felix Alfred – Breslau: Maruschke & Berendt 1935 [mf ed 1990] – (= ser Deutschkundliche arbeiten. b, schlesische reihe 5) – 1r – 1 – (filmed with: gerhart hauptmann : kritische studien / artur kutsche et al) – mf#2702 – us Wisconsin U Libr [430]
Antike und christentum / Doelger, Frans J – Muenster. v1-6. 1930-1940/50 – 6v on 50mf – 8 – €95.00 – ne Slangenburg [230]
Antikhrist : petr i aleksei / Merezhkovskii, D – 1905 – 9mf – 8 – mf#R-748 – ne IDC [947]
Antikhrist : petr i aleksei / Merezhkovskii, D – 1906 – 11mf – 8 – mf#R-749 – ne IDC [947]
Antikomintern – Berlin DE, 1936-39 [gaps] – 1 – gw Misc Inst [320]
Die antikriegsaktion – Paris (F), 1933 aug-sep – 1r – 1 – gw Misc Inst [934]
Antilaicismo / Goma Tomas, Isidro – Madrid: Razon y Fe, 1935 – 1 – (2v barcelona, 1935) – sp Bibl Santa Ana [946]
Antilegomena : die reste der ausserkanonischen evangelien und urchristlichen ueberlieferungen – Giessen: J Ricker, 1901 [mf ed 1989] – 1mf – 9 – 0-7905-1790-6 – (in greek, german & latin. incl bibl & ind) – mf#1987-1790 – us ATLA [225]
Antilia / Balen, Willem Julius Van – Amsterdam, Netherlands. 1935 – 1r – us UF Libraries [972]
Antillas / Corton, Antonio – Barcelona, Spain. 1898 – 1r – us UF Libraries [972]
Antilles : filles de france / Oulie, Marthe – Paris, France. 1935 – 1r – us UF Libraries [972]
Antilles : la france, le monde francais / Lasserre, Guy – Caen, France. 1961 – 1r – us UF Libraries [972]
Les antilles – Saint Pierre, Martinique. 1872-1901 (1) – mf#67949 – us UMI ProQuest [079]
Les antilles francaises, particulierement la guadeloupe : depuis leur decouverte jusqu'au 1er janvier 1823 / Boyer-Peyreleau, Eugene – Paris 1823 [mf ed Hildesheim 1995-98] – 3v on 10mf – 9 – €100.00 – ISBN-10: 3-487-26938-4 – ISBN-13: 978-3-487-26938-2 – gw Olms [918]
Antillon, A *see* Antro fuego
Antillon, Isidoro de *see* Geographie physique et politique de l'espagne et du portugal
Antilogia papae : hoc est, de corrupto ecclesiae statu, et totius cleri papistici peruersitate / [Flacius Illyricus d A, M] – Basileae, [1555] – 9mf – 9 – mf#TH-1 mf 453-461 – ne IDC [242]
Antilutherus...tres libros complectens / Clichtove, J – Parisiis, 1524 – 4mf – 9 – mf#CA-80 – ne IDC [241]

The anti-lynching campaign, 1912-1955 – 2ser – (= ser Papers of the naacp 7) – 1 – (ser a: anti-lynching investigative files, 1912-53 30r isbn 0-89093-971-3 $5810. ser b: anti-lynching legislative & publicity files, 1916-55 35r isbn 0-89093-972-1 $6775. with p/g) – us UPA [322]

Antilynching: hearings:...january 29, 1920 / U.S. Congress. House. Committee on the Judiciary – Washington, Govt. Print. Off., 1920. 65 p. LL-1359 – 1 – us L of C Photodup [340]

Anti-masonic review and magazine – La Salle IL 1828-30 – 1 – mf#3717 – us UMI ProQuest [366]

Anti-methodist publications issued during the eighteenth century : a chronologically arranged and annotated bibliography of all known books and pamphlets written in opposition to the methodist revival during the life of wesley / Green, Richard – London: pub...by CH Kelly, 1902 [mf ed 1990] – 1mf – 9 – 0-7905-8035-7 – mf#1988-6016 – us ATLA [242]

Antimicrobial agents and chemotherapy – Washington DC 1972+ – 1,5,9 – ISSN: 0066-4804 – mf#6608 – us UMI ProQuest [576]

Anti-modernisteneid, freie forschung und theologische fakultaeten : mit anhang, der anti-modernisteneid, lateinisch und deutsch, nebst aktenstuecken / Mulert, Hermann – Halle (Saale): Verlag des Evangelischen Bundes, 1911 [mf ed 1990] – 1mf – 9 – 0-7905-6308-8 – mf#1988-2308 – us ATLA [241]

Antimon auf si (113) : ein surfaktant auf einer thermisch stabilen oberflaeche / Wolff, Gunter – (mf ed 1997) – 1mf – 9 – €30.00 – 3-8267-2436-4 – mf#DHS 2436 – gw Frankfurter [530]

Anti-monarchist and republican watchman – 1808 dec 21; 1809 jan 25, oct 11 – 1 – mf#857194 – us WHS [071]

Anti-monopolist – 1873 oct – 1 – mf#3177689 – us WHS [071]

The anti-monopolist – Grand Island, NE:Grand Island Pub Co. v1 n1. jan 3-nov 1883// (wkly) [mf ed jan 3-oct 31 1883 (gaps)] – 1r – 1 – (absorbed by: grand island independent. some articles in german) – us NE Hist [071]

Anti-napoleonische pamphlete : politische schriften aus den freiheitskriegen 1813-15 / ed by Schoewerling, Rainer & Steinecke, Hartmut – Hildesheim 1996 [mf ed Hildesheim 1995-98] – 417mf – 9 – diazo €1640.00 silver €1798.00 – gw Olms [943]

Antinationaux / Janvier, Louis Joseph – Paris, France. 1884 – 1r – us UF Libraries [972]

Der anti-neckerj h mercks und der minister fr k v moser : ein beitrag zur beurteilung j h mercks / Loebell, Richard – Darmstadt: A Klingelhoeffer 1896 [mf ed 1990] – 1r – 1 – (filmed with: f l w meyer / curt zimmermann & other titles) – mf#2834p – us Wisconsin U Libr [943]

Antingen-eller : en roest till det swenska zion med anledning af striden om gud, christus och foersoningen / Moeller, Christian – Chicago: Engberg & Holmberg 1877 [mf ed 1993] – 1mf – 9 – 0-524-06433-4 – mf#1991-2555 – us ATLA [240]

Antinomianism explained, exposed, and exploded / Hopwood, W – London, England. 1822 – 1r – us UF Libraries [240]

Antinomianism in the colony of massachusetts bay, 1636-1638 : including the short story and other documents / ed by Adams, Charles Francis – Boston: Prince Society, 1894. Chicago: Dep of Photodup, U of Chicago Lib, 1968 (1r); Evanston: American Theol Libr Assoc, 1984 (1r) – (= ser Publications Of The Prince Society) – 1 – 0-8370-0507-8 – (incl bibl ref and ind) – mf#1984-B079 – us ATLA [975]

Antioch baptist church – Darlington Co, SC. 1830-1895, 1956-1975 – 1r – 1 – $17.01 – (378p) – us Southern Baptist [242]

Antioch baptist church – Cherokee Co, SC. 1815-1959 – 1 – $29.61 – us Southern Baptist [242]

Antioch baptist church – Chicago. 1967-1979 (1) 1971-1979 (5) 1976-1979 (9) – 1 – $17.01 – mf#6499 – us Southern Baptist [242]

Antioch baptist church – Chri Bibb Co, AL. 1833-69 – 1 – $15.03 – us Southern Baptist [242]

Antioch baptist church – Enoree, SC. 1835-1970 – 1 – $41.49 – us Southern Baptist [242]

Antioch baptist church – Fairfield, TX. 1906-1972 – 1 – $25.56 – us Southern Baptist [242]

Antioch baptist church – Edgefield Co, SC. 520p. 1830-1982 – 1 – $23.40 – (incomplete) – us Southern Baptist [242]

Antioch baptist church – Lafayette, AL. 1835-1951 – 1 – $42.30 – us Southern Baptist [242]

Antioch baptist church – Johnson City, TN. 1875-1987 – 1 – $40.37 – (minutes, 1952-87. history/other, 1875-1986) – us Southern Baptist [242]

Antioch baptist church – Monroe, NC. oct 1892-nov 1916 – 1 – $7.38 – us Southern Baptist [242]

Antioch baptist church – Orangeburg Co, SC 1867-80 – 1 – $11.07 – us Southern Baptist [242]

[Antioch-] ledger dispatch – CA. 1870-oct 1998 – 321r – 1 – $19,260.00 – (cont: ledger; weekly ledger; became pt of: the contra costa times in 1998) – mf#BC02019 – us Library Micro [071]

Antioch missionary baptist church – ROSEBUD, IL. 28 May 1864-Mar 1972 – 1 – $30.69 – us Southern Baptist [242]

Antioch record series / Greene Co. Yellow Spring – jul 1964-jun 1979 [wkly] – 10r – 1 – mf#B2244-2253 – us Ohio Hist [378]

Antioch review – Yellow Springs OH 1941+ – 1,5,9 – ISSN: 0003-5769 – mf#1041 – us UMI ProQuest [410]

[Antioch-] the county paper – Antioch, CA. 1898-99 – 1r – 1 – $60.00 – mf#B02018 – us Library Micro [071]

Antioche paienne et chretienne / Festugiere, A J – Paris, 1959 – 14mf – 8 – €27.00 – ne Slangenburg [240]

Anti-opium : or, things to think on for his royal highness the prince of wales and every british senator, on india, opium, and china / lota – London, 1873 – (= ser 19th c british colonization) – 1mf – 9 – mf#1.1.7077 – uk Chadwyck [330]

Antioquia medica – Medellin, Colombia 1972-73 – 1 – ISSN: 0044-8389 – mf#8112 – us UMI ProQuest [610]

Anti-papa / Jack, Thomas Godfrey – London, England. 189-? – 1r – us UF Libraries [240]

The antipapal tracts of the fourteenth century / Schaff, David S – [S.l.: s.n.], 1901 – 1mf – 9 – 0-8370-7826-1 – mf#1986-1826 – us ATLA [240]

Antipas, F D *see* Coming king

Antipas, son of chuza : and others whom jesus loved / Houghton, Louise Seymour – New York: Anson D F Randolph, c1895 [mf ed 1985] – 1mf – 9 – 0-8370-3673-9 – mf#1985-1673 – us ATLA [830]

The anti-pelagian works of saint augustine, bishop of hippo – Edinburgh: T & T Clark, 1872-76 [mf ed 1985] – (= ser The works of aurelius augustine, bishop of hippo 4,12,15) – 3v on 4mf – 9 – 0-8370-2516-8 – (in english) – mf#1985-0516 – us ATLA [240]

Antiphonae, sev sacrae cantiones : qvae in totivs anni vesperarvm ac completorii solemnitatibus decantari solent / Anerio, Giovanni Francesco – Romae: Io. Baptistam Roblectum 1613 [mf ed 1989] – 3v on 1r – 1 – mf#pres. film 68 – us Sibley [780]

Antiphonale diurnum – 1625 – (= ser Mssa) – 10mf – 9 – €110.00 – mfchl 82 – gw Fischer [780]

Antiphonale missarum sextuplex / ed by Hesbert, R J – Bruxelles, 1935 – 14mf – 8 – €27.00 – ne Slangenburg [241]

Antiphonarium (cima37) : farbmikrofiche-edition der handschrift karlsruhe, badische landesbibliothek, aug perg 60 – (mf ed 1995) – (= ser Codices illuminati medii aevi (cima) 37) – 10 color mf – 15 – €385.00 – 3-89219-037-2 – (int & description by hartmut moeller. with app: verzeichnis der gesangsinitien) – gw Lengenfelder [090]

Antiphonarium juxta breviarium romanum : ex decreto sacro-sancte concilii tridentini – Parisus: apud Ludovicum Sevestre, 1668 [mf ed 1988] – 1r – 1 – (with ind) – mf#SEM35P291 – cn Bibl Nat [241]

Antiphonarium mozarabicum de la catedral de leon – Leon, 1928 – 11mf – 8 – €21.00 – ne Slangenburg [241]

Antiphonarium proprium : no[n]nullaq[ue] quoru[m]dam sanctoru[m] noua officia... = Antiphonary – [Venetijs impressu[s] sumtib[us] nobilis viri d[omi]ni. luce antonij de giu[n]ta flore[n]tini...1523] [mf ed 19–] – 1r – 1 – mf#film 1895 – us Sibley [780]

Antiphonarium romanum de tempore et sanctis – 1695 – (= ser Mssa) – 9mf – 9 – €105.00 – mfchl 83 – gw Fischer [780]

Antiphonarium seu magnus liber organi de gradali et antiphonario (cima45) : color microfiche edition of the manuscript firenze, biblioteca medicea laurenziana, plut 29.1 – (mf ed 1996) – (= ser Codices illuminati medii aevi (cima) 45) – 42p on 15 color mf – 15 – €490.00 – 3-89219-045-3 – (int by edward h roesner) – gw Lengenfelder [090]

[Antiphonarium speciale augustense] antiphonarii opusculum speciale impensis christophori thum civis august – Basilea: Jacob de Pforzheim 1511 – (= ser Hqab. literatur des 16. jahrh.) – 2mf – 9 – €30.00 – mf#1511 – gw Fischer [780]

The antiphonary of bangor (hbs4) : an early irish ms in the ambrosian library at milan, pt 1 / Warren, F E – 1893 – (= ser Henry bradshaw society (hbs)) – 6mf – 8 – €14.00 – ne Slangenburg [241]

The antiphonary of bangor, pt 2 (hbs10) / Warren, G – 1895 – (= ser Henry bradshaw society (hbs)) – 6mf – 8 – €14.00 – ne Slangenburg [241]

Antipistorius : order widerlegung des calvinischen politici simonis ulrich pistoris i seuselitz / Gedik, S – Leipzig, 1620 – 5mf – 9 – mf#TH-1 mf 521-525 – ne IDC [242]

Antipode – Oxford, England 1986+ – 1,5,9 – ISSN: 0066-4812 – mf#17385 – us UMI ProQuest [900]

Antipologia breve en que se prueba el verdadero temperamento que la nieve posee... / Mirez Carvajal, C – Granada, 1652 – 2mf – 9 – sp Cultura [610]

Antipriscilliana : dogmengeschichtliche untersuchungen und texte aus dem streite gegen priscillians irrlehre / Kuenstle, Karl – Freiburg im Breisgau; St Louis, MO: Herder, 1905 [mf ed 1991] – 1mf – 9 – 0-7905-9405-6 – (discussion in german. text in latin) – mf#1989-2630 – us ATLA [240]

Antiquaries journal – London, England 1921+ – 1,5,9 – ISSN: 0003-5815 – mf#1273 – us UMI ProQuest [930]

Antiquaries journal – London. v1-26. 1921-1946 – 258mf – 8 – mf#H-803c – ne IDC [900]

Antiquarius, Jacobus *see* Exposite in terentium...

Antiquarum statuarum urbis romae / Calcagni, F – Roma, 1668 – 2mf – 9 – mf#GDI-5 – ne IDC [700]

Antiquary [1871] : a medium of intercommunication for men of letters, the archaeologist, and the reading public – La Salle IL 1871-73 – 1 – mf#2778 – us UMI ProQuest [930]

Antiquary [1880] : a magazine devoted to the study of the past – London, England 1880-1915 – 1 – mf#5232 – us UMI ProQuest [930]

Antique automobile – Hershey PA 1971+ – 1,5,9 – ISSN: 0003-5831 – mf#5950 – us UMI ProQuest [629]

Antique collector – Harrow, England 1976-96 – 1,5,9 – ISSN: 0003-5858 – mf#11131 – us UMI ProQuest [740]

Antique Doorknob Collectors of America *see* Doorknob collector

Antique jewellery and its revival / Castellani, Alessandro – London [1862] – (= ser 19th c art & architecture) – 1mf – 9 – mf#4.2.1233 – uk Chadwyck [730]

Antique monthly – Atlanta GA 1967-93 – 1,5,9 – ISSN: 0003-5882 – mf#7510 – us UMI ProQuest [740]

Antique phonograph monthly – 1973-81 – 1 – mf#642634 – us WHS [780]

Antique point and honiton lace / Treadwin [Mrs] – London [1873?] – (= ser 19th c art & architecture) – 2mf – 9 – mf#4.2.112 – uk Chadwyck [740]

Antique price report – 1973 jun-1978 sep; 1978 oct-1980 sep – 1 – mf#498180 – us WHS [745]

Antique radio classified – 1984 jan-1988 mar; 1988 apr-1990 mar; 1990 apr-1991 jun; 1991 jul-1992 dec; 1993 – 1 – mf#2929633 – us WHS [745]

Antique trader price guide to antiques and collectors' item – 1975 winter-1979 winter; 1980 spring-1982 winter; 1983 spring-1985 dec; 1986 feb-1988 jun – 1 – mf#515938 – us WHS [745]

Antique Wireless Association *see* Old timer's bulletin

Antiques : architecture, applied arts, studio arts – (= ser Art exhibition catalogues on microfiche) – 15 catalogues on 21mf – 9 – £176.00 – (individual titles not listed separately) – uk Chadwyck [740]

Antiques ceremonies dans l'abbaye de saint-evroult / Guery, Ch – Alencon, 1916 – €3.00 – ne Slangenburg [241]

Antiques & collecting hobbies – Chicago IL 1986-93 – 1,5,9 – (cont: hobbies; cont by: antiques and collecting magazine) – ISSN: 0884-6294 – mf#2541.02 – us UMI ProQuest [790]

Antiques & collecting magazine – Chicago IL 1994+ – 1,5,9 – (cont: antiques & collecting hobbies) – ISSN: 1084-0818 – mf#2541,02 – us UMI ProQuest [790]

Antiques gazette – v1 n1-v2 n12 [1974 aug 1-1976 jul] – 1 – mf#345254 – us WHS [745]

Antiques journal – Dubuque IA 1977-81 – 1,5,9 – ISSN: 0003-5963 – mf#11399 – us UMI ProQuest [740]

Antiques usa – [v8 n4]-v9 n4 [[52]-56] [1980 oct/nov; 1981 sep/oct] – 1 – mf#669580 – us WHS [071]

Antiques world – New York NY 1979-81 – 1,5,9 – ISSN: 0163-0911 – mf#11871 – us UMI ProQuest [740]

Antiquitates italicae medii aevi / Muratori, Lodovico A – Mediolani. v1-6. 1738-1742 – 6v on 171mf – 9 – €324.00 – ne Slangenburg [931]

L'antiquite de la terre et de l'homme : memoire / Baillairge, Charles P Florent – S.l: s.n, 1899? – 1mf – 9 – mf#00084 – cn Canadiana [900]

L'antiquite erotique / Tennordrac, M J – Paris: Editions & publ de Lutece, 1952 – (= ser Les femmes [coll]) – 2mf – 9 – mf#11411 – fr Bibl Nationale [306]

Antiquites anglo-normandes de ducarel / Thieullier, Smart le – Caen 1824 [mf ed Hildesheim 1995-98] – 1v on 4mf – 9 – €120.00 – 3-487-25957-5 – gw Olms [944]

Antiquities : including large collections from the louvre & egyptian museum in cairo / Caisse Nationale des Monuments Historiques et des Sites. Paris – (= ser Fine and decorative arts in france) – 98mf – 9 – $670.00 – 0-907006-85-X – (over 5500 reproductions) – uk Mindata [930]

Antiquities from san thome and mylapore : [the traditional site of the martyrdom and tomb of st thomas, the apostle] / Hosten, Henry – Madras: Diocese of Mylapore 1936 [mf ed 1985] – 1r [ill] – 1 – (incl bibl ref & ind; foreword by p j thomas. filmed with: english and latin / ogle, marbury b & other titles) – mf#8110 – us Wisconsin U Libr [930]

Antiquities from the city of benin : and from other parts of west africa / British Museum, London. Dept of British and Mediaeval Antiquities – London 1899 – (= ser 19th c art & architecture) – 4mf – 9 – mf#4.1.394 – uk Chadwyck [700]

Antiquities, historical and monumental of the county of cornwall / Borlase, William – 6mf – 7 – mf#87014 – uk Microform Academic [941]

Antiquities of athens : and other places in greece sicily etc / Cockerell, Charles Robert – London 1830 – (= ser 19th c art & architecture) – 8mf – 9 – mf#4.2.863 – uk Chadwyck [700]

Antiquities of bhimbar and rajauri / Kak, Ram Chandra – Calcutta: Supt Govt Print, India, 1923 – (= ser Samp: indian books) – us CRL [930]

Antiquities of free-masonry [the] : comprising illustrations of the five grand periods of masonry, from the creation of the world to the dedication of king solomon's temple / Oliver, George – new ed. London :Richard Spencer 1843 – 4mf – 9 – (with large additions & improvements by aut) – mf#vrl-88 – ne IDC [366]

The antiquities of heraldry...from literature, coins, gems, vases, and other monuments of pre-christian and mediaeval times. / Ellis, William Smith – London: J.R. Smith, 1869. xxiv,276p – 1 – us Wisconsin U Libr [920]

Antiquities of india : an account of the history and culture of ancient hindustan / Barnett, Lionel David – London: Philip L Warner, 1913 [mf ed 1992] – (= ser Handbooks to ancient civilizations series) – 1mf – 9 – 0-524-02415-4 – mf#1990-2999 – us ATLA [930]

The antiquities of israel = Die altherthumer des volkes israel / Ewald, Heinrich – London: Longmans, Green, 1876 [mf ed 1988] – 1mf – 9 – 0-7905-0009-4 – (english by henry shaen solly. incl bibl ref & ind) – mf#1987-0009 – us ATLA [221]

The antiquities of magna graecia / Wilkins, William – Cambridge 1807 – (= ser 19th c art & architecture) – 7mf – 9 – mf#4.2.1207 – uk Chadwyck [930]

The antiquities of sind : with historical outline / Cousens, Henry – Calcutta: Govt of India, Central Publication Branch, 1929 – (= ser Samp: indian books) – us CRL [930]

The antiquities of tell el yahudiyeh *see* The mound of the jew and the city of onias

The antiquities of the christian church = Handbuch der christlichen archaeologie. Selections / Augusti, Johann Christian Wilhelm – Andover [Mass]: Gould, Newman & Saxton, 1841 – 2mf – 9 – 0-524-08415-7 – (in english) – mf#1993-1025 – us ATLA [930]

Antiquities of the inns of court and chancery : containing historical and descriptive sketches relative to their original foundation, customs, ceremonies, buildings, government, etc... / Herbert, William – London: Vernor & Hood, 1804 – 5mf – 9 – $7.50 – mf#LLMC 84-295 – us LLMC [347]

Antiquities of the mesa verde national park, cliff palace / Fewkes, Jesse Walter – Washington, DC. 1911 – 1r – us UF Libraries [790]

Antiquity – York, England 1976+ – 1,5,9 – ISSN: 0003-598X – mf#15199 – us UMI ProQuest [930]

Antiquity *see* Rock-pictures and archaeology in the libyan desert

Antiquity and survival : vol 2: the holy land: new light on the prehistory and early history of israel / ed by Ruysch, W A – Israel Exploration Society, The Hague & Jerusalem, 1957 – 9 – $10.00 – us IRC [930]

"The antiquity of man" : an examination of sir charles lyell's recent work / Pattison, Samuel Rowles – London, 1863 – (= ser 19th c evolution & creation) – 1mf – 9 – mf#1.1.6351 – uk Chadwyck [210]

Antiquity of man : as deduced from the discovery of a human skeleton during the excavations of the east and west india dock-extensions at tilbury, north bank of the thames / Owen, Richard – London, 1884 – (= ser 19th c evolution & creation) – 1mf – 9 – mf#1.1.1159 – uk Chadwyck [573]

Antiquity of man : historically considered / Rawlinson, George – London, England. 1883? – 1r – us UF Libraries [240]

Antiquity of the church of england / Foye, Martin Wilson – Birmingham, England. 1836 – 1r – us UF Libraries [241]

Antiqvæ mvsicae avctores septem / Meibom, Marcus – Amstelodami: apud Lucovicum Elzevirium 1652 [mf ed 19–] – 2v on 14mf – 9 – (in greek & latin) – mf#fiche 521 – us Sibley [780]

Antiqvissima fides et vera religio / Bullinger, Heinrich – [Tigvri, Christoph Froschouer, 1544] – 2mf – 9 – mf#PBU-133 – ne IDC [240]

Antiqvitates literatvrae hvngaricae / Revai, Miklos – Pestini: Trattner 1803 – (= ser Whsb) – 4mf – 9 – €50.00 – mf#Hu 165 – gw Fischer [490]

Antiqvitatum convivialivm libri 3 in qvibvs hebraeorvm, graecorvm, romanorvm aliarvmqve nationvm antiqva conviviorvm genera... / Stucki, J W – Tigvri: Ioannes Wolph, 1582 – 9mf – 9 – mf#PBU-503 – ne IDC [240]

Antiqvitatum convivialivm libri 3 in qvibvs hebraeorvm, graecorvm, romanorvm aliarvmqve nationvm antiqva conviviorvm genera... / Stucki, J W – Tigvri, Ioannes Wolph, 1597 – 15mf – 9 – mf#PBU-622 – ne IDC [240]

Anti-Racist Action [Organization] *see* Fighting words

Anti-revolutionary tracts *see* Political tracts and pamphlets... 19th c

Anti-Seigniorial Convention (1854 : Montreal, Quebec) *see* La convention anti-seigneuriale de montreal au peuple

Antisemitisches volksblatt *see* Reichs-geldmonopol

Antisemitism i pogromy na ukraini / Tcherikower, Elias – Berlin, Germany. 1923 – 1r – us UF Libraries [939]

Anti-semitism in west germany / Seydewitz, Ruth – Berlin, Germany. 1956 – 1r – us UF Libraries [939]

Antisemitismo (version del aleman) y el antisemitismo / Coudenhove-Kalergi, Richard Nicolaus – Mexico City?, Mexico. 1939 – 1r – us UF Libraries [939]

Antiseptic – Tamil Nadu, India 1951-53 – 1 – ISSN: 0003-5998 – mf#606 – us UMI ProQuest [610]

Antiseptic surgery / Fenwick, George Edgeworth – S.l: s.n, 1881? – 1mf – 9 – mf#44720 – cn Canadiana [617]

AntiShyster : a critical examination of the american – v1 n1 [1990 dec]; v1 n2-4, 6-8 [1991 feb-may, aug-nov]; v2 n1-6 [1992: jan-nov/dec]; v3 n1 [1993] – 1 – mf#2682576 – us WHS [071]

Anti-slavery advocate – London, England oct 1852-1 may 1863 – 1 1/2r – 1 – uk British Libr Newspaper [320]

Anti-slavery bugle / Columbiana Co. Salem – v1 n1. jun 1845-apr 1861// [wkly] – 4r – 1 – mf#B4336-4339 – us Ohio Hist [976]

Anti-slavery bugle – Jan 1851-May 1861 – 1 – 74.00 – us L of C Photodup [976]

Anti-slavery bugle – Killen TX 1845-61 – 1 – mf#3084 – us UMI ProQuest [976]

Anti-slavery collection : 1795-1880 / TheRhodes House Library. Oxford – 59r – 1 – £2450.00 – mf#RHL – uk World [976]

Anti-slavery collection : 18th-19th centuries / Friends House Library. The Religious Society of Friends – 25r – 1 – £1250.00 – mf#ASL – uk World [976]

Anti-slavery examiner / American Anti-Slavery Society – New York. n1-14. 1836-45 [all publ] – (= ser Black journals, series 1) – 17mf – 9 – $165.00 – us UPA [976]

Anti-slavery examiner – La Salle IL 1836-45 – 1 – mf#3934 – us UMI ProQuest [976]

Anti-Slavery International *see* Anti-slavery reporter 1825-1994

Anti-slavery international : anti-slavery material from the worlds oldest human rights organization – 1767 et seq [mf ed Academic Microforms Ltd] – 45r – 1 – 1-897955-44-8 – (with ind) – uk Academic [322]

Anti-slavery international – 2pt – 1 – (pt1: annual reports, 1800-2000, submissions to unchr, ephemera and publications of anti-slavery international, 1980-2000 [13r] £1200; pt2: publications & reports of anti-slavery international and predecessors, 1880-1979 [5r] £475; with d/g) – uk Matthew [341]

Anti-slavery materials : regional records and other pamphlets 18th-19th centuries / John Rylands University Library. Manchester – 19r – 1 – £980.00 – mf#MUA – uk World [976]

Anti-slavery notes / Gregg, Frank M – 1r – 1 – mf#B41437 – us Ohio Hist [976]

Anti-slavery propaganda collection, oberlin college, 1835-1863 – 7235mf – 9 – (coll includes american anti-slavery propaganda publ before jan 1 1863, the date of the emancipation proclamation) – mf#C39-23100 – us Primary [976]

Anti-slavery record / American Anti-Slavery Society – New York. v1-3. 1835-37 [all publ] – (= ser Black journals, series 1) – 6mf – 9 – $85.00 – us UPA [976]

Anti-slavery record – La Salle IL 1835-37 – 1 – mf#3935 – us UMI ProQuest [976]

Anti-slavery reporter – La Salle IL 1833 – 1 – mf#3936 – us UMI ProQuest [976]

Anti-slavery reporter 1825-1994 : campaigning for the abolition of slavery / Anti-Slavery International – [mf ed Academic Microforms Ltd] – 18r – 1 – 1-897955-39-1 – (formerly: anti-slavery reporter and aborigines friend) – uk Academic [306]

Anti-slavery reporter and aborigines' friend – La Salle IL 1840-1915 – 1 – mf#4745 – us UMI ProQuest [976]

Anti-slavery society papers: trinidad, 1836-1842 : from rhodes house library – 1836-42 – 1r – 1 – mf#96659 – uk Microform Academic [972]

The antislavery struggle and triumph in the methodist episcopal church / Matlack, Lucius C – New York: Phillips & Hunt; Cincinnati: Walden & Stowe, 1881 [mf ed 1990] – 1mf – 9 – 0-7905-5432-1 – (incl bibl ref. int by d d whedon) – mf#1988-1432 – us ATLA [242]

Anti-slavery tracts / American Anti-Slavery Society – 2ser. New York, 1855-61 [all publ] – (= ser Black journals, series 1) – 16mf – 9 – $165.00 – (ser 1: n1-20 [all publ] 1855-56. ser 2: n1-25 [all & last publ] 1860-61) – us UPA [976]

Anti-slavery tracts – Westport, CT: Negro Universities Press. 2v. 1970 [mf ed Westport, CT: Greenwood Pub Corp [1970?]] – 16mf – 9 – mf#Sc Micro F-166 – Located: NYPL – us Misc Inst [976]

Anti-socialist organisations in britain : anti-socialist journals, 1874-1914 – 19r – 1 – (part 1: anti-socialist journals, 1874-1914) – us Primary [325]

Antisofisma – 1787 – 9 – sp Bibl Santa Ana [946]

Anti-soviet newspapers – 493 titles on 91r – 1 – (with finding aids; titles also listed individually) – ne IDC [947]

Anti-strauss : ernstes zeugniss fuer die christliche wahrheit wider die alte und neue unglaubenslehre / Kratander – Stuttgart: JF Steinkopf, 1841 [mf ed 1993] – 1mf – 9 – 0-524-08086-0 – mf#1992-1146 – us ATLA [240]

The anti-sweater – A journal devoted to the exposure of the sweating system and for the organization of the journeyman tailors and machinists. London. -m. Jul 1886-Feb 1887. (5 ft) – 1 – uk British Libr Newspaper [331]

Anti-teapot review – La Salle IL 1864-69 – 1 – mf#4709 – us UMI ProQuest [073]

Anti-terrorism explosives act of 2002 : hearing...house of representatives, 107th congress, 2nd session on h r 4864, june 11 2002. / United States. Congress. House. Committee on the Judiciary. Subcommittee on Crime, Terrorism, and Homeland Security. – Washington: US GPO 2002 [mf ed 2002] – 1mf – 9 – 0-16-068809-4 – us GPO [344]

Anti-theatre *see* Eighteenth century journals

Anti-theistic theories / Flint, Robert – 4th ed. Edinburgh: W Blackwood, 1889 [mf ed 1990] – (= ser Baird lecture 1877) – 2mf – 9 – 0-7905-7819-0 – mf#1989-1044 – us ATLA [210]

The antithesis between symbolism and revelation : lecture delivered before the historical presbyterian society in philadelphia, pa / Kuyper, Abraham – Amsterdam: Hoeveker & Wormser, [1898?] [mf ed 1992] – 1mf – 9 – 0-7905-3351-0 – mf#1987-3351 – us ATLA [230]

Antithesis et compendivm evangelicae et papisticae doctrinae / Bullinger, Heinrich – [Zuerich], Christoph Froschouer, 1551 – 1mf – 9 – mf#PBU-166 – ne IDC [240]

Antitrinitarian biography : or, sketches of the lives and writings of distinguished antitrinitarians. exhibiting a view of the state of the unitarian doctrine and worship in the principal nations of europe... / Wallace, Robert – London: E T Whitfield, 1850 [mf ed 1990] – 3v on 5mf – 9 – 0-7905-8098-5 – mf#1988-8034 – us ATLA [243]

L'antitrinitarisme a geneve au temps de calvin : etude historique / Cologny, L – Geneve: Taponnier & studer, 1873 [mf ed 1993] – (= ser Presbyterian coll) – 2mf – 9 – 0-524-07403-8 – mf#1991-3063 – us ATLA [242]

Antitrust, 1890-1990 / ed by Wood, Diane P – 509mf – 9 – $4925.00 – 1-55655-433-8 – (p/g only $500) – us UPA [332]

Antitrust bulletin – New York NY 1955+ – 1,5,9 – ISSN: 0003-603X – mf#6476 – us UMI ProQuest [340]

Antitrust law journal – v1-68. 1952-2000 – 9 – $971.00 set – (v1-60 1952-92 in reel or mf $731. v61-68 1993-2000 in mf $240. title varies: v1-31 1952-66 as american bar association. section of antitrust law proceedings) – ISSN: 0003-6056 – mf#100201 – us Hein [340]

Anti-trust laws with special reference to the mennen co decision, the hardwood lumber decision and the edge resolution : an address / Levy, Felix Holt – [New York: Beacon Press [1922?] (mf ed 1992) – 1r – 1 – mf#21809 – us Harvard [346]

Antiviral chemistry & chemotherapy – 1990+ – 1,5,9 – ISSN: 0956-3202 – mf#17754 – us UMI ProQuest [576]

Antiviral research – Oxford, England 1981+ – ISSN: 0166-3542 – mf#42240 – us UMI ProQuest [576]

Anti-war viewpoints – 1966 may?-1973 feb – 1 – mf#1052195 – us WHS [071]

Anti-xenien : in auswahl / ed by Stammler, Wolfgang – Bonn: A Marcus & E Weber 1911 [mf ed 1990] – (= ser Kleine texte fuer vorlesungen und uebungen 81) – 1r – 1 – (incl bibl ref. filmed with: goethes romische elegien / albert leitzmann) – mf#7371 – us Wisconsin U Libr [430]

Anti-zarathustra : gedanken ueber friedrich nietzsches hauptwerke / Henne am Rhyn, Otto – Altenburg, S.-A: Alfred Tittel 1899 [mf ed 1995] – 1r – 1 – (incl bibl ref & ind. filmed with: credit / august niemann) – mf#3699p – us Wisconsin U Libr [190]

Antler american : [official county and city paper 1911] – Antler, ND: A J Drake. v1 n9 jul 22 1905-v15 n6 jul 24 1919 (wkly) – (= ser Kuroki booster) – 1 – (v1 n1 apr 20 1916-v1 n52 apr 19 1917 of kuroki booster publ as back page of antler american; cont: american (antler, nd); absorbed by: the westhope standard) – mf#03941-03944 – us North Dakota [071]

Antlitz der zeit : sinfonie moderner industriedichtung: selbstbildnis und eigenauswahl der autoren / ed by Haas, Wilhelm – Berlin: Wegweiser-Verlag, [1929?] – (= ser Volksverband der buecherfreunde) – 234p/[8]pl – 1 – (incl bibl ref and ind) – mf#8361 – us Wisconsin U Libr [810]

Antoine, A *see* Histoire des emigres francais

Antoine Francois Prevost d'Exiles *see* Prevosts "manon lescaut" in deutschen uebersetzungen des 18., 19. und 20. jahrhunderts

Antoine goulet, de la societe des poetes canadiens-francais : bio-bibliographie analytique / Jacques, Marthe – 1961 [mf ed 1978] – (= ser Bibliographies du cours...1947-66) – 1mf – 9 – mf#SEM105P4 – cn Bibl Nat [440]

Antoinette bourignon, quietist / Macewen, Alexander Robertson – London: Hodder & Stoughton, 1910 [mf ed 1990] – 1mf – 9 – 0-7905-5186-1 – (incl bibl ref) – mf#1988-1186 – us ATLA [240]

Antologia / Caro, Jose Eusebio – Bogota, Colombia. 1951 – 1r – us UF Libraries [972]

Antologia / Dario, Ruben – Mexico City?, Mexico. 1958 – 1r – us UF Libraries [972]

Antologia / Garcia Godoy, Federico – Ciudad Trujillo, Dominican Republic. 1951 – 1r – us UF Libraries [800]

Antologia / Gomez De Avellaneda Y Arteaga, Gertrudis – Buenos Aires, Argentina. 1945 – 1r – us UF Libraries [972]

Antologia / Henriquez Urena, Pedro – Ciudad Trujillo, Dominican Republic. 1950 – 1r – us UF Libraries [972]

Antologia / Hostos, Eugenio Maria De – Madrid, Spain. 1952 – 1r – us UF Libraries [972]

Antologia / Mieses Burgos, Franklin – Ciudad Trujillo, Dominican Republic. 1952 – 1r – us UF Libraries [972]

Antologia / Moreno Jimenes, Domingo – Ciudad Trujillo, Dominican Republic. 1949 – 1r – us UF Libraries [800]

Antologia / Rosales Y Rosales, Vicente – San Salvador, El Salvador. 1959 – 1r – us UF Libraries [972]

Antologia : seleccion / Lugo, Americo – Ciudad Trujillo, Dominican Republic. 1949 – 1r – us UF Libraries [800]

Antologia : verso y prosa / Molina, Juan Ramon – San Salvador, El Salvador. 1959 – 1r – us UF Libraries [800]

Antologia americana / Ghiraldo, Alberto – Madrid, Spain. v1-5. 1920- – 1r – us UF Libraries [972]

Antologia brasileira / Werneck, Eugenio – Rio de Janeiro, Brazil. 1941 – 1r – us UF Libraries [972]

Antologia chilena / Dario, Ruben – Santiago, Chile. 1941 – 1r – us UF Libraries [972]

Antologia comentada de textos espanoles e hispanoa / Remos Y Rubio, Juan Nepomuceno Jose – Habana, Cuba. 1926 – 1r – us UF Libraries [972]

Antologia contemporanea / Brandao, Claudio – Rio de Janeiro, Brazil. 1939 – 1r – us UF Libraries [972]

Antologia critica de jose marti / Gonzaliz, Manuel Pedro – Mexico City?, Mexico. 1960 – 1r – us UF Libraries [440]

Antologia critica del modernismo hispanoamericano / Silva Castro, Raul – New York, NY. 1963 – 1r – us UF Libraries [972]

Antologia da poesia mineira, fase modernista / Guimmaraens Filho, Alphonsus De – Belo Horizonte, Brazil. 1946 – 1r – us UF Libraries [440]

Antologia de autores clasicos extranjeros / Tamayo Zamora, B – Badajoz: Tip. Lib. A. Arqueros, 3rd ed 1909 – 1 – sp Bibl Santa Ana [440]

Antologia de autores latinos-cristianos – Toledo: Biblioteca Latina, vol 3. 1944 – 1 – sp Bibl Santa Ana [240]

Antologia de cuentistas brasilenos / Orico, Osvaldo – Santiago, Chile. 1946 – 1r – us UF Libraries [972]

Antologia de cuentos puertorriquenos – Godfrey, IL. 1956 – 1r – us UF Libraries [972]

Antologia de cuentos puertorriquenos – Mexico City?, Mexico. 1954 – 1r – us UF Libraries [972]

Antologia de fundamentos de filosofia – Ciudad Universitaria, Costa Rica. 1961 – 1r – us UF Libraries [100]

Antologia de la literatura dominicana – Santiago, Dominican Republic. v1-2. 1944 – 1r – us UF Libraries [440]

Antologia de la novela cubana – Habana, Cuba. 1960 – 1r – us UF Libraries [830]

Antologia de la nueva poesia colombiana – Bogota, Colombia. 1949 – 1r – us UF Libraries [810]

Antologia de la poesia cubana / Lezama Lima, Jose – Habana, Cuba. v1-3. 1965 – 1r – us UF Libraries [810]

Antologia de la poesia hispanoamericana / Panero, Leopoldo – Madrid, Spain. v1-2. 1944 – 1r – us UF Libraries [810]

Antologia de la poesia moderna hispanoamericana / Abril, Xavier – Montevideo, Uruguay. 1956 – 1r – us UF Libraries [810]

Antologia de lendas do indio brasileiro / Silva, Alberto Da Costa E – Rio de Janeiro, Brazil. 1957 – 1r – us UF Libraries [810]

Antologia de mis antologias de mis doce mil sonetos / Sanchez Arjona, Vicente – Sevilla: Imprenta Zambrano, 1959 – 1 – sp Bibl Santa Ana [810]

Antologia de mis cantares / Sanchez Arjona, Vicente – Sevilla: Imprenta Carlos Acuna, 1944 – 1 – sp Bibl Santa Ana [780]

Antologia de mis penas cortas, pensamientos y madrigales / Sanchez Arjona, Vicente – Sevilla: Imprenta Carlos Acuna, 1945 – 1 – sp Bibl Santa Ana [800]

Antologia de mis sonetos / Sanchez Arjona, Vicente – Sevilla: Imp. Carlos A., Tomo 3. 1954. Tomo 4-10, anos 1956, 1958, 1959 – 1 – sp Bibl Santa Ana [810]

Antologia de mis sonetos / Sanchez Arjona, Vicente – Sevilla: Imprenta Carlos Acuna, Tomo 2. 1952 – 1 – sp Bibl Santa Ana [810]

Antologia de mis ultimos sonetos / Sanchez Arjona, Vicente – Sevilla: Imprenta Alvarez, Tomo 7. 1959 – 1 – sp Bibl Santa Ana [810]

Antologia de panama / Korsi, Demetrio – Barcelona, Spain. 1926 – 1r – us UF Libraries [972]

Antologia de periodistas cubanos / Soto Paz, Rafael – Habana, Cuba. 1943 – 1r – us UF Libraries [073]

Antologia de poesia antioquena – Lima, Peru. 1961? – 1r – us UF Libraries [810]

Antologia de poetas americanos / Morales, Ernesto – Buenos Aires, Argentina. 1941 – 1r – us UF Libraries [440]

Antologia de poetas brasileiras bissextos contempo... / Bandeira, Manuel – Rio de Janeiro, Brazil. 1946 – 1r – us UF Libraries [440]

Antologia de poetas contemporaneos de puerto rico / Labarthe, Pedro Juan – Mexico City?, Mexico. 1946 – 1r – us UF Libraries [440]

Antologia de poetas costarricenses / Padilla, Rosario De – San Jose, Costa Rica. 1946 – 1r – us UF Libraries [440]

Antologia de poetas hispano-americanos – Madrid, Spain. v1-4. 1927- – 2r – us UF Libraries [810]

Antologia de poetas hondurenos / Castro, Jesus – Tegucigalpa, Mexico. 1939 – 1r – us UF Libraries [440]

Antologia de poetas jovenes de honduras desde 1935 / Barrera, Claudio – Tegucigalpa?, Honduras. 1950 – 1r – us UF Libraries [440]

Antologia de poetas precursores del modernismo / Torres-Rioseco, Arturo – Washington, DC. 1949 – 1r – us UF Libraries [440]

Antologia de poetas y prosistas hispanoamericanos / Monterde, Francisco – Mexico City?, Mexico. 1931 – 1r – us UF Libraries [440]
Antologia de prosistas guatemaltecos / Echeverria, Amilcar – Guatemala, 1957 – 1r – us UF Libraries [972]
Antologia de sus obras / Fernandez Juncos, Manuel – Mexico City?, Mexico. 1965 – 1r – us UF Libraries [972]
Antologia del cuento antioqueno / Mejia Vallejo, Manuel – Lima, Peru. 196-? – 1r – us UF Libraries [440]
Antologia del cuento colombiano / Pachon Padilla, L'duardo – Bogota, Colombia. 1959 – 1r – us UF Libraries [440]
Antologia del cuento en cuba (1902-1952) / Bueno, Salvador – Habana, Cuba. 1963 – 1r – us UF Libraries [440]
Antologia del cuento hispanoamericano / Garcini, Maria Del Carmen – Habana, Cuba. 1963 – 1r – us UF Libraries [440]
Antologia del cuento salvadoreno, 1880-1955 / Barba Salinas, Manuel – San Salvador, El Salvador. 1959 – 1r – us UF Libraries [972]
Antologia del ilustrisimo senor manuel jose mosque – Bogota, Colombia. 1954 – 1r – us UF Libraries [972]
Antologia del soneto – Habana, Cuba. 1942 – 1r – us UF Libraries [972]
Antologia do folclore brasileiro / Cascudo, Luis Da Camara – Sao Paulo, Brazil. 1943 – 1r – us UF Libraries [390]
Antologia do pensamento social e politico no brasil / Vita, Luis Washington – Sao Paulo, Brazil. 1968 – 1r – us UF Libraries [300]
Antologia dos poetas brasileiros da fase romantica / Bandeira, Manuel – Rio de Janeiro, Brazil. 1949 – 1r – us UF Libraries [440]
Antologia ecuatoriana : cantares del pueblo ecuatoriano – Quito: Universidad Central del Ecuador 1892 [mf ed Bloomington IN: Indiana Uni Lib, Preservation Dept 1984] – xxvi 504p on 1r – 1 – us Indiana Preservation [390]
Antologia escolar brasileira / Rebelo, Marques – Rio de Janeiro, Brazil. 1967 – 1r – us UF Libraries [972]
Antologia euclidiana / Cunha, Eucyldes Da – Sao Paulo, Brazil. 1967 – 1r – us UF Libraries [972]
Antologia general del cuento puertorriqueno / Rosa-Nieves, Cesareo – San Juan, Puerto Rico. v1-2. 1959 – 1r – us UF Libraries [972]
Antologia guajira / Riveron Hernandez, Francisco – Habana, Cuba. 1958 – 1r – us UF Libraries [972]
Antologia herediana / Heredia, Jose Maria – Habana, Cuba. 1939 – 1r – us UF Libraries [972]
Antologia hispano-americana / Campos, Jorge – Madrid, Spain. 1950 – 1r – us UF Libraries [972]
Antologia latina / Franco y Lozano, Francisco – Badajoz: Uceda Hermanos, 1907 – 1 – sp Bibl Santa Ana [946]
Antologia lirica / Caparroso, Carlos Arturo – Bogota, Colombia. 1951 – 1r – us UF Libraries [972]
Antologia mariana / Trujillo Gutierrez, Eduardo – Bogota, Colombia. 1954 – 1r – us UF Libraries [972]
Antologia mayor / Guillen, Nicolas – Habana, Cuba. 1964 – 1r – us UF Libraries [972]
Antologia poetica / Alvarez Magana, Manuel – San Salvador, El Salvador. 1961 – 1r – us UF Libraries [810]
Antologia poetica / Cadilla De Ruibal, Carmen Alicia – San Juan, Puerto Rico. 1941 – 1r – us UF Libraries [810]
Antologia poetica : con un poema de rafael alberti / Rugeles, Manuel Felipe – Buenos Aires, Argentina. 1952 – 1r – us UF Libraries [810]
Antologia poetica / Dario, Ruben – Habana, Cuba. 1962 – 1r – us UF Libraries [810]
Antologia poetica / Dario, Ruben – Santiago, Chile. 1946 – 1r – us UF Libraries [810]
Antologia poetica / Lainez, Daniel – Tegucigalpa, Mexico. 1950 – 1r – us UF Libraries [810]
Antologia poetica / Miro, Ricardo – Guatemala, 1951 – 1r – us UF Libraries [810]
Antologia poetica / Pombo, Rafael – Bogota, Colombia. 1952 – 1r – us UF Libraries [810]
Antologia poetica / Ribera Chevremont, Evaristo – San Juan, Puerto Rico. 1957 – 1r – us UF Libraries [810]
Antologia poetica : seleccion / Dario, Ruben – Guatemala, 1948 – 1r – us UF Libraries [810]
Antologia poetica (1907-1937) / Miro, Ricardo – Panama, 1937 – 1r – us UF Libraries [810]
Antologia poetica (1918-1938) / Pedros, Regino – Habana, Cuba. 1939 – 1r – us UF Libraries [810]
Antologia poetica (1924-1950) / Ribera Chevremont, Evaristo – Madrid, Spain. 1954 – 1r – us UF Libraries [810]
Antologia poetica, 1934-1954 / Barcena, Lucas – Panama, 1959 – 1r – us UF Libraries [810]
Antologia poetica dominicana / Contin Aybar, Pedro Rene – Ciudad Trujillo, Dominican Republic. 1943 – 1r – us UF Libraries [810]
Antologia poetica dominicana / Contin Aybar, Pedro Rene – Ciudad Trujillo, Dominican Republic. 1951 – 1r – us UF Libraries [810]
Antologia poetica guadalupense *see* En el alcazar de la reina. antologia poetica guadalupense
Antologia poetica hispano-americana / Oyuela, Calixto – Buenos Aires, Argentina. v1-v3 pt2. 1919 – 2r – us UF Libraries [810]
Antologia poetica moderna / Saz Sanchez, Agustin Del – Barcelona, Spain. 1948 – 1r – us UF Libraries [810]
Antologia poetica trujillista / Castro Noboa, H B De – Santiago, Dominican Republic. 1946 – 1r – us UF Libraries [810]
Antologia poetilor tineri ' zaharia stancu – Bucurejsti, Romania. 1934 – 1r – us UF Libraries [810]
Antologia puertorriquena / Silva De Quinones, Rosita – San Juan, Puerto Rico. no date – 1r – us UF Libraries [972]
Antologia rota / Leon Felipe – Buenos Aires, Argentina. 1957 – 1r – us UF Libraries [972]
Antologie fun der yidisher literatur in argentina / Comite de Homenaje a "El Diario Israelita", Buenos Aires – Buenos Aires, Argentina. 1944 – 1r – us UF Libraries [470]
Antologio de mis -hasta ahora- seiscientos sonetos / Sanchez Arjona, Vicente – Sevilla: Imprenta Carlos Acuna, 1945 – 1 – sp Bibl Santa Ana [810]
Antologyah shel ha-sifrut ha-lita'it – Kovnah, Lithuania. 1932 – 1r – us UF Libraries [939]
Antommarchi, Francesco *see* Memoires du docteur f antommarchi
Anton auerspergs (anastasius gruens) politische reden und schriften / ed by Hock, Stefan – Wien: Literarischer Verein 1906 [mf ed 1993] – (= ser Schriften des literarischen vereins in wien 5) – 1r – 1 – (incl bibl ref; filmed with: schriften des literarischen vereins in wien) – mf#3333p – us Wisconsin U Libr [850]
Anton Crespo, Emilio *see*
– Un ano de accion sindical en la provincia
– Un ano de accion sindical en la provincia de badajoz
Anton fabers neue europaeische staats-canzley, welche die wichtigsten oeffentlichen angelegenheiten, vornehmlich des deutschen reiches in sich fasset – Ulm, Frankfurt, Leipzig 1761-82 – (= ser Dz. historisch-politische abt) – 174mf – 9 – €1044.00 – mf#k/n1676 – gw Olms [323]
Anton, Fernando De *see* Cuestion social
Anton, Ludwig *see* Wirrwarr
Anton reiser : ein psychologischer roman / Moritz, Karl Philipp; ed by Geiger, Ludwig – Heilbronn: Gebr Henninger, 1886 [mf ed 1993] – (= ser Deutsche litteraturdenkmale des 18. und 19. jahrhunderts 23) – 443p – 1 – (repr of an ed in 4pts, berlin 1785-1790. incl bibl ref) – mf#8676 reel 3 – us Wisconsin U Libr [830]
Anton tuchers haushaltsbuch : 1507 bis 1517 / ed by Loose, Wilhelm – Stuttgart: Litterarischer Verein, 1877 (Tuebingen: H Laupp) [mf ed 1993] – (= ser Blvs 134) – 220p – 1 – mf#8470 reel 28 – us Wisconsin U Libr [640]
Anton wildgans : ein leben in briefen / ed by Wildgans, Lilly – Wien: W Frick, 1947 [mf ed 1992] – 3v on 1r – 1 – (v1: 1900-16. v2: 1917-24. v3: 1925-32) – mf#7956 – us Wisconsin U Libr [860]
Antoniades, M *see* Ekphrasis tes hagias sophias
Antonianum – 1(1926)-35(1960) – 465mf – 9 – €888.00 – ne Slangenburg [073]
Antonie van leeuwenhoek – Dordrecht, Netherlands 1991+ – 1,5,9 – ISSN: 0003-6072 – mf#16763 – us UMI ProQuest [576]
Antonievich, V I *see* Nasha mysl' [Kharbin: 1919]
Antoniewicz, Johann von *see* Johann elias schlegels aesthetische und dramaturgische schriften
Antonii sanderi presbyteri gandavum sive gandavensium rerum libri sex – Bruxellis: Apud Ioannem Pepermanum, 1627 [mf ed 1988] – 6mf – 9 – mf#SEM105P947 – cn Bibl Nat [949]
Antonij sucquet e societate iesu via vitae aeternae iconibus... – Antwerpiae: Apud Henricum Aertssium, 1625 – 14mf – 9 – mf#0-1919 – ne IDC [090]
Antonij sucquet et societate iesu via vitae aeternae iconibus... – Antverpiae: Typis Martini Nutij, 1620 – 11mf – 9 – mf#0-769 – ne IDC [090]
Antonil, Andre Joao *see* Cultura e opulencia do brasil
The antonin genizah in the saltykov-shchedrin public library in leningrad / Katsh, A I – New York, 1963 – 1mf – 9 – mf#R-10882 – ne IDC [956]
Antoninus de Florentia *see* Summa theologica
Antonio alimundo – Minorca, Spain. v938. 1763-1772 – 2r – us UF Libraries [920]
Antonio averli filarete's tractat ueber die baukunst : nebst seinen buechern von der zeichenkunst und den bauten der medici / [Filarete] Oettingen, W von – Wien. v3. 1890 – 10mf – 9 – mf#0-517 – ne IDC [720]
Antonio brasio...monumenta... / Barrado Manzano, Arcangel – Madrid: Archivo Ibero-Americano, 1959 – 1 – sp Bibl Santa Ana [240]
Antonio conselheiro / Macedo, Nertan – Rio de Janeiro, Brazil. 1969 – 1r – us UF Libraries [920]
Antonio da silva riego : documentao... / Barrado Manzano, Arcangel – Madrid: Archivo Ibero-Americano, 1959 – 1 – sp Bibl Santa Ana [946]
Antonio de Beatis *see* Voyage du cardinal d'aragon en allemagne, hollande, belgique, france et italie (1517-1518)
Antonio de santa maria...en custodios y provinciales de la provincia de san jose / Perez, Lorenzo – Madrid: Archivo Ibero Americano, 1924 – 1 – sp Bibl Santa Ana [946]
Antonio de Trujillo *see* San marcos defendido en el milagro que obra dios
Antonio flaquer – Minorca, Spain. v370-364. 1745-1775 – 5r – us UF Libraries [920]
Antonio hurtado / Blanco Garcia, Francisco – Madrid: Saenz de Jubera, 1909 – sp Bibl Santa Ana [440]
Antonio jose : o judeu / Juca Filho, Candido – Rio de Janeiro, Brazil. 1940 – 1r – us UF Libraries [972]
Antonio jose de sucre / Sherwell, Guillermo Antonio – Washington, DC. 1924 – 1r – us UF Libraries [972]
Antonio labriola : la vita e il pensiero / Pane, Luigi dal – Bologna: Forni 1968 [mf ed 1980] – 1r – 1 – (repr of rome 1935 ed; incl ind & bibl ref; pref by giovacchino volpe) – mf#101 – us Wisconsin U Libr [945]
Antonio leila de fermo y la condenacion del indianum jure... / Leturia, Pedro S – Madrid: Missionalis Hispanica, 1949 – 1 – sp Bibl Santa Ana [240]
Antonio, Nicolas *see* Bibliotheca hispana vetus
Antonio pons – Minorca, Spain. v913. 1761-1771 – 1r – us UF Libraries [920]
Antonio pons y pons – Minorca, Spain. v306 and v317. 1767? – 1r – (gaps) – us UF Libraries [920]
Antonio s pedreira : buceador de la personalidad p... / Sierra Berdecia, Fernando – San Juan, Puerto Rico. 1942 – 1r – us UF Libraries [920]
Antonio scialoja : memorie e documenti, 1845-1877 / Cesare, Raffaele de – Citta di Castello: S Lapi 1893 [mf ed 1991] – 1r – 1 – mf#c1738 – us Harvard [330]
Antonio y Hernandez, Pedro de A *see* Aritmetica y sistema legal de pesas
Antoniotto, Giorgio *see* L'arte armonica
Antonius corvinus : ein maertyrer des evangelisch-lutherischen bekenntnisses / Uhlhorn, Gerhard – Halle: Verein fuer Reformationsgeschichte, 1892 [mf ed 1990] – (= ser Schriften des vereins fuer reformationsgeschichte 9/4/37) – 1mf – 9 – 0-7905-4717-1 – (incl bibl ref) – mf#1988-0717 – us ATLA [242]
Antonius de arena provencalis, de bragardissima villa de soleriis : ad suos compagnones studiantes, qui sunt de persona frigantes, bassas dansas & branlos practicantes, nouvellos quamplurimos mandat... / Arena, Antoine – [s.l]: Stampatus in Stampatura Stampatorum 1670 [mf ed 19-] – 2mf – 9 – mf#fiche6 – us Sibley [790]
Anton-Maria da Vicenza *see* Lexicon bonaventurianum
Antonov, Ark *see* Syzranskii vestnik
La antorcha – Managua: Convencion Bautista de Nicaragua [mayo 1982-oct/dic 1992] (mthly) – 1r – 1 – us CRL [242]
La antorcha catolica – Almendralejo, 1901 – 5 – sp Bibl Santa Ana [241]
Antrakt – Moscow, 1866-68 [wkly] – 19mf – 9 – (cont: artiste russe) – us UMI ProQuest [780]
Antrakt – St Petersburg, Russia 1882 [mf ed Norman Ross] – 1 – mf#nrp-1578 – us UMI ProQuest [077]
Antrieb – Magdeburg DE, 1956 18 jan-1974 nov [gaps], 1975 jan-nov, 1976 jan-nov [gaps], 1977-1984 nov [gaps], 1986-1988 nov, 1989 jan-nov – 6r – 1 – (schwermaschinenbau) – gw Misc Inst [620]
Der antrieb – Dessau DE, 1953 feb-1985 nov [gaps], 1986-1989 13 dec [gaps] – 6r – 1 – (notes: elektromotorenwerk) – gw Misc Inst [621]
Antrim guardian – Antrim, Ireland 6 dec 1973-[mf 1986-] – 1 – uk British Libr Newspaper [072]
Antrim times and ballymena observer – Ballymena, Ireland 13 jun 1985-30 dec 1986, 13 may 1987-4 sep 1991 – 1 – (cont by: antrim times [11 sep 1991-]) – uk British Libr Newspaper [072]
Antro fuego / Antillon, A – San Jose, Costa Rica. 1955 – 1r – us UF Libraries [920]
Antrobus, st mark – (= ser Cheshire monumental inscriptions) – 1mf – 9 – £2.50 – mf#1 – uk CheshireFHS [929]
Antropologia filosofica 1. preliminares y cuestiones basicas / Frutos Cortes, Eugenio – Zaragoza: Facultad de Filosofia y Letras, 1971 – sp Bibl Santa Ana [100]
Antropologia filosofica 2. dimensiones entitativas del hombre / Frutos Cortes, Eugenio – Zaragoza: Facultad de Filosofia y Letras, 1972 – sp Bibl Santa Ana [100]
Antropov, P A *see* Finansovo-statisticheskii atlas rossii
Antshel, Kevin M *see* The effect of time of season on the athletic identity in collegiate swimmers
Antsyferov, A N *see*
– Kooperativnyi institut i muzei
– Kooperativnyi kredit
– Ocherki po kooperatsii
– Sovet vserossiiskikh kooperativnykh sezdov
– Tsentralnye banki kooperativnogo kredita
Antun, John M *see* Journal of culinary science and technology
Antuna, Jose Gervasio *see* Perspectivas de america
Antuna, Rosario *see* Son de otros
Antunes, De Paranhos *see* Passado e presente da economia brasileira
Antunez Toriblo, Manuel *see* La castellana de ribera del fresno. leyenda.
Antung ved jalufloden / Ellerbek, Soren Anton – Kobenhavn: Kirkelig forening for den indre mission i Danmark, 1910 [mf ed 1995] – (= ser Yale coll) – 34p (ill) – 1 – 0-524-09501-9 – (in danish) – mf#1995-0501 – us ATLA [951]
Antwerp, 1477-1559 : from the battle of nancy to the treaty of cateau cambresis / Wegg, Jervis – London: Methuen [1916] [mf ed 1987] – 1r – 1 – (filmed with: the dravidian element in indian culture / slater, g) – mf#6833 – us Wisconsin U Libr [949]
Antwerp, E I van *see* Augustine
Antwerp, Gertruida Cornelia van *see* Fantasie en verbeelding as moontlikheidsvoorwaardes vir kreatiewe denke gedurende aanvangsonderwys
Antwerp, William Clarkson van *see* The war and wall street
Antwort auf die streitschrift d cremers : zum kampf um das apostolikum / Harnack, Adolf von – Leipzig: Grunow, 1892 [mf ed 1990] – (= ser Hefte zur "christlichen welt" 3) – 1mf – 9 – 0-7905-5895-5 – mf#1988-1895 – us ATLA [240]
Antwort auff d christophori pezelii predigers zu bremen falsch gebrauchte gruende / Hoffmann, D – Helmstadt, 1589 – 2mf – 9 – mf#TH-1 mf 689-690 – ne IDC [242]
Antwort auff das buch des osiandrischen schwermers in preussen, m vogels, / Moerlin, J – [Magdeburg, 1557] – 1mf – 9 – mf#TH-1 mf 1173 – ne IDC [242]
Antwort der dieneren der kyrchen zuo zueryсh vff d. jacoben anderesen...widerlegen...antwort...vff d. jacoben andresen...erinnerung... / Bullinger, Heinrich – [Zuerych, Christoffel Froschower, 1575] – 6mf – 9 – mf#PBU-256 – ne IDC [240]
Antwort, glaub vnd bekentnis auff das schoene vnd liebliche interim / Amsdorff, N von – [Magdeburg], 1548 – 1mf – 9 – mf#TH-1 mf 11 – ne IDC [242]
Antwort matthiae flacii illirici, auff das stenckfeldische buechlein iudicium das genant / Flacius Illyricus d A, M – [Nuernberg, 1555] – 1mf – 9 – mf#TH-1 mf 462 – ne IDC [242]
Eyn antwurt huldrychs zuinglins uff die epistel joannis pugenhag...das nachtmal christi betreffende / Zwingli, H – Zuerich: Christoffel Froschouer, 1526 – 1mf – 9 – mf#PBU-523 – ne IDC [242]
Antysemityzm w literaturze polskiej 15-17 w... / Bartoszewicz, Kazimierz – Warszawa, Poland. 1914 – 1r – us UF Libraries [939]
Anuario 1972-1974 / Cuba. Direccion Central de Estadistica – (= ser Latin american & caribbean...1821-1982) – 20mf – 9 – uk Chadwyck [318]
Anuario de estadistica 1963/1968-1966/1971 / Ecuador. Instituto Nacional de Estadistica – (= ser Latin american & caribbean...1821-1982) – 13mf – 9 – (1966/71 not available) – uk Chadwyck [318]
Anuario de estadistica de la ciudad de buenos aires...ano 1883-1895 – Buenos Aires, 1885-1896 – 100mf – 9 – sp Cultura [318]
Anuario de estadistica de la provincia de tucuman...ano 1895, tomo 1 – Buenos Aires, 1896 – 9mf – 9 – sp Cultura [318]
Anuario de estadistica de la provincia de tucuman...ano 1897-1898 – Buenos Aires, 1898-1899 – 14mf – 9 – sp Cultura [317]
Anuario de la direccion general de estadistica 1894 – Buenos Aires, 1895 – 10mf – 9 – sp Cultura [310]

Anuario de la direccion general de estadistica 1897 : tomo 1-2 – Buenos Aires, 1898 – 18mf – 9 – sp Cultura [310]

Anuario de la direccion general de estadistica 1898 / Guatemala. Direccion General de Estadistica – (= ser Latin american & caribbean...1821-1982) – 3mf – 9 – uk Chadwyck [318]

Anuario de la direccion general de estadistica 1892-1914 / Argentine Republic. Direccion General de Estadistica – (= ser Latin american & caribbean...1821-1982) – 215mf – 9 – uk Chadwyck [318]

Anuario de psicologia / Universidad De San Carlos De Guatemala Facultad De Humanidades – Guatemala. v1. 1962 – 1r – us UF Libraries [150]

Anuario deportivo 1969 / Junta Provincial de Educacion Fisica y Deportes – Badajoz: Graf. Nemesio Jimenez, 1970 – sp Bibl Santa Ana [946]

Anuario estadistico 1970 / Guatemala. Direccion General de Estadistica – (= ser Latin american & caribbean...1821-1982) – 3mf – 9 – uk Chadwyck [318]

Anuario estadistico 1848/1858-1937 / Chile. Servicio Nacional de Estadistica y Censos – (= ser Latin american & caribbean...1821-1982) – 512mf – 9 – (several missing vols & pts) – uk Chadwyck [318]

Anuario estadistico 1848/1949-1970 = Statistical yearbook 1848/1949-1970 / Puerto Rico. Bureau of Economics and Statistics – (= ser Latin american & caribbean...1821-1982) – 65mf – 9 – (1965 not available) – uk Chadwyck [318]

Anuario estadistico 1877-1969 / Venezuela. Direccion General de Estadistica y Censos Nacionales – (= ser Latin american & caribbean...1821-1982) – 228mf – 9 – (some iss between 1879-1937 not publ. 1878, 1884, 1887, 1889, 1891, 1941 not available) – uk Chadwyck [318]

Anuario estadistico 1883-1969 / Costa Rica. Direccion General de Estadistica y Censo – (= ser Latin american & caribbean...1821-1982) – 198mf – 9 – (1892, 1894-1906 not publ. 1883-87, 1890-93, 1908, 1910, 1917, 1939, 1948-50 not available) – uk Chadwyck [318]

Anuario estadistico 1884-1967/1969 / Uruguay. Direccion General de Estadistica – (= ser Latin american & caribbean...1821-1982) – 466mf – 9 – (1890, 1943-44 not available) – uk Chadwyck [318]

Anuario estadistico 1886-1969 / Paraguay. Direccion General de Estadistica – (= ser Latin american & caribbean...1821-1982) – 45mf – 9 – (1886-1913, 1914-15, 1918-24, 1925-26, 1927, 1930-34 not available) – uk Chadwyck [318]

Anuario estadistico 1893-1968/1969 / Mexico. Direccion General de Estadistica – (= ser Latin american & caribbean...1821-1982) – 255mf – 9 – (1908-22, 1925-29, 1931-37 not publ. 1905, 1923-24, 1946-50 not available) – uk Chadwyck [318]

Anuario estadistico 1911-1965 / El Salvador. Direccion General de Estadistica – (= ser Latin american & caribbean...1821-1982) – 288mf – 9 – (1915, 1918, 1923-24 estadistica comercial; 1936, 1945 v1, 1963 v1 not available) – uk Chadwyck [318]

Anuario estadistico 1936-1954 / Dominican Republic. Direccion General de Estadistica y Censos – (= ser Latin american & caribbean... 1821-1982) – 199mf – 9 – uk Chadwyck [318]

Anuario estadistico 1938-1947 / Nicaragua. Direccion General de Estadistica – (= ser Latin american & caribbean...1821-1982) – 17mf – 9 – (1938, 1946 not available) – uk Chadwyck [318]

Anuario estadistico 1944-1957 / Argentine Republic. Direccion Nacional de Estadistica y Censos – (= ser Latin american & caribbean... 1821-1982) – 52mf – 9 – (1951-56 not publ) – uk Chadwyck [318]

Anuario estadistico 1948-1950 / Spanish Sahara. Secretario General – (= ser African official statistical serials, 1867-1982) – 8mf – 9 – uk Chadwyck [316]

Anuario estadistico 1952-1969 / Honduras. Direccion General de Estadistica y Censos – (= ser Latin american & caribbean...1821-1982) – 66mf – 9 – (1969 not available) – uk Chadwyck [318]

Anuario estadistico 1968-1969 / Nicaragua. Ministerio de Economia, Industria y Commercio and Banco Central de Nicaragua – (= ser Latin american & caribbean...1821-1982) – 4mf – 9 – uk Chadwyck [318]

Anuario estadistico de cuba 1952, 1956-1957 / Cuba. Direccion General de Estadistica – (= ser Latin american & caribbean...1821-1982) – 22mf – 9 – uk Chadwyck [318]

Anuario estadistico de espana / Instituto Geografico – Madrid, 1858-1867 – 76mf – 9 – sp Cultura [314]

Anuario estadistico de espana 1858-1867, 1912-1934, 1943-1970 / Spain. Instituto Nacional de Estadistica – (= ser European official statistical serials, 1841-1984) – 504mf – 9 – (1970 not available) – uk Chadwyck [314]

Anuario estadistico de la republica de paraguay, ano 1887 – Asuncion, 1889 – 9mf – 9 – sp Cultura [318]

Anuario estadistico de la republica oriental de uruguay, ano 1884-1896 – Montevideo, 1885-1898 – 64mf – 9 – sp Cultura [318]

Anuario estadistico de peru 1944/1945-1958/1966 / Peru. Direccion de Estadistica – (= ser Latin american & caribbean...1821-1982) – 104mf – 9 – (1958-66 not available) – uk Chadwyck [318]

Anuario estadistico republica de chile, contralor – Santiage de Chile, Chile. 1914-1926 – 7r – (gaps) – us UF Libraries [079]

Anuario estatistico 1926-1973 = Statistical yearbook 1926-1973 / Mozambique. Reparticao Tecnica de Estatistica – (= ser African official statistical serials, 1867-1982) – 290mf – 9 – uk Chadwyck [316]

Anuario estatistico 1933-1952 / Cape Verde Islands. Seccao de Estatistica – (= ser African official statistical serials, 1867-1982) – 28mf – 9 – uk Chadwyck [316]

Anuario estatistico 1933-1973 / Angola. Reparticao de Estatistica Geral – (= ser African official statistical serials, 1867-1982) – 147mf – 9 – uk Chadwyck [316]

Anuario estatistico 1947-1958 / Guinea-Bissau (formerly Portuguese Guinea). Reparticao Provincial dos Servicos de Economia e Estatistica Geral – (= ser African official statistical serials, 1867-1982) – 19mf – 9 – uk Chadwyck [316]

Anuario estatistico de brasil 1908/1912-1969 / Brazil. Instituto Brasileiro de Geografia e Estatistica – (= ser Latin american & caribbean...1821-1982) – 207mf – 9 – (1946 not available) – uk Chadwyck [318]

Anuario estatistico de portugal 1875-1970 = Annuaire statistique 1875-1970 / Portugal. Instituto Nacional de Estatistico – (= ser European official statistical serials, 1841-1984) – 344mf – 9 – uk Chadwyck [314]

Anuario general de estadística 1905-1969/1970 / Colombia. Departamento Administrativo Nacional de Estadistica – (= ser Latin american & caribbean...1821-1982) – 243mf – 9 – (1906-14 not publ. 1918-25, 1926-28, 1935, 1955, 1958, 1963, 1965 v4 not available) – uk Chadwyck [318]

Anuario geografico y estadistico de la republica de bolivia 1919 / Bolivia. Direccion General de Estadistica y Estudios Geograficos – (= ser Latin american & caribbean...1821-1982) – 8mf – 9 – uk Chadwyck [318]

Anuario legislativo de instruccion publica... 1885-1909 – Madrid, 1890-1910 – 200mf – 9 – sp Cultura [340]

Anuario meteorologia / Ministerio De Transportes Y Comunicaciones – La Paz, Bolivia. 1975 – 1r – us UF Libraries [550]

Anuario nacional estadistico y geografico de bolivia 1917 / Bolivia. Direccion General de Estadistica y Estudios Geograficos – (= ser Latin american & caribbean...1821-1982) – 7mf – 9 – uk Chadwyck [318]

Anuario para o ano de...publicado pelo observatorio nacional rio de janeiro / Observatorio Nacional [Brazil] – Rio de Janeiro (annual) [mf ed 2001] – anno 6-66(1890-1950) on 10r [ill] – 1 – (cont: annuario [imperial observatorio do rio de janeiro]; cont by: efemerides astronomicas; ceased in 1976?; imprint varies) – mf#film mas c5052 – us Harvard [520]

Anuarul statistic al romaniei 1904-1939/40 / Romania. Institut Central de Statistica – (= ser European official statistical serials, 1841-1984) – 93mf – 9 – (irregular. 1905-08, 1910-11 not publ. 1930 not available) – uk Chadwyck [314]

Anuarul statistic al rpr 1957-1970 / Romania. Directiunea Centrala de Statistica – (= ser European official statistical serials, 1841-1984) – 125mf – 9 – uk Chadwyck [314]

Anucasana parva – Calcutta: Bharata Press, 1893 [mf ed 1993] – (= ser The mahabharata of krishna-dwaipayana vyasa) – 2mf – 9 – 0-524-08008-9 – (trans by kesari mohan ganguli) – mf#1991-0230 – us ATLA [490]

El anunciador – Trinidad, colo: la compania publicista de "el anunciador" [apr 1918-nov 18 1922] – 2r – 1 – us CRL [079]

El anunciador *see* Miscellaneous newspapers of las animas county, reel 2

Anunciar : revista para catequistas – Quito: Imprenta del Colegio Tecnico Don Bosco, [ano 1 n1-ano 9 n108 (1984-1992)] (mthly) – 2r – 1 – us CRL [241]

Anup Singh *see* Nehru, the rising star of india

anuruddha *see* Tan chve e sangruih path

Anuvario estatistico de angola / Angola. Reparticao de Estatistica Geral – 1 – us L of C Photodup [960]

Anvar, 'Isharat Hasan *see* The metaphysics of iqbal

The anvar-i suhaili, or the lights of canopus / ed by Kashifi, Husayn Vaiz – Hertford: Stephen Austin 1854 [mf ed Bloomington IN: Indiana Uni Lib, Preservation Dept 1984] – 650p on 1r – 1 – (moral fables of bidpai, trans into english) – us Indiana Preservation [390]

Anvers : ou, la prise de la citadelle / Ces-Caupenne, Octave – Paris, France. 1833 – 1r – us UF Libraries [440]

Anvil – 1934 mar-aug – 1 – mf#147048 – us WHS [071]

Anvil – Killen TX 1949-60 – 1 – ISSN: 0003-6226 – mf#2487 – us UMI ProQuest [410]

The anvil / Frenssen, Gustav – Boston; New York: Houghton Mifflin 1930 [mf ed 1989] – 1r – 1 – (trans of: otto babendiek by huntley paterson. filmed with: moewen und maeuse) – mf#7266 – us Wisconsin U Libr [830]

The anvil *see* Miscellaneous newspapers of pueblo county

Anvil chorus – 1949 nov-51 sep – 1 – mf#1109769 – us WHS [071]

The anvil: the proletarian fiction magazine – Moberly, Mo. etc. v. 1-3 no. 13. May 1933-Oct Nov 1935 – 1 – us NY Public [335]

Anville, J B B d' *see*
– Memoire de m. d'anville, premier geographe du roy, des academies royales des belles-lettres, & des sciences. svr la chine
– Nouvel atlas de la chine.

Anville, Jean-Baptiste B d' *see* Memoires sur l'egypte ancienne et moderne

Anwand, Oscar *see* Beitraege zum studium der gedichte von j.m.r. lenz

Anwander, Georg *see* Christliche predigt von der vocal und instrumentalischen music

Anweisung fuer ansiedler an die ottawa und opeongo strasse und umgegend / French, Thomas P – Toronto: [s.n, 1857?] [mf ed 1993] – 1mf – 9 – 0-665-91891-7 – (also available in english) – mf#91891 – cn Canadiana [917]

Die anwendung der lippenbluetler in der zahnheilkunde von der antike bis heute / Kolek, Iveta – (mf ed 2000) – 2mf – 9 – €40.00 – 3-8267-2709-6 – mf#DHS 2709 – gw Frankfurter [617]

Die anwendung des buches hiob in der rabbinischen agadah : 1. theil, die tannaitische interpretation von hillel bis chija nach schulen / Kaufmann, Herman Ezechiel – Frankfurt a. M: J Kauffmann, 1893 – 1mf – 9 – 0-8370-3849-9 – (incl bibl ref) – mf#1985-1849 – us ATLA [221]

Die anwendung von asteraceae (korbbluetler) in der zahnheilkunde von der antike bis heute / Hammerich, Angelika – (mf ed 2000) – 221p on 3mf – 9 – €49.00 – 3-8267-2700-2 – mf#DHS 2700 – gw Frankfurter [617]

Anwendungsorientierte modellierung der bodenerosionsgefahr auf alpweiden am beispiel des gunzesrieder tals (oberallgaeu) / Proswitz, Elisabeth – (mf ed 1998) – 2mf – 9 – €40.00 – 3-8267-2529-8 – mf#DHS 2529 – gw Frankfurter [550]

Anwyl, Edward *see* Celtic religion in pre-christian times

The anxious bench / Nevin, John Williamson – 2nd rev enl ed. Chambersburg, PA: printed... the German Reform Church, 1844 [mf ed 1990] – 1mf – 9 – 0-7905-6418-1 – (1st publ 1843) – mf#1988-2418 – us ATLA [240]

The anxious inquirer after salvation, directed and encouraged / James, John Angell – Toronto: repr fr London ed by Lovell & Gibson, 1850 [mf ed 1994] – 2mf – 9 – 0-665-94673-2 – mf#94673 – cn Canadiana [240]

An anxious moment etc / Hungerford, Margaret Wolfe (Hamilton) – London: Chatto & Windus, 1897 – (= ser 19th c women writers) – 4mf – 9 – mf#5.1.95 – uk Chadwyck [830]

Anythony's lagoon mortuary book 1890-1948 *see* Borooloola inquest book, 28 december 1889 to 10 november 1930

Anz, Henricus *see* Subsidia ad cognoscendum graecorum sermonem vulgarem e pentateuchi versione alexandrine repetita

Anz journal of surgery – Oxford, England 2001+ – 1,5,9 – (cont: australian and new zealand journal of surgery) – ISSN: 1445-1433 – mf#2731,01 – us UMI ProQuest [617]

Anz, W *see* Zur frage nach dem ursprung des gnostizismus

Anz, Wilhelm *see* Zur frage nach dem ursprung des gnostizismus

Anzaas congress papers / Australian and New Zealand Association for the Advancement of Science – 1970-97 – 9 – price varies – ISSN: 0 – at UNSW Lib [500]

Anzanische inschriften und vorarbeiten zu ihrer entzifferung / Weissbach, Franz Heinrich – Leipzig: S Hirzel, 1891 [mf ed 1986] – (= ser Abhandlungen der philologisch-historischen classe der koenigl saechsischen gesellschaft der wissenschaften 12/2) – 1mf – 9 – 0-8370-7751-6 – (comm in german; text in elamite) – mf#1986-1751 – us ATLA [470]

Anzano, T *see* Elementos preliminares para poder formar un systema de gobierno del hospicio general

Anzeige der leipziger oekonomischen gesellschaft (societaet) – Dresden 1764-1814 – (= ser Dz) – 62iss on 72mf – 9 – €720.00 – mf#k/n2878 – gw Olms [330]

Anzeige-blatt der kreishauptstadt speyer *see* Speyerer anzeige-blatt 1811

Anzeigeblatt der staedt behoerden zu frankfurt – Frankfurt/M DE, 1874, 1882, 1886, 1890, 1898, 1900, 1902 – 7r – 1 – gw Misc Inst [350]

Anzeige-blatt des kreises zweybruecken *see* Zweybrueckisches wochenblatt

Anzeige-blatt fuer den kreis biedenkopf und bezirk voehl – Biedenkopf DE, 1983- – ca 7r/yr – 1 – (bezirksausgabe von wetzlarer neue zeitung; title varies: 3 jan 1849: der hinterlaender bote /.../; 6 jan 1869: kreisblatt; 3 jan 1877: hinterlaender anzeiger, later regional ed of: wetzlarer neue zeitung; filmed by other misc inst: 1841 9 jan-25 dec, 1843, 1846, 1847, 1849, 1851, 1854, 1855, 1861, 1863, 1865-69 [several gaps], 1872-75 [gaps], 1877-1943, 1949 1 aug-1969 24 oct; with suppls) – gw Misc Inst [074]

Anzeige-blatt fuer den markt redwitz *see* Wochen-blatt fuer den markt redwitz und umgegend

Anzeigeblatt fuer die stadt giessen – Giessen, Lahn DE, 1848-49 – 1r – 1 – (title varies: 2 jan 1868: giessener anzeiger; filmed by misc inst: 1969- [ca 11r/yr]) – gw Misc Inst [074]

Anzeigen fuer das fuerstenthum luebeck *see* Eutinische woechentliche anzeigen

Anzeigen fuer den landdrosteibezirk stade *see* Intelligenz-blatt des nord-departements

Anzeigen fuer tanga – Tanga (EAT), 1912 6 jan-23 nov – 1r – 1 – (filmed by misc inst: 1901/02, 1904-16 [gaps]) – gw Misc Inst [079]

Anzeigen fuer tanga – Tanga (EAT), 1912 6 jan-23 nov – 1r – 1 – (filmed by other misc inst: 1901/02, 1904-16 [gaps]) – gw Misc Inst [079]

Anzeigen fuer tanga : veroeffentlichungsstelle fuer bekanntmachungen der kaiserlichen behoerden – Tanga (EAT), 1912 6 jan-23 nov – 1r – 1 – (filmed by other misc inst: 1901/02, 1904-16 [gaps]; later: usambara-post) – gw Misc Inst [079]

Anzeigenaushang fuer den kreis herzogtum lauenburg – Ratzeburg DE, 1946 14 may-1949 30 sep [gaps] – 1r – 1 – gw Misc Inst [943]

Anzeigenblatt fuer den kreis rendsburg – Rendsburg DE, 1948 23 jul-1949 29 mar 29 – 1r – 1 – gw Misc Inst [074]

Anzeiger / Hamilton Co. Cincinnati – jan 2 1881-oct 20 1901 – 77 – 1 – (in german) – mf#B36996-37072 – us Ohio Hist [071]

Anzeiger – Cottbus DE, 1900, 1901 jul/dec-1944 jan-jun – 89r – 1 – (title varies: 1 jul 1871: cottbuser anzeiger) – gw Misc Inst [074]

Anzeiger *see* Dortmunder wochenblatt

Der anzeiger – Gotha DE, 1794, 1795 [gaps], 1796-97, 1798 [gaps], 1800-09, 1826-37 – 45r – 1 – (title varies: 1 jul 1793: der reichsanzeiger, 19 sep1806: allgemeiner anzeiger der deutschen, 2 jan 1830: allgemeiner anzeiger und national-zeitung der deutschen, 2 jan 1850: reichsanzeiger der deutschen) – gw Misc Inst [074]

Der anzeiger – Goerlitz DE, 1870 n1-149 – 1r – 1 – (title varies: 6 jan 1803: neuer goerlitzer anzeiger, 14 jan 1808: goerlitzer anzeiger, 14 jan 1876: goerlitzer nachrichten und anzeiger, 1 feb 1929: vereinigte goerlitzer nachrichten und niederschlesische zeitung, 1 jan 1932: goerlitzer nachrichten; filmed by other misc inst: 1799 3 jan-1943 31 mar [238r]) – gw Misc Inst [074]

Anzeiger der bibliothekwissenschaft *see* Anzeiger fuer literatur der bibliothekwissenschaft 1840-1844

Anzeiger der verordnungen der landesverwaltungen und gerichte – Bayreuth DE, 1808 4 oct-1809 [gaps], 1810 3 jul-3 aug, 1811 – 2r – 1 – gw Misc Inst [350]

Anzeiger des siegkreises – Siegburg DE, 1958-1962 12 dec – 1 – (filmed by other misc inst: 1855 31 dec-1868, 1871-73, 1874 15 feb-1875 19 dec, 1876-1937 27 feb, 1949 29 oct-1967 13 aug; title varies: 1861?: siegburger kreisblatt; 3 jan 1866: kreisblatt des rhein-siegkreises; 1868: siegburger zeitung; 1885: siegburger kreisblatt; 1925: siegburger zeitung; 1935: neue siegburger zeitung; 29 oct 1949: siegburger zeitung) – gw Misc Inst [074]

Anzeiger fuer barr und umgebung – Barr, F. 1900-16 – 1 – fr ACRPP [944]

Anzeiger fuer den kreis paderborn *see* Paderborner kreisblatt ueber politik, handel und gewerbe

Anzeiger fuer den kreis pless – Pless (Pszczyna PL), 1923 15 apr-30 dec, 1924 feb-1931, 1932 10 aug-30 sep – 6r – 1 – (aka: nikolaier anzeiger and: plesser stadtblatt) – gw Misc Inst [077]

Anzeiger fuer den landkreis eutin *see* Eutinische woechentliche anzeigen
Anzeiger fuer die gesamte kinematografen-industrie – Bruenn (Brno CZ), 1907 sep-1908 21 feb – 1r – 1 – gw Mikrofilm [790]
Anzeiger fuer die landrathlichen kreise aschersleben, calbe, mansfeld – Aschersleben, Calbe S, Mansfeld DE, 1855 3 jan-1880, 1882-84, 1886-88, 1890, 1892-1900 – gw Misc Inst [074]
Der anzeiger fuer goldingen und windau – Goldingen (Kuldiga LV), 1927 5 nov-1929 [gaps] – 1 – gw Misc Inst [077]
Anzeiger fuer harlingerland – Wittmund DE, 1988- – 7r/yr – 1 – gw Misc Inst [074]
Der anzeiger fuer hemelingen : und die bremer suedoestlichen vororte hastedt, sebaldsbrueck und osterholz – Bremen DE, 1938 1 oct-1940 30 sep – 4r – 1 – gw Misc Inst [943]
Anzeiger fuer literatur der bibliothekswissenschaft 1840-1844 – [mf ed 1990] – 83mf – 9 – €500.00 – 3-89131-034-X – (filmed with: anzeiger der bibliothekswissenschaft, dresden und leipzig sp halle 1845-49, anzeiger fuer bibliographie und bibliothekswissenschaft, halle 1850-55, neuer anzeiger fuer bibliographie und bibliothekswissenschaft, dresden sp berlin u stuttgart 1856-1886) – gw Fischer [020]
Anzeiger fuer lommatzsch und umgegend – Lommatzsch DE, 1854-1943 15 apr – 59r – 1 – (title varies: 1903?: lommatzscher anzeiger; 1 jul 1926: lommatzscher anzeiger und tageblatt) – gw Misc Inst [074]
Anzeiger fuer oberhessen *see* Intelligenzblatt fuer die provinz oberhessen
Anzeiger fuer perleberg und umgegend – Perleberg DE, 1844 11 feb-1848 1 apr – 1 – gw Misc Inst [074]
Anzeiger fuer rosswein und umgegend – Rosswein DE, 1834-1945? – ca 116r – 1 – (title varies: 1 jul 1882: rossweiner tageblatt) – gw Misc Inst [074]
Anzeiger fuer rosswein, waldheim und die umliegenden orte – Rosswein DE, 1839 11 jan-1897 – 29r – 1 – (title varies: 23 sep 1848: anzeiger und unterhaltungsblatt fuer waldheim, rossheim, hartha und die umgegend, 1852: anzeiger und unterhaltungsblatt fuer doebeln, waldheim, hartha, rosswein und die umliegenden ortschaften, 2 jan 1858: anzeiger und amtsblatt fuer die koeniglichen gerichtsaemter und stadtraethe zu doebeln, hartha, waldheim, 3 jan 1863: anzeiger fuer doebeln, hartha und waldheim, 2 jan 1876: anzeiger fuer waldheim und hartha, 1888: anzeiger und tageblatt fuer waldheim und hartha) – gw Misc Inst [074]
Anzeiger fuer schweizerische altertumskunde – Zuerich, 1869-1920 – (= ser Architectural periodicals at avery library, columbia university) – 9r – 1 – $1240.00 – us UPA [930]
Anzeiger fuer sobernheim, kirn und umgegend 1859 – Sobernheim DE, 1859-60 [gaps], 1862 7 jan-1866 [gaps], 1867 3 jul-29 dec – 1 – (title varies: 21 mar 1860: anzeiger fuer sobernheim und umgegend, 27 mar 1862: anzeiger fuer kirn und umgegend, 30 mar 1862: anzeiger fuer sobernheim und umgegend, 2 apr 1862: anzeiger fuer sobernheim, kirn und umgegend, 3 jul 1862: sobernheimer und kirner intelligenzblatt, 3 jul 1864: sobernheim-kirner intelligenzblatt) – gw Misc Inst [074]
Anzeiger fuer stadt und kreis schluechtern – Schluechtern DE, 1902 11 jan-1904 30 nov – 1 – (title varies: 15 nov 1902: schluechterner anzeiger fuer stadt und kreis) – gw Misc Inst [074]
Anzeiger fuer weisswasser – Weisswasser DE, 1904 6 jan-31 mar & 1 oct-25 dec, 1909-1910 30 mar, 1913 1 jan-29 jun, 1914 3 jan-31 mar – 3r – 1 – (tw. auch: rietschener tageblatt) – gw Misc Inst [074]
Anzeiger und amtsblatt fuer das koenigliche gerichtsamt und den stadtrath zu leisnig *see* Leisniger wochenblatt
Anzeiger und post – Lawrence MA (USA), 1934 6 jan-1939 25 nov – 3r – 1 – gw Misc Inst [071]
Anzeiger und unterhaltungsblatt fuer waldheim, rossheim, hartha und die umgegend *see* Anzeiger fuer rosswein, waldheim und die umliegenden orte
Anzeiger und wochenblatt fuer hoerde, schwerte, aplerbeck, westhofen und umgebung – Dortmund DE, 1957 1 apr-1859, 1884-85, 1888-95, 1897-1900, 1902, 1905 jan-jun, 1909-1910 mar, 1910 jun-1911 1 jan, 1912-1914 19 sep, 1915 1 jul-1917 29 sep, 1918-1919 30 aug, 1920-1926 30 sep, 1927 3 jan-31 mar, 1927 1 jul-1932 jun, 1933-1941 31 may [gaps], 1949 1 nov-1950, 1951 18 jun-1955 14 mar – 1 – (missing: 1914 jul-12 sep; title varies: 1860: hoerder volksblatt, 22 mar 1934: volksblatt, 1955?: westdeutsche allgemeine / hoerder volksblatt; with suppl: auftragsbuch der anzeigen-redaktion 1918 18 mar-1919 16 may) – gw Misc Inst [074]
Anzeiger von oberkotzau – Oberkotzau DE, 1908 18 jul, 1910-14, 1919 16 oct-1927 12 sep, 1929-39, 1951 24 aug-1970 – 22r – 1 – (title varies: 1 oct 1910: oberkotzauer zeitung) – gw Misc Inst [074]
Anzeiger von wurzach – Bad Wurzach DE, 1896 12 may-1907 – 1 – (aka: der bote vom allgaeu) – gw Misc Inst [074]
Anzeigung was der gebrauch vnd gewonhait in des turcken land ist ... – n.p, 1526 – 1mf – 9 – mf#H-8139 – ne IDC [956]
Anzelc-Spesia, Meredith L *see* The effects of exercise on premenstrual syndrome and progesterone concentrations
Anzengruber / David, Jakob Julius – Berlin: Schuster & Loeffler [1920?] [mf ed 1988] – (= ser Die dichtung 2) – 1r [ill] – 1 – (Filmed with: Das vierte Gebot / Ludwig Anzengruber) – mf#6951 – us Wisconsin U Libr [430]
Anzengruber, Ludwig *see*
- Allerhand humore
- Anzengrubers werke in vierzehn teilen
- Aus'm gewohntem g'leis
- Brave leut' vom grund
- Briefe von ludwig anzengruber
- Doppelselbstmord
- Dorfgaenge
- Elfriede
- Ein faustschlag
- Gesammelte werke
- Der g'wissenswurm
- Hand und herz
- Heimg'funden!
- Kleiner markt
- Der ledige hof
- Letzte dorfgaenge
- Ludwig anzengrubers ausgewaehlte werke
- Ludwig anzengrubers gesammelte werke
- Der meineidbauer
- Der pfarrer von kirchfeld
- 'S jungferngift
- Stahl und stein
- Der sternsteinhof
- Die tochter des wucherers
- Das vierte gebot

Anzengrubers werke in vierzehn teilen / ed by Bettelheim, Anton – Berlin: Bong, [1918] [mf ed 1988] – 14v in 7 (ill) – 1 – mf#6946 – us Wisconsin U Libr [802]
Anzuelo de dios / Lindo, Hugo – San Salvador, El Salvador. 1962 – 1r – us UF Libraries [972]
Ao Bha Sa Bhi Vam Sa, Ther, A Rhan *see* Mran ma a bhi dhan
Ao la ri ka su khu ma dhat kyam / Aon Dan, U – Ran Kun: U Khan Mon 1977 [mf ed 1990] – 1r with other items – 1 – (in burmese) – mf#mf-10289 seam reel 165/5 [§] – us CRL [615]
Ao nagas / Majumder, Surendra Nath – Calcutta: Sailen Majumdar, 1925 – (= ser Samp: indian books) – us CRL [305]
Ao som da viola / Barroso, Gustavo – Rio de Janeiro, Brazil. 1949 – 1r – us UF Libraries [972]
Aoba local council, new hebrides : minute book – 16 nov 1962-9 jan 1969 – 1r – 1 – mf#pmb48 – at Pacific Mss [350]
Aon Chve, Mon, Takkasuil *see*
- A khyac Ivan su-le phru phru
- Ka le lu nay ca pe

Aon Dan, U *see* Ao la ri ka su khu ma dhat kyam
Aon Kyo Cann *see* A dhi pa ti lam
Aon Prann *see* Le thai chok tai tuik ka le
Aon Puin Phrui et al *see* Mran ma ca ka pum phrac can vatthu mya
Aone's leadership prospectives – Philadelphia PA 1993-96 – 1,5,9 – (cont: nursing scan in administration) – ISSN: 1072-5067 – mf#20629 – us UMI ProQuest [610]
Aonio paleario : a chapter in the history of the italian reformation / Bonnet, Jules – London: Religious Tract Society, 1884 [mf ed 1990] – 1mf – 9 – 0-7905-5633-2 – (incl bibl ref. in english) – mf#1988-1633 – us ATLA [242]
Aonio paleario and his friends : with a revised edition of the benefit of christ's death / Blackburn, William Maxwell & Benedetto da Mantova – Philadelphia: Presbyterian Board of Pub, 1866 [mf ed 1992] – 1mf – 9 – 0-524-04607-7 – mf#1990-1267 – us ATLA [242]
Aontas gaedheal weekly post – Dublin, 31 may-28 jun 1935 – 0.5r – 1 – ie National [072]
Aos aspirantes da escolar militar / Monteiro, Goes – Rio de Janeiro, Brazil. 1942 – 1r – us UF Libraries [355]
Aotea news – Tryphena, NZ. 1985-89 – 2r – 1 – mf#11.62 – nz Nat Libr [079]
Aotearoa he nupepa ma nga tangata maori – Napier, NZ. 1892 – 1r – 1 – mf#31.5 – nz Nat Libr [079]
Ap world – v1 n1-v14 n3 [1945 jan/feb-59: autumn] – 1 – mf#604904 – us WHS [071]
APA del Colegio "Santisima Trinidad" de Plasencia *see* Estatutos
Apa monitor / American Psychological Association – Washington DC 1970-99 – 1 – (cont by: monitor on psychology) – ISSN: 0001-2114 – mf#6088.01 – us UMI ProQuest [150]
The apa movement : a sketch / Desmond, Humphrey Joseph – Washington: New Century Press, 1912 [mf ed 1990] – 1mf – 9 – 0-7905-5818-1 – mf#1988-1818 – us ATLA [360]
Apacible, G *see* To the american people
Apalachicola commercial advertiser – Apalachicola, FL. 1844 jan-1848 [incomplete] – 1r – us UF Libraries [071]
Apalachicola (Fla) Ordinances, Etc *see* Code of ordinances of the city of apalachicola, fl...
Apalachicola, florida – s.l, s.l? . 193-? – 1r – us UF Libraries [978]
Apalachicola gazette – Apalachicola, FL. 1836 mar 10-1839 dec 21 – 1r – us UF Libraries [071]
Apalachicola times – Apalachicola, FL. 1946 sep-1992 – 29r – (gaps) – us UF Libraries [071]
Les apaotres : essai d'histoire religieuse d'apres la methode des sciences naturelles / Ferriere, Emile – Paris: Germer Bailliere, 1879 – 2mf – 9 – 0-7905-1324-2 – (incl bibl ref) – mf#1987-1324 – us ATLA [240]
Les apaotres / Renan, Ernest – Paris: Michel Levy, 1866 – 2mf – 9 – 0-8370-9412-7 – (incl bibl ref) – mf#1986-3412 – us ATLA [240]
Aparato bibliografico...extremadura / Barrantes Moreno, Vicente – 1875. 3 tomos – 9 – sp Bibl Santa Ana [946]
Aparencia do rio de janeiro / Cruls, Gastao – Rio de Janeiro, Brazil. v1-2. 1965 – 1r – us UF Libraries [972]
Aparicio, Raul *see* Hijos del tiempo
Aparta de tus ojos / Socorro De Tinoco, Maria Del – San Jose, Costa Rica. 1947 – 1r – us UF Libraries [972]
Apartment ideas – New York NY 1969-73 – 1 – (cont by: apartment life) – ISSN: 0003-6366 – mf#9299.02 – us UMI ProQuest [640]
Apartment life – New York NY 1974-80 – 1,5,9 – (cont by: metropolitan home) – ISSN: 0092-0444 – mf#9299.02 – us UMI ProQuest [640]
The apartments of the house : their arrangement furnishing and decoration / Crouch, Joseph & Butler, Edmund – London: At the Sign of the Unicorn, 1900 – (= ser 19th c art & architecture) – 3mf – 9 – mf#4.1.82 – uk Chadwyck [740]
Apastamba, yagna-paribhasha-sutras *see* The grihya-sutras (stbe30)
The apatite deposits of canada / Hunt, Thomas Sterry – S.l: s.n, 1884? – 1mf – 9 – mf#07733 – cn Canadiana [622]
Apco horizons : a publication of the african peoples' christian organization – 1985 jan; 1987 may – 1 – mf#4877706 – us WHS [071]
Apea reporter – 1976 jul, v4-5 (1977) [1]; 1980 oct-1987 jun/aug [2] – 1 – mf#600118 [1]; 1520602 [2] – us WHS [071]
Apelles symbolicus exhibens seriem amplissimam symbolorum, poetisque, oratoribus ac verbi dei praedicatoribus conceptus subministrans varios / Ketten, J M von der – Amstelaedami & Gedani: Apud Janssonio-Waesbergios, 1699. 2v – 13mf – 9 – mf#0-320 – ne IDC [090]
Apendice a la...salida de don quixote / Habela Patino, Eugenio – 1789 – 9 – sp Bibl Santa Ana [830]
Apendice de la memoria historica / Gonzalez, Manuel Dionisio – Santa Clara, Cuba. 1925 – 1r – us UF Libraries [972]
Apenrader tageblatt – Apenrade (Aabenraa DK), 1920 21 dec-1929 31 jan – 1 – gw Misc Inst [074]
Apercu de l'origine des diverses ecritures de l'ancien monde / Klaproth, Heinrich Julius – Paris [1832] – (= ser Whsb) – 2mf – 9 – €30.00 – (fr: courtin, eustache marie pierre marc antoine: encyclopedie moderne) – mf#Hu 016 – gw Fischer [410]
Apercu du plan d'etudes et de la methode d'enseignement / Universite d'Ottawa – Ottawa: [s.n.], 1893 [mf ed 1987] – 1mf – 9 – 0-665-34054-0 – mf#34054 – cn Canadiana [378]
Apercu du plan d'etudes et de la methode d'enseignement suivis : au college d'ottawa – Ottawa: [s.n.], 1882 [mf ed 1980] – 1mf – 9 – 0-665-00711-6 – mf#00711 – cn Canadiana [370]
Apercu general du voyage de m brosset dans la transcaucasie – Spb, 1850. v7 – 2mf – 9 – mf#R-1702 – ne IDC [910]
Apercu historique et statistique sur la regence d'alger : intitule en arabe le miroir / Hamdan ibn Uthman Khawajah – Paris 1833 [mf ed Hildesheim 1995-98] – 1v on 3mf – 9 – €90.00 – 3-487-27352-7 – gw Olms [960]
Apercu statistique de l'ile de cuba : precede de quelques lettres sur la havane, et suivi de tableaux synoptiques / Huber, B – Paris 1826 [mf ed Hildesheim 1995-98] – 1v on 3mf – 9 – €90.00 – 3-487-26942-2 – gw Olms [318]
Apercu sur la formation historique de la nation ha... / Charlier, Etienne D – Port-Au-Prince, Haiti. 1954 – 1r – us UF Libraries [972]
Apercu sur les structures grammaticales des langues / Houis, Maurice – Lyon, France. 1967 – 1r – us UF Libraries [440]
Apercu sur quelques contemporains – [S.l: s.n, 18–] [mf ed 1980] – 1mf – 9 – 0-665-04201-9 – mf#04201 – cn Canadiana [355]
Apercu sur quelques contemporains – [S.l: s.n, 18–?] [mf ed 1984] – 1mf – 9 – 0-665-47691-4 – mf#47691 – cn Canadiana [355]
Apercus de taxinomie generale / Durand, Joseph-Pierre – Paris: F Alcan, 1899 – 1r – 1 – us CRL [560]
Apercus sur la biscaye, les asturies et la galice : precis de la defense des frontieres du guipuscoa et de la navarre; par le general don ventura caro, en 1793 et 1794 / Marcillac, Pierre L de – Paris 1807 [mf ed Hildesheim 1995-98] – 2mf – 9 – €60.00 – 3-487-29859-7 – gw Olms [946]
Apercus sur l'espagne chretienne du 4th siecle : ou le "de lapso" de bachiarius / Duhr, J – Louvain, 1934 – 3mf – 8 – €7.00 – ne Slangenburg [240]
Apercus sur linstitution communale / Price, Hannibal – Port-Au-Prince, Haiti. 1902 – 1r – us UF Libraries [972]
Aperture – Millerton NY 1952+ – 1,5,9 – ISSN: 0003-6420 – mf#2156 – us UMI ProQuest [770]
Apes, William *see* The experiences of five christian indians of the pequod tribe
Apfelbaum, Abe *see* Mosheh zakuth
Apg news – 1980 oct 22-1981 jun-1993 jan-dec – 1 – mf#565089 – us WHS [071]
Apg newsletter – 1979 jul-1985 dec – 1 – mf#823455 – us WHS [071]
Apha letter – n1-1968 [1974 nov-1985 nov/dec] – 1 – mf#378533 – us WHS [071]
Aphasiology – Abingdon, Oxfordshire 1991+ – 1,5,9 – ISSN: 0268-7038 – mf#17294 – us UMI ProQuest [616]
Aphorismen / Euringer, Richard – Hamburg: Hanseatische Verlagsanstalt c1943 [mf ed 1989] – 1r – 1 – (filmed with: die arbeitslosen) – mf#7226 – us Wisconsin U Libr [880]
Aphorismi confessariorum / Sa, Emm – Duaci, 1623 – 8mf – 8 – €14.00 – ne Slangenburg [240]
Aphorismi doctrinae christianae / Piscator, J – Herbornae, 1605 – 5mf – 9 – mf#PBA-296 – ne IDC [240]
Aphorismi urbigerani : or certain rules, clearly demonstrating the three infallible ways of preparing the grand elixir... / Urbigerus, Baro – London: printed for H Faitborne 1690 [mf ed 1984] – 1r – 1 – mf#1231 – us Wisconsin U Libr [540]
Aphorismorum libri sex de consideratione eucharistiae... / Vadian, J – Zurch, Cristof Froshower, [1536] – 5mf – 9 – mf#PBU-401 – ne IDC [240]
Aphorismos sacados de la historia de p.c. tacito / Arias Montano, Benito – 1614 – 9 – sp Bibl Santa Ana [450]
Aphorisms and reflections : conduct, culture and religion / Spalding, John Lancaster – Chicago: AC McClurg, 1901 [mf ed 1990] – 1mf – 9 – 0-7905-5968-4 – mf#1988-1968 – us ATLA [100]
The aphorisms of sandilya : with the commentary of swapneswara, or, the hindu doctrine of faith – Calcutta: Asiatic Society of Bengal, 1878 [mf ed 1993] – (= ser Bibliotheca indica) – 1mf – 9 – 0-524-07145-4 – (english by edward byles cowell) – mf#1991-0075 – us ATLA [280]
Aphra – Brooklyn NY 1969-76 – 1,5,9 – ISSN: 0003-6447 – mf#6557 – us UMI ProQuest [320]
Aphraates, frere *see* Arithmetique
Aphraetes : patrologia syriaca 1 / ed by Graffin, B – Paris, 1894 – 2v on 24mf – 8 – €46.00 – ne Slangenburg [240]
Aphrahat's des persischen weisen homilien : die akten des karpus, des papylus und der agathonike: eine urkunde aus der zeit marc aurel's / Bert, Georg & Harnack, Adolf von – Leipzig: J C Hinrichs, 1888 [mf ed 1989] – (= ser Tugal 3/3) – 2mf – 9 – 0-7905-4010-X – (in german & greek) – mf#1988-0010 – us ATLA [240]
Aphrahat's des persischen weisen homilien / Bert, Georg – Leipzig, 1888 – (= ser Tugal 1-3/4a) – 7mf – 9 – €15.00 – ne Slangenburg [240]

Aphroditographische fragmente, zur genauern kenntniss des planeten venus : sammt beygefuegter beschreibung des lilienthalischen 27 fuessigen telescops, mit practischen bemerkungen und beobachtungen ueber die groesse der schoepfung / Schroeter, Johann Hieronymus – Helmstedt: C G Fleckeisen 1796 [mf ed 1998] – 1r [pl/ill] – 1 – (incl bibl ref) – mf#film mas 28419 – us Harvard [520]

Aphum pan khyi u cam tui *see* Cac kuin u bhui san / aphum pan khyi u cam tui

Api – Djakarta: Masa Merdeka (daily) [mf ed Jakarta, Indonesia: Perpustakaan Nasional 1987] – 1r [oct 10-15 1965] – 1 – mf#mf-12762 seam – us CRL [079]

Apiacta – Bucharest, Romania 1975 – 1,5,9 – ISSN: 0003-6455 – mf#9718 – us UMI ProQuest [630]

Apic – Oxford, England 1977 – 1,5,9 – ISSN: 0161-8717 – mf#12232.03 – us UMI ProQuest [610]

Apic keynoter : news of the american political items – 1970 spring-1978 spring – 1 – mf#1496170 – us WHS [321]

Apices juris : and other legal essays in prose and verse / Morse, Charles – Toronto: Canadian Law Book Co, 1906 – 4mf – 9 – 0-665-77607-1 – mf#77607 – cn Canadiana [340]

Apis und este : [so fing es an] / Brehm, Bruno – Muenchen: R Piper c1931 [mf ed 1989] – 1r – 1 – (filmed with other titles) – mf#7066 – us Wisconsin U Libr [830]

Apla quarterly journal *see* Aipla quarterly journal

Aplerbecker zeitung – Dortmund DE, 1919 9 dec-1922 31 aug [gaps] – 1r – 1 – gw Misc Inst [074]

Apo (african political organisation) – [Cape Town: The Organisation. [v1 n2-v8 n219]. jun 5 1901-apr 8 1922 – 5r – 1 – us CRL [325]

Apocalipseos interpretatio litteralis : ejusque cum aliis libris sacris concordantia / Eyzaguirre, Raphaele – Romae: Ex Officina Unionis Editricis, 1911 [mf ed 1993] – 2mf – 9 – 0-524-06834-8 – mf#1992-0976 – us ATLA [225]

Apocalypse / Huntington, William Reed – London, England. 1892 – 1r – us UF Libraries [240]

The apocalypse : an introductory study of the revelation of st. john the divine: being a presentment of the structure of the book and of the fundamental principles of its interpretation / Benson, Edward White – London: Macmillan, 1900 – 1mf – 9 – 0-8370-2275-4 – (includes appendix) – mf#1985-0275 – us ATLA [221]

The apocalypse : its structure and primary predictions / Brown, David – New York: Christian Literature Co, 1891 [mf ed 1985] – 1mf – 9 – 0-8370-2474-9 – mf#1985-0474 – us ATLA [225]

The apocalypse : or, revelation of s john the divine: six lectures / Scott, Joseph John – London: J Murray, 1909 [mf ed 1993] – 1mf – 9 – 0-524-05632-3 – mf#1992-0487 – us ATLA [225]

The apocalypse : viewed under the light of the doctrines of the unfloding ages and the restitution of all things / Waller, Charles B – London: C Kegan Paul, 1878 [mf ed 1985] – 1mf – 9 – 0-8370-5689-6 – mf#1985-3689 – us ATLA [225]

The apocalypse : with a commentary and an introduction on the reality of prediction, the history of christendom... / Huntingford, Edward – London: Kegan Paul, Trench, 1881 [mf ed 1985] – 1mf – 9 – 0-8370-3696-8 – mf#1985-1696 – us ATLA [225]

The apocalypse : with notes and reflections / Williams, Isaac – London: Francis & John Rivington, 1852 [mf ed 1989] – 2mf – 9 – 0-7905-2698-0 – mf#1987-2698 – us ATLA [225]

L'apocalypse de jean / Loisy, Alfred Firmin – Paris: Emile Nourry, 1923 [mf ed 1985] – 1mf – 9 – 0-8370-4657-2 – (in french. incl bibl ref and ind) – mf#1985-2657 – us ATLA [225]

L'apocalypse de s jean : ordonnance et interpretation des visions allegoriques et prophetiques de ce livre / Gallois, M-Aug – Paris: P Lethielleux, 1895 [mf ed 1993] – 1mf – 9 – 0-524-06333-8 – (in french and latin) – mf#1992-0871 – us ATLA [225]

Apocalypse explained / Swedenborg, Emanuel – New York, NY. v1-6. 1897 – 2r – us UF Libraries [240]

The apocalypse explained : light for the times / Collins, George – Ottawa: Hunter, Rose, 1869 [mf ed 1985] – 1mf – 9 – 0-665-03013-4 – mf#03013 – cn Canadiana [225]

The apocalypse of baruch / ed by Charles, Robert Henry – London: A & C Black, 1896 – 1r – 1 – 0-8370-0505-1 – (transl from the syriac) – mf#1984-B278 – us ATLA [221]

The apocalypse of jesus christ : an exposition / Mead, Willis W – New York: WW Mead, 1909 [mf ed 1985] – 1mf – 9 – 0-8370-4370-0 – (incl bibl ref) – mf#1985-2370 – us ATLA [225]

The apocalypse of st john : the greek text / Swete, Henry Barclay – 3rd ed. London, New York: Macmillan, 1909 [mf ed 1989] – 2mf – 9 – 0-7905-2876-2 – (in english & greek. incl bibl ref, int, notes & ind) – mf#1987-2876 – us ATLA [225]

The apocalypse of st john 1-3 : the greek text with introduction, commentary, and additional notes / Hort, Fenton John Anthony – London: Macmillan, 1908 [mf ed 1985] – 1mf – 9 – 0-8370-3660-7 – (incl ind) – mf#1985-1660 – us ATLA [225]

The apocalypse of st john 1-3 / Hort, Fenton John Anthony – 1908 – 9 – $10.00 – us IRC [240]

Apocalypse revealed / Swedenborg, Emanuel – Boston, MA. v1-2. 1907 – 1r – us UF Libraries [240]

Apocalypse unveiled and a fight with death and slander / Gow, William – Perth, Australia. 1888 – 1r – us UF Libraries [240]

Apocalypses apocryphae : mosis, esdrae, pauli, iohannis, item, mariae dormitio / ed by Tischendorf, Constantin von – Lipsiae: H Mendelssohn, 1866 [mf ed 1990] – 3mf – 9 – 0-8370-1780-7 – (text in greek & latin) – mf#1987-6168 – us ATLA [221]

Les apocalypses juives : essai de critique litteraire et theologique / Faye, Eugene de – Paris: Fischbacher, 1892 [mf 1989] – 1mf – 9 – 0-7905-0882-6 – (incl bibl ref) – mf#1987-0882 – us ATLA [221]

Apocalypsis alfordiana : or, five letters to the very rev h alford, dean of canterbury... / Elliott, Edward Bishop – London: Seeley, Jackson & Halliday, 1862 [mf ed 1985] – 1mf – 9 – 0-8370-3059-5 – (incl bibl ref) – mf#1985-1059 – us ATLA [225]

Apocalypsis et actus apostolorum : cum quarti maccabaeorum libri fragmento / ed by Tischendorf, Constantin von – Lipsiae: JC Hinrichs, 1869 [mf ed 1986] – (= ser Monumenta sacra inedita. nova collectio 6) – 4mf – 9 – 0-8370-9427-5 – mf#1986-3427 – us ATLA [090]

Apocalypsis et actus cum fragmentis evangelicis (msi6) / ed by Tischendorf, G F C – Lipsiae, 1869 – (= ser Monumenta sacra inedita. nova collectio v6) – €52.00 – ne Slangenburg [220]

Apocalyptic sketches : lectures on the seven churches of asia minor / Cumming, John – Philadelphia: Lindsay & Blakiston, 1854 [mf ed 1992] – 2mf – 9 – 0-524-04090-7 – mf#1992-0048 – us ATLA [225]

Apocalyptical key : an extraordinary discourse on the rise and fall of papacy... / Fleming, Robert – New York: American & Foreign Christian Union, 1855 [mf ed 1986] – 1mf – 9 – 0-8370-8338-9 – mf#1986-2338 – us ATLA [225]

Apocalyptical key / Fleming, Robert – London, England. 1793 – 1r – us UF Libraries [240]

The apocrypha : greek and english in parallel columns – 2nd ed. London: Samuel Bagster, 1906 – 1mf – 9 – 0-8370-1269-4 – (greek and english in parallel columns) – mf#1987-6035 – us ATLA [221]

The apocrypha : translated out of the greek and latin tongues, being the version set forth a.d. 1611, compared with the most ancient authorities and revised a.d. 1894 – New York: Thomas Nelson, [1894?] – 1mf – 9 – 0-524-08068-2 – mf#1992-1128 – us ATLA [221]

Apocrypha and pseudepigrapha of the old testament / Charles, Robert Henry – 1913 – 9 – $36.00 – us IRC [221]

The apocrypha and pseudepigrapha of the old testament in english : with introductions and critical and explanatory notes to the several books / ed by Charles, Robert Henry – Oxford: Clarendon Press, 1913 – 15mf – 9 – 0-8370-1850-1 – mf#1987-6237 – us ATLA [221]

Apocrypha anecdota : a collection of 13 apocryphal books and fragments / ed by James, Montague Rhodes – Cambridge: University Press; New York: Macmillan [dist] 1893 [mf ed 1989] – (= ser Texts and studies (cambridge, england) 2/3) – 1mf – 9 – 0-7905-1333-1 – (in latin & greek. int in english. incl ind) – mf#1987-1333 – us ATLA [220]

Apocrypha anecdota / ed by James, Montague Rhodes – Cambridge: University Press 1897 [mf ed 1989] – (= ser Texts and studies (cambridge, england) 5/1) – 1mf – 9 – 0-7905-1900-3 – (in greek & english; int in english, german, greek & latin; incl ind) – mf#1987-1900 – us ATLA [220]

Apocrypha anecdota (ts5/1) : second series / ed by James, M R – 1897 – (= ser Texts and studies (ts)) – 5mf – 9 – €12.00 – ne Slangenburg [220]

Apocrypha anedocta (ts2/3) / ed by James, M R – 1893 – (= ser Texts and studies (ts)) – 4mf – 9 – €11.00 – ne Slangenburg [220]

Apocrypha arabica / ed by Gibson, Margaret Dunlop – London: C.J. Clay; New York: Macmillan [dist] 1901 [mf ed 1990] – (= ser Studia sinaitica 8) – 1mf – 9 – 0-8370-1667-3 – (english trans by ed) – mf#1987-6097 – us ATLA [225]

Apocrypha controversy : aberdeenshire auxiliary bible society – Edinburgh, Scotland. 1827 – 1r – us UF Libraries [240]

Apocrypha controversy – Edinburgh, Scotland. 1826 – 1r – us UF Libraries [240]

Apocrypha controversy – Edinburgh, Scotland. 1829 – 1r – us UF Libraries [240]

Apocrypha controversy : review of the statement by the glasgow diss... – Edinburgh, Scotland. 1826 – 1r – us UF Libraries [240]

The apocrypha of the old testament : with historical introductions, a revised translation, and notes critical and explanatory / Bissell, Edwin Cone – New York: Charles Scribner, 1880 [mf ed 1986] – (= ser A commentary on the holy scriptures. old testament 15) – 2mf – 9 – 0-8370-6723-5 – (incl app) – mf#1986-0723 – us ATLA [221]

Apocrypha sinaitica / ed by Gibson, Margaret Dunlop – London: C J Clay, 1896 [mf ed 1990] – (= ser Studia sinaitica 5) – 1v on 1mf – 9 – 0-8370-1834-X – (english trans by ed) – mf#1987-6222 – us ATLA [225]

The apocryphal acts of paul, peter, john, andrew and thomas – Chicago: Open Court, 1909 – 1mf – 9 – 0-8370-1912-5 – (includes bibliographic references) – mf#1987-6299 – us ATLA [226]

The apocryphal and legendary life of christ : being the whole body of the apocryphal gospels and other extra canonical literature which pretends to tell of the life and words of jesus christ... / by Donehoo, James DeQuincey – New York: Macmillan, 1903 – 2mf – 9 – 0-8370-1992-3 – mf#1987-6379 – us ATLA [220]

Apocryphal gospel of peter – London, England. 1892 – 1r – us UF Libraries [226]

Apocryphal gospels : a lecture delivered in the new hall of science / Cowper, Benjamin Harris – London, England. 1874 – 1r – us UF Libraries [240]

The apocryphal gospels and other documents relating to the history of christ – 6th ed. London: D. Nutt, 1897. Chicago: U of Chicago Lib, 1975 (1r); Evanston: American Theol Lib Assoc, 1984 (1r) – 1 – 0-8370-1550-2 – mf#1984-B494 – us ATLA [240]

The apocryphal new testament / James, M R – Oxford, 1924 – 9 – $21.00 – us IRC [240]

The apocryphal new testament... – London: W. Hone, 1820,1888 printing. xv,271p – 1 – us Wisconsin U Libr [240]

Apocryphes coptes du nouveau testament / ed by Revillout, Eugene – Paris: F Vieweg, 1876 [mf ed 1990] – (= ser Etudes egyptologiques 7) – 1mf – 9 – 0-8370-1842-0 – (no more publ?) – mf#1987-6230 – us ATLA [225]

Les apocryphes de l'ancien testament / Andre, Louis Edouard Tony – Florence: Osvaldo Paggi, 1903 [mf ed 1985] – 1mf – 9 – 0-8370-2098-0 – (incl ind. also available in reels) – mf#1985-0098 – us ATLA [221]

Apogryphal acts of the apostles edited from syriacs mss / Wright, W – London. v1-2. 1871 – €23.00 – ne Slangenburg [226]

Apokalipsis v russkoi literature / Kruchenykh, A – 1923 – 46p 1mf – 8 – mf#R-951 – ne IDC [243]

Die apokalypse / Hesler, Heinrich von; ed by Helm, Karl – Berlin: Weidmann, 1907 [mf ed 1993] – (= ser Deutsche texte des mittelalters 8; Dichtungen des deutschen ordens 1) – 1 – (incl bibl ref & ind) – mf#8623 reel 3 – us Wisconsin U Libr [810]

Apokalypse / antichrist / ars memorandi / canticum canticorum / defensorium inviolatae virginitatis b. mariae / biblia pauperum / ars moriendi / speculum humanae salvationis / kalender des regiomontanus (mxt5) : farbmikrofiche-edition der blockbuecher der universitaetsbibliothek muenchen, cim.45-45a, 46-47a, 48-52, 40 – (mf ed 2002) – (= ser Monumenta xylographica et typographica (mxt) 5) – ca 40p on ca 7 color mf – 15 – ca €350.00 – 3-89219-405-X – gw Lengenfelder [090]

Apokalypse / ars moriendi / biblia pauperum / antichrist / fabel vom kranken loewen / kalendarium und planetenbuecher / historia david (mxt2) : die lateinisch-deutschen blockbuecher des berlin-breslauer sammelbandes. berlin, staatliche museen preussischer kulturbesitz, kupferstichkabinett, cim 1, 2, 5, 7, 9, 10, 12. farbmikrofiche-edition – (mf ed 1992) – (= ser Monumenta xylographica et typographica (mxt) 2) – 98p/5pl on 4 color mf – 15 – €335.00 – 3-89219-402-5 – (int & description by nigel f palmer) – gw Lengenfelder [090]

Apokalypse / ars moriendi / medizinische traktate / tugend- und lasterlehren (cima39) : die erbaulich-didaktische sammelhandschrift london, wellcome institute for the history of medicine, ms 49. farbmikrofiche-edition – (mf ed 1995) – (= ser Codices illuminati medii aevi (cima) 39) – 76p on 3 color mf – 15 – €290.00 – 3-89219-039-9 – (int, catalogue & ind by almuth seebohm) – gw Lengenfelder [090]

Die apokalypse des elias / Steindorff, G – Leipzig, 1899 – (= ser Tugal 2-17/3a) – 3mf – 9 – €7.00 – ne Slangenburg [221]

Die apokalypse des elias : eine unbekannte apokalypse und bruchstuecke der sophonias-apokalypse / Steindorff, Georg – Leipzig: J C Hinrichs, 1899 [mf ed 1989] – (= ser Tugal 17/3a) – 1mf – 9 – 0-7905-1851-1 – (in german & coptic) – mf#1987-1851 – us ATLA [221]

Apokalypse / koenigsberger apokalypse (cima27) : mikrofiche-edition der handschriften torun, biblioteka uniwersytetu mikolaja kopernika, ms rps 64 und ms rps 44 / Hesler, Heinrich von – (mf ed 2000) – (= ser Codices illuminati medii aevi (cima) 27) – 64p on 3 color+3 b/w mf – 15 – €260.00 – 3-89219-027-5 – (int & description by volker honemann) – gw Lengenfelder [090]

Die apokalypse (mxt1) : blockbuch-ausgabe 4 e. farbmikrofiche-edition des exemplars mainz, gutenberg-museum, ink 131 – (mf ed 1991) – (= ser Monumenta xylographica et typographica (mxt) 1) – 34p on 1 color mf – 15 – €135.00 – 3-89219-401-7 – (int by elke purpus) – gw Lengenfelder [090]

Die apokalypse und ihre neueste kritik / Hirscht, Arthur – Leipzig: August Neumann, 1895 – 1mf – 9 – 0-8370-3592-9 – (incl bibl ref) – mf#1985-1592 – us ATLA [221]

Apokrificheskie teksty / Lavrov, P A – 1899. v6(3) – 4mf – 8 – mf#R-4079 – ne IDC [243]

Apokrificheskiia skazaniia o novozavetnykh litsakh i sobytiiakh, po rukopisiam soloveckoi biblioteki / Porfirev, I I – 1890 – 471p 9mf – 8 – (sbornik otdeleniia russkago iazyka i slo vesnosti imp akademii nauk,52:4) – mf#R-4072 – ne IDC [243]

Apokrificheskiia skazaniia o vetkhozavetnykh litsakh i sobytiiakh po rukopisiam solovetskoi biblioteki / Porfirev, I I – 1877 – 5mf – 8 – (sbornik otdeleniia ruskago iazyka i slo vesnosti ak nauk, tom 17 n1) – mf#R-4067 – ne IDC [243]

Apokrify i lehendy z ukrainskykh rukopisiv : zibrav, uporiadkovav i poiasnyv i franko – U Lvovi, 1896-1910. 5v – 37mf – 8 – mf#R-4057 – ne IDC [243]

Apokrisis / Filalet Khristofor – [Ostrog, 1598] – 9mf – 9 – mf#RHB-38 – ne IDC [460]

Die apokryphen : vertheidigung ihres althergebrachten anschlusses an die bibel / Stier, R – Braunschweig: C A Schwetschke, 1853 – 1mf – 9 – 0-7905-3171-2 – (incl bibl ref) – mf#1987-3171 – us ATLA [220]

Die apokryphen apostelgeschichten und apostellegenden : ein beitrag zur altchristlichen literaturgeschichte / Lipsius, Richard Adelbert – Braunschweig: C A Schwetschke, 1883-1887. Chicago: Dep of Photodup, U of Chicago Lib, 1971 (1r); Evanston: American Theol Lib Assoc, 1984 (1r) – 1 – 0-8370-0539-6 – (incl ind) – mf#1984-B255 – us ATLA [225]

Die apokryphen briefe des paulus an die laodicener und korinther / ed by Harnack, Adolf von – Bonn: A Marcus & E Weber 1905 [mf ed 1992] – (= ser Kleine texte fuer theologische vorlesungen und uebungen 12/4) – 1mf – 9 – 0-524-04751-0 – (incl bibl ref; text in latin & greek; notes in german) – mf#1992-0193 – us ATLA [227]

Die apokryphen des alten testaments : ein zeugniss wider dieselben auf grund des wortes gottes / Keerl, Philipp Friedrich – Leipzig: Gebhardt und Reisland, 1852 – 1mf – 9 – 0-7905-3143-7 – mf#1987-3143 – us ATLA [221]

Die apokryphen und pseudepigraphen des alten testaments / ed by Kautzsch, Emil – Tuebingen: J C B Mohr, 1900 – 3mf – 9 – 0-8370-1776-9 – (incl ind) – mf#1987-6164 – us ATLA [221]

Die apokryphen und pseudepigraphen des alten testaments / Kautzsch, Emil F – Tuebingen, 1900. neudr. 1921 – 28mf – 8 – €54.00 – ne Slangenburg [221]

Apolineo caduceo...concordia entre opiniones... sobre consultas de los medicos... / Luque, C – Sevilla, 1694 – 7mf – 9 – sp Cultura [610]

Apollinarios von laodicea / Draeseke, J – Leipzig, 1892 – (= ser Tugal 1-7/3.4) – 8mf – 9 – €17.00 – ne Slangenburg [920]

Apollinarios von laodicea : sein leben und seine schriften / Draeseke, Johannes – Leipzig: JC Hinrichs, 1892 [mf ed 1989] – (= ser Tugal 7/3-4) – 2mf – 9 – 0-7905-4031-2 – (in german, greek & latin. incl bibl ref) – mf#1988-0031 – us ATLA [240]

Apollo / Allen, Wilkes – 1790. Cover title. Manuscript collection of vocal and instrumental music in 1, 2, 3, and 4 parts, playable by solo instruments or keyboard and 1 or 2 solo instruments. Six of the tunes are the compiler's originals. MUSIC 1971 – 1 – us L of C Photodup [780]

Apollo – London, England 1925+ – 1,5,9 – ISSN: 0003-6536 – mf#1382 – us UMI ProQuest [700]

Apollo : eine monatsschrift / ed by Meissner, August Gottlieb – Prag, Leipzig 1793-97 – (= ser Dz. abt literatur) – 27mf – 9 – €270.00 – mf#k/n4594 – gw Olms [400]

Apollo 11 moon landing newspaper selections, may-aug 1969 – 8r – 1 – mf#B29350-29357 – us Ohio Hist [355]

Apollo e dafne : balletto anacreontico, composto e diretto dal sig. pietro hus. rappresentato per la prima volta in palermo nel real teatro santa cecilia nell'autunno del 1825 / Hus, Pierre – Palermo: Dalla Società tip., 1825 – 1 – (scenery designed by giovanni li volsi and gaetano riolo) – mf#*ZBD-*MGTZ pv 7-Res – Located: NYPL – us Misc Inst [790]

Apollodorus *see* Epitoma vaticana ex apollodori bibliotheca

Apollon – Washington. 1957-1975 (1) 1971-1973 (5) – 213mf – 9 – mf#1105 – ne IDC [077]

L'apollon moderne : ou, le developpement intellectuel par les sons de la musique... / Brijon, C R – 2. ed, Lyon: [s.n.] 1782 [mf ed 19–] – 6mf – 9 – mf#fiche 699 – us Sibley [780]

Apollonios rhodios : interpretationen zur erzaehlungskunst und quellenverwertung / Stoesel, Franz – Bern: P Haupt 1941 [mf ed 1979] – 1r – 1 – (incl bibl ref) – mf#film mas 8407 – us Harvard [450]

Apollonius, Dyscolus *see* De pronominibus, pars generalis

Apollonius of tyana : the pagan christ of the third century: an essay = Le christ paien au 3e siecle / Reville, Albert – London: John Camden Hotten 1866 [mf ed 1990] – 2mf – 9 – 0-7905-7457-8 – (in english) – mf#1989-0682 – us ATLA [240]

Apollonius von tyana und christus : oder, das verhaeltniss des pythagoreismus zum christenthum / Baur, Ferdinand Christian – Tuebingen: LF Fues, 1832 [mf ed 1990] – 1mf – 9 – 0-7905-7043-2 – mf#1988-3043 – us ATLA [180]

Apollonius von tyrland (cima49) : farbmikrofiche-edition der handschrift chart a 689 der forschungs- und landesbibliothek gotha / Neustadt, Heinrich von – (mf ed 1998) – (= ser Codices illuminati medii aevi (cima) 49) – 39p on 6mf – 15 – €335.00 – 3-89219-049-6 – (int by wolfgang achnitz) – gw Lengenfelder [090]

Apollonius, W *see* Jus majestatis circa sacra...

Apollos : or, studies in the life of a great layman of the first century / Wynne, George Robert – London: SPCK; New York: E S Gorham, 1912 [mf ed 1989] – 1mf – 9 – 0-7905-0539-8 – (incl bibl ref) – mf#1987-0539 – us ATLA [225]

Apolo ne kama – Kasempa, Zambia. 1955 – 1r – us UF Libraries [960]

Apolo y coatlicue / Cardoza Y Aragon, Luis – Mexico City?, Mexico. 1944 – 1r – us UF Libraries [972]

Der apologet aristides : der text seiner uns erhaltenen schriften nebst einleitenden untersuchungen ueber dieselben = Apology for the christian faith / Aristides – Erlangen: A Deichert, 1894 [mf ed 1991] – 1mf – 9 – 0-7905-9116-2 – (incl bibl ref; in german & greek) – mf#1989-2341 – us ATLA [240]

Apologetic lectures on the fundamental truths of christianity : delivered in leipsic...winter 1864 = Apologetische vortraege ueber die grundwahrheiten des christenthums / Luthardt, Christoph Ernst – 7th ed. Edinburgh: T & T Clark, 1888 [mf ed 1985] – 2mf – 9 – 0-8370-4203-8 – (trans by sophia taylor; incl ind) – mf#1985-2203 – us ATLA [240]

Apologetic lectures on the fundamental truths of christianity : delivered in leipzig in the winter of 1864 = Apologetische vortraege ueber die grundwahrheiten des christenthums / Luthardt, Christoph Ernst – 6th ed. Edinburgh: T & T Clark, 1882 [mf ed 1984] – 6mf – 9 – 0-8370-0856-5 – (english by sophia taylor. incl ind) – mf#1984-4212 – us ATLA [240]

Apologetic lectures on the moral truths of christianity / Luthardt, Christoph Ernst – 3rd ed. Edinburgh: T & T Clark, 1881 [mf ed 1984] – 5mf – 9 – 0-8370-0858-1 – (incl ind) – mf#1984-4210 – us ATLA [240]

Apologetic lectures on the saving truths of christianity / Luthardt, Christoph Ernst – 4th ed. Edinburgh: T & T Clark, 1880 [mf ed 1984] – 5mf – 9 – 0-8370-0857-3 – (english by sophia taylor. incl ind) – mf#1984-4211 – us ATLA [240]

The apologetic of modern missions : eight outline studies / Murray, John Lovell – New York: Student Volunteer Movement, c1909 – 1mf – 9 – 0-8370-6824-X – mf#1986-0824 – us ATLA [240]

The apologetic of the new testament / Scott, Ernest Findlay – London: Williams & Norgate; New York: G P Putnam, 1907 [mf ed 1985] – (= ser Crown theological library 22) – 1mf – 9 – 0-8370-5189-4 – (incl ind) – mf#1985-3189 – us ATLA [240]

Apologetic postscript to the rhapsody / Barton, E – Dublin, Ireland. 1823 – 1r – us UF Libraries [240]

Apologetica expositio / Bullinger, Heinrich – Tigvri, Andreas et Iacobus Gesner, [1556] – 2mf – 9 – mf#PBU-190 – ne IDC [240]

Apologetica ioann oecolampadii de dignitate evcharistiae sermones duo / Oecolampadius, J – Zuerich, Christof Froschover, 1526 – 4mf – 9 – mf#PBU-367 – ne IDC [240]

Apologetico discurso...en que se prueba que los polvos de quarango se deben usar... / Salado Garces de Leon, D – Sevilla, 1687 – 1mf – 9 – sp Cultura [615]

Apologetics : a course of lectures / Smith, Henry Boynton; ed by Karr, William Stevens – New York: AC Armstrong, 1882, c1881 [mf ed 1985] – 1mf – 9 – 0-8370-5289-0 – mf#1985-3289 – us ATLA [230]

Apologetics : or, a system of christian evidence / Lindberg, Conrad Emil – Rock Island IL: Augustana Book Concern 1917 [mf ed 1992] – 1mf – 9 – 0-524-05013-9 – (incl bibl ref) – mf#1991-2183 – us ATLA [240]

Apologetics : or, christianity defensively stated / Bruce, Alexander Balmain – New York: Scribner, 1892 [mf ed 1989] – (= ser The international theological library 3) – 2mf – 9 – 0-7905-3314-6 – (incl bibl ref & ind) – mf#1987-3314 – us ATLA [240]

Apologetics : or, the rational vindication of christianity in three volumes. vol 1: fundamental apologetics / Beattie, Francis Robert – Richmond, VA: Presbyterian Comm of Publ, c1903 [mf ed 1985] – 2mf – 9 – 0-8370-2223-1 – (incl ind. no more publ?) – mf#1985-0223 – us ATLA [210]

Apologetics : or, the scientific vindication of christianity / Ebrard, Johannes Heinrich August – [2d ed.] Edinburgh: T & T Clark 1886-87 [mf ed 1984] – (= ser Clark's foreign theological library. new series 26,29,31) – 3v on 1r – 1 – 0-8370-0754-2 – (incl bibl ref & ind) – mf#1984-t108 – us ATLA [240]

Apologeticum scriptum / Micronius, M – n.p, 1557 – 2mf – 9 – mf#PBA-266 – ne IDC [240]

Apologetique et raison dans les pensees de pascal / Bouchilloux, Helene – 2mf – 9 – (10064) – fr Atelier National [190]

Apologetische, dogmatische und montanistische schriften, 2.bd (bdk24) / Tertullian – (= ser Bibliothek der kirchenvaeter. 1. reihe (bdk)) – €19.00 – ne Slangenburg [240]

Das apologetische schreiben des josua lorki an den abtruennigen don salomon ha-lewi (paulus de santa maria) *see* Igeret r yehoshua ha-lorki

Apologetische vortraege / Fuchs, M et al – Barmen: H Klein, [18–?] [mf ed 1992] – 1mf – 9 – 0-524-02062-0 – mf#1990-0559 – us ATLA [240]

Apologetische vortraege : von den bernischen geistlichen bernard, dubuis, von greyerz, gueder...gehalten im winter 1869 auf 1870 / Bernard, Auguste et al – Bern: Haller, 1870 [mf ed 1985] – 1mf – 9 – 0-8370-2107-3 – mf#1985-0107 – us ATLA [240]

Apologetische vortraege ueber die grundwahrheiten des christenthums / Luthardt, Christoph Ernst – Leipzig: Doerffling und Franke, 1864 [mf ed 1985] – 1mf – 9 – 0-8370-4205-4 – mf#1985-2205 – us ATLA [240]

Apologetische vortraege ueber die grundwahrheiten des christenthums *see* Apologetic lectures on the fundamental truths of christianity

Apologetische vortraege ueber die heilswahrheiten des christenthums : im winter 1867 zu leipzig gehalten / Luthardt, Christoph Ernst – 6. durchgesehene Aufl. Leipzig: Doerffling und Franke, 1890. Chicago: Dep of Photodup, U of Chicago Lib, 1975 (1r); Evanston: American Theol Lib Assoc, 1984 (1r) – (= ser HIS apologie des christentums) – 1 – 0-8370-1274-0 – (incl ind) – mf#1984-6025 – us ATLA [240]

Apologetische zeitstimmen / Oosterzee, Johannes Jacobus van – Guetersloh: C Bertelsmann, 1868 [mf ed 1985] – 1mf – 9 – 0-8370-3937-1 – (trans fr dutch into german by friedrich & ludwig meyeringh) – mf#1985-1937 – us ATLA [240]

Apologi creaturarum : g de jode excu / [Moerman, J] – Antverpiae: Excudebat Gerardo Judeae Christophorus Plantinus, 1584 – 2mf – 9 – mf#O-698 – ne IDC [090]

Apologia : an explanation and defence / Abbott, Edwin Abbott – London: Adam & Charles Black, 1907 [mf ed 1985] – 1mf – 9 – 0-8370-2011-5 – mf#1985-0011 – us ATLA [226]

Apologia : oder verantwortung dess christlichen concordienbuchs jn welcher die wahre christliche lehre vertheydiget / Kirchner, T – Heydelberg, 1583 – 5mf – 9 – mf#TH-1 mf 806-810 – ne IDC [242]

Apologia auff die vermeinte widerlegung des osiandrischen schwermers in preussen : m vogels sampt gruendlichem kurtzen bericht, was der haubtstreit vnd die lere osiandri gewesen sey / Moerlin, J – [Magdeburg], 1557 – 1mf – 9 – mf#TH-1 mf 1174 – ne IDC [242]

Apologia catolica... / Sallaberry, Juan Faustino – Madrid: Razon y Fe, 1930 – 1 – sp Bibl Santa Ana [241]

Apologia confessionis de coena domini, contra corrvptelas calumnias ionnis caluini / Westphal, J aus Hamburg – Vrsellis, 1558 – 5mf – 9 – mf#TH-1 mf 1469-1473 – ne IDC [242]

Apologia danielis hofmanni : missa ad theodorvm bezam qua to reton in verbis coenae sacrae / Hoffmann, D – Helmstadii, 1586 – 7mf – 9 – mf#TH-1 mf 691-697 – ne IDC [242]

Apologia de la lengua bascongada : o ensayo critico filosofico de su perfeccion y antigueedad / Astarloa y Aguirre, Pablo Pedro de – Madrid: Ortega 1803 – (= ser Whsb) – 5mf – 9 – €60.00 – mf#Hu 049 – gw Fischer [410]

Apologia de la pequena nacion / Picon-Salas, Mariano – Rio Piedras, Puerto Rico. 1946 – 1r – us UF Libraries [972]

Apologia de las 7 de la manana / Entralgo, Elias Jose – Habana, Cuba. 1950 – 1r – us UF Libraries [972]

Apologia de las 7 de la manana / Entralgo, Elias Jose – Habana, Cuba. 1959 – 1r – us UF Libraries [972]

Apologia de los banos de la muy noble y leal ciudad de alhama / Vergara Cabezas, F – Granada, 1636 – 2mf – 9 – sp Cultura [615]

Apologia del doctor... / Luna Vega, J – Sevilla, 1605 – 1mf – 9 – sp Cultura [610]

Apologia del presidente roosevelt y un poema / Brenes Mesen, Roberto – San Jose, Costa Rica. 19– – 1r – us UF Libraries [972]

Apologia m casparis aqvilae / Adler, K – [Magdeburg], 1548 – 1mfmf – 9 – mf#TH-1 mf 1 – ne IDC [242]

Apologia matthiae flacij illyrici ad scholam vitebergensem in adiaphororum causa / Flacius Illyricus d A, M – [Magdeburgi, 1549] – 1mf – 9 – (missing title pg) – mf#TH-1 mf 463 – ne IDC [242]

Apologia medicinalis acedunt egregiae censurae de venae sextione in febribus... / Vaez, P – Barcelona, 1593 – 5mf – 9 – sp Cultura [615]

Apologia oder verantwortung dess christlichen concordienbuchs / Chemnitz d A, M – Heydelberg, 1583 – 5mf – 9 – mf#TH-1 mf 199-203 – ne IDC [242]

Apologia pro consilio medicinali in diminute visiones adversus duas epitolas... / Aguiar, T – Marchena, 1621 – 6mf – 9 – sp Cultura [610]

Apologia pro justificatione adversus lescalium / Beza, Theodor de – Geneve, Le Preux, 1592 – 4mf – 9 – mf#PFA-116 – ne IDC [240]

Apologia pro reverendis et illvstris principibvs catholicis : ac alijs ordinibus imperij aduersus mucores & calumnias buceri, super actis comitiorum ratisponae. apologia pro reuerendiss se ap legato & cardinale, caspare contareno / Eck, Johann – Coloniae: Novesiana, 1542. Chicago: Dep of Photodup, U of Chicago Lib, 1971 (1r); Evanston: American Theol Lib Assoc, 1984 (1r) – 1 – 0-8370-0490-X – mf#1984-B250 – us ATLA [241]

Apologia pro sanctissima virgine maria, matre domini / Rivetus, Andr – Lugd Batavorum, 1639 – 8mf – 8 – €17.00 – ne Slangenburg [241]

Apologia pro vita sua : being a reply to a pamphlet entitled "what, then, does dr newman mean?" / Newman, John Henry – London: Longman, Green, Longman, Roberts & Green, 1864 [mf ed 1990] – 2mf – 9 – 0-7905-7430-6 – mf#1989-0655 – us ATLA [241]

Apologia sive excusatio atque etiam assertio veritatis / Occam, Guillelmus de (Ockham, William of) – Lugduni, 1495 – €5.00 – ne Slangenburg [241]

Apologia...del agua de la vida...en que se hace examen y juicio de los papeles... / Amigo y Bertran, L – Zaragoza, 1682 – 1mf – 9 – sp Cultura [610]

Apologie / Moded, H – Haarlem, 1879 – 1mf – 9 – mf#PBA-273 – ne IDC [240]

Apologie d'apulee / Vallette, Paul – Paris, France. 1908 – 1r – us UF Libraries [960]

Apologie der selbstausloesung : ethik und metaphysik in philipp mainlaenders theorie des zerfalls / Mueller, Winfried H – (mf ed 1996) – 3mf – 9 – €49.00 – 3-8267-2328-7 – mf#DHS 2328 – gw Frankfurter [110]

Die apologie des aristides : recension und rekonstruktion des textes = Apology for the christian faith / Hennecke, Edgar – Leipzig: JC Hinrichs, 1893 [mf ed 1989] – (= ser Tugal 4/3) – 1mf – 9 – 0-7905-4011-8 – mf#1988-0011 – us ATLA [240]

Die apologie des aristides / Raabe, R – Leipzig, 1893 – (= ser Tugal 1-9/1b) – 2mf – 9 – €5.00 – ne Slangenburg [240]

Apologie des christenthums / Luthardt, Christoph Ernst – Leipzig: Doerffling & Franke, 1880-1898 – 2r – 1 – mf#1984-B406 – us ATLA [240]

Apologie des christenthums *see* Christian apology

Apologie des eglises reformees... / Daille, Jean – Charenton, 1641 – 3mf – 9 – mf#PRS-135 – ne IDC [240]

Apologie du gout francois, relativement a l'opera : poeme, avec un discours apologetique, et des adieux aux bouffons / Caux de Cappeval – [Paris 1754] [mf ed 19–] – 2mf – 9 – mf#fiche 388 – us Sibley [780]

L'apologie d'un incredule *see* Reasons for unbelief

Een apologie of verandtwoordinghe : op 20 verscheyden artikelen... / Micronius, M – [Embden], 1558 – 5mf – 9 – mf#PBA-2 – ne IDC [240]

Een apologie of verandtwoordinghe : op 20 verscheyden artikelen die menno symons... / Micronius, M – Embden, 1557 – 2mf – 9 – mf#PBA-264 – ne IDC [240]

Apologie pour la morale des reformez... / [Jurieu, P] – Quevilly, 1675 – 7mf – 9 – mf#PRS-149 – ne IDC [240]

Apologie pour la reformation, pour les reformateurs, et pour les reformez / Jurieu, P – Rotterdam, 1683 – 11mf – 9 – mf#CA-134 – ne IDC [242]

Apologie pour les catholiques contre les faussetez et les calomnies d'un livre intitule : la politique du clerge de france / Arnauld, A – Liege, 1681-1682. 2v – 13mf – 9 – mf#CA-114 – ne IDC [944]

Apologie pour l'ordre des francsmacons : [avec deux chansons composees par le frere americain] – A la Haye: Chez Pierre Gosse 1742 – 2mf – 9 – mf#vrl-7 – ne IDC [366]

Apologie tegen : een uitgegeven vraagboekje door ds h van der werp, bij de h chr ger gemeente te roseland, ill / Meinders, E L – Holland, MI: De Grondwet en News Stoomdrukkerij, 1889 [mf ed 1993] – (= ser Reformed church coll) – 1mf – 9 – 0-524-06644-2 – mf#1991-2699 – us ATLA [242]

Apologie...en laquelle est demonstré... / Bullinger, Heinrich – [Geneve], Matthieu de la Roche, 1558 – 2mf – 9 – mf#PBU-192 – ne IDC [240]

The apologies of justin martyr : to which is appended the epistle to diognetus – New York: Harper, 1877 [mf ed 1985] – (= ser Douglass series of christian greek and latin writers 5) – 1mf – 9 – 0-8370-5922-4 – (incl ind. int and notes by basil lanneau gildersleeve) – mf#1985-3922 – us ATLA [240]

The apologies of justin martyr. to which is appended the epistle to diognetus / Justin the Martyr, Saint – Introd. and Notes by Basil L. Gildersleeve. New York: Harper, 1877. xli,289p. Greek or English text – 1 – us Wisconsin U Libr [240]

Les apologistes chretiens au 2e siecle / Freppel, Charles – 2e ed. Paris: Bray et Retaux, 1870 [mf ed 1986] – 1mf – 9 – 0-8370-6902-5 – (incl bibl ref) – mf#1986-0902 – us ATLA [240]

Les apologistes grecs du 2e siecle de notre ere / Puech, Aime – Paris: Hachette, 1912 [mf ed 1991] – 1mf – 9 – 0-7905-9072-7 – (incl bibl ref) – mf#1989-2297 – us ATLA [240]

An apology for a work entitled "contrasts" : being a defence of the assertions advanced in that publication, against the various attacks lately made upon it / Pugin, Augustus Welby Northmore – Birmingham: printed for aut, by R P Stone & Son, 1837 – (= ser 19th c art & architecture) – 1mf – 9 – mf#4.1.199 – uk Chadwyck [700]

An apology for actors (1612) by thomas heywood : bound with a refutation of the apology for actors (1615) by i.g. / Heywood, Thomas – Introds. and bibliog. notes by Richard H. Perkinson. 1941. 156p – 9 – us Scholars Facs [790]

Apology for christmas-day / Manning, James – Exeter, England. 1822 – 1r – us UF Libraries [240]

An apology for great britain : in allusion to a pamphlet intituled "considerations etc par un canadien, mpp" / Cuthbert, Ross – Quebec: J. Neilson, 1809 [mf ed 1971] – 1r – 5 – mf#SEM16P18 – cn Bibl Nat [971]

Apology for lollard doctrines : attributed to wycliffe / Wycliffe, John – London: Camden Society 1842 [mf ed 1984] – 1r – 1 – 0-8370-1560-X – mf#1984-b495 – us ATLA [242]

An apology for mohammed and the koran / Davenport, John – London: J Davy, 1869 [mf ed 1991] – 1mf – 9 – 0-524-01268-7 – mf#1990-2304 – us ATLA [260]

Apology for slavery : or, six cogent arguments against the immediate abolition of the slave-trade / Geddes, Alexander – London: Printed for J Johnson & R Faulder c1792 [mf ed 2004] – (= ser Selected americana from sabin's dictionary of books relating to america, from its discovery to the present time) – 1mf – 9 – us Primary [306]

An apology for the architectural monstrosities of london / Hakewill, Arthur William – London 1835 – (= ser 19th c art & architecture) – 1mf – 9 – mf#4.2.1259 – uk Chadwyck [720]

Apology for the bible / Watson, Richard – London, England. 1796 – 1r – us UF Libraries [220]

An apology for the book of psalms in five letters : addressed to the friends of union in the church of god / McMaster, Gilbert – 4th ed. Philadelphia: Daniels & Smith, 1852 [mf ed 1993] – (= ser Presbyterian coll) – 1mf – 9 – 0-524-07442-9 – mf#1991-3102 – us ATLA [220]

Apology for the christian faith *see*
– Der apologet aristides
– Die apologie des aristides

An apology for the colouring of the greek court / Jones, Owen – London 1854 – (= ser 19th c art & architecture) – 1mf – 9 – mf#4.2.148 – uk Chadwyck [700]

An apology for the common english bible : and a review of the extraordinary changes made in it by managers of the american bible society / Coxe, Arthur Cleveland – 3rd ed Baltimore: J Robinson, 1857 [mf ed 1990] – 1mf – 9 – 0-8370-1876-5 – mf#1987-6263 – us ATLA [220]

Apology for the disuse of alcoholic drinks / Macdonald, G B – London, England. 1841 – 1r – us UF Libraries [240]

Apology for the freedom of the press and for general liberty / Hall, Robert – London, England. 1822 – 1r – us UF Libraries [240]

Apology for the more frequent administration of the lord's supper / Brown, John – Edinburgh, Scotland. 1804 – 1r – us UF Libraries [240]

Apology for the plain sense of the doctrine of the prayer book on h... / Watson, Alexander – London, England. 1850 – 1r – us UF Libraries [240]

Apology for the religious orders : being a translation from the latin of two of the minor works of the saint = Contra impugnantes dei cultum et religionem / Thomas, Aquinas, Saint; ed by Procter, John – London: Sands, 1902 [mf ed 1991] – 2mf – 9 – 0-7905-9712-8 – (in english; int by ed) – mf#1989-1437 – us ATLA [241]

An apology for the septuagint : in which its claims to biblical and canonical authority are briefly stated and vindicated / Grinfield, Edward William – London: William Pickering, 1850 [mf ed 1989] – 1mf – 9 – 0-7905-1054-5 – (in english, greek & latin. incl bibl ref) – mf#1987-1054 – us ATLA [221]

Apology for the study of divinity / Rose, Hugh James – London, England. 1834 – 1r – us UF Libraries [240]

Apology for the true christian divinity : as the same is held forth and preached by the people, in scorn, called quakers... = Theologiae ver e christianae apologia / Barclay, Robert – 13th corr ed. Manchester: W Irwin [mf ed 1993] – (= ser Society of friends (quakers) coll) – 1mf – 9 – 0-524-06977-8 – (incl bibl ref; in english) – mf#1991-2830 – us ATLA [243]

The apology of al kindy : written at the court of al maamaun (circa a h 215, a d 830), in defence of christianity against islam / al-Kindi, Abd al-Masih – 2nd ed. London: SPCK, 1911 [mf ed 1991] – 1mf – 9 – 0-524-01540-6 – mf#1990-2494 – us ATLA [260]

The apology of aristides / Aristides – London, New York: W Scott, 1909 [mf ed 1990] – 1mf – 9 – 0-7905-3630-7 – (trans fr greek by w s walford) – mf#1989-0123 – us ATLA [240]

The apology of aristides on behalf of the christians (ts1/1) : from a syriac ms / ed by Harris, J R – 1891 – (= ser Texts and studies (ts)) – 3mf – 9 – €7.00 – (app by j a robinson) – ne Slangenburg [230]

The apology of origen in reply to celsus : a chapter in the history of apologetics / Patrick, John – Edinburgh: W Blackwood, 1892 [mf ed 1991] – 1mf – 9 – 0-7905-9835-3 – mf#1989-1560 – us ATLA [240]

Apology of rejoicing christmas – Edinburgh, Scotland. 1828 – 1r – us UF Libraries [240]

The apology of the christian religion : historically regarded with reference to supernatural revelation and redemption / Macgregor, James – Edinburgh: T & T Clark, 1891 [mf ed 1990] – 1mf – 9 – 0-7905-5112-8 – mf#1988-1112 – us ATLA [240]

Apontamentos historicos especialementes eclesiasticos sobre as ilhas e diocese de s thome e principe / Lima, Jose Joaquim Lopes de – [S.l, s.n, 19–?] – 1 – us CRL [241]

Apontamentos para a historia da administracao da diocese e da organiscao do seminario lyceu / Ferreira da Silva, Francisco – Lisboa: Typ Minerva Central 1899 [mf ed 1987] – 1r – 1 – (filmed with: o estado de sao paulo / amaral, t & other titles) – mf#1866 – us Wisconsin U Libr [241]

Apontamentos para a historia da colonizacao de blu... / Ferraz, Paulo Malta – Sao Paulo, Brazil. 1949 – 1r – us UF Libraries [972]

Apontamentos para a historia da republica / Dornas, Joao – Curitiba, Brazil. 1941 – 1r – us UF Libraries [972]

Apontamentos para a historia da republica dos esta / Campos Porto, Manuel Ernesto De – Rio de Janeiro, Brazil. 1890 – 1r – us UF Libraries [972]

Apontamentos para a historia d'angola / Carvalho E Menezes, Vasco Guedes De – Funchal, Portugal. 1882 – 1r – us UF Libraries [960]

Apontamentos para a historia de guerra de zambezia, 1871-1875 / De Silva Barahona e Costa, Henrique Cesar – Lisbon: P Aurea, 1895 – 1 – us CRL [960]

Apophtegmata : studien zur geschichte des aeltesten moenchtums / Bousset, Wilhelm – Tuebingen, 1923 – 9mf – 8 – €18.00 – ne Slangenburg [241]

Apophtegmata symbolica per moralia et ethica dogmata, rythmice constructa... / Redel, A C – Augustae Vindelicorum: Apud Johann Philippum Steudner, n.d. – 2mf – 9 – mf#O-1802 – ne IDC [090]

Apopka chief – Apopka, FL. 1988 apr 22-1996 – 17r – (gaps) – us UF Libraries [071]

Aportacion al estudio de los acidos l(-) malico, d(-) y l(+) lactico en mostos y vinos de tierra de barros / Pinto Corraliza, Maria del Carmen – Badajoz: Universidad de Extremadura, 1980 – 1 – sp Bibl Santa Ana [946]

Aportacion al estudio de los compuestos nitrogenados en mostos y vinos de tierra de barros / Macias Laso, Pedro – Badajoz: Universidad de Extremadura, 1980 – 1 – sp Bibl Santa Ana [946]

Aportacion al estudio de los vinos de la zona de "tierra de barros" (badajoz) / Maynar Marino, Juan – Badajoz: Universidad de Extremadura. Facultad de Ciencias. Departamento de Quimica Fisica, 1975 – 1 – sp Bibl Santa Ana [550]

Aportacion al estudio del fuero de baylio / Cerro Sanchezherrera, Eduardo – Madrid: Editorial Revista de Derecho Privado, 1964 – 1 – sp Bibl Santa Ana [370]

Aportacion al vocabulario / Rodriguez Perera, Francisco – Badajoz: Imprenta de la Diputacion Provincial, 1959 – sp Bibl Santa Ana [440]

Aportaciones al lenguaje de torres naharro / Segura Covarsi, Enrique – Badajoz: Imp. Dip. Provincial, 1944 – 1 – sp Bibl Santa Ana [946]

Aportaciones para una politica economica cubana / Cuba Ministerio De Hacienda – Habana, Cuba. 1937 – 1r – us UF Libraries [300]

El aposento alto – Nashville TN: Board of Missions, Methodist Episcopal Church, South (jan/feb/mar 1938-sep/oct 2006) [mf ed 2007] – (= ser Religious periodical literature of the hispanic and indigenous peoples of the americas, 1850-1985) – 13r – 1 – (imprint varies; many iss lacking; spanish ed of: upper room) – ISSN: 0003-6552 – mf#2006h-s020 – us ATLA [242]

Apostacy in perilous times / Wallace, Robert – London, England. 1837 – 1r – us UF Libraries [240]

Apostasy developed – London, England. 1846 – 1r – us UF Libraries [240]

Apostasy of the roman catholic church clearly demonstrated / Gregg, Tresham Daines – Sheffield, England. 18– – 1r – us UF Libraries [241]

Ein apostel der wiedertaeufer / Keller, Ludwig – Leipzig: S Hirzel, 1882 – 1mf – 9 – 0-8370-8913-1 – mf#1986-2913 – us ATLA [243]

Der apostel johannes / Krenkel, Max – Berlin: F Henschel, 1871 – 1mf – 9 – 0-7905-3348-0 – (incl bibl ref) – mf#1987-3348 – us ATLA [240]

Der apostel paulus / Bousset, Wilhelm – Halle a S: Gebauer-Schwetschke, [1906?] – 1mf – 9 – 0-524-05030-9 – mf#1992-0283 – us ATLA [240]

Der apostel paulus und sein evangelium als autoritaet fuer den glauben / Oehler, Theodor – Basel: Basler Missionsbuchh, 1907 – 1mf – 9 – 0-524-06150-5 – mf#1992-0817 – us ATLA [225]

Das aposteldecret nach seiner aussercanonischen textgestalt / Resch, A – Leipzig, 1905 – (= ser Tugal 2-28/3) – 3mf – 9 – €7.00 – ne Slangenburg [240]

Das aposteldecret nach seiner ausserkanonischen textgestalt / Resch, Gotthold – Leipzig: J C Hinrichs, 1905 – (= ser Tugal) – 1mf – 9 – 0-7905-1731-0 – (incl bibl ref and ind) – mf#1987-1731 – us ATLA [225]

Apostelgeschichte *see* The acts of the apostles

Die apostelgeschichte : eine ausgleichung des paulinismus und des judenthums innerhalb der christlichen kirche / Bauer, Bruno – Berlin: Gustav Hempel, 1850 – 1mf – 9 – 0-7905-0904-0 – (incl bibl ref) – mf#1987-0904 – us ATLA [225]

Die apostelgeschichte / Belser, Johannes Evangelist – 1. & 2. aufl. Muenster i W: Aschendorff 1908 [mf ed 1992] – (= ser Biblische zeitfragen 1/7) – 1mf – 9 – 0-524-05576-9 – mf#1992-0436 – us ATLA [226]

Die apostelgeschichte / Belser, Johannes Evangelist – Wien: Mayer, 1905 – (= ser Kurzgefasster wissenschaftlicher Kommentar zu den Heiligen Schriften des Neuen Testamentes) – 1mf – 9 – 0-8370-2258-4 – mf#1985-0258 – us ATLA [225]

Die apostelgeschichte / Felten, Joseph – Freiburg i B: Herder, 1892 – 2mf – 9 – 0-524-05722-2 – (incl bibl ref) – mf#1992-0565 – us ATLA [225]

Die apostelgeschichte : oder der entwicklungsgang der kirche von jerusalem bis rom / Baumgarten, M – 2. aufl. Braunschweig, 1859 – 2v on 20mf – 8 – €38.00 – ne Slangenburg [225]

Die apostelgeschichte / Preuschen, Erwin – Tuebingen: J C B Mohr, 1912 – (= ser Handbuch zum neuen testament) – 1mf – 9 – 0-7905-3465-7 – (incl bibl ref) – mf#1987-3465 – us ATLA [225]

Die apostelgeschichte : textkritische untersuchungen und textherstellung / Weiss, Bernhard – Leipzig: J C Hinrichs, 1893 [mf ed 1986] – (= ser Das neue testament 1/1; Texte und untersuchungen zur geschichte der altchristlichen literatur (tugal) 9/3-4) – 1mf – 9 – 0-8370-9588-3 – mf#1986-3588 – us ATLA [226]

Die apostelgeschichte : untersuchungen / Harnack, Adolf von – Leipzig, 1908 – 4mf – 8 – €11.00 – ne Slangenburg [226]

Die apostelgeschichte / Weiss, Bernhard – Leipzig, 1893 – (= ser Tugal 1-9/3.4) – 5mf – 9 – €12.00 – ne Slangenburg [225]

Die apostelgeschichte bei dewette-overbeck und bei adolf harnack / Schmidt, Paul Wilhelm – Basel: Helbing & Lichtenhahn, 1910 – 1mf – 9 – 0-8370-9576-X – (incl bibl ref) – mf#1986-3576 – us ATLA [225]

Die apostelgeschichte im lichte der neueren text-, quellen- und historisch-kritischen forschungen : ferienkurs-vortraege / Clemen, Carl – Giessen: Alfred Toepelmann, 1905 – 1mf – 9 – 0-8370-2681-4 – mf#1985-0681 – us ATLA [225]

Die apostelgeschichte in bibelstunden / Gerok, Karl – Stuttgart: S G Liesching 1868 [mf ed 1991] – 2v on 3mf – 9 – 0-7905-8345-3 – mf#1987-6444 – us ATLA [226]

Die apostelgeschichte s lucae / Hoffmann, Heinrich – Leipzig: A Deichert, 1903 – 1mf – 9 – 0-8370-9550-6 – mf#1986-3550 – us ATLA [225]

Die apostelgeschichte uebersetzt und erklaert / Felten, J – Freiburg im Breisgau, 1892 – 9mf – 8 – €18.00 – ne Slangenburg [225]

Die apostelgeschichte und ihr geschichtlicher wert / Hadorn, W – Gr Lichterfelde-Berlin: E Runge 1906 [mf ed 1990] – (= ser Biblische zeit- und streitfragen 2/6) – 1mf – 9 – 0-7905-3339-1 – mf#1987-3339 – us ATLA [226]

Die apostellehre und die juedischen beiden wege / Harnack, Adolf von – Leipzig: J C Hinrichs, 1886 [mf ed 1989] – 1mf – 9 – 0-7905-1327-7 – (comm in german. text in greek. notes in greek & latin) – mf#1987-1327 – us ATLA [240]

Das apostilische zeitalter der christlichen kirche / Weizsaecker, Carl – Zweite, neu bearbeitete Aufl. Freiburg, i. B.: Mohr, 1892. Chicago: Dep of Photodup, U of Chicago Lib, 1970 (1r); Evanston: American Theol Lib Assoc, 1984 (1r) – 1 – 0-8370-0317-2 – (incl ind) – mf#1984-B129 – us ATLA [240]

Apostillas / Posada, Eduardo – Bogota, Colombia. 1926 – 1r – us UF Libraries [240]

Apostillas... / Posada, Eduardo – Madrid: Razon y Fe, 1927 – 1 – sp Bibl Santa Ana [200]

The apostle of alaska : the story of william duncan of metlakahtla / Arctander, John William – New York: Fleming H Revell, c1909 [mf ed 1986] – 1mf – 9 – 0-8370-6241-1 – (incl ind) – mf#1986-0241 – us ATLA [240]

The apostle of burma : a missionary epic: in commemoration of the centennial of the birth of adoniram judson / Richards, William Carey – Boston: Lee & Shepard; New York: C T Dillingham, 1889, c1888 [mf ed 1986] – 1mf – 9 – 0-8370-6344-2 – mf#1986-0344 – us ATLA [810]

An apostle of personal harmonizing / Carman, Bliss – [New Canaan, CT?: s.n, 1911?] – 1mf – 9 – 0-665-77808-2 – mf#77808 – cn Canadiana [613]

The apostle of ryo-u – Philadelphia: Board of foreign missions, Reformed church in the US, 1917 [mf ed 1995] – (= ser Yale coll) – 124p (ill) – 1 – 0-524-09873-5 – mf#1995-0873 – us ATLA [920]

Apostle of the gentiles, and his glorying / Melson, John Barritt – London, England. 1850 – 1r – us UF Libraries [240]

An apostle of the north : memoirs of the right reverend william carpenter bompas... / Cody, Hiram Alfred – New York: E P Dutton, 1908 [mf ed 1990] – 1mf – 9 – 0-7905-4786-4 – mf#1988-0786 – us ATLA [240]

The apostle of the north, rev james evans / Young, Egerton Ryerson – London: Marshall, 1899 – 4mf – 9 – mf#30583 – cn Canadiana [242]

The apostle of the north, rev. james evans / Young, Egerton Ryerson – New York: Fleming H Revell, c1899 – 1mf – 9 – 0-8370-6637-9 – mf#1986-0637 – us ATLA [240]

An apostle of the western church : memoir of the right reverend jackson kemper, doctor of divinity, first missionary bishop of the american church / White, Greenough – New York: T Whittaker, 1900 [mf ed 1992] – (= ser Anglican/episcopal coll) – 1mf – 9 – 0-524-04825-8 – mf#1992-2054 – us ATLA [240]

An apostle of the wilderness : james lloyd breck...his missions and his schools / Holcombe, Theodore Isaac – New York: T Whittaker, 1903 [mf ed 1991] – 1mf – 9 – 0-524-00558-3 – mf#1990-0058 – us ATLA [240]

The apostle paul : a sketch of the development of his doctrine = Apaotre paul / Sabatier, Auguste; ed by Findlay, George Gillanders – 3rd ed. New York: James Pott, 1896 – 1mf – 9 – 0-8370-5012-X – (incl bibl ref. in english) – mf#1985-3012 – us ATLA [225]

The apostle paul / Whyte, Alexander – Edinburgh: Oliphant Anderson and Ferrier, 1903 – 1mf – 9 – 0-7905-2214-4 – mf#1987-2214 – us ATLA [240]

Apostle paul an unitarian / Mardon, Benjamin – London, England. 1826 – 1r – us UF Libraries [243]

Apostlernes gjerninger forklaret i bibellaesninger / Besser, Wilhelm Friedrich – Christiania [Oslo]: Wm Grams, 1863-65 [mf ed 1992] – 2v on 3mf – 9 – 0-524-05392-8 – mf#1992-0402 – us ATLA [226]

The apostles / Renan, Ernest – New York: Carleton, 1870 – 1mf – 9 – 0-524-07540-9 – mf#1992-1083 – us ATLA [240]

The apostles' creed = Apostolisches symbolum / Harnack, Adolf von; ed by Saunders, Thomas Bailey – London: A. and C. Black, 1901 – 1mf – 9 – 0-7905-4806-2 – (incl bibl ref. in english) – mf#1988-0806 – us ATLA [240]

The apostles' creed / Burn, Andrew Ewbank – 3rd ed. London: Rivingtons, 1914 – (= ser Oxford church text books) – 1mf – 9 – 0-524-02975-X – (incl bibl ref) – mf#1990-0762 – us ATLA [226]

The apostles' creed : its origin, its purpose, and its historical interpretation / McGiffert, Arthur Cushman – New York: Scribner, 1902 – 1mf – 9 – 0-7905-4836-4 – mf#1988-0836 – us ATLA [226]

The apostles' creed : its relation to primitive christianity / Swete, Henry Barclay – 3rd ed. Cambridge: University Press, 1899 – 2mf – 9 – 0-7905-6635-4 – (incl bibl ref) – mf#1988-2635 – us ATLA [226]

The apostles' creed : a vindication of the apostolic authorship of the creed on the lines of scripture and tradition, together with some account of its development and critical analysis of its contents / MacDonald, Alexander – 2nd rev and enl ed. London: K Paul, Trench, Truebner, 1925 – 1mf – 9 – 0-524-08117-4 – (incl bibl ref) – mf#1993-9023 – us ATLA [240]

The apostles' creed and the new testament = Apostolische glaubensbekenntnis und das neue testament / Kunze, Johannes – New York: Funk & Wagnalls, 1912 – 1mf – 9 – 0-7905-4992-1 – (in english) – mf#1988-0992 – us ATLA [240]

The apostles' creed to-day / Drown, Edward Staples – New York: Macmillan, 1917 – (= ser Church Principles for Lay People) – 1mf – 9 – 0-524-04371-X – mf#1991-2075 – us ATLA [240]

Apostles' doctrine and fellowship – London, England. 1871 – 1r – us UF Libraries [240]

Apostle's faith – v1 n1 [194-?] – 1 – mf#4026913 – us WHS [225]

Apostles of freedom / Vaswani, Thanwardas Lilaram – Madras: Ganesh & Co, 1922 – (= ser Samp: indian books) – us CRL [954]

Apostles of mediaeval europe / Maclear, George Frederick – London: Macmillan, [1869?] [mf ed 1986] – 1mf – 9 – 0-8370-6148-2 – (incl bibl ref) – mf#1986-0148 – us ATLA [240]

Apostles of the lord : being six lectures on pastoral theology / Newbolt, William Charles Edmund – London, New York: Longmans, Green, 1901 [mf ed 1991] – 1mf – 9 – 0-7905-8533-2 – mf#1989-1758 – us ATLA [240]

The apostles' school of prophetic interpretation : with its history down to the present time / Maitland, Charles – London: Longman, Brown, Green, and Longmans, 1849 – 2mf – 9 – 0-7905-1355-2 – (incl bibl ref and indexes) – mf#1987-1355 – us ATLA [225]

The apostleship of prayer : a holy league of christian hearts united with the heart of jesus = Apostolat de la priere / Ramiere, Henri – Baltimore: John Murphy, 1866 – 1mf – 9 – 0-8370-7186-0 – (in english) – mf#1986-1186 – us ATLA [240]

Apostol – Ed 2. Vil'no: Mamonich Printing House, [1592] – 10mf – 9 – mf#RHB-29 – ne IDC [460]

Apostol – [Ed 3]. Vil'no: Mamonich Printing House, [1595] – 10mf – 9 – mf#RHB-35 – ne IDC [460]

Apostol – L'vov: Ivan Fedorov Printing House, 1574 – 10mf – 9 – mf#RHB-3 – ne IDC [460]

Apostol – M: Andronik Timofeev Nevezha, 1597 – 12mf – 9 – mf#RHB-2 – ne IDC [460]

[Apostol] – Vil'no: Mamonich Printing House, 1591 – 9mf – 9 – mf#RHB-13 – ne IDC [460]

Apostol bautista en la perla antillana : biography of dr m n mccall / Munoz, A Lopez – 1 – $14.42 – us Southern Baptist [242]

Un apostol con temple de martin. la beata filipina duchense / Bayle, Constantino – Barcelona, Madrid: Razon y Fe, 1944 – 1 – sp Bibl Santa Ana [240]

El apostol mariano representado en la vida del v p juan maria de salvatierra, de la compania de jesus : fervoroso missionero en la provincia de nueva-espana, y conquistador apostolico de las californias / Venegas, Miguel – en Mexico: impr Dona Maria de Ribera 1754 – (= ser Books on religion...1543/44-c1800: jesuitas) – 4mf – 9 – mf#crl-230 – ne IDC [241]

L'apostolat de la presse : vade-mecum des propagateurs de la "croix" – Montreal: La Maison de la bonne Presse, 1894 – 1mf – 9 – mf#25324 – cn Canadiana [241]

Apostolat des bons livres. Bibliotheque *see*
– Catalogue de la bibliotheque de l'apostolat des bons livres
– Deuxieme supplement au catalogue de la bibliotheque de l'apostolat des bons livres
– Supplement au catalogue de la bibliotheque de l'apostolat des bons livres

Apostolat des bons livres, oeuvre annexe de l'apostolat de la priere : catalogue des ouvrages contenus dans la bibliotheque de cette association – Quebec: Leger Brousseau, impr, 1895 [mf ed 1998] – 9 – cn Bibl Nat [020]

Apostolat des oblats de marie immaculee – 1-50. 1929-79 – 1 – us CRL [240]

Apostolat en haiti / Bonnaud, L – Priziac, France. 1938 – 1r – us UF Libraries [972]

Das apostolat und martirium de gesellschaft jesu in japan / Patiss, Georg – Wien: Ludwig Mayer, 1863 [mf ed 1995] – (= ser Yale coll) – viii/461p – 1 – 0-524-09590-6 – (in german) – mf#1995-0590 – us ATLA [241]

Apostolate of Christian Action *see* Divine love

The apostolic age = Apostolische zeitalter / Dobschuetz, Ernst von – London: Philip Green, 1909 – 1mf – 9 – 0-524-02852-4 – (incl bibl ref. in english) – mf#1990-0709 – us ATLA [240]

The apostolic age : its life, doctrine, worship and polity / Bartlet, James Vernon – New York: Scribner, 1899 – (= ser Ten Epochs Of Church History) – 2mf – 9 – 0-7905-4429-6 – mf#1988-0429 – us ATLA [240]

The apostolic age of the christian church = Apostolische zeitalter der christlichen kirche / Weizsaecker, Carl – 3rd ed. London: Williams and Norgate; New York: Putnam, 1907-1912 – (= ser Theological Translation Library) – 2mf – 9 – 0-7905-4059-2 – (in english) – mf#1988-0059 – us ATLA [240]

Apostolic and modern missions / Martin, Chalmers – New York: Fleming H Revell, 1898 [mf ed 1986] – (= ser Students' lectures on missions 1895) – 1mf – 9 – 0-8370-6273-X – mf#1986-0273 – us ATLA [240]

The apostolic and post-apostolic times : their diversity and unity in life and doctrine / Lechler, Gotthard Victor – 3d ed., thoroughly revised and re-written. Edinburgh: T. & T. Clark, 1886. Beltsville, Md: NCR Corp, 1978 (9mf); Evanston: American Theol Lib Assoc, 1984 (9mf) – 9 – 0-8370-0971-5 – (incl bibl ref and index) – mf#1984-4345 – us ATLA [240]

Apostolic christianity : notes and inferences mainly based on s. paul's epistles to the corinthians / Henson, Hensley – London: Methuen, 1898 [mf ed 1990] – 1mf – 9 – 0-7905-6180-8 – mf#1988-2180 – us ATLA [227]

Apostolic christianity, a d 1-100 / Schaff, Philip – new rev enl ed. New York: Scribner, 1886 [mf ed 1991] – (= ser History of the christian church 1) – 2mf – 9 – 0-524-01890-1 – (incl bibl ref) – mf#1990-0517 – us ATLA [225]

The apostolic church / Simpson, Albert B – Nyack, NY: Christian Alliance Pub Co, [1898?] [mf ed 1992] – (= ser Christian & missionary alliance coll) – 1mf – 9 – 0-524-02497-9 – mf#1990-4356 – us ATLA [227]

The apostolic faith : official publication of the azusa street mission / ed by Seymour, William Joseph – Foley AL: Together with the Harvest Publ 1997 [v1-2 (1906-1908)] [mthly] [mf ed 2005] – 1v on 1r – 1 – (lacks: pref material, glos, ind; publ by: apostolic faith movement of los angeles, sep-oct 1906; pacific apostolic faith movement, nov 1906-jan 1907; apostolic faith mission, feb/mar 1907-may 1908; originally publ: los angeles: apostolic faith movement of los angeles 1906-08; continued by: apostolic faith (portland, or: 1908), some pgs damaged) – mf#2005-s103 – us ATLA [242]

Apostolic faith and the lower light – Portland OR: The Apostolic Faith, 1955-56 [bimthly] – 1r – 1 – (merger of: the apostolic faith (portland, or: 1908) and: the lower light; cont by: apostolic faith (portland, or: 1956); filmed with: the apostolic faith, the lower light, the light of hope, the armour bearer, the convict's hope and: the prisoner's hope) – us Oregon Lib [240]

Apostolic faith and the lower light *see* Apostolic faith (portland, or: 1908)

The apostolic faith and the lower light *see* Lower light

Apostolic Faith Church *see* Light of hope

Apostolic faith (portland, or: 1908) – Portland OR: The Apostolic Faith [irreg] – 1r – 1 – (ceased in 1954; merged with: the lower light, to form: the apostolic faith and the lower light) – us Oregon Lib [071]

The apostolic faith (portland, or: 1908) *see* Lower light

Apostolic faith (portland, or: 1956) – Portland OR: The Apostolic Faith, 1956- [bimthly] – 1r – 1 – (ceased with v58 n6 (nov-dec 1965)? cont: apostolic faith and the lower light; cont by: light of hope) – us Oregon Lib [071]

The apostolic faith restored / Lawrence, Bennett Freeman – St Louis, Mo: Gospel Pub House, c1916 – 1mf – 9 – 0-524-00279-7 – mf#1989-2979 – us ATLA [240]

The apostolic fathers – [Shanghai?]: Church Literature Committee...by the help of SPCK, London 1918 – (= ser Yale coll) – 128p – 1 – 0-524-10183-3 – (chinese trans by montgomery hunt throop) – mf#1995-1183 – us ATLA [240]

The apostolic fathers : comprising the epistles (genuine and spurious) of clement of rome, the epistles of s. ignatius, the epistle of s. polycarp, the martyrdom of s. polycarp, the teaching of the apostles... / ed by Harmer, J R – London: Macmillan, 1891 – 2mf – 9 – 0-8370-9555-7 – (texts in english, greek and latin; notes in english. incl ind) – mf#1986-3555 – us ATLA [241]

The apostolic fathers / Holland, Henry Scott – London: SPCK; New York: Pott, Young [1878?] [mf ed 1990] – (= ser The fathers for english readers) – 1mf – 9 – 0-7905-5710-X – mf#1988-1710 – us ATLA [240]

The apostolic fathers / Lichtfoot, J – London, 1890-1889 – 51mf – 8 – €98.00 – (pt1: s clement of rome. pt2: s ignatius and polycarp) – ne Slangenburg [241]

The apostolic fathers / Lightfoot, Joseph Barber – London; New York: Macmillan, 1885-1890. Beltsville, Md: NCR Corp, 1978 (33mf); Evanston: American Theol Lib Assoc, 1984 (33mf) – (= ser Biblical crit – us & gb) – 9 – 0-8370-1209-0 – (incl bibl ref and ind) – mf#1984-1073 – us ATLA [240]

The apostolic fathers : pt 1: s clement of rome. pt 2: s ignatius and s polycarp / Lightfoot, J – London, 1890-1889 – €98.00 – ne Slangenburg [240]

The apostolic fathers and the apologists of the second century / Jackson, George Anson – New York: D Appleton, 1879 – (= ser Early Christian Literature Primers) – 1mf – 9 – 0-524-05148-8 – mf#1990-1404 – us ATLA [240]

The apostolic gospel : with a critical reconstruction of the text – London: Smith, Elder, 1896 – 1mf – 9 – 0-524-05792-3 – mf#1992-0619 – us ATLA [226]

Apostolic hymns / Kirkland, J V & Kirkland, R S – 1898 – 1 – $10.18 – us Southern Baptist [242]

Apostolic life as revealed in the acts of the apostles / Parker, Joseph – New York: Funk & Wagnalls, 1883-84 [mf ed 1993] – 3v on 3mf – 9 – 0-524-08510-2 – mf#1993-0035 – us ATLA [226]

The apostolic liturgy and the epistle to the hebrews : being a commentary on the epistle in its relation to the holy eucharist: with appendices on the liturgy of the primitive church / Field, John Edward – London: Rivingtons, 1882 – 2mf – 9 – 0-7905-0079-5 – (incl bibl ref and indexes) – mf#1987-0079 – us ATLA [227]

Apostolic Lutheran Church of America *see* Christian monthly

Apostolic ministry : compared with the pretensions of spurious relig... / Smith, John Pye – London, England. 1810 – 1r – us UF Libraries [240]

Apostolic ministry : sermons and addresses / Lidgett, John Scott – London: CH Kelly, [1909?] [mf ed 1991] – 1mf – 9 – 0-7905-9308-4 – mf#1989-2533 – us ATLA [242]

The apostolic ministry : a discourse. delivered in rochester, n.y., before the new york baptist union for ministerial education... / Wayland, Francis – Rochester: Sage & Bro, 1853 – 1mf – 9 – 0-524-07597-2 – mf#1991-3217 – us ATLA [242]

Apostolic ministry in the scottish church / Story, Robert Herbert – Edinburgh: W Blackwood, 1897 – (= ser Baird lecture) – 1mf – 9 – 0-7905-9687-3 – mf#1989-1412 – us ATLA [240]

Apostolic order and unity / Bruce, Robert – Edinburgh: T & T Clark, 1903 [mf ed 1992] – 1mf – 9 – 0-524-02790-0 – mf#1990-0694 – us ATLA [240]

The apostolic preaching and its development / Dodd, C H – 1936 – 9 – $10.00 – us IRC [240]

The apostolic rite of confirmation : being the substance of two sermons preached before his congregation on sunday, january 27, 1867 / Bedford-Jones, T – Ottawa?: G E Desbarats, 1867 – 1mf – 9 – mf#07665 – cn Canadiana [240]

Apostolic succession : a discourse / Noyes, Eli – Pawtucket, RI: AW Pearce, 1851 [mf ed 1991] – 1mf – 9 – 0-7905-9045-X – mf#1989-2270 – us ATLA [240]

Apostolic succession : a review of the claims of the established church, considered as an apostolical institution, and especially as an authorised interpreter of holy scripture – [Brockville UC?: s.n.] 1840 [mf ed 1987] – 1mf – 9 – 0-665-55438-9 – (incl bibl ref) – mf#55438 – cn Canadiana [242]

Apostolic succession : a sermon preached on the feast of st matthias, 1897, at the episcopal consecration of right rev. edmond f. prendergast, d.d., bishop of scillio and bishop auxiliary of philadelphia / Loughlin, James F – Philadelphia: H L Kilner, [1897?] [mf ed 1986] – 1mf – 9 – 0-8370-7305-7 – mf#1986-1305 – us ATLA [241]

Apostolic succession in the church of sweden / Nicholson, Aldwell – London: Rivingtons 1880 [mf ed 1993] – 1mf – 9 – 0-524-06652-3 – (incl bibl ref) – mf#1991-2707 – us ATLA [242]

Apostolic Union of Secular Priests *see* General rule of the apostolic union of secular priests

Apostolic voice – v1 n11; 1949? – 1 – mf#4025150 – us WHS [240]

Apostolic woman – 1984 jan, apr; 1985 dec/jan -feb/mar, apr/may – 1 – mf#4026001 – us WHS [071]

Apostolic woman's newsletter – 1988 jun, oct/ nov – 1 – mf#4026011 – us WHS [071]

The apostolical and primitive church : popular in its government, informal in its worship / Coleman, Lyman – [rev ed] Philadelphia: J B Lippincott, 1869 [mf ed 1989] – 1mf – 9 – 0-7905-0875-3 – (incl bibl ref and ind) – mf#1987-0875 – us ATLA [240]

Apostolical christianity : its history and development / Row, Charles Adolphus – London: Church of England Sunday School Institute, [1879?] [mf ed 1989] – 1mf – 9 – 0-7905-3214-X – mf#1987-3214 – us ATLA [240]

Apostolical commission / Thirlwall, Connop – London, England. 1852 – 1r – us UF Libraries [240]

The apostolical constitutions, and cognate documents : with special reference to their liturgical elements / O'Leary, De Lacy – London: SPCK 1906 [mf ed 1992] – (= ser Early church classics) – 1mf – 9 – 0-524-04686-7 – mf#1990-1313 – us ATLA [240]

Apostolical institution of episcopacy – London, England. 1853 – 1r – us UF Libraries [240]

Apostolical method of preaching the gospel / Birt, John – Hull, England. 1814 – 1r – us UF Libraries [240]

Apostolical ministry / Wilberforce, Samuel – London, England. 1833 – 1r – us UF Libraries [240]

Apostolical succession / Elrington, Charles K Richard – Dublin, Ireland. 1840 – 1r – us UF Libraries [240]

Apostolical succession / Hawkins, Edward – London, England. 1842 – 1r – us UF Libraries [240]

Apostolical succession / Weir, John – London, England. 1848 – 1r – us UF Libraries [240]

Apostolical succession and canon 15 : a reply to the rev w goode's tract... / Scott, William R – London: Joseph Masters, 1852 [mf ed 1993] – (= ser Anglican/episcopal coll) – 1mf – 9 – 0-524-05775-3 – mf#1991-2331 – us ATLA [242]

Apostolical succession in the church of england / Haddan, A W – London, 1869 – €23.00 – ne Slangenburg [242]

Apostolical succession in the church of england / Haddan, Arthur West – new ed. London: Rivingtons, 1883 [mf ed 1990] – 1mf – 9 – 0-7905-5601-4 – mf#1988-1601 – us ATLA [242]

Apostolical succession in the light of history and fact : the congregational union lecture for 1897 / Brown, John – London: Congregational Union of England and Wales, 1898 [mf ed 1989] – 2mf – 9 – 0-7905-0817-6 – (incl bibl ref) – mf#1987-0817 – us ATLA [240]

The apostolical system of the church defended : in a reply to dr. whately on the kingdom of christ / Buel, Samuel – Philadelphia: H Hooker, 1844 – 1mf – 9 – 0-524-00008-5 – mf#1989-2708 – us ATLA [240]

Apostolicarum epistolarum libri quinque / Pii quinti pont max; ed by Goubau, Francisci – Antverpiae, 1640 – €18.00 – (nunc primum in lucem editi opera et cura francisci goubau) – ne Slangenburg [226]

Das apostolicum : sein ursprung und seine biblische begruendung / Werther, Richard – Rathenow: A Haase, 1875 – 1mf – 9 – 0-8370-5799-X – (incl bibl ref) – mf#1985-3799 – us ATLA [220]

Het apostolisch vicariaat van zuid-shansi in de eerste vijf-en-twintig jaren van zijn bestaan (1890-1915) : gedenkschrift / Timmer, Odoricus – Leiden: G F Theonville [1915] [mf ed 1995] – (= ser Yale coll) – 111p (ill) – 1 – 0-524-09710-0 – (in dutch) – mf#1995-0710 – us ATLA [241]

Das apostolische glaubensbekenntnis und das neue testament / Kunze, Johannes – Berlin: Edwin Runge 1911 [mf ed 1989] – (= ser Biblische zeit- und streitfragen 7/6-7) – 1mf – 9 – 0-7905-2598-4 – (incl bibl ref) – mf#1987-2598 – us ATLA [240]

Das apostolische glaubensbekenntniss : eine apologetisch-geschichtliche studie / Blume, Clemens – Freiburg im Breisgau; St Louis, Mo: Herder, 1893 – 1mf – 9 – 0-7905-7272-2 – (incl bibl ref and index) – mf#1989-0495 – us ATLA [240]

Das apostolische glaubensbekenntniss : ein geschichtlicher bericht nebst einem nachwort / Harnack, Adolf von – 13. durch Zusaetze verm Aufl. Berlin: A Haack, 1892 – 1mf – 9 – 0-524-05145-3 – mf#1990-1401 – us ATLA [240]

Das apostolische symbol : seine entstehung, sein geschichtlicher sinn, seine urspruengliche stellung im kultus und in der theologie der kirche / Kattenbusch, Ferdinand – Leipzig: J C Hinrichs, 1894-1900 – 4mf – 9 – 0-7905-4883-6 – (incl bibl ref) – mf#1988-0883 – us ATLA [240]

Das apostolische symbol im mittelalter / Wiegand, F – Giessen, 1904 – 1mf – 8 – €3.00 – ne Slangenburg [240]

Das apostolische symbolum : vortrag / Zoeckler, Otto – Guetersloh: C Bertelsmann, 1872 – 1mf – 9 – 0-7905-8991-5 – (incl bibl ref) – mf#1989-2216 – us ATLA [240]

Das apostolische symbolum / Zahn, Theodor – Erlangen 1893 – €5.00 – ne Slangenburg [240]

Die apostolische vaeter (bdk35 1.reihe) – (= ser Bibliothek der kirchenvaeter. 1. reihe (bdk 1.reihe)) – €14.00 – ne Slangenburg [240]

Die apostolische vaeter (gcsej2) / ed by Whittaker, M – 1956 – (= ser Griechische christlichen schriftsteller der ersten jahrhunderte (gcsej)) – €7.00 – ne Slangenburg [240]

Die apostolische vollmacht des papstes in glaubens-entscheidungen / Weninger, Francis Xavier – Innsbruck: Felician Rauch, 1841 – 1mf – 9 – 0-8370-6853-3 – (incl bibl ref) – mf#1986-0853 – us ATLA [241]

Das apostolische zeitalter / Dobschuetz, Ernst von – Halle a. S: Gebauer-Schwetschke 1904 [mf ed 1989] – (= ser Religionsgeschichtliche volksbuecher fuer die deutsche christliche gegenwart 1/9) – 1mf – 9 – 0-7905-3127-5 – (incl bibl ref) – mf#1987-3127 – us ATLA [240]

Das apostolische zeugniss von christi person und werk : nach seiner geschichtlichen entwicklung / Gess, Wolfgang Friedrich – Basel: Bahnmaier. 2v. 1878-79 – 2mf – 9 – 0-7905-0837-0 – mf#1987-0837 – us ATLA [220]

Die apostolischen konstitutionen : eine litterar-historische untersuchung / Funk, Franz Xaver von – Rottenburg am Neckar: W Bader, 1891 [mf ed 1990] – 1mf – 9 – 0-7905-5880-7 – (incl bibl ref) – mf#1988-1880 – us ATLA [240]

Die apostolischen vaeter : untersuchungen ueber inhalt und ursprung der unter ihrem namen erhaltenen schriften / Hilgenfeld, Adolf – Halle: C.E.M. Pfeffer, 1853 – 1mf – 9 – 0-7905-4922-0 – (incl bibl ref) – mf#1988-0922 – us ATLA [240]

Apostol...to est' deianiia i poslaniia apostol'skiia : perevod s grecheskago perevoda semidesiati dvukh bogomudrykh tolkovnikov... – M: Ivan Fedorov and Petr Timofeev Mstislavets, 1564 – 10mf – 9 – mf#RHB-1 – ne IDC [460]

L'apostrophe : journal etudiant / Association Generale des Etudiants du Cegep Andre-Laurendeau – Lasalle: AGECAL. v1 n1 sep 1976-v2 n3 dec 1977 [mf ed 1988] – 9 – (cont by: la meche (la salle, quebec 1978)) – mf#SEM105P977 – cn Bibl Nat [378]

Apoteck fuer den gemainen man : der die ertzte zu eruchen, am gut nicht vermuegens, oder sonst jn der not, allwege nicht erraichen kan / Brunschwig, Hieronymus – [Nuermberg: Fryderich Peypus 1529] [mf ed 19–] – (= ser German books before 1601) – 1r – 1 – us OmniSys [615]

L'apotre des indes et du japon : saint francois xavier / Bellessort, Andre – Paris: Perrin, 1917 [mf ed 1995] – (= ser Yale coll) – ii/344p – 1 – 0-524-09564-7 – (in french) – mf#1995-0564 – us ATLA [241]

L'apotre du peuple : journal socialiste, politique, litteraire et artistique, paraissant le mardi, le jeudi et le samedi de chaque semaine – Montmartre [Paris]: Pilloy freres [jun 3-6 1848] (3 times/wk) – (= ser French revolution of 1848. newspapers) – 1r – 1 – us CRL [074]

L'apotre paul : esquisse d'une histoire de sa pensee / Sabatier, Auguste – Paris: Librairie Fischbacher, 1896 [mf ed 1985] – 1mf – 9 – 0-8370-5013-8 – (in french. incl ind) – mf#1985-3013 – us ATLA [225]

L'apotre paul et jesus-christ / Goguel, Maurice – Paris: Fischbacher, 1904 [mf ed 1985] – 1mf – 9 – 0-8370-3327-6 – (in french. incl ind of biblical citations and bibl ref) – mf#1985-1327 – us ATLA [225]

Appa standard – 1973 autumn/winter-1978 spring/summer – 1 – mf#642467 – us WHS [071]

Appadorai, Angadipuram *see*
- Democracy in india
- Dyarchy in practice
- Revision of democracy
- The substance of politics

Appaji Bapuji *see* Short memoir of the late rev hari ramchandra khisti

Appalachia : an economic report / Appalachian Regional Commission; ed by Maher, Judith F – Washington: Appalachian Regional Commission, 1977 [mf ed 1983] – 1r – 1 – (updates the 1973 suppl to appalachia – an economic report. incl bibl ref) – mf#*ZT-1381 n14 – us NY Public [331]

Appalachia – Washington DC 1967+ – 1,5,9 – ISSN: 0003-6595 – mf#11415 – us UMI ProQuest [370]

Appalachia medicine – Lexington KY 1972 – 1,5,9 – ISSN: 0003-6609 – mf#7072 – us UMI ProQuest [610]

Appalachian Committee for Full Employment *see* Voice for jobs and justice

Appalachian florida / Satsumaland Fruit Growers – Round Lake, FL. 1925 – 1r – us UF Libraries [634]

Appalachian heritage – Berea KY 1973+ – 1,5,9 – ISSN: 0363-2318 – mf#9124 – us UMI ProQuest [978]

Appalachian notes – 1977 jan-apr – 1 – mf#535249 – us WHS [071]

Appalachian Oral History Project *see* Mountain memories

Appalachian proutist – 1980 oct 17-nov25 – 1 – mf#1313568 – us WHS [071]

Appalachian Regional Commission *see* Appalachia

Appalachian renaissance – v1 n5, 7-8 [1981 jan, mar-apr] – 1 – mf#1313569 – us WHS [071]

Appalachian review – Morgantown WV 1966-68 – 1 – mf#2356 – us UMI ProQuest [301]

Appalachian State University *see* Watauga county times past

Appaloosa news – Moscow ID 1973+ – 1,5,9 – ISSN: 0003-665X – mf#7322.01 – us UMI ProQuest [636]

Appanoose Union Sunday School, Franklin County, KS *see* Record book

Appantampuran *see* Bhaskara menon

Apparaat voor de studie der geschiedenis / Romein, J M – Groningen, 1960 – €5.00 – ne Slangenburg [240]

An apparatus criticus to chronicles in the peshitta version : with a discussion of the value of the codex ambrosianus / Barnes, William Emery – Cambridge: University Press, 1897 [mf ed 1988] – 1mf – 9 – 0-7905-0249-6 – (pref & int in english; crit app in syriac & latin. incl ind) – mf#1987-0249 – us ATLA [221]

Ein apparatus criticus zur pesitto zum propheten jesaia / Diettrich, Gustav – Giessen: Alfred Toepelmann, 1905 – 1mf – 9 – 0-8370-2913-9 – mf#1985-0913 – us ATLA [221]

Apparatus hostiensis in decretales, libros 3-5 (siecle 14) – Calahorra – 1r – 5,6 – sp Cultura [240]

Apparatus in librium 6 (siecle 14) / Baisio, Guido de – Barcelona – 2r – 5,6 – sp Cultura [240]

Apparatus musico-organisticus invictissimo leopoldo 1 : imperatori semper augusto ad coronationem auspicatissimam coniugis... / Muffat, Georg – ed prima, [Wien: between 1704 & 1709] [mf ed 19–] – 1r – 1 – mf#film 607 – us Sibley [780]

Apparatus work for boys and girls : a course of graded instruction...in the use of horizontal bars, parallel bars, horses, rings, ladders... / Zwarg, Leopold Fredrick – Philadelphia: J J McVey [c1923] – 1 – (filmed with: apontamentos para a historia / ferreira da silva, f) – mf#1866 – us Wisconsin U Libr [790]

Apparel industry magazine – New York NY 1995+ – 1,5,9 – ISSN: 0192-1878 – mf#20016 – us UMI ProQuest [680]

Apparel merchandising – New York NY 1985-90 – 1,5,9 – ISSN: 0746-889X – mf#15624 – us UMI ProQuest [680]

Apparel world – New York NY 1981-86 – 1,5,9 – mf#12754 – us UMI ProQuest [680]

Apparitions : or, the mystery of ghosts, hobgoblins, and haunted houses developed / Taylor, Joseph – London : Lackington, Allen 1814 [mf ed Bloomington IN: Indiana Uni Lib, Preservation Dept 1984] – xi 223p on 1r [ill] – 1 – us Indiana Preservation [390]

Appayya Diksita *see*
- Kuvalayananda karikas
- Sivadvaita nirnaya

Appeal – Arlington OR: S A Thomas, 1903-05 [wkly] [mf ed 1968] – 1r – 1 – (cont: arlington appeal (1903-03)) – us Oregon Lib [071]

Appeal – St Paul, MN: Northwestern Pub Co. 1889-v39 n47. nov 24 1923 (wkly) [mf ed 1947] – (= ser Negro newspapers on microfilm) – 6r – 1 – (cont: western appeal (st paul, mn: 1885)) – us L of C Photodup [071]

Appeal : on behalf of the committee of united dissenters of manchester / Johns, W – Manchester, England. 1834 – 1r – us UF Libraries [240]

Appeal : the scottish civil disabilities of 1792 – Edinburgh, Scotland. 1862 – 1r – us UF Libraries [240]

Appeal and a defiance : an appeal to the good faith of a protestant by birth, a defiance to the reason of a rationalist by profession = Appel et defi / Dechamps, Victor Auguste – New York: Benziger, 1883 [mf ed 1991] – 1mf – 9 – 0-7905-8781-5 – (in english) – mf#1989-2006 – us ATLA [240]

Appeal and remonstrance to his holiness pope pius 7 / O'conor, Charles – London, England. 1824 – 1r – us UF Libraries [240]

An appeal by a lancashire liberal against the unjust operation of the irish land act / Anderson, Christopher – Liverpool, 1882 – (= ser 19th c ireland) – 1mf – 9 – mf#1.1.9171 – uk Chadwyck [333]

Appeal by the incumbent and churchwardens of the church of st john – Aberdeen, Scotland. 1862 – 1r – us UF Libraries [240]

An appeal for the ancient doctrines of the religious society of friends – Philadelphia: J Kite, 1847 [mf ed 1993] – (= ser Society of friends (quakers) coll) – 1mf – 9 – 0-524-06701-5 – mf#1991-2731 – us ATLA [243]

Appeal for the native race, settlers and miners of british columbia – S.l: s.n, 1871?] [mf ed 1987] – 1mf – 9 – 0-665-56230-6 – mf#56230 – cn Canadiana [971]

Appeal for the sustentation fund / Free Church Of Scotland – Edinburgh, Scotland. 18– – 1r – us UF Libraries [240]

An appeal for unity in faith : being an appeal to anglicans (protestant episcopalians) and protestants of other denominations to return to the unity of the faith / Phelan, John – 2nd rev enl ed. Chicago: MA Donohue, c1911 [mf ed 1992] – 1mf – 9 – 0-524-02797-8 – (no more publ) – mf#1990-0701 – us ATLA [240]

Appeal from tradition to scripture and common sense : or, an answer to the question, what constitutes the divine rule of faith and practice / Peck, George – New York:...for the Methodist Episcopal Church, 1844 [mf ed 1984] – 6mf – 9 – 0-8370-1125-6 – (incl bibl ref & ind) – mf#1984-4116 – us ATLA [220]

Appeal in behalf of church government – London, England. 1840 – 1r – us UF Libraries [240]

An appeal in behalf of the further endowment of the divinity school of harvard university / Bellows, Henry Whitney – Cambridge: John Wilson, 1879 [mf ed 1993] – (= ser Unitarian/universalist coll) – 1mf – 9 – 0-524-07809-2 – mf#1991-3356 – us ATLA [378]

The appeal in indian music / Sahukar, Mani – Bombay: Thacker & Co, 1943 – (= ser Samp: indian books) – us CRL [780]

The appeal of india : a report of visits to the british india mission fields of the american baptist foreign mission society... / Robbins, Joseph Chandler – Philadelphia, Boston: American Baptist Publ Society, 1919 [mf ed 1995] – (= ser Yale coll) – x/90p (ill) – 1 – 0-524-09303-2 – mf#1995-0303 – us ATLA [242]

The appeal of medical missions / Moorshead, Robert Fletcher – Edinburgh: Oliphant, Anderson & Ferrier, 1913 – 1mf – 9 – 0-524-07758-4 – mf#1991-3326 – us ATLA [240]

The appeal of romanism to educated protestants : a paper read before the evangelical alliance, new york, oct 8, 1873 / Storrs, Richard Salter – New York: Harper, 1874 – 1mf – 9 – 0-8370-7992-6 – mf#1986-1992 – us ATLA [242]

Appeal on behalf of the house of mercy at bussage – Gloucester, England. 1853? – 1r – us UF Libraries [240]

Appeal on the common school law : its incongruity and maladministration... / Dallas, Angus – Toronto: Printed and publ..."Catholic Citizen", 1858 – 1mf – 9 – mf#10786 – cn Canadiana [370]

Appeal to all classes : on the subject of church patronage in scotland – Glasgow, Scotland. 1824 – 1r – us UF Libraries [240]

Appeal to all that doubt or disbelieve the truths of the gospel / Law, William – London, England. 1845 – 1r – us UF Libraries [240]

Appeal to british protestants / Cooke, William – London, England. 18– – 1r – us UF Libraries [242]

An appeal to capitalists : and the rest of the community of the british empire, on the state of its trading and commercial interests, and submitting a remedy for the evils to which they are subjected / Amator Patriae – London: printed...& sold by Holdsworth & Ball, 1829 – (= ser 19th c economics) – 1mf – 9 – mf#1.1.429 – uk Chadwyck [330]

An appeal to irishmen to unite in supporting measures formed on principles of common justice and common sense / Naper, James Lenox William – London, 1848 – (= ser 19th c ireland) – 1mf – 9 – mf#1.1.1995 – uk Chadwyck [339]

Appeal to liberal christians for the cause of christianity in india *see* Correspondence relative to the prospects of christianity

Appeal to protestant charity and english justice – London, England. 1813 – 1r – us UF Libraries [242]

Appeal to reason – 1895 aug 31-1899; 1900-1917 dec 15; n603-995 [1909 jan 2-14 dec 26]; n944-945 [1914 jan 3-10] – 1 – mf#625705 [1]; 1269507 [2]; 345719 [3]; 964079 [5]; 964084 [6] – us WHS [071]

Appeal to the american people on behalf of cuba / Cisneros Y Betancourt, Salvador – New York, NY. 1900 – 1r – us UF Libraries [972]

Appeal to the british churches : in reply to the british banner / Urwick, William – Dublin, Ireland. 1854 – 1r – us UF Libraries [240]

An appeal to the british government, in behalf of the british colony and province of ceylon : with an appendix containing various notices of the island by authors and travellers of the early and middle ages / Peter, William – 2nd ed. Frankfurt a.Main, 1836 – (= ser 19th c economics) – 2mf – 9 – mf#1.1.369 – uk Chadwyck [915]

An appeal to the british government, in behalf of the british colony and province of ceylon : with an appendix containing various notices of the island by authors and travellers of the early and middle ages / Peter, William – Francfort O.M. 1836 – (= ser 19th c british colonization) – 2mf – 9 – mf#1.1.369 – uk Chadwyck [339]

An appeal to the british nation on the treatment experienced by napoleon buonaparte in the island of st helena = Appel a la nation anglaise... / Doris, Charles, de Bourges [M Santini pseud] – 3rd augm ed. London: Ridgways 1817 [mf ed 1988] – 1r – 1 – (french & english parallel texts; with pref. filmed with: waterloo: the campaign and the battle / depeyster, j w) – mf#2210 – us Wisconsin U Libr [941]

An appeal to the canadian institute on the rectification of parliament / Fleming, Sandford – Toronto: Copp, Clark, 1892 [mf ed 1980] – 2mf – 9 – 0-665-03126-2 – mf#03126 – cn Canadiana [325]

Appeal to the christian public on the evils of theatrical amusement / Hogg, R – Whitehaven, England. 1823 – 1r – us UF Libraries [240]

Appeal to the christian women of the south / Grimke, Angelina Emily – [New York: American Anti-slavery Soc, 1836] [mf ed 1984] – (= ser Women & the church in america 16) – 1mf – 9 – 0-8370-0235-4 – mf#1984-2016 – us ATLA [976]

An appeal to the citizens of michigan : showing the necessity of the early completion of the great western rail way from detroit to the niagara river – [Detroit?: s.n.] 1851 [mf ed 1987] – 1mf – 9 – 0-665-55439-7 – mf#55439 – cn Canadiana [380]

Appeal to the clergy and laity of the church of england to combine / Denison, George Anthony – London, England. 1850 – 1r – us UF Libraries [241]

Appeal to the clergy of the church of scotland – Edinburgh, Scotland. 1875 – 1r – us UF Libraries [240]

Appeal to the consciences of protestant members of parliament again – London, England. 1855 – 1r – us UF Libraries [242]

Appeal to the episcopal synod of the church in scotland – Aberdeen, Scotland. 1865 – 1r – us UF Libraries [240]

Appeal to the evangelical clergy against their concurrence in the d... / Jordan, J – London, England. 1850 – 1r – us UF Libraries [242]

An appeal to the imperial parliament : upon the claims of the ceded colony of trinidad, to be governed by a legislature and judicature, founded on principles sanctioned by colonial precedents and long usage / Sanderson, John – London 1812 – (= ser 19th c british colonization) – 3mf – 9 – mf#1.1.3800 – uk Chadwyck [323]

An appeal to the inhabitants of lower canada, on the use of ardent spirits – [Montreal?: s.n.] 1828 [mf ed 1987] – 1mf – 9 – 0-665-55440-0 – mf#55440 – cn Canadiana [362]

Appeal to the members of the two universities presenting ten reason... / Campian, Edmond – London, England. 1827 – 1r – us UF Libraries [240]

An appeal to the methodist episcopal church : concerning what its next general conference should do on the question of slavery / Stevens, Abel – New York: printed by John F Trow, 1859 [mf ed 1990] – 1mf – 9 – 0-7905-6507-2 – (incl discourse) – mf#1988-2507 – us ATLA [242]

An appeal to the methodists in opposition to the changes proposed in their church government / Bond, Thomas E – Baltimore: Armstrong & Plaskitt 1827 – 1r – 1 – $35.00 – mf#um-15 – us Commission [242]

An appeal to the montreal conference and the methodist church generally : from a charge by rev william scott, in which is shown his charge to be invalid, and his defence of the seminary of st sulpice against the indians of oka to be baseless / Borland, John – [Montreal?: s.n.], 1883 [mf ed 1979] – 1mf – 9 – 0-665-00183-5 – mf#00183 – cn Canadiana [242]

Appeal to the parishioners of mortimer / Vaughan, John J – Salisbury, England. 1833? – 1r – us UF Libraries [240]

Appeal to the preachers of all the creeds / Brown, Gamaliel – Ramsgate, England. 1871 – 1r – us UF Libraries [240]

An appeal to the public : from the charges contained in the "reply of the rev james smith, to the strictures of the rev alexander m'caine;" accompanied with remarks upon the government of the methodist episcopal church church / M'Caine, Alexander – Baltimore: Matchett 1826 – 1r – 1 – $35.00 – mf#um-15 – us Commission [242]

An appeal to the public : occasioned by the suspension of the architectural lectures in the royal academy / Soane, John – London 1812 – (= ser 19th c art & architecture) – 2mf – 9 – mf#4.2.1322 – uk Chadwyck [720]

Appeal to the reason and good feeling of the enlgish people on the... / Wiseman, Nicholas Patrick Stephen – London, England. 1850 – 1r – us UF Libraries [240]

An appeal to the right hon w e gladstone, mp, her majesty's prime minister : respecting the suppression of certain papers by the government, the "red river rebellion," and the illegal transfer of the north-west territories... / Corbett, Griffith Owen – London, [1870] – (= ser 19th c books on british colonization) – 1mf – 9 – mf#1.1.3699 – uk Chadwyck [330]

An appeal to the scottish bishops and clergy : and generally to the church of their communion / [Palmer, William] – Edinburgh: Alex. Laurie, 1849 [mf ed 1990] – 2mf – 9 – 0-7905-6609-5 – mf#1988-2609 – us ATLA [243]

An appeal to the senate on the...plans for the university library [of the university of cambridge] / Wilkins, William – Cambridge 1831 – (= ser 19th c art & architecture) – 1mf – 9 – mf#4.1.371 – uk Chadwyck [720]

An appeal to the sense of the people on the present posture of affairs : wherein the nature of the late treaties are inquired into, and the conduct of the m–i–y with regard to m–n–ca, a–r–ca, etc is considered... – London: printed for David Hookham...1756 [mf ed 1984] – 1mf – 9 – 0-665-20195-8 – mf#20195 – cn Canadiana [971]

An appeal to the sons of africa / Jones, Charles P – 1902 – $20.00 – us ABHS [240]

An appeal to the women of the nominally free states – 2nd ed. Boston: Isaac Knapp, 1838 [mf ed 1984] – (= ser Women & the church in america 79) – 1mf – 9 – 0-8370-1205-8 – mf#1984-2079 – us ATLA [976]

Appeal to truth : a letter: addressed to the cardinals, archbishops and bishops of germany, bavaria, and austria-hungary = Lettre de l'episcopat belge aux cardinaux et aux eveques d'allemagne, de baviere et d'autriche / Mercier, Desire et al – London: Hodder & Stoughton [1916?] [mf ed 1992] – 1mf – 9 – 0-524-04019-2 – mf#1990-1191 – us ATLA [933]

Appeal to unionists / Gordon, James – Glasgow, Scotland. 1870 – 1r – us UF Libraries [240]

An appeal to unitarians : being a record of religious experiences – London: Longmans, Green, 1890 [mf ed 1992] – 1mf – 9 – 0-524-05311-1 – mf#1990-1429 – us ATLA [243]

Appeals expediting systems : an evaluation of 2nd and 8th circuit procedures / Farmer, Larry C – Washington: FJC, Sept 1981 – 1mf – 9 – $1.50 – mf#LLMC 95-818 – us LLMC [340]

Appearance and reality / Bradley, Francis Herbert – London, England. 1893 – 1r – us UF Libraries [960]

Appearance and reality : a metaphysical essay / Bradley, Francis Herbert – London: S Sonnenschein; New York: Macmillan, 1893 [mf ed 1990] – 2mf – 9 – 0-7905-3808-3 – mf#1989-0301 – us ATLA [110]

The appearances of our lord after the passion : a study in the earliest christian tradition / Swete, Henry Barclay – London: Macmillan, 1907 [mf ed 1985] – 1mf – 9 – 0-8370-5472-9 – (incl bibl ref & ind) – mf#1985-3472 – us ATLA [240]

L'appel – Port-au-Prince: Imp Mme F Smith, jun 1-14,21,30, jul 9-23 1902 – us CRL [079]

L'appel – Prisyv – Paris. n3-60. 17 oct 1915-31 mars 1917 – 1 – (in Russian. mq n1-2, 7, 28, 36, 41, 45, 54, 59) – fr ACRPP [073]

L'appel – Paris, France 26 jun, 10 jul, 2 oct 1941; 22 jan 1942-13 jul 1944 – 1 – (wanting n48,51-59,64,79-88,95-97,116,170) – uk British Libr Newspaper [074]

Appel au parlement imperial et aux habitants des colonies angloises, dans l'amerique du nord : sur les pretentions exorbitantes du gouvernement executif et du conseil legislatif de la province du bas-canada / Blanchet, Francois Xavier – Quebec: Flavien Vallerand, 1824 [mf ed 1982] – 1mf – 9 – mf#SEM105P148 – cn Bibl Nat [323]

L'appel au peuple – Tours. no. spec., no. 1-8. 20 sept-21 oct 1888 – 1 – fr ACRPP [073]

Appel au peuple : sauvons la france, sauvons la liberte – Paris [1849?] – (= ser Pamphlets and periodicals relating to the French Revolution of 1848) – 1r – 1 – us CRL [944]

Appel au peuple francais : sauvons la france, sauvons la liberte – [Paris, 1849?] – (= ser Pamphlets and periodicals relating to the French Revolution of 1848) – 1r – 1 – us CRL [944]

Appel au tribunal de l'opinion publique du rapport de m chabroud : et du decret rendu par l'assemblee nationale le 2 octobre 1790; examen du memoire du duc d'orleans, et du plaidoyer du comte de mirabeau, et nouveaux eclaircissemens sur les crimes du 5 et 6 oct / Mounier, Jean J – Geneve 1790 [mf ed Hildesheim 1995-98] – 1v on 3mf – 9 – €90.00 – 3-487-26287-8 – gw Olms [944]

Appel aux armes : sermon preche a l'eglise saint-andre, ottawa, le dimanche, 27 juin 1915 = The call of the war: recruiting sermon preached in st andrew's church, ottawa...june 27, 1915 / Herridge, William Thomas – Ottawa: [s.n.], 1915 – 1mf – 9 – 0-665-74553-2 – mf#74553 – cn Canadiana [240]

Appel du clerge en faveur de la colonisation : rapport du comite de direction – Montreal: Plinguet & Laplante, 1865 – 1mf – 9 – mf#47718 – cn Canadiana [304]

Appel d'une femme du peuple, sur l'affranchissement de la femme / Demar, Claire – Paris, 1833 – (= ser Les femmes [coll]) – 1mf – 9 – (filmed with: ma loi d'avenir 1833; paris: bureau de la tribune des femmes 1834) – mf#6908 – fr Bibl Nationale [305]

Appel, Ernst *see* Leone medigos lehre vom weltall und ihr verhaeltnis zu griechischen und zeitgenoessischen anschauungen

Appel et defi *see* Appeal and a defiance

Appel, Heinrich *see* Die komposition des aethiopischen henochbuches

Appel, Theodore *see*
- The beginnings of the theological seminary of the reformed church in the united states, from 1817 to 1832
- Letters to boys and girls about the holy land
- The life and work of john williamson nevin, d.d., ll.d

Appelbaum, Theodore *see* Johannes brahms

Appelius, Erhard W et al *see* Zur ostdeutschen agrargeschichte

Appelius, Karl Theodor *see* Geistliche selbstbekenntnisse

Appell : die geschichte einer frontkameradschaft / Blasius, Richard – Berlin: K Curtius, 1943 [mf ed 1989] – 160p – 1 – mf#7030 – us Wisconsin U Libr [830]

Appell – Stockholm, Sweden. 1907-08, 1912-25 – 4r – 1 – sw Kungliga [078]

Appell, Johann *see* Kurhessen in einer geographisch-statistisch-historischen uebersicht

Appellate case files of the supreme court of the united states 1792-1831 / U.S. Supreme Court – (= ser Records of the supreme court of the united states) – 96r – 1 – (with printed guide) – mf#M214 – us Nat Archives [347]

Appellate case files of the us circuit court for the southern district of us, 1793-1845 / U.S. Circuit and District Courts – (= ser Records of district courts of the united states) – 8r – 1 – (with printed guide) – mf#M855 – us Nat Archives [347]

Appellate jurisdiction of the crown in matters spiritual / Manning, Henry Edward – London, England. 1850 – 1r – us UF Libraries [240]

The appellate jurisdiction of the house of lords in scotch causes : illustrated by the litigation relating to the custody of the marquis of bute / Macpherson, Norman – Edinburgh, Clark, 1861. 95 p. LL-2346 – 1 – us L of C Photodup [340]

Appellate opinion writing : presented at a seminar for federal appellate judges, march 11-14, 1975 / Re, Edward D – Washington: FJC, 1975 – 1mf – 9 – $1.50 – mf#LLMC 95-300 – us LLMC [340]

Appellate practice and procedure in the supreme court of the united states / Robertson, Reynolds – Indianapolis: Bobbs-Merrill, 1905. 878p. LL-1495 – 1 – us L of C Photodup [347]

Appellatio flaviani : the letters of appeal from the council of ephesus, a d 449, addressed by flavian and eusebius to st leo of rome / Flavian, Saint, Patriarch of Constantinople; ed by Lacey, Thomas Alexander – London: SPCK 1903 [mf ed 1993] – (= ser Church historical society (series) 70) – 1mf – 9 – 0-524-05502-5 – (in english & latin. int by ed) – mf#1990-1497 – us ATLA [240]

Appellation fuer die 12. ort einer lobl. eydtgnoschafft wider die vermeinte disputation zu bern gehalten / Eck, J et al – Luzern, 1528 – 1mf – 9 – mf#ZWI-25 – ne IDC [240]

L'appello – Cleveland OH, 1917* – 1r – 1 – (italian periodical) – us IHRC [073]

Appelt, E P *see* Modern german prose

Appelt, Ewald Paul *see* Die haeuser von ohlenhof

Appendice au rapport du commissaire des terres de la couronne – Toronto: John Lovell, 1859 [mf ed 1993] – 1mf – 9 – mf#SEM105P1857 – cn Bibl Nat [333]

Appendice du premier rapport, 1849 = Appendix to first report, 1849 – Montreal: impr par Stewart Derbishire & George Desbarats, 1849 [mf ed 2000] – 8mf – 9 – mf#SEM105P3226 – cn Bibl Nat [317]

Appendices ad hainii – copingeri repertorium bibliographeum / Reichling, Dietrich – 1905-11. vWith 1914 supplementum. 8 v – 1,9 – us AMS Press [010]

The appendices to the gospel according to mark : a study in textual transmission / Williams, Clarence Russell – New Haven, Conn.: Yale University Press, 1915 – (= ser Transactions Of The Connecticut Academy Of Arts And Sciences) – 1mf – 9 – 0-7905-3419-3 – (incl bibl ref) – mf#1987-3419 – us ATLA [226]

Appendices to the sermon preached by the rev e b puseym... – Oxford, England. 1838 – 1r – us UF Libraries [240]

Appendices to votes and proceedings, 1817-1890 and reports of the select committees on public petitions, 1833-1900 : british public petitions of the 19th century / ed by Torrington, F W – 1817-1900 [mf ed Chadwyck-Healey, 1981] – (= ser House of commons parliamentary papers, 1801-1900) – 1220mf – 9 – (incl ind for 1833-52) – uk Chadwyck [324]

Appendix ad theologiam pacificam : sive modesta responsio ad...s maresii indiculum controversiarum... / Wittichius, C – Lugduni Batavorum, 1672 – 2mf – 9 – mf#PBA-410 – ne IDC [240]

Appendix (b) to report on the affairs of british north america : from the earl of durham, her majesty's high commissioner etc etc / Durham, John George Lambton, Earl of – [London, England: s.n, 1839] (mf ed 1998) – 3mf – 9 – mf#SEM105P2901 – cn Bibl Nat [971]

Appendix bibliothecae conradi gesneri / [Gessner, K] – Zuerich, Christoph Froschauer, 1555 – 3mf – 9 – mf#PBU-408 – ne IDC [240]

Appendix libelli adversus interim adultero-germanum : in qua refutat joannes calvinus censuram quandam typographi ignoti de parvulorum sanctificatione, et muliebri baptismo / Calvin, John – [Genevae: Jean Girard] – 1mf – 9 – mf#CL-7 – ne IDC [242]

Appendix no 26 to the honorable the president, and the honorable the members of the legislative council – [Quebec]: [s.n.], [1835] (mf ed 1989) – 1mf – 9 – mf#SEM105P1130 – cn Bibl Nat [324]

Appendix sive vol 9 : codex actuum laudianus – Lipsiae, 1870 – (= ser Monumenta sacra inedita. nova collectio v9) – €39.00 – ne Slangenburg [220]

Appendix to an inquiry into the prophetic numbers contained in the... / Mason, Archibald – Glasgow, Scotland. 1818 – 1r – us UF Libraries [240]

An appendix to cowen's treatise on the civil jurisdiction of justices of the peace in the state of new york. / Hayden, Chester – Albany: Gould, Banks and Gould, 1848. 142p. LL-646 – 1 – us L of C Photodup [340]

Appendix to five lectures on attrition, contrition, and sovereign love / Ward, William George – [S.l: s.n, 1858?] [mf ed 1991] – 1mf – 9 – 0-7905-8963-X – mf#1989-2188 – us ATLA [241]

Appendix to memoranda of june, 1892 : on the subject of free tuitions in the faculty of arts: exemptions from fees in favour of students of affiliated theological colleges / Dawson, John William – [S.l: s.n, 1892?] [mf ed 1980] – 1mf – 9 – mf#03663 – cn Canadiana [378]

Appendix to minutes of evidence taken before select committee of the house of lords on colonization from ireland / Grande-Bretagne. Parliament. House of Lords – [s.l.]: [s.n.], 1847 [mf ed 1983] – 3mf – 9 – mf#SEM105P158 – cn Bibl Nat [324]

Appendix to the report of the british commissioners appointed in july 1839 : to explore and survey the territory in dispute between the governments of great britain and the united states of america, under the 2nd article of the treaty of ghent – [S.l: s.n, 1839?] (mf ed 1992) – 1mf – 9 – mf#SEM105P1385 – cn Bibl Nat [970]

Appendix to the report of the commissioner of crown lands – Toronto: printed by John Lovell, 1859 [mf ed 1993] – 1mf – 9 – mf#SEM105P1856 – cn Bibl Nat [324]

Appendix to the...annual report of the council of the indian institute of science, bangalore / Indian Institute of Science. Bangalore – Bangalore: The Institute, 1918-1939 – 1r – 1 – (summary of research done at the institute during the year) – us CRL [500]

Apperson, M M *see* Victory

Appert, Benjamin Nicolas Marie *see* Dix ans a la cour du roi louis philippe et souvenirs du tems de l'empire et de la restauration

Appia, Adolphe *see* La mise en scene du drame wagnerien

Appian of Alexandria *see* Livius, books 31-40/dictys...

Appita journal – Carlton, Australia 1978+ – 1,5,9 – ISSN: 1038-6807 – mf#8930 – us UMI ProQuest [670]

Apple – Springfield IL 1975-76 – 1,5,9 – ISSN: 0003-6765 – mf#7480 – us UMI ProQuest [810]

The apple – London, 1920-22 [mf ed Chadwyck-Healey] – (= ser Art periodicals on microform) – 1r – 1 – uk Chadwyck [760]

Apple, James G et al *see* Manual for cooperation between state and federal courts

The apple of discord, or, temporal power in the catholic church / Zurcher, George – Buffalo, NY: Apple of Discord, 1905 – 2mf – 9 – 0-7905-6979-5 – (incl bibl ref) – mf#1988-2979 – us ATLA [241]

Apple pie – iss n1-1915 [1973 may 7-1975 fall] – 1 – mf#622513 – us WHS [071]

Apple river journal – 1975 jan 5-1977 may 7 – 1 – mf#931888 – us WHS [071]

Applebee, John Henry et al *see* West roxbury magazine

Appleby newsletter – v1 n1-2; v2 n1-v5 n2; v6 n1-2; v7 n1; [1981 jul-oct; 1982 apr-1985 oct; 1987 apr-oct; 1988 apr] – 1 – mf#1052217 – us WHS [071]

Applegarth, Margaret Tyson *see* Fifty-two primary missionary stories

Applegate, Michael T *see* The economic impact of dean e. smith activities center events on chapel hill, north carolina

Applegate, Thomas *see* The voice of sacred triples

Applegate's mineral springs / Briggs, John C – s.l, s.l? . 1936 – 1r – us UF Libraries [978]

Appleland bulletin – v7 n1-v15 n4 [1978 fall-1987 summer] – 1 – mf#1532794 – us WHS [071]

Apples of gold in pictures of silver : or, good words and comfortable words / Fordyce, Alexander Dingwall – Fergus, Ont?: s.n, 1881 – 1mf – 9 – mf#33729 – cn Canadiana [240]

Appleton city times – 1870 feb 26-1871 feb 25; 1873 apr 3 [v7 n6] – 1 – mf#917756 – us WHS [071]

Appleton crescent – 1853 feb 10-1855 oct 27; 1905 jan 7-1906 dec 29 [with gaps] – 1 – mf#890517 – us WHS [071]

Appleton daily post – 1885 mar 19; 1887 may 26; 1899 aug 11-1900 feb 21; 1900 apr 21; 1919 nov 3-dec 31 [with gaps] – 1 – mf#931568 – us WHS [071]

Appleton, Elizabeth *see*
- A guide to the french language
- The spring bud

Appleton evening crescent – 1897 apr 16-oct 18; 1919 dec 4-dec 31 [with gaps] – 1 – mf#918046 – us WHS [071]

Appleton, hill cliffe baptist chapel – (= ser Cheshire monumental inscriptions) – 3mf – 9 – £4.50 – mf#414 – uk CheshireFHS [929]

Appleton motor – 1859 aug 18-1925 – 1 – mf#918394 – us WHS [071]

Appleton Papers, Inc *see*
- Hi-lites
- Hi-lites, appleton plant
- Hi-lites, corporate
- Hi-lites, harrisburg plant
- Hi-lites, locks mill
- Hi-lites, spring mill
- Hi-lites, west carrollton mill
- News break

Appleton post – 1866 sep 20-1870 jan 6-1888 mar 1-1889 apr 25 [with few gaps] – 1 – mf#928612 – us WHS [071]

Appleton post-crescent – 1933 jan 2-nov 1957 [with gaps] – 1 – mf#1165526 – us WHS [071]

Appleton, R J *see* Testing the validity of the near infrared body composition technique in children

Appleton review – 1930 jan 16-dec 23 – 1 – mf#917749 – us WHS [071]

Appleton thorn, st cross – (= ser Cheshire monumental inscriptions) – 1mf – 9 – £2.50 – mf#2 – uk CheshireFHS [929]

Appleton, Victor *see*
- Tom swift and his sky train
- Tom swift and his television detector

Appleton volksfreund – 1874 oct 23-1929 [with gaps] – 1 – mf#959838 – us WHS [071]

Appleton volksfreund – Appleton WI (USA), 1925 20 aug-1926, 1931 19 nov-1932 18 may – 1r – 1 – gw Misc Inst [071]

Appleton wecker – 1887 apr 28 – 1 – mf#916101 – us WHS [071]

Appleton weekly post – 1889 may 2-1912 dec 26 [with gaps] – 1 – mf#958914 – us WHS [071]

Appleton's general guide to the united states and canada, 1879, 1882-1901 – 1879-1901 – 1 – us L of C Photodup [917]

Appletons' guide-book to alaska and the northwest coast : including the shores of washington, british columbia, southeastern alaska, the aleutian and the seal islands, the bering and the arctic coasts, the yukon river and klondike district / Scidmore, Eliza Ruhamah – new ed. New York: D Appleton, 1899 [mf ed 1981] – 3mf – 9 – (with chap on the klondike) – mf#16106 – cn Canadiana [917]

Appleton's journal : a magazine of general literature – La Salle IL 1869-81 – 1 – mf#5234 – us UMI ProQuest [410]

Appleton's magazine – Killen TX 1903-09 – 1 – mf#2861 – us UMI ProQuest [400]

Appleyard, J W *see* Kafir-english dictionary

Appleyard, Lula Dee Keith *see* Plantation life in middle florida, 1821-1845

Appliance – Oak Brook IL 1944+ – 1,5,9 – ISSN: 0003-6781 – mf#1488 – us UMI ProQuest [690]

Appliance engineer – Oak Brook IL 1967-74 – 1,5,9 – ISSN: 0003-6773 – mf#8435 – us UMI ProQuest [620]

Appliance manufacturer – Troy MI 1984+ – 1,5,9 – ISSN: 0003-679X – mf#14864.01 – us UMI ProQuest [680]

Application a la geographie des methodes d'etude d... / Cailleux, Andre – Rio de Janeiro, Brazil. 1961 – 1r – us UF Libraries [972]

Application de la geographie a l'histoire : ou etude elementaire de geographie et d'histoire generales comparees / Braconnier, Edouard – Paris: Simon. 1845 [mf ed 1985] – 2v on 1mf – 9 – 0-665-51981-8 – mf#51981 – cn Canadiana [910]

Application for assurance in the federal life assurance company of ontario : head office in hamilton / Federal Life Assurance Co – [s.l: s.n. 189-?] [mf ed 1987] – 1mf – 9 – 0-665-41046-8 – mf#41046 – cn Canadiana [368]

The application of affirmative action policies in the south african correctional services department / Makgoba, Matsemela Johannes – Uni of South Africa 2001 [mf ed Johannesburg 2001] – 4mf – 9 – (incl bibl) – mf#mfm14675 – sa Unisa [365]

An application of bowen family system theory to the pastoral ministry of the sharon baptist church / Robertson, James Errol – 1982 – 1 – $5.36 – us Southern Baptist [242]

The application of federal laws in american samoa, guam, the northern mariana islands, and the virgin islands / American Samoa. Dept of Interior – Washington: Office of the Solicitor. v1-3. oct 1993 – 17mf – 9 – $25.50 – mf#LLMC 95-036 – us LLMC [327]

The application of human motor control principles to a collective robotic arm / Harty, Tyson H – 2000 – 96p on 1mf – 9 – $5.00 – mf#PSY 2151 – us Kinesology [629]

An application of item response theory to the rest of gross motor development / Cole, E L – 1990 – 2mf – 9 – $8.00 – us Kinesology [150]

Application of john galbraith...for the chair of civil engineering, in the school of practical science, province of ontario : together with copies of testimonials and recommendations – Toronto?: Rowsell & Hutchison, 1878 – 1mf – 9 – mf#26525 – cn Canadiana [378]

Application of lma principles in ethnic dance training / Christopher, Tara L – 2000 – 242 on 3mf – 9 – $5.00 – mf#PE 4133 – us Kinesology [790]

The application of logic / Sidgwick, Alfred – London: Macmillan, 1910 – 1mf – 9 – 0-7905-7371-7 – mf#1989-0596 – us ATLA [160]

The application of ngara's linguistic format in n f mbhele's short stories, "amayezi namathunzi" / Malinga, Sizakele Sellina – Pretoria: Vista University 2000 [mf ed 2000] – 3mf – 9 – (incl bibl) – mf#mfm15154 – sa Unisa [470]

The application of ornament / Day, Lewis Foreman – 2nd ed. London 1891 – (= ser 19th c art & architecture) – 2mf – 9 – mf#4.2.115 – uk Chadwyck [740]

Application of outcomes-based education in the teaching of sesotho : in the general education and training band at the foundation phase (grade one) / Ramabenyane, Mamosebatho Julia – Vista University 2000 [mf ed Johannesburg 2000] – 5mf [ill] – 9 – (incl bibl ref) – mf#mfm14743 – sa Unisa [370]

The application of the christian faith by small college christian american athletes within the sport of baseball / Wendt, Vernon E – 2000 – 569p on 6mf – 9 – $30.00 – mf#PSY 2164 – us Kinesology [303]

Application of the credit valley railway for right of way and crossings at the city of toronto : second interview of the railway delegation with the railway comittee of the privy council, ottawa, thursday, june 3rd, 1879 / Holland, A & Holland, George C [comp] – [Ottawa?: s.n, 1879?] [mf ed 1980] – 1mf – 9 – 0-665-04186-1 – mf#04186 – cn Canadiana [380]

The application of the polygraph in the criminal justice system / Martin, Raymond Charles – Uni of South Africa 2001 [mf ed Johannesburg 2001] – 7mf – 9 – (incl bibl ref) – mf#mfm15041 – sa Unisa [365]

The application of the roman alphabet to all the oriental languages : contained in a series of papers / Trevelyan, Charles Edward et al – [Serampore?]: Serampore Press, 1834 – 1mf – 9 – 0-524-03109-6 – mf#1990-0834 – us ATLA [490]

Application of the transtheoretical model of behavior change to physical activity behavior in a college education course / Vogler, Dawn R – 1999 – 1mf – 9 – $4.00 – mf#HE 644 – us Kinesology [378]

Application of the transtheoretical model to exercise adherence / Murphy, Debra & Roberts, John A – 1992 – 2mf – 9 – $8.00 – us Kinesology [150]

Applications / Society of Colonial Wars in the State of South Carolina – [mf ed Summerville SC: Charleston Micrographics, 1998] – 1r – 1 – mf#45-359 – us South Carolina Historical [978]

Applications for certificates of necessity, 1941-1945 / U.S. War Production Board – (= ser Records of the war production board) – 1095r – 1 – mf#M1200 – us Nat Archives [934]

Applications for enrollment and allotment of washington indians, 1911-1919 / U.S. Bureau of Indian Affairs – (mf ed 1984) – (= ser Records relating to census rolls and other enrollments; National archives microfilm publications) – 6r – 1 – mf#M1343 – us Nat Archives [317]

Applications for enrollment of the commission to the five civilized tribes, 1898-1914 / U.S. Bureau of Indian Affairs – (= ser Records relating to census rolls and other enrollments) – 468r – 5 – mf#M1301 – us Nat Archives [317]

Applications from the bureau of indian affairs, muskogee area office, relating to enrollment in the five civilized tribes under the act of 1896 / U.S. Bureau of Indian Affairs – (= ser Records relating to census rolls and other enrollments) – 54r – 1 – mf#M1650 – us Nat Archives [317]

Applications to the land court / Government of Niue, Justice, Lands and Survey Department, Land Court – 1985-2003 – 1r – 1 – mf#pmb1259 – at Pacific Mss [347]

Applied acoustics = Acoustique applique – Oxford, England 1968+ – 1,5,9 – ISSN: 0003-682X – mf#42073 – us UMI ProQuest [621]

Applied and environmental microbiology – Washington DC 1964+ – 15,9 – (cont: applied microbiology) – ISSN: 0099-2240 – mf#1713,01 – us UMI ProQuest [576]

Applied animal behaviour science – Oxford, England 1975+ – 1,5,9 – ISSN: 0168-1591 – mf#42213 – us UMI ProQuest [590]

Applied artificial intelligence – Oxford, England 1987+ – 1,5,9 – ISSN: 0883-9514 – mf#16652 – us UMI ProQuest [000]

Applied biochemistry and microbiology – Dordrecht, Netherlands 1965+ – 1,5,9 – ISSN: 0003-6838 – mf#10900 – us UMI ProQuest [576]

Applied catalysis – Oxford, England 1981-91 – 1,5,9 – ISSN: 0166-9834 – mf#42214 – us UMI ProQuest [660]

Applied catalysis a : general – Oxford, England 1992+ – 1,5,9 – mf#42658 – us UMI ProQuest [660]

Applied catalysis b : environmental – Oxford, England 1993+ – 1,5,9 – ISSN: 0926-3373 – mf#42659 – us UMI ProQuest [660]

Applied christianity : moral aspects of social questions / Gladden, Washington – Boston: Houghton, Mifflin, 1886 – 1r – 1 – 0-8370-0348-2 – mf#1984-B264 – us ATLA [240]

Applied christianity in the hokkaido : an attempt at prison reform in japan / Curtis, William Willis – Boston: American Board of Commissioners for Foreign Missions, [189-] [mf ed 1995] – (= ser Yale coll) – 12p – 1 – 0-524-09692-9 – mf#1995-0692 – us ATLA [230]

Applied cognitive psychology – Hoboken NJ 1987+ – 1,5,9 – (cont: human learning) – ISSN: 0888-4080 – mf#16095 – us UMI ProQuest [150]

Applied developmental science – Mahwah NJ 2001+ – 1 – mf#28516 – us UMI ProQuest [150]

Applied economics – Abingdon, Oxfordshire 1983-98 – 1,5,9 – ISSN: 0003-6846 – mf#14398 – us UMI ProQuest [338]

Applied energy – Oxford, England 1975+ – 1,5,9 – ISSN: 0306-2619 – mf#42215 – us UMI ProQuest [333]

Applied ergonomics – Oxford, England 1969+ – 1,5,9 – ISSN: 0003-6870 – mf#13320 – us UMI ProQuest [620]

Applied geochemistry – Oxford, England 1986+ – 1,5,9 – ISSN: 0883-2927 – mf#49481 – us UMI ProQuest [550]

Applied geography – Oxford, England 1981+ – 1,5,9 – ISSN: 0143-6228 – mf#17211 – us UMI ProQuest [910]

Applied imagination / Osborn, Alexander Faickney – New York, NY. 1963 – 1r – us UF Libraries [150]

Applied linguistics – Oxford, England 1983+ – 1,5,9 – ISSN: 0142-6001 – mf#14049 – us UMI ProQuest [400]

Applied marketing research – Rocky Hill CT 1961-91 – 1,5,9 – ISSN: 1064-1157 – mf#5095,03 – us UMI ProQuest [650]

Applied mathematical modelling – Oxford, England 1976+ – 1,5,9 – ISSN: 0307-904X – mf#13321 – us UMI ProQuest [510]

Applied mathematics and computation – Oxford, England 1975+ – 1,5,9 – ISSN: 0096-3003 – mf#42072 – us UMI ProQuest [510]

Applied mathematics and optimization – Dordrecht, Netherlands 1974+ – 1,5,9 – ISSN: 0095-4616 – mf#13130 – us UMI ProQuest [510]

Applied mathematics letters – Oxford, England 1988+ – 1,5,9 – ISSN: 0893-9659 – mf#49514 – us UMI ProQuest [510]

Applied measurement in education – Mahwah NJ 1998+ – 1,5,9 – ISSN: 0895-7347 – mf#25211 – us UMI ProQuest [370]

Applied mechanics / Poorman, Alfred Peter – New York, NY. 1930 – 1r – us UF Libraries [621]

Applied mechanics reviews – New York NY 1948+ – 1,5,9 – ISSN: 0003-6900 – mf#1015 – us UMI ProQuest [621]

Applied microbiology and biotechnology – Dordrecht, Netherlands 1984+ – 1,5,9 – (cont: european journal of applied microbiology and biotechnology) – ISSN: 0175-7598 – mf#13165,02 – us UMI ProQuest [576]

Applied numerical mathematics – Oxford, England 1985+ – 1,5,9 – ISSN: 0168-9274 – mf#42483 – us UMI ProQuest [510]

Applied nursing research [anr] – Philadelphia PA 1992+ – 1,5,9 – ISSN: 0897-1897 – mf#21092 – us UMI ProQuest [610]

Applied organometallic chemistry – Hoboken NJ 1991+ – 1,5,9 – ISSN: 0268-2605 – mf#17160 – us UMI ProQuest [540]

Applied physics – Dordrecht, Netherlands 1973-81 – 1,5,9 – ISSN: 0340-3793 – mf#13252.02 – us UMI ProQuest [621]

Applied physics a and b – Dordrecht, Netherlands 1984-96 – 1,5,9 – mf#13254 – us UMI ProQuest [621]

Applied physics a. materials science and processing – Dordrecht, Netherlands 1995-96 – 1,5,9 – (cont: applied physics a: solids and surfaces) – ISSN: 0947-8396 – mf#13252,02 – us UMI ProQuest [621]

Applied physics a. solids and surfaces – Dordrecht, Netherlands 1981-94 – 1,5,9 – (cont by: applied physics a: materials science and processing) – ISSN: 0721-7250 – mf#13252.02 – us UMI ProQuest [621]

Applied physics b. lasers and optics – Dordrecht, Netherlands 1994+ – 1,5,9 – (cont: applied physics b: photophysics and laser chemistry) – ISSN: 0946-2171 – mf#13253,01 – us UMI ProQuest [621]

Applied physics b. photophysics and laser chemistry – Dordrecht, Netherlands 1981-93 – 1,5,9 – (cont by: applied physics b: lasers and optics) – ISSN: 0721-7269 – mf#13253.01 – us UMI ProQuest [621]

Applied physics letters – v1- 1962- – 1,5,6,9 – us AIP [621]

Applied psycholinguistics – Cambridge, England 1980+ – 1,5,9 – ISSN: 0142-7164 – mf#13014 – us UMI ProQuest [400]

Applied psychological measurement – Newbury Park CA 1977+ – 1,5,9 – ISSN: 0146-6216 – mf#12872 – us UMI ProQuest [150]

Applied psychophysiology and biofeedback – Dordrecht, Netherlands 1997+ – 1,5,9 – (cont: biofeedback and self-regulation) – ISSN: 1090-0586 – mf#17657,01 – us UMI ProQuest [612]

Applied radiation and isotopes – Oxford, England 1993+ – 1,5,9 – (cont: international journal of radiation applications & instrumentation pt a: applied radiation & isotopes) – ISSN: 0969-8043 – mf#49091,01 – us UMI ProQuest [530]

Applied Research Center *see* Racefile

Applied research in mental retardation – Oxford, England 1980-86 – 1, 5,9 – ISSN: 0270-3092 – mf#49389 – us UMI ProQuest [370]

Applied scientific research – Dordrecht, Netherlands 1991-94 – 1,5,9 – (cont by: flow, turbulence & combustion) – ISSN: 0003-6994 – mf#16767.01 – us UMI ProQuest [500]

Applied stochastic models and data analysis – Hoboken NJ 1985-94 – 1,5,9 – (cont by: applied stochastic models in business & industry) – ISSN: 8755-0024 – mf#16096.01 – us UMI ProQuest [510]

Applied stochastic models in business and industry – Hoboken 2001+ – 1,5,9 – (cont: applied stochastic models & data analysis) – ISSN: 1524-1904 – mf#16096,01 – us UMI ProQuest [510]

Applied surface science – Oxford, England 1977+ – 1,5,9 – ISSN: 0169-4332 – mf#42015 – us UMI ProQuest [540]

An applied therapeutic magnet has no effect on grip strength / Perkins, Kelli L – 2000 – 1mf – 9 – $4.00 – mf#PH 1692 – us Kinesology [612]

Applied thermal engineering – Oxford, England 1996+ – 1,5,9 – (cont: heat recovery systems and chp) – ISSN: 1359-4311 – mf#49386,02 – us UMI ProQuest [530]

Applied thermal sciences – Hoboken NJ 1988-89 – 1,5,9 – ISSN: 1042-0959 – mf#18116 – us UMI ProQuest [621]

The appointed time : being scriptural, historical, and astronomical proofs of the end of the gentile times in 1898 1/4 and the coming of the lord / Dimbleby, Jabez Bunting – 2nd ed. London: E Nister, 1896 [mf ed 1992] – 1mf – 9 – 0-524-03968-2 – mf#1992-0011 – us ATLA [220]

Appointment and promise of messiah / Craig, Edward – Edinburgh, Scotland. 1821 – 1r – us UF Libraries [240]

Appointment book of president kennedy (1961-1963) – (= ser Presidential documents series) – 3r – 1 – $500.00 – 0-89093-357-X – (with p/g) – us UPA [977]

Appointment congo / Law, Virginia W – Chicago, IL. 1966 – 1r – us UF Libraries [960]

Appointment of popish bishops : speech – London, England. 1850? – 1r – us UF Libraries [240]

Apponius *see* In canticum canticorum expositio

Appraisal journal – Chicago IL 1932+ – 1,5,9 – ISSN: 0003-7087 – mf#6524 – us UMI ProQuest [333]

An appraisal model for academic staff performance at technical colleges in south africa / Litheko, Segano R S – Pretoria: Vista University 2001 [mf ed 2001] – 5mf – 9 – (incl bibl ref) – mf#mfm15319 – sa Unisa [650]

Appraisal of educators in schools in the bloemfontein area / Toolo, Lynnette Lineo – Vista University 2000 [mf ed Johannesburg 2000] – 2mf – 9 – (incl bibl ref) – mf#mfm14748 – sa Unisa [370]

Appraisal of the development of kicking behavior of preschool age children / Paula, E A – 1991 – 1mf – 9 – $4.00 – us Kinesology [150]

Appraising physical status : methods and norms / McCloy, Charles H – 1938 – 10mf – 9 – $30.00 – us Kinesology [612]

Appraising physical status : the selection of measurements / McCloy, Charles H – 1936 – 4mf – 9 – $12.00 – us Kinesology [612]

The appreciation of art / Overton, Alfred C – Allahabad, India: Kitab Kutir, 1949 – (= ser Samp: indian books) – (foreword by nandalal bose) – us CRL [700]

Appreciations and criticisms of the works of charles dickens / Chesterton, G K – London, England. 1911 – 1r – us UF Libraries [420]

Apprenti gabriel / Deyrieux, L – Lyon, France. 18–? – 1r – us UF Libraries [440]

Apprentice's companion – Killen TX 1835 – 1 – mf#4785 – us UMI ProQuest [073]

Apprentices Library Society. Charleston, South Carolina *see* Records of the apprentices' and minors' library society

Approach – 1988 oct-1993 dec – 1 – mf#5486805 – us WHS [071]

Approach : the mission/education newsweekly – 1968 apr 22-1969 jun 16 – 1 – mf#1052226 – us WHS [071]

Approach [1955] – Washington DC 1955-95 – 1,5,9 – ISSN: 0570-4979 – mf#6290 – us UMI ProQuest [629]

Approach [1997] – Washington DC 1997+ – 1,5,9 – ISSN: 1094-0405 – mf#26513 – us UMI ProQuest [629]

Approach mech – Washington DC 1995-96 – 1,5,9 – ISSN: 1086-928X – mf#21883 – us UMI ProQuest [629]

The approach of christ to modern india / Farquhar, John Nicol – Calcutta: Association Press, 1913 – 1mf – 9 – 0-7905-6057-7 – mf#1988-2057 – us ATLA [240]

Approach to a practical pedagogy of piano sight-reading / Hart, Lawrence Elbert – U of Rochester 1958 [mf ed 19–] – 5mf – 9 – mf#fiche 243, 535 – us Sibley [780]

Approach to the artistic playing of wind instruments / Peters, Harry B – U of Rochester 1940 [mf ed 19–] – 2mf – 9 – (with bibl) – mf#fiche 1133 – us Sibley [780]

The approach to the gospel : addresses delivered to the annual conference of the american presbyterian mission of western india at panhala, kolhapur / Hooper, J S M – Kolhapur: A P Mission Press, 1918 [mf ed 1995] – (= ser Yale coll) – 44p – 1 – 0-524-09540-X – mf#1995-0540 – us ATLA [242]

The approach to the social question : an introduction to the study of social ethics / Peabody, Francis Greenwood – New York: Macmillan, 1912 – (= ser Earl Lectures) – 1mf – 9 – 0-524-04846-0 – (incl bibl ref) – mf#1990-1338 – us ATLA [360]

Approche ethnolinguistique de la tradition orale wolof : contes et taasu / Keita, Abdoulaye – 1986 – (= ser Kesteloot coll) – 6mf – 9 – us CRL [390]

Appropriate technology : its importance for african women / Carr, Marilyn. – 1977 – (= ser African training and research centre for women publications on microfilm) – 2mf – 9 – us CRL [305]

Appropriate technology for african women / Carr, Marilyn – [Addis Ababa]: African Training & Research Ctr for Women, Economic Commission for Africa, UN 1978 – (= ser African training and research centre for women publications on microfilm) – 2mf – 9 – us CRL [305]

Approval of agressive acts in wrestling : individual and contextual variables / DeVries, Steven N – 1998 – 242p on 3mf – 9 – $15.00 – mf#PSY 2176 – us Kinesology [150]

Apr – Sydney, Australia 1950-56 – 1 – mf#579 – us UMI ProQuest [770]

Apraes le concile : ou, hyacinthe et doellinger – Lyon: Denis, 1872 [mf ed 1986] – 1mf – 9 – 0-8370-8375-3 – mf#1986-2375 – us ATLA [241]

Aprendiz de la exercitation 36 / Luna Vega, J – SL, 1618 – 1mf – 9 – sp Cultura [610]

Aprent, Johannes *see* Erzaehlungen

Apres de Mannevillette, J B *see* Instructions sur la navigation des indes orientales et de la chine, pour servir au neptune oriental

D'apres les paraboles histoires vraies... / Debout, Jacques; ed by Bayle, Constantino – Madrid: Razon y Fe, 1928 – 9 – sp Bibl Santa Ana [240]

Apresentacao da poesia brasileira / Bandeira, Manuel – Rio de Janeiro, Brazil. 1946 – 1r – us UF Libraries [972]

Apri news release – 1983 jul 8-1985 jun 6 – 1 – mf#1231756 – us WHS [071]

April airs : a book of new england lyrics / Carman, Bliss – Boston: Small, Maynard, 1916 – 2mf – 9 – 0-665-77893-7 – mf#77893 – cn Canadiana [810]

April, Ewart Zolile *see* The effectiveness of the cascade model in the training of educators for implementing outcomes-based education

April's lady : a novel / Hungerford, Margaret Wolfe – London: F V White & Co. 3v. 1891 – (= ser 19th c women writers) – 9mf – 9 – mf#5.1.113 – uk Chadwyck [830]

Aprobacion y confirmacion que dio el...al parecer y adicion que hizo. fray juan de los reyes... / Colegio Mayor del Conde Duque – S.l., s.i., s.a. 1640 – 1 – sp Bibl Santa Ana [240]

Aprokos gospels [complete] – 1220s-1230s – 9mf – 9 – (old russian version) – us UMI ProQuest [090]

Aprokos gospels [complete] – 1300s, 1800s – 9mf – 9 – (russian version) – us UMI ProQuest [090]

Aprokos gospels [complete] – 1350s-90s – 7mf – 9 – (russian version) – us UMI ProQuest [090]

Apropos [sic] : au public du montreal pour les adieux de la compagnie / Achintre, Auguste – [Montreal?: s.n, 1867?] [mf ed 1980] – 1mf – 9 – mf#06573 – cn Canadiana [780]

Aproximacion a la evolucion socio-economica de la provincia de badajoz en 1972 / Secretariado de Asuntos Economicos de Badajoz – Badajoz: Graf. Jimenez, s.a. – sp Bibl Santa Ana [330]

Aproximacion a un estudio de antonio hurtado como poeta / Garcia Camino, Victor Gerardo & Garcia Caminos Burgos, Luis F – Badajoz: Dip. Provincial, 1958 – 1 – sp Bibl Santa Ana [440]

Aps news – 1,5,6,9 – us AIP [530]

Apsley house and walmer castle / Ford, Richard – London 1853 – (= ser 19th c art & architecture) – 3mf – 9 – mf#4.2.1732 – uk Chadwyck [720]

Apt to teach / Dickson, William – Edinburgh, Scotland. 1875 – 1r – us UF Libraries [240]

The aptos voice *see* [Santa cruz-] miscellaneous titles

Aptowitzer, V *see* Kain und abel in der agada der apokryphen, der hellenistischen, christlichen und muhammedanischen literatur

Apuntaciones criticas sobre el lenguaje bogotano / Cuervo, Rufino Jose – Bogota, Colombia. 1939 – 1r – us UF Libraries [972]

Apuntaciones criticas sobre el lenguaje bogotano / Cuervo, Rufino Jose – Paris, France. 1914 – 1r – us UF Libraries [972]

Apuntaciones literarias / Vitier, Medardo – Habana, Cuba. 1935 – 1r – us UF Libraries [972]

Apuntaciones para la historia natural de las aves...tomo 1-2, asuncion / Azara, Felix – 9mf – 9 – sp Cultura [590]

Apuntameintos sobre la topografia fiscia / Guzman, David Joaquin – San Salvador, El Salvador. 1883 – 1r – us UF Libraries [972]

Apuntamiento de indias, ms. 1568-1637 – 7mf – 9 – sp Cultura [340]

Apuntamiento legal de la o. santiago / Chaves, Bernabe de – 1 – sp Bibl Santa Ana [340]

Apuntamientos sobre el adelantamiento de yucatan, de amalio huarte y echenique / Beltran y Rozpide, Ricardo – Madrid: Fortanet, 1920. B.R.A.H. 76. pp. 5-6 – 1 – sp Bibl Santa Ana [946]

Apuntamientos...de como se deben reformar... las doctrinas y la manera de... / Abril, P – Madrid, 1589 – 1mf – 9 – sp Cultura [610]

Apunte descriptivo de la serena / Hidalgo, Juan Francisco – Castuera: Imp. La Puritana, 1924 – 1 – sp Bibl Santa Ana [946]

Apuntes / Weber, Delia – Ciudad Trujillo, Dominican Republic. 1949 – 1r – us UF Libraries [972]

Apuntes bibliograficos de la prensa periodica de la baja extremadura. 1 y 2 / Guerra Guerra, Arcadio – Badajoz: Imp. Dip. Provincial, 1974. Sep. REE – sp Bibl Santa Ana [073]

Apuntes bibliographicos / Rodriguez Y Exposito, Cesar – Habana, Cuba. 1947 – 1r – us UF Libraries [972]

Apuntes bilogicos sobre el pojo de las habas / Moreno Marquez, Victor – Madrid: Rev. Fitopatologia, 1944 – 1 – sp Bibl Santa Ana [574]

Apuntes biograficas del m.i. sr. d. francisco de paula soto y mancera, arcipreste de la santa y apostolica iglesia de santiago de compostela, natural de zafra, provincia de badajoz / Calderon, Cesareo – Badajoz: Uceda Hermanos, 1905 – 1 – sp Bibl Santa Ana [240]

Apuntes biograficos de don juan carrillo sanchez / Anguita Valdivia, Jose – Madrid: Razon y Fe, 1929 – 1 – sp Bibl Santa Ana [920]

Apuntes biograficos de emiia casanova de villaverde escritos / Asonova De Villaverde, Emilia – New York, NY. 1874 – 1r – us UF Libraries [972]

Apuntes biograficos en torno a la vida / Figueroa De Cifredo, Patria – San Juan, Puerto Rico. 1965 – 1r – us UF Libraries [972]

Apuntes de cancerolojia, para el prontuario / Pieter, Heriberto – Ciudad Trujillo, Dominican Republic. 1950 – 1r – us UF Libraries [972]

Apuntes de esparragalejo / Parejo Gonzalez, Jose – Badajoz: Tip. y Libr. Bernardo Vadillo Serrano, 1946 – 1 – sp Bibl Santa Ana [946]

Apuntes de haiti / Monclus, Miguel Angel – Ciudad Trujillo, Dominican Republic. 1952 – 1r – us UF Libraries [972]

Apuntes de heraldica cacerena / Rueda Sanchez Malo, Jose Miguel – Badajoz: Institucion Pedro de Valencia, 1971 – 1 – sp Bibl Santa Ana [946]

Apuntes de historia eclesiastica de venezuela. caracas, 1929 / Talavera y Garces, Mariano – Madrid: Razon y Fe, 1930 – 1 – sp Bibl Santa Ana [240]

Apuntes de historia natural y mamiferos de guatemala / Ibarra, Jorge A – Guatemala, 1959 – 1r – us UF Libraries [972]

Apuntes de la delegacion de hacienda / Lateulade, Emilio – Guantanamo, Cuba. 1930 – 1r – us UF Libraries [972]

Apuntes de la sublevacion fascista : impresiones de un militar republicano / Romero, Luis – Barcelona: Oficinas de Propaganda CNT FAI [1937] – 9 – mf#w1149 – us Harvard [946]

Apuntes de matematicas / Llinas Estevez, Jeronimo – Badajoz: La minerva extremena, s.a. – 1 – sp Bibl Santa Ana [510]

Apuntes de ortografia para ingreso (colegio de san jose) – Villafranca de los Barros: Imp. Rodriguez, s.a. – 1 – sp Bibl Santa Ana [946]

Apuntes de pedagogia / Maillo, Adolfo – Caceres, s.i. 1953 – 1 – sp Bibl Santa Ana [370]

Apuntes de pedagogia deportiva / Fernandez Santana, Ezequiel – Badajoz: Tip. Joaquin Sanchez, 1922 – sp Bibl Santa Ana [946]

Apuntes de psicologia y logica – Villafranca de los Barros: Tip. F. Rodriguez, 1923 – 1 – sp Bibl Santa Ana [150]

Apuntes de recuerdos / Guel, Conde de (Marques de Comillas); ed by Bayle, Constantino – Madrid: Razon y Fe, 1928 – 9 – sp Bibl Santa Ana [920]

Apuntes de un turista tropical / Iraizoz Y De Villar, Antonio – Habana, Cuba. 1931 – 1r – us UF Libraries [972]

Apuntes historico-biograficos / Martinez Delgado, Luis – Bogota, Colombia. 1940 – 1r – us UF Libraries [972]

Apuntes historicos de la que fue sede arzobispal de merida del 507 al 910 (403 anos) / Munoz Gallardo, Juan Antonio – Badajoz: Imprenta de la Diputacion Provincial, 1971. Separata Revista Estudios Extremenos – 1 – sp Bibl Santa Ana [240]

Apuntes historicos...villa de fuente del maestre desde... / Gomez Jara y Herrera, Juan de la Cruz – 1873 – 9 – sp Bibl Santa Ana [946]

Apuntes ineditos / Marti, Jose – Habana, Cuba. 1951 – 1r – us UF Libraries [972]

Apuntes para el presente y porvenir de cuba / Pujol Y De Camps, Marcelo – Habana, Cuba. 1885 – 1r – us UF Libraries [972]

Apuntes para el tiempo / Diaz Castro, Tania – Habana, Cuba. 1964 – 1r – us UF Libraries [972]

Apuntes para la h de pacora / Gutierrez, Gonzalo – Armenia, Colombia. 1942 – 1r – us UF Libraries [972]

Apuntes para la historia de serradilla (caceres) / Sanchez Rodrigo, Agustin – Serradilla (Caceres): Imprenta de Sanchez Rodrigo, 1930 – 1 – sp Bibl Santa Ana [946]

Apuntes para la historia de villafranca de los barros / Asensio, Jose Maria – Madrid: Fortanet, 1904. B.R.A.H. XLIV, pp. 246-249 – sp Bibl Santa Ana [946]

Apuntes para la historia de villafranca de los barros (badajoz) / Cascales Munoz, Jose – Madrid: Est. Tip. Fortanet, 1904 – sp Bibl Santa Ana [946]

Apuntes para la historia de villafranca de los barros de jose cascales munoz... / Asensio, Jose Maria – Madrid: Tip. de Fortanet, 1899 – sp Bibl Santa Ana [946]

Apuntes para la historia de zafra / Osuna Lara, Antonio J – Badajoz: Imp. de la Diputacion Provincial, 1976 – sp Bibl Santa Ana [946]

Apuntes para la historia del clero de caldas / Duque Botero, Guillermo – Medellin, Colombia. 1957 – 1r – us UF Libraries [972]

Apuntes para la historia del derecho de mejico / Esquivel Obregon, Toribio – Madrid: Missionalia Hispanica, 1945. 3v – 1 – sp Bibl Santa Ana [972]

Apuntes para la historia del origen y desenvolvimiento del regio patronato indiano hasta 1857 / Garcia Gutierrez, Jesus – Mexico, 1941; Madrid: Missionalia Hispanica, 1944 – 1 – sp Bibl Santa Ana [240]

Apuntes para la historia literaria de puerto rico / Cabrera, Francisco Manrique – San Juan, Puerto Rico. 1957 – 1r – us UF Libraries [972]

Apuntes para la historia...plasencia / Barrio y Rufo, Jose – 1851 – 9 – sp Bibl Santa Ana [946]

Apuntes para un diccionario de escritores del s. 19 / Ossorio, Bernard M – 1889 – 9 – sp Bibl Santa Ana [440]

Apuntes para una sociologia costarricense / Rodriguez Vega, Eugenio – San Jose, Costa Rica. 1953 – 1r – us UF Libraries [972]

Apuntes para...topografico...burguillos / Martinez Martinez, Matias Ramon – 1884 – 9 – sp Bibl Santa Ana [910]

Apuntes sobre el movimiento de poblacion de buenos aires de 1879 – Buenos Aires, 1879 – 2mf – 9 – sp Cultura [318]

Apuntes sobre la escenificacion de los romances. manuscrito / Garcia Lorca, Federico – 1mf – 9 – sp Cultura [820]

Apuntes sobre la provincia misional del orinoco...caracas, 1933 / Bueno, Ramon – Madrid: Razon y Fe, 1934 – 1 – sp Bibl Santa Ana [946]

Apuntes sobre los sucesos ocurridos en... miajadas – 1873 – 9 – sp Bibl Santa Ana [946]

Apuntes sobre los urbach / Portuondo, Jose Antonio – Habana, Cuba. 1953 – 1r – us UF Libraries [972]

Apuntes sobre poesia popular y poesia negra en las... / Hernandez Franco, Tomas Rafael – San Salvador, El Salvador. 1942 – 1r – us UF Libraries [440]

Apuntes y documentos contra la orden dominicana en colombia 1680-1930. caracas, 1936 / Mesanza, Andres – Burgos: Razon y Fe, 1938 – 1 – sp Bibl Santa Ana [240]

Apuntes y documentos sobre la orden dominicana en... / Mesanza, Andres – Caracas, Venezuela. 1936 – 1r – us UF Libraries [240]

Apuntes y documentos...administrativas / Bravo Murillo, Juan – 1858 – 9 – sp Bibl Santa Ana [946]

Apuntes y materiales para la biografia de don jose de espronceda / Cascales Munoz, Jose – Extrait de la Revue Hispanique. Tome 23. New York, Paris. 1910 – 1 – sp Bibl Santa Ana [920]

Apuntes y pinchazos / Jimenez Lugo, Angel – San Juan, Puerto Rico. 1959 – 1r – us UF Libraries [972]

Apuntes...archivo general de simancas / Romero de Castillay Perosso, Francisco – 1873 – 9 – sp Bibl Santa Ana [025]

Apuntes...historia...amigos del pais de badajoz / Merino de Torres, Alberto – 1898 – 9 – sp Bibl Santa Ana [830]

Apuntes...villa de gata / Guerra Hontiveros, Marcelino – 1897 – 9 – sp Bibl Santa Ana [946]

Apwu review / Montgomery Co. Dayton – (oct 1971-sep 1972), apr 1973-74 [irreg, mthly] – 1r – 1 – mf#B10192 – us Ohio Hist [331]

Apwu seattle news-report – v1 n1-3 1971 oct-dec – 1 – mf#633277 – us WHS [071]

Aqmulla – Troitsk, Russia jul 191-oct 1917 [mf ed Norman Ross] – 1r – 1 – mf#nrp-1831 – us UMI ProQuest [077]

Aqua [1983] – Oxford, England 1983-88 – 1,5,9 – (cont by: aqua [1988]) – ISSN: 0003-7214 – mf#49440 – us UMI ProQuest [333]

Aqua [1989] – Oxford, England 1989+ – 1,5,9 – (cont: aqua [1983]) – ISSN: 0003-7214 – mf#17143 – us UMI ProQuest [333]

Aqua pictura / Hassell, John – 2nd ed. London [1813] – (= ser 19th c art & architecture) – 2mf – 9 – mf#4.2.879 – uk Chadwyck [750]

Aq-ua-chamine = Menominee talking – 1974 sep 15-1976 jun 30 – 1 – mf#345246 – us WHS [071]

Aquachamine / Menominee Restoration Committee – 1974-76 – (= ser American indian periodicals... 2) – 3mf – 9 – $95.00 – us UPA [305]

Aquacultural engineering – Oxford, England 1982+ – 1,5,9 – ISSN: 0144-8609 – mf#42402 – us UMI ProQuest [620]

Aquaculture – Oxford, England 1972+ – 1,5,9 – ISSN: 0044-8486 – mf#42074 – us UMI ProQuest [639]

Aquaculture and fisheries management – Oxford, England 1985-94 – 1,5,9 – (cont: fisheries management; cont by: aquaculture research) – ISSN: 0266-996X – mf#15503.03 – us UMI ProQuest [639]

Aquaculture research – Oxford, England 1995+ – 1,5,9 – (cont: aquaculture and fisheries management) – ISSN: 1355-557X – mf#15503,02 – us UMI ProQuest [639]

Aquapreneurship : characteristics and business management practices of current and potential swim school owners / Mackey, Marcia J & Parkhouse, Bonnie L – 1992 – 2mf – 9 – $8.00 – us Kinesology [790]

Aquarium hobbyist – Riverside CT 1971-73 – 1,5,9 – ISSN: 0044-8532 – mf#7606 – us UMI ProQuest [639]

Aquatic botany – Oxford, England 1975+ – 1,5,9 – ISSN: 0304-3770 – mf#42066 – us UMI ProQuest [580]

Aquatic conservation : marine and freshwater ecosystems – Hoboken NJ 1991+ – 1,5,9 – ISSN: 1052-7613 – mf#18156 – us UMI ProQuest [574]

Aquatic toxicology – Oxford, England 1981+ – 1,5,9 – ISSN: 0166-445X – mf#42241 – us UMI ProQuest [574]

Aquaviva / Gomez Bravo, Vicente – Villafranca de los Barros (Badajoz): Imp. Bolanos Iglesias, 1941 – 1 – sp Bibl Santa Ana [946]

Aquayo Spencer, Rafael *see* Don vasco de quinoga...

The aqueduct, quebec : september 1885 / Baillairge, Charles P Florent – Quebec?: s.n, 1885? – 1mf – 9 – mf#04766 – cn Canadiana [624]

Aquellos tiempos / Garcia Rios, Miguel A – San Juan, Puerto Rico. 1963 – 1r – us UF Libraries [972]

Aqui la codosera / Corredor Garcia, Antonio – Caceres: Ediciones Gruzada Mariana. Imp. Rodriguez, 1973 – 1 – sp Bibl Santa Ana [946]

Aqui se cuetan cuentos / Lindo, Hugo – Bogota, Colombia. 1959 – 1r – us UF Libraries [972]

Aquila *see* Serious thoughts on the fall and restoration of man, with some remarks on the doctrines of...

Aquila, Dominic Anthony *see* The maieutic art of paul rosenfeld

Aquila grandis magnarum alarum, gentilibus gentis rabattae typus, celsissimo ac reverendissimo principi...raymundo ferdinando – Passauy: Apud Mariam Margaretam Hoellerin, 1714 – 1mf – 9 – mf#0-2038 – ne IDC [090]

Aquileiensis, Paulinus *see* Contra felicem

Aquileo j echeverria / Ibarra Bejarano, Georgina – San Jose, Costa Rica. 1946 – 1r – us UF Libraries [972]

Aquililla, Araceli De *see* Primeros recuerdos

Aquinas ethicus : or, the moral teaching of st thomas – London: Burns & Oates; New York: Benziger Bros, 1896 [mf ed 1991] – (= ser Quarterly series 79-80) – 2v on 3mf – 9 – 0-524-00170-7 – (trans by joseph rickaby) – mf#1989-2870 – us ATLA [230]

Aquinas, Thomas, Saint *see*
- Opera omnia
- Quaestiones de duodecim quodlibet
- Summa theologica

Aquino, C de *see* Sacra exequialia in funere jacobi 2

Ar [american review] – New York NY 1967-77 – 1,5,9 – mf#8907 – us UMI ProQuest [400]

Ar vro : revue culturelle independante – Concarneau, Paris. 1959-66 [bimnthly] – 1 – fr ACRPP [073]

Ara coeli : an essay in mystical theology / Chandler, Arthur – London: Methuen, c1916 [mf ed 1993] – (= ser Anglican/episcopal coll) – 1mf – 9 – 0-524-06480-6 – mf#1991-2580 – us ATLA [230]

Ara romana de barcarrota / Fita, Fidel – 1900 – 9 – sp Bibl Santa Ana [930]

Ara romana de barcarrota / Fita, Fidel – Madrid: Tip de Fortanet, 1900 – 1 – sp Bibl Santa Ana [946]

Ara y canta – Badajoz, 1927-1929 – 5 – sp Bibl Santa Ana [073]

Arab builders of zimbabwe / Mullan, James E – Salisbury, Zimbabwe. 1969 – 1r – us UF Libraries [960]

The arab civilization / Hell, Joseph – Cambridge, England: W Heffer & Sons, 1926 – (= ser Samp: indian books) – (trans fr german of joseph hell by s khuda bukhsh) – us CRL [956]

Arab confederation and other issues, 1950-1959 – (= ser Confidential u s state department central files) – 27r – 1 – $5225.00 – 1-55655-382-X – (with p/g) – us UPA [327]

The arab conquest of egypt and the last thirty years of the roman dominion / Butler, Alfred Joshua – Oxford: Clarendon Press, 1902 – 1mf – 9 – 0-524-08155-7 – (incl bibl ref) – mf#1991-0285 – us ATLA [960]

The arab conquests in central asia / Gibb, H A R – 1923 – (= ser Royal asiatic society. j g forlong fund) – 1r – 1 – mf#2155 – uk Microform Academic [900]

The arab conquests in central asia / Gibb, H A R – London, 1923 – 2mf – 8 – mf#U-608 – ne IDC [956]

Arab economic prospects in the 1980's = Al-sira' al-'arabi al-isra'ili wa-al-tahaddiyat al-iqtisadiyah lil-duwal al-'arabiyah fi al-thamanihat / Kubursi, A A – Beirut: Institute for Palestine Studies 1980 [mf ed Chicago IL: University of Chicago Library Photoduplication Lab for the Middle Eastern Microfilm Project at CRL 1993] – 1r – 1 – (also publ in arabic; incl bibl ref) – us CRL [330]

Arab filolog o turetskom iazyke / Melioranskii, P M – Spb, 1900 – 6mf – 8 – mf#U-358 – ne IDC [956]

Arab geographers' knowledge of southern india / Nainar, S Muhammad Husayn – Madras: University of Madras, 1942 – (= ser Samp: indian books) – us CRL [915]

The arab higher committee : its origins, personnel and purposes – New York, 1947 – 1mf – 9 – mf#J-28-150 – ne IDC [956]

The arab kingdom and its fall / Wellhausen, Juliu – [Calcutta]: University of Calcutta, 1927 – (= ser Samp: indian books) – (trans by margaret graham weir) – us CRL [956]

Arab law quarterly – v1-15. 1985-2000 – 9 – $850.00 set – ISSN: 0268-0556 – mf#111801 – us Hein [340]

Arab liberation front publications – Chicago, IL: filmed...for the Middle Eastern Microfilm Project at the Center for Research Libraries, 1994 – 1 – us CRL [327]

Arab newspapers : a contemporary record of 40 years of crucial change and development in the arab world – 1950-88 [mf ed Chadwyck-Healey] – 1 – (al-anwar, 1960-88 [lebanon] 99r. al-hayat, 1950-76 [lebanon] 53r) – uk Chadwyck [079]

The Arab Palestine Office *see* Commentary on water development in the jordan valley region

Arab report – Washington: Arab Information Center. v2 n1-21,23-24 dec 1975-oct 1 1976; nov 23-24 1976. v3 n1-21,23-24 dec 1976-oct 21 1977; nov 1-nov 15/dec 15 1977. v4 n1-19,21,23-24 jan-oct 1, nov 1, dec 1978). v5 (1979). v6 n1-3, 5-8 (jan-feb 1, mar-apr 1980) – 1r – 1 – us CRL [956]

Arab republic of egypt : [country report] – [197-?] – (= ser African training and research centre for women publications on microfilm) – 1mf – 9 – us CRL [956]

Arab studies quarterly – New York NY 1992+ – 1,5,9 – ISSN: 0271-3519 – mf#19174 – us UMI ProQuest [956]

The arab world – New York: Arab Information Center [v1 n4-v18 n5/6 (jul 1955-may/jun 1972)] (bimthly) – 2r – 1 – us CRL [079]

Arab-english dictionary / Hava, J – 1915 – 9 – $33.00 – us IRC [040]

Arabia / Conder, Josiah – London 1825 [mf ed Hildesheim 1995-98] – 1v on 3mf – 9 – €90.00 – 3-487-27666-6 – gw Olms [956]

Arabia and the bible / Montgomery, James A – 1934 – 9 – $10.00 – us IRC [220]

Arabia, egypt, india : a narrative of travel / Burton, Isabel, Lady – London: W Mullan & Son 1879 [mf ed 1987] – 1r [ill] – 1 – (filmed with: the power and beauty of superb womanhood / macfadden, b) – mf#10645 – us Wisconsin U Libr [910]

Arabia petraea / Musil, A – Wien, 1907-1908. 3v – 22mf – 9 – mf#H-2854 – ne IDC [915]

Arabian journal for science and engineering – Hoboken NJ 1979+ – 1,5,9 – ISSN: 0377-9211 – mf#11995 – us UMI ProQuest [073]

Arabian nights – London, England. no date – 1r – us UF Libraries [470]

Arabian poetry for english readers / ed by Clouston, William Alexander – Glasgow: priv print, 1881 [mf ed 1993] – 2mf – 9 – 0-524-08156-5 – mf#1991-0286 – us ATLA [470]

Arabian society at the time of muhammad / Kennedy, Pringle – Calcutta: Thacker, Spink & Co, 1926 – (= ser Samp: indian books) – us CRL [956]

Arabian tales and anecdotes : being a selection from the notes to the new translation of 'the thousand and one nights' / Lane, E W – London, 1845 – 3mf – 9 – mf#HT-292 – ne IDC [470]

Arabic and chinese trade in walrus and narwhal ivory / Laufer, B – Leide, 1913 – 1mf – 8 – mf#U-523 – ne IDC [380]

[Arabic manuscripts from ghana and adjacent territories] – Chicago, IL: U of Chicago, Photoduplication Dept, 1974 – 1 – us CRL [960]

Arabic newspapers and periodicals – 3pt, 65 titles on 4276mf, 4727r – 9,1 – €490,422.00 – (3pt consist of: palestine newspapers [55 titles on 3831mf, 4727r], egyptian oppositional periodicals [5 titles on 233mf], and early arabic periodicals [5 titles on 212mf]) – ne IDC [956]

Arabic sciences and philosophy – Cambridge, England 1991+ – 1,5,9 – ISSN: 0957-4239 – mf#17641 – us UMI ProQuest [073]

An arabic version of the acts of the apostles and the seven catholic epistles : from an 8th or 9th century ms in the convent of st catharine on mount sinai / ed by Gibson, Margaret Dunlop – London: C J Clay 1899 [mf ed 1990] – (= ser Studia sinaitica 7) – 1v on 1mf – 9 – 0-8370-1786-6 – mf#1987-6174 – us ATLA [226]

An arabic version of the epistles of st paul to the romans, corinthians, galatians : with part of the epistle to the ephesians / ed by Gibson, Margaret Dunlop – London: C J Clay 1894 [mf ed 1990] – (= ser Studia sinaitica 2) – 1mf – 9 – 0-8370-1787-4 – mf#1987-6175 – us ATLA [227]

An arabic vocabulary and index for richardson's arabic grammar : in which the words are explained according to the parts of speech, and the derivatives are traced to their originals in the hebrew, chaldee, and syriac languages / Noble, James – Edinburgh, 1820 – (= ser 19th c books on linguistics) – 2mf – 9 – (ind to grammar of the arabick language by john richardson) – mf#2.1.50 – uk Chadwyck [470]

Arabisch-deutsches lexikon zum sprachgebrauch des maimonides / Friedlaender, I – Frankfurt a.M., 1902 – 3mf – 9 – mf#J-412-5 – ne IDC [470]

Das arabische reich und sein sturz / Wellhausen, Julius – Berlin: Georg Reimer, 1902 – 1mf – 9 – 0-7905-3114-3 – (incl bibl ref) – mf#1987-3114 – us ATLA [260]

Die arabischen uebersetzungen von aristoteles' schrift de caeolo / Endress, Gerhard – Frankfurt a.M., 1966 – 3mf – 9 – 3-89349-667-X – gw Frankfurter [180]

Arab-islamic biographical archive (aiba) = Arabisch-islamisches biographisches archiv (aiba) / Kramme, Ulrike [comp] – [mf ed 1995-2002] – 560mf (1:24) in 12 installments – 9 – diazo €10,060.00 (silver €11,080 isbn: 978-3-598-33881-6) – ISBN-10: 3-598-33880-5 – ISBN-13: 978-3-598-33880-9 – (with printed ind) – gw Saur [956]

Arab-islamic biographical archive. series 2 (aiba2) = Arabisch-islamisches biographisches archiv. neue folge (aiba) / Cikar, Jutta [comp] – [mf ed 2004-07] – 392mf (1:24) in 12 installments – 9 – diazo €10,060.00 (silver €11,080 isbn: 978-3-598-35471-7) – ISBN-10: 3-598-35470-3 – ISBN-13: 978-3-598-35470-0 – (with printed ind) – gw Saur [956]

Arab-jewish unity : testimony before the anglo-american inquiry commission for the ihud (union) association / Buber, Martin & Magnes, L – London, 1947 – 2mf – 9 – mf#J-28-184 – ne IDC [956]

The arabs and the turks : their origin and history, their religion, their imperial greatness in the past, and their condition at the present time / Clark, Edson Lyman – Boston: Congregational Pub Society, 1876, c1875 – 1mf – 9 – 0-8370-9769-X – (incl bibl ref) – mf#1986-3769 – us ATLA [260]

Arabskie i persidskie fizikomatematicheskie rukopisi v bibliotekakh sovetskogo soiuza : fiziko-matematicheskie nauki v stranakh vostoka / Rozenfeld, B A – Sbornik statei i publikatsii. M, 1925 v1 – 1mf – 9 – mf#R-10665 – ne IDC [956]

Arabskie rukopisi sobraniia leningradskogo gosudarstvennogo universiteta / Beliaev, V I & Bulgakov, P G – (Pamiati akademika Ignatiia Iulianovicha Krachkovskogo. Sbornik statei, [L., 1958] – 1mf – 9 – mf#R-10921 – ne IDC [956]

Arabyazdi, Behjat *see* Determination of occupational stress and coping strategies of mediators utilizing the delphi technique

Arachne : historischer roman / Ebers, Georg – 5. aufl. Stuttgart: Deutsche Verlags-Anstalt, 1898 [mf ed 1989] – 502p – 1 – mf#7193 – us Wisconsin U Libr [830]

Die arachniden australiens nach der natur beschrieben und abgebildet / Koch, L & Keyserling, E – Nuernberg, 1871-1889. 2v – 30mf – 9 – mf#Z-2251 – ne IDC [590]

Les arachnides de france / Simon, E – Paris, 1874-1884, v1-5, 7; 1914-1937, v6 – 36mf – 9 – mf#Z-2239 – ne IDC [590]

The arachnological library : key works in spider systematics. easy access to the classic reference works – [mf ed Pergamon] – 4 sect on 395mf – 9 – (with p/g & int by j a l cooke. sect a: early classical works 74mf. sect b: european & general studies 178mf. sect c: new world studies 75mf. sect d: indian, australian & far eastern works 68mf. may be purchased separately) – us UMI ProQuest [590]

Arader zeitung – Arad (RO), 1921-1943 29 dec – 16r – 1 – gw Misc Inst [077]

Aradillas Agudo, Antonio *see*
- Amen sugerencias liturgicas
- Ante todo, esposos
- Bendicenos senor
- El beso?
- Cartas a la novia
- Coeducacion
- Como ensenar a los hijos a vivir con alegria
- Concilio y vida cristiana
- Los curas
- David, hoy
- El dialogo sexual
- Divorciarse en espana mercado negro y corrupcion
- Divorcio en espana
- En los matrimonios rotos que hacemos con los hijos?
- La familia en directo
- Firmes
- Fraude nos tribunais eclesiasticos
- Gozoy liturgia de la santa misa
- Iglesia
- Iglesia ano 2000
- Igreja 2001
- Impacto. meditaciones para militantes
- Matrimonios rotos
- Nosotros...libro de preces de la militante deaccion catolica
- La oracion de todas las noticias
- Orad hermanos
- Papeles prohibidos
- Proceso a los tribunales
- Si, mujer

Arago, Jacques *see* Promenade autour du mon dependant les annees 1817, 1818, 1819 et 1820

Arago, Jacques Etienne Victor *see* Mon ami cleobul

Aragon Fernandez, Antonio *see* Las ermitas de cordoba

Aragua (Venezuela : State) *see* Leyes del estado aragua

Araguary – Araguari, MG: Typ Progredior-Araguary, 21 abr-maio 1894; set 1895; jan 1896; out 1898; maio 1900; out 1908-maio 1909; dez 1910; jan 1911; nov 1912; nov 1915; nov 1921; mar 1925; abr-maio 1926; abr, nov 1932; 12 mar 1933 – (= ser Ps 19) – 1,5,6 – bl Biblioteca [079]

'Arakhim / Brenner, Joseph Hayyim – Tel-Aviv, Israel. 1934 – 1r – 1 – us UF Libraries [939]

L'araldo / Cuyahoga Co. Cleveland – dec 1942-jan 1952, mar 52-nov 1953 [wkly] – 7r – 1 – (in italian) – mf#B8506-8512 – us Ohio Hist [071]

L'araldo. : organo in lingua italiana del partito comunista francese (s.f.i.c.) / Communist Party. France – Paris. 4 mars 1922-1er dec 1923, inc – 1 – fr ACRPP [335]

Die aramaeer : historisch-geographische untersuchungen / Schiffer, Sina – Leipzig: J C Hinrichs, 1911 – 1mf – 9 – 0-7905-2035-4 – (incl bibl ref and indexes) – mf#1987-2035 – us ATLA [956]

Aramaeische papyrus aus elephantine / Ungnad, Arthur – Leipzig: JC Hinrichs, 1911 [mf ed 1989] – (= ser Hilfsbuecher zur kunde des alten orients 4) – 1mf – 9 – 0-7905-2562-3 – (in german & aramaic. incl bibl ref) – mf#1987-2562 – us ATLA [470]

Aramaeische pflanzennamen / Loew, Immanuel – Leipzig: Wilhelm Engelmann, 1881 [mf ed 1986] – 2mf – 9 – 0-8370-8125-4 – (incl ind) – mf#1986-2125 – us ATLA [580]

Aramaeische sprichwoerter und volkssprueche : ein beitrag zur kenntnis eines ostaramaeischen dialekts sowie zur vergleichenden paroemiologie / Lewin, Moses – Berlin: H Itzkowski 1895 [mf ed 1986] – 1mf – 9 – 0-8370-7302-2 – (discussion in german; texts in aramaic. incl ind) – mf#1986-1302 – us ATLA [470]

Aramaeische urkunden zur geschichte des judentums : im 6 und 5 jahrhundert vor chr / Staerk, Willy – Bonn: A Marcus & E Weber 1908 [mf ed 1986] – (= ser Kleine texte fuer theologische und philologische vorlesungen und uebungen 32) – 1mf – 9 – 0-8370-7343-X – (comm in german, text in aramaic) – mf#1986-1343 – us ATLA [939]

An aramaic method : a class book for the study of the elements of aramaic: from bible and targums / Brown, Charles Rufus – Chicago: American Publ Society of Hebrew, 1884-86 [mf ed 1986] – 1mf – 9 – 0-8370-7047-3 – mf#1986-1047 – us ATLA [470]

Aramaische sprichwoerter und volksspruche / Lewin, Moses – Berlin, Germany. 1895 – 1r – us UF Libraries [470]

Die aramaismen im alten testament untersucht : 1. lexikalischer teil / Kautzsch, Emil – Halle a. S: M Niemeyer, 1902. Chicago: Dep of Photodup, U of Chicago Lib, 1964 (1r); Evanston: American Theol Lib Assoc, 1984 (1r) – 1 – 0-8370-0107-2 – (incl bibl ref) – mf#1984-B016 – us ATLA [221]

Aramayo, Avelino *see* Proyecto de una nueva via entre bolivia y el oceano pacifico

Aramayo-francke archives – [Cochabamba, Bolivia; Chicago, IL: microfilmed by Microcentro for Latin American Microform Project at Center for Research Libraries, 1997] – 1 – us CRL [972]

Aramburo Y Machado, Mariano *see* Impresiones y juicios

Aramburo y Machado, Mariano *see* Discursos

Arami, M M *see* Vive tu vida

The aran islands / Synge, John Millington – Drawings by Jack B. Yeats. Boston, 1911 – 1 – us Wisconsin U Libr [840]

Arana, Felipe N *see* Sementera

Arana Soto, Salvador *see* Diccionario de temas regionalistas en la poesia pu...

Aranceles de aduanas para los puertos de la isla d... / Cuba Laws, Statutes, Etc – Habana, Cuba. 1902 – 1r – us UF Libraries [972]

Aranda y Marzo, J *see* Description tripartita medico-astronomica que toca... sobre la constitucion epidemica... de espana, con especialidad en la villa de orgaz en 1735 y 1737

Arango, Angel *see* Adonde van los cefalomos?

Arango Bueno, Teresa *see*
- Precolombia

Arango Cano, Jesus *see*
- Geografia fisica y economica de colombia
- Inmigracion y colonizacion en la grancolombia

Arango Ferrer Javier *see* Literatura de colombia

Arango Ferrer, Javier *see* Dos horas de literatura colombiana

Arango H, Ruben *see* Mi literatura

Arango Uribe, Arturo *see* 180 [i.e. ciento ochenta] dias en el frente

Aranguez Sanz, Bibiano *see* Industrias carnicas

Aranguren Martinez, Benito De *see* Recuerdos

Aranha, Graca *see*
- Canaan
- Chanaan
- Viagem maravilhosa

Aranha, Jose Pereira Da Graca *see* Canaan

Aranha, Oswaldo *see* Revolucao e a america

La aranya / Guimera, Angel – Barcelona. 1908 – 1 – us CRL [830]

Aranzaes, Nicanor *see* Diccionario historico del departamento de la paz

Arapahoe county miscellaneous newspapers – Englewood, CO (mf ed 1991) – 1r – 1 – (englewood messenger (1924-27), englewood news (1964, 1966, jun 6 1974-feb 26 1975), the tabloid (mar 12 1918-apr 1 1918)) – mf#MF Z99 Ar14e – us Colorado Hist [071]

Arapahoe pioneer – Arapahoe, NE: [Fred Boehner] v1 n1. jul 3 1879-aug 24 1911// (wkly) – 6r – 1 – us Bell [071]

Arapahoe public mirror – Arapahoe, NE: T M Gill. v95 n23. jun 6 1974- (wkly) – 14r – 1 – (absorbed: holbrook observer; cont: public mirror; iss for apr-aug 1979, 1980 accompanied by a separately numbered suppl: laker (elmwood, ne)) – us NE Hist [071]

Arapahoe road baptist church – LITTLETON, CO. 1965-75 – 1 – $39.33 – (cherry hills baptist church. church records. 1960-65) – us Southern Baptist [242]

Arapov, P *see* Letopis russkogo teatra

Araquistain, Luis *see* Agonia antillana

Ararat – New York NY 1972+ – 1,5,9 – ISSN: 0003-7583 – mf#6614 – us UMI ProQuest [400]

Ararat : roman / Ulitz, Arnold – Muenchen: A Langen 1920 [mf ed 1993] – 1r – 1 – (filmed with: der grosse janja / arnold ulitz & other titles) – mf#2932p – us Wisconsin U Libr [830]

Ararat baptist church – Jackson, TN – 1 – $10.00 – (church minutes, oct 1850-sep 1874 (missing 1857-66); church history 1850-1991. 100p) – mf#7077 – us Southern Baptist [242]

O ararigboia – Rio de Janeiro, RJ: Typ de Silva Santos & Cia, 20 abr-20 ago 1853 – (= ser Ps 19) – mf#P01B,05,12 – bl Biblioteca [321]

O arassuahy : orgao do governo municipal – Arassuahy, MG. 10 mar 1897 – (= ser Ps 19) – bl Biblioteca [350]

Arata, Alan W *see* Kinematic and kinetic evaluation of high speed backward running

Aratapu gazette – may 1884-mar 1885 – 1r – 1 – mf#12.22 – nz Nat Libr [079]

Aratuhype : periodico noticioso, commercial e agricola – Aldeia, BA: Typ do Aratuhype, 06 maio,11 nov 1883 – (= ser Ps 19) – 1,5,6 – mf#P11,2,5 – bl Biblioteca [630]

La araucana : primera, segunda, y tercera parte / Ercilla y Zuniga, Alonso de – Madrid: Abad 1733 – (= ser Whsb) – 3mf – 9 – €40.00 – (incl: la araucana: quarta, y quinta parte en que se prosique y acaba, la historia de d alonso de ercilla...by diego de santistevan osorio) – mf#Hu 032 – gw Fischer [410]

La araucana : quarta, y quinta parte en que se prosique y acaba, la historia de d. alonso de ercilla...by diego de santistevan osorio / Santistevan Osorio, Diego de – Madrid: Abad 1735 – (= ser Whsb) – 2mf – 9 – €30.00 – mf#Hu 032a – gw Fischer [410]

Araujo, Alceu Maynard *see*
- 100 melodias folcloricas
- Medicina rustica

Araujo De Figueroa, Cayita *see* Poesias revolucionarias para la ninez cubana

Araujo Filho, Jose Ribeiro De *see* Santos, o porto do cafe

Araujo Jorge, Arthur Guimaraes De *see* Introducao as obras do barao do rio-branco

Araujo Jorge, Arthur Guimaraes de *see* Ensaios de historia e critica

Araujo, Oscar Egidio De *see* Uma pesquisa de padrao de vida

Arauto – Ouro Preto, MG: [s.n.] 13 maio 1894 – (= ser Ps 19) – mf#P31,03,48 – bl Biblioteca [079]

Arauto – Rio Novo, MG. 29 ago 1897 – (= ser Ps 19) – bl Biblioteca [079]

O arauto : noticioso e litterario – Itajai, SC. 19 jul, set, 15 nov 1903 – (= ser Ps 19) – mf#UFSC/BPESC – bl Biblioteca [410]

O arauto : orgao hebdomadario – Cataguazes, MG. 21 dez 1902 – (= ser Ps 19) – mf#P11B,03,86 – bl Biblioteca [079]

O arauto : periodico evangelico – Sao Paulo, SP: Typ a Vapor da Casa Ecletica, 15 abr 1898 – (= ser Ps 19) – mf#P17,02,222 – bl Biblioteca [240]

Aravamuthan, T G *see*
- The kaveri, the maukharis and the sangam age
- Portrait sculpture in south india
- Some survivals of the harappa culture

The aravidu dynasty of vijayanagara / Heras, Henry – Madras: BG Paul & Co, 1927- – (= ser Samp: indian books) – us CRL [954]

O araxaense – Araxa, MG. 24-30 ago 1891 – (= ser Ps 19) – mf#P17,02,69 – bl Biblioteca [079]

Arazola Gil, Luis Enrique *see* Contribucion a la historia de la colonia del sacramento. la epopeya de manuel lobo...

Arbaces und panthea : oder die geschwister: schauspiel nach francis beaumont in fuenf aufzuegen von leo greiner / Beaumont, Francis – Berlin: Reiss, [1912] [mf ed 1990] – 140p – 1 – mf#7409 – us Wisconsin U Libr [820]

Der arbaiter : organ fun di poilishe sotsialistishe partai – London, England dec 1898; dec 1900; apr, aug 1901; jul, nov 1902 – 1 – (in yiddish) – uk British Libr Newspaper [335]

L'arbalete – Lyon. n1-13. fevr 1940-1948 – 1 – fr ACRPP [073]

Arbanere, Etienne G *see* Tableau des pyrenees francaises

Den arbeid van mars... / Mallet, A M – Amsterdam, 1672 – 9mf – 9 – mf#OA-154 – ne IDC [720]

Die arbeider en arm boer : [afrikaanse maandblad van die kommunistiese party van suid afrika] – Johannesburg: Kommunistiese Party -[1935] [mthly] – 1r – 1 – (began with n1 [jan 1935]; iss for jun 13-aug 8 1936 filmed with: umsebenzi [cape town, south africa: 1930], 1933-jun 6 1936 [filmed between jun 22 & jun 29 1935 iss]; and: south african worker [johannesburg, south africa: 1936], jun 13-aug 8 1936) – mf#mf-753 reel 3 – us CRL [335]

Die arbeit – Eupen (B), 1923 3 feb-1924 30 aug, 1926-27, 1931 3 jan-1936 6 jun – 1 – gw Misc Inst [331]

Die arbeit – Eupen (B), 1923 3 feb-1924 30 aug, 1926-27, 1931 3 jan-1936 6 jun – 1 – gw Misc Inst [331]

Die arbeit : gewerkschafts-zeitung – London (GB), 1941 15 mar-15 nov – 1r – 1 – gw Misc Inst [331]

Die arbeit / ed by Leipart, Theodor – Zeitschrift fuer gewerkschaftspolitik und wirtschaftskunde.... v1-10. 1924-33. -m. (Serial publications of German trade unions in the Memorial Library, University of Wisconsin-Madison) – 1 – us Wisconsin U Libr [330]

Die arbeit : organ der zionistischen volkssozialistischen partei hapoel-hazair / ed by Landauer, Georg – Berlin. v1-5. 1919-1924. 1928 (special iss) [complete] – (= ser German-jewish periodicals...1768-1945, pt 1) – 1r – 1 – $125.00 – mf#B15 – us UPA [325]

Die arbeit – Organ fuer die Sozialen Reformbestrebungen, hrsg. von Eduard Pfeiffer. v1, nos1-5, 7-8. 1866. Frankfurt am Main. Place of publ. varies. (Serial publications of German trade unions in the Memorial Library, University of Wisconsin-Madison) – 1 – us Wisconsin U Libr [331]

Die arbeit : zeitschrift fuer gewerkschaftspolitik und wirtschaftskunde – Barmen (Wuppertal), Bochum, Duisburg DE, 1907 14 apr-1917 22 dec – 4r – 1 – (title varies: 1919 n27: die wacht) – mf#3257 – gw Mikropress [074]

Arbeit und arbeitsrecht : monatsschrift fuer die betriebliche praxis – Berlin: Verlag die Wirtschaft 1963- [mf ed 2-] – 1r – 1 – (mthly 1978-1991, semimthly 1963-77, publ by: aua gmbh 1991-; jahrgang numbering cont numbering of arbeit und sozialfuersorge, vols for 1968- incl suppl: verfuegungen und mitteilungen des staatlichen amtes fuer arbeit und loehne beim ministerrat) – mf#816 – us Wisconsin U Libr [344]

Arbeit und sitte in palaestina / Dalman, Gustaf – Guetersloh, 1928-1932. 3v – 19mf – 9 – mf#H-2886 – ne IDC [956]

Arbeit und sozialfuersorge : amtliches organ der deutschen verwaltung fuer arbeit und sozialfuersorge der sowjetischen besatzungszone in deutschland – Berlin: Zentralverwaltung fuer Arbeit und Sozialfuersorge 1946-62 [mf ed 1983] – 17v on 7r – 1 – (ceased in 1962; cont by: arbeit und arbeitsrecht) – mf#816 – us Wisconsin U Libr [344]

Arbeit und wehr – Berlin DE, 1937 n48, 1938 [gaps], 1940-41 [gaps] – 1 – gw Misc Inst [331]

Arbeiten der kurlaendischen gesellschaft fuer literatur und kunst – Mitau 1847-51 – v1-10 on 31mf – 9 – mf#kr-1654/2 – ne IDC [410]

Die arbeiten des vatikanischen concils / Martin, Konrad – 2. unveraend Aufl. Paderborn: Ferdinand Schoeningh, 1873 – 1mf – 9 – 0-8370-8767-8 – (incl ind) – mf#1986-2767 – us ATLA [241]

Arbeiten einer vereinigten gesellschaft in der oberlausitz zu den geschichten und der gelahrtheit ueberhaupt gehoerende – Leipzig, Lauban 1750-56 – (= ser Dz. abt literatur) – 6v on 20mf – 9 – €200.00 – mf#k/n4403 – gw Olms [400]

Arbeiten zu film und fernsehen / Boll, Uwe – (mf ed 1992) – 3mf – 9 – €49.00 – 3-89349-460-X – mf#DHS 460 – gw Frankfurter [790]

Arbeiten zur deutschen literatur, 1750-1850 / Sengle, Friedrich – Stuttgart: Metzler, c1965 [mf ed 1993] – 243p – 1 – (incl bibl ref and ind) – mf#8207 – us Wisconsin U Libr [430]

Die arbeitende jugend – Berlin DE, 1905 1 feb-1908 1 dec, 1909-1933 feb – 5r – 1 – (title varies: 1909: arbeiter-jugend) – mf#6407 – gw Mikropress [331]

Der arbeiter – Budapest (H), 1893-94 – 1r – 1 – gw Misc Inst [331]

Der arbeiter – Muenchen DE, 1895 5 jan-1907, 1909-18 – 12r – 1 – mf#2163 – gw Mikropress [331]

Der arbeiter – New York. N.Y. The workman. 1904-11 – 1 – us AJPC [071]

Der arbeiter – New York. Oct 8 1904-Aug 12 1911. Incomplete – 1 – us NY Public [071]

Der arbeiter – New York. v. 1-11. n7. sept. 15, 1927-feb. 13, 1937 – 1 – us NY Public [325]

Die arbeiter : drama in vier aufzuegen / Bulthaupt, Heinrich Alfred – Leipzig: P Reclam 1893 [mf ed 1989] – 1r – 1 – (filmed with: der philister vor, in und nach geschichte / clemens brentano) – mf#7094 – us Wisconsin U Libr [820]

Der arbeter fraind = Worker's friend – London, England 15 jul 1885-26 mar 1897; 14 oct 1898-21 jul 1916; apr 1920-23 dec 1923 – 1 – (not publ between 21 jul 1916 & apr 1920; publ by the international workingmen's educational club; discontinued) – uk British Libr Newspaper [331]

Arbeiter illustrierte zeitung – Berlin. 5-15 no. 33. Jan. 1926-Aug. 12, 1936. Incomplete – 1 – us NY Public [331]

Arbeiter in der gegenwartsliteratur / Roehner, Eberhard – 1. aufl. Berlin: Dietz 1967 [mf ed 1992] – 1r – 1 – (incl bibl ref & ind. filmed with: the era of expressionism / ed & ann by paul raabe) – mf#3301p – us Wisconsin U Libr [430]

Arbeiter socialtische zeitung – 1900 dec – 1 – mf#3177686 – us WHS [071]

Der arbeiter und sein arzt : bemerkungen zur wiederkehr des jahrestages der herausgabe des befehls 234 zur verbesserung der aerztlichen betreuung der arbeiter und angestellten in betrieben der sowjetischen zone / ed by Pressestelle der Deutschen Zentralverwaltung fuer das Gesundheitswesen in der sowjetischen Besatzungszone – Dresden: Verlag des Deutschen Hygiene-Museums 1948 [mf ed 1989] – 1 – mf#7136 – us Wisconsin U Libr [610]

Arbeiter- und soldatenrat – Sitzungsprotokolle. Bremen, 1918-19 – 1 – gw Mikropress [943]

Arbeiter union / National Arbeiter Union – New York. v. 1 no. 1-49. June 13 1868-May 15 1869. Weekly; v. 1-2 no. 101. May 22 1869-Sept 17 1870. Daily – 1 – us NY Public [072]

Arbeiter zeitung – Vienna. Oct 1909-Feb 1934; Aug 1945-Dec 1948 – 67r – 1 – us L of C Photodup [074]

Die arbeiter zeitung – New York. 1890-1902 – 1 – us NY Public [071]

Die arbeiter zeitung – New York. N.Y. The workman's paper. 1890-1902 – 1 – us AJPC [071]

Arbeiterbewegung und klassik : ausstellung im goethe- und schiller-archiv der nationalen forschungs- und gedenkstaetten der klassischen deutschen literatur in weimar, 1964-1966 / Holtzhauser, Helmut – 1. aufl. Weimar: Aufbau-Verlag 1964 [mf ed 1992] – 1r [ill] – 1 – (incl bibl ref. filmed with: die deutsche treue in sage und poesie / oscar dolch) – mf#3208p – us Wisconsin U Libr [430]

Arbeiterblatt – Vienna, Austria jul-dec 1868 [mf ed Norman Ross] – 1r – 1 – mf#nrp-1930 – us UMI ProQuest [331]

Arbeiter-chronik – Nuernberg DE, 1888-1889 16 mar – 1r – 1 – gw Misc Inst [331]

Die arbeiterclassen-bewegung in england / Marx-Aveling, Eleonore – Nuernberg, 1895 – 1 – gw Mikropress [335]

Arbeiterdichtung : analysen, bekenntnisse, dokumentationen / ed by Oesterreichischen Gesellschaft fuer Kulturpolitik – Wuppertal: Hammer, c1973 [mf ed 1993] – 324p – 1 – (incl bibl ref) – mf#8265 – us Wisconsin U Libr [430]

Die arbeiterdichtung in frankreich : ausgewaehlte lieder franzoesischer proletarier – London: Truebner; Hamburg: J P F E Richter, [18–?] (mf ed 1990) – 1 – (filmed with: goethes faust in ursprueglicher gestalt) – us Wisconsin U Libr [810]

Der arbeiter-fotograf – Berlin, Halle S DE, 1926 aug-1932 feb – 1r – 1 – (filmed by misc inst: 1926-1932 feb; 1926 aug-1932 nov) – gw Mikrofilm; gw Misc Inst [770]

Die arbeiterfrage und das christenthum / Ketteler, Wilhelm Emmanuel, Freiherr von – Mainz: F. Kirchheim, 1864 – 1mf – 9 – 0-7905-6001-1 – mf#1988-2001 – us ATLA [240]

Arbeiterfreund – Berlin. 1-52. 1863-1914 – 1 – us NY Public [331]

Der arbeiterfreund : zeitschrift des centralvereins in preussen fuer das wohl der arbeitenden klassen – Berlin DE, 1863-83, 1884 [gaps], 1885-1911, 1912 [gaps], 1913-14 – 12r – 1 – mf#2207 – gw Mikropress [331]

Der arbeiterfreund : zeitschrift des centralvereins in preussen fuer das wohl der arbeitenden klassen – Berlin: Verlag von Otto lanke & Co 1863-[1914] [mf ed 1981] – 1r – 1 – (subtitle varies) – mf#7703 reel 13 – us Wisconsin U Libr [331]

Arbeiterfunk *see* Der neue rundfunk

Arbeiter-illustrierte-zeitung aller laender *see* Sowjetrussland im bild

Die arbeiterin – Stuttgart, Berlin DE, 1892 11 jan-1922 15 aug – 8r – 1 – (filmed by misc inst: 1892-1904, 1909-18, 1920. title varies: 1892: die gleichheit; fr 5 jul 1919 publ in berlin) – gw Mikropress; gw Misc Inst [331]

Die arbeiterin – Hamburg DE, 1890-91 [gaps] – 1r – 1 – gw Misc Inst [331]

Arbeiterinnen-zeitung – Vienna, Austria jan 1892-dec 1934 [mf ed Norman Ross] – 7r – 1 – (aka: die frau after 1926) – mf#nrp-1931 – us UMI ProQuest [331]

Arbeiter-jugend – Berlin: Vorwaerts. v1-25 n4 [1909-apr 1933] [mf ed 1978] – 25v on 5r [ill] – 1 – (incl suppl) – mf#film mas c351 – us Harvard [331]

Arbeiter-jugend : organ fuer die geistigen und wirtschaftlichen interessen der jungen arbeiter und arbeiterinnen – Berlin, 1909-Feb 1933 – 4r – 1 – gw Mikropress [331]

Arbeiter-jugend *see* Die arbeitende jugend

Die arbeiter-kolonie : correspondenzblatt fuer die interessen der deutschen arbeiter-kolonien – Wustrau: Central-Vorstand der Deutschen Arbeiter-Kolonien. 1-13 jahrg apr 1884- [mthly] [mf ed 1981] – 13v on 1r – 1 – (cont by: wanderer; iss by: central-vorstand deutscher arbeiter-kolonien. organ of: gesamt-verband der deutschen natural-verpflegungs-stationen, and: deutscher herbergsverein) – mf#7703 reel 21 – us Wisconsin U Libr [630]

Arbeiterlesebuch : nicht nur fuer arbeiter / ed by Werkkreis Literatur der Arbeitswelt. Werkstatt Bremen – Frankfurt/Main: Fischer Taschenbuch Verlag, 1981 [mf ed – 164p (ill) – 1 – (incl bibl ref) – mf#8549 – us Wisconsin U Libr [430]

Arbeiterlesebuch : rede lassalle's zu frankfurt am main am 17. und 19. mai 1863, nach dem stenographischen bericht – 4. aufl. Chicago : Charles Ahrens, 1872 [mf ed 1984] – (= ser Sammlung social-politischer schriften v2 n6) – 72p – 1 – mf#8479 reel 1 v2 n6 – us Wisconsin U Libr [331]

Arbeiteroeffentlichkeit und literatur : zur theorie des werkkreises literatur der arbeitswelt juergen alberts / Alberts, Juergen – Hamburg: VSA, 1977 [mf ed 1992] – 126p – 1 – (incl bibl ref. aft by horst hensel) – mf#7998 – us Wisconsin U Libr [331]

Arbeiter-philosophen und- dichter / ed by Levenstein, Adolf – Berlin: Eberhard Frowein 1908 [mf ed 1993] – 1r – 1 – (no more publ? filmed with: "in uns ist alles" / ed by karl cerff) – mf#3334p – us Wisconsin U Libr [800]

Arbeiterpolitik – Asch (CZ), 1935 dec, 1936 jan, 1937 nov, 1938 jan-sep – 1 – gw Misc Inst [331]

Arbeiterpolitik : organ der kommunistischen partei-opposition, elsass – Strassburg (Strasbourg F), 1934-39 [gaps] – 4r – 1 – gw Misc Inst [331]

Arbeiterpolitik – Prag (CZ), 1929 22 jun-1930 23 aug – 1r – 1 – gw Misc Inst [331]

Arbeiterpolitik : wochenschrift fuer den sozialismus bremen – Berlin, 1917 – 1r – 1 – gw Mikropress [325]

Arbeiterpolitik : wochenschrift fuer wissenschaftlichen sozialismus – Bremen DE, 1916 24 jun-1919 8 mar [gaps] – 1r – 1 – gw Misc Inst [335]

Arbeiterpolitik : wochenschrift fuer wissenschaftlichen sozialismus – Bremen DE, 1917 – 1r – 1 – mf#3349 – gw Mikropress [335]

Arbeiterpresse *see* Arbeiter-wochen-chronik

Der arbeiter-rat : organ der arbeiterraete deutschlands – Berlin DE, 1919 feb-1920 – 1r – 1 – mf#4188 – gw Mikropress [331]

Arbeiterrat Gross-Hamburg *see* Jahrbuch

Arbeiter-Sekretariat. Bremen *see* Jahresbericht...

Arbeiter-Sekretariat. Halle *see* Geschaeftsbericht...

Arbeiter-Sekretariat. Muenchen *see* Jahresbericht...

Arbeiter-Sekretariat. Nuremberg *see* Jahresbericht...

Arbeitersender *see* Unser sender

Arbeiterstimme – Dresden DE, 1925 apr-jun, 1926-30, 1931 mai-1932 apr, 1933 jan-feb – 17r – 1 – gw Misc Inst [331]

Arbeiterstimme – Wroclaw, Poland. Jul 1952-Jan 1956; Sept 1956-Apr 1958 – 7r – 1 – us L of C Photodup [077]

Arbeitertag fuer braunschweig und weiterere umgegend : abgeschalten zu braunschweig, sonntag, den 21 jul 1867 – [Braunschweig: Berglein & Limbach 1867] [mf ed 1981] – 1r – 1 – (filmed from poor condition original. filmed with: hauptergebnisse der amtlichen lohnerhebung in der schuhindustrie) – mf#7703 reel 114 n3 – us Wisconsin U Libr [331]

Arbeiter-tribuene – Stuttgart DE, 1930 – 1r – 1 – (filmed by other misc inst: 1929 n4-1930 [2r]) – gw Misc Inst [331]

Arbeitertum : blaetter fuer theorie und praxis der nsbo – Berlin DE, 1931 1 mar-1940 31 mar – 3r – 1 – mf#5944 – gw Mikropress [331]

Arbeiter-turn-und-sportzeitung *see* Arbeiter-turn-zeitung

Arbeiter-turn-zeitung – Leipzig DE, 1893 15 jul-1905, 1908-1933 22 mar – 10r – 1 – (fr 1931: arbeiter-turn-und-sportzeitung; with suppls) – mf#4983 – gw Mikropress [790]

Arbeiter-wochen-chronik : sozialdemokratisches volksblatt – Budapest (H), 1873-94 – 6r – 1 – (1880-81: numerous disguised ed with individual titles & numbering e.g. telephon, 1891: arbeiterpresse) – gw Misc Inst [331]

Arbeiterwohl – Koeln/Moenchengladbach DE, 1898-1903, 1905 – 2r – 1 – (title varies: 1905: soziale kultur) – gw Misc Inst [301]

Arbeiterwohlfahrt – [1] 1926-[8] 1933 [mf ed 2005] – (= ser Freie wohlfahrtspflege 11) – 38mf – 9 – €390.00 – 3-89131-474-4 – gw Fischer [335]

Arbeiterwohlfahrt Hauptausschuss *see* Geschaeftsbericht...

Das arbeiterwort – Zuerich (CH), 1961 oct, 1962 may-1969 feb [gaps] – 1r – 1 – gw Mikrofilm [331]

Arbeiter-Zeitung – St-Louis Elsass (F), 1898, 1899-1900 [gaps], 1901-16 – 6r – 1 – gw Misc Inst [331]

Arbeiter-zeitung – Bern. v1-2 n1-33 jul 15 1876-oct 13 1877 – 1 – us CRL [074]

Arbeiter-zeitung – Wien (A), 1963-66 – 1 – (filmed by other misc inst: 1931-1934 feb [9r], 1945 5 aug-1989 31 mar [156r]) – gw Misc Inst [331]

Arbeiter-zeitung – Ludwigshafen DE, 1924 1 aug-30 sep, 1928 2 jan-1933 28 feb – 12r – 1 – mf#6048 – gw Mikropress [331]

Arbeiter-zeitung – New York NY (USA), 1873 8 feb-1875 13 mar – 1r – 1 – gw Misc Inst [071]

Arbeiter-zeitung – Reichenberg (Liberec CZ), 1929 20 jul-28 dec, 1930 4 jan-1 mar – 1r – 1 – gw Misc Inst [331]

Arbeiter-zeitung – Bratislava, Czechoslovakia. Jun 1935-Jan 1937 – 1r – 1 – (semi-monthly ed) – us L of C Photodup [077]

Arbeiter-zeitung / Sozialdemokratische Arbeiter-partei OEsterreichs/RSOE/SPOE – Chicago, 1949- – 1 – us CRL [331]

Arbeiter-zeitung – St Louis MT (USA), 1922 1 apr-1929 28 dec – 3r – 1 – gw Misc Inst [331]

Arbeiter-zeitung – Essen DE, 1907 26 oct-1910, 1911 jul-dez, 1912 jul-1914, 1916-22, 1923 apr-jun, 1924-1925 sep, 1926-1927 mar, 1927 jul-1928 sep, 1929 jan-sep 1930, 1931 apr-1932 – 50r – 1 – (title varies: 1 nov 1919: essener arbeiter-zeitung, 1 may 1926: volkswacht) – mf#3475 – gw Mikropress [331]

Arbeiter-zeitung – Coburg DE, 1863-65, 1866 [gaps] – 2r – 1 – (title varies: 8 apr 1863: allgemeine deutsche arbeiter-zeitung) – gw Misc Inst [331]

Arbeiter-zeitung – Bruenn (Brno CZ), 1934 18 mar-1935 20 oct [gaps], 1936 5 jan-22 nov [gaps] – 1r – 1 – (title varies: organ der oesterreichischen sozialdemokratie) – gw Misc Inst [331]

Arbeiter-zeitung – Bratislava, Czechoslovakia. Sept 1934-Jul 1936 – 1r – 1 – (weekly ed. scattered issues) – us L of C Photodup [077]

Arbeiter-zeitung – Wien: R Pokorny, 1949-oct 14 1985 – 1 – us CRL [074]

Arbeiter-zeitung *see*
- Buffaloer arbeiter-zeitung
- Schlesische arbeiter-zeitung
- Westfaelische freie presse 1890

Arbeiterzeitung – Vienna, Austria dec 1886-feb 1934 [mf ed Norman Ross] – 131r – 1 – (social democrat) – mf#nrp-1932 – us UMI ProQuest [074]

Arbeiterzeitung – Temeschburg (Timisoara RO), 1926 3 nov-1930 11 dec – 4r – 1 – gw Misc Inst [331]

Die arbeiter-zeitung – Essen DE, 1907 26 oct-1910, 1911 jul-dec, 1912 jul-1914, 1916-22, 1923 apr-jun, 1924-1925 sep, 1926-1927 mar, 1927 jul-1928 sep, 1929 jan-sep, 1930, 1931 apr-1932 – 51r – 1 – (title varies: 1 nov 1919: essener arbeiter-zeitung; 1 may 1926: volkswacht. with suppls: kinderfreund 1907-12 [gaps], 1914 n1-8) – mf#3475 – gw Mikropress [331]

Arbeiter-zeitung fuer gelsenkirchen und umgebung – Gelsenkirchen DE, 1923 2-14 & 30 aug-10 sep [gaps] – 1r – 1 – (bezirksausgabe von ruhr-echo, essen; filmed by other misc inst: 1922 20 sep-1923 7 sep) – gw Misc Inst [331]

Arbeiter-zeitung fuer hessen-waldeck und suedhannover – Kassel DE, 1920 27 oct-1922 31 mar – 3r – 1 – gw Misc Inst [331]

Der arbeitgeber – Frankfurt/M DE, 1857-66 – 2r – 1 – mf#4404 – gw Mikropress [331]

Der arbeitgeber – Frankfurt/M DE, 1869-79 [gaps] – 1 – gw Misc Inst [331]

Arbeitsausschuss Freigewerkschaftlicher Bergarbeiter *see* Bergarbeiter-mitteilungen

Der arbeitseinsatz im deutschen reich / Germany. Reichsarbeitsministerium. Hauptabteilung – 1927-42. 1939. No. 17-24 and 1940 wanting – 1 – $195.00 – us L of C Photodup [943]

Arbeitsgruppe der Plattdeutschen Gilde zu Rostock *see* John brinckmans plattdeutsche werke

Die arbeitslosen : roman aus der gegenwart / Euringer, Richard – Hamburg: Hanseatische Verlagsanstalt c1930 [mf ed 1989] – 1r – 1 – (filmed with: aphorismen & other titles) – mf#7226 – us Wisconsin U Libr [830]
Arbeitslosigkeit zwischen lohn und effizienz : eine darstellung und kritik der effizienzlohntheorien / Kaufmann, Frank – (mf ed 1993) – 2mf – 9 – €49.00 – 3-89349-656-4 – mf#DHS 656 – gw Frankfurter [331]
Arbeitsmann – Berlin. v. 1 no. 1-v. 7 no. 16. Oct 5 1935-Apr 19 1941. Incomplete – 1 – us NY Public [074]
Arbeitsplan fur chanukka / Ehrmann, Eliezer L – Berlin, Germany. 1937 – 1r – us UF Libraries [939]
Arbeitsplatznahe weiterbildung : betriebspaedagogische konzepte und betriebliche umsetzungsstrategien / Severing, Eckhart – Neuwied, Kriftel, Berlin: Luchterhand 1994 (mf ed 1996) – (= ser Grundlagen der weiterbildung) – 3mf – 9 – €38.00 – 3-8267-9692-6 – mf#DHS 9692 – gw Frankfurter [374]
Arbejderen – 1898 oct 20; 1899 jun 29; 1900 jan 4 – 1 – mf#868586 – us WHS [071]
Arbelaez, Tulio *see* Episodios de la guerra de 1899 a 1903
Arbelaez Urdaneta, Carlos *see* Biografia del general rafael urdaneta
d'Arbelles, Salvador *see* Corinto a traves de la historia (1514-1933). corinto (nicaragua)
Arbenz, E *see* Die vadianische briefsammlung der stadtbibliothek st gallen
Arbeonis epsicopi frisingensis viae sanctorum haimhrammi et corbiniani (mgh7:13.bd) – 1920 – (= ser Monumenta germaniae historica 7: scriptores rerum germanicarum in usum scholarum (mgh7)) – €11.00 – ne Slangenburg [240]
Arber, Agnes Robertson *see* Herbals
Arber, Edward *see*
– Seven sermons before edward 6
– The story of the pilgrim fathers, 1606-1623 a.d
– A supplication for the beggars
Arberry, Arthur John *see* An introduction to the history of sufism
Arbetarbladet – Gavle, Sweden. 1902-78 – 444r – 1 – sw Kungliga [078]
Arbetarbladet – Gavle, Sweden. 1979- – 1 – sw Kungliga [078]
Arbetarebladet – Gaevle, 1869-88 – 9 – sw Kungliga [078]
Arbetaren – Goeteborg, Sweden. 1869-70 – 1 – sw Kungliga [078]
Arbetaren – Stockholm, Sweden. 1922-78 – 171r – 1 – (previous title: syndikalisten) – sw Kungliga [078]
Arbetaren – Stockholm, Sweden. 1902-06 – 1 – sw Kungliga [078]
Arbetaretidningen – Stockholm, Sweden. 1900-12, 1914 – 2r – 1 – sw Kungliga [078]
Arbetarevannen – Lulea, Sweden. 1863-66 – 1 reel – 1 – sw Kungliga [078]
Arbetarpolitiken – Borlaenge, 1921-23 – 2r – 1 – sw Kungliga [078]
Arbetartidningen ny dag – Stockholm, Sweden. 1974-82 – 1 – sw Kungliga [078]
Arbetartidningen ny dag – Stockholm, Sweden. 1979-82 – 1 – sw Kungliga [078]
Arbetartidningen ny dag *see* Ny dag
Arbeter in der yidisher literatur – Moskve, Russia. 1931 – 1r – us UF Libraries [470]
Arbetet – Malmo, Sweden. 1979-95 – 1 – sw Kungliga [078]
Arbetet – Malmoe, Sweden. 1999-2000 – 35r – 1 – sw Kungliga [078]
Arbetet – Malmo, Sweden. 1887-1995 – 79r – 1 – (title changes to: arbetet nyheterna from 1995; vastsv ed 1966-78 152r; editorial pages, 1952-78 24r) – sw Kungliga [078]
Arbetet ny tid – Malmoe, Sweden. 1999-2000 – 11r – 1 – sw Kungliga [078]
Arbetet nyheterna – Malmoe, 1995- – 9 – (previous title: arbetet) – sw Kungliga [078]
Arbetet nyheterna – Goeteborg, 1991- – 1 – (previous title: arbetet väst) – sw Kungliga [078]
Arbetet nyheterna *see* Arbetet
Arbetet vaest – Goeteborg, 1979-91 – 146r – 1 – sw Kungliga [078]
Der arbeyter = The workman – New York [NY]: Socialist Labor Club of New York. v1 n1. oct 8 1904- (wkly) [mf ed [197-?]] – 1 – (publ by: the jewish socialist labor federation, apr 24 1909-aug 12 1911. yiddish, with occasional advertisements in english) – mf#*ZAN-*P881 – us NY Public [331]
Arbiter of elegance / Bagnani, Gilbert – Toronto, ON. 1954 – 1r – us UF Libraries [960]
Arbitrage du tres saint-pere le pape entre la repu... – Paris, France. 1896 – 1r – us UF Libraries [972]
Arbitraje de limites entre honduras y guatemala / Honduras – Washington, DC. 1932 – 1r – us UF Libraries [972]
Arbitraje en el derecho privado / Briseno Sierra, Humberto – Mexico City?, Mexico. 1963 – 1r – us UF Libraries [972]
Arbitraje entre honduras y nicaragua / Ramirez Y Fernandez Fontecha, Antonio Abad – New York, NY. 1938 – 1r – us UF Libraries [972]
Arbitration engagements now existing in treaties, treaty provisions and national constitutions / Myers, Denys Peter [comp] – Boston: World Peace Foundation 1915 [mf ed 1992] – (= ser World peace foundation pamphlet series 5/5.3) – 1mf – 9 – 0-524-03240-8 – mf#1990-0868 – us ATLA [341]
Arbitration international – v1-16. 1985-2000 – 9 – $944.00 set – ISSN: 0957-0411 – mf#112961 – us Hein [341]
Arbitration journal – New York NY 1937-92 – 1,5,9 – (cont by: dispute resolution journal) – ISSN: 0003-7893 – mf#2498.01 – us UMI ProQuest [303]
Arbitrator *see* Clear creek county miscellaneous newspapers
Arbman, Ernst *see* Rudra
Arbman, Per Theodor *see* Vad ar evangelium?
Arbog... : for kirkelig forening den indre mission i danmark – 1990-93 [complete] – 1r – 1 – mf#ATLA S0900 – us ATLA [073]
Arboga – Arboga, Sweden. 1888-89 – 1r – 1 – sw Kungliga [078]
Arboga tidning – Arboga, 1851-55 – 1r – 1 – sw Kungliga [078]
Arboga tidning – Arboga, Sweden. 1851-55 – 1 reel – 1 – sw Kungliga [078]
Arboga tidning – Arboga, Sweden. 1858-80 – 9r – 1 – sw Kungliga [078]
Arboga tidning – Arboga, Sweden. 1881-1970 – 141r – 1 – sw Kungliga [078]
Arboga weckoblad – Arboga, Sweden. 1855-58 – 1r – 1 – (nya weckobladet i arboga, 1853-55) – sw Kungliga [078]
Arbogabladet – Arboga, Sweden. 1848-49 – 1r – 1 – (nya arbogabladet, 1850) – sw Kungliga [078]
Arbogaposten – Arboga, Sweden. 1880-81 – 1r – 1 – sw Kungliga [078]
Arbois de Jubainville, M H d' *see* Etudes sur l'etat interieur des abbayes cisterciennes
Arbol criollo / Jimenez-Quiros, Otto – Cartago, Costa Rica. 1964 – 1r – us UF Libraries [972]
Arbol cronologico que manifiesta los comisarios generales de indias del orden de san francisco y plan de todas las provincias... / Gonzalez de Agueros, Pedro – [Madrid]: impr Bento Cano, ano de 1789 – (= ser Books on religion...1543/44-c1800: franciscanos) – 1mf – 9 – mf#crl-212 – ne IDC [241]
Arbol de la noche alegre / Blanco, Andres Eloy – Caracas, Venezuela. 1960 – 1r – us UF Libraries [972]
Arbol de las veras / Lopez de Haro, Alonso – 1636 – 9 – sp Bibl Santa Ana [810]
Arbol de las veras y...elogios / Mogroveio de Cerda, Ivan – 1636 – 9 – sp Bibl Santa Ana [810]
Arbol lleno de cantos / Miranda, Luis Antonio – San Juan, Puerto Rico. 1946 – 1r – us UF Libraries [972]
Arbol y luego bosque / Fernandez, David – La Habana, Cuba. 1964 – 1r – us UF Libraries [972]
Arboleda, Gustavo *see* Historia de cali
Arboleda, Julio *see* Poesias
Arboleda Llorente, Jose Maria *see*
– Indio en la colonia
– Vida del illmo senor manuel joe mosquera
Arboleda R, J Vicente *see* Study of the value of starter solutions for transplanting certain v...
Arboleda, Sergio *see* Constitucion politica
Arboles / Velez, Clemente Soto – New York, NY. 1955 – 1r – us UF Libraries [972]
Arboles mios / Palma, Marigloria – Barcelona, Spain. 1965 – 1r – us UF Libraries [972]
Arboles sin raices / Gonzalez De Cascorro, Raul – Santa Clara, Cuba. 1960 – 1r – us UF Libraries [972]
Arbor day : a few advices to farmers on the planting of forest and ornamental trees / Chapais, Jean Charles – Montreal: E Senecal, 1884 [mf ed 1980] – 1mf – 9 – 0-665-02870-9 – mf#02870 – cn Canadiana [634]
Arbor day : programme for its celebration in the year 1885 and advice on the planting and sowing of forest trees / Chapais, Jean-Charles – Quebec: [s.n.], 1885 [mf ed 1981] – 1mf – 9 – mf#12007 – cn Canadiana [634]
Arbor day, ontario : suggestions and regulations in regard to its observance by school trustees, teachers and pupils in ontario – [Toronto?: s.n.], 1887 [mf ed 1986] – 1mf – 9 – 0-665-54744-7 – mf#54744 – cn Canadiana [634]
Arbor day, province of quebec : proclamations, etc, instructions for planting trees – [S.l: s.n, 1883?] [mf ed 1980] – 1mf – 9 – 0-665-02479-7 – mf#02479 – cn Canadiana [634]
Arbor hills association newsletter – 1980 nov-1983 oct; 1980 nov-1995 dec – 1 – mf#693432 – us WHS [071]
Arbor Hills Neighborhood Association *see* Newsletter
Arbor state – Wymore, NE: Wymore Arbor State Inc. 8v. v94 n10. apr 29 1976-v101 n17. dec 9 1982 (wkly) [mf ed filmed 1978-84] – 8r – 1 – (cont: wymore arbor state. absorbed: cortland news. cont by: wymore arbor state (1982). publ as wymore arbor state jul 9-30 1981 and jul 1-15 1982) – us NE Hist [071]
Arbor vitae crucifixae jesu / Ubertinus de Casali (Ubertino of Casale) – Venetiis, 1485 – €40.00 – ne Slangenburg [241]
L'arbore di diane : der baum der diana: eine comische oper in 2 acten... / Marti'n y Soler, Vicente – Bonn: bei N Simrock [1790?] [mf ed 19–] – 1r – 1 – (arr for piano by c g neefe; italian & german words; libretto by lorenzo da ponte) – mf#pres. film 180 – us Sibley [780]
Arboreus, Ioan *see* Theosophia
Arbousset, Jean Thomas *see*
– Relation d'un voyage d'exploration au nord-est de la colonie...
– Relation d'un voyage d'exploration au nord-est de la colonie du cap du bonne-esperance en 1836
– Voyage d'exploration aux montagnes bleues
Arbre historique des dynasties francaises / Soeurs de la charite de Quebec Academie – [Levis, Quebec?: s.n.], 1894 [mf ed 1980] – 1mf – 9 – 0-665-03985-9 – mf#03985 – cn Canadiana [929]
Arbroath argus : or forfarshire political and critical review – [Scotland] Angus, Arbroath: J Bremnar 1 feb 1836-19 apr 1837 (mthly) [mf ed 2004] – 1r – 1 – (cont by: arbroath journal and forfarshire political and critical review) – uk Newsplan [072]
Arbroath argus redivivus – [Scotland] Angus, Arbroath: J Bremnar jan 1842 (mthly) [mf ed 2004] – 2v on 1r – 1 – uk Newsplan [072]
Arbroath guide and county of forfar advertiser – [Scotland] Angus, Arbroath: B M Kennedy 26 mar 1842-30 dec 1843 (wkly) [mf ed 2004] – 1r – 1 – (absorbed: carnoustie times; title varies) – uk Newsplan [072]
Arbroath herald and angus county advertiser – Arbroath: Arbroath Herald Ltd 1940- (wkly) [mf ed 3 jan 1992-] – 1 – (cont: arbroath herald and advertiser for the montrose burghs; foll suppls are currently entered separately: arbroath & angus today) – uk Scotland NatLib [072]
Arbroath herald and angus-shire, political, literary, commercial, and agricultural advertiser – [Scotland] Angus, Arbroath: J Duff 30 nov 1838-27 dec 1839 (fortnightly) [mf ed 2004] – 57v on 1r – 1 – uk Newsplan [072]
Arbroath journal : and forfarshire political and critical review – [Scotland] Angus, Arbroath: J Daniel, 9 oct 1841-19 mar 1842 (wkly) [mf ed 2004] – 1r – 1 – (cont by: arbroath argus, or, forfarshire political and critical review) – uk Newsplan [072]
Arbroath journal : a local advertiser and miscellany of amusement and instruction – [Scotland] Angus, Arbroath: J Bremnar may 1854 [mf ed 2004] – 1r – 1 – uk Newsplan [072]
Arbroath miscellany of instruction & amusement – [Scotland] Angus, Arbroath: J Bremnar mar 1838-7 jul 1838 (mthly) [mf ed 2004] – 6v on 1r – 1 – uk Newsplan [072]
[Arbuckle-] arbuckle american – CA. 1927; 1929-31; 1933-34; 1948-65; 1967-69 – 10r – 1 – $600.00 – mf#B03137 – us Library Micro [071]
O arbusto : jornal critico, litterario e noticioso – Teresina, Pl. 05 set 1878 – (= ser Ps 19) – mf#P17,02,126 – bl Biblioteca [410]
Arbustum vel arboretum augustaeum : aeternitati ac domui augustae seleniae sacrum... / Gosky, Martino – Wolfenbuettel: Typis Johan et Henr. Stern, 1650 – 15mf – 9 – mf#O-1873 – ne IDC [090]
Arbuthnot, Alexander John *see* Memories of rugby and india
L'arc – Aix-en-Provence. n1-12. 1958-oct 1960 – 1 – fr ACRPP [073]
Arc – 1980 mar/apr-aug/sep – 1 – mf#615530 – us WHS [071]
Arc – Arlington VA 1980-91 – 1,5,9 – (cont: mental retardation news; cont by: arc today) – ISSN: 0199-9435 – mf#8620.02 – us UMI ProQuest [370]
Arc in wisconsin news – v23 n1-v25 n1 [1978 mar-1979 dec/1980 jan] – 1 – mf#615529 – us WHS [071]
Arc news – 1980 dec-1982 sep – 1 – mf#615531 – us WHS [071]
Arc newsletter – v1 n1-v6 n2 [1977 mar-1982 jun] – 1 – mf#656646 – us WHS [071]
Arc today – Arlington VA 1992-96 – 1,5,9 – (cont: arc) – mf#8620,02 – us UMI ProQuest [370]
Arca noe : thesaurus linguae sanctae novus / Marinus, Marcus – Venetiis. pars 1+2. 1593 – 71mf – 8 – €136.00 – ne Slangenburg [225]
Arcadia : [devoted exclusively to music, art and literature] – Montreal. v1 n1-21. may 2 1892-mar 1 1893// – 1r – 1 – Can$110.00 – cn McLaren [071]
Arcadia / Sannazarius, Jacobus [Sannazaro, Jacopo] – 15th c – (= ser Holkham library manuscript books 522) – 1r – 1 – mf#2723 – uk Microform Academic [810]
Arcadia anzeiger – 1911 jan 13; 1912 nov 1-14 jul 24; 1914 aug 7-1916 feb 4 – 1 – mf#918028 – us WHS [071]
[Arcadia-] arcadia tribune – CA. 1977-80 – 8r – 1 – $480.00 – mf#R02020 – us Library Micro [071]
Arcadia champion – Arcadia, NE: Clarence L Day. -v31 n4. jul 15 1926 (wkly) – 9r – 1 – (cont: arcadia courier; merged with: arcadia tribune to form: arcadia champion and the arcadia tribune; iss for may 29 1896- called v2 n4-) – us Bell [071]
Arcadia champion and the arcadia tribune – Arcadia, NE: Champion Pub Co. v31 n5. jul 22 1926- (wkly) – 1r – 1 – (formed by the union of: arcadia champion and: arcadia tribune) – us Bell [071]
Arcadia courier – Arcadia, NE: O D Crane (wkly) – 1r – 1 – (cont by: arcadia champion) – us Bell [071]
Arcadia daily news – Arcadia, FL. 1914 mar 17-1916 jun – 3r – (gaps) – us UF Libraries [071]
Arcadia enterprise – Arcadia, FL. 1912-1923 – 6r – (gaps) – us UF Libraries [071]
The arcadia guide – Arcadia, NE: Mrs A Rasmussen. v1 n1. aug 26 1948- (biwkly) [mf ed jan 6 1955- (gaps)] – 4r – (suspended foll dec 16 1971 issue; resumed with mar 2 1972 issue) – us NE Hist [071]
Arcadia leader – [1875 jul 1-1876 mar 10]; 1875 jul 1-1877 nov 29; 1876 jan 14-1877 nov 29 – 1 – mf#918400 – us WHS [071]
Arcadia news-leader – 1939 nov 2-1995 – 1 – mf#968077 – us WHS [071]
Arcadia record – 1911 nov 24-1913 may 9 – 1 – mf#958909 – us WHS [071]
Arcadian – 1895 may 23-1898 apr 14; 1898 apr 21-1899 nov 24; 1899 dec 1-1901 may 24; 1901 may 31-1902 nov 28; 1902 dec 5-1904 apr 22; 1904 apr 29-1905 sep 15; 1905 sep 22-1907 aug 9 – 1 – mf#918026 – us WHS [071]
Arcadian – Arcadia, NE: S B Warden. -v15 n21. feb 29 [ie 25] 1943 (wkly) – (= ser Ord Quiz) – 2r – 1 – (absorbed by: ord quiz) – us Bell [071]
Arcadian – Arcadia, FL. 1925-1996 – 66r – (gaps) – us UF Libraries [071]
Arcaismo vulgar en el espanol de puerto rico / Alvarez Nazario, Manuel – Mayaguez, Puerto Rico. 1957 – 1r – us UF Libraries [440]
Arcaismo y modernidad en la explotacion agraria de valdeburon (leon) / Martin Galindo, Jose Luis – 1 – sp Bibl Santa Ana [630]
Arcana historia (cbh3,3) / Procopii Caesariensis; ed by Maltret, Cl – Parisiis, 1663 – (= ser Corpus byzantinae historiae (cbh)) – €27.00 – ne Slangenburg [243]
Arcangelo carradori's ditionario della lingua italiana e nubiana – [Uppsala, Sweden, A.B. Lundequistska Bokhandeln, 1931] – 1r – 1 – us CRL [440]
Arcani musicali suelati dalla vera amicitia : ne' quali appariscono diversi studii artificiosi, molte osservationi... / Berardi, Angelo – Bologna: Per P-M Monti 1690 [mf ed 19–] – 1mf – 9 – mf#fiche 355 – us Sibley [780]
[Arcata-] arcata coop newsletter – CA. mar 1976-nov 1985; dec 1990-jan 1995 – 3r – 1 – $180.00 – mf#B03139 – us Library Micro [071]
[Arcata-] daily evening telephone – CA. dec 1881-dec 1882 – 2r – 1 – $120.00 – (aka: weekly telephone) – mf#B02021 – us Library Micro [071]
[Arcata-] drift dodger – CA. 1982-88 – 1r – 1 – $60.00 – mf#B05026 – us Library Micro [071]
[Arcata-] econews – CA. may 1971-dec 1995 – 4r – 1 – $240.00 – mf#B03140 – us Library Micro [071]
[Arcata-] osprey – CA. 1973-90 – 1r – 1 – $60.00 – mf#B05027 – us Library Micro [071]
[Arcata-] the lumberjack – CA. 1929-99 – 21r – 1 – $1260.00 – (aka: hstc hooter humboldt state university newspaper oct 30 1929-jun 5 1989) – mf#B03141 – us Library Micro [071]
[Arcata-] the union – CA. jul 31 1886- (wkly) – 86r – 1 – $5160.00 (subs $90/y) – mf#B02022 – us Library Micro [071]
Arcata union *see* [Mckinleyville-] mckinleyville journal
Arcaya, Pedro Manuel *see*
– Estudios sobre personajes y hechos de la historia
– Venezuela y su actual regimen
Arce, Antonio M *see* Sociologia y desarrollo rural
Arce, David N *see* Etica y estetica en la danza
Arce De Vazquez, Margot *see* Impresiones
Arce, Joan C *see*
– Dificultades vencidas...para la limpieza y aseo de las calles de esta corte
– Testosterone and physical activity

Arce, Jose M *see* Manuel gonzalez zeledon
Arce, Luis A *see* Jose antonio cortina
Arce, Manuel Jose *see* En el nombre del padre
Arce Y Valladares, Manuel Jose *see* Romacero de yndias
Arce y Valladares, Manuel Jose *see* 7 sonetos de ausencia
Arcebispo de cangranor / Meyrelles de Souto, A – 1960 – 1 – sp Bibl Santa Ana [240]
O arcebispo de goa e a congregacao de propaganda fide / [Rivara, Joaquim Heliodoro da Cunha] – Nova-Goa: Imprensa Nacional, 1862 [mf ed 1995] – (= ser Yale coll) – 102p – 1 – 0-524-10108-6 – (in portuguese) – mf#1995-1108 – us ATLA [241]
L'arcenal de chirurgie... / Scultet, J – Lyon, 1675 – 8mf – 9 – sp Cultura [617]
Arch notes – 1983/1-1987/6 [1982 nov/dec-1987 nov/dec] – 1 – mf#1109781 – us WHS [071]
The arch of titus and the spoils of the temple / Knight, William – New York: Fleming H Revell [1896?] [mf ed 1992] – (= ser By-paths of bible knowledge 22) – 1mf – 9 – 0-524-05225-5 – (int by lord bishop of durham) – mf#1992-0358 – us ATLA [930]
Archadelt, Jacobus *see* Missae tres...
Archaeologia : or, miscellaneous tracts relating to antiquity – London, England 1770-1992 – 1 – ISSN: 0261-3409 – mf#6173 – us UMI ProQuest [930]
Archaeologia : or miscellaneous tracts relating to antiquity – London, Oxford, 1779-1945. v1-91+ind v1-50 – 1084mf – 9,8 – mf#H-805c – ne IDC [930]
Archaeologia americana : transactions and collections of the american antiquarian society / American Antiquarian Society – Worcester MA: Manning 1820 – (= ser Whsb) – 5mf – 9 – €60.00 – mf#Hu 496 – gw Fischer [975]
Archaeologia americana : transactions and collections of the american antiquarian society – v2 – 1 – mf#1058107 – us WHS [071]
Archaeologia cambrensis – La Salle IL 1846-1905 – 1 – mf#5236 – us UMI ProQuest [930]
Archaeologia or miscellaneous tracts relating to antiquity – v. 1-98. 1773-1961. Index, v. 1-50 – 1 – 959.00 – us L of C Photodup [930]
Archaeological atlas of ohio, 1914 / Mills, William – 1r – 1 – mf#B26300 – us Ohio Hist [933]
Archaeological Dept. Southern Circle. Madras *see* Annual report...for the year...
Archaeological institute of america bulletin – Boston MA 1975-96 – 1,5,9 – mf#10306 – us UMI ProQuest [930]
The Archaeological Museum. Naples *see* National archaeological museum, naples
Archaeological reconnaissance of northwestern honduras / Yde, Jens – Copenhagen, Denmark. 1938 – 1r – us UF Libraries [930]
Archaeological reports published under official authority – Simla: Govt Central Printing Office, 1904 – (= ser Lists and guides to official Indian publications, 1892-1970) – 1r – 1 – us CRL [930]
Archaeological researches in palestine during the years 1873-1874 / Clermont-Ganneau, Charles – London: publ for the Cttee of the Palestine Exploration Fund 1896-99 [mf ed 1990] – 2v on 3mf [ill] – 9 – 0-8370-1662-2 – (trans by aubrey stewart, v2 trans by john macfarlane; ill by a lecomte du noiiy) – mf#1987-6092 – us ATLA [933]
Archaeological review – La Salle IL 1888-90 – 1 – mf#5237 – us UMI ProQuest [930]
Archaeological Society of British Columbia *see*
– Midden
– Newsletter of the archaeological society of b c
Archaeological Society of Connecticut *see* Connecticut news
Archaeological society of delaware. bulletin – Wilmington DE 1977-78 – 1,5,9 – ISSN: 0003-8067 – mf#10729 – us UMI ProQuest [930]
Archaeological Society of New Jersey *see* Newsletter of the archaeological society of new jersey
Archaeological Survey of India *see* Hampi ruins
Archaeological survey of india : photographs in the india office collections, in the british library, london – 230mf – 9 – £1550 – (with d/g) – uk Matthew [930]
Archaeological Survey of India. Northern Circle *see* Annual progress report of the superintendent, archaeological survey of india, northern circle, muhammadan and british monuments...
Archaeological Survey of India. Northern Circle. Muhammadan and British Monuments *see* Annual progress report of the superintendent... for the year ending...
An archaeological tour in gedrosia / Stein, Aurel – Calcutta: Govt of India, Central Publ Branch, 1931 – (= ser Samp: indian books) – us CRL [930]
An archaeological tour in upper swat and adjacent hill tracts / Stein, Aurel – Calcutta: Govt of India, Central Publ Branch, 1930 – (= ser Samp: indian books) – us CRL [930]
An archaeological tour in waziristan and northern baluchistan / Stein, Aurel – Calcutta: Govt of India, Central Publication Branch, 1929 – (= ser Samp: indian books) – us CRL [930]
Archaeologische entdeckungen des neunzehnten jahrhunderts *see* Century of archaeological discoveries
Archaeology – Boston MA 1948+ – 1,5,9 – ISSN: 0003-8113 – mf#11939 – us UMI ProQuest [930]
Archaeology : progress report of the archaeological survey of india, western circle... – Bombay: The Survey, 1906-13 – 1r – 1 – us CRL [930]
Archaeology and physical anthropology in oceania – Sydney, Australia 1966-80 – 1,5,9 – (cont by: archaeology in oceania) – ISSN: 0003-8121 – mf#5968.01 – us UMI ProQuest [930]
Archaeology and the religion of israel / Albright, W F – Johns Hopkins Press, 1953 – 9 – $10.00 – us IRC [270]
Archaeology in india / India. Ministry of Education, Department of Archaeology – Delhi: Manager of Publ, 1950 – (= ser Samp: indian books) – us CRL [930]
Archaeology in oceania – Sydney, Australia 1981+ – 1,5,9 – (cont: archaeology and physical anthropology in oceania) – ISSN: 0728-4896 – mf#5968,01 – us UMI ProQuest [930]
The archaeology of baptism / Cote, Wolfred Nelson – London: Yates and Alexander, 1876 – 4mf – 9 – 0-524-07407-0 – (incl bibl ref) – mf#1991-3067 – us ATLA [242]
The archaeology of gujarat : including kathiawar / Sankalia, Hasmukhlal Dhirajlal – Bombay: Natwarlal & Co, 1941 – (= ser Samp: indian books) – us CRL [930]
The archaeology of palestine / Albright, W F – Penguin Books, 1949 – 9 – $10.00 – us IRC [930]
The archaeology of palestine and the bible / Albright, W F – Revell, 1932 – 9 – $10.00 – us IRC [930]
The archaeology of the cuneiform inscriptions / Sayce, Archibald Henry – 2nd ed., rev. London: S.P.C.K.; New York: E.S. Gorham, 1908 – (= ser Rhind Lectures) – 1mf – 9 – 0-7905-3407-X – (incl bibl ref) – mf#1987-3407 – us ATLA [930]
The archaeology of the napa, california region / Heiser, Robert F – 1953 – (= ser Volcanology-archaelogy coll) – 1r – 1 – $50.00 – mf#B70062 – us Library Micro [930]
Archaeology of the napa region – Napa Co, CA: Robert F Heizer, 1953 – 1r – 1 – $50.00 – mf#B40236 – us Library Micro [930]
Archaeology of the old testament / Naville, Edouard – London: Robert Scott 1913 [mf ed 1989] – (= ser Library of historic theology) – 1mf – 9 – 0-7905-1313-7 – (incl ind) – mf#1987-1313 – us ATLA [221]
Archaia : or, studies of the cosmogony and natural history of the hebrew scriptures / Dawson, John William – Montreal: B Dawson; London: Sampson Low 1860 [mf ed 1989] – 1mf – 9 – 0-7905-0754-4 – (incl bibl ref & ind) – mf#1987-0754 – us ATLA [221]
Archaiologike ephemeris – Athens, 1837-1982 – (= ser Architectural periodicals at avery library, columbia university) – 24r – 1 – $3230.00 – us UPA [930]
Archaische texte aus uruk / Falkenstein, A – Berlin, 1936 – (= ser Ausgrabungen der deutschen forschungsgemeinschaft in uruk-warka 2) – 7mf – 9 – mf#NE-444 – ne IDC [490]
Archambault, Joseph-Papin *see* Les familles au sacre-coeur
Archambault, Pedro Maria *see* Historia de la restauracion
Archambault, Urgel Eugene *see*
– Ecole polytechnique de montreal
– Prospectus, ecole polytechnique de montreal
Archbald citizen – Archbald, PA. v21 n1138. apr 1-june 17, 1916 [wkly] – 1r – 1 – (other titles: citizen [archbald, pa]) – mf#11 L1.1 – us Western Res [071]
[Archbald, pa-] citizen *see* Archbald citizen
Archbell, James *see*
– A grammar of the bechuana language
Archbishop maclagan : being a memoir of the most reverend the right honourable william dalrymple maclagan, d.d., archbishop of york and primate of england / How, Frederick Douglas – London: W Gardner, Darton 1911 [mf ed 1990] – 2mf [ill] – 9 – 0-7905-4927-1 – mf#1988-0927 – us ATLA [241]
Archbishop of canterbury's assyrian mission. report – London: SPCK, 1894-1915 [mf ed 2001] – (= ser Christianity's encounter with world religions, 1850-1950) – 1r – 1 – mf#2001-s000 – us ATLA [240]
The archbishop of goa and the congregation de propaganda fide / Rivara, Joaquim Heliodoro da Cunha – New-Goa: National Press, 1862 [mf ed 1995] – (= ser Yale coll) – 92p – 1 – 0-524-10032-2 – mf#1995-1032 – us ATLA [241]
Archbishop purcell and the archdiocese of cincinnati : a study based on original sources / McCann, Mary Agnes – 1918 [mf ed 1993] – 1mf – 9 – 0-524-06268-4 – mf#1991-2459 – us ATLA [241]
Archbishop secker's five sermons against popery / Porteus, Beilby – London, England. 1835 – 1r – us UF Libraries [241]
Archbishop secker's lectures on the creed / Secker, Thomas – Dublin, Ireland. 1854 – 1r – us UF Libraries [241]
Archbishop thomas bradwardine : a fourteenth century augustinian / Obermann, H A – Utrecht, 1958 – 5mf – 8 – €12.00 – ne Slangenburg [241]
Archbishop wake and the project of union (1717-1720) : between the gallican and anglican churches / Lupton, Joseph Hirst – London: G Bell; Cambridge: Deighton, Bell 1896 [mf ed 1990] – 1mf – 9 – 0-7905-5482-8 – (incl bibl ref) – mf#1988-1482 – us ATLA [242]
Archbishop's champion brought to book / Stearns, Edward Josiah – New York: T Whittaker 1881 [mf ed 1986] – 1mf – 9 – 0-8370-8228-5 – mf#1986-2228 – us ATLA [241]
Archbold, William Arthur Jobson *see*
– Bengal haggis
– Essays on the teaching of history
– Outlines of indian constitutional history
Archdiocese of Quebec. Catholic Church *see*
– Circulaire au clerge
Archdiocese of St. Paul and Minneapolis *see* Parish questionnaires and related materials
Arche, Jose Vicente *see* Castilla agricola para la ensenanza de la agricultura...caceres
L'arche sainte : guide du franc-macon destine a perfectionner l'instruction des recipiendaires a tous les degres, et contenant l'origine, les principes, l'appreciation des rites, grades, ceremonies, fetes, usages, etc, de la maconnerie... – 6e enl ed. Lyon: C Jaillet 1865 – 3mf – 9 – mf#vrl-9 – ne IDC [366]
Archeley : das ist gruendlicher und...von geschuetz / Ufa, D – Zutphen, 1630 – 4mf – 9 – mf#OA-181 – ne IDC [720]
Archenhold-Sternwarte Berlin-Treptow *see* Mitteilungen der archenhold-sternwarte berlin-treptow
Archenholtz, Johann W von *see* England und italien
Archenholtz, Johann Wilhelm von *see*
– Annalen der brittischen geschichte
– Litteratur und voelkerkunde
– Neue litteratur und voelkerkunde
Archenholz, J W von *see* England und italien
Archenholz, Johann Wilhelm von *see* Minerva
Archeological material from saba and st eustatius / Josselin De Jong, Jan Petrus Benjamin De – Leiden, Netherlands. 1947 – 1r – us UF Libraries [930]
Archeological Museum of Merida *see* A brief guide to the museum
Archeological Society of Virginia *see* Quarterly bulletin
L'archeologie egyptienne / Maspero, G – Paris, 1907 – 4mf – 9 – mf#NE-20409 – ne IDC [956]
Archeologie mesopotamienne : les etapes / Parrot, A – Paris, 1946-1953. 2v – 12mf – 9 – mf#NE-418 – ne IDC [956]
Archeologie paleochretienne et culte chretien / Nedoncelle, M et al – Straszbourg, 1962 – 4mf – 8 – €11.00 – ne Slangenburg [930]
Archeologie religieuse du diocese de montreal, 1850 / Viger, Jacques – Montreal: impr par Lovell & Gibson, 1850 [mf ed 1974] – 1r – 5 – mf#SEM16P217 – cn Bibl Nat [241]
Archeologie religieuse du diocese de montreal, 1850 / Viger, Jacques – [Montreal?: s.n.] 1850 [mf ed 1983] – 1mf – 9 – 0-665-41723-3 – mf#41723 – cn Canadiana [241]
The archeology of crete / Pendlebury, J D S – Methuen. 1939 – 9 – $15.00 – us IRC [930]
Archeology of the northern san joaquin valley / Schenck, William Egbert – Berkeley, CA. 1929 – 1r – us UF Libraries [930]
The archeology of the white buffalo robe site / Ahler, Stanley A et al; ed by Chung, Ho Lee – Grand Forks ND: University of North Dakota, 1980 [mf ed 1982] – 2v on 1r – 1 – (incl bibl ref) – mf#n82-298 – us Wisconsin U Libr [975]
Archer, Andrew *see*
– Canada
– Canada, a short history of the dominion of canada
– A history of canada
– History of canada
Archer, Edward C *see* Tours in upper india, and in parts of the himalaya mountains
Archer, Fred Palmer *see* Papers relating to plantations in wuvulu, bougainville and buka, papua new guinea
Archer, Harry Glasier *see*
– The burial service
– The psalter and canticles pointed for chanting to the gregorian psalm tones
Archer, John Clark *see* The sikhs in relation to hindus, moslems, christians, and ahmadiyyas
Archer, Mildred *see* Patna painting
Archer, Thomas *see* India and the gospel
Archer, William *see*
– India and the future
– Pirate's progress
Archer, William George *see*
– Bazaar paintings of calcutta
– The dove and the leopard
– Indian painting in the punjab hills
– The vertical man
Archer, William Henry *see* Statistical register of victoria, from the foundation of the colony
Archer's craft / Hodgkin, Adrian Eliot – New York, NY. 1951? – 1r – us UF Libraries [025]
Archery world – Plymouth MN 1974-89 – 1,5,9 – (cont by: bowhunting world) – ISSN: 0003-827X – mf#10676.01 – us UMI ProQuest [790]
Arches – 1981 sep 25-1987 nov 13 – 1 – mf#654534 – us WHS [071]
Archibald, Adams George *see* Address delivered on the 30th day of october, ad 1883
Archibald, Alexander *see* Experimental religion exemplified
Archibald, Andrew Webster *see* The bible verified
Archibald, F A *see* Methodism and literature
Archibald, Francis A *see* Methodism and literature
Archibald, John *see* The historic episcopate in the columban church and in the diocese of moray
Archibald, Smith *see* Business ledger 1827-1874
Archibishops' answer to the pope – London, England. 1897? – 1r – us UF Libraries [240]
Archiconfrerie de sainte-anne de beaupre : manuel du directeur – [Quebec?: s.n.] 1868 [mf ed 1985] – 1mf – 9 – 0-665-10181-3 – (incl text in latin) – mf#10181 – cn Canadiana [240]
Archidiaconal functions / Fry, Lucius G – London, England. 1900 – 1r – us UF Libraries [240]
Archie – iss n1-50. win 1942-jun 1951 – 15 – mf#001AR-010AR – us MicroColour [740]
Archief geschiedenis aartsbisdom utrecht – 1(1875)-75(1957) – 420mf – 9 – €801.00 – ne Slangenburg [949]
Archief van kerkelijke geschiedenis – 1(1829)-11(1840) – 54mf – 9 – €103.00 – (1841-1849: nederl archief v kerkel gesch 1852-1854: nieuw archief v kerkel gesch) – ne Slangenburg [242]
Het archief van prof dr w h de vriese betreffende zijn onderzoek naar de kultures in nederlands indie, 1857-1862 = the archives of prof dr w h de vriese concerning his investigations into the netherlands indies cultivations / Netherlands. General State Archives – 204mf – 9 – €1,270.00 – mf#M106 – ne MMF Publ [949]
Archief voor de geschiedenis der oude hollandsche zending / Grothe, J A – Utrecht 1884-91 [mf ed 2004] – (= ser Rare printed sources and reference works for the history of dutch colonialism) – 6v on 18mf – 9 – €225.00 – mf#mmp114 – ne Moran [959]
Archief voor de geschiedenis der oude hollandsche zending – Utrecht. deel 1-3. 1884-86 – 16mf – 9 – €31.00 – ne Slangenburg [949]
Archief voor de koffiecultuur in nederlandsch indie – Batavia, Djakarta, 1927-1950. v1-17 – 72mf – 8 – (missing: 1947 v16) – mf#SE-23 – ne IDC [959]
Archief voor de rubbercultuur in nederlandsch indie – Bogor, 1917-1958. v1-35 – 448mf – 8 – mf#SE-24 – ne IDC [959]
Archief voor nederlandsche kerkgeschiedenis – 1(1885)-7(1889) – 42mf – 9 – €80.00 – (1857-1866: kerkhistorisch archief. 1870-1880: studien en bijdragen op 't gebied der historische theologie) – ne Slangenburg [242]
Archiepiscopatus parisiensis (gc7) – Parisiis, 1744 – (= ser Gallia christiana (gc)) – €76.00 – ne Slangenburg [240]
Archiepiscopi ravennatis homiliae / Petri Chrysologi (Peter Chrysologus, Saint) – Coloniae, 1541 – €29.00 – ne Slangenburg [241]
Archilochus : griechisch und deutsch – Muenchen, Germany. 1959 – 1r – us UF Libraries [450]
Archimedes *see* Archimedis opera non nvlla
Archimedes in alexandrien : erzaehlung / Colerus, Egmont – Berlin: P Zsolnay, 1941, c1939 [mf ed 1989] – 196p – 1 – mf#7156 – us Wisconsin U Libr [830]
Archimedis opera non nvlla : a federico commandino vrbinate, nvper in latinvm conversa, et commentariis illvstrata, quorum nomina in sequenti pagina leguntur – Venetiis: Apud Paulum Manutium, 1558 [mf ed 1982] – 2v (ill) – 1 – mf#606 – us Wisconsin U Libr [510]
Archinard, Andre *see* Les edifices religieux de la vieille geneve
L'archipel des comores / Manicacci, Jean – Tananarive: Impr Officielle, 1939 – 1 – us CRL [960]
Archipel lenoir / Salacrou, Armand – Paris, France. 1948 – 1r – us UF Libraries [440]

Architect – London, England 1971-78 – 1,5,9 – ISSN: 0003-8415 – mf#6265 – us UMI ProQuest [720]

Architect and building news – London, England 1950-68 – 1 – ISSN: 0570-6416 – mf#666 – us UMI ProQuest [690]

The architect and his artists / White, William Henry – London 1892 – (= ser 19th c art & architecture) – 1mf – 9 – mf#4.2.150 – uk Chadwyck [720]

Architect, engineer and surveyor – La Salle IL 1840-43 – 1 – mf#5239 – us UMI ProQuest [624]

The architect of the new palace at westminster / Barry, Alfred, Bishop of Sydney – London 1868 – (= ser 19th c art & architecture) – 2mf – 9 – mf#4.2.177 – uk Chadwyck [720]

Architect [overseas ed] – London UK 1986-87 – 1,5,9 – (cont: riba journal; cont by: journal / royal institute of british architects [overseas ed]) – ISSN: 0950-8902 – mf#1383.03 – us UMI ProQuest [720]

Architectonisches alphabet bestehend aus dreyssig rissen... / Steingruber, J D – Schwabach, 1773 – 2mf – 9 – mf#OA-111 – ne IDC [720]

An architect's experiences : professional, artistic, and theatrical / Darbyshire, Alfred – Manchester: J E Cornish, 1897 – (= ser 19th c art & architecture) – 4mf – 9 – mf#4.1.104 – uk Chadwyck [720]

Architects' journal – London UK 1924+ – 1,5,9 – ISSN: 0003-8466 – mf#1227 – us UMI ProQuest [720]

Architects law reports – v1-4. 1904-08 (all publ) – 9 – mf#LLMC 84-704 – us LLMC [340]

Architects Renewal Committee in Harlem *see* Partisan planning

Architects' Renewal Committee in Harlem (ARCH) *see* Harlem news

Architectura : klare en duydelijcke demonstration der vijf ordens, uyt...vincent scamozzi / Anhaltin, C M – Amsterdam, 1661 – 1mf – 9 – mf#OA-284 – ne IDC [720]

Architectura / Vignola, J – SL, 1582 – 2mf – 9 – sp Cultura [720]

La architectura / Alberti, L B – SL, 1575 – 7mf – 9 – sp Cultura [720]

Architectura chivilis : vertoonende verscheyde treffelijcke cappen soo van toorens, kercke, als mede...huysen en eenige wenteltrappe... / Danckers, J – Amsterdam, n d – 3mf – 9 – mf#O-1143 – ne IDC [720]

Architectura civilis : oder beschreibung und vorreisung vieler vornehmer dachwerck / Wilhelm, J – Nuernberg, 1649. 2v – 2mf – 9 – mf#OA-125 – ne IDC [720]

Architectura civilis... / Furttenbach, J – Ulm, 1628 – 5mf – 9 – mf#O-1155 – ne IDC [720]

Architectura curiosa va... / Boeckler[n], G A – Rimbergae, [1664] – 14mf – 9 – mf#O-1147 – ne IDC [720]

Architectura ecclesiastica londini : or graphical survey of...churches, in london / Clarke, Charles – London 1820 – (= ser 19th c art & architecture) – 5mf – 9 – mf#4.2.1281 – uk Chadwyck [720]

Architectura et perspectiva des fortifications... / Perret, I – Francfort sur le Mein, 1602 – 3mf – 9 – mf#O-1144 – ne IDC [720]

Architectura hydraulica en las fabricas de puentes methodo de proyectarlos y reparlos / Pontones, P – SL, 1759 – 9mf – 9 – sp Cultura [620]

Architectura libri dece... / Vitruvius Pollio, M – Como, 1521 – 7mf – 9 – mf#OA-5 – ne IDC [720]

Architectura martialis... / Furttenbach, J – Ulm, 1630 – 6mf – 9 – mf#O-1140 – ne IDC [720]

Architectura militaris... / Freitag, A – Amsterdam, 1665 – 4mf – 9 – mf#OA-216 – ne IDC [720]

Architectura militaris... / Freitag, A – Leyden, 1642 – 4mf – 9 – mf#OA-215 – ne IDC [720]

Architectura militaris... / Sturm, L C – Nuernberg, 1736 – 3mf – 9 – mf#OA-218 – ne IDC [720]

Architectura militaris moderna / Doegen, M – Amstelodami, 1647 – 9mf – 9 – mf#OA-145 – ne IDC [720]

Architectura moderna ofte bouwinge van onsen tyt / Keyser, H de & Danckerts, C – Amstelredam, 1631 – 4mf – 9 – mf#O-1164 – ne IDC [720]

Architectura numismatica : or, architectural medals of classic antiquity / Donaldson, Thomas Leverton – London 1859 – (= ser 19th c art & architecture) – 6mf – 9 – mf#4.2.1620 – uk Chadwyck [720]

Architectura privata... / [Furttenbach, J] – Augspurg, 1641 – 4mf – 9 – mf#O-1157 – ne IDC [720]

Architectura recreationis... / Furttenbach, J – Augspurg, 1640 – 7mf – 9 – mf#O-1156 – ne IDC [720]

Architectura universalis... / Furttenbach, J – Ulm, 1635 – 9mf – 9 – mf#O-1139 – ne IDC [720]

Architectura von ausstheilung / symmetria und proportion der fuenff seulen... / Dietterlin, W – [Nuernberg], 1598 – 11mf – 9 – mf#O-1020 – ne IDC [720]

Architectura von vestungen... / Speckle, D – Dresden, 1710 – 5mf – 9 – mf#OA-268 – ne IDC [720]

Architectura von vestungen... / Speckle, D – Straszburg, 1608 – 6mf – 9 – mf#OA-217 – ne IDC [720]

Architectural and design history *see* The life and work of a w n pugin

Architectural and engineering news – New York NY 1958-70 – 1,5,9 – ISSN: 0003-8482 – mf#1938 – us UMI ProQuest [720]

An architectural and general description of the town hall, manchester / Axon, William Edward Armytage – Manchester 1878 – (= ser 19th c art & architecture) – 3mf – 9 – mf#4.2.1056 – uk Chadwyck [720]

The architectural antiquities of great britain / Britton, John – London 1807-26 – (= ser 19th c art & architecture) – 18mf – 9 – mf#4.2.815 – uk Chadwyck [720]

Architectural antiquities of normandy / Cotman, John Sell – London [1821-22] – (= ser 19th c art & architecture) – 12mf – 9 – mf#4.2.269 – uk Chadwyck [720]

The architectural antiquities of rome / Taylor, George Ledwell – London 1821,22 – (= ser 19th c art & architecture) – 11mf – 9 – mf#4.2.1220 – uk Chadwyck [720]

The architectural antiquities of western india / Cousens, Henry – London: India Society, 1926 – (= ser Samp: indian books) – us CRL [720]

Architectural association journal, 1936-1959 – 7r – 1 – mf#537 – uk Microform Academic [720]

Architectural Association, London *see* A visit to the domed churches of charente

Architectural design cost and data – Tampa FL 1974-78 – 1,5,9 – (cont by: design cost and data for the construction industry) – ISSN: 0003-8512 – mf#9904.05 – us UMI ProQuest [720]

Architectural designs : manufactured in imperishable terra cotta / Doulton and Co Ltd – [London? 1872?] – (= ser 19th c art & architecture) – 1mf – 9 – mf#4.2.861 – uk Chadwyck [720]

Architectural digest – New York NY 1994+ – 1,5,9 – ISSN: 0003-8520 – mf#18401 – us UMI ProQuest [720]

Architectural drawing / Spiers, Richard Phene – London 1887 – (= ser 19th c art & architecture) – 2mf – 9 – mf#4.2.1197 – uk Chadwyck [720]

Architectural drawings from the victoria and albert museum – 23r – 1 – (incl drawings by sir christopher wren, nicholas hawksmoor, robert adam and sir gilbert scott. with ind) – mf#96742 – uk Microform Academic [720]

Architectural forum – 1892-1974 – 1,5,9 – ISSN: 0003-8539 – mf#1164 – us UMI ProQuest [720]

The architectural history of canterbury cathedral / Willis, Robert – London 1845 – (= ser 19th c art & architecture) – 2mf – 9 – mf#4.2.817 – uk Chadwyck [720]

The architectural history of chichester cathedral / Willis, Robert – Chichester 1861 – (= ser 19th c art & architecture) – 1mf – 9 – mf#4.2.1262 – uk Chadwyck [720]

An architectural history of the cathedral church of manchester / Crowther, Joseph Stretch – Manchester 1893 – (= ser 19th c art & architecture) – 3mf – 9 – mf#4.2.1493 – uk Chadwyck [720]

The architectural history of the university of cambridge / Willis, Robert – Cambridge 1886 – (= ser 19th c art & architecture) – 29mf – 9 – mf#4.2.1018 – uk Chadwyck [720]

The architectural history of the...holy sepulchre at jerusalem / Willis, Robert – London 1849 – (= ser 19th c art & architecture) – 3mf – 9 – mf#4.2.301 – uk Chadwyck [720]

The architectural history of winchester cathedral / Willis, Robert – London 1846 – (= ser 19th c art & architecture) – 1mf – 9 – mf#4.2.816 – uk Chadwyck [720]

Architectural illustrations and account of the temple church / Billings, Robert William – London 1838 – (= ser 19th c art & architecture) – 3mf – 9 – mf#4.2.1487 – uk Chadwyck [720]

Architectural illustrations and description of the cathedral church at durham / Billings, Robert William – London 1843 – (= ser 19th c art & architecture) – 3mf – 9 – mf#4.2.1378 – uk Chadwyck [720]

Architectural illustrations of kettering church, northamptonshire / Billings, Robert William – London 1843 – (= ser 19th c art & architecture) – 1mf – 9 – mf#4.2.1490 – uk Chadwyck [720]

Architectural illustrations of windsor castle / Gandy, Michael & Baud, Benjamin – London 1842 – (= ser 19th c art & architecture) – 6mf – 9 – mf#4.2.1434 – uk Chadwyck [720]

Architectural magazine and journal of improvement in architecture, building, and furnishing – La Salle IL 1834-39 – 1 – mf#5238 – us UMI ProQuest [720]

Architectural nomenclature of the middle ages / Willis, Robert – Cambridge 1844 – (= ser 19th c art & architecture) – 2mf – 9 – mf#4.2.1590 – uk Chadwyck [720]

Architectural notes on german churches / Whewell, William – [3rd ed]. Cambridge 1842 – (= ser 19th c art & architecture) – 4mf – 9 – mf#4.2.95 – uk Chadwyck [720]

Architectural ornaments : or a collection of capitals, friezes, roses, entablatures, mouldings / Aglio, Augustine – London 1820 – (= ser 19th c art & architecture) – 1mf – 9 – mf#4.2.1326 – uk Chadwyck [720]

Architectural periodicals at avery library, columbia university – Clearwater Publ Co – 286r – 1 – $34,795.00 coll – (individual titles listed separately) – us UPA [720]

Architectural precedents : consisting of plans, elevations, sections and details / Davy, Christopher – 3rd ed. London 1841 – (= ser 19th c art & architecture) – 7mf – 9 – mf#4.2.1586 – uk Chadwyck [720]

Architectural principles in the age of humanism / Wittkower, Rudolf – London, England. 1952 – 1r – us UF Libraries [720]

Architectural Publication Society *see* The dictionary of architecture

Architectural record – New York NY 1891+ – 1,5,9 – ISSN: 0003-858X – mf#1042 – us UMI ProQuest [720]

Architectural record – New York. v1-48. 1891-1920 – 497mf – 9 – mf#O-1228 – ne IDC [720]

Architectural review – London UK 1896+ – 1,5,9 – ISSN: 0003-861X – mf#1228 – us UMI ProQuest [720]

Architectural review for the artist and craftsman – London, 1896/1897-1902. v1-11 – 359mf – 9 – (cont as: architectural review. london, 1902-1920. v12-48) – mf#O-491 – ne IDC [720]

Architectural science review – Sydney, Australia 1989+ – 1,5,9 – ISSN: 0003-8628 – mf#14797 – us UMI ProQuest [720]

Architectural sketches from the continent / Shaw, Richard Norman – London [1858] – (= ser 19th c art & architecture) – 6mf – 9 – mf#4.2.865 – uk Chadwyck [720]

Architectural Society, London *see* Essays of the london architectural society

Architectural studies in france / Davie, W Galsworthy – [London? 1877?] – (= ser 19th c art & architecture) – 5mf – 9 – mf#4.2.1540 – uk Chadwyck [720]

An architectural survey of the churches in... lindisfarne / Wilson, Frederick Richard – Newcastle-upon-Tyne 1870 – (= ser 19th c art & architecture) – 5mf – 9 – mf#4.2.1302 – uk Chadwyck [720]

An architectural tour of normandy / Knight, Henry Gally – London 1836 – (= ser 19th c art & architecture) – 3mf – 9 – mf#4.1.376 – uk Chadwyck [720]

Architecture / Allen, L – s.l, s.l? . 193-? – 1r – us UF Libraries [720]

Architecture / Blanton, Kelsey – s.l, s.l? . 1936 – 1r – us UF Libraries [720]

Architecture / Brooks, Alfred Mansfield – Boston, MA. 1924 – 1r – us UF Libraries [720]

Architecture : clearwater, florida / Walk Chas E – s.l, s.l? . 1936 – 1r – us UF Libraries [978]

Architecture : especially in relation to our parish churches / Bishop, Henry Halsall – London 1886 – (= ser 19th c art & architecture) – 3mf – 9 – mf#4.2.506 – uk Chadwyck [720]

Architecture : gothic and renaissance / Smith, Thomas Roger – London, England. 1884 – 1r – us UF Libraries [720]

Architecture : a profession or an art / Shaw, Richard Norman & Jackson, Thomas Graham, Baronet – London 1892 – (= ser 19th c art & architecture) – 3mf – 9 – mf#4.2.174 – uk Chadwyck [720]

Architecture – s.l, s.l? . 1937 – 1r – us UF Libraries [720]

Architecture... / De l'Orme, Ph – Paris, 1568 – 11mf – 9 – mf#OA-35 – ne IDC [720]

Architecture: a monthly magazine of architectural art – La Salle IL 1896-98 – 1 – mf#5240 – us UMI ProQuest [720]

Architecture and early photography in france / Caisse Nationale des Monuments Historiques et des Sites. Paris. Archives Photographiques – 255mf coll – 9 – 0-907006-74-4 – (architecture and monuments in france 137mf $870 isbn: 0-907006-64-7. paris views and early photography in france 118mf $760 isbn: 0-907006-69-8) – uk Mindata [700]

Architecture and environment : architecture, applied arts, studio arts – (= ser Art exhibition catalogues on microfiche) – 53 catalogues on 64mf – 9 – £470.00 – (individual titles not listed separately) – uk Chadwyck [720]

Architecture and monuments in france – 137mf – 9 – $1000.00 – 0-907006-64-7 – (unique photographic record fr french state photographic archives by the photographer felix martin-sabon; photos are in topographical order & incl full ind; over 17,000 reproductions) – uk Mindata [720]

Architecture and naive art *see* The index of american design (tiam)

Architecture and other arts... : in northern central syria and the djebel hauran / Butler, H C – New York, 1904 – 8mf – 9 – mf#H-2831 – ne IDC [956]

Architecture and public buildings / White, William Henry – London 1884 – (= ser 19th c art & architecture) – 3mf – 9 – mf#4.2.564 – uk Chadwyck [720]

Architecture and sculpture collection / Victoria and Albert Museum. London – (= ser Decorative art in the victoria and albert museum) – 152mf – 9 – $1125.00 – 0-907006-25-6 – (over 9000 reproductions) – uk Mindata [720]

Architecture d'aujourd'hui – Paris, France 1935+ – 1,5,9 – ISSN: 0003-8695 – mf#6983 – us UMI ProQuest [720]

L'Architecture et art de bien bastir du... / [Alberti, L B] – Paris, 1553 – 9mf – 9 – mf#OA-32 – ne IDC [720]

L'architecture francaise – n1-280. Paris. nov 1940-65 – 5 – fr ACRPP [720]

Architecture hydraulique... / Belidor, [B F] – Paris, 1737-1753 – 7mf – 9 – mf#OA-250 – ne IDC [720]

Architecture in dharwar and mysore / Taylor, Meadows [i.e. Philip Meadows] – London 1866 – (= ser 19th c art & architecture) – 10mf – 9 – mf#4.2.1512 – uk Chadwyck [720]

Architecture in italy from the sixth to the eleventh century / Cattaneo, Raffaele – London 1896 – (= ser 19th c art & architecture) – 4mf – 9 – mf#4.2.1192 – uk Chadwyck [720]

Architecture militair : waar by de versterckinge des vyfhoecx, vande heer...m: van coehoorn... wert verbroken / Paen, L – Leeuwarden, 1682 – 1mf – 9 – mf#OA-274 – ne IDC [720]

Architecture moderne o- l'art de bien batir... / [Briseux, C E] – Paris, 1728-1729. 2v – 16mf – 9 – mf#O-1160 – ne IDC [720]

The architecture of ancient delhi / Cole, Henry Hardy – London 1872 – (= ser 19th c art & architecture) – 3mf – 9 – mf#4.2.513 – uk Chadwyck [710]

Architecture of machinery : an essay on propriety of form and proportion, with a view to assist and improve design / Clegg, Samuel – London: Architectural Library, 1842 – (= ser 19th c art & architecture) – 2mf – 9 – mf#4.1.38 – uk Chadwyck [680]

The architecture of russell warren / Alexander, Robert L – [New York: New York Uni] 1952 [mf ed 1981] – 5mf – 9 – (incl bibl) – mf#50-01 – us South Carolina Historical [720]

Architecture of seattle, washington : a selected bibliography / White, Anthony G – Monticello, IL: Vance Bibliographies, [1982] (mf ed 1982) – 1mf – 9 – mf#*XMC-430 – us NY Public [720]

The architecture of the churches of denmark / Heales, Alfred Charles – London 1892 – (= ser 19th c art & architecture) – 2mf – 9 – mf#4.2.236 – uk Chadwyck [720]

The architecture of the heavens / Nichol, John Pringle – 9th rev enl ed. London: H Bailliere 1851 – (= ser Library of illustrated standard scientific works 9) – 1r [pl/ill] – 1 – mf#film mas 28416 – us Harvard [520]

The architecture of the intelligible universe in the philosophy of plotinus / Armstrong, A H – Cambridge, 1940 – €7.00 – ne Slangenburg [110]

Architecture of the middle ages in italy / Cresy, Edward & Taylor, George Ledwell – London 1829 – (= ser 19th c art & architecture) – 3mf – 9 – mf#4.2.758 – uk Chadwyck [720]

The architecture of the park : a series of designs comprising plans, elevations, perspective views, and details for buildings... / Starforth, John – [Edinburgh]: Banks & Co, 1890 – (= ser 19th c art & architecture) – 3mf – 9 – mf#4.1.152 – uk Chadwyck [720]

Architecture of the renaissance in england / Gotch, John Alfred – London 1891-94 – (= ser 19th c art & architecture) – 15mf – 9 – mf#4.2.271 – uk Chadwyck [720]

Architecture of the renaissance in italy / Anderson, William James – London, England. 1927? – 1r – us UF Libraries [720]

The architecture of the renaissance in italy / Anderson, William James – [2nd ed] London 1898 – (= ser 19th c art & architecture) – 4mf – 9 – mf#4.1.279 – uk Chadwyck [720]

Architecture of the seventeenth century / Hakewill, Arthur William – [London?] 1856 – (= ser 19th c art & architecture) – 1mf – 9 – mf#4.2.1607 – uk Chadwyck [720]
Architecture plus – New York NY 1973-74 – 1,5,9 – ISSN: 0570-6556 – mf#8218 – us UMI ProQuest [720]
L'architecture pratique : qui comprend le detail du toise, et du devis des ouvrages de massonerie... / Bullet, P – Paris, 1691 – 5mf – 9 – mf#OA-45 – ne IDC [720]
Architecture: the aia journal – New York NY 1983+ – 1,5,9 – (cont: american institute of architects aia journal) – ISSN: 0746-0554 – mf#114,01 – us UMI ProQuest [720]
Architecture von den funf seulen sambt iren ornamenten und zierden... / Krammer, G – [Koeln, 1610] – 2mf – 9 – mf#OA-61 – ne IDC [720]
Architecture.west – Seattle WA 1965-69 – 1,5,9 – mf#1933 – us UMI ProQuest [720]
Der architekt *see* Art and decoration
Architektenwettbewerbe in deutschland : geschichte ihrer entwicklung von 1860-1914 / Skiba, Petra – (mf ed 1997) – 6mf – 9 – €62.50 – 3-8267-2486-0 – mf#DHS 2486 – gw Frankfurter [720]
L'architettura... / Cataneo, P – [Venezia, 1567] – 6mf – 9 – mf#O-1007 – ne IDC [720]
Architettura campestre... : simple and economical forms in the modern...style / Hunt, Thomas Frederick – London 1827 – (= ser 19th c art & architecture) – 1mf – 9 – mf#4.2.1128 – uk Chadwyck [720]
L'architettura civile preparata su la geometria, e ridotta alle prospettive : considerazioni pratiche / Bibiena, F G da – Parma, 1711 – 9mf – 9 – mf#OA-17 – ne IDC [720]
L'architettura di leon batista alberti tradotta in lingua fiorentina da c bartoli / Alberti, L B – Venetia, 1565 – 9mf – 9 – mf#O-1018 – ne IDC [720]
Archiv aller buergerlichen wissenschaften zum nutzen und vergnuegen [...] – Hamburg DE, 1804-06 – 3r – 1 – gw Misc Inst [330]
Archiv Bibliographia Judaica *see* Dokumentation zur juedischen kultur in deutschland 1840-1940, abt 1
Archiv Bibliographia Judaica e.V. *see*
– Dokumentation zur juedischen kultur in deutschland 1840-1940, abt 2
– Dokumentation zur juedischen kultur in deutschland 1840-1940, abt 3
– Dokumentation zur juedischen kultur in deutschland 1840-1940, abt 3. neue folge
– Dokumentation zur juedischen kultur in deutschland 1840-1940, abt 4
– Dokumentation zur juedischen kultur in deutschland 1840-1940, abt 5
– Dokumentation zur juedischen kultur in deutschland 1840-1940, abt 6
– Dokumentation zur juedischen kultur in deutschland 1840-1940, abt 8
Archiv der deutschen Frauenbewegung Kassel *see*
– Die frau im staat
– Die frauenbewegung
Archiv der erziehungskunde fuer deutschland / ed by Stephani, Heinrich – Weissenfels, Leipzig 1791-94 – (= ser Dz) – 4v on 7mf – 9 – €140.00 – mf#k/n722 – gw Olms [370]
Archiv der forst und- jagd-gesetzgebung der deutschen bundesstaaten / Behlen, Stephan – Freiburg im Breisgau: Fr Wagner'schen Buchh 1834-46 (irreg) [1.-20.bd nf: 1.-6.bd(1835-46) [mf ed 2005] – 4r [ill] – 1 – (publ: frankfurt a m: j d sauerlaender 1845-46) – mf#film mas c6051 – us Harvard [346]
Archiv der gegenwart – Bonn-Bad Godesberg, 1946-1993 – 47r – 1 (Subsc DM150.00) – gw Mikropress [943]
Archiv der gegenwart *see* Keesing's archiv der gegenwart
Archiv der gesellschaft fuer aeltere deutsche geschichtskunde – 1(1819-20)-12(1872) – 166mf – 9 – €317.00 – ne Slangenburg [930]
Archiv der mathematik = Archives of mathematics – Dordrecht, Netherlands 1992+ – 1,5,9 – ISSN: 0003-889X – mf#13940 – us UMI ProQuest [510]
Archiv der mathematik und physik – 115v. 1841-1920 – 1 – $1296.00 – (in german) – mf#0058 – us Brook [500]
Archiv der mathematik und physik – Greifswald sp Leipzig 1841-1920 [mf ed 1994] – 535mf – 9 – €2870.00 – 3-89131-163-X – (reel 1: v1-70 1841-84; reel 2: v1-17 1884-1900; reel 3: v1-28 1901-20; incl: literarische berichte, bibliographische mitteilungen, register) – gw Fischer [500]
Archiv der saechsischen geschichte / ed by Arndt, Gottfried August – Leipzig, 1784-86 – (= ser Dz. historisch-geographische abt) – 3pt on 9mf – 9 – €180.00 – mf#k/n1148 – gw Olms [943]
Archiv des historischen vereins fuer niedersachsen – 1(1845)-5(1849) – 36mf – 9 – €68.00 – ne Slangenburg [943]
Archiv des historischen vereins von unterfranken und aschaffenburg – Wuerzburg. v1-77 [[1832/33]-1937/38] – 1 – (cont by: mainfraenkisches jahrbuch fuer geschichte und kunst) – us Harvard [943]
Archiv des vereins fuer geschichte und alterthuemer der herzogthuemer bremen und verden und des landes hadeln zu stade – Stade [Germany]: In Commission der A Pockwitz'schen Buchhandlung 1863- v1-11 [1862-96] [mf ed 1978] – 11v on 2r [ill] – 1 – (cont in 1891 by: zeitschrift des historischen vereins fuer niedersachsen; ceased in 1886; iss by the society under its earlier name: verein fuer geschichte und des landes hadeln zu stade) – mf#film mas c290 – us Harvard [943]
Archiv fuer anthropologie – 1 – (v1-32 1866-1906 $660 [0059]. v33-54 1906-40 $144 [0060]. in german) – us Brook [301]
Archiv fuer asiatische literatur, geschichte und sprachkunde – Spb., 1810. v1 – 8mf – 9 – mf#R-1659 – ne IDC [077]
Archiv fuer buergerliches recht / ed by Kohler, Josef et al – Berlin: C Heymann. v1-43.1889-1919 – (= ser Civil law 3 coll) – 214mf – 9 – (cont: archiv fuer theorie und praxis des allgemeinen deutschen handels- und wechselrecht. absorbed by: archiv für die civilistische praxis) – mf#LLMC 96-571 – us LLMC [340]
Archiv fuer chemie und meteorologie *see* Archiv fuer die gesammte naturlehre
Archiv fuer das studium der neueren sprachen / ed by Herrig, L & Bischoff, H – Elberfeld. v1-43 1846-1868; v134-137 916-1918 – 364mf – 8 – mf#H-399 – ne IDC [410]
Archiv fuer das studium der neueren sprachen und literaturen – v1. 1848- – 1 – us AMS Press [400]
Archiv fuer den menschen und buerger in allen verhaeltnissen... / ed by Schlettwein, Joh August – Leipzig 1780-84 – (= ser Dz) – 8v on 30mf – 9 – €300.00 – mf#k/n2728 – gw Olms [320]
Archiv fuer die ausuebende erziehungskunst / ed by Heyler, K Chr (spaeter J F Ross) – Giessen 1777-84 – (= ser Dz) – 12pt on 25mf – 9 – €250.00 – mf#k/n633 – gw Olms [370]
Archiv fuer die gesammte naturlehre – Nuernberg 1824-35 [mf ed 1993] – 27v on 108mf – 9 – €710.00 – 3-89131-161-3 – (with v19 also under the title: archiv fuer chemie und meteorologie) – gw Fischer [500]
Archiv fuer die geschichte der arzneykunde in ihrem ganzen umfang / ed by Wittwer, P L – Nuernberg 1790 – (= ser Dz) – v1 st1 on 2mf – 9 – €60.00 – mf#k/n3660 – gw Olms [615]
Archiv fuer die geschichte der philosophie – 1(1888)-30(1917) – 305mf – 9 – €582.00 – ne Slangenburg [100]
Archiv fuer die geschichte des niederrheins – 1(1831)-7(1870) – 57mf – 9 – €109.00 – ne Slangenburg [943]
Archiv fuer die geschichte des sozialismus und der arbeiterbewegung – v1-15. 1911-30 – 1 – $144.00 – (in german) – mf#0064 – us Brook [335]
Archiv fuer die saechsische geschichte – Leipzig: B Tauchnitz. bd1-12 [1863-1873/74] 2r (film mas c301); nf: bd1-6 [1874/75-1879/1880] 1r (film mas c156) [annual] [mf ed 1978] – 1 – (merged with: mittheilungen des k saechsischen alterthumsvereins and: neues archiv fuer saechsische geschichte und altertumskunde; publ with subvention of the royal government of saxony) – us Harvard [943]
Archiv fuer die theologie und ihre neueste literatur – Tuebingen, 1(1815)-8(1826) – 99mf – 9 – €189.00 – ne Slangenburg [200]
Archiv fuer die theoretische und praktische rechtsgelehrsamkeit / ed by Hagemann, Theodor et al – Helmstedt 1788-92 – (= ser Dz) – 6pt on 13mf – 9 – €130.00 – mf#k/n2591 – gw Olms [340]
Archiv fuer elektrotechnik – Heidelberg, Germany 1981-82 – 1,5,9 – ISSN: 0003-9039 – mf#13131 – us UMI ProQuest [621]
Archiv fuer entscheidungen der obersten gerichte in den deutschen staaten / ed by Seuffert, Johann Adam et al – os: v1-30 1847-75 ns: v1-22 1987-97. Muenchen: I G Cotta, 1847-67; R Oldenbourg, 1870-95 – (= ser Civil law 3 coll) – 228mf – 9 – (series cont to v98 1944. in the portion of this series, some vols are repr, & os v12-14, ind 21-25, & ind 26-30 lack title pp; all ns annuals have their own ind) – mf#llmc 96-567, 97-567 b-d – us LLMC [342]
Archiv fuer frauenarbeit / ed by Silbermann, J – 1913-22 [mf ed 2000] – (= ser Hq 43) – 10v on 32mf – 9 – €220.00 – 3-89131-362-4 – (im auftrage des kaufmaennischen verbandes fuer weibliche angestellte) – gw Fischer [305]
Archiv fuer frauenarbeit *see* Jahrbuch fuer frauenarbeit
Archiv fuer frauenkunde und eugenik / ed by Hirsch, Max – v1-5. 1914-19 [mf ed 1996] – (= ser Hq 19) – 85mf – 9 – €640.00 – 3-89131-131-1 – (filmed with: archiv fuer frauenkunde und eugenetik, sexualbiologie und vererbungslehre [v6-8 1920-22], archiv fuer frauenkunde und eugenetik, sexualbiologie und konstitutionsforschung [v9 1923], archiv fuer frauenkunde und konstitutionsforschung [v10-19 1924-33]) – gw Fischer [618]
Archiv fuer frauenkunde und konstitutionsforschung *see* Archiv fuer frauenkunde und eugenik
Archiv fuer funkrecht – Berlin DE, 1928-1944 apr, 1944 sep – 5r – 1 – gw Mikrofilm [343]
Archiv fuer geschichte von oberfranken / Historischer Verein fuer Oberfranken zu Bayreuth – Bayreuth. v13-34 [1875-1941] (gaps) [mf ed 1978] – 3r [ill] – 1 – (cont: archiv fuer geschichte und alterthumskunde des ober-main-kreises; title varies: archiv fuer geschichte und altertumskunde von oberfranken (varies slightly), 1889-1919; archiv fuer die geschichte von oberfranken) – ISSN: 0066-6335 – mf#film mas c319 – us Harvard [943]
Archiv fuer hessische geschichte und altertumskunde / Historischer Verein fuer das Grossherzogtum Hessen – Darmstadt: Im Selbstverlag des Historischen Vereins fuer das Grossherzogthum Hessen 1835- nf: v1-22 (1894-1942) [mf ed 1978] – 4r – 1 – ISSN: 0066-636X – mf#film mas c410 – us Harvard [943]
Archiv fuer katholisches kirchenrecht – 1(1857)-22(1869) – 217mf – 9 – €414.00 – ne Slangenburg [241]
Archiv fuer kriminologie – v1-100. 1899-1937 – 9 – $780.00 – (in german) – mf#0061 – us Brook [364]
Archiv fuer kriminologie, kriminalanthropologie und kriminalistik – Berlin DE, 1932 jul-aug, 1933-44 – 3r – 1 – mf#12913 – gw Mikropress [364]
Archiv fuer litteratur- und kirchengeschichte des mittelalters – 1(1885)-7(1900) – 95mf – 9 – €181.00 – ne Slangenburg [931]
Archiv fuer litteraturgeschichte – Leipzig: B G Teubner. v1-15 [1870-1887] [mf ed 1978] – 15v on 4r [ill] – 1 – (cont: jahrbuch fuer litteraturgeschichte; cont by : vierteljahrschrift fuer litteraturgeschichte) – mf#film mas c284 – us Harvard [400]
Archiv fuer mikroskopische anatomie [1889] – Dordrecht, Netherlands 1889-94 – 1 – (cont by: archiv fuer mikroskopische anatomie und entwicklungsgeschichte) – mf#13232.07 – us UMI ProQuest [574]
Archiv fuer mikroskopische anatomie [1911] – Dordrecht, Netherlands 1911-23 – 1 – (cont: archiv fuer mikroskopische anatomie und entwicklungsgeschichte) – mf#13232.07 – us UMI ProQuest [574]
Archiv fuer mikroskopische anatomie und entwicklungsgeschichte – Dordrecht, Netherlands 1895-1911 – 1 – (cont: archiv fuer mikroskopische anatomie; cont by: archiv fuer mikroskopische anatomie) – mf#13232.07 – us UMI ProQuest [574]
Archiv fuer mikroskopische anatomie und entwicklungsmechanik – Dordrecht, Netherlands 1923-25 – 1 – (cont by: wilhelm roux' archiv fuer entwicklungsmechanik der organismen) – mf#13232.07 – us UMI ProQuest [574]
Archiv fuer musikwissenschaft – Trossingen. 1918-27. 2 reels – 1 – us L of C Photodup [780]
Archiv fuer neutestamentliche zeitgeschichte und kulturkunde / ed by Leipoldt, J – Leipzig, 1925-1926. v1-2 – 13mf – 8 – mf#H-227c – ne IDC [240]
Archiv fuer oesterreichische geschichte – v1-116. 1848-1944 – 1 – $1140.00 – (in german) – mf#0062 – us Brook [943]
Archiv fuer physiologische heilkunde – v1-15 1842-56; nf: v1-3 1857-59 [mf ed 1994] – (= ser Medizinische zeitschriften) – 62mf – 9 – €1080.00 – 3-89131-189-3 – gw Fischer [615]
Archiv fuer presserecht *see* Der zeitungs-verlag
Archiv fuer protistenkunde – v1-96. 1902-43 – 1 – $960.00 – (in german) – mf#0063 – us Brook [574]
Archiv fuer psychiatrie und nervenkrankheiten = Archives of psychiatry and neurological sciences – Dordrecht, Netherlands 1981-82 – 1,5,9 – (cont by: european archives of psychiatry and neurological sciences) – ISSN: 0003-9373 – mf#13132.02 – us UMI ProQuest [616]
Archiv fuer rassen- und gesellschafts-biologie : einschliesslich rassen- und gesellschafts-hygiene – Berlin. v1-37. 1904-44 – 1 – us L of C Photodup [943]
Archiv fuer rassen- und gesellschaftsbiologie – Muenchen DE, 1904-1916/18, 1921-39 – 1 – gw Misc Inst [573]
Archiv fuer reformationsgeschichte – 1(1903)-52(1961) – 297mf – 9 – €566.00 – ne Slangenburg [242]
Archiv fuer religionswissenschaft – 1(1898)-37(1941) – 317mf – 9 – €605.00 – ne Slangenburg [200]
Archiv fuer saechsische geschichte – Leipzig DE, 1863-80 – 8r – 1 – gw Misc Inst [943]
Archiv fuer schweizerische geschichte – Zuerich, 1(1843)-20(1875) – 153mf – 9 – €292.00 – ne Slangenburg [949]
Archiv fuer slavische philologie – Berlin. v1-42. 1876-1929 – 480mf – 8 – mf#652 – ne IDC [460]
Archiv fuer wissenschaftliche erforschung des alten testaments – Halle, 1(1867-69) – 13mf – 9 – €25.00 – ne Slangenburg [221]
Archiv fur die naturkunde liv-, esth- und kurlands – Portland. 1972-1981 (1) 1977-1981 (5) 1977-1981 (9) – 154mf – 9 – mf#8605 – ne IDC [077]
Archiv gemeinnuetziger physischer und medicinischer kenntnisse / ed by Rahn, Johann Heinrich – Zuerich 1787-91 – (= ser Dz) – 6v on 16mf – 9 – €160.00 – mf#k/n3634 – gw Olms [610]
Archiv zur laender- und geschicht-kunde unsrer zeit / ed by Stoever, Dietrich Heinrich – Schwerin 1790 – (= ser Dz. historisch-geographische abt) – 1v on 2mf – 9 – €60.00 – mf#k/n1228 – gw Olms [900]
Archival-urkunden : documenta und probationes in causa monasterii augiae majoris – 1750 – €27.00 – ne Slangenburg [241]
Archive for history of exact sciences – Dordrecht, Netherlands 1960+ – 1,5,9 – ISSN: 0003-9519 – mf#13133 – us UMI ProQuest [500]
Archive for rational mechanics and analysis – Dordrecht, Netherlands 1980+ – 1,5,9 – ISSN: 0003-9527 – mf#13134 – us UMI ProQuest [510]
Archive of c o van der plas *see* War and decolonization in indonesia, 1940-1950
Archive of count johannes van den bosch *see* Sources for the study of colonial indonesia and dutch colonial policy
Archive of dr p j koets *see* War and decolonization in indonesia, 1940-1950
Archive of gerard jan chretien schneither *see* Sources for the study of colonial indonesia and dutch colonial policy
Archive of h j van mook *see* War and decolonization in indonesia, 1940-1950
Archive of j h van royen *see* War and decolonization in indonesia, 1940-1950
Archive of jean chretien baron baud *see* Sources for the study of colonial indonesia and dutch colonial policy
Archive of nicolaus engelhard *see* Sources for the study of colonial indonesia and dutch colonial policy
Archive of professor dr w h de vriese : concerning his investigation into the netherlands indies cultivations, 1857-1862 – 204mf – 9 – €1395.00 – mf#m106 – ne MMF Publ [580]
Archive of rudolf rocker, 1894-1958 : theoretician of anarchosyndicalism / Amsterdam. International Institute of Social History – [mf ed 2001] – 464mf – 9 – €3130.00 – (p/g in german with int in english) – mf#m490 – ne MMF Publ [320]
Archive of the amsterdam booksellers guild, 1662-1812 – 141mf – 9 – €1570.00 – (with p/g, inventory & int in english) – mf#m420 – ne MMF Publ [070]
Archive of the colonial school for girls and women, the hague, 1920-1949 – [mf ed 2004] – (= ser Women in the netherlands east indies 1) – 12r – 1 – €1320.00 – (printed inventory in dutch; int in english) – mf#mmp110 – ne Moran [376]
Archive of the conseil des troubles, 1567-76 *see* The inquisitions
Archive of the dutch consulate at nagasaki 1860-1915 *see* Japan and the west
Archive of the dutch consulate at yokohama, 1860-1870 *see* Japan and the west
Archive of the dutch legation in japan, 1870-1890 *see* Japan and the west
Archive records / Trade Union Council of South Africa – Chicago, IL: Coop Africana Microform Project CRL, 1914-69 [mf ed 19–?] – 1 – us CRL [331]
Archives / Fiji Independent News Service – 1987-1992 – (filming in progress) – mf#PMB1079 – at Pacific Mss [070]
Archives / Fiji Trades Union Congress – 1959-1995 – 26r – 1 – mf#PMB1085 – at Pacific Mss [331]
Archives : five sections covering the period 1754 – c. 1800 / Royal society of arts archives – 5 sects – 31r – 1 – £1400.00 – mf#RSA – uk World [700]
Archives / Honiara. Catholic Archdiocese – 1905-82 – 6r – 1 – mf#PMB1120 – at Pacific Mss [980]
Archives / Maryland – v1-72. 1883-1972 – 1 – us MD Archives [324]
Archives, 1920s-1974 / Losuia District Administration, Kiriwina, Trobriand Islands, Papua New Guinea – r1-2 – 1 – (available for ref) – mf#pmb1177 – at Pacific Mss [980]

Archives, 1792-1914 / Baptist Missionary Society. London – 165,931p – 1 – $6637.24 – (minutes, 1792-1914; committees, 1793-1914; home office correspondence, 1792-1914; missionary journals & correspondence, 1792-1914) – us Southern Baptist [242]

Archives, 1902-1992 / Levers Pacific Plantations Pty Ltd & Lever Solomons Ltd – r1-6 – 1 – (access under negotiation) – mf#PMB1121 – at Pacific Mss [980]

Archives, 1927-1994 / Losuia District Administration, Kiriwina, Trobriand Islands, Papua New Guinea – 6r – 1 – (available for ref) – mf#pmb1165 – at Pacific Mss [980]

Archives, 1963-2000 / Young Women's Christian Association of Fiji – r1-2 – 1 – (closed till jan 2005 then available for ref) – mf#pmb1211 – at Pacific Mss [240]

Archives, 1969-95 / Papua New Guinea Trades Union Congress – 5r – 1 – mf#PMB1117 – at Pacific Mss [331]

Archives, 1975-1999 / Solomon Islands National Union of Workers – r1-4 – 1 – (available for ref) – mf#pmb1187 – Available for reference – at Pacific Mss [331]

Archives, 1989-1999 / South Pacific and Oceania Council of Trade Unions – r1-5 – 1 – (available for ref) – mf#pmb1166 – at Pacific Mss [331]

Archives and history news – 1970 jan-1982 winter/spring – 1 – mf#615763 – us WHS [071]

Archives annuelles de la normandie : historiques, monumentales, litteraires et statistiques – Caen 1824 [mf ed Hildesheim 1995-98] – 1v on 2mf – 9 – €60.00 – gw Olms [944]

Archives annuelles de la normandie historiques, monumentales, litteraires et statistiques – Caen 1824 [mf ed Hildesheim 1995-98] – 1v on 2mf – 9 – €60.00 – 3-487-25955-9 – gw Olms [944]

Archives authority : convict registers etc – 9 – (apply to publ for details) – at Pascoe [920]

Archives Berberes *see* Publication du comite d'etudes berberes de rabat

Les archives berberes / Comite d'Etudes Berberes de Rabat – Paris. 1915-20 (1-4) – 1 – fr ACRPP [073]

Les archives berberes – Paris: Comite d'etudes berberes de Rabat [etc]. v1-4 n1/2. 1915/16-1919/20 – 1 – us CRL [960]

Archives berberes et bulletin de l'institut des hautes-etudes marocaines – Paris, 1921-1946. v1-33 – 300mf – 8 – mf#H-512c – ne IDC [956]

Archives biographiques francaises (abf1) = French biographical archive (abf1) / ed by Bradley, Susan – [mf ed 1989-91] – 1065mf (1:24) – 9 – diazo €10,060.00 (silver €11,080 isbn:978-3-598-32579-3) – ISBN-10: 3-598-32564-9 – ISBN-13: 978-3-598-32564-9 – (with printed ind) – gw Saur [944]

Archives biographiques francaises. deuxieme serie (abf2) = French biographical archive. series 2 (abf2) / Nappo, Tommaso [comp] – [mf ed 1993-96] – 664mf (1:24) in 12 installments – 9 – diazo €10,060.00 (silver €11,080 isbn: 978-3-598-33568-6) – ISBN-10: 3-598-33555-5 – ISBN-13: 978-3-598-33555-6 – (with printed ind) – gw Saur [944]

Archives biographiques francaises. deuxieme serie (abf2) supplement = French biographical archive. series 2 (abf2) supplement / Nappo, Tommaso [comp] – [mf ed 1999] – 108mf in 2 installments – 9 – diazo €2030.00 (silver €2460 isbn: 978-3-598-33504-4) – ISBN-10: 3-598-33503-2 – ISBN-13: 978-3-598-33503-7 – gw Saur [944]

Archives biographiques francaises jusqu a 1999 (abf3) = French biographical archive to 1999 (abf3) / Nappo, Tommaso [comp] – [mf ed 2001-02] – 481mf (1:24) – 9 – diazo €10,060.00 (silver 11,080 isbn: 978-3-598-34751-1) – ISBN-10: 3-598-34750-2 – ISBN-13: 978-3-598-34750-4 – (with printed ind) – gw Saur [944]

Archives bulletin – 1975 aug-1979 jun – 1 – mf#639179 – us WHS [071]

Archives d'anatomie microscopique – v1-16. 1897-1915 – 9 – $420.00 – mf#0070 – us Brook [576]

Archives d'anatomie microscopique et de morphologie experimentale – Paris, France 1968-81 – 1,5,9 – (cont by: biological structures and morphogenesis) – ISSN: 0003-9594 – mf#3419 – us UMI ProQuest [578]

Archives d'anthropologie criminelle de medecine legale et de psychologie normale et pathologique – v. 1-29. 1886-1915 – 1 – us L of C Photodup [360]

Archives de biologie – v1-25. 1880-1909 – 9 – $660.00 – (in french) – mf#0071 – us Brook [574]

Archives de la france monastique (afm) – Paris. v1-50 – 9 – €631.00 set – (vols also listed individually) – ne Slangenburg [241]

Archives de la franc-maconnerie : ou, les secrets et travaux de tous les grades: jusqu'a celui de rose-croix, y compris les grades ecossais – Paris: J G Dentu 1821 – 3mf – 9 – mf#vrl-10 – ne IDC [366]

Archives de la province de Quebec *see*
– Rapport de l'archiviste de la province de quebec pour 1929-1930
– Rapport de l'archiviste de la province de quebec pour 1936-1937

Archives de la theologie catholique – 1(1861)-8(1864) – 72mf – 9 – €137.00 – ne Slangenburg [241]

Archives de l'hotel-dieu saint-michel de roberval 1917-1922 : bibliographie / Marie des Anges, soeur – 1962 [mf ed 1978] – (= ser Bibliographies du cours...1947-66) – 2mf – 9 – (pref by pere hilaire de la perade) – mf#SEM105P4 – cn Bibl Nat [360]

Archives de l'institut pasteur d'algerie / L'Institut Pasteur d'Algerie – Algiers 1950-71 – 1,5 – ISSN: 0020-2460 – mf#565 – us UMI ProQuest [616]

Archives de l'orient latin – Paris, 1881-1884. 2v – 32mf – 9 – mf#H-2503 – ne IDC [915]

Archives de pediatrie – Oxford UK 1994+ – 1,5,9 – ISSN: 0929-693X – mf#42745 – us UMI ProQuest [618]

Les archives de thalie – ou Observations sur les sciences, les arts et la litterature, publiees par Ricord aine; pour faire suite au Journal des theatres. Paris. avr 1818-fevr 1819 – 1 – fr ACRPP [790]

Archives des maitres de l'orgue des 16e, 17e, et 18e siecles = Organ masters of the 16th, 17th and 18th centuries / ed by Guilmant, A & Pirro, A – Paris. 10v. 1898-1910 – 11 – $110.00 – us Univ Music [780]

Archives des maladies professionnelles de medecine du travail et de securite sociale – Paris, France 1968-79 – 1,5,9 – ISSN: 0003-9691 – mf#3417 – us UMI ProQuest [610]

Archives des missions scientifiques et litteraires : choix de rapports et instructions pub... – Paris: Impr nationale [mf ed 1988] – 1r – 1 – (cont by: nouvelles archives des missions scientifiques et litteraires) – mf#6845 – us Wisconsin U Libr [073]

Archives des sciences – Geneva, Switzerland 1993+ – 1,5,9 – ISSN: 0003-9705 – mf#10138 – us UMI ProQuest [500]

Archives d'histoire doctrinale et litteraire du moyen-age – 1(1926)-29(1962) – 331mf – 9 – €631.00 – ne Slangenburg [931]

Archives d'histoire du droit oriental – 1937-51 [mf ed 2001] – (= ser Christianity's encounter with world religions, 1850-1950) – 1r – 1 – (in french) – mf#2001-s044 – us ATLA [340]

Archives diplomatiques – Series 1-4. 1861-1914 – 1 – 868.00 – us L of C Photodup [025]

Archives d'ophtalmologie – Paris, France 1976-77 – 1,5,9 – (cont: archives d'ophtalmologie et revue generale d'ophtalmologie) – ISSN: 0399-4236 – mf#3397,01 – us UMI ProQuest [617]

Archives d'ophtalmologie et revue generale d'ophtalmologie – Paris, France 1881-1975 – 1,5,9 – (cont by: archives d'ophtalmologie) – ISSN: 0003-973X – mf#3397.01 – us UMI ProQuest [617]

Archives europeennes de sociologie = European journal of sociology – Cambridge UK 1977+ – 1,5,9 – ISSN: 0003-9756 – mf#11556 – us UMI ProQuest [301]

Archives francaises de pediatrie – Velizy, France 1968-93 – 1,5,9 – ISSN: 0003-9764 – mf#3418 – us UMI ProQuest [618]

Archives historiques du poitou – v1-50. 1872-1938 – 1 – $594.00 – (v51-61 1939-82 $114 [0073]) – mf#0072 – us Brook [940]

Archives historiques et litteraires du nord de la france et du midi de la belgique – 18v. 1829-57 – 1 – $264.00 – (in french) – mf#0074 – us Brook [944]

Archives, history, records, annual guides / U.S. Volleyball Association – 1916-75. 126 fiches – 9 – $115.00 – (reviews (1940-80) and official guide (1976-80) 67mf) – us Kinesology [790]

Archives information bulletin – v1 n3-v2 n4 [1979 jul-1980 oct] – 1 – mf#630966 – us WHS [071]

Archives internationales de pharmacodynamie et de therapie – Ghent, Belgium 1977-96 – 1,5,9 – ISSN: 0003-9780 – mf#11417 – us UMI ProQuest [615]

Archives israelites – Paris. 1841-1935 – 1 – fr ACRPP [939]

Archives israelites – Paris. v. 1-98. 1840-1935. Incomplete – 1 – us NY Public [939]

Archives marocaines / Mission scientifique du Maroc – Paris. 1904-34 – 1 – fr ACRPP [956]

Archives Nationales, Paris *see* Plans of paris from the archives nationales

Archives neerlandaises de physiologie de l'homme et des animaux – v1-23. 1918-38 – 9 – $360.00 – (in french) – mf#0065 – us Brook [612]

Archives of andrology – Abingdon, Oxfordshire 1989+ – 1,5,9 – ISSN: 0148-5016 – mf#14240 – us UMI ProQuest [612]

Archives of archaeology – v1-29 – 9 – $480.00 – mf#0075 – us Brook [930]

The archives of canada / Anderson, William James – S.l: s.n, 1872? – 1mf – 9 – mf#01153 – cn Canadiana [025]

Archives of clinical neuropsychology – Oxford UK 1986+ – 1,5,9 – ISSN: 0887-6177 – mf#49494 – us UMI ProQuest [616]

Archives of dermatological research = Archiv fuer dermatologische forschung – Dordrecht, Netherlands 1981+ – 1,5,9 – ISSN: 0340-3696 – mf#13135.06 – us UMI ProQuest [616]

Archives of dermatology – Chicago IL 1920+ – 1,5,9 – ISSN: 0003-987X – mf#67 – us UMI ProQuest [616]

Archives of disease in childhood – London UK 1926+ – 1,5,9 – ISSN: 0003-9888 – mf#1328 – us UMI ProQuest [618]

Archives of elkin matthews, 1811-1938 – [mf ed Chadwyck-Healey] – (= ser Archives of british and american publishers) – 1r – 1 – (incl catalogue) – uk Chadwyck [070]

Archives of emergency medicine – London UK 1984-1993 – 1,5,9 – (cont by: journal of accident and emergency medicine) – ISSN: 0264-4924 – mf#15504.02 – us UMI ProQuest [610]

Archives of environmental contamination and toxicology – Dordrecht, Netherlands 1973+ – 1,5,9 – ISSN: 0090-4341 – mf#13136 – us UMI ProQuest [362]

Archives of environmental health – Washington DC 1960+ – 1,5,9 – ISSN: 0003-9896 – mf#5341.01 – us UMI ProQuest [614]

Archives of family medicine – Chicago IL 1992-2000 – 1,5,9 – ISSN: 1063-3987 – mf#20762 – us UMI ProQuest [610]

Archives of general psychiatry – Chicago IL 1959+ – 1,5,9 – ISSN: 0003-990X – mf#1182 – us UMI ProQuest [616]

Archives of george allen and company, 1893-1915 – (= ser Archives of british and american publishers) – 27r – 1 – (with ind [2mf]) – uk Chadwyck [070]

Archives of george routledge and company, 1853-1902 – [mf ed Chadwyck-Healey] – (= ser Archives of british and american publishers) – 6r – 1 – (incl ind) – uk Chadwyck [070]

Archives of gerontology and geriatrics – Oxford UK 1985+ – 1,5,9 – ISSN: 0167-4943 – mf#42403 – us UMI ProQuest [618]

Archives of grant richards, 1897-1948 – [mf ed Chadwyck-Healey] – (= ser Archives of british and american publishers) – 72r – 1 – (with ind) – uk Chadwyck [070]

Archives of gynecology – Dordrecht, Netherlands 1981-86 – 1,5,9 – ISSN: 0170-9925 – mf#13137.02 – us UMI ProQuest [618]

Archives of harper and brothers, 1817-1914 – [mf ed Chadwyck-Healey] – (= ser Archives of british and american publishers) – 58r – 1 – (with ind) – uk Chadwyck [070]

Archives of internal medicine – Chicago IL 1908+ – 1,5,9 – ISSN: 0003-9926 – mf#68 – us UMI ProQuest [616]

Archives of kegan paul, trench, trubner and henry s king, 1858-1912 – [mf ed Chadwyck-Healey] – (= ser Archives of british and american publishers) – 27r – 1 – (with ind [2mf]) – uk Chadwyck [070]

Archives of macmillan and company, 1854-1924 : pts 1 and 2 – [mf ed Chadwyck-Healey] – (= ser Archives of british and american publishers) – 73r – 1 – (pt1: readers' report 1867-1934 8r. pt2: publishing records 1860-1921 65r. with ind) – uk Chadwyck [070]

Archives of maryland – 3,5,8,11,18,20,21,22 – 1 – mf#1518973 – us WHS [978]

Archives of maryland / Maryland Historical Society – 72v 1883-1972 – 1 – us AMS Press [978]

Archives of maryland – v1-72 – 9 – $900.00 – mf#0076 – us Brook [978]

Archives of medical research – Oxford UK 1999+ – 1,5,9 – (cont: archivos de investigacion medica) – ISSN: 0188-4409 – mf#42829 – us UMI ProQuest [616]

Archives of microbiology – Dordrecht, Netherlands 1980+ – 1,5,9 – (cont: archiv fuer mikrobiologie) – ISSN: 0302-8933 – mf#13105,01 – us UMI ProQuest [576]

Archives of neurology – Chicago IL 1959+ – 1,5,9 – ISSN: 0003-9942 – mf#1181 – us UMI ProQuest [616]

The archives of nova scotia *see* "Evangeline" and "the archives of nova scotia"

Archives of ophthalmology / American Medical Association – Chicago IL 1869+ – 1,5,9 – ISSN: 0003-9950 – mf#70 – us UMI ProQuest [617]

Archives of oral biology – Oxford UK 1959+ – 1,5,9 – ISSN: 0003-9969 – mf#49014 – us UMI ProQuest [617]

Archives of orthopaedic and trauma surgery [1989] – Dordrecht, Netherlands 1989+ – 1,5,9 – (cont: archives of orthopaedic and traumatic surgery) – ISSN: 0936-8051 – mf#13138,03 – us UMI ProQuest [617]

Archives of orthopaedic and traumatic surgery [1981] = Archiv fuer orthopaedische und unfallchirurgie – Dordrecht, Netherlands 1981-85 – 1,5,9 – (cont by: archives of orthopaedic and trauma surgery) – ISSN: 0344-8444 – mf#13138.03 – us UMI ProQuest [617]

Archives of otolaryngology – Chicago IL 1925-85 – 1,5,9 – (cont by: archives of otolaryngology – head and neck surgery; cont: archives of otolaryngology) – ISSN: 0003-9977 – mf#71.01 – us UMI ProQuest [617]

Archives of otolaryngology – head and neck surgery – Chicago IL 1986+ – 1,5,9 – (cont: archives of otolaryngology) – ISSN: 0886-4470 – mf#71,01 – us UMI ProQuest [617]

Archives of otology – Chicago IL 1869-1908 – 1,5,9 – mf#1872 – us UMI ProQuest [617]

Archives of oto-rhino-laryngology – Dordrecht, Netherlands 1981-82 – 1,5,9 – ISSN: 0302-9530 – mf#13139.05 – us UMI ProQuest [617]

Archives of pathology – Chicago IL 1926-75 – 1,5,9 – (cont by: archives of pathology and laboratory medicine) – ISSN: 0363-0153 – mf#72.01 – us UMI ProQuest [619]

Archives of pathology and laboratory medicine – Chicago IL 1976+ – 1,5,9 – (cont: archives of pathology) – ISSN: 0003-9985 – mf#72,01 – us UMI ProQuest [619]

Archives of pediatrics – Baltimore MD 1949-62 – 1 – ISSN: 0096-6630 – mf#119 – us UMI ProQuest [618]

Archives of pediatrics and adolescent medicine – Chicago IL 1994+ – 1,5,9 – (cont: american journal of diseases of children) – ISSN: 1072-4710 – mf#408,01 – us UMI ProQuest [618]

Archives of physical medicine and rehabilitation – Philadelphia PA 1920+ – 1,5,9 – ISSN: 0003-9993 – mf#392 – us UMI ProQuest [617]

Archives of plaid cymru : 1926-99+ / Plaid Cymru – 45r – 1 – £2150.00 – (the archives consist of: pamphlets, leaflets etc. publ since 1926; complete runs of both party newspapers – ddraid goch and weish nation; books publ by, and on the behalf of, plaid cymru; programs and minutes of the party conference; minutes of the meetings of the national council of plaid cymru and the internal memoranda and publ of the plaid cymru research group) – mf#APC – uk World [941]

Archives of psychiatric nursing – Philadelphia PA 1992+ – 1,5,9 – ISSN: 0883-9417 – mf#21093 – us UMI ProQuest [610]

Archives of richard bentley and son, 1829-1898 – [mf ed Chadwyck-Healey] – (= ser Archives of british and american publishers) – 116r – 1 – (with ind) – uk Chadwyck [070]

Archives of sexual behavior – Dordrecht, Netherlands 1971+ – 1,5,9 – ISSN: 0004-0002 – mf#10848 – us UMI ProQuest [150]

Archives of surgery – Chicago IL 1920+ – 1,5,9 – ISSN: 0004-0010 – mf#73 – us UMI ProQuest [617]

Archives of swan sonnenschein and company, 1878-1911 – [mf ed Chadwyck-Healey] – (= ser Archives of british and american publishers) – 25r – 1 – (with ind [1mf]) – uk Chadwyck [070]

Archives of the british conservative party – 53r 2006mf (coll) – 1,9 – (pamphlets & leaflets 1093mf c39-27713. executive committe minutes of the national union of conservative associations, 1897-1956, together with central committee minutes and annual reports 150mf c39-28970. minutes and reports of conservative party conferences, 1867-1946 134mf c39-27711. british general election campaign guides, 1885-1950 81mf c39-28971. national union gleanings and successors, 1893-1968 47r c39-27712. conservative party committee minutes, 1909-64 6r c39-28972. conference reports, 1947-63 27mf c39-28973. campaign guides, 1951-74 31mf c39-28974. conservative agents' journal, 1902-83 278mf c39-28975. conservative party conference reports, 1965-91; british general election guides, 1975-91 120mf c39-28976) – mf#C39-27710 – us Primary [941]

Archives of the british labour party – 156r 3041mf (coll) – 1,9 – (general correspondence and political records (covers much of the period fr 1873-1968) 141r 38mf c39-27691. pamphlets and leaflets (1900-69) 581mf c39-27692. national executive committee minutes, 1900-83 2052mf c39-27693. speeches and press statements (1964-73) 197mf c39-27694. the fiche nos for each release are cumulative, hence the 1st fiche in pt 6 of this coll is no 582. with guide) – mf#C39-27690 – us Primary [941]

Archives of the british liberal party – 27r 481mf (coll) – 1,9 – (pamphlets & leaflets: pt1: 1885-1911 107mf c39-27701; pt2: 1912-39 101mf c39-27702; pt3: 1940-63 124mf c39-27703; pt4: 1964-74 96mf c39-27704; national liberal federation annual reports, 1877-1936 53mf c39-27710; the liberal magazine, 1893-1950 27r c39-27720) – mf#C39-27700 – us Primary [325]

Archives of the british trades union congress – 63r 533mf (coll) – 1,9 – (the mining crisis and the general strike, 1925-26: the documentary record 22r c39-20010. trade union congress committee minutes and papers, 1922-50 25r c39-20020 – pt 1: economic committee, finance & general purposes committe etc 12r c39-20021 pt 2: colonial advisory committee, industrial welfare committee etc 13r c39-20022. trade union congress general council minute books 88mf c39-20030 – pt 1: 1921-32 36mf c39-20031 pt 2: sep 1932-dec 1946 52mf c39-20032. pamphlets and leaflets of the british tuc 445mf c39-20040 – pt 1: 1887-1930 101mf c39-20041 pt 2: 1931-47 97mf c39-20042 pt 3 1948-66 135mf c39-20043 pt 4: 1967-72 112mf c39-20044. tuc periodicals and serial publ 16r c39-20050) – mf#C39-20000 – us Primary [331]
Archives of the cambridge university press, 1669-1902 – [mf ed Chadwyck-Healey] – (= ser Archives of british and american publishers) – 11r – 1 – (with p/g by e s leedham-green) – uk Chadwyck [070]
Archives of the campaign for nuclear disarmament *see* Nuclear disarmament after the cold war
Archives of the destruction : a photographic record of the holocaust – 245mf – 9 – (15,000 photographs of the holocaust. with ind) – mf#C39-27930 – us Primary [943]
Archives of the english province of the society of jesus : from the society of jesus, london – 2r – 1 – mf#96729 – uk Microform Academic [240]
Archives of the fabian society – Hassocks, Sussex: Harvester Press, 1975-85 (mf ed) – 84r 141mf (coll) – 1,9 – (pt 1: minute books and records 1884-1918 141mf c39-27601. pt 2: minutes of the executive committee and lectures 1919-60 10r c39-27602. pt 3: correspondence of eminent persons and early material and memorials 1881-1952 12r c39-27603. pt 4: papers and records of the finance and general purposes committee 1919-64, and the fabian local societies 1919-64 15r c39-27604. pt 5: papers and records of the fabian women's group 1919-51; the society for socialist inquiry and propaganda 1931-32; the new fabian research bureau 1931-39, and other bodies 15r c39-27605. pt 6: home research committee minutes 1943-64, and papers, sect a 1930-49 18r c39-27606. pt 7: home research committee papers, sect b, 1950-64; international and commonwealth bureau minutes and papers 1940-64; london labour party and fabian regional councils 1945-62 14r c39-27607. incl printed guide) – mf#C39-27600 – us Primary [025]
Archives of the federal writers' project : printed and mimeograph publications in the surviving federal writers' project files, 1933-1943 (excludimg state guides) – 35r – 1 – mf#C35-28290 – us Primary [025]
Archives of the feltrinelli institute – 20r – 1 – (coll incl rare political periodicals and vols wh are now out-of-print. comprising bollettino dell' opposizione comunista italian (1931-33); bollettino di partito (1944-45); il domani d'italia (1901-03); il domani d'italia (1922-24); la nostra lotta (1943-45); l'ordine nuovo (1919-25); pagine rosse (1923-24); politica socialista (1933-35); prometeo (1924); rassegna communista (1921-22); il soviet (1918-22); la stato operaio (1927-39); l'unita (1924-26)) – mf#C39-20100 – us Primary [945]
The archives of the french protestant church, 1560-1889 : manuscripts held at l'eglise protestante francaise de londres – 37r – 1 – £1850.00 – mf#FPC – uk World [242]
Archives of the german embassy at washington : (american historical association project 1) / Germany. Embassy at Washington, DC – (= ser National archives coll of foreign records seized, 1941-) – 52r – 1 – mf#T290 – us Nat Archives [327]
Archives of the house of longman, 1794-1914 – [mf ed Chadwyck-Healey] – (= ser Archives of british and american publishers) – 73r – 1 – (with ind) – uk Chadwyck [070]
Archives of the independent labour party – 10pt-coll – 46r 688mf (coll) – 1,9 – (pt1-5: pamphlets & leaflets 1893-1975 512mf c39-27721. pt6: minutes & related records, national administrative council minutes & related records, 1894-1950 76mf c39-27726. pt7: minutes & related records, bracnh minutes & related records, 1892-1950 100mf c39-27727. pts8-9: the francis johnson correspondence, 1888-1950 21r c39-27728. pt10: organizational & regional records of the independent labour party) – mf#C39-27720 – us Primary [941]
Archives of the inner temple library : 1547-1970 / Inner Temple. Library – 55r – 1 – £2600.00 – mf#LIU – uk World [340]
Archives of the international institute of social history – 1r – 1 – us Primary [302]
Archives of the moravian church, bristol, 1756-1806 – 5r – 1 – mf#635 – uk Microform Academic [025]
Archives of the parliamentary labour party – 264mf – 9 – (incl minutes and records of the administrative committee, 1941-45; executive committee, 1923-27; liaison committee, 1945-68; parliamentary committee, 1951-64; parliamentary labour party, 1906-68) – mf#C39-20300 – us Primary [941]
The archives of the race relations department of the united church board for homeland ministries – 1942-76 [mf ed 2002] – 58r – 1 – $7540.00 set – (guide may be purchased separately $40) – mf#d3625 – us Amistad [305]
Archives of the royal college of physicians, 1518-1988 – 3pt – 9 – (pt1: annals of the royal college of physicians 1518-1915 [417mf] £2600; pt2: annals...1916-88 [602mf] £3550; pt3: council minutes...1836-1978 [333mf] £2000; with d/g) – uk Matthew [610]
Archives of the royal literary fund : 1790-1918 / Cross, Nigel [comp] – 145r – 1 – £5950.00 – (with printed guide) – mf#RLF – uk World [420]
Archives of the settlement movement : archives of the national federation of settlements and successors, 1899-1958 – 5pt-coll – 74r – 1 – (pt 1: minutes, reports and proceedings of central policy making groups 20r. pt 2: domestic programmes, project files on public policy and social issues 1911-61 22r. pt 3: national federation of settlements and successors, domestic programmes, selected files on nfs member houses and city federations 1800-1961 15r. pt 4: international activities of the national federation of settlements and successors, c1920-60 10r. pt 5: major figures of the settlement movement, correspondence, speeches and articles c1899-1958 7r. includes a printed guide) – mf#C36-27430 – us Primary [975]
Archives of the soviet communist party and soviet state : from the state archive of the russian federation (garf – 2 sites), the russian centre for the preservation and study of documents of most recent history (rtskhidni), and the centre for the preservation of contemporary documentation (tskhsd) – [mf ed Chadwyck-Healey] – 10,534r – 455r of opisi [finding aids] 10,079r of dela [files of docs] – 1 – (enquire for further details; tskhsd has been renamed russian state archive of contemporary history (rossiiskii gosudarstvennyi arkhiv noveishei istorii – rgani). rtskhidni has been renamed russian state archives of social and political history (rossiiskii gosudarstvennyi arkhiv sotsialno-politicheskoi istorii – rgaspi)) – State Archival Service of Russia (Rosarkhiv) and the Hoover Institution on War, Revolution and Peace – uk Chadwyck [947]
Archives of the spanish government of west florida, 1782-1816 – 7r – 1 – mf#T1116 – us Nat Archives [978]
Archives of the tongan judiciary / Tonga. Ministry of Justice – 1905-1995 – 21r – 1 – mf#PMB1088 – at Pacific Mss [340]
Archives of the work projects administration and predecessors, 1933-1943 : the final state reports, 1943 – 13r (coll) – 1 – (pt 1: final reports of the state program, 6r. pt 2: final state reports for the federal music program, the federal art program, the federal crafts program, the museum and visual aids program, the federal theater program and the federal writers program, 7r) – us Primary [350]
Archives of toxicology = Archiv fuer toxikologie – Dordrecht, Netherlands 1930+ – 1,5,9 – (cont: archiv fuer toxikologie) – ISSN: 0340-5761 – mf#13140,02 – us UMI ProQuest [615]
Archives of useful knowledge – La Salle IL 1810-13 – 1 – mf#4419 – us UMI ProQuest [630]
Archives of virology – Dordrecht, Netherlands 1983+ – 1,5,9 – ISSN: 0304-8608 – mf#13263,01 – us UMI ProQuest [576]
Archives ou correspondance inedite de la maison d'orange-nassau – v1-23 – 9 – $498.00 – mf#0077 – us Brook [940]
Archives parlementaires de 1787 a 1860 : recueil complet des debats legislatifs et politiques des chambres francaises – v1-106. 1800-37 – 9 – $2484.00 – (v1-82 1787-97 $1980 [0067]. v82S-88 1794 $267 [0068]. v107-127 1837-39 $873 [0069]. in french) – mf#0066 – us Brook [342]
Archives parlementaires de 1787-1860 / France. Chambres francaises – Paris: premiere serie: 1787-99. 4 jan 1794 (1-82) – 1 – (deuxieme serie: 1799-1860. 13 dec 1799-juil 1839 (1-126)) – fr ACRPP [323]
Archives philosophiques, politiques et litteraires – Paris. juil. 1817-18 – 1 – fr ACRPP [073]
Archives pour servir... : l'etude de l'histoire, des langues, de la geographie et de l'ethnographie de l'asie orientale / T'oung, Pao – Leiden, 1890-1899, v1-10; S 2, 1900-1944, v1-37 – 475mf – 9 – mf#391c – ne IDC [915]
Archives, records, reference material and conference reports / Council for National Cooperation in Aquatics – 1951-72.32 fiches – 9 – (1974-80: biennial conference reports 8mf) – us Kinesology [790]
Archives royales de mari – Paris. v1-6. 1946-1953 – 8 – €42.00 – (1 lettres publ par g dossin, paris 1946 5mf. 2 lettres publ par ch f jean, paris 1941 5mf. 3 lettres publ par j r kupper, paris 1948 3mf. 4 lettres publ par g dossin, paris 1951 3mf. 5 lettres publ par g dossin, paris 1951 3mf. 6 lettres publ par j r kupper, paris 1953 3mf) – ne Slangenburg [240]
Archivio biografico italiano (abi1) = Italian biographical archive (abi1) / Nappo, Tommaso [comp] – [mf ed 1987-90] – 1046mf (1:24) – 9 – diazo €10,060.00 (silver €11,080 ISBN: 978-3-598-31520-6) – ISBN-10: 3-598-31540-6 – ISBN-13: 978-3-598-31540-4 – (with printed ind) – gw Saur [945]
Archivio biografico italiano. nuova serie (abi2) = Italian biographical archive. series 2 (abi2) / Nappo, Tommaso [comp] – [mf ed 1991-94] – 710mf (1:24) – 9 – diazo €10,060.00 (silver €11,080 isbn: 978-3-598-33154-1) – ISBN-10: 3-598-33140-1 – ISBN-13: 978-3-598-33140-4 – (with printed ind) – gw Saur [945]
Archivio biografico italiano. nuova serie (abi2). supplemento = Italian biographical archive. new series (abi2) supplement / Nappo, Tommaso [comp] – [mf ed 1997] – 95mf (1:24) – 9 – diazo €2030.00 (silver €2460 isbn: 978-3-598-33331-6) – ISBN-10: 3-598-33330-7 – ISBN-13: 978-3-598-33330-9 – (with printed ind) – gw Saur [945]
Archivio biografico italiano sino al 1996 (abi3) = Italian biographical archive to 1996 (abi3) / Nappo, Tommaso [comp] – [mf ed 1998-2000] – 458mf (1:24) in 12 installments – 9 – diazo €10,060.00 (silver €11,080 isbn: 978-3-598-34301-8) – ISBN-10: 3-598-34300-0 – ISBN-13: 978-3-598-34300-1 – (with printed ind) – gw Saur [945]
Archivio biografico italiano sino al 2001 (abi4) = Italian biographical archive to 2001 (abi4) / Nappo, Tommaso [comp] – [mf ed 2002-04] – 518mf (1:24) in 12 installments – 9 – diazo €10.060.00 (silver €11,080 isbn: 978-3-598-35081-8) – ISBN-10: 3-598-35080-5 – ISBN-13: 978-3-598-35080-1 – (with printed ind) – gw Saur [945]
Archivio glottologico italiano – v1-65. 1873-1980 – 9 – $720.00 – (in italian) – mf#0078 – us Brook [440]
Archivio per lo studio delle tradizioni popolari – v1-24. 1882-1909 – 1 – $360.00 – (in italian) – mf#0079 – us Brook [390]
Archivio storico dell' arte – Roma. v1-7 1888-1894; v1-3 1895-1897 – 157mf – 9 – (with ind) – mf#0-1210 – ne IDC [700]
Archivio storico dell'arte – Rome, 1888-1897 [mf ed Chadwyck-Healey] – (= ser Art periodicals on microform) – 5r – 1 – uk Chadwyck [700]
Archivio storico per la sicilia orientale – v1-35 1904-39 – 1 – $240.00 – mf#0080 – us Brook [945]
O archivo : revista destinada a vulgarizacao de documentos geographicos e... – Cuiaba, MT. abr 1906 – (= ser Ps 19) – mf#P17,02,118 – bl Biblioteca [900]
Archivo biografico de espana, portugal e iberoamerica 1960-1995 [abepi3] = Spanish, portuguese and latin-american biographical archive 1960-1995 [abepi3] / Herrero Mediavilla, Victor [comp] – [mf ed 1996-98] – 471mf (1:24) in 12 installments – 9 – diazo €10,060.00 (silver €11,080 isbn: 978-3-598-34031-4) – ISBN-10: 3-598-34030-3 – ISBN-13: 978-3-598-34030-7 – (with printed ind) – gw Saur [946]
Archivo biografico de espana, portugal e iberoamerica [abepi1] = Spanish, portuguese and latin american biographical archive [abepi1] / Herrero Mediavilla, Victor [comp] – [mf ed 1986-89] – 1144mf (1:24) – 9 – diazo €10,060.00 (silver €11,080 isbn: 978-3-598-32045-3) – ISBN-10: 3-598-32030-2 – ISBN-13: 978-3-598-32030-9 – (with printed ind) – gw Saur [946]
Archivo biografico de espana, portugal e iberoamerica [abepi4] = Spanish, portuguese and latin-american biographical archive to 2001 [abepi4] / Herrero Mediavilla, Victor [comp] – [mf ed 2002-04] – 700mf (1:24) in 12 installments – 9 – diazo €10,060.00 (silver €11,080 isbn: 978-3-598-35051-1) – ISBN-10: 3-598-35050-3 – ISBN-13: 978-3-598-35050-4 – (with printed ind) – gw Saur [946]
Archivo biografico de espana, portugal e iberoamerica. nueva serie [abepi2] = Spanish, portuguese and latin-american biographical archive. series 2 / Herrero Mediavilla, Victor [comp] – [mf ed 1991-93] – 1018mf (1:24) – 9 – diazo €10,060.00 (silver €11,080 isbn: 978-3-598-32964-7) – ISBN-10: 3-598-32977-6 – ISBN-13: 978-3-598-32977-7 – (with printed ind) – gw Saur [946]
Archivo de historia y variedades / Febres Cordero, Julio – Caracas, 1930-31; Madrid: Razon y Fe, 1932. 2v – 1 – sp Bibl Santa Ana [946]
Archivo de la corazon de aragon – Minorca, Spain. v1-3. no date – 2r – us UF Libraries [324]
Archivo de la nacion : correspondencia de lord strangford y de la estacion naval en el rio de la plata con el gobierno de buenos aires, 1810-1812 / Bayle, Constantino – Madrid: Razon y Fe, 1943 – 1 – sp Bibl Santa Ana [355]
El archivo de los condes de canilleros : sep de la revista hidalguia / Munoz de San Pedro, Miguel – enero-marzo 1954 n4 – sp Bibl Santa Ana [020]
Archivo de ruben dario / Ghiraldo, Alberto – Buenos Aires, Argentina. 1943 – 1r – us UF Libraries [440]
Archivo del general miranda / Bayle, Constantino – Caracas, Madrid: Razon y Fe, 1931. 6v – 1 – sp Bibl Santa Ana [355]
Archivo del general miranda : revolucion francesa, tomo 9 a 12 / Bayle, Constantino – Caracas, 1931-32; Madrid: Razon y Fe, 1933 – 1 – sp Bibl Santa Ana [944]
Archivo diplomatico da independencia / Brazil. Ministerio das Relacoes Exteriores – Rio de Janeiro, Brazil. v1-6. 1922 – 2r – us UF Libraries [972]
Archivo documental espanol / Academia de la Historia Madrid – v1-10. 1950-59 – 1 – $120.00 – (in spanish) – mf#0003 – us Brook [946]
Archivo Extremeno *see* Documentos historicos referentes a extremadura
Archivo extremeno – v. 1-4. 1908-11 – 9 – sp Bibl Santa Ana [010]
Archivo General De Indias *see*
- Catalogo de los fondos cubanos
- Catalogo de pasajeros a indias durante los siglos 16, 17 y 18

El archivo general de indias de sevilla / Torre Revello, Jose – Buenos Aires, 1929; Madrid: Razon y Fe, 1931 – 1 – sp Bibl Santa Ana [305]
Archivo General de la Nacion *see* Acuerdos del extinguido cabildo de buenos aires
Archivo general de la nacion : acuerdos del extinguido cabildo de buenos aires. serie 2, tomo 9 / Bayle, Constantino – Buenos Aires, 1931; Madrid: Razon y Fe, 1932 – 1 – sp Bibl Santa Ana [972]
Archivo general de la nacion : universidad autonoma de mexico. nuevos documentos relativos a los bienes de hernando cortes, 1547-1947. tomo 2. mexico, 1946 / Bayle, Constantino – Madrid: Razon y Fe, 1948 – 1 – sp Bibl Santa Ana [972]
Archivo Historico Nacional *see* Consejo de castilla
Archivo leonessano : documenti riguardanti la vita e il culto di san giuseppe de leonessa, roma 1965 / Chiaretti, Giuseppe – Madrid: Graf. Calleja, 1966 – 1 – sp Bibl Santa Ana [240]
Archivo litterario – Sao Paulo, SP: Typ Imparcial de Joaquim Roberto de Azevedo Marques, ago-set 1865; mar-abr 1866; set 1867; maio-jun, out 1868 – (= ser Ps 19) – mf#P17,02,240 – bl Biblioteca [440]
Archivo Ministerio de Hacienda MS. *see* Explicacion de los estados que forman la balanza del comercio reciproco que hizo espana...en 1795
Archivo Municipal, Serradilla *see* Carta real por la que se exime a serradilla de la jurisdiccion de plasencia (24 de noviembre de 1557)
Archivo Nacional De Cuba *see*
- Catalogo de la exposicion fortografica vida de mat...
- Catalogo de los fondos
- Nuevos papeles sobre la toma de la habana por los...
- Papeles sobre la toma de la habana por los inglese...

Archivo santander – Bogota, Colombia. v1-24. 1913-1932 – 5r – us UF Libraries [972]
Archivos da palestra scientifica do rio de janeiro – Rio de Janeiro, RJ. 1858 – (= ser Ps 19) – bl Biblioteca [079]
Archivos de investigacion medica – Mexico City 1973-80 – 1,5,9 – ISSN: 0066-6769 – mf#8934 – us UMI ProQuest [610]
Archivos de medicina – Rio de Janeiro, RJ: Typ Commercial, maio-jun 1874 – (= ser Ps 19) – mf#P17,01,80 – bl Biblioteca [610]
Los archivos de salta y jujuy / Olguin, Eduardo Fernandez – Madrid: Razon y Fe, 1927 – 1 – sp Bibl Santa Ana [025]
Archivos del folklore cubano – Habana, Cuba. v1-5. 1930 – 1r – us UF Libraries [390]
Archivos del folklore cubano – Havana. 5v. 1924-30 – 1 – us L of C Photodup [025]
Archivos historicos de puerto rico / Canedo, Lino Gomez – San Juan, Puerto Rico. 1964 – 1r – us UF Libraries [972]
Archivos venezolanos de puericultura y pediatria – Caracas, Venezuela 1950-52 – 1 – ISSN: 0004-0649 – mf#621 – us UMI ProQuest [618]
Archivum franciscanum historicum – 1(1908)-40(1947) – 585mf – 9 – €1115.00 – ne Slangenburg [241]

Archivum historicum societatis iesu – Rome, Italy 1932-84 – 1,5,9 – mf#7071 – us UMI ProQuest [940]
Archivum romanicum / ed by Bertoni, G – Geneve. v1-25. 1917-1941 – 251mf – 8 – mf#H-301 – ne IDC [460]
Archon – v17 n2-v32 n2 (1951 dec-1978), v40 n1 (1990 spring/summer), v41 n1 (1991 spring/summer), v44-v45 (1993 spring-1997 spring/summer), 1993/94 [1]; v2 n1-v3 n32 (1896 sep 15-1898 may 31) [2] – 1 – mf#498179 [1]; 710162 [2] – us WHS [071]
The arch-satirist : [novel] / Williams, Frances Fenwick – Toronto: McLeod & Allen, 1910 – 5mf – 9 – 0-665-65377-8 – (ill by charles copeland) – mf#65377 – cn Canadiana [830]
Archys life of mehitabel : verse / Marquis, Don – 1st ed. Garden City NY: Doubleday, Doran & Co 1933 [mf ed 1986] – 1r – 1 – (filmed with: fear / mosso, a) – mf#1670 – us Wisconsin U Libr [810]
Arcieri, Giovanni P *see* The circulation of the blood
Arciero, Paul J *see* Influence of age and caffeine on resting metabolic rate, blood pressure, and mood state in younger and older individuals
Arcila Farias, Eduardo *see*
– Economia colonial de venezuela
– Regimen de la ecomienda de venezuela
Arcila Robledo, Gregorio *see* Constelacion de celebres terciarios
Arcilla y pajaro – Caceres, 1953 – 5 – sp Bibl Santa Ana [073]
Arcimegas, German *see* Los demanes en la conquista de america, buenos aires, 1943
Arcin, Andre *see* La guinee francaise
Arciniega, Rosa *see*
– Don pedro de valdivia conquistador de chile
– Francisco pizarro. biografia del conquistador del peru
– Pizarro biografia del conquistador del peru
Arciniegas, German *see*
– Biografia del caribe
– Caballero de el dorado
– Caribbean
– Germans in the conquest of america
Arciniegas, Ismael Enrique *see* Ramancero de la conquista y la colonia...
Arco iris / Dominguez, Blanca – New York, NY. 1964 – 1r – us UF Libraries [972]
d'Arco, Patrick H *see* Clinical, functional, and radiographic assessment of the conventional and modified boyd-anderson surgical procedures for repair of distal biceps tendon ruptures
Arcocha, Juan *see* Muertos andan solos
Arcoiris / Fina Garcia, Francisco – Santiago, Cuba. 1961 – 1r – us UF Libraries [972]
Arconada, Mariano *see* Benito arias montano y aubrey f.g. beel
Arcos, Francisco de *see* Vida de la venerable maria de jesus
Arcos, Marcos De Noronha E Brito, Conde De, *see* Ultimo vice-rei do brasil
Arctander, John William *see* The apostle of alaska
Arctic – Calgary AB 1948+ – 1,5,9 – ISSN: 0004-0843 – mf#1828 – us UMI ProQuest [919]
Arctic and alpine research – Boulder CO 1972-97 – 1,5,9 – (cont by: arctic, antarctic, and alpine research) – ISSN: 0004-0851 – mf#6917.01 – us UMI ProQuest [550]
Arctic, antarctic, and alpine research – Boulder CO 1999+ – 1,5,9 – (cont: arctic and alpine research) – ISSN: 1523-0430 – mf#6917,01 – us UMI ProQuest [550]
Arctic anthropology – Madison WI 1962+ – 1,5,9 – ISSN: 0066-6939 – mf#2374 – us UMI ProQuest [301]
Arctic bibliography – Calgary AB 1953-75 – 1 – ISSN: 0066-6947 – mf#5174 – us UMI ProQuest [990]
An arctic boat journey in the autumn of 1854 / Hayes, Isaac Israel – Boston: Brown & Taggard, 1860 [mf ed 1984] – 5mf – 9 – 0-665-44954-2 – (a partial account of the second grinnell expedition) – mf#44954 – cn Canadiana [919]
Arctic Club of America *see* Bulletin
Arctic experiences : containing capt george e tyson's wonderful drift on the ice-floe / ed by Blake, Euphemia Vale – New York: Harper, 1874 [mf ed 1979] – 6mf – 9 – 0-665-00172-X – (incl ind) – mf#00172 – cn Canadiana [919]
Arctic exploration : with information respecting sir john franklin's missing party / Rae, J – London, 1855. v25 – 1mf – 9 – mf#N-357 – ne IDC [919]
Arctic explorations : the second grinnell expedition in search of sir john franklin, 1853, 1854, 1855 / Kane, E K – Philadelphia, 1856. 2v – 18mf – 9 – mf#N-279 – ne IDC [919]
The arctic home in the vedas : being also a new key to the interpretation of many vedic texts and legends / Tilak, Bal Gangadhar – Poona City: Kesari, [1903?] – (= ser Samp: indian books) – us CRL [490]
Arctic Missions, Inc *see*
– Arctic news
– Arctic voice
– Family news
Arctic news / Arctic Missions, Inc – Portland OR: Arctic Missions, Inc [irreg] [mf ed 2006] – 1964-66 on 1r – 1 – (lacks iss?]; publ in: portland, or, 1964-65; in gresham, or, 1965-66; cont: family news (portland, or) [2006i-s002]; cont by: arctic voice [2006i-s004]) – mf#2006i-s003 – us ATLA [240]
The arctic news – 1941-76 – (= ser American indian periodicals... 1) – 13mf – 9 – $115.00 – us UPA [305]
Arctic refueler – 1992 apr-1993 oct – 1 – mf#2713517 – us WHS [071]
Arctic researches and life among the esquimaux *see* Life with the esquimaux
Arctic searching expedition : a journal of a boat-voyage through rupert's land and the arctic sea, in search of the discovery ships under command of sir john franklin / Richardson, J – London: Longman, Brown, Green and Longmans, 1851. 2v – 17mf – 9 – mf#N-369 – ne IDC [919]
Arctic soldier : the alaskan military magazine – 1984 winter-1993 spring – 1 – mf#604686 – us WHS [071]
Arctic sounder – v1 n22-26; v2 n1-7,9-11,21-1925; v3 n1-14; [1987 jan; 21-mar 18, apr 1-jun 24, jul 22-aug 19; 1988 jan 20-mar 16, 30-sep] – 1 – mf#1670001 – us WHS [071]
Arctic star – 1992 jan 10-1993 sep 17 – 1 – mf#1726461 – us WHS [071]
Arctic temperatures and exploration / Jenkins, Stuart – [s.l: s.n, 1888?] [mf ed 1984] – 1mf – 9 – 0-665-44935-6 – mf#44935 – cn Canadiana [919]
Arctic voice / Arctic Missions, Inc – Gresham OR: Arctic Missions [qrtly irreg] [mf ed 2006] – v16 n4-v18 n2 (dec 1967-jun/jul 1969); v3-5 n2 (1970-fall 1973); [n1] 1974-aug 1988) [gaps] on 1r – 1 – (some iss lacking; damaged: (1975) [cover of summer iss]; iss fr dec 1967-jun/jul 1969 called v16-18; fr 1970-73 called v3-5; vol designation dropped fr 1974-88; cont: arctic news (portland, or) [2006i-s003]; cont by: interaction (boring, or) [2006i-s005]) – mf#2006i-s004 – us ATLA [240]
Arcturus : a canadian journal of literature and life – Toronto: J C Dent. v1 n1-24. jan 15-jun 25 1887// (wkly) – 1r – 1 – Can$110.00 – cn McLaren [071]
Arcturus : a journal of books and opinion – v1-3. 1840-42 – 1 – us AMS Press [800]
Arcturus – La Salle IL 1840-42 – 1 – mf#4153 – us UMI ProQuest [420]
Arcudius, P *see*
– De concordia ecclesiae occidentalis et orientalis
– Opuscula aurea theologica...circa processionem spiritus sancti
Arcus aliquot triumphal : et monimenta victor classicae, in honor invictissimi illustriss jani austriae victoris non quieturi / Sambucus, J – Antverpiae: Apud Philippum Gallaeum, 1572 – 1mf – 9 – mf#O-745 – ne IDC [700]
d'Arcy, Charles Frederick *see*
– Christian ethics and modern thought
– God and freedom in human experience
– Idealism and theology
Arcy, Charles Frederick d' *see* A short study of ethics
Ard news letter – 1979 apr/may – 1 – mf#4877647 – us WHS [071]
Ardagh, Alice Maud *see* Tangled ends
Ardant du Picq, Charles Pierre *see* La langue songhay, dialecte dyerma
Ardelian, Elena Lukinichna *see* Dva goda v kambodzhe
Ardelt, Margaret E *see* Ventilatory responsiveness to acetazolamide during normoxic and hypoxic rest and exercise
Arden, A *see* Banning and arden's reports of patent cases in the u.s. circuit courts
Arden of feversham – London, England. 1887 – 1r – us UF Libraries [025]
Ardener, Edwin *see* Coastal bantu of the cameroons
Ardenne de Tizac, Andree Francoise Cardine d' *see* French writer andree viollis speaks in paris about the admirable defense of the spanish capital
Ardennes campaign statistics : 16 dec 1944-19 jan 1945 / U.S. Army. Office of the Chief of Military History – 1952 – 1 – us L of C Photodup [977]
The ardent pilgrim : an introduction to the life and work of mohammad iqbal / Ikabala Singha – Bombay: Orient Longmans; New York: Longmans, Green & Co, 1951 – (= ser Samp: indian books) – us CRL [920]
Ardila B, Jose Joffre *see* Desarrollo del sistema de transportes en colombia
Ardlethan beckom times – Ardlethan, jan 1964-dec 1968 – 2r – A$124.30 vesicular A$135.30 silver – at Pascoe [079]
Ardovino, Patricia S *see* The meaning of leisure experience in the lives of adult male offenders and former offenders with mental retardation
Ardrossan and saltcoats herald – Ardrossan, Scotland 2 jan 1854- [mf 1986-] – 1 – uk British Libr Newspaper [072]
Ardrossan & saltcoats herald (ardrossan, scotland : 1993) – Ardrossan: Guthrie Newspaper Group 1993- (wkly) [mf ed 7 jan 1994-] – 1 – (cont: ardrossan & saltcoats herald, arran and west coast advertiser) – ISSN: 0963-4088 – uk Scotland NatLib [072]
Ardrossan, saltcoats & stevenston standard – [Scotland] North Ayrshire, Ardrossan: J Cran 1 jun 1900-10 aug 1901 (wkly) [mf ed 2004] – 1r – 1 – uk Newsplan [072]
Arduino terzi : memoire franciscane nella valle... / Barrado Manzano, Arcangel – Madrid: Archivo Ibero-Americano, 1959 – 1 – sp Bibl Santa Ana [240]
Arduino terzi : san fabiano de la foresta... / Barrado Manzano, Arcangel – Madrid: Archivo Ibero-Americano, 1959 – 1 – sp Bibl Santa Ana [240]
Arduino terzi : san francisco d'assisi a roma... / Barrado Manzano, Arcangel – Madrid: Archivo Ibero-Americano, 1959 – 1 – sp Bibl Santa Ana [240]
Are anglican orders valid? / MacDevitt, John – Dublin: Sealy, Bryers and Walker; New York: Benziger 1896 [mf ed 1986] – 1mf – 9 – 0-8370-6998-X – mf#1986-0998 – us ATLA [242]
Are cathedral institutions useless? – Eton, England. 1838 – 1r – us UF Libraries [240]
Are christ and belial united? are the church and the world agreed? : a sermon preached january 22nd, 1888 / Adams, Henry – Yarmouth, NS?: C Carey, 1888? – 1mf – 9 – mf#09101 – cn Canadiana [242]
Are foreign missions doing any good? – London, England. 1887 – 1r – us UF Libraries [240]
Are health educators socialized to perceive role modeling as a professional responsibility / Scott, Lisa A – Purdue University, 1996 – 1mf – 9 – mf#HE 570 – us Kinesology [613]
Are pre-millennialists right? : or, reasons for believing in the pre-millenial advent of christ / Kellogg, Samuel Henry – Chicago: F H Revell [1885?] [mf ed 1989] – 1mf – 9 – 0-7905-2230-6 – mf#1987-2230 – us ATLA [240]
Are roman catholics forbidden to read the holy scriptures – London, England. 18– – 1r – us UF Libraries [241]
Are secret societies a blessing or a curse? : an address / Carradine, Beverly – Chicago: National Christian Assoc, 1891 [mf ed 1992] – 1mf – 9 – 0-524-02976-8 – mf#1990-0763 – us ATLA [242]
Are the critics right? : historical and critical considerations against the graf-wellhausen hypothesis = Historisch-kritische bedenken gegen die graf-wellhausensche hypothese von einem frueheren anhaenger / Moeller, Wilhelm – 2nd ed. London: Religious Tract Soc 1903 [mf ed 1985] – 1mf – 9 – 0-8370-4457-X – (incl ind. english trans fr german by clarke huston irwin; int by conrad von orelli) – mf#1985-2457 – us ATLA [221]
Are the effects of use and disuse inherited? : an examination of the view held by spencer and darwin / Ball, William Platt – London, 1890 – (= ser 19th c evolution & creation) – 2mf – 9 – mf#1.1.4249 – uk Chadwyck [575]
Are there south africans? / Hancock, William Keith – Johannesburg, South African Institute of Race Relations, 1966 – (= ser Alfred and winifred hoernle memorial lecture) – 1mf – 9 – (incl bibl ref) – mf#Sc Micro F-1429 – Located: NYPL – us Misc Inst [321]
Are we immortal? / Emberson, Frederick C – Montreal: [s.n, 189-?] [mf ed 1993] – 2mf – 9 – 0-665-91439-3 – mf#91439 – cn Canadiana [870]
Are we justified in distinguishing between an altered and an unaltered augustana as the confession of the lutheran church? / Neve, Juergen Ludwig – Burlington IA: German Literary Board 1911 [mf ed 1993] – 1mf – 9 – 0-524-06651-5 – mf#1991-2706 – us ATLA [242]
Are you afraid to die? – Dublin, Ireland. 18– – 1r – us UF Libraries [240]
Are you forgiven? / Ryle, J C – Ipswich, England. 1854 – 1r – us UF Libraries [240]
Are you going to heaven? / White, J Metcalfe – Dublin, Ireland. 18– – 1r – us UF Libraries [240]
Are you holy? / Ryle, J C – Ipswich, England. 1855 – 1r – us UF Libraries [240]
Are zionism and reform judaism incompatible? – New York, 1943 – 1mf – 9 – mf#J-28-1 – ne IDC [270]
Area development site and facility planning – Westbury NY 1966+ – 1,5,9 – ISSN: 1048-6534 – mf#8230 – us UMI ProQuest [710]
Area editions: asia and pacific / U.S. Foreign Broadcast Information Service – 1 – us L of C Photodup [950]
Area editions: communist china / U.S. Foreign Broadcast Information Service – 1 – us L of C Photodup [951]
Area editions: eastern europe / U.S. Foreign Broadcast Information Service – 1 – us L of C Photodup [949]
Area editions: latin america / U.S. Foreign Broadcast Information Service – 1 – us L of C Photodup [972]
Area editions: middle east and africa / U.S. Foreign Broadcast Information Service – 1 – us L of C Photodup [956]
Area editions: soviet union / U.S. Foreign Broadcast Information Service – 1 – us L of C Photodup [947]
Area editions: western europe / U.S. Foreign Broadcast Information Service – 1 – us L of C Photodup [940]
Area file of the naval records collection, 1775-1910 / U.S. Navy – (= ser Naval records coll of the office of naval records and library) – 414r – 1 – (with printed guide) – mf#M625 – us Nat Archives [355]
Area handbook for angola / Herrick, Allison Butler – Washington, DC. 1967 – 1r – us UF Libraries [960]
Area handbook for burundi / Mcdonald, Gordon C – Washington, DC. 1969 – 1r – us UF Libraries [960]
Area handbook for colombia / American University (Washington, DC) Foreign Area Studies Division – Washington, DC. 1964 – 1r – us UF Libraries [972]
Area handbook for mozambique / Herrick, Allison Butler – Washington, DC. 1969 – 1r – us UF Libraries [960]
Area handbook for rwanda / Nyrop, Richard F – Washington, DC. 1969 – 1r – us UF Libraries [960]
Area handbook for tanzania / Herrick, Allison Butler – Washington, DC. 1968 – 1r – us UF Libraries [960]
Area handbook for venezuela / American University (Washington, DC) Foreign Area Studies Division – Washington, DC. 1964 – 1r – us UF Libraries [972]
Area news – Griffith, sep 1929-jun 1932, jan 1934-dec 1968, jan 1969-jun 1997 – A$808.50 vesicular A$924.00 silver – (incorporated in: riverina daily news) – at Pascoe [079]
Area studies program / Human Relations Area Files – 1980. Five major groups: Africa; the Americas; Asia; Europe and Oceania. At least 3 modules per major group – 9 – us HRAF [301]
Area trends in employment and unemployment – Washington DC 1974+ – 1,5,9 – ISSN: 0004-0916 – mf#9152 – us UMI ProQuest [331]
Areal-anzeiger *see* Duesseldorfer lokal-zeitung
Areas of concern – Bryn Mawr PA 1973-74 – 1 – ISSN: 0044-8788 – mf#7484 – us UMI ProQuest [333]
Arechederreta y Escalada, Juan Bautista de *see* Catalogo de los colegiales del insigne
Arede, Joao Domingues *see* Estudos regionaes
Arelatensis episcopus, regula sanctarum virginum aliaque opuscula ad sanctimoniales directa / Caesarius, S – Bonn, 1933 – 2mf – 8 – €5.00 – ne Slangenburg [240]
Areler volkszeitung – Arel (B) (Arlon, Aarlen) 1941 9 aug-1944 19 aug [gaps] [mf ed 2005] – 1 – gw Mikrofilm [074]
Arellano Moreno, Antonio *see* Guia de historia de venezuela, 1492-1945
Arena – 1898 jul/dec [1]; 1981 jul-1985 oct [2] – 1 – mf#1167773 [1]; 963369 [2] – us WHS [071]
Arena – La Salle IL 1889-1909 – 1 – mf#3874 – us UMI ProQuest [240]
The arena – v1-41. dec 1889-aug 1909 – 1 – us L of C Photodup [073]
The arena and the throne / Townsend, Luther Tracy – Boston: Lee and Shepard, 1874 – 1mf – 9 – 0-524-05639-0 – mf#1992-0494 – us ATLA [210]
Arena, Antoine *see* Antonius de arena provencalis, de bragardissima villa de soleriis
Arena, Antonius *see* ...Bassas dansas...augmentatus...
Arena star – 1874 aug 14 [1]; 1878 jun 21-1880 apr 2; 1880 apr 9-1883 jun 29; 1883 jul 6-nov 23 [2] – 1 – mf#1139530 [1]; 962726 [2] – us WHS [071]
Arenal de Garcia Carrasco, Concepcion *see* Obras completas
Arenal, Humberto *see*
– Tiempo ha descendido
– Vuelta en redondo
Arenales / Figueroa, Loida – Barcelona, Spain. 1961 – 1r – us UF Libraries [972]
Arenas, Braulio *see* La promesa en blanco
Arenas del uruguay / Fajardo, Heraclio C – Buenos Aires, Argentina. 1862 – 1r – us UF Libraries [972]
Arenas Lopez, Anselmo *see*
– Curso de historia de espana
– Curso de historia de espana, tomo 1
– Curso de historia general
– La lusitania celtiberica
– Programa de examen de geografia
– Resumen de geografia
– Resumen de historia de espana

Arendt, Erich *see* Heroes: narraciones para soldados

Arens, Bernard *see* Das katholische zeitungswesen in ostasien und ozeanien

Arens, Eduard *see* Das geistliche jahr / geistliche lieder

Arensohn, Moses Solomon *see* Moreh nevukhe ha-dor

Arent de gelder : sein leben und seine kunst – Haag, v4. 1914 – 4mf – 9 – mf#O-518 – ne IDC [700]

Arentz, Erik *see* Angkor, fransk indo-kina

Arenz, Karl *see* Die entdeckungsreisen in nord- und mittel-afrika von richardson, overweg, barth und vogel

L'areopage / Cuthbert, Ross – Quebec: J Neilson, 1803 [mf ed 1971] – 1r – 5 – mf#SEM16P19 – cn Bibl Nat [071]

Areopagus – Shatin, China 1987-97 – 1,5,9 – (cont: update: a quarterly journal on new religious movements) – ISSN: 1011-8101 – mf#16749 – us UMI ProQuest [290]

Aresi, P *see* Delle imprese sacre con utili e dilettevoli discorsi accompagnate, libro prima

Aretas 4, koenig der nabataeer : eine historisch-exegetische studie zu 2 kor 11, 32 f / Steinmann, Alphons – Freiburg i.B, St Louis MO: Herder 1909 [mf ed 1989] – 1mf – 9 – 0-7905-0441-3 – (incl bibl ref) – mf#1987-0441 – us ATLA [227]

Der arethascodex paris gr 41... *see* Die altercatio simonis iudaei et theophili christiani

Arethusa – Baltimore MD 1968+ – 1,5,9 – ISSN: 0004-0975 – mf#6613 – us UMI ProQuest [450]

Aretino : oder dialog ueber malerei von lodovico dolce. nach der ausgabe vom jahre 1557 aus dem italienischen uebersetzt von cajetan cerri / [Dolce, Lodovico]; ed by Eitelberger von Edelberg, R – Wien, 1871. v2 – 2mf – 9 – mf#O-517 – ne IDC [700]

Aretino, Leonardo *see*
- Bellum punicum 1...
- Opuscula 30...

Aretino, Pietro *see* Coloquio de las damas

Arets / Brawer, A J – Tel-Aviv, Israel. 1927 – 1r – us UF Libraries [939]

Arets / Saphir, Elijah – Jaffa, Israel. 1911 – 1r – us UF Libraries [939]

Arets hogtider / ed by Nilsson, Martin P – Stockholm: A Bonnier 1938 [mf ed Bloomington IN: Indiana Uni Lib, Preservation Dept 1984] – 153p on 1r – 1 – us Indiana Preservation [390]

Arets veha-'avodah – Yafo, Israel. n1-5. 1918-1919 – 1r – us UF Libraries [939]

Arevalo, Faustino *see*
- Caelii sedulii opera
- Draconti carmina
- Hymnodia hispanica
- Iuvenci carmina
- M aureli clementis prudenti carmina
- Maurelio clementis prudenti carmina
- Opera omnia...recensente

Arevalo, Juan Jose *see*
- La adolescencia como evasion y retorno
- Discursos en la presidencia
- Escritos politicos
- Fabula del tiburon y las sardinas
- Guatemala
- Presidente electo al pueblo de la republica

Arevalo Martinez, Rafael *see*
- Duques de endor
- Ecce pericles
- Hombre que parecia un caballo, y otros cuentos
- Hombre que parecia un cabillo, y las rosas de enga
- Llama
- Manuel aldano
- Mundo de los maharachias
- Noches en el palacio de la nunciatura
- Obras escogidas
- Oficina de paz de orolandia, novela del imperialis
- Poemas
- Rafael arevalo martinez

Arevalo, Rafael *see* Derecho penal islamico, escuela malekita

Arex : declaracion programatica / Accion Regional Extremena – Caceres: Tip. Extremadura, 1977 – 1 – sp Bibl Santa Ana [946]

Areyto / Belaval, Emilio S – San Juan, Puerto Rico. 1948 – 1r – us UF Libraries [972]

Arfe y Villafane, J *see* De varia commesuracion para la esculptura y architectura

Argelander, Friedrich *see*
- Neue uranometrie
- Uranometria nova

Argens, Jean-Baptiste de B d' *see* Memoires du marquis d'argens, chambellan de frederic-le-grand, roi de prusse, et directeur de l'academie royale de berlin

Argens, Olivier d' *see* Memoires d'olivier d'argens et correspondances des generaux charette, stofflet, puisaye, d'autichamp, frotte, cormatin, botherel

Argensola, B L de *see*
- Conquista de las islas malucas al rey felipe 3
- Histoire de la conque'te des isles moluques par les espagnols, par les portugais, & par les hollandois
- Two letters taken out of...his treatise, called conquista de las islas malucas

Argent, Sophie *see* Settling day

Argenterie orientale / Smirnov, P – St. Petersburg, 1909 – 1r – 1 – mf#4614 – uk Microform Academic [740]

Argentina : internal affairs and foreign affairs, 1945-1959 / U.S. State Dept – (= ser Confidential u s state department special files) – 1 – $20,020.00 coll – (internal affairs, 1945-49: pt1: political, governmental, & national defense affairs 28r isbn 0-89093-538-6 $5410; pt2: social, economic, & industrial affairs 19r isbn 0-89093-539-4 $3675. foreign affairs, 1945-49 5r isbn 0-89093-537-8 $970. internal affairs & foreign affairs, 1950-54 29r isbn 0-89093-954-3 $5610. 1955-59 28r isbn 0-89093-872-5 $5410. with p/g) – us UPA [327]

Argentina *see* Conferencia de ministros de hacienda, seccion asun...

Argentina, 1918-1941 – (= ser U s military intelligence reports) – 4r – 1 – $710.00 – 0-89093-642-0 – (with p/g) – us UPA [355]

Argentina, brazil and chile since independence / George Washington University Seminar Conference – Washington, DC. 1935 – 1r – us UF Libraries [972]

Argentina. Contaduria General de la Nacion *see* Memoria...

Argentina Departamento Nacional de Agricultura *see* Informe del departamento nacional de agricultura

Argentina. Ministerio de Agricultura *see*
- Memoria...
- Memorias de las direcciones de comercio e industrias, tierras y colonias, agricultura y ganaderia e inmigracion y recopilacion de mensajes al honorable congreso, decretos, notas y otros documentos...

Argentina. Ministerio de Comercio e Industria *see* Memoria correspondiente al ano

Argentina. Ministerio de Finanzas de la Nacion *see* Memoria...

Argentina. Ministerio de Hacienda de la Nacion *see* Memoria...

Argentina. Ministerio de Obras Publicas *see* Memoria...

Argentina. Ministerio de Relaciones Exteriores *see* Memoria...

Argentina. Ministerio de Relaciones Exteriores y Culto *see* Memoria...

Argentina. Ministerio de Trabajo y Prevision *see* Memoria...

Argentina. Secretaria de Estado de Hacienda *see* Memoria...

Argentina, Thomas d' *see* Scripta super 4 libros sententiarum

Argentina y conquista del rio de la plata con otros acae cimientos de los reinos del peru. tucuman y estado de brasil. notas bibliograficas y biograficas de carlos navarro y lamarca / Barco Centenera, Arcediano – Buenos Aires: Angel Estrada. Cia. Edit., 1912 – sp Bibl Santa Ana [946]

Argentina...y conquista del rio de la plata con otros acaecimientos de los reynos... / Barco de Centenera, M – Liboa, 1602 – 8mf – 9 – sp Cultura [972]

Argentine / Gabriel, M – Paris, France. 1839 – 1r – us UF Libraries [440]

Argentine literature / Leavitt, Sturgis Elleno – Chapel Hill, North Carolina. 1924 – 1r – us UF Libraries [440]

Argentine Republic *see*
- Boletin oficial
- Informes de los consejeros legales del poder ejecutivo
- Registro nacional.
- Registro nacional...que comprende los documentos desde 1810 hasta 1891

Argentine republic. comision nacional del censo – v1-10. 1914 – 1 – $168.00 – mf#0082 – us Brook [318]

Argentine Republic. Contaduria General de la Nacion *see* Memoria

Argentine Republic. Corte Suprema de Justicia de la Nacion *see* Fallos...: con la relacion de sus respectivas causas

Argentine Republic. Courts *see* Boletin judicial de la republica argentina

Argentine Republic. Direccion de Economica Rural y Estadistica *see* Datos estadisticos

Argentine Republic. Direccion General de Estadistica *see*
- Anuario de la direccion general de estadistica 1892-1914
- Extracto estadistico de la republica argentina 1915

Argentine Republic. Direccion general de ferrocarriles *see* Estadistica de los ferrocarriles en exploitacion

Argentine Republic. Direccion Nacional de Estadistica y Censos *see* Anuario estadistico 1944-1957

Argentine Republic. Junta de Administracion del Credito Publico Nacional *see* Informe del presidente del credito publico nacional pedro agote sobre le deudo publica

Argentine Republic. Laws, Statutes, etc *see*
- Coleccion completa de leyes nacionales sancionadas por el honorable congreso
- Leyes nacionales clasificadas y sus decretos reglamentarios
- Leyes nacionales sancionadas en el periodo lejislativo de 1883-

Argentine Republic. Presidente, 1932-38 (Justo) *see* Poder ejecutivo nacional periodo 1932-38

Argentine Republic. Secretaria de Comunicaciones *see* Boletin

The argentine review – Buenos Aires: [s.n, [v1 n4-v6 n10 (oct 1924-oct 1929)] (mthly) – 8r – 1 – us CRL [073]

Argentine weekly – Buenos Aires, Argentina 1 jun 1923-30 jun 1932 – 18r – 1 – (discontinued) – uk British Libr Newspaper [079]

Argentiner magazin – Buenos Aires. v3-23. 1937-1957.(incomplete) – 1 – us NY Public [073]

Argentinisches tageblatt – Buenos Aires (RA), 1914 1 aug-1919 1 sep [gaps] – 12r – 1 – (filmed by other misc inst: 1914 1 aug-1919 1 sep [gaps], 1919 12 dec-1933 8 mar [gaps], 1933 4 apr, 14 jun & 21 jun, 1972- [48r until 1933]; 1981-1983 23 apr) – gw Misc Inst [079]

Argentinisches wochenblatt – Buenos Aires (RA), 1897 14 jul-1906 30 jun, 1906 6 oct-1909 27 mar, 1909 3 jul-25 sep, 1910, 1911 1 apr-1913 27 sep – 47r – 1 – (weekend ed of argentinisches tageblatt; filmed by other misc inst: 1897 14 jul-1906 30 jun, 1906 6 oct-1909 27 mar, 1909 3 jul-25 sep, 1910, 1911 1 apr-1913 27 sep, 1914 15 aug-1915 24 jun, 1916 7 oct-1919 27 dec, 1920 17 jan-1924 27 sep [gaps], 1925-28 [gaps], 1929 19 jan-1939 24 jun (gaps) [67r]. with suppl: hueben und drueben 1904-39 [gaps] and: der kolonist 1910-18) – gw Misc Inst [079]

Argentores: revista teatral – Buenos Aires. v1-31. Apr 1934-15 Aug 1946 – 1 – us L of C Photodup [790]

Arger, Jane *see* Les agrements et le rythme

Agriculture au katanga / Hock, A – Bruxelles, Belgium. 1912 – 1r – us UF Libraries [630]

Argivale inligtingontsluiting en -herwinning vir die historiese navorser / Ingram, Annette – Uni of South Africa 2000 [mf ed Johannesburg 2000] – 7mf – 9 – (text in afrikaans and english; abstract in afrikaans and english; incl bibl ref) – mf#mfm14847 – sa Unisa [960]

Argo – 1886 dec 1 – 1 – mf#851199 – us WHS [071]

O argonauta : periodico litterario, critico e chistoso – Teresina, Pl. 26 jul 1877 – (= ser Ps 19) – mf#P17,02,125 – bl Biblioteca [410]

Los argonautas ingleses de ultima hora / Bayle, Constantino – Madrid: Razon y Fe, 1928 – 9 – sp Bibl Santa Ana [999]

Les argonautes – Paris. avr 1908-10 – 1 – fr ACRPP [073]

Argonne post weekly – 1920 aug 27-1921 feb 25 – 1 – mf#955529 – us WHS [071]

Das argon-resonanzkontinuum im vakuum-ultravioletten spektralbereich zwischen 50 mm und 78.7nm / Trommer, Gert F – (mf ed 1995) – 2mf – 9 – €40.00 – 3-8267-2159-4 – mf#DHS 2159 – gw Frankfurter [621]

Argos – Manaus, AM: Typ Liberal, 21 abr 1872 – (= ser Ps 19) – mf#P11B,06,10 – bl Biblioteca [321]

O argos : da provincia de santa catarina – Desterro, SC: Typ de Jose Joaquim Lopes, 04 jan 1856-set, dez 1857; 02 jan 1858-30 dez 1861; jan-fev, abr, 17 jun 1862 – (= ser Ps 19) – bl Biblioteca [079]

O argos cearense : jornal politico e liberal – Fortaleza, CE: Typ Fidelissima de Francisco Luis de Vasconcellos, 07 set, nov-dez 1850; 04 set 1851 – (= ser Ps 19) – mf#P18B,03,74 – bl Biblioteca [320]

Argos oder der mann mit den hundert augen – Strassburg (Strasbourg F), 1792 3 jul-1794 16 jun, 1796 20 apr-30 jun – 1 – fr ACRPP [073]

Argosy – Ansley, NE: A H Barks. v27 n16. jul 21 1910- (wkly) – 2r – 1 – (cont: argosy and the chronicle-citizen (1909)) – us Bell [071]

Argosy – Ansley, NE: A H Barks. 1v. v7 n6-n9. oct 8-29 1909 (wkly) – 1r – 1 – (cont: argosy and the chronicle-citizen. merged with: chronicle-citizen (ansley ne 1909) to form: argosy and the chronicle-citizen (ansley ne 1909)) – us Bell [071]

Argosy – New York. 1882-1905. – 1 – (scattered issues lacking.) – us L of C Photodup [073]

Argosy [1866] – La Salle IL 1866-1901 – 1 – mf#3901 – us UMI ProQuest [073]

Argosy [1973] – Newark NJ 1973-78 – 1,5,9 – ISSN: 0191-426X – mf#8306 – us UMI ProQuest [790]

Argosy and the chronicle-citizen – Ansley, NE: A H Barks. v26 n32. nov 4 1909-jul 14 1910// (wkly) – 1r – 1 – (formed by the union of: argosy (1909) and: chronicle-citizen (1909). cont by: argosy (1910)) – us Bell [071]

Argosy and the chronicle-citizen – Ansley, NE: A H Barks. 3v. v24 n27. oct 3 1907-v26 n27. sep 30 1909 (wkly) – 1r – 1 – (formed by the union of: argosy (ansley ne) and: chronicle-citizen. split into: argosy (1909) and: chronicle-citizen (1909). cont numbering of: chronicle-citizen) – us Bell [071]

The argosy. (weekly argosy) – Georgetown, Guyana. Jan 1887-Oct 1908.-w. 28 reels – 1 – uk British Libr Newspaper [072]

Arguello, Agenor *see*
- Jardin de liliana
- Precursores de la poesia nueva en nicaragua

Arguello Castrillo, A *see*
- Discurso sobre el charlatanismo medico y quirurgico...
- Disertacion chirurgica relativa al gobierno politico en la que se proponen los danos de la castracion vulgar segun se practica para curar ninos quebrados
- Methodo exemplar del doctor mejano para el estudio de la medicina...

Arguello, Manuel de *see*
- Sermon de la dominica septuagessima
- Sermon moral al real acuerdo de mexico al tiempo que omo posession cut publica entrada el ex[cellentissi]mo senor d joseph sarmiento valladares...
- Sermon panegyrico: que en la celebridad de la dedicacion del templo nuevo de san bernardo

Arguello Mora, Manuel *see* Obras literarias e historicas

Arguello, Santiago *see*
- Libro de los apologos y de otras cosas espirituale
- Mi mensaje a la juventud
- Modernismo y modernistas
- Poesias escogidas y poesias nuevas

The argument : a priori, for the being and the attributes of the lord god, the absolute one, and first cause / Gillespie, William Honyman – 6th ed. Edinburgh: T & T Clark, 1906 [mf ed 1985] – 1mf – 9 – 0-8370-3288-1 – mf#1985-1288 – us ATLA [210]

Argument before the interstate commerce commission, washington, dc : in behalf of the national association of owners of railroad securities, june 11 1917 / Warfield, Solomon Davies – [S.l: s.n, 1917?] (mf ed 19–) – [6]p – mf#ZV-TPG pv128 n3 – us Harvard [380]

The argument delivered before the judicial committee of the privy council in the case of ridsdale v. clifton and others : together with the proceedings in the case, the judgment of lord penzance, and the report of the judicial committee / Stephen, James Fitzjames et al – London: C Kegan Paul, 1878 – 2mf – 9 – 0-524-03545-8 – mf#1990-4740 – us ATLA [240]

Argument for a church-establishment / Scholefield, James – Cambridge, England. 1833 – 1r – us UF Libraries [240]

The argument for christianity / Lorimer, George Claude – Philadelphia: American Baptist Pub Soc, 1894 [mf ed 1991] – 2mf – 9 – 0-7905-7905-7 – (incl bibl ref) – mf#1989-1130 – us ATLA [240]

Argument from christian baptism for christian education / Trevor, George – Oxford, England. 1836 – 1r – us UF Libraries [242]

The argument from prophecy / Maitland, Brownlow – London: The Christian Evidence Cttee of the SPCK, 1877 [mf ed 1984] – (= ser Biblical crit – us & gb 25) – 3mf – 9 – 0-8370-0169-2 – (incl bibl ref) – mf#1984-1025 – us ATLA [221]

An argument in defence of the exclusive right claimed by the colonies to tax themselves : with a review of the laws of england, relative to representation and taxation – London: printed...by Brotherton and Sewell...T Evans...and W Davis...1774 – 2mf – 9 – mf#20461 – cn Canadiana [336]

Argument of adam crooks, qc : against the great western application for a railway line from glencoe to the niagara river: and in favour of the amendments to the charter of the erie and niagara extension railway company... – Toronto: Hunter, Rose, 1869 – 1mf – 9 – mf#03627 – cn Canadiana [380]

Argument of mr joseph s auerbach : before the judiciary committee of the senate in opposition to the so-called anti-trust bills, april 8, 1897 / Auerbach, Joseph Smith – [S.l: s.n, 1897?] (mf ed 19–) – 59p – mf#ZT-TN pv40 n9 – us Harvard [340]

The argument of the book of job unfolded / Green, William Henry – New York: Robert Carter, 1874, c1873 – 1mf – 9 – 0-8370-9388-0 – mf#1986-3388 – us ATLA [221]

Argument of the rev henry preserved smith before the presbytery of cincinnati / Smith, Henry Preserved – Cincinnati: R Clarke 1892 [mf ed 1990] – 1mf – 9 – 0-7905-6318-5 – mf#1988-2318 – us ATLA [242]

Argumentation – Dordrecht, Netherlands 1987+ – 1,5,9 – ISSN: 0920-427X – mf#15256 – us UMI ProQuest [300]
Argumentation and advocacy – River Falls WI 1989+ – 1,5,9 – (cont: journal of the american forensic association) – ISSN: 1051-1431 – mf#6485,01 – us UMI ProQuest [340]
Argumento de la nueva espana – [Buenos Aires 1937] – 9 – mf#w869 – us Harvard [946]
El argumento de un drama / Hurtado, Antonio – 1867 – 9 – sp Bibl Santa Ana [410]
Argumentorum et objectionum : de praecipuis articulis doctrinae christianae... / Pezelius, C – Neapoli Nemetum, 1583-96 – 50mf – 9 – mf#PBA-291 – ne IDC [240]
Arguments – n1-27 28. Paris. 1957-62. – 1 – (lacking: n16) – fr ACRPP [073]
Arguments for and against a baptist theological school at williamsburg, va / Jones, Scervant – 1837 – 1 – $5.00 – us Southern Baptist [242]
Arguments in behalf of the united states, with supplement and appendix : presented...under the treaty between great britain and the united states for the final settlement of the claims of the hudson's bay and puget's sound agricultural companies / Cushing, Caleb, 1800-1879 – [Washington?: s.n.] 1868 [mf ed 1983] – 2mf – 9 – 0-665-14503-9 – mf#14503 – cn Canadiana [341]
Arguments in favour of lay representation in ecclesiastical synods / Farquhar, William – Edinburgh, Scotland. 1853 – 1r – us UF Libraries [240]
The arguments of romanists : from the infallibility of the church and the testimony of the fathers in behalf of the apocrypha / Thornwell, James Henley – New-York: Leavitt, Trow, 1845 – 5mf – 9 – 0-524-08820-9 – (incl bibl ref) – mf#1993-3312 – us ATLA [220]
Arguments on behalf of the complainants in the matter of the complaint against egbert c smyth, brown professor of ecclesiastical history : heard dec 28, 29, 30, 31, 1886, before the board of visitors of andover theological seminary – Boston: Rand Avery 1887 [mf ed 1992] – 1mf – 9 – 0-524-02753-6 – mf#1990-4428 – us ATLA [242]
Arguments to prove the policy and necessity of granting to newfoundland a constitutional government : in a letter to the right honourable w huskisson... / Morris, Patrick – London: Hunt & Clarke 1828 [mf ed 1987] – 2mf – 9 – 0-665-67801-0 – mf#67801 – cn Canadiana [320]
Argumenty – Moscow, Russia 1985-86 – 1 – mf#16316 – us UMI ProQuest [200]
Argus – 1899 dec 23-17 dec 15 – 1 – mf#1109818 – us WHS [071]
Argus – Ashfield, feb 1924-dec 1942; jan1958-dec1963 – 2r – 1 – A$110.79 vesicular A$121.79 silver – at Pascoe [079]
Argus – Cape Town: Argus Group, 1 dec 1969- (daily) [mf ed Johannesburg: Microfile 1969-] – 1 – sa Misc Inst [079]
Argus – [London & SE] East Sussex, Brighton Ref Lib 1880-96 – 1 – (cont as: evening argus (brighton) 1896-; blnl: jul-dec 1897, 1899, jul-dec 1910) – uk Newsplan [072]
Argus – South Sioux City, NE: E B Wilbur. v12 n25. nov 20 1891-1902// (wkly) [mf ed 1891-1902 (gaps) filmed 1958-[1974?]] – 2r – 1 – (cont: north nebraska argus. absorbed by: dakota county record) – us NE Hist [071]
Argus / Corporation des Bibliothecaires Professionnels du Quebec – Montreal: la Corporation. v1 n[1] [nov/dec 1971]- (bimthly) [mf ed 1977-92] – 2r – 5 – (filmed with: bulletin de nouvelles = news bulletin, v1 n1 janv 1966-v1 n24 sep 1971) – mf#SEM16P290 – cn Bibl Nat [020]
Argus / Corporation des bibliothecaires professionnels du Quebec – Montreal: la Corporation. v1 n[1] [nov./dec 1971]- (bimthly) [mf ed 1992-] – 9 – mf#SEM105P1705 – cn Bibl Nat [020]
Argus – Melbourne jun 1846-jan 1957 – 465mf – 9 – at Pascoe [079]
Argus / Montgomery Co. Englewood – feb 1975-may 1976 [wkly] – 2r – 1 – mf#B33778-33779 – us Ohio Hist [071]
Argus / Montgomery Co. Englewood – may 1976-jan 1981 [wkly] – 7r – 1 – mf#B33685-33691 – us Ohio Hist [071]
Argus – Paris, oct 1802-sept 1803 – 1 – fr ACRPP [073]
Argus / Paulding Co. Antwerp – may 1885-jun 1887 [wkly] – 1r – 1 – mf#B920 – us Ohio Hist [071]
Argus / Tuscarawas Co. Newcomerstown – oct 1873-dec 1876 [wkly] – 1r – 1 – mf#B34609 – us Ohio Hist [071]
Argus / Union Co. Marysville – may 1844-may 1845 [wkly] – 1r – 1 – mf#B5536 – us Ohio Hist [071]
The argus – London, UK. 3 feb 1839-12 sep 1846 – 8r – 1 – (english gentleman: 3 jan-12 sep 1846) – uk British Libr Newspaper [072]
The argus – Melbourne, Australia. -d. Jan 1941-Dec 1952; Nov 1953-Dec 1956. 218 reels – 1 – uk British Libr Newspaper [072]
The argus – Monaghan, Ireland. -w. 15 Jan 1875-23 July 1881 (1875 very imperfect.) 2 1 2 reels – 1 – uk British Libr Newspaper [072]
The argus *see* Miscellaneous newspapers of larimer county
[Adin-] argus – CA. may 4 1882-dec 20 1883; jan 20 1887; jan 25-dec 1894 – 6r – 1 – $360.00 – (may 1882-dec 1947 (incomplete)) – mf#B02001 – us Library Micro [071]
Argus: a weekly journal – Bombay, India [6 apr 1872-27 dec 1873] – 1 – uk British Libr Newspaper [079]
Argus [brighton] – Brighton, England 30 mar 1880-24 aug 1896 – 1 – (cont by: evening argus [25 aug 1896-24 jan 2004]; argus [26 jan 2004-]) – uk British Libr Newspaper [072]
Argus [bucharest] – Bucharest, Romania [ns] 29 jan-19 may 1920, 4,7,25 nov 1942, 1 jan 1944-16 may, 15-29 jun, 2 aug-27 sep 1948 (very imperfect) – 1 – uk British Libr Newspaper [077]
Argus [cape town] – Cape Town, South Africa dec 1969- (imperfect) [1857-jun 1860, 1869, jan-apr mar, nov-dec 1901, jan-21 mar 1905, 1971-] – 1 – (cont: cape argus [3 jan 1857-2 nov 1969]) – uk British Libr Newspaper [079]
L'Argus des Revues *see* N S
Argus [drogheda] – Drogheda 1936, 1943-47 – 5r – 1 – ie National [072]
Argus [dundalk] – Dundalk, Ireland 18 nov 1977- [mf 1986-] – 1 – (cont: dundalk argus [20 sep 1974-11 nov 1977]) – uk British Libr Newspaper [072]
Argus (hastings ed) – [London & SE] PO 30 mar 1880-dec 1950 – 326r – 1 – (cont by: evening argus [sep 1896-dec 1950]) – uk Newsplan [072]
Argus (hillsboro, or) – Hillsboro OR: Argus Co, -1895 [wkly] – 1 – (cont by: hillsboro argus (hillsboro, or)) – us Oregon Lib [071]
L'argus indochinois – Hanoi. fevr 1922-juil 1930 – 1 – fr ACRPP [073]
Argus journal / Corporation des bibliothecaires professionnels du Quebec – [Montreal]: la Corporation. n1 aout 1975-n72 sep/oct 1985 (mthly) [mf ed 1984-] – 5 – (cont by: bulletin argus) – mf#SEM16P345 – cn Bibl Nat [020]
Argus leader *see* Grey river argus
Argus leader (greymouth) – jan-dec 1904, jan-jun 1907, jan-dec 1939, jan 2- mar 27 1954, jun 24-sep 27 1954, sep 18-dec 31 1954, sep 26-dec 31 1955, apr 6-jun 30 1956, jul 2-sep 29 1956, jan 2 1957-jan 21 – 1 – (title changed fr: grey river argus) – mf#60.1 – nz Nat Libr [079]
Argus [liverpool] : a weekly review of politics, literature and social science – Liverpool. England 21 oct 1876-25 dec 1880 (wkly) – 4r – 1 – uk British Libr Newspaper [072]
Argus [melbourne] – Melbourne, Australia 15 sep 1848-dec 1952, 29 oct 1953-16 jan 1957 – 202r – 1 – (cont: melbourne argus [2 jun 1846-12 sep 1848]) – uk British Libr Newspaper [079]
Argus [monaghan] – Monaghan, 14th aug 1954-1959 – 5r – 1 – ie National [072]
Argus, monaghan, armagh, cavan, fermanagh, louth, meath & tyrone advertiser – Monaghan, Ireland 15 jan, 5,12 mar, 9,16,23 apr, 28 may, 4 jun, 1 oct 1875-2 nov 1877; 1 feb 1878-23 jul 1881 – 1 – (discontinued) – uk British Libr Newspaper [072]
Argus northmont / Montgomery Co. Englewood – jan 1969-feb 1975 [wkly] – 8r – 1 – mf#B33850-33857 – us Ohio Hist [071]
Argus [norwich] – Norwich, England 6 jan 1877-24 jan 1893 – 1 – (wanting: 1872,1895,1896; cont: norwich argus [10 jan 1863-30 dec 1876]; cont by: norfolk weekly standard & argus [28 jan 1893-26 sep 1913]) – uk British Libr Newspaper [071]
Argus observer – Ontario OR: Malheur Pub Co, 1986- [daily ex sat] – 1 – (cont: daily argus observer (1970-86)) – us Oregon Lib [071]
Argus observer *see* Ontario argus
Argus [pseud] *see* A mild remonstrance against the taste-censorship
Argus (rogue river, or) – Rogue River OR: W R Brower, [wkly] – 1 – us Oregon Lib [071]
Argus sentinel / Montgomery Co. Englewood – jan 1981-jan 1984 [wkly] – 4r – 1 – mf#B34017-34020 – us Ohio Hist [071]
Argus-journal – 1966 oct-1972 jun; 1972 jul-1976 dec; 1977-82; 1983-87 – 1 – mf#205540 – us WHS [071]
Argyle agenda – 1979 nov 1-2000 jan-may [1]; 1970 jan 21-1971 dec 30, 1972 jan 6-1973 mar 15 [2] – 1 – mf#999741 [1]; 999736 [2] – us WHS [071]
Argyle atlas – 1884 dec 16-1968 may 2 [with gaps] – 1 – mf#961326 – us WHS [071]
Argyle Co-operative House *see* Canadian free press
Argyle liberal – Crookwell, feb 1910-mar 1930 (misc. issues) – 1r – A$32.12 vesicular A$37.62 silver – at Pascoe [079]
Argyle liberal – Crookwell, oct 1903-dec 1907 – 2r – A$135.70 vesicular A$146.70 silver – at Pascoe [079]
Argyll, George Douglas Campbell, 8th duke of *see*
- Organic evolution cross-examined
- Primeval man
- The reign of law
- Speech of the duke of argyll
- The unity of nature

Argyll, George Douglas Campbell, Duke of *see* What is truth?
Argyll, John Douglas Sutherland Campbell, Duke of *see*
- Canadian life and scenery
- The canadian north west
- Guido and lita
- Imperial federation
- Love and peril
- Memories of canada and scotland
- Yesterday and to-day in canada

Argyll, John Douglas Sutherland Campbell, duke of *see*
- Canadian life and scenery
- Canadian pictures
- Canadian pictures, drawn with pen and pencil

Argyllshire advertiser – Lochgilphead: Argyllshire Advertiser 1948- (wkly) [mf ed 2000-] – 1 – (cont: argyllshire advertiser and lochfyneside echo) – uk Newsplan; uk Scotland NatLib [072]
Argyllshire advertiser and west coast journal – [Scotland] Argyll & Bute, Lochgilphead: R Paterson 4 feb 1887-dec 1950 (wkly) [mf ed 2003] – 28r – 1 – (missing: nov 1888-oct 1895; cont by: argyllshire advertiser and lochfyneside echo [nov 1895-dec 1948]; argyllshire advertiser [jan 1949-dec 1950]) – uk Newsplan [072]
Argyllshire herald and campbeltown advertiser – [Scotland] Argyll & Bute 8 sep 1855-9 mar 1918 [mf ed 2003] – 32r – 1 – (missing: 1877; cont by: argyllshire herald [11 jan 1861-mar 1918]) – uk Newsplan [072]
Argyllshire leader and western isles gazette – [Scotland] Argyll & Bute 31 jan 1929-10 nov 1934 [mf ed 2003] – 5r – 1 – uk Newsplan [072]
Arhe pui karan / Thin Naung, Man – Yan Kun: Chave Phru 1978 [mf ed 1990] – 1r with other items – 1 – (in burmese) – mf#mf-10289 seam reel 137/7 [§] – us CRL [305]
Ari – Buffalo, NY. 1972-86 – 1 – us AJPC [071]
Aria : consolata amato bene, cavata nel opera la cosa rara / Viotti, Giovanni Battista – [179-] [mf ed 19–] – 1r – 1 – (for soprano & orchestra; composed in 1791 for int into martin y soler's opera, la cosa rara) – mf#pres. film 33 – us Sibley [780]
Ariadne : the story of a dream / Ouida – Toronto: Belford, 1877 – 5mf – 9 – mf#11663 – cn Canadiana [880]
Ariadne auf naxos : oper in einem aufzuge von hugo von hofmannsthal / Strauss, Richard & Hofmannsthal, Hugo von – Berlin, Paris: A Fuerstner 1912 – 1 – mf#2728p – us Wisconsin U Libr [780]
[Ariadne auf naxos] : ein duodram / Benda, Georg – [Leipzig? 1785?] [mf ed 19–] – 1r – 1 – mf#film 664 – us Sibley [780]
Ariah park news – Ariah Park, 1923-42 – 5r – A$355.10 vesicular A$362.60 silver – at Pascoe [079]
The arian controversy / Gwatkin, Henry Melvill – 2nd ed. London: Longmans, Green, 1891 – (= ser Epochs Of Church History) – 1mf – 9 – 0-7905-5765-7 – (incl ind) – mf#1988-1765 – us ATLA [240]
The arian controversy / Gwatkin, Henry Melvill – New York: Anson D.F. Randolph & Co., 1889. 176p – 1 – us Wisconsin U Libr [240]
The arian movement in england / Colligan, James Hay – Manchester: University Press; New York: Longmans, Green [distributor], 1913 – 1mf – 9 – 0-7905-4213-7 – (incl bibl ref) – mf#1988-0213 – us ATLA [240]
The arian witness : or, the testimony of arian scriptures / Banerjea, Krishna Mohan – Calcutta: Thacker, Spink, 1875 – 1mf – 9 – 0-524-05836-9 – (incl bibl ref) – mf#1990-3500 – us ATLA [230]
Ariane / Corneille, Thomas – Paris, France. 1803 – 1r – 1 – us UF Libraries [440]
L'arianna tragedia : tragedia del signor ottavio rinuccini... / Monteverdi, Claudio – Venetia: Apresso Ghirardo & Ileppo Imberti, Fratelli 1622 [mf ed 19–] – 1mf – 9 – mf#fiche 837, 655 – us Sibley [780]
Arianoff, A D' *see* Histoire des bagesera, souverains du gisaka
The arians of the fourth century / Newman, John Henry – 4th ed. London: Basil Montague Pickering, 1876 – 2mf – 9 – 0-7905-7124-2 – (incl bibl ref) – mf#1988-3124 – us ATLA [240]
Arias Corrales, Juan *see* Cuatro leyendas cacerenas
Arias de Quintanaduenas, Jacinto *see* Antiguedades y santos...de alcanatara
[Arias, duets, trios, and choruses in italian] : vocal or orchestral scores – London: [various imprints] c1790-1810 [mf ed 1991] – 1v on 1r – 1 – (in italian) – mf#pres. film 109 – us Sibley [780]
Arias, Fernando de *see* El predicador fr fernando de arias prior provincial de esta provincia del santissimo nombre de jesus de nueva-espana de la regular observancia de los hermanitanos de ntro p s augustin
Arias, Juan De Dios *see*
- Institucion cultural santandereana
- Letras santandereanas
- Practicas eclesiatica para el uso y ejercicios de notarios publicos

Arias Larreta, Abraham *see* From columbus to bolivar
Arias Madrid, Arnulfo *see* Discursos pronunciados
Arias montano : humanista / Gonzalez de la Calle, Pedro Urbano – Badajoz: Imp. del Hospicio Provincial, 1928 – 1 – sp Bibl Santa Ana [946]
Arias montano / Vazquez, Jose Andr'es – Madrid: Biblioteca Nueva, 1943 – 1 – sp Bibl Santa Ana [780]
Arias Montano, Benito *see*
- Aphorismos sacados de la historia de p.c. tacito
- Benito arias montano
- Biblia sacra regia...
- Comentario in profetas mimos
- Commentaria in duodecim prophetas
- Commentaria in isiae prophetae
- Davidis regis...psalmi
- De optimo imperio sive josuae
- De varia republica sive commentaria in librum judicum
- Dictatum christianum
- Elucidationes in omnia sanctorum apostolorum scripta
- Elucidationes in quator evangelia metthaei, marci, lucae, iohannis...
- Hymni et saecula
- Hymni et secula
- In 31 davidis psalmos
- Liber generationis et regenerationis adae sive de historia
- Monumentos sagrados de la salud del hombre
- Parafrasis del maestro sobre el cantar de los cantares en tono pastoril
- Q biblia sacra (regia). tomo 1
- Rey de nuestros escriturarios
- Rhetoricorum libri 4

Arias montano escribe a justo lipsio y a juan moreto / Lopez de Toro, Jose – Madrid: Rev. de Archivos, Bibliotecas y Museos, 1954. pp. 533-543 – 1 – sp Bibl Santa Ana [780]
Arias montano y el monumento al duque de alba / Schubart, Herta – Madrid: Cruz y Raya, 1933 – 1 – sp Bibl Santa Ana [946]
Arias montano y la politica de felipe 2 : en flandes por luis morales... / Garcia Garcia, Rafael – Malaga: Revista Espanola de Estudios Biblicos, 1928 – 1 – sp Bibl Santa Ana [320]
Arias montano y la politica de felipe 2nd en flandes : por luis morales oliver / Maura Gamazo, Gabriel – Madrid: Tip. R.Bib. Archivos y Museos, 1929 – 1 – sp Bibl Santa Ana [320]
Arias montano y los jesuitas / Perez Goyena, A – Madrid: Estudios Biblicos, 1928 – 1 – sp Bibl Santa Ana [241]
Arias montano y su tratado "de optimo imperio" / Duran Ramas, Maria de los Angeles – Sevilla: Universidad, 1981 – 1 – sp Bibl Santa Ana [946]
Arias Montanus, B *see*
- David
- Humanae salutis monumenta b. ariae montani studio constructa et decantata

Arias, Paolo Enrico *see* Skopas
Arias Ramirez, Fernando *see* Colombia y su pueblo
Arias Regodon, Publio *see* Novena de san gregorio obispo de ostia patrono popular de la villa de ruanes por...
Arias Trujillo, Bernardo *see* Risaralda
Ariate, Nicolas *see*
- An cadanayan na lacao nin quinaban napipinta sa buhay ni nicolas asin crispina na mag-amang macalolodoc
- Comedia can panahon nin paghadit
- Historia nin hadeng salomon asin ni reina saba
- An nagpacombaba no principe igmedio asin an namoot na princesa cloriana
- An pagcabuhay ni d pedro tabios asin ni manuela lilay

Arid soil research and rehabilitation – Abingdon, Oxfordshire 1987-96 – 1,5,9 – ISSN: 0890-3069 – mf#17306.01 – us UMI ProQuest [630]
Ariel – 1902 oct, 1903 apr – 1 – mf#3177684 – us WHS [071]
Ariel : journal du monde elegant – n1-20. Paris. 2 mars-7 mai 1836 – 5 – (lacking: n17) – fr ACRPP [073]
Ariel – La Salle IL 1827-32 – 1 – mf#3846 – us UMI ProQuest [000]
Ariel – Stockholm, Moelnlycke. 1938-65 – 9 – sw Kungliga [078]

Ariel, Pablo *see* Mi amigo pedro
Ariel's offenbarungen : [roman] / Arnim, Ludwig Achim, Freiherr von; ed by Minor, Jacob – Weimar: Gesellschaft der Bibliophilen, 1912 [mf ed 1988] – 324p – 1 – (incl bibl ref) – mf#6956 – us Wisconsin U Libr [830]
Ariel's offenbarungen *see* Die beziehungen des dramatikers achim von arnim zur altdeutschen litteratur
Ariets / Soiuz Slavian – Riazan', Russia. n1[1](mar '98)-n5(dek '98) – 1 – mf#mf-12248 (reel 2) – us CRL [077]
'Arif, Kethudazade *see* The divan project
Arif, Nail *see* Eine untersuchung ueber wirkungsunterschiede zwischen fermentierten und unfermentiertem iscador bei hiv-positiven und gesunden probanden
'Arif (Seyhuelislam), Hikmet *see* The divan project
Arifauna y flora : nos costumes, supersticoes e lendas brasileiras e americanas... / Teschaner, C – Madrid: Razon y Fe, 1926 – 1 – sp Bibl Santa Ana [580]
Arifin, Hassan Noel *see*
– Almanak dai toa (asia timoer raja) disoesoen oleh hassan noel 'arifin
– Poelau darah
Arinez, Agustin Maria De *see* Diccionario hispano-kanaka
Arinos De Melo Franco, Afonso *see*
– Desenvolvimento da civilizacao material no brasil
– Estadista da republica
– Historias y paizagens
– Homens e temas do brasil
– Indio brasileiro e a revolucao francesa
– Lendas e tradicoes brasileiras
– Notas do dia
– Pelo sertao
– Terra do brasil
Arioaldo, re de' longobardi : ballo tragico in cinque atti, inventato e diretto da tomaso casati, da rappresentarsi nell'i r teatro della canobbiana il carnevale 1841 / Casati, Tomaso – Milano: G Truffi, 1841 – 1 – mf#*ZBD-*MGTZ pv 3-Res – Located: NYPL – us Misc Inst [790]
Arion – Boston MA 1962+ – 1,5,9 – ISSN: 0095-5809 – mf#12023 – us UMI ProQuest [450]
Arion : a canadian journal of art... – Toronto. v1 n1-12. oct 1880-sep 1881// (mthly) – 1r – 1 – Can$90.00 – cn McLaren [700]
Arion – Recife, PE: [s.n.] 05 set-nov 1891; out-nov 1892 – (= ser Ps 19) – mf#P16,01,03 – bl Biblioteca [440]
Ariosto, Ludovico *see* Die kosmographie in ariosts orlando furioso
Aris, Reinhold *see* Die staatslehre adam muellers in ihrem verhaeltnis zur deutschen romantik
Arische freiheit – Dinkelsbuehl DE, 1927 – 1r – 1 – gw Misc Inst [074]
Arische religion / Schroeder, Leopold von – Leipzig: H Haessel 1914-16 [mf ed 1992] – 2v on 4mf – 9 – 0-524-04167-9 – (incl bibl ref) – mf#1990-3297 – us ATLA [290]
O arisdarcho – Rio de Janeiro, RJ: Typ do Diario de N L Vianna, 09 maio-02 jun 1840 – (= ser Ps 19) – mf#P15,01,70 n02 – bl Biblioteca [321]
Arise, o lord / Hawkins, James – [171-?] [mf ed 19–] – 1r – 1 – mf#pres. film 31 – us Sibley [780]
Aris's birmingham gazette – Birmingham, England [mf jan-aug 1871; jan-dec 1876] – 1 – (incorp with: birmingham daily gazette; wanting: sep-dec 1871; cont: birmingham gazette; or, the general correspondent [16 nov 1741]) – uk British Libr Newspaper [072]
O aristarcho : orgao quinzenal, litterario e noticioso – Cidade de Santo Antonio do Monte, MG, 15 jun 1885 – (= ser Ps 19) – bl Biblioteca [073]
O aristarcho : propriedade de uma associacao – Manaus, AM: Typ do Jornal do Amazonas, 25 fev-03 mar 1884 – (= ser Ps 19) – mf#P11B,06,11 – bl Biblioteca [073]
Aristarco...padre san francisco / Trujillo, Antonio de – 1683 – 9 – (1685) – sp Bibl Santa Ana [240]
Aristas / Betancourt, Gaspar – Habana, Cuba. 1935 – 1r – us UF Libraries [972]
Aristeguieta Rojas, Francisco De Paula *see* Grano de arena
Aristeguieta Silva, F *see* Espana moscovita y sus consecuencias
Aristide, Achille *see* Problemes haitiens
Aristidean : a magazine of review, politics and light literature – La Salle IL 1845 – 1 – mf#3937 – us UMI ProQuest [320]
Aristides *see*
– Der apologet aristides
– The apology of aristides
– Supplementary observations upon the proceedings of the house of assembly in this province
Aristides lobo / Moreno Brandao – Rio de Janeiro, Brazil. 1938 – 1r – us UF Libraries [972]
Aristippe / Kreutzer, Rodolphe – Paris, France. 1810 – 1r – us UF Libraries [440]
Aristo / Tawiow, Israel Hayyim – Warsaw, Poland. 1898 – 1r – us UF Libraries [939]
Aristocracy and evolution : a study of the rights, the origin and the social functions of the wealthier classes / Mallock, William Hurrell – London: A & C Black, 1898 – 5mf – 9 – mf#09520 – cn Canadiana [305]
Aristocracy, the state, and the local community : the hastings correspondence from the huntingdon library, san marino, california – 39r – 1 – (pt 1: 1477-1701 19r. pt 2: 1702-1828 20r) – us Primary [941]
The aristocracy, the state and the local community : from the huntington library, san marino, california – 39r coll – 1 – (pt 1: the hastings correspondence, 1477-1701 19r c39-16501. pt 2: 1701-1828 20r c39-16502. material on military affairs, taxation, politics, jacobite politics, plantations in america, the american revolution, the napoleonic wars and other issues between 1477 to 1828) – mf#C39-16500 – us Primary [941]
Aristocratic women = the social, political and cultural history of rich and powerful women – [mf ed Marlborough 1994, 1997] – 2pt – 1 – (pt1: the correspondence of jemima, marchioness grey (1722-97) & her circle, fr bedfordshire county record office [10r] £900; pt2: correspondence & diaries of charlotte georgiana, lady bedingfeld (formerly jerningham) c1779-1833, together with the letters of anna seward c1791-1804, & lady stafford c1774-1837, fr birmingham university library [15r] £1350; with d/g) – uk Matthew [860]
Aristophanes *see* The comedies of aristophanes
Aristoteles : politicorum libri – Zaragoza, Barcelona SP. 1480? – 1,5 – sp Cultura [180]
Aristoteles [Aristotle] *see* Ethica, politica, oeconomica
Aristotelianism / Smith, Isaac Gregory – [3rd ed] London: SPCK; New York: E & JB Young, 1889 [mf ed 1991] – (= ser Chief ancient philosophies) – 1mf – 9 – 0-7905-9666-0 – (incl bibl ref) – mf#1989-1391 – us ATLA [180]
Aristotelis *see* Metaphysicorum libricum averrois commentariis et epitome venetiis
Aristotelis ethica nicomachea recognovit franciscus susemihl – Lipsiae, Germany. 1903 – 1r – us UF Libraries [180]
Aristotelis metaphysica – Bonnae, Germany. v1-2. 1848 – 1r – us UF Libraries [180]
Aristotelis quae feruntur de plantis – Lipsiae, Germany. 1888 – 1r – us UF Libraries [180]
Aristotle *see*
– Athenaion politeia
– Athenian constitution
– Epistola ad quintum fratrem...
– The ethics of aristotle
– Filosofia moral
– Metaphysics of aristotle
Aristotle et al *see* The classical psychologists
Aristotle's criticism of plato and the academy / Cherniss, Harold Frederick – Baltimore, MD. 1944 – 1r – us UF Libraries [180]
Aristov, N *see* Pervyia vremena khristianstva v rossii po tserkovno-istoricheskomu soderzhaniiu...
Arithmetic for the use of schools / Liebich, Max – Montreal: E M Renouf, c1901 [mf ed 1996] – (= ser Renouf's mathematical series) – 1mf – 9 – 0-665-81348-1 – mf#81348 – cn Canadiana [510]
Arithmetic teacher – Reston VA 1954-94 – 1,5,9 – (cont by: teaching children mathematics) – ISSN: 0004-136X – mf#1564 – us UMI ProQuest [510]
Arithmetical tables compiled for the use of schools : including various useful tables etc – 13th ed. Montreal: R Miller, 1869 [mf ed 1995] – 1mf – 9 – 0-665-94773-9 – mf#94773 – cn Canadiana [510]
Arithmetique : cours elementaire: livre de l'eleve – Montreal: [s.n, 1883?] [mf ed 1980] – 2mf – 9 – 0-665-04353-8 – mf#04353 – cn Canadiana [510]
Arithmetique : cours elementaire: livre du maitre / Aphraates, frere – Montreal: [freres des ecoles chretiennes, entre 1893 et 1927] (mf ed 1993) – 2mf – 9 – mf#SEM105P2037 – cn Bibl Nat [510]
Arithmetischer tausendkuenstler / Faulhaber, Johannes – Ulm: Auf Kosten der Gaumischen Handlung 1762 [mf ed 1979] – 1r – 1 – (filmed with: numerus figuratus, [s.l. 1614?]) – mf#9102 – us Wisconsin U Libr [510]
Aritmetica / Santos Redondo, Ignacio – 1890 – 9 – sp Bibl Santa Ana [510]
Aritmetica delle nazioni e divisione del tempo fra l'orientali : opera / Hervas y Panduro, Lorenzo – Cesena: Biasini 1786 – (= ser Whsb) – 3mf – 9 – €40.00 – mf#Hu 003 – gw Fischer [410]
Aritmetica elemental / Montero y Santaren, Eulogio – 1892 – 9 – sp Bibl Santa Ana [510]
Aritmetica, las cuatro operaciones fundamentales / Crespo, Nicasio – Badajoz: Tip. La Economica, 1905 – sp Bibl Santa Ana [510]
Aritmetica para los alumnos / Botello del Castillo, Carlos – 1880 – 9 – sp Bibl Santa Ana [510]
Aritmetica y sistema legal de pesas / Antonio y Hernandez, Pedro de A – 1891. Incompleto – 9 – sp Bibl Santa Ana [510]
Arius the libyan : a romance of the primitive church / Kouns, Nathan Chapman – San Francisco: John Howell 1914 [mf ed 1992] – 1mf – 9 – 0-524-02404-9 – (incl bibl ref) – mf#1990-0607 – us ATLA [830]
A-rivista anarchica – Milan, Italy 1971-73 – 1 – ISSN: 0044-5592 – mf#8536 – us UMI ProQuest [320]
Ariyasaccakatha / Vimalappanna, Brah Gru – rev corr ed. Phnom-Penh: Impr nouvelle A Portail [mf ed 1989] – 1r with other items – 1 – (title & text in khmer; added t.p. in french) – mf#mf-10289 seam reel 113/9 [§] – us CRL [280]
Ariza, Sander *see* Trujillo
Arizona : arizona revised statutes annotated – St. Paul: West Pub Co, 1956-Jun 2002 update – 9 – $3307.00 set – mf#401590 – us Hein [348]
Arizona : session laws of american states and territories – 1864-2001 – 9 – $1728.00 set – mf#402500 – us Hein [348]
Arizona *see* Reports and opinions (a-g)
The Arizona Academy of Science *see* Journal
Arizona and the west – Tucson AZ 1959-86 – 1,5,9 – (cont by: journal of the southwest) – ISSN: 0004-1408 – mf#2004.01 – us UMI ProQuest [975]
Arizona architect – Tucson AZ 1957-74 – 1,5,9 – ISSN: 0004-1416 – mf#8001 – us UMI ProQuest [720]
Arizona attorney – v1-37. 1965-2 – 9 – $602.00 – (cont: arizona bar journal v1-23 1965-88) – ISSN: 0004-1424 – mf#108571 – us Hein [340]
Arizona attorney general reports and opinions – 1915-2001 – 9 – $396.00 set – (1915-77 on reel $70. 1978-2001 on mf $326) – mf#408120 – us Hein [340]
Arizona baptist beacon – Phoenix, 1937-90 – 1 – $804.08 – us Southern Baptist [242]
Arizona business – Tempe AZ 1967-94 – 1,5,9 – (cont by: azb: arizona business) – ISSN: 0093-0717 – mf#6383.01 – us UMI ProQuest [338]
Arizona Civil Liberties Union *see* Civil liberties in arizona
Arizona Commission of Indian Affairs *see* Capitol drumbeat
Arizona Education Association *see* Aea advocate
Arizona. Fort Verde Headquarters *see* Headquarters records of fort verde, arizona, 1886-1891
Arizona heritage news – v8 n6/7; v1 n1-2 [1978 jul; 1979 jan-feb] – 1 – mf#668952 – us WHS [071]
Arizona indian monthly – v3 iss 8-1912 [1980 dec-1981 apr] – 1 – mf#674422 – us WHS [071]
Arizona indian now – v3 iss 7 [1980 nov] – 1 – mf#674419 – us WHS [071]
Arizona informant – 1979 mar 14-2001 jan-jun [with gaps] – 1 – mf#627949 – us WHS [071]
Arizona journal of international and comparative law review – 1982- v18. 2001 – 9 – $308.00 set – (none publ 1983, 1986. v numbering began with v6 1989) – ISSN: 0743-6963 – mf#101651 – us Hein [341]
Arizona law review – Tucson AZ 1959+ – 1,5,9 – ISSN: 0004-153X – mf#10074 – us UMI ProQuest [340]
Arizona librarian – Tucson AZ 1940-71 – 1,5 – ISSN: 0004-1548 – mf#2524 – us UMI ProQuest [020]
Arizona medicine – Phoenix AZ 1944-85 – 1,5,9 – ISSN: 0004-1556 – mf#2387 – us UMI ProQuest [610]
Arizona nurse – Tempe AZ 1972+ – 1,5,9 – ISSN: 0004-1599 – mf#7328 – us UMI ProQuest [610]
Arizona. Presbytery (Pres. Church in the U.S.A.) *see* Minutes, 1888-1906
Arizona preservation news – v1 n1-v8 n5 [1970 dec-1978 may] – 1 – mf#668953 – us WHS [071]
Arizona prince hall masonic journal – 1963 may – 1 – mf#5026221 – us WHS [071]
Arizona public employee – 1976 sep-1979 nov – 1 – mf#630430 – us WHS [071]
Arizona quarterly – Tucson AZ 1945+ – 1,5,9 – ISSN: 0004-1610 – mf#263 – us UMI ProQuest [071]
Arizona reports – v1-21. 1866-1920 – 167mf – 9 – $565.00 – (updates planned) – mf#LLMC 84-123 – us LLMC [340]
Arizona republican – 1900 nov 1-1901 mar 7; 1901 mar 14-apr 25 – 1 – mf#854114 – us WHS [071]
Arizona. State Bar Association *see* Proceedings
Arizona State Genealogical Society *see* Copper state bulletin
Arizona state law journal – 1969-v32. 1969-2000 – 5,6,9 – $652.00 set – (1969-84 in reel $280; 1985-v32 1985-2000 in mf $372; title varies: 1969-74 as law and the social order; vol numbering began with v19 1987) – ISSN: 0164-4297 – mf#100681 – us Hein [348]
Arizona. Supreme Court *see* Arizona supreme court reports
Arizona supreme court reports / Arizona. Supreme Court – v1-27. 1866-1925 – 26mf (1:42) 109mf (1:24) – 9 – $280.00 – (no pre-nrs vols. add vols planned) – mf#LLMC 84-123 – us LLMC [347]
Arizona teacher – Phoenix AZ 1949-74 – 1,5 – mf#215 – us UMI ProQuest [370]
Arizona. Territory. Board of Control *see* Report
Arizona. Territory. Prisons *see* Biennial report
Arizonans for National Security. Committee for National Security [AZ] *see* Eagle's eye
Arjanie polscy : zma rycinami / Morawski, Szczesny – We Lwowie: Naknadem autora 1906 – 2mf [ill] – 9 – 0-524-01231-8 – mf#1990-0370 – us ATLA [943]
Arjo, Jose *see* Carta del p joseph de arjo, de la compania de jesus, preposito de la casa professa de esta ciudad de mexico
Arjun, Guru *see* The psalm of peace
Ark – Cincinnati. 1911-23 – 1 – us AJPC [073]
Ark of god / Pinder, John H – London, England. 1840? – 1r – us UF Libraries [240]
Arkadas – Istanbul: Cumhuriyet Matbaasi, 1928-29. Sahibi ve Mueduer-i Mesul: Sedat [Simavi]. n1-35 (27 Haziran 1928-20 Subat 1929) – (= ser O & t journals) – 10mf – 9 – $165.00 – us MEDOC [079]
Arkamistuuled kodumaal = Winds of revival in the homeland / Laks, Johannes – Toronto: Toronto Vabakoguduse Kirjastus (Toronto Free Church Publishing House), 1966. 137p. Publ. No. 6295 b. One item of four on reel – 1 – us Southern Baptist [242]
Arkansas – (= ser General education board: the early southern program) – 6r – 1 – $780.00 – us Scholarly Res [370]
Arkansas : code of 1987 annotated – Charlottesville: Michie Company, 1947-Jul 2002 update – 9 – $3709.00 set – mf#401740 – us Hein [348]
Arkansas : session laws of american states and territories – 1818-1997 – 9 – $2,384.00 set – mf#402510 – us Hein [348]
Arkansas *see*
– Reports and opinions
– Reports, pre-nrs
– State reports, post-nrs
Arkansas advocate – v1-v4 n11 [1972 jun-1976 oct] – 1 – mf#345712 – us WHS [071]
Arkansas and mississippi superintendents' attitudes toward k-6 physical education in the public school / Williams, Lisa G – 1998 – 1mf – 9 – $4.00 – mf#PE 4016 – us Kinesology [790]
Arkansas attorney general reports and opinions – 1877-2001 – 6,9 – $1299.00 set – (1877-1943, 1953-79 on reel $315. 1980-2001 on mf $984. 1944-52 not available) – mf#408130 – us Hein [340]
Arkansas baptist – 1890-1991 – 1 – $3,456.48 – us Southern Baptist [242]
Arkansas baptist materials – 76p – 1 – $5.00 – ((1) northwestern arkansas associations; (2) ben m bogard, an old landmark made plain; (3) j l chastain, which of the two organizations, claiming it, is rightly entitled to the name, the benton county association?; (4) history of missionary baptist associations in benton county, ar, 1810-1940) – us Southern Baptist [242]
Arkansas bar association reports – 1900-47 (all publ) – 105mf – 9 – $157.00 – mf#LLMC 84-404 – us LLMC [340]
Arkansas business – Little Rock AR 1996+ – 1,5,9 – ISSN: 1053-6582 – mf#17830 – us UMI ProQuest [650]
Arkansas business and economic review – Fayetteville AR 1991-2000 – 1,5,9 – ISSN: 0004-1742 – mf#15708 – us UMI ProQuest [338]
Arkansas Central Baptist Association *see* Miscellaneous new bulletins
Arkansas Central Baptist Associations *see* Annuals
Arkansas dental journal – North Little Rock AR 1973-90 – 1,5,9 – (cont by: arkansas dentistry) – ISSN: 0004-1769 – mf#8366.01 – us UMI ProQuest [617]
Arkansas dentistry – North Little Rock AR 1991+ – 1,5,9 – (cont: arkansas dental journal) – ISSN: 1056-4764 – mf#8366,01 – us UMI ProQuest [617]
Arkansas echo – Little Rock AR (USA), 1922 7 dec-1932 24 aug [gaps] – 4r – 1 – gw Misc Inst [071]
Arkansas evangel – 714p. 1881-25 Mar 1886 – 1 – $24.99 – us Southern Baptist [242]
Arkansas folklore – 1950 jul 6-1958 feb – 1 – mf#234166 – us WHS [390]
Arkansas gazette – 1819 nov 20-1836 oct 4 [1]; 1919 jun 6-1940 (with gaps) [2] – 1 – mf#2738738 [1]; 846032 [2] – us WHS [071]

Arkansas genealogical register – 1971 mar-1974 jun – 1 – mf#1052328 – us WHS [929]
Arkansas guard – 1981 jun-1982 jul/aug; 1982 nov/dec – 1 – mf#1051443 – us WHS [071]
Arkansas law journal – Fort Smith. v1. 1877 – (= ser Historical legal periodical series) – 1 – $45.00 – mf#408830 – us Hein [340]
Arkansas law review – v1-53. 1946-2000 – 5,6,9 – $710.00 set – (v1-38 1946-85 in reel $430. v39-53 1985-2000 in mf $280) – ISSN: 0004-1831 – mf#100691 – us Hein [340]
Arkansas lawyer – v1-36. 1967-2001 – 9 – $377.00 set – ISSN: 0571-0502 – mf#401300 – us Hein [340]
Arkansas legionnaire – 1940 jan 5-1942 dec 25 – 1 – mf#1052330 – us WHS [071]
Arkansas libraries – Little Rock AR 1944+ – 1,5,9 – ISSN: 0004-184X – mf#2246 – us UMI ProQuest [020]
Arkansas materials / Baptist Associations – 1810-1940 – 1 – $5.00 – ((1) northwestern arkansas associations; (2) ben m bogard, an old landmark made plain; (3) j l chastain, which of the two organizations, claiming it, is rightly entitled to the name, the benton county association?; (4) history of missionary baptist associations in benton county, ar) – us Southern Baptist [242]
Arkansas Radical Media Co-op *see* Different drummer
Arkansas. State Bar Association *see* Proceedings
Arkansas State Federation of Labor *see* Proceedings of the...convention of the arkansas state federation of labor
Arkansas state gazette – 1836 oct 11-1848 apr 28; 1848 may 5-1850 feb 1 – 1 – mf#854232 – us WHS [071]
Arkansas state press – jul 1-dec 30; 1994 jan 6-dec 29; 1995 jan 5-dec 28; 1996 jan 18-dec 26; 1997 jan 2-dec 25; 1998 jan 8-aug 27 [1]; 1946 may 3 [2] – 1 – mf#2792787 [1]; 870669 [2] – us WHS [071]
Arkansas. Supreme Court *see* Arkansas supreme court reports
Arkansas supreme court reports / Arkansas. Supreme Court – v1-170. 1837-1926 – 1275mf – $1912.00 – (pre-nrs: v1-46 1837-85 359mf $538. updates planned) – mf#LLMC 82-981 – us LLMC [347]
Arkansas times and advocate – 1838 may 7, jul 23 – 1 – mf#846022 – us WHS [071]
Arkansas traveler – Chicago.v12, no.12-v22, no.12.11 Feb 1888-12 Aug 1893 – 1 – us L of C Photodup [978]
Arkell, Herbert Samuel *see* Production and markets
Arkell, W J *see*
– Prehistoric survey of egypt and western asia
– Prehistoric survey of egypt and western asia, vol 2
– Prehistoric survey of egypt and western asia, vol 3
– Prehistoric survey of egypt and western asia, vol 4
Arkhangel'skaia gub ispolnitel'nyj komitet sovetov *see* Izvestiia arkhangel'skogo gubernskogo ispolnitel'nogo komiteta sovetov rabochikh i krest'ianskikh deputatov
Arkhangel'skie gubernskie vedomosti – Arkhangel'sk, Russia 1859-79 [mf ed Norman Ross] – 22r – 1 – mf#nrp-194 – us UMI ProQuest [077]
Arkhangelskii, A S *see* Tvoreniia ottsov tserkvi v drevne-russkoi pismennosti
Arkhangel'skii Gorodskoi Bank *see* Otchet za 2-i operatsionnyi god 1-go oktiabria 1924 g po 1-e oktiabria 1925 g
Arkhangel'skij sovet rabochikh i krest'ianskikh deputatov *see* Izvestiia arkhangel'skogo soveta rab i sold deputatov
Arkheologicheskaia letopis iuzhnoi rossii – Kiev, 1899-1904. v1-6 – 28mf – 9 – mf#R-3389 – ne IDC [077]
Arkheologicheskie izvestiia i zametki, izdavaemye imperatorskim moskovskim arkheologicheskim obshchestvom – Providence. 1954+ (1) 1975+ (5) 1975+ (9) – 88mf – 9 – mf#1679 – ne IDC [077]
Arkhipova, M I *see* Golos sibiri
Arkhitekturnyi muzei imperatorskoi akademii khudozhestv – Spb., 1902-1903 – 21mf – 9 – (missing: 1903(2, 4-10)) – mf#R-3390 – ne IDC [077]
Arkhiv biologicheskikh nauk – Spb., Pg., L., 1892-1940. v1-58 – 899mf – 9 – (missing: 1934, v36b; 1938, v49-51(2)) – mf#R-1517 – ne IDC [077]
Arkhiv gosudarstvennogo soveta – Spb., 1869-1904. v1-5 (1) – 258mf – 9 – mf#R-4242 – ne IDC [077]
Arkhiv i biblioteka sv sinoda i konsistorskie arkhivy / Zdravomyslov, K I – 1906 – 61p 1mf – 9 – mf#R-9891 – ne IDC [243]
Arkhiv istoricheskikh i prakticheskikh svedenii, otnosiashchikhsia do rossii – Cambridge. 1962-1968 (1) 1966-1968 (5) 1966-1968 (9) – 89mf – 9 – mf#1680 – ne IDC [077]
Arkhiv istoriko-iuridicheskikh svedenii, otnosiashchikhsia do rossii / ed by Kalachov, N V – M., 1850-1876. 3 v – 16mf – 9 – mf#R-8213 – ne IDC [077]
Arkhiv iugo-zapadnoi rossii – Kiev: Vremennaia komissiia dlia razbora drevnikh aktov. 1859-1914. 36 v – 311mf – 9 – mf#R-14936 – ne IDC [077]
Arkhiv polotskoi dukhovnoi konsistorii / Sapunov, A P – 1898 – 102p 2mf – 9 – mf#R-14,220 – ne IDC [243]
Arkhiv pravitelstvuiushchego senata – 3v 24mf – 9 – mf#R-10813 – ne IDC [947]
Arkhiv Radians'koi Ukrainy *see* Istorychno-arkhivoznavchyi zhurnal
Arkhiv russkoi artillerii / ed by Strukov, D P – Spb., 1889. v1(1700-1718) – 12mf – 9 – mf#R-10934 – ne IDC [077]
Arkhiv veterinarnykh nauk – Spb., 1876-1917 – 1206 – 9 – (missing: 1899 v29(10); 1906 v36; 1907 v37(1, 8); 1917 v47) – ne IDC [077]
Arkhivnoe delo – Moscow. 58v. 1923-41 – 1 – us L of C Photodup [025]
Arkite worship / Balgarnie, Robert – London: James Nisbet 1881 [mf ed 1993] – 1mf – 9 – 0-524-05787-7 – mf#1992-0614 – us ATLA [221]
Arkley, Patrick *see* Letter to the reverend alexander beith, stirling
Arklow reporter – Arklow, Ireland 30 aug 1890-10 jun 1893 – 1 1/4r – 1 – (incorp with: bray herald) – uk British Libr Newspaper [072]
Arlegui, Jose de *see* Chronica de la provincia de n p s francisco de zacatecas
O arlequim – Rio de Janeiro, RJ: Typ do Arlequim, 05 maio-29 dez 1867 – (= ser Ps 19) – mf#P03,01,14 – bl Biblioteca [321]
Arlequin sauvage / Lisle de la Drevetiere, Louis-Francois de – comedie en prose et en 3 actes. Paris. 1783 – 1 – fr ACRPP [440]
Arlesey – (= ser Bedfordshire parish register series) – 3mf – 9 – £7.50 – uk BedsFHS [929]
Arlincourt, Charles V d' *see*
– Dieu le veut
– Place au droit
Arlington 1754-1849 – Oxford MA (mf ed 1994) – (= ser Massachusetts vital record transcripts to 1850) – 1v on 4mf – 9 – 0-87623-201-2 – (mf 1t: births 1754-1834; marriages 1838-40, 1845; deaths 1766-1835; intentions 1840-45. mf 2t: intentions 1807-40; marriages 1807-22. mf 3t: marriages 1818-44. mf 3t-4t: births 1843-49. mf 4t: marriages 1844-49; deaths 1845-49) – us Archive [978]
Arlington appeal – Arlington OR: S A Thomas, 1903 [wkly] [mf ed 1968] – 1r – 1 – (cont by: appeal (1903-05)) – us Oregon Lib [071]
[Arlington-] arlington times – CA. sep 24 1908-sep 17 1977 – 17r – 1 – $1020.00 – mf#R03142 – us Library Micro [071]
Arlington bulletin – Jefferson OR: W E & J W Burton, [wkly] [mf ed 1966] – 6r – 1 – (ceased in 1942; absorbed: boardman mirror (1921-25)) – us Oregon Lib [071]
Arlington bulletin *see* Boardman mirror
The arlington citizen – Blair, NE: J Hilton Rhoades, mar 1966- (wkly) [mf ed mar 2 1967] – 8r – 1 – (issues for nov 21 1974-called v20 n39-) – us NE Hist [071]
Arlington first baptist church : church records – ARLINGTON, TX. 1911-41 – 1 – $65.43 – us Southern Baptist [242]
Arlington herald – Arlington, NE: Geo F Goodell, 1885-v20 n35. jul 5 1902 (wkly) – 1r – 1 – (cont by: herald (arlington ne)) – us Bell [071]
Arlington independent – Arlington OR: H W Lang, 1913- [wkly] [mf ed 1977] – 1r – 1 – us Oregon Lib [071]
Arlington record – Arlington OR: J M Johns, [wkly] [mf ed 1966] – 2r – 1 – us Oregon Lib [071]
Arlington review-herald – Arlington, NE: Fassett Print Co, 1904-v64 n4. nov 25 1948 (wkly) – 13r – 1 – (formed by the union of: arlington review and herald (arlington ne). cont by: washington county review-herald) – us Bell [071]
Arlington times – Arlington, NE: R O Willis & Co, 1892 (wkly) – 2r – 1 – (cont: people's defender (arlington ne)) – us Bell [071]
[Arlington-] usa today – VA. oct 18 1989-Loma Prieta Earthquake – $110.00 – mf#B06071 – us Library Micro [071]
Arlington weekly news – Arlington, NE: B C Maynard (wkly) – 1r – 1 – us Bell [071]
Arlis/north america newsletter / Art Libraries Society of North America – Kanata ON 1972-81 – 1,5,9 – (cont by: art documentation: bulletin of the art libraries society of north america) – ISSN: 0090-3515 – mf#10830 – us UMI ProQuest [700]
Arliss, Jean *see* The quick years
Arlosoroff, C *see* Der juedische volkssozialismus
Art, Gustave Otto *see* Trutznachtigall
Arm – 1970 may-jun – 1 – mf#1051476 – us WHS [071]
Arm and hammer – 1898 dec – 1 – mf#3177680 – us WHS [071]
The arm chair – New York. v1-5 1878-84 (wanting v3 no 134) – 1 – $47.00 – us L of C Photodup [640]
Arm report – n72-76 [1974 jun ?-1976 jun] – 1 – mf#379634 – us WHS [071]
Arm the masses – 1991 sep-1995 may/jun – 1 – mf#4712809 – us WHS [071]
Arma records management quarterly / Association of Records Managers and Administrators – Lenexa KS 1967-98 – 1,5,9 – (cont by: information management journal) – ISSN: 1050-2343 – mf#6778.01 – us UMI ProQuest [020]
The armada watchman – Armada, NE: R A Reid. v1 n47. apr 11 1889 (wkly) [mf ed Apr.11,1889-July 10,1890 (gaps) filmed 1976] – 1r – 1 – us NE Hist [071]
Armaes do museu paulista, tomos 1 y 2 / ed by Bayle, Constantino – Madrid: Razon y Fe, 1927 – 1 – sp Bibl Santa Ana [355]
Armagh gazette – Armagh, Ireland 12 jan-12 oct 1850 – 1 – (cont: ulster gazette, agricultural & sporting chronicle, etc [7 oct 1844-31 dec 1849]; cont by: ulster gazette [23 nov 1850-5 jun 1909]) – uk British Libr Newspaper [072]
Armagh guardian – Ireland, 3 dec 1844-1971; 1973-15 oct 1982 – 119 1/2r – 1 – uk British Libr Newspaper [072]
Armagh observer – Armagh, Ireland. 8 jun 1935-43; 1986-93 – 29r – 1 – uk British Libr Newspaper [072]
Armagh observer – [Northern Ireland] Belfast jan 1944-dec 1950 [mf ed 2002] – 5r – 1 – uk Newsplan [072]
Armagh standard – Armagh, Ireland 11 apr 1884-4 jun 1909 [mf 1885-87, 1890-96] – 1 – (incorp with: ulster gazette fr jun 1909) – uk British Libr Newspaper [072]
Arman – Tehran: Sazman-i Javanan va Danishjuyan-i Dimukrat-i Iran. dawrah-'i 2, sal-i 1, shumarah-'i 1-22. 16 isfand 1357-17 murdad 1358 [6 mar 1979-8 aug 1980] – 1r – 1 – $53.00 – us MEDOC [079]
Armana prouvencau – Avignon. 39v. 1855-1893 – 117mf – 8 – mf#H-1377 – ne IDC [440]
Armand, Joseph *see* Plaies sociales au dix-neuvieme siecle
Armand, L M *see*
– Dusha kooperatsii
– Kooperativnaia chainaia
– Narodnyi teatr i kooperatsiia
Arman-i mustaz'afin – Tehran, 1979-80. shumarah-'i 1-32 [apr 1979-mar 1980] – (= ser Persian journals and periodicals) – 1r – 1 – $53.00 – (missing: n19-28, 30-31) – us MEDOC [956]
Armas Chitty, Jose Antonio De *see* Zaraza
Armas dominicanas / Morel, Emilio A – Ciudad Trujillo, Dominican Republic. 1939 – 1r – us UF Libraries [972]
Armas, Gabriel *see*
– Donoso cortes
– Donoso cortes en la problematica de la espiritualidad (esbozo de biografia mistica)
– Fama, eclipse y resurreccion de donoso
– Por que volvemos a donoso cortes
Armas i triunfos...de los hijos de galicia... / Gandara, F – Madrid, 1662 – 13mf – 9 – sp Cultura [946]
Armas Medina, Fernando *see* Pizarro
Armas para ganar una nueva batalla / Guinea, Gerardo – Guatemala, 1957 – 1r – us UF Libraries [972]
Armas Y Cardenas, Jose De *see*
– Historia y literatura
– Perfidia espanola ante la revolucion de cuba
Armas Y Cardenas, Susini De... *see* Seleccion de trabajos
Armas Y Cespedes, Jose De *see* Frasquito
Armbrust, L *see* Die territoriale politik der paepste von 500 bis 800
Armbruster, Joh Mich *see* Schwaebisches museum
Der arme heinrich / Aue, Hartmann von der; ed by Paul, Hermann – 6. aufl. Halle (Saale): M Niemeyer 1921 [mf ed 1993] – (= ser Altdeutsche textbibliothek 3) – 6r – 1 – (incl bibl ref) – mf#3244p – us Wisconsin U Libr [800]
Der arme heinrich hartmanns von aue : eine interpretation / Nagel, Bert – Tuebingen: M Niemeyer 1952 [mf ed 1993] – 1r – 1 – (incl bibl ref. filmed with: hartmann von aue als lyriker / f saran & other titles) – mf#3395p – us Wisconsin U Libr [430]
Der arme heinrich nebst dem inhalte des "erek" und "iwein" / Aue, Hartmann von der & Wernher der Gartenaere – 9. aufl. Halle (Saale): Waisenhauses, 1925 [mf ed 1996] – (= ser Denkmaeler der aelteren deutschen literatur 2/2) – vi/126p – 1 – (trans by gotthold boetticher. incl bibl ref) – mf#9734 – us Wisconsin U Libr [430]
Der arme heinrich nebst dem inhalte des 'erek' und 'iwein' / Aue, Hartmann von – Halle/S: Verlag der Buchhandlung des Waisenhauses, 1891 [mf ed 1993] – (= ser Denkmaeler der aelteren deutschen literatur 2/2; Die hoefische dichtung des mittelalters) – 1 – (incl bibl ref) – mf#8185 – us Wisconsin U Libr [810]
Die arme kleine; stille welt / Ebner-Eschenbach, Marie von – Leipzig: H Fikentscher, H Schmidt & H Guenther, [1928] – 2r – 1 – us Wisconsin U Libr [430]
Der arme konrad – Berlin DE, 1896-97 [gaps] – 1r – 1 – gw Misc Inst [074]
Der arme konrad : kalender fuer das arbeitende volk – Muenchen DE, 1903, 1906 – 1r – 1 – gw Misc Inst [331]
Der arme mann im tockenburg / Braeker, Ulrich; ed by Buelow, Eduard – Leipzig: G Wigand, 1852 [mf ed 1989] – x/411p (ill) – 1 – mf#7062 – us Wisconsin U Libr [830]
Der arme narr : schauspiel in einem akt / Bahr, Hermann – 2. aufl. Wien: C Konegen (E Stuepnagel), 1906 [mf ed 1998] – 92p – 1 – mf#9960 – us Wisconsin U Libr [810]
Der arme teufel – Berlin DE, 1902-04 – 1 – gw Misc Inst [074]
Der arme teufel – Detroit MI (USA), 1884 6 dec-1894 17 nov [gaps] – 4r – 1 – gw Misc Inst [071]
Armed citizen news – 1974 feb-1978 dec/1980 jan – 1 – mf#1051461 – us WHS [071]
Armed force – 1945 oct 13-1952 nov 8 – 1 – mf#1052343 – us WHS [071]
Armed forces and society – Piscataway NJ 1978+ – 1,5,9 – ISSN: 0095-327X – mf#11904 – us UMI ProQuest [306]
Armed forces chemical journal – Arlington VA 1946-64 – 1 – mf#1545 – us UMI ProQuest [355]
Armed forces chemical journal – v. 1-16. Oct 1946-Dec 1962 – 1 – 93.00 – us L of C Photodup [660]
Armed Forces Communications Association *see* Signal
Armed forces comptroller – Alexandria VA 1956+ – 1,5,9 – ISSN: 0004-2188 – mf#5805 – us UMI ProQuest [355]
Armed forces journal – Springfield VA 1863-73 – 1,5,9 – (cont by: armed forces journal international) – ISSN: 0004-220X – mf#1773.02 – us UMI ProQuest [355]
Armed forces journal international – Springfield VA 1974+ – 1,5,9 – (cont: armed forces journal) – ISSN: 0196-3597 – mf#1773.02 – us UMI ProQuest [355]
Armed forces management – v. 1-9. Oct 1954-Sep 1963 – 1 – us L of C Photodup [355]
Armed forces talk – 1944 jul 31-1945 sep 15 – 1 – mf#450049 – us WHS [335]
Armed services ymca : [newsletter] – v5 n2-3 [1986 summer-fall]; 1992 spring, summer, winter – 1 – mf#1051802 – us WHS [071]
The armed struggle and life of the khmer people in the liberated areas in pictures – [n.p.] NUFC Press [National United Front of Cambodia] [mf ed 1989] – [ill] 1r with other items – 1 – mf#mf-10289 seam reel 017/12 [§] – us CRL [959]
Armed struggle in southern africa / Africa Research Group – Ithaca, NY. 1969? – 1r – us UF Libraries [960]
L'armee d'afrique – n1-49. Alger. 1924-nov 1928 – 1 – fr ACRPP [355]
L'armee de la republique espagnole qui defend la democratie et la paix – Paris, 1937? – (= ser Blodgett coll) – 9 – mf#fiche w 727 – us Harvard [946]
Armee und marine – Berlin DE, 1900 1 oct-1904/05 – 2mf=4df – 9 – (aufgabe in ueberall) – gw Mikrofilm [074]
Armeiskii sbornik – 1921-96 – 9 – (ceased publ) – mf#m0028 – us East View [077]
Armeleutslieder / Kamp, Otto – 3. durchg aufl. Frankfurt a.M: Knauer 1888 [mf ed 1991] – 1r – 1 – (filmed with: ernst junger / wulf dieter muller) – mf#2749p – us Wisconsin U Libr [810]
Armellada, Cesareo De *see* Tauron panton
Armellini, Giuseppe *see* Astronomia e geodesia
Armen, Eric V *see* The caseload experiences of the district courts from 1972 to 1983
Armenia : a year at erzurum / Curzon, R – London, 1854 – 4mf – 9 – mf#AR-1425 – ne IDC [915]
Armenia and the near east / Nansen, F – London, 1928 – 4mf – 9 – mf#AR-1445 – ne IDC [956]
[Armenia-] kommunist armenia – USSR. 1963; 1964-1966; 1968-1974 – 11r – 1 – $550.00 – mf#B63585 – us Library Micro [320]
Armenia weekly – Sydney – at Pascoe [079]
Armeniaca *see* Zeitschrift fuer die erforschung der sprache und kultur armeniens
The armenian apology and acts of apollonius : and other monuments of early christianity / ed by Conybeare, Frederick Cornwallis – 2nd ed. London: S Sonnenschein; New York: Macmillan, 1896 – 1mf – 9 – 0-7905-8023-3 – (incl bibl ref) – mf#1988-6004 – us ATLA [243]
The armenian awakening : a history of the armenian church, 1820-60 / Arpee, Leon – Chicago: University of Chicago Press; London: T Fisher Unwin, 1909 – 1mf – 9 – 0-8370-7601-3 – (incl ind) – mf#1986-1601 – us ATLA [240]
The armenian church / Dowling, Theodore Edward – London: S.P.C.K.; New York: E.S. Gorham, 1910 – 1mf – 9 – 0-8370-7624-2 – mf#1986-1624 – us ATLA [240]
Armenian Church of America *see* Bema

Armenian General Benevolent Union *see* Hoosharar

The armenian genocide in the us archives, 1915-1918 / U.S. National Archives and Library of Congress – [mf ed Chadwyck-Healey] – 396mf – 9 – (with p/ind) – uk Chadwyck [956]

Armenian kingdom of cilicia / Kurkjian, Vahan M – New York: V M Kurkjian 1919 [mf ed 1999] – 1r – 1 – mf#28641 – us Harvard [933]

Armenian mirror-spectator – 1979 feb 24-dec; 1980; 1981-1982 jun; 1982 jul-1983 dec; 1984 jan-1985 apr; 1985 may 4-1986 dec 27; 1987 jan-1988 jun; 1988 jul-1989 sep – 1 – mf#1109626 – us WHS [071]

Armenian observer – 1971 jul 14-1973 jul 25; 1973 aug 1-1975 jun 25; 1975 jul-1977jun; 1977 jul 6-1978 dec 27; 1979 jan 3-1980 dec; 1981-82; 1983-84; 1985-1986 aug; 1986 sep 3-1988 aug 31 – 1 – mf#1052346 – us WHS [071]

Armenian reporter – 1971 jan-1972 jun; 1972 jul-1973 nov; 1973 dec-1975 jun; 1977 feb-1978 jul; 1978 aug-1979; 1980; 1981-1982 apr; 1982 jul-1983 dec; 1984-85 feb 21; 1985 feb 28-1986 may; 1986 jun-1987 sep; 1987 sep 24-1988; 1989-92 – 1 – mf#801088 – us WHS [071]

Armenian review – Watertown MA 1948+ – 1,5,9 – ISSN: 0004-2366 – mf#7461 – us UMI ProQuest [073]

Armenian weekly – 1986 oct 4-1987; 1988-1989 mar 25; 1989 apr-1990 mar – 1 – mf#289405 – us WHS [071]

Armenian weekly – Watertown MA 1974+ – 1,5,9 – ISSN: 0148-2971 – mf#8736 – us UMI ProQuest [305]

The armenians : a tale of constantinople / Mac Farlane, C – London, 1830. 3v – 12mf – 9 – mf#AR-2027 – ne IDC [956]

Armenians in india : from the earliest times to the present day: a work of original research / Seth, Mesrovb Jacob – Calcutta: M J Seth, 1937 – (= ser Samp: indian books) – us CRL [954]

L'armenie chretienne et sa litterature / Neve, Felix – Louvain: Charles Peeters, 1886 [mf ed 1986] – 1mf – 9 – 0-8370-7894-6 – (incl bibl ref) – mf#1986-1894 – us ATLA [240]

Armenien einst und jetzt / Lehman-Haupt, C F – Berlin, 1910-1931. 2v – 14mf – 9 – (missing: v1) – mf#AR-1429 – ne IDC [915]

Armenien unter der arabischen herrschaft... / Ghazarian, M – Marburg, 1903 – 1mf – 9 – mf#AR-1428 – ne IDC [950]

Armenien-index : bilddokumentation zur kunst in armenien / ed by Bildarchiv Foto Marburg – Deutsches Dokumentationszentrum fuer Kunstgeschichte Philipps- Universitaet Marburg – [mf ed 2000] – 89mf (1:24) – 9 – diazo €1896.00 – ISBN-10: 3-598-34535-6 – ISBN-13: 978-3-598-34535-7 – gw Saur [700]

Armenini, G B *see* De veri precetti della pittura...

Armenische irenaeusfragmente / Irenaeus – Leipzig: J C Hinrichs 1913 [mf ed 1989] – (= ser Tugal 6/3) – 1mf – 9 – 0-7905-1899-6 – (in german & armenian; incl bibl ref & ind) – mf#1987-1899 – us ATLA [490]

Armenische irenaeusfragmente / Jordan, H – Leipzig, 1913 – (= ser Tugal 3-36/3) – 4mf – 9 – €11.00 – ne Slangenburg [240]

Die armenische kirche in ihren beziehungen zu den syrischen kirchen bis zum ende des 13. jahrhunderts / Ter Minassiantz, E – Leipzig, 1904 – (= ser Tugal 2-26/4) – 4mf – 9 – €11.00 – ne Slangenburg [240]

Die armenische kirche in ihren beziehungen zur byzantinischen : (vom 4 bis 13 jahrhundert) / Ter-Mikelian, A – Leipzig, 1892 – 2mf – 9 – mf#AR-1825 – ne IDC [243]

Armenische Vaeter *see*
- Ausgewaehlte schriften, 1. bd (bdk57 1.reihe)
- Ausgewaehlte schriften, 2. bd (bdk58 1.reihe)

Die armennot *see* Die wassernot im emmental / die armennot / eines schweizers wort

Die armennot / ein sylvestertraum / eines schweizers wort / Gotthelf, Jeremias [Albert Bitzius]; ed by Vetter, Ferdinand – Bern: Schmid & Francke, 1899 [mf ed 1993] – (= ser Volksausgabe seiner werke im urtext 7) – 357p – 1 – mf#8532 reel 2 – us Wisconsin U Libr [830]

Armeno, Christoforo *see* Die reise der soehne giaffers

Armenpflege und wohltätigkeit in Zuerich zur zeit Ulrich Zwinglis / Koehler, W – Zuerich, 1919 – 1mf – 9 – mf#ZWI-74 – ne IDC [242]

Armenth-Brothers, Francine R *see* Freshmen athletes' perceptions of adjustment to intercollegiate athletics

Armes for red spain / Bayle, Constantino & Hericurt, Pierre – London, 1938; Burgos: Razon y Fe, 1938 – 1 – sp Bibl Santa Ana [946]

Les armes romaines / Couissin, P – Paris, 1926 – €21.00 – ne Slangenburg [930]

Armfield, H T *see* The three witnesses

Armidale argus – Armidale, apr 1899-dec 1907 – 3r – A$195.32 vesicular A$212.06 silver – at Pascoe [079]

Armidale chronicle – Armidale, 1872-76, 1910, 1917-29 – at Pascoe [079]

Armidale express – Armidale, jan 1964-dec 1968, jan 1969-aug 1997 – at Pascoe [079]

Armidale express, and new england general advertiser – Armidale, Australia 8 aug 1893-30 jun 1922 (very imperfect) – 33r – 1 – uk British Libr Newspaper [079]

Armidale independent – Armidale – at Pascoe [079]

Armide : drame-heroique, en cinq actes / Gluck, Christophe Willibald – Paris, France. 1814 – 1r – us UF Libraries [440]

Armide : tragedie mise en musique / Lully, Jean Baptiste – Paris: C Ballard 1686 [mf ed 19–] – 1r – 1 – (libretto by philippe quinault) – mf#film 617 – us Sibley [780]

The armies of the native states of india – London, 1884 – (= ser 19th c books on british colonization) – 2mf – 9 – mf#1.1.9018 – uk Chadwyck [355]

Armiia 7-aia *see* Izvestiia armejskogo komiteta 7-j armii

Armiia 8-aia : izvestiia armejskogo komiteta 8-oj armii – Moscow, Russia 1917 [mf ed Norman Ross] – 1r – 1 – mf#nrp-81 – us UMI ProQuest [077]

Armiia 8-aia *see* Izvestiia 2-go armejskogo s"ezda 8-oj armii

Armiia 9-aia *see* Izvestiia armejskogo komiteta 9-oj armii

Armiia 11-aia *see* Izvestiia shtaba 11-oj armii

Armiia chetvertaia *see*
- Golos soldata
- Izdaetsia armejskim komitetom

Armiia i narod – Ufa: Biuro pechati Shtaba voisk Samar armii 1918 [1918 1 sent-] – (= ser Asn 1-3) – n1-89 [1918] item 14, on reel n6 – 1 – mf#asn-1 014 – ne IDC [077]

Armiia piataia *see* Izvestiia ariejskogo ispolnitel'nogo komiteta 5-j armii

Arminia – jahrg 5 hft 18 [1886 jul 11] – 1 – mf#1002920 – us WHS [071]

Arminian inconsistencies and errors : in which it is shown that all the distinctive doctrines of the presbyterian confession of faith are taught by standard writers of the methodist episcopal church / Brown, Henry – Philadelphia: W S & A Martien 1856 [mf ed 1992] – 1mf – 9 – 0-524-04252-7 – (incl bibl ref) – mf#1991-2036 – us ATLA [242]

Arminian magazine : consisting of extracts and original treatises on general redemption – La Salle 1789-90 – 1 – mf#3511 – us UMI ProQuest [200]

Arminian magazine: : consisting of extracts and original treatises on universal redemption – London 1778-97 – 20v on 1089mf – 9 – (cont as: the methodist magazine. london, 1798-1821 v21-44;the eesleyan-methodist magazine. london, 1822-1894. v45-117;1820,(v43:aug,p561-640),1821(v44:jun/jul,p401-560),1856(v79:pt2),1878-1891(v101-114)) – mf#H-2753 – ne IDC [242]

The arminian magazine... – Stoke-Damarel, 1822-1828. v1-7 – 100mf – 9 – (cont as: the bible christian magazine...shebbear,1829-1850 n.s.,v1-7;s3,v1-15; missing: 1822(v1),1823(v2:p37-72),1825(v4),1828(v7:n.s.),1829-1832(v1-3),1832(v4:p195-198),1833(v5:p105-108),s3,1836-1838,v1-3) – mf#MP-350 – ne IDC [240]

Arminianism in history : or, the revolt from predestinationism / Curtiss, George Lewis – Cincinnati: Cranston & Curts; New York: Hunt & Eaton 1894 [mf ed 1989] – 1mf – 9 – 0-7905-4114-9 – mf#1988-0114 – us ATLA [242]

Arminius – Muenchen DE, 1926 13 jan-1927 11 sep – 1r – 1 – gw Misc Inst [074]

Arminius, Iac *see* Opera theologica

Arminius, Jacobus *see*
- Iacobi arminii...disputationes...publicae et privatae
- Opera theologica
- Verclaringhe iacobi arminii...

Arminjon, P *see*
- De la nationalite dans l'empire ottoman specialement en egypte
- Les societes anonymes etrangeres en egypte...

Armitage, John *see* Historia do brasil

Armitage, Maria T *see* Historical overview of the national baseball library

Armitage, Merle *see* George gershwin

Armitage, Thomas *see*
- Christian union
- A history of the baptists
- Jesus
- Opening sermon before the hudson river association south, the

Armitage-Smith, G *see* The citizen of england

Armley and wortley news – Armley, England 6 sep1889-30 sep 1932, 8 aug 1947-4 apr 1958 – 1 – (discontinued; between 30 sep 1932 & 8 aug 1947 incorp with: leeds guardian; wanting: 1894, 1897, 1900, 1909) – uk British Libr Newspaper [072]

Armonia / Melara Berrocal, Isidro – Caceres: (Tip. El Noticiero), 1948 – 1 – sp Bibl Santa Ana [946]

La armonia del parnas, mes nvmerosa en las poesias varias del atlant del cel poetic : recopiladas, y emendadas per dos ingenis de la molt illustre / Garcia, Francesc V – Barcelona: Figuero 1700 – (= ser Whsb) – 3mf – 9 – €40.00 – mf#Hu 040 – gw Fischer [410]

Armonia ecclesiasticorum concertuum / Vernizzi, Ottavio – 1604 – (= ser Mssa) – 1mf – 9 – €20.00 – mfchl 434 – gw Fischer [780]

Armor – Fort Knox KY 1888+ – 1,5,9 – ISSN: 0004-2420 – mf#2361 – us UMI ProQuest [621]

Armour and weapons / Ffoulkes, Charles John – Oxford, England. 1909 – 1r – us UF Libraries [355]

Armour, Edward Douglas *see*
- Division courts and small credits
- Essays on the devolution of land upon the personal representative
- A treatise on the investigation of titles to real estate in ontario
- A treatise on the law of real property

Armour, John M *see* Atonement and law

The armour of light *see* Kuang ming ti chuang pei (ccm276)

Armoury : a magazine of weapons for christian warfare – La Salle IL 1873-81 – 1 – mf#4746 – us UMI ProQuest [355]

Arms control and disarmament – 1964 65-1970 – 1 – $88.00 – us L of C Photodup [327]

Arms control and disarmament / U.S. Dept of State – n1-11. 1963-68 [all publ] – 11mf – 9 – $16.50 – mf#llmc 81-912 – us LLMC [327]

Arms control and disarmament / U.S. Library of Congress – Library of Congress. v1-9. 1965-73 (all publ) – 65mf – 9 – $97.00 – mf#LLMC 79-450 – us LLMC [327]

Arms control and disarmament – Washington DC 1964-73 – 1,5,9 – ISSN: 0000-0272 – mf#1757 – us UMI ProQuest [327]

Arms control and disarmament bibliography – v1 n1-v9 n2. 1964-73 [complete] – (= ser The library of world peace studies) – 44mf – 9 – $270.00 – us UPA [327]

Arms for red spain / Hericourt, Pierre – London: Burns, Oates & Washbourne [1938] [mf ed 1977] – (= ser Blodgett coll) – 1mf – 9 – mf#w939 – us Harvard [946]

Die armseligen besenbinder : altes maerchen in fuenf akten / Hauptmann, Carl – Leipzig: K Wolff, 1913 – 1r – 1 – us Wisconsin U Libr [820]

Armson, Thomas *see* Remarks and animadversions on the roman catholic religion

Armstong, Richard Acland *see* Agnosticism and theism in the nineteenth century

Armstrong, A H *see* The architecture of the intelligible universe in the philosophy of plotinus

Armstrong, Alexander *see* Shantung (china)

Armstrong, Charles Newhouse *see* Canada and her resources

Armstrong, E S *see* The history of the melanesian mission

Armstrong, Edward J *see* The sportsman's and tourist's guide to the hunting, fishing and pleasure resorts of new brunswick

Armstrong, Frances Charlotte *see* Old caleb's will

Armstrong, George *see* Names and places in the old and new testament and apocrypha

Armstrong, George Dodd *see*
- A half hour with robert elsmere
- The sacraments of the new testament
- The theology of christian experience

Armstrong, George E *see*
- Angioma of the head
- Clinical lecture on the surgical treatment of perforated gastric ulcer
- Cystic tumors of the brain following traumatism – jackson epilepsy – operation – perfect recovery
- Excision of half the tongue
- Gall-stone surgery
- Hospital abuse
- The pathology, diagnosis and treatment of perforated gastric ulcer
- The surgical treatment of typhoid fever
- Tuberculous disease of the spine
- The wisdom of surgical interference in haematemesis and melaena from gastric and duodenal ulcer

Armstrong, George Frederick *see* Inaugural lecture of the department of practical science in mcgill university, montreal

Armstrong, George Gilbert *see* Richard acland armstrong

Armstrong, Isabel *see* Nineteenth-century british periodicals

Armstrong, J A *see* Jubilee of "christ church", newbury ont

Armstrong, J S *see* Specification

Armstrong, James *see* A treatise on the law relating to marriages in lower canada

Armstrong, Jessie F *see*
- Celestine and sallie
- Ernest and ida
- Little phil's christmas gifts
- Through rosamund's eyes

Armstrong, John *see*
- Church's office towards the young
- The eclectic almanac for the year 1839
- Histoire naturelle et civile de l'isle de minorque
- Pattern of church building
- Reminiscences

Armstrong, John Gilbert *see*
- Separate schools
- The supremacy of the sovereign

Armstrong, John Simeon *see* Schemes showing the possibilites of st john, nb

Armstrong, John Simpson *see* A manual of the law and practice at elections in ireland

Armstrong, Joseph H *see* Lyrics, idyls and fragments

Armstrong, Louis Olivier *see*
- A canoe trip through temagaming the peerless in the land of hiawatha
- For actual settlers
- Southern manitoba and turtle mountain country

Armstrong, Orland Kay *see* Life and work of dr a a murphree

Armstrong, Paul *see* Mysterieux jimmy

Armstrong, Richard Acland *see*
- Discourses
- Faith and doubt in the century's poets
- God and the soul
- Latter-day teachers
- Man's knowledge of god
- Martineau's "study of religion"
- Modern review
- The trinity and the incarnation

Armstrong, Richard Acland et al *see* The triumph of faith

Armstrong, Robert *see* Linear phonography

Armstrong, Robert Cornell *see*
- Just before the dawn
- Light from the east

Armstrong, Robert G *see*
- Memoir of hannah hobbie
- Study of west african languages

Armstrong, Samuel Chapman *see*
- Education for life
- Founding of the hampton institute

Armstrong surname bulletin – 1969-1979 oct – 1 – mf#498177 – us WHS [929]

Armstrong, T B *see* Journal of travels in the seat of war, during the last two campaigns of russia and turkey

Armstrong, Walter *see*
- Alfred stevens
- The art annual for 1891 briton riviere royal academician
- The art of velazquez
- The art of william quiller orchardson
- El arte en la gran bretana e irlanda
- Celebrated pictures exhibited at the glasgow international exhibition
- The life of velazquez
- Memoir of peter de wint
- Scottish painters

Armstrong, William *see* Five-minute sermons to children

Armstrong, William Dunwoodie *see* Inaugural address delivered before knox college metaphysical and literary society

Armstrong, William Dunwoodie [i.e. Vindex] *see*
- Criticism of mr lesueur's pamphlet, entitled defence of modern thought
- Reply to the appendix of mr lesueur's criticism no 2

Armstrong, William Jackson *see* The masses and the millionaires

Armstrong, William Reginald *see*
- Essay on the times
- Essays on the times
- Romanism

Armstrong's contested election cases / New York. (State) – 1v. 1777-1871 – 3mf – 9 – $4.50 – mf#LLMC 80-014 – us LLMC [340]

Armut : ein trauerspiel / Wildgans, Anton – Leipzig: L Staackmann 1927, c1914 [mf ed 1991] – 1r – 1 – (filmed with: kirbisch & other titles) – mf#3052p – us Wisconsin U Libr [820]

Army – Arlington VA 1950+ – 1,5,9 – ISSN: 0004-2455 – mf#2717 – us UMI ProQuest [355]

Army – v. 1-9. Aug 1954-Jul 1963 – 1 – us L of C Photodup [355]

Army al&t – Washington DC 2000+ – 1,5,9 – (cont: army rd and a) – ISSN: 1529-8507 – mf#2118,04 – us UMI ProQuest [355]

Army and navy : coast guard / Goebel, Rubye K – s.l, s.l? . 1936 – 1r – us UF Libraries [355]

Army and navy : (daytona beach) / Goebel, Rubye K – s.l, s.l? . 1936 – 1r – us UF Libraries [355]

Army and navy chronicle – La Salle IL 1835-42 – 1 – mf#3938 – us UMI ProQuest [355]

Army and navy chronicle and scientific repository – La Salle IL 1843-44 – 1 – mf#3939 – us UMI ProQuest [355]

Army and navy journal – v. 1-83. 1863-1946 – 1 – us L of C Photodup [355]

Army and navy posts : fort matanzas / Keleher, M R – s.l, s.l? . 1935 – 1r – us UF Libraries [355]
Army Armament Research and Development Center [US] *see* Voice
Army desert training – Historical Archive, 1946 – 1r – 1 – $50.00 – mf#R60010 – us Library Micro [355]
An army doctor's romance / Allen, Grant – London: Tuck, [1893] – (= ser Breezy library series) – 2mf – 9 – 0-665-94433-0 – mf#94433 – cn Canadiana [830]
Army dollar – v9 n13,17,19-21,23-1925 [1983 jun 30, aug 17, sep 19-oct 26; nov 17-dec 15]; v10 n5-6,9-[1984 mar 8-1922 apr 19-sep 20]; v11 n17-24 [1985 aug 24-sep 29, dec 5]; v12 n1,5,9-12,14-1925 n26 [1986 jan 9, mar 6, may 1-jun 12, jul 10] – 1 – mf#1052351 – us WHS [071]
Army families – 1987 sep-1993 sep – 1 – mf#1551557 – us WHS [355]
Army flier – 1980 aug 21-1992 mar-sep [with gaps] – 1 – mf#1002310 – us WHS [355]
The army historical program in the european theater and command, 8 may 1943-31 dec 1950 / U.S. Army. Historical Division. European Command – v. 1-4. 1951 – 7 – us L of C Photodup [977]
Army in europe – 1967 feb-1972 dec – 1 – mf#1532601 – us WHS [355]
Army information digest – v. 1-17. 1946-62 – 1 – us L of C Photodup [355]
Army lawyer – Washington DC 1976+ – 1,5,9 – ISSN: 0364-1287 – mf#11232 – us UMI ProQuest [355]
The army lawyer – v1-4. 1971-74; 1975-84; 1987-95 – (= ser Military law coll) – 225mf – 9 – $337.00 – (none publ between 1985-86. updates planned) – mf#LLMC 84-238 – us LLMC [340]
Army lists, 1740-1784 – London: Royal Artillery Institution Library – (= ser BRRAM series) – 81mf – 9 – £648 / $1296 – (with p/g; int by ivor burton) – mf#87300 – uk Microform Academic [355]
Army logistician – Washington DC 1969+ – 1,5,9 – (cont by: alog: army logistician) – ISSN: 0004-2528 – mf#5713.02 – us UMI ProQuest [355]
Army manuals and regulations index : (consolidated index of army publications and blank forms) / U.S. National Technical Information Service – Quarterly. Items listed by title and Army number – 9 – us NTIS [000]
Army museum newsletter – 1969 sep-1974 oct – 1 – mf#225062 – us WHS [355]
Army, navy : coast guard / Scoville, Dorothy R – s.l, s.l? . 1936 – 1r – us UF Libraries [355]
Army news – Darwin, 1941-45 – 3r – A$115.50 vesicular A$132.00 silver – at Pascoe [079]
The army of india question / Porter, Neale – London 1860 – (= ser 19th c british colonization) – 1mf – 9 – mf#1.1.3662 – uk Chadwyck [355]
Army orders, vouchers, returns, 1792-1793 / Torrence, Aaron – 1r – 1 – mf#B25933 – us Ohio Hist [355]
Army quarterly and defence journal, 1920-1983 – 26r – 1 – mf#C39-27670 – us Primary [355]
Army r, d and a – Washington DC 1978-85 – 1,5,9 – (cont: army research and development; cont by: army rd and a magazine) – ISSN: 0162-7082 – mf#2118.04 – us UMI ProQuest [355]
Army rd and a – Washington DC 1995-99 – 1,5,9 – (cont by: army al&t) – ISSN: 0892-8657 – mf#2118.04 – us UMI ProQuest [355]
Army rd and a magazine – Washington DC 1986 – 1,5,9 – (cont: army r, d and a) – ISSN: 0895-111X – mf#2118.04 – us UMI ProQuest [355]
Army re-organization : with special reference to the british soldier in india / Mouat, Frederic John – London 1881 – (= ser 19th c british colonization) – 1mf – 9 – mf#1.1.3750 – uk Chadwyck [355]
Army reporter – v6 n23-v8 n13 [1970 jun 8-1972 apr 24] – 1 – mf#630837 – us WHS [355]
Army research and development – Washington DC 1960-94 – 1,5,9 – (cont by: army r, d and a) – ISSN: 0004-2560 – mf#2118.04 – us UMI ProQuest [355]
Army reserve magazine – Washington DC 1979+ – 1,5,9 – ISSN: 0004-2579 – mf#12111,01 – us UMI ProQuest [355]
Army times – 1944 mar 25-v5 n47 [1945 jun 30]; v2 n20 [1941 dec 27]-1944 mar 18 – 1 – mf#764665 – us WHS [355]
Army times – Springfield VA 1940+ – 1,5,9 – ISSN: 0004-2595 – mf#1759 – us UMI ProQuest [355]
Arnason, Jon *see* Islandske folkesagn og aeventyr
Arnau, J *see*
– Certamen pharmaceutico-galenico in quo tres continentur dissertationes...
– Opus neotericum medicum theorico practicum de laxo et estricto, justa divini. hippocratis mentem
Arnaud, E *see*
– La palestine ancienne et moderne, ou, geographie historique et physique de la terre sainte
– Le pentateuque mosaique defendu contre les attaques de la critique negative
Arnaud, Eugene *see*
– Memoires historiques sur l'origine, les moeurs, les souffrances et la conversion au protestantisme des vaudois du dauphine
– Notice historique et bibliographique
– Recherches critiques sur l'epitre de jude
Arnaud, Henri *see* Glorious recovery by the vaudois of their valleys, from the original with a compendious history of that people, previous and subsequent to that event, by hugh dyke acland
Arnaud, Pierre *see* Trois quatuors a deux violons, alto et basse, oeuvre 1er
Arnaud, Robert *see*
– L'homme qui rit jaune
– L'islam et la politique musulmane francaise en afrique occidentale francaise
Arnauld, A *see*
– Apologie pour les catholiques contre les faussetez et les calomnies d'un livre intitule
– La perpetuite de la foy de l'eglise catholique touchant l'eucharistie
– Le renversement de la morale de jesus-christ par les erreurs des calvinistes, touchant la justification
Arnault, Antoine V *see* Souvenirs d'un sexagenaire
Arnault, Antoine-Vincent *see* Germanicus
Arnault, Lucien Emile *see* Regulus
Arnault, Lucien-Emile *see* Pierre de portugal
Arndt, Augustin *see* Nicolai lancicii de praestantia instituti societatis jesu
Arndt, Ernst M *see*
– Bruchstuecke aus einer reise durch einen theil italiens
– Bruchstuecke aus einer reise von baireuth bis wien im sommer 1798
– Bruchstuecke einer reise durch frankreich im fruehling und sommer 1799
Arndt, Ernst Moritz *see*
– Anthologie aus den werken von ernst moritz arndt
– Blaetter der erinnerung
– Du mein vaterland
– Ernst moritz arndt
– Ernst moritz arndts briefe an eine freundin
– Ernst moritz arndt's saemmtliche werke
– Fragmente ueber menschenbildung
– Der waechter
Arndt, Erwin *see* Fortunatus
Arndt, Friedrich *see* Die vier temperamente
Arndt, Gottfried August *see* Archiv der saechsischen geschichte
Arndt, J *see*
– Paradisz gaertlein
– Postilla
– Vier buecher von wahrem christenthumb
[Arndt, J] *see* Anatomia et laboratorium veri christiani
Arndt, Johann *see*
– Des hocherleuchteten lehrers, herrn johann arndts, weiland general-superintendenten des fuerstenthums lueneburg, sechs buecher vom wahren christenthum
– Johann arndts vier buecher vom wahren christenthum
– True christianity
Arndt, Theodor *see* Die stellung ezechiels in der alttestamentlichen prophetie
Arndts, L et al *see* Kritische ueberschau der deutschen gesetzgebung und rechtswissenschaft
Arndts von Arnesberg, Karl Ludwig, Ritter *see*
– Gesammelte civilistische schriften
– Lehrbuch der pandekten
Arne, Thomas Augustine *see*
– Artaxerxes
– Artaxerxes, an english opera as it is performed at the theatre royal in covent garden
– Comic tunes in the celebrated entertainment call'd harlequin sorcerer as they are perform'd at the theatre-royal in covent-garden
– Soldier tir'd of war's alarms
– Trip to portsmouth
– Where the bee sucks
Arnes Luna, Alfredo *see* Ensayos literarios
Arneth et al *see* Neue encyclopaedie der wissenschaften und kuenste (ael1/24)
Arneth, Franz Hektor, Ritter von *see* Das classische heidenthum und die christliche religion
Arnett, Edward John *see*
– Gazetteer of sokoto province
– Gazetteer of zaria province
Arnett, John Andrews *see* Bibliopegia
Arnett, Thomas *see* Journal of the life, travels and gospel labors of thomas arnett
Arnhard, C von *see* Liturgie zum tauf-fest der aethiopischen kirche
Arnheim, Rudolf et al *see* Jugend und welt
Arnhold, Erna *see* Goethes berliner beziehungen
Arniches, Carlos *see* Sorrow of the writer arniches for the ruin of madrid
Arnigio, B *see* Rime de gli academici occulti con le loro imprese et discorsi
Arnim, Bettina von *see*
– Bettina von arnim
– Clemens brentanos fruehlingskranz
– Correspondence of fraeulein guenderode and bettine von arnim
– Dies buch gehoert dem koenig
– Gespraeche mit daemonen
– Goethes briefwechsel mit einem kinde
– Die guenderode
– Jlius [sic] pamphilius und di ambrosia
– Saemtliche werke
Arnim, H von *see*
– Die entstehung der gotteslehre des aristoteles
– Eudemische ethik und metaphysik
Arnim, Karl O von *see* Reise ins russische reich im sommer 1846
Arnim, Ludwig Achim, Freiherr von *see*
– Achim von arnims werke
– Ariel's offenbarungen
– Arnims troest einsamkeit
– Arnims werke
– Contes bizarres
– Des knaben wunderhorn
– Erzaehlungen
– Fuerst ganzgott und saenger halbgott
– Hollin's liebeleben
– Novellen
– Unbekannte aufsaetze und gedichte
Arnims troest einsamkeit / ed by Pfaff, Fridrich – 2. ausg. Freiburg i.B: J C B Mohr, 1890 [mf ed 1988] – xcvi/412p (ill) – 1 – (incl ind) – mf#6956 – us Wisconsin U Libr [880]
Arnims werke / ed by Schier, Alfred – krit durchges erl ausg. Leipzig: Bibliographisches Institut, [1925?] [mf ed 1988] – (= ser Meyers klassiker-ausgaben) – 3v – 1 – mf#6955 – us Wisconsin U Libr [800]
Arno holz und die deutsche presse / Ress, Robert – Dresden: C Reissner 1913 [mf ed 1991] – 1r – 1 – (incl bibl ref. filmed with: einer baut einen dom / vcarl maria holzapfel) – mf#2733p – us Wisconsin U Libr [070]
Arno holz und die juengstdeutsche bewegung / Strobl, Karl Hans – Berlin: Gose & Tetzlaff c1902 [mf ed 1990] – 1r – 1 – (incl bibl ref. filmed with: der dichter vor der geschichte : holderlin, novalis / reinhold schneider) – mf#2732p – us Wisconsin U Libr [430]
Arnobius lunior *see*
– Opera minora
– Praedestinatus (ccsl25b
Arnobius maior *see* Thesaurus arnobii maioris
Arnold, Albert Nicholas *see*
– Baptist pamphlets
– Commentary on the epistle to the romans
– The scriptural terms of admission to the lord's supper
Arnold, Byron *see* Personality traits of music students.
Arnold, Carl Franklin *see*
– Die ausrottung des protestantismus in salzburg unter erzbischof firmian und seinen nachfolgern
– Caesarius von arelate und die gallische kirche seiner zeit
– Gemeinschaft der heiligen und heiligungs-gemeinschaften
– Die neronische christenverfolgung
Arnold, Charles D *see* Analysis of beginning string ensemble methods
Arnold, Charles Edward *see*
– Chart of christ's journeyings
– Normal studies on the life and ministry of christ
Arnold de Jesus, frere *see* Aux honorables membres du comite catholique du conseil de l'instruction publique
Arnold, Edward Vernon *see* Roman stoicism
Arnold, Edward W *see* One thousand legal facts.
Arnold, Edwin *see*
– Education in india
– Indian idylls
– Indian poetry
– The light of asia
– The light of the world
– The marquis of dalhousie's administration of british india
– Pearls of the faith
Arnold, Friedrich Christian von *see* Beitraege zum teutschen privat-rechte
Arnold, Gary J *see* The lloyd papers
Arnold, Georges-Daniel *see*
– Der pfingstmontag
Arnold, Gottfried *see*
– Die abwege, oder irrungen und versuchungen
– Das eheliche und unverehelichte leben der ersten christen, nach ihren eigenen zeugnissen und exempeln
– Die geistliche gestalt eines evangelischen lehrers
– Gottfried arnolds auserlesene send-schreiben derer alten
– Die verklaerung jesu christi in der seele
Arnold, Hans *see* Lebensdrang und todesverlangen in der deutschen literatur von 1850-1880 im zusammenhang mit der philosophie schopenhauers
Arnold, Heinz Ludwig *see* Als schriftsteller leben
Arnold houbraken und seine "groote schovwburgh" / [Houbraken, A] Hofstede de Groot, C – Haag, 1893. v1 – 7mf – 9 – mf#O-518 – ne IDC [700]
Arnold houbraken's grosse schouburgh der niederlaendische maler und malerinnen / [Houbraken, A] Wurzbach, A von – Wien, 1880. v14 – 7mf – 9 – mf#O-517 – ne IDC [700]
Arnold, James N *see* Vital records of rehoboth massachusetts to 1850
Arnold, Johannes et al *see* Uns blaest der wind nicht ins gesicht
Arnold, John Muehleisen *see*
– Genesis and science
– Islam
Arnold, Matthew *see*
– Corydon
– Discourses in america
– God and the bible
– Irish essays
– Isaiah 40-66
– Letters of matthew arnold, 1848-1888
– Literature and dogma
– St paul and protestantism
Arnold, P T *see*
– Feeding value and nutritive properties of citrus by-products 2
– Management of dairy cattle in florida
Arnold, Paul Johannes *see* Talib
Arnold, Robert Arthur *see*
– The land and the people
Arnold, Robert E *see* An investigation into the grief process and the emotional restabilization of the divorcee with some possible implications for the minister as a therapeutic agent
Arnold, Robert Franz *see* Achtzehnhundertneun
Arnold, Ruth A *see* Quality elementary physical education programs
Arnold, Samuel *see*
– Inkle and yarico
– Piper o'er the meadows straying
– Shipwreck
Arnold sentinel – Arnold, NE: H J Bedford, jul 20 1911 (wkly) – 13r – 1 – us Bell [071]
Arnold sentinel – Arnold, NE: H J Bedford, jul 20 1911 (wkly) [mf ed jun 2 1955 (gaps) filmed 1969] – 17r – 1 – us NE Hist [071]
Arnold, T *see*
– Henrici huntenduniensis historia anglorum
– Memorials of st edmunds abbey
– Symeonis monachi opera omnia
Arnold, Thomas *see*
– Christian duty of granting the claims of the roman catholics
– The christian life
– Postscript to principles of church reform
– Sermons
Arnold, Thomas Kerchever *see*
– Examination of some portions of the rev w goode's "letter to the...
– Remarks on the rev gs faber's primitive doctrine
Arnold, Walter *see* The life and death of the sublime society of beef steaks
Arnold, Wilhelm *see* Wormser chronik
Arnold, William R *see* Ancient-babylonian temple records in the columbia university library
Arnold-Forster, Hugh Oakeley *see* The truth about the land league
Arnoldi chronica slavorum (mgh7:14.bd) – 1886 – (= ser Monumenta germaniae historica 7: scriptores rerum germanicarum in usum scholarum (mgh7)) – €14.00 – ne Slangenburg [240]
Arnold's magazine of the fine arts – La Salle IL 1831-34 – 1 – mf#5242 – us UMI ProQuest [700]
Arnoldsville baptist church – Arnoldsville, GA 1908-94 – 1 – $25.70 – (articles of faith 1993 constitution, church minutes 1908-1994 571p) – mf#6847 – us Southern Baptist [242]
Arnoldus buchelius "res pictoriae" : aanteekeningen over kunstenaars en kunstwerken voorkomende in zijn diarium, res pictoriae, notae quotidianae en descriptio urbis ultrajectinae / Hoogewerff, G J & Regteren Altena, J Q van – 's-Gravenhage, 1928 – 2mf – 9 – mf#O-518 – ne IDC [700]
Arnolphus, B *see* Oratio habita ad sanctissimu dum nostru leone 10 pont. max...
Arnot, Frederick Stanley *see*
– Bihe and garenganze
– Garenganze
– Piloted into port
Arnot, J G *see* The prisoners of the forty-five
Arnot, Sandford *see*
– Clavis orientalis, pt 1
– Clavis orientalis, pt 2
– An essay on the origin and structure of the hindoostanee tongue, or general language of british india
– Letter to the right honourable the president of the india board
– A new persian grammar
– A new self-instructing grammar of the hindustani tongue, the most useful and general language of british india, in the oriental and roman character

Arnot, William *see*
- Grounds of legislative restriction applied to public-houses
- Laws from heaven for life on earth
- The lesser parables of our lord and lessons of grace in the language of nature
- The parables of our lord
- Sabbath school teaching, in its principles and practice

Arnott, G A W *see*
- The botany of captain beechey's voyage
- Notice of a journal of a voyage from rio de janeiro to the coast of peru

Arnott, Peter *see* More impertinence

Arnould, Auguste Jean Francois *see*
- Amant malheureux
- Fete des fous
- Secret

Arnoux, Alexandre *see* Huon de bordeaux

Arnsberger zeitung – Arnsberg 1886, 1889, 1891, 1895, 1898 20 jan-31 mar, 1898 8 mai-1900, 1904-05, 1907, 1908 1 jul-1910 30 jun, 1911, 1912 1 jul-1913 [mf ed 2004] – 10r – 1 – (12 dec 1906: westfaelische tageszeitung) – gw Mikrofilm [074]

Arnsperger, Walther *see* Lessings seelenwanderungsgedanke kritisch beleuchtet

Arnstaedtische woechentliche anzeigen und nachrichten – Arnstadt DE, 1909 [gaps] – 1mf=2df – 1 – (title varies: 1828: der beobachter; 1869: arnstaedtisches nachrichts- und intelligenzblatt) – gw Misc Inst [350]

Arnton, William H *see* Catalogue of the law library of the late r a ramsay, esq, advocate

Arntzenius, Louis Marie George *see* Balletmuziek

Aro citizen – v3 n36 [1881 jan 6] – 1 – mf#852757 – us WHS [071]

Arocha, Jose Ignacio *see* Diccionario geografico, estadistico e historico de...

Arocho Rivera, Minerva *see*
- Paisajes de oro y soledad
- Quimeras e inquietudes
- Sinfonia en negro

Aromas de mi huerto / Negron, Virgilio – Santurce?, Puerto Rico. 1954 – 1r – us UF Libraries [972]

Die aromata in ihrer bedeutung fuer religion, sitten, gebraeuche, handel und geographie des alterthums : bis zu den ersten jahrhunderten unserer zeitrechnung / Sigismund, Reinhold – Leipzig: CF Winter, 1884 [mf ed 1992] – 1mf – 9 – 0-524-02371-9 – (incl bibl ref) – mf#1990-2982 – us ATLA [390]

Aron, Joseph *see* Canada / transvaal

Aron, Pietro *see*
- Compendiolo di molti dvbbi
- Libri tres de institutione harmonica..
- Toscanello
- Toscanello in mvsica di messer piero aron... nvovamente stampato con laggivnta da lvi fatta et con diligentia corretto...
- Trattato della natura et cognitione di tutti gli tuoni di canto figurato

L'arondelle *see* La vie ecoliere

Aroni, Julius *see* Futures

Aronson, Alex *see*
- Rabindranath through western eyes
- Rolland and tagore
- Romain rolland

Aronson, Robert H *see* Attorney-client fee arrangement

Aropagitica / Milton, John – London, England. 1840 – 1r – us UF Libraries [240]

Aros – Vaesteras, Sweden. 1865-69 – 1 – sw Kungliga [078]

Arosemena G, Diogenes A *see* Historia documental del canal de panama

Arosemena, Pablo *see* Escritos

Around the bend – 1978 mar-dec [1]; 1981 spring-1985 fall [2] – 1 – mf#620630 [1]; 1095484 [2] – us WHS [071]

Around the caribbean and across panama / Nicholas, Francis Child – Boston, MA. 1903 – 1r – us UF Libraries [918]

Around the creek – 1948 dec 7-1949 mar 24 – 1 – mf#681689 – us WHS [071]

Around the cross : some of the first principles of the doctrine of christ / Aitken, William Hay Macdowall Hunter – new ed. London: John F Shaw [1884?] [mf ed 1992] – 1mf – 9 – 0-524-04366-3 – mf#1991-2070 – us ATLA [242]

Around the home table / Jacoby, James Calvin – [2nd ed]. Philadelphia PA: Lutheran Publ Soc c1911 [mf ed 1992] – 1mf – 9 – 0-524-04552-6 – mf#1991-2116 – us ATLA [242]

Around the wicket gate : or, a friendly talk with seekers concerning faith in the lord jesus christ / Spurgeon, Charles Haddon – New York: American Tract Society c1890 [mf ed 1985] – 1mf [ill] – 9 – 0-8370-5510-5 – mf#1985-3510 – us ATLA [240]

Around the world studies and stories of presbyterian foreign missions / Bradt, Charles Edwin et al – Wichita KS: Missonary Press c1912 [mf ed 1992] – 2mf [ill] – 9 – 0-524-04368-X – mf#1991-2072 – us ATLA [242]

Aroysgevorfene reyd / Gorodiski, Jonah – Buenos Aires, Argentina. 1948 – 1r – us UF Libraries [939]

Aroz Pascual, L *see* Toledo. archivo de la catedral. sellos eclesiasticos del archivo de la catedral de toledo (1099-1792)

Aroza, D *see* Tesoro de las excelencias y utilidades de la medicina y espejo del prudente y sabio medico...

Arozteguy, Abdon *see* Revolucion oriental de 1870

Arpas y clarines / Silva Munoz Del Canto, Oscar – Camaguey, Cuba. 1951 – 1r – us UF Libraries [972]

Arpee, Leon *see* The armenian awakening

Arpegios / Lopez Ortiz de Leon, Angel – Badajoz: Arqueros, 1907 – 1 – sp Bibl Santa Ana [946]

Arpentigny, Casimir S d' *see* Voyage en pologne et en russie

Arqueologia agustiniana / Perez De Barradas, Jose – Bogota, Colombia. 1943 – 1r – us UF Libraries [930]

Arqueologia de magacela / Jimenez Navarro, E et al – Badajoz: Dip. Provincial, 1951. Sep. REE – 1 – sp Bibl Santa Ana [930]

La arqueologia de norba cesarina / Callejo Serrano, Carlos – Madrid: Diana Artes Graficas, 1969 – 1 – sp Bibl Santa Ana [930]

Arqueologia venezolana / Rouse, Irving – New Haven, CT. 1963 – 1r – us UF Libraries [930]

Arqueologia y antropologia de la tierra / Perez de Barrados, Jose – Madrid: Razon y Fe, 1940 – 1 – sp Bibl Santa Ana [930]

Arquitectura – Mexico City 1949-54 – 1 – ISSN: 0004-2684 – mf#607 – us UMI ProQuest [720]

Arquitectura colonial en venezuela / Gasparini, Graziano – Caracas, Venezuela. 1965 – 1r – us UF Libraries [720]

La arquitectura naval espanola (en madera) – Madrid: Razon y Fe, 1926 – 1 – sp Bibl Santa Ana [355]

Arquitectura precolombina en mexico / Amabilis Dominguez, Manuel – Mexico City?, Mexico. 1956 – 1r – us UF Libraries [720]

Arquitectura religiosa del s. 16 de la tierra de barros / Garrido Santiago, Manuel – Limanas, Tomo 1 (Aceuchal, Fuente del Maestre). Caceres: Univ. de Extremadura. Facult. de Filosofia y Letras, y Tomo 2. 1980 – 1 – sp Bibl Santa Ana [720]

Arquivo Nacional (Brazil) *see*
- Colecao de portugal
- Perfil de cayru

Arraches aux tenebres / Graeme, Bruce – Paris, France. v1-2. 1950 – 2r – us UF Libraries [972]

Arradcom voice – 1981 jan 19-1983 jun 20 – 1 – mf#671455 – us WHS [071]

Arrangement of parish churches considered / Hewett, John William – Cambridge, England. 1848 – 1r – us UF Libraries [240]

An arrangement of the psalms and spiritual songs of the rev isaac watts / Winchell, James M – 1832 – 1 – $29.40 – us Southern Baptist [780]

Arras de cristal y clara lair / Cuchi Coll, Isabel – Ciudad Trujillo, Dominican Republic. 1938 – 1r – us UF Libraries [972]

Arras, Louisa Augusta d' (Lechmere) *see* The two friends

Ar-rasad – Alger. n1-56. 1938-39 – 1 – fr ACRPP [073]

Arratibel, Juan *see* Manual de las cuarenta horas...

Arraz, Antonio *see* Damaso velazquez

Arrebol : jornal academico – Sao Paulo, SP: Typ Liberal, jul 1849 – (= ser Ps 19) – mf#P17,02,217 – bl Biblioteca [073]

Arredondo, Alberto *see* Negro en cuba

Arredondo, Martin *see*
- Obras de albeyteria

Arrendamientos rusticos protegidos : (comentarios a la ley de 15 de julio de 1954) / Fernandez de Soria y Villanueva, Fernando – Badajoz, 1954 – sp Bibl Santa Ana [946]

Arreola, Eduardo *see* Centroamerica

Arrest du conseil d'estat du roi, qui ordonne que les proprietaires anglois de papiers du canada... – Poitiers: Chez Jean Faulcon... [1766?] [mf ed 1983] – 1mf – 9 – 0-665-44549-0 – mf#44549 – cn Canadiana [332]

Arrest [d]u conseil d'estat du roy : [pou]r la prise de possession du bail de la ferme generale des domaines d'occident...du 10. septembre 1726 – Paris: De l'Impr royale, 1726 [mf ed 1984] – 1mf – 9 – 0-665-44553-9 – mf#44553 – cn Canadiana [380]

Arrest du conseil d'estat du roy : ordonne l'execution de l'edit du present mois, qui accorde a la compagnie des indes...du 21. juillet 1720 – Paris: De l'Impr royale, 1720 [mf ed 1983] – 1mf – 9 – 0-665-44554-7 – mf#44554 – cn Canadiana [380]

Arrest du conseil d'estat du roy : portant defenses d'exposer ou recevoir dans les provinces de l'obeissance de sa majeste en europe...du 20. mars 1728 – [Lille, France?]: De l'Impr de C M Crame...[1728?] [mf ed 1983] – 1mf – 9 – 0-665-44556-3 – mf#44556 – cn Canadiana [380]

Arrest du conseil d'estat du roy : qui nomme les directeurs de la compagnie d'occident: du 12 septembre 1717 – Paris: Chez la veuve Saugrain, & Pierre Prault...1720 [mf ed 1983] – 1mf – 9 – 0-665-44509-1 – mf#44509 – cn Canadiana [338]

Arrest du conseil d'estat du roy : qui ordonne que les pelleteries et denrees provenant du cru...du 21. may 1721 – [s.l: s.n, 1721?] [mf ed 1984] – 1mf – 9 – 0-665-44550-4 – mf#44550 – cn Canadiana [380]

Arrest du conseil d'estat du roy : qui proroge pendant un an, a compter du 23. octobre prochain...du 27 juillet 1728 – [Lille?]: De l'Impr de C M Crame...1728 [mf ed 1983] – 1mf – 9 – 0-665-44511-3 – mf#44511 – cn Canadiana [380]

Arrest du conseil d'estat du roy, du 12. fevrier 1726 : qui casse une ordonnance de m l'intendant du canada... – Paris: De l'Impr de la veuve & M-G Jouvenel...1726 [mf ed 1983] – 1mf – 9 – 0-665-44557-1 – mf#44557 – cn Canadiana [380]

Arrest du conseil d'estat du roy, qui fait deffenses a tous armateurs et negocians...du 9 may 1733 – Paris: De l'Impr royale, 1733 [mf ed 1983] – 1mf – 9 – 0-665-44552-0 – mf#44552 – cn Canadiana [380]

Arrest du conseil d'estat du roy, qui permet pendant une annee seulement...du 23 decembre 1727 – [s.l: s.n, 1727?] [mf ed 1983] – 1mf – 9 – 0-665-44548-2 – mf#44548 – cn Canadiana [380]

Arrest du conseil d'estat, servant de reglement : pour restablir la regularite dans un monastere de s benoi (saint victor de marseille) – 1668 – €3.00 – ne Slangenburg [241]

Arrest du conseil d'etat du roi : concernant les interets des reconnoissances donnees en echange des papiers du canada...du 29 decembre 1765 – Paris: De l'Impr royale, 1766 [mf ed 1983] – 1mf – 9 – 0-665-44545-8 – mf#44545 – cn Canadiana [332]

Arrest du conseil d'etat du roi : portant prorogation pendant la presente guerre, de l'entrepot des marchandises et denrees destinees pour le commerce des isles et colonies francoises, du 4 may 1745... – [s.l.]: De l'impr de la veuve de C M Crame [1745?] [mf ed 1983] – 1mf – 9 – 0-665-39998-7 – mf#39998 – cn Canadiana [320]

Arrest du conseil d'etat du roi : qui ordonne qu'a l'avenir les martres, autres que zibelines... du 19 janvier 1767 – Paris: De l'Impr royale, 1767 [mf ed 1983] – 1mf – 9 – 0-665-44510-5 – mf#44510 – cn Canadiana [380]

Arrest du conseil d'etat du roy, qui ordonne que le commerce du castor, demeura libre...du 16 may 1720 – Paris: Chez la veuve Saugrain, & Pierre Prault...1720 [mf ed 1983] – 1mf – 9 – 0-665-44514-8 – mf#44514 – cn Canadiana [380]

Arrest du conseil d'etat du roy...a compter du premier janvier 1746...ordonne par la declaration du 10 novembre 1727... – [s.l]: De l'Impr de la veuve de C M Crame...[1746?] [mf ed 1983] – 1mf – 9 – 0-665-44513-X – mf#44513 – cn Canadiana [380]

Arrest du conseil d'etat du roy...a compter su premier janvier 1749...ordonne par la declaration du 10 novembre 1727... – [s.l]: De l'Impr de la veuve de C M Crame...[1748?] [mf ed 1983] – 1mf – 9 – 0-665-44512-1 – mf#44512 – cn Canadiana [380]

Arrest law bulletin – Boston MA 1985+ – 1,5,9 – ISSN: 8755-8300 – mf#13090 – us UMI ProQuest [345]

Arrests du conseil d'estat du roy : le premier ordonne sans s'arreter a l'ordonnance du sieur de fontainieu, intendant de dauphine, du premier septembre 1733... – Paris: Chez Pierre Prault...1735 [mf ed 1983] – 1mf – 9 – 0-665-44555-5 – mf#44555 – cn Canadiana [346]

Arrests summarized by assisting personnel – 1989 jul 31-1991 jun 30 – 1 – mf#550515 – us WHS [071]

Arret du conseil d'etat du roi : qui fixe les droits que doivent payer par douzaine...du 12 decembre 1781 – Paris: De l'Impr royale, 1782 [mf ed 1983] – 1mf – 9 – 0-665-44551-2 – mf#44551 – cn Canadiana [380]

Arret du conseil d'etat du roi : qui ordonne la liquidation des lettres de change et billets de monnoie du canada du 29 juin 1764... – Lyon: De l'Impr de P Valfray...1764 [mf ed 1994] – 1mf – 9 – 0-665-94710-0 – mf#94710 – cn Canadiana [332]

Arrete pour l'application au cambodge de la reglementation des concessions dominales du 19 septembre 1926 – Phnom-Penh: Impr du Gouvernement [mf ed 1989] – 1r with other items – 1 – mf#mf-10289 seam reel 026/18 [§] – us CRL [343]

Arrets de la chambre des comptes de paris / France. Ancien Regime – 1718-90 – 1 – fr ACRPP [323]

Arrets de la cour des aides / France. Ancien Regime – 1763-76 – 1 – fr ACRPP [324]

Arrets de la cour des monnaies / France. Ancien Regime – 1759-75 – 1 – fr ACRPP [324]

Arrets de la cour du parlement / France. Ancien Regime – 1770-71 – 1 – fr ACRPP [324]

Arrets du conseil d'etat du roi / France. Ancien Regime – 1703, 1770-71 – 1 – fr ACRPP [324]

Arrets du grand conseil du roi / France. Ancien Regime – 1748-87 – 1 – fr ACRPP [324]

Arrets royaux / France. Ancien Regime – 1770-71 – 1 – fr ACRPP [324]

Arria und messalina : trauerspiel in fuenf aufzuegen / Wilbrandt, Adolf von – Wien: L Rosner 1874 [mf ed 1995] – 1r – 1 – (filmed with: wieland und die schweiz / emil ermatinger) – mf#3761p – us Wisconsin U Libr [820]

Arriaga, Pablo Jose de *see* Extirpacion de la idolatria del piru

Arriaga, R de *see* Cursus philosophicus

Arrian *see* Ancient india as described by megasthenaes and arrian

Arriani historici et philosophi ponti euxini et maris erythraei periplus... / Arrianus & Flavius – Lvgdvni, apvd Bartholomaevm Vincentivm, 1577 – 8mf – 9 – mf#PBU-646 – ne IDC [240]

Arrianus *see* Arriani historici et philosophi ponti euxini et maris erythraei periplus...

Arriaza, Juan Bautista *see* Poesias o rimas juveniles

Arriba – Madrid, Spain 27 mar 1940-31 aug 1946, 16 nov 1952-30 dec 1956, 1 jan 1958-31 dec 1960 (imperfect) – 87r – 1 – uk British Libr Newspaper [074]

Arriba espana – Pamplona, Spain 16 jan-13 oct 1937, 2 sep 1941-27 may 1945 (imperfect) – 11r – 1 – uk British Libr Newspaper [074]

Arriba. Madrid *see* Siete editoriales de arriba y su comentario

Arrigo, Bruce A *see* Journal of forensic psychology practice

Arrillaga, Enrique de *see* Un general espanol del siglo 27, don jose de garro

Arrington 1528-1950 – (= ser Cambridgeshire parish register transcript) – 4mf – 9 – £5.00 – uk CambsFHS [929]

Arriola, Jorge Luis *see* Galvez en la encrucijada

L'arrivee apostologique aux eglises representee par celles de l'apostre saint paul, aux eglises de rome et de corinthe / Labadie, Jean de – Middelbourg, 1667 – 2mf – 9 – mf#PPE-163 – ne IDC [225]

Arrivi, Francisco *see*
- Bolero y plena
- Ciclo de lo ausente
- Club de solteros
- Entrada por las raices
- Escultor de la sombra
- Fronteras
- Isla y nada
- Maria soledad
- Sirena
- Sombra menos
- Vejigantes

Arrocha Graell, C *see* Historia de la independencia de panama

Arrom, Jose Juan *see*
- Certidumbre de america
- Estudios de literatura hispanoamericana

Arros, Jean d' *see* Leon 13 d'apres ses encycliques

Arrow – 1967 n2-1981 n3; 1981 n4-1988 n4 – 1 – mf#1826591 – us WHS [071]

Arrow – 1979 may 24, 1980 dec-1982 apr/may – 1 – mf#656638 – us WHS [071]

Arrow – Redditch, England 18 jul 1889-2 jun 1892 – 1 – (cont by: redditch news & east worcestershire advertiser [9 jun 1892-5 apr 1894]) – uk British Libr Newspaper [072]

Arrow = Flecha – v2 n3 [1979 fall]; v3 n1-v11 n4 – 1 – mf#1549311 – us WHS [071]

Arrow : a weekly sporting newspaper – Sydney, Australia 19 mar 1920-7 jan 1922 (wkly) – 1r – 1 – uk British Libr Newspaper [790]

The arrow – 1904-08 – (= ser American indian periodicals... 1) – 1r – 1 – $125.00 – us UPA [305]

The arrow – Dublin. oct 20 1906-aug 25 1909; summer 1939 – 1 – us NY Public [073]

The arrow : an illustrated journal of canadian wit and humour – Toronto: Crawford & Hunter, [1886] – 9 – ISSN: 1190-7193 – mf#P04323 – cn Canadiana [870]

The arrow – Morrinsville, NZ. mar 1972-nov 1977 – 1 – (incorp in: piako post) – mf#15.24 – nz Nat Libr [079]

The arrow – Quebec: [s.n, 1864] – 9 – mf#P04036 – cn Canadiana [071]

The arrow and morrinsville star *see* Piako post

The arrow of gold : a story between two notes / Conrad, Joseph – Toronto: Ryerson Pres, [1919?] – 4mf – 9 – 0-665-73296-1 – mf#73296 – cn Canadiana [830]

Arrowood baptist church : church records – CHESNEE, SC. 1843-1955 – 1 – $40.41 – us Southern Baptist [242]

Arrowsmith, A *see* The house decorator and painter's guide

Arrowsmith, H W *see* The house decorator and painter's guide

Arrowsmith, James *see* The paper-hanger's and upholsterer's guide

Arrowsmith's dictionary of bristol – 2nd ed. Bristol: J W Arrowsmith 1906 [mf ed 1987] – 1r [ill] – 1 – (revision & enl ed of the original arrowsmith's dictionary of bristol, publ 1884- . filmed with: the life and works of john arbuthnot / aitken, g a) – mf#2038 – us Wisconsin U Libr [059]

Arroyave Velez, Eduardo *see* Naipes de antioquia

Arroyo, Angel Manuel *see*
– Cenizas del alma
– Laminas de mi infinito

Arroyo, Anita *see* Caballito verde

Arroyo, Augusto *see* Tierras comunarias y sucesion hereditaria en la reforma agraria

[Arroyo grande-] 5 cities times press recorder – CA. 1970-1979 – 36r – 1 – $2160.00 – mf#B02023 – us Library Micro [071]

[Arroyo grande-] arroyo grande valley herald recorder – CA. apr 1950-dec 1969 – 14r – 1 – $840.00 – (cont by: 5 cities times press recorder) – mf#B03143 – us Library Micro [071]

[Arroyo grande-] herald recorder – CA. 1940-1950 – 5r – 1 – $300.00 – (cont by: arroyo grande valley herald recorder) – mf#B03144 – us Library Micro [071]

[Arroyo grande-] nipoma's adobe press – Arroyo Grande, CA. 1983- – 13r – 1 – $780.00 (subs $50/y) – mf#R03145 – us Library Micro [071]

[Arroyo grande-] times press recorder – CA. jan 1983- – 71r – 1 – $4260.00 (subs $325/y) – mf#R04005 – us Library Micro [071]

Arroyo, Jaime *see* Historia de la gobernacion de popayan seguida de l...

Arroyo, Leonardo *see* Igrejas de sao paulo

Arrufat, Anton *see*
– En claro
– Mi antagonista, y otras observaciones
– Repaso final
– Teatro

Ar-ruh – Blida. n1-24. nov 1937-fev 1939 – (= ser Ruh) – 1 – fr ACRPP [073]

Ars ambrosiana – 1986 – (= ser ILL – ser b; Ccsl 133c) – 4mf+27p – 9 – €30.00 – 2-503-71332-7 – be Brepols [400]

Ars ambrosiana : commentum anonymum in donati partes maiores. formae tplila 6 – 1982 – (= ser ILL – ser a; Ccsl 133c) – 5mf+42p – 9 – €20.00 – 2-503-61338-1 – be Brepols [400]

Ars cantandi : das ist, richtiger und ausfuehrlicher weg, die jugend aus dem rechten grund in der sing-kunst zu unterrichten... / Carissimi, Giacomo – Augspurg: Drucks & Verlegts Jacob Koppmayer 1693 [mf ed 1983?] – 1mf – 9 – mf#fiche 385 – us Sibley [780]

Ars cisterciensi : buchmalereien aus mittel- und ostdeutschen klosterbibliotheken / Libor, Reinhard Maria – Wuerzburg: Holzner Verlag 1967 [mf ed 1993] – (= ser Ostdeutsche beitraege aus dem goettinger arbeitskreis 41) – 10r [ill] – 1 – (incl bibl ref. filmed with: ostdeutsche beitraege aus dem goettinger arbeitskreis) – mf#3180p – us Wisconsin U Libr [090]

Ars (gencligin sesi) – Sahibi ve Yazi Isleri Mueduerue: Cetin Evren. n1. 10 kasim 1949 – (= ser O & t journals) – 1mf – 9 – $25.00 – us MEDOC [956]

Ars generalis ultima : formae tplila 3 / Lullus, Raymond – 1986 – (= ser ILL – ser a; Cccm 75) – 14mf+49p – 9 – €40.00 – 2-503-63752-3 – be Brepols [400]

Ars gramatica japonicae linguae / Collado, Fr. Diego – 1632 – 9 – sp Bibl Santa Ana [480]

Ars grammatica : formae tplila 10 / Donatus Ortigraphus – 1982 – (= ser ILL – ser a; Cccm 40d) – 5mf+45p – 9 – €20.00 – 2-503-60400-5 – be Brepols [400]

Ars grammaticae japonicae linguae : in gratiam et adiutorium eorum, qui praedicandi evangelii causa ad iaponiae regnum se voluerint conferre / Collado, Diego – Romae: Typ Sac Congreg de propag fide 1632 – (= ser Whsb) – 1mf – 9 – €20.00 – (incl: additiones ad dictionarium japonicum) – mf#Hu 327 – gw Fischer [490]

Ars habendi et audiendi conciones sacras / Zepperus, W – Sigenae, 1598 – 5mf – 9 – mf#PBA-419 – ne IDC [240]

Ars ignatiana animorum ad deum per christum adducendorum : quae latet in libro exercitiorum spiritualium / Nonell, Iacobus – Barcinone [Barcelona]: Franciscus Rosalius 1888 [mf ed 1986] – 1mf – 9 – 0-8370-7089-9 – mf#1986-1089 – us ATLA [241]

Ars moriendi *see*
– Apokalypse / ars moriendi / biblia pauperum / antichrist / fabel vom kranken loewen / kalendarium und planetenbuecher / historia david
– Apokalypse / ars moriendi / medizinische traktate / tugend- und lasterlehren

Arsdekin, R *see* Theologia tripartita universa...

Arsenal accents – 1980 nov 7-1991 – 1 – mf#1052186 – us WHS [071]

Arsenev, K K *see* Zakonodatelstvo o pechati

Arsenne thiebaut's von berneaud...schilderung der insel elba : nebst notizen von den uebrigen kleinen inseln des tyrrhenischen meeres, meist nach eigener ansicht entworfen – Weimar 1809 [mf ed Hildesheim 1995-98] – 1v on 1mf – 9 – €40.00 – 3-487-26548-6 – (trans fr french) – gw Olms [914]

Arsfundr hins... / Evangeliska luterska kirkjufelag islendinga i vesturheimi – [Winnipeg MB?]: s.n.] n1-3. 1885-87 [annual] [mf ed 2003] – 3v on 1r – 1 – (3rd report pub in jul/aug 1887 iss of: sameiningin; filmed with later titles: arsping hins...and: gjorabok...arsping hins...) – mf#2003-s502a – us ATLA [242]

Arsinoe, regina di cassandrea : azione tragica in sei atti, composta e diretta dal coreografo giacomo serafini, da rappresentarsi nel nobile teatro di apollo nel carnevale dell' anno 1839 / Serafini, Giacomo – Roma: Tip Puccinelli [1839?] – 1 – (scenery by giuseppe badiali; costumes by antonio ghelli) – mf#*ZBD-*MGTZ pv 3-Res – Located: NYPL – us Misc Inst [790]

Arskii, P A *see* Zagadka aavinkova

Arslanian, Sharon P *see* The history of tap dance in education

Arslanian, Sharon Park *see* Dance concert

Arsping hins *see* Arsfundr hins...

Arsping hins... / Evangeliska luterska kirkjufelag islendinga i vesturheimi – Winnipeg: Prentsmidja Loegbergs, 1888-1909 [annual] [mf n4-25 1888-1909] – 22v on 1r – 1 – (lacks: 6th p81-84. 4th-16th report pub each jul or jul/aug 1888-1900 of: sameiningin; filmed with earlier and later titles: arsping hins...and: gjorabok...arsping hins...) – mf#2003-s502b – us ATLA [242]

L'art : ou les principes philosophiques du chant / Blanchet, Joseph – 2. corr aug ed, Paris: A M Lottin [etc] 1756 [mf ed 19–] – 3mf – 9 – mf#fiche 364 – us Sibley [780]

L'art – Paris. n1-7. mai-juin 1868 – 1 – fr ACRPP [700]

Lart – no. 1-10. Paris. nov 1865-janv 1866 – 1 – fr ACRPP [700]

L'art *see* The artist 1880-82 – l'artist et courier de l'art

Art : clearwater / Walk, Charles E – s.l, s.l? . 1936 – 1r – us UF Libraries [978]

Art : tarpon springs / Walk, Charles E – s.l, s.l? . 1936 – 1r – us UF Libraries [978]

Art amateur : a monthly journal devoted to art in the household – La Salle IL 1879-1903 – 1 – mf#5243 – us UMI ProQuest [700]

Art and antiques – New York NY 1984-92 – 1,5,9 – ISSN: 0195-8208 – mf#12626,01 – us UMI ProQuest [700]

Art and archaeology – Boston MA 1914-34 – 1 – mf#10307 – us UMI ProQuest [930]

Art and archaeology abroad : a report intended primarily for indian students desiring to specialize in those subjects in the research centres of europe and america / Naga, Kalidasa – Calcutta: University of Calcutta, [1937] – (= ser Samp: indian books) – us CRL [700]

Art and archaeology newsletter – New York NY 1973-75 – 1 – ISSN: 0004-2986 – mf#9197 – us UMI ProQuest [930]

Art and architecture of ancient egypt / Smith, William Stevenson – Harmondsworth, England. 1958 – 1r – us UF Libraries [720]

The art and architecture of bikaner state / Goetz, Hermann – Oxford: Published for the Government of Bikaner State and the Royal India and Pakistan Society by Bruno Cassirer, 1950 – (= ser Samp: indian books) – us CRL [700]

The art and architecture of india : buddhist, hindu, jain / Rowland, Benjamin – London; Baltimore, USA: Penguin Books, 1953 – (= ser Samp: indian books) – us CRL [700]

Art and art industries in japan / Alcock, Rutherford – London 1878 – (= ser 19th c art & architecture) – 4mf – 9 – mf#4.2.80 – uk Chadwyck [700]

Art and artists of the capitol of the united states of america / Fairman, Charles E – Washington: GPO, 1927 – 6mf – 9 – $9.00 – mf#LLMC 96-038 – us LLMC [700]

Art and craft – Leamington Spa UK 1979-99 – 1,5,9 – (cont: art and craft in education; cont by: art and design) – ISSN: 0262-7035 – mf#7630.02 – us UMI ProQuest [740]

Art and crafts in our schools / Gaitskell, Charles D – Toronto, ON. 1949 – 1r – us UF Libraries [700]

Art and decoration / National Art Library – (= ser The art periodicals coll at the v and a museum, 1750-1920, pt 3) – 28r (3 col) – 1 – £1550.00 – (art and decoration 28r £1550; la decoration 1893-1905 2r £180; bulletin des metiers d'art 1r £100; la chronique des arts 1862-1922 12r £580; der architekt 1895-1916 6r £320; les arts 1902-20 7r £360) – mf#VAT – uk World [700]

Art and decoration *see* The artist 1880-82 – l'artist et courier de l'art

Art and design – Leamington Spa UK 1972+ – 1,5,9 – (cont: art and craft) – ISSN: 1470-9724 – mf#7630,02 – us UMI ProQuest [740]

The art and design of utopian and religious communities *see* The index of american design (tiam)

Art and hand work for the people...three papers / Tuckwell, William et al – Manchester 1885 – (= ser 19th c art & architecture) – 1mf – 9 – mf#4.1.291 – uk Chadwyck [700]

Art and handicraft / Sedding, John Dando – London 1893 – (= ser 19th c art & architecture) – 2mf – 9 – mf#4.2.166 – uk Chadwyck [740]

Art and history : subject collections – (= ser Art exhibition catalogues on microfiche) – 86 catalogues on 146mf – 9 – £920.00 – (individual titles not listed separately) – uk Chadwyck [700]

Art and letters : an illustrated review – La Salle IL 1888-89 – 1 – mf#5244 – us UMI ProQuest [700]

Art and life : and the building and decoration of cities – London 1897 – (= ser 19th c art & architecture) – 3mf – 9 – mf#4.2.499 – uk Chadwyck [720]

Art and life / Moore, Thomas Sturge – London: Methuen [1910] [mf ed 1985] – 1r [ill] – 1 – (with bibl footnotes. filmed with: the mythology of greece and rome / seemann, o) – mf#6844 – us Wisconsin U Libr [700]

Art and life : snippets, essays, and essayettes / Krishna, Roop – Lahore: Rama Krishna and Sons, 1940 – (= ser Samp: indian books) – us CRL [700]

Art and literature *see* The artist 1880-82 – l'artist et courier de l'art

Art and man – New York NY 1970-92 – 1,5,9 – (cont by: scholastic art) – ISSN: 0004-3052 – mf#8027.01 – us UMI ProQuest [700]

Art and poetry : being thoughts towards nature – La Salle 1850 – 1 – mf#4177 – us UMI ProQuest [410]

Art and science / Stokes, Adrian Durham – London, England. 1949 – 1r – us UF Libraries [700]

Art and the formation of taste : six lectures / Crane, Lucy – London 1882 – (= ser 19th c art & architecture) – 4mf – 9 – mf#4.2.457 – uk Chadwyck [700]

Art and the law *see* Columbia-vla journal of law and the arts

Art and tradition / Haladara, Asitakumara – Agra: Educational Publ, [1938] – (= ser Samp: indian books) – us CRL [700]

Art and work / Davis, Owen William – London 1885 – (= ser 19th c art & architecture) – 3mf – 9 – mf#4.2.1054 – uk Chadwyck [700]

Art annual : the life and work of sir f leighton...sir j e millais...l alma tadema...and j l meissonier – London 1887 – (= ser 19th c art & architecture) – 4mf – 9 – mf#4.2.834 – uk Chadwyck [750]

The art annual for 1889 rosa bonheur her life and work by rene peyrol / Peyrol, Rene – London: J S Virtue & Co Ltd, [1889] – (= ser 19th c art & architecture) – 1mf – 9 – mf#4.1.113 – uk Chadwyck [750]

The art annual for 1890 birket foster his life and work by marcus b huish : xmas number of the art journal / Huish, Marcus Bourne – London: J S Virtue & Co Ltd, [1890] – (= ser 19th c art & architecture) – 1mf – 9 – mf#4.1.114 – uk Chadwyck [700]

The art annual for 1891 briton riviere royal academician : his life and work by w armstrong / Armstrong, Walter – London: J S Virtue & Co Ltd, [1891] – (= ser 19th c art & architecture) – 1mf – 9 – mf#4.1.115 – uk Chadwyck [700]

The art annual for 1892 professor hubert herkomer royal academician his life and work by w.l. courtney / Courtney, William Leonard – London: J S Virtue & Co Ltd, [1892] – (= ser 19th c art & architecture) – 1mf – 9 – mf#4.1.116 – uk Chadwyck [700]

The art annual for 1894 sir edward burne-jones, bart : his life and work by julia cartwright (mrs ady). xmas number of the art journal / Ady, Julia Mary – London: J S Virtue & Co Ltd, [1894] – (= ser 19th c art & architecture) – 1mf – 9 – mf#4.1.118 – uk Chadwyck [750]

The art annual of 1893 william holman hunt : his life and work by the venerable archdeacon farrar...and mrs meynell / Farrar, Frederic William & Thompson, Alice C – London: J S Virtue & Co Ltd, [1893] – (= ser 19th c art & architecture) – 1mf – 9 – mf#4.1.117 – uk Chadwyck [750]

Art applied to industry : a series of lectures / Burges, William – Oxford 1865 – (= ser 19th c art & architecture) – 2mf – 9 – mf#4.2.7 – uk Chadwyck [740]

Art, architecture and photography with art periodicals 1895-1972 *see* Publications of the venice biennale, 1895-1977 (pvb)

Art as applied to dress : with special reference to harmonious colouring / Higgin, Louis – London 1885 – (= ser 19th c art & architecture) – 2mf – 9 – mf#4.2.1013 – uk Chadwyck [740]

Art Association of Montreal *see*
– 21st loan exhibition of paintings in the art gallery, phillips square
– The catalogue of first annual loan and sale exhibition of the newspaper artists' association
– First exhibition of works of art in black and white, february, 1881

Art bulletin – New York NY 1913+ – 1,5,9 – ISSN: 0004-3079 – mf#1476 – us UMI ProQuest [700]

Art, Business & Culture Exchange International *see* Globescope

The art collector – 1889-99 [mf ed Chadwyck-Healey] – (= ser Rare 19th century american art journals) – 31mf – 9 – uk Chadwyck [700]

L'art dans le parler – [Phnom Penh? 1968?] [mf ed 1990] – 1r with other items – 1 – (Title & text in khmer; added title in french) – mf#mf-10289 seam reel 123/3 [§] – us CRL [959]

Art dans les deux mondes – Killen TX 1890-91 – 1 – mf#3153 – us UMI ProQuest [700]

L'art de charpenterie de mathurin jousse : corrige et augmente de ce qu'il y a de plus curieux dans cet art, et des machines les plus necessaires a un charpentier – Paris: Chez Thomas Moette, 1702 [mf ed 1975] – 1r – 1 – mf#SEM35P120 – cn Bibl Nat [690]

Art de deplaire / Melesville, M – Paris, France. 1855 – 1r – us UF Libraries [700]

L'art de jetter les bombes / Blondel, F – Amsterdam, 1690 – 7mf – 9 – mf#OA-185 – ne IDC [720]

L'art de jetter les bombes / Blondel, F – Paris, 1683 – 5mf – 9 – mf#OA-251 – ne IDC [720]

L'art de jouer le violon : contenant les regles necessaires a la perfection de cet instrument... opera 9 = Art of playing on the violin / Geminiani, Francesco – Paris: De la Chevardiere [175-] [mf ed 19–] – 4mf – 9 – mf#fiche 472 – us Sibley [780]

L'art de la guerre... / Gaya, L de – ed 4. La Haye, 1689 – 3mf – 9 – mf#OA-260 – ne IDC [720]

L'art de la musique enseigne et pratique par la nouvelle methode du bureau typografique etablie sur une seule cle : sur un seul ton, et sur un seul signe de mesure... / Dumas, Antoine Joseph – Paris: chez l'auteur [1753] [mf ed 19–] – 17mf – 9 – mf#fiche 447 – us Sibley [780]

L'art de la poesie francoise et latine : avec une idee de la musique sous une nouvelle methode, omnia in pondere, numero & mensura. en trois parties / La Croix, A Pherotee de – A Lyon: Chez Thomas Amaulry 1694 [mf ed 19–] – 10mf – 9 – mf#fiche 589 – us Sibley [780]

L'art de richard wagner : l'oeuvre poetique / Ernst, Alfred – Paris: E Plon, Nourrit 1893 – 6mf – 9 – mf#wa-24 – ne IDC [780]

L'art de se perfectionner dans le violon : ou l'on donne a etudier des lecons sur toutes les positions des quatre cordes du violon et lesdifferens coups d'archet... / Corrette, Michel – Paris: l'auteur [1783?] [mf ed 1978] – 1mf – 9 – mf#fiche 397 – us Sibley [780]

L'art de toucher le clavecin...organish [!] du roi... / Couperin, Francois – Paris: chez l'auteur...1717 [mf ed 19–] – 6mf – 9 – mf#fiche 418 – us Sibley [780]

L'art decoratif *see* The artist 1880-82 – l'artist et courier de l'art

L'art decoratif francais : catalogue of the first exhibition of french artists in decorative art / Grafton Galleries, London – London [1893] – (= ser 19th c art & architecture) – 1mf – 9 – mf#4.2.992 – uk Chadwyck [740]

L'art des emblemes : ou s'enseigne la morale par les figures de la fable... / Menestrier, C F – Paris: RJB de la Caille, 1684 – 5mf – 9 – mf#O-690 – ne IDC [090]

L'art des emblemes... / Menestrier, C F – Lyon: Benoist Coral, 1662 – 3mf – 9 – mf#O-46 – ne IDC [090]

Art direction – Stamford CT 1953-75 – 1,5 – ISSN: 0004-3109 – mf#1802 – us UMI ProQuest [700]

Art documentation : bulletin of the art libraries society of north america – Kanata ON 1989-99 – 1 – (cont: arlis/north america arlis/na newsletter) – ISSN: 0730-7187 – mf#13968 – us UMI ProQuest [700]

L'art du chant : dedie a madame de pompadour / Berard, Jean-Antoine – Paris: Chez Dessaint & Saillant [etc] 1755 [mf ed 19–] – 4mf – 9 – mf#fiche 354, 917 – us Sibley [780]

Art du cirier / Duhamel du Monceau, Henri Louis – [Paris]: Impr de H L Guerin & L F Delatour 1762 [mf ed 1979] – 1r [ill] – 1 – (incl ind) – mf#39 – us Wisconsin U Libr [660]

Art du couvreur / Duhamel du Monceau, Henri Louis – [S.l.]: [s.n.], 1766 ([Paris]: de l'impr de L F Delatour) [mf ed 1992] – 5mf – 9 – (with ind) – mf#SEM105P1493 – cn Bibl Nat [690]

L'art du menuisier / Roubo, Andre J – [Paris?]: [de l'impr de L F Delatour] 4v in 5. 1769-1775 [mf ed 1977] – 1r – 1 – mf#SEM35P146 – cn Bibl Nat [690]

L'art du peintre, doreur et vernisseur : ouvrage utile aux proprietaires ou locataires qui veulent decorer eux-memes leur sejour, ainsi qu'a ceux qui se destinent a la profession de peintre, doreur et vernisseur / Watin, Jean Felix – 9e ent ref augm ed. Paris: Berlin-Leprieur...1823 [mf ed 1979] – 1r – 5 – mf#SEM16P278 – cn Bibl Nat [760]

L'art du plein-chant : ou traite theorico-pratique sur la facon de le chanter... – Villefranche-de-Rouergue: P Vedeilhie 1764 [mf ed 19–] – 7mf – 9 – mf#fiche331 – us Sibley [780]

L'art du tuilier et du briquetier / Duhamel du Monceau, Henri Louis – P [Paris]: [s.n.], 1763 [mf ed 1992] – 7mf – 9 – mf#SEM105P1492 – cn Bibl Nat [690]

Art education – Reston VA 1948+ – 1,5,9 – ISSN: 0004-3125 – mf#6520 – us UMI ProQuest [700]

Art education at home and abroad / Yapp, George Wagstaffe – [2nd ed]. London 1853 – (= ser 19th c art & architecture) – 1mf – 9 – mf#4.2.962 – uk Chadwyck [700]

Art embroidery : a treatise on the revived practice of decorative needlework / Lockwood, Mary (Smith) & Glaister, Elizabeth – London: Marcus Ward & Co; Belfast: Strand & Royal Ulster Works, 1878 – (= ser 19th c art & architecture) – 2mf – 9 – mf#4.1.146 – uk Chadwyck [740]

Art et critique – no. 1-95. Paris. juin 1889-janv 1891, janv-mars 1892 – 1 – fr ACRPP [700]

L'art et la vie – Paris. 1892-97 [mnthly] – 1 – (revue mensuelle, litteraire, artisitque et sociale) – fr ACRPP [073]

The art exemplar / Stannard, William J – [London? 1859?] – (= ser 19th c art & architecture) – 7mf – 9 – mf#4.2.196 – uk Chadwyck [700]

Art exhibition catalogs on microfiche / North America. Museums and Galleries – 1935-76 – 252mf – 9 – (among these sets are the complete coll of arts council of great britain catalogues [1942-78], the french salon catalogues [1673-1925] as well as several american colls incl catalogues fr the sidney janis gallery) – uk Chadwyck [700]

Art exhibition catalogs subject index, 1977-1990 / Santa Barbara. Arts Library at the University of California – 195mf – 9 – £990.00 – (former title: the catalogs of the art exhibition catalog collection of the arts library, university of california at santa barbara) – uk Chadwyck [700]

Art exhibition catalogues on microfiche : catalogues of major museums and galleries; major subject collections; subject collections; architecture, applied arts, studio arts – 5,687mf – 9 – £14,000.00 coll – (some titles listed separately. categories as provided: catalogues of major museums and galleries 3633mf. major subject collections 1431mf. subject collections 3551mf. architecture, applied arts, studio arts 922mf) – uk Chadwyck [700]

L'art flamand et hollandais – Anvers. v1-25. 1904-1925 – 123mf – 9 – mf#O-1212 – ne IDC [700]

Art foliage : for sculpture and decoration / Colling, James Kellaway – [2nd ed] London 1878 – (= ser 19th c art & architecture) – 4mf – 9 – mf#4.2.98 – uk Chadwyck [700]

Art for art's sake / Van Dyke, John Charles – London 1893 – (= ser 19th c art & architecture) – 4mf – 9 – mf#4.2.518 – uk Chadwyck [700]

"L'art francais" presente...[...du nouveau avec marc-aurele fortin, arca] = "L'art francais" presents...[something new...by marc-aurele fortin, arca] / Galerie l'Art francais – [Montreal: l'art francais, 1946?] (mf ed 1993) – 1mf – 9 – mf#SEM105P1895 – cn Bibl Nat [700]

Art furniture : from designs by e w godwin...and others / Watt, William – London 1877 – (= ser 19th c art & architecture) – 1mf – 9 – mf#4.2.48 – uk Chadwyck [740]

Art guides : a guide to the painting and sculpture in the justice building / U.S. Dept of Justice – Washington: Art in Federal Buildings Inc, 1938 – 1mf – 9 – $1.50 – mf#LLMC 91-080 – us LLMC [700]

Art history – Oxford UK 1988+ – 1,5,9 – ISSN: 0141-6790 – mf#17386 – us UMI ProQuest [700]

Art impressions of dresden, berlin, and antwerp / Wilkins, William Noy – London 1860 – (= ser 19th c art & architecture) – 3mf – 9 – mf#4.2.1783 – uk Chadwyck [700]

Art in america – New York, 1913-1920. v1-8 – 49mf – 9 – mf#O-1213 – ne IDC [700]

Art in america – New York NY 1913+ – 1,5,9 – ISSN: 0004-3214 – mf#2136 – us UMI ProQuest [700]

Art in america – v1-38. 1913-50 – 1 – us AMS Press [700]

Art in ancient rome / Strong, Eugenie Sellers – New York, NY. v1-2. 1928 – 1r – us UF Libraries [700]

Art in everything / Fawcett, Henry – London 1882 – (= ser 19th c art & architecture) – 2mf – 9 – mf#4.2.594 – uk Chadwyck [700]

Art in needlework : a book about embroidery / Day, Lewis Foreman & Buckle, Mary – London: B T Batsford, 1900 – (= ser 19th c art & architecture) – 4mf – 9 – mf#4.1.202 – uk Chadwyck [740]

Art in ornament and dress = L'art dans la parure et dans le vetement / Blanc, Charles – London: Chapman & Hall, 1876 – (= ser 19th c art & architecture) – 3mf – 9 – mf#4.1.143 – uk Chadwyck [740]

Art in provincial france...1882 / Carr, Joseph William Comyns – London 1883 – (= ser 19th c art & architecture) – 2mf – 9 – mf#4.2.1150 – uk Chadwyck [700]

Art in the modern state / Dilke, Emilia Frances (Strong) – London 1888 – (= ser 19th c art & architecture) – 4mf – 9 – mf#4.2.446 – uk Chadwyck [700]

Art in the school – [Scotland] [Edinburgh: Educational Institute of Scotland 8 jun, 14 dec 1934 (semiannual) [mf ed 2004] – 2v on 1r – 1 – (suppl to: scottish educational journal (edinburgh, scotland : 1918)) – uk Newsplan [072]

Art industry metal-work : illustrating the chief processes of art-work / Yapp, George Wagstaffe – London [1877?] – (= ser 19th c art & architecture) – 13mf – 9 – mf#4.2.954 – uk Chadwyck [730]

Art Institute of Chicago *see*
– French drawings and sketchbooks of the nineteenth century, vol 1
– French drawings and sketchbooks of the nineteenth century, vol 2
– French drawings of the sixteenth and seventeenth centuries
– Italian drawings of the 18th and 19th centuries and spanish drawings of the 17th through 19th centuries
– Twentieth-century european paintings

Art institute of chicago. bulletin – Chicago IL 1907-82 – 1,5,9 – ISSN: 0094-3312 – mf#1768 – us UMI ProQuest [700]

Art instruction in england / Hulme, Frederick Edward – London 1882 – (= ser 19th c art & architecture) – 2mf – 9 – mf#4.2.70 – uk Chadwyck [700]

Art international [1957] – Paris, France 1957-84 – 1,5,9 – (cont by: art international [1987]) – ISSN: 0004-3230 – mf#1815 – us UMI ProQuest [700]

Art international [1987] – Paris, France 1987-91 – 1,5,9 – (cont: art international [1957]) – mf#17358 – us UMI ProQuest [700]

Art international aujourd'hui – Paris, 1929-30 [mf ed Chadwyck-Healey] – (= ser Art periodicals on microform) – 2r – 1,14 – uk Chadwyck [720]

Art journal – New York NY 1941+ – 1,5,9 – ISSN: 0004-3249 – mf#1477 – us UMI ProQuest [700]

The art journal – 1875-87 [mf ed Chadwyck-Healey] – (= ser Rare 19th century american art journals) – 65mf – 9 – uk Chadwyck [700]

Art journal, london : the illustrated catalogue of the universal exhibition – London [1868] – (= ser 19th c art & architecture) – 6mf – 9 – mf#4.2.915 – uk Chadwyck [700]

Art Libraries Society of North America *see* Arlis/north america newsletter

Art, literature, music, drama – s.l, s.l? . 193-? – 1r – us UF Libraries [700]

L'art litteraire : bulletin d'art, de critique et bibliographie – Paris. n1-13; ns: n1-12. oct 1892-94 – 1 – fr ACRPP [400]

Art looting and nazi germany : records of the fine arts and monuments adviser, ardelia hall, 1945-1961 – 2pt – 1 – (pt1: country files for austria, italy, & germany 6r isbn 1-55655-860-0 $1165. pt2: subject files 14r isbn 1-55655-891-0 $2550. with p/g) – us UPA [934]

Art material trade news – Shawnee Mission KS 1991-92 – 1,5,9 – ISSN: 0004-3265 – mf#17086 – us UMI ProQuest [700]

Art meets labor : [newsletter] – 1986 jan/feb-1988 sep/oct – 1 – mf#1520123 – us WHS [071]

L'art moderne – Paris. dec 1882-83 – 1 – fr ACRPP [700]

L'art moderne : revue de critique des arts et de la litterature – Bruxelles, 1881-93; 1895-1913 – 1 – fr ACRPP [700]

Art news annual – New York NY 1970-72 – 1,5,9 – ISSN: 0066-7994 – mf#6025 – us UMI ProQuest [700]

Art no 663 : new smyrna – s.l, s.l? . 1936 – 1r – us UF Libraries [978]

Art noveau : subject collections – (= ser Art exhibition catalogues on microfiche) – 18 catalogues on 28mf – 9 – £235.00 – (individual titles not listed separately) – uk Chadwyck [700]

The art of amrita sher-gil – Allahabad: Roerich Centre of Art & Culture, 1937 – (= ser Samp: indian books) – (int by r c tandan) – us CRL [700]

The art of beauty / Haweis, Mary Eliza (Joy) – London 1878 – (= ser 19th c art & architecture) – 4mf – 9 – mf#4.2.299 – uk Chadwyck [700]

The art of being alive : success through thought / Wilcox, Ella Wheeler – New York: Harper, c1914 [mf ed 1998] – 1r – 1 – (filmed with: boy life on the prairie / hamlin garland) – mf#4390 – us Wisconsin U Libr [840]

The art of bernard shaw / Sen Gupta, Subodh Chandra – Calcutta: A Mukherjee & Co, 1950 – (= ser Samp: indian books) – us CRL [420]

Art of chorale-preluding and chorale accompaniment as presented in kittel's der angehende praktische organist / Brown, Charles Stagmaier – U of Rochester 1970 [mf ed 19–] – 2v on 12mf – 9 – mf#fiche378 – us Sibley [780]

The art of controversy, and other posthumous papers = Selections. 1896 / Schopenhauer, Arthur – London: S Sonnenschein; New York: Macmillan, 1896 – (= ser The Philosophy at Home Series) – 1mf – 9 – 0-7905-7369-5 – (in english) – mf#1989-0594 – us ATLA [190]

Art of courtly love / Andre, Le Chapelain – New York, NY. 1957 – 1r – us UF Libraries [025]

The art of cross-examination / Wellman, Francis Lewis – New and enl. ed. New York, Macmillan, 1904. 404 p. LL-1183 – 1 – us L of C Photodup [340]

The art of decoration / Haweis, Mary Eliza (Joy) – London 1881 – (= ser 19th c art & architecture) – 5mf – 9 – mf#4.1.181 – uk Chadwyck [740]

The art of decorative design / Dresser, Christopher – London 1862 – (= ser 19th c art & architecture) – 4mf – 9 – mf#4.2.1247 – uk Chadwyck [740]

The art of dress / Haweis, Mary Eliza (Joy) – London 1879 – (= ser 19th c art & architecture) – 2mf – 9 – mf#4.2.297 – uk Chadwyck [740]

The art of dress : or, guide to the toilette / Howard, Frank – London 1839 – (= ser 19th c art & architecture) – 1mf – 9 – mf#4.2.926 – uk Chadwyck [740]

The art of dress : or, guide to the toilette – London 1839 – (= ser 19th c art & architecture) – 1mf – 9 – mf#4.1.454 – uk Chadwyck [740]

The art of dressmaking at home and in the workroom, vol 1 : select lessons in cutting and fitting ladies' garments... / Boudet, Marie – Montreal: E Boulet, 1903 – 2mf – 9 – 0-659-92124-3 – mf#9-92124 – cn Canadiana [640]

Art of enamelling on metal / Brown, William Norman – London 1900 – (= ser 19th c art & architecture) – 1mf – 9 – mf#4.2.157 – uk Chadwyck [730]

The art of engraving, with the various modes of operation, under the following different divisions : etching. soft-ground etching. line engraving. chalk and stipple. aquatint. mezzotint. lithography. wood engraving. medallic engraving. electrography. and photography / Fielding, Theodore Henry Adolphus – London: Ackermann & Co, 1841 – (= ser 19th c art & architecture) – 2mf – 9 – mf#4.1.128 – uk Chadwyck [760]

The art of extempore speaking : hints for the pulpit, the senate, and the bar = Etude sur l'art de parler en public / Bautain, Louis – 6th ed. New York: Charles Scribner, 1858 [mf ed 1993] – 1mf – 9 – 0-524-08436-X – (in english) – mf#1993-2041 – us ATLA [400]

Art of fingering : or the easiest and surest method how to learn to play on the harpsichord with propriety and expedition...: to which is added a table of all the different keys... / Heck, Johann Caspar – London: printed & sold by W Randall & I Abell...[1766?] [mf ed 19–] – 2mf – 9 – mf#fiche 548 – us Sibley [780]

The art of flower painting / Duffield, Mary Elizabeth (Rosenberg) – London: Winsor & Newton, 1856 – (= ser 19th c art & architecture) – 2mf – 9 – mf#4.1.135 – uk Chadwyck [750]

The art of fresco painting, as practised by the old italian and spanish masters : with a preliminary inquiry into the nature of the colours used in fresco painting, with observations and notes / Merrifield, Mary Philadelphia – London: Publ...by Charles Gilpin; Brighton: Arthur Wallis, 1846 – (= ser 19th c art & architecture) – 3mf – 9 – mf#4.1.42 – uk Chadwyck [750]

The art of furnishing on rational and aesthetic principles / Cooper, H J of South Hampstead – London 1876 – (= ser 19th c art & architecture) – 2mf – 9 – mf#4.2.35 – uk Chadwyck [740]

The art of garnishing churches at christmas and other festivals / Cox, Edward Young – London: Cox & Son, [1868] – (= ser 19th c art & architecture) – 2mf – 9 – (with photographs, lithographs, & wood engravings, ill the original designs) – mf#4.1.57 – uk Chadwyck [740]

The art of good living and good dying : emmanuel college, cambridge, ms. 4.1.16 / Verard, A – 1r – 1 – (int by f h stubbings) – mf#96593 – uk Microform Academic [240]

The art of hindu dance / Bhadury, Manjulika & Chatterjee, Santosh – Calcutta: SK Chatterjee: Sole distributor, Bankim Chandra Chatterjea, 1945 – (= ser Samp: indian books) – us CRL [790]

The art of illuminating / Tymms, William Robert – London 1860 – (= ser 19th c art & architecture) – 4mf – 9 – mf#4.1.265 – uk Chadwyck [740]

The art of illumination and missal painting : a guide to modern illuminators / Humphreys, Henry Noel – London: H G Bohn, 1849 – (= ser 19th c art & architecture) – 2mf – 9 – mf#4.1.188;c.4.1.235 – uk Chadwyck [740]

The art of illustration / Blackburn, Henry – London 1894 – (= ser 19th c art & architecture) – 3mf – 9 – mf#4.2.77 – uk Chadwyck [740]

Art of india : paintings and manuscripts from the victoria & albert museum, london – 71mf – 9 – £1800 – (with d/g) – uk Matthew [700]

The art of india and pakistan : a commemorative catalogue of the exhibition held at the royal academy of arts, london, 1947-8 / ed by Ashton, Leigh – London: Faber and Faber, 1950 – (= ser Samp: indian books) – us CRL [700]

The art of invigorating and prolonging life, by food, clothes, air, exercise, wine, sleep, etc : or, the invalid's oracle / Kitchiner, William – 6th ed. London: Printed for Geo. B. Whittaker by J. Moyes, 1828. 337p – 1 – us Wisconsin U Libr [615]

The art of iron moulding, in all its various branches – Boston, 1853 – 1 – us CRL [740]

The art of java / Gangoly, Ordhendra Coomar – Calcutta: Rupam, [19–] – (= ser Samp: indian books) – us CRL [700]

The art of judging the character of individuals from their handwriting and style – Edward Lumley, editor. London: John Russell Smith, 1875. viii,177p. 35 leaves of plates, facsims – 1 – us Wisconsin U Libr [150]

The art of kathakali / Pandeya, Avinash C – Allahabad: Kitabistan, 1943 – (= ser Samp: indian books) – (int by his highness maharana shree vijayadevji rana; foreword by gopi nath) – us CRL [790]

Art of literature / Schopenhauer, Arthur – London, England. 1900 – 1r – us UF Libraries [400]

The art of making devises : treating of hieroglyphicks, symboles, emblemes... / Estienne, H – London: W.E. and J.G., 1646 – 2mf – 9 – (transl into english by thomas blount) – mf#O-606 – ne IDC [090]

The art of marine painting in water-colours / Carmichael, James Wilson – London 1859 – (= ser 19th c art & architecture) – 2mf – 9 – mf#4.2.939 – uk Chadwyck [750]

Art of modulating illustrated in one grand lesson and two preludes : for the pianoforte, harpsichord or organ / Bemetzrieder, Anton – London: printed by T Skillern [1796] [mf ed 19–] – 1mf – 9 – mf#fiche 346 – us Sibley [780]

The art of music : A comprehensive library of information for music lovers and musicians / Mason, Daniel G – New York. v1-14. 1915-17 – 1 – $120.00 – mf#0351 – us Brook [780]

Art of musick / Lampe, John Frederick – London: C Corbett 1740 [mf ed 19–] – 1mf – 9 – mf#fiche 591, 751 – us Sibley [780]

The art of painting in the queen's reign : being a glance at some of the painters and paintings of the british school during the last sixty years / Temple, Alfred George – London: Chapman & Hall Ltd, 1897 – (= ser 19th c art & architecture) – 6mf – 9 – mf#4.1.89 – uk Chadwyck [750]

Art of playing thorough bass with correctness according to the true principles of composition : fully explained by a great variety of examples in various stiles; to which are added by way of suppliment six lessons of accompaniment... / Heck, Johann Caspar – London: J Preston [1793] [mf ed 19–] – 9mf – 9 – mf#fiche 549 – us Sibley [780]

The art of preaching *see* Cheng tao i chu [ccm8]

Art of questioning / Fitch, Joshua Girling – London, England. 18– – 1r – us UF Libraries [240]

The art of questioning / Bryan, Joseph Harris – St Louis, MO: Christian Pub Co, c1909 – 1mf – 9 – 0-524-06084-3 – mf#1991-2397 – us ATLA [240]

Art of reading latin / Hale, William Gardner – New York, NY. 1887 – 1r – us UF Libraries [450]

The art of school management : a text-book for normal schools and normal institutes, and a reference book for teachers, school officers and parents / Baldwin, Joseph – Toronto: Warwick, 1886 – 4mf – 9 – (with app) – mf#25082 – cn Canadiana [370]

Art of securing attention in a sunday school class / Fitch, Joshua Girling – London, England. 18– – 1r – us UF Libraries [240]

The art of simpling / Coles, William – An introduction to the knowledge and gathering of plants. London. 1656 – 1 – us Wisconsin U Libr [631]

The art of sketching from nature / Delamotte, Philip Henry – London 1871 – (= ser 19th c art & architecture) – 2mf – 9 – mf#4.2.1244 – uk Chadwyck [740]

The art of teaching : a manual for the use of teachers and school commissioners / Emberson, Frederick C – Montreal: Dawson, 1877 – 3mf – 9 – mf#27112 – cn Canadiana [370]

The art of the house / Watson, Rosamund Marriott (Tomson) – London 1897 – (= ser 19th c art & architecture) – 3mf – 9 – mf#4.2.51 – uk Chadwyck [720]

The art of the old english potter / Solon, Louis Mark Emanuel – London 1883 – (= ser 19th c art & architecture) – 6mf – 9 – mf#4.2.267 – uk Chadwyck [730]

The art of the pal empire / French, John Calvin – London: Oxford University Press, 1928 – (= ser Samp: indian books) – us CRL [700]

Art of thought / Wallas, Graham – New York, NY. 1926 – 1r – us UF Libraries [100]

The art of transparent painting on glass / Groom, Edward – London 1855 – (= ser 19th c art & architecture) – 1mf – 9 – mf#4.2.189 – uk Chadwyck [740]

The art of using the china missionary survey / Clark, Sidney J W – us ATLA [240]

The art of using the china missionary survey / Clark, Sidney James Wells – [S.l.: s.n., 1922?] (Shanghai: Shanghai Mercury). Chicago: Dep of Photodup, U of Chicago Lib, 1971 (1r); Evanston: American Theol Lib Assoc, 1984 (1r) – 1 – mf#1984-6294 – us ATLA [240]

The art of velazquez / Armstrong, Walter – London 1896 – (= ser 19th c art & architecture) – 2mf – 9 – mf#4.1.347 – uk Chadwyck [750]

The art of water drawing, 1659-60 / D'Acres, R – [mf ed 1930] – (= ser Newcomen society extra publication 2) – 3mf – 7 – (int by rhys jenkins) – mf#86574 – uk Microform Academic [550]

The art of william quiller orchardson / Armstrong, Walter – London 1895 – (= ser 19th c art & architecture) – 2mf – 9 – mf#4.2.377 – uk Chadwyck [750]

L'art paien sous les empereurs chretiens / Allard, Paul – Paris: Didier, 1879 [mf ed 1990] – 1mf – 9 – 0-7905-4601-9 – (in french. incl bibl ref) – mf#1988-0601 – us ATLA [700]

Art papers – Atlanta GA 1981+ – 1,5,9 – ISSN: 0278-1441 – mf#15200 – us UMI ProQuest [700]

Art photography in short chapters / Robinson, Henry Peach – London: Hazell, Watson, & Viney Ltd, 1890 – (= ser 19th c art & architecture) – 2mf – 9 – mf#4.1.102 – uk Chadwyck [770]

Art Place/Center Gallery [Madison WI] *see* Center gallery newsletter

L'art pour tous *see* The artist 1880-82 – l'artist et courier de l'art

Art precolombien d'haiti – Port-Au-Prince, Haiti. 1941 – 1r – us UF Libraries [700]

The art press / Victoria and Albert Museum. London – [mf ed Chadwyck-Healey, 1976] – 21mf – 9 – (incl ind) – uk Chadwyck [700]

Art psychotherapy – Oxford UK 1973-79 – 1,5,9 – (cont by: arts in psychotherapy) – ISSN: 0090-9092 – mf#49015 – us UMI ProQuest [150]

Art quarterly – Detroit MI 1938-74 – 1,5 – ISSN: 0004-3303 – mf#356 – us UMI ProQuest [700]

The art quarterly – Detroit, 1938-1946. v1-9 – 60mf – 9 – mf#0-493c – ne IDC [700]

L'art religieux au caucase / Mourier, J – Paris, 1887 – 2mf – 9 – mf#AR-1850 – ne IDC [243]

Art review – London UK 1994-1997 – 1,5,9 – (cont: arts review) – mf#8170,01 – us UMI ProQuest [700]

Art sales : a history of sales of pictures and other works of art / Redford, George – London, 1888 – (= ser 19th c art & architecture) – 2v on 16mf – 9 – mf#4.1.52 – uk Chadwyck [700]

The art student : an illustrated magazine conducted by members of the birmingham school of art – Birmingham 1885-87 – (= ser 19th c art & architecture) – 3mf – 9 – mf#4.2.1689 – uk Chadwyck [700]

Art teacher – Reston VA 1971-80 – 1,5,9 – mf#6522 – us UMI ProQuest [700]

The art teaching of john ruskin / Collingwood, William Gershom – London 1891 – (= ser 19th c art & architecture) – 5mf – 9 – mf#4.2.71 – uk Chadwyck [700]

An art tour to northern capitals of europe / Atkinson, John Beavington – London 1873 – (= ser 19th c art & architecture) – 5mf – 9 – mf#4.2.138 – uk Chadwyck [700]

Art treasures of the united kingdom... : from the manchester art treasures exhibition, 1857 / Waring, John Burley – London [1858] – (= ser 19th c art & architecture) – 9mf – 9 – mf#4.1.241 – uk Chadwyck [700]

L'art universel des fortifications... / Brueil, J du – Paris, 1665 – 4mf – 9 – mf#OA-253 – ne IDC [720]

L'art vivant – Paris, 1925-39 [mf ed Chadwyck-Healey] – (= ser Art periodicals on microform) – 10r – 1 – uk Chadwyck [740]

The art wealth of england : a series of photographs representing fifty of the most remarkable works of art contributed on loan to the special exhibition at the south kensington museum, 1862 / Robinson, John Charles & Thompson, C Thurston – [London]: publ by the authority of the Science and Art Dept...by...Scott & Co, 1862 – (= ser 19th c art & architecture) – 3mf – 9 – mf#4.1.182 – uk Chadwyck [700]

Art work on british columbia, canada / Carre, William H – [S.l.]: W Carre, 1900 [mf ed 1980] – 2mf – 9 – 0-665-02463-0 – mf#02463 – cn Canadiana [917]

Art work on city of saint john, new brunswick – [S.l.]: W H Carre, 1899 [mf ed 1980] – 2mf – 9 – 0-665-01003-6 – mf#01003 – cn Canadiana [720]

Art work on hamilton, canada / Carre, William H – [S.l: s.n.], 1899 [mf ed 1979] – 2mf – 9 – 0-665-00025-1 – mf#00025 – cn Canadiana [700]

Art work on montreal, canada / Carre, William H – [S.l: s.n.], 1898 [mf ed 1979] – 3mf – 9 – 0-665-00027-8 – (text by arthur weir) – mf#00027 – cn Canadiana [700]

Art work on ottawa, canada / Carre, William H – [S.l.]: W Carre, 1898 [mf ed 1979] – 2mf – 9 – 0-665-00029-4 – mf#00029 – cn Canadiana [700]

Art work on toronto, canada – [Toronto?]: W H Carre, 1898 [mf ed 1980] – 3mf – 9 – 0-665-01086-9 – mf#01086 – cn Canadiana [720]

Art work on winnipeg, manitoba, canada – [Winnipeg?]: W H Carre, 1900 [mf ed 1983] – 4mf – 9 – mf#30221 – cn Canadiana [700]

Art work – quebec, canada / Carre, William H – [S.l.]: W Carre, [18–?] [mf ed 1980] – 2mf – 9 – 0-665-02500-9 – mf#02500 – cn Canadiana [700]

The art workmanship of the maori race in new zealand / Hamilton, Augustus – Dunedin 1896-1901 – (= ser 19th c art & architecture) – 6mf – 9 – mf#4.1.343 – uk Chadwyck [700]

Artamonov, P I *see* Novosti vladivostoka

L'artaserse : drama in tre atti di pietro metastasio / Vinci, Leonardo – [n.p. 175-?] [mf ed 19–] – 3v on 12mf – 9 – mf#fiche 1131 – us Sibley [780]

Artault, Thibault *see* Tres ample et vraye exposition de la regle de monsieur sainct benoist

Artaxerce / Delrieu, Etienne Joseph Bernard – Paris, France. 1808 – 1r – us UF Libraries [440]

Artaxerxes : soldier tir'd of war's alarms / Arne, Thomas Augustine – [London: between c1804 & 1812] [mf ed 19–] – 1r – 1 – mf#pres. film 101 – us Sibley [780]

Artaxerxes 3 ochus and his reign : with special consideration of the old testament sources bearing upon the period / Hirschy, Noah Calvin – Chicago: University of Chicago 1909 [mf ed 1989] – 1mf – 9 – 0-7905-1103-7 – (incl bibl) – mf#1987-1103 – us ATLA [930]

Artaxerxes, an english opera as it is performed at the theatre royal in covent garden / Arne, Thomas Augustine – London: printed for John Johnson...[176-?] [mf ed 19–] – 3v in 1 on 1r – 1 – (libretto trans fr pietro metastasio's artaserse [1729] by the composer) – mf#film 1396 – us Sibley [780]

L'arte – Rome, 1898-1971 [mf ed Chadwyck-Healey] – (= ser Art periodicals on microform) – 22r – 1 – uk Chadwyck [700]

A arte : orgam defensor do theatro nacional – Sao Joao d'El Rey, MG: Companhia Luso-Brasileira, 26 ago-2 set 1905 – (= ser Ps 19) – 1,5,6 – bl Biblioteca [079]

L'arte armonica : or, a treatise on the composition of musick, in three books; with an introduction, on the history, and progress of musick, from it's [sic.] beginning to this time / Antoniotto, Giorgio – London: printed by J Johnson 1760 [mf ed 19–] – 2v on 4mf – 9 – mf#fiche763 – us Sibley [780]

Arte, bocabulario : tesoro y catecismo de la lengva gvarani / Ruiz De Montoya, Antonio – Leipzig, Germany. v1-4. 1876 – 1r – us UF Libraries [025]

Arte colonial en santo domingo, siglos 16-18 / Universdad De Santo Domingo – Ciudad Trujillo, Dominican Republic. 1950 – 1r – us UF Libraries [700]

Arte de furtar : e o seu autor / Pena, Afonso – Rio de Janeiro, Brazil. v1-2. 1946 – 1r – us UF Libraries [972]

Arte de hablar en prosa y verso / Gomez Hermosilla, Jose – Buenos Aires, Argentina. 1943 – 1r – us UF Libraries [972]

Arte de hablar...verso / Gomez Hermosilla, Jose – 1876 – 9 – sp Bibl Santa Ana [810]

Arte de la lengua general del reyno de chile, con un dialogo chileno-hispano muy curioso : a que se anade la doctrina christiana, esto es, rezo, catecismo, coplas, confesionario, y platicas; lo mas en lengua chilena y castellana... / Febres, Andres – en Lima: En la calle de la Encarnacion, ano de 1765 – (= ser Books on religion...1543/44-c1800: catecismos) – 8mf – 9 – mf#crl-15 – ne IDC [241]

Arte de la lengua general del reyno de chile, con un dialogo chileno-hispano muy curioso : a que se anade la doctrina christiana...y por fin un vocabulario hispano-chileno... / Febres, Andres – Lima: En la calle de la Encarnacion 1765 – (= ser Whsb) – 8mf – 9 – €80.00 – mf#Hu 385 – gw Fischer [440]

Arte de la lengua hiliguayna de la isla de panay / Mentrida, Alonso de – Manila. 1818 – 1 – us CRL [490]

Arte de la lengua mexicana : y breves platicas de los mysterios de n santa fee catholica, y otras para exortacion de su obligacion a los indios / Avila, Francisco de – en Mexico: ...Miguel de Ribera Caldero[n] en el Empedradillo, ano de 1717 – (= ser Books on religion...1543/44-c1800: doctrina cristiana, obras de devocion) – 2mf – 9 – mf#crl-54 – ne IDC [241]

Arte de la lengua moxa : con su vocabulario, y cathecismo / Marban, Pedro – [Lima]: impr Joseph de Contreras [1701 or 1702] – (= ser Books on religion...1543/44-c1800: catecismos) – 10mf – 9 – mf#crl-8 – ne IDC [241]

Arte de la lengua moxa, con su vocabulario, y cathecismo / Marban, Pedro – [Lima]: Contreras [1701] – (= ser Whsb) – 11mf – 9 – €95.00 – (incl: cathecismo menor en lengua espanola, y moxa) – mf#Hu 387 – gw Fischer [440]

Arte de la lengua pampanga : dedicale al m.r.p.p. fr. francisco zenzano / Bergano, Diego – [Manila]: Impr de la Compania de Jesus por S L Sabino, 1729 – 1 – us CRL [490]

Arte de la lengua quichua / Torres Rubio, Diego de – en Lima: Por Francisco Lasso, ano de 1619 – (= ser Books on religion...1543/44-c1800: catecismos) – 3mf – 9 – mf#crl-3 – ne IDC [241]

Arte de la lengua quichua : y nuevamente van anadidos los romances, el cathecismo pequeno, todas las oraciones, los dias de fiesta... / Torres Rubio, Diego de – en Lima: A costa de Francisco Farfan de los Godos...; por Joseph de Contreras, y Alvarado...de la Santa Cruzada [1700?] – (= ser Books on religion...1543/44-c1800: catecismos) – 3mf – 9 – mf#crl-7 – ne IDC [241]

Arte de la lengua tagala : y manual tagalog, para la administracion de los santos sacramentos que de orden de sus superiores compuso sebastian de totanes – Sampaloc: [s. n.] 1745 – (= ser Whsb) – 2mf – 9 – €30.00 – (incl: manual tagalog, para auxilio a los religiosos de esta santa provincia de s gregorio magno) – mf#Hu 322 – gw Fischer [490]

Arte de la lengua tagala, y manual tagalog : para la administracion de los santos sacramentos / Totanes, Sebastian de – Sampaloc: Argueelles de la Concepion 1796 – (= ser Whsb) – 5mf – 9 – €60.00 – mf#Hu 325 – gw Fischer [490]

Arte de la lengua tagala; y, manual tagalog : para la administracion de los santos sacramentos / De Totanes, Sebastian – Manila: Estab tip del Colegio de Sto Tomas, 1850 – 1 – us CRL [490]

Arte de la lengua totonaca : conforme a el arte de antonio nebrija / Zambrano Bonilla, Jose – en la Puebla: impr Miguel de Ortega, ano de 1752 – (= ser Books on religion... 1543/44-c1800: doctrina cristiana, obras de devocion) – 3mf – 9 – mf#crl-49 – ne IDC [241]

Arte de la lengua totonaca : conforme a el arte de antonio nebrija, compuesto por d. joseph zambrano bonilla...dedicado a el illmo. sr. dr. d. domingo panlaleon alvarez de abreu... / Zambrano Bonilla, Jose – Puebla: En la impr de la viuda de m de Ortega, 1752 – (= ser Microfilm coll of manuscripts on cultural anthropology) – 22p (ill) – us Chicago U Pr [490]

Arte del barbero-peluquero-banero : que contiene el modo de hacer la barba, construccion de pelucas... modos de peinados / Garsault – Madrid, 1771 – 5mf – 9 – sp Cultura [640]

L'arte del contraponto / Artusi, Giovanni Maria – 1598 – (= ser Mssa) – 1mf – 9 – €20.00 – mfchl 48a – gw Fischer [780]

L'arte del contraponto ridotta in tavole : dove brevemente si contiene i precetti a quest' arte necessarij / Artusi, Giovanni Maria – In Venetia Presso Giacomo Vincenzi, & Ricciardo Amadino, co 1586 [mf ed 19–] – 2mf / 1r – 9,1 – (bound with: seconda parte dell' arte del contraponto [venetia 1589]) – mf#fiche63 / film 609 – us Sibley [780]

Arte del cuento en puerto rico / Melendez, Concha – New York, NY. 1961 – 1r – us UF Libraries [972]

L'arte del navegar : in laqval si contengono le regole, dechiarationi, secreti, e auisi, alla bona nauegation necessarij / Medina, Pedro de – Vinetia: G Pedrezato, 1554 [mf ed 1988] – 4mf – 9 – mf#SEM105P877 – cn Bibl Nat [520]

Arte dentaria : revista mensal da cirugia e da prothese dentarias – Rio de Janeiro, RJ: Typ Imperial e Constitucional de J Villeneuve & C, set 1869 – (= ser Ps 19) – mf#P17,01,70 – bl Biblioteca [617]

L'arte di ordinare i giardini... / Marulli, V – Napoli, 1804. 2v – 2mf – 9 – mf#GDI-17 – ne IDC [700]

L'arte d'inventer a l'improviste des fantaisies et cadences pour le violon : formant un recueil de 246 pieces amusantes et utiles en tous les tons majeurs et mineurs, oeuvre 17 / Campagnoli, Bartolomeo – Leipzig: Breitkopf & Hartel [c1812] [mf ed 1992] – 1r – 1 – mf#pres. film 117 – us Sibley [780]

El arte dramatico en lima durante el virreinato. madrid, 1945 / Lohmann Villena, Guillermo – Madrid: Razon y Fe, 1946 – 1 – sp Bibl Santa Ana [790]

Arte en america y filipinas : cuaderno 1. sevilla, 1935 / Bayle, Constantino – Madrid: Razon y Fe, 1936 – 1 – sp Bibl Santa Ana [700]

El arte en la gran bretana e irlanda / Diez Canedo, Enrique & Armstrong, Walter – Madrid: libreria gutenberg de jose ruiz, 1909 – sp Bibl Santa Ana [700]

El arte en la revolucion : conferencia pronunciada en el cine coliseum de barcelona, el dia 21 de marzo de 1937 / Noja Ruiz, Higinio – Barcelona? 1937? – (= ser Blodgett coll) – 9 – mf#fiche w1079 – us Harvard [946]

El arte explicado y gramatico / Marquez de Medina, Marcos – 1804 – 9 – sp Bibl Santa Ana [700]

El arte extremeno actual / Martin Gil, Thomas – Caceres: Tip. Extremadura, 1929 – 1 – sp Bibl Santa Ana [700]

L'arte (gi...archivio storico dell 'arte) / ed by Venturi, A – Roma. v1-2. 1898-99 – 490mf – 9 – (cont as: l'arte (periodico di storia dell'arte medievale e moderna e d'arte decorativa). roma, 1900-43. v3-47) – mf#0-494c – ne IDC [700]

Arte legal para estudiar jurisprudencia con la paratitla y exposicion... / Bermudez de Pedraza, F – Salamanca, 1612 – 5mf – 9 – sp Cultura [340]

Arte monumental prehistorico / Preuss, Konrad Theodor – Bogota, Colombia. v1-2. 1931 – 1r – us UF Libraries [720]

L'arte musicale in italia / Torchi, Luigi – Milan. 1897. 7v – 1 – 69.00 – us L of C Photodup [780]

L'arte musicale in italia / ed by Torchi, Luigi – Rome, Milan. 7v. 1897-1908 – 11 – $115.00 – us Univ Music [780]

L'arte naive – Reggio Emilia, 1974-75 [mf ed Chadwyck-Healey] – (= ser Art periodicals on microform) – 1r – 1 – uk Chadwyck [700]

Arte novissima de lengua mexicana / Tapia Zenteno, Carlos de – Mexico: de Hogal 1753 – (= ser Whsb) – 1mf – 9 – €20.00 – mf#Hu 407 – gw Fischer [440]

Arte, o compendio general del canto-llano, figurado, y organo / Marcos y Navas, Francisco – Madrid: por D Joachin Ibarra...1777 [mf ed 198-] – 12mf – 9 – mf#fiche 875 – us Sibley [780]

The arte of rhetorique / Wilson, Thomas – 1553 – 9 – us Scholars Facs [410]

L'arte organica / Antegnati, Costanzo – 1608 – (= ser Mssa) – 1mf – 9 – €20.00 – mfchl 47 – gw Fischer [780]

Arte pratica di contrappunto dimostrata con esempj di varj autori e con osservazioni / Paolucci, Giuseppe – Venezia: A de Castro 1765-72 [mf ed 19–] – 3v on 17mf / 1mf – 9 – mf#fiche 635 / fiche 924 – us Sibley [780]

Arte prattica et poetica : das ist: ein kurtzer unterrricht wie man einen contrapunct machen und componiren sol lernen (in zehen buecher abgetheilet) sehr kuertz- und leichtlich zu begreiffen / Herbst, Johann Andreas – Franckfurt: Anthonio Hummen. In Verlegung Thomae Matthiae Goetzens 1653 [mf ed 19–] – 1mf – 9 – mf#fiche 872 – us Sibley [780]

Arte precolombino en mexico y en la america central / Toscano, Salvador – Madrid: Missionalia Hispanica, 1948 – 1 – sp Bibl Santa Ana [700]

Arte, vocabulario y confessionario en el idioma mexicano : como se usa en el obispado de guadalaxara / Cortes y Zedeno, Jeronimo Tomas de Aquino – [Puebla]: impr Colegio Real de San Ignacio de la Puebla de los Angeles, ano de 1765 – (= ser Books on religion...1543/44-c1800: confesionarios) – 3mf – 9 – mf#crl-27 – ne IDC [241]

Arte y el amor en montparnasse / Maribona, Armando R – Mexico City?, Mexico. 1950 – 1r – us UF Libraries [700]

Arte, y gramatica general de la lengva qve corre en todo el reyno de chile, con vn vocabulario, y confessionario / Valdivia, Luis de – en Lima: Por Francisco del Canto, ano 1606 – (= ser Books on religion...1543/44-c1800: confesionarios) – 3mf – 9 – mf#crl-21 – ne IDC [241]

El arte y la pintura de adelardo covarsi / Vaca Morales, Francisco – Badajoz: Diputacion Provincial, 1944 – 1 – sp Bibl Santa Ana [750]

Arte y uso de architectura... / Lorenzo de San Nicolas, Fray – SL, SA – 6mf – 9 – sp Cultura [700]

Arte y uso de la arquitectura : con el primer libro de euclides... / Lorenzo de San Nicolas, Fray – Madrid, 1796 – 13mf – 9 – sp Cultura [700]

Arte, y vocabulario de la lengua quichua general de los indios de el peru : ahora nuevamente corregido, y aumentado en machos vocablos, y varias advertencias, notas... / Torres Rubio, Diego de – Lima: Impr de la Plazuela de San Christoval 1754 – (= ser Whsb) – 6mf – 9 – €70.00 – mf#Hu 391 – gw Fischer [440]

Arteaga, Rolando *see* Manifiesto del hombre reciente

Arteaga, Stefano *see* Le rivoluzioni del teatro musicale italiano

Arteau, Jean-Marie *see* Bio-bibliographie de monsieur carl faessler

Artefactos symmetriacos, e geometricos, advertidos, e descobertos pela industriosa perfeicao das artes, esculturaria, architectonica, e da pintura / Vasconcellos da Piedade, I – Lisboa, 1733 – 14mf – 9 – mf#O-1170 – ne IDC [700]

Artefizieller sphinkter 'as 800' am blasenhals : methode der wahl bei maennlicher stressinkontinenz unter schonung der erektion und ejakulation / Borkowski, Jerzy Roman – (mf ed 1996) – 2mf – 9 – €40.00 – 3-8267-2280-9 – mf#DHS 2280 – gw Frankfurter [616]

Arteli rabochikh dlia osnovaniia fabrik ili masterskikh : assotsiatsii / Miloradovich, L – 1962 – 16p 1mf – 9 – mf#COR-75 – ne IDC [335]

Arteli v drevnei i nyneshnei rossii / Kalachov, N V – 1864 – 93p 2mf – 9 – mf#COR-41 – ne IDC [335]

Arteli v rossii / Isaev, A A – Iaroslavl, 1881 – 336p 4mf – 9 – mf#COR-35 – ne IDC [335]

Artelnoe delo – Pg., 1916-1917(10) – 12mf – 9 – (cont as: trudovoe edinenie. missing: 1918-1919(2); 1916(10); 1918(10-12)) – mf#COR-546 – ne IDC [077]

Artelnoe delo / ed by Izdanie obshchestva dlia sodeistviia artelnomu delu v rossii – 1916-1917, 1918-1919 – 12mf – 9 – (cont as:trudovoe edinenie.missing:1916(10),1918(10-12)) – mf#COR-546 – ne IDC [335]

Artelnyi mir – 1913-1914(12) – 9mf – 9 – mf#COR-548 – ne IDC [335]

The artemas ward papers, 1721-1953 – [mf ed 1967] – 5r – 1 – (with p/g) – us MA Hist [355]

Artemev, E A *see* Sputnik kustaria i remeslennika

Artem'ev, V Ia *see* Statisticheskii ezhegodnik

L'artemisia / Cimarosa, Domenico – [n.p. 180-?] [mf ed 19–] – 1r – 1 – (in italian) – mf#pres. film 21 – us Sibley [780]

Arterial hypoxemia and performance during intense exercise / Koskolou, Maria D & McKenzie, Donald C – 1991 – 1mf – 9 – $4.00 – us Kinesology [613]

Arteriosclerosis : an official journal of the american heart association, inc / American Heart Association, Inc – Baltimore MD 1981-90 – 1,5,9 – (cont by: arteriosclerosis and thrombosis) – ISSN: 0276-5047 – mf#13359.02 – us UMI ProQuest [616]

Arteriosclerosis and thrombosis – Baltimore MD 1991-94 – 1,5,9 – (cont: arteriosclerosis: an official journal of the american heart association, inc; cont by: arteriosclerosis, thrombosis and vascular biology) – ISSN: 1049-8834 – mf#13359.02 – us UMI ProQuest [616]

Arteriosclerosis, thrombosis and vascular biology – Baltimore MD 1995+ – 1,5,9 – (cont: arteriosclerosis and thrombosis) – ISSN: 1079-5642 – mf#13359,02 – us UMI ProQuest [616]

Artes e letras – Florianopolis, SC. 03 fev 1924 – (= ser Ps 19) – mf#UFSC/BPESC – bl Biblioteca [079]

Artes plasticas na semana de 22 / Amaral, Aracy A – Sao Paulo, Brazil. 1970 – 1r – us UF Libraries [972]

Artes praedicandi : contribution a l'histoire de la rhetorique au moyen age / Charland, T-M – Paris, 1936 – 7mf – 8 – €15.00 – ne Slangenburg [400]

Artesanato e desenvolvimento / Rios, Jose Arthur – Rio de Janeiro, Brazil. 1969? – 1r – us UF Libraries [972]

The artesian and other deep wells on the island of montreal / Adams, Frank Dawson & LeRoy, Osmond Edgar – Montreal: [s.n.], 1906 – 2mf – 9 – 0-665-72208-7 – (incl bibl ref) – mf#72208 – cn Canadiana [550]

Artforum – New York NY 1962+ – 1,5,9 – ISSN: 1086-7058 – mf#3376 – us UMI ProQuest [700]

Arthington, Maria *see* Poetry of bye-gone days

Arthritis and rheumatism – Hoboken NJ 1958+ – 1,5,9 – ISSN: 0004-3591 – mf#10528 – us UMI ProQuest [616]

Arthropod structure and development – Oxford UK 2000+ – 1,5,9 – (cont: international journal of insect morphology and embryology) – ISSN: 1467-8039 – mf#49095,01 – us UMI ProQuest [590]

Arthur A Schomburg I S 201 Educational Complex *see* Kweli, verdad, truth

The arthur a schomburg papers, 1724-1938 – (= ser Manuscript colls from the schomburg center for research in black culture, the new york public library) – 12r – 1 – $2145.00 – 1-55655-376-5 – (with p/g) – us UPA [305]

Arthur a shurcliff collection of glass lantern slides – [mf ed 1985] – 2r – 1 – (with p/g; 879 glass lantern slides illus 19th- and early 20th-c urban and landscape planning in boston area) – us MA Hist [710]

The arthur advocate – Arthur, C W [Ont]: T G Greenham, [186–18–?] – 9 – mf#P06136 – cn Canadiana [071]

Arthur boyd houghton : a selection from his work in black and white / Housman, Laurence – London 1896 – (= ser 19th c art & architecture) – 3mf – 9 – mf#4.2.1764 – uk Chadwyck [740]

Arthur enterprise – Arthur, NE: H E Roush, 1914 (wkly) – 9r – 1 – (cont: hustler (read ne); v7 n1-10 misnumbered v6 n1-10) – us Bell [071]

The arthur enterprise – Arthur, NE: H E Roush, 1914 (wkly) [mf ed oct 21 1954-jun 27 1991] – 11r – 1 – (cont: hustler. v7 n1-v7 n10 misnumbered v6 n1-v6 n10) – us NE Hist [071]

Arthur family newsletter – 1976 nov-1984 oct – 1 – mf#950380 – us WHS [071]

Arthur fitger : sein leben und schaffen / Wocke, Helmut – Stuttgart: Metzler, 1913 [mf ed 1992] – (= ser Breslauer beitraege zur literaturgeschichte. neue folge 36) – x/152p – 1 – (incl bibl ref and ind) – mf#8014 reel 4 – us Wisconsin U Libr [430]

Arthur foote : american composer and theorist / Kopp, Frederick Edward – U of Rochester 1957 [mf ed 19–] – 2v on 1r / 11mf – 1 – mf#film 982 / fiche 1014 – us Sibley [780]

Arthur J Jones, Son and Co *see* Description of a suite of sculptured decorative furniture

Arthur nortje in port elizabeth : a regional reconstruction of his world in the eastern cape: 1942-1965: through interviews, letters, diaries, poems and photographs / Hendricks, Shaheed – Pretoria: Vista University 2002 [mf ed 2002] – 5mf [ill] – 9 – (incl bibl ref) – mf#mfm15349 – sa Unisa [800]

Arthur schopenhauer : his life and philosophy / Zimmern, Helen – London: Longmans, Green 1876 [mf ed 1991] – 1mf [ill] – 9 – 0-7905-8989-3 – mf#1989-2214 – us ATLA [120]

Arthur shepherd / Loucks, Richard – U of Rochester 1960 [mf ed 19–] – 4v on 24mf – 9 – (with bibl & app) – mf#fiche 70, 71 – us Sibley [780]

Arthur stanton : a memoir / Russell, George William Erskine – London: Longmans, Green 1917 [mf ed 1992] – 1mf [ill] – 9 – 0-524-04942-4 – mf#1992-2063 – us ATLA [241]

Arthur, William *see*
- Addresses
- French revolution of 1848
- Italy in transition
- On the difference between physical and moral law
- The pope, the kings and the people
- Revival in ballymena and coleraine
- Shall the loyal be deserted and the disloyal set over them?
- The tongue of fire

Arthur young's tour in ireland [1776-1779] / ed by Hutton, Arthur Wollaston – London, New York: G Bell & Sons 1892 [mf ed 1985] – 2v on 1r – 1 – (int & notes by ed, bibl by john p anderson; filmed with: la belgique sous la domination etrangere / pollet, ch) – mf#6569 – us Wisconsin U Libr [914]

Arthurian legends and the influence of french prose romance : the grail, lancelot, tristan and related manuscripts from the british library – 17r – 1 – £1600.00 – (with d/g) – uk Matthew [410]

Arthurian romances / Chretien, De Troyes – London, England. 1913 – 1r – us UF Libraries [390]

Arthurian tales : the greatest of romances, which recount the noble and valorous deeds of king arthur and the knights of the round table = Morte d'arthur. Selections / Malory, Thomas; ed by Rhys, Ernest – London: Norroena Society, 1907 – (= ser Anglo saxon classics) – 5mf – 9 – 0-524-08193-X – mf#1991-0306 – us ATLA [390]

Arthur's home magazine – La Salle IL 1852-97 – 1 – mf#5246 – us UMI ProQuest [640]

Arthur's magazine – La Salle IL 1844-46 – 1 – mf#3940 – us UMI ProQuest [640]

Die arthur-sage und die maehrchen des rothen buches von hergest / Schulz, Albert – Quedlinburg; Leipzig: G Basse, 1842 – 10r – 1 – (incl bibl ref) – us Wisconsin U Libr [430]

Le arti di bologna disegnate da annibale caracci ed intagliate da simone guillini coll'assistenza di alessandro algardi / Caracci, A – Roma, 1740 – 5mf – 9 – mf#O-1094 – ne IDC [700]

Artibus asiae – Hellerau, Dresden, 1925. v1 – 7mf – 8 – mf#CH-857c – ne IDC [956]

Artickel : deren sich die bischoff und gleerten des koenigreychs engelland in eine synodo im jar des herren mdlii zu london gehalten, vereiniget habed... – Zuerych, Andreas Gessner, [1553] – 1mf – 9 – mf#PBU-664 – ne IDC [240]

Article 29 considered... / Grueber, Charles Stephen – London, England. 1855 – 1r – us UF Libraries [240]

The articled clerk's journal and examiner – London. v1-3. 1879-81 (all publ) – 4mf – 9 – $6.00 – mf#LLMC 84-406 – us LLMC [340]

Articles d'association de la compagnie d'assurance de montreal contre les accidents du feu / Compagnie d'assurance de Montreal contre les accidents du feu – Montreal: De l'imprimerie de C B Pasteur, 1819? – 1mf – 9 – mf#21074 – cn Canadiana [368]

Articles d'association etablissant une compagnie d'assurance contre les accidens du feu dans la cite de quebec / Compagnie d'assurance de Quebec contre les accidens du feu – Quebec: Impr par John Neilson i.e. Neilson...1818 – 1mf – 9 – mf#21052 – cn Canadiana [368]

Les articles de la sacree faculte de theologie de paris, concernans nostre foy et religion chrestienne, et forme de prescher : avec le remede contre la poison / [Calvin, J] – [Geneva: Jean Girard], 1544 – 2mf – 9 – mf#CL-47 – ne IDC [240]

Articles enacted in the act intituled "an act to repeal a certain act therein-mentioned : and to provide for the police of the borough of william-henry, and certain other villages in this province" (9th march, 1824) passed in the fourth session of the eleventh provincial parliament of lower-canada = Articles statues dans l'acte intitule "acte pour rappeler un certain acte y mentionne et pour pourvoir a la police du bourg de William Henry... / Bas-Canada. Laws, Statutes etc – Quebec: printed by P E Desbarats, 1824 [mf ed 1991] – 1mf – 9 – mf#SEM105P1221 – cn Bibl Nat [350]

Articles, letters and miscellaneous papers, 1873-1907 / Fison, Lorimer – 1r – mf#PMB1042 – at Pacific Mss [980]

Articles of agreement – Edinburgh, Scotland. 1870 – 1r – us UF Libraries [240]

Articles of association of the montreal bank / Bank of Montreal – Montreal: Printed by N Mower, 1818 – 1mf – 9 – mf#55036 – cn Canadiana [332]

Articles of association of the [sic] quebec bank – [Quebec?]: J Neilson, Printer, [1820?] [mf ed 1993] – 1mf – 9 – 0-665-91319-2 – mf#91319 – cn Canadiana [332]

Articles of association, subscribers list / Ohio Company – 1r – 1 – mf#B26298 – us Ohio Hist [338]

The articles of the faith : approved by the synod of the presbyterian church of england, 1st may, 1890 – London: Publication Committee of the Presbyterian Church of England, [1890?] – 1mf – 9 – 0-524-07262-0 – mf#1991-3003 – us ATLA [242]

The articles of war : historical texts – repr of appendix to 2nd ed 1896, of William Winthrop's classic treatise on military law. Washington: GPO, 1920 – 2mf – 9 – $3.00 – mf#LLMC 88-029 – us LLMC [340]

Articles on romanism : monsignor capel, dr littledale / Hopkins, John Henry – New York: Thomas Whittaker 1890 [mf ed 1986] – 1mf – 9 – 0-8370-8522-5 – (incl ind) – mf#1986-2522 – us ATLA [241]

Articles on the solomon islands / Metcalfe, John R – 1350-c1961 – 1r – mf#pmb67 – at Pacific Mss [980]

Articles published in "realites cambodgiennes" june 22-july 27 1962 / Norodom Sihanouk, Prince – Washington DC: Royal Cambodian Embassy [1962?] [mf ed 1989] – 1r with other items – 1 – mf#mf-10289 seam reel 015/01 [§] – us CRL [327]

Articuli a facultate sacrae theologiae parisiensi determinati super materiis fidei nostrae hodie controversis : cum antidoto / [Calvin, J] – [Genevae: Jean Girard], 1544 – 1mf – 9 – mf#CL-6 – ne IDC [240]

Articuli ecclesiae anglicanae : or, the several editions of the articles of the church of england / ed by Davey, William Harrison – Oxford: J H & Jas Parker 1861 [mf ed 1986] – 1mf – 9 – 0-8370-8801-1 – (pref in english; texts in english & latin) – mf#1986-2801 – us ATLA [242]

Articulos de costumbres / Bentancourt, Luis Victoriano – Habana, Cuba. 1929 – 1r – us UF Libraries [972]

Articulos de costumbres / Cavanillas y Munoz, Juan Alonso – 1879 – 9 – sp Bibl Santa Ana [390]

Articulos periodisticos / Bobadilla, Emilio – Havana, Cuba. 1952 – 1r – us UF Libraries [972]

Articulos periodisticos / Varona, Enrique Jose – Habana, Cuba. 1949 – 1r – us UF Libraries [972]

Articulos politico-humoristicos y literarios / Geigel Y Zenon, Jose – Barcelona, Spain. 1936 – 1r – us UF Libraries [972]

Articulos varios de jose y giullerma camancho carrizosa – Bogota, Colombia. 1936 – 1r – us UF Libraries [972]

Articulus de audientia confessionum *see* Tractatus de causa immediata ecclesiasticae potestatis. articulus de audientia confessionum

Articvlvs de libero arbitrio, sev hvmani arbitrii viribvs, ex scriptvrae / Hunnius, A – Francofvrti ad Moenvm, 1597 – 2mf – 9 – mf#TH-1 mf 750-751 – ne IDC [242]

Articvlvs de persona christi : dvarvm in ea natvrarvm vnione hypostatica / Hunnius, A – [Vrsellis, 1585] – 6mf – 9 – mf#TH-1 mf 752-757 – ne IDC [242]

Artifacts / Shepaug Valley Archaeological Society & American Indian Archeological Institute – 1972 sep-1986 winter – 1r – 1 – (cont by: netop) – mf#1152845 – us WHS [930]

Les artifices des heretiques / Rapin, R – Paris, 1681 – 5mf – 9 – mf#CA-143 – ne IDC [240]

Artificial intelligence – Oxford UK 1970+ – 1,5,9 – ISSN: 0004-3702 – mf#42071 – us UMI ProQuest [000]

Artificial intelligence and the law – v1-7. 1992-99 – 9 – $403.00 – ISSN: 0197-1093 – mf#114041 – us Hein [340]

Artificial intelligence in medicine – Oxford UK 1992+ – 1,5,9 – ISSN: 0933-3657 – mf#42705 – us UMI ProQuest [000]

Artificial intelligence review – Dordrecht, Netherlands 1986+ – 1,5,9 – ISSN: 0269-2821 – mf#15613 – us UMI ProQuest [000]

Artificial limbs – Washington DC 1954-72 – 1,5,9 – ISSN: 0004-3729 – mf#2489 – us UMI ProQuest [617]

The artificial propagation of marine food fishes and edible crustaceans / Harvey, Moses – [Ottawa?: s.n, 1892?] – 1mf – 9 – 0-665-93951-5 – mf#93951 – cn Canadiana [639]

Artificialis introductio iacobi fabri stapule[n]sis : in dece[m] ethicoru[m] libros aristotelis; adiuncto familiari comme[n]tario iudoci clichtovei declarata; leonardi aretini dialogus de moribus ad galeotum amicum dialogo paruoru[m] moralium aristotelis ad eudemium respondens... / Lefevre d'Etaples, Jacques – Argentorati: Ex officina sua impressoria publicauit, anno...1511 mense Martio – (= ser Ethics in the early modern period) – [1]lea on 3mf – 9 – mf[u]-153 – ne IDC [170]

Artigas, Miguel *see* Las cien mejores poesias (liricas) de la lengua castellana

Artiles Rodriguez, Jenaro *see* Habana de velazquez

Artilleria colombiana / Centro De Artilleria (Bogota, Colombia) – Bogota, Colombia. 1960? – 1r – us UF Libraries [972]

L'artillerie au maroc : campagnes en chaoufa / Feline, Marie Charles – Paris: Berger-Levrault, 1912 – 1 – us CRL [960]

Artillery for the us land service with plates : by brevet major alfred mordecai (washington 1848-1849) / U.S. War Dept. Adjutant General's Office – (= ser Records of the adjutant general's office, 1780's-1917) – 1r – 1 – mf#T1104 – us Nat Archives [355]
Artime Bueen, Manuel Francico *see* Marches de guerra y cantos de presidio por manuel
Artinano y Zuricalday, Aristides de *see* Vida del beato valentin de berrio-ochoa y aristi
Artis cabbalisticae : hoc est reconditae theologiae et philosophiae scriptores / Ricius, P et al – Basileae, 1587 – €84.00 – ne Slangenburg [240]
Artis mvsicae legibvs logicis methodice informatae libri dvo : ad totum musicae artificium & comprimis solidum sonorum, modorumque musicorum fundamentum... / Magirus, Johann – [Brunsvigael] sumptibus avtoris 1611 [mf ed 19–] – 2mf – 9 – mf#fiche 155, 791 – us Sibley [780]
L'artisan : journal de la classe ouvriere – prosp., n1-4. Paris. sept-oct 1830 – 1 – fr ACRPP [073]
Artisan – Toronto. v1 n4 oct 12 1848; v1 n9-13 nov 16-dec 14 1848 (wkly) – 1r – 1 – Can$110.00 – cn McLaren [071]
The artisan – Freetown. Sierra Leone. -m. May 1884-Dec 1888. (28 ft) – 1 – uk British Libr Newspaper [072]
An artisan missionary on the zambesi : being the life story of william thomson waddell, largely drawn from his letters and journals / MacConnachie, J – Edinburgh, London, [1901] – 2mf – 9 – mf#HTM-107 – ne IDC [920]
L'artisanat au canada francais (1900-1950) : bibliographie analytique / Falardeau, Edith – 1956 [mf ed 1978] – 1r – 1 – (incl ind; pref by d'Emile Asselin) – mf#SEM105P4 – cn Bibl Nat [740]
Artishchev, R T [comp] *see* Statisticheskii ezhegodnik za 1924 g
Artist *see* Teatralnyi, muzykalnyi i khudozhestvennyi zhurnal
Der artist – Duesseldorf DE, 1898-1902, 1904-71 – 130r – 1 – (title varies: 1936: die unterhaltungsmusik, 1941: das podium der unterhaltungsmusik) – gw Misc Inst [780]
The artist : or young ladies' instructor in ornamental painting, drawing, etc / Gandee, B F – London 1835 – (= ser 19th c art & architecture) – 4mf – 9 – mf#4.2.1680 – uk Chadwyck [740]
The artist *see* The artist 1880-82 – l'artist et courier de l'art
Artist [1972] – New York NY 1972-91 – 1,5,9 – ISSN: 0004-3877 – mf#6626 – us UMI ProQuest [700]
The artist 1880-82 – l'artist et courier de l'art : and other periodicals / National Art Library – (= ser The art periodicals coll at the v and a museum, 1750-1920, pt 2) – 56r (19col) – 1 – £3300.00 – (incl: the artist 1880-82 – l'artist et courier de l'art 56r £3300. the chromolithograph 1867-69 2r £180. l'art pour tous 1861-1906 9r £800. l'art decoratif 1900-13 8r £710. the artist 1880-1902 9r £470. l'art 1875-1907 23r £1200. courier de l'art 1881-90 4r £220. art and decoration 1885-86 – art and literature 1886-89 1r £65. with printed guide) – mf#VAS – uk World [700]
Artist: a monthly lady's book – La Salle IL 1842-43 – 1 – mf#4556 – us UMI ProQuest [700]
Artist and amateur's magazine – La Salle IL 1843-44 – 1 – mf#5248 – us UMI ProQuest [700]
Artist files / New York. Museum of Modern Art – (mf ed 1992 A-F) – 5697mf – 9 – £18,900.00 coll – uk Chadwyck [700]
Artist in unknown india / Milward, Marguerite – London: T Werner Laurie, 1948 – (= ser Samp: indian books) – us CRL [306]
Artist muzykant – Moscow, 1918-19 [3 iss publ] – 1mf – 9 – us UMI ProQuest [780]
Artista : artes commercio e agricultura – Salvador, BA: Lith-Typ de Y G Tourinho, 06 maio 1876 – (= ser Ps 19) – mf#P18B,02,19 – bl Biblioteca [079]
Artista : jornal politico, litterario e noticioso – Rio Grande, RS: Typ do Artista, 26 nov 1867 – (= ser Ps 19) – bl Biblioteca [073]
O artista : orgao industrioso e artistico da provincia de santa catharina – Desterro, SC: Typ de Alex Margarida, 24 nov-dez 1878; fev 1879-17 mar 1880 – (= ser Ps 19) – mf#P16,02,53 – bl Biblioteca [321]
O artista : periodico dedicado a industria e principamente as artes – Rio de Janeiro, RJ: Typ de Aranha & Guimaraes, 27 nov 1870-12 mar 1871 – (= ser Ps 19) – mf#P05,04,28 – bl Biblioteca [338]
O artista brasileiro – Rio de Janeiro, RJ. 16 abr-23 maio 1949 – (= ser Ps 19) – mf#DIPER – bl Biblioteca [079]
Artistarkos (S van Mierlo) *see* Het voornemen der eeuwen en de gemeente der verborgenheid
Los artistas pintores de la expedicion malaspina. buenos aires, 1944 / Torre Revello, Jose – Madrid: Razon y Fe, 1946 – 1 – sp Bibl Santa Ana [700]
L'artiste : journal de la litterature et des beaux-arts – Paris. 1831-99, 1901, dec 1904 (incomplete) – 1 – fr ACRPP [073]
L'artiste – Montreal: impr Pour les Proprietaires par J Lovell, 1860 – 9 – mf#P04035 – cn Canadiana [700]
L'artiste *see* Revue de paris
Artiste russe – St Petersburg, 1846-48 [bimthly] – 24mf – 9 – us UMI ProQuest [780]
Artistes et artisans du canada / Falardeau, Emile – Montreal: G Ducharme, 1940-1969 [mf ed 1990] – 7mf – 9 – mf#SEM105P1267 – cn Bibl Nat [700]
Artistic conservatories : and other horticultural buildings / Godwin, Edward William et al – London 1880 – (= ser 19th c art & architecture) – 1mf – 9 – mf#4.2.1004 – uk Chadwyck [720]
The artistic evolution of the english home / Waring and Gillow Ltd – London [1900?] – (= ser 19th c art & architecture) – 2mf – 9 – mf#4.2.1763 – uk Chadwyck [720]
Artistic homes : or, how to furnish with taste – London [1880] – (= ser 19th c art & architecture) – 2mf – 9 – mf#4.2.36 – uk Chadwyck [740]
Artistic japan *see* Ackermann's 'repository of arts'
Artistic pedigree – 1993 feb 18/mar 4-1995 may 12 – 1 – mf#2680355 – us WHS [071]
Artistic Supply Co Ltd *see* Catalogue of illustrations of the artistic supply company
Artistic-country seats – New York, NY. 1886 – 1r – us UF Libraries [025]
Artist-muzykant – Moscow, 1918 – 2mf – 9 – us UMI ProQuest [790]
Artists : lee county / Hanson, W Stanley – s.l, s.l? . 1936 – 1r – us UF Libraries [978]
Artists' and writers' chap book – New York. Dec 15 1933; May 3 1935 – 1 – us NY Public [420]
Artists at home : photographed by j p mayall / Stephens, Frederic George – London 1884 – (= ser 19th c art & architecture) – 3mf – 9 – mf#4.2.125 – uk Chadwyck [770]
Artists' homes : a portfolio of drawings / Adams, Maurice Bingham – London 1883 – (= ser 19th c art & architecture) – 2mf – 9 – mf#4.1.328 – uk Chadwyck [740]
Artist's proof – New York NY 1961-71 – 1,5,9 – ISSN: 0571-2149 – mf#2536 – us UMI ProQuest [700]
Artists review – Toronto: Artists Cooperative Toronto. v1-4 n1. oct 11 1977-oct 1980// – 1r – 1 – Can$110.00 – cn McLaren [700]
Artists scrapbooks : anni albers to frank lloyd wright / New York. Museum of Modern Art – 642mf – 9 – £2,975.00 coll – (126 vols on 42 artists) – uk Chadwyck [700]
Artists' sketchbooks in the british museum – [mf ed 1996] – 12r – 1 – $1600.00 – 1-900853-50-7 – (with printed ind of artists) – uk Mindata [700]
Artitudes international – St Jeannet, 1972-74 [mf ed Chadwyck-Healey] – (= ser Art periodicals on microform) – 1r – 1 – uk Chadwyck [700]
The artizan's guide and everybody's assistant : embracing nearly four thousand new and valuable receipts tables, etc in almost every branch of business connected with civilized life, from the household to the manufactory / Moore, Richard – Montreal: J Lovell, 1875 – 6mf – 9 – mf#54941 – cn Canadiana [640]
Art-journal – Killen TX 1839-1912 – 1 – mf#5245 – us UMI ProQuest [700]
The art-journal / Dublin. Exhibition of Art and Art-industry, 1853 – London 1853 – (= ser 19th c art & architecture) – 10mf – 9 – mf#4.2.916 – uk Chadwyck [700]
The art-manufactures of birmingham and midland counties / Wallis, George – London [1862] – (= ser 19th c art & architecture) – 2mf – 9 – mf#4.2.911 – uk Chadwyck [700]
Artner, Maria T von *see* Briefe ueber einen theil von croatien und italien an caroline pichler
Artnews – New York NY 1902+ – 1,5,9 – ISSN: 0004-3273 – mf#923 – us UMI ProQuest [700]
Arto-Haumacher, Rafael *see* C F gellerts briefstilreform
Artois, Armand D' *see*
– Maris ont tort
– Suites d'un mariage de raison
– Valentine
Arton's Cultural Affairs Society & Publishing *see* Fuse
Artrage newsletter – 1994 jul/aug – 1 – mf#4864026 – us WHS [071]
Arts : beaux-arts, litterature, spectacles. – Paris. 31 janv 1945-aout 1966 – 1 – fr ACRPP [073]
Arts – Paris, France 9 mar 1945-jul 1967 – 28 1/2r – 1 – uk British Libr Newspaper [700]
Les arts *see* Art and decoration
Arts anciens de flandre – Bruges. v1-6. 1905-1913 – 48mf – 9 – mf#0-1215 – ne IDC [700]
Arts and activities – San Diego CA 1939+ – 1,5,9 – ISSN: 0004-3931 – mf#2199 – us UMI ProQuest [370]
The arts and artists...of the schools of painting, sculpture and architecture / Elmes, James – London 1825 – (= ser 19th c art & architecture) – 12mf – 9 – mf#4.2.1151 – uk Chadwyck [700]
Arts and crafts : orlando, florida / Leonard, Agnes Mckenna – s.l, s.l? . 1936 – 1r – us UF Libraries [978]
Arts and Crafts Exhibition Society, London *see*
– [Exhibition catalogue. 1888]
– [Exhibition catalogue. 1889]
– [Exhibition catalogue. 1890]
– [Exhibition catalogue. 1893]
– [Exhibition catalogue. 1896]
– [Exhibition catalogue. 1899]
The arts and crafts of travancore / Kramrisch, Stella et al – London: Royal India Society and Govt of Travancore, 1948 – (= ser Samp: indian books) – us CRL [700]
The arts and the artistic manufactures of denmark / Boutell, Charles – London 1874 – (= ser 19th c art & architecture) – 2mf – 9 – mf#4.1.339 – uk Chadwyck [740]
Arts, antiquities, and chronology of ancient egypt / Wathen, George Henry – London 1843 – (= ser 19th c art & architecture) – 4mf – 9 – mf#4.2.1738 – uk Chadwyck [930]
Arts & architecture – Santa Monica CA 1911-67 – 1 – (cont by: arts + architecture) – ISSN: 0730-9481 – mf#1033 – us UMI ProQuest [720]
Arts + architecture – Santa Monica CA 1983-84 – 1,5,9 – (cont: arts & architecture) – ISSN: 0730-9481 – mf#13831 – us UMI ProQuest [720]
Arts council of great britain catalogue – London, 1942-1978 – (= ser Art exhibition catalogues on microfiche) – 892 catalogues on 1022mf – 9 – £5,365.00 – (individual titles not listed separately) – uk Chadwyck [700]
Arts education policy review – Washington DC 1993+ – 1,5,9 – (cont: design for arts in education) – ISSN: 1063-2913 – mf#824,02 – us UMI ProQuest [700]
Les arts en portugal / Raczynski, A – Paris, 1846 – 6mf – 9 – mf#0-1058 – ne IDC [700]
Arts et metiers / Societe des Anciens Eleves des Ecoles nationales d'Arts et Metiers – Paris. oct 1920-41 – 1 – fr ACRPP [740]
Arts et metiers graphiques – Paris, 1927-39 [mf ed Chadwyck-Healey] – (= ser Art periodicals on microform) – 6r – 1 – uk Chadwyck [760]
Arts, heraldique, archeologie / Le Beffroi – Bruges, 1863-1876. v1-4 – 28mf – 9 – mf#0-1216 – ne IDC [700]
Arts in psychotherapy – Oxford UK 1980+ – 1,5,9 – (cont: art psychotherapy) – ISSN: 0197-4556 – mf#49534 – us UMI ProQuest [700]
Arts in society – Madison WI 1958-76 – 1,5,9 – ISSN: 0004-4024 – mf#1952 – us UMI ProQuest [700]
Arts magazine – Forest Hills NY 1926-92 – 1,5,9 – ISSN: 0004-4059 – mf#8000 – us UMI ProQuest [700]
Arts management – New York NY 1962+ – 1,5,9 – ISSN: 0004-4067 – mf#8555 – us UMI ProQuest [700]
The arts of the hausa : an aspect of islamic culture in northern nigeria / Commonwealth Institute – 1977 – 2 color mf – 15 – $45.00f – 0-226-68899-2 – (62p accompanying text) – us Chicago U Pr [700]
Arts review – London UK 1973-92 – 1,5,9 – (cont by: art review) – ISSN: 0004-4091 – mf#8170.01 – us UMI ProQuest [700]
Arts under arms / Fitzgibbon, Maurice – New York, NY. 1901 – 1r – us UF Libraries [700]
Artscanada – Toronto ON 1943-82 – 1,5,9 – ISSN: 0004-4113 – mf#2253 – us UMI ProQuest [700]
The art-union exhibition, for 1843 : a hand-book guide for visitors / Clarke, Henry Green – London: H G Clarke & Co, 66, 1843 – (= ser 19th c art & architecture) – 1mf – 9 – mf#4.1.43 – uk Chadwyck [700]
Artus, Gaston Andre *see* Gijig-anang mekateokonaie, s j o gagikwewinan
Artus, Wilfrido *see* Los reformadores espanoles del siglo 16
L'artusi : ouero delle imperfettioni della moderna mvsica ragionamenti dui / Artusi, Giovanni Maria – Nouamente stampato Venetia: G Vincenti 1600 [mf ed 19–] – 1r – 1 – us UMI ProQuest [780]
Artusi, Giovanni Maria *see*
– L'arte del contraponto
– L'arte del contraponto ridotta in tavole
– L'artusi
– L'artusi overo delle imperfettioni...
L'artusi overo delle imperfettioni... / Artusi, Giovanni Maria – 1600 – (= ser Mssa) – 4mf – 9 – €60.00 – mfchl 48 – gw Fischer [780]
Artvin vilayeti hakkinda malumat-i umumiye / Zeki, Muvahhid – [Istanbul]: Sikret-i Mertebiye Matbaasi, 1927 – (= ser Ottoman histories and historical sources) – 3mf – 9 – $55.00 – us MEDOC [956]
Artweek – Palmyra WI 1970+ – 1,5,9 – ISSN: 0004-4121 – mf#7152 – us UMI ProQuest [700]
The art-workman's position : a lecture delivered in behalf of the architectural museum / Beresford-Hope, Alexander James Beresford – London 1864 – (= ser 19th c art & architecture) – 1mf – 9 – mf#4.2.960 – uk Chadwyck [720]
Artz – 1991 oct/nov; 1992 mar-apr, aug-1993 jan/feb; summer – 1 – mf#4879190 – us WHS [071]
Aruanne = Report / Estonian Baptist Union – 62p. 1922, 19th Report; 1939, 36th Report. – 1 – $5.00 – us Southern Baptist [242]
Aruban annals / Rings, William Refus – Plain City, OH. 1943 – 1r – us UF Libraries [972]
Aruchas bas-ammi : israels heilung / Ruelf, I – Frankfurt a M, 1883 – 1mf – 9 – mf#J-28-60 – ne IDC [956]
'Arukh / Nathan Ben Jehiel – Lemberg, Ukraine. 1865 – 1r – 1 – us UF Libraries [090]
Arun *see* Testament of subhas bose
Arunachalam, S *see* The history of the pearl fishery of the tamil coast
Arunanti Civacariyar *see* Sivajnana siddhiyar of arunandi sivacharya
Arundale, Francesca *see* The idea of re-birth
Arundale, Francis *see*
– Examples and designs of verandahs
– Gallery of antiquities selected from the british museum
Arundale, George Sydney *see* Freedom and friendship
Arundel family papers – 1803-1935 – 2r – 1 – mf#pmb1227 – at Pacific Mss [929]
Arundel, John *see* Causes of declension in christian churches
Arundel, John T *see*
– Correspondence
– Diaries
– Miscellaneous correspondence
– Miscellaneous papers on the phosphatre industry
– 'Sundry data of my life'
Arundel Society, London *see*
– A classified list of photographs of drawings, paintings, and sculpture, precious metals and enamels
– The cloisters of monreale in sicily
– Decorative furniture english, italian, german, flemish, etc
– Ecclesiastical metal work of the middle ages
– The sculptured ornament of the monastery of batalha
– The treasure of petrossa
Arundell, Francis V *see*
– Discoveries in asia minor
– A visit to the seven churches of asia with an excursion into pisidia
Arundell of Wardour, John Francis Arundell, Baron *see*
– The scientific value of tradition
– Tradition
Arunodaya : the autobiography of baba padmanji: containing a description of his former life as a hindu; and the causes which led to his conversion / Padmanji, Baba – 2nd rev ed. Bombay: Bombay Tract & Book Society, 1908 [mf ed 1995] – (= ser Yale coll) – 15p/252p (ill) – 1 – 0-524-09915-4 – (in marathi) – mf#1995-0915 – us ATLA [920]
Arusmont, Frances d' *see* Views of society and manners in america
Arvelo Larriva, Alfredo *see* Sones y canciones, y otros poemas
Arvelo, Teresa *see* Emilia
Arvendel : or, sketches in italy and switzerland / Noel, Gerard T – London 1826 [mf ed Hildesheim 1995-98] – (= ser Fbc) – 1v on 1mf – 9 – €40.00 – 3-487-27783-2 – gw Olms [914]
Arvika allehanda – Arvika, 1890-94 – 9 – sw Kungliga [078]
Arvika nyheter – Arvika, Sweden. 1895- – 1 – sw Kungliga [078]
Arvika tidning – Kristinehamn, Arvika, 1884-1962 – 116r – 1 – sw Kungliga [078]
Arvikakuriren – Arvika, 1905 – 1r – 1 – sw Kungliga [078]
Arvin, Neil Cole *see* Eugene scribe and the french theatre, 1815-1860
Arvin, Newton *see* Herman melville
Arwed : ou, les represailles / Etienne, Charles Guillaume – Bruxelles, Belgium. 1830 – 1r – us UF Libraries [440]
Arx, Walther von *see* Gottfried keller
Arya : a philosophical review = Revue de grande synthese – Pondicherry: All India Books, Sri Aurobindo's Ashram, 1990 [mthly] [mf v1-7 1914-21 filmed 2003] – 7v on 3r – 1 – (in english) – mf#2003-s013 – us ATLA [280]

The arya dharma of sakya muni, gautama, buddha : or, the ethics of self discipline / Dharmapala, Anagarika – Calcutta: Maha Bodhi Society, 1917 [mf ed 1995] – (= ser Yale coll) – 232p – 1 – 0-524-09170-6 – mf#1995-0170 – us ATLA [280]

The arya samaj : an account of its origin, doctrine and activities, with a biographical sketch of the founder / Lajpat Rai, Lala – Lahore: Uttar Chand Kapur & Sons, 1932 – (= ser Samp: indian books) – us CRL [280]

The arya samaj : an account of its origin, doctrines, and activities / Lajpat Rai, Lala – London: Longmans, Green, 1915 – 1mf – 9 – 0-524-01191-5 – (incl bibl ref) – mf#1990-2267 – us ATLA [280]

The arya samaj and its detractors : a vindication / Sraddhananda, swami – 1st ed. Dayanandab[a]d: [s.n.], 1910 [mf ed 1992] – 2mf – 9 – 0-524-05463-0 – mf#1990-3489 – us ATLA [280]

Aryan – 1977/1978 winter-1980 dec – 1 – mf#597186 – us WHS [071]

Aryan, C Leon de *see*
– The aryan sun-work-shop and the broom
– The broom

The aryan home : a thesis on the location of the original aryan home and other early aryan settlements: historicogeographical solution of the problem / Pithawalla, Maneck B – Karachi: [sn], 1946 – (= ser Samp: indian books) – us CRL [930]

Aryan Nations-Teutonic Unity Publ et al *see* Calling our nation

Aryan sun-myths the origin of religions / Titcomb, Sarah Elizabeth – [s.l]: SE Titcomb, c1890 [mf ed 1991] – 1mf – 9 – 0-524-01318-7 – (incl bibl ref) – mf#1990-2354 – us ATLA [230]

The aryan sun-work-shop and the broom / ed by Aryan, C Leon de – San Diego, CA: C Leon de Aryan [v23 n49-v26 n51(easter 1954-apr 1965)] (mthly) – 5r – 1 – us CRL [073]

Aryan Youth Movement [US] *see* White student union

The aryanisation of india / Dutt, Nripendra Kumar – Calcutta: Nripendra Kumar Dutt, 1925 – (= ser Samp: indian books) – us CRL [954]

Aryas, semites and jews : jehovah and the christ / Burge, Lorenzo – Boston: Lee & Shepard, 1889, c1888 [mf ed 1989] – 1mf – 9 – 0-7905-0869-9 – mf#1987-0869 – us ATLA [939]

Aryavarta – Patna, India. 2 May 1949-1951; Jul 1952-1987 – 113r – 1 – us L of C Photodup [079]

Aryo-semitic speech : a study in linguistic archaeology / McCurdy, James Frederick – Andover: Warren F Draper 1881 [mf ed 1986] – 1mf – 9 – 0-8370-8202-1 – (incl bibl ref & ind) – mf#1986-2202 – us ATLA [490]

Arzobispo de bogota / Leon, Eugenio – Medellin, Colombia. 1950 – 1r – us UF Libraries [972]

Arzobispo valera / Henriquez Urena, Max – Rio de Janeiro, Brazil. 1944 – 1r – us UF Libraries [972]

Arzt, Frederick Karl *see* History and outline of laws relating to vessel inspection

Der arzt im spiegelbild der deutschen schoengeistigen literatur seit dem beginn des naturalismus / Wittmann, Fritz – Berlin: E Ebering, 1936 [mf ed 1993] – (= ser Abhandlungen zur geschichte der medizin und der naturwissenschaften 18) – 133p – 1 – (incl bibl ref) – mf#8146 – us Wisconsin U Libr [430]

Arzu, Jose *see* Pepe batred intimoo

As a fire / Latham, Henry Jepson [comp] – Brooklyn NY: H J Latham 1907 [mf ed 1985] – 1mf – 9 – 0-8370-4350-6 – mf#1985-2350 – us ATLA [240]

As aliancas / Ivo, Ledo – Rio de Janeiro, Brazil. 1947 – 1r – us UF Libraries [972]

As chulipas : cronica quinzenal das lettras, artes, costumes e politica – Rio de Janeiro, RJ: Typ Fluminense, 15-30 jul 1876 – (= ser Ps 19) – mf#P17,01,102 – bl Biblioteca [079]

As elites de cor / Azevedo, Thales De – Sao Paulo, Brazil. 1955 – 1r – us UF Libraries [972]

As gavetas da torre fombo 4, lisboa 1964 / Barrado Manzano, Arcangel – Madrid: Graf. Calleja, 1967 – 1 – sp Bibl Santa Ana [946]

As happy as a prince / Smith, James – London, England. 18– – 1r – us UF Libraries [240]

"As i remember kansas city from my boyhood and its townhood days." / Hymer, Julian B – 1 – ("Reminiscence of 64 years of Railroading." 1) – us Kansas [978]

As it is – 1979 nov-1985 – 1 – mf#1221359 – us WHS [071]

As it was in the beginning : or, the historic principle applied to the mosaic scriptures / Cridge, Edward – Chicago: F H Revell c1900 [mf ed 1993] – 1mf – 9 – 0-524-05605-6 – mf#1992-0460 – us ATLA [221]

As lettras : revista quinzenal do gremio litterario "amor e progesso" – Rio de Janeiro, RJ: Typ de Machado & C, 15 nov 1880 – (= ser Ps 19) – mf#P17,01,174 – bl Biblioteca [440]

As missoes... / Silva Rego, Antonio da – Madrid: Archivo Ibero Americano, 1960 – 1 – sp Bibl Santa Ana [240]

As moedas visigodas da lusitania / Elias Garcia, A – Guimaraes, 1950 – 1 – sp Bibl Santa Ana [946]

As noch de tankruesel brenn' : mit biller ut theodor herrmann sin warkstaed / Frahm, Ludwig – Hamborg: R Hermes 1918 [mf ed 1989] – (= ser Nedderduetsch boekeri 4) – 1r [ill] – 1 – (filmed with: theodor fontane / paul von szczepanski) – mf#7253 – us Wisconsin U Libr [880]

As novidades – Fall River, MA: As novidades Pub Co, dec 1917-1948] – 15r – us CRL [946]

As others saw him : a retrospect, a d 54 / Jacobs, Joseph – New York: Funk & Wagnalls 1903 [mf ed 1985] – 1mf – 9 – 0-8370-3745-X – (with int, aft & notes; first publ anonymously in 1895) – mf#1985-1745 – us ATLA [830]

As others see us : a study of progress in the united states / Brooks, John Graham – New York: Macmillan 1908 [mf ed 1990] – 1mf – 9 – 0-7905-5638-3 – (incl bibl ref) – mf#1988-1638 – us ATLA [301]

As others see us, and as we are : the plea and position of the disciples of christ, as they are, presented in contrast with the erroneous views usually held of them by the denominational world / Hill, John Louis – Cincinnati, O[hio]: Standard Pub Co c1908 [mf ed 1992] – 2mf – 9 – 0-524-02256-9 – mf#1990-4263 – us ATLA [240]

A.S Pratt and Sons. Washington, DC *see* Duties, powers, and liability of national bank directors

As sirat : organe de l'association des ulamas musulmans algeriens. – n1-15. Constantine. sept-dec 1933 – 1 – fr ACRPP [260]

As to roger williams and his 'banishment' from the massachusetts plantation : with a few further words concerning the baptists, the quakers, and religious liberty / Dexter, Henry Martyn – Boston: Congregational Pub Society 1876 [mf ed 1990] – 2mf – 9 – 0-7905-7219-2 – (incl bibl ref) – mf#1988-3219 – us ATLA [242]

As to sharing fairly / Wheeler, Everett Pepperrell – New York: H Holt & Co, c1920 (mf ed 19–) – 15p – (repr fr: unpartizan review, mar-apr 1920) – mf#ZT-TB pv144 n11 – us Harvard [338]

Asa journal : the journal of the archaeological survey association of southern california – 1977 spring/summer-1988 spring/summer – 1 – mf#1544110 – us WHS [930]

Asa turner : a home missionary patriarch and his times / Magoun, George Frederick – Boston: Congregational Sunday-School & Pub Soc c1889 [mf ed 1990] – 1mf – 9 – 0-7905-5430-5 – mf#1988-1430 – us ATLA [242]

Asabari / Banaphula – Kalakata, 1381 [1974] – 1r – 1 – us CRL [954]

Asad Sulayman Abduh *see* Bad awjuh al-ikhtilaf fi rasm ism al-makan al-wahid bi-huruf al-lughah al arabiyah fi al-mamlakah al-arabiyah al-saudiyah

Asahi evening news – Tokyo, Japan 1 jul 1965-30 dec 1978 – 1 – uk British Libr Newspaper [079]

Asahi shimbun – 1888- mthly updates – (japanese daily newspaper) – us Primary [079]

Asaio journal – Baltimore MD 1992-96 – 1,5,9 – (cont: asaio transactions) – ISSN: 1058-2916 – mf#16011,02 – us UMI ProQuest [610]

Asaio transactions – Baltimore MD 1986-91 – 1,5,9 – (cont by: asaio journal) – ISSN: 0889-7190 – mf#16011.02 – us UMI ProQuest [610]

Asalh update – 1991 spring – 1 – mf#4851377 – us WHS [071]

Asam bani – Gauhati, India. 1962-Jun 1965 – 4r – 1 – (assamese language) – us L of C Photodup [079]

Asamblea asistencial de la c.n.s. de caceres : plan asistencial provincial en su alcance global / Central Nacional Sindicalista. Caceres – Caceres: Tip. El Noticiero, 1948 – sp Bibl Santa Ana [330]

Asamblea filipina – Manila, Philippines 18 oct-11 nov 1907, 27 jul-26 sep 1908 (imperfect) – 1 – uk British Libr Newspaper [079]

Asamblea nacional constituyente de 1885 / Garcia, Miguel Angel – San Salvador, El Salvador. 1936 – 1r – us UF Libraries [972]

Asamblea plenaria del consejo economico provincial / Wulff Martin, Enrique – Caceres, s.i., 1953 – sp Bibl Santa Ana [330]

Asamblea plenaria del consejo economico sindical / Solano Pedrero, Carlos – Caceres, s.i., 1947 – sp Bibl Santa Ana [330]

Asambleas constituyentes argentinas... / Raviguani, Emilio – Madrid: Razon y Fe, 1940 – 1 – sp Bibl Santa Ana [972]

Die asaph-psalmen : historisch-kritisch untersucht / Kopfstein, Marcus – Marburg: In Commission von Oscar Ehrhardt's Universitaets-Buchhandlung, 1881 – 1mf – 9 – 0-8370-3984-3 – (incl bibl ref) – mf#1985-1984 – us ATLA [220]

Asar-i hamide-i aklam – Osman Rasih, 1290 [1873] – (= ser Ottoman histories and historical sources) – 1mf – 9 – $25.00 – us MEDOC [956]

Asatkin, O M *see*
– Narodnoe khoziaistvo ukssr
– Ukraina v tsifrakh

Asbarez – 1978 aug 16; 1979 jan 31-1980 jun; 1980 jul-1981; 1982-1983 jun; 1983 jul-1984; 1985-1986 jun; 1986 jul-1987; 1988-1989 jun; 1998-2000 – 1 – mf#1012153 – us WHS [071]

Asbeck, M d' *see* La mystique de ruysbroeck l'admirable. un echo du neoplatonisme au 14th siecle

Asbeck, Wilhelm Ernst *see* Der geheimnisvolle hof

Asbestos case management : pretrial and trial procedures / Willging, Thomas E – Washington: FJC, 1985 – 1mf – 9 – $1.50 – mf#LLMC 95-321 – us LLMC [340]

Asbjornsen, Peter Christen *see* Round the yule log

Asbury and his coadjutors / Larrabee, William Clark; ed by Clark, Davis Wasgatt – Cincinnati: publ...for the Methodist Episcopal Church 1853 [mf ed 1984] – 2v on 8mf – 9 – 0-8370-0801-8 – mf#1984-4153 – us ATLA [242]

Asbury, Samuel Ralph *see* The book of the prophet jeremiah

Asbury seminarian – Wilmore KY 1946-85 – 1,5,9 – (cont by: asbury theological journal) – ISSN: 0004-4253 – mf#8932.01 – us UMI ProQuest [240]

Asbury theological journal – Wilmore KY 1986+ – 1,5,9 – (cont: asbury seminarian) – mf#8932,01 – us UMI ProQuest [240]

ASCE Technical Council on Codes and Standards *see* Journal of technical topics in civil engineering

Ascendance – 1988 nov; 1989 – 1 – mf#4848549 – us WHS [071]

Ascendancy of popery fatal to the truth of the gospel / Bridge, Stephen – London, England. 1850 – 1r – us UF Libraries [240]

The ascended christ : a study in the earliest christian teaching / Swete, Henry Barclay – London, New York: Macmillan, 1910 [mf ed 1988] – 1mf – 9 – 0-7905-0353-0 – (incl bibl ref & ind) – mf#1987-0353 – us ATLA [240]

La ascendencia espanola el inca garcilaso de la vega precisiones genealogicas / Lohmann Villena, Guillermo – Madrid, s.i. 1958 – 1 – sp Bibl Santa Ana [920]

Ascendientes y descendientes de hernando cortes : linea de medina-sidonia y otros / Valgoma y Diaz-Varela, Dalmiro de la – Madrid: Cultura Hispanica, 1951 – 1 – sp Bibl Santa Ana [920]

Ascensao e queda de miguel arraes / Barros, Adirson De – Rio de Janeiro, Brazil. 1965 – 1r – us UF Libraries [972]

The ascension: a sacred oratorio / Hook, James – 1776. Copyist's manuscript in ink with holograph corrections. Vocal score. MUSIC 1873, Item 1 – 1 – us L of C Photodup [780]

The ascension and heavenly priesthood of our lord / Milligan, William – London New York: Macmillan 1892 [mf ed 1985] – (= ser The baird lecture 1891) – 1mf – 9 – 0-8370-4438-3 – (incl bibl ref & ind) – mf#1985-2438 – us ATLA [240]

Ascension d'isaie = Ascension of isaiah / Tisserant, Eugene – Paris: Letouzey, 1909 [mf ed 1988] – (= ser Documents pour l'etude de la bible) – 1mf – 9 – 0-7905-0401-4 – (in french and latin. int and notes by eugene tisserant. incl ind) – mf#1987-0401 – us ATLA [220]

The ascension of isaiah / ed by Charles, Robert Henry – London, A and C Black, 1900. Chicago: Dep of Photodup, U of Chicago Lib, 1971 (1r); Evanston: American Theol Lib Assoc, 1984 (1r) – 1 – 0-8370-0513-2 – mf#1984-B301 – us ATLA [221]

Ascensioni sul monte kenya / Devalle, Giovanni – Firenze: Istituto geografico militare 1938 [mf ed [Nairobi]: Kenya National Archives Photographic Service 1970] – 16p [ill] on 1r with other items – 1 – (offprint fr: universo. anno 19 n8 [agosto 1938]) – mf#mf-1550 r43 – us CRL [960]

The ascent of faith : or, the grounds of certainty in science and religion / Harrison, Alexander James – New York: Thomas Whittaker, 1894 [mf ed 1985] – (= ser The boyle lectures 1892, 1893) – 1mf – 9 – 0-8370-3499-X – (incl ind) – mf#1985-1499 – us ATLA [210]

The ascent of nanda devi / Tilman, Harold William – Cambridge: University Press, 1937 – (= ser Samp: indian books) – (foreword by t g longstaff) – us CRL [280]

The ascent of olympus / Harris, James Rendel – Manchester: University Press, 1917 – 1mf – 9 – 0-524-00887-6 – (incl bibl ref) – mf#1990-2110 – us ATLA [250]

The ascent of the soul / Bradford, Amory Howe – New York: Outlook, 1903 – 1mf – 9 – 0-8370-3391-8 – (incl ind) – mf#1985-1391 – us ATLA [240]

The ascent through christ : a study of the doctrine of redemption in the light of the theory of evolution / Griffith-Jones, Ebenezer – New York: James Pott, 1900 – 2mf – 9 – 0-7905-9383-1 – mf#1989-2608 – us ATLA [240]

L'ascesa del proletario – Wilkes-Barre, PA. oct 1 1908-sep 15 1910 – 1r – 1 – (italian newspaper) – us IHRC [071]

Das ascetentum der drei ersten christl jahrhunderte und das egyptische moenchtum *see* Das morgenlaendische moenchtum

Ascetical treatise on the sacrifice of the mass / a letter on the great importance of the divine = De sacrificio missae tractatus asceticus / Bona, Giovanni – London: John Philp [1871?] [mf ed 1993] – 2v on 1mf – 9 – 0-524-06286-2 – (in english) – mf#1990-5215 – us ATLA [240]

Asch, Sholem *see*
– Amerika
– A passage in the night
– Petersburg
– Reb shloyme nogid

Aschaffenburger anzeiger *see* Privilegirte kur-mainzische landes-zeitung

Aschaffenburger zeitung *see* Privilegirte kur-mainzische landes-zeitung

Ascham, Roger *see* The schoolmaster

Aschenbrenner, Michael *see* Lehrbuch der metaphysik

Aschenbroedel : dramatisches maehrchen / Grabbe, Christian Dietrich – Duesseldorf: J H C Schreiner, 1835 [mf ed 1995] – 99p – 1 – mf#8749 – us Wisconsin U Libr [390]

Ascher, Fritz *see* Palaestina nachrichten

Ascher tagesbote *see* Ascher zeitung

Ascher zeitung – Asch (CZ), 1929 16 apr-1938 – 22r – 1 – (title varies: 1933-34: ascher tagesbote, 1 dec 1933-5 feb 1934 wkly) – gw Misc Inst [077]

Aschera und astarte : ein beitrag zur semitischen religionsgeschichte / Torge, Paul – Leipzig: J C Hinrichs 1902 [mf ed 1989] – 1mf – 9 – 0-7905-2558-5 – (in german & hebrew; incl bibl ref) – mf#1987-2558 – us ATLA [220]

Aschoff, Wiebke *see* Studien zu niccolo tribolo

Asconius [Tiberius Catius Asconius Silius Italicus] *see* In orationes quasdam ciceronis...

Ascorbic acid content of some florida-grown guavas / Mustard, Margaret J – Gainesville, FL. 1945 – 1r – us UF Libraries [634]

Ascot times – Ascot, England 7 jun 1984-19 jul 1990 [mf 1984-90] – 1 – (variant ed of: bracknell times) – uk British Libr Newspaper [072]

Asdonk, Ben *see* Zum verhaeltnis von religion und moderner schule

Asea news – 1977 feb-1982 jul; 1982 aug-1988 – 1 – mf#615222 – us WHS [071]

Ased y Latorre, A *see*
– Historia de la epidemia...de barbastro en el ano de 1748...
– Memoria instructiva de los medios de precaver los males resultas de un temporal excesiva mente humedo

Asedio de huesca, 18 julio 1936, 25 marzo 1938 / Gode, Antonio – Zaragoza: Excmo Ayuntamiento de Huesca 1938 – 9 – mf#w921 – us Harvard [946]

Asedio, y otros cuentos / Diaz Valcarcel, Emilio – Mexico City?, Mexico. 1958 – 1r – us UF Libraries [972]

Asee prism – Washington DC 1993+ – 1,5,9 – ISSN: 1056-8077 – mf#20931 – us UMI ProQuest [370]

Aseev, Nikolai Nikolaevich *see* Estafeta

Asencio-Camacho, Fernando *see* Problemas actuales de sociologia medica del puerto

Asenjo, Conrado *see* Geografia de la isla de puerto rico

Asensio, Jose Maria *see*
– Apuntes para la historia de villafranca de los barros
– Apuntes para la historia de villafranca de los barros de jose cascales munoz...

Asensio Menendez, Jose *see* Batalla de san pedro perulapan

Asenso, Antonio *see* La prensa madrilena a traves de los siglos

El asesinato de don francisco pizarro / Fernandez-Davila, Guillermo – Lima: imp. lux, 1945 – sp Bibl Santa Ana [350]

El asesinato del conquistador del peru : don francisco pizarro (26 de junio de 1541 / Fernandez-Davila, Guillermo – Lima: imp. lux. de l.e. castro, 1941 – sp Bibl Santa Ana [350]

El asesinato del regato de los avellanos / Ibarrola, Jose – Caceres: Tip. El Noticiero, s.a. – 1 – sp Bibl Santa Ana [946]

Asesinos de espana esta es vuestra obra – [Madrid? 193-] – (= ser Blodgett coll) – 9 – mf#w731 – us Harvard [946]

Asf scan – v24 n1-v31 n6 [1975 jan-1983 jan] – 1 – mf#1519027 – us WHS [071]

Asfeld, L T d' *see* Haslam-gherai, sultan de crimee

The asg report : an official government periodical / American Samoa. Government – Pago Pago: Govt of American Samoa, 28 Jul 1986-20 May 1988 – 6mf – 9 – $9.00 – mf#LLMC 82-100C Title 48 – us LLMC [324]

Asgard and the gods / Wagner, Wilhelm – New York, NY. 1917 – 1r – us UF Libraries [025]

Asghar aga – London, 1980- . sal-i 1, shumarah-'i 23-sal-i 4, shumarah-'i 122. 22 day 1358-30 murdad 1361 [12 jan 1980-21 aug 1982] – 1r – 1 – $53.00 – (cont: taghut; incl on r) – us MEDOC [079]

Ash, Edward *see*
- The christian profession of the society of friends
- An inquiry into some parts of christian doctrine and practice

Ash, Rodney Philip *see* Techniques of piano transcribing, 1800-1954

Ashanti and the gold coast : and what we know of it / Hay, John Charles Dalrymple – London: E Stanford, 1874 – 1 – us CRL [960]

Ashanti heroes / Bonsu Kyeretwie, K – Accra, Ghana. 1964 – 1r – us UF Libraries [960]

Ashanti pioneer – Kumasi, Ghana 21 nov 1939-30 dec 1950 (imperfect) – 1 – uk British Libr Newspaper [079]

Ashanti-danish relations : 1780-1831 / Kea, Ray A – 1967 – 1r – 1 – us CRL [960]

Ashbee, Charles Robert *see*
- A few chapters in work-shop re-construction and citizenship
- The manual of the guild and school of handicraft
- Transactions of the guild and school of handicraft, vol 1

Ashburner, John *see* Notes and studies in the philosophy of animal magnetism and spiritualism

Ashburnham 1735-1900 – Oxford, MA (mf ed 1993) – (= ser Massachusetts vital records) – 53mf – 9 – 0-87623-176-8 – (mf 1-3: proprietors 1735-1801. mf 4-18: town records 1801-56. mf 19-20: births 1750-1847. mf 20: marriages 1800-43. mf 21: deaths 1760-99. mf 22-28: town records 1765-1801. mf 25: intentions 1765-69. mf 25-26: births 1752-1800. mf 27: deaths 1767-99. mf 28: marriages & intents 1768-1801. mf 29,31: intentions 1801-1847. mf 29-31: births 1775-1847. mf 30-34: town records 1797-1846. mf 31: deaths 1784-1844. mf 31-34: marriages 1800-43. mf 35,37: militia 1847-62. mf 35-36: intentions 1847-77. mf 36-38: town records 1844-76. mf 38: dog licenses 1859-76. mf 39-40: birth index 1843-1900. mf 40-41: marriage index 1843-1900. mf 42-43: death index 1843-1900. mf 44-45: vital records 1843-50. mf 46-48: births 1850-1900. mf 48-51: marriages 1851-1900. mf 51-53: deaths 1850-1900) – us Archive [978]

Ashburnham 1752-1849 – Oxford, MA (mf ed 1994) – (= ser Massachusetts vital record transcripts to 1850) – 10mf – 9 – 0-87623-202-0 – (mf 1t: intentions 1765-69; births 1752-1800. mf 2t: births 1755-1800; deaths 1767-99. mf 2t-3t: marriages & intentions 1769-1802. mf 3t-4t: intentions 1801-39. mf 4t-5t: births 1783-1847. mf 6t: intentions of marrige 1839-47; deaths 1784-1844; births 1834-41. mf 6t-7t: marriages 1800-43. mf 7t: out-of-town marriages 1756-99. mf 7t-8t: births 1843-49. mf 9t: marriages 1843-49. mf 10t: deaths 1843-49) – us Archive [978]

Ashburton guardian – nov 1974-oct 1976; jan 1977-feb 2002 – 1 – mf#70.4 – nz Nat Libr [079]

Ashburton western guardian – [SW England] Devon 1902, jan 1904-dec 1907 [mf ed 2003] – 5r – 1 – uk Newsplan [072]

Ashburton's the courier – sep 1985-dec 1988 – 3r – 1 – mf#75.11 – nz Nat Libr [079]

Ashby 1754-1891 – Oxford, MA (mf ed 1983) – (= ser Massachusetts vital records) – 25mf – 9 – 0-931248-42-6 – (mf 1-4: births 1754-1876. mf 5-8: marriages 1768-1859. mf 9-11: deaths 1755-1862. mf 12-14: births 1754-1876. mf 15-16: marriages 1768-1859. mf 17: town records 1795-1821. mf 18-19: deaths 1755-1862. mf 20-22: marriages 1768-1859. mf 23-25: deaths 1851-91) – us Archive [978]

Ashby, Caroline W *see* Lilies and shamrocks

Ashby, Irene M *see* Elizabeth fry

Ashby, John W *see* Alachua, the garden county of floridia, its resources and advantage

Ashby, Lillian Luker *see* My india

Ashcraft Family *see* Materials from the scrapbook of the ashcraft family

Ashcraft news – n1-[1983 jan-1988 sep] – 1 – mf#1336721 – us WHS [071]

Ashcroft, Frank *see* Story of our rajputana mission

Ashcroft herald *see* Miscellaneous newspapers of pitkin county

Ashcroft, Robert *see* The scriptures opened

Ashe, R P *see* Chronicles of uganda

Ashe, Robert Hoadly *see* Letter to the rev john milner

Ashe, Thomas *see*
- History of the azores, or western islands
- Travels in america

Asher Ben Jehiel *see* Perush rabenu asher

Asher, John A *see* Der guote gerhart

Asher, Robert *see* The worker and technological change, 1930-80

The ashes of a god – London: Medici Society, 1914 – (= ser Samp: indian books) – (trans fr original mss by f w bain) – us CRL [280]

Asheville advocate – 1992 feb 21/28, jun 26/jul 3-10/17, sep 18/25, oct 2/9; 1993 jan 29/feb 5, 26/mar 5-5/12, 19/26-26/apr 2, jun 4/11-11/18, sep 3/9-16/23, oct 8/14-15/21, 29/nov 5-5/12, 19/26; 1994 jan 7/14-21/28, aug 5/19-nov 12/26, dec 11/21-1995 mar 20/apr 3, 17/may – 1 – mf#2504749 – us WHS [071]

Ashfield 1740-1849 – Oxford, MA (mf ed 1994) – (= ser Massachusetts vital record transcripts to 1850) – 7mf – 9 – 0-87623-203-9 – (mf 1t: marriages & intentions 1762-64. mf 1t-2t: births 1750-1847. mf 2t: deaths 1763-1843. mf 2t-5t: marriages 1762-1843. mf 3t-4t: intentions 1787-1849. mf 5t: out-of-town marriages 1768-1799. mf 5t-6t: births & deaths 1740-1849. mf 6t-7t: births 1843-49. mf 7t: marriages 1844-49; deaths 1843-49) – us Archive [978]

Ashfield 1750-1895 – Oxford, MA (mf ed 1987) – (= ser Massachusetts vital records) – 26mf – 9 – (mf 1-3: index: birth 1750-1895; marriage, death 1843-95. mf 4: index to deaths 1750-1895. mf 5: index to births 1896-1911. mf 6: index to marriages 1896-1911. mf 7: index to deaths 1896-1911. mf 8-11: index: births, marriages, deaths 1912-60. mf 12-15: town & vital records 1750-1857. mf 16-18: town & vital records 1763-1876. mf 19-20: births, marriages, deaths 1843-56. mf 21-26: births, marriages, deaths 1857-95) – us Archive [978]

Ashford and alfred news and general advertiser – Ashford, England 17 jul 1855-12 jun 1858 – 1 – (cont by: kentish express (all ed mf), ashford & alfred news, hythe gazette [19 jun-20 nov 1858]) – uk British Libr Newspaper [072]

Ashhurst action – jun 1973-1983 – 3r – 1 – mf#45.9 – nz Nat Libr [079]

Ashkenazi, Bezalel Ben Abraham *see* Shitah mekubetset

Ashland 1840-1895 – Oxford, MA (mf ed 1989) – (= ser Massachusetts vital records) – 17mf – 9 – 0-87623-101-6 – (mf 1-2: index to births 1840-95. mf 2-3: index to marriages 1846-97. mf 3-4: index to deaths 1846-97. mf 5-7: births 1840-95. mf 7-9: marriages 1846-97. mf 10-11: deaths 1846-97. mf 12-13: index to deaths 1898-1949. mf 14-15: index to marriages 1898-1949. mf 16-17: index to births 1896-1949) – us Archive [978]

Ashland advertiser – Ashland OR: W Y Crowson, -1898 [wkly] [mf ed 1978] – 1r – 1 – (began in 1892? suspended in 1894. resumed with v3 n4 (jun 12 1895)) – us Oregon Lib [071]

Ashland advocate – Ashland, PA. -w 1889-1901; 1902-1907 – 13 – $25.00r – us IMR [071]

Ashland american – Ashland OR: P Robinson, 1927 [wkly] – 1 – (cont by: ashland register (1927-); cont: central point american (1925-27)) – us Oregon Lib [071]

Ashland appeal – 1894 aug 15-1895 jan 12 – 1 – mf#916293 – us WHS [071]

Ashland bladet och ashland posten – 1903 feb 7-1905 may 27; 1905 jun 3-1906 jul 20; 1906 jul 27-1907 sep 13 – 1 – mf#958905 – us WHS [071]

Ashland [city directory listing] – 1888; 1893 – 1 – mf#3059374 – us WHS [917]

Ashland Co. Ashland *see*
- Press
- Times
- Union series

Ashland Co. Hayesville *see* Journal

Ashland Co. Loudonville *see*
- Advocate
- Democrat
- Loudenville times
- Times

Ashland county atlas, 1874 – 1r – 1 – mf#B27423 – us Ohio Hist [978]

Ashland county herold – 1906 mar 22-1907 mar 16 – 1 – mf#1221727 – us WHS [071]

Ashland County & Wayne County. Ohio *see* Deeds, ms 3193

Ashland daily evening tidings – Ashland OR: [s.n] 1890 [daily ex sun] – 1 – (variant ed of: ashland tidings (1876-1919)) – us Oregon Lib [071]

Ashland daily evening tidings *see* Ashland tidings

Ashland daily news – 1887 sep 25-1888 sep 30; 1888 oct 1-1889 dec 31; 1890; 1891; 1892 jan 1-1893 mar 30; 1895 jan-may 6 – 1 – mf#961923 – us WHS [071]

Ashland daily press – 1888 jun 1-1966 apr 30 [with gaps] – 1 – mf#1137350 – us WHS [071]

Ashland daily tidings (ashland, or: 1919) – Ashland OR: Ashland Printing Co, 1919-70 [daily ex sun] – 1 – (related to: ashland weekly tidings; cont: ashland tidings; cont by: daily tidings) – us Oregon Lib; us Oregon Hist [071]

Ashland daily tidings (ashland, or: 1993) – Ashland OR: Capital Cities/ABC, 1993- [daily ex sun] – 1 – (cont: daily tidings (ashland, or)) – us Oregon Lib; us Oregon Hist [071]

Ashland gazette – Ashland, NE: [T J Pickett, Jr] (wkly) – 40r – 1 – (cont: saunders county reporter) – us Bell [071]

Ashland gazette – Ashland, NE: [T J Pickett, Jr] v6 n11. jun 22 1883 (wkly) [mf ed dec 26 1884- (gaps) filmed 1969] – 27r – 1 – (cont: saunders county reporter; other ed available: daily gazette, sep 1881) – us NE Hist [071]

Ashland journal – Ashland, NE: Phil R Wilmarth. 2v. v9 n45. nov 10 1905-v2 n20. mar 15 1907 (wkly) – 1r – 1 – (cont: saunders county journal; publ in omaha feb 3-mar 15 1907; iss for nov 24 1905-mar 15 1907 called v1 n4-v2 n20) – us Bell [071]

Ashland, Maine. Baptist Church *see* Records

Ashland news – 1895 may 7-1910 dec 31 [1]; 1885 may 13-1887 sep 21 [2] – 1 – mf#961917 [1]; 919952 [2] – us WHS [071]

The ashland news – Ashland, NE: George B Pickett. -3rd yr n37. sep 25 1896 (wkly) – 1r – 1 – us Bell [071]

Ashland press – 1872 jun 22-1874 mar 28; 1874 apr 4-1877 may 12; 1877 may 19-1880 aug 7; 1880 aug 14-1883 dec 14; 1883 dec 22-1887 mar 19; 1887 mar 26-1889 jun 8; 1889 jun 15-1892 aug 13; 1892 aug 20-1893 oct 14 – 1 – mf#918815 – us WHS [071]

Ashland record – Ashland OR: Charles B Wolf, [wkly] – 1 – (began in 1911; ceased in 1919; cont: valley record (1888-1911)) – us Oregon Lib [071]

Ashland register – Ashland OR: C J Read, 1927- [semiwkly] – 1 – (cont: ashland american (1927)) – us Oregon Lib [071]

Ashland tidings – Ashland OR : J M Sutton, 1876-1919 [freq varies] – 1 – (cont by: ashland daily tidings (1919); other ed available: ashland daily evening tidings) – us Oregon Hist [071]

Ashland tidings – Ashland OR: J M Sutton, 1876-1919 [semiwkly] – 1 – (related to: ashland daily evening tidings (1890-90); cont by: ashland daily tidings (1919-70)) – us Oregon Lib [071]

Ashland tidings *see* Ashland daily evening tidings

Ashland times – Ashland, NE: Orin H Mathews (wkly) [mf ed v1 n41. jan 20 1871,1872,1876 (gaps)] – 1r – 1 – (cont: weekly ashland times) – us NE Hist [071]

Ashland weekly news – 1887 sep 28-1888 may 2; 1888 may 9-1890 apr 30; 1897 jan 6-1898 oct 12; 1898 oct 19-1901 dec 25; 1903-05 – 1 – mf#919961 – us WHS [071]

Ashland weekly press – 1893 oct 21-1894 jun 2; 1894 jun 9-1896 mar 21; 1896 mar 28-1897 nov 6; 1897 nov 13-1899 aug 5; 1899 aug 12-1901 jun 8; 1901 jun 15-1902 dec 13; 1902 dec 20-1904 sep 17; 1904 sep 24-1906 aug 25; 1906 sep 1-1908 apr 11; 1908 apr 18-1909 sep 11; 1909 sep 18-11 apr 22; 1911 apr 29-1912 dec 21; 1912 dec 28-14 aug 18; 1914 aug 15-1916 oct 7 – 1 – mf#918823 – us WHS [071]

Ashland weekly tidings – Ashland OR: Ashland Printing Co, 1919-24// [wkly] – 1 – (variant ed of: ashland daily tidings (1919-70)) – us Oregon Lib; us Oregon Hist [071]

Ashland-posten – 1900 dec 22 – 1 – mf#1221712 – us WHS [071]

Ashley, Barnas Freeman *see*
- Air castle don
- Dick and jane's adventures on sable island
- Tan pile jim

Ashley, R K *see* Glances over the field of faith and reason, or, christianity in its idea and development

Ashley river baptist church – CHARLESTON, SC. 1736-69, 1943-59, 1962-72 – 1 – $26.65 – us Southern Baptist [242]

Ashley, William James *see*
- The character of villein tenure
- Nine lectures on the earlier constitutional history of canada

Ashley-cum-silverley 1630-1950 – (= ser Cambridgeshire parish register transcript) – 5mf – 9 – £6.25 – uk CambsFHS [929]

Ashman, Louis S *see* Law and forms of prayers and instructions

Ashman, Mary *see* Pathways and clearingsh

Ashmead, William Harris *see*
- A monograph of the north american proctotrypidae
- Orange insects

Ashmodai / Andreyev, Leonid – New York, NY. 1909 – 1r – us UF Libraries [939]

The ashmole bestiary : ashmole ms. 1511 / Ashmolean Museum – 13th c – 1r – 14 – mf#C502 – uk Microform Academic [240]

Ashmolean Museum *see*
- The ashmole bestiary
- Drawings of raphael in the ashmolean museum

Ashmun, Jehudi *see* History of the american colony in liberia, 1821-1823

Ashpitel, Francis *see* The increase of the israelites in egypt shewn to be probable from the statistics of modern populations

Ashrae journal / American Society of Heating, Refrigerating and Air Conditioning Engineers – Atlanta GA 1959+ – ISSN: 0001-2491 – mf#C1523 – us UMI ProQuest [621]

Ashrae transactions – Atlanta GA 1895+ – 1,5,9 – ISSN: 0001-2505 – mf#6586 – us UMI ProQuest [690]

Ashre ha-ish / Margaliyot, Yesha'y Asher Zelig – Yerushalayim, Israel. 1927 – 1r – us UF Libraries [939]

Ashtabula Co. Andover *see* Citizen

Ashtabula Co. Ashtabula *see*
- Daily telegraph
- Democratic free press
- Harbor journal series
- News
- Sentinel
- Star
- Star beacon
- Star-beacon
- Star-beacon – morning edition
- Telegraph
- Telegraph series
- Weekly telegraph

Ashtabula Co. Conneaut *see*
- Ashtabula county advance
- Gazette
- News-herald
- Reporter

Ashtabula Co. Geneva *see*
- Free press
- Free press-times
- Free press-times series
- Times

Ashtabula Co. Jefferson *see*
- Ashtabula sentinel
- Gazette

Ashtabula Co. Orwell *see* Weekly welcome

Ashtabula Co. Rock Creek *see* Banner

Ashtabula county advance / Ashtabula Co. Conneaut – 3/1914-1/1915 [daily, wkly] – 1r – 1 – (a socialist newspaper) – mf#B306 – us Ohio Hist [071]

Ashtabula county sentinel – Jefferson, OH: J A Howells & Co, 1900-1910 (wkly) – 1 – us CRL [071]

Ashtabula sentinel / Ashtabula Co. Jefferson – jan 1853-dec 1856 [wkly] – 2r – 1 – mf#B570-571 – us Ohio Hist [071]

Ashtabula sentinel / Ashtabula Co. Jefferson – jan 1867-dec 1867 [wkly] – 1r – 1 – mf#B5640 – us Ohio Hist [071]

Ashtabula sentinel – Ashtabula, OH: O H Fitch, jan 21 1832-oct 17 1878 – 13r – 1 – (issues for aug 16 1877-oct 17 1878 filmed with: semi-weekly ashtabula sentinel, oct 23 1878-apr 24 1880) – us CRL [071]

Ashtabula sentinel – Jefferson, OH: J A Howells & Co [1884-1899] (wkly) – 1r – 1 – us CRL [071]

Ashtavakra samhita : text with word-for-word trans, english rendering and comments – Mayavati, Almora: Advaita Ashrama, 1940 – (= ser Samp: indian books) – us CRL [140]

Ashton and audenshaw reporter – Ashton-under-Lyne, England 29 jul 1993-9 oct 1997 [mf 1986-] – 1 – (cont: ashton-under-lyne reporter (main ed) 14 nov 1885-22 jul 1993; cont by: reporter. ashton, audenshaw, etc [16 oct 1997-24 sep 1998]) – uk British Libr Newspaper [072]

Ashton and stalybridge guardian and dukinfield, mossley and droylsden sentinel – [NW England] Ashton, Stalybridge Lib 2 nov 1867-18 jul 1868 – 1 – uk MLA; uk Newsplan [072]

Ashton and stalybridge reporter – Ashton-under-Lyne, England 10 apr 1858-20 oct 1866 – 1 – (cont: ashton weekly reporter, & stalybridge & dukinfield chronicle [14 apr 1855-3 apr 1858]; cont by: ashton reporter [27 oct 1866-7 nov 1885]) – uk British Libr Newspaper [072]

Ashton chronicle and district advertiser – [NW England] Ashton, Stalybridge Lib mar 1848-3 nov 1849 – 1 – uk MLA; uk Newsplan [072]

Ashton, Douglas F *see* Temperature rise in human muscle during ultrasound treatments utilizing flex-all as a coupling agent

Ashton, dukinfield and stalybridge news – [NW England] Ashton-under-Lyne 11 may 1857 [mf ed 2004] – 1r – 1 – uk Newsplan [072]

Ashton, Edmund Hugh *see*
- Basuto

Ashton guardian – [NW England] Ashton, Stalybridge Lib 27 jan 1877-25 may 1878* – 1 – (wanting: 19 jan, 9 feb, 9-23 mar 1878) – uk MLA; uk Newsplan [072]

Ashton hayes, st john the evangelist – (= ser Cheshire monumental inscriptions) – 1mf – 9 – £2.50 – mf#4 – uk CheshireFHS [929]
Ashton herald – Ashton, NE: J R Gardiner, -nov 1934// (wkly) – (= ser Sherman County Times) – 3r – 1 – (absorbed by: sherman county times (loup city ne: 1915)) – us Bell [071]
Ashton, Hugh *see* Problem territories of southern africa
Ashton, John *see* The devil in britain and america
Ashton, Leigh *see* The art of india and pakistan
Ashton reporter – [NW England] Ashton 2 jul 1971 – 1 – (Title change: Ashton & Haydock Reporter [9 jul 1971-11 Jul 1979]) – uk MLA; uk Newsplan [072]
Ashton reporter (denton edition) – [NW England] Ashton, Stalybridge Lib 1964 – 1 – (title change: denton reporter) – uk MLA; uk Newsplan [072]
Ashton reporter (droylsden ed) – [NW England] Droylsden, Stalybridge Lib 1962 – 1 – (title change: droylsden reporter [1962-70, 1973, 1978]) – uk MLA; uk Newsplan [072]
Ashton reporter (dukinfield ed) – [NW England] Dukinfield, Stalybridge Lib 1948-67 – 1 – (title change: dukinfield reporter [1967-70, 1978-mar 1986]) – uk MLA; uk Newsplan [072]
Ashton reporter (hyde edition) – [NW England] Tameside jan 1876-dec 1894 [mf ed 2003] – 15r – 1 – (missing: 1878, 1883, 1885-88, 1890; cont as: hyde, marple and glossop reporter [jan 1889-dec 1893]; hyde reporter and telegraph [1894]) – uk Newsplan [072]
Ashton ricker and municipal mirror – [NW England] Tameside dec 1852, apr 1854, jul 1855, 5 jul 1856 [mf ed 2004] – 1r – 1 – uk Newsplan [072]
Ashton standard – [NW England] Ashton-under-Lyne 2 jan 1858-16 feb 1867, 16 oct 1869-29 dec 1900 [mf ed 1896-nov 1897] – 1 – (wanting: dec 1897) – uk MLA; uk Newsplan [072]
Ashton times and oldham visitor – [NW England] Tameside mar, apr, jun, nov 1851; apr, jul, nov 1852; jan 1853 [mf ed 2003] – 1r – 1 – uk Newsplan [072]
Ashton weekly reporter, and stalybridge and dukinfield chronicle – [North West] Ashton, Stalybridge Lib 14 apr 1855-3 apr 1858 [mf 1855-] – 1 – (cont by: ashton & stalybridge reporter, dukinfield, mossley, glossop, hyde, mottram & droylsden chronicle [10 apr 1858-20 oct 1866]; ashton reporter [27 oct 1866-7 nov 1885]; ashton-under-lyne reporter (& the ashton-under-lyne herald) [14 nov 1885]) – uk MLA; uk Newsplan [072]
Ashtonian – [NW England] Tameside may 1847, feb, jun 1849 [mf ed 2003] – 1r – 1 – uk Newsplan [072]
Ashton-under-lyne and district advertiser – [NW England] Ashton, Stalybridge Lib 11 mar-30 dec 1909 – 1 – uk MLA; uk Newsplan [072]
Ashton-under-lyne herald – [NW England] Ashton-under-Lyne, Stalybridge Lib 6 jul 1889-13 april 1901 [mf jul-dec 1889, jan-dec 1902] – 1 – (wanting 1896,1897.1912; cont by: weekly herald [20 apr 1901-30 apr 1910]; incorp with: ashton-under-lyne reporter) – uk MLA; uk Newsplan [072]
Ashton-under-lyne news – [NW England] Ashton-under-Lyne, Stalybridge Lib 18 jan 1868-14 feb 1874 – 1 – (discontinued) – uk MLA; uk Newsplan [072]
Ashton-under-lyne reporter (audenshaw edition) – [NW England] Ashton-underLlyne, Stalybridge Lib 1973-mar 1986 – 1 – (discontinued) – uk MLA; uk Newsplan [072]
Ashton-under-lyne reporter denton edition – [NW England] Tameside jan 1888-dec 1898 [mf ed 2003] – 8r – 1 – (missing: 1891-92, 1895) – uk Newsplan [072]
Ashton-under-lyne reporter [dukinfield ed] – Ashton-under-Lyne, England 22 nov 1963-14 jul 1967 – 1 – (cont as: dukinfield reporter) – uk British Libr Newspaper [072]
Ashton-under-lyne reporter [main ed] – Ashton-under-Lyne, England 14 nov 1885-22 jul 1993 – 1 – (cont: ashton reporter [27 oct 1866-7 nov 1885]; cont by: ashton & audenshaw reporter [29 jul 1993-9 oct 1997]) – uk British Libr Newspaper [072]
Ashton-upon-mersey, st martin – (= ser Cheshire monumental inscriptions) – 4mf – 9 – £5.00 – mf#92 – uk CheshireFHS [929]
Ashton-upon-mersey, st martin: baptisms 1605-1880 – [North Cheshire FHS] – (= ser Cheshire church registers) – 4mf – 9 – £5.50 – mf#421 – uk CheshireFHS [929]
Ashton-upon-mersey, st martin: baptisms/ marriages 1701-1760 – (= ser Cheshire church registers) – 1mf – 9 – £2.50 – mf#374 – uk CheshireFHS [929]
Ashton-upon-mersey, st martin: burials 1608-1731 – [North Cheshire FHS] – (= ser Cheshire church registers) – 1mf – 9 – £3.00 – mf#419 – uk CheshireFHS [929]
Ashton-upon-mersey, st martin: burials 1731-1890 – [North Cheshire FHS] – (= ser Cheshire church registers) – 6mf – 9 – £7.00 – mf#251 – uk CheshireFHS [929]
Ashton-upon-mersey, st martin: marriages 1605-1730 – [North Cheshire FHS] – (= ser Cheshire church registers) – 1mf – 9 – £3.00 – mf#420 – uk CheshireFHS [929]
Ashurbanipal, King of Assyria *see*
– The annals of ashurbanapal (5 rawlinson pl 1-10)
– History of assurbanipal
Ashurnasirpal 2 *see* Die annalen asurnazirpals (884-860 v. chr.)
Ashwaubenon allouez howard-suamico press – 1977 sep 16-1978 nov 10 [1]; 1978 nov 17-1997 [2] – 1 – mf#1223692 [1]; 944284 [2] – us WHS [071]
Ashwaubenon press – 1976 feb 20-aug 6 – 1 – mf#1223684 – us WHS [071]
Ashwaubenon-allouez press – 1976 aug 23-1977 sep 9 – 1 – mf#1223686 – us WHS [071]
Ashwell, A R *see*
– Life of the right reverend samuel wilberforce
Ashwood prospector – Ashwood OR: M. Luddemann, [wkly] [mf ed 1971] – 1r – 1 – us Oregon Lib [071]
Ashworth, James *see* Safety lamps and colliery explosions
Ashworth, Robert A *see* The union of christian forces in america
Asi era el hermano agustin. un jesuita desconocido. cincuenta anos de heroismo coronados por el moctivio. vida heroica de agustin maria diaz zapata...1869-1936 / Mateos, Francisco & Stachlin, Carlos Maria – Madrid: Razon y Fe, 1944 – 1 – sp Bibl Santa Ana [241]
Asi es costa rica / Reyes H, Alfonso – San Jose, Costa Rica. 1945 – 1r – us UF Libraries [972]
Asi es la selva : estudio geografico y etnologia de la provincia de bajo amazonas. lima, 1943 / Villarejo, Avencio – Madrid: Razon y Fe, 1947 – 1 – sp Bibl Santa Ana [900]
Asi fue la revolucion / Estrada Monsalve, Joaquin – Bogota, Colombia. 1950 – 1r – us UF Libraries [972]
Asi happenings – v9 n6-v13 n10 [1977 jun/ aug-1981 dec] – 1 – mf#603330 – us WHS [071]
Asi microfiche library : retrospective / U.S. Government – 1960s-1996 – 9 – apply for price – us CIS [324]
Asi murio el insigne bibliografo don bartolome j gallardo / Martinez, Ildefonso – Badajoz: La Alianza, 1935 – 1 – sp Bibl Santa Ana [946]
Asi paga el diablo... / Trigo, Felipe – Madrid: Renacimiento, 1911 – sp Bibl Santa Ana [946]
Asi progresa un pueblo / Venezuela Direccion Nacional De Informacion – Caracas, Venezuela. 1956 – 1r – us UF Libraries [972]
Asia [1964] – New York NY 1964-73 – 1,5,9 – ISSN: 0161-4355 – mf#2331 – us UMI ProQuest [327]
Asia [1978] – New York NY 1978-83 – 1,5,9 – (cont: asia bulletin) – ISSN: 0161-4355 – mf#11979 – us UMI ProQuest [321]
Asia bulletin – New York NY 1974-78 – 1 – (cont by: asia) – ISSN: 0161-4355 – mf#10375 – us UMI ProQuest [321]
Asia calling – Pacific Palisades CA 1977-78 – 1,5,9 – ISSN: 0004-4431 – mf#7727 – us UMI ProQuest [320]
Asia in the modern world / Venkatasubbiah, H – New Delhi: Asian Relations Conference, Indian Council of World Affairs, 1947 – (= ser Samp: indian books) – us CRL [321]
Asia in the twentieth century / Whyte, Alexander Frederick – New York: Charles Scribner's Sons, 1926 – (= ser Samp: indian books) – us CRL [321]
Asia Information Group *see* Indochina bulletin
Asia journal of theology – Bangalore, India 1987+ – 1,5,9 – (cont: east asia journal of theology) – ISSN: 0217-1244 – mf#16183 – us UMI ProQuest [200]
Asia mail – Alexandria VA 1976-82 – 1,5,9 – mf#10809 – us UMI ProQuest [327]
Asia major : a british jouranl of far eastern studies – v1-15 new series. 1949-69 – 1 – us AMS Press [950]
Asia major *see* Volkskundliches aus altturkestan
Asia money and finance – London UK 1991-93 – 1,5,9 – (cont by: asiamoney) – mf#19518.01 – us UMI ProQuest [332]
Asia pacific journal of management [apjm] – Dordrecht, Netherlands 1991+ – 1,5,9 – ISSN: 0217-4561 – mf#18139 – us UMI ProQuest [650]
Asia Pacific Regional Office *see* Pacific unionist
Asia pacific viewpoint – Oxford UK 1996+ – 1,5,9 – (cont: pacific viewpoint) – ISSN: 1360-7456 – mf#1903,01 – us UMI ProQuest [900]
Asia polyglotta / Klaproth, J [H von] – Paris: A Schubart, 1823 – 5mf – 9 – mf#AR-1598 – ne IDC [915]
Asia today international – Sydney, Australia 2001+ – 1,5,9 – mf#27159,01 – us UMI ProQuest [341]
L'asia...consigliero del christianissimo re di portogallo / Barros, J de – Venetia, 1561-1562. 2pts – 10mf – 9 – mf#H-8299 – ne IDC [956]
Asiamoney – London UK 1993+ – 1,5,9 – (cont: asia money and finance) – mf#19518,01 – us UMI ProQuest [332]
Asian affairs : an american review – Washington DC 1979+ – 1,5,9 – ISSN: 0092-7678 – mf#12619 – us UMI ProQuest [327]
Asian american advertiser – v1 n1-11 [1980 jul 15-1981 mar 15] – 1 – mf#637644 – us WHS [071]
Asian american journey – 1977 dec [1]; v1 n2-v5 n5 (1978 feb-1982 may) [2] – 1 – mf#655304 [1]; 655299 [2] – us WHS [071]
Asian and african studies – Piscataway NJ 1976-93 – 1,5,9 – ISSN: 0066-8281 – mf#11100 – us UMI ProQuest [900]
Asian and pacific quarterly of cultural and social affairs – Seoul, South Korea 1990-92 – 1 – (cont by: asian pacific quarterly) – ISSN: 0251-3110 – mf#17727.03 – us UMI ProQuest [079]
Asian art – Oxford UK 1992-93 – 1 – (cont by: asian art and culture) – ISSN: 0894-234X – mf#17035.01 – us UMI ProQuest [700]
Asian art and culture – Oxford UK 1994-96 – 1 – (cont: asian art) – ISSN: 1352-2744 – mf#17035,01 – us UMI ProQuest [700]
Asian books in the russian language : from the biblioteka akademii nauk (ban), st. petersburg / Biblioteka Akademii Nauk (BAN), St Petersburg – 697mf – 9 – $3,200.00 coll $25.00t – (china: 350mf $1600. japan: 200mf $900. manchuria: 102mf $450. mongolia: 281mf $1300. tibet: 89mf $400) – us UMI ProQuest [480]
Asian books newsletter – Calcutta, India 1966-74 – 1 – ISSN: 0004-4547 – mf#2309 – us UMI ProQuest [020]
Asian business – North Point HK 1987+ – 1,5,9 – ISSN: 0254-3729 – mf#15712,01 – us UMI ProQuest [338]
Asian business and community news – 1981 dec, 1982 mar-jul, oct/nov-1985 sep – 1 – mf#855239 – us WHS [071]
Asian case research journal – Hoboken NJ 2001+ – 1,5,9 – ISSN: 0218-9275 – mf#25569 – us UMI ProQuest [338]
Asian culture, 1845-1949 : the periodical perspective – ca 52r – 1 – (previous title: asian periodicals, 1845-1949. 11 periodicals detailing asian culture and reactions of both western observers and asians to that culture) – mf#C39-27870 – us Primary [950]
Asian development bank release – n1. 18 dec 1967 (all publ) – (= ser The sec release series preceding the sec docket) – 1mf – 9 – $1.50 – mf#LLMC 89-007 – us LLMC [346]
Asian economic history series, series 1 : the opium trade and the united nations commission on narcotic drugs, 1945-48 [public record office class fo 371/50647-50654, 57020-57024, 67641-67644, 72907-72915] – [mf ed Marlborough 1991] – 4r – 1 – £380.00 – (with d/g) – uk Matthew [380]
Asian economic history series, series 2 : economic development in brunei, hong kong, malaysia, singapore, south korea and taiwan, 1950-80 [public record office files from the foreign office, colonial office, treasury, dominions office, board of trade and cabinet committees] – 4pt – 1 – (pt1: files for 1950-54 [24r] £2275; pt2: files for 1955-58 [24r] £2275; pt3: files for 1959-62 [36r] £3400 [mf ed 2004]; pt4: files for 1963-66 [24r] £2250; with d/g) – uk Matthew [338]
Asian family affair – v6 n5; v7 n3-v13 n7 [1977 aug/sep; 1978 sep-1984 nov] – 1 – mf#998606 – us WHS [305]
Asian folklore studies – Nagoya, Japan 1989+ – 1,5,9 – ISSN: 0385-2342 – mf#18194,01 – us UMI ProQuest [390]
Asian forum – Washington DC 1969-81 – 1,5,9 – ISSN: 0004-4563 – mf#8913 – us UMI ProQuest [950]
Asian immigration and exclusion, 1906-1913 – (= ser Records of the immigration and naturalization service, series a: subject correspondence files 1; Research colls in american immigration) – 30r – 1 – $5365.00 – 1-55655-160-6 – (pt1 suppl 1898-1941 16r $3115 isbn 1-55655- 605-5. with p/g) – us UPA [342]
Asian law journal – v1-8. 1994-2001 – 9 – $119.00 set – mf#117331 – us Hein [342]
The asian mystery illustrated in the history, religion, and present state of the ansaireeh or nusairis of syria / Lyde, Samuel – London: Longman, Green, Longman, and Roberts, 1860 – 1mf – 9 – 0-524-01843-X – mf#1990-2678 – us ATLA [260]
Asian pacific quarterly – Seoul, South Korea 1993-94 – 1 – (cont: asian and pacific quarterly of cultural and social affairs) – mf#17727,03 – us UMI ProQuest [079]
Asian perspectives – Honolulu HI 1986+ – 1,5,9 – ISSN: 0066-8435 – mf#1581 – us UMI ProQuest [079]
Asian reporter – Portland OR: Asian Reporter, 1991- [wkly] – 1 – us Oregon Lib [071]
Asian studies – v1-3. 1963-65 – 1 – (v4. 1966- in prep) – us AMS Press [950]
Asian studies professional review – Ann Arbor MI 1971-76 – 1 – ISSN: 0044-9245 – mf#8375 – us UMI ProQuest [950]
Asian survey – Berkeley CA 1971+ – 1,5,9 – ISSN: 0004-4687 – mf#6045 – us UMI ProQuest [321]
Asian textile business – Osaka, Japan 2003+ – 1,5,9 – mf#27500,04 – us UMI ProQuest [670]
Asian theatre journal [atj] – Honolulu HI 1989+ – 1 – ISSN: 0742-5457 – mf#16284 – us UMI ProQuest [790]
Asian thought and society – Oneonta NY 1989+ – 1,5,9 – ISSN: 0361-3968 – mf#17611 – us UMI ProQuest [306]
Asianadian – 1978 spring-1985 summer – 1 – mf#817752 – us WHS [071]
Asian-South Pacific Bureau of Adult Education *see* Aspbae journal
Asianweek – 1983; 1984; 1985 jan-1995 dec 22 [with gaps] – 1 – mf#1118269 – us WHS [071]
Asia-pacific development journal – New York NY 1994+ – 1,5,9 – (cont: economic bulletin for asia and the pacific) – ISSN: 1020-1246 – mf#21287 – us UMI ProQuest [338]
Asia-raya : 1 tahoen nimer peringatan – (Djakarta, 2603) 1v – 108p 4mf – 9 – mf#SE-2002 mf13-16 – ne IDC [959]
Asia-raya : oentoek memperingati enam boelan balatentara dai-nippon melindoengi indonesia – Djakarta, 2602. 1v – 162p 5mf – 9 – (nomer istimewa ini diselenggarakan oleh winarno, andjar asmara dan kamadjaja; on spine: nomer istimewa, 9 sep 2602 (=1942)) – mf#SE-2002 mf8-12 – ne IDC [959]
Asiatic and colonial quarterly journal – London: James Madden. v1-6 n1-12 [dec ? 1846-sep 1849] – 1 – us CRL [950]
Asiatic annual register, 1799-1811 – London, 1801-1812. v1-12 – 172mf – 8 – mf#I-201 – ne IDC [956]
The asiatic dionysos / Davis, Gladys Mary Norman – London: G Bell, 1914 – 1mf – 9 – 0-524-01691-7 – (incl bibl ref) – mf#1990-2593 – us ATLA [250]
Asiatic fields : addresses delivered before the eastern missionary convention of the methodist episcopal church, philadelphia, pa, oct 13-15 1903 – New York: Eaton & Mains; Cincinnati: Jennings & Pye, [1904] [mf ed 1995] – (= ser Yale coll; Philadelphia convention addresses) – 1 – 0-524-09267-2 – mf#1995-0267 – us ATLA [242]
The asiatic islands and new holland : being a description of the manners, customs, character, and state of society of the various tribes by which they are inhabited / ed by Shoberl, Frederick – London [1824] [mf ed Hildesheim 1995-98] – 2v on 4mf – 9 – €120.00 – 3-487-27435-3 – gw Olms [915]
Asiatic journal and monthly review – London, 1816-1829 v1-28; n.s. 1830-1843 v1-40; s3 1843-1845, v1-4 – 900mf – 8 – mf#I-202 – ne IDC [950]
Asiatic quarterly review – London 1886-90 – 10v on 944mf – 8 – (cont as: the imperial and asiatic quarterly review, s2 london 1891-1895 v1-10. s3 woking 1896-1912 v1-34; the asiatic quarterly review n.s. woking 1913 v1-2; the asiatic review n.s. london 1914-46 v3-42) – mf#I-200c – ne IDC [079]
Asiatic researches – Calcutta, 1788-1836. v1-20. 1835, ind v1-18 – 404mf – 9 – mf#I-102 – ne IDC [915]
Asiatic society monographs of the royal asiatic society of great britain and ireland – London, 1899-1909. v1-9 – 48mf – 8 – (missing: 1904 v6) – mf#I-544 – ne IDC [956]
Asiatic society of bengal, calcutta. journal and proceedings – ns: v1-30 1905-34 – 9 – $660.00 – mf#0083 – us Brook [954]
Asiatic society of bengal. journal – v1-38. 1832-69 – 9 – $840.00 – mf#0084 – us Brook [950]
Asiatic society of japan. transactions – Tokyo. v1-50. 1872-1922 – 1 – $390.00 – (ser2: v1-19 1924-40 $108 [0086]) – mf#0085 – us Brook [950]
Asiatic studies : religious and social / Lyall, Alfred Comyn – 2nd ed. London John Murray, 1907 [mf ed 1995] – (= ser Yale coll) – 2v – 1 – 0-524-09062-9 – mf#1995-0062 – us ATLA [950]
Asiatisches magazin : oder nachrichten von den sitten und gebraeuchen, den wissenschaften und kuensten, den handwerken und gewerben, der denkart und der religion der asiaten, von den thieren, den pflanzen, den mineralien, dem boden und dem clima – Leipzig 1806/07-1811 [mf ed Hildesheim 1995-98] – 3v on 7mf – 9 – €140.00 – 3-487-27601-1 – gw Olms [950]
Asie / Lenormand, Henri-Rene – Paris, France. 1931 – 1r – us UF Libraries [440]
L'asie francaise – v1-40. 1901-40 – 1 – us L of C Photodup [944]
Asie francaise / Comite de l'Asie francaise – no1-378. Paris. avr 1901-avr 1940 – 1 – fr ACRPP [959]

ASPEKTE

Asien, afrika, lateinamerika – 1973-1989 – 411mf – 1 – gw Mikropress [900]
Asif – Tel-Aviv, Israel. 1942? – 1r – us UF Libraries [939]
Asikane / Rauf, Mehmet – Istanbul: Hilal Matbaasi, 1325 [1909] – (= ser Ottoman literature, writers and the arts) – 3mf – 9 – $55.00 – us MEDOC [470]
Asikin widjaja kusumah, D Raden *see* Diagnosa-kimia dan tafsir-kliniknja
Asikpasazade *see*
– Tevarih-i al-i osman [asikpasazade tarihi]
Asile d'alienes de Quebec *see* Report of the quebec lunatic asylum
Asile d'alienes de quebec : reglement – [Levis?, Quebec: s.n.] 1875 [mf ed 1984] – 1mf – 9 – 0-665-44848-1 – mf#44848 – cn Canadiana [360]
Asile de nuit / Maurey, Max – Paris, France. 1905 – 1r – us UF Libraries [440]
Asile (hospice) st-jean de dieu, longue-pointe, pq, canada – Montreal: [s.n.], 1892 [mf ed 1979] – 1mf – 9 – 0-665-00035-9 – (in french and english) – mf#00035 – cn Canadiana [616]
Asile Sainte-Darie (Montreal, Quebec) *see* Annales du monastere de notre-dame de charite du bon pasteur d'angers, dit asile sainte darie a montreal 1870-1900
Les asiles d'alienes de la province de quebec et leurs detracteurs / Tache, Joseph-Charles – Hull, Quebec?: La Vallee d'Ottaoua, 1885 – 1mf – 9 – mf#24430 – cn Canadiana [360]
Asilo diplomatico / Corpeno V, Roberto S – Guatemala, 1963 – 1r – us UF Libraries [972]
Asils international law journal *see* Ilsa journal of international law
Asim, Salih *see* Ueskub tarihi ve civari
Asim tarihi / Efendi, Ahmed Asim – Istanbul: Ceride-i Havadis Matba'asi, 1274 [1857] – (= ser Ottoman histories and historical sources) – 30mf – 9 – $500.00 – us MEDOC [959]
Asim, Tuhfe-i *see* The divan project
Asimov, Isaac *see* Stars, like dust
Asimov's science fiction – New York NY 1977+ – 1,5,9 – (cont: isaac asimov's science fiction magazine) – ISSN: 1065-2698 – mf#11672,01 – us UMI ProQuest [420]
Asin Palacios, Miguel *see* Comentarios de don garcia de silva y figueroa de la embajada que de parte del rey de espana don felipe 3rd hizo al rey xa abas de persia
L'asino – New York NY, jul 5 1908-feb 27 1910 – 1r – 1 – (italian newspaper) – us IHRC [071]
Asins, Manuel *see* Monologo en prosa. musica de don joaquin perez
Asir [yeni asir] – Selanik: Asir Matbaasi, Yeni Asir Matbaasi, 1895-1928. Sahib-i Imtiyaz: Abdurrahman Nafiz. n1537,1644,1743,6784. 21 eyluel 1927 – (= ser O & t journals) – 1mf – 9 – $25.00 – us MEDOC [956]
Asiri, Fazl Mahmud *see* Studies in urdu literature
Asistencia social / Henriquez Almanzar, Carmen Adolfina – Ciudad Trujillo, Dominican Republic. 1947 – 1r – us UF Libraries [972]
Asit kumar haldar / Cousins, James Henry – Calcutta: Harimohan Mukhurji: Sold by Rupam, [1924] – (= ser Samp: indian books) – (with ann on the plates by ordhendra coomar gangoly) – us CRL [954]
Asiyan – Istanbul. 1-2. sene n1-26. 28 agustos 1324-29; subat 1325 [28 aug 1906-27 feb 1907] [all publ] – (= ser O & t journals) – 13mf – 9 – $210.00 – us MEDOC [956]
Asj-Sju'llah *see* Beberapa penggalan dari sedjarah perdjoeangan oemmat islam
Ask for beck's 'zingib,' the favorite drink – S.l: s.n, 1900? – 1mf – 9 – mf#60032 – cn Canadiana [650]
Ask newsletter – Vancouver: Assoc for Social Knowledge. v1-5. apr 1964-feb 1968// (mthly) – 1r – 1 – Can$150.00 – (canada's first periodical devoted to gay liberation) – cn McLaren [305]
Aska weint : eine mythe aus urfernen tagen / Lettenmair, Josef Guenther – Berlin: Nordland Verlag [c1942] [mf ed 1992] – 1r [ill] – 1 – (filmed with: deutschland muss leben! / heinrich lersch) – mf#2820p – us Wisconsin U Libr [830]
Askersunds tidning – Askersund, Sweden. 1857-68; 1956-58 – 5r – 1 – sw Kungliga [078]
Askersunds tidning – Askersund, Sweden. 1857-78 – 7r – 1 – sw Kungliga [078]
Askersunds tidning – Askersund, Sweden. 1917-56 – 60r – 1 – sw Kungliga [078]
Askersunds veckoblad – Askersund, Sweden. 1879-1917 – 21r – 1 – sw Kungliga [078]
Askese und moenchtum / Zoeckler, Otto – 2 rev enl ed. Frankfurt a M: Heyder & Zimmer 1897 [mf ed 1994] – 2v in 1 on 2mf – 9 – 0-524-08844-6 – mf#1993-1103 – us ATLA [240]
Askhabad – Ashkhabad, Turkmenistan 1900 [mf ed Norman Ross] – 1 – mf#nrp-202 – us UMI ProQuest [077]
Askim, Karen L *see* Validity of whole-body bioelectrical impedance analysis in the prediction of percent body fat in women
Askwith, Edward Harrison *see*
– The christian conception of holiness
– The epistle to the galatians
– The historical value of the fourth gospel
– An introduction to the thessalonian epistles
Asm del rey don alfonso 13 : recuerdo de sumpaso regio por extremadura...serenata del primer cuarteto / Mora, Angel – Transcrita...por A.Voadmirll. Merida J. Joaquin Soler Segura, 1905 – 1 – sp Bibl Santa Ana [946]
Asmodee – Amsterdam, Netherlands 3 may 1854-25 dec 1861, 7 jan 1864-31 dec 1874, 4 jan-27 dec 1877 – 1 – uk British Libr Newspaper [074]
Asmonean – New York. 1849-58 – 1 – us AJPC [073]
Asmus, Georg *see*
– Amerikanisches skizzebuechelche
– Gedichtbuechelchen
Asmus, Paul *see*
– Das absolute und die vergeistigung der einzelnen indogermanischen religionen
– Indogermanische naturreligion
Asmusson, Erling *see* Body temperature and capacity for work
Asnad-i nahat-i azadi-i iran *see* Ba hashiyah va bi hashiyah
Asnad-i tarikhi-i jubnish-i kargari, susiyal-dimukrasi va kumunisti-i iran = Historical documents of the workers', social-democratic, and communist movement in iran / ed by Chaqueri, Cosroe – 23v – 3r – 1 – $200.00 – us MEDOC [320]
El asno erudito / Forner Segarra, Juan Pablo – Valencia: editorial castalia, 1948 – sp Bibl Santa Ana [946]
Asociacion Amigos de Guadalupe *see*
– Extremadura y el mar
– Memoria y actas de las reuniones pro-hispanidad celebradas en el real monasterio de guadalupe...mayo de 1948
Asociacion Benefica Ntra. Sra. de la Luz. Malpartida de Plasencia *see* Ala santisima virgen de la luz 1978
Asociacion Cultural Chambra *see* Fiestas del risco 1979
Asociacion Cultural "Pedro de Trejo" *see* Ferias y fiestas 1973
Asociacion Cultural Placentina Pedro de Trejo *see* Ferias y fiestas de plasencia 1975
Asociacion de Amigos de Guadalupe *see* Estatutos provisionales. octubre, 1946
Asociacion de Hijas de la Purisima e Inmaculada...Viregen Maria. Spain *see*
– Oraciones
Asociacion de Medicina Extremena *see* Estatutos y reglamento
Asociacion de Padres de Alumnos y Amigos de la Esuela de EGB *see* Estatutos
Asociacion Empresarial de panaderos de la Provincia de Caceres *see* Estatutos
Asociacion Empresarial Harino-Panadera de la provincia de Caceres *see* Estatutos de la...
Asociacion Espanola de Hematologia y Hemoterapia, 12 Reunion *see* 1 reunion hispano portuguesa de hematologia
Asociacion familiar "Los Alamos" Casas de Don Antonio *see* Estatutos
Asociacion Geofisica de Mexico *see* Boletin
Asociacion Morala de Padres de Alumnos de Educ. General Basica. Navlamoral de la Mata *see* Estatutos
Asociacion Provincial de amas de casa *see* Estatutos de la...
Asociacion Provincial del Magisterio de Caceres *see* Reglamento de...reformado en 1940
Asociacion Santa Eulalia. Spain *see* Corona poetica
Asoka / Bhandarkar, Devadatta Ramakrishna – Calcutta: University of Calcutta, 1932 – (= ser Samp: indian books) – us CRL [954]
Asoka : the buddhist emperor of india / Smith, Vincent Arthur – 2nd rev enl ed. Oxford: Clarendon Press, 1909 [mf ed 1995] – (= ser Yale coll) – 252p (ill) – 1 – 0-524-09854-9 – mf#1995-0854 – us ATLA [954]
Asola, Giovanni Matteo *see* Madrigali a due voci
Asonada / Mancisidor, Jose – Jalapa, Mexico. 1931 – 1r – us UF Libraries [972]
Asonante final / Florit, Eugenio – Habana, Cuba. 1955 – 1r – us UF Libraries [972]
Asonova De Villaverde, Emilia *see* Apuntes biograficos de emiia casanova de villaverde escritos
Asp – 1983 mar-apr, aug-oct, dec; 1984 mar, may-aug – 1 – mf#1477072 – us WHS [071]
Aspaklariya ha-me'irah / Slivkin, Hayyim Shalom – s.l, s.l? . 1903 – 1r – us UF Libraries [939]
Asparagus caterpillar / Wilson, J W – Gainesville, FL. 1934 – 1r – us UF Libraries [634]
Aspbae journal / Asian-South Pacific Bureau of Adult Education – New Delhi, India 1975 – 1,5,9 – ISSN: 0001-2602 – mf#10327 – us UMI ProQuest [374]
Aspdin, James *see* Emigration
Aspect of prophecy respecting the present and future state of the j... / Collyer, William Bengo – London, England. 1829 – 1r – us UF Libraries [240]
Aspectos da economia brasileira / Sa, Jayme Margrassi De – Sao Paulo, Brazil. 1970 – 1r – us UF Libraries [972]
Aspectos da historia e da cultura do brasil – Lisboa, Portugal. 1923 – 1r – us UF Libraries [972]
Aspectos da industrializacao brasileira – Sao Paulo, Brazil. 1969? – 1r – us UF Libraries [972]
Aspectos da literatura brasileira / Andrade, Mario De – Rio de Janeiro, Brazil. 1943 – 1r – us UF Libraries [972]
Aspectos da litteratura colonial brazileira / Lima, Oliveira – Leipzig, Germany. 1896 – 1r – us UF Libraries [972]
Aspectos de historia e da cultura do brasil / Lima, Oliveira – Lisboa, Portugal. 1923 – 1r – us UF Libraries [972]
Aspectos do brasil / Magalhaes, Symphronio De – Rio de Janeiro, Brazil. 1930 – 1r – us UF Libraries [972]
Aspectos do nacionalismo economico brasileiro / Luz, Nicia Villela – Sao Paulo, Brazil. 1959 – 1r – us UF Libraries [330]
Aspectos do padre antonio vieira / Lins, Ivan Monteiro De Barros – Rio de Janeiro, Brazil. 1962 – 1r – us UF Libraries [972]
Aspectos do romance brasileiro / Castello, J Aderaldo – Rio de Janeiro, Brazil. 1960 – 1r – us UF Libraries [972]
Aspectos e perspectivas da economia nacional / Dias Rollemberg, Luiz – Rio de Janeiro, Brazil. 1941 – 1r – us UF Libraries [330]
Aspectos e trachos escolhidos dos sermoes e cartas / Lins, Ivan Monteiro De Barros – Rio de Janeiro, Brazil. 1966 – 1r – us UF Libraries [972]
Aspectos economicos de nuestra revolucion / Cardona Rossell, Mariano – [Barcelona?]: Oficinas de Propaganda CNT, FAI [1937?] – 9 – mf#w773 – us Harvard [946]
Aspectos geograficos de la colonizacion agricola e... / Sandner, Gerhard – San Jose, Costa Rica. 1961 – 1r – us UF Libraries [972]
Aspectos gerais de pelotas / Pimentel, Fortunato – Porto Alegre, Brazil. 1940 – 1r – us UF Libraries [972]
Aspectos legais e economicos da pequena empresa br... / Bouzan, Ary – Rio de Janeiro, Brazil. 1968 – 1r – us UF Libraries [330]
Aspectos naiconales / Velasco Y Perez, Carlos De – Habana, Cuba. 1915 – 1r – us UF Libraries [972]
Aspectos sociais de luanda inferidos dos anuncios publicados / Mario Antonio – Coimbra, Portugal. 1965 – 1r – us UF Libraries [960]
Aspectos socio-geograficos do amazonas / Jobim, Anisio – Manaos, Brazil. 1950 – 1r – us UF Libraries [972]
Aspects articulatoires de la labiale vocalique en francais. contribution a la modelisation a partir de labiophotographies : labiofilms et films radiologiques. etude statique, dynamique et contrastive / Zerling, Jean Pierre – 2mf – 9 – (10218) – fr Atelier National [440]
Aspects de la france – Paris. 10 juin 1947-86 – 1 – fr ACRPP [073]
Aspects du genie d'israel – Paris, France. 1950 – 1r – us UF Libraries [956]
Aspects of abul kalam azad : essays on his literary, political and religious activities / ed by Butt, Abdullah – Lahore: Maktaba-I-Urdu, 1942 – (= ser Samp: indian books) – us CRL [920]
Aspects of adjudication / Warden, Robert Bruce – Washington, Ernest Institute, 1886. 22 p. LL-1368 – 1 – us L of C Photodup [340]
Aspects of authority in the christian religion / Robins, Henry Burke – Philadelphia: Griffith & Rowland Press c1911 [mf ed 1991] – 1mf – 9 – 0-7905-9851-5 – (incl bibl ref) – mf#1989-1576 – us ATLA [240]
Aspects of bengali society from old bengali literature / Das Gupta, Tamonash Chandra – Calcutta: University of Calcutta 1935 [mf ed 1996] – (= ser Samp: indian books 09269) – 1r – 1 – (filmed with other items; incl bibl ref & ind) – mf#mf-10881 r027 – us CRL [301]
Aspects of central african history – Evanston, IL. 1968 – 1r – us UF Libraries [960]
Aspects of christ / Selbie, William Boothby – London: Hodder & Stoughton 1909 [mf ed 1985] – 1mf – 9 – 0-8370-5608-X – (incl bibl ref & ind) – mf#1985-3608 – us ATLA [240]
Aspects of christian experience / Merrill, Stephen Mason – Cincinnati: Walden & Stowe 1882 [mf ed 1992] – 1mf – 9 – 0-524-06188-2 – mf#1991-2444 – us ATLA [242]
Aspects of christian mysticism / Scott, William Major – New York: E P Dutton 1907 [mf ed 1992] – 1mf – 9 – 0-524-04852-5 – mf#1990-1344 – us ATLA [230]
Aspects of early assamese literature / ed by Kakati, Banikanta – Gauhati: Gauhati University, 1953 – (= ser Samp: indian books) – us CRL [490]
Aspects of education : a study in the history of pedagogy / Browning, Oscar; ed by Butler, Nicholas Murray – New York: Industrial Education Assoc, 1888 [mf ed 1986] – 1mf – 9 – 0-8370-7773-7 – mf#1986-1773 – us ATLA [370]
Aspects of internet payment instruments / Lawack-Davids, Vivienne Antoinette – Uni of South Africa 2000 [mf ed Johannesburg 2000] – 14mf – 9 – (incl bibl ref) – mf#mfm15019 – sa Unisa [000]
Aspects of islam / Macdonald, Duncan Black – New York: Macmillan 1911 [mf ed 1991] – (= ser Hartford-lamson lectures on the religions of the world) – 1mf – 9 – 0-524-01621-6 – mf#1990-2560 – us ATLA [260]
Aspects of mexican civilization / Vasconcelos, Jose – Chicago, IL. 1926 – 1r – us UF Libraries [972]
Aspects of professional career success and the implications for life skills education / Villiers, Sarah Leone de – Uni of South Africa 2001 [mf ed Pretoria: UNISA 2000] – 8mf – 9 – (incl bibl ref) – mf#mfm14732 – sa Unisa [370]
Aspects of religious and scientific thought / Hutton, Richard Holt; ed by Roscoe, Elizabeth Mary – London, New York: Macmillan 1899 [mf ed 1985] – 1mf – 9 – 0-8370-4847-8 – mf#1985-2847 – us ATLA [210]
Aspects of religious belief and practice in babylonia and assyria / Jastrow, Morris – New York: G P Putnam 1911 [mf ed 1989] – (= ser American lectures on the history of religions 9th ser) – 2mf – 9 – 0-7905-1164-9 – mf#1987-1164 – us ATLA [290]
Aspects of revelation / Brewster, Chauncey Bunce – New York: Longmans, Green 1901 [mf ed 1991] – (= ser Baldwin lectures 1900) – 1mf – 9 – 0-7905-7691-0 – mf#1989-0916 – us ATLA [240]
Aspects of scepticism : with special reference to the present time / Fordyce, John – New York: T Whittaker 1884 [mf ed 1984] – 1r – 1 – (incl bibl ref. filmed with: meteor / behrman, s n) – mf#1096 – us Wisconsin U Libr [140]
Aspects of sexual reproduction and their effects on population genetic structure in creeping thistle (cirsium arvense l scop) / Heimann, Bettina – (mf ed 1997) – 2mf – 9 – €40.00 – 3-8267-2490-9 – mf#DHS 2490 – gw Frankfurter [580]
Aspects of spanish-american literature / Torres-Rioseco, Arturo – Seattle, WA. 1963 – 1r – us UF Libraries [440]
Aspects of the atonement : the atoning sacrifice illustrated from the various sacrificial types of the old testament, and from the successive ages of christian thought / Ragg, Lonsdale – London: Rivingtons 1904 [mf ed 1989] – 1mf – 9 – 0-7905-3163-1 – mf#1987-3163 – us ATLA [240]
Aspects of the collateral source rule : with special reference to pension benefits / Manamela, Makwena Ernest – Uni of South Africa 2000 [mf ed Johannesburg 2000] – 1mf – 9 – (incl bibl ref) – mf#mfm14956 – sa Unisa [360]
Aspects of the infinite mystery / Gordon, George Angier – Boston: Houghton Mifflin 1916 [mf ed 1991] – 1mf – 9 – 0-7905-7743-7 – mf#1989-0968 – us ATLA [240]
Aspects of the old testament : considered in eight lectures. delivered before the university of oxford / Ottley, Robert L – London, New York: Longmans, Green 1897 [mf ed 1986] – (= ser Bampton lectures 1897) – 2mf – 9 – 0-8370-9497-6 – (incl bibl ref & ind) – mf#1986-3497 – us ATLA [221]
Aspects of the rural occupance of namaqualand, south africa / Slade, Donald Graeme Bartlett – U of Liverpool 1974 [mf ed S.l: Mille Rand c1984] – 7mf – 9 – sa Misc Inst [307]
Aspects of the spiritual / Brierley, Jonathan – New York: Thomas Whittaker 1909 [mf ed 1985] – 1mf – 9 – 0-8370-2831-0 – mf#1985-0831 – us ATLA [240]
Aspects of the vedanta / Avergal, N Vythinatha Aiyar et al – 3rd ed. Madras: G A Natesan [1903?] [mf ed 1993] – 3mf – 9 – 0-524-07932-3 – mf#1991-0182 – us ATLA [280]
Aspects of theism / Knight, William Angus – London, New York: Macmillan 1893 [mf ed 1985] – 1mf – 9 – 0-8370-3944-4 – mf#1985-1944 – us ATLA [210]
Aspe-Fleurimont, Lucien Auguste *see* La guinee francaise, conakry et rivieres du sud
Aspek as uitdrukkingsmiddel van handeling – Pretoria, South Africa. 1958 – 1r – us UF Libraries [960]
Aspekte / Wachsmann, Konrad – Wiesbaden: Krausskopf 1961 [mf ed 1984] – 1r [ill] – 1 – mf#965 – us Wisconsin U Libr [770]

Aspekte der selbstbestimmungsproblematik in den vereinten nationen : fallstudien zu zypern und puerto rico / Nikitopoulos, Ingeborg – Heidelberg, 1970 – 6mf – 9 – 3-89349-759-5 – gw Frankfurter [327]

Aspekte des funktional-semantischen feldes der art und weise im modernen englisch / Biederstaedt, Birgit – (mf ed 1999) – 3mf – 9 – €49.00 – 3-8267-2648-0 – mf#DHS 2648 – gw Frankfurter [420]

Aspekte des personalmanagements bei der einfuehrung und durchsetzung von ganzheitlich orientierten qualitaetssicherungssystemen nach din iso 9000ff / Natschke, Birgit – (mf ed 1994) – 1mf – 9 – €30.00 – 3-8267-2010-5 – mf#DHS 2010 – gw Frankfurter [650]

Aspekte einer provokativen tschechischen germanistik / Preisner, Rio – Wuerzburg: Jal-Verlag 1977-81 [mf ed 1993] – (= ser Colloquium slavicum 8) – 2v on 1r – 1 – (incl bibl ref. filmed with: gestaltung, umgestaltung / ed by joachim mueller) – mf#3176p – us Wisconsin U Libr [430]

Aspekte van die onafhanklikheid van die strafhowe : 'n regsvergelykende ondersoek / Nel, Susanna Sophia – Uni of South Africa 2000 [mf ed Johannesburg 2000] – 16mf – 9 – (summary in english; text in afrikaans; incl bibl ref & ind) – mf#mfm14919 – sa Unisa [380]

Aspekte van die xhosakortverhaal / Botha, Christoffel Rudolph – [Stellenbosch]: U van Stellenbosch 1978 [mf ed 1978] – 5mf – 9 – (incl bibl) – mf#mf.256 – sa Stellenbosch [470]

Aspen times *see* Miscellaneous newspapers of pitkin county

Aspen weekly press *see* Miscellaneous newspapers of pitkin county

Asphalt : a quarterly publication of the asphalt institute – Lexington KY 1949-76 – 1,5,9 – ISSN: 0004-4954 – mf#1107 – us UMI ProQuest [624]

Asphalt block pavement / Baillairge, Charles P Florent – S.l: s.n, 1899? – 1mf – 9 – (repr fr: canadian engineer, sept, 1899) – mf#10184 – cn Canadiana [625]

Aspillera, Paraluman S *see* Improve your tagalog

Aspin procurement report – 1988 feb, jun, sep, dec; 1989 may, sep; 1990 winter, summer; 1991 winter – 1 – mf#1110529 – us WHS [071]

Aspinall, Algernon Edward *see*
– Pocket guide to the west indies
– Pocket guide to the west indies and british guiana
– Wayfarer in the west indies

Aspinall's reports of maritime cases – v1-18. 1870-1936 – 9 – $540.00 – mf#0087 – us Brook [341]

Aspinion, Robert *see* Contribution a l'etude du droit coutumier berbere marocain

O aspirante – Ouro Preto, MG: Typ Silvia Cabral, 05 maio 1894 – (= ser Ps 19) – mf#P31,03,50 – bl Biblioteca [079]

Les aspirations : poesies canadiennes / Chapman, William – Paris: Librairies-imprimeries reunies, 1904 – 4mf – 9 – 0-665-75957-6 – mf#75957 – cn Canadiana [810]

Aspirations of nature / Hecker, Isaac Thomas – 4th ed. New York: Catholic Publ House 1869, c1857 – 1mf – 9 – 0-8370-6981-5 – (incl bibl ref) – mf#1986-0981 – us ATLA [230]

Aspland, Robert *see* Reunion of the wise and good in a future state

Aspley guise – (= ser Bedfordshire parish register series) – 2mf – 9 – £5.00 – uk BedsFHS [929]

Asplund, John *see* The annual register of the baptist denomination in north america, 1790-1794

Asq six sigma forum magazine – Milwaukee WI 2002+ – 1,5,9 – ISSN: 1539-4069 – mf#31895 – us UMI ProQuest [338]

Asqueroso : orgam arrecadante – Fortaleza, CE. 06 jan 1896 – (= ser Ps 19) – mf#P18B,03,72 – bl Biblioteca [870]

'Asr-i 'Amal – West Germany: Intisharat-i Asr-i 'Amal. shumarah-'i 1-7 – 1r – 1 – $53.00 – us MEDOC [079]

Assab e i dan...chili. viaggio e studii / Licata, G B – Milano, 1885 – 4mf – 9 – mf#NE-20205 – ne IDC [918]

Assab e i suoi critici : con la carta della baja d'assab e regioni adiacenti / Sapeto, G – Genova, 1879 – 3mf – 9 – mf#NE-20297 – ne IDC [956]

L'assaba : essai monographique / Munier, Pierre Marie – Saint-Louis, Senegal. 1952 – 1 – us CRL [306]

Assabghy, A *see* Les questions de nationalite en egypte

The assam gazette / Assam. India – 1963-1966. Incomplete – 1 – us NY Public [324]

Assam. India *see* The assam gazette

Assam planter : tea planting and hunting in the assam jungle / Ramsden, A R – London: John Gifford, 1945 – (= ser Samp: indian books) – us CRL [630]

Assam tribune – Gauhati, India. May 1944-Dec 1994 – 146r – 1 – us L of C Photodup [079]

Assam valley : beliefs and customs of the assamese hindus / Muirhead-Thomson, R C – London: Luzac & Co, 1948 – (= ser Samp: indian books) – us CRL [390]

Assamese : its formation and development / Kakati, Banikanta – Gauhati, Assam: Govt of Assam, 1941 – (= ser Samp: indian books) – us CRL [490]

Assamese literature / Barua, Birinchi Kumar – Bombay: For the PEN All-India Centre by International Book House, 1941 – (= ser Samp: indian books) – us CRL [490]

L'assassinat de andres nin : ses causes, ses auteurs – Paris, 1939 – (= ser Blodgett coll) – 9 – mf#fiche w 732 – us Harvard [946]

L'assassinat maconnique / le crime rituel / la trahison juive – Paris: Renaissance francaise 1905 – 2mf – 9 – mf#vrl-122 – ne IDC [366]

Assassination Information Bureau *see* People and the pursuit of truth

Assassination of catholic priests in the diocese of barcelona, spain : under the so-called spanish republic, now the spanish republic in exile / Spain. Embajada (United States). Office of Cultural Relations – Washington DC [1946?] – 9 – mf#w733 – us Harvard [946]

L'assault / Front national-syndicaliste – Paris. n1-3. mai 1933-mars 1934 – 1 – fr ACRPP [325]

L'Assaut *see* Haiti et les problemes panamericaines

The assay of gold and silver wares / Ryland, Arthur – London: Smith, Elder, 1852. 212p. LL-4084 – 1 – us L of C Photodup [540]

Asschepoester, groot toover-ballet in drie bedrijven : gemonteerd door den balletmeester rives / Albert – Amsterdam: M Westerman, 1824 – 1 – mf#*ZBD-*MGTZ pv7-Res – Located: NYPL – us Misc Inst [790]

Asselin, Benoit, Boucher, Ducharme, Lapointe, inc *see* Report on the economic studies for localizing the powerhouse manicouagan 5

Asseline, Louis *see* Mary alacoque and the worship of the sacred heart of jesus

Assemani, B *see* Opera omnia, graece, syriace et latine

Assemani, E *see* Bibliothecae mediceae laurentianae et palatinae codicum mms

Assemani, J A *see*
– Commentarius theologico-canonico-criticus de ecclesiis
– De catholicis seu patriarchis chaldaeorum et nestorianorum

Assemanius, S E *see* Acta sanctorum martyrum orientalium et occidentalium

Assemanus, J A *see* Codex liturgicus ecclesiae universae

Assemanus, J S *see*
– Bibliotheca orientalis clementino-vaticana
– Bibliothecae apostolicae vaticanae codicum manuscriptorum catalogus, vol 1
– Kalendaria ecclesiae universae. kalendaria ecclesiae slavicae sive graeco-moschae

Assemanus, S E *see* Bibliothecae apostolicae vaticanae codicum manuscriptorum catalogus, vol 1

Assemblea dos membros da communidade portugueza (1888: Bombay) *see* Acta da assemblea dos membros da communidade portugueza de bombaim

Assemblee a saint-hyacinthe le 8 decembre 1885 pour protester contre l'execution de riel : discours de l'hon m bellerose – S.l: s.n, 1885? – 1mf – 9 – mf#30256 – cn Canadiana [971]

A une assemblee des electeurs de la ville et des faubourgs de quebec : qui approuvent la conduite de la chambre d'assemblee, tenue a l'hotel de malhiot, 13 nov 1827 / Lagueux, Louis Abraham – S.l: s.n, 1827? – 1mf – 9 – mf#35391 – cn Canadiana [323]

Assemblee generale... / Societe pour le patronage des jeunes detenus et des jeunes liberes du departement de la Seine – Paris, 1841 [mf ed 1969] – 1r – 1 – mf#SLL pv2 – us NY Public [366]

Assemblee legislative de quebec : seance du 13 decembre 1876 / Angers, Auguste Real – [S.l: s.n, 1876?] [mf ed 1979] – 1mf – 9 – 0-665-00818-X – mf#00818 – cn Canadiana [323]

L'assemblee nationale – Paris, 1849-50, 1852-53 – 1 – fr ACRPP [323]

Les assemblees du clergue et le jansenisme / Bourlon, I – Paris: Bloud, 1909 – 1mf – 9 – 0-8370-8407-5 – (incl bibl ref) – mf#1986-2407 – us ATLA [240]

Assemblies of God *see* Christ for all

Assemblies of god home missions – 1978 sep/oct-1983 jul/aug – 1 – mf#351905 – us WHS [071]

Assembly – Troy MI 1991+ – 1,5,9 – (cont: assembly engineering) – ISSN: 1050-8171 – mf#8698,01 – us UMI ProQuest [620]

Assembly engineering – Troy MI 1958-89 – 1,5 – (cont by: assembly) – ISSN: 0004-5063 – mf#8698.01 – us UMI ProQuest [620]

Assembly of 1881 and the case of professor robertson smith / Innes, Alexander Taylor – Edinburgh: John Maclaren, [188?] Princeton: Speer Lib, and Dep of Photodup, U of Chicago Lib, 1978 (1r); Evanston: American Theol Lib Assoc, 1984 (1r) – (= ser Case of william robertson smith in the free church of scotland) – 1 – 0-8370-0635-X – (incl bibl ref) – mf#1984-6275 – us ATLA [242]

Assembly of Governmental Employees [Washington DC] *see* Coverage

Assembly Of Hebrew Orthodox Rabbis Of America *see* Sefer keneset ha-rabanim ha-ortodoksim ba-'amerika

The assembly order books : at sutton's hospital, charterhouse, 1613-1982 – 9r – 1 – £430.00 – (documents a full spectrum of social and economic history. a record of all that the governors 'ordered and appointed' to be done for nomination and admission of 'brothers') – mf#AOB – uk World [941]

Assembly proceedings / East Bengal (Pakistan). Legislative Assembly – Dacca, East Bengal Govt Press. v1 n3- v12 (mar 29 1948-aug 5 1955) (irreg) – 5r – 1 – us CRL [323]

Assembly proceedings / East Pakistan (Pakistan). Assembly – Dacca, East Pakistan Govt Press [v13-20 may 22 1956-jun 25 1958] (irreg) – 4r – 1 – us CRL [323]

Asser, Bishop of Sherborne *see* Asser's life of king alfred

Asser, John *see* Asser's life of king alfred

Asser's life of king alfred : together with the annals of saint neots erroneously ascribed to asser / Asser, Bishop of Sherborne; ed by Stevenson, William Henry – Clarendon Press, 1904 [mf ed 1986] – cxxx/386p – 1 – (int & comm by ed) – mf#6976 – us Wisconsin U Libr [941]

Asserta aphoristica et chirurgica ex libris... / Gasco y Navarro, J M – Valencia, 1745 – 1mf – 9 – sp Cultura [617]

Asserta theo-subtitulia...efficacia / Gil Becerra, Benito – 1737. 2v – 9 – sp Bibl Santa Ana [240]

Assertationes sacrae theologiae juxta mentem seraphici, subtilisque doctorum propugnandae / Ayzinena, Jose de – [Mexico City]: Apud viduam D Sebastiani de Arevolo, anno 1776 – (= ser Books on religion...1543/44-c1800: teologia "culta"; derecho canonico) – 1mf – 9 – mf#crl-277 – ne IDC [241]

Assertio contra scriptum de adoratione carnis christi / Daneau, Lambert – Geneve, E Vignon, 1585 – 1mf – 9 – mf#PFA-130 – ne IDC [240]

Assertio orthodoxae doctrinae de duabus naturis christi... / Simler, J – Tigvri, Christoph Froschover, 1575 – 2mf – 9 – mf#PBU-332 – ne IDC [240]

Assertio sanae et orthodoxae doctrinae de persona et maiestate domini nostri iesv christi / Hunnius, A – Francofvrti ad Moenvm, 1592 – 5mf – 9 – mf#TH-1 mf 744-748 – ne IDC [242]

Assertio septem sacramentorum : or, defence of the seven sacraments / ed by O'Donovan, Louis – New York: Benziger 1908 [mf ed 1990] – 2mf – 9 – 0-7905-7050-5 – (incl bibl ref; in english & latin; int by ed) – mf#1988-3050 – us ATLA [241]

Assessing america's health risks : how well are medicare's clinical preventive benefits serving america's seniors?: hearing...house of representatives, 107th congress, 2nd session, may 23 2002. / United States. Congress. House. Committee on Energy and Commerce. Subcommittee on Oversight and Investigations – Washington: US GPO 2002 [mf ed 2002] – 1mf – 9 – 0-16-068793-4 – (incl bibl ref) – us GPO [362]

Assessing retiree health legacy costs : is america prepared for a healthy retirement? : hearing...house of representatives, 107th congress, 2nd session...washington dc, may 16 2002 / United States. Congress. House. Committee on Education and the Workforce. Subcommittee on Employer-Employee Relations – Washington: US GPO 2002 [mf ed 2002] – 2mf – 9 – 0-16-068964-3 – us GPO [362]

Assessing the child care and development block grant : hearing...house of representatives, 107th congress, 2nd session...washington dc, feb 27 2002 / United States. Congress. House. Committee on Education and the Workforce. Subcommittee on 21st Century Competitiveness – Washington: US GPO 2002 [mf ed 2002] – 2mf – 9 – 0-16-069238-5 – (incl bibl ref & ind) – us GPO [360]

Assessment and comparison of the stress experienced by international and american students at the university of north texas / Islam, Nehalul – 2001 – 50p on 1mf – 9 – $5.00 – mf#HE 686 – us Kinesiology [150]

Assessment and evaluation in higher education – Abingdon, Oxfordshire 1981+ – 1,5,9 – ISSN: 0260-2938 – mf#12905,01 – us UMI ProQuest [378]

Assessment for effective intervention – Austin TX 1980+ – 1,5,9 – ISSN: 1534-5084 – mf#12743,01 – us UMI ProQuest [370]

Assessment journal – Chicago IL 1994+ – 1,5,9 – ISSN: 1073-8568 – mf#20656 – us UMI ProQuest [333]

Assessment of a marketing order prorate suspension : a study of california – arizona navel oranges / Powers, Nicholas John et al – Washington DC: US Dept of Agriculture, Economic Research Service...1986 – (= ser Agricultural economic report 557) – 9 – (incl bibl ref) – us GPO [634]

An assessment of cuba broadcasting : the voice of freedom: hearing...house of representatives, 107th congress, 2nd session, june 6, 2002 / United States. Congress. House. Committee on International Relations. Subcommittee on International Operations and Human Rights – Washington: US GPO 2002 [mf ed 2002] – 1mf – 9 – 0-16-068889-2 – us GPO [322]

The assessment of destination awareness of indiana's tourism potential / Yen, J – 1990 – 1mf – 9 – $4.00 – us Kinesology [338]

Assessment of factors which influence college students to participate in regular physical activity : a precede approach / Brawley, Jodi – 1999 – 1mf – 9 – $4.00 – mf#HE 655 – us Kinesology [613]

An assessment of fear of failure as related to gender, athletic participation, level of athletic competition, and sport type / NiiLampti, Nyaka – 2000 – 102p on 2mf – 9 – $10.00 – mf#PSY 2156 – us Kinesology [150]

The assessment of learning programmes for the senior phase at environmental education centres in mpumalanga / Maila, Mago William – Uni of South Africa 2001 [mf ed Johannesburg 2001] – 3mf – 9 – (incl bibl ref) – mf#mfm15052 – sa Unisa [370]

An assessment of selected risk management practices in local indiana park and recreation departments / Mukundan, V – 1991 – 2mf – 9 – $8.00 – us Kinesology [650]

Assessment of technology infrastructure in native communities / Riley, Linda Ann – Washington DC: Economic Devt Administration, US Dept of Commerce [1999?] [mf ed 1999] – 2mf – 9 – (incl bibl ref & ind) – us GPO [338]

An assessment of the attitudes of college students : regarding selected health issues and pregnancy / Hunt, Amy R – 2000 – 1mf – 9 – $4.00 – mf#HE 656 – us Kinesology [150]

An assessment of the effectiveness of the cool cape on the rapid reduction of exercise-induced, elevated body core temperature / Peterson, Paul A & Prentice, William E – 1992 – 1mf – 9 – $4.00 – us Kinesology [613]

Assessment of the extent and influence of indoor radon exposures in south africa / Atomic Energy Corporation of South Africa – Pretoria: Atomic Energy Corporation of South Africa 1991 [mf ed Pretoria, RSA: State Library [199-]] – 46lea on 1r with other items – 5 – mf#a 92-0508 r26 – us CRL [360]

An assessment of the factor validity of the precompetitive stress inventory / Finch, Laura M – 1988 – 143p 2mf – 9 – $8.00 – us Kinesology [150]

An assessment of the health habits and counseling practices of physicians / Kelley, Kristi S – University of North Carolina at Charlotte, 1996 – 2mf – 9 – $8.00 – mf#HE 566 – us Kinesology [613]

Assessment of the impact of tobacco enforcement citation on oregon tobacco retailers' knowledge, attitudes, practices and policies towards minors' access / Street-Muscato, Louise – 1997 – 2mf – 9 – $8.00 – mf#HE 596 – us Kinesology [360]

An assessment of the marketing and promotions of women's lacrosse in ncaa division 1 / Ervin, James R – 1998 – 1mf – 9 – $4.00 – mf#PE 3871 – us Kinesology [790]

An assessment of the nature and prevalence of sport psychology service provision in professional sports / Dunlap, Erik M – 1999 – 2mf – 9 – $8.00 – mf#PE 4081 – us Kinesology [150]

Assessment of the need for certified athletic trainers in new york state high schools / Koabel-Bagley, Patricia – 1994 – 1mf – $4.00 – us Kinesology [617]

Assessment of the planning and implementation process of worksite health promotion programs / Underwood, Lisa S – 1994 – 1mf – $4.00 – us Kinesology [613]

An assessment of the possessed qualifications and important qualifications of aquatic administrators / Wydan, Mary E – 1989 – 73p 1mf – 9 – $4.00 – us Kinesology [790]

An assessment of the relationship between participation in intercollegiate athletics and the dynamics of romantic relationships / Goldman, Cheryl L – 1997 – 2mf – 9 – $8.00 – mf#PSY 1985 – us Kinesology [150]
An assessment of the use of union dues for political purposes : is the law being followed or violated?: hearing...house of representatives, 107th congress, 2nd session...washington dc, june 20 2002 / United States. Congress. House. Committee on Education and the Workforce. Subcommittee on Workforce Protections – Washington: US GPO 2002 [mf ed 2002] – 2mf – 9 – 0-16-068898-1 – (incl ind) – us GPO [331]
Assessment of the VISA-A questionnaire for Achilles tendinopathy 109=and its correlation with imaging / Robinson, Jennifer M – 2000 – 89p on 1mf – 9 – $5.00 – us Kinesology [617]
Assessment tools used by elementary level adapted physical educators in wisconsin / Steinbrunner, Pamela J – University of Wisconsin-La Crosse, 1995 – 1mf – 9 – $4.00 – mf#PE3620 – us Kinesology [370]
Assessment update – Hoboken NJ 1989+ – 1,5,9 – ISSN: 1041-6099 – mf#17614 – us UMI ProQuest [378]
Assessor lankens verlobung : novellen / Kretzer, Max – Berlin: Phoenix-Verlag C Siwinna c1920 [mf ed 1995] – 1r – 1 – (filmed with: die tuerken vor wien / richard kralik) – mf#3909p – us Wisconsin U Libr [830]
The assessor's guide : a manual of the duties of assessors pursuant to the statutes of the legislature of ontario relating thereto – Toronto: N Ure, 1882 – 1mf – 9 – mf#06698 – cn Canadiana [336]
Assessors journal – Chicago IL 1966-81 – 1,5,9 – ISSN: 0004-5071 – mf#6995 – us UMI ProQuest [333]
Assessors recorders / Nye. Nevada – Official records – 1 – us Library Micro [317]
Assessors recorders / Ormsby. Nevada – Official records – 1 – us Library Micro [317]
Asset securitization report – New York NY 2001+ – 1,5,9 – ISSN: 1547-3422 – mf#32307 – us UMI ProQuest [332]
Assets – v1 n1 [1983] – 1 – mf#5294727 – us WHS [071]
Assets protection – Madison WI 1978-84 – 1,5,9 – (cont by: data processing and communications security) – ISSN: 0098-9169 – mf#11931.03 – us UMI ProQuest [000]
Assfalg, J *see* Die ordnung des priestertums
Assid Door *see* De eeuwige cirkel leven en strijd van de indianen en marrons in suriname
As-siddiq – n1-54. Alger. aout 1920-mars 1922 – 1 – (lacking: n11) – fr ACRPP [073]
Assignment africa / Swanson, Donald – Cape Town, South Africa. 1965 – 1r – us UF Libraries [960]
Assimilacao e mobilidade / Durham, Eunice Ribeiro – Sao Paulo, Brazil. 1966 – 1r – us UF Libraries [972]
Assis Brazil, Joaquim Francisco De *see* Democracia representativa
Les assises du temple : poesies maconniques / Malvesin, Louis – 2e ed. Bordeaux: Metreau et Cie 1857 – 2mf – 9 – mf#vrl-72 – ne IDC [366]
Assistance due aux parents – nouv augm rev ed. Montreal: Cadieux & Derome, 1883 [mf ed 1985] – 2mf – 9 – 0-665-08802-7 – mf#08802 – cn Canadiana [170]
Assistance publique et privee en haiti / Mathurin, Augustin – Port-Au-Prince, Haiti. 1944 – 1r – us UF Libraries [972]
Assistant librarian (al) – Birmingham UK 1898-1997 – 1,5,9 – (cont by: impact) – ISSN: 0004-5152 – mf#2285 – us UMI ProQuest [020]
Assistant secretary for labor-management relations decisions / U.S. Dept of Labor – v1-8. 1970-78 – 101mf – 9 – $151.00 – (with ind/digest 1970-78) – mf#llmc 82-602 – us LLMC [344]
Assisted reproduction reviews – v1-6. 1991-1996 – 6r – 1,5,6,9 – $90.00r – us Lippincott [618]
Assistencia tecnica / Mancini, Luiz Carlos – Rio de Janeiro, Brazil. 1956 – 1r – us UF Libraries [600]
Assistencia tecnica aos produtores de borracha – Rio de Janeiro, Brazil. 1970 – 1r – us UF Libraries [600]
Assisting teachers to support mildly intellectually disabled learners in the foundation phase : in accordance with the policy of inclusion / Sethosa, Mosima Francisca – Uni of South Africa 2001 [mf ed Johannesburg 2001] – 7mf – 9 – (incl bibl ref) – mf#mfm15067 – sa Unisa [370]
Assmann, Elisabeth *see* Die entwicklung des lyrischen stils bei detlev von liliencron
Associacao Brasileira De Enfermagem *see* Survey of needs and resources of nursing in brazil
Associacao Do Comercio E Industria De Luanda *see* Consideracoes sobre o problema das transferencias de angola
Associacao Industrial De Angola *see* Guia industrial de angola
Associacoes secretas entre os ind'igenas de angola / Serra Frazao – Lisboa, Portugal. 1946 – 1r – us UF Libraries [960]
The associate creed of andover theological seminary / Park, Edwards Amasa – Boston: Franklin Press, 1883 – 1mf – 9 – 0-7905-7179-X – mf#1988-3179 – us ATLA [240]
Associate dispatch – 1983 jul-1993 nov/dec – 1 – mf#1052456 – us WHS [071]
Associate Presbyterian Church of North America *see* Book of discipline
Associate Reformed Church in North America. General Synod *see* Minutes, 1782-1822
Associate reformed presbyterian – Greenville SC 1989+ – 1,5,9 – ISSN: 0362-0816 – mf#15956 – us UMI ProQuest [242]
Associate Synod of North America *see* Minutes, 1801-1821
Associated Charities of Milwaukee *see* Annual report
Associated Committee of Friends on Indian Affairs *see* Indian progress
Associated General Contractors of America *see*
– Labor letter
– Newsletter
– Open shop
– Safety slants
Associated General Contractors of America et al *see* Legislative bulletin
The Associated Negro Press *see*
– The claude a barnett papers
Associated Press *see* Wisconsin a p log
Associated press clipping file – New York NY 1937-74 – 1 – mf#3173 – us UMI ProQuest [070]
Associated press clippings file. europe disorders – New York NY 1937-63 – 1 – mf#3222 – us UMI ProQuest [070]
Associated Students of UCLA *see*
– Gente
– Gente de aztlan
Association : an essay analytic and experimental / Calkins, Mary Whiton – New York: Macmillan 1896 [mf ed 1993] – (= ser The psychological review. monograph supplement 2) – 1mf – 9 – 0-524-08439-4 – mf#1993-2044 – us ATLA [150]
The association between variables obtained using velocity- and load-regulated squats / Murlasits, Zsolt – 2000 – 1mf – 9 – $4.00 – mf#PE 4074 – us Kinesology [611]
Association bienveillante des pompiers de Montreal *see* Constitution et reglements de l'association bienveillante des pompiers de montreal
Association Canada-Normandie. Section de Montreal *see* Canada-normandie
Association canadienne des bibliothecaires de langue francaise *see* Nouvelles de l'acblf
L'Association canadienne des parents des prisonniers de guerre *see* Bulletin
Association canadienne d'histoire du chemin de fer *see* News report
Association catholique de bienfaisance mutuelle du Canada. Grand conseil *see* Acte d'incorporation et constitution et reglements du grand conseil de l'association catholique de bienfaisance mutuelle du canada et de ses succursales, revisee en août 1896
Association concordia of japan : report – n1-2. 1913-14; n15 [complete] – 1r – 1 – mf#ATLA B0141 – us ATLA [950]
Association d'annexion de Montreal *see*
– Addresse de l'association d'annexion de montreal au peuple du canada
– The annexation manifesto of 1849
– Circulaire de l'association d'annexion de montreal
– Circulaire du comite de l'association d'annexion de montreal
– Circular of the committee of the annexation association of montreal
Association de la jeunesse canadienne-francaise *see* Memoire de l'association de la jeunesse canadienne-francaise
Association de la propagation de la foi (Diocese de Montreal) *see*
– Rapport de l'association de la propagation de la foi
– Rapport de l'association de la propagation de la foi pour le diocese de montreal
Association de la propagation de la foi (Diocese de Quebec) *see* Rapport sur les missions du diocese de quebec
Association de l'unite de Shipshaw-Valin *see* Rapport sur l'agriculture
Association des Amis de Romain Rolland *see* Bulletin
L'Association des Anciens Eleves de l'enseignement Colonial. Chambre de Commerce. Lyon *see* Lyon colonial
L'association des apiculteurs de quebec – [Quebec?: Association des apiculteurs de Quebec, 1930?] (mf ed 1993) – 1mf – 9 – mf#SEM105P2014 – cn Bibl Nat [630]
Association des architectes-paysagistes et urbanistes du Canada *see* Memoire presente a la commission parent...
Association des bibliothecaires du Quebec *see*
– Bulletin de l'abq
– Bulletin de l'association des bibliothecaires du quebec
– Bulletin de nouvelles
Association des denturologistes du Quebec *see*
– Le denturo
Association des familles Berube *see* Info Berube
Association des hommes d'affaires de l'Ile Jesus *see* Memoire de l'association des hommes d'affaires de l'Ile jesus inc a l'honorable jean lesage, premier ministre du gouvernement de la province de quebec
Association des instituteurs de la circonscription de l'Ecole normale Jacques-Cartier *see* Constitution et reglements de l'association des instituteurs en rapport avec l'ecole normale jacques-cartier
Association des professeurs du Conservatoire de musique et d'art dramatique de la province de Quebec *see* Memoire presente a la commission d'enquete sur l'enseignement des arts
Association des psychiatres du Canada *see* Canadian psychiatric association journal
Association des recherches sur les sciences religieuses et profanes au Canada *see* Culture
L'Association Emile-Zola *see* Bulletin
Association "filipino sailors home," manila p i = palatuntunan ng kapisanan "filipino sailors home" – Manila: I R Morales, 1921 [mf ed 1985] – (= ser Philippine labor publications 1/2) – 1v – (in tagalog) – mf#6580 reel 1 n2 – us Wisconsin U Libr [366]
Association for Business Communication (US) *see*
– Bulletin of the association for business communication
– Business communication quarterly
Association For Childhood Education International Literature *see* Told under the magic umbrella
Association for Communication Administration *see* Jaca
Association for communication administration. aca bulletin – Annandale VA 1977-92 – 1,5,9 – (cont by: jaca: journal of the association for communication administration) – ISSN: 0360-0939 – mf#11714.02 – us UMI ProQuest [400]
Association for computing machinery. communications of the acm – New York NY 1959+ – 1,5,9 – (cont: communications of the association for computing machinery) – ISSN: 0001-0782 – mf#12688,01 – us UMI ProQuest [000]
Association for computing machinery. journal – New York NY 1954+ – 1,5,9 – ISSN: 0004-5411 – mf#12690 – us UMI ProQuest [000]
Association for Economic Studies et al *see* Labor today
Association for Educational Communications and Technology *see* Journal of instructional development
Association for Gravestone Studies *see*
– Newsletter
– Newsletter of the association for gravestone studies
Association for library service to children. alsc newsletter – Chicago IL 1989-92 – 1 – ISSN: 0162-6612 – mf#12515 – us UMI ProQuest [020]
Association for obtaining an official inquiry into the pauperism of... – Edinburgh, Scotland. 1840? – 1r – us UF Libraries [240]
Association for Preservation Technology *see*
– Communique
– Newsletter
– Newsletter of the association for preservation technology
Association for Promoting University Consolidation *see*
– A short statement of the advantages of university consolidation
Association for promotion of canadian industry : its formation, by-laws, etc – Toronto: [s.n.] 1866 [mf ed 1984] – 1mf – 9 – 0-665-32219-4 – mf#32219 – cn Canadiana [338]
Association for Report on Confederation *see*
– Report on confederation
– Report: the magazine of public affairs
Association for Study of American Indian Literatures [US] *see*
– Newsletter of the association for study of american indian literatures
– Studies in american indian literatures
Association for Study of Connecticut History *see*
– Connecticut history
– Connecticut history newsletter
– Connecticut history newsletter of the...
Association for supervision and curriculum development yearbook – Alexandria VA 1928+ – 1,5,9 – ISSN: 1042-9018 – mf#6587 – us UMI ProQuest [370]
Association for the Advancement of Women *see*
– Papers and reports read before the association for the advancement of women at its annual congress
– Papers read at the...congress of women
– Papers read before the association for the advancement of womenat its annual congress
– Report of the association for the advancement of women
Association for the care of children in hospitals. journal – Mahwah NJ 1979-81 – 1,5,9 – ISSN: 0145-3351 – mf#12161.01 – us UMI ProQuest [366]
Association for the Care of Coloured Orphans (Cheyney, PA) *see*
– ...Annual report of the association for the care of coloured orphans
– Annual report of the shelter for colored children
– Annual report of the shelter for colored girls
– Annual report of the shelter for colored orphans
– Report of the association for the care of coloured orphans
Association for the Preservation of Political Americana *see* Newsletter
Association for the Protection of Our Water Supply *see* National fluoridation news
Association for the Study of Negro Life and History, Inc *see* Annual report of the director...
Association for the study of perception. international journal – Dekalb IL 1966-89 – 1,5,9 – ISSN: 0004-5454 – mf#8381 – us UMI ProQuest [150]
Association for Union Democracy *see* Union democracy review
L'Association francaise de regulation et d'automatisme *see* Automatisme
Association francaise des observateurs d'etoiles variables *see*
– Bulletin bibliographique
– Bulletin de l'association francaise des observateurs d'etoiles variables
Association generale des etudiants de la Faculte de l'education permanente de l'Universite de Montreal *see*
– Ageefep
– Cite educative
Association Generale des Etudiants du Cegep Andre-Laurendeau *see* L'apostrophe
L' Association Generale des Etudiants Socialistes. Flenu, Jupille, Liege, etc *see* L'etudiant socialiste
Association international des Travailleurs. Section de la Suisse Romande *see* Journal
L'Association internationale des Travailleurs. La Federation jurassienne *see* Bulletin de la federation jurassienne de l'association internationale des travailleurs
Association internationale du Haut-Congo Vivi Station *see* Records of the vivi station, 1881-1885
Association internationale du Haut-Congo. Vivi Station *see* Records of the vivi station, 1881-1885
L'association journal d'economie sociale – Quebec: L'Association, [1890-1891] – 9 – ISSN: 1190-7657 – mf#P04037 – cn Canadiana [366]
Association letter – n11-1921 [1941 apr-1942 apr] – 1 – mf#676621 – us WHS [071]
Association libertiste : ou, embrigadement moral de la societe / ed by Citoyen Pinto – Paris: E Proux, [n1 (1848)] – (= ser French revolution of 1848. newspapers) – 1r – 1 – us CRL [074]
Association management / American Society of Association Executives – Washington DC 1980+ – 1,5,9 – ISSN: 0004-5578 – mf#12800,01 – us UMI ProQuest [650]
Association news – v1 n1 [1973 sep/oct] – 1 – mf#624405 – us WHS [071]
Association of afrikan historians newsletter – 1975 sep-oct; 1976 sep-oct – 1 – mf#4990669 – us WHS [071]
Association of American Dancing *see* Acrobatic enchainements and hints on presentation
Association of american geographers. aag newsletter – Washington DC 1975+ – 1,5,9 – ISSN: 0275-3995 – mf#10801 – us UMI ProQuest [900]
Association of american geographers. annals – Oxford UK 1911+ – 1,5,9 – ISSN: 0004-5608 – mf#10808 – us UMI ProQuest [910]
Association of american geographers. proceedings – Washington DC 1969-76 – 1,5,9 – ISSN: 0572-4295 – mf#10807 – us UMI ProQuest [900]
Association of American Indian Affairs *see* We shake hands
Association of American Indian Physicians *see* Newsletter
Association of American Law Schools *see* Select essays in anglo-american legal history
Association of american law schools : handbook and proceedings of the annual meetings – 1st to 63rd. 1902-64 – 168mf – 9 – $252.00 – (lacking: 1906) – mf#LLMC 84-407 – us LLMC [378]
Association of american law schools. proceedings – Washington DC, 1900-83 – 9 – $360.00 – mf#0088 – us Brook [340]

Association of american physicians. transactions – Philadelphia. v1-25. 1886-1910 – 1 – $378.00 – mf#0089 – us Brook [610]

Association of American Railroads *see* Rail news update

Association of Baptists for Evangelism in the Orient *see*
- The message
- Report of association of baptists for evangelism in the orient, inc

Association of baptists for evangelism in the orient, inc : [papers] – [s l]: Association of Baptists for Evangelism in the Orient [n1[1929]; feb, dec 1930; jan feb/mar 1931]] [mf ed 2005] – 1r – 1 – (iss lack printed dates & numbering; cont by: message (association of baptists for evangelism in the orient)) – mf#2005c-s065 – us ATLA [242]

Association of Black Anthropologists *see* Notes from the aba

Association of Canadian Etchers *see* Catalogue of the first annual exhibition of the association of canadian etchers

Association of Casualty and Surety Companies. Law Dept *see* Chart analysis of the automobile liability security laws of the united states and canada.

Association of Catholic Trade Unionists *see* Labor leader

Association of Civilian Technicians [US] *see* Technician

Association of collegiate schools of planning. bulletin – Cincinnati OH 1973-79 – 1,5,9 – ISSN: 0004-5675 – mf#8707 – us UMI ProQuest [710]

Association of departments of english. ade bulletin – New York NY 1975+ – 1,5,9 – ISSN: 0001-0898 – mf#10580 – us UMI ProQuest [420]

Association of departments of foreign languages. adfl bulletin – New York NY 1977-2000 – 1,5,9 – ISSN: 0148-7639 – mf#10993 – us UMI ProQuest [400]

Association of Dominion Land Surveyors *see* Memorandum...fifth annual meeting of the dominion land surveyors association

Association of Food and Drug Officials of the US *see* Quarterly bulletin

Association of Forest Service Employees for Environmental Ethics *see* Inner voice

Association of Full Gospel Women Clergy *see* Women of the word

Association of governing boards of universities and colleges. [agb] reports – Washington DC 1977-92 – 1,5,9 – ISSN: 0044-961X – mf#11647 – us UMI ProQuest [378]

Association of hospital and institution libraries. quarterly – Chicago IL 1960-74 – 1,5,9 – ISSN: 0090-3116 – mf#2003 – us UMI ProQuest [020]

Association of Iron and Steel Engineers *see* Aise steel technology

Association of Lake Underwriters. Board of Marine Inspectors *see* Proceedings...held at buffalo, august 1856

Association of Latvian Engineers Abroad *see* Technikas apskats

Association of life insurance counsel papers – v1-13. 1913-57 – 110mf – 9 – $165.00 – (lacking: v3 p350-399. v12. after 1949 were called: proceedings) – mf#LLMC 84-409 – us LLMC [368]

Association of life insurance medical directors of america. transactions : annual meeting – Hartford CT 1889-1978 – 1,5,9 – ISSN: 0066-9598 – mf#2447 – us UMI ProQuest [368]

Association of medical officers of the militia of canada : inaugural address by col g sterling ryerson...president, knight of grace of the order of st john of jerusalem in england / Ryerson, George Sterling – Toronto: [s.n.], 1908 – 1mf – 9 – 0-665-87713-7 – (repr fr: the canada lancet, august 1908) – mf#87713 – cn Canadiana [355]

Association of Metis and Non-Status Indians of Saskatchewan *see* New breed

Association of Municipal Corporations. London *see* Municipal review

Association of Neighborhood Housing Developers *see* City limits

Association of Ontario Land Surveyors *see*
- By-laws and rules as revised 1899
- By-laws of the association of ontario land surveyors
- Proceedings of the...
- Report of committee on topographical surveying

Association of operating room nurses. aorn journal – Denver CO 1963+ – 1,5,9 – ISSN: 0001-2092 – mf#9805 – us UMI ProQuest [610]

Association of pacific coast geographers. yearbook – Corvallis OR 1935-86 – 1,5,9 – mf#8615 – us UMI ProQuest [910]

Association of Political Items Collectors *see* Bull moose

Association of practitioners before the icc : reports of annual meetings – v1-3. 1930-32 (all publ) – 9mf – 9 – $13.50 – mf#LLMC 84-410 – us LLMC [340]

Association of protestant teachers of the province of quebec : montreal meeting, 1886 / Dawson, John William – [Montreal?: s.n, 1886?] [mf ed 1980] – 1mf – 9 – mf#03662 – cn Canadiana [377]

Association of Records Managers and Administrators *see* Arma records management quarterly

Association of rehabilitation nurses. arn journal – Glenview IL 1975-80 – 1,5,9 – (cont by: rehabilitation nursing) – ISSN: 0362-3505 – mf#12251.01 – us UMI ProQuest [617]

Association of Research Libraries. Foreign Newspaper Microfilm Project *see* Circular letter

Association of Residents and Internes of British Columbia *see* Pariscope

Association of southeastern biologists. asb bulletin – Morehead City NC 1974+ – 1,5,9 – ISSN: 0001-2386 – mf#9103.01 – us UMI ProQuest [574]

Association of teachers of japanese. journal – Boulder CO 1975-95 – 1,5,9 – ISSN: 0885-9884 – mf#10616.01 – us UMI ProQuest [480]

Association of Teachers of Russian (Great Britain) *see* Journal of russian studies

Association of the bar of the city – 1920-23 – 9 – $20.00 set – mf#100731 – us Hein [340]

Association of the Bar of the City of New York *see*
- In memoriam, marshall s bidwell
- Yearbooks-annual reports

Association of the bar of the city of new york. record – New York NY 1946+ – 1,5,9 – ISSN: 0004-5837 – mf#7144 – us UMI ProQuest [340]

Association of the bar of the city of new york yearbooks – 1870-1977 – 131mf – 9 – $196.00 – (lacking: 1976. before 1909 were called: annual reports) – mf#LLMC 84-408 – us LLMC [340]

Association of the Third Regiment Wisconsin Infantry Veteran Volunteers *see* Proceedings of the...annual reunion of the Association of the Third Regiment Wisconsin Infantry Veteran

Association of trial lawyers of america. atla law reporter – Washington DC 1979-87 – 1,5,9 – (cont by: law reporter) – ISSN: 0364-8125 – mf#12072.05 – us UMI ProQuest [340]

Association of United Ukrainian Canadians *see* Zvit i rezoliutsii z'izdu

Association of Western Pulp and Paper Workers *see* Rebel

Association of Wisconsin School Administrators *see* Bulletin of the association...

Association of workers of revolutionary cinematography : form the russian state archive of literature and art – (= ser The Russian Archives) – 14r – 1 – (coll includes correspondence, memoranda, notes and minutes of meetings, and aarc's periodicals) – us Primary [790]

Association on American Indian Affairs *see*
- The american indian
- Indian affairs
- Indian family defense
- Indian natural resources

Association pour l'avancement des sciences et des techniques de la documentation *see*
- Documentation et bibliotheques

Association professionnelle des industriels. Congres patronal (2e : 1946 : Montreal, Quebec) *see* Organisation et reforme de l'industrie moderne

Association professionnelle des industriels. Congres patronal (4e : 1949 : Montreal, Quebec) *see* Ou va l'industrie

Association progress *see* Ch'ing-nien chin-pu [ccs23]

Association quebecoise des professeurs de francais *see*
- Quebec-francais

Association record / Young Men's Christian Association of Montreal – Montreal: Young Men's Christian Association of Montreal, [18–?-18– or 19–] – 9 – (ceased 188-?) – ISSN: 1190-7002 – mf#P04039 – cn Canadiana [366]

Association Saint-Antoine de Montreal *see* Constitution et reglements de l'association saint antoine de montreal

Association Saint-Jean-Baptiste de Montreal. Caisse nationale d'economie *see* Caisse nationale d'economie, fondee le 1er janvier 1899

Association St. Antoine de Montreal *see* Constitution et reglements de...

Association st jean-baptiste de montreal, fondee en 1834 : statuts et reglements – [Montreal?: s.n.] 1868 [mf ed 1984] – 1mf – 9 – 0-665-44283-1 – mf#44283 – cn Canadiana [366]

Association to Preserve the Eatonville Community *see* Hurston herald

Les associations bambara et leurs chants recreatifs, tome 1 / Couloubaly, Pascal Baba F – [Dakar]: Universite de Dakar, IFAN, Departement de litterature africaine, 1984 – (= ser Kesteloot coll) – 2mf – 9 – us CRL [470]

Associative thesaurus of english, an... / Kiss, G R et al [comp] – 1r – 5 – (avail in edited 96892 and unedited 96975 form) – uk Microform Academic [420]

Associazione Degli Africanisti Italiani *see* Bollettino della associazione degli africanisti italiani

Assommoir / Busnach, William – Paris, France. 1881 – 1r – us UF Libraries [440]

Assomption de hannele mattern / Hauptmann, Gerhart – Paris, France. 1894 – 1r – us UF Libraries [440]

Assorted materials dealing with india – London: British Museum Photographic Service, [19–] – 1 – us CRL [954]

Assorted rhodesian and south african pamphlets / Boston University. African Studies Library – Boston: [The Library, 197-] – 1 – us CRL [960]

Assoziationsversuche mit jugendlichen rauchern und nichtrauchern / Vockrodt-Scholz, Viola – (mf ed 1995) – 5mf – 9 – €59.00 – 3-8267-2205-1 – mf#DHS 2205 – gw Frankfurter [150]

Assu, Jacare [pseud] *see* Brazilian colonization

O assuense : periodico politico, moral e noticioso – Assu, RN: Typ Liberal Assuense, 23-30 mar, maio-jul, set, nov 1867; mar-maio 1868; ago-out 1870; jun-jul, out 1871; mar-abr, 08 jul 1872 – (= ser Ps 19) – bl Biblioteca [079]

The assumption of moses : translated from the latin sixth century ms / ed by Charles, Robert Henry – London: Adam and Charles Black, 1897. Chicago: Dep of Photodup, U of Chicago Lib, 1971 (1r); Evanston: American Theol Lib Assoc, 1984 (1r) – 1 – 0-8370-0541-8 – (transl from the latin 6th century mss) – mf#1984-B277 – us ATLA [221]

Assumption of risk in sport activities : an analysis of contributing factors to legal outcomes in reported cases / Spengler, John O – 1999 – 2mf – 9 – $8.00 – mf#PE 4093 – us Kinesology [790]

The assumptions of the seminary of st sulpice to be the owners of the seigniory of the lake of two mountains and the one adjoining examined and refuted : and their treatment of the indians of the lake of two mountains... / Borland, John – Montreal?: The "Gazette", 1872 – 1mf – 9 – mf#23792 – cn Canadiana [971]

Assunta leoni : schauspiel in fuenf aufzuegen / Wilbrandt, Adolf von – Wien: L Rosner, 1883 [mf ed 1995] – 1 – mf#8910 – us Wisconsin U Libr [820]

L'assunta nell'odierna teologia cattolica : studio pubblicato sul periodico la scuola cattolica, organo della facolt a teologica pontificia di milano / Crosta, Clino – Monza: Artigianelli, 1903 [mf ed 1986] – 1mf – 9 – 0-8370-8415-6 – (in italian) – mf#1986-2415 – us ATLA [241]

Assuntos insulanos / Cabral, Oswaldo R – Florianopolis, Brazil. 1948 – 1r – us UF Libraries [972]

Assurance / Ryle, J C – Ipswich, England. 1850 – 1r – us UF Libraries [240]

Assurance, banque et stocks : chiffres et documents / Filiatreault, Aristide – Montreal: [s.n.] 1905 [mf ed 1995] – 1mf – 9 – 0-665-76964-4 – mf#76964 – cn Canadiana [368]

The assurance of faith / Guth, William Westley – Cincinnati: Jennings and Graham; New York: Eaton and Mains, c1911 – 1mf – 9 – 0-7905-7746-1 – mf#1989-0971 – us ATLA [240]

The assurance of immortality / Fosdick, Harry Emerson – New York: Macmillan, 1913 – 1mf – 9 – 0-7905-3839-3 – mf#1989-0332 – us ATLA [240]

Assurance of salvation – Dublin, Ireland. 18– – 1r – us UF Libraries [240]

Assurance of salvation practically considered / Davidson, James – Edinburgh, Scotland. 18– – 1r – us UF Libraries [240]

Les assurances au canada : projet d'agence d'une compagnie francaise d'assurance contre l'incendie, sur la vie, et contre les risques maritimes / Fournier, Jules – Montreal: J Lovell, 1865 – 1mf – 9 – mf#47459 – cn Canadiana [368]

Assurbanipal und die letzten assyrischen koenige... / Streck, M – Leipzig, 1916 – 16mf – 9 – mf#NE-412 – ne IDC [956]

Assynt news – Lochinver: Assynt News 1980- – 1 – uk Scotland NatLib [072]

Assyria : from the earliest times to the fall of nineveh / Smith, George – new rev ed. London: SPCK 1886 [mf ed 1992] – (= ser Ancient history from the monuments) – 1mf – 9 – 0-524-05421-5 – (new rev ed by archibald henry sayce) – mf#1992-0431 – us ATLA [930]

Assyria : its princes, priests, and people / Sayce, Archibald Henry – London: Religious Tract Soc [1885?] [mf ed 1988] – (= ser By-paths of bible knowledge 7) – 1mf – 9 – 0-7905-0283-6 – (incl ind) – mf#1987-0283 – us ATLA [930]

Assyriaca : eine nachlese auf dem gebiete der assyriologie / Hilprecht, Hermann Vollrat – Boston: Ginn; Halle (Saale): Max Niemeyer 1894 [mf ed 1986] – (= ser Publications of the university of pennsylvania. series in philology, literature, and archaeology 3/1) – 1mf [ill] – 9 – 0-8370-8434-2 – (no more publ?) – mf#1986-2434 – us ATLA [470]

Assyrian and babylonian literature : selected translations / Harper, Robert Francis – New York: D Appleton 1904 [mf ed 1986] – 2mf [ill] – 9 – 0-8370-8823-2 – (trans fr akkadian; crit int by robert francis harper) – mf#1986-2823 – us ATLA [470]

Assyrian and babylonian religious texts : being prayers, oracles, hymns etc – Leipzig: J C Hinrichs 1895-97 [mf ed 1986] – (= ser Assyriologische bibliothek 13) – 2v on 2mf – 9 – 0-8370-9050-4 – (v3 never publ?) – mf#1986-3050 – us ATLA [470]

Assyrian and babylonian religious texts / Craig, J A – Leipzig, 1895-1897. 2v – (= ser Assyriologische bibliothek) – 4mf – 9 – (assyriologische bibliothek v13) – mf#NE-427 – ne IDC [956]

An assyrian doomsday book : or, liber censualis of the district round harran, in the 7th century b c / Johns, Claude Hermann Walter – Leipzig: JC Hinrichs, 1901 [mf ed 1986] – (= ser Assyriologische bibliothek 17) – 1mf – 9 – 0-8370-7709-5 – (text in english & akkadian. comm in english) – mf#1986-1709 – us ATLA [470]

Assyrian echoes of the word / Laurie, Thomas – New York: American Tract Society c1894 [mf ed 1986] – 1mf – 9 – 0-8370-9398-8 – mf#1986-3398 – us ATLA [220]

Assyrian grammar : with paradigms, exercises, glossary, and bibliography / Delitzsch, Friedrich – Berlin: H Reuther; New York: B Westermann 1889 [mf ed 1986] – (= ser Porta linguarum orientalium 10) – 2mf – 9 – 0-8370-8567-5 – (english trans fr german by archibald robert stirling kennedy) – mf#1986-2567 – us ATLA [470]

The assyrian laws / Driver, G R – 1935 – 9 – $18.00 – us IRC [348]

Assyrian life and history / Harkness, Margret Elise – London: Religious Tract Society [1883?] [mf ed 1993] – (= ser By-paths of bible knowledge 2) – 2mf – 9 – 0-524-07963-3 – mf#1992-1118 – us ATLA [930]

An assyrian manual : for the use of beginners in the study of the assyrian language / Lyon, David Gordon – Chicago: American Publ Society of Hebrew, 1886 [mf ed 1986] – 1mf – 9 – 0-8370-8447-4 – (grammar in english. texts in akkadian) – mf#1986-2447 – us ATLA [470]

The assyrian monuments illustrating the sermons of isaiah / Kellner, Maximilian – Boston: Damrell & Upham, 1900 – 1mf – 9 – 0-8370-3872-3 – mf#1985-1872 – us ATLA [221]

Assyrian texts : being extracts from the annals of shalmaneser 2., sennacherib, and assur-bani-pal: with philological notes / Budge, Ernest Alfred Wallis – London: Truebner: Samuel Bagster 1880 [mf ed 1986] – 1mf – 9 – 0-8370-7048-1 – mf#1986-1048 – us ATLA [470]

Assyrien und babylonien nach den neuesten entdeckungen / Kaulen, Franz – 5. aufl. Freiburg i B, St Louis MO: Herder 1899 [mf ed 1989] – (= ser Illustrierte bibliothek der laender und voelkerkunde) – 1mf [ill] – 9 – 0-7905-2721-9 – mf#1987-2721 – us ATLA [930]

Assyriology : its use and abuse in old testament study / Brown, Francis – New York: Charles Scribner 1885 [mf ed 1986] – 1mf – 9 – 0-8370-9846-7 – mf#1986-3846 – us ATLA [221]

Assyrisch-babylonische briefe : religioesen inhalts aus der sargonidenzeit / Behrens, Emil – Leipzig: August Pries 1905 [mf ed 1986] – 1mf – 9 – 0-8370-7442-8 – (text in akkadian & german; comm in german) – mf#1986-1442 – us ATLA [470]

Assyrisch-babylonische chrestomathie : fuer anfaenger – Leiden: E J Brill 1895 [mf ed 1986] – 2mf – 9 – 0-8370-8596-9 – (incl glos; text in akkadian & german, discussion in german) – mf#1986-2596 – us ATLA [470]

Assyrisch-babylonische mythen und epen / Jensen, Peter – Berlin: Reuther & Reichard 1900 [mf ed 1989] – (= ser Keilinschriftliche bibliothek 6/1) – 2mf – 9 – 0-7905-2718-9 – mf#1987-2718 – us ATLA [470]

Die assyrisch-babylonischen keilinschriften : kritische untersuchung der grundlagen ihrer entzifferung: nebst dem babylonischen texte der trilinguen inschriften in transcription sammt uebersetzung und glossar / Schrader, Eberhard – Leipzig: In Commission bei F A Brockhaus, 1872 – 1mf – 9 – 0-8370-8381-8 – (texts in akkadian and german. incl bibl ref) – mf#1986-2381 – us ATLA [470]

Die assyrische beschwoerungssammlung maql / Meier, G – Berlin, 1937 – (= ser Archiv orientforschung) – 1mf – 9 – (archiv orientforschung, beiheft 2) – mf#NE-468 – ne IDC [956]
Assyrische gebete an den sonnengott fuer staat und koenigliches haus : aus der zeit asarhaddons und asurbanipals / ed by Knudtzon, J A – Leipzig: Eduard Pfeiffer 1893 [mf ed 1986] – 2v on 2mf – 9 – 0-8370-7074-0 – (incl bibl ref & glos. filmed by idc: 8mf order#NE-20013) – mf#1986-1074 – us ATLA; ne IDC [470]
Der assyrische gott / Tallqvist, K – (= ser Studia Orientalia) – 3mf – 9 – (studia orientalia, helsingforsiae 1932. v4) – mf#NE-410 – ne IDC [956]
Assyrische grammatik : mit uebungsstuecken und kurzer literatur-uebersicht / Delitzsch, Friedrich – 2. aufl. Berlin: Reuther & Reichard; New York: Lemcke & Buechner 1906 [mf ed 1986] – (= ser Porta linguarum orientalium 10) – 1mf – 9 – 0-8370-8498-9 – (discussion in german; exercises in akkadian) – mf#1986-2498 – us ATLA [470]
Assyrische jagden : auf grund alter berichte und darstellungen / Meissner, Bruno – Leipzig: J C Hinrichs 1911 [mf ed 1989] – (= ser Der alte orient 13/2) – 1mf – 9 – 0-7905-2030-3 – (incl bibl ref) – mf#1987-2030 – us ATLA [930]
Assyrische lesestuecke : mit den elementen der grammatik und vollstaendigem glossar / Delitzsch, Friedrich – 5th rev ed. Leipzig: J C Hinrichs 1912 [mf ed 1986] – (= ser Assyriologische bibliothek 16) – 1mf – 9 – 0-8370-8978-6 – mf#1986-2978 – us ATLA [470]
Assyrische rechtsurkunden / David, M & Ebeling, E – Stuttgart, 1929 – 1mf – 9 – mf#NE-420 – ne IDC [930]
Assyrische thiernamen : mit vielen excursen und einem assyrischen und akkadischen glossar / Delitzsch, Friedrich – Leipzig: J C Hinrichs 1874 [mf ed 1986] – 1mf – 9 – 0-8370-9054-7 – (in german, akkadian, sumerian; incl bibl ref) – mf#1986-3054 – us ATLA [590]
Die assyrische verbtafel : die assyrische zeichenordnung auf grund von sa und v. rawl. 45 / Peiser, Felix Ernst – Muenchen: F Straub, 1886 – 1mf – 9 – 0-8370-8850-X – mf#1986-2850 – us ATLA [470]
Das assyrische weltreich im urteil der propheten / Staerk, Willy – Goettingen: Vandenhoeck und Ruprecht, 1908 – 1mf – 9 – 0-8370-5356-0 – (incl ind of biblical texts cited) – mf#1985-3356 – us ATLA [221]
Assyrisches beamtentum nach briefen aus der sargonidenzeit / Klauber, E – Leipzig, 1910 – (= ser Leipziger semitistische Studien) – 2mf – 9 – (leipziger semitistische studien, leipzig 1914 v5 pt3) – mf#NE-20117 – ne IDC [956]
Assyrisches handwoerterbuch / Delitzsch, Friedrich – Leipzig: J C Hinrichs; Baltimore: Johns Hopkins 1896 [mf ed 1986] – 2mf – 9 – 0-8370-8979-4 – (supersedes: assyrisches woerterbuch zur gesamten bisher veroeffentlichten keilschriftliteratur 1887-90, of wh only 3 fasc wer publ; iss in pts. filmed by idc: 8mf order#ne-473) – mf#1986-2979 – us ATLA; ne IDC [470]
Assyrisches syllabar : fuer den gebrauch in seinen vorlesungen / ed by Schrader, Eberhard – Berlin: Buchdr der koenigl Akademie der Wissenschaften 1880 [mf ed 1986] – 1mf – 9 – 0-8370-8616-7 – (in german & akkadian) – mf#1986-2616 – us ATLA [470]
Assyrisches und talmudisches : kulturgeschichtliche und lexikalische notizen / Pick, Hermann – Berlin: S Calvary 1903 [mf ed 1985] – 1r – 1 – 0-8370-4741-2 – mf#1985-2741 – us ATLA [470]
Astarloa y Aguirre, Pablo Pedro de *see* Apologia de la lengua bascongada
Astarte : ein beitrag zur mythologie des orientalischen alterthums / Mueller, Alois – Wien: Aus der KK Hof- und Staatsdruckerei 1861 [mf ed 1991] – 1mf – 9 – 0-524-01853-7 – mf#1990-2688 – us ATLA [290]
Astbury, st mary – (= ser Cheshire monumental inscriptions) – 4mf – 9 – £5.00 – mf#3 – uk CheshireFHS [929]
Astell, Mary *see* The christian religion as profes'd by a daughter of the church of england
Aster, Ernst von *see* Goethes faust
Asterius of Amasea, Bishop of Amasea *see* Ancient sermons for modern times
O asteroide : orgam d'instruccao e defeza do povo – Cachoeira, BA: [s.n.] 23 set 1887-13 maio 1889 – (= ser Ps 19) – mf#P18B,02,44 – bl Biblioteca [079]
Astghaditaran, Byurakani *see* Soobshcheniia biurakanskoi observatorii. byurakani astghaditarani haghordumner
The asthmatic athlete : metabolic and ventilatory responses during exercise with and without pre-exercise medication / Ienna, Tiziana M – 1994 – 2mf – $8.00 – us Kinesology [612]
Astianatte : [opera rappresentata l'anno 1741 a torre argentina] / Jommelli, Nicolo – [178-?] [mf ed 1988] – 1r – 1 – (text in italian) – mf#pres. film 15 – us Sibley [780]
Asticou : organe de la societe historique de l'ouest du – 1968 jun 24-1985 jul – 1 – mf#296958 – us WHS [071]
Astie, Jean-Frederic *see*
– Les deux theologies nouvelles dans le sein du protestantisme francais
– Explication de l'evangile selon saint jean
– Histoire de la republique des etats-unis depuis l'etablissement des premieres colonies jusqu' a l'election du president lincoln
– La theologie allemande contemporaine
Astilla, Michael J *see* Kinesthetic sense and consistency in multijoint movement sequences
Astley, Hugh John Dukinfield *see* Prehistoric archaeology and the old testament
Astley, Hugh John Dunkinfield *see* Biblical anthropology compared with and illustrated
Astley, T *see* A new general collection of voyages and travel
Astm bulletin – Conshohocken PA 1921-60 – 1 – mf#976 – us UMI ProQuest [620]
Astolfi, G-F *see* Scelta curiosa
Aston, Louise *see* Freischaerler-reminiscenzen
Aston, Thomas H *see* John rogers
Aston, William George *see* Shinto, the way of the gods
Aston-by-sutton, st peter – (= ser Cheshire monumental inscriptions) – 2mf – 9 – £4.00 – mf#5 – uk CheshireFHS [929]
Astonishing *see* Marvel boy / astonishing
Astor, Nancy *see* Women, politics and welfare
Astoria daily budget – Astoria OR: Budget Pub Co, -1914 [daily ex sun] – 1 – (cont by: astoria evening budget (1914-30); related to: astoria weekly budget) – us Oregon Lib [071]
Astoria daily budget *see* Astoria weekly budget
Astoria daily news – Astoria OR: [s.n.] [daily ex sun] – 1 – (ceased in 1903; absorbed by: morning astorian (1899-1930)) – us Oregon Lib [071]
Astoria evening budget – Astoria OR: [s.n.], 1914-1930 [daily ex sun] – 1 – (related to: astoria weekly budget. cont: astoria daily budget (-1914). merged with: morning astorian (1899-1930) to form: evening astorian-budget) – us Oregon Lib [071]
Astoria evening budget *see* Astoria weekly budget
Astoria herald – Astoria OR: Herald Pub Co [wkly] – 2r – 1 – us Oregon Lib [071]
Astoria oder geschichte einer handelsexpedition jenseits der rocky mountains / Irving, Washington – Stuttgart [Germany]: J G Cotta, 1838 [mf ed 1984] – 5mf – 9 – 0-665-45446-5 – (trans fr english. original iss in ser: reisen und landerbeschreibungen der alteren und neusten zeit) – mf#45446 – cn Canadiana [917]
Astoria weekly budget – Astoria OR: [s.n.] [wkly] – 1 – (related to: astoria daily budget (-1914); astoria evening budget (1914-30)) – us Oregon Lib [071]
Astoria weekly budget *see* Astoria daily budget
Astorian – Astoria OR: J S Dellinger Co, [wkly] – 1 – (related to: morning astorian (1899-1930)) – us Oregon Lib [071]
Astorian-budget – Astoria OR: Astorian-Budget Pub Co, 1960 [daily ex sun & hols] – 1 – (related to: weekly astorian. cont: evening astorian-budget (1930-60); cont by: daily astorian evening budget (1961-61)) – us Oregon Lib [071]
Astorian-budget *see* Weekly astorian (astoria, or)
Astorpsposten skane-smaland – Angelholm, Sweden. 1899-1901 – 2r – 1 – sw Kungliga [078]
Astouacasounc matean hin eu nor ktakaranac est csgrit thargmanouthean nahneac meroc i hellenakann hauatarmagojn bnagre i hajkakans barbar : norogapes i lojs encajeal jentrelagojn grcagir galaphare hamematoutheamb ajl eu gjlorinakac... – Venedig: Lazar 1805 – (= ser Whsb) – 10mf – 9 – €90.00 – (die heilige schrift d. alten u. neuen testaments nach d. genauen uebersetzung unserer vorvaeter aus d. griechischen, getreuen originale ins armenische [armen.]) – mf#Hu 230 – gw Fischer [490]
Astounding science fiction – Norwalk CT 1930-60 – 1 – mf#6220 – us UMI ProQuest [410]
Astra : roman / Sylva, Carmen – Bonn: E Strauss 1886 [mf ed 1989] – 1r – 1 – (filmed with: es klopft & other titles) – mf#7214 – us Wisconsin U Libr [830]
Astrakhanskie gubernskie vedomosti – Astrakhan', Russia 1856-75 [mf ed Norman Ross] – 12r – 1 – mf#nrp-212 – us UMI ProQuest [077]
Astrakhanskii vestnik – Astrakhan', Russia 1889-92 [mf ed Norman Ross] – 1 – mf#nrp-213 – us UMI ProQuest [077]
The astral plane : its scenery, inhabitants and phenomena / Leadbeater, Charles Webster – 4th rev ed. London: Theosophical Pub Society, 1905 – (= ser Theosophical Manuals (London, England)) – 1mf – 9 – 0-524-02310-7 – mf#1990-2933 – us ATLA [290]
Astral projection – Santa Fe NM 1968-72 – 1 – ISSN: 0004-6116 – mf#7317 – us UMI ProQuest [130]
Astral projection – v4 n1=13 [1971 dec 15] – 1 – mf#1582965 – us WHS [071]
Die astralmythologische weltanschauung und das alte testament / Wilke, Fritz – Berlin: Edwin Runge 1907 [mf ed 1989] – (= ser Biblische zeit- und streitfragen 3/10) – 1mf – 9 – 0-7905-2339-6 – mf#1987-2339 – us ATLA [221]
Astrana Marin, Luis *see* Cervantines
Astray – Toronto: W Chewett, [18–] [mf ed 1980] – 1mf – 9 – 0-665-02454-1 – mf#02454 – cn Canadiana [810]
Astrea / Rivera Hernandez, Alejandro – Mexico City?, Mexico. 1963 – 1r – us UF Libraries [972]
Astreia : semanario imparcial – Itauna, MG. 31 maio 1896 – (= ser Ps 19) – bl Biblioteca [079]
Astrid Nypan, Astrid *see* Market trade
Astro do seculo – Sao Joao del Rei, MG: [s.n.] 17, 31 ago 1893 – (= ser Ps 19) – mf#P11B,03,84 – bl Biblioteca [079]
Astrodynamics 1983 [aasms45] – 1984 – (= ser Aasms 1968) – 33papers on 13mf – 9 – $40.00 – 0-87703-192-4 – (suppl to v54, advances) – us Univelt [629]
Astrodynamics 1985 [aasms51] – 1986 – (= ser Aasms 1968) – 55papers on 22mf – 9 – $60.00 – 0-87703-247-5 – (suppl to vol 58, advances) – us Univelt [629]
Astrodynamics 1987 [aasms55] – 1988 – (= ser Aasms 1968) – 48papers on 20mf – 9 – $70.00 – 0-87703-287-4 – (suppl to vol 65, advances) – us Univelt [629]
Astrodynamics 1989 [aasms59] – 1990 – (= ser Aasms 1968) – 25papers on 12mf – 9 – $50.00 – 0-87703-319-6 – (suppl to vol 71, advances) – us Univelt [629]
Astrodynamics 1991 [aasms63] – 1992 – (= ser Aasms 1968) – 29papers on 14mf – 9 – $75.00 – 0-87703-348-X – (suppl to vol 76, advances) – us Univelt [629]
Astrodynamics 1993 [aasms69] – 1994 – (= ser Aasms 1968) – 9papers on 5mf – $30.00 – 0-87703-381-1 – (suppl to vol 85, advances) – us Univelt [629]
Astrodynamics 1995 [aasms72] – 1996 – (= ser Aasms 1968) – 6papers on 4mf – $15.00 – 0-87703-408-7 – (suppl to vol 90, advances) – us Univelt [629]
Astrodynamics specialist conference [aasms7] – 1968 – (= ser Aasms 1968) – 69 papers on 21mf – 9 – $30.00 – 0-87703-227-0 – us Univelt [629]
Astrodynamics specialist conference [aasms20] – 1971 – (= ser Aasms 1968) – 91papers on 66mf – 9 – $70.00 – 0-87703-237-8 – us Univelt [629]
L'Astrolabe et la Zelee *see* Voyage au pole sud et dan l'oceanie sur les corvettes l'astrolabe et la zelee
An astrologer's day and other stories / Narayan, R K – London: Eyre & Spottiswoode, 1947 – (= ser Samp: indian books) – us CRL [130]
Astrologica et divinatoria : formae tplila 134 / Hermes Trismegistus – [mf ed 2003] – (= ser ILL – ser a; Cccm 144c) – 3mf+ix/35p – 9 – €33.00 – 2-503-64446-5 – be Brepols [400]
Astrological bulletina – 1919 apr – 1 – mf#3910367 – us WHS [130]
Astrological-astronomical texts – Leipzig: J C Hinrichs 1899 [mf ed 1986] – 1mf – 9 – 0-8370-7053-8 – mf#1986-1053 – us ATLA [520]
Astrologische bibliothek – Leipzig: Theosophisches Verlagshaus [1910- [v1-21 1919-27] (annual) [mf ed 1978] – 2r – 1 – mf#film mas c640 – us Harvard [130]
Astrology and religion among the greeks and romans / Cumont, Franz Valery Marie – New York: G P Putnam 1912 [mf ed 1989] – (= ser American lectures on the history of religions 9th ser) – 1mf – 9 – 0-7905-4273-0 – (incl bibl ref) – mf#1988-0273 – us ATLA [250]
The astrology of personality / Rudhyar, Dane – A reformulation of astrological concepts and ideals, in terms of contemporary psychology and philosophy. Garden City, NY: (Doubleday, 1970). xviii,500p. illus – 1 – us Wisconsin U Libr [150]
Astrology-your daily horoscope – New York NY 1975-80 – 1,5,9 – ISSN: 0195-0851 – mf#10086 – us UMI ProQuest [130]
Astronautica acta – Oxford UK 1955-73 – 1,5,9 – (cont by: acta astronautica) – ISSN: 0004-6205 – mf#49492 – us UMI ProQuest [520]
Astronautics – Reston VA 1957-63 – 1 – ISSN: 0097-7152 – mf#5077 – us UMI ProQuest [629]
Astronautics and aeronautics – Reston VA 1963-83 – 1,5,9 – (cont by: aerospace america) – ISSN: 0004-6213 – mf#1605.01 – us UMI ProQuest [629]
Astronautics international [aasms6] – 1968 – (= ser Aasms 1968) – 16 papers on 10mf – 9 – $15.00 – 0-87703-226-2 – us Univelt [629]
Astronomia e geodesia / Armellini, Giuseppe – Varese: V Bompiani 1941 [mf ed 2004] – (= ser Enciclopedia scientifica monografica italiana del ventesimo secolo. serie 1 5) – 1r – 1 – (incl ind; fr: "bibliografia astronomica e geodetica italiana degli ultimi trenta anni" p[267]-292) – mf#film mas 36158 – us Harvard [520]
Astronomical and magnetical and meteorological observations made at the royal observatory, greenwich, in the year... / Airy, George Biddell et al – London: J Barker 1840-1926 (annual) [mf ed 2000] – 1839-1924 on 87r [ill] – 1 – (cont: astronomical observations made at the royal observatory at greenwich...; cont by: observations made at the royal observatory, greenwich in the year...in astronomy, magnetism and meteorology; aka: greenwich astronomical observations, greenwich observations) – mf#film mas c4560 – us Harvard [520]
Astronomical and magnetical and meteorological observations made at the royal observatory, greenwich, in the year ... 1868 [supplement 1] *see* On the corrections of bouvard's elements of the orbits of jupiter and saturn
Astronomical and meteorological observations made at the sydney observatory in the year... / Scott, W – Sydney: T Richards, Govt Printer 1861- (annual) [mf ed 2001] – 1860 on 1r – 1 – (cont: astronomical observations made at the sydney observatory in the year...) – mf#film mas c5056 – us Harvard [520]
Astronomical and meteorological observations made at the united states naval observatory during the year... – Washington: GPO 1862-85 (annual) [mf ed 1998] – 23v on 18r [ill] – 1 – (cont: astronomical observations made..., issn 1061-2742; cont by: observations [united states naval observatory]; vols for 1853-60 not iss except in a combined vol with title: results of observations made at the united states naval observatory with the transit instrument and mural circle; title varies slightly) – ISSN: 1061-3110 – mf#film mas c4123 – us Harvard [520]
Astronomical circular [tashkent, uzbekistan] / Toshkent astronomik observatoriiasi – Tashkent: The Observatory 1932-[1946] [irreg] [mf ed 2003] – 1r – 1 – (cont by: tsirkuliar taskentkoi astronomicheskoi observatorii; chiefly in english with some russian; with ind) – mf#film mas c5598 – us Harvard [520]
Astronomical instruments in the delhi museum / Kaye, George Rusby – Calcutta: Supt, Govt Print, India, 1921 – (= ser Samp: indian books) – us CRL [520]
Astronomical journal – Chicago IL 1995+ – 1,5,9 – ISSN: 0004-6256 – mf#26918 – us UMI ProQuest [520]
The astronomical journal – v1-. 1849- – 1,5,6 – us AIP [520]
Astronomical, magnetic and meteorological observations made during the year... : at the united states naval observatory – Washington: GPO 1895-99 (annual) [1890-1892] [mf ed 1998] – 3v on 2r [ll] – 1 – (some app have added title: washington observations) – ISSN: 1061-3137 – mf#film mas c4125 – us Harvard [520]
Astronomical observations and researches made at dunsink, the observatory of trinity college, dublin – Dublin: Hodges, Foster & Co 1870-1900 (irreg) [mf ed 1999] – 9v on 1r [ill] – 1 – mf#film mas c4258 – us Harvard [520]
Astronomical observations made... / United States Naval Observatory – Washington DC: publ by authority of the Secretary of the Navy 1846-67 (annual) [v1 (1845)-1851/1852] [mf ed 1998] – 6v on 3r [ill] – 1 – (iss for 1845-1849/50 called v1-5; title varies slightly) – ISSN: 1061-2742 – mf#film mas c4122 – us Harvard [520]
Astronomical observations made at the honorable, the east india company's observatory at madras in the years... / =Taylor, Thomas Glanville et al – Madras: Printed by order of the Madras Govt, at the American Mission & C K S Presses 1844-54 (irreg) [mf ed 2000] – [v6-8](1830/43-1848/52) on 1r – 1 – (cont: result of astronomical observations made at the honorable, the east india company's observatory at madras) – mf#film mas c4642 – us Harvard [520]
Astronomical observations made at the royal observatory at greenwich... / Bradley, James et al – Oxford: Clarendon Press 1776-1840 (irreg, qrtly, annual) [mf ed 2000] – 1750/1762-1838 on 12r [ill] – 1 – (cont by: astronomical and magnetical and meteorological observations made at the royal observatory, greenwich, in the year...; some vols iss out of chronological sequence; aka: greenwich astronomical observations, greenwich observations) – mf#film mas c4559 – us Harvard [520]

Astronomical observations made at the sydney observatory in the year... / Scott, W – Sydney: T Richards, Govt Printer -1860 (annual) [mf ed 2001] – 1859 on 1r [ill] – 1 – (cont by: astronomical and meteorological observations made at the sydney observatory in the year...) – mf#film mas c5056 – us Harvard [520]

Astronomical observations made at the university observatory, oxford / Pritchard, Charles – Oxford: Clarendon Press 1878- (irreg) [mf ed 2000] – n1-4(1878-92) on 1r – 1 – (v2 lacks numbering; v3-4 called fasc 3-4) – mf#film mas c4555 – us Harvard [520]

Astronomical observations made at the williamstown observatory in the years... / Williamstown Observatory [Williamstown, Victoria] – Melbourne: J Ferres, govt printer 1869 (irreg) [mf ed 2000] – 1861/1863 on 1r – 1 – (cont by: results of astronomical observations made at the melbourne observatory in...) – mf#film mas c4488 – us Harvard [520]

The astronomical observatories of jai singh / Kaye, George Rusby – Calcutta: Supt, Govt Print, 1918 – (= ser Samp: indian books) – us CRL [520]

Astronomical Observatory of Harvard College *see* Miscellaneous papers

Astronomical register : a medium of communication for amateur observers and all others interested in the science of astronomy – London: J D Potter 1864-86 (mthly) [mf ed 2000] – v1-24(1863-86) on 4r – 1 – (with ind) – mf#film mas c4470 – us Harvard [520]

Astronomical Society of South Africa *see* Monthly notes [astronomical society of south africa]

Astronomical Society of South Africa, Cape Centre *see* Monthly notes [astronomical society of south africa. cape centre]

Astronomical Society of South Africa. Cape Centre *see* Astronomical society of south africa, cape centre

Astronomical society of south africa, cape centre : [monthly notes] / Astronomical Society of South Africa. Cape Centre – [Cape of Good Hope]: Astronomical Society of South Africa, Cape Centre [1940] [mthly] [mf ed 2003] – 7v on 1r – 1 – (cont by: monthly notes [astronomical society of south africa. cape centre]; with ind) – mf#film mas c5541 – us Harvard [520]

Astronomical society of the pacific. publications – Chicago IL 1988+ – 1,5,9 – ISSN: 0004-6280 – mf#14275 – us UMI ProQuest [520]

Astronomicheskaia Engel'gardtovskaia observatoriia *see* Biulleten' astronomicheskoi observatorii im vp engel'gardta

Astronomicheskii tsirkuliar = Astronomical circulars / Akademiia nauk SSSR. Astronomicheskii institut – [Leningrad]: Astronomicheskii institut Akademii nauk SSSR: Vsesoiuznoe astronomo-geodezicheskoe obshchestvo pri AN SSSR [1940]- [irreg] [mf ed 2003] – 1r – 1 – (publ: [kazan']: biuro astronomicheskikh soobshchenii akademii nauk sssr [1943 feb 12]-[1954 iiun' 11]; moskva: uop ila an sssr [dec 1989-]; with ind) – ISSN: 0373-191X / 0236-2457 – mf#film mas c5594 – us Harvard [520]

Astronomicon liber 5 / Manilius, Marcus – Parisiis: Edibus Serpentinis 1786 [mf ed 1998] – 2v on 1r – 1 – (int in french; latin & french parallel text) – mf#film mas 28300 – us Harvard [520]

Astronomie : les astres, l'univers / Rudaux, Lucien et al – Paris: Larousse c1948 [mf ed 1998] – (= ser Coll in-quarto larousse) – 1r [ill] – 1 – mf#film mas 28404 – us Harvard [520]

Astronomie indienne d'apres la doctrine et les livres anciens et modernes des brammes sur l'astronomie, l'astrologie et la chronologie : suivie de l'examen de l'astronomie des anciens peuples de l'orient et de l'explication des principaux monuments astronomico-astrologiques de l'egypte et de la perse / Guerin, J M F – Paris: Impr Royale 1847 [mf ed 1998] – 1r [pl/ill] – 1 – mf#film mas 28218 – us Harvard [520]

Astronomische abhandlungen der hamburger sternwarte / Hamburger Sternwarte – Hamburg: Sternwarte 1909-68. 7v (irreg) [mf ed 2000] – v1-v5 n11(1909-62) on 2r [ill] – 1 – (cont by: abhandlungen aus der hamburger sternwarte; some iss publ out of chronological sequence; v1-5 n7 have title: astronomische abhandlungen der hamburger sternwarte in bergedorf) – mf#film mas c4619 – us Harvard [520]

Astronomische beobachtungen angestellt auf der k sternwarte zu bogenhausen / K Sternwarte [Munich, Germany] – Muenchen: F S Huebschmann [1835?]-1838 (irreg) [mf ed 2000] – t1-5(1821-27) on 1r – 1 – (cont by: observationes astronomicae in specula regia monachiensi; iss under following name variants: k sternwarte zu bogenhausen 1820/1821; by: koenigl sternwarte zu bogenhausen 1822-1827) – mf#film mas c4508 – us Harvard [520]

Astronomische beobachtungen auf der koeniglichen sternwarte zu berlin / Koenigliche Sternwarte zu Berlin – Berlin: F. Duemmler 1840- (irreg) [v1-5; 2.ser: v1-3 (1840-1904)] [mf ed 1999] – 4r [ill] – 1 – (companion publ to: beobachtungs-ergebnisse der koeniglichen sternwarte zu berlin; absorbed by: veroeffentlichungen der koeniglichen sternwarte zu berlin-babelsberg; suspended 1858-83, 1893-1903; ed by johann franz encke 1840-1848; wilhelm julius foerster 1884) – mf#film mas c4283 – us Harvard [520]

Astronomische mittheilungen von der koeniglichen sternwarte zu goettingen / Koenigliche Sternwarte zu Goettingen et al – Goettingen: A Rente 1869-1919 (irreg) [mf ed 1999] – t1-20(1869-1919) on 2r [ill] – 1 – (cont by: veroeffentlichungen der universitaets-sternwarte zu goettingen; imprint varies) – mf#film mas c4492 – us Harvard [520]

Astronomische tafeln zur bestimmung der zeit : aus der beobachteten gleichen obwohl unbekannten hoehe zweyer fixsterne: vorzueglich zum nutzen der schiffahrt / Koch, Julius August – Berlin, Stralsund: Bei Gottlieb August Lange 1797 [mf ed 1998] – 1r – 1 – (incl: ein anhang zu bodens astronomischen jahrbuch fuer 1799) – mf#film mas 28219 – us Harvard [520]

Astronomisches aus babylon : oder, das wissen der chaldaeer ueber den gestirnten himmel / Epping, Joseph – Freiburg i.B, St Louis: Herder 1889 [mf ed 1986] – 1mf – 9 – 0-8370-7060-0 – (incl bibl ref) – mf#1986-1060 – us ATLA [520]

Astronomisches Rechen-Institut zu Berlin-Dahlem *see* Veroeffentlichungen des astronomischen rechen-instituts zu berlin-dahlem

Astronomiska iakttagelser och undersoekningar...anstaelda pa stockholms observatorium / Stockholms observatorium – Stockholm: Kongl boktryckeriet. 13v. -1934 (irreg) [mf ed 1999] – v1-11(1880-1934) on 4r [ill] – 1 – (lacks v2; cont by: stockholms observatoriums annaler; in swedish & french; imprint varies; publ: almqvist & wiksell boktryckeri 1908-1934) – mf#film mas c4317 – us Harvard [520]

Astronomska opservatorija u Beogradu *see*
- Bulletin de l'observatoire astronomique de beograd
- Memoires [Astronomska opservatorija u Beogradu]

Astronomy / Fisher, Clyde et al – New York: J Wiley & sons; London: Chapman & Hall c1940 [mf ed 1998] – (= ser The sciences, a survey course for colleges) – 1r [ill] – 1 – (incl bibl ref & ind) – mf#film mas 28211 – us Harvard [520]

Astronomy – Waukesha WI 1973+ – 1,5,9 – ISSN: 0091-6358 – mf#11470 – us UMI ProQuest [520]

Astronomy and astro-physics – Killen TX 1882-94 – 1 – mf#2955 – us UMI ProQuest [520]

Astronomy and astrophysics – Heidelberg, Germany 1974-97 – 1,5,9 – ISSN: 0004-6361 – mf#13106 – us UMI ProQuest [520]

Astronomy and general physics considered with reference to natural theology / Whewell, William – London, 1833 – (= ser 19th c evolution & creation) – 5mf – 9 – mf#1.1.10850 – uk Chadwyck [210]

Astronomy and geophysics – Oxford UK 2001+ – 1,5,9 – (cont: quarterly journal of the royal astronomical society) – ISSN: 1366-8781 – mf#15540,01 – us UMI ProQuest [520]

Astronomy and meteorology – Montreal: W H Smith. n1 apr 1887-n7 oct 1887 (mthly) [mf ed 1989] – 6mf – 9 – mf#P04080 – cn Canadiana [520]

Astronomy in florida – s.l, s.l? . 193-? – 1r – us UF Libraries [520]

Astronomy, in infancy, youth and maturity / Harvey, Arthur – [Toronto? s.n, 1900?] – 1mf – 9 – 0-665-91570-5 – mf#91570 – cn Canadiana [520]

Astronomy in the old testament = Astronomia nell'antico testamento / Schiaparelli, Giovanni Virginio – Oxford: Clarendon Press 1905 [mf ed 1985] – 1mf – 9 – 0-8370-5088-X – (english ed with many corr & additions by aut; incl bibl ref) – mf#1985-3088 – us ATLA [520]

Astronomy letters – v1- 1975- – 1,5,6 – us AIP [520]

The astronomy of the bible / Mitchel, Ormsby MacKnight – New York:Blakeman and Mason, 1863 – 1mf – 9 – 0-8370-4450-2 – mf#1985-2450 – us ATLA [210]

[Astronomy pamphlets] / Herschel, William – 1788-1805 [mf ed 1998] – 6 pcs on 1r – 1 – (coll of off-prints fr: philosophical transactions iss by the royal astronomical society [london] in 1788-1805) – mf#film mas 28422 – us Harvard [520]

Astronomy quarterly – Oxford UK 1988-91 – 1,5,9 – ISSN: 0364-9229 – mf#49585 – us UMI ProQuest [520]

Astrophysica norvegica / ed by Universitetet i Oslo. Institutt for teoretisk astrofysikk [Institute of Theoretical Astrophysics, University of Oslo] – Oslo: Universitetsforlaget 1936- (irreg) [mf ed 2000] – v1-5(1934-57) on 1r [ill] – 1 – (in norwegian, english, french & german) – ISSN: 0067-0030 – mf#film mas c4656 – us Harvard [520]

Astrophysical journal – Chicago IL 1895+ – 1,5,9 – ISSN: 0004-637X – mf#134 – us UMI ProQuest [520]

The astrophysical journal : supplement series / American Astronomical Society – Chicago: Publ for the American Astronomical Society. v1-mar 1954- [mthly] – 9 – (with ind) – ISSN: 0004-637X – us Chicago U Pr [520]

Astrophysics – Dordrecht, Netherlands 1965+ – 1,5,9 – ISSN: 0571-7256 – mf#10901 – us UMI ProQuest [520]

Astrophysics and space science – Dordrecht, Netherlands 1989+ – 1,5,9 – ISSN: 0004-640X – mf#14740 – us UMI ProQuest [520]

Astrophysikalisches Observatorium Potsdam *see* Publicationen des astrophysikalischen observatoriums zu potsdam

Astrov, N I et al *see* Zakonodatelnye proekty i predlozheniia partii narodnoi svobody:

Astrum alberti – Belleville [Ont]: Students of Albert College, [1883?-18– or 19–] [mf ed v1 n1 jan 1883-v1 n6 jun 1883] – 9 – mf#P04389 – cn Canadiana [378]

Astuatsashunch girk hnots ew norots ktakaranats – 1817 – 20mf – 9 – mf#AR-1461 – ne IDC [243]

Astuatsashunch hnots ew norots ktakaranats – K Polis, 1705 – 13mf – 9 – mf#AR-1460 – ne IDC [243]

Asturias, Francisco *see* Belice

Asturias, Miguel Angel *see*
- Alhajadito
- Audiencia de los confines
- Leyendas de guatemala
- Mulata de tal
- Papa verde
- Poesia
- Senor presidente
- Teatro

Astwick – (= ser Bedfordshire parish register series) – 1mf – 9 – £3.00 – uk BedsFHS [929]

Astwick, st guthlac monumental inscriptions – Bedfordshire Family HS 1985 – (= ser Bedfordshire parish register series) – 1mf – 9 – £1.25 – uk BedsFHS [929]

[Asuncion-] al cor – PY. 1964-69 – 1r – 1 – $50.00 – mf#R63578 – us Library Micro [079]

[Asuncion-] la republica – PY. oct-dec 1981 – 1r – 1 – $50.00 – mf#R63619 – us Library Micro [079]

El asunto de plasencia / Diaz de la Cruz, Felipe – 1888 – 9 – sp Bibl Santa Ana [946]

The asuri-kalpa : a witchcraft practice of the atharva-veda / Atharvaparisishta – Baltimore: I Friedenwald, 1889 – 1mf – 9 – 0-524-07492-5 – mf#1991-0113 – us ATLA [280]

Asvaghosa / Law, Bimala Churn – Calcutta: Royal Asiatic Society of Bengal, 1946 – (= ser Samp: indian books) – us CRL [490]

Asvaghosa *see* Acvaghosha's discourse on the awakening of faith in the mahaayaana

Aswegen, Suzanne L van *see* Success criteria of facilitators involved in community-based satellite distance education

Asylum for the Colored Insane of North Carolina *see* Report of the board of directors and superintendent of the asylum for the colored insane of north carolina

Asymmetrische 1,3-dipolare cycloadditionen und hetero-diels-alder-reaktionen unter verwendung von (s)-prolinestern als chirale auxiliare / Blaeser, Edwin – (mf ed 1996) – 2mf – 9 – €40.00 – 3-8267-2388-0 – mf#DHS 2388 – gw Frankfurter [540]

At a court of general sessions of the peace holden at the session-house in the city of montreal...tenth day of january, one thousand, eight hundred and nine... = A une cour des sessions generales de la paix tenue a la maison d'audience dans la ville de montreal...dixieme jour de janvier, mil huit cent neuf... – [Montreal?: s.n, 1809?] [mf ed 1983] – 1mf – 9 – 0-665-42552-X – mf#42552 – cn Canadiana [343]

At a special meeting of the emigrants' society : held in the grand jury room of the court-house at quebec, the 11th oct 1819... / Quebec Emigrant Society – [Quebec?: s.n, 1819?] [mf ed 1993] – 1mf – 9 – 0-665-91314-1 – mf#91314 – cn Canadiana [320]

AT and T Bell Laboratories *see* Record

At and t -selected commission decisions : i.c.c. activities affecting telecommunications – American Telephone and Telegraph Co.-Legal department. nos 1-254 46 bks. 1912-16 (all publ) – 229mf – 9 – $1030.00 – mf#LLMC 84-403 – us LLMC [380]

At grips : talks with the telugus of south india / Goffin, Herbert J – [London]: London Missionary Society, 1913 [mf ed 1995] – (= ser Yale coll) – 153p (ill) – 1 – 0-524-09022-X – mf#1995-0022 – us ATLA [954]

At home and abroad : a description of the english and continental missions of the london society for promoting christianity amongst the jews / Gidney, Williams Thomas – London: Operative Jewish Converts' Institution, 1900 [mf ed 1986] – 1mf – 9 – 0-8370-7142-9 – mf#1986-1142 – us ATLA [240]

At home and abroad : a magazine of home and foreign missions for young helpers in the work – London: Wesleyan Mission-House 1879-1914 (qrterly, mthly) [mf ed 2003] – 4r – 1 – (began in 1879; latest iss consulted: spring 1973; several iss lacking) – mf#2003-s075 – us ATLA [242]

At home with god : priedieu papers on spiritual subjects / Russell, Matthew – London, New York: Longmans, Green 1910 [mf ed 1986] – 1mf – 9 – 0-8370-7013-9 – mf#1986-1013 – us ATLA [241]

At home with the patagonians : a year's wanderings over untrodden ground from the straits of magellan to the rio negro / Musters, George Chaworth – London: John Murray 1871 [mf ed 1988] – 1r [ill] – 1 – (filmed with: universal biography / lempriere, j) – mf#2145 – us Wisconsin U Libr [918]

At, Jean Antoine *see* La loi

At kalahari's brink – Plumtree, Zimbabwe. 1964 – 1r – us UF Libraries [960]

At last : a christmas in the west indies / Kingsley, Charles – London, England. 1873 – 1r – us UF Libraries [880]

At last / Kingsley, Charles – London, England. 1889 – 1r – us UF Libraries [880]

At market value : a novel / Allen, Grant – Chicago, New York: F T Neely, 1894 – (= ser Neely's international library) – 4mf – 9 – mf#26646 – cn Canadiana [830]

At michaelmas : a lyric / Carman, Bliss – [S.l: s.n, 1895] [mf ed 1980] – 1mf – 9 – 0-665-00472-9 – mf#00472 – cn Canadiana [780]

At onement : or, reconciliation with god / Workman, George Coulson – New York: Fleming H Revell c1911 [mf ed 1989] – 1mf – 9 – 0-7905-2457-0 – mf#1987-2457 – us ATLA [240]

At our own door : a study of home missions with special reference to the south and west / Morris, Samuel Leslie – New York: Fleming H Revell c1904 [mf ed 1986] – 1mf – 9 – 0-8370-6222-5 – (incl ind) – mf#1986-0222 – us ATLA [242]

At sea and in port : or, life and experience of william s fletcher: for thirty years seaman's missionary in portland, oregon / Fletcher, William S – Portland OR: J K Gill 1898 [mf ed 1992] – 1mf [ill] – 9 – 0-524-04123-7 – mf#1992-2009 – us ATLA [242]

At & t bell laboratories technical journal : a journal of the at&t companies – New Providence NJ 1984 – 1,5,9 – (cont: bell system technical journal; cont by: at and t technical journal) – ISSN: 0748-612X – mf#58.02 – us UMI ProQuest [621]

At & t technical journal – New Providence NJ 1985-96 – 1,5,9 – (cont: at and t bell laboratories technical journal : a journal of the at&t companies) – ISSN: 8756-2324 – mf#58,02 – us UMI ProQuest [621]

At & t technology – Murray Hill NJ 1986-96 – 1,5,9 – (cont: record / at and t bell laboratories) – ISSN: 0889-8979 – mf#15465 – us UMI ProQuest [621]

At the back of the black man's mind : or, notes on the kingly office in west africa / Dennett, Richard Edward – London: Macmillan 1906 [mf ed 1992] – 1mf – 9 – 0-524-03365-X – mf#1990-3199 – us ATLA [305]

At the center – 1991 jun, sep – 1 – mf#4867188 – us WHS [071]

At the crossroads – 1997 winter – 1 – mf#3843831 – us WHS [071]

At the cross-roads, 1885-1946 : the autobiography of nripendra chandra banerji – Calcutta: A Mukherjee & Co, [1950] – (= ser Samp: indian books) – us CRL [920]

At the feet of the master / Krishnamurti, Jiddu – amer ed. Chicago: Rajput Press 1911 [mf ed 1991] – 1mf [ill] – 9 – 0-524-01775-1 – mf#1990-2623 – us ATLA [290]

At the green dragon, a bird of passage, and the umbrella mender / Harradan, Beatrice – Chicago, IL. no date – 1r – us UF Libraries [960]

At the sign of the brush and pen : being notes on some black and white artists of to-day / Reid, J G – Aberdeen: A Brown & Co; Edinburgh...London...1898 – (= ser 19th c art & architecture) – 2mf – 9 – mf#4.1.90 – uk Chadwyck [700]

At the terminals : an act in the holy drama of israelism / Kahan, Louis – Seattle, WA: Beacon Press, c1926 – 1r – 1 – (incl bibl ref) – mf#*ZP-1500 – us NY Public [220]

At work : letters of marie elizabeth hayes, mb, missionary doctor, delhi, 1905-8 – London: Marshall Bros, [1909] [mf ed 1995] – (= ser Yale coll) – xii/263p (ill) – 1 – 0-524-09494-2 – (int by george robert wynne) – mf#1995-0494 – us ATLA [610]
Ata magazine / Alberta Teachers' Association – Edmonton AB 1920+ – 1,5,9 – ISSN: 0380-9102 – mf#10092 – us UMI ProQuest [370]
Ata news – 1983 may 24-1988 dec 5 – 1 – mf#1109423 – us WHS [071]
Ata, Tayyarzade Ahmed *see* Tarih-i 'ata
Atabalipa degli incas : o pizzarro alla scoperta delle indie, ballo storico diviso in 7 quadri del coreografo lodovico pedoni, da rappresentarsi al r. teatro apollo la stagione d'inverno 1871 in 1872 / Pedoni, Lodovico – Roma, Tip. Olivieri [1872?] – mf#*ZBD-*MGTZ pv1-Res – Located: NYPL – us Misc Inst [790]
Atabeyoglu, Salahattin Enis *see* Sara
Ataide, antonio de : viajens do reino para a india... / Alvarez, Arturo – Madrid: Archivo Ibero Americano, 1961 – 1 – sp Bibl Santa Ana [970]
Atalaia – Rio de Janeiro, RJ: Typ Nacional, 31 maio-02 set 1823 – (= ser Ps 19) – mf#P01,04,07 – bl Biblioteca [321]
O atalaia : litterario, critico e noticioso – Natal, RN: Typ Independente, 02 dez 1876 – (= ser Ps 19) – bl Biblioteca [079]
Atalaia da liberdade – Rio de Janeiro, RJ: Typ de Plancher, 04 fev-17 mar 1826 – (= ser Ps 19) – mf#P15,01,68 – bl Biblioteca [321]
Atalanta fugiens : hoc est emblemata nova de secretis naturae chymica... / Maier, M – Oppenheimii: Ex typgraphia Hieronymi Galleri, sumptibus Joh. Theodori de Bry, 1618 – 3mf – 9 – mf#0-357 – ne IDC [090]
El atalaya bautista – 1908-24. Name changes: El Atalaya Bautista, 1908 – May 1910; El Bautista, May 1910 – Dec 1912; El Foro Cristiana, Jan 1915 – May 1915; El Alalaya Bautista, Dec 1917-24 – 1 – us Southern Baptist [242]
Atalaya de la mancha – Ano 1813-1815 (16-VII/26-IV) – 53mf – 9 – sp Cultura [946]
Ataque a manzanillo por dos buques corsarios en el... / Tamayo Y Lastres, Jose – Habana, Cuba. 1909 – 1r – us UF Libraries [972]
[Atascadero-] atascadero news – CA. 1916-24; 1928- – 122r – 1 – $7320.00 (subs $240/y) – mf#BC02024 – us Library Micro [071]
Atbasarskaia zhizn' : Organ bespartiin obshchestv-polit mysli / ed by Belov, S I – Atbasar [Akmol obl]: Izd t-vo "Raketa" 1918-19 [1918 1 dek-1919, 9 avg – (= ser Asn 1-3) – n1 [1918]-n199 [1919] [gaps] item 15, on reel n6 – 1 – mf#asn-1 015 – ne IDC [077]
Atbilde : "pret baptistu sekti" = A reply: "against the sect of the baptists" – 1 – (riga: aleksander stahl, 1882 publ n6298 d. one of five items on a reel) – us Southern Baptist [242]
Atchara Chiwaphan *see* Khumu kanson phasa thai
Atchison County. Kansas. School District 9 *see* Minutes of the annual meetings of the board of trustees. 1861-70
Atchison daily champion – 1885 sep 29-1892 jun 12 [with gaps] – 1 – mf#846326 – us WHS [071]
Atchison. Kansas. Trinity Episcopal Church *see* Records
Atchison, Thomas *see* Petition to the king
Atchison, Topeka and Santa Fe Railroad Company *see*
– Correspondence
– Daily construction journal
– List of buildings of at and sf rr and leased lines
– Payroll records
– Records and correspondence
– Santa fe rate schedules
Atchison, Topeka and Santa Fe Railway Company *see*
– Records
– "Splinters."
Atchley, Edward Godfrey Cuthbert Frederic *see* History of the use of incense in divine worship
Atchley, Edward Godfrey Cuthbert Frederic et al *see* Some principles and services of the prayer-book historically considered
Ateitis – Pittsburgh PA, 1900-01 – 1r – 1 – (lithuanian newspaper) – us IHRC [071]
Ateitis – South Boston, MA: Leidzia Ateities Kooperacijos Draugija, Ink., Petnyciomis, jan-apr 1918 – 15 – 1 – us CRL [071]
L'atelier – Paris. dec 1940-aout 1944 – 1 – fr ACRPP [073]
L'atelier – Paris. mars 1920-23 [wkly] – 1 – fr ACRPP [073]
L'atelier – Paris. sept 1840-juil 1850 – 1 – fr ACRPP [073]
Das atelier – Malmedy (B), 1922/23 1 nov-1923/24 apr – 1 – gw Misc Inst [074]
Atelier national de reproduction des theses de lille – 1971- – 24,500 titles – 9 – (all french dissertations literature, human science, law, politics, sociality, economics) – ISSN: 0 – fr Atelier National [378]
L'atelier pour le plan / Confederation Generale du Travail – Paris. n1-31. mai 1935-37 – 1 – fr ACRPP [330]
Atelier / seikatsu bujutsu : an art journal founded by kanae yamamoto, 1924-1943 – Atelier v1 n1-v18 n8(1941 feb-1941 aug) 218 iss, seikatsu bijutsu v1 n1-v3 n11 (1941 aug-1943 nov) 27 iss – 50r – 1 – Y750,000 – (title change to: seikatsu bijutsu; with 180p guide; in japanese) – ja Yushodo [700]
Atem einer floete : gedichte / Baumann, Hans – 15.-24. tausend. Jena: E Diederichs, 1942 [mf ed 1989] – 61p – 1 – mf#6983 – us Wisconsin U Libr [810]
ATEN *see* Bulletin d'information de l'a.t.e.n
Atenas news – Miami, FL. 1982 nov 10-1993 may 20 – 12r – (gaps) – us UF Libraries [071]
Atencion, guatemala / Heredia, Manuel De – Madrid, Spain. 1962 – 1r – us UF Libraries [972]
Atenea: revista mensual de ciencias, letras y artes – Santiago. Issued by Universidad of Concepcion, Chile. v. 1-92. Apr 1924-Mar 1949 – 1 – us L of C Photodup [800]
Atenei – New York. 1959-1972 (1) 1964-1972 (5) 1970-1972 (9) – 118mf – 9 – mf#1192 – ne IDC [077]
Atenei: istoriko-literaturnyi vremennik – Leningrad. v. 1-3. 1924-1926 – 1 – us NY Public [947]
Ateneo *see*
– Cervantes en el ateneo de badajoz
– Primer concurso fotografico catalogo mayo 1927
Ateneo De La Habana Comite 'Pro-Zenea' *see* Juan clemente zenea
Ateneo Dominicano *see* Album simbolico
'Ateret Hakhamim / Taxin, Menahem Zevi – Warsaw, Poland. 1907 – 1r – 1 – us UF Libraries [939]
'Ateret Ha-Leviyim / Pesis, Phinehas – Varsha, Poland. 1902 – 1r – 1 – us UF Libraries [939]
'Ateret Tif'eret / Schorr, Levi Isaac Dov – Brody, Ukraine. 1906 – 1r – 1 – us UF Libraries [939]
'Ateret Tsevi / Taxin, Menahem Zevi – Vilna, Lithuania. 1910 – 1r – 1 – us UF Libraries [939]
'Ateret Tsevi / Taxin, Menahem Zevi – Warsaw, Poland. 1901 – 1r – 1 – us UF Libraries [939]
'Ateret Yitshak / Savitski, Yitshak Ayzik – Boston, MA. 1916 – 1r – 1 – us UF Libraries [939]
Athalye, D V *see*
– The life of lokamanya tilak
– The life of mahatma gandhi
Athalye, S B *see* Vikramorvasiya
Athanase *see* Chronicles of florida
Athanasian creed : its use in the services of the church / Ommanney, George Druce Wayne – London, England. 1872 – 1r – us UF Libraries [240]
Athanasian creed : a plea for its disuse in the public worship / Lake, John – London, England. 1875 – 1r – us UF Libraries [240]
The athanasian creed : by whom written and by whom published / Ffoulkes, Edmund Salisbury – London: J T Hayes, [1871?] – 1mf – 9 – 0-7905-5466-6 – mf#1988-1466 – us ATLA [240]
The athanasian creed and its early commentaries / Burn, Andrew Ewbank – Cambridge: University Press 1896 [mf ed 1989] – (= ser Texts and studies (cambridge, england) 4/1) – 1mf – 9 – 0-7905-1808-2 – (incl bibl ref & ind) – mf#1987-1808 – us ATLA [240]
The athanasian creed and its early commentaries (ts4/1) / Burn, A E – 1986 – (= ser Texts and studies (ts)) – 3mf – 9 – €7.00 – ne Slangenburg [240]
Athanasian creed and the theology of nature compared / Marsden, T – London, England. 1872 – 1r – us UF Libraries [240]
Athanasian creed vindicated and explained / Dodwell, William – London, England. 1819 – 1r – us UF Libraries [240]
The athanasian origin of the athanasian creed / Brewer, John Sherren – London: Rivingtons, 1872 – 1mf – 9 – 0-7905-4099-1 – mf#1988-0099 – us ATLA [240]
The athanasian warnings / Sparrow-Simpson, William John – London; New York: Longmans, Green, 1911 – 1mf – 9 – 0-524-00139-1 – mf#1989-2839 – us ATLA [240]
Athanasiana : litterar- und dogmengeschichtliche untersuchungen / Stuelcken, Alfred – Leipzig: J C Hinrichs 1899 [mf ed 1989] – (= ser Tugal 19/4) – 1mf – 9 – 0-7905-1796-5 – (in german & greek, also iss in pt under title: beitraege zur athanasius) – mf#1987-1796 – us ATLA [240]
Athanasiana / Stuelcken, A – Leipzig, 1899 – (= ser Tugal 2-19/4) – 3mf – 9 – €7.00 – ne Slangenburg [240]
Athanasii kircheri prodromus coptus sive aegyptiacus : in quo cum linguae coptae sive aegyptiacae, quondan pharaonicae, origo, aetas, vicissitudo, inclinatio, tum hieroglyphicae literaturae instauratio...exhibentur / Kircher, Athanasius – Romae: Ty. Sac Congreg de propag fide 1636 – (= ser Whsb) – 4mf – 9 – €50.00 – mf#Hu 377 – gw Fischer [490]
[Athanasii kircheri]...mvsvrgia vniversalis : sive ars magna consoni et dissoni in 10. libros digesta / Kircher, Athanasius – Romae: ex typographia haeredum Francisci Corbelletti 1650 [mf ed 19–] – 2v on 1r / 22mf – 1,9 – mf#film 95a, 95b / fiche147-148 – us Sibley [780]
Athanasius / Goerres, Joseph von – Regensburg: G J Manz 1838 [mf ed 1990] – 1mf – 9 – 0-7905-6228-6 – mf#1988-2228 – us ATLA [240]
Athanasius : his life and life-work / Reynolds, Henry Robert – London: Religious Tract Soc 1889 [mf ed 1989] – (= ser The church history series 5) – 1mf – 9 – 0-7905-4478-4 – mf#1988-0478 – us ATLA [240]
Athanasius *see*
– The festal letters of athanasius in an ancient syriac version
– Gegen die arianer, 1. bd (bdk13 1.reihe)
– Gegen die heiden / ueber die menschwerdung / leben des hl antonius und pachomius, 2. bd (bdk31 1.reihe)
Athanasius der grosse und die kirche seiner zeit : besonders im kampfe mit den arianismus / Moehler, Johann Adam – 2. veraend. aufl. Mainz: F Kupferberg, 1844 [mf ed 1990] – 2mf – 9 – 0-7905-5773-8 – (1st printed 1827. incl bibl ref) – mf#1988-1773 – us ATLA [240]
Athanasius, Saint *see* The orations of s athanasius against the arians
Athanasius, Saint, Bishop Alexander of Alexandria *see* Opera omnia
Athanasius, Saint, Patriarch of Alexandria *see*
– A discourse
– A discourse of s athanasius on the incarnation of the word of god
– Select treatises of st. athanasius in controversy with the arians
Atharva pratisakhya : edited for the first time together with an introduction, english translation, notes, and indices / ed by Kanta, Surya – Lahore: Mehar Chand Lachhman Das, 1939 – (= ser Samp: indian books) – us CRL [490]
Atharvaparisishta *see* The asuri-kalpa
The atharva-veda and the gopatha-brahmana / Bloomfield, Maurice – [Strassburg: K J Truebner 1899] [mf ed 1993] – (= ser Grundriss der indo-arischen philologie und altertumskunde 2/1/b) – 1mf – 9 – 0-524-08038-0 – (incl bibl ref) – mf#1991-0254 – us ATLA [280]
Atharva-veda samhita / ed by Lanman, Charles Rockwell – Cambridge MA: Harvard University 1905 [mf ed 1993] – (= ser Harvard oriental series 7,8) – 13mf – 9 – 0-524-07385-6 – mf#1991-0105 – us ATLA [280]
Atharva-veda-samhita – Cambridge MA: Harvard University, 1905 [mf ed 1995] – (= ser Yale coll; Harvard oriental series 7-8) – 2v (ill) – 1 – 0-524-09071-8 – (trans with crit & exegetical comm by william dwight whitney. rev and brought nearer to completion and ed by charles rockwell lanman. incl ind. in english) – mf#1995-0071 – us ATLA [490]
Athearn, Walter Scott *see*
– The beginners' department of the church school
– The church school
– The junior department of the church school
– The primary department of the church school
– The senior department of the church school
Atheism and arithmetic / Hastings, H L – Boston, MA. 1885 – 1r – us UF Libraries [240]
Atheism and arithmetic : mathematical law in nature / Hastings, Horace Lorenzo – Boston, MA: H L Hastings; London: S Bagster, 1885 [mf ed 1986] – 1mf – 9 – 0-8370-7386-3 – mf#1986-1386 – us ATLA [210]
Atheism and the value of life : five studies in contemporary literature / Mallock, William Hurrell – London: Richard Bentley, 1884 – 1mf – 9 – 0-8370-4180-5 – mf#1985-2180 – us ATLA [420]
Atheism in philosophy : and other essays / Hedge, Frederic Henry – Boston: Roberts Bros 1884 [mf ed 1991] – 1mf – 9 – 0-524-00035-2 – mf#1989-2735 – us ATLA [140]
Atheisms in geology : sir c lyell, hugh miller etc confronted with the rocks / Smith, James Alexander – London, 1857 – (= ser 19th c evolution & creation) – 1mf – 9 – mf#1.1.9057 – uk Chadwyck [550]
Der atheismus / Schaarschmidt, Carl von – [S.l: s.n., 18–?] – (= ser Sammlung von Vortraegen fuer das deutsche Volk) – 1mf – 9 – 0-524-03295-5 – mf#1990-0906 – us ATLA [210]
Der atheist – Vienna, Austria jan 1927-dec 1932 [mf ed Norman Ross] – 1r – 1 – mf#nrp-1939 – us UMI ProQuest [210]
Athena press (athena, or: 1893) – Athena OR: J W Smith, -1942 [wkly] [mf ed 1973] – 7r – 1 – (absorbed by: milton eagle) – us Oregon Lib [071]
Athena press (athena, or: 1946) – Athena OR: R C Cooke, 1946-1985 [wkly] – 8r – 1 – (related to: milton eagle (1887-1951)) – us Oregon Lib [071]
Athenaeum : philosophische zeitschrift / ed by Frohschammer, J – bd1-3(1862-1864) – 3v on 36mf – 9 – €69.00 – ne Slangenburg [100]
Athenaeum / ed by Schlegel, August Wilhelm et al – Braunschweig 1798, Berlin 1799-1800 – (= ser Dz. abt literatur) – 3v on 13mf – 9 – €130.00 – mf#k/n4619 – gw Olms [430]
Athenaeum [1807] : a magazine of literary and miscellaneous information – La Salle IL 1807-09 – 1 – mf#4196 – us UMI ProQuest [420]
Athenaeum [1814] – La Salle IL 1814 – 1 – mf#3552 – us UMI ProQuest [420]
Athenaeum [1828] – London UK 1828-1921 – 1 – mf#2855 – us UMI ProQuest [400]
Athenaeum – eine zeitschrift / ed by Schlegel, August Wilhelm & Schlegel, Friedrich – Berlin 1798-1800 [mf ed Hildesheim 1977] – (= ser Allgemeinwissenschaftliche und literarische zeitschriften des 17. und 18. jahrhunderts) – 13mf – 9 – diazo €130.00 – gw Olms [074]
L'Athenaeum francais *see* Journal universel de la litterature, de la science et des beaux-arts
Athenagorae libellus pro christianis / Schwartz, Eduard – Leipzig, 1891 – (= ser Tugal 1-4/2) – 3mf – 9 – €7.00 – ne Slangenburg [230]
Athenagorae libellus pro christianis / oratio de resurrectione cadaverum / ed by Schwartz, Eduard – Leipzig: J C Hinrichs 1891 [mf ed 1989] – (= ser Tugal 4/2) – 1mf – 9 – 0-7905-4014-2 – (in greek & latin) – mf#1988-4 – us ATLA [240]
Athenai – Athens, Greece 9 feb 1917-12 feb, 15-28 jun, 16 jul 1919 (imperfect) – 1 – (in greek) – uk British Libr Newspaper [074]
Athenaion politeia / Aristotle – London, England. 1891 – 1r – us UF Libraries [180]
Athene – Corsham UK 1975-78 – 1,5,9 – ISSN: 0004-6582 – mf#8897 – us UMI ProQuest [700]
Atheneo bahiano – Bahia: Imprensa Economica, 09 abr 1878-jul 1879 – (= ser Ps 19) – mf#P17,01,15 – bl Biblioteca [079]
O atheneu : orgam dos alumnos da 3a serie da escola de pharmacia – Ouro Preto, MG: Imprensa Official da Minas Geraes, 15 dez 1893 – (= ser Ps 19) – mf#P11B,03,83 – bl Biblioteca [500]
Atheneum : or, spirit of the english magazine – La Salle IL 1817-33 – 1 – mf#4420 – us UMI ProQuest [073]
The atheneum – Toronto: T Bengough, [1883-18– or 19–] – 9 – mf#P04340 – cn Canadiana [370]
The atheneum or spirit of the english magazine – Boston. v1-32. 1817-33 – 1 – us L of C Photodup [073]
Athenian constitution / Aristotle – Cambridge, MA. 1935 – 1r – us UF Libraries [323]
Athenian gazette : or casuistical mercury – La Salle IL 1691-97 – 1 – mf#4201 – us UMI ProQuest [073]
Athenian news : or, dunton's oracle – La Salle IL 1710 – 1 – mf#4202 – us UMI ProQuest [420]
Athenian oracle – La Salle IL 1703-10 – 1 – mf#4203 – us UMI ProQuest [900]
Atheniensis historiarum libri 10 (cshb45) / Laonici Chalcocondylae; ed by Bekkeri, Imm – Bonnae, 1843 – (= ser Corpus scriptorum historiae byzantinae (cshb)) – €19.00 – ne Slangenburg [243]
Athens and attica : journal of a residence there / Wordsworth, C – London, 1836 – 4mf – 9 – mf#HT-160 – ne IDC [914]
Athens Co. Athens *see*
– County gazette
– Herald
– Messenger
Athens Co. Nelsonville *see*
– Athens county press and nelsonville tribune
– Hocking valley press
– Valley register
Athens county atlas, 1905 – 1r – 1 – mf#B27425 – us Ohio Hist [978]
Athens county press and nelsonville tribune / Athens Co. Nelsonville – jun-oct 1971 [wkly] – 1r – 1 – mf#B33694 – us Ohio Hist [071]
The athens daily news – Sayre, PA., 1889-1890 – 13 – $25.00r – us IMR [071]
Athens first baptist church – ATHENS, TN. 1871-1904 – 1 reel – 1 – $11.25 – (250p) – us Southern Baptist [242]
Athens first baptist church – ATHENS, TX. 1887-1945, 1966-77 – 1 – $45.45 – us Southern Baptist [242]
The athens news – Sayre, PA., 1889-1890 – 13 – $25.00r – us IMR [071]
Athens. Presbytery (Cum. Pres. Ch.) *see* Minutes, 1854-1907
Athens record – 1901 oct 10-1965 jan 28 [with gaps] – 1 – mf#918628 – us WHS [071]

O atheo : jornal critico-theatral – Rio de Janeiro, RJ: Typ Brasiliense de Francisco Manoel Ferreira, 10 ago-14 set 1851 – (= ser Ps 19) – mf#P15,01,75 – bl Biblioteca [790]

Atherosclerosis – Oxford UK 1961+ – 1,5,9 – ISSN: 0021-9150 – mf#42075 – us UMI ProQuest [616]

Atherton, Gertrude *see* American wives and english husbands

Atherton, William Henry *see* Old montreal in the early days of british canada, 1778-1788

Athey, Irene et al *see* Language, reading and deafness

O athleta : jornal medico-homeopathico – Rio de Janeiro, RJ: Typ Brasiliense de Francisco Manoel Ferreira, 14 jan-07 ago 1852 – (= ser Ps 19) – mf#P15,01,55 – bl Biblioteca [615]

O athleta : periodico litterario e noticioso – Pilar, AL. 25 maio, out 1902; fev 1903; set-20 dez 1908 – (= ser Ps 19) – mf#P18B,01,55 – bl Biblioteca [410]

Athletes' perceptions of coaching performance among ncaa division 3 and naia head football coaches in the state of mississippi / Jubenville, Colby B – 1999 – 1mf – 9 – $4.00 – mf#PE 3962 – us Kinesology [790]

Athletes' perceptions of social support provided by their head coach, and athletic trainer, pre-injury and during rehabilitation / Robbins, Jamie E – 2000 – 108p on 2mf – 9 – $10.00 – mf#PE 4110 – us Kinesology [150]

Athletic department practices that relate to the academic performance and persistence of students-athletes / Unruh, Richard L – 1999 – 199p on 3mf – 9 – $15.00 – mf#PE 4179 – us Kinesology [370]

Athletic director – Reston VA 1969-88 – 1,5,9 – ISSN: 0004-6647 – mf#9287 – us UMI ProQuest [790]

Athletic directors' perceived prevalence of north carolina high school athletes' drug and substance use / Moose, Karen S – University of North Carolina at Chapel Hill, 1995 – 1mf – 9 – $4.00 – mf#HE557 – us Kinesology [360]

Athletic fund-raising : exploring the motives behind private donations / Smith, Joseph C, Jr. – 1989 – 91p 1mf – 9 – $4.00 – us Kinesology [790]

Athletic Institute *see* Volleyball

Athletic journal – New York NY 1921-87 – 1,5,9 – ISSN: 0004-6655 – mf#2137 – us UMI ProQuest [790]

Athletic lancet : a sporting journal for dumbartonshire – [Scotland] Dumbarton: Bennett & Thomson 29 sep-20 oct 1887 (wkly) [mf ed 2004] – 4v on 1r – 1 – uk Newsplan [790]

Athletic leaves – Montreal: [Montreal Amateur Athletic Association, 1888-188-?] [mf ed v1 n1 sep 25, 1888] – 9 – mf#P04022 – cn Canadiana [790]

Athletic news – [NW England] Manchester ALS 17 feb 1896-apr 1917, 9 dec 1918-20 jul 1931 – 1 – uk MLA; uk Newsplan [790]

The athletic organizational structure and administrative views of university and athletic governing personnel in the southwest conference / Cheatham, Tina R & Myers, Bettye – 1992 – 3mf – $12.00 – us Kinesology [790]

Athletic training – Dallas TX 1956-91 – 1,5,9 – (cont by: journal of athletic training) – ISSN: 0160-8320 – mf#10829.01 – us UMI ProQuest [790]

Athlone news – Athlone, Westmeath. 2 dec 1961-23 mar 1962 – 0.5r – 1 – ie National [072]

Athlone observer – Athlone, Westmeath. 24 may 1985-1995 – 11r – 1 – ie National [072]

Athlone sentinel – Athlone, Ireland 21 nov 1834-31 jul 1861 (imperfect) [mf 1851-61] – 1 – uk British Libr Newspaper [072]

Athlone times – Athlone, Ireland 4 may 1889-25 jan 1902 [mf 1889-97] – 6r – 1 – uk British Libr Newspaper [072]

Athlone times – Athlone, Westmeath. nov 1887, 1899 – 0.5r – 1 – ie National [072]

Athol 1734-1905 – Oxford, MA (mf ed 1999) – (= ser Massachusetts vital records) – 138mf – 9 – 0-87623-406-6 – (mf 1-9, 20-22: proprietors & land 1734-72. mf 10-19, 23-40: town records 1762-1864. mf 22, 25-27: births & deaths 1737-1846. mf26-27,32: marriages/intentions 1791-1851. mf 39: births & deaths 1737-93. mf 41: marriages & intentions 1751-93. mf 42-47: town records 1793-1831. mf 48-51: births & deaths 1737-1844. mf 49-61: town records 1792-1871. mf 50-51: marriages & intentions 1791-1843. mf 54-58: militia lists 1840-52. mf 59-60: marriage intentions 1836-51. mf 62-70: town meetings & militia 1853-71. mf 71-72: selectmen records 1844-88. mf 73-83: civil war records 1860-1928. mf 80: naturalizations 1885-1900. mf 84-90: voters 1884-1905. mf 91: out-of-town marriages 1746-99. mf 91-92: births 1843-51. mf 92: marriages 1843-49. mf 92,97: deaths 1843-49. mf 93-105: births 1850-1905. mf 106-117: marriages 1850-1904. mf 118-125: marrriage intentions 1850-1905. mf 126-137: deaths 1850-1908. mf 138: non-resident burials 1898-1920+) – us Archive [978]

Athol 1737-1849 – Oxford, MA (mf ed 1994) – (= ser Massachusetts vital record transcripts to 1850) – 8mf – 9 – 0-87623-204-7 – (mf 1-4t: births & deaths 1737-1847. mf 4t: intentions 1762-93. mf 4t-6t: marriages 1750-1843. mf 6t-7t: intentions 1793-1836. mf 7t: births 1843-46. mf 8t: deaths 1845-49) – us Archive [978]

Atholl and breadalbane times and aberfeldy, pitlochry, dunkeld, and killin advertiser – [Scotland] Perth: A Wright 13 jun-5 sep 1896 (wkly) [mf ed 2003] – 13v on 1r – 1 – uk Newsplan [072]

Atholl, Katharine Marjory *see*
- My impressions of spain
- Report of our visit to spain

Athos : or, the mountain of the monks / Riley, Athelstan – London: Longmans, Green 1887 [mf ed 1992] – 1mf – 9 – 0-524-04146-6 – mf#1990-1216 – us ATLA [243]

Athribis / Petrie, W M – London, 1908 – 3mf – 9 – mf#NE-20356 – ne IDC [956]

Aticismos tropicales / Vincenzi, Moises – Habana, Cuba. 1919 – 1r – us UF Libraries [972]

'Atidot Yisra'el / Deinard, Ephraim – Newark, New Jerseyusa. v1-2. 1891 – 1r – 1 – us UF Libraries [939]

Atif, Mehmet *see* Nazar-i seriatte kuvvet-i beriye ve bahriye'nin ehemmiyet ve vucubu

Atirador franco – Rio de Janeiro, RJ. 01 jan-29 abr 1881 – (= ser Ps 19) – mf#P05,04,32 – bl Biblioteca [321]

Atividades do inps, em 1970 / Instituto Nacional De Previdencia Social Diretori – Rio de Janeiro, Brazil. 1970 – 1r – us UF Libraries [972]

Atjeh press service : almanak umum – Kutaradja, 1959 – 8mf – 9 – mf#SE-610 – ne IDC [950]

Atkins, Frederick Anthony *see* Bible difficulties and how to meet them

Atkins, Gaius Glenn *see*
- Pilgrims of the lonely road
- Procession of the gods

Atkins, Henry Gibson *see* German literature through nazi eyes

Atkins, J *see* A voyage to guinea, brasil, and the west-indies

Atkins, J Alston *see* The texas negro and his political rights

Atkins, James *see* The kingdom in the cradle

Atkins, Sarah *see* Memoirs of john frederic oberlin, pastor of waldbach, in the ban de la roche

Atkins, Thomas B *see* Selections from the public documents of the province of nova scotia

Atkins, Wilfred Guy *see* Suggestions for an amended spelling and word division of nyanja

Atkinson, A *see* Ireland exhibited to england in a political and moral survey of her population

Atkinson, Abraham Fuller *see* The leading doctrines of the gospel

Atkinson, Archibald *see* Speech of mr atkinson, of virginia, on the oregon question

[Atkinson art gallery exhibition handbook 1879] – London 1879 – (= ser 19th c art & architecture) – 1mf – 9 – mf#4.2.775 – uk Chadwyck [700]

Atkinson, Christopher William *see*
- The emigrant's guide to new brunswick, british north america
- A guide to new brunswick, british north america etc
- A historical and statistical account of new-brunswick, bna
- Interesting extracts, etc on religious and moral subjects

Atkinson, Dorothy *see*
- Cherry lake farms
- History of dixie county
- History of franklin county
- History of gadsden county
- History of lafayette county
- History of liberty county
- History of madison county
- Lafayette county
- St george, st vincent, dog island

Atkinson, Edwin Thomas *see* Notes on the history of religion in the himalaya of the n(orth)-w(est) provinces, india

Atkinson enterprise – Atkinson, NE: Ragon & Woods, 1890 (wkly) [mf ed v2 n11. mar 20 1891-may 6 1892 (gaps) filmed 1998] – 1r – 1 – us NE Hist [071]

Atkinson, Ernest Edwin *see* A selected bibliography of hispanic baptist history

Atkinson, Geoffroy *see*
- Les nouveaux horizons de la renaissance francaise
- Nouveaux horizons de la renaissance francaise
- Relations de voyages du 17e siecle et l'evolution des idees

Atkinson, George E *see*
- The game birds of manitoba
- Manitoba birds of prey

Atkinson, George Henry *see*
- Address delivered by rev g h atkinson, dd, before the chamber of commerce of the state of new-york
- The northwest coast including oregon, washington and idaho

Atkinson, George M *see* Ogam inscribed monuments of the gaedhil in the british island

Atkinson graphic – Atkinson, NE: A M Church, dec 1901 (wkly) – 29r – 1 – (cont: atkinson plain dealer, atkinson graphic and holt county republican-consolidated) – us Bell [071]

Atkinson graphic – Atkinson, NE: A M Church, dec 1901 (wkly) [mf ed may 31 1957-] – 19r – 1 – (cont: atkinson plain dealer, atkinson graphic and holt county republican-consolidated) – us NE Hist [071]

Atkinson graphic – Atkinson, Holt Co, NE: H W Mathews. v1 n1. aug 10 1882-v15 n43. jun 10 1897 (wkly) [mf ed sep 28 1882-jan 24 1895 (gaps) filmed 1970] – 1r – 1 – (merged with: atkinson plain dealer to form: atkinson plain dealer and graphic-consolidated) – us NE Hist [071]

Atkinson graphic – Atkinson, NE: H W Mathews. v1 n1. aug 19 1882-v15 n43. jun 10 1897 (wkly) – 2r – 1 – (merged with: atkinson plain dealer to form: atkinson plain dealer and graphic-consolidated) – us Bell [071]

Atkinson, Henry A *see* The church and the people's play

Atkinson, Henry George *see* Letters on the laws of man's nature and development

Atkinson, J Augustus *see* Paper on the best way to make the sunday school a preparation for c...

Atkinson, John *see*
- The beginnings of the wesleyan movement in america and the establishment therein of methodism
- Centennial history of american methodism

Atkinson, John Beavington *see*
- An art tour to northern capitals of europe
- English painters of the present day

Atkinson, Joseph Beavington *see*
- Overbeck
- The schools of modern art in germany
- Studies among the painters

Atkinson, Louise Warren *see*
- The story of paul of tarsus

The atkinson memorial : discourses / Silloway, Thomas William – Boston: James M Usher, 1861 – 1mf – 9 – 0-524-08585-4 – mf#1993-3170 – us ATLA [240]

Atkinson plain dealer – Atkinson, NE: Edwin S Eves. v16 n18. dec 15 1897-v16 n47. feb 9 1899 (wkly) – 2r – 1 – (cont: atkinson plain dealer and graphic-consolidated. cont by: atkinson plain dealer and graphic-consolidated (1899)) – us Bell [071]

Atkinson plain dealer – Atkinson, NE: O C Bates & E S Eves, 1893-v4 n43. jun 9 1897 (wkly) – 1r – 1 – (merged with: atkinson graphic to form: atkinson plain dealer and graphic-consolidated) – us Bell [071]

Atkinson plain dealer – Atkinson, NE: O C Bates & E S Eves, 1893-v4 n43. jun 9 1897 (wkly) [mf ed jun 5 1895-dec 9 1896 gaps) filmed 1993] – 1r – 1 – (merged with: atkinson graphic to form: atkinson plain dealer and graphic-consolidated) – us NE Hist [071]

Atkinson plain dealer and graphic-consolidated – Atkinson, NE: E S Eves. v16 n48. feb 16 1899-v18 n18. jul 26 1900 (wkly) – 1r – 1 – (cont: atkinson plain dealer (1897). merged with: holt county republican to form: atkinson plain dealer, atkinson graphic, and holt county republican-consolidated) – us Bell [071]

Atkinson plain dealer and graphic-consolidated – Atkinson, NE: E S Eves. v15 n44-v16 n17. jun 16-dec 8 1897 (wkly) – 1v on 2r – 1 – (formed by the union of: atkinson plain dealer and: atkinson graphic. cont by: atkinson plain dealer (1897). cont numbering of: atkinson graphic) – us Bell [071]

Atkinson plain dealer and graphic-consolidated *see* The holt county republican

Atkinson plain dealer, atkinson graphic and holt county republican-consolidated – Atkinson, NE: Lee W Henry. v18 n19. aug 2 1900-dec 1901// (wkly) – 2r – 1 – (formed by the union of: atkinson plain dealer and graphic-consolidated (1899) and: holt county republican. cont by: atkinson graphic (1901)) – us Bell [071]

Atkinson plain dealer, atkinson graphic and holt county republican-consolidated *see* The holt county republican

Atkinson, R *see*
- The irish liber hymnorum
- On a new and cheap method of dressing car wheels, axles, etc etc

Atkinson, Solomon *see* A second letter to the right hon w huskisson

Atkinson, T *see* Christian unity

Atkinson, T [comp] *see* Church psalmody

Atkinson, Theresa S *see* Experimental and analytical development of a poroelastic finite element model for tendon

Atkinson, Timothy *see* The fundamental principle of the word of god, the legitimate basis of temperance societies

Atkinson, William *see* Views of picturesque cottages with plans

Atkinson, William Walker *see*
- Mental fascination
- New thought, its history and principles, or, the message of the new thought

Atla law reporter *see* Law reporter (atla)

Atlanta – Atlanta GA 1961+ – 1,5,9 – ISSN: 0004-6701 – mf#3039 – us UMI ProQuest [350]

Atlanta / Cox, Jacob Dolson – New York: C Scribner's Sons, 1882 – (= ser Campaigns of the civil war 9) – 4mf – 9 – (incl ind) – mf#05307 – cn Canadiana [976]

Atlanta age – Atlanta, GA: W A Pledger & A M Hill, 1898 (mf ed 1947) – (= ser Negro Newspapers on Microfilm) – 1r – 1 – us L of C Photodup [071]

The atlanta black crackers / Joyce, Allen E – 1975 – 3mf – 9 – $12.00 – mf#PE 4036 – us Kinesology [790]

Atlanta business chronicle – Charlotte NC 1988+ – 1,5,9 – ISSN: 0164-8071 – mf#16669 – us UMI ProQuest [650]

Atlanta Clergy and Laity Concerned *see* Georgia peace and justice report

Atlanta Cooperative News Project et al *see* Great speckled bird

Atlanta daily register – 1864 mar 19, 23, apr 3 – 1 – mf#859860 – us WHS [071]

Atlanta economic review – Atlanta GA 1951-78 – 1,5,9 – (cont by: business) – ISSN: 0004-671X – mf#7592.01 – us UMI ProQuest [338]

Atlanta Federation of Trades et al *see* Journal of labor

Atlanta five – 1980 dec 5-1982 jun 26 – 1 – mf#656368 – us WHS [071]

Atlanta Friends National Legislative Committee et al *see* News/views

[Atlanta-] great speckled bird – GA. 1972-1974 – 10r – 1 – $600.00 – mf#R04143 – us Library Micro [071]

Atlanta inquirer – [1979 feb 10-1980 dec 27]; 1965 apr 24, may 1,22,29, jun 5,12,19; 1981 jan 3-sep 26; 1981 oct 3-1982 jun 26; 1982 jul 3-1983 mar 26; 1983 apr-dec; 1984-1999 dec 25 [with gaps] – 1 – mf#565085 – us WHS [071]

Atlanta metro – 1994 feb, mar; 1997 mar-dec; 1998 – 1 – mf#3067941 – us WHS [071]

Atlanta news – 1978 feb-1980 aug – 1 – mf#666353 – us WHS [071]

Atlanta news leader – 1997 jan 16-jun; 1997 jul 10-dec 31 – 1 – mf#3067948 – us WHS [071]

Atlanta report – 1979 apr 10-1982 sep 24, nov 15-1984 summer – 1 – mf#670095 – us WHS [071]

[Atlanta-] soul force – CA. 1969-1973 – 1r – 1 – $60.00 – mf#R04144 – us Library Micro [071]

Atlanta star report – 1994 sep – 1 – mf#4026888 – us WHS [071]

Atlanta tribune – 1991 apr-1992 dec; 1993 jan-dec; 1994 jan 1-dec 15; 1995 jan 1-jun 19; 1995 jul 3-dec 30; 1996 jan 1/15-dec 15/30; 1997 jan 1-dec 15/31 – 1 – mf#1878024 – us WHS [071]

Atlanta University Center (GA) *see* Black college radio news

Atlanta voice – 1993 jan 16/22-dec 25/1994 jan 7; 1994 jan 8/14-dec 31/1995 jan 6; 1995 jan 7/13-dec 30/1996 jan 5; 1996 jan 6/12-dec 21/28; 1997 jan 4/17-jun 28/jul 4; 1997 jul 5/11-dec 27/1998 jan 9; 1998 jan 6/10-jun 27/jul 3; 1998 jul 4/10-dec 26/jan 1; 1999 jan 2/8-jun 26/jul 2; 1999 jul 3/9-dec 25/31; 2000 jan 1/7-jun 24/30; 2000 jul 1/7-dec 30/jan 5; 2001 jan-jun – 1 – mf#804980 – us WHS [071]

Atlanta-Fulton Public Library *see* Catalyst

Atlante / Vincenzi, Moises – Quito, Ecuador. 1924 – 1r – us UF Libraries [972]

Atlantes extremenos? / Roso de Luna, Mario – Caceres: Tip. Enc. y Lib. Jimenez, 1905 – 1 – sp Bibl Santa Ana [946]

Atlantic – Washington DC 1982-92 – 1,5,9 – (cont: atlantic monthly; cont by: atlantic monthly) – ISSN: 0276-9077 – mf#158.02 – us UMI ProQuest [073]

Atlantic and american notes : a paper read at euston station, london, on monday, march 13, 1882 / Neele, George P – London: McCorquodale, 1882 [mf ed 1981] – 1mf – 9 – mf#16790 – cn Canadiana [380]

Atlantic and northwest railway : general specification / Ross, James – [Sherbrooke, Quebec?: s.n, 1887?] [mf ed 1994] – 1mf – 9 – 0-665-94685-6 – mf#94685 – cn Canadiana [380]

Atlantic and St Lawrence Railroad Company. Provisional Committee *see* Report of the... appointed 2nd october, 1845, at halifax

The atlantic cables : a review of recent telegraphic legislation in canada / Chesson, Frederick William – London: E Wilson, 1875 – 1mf – 9 – mf#00601 – cn Canadiana [343]

Atlantic Cape May Central Labor Council et al *see* Labor's voice

Atlantic charter and africa from an american standpoint / Committee On Africa, The War, And Peace Aims – New York, NY. 1942 – 1r – us UF Libraries [960]

Atlantic Coast Line *see* Vestibuled train to florida

Atlantic community news – Washington DC 1962-87 – 1,5,9 – (cont by: atlantic council news) – ISSN: 0571-7744 – mf#7023.01 – us UMI ProQuest [350]

Atlantic community quarterly – Washington DC 1963-88 – 1,5,9 – ISSN: 0004-6760 – mf#1621 – us UMI ProQuest [350]

Atlantic council news – Washington DC 1988 – 1,5,9 – (cont: atlantic community news) – ISSN: 1046-0233 – mf#7023,01 – us UMI ProQuest [350]

Atlantic economic journal [aej] – St Louis MO 1978+ – 1,5,9 – ISSN: 0197-4254 – mf#14380 – us UMI ProQuest [338]

The atlantic express and the future british port of arrival – [S.l: s.n, 1893?] – 1mf – 9 – 0-665-02203-4 – mf#02203 – cn Canadiana [380]

Atlantic guardian – Fredericton NB 1950-54 – 1 – mf#594 – us UMI ProQuest [071]

The atlantic islands as resorts of health and pleasure / Benjamin, Samuel Greene Wheeler – New York: Harper, 1878 – 3mf – 9 – mf#06708 – cn Canadiana [917]

The atlantic islands as resorts of health and pleasure / Benjamin, Samuel Greene Wheeler – New York: Harper & Bros, 1878 [mf ed 2003] – [7]-274p (ill) – 1 – mf#5295 – us Wisconsin U Libr [910]

Atlantic journal and friend of knowledge – La Salle IL 1832-33 – 1 – mf#3847 – us UMI ProQuest [900]

Atlantic magazine – La Salle IL 1824-25 – 1 – mf#3679 – us UMI ProQuest [071]

Atlantic monthly [1857] – Washington DC 1857-1980 – (cont by: atlantic) – ISSN: 0004-6795 – mf#158.02 – us UMI ProQuest [073]

Atlantic monthly [1994] – Washington DC 1994+ – 1,5,9 – (cont: atlantic) – ISSN: 1072-7825 – mf#158,02 – us UMI ProQuest [073]

The atlantic naturalist – v1-25. 1946-1970 – 1 – us AMS Press [574]

Atlantic provinces economic council. apec newsletter – Halifax NS 1973 – 1 – ISSN: 0044-989X – mf#7850.01 – us UMI ProQuest [330]

Atlantic provinces library association. apla bulletin – Halifax NS 1936+ – 1,5,9 – ISSN: 0001-2203 – mf#2254 – us UMI ProQuest [020]

Atlantic reporter – 1st series: v1-129. 1885-1925 – (= ser National reporter system, 1879 thru 1919) – 1588mf – 9 – $2382.00 – (add vols planned) – mf#LLMC 79-404C – us LLMC [340]

Atlantic souvenir : a christmas and new year's offering – Ann Arbor MI 1826-32 – 1 – mf#4862 – us UMI ProQuest [073]

Atlantic steam navigation / Fry, Henry – Quebec; Bristol: s.n, 1883 – 1mf – 9 – mf#16919 – cn Canadiana [380]

Atlantic-gulf ship canal / United States Congress House Committee On River – Washington, DC. 1937 – 1r – us UF Libraries [380]

Atlantida / Decoud, Diogenes – Buenos Aires, Argentina. 1901 – 1r – us UF Libraries [025]

Atlantide = Atlantida – Port-au-Prince: Impr Nemours Telhomme. v1 n1-7. oct 1932-apr 1933 – 5 sheets – us CRL [079]

Die atlantik vokhenblatt – Atlantic City. N.J. 1922-23 – 1 – us AJPC [071]

Atlantis – London UK 1975-76 – 1,5,9 – (cont: atlantis incorporating uranus) – mf#7127,01 – us UMI ProQuest [629]

Atlantis – New York NY 2 mar 1895-31 aug 1916, 3 mar 1917-21 nov 1944 – 1 – (in greek; 1942-44: imperfect) – uk British Libr Newspaper [071]

Atlantis – ns: v6 n1-v7 n6 [1857 jan-dec] – 1 – mf#1078956 – us WHS [071]

Atlantis : untersuchungen der platonischen schriften timaios und kritias / Schmeck, Alfred – (mf ed 2000) – 1mf – 9 – €30.00 – 3-8267-2715-0 – mf#DHS 2715 – gw Frankfurter [949]

Atlantis incorporating uranus – London UK 1948-72 – (cont by: atlantis) – ISSN: 0004-6906 – mf#7127.01 – us UMI ProQuest [629]

Atlantis journal des neuesten und wissenswuerdigsten : aus dem gebiete der politik, geschichte, geographie, statistik, culturgeschichte und literatur der nord- und suedamerikanischen reiche mit einschluss des westindischen archipelagus – Leipzig 1826-27 [mf ed Hildesheim 1995-98] – 4v on 10mf – 9 – €100.00 – 3-487-27184-2 – gw Olms [073]

Der atlantische sklavenhandel von dahomey (1740-1797) : eine untersuchung zur wirtschafts- und sozialgeschichte afrikas und wirtschaftsanthropologie / Peukert, Werner – Frankfurt, 1975 – 1mf – 9 – 3-89349-380-8 – gw Frankfurter [380]

Atlantische studien von deutschen in amerika – bd 1-5 hft 3 [1853-54]; bd 6-8 [1855-57] – 1 – mf#1265595 – us WHS [305]

Atlantische studien von deutschen in amerika – Goettingen 1853-57 [mf ed Hildesheim 1995-98] – 8v on 16mf – 9 – €160.00 – 3-487-27179-6 – gw Olms [305]

Atlas – 1856 nov 15-1857 oct 17, 1858 nov 29-1859 jun 3, 1859 jun 4-dec 15, 1859 dec 16-1860 jun 1, 1860 jun 2-nov 28 [1]; 1858 nov 29-1859 jun 3, 1859 jun 4-dec 15, 1859 dec 16-1860 jun 1, 1860 jun 2-nov 28 [2] – 1 – mf#1159109 [1]; 1159107 [2] – us WHS [071]

Atlas : the magazine of the world press – New York NY mar 1961-apr 1972 – 1 – (cont by: atlas: world press review [may 1974-dec 1979]) – uk British Libr Newspaper [070]

Atlas / Newson – Spencer County, KY. 36p. 1882 – 1 – $5.00 – us Southern Baptist [910]

Atlas – Sydney, nov 1844-dec 1848 – 2r – A$133.67 vesicular A$144.67 silver – at Pascoe [079]

Atlas : sydney weekly journal of politics, commerce, and literature – Sydney, Australia 30 nov 1844-26 dec 1846 – 2r – 1 – uk British Libr Newspaper [079]

Atlas : world press review – New York NY may 1974-dec 1979 – 1 – (cont: atlas [the magazine of the world press] mar 1961-apr 1972; cont by: world press review [jan 1980-dec 1995) – uk British Libr Newspaper [071]

Atlas, 1875 / Fairfield County, OH – 1r – 1 – mf#B27423 – us Ohio Hist [978]

Atlas, 1875 / Fayette County, OH – 1r – 1 – mf#B27425 – us Ohio Hist [978]

Atlas archeologique de la bible : d'apres les meilleurs documents, soit anciens... / Fillion, Louis-Claude – Lyon: Delhomme & Briguet 1883 [mf ed 1993] – 5mf – 9 – 0-524-07479-8 – mf#1992-1111 – us ATLA [220]

Atlas biblicus : continens duas et viginti tabulas quibus accedit index topographicus in universam geographiam bibliam / ed by Hagen, Martin – Paris: Lethielleux 1907 [mf ed 1992] – (= ser Cursus scripturae sacrae) – 1mf [ill] – 9 – 0-524-03881-3 – mf#1987-6494 – us ATLA [220]

Atlas (buenos aires, argentina) / ed by Agrupacion Trabajadores Latinoamericanos Sindicalistas – Buenos Aires: Atlas [mf ed 1984] – 1 – mf#1164 – us Wisconsin U Libr [331]

[Atlas consisting of 43 maps of the counties of lower canada and 42 maps of upper canada] / Devine, Thomas – [Montreal]: [Matthew's Lith], [s.d] (mf ed 1974) – 1r – 1 – mf#SEM35P90 – cn Bibl Nat [917]

Atlas da camara da cidade de sao paulo / Sao Paulo. Divisao do Arquivo Historico do Departamento de Cultura – [v1-72. 1562/96-1886] – 1 – us CRL [972]

Atlas de sanson *see* [Atlas nouveau]

Atlas der evangelischen missions-gesellschaft zu basel : nach den angaben der missionare locher, plessing, kies, albrecht, wiegle, dr. gundert, lechler & winnes / Josenhans, Joseph – 2. Aufl. Basel: Verlag des Comptoires des Evangelischen Missions-Gesellschaft in Basel, 1859. Chicago: Dep of Photodup, U of Chicago Lib, 1973 (1r); Evanston: American Theol Lib Assoc, 1984 (1r) – 1 – 0-8370-0422-5 – mf#1984-B437 – us ATLA [242]

Atlas des gestirnten himmels : fuer freude der astronomie / Littrow, Joseph Johann; ed by Littrow, Karl Ludwig – 2nd rev enl ed. Stuttgart: Hoffmann 1854 [mf ed 1998] – 1r [pl/ill] – 1 – mf#film mas 28218 – us Harvard [520]

Atlas d'histoire naturelle de la bible : d'apres les monuments anciens et les meilleures sources modernes et contemporaines / Fillion, Louis-Claude – Lyon: Briday 1884 [mf ed 1990] – 4mf – 9 – 0-7905-3436-3 – (incl bibl ref) – mf#1987-3436 – us ATLA [220]

Atlas ethnographique du globe : ou classification des peubles anciens et modernes d'apres leur langue: partie historique et litteraire / Balbi, Adriano – Paris: Rey et Gravier 1826 – (= ser Whsb) – 6mf – 9 – €70.00 – (tome 1: discours preliminaire et introduction) – mf#Hu 010 – gw Fischer [410]

Atlas geographique / l'Isle, Guillaume de et al – [Paris etc]: [s.n.], 169 -1715?] (mf ed 1975) – 1r – 1 – mf#SEM35P123 – cn Bibl Nat [910]

Atlas hierarchicus : descriptio geographica et statistica s. romanae ecclesiae tum occidentis tum orientis juxta statum praesentem / Streit, Karl – Paderbornae in Guestfalia: Sumptibus typographiae Bonifacianae, 1913 – 1r – 1 – 0-524-06346-X – mf#1990-B002 – us ATLA [240]

Atlas le-toldot yisra'el ba-'arets uva-golah / Theilhaber, Felix A – Tel-Aviv, Israel. 1946 – 1r – us UF Libraries [939]

Atlas, M *see* Razvitie gosudarstvennogo banka sssr

Atlas, M S *see* Natsionalizatsiia bankov v sssr

[Atlas nouveau] : contenant toutes les parties du monde... / Sanson, Nicolas – [S.l.]: [s.n.], [s.d.] (mf ed 1981) – 1r – 5 – mf#SEM35P173 – cn Bibl Nat [914]

Atlas of africa / Horrabin, James Francis – New York, NY. 1960 – 1r – us UF Libraries [960]

Atlas of african affairs / Boyd, Andrew – New York, NY. 1962 – 1r – us UF Libraries [960]

An atlas of astronomy : a series of seventy-two plates, with introduction and index / Ball, Robert Stawell – London: Philip 1892 [mf ed 1998] – 1r [pl/ill] – 1 – mf#film mas 28251 – us Harvard [520]

Atlas of fresno county – Fresno Co, CA: Wm Harvey, Sr, 1907 – 1r – 1 – $50.00 – mf#B40214 – us Library Micro [917]

The atlas of john speed, entitled "the theatre of the empire of great britaine" : from the bodleian library, oxford – 1611-12 – 1r – 1 – (int by r a skelton) – mf#589 – uk Microform Academic [912]

Atlas of the chinese empire : containing separate maps of the eighteen provinces of china proper...together with an index to all the names on the maps and a list of all protestant missions stations etc / Stanford, Edward – London, Philadelphia: China Inland Mission; Morgan & Scott, [1908] [mf ed 1995] – (= ser Yale coll) – xii/16p – 1 – 0-524-10239-2 – mf#1996-1239 – us ATLA [915]

Atlas of the city and island of montreal : including the counties of jacques cartier and hochelaga, from actual surveys, based upon the cadastral plans deposited in the office of the department of crown lands / Hopkins, Henry Whitmer – [Montreal]: Provincial Surveying & Pub Co, 1879 [mf ed 1973] – 1r – 1 – mf#SEM35P50 – cn Bibl Nat [917]

Atlas of the city of montreal : from special survey and official plans, showing all buildings and names of owners / Goad, Chas E – rev ed. Montreal: the author. 2v. 1890 [mf ed 1973] – 1r – 1 – mf#SEM35P49 – cn Bibl Nat [917]

Atlas of the city of montreal and vicinity : in four volumes, from official plans – special surveys showing cadastral numbers, buildings and lots / Chas E Goad Co – Montreal [etc]: Chas E Goad, Co, 1912-1914 [mf ed 1973] – 1r – 1 – mf#SEM35P48 – cn Bibl Nat [917]

Atlas of the historical geography of the holy land / ed by Smith, George Adam – London: Hodder and Stoughton, 1915 – 1r – 1 – 0-524-02758-7 – (incl bibl ref) – mf#1987-B008 – us ATLA [912]

Atlas of the island and city of montreal and ile bizard : a compilation of the most recent cadastral plans from the book of reference / Pinsoneault, A R – [S.l.]: The Atlas Publ Co, [ca 1907] (mf ed 1975) – 1r – 1 – mf#SEM35P122 – cn Bibl Nat [917]

Atlas of the presbytery of st john, n b / Fotheringham, Thomas Francis – St John, NB: s.n, 1886 – 1mf – 9 – (incl ind) – mf#06024 – cn Canadiana [242]

Atlas of the town of sorel and county of richelieu, province of quebec : from actual surveys, based upon the cadastral plans deposited in the office of the department of crown lands / Hopkins, Henry Whitmer – [Quebec]: Provincial surveying & Pub, 1880 [mf ed 1984] – 1r – 1 – mf#SEM35P197 – cn Bibl Nat [917]

Atlas sedjarah / Yamin, M – Amsterdam, Djakarta, 1956 – 3mf – 9 – mf#SE-823 – ne IDC [959]

Atlas to accompany the official records of the union and confederate armies – 1r – 1 – $50.00 – mf#B40145 – us Library Micro [976]

Atlas, Z V *see*
– Denezhnoe obrashchenie i kredit sssr
– Ocherki po istorii denezhnogo obrashcheniia v sssr

Atlas-geographie : etude physique, politique, economique des cinq parties du monde – Montreal: Librairie Granger freres, limitee, 1954 [mf ed 1993] – 3mf – 9 – mf#SEM105P1938 – cn Bibl Nat [910]

Atlas-hotel / Salacrou, Armand – Paris, France. 1931 – 1r – us UF Libraries [440]

Atm & debit news – New York NY 2001+ – 1,5,9 – mf#32202 – us UMI ProQuest [332]

Atmadja, A *see* Sipatahoean, koran dina basa sunda, 30 tahun

Atmoda – Riga, USSR. 1989-1990; Apr 25 1991-Dec 1992 – 3r – 1 – us L of C Photodup [077]

Atmospheric environment – Oxford UK 1967+ – 1,5,9 – ISSN: 1352-2310 – mf#49017 – us UMI ProQuest [333]

Atmospheric environment pt b : urban atmosphere – Oxford UK 1990-93 – 1,5,9 – ISSN: 0957-1272 – mf#49515 – us UMI ProQuest [333]

Atmospheric research – Oxford UK 1986+ – 1,5,9 – ISSN: 0169-8095 – mf#42442,01 – us UMI ProQuest [550]

Atnometev, Nijat Baku *see* Bukvar' tatarskago i arabskago pis'ma: s priloz. slov so znakami, pokazuvajuscimi ich vygovor

Ato e o fato / Cony, Carlos Heitor – Rio de Janeiro, Brazil. 1964 – 1r – us UF Libraries [972]

Atoll research bulletin – Washington DC: Smithsonian Institute. v1-40. 1951-55 – 1 – $60.00 – mf#0090 – us Brook [590]

Atom – Los Alamos NM 1972-81 – 1,5,9 – ISSN: 0004-7023 – mf#6648 – us UMI ProQuest [530]

The atom – Bombay: R K Karanjia, feb 1 1950-sep 20 1950 – 1r – 1 – us CRL [079]

The atom – Bombay: R K Karanjia, feb 1-sep 21 1950; jan 10 1951-aug 27 1952 – 3r – 1 – us CRL [950]

The atom – Bombay: R K Karanjia, jan 10 1951-aug 27 1952 – 2r – 1 – us CRL [079]

Atom und strom – Frankfurt, Germany 1977-80 – 1,5,9 – ISSN: 0004-7066 – mf#10099 – us UMI ProQuest [621]

A'tome – 1974 apr 25-1975 apr 10 – 1 – mf#626719 – us WHS [071]

Atomic bomb / Johnsen, Julie Emily – New York, NY. 1946 – 1r – us UF Libraries [320]

Atomic energy act of 1954 : legislative history / U.S. Atomic Energy Commission – Washington: GPO. 3v. 1955 – 45mf – 9 – $67.00 – mf#llmc 81-104 – us LLMC [348]

Atomic energy commission annual reports / U.S. Nuclear Regulatory Commission – 1956-74 (all publ) – 82mf – 9 – $123.00 – (lacking: 1956-58) – mf#LLMC 80-500 – us LLMC [324]

Atomic energy commission reports / U.S. Nuclear Regulatory Commission – v1-8. 1956-75 (all publ) – 91mf – 9 – $136.00 – (cont by: atomic energy commission annual reports) – mf#LLMC 80-499 – us LLMC [344]

Atomic Energy Corporation of South Africa *see* Assessment of the extent and influence of indoor radon exposures in south africa

Atomic energy newsletter – New York NY 1949-61 – 1 – mf#551 – us UMI ProQuest [530]

Atomic energy review – Vienna, Austria 1963-65 – 1 – ISSN: 0004-7112 – mf#2126 – us UMI ProQuest [621]

Atomic spectroscopy – Shelton CT 1985+ – 1,5,9 – ISSN: 0195-5373 – mf#14932,01 – us UMI ProQuest [530]

Atomic submarine and admiral rickover / Blair, Clay – New York, NY. 1954 – 1r – us UF Libraries [500]

Atomism : dr tyndall's atomic theory of the universe: an irenicum, or, plea for peace and co-operation between science and theology / Watts, Robert – Belfast: William Mullan 1874 [mf ed 1985] – 1mf – 9 – 0-8370-5728-0 – mf#1985-3728 – us ATLA [210]

Atomos que nuevamente se han descubierto con las luces de apolo, en la controversia... / Tenorio de Leon, A – SL, SA – 2mf – 9 – sp Cultura [610]

Atomwirtschaft, atomtechnik – Duesseldorf, Germany 1975-84 – 1,5,9 – ISSN: 0365-8414 – mf#9302 – us UMI ProQuest [530]

Atomzeitalter – Koeln DE, 1960-66 – 1 – gw Misc Inst [074]

Atonement / Bradlaugh, Charles – London, England. 1884 – 1r – us UF Libraries [240]

Atonement : the fundamental fact of christianity / Hall, Newman – New York: F H Revell [1893?] [mf ed 1984] – 2mf – 9 – 0-8370-0903-0 – (incl bibl ref & ind) – mf#1984-4268 – us ATLA [240]

Atonement / Haldane, J A – London, England. 18– – 1r – us UF Libraries [240]

Atonement : the only efficient exponent of god's love to man, and the source and motive of man's love to god / Maxwell, Somerset Richard – London: William Yapp 1866 [mf ed 1991] – 1mf – 9 – 0-7905-8847-1 – mf#1989-2072 – us ATLA [240]

Atonement : a play of modern india, four acts / Thompson, Edward John – London: Ernest Benn Ltd, 1924 – (= ser Samp: indian books) – us CRL [820]

Atonement / Tucker, William – Boston: Universalist Pub House 1893 [mf ed 1993] – (= ser Manuals of faith and duty 10) – 1mf – 9 – 0-524-06504-7 – mf#1991-2604 – us ATLA [240]

The atonement / Bowne, Borden Parker – Cincinnati: Curts & Jennings; New York: Eaton & Mains c1900 [mf ed 1985] – 1mf – 9 – 0-8370-2811-6 – mf#1985-0811 – us ATLA [240]

The atonement : the congregational union lecture for 1875 / Dale, Robert William – London: Congregational Union of England and Wales 1909 [mf ed 1989] – 2mf – 9 – 0-7905-1749-3 – mf#1987-1749 – us ATLA [220]

The atonement : discourses and treatises / Edwards, Smalley et al – Boston: Congregational Board of Publ 1859 [mf ed 1989] – 2mf – 9 – 0-7905-2949-1 – (with int essay by edwards a park) – mf#1987-2949 – us ATLA [240]

The atonement / Hodge, Archibald Alexander – Philadelphia:Presbyterian Board of Publ c1867 [mf ed 1985] – 1mf – 9 – 0-8370-4511-8 – (incl bibl ref & ind) – mf#1985-2511 – us ATLA [240]

The atonement : in its relations to the covenant, the priesthood, the intercession of our lord / Martin, Hugh – Philadelphia: Smith, English 1871 [mf ed 1985] – 1mf – 9 – 0-8370-4150-3 – mf#1985-2150 – us ATLA [240]

The atonement : its efficacy and extent / Candlish, Robert Smith – Edinburgh: Adam & Charles Black 1867 [mf ed 1985] – 1mf – 9 – 0-8370-4671-8 – mf#1985-2671 – us ATLA [240]
The atonement / Pullan, Leighton – London, New York: Longmans, Green 1906 [mf ed 1989] – (= ser The oxford library of practical theology) – 1mf – 9 – 0-7905-2308-6 – (incl ind) – mf#1987-2308 – us ATLA [240]
The atonement / Stalker, James – New York: A C Armstrong 1909 [mf ed 1985] – 1mf – 9 – 0-8370-5524-5 – (incl bibl ref) – mf#1985-3524 – us ATLA [240]
The atonement : viewed as assumed divine responsibility, traced as the fact attested in divine revelation... / Samson, George Whitefield – Philadelphia: J B Lippincott 1878 [mf ed 1985] – 1mf – 9 – 0-8370-5397-8 – (incl ind) – mf#1985-3397 – us ATLA [240]
The atonement and intercession of christ / Davies, David Charles; ed by Jenkins, D E – Edinburgh: T & T Clark; New York: Charles Scribner [dist] 1901 [mf ed 1985] – 1mf – 9 – 0-8370-3366-7 – (incl ind) – mf#1985-1366 – us ATLA [240]
Atonement and law : or, redemption in harmony with law as revealed in nature / Armour, John M – 3rd ed. Philadelphia: Christian Statesman Publ Co: H B Garner 1886, c1885 [mf ed 1985] – 1mf – 9 – 0-8370-2245-2 – (incl ind) – mf#1985-0245 – us ATLA [240]
The atonement and modern thought : being the donnellan lectures / Hitchcock, Francis Ryan Montgomery – London: Wells Gardner, Darton 1911 [mf ed 1989] – (= ser Donnellan lectures 1911?) – 1mf – 9 – 0-7905-1104-5 – (incl ind) – mf#1987-1104 – us ATLA [240]
The atonement and modern thought / Remensnyder, Junius Benjamin – Philadelphia PA: Lutheran Publ Soc c1905 [mf ed 1985] – 1mf – 9 – 0-8370-5273-4 – (incl bibl ref & ind; int by benjamin b warfield) – mf#1985-3273 – us ATLA [240]
Atonement and personality / Moberly, Robert Campbell – New York: Longmans, Green 1901 [mf ed 1989] – 1mf – 9 – 0-7905-2927-0 – mf#1987-2927 – us ATLA [240]
Atonement and progress / Marshall, Newton Herbert – London: James Clarke 1908 [mf ed 1991] – 1mf – 9 – 0-7905-8844-7 – mf#1989-2069 – us ATLA [240]
The atonement and the at-one-maker / Blunt, John Henry – London: Joseph Masters 1855 [mf ed 1985] – 1mf – 9 – 0-7905-2834-7 – mf#1987-2834 – us ATLA [240]
The atonement and the living christ : notes of last lectures and addresses / Body, George – London: Mowbray [19–?] [mf ed 1992] – 1mf – 9 – 0-524-05396-0 – mf#1992-0406 – us ATLA [240]
The atonement and the modern mind / Denney, James – New York: A C Armstrong 1903 [mf ed 1985] – 1mf – 9 – 0-8370-3535-X – mf#1985-1535 – us ATLA [240]
Atonement as set forth in the old testament / Stuart, C E – London, England. 18– – 1r – us UF Libraries [221]
The atonement in christ / Miley, John – New York: Phillips & Hunt; Cincinnati: Hitchcock & Walden 1879 [mf ed 1991] – 1mf – 9 – 0-7905-9816-7 – mf#1989-1541 – us ATLA [240]
The atonement in its relations to law and moral government / Barnes, Albert – Philadelphia: Parry & McMillan 1859 [mf ed 1989] – 1mf – 9 – 0-7905-0853-2 – (incl bibl ref) – mf#1987-0853 – us ATLA [240]
Atonement in literature and life / Dinsmore, Charles Allen – Boston: Houghton, Mifflin 1906 [mf ed 1985] – 1mf – 9 – 0-8370-3591-0 – mf#1985-1591 – us ATLA [230]
The atonement in modern religious thought : a theological symposium / Godet, Frederic Louis – London: James Clarke 1900 [mf ed 1988] – 1mf – 9 – 0-7905-0370-0 – mf#1987-0370 – us ATLA [240]
The atonement of christ / Pendleton, James Madison – Philadelphia: American Baptist Publ Soc c1885 [mf ed 1993] – 1mf – 9 – 0-524-06494-6 – mf#1991-2594 – us ATLA [240]
The atonement of christ : six lectures...hereford cathedral during holy week, 1871 / Barry, Alfred – London, New York: Macmillan 1871 [mf ed 1985] – 1mf – 9 – 0-8370-5605-5 – mf#1985-3605 – us ATLA [240]
The atonement of christ and the justification of the sinner / Fuller, Andrew – New York: American Tract Soc c1854 [mf ed 1991] – 1mf – 9 – 0-7905-9931-7 – mf#1989-1656 – us ATLA [240]
The atonement shown to be an absolute necessity / Varley, Henry – London: A Holness [1901?] [mf ed 1991] – 1mf – 9 – 0-7905-8952-4 – mf#1989-2177 – us ATLA [240]
The atonement viewed in the light of certain modern difficulties : being the hulsean lectures for 1883, 1884 / Lias, John James – 2nd ed. London: James Nisbet 1888 [mf ed 1991] – 1mf – 9 – 0-7905-7899-9 – (incl bibl ref) – mf#1989-1124 – us ATLA [240]
The atoning life / Nash, Henry Sylvester – New York: Macmillan, 1908 – 1mf – 9 – 0-7905-9532-X – mf#1989-1237 – us ATLA [240]
The atoning work of christ viewed in relation to some current theories : in eight sermons / Thomson, William – London: Longman, Brown, Green and Longmans, 1853 – 1mf – 9 – 0-7905-0355-7 – (incl bibl ref) – mf#1987-0355 – us ATLA [240]
Atorney general opinions : from the territory to date / Hawaii – 1904-94 – (= ser Hawaii appellate reports) – 180mf – 9 – $270.00 – (vols after 1994 planned) – mf#LLMC 77-104 – us LLMC [340]
Atp maintenance library for avionics – 9 – (includes navigation, communication, radar and electronic equipment for bendix, king, collins, foster airdata, global wulfsberg, narco, avionic instruments, 3-m stormscope, flite-tronics and bonzer. 32 sublibraries available separately. biweekly revision service.) – us Aircraft Tech [600]
Atp maintenance library for helicopters – 9 – (service and maintenance incl aerospatiale, robinson, sikorsky agusta, bell, mbb, enstrom, hiller, mcdonnell douglas eurocopter france, allison gas turbine, textron lycoming, pratt & whitney of canada, turbomeca. service manuals, parts catalogs & service information. with 12-months revision service. 62 sub-libraries available. biwkly revision service separately) – us Aircraft Tech [629]
Atp maintenance library for jet aircraft over 12,500 lbs – 9 – (incl avions marcel dassault falcon jet, beechjet, british aerospace, cessna citation, learjet, gulfstream, raytheon hawker, lockheed jetstar, mitsubishi, sabreliner, garrett, general electric, rolls royce, pratt & whitney of canada and pratt & whitney with thrust reversers and apus. 54 sublibraries available. biwkly revision service) – us Aircraft Tech [600]
Atp maintenance library for propeller aircraft – 9 – (non-jet, fixed wing under 12,500 lbs. airframes, engines and propellers. contains maintenance manuals, wiring diagrams, overhaul manuals, parts catalogs, parts price lists, structural repair manuals, service information and airworthiness directives. 110 sub-libraries available separately. biweekly revision service.) – us Aircraft Tech [629]
Atr [australian telecommunication research] – Melbourne, Australia 1977 – 1,5,9 – ISSN: 0001-2777 – mf#10490 – us UMI ProQuest [380]
Atras hay dios / Palomino, Martin Alfonso – Caceres, 1973 – 1 – sp Bibl Santa Ana [946]
Atras los invasores : el pueblo entero tiene que movilizarse al llamamiento de la partria / Hernandez, Jesus – Barcelona: Sociedad General de Publicaciones [1938] [mf ed 1977] – (= ser Blodgett coll) – 1mf – 9 – mf#w941 – us Harvard [946]
Atraves da bahia / Spix, Johann Baptist Von – Sao Paulo, Brazil. 1938 – 1r – us UF Libraries [972]
Atraves da historia naval brasileira / Prado Maia, Joao Do – Sao Paulo, Brazil. 1936 – 1r – us UF Libraries [355]
Atraves do sertao do brasil / Roosevelt, Theodore – Sao Paulo, Brazil. 1944 – 1r – us UF Libraries [972]
Atreya, Bhikhan Lal *see* The philosophy of the yoga-vasistha
Atrocities of the pirates : being a faithful narrative of the unparalleled sufferings endured by the author during his captivity among the pirates of the island of cuba / Smith, Aaron – London 1824 [mf ed Hildesheim 1995-98] – 1v on 2mf – 9 – €60.00 – ISBN-10: 3-487-26936-8 – ISBN-13: 978-3-487-26936-8 – gw Olms [920]
Atta – inuvialuit – 1981 jul/aug-1982 jul/aug – 1 – mf#656629 – us WHS [071]
The attache : or, sam slick in england / Haliburton, Thomas Chandler – new rev ed. New York: Dick & Fitzgerald, [187-?] [mf ed 1984] – 4mf – 9 – 0-665-45545-3 – mf#45545 – cn Canadiana [830]
Attache magazine – v1 n6 [1984 [apr?]] – 1 – mf#4877625 – us WHS [071]
Attachment to life / Hughes, Joseph – London, England. 1822 – 1r – us UF Libraries [240]
Attachment to the church of christ / Skinner, William – Aberdeen, Scotland. 1833 – 1r – us UF Libraries [240]
Attachments to recreation settings : the case of rail-trail users / Moore, R L – 1991 – 2mf – 9 – $8.00 – us Kinesology [790]
Attack! – 1969 fall-1978 feb – 1 – mf#384128 – us WHS [071]
Attack upon the university of oxford / Sewell, William – London, England. 1834 – 1r – us UF Libraries [378]
Die attacke : eine tragische komoedie in drei aufzuegen mit einem vor- und einem nachspiel / Dominik, Heinrich – 1. u 2. Aufl. Berlin: S Fischer, 1919 – 1r – 1 – us Wisconsin U Libr [820]
Attah – Madison, WI. 1971-77 – 1 – us AJPC [071]
Attah = atah [romanized] – 1971 apr-1978 mar – 1 – mf#492770 – us WHS [071]
At-Tai *see* The adventures of hatim tai
Attakaddoum – Alger. mai 1923-juil 1931 – 1 – fr ACRPP [073]
Attakapas register – 1861 jan 17-jun 27, jul 18-nov 14 – 1 – mf#860075 – us WHS [071]
L'attawadod d'abou naddara – Paris. 1888-89, avr 1894-nov 1898 – 1 – fr ACRPP [073]
Attempt at the isolation of an organic toxcant in an everglade / Hazard, John Beach – s.l, s.l? . 1925 – 1r – us UF Libraries [630]
An attempt to approximate to the antiquity of man by induction from well established facts / Denison, William Thomas – Madras, 1865 – (= ser 19th c evolution & creation) – 1mf – 9 – mf#1.1.7899 – uk Chadwyck [573]
Attempt to ascertain the meaning of a passage in the twenty-second c... / Gyles, J F – Bath, England. 1826 – 1r – us UF Libraries [240]
An attempt to define geometric proportions of gothic architecture / Billings, Robert William – London 1840 – (= ser 19th c art & architecture) – 1mf – 9 – mf#4.2.1320 – uk Chadwyck [720]
An attempt to define the principles...in the decorative arts / Wyatt, Matthew Digby – London 1852 – (= ser 19th c art & architecture) – 1mf – 9 – mf#4.2.320 – uk Chadwyck [740]
An attempt to elucidate the principles of malayan orthography / Robinson, William – [Fort Marlborough]: printed at the Mission Press, 1823 – (= ser 19th c books on linguistics) – 4mf – 9 – mf#2.1.37 – uk Chadwyck [490]
Attempt to investigate the true principles of cathedral reform / Selwyn, William – Cambridge, England. 1839 – 1r – us UF Libraries [240]
Attempt to point out the duty which the church owes to the people... / Chalmers, Thomas – Edinburgh, Scotland. 1836 – 1r – us UF Libraries [240]
Attempt to promote the peace and edification of the church by uniti... / Mortimer, Thomas – Cambridge, England. 1838 – 1r – us UF Libraries [240]
Attempt to promote true brotherly affection and christian union / Melson, Robert – York, England. 1818 – 1r – us UF Libraries [240]
An attempt to remove error : designed as a letter to a friend on the important subject of the sabbath / Stillman, William – 1 – $5.00 – us Southern Baptist [242]
Atten opbyggelige taler / Kierkegaard, Soeren – Kobenhavn: P G Philipsen 1843-45 [mf ed 1990] – 2mf – 9 – 0-7905-7412-8 – (iss in pts) – mf#1989-0637 – us ATLA [130]
Attenberger, Toni *see* Der endlose wald
Attenborough, David *see* Zoo quest to guiana
Attendance of protestant children at roman catholic schools / Carteret-Hill, P – London, England. 1893? – 1r – us UF Libraries [240]
Attention in young soccer players : the development of an attentional focus training program / Papanikolaou, Zissis & Oglesby, Carole – 1992 – 2mf – 9 – $8.00 – us Kinesology [150]
The attention value of advertisements in a leading periodical : an experiment in measuring the relative attention secured by the various advertisements... / Hotchkiss, George Burton & Franken, Richard B – New York: New York University, c1920 (mf ed 19–) – 32p – mf#ZT-TB+ pv190 n8 – us Harvard [650]
Attentional style differences of injured and non-injured athletes / Noun, Holly A – 1996 – 2mf – 9 – $8.00 – mf#PSY 1954 – us Kinesology [612]
Atterbury, Anson Phelps *see* Islam in africa
Atteridge, Andrew Hilliard *see* Famous modern battles
L'attesa – Agen. n1-17. 21 nov 1926-16 mars 1927 – 1 – (lacking: n16) – fr ACRPP [073]
Attestations de six cures : au sujet de la conduite en 1837-38, du colonel gugy – S.l: s.n, 1841? – 1mf – 9 – mf#24911 – cn Canadiana [971]
Atthill, Lombe *see* Manuel des maladies des femmes
Atti / Accademia nazionale Luigi Cherubini di Musica. Lettere e Arti figurative – 1 30, 1863-92; 39-50, 1911-13; 61, 1941 – 1 – $23.00 – us L of C Photodup [780]
Atti – Venezia. v50. 1891/92 – us CRL [074]
Gli atti dei ss montano : lucio e compagni / Cavalieri, Pio Franchi de – Roma, 1898 – €7.00 – ne Slangenburg [241]
Atti del terzo congresso di studi coloniali : firenze-roma, 12-17 aprile 1937-15 – Firenze 1937 – us CRL [945]
Atti della pontificia accademia romana di archeologia – Roma. v1-4. 1821-1831 – 317mf – 9 – (cont as: dissertazioni della pontificia...roma, 1835-1864 v5-15; 1881-1921 v1-15) – mf#O-1205 – ne IDC [930]
Atti della societa italiana di scienze naturali – Milano: La Societa 1860-1895. v11. 1868. – 1r – us CRL [500]
Atti parlamentari / Italy. Parlamento. Legislatura – 1870-1943 – 1 – us L of C Photodup [945]
Attias, Moshe *see* Keneset yisra'el be-erets yisra'el, yisudah ve-irgunah
The attic theatre / Haigh, Arthur Elam – Oxford, 1889. 13 plus 341p. Illus. Plates. With: Occult Japan by P. Lowell. 1 reel. 1305 – apply for price – us Wisconsin U Libr [000]
Attica news – v2 n?-1910 [1974 ?-1974 aug 21] – 1 – mf#3422114 – us WHS [071]
Attica news service – 1973 mar-? – 1 – mf#774824 – us WHS [071]
Attila : historischer roman aus der voelkerwanderung / Dahn, Felix – Leipzig: Breitkopf & Haertel 1888 [mf ed 1979] – (= ser Kleine romane aus der voelkerwanderung 6) – 2r – 1 – (filmed with: bissula, chlodovech, vom chiemgau & stilicho) – mf#film mas c494 – us Harvard [830]
Attila : historischer roman aus der voelkerwanderung (a 453 n chr) / Dahn, Felix – Philadelphia: Morwitz [18–?] [mf ed 1993] – 1r – 1 – (filmed with: sein bauermaedchen / marianne fleischhack) – mf#8538 – us Wisconsin U Libr [830]
Attilas ende : erzaehlung / Zillich, Heinrich – Muenchen: A Langen, G Mueller c1938 [mf ed 1993] – 1r – 1 – (filmed with: conversations of goethe with eckermann and soret / trans by john oxenford) – mf#8551 – us Wisconsin U Libr [830]
Attilio regolo [dramma per musica si] sig gio adolfo hasse [la poesia e del sig abbate pietro metastasio / Hasse, Johann Adolf – [n.p. 175-?] [mf ed 19–] – 1r – 1 – mf#film 705 – us Sibley [780]
Le attioni d'arrigo terzo re di francia : et quarto di polonia, descritte in dialogo / [Porcacchi, T] – Vinetia: Appresso Giorgio Angelieri, 1574 – 1mf – 9 – mf#O-1944 – ne IDC [090]
Attis : seine mythen und sein kult / Hepding, Hugo – Gieszen [Giessen]: J Ricker 1903 [mf ed 1992] – (= ser Religionsgeschichtliche versuche und vorarbeiten 1) – 1mf – 9 – 0-524-03478-8 – (incl bibl ref) – mf#1990-3220 – us ATLA [250]
The attis of caius valerius catallus : translated into english verse, with dissertations on the myth of attis, on the origin of tree-worship, and on the gallambic metre / Allen, Grant – London: D Nutt, 1892 – (= ser Bibliotheque de carabas) – 2mf – 9 – mf#03848 – cn Canadiana [450]
Attisches museum / ed by Wieland, Christoph Martin – 1796-1809 – (= ser Dz) – 4v on 14mf – 9 – €140.00 – mf#k/n890 – gw Olms [060]
Attitude changes on physical fitness after completing a fitness walking class / Shunk, Anna L – 1997 – 1mf – 9 – $4.00 – mf#PSY 1958 – us Kinesology [790]
Attitude check – v1 n1; v2 n1-3 [1969 nov 1; 1970 feb 1-apr] – 1 – mf#721047 – us WHS [071]
L'attitude internationale – Madrid, 1936? – (= ser Blodgett coll) – 9 – mf#fiche w 736 – us Harvard [946]
Attitude of american courts in labor cases : a study in social legislation / Groat, George Gorham – New York, Longmans, Green, 1911. 400 p. LL-1054 – 1 – us L of C Photodup [344]
The attitude of community health nurses towards integration of traditional healers in primary health care in north west province / Peu, Mmapheko Doriccah – Uni of South Africa 2000 [mf ed Johannesburg 2000] – 4mf – 9 – (incl bibl ref) – mf#mfm14946 – sa Unisa [610]
The attitude of physical education teachers in ireland toward the assessment in second level teaching program / Murphy, G – 1990 – 2mf – 9 – $8.00 – us Kinesology [150]
Attitude of the episcopal church towards non-episcopal churches : from the church standard, reprinted by request – [s.l: s.n 1904?] [mf ed 1993] – 9 – 0-524-07187-X – mf#1990-5345 – us ATLA [242]
The attitude of the texas banker to texas railroads / Duff, Robert C – Houston: Rein & Sons, Printers, [191-?] (mf ed 19–) – 18p – (fr april number of texas bankers journal) – mf#ZV-TPR pv14 n9 – us Harvard [380]
Attitudes and behaviors toward weight, body shape and eating in male and female college students / Lofton, Stacy L – 2000 – 95p on 1mf – 9 – $5.00 – mf#PSY 2130 – us Kinesology [150]
Attitudes associated with competitive age-group swimming / Lauber, Russell L – 1980 – 1mf – 9 – $4.00 – us Kinesology [790]

Attitudes of children in integrated and segregated physical education programs toward peers with disabling conditions / Tripp, April – 1989 – 136p 2mf – 9 – $8.00 – us Kinesology [360]

Attitudes of korean national athletes and coaches toward athletics participation / Cho, Kwang M et al – 1990 – 2mf – 9 – $8.00 – us Kinesology [150]

Attitudes of olympic sport student-athletes and coaches toward ncaa restrictions on practice time-in season / Whitfield, Davis – University of North Carolina at Chapel Hill, 1995 – 2mf – 9 – $8.00 – mf#PSY1873 – us Kinesology [150]

The attitudes of secondary school learners towards biology and implications for curricula development / Manganye, Hlengani Thomas – Uni of South Africa 2001 [mf ed Johannesburg 2001] – 6mf – 9 – (incl bibl ref) – mf#mfm14820 – sa Unisa [373]

Attitudes of therapeutic recreation professionals toward persons with aids and the relationship of their attitude to their knowledge of aids / Glenn, Cherie A & Datillo, John P – 1992 – 2mf – 9 – $8.00 – us Kinesology [150]

Attitudes, perceptions and coping skills of long-term breast cancer survivors / Baskerville, LF – 1990 – 3mf – 9 – $12.00 – us Kinesology [150]

The attitudes toward mormonism in illinois as recorded by the press / Snider, Cecil A – 1831-1849. Warsaw, Illinois. Collected 1933-1934 – 1 – us NY Public [243]

Attitudes toward physical activity of obese and non-obese children and their parents / Steininger, Wanda K – 1982 – 2mf – 9 – $8.00 – us Kinesology [790]

Attitudes towards physical activity among american and german senior citizens / Eckl, C – 1990 – 2mf – 9 – $8.00 – us Kinesology [150]

Attleboro 1692-1849 – Oxford, MA (mf ed 1994) – (= ser Massachusetts vital record transcripts to 1850) – 22mf – 9 – 0-87623-205-5 – (mf 1t-4t: vitals 1692-1796. mf 4t: marriages 1741-56; marriages 1741-56. mf 5t-9t: intentions 1797-1849. mf 9t-10t: vitals 1732-1837. mf 11t: deaths 1753-1835; marriages 1778-1805. mf 11t-14t: vitals 1709-1858. mf 14t-16t: marriages 1695-1845. mf 16t-19t: intentions 1724-97. mf 19t-20t: births 1843-49. mf 21t: marriages 1695-1799; marriages 1843-49. mf 22t: deaths 1843-49) – us Archive [978]

Attleboro 1694-1890 – Oxford, MA (mf ed 1986) – (= ser Massachusetts vital records) – 99mf – 9 – 0-87623-008-7 – (mf 1-13: births, marriages, deaths 1694-1844. mf 14-17: vital statistics 1715-1803. mf 18-23: town meeting records 1723-34. mf 24-27: bounds of lands 1735-95. mf 28-34: precinct records 1745-70. mf 35-42: town meeting records 1757-78. mf 43-49: town records 1782-91. mf 50-54: births, marriages, deaths 1844-70. mf 55: index to b,m,d 1844-70. mf 56-64: births 1856-90. mf 65-72: index to births 1844-1910. mf 73-80: marriages 1856-90. mf 81-87: index to marriages 1844-1910. mf 88-93: deaths 1856-90. mf 94-99: index to deaths 1844-1904) – us Archive [978]

Attorney fee petitions : suggestions for administration and management / Willging, Thomas E & Weeks, Nancy – 2mf – 9 – $3.00 – mf#LLMC 95-322 – us LLMC [340]

Attorney general. official opinions of...us department of justice – v1-43. Digest 1-4 + Annual Reports 1980-1994 – 6,9 – $532.00 set – (v1-41 digest 1-3 on reel $455. v42-43 + digest 4 on mf $38.50. annual reports 1980-1994 $39) – mf#200051 – us Hein [340]

The attorney general's committee on administrative procedure : monographs / U.S. Dept of Justice – 28pts. 1940-41 (all publ) – 56mf – 9 – $84.00 – mf#LLMC 81-214 – us LLMC [340]

The attorney general's survey on release procedures / U.S. Dept of Justice – 5v. 1939-40 (all publ) – 36mf – 9 – $54.00 – mf#LLMC 81-213 – us LLMC [340]

Attorney rolls, 1790-1951 / U.S. Supreme Court – (= ser Records of the supreme court of the united states) – 4r – 1 – (with printed guide) – mf#M217 – us Nat Archives [347]

Attorney-client fee arrangement : regulation and review / Aronson, Robert H – Washington: n.p., 1980? – 2mf – 9 – $3.00 – mf#LLMC 95-305 – us LLMC [348]

Attorneys' ethics collection : publications of the office of disciplinary counsel and the disciplinary board of the hawaii supreme court; comprising the formal and informal opinions of the board / Hawaii – 1974-83 – (= ser Hawaii appellate reports) – 24mf – 9 – $36.00 – (with ind. also summaries for 1857-1983 and ind. incl text of all decisions dealing with attorney ethics between 1857-1983) – mf#LLMC 81-110 – us LLMC [170]

Attorneys' fees in class actions : a report to the federal judicial center / Miller, Arthur R – Washington: FJC, July 1980 – 5mf – 9 – $7.50 – mf#LLMC 95-824 – us LLMC [340]

The Attorneys' Mutual Association *see* List of selected domestic and foreign attorneys, and a telegraphic code.

Attorney's record / Leavenworth County. Kansas. District Court – 1855-1976 – 1 – us Kansas [324]

Attorneys' views of local rules limiting interrogatories / Shapard, John & Seron, Carroll – Washington: FJC, 1986 – 1mf – 9 – $1.50 – mf#LLMC 95-329 – us LLMC [340]

Attraction of the cross / James, J A – London, England. 1819? – 1r – us UF Libraries [240]

The attraction of the cross : designed to illustrate the leading truths, obligations and hopes of christianity / Spring, Gardiner – 3rd ed. New-York: M W Dodd, 1846, c1845 – 1mf – 9 – 0-7905-0385-9 – mf#1987-0385 – us ATLA [240]

Attractions of an excursion upon the great lakes : routes and rates for summer tours – Buffalo?: s.n, 1880? – 1mf – 9 – (incl ind) – mf#37123 – cn Canadiana [917]

The attractive christ : and other sermons / MacArthur, Robert Stuart – Philadelphia: American Baptist Publ Society, 1898 – 1mf – 9 – 0-8370-7484-3 – mf#1986-1484 – us ATLA [240]

Attraverso il benadir / Carletti, Tomaso – Viterbo: Agnesotti, 1910 – 1 – us CRL [910]

Die attribute der heiligen : ein alphabetisches nachschlagebuch zum verstaendnis kirchlicher kunstwerke / Pfleiderer, R – Ed 2. Ulm: Heinrich Kerler, 1920 – 3mf – 9 – mf#O-1255 – ne IDC [700]

The attributes of christ : or, christ the wonderful, the counsellor, god the mighty, the father of the world to come, the prince of peace / Gasparini, Joseph – New York: P O'Shea, 1870 – 1mf – 9 – 0-8370-7461-4 – mf#1986-1461 – us ATLA [240]

Attributs et symboles dans l'art profane 1450-1600 / Tervarent, G de – Geneve, 1958 – €18.00 – ne Slangenburg [700]

The attorney general's investigation of government patent practices and policies / U.S. Dept of Justice – Washington, DC: GPO. v1-3. 1947 (all publ) – 12mf – 9 – $18.00 – mf#LLMC 81-215 – us LLMC [346]

Attwood, Peter Harold *see* A jubilee essay on imperial confederation as affecting manitoba and the northwest

Attwood, Thomas *see* Adopted child

Atualizacao : revista e divulgacao teologica para o cristao de hoje – Belo Horizonte: Editora "O Lutador". v1 n1-v21 n240. dec 1969-nov/dec 1992 – 8r – us CRL [200]

Atwater, Albert William *see* Some remarks on advocacy in civil cases

Atwater, Anna Robison *see* Jehovah's war against false gods

[Atwater-] atwater's new times – CA. 1979- – 10r – 1 – $600.00 (subs $50/y) – mf#B02026 – us Library Micro [071]

Atwater, Edward Elias *see* History and significance of the sacred tabernacle of the hebrews

Atwater, John Milton *see* Jehovah's war against false gods

[Atwater-] the signal – CA. 1911-aug 1918; jan 28 1927-dec 1931; 1933- – 42r + – 1 – $2520.00 (subs $50/y) – mf#B02025 – us Library Micro [071]

Atwood, A *see* Glimpses in pioneer life on puget sound

Atwood, Harry *see* The constitution explained

Atwood, Harry F *see* Back to the republic

Atwood, Henry C [comp] *see* The master workman

Atwood, Isaac Morgan *see*
- Glance at the religious progress of the country in a hundred years
- Revelation

Atys : tragedie mise en musique / Lully, Jean Baptiste – Paris: C Ballard 1689 [mf ed 19–] – 1r – 1 – (libretto by philippe quinault) – mf#film 507 – us Sibley [780]

Atzberger, Leonhard *see*
- Die christliche eschatologie in den stadien ihrer offenbarung im alten und neuen testaments
- Geschichte der christlichen eschatologie innerhalb der vornicaenischen zeit
- Die logoslehre des hl. athanasius
- Die unsuendlichkeit christi

Atzinger, Elizabeth Benjamin *see* Provision of education to minority groups in austria

Au bord de l'abime : ou, un roman a la mode / Fournier, Narcisse – Paris, France. 1844? – 1r – us UF Libraries [440]

Au bresil / Walle, Paul – Paris, France. 1910 – 1r – us UF Libraries [972]

Au cambodge / Agostini, Jules – Paris: [Plon, Nourrit 1898] [mf ed 1989] – (= ser Bibliotheque illustree des voyages autour du monde par terre & par mer) – 1r with other items – 1 – mf#mf-10289 seam reel 014/02 [§] – us CRL [915]

Au cambodge et en annam : voyage pittoresque / Lagrilliere-Beauclerc, Eugene – Paris: La Librairie Mondiale [189-?] [mf ed 1989] – [ill] 1r with other items – 1 – (35 ill & photos by aut) – mf#mf-10289 seam reel 007/06 [§] – us CRL [915]

Au ciel! au ciel! : ou, un chemin court et facile pour aller au ciel / Liguori, Alfonso Maria de', Saint – Sainte-Anne de Beaupre, Que[bec: s.n.] 1920 [mf ed 1995] – 1mf – 9 – 0-665-75375-6 – mf#75375 – cn Canadiana [241]

Au coeur de l'afrique / Villelune, E de – Paris: G Beauchesne, 1909 – 1 – us CRL [960]

Au coin du feu / Marjolaine – Montreal: Librairie d'Action canadienne-francaise, limitee, 1931 [mf ed 1991] – 2mf – 9 – (ill by j mcisaac) – mf#SEM105P1333 – cn Bibl Nat [971]

Au coin du feu : contes = Traeumereien an franzoesischen kaminen / Volkmann, Richard von – Paris: Librairie Fischbacher 1889 [mf ed 1995] – 1 – (french trans fr german. filmed with: kegsardoemets gvinnor / hans wachenhusen & other titles) – mf#3758p – us Wisconsin U Libr [390]

Au congo : comment les noirs travaillent / Lemaire, Charles Francois Alexandre – Bruxelles: Impr Ch Bulens, 1895 – 1 – us CRL [960]

Au congo belge : avec des notes et documents recents relatifs au congo francais / Mille, Pierre – Paris: Colin, 1899 – 1 – us CRL [960]

Au congo belge / Calmeyn, Maurice – Bruxelles, Belgium. 1912 – 1r – us UF Libraries [960]

Au congo et aux indes – Bruxelles, Belgium. 1906 – 1r – us UF Libraries [960]

Au congres eucharistique de malte / Emard, Joseph-Medard – Valleyfield [Quebec: s.n.] 1913 [mf ed 1994] – 5mf – 9 – 0-665-73177-9 – (incl bibl ref) – mf#73177 – cn Canadiana [240]

Au dahomey, 1892 / Masse, Daniel – from La revue de Paris. Paris. 1899 – 1 – us CRL [960]

Au dela des forces / Bjornson, Bjornstjerne – Paris, France. 1910 – 1r – us UF Libraries [440]

Au dela du jourdain : souvenirs d'une excursion / Gautier, Lucien – 2e ed. Geneve: Ch Eggimann; Paris: Librairie Fischbacher 1896 [mf ed 1989] – 1mf – 9 – 0-7905-1397-8 – mf#1987-1397 – us ATLA [915]

Au foyer de mon presbytere : poemes et chansons / Gingras, Apollinaire – Quebec?: A Cote, 1881 – 3mf – 9 – mf#52929 – cn Canadiana [810]

Au grand soleil d'afrique / Gouzy, Rene – Geneve: A Jullien, 1923 – 1 – us CRL [960]

Au gre du souvenir / Marcelin, Frederic – Paris, France. 1913 – 1r – us UF Libraries [972]

Au, Hans von der *see* Deutsche volkstaenze aus der dobrudscha

Au jeudi saint : meditation sacerdotale / Emard, Joseph-Medard – Valleyfield [Quebec]: Bureaux de la Chancellerie, 1915 [mf ed 1995] – 1mf – 9 – 0-665-74168-5 – mf#74168 – cn Canadiana [241]

Au jour de l'an / Emard, Joseph-Medard – Valleyfield [Quebec: s.n.] 1913 [mf ed 1994] – 2mf – 9 – 0-665-73182-5 – mf#73182 – cn Canadiana [241]

Au maroc, 1911-1914 : souvenirs d'un africain / Goureaud, P H – Paris: Plon, [1949] – 1 – us CRL [960]

Au niger : recits de campagnes 1891-1892 / Peroz, Marie Etienne – Paris: C Levy, 1895 – 1 – us CRL [960]

Au nord : brochure accompagnee d'une carte geographique des cantons a coloniser dans les vallees de la riviere rouge et du lievre et dans partie des vallees de la mattawin et de la gatineau – Saint-Jereme [Quebec: s.n.], 1883 [mf ed 1979] – 1mf – 9 – 0-665-00037-5 – mf#00037 – cn Canadiana [971]

Au pays de l'esclavage : moeurs et coutumes de l'afrique centrale d'apres des notes recueillies / Behagle, Ferdinand de – Paris: J Maisonneuve, 1900 – 1 – us CRL [305]

Au pays de l'or rouge / Walle, Paul – Paris, France. 1921 – 1r – us UF Libraries [972]

Au pays des castes : voyage a la cote de la pecherie / Coube, Stephen – nouv ed. Paris: Victor Retaux, 1901 [mf ed 1995] – (= ser Yale coll) – 274p – 1 – 0-524-09857-3 – (in french) – mf#1995-0857 – us ATLA [241]

Au pays des etapes : notes d'un legionnaire / Ecorres, Charles des – Paris, Limoges France: H Charles-Lavauzelle, 1892 – 4mf – 9 – mf#10435 – cn Canadiana [355]

Au pays des fetiches / Vigne d'Octon, Paul – Paris: A Lemerre, 1891 – 1 – us CRL [960]

Au pays des massai / Thomson, Joseph – Paris, France. 1886 – 1r – us UF Libraries [960]

Au pays des pardons – Paris: Calmann-Levy [191?] [mf ed Bloomington IN: Indiana Uni Lib, Preservation Dept 1984] – xv 369p on 1r – 1 – us Indiana Preservation [390]

Au pays du soleil et de l'or / Mevil, Andre – Paris: Didot, [1897?] – 1 – us CRL [960]

Au pays ghimirra / Montandon, G – Paris, 1913 – 6mf – 9 – mf#NE-20227 – ne IDC [916]

Au pays tsimihety : feuilles de route d'un missionaire / Rusillon, Henri – Paris: Societe des Missions Evangeliques, 1923 – (= ser Les Cahiers missionnaires) – 1r – 1 – 0-8370-0498-5 – mf#1984-B224 – us ATLA [240]

Au peuple de la haute ville de quebec et du fauxbourg st jean : c'est le onze de ce mois que nous devons choisir un representant... – S.l: s.n, 1805? – 1mf – 9 – mf#58507 – cn Canadiana [325]

Au pied de l'autel / Sylvain, Adrien – [Quebec?: s.n, 1879?] [mf ed 1984] – 1mf – 9 – 0-665-46399-5 – mf#46399 – cn Canadiana [240]

Au pilori – Paris, France 31 jan, 12,26 jun, 10 jul 1941; 15 jan 1942-12 jul 1944 – 1 – uk British Libr Newspaper [074]

Au portique des laurentides : une paroisse moderne; le cure labelle / Buies, Arthur – Quebec: impr par C Darveau, 1891 [mf ed 1982] – 2mf – 9 – mf#SEM105P74 – cn Bibl Nat [830]

Au ruanda sur les bords du lac kivu (congo belge) : un royaume hamite au centre de l'afrique / Pages, G – Bruxelles: G van Campenhout 1933 – us CRL [960]

Au service de l'enfance : l'association quebecoise de la goutte de lait, 1915-1965 / Fortier, de la Broquerie – Quebec: edition Garneau, 1966 [mf ed 1995] – 2mf – 9 – mf#SEM105P2520 – cn Bibl Nat [366]

Au service d'une cause – Port-Au-Prince, Haiti. 1948 – 1r – us UF Libraries [972]

Au tchad : trois ans chez les senoussistes, les ouaddaiens et les kirdis / Cornet, Charles Joseph Alexandre – 2. ed. Paris: Plon-Nourrit, 1910 – 1 – us CRL [916]

Au temps des "petits chars" / Grenon, Hector – Montreal: Editions internationales Alain Stanke, 1975 [mf ed 2001] – 6mf – 9 – mf#SEM105P3306 – cn Bibl Nat [920]

Au temps des pharaons / Moret, A – Paris, 1925 – 4mf – 9 – mf#NE-20399 – ne IDC [956]

Au temps des pharaons *see* In the time of the pharaohs

Au travail : hebdomadaire populaire syndicaliste – Chambery, France 13 dec 1941-1 jan 1944 – 1 – (wanting: n53,54,59,75-77,83,91,103,105, 111,147) – uk British Libr Newspaper [074]

Au travers des forets vierges de la guyane holland / Cappelle, Herman Van – Baarn, Surinam. 1905 – 1r – us UF Libraries [972]

Aua : novela negra / Duarte, Fausto – Lisbon, 1945 – 1 – us CRL [830]

Aua report – 1946 jul-1952 jan – 1 – mf#1051489 – us WHS [071]

Aua reporter – 1945 jun-jul – 1 – mf#1051490 – us WHS [071]

Au-authm action news – 1978 apr 24, dec 25; 1979 jan 29-1982 jun 22 – 1 – mf#624453 – us WHS [071]

L'aube – Paris, France 9 feb-11 mar, 27 may-3 jun 1940; 5 sep 1944-29 apr 1947; 2 aug 1950-20 oct 1951 [wkly] – 1 – uk British Libr Newspaper [074]

Aube, Benjamin *see*
- Les chretiens dans l'empire romain
- De constantino imperatore, pontificio maximo
- L'eglise et l'etat dans la seconde moitie du 3e siecle
- Histoire des persecutions de l'eglise, la polemique paienne a la fin du 2e siecle
- Polyeucte dans l'histoire

Aube. France (Dept) *see* Projet de budget des recettes et des depenses departementales and decision modificative

Aube, Theophile *see* Martinique

Aubel, Hermann *see* Ein polarsommer reise nach lappland und kanin

Auber, D F E *see*
- Fra diavolo, oder gasthaus von terracino, romantische oper in drey aufzuegen
- La part du diable

Auber, Jacques *see*
- Francais, malgaches, bantous, arabes, turcs, chinois, canaques...parlons-nous une meme langue?
- La langue malgache en 30 familles de mots

Auber, Peter *see*
- China

Auberge d'auray / Moreau, Charles Francois Jean Baptiste – Paris, France. 1830 – 1r – us UF Libraries [440]

Auberge du grand frederic / Lafontaine, W – Paris, France. 1821 – 1r – us UF Libraries [440]

Auberlen, Carl August *see*
- The divine revelation
- Schleiermacher

Auberlen, Carl August et al *see*
- The foundations of our faith
- The two epistles of paul to the thessalonians

Aubert, Alexandre *see* Les experiences religieuses et morales du prophete amos

Aubert de Gaspe, Philippe *see*
- Les anciens canadiens
- Le chercheur de tresors

Aubert de la Rue, Edgar *see* La somalie francaise

Aubert, G *see* L'histoire des gverres faictes par les chrestiens contre les tvrcs

Aubert, Georges *see* L'afrique du sud

Aubert, Hermann *see* Grundzuege der physiologischen optik

Aubertin, John James *see* A fight with distances

Aubigne, T A d' *see*
- Histoire universelle [maille, 1616-1620]
- Memoires...

Aubignosc, L P d' *see* Conjuration du general malet contre napoleon

Aubin, Napoleon *see* La chimie agricole mise a la portee de tout le monde

Aublet, Edouard Edmond *see* La guerre au dahomey, 1888-1893

Auboin, Elie *see* Technique et psychologie du comique

Aubrey beardsley / Symons, Arthur – London 1898 – (= ser 19th c art & architecture) – 1mf – 9 – mf#4.2.1723 – uk Chadwyck [740]

Aubrey de vere : a memoir based on his unpublished diaries and correspondence / Ward, Wilfred Philip – London, New York: Longmans, Green 1904 [mf ed 1986] – 2mf [ill] – 9 – 0-8370-7210-7 – (incl ind) – mf#1986-1210 – us ATLA [420]

Aubrey, John *see* Remaines of gentilisme and judaisme

Aubry, Charles *see* Cours de droit civil francais, d'apres l'ouvrage allemand de c.-s. zachariae

Aubry, Jean-Baptiste *see* Les chinois chez eux

Aubry, P *see* Les proses d'adam de saint-victor

Auburn – 1992- – (= ser California telephone directory coll) – 3r – 1 – $150.00 – mf#P00006 – us Library Micro [917]

Auburn 1704-1900 – Oxford MA (mf ed 1994) – (= ser Massachusetts vital records) – 15v on 54mf – 9 – 0-87623-375-2 – (mf1-4: vital records 1753-1844. mf5-6: births & deaths 1824-45. mf7-9: births & deaths 1704-1844. mf10-11: intentions 1804-48. mf11: marriages 1803-43. mf12-15: town records 1773-86. mf15: intentions 1778. mf16-20: tax lists 1786-1800. mf20-24: accounts 1783-1812. mf25-30: town records 1786-1802. mf31-36: town meetings 1802-23. mf37-40: mortgages 1832-49. mf41: intentions index: 1848-99. mf42-44: intentions 1848-99. mf44: women voters 1914-15. mf45-46: birth index 1844-1909. mf46-47: marriage index 1844-1909. mf47-48: death index 1844-1909. mf49-50: births 1844-74. mf50: marriages 1843-66. mf50: out-town marriages 1778-96. mf51: deaths 1843-66. mf52: births 1873-1902. mf53: marriages 1867-1902. mf54: deaths 1867-1903) – us Archive [978]

Auburn baptist church – AUBURN, KY. 1929-59 – 1 – $23.76 – us Southern Baptist [242]

Auburn daily advertiser – Auburn, [NE]: George R Peck (daily) – 1r – 1 – us Bell [071]

Auburn daily tribune – Auburn, NE: John Stuart. 2v. v3 n57. aug 27 1953-v4 n1. jan 2 1954 (daily ex sun and mon) – 1r – 1 – (cont: auburn tribune. merged with: stella press to form: auburn press-tribune) – us NE Hist [071]

Auburn evening post – Auburn, NE: Rush O Fellows, apr 4 1887 (daily ex sun) – 1r – 1 – us Bell [071]

Auburn journal – Auburn, NE: Melvin V Wilkins. v2 n38. jun 18 1936- (daily ex sun) – 1r – 1 – (cont: peru enterprise) – us Bell [071]

[Auburn-] journal – CA. jul 1914-feb 1988 – 180r – 1 – $10,800.00 – mf#BC02027 – us Library Micro [073]

[Auburn-] placer county leader – CA. may 1898-1902 (wkly) – 1r – 1 – $60.00 – mf#CB02029 – us Library Micro [071]

[Auburn-] placer county republican – CA. may 1885-dec 1898; aug 1903-nov 7 1918 (wkly) – 15r – 1 – $900.00 – mf#CB02030 – us Library Micro [071]

[Auburn-] placer press – CA. may 30 1857-sep 4 1858 – 1r – 1 – $60.00 – mf#C03146 – us Library Micro [071]

Auburn post – Auburn, NE: Rush O Fellows, -sep 1904// (wkly) – 4r – 1 – (cont: sheridan post; cont by: nemaha county republican (auburn ne)) – us Bell [071]

Auburn press-tribune – Auburn, NE: John Vonnes. 72nd yr n28. jan 5 1954- (wkly) – 2r – 1 – (absorbed: peru pointer. formed by the union of: stella press and: auburn daily tribune. cont numbering of: stella press) – us Bell [071]

Auburn press-tribune – Auburn, NE: John Vonnes. 72nd yr n28. jan 5 1954- (wkly) [mf ed apr 10 1956- (gaps)] – 21r – 1 – (absorbed: peru pointer. formed by the union of: stella press and: auburn daily tribune. cont numbering of stella press) – us NE Hist [071]

[Auburn-] republican argus – CA. 1899-jul 1903 (wkly) – 3r – 1 – $180.00 – mf#B02031 – us Library Micro [071]

Auburn seminary record – v1-27. 1905-32 [complete] – 7r – 1 – mf#ATLA 1993-S508 – us ATLA [200]

[Auburn-] stars and stripes – CA. jul 3 1867-jun 26 1871 (wkly) – 1r – 1 – $60.00 – mf#B02033 – us Library Micro [071]

[Auburn-] the placer argus – CA. sep 1872-dec 1888; aug 1889-1897 (wkly) – 14r – 1 – $840.00 – mf#B02028 – us Library Micro [071]

Auburn times – 1903 mar 20-1904 mar 24 – 1 – mf#1093480 – us WHS [071]

Auburn tribune – Auburn, NE: [John Stuart] 3v. v1 n1. aug 23 1951-v3 n56. aug 26 1953 (daily ex sun and mon) – 2r – 1 – (cont by: auburn daily tribune) – us NE Hist [071]

[Auburn-] union advocate – CA. sep 1862-aug 1863 (wkly) – 1r – 1 – $60.00 – mf#B02032 – us Library Micro [071]

[Auburn-] weekly placer herald – CA. sep 1852-1947 (wkly) – 53r – 1 – $3180.00 – (see: placer herald-rocklin) – mf#BC02034 – us Library Micro [071]

Auburndale, florida : permanently substantial – Auburndale, FL. 1926 – 1r – us UF Libraries [978]

Auc digest : the newspaper serving the atlanta university center – 1979 apr 16; 1999 jan 11-nov 22; 1994 sep 19-1995 may 8 [v21 n30-v22 n26]; 1995 aug 21-1998 may 11 [v22 n27-v25 n26]; 1998 aug 31-1999 dec 6; 2000 – 1 – mf#2699170 – us WHS [071]

Auch eine denkschrift ueber den gegenwaertigen zustand von deutschland : oder wuerdigung der denkschrift des herrn von stourdza in juridischer, moralischer, politischer und religioser hinsicht / Krug, Wilhelm Traugott – Leipzig 1819 [mf ed Hildesheim 1995-98] – 1v on 1mf – 9 – €40.00 – 3-487-26349-1 – gw Olms [943]

Aucher-Eloy, P M R *see* Relations de voyages en orient de 1830...1838...

Auchincloss' chronology of the holy bible / Auchincloss, William Stuart – New York: D van Nostrand 1909 [mf ed 1985] – 1mf – 9 – 0-8370-2125-1 – (incl chronological ind; int by archibald henry sayce) – mf#1985-0125 – us ATLA [221]

Auchincloss, William Stuart *see*
- Auchincloss' chronology of the holy bible
- Bible chronology from abraham to the christian era
- The book of daniel unlocked
- How to read josephus
- The only key to daniel's prophecies

Auchterarder chronicle and strathearn advertiser – [Scotland] Perthshire, Auchterarder: Tovani & Co 20 feb 1892 (wkly) [mf ed 2004] – 1r – 1 – uk Newsplan [072]

Auckland chronicle – Bishop Auckland, England 6 jan 1866-9 jul 1869, 1 may 1874-27 sep 1906 [mf 1899] – 1 – (wanting: 1877,1881,1896,1897; cont by: south durham & auckland chronicle [oct 1906-26 dec 1912, 1 jan-7 may 1914]) – uk British Libr Newspaper [079]

Auckland papers : material relating to the american revolution. from the british library – (= ser British records relating to america in microform) – 5r – 1 – (with guide. int by g c bolton) – mf#96662 – uk Microform Academic [975]

Auckland star – [Auckland NZ]: 16 jan 1979-31 oct 1982 [daily] – 91r – 1 – (began with apr 13 1887...in new zealand; ceased with 16 aug 1991) – ISSN: 0004-7449 – mf#11.28 – nz Nat Libr [079]

Auckland sun – 10 aug 1987-8 jul 1988// – 22r – 1 – (aka: the sun; ceased publ 8 jul 1988) – mf#11.53 – nz Nat Libr [079]

Auckland times and herald – 1 – (incorp with: south durham & auckland chronicle [3 jan 1866-27 dec 1895, 5 jan 1900-apr 1910] [mf 1887]) – uk British Libr Newspaper [072]

Auckland truth – jul 1913-jun 1914; jan 1915-dec 1924 – 17r – 1 – mf#11.30 – nz Nat Libr [079]

Auckland weekly news – Auckland, New Zealand 22 dec 1877-2 may 1934 – 1 – (cont: weekly news [3 dec 1864-15 dec 1877]; cont by: weekly news [9 may 1934-29 may 1940; 19 feb 1941, 13 jan 1954; 13 feb, 24 apr, 1 may 1963]) – uk British Libr Newspaper [079]

Auclair, Elie-Joseph *see*
- Cause masson-prevost
- La foi catholique dans ses relations avec la raison et la volonte

Auclair, Joseph *see*
- Le congres
- Le congres de la baie saint paul
- Message du grand chef au congres de 1884
- La sainte enfance dans le diocese de quebec

AuCoin, Rhonda B G *see* Site specific motor unit recruitment during fatigue in human soleus muscle

Aucouturier, Michel *see* Don juan

Auctarium chartularii universitatis parisiensis / Denifle, Heinrich – Parisiis. v1-5. 1937-42 – 5v on 102mf – 8 – €195.00 – ne Slangenburg [378]

Auctarium codicis apocryphi n t fabriciani – Havinae: Arntzen & Hartier, 1804 – 1r – 1 – 0-8370-1097-7 – mf#1984-6241 – us ATLA [225]

Auctarium d c de visch ad bibliothecam scriptorum s o cisterciensis / ed by Canivez, J-M – Brigantii, 1927 – 2mf – 8 – €9.00 – ne Slangenburg [241]

Auction sale of household furniture : there will be sold by public auction on the market square ingersoll, on saturday, may 30th, '91... – Ingersoll, ON?: s.n, 1891? – 1mf – 9 – mf#54241 – cn Canadiana [680]

Auction sale of timer limits, saw mill, lumbering plant, etc – Toronto?: s.n, 1892? – 1mf – 9 – mf#11091 – cn Canadiana [670]

Auctor vetus de beneficiis (mgh leges 3: 2.bd. pars 1a) : textus latini – 1964 – (= ser Monumenta germaniae historica leges 3. fontes iuris germanici antiqui, nova series (mgh leges 3)) – €11.00 – ne Slangenburg [240]

Auctor vetus de beneficiis (mgh leges 3: 2.bd. pars 2a) : archetypus und goerlitzer rechtsbuch – 1966 – (= ser Monumenta germaniae historica leges 3. fontes iuris germanici antiqui, nova series (mgh leges 3)) – €12.00 – ne Slangenburg [342]

L'audace litteraire – Paris. n1-17. 1909-mai juin 1911 – 1 – fr ACRPP [400]

Audacias literarias de una pobre pluma / Rodriguez Romero, Manuel – Montevideo, Uruguay. 1927 – 1r – us UF Libraries [972]

Audain, Leon *see* Choses d'haiti

Audain, Louis *see* Quelques fragments inedits de notre histoire conte

Aude, Joseph *see* Madame angot au malabar

Audecibel – Livonia MI 1952-99 – 1,5,9 – (cont by: hearing professional) – ISSN: 0004-7473 – mf#9865.01 – us UMI ProQuest [616]

Audelco newsletter – 1973 nov; 1974 apr, sep/oct; 1975 jan/feb – 1 – mf#4852619 – us WHS [071]

Audet, Diane *see* Metabolic cost of downhill ski ergometry in males

Audet, Francis-Joseph *see*
- Canadian historical dates and events
- Le clerge protestant du bas-canada de 1760 a 1800
- Gouverneurs, lieutenants-gouverneurs, et administrateurs de la province de quebec, des bas et haut canadas, du canada sous l'union et de la puissance du canada, 1763-1908
- Jean-daniel dumas, le heros de la monongahela

Audet, Louis *see* Lettres d'un etudiant

Audet, Maurice *see*
- Haiti, le reveil d'une race
- Phare dans les antilles

Audet, M-R *see* Bibliographie analytique du docteur emile gaumond chef du service de dermato-syphiligraphie, hotel-dieu, quebec

Audette, Louis Arthur *see* Almanach judiciaire de la province de Quebec

Audi alteram partem / Alford, Henry – London, England. 1851 – 1r – us UF Libraries [240]

Audible – v8 n9; v10 n2-v15 n6 [1976 dec; 1978 feb-1983 dec] – 1 – mf#965472 – us WHS [071]

L'audience – Paris. 10 oct 1873-24 aout 1884, 1883 inc., mq no. 28, 1884 inc., mq no. 15-16 – 1 – fr ACRPP [074]

Audience [1968] – Van Nuys CA 1968-76 – 1,5,9 – ISSN: 0004-7503 – mf#7799 – us UMI ProQuest [790]

Audience [1971] – Boston MA 1971-73 – 1,5,9 – ISSN: 0004-749X – mf#6379 – us UMI ProQuest [700]

Audience Development Committee *see* Intermission

Audience enjoyment of dance performance improvisation as affected by improvisational structures and audience education / Ahlander, Julie D – 1996 – 1mf – 9 – $4.00 – mf#PSY 1973 – us Kinesology [150]

Audience memorable au tribunal de cassation / Bouchereau, Paul – Port-Au-Prince, Haiti. 1941 – 1r – us UF Libraries [972]

Audience reactions to the portrayal of blacks in athletic apparel commercials / Wilson, Brian – University of British Columbia, 1995 – 2mf – 9 – $8.00 – mf#PSY1875 – us Kinesology [302]

Audiencia casa de contratacion : juicios de residencia – Sevilla, SP. 1544-45 – 1,5 – sp Cultura [340]

Audiencia de canarias : juicios de residencia – Sevilla, SP. 1567-1573 – 1,5 – sp Cultura [340]

Audiencia de charcas : juicios de residencia – Sevilla, SP. 1564-80 – 1,5 – sp Cultura [340]

Audiencia de los confines / Asturias, Miguel Angel – Buenos Aires, Argentina. 1957 – 1r – us UF Libraries [972]

Audiencia de mejico : juicios de residencia – Sevilla, SP. 1529-1675 – 1,5 – sp Cultura [340]

Audin, Jean M *see*
- Guide du voyageur en france
- Histoire de la saint-barthelemy d'apres les chroniques, memoires et manuscrits du 16e siecle
- Merveilles et beautes de la nature en suisse

Audin, Jean Marie Vincent *see* History of the life, writings, & doctrines of luther

Audin, M *see* Favole heroiche contenenti le vere massime della politica, et della morale...parte prima

Audio – New York NY 1920-2000 – 1,5,9 – ISSN: 0004-752X – mf#400 – us UMI ProQuest [621]

Audio amateur – Peterborough NH 1970-95 – 1,5,9 – (cont by: audio electronics) – ISSN: 0004-7546 – mf#9181.01 – us UMI ProQuest [621]

Audio electronics – Peterborough NH 1997-2000 – 1,5,9 – (cont: audio amateur) – ISSN: 1092-552X – mf#9181,01 – us UMI ProQuest [621]

Audio engineering society. journal – New York NY 1953-85 – 1,5,9 – (cont by: aes: journal of the audio engineering society, audio/acoustics/applications) – ISSN: 0004-7554 – mf#6175.01 – us UMI ProQuest [621]

Audio scene canada – Toronto ON 1975-81 – 1,5,9 – (cont by: audio video canada) – ISSN: 0315-1182 – mf#10762.01 – us UMI ProQuest [621]

Audio video canada – Toronto ON 1981-82 – 1,5,9 – (cont: audio scene canada) – ISSN: 0710-4413 – mf#10762,01 – us UMI ProQuest [621]

Audio visual – Croydon UK 1972+ – 1,5,9 – ISSN: 0305-2249 – mf#9967 – us UMI ProQuest [370]

Audio visual directions – White Plains NY 1981-84 – 1,5,9 – (cont by: av video) – ISSN: 0746-8989 – mf#12771.04 – us UMI ProQuest [621]

Audiology – Hamilton ON 1971-74 – 1,5,9 – ISSN: 0020-6091 – mf#5928 – us UMI ProQuest [617]

Audiovisual instruction – Bloomington IN 1956-78 – 1,5,9 – (cont by: audiovisual instruction / instructional resources) – ISSN: 0004-7635 – mf#1475.03 – us UMI ProQuest [621]

Audiovisual instruction / instructional resources – Bloomington IN 1978-79 – 1,5,9 – (cont: audiovisual instruction; cont by: instructional innovator) – ISSN: 0191-3417 – mf#1475.03 – us UMI ProQuest [621]

Audiovisual librarian – Milton Keynes UK 1984-98 – 1,5,9 – ISSN: 0302-3451 – mf#14583 – us UMI ProQuest [020]

AudioXpress – Peterborough NH 2001+ – 1,5,9 – mf#31265 – us UMI ProQuest [790]

Audit bureau of circulations ltd : summarised list of net sales of all publications in membership of the bureau... – London, England 1 jul/31 dec 1936- – 1 – (wanting: 1940-47) – uk British Libr Newspaper [650]

Auditing – Sarasota FL 1989+ – 1,5,9 – ISSN: 0278-0380 – mf#18253 – us UMI ProQuest [650]

Auditing theory and practice / Montgomery, Robert H – New York: The Ronald Press, 1913 – 8mf – 9 – $12.00 – mf#LLMC 92-184 – us LLMC [340]

Auditor switching / Garach, Hematlal – Uni of South Africa 2001 [mf ed Johannesburg 2001] – 4mf – 9 – (incl bibl ref) – mf#mfm14989 – sa Unisa [650]

Auditors' report – 1929 dec 1/1930 jun 1 – 1 – mf#3314760 – us WHS [650]

Audorf, Jacob et al *see* Leuchtkugeln

Audouin de Geronval, Maurice E *see* Lettres sur la champagne

Audric, John *see* Angkor and the khmer empire

Audsley, George Ashdown *see*
- Descriptive catalogue of art works in japanese lacquer
- Guide to the art of illuminating and missal painting
- Keramic art of japan
- Notes on japanese art
- The ornamental arts of japan
- Outlines of ornament in the leading styles
- The practical decorator and ornamentalist
- Taste versus fashionable colours

Audsley, Maurice Ashdown *see* The practical decorator and ornamentalist

Audsley, William James *see*
- Guide to the art of illuminating and missal painting
- Outlines of ornament in the leading styles
- Taste versus fashionable colours

Audubon and his journals – New York, NY. v1-2. 1897 – 1r – us UF Libraries [590]

Audubon, John James *see*
- Delineations of american scenery and character
- Scenes de la nature dans les etats unis et le nord de l'amerique

Aue, Hartmann von *see* Der arme heinrich nebst dem inhalte des 'erek' und 'iwein'

Aue, Hartmann von der *see*
- Der arme heinrich
- Der arme heinrich nebst dem inhalte des "erek" und "iwein"
- Erec
- Gregorius

Auer, J *see* Die menschliche willensfreiheit im lehrsystem des thomas von aquin und johannes duns scotus

Auer, Wilhelm *see* Johannes calvins leben und seine stellung innerhalb der gesamtkirche
Auerbach, Berthold *see*
- Andree hofer
- Auf wache
- Barfuessele
- Brigitta
- Dramatische eindruecke
- Joseph im schnee
- Neues leben
- On the heights
- Saemmtliche schwarzwaelder dorfgeschichten
- Schwarzwaelder dorfgeschichten
- Storbonden og hand sonner

Auerbach, Ephraim *see* Shenot reshit
Auerbach, Erich *see* Vier untersuchungen zur geschichte der franzosischen bildung
Auerbach, Guenter *see* Sachgehalt und wahrheitsgehalt in shakespeares "the tempest"
Auerbach, Joseph Smith *see*
- Argument of mr joseph s auerbach
- The bible and modern life – bible words and phrases

Auerbach, Sigmund *see* Ueber die bildende nachahmung des schoenen
Auernheimer, Raoul *see*
- Casanova in wien
- Die dame mit der maske

Auerswald, Annmarie von *see*
- Die ewige ordnung
- Heresgast
- Kampf um irland
- Das radkreuz
- Sonnwill

Auf abschuessiger bahn : roman / Byr, Robert [Bayer, Robert von] – Berlin: Hausfreund-Expedition, [19–?] [mf ed 1989] – 4v in 2 – 1 – (each vol has separate t p) – mf#6992 – us Wisconsin U Libr [830]
Auf, auf ihr christen / Abraham a Sancta Clara – Wien: C Konegen, 1883 [mf ed 1988] – (= ser Wiener neudrucke 1) – xiv/135p – 1 – mf#6934 n2 – us Wisconsin U Libr [240]
Auf biegen und brechen : den weltkrieg begreifen heisst erleben und reifen / Zindler, Erwin – Leipzig: K F Koehler c1929 [mf ed 1992] – 1r – 1 – (filmed with: der bildhauer / hanns von zobeltitz; der befehl des gewissens / hans zoeberlein) – mf#3069p – us Wisconsin U Libr [830]
Auf dem heimweg : neue gedichte / Fischer, Johann Georg – Stuttgart: J G Cotta 1891 [mf ed 1993] – 1r – 1 – (filmed with: aus frischer luft / j g fischer & other titles) – mf#8576 – us Wisconsin U Libr [810]
Auf dem kriegspfad gegen die massai : eine fruehlingsfahrt nach deutsch-ostafrika / Kallenberg, Friedrich – Muenchen 1892 [mf ed Hildesheim 1995-98] – 1v on 2mf – 9 – €60.00 – 3-487-27319-5 – gw Olms [960]
Auf dem schlachtfelde von custozza : [a poem] / Spindler, William – [s.l.: s.n.] 1870 [mf ed 1993] – 1r – 1 – (filmed with: ein beitrag zu theodor storm's stimmungskunst / hermann stamm & other titles) – mf#2905p – us Wisconsin U Libr [810]
Auf dem schnittpunkt zweier gattungen : dramatisierungen englischer romane des 18. jahrhunderts / Sebastian, Astrid – (mf ed 1994) – 3mf – 9 – €49.00 – 3-89349-856-7 – mf#DHS 856 – gw Frankfurter [420]
Auf dem wege zum monotheismus : rektoratsrede / Budde, Karl – Marburg: NG Elwert, 1910 [mf ed 1992] – (= ser [Marburger akademische reden] 24) – 1mf – 9 – 0-524-04694-8 – (incl bibl ref) – mf#1990-3403 – us ATLA [210]
Auf den kargen huegeln der neumark : zur geschichte eines schaefer- und bauerngeschlechts im warthebruch / Kuenkel, Hans – Wuerzburg: Holzner Verlag 1962 [mf ed 1993] – (= ser Ostdeutsche beitraege aus dem goettinger arbeitskreis 21) – 10r [ill] – 1 – (filmed with: ostdeutsche beitraege aus dem goettinger arbeitskreis) – mf#3180p – us Wisconsin U Libr [920]
Die auf den morgen warten! : roman / Brock, Paul – Berlin: F Eher, 1939 [mf ed 1989] – (= ser Deutsche kulturbuchreihe) – 333p – 1 – mf#7088 – us Wisconsin U Libr [830]
Der auf den parnass versetzte gruene hut : 1767 [ein lustspiel in drey aufzuegen] / Klemm, Christian Gottlob – Wien: C Konegen 1883 [mf ed 1988] – 1r – 1 – (filmed with: ueber die schiftstellerische thaetigkeit thomas abbt's / dr geisler) – mf#6934 n5 – us Wisconsin U Libr; us Primary [820]
Auf der feuerstaette : roman / Jensen, Wilhelm – Leipzig: Carl Reissner, 1893 [mf ed 1995] – 2v (ill) – 1 – mf#8795 – us Wisconsin U Libr [830]
Auf der maerchensuche : die entstehung meiner maerchensammlung / Wisser, Wilhelm – Hamburg: Hanseatische Verlagsanstalt [1926] [mf ed 1992] – 1r [ill] – 1 – (filmed with: dichterische arbeiten / eugen gottlob winkler) – mf#3058p – us Wisconsin U Libr [390]
Auf der station : skizzen und novellen aus dem soldaten-leben / Byr, Robert [Bayer, Robert von] – Berlin: L Gerschel, 1865 [mf ed 1989] – 181p – 1 – mf#6992 – us Wisconsin U Libr [830]
Auf der walz vor fuenfzig jahren / Krebs, Werner – Bern: Verein fuer Verbreitung guter Schriften 1927 [mf ed 1990] – (= ser Berner schriften 147) – 1r – 1 – (filmed with: kotzebue in england / walter sellier) – mf#2776p – us Wisconsin U Libr [920]
Auf geht's! humor aus der kampfzeit der nationalsozialistischen bewegung *see* Nsdap (national socialist german workers party) nazi publications
Auf glaubenspfaden : drei erzaehlungen / Dorn, Kaethe – Reutlingen: Ensslin & Laiblin, 1926 [mf ed 1995] – 160p/1pl (ill) – 1 – mf#9032 – us Wisconsin U Libr [880]
Auf grosser fahrt : eine seefahrergeschichte / Gerstner, Hermann – Muenchen: Zentralverlag der NSDAP, F Eher 1942 [mf ed 1990] – (= ser Soldaten-kameraden! 46) – 1r – 1 – (filmed with: die regulatoren in arkansas / friedrich gerstacker) – mf#2609p – us Wisconsin U Libr [830]
Auf gut deutsch – Muenchen DE, 1918-19 – 1r – 1 – gw Misc Inst [074]
Auf halbem wege : [historical novel] / Dwinger, Edwin Erich – Jena: E Diederichs, 1939 [mf ed 1989] – 571p – 1 – mf#7192 – us Wisconsin U Libr [830]
Auf leben und tod : zwei novellen / Storm, Theodor – Bayreuth: Der Gauverlag 1944 [mf ed 1993] – (= ser Bayreuther feldpostausgaben) – 1r – 1 – (filmed with: es klingt wie eine sagen & other titles) – mf#2903p – us Wisconsin U Libr [830]
Auf missionspfaden in japan / Dalton, Hermann – Bremen: C Ed Meuller, 1895 [mf ed 1995] – (= ser Yale coll) – xv/446p – 1 – 0-524-09542-6 – (in german) – mf#1995-0542 – us ATLA [950]
Auf neuer scholle / Gerlach, Fritz – Berlin, Germany. 1941 – 1r – us UF Libraries [943]
Auf spuren des jungen goethe : bilder aus dem alten frankfurt / ed by Sutter, Otto Ernst – Frankfurt/Main: [Niemeyer], 1932 [mf ed 1992] – 37p (ill) – 1 – mf#7984 – us Wisconsin U Libr [750]
Auf stillen wegen : dichtungen / Hammer, Julius – 3. aufl. Leipzig: F A Brockhaus 1878 [mf ed 1990] – 1r – 1 – (filmed with: albrecht von haller / stephen d'irsay) – mf#2696p – us Wisconsin U Libr [810]
Auf vorposten – Berlin DE, 1912-25 – 1r – 5 – gw Misc Inst [074]
Auf vorposten in china : aus dem tagebuche einer missionarsfrau / Leuschner, W – Berlin: Berliner evang. Missionsgesellschaft, 1913 [mf ed 1995] – (= ser Yale coll) – 148p – 1 – 0-524-09163-3 – (in german) – mf#1995-0163 – us ATLA [951]
Auf wache : novelle / Auerbach, Berthold – New York: H Holt, [18–?] [mf ed 1991] – 126p – 1 – (incl: der gefrorene kuss: novelle by otto roquette; ed with int & notes by a a macdonell; incl glos) – mf#7532 – us Wisconsin U Libr [830]
Auf wiedersehn, susanne! : roman / Brehm, Bruno – Muenchen: R Piper c1939 [mf ed 1989] – 1r – 1 – (filmed with other titles) – mf#7066 – us Wisconsin U Libr [830]
Auf zum werk = To the work – Moundridge, KA. aug 1921-jan 1923 – 1 – $10.78 – (organ of the mennonite russian bible society) – us Southern Baptist [242]
Auf zur wolga : schicksale deutscher auswanderer / Ponten, Josef – Feldpostausg. Koeln: H Schaffstein 1944 [mf ed 1992] – (= ser Blaue baendchen 203) – 1r – 1 – (filmed with: liebe ist ewig / wilhelm von polenz) – mf#2864p – us Wisconsin U Libr [830]
Der aufbau – New York NY (USA), 1934- – 1 – (filmed by misc inst: 1972-) – gw Mikrofilm; gw Misc Inst [071]
Der aufbau : gedichte / Anacker, Heinrich – Muenchen: Zentralverlag der NSDAP, F Eher, 1936 [mf ed 1988] – 114p – 1 – mf#6939 n10 – us Wisconsin U Libr [810]
Aufbau, aufgaben und ergebnisse der internationalen arbeitsorganisation / Dietz, Theodor – Wuerzburg, 1934 (mf ed 1995) – 1mf – 9 – €24.00 – 3-8267-3172-7 – mf#DHS 3172 – gw Frankfurter [343]
Der aufbau der amosreden / Baumann, Eberhard – Giessen: J Ricker, 1903 [mf ed 1985] – (= ser Beihefte zur zeitschrift fuer die alttestamentliche wissenschaft 7) – 1mf – 9 – 0-8370-2210-X – mf#1985-0210 – us ATLA [221]
Aufbau und evaluation eines messplatzes zur quantifizierenden erfassung von spastik / Edelhaeuser, Friedrich – (mf ed 1997) – 3mf – 9 – €49.00 – mf#DHS 2440 – gw Frankfurter [612]
aufbau und frieden *see* Mansfeld-echo
Aufbereitung und interpretation von ergebnissen externer qualitaetssicherungsmassnahmen als einstieg in die interventionen von qualitaetsproblemen : konzepte und loesungsvorschlaege am beispield der externen qualitaetssicherung in krankenhaeusern schleswig-holsteins / Niemann, Frank-Michael – (mf ed 1995) – 2mf – 9 – €40.00 – 3-8267-2184-5 – mf#DHS 2184 – gw Frankfurter [360]
Der aufbruch : kampfblatt im sinne des leutnant a d scheringer. zeitschrift fuer wehrfragen, kriegsprobleme und kampf gegen den faschismus – Berlin DE, 1931-1933 n1 – 1r – 1 – gw Misc Inst [355]
Der aufbruch – Lodz (PL), 1935 2 may-21 dec – 2r – 1 – gw Misc Inst [077]
Der aufbruch : stimmen junger deutscher – Kattowitz (Katowice PL), 1933 11 mar-1934 29 sep, 1936, 1937 apr-dec, 1938 25 nov-31 dec – 1 – gw Misc Inst [077]
Der aufenthalt israels in aegypten im lichte der aegyptischen monumente / Spiegelberg, Wilhelm – 3. Aufl. Strassburg: Schlesier & Schweikhardt, 1904 – 1mf – 9 – 0-8370-9905-6 – mf#1986-3905 – us ATLA [930]
Aufenthalt und reisen in mexico in den jahren 1825 bis 1834 : bemerkungen ueber land, produkte, leben und sitten der einwohner und beobachtungen aus dem gebiete der mineralogie, geognosie, bergbaukunde, meteorologie, geographie etc / Burkart, Joseph – Stuttgart 1836 [mf ed Hildesheim 1995-98] – 2v on 6mf – 9 – €120.00 – 3-487-26985-6 – gw Olms [918]
Die auferstehung christi : die berichte ueber auferstehung, himmelfahrt und pfingsten:ihre entstehung, ihr geschichtlicher hintergrund und ihre religioese bedeutung / Meyer, Arnold – Tuebingen:J C B Mohr (Paul Siebeck), 1905 – 1mf – 9 – 0-8370-4399-9 – mf#1985-2399 – us ATLA [240]
Die auferstehung christi und die radikale theologie : die feststellung und deutung der geschichtlichen tatsachen der auferstehung des herrn durch die fortgeschrittene moderne theologie (arnold meyer und h. holtzmann) in kritischer beleuchtung / Korff, Theodor – Halle a. S.: Eugen Strien, 1908 – 1mf – 9 – 0-8370-5855-4 – (incl bibl ref) – mf#1985-3855 – us ATLA [240]
Die auferstehung jesu / Spitta, Friedrich – Goettingen: Vandenhoeck & Ruprecht, 1918 – 1mf – 9 – 0-524-06220-X – mf#1992-0858 – us ATLA [220]
Die auferstehung jesu christi nach den berichten des neuen testamentes / Dentler, Eberhard – Muenster i W: Aschendorff 1908 [mf ed 1992] – (= ser Biblische zeitfragen 1/6) – 1mf – 9 – 0-524-05608-0 – mf#1992-0463 – us ATLA [225]
Die auferstehung jesu in ihrer bedeutung fuer den christlichen glauben / Krueger, Wilhelm – Bremen: C Ed Mueller, 1867 – 1mf – 9 – 0-8370-4005-1 – (incl bibl ref) – mf#1985-2005 – us ATLA [240]
Die auferstehung und ihre neueste bestreitung : vortrag gehalten zu stettin den 24. januar 1865 / Beyschlag, Willibald – Berlin: Ludwig Rauh, [1865] – 1mf – 9 – 0-8370-5986-0 – mf#1985-3986 – us ATLA [240]
Die auferstehungsgeschichte unsers herrn jesu christi nach den vier evangelien / Nebe, August – Wiesbaden: J Niedner, 1882 – 1mf – 9 – 0-524-04411-2 – mf#1992-0104 – us ATLA [225]
Auff den bericht vnd radtschlag : so vnter dem namen des herrn philippi melanthonis zu heidelberg gedruckt vnd ausgangen ist / Moerlin, J – [Magdeburg], 1560 – 1mf – 9 – mf#TH-1 mf 1172 – ne IDC [242]
Auffahrt : neue gedichte / Pulver, Max – Leipzig: Insel-Verlag 1919 [mf ed 1991] – 1r – 1 – (filmed with: heiterer guckkasten / bruno wolfgang) – mf#2865p – us Wisconsin U Libr [810]
Die auffassung des hohenliedes bei den abessiniern : ein historisch-exegetischer versuch / Euringer, Sebastian – Leipzig: J C Hinrichs, 1900 – 1mf – 9 – 0-8370-3076-5 – mf#1985-1076 – us ATLA [220]
Die auffassung des jungen herder vom mittelalter : ein beitrag zur geschichte der aufklaerung / Stolpe, Heinz – Weimar: H Boehlaus Nachfolger, 1955 – 1 – (incl bibl ref and index) – us Wisconsin U Libr [430]
Aufforderung zum laecheln : eine auslese heiterer erzaehlungen / Frenzel, Herbert Alfred & Schmidt, Dietmar – 1.-4.aufl. Berlin: H Reichel, [1943?] [mf ed 1989] – 260p – 1 – mf#7269 – us Wisconsin U Libr [830]
Die auffuehrung des ganzen faust auf dem wiener hofburgtheater : nach dem ersten eindruck besprochen / Schroeer, Karl Julius – Heilbronn: G Henninger, 1883 [mf ed 1990] – xii/58p – 1 – mf#7360 – us Wisconsin U Libr [790]
Die auffuehrung von beethovens neunter symphonie unter richard wagner in bayreuth : 22. mai 1872 / Porges, Heinrich – Leipzig: Kahnt 1872 – 1mf – 9 – mf#wa-85 – ne IDC [780]
Die aufgabe der geschichte der alttestamentlichen auslegung in der gegenwart : akademische antrittsrede / Siegfried, Carl – Jena: Hermann Dufft 1876 [mf ed 1989] – 1mf – 9 – 0-7905-2196-2 – mf#1987-2196 – us ATLA [221]
Aufgaben aus "die jungfrau von orleans" / Schroeder, W – 6. verm. aufl. Leipzig: W Engelmann 1908 [mf ed 1993] – (= ser Aufgaben aus klassischen dramen, epen und romanen 2) – 1r – 1 – (filmed with: don karlos in der geschichte und in der poesie / richard pappritz) – mf#2873p – us Wisconsin U Libr [430]
Aufgaben aus "maria stuart" / Heinze, Hermann – 3. umgearb aufl. Leipzig: W Engelmann 1905 [mf ed 1993] – (= ser Aufgaben aus klassischen dramen, epen und romanen) – 1r – 1 – (incl bibl ref. filmed with: don karlos in der geschichte und in der poesie / richard pappritz) – mf#2873p – us Wisconsin U Libr [430]
Aufgaben aus wallenstein / Heinze, Hermann – 6. verbess ausg. Leipzig: W Engelmann 1907 [mf ed 1995] – (= ser Aufgaben aus klassischen dramen, epen und romanen) – 1r – 1 – (incl bibl ref. filmed with: schiller, don carlos / rudolf ibel) – mf#3731p – us Wisconsin U Libr [430]
Die aufgaben der missionspredigt in indien : hauptsache nach ein vortrag, den der verfasser auf dem missionskursus in gernsbach, baden, am 30. mai 1901 gehalten hat / Hoch, Mark – Basel: Verlag der Missionsbuchhandlung, 1901 [mf ed 1995] – (= ser Yale coll; Basler missionsstudien 8) – 27p – 1 – 0-524-09833-6 – (in german) – mf#1995-0833 – us ATLA [240]
Die aufgaben der neutestamentlichen forschung in der gegenwart / Fiebig, Paul – Leipzig: J C Hinrichs 1909 [mf ed 1985] – 1mf – 9 – 0-8370-3123-0 – (incl bibl ref) – mf#1985-1123 – us ATLA [225]
Die aufgaben der neutestamentlichen wissenschaft in der gegenwart / Weiss, Johannes – Goettingen: Vandenhoeck & Ruprecht 1908 [mf ed 1986] – 1mf – 9 – 0-8370-6454-6 – mf#1986-0454 – us ATLA [225]
Aufgaben der protestantischen theologie / Roehm, Johann Baptist – Augsburg: Max Huttler 1882 [mf ed 1985] – 1mf – 9 – 0-8370-4946-6 – (incl bibl ref) – mf#1985-2946 – us ATLA [242]
Aufg'setzt : eine baierische bauerngeschichte / Schmid, Herman – Leipzig: Keil [18–?] – (= ser Gesammelte schriften. volks- und familien-ausgabe 47 nf15) – 1 – (bound with: ledige kinder) – mf#film mas c438 – us Harvard [830]
Aufhauser, Johannes Baptist *see*
- Antike jesus-zeugnisse
- Die heilslehre des hl. gregor von nyssa
- Konstantins kreuzesvision in ausgewaehlten texten

Die aufhebung des ediktes von nantes im oktober 1685 / Schott, Theodor – Halle: Verein fuer Reformationsgeschichte, 1885 – (= ser [Schriften des Vereins fuer Reformationsgeschichte]) – 1mf – 9 – 0-7905-4659-0 – (incl bibl ref) – mf#1988-0659 – us ATLA [944]
Die aufhebung des jesuiten-ordens : eine beleuchtung der alten und neuen anklagen wider denselben / Riffel, Caspar – Mainz: Kirchheim, Schott und Thielmannn, 1845 – 1mf – 9 – 0-524-08504-8 – mf#1993-3149 – us ATLA [241]
Der aufklaerer, friedrich nicolai / Aner, Karl – Giessen: A. Toepelmann, 1912 – (= ser Studien zur Geschichte des neueren Protestantismus) – 1mf – 9 – 0-7905-4370-2 – (incl bibl ref) – mf#1988-0370 – us ATLA [242]
Aufklaerung – Gelsenkirchen DE, 1951-52 n6 [gaps] – 1 – gw Misc Inst [074]
Aufklaerung von elektrodenprozessen der positiven masse einer li/licoo$_2$ sekundaerbatterie mittels elektrochemischer impedanzspektroskopie / Doege, Volker – (mf ed 1997) – 2mf – 9 – €40.00 – 3-8267-2456-9 – mf#DHS 2456 – gw Frankfurter [540]
Aufrecht, Louis *see* Lelamed bene yehudah
Aufrichtig-deutsche volks-zeitung – Gera DE, 1795-99 – 5r – 1 – (title varies: 1796: aufrichtigdeutsche volks-zeitung, 1798: aufrichtig-teutsche volks-zeitung, 15 jul 1800: neue privilegirte geraische zeitung, 1812: geraische zeitung, 1 jul 1848: fuerstlich reuss-geraische zeitung, 2 oct 1866: fuerstlich reuss geraer zeitung, 12 nov 1918: geraer zeitung, 1 feb 1938: geraer zeitung – geraer beobachter) – gw Misc Inst [074]
Aufruf : streitschrift fuer menschenrechte – Prag (CZ), 1933 15 apr-1934 15 sep – 1r – 1 – gw Misc Inst [322]
Aufruf *see* Europaeische hefte
Der aufruhr um den junker ernst : erzaehlung / Wassermann, Jakob – 1.-15. aufl. Berlin: S Fischer c1926 [mf ed 1991] – 1r [ill] – 1 – (filmed with: christian wahnschaffe) – mf#3025p – us Wisconsin U Libr [880]
Aufsaetze das numinose betreffend / Otto, R – Stuttgart-Gotha, 1923 – €17.00 – ne Slangenburg [240]

Aufsaetze ueber goethe / Scherer, Wilhelm – Berlin: Weidmann, 1886 [mf ed 199] – vi/355p – (coll of articles publ in various magazines from 1874-85, ed with "vorwort" and notes by erich schmidt. incl bibl ref) – mf#10162 – us Wisconsin U Libr [430]

Aufsaetze und vortraege / Reischle, Max; ed by Haering, Theodor & Loofs, Friedrich – Tuebingen: J C B Mohr (Paul Siebeck) 1906 [mf ed 1986] – 1mf – 9 – 0-8370-6311-6 – mf#1986-0311 – us ATLA [240]

Aufsaetze zur deutschen literaturgeschichte / Mehring, Franz; ed by Koch, Hans – Leipzig: P Reclam 1966 [mf ed 1993] – (= ser Reclams universal-bibliothek; Sprache und literatur) – 1 – (incl bibl ref & ind) – mf#8137 – us Wisconsin U Libr [430]

Aufschlager, Johann F *see*
- L'alsace
- Das elsass

Aufstand / Brant, Stefan – Stuttgart, Germany. 1954 – 1r – us UF Libraries [943]

Aufstand der fischer von st barbara / Seghers, Anna – Berlin: G Kiepenheuer 1929 [mf ed 1991] – 1r – 1 – (filmed with: charles sealsfield (carl postl) / albert b faust) – mf#2941p – us Wisconsin U Libr [830]

Der aufstand im warschauer ghetto : entstehung und verlauf = Powstanie w getcie warszawskim / Mark, Bernard – Berlin: Dietz, 1957 (mf ed 1995) – 1r – 1 – (in german. incl bibl ref and ind) – mf#ZZ-34402 – us NY Public [943]

Aufstieg – Reval (Tallinn EW), 1932 20 mar-1933 3 dec – 1r – 1 – gw Misc Inst [077]

Der aufstieg – Berlin DE, Wien (A), 1930-32 – 1r – 1 – gw Misc Inst [074]

Der aufstieg : eine juedische monatsschrift – Berlin-Vienna. v1. 1930-32 [complete] – (= ser German-jewish periodicals...1768-1945, pt 2) – 1r – 1 – $125.00 – mf#B22 – us UPA [939]

Der aufstieg der muttersprache im deutschen denken des 15. und 16. jahrhunderts / Daube, Anna – Frankfurt am Main: M Diesterweg 1940 [mf ed 1993] – (= ser Deutsche forschungen 34) – 1r – 1 – (incl bibl ref) – mf#8023 reel 5 – us Wisconsin U Libr [430]

Der aufstieg zur klassik in der kritik der zeit : die wesentlichen und die umstrittenen rezensionen aus der periodischen literatur von 1750 bis 1795, begleitet von den stimmen der umwelt, in einzeldarstellungen / ed by Fambach, Oscar – Berlin: Akademie-Verlag, 1959 [mf ed 1993] – (= ser Jahrhundert deutscher literaturkritik, 1750-1850 v3) – xxii/685p – 1 – (incl bibl ref) – mf#8223 reel 1 – us Wisconsin U Libr [430]

Der auftakt; musikblaetter – v. 1-18, no. 3 4. 1920-38 – 1 – 50.00 – us L of C Photodup [780]

Aufwaerts – Grosskayna DE, 1952-1968 21 mar [gaps] – 4r – 1 – (braunkohlenwerk) – gw Misc Inst [622]

Aufwaerts – Halle S DE, 1948 5 nov-1995 5 oct [gaps] – 22r – 1 – (chemische werke buna) – gw Misc Inst [660]

Aufwaerts – Ruebeland DE, 1950 9 dec-1952 31 jul [gaps] – 1r – 1 – (chemische werke buna, vhk ruebeland) – gw Misc Inst [660]

Aufwaerts : jugendzeitschrift des dgb (deutscher gewerkschaftsbund) – Koeln DE, 1948 16 jun-1966 15 dec, 1968 15 mar 1968, 1969 15 jan-15 nov – 6r – 1 – mf#7223 – gw Mikropress [331]

Aufwaerts : soziale wochenzeitung fuer die hand- und kopfarbeit – mitteilungsblatt des deutschen gewerkschaftsbundes – Duesseldorf DE, 1920 18 sep-1926 11 jul – 9r – 1 – (with suppl: an rhein und ruhr 1924 12 jan-1925 1 jul [1r], chronik der arbeit 1924 12 jan-1932 10 nov [4r], heimat und welt 1926 7 feb-4 jul & 7 nov-25 dec [1r publ in essen & koeln], ringende jugend (later: werkblatt fuer jugendarbeit in duesseldorf-stadt und -land) 1924 12 jan-15 aug, 31 jan-1925 31 mar 31 [2r]) – gw Misc Inst [331]

Die aufwertung des rhythmus in der neuen musik des fruehen 20. jahrhunderts / Schmidt, Steffen Alexander – (mf ed 2000) – 3mf – 9 – €49.00 – 3-8267-2713-4 – mf#DHS 2713 – gw Frankfurter [780]

Aufzeichnungen des schweizerischen reformators heinrich bullinger... / Krafft, C – Elberfeld: S Lucas, 1870 – 2mf – 9 – mf#PBU-432 – ne IDC [242]

Augar, F *see* Die frau im roemischen christenprocess

Auge, Claude *see* Petit larousse illustre (ael2/11)

El auge del imperio espanol en america / Madariaga, Salvador – Buenos Aires: editorial sudamerica, 1955 – 527p (ill) – 1 – us Wisconsin U Libr [972]

Die augen der erinnerung und anderes / Seidel, Heinrich – Leipzig: A G Liebeskind 1897 – 1 – mf#film mas c419 – us Harvard [830]

Die augen des ewigen bruders : eine legende / Zweig, Stefan – Leipzig: Insel-Verlag, [1933] [mf ed 1992] – (= ser Insel-buecherei n349) – 63p – 1 – mf#7801 – us Wisconsin U Libr [390]

Der augenblick des gluecks : aus den memoiren eines fuerstlichen hofes / Hacklaender, Friedrich Wihelm – Philadelphia: Philadelphia Demokrat Publishing, [18–?] [mf ed 1993] – 427p (ill) – 1 – mf#8668 – us Wisconsin U Libr [880]

Auger, Cleophas *see* Le pilotage du saint-laurent de quebec a montreal

Augias, Carlo *see* Elementi scientifici di etica civile e diritto

Augier, Angel I *see*
- Breve antologia
- Canciones para tu historia
- Isla en el tacto

Augier, Emile *see*
- Effrontes
- Jeunesse
- Mariage d'olympe
- Paul forestier
- Post-scriptum

Augier, Ernest *see* Du tac au tac

Auglaize county atlas, 1880 – 1r – 1 – mf#B7070 – us Ohio Hist [978]

Auglaize county democrat / Auglize Co. Wapakoneta – 1886-II/92, 12/93-10/1923 (poor quality) [wkly] – 17r – 1 – mf#B6655-6671 – us Ohio Hist [071]

Auglaize republican / Auglize Co. Wapakoneta – (aug 1886-jul 1891) poor quality [wkly] – 2r – 1 – mf#B5630-5631 – us Ohio Hist [071]

Auglize Co. Cridersville *see* Press

Auglize Co. Jackson Centr *see* Record

Auglize Co. Minster *see* Community post

Auglize Co. New Bremen *see*
- Stern des westlichen
- Sun

Auglize Co. Saint Marys *see* Evening leader

Auglize Co. Shawnee-Cride *see* Press

Auglize Co. Wapakoneta *see*
- Auglaize county democrat
- Auglaize republican
- Daily news
- Democratic times
- Shelby review series

Auglize Co. Waynesfield *see* Journal

Augsburg confession / Krauth, Charles Porterfield – Philadelphia: Tract & Book Society of St John's Evangelical Lutheran Church 1869 [mf ed 1993] – 1mf – 9 – 0-524-06397-4 – (trans fr original latin, with most important additions of german text incorporated; incl bibl ref; int, notes & ind by charles p krauth) – mf#1991-2519 – us ATLA [242]

The augsburg confession : and formula for the government and discipline of the evangelical lutheran church of the general synod of the united states – Philadelphia: Lutheran Publ Soc 1890 [mf ed 1993] – 1mf – 9 – 0-524-06605-1 – mf#1991-2660 – us ATLA [242]

The augsburg confession : a brief review of its history and an interpretation of its doctrinal articles with introductory discussions on confessional questions / Neve, Juergen Ludwig – Philadelphia PA: Lutheran Publ Soc c1914 [mf ed 1992] – 1mf – 9 – 0-524-03046-4 – mf#1990-0803 – us ATLA [242]

The augsburg confession : an introduction to its study and an exposition of its contents / Loy, Matthias – Columbus OH: Lutheran Book Concern 1908 [mf ed 1990] – 3mf – 9 – 0-7905-5107-1 – mf#1988-1107 – us ATLA [242]

The augsburg confession : presented at the diet of augsburg, a.d. 1530 – Philadelphia PA: United Lutheran Publ House [1913?] – 1mf – 9 – 0-524-06606-X – mf#1991-2661 – us ATLA [242]

Augsburger abendzeitung – Augsburg, Germany. 1858-Sept 1875; Apr 1882-Jun 1884; Apr-Jun 1895; 1901-Sept 1912 – 46r – 1 – us L of C Photodup [074]

Augsburger allgemeine *see* Schwaebische landeszeitung

Die augsburger "allgemeine zeitung" 1798-1866 : mikrofiche-edition mit handbuch. nach dem redaktionsexemplar im cotta-archiv (stiftung der "stuttgarter zeitung") / Fischer, Bernhard [comp] – [mf ed 2002-05] – 1797mf (1:24) 3pt in 9 installments + suppl – 9 – silver €16,000.00 [excl suppl] – ISBN-10: 3-598-34960-2 – ISBN-13: 978-3-598-34960-7 – (pt1: 1798-1832 [mf ed 2002-03] 468mf [€3990] isbn: 978-3-598-34961-4; pt2: 1833-49 [mf ed 2003-04] 565mf [€4500] isbn: 978-3-598-34965-2; pt3: 1850-66 [mf ed 2004-05] 580mf [€5550] isbn: 978-3-598-34969-0, suppl 1867-71 [mf ed 2005] 183mf [€2772] isbn: 978-3-598-34976-8; all pt with guidebooks) – gw Saur [074]

Der augsburger patriciers philipp hainhofer reisen nach innsbruck und dresden / Doering, O – Wien. v10. 1901 – 4mf – 9 – mf#O-517 – ne IDC [915]

Augsburger postzeitung – Augsburg, 1918-21 – 7r – 1 – (mit literarischer beilage, 1917-21) – gw Mikropress [074]

Augsburger tagblatt – Augsburg DE, 1848-49 – 2r – 1 – gw Misc Inst [074]

Augsburger tagespost – Wuerzburg DE, 1948 28 aug-30 dec, 1952-54 – 4r – 1 – (filmed by misc inst: 1953 1 jun-1983, 1967-2002 19 mar (gaps) [ca 2r/yr]; title varies: 12 dec 1948: die tagespost, 1 jan 1950: deutsche tagespost, 3 apr 1999: die tagespost, publ in augsburg, fr 1 mar 1951 in regensburg, fr 1 jul 1955 in wuerzburg. incl suppls: roemische warte 1960 23 aug-1971 28 sep, voelker im aufbruch 1961 27 jan-1981 25 may [2r]) – gw Mikrofilm; gw Misc Inst [074]

Augsburger volkszeitung *see* Volkszeitung

Die augsburgische confession / Vilmar, August Friedrich Christian; ed by Piderit, Karl Wilhelm – Guetersloh: C Bertelsmann, 1870 – 1mf – 9 – 0-7905-9730-6 – mf#1989-1455 – us ATLA [240]

Die augsburgische confession als symbolische lehrgrundlage der deutschen reformationskirche / Zoeckler, Otto – Frankfurt a M: Heyder & Zimmer, 1870 – 1mf – 9 – 0-7905-9654-7 – (incl bibl ref) – mf#1989-1379 – us ATLA [240]

Den augsburgske konfession : eller, den troesbekjendelse, som blev overrakt kejser karl den 5te paa rigsdagen i augsburg 1530 – Decorah IA: Norske synodes forlag 1876 [mf ed 1993] – 1mf – 9 – 0-524-08746-6 – mf#1993-3251 – us ATLA [242]

Augspurg, Anita *see* Frauenstimmrecht

Augspurgische ordinari-zeitung – Augsburg, Muenchen DE, 1848-49 – 2r – 1 – (title varies: 5 jan 1735: augspurger ordinari-zeitung, 12 jan 1735: augspurgische ordinari-zeitung, 1743: augspurger ordinari-zeitung, 1746: augspurgische ordinari-zeitung, 1775: augsburgische staats- und gelehrte zeitung, 1784: augspurger ordinaere zeitung, 1796: augsburgische ordinaere zeitung, 1800: augsburgische ordinari zeitung, 1802: augsburgische ordinaere deutsche zeitung, 1804: ordinaere augsburgische zeitung, 1809: augsburgische politische zeitung, 1818: augsburger politische abendzeitung, 1 jul 1826: augsburger abendzeitung, 2 sep 1912: muenchen-augsburger abendzeitung. filmed by other misc inst: 1742 29 jan, 1771, 1773-85, 1792-1798 30 jun, 1801-07, 1810 22 feb-1812, 1814-17, 1819-25, 1827-1829 jul, 1832, 1835-1934) – gw Misc Inst [074]

Augspurgisch-wochentlicher kern der curiositaeten – Augsburg DE, 1702 4 jan-26 apr – 1r – 1 – gw Misc Inst [074]

Augur – 1970 sep 24/oct 7-1972 may 12 – 1 – mf#701364 – us WHS [071]

Augur – Eugene OR: Augur Pub Co, 1969 [semimthly] – 1 – (cont by: eugene augur (1970)) – us Oregon Lib [071]

Augur (eugene, or) – Eugene OR: Augur Pub Co, 1970-73 [mthly] – 1 – (cont: augur from eugene; cont by: eugene augur (eugene, or)) – us Oregon Lib [071]

The august 1983 amendments to the federal rules of civil procedure : promoting effective case management and lawyer responsibility / Miller, Arthur R – Washington: FJC, 1984 – 1mf – 9 – $1.50 – mf#LLMC 95-810 – us LLMC [347]

August der starke : tragoedie in fuenf akten / Buechler, Franz – Berlin: A Langen/G Mueller 1937 [mf ed 1989] – (= ser Buecherei der dramatischen dichtung) – 1r – 1 – (filmed with: hofische spuren im protestantischen schuldrama um 1600 / hildegard schaefer) – mf#7093 – us Wisconsin U Libr [820]

August dillman / Baudissin, Wolf Wilhelm – Leipzig: S Hirzel 1895 [mf ed 1989] – 1mf – 9 – 0-7905-3240-9 – mf#1987-3240 – us ATLA [920]

August gottlieb spangenberg : bischof der bruederkirche / Reichel, Gerhard – Tuebingen: J C B Mohr 1906 [mf ed 1990] – 1mf – 9 – 0-7905-6549-8 – (incl bibl ref) – mf#1988-2549 – us ATLA [240]

August graf von platens saemtliche werke in zwoelf baenden : historisch-kritische ausgabe mit einschluss des handschriftlichen nachlasses / ed by Koch, Max & Petzet, Erich – Leipzig: Max Hesse [1909?] [mf ed 1995] – 12v in 2r – 1 – (incl bibl ref & ind) – mf#3701p – us Wisconsin U Libr [802]

August graf von platens saemtliche werke in zwoelf baenden : historisch-kritische ausgabe mit einschluss des handschriftlichen nachlasses / ed by Koch, Max & Petzet, Erich – Leipzig: Max Hesse, [1909?] [mf ed 1995] – 12v in 4 on 2r – 1 – (incl bibl ref & ind) – mf#8829 – us Wisconsin U Libr [802]

August hermann francke : ein lebensbild / Kramer, Gustav – Halle a. S: Buchh des Waisenhauses 1880-82 [mf ed 1990] – 2v on 2mf [ill] – 9 – 0-7905-5659-6 – (incl bibl ref) – mf#1988-1659 – us ATLA [242]

August, Jonathan A *see* Teaching, learning and evaluating clinical skills in athletic training

August neander : ein beitrag zu seiner charakteristik / Krabbe, Otto – Hamburg: Rauhen 1852 [mf ed 1991] – 1mf – 9 – 0-524-00997-X – mf#1990-0274 – us ATLA [240]

August von platen in italia e giosue carducci / Saracco, Maria – Torino: Cesare Valentino 1930 [mf ed 1995] – 1r – 1 – (incl bibl ref. filmed with: neue marksteine / adolf pichler) – mf#3704p – us Wisconsin U Libr [410]

August wilhelm ifflands schauspielkunst bis zum abschluss der mannheimer zeit (1796) / David, Siegfried – Heidelberg, 1933 (mf ed 1994) – 1mf – 9 – €24.00 – 3-8267-3011-9 – mf#DHS-AR 3011 – gw Frankfurter [790]

August wilhelm schlegel als lyriker : kapitel 1: fruehzeit / Wulf, Erich – Berlin, 1913 (mf ed 1995) – 1mf – 9 – €24.00 – 3-8267-3118-2 – mf#DHS-AR 3118 – gw Frankfurter [430]

Augusta / Fauchois, Rene – Paris, France. 1949, c1936 – 1r – us UF Libraries [440]

Augusta area times – 1964 jan-dec; 1965 jan 7-1999 [with gaps] – 1 – mf#1026843 – us WHS [071]

Augusta baptist church : church records – AUGUSTA, KY. 1907-62 – 1 – $24.84 – us Southern Baptist [242]

Augusta chronicle – Augusta, GA. 175th anniv – 1r – us UF Libraries [071]

Augusta chronicle and georgia advertiser – 1829 jul 11 – 1 – mf#845734 – us WHS [071]

Augusta eagle – 1915 nov 5, 1915 nov 12-17 mar 16, 1917 mar 23-1918 jul 19, 1918 jul 26-oct 29 [1]; 1874 jul 11-1877 may 19, 1877 may 26-1880 apr 24, 1880 may 1-1883 jun 30, 1883 jul 7-1886 oct 23, 1886 oct 30-1890 jan 18, 1890 jan 25-1893 may 6, 1893 may 13-1896 jul 25, 1896 aug 1-1899 feb 25 [2] – 1 – mf#1044329 [1]; 1044336 [2] – us WHS [071]

Augusta eagle times – 1927 jan 13-mar 3 – 1 – mf#1044308 – us WHS [071]

Augusta Educational News Cooperative *see* Polylogue

Augusta first baptist church : church records – AUGUSTA, GA. Newsletter, v1-5. March 1887-Jan 1893 – 1 reel – 1 – $9.81 – us Southern Baptist [242]

Augusta focus – 1991 mar 28-1999 dec 30/2000 jan 5 [with gaps] – 1 – mf#1871666 – us WHS [071]

Augusta herald – 1871 sep 23-1872 mar 30 – 1 – mf#923893 – us WHS [071]

Augusta times – 1886 aug 14-1897 may 14; 1904 may 13-1904 dec 29; 1906 jan 5-1907 jan 4; 1907 jan 11-1908 jan 3; 1908 jan 3-1909 jul 30; 1909 aug 6-1909 dec 31; 1910 jan 7-1911 jun 30; 1911 jul 7-1912 dec 27; 1913 jan 3-1913 dec 26; 1914 jan 2-14 dec 25; 1915 jan 1-1916 dec 26; 1917 jan 5-17 dec 28; 1918 jan 4-1918 dec 27; 1919 jan 3-19, nov 28 – 1 – mf#959668 – us WHS [071]

Augusta union – 1928 jan 19-29 apr 11; 1929 apr 18-1930 jun 26; 1930 jul 3-1931 sep 17; 1931 sep 24-1932 dec 29; 1933 jan 5-1934 jun 14; 1934 jun 21-35 dec 12; 1935 dec 19-1937 aug 5; 1937 aug 12-1939 mar 16; 1939 mar 23-1939 dec 28; 1940-64; 1965-97 [with gaps] – 1 – mf#787164 – us WHS [071]

The augusta union – Augusta, GA: [A W Wimberly], 1889 (wkly) [mf ed 1947] – (= ser Negro Newspapers on Microfilm) – 1r – 1 – us L of C Photodup [071]

Augustan review : a monthly production – La Salle IL 1815-16 – 1 – mf#4204 – us UMI ProQuest [420]

Augustan Society *see* Genealogical library quarterly

Augustana *see* Lutheran companion

Augustana evangelical lutheran church : augustana lutheran church women. constitution and by-laws, 1926-1956, 1941-1956 / Augustana Lutheran Church Women – 1r – 1 – (with ind; forms pt of: record group aug 40 augustana lutheran church women; reel also incl xa0100r) – mf#xa0099r – us ATLA [242]

Augustana evangelical lutheran church : augustana lutheran church women. convention material, 1895, 1906-1960 / Augustana Lutheran Church Women – 1r – 1 – (with ind; forms pt of: subgroup aug 40/4 general convention) – mf#xa0102r – us ATLA [242]

Augustana evangelical lutheran church : augustana lutheran church women. corporation documents, 1899, 1944, 1949, 1958 / Augustana Lutheran Church Women – 1r – 1 – (with ind; forms pt of: record group aug 40 augustana lutheran church women; reel also incl xa0099r) – mf#xa0100r – us ATLA [242]

Augustana evangelical lutheran church : augustana lutheran church women. minutes, 1892-1893, 1918-1946 / Augustana Lutheran Church Women – 1r – 1 – (with ind; forms pt of: subgroup aug 40/4 general convention) – mf#xa0101r – us ATLA [242]

Augustana evangelical lutheran church : augustana lutheran church women. minutes, 1895-1962 / Augustana Lutheran Church Women – 5r – 1 – (with ind; forms pt of: subgroup aug 40/5 executive board/board of directors) – mf#xa0098r – us ATLA [242]

Augustana evangelical lutheran church : minnesota conference minutes – v1-21. 1858-1962 [complete] – (= ser Minnesota conferensens protokoll) – 10r – 1 – (title varies) – mf#ATLA S0216 – us ATLA [242]
Augustana evangelical lutheran church : report of the synod – n1-103. 1860-1962 – 20r – 1 – (lacks some pg) – mf#ATLA S0093 – us ATLA [242]
Augustana Evangelical Lutheran Church. Board of Foreign Missions *see* Lutheran theological seminary, annual reports and minutes 1913-1948
Augustana Lutheran Church Women *see*
– Augustana evangelical lutheran church
Augustana observer – 1921 sep 9-1928 sep 27; 1928 oct 4-1937 nov 18; 1937 dec 2-1946 dec 18 – 1 – mf#945784 – us WHS [071]
Augustana quarterly : the church quarterly of the augustana lutheran church – v1-27. 1922-48 [complete] – 7r – 1 – mf#ATLA S0021 – us ATLA [242]
The augustana synod : a brief review of its history, 1860-1910 / Petri, Carl Johan et al – Rock Island, IL: Augustana Book Concern, 1910 – 1mf – 9 – 0-524-03734-5 – mf#1990-4839 – us ATLA [240]
Augustana Synod Mission. Conference *see*
– Minutes of the...annual conference of the augustana synod mission
– Proceedings of the annual conference of the augustana synod mission
Augustana synod mission. conference. minutes of the...annual conference – [China: s.n. 19–] (Hankow: Central China Post) (1925-1948) [mf ed 2005-07] – (= ser Christianity's encounter with world religions, 1850-1950) – 2r – 1 – (mf lacks 34th,37th; 15th, 17th-18th, 21st & 38th reports also contain summer conference minutes; 19th & 39th reports contain emergency conference minutes; cont: augustana synod mission. conference. proceedings of the annual conference of the augustana synod mission) – mf#2005c-s024 – us ATLA [242]
Augustana synod mission. conference. proceedings of the annual conference – Juchow, Honan, China: [s.n.] 1922-23 [mf ed 2005-07] – (= ser Christianity's encounter with world religions, 1850-1950) – 2r – 1 – (lacks: 1923 [p69-70]; cont: augustana synod mission. summer conference. proceedings of two conferences of the augustana synod mission; cont by: augustana synod mission. conference. minutes of the...annual conference of the augustana synod mission) – mf#2005c-s023 – us ATLA [242]
Auguste, Charles A *see* Pour une education haitienne
Auguste comte and positivism / Mill, John Stuart – London: Truebner 1865 [mf ed 1991] – 1mf – 9 – 0-7905-9817-5 – (repr fr the westminster review) – mf#1989-1542 – us ATLA [140]
Auguste, M *see* M des chalumeaux
Augusti, Johann Christian Wilhelm *see*
– The antiquities of the christian church
– Beytraege zur geschichte und statistik der evangelischen kirche
Augustin / Hertling, Georg, Graf von – Mainz: F Kirchheim 1902 [mf ed 1990] – (= ser Weltgeschichte in karakterbildern 1.abt) – 1mf – 9 – 0-7905-5603-0 – mf#1988-1603 – us ATLA [240]
Augustin, Caspar *see* Der newen cornet und fahnen
Augustin, Catherine *see* Deixis
Augustin, Hermann *see* Goethes und stifters nausikaa-tragoedie
Augustin tuengers facetiae / ed by Keller, Adelbert von – Stuttgart: Litterarischer Verein, 1874 (Tuebingen: H Laupp) [mf ed 1993] – (= ser Blvs 118) – 163p – 1 – (latin & german text) – mf#70 – us Wisconsin U Libr [430]
Augustin tuengers facetiae / ed by Keller, Adelbert von – Stuttgart: Litterarischer Verein, 1874 (Tuebingen: H Laupp) [mf ed 1993] – (= ser Blvs 118) – 163p – 1 – (latin & german text) – mf#8470 reel 25 – us Wisconsin U Libr [450]
Augustin und Luther : ein historisch-apologetischer versuch / Roos, Johannes – Guetersloh: C Bertelsmann 1876 [mf ed 1992] – 1mf – 9 – 0-524-03911-9 – mf#1990-1170 – us ATLA [240]
Augustine : the divination of demons and care for the dead / Antwerp, E I van – Washington DC, 1955 – 2mf – 8 – €5.00 – ne Slangenburg [240]
Augustine *see* Confessions of st augustine
Augustine and his companions : four lectures / Browne, George Forrest – 3rd ed. London: SPCK 1906 [mf ed 1989] – 1mf – 9 – 0-7905-4161-0 – mf#1988-0161 – us ATLA [242]
Augustine and his companions : four lectures / delivered at st paul's in january, 1895 / Browne, George Forrest – London: S.P.C.K., 1906 – 1mf – us ATLA [240]
Augustine of canterbury / Cutts, Edward Lewes – London: Methuen, 1895. (Leaders of religion) – 1mf – us ATLA [241]
Augustine of canterbury / Cutts, Edward Lewes – London: Methuen 1895 [mf ed 1989] – (= ser Leaders of religion) – 1mf – 9 – 0-7905-4221-8 – mf#1988-0221 – us ATLA [241]
Augustine, Saint *see* Sermons
Augustine, Saint, Bishop of Hippo *see*
– The anti-pelagian works of saint augustine, bishop of hippo
– De catechizandis rudibus
– The city of god
– The confessions of augustine
– The confessions of st augustine
– Fuenf festpredigten augustins in gereimter prosa
– Letters of saint augustine, bishop of hippo
– On christian doctrine
– Preaching and teaching according to s augustine
– S aurelii augustini confessiones
– La sagrada regla del gran doctor de la iglesia n p san agustin
– Select anti-pelagian treatises of st augustine
– Selections
– The sermon on the mount expounded
– The soliloques of saint augstine
– The soliloquies of st. augustine
– The confessions of st augustine, bishop of hippo
– Thirteen homilies of st augustine on st john 14
– Treatise of saint aurelius augustine, bishop of hippo
Augustine synthesis – New York, NY. 1945 – 1r – us UF Libraries [240]
Der augustinermoench johannes hoffmeister : ein lebensbild aus der reformationszeit / Paulus, Nikolaus – Freiburg i B: Herder, 1891 – 1mf – 9 – 0-524-01586-4 – (incl bibl ref) – mf#1990-0452 – us ATLA [240]
Augustines lehre von der einheit und dreieinheit in ihrer bedeutung fuer sein und erkennen / Mueller, Eberhard – Erlangen [Germany]: K. Doerres, 1929 – 1r – 1 – 0-8370-1489-1 – mf#1984-B040 – us ATLA [240]
The augustinian revolution in theology : illustrated by a comparison with the teaching of the antiochene divines of the fourth and fifth centuries / Allin, Thomas; ed by Lias, John James – London: J Clarke, 1911 – 1mf – 9 – 0-7905-3517-3 – mf#1989-0010 – us ATLA [240]
Augustinis, Aemilius M de *see* The true faith of our forefathers
Der augustinismus : eine dogmengeschichtliche studie / Rottmanner, Odilo – Muenchen: JJ Lentner, 1892 – 1mf – 9 – 0-524-03913-5 – (incl bibl ref) – mf#1990-1172 – us ATLA [240]
Augustinus : sein theologisches system und seine religionsphilosophische anschauung / Dorner, August – Berlin: W Hertz, 1873. Chicago: Dep of Photodup, U of Chicago Lib, 1969 (1r); Evanston: American Theol Lib Assoc, 1984 (1r) – 1 – 0-8370-0405-5 – (incl bibl ref and ind) – mf#1984-B098 – us ATLA [240]
Augustinus (Augustine, Saint, Bishop of Hippo) *see*
– 15 buecher ueber die dreieinigkeit, 11. bd (bdk13 2.reihe)
– 15 buecher ueber die dreieinigkeit, 12. bd (bdk14 2.reihe)
– Ausgewaehlte briefe, 9. bd 9 1. teil (bdk29 1.reihe)
– Ausgewaehlte briefe, 9. bd 9 2. teil (bdk30 1.reihe)
– Ausgewaehlte praktische schriften homiletischen und katechetischen inhalts, 8. bd (bdk49 1.reihe)
– Bekenntnisse, 7. bd (bdk18 1.reihe)
– Gottesstaat, 1. bd (bdk1 1.reihe)
– Gottesstaat, 2. bd (bdk16 1.reihe)
– Gottesstaat, 3. bd (bdk28 1.reihe)
– Liber soliloquiorum
– Vortraege ueber das evangelium des hl johannes, 4. bd (bdk8 1.reihe)
– Vortraege ueber das evangelium des hl johannes, 5. bd (bdk11 1.reihe)
– Vortraege ueber das evangelium des hl johannes, 6. bd (bdk19 1.reihe)
Augustinus (Augustine, Saint, Bishop of Hippo) [comp] *see* Duo tractatus, quorum alter vocatur florigerus
Augustinus, Orosius *see* Contra adversarium legis et prophetarum. contra priscillianistas et orienistas. de errore priscillianistarum et origenistarum
Augustinus, St *see*
– Confessiones
– De doctrina christiana
– In psalmos (siecle 7)
– Retractationes
Augustinus seu doctrina sancti augustini de humanae naturae sanitate / Cornelius Jansen, the Elder – Rothomagi. v1-3. 1652 – 56mf – 8 – €107.00 – ne Slangenburg [241]
Augustinus-blatt – Duesseldorf DE, 1918-1934 feb – 2r – 1 – gw Misc Inst [074]
Augustiny, Waldemar *see*
– Die tochter tromsees
– Die wiederkehr des novalis
Augusto, Jose *see* Presidencialismo versus parlamentarismo
Augusto leverger : almirante barao de melgaco / Taunay, Alfredo D'escragnolle Taunay – Sao Paulo, Brazil. 1931? – 1r – us UF Libraries [972]
Augustus bohse genannt talander : ein beitrag zur geschichte der galanten zeit in deutschland / Schubert, Ernst – Breslau: F Hirt, 1911 [mf ed 1992] – (= ser Breslauer beitraege zur literaturgeschichte. neue folge 17) – 110p – 1 – (incl bibl ref) – mf#8014 reel 3 – us Wisconsin U Libr [430]
Augustus caesar and the organization of the empire of rome / Firth, John Benjamin – New York: G P Putnam 1903, c1902 [mf ed 1990] – (= ser Heroes of the nations) – 1mf – 9 – 0-7905-5531-X – mf#1988-1531 – us ATLA [930]
Augustus m toplady and contemporary hymn-writers / Wright, Thomas – London: Farncombe 1911 [mf ed 1993] – (= ser Lives of the british hymn writers 2) – 1mf – 9 – 0-524-06506-3 – mf#1991-2606 – us ATLA [242]
Augustus velleris aurei ordo per emblemata, ectheses politicas et historiam demonstratus / Erath, A – Ratisbonae: Sumptibus JZ Seidelii, typis JG Hofmanni, 1697 – 3mf – 9 – mf#O-1568 – ne IDC [090]
Aujourd' hui – Paris, France 9 mar 1942-17 jul 1944 – 4r – 1 – (wanting: n1191,1215-1223) – uk British Libr Newspaper [074]
Aujourd'hui et demain : les evenements devoiles / Bertrand, Isidore – Paris: Bloud et Barral [18–] – 2mf – 9 – mf#vrl-11 – ne IDC [366]
Aujourd'hui-quebec : mensuel d'idees et d'information – Montreal: [s.n.] v1 n1 mars 1965-v3 n9 nov 1967 [mf ed 1976] – 1r – 5 – mf#SEM16P274 – cn Bibl Nat [073]
Auk – Champaign IL 1884+ – 1,5,9 – ISSN: 0004-8038 – mf#467 – us UMI ProQuest [590]
Aukers, Steven M *see*
– The development of a decision precess map
– The development of an empirically grounded set of salient ski resort attributes
Aul, Joachim *see*
– Bibliographischer zugang zur griechischen philosophie
– Schopenhauer bibliographie
– Schopenhauers begruendungstheorie im lichte der ergebnisse der modernen wissenschaftstheorie
Aulae turcicae, othomanniciq've imperii descriptio...pars 1. solymanni 12 and selymi 13 tvrcar impp contra christianos...pars 2 / Geuffroy, A – Basileae, 1577 – 7mf – 9 – mf#H-8246 – ne IDC [950]
Aulaea romana : contra peristromata turcica expansa / [Harsdoerffer, G P] – n.p, 1642 – 1mf – 9 – mf#O-1599 – ne IDC [090]
Aulard, Francois Victor Alphonse *see* La societe des jacobins
Aulard, Francois-Alphonse *see* Les orateurs de la revolution
The auld kirkyard, fergus : in it, and about it / Fordyce, Alexander Dingwall – S.l: s.n, 1882? – 1mf – 9 – mf#05582 – cn Canadiana [929]
Aulen, Gustaf *see*
– Brev till henrik reuterdahl
– Dogmhistoria
– Evangelisk kyrklighet
– H reuterdahls teologiska askadning
– Den kristna tankens tolkning af jesu person
– Syndernas foerlatelse
– Till belysning af den lutherska kyrkoiden
Aulia, dr *see* "Makanan jang sehat" dan beberapa ichtiar jang lain boeat memeliharakan kesehatan dan menolong menjemboehkan penjakit
Aulio persio flaco-saturnae...et scholiis (sic) / Sanchez de las Brozas, Francisco – 1599 – 9 – sp Bibl Santa Ana [450]
L'Aulnaye, M de [Francois Henri Stanislas] *see* Thuileur des trente-trois degres de l'ecossisme du rit ancien, dit accepte
Aulnoy, Marie-Catherine d' *see* Contes des fees
Aum, the cosmic light – v4 n53-1958 [1978 apr-sep] – 1 – mf#361934 – us WHS [071]
Aum, the cosmic light newsletter – 1978 nov-dec; 1979 easter; 1979 dec – 1 – mf#667904 – us WHS [071]
Aumsville advance – Aumsville OR: J A Seabury, [wkly] [mf ed 1967] – 1r – 1 – us Oregon Lib [071]
Aumsville star – Aumsville OR: C S Clark, 1923- [wkly] [mf ed 1967] – 2r – 1 – (cont: weekly record (aumsville, or)) – us Oregon Lib [071]
Aun mhut nan nann lam nvhan : san kyan ma re a tvak / Tin Win – Ran kun: Kyui pyo ca pe 1979 [mf ed 1990] – 1r with other items – 1 – (in burmese; with bibl) – mf#mf-10289 seam reel 177/3 [§] – us CRL [615]
L'aunae insectorum germanicae initia / ed by Panzer, G Vd E – Nuernberg, Regensburg 1792-1844 – pt1-190 on 174mf – 9 – mf#2455/2 – ne IDC [590]
Aung Htu, Saw *see* Karan nhac sac ku pvai
An aunsweare unto certaine assertions : tending to maintaine the churche of rome, to bee the true and catholique churche / Knewstub, J – London: Thomas Dawson, 1579 – 3mf – 9 – mf#PW-72 – ne IDC [240]
Auquier, Philippe *see* Pierre puget
Aura : periodico litterario e recreativo – Rio de Janeiro, RJ. 25 set-18 out 1881 – (= ser Ps 19) – mf#P17,01,67 – bl Biblioteca [440]
Aurand, Charles Monroe *see* Rays of light
Aurangzeb and his times / Faruki, Zahiruddin – Bombay: DB Taraporevala Sons & Co, 1935 – (= ser Samp: indian books) – us CRL [954]
Auras de argeme / Villalobos Bote, Rufino – Coria: Edit. Fernandez, 1953 – 1 – sp Bibl Santa Ana [946]
Aureli, Willy *see*
– Roncador
– Sertoes bravios
Aurelino leal : sua vida, sua repoca, sua obra hami / Leal, Hamilton – Rio de Janeiro, Brazil. 1968 – 1r – us UF Libraries [972]
Aurelius, Marcus *see* The meditations of marcus aurelius antonius
Aureville, J A d' *see* De la passion du jeu, de l'infidelite des joueurs, et de leurs ruses
Auricular confession / Hook, Walter Farquhar – London, England. 1848? – 1r – us UF Libraries [240]
Auricular confession / Lowe, Thomas Hill – Exeter, England. 1852 – 1r – us UF Libraries [240]
Auricular confession and popish nunneries / Hogan, William – London, England. 1848 – 1r – us UF Libraries [240]
Auricular confession in the protestant episcopal church / Hawks, Francis Lister – New York: Geo P Putnam 1850 [mf ed 1990] – 1mf – 9 – 0-7905-7238-9 – mf#1988-3238 – us ATLA [242]
L'aurora – Paterson NJ, sep 15 1899-may 1930 – 1r – 1 – (italian newspaper) – us IHRC [071]
L'aurora – Utica NY, 1928-29* – 1r – 1 – (italian periodical) – us IHRC [073]
A aurora – Cameta, PA. 26 maio 1887 – (= ser Ps 19) – bl Biblioteca [079]
A aurora : periodico litterario e critico – Rio de Janeiro, RJ: Typ de F A de Almeida, 15 jun-17 ago 1851 – (= ser Ps 19) – mf#P15,01,49 – bl Biblioteca [440]
Aurora – 1899-1927 – 1 – sw Kungliga [070]
Aurora – Joseph OR: J A Burleigh, [wkly] – 1 – (ceased in 1897; place of publ moved to enterprise, or 1895; absorbed by: wallowa herald) – us Oregon Lib [071]
Aurora / Columbiana Co. New Lisbon – (mar 1838-jun 1856) very scattered [wkly] – 1r – 1 – mf#B30147 – us Ohio Hist [071]
Aurora – La Salle IL 1834-35 – 1 – mf#3718 – us UMI ProQuest [420]
Aurora – v1 n1-v2 n8 (1986 oct-1987 dec) [1]; 1983 spring-1984 spring; 1986 fall-1987 spring [2] – 1 – mf#1611809 [1]; 1049418 [2] – us WHS [071]
Aurora : eine zeitschrift aus dem suedlichen deutschland / ed by Joh Chr, Freiherr von Aretin et al – Muenchen 1804-05 – (= ser Dz. abt literatur) – 2jge on 8mf – 9 – €160.00 – mf#k/n4664 – gw Olms [430]
La aurora – Albuquerque, NM. v. 1-14. 1900-1915. (incomplete file) – 1 – $50.00 – us Presbyterian [071]
La aurora – [Habana]: Imprenta de la viuda Barcina y compa. v1 n1-v3 n1. oct 22 1865-may 3 1868 – us CRL [079]
The aurora – Freetown, Sierra Leone. -w. 16 April-31 Dec 1921. 1 reel – 1 – uk British Libr Newspaper [072]
The aurora : a monthly for the mothers and daughters of the south and west – Murfreesboro, TN. 2237p. jan 1858-jun 1861 – 2r – 1 – mf#6984 – us Southern Baptist [242]
The aurora : monthly magazine...as a monthly record of our work, and of indian education and progress – Middle Church, Man: Rupert's Land Industrial School, [1893?-189- or 19–] – 9 – mf#P04339 – cn Canadiana [370]
[Aurora-] aurora daily times – NV. 27-28 nov and 12 dec 1863 (3 issues only) – 1r – 1 – $60.00 – mf#U04400 – us Library Micro [071]
Aurora banner – Ontario, CN. jan 1900-dec 1975 – 49r – 1 – cn Commonwealth Imaging [071]
Aurora borealis – 1833 – (= ser English gift books and literary annuals, 1823-1857) – 4mf – 9 – uk Chadwyck [800]
Aurora borealis – 1898 dec 31-1899 mar 1 – 1 – mf#853292 – us WHS [071]
Aurora borealis – Aurora OR. Dixon & Hoskinson, [wkly] – 1r – 1 – (ceased in 1909; absorbed by: tribune (1909)) – us Oregon Lib [071]

[Aurora-] borealis – NV. 23 dec 1905 – 1r – 1 – $60.00 – mf#U04402 – us Library Micro [071]
Aurora commercial – 1863 jan 1; 1864 sep 8; 1967 mar 16-23; apr 20, may 18, oct 26 – 1 – mf#855966 – us WHS [071]
Aurora commercial advertiser – 1867 spring – 1 – mf#855968 – us WHS [071]
Aurora daily beacon-news – Aurora IL, 1940 oct 20 – 1r – 1 – mf#1145518 – us WHS [074]
la aurora en copacavana / Calderon de la Barca, Pedro – Madrid: Rivadeneyra, 1850 – 1 – sp Bibl Santa Ana [946]
[Aurora-] esmeralda Star – NV. jul 1862 – 1r – 1 – $60.00 – (Suppl to 30 Dec 1863) – mf#U04401 – us Library Micro [071]
Aurora fluminense – Rio de Janeiro, RJ: Typ do Republico, 26 maio-22 ago 1855 – (= ser Ps 19) – bl Biblioteca [320]
Aurora [greenock, scotland] – [Scotland] Inverclyde, Greenock: J Lennox 16 aug 1848 [mf ed 2004] – 1r – 1 – uk Newsplan [072]
The aurora karakul sheep co : breeders of karakul sheep – Aurora [Ont: Grand & Toy, 1915?] – 1mf – 9 – 0-659-90805-0 – mf#9-90805 – cn Canadiana [636]
Aurora news – Aurora, NE: A L Burr. v1 n1. apr 19 1929-v14 n29. nov 6 1942=sun ser: v44 n2185-v59 n18 (wkly) – 9r – 1 – (cont: aurora sun. merged with: republican-register (aurora ne) to form: aurora news-register) – us Bell [071]
Aurora news – Aurora, NE: A L Burr. v1 n1. apr 19 1929-v14 n29. nov 6 1942=sun ser. v44 n2185-sun ser. v59 n18 (wkly) [mf ed apr 19 1929-nov 6 1942 (gaps)] – 1 – (cont: aurora sun. merged with: republican-register, to form: aurora news-register) – us NE Hist [071]
Aurora news – Aurora, NE: Hellings & Stone. 3v. old ser: v12 n29. aug 21 1885-v14 n22. jun 22 1887 (wkly) – 1r – 1 – (cont: hamilton county news; absorbed by: aurora republican; iss for aug 21 1885-aug 27 1886 called also new ser: v1 n1-v2 n2) – us Bell [071]
Aurora news – Aurora, NE: Hellings & Stone. 3v. os: v12 n29. aug 21 1885-v14 n22. jun 22 1887 (wkly) [mf ed aug 21 1885-aug 27 1887 filmed 2000] – 2r – 1 – (cont: hamilton county news. absorbed by: aurora republican. issues for aug 21 1885-aug 27 1886 called also new ser. v1 n1-v2 n2) – us NE Hist [071]
Aurora news-register – Aurora, NE: Bremer Pub Co. v70 n34. nov 13 1942- (wkly) – 12r – 1 – (formed by the union of: aurora news (1929) and: republican-register (aurora ne). issues for nov 13 1942-feb 26 1965 called also v14 n30-v36 n49. issues for nov 20 1942-nov 23 1967 called also v1 n2-v26 n1) – us Bell [071]
Aurora news-register – Aurora, NE: Bremer Pub Co. v70 n34. nov 13 1942 (wkly) [mf ed aug 10 1956] – 55r – 1 – (formed by the union of: aurora news (1929), and: republican-register. issues for nov 13 1942-feb 26 1965 called also v14 n30-v36 n49. issues for nov 20 1942-nov 23 1967 called also v1 n2-v26 n1) – us NE Hist [071]
Aurora observer – Aurora OR: N C Westcott, -1940 [wkly] [mf ed 1967] – 4r – 1 – (cont by: north marion county observer (1940-19-?)) – us Oregon Lib [071]
Aurora of the valley – Newbury VT 4 mar 1848-26 dec 1868 – 1 – uk British Libr Newspaper [071]
Aurora paulistana : folha litteraria, industrial e politica – Sao Paulo, SP: Typ Commercial, 28 ago 1851-07 set 1852 – (= ser Ps 19) – mf#P19,03,07 – bl Biblioteca [079]
Aurora republican – Aurora, NE: L W Hastings, 1873-v56 n44. mar 22 1929 (wkly) – 27r – 1 – (absorbed: aurora news. merged with: hamilton county register to form: hamilton county republican-register. publ as: aurora daily republican sep 16-18 1891) – us Bell [071]
Aurora republican – Aurora, NE: L W Hastings, 1873-v56 n44. mar 22 1929 (wkly) [mf ed apr 25 1877-mar 22 1929 (gaps) filmed 2000] – 1 – (absorbed: aurora news. merged with: hamilton county register to form: hamilton county republican-register. publ as: aurora daily republican sep 16-18 1891) – us NE Hist [071]
Aurora Snow Shoe Club *see* Constitution and by-laws of the aurora snow shoe club
The aurora sun – Aurora, NE: E W Hurlbut. v1 n1. aug 8 1885-apr 19 1929// (wkly) [mf ed aug 8 1885-jan 17 1929 (gaps)] – 13r – 1 – (cont by: aurora news. suspended foll nov 26 1926 issue; resumed on may 3 1929 carrying two numberings: v1 n1, and old ser. v42 n2154) – us NE Hist [071]
The aurora telegraph – Aurora, NE: Sheppard & Fritz. 2v. -v2 n8. feb 18 1879 (wkly) – 1r – 1 – us Bell [071]
Aurora trade guide and advertiser! – [Aurora, Ont?]: J R Beden, [1865-18– or 19–] [mf ed v1 n1 apr 1865] – 9 – mf#P06074 – cn Canadiana [071]
Aurora und christliche woche – Buffalo NY (USA), 1921 28 oct-1924, 1926-27, 1929-1933 14 apr, 1933 15 sep-1936 25 dec [gaps] – 5r – 1 – gw Misc Inst [240]
Aurora volksfreund – Aurora, IL: Peter Klein, jul 2 1917-jun 16 1922 – 10r – 1 – us CRL [071]
Aurora weekly herald – 1869 jun 22 – 1 – mf#874017 – us WHS [071]
Aurora weekly standard – 1853 oct 6 – 1 – mf#856267 – us WHS [071]
Auroras, poetias / Morandeyra, Mary – Havana, Cuba. 1929 – 1r – us UF Libraries [972]
L'aurore – Fort-de-France. nov 1932-avr 1935 – 1 – fr ACRPP [079]
L'aurore – Paris. 1er oct 1897-2 aout 1914 – 1 – fr ACRPP [074]
L'aurore – Paris. sept 1944-1986 – 1 – fr ACRPP [073]
L'aurore de la republique – Vaugirard [Paris]: Impr de Moncheny. n1. feb 27 1848 – 1r – 1 – us CRL [074]
Aus alten arten / Lehmann, Emil – Dresden, Germany. 1886 – 1r – us UF Libraries [939]
Aus alten tagen – Lindlar DE, 1906-1907 9 mar – 1r – 1 – (suppl to: bergischer tuermer) – gw Misc Inst [074]
Aus amerika : erfahrungen, reisen und studien / Froebel, Julius – Leipzig 1857-58 [mf ed Hildesheim 1995-98] – 2v on 8mf – 9 – €160.00 – 3-487-27021-8 – gw Olms [917]
Aus amerika : reisebriefe / Herzog, Karl – Berlin 1884 [mf ed Hildesheim 1995-98] – 2v on 6mf – 9 – €120.00 – 3-487-27053-6 – gw Olms [880]
Aus amerika / Wislicenus, Gustav A – Leipzig 1854 [mf ed Hildesheim 1995-98] – 2v on 2mf – 9 – €60.00 – 3-487-27029-3 – gw Olms [917]
Aus bimbos seelenwanderungen *see* Das judengrab / aus bimbos seelenwanderungen
Aus carmen sylva's leben / Stackelberg, Natalie, Freiin von – 4. aufl. Heidelberg: Carl Winter 1886 [mf ed 1989] – 1r – 1 – (filmed with: astra) – mf#7214 – us Wisconsin U Libr [920]
Aus chamissos fruehzeit : ungedruckte briefe nebst studien / Chamisso, Adelbert von; ed by Geiger, Ludwig – Berlin: Paetel 1905 [mf ed 1989] – 1r – 1 – mf#7151 – us Wisconsin U Libr [860]
Aus dantes verbannung : literarhistorische studien / Scheffer-Boichorst, Paul – Strassburg, 1882 (mf ed 1992) – 2mf – 9 – €24.00 – 3-89349-066-3 – mf#DHS-AR 29 – gw Frankfurter [450]
Aus dem altbabylonischen recht / Meissner, Bruno – Leipzig: J C Hinrichs 1905 [mf ed 1989] – (= ser Der alte orient 7/1) – 1mf – 9 – 0-7905-2053-2 – mf#1987-2053 – us ATLA [340]
Aus dem alten wien / Vogl, Johann – Wien 1865 [mf ed Hildesheim: 1995-98] – (= ser Fbc) – 270p on 2mf – 9 – €60.00 – 3-487-29462-1 – gw Olms [914]
Aus dem archiv der deutschen schillerstiftung *see* Veroeffentlichungen aus dem archiv der deutschen schillerstiftung, weimar
Aus dem belagerten tsingtau : tagebuchblaetter / Voskamp, Carl John – Berlin: Berliner evang Missionsgesellschaft, 1915 [mf ed 1995] – (= ser Yale coll) – 143p – 1 – 0-524-09534-5 – (in german) – mf#1995-0534 – us ATLA [880]
Aus dem belagerten tsingtau : tagebuchblaetter von c j voskamp – Berlin: Berliner evang. Missionsgesellschaft, 1915 (mf ed 19–) – (= ser Day missions monographs on microfilm) – 143p – mf#Z-BTZE pv136 n7 – us NY Public [951]
Aus dem boehmerwalde / Rank, Josef – Prag: J G Calve (Robert Lerche) 1917 [mf ed Bloomington IN: Indiana Uni Lib, Preservation Dept 1984] – 1r – 1 – us Indiana Preservation [390]
Aus dem dunkelsten berlin – Berlin DE, 1898-1913 – 1 – gw Misc Inst [074]
Aus dem felde : [poems] / Wolff, Julius – Berlin: G Grote 1895 [mf ed 1995] – (= ser Grote'sche sammlung von werken zeitgenoessischer schriftsteller 55) – 1r – 1 – (filmed with: ottilie wildermuths gesammelte werke / ed by ida lackowitz) – mf#3764p – us Wisconsin U Libr [810]
Aus dem geistesleben der thiere : oder staaten und thaten der kleinen / ed by Roser, Andreas – Leipzig: 1880 (mf ed 1997) – (= ser Passauer texte zur philosophie) – 5mf – 9 – €59.00 – 3-8267-3211-1 – mf#DHS 3211 – gw Frankfurter [110]
Aus dem goethe-national-museum / ed by Ruland, Carl – Weimar: Goethe-Gesellschaft, 1895-1904 [mf ed 1993] – (= ser Schriften der goethe-gesellschaft 10,12,19) – 3 portfolios (ill) – 1 – (incl bibl ref) – mf#8657 reel 3 – us Wisconsin U Libr [060]
Aus dem harze : skizzen und sagen / Proehle, Heinrich – Leipzig 1851 [mf ed Hildesheim: 1995-98] – (= ser Fbc) – viii/120p on 1mf – 9 – €40.00 – 3-487-29586-5 – gw Olms [880]
Aus dem indischen leben / Hoevell, Walter R van – Leipzig 1868 [mf ed Hildesheim 1995-98] – 1v on 2mf – 9 – €60.00 – 3-487-27485-X – gw Olms [954]
Aus dem inneren leben der katholischen kirche im 19. jahrhundert : erster band / Nielsen, Fredrik – Karlsruhe: H Reuther 1882 [mf ed 1986] – 1mf – 9 – 0-8370-7895-4 – (no more publ; incl bibl ref) – mf#1986-1895 – us ATLA [241]
Aus dem jahrhundert des grossen krieges / Freytag, Gustav – New York: Maynard, Merrill, & Co 1894 [mf ed 1989] – (= ser Maynard's german texts 15) – 1r – 1 – (advanced text with [english] int & notes by r j morich. filmed with: ratsel um herta / hermann freyberg) – mf#7273 – us Wisconsin U Libr [840]
Aus dem josephinischen wien : geblers und nicolais briefwechsel waehrend der jahre 1771-1786 / Gebler, Tobias Philipp, Freiherr von; ed by Werner, Richard Maria – Berlin: W Hertz (Bessersche Buchhandlung) 1888 [mf ed 1989] – 1r – 1 – (incl ind. filmed with: so war das / rudolf geck; emanuel geibels gesammelte werke) – mf#7285 – us Wisconsin U Libr [860]
Aus dem kampf der schwaermer gegen luther : drei flugschriften (1524, 1525) / ed by Enders, Ludwig – Halle a. S: M Niemeyer, 1893 [mf ed 1993] – (= ser Neudrucke deutscher literaturwerke des 16. und 17. jahrhunderts 118; Flugschriften aus der reformationszeit 10) – xviii/55p – 1 – (incl bibl ref) – mf#8413 reel 5 – us Wisconsin U Libr [943]
Aus dem kaukasus : reisen und studien / Hahn, C von – Leipzig, 1892 – 4mf – 9 – mf#AR-1596 – ne IDC [914]
Aus dem klassenkampf : soziale gedichte / ed by Fuchs, Eduard et al – Berlin: Akademie-Verlag, 1978 [mf ed 1993] – (= ser Textausgaben zur fruehen sozialistischen literatur in deutschland v18) – xxxvii/89p – 1 – (incl bibl ref) – mf#8367 – us Wisconsin U Libr [810]
Aus dem lager der goethe-gegner / Holzmann, Michael – Berlin: B Behr, 1904 [mf ed 1993] – (= ser Deutsche litteraturdenkmale des 18. und 19. jahrhunderts 129, 3. folge n9) – 224p – 1 – (incl bibl ref) – mf#8676 reel 7 – us Wisconsin U Libr [430]
Aus dem leben : skizzen / Christen, Ada – Leipzig: E J Guenther 1876 [mf ed 1993] – 1r – 1 – (filmed with: arbeiterlesebuch, nicht nur fuer arbeiter / ed by werkstatt bremen) – mf#3471p – us Wisconsin U Libr [880]
Aus dem leben der arabischen bevoelkerung in sfax (regentschaft tunis) / Narbeshuber, Karl – Leipzig: R Voigtlaender 1907 – (= ser Veroeffentlichungen des staedtischen museums fuer voelkerkunde zu leipzig 2) – 1mf – 9 – 0-524-01857-X – (incl selections in arabic, roman transcr & german trans) – mf#1990-2692 – us ATLA [390]
Aus dem leben der juden deutschlands im mittelalter / Berliner, Abraham; ed by Ellbogen, Ismar – Berlin, 1937 (mf ed 1996) – (= ser Monographien zur wissenschaft des judentums) – 2mf – 9 – €31.00 – 3-8267-3195-6 – mf#DHS 3195 – gw Frankfurter [270]
Aus dem leben eines dorpater universitaetslehrers : erinnerungen des mediziners prof dr friedrich v bidder, 1810-1894 – Wuerzburg: Holzner-Verlag 1959, c1958 [mf ed 1992] – (= ser Ostdeutsche beitraege aus dem goettinger arbeitskreis 11) – 10r [ill] – 1 – (incl bibl ref & ind. filmed with: ostdeutsche beitraege aus dem goettinger arbeitskreis) – mf#3180p – us Wisconsin U Libr [370]
Aus dem leben eines reformierten pastors / Zahn, Adolf – 2. veraend aufl. Barmen: H Klein [1885?] [mf ed 1991] – 1mf – 9 – 0-524-01036-6 – (first printed anonymously in 1881) – mf#1990-0313 – us ATLA [242]
Aus dem leben und der arbeit eines china-missionaers / Leuschner, F W – Berlin: Berliner evangelischen Missionsgesellschaft, [1902] [mf ed 1995] – (= ser Yale coll) – 128p (ill) – 1 – 0-524-09497-7 – (in german) – mf#1995-0497 – us ATLA [920]
Aus dem leben von niklaus bolt / Teuteberg, Rene et al – Basel: F Reinhardt, [194-?] [mf ed 1989] – 33p – 1 – mf#7047 – us Wisconsin U Libr [920]
Aus dem letzten jahrzehnt vor dem vatikankonzil / Nippold, Friedrich – Jena: Hermann Costenoble 1899 [mf ed 1986] – 2mf – 9 – 0-8370-9088-1 – (incl bibl ref) – mf#1986-3088 – us ATLA [241]
Aus dem missionsleben draussen feur die arbeit daheim / Witte, Johannes – Berlin: Hutten-Verlag, 1919 [mf ed 1995] – (= ser Yale coll) – xiv/378p – 1 – 0-524-09583-3 – (in german) – mf#1995-0583 – us ATLA [920]
Aus dem nachlass / Fontane, Theodor; ed by Ettlinger, Josef – 4.aufl. Berlin: F Fontane, 1908 [mf ed 1989] – xviii/316p/[1pl] (ill) – 1 – mf#7074 – us Wisconsin U Libr [800]
Aus dem nachlass varnhagen's von ense *see* Tagebuecher
Aus dem natur- und voelkerleben im tropischen amerika : skizzenbuch / Scherzer, Karl von – Leipzig 1864 [mf ed Hildesheim 1995-98] – 1v on 3mf – 9 – €90.00 – 3-487-26997-X – gw Olms [880]
Aus dem ostlande – Lissa i.P: O Eulitz 1916-[jahrg11-13(1916-18)] [mf ed 198-] – 4v [ill] – 1 – (cont: aus dem posener lande; ceased in 1919 with v14 n6; incl suppl: dies und das aus dem ostlande; imprint varies) – mf#film mas c705 – us Harvard [943]
Aus dem palmenlande : selbsterlebtes aus ost- und westindien / Flex, Oscar Theodor – Geutersloh: C Bertelsmann, 1907 [mf ed 1995] – (= ser Yale coll) – vi/282p (ill) – 1 – 0-524-09944-8 – (in german) – mf#1995-0944 – us ATLA [306]
Aus dem persoenlichen verkehre mit franz grillparzer / Littrow-Bischoff, Auguste von – Wien: L Rosner 1873 [mf ed 1990] – 1r – 1 – (filmed with: franz grillparzer / adalbert faulhammer) – mf#2689p – us Wisconsin U Libr [430]
Aus dem reich der todten *see* Politische gespraeche der todten
Aus dem sozialen und politischen kampf : die zwoelf artikel der bauern, 1525 / ed by Goetze, A – Halle (Saale): M Niemeyer, 1953 [mf ed 1993] – (= ser Neudrucke deutscher literaturwerke des 16. und 17. jahrhunderts 322) – 64p – 1 – (incl bibl ref) – mf#8413 reel 11 – us Wisconsin U Libr [943]
Aus dem tagewerk eines assyrischen zauberpriesters / Ebeling, E – Leipzig, 1931 – (= ser Mitteilungen der altorientalischen Gesellschaft) – 1mf – 9 – (mitteilungen der altorientalischen gesellschaft. v5, pt 3) – mf#NE-20104 – ne IDC [956]
Aus dem walde / Burckhardt Heinrich – Hannover: C Ruempler 1865- (irreg) [1.-10.heft (1865-81)] [mf ed 2005] – 2r [ill] – 1 – (ceased in 1900?) – mf#film mas c6052 – us Harvard [634]
Aus dem weichseldelta : reiseskizzen / Passarge, Louis – Berlin 1857 [mf ed Hildesheim 1995-98] – 3mf – 9 – €90.00 – 3-487-29539-3 – gw Olms [914]
Aus den anfaengen der sozialistischen dramatik / ed by Muenchow, Ursula – Berlin: Akademie-Verlag, 1965-87, c1965-72 [mf ed 1993] – (= ser Textausgaben zur fruehen sozialistischen literatur in deutschland 3, 5, 11) – 3v – 1 – (incl bibl ref) – mf#8192 – us Wisconsin U Libr [430]
Aus den archiven des belgischen kolonialministeriums berlin, 1916 / Belgium. Ministere des Colonies – 1Folge. Berlin, 1918 – 1 – us CRL [960]
Aus den briefen der herzogin elisabeth charlotte von orleans an etienne polier de bottens / ed by Hellmann, S – Stuttgart: Litterarischer Verein, 1903 (Tuebingen: H Laupp, Jr) [mf ed 1993] – (= ser Blvs 231) – xviii/131p – 1 – (incl bibl ref and ind. french text. int in german) – mf#8470 reel 47 – us Wisconsin U Libr [860]
Aus den briefen der herzogin elisabeth charlotte von orleans an etienne polier de bottens / Orleans, Charlotte-Elisabeth, duchesse d'; ed by Hellmann, S – Stuttgart: Litterarischer Verein 1903 (Tuebingen: H Laupp, Jr) [mf ed 1993] – 58r – 1 – (incl bibl ref & ind; french text, int in german. filmed with: bibliothek des literarischen vereins in stuttgart) – mf#3420p – us Wisconsin U Libr [860]
Aus den briefen des paulus nach korinth / Aner, Karl – Tuebingen: J C B Mohr (Paul Siebeck) 1913 [mf ed 1986] – (= ser Religionsgeschichtliche volksbuecher fuer die deutsche christliche gegenwart 6/1) – 1mf – 9 – 0-8370-9523-9 – mf#1986-3523 – us ATLA [227]
Aus den erinnerungen eines achtundvierzigers = Recollections of a 48'er / Mueller, Jakob – Cleveland, OH: Rud, Schmidt Printing Co, 1896 – 1r – 1 – (german language history of immigration, and the early german community in cleveland) – mf#F34ZSL G1M9 Vault – us Western Res [304]
Aus den hochgebirgen von granada : naturschilderungen, erlebnisse und erinnerungen; nebst granadinischen volkssagen und maerchen / Willkomm, Moritz – Wien 1882 [mf ed Hildesheim 1995-98] – 3mf – 9 – €90.00 – 3-487-29852-X – gw Olms [390]
Aus den noerdlichen kalkalpen : ersteigungen und erlebnisse in den gebirgen berchtesgadens, des algaeu, des innthales, des isar-quellengebietes und des wettersteins; mit erlaeuternden beitraegen zur orographie und hypsometrie der noerdlichen kalkalpen / Barth, Hermann von – Gera 1874 [mf ed Hildesheim: 1995-98] – (= ser Fbc) – xxiv/637p on 5mf – 9 – €100.00 – 3-487-29476-1 – gw Olms [914]

Aus den tagen bonifaz 8 : funde und forschungen / Finke, Heinrich – Muenster i.W: Aschendorff 1902 [mf ed 1986] – (= ser Vorreformationsgeschichtliche forschungen 2) – 2mf – 9 – 0-8370-7860-1 – (incl bibl ref & ind) – mf#1986-1860 – us ATLA [241]

Aus den tagen der hansa : drei novellen / Jensen, Wilhelm – Freiburg i B: Kiepert & von Bolschwing, 1885 [mf ed 1996] – 3v – 1 – mf#9698 – us Wisconsin U Libr [830]

Aus den tagen der occupation : eine osterreise durch nordfrankreich und elsass-lothringen 1871 / Fontane, Theodor – Berlin 1871 [mf ed Hildesheim 1995-98] – 2v on 5mf – 9 – €100.00 – 3-487-29722-1 – gw Olms [914]

Aus den vorbergen : novellen / Heyse, Paul – 2. aufl. Berlin: W Hertz 1893 [mf ed 1995] – 1r – 1 – (filmed with: der zerbrochene franz / ludwig hevesti) – mf#3637p – us Wisconsin U Libr [830]

Aus der antiken schule : sammlung griechischer texte auf papyrus, holztafeln, ostraka / Ziebarth, Erich – Bonn: A Marcus & E Weber 1910 [mf ed 1992] – (= ser Kleine texte fuer theologische und philologische vorlesungen und uebungen 65) – 1mf – 9 – 0-524-04321-3 – (incl bibl ref) – mf#1990-1247 – us ATLA [370]

Aus der arbeit des heims des judischen frauenbundes isenburg / Pappenheim, Bertha – Frankfurt am Main, Germany. 1926 – 1r – us UF Libraries [939]

Aus der barockbibliothek nuenning : sammlung von seltenen werken zur kulturgeschichte = From the baroque library nuenning – [mf ed 2001] – 510mf – 9 – €2850 diazo €3420 silver – 3-89131-376-4 – (incl unimarc-files & catalogue) – gw Fischer [020]

Aus der briefmappe eines burgtheaterdirektors / Dingelstedt, Franz, Freiherr von – Wien: Schroll 1925 [mf ed 1993] – 1r [ill] – 1 – (with biogr sketch & ann by karl glossy; incl bibl ref & ind. filmed with: gedichte / deinhardstein) – mf#8539 – us Wisconsin U Libr [790]

Aus der burschenzeit : ein idyll / Leander, Richard – Halle a/S: M Niemeyer 1876 [mf ed 1993] – 1r – 1 – (filmed with: die woelfe / herbert volck) – mf#2946p – us Wisconsin U Libr [810]

Aus der ecke : sieben neue novellen / Riehl, Wilhelm Heinrich – Bielefeld: Velhagen und Klasing 1874 [mf ed 1995] – 1r [ill] – 1 – (filmed with: jean paul / richard benz) – mf#3715p – us Wisconsin U Libr [830]

Aus der ferne : [arie fuer singstimme mit pianofortebegleitung] / Righini, Vincenzo – [Berlin: Concha 180-?] [mf ed 1989] – 1r – 1 – mf#pres. film 51 – us Sibley [780]

Aus der heimat : neue gedichte / Prutz, Robert Eduard – Leipzig: F A Brockhaus 1858 [mf ed 1991] – 1r – 1 – (filmed with: liebe ist ewig / wilhelm von polenz) – mf#2864p – us Wisconsin U Libr [810]

Aus der heimat – Zittau DE, 1899-1901 – 1r – 1 – gw Misc Inst [074]

Aus der jugendzeit : lebenserinnerungen / Stahr, Adolf Wilhelm Theodor – Schwerin i. M: A Hildebrand 1870-77 [mf ed 1979] – 2v in 1 on 1r – 1 – mf#film mas 8405 – us Harvard [920]

Aus der knabenzeit / Gutzkow, Karl – Frankfurt/Main: Literarische Anstalt, 1852 [mf ed 2001] – xii/305p – 1 – mf#10526 – us Wisconsin U Libr [920]

Aus der rechtsgeschichte benediktinischer verbaende / Molitor, R – Muenster i.W. v1-3. 1928-33 – 3v on 37mf – 8 – €71.00 – ne Slangenburg [241]

Aus der schule des wulfila : auxenti dorostorensis epistula de fide vita et obitu wulfilae / ed by Kauffmann, Friedrich – Strassburg: K J Truebner 1899 [mf ed 1992] – (= ser Texte und untersuchungen zur altgermanischen religionsgeschichte 1) – 2mf – 9 – 0-524-02787-0 – (incl bibl ref) – mf#1987-6481 – us ATLA [243]

Aus der tiefe *see* Unser grundstoff

Aus der waffenkammer des sozialismus – Frankfurt/M DE, 1903-10 – 1r – 1 – mf#6080 – gw Mikropress [335]

Aus der welt der papyri / Wessely, Carl – Leipzig: H Haessel 1914 [mf ed 1989] – 1mf – 9 – 0-7905-3116-X – (with bibl app) – mf#1987-3116 – us ATLA [930]

Aus der werdezeit des christentums : studien und charakteristiken / Geffcken, Johannes – 2. aufl. Leipzig: B G Teubner 1909 [mf ed 1990] – (= ser Aus natur und geisteswelt 54) – 1mf – 9 – 0-7905-5213-2 – mf#1988-1213 – us ATLA [240]

Aus der werkstatt : studien und anregungen / Fulda, Ludwig – Stuttgart, Berlin: J G Cotta 1904 [mf ed 1989] – 1r – 1 – (filmed with: gedichte & other titles) – mf#7280 – us Wisconsin U Libr [880]

Aus der zeit, gegen die zeit : gesammelte essays / Berg, Leo – Leipzig: Huepeden & Merzyn, 1905 [mf ed 1989] – vii/453p – 1 – mf#7008 – us Wisconsin U Libr [840]

Aus deutschen bussbuechern : ein beitrag zur deutschen culturgeschichte / Friedberg, Emil – Halle: Verlag der Buchh. des Waisenhauses 1868 [mf ed 1990] – 1mf – 9 – 0-7905-5874-2 – (in german & latin; incl bibl ref) – mf#1988-1874 – us ATLA [240]

Aus deutscher seele : ein buch volkslieder / Jacobowski, Ludwig – Minden/Westf: J C C Brun [1899] [mf ed 1993] – 1r – 1 – (incl bibl ref. filmed with: die ammen-uhr / dresdener kuenstlern) – mf#8359 – us Wisconsin U Libr [780]

Aus eigener kraft : gedanken und erfahrungen eines versehrten... / Kugelgen, Carlo Von – Nurnberg, Germany. 1943 – 1r – us UF Libraries [920]

Aus eigener kraft – Teutschenthal DE, 1951 1 apr-1956 16 nov [gaps] – 2r – 1 – (kaliwerk) – gw Misc Inst [622]

Aus eigener kraft : roman in drei baenden / Hillern, Wilhelmine von – 2. aufl. Leipzig: E Keil's Nachfolger [189-?] [mf ed 1994] – 3v on 1r – 1 – mf#3639p – us Wisconsin U Libr [830]

Aus einem griechischen zauberpapyrus / Wuensch, Richard – Bonn: A Marcus & E Weber, 1911 [mf ed 1992] – (= ser Kleine texte fuer vorlesungen und uebungen 84) – 1mf – 9 – 0-524-04709-X – (text in greek. notes in german) – mf#1990-3418 – us ATLA [130]

Aus einer ganz kleinen garnison : [a novel] – Dresden-Niedersedlitz: H G Muenchmeyer [18–?] [mf ed 1988] – 1r – 1 – (filmed with: anastasius groun's gesammelte werke / ed by ludwig august frankl) – mf#6970 – us Wisconsin U Libr [830]

Aus einer ganz kleinen garnison – baal *see* Anastasius gruen's gesammelte werke

Aus einer kleinen garnison : ein militaerisches zeitbild / Bilse, Fritz Oswald – [Vienna]: Wiener Verlag, 1904 [mf ed 1995] – 269p – 1 – mf#9161 – us Wisconsin U Libr [830]

Aus einer kleinen stadt / Freytag, Gustav – Leipzig: S Hirzel 1880 [mf ed 1993] – 1r – 1 – (filmed with: jenny / fanny lewald) – mf#8579 – us Wisconsin U Libr [880]

Aus england : studien und briefe ueber londoner theater, kunst und presse / Fontane, Theodor – Stuttgart: Ebner & Seubert, 1860 [mf ed 1989] – viii/325p – 1 – mf#7248 – us Wisconsin U Libr [790]

Aus englischen bibliotheken / Levison, Wilhelm – Hannover: Hahn [1910?] [mf ed 1986] – 1mf – 9 – 0-8370-7808-3 – (in german & latin; incl bibl ref) – mf#1986-1808 – us ATLA [241]

Aus friedrich hebbels werdezeit / Neumann, Alfred, of Zittau – Zittau: M Boehme 1899 [mf ed 1992] – (= ser Jahresbericht des koeniglichen realgymnasiums in zittau; ostern 1899 587) – 1 – (incl bibl ref. filmed with: die goethe-bildnesse / hermann rollett) – mf#3079p – us Wisconsin U Libr [430]

Aus friedrichs hebbels korrespondenz : ungedruckte briefe von und an den dichter nebst beitraegen zur textkritik einzelner werke / ed by Hirth, Friedrich – Muenchen, Leipzig, 1913 (mf ed 1994) – 2mf – 9 – €31.00 – 3-8267-3021-6 – mf#DHS-AR 3021 – gw Frankfurter [430]

Aus frischer luft : gedichte / Fischer, Johann Georg – 2. aufl. Stuttgart: C Grueninger 1873 [mf ed 1993] – 1r – 1 – (filmed with: auf dem heimweg) – mf#8576 – us Wisconsin U Libr [810]

Aus fuenf jahrtausenden morgenlaendischer kultur : festschrift max freiherrn von oppenheim zum 70 geburtstage gewidmet von freunden und mitarbeitern / [Oppenheim, M von] – 5mf – 9 – (archiv fuer orientforschung, berlin 1933 beiband 1) – mf#NE-20055 – ne IDC [956]

Aus goethes archiv : die erste weimarer gedichtsammlung in facsimile-wiedergabe / ed by Suphan, Bernhard & Wahle, Julius – Weimar: Goethe-Gesellschaft, 1908 [mf ed 1993] – (= ser Schriften der goethe-gesellschaft 23) – 1 portfolio/26p/24pl – 1 – (incl bibl ref) – mf#8657 reel 6 – us Wisconsin U Libr [430]

Aus goethe's freundeskreise : darstellungen aus dem leben des dichters / Duentzer, Heinrich – Braunschweig: F Vieweg, 1868 [mf ed 1991] – xi/552p – 1 – (incl bibl ref) – mf#7544 – us Wisconsin U Libr [920]

Aus goethes lebenskreis : drei essays / Kahn-Wallerstein, Carmen – Bern: A Francke, 1946 [mf ed 1990] – 107p – 1 – mf#7373 – us Wisconsin U Libr [430]

Aus goethes tagebuechern / Graef, Hans Gerhard [comp] – Leipzig: Insel-Verlag 1908 [mf ed 1990] – 1r [ill] – 1 – (incl bibl ref, int by comp) – mf#7366 – us Wisconsin U Libr [880]

Aus grossen hoehen : alpenroman / Ompteda, Georg, Freiherr von – Berlin: F Fontane 1903 [mf ed 1991] – 1r – 1 – (filmed with: das passions-schauspiel in oberammergau) – mf#2857p – us Wisconsin U Libr [830]

Aus grosser zeit *see* Die post

Aus hellas, rom und thule : cultur- und litteraturbilder / Poestion, Josef Calasanz – 2. aufl. Leipzig: W Friedrich [1884] [mf ed 1991] – 1r – 1 – (filmed with: mutter marie / heinrich mann) – mf#2844p – us Wisconsin U Libr [410]

Aus herders nachlass / Herder, Johann Gottfried; ed by Duentzer, Heinrich & Herder, Ferdinand Gottfried von – Frankfurt a.M: Meidinger, 1856-57 [mf ed 1992] – 3v – 1 – (incl bibl ref) – mf#7636 – us Wisconsin U Libr [430]

Aus indien : reisebriefe eines missionaers / Noti, Severin – 1. aufl. Einsiedeln; New York: Benziger, 1908 [mf ed 1995] – (= ser Yale coll) – 370p (ill) – 1 – 0-524-09916-2 – (in german) – mf#1995-0916 – us ATLA [241]

Aus israels propheten : amos, hosea, jesaja, jeremia, deuterojesaja – Tuebingen: J C B Mohr 1914 [mf ed 1993] – (= ser Religionsgeschichtliche volksbuecher fuer die deutsche christliche gegenwart 6/5) – 1mf – 9 – 0-524-05793-1 – mf#1992-0620 – us ATLA [221]

Aus jerusalem / Hahn-Hahn, Ida M – Mainz 1851 [mf ed Hildesheim 1995-98] – 1v on 2mf – 9 – €60.00 – 3-487-27685-2 – gw Olms [915]

Aus joh jac winckelmanns briefen / ed by Meszlenyi, Richard – Berlin: B Behr (F Feddersen), 1913- [mf ed 1993] – (= ser Deutsche litteraturdenkmale des 18. und 19. jahrhunderts 145, 3. folge n25) – 1 – (no more publ?) – mf#8676 reel 9 – us Wisconsin U Libr [860]

Aus kampfgewuehl und einsamkeit : gedichte / Seidel, Robert – Stuttgart: J H W Dietz 1895 [mf ed 1991] – 1r – 1 – (filmed with: heinrich seidel und der deutsche humor / alfred biese) – mf#2942p – us Wisconsin U Libr [810]

Aus kaukasischen laendern : reisebriefe / Abich, H – Wien, 1896. 2v – 11mf – 9 – mf#AR-1583 – ne IDC [915]

Aus klassischer zeit : wieland und reinhold: original-mittheilungen als beitraege zur geschichte des deutschen geisteslebens im 18. jahrhundert / ed by Keil, Robert – new ed. Leipzig: W Friedrich [1890] [mf ed 1991] – 1r – 1 – (1st ed publ in 1885 under title: wieland und reinhold; incl ind. filmed with: the graces / christoph martin wieland) – mf#3049p – us Wisconsin U Libr [860]

Aus kunst und leben / Keppler, Paul Wilhelm von – 2. unver aufl. Freiburg i.B, St Louis MO: Herder 1905 [mf ed 1986] – 1mf – 9 – 0-8370-6818-5 – mf#1986-0818 – us ATLA [700]

Aus lavaters brieftasche : ungedruckte handschrifte nebst lavater-erinnerungen mit facsimiles / Lavater, Johann Caspar; ed by Mueller, Gustav Adolf – Muenchen: Seitz & Schauer 1897 [mf ed 1990] – 1r [ill] – 1 – (filmed with: katenlud / fritz lau) – mf#2818p – us Wisconsin U Libr [800]

Aus leben und arbeit / Schuchhardt, Karl – Berlin: W de Gruyter 1944 [mf ed 1986] – 1r [ill] – 1 – (filmed with: biblical literature and its backgrounds / macarthur, j r) – mf#7240 – us Wisconsin U Libr [930]

Aus lichtenbergs nachlass : aufsaetze, gedichte, tagebuchblaetter, briefe: zur hundertsten wiederkehr seines todestages (24. februar 1799) / ed by Leitzmann, Albert – Weimar: H Boehlau Nachfolger, 1899 [mf ed 1996] – xxiii/272p/1pl – 1 – mf#9708 – us Wisconsin U Libr [800]

Aus luise hensels jugendzeit : neue briefe und gedichte, zum jahrhunderttag ihrer konversion (8 dezember 1818) / Cardauns, Hermann – Freiburg i.B: Herder 1918 [mf ed 1990] – 1r – 1 – (incl bibl ref. filmed with: j j wilhelm heinse und die asthetik zur zeit der deutschen aufklarung / von emil utitz) – mf#2721p – us Wisconsin U Libr [920]

Aus lydien : epigraphisch-geographische reisefruechte / Buresch, K – Leipzig, 1898 – 6mf – 9 – mf#NE-110 – ne IDC [915]

Aus masorah und talmudkritik : exegetische studien / Koenigsberger, Bernhard – Berlin: Mayer & Mueller 1892 [mf ed 1985] – 1mf – 9 – 0-8370-3977-0 – (incl bibl ref) – mf#1985-1977 – us ATLA [221]

Aus meinem leben : erinnerungen des tamulenpastors nj dewasagajam b a – Leipzig: Verlag der Evangelisch-lutherischen Mission, 1919 [mf ed 1995] – (= ser Yale coll) – 120p (ill) – 1 – 0-524-09954-5 – (in german) – mf#1995-0954 – us ATLA [920]

Aus meinem leben : erinnerungsblaetter / Bodenstedt, Friedrich M von – Leipzig 1879 [mf ed Hildesheim: 1995-98] – (= ser Fbc) – 1v on 2mf – 9 – €60.00 – 3-487-29475-3 – gw Olms [880]

Aus meinem leben / Felder, Franz Michael; ed by Schoenbach, Anton E – Wien: Literarischer Verein 1904 [mf ed 1993] – (= ser Schriften des literarischen vereins in wien 2) – 5r – 1 – mf#3333p – us Wisconsin U Libr [430]

Aus meinem leben : selbstbiographie / Bretschneider, Karl Gottlieb; ed by Bretschneider, Horst – 2. ausg. Gotha: J G Mueller 1852 [mf ed 1990] – 1mf [ill] – 9 – 0-7905-4845-3 – (incl bibl ref) – mf#1988-0845 – us ATLA [240]

Aus meiner jugend : autobiographie / Lazarus, M[oritz] – Frankfurt a.M. 1913 (mf ed 1996) – (= ser Monographien zur wissenschaft des judentums) – 2mf – 9 – €31.00 – 3-8267-3188-3 – (vorwort und anhang herausgegeben von nahida lazarus) – mf#DHS 60 – gw Frankfurter [920]

Aus meiner liedermappe : gedichte / Zeise, Heinrich – 2. verm veraend aufl. Hannover: A Weichelt 1883 [mf ed 1992] – 1r – 1 – (filmed with: feldmuenster / franz graf zedtwitz) – mf#3066p – us Wisconsin U Libr [810]

Aus morgenland und abendland : neue gedichte und sprueche / Bodenstedt, Friedrich Martin von – 2. aufl. Leipzig: F A Brockhaus, 1884 [mf ed 1989] – x/284p – 1 – mf#7040 – us Wisconsin U Libr [810]

Aus nachgelassenen schriften eines fruehvollendeten / Braun, Otto; ed by Vogelstein, Julie – Berlin: B Cassirer, 1920 [mf ed 1989] – 307p (ill) – 1 – mf#7063 – us Wisconsin U Libr [880]

Aus nacht zum licht : drei erzaehlungen von hueben und drueben / Graepp, L W et al – Milwaukee, WI: G Brumder, [1889?] [mf ed 1993] – 363p – 1 – mf#8363 – us Wisconsin U Libr [830]

Aus nord und sued : missionsblatt der bruedergemeine fuer die jugend – Herrnhut: Missionsbuchhandlung 1900-39 [bimthly, mthly] [mf ed 2005] – v1-40 (1900-39) on 1r – 1 – (subtitle varies; several iss missing; some pgs damaged) – mf#2005c-s043 – us ATLA [242]

Aus ottilie von goethes nachlass / ed by Oettingen, Wolfgang von – Weimar: Goethe-Gesellschaft, 1912-13 [mf ed 1993] – (= ser Schriften der goethe-gesellschaft 27-28) – 2v – 1 – (incl bibl ref) – mf#8657 reel 7 – us Wisconsin U Libr [860]

Aus paris : beitraege zur charakteristik des gegenwaertigen frankreichs / Lindau, Paul – Stuttgart 1865 [mf ed Hildesheim 1995-98] – 2mf – 9 – €60.00 – 3-487-29634-9 – gw Olms [944]

Aus politik und zeitgeschichte – Bonn: Bundeszentrale fuer politische Bildung etc 1945- [wkly] [mf ed 1984-90] – 7r – 1 – (suppl to: wochen zeitung, das parlament; 1956-58 also suppl to: deutsche studentenzeitung) – mf#8643 – us Wisconsin U Libr [321]

Aus schleiermachers hause : jugenderinnerungen seines stiefsohnes / Willich, Ehrenfried von – Berlin: G Reimer 1909 [mf ed 1991] – 1mf [ill] – 9 – 0-524-00215-0 – mf#1989-2915 – us ATLA [240]

Aus schleiermachers leben *see* The life of schleiermacher

Aus schrift und geschichte : theologische abhandlungen und skizzen – Basel: R Reich 1898 [mf ed 1989] – 2mf – 9 – 0-7905-1550-4 – (incl bibl ref) – mf#1987-1550 – us ATLA [240]

Aus schwerer vergangenheit : ein geschichten-cyklus / Jensen, Wilhelm – 4. aufl. Leipzig: B Elischer Nachf, [191-?] [mf ed 1996] – 381p – 1 – mf#9698 – us Wisconsin U Libr [830]

Aus sibirien : lose blaetter aus dem tagebuche eines reisenden linguisten / Radlov, V V [Radloff, W] – Leipzig, 1884. v1-2 – 20mf – 9 – mf#U-331 – ne IDC [915]

Aus spaetherbsttagen : erzaehlungen / Ebner-Eschenbach, Marie von – Berlin: Gebrueder Paetel 1902 [mf ed 1993] – 2v on 1r – 1 – (filmed with: erzaehlungen / marie von ebner-eschenbach) – mf#8572 – us Wisconsin U Libr [830]

Aus tirol : berg- und gletscher-reisen in den oesterreichischen hochalpen / Ruthner, Anton von – Wien 1869 [mf ed Hildesheim: 1995-98] – (= ser Fbc) – viii/464p on 3mf – 9 – €90.00 – 3-487-29430-3 – gw Olms [914]

Aus transkaukasien und armenien : reisebriefe / Petersen, W – Leipzig, 1885 – 2mf – 9 – mf#AR-1982 – ne IDC [915]

Aus vergangenheit und gegenwart der juden und den juedischen gemeinden in den posener laendern – Koschmin, Wroclaw PL, 1904-29 – 1r – 1 – us UMI ProQuest [939]

Aus vier welttheilenein : reise-tagebuch in briefen / Wichura, Max – Breslau 1868 [mf ed Hildesheim 1995-98] – 1v on 3mf – 9 – €90.00 – 3-487-26633-4 – gw Olms [910]

Aus wald und heide : geschichten und schilderungen / Loens, Hermann – Hannover: A Sponholtz [193-?] [mf ed 1995] – 1r [ill] – 1 – (filmed with: bert brecht / willy haas) – mf#3941p – us Wisconsin U Libr [880]

Aus wissenschaft und leben / Harnack, Adolf von – Giessen: A Toepelmann 1911 [mf ed 1989] – (= ser Reden und aufsaetze: neue folge 1-2) – 2v on 2mf – 9 – 0-7905-1088-X – mf#1987-1088 – us ATLA [240]

Aus zwei seelen : neue gedichte / Presber, Rudolf – 3. & 4. stark verm aufl. Stuttgart: Deutsche Verlags-Anstalt 1919 [mf ed 1996] – 1r – 1 – (filmed with: wurzellocker / wilhelm von polenz) – mf#3986p – us Wisconsin U Libr [810]
Ausbreitung und verfall der romantik / Huch, Ricarda Octavia – Leipzig: H Haessel, 1902 [mf ed 1993] – 365p/[3]p – 1 – (incl bibl ref) – mf#8360 – us Wisconsin U Libr [430]
Ausbreitungsbiologische merkmalstypenspektrenanalyse von bryophytengesellschaften am beispiel eines ariden und eines tropischen standortes / Bachmann, Cordula – (mf ed 2000) – 3mf – 9 – €49.00 – 3-8267-2688-X – mf#DHS 2688 – gw Frankfurter [574]
Ausdrucksbewegungen in heinrich von kleists werk / Blau, Georg – (mf ed 1995) – 7mf – 9 – €65.00 – 3-8267-2093-8 – mf#DHS 2093 – gw Frankfurter [430]
Ausdrucksformen der lateinischen liturgiesprache bis ins elfte jahrhundert *see* Das irische palimpsest-sakramentar in clm 14429 [tab53-54]
Der ausdrucksgehalt des menschlichen ganges / Kietz, Gertraud – 2nd enl ed. Leipzig: J A Barth, 1952 [mf ed 2002] – (= ser Beihefte...zur zeitschrift fuer angewandte psychologie und charakterkunde 93) – 1r – 1 – (incl bibl ref) – mf#5175 – us Wisconsin U Libr [150]
Der ausdrucksgehalt des menschlichen ganges / Kietz, Gertraud – Leipzig: J A Barth, 1948 [mf ed 2002] – (= ser Beihefte...zur zeitschrift fuer angewandte psychologie und charakterkunde 93) – 1r – 1 – (incl bibl ref. filmed with (reel 12): beitraege zur typologie und symptomatologie der arbeitskurve / von heinz remplein (v91 1942)) – mf#5153 reel 12 – us Wisconsin U Libr [150]
Ausdruckswelt : essays und aphorismen / Benn, Gottfried – Wiesbaden: Limes, 1949 [mf ed 1989] – 112p – 1 – mf#7006 – us Wisconsin U Libr [840]
Die auseinandersetzung mit der franzoesischen revolution in der geschichtsschreibung der "kleindeutschen" schule / Voelker, Monika – Frankfurt a.M., 1978 – 3mf – 9 – 3-89349-382-4 – gw Frankfurter [370]
Ausencia pura / Gonzalez Y Contreras, Gilberto – Mexico City?, Mexico. 1946 – 1r – us UF Libraries [972]
Ausencia y presencia de jose matial delgado / Duran, Miguel Angel – San Salvador, El Salvador. 1961 – 1r – us UF Libraries [440]
Auserlesene anmerckungen ueber allerhand wichtige materien und schriften / ed by Thomasius, Christian – Frankfurt/Leipzig 1704-07 [mf ed 1993] – (= ser Aus den anfaengen des zeitschriftenwesens: fruehe deutsche zeitschriften) – 5v on 9mf – 9 – €140.00 – 3-89131-093-5 – gw Fischer [074]
Auserlesene bibliothek der neuesten deutschen litteratur / ed by Helwing, Christian Friedrich et al – Lemgo 1772-81 – (= ser Dz) – 20v on 98mf – 9 – €980.00 – mf#k/n291 – gw Olms [430]
Auserlesene gedichte / Opitz, Martin; ed by Mueller, Wilhelm – Leipzig: F A Brockhaus, 1822 [mf ed 1993] – xx/220p – 1 – mf#8455 – us Wisconsin U Libr [810]
Auserlesene gedichte deutscher poeten / Zinkgref, Julius Wilhelm [comp]; ed by Braune, Wilhelm – Halle a. S: M Niemeyer 1879 [mf ed 1993] – (= ser Neudrucke deutscher literaturwerke des 16. und 17. jahrhunderts 15) – 11r – 1 – (incl bibl ref. filmed with: neudrucke deutscher literaturwerke des 16. und 17. jahrhunderts) – mf#3387p – us Wisconsin U Libr [810]
Auserlesene litteratur des katholischen deutschlands / ed by Schwarz, Ildefons – Koburg 1788-90 – (= ser Dz. abt literatur) – 3v[zu je 4st] on 13mf – 9 – €130.00 – mf#k/n4569 – gw Olms [241]
Auserlesene sammlung vermischter oeconomischer schriften / ed by Riem, J – Dresden 1790-92 – (= ser Dz) – 2v+2 suppl on 8mf – 9 – €160.00 – mf#k/n2976 – gw Olms [330]
Ausfahrt und landung : festgabe fuer bibliotheksdirektor wolfgang van der briele, zum 65. geburtstag am 16. mai 1959 / ed by Stadtbibliothek Wuppertal – Wuppertal: J H Born 1960 [mf ed 1992] – 1r [ill] – 1 – (incl bibl ref; filmed with: studienausgaben zur neueren deutschen literatur ed by deutschen akademie der wissenschaften zu berlin...) – mf#3137p – us Wisconsin U Libr [020]
Ausflucht an den rhein und dessen naechste umgebungen im sommer des ersten friedlichen jahres / Schopenhauer, Johanna – Leipzig 1818 [mf ed Hildesheim: 1995-98] – (= ser Fbc) – 296p on 2mf – 9 – €60.00 – 3-487-29547-4 – gw Olms [914]
Ausflug an den niederrhein und nach belgien im jahr 1828 / Schopenhauer, Johanna – Leipzig 1831-32 [mf ed Hildesheim: 1995-98] – (= ser Fbc) – 2v on 4mf – 9 – €120.00 – 3-487-29548-2 – gw Olms [914]
Ein ausflug nach comacchio / Jacoby, Leopold – Triest: J Dase, 1881 – 1r – 1 – us Wisconsin U Libr [800]
Ein ausflug nach england / Heeringen, Gustav von – Gotha 1841 [mf ed Hildesheim 1995-98] – (= ser Fbc) – 3mf – 9 – €90.00 – 3-487-28817-6 – gw Olms [914]
Ausflug nach portugal im sommer 1863 : mit einer abhandlung ueber die portugiesische sprache / Brandes, Heinrich K – Lemgo [u.a.] 1864 [mf ed Hildesheim 1995-98] – 2mf – 9 – €60.00 – 3-487-29821-X – gw Olms [914]
Ausflug ueber constantinopel nach taurien im sommer 1831 / Brunner, Samuel – St Gallen [u. a.] 1833 [mf ed Hildesheim 1995-98] – (= ser Fbc) – 3mf – 9 – €90.00 – 3-487-28987-3 – gw Olms [915]
Ausflug von lissabon nach andalusien : und in den norden von marokko im fruehjahr 1845 / Loewenstein, Wilhelm zu – Dresden [u a] 1846 [mf ed Hildesheim 1995-98] – 2mf – 9 – €60.00 – 3-487-29860-0 – gw Olms [914]
Ausfuehrliche anleitung zu der gantzen civil-bau-kunst... : nebst denen lebens-beschreibungen und den fuenf ordnungen von j bar de vigla wie auch dessen und des beruehmten michangelo vornemsten gebaeuden / Daviler, A C – Amsterdam, 1699 – 6mf – 9 – mf#OA-101 – ne IDC [720]
Ausfuehrliche anleitung zur buergerlichen baukunst / Penther, J F – Augsburg, 1744-1748. 4v – 24mf – 9 – mf#OA-106 – ne IDC [720]
Ausfuehrliche beschreibung des meissnischen ober-ertzgebuerges : nach seiner lage, gestalt, bergen, thaelern, felsen, fluessen...wie auch angemerckten zustande der elemente, himmels-zeichen, witterung und allerhand curioesen begebenheiten gefertiget / Lehmann, Christian – Leipzig: Bey Friedrich Lanckischens Erben 1747 [mf ed 1979] – 1r [ill/pls] – 1 – mf#film mas 8986 – us Harvard [914]
Ausfuehrliche encyklopaedie der gesammten staatsarzneikunde (ael3/23) / Most, Georg Friedrich – Leipzig 1838, 1840 [mf ed 1995] – (= ser Archiv der europaeischen lexikographie: fach-enzyklopaedien) – 2v+1 suppl vol on 29mf – 9 – €210.00 – 3-89131-219-9 – (int by michael stolberg) – gw Fischer [610]
Ausfuehrliche geographisch-statistisch-topographische beschreibung des regierungsbezirks erfurt : auf anordnung der koeniglichen regierung nach amtlichen und andern zuverlaessigen quellen, so wie nach den vom professor voelker hinterlassenen materialien / Noback, Carl A – Erfurt 1840 [mf ed Hildesheim: 1995-98] – (= ser Fbc) – 3mf – 9 – €90.00 – 3-487-29585-7 – gw Olms [914]
Ausfuehrliche grammatik der griechischen sprache : zweiter teil: satzlehre / Kuehner, Raphael – 3. aufl. Hannover: Hahn 1898-1904 [mf ed 1994] – 2v on 4mf – 9 – 0-524-08610-9 – (incl bibl ref) – mf#1993-0045 – us ATLA [450]
Ausfuehrliche volks-gewerbslehre : oder, allgemeine und besondere technologie zur belehrung und zum nutzen fur alle staende / Poppe, Johann Heinrich Moritz von – Stuttgart: C Hoffmann 1833 [mf ed 1980] – (= ser Goldsmiths'-kress library of economic literature) – 2v on 1r [ill] – 1 – (incl ind in v2) – us Primary [331]
Ausfuehrlicher bericht von allerhand neuen buechern und anderen dingen so zur heutigen historie der belehrsamkeit gehoerig / ed by Woltereck, C – Frankfurt/Leipzig 1708-10 [mf ed 1993] – (= ser Aus den anfaengen des zeitschriftenwesens: fruehe deutsche zeitschriften) – 5mf – 9 – €120.00 – 3-89131-088-9 – (cont: monatliche unterredungen and: curieuse bibliothec) – gw Fischer [070]
Ausfuehrliches lehrbuch der hebraeischen sprache / Boettcher, Friedrich – Leipzig: J A Barth 1866-68 [mf ed 1989] – 2v on 4mf – 9 – 0-8370-1313-5 – (incl ind) – mf#1987-6046 – us ATLA [470]
Ausfuehrliches verzeichnis der aegyptischen altertuemer und gipsabguesse / Erman, A – Berlin, 1899 – 6mf – 9 – mf#NE-20393 – ne IDC [930]
Die ausfuehrungsgesetze zum buergerlichen gesetzbuche : sammlung der von den bundesstaaten zur ausfuehrung des buergerlichen gesetzbuchs und seiner nebengesetze erlassenen gesetze und mit gesetzeskraft versehenen verordnungen / ed by Becher, Heinrich – Muenchen: J Schweitzer. v1-2. 1901 – (= ser Civil law 3 coll) – 36mf – 9 – (incl bibl ref and index) – mf#LLMC 96-536 – us LLMC [348]
Ausfuhrliche volks-gewerbslehre : oder allgemeine und besondere technologie / Poppe, G H M von – Stuttgart 1859 – 1 – gw Mikropress [600]
Ausg. nuernberg – 1933, 19.4.-31.5. – 1 – gw Mikropress [943]
Der ausgang der prophetie / Haller, Max – Tuebingen: J C B Mohr 1912 [mf ed 1989] – (= ser Religionsgeschichtliche volksbuecher fuer die deutsche christliche gegenwart 2/12) – 1mf – 9 – 0-7905-1605-5 – mf#1987-1605 – us ATLA [221]
Ausgewaehlt werke / Rosegger, Peter – Wien [etc]: A Hartleben 1888-91 [mf ed 1979] – 6v on 2r [ill] – 1 – (ill by a greil & a schmidhammer) – mf#film mas c568 – us Harvard [802]
Ausgewaehlte akademische reden und abhandlungen / Stade, Bernhard – Giessen: J Ricker 1899 [mf ed 1985] – 1mf – 9 – 0-8370-5353-6 – (incl bibl ref) – mf#1985-3353 – us ATLA [221]
Ausgewaehlte akten persischer martyrer (bdk22 1.reihe) – (= ser Bibliothek der kirchenvaeter. 1. reihe (bdk 1.reihe)) – €15.00 – ne Slangenburg [240]
Ausgewaehlte analytische loesungsmethoden fuer komplexe konvektive waermeuebertragungsprobleme / Weigand, Bernhard – (mf ed 1999) – 3mf – 9 – €49.00 – 3-8267-2624-3 – mf#DHS 2624 – gw Frankfurter [621]
Ausgewaehlte aspekte rechnergestuetzter informations- und kommunikationstechnologien / Schoop, Michael – (mf ed 1994) – 1mf – 9 – €30.00 – 3-8267-2021-0 – mf#DHS 2021 – gw Frankfurter [000]
Ausgewaehlte briefe / Strauss, David Friedrich; ed by Zeller, Eduard – Bonn: Emil Strauss 1895 [mf ed 1993] – 1mf [ill] – 9 – 0-524-07661-8 – mf#1992-1102 – us ATLA [190]
Ausgewaehlte briefe / Winckelmann, Johann Joachim; ed by Uhde-Bernays, Hermann – Leipzig: Insel-Verlag 1925 [mf ed 1991] – 1r [ill] – 1 – (filmed with: winckelmanns kleine schriften zur geschichte der kunst des altertums / ed by hermann uhde-bernays) – mf#3055p – us Wisconsin U Libr [700]
Ausgewaehlte briefe, 1. bd (bdk46 1.reihe) / Basilius des Grossen (Basil The Great, Saint (Basil of Caesarea)) – (= ser Bibliothek der kirchenvaeter. 1. reihe (bdk 1.reihe)) – €15.00 – ne Slangenburg [240]
Ausgewaehlte briefe, 9. bd 9 1. teil (bdk29 1.reihe) / Augustinus (Augustine, Saint, Bishop of Hippo) – (= ser Bibliothek der kirchenvaeter. 1. reihe (bdk 1.reihe)) – €18.00 – ne Slangenburg [241]
Ausgewaehlte briefe, 9. bd 9 2. teil (bdk30 1.reihe) / Augustinus (Augustine, Saint, Bishop of Hippo) – (= ser Bibliothek der kirchenvaeter. 1. reihe (bdk 1.reihe)) – €14.00 – ne Slangenburg [241]
Ausgewaehlte briefe aus den jahren 1883 bis 1902 / Dehmel, Richard – Berlin: S Fischer 1922 [mf ed 1989] – 1r [ill] – 1 – (incl ind. filmed with: schone wilde welt) – mf#7209 – us Wisconsin U Libr [860]
Ausgewaehlte briefe aus den jahren 1902 bis 1920 / Dehmel, Richard – Berlin: S Fischer 1923 [mf ed 1989] – 1r [ill] – 1 – (incl ind. filmed with: schone wilde welt) – mf#7209 – us Wisconsin U Libr [860]
Ausgewaehlte briefe (bdk18 2.reihe) / Hieronymus – (= ser Bibliothek der kirchenvaeter. 2. reihe (bdk 2.reihe)) – €18.00 – ne Slangenburg [240]
Ausgewaehlte briefen, 2. bd (bdk16 2.reihe) / Hieronymus – (= ser Bibliothek der kirchenvaeter. 2. reihe (bdk 2.reihe)) – €17.00 – ne Slangenburg [240]
Ausgewaehlte dichtungen / Klopstock, Friedrich Gottlieb; ed by Heinemann, K – Bielefeld: Velhagen & Klasing 1912 [mf ed 1995] – 1r [ill] – 1 – (incl bibl ref. filmed with: "penthesilea" in der kleistliteratur / werner schmidt) – mf#3676p – us Wisconsin U Libr [810]
Ausgewaehlte dichtungen / Vierordt, Heinrich – Heidelberg: C Winter 1906 [mf ed 1991] – 1r – 1 – (pref by ludwig fulda. filmed with: die wanderung des herrn ulrich von hutten / will vesper) – mf#2945p – us Wisconsin U Libr [810]
Ausgewaehlte erzaehlungen / Ebner-Eschenbach, Marie von – Berlin: Paetel 1910 [mf ed 1993] – 3v on 1r – 1 – mf#8571 – us Wisconsin U Libr [830]
Ausgewaehlte erzaehlungen / Keller, Gottfried – Wien: W Frick 1945 [mf ed 1995] – (= ser Deutsche erzaehler) – 1r – 1 – (filmed with: goethes faust / robert petsch) – mf#3647p – us Wisconsin U Libr [830]
Ausgewaehlte fabeln und gedichte / Pfeffel, Gottlieb Konrad – Stuttgart: Kroener [18–?] [mf ed 1991] – 1r – 1 – (filmed with: die prufungen der baptisten zu littleville / heinrich pfitzner) – mf#2863p – us Wisconsin U Libr [800]
Ausgewaehlte gedichte / Falke, Gustav – Hamburg: A Janssen 1908 [mf ed 1989] – (= ser Hamburgische hausbibliothek) – 1r – 1 – (filmed with: blut und eisen / max eyth) – mf#7228 – us Wisconsin U Libr [810]
Ausgewaehlte gedichte / Hesse, Hermann – 1.-5. aufl. Berlin: S Fischer 1921 [mf ed 1990] – 1r – 1 – (filmed with: das problem "volkstum und dichtung" bei herder / reta schmitz) – mf#2724p – us Wisconsin U Libr [810]
Ausgewaehlte historische, homiletische und dogmatische schriften, 1. bd (bdk15 1.reihe) / Hieronymus – (= ser Bibliothek der kirchenvaeter. 1. reihe (bdk 1.reihe)) – €19.00 – ne Slangenburg [240]
Ausgewaehlte homilien und predigten, 2. bd (bdk47 1.reihe) / Basilius des Grossen (Basil The Great, Saint (Basil of Caesarea)) – (= ser Bibliothek der kirchenvaeter. 1. reihe (bdk 1.reihe)) – €17.00 – ne Slangenburg [240]
Ausgewaehlte kapitel zu einer hans-sachs-grammatik / Albrecht, Julius – Freiburg, 1896 (mf ed 1995) – 1mf – 9 – €24.00 – 3-8267-3141-7 – mf#DHS-AR 3141 – gw Frankfurter [430]
Ausgewaehlte kleine schriften / Forster, Georg; ed by Leitzmann, Albert – Stuttgart: G J Goeschen, 1894 [mf ed 1993] – (= ser Deutsche litteraturdenkmale des 18. und 19. jahrhunderts 46-47) – xx/165p – 1 – mf#8676 reel 4 – us Wisconsin U Libr [800]
Ausgewaehlte kleine schriften / Gelzer, Heinrich – Leipzig: B G Teubner 1907 [mf ed 1990] – 1mf – 9 – 0-7905-7229-X – (incl bibl ref) – mf#1988-3229 – us ATLA [240]
Ausgewaehlte kritische schriften / Tieck, Ludwig – Tuebingen: M. Niemeyer c1975 [mf ed 1992] – 1r – 1 – (incl bibl ref & ind; int by ernst ribbat. filmed with: die zeit als einbildungskraft des dichters / emil staiger) – mf#3179p – us Wisconsin U Libr [410]
Ausgewaehlte maerchen und gedichte / Baumbach, Rudolf; ed by Manley, Edward – Boston; New York: Ginn, c1910 [mf ed 1989] – (= ser International modern language series) – xiii/209p – 1 – (incl int and notes) – mf#6983 – us Wisconsin U Libr [800]
Ausgewaehlte maertyreracten *see* Acta martyrum selecta
Ausgewaehlte novellen / Storm, Theodor – Weimar: Alexander Duncker [1919?] [mf ed 1995] – 1r [ill] – 1 – (filmed with: novellen; novellen der liebe) – mf#3749p – us Wisconsin U Libr [830]
Ausgewaehlte praktische schriften homiletischen und katechetischen inhalts, 8. bd (bdk49 1.reihe) / Augustinus (Augustine, Saint, Bishop of Hippo) – (= ser Bibliothek der kirchenvaeter. 1. reihe (bdk 1.reihe)) – €18.00 – ne Slangenburg [241]
Ausgewaehlte predigten (bdk43 1.reihe) / Petrus Chrysologus (Peter Chrysologus, Saint) – (= ser Bibliothek der kirchenvaeter. 1. reihe (bdk 1.reihe)) – €15.00 – ne Slangenburg [240]
Ausgewaehlte predigten johann taulers / Tauler, Johannes; ed by Naumann, Leopold – Bonn: A Marcus & E Weber 1914 [mf ed 1992] – (= ser Kleine texte fuer vorlesungen und uebungen 127) – 1mf – 9 – 0-524-04689-1 – mf#1990-1316 – us ATLA [240]
Ausgewaehlte predigten johann taulers (kit127) / ed by Naumann, L – Bonn, 1914 – €5.00 – ne Slangenburg [240]
Ausgewaehlte predigten und reden / Einhorn, David; ed by Kohler, Kaufmann – New York: E Steiger 1880 – viii/399p – 1 – mf#1247 – us Wisconsin U Libr [270]
Ausgewaehlte psalmen / Gunkel, Hermann – 4. verb aufl. Gottingen: Vandenhoeck & Ruprecht 1917 [mf ed 1985] – 1mf – 9 – 0-8370-3420-5 – (incl ind) – mf#1985-1420 – us ATLA [221]
Ausgewaehlte psalmen / Stoeckhardt, George – St Louis MO: Concordia 1915 [mf ed 1992] – 1mf – 9 – 0-524-04114-8 – mf#1992-0072 – us ATLA [221]
Ausgewaehlte reden und lieder / nisibenische hymnen, bd. 1 (bdk37 1.reihe) / Ephraem der Syrer (Ephraem Syrus, Saint) – (= ser Bibliothek der kirchenvaeter. 1. reihe (bdk 1.reihe)) – €15.00 – ne Slangenburg [240]
Ausgewaehlte schriften / Guenzburg, Johann Eberlin von; ed by Enders, Ludwig – Halle a.S: Max Niemeyer 1896-1902 [mf ed 1993] – (= ser Neudrucke deutscher literaturwerke des 16. und 17. jahrhunderts) – 11r – 1 – (incl bibl ref. filmed with: neudrucke deutscher literaturwerke des 16. und 17. jahrhunderts) – mf#3387p – us Wisconsin U Libr [800]
Ausgewaehlte schriften / Mueller, Otto – Gera: Griesbach 18-? [mf ed 1979] – (= ser 1r) – 12v on 1r – 1 – mf#film mas 8547 – us Harvard [802]
Ausgewaehlte schriften / Saphir, Moritz Gottlieb – 4 aufl. Bruenn etc: F Karafiat 1870 [mf ed 1979] – 10v in 5, 1r – 1 – mf#film mas 8494 – us Harvard [802]
Ausgewaehlte schriften / Saphir, Moritz Gottlieb – 6. aufl. Bruenn: Fr Karafiat 1871 [mf ed 1995] – 1r – 1 – mf#3719p – us Wisconsin U Libr [800]
Ausgewaehlte schriften / Speidel, Ludwig; ed by Radecki, Sigismund von – 1. aufl. Wedel in Holstein: Alster Verlag C Brauns 1947 [mf ed 1995] – (= ser Meister der kleinen form 2) – 1r – 1 – mf#3745p – us Wisconsin U Libr [800]

Ausgewaehlte schriften, 1. bd (bdk5 2.reihe) : zwoelf buecher ueber die dreieingkeit, buch 1-7 / Hilarius von Poitiers (Hilary of Poitiers, Saint) – (= ser Bibliothek der kirchenvaeter. 2. reihe (bdk 2.reihe)) – €15.00 – ne Slangenburg [240]

Ausgewaehlte schriften, 1. bd (bdk6 2.reihe) : zwoelf buecher ueber die dreieingkeit, buch 8-12 / Hilarius von Poitiers (Hilary of Poitiers, Saint) – (= ser Bibliothek der kirchenvaeter. 2. reihe (bdk 2.reihe)) – €14.00 – ne Slangenburg [240]

Ausgewaehlte schriften, 1. bd (bdk7 2.reihe) : mahnrede an die heiden. der erzieher 1 / Klemens von Alexandrien (Clement of Alexandria, Saint) – (= ser Bibliothek der kirchenvaeter. 2. reihe (bdk 2.reihe)) – €12.00 – ne Slangenburg [240]

Ausgewaehlte schriften, 1. bd (bdk57 1.reihe) / Armenische Vaeter – (= ser Bibliothek der kirchenvaeter. 1. reihe (bdk 1.reihe)) – €14.00 – ne Slangenburg [240]

Ausgewaehlte schriften, 2. bd (bdk8 2.reihe) : der erzieher 2-3 / Klemens von Alexandrien (Clement of Alexandria, Saint) – (= ser Bibliothek der kirchenvaeter. 2. reihe (bdk 2.reihe)) – €15.00 – ne Slangenburg [240]

Ausgewaehlte schriften, 2. bd (bdk58 1.reihe) / Armenische Vaeter – (= ser Bibliothek der kirchenvaeter. 1. reihe (bdk 1.reihe)) – €14.00 – ne Slangenburg [240]

Ausgewaehlte schriften, 2.bd (bdk1 2.reihe) : kirchengeschichte / Eusebius von Caesarea – (= ser Bibliothek der kirchenvaeter. 2. reihe (bdk 2.reihe)) – €18.00 – ne Slangenburg [240]

Ausgewaehlte schriften, 2.bd (bdk3 2.reihe) : vier buecher dialoge / Gregor der Grosse – (= ser Bibliothek der kirchenvaeter. 2. reihe (bdk 2.reihe)) – €14.00 – ne Slangenburg [240]

Ausgewaehlte schriften (bdk2 2.reihe) : angebliche schriften uber "gottliche namen". angeblicher brief an den moench demophilus / Dionysius Areopagita (Dionysius the Areopagite) – (= ser Bibliothek der kirchenvaeter. 2. reihe (bdk 2.reihe)) – €11.00 – ne Slangenburg [240]

Ausgewaehlte schriften (bdk9 2.reihe) / Fulgentius von Ruspe (Fulgentius of Ruspe, Saint) – (= ser Bibliothek der kirchenvaeter. 2. reihe (bdk 2.reihe)) – €11.00 – ne Slangenburg [241]

Ausgewaehlte schriften (bdk12 2.reihe) / Cyrillus von Alexandrien (Cyril of Alexandria, Saint) – (= ser Bibliothek der kirchenvaeter. 2. reihe (bdk 2.reihe)) – €12.00 – ne Slangenburg [240]

Ausgewaehlte schriften der armenischen kirchenvaeter / ed by Weber, S – Muenchen, 1927. 2v – 8mf – 9 – mf#AR-1572 – ne IDC [243]

Ausgewaehlte schriften der syrischen dichter cyrillonas, balaus, isaak von antiochien und jakob von sarug (bdk6 1.reihe) – (= ser Bibliothek der kirchenvaeter. 1. reihe (bdk 1.reihe)) – €17.00 – ne Slangenburg [240]

Ausgewaehlte schriften des heiligen gregorius des grossen, papstes und kirchenlehrers / Gregory 1, Pope – Kempten: Jos Koesel 1873-74 [mf ed 1986] – (= ser Bibliothek der kirchenvaeter) – 2v on 4mf – 9 – 0-8370-8514-4 – (trans fr latin into german by theodor kranzfelder) – mf#1986-2514 – us ATLA [240]

Ausgewaehlte schriften und briefe / Winckelmann, Johann Joachim; ed by Rehm, Walter – Wiesbaden: Dieterich 1948 [mf ed 1991] – 1r – 1 – (incl bibl ref & ind. filmed with: winckelmanns kleine schriften zur geschichte der kunst des altertums / ed by hermann uhde-bernays) – mf#3055p – us Wisconsin U Libr [700]

Ausgewaehlte schriften von columban, alkuin, dodana, jonas, hrabanus maurus, notker balbulus, hugo von sankt viktor und peraldus – Freiburg, 1890 – 6mf – 8 – €14.00 – (pref and trans by p gabriel meier) – ne Slangenburg [240]

Ausgewaehlte und neue gedichte / Binding, Rudolf Georg – Frankfurt/M: Ruetten & Loening, 1930 [mf ed 1989] – 204p – 1 – mf#7024 – us Wisconsin U Libr [810]

Ausgewaehlte werke / Bjornson, Bjornstjerne; ed by Schaefer, Thomas – Berlin: P J Oestergaard, [1910] – 3v – 1 – mf#1648 – us Wisconsin U Libr [802]

Ausgewaehlte werke / Brachvogel, Albert Emil – new rev ed. Berlin: [O] Janke [1872-74?] [mf ed 1979] – 4v on 1r – 1 – mf#film mas 8279 – us Harvard [802]

Ausgewaehlte werke / Brentano, Franz Clemens; ed by Morris, Max – Leipzig: M Hesse [1904?] [mf ed 1990] – 4v on 1r – 1 – (int by ed. filmed with: clemens brentano / wolfgang pfeiffer-belli; vom leichten lacheln / e c christophe) – mf#7071 – us Wisconsin U Libr [800]

Ausgewaehlte werke / Hoelderlin, Friedrich; ed by Schwab, Christoph Theodor – Stuttgart: J G Cotta 1874 [mf ed 1994] – 1r – 1 – (incl bibl ref. filmed with: hoelderlin : choix de textes, bibliographie, dessins.../ rudolf leonhard & robert rovini) – mf#3641p – us Wisconsin U Libr [800]

Ausgewaehlte werke / Vischer, Friedrich Theodor; ed by Keyssner, Gustav – Stuttgart: Deutsche Verlags-Anstalt 1918 [mf ed 1991] – 3v on 1r – 1 – mf#2952p – us Wisconsin U Libr [800]

Ausgewaehlte werke / Weerth, Georg; ed by Kaiser, Bruno – Berlin: Verlag Volk & Welt 1948 [mf ed 1995] – 1r – 1 – (incl bibl ref & ind. filmed with: au coin du feu / richard leander) – mf#3758p – us Wisconsin U Libr [800]

Ausgewaehlte werke / Wildenbruch, Ernst von; ed by Elster, Hanns Martin – Berlin: G Grote 1919 [mf ed 1996] – 4v on 1r – 1 – (int by ed) – mf#4062p – us Wisconsin U Libr [802]

Ausgewahlte predigten / Hess, Mendel – Hersfeld, Germany. 1871 – 1r – us UF Libraries [939]

Ausgewahlte sonaten : sonata 1, 2, 3, 4, 5, 6, per flauto traverso con cembalo o con basso / Quantz, Johann Joachim – Leipzig. 1921. 7v – 1 – us L of C Photodup [780]

Die ausgrabungen in palaestina und das alte testament / Gressmann, Hugo – Tuebingen: J C B Mohr (Paul Siebeck) 1908, c1905 [mf ed 1988] – (= ser Religionsgeschichtliche voelksbuecher fuer die deutsche christliche gegenwart 3/10) – 1mf – 9 – 0-7905-0129-5 – mf#1987-0129 – us ATLA [930]

Die ausgrabungen und entdeckungen im zweistroemeland / Witzel, Theophilus – 1. & 2. aufl. Muenster i W: Aschendorff 1911 [mf ed 1993] – (= ser Biblische zeitfragen 4/3-4) – 1mf – 9 – 0-524-05644-7 – (incl bibl ref) – mf#1992-0499 – us ATLA [930]

'Auslaenderinnen reichen die hand.' : britische und amerikanische frauenpolitik in deutschland im rahmen der demokratischen reedukation nach 1945 / Teller, Gustav – (mf ed 1999) – 11mf – 9 – €77.50 – 3-8267-2595-6 – mf#DHS 2595 – gw Frankfurter [943]

Das ausland : ein tageblatt fuer kunde des geistigen und sittlichen lebens der voelker mit besonderer ruecksicht auf verwandte erscheinungen in deutschland – Johann Friedrich Cotta 1828-93 [mf ed 1999] – 66v on 714mf – 9 – €2150.00 – 3-89131-351-9 – gw Fischer [074]

Das ausland (1828-1893) : tagblatt/wochenschrift fuer kunde des geistigen und sittlichen lebens der voelker = Foreign countries / ed by Deutsches Literaturarchiv. Marbach am Neckar – [mf ed 1999] – 547mf (1:24) – 9 – silver €2968.00 – ISBN-10: 3-598-32540-1 – ISBN-13: 978-3-598-32540-3 – (based on the editorial copy fr the cotta archive; with handbook) – gw Saur [074]

Das ausland: ueberschau der neuesten forschungen auf dem gebiete der natur-, erd-, und voelkerkunde – Stuttgart. v49-66. 1876-93 – 1 – us L of C Photodup [910]

Der auslandsdeutsche – Stuttgart DE, 1937-38 – 1 – gw Misc Inst [074]

Auslands-Organisation der NSDAP *see*
– N s report
– New order

Auslandsvertretungen der deutschen gewerkschaften in stockholm : rundbrief – Stockholm (S), 1942 dec-1945 – 1r – 1 – (several title changes: fr sep 1944: mitteilungsblatt/landesgruppe deutscher gewerkschafter in schweden) – gw Misc Inst [331]

Auslegung der epistel s pauli an die epheser in zehen predigt den doerffpfarherrn vnd hausvetern zu dienst einfeltiglich verfassel / Major, G – Wittemberg, 1559 – 9mf – 9 – mf#TH-1 mf 908-916 – ne IDC [242]

Auslegung des 129 psalmen dauids auff gegenwertigen zustand der kirchen zu wittenberg / Huber, S – Wittemberg, 1593 – 1mf – 9 – mf#TH-1 mf 709 – ne IDC [242]

Die auslegung des hohenliedes in der juedischen gemeinde und der griechischen kirche / Riedel, Wilhelm – Leipzig: A Deichert (Georg Boehme) 1898 [mf ed 1985] – 1mf – 9 – 0-8370-4896-6 – (incl bibl ref & ind) – mf#1985-2896 – us ATLA [221]

Auslegung von schriftabschnitten zum zweck des hoeheren schulunterrichts / Lange, H – Breslau: Grass, Barth 1867 [mf ed 1986] – 1mf – 9 – 0-8370-8586-1 – mf#1986-2586 – us ATLA [225]

Der auslessneste und nach den regeln der antiquen bau-kunst... / Sturm, L C – Augsburg, 1721 – 5mf – 9 – mf#OA-112 – ne IDC [720]

Aus'm gewohntem g'leis : posse mit gesang in fuenf abtheilungen / Anzengruber, Ludwig – Stuttgart: J G Cotta, [1879] [mf ed 1988] – 83p – 1 – mf#6947 – us Wisconsin U Libr [820]

Ausonia – v1-24. 1946-69 – 1 – us AMS Press [800]

Ausonius *see* Metamorphoses...

Die ausrottung des protestantismus in salzburg unter erzbischof firmian und seinen nachfolgern : ein beitrag zur kirchengeschichte des achtzehnten jahrhunderts / Arnold, Carl Franklin – Halle: Verein fuer Reformationsgeschichte, 1900-1901 – (= ser [Schriften des Vereins fuer Reformationsgeschichte]) – 1mf – 9 – 0-7905-4721-X – (incl bibl ref) – mf#1988-0721 – us ATLA [242]

Auss glareani musick ein vsszug : mit verwillung vn hilff glareani alle christenlichen kirchen alt vn goetlich gsang zuelerne auch zueverstan ganz nutzlich vn denen zu hilff so der mathematick vn vyllicht der latinischen sprachnitt gantz vnderricht / Glareanus, Henricus – Basel: H P Mertz 1559 [mf ed 19–] – 1r / 2mf – 1,9 – (film version is in latin & german; fiche in german only) – mf#film 233 / fiche 483 – us Sibley [780]

Aussaat und ernte : gedichte / Wills, Franz Hermann – Berlin: E Schmidt 1946 [mf ed 1991] – 1r – 1 – (filmed with: christoph marlow / ernst von wildenbruch) – mf#2963p – us Wisconsin U Libr [810]

Aussbund schoener teutscher liedlein – 1552 – (= ser Mssa) – 7mf – 9 – €90.00 – mfchl 97 – gw Fischer [780]

Aussenohrcharakteristik bei verschiedenen nagern / Blaheta, Roman – (mf ed 1994) – 2mf – 9 – €40.00 – 3-8267-2076-8 – mf#DHS 2076 – gw Frankfurter [574]

Aussenpolitik zwischen machtpolitik und dogma : die deutsch-italienischen beziehungen von der jahreswende 1932/33 bis zur stresa-konferenz / Poulain, Marc – Frankfurt a.M., 1971 – 2mf – 3-89349-997-0 – gw Frankfurter [327]

Die ausserchristlichen religionen und die religion jesu christi / Malapert-Neufville, Marie Constanze, Freifrau von – Leipzig: A Deichert, 1914 – 1mf – 9 – 0-524-01792-1 – mf#1990-2640 – us ATLA [230]

Ausserer, Alois *see* De clausulis minucianis et de ciceronianis quae quidem

Die aussermasorethischen uebereinstimmungen zwischen der septuaginta und der peschittha in der genesis / Haenel, Johannes – Giessen: Alfred Toepelmann, 1911 – 1mf – 9 – 0-7905-0497-9 – (incl bibl ref) – mf#1987-0497 – us ATLA [221]

Aussichten fuer die evangelische kirche deutschlands : in folge der beschluesse der reichsversammlung in frankfurt / Hoffmann, C – Stuttgart: J F Steinkopf 1849 [mf ed 1992] – 1mf – 9 – 0-524-05436-3 – mf#1990-1468 – us ATLA [242]

Aussiger tagblatt – Aussig (Usti nad Labem CZ), 1928 2 nov-1938 – 31r – 1 – (title varies: 1 sep 1941: elbetal-zeitung; filmed by misc inst: 1939 1 nov-1943 30 jun, 1944 3 jan-30 jun) – gw Misc Inst [077]

Ausslegung : das ist, erklaerung der deutschen geistlichen lieder, so von herrn doctore martino lvthero, vnd andern gottseligen christen gemacht / Pauli, S – (Magdeburg, 1588) – 7mf – 9 – mf#TH-1 mf 1259-1265 – ne IDC [242]

Ausslegung der episteln vnd euangelien / Musaeus, S – Franckfurt am Main, 1590 – 4mf – 9 – mf#TH-1 mf 1192-1195 – ne IDC [242]

Ausslegung der ersten acht capitel der episteln s pavli an die roemer / Spangenberg, C – Strassburg, 1566 – 8mf – 9 – mf#TH-1 mf 1429-1436 – ne IDC [242]

Ausslegung der euangelien von den fuernembsten festen vom advent biss auff ostern [3 und 4] / Mathesius, J – Nuernberg, 1571 – 10mf – 9 – mf#TH-1 mf 981-990 – ne IDC [242]

Ausslegung der letsten acht capitel : der episteln s pavli an die roemer / Spangenberg, C – Strassburg, 1569 – 8mf – 9 – mf#TH-1 mf 1392-1399 – ne IDC [242]

Ausstellungs-tageblatt – Duesseldorf DE, 1902 1 may-21 oct – 1r – 1 – gw Misc Inst [700]

Ausstellungs-zeitung – Duesseldorf DE, 1880 9 may-1 oct – 1r – 1 – gw Misc Inst [700]

Aust, Emil *see* Die religion der roemer

Austen, Ernest Edward *see* Tsetse-flies

Austen, Ralph A *see* Northwest tanzania under german and british rule

Austin, Alfred *see*
– Hibernian horrors
– Russia before europe

Austin, Benjamin Fish *see*
– Glimpses of the unseen
– The gospel to the poor versus pew rents
– The higher christian education of women
– The jesuits
– The methodist episcopal church pulpit
– "The plebiscite"
– Popular sins
– The prohibition leaders of america
– Rational memory training
– Woman
– Woman, her character, culture and calling

Austin, Dennis *see* Britain and south africa

Austin employee news – 1983 mar-1987 nov – 1 – mf#1288331 – us WHS [331]

Austin, F *see* Collection of ornaments at austin's artificial stone works

Austin, Mary Hunter *see* The man jesus

[Austin-] nevada progressive – NV. 1924-26 – 1r – 1 – $60.00 – mf#U04403 – us Library Micro [071]

Austin phelps : a memoir / Phelps, Elizabeth Stuart – New York: Scribner 1892 [mf ed 1993] – 1mf [ill] – 9 – 0-524-06912-3 – mf#1991-2825 – us ATLA [242]

Austin prout news – v1 n1-6 [1980 dec-81 may] – 1 – mf#1304010 – us WHS [071]

Austin, Rachel A *see* Negro churches

[Austin-] reese river reveille – NV. 1863-64; may 1867; 1872-1949 (broken series); 1952-93 – 89r – 1 – $5340.00 – mf#U04404 – us Library Micro [071]

Austin, Sarah *see* A memoir of the reverend sydney smith

Austin sun – 1974 jul 24, nov 21-1975 nov 13; 1975 dec 4-1977 feb 25; 1977 mar 4-1978 jan 27; 1978 feb 3-jun 29 – 1 – mf#333764 – us WHS [071]

[Austin-] sun – NV. 1933-34 [wkly] – 1r – 1 – $60.00 – mf#U04405 – us Library Micro [071]

Austin, Thomas *see* Two fifteenth century cookery-books

Austin weekly news – 1994 jan 7-1995 aug 18 – 1 – mf#2898854 – us WHS [071]

The austinian theory of law : being an edition of lectures 1, 5 and 6 of austin's "jurisprudence" and of austin's "essay on the uses of the study of jurisprudence" with critical notes and excursus / Brown, W Jethro – London: John Murray, 1906 – 5mf – 9 – $7.50 – (llmc's copy lacks 135-136p & 335-336p) – mf#LLMC 95-188 – us LLMC [340]

Austral africa / Mackenzie, John – New York, NY. v1-2. 1969 – 1r – us UF Libraries [960]

Austral ecology – Oxford UK 2002+ – 1,5,9 – (cont: australian journal of ecology) – ISSN: 1442-9985 – mf#15506,01 – us UMI ProQuest [574]

Australasia *see* Dalgetys review [weekly]

Australasia and the world's evangelisation : addresses delivered...melbourne, australia, apr 10-12 1903, and christchurch, new zealand, may 2-3 1903 – Sydney: Australasian Student Union [1903?] [mf ed 1986] – 1mf – 9 – 0-8370-7194-1 – mf#1986-1194 – us ATLA [242]

Australasian – Melbourne, Australia 1 oct 1864-26 jun 1869, 7 jan 1871-24 apr 1880, 5 jan 1884-21 oct 1939, 4 jan 1941-30 mar 1946 – 1 – (cont by: australasian post [6 jun 1946-31 jan 1952]) – uk British Libr Newspaper [079]

Australasian – Melbourne, jul 1902-dec 1902 – 2r – A$77.00 vesicular A$88.00 silver – at Pascoe [079]

Australasian annals of medicine – Sydney, Australia 1952-70 – 1 – mf#2164 – us UMI ProQuest [610]

Australasian biographical archive (anzo-ba) = Australasiatisches biographisches archiv (anzo-ba) / Herrero Mediavilla, Victor [comp] – [mf ed 1990-95] – 423mf (1:24) – 9 – diazo €10,060.00 (silver €11,080 isbn: 978-3-598-32930-2) – ISBN-10: 3-598-32944-X – ISBN-13: 978-3-598-32944-9 – (with printed ind) – gw Saur [980]

Australasian biographical archive (anzo-ba). supplement = Australasiatisches biographisches archiv (anzo-ba). supplement / Herrero Mediavilla, Victor [comp] – [mf ed 1995] – 119mf – 9 – diazo €2030.00 (silver €2460 isbn: 978-3-598-32925-8) – ISBN-10: 3-598-32924-5 – ISBN-13: 978-3-598-32924-1 – (with printed ind) – gw Saur [980]

Australasian builder and contractors news – Sydney, apr 1887-apr 1895 – 5r – A$305.14 vesicular A$332.64 silver – at Pascoe [079]

Australasian decorator and painter – Sydney, Australia 1 oct 1908-1 may 1924 – 1 – (cont by: decorator & painter for australia & new zealand [1 jun 1924-30 dec 1946]) – uk British Libr Newspaper [640]

Australasian engineer – Sydney, Australia 31 jul 1925-oct 1972, jan 1973 – 56 1/2r – 1 – uk British Libr Newspaper [620]

Australasian hebrew – Sydney, nov 1895-nov 1896 – 1r – A$58.04 vesicular A$63.54 silver – at Pascoe [079]

The australasian hebrew – Sydney. v. 1, no. 1-2, no. 26. 22 Nov 1895-13 Nov 1896 – 1 – us NY Public [939]

Australasian insurance and banking record – Melbourne, Australia apr 1886-dec 1921 – 68r – 1 – uk British Libr Newspaper [332]

Australasian manufacturer – Sydney, Australia 15 nov 1919-24 jun 1922, 21 jun 1924-14 jan, 25 nov 1967 – 158r – 1 – uk British Libr Newspaper [670]

Australasian manufacturer – Sydney, Australia 1950-51 – 1 – ISSN: 0004-8410 – mf#633 – us UMI ProQuest [670]

Australasian pastoralists' review – Melbourne, Australia 16 mar 1891-15 feb 1901 – 1 – (cont by: pastoralists' review [15 mar 1901-15 dec 1912]) – uk British Libr Newspaper [240]

Australasian photo-review *see* Apr

Australasian post – Melbourne, Australia 6 jun 1946-31 jan 1952 – 1 – (cont: australasian [1 oct 1864-26 jun 1869, 7 jan 1871-24 apr 1880, 5 jan 1884-21 oct 1939, 4 jan 1941-30 mar 1946]; cont by: new australasian post [17 jul-30 oct 1952]; cont by: new australasian post [17 jul-30 oct 1952]) – uk British Libr Newspaper [079]

Australasian science [1998] : incorporating search – East Malvern, Australia 1998-99 – 1,5,9 – (cont by: australasian science [2000]) – ISSN: 1440-3919 – mf#26868.01 – us UMI ProQuest [500]

Australasian science [2000] – East Malvern, Australia 2000+ – 1,5,9 – (cont: australasian science, incorporating search) – ISSN: 1442-679X – mf#26868,01 – us UMI ProQuest [500]

Australasian sketcher with pen and pencil – Melbourne, Australia 19 apr 1873-24 apr 1880, 1 jan-18 jun 1881, 14 jan-16 dec 1882, 16 jan 1884-26 dec 1889 – 1 – (lacking: n154, 157) – uk British Libr Newspaper [740]

Australasian traveller – Melbourne, Australia 5 jul 1913-6 dec 1921 – 1 – (missing: sep-nov 1913) – uk British Libr Newspaper [919]

Australasian traveller's gazette – Melbourne, Australia jan 1918-dec 1921 – 1 – (cont: cook's australasian traveller's gazette [may, jun, aug 1911, jul 1914-dec 1915]) – uk British Libr Newspaper [918]

Australasian travellers gazette *see* Cooks australasian travellers gazette

Australia / David, Arthur Evan – London: Mowbray 1908 [mf ed 1992] – (= ser Handbooks of english church expansion 4) – 1mf [ill] – 9 – 0-524-03444-3 – mf#1990-4704 – us ATLA [242]

Australia : the making of a nation / Fraser, John Foster – London, Toronto: Cassell, 1910 – 5mf – 9 – 0-665-73166-3 – (incl ind) – mf#73166 – cn Canadiana [980]

Australia : or facts and features, sketches and incidents of australia and australian life / Morison, John – London 1867 – (= ser 19th c british colonization) – 4mf – 9 – (with notices of new zealand by a clergyman) – mf#1.1.5632 – uk Chadwyck [980]

Australia : session laws of australia – 1891-99 – 9 – $4755.00 set – mf#408060 – us Hein [348]

Australia and cambodia : a model relationship / Jaspan, Meryn Aubrey – [Perth] University of Western Australia, Dept of Anthropology [1965] [mf ed 1989] – 1 – mf#mf-10289 seam reel 020/04 [§] – us CRL [327]

Australia and its gold fields : a historical sketch of the progress of the australian colonies, from the earliest times to the present day... / Hargraves, Edward Hammond – London, 1855 – (= ser 19th c books on british colonization) – 3mf – 9 – mf#1.1.5690 – uk Chadwyck [980]

Australia and new zealand bulletin / London Missionary Society – Auckland, jan 1952-jul 1970 – 1r – 1 – mf#PMB Doc413 – at Pacific Mss [980]

Australia and new zealand journal of developmental disabilities – Abingdon, Oxfordshire 1982-95 – 1,5,9 – (cont: australian journal of developmental disabilities;; cont by: journal of intellectual and developmental disability) – ISSN: 0726-3864 – mf#10741.03 – us UMI ProQuest [150]

Australia and the empire / Martin, Arthur Patchett – Edinburgh 1889 – (= ser 19th c british colonization) – 4mf – 9 – mf#1.1.3781 – uk Chadwyck [337]

Australia and the united states – Boston, MA. 1941 – 1r – us UF Libraries [500]

Australia. Army. Military Court *see* Proceedings of a military court held at manus island, june 1950-april 1951

Australia. Bureau of Census and Statistics *see* Overseas trade, statistics of overseas imports and exports and customs and excise revenue

Australia. bureau of census and statistics. official yearbook – v1-30. 1908-37 – 9 – $990.00 – mf#0092 – us Brook [319]

Australia china times – Sydney, oct 1955-jun 1957 – 1r – A$39.47 vesicular A$44.97 silver – at Pascoe [079]

Australia: colonial life and settlement : the colonial secretary's papers, 1788-1825 from the state records authority of new south wales – 3pt – 1 – (pt1: letters sent, 1808-25 [19r] £1750; pt2: special bundles (topic coll), proclamations, orders and related records, 1789-1825 [21r] £1950; pt3: letters received, 1788-1825 [32r] £2950; with d/g) – uk Matthew [980]

Australia Commonwealth *see*
- Report of the royal commission inquiry into the present conditions
- Selected ministerial statements on papua and new guinea

Australia Commonwealth Dept of External Territories *see* Ministerial press statements, speeches, addresses etc

Australia Commonwealth Dept of Territories *see* Papua and new guinea newsletter

Australia e ceylan / Balangero, Giovanni Battista – Torino: G B Paravia, [1897] [mf ed 1995] – (= ser Yale coll) – xiv/386p (ill) – 1 – 0-524-09719-4 – (in italian) – mf#1995-0719 – us ATLA [240]

Australia. Federal Council for Aboriginal Advancement. Sub-committee on Legislative Reform *see* Government legislation and the aborigines

Australia. Northern Territory *see* Government gazette

Australia. Western *see* Government gazette

Australian – Braddon, Australia. 1 jul-dec 1971 – 1 1/2r – 1 – uk British Libr Newspaper [072]

Australian – Canberra, 1964-96 – 373r – 1 – at Pascoe [079]

Australian – Canberra, Australia 15 jul 1964-31 dec 1992 – 1 – uk British Libr Newspaper [079]

Australian – Sydney, Australia 14 oct 1824-7 may 1828, 16 jan 1829, 3 aug 1838, 29 aug, 7,26,28 sep, 3,15,22 oct, 2 nov 1839 – 1 – (cont by: australian journal of commerce, agriculture & politics [ns:] 7,28 jan, 17 mar, 30 jun 1848) – uk British Libr Newspaper [079]

Australian – Sydney, 1824-48 – 10r – 1 – A$385.00 vesicular A$440.00 silver – at Pascoe [079]

The australian – Canberra, Australia. -d. 15 July 1964-Dec 1971. 84 reels – 1 – uk British Libr Newspaper [072]

The australian – Sydney, Australia. -w. Oct 1824-Jan 1829; Aug 1838-June 1848. 3 reels – 1 – uk British Libr Newspaper [072]

The australian aboriginal and the christian church / Pitts, Herbert – London: SPCK; New York: E S Gorham, 1914 [mf ed 1995] – (= ser Yale coll) – x/133p (ill) – 1 – 0-524-09201-X – mf#1995-0201 – us ATLA [230]

Australian accountant – Melbourne, Australia 1936-97 – 1,5,9 – (cont by: australian cpa) – ISSN: 0004-8631 – mf#7333.02 – us UMI ProQuest [650]

Australian and chinese herald – Sydney, sep 1894-oct 1897 – 1r – A$69.17 vesicular A$74.67 silver – at Pascoe [079]

Australian and New Zealand Association for the Advancement of Science *see* Anzaas congress papers

Australian and new zealand journal of medicine – Oxford UK 1971-2000 – 1,5,9 – ISSN: 0004-8291 – mf#6645.01 – us UMI ProQuest [610]

Australian and new zealand journal of mental health nursing – Oxford UK 1995-2001 – 1,5,9 – ISSN: 1324-3780 – mf#21590.02 – us UMI ProQuest [610]

Australian and new zealand journal of surgery – Oxford UK 1931-2000 – 1,5,9 – ISSN: 0004-8682 – mf#2731.01 – us UMI ProQuest [617]

Australian band news – Sydney – 5r – A$333.65 vesicular A$361.15 silver – (aka: australasian band & orchestra news) – at Pascoe [079]

Australian bankruptcy bulletin – Sydney, Australia 1973 – 1 – ISSN: 0045-0286 – mf#8247 – us UMI ProQuest [332]

Australian baptists : misc. historical items – 38p. 1868, 1870-71 – 1 – $5.00 – us Southern Baptist [242]

Australian biblical review – 1951-92 [complete] – 2r – 1 – mf#ATLA S0658 – us ATLA [220]

Australian board of missions review *see* Abm review

Australian brewers' journal – Melbourne, Australia 20 oct 1882, 20 dec 1910-20 apr 1921 – 1 – (cont by: australian brewing & wine journal [20 may-20 dec 1921]) – uk British Libr Newspaper [660]

Australian brewing and wine journal – Melbourne, Australia 20 may-20 dec 1921 – 1 – (cont: australian brewers' journal [20 oct 1882, 20 dec 1910-20 apr 1921]) – uk British Libr Newspaper [660]

Australian business law review – Rozelle, Australia 1987-95 – 1,5,9 – ISSN: 0310-1053 – mf#15017 – us UMI ProQuest [346]

Australian churchman – Sydney, 1867-70 – 6r – 1 – A$443.17 vesicular A$476.17 silver – at Pascoe [079]

Australian civil engineering – Chippendale, Australia 1969-70 – 1 – ISSN: 0572-0826 – mf#3431 – us UMI ProQuest [624]

Australian clinical review – Oxford UK 1991-93 – 1,5,9 – (cont by: journal of quality in clinical practice) – ISSN: 0726-3139 – mf#18092.01 – us UMI ProQuest [610]

Australian commonwealth : (new south wales, tasmania, victoria, western australia, south australia, queensland, new zealand) / Tregarthen, Greville Philipps – London 1893 – (= ser 19th c british colonization) – 6mf – 9 – mf#1.1.4473 – uk Chadwyck [980]

Australian communist / communist / workers weekly / tribune – 24r – 1 – A$938.48 vesicular A$1070.48 silver – (communist (may 1921); workers weekly (22 jun 1923); tribune (sep 1939)) – at Pascoe [079]

Australian computer journal – Darlinghurst, Australia 1974-96 – 1,5,9 – (cont by: journal of research and practice in information technology) – ISSN: 0004-8917 – mf#2758.01 – us UMI ProQuest [000]

Australian Council for Health, Physical Education and Recreation *see* Achper healthy lifestyles journal

Australian courier – Sydney, jan 1899-dec 1903 – 1r – A$93.72 vesicular A$99.22 silver – at Pascoe [079]

Australian cpa – Melbourne, Australia 1999+ – 1,5,9 – (cont: australian accountant) – mf#7333.02 – us UMI ProQuest [650]

Australian dance band news / music maker – Sydney, jun 1932-dec 1950 – (= ser Music maker) – 4r – 1 – A$272.62 vesicular A$294.62 silver – at Pascoe [079]

Australian dental journal – St Leonards, Australia 1973+ – 1,5,9 – ISSN: 0045-0421 – mf#8114 – us UMI ProQuest [617]

Australian dictionary of dates and men of the time : containing the history of australasia from 1542 to may, 1879 / Heaton, John Henniker – Sydney: G Robertson 1879 [mf ed 1990] – 1v on 6mf – 9 – 0-7905-8264-3 – mf#1988-6142 – us ATLA [059]

Australian economic history review [aehr] – Oxford UK 1977+ – 1,5,9 – ISSN: 0004-8992 – mf#11431 – us UMI ProQuest [330]

Australian economic review – Oxford UK 1990+ – 1,5,9 – ISSN: 0004-9018 – mf#18354 – us UMI ProQuest [330]

Australian education review – Camberwell, Australia 1974-85 – 1,5,9 – ISSN: 0311-6875 – mf#8095,01 – us UMI ProQuest [370]

The australian emigrant's manual : or, a guide to the gold colonies of new south wales and port phillip / Lang, John Dunmore – London, 1852 – (= ser 19th c british colonization) – 2mf – 9 – mf#1.1.4233 – uk Chadwyck [304]

Australian evangel – Sydney, sep 1929-may 1961 – 2r – A$157.43 vesicular A$168.43 silver – at Pascoe [079]

Australian Federal Convention *see* The draft bill to constitute the commonwealth of australia

Australian field : a journal for squatters, sportsmen, farm & fireside – Sydney, Australia 1 jun 1901-2 mar 1902, 12 oct 1907-22 feb 1912 (imperfect) – 1 – uk British Libr Newspaper [079]

Australian field – Sydney, dec 1894-feb 1912 – 14r – A$980.76 vesicular A$1057.76 silver – at Pascoe [079]

Australian financial review – Sydney, Australia 2 jun 1971-28 jun 1996 (imperfect) – 1 – uk British Libr Newspaper [332]

The australian forestry journal – Sydney: New South Wales Forestry Commission [1918-31] [mf ed 2004] – 14v on 4r [ill] – 1 – (qrtly [1918], mthly 1919-27; qrtly 1928-31) – mf#film mas c5698 – us Harvard [634]

Australian friend : the organ of the religious society of friends (quakers) in australia – [Sydney NSW] 1947-71 (bimthly) (incomplete) [mf ed 1989-] – 1 – (cont: friend of australia and new zealand) – mf#S0863 – us ATLA [243]

Australian g p – Penrith, Australia 1976-78 – 1,5,9 – ISSN: 0045-0499 – mf#7672 – us UMI ProQuest [610]

Australian government gazette – Canberra, jul 2 1973-jun 1977 – 16r – 1 – (includes ed: general and public service) – us CRL [079]

Australian graphic – Sydney, nov 1883-apr 1884 – 1r – 1 – A$29.79 vesicular A$35.29 silver – at Pascoe [079]

Australian Institute of Aboriginal Studies *see* Newsletter

Australian Institute of Agricultural Science *see* Journal of the australian institute of agricultural science

Australian israelite – Melbourne, jun 1871-may 1875 – 1r – A$72.60 vesicular A$78.10 silver – at Pascoe [079]

The australian israelite – Melbourne. v. 1, no. 1-4, no. 44. 30 Jun 1871-7 May 1875. Incomplete – 1 – us NY Public [939]

Australian jazz quarterly : a magazine for the connoisseur of hot music – n1-31. may 1946-apr 1957 (irreg) [all publ] – (= ser Jazz periodicals, 1914-1977) – 1r – 1 – $125.00 – us UPA [780]

Australian jewish chronicle – Sydney, mar 1922-feb 1931 – 3r – A$218.37 vesicular A$234.87 silver – at Pascoe [079]

Australian jewish forum – Sydney. v. 1-9. no. 77. Feb 1941-Sept 1949 – 1 – us NY Public [939]

Australian jewish herald – Melbourne, jan 1947-dec 1958 – 5r – A$328.81 vesicular A$356.31 silver – at Pascoe [079]

Australian jewish times – Sydney. oct 1953-jun 1978 – 31r – 1 – A$2232.03 vesicular A$2402.53 silver – (aka: australian jewish news) – at Pascoe [939]

Australian journal – Sydney, oct 1824-sep 1848 – 30r – 1 – A$1155.00 vesicular A$1320.00 silver – at Pascoe [079]

Australian journal for health, physical education and recreation (ajhper) – Hindmarsh, Australia 1975-83 – 1,5,9 – (cont by: achper national journal) – ISSN: 0004-9492 – mf#10250.03 – us UMI ProQuest [613]

Australian journal of adult and community education – Canberra, Australia 1990-99 – 1,5,9 – (cont: australian journal of adult education) – ISSN: 1035-0462 – mf#7581.02 – us UMI ProQuest [374]

Australian journal of adult education – Canberra, Australia 1972-89 – 1,5,9 – (cont by: australian journal of adult and community education) – ISSN: 0004-9387 – mf#7581.02 – us UMI ProQuest [374]

Australian journal of adult learning – Canberra, Australia 2000+ – 1,5,9 – (cont: australian journal of adult and community education) – mf#7581,02 – us UMI ProQuest [374]

Australian journal of agricultural and resource economics – Oxford UK 2001+ – 1,5,9 – ISSN: 1364-985X – mf#25719 – us UMI ProQuest [630]

Australian journal of agricultural research – Collingwood, Australia 1979+ – 1,5,9 – ISSN: 0004-9409 – mf#11674 – us UMI ProQuest [630]

Australian journal of anthropology – Sydney, Australia 1990+ – 1,5,9 – ISSN: 1035-8811 – mf#18295,01 – us UMI ProQuest [301]

Australian journal of biological sciences – Collingwood, Australia 1977-88 – 1,5,9 – ISSN: 0004-9417 – mf#11675 – us UMI ProQuest [574]

Australian journal of botany – Collingwood, Australia 1977+ – 1,5,9 – ISSN: 0067-1924 – mf#11676 – us UMI ProQuest [580]

Australian journal of chemistry – Collingwood, Australia 1953+ – 1,5,9 – ISSN: 0004-9425 – mf#11677 – us UMI ProQuest [540]

Australian journal of chinese affairs = Ao chung – Canberra, Australia 1994 – 1 – (cont by: china journal=chung kuo yen chiu) – ISSN: 0156-7365 – mf#17530.01 – us UMI ProQuest [951]

Australian journal of commerce, agriculture and politics – Sydney, Australia 7,28 jan, 17 mar, 30 jun 1848 – 1 – (cont: australian [14 oct 1824-7 may 1828, 16 jan 1829, 3 aug 1838, 29 aug, 7,26,28 sep, 3,15,22 oct, 2 nov 1839]) – uk British Libr Newspaper [079]

Australian journal of dairy technology – Glen Iris, Australia 1992+ – 1,5,9 – ISSN: 0004-9433 – mf#14420 – us UMI ProQuest [630]

Australian journal of developmental disabilities – Abingdon, Oxfordshire 1980-81 – 1,5,9 – (cont: australian journal of mental retardation; cont by: australia and new zealand journal of developmental disabilities) – ISSN: 0159-9011 – mf#10741.03 – us UMI ProQuest [150]

Australian journal of early childhood – Watson, Australia 1982+ – 1,5,9 – ISSN: 0312-5033 – mf#12906 – us UMI ProQuest [640]

Australian journal of earth sciences – Oxford UK 1984+ – 1,5,9 – (cont: journal of the geological society of australia) – ISSN: 0812-0099 – mf#15505,01 – us UMI ProQuest [550]

Australian journal of ecology – Oxford UK 1980-93 – 1,5,9 – (cont by: austral ecology) – ISSN: 0307-692X – mf#15506.01 – us UMI ProQuest [574]

Australian journal of education – Camberwell, Australia 1983+ – 1,5,9 – ISSN: 0004-9441 – mf#14054 – us UMI ProQuest [370]

Australian journal of education – Sydney, Australia 1 jul 1903-15 dec 1911 – 2 1/2r – 1 – uk British Libr Newspaper [370]

Australian journal of experimental agriculture – Collingwood, Australia 1985-96 – 1,5,9 – (cont: australian journal of experimental agriculture and animal husbandry) – ISSN: 0816-1089 – mf#12122,01 – us UMI ProQuest [630]

Australian journal of experimental agriculture and animal husbandry – Collingwood, Australia 1980-84 – 1,5,9 – (cont by: australian journal of experimental agriculture) – ISSN: 0045-060X – mf#12122.01 – us UMI ProQuest [636]

Australian journal of experimental biology and medical science – 1924-86 – 1,5,9 – (cont by: immunology and cell biology) – ISSN: 0004-945X – mf#10592.01 – us UMI ProQuest [574]

Australian journal of international affairs – Abingdon, Oxfordshire 1990+ – 1,5,9 – (cont: australian outlook) – ISSN: 1035-7718 – mf#8353,01 – us UMI ProQuest [327]

Australian journal of language and literacy – Norwood 1992+ – 1.5.9 – (cont: australian journal of reading) – ISSN: 1038-1562 – mf#14135,01 – us UMI ProQuest [370]

Australian journal of linguistics – Abingdon, Oxfordshire 1989 – 1 – ISSN: 0726-8602 – mf#16518 – us UMI ProQuest [400]

Australian journal of marine and freshwater research – Collingwood, Australia 1977-94 – 1,5,9 – (cont by: marine and freshwater research) – ISSN: 0067-1940 – mf#11678.01 – us UMI ProQuest [574]

Australian journal of mental retardation – Abingdon, Oxfordshire 1974-79 – 1,5,9 – (cont by: australian journal of developmental disabilities) – ISSN: 0045-0634 – mf#10741.03 – us UMI ProQuest [616]

Australian journal of music education / Australian Society for Music Education – Parkville, Australia 1972-82 – 1,5,9 – ISSN: 0004-9484 – mf#7225 – us UMI ProQuest [780]

Australian journal of pharmacy [ajp] – Hawthorn, Australia 1952-90 – 1,5,9 – ISSN: 0311-8002 – mf#755 – us UMI ProQuest [615]

Australian journal of physics – Collingwood, Australia 1977+ – 1,5,9 – ISSN: 0004-9506 – mf#11679 – us UMI ProQuest [530]

Australian journal of plant physiology – Collingwood, Australia 1977-96 – 1,5,9 – ISSN: 0310-7841 – mf#11680.01 – us UMI ProQuest [580]

Australian journal of political science – Abingdon, Oxfordshire 1991+ – 1,5,9 – (cont: politics) – ISSN: 1036-1146 – mf#9163,01 – us UMI ProQuest [320]

Australian journal of public health – Curtin, Australia 1993-95 – 1,5,9 – (cont by: australian and new zealand journal of public health) – ISSN: 1035-7319 – mf#14280.02 – us UMI ProQuest [614]

Australian journal of reading – Norwood SA 1983-91 – 1,5,9 – (cont by: australian journal of language and literacy) – ISSN: 0156-0301 – mf#14135.01 – us UMI ProQuest [370]

Australian journal of social issues – Strawberry Hills, Australia 1961-96 – 1,5,9 – ISSN: 0157-6321 – mf#8600 – us UMI ProQuest [360]

Australian journal of soil research – Collingwood, Australia 1979-90 – 1,5,9 – ISSN: 0004-9573 – mf#11682 – us UMI ProQuest [630]

Australian journal of zoology – Collingwood, Australia 1977+ – 1,5,9 – ISSN: 0004-959X – mf#11683 – us UMI ProQuest [590]

Australian law librarian – v1-9. 1993-2 – 9 – $197.00set – (cont: australian law librarians group newsletter) – ISSN: 1039-6626 – mf#117051 – us Hein [340]

Australian law librarians' group newsletter – v1-3 n1-113. 1973-92 – 9 – $66.00 set – (cumulative index 1973-90; cont by: australian law librarian) – ISSN: 0311-5984 – mf#307661 – us Hein [020]

Australian law news – Melbourne, Australia 1977 – 1,5,9 – (cont: law council newsletter) – ISSN: 0159-7531 – mf#9288.02 – us UMI ProQuest [340]

Australian lawyer – Chatswood, Australia 1958-67 – 1 – mf#2153 – us UMI ProQuest [340]

Australian leather journal – Melbourne, Australia 15 dec 1910-15 jul 1915, 15 jan 1916-15 dec 1921 – 10r – 1 – uk British Libr Newspaper [670]

Australian library journal – Kingston, Australia 1951+ – 1,5,9 – ISSN: 0004-9670 – mf#2087 – us UMI ProQuest [020]

Australian literary studies – St Lucia, Australia 1963+ – 1,5,9 – ISSN: 0004-9697 – mf#8108 – us UMI ProQuest [400]

Australian mining and engineering review – Melbourne, Australia 5 dec 1908-5 dec 1910 (very imperfect) – 1 – (cont by: mining & engineering review [5 jan 1911-sep 1917]) – uk British Libr Newspaper [622]

Australian mining standard – Melbourne, Australia 12 mar 1890-13 sep 1900, 1 jan 1901-19 nov 1914 – 1 – (cont by: australian statesman & mining standard [26 nov 1914-28 dec 1916]) – uk British Libr Newspaper [622]

Australian monthly and colonial monthly – Sydney, 1865-70 – 3r – 1 – A$115.50 vesicular A$132.00 silver – at Pascoe [073]

Australian nation – Sydney, feb 1899-sep 1900 – 1r – A$40.70 vesicular A$46.20 silver – at Pascoe [079]

Australian native policy : its history especially in victoria / Foxcroft, Edmund John Buchanan – Melbourne, London: Melbourne UP in assoc with Oxford UP 1941 [mf ed 1991] – 1r – 1 – (filmed with: why men hate / tenenbaum, s & other titles) – mf#1867 – us Wisconsin U Libr [322]

Australian nurses' journal – Melbourne, Australia 1971-93 – 1,5,9 – (cont by: australian nursing journal: anj) – ISSN: 0045-0758 – mf#7591 – us UMI ProQuest [610]

Australian nursing journal [anj] – Melbourne, Australia 1993+ – (cont: australian nurses' journal) – ISSN: 1320-3185 – mf#20867 – us UMI ProQuest [610]

Australian occupational therapy journal – Oxford UK 1994+ – 1,5,9 – ISSN: 0045-0766 – mf#20743 – us UMI ProQuest [615]

Australian office correspondence files / J T Arundel & Co and Pacific Islands Co Ltd – 1892-1904 – 8r – 1 – mf#pmb1174 – at Pacific Mss [338]

Australian outlook – Abingdon, Oxfordshire 1947-89 – 1,5,9 – (cont by: australian journal of international affairs) – ISSN: 0004-9913 – mf#8353.01 – us UMI ProQuest [327]

Australian paediatric journal – Oxford UK 1965-89 – 1,5,9 – (cont by: journal of paediatrics and child health) – ISSN: 0004-993X – mf#8744.01 – us UMI ProQuest [618]

Australian plants – Picnic Point, Australia 1959-80 – 1,5,9 – ISSN: 0005-0008 – mf#8937 – us UMI ProQuest [580]

The australian presbyterian : the magazine of the presbyterian church in australia – Melbourne, Australia: National Journal Cttee of the Presbyterian Church of Australia 1998- (mthly ex jan) [mf ed 2000-] – 1 – (subtitle varies slightly) – mf1003 – us ATLA [242]

Australian Railways Union *see* Railroad

An australian ramble : or a summer in australia / Ritchie, James Ewing – London 1890 – (= ser 19th c british colonization) – 3mf – 9 – mf#1.1.3939 – uk Chadwyck [880]

Australian school librarian – East Melbourne, Australia 1972-85 – 1,5,9 – ISSN: 0005-0199 – mf#8075 – us UMI ProQuest [020]

Australian School of Pacific Administration *see* Reports and related papers

Australian school of pacific administration : annual reports – 1955-70 – 1r – 1 – mf#pmb doc27 – at Pacific Mss [350]

Australian school of pacific administration : reports, correspondence and related papers – 1946-92 – 2r – 1 – mf#pmb1158 – at Pacific Mss [350]

Australian science teachers journal – Deakin West, Australia 1975+ – 1,5,9 – ISSN: 0045-0855 – mf#10529 – us UMI ProQuest [370]

Australian Society for Music Education *see* Australian journal of music education

Australian star – Sydney, jul 1888-jun 1910 – 45r – 9 – A$3143.71 vesicular A$3391.21 silver – at Pascoe [079]

Australian state session laws – Backfile 1980-2002 update n1 – 9 – $10,155.00 set – 0-89941-414-1 – mf#400460 – us Hein [323]

Australian statesman and mining standard – Melbourne, Australia 26 nov 1914-28 dec 1916 – 1 – (cont: australian mining standard [12 mar 1890-13 sep 1900, 1 jan 1901-19 nov 1914]; cont by: industrial australian [15 oct 1932-15 feb 1933]) – uk British Libr Newspaper [622]

Australian stock exchange intelligence – Melbourne, Australia aug 1895-jul 1912 – 1 – uk British Libr Newspaper [332]

Australian sugar journal – Brisbane, Australia 7 jul 1910-9 dec 1921 – 11r – 1 – uk British Libr Newspaper [660]

Australian town and country journal – Sydney, Australia 8 jan 1870-25 jun 1919 (imperfect) – 1 – (discontinued) – uk British Libr Newspaper [079]

Australian town and country journal – Sydney, jan 1870-jun 1919 – 95r – 1 – A$6267.71 vesicular A$6790.21 silver – at Pascoe [073]

Australian transport – Sydney, Australia 1972-78 – 1,5,9 – ISSN: 0005-0385 – mf#7175 – us UMI ProQuest [380]

Australian university – Canberra City, Australia 1963-76 – 1,5,9 – ISSN: 0005-0415 – mf#7160 – us UMI ProQuest [378]

Australian, windsor richmond and hawkesbury advertiser – Windsor. 1873-1883, 2 jul 1896 – 4r – 9 – A$168.50 vesicular A$196.02 silver – at Pascoe [079]

Australian, windsor richmond and hawkesbury advertiser – Windsor, 1899 – 1r – A$27.50 vesicular A$33.00 silver – at Pascoe [079]

Australian womens weekly – Sydney. 1933- – 1 – at Pascoe [079]

Australian worker – Sydney, Australia 7 jan 1915-13 apr 1960 – 1 – (cont: worker [5 jan 1895-31 dec 1898]) – uk British Libr Newspaper [331]

Australian worker – Sydney, 1891-dec 1950 – 50r – 1 – A$2602.64 vesicular A$2877.64 silver – at Pascoe [073]

Australian workman – Sydney, 1890-97 – 1r – 1 – A$77.00 vesicular A$88.00 silver – at Pascoe [073]

Australian yearbook of international law – v1-19. 1965-98 – 9 – $702.00 set – (numbering began with v6 1974-75) – ISSN: 0084-7658 – mf#110001 – us Hein [341]

Australia's first preacher : the rev richard johnson, first chaplain of new south wales / Bonwick, James – London: Sampson Low, Marston 1898 [mf ed 1989] – 1mf – 9 – 0-7905-4381-8 – mf#1988-0381 – us ATLA [242]

The australiasian – Melbourne, Australia. Australasian Post New Australasian Post. - w. Oct 1864-26 June 1869; 7 Jan 1871-24 April 1880; 5 Jan 1884-21 Oct 1939; 4 Jan 1941-30 March 1946; 6 June 1946-31 Jan 1952; 17 July-23 Oct 1952. 219 reels – 1 – uk British Libr Newspaper [079]

Australien : geschichte der entdeckung und kolonisation; bilder aus dem leben der ansiedler in busch und stadt / ed by Christmann, Friedrich – Leipzig 1880 [mf ed Hildesheim 1995-98] – 1v on 4mf – 9 – €120.00 – 3-487-26816-7 – gw Olms [980]

Das australische abenteuer : ein roman vom leben, vom gold und von der geschichte des fuenften kontinents / Becker, Otto Eugen Hasso – Leipzig: H H Kreisel, c1939 [mf ed 1989] – 268p – 1 – mf#7004 – us Wisconsin U Libr [830]

Austria : containing a description of the manners, customs, character, and costumes of the people of that empire – London 1823 [mf ed Hildesheim 1995-98] – (= ser Fbc) – 2v on 4mf – 9 – €120.00 – 3-487-29452-4 – gw Olms [943]

Austria : including hungary, transylvania, dalmatia, and bosnia / Karl Baedeker (Firm) – Leipsic, Germany. 1900 – 1r – us UF Libraries [025]

Austria as it is : or, sketches of continental courts / Sealsfield, Charles – London 1828 [mf ed Hildesheim 1995-98] – 2mf – 9 – €60.00 – 3-487-29445-1 – gw Olms [914]

Austria. Bundesministerium fuer soziale Verwaltung *see* Amtliche nachrichten

Austria. Bundesministerium fuer Unterricht *see* Verordnungsblatt

Austria. Bundespolizeidirektion, Vienna *see* Fahndungsverzeichnis zu dem zentralpolizeiblatte, dem wiener, grazer und innsbrucker taeglichen fahndungsblatte

Austria. Finanz-Ministerium *see* Staats-voranschlag fuer die im reichsrathe vertretenen koenigreiche und laender

Austria. Herrenhaus *see* Stenographische protokolle ueber die sitzungen des oesterreichischen reichsrathes

Austria. (Lower) *see*
- Amtliche nachrichten
- Stenographische protokolle.

Austria. main government archives : records = (Allgemeines verwaltungs-archiv) (vienna) – 1910-13 – 1r – 1 – (records consist of reports & correspondence fr austro-hungarian consulates in the us pertaining to immigration. in german. inventory available (in german only)) – us IHRC [324]

Austria. Reichsrat. Abgeordnetenhaus *see* Stenographische protokolle des abgeordnetenhauses des reichsrathes

Austria. state archives : records = (Staatsarchiv) (vienna) – 1848-1919 – 32r – 1 – (records consist of consular dispatches, telegrams, police reports & correspondence pertaining to emigrants fr austro-hungarian empire. in german. partial inventory available (in german and english)) – us IHRC [324]

Austria. Statistische Zentralkommission *see*
- Oesterreichisches statistisches handbuch 1882-1917
- Statistisches jahrbuch der oesterreichischen monarchie 1863-1881
- Tafeln zur statistik der oesterreichischen monarchie 1842-1859

Austria. Statistisches Zentralamt *see* Statistisches handbuch fuer die republik oesterreich 1920-1938, 1950-1965

Austria to-day *see* Germany to-day

...Austriaci viennensis e scholis piis : vertumnus vanitatis / Martinus...S Brunone – Augsburg: Typis Augustanis per Joan. Jacob Lotter, 1725 – 4mf – 9 – mf#0-36 – ne IDC [090]

The austrian court from within / Radziwill, Ekaterina Rzewuska kniagina – New York: Frederick A. Stokes, 1916. 235p. illus – 1 – us Wisconsin U Libr [943]

Austrian information – Washington DC 1948+ – 1,5,9 – ISSN: 0005-0520 – mf#1489 – us UMI ProQuest [327]

Der austritt deutschlands aus dem volkerbund, seine vorgeschichte und seine nachwirkungen / Fraser, Christine – Bonn, 1969 – 1 – gw Mikropress [943]

Austro-Hungarian Monarchy *see* Voranschlag

Austro-Hungarian Monarchy. Ministerium des K. und K. Hauses und des aussern *see* Oesterreich-ungarns aussenpolitik

Austro-Hungarian Monarchy. Reichskriegsministerium *see* Verordnungsblatt fuer das k. u. k. heer

Die ausubung des verordnungsrechts im freistaate hessen seit der revolution / Schmahl, Ludwig – Giessen 1921 – 1 – gw Mikropress [943]

Auswaertige politik – Berlin DE, 1939 n7-1944 n4 – 1 – gw Misc Inst [327]

Auswahl / Herder, Johann Gottfried; ed by Kuehnemann, Eugen – 3. aufl. Leipzig: Duerr 1910 [mf ed 1991] – 1r – 1 – (filmed with: eck segge man bloss / wilhelm henze) – mf#7472 – us Wisconsin U Libr [800]

Eine auswahl aus dem dichterischen werk / Blunck, Hans Friedrich – Bielefeld: Velhagen & Klasing, 1943 [mf ed 1989] – (= ser Velhagen und klasings deutsche lesebogen 249) – 63p – 1 – mf#7040 – us Wisconsin U Libr [810]

Auswahl aus den iliasscholien : zur einfuehrung in die antike homerphilologie – Bonn: A Marcus & E Weber 1912 [mf ed 1992] – (= ser Kleine texte fuer vorlesungen und uebungen 111) – 1mf – 9 – 0-524-05526-2 – mf#1990-3496 – us ATLA [450]

Eine auswahl aus seinen schriften auf das vierhundertjaehrige jubilaeum der zuericher reformation / Zwingli, H – Zuerich, 1918 – 9mf – 9 – mf#ZWI-71 – ne IDC [242]

Auswahl der minnesaenger : fuer vorlesungen und zum schulgebrauch; mit einem woerterbuche und einem abrisse der mhd formenlehre / ed by Volckmar, Karl – Quedlinburg; Leipzig: G Basse, 1845 [mf ed 1993] – (= ser Bibliothek der gesammten deutschen national-literatur von der aeltesten bis auf die neuere zeit sect1/15) – xxiv/216p – 1 – mf#8438 reel 4 – us Wisconsin U Libr [810]

Auswahl deutscher prosa der gegenwart : mit lebensbeschreibungen der verfasser und anmerkungen / ed by Hein, Gustav – 3. ausg. Oxford: Universitaetsverlag, 1914 [mf ed 1993] – 208p – 1 – mf#8373 – us Wisconsin U Libr [830]

Auswahl von schraubwerkzeugen / Fischer, Thomas – (mf ed 1996) – 2mf – 9 – €40.00 – 3-8267-2297-3 – mf#DHS 2297 – gw Frankfurter [621]

Auswanderung der saechsischen lutheraner im jahre 1838 : ihre niederlassung in perry-co, mo., und damit zusammenhaengende interessante nachrichten / Koestering, Johann Friedrich & Walter, Carl Ferdinand Wilhelm – 2. aufl. St Louis MO: A Wiebusch 1867 [mf ed 1992] – 1mf [ill] – 9 – 0-524-02830-3 – mf#1990-4451 – us ATLA [242]

Auswanderungs-katechismus : ein rathgeber fuer auswanderer / Wander, Karl F – Glogau 1852 [mf ed Hildesheim 1995-98] – 1v on 3mf – 9 – €90.00 – 3-487-27003-X – gw Olms [917]

Der ausweg – Paris (F), 1934-35 [gaps] – 1 – (later publ in zuerich) – gw Misc Inst [074]

Die auswirkungen der fabrikarbeit auf das traditionale rollenverhalten der frau in megara, griechenland : eine vergleichende untersuchung / Kalfelis, Petros – Heidelberg, 1974 – 4mf – 9 – 3-89349-755-2 – gw Frankfurter [331]

Die auswirkungen des bilingualismus in algerien (franzoesisch/arabisch) auf das erlernen des deutschen als fremdsprache / Boudaa, Azzedine – (mf ed 1992) – 3mf – 9 – €49.00 – 3-89349-598-3 – mf#DHS 598 – gw Frankfurter [410]

Die auswirkungen des europaeischen binnenmarktes auf die wirtschaftsbeziehungen zur dritten welt / Pennekamp, Markus – (mf ed 1995) – 1mf – 9 – €30.00 – 3-8267-2274-4 – mf#DHS 2274 – gw Frankfurter [337]

Auswirkungen von ausleitungen zur wasserkraftnutzung auf die besiedlung durch makroobenthon in gewaesserstrecken des nordschwarzwaldes / Jehle, Robert – (mf ed 1996) – 2mf – 9 – €40.00 – 3-8267-2292-2 – mf#DHS 2292 – gw Frankfurter [574]

Ein auszug aus dem reise-journal eines unterrichteten maurers / Die Loge zu Z – 1mf – 9 – mf#VR-16.20 – ne IDC [910]

Auszug aus der vorderasiatischen geschichte / ed by Winckler, Hugo – Leipzig: J C Hinrichs 1905 [mf ed 1989] – (= ser Hilfsbuecher zur kunde des alten orients 2) – 1mf – 9 – 0-7905-2812-6 – mf#1987-2812 – us ATLA [930]

Auszug der neuesten weltbegebenheiten – Stralsund DE, 1811 – 1r – 1 – (filmed by other misc inst: 1799-1804, 1815 (single iss) [6r]. title varies: 2 jan 1772: auszug der neuesten weltbegebenheiten. with suppls: mode und heim 1895-1909, sonntags-beilage 1892 2 oct-1896, 1898-1929, zick-zack 1895-1900 (berlin/schwerin)) – gw Misc Inst [074]

Auszug der neuesten weltgeschichte *see* Christian-erlangischer zeitungs-extract

Auszug der neuesten zeitungen 1770 – Rostock DE, 1770 & 1774 [single iss], 1775 & 1782 [single iss], 1786 nov-1788 mar [gaps], 1789-90, 1792 [single iss], 1793, 1797-1806, 1807 jan-jun, 1808-43, 1846-1921 – 202r – 1 – (title varies: 2 feb 1812: Auszug aus den Neuesten Zeitungen, 1 jan 1815: Auszug der Neuesten Zeitungen, 12 nov 1846: Rostocker Zeitung. incl suppls: frauen rundschau 1908-, haus und landwirtschaft 1908-14, illustrirte rundschau 1908-09, mecklenburgischer generalanzeiger 1899-1901, mecklenburgisches neues wochenblatt 1908-14, officielle beilage fuer amtliche bekanntmachungen 1857, 1859, 1879 [single iss], 1880-81, 1887, 1900, 1920 [gaps], rostocker sonntagsbote 1909-, rundschau 1890-93 [gaps]) – gw Misc Inst [074]

Auszug vnd kurtzer bericht : von der gerechtigkeit der christen fur gott aus einer predig vber die wort johannis / Funck, J – Kuenigsberg, 1552 – 1mf – 9 – mf#TH-1 mf 477 – ne IDC [242]

Auszuge aus originalbriefen : geschrieben in franzoesischer sprache von den apostolischen vikarien und missionarien in china, tunkin, cochinchina, etc euber den zustand jener missionen – Wien: Mathias Andreas Schmidt, 1811 [mf ed 1995] – (= ser Yale coll) – 3v in 1 (ill) – 1 – 0-524-09245-1 – (in german) – mf#1995-0245 – us ATLA [240]

Aut ta khyap sai ta pvan : [short stories] / Yan Lhuin, Mon – Ran Kun: Ca pe Biman 1980 [mf ed 1990] – (= ser Prann su lak cvai ca can) – 1r with other items – 1 – (in burmese) – mf#mf-10289 seam reel 172/3 [$] – us CRL [830]

Autenrieth, Georg *see* Homeric dictionary for use in schools and colleges

El autentico espronceda pornografico y el apocrifo en general / Cascales Munoz, Jose – Toledo, 1932 – 1 – sp Bibl Santa Ana [946]

Les auteurs hindoustanis et leurs ouvrages : d'apres les biographies originales / Tassy, M Garcin de – 2. ed. Paris: E Thorin 1868 – us CRL [920]

An authentic account of an embassy from the king of great britain to the emperor of china : including cursory observations made, and information obtained... / Staunton, G L – London: W Bulmer and Co, 1797. 2v – 24mf – 9 – mf#H-6150 – ne IDC [915]

An authentic account of an embassy from the king of great britain to the emperor of china : including cursory observations made, and information obtained, in travelling through that ancient empire, and a small part of chinese tartary / Staunton, George L – London 1798 [mf ed Hildesheim 1995-98] – 3v on 13mf – 9 – €130.00 – 3-487-27251-2 – gw Olms [951]

An authentic account of the embassy of the dutch east-india company : to the court of the emperor of china, in the years 1794 and 1795... / Braam, A E van – London: R Phillips. 2v. 1798 – 8mf – 9 – mf#HT-779 – ne IDC [915]

Authentic account of the enthronement of his eminence cardinal wise / Wiseman, Nicholas Patrick – London, England. 1850? – 11r – us UF Libraries [240]

Authentic account of the late unfortunate death of lord camelford / Cockburne, William – London, England. 1804 – 1r – us UF Libraries [240]

Authentic and useful history of issac jenkins, his wife... / Beddoes, Thomas – London, England. 1826 – 1r – us UF Libraries [240]

Authentic copies of the preliminary articles of peace : between his britannic majesty and the most christian king, his most catholic majesty and the united states of america: signed at versailles, the 20th of jan 1783 – London: printed for J Debrett...Montreal; repr by F Mesplet, 1783 [mf ed 1992] – 1mf – 9 – 0-665-94702-X – (also available in french) – mf#94702 – cn Canadiana [341]

An authentic history of the missions : under the care of the missionary society of the methodist episcopal church / Bangs, Nathan – New York:...for the Methodist Episcopal Church, 1832 [mf ed 1990] – 1mf – 9 – 0-7905-5804-1 – mf#1988-1804 – us ATLA [242]

An authentic history of the prayer book, its five revisions, and the periods at which they were made : showing from whence each of the present conflicting parties in the church presume to derive their authorities / Amicus – Montreal: J Starke & Co, printers, 1874 [mf ed 1980] – 1mf – 9 – 0-665-05841-1 – (a repr of int to the book of "common prayer", ed by richard mant...publ at oxford, london, ad 1820) – mf#05841 – cn Canadiana [240]

Authentic letters from upper canada : with an account of canadian field sports / Magrath, Thomas W – Dublin 1833 [mf ed Hildesheim 1995-98] – 1v on 3mf – 9 – €90.00 – 3-487-27088-9 – gw Olms [790]

An authentic narrative of four years' residence at tongataboo : one of the friendly islands, in the south-sea, with an appendix, by an eminent writer / Vason, George – London 1810 [mf ed Hildesheim 1995-98] – 1v on 2mf – 9 – €60.00 – 3-487-26781-0 – gw Olms [980]

Authentic report of the discussion held in rome on the evenings – London, England. 1873 – 1r – us UF Libraries [240]

Authentic report of the discussion on the unitarian controversy / Porter, John Scott – London, England. 1834 – 1r – us UF Libraries [243]

Authentic report of the discussion which took place between the rev... / Burnet, J – Birmingham, England. 1827 – 1r – us UF Libraries [240]

Authenticated report of the discussion which took place between the rev messrs maguire and gregg : in the rotunda, dublin, in may, 1838 / Maguire, Thomas – [Montreal?: s.n.] 1839 [mf ed 1984] – 3mf – 9 – 0-665-46275-1 – mf#46275 – cn Canadiana [241]

L'authenticite mosaique du pentateuque / Mangenot, Eugene – Paris: Letouzey & Ane, 1907 [mf ed 1986] – 1mf – 9 – 0-8370-7000-7 – (incl text of de mosaica authentia pentateuchi of 1906 in latin and french) – mf#1986-1000 – us ATLA [221]

The authenticity and messianic interpretation of the prophecies of isaiah : vindicated in a course of sermons preached before the university of oxford / Payne Smith, Robert – Oxford: John Henry and James Parker, 1862 – 4mf – 9 – 0-8370-1223-6 – mf#1984-1081 – us ATLA [221]

The authenticity and messianic interpretation of the prophecies of isaiah : vindicated in a course of sermons preached before the university of oxford / Payne Smith, Robert – Oxford: John Henry and James Parker, 1862 – 1r – 1 – mf#1984-B313 – us ATLA [221]

The authenticity of the drive-in worship ministry / McCormick, Bert Edward – Princeton, NJ: [s.n.], 1978. Chicago: Dep of Photodup, U of Chicago Lib, 1979 (1r); Evanston: American Theol Lib Assoc, 1984 (1r) – 1 – 0-8370-1353-4 – mf#1984-T200 – us ATLA [210]

The authenticity of the gospel of st luke : its bearing upon the evidences of the truth of christianity / Hervey, Arthur Charles – 2nd ed. London: SPCK; New York: E & J B Young, 1892 – 1mf – 9 – 0-8370-3576-7 – mf#1985-1576 – us ATLA [226]

Authentick memoirs of the christian church in china : being a series of facts to evidence the causes of the declension of christianity in that empire / Mosheim, Johann Lorenz – London: printed for J & R Tonson and S Draper, 1750 [mf ed 1995] – (= ser Yale coll) – 60p – 1 – 0-524-09522-1 – (trans fr german) – mf#1995-0522 – us ATLA [240]

Die authentie des pentateuches *see* Dissertations on the genuineness of the pentateuch

Authentische berichte ueber luthers letzte lebensstunden / Jonas, Justus et al; ed by Strieder, Jacob – Bonn: A Marcus & E Weber 1912 [mf ed 1992] – (= ser Kleine texte fuer vorlesungen und uebungen 99) – 1mf – 9 – 0-524-04626-3 – mf#1990-1286 – us ATLA [242]

Author – London UK 1973-90 – 1,5,9 – ISSN: 0005-0628 – mf#8898 – us UMI ProQuest [400]

The author, 1890-1960 : the journal of the society of authors – 10r – 1 – mf#95725 – uk Microform Academic [800]

Author: a monthly magazine for literary workers – Killen TX 1889-92 – 1 – mf#3166 – us UMI ProQuest [070]

Author and journalist – Washington DC 1916-69 – 1 – ISSN: 0005-0636 – mf#264 – us UMI ProQuest [070]

Author and title catalogue of transmitted drama and features, 1936-1975 : with chronological list of transmitted plays (television) / BBC (British Broadcasting Corporation) Television – [mf ed Chadwyck-Healey] – 63mf – 9 – uk Chadwyck [790]

Author and title catalogue of transmitted drama, poetry and features 1929-1975 / BBC (British Broadcasting Corporation) Radio – [mf ed Chadwyck-Healey] – 125mf – 9 – uk Chadwyck [302]

Author catalogues / Biblioteca Nacional, Madrid – [mf ed Chadwyck-Healey] – 2 catalogues on 4982mf – 9 – (in spanish; the national library of spain contains the largest & most comprehensive coll of spanish books in the world; mf ed divided into 2pts: catalogo general de libros impresos [hasta 1981] 4409mf; catalogo general de libros impresos [1982-87] 573mf; with p/g) – uk Chadwyck [020]

The author catalogues of printed books / Bibliotheque Nationale. France – [mf ed Chadwyck-Healey] – 3 catalogues on 4765mf – 9 – (catalogue general: auteurs 1897-1959 1445mf. catalogue general: auteurs, collectivites-auteurs, anonymes 1960-69 430mf. catalogue general: auteurs, 1897-1959 suppl 2890mf. with p/g) – uk Chadwyck [020]

Author headings for the official publications of oklahoma / Cramer, Rose Fulton – 1944 – (filmed with: author headings for the official publications of the state of wisconsin by ruth lillian whitlock) – us CRL [324]

Author headings for the official publications of the state of wisconsin / whitlock, ruth lillian – 1941 – (filmed with: author headings for the official publications of oklahoma by rose fulton cramer) – us CRL [324]

Authoritative christianity : the first ecumenical council, that is, the first council of the whole christian world, which was held a.d. 325 at nicaea in bithynia / Chrystal, James – Jersey City, N.J., U.S.A.: J. Chrystal, 1891 – 2mf – us ATLA [240]

Authorities, deductions and notes in contracts / Pattee, William Sullivan – Minneapolis: University, 1900. 171p. LL-999 – 1 – us L of C Photodup [346]

Authorities, deductions and notes in real property / Pattee, William Sullivan – Minneapolis: University, 1902. 148p. LL-1036 – 1 – us L of C Photodup [346]

Authorities, deductions, and notes in the elements of equity / Pattee, William Sullivan – Minneapolis: University, 1900. 131p. LL-1016 – 1 – us L of C Photodup [340]

Authority : the function of authority in life and its relation to legalism in ethics and religion / Huizinga, Arnold van Couthen Piccardt – Boston: Sherman, French 1911 [mf ed 1990] – 1mf – 9 – 0-7905-7341-5 – mf#1989-0566 – us ATLA [170]

Authority : or, a plain reason for joining the church of rome / Rivington, Luke – 7th rev ed. London: Catholic Truth Soc 1897 [mf ed 1986] – 1mf – 9 – 0-8370-8465-2 – mf#1986-2465 – us ATLA [241]

Authority and archaeology, sacred and profane : essays on the relation of monuments to biblical and classical literature / Driver, Samuel Rolles et al; ed by Hogarth, David George – New York: Charles Scribner; London: John Murray 1899 [mf ed 1989] – 2mf – 9 – 0-7905-0429-4 – (incl bibl ref & ind) – mf#1987-0429 – us ATLA [930]

The authority and person of our lord / Hutton, John Alexander – New York: Fleming H Revell c1910 [mf ed 1989] – 1mf – 9 – 0-7905-1009-X – mf#1987-1009 – us ATLA [240]

Authority and the light within / Grubb, Edward – London: James Clarke 1908 [mf ed 1985] – 1mf – 9 – 0-8370-3413-2 – (incl bibl ref & ind) – mf#1985-1413 – us ATLA [210]

Authority, ecclesiastical and biblical / Hall, Francis Joseph – New York: Longmans, Green 1908 [mf ed 1990] – (= ser Dogmatic theology 2) – 1mf – 9 – 0-7905-3888-1 – (incl bibl ref) – mf#1989-0381 – us ATLA [240]

Authority in matters of faith / Robertson, Alexander et al – 2nd rev ed. London: SPCK 1897 [mf ed 1993] – (= ser Church historical society [series] 9-13) – 1mf – 9 – 0-524-06442-3 – mf#1991-2564 – us ATLA [210]

Authority in religion / Leckie, Joseph Hannay – Edinburgh: T & T Clark 1909 [mf ed 1985] – 1mf – 9 – 0-8370-4074-4 – (incl ind) – mf#1985-2074 – us ATLA [200]

Authority in the church / Strong, Thomas Banks – London, New York: Longmans, Green 1903 [mf ed 1985] – (= ser Handbooks for the clergy) – 1mf – 9 – 0-8370-5456-7 – mf#1985-3456 – us ATLA [240]

The authority of christ / Forrest, David William – Edinburgh: T & T Clark 1906 [mf ed 1990] – 2mf – 9 – 0-7905-3837-7 – (incl bibl ref) – mf#1989-0330 – us ATLA [240]

Authority of christ over the individual, the church, and the nation / Dick, James – Belfast, Northern Ireland. 1893 – 1r – us UF Libraries [240]

The authority of god : or, the true barrier against romish and infidel aggression. four discourses = Autorite des ecritures inspirees de dieu / Merle d'Aubigne, Jean Henri – aut's complete ed. New York: R Carter 1851 [mf ed 1991] – 1mf – 9 – 0-7905-7987-1 – (trans fr french into english) – mf#1989-1272 – us ATLA [220]

The authority of holy scripture : an inaugural address / Briggs, Charles Augustus – 2nd ed. New York: Scribner 1891 [mf ed 1984] – (= ser Biblical crit – us & gb 5) – 2mf – 9 – 0-8370-0250-8 – (incl bibl ref) – mf#1984-1005 – us ATLA [220]

The authority of scripture : a re-statement of the argument / Redford, Robert Ainslie – London: Religious Tract Soc [1883?] [mf ed 1989] – 1mf – 9 – 0-7905-2790-1 – mf#1987-2790 – us ATLA [220]

Authority of scripture considered in relation to christian union / Craik, Henry – London, England. 1863? – 1r – us UF Libraries [240]

The authority of the archbishop in the lincoln case – London: W Knott 1891 [mf ed 1992] – 1mf – 9 – 0-524-04170-9 – mf#1990-4974 – us ATLA [242]

The authority of the church : as set forth in the book of common prayer, articles and canons / Dix, Morgan – London: Wells Gardner, Darton; New York: E & J B Young [1891?] [mf ed 1990] – 1mf – 9 – 0-7905-7623-6 – mf#1989-0848 – us ATLA [242]

Authority of the pope in england – York, England. 18– – 1r – us UF Libraries [240]

Authorized and authentic life and works of t de witt talmage / Banks, Charles Eugene – [s.l: s.n.] c1902 [mf ed 1991] – 2mf – 9 – 0-524-01641-0 – mf#1990-0462 – us ATLA [240]

The authorized catalogue of the first annual exhibition of the agricultural and industrial exhibition association of toronto : held in the new exhibition park, in the city of toronto: open from september 1st to september 19th, 1879 / Agricultural and Industrial Exhibition (1st : 1879 : Toronto, Ont) – Toronto: Copp, Clark, [1879?] [mf ed 1981] – 2mf – 9 – 0-665-09647-X – mf#09647 – cn Canadiana [630]

The authorized edition of the english bible (1611) : its subsequent reprints and modern representatives / Scrivener, Frederick Henry Ambrose – Cambridge: University Press 1884 [mf ed 1985] – 1mf – 9 – 0-8370-5199-1 – (incl bibl ref & ind of persons & subjects) – mf#1985-3199 – us ATLA [220]

Authorized or revised? : sermons on some of the texts in which the revised version differs from the authorized / Vaughan, Charles John – London: Macmillan 1882 [mf ed 1986] – 2mf – 9 – 0-8370-9992-7 – mf#1986-3992 – us ATLA [242]

Authorized record of proceedings / Baptist World Congress (1905: London, England) – London: Baptist Union Publ Dept 1905 [mf ed 1986] – 2mf – 9 – 0-8370-8964-6 – (incl bibl ref & ind; int by j h shakespeare) – mf#1986-2964 – us ATLA [242]

Authorized report of the proceedings of the first congress of the protestant episcopal church in the united states : held in the city of new york, oct 6th and 7th, 1874 / Wildes, George Dudley – New York: T Whittaker 1875 [mf ed 1993] – 1mf – 9 – 0-524-06280-3 – mf#1991-2471 – us ATLA [242]

The authorized version of the bible and its influence / Cook, Albert Stanburrough – New York: G P Putnam 1910 [mf ed 1986] – 1mf – 9 – 0-8370-9214-0 – mf#1986-3214 – us ATLA [220]

An author's adventures : or, personal reminiscences in book-making / Ballantyne, Robert Michael – London: J Nisbet, 18–? – 3mf – 9 – mf#50418 – cn Canadiana [070]

Authors and their public in ancient times : a sketch of literary conditions and of the relations with the public of literary producers, from the earliest times to the invention of printing / Putnam, George Haven – New York: Putnam, 1894 [mf ed 1990] – 1mf – 9 – 0-7905-7190-0 – mf#1988-3190 – us ATLA [070]

The author's apology for protesting against the methodist episcopal government / O'Kelly, James – Richmond: Printed for the author, 1798. Chicago: Dep of Photodup, U of Chicago Lib, 1968 (1r); Evanston: American Theol Lib Assoc, 1984 (1r) – 1 – 0-8370-0475-6 – mf#1984-B086 – us ATLA [242]

Authors in miami / Garcia, Helen M – s.l, s.l? . 193-? – 1r – us UF Libraries [978]

Authorship and date of the books of moses considered : with special reference to professor smith's views / Paul, William – Aberdeen: Lewis Smith, 1878. Princeton: Speer Lib, and Dep of Photodup, U of Chicago Lib, 1978 (1r); Evanston: American Theol Lib Assoc, 1984 – (= ser Case of william robertson smith in the free church of scotland) – 1 – 0-8370-0616-3 – mf#1984-6267 – us ATLA [240]

The authorship of a journal of the siege of quebec : in the year 1759 / Quebec Literary and Historical Society. Associate member – [S.l: s.n, 1872?] [mf ed 1985] – 1mf – 9 – 0-665-10402-2 – mf#10402 – cn Canadiana [971]

The authorship of the de imitatione christi : with many interesting particulars about the book / Kettlewell, Samuel – London: Rivingtons, 1877 – 2mf – 9 – 0-7905-6653-2 – mf#1988-2653 – us ATLA [240]

Authorship of the dialogus de vita crysostomi / Butler, Edward Cuthbert – Roma: Tipografia Poliglotta 1908 [mf ed 1992] – 1mf – 9 – 0-524-05138-0 – mf#1990-1394 – us ATLA [240]

The authorship of the fourth gospel : and other critical essays / Abbot, Ezra; ed by Thayer, J H – Boston: G H Ellis, 1888 – 2mf – 9 – 0-8370-2004-2 – (incl ind) – mf#1985-0004 – us ATLA [226]

The authorship of the fourth gospel : external evidence / Abbot, Ezra – Boston: G H Ellis, 1880 – 1mf – 9 – 0-8370-2005-0 – mf#1985-0005 – us ATLA [226]

The authorship of the west saxon gospels / Drake, Allison Emery – New York: E Scott, 1894 – 1mf – 9 – 0-8370-2961-9 – mf#1985-0961 – us ATLA [226]

The author/title and subject catalogue of books / British Architectural Library – 1132 mf – 9 – £3750.00 – (pt 1: aut/title catalogue alphabetically arranged. pt 2: subject catalogues. a: alphabethical ind to classified catalogue.s b: classified subjects catalogue to 1955. c: classified subjects catalogue 1956-83. d: classified subjects catalogue to 1983 (class 91; Topography). e: classified subjects catalogue to 1983 (class 92: biography)) – mf#RCC – uk World [720]

Autio, Wesley R *see* Journal of tree fruit production

Autissiodorensis, Heiricus *see* Homiliae per circulum anni

L'auto – Paris, France 10 aug-31 dec 1916, 3 mar 1941-31 aug 1942 (imperfect) – 2r – 1 – uk British Libr Newspaper [629]

Auto age dealer business – Shawnee Mission KS 1992-95 – 1,5,9 – (cont by: ward's dealer business) – ISSN: 1070-8294 – mf#18406.03 – us UMI ProQuest [380]

Auto de fe and jew / Adler, Elkan Nathan – London, New York: OUP 1908 [mf ed 1990] – 1mf [ill] – 9 – 0-7905-5500-X – (incl bibl ref) – mf#1988-1500 – us ATLA; us UF Libraries [946]

Auto del nuncio contra una asserta sentencia impressa... / Fachenetti, Cesar – S.l., s.i., s.a. 1642 – 1 – sp Bibl Santa Ana [946]
Auto general de la fee : celebrado por los senores, el il[lustrissi]mo y r[everendiss]mo senor don iuan de manozca, arcobispo de mexico... / Bocanegra, Matias de – en Mexico: Por Antonio Calderon [1649] – (= ser Books on religion...1543/44-c1800: inquisicion) – 2mf – 9 – mf#crl-255 – ne IDC [241]
Auto general de la fee que assistio presidiendo en nombre : y representacion de la catholica magestad del rey n senor d felipe quarto [que dios guarde]... / Ruiz de Zepeda Martinez, Rodrigo – en Mexico: impr de Bernardo de Calderon [1660?] – (= ser Books on religion...1543/44-c1800: inquisicion) – 2mf – 9 – mf#crl-256 – ne IDC [241]
Auto racing digest – Evanston IL 1985+ – 1,5,9 – ISSN: 0090-8029 – mf#11613 – us UMI ProQuest [790]
Auto repair books : domestic and imported car, light truck and van – Over 100,000 pages. Fully indexed. Over 700mf. Sections available separately – 9 – $1,777.00; $2,363.00 in Canada – (annual update) – us Mitchell Int [629]
Auto repair books : domestic car repair – 10 years of domestic car model information – 9 – $750.00; $998.00 in Canada – (annual update: $180 $239.00 in canada) – us Mitchell Int [629]
Auto repair books : domestic light truck and van repair – 10 years of domestic light truck and van information – 9 – $500.00; $665.00 in Canada – (annual update: $125 $166.00 in canada) – us Mitchell Int [629]
Auto repair books : imported car and light truck repair – 10 years of imported car and light truck service and repair information – 9 – $800.00; $1,064.00 in Canada – (annual update: $195 $259.00 in canada) – us Mitchell Int [629]
Auto repair books : technical service bulletins – Published yearly – 9 – $50.00; $67.00 in Canada – us Mitchell Int [629]
Auto sacramental nuebo [sic] de las pruebas del linaje umanO [sic] – Madrid, Spain. 1897 – 1r – us UF Libraries [960]
Auto worker – 1919 may-1924 dec – 1 – mf#1052677 – us WHS [629]
Auto worker, 1919-24 / the spark plug, 1917 / United Automobile, Aircraft and Vehicle Workers of America – (= ser Labor union periodicals, pt 1: the metal trades) – 1r – 1 – $210.00 – 1-55655-227-0 – us UPA [331]
Auto workers news / Auto Workers Union – 1927-1934 – (= ser Labor union periodicals, pt 1: the metal trades) – 1r – 1 – $210.00 – 1-55655-228-9 – us UPA [331]
Auto workers news – v1-3 [1927 may-1930 may]; v8 n1, 4-16 [1934 jan 13, feb 24-1934 aug 18] – 1 – mf#1052678 – us WHS [629]
Auto Workers Union see Auto workers news
Auto y sentencia en favor del r. padre fr. sebastian de moratilla vicario paterno del monaterio de guadalupe... / Nunciatura Apostolica – S.l., s.i., s.a. 1641 – 1 – sp Bibl Santa Ana [946]
Autoaggression und pathologische informationsverarbeitung bei geistigbehinderten mit autistischen zuegen / Elbing, Ulrich – (mf ed 1992) – 3mf – 9 – €49.00 – 3-89349-468-5 – mf#DHS 468 – gw Frankfurter [616]
Auto-aircraft news – 1946 sep 23-1947 may 8; 1949 feb 28-1954 dec – 1 – mf#1110031 – us WHS [629]
Autobiografia : cartas y versos / Manazano, Juan Francisco – Habana, Cuba. 1937 – 1r – us UF Libraries [972]
Autobiografia / Dario, Ruben – Buenos Aires, Argentina. 1947 – 1r – us UF Libraries [920]
Autobiografia : 'exposicao aos credores e ao public... / Maua, Irineo Evangelista De Souza – Rio de Janeiro, Brazil. 1942 – 1r – us UF Libraries [972]
Autobiographic memoirs / Harrison, Frederic – London: Macmillan 1911 [mf ed 1992] – 2v on 2mf – 9 – 0-524-04297-7 – (incl bibl ref) – mf#1992-2017 – us ATLA [920]
Autobiographical material : miss willie kelley / Barton, L E – 148p – 1 – $5.18 – us Southern Baptist [920]
Autobiographical memoirs, 1821-52 / Orchard, George Herbert – 96p – 1 – $5.00 – us Southern Baptist [242]
Autobiographical notes / Clark, Marona M (Still) – undated, Miscellaneous titles – 1 – us Kansas [920]
[Autobiographical pamphlets] / Mullen, Mary B et al – [s.l: s.n. 1898?-1908?] [mf ed 1992] – (= ser Christian & missionary alliance coll) – 1mf – 9 – 0-524-03729-9 – (together with: musgrove, sara m c: history of twenty-five years' four-fold gospel work in troy [new york 1908]; senft, frederic herbert [mrs]: jesus, my physician [1898]) – mf#1990-4834 – us ATLA [920]
Autobiographical record : adventures of a guano digger in the eastern pacific / Chave, Richard Branscombe – 1871 – 1r – 1 – mf#pmb20 – at Pacific Mss [920]
An autobiographical sketch of the services of the late captain andrew bulger : of the royal newfoundland fencible regiment / Bulger, Andrew H – Bangalore, India: Regimental Press, 1865 – 1mf – 9 – mf#48336 – cn Canadiana [975]
Autobiographical sketches and recollections : during a thirty-five years' residence in new orleans / Clapp, Theodore – 3rd ed. Boston: Phillips, Sampson 1858 [mf ed 1992] – 1mf – 9 – 0-524-05125-9 – mf#1992-2078 – us ATLA [242]
Autobiographies from men of all ranks : sources from the british library, london – 1 – (pt1: autobiographies c1760-1820 [20r] £1850; with d/g) – uk Matthew [920]
An autobiography : the story of the lord's dealings with mrs amanda smith, the colored evangelist / Smith, Amanda – Chicago: Meyer & Brother, 1893 [mf ed 1984] – (= ser Women & the church in america 130) – 2mf – 9 – 0-8370-1393-3 – mf#1984-2130 – us ATLA [305]
Autobiography / Dole, Artemus Wood – 1856-67 – 1 – us Kansas [920]
Autobiography : every goose a swan, vol 2 / Langdon, Robert Adrian – 1993 – 1r – 1 – mf#pmb1230 – at Pacific Mss [920]
Autobiography / Harris, Joseph S – 1 – $50.00 – us Presbyterian [920]
Autobiography / Hoshour, Samuel Klinefelter – St Louis: John Burns, 1884 [mf ed 1993] – (= ser Christian church (disciples of christ) coll) – 1mf – 9 – 0-524-07010-5 – (int by isaac errett, app by ryland t brown) – mf#1991-2863 – us ATLA [920]
Autobiography / Mill, John Stuart; ed by Taylor, Helen – New York: H Holt, 1873 [mf ed 1991] – 1mf – 9 – 0-7905-9034-4 – mf#1989-2259 – us ATLA [190]
Autobiography / Reiley, James – 1950 – 1r – 1 – $35.00 – mf#um-140 – us Commission [242]
Autobiography / Rennolds, Edwin Hansford – Paris, Tennessee. 104p. 2 Sep 1897 – 1 – $5.00 – (also clippings, pictures and paper, ythe baptist reaper. containing picture of grandfather asa cox) – us Southern Baptist [242]
Autobiography / Schwarz, Conrad Herman Alfred – 1 – us Kansas [978]
Autobiography / Sinclair, Addie M – 1 – us Kansas [978]
Autobiography / Tregurtha, Edward Primrose – 1803-52 – 1r – 1 – mf#pmb12 – at Pacific Mss [920]
Autobiography / Wilkes, Charles – 1798-1877 – 1 – $59.00 – us L of C Photodup [920]
Autobiography... : for forty-nine years a missionary in the orient... / Schauffler, W G – New York, 1887 – 4mf – 9 – mf#HTM-170 – ne IDC [920]
The autobiography : a critical and comparative study / Burr, Anna Robeson Brown – Boston: Houghton Mifflin, 1909 [mf ed 1990] – 2mf – 9 – 0-7905-4664-7 – (incl bibl ref) – mf#1988-0664 – us ATLA [410]
Autobiography 1764 / Cherbury, Edward, Lord Herbert of – 2nd rev ed. London, 1906 – €7.00 – ne Slangenburg [920]
Autobiography, 1880-1945 / King, Spencer B – 166p – 1 – $5.81 – us Southern Baptist [242]
Autobiography, 1881-1891 / Bascom, Flavel – 2r – 1 – us Amistad [242]
Autobiography and conversion of w r bradlaugh / Bradlaugh, William Robert – London, England. 1884? – 1r – us UF Libraries [240]
Autobiography and correspondence / Nassau, Robert Hamill – 1 – $100.00 – us Presbyterian [920]
The autobiography and diary of samuel davidson : with a selection of letters from english and german divines, and an account of the davidson controversy of 1857 / Picton, James Allanson; ed by Davidson, Anne Jane – Edinburgh: T & T Clark, 1899 [mf ed 1984] – (= ser Biblical crit – us & gb 14) – 5mf – 9 – 0-8370-0672-4 – (incl bibl ref & ind) – mf#1984-1014 – us ATLA [240]
Autobiography and jefferson lodge minutes / Steele, Oliver Hazard Perry – undated, Records of the lodge in Franklin County, Ohio, and lengthy autobiography of Steele, a resident of Columbus, Ohio, and Ogle County, Illinois – 1 – us Kansas [920]
Autobiography and other materials by him / Dagg, John L – 1 – $60.90 – us Southern Baptist [242]
Autobiography and personal recollections of john b gough : with twenty-six years' experience as a public speaker / Gough, John Bartholomew – Springfield, MA: Bill, Nichols; Chicago, IL: Bill & Heron, 1870, c1869 [mf ed 1990] – 2mf – 9 – 0-7905-4794-5 – mf#1988-0794 – us ATLA [920]
Autobiography and work of bishop m f jamison... : a narration of his whole career... / Jamison, Monroe Franklin – Nashville, TN: printed for aut...1912 [mf ed 1993] – (= ser Methodist coll) – 1mf – 9 – 0-524-07011-3 – mf#1991-2864 – us ATLA [242]
Autobiography, correspondence etc of lyman beecher / ed by Beecher, Charles – New York: Harper, 1865 [mf ed 1984] – 13mf – 9 – 0-8370-0222-2 – (incl bibl ref) – mf#1984-0056 – us ATLA [242]
Autobiography, intellectual, moral, and spiritual / Mahan, Asa – London: T Woolmer, 1882 [mf ed 1991] – 2mf – 9 – 0-7905-8695-9 – mf#1989-1920 – us ATLA [920]
The autobiography, longtime missionary to brazil / Bratcher, Lewis M – 1888-1953 – 1 – 5.67 – us Southern Baptist [242]
Autobiography, memories and experiences of Moncure Daniel Conway – Boston: Houghton, Mifflin, 1904 [mf ed 1990] – 2v on 3mf – 9 – 0-7905-8123-X – mf#1988-8040 – us ATLA [920]
Autobiography of a pioneer printer : together with sketches of the war of 1812 along the niagara frontier / Howe, Eber D – Painesville, OH; Telegraph Steam Printing Co, 1878 – 1r – 1 – us Western Res [680]
Autobiography of a shaker : and revelation of the apocalypse / Evans, Frederick William – new enl ed. Glasgow: United Publ Co; New York: American News Co, 1888 [mf ed 1990] – 1mf – 9 – 0-7905-5464-X – (with app) – mf#1988-1464 – us ATLA [243]
Autobiography of a soldier of the civil war / Snyder, Edward P – 1r – 1 – mf#B41471 – us Ohio Hist [976]
Autobiography of abraham snethen : the barefoot preacher / Lamb, N E [comp] – Dayton, OH: Christian Pub Assoc, 1909 [mf ed 1990] – 1mf – 9 – 0-7905-8016-0 – (corr & rev by john franklin burnett) – mf#1988-8016 – us ATLA [240]
Autobiography of adin ballou, 1803-1890 : containing an elaborate record and narrative of his life from infancy to old age / Ballou, Adin; ed by Heywood, William Sweetzer – Lowell, MA: Vox Populi Press, 1896 [mf ed 1990] – 2mf – 9 – 0-7905-3533-5 – mf#1989-0026 – us ATLA [920]
Autobiography of allen jay : born 1831, died 1910 – Philadelphia: John C Winston, c1910 [mf ed 1992] – (= ser Society of friends (quakers) coll) – 2mf – 9 – 0-524-02692-0 – mf#1990-4399 – us ATLA [243]
The autobiography of an indian princess / Sunity Devee, Maharani of Cooch Behar – London: John Murray, 1921 – (= ser Samp: indian books) – 1r – us CRL [920]
Autobiography of andrew dickson white – New York: Century, c1905 [mf ed 1990] – 2v on 3mf – 9 – 0-7905-8099-3 – mf#1988-8035 – us ATLA [920]
The autobiography of archbishop ullathorne : with selections from his letters / Ullathorne, William Bernard, Archbishop of – 2nd ed. London: Burns & Oates; New York: Benziger, [1892?] [mf ed 1990] – 1mf – 9 – 0-7905-6379-7 – mf#1988-2379 – us ATLA [241]
Autobiography of benjamin franklin – New York, NY. 190-? – 1r – us UF Libraries [920]
Autobiography of benjamin hallowell – Philadelphia: Friends' Book Assoc, 1883 [mf ed 1993] – (= ser Society of friends (quakers) coll) – 1mf – 9 – 0-524-06540-3 – mf#1991-2624 – us ATLA [920]
Autobiography of bishop isaac lane : with a short history of the c m e church in america and of methodism – Nashville, TN: Pub House of the ME Church, South, 1916 [mf ed 1993] – (= ser Methodist coll) – 1mf – 9 – 0-524-07016-4 – mf#1991-2869 – us ATLA [242]
The autobiography of charles h spurgeon – Chicago: Fleming H. Revell, 1898-1900 [mf ed 1990] – 4mf – 9 – 0-7905-8093-4 – mf#1988-8029 – us ATLA [920]
Autobiography of dan young : a new england preacher of the olden time / ed by Strickland, William Peter – New York: Carlton & Porter, 1860 [mf ed 1990] – 1mf – 9 – 0-7905-8196-5 – mf#1988-8079 – us ATLA [242]
Autobiography of dean merivale : with selections from his correspondence / Merivale, Charles; ed by Merivale, Judith Anne – London: E Arnold 1899 [mf ed 1984] – 1r – 1 – (filmed with: shelley memorials / shelley, j g & other titles) – mf#8664 – us Wisconsin U Libr [930]
Autobiography of dean merivale : with selections from his correspondence / ed by Merivale, Judith Anne – London: E Arnold, 1899 [mf ed 1990] – 1mf – 9 – 0-7905-5715-0 – mf#1988-1715 – us ATLA [920]
Autobiography of edward gibbon / ed by Sheffield, John Baker Holroyd, Earl of – London, New York: Oxford UP [1907?] [mf ed 1991] – (= ser The world's classics 139) – 1mf – 9 – 0-524-00636-9 – mf#1990-0136 – us ATLA [900]
Autobiography of elder jacob knapp / Knapp, Jacob – New York: Sheldon, 1868 [mf ed 1984] – (= ser Revivalism and revival preachers in america 5) – 4mf – 9 – 0-8370-0693-7 – (incl bibl ref) – mf#1984-3005 – us ATLA [242]
Autobiography of erastus o haven : one of the bishops of the methodist episcopal church / Haven, Erastus Otis; ed by Stratton, Charles Carroll – New York: Phillips & Hunt; Cincinnati: Walden & Stowe, 1883 [mf ed 1991] – 1mf – 9 – 0-524-00555-9 – mf#1990-0055 – us ATLA [242]
Autobiography of george mueller – London: J Nisbet, 1905 [mf ed 1991] – 2mf – 9 – 0-524-01657-7 – mf#1990-0478 – us ATLA [240]
Autobiography of george tyrrell, 1861-1884 – London: E Arnold, 1912 [mf ed 1992] – (= ser Anglican/episcopal coll) – 1mf – 9 – 0-524-04824-X – mf#1992-2053 – us ATLA [241]
The auto-biography of goethe : truth and poetry: from my own life = Aus meinem leben – London: George Bell & Sons, 1874 [mf ed 1999] – 2v – 1 – (v1 trans by by john oxenford. v2 trans by alexander james william morrison) – mf#8631 – us Wisconsin U Libr [920]
Autobiography of hollington k tong *see* Tung hsien-kuang tzu chuan (ccm302)
Autobiography of john j cornell : containing an account of his religious experiences and travels in the ministry – Baltimore, MD: Lord Baltimore Press, 1906 [mf ed 1993] – (= ser Society of friends (quakers) coll) – 2mf – 9 – 0-524-06711-2 – mf#1991-2741 – us ATLA [243]
The autobiography of judas iscariot : a character study / Hart, James W T – London: Kegan Paul, Trench, 1884 [mf ed 1985] – 1mf – 9 – 0-8370-3504-X – mf#1985-1504 – us ATLA [830]
Autobiography Of Lydia Sexton / Sexton, Lydia – Dayton, OH: United Brethern Publ House, 1882 – 1r – 1 – (autobiography of an female evangelic protestant minister) – us Western Res [240]
The autobiography of maharshi devendranath tagore – London: Macmillan 1914 [mf ed 1995] – (= ser Yale coll) – 1mf – 1 – 0-524-09439-X – (trans original bengali by satyendranath tagore & indira devi) – mf#1995-0439 – us ATLA [290]
Autobiography of peter cartwright : the backwoods preacher / ed by Strickland, William Peter – New York: Carlton & Porter, 1857, c1856 [mf ed 1990] – 2mf – 9 – 0-7905-4725-2 – mf#1988-0725 – us ATLA [240]
Autobiography of rev alvin torry : first missionary to the six nations and the northwestern tribes of british north america / Torry, Alvin; ed by Hosmer, William – Auburn: WJ Moses, 1861 [mf ed 1993] – (= ser Methodist coll) – 1mf – 9 – 0-524-07047-4 – mf#1991-2900 – us ATLA [242]
Autobiography of rev james b finley : or, pioneer life in the west / ed by Strickland, William Peter – Cincinnati: Methodist Book Concern for the author, 1853 [mf ed 1990] – 2mf – 9 – 0-7905-4566-7 – mf#1988-0566 – us ATLA [240]
The autobiography of sir henry morton stanley / ed by Stanley, Dorothy – Boston: Houghton Mifflin, 1909 [mf ed 1993] – 2mf – 9 – 0-524-06937-9 – mf#1990-3563 – us ATLA [920]
Autobiography of thaddeus lewis : a minister of the methodist episcopal church in canada – [Picton, Ont?: s.n.] 1865 [mf ed 1984] – 3mf – 9 – 0-665-45533-X – mf#45533 – cn Canadiana [242]
The autobiography of the australian and new zealand secretary of the london missionary society *see* Struts and frets his hour
Autobiography of the first forty-one years of the life of sylvanus bobb : to which is added a memoir – Boston: Universalist Pub House, 1867 [mf ed 1993] – (= ser Unitarian/universalist coll) – 2mf – 9 – 0-524-07309-0 – mf#1991-3024 – us ATLA [240]
Autobiography of the late donald fraser : and a selection from his sermons – London: James Nisbet & Co, 1892 [mf ed 1980] – 3mf – 9 – 0-665-03175-0 – (pref by j oswald dykes) – mf#03175 – cn Canadiana [242]
Autobiography of the late rev nelson burns : a new study of the christ life, etc – S.l: Christian Association, 1904? – 3mf – 9 – mf#35715 – cn Canadiana [242]
The autobiography of the rev charles freshman : late rabbi of the jewish synagogue at quebec... – Toronto: S Rose, 1868 – 4mf – 9 – mf#03258 – cn Canadiana [920]
The autobiography of the rev charles freshman : late rabbi of the jewish synagogue at quebec... – Toronto: Samuel Rose, 1868 [mf ed 1986] – 1mf – 9 – 0-8370-6664-6 – mf#1986-0664 – us ATLA [242]

The autobiography of the rev enoch pond : for fifty years professor in bangor theological seminary – Boston: Congregational Sunday-School & Pub Soc, c1883 [mf ed 1990] – 1mf – 9 – 0-7905-5557-3 – mf#1988-1557 – us ATLA [378]

Autobiography of the rev joseph townend : with reminiscences of his missionary labours in australia – 2nd ed London: W Reed, 1869 [mf ed 1990] – 1mf – 9 – 0-7905-8183-3 – mf#1988-8066 – us ATLA [240]

Autobiography of the rev luther lee – New York: Phillips & Hunt, 1882 [mf ed 1984] – 4mf – 9 – 0-8370-1623-1 – mf#1984-4152 – us ATLA [242]

Autobiography of the rev william jay – New York: R Carter, 1856 [mf ed 1984] – 9mf – 9 – 0-8370-1652-5 – mf#1984-6246 – us ATLA [240]

The autobiography of theophilus waldmeier, missionary : being an account of ten years' life in abyssinia, and sixteen years in syria – London: SW Partridge, [1886?] [mf ed 1993] – (= ser Society of friends (quakers) coll) – 1mf – 9 – 0-524-06782-1 – mf#1991-2789 – us ATLA [240]

The autobiography of theophilus waldmeier, missionary – London, 1886 – 4mf – 9 – mf#NE-20217 – ne IDC [910]

Autobiography of thomas guthrie... : and memoir / Guthrie, David Kelly & Guthrie, Charles John Guthrie, Lord – New York: R Carter, 1876 [mf ed 1991] – 2v on 3mf – 9 – 0-524-00553-2 – mf#1990-0053 – us ATLA [242]

Autobiography of william g schauffler : for forty-nine years a missionary in the orient – New York: ADF Randolph, c1887 [mf ed 1990] – 1mf – 9 – 0-7905-8177-9 – (int by e a park) – mf#1988-8060 – us ATLA [240]

Autobiography, poems and prayers / Parker, Theodore; ed by Leighton, Rufus – centenary ed. Boston: American Unitarian Assoc [1911?] [mf ed 1993] – (= ser Unitarian/universalist coll) – 6mf – 9 – 0-524-07449-6 – mf#1991-3109 – us ATLA [240]

Autobiography...including his voyages, travels, adventure, speculations, successes and failures, faithfully and frankly narrated / Buckingham, James Silk – London: Longman, Brown, Green & Longmans 1855 [mf ed 1987] – 2v on 1r – 1 – (no more publ. filmed with: my life / duncan, i) – mf#1980 – us Wisconsin U Libr [920]

Autocar – London. v30-95.1913-49 – 1 – $4252.00 – mf#0095 – us Brook [629]

Autocrata / Wyld Ospina, Carlos – Guatemala, 1929 – 1r – us UF Libraries [972]

Auto-enseignement en histologie humaine : 800 questions a choix multiples, avec reponses commentees / Messier, Bernard – St-Hyacinthe: Edisem; Paris: Maloine, 1981 [mf ed 1995] – 2mf – 9 – mf#SEM105P2373 – cn Bibl Nat [611]

Autogestao / Nogueira, Paulo – Rio de Janeiro, Brazil. 1969 – 1r – us UF Libraries [972]

Unos autografos de don bartolome jose gallardo / Llanos y Torrigua, Felix de – Madrid: Tip. Revista de Arch. Bibliot. y Museos, 1924 – 1 – sp Bibl Santa Ana [920]

Autograph collector's magazine – 1986 jul/aug-1989 dec – 1 – mf#1700778 – us WHS [073]

Autographs of the signers of the declaration / Jenkins, Charles Francis – Philadelphia, 1926 (mf ed 19–) – 1 – mf#*ZH-IAG pv351 n5 – us NY Public [975]

Autologe fibroblasten in der endoskopischen therapie des vesikorenalen refluxes / Zoeller, Gerhard Maximilian – (mf ed 1997) – 2mf – 9 – €40.00 – 3-8267-2500-X – mf#DHS 2500 – gw Frankfurter [616]

Automat union news – n1-7 [1942 may 20-aug 15] – 1 – mf#630133 – us WHS [331]

Automated builder – Ventura CA 1989+ – 1,5,9 – (cont: automation in housing and manufactured home dealer) – ISSN: 0899-5540 – mf#1826,03 – us UMI ProQuest [690]

Automated times – v1 n1-v2 n2 [1984/1985 dec/jan-1986 fall] – 1 – mf#1533927 – us WHS [071]

Automates vivants : ou, l'atelier de vaucanson / Sonat, G – Paris, France. 18– – 1r – us UF Libraries [440]

Automatic data processing – London UK 1958-61 – 1 – ISSN: 0572-2217 – mf#1302 – us UMI ProQuest [000]

Automatica – Oxford UK 1963+ – 1,5,9 – ISSN: 0005-1098 – mf#49018 – us UMI ProQuest [621]

Automatik – Magdeburg DE, 1960, 17 nov -1989 nov [gaps] – 5r – 1 – (werkzeugmaschinenfabrik) – gw Misc Inst [620]

Automation – London UK 1975-81 – 1,5,9 – ISSN: 0005-1152 – mf#9977 – us UMI ProQuest [629]

Automation [1954] – Cleveland OH 1954-76 – 1,5,9 – (cont by: production engineering) – ISSN: 0005-1160 – mf#1077.03 – us UMI ProQuest [629]

Automation [1987] – Cleveland IL 1987-91 – 1,5,9 – (cont: production engineering; cont by: penton's controls and systems) – ISSN: 0896-6052 – mf#1077.03 – us UMI ProQuest [620]

Automation and remote control – Dordrecht, Netherlands 1956+ – 1,5,9 – ISSN: 0005-1179 – mf#1533 – us UMI ProQuest [629]

Automation in housing – Ventura CA 1964-74 – 1,5,9 – ISSN: 0005-1217 – mf#1826.03 – us UMI ProQuest [690]

Automation in housing and manufactured home dealer – Ventura CA 1984-87 – 1,5,9 – (cont: automation in housing and systems building news; cont by: automated builder) – ISSN: 0740-3534 – mf#1826.03 – us UMI ProQuest [690]

Automation in housing and systems building news – Ventura CA 1975-82 – 1,5,9 – (cont by: automation in housing and manufactured home dealer) – ISSN: 0362-0395 – mf#1826.03 – us UMI ProQuest [690]

Automatische extraktion von mitochondrien aus mikroskopischen bildern von herzmuskelzellen / Oth, Volker & Wirth, Georg – 2000 – 1mf – 9 – 3-8267-2676-6 – mf#DHS 2676 – gw Frankfurter [621]

Automatisme / L'Association francaise de regulation et d'automatisme – Paris. 1956-72 – 5 – fr ACRPP [670]

Automobile Association Of South Africa *see* Trans-african highways

Automobile engineer – Sutton UK 1950-72 – 1,5,9 – ISSN: 0005-1381 – mf#654 – us UMI ProQuest [629]

Automobile magazine – Killen TX 1899-1907 – 1 – mf#4796 – us UMI ProQuest [629]

Automobile year = Annee automobile / Auto jahr – Lausanne, Switzerland 1975-80 – 1,5,9 – ISSN: 0084-7674 – mf#8010 – us UMI ProQuest [629]

Automotive daily news – 1937 feb 8,10 – 1 – mf#3910377 – us WHS [071]

Automotive design and production [ad&p] – Cincinnati OH 2002+ – 1,5,9 – (cont: automotive manufacturing and production (amp)) – mf#25565,01 – us UMI ProQuest [629]

Automotive engine testing / Gruber, Foster M – New York, NY. 1940 – 1r – us UF Libraries [629]

Automotive engineer – London UK 1975+ – 1,5,9 – ISSN: 0307-6490 – mf#11215 – us UMI ProQuest [629]

Automotive executive – McLean VA 1989-94 – 1 – (cont by: nada's automotive executive) – ISSN: 0195-1564 – mf#12284.01 – us UMI ProQuest [629]

Automotive industries [1899] – London UK 1899-1975 – 1,5,9 – (cont by: chilton's automotive industries) – ISSN: 0886-4675 – mf#45.02 – us UMI ProQuest [629]

Automotive industries [1995] – London UK 1995+ – 1,5,9 – (cont: chilton's automotive industries) – ISSN: 1099-4130 – mf#45,02 – us UMI ProQuest [629]

Automotive manufacturing and production [amp] – Cincinnati OH 1997-2000 – 1,5,9 – mf#25565.01 – us UMI ProQuest [338]

Automotive marketing – Cleveland OH 1998-2001 – 1,5,9 – (cont: chilton's automotive marketing) – mf#11779,04 – us UMI ProQuest [380]

Automotive news – Detroit MI 1925+ – 1,5,9 – ISSN: 0005-1551 – mf#6116 – us UMI ProQuest [629]

Automotive plastics – Dearborn MI 2000-01 – 1,5,9 – (cont: molding systems) – ISSN: 1531-6815 – mf#14875,02 – us UMI ProQuest [660]

Automotive production – Cincinnati OH 1996 – 1,5,9 – (cont: production) – ISSN: 1086-9298 – mf#8949,01 – us UMI ProQuest [338]

Automotive service digest – Chicago IL 1958-61 – 1 – mf#1137 – us UMI ProQuest [629]

The automotive trade magazine *see* The horseless age

Autonomia universitaria / Febres Cordero, Focion – Caracas, Venezuela. 1959 – 1r – us UF Libraries [972]

Autonomic and autacoid pharmacology – Oxford UK 2002+ – 1,5,9 – ISSN: 1474-8665 – mf#16734,01 – us UMI ProQuest [615]

Autonomic neuroscience : basic and clinical – Oxford UK 2002+ – 1 – (cont: journal of the autonomic nervous system) – ISSN: 1566-0702 – mf#42212,01 – us UMI ProQuest [612]

Autonomie – London. v1-8 n212. nov 6 1886-apr 22 1893 – 1 – us CRL [073]

Autonomie : victoires autonomistes de sir wilfrid laurier: la nation canadienne – [Ottawa?: s.n, 1911?] – 1mf – 9 – 0-665-76290-9 – mf#76290 – cn Canadiana [320]

Autonomie d'haiti / Johnson, James Weldon – Port-Au-Prince, Haiti. 1921 – 1r – us UF Libraries [972]

Autoperfusionsballonkatheder behandlung : akut-ischaemische komplikationen bei koronarintervention / Piepenbrink, Thomas – (mf ed 2000) – 2mf – 9 – €40.00 – 3-8267-2717-7 – mf#DHS 2717 – gw Frankfurter [617]

Autoproducts – Torrance CA 1973-74 – 1 – (cont by: ad and d: automotive design and development) – ISSN: 0005-1675 – mf#7934.01 – us UMI ProQuest [629]

L'autore del quarto evangelo rivendicato / Polidori, Eugenio – 3. ed, migliorata. Roma: Civit a Cattolica, 1906 [mf ed 1993] – 1mf – 9 – 0-524-06152-1 – (in italian. incl bibl ref) – mf#1992-0819 – us ATLA [225]

Autores contemporaneos / Ribeiro, Joao – Rio de Janeiro, Brazil. 1937 – 1r – us UF Libraries [440]

Autores germanos en el peru : florilegio de la poesia alemana en versiones peruanas / Estuardo, Nunez – Lima: Ministerio de Educacion Publica 1953 [mf ed 1993] – (= ser Concurso de fomento de la cultura. 2a epoca 1) – 1r – 1 – (filmed with: deutsch-indische geistesbeziehungen / ludwig alsdorf) – mf#8141 – us Wisconsin U Libr [410]

Autores pre-romanticos alemaes / Rosenfeld, Anatol [comp] – Sao Paulo: Herder, 1965 [mf ed 1993] – (= ser Colecao pensamento estetico) – 129p – 1 – (int and ann by comp. incl bibl ref) – mf#8210 – us Wisconsin U Libr [430]

La autoridad doctrinal de la iglesia catolica y la libertad de pensamiento / Vazquez Camarasa, Enrique – Madrid: Imprenta del Asilo de Hurfanos del S.C. de Jesus, 1916 – 1 – sp Bibl Santa Ana [241]

Die autoritaet der hlg schrift und die kritik : nach der schrift und den grundsaetzen luthers / Haug, Karl – Strassburg: Strassburger Druckerei & Verlagsanstalt, 1891 – 1mf – 9 – 0-8370-3523-6 – mf#1985-1523 – us ATLA [220]

Die autoritaet des alten testamentes fuer den christen / Oettli, Samuel – Berlin: Edwin Runge, 1906 – 1mf – 9 – 0-7905-0507-X – mf#1987-0507 – us ATLA [221]

L'autorite – Paris. 25 fev 1886-6 sept 1914, 24 mars-19 sept 1928, 16 fevr-14 sept 1929 – 1 – fr ACRPP [073]

L'autorite – Paris: Schneider, may 1849 – 1 – us CRL [074]

L'autorite de jesus : envisagee du point de vue de ses disciples immediats d'apres le temoignage des ecrivains synoptiques... / Wuarin, Louis – Genever: B Soullier, 1875 [mf ed 1985] – 1mf – 9 – 0-8370-5929-1 – (incl bibl ref) – mf#1985-3929 – us ATLA [240]

Autorite des ecritures inspirees de dieu *see* The authority of god

L'autorite humaine des livres saints / Mechineau, Lucien – Paris: Librairie Bloud, 1903 [mf ed 1993] – (= ser Science et religion) – 1mf – 9 – 0-524-05813-X – mf#1992-0640 – us ATLA [220]

Autos acordados de la real audiencia / Puerto Rico Real Audiencia – Puerto Rico, Puerto Rico. 1857 – 1r – us UF Libraries [972]

Autos de devassa da inconfidencia mineira – Rio de Janeiro, Brazil. v1-7. 1936-38 – 1r – us UF Libraries [972]

Autour de kita : etude soudanaise / Tellier, G – Paris: H Chales-Lavauzell, [1898] – 1 – us CRL [960]

Autour de la docte ignorance : une controverse sur la theologie mystique au 15e siecle / Vansteenberghe, E – 1915 – (= ser Bgphma 14/2-4) – €11.00 – ne Slangenburg [200]

Autour de la maison / Normand, Michelle le – Montreal: edition du Devoir, 1918 [mf ed 1999] – 2mf – 9 – 0-659-90206-0 – mf#9-90206 – cn Canadiana [830]

Autour de la question biblique : une nouvelle ecole d'exegese et les autorites qu'elle invoque / Delattre, Alphonse J – Liege: H Dessain [1904?] [mf ed 1986] – 1mf – 9 – 0-8370-7133-X – mf#1986-1133 – us ATLA [220]

Autour de l'isthme de panama / Justin, Joseph – Port-Au-Prince, Haiti. 1913 – 1r – us UF Libraries [970]

Autour de mandera : notes sur l'ouzigoua, l'oukwere et l'oudoe / Picarda, R P – In Missions Catholiques. Lyons. 1886 – 1 – us CRL [306]

Autour de platon / Dies, Auguste – Paris, France. v1-2. 1927 – 1r – us UF Libraries [180]

Autour du concile : souvenirs et croquis d'un artiste a rome / Yriarte, Charles – Paris: J Rothschild 1887 [mf ed 1986] – 1mf [ill] – 9 – 0-8370-8639-6 – mf#1986-2639 – us ATLA [914]

Autour du proces de p.o.u.m. : des revolutionnaires en danger de mort: julian gorkin, juan andrade, gironella, jose rovira, jordi arquer, daniel rebull, pedro bonet, jose escuder – [Paris?]: Independent News 1938 – 9 – mf#w737 – us Harvard [335]

Autour d'un petit livre / Loisy, Alfred Firmin – 2e ed. Paris: Alphonse Picard, 1903 [mf ed 1989] – 1mf – 9 – 0-7905-1186-X – mf#1987-1186 – us ATLA [240]

Autour d'un tiare / Gebhart, Emile – Paris: Georges Cres [18–?] [mf ed 1986] – (= ser Coll gallia 12) – 1mf [ill] – 9 – 0-8370-7863-6 – mf#1986-1863 – us ATLA [440]

Autour d'une carriere politique : joseph israel tarte, 1880-1897, dix-sept ans de contradictions – [Montreal?: s.n, 1897?] [mf ed 1980] – 2mf – 9 – 0-665-00068-5 – mf#00068 – cn Canadiana [325]

Autour d'une femme sous les tropiques / Poulet, Georges – Paris: Albin Michel, 1927 – (= ser Les femmes [coll]) – 3mf – 9 – mf#12696 – fr Bibl Nationale [305]

L'auto-velo – Paris. 16 oct 1900-17 Aout 1944 – 1 – (puis l'auto.) – fr ACRPP [629]

AutoWeek – Detroit MI 1986+ – 1,5,9 – ISSN: 0192-9674 – mf#15195 – us UMI ProQuest [629]

Autre messie / Soumagne, Henry – Paris, France. 1924 – 1r – us UF Libraries [440]

Autre part du diable : ou, le talisman du mari / Varner, Antoine-Francois – Paris, France. 1843 – 1r – us UF Libraries [440]

L'autriche, ou moeurs, usages et costumes des habitans de cet empire : suivi d'un voyage en baviere et au tyrol / Serres, Marcel de – Paris 1821 [mf ed Hildesheim 1995-98] – 6v on 12mf – 9 – €120.00 – 3-487-29449-4 – gw Olms [914]

L'autriche slave et roumaine – Paris. n1-29. 8 nov 1887-30 mai 1888 [wkly] – 1 – (journal politique) – fr ACRPP [320]

Autumn and winter season sketchbook / Foster and Co, Allen – [London? 1900?] – (= ser 19th c art & architecture) – 1mf – 9 – mf#4.2.1002 – uk Chadwyck [740]

Autumn gatherings : mabel ashton: a tale of the crimean war – and the recluse of rutherford manor / Allnatt, Elizabeth – London: S W Partridge & Co, 1879 – (= ser 19th c women writers) – 2mf – 9 – mf#5.1.115 – uk Chadwyck [830]

An autumn in greece : comprising sketches of the character, customs, and scenery of the country, with a view of its present critical state / Bulwer, William H – London 1826 [mf ed Hildesheim 1995-98] – (= ser Fbc) – 3mf – 9 – €90.00 – 3-487-29054-5 – gw Olms [914]

An autumn in italy : being a personal narrative of a tour in the austrian, tuscan, roman and sardinian states, in 1827 / Sinclair, J D – Edinburgh 1829 [mf ed Hildesheim 1995-98] – (= ser Fbc) – 2mf – 9 – €60.00 – 3-487-29271-8 – gw Olms [914]

Autumn leaves : articles and news items on the pacific islands – v1-21 [1888-1908] – 1r – 1 – (autumn leaves is a publ of the reorganized church of jesus christ of latter day saints; articles mainly concern the work of missionaries of the reorganized church in hawaii, tahiti & the tuamotu archipelago) – mf#pmb94 – at Pacific Mss [243]

Autumn leaves : articles on the pacific islands – v22-44 [1909-1931] – 1r – 1 – (autumn leaves, renamed vision fr v42 is a publ of the reorganized church of jesus christ of latter day saints; articles mainly concern the work of missionaries of the reorganized church in hawaii, tahiti & the tuamotu archipelago) – mf#pmb109 – at Pacific Mss [243]

An autumn near the rhine : or, sketches of courts, society, scenery etc in some of the german states bordering on the rhine / Dodd, Charles E – London 1818 [mf ed Hildesheim 1995-98] – 4mf – 9 – €120.00 – 3-487-29556-3 – gw Olms [914]

Auvergne et provence : album pittoresque – Paris [1833] [mf ed Hildesheim 1995-98] – 2mf – 9 – €60.00 – 3-487-29711-6 – gw Olms [914]

Aux antilles / Meignan, Victor – Paris, France. 1878 – 1r – us UF Libraries [972]

Aux bambins canadiens / Marjolaine – Montreal: editions Albert Levesque, 1934 [mf ed 1991] – 2mf – 9 – (ill by J McIsaac) – mf#SEM105P1332 – cn Bibl Nat [971]

Aux directeurs de la compagnie du chemin de fer de phillipsburg, farnham et yamaska / Foster, John – [S.l: s.n, 1872?] [mf ed 1980] – 1mf – 9 – 0-665-04334-1 – mf#04334 – cn Canadiana [380]

Aux ecoutes – Paris. 1925-aout 1940 – 1 – fr ACRPP [073]

Aux electeurs du comte de laprairie et napierville : documents et faits – [S.l: s.n, 1896?] [mf ed 1980] – 1mf – 9 – 0-665-00843-0 – mf#00843 – cn Canadiana [370]

Aux electeurs du comte de laprairie et napierville : documents et faits – [S.l: s.n, 1896?] [mf ed 1985] – 1mf – 9 – 0-665-52164-2 – mf#52164 – cn Canadiana [370]

Aux electeurs du comte de l'assomption : une autre voix episcopale: categoriques declarations de mgr cameron: lettre circulaire au clerg d'antigonish – [S.l: s.n, 1896?] [mf ed 1980] – 1mf – 9 – 0-665-04087-3 – mf#04087 – cn Canadiana [320]

Aux electeurs du comte de l'assomption : documents et faits – [S.l.]: [s.n.], [1896?] [mf ed 1979] – 9 – 0-665-00816-3 – mf#00816 – cn Canadiana [325]

Aux electeurs du comte de l'assomption : protection, faits et questions d'ecoles – [S.l.]: [s.n.], [1896?] [mf ed 1981] – 1mf – 9 – 0-665-11647-0 – mf#11647 – cn Canadiana [325]

Aux electeurs du comte de montmorency / Casgrain, Thomas Chase – Quebec: s.n, 1896? – 1mf – 9 – mf#04026 – cn Canadiana [377]

Aux electeurs du comte de quebec / Caron, Adolphe – Quebec?: s.n, 1874 – 1mf – 9 – mf#23964 – cn Canadiana [325]

Aux electeurs du comte de quebec : programme de m g e amyot, candidat liberal ministeriel, 1er octobre 1906 – [Quebec?: s.n, 1906?] – 1mf – 9 – 0-665-76000-0 – mf#76000 – cn Canadiana [323]

Aux etats-unis et dans ontario – Montreal?: A-T Lepine, 1892 – 1mf – 9 – mf#02449 – cn Canadiana [305]

Aux femmes / Jacob, mlle [Victoire, Jeanne] – n.p. 1832 – (= ser Les femmes [coll]) – 1mf – 9 – mf#6910 – fr Bibl Nationale [305]

Aux femmes priviligiees. jeanne-desiree, proletaire *see* Lettre au roi

Aux fillettes canadiennes / Marjolaine – Montreal: editions Albert Levesque, 1933 [mf ed 1993] – 1mf – 9 – (ill by james mcisaac) – mf#SEM105P1820 – cn Bibl Nat [971]

Aux fillettes canadiennes / Marjolaine – Montreal: editions Albert Levesque, 1936 [mf ed 1993] – 2mf – 9 – (ill by james mcisaac) – mf#SEM105P1819 – cn Bibl Nat [971]

Aux honorables chevaliers, citoyens et bourgeois de la province du bas-Canada, assembles en parlement provincial : la petition des soussignes, electeurs dument qualifiee a choisir des membres... – [S.l: s.n, 1818?] [mf ed 1986] – 1mf – 9 – 0-665-58511-X – mf#58511 – cn Canadiana [323]

Aux honorables chevaliers, citoyens et bourgeois, les communes du royaume-uni de la grande bretagne et d'irlande, assemblees en parlement / Papineau, Louis Joseph – [Quebec]: [Chambre d'Assemblee], [1834] – 1mf – 9 – mf#SEM105P1129 – cn Bibl Nat [323]

Aux honorables membres du comite catholique du conseil de l'instruction publique / Arnold de Jesus, frere – Montreal: [s.n.], 1884 [mf ed 1980] – 1mf – 9 – 0-665-02200-X – (incl bibl ref) – mf#02200 – cn Canadiana [377]

Aux honorables membres du comite catholique du conseil de l'instruction publique / Reticius, frere – [Montreal?: s.n, 1884?] [mf ed 1981] – 1mf – 9 – mf#12338 – cn Canadiana [377]

[Aux honorables] membres du conseil executif, du conseil legislatif [et de l'as]semblee legislative de la province de Quebec / Bureau des commissaires d'ecoles catholiques romains de la cite de Montreal – [S.l: s.n, 1879?] [mf ed 1986] – 1mf – 9 – 0-665-60830-6 – mf#60830 – cn Canadiana [350]

Aux iles caraibes / Jacquemin, Charles – Havre, France. 1936 – 1r – us UF Libraries [972]

Aux jeunes gens qui veulent reussir / Clement, Alex – [Montreal?: s.n.], 1898 [mf ed 1980] – 1mf – 9 – mf#00682 – cn Canadiana [650]

Aux libres et intelligents electeurs de la province de quebec / Joly de Lotbiniere, Henri Gustave – S.l: s.n, 1878 – 1mf – 9 – mf#04505 – cn Canadiana [320]

Aux meres canadiennes : le lait pour l'alimentation dans les villes / Barre, Stanislas Morrier – [Montreal: s.n, 1905?] – 1mf – 9 – 0-665-97002-1 – mf#97002 – cn Canadiana [630]

Aux mines d'or du klondike : du lac bennett a dawson city / Boillet, Leon – Paris: Hachette, 1899 [mf ed 1979] – 9 – 0-665-00163-0 – mf#00163 – cn Canadiana [622]

Aux origines de la sorbonne : 1: robert de sorbon / Glorieux, P – Paris, 1966 – €15.00 – (2: le cartulaire, Paris 1965 €25) – ne Slangenburg [378]

Aux ruines des grandes cites soudanaises / Peyrissac, Leon – Paris: A Challamel, 1910 – 1 – us CRL [930]

Aux urnes, citoyennes! / Dunord, Charles – Paris, France. 1924 – 1r – us UF Libraries [440]

Aux zouaves : dernier adieu / Verreau, Hospice Anthelme – Montreal: s.n, 1868 (Montreal: Typ de la Minerve) – 1mf – 9 – mf#07067 – cn Canadiana [320]

O auxiliador – Rio de Janeiro, RJ: Typ Franceza, 24 ago-03 set 1841 – (= ser Ps 19) – mf#P17,01,66 – bl Biblioteca [321]

O auxiliador da administracao do correio da corte – Rio de Janeiro, RJ: Typ de N. Lobo Vianna & Filhos, 1856-1857 – (= ser Ps 19) – mf#P12,05,29-30 – bl Biblioteca [350]

O auxiliador da industria nacional – Rio de Janeiro, RJ: Typ de I F Torres, 15 jan 1833-31 dez 1892 – (= ser Ps 19) – mf#P19A,01,1-30P19A,02,01-29 – bl Biblioteca [079]

Auxiliaries newsletter – 1960 oct-1962 apr – 1 – mf#3579688 – us WHS [071]

Auxiliary bulletin. solar / carothers observatory (private astronomical) / Carothers Observatory; ed by Carothers, Warren Fay – Houston TX: Carothers Observatory, n1 [mar 1916]- (mthly) [mf ed 2007] – 1r – 1 – (companion to: auxiliary bulletin. weather) – mf#film mas 37496 – us Harvard [520]

Auxiliary bulletin. weather / carothers observatory (private astronomical) / Carothers Observatory; ed by Carothers, Warren Fay – Houston TX: Carothers Observatory, n1 [mar 1916]- (mthly) [mf ed 2007] – 1r – 1 – (companion to: auxiliary bulletin. solar) – mf#film mas 37496 – us Harvard [520]

El auxilio de america para la reconstruccion de espana : texto taquigrafico de la conferencia pronunciada en la sala studium de barcelona, el dia 9 de octubre de 1938 / Prieto, Indalecio – Barcelona, 1938 – (= ser Blodgett coll) – 9 – mf#fiche w1117 – us Harvard [946]

Auxilio Social *see* Normas de funcionamiento en los centros de alimentacion infantil

Auxilio social – [Valladolid?: s.n. 1937?] – 9 – mf#w738 – us Harvard [946]

Auzary, Bernadette *see* Fluctuat nec mergitur. la prevote des marchands et l'urbanisme parisien au 15 siecle d'apres la jurisprudence du parlement (1380-1500)

Auzias-Turenne, Raymond *see* Voyage au pays des mines d'or

Auziere, Louis *see* Essai historique sur les facultes de theologie de saumur et de sedan

Av communication review – Bloomington IN 1953-77 – 1,5,9 – (cont by: ectj: educational communication and technology) – ISSN: 0001-2890 – mf#1466.01 – us UMI ProQuest [370]

Av guide – Des Plaines IL 1922+ – 1,5,9 – ISSN: 0091-360X – mf#384 – us UMI ProQuest [370]

Av video [1984] – White Plains NY 1984-96 – 1,5,9 – (cont: audio visual directions) – ISSN: 0747-1335 – mf#12771.04 – us UMI ProQuest [380]

Av video and multimedia producer – Rockville MD 1996+ – 1,5,9 – ISSN: 1090-7459 – mf#26293.01 – us UMI ProQuest [380]

Ava *see* Die dichtungen der frau ava

Avadana kalpalata : a collection of legendary stories about the bodhisattvas; with its tibetan version / Ksmendra – Calcutta: printed by W Carey 1888-1913 [mf ed 1986] – 2v on 1r – 1 – mf#1729 – us Wisconsin U Libr [280]

Avadana-cataka, cent legendes bouddhiques – Paris: E Leroux 1891 [mf ed 1993] – (= ser Annales du musee guimet 18) – 5mf – 9 – 0-524-08029-1 – (trans fr sanskrit into french by leon feer) – mf#1991-0251 – us ATLA [280]

Availability of the phosphorus of various types of phosphates added to everglades peat land / Neller, J R – Gainesville, FL. 1945 – 1r – us UF Libraries [630]

Avalanche in central africa / Gibbs, Peter – London, England. 1961 – 1r – us UF Libraries [630]

Avaliani, S *see* Zemelnyi vopos v rossii i kooperatsiia

Avalon, Arthur *see* Principles of tantra

[Avalon-] catalina islander – CA. 1934-56 – 9r – 1 – $540.00 – mf#C02035 – us Library Micro [071]

Avalon news – Avalon, dec 1950-aug 1967 – 2r – A$121.75 vesicular A$132.75 silver – at Pascoe [079]

Avance – Plasencia 1899-1935 – 5 – sp Bibl Santa Ana [074]

Avance para la bibliografia de obras impresas del d. benito arias montano / Morales Oliver, Luis – Badajoz: Imprenta del Hospicio Princial, 1928. Separata R.C.Est.Ex. – 1 – sp Bibl Santa Ana [010]

Avances arqueologicos en santa amalia / Roso de Luna, Mario – Madrid: Fortanet, 1912. B.R.A.H. 60, p. 260 – sp Bibl Santa Ana [930]

Avangard – Avgustov, Poland 1939-41 [mf ed Norman Ross] – 1 – mf#nrp-220 – us UMI ProQuest [934]

L'avant garde : organe de l'union marocaine du travail – Casablanca: L'Union, 1963-feb 19 1971 – 4r – 1 – us CRL [079]

l'avant-coureur – Paris. 1760-1773 – 1 – (Suite de: la feuille) – fr ACRPP [073]

Avante : orgam noticioso e independente – Ouro Verde, SC. 21 jan 1930; 21 jan 1931 – (= ser Ps 19) – bl Biblioteca [079]

L'avant-garde – Paris: Imp A Vallee, mar 26-31, apr 1-3,5-7,9,11-12,14-15,17-23,25-30, may 1-11,13-19,21-22 – 2r – 1 – (Filmed as pt of: Commune de Paris newspapers. Newspapers on these reels are filmed chronologically, not alphabetically) – us CRL [074]

L'avant-garde – Paris. avr 1905-mars 1906 – 1 – fr ACRPP [073]

L'avant-garde – Port-au-Prince: Impr de l'Oeil, 1ere annee n1-2e annee n77. 12 janv 1882-28 juin 1883 – 6 sheets – us CRL [079]

Avant-garde – New York NY 1968-71 – 1 – ISSN: 0005-1918 – mf#9673 – us UMI ProQuest [073]

L'avant-garde de normandie – Rouen. 1913-14 – 1 – fr ACRPP [073]

L'avant-garde ouvriere et communiste : organe de defense des jeunes travailleurs – Paris. n1-687. sep 1920-36 – 1 – (puis ouvriere et paysanne) – fr ACRPP [335]

L'avant-garde republicaine et socialiste – Toulouse. sept 1891-juin 1892 – 1 – fr ACRPP [320]

L'avant-garde voltaique : organe de la jeunesse voltaique, rda – [Ouagadougou: s.n. v1 n1 (undated). v1 n2 jul 22 1958 – us CRL [079]

L'avanti! – Chicago, [oct 1918-oct 1 1921] – 1 – us CRL [071]

Avanti! : bulletin psi / Partito Socialisto Italiano – Paris, Milan, 1956- – 1 – us CRL [074]

Avanti! : organ of the italian socialist party – Clearwater Publ Co, 1896-1955 – 58r – 1 – $6125.00 – us UPA [335]

Avanti! – Roma: Typ de l'Avanti [1896-[dec25,1896-1925;1956-] (daily) – 160r – 1 – us CRL [074]

Avanti! – Roma: Typ de l'Avanti, [1896-dec 25 1896-1991] – 1 – us CRL [074]

Avanti! [milan] : giornale socialista. quotidiano del partito socialista – Milan, Italy 23 dec 1896-29 oct 1926, 1 may 1935-1 may 1940, 1 aug 1943-31 dec 1955, 29 sep 1959-22 apr 1962 – 1 – (fr 23 dec 1896-8 oct 1911, 15 jun-31 dec 1944, 1 aug 1950-30 jun 1951 publ in rome; may 1935-may 1940 publ in paris) – uk British Libr Newspaper [074]

Avanti [zuerich] : l'avvenire de lavoratore – Zuerich, Switzerland 29 mar 1930-12 may 1935 – 1 – (cont: l'avvenire de lavoratore: settimanale del partito socialista italiano, nella suizzera [4 jan-15 mar 1930]; cont by: nuovo avanti [19 may 1935-8 jun 1940]) – uk British Libr Newspaper [335]

L'avant-poste : revue de litterature et de critique – Paris. n1-3. juin-oct nov 1933 – 1 – fr ACRPP [410]

L'avant-scene : theatre – Paris. n1-327. mars 1949-fevr 1965 – 1 – (suite de: opera: hebdomadaire du theatre, du cinema des lettres et des arts) – fr ACRPP [790]

Avant-scene – Theatre, litterature, beaux-arts, modes. Red. en chef Charles de Sarlat. 43 no. Paris. 25 janv 1867-28 mai 1872 – 1 – fr ACRPP [800]

Les avantures de monsieur robert chevalier, dit de beauchene : capitaine de flibustiers dans la nouvelle-france / Sage, Alain Rene le – Paris: Chez Etienne Ganeau...1733 [mf ed 1984] – 2v on 1mf – 9 – 0-665-12620-4 – mf#12620 – cn Canadiana [830]

Avantures du sr c lebeau, avocat en parlement : ou voyage curieux et nouveau, parmi les sauvages de l'amerique septentrionale dans le quel on trouvera une description du canada... / Beau, Claude le – Amsterdam 1738 [mf ed Hildesheim 1995-98] – 2v on 10mf – 9 – €100.00 – 3-487-27107-9 – gw Olms [971]

El avaro / Moreno Torrado, Luis – Badajoz: tipografia y encuadernacion la minerva extremena, 1907 – 1 – sp Bibl Santa Ana [810]

Avatar – 1967 jun 23/jul 7-1968 aug 15 – 1 – mf#1052705 – us WHS [071]

Avataras / Besant, Annie Wood – London: Theosophical Pub Society, 1900 – (= ser Samp: indian books) – us CRL [180]

Avaux, Jean A d' *see* Negociations de monsieur le comte d'avaux en hollande

Avc bulletin – 1946 feb 1-1965; 1966 jan-1969 dec; 1970 apr-1988 fall – 1 – mf#668869 – us WHS [071]

Ave eva : erzaehlung / Johst, Hanns – Muenchen: A Langen 1932 [mf ed 1990] – 1r – 1 – (filmed with: eddystone / wilhelm jensen) – mf#2743p – us Wisconsin U Libr [830]

Ave maria – 1966 may 14-1968 jul 27; 1968 aug 3-1970 mar 21 – 1 – mf#585386 – us WHS [071]

Ave maria : cuaresma / Gil Becerra, Benito – 1733 – 9 – sp Bibl Santa Ana [240]

Ave maria : paraiso / Gil Becerra, Benito – 1739 – 9 – sp Bibl Santa Ana [240]

Ave maria : paraiso de oraciones sagradas en... / Gil Becerra, Benito – Madrid: Thomas Rodriguez Frias, s.a. – 1 – sp Bibl Santa Ana [240]

Ave maria ofrecimiento del santisimo rosario de la virgen nuestra senora – Manila: Fajardo 1907 [mf ed Bloomington IN: Indiana Uni Lib, Preservation Dept 1984] – (= ser Coll of spanish devotional literature) – 1r [ill] – 1 – us Indiana Preservation [241]

Ave maris stella *see* Missarum josquin liber secundus

Avebury, John Lubbock, 1st Baron *see*
- On the senses, instincts, and intelligence of animals
- Pre-historic times

Avedisian, Lori-Ann *see* The effect of selected buffering agents on performance in the competitive 1600 meter run

Aveling, Edward Bibbins *see* The religious views of charles darwin

Aveling, Francis *see*
- On the consciousness of the universal and the individual
- The philosophers of the smoking-room

Avellaneda y sus obras / Cotarelo Y Mori, Emilio – Madrid, Spain. 1930 – 1r – us UF Libraries [972]

[Avenal-] avenal progress – CA. mar 5 1986- – 3r – 1 – $180.00 (subs $50/y) – mf#B03582 – us Library Micro [071]

[Avenal-] avenal times – CA. 1980-jun 30 1982 – 3r – 1 – $180.00 – mf#B02036 – us Library Micro [071]

Avenarius and the standpoint of pure experience / Bush, Wendell T – New York: Science Press 1905 [mf ed 1990] – (= ser Columbia university contributions to philosophy and psychology 10/4; Archives of philosophy, psychology and scientific methods 2) – 1mf – 9 – 0-7905-7325-3 – (incl bibl ref) – mf#1989-0550 – us ATLA [120]

Avenarius, Ferdinand *see*
- Baal
- Balladenbuch
- Faust
- Jesus

Avenarius, Ferdinand [comp] *see* Das froehliche buch

Avenarius, Richard *see* Ueber die beiden ersten phasen des spinozischen pantheismus

Avencebrolis (ibn gebirol fons vitae / Baeumker, C – Muenster, 1892/95 – (= ser Bgphma 1/2-4) – 10mf – 9 – €19.00 – ne Slangenburg [100]

Avendano, Fernando de *see* Sermones de los misterios de nuestra santa fe catolica, en lengua castellana, y la general del inca

Avendano Suarez de Soussa, Pedro de *see*
- Sermon de la esclarecida virgen, y inclita martyr de christo sta barbara
- Sermon del glorioso abbad s bernardo
- Sermon del primer dia de pasqua del espiritu santo
- Sermon que en la fiesta titular que celebra la compania de bethlem en su hospital de convalecientes de aquesta ciudad de mexico

Avenement du general fabre nicolas geffrard / Michel, Antoine – Port-Au-Prince, Haiti. 1932 – 1r – us UF Libraries [972]

L'avenement du peuple – Paris. No.1-75.19 sept-1 dec 1851 – 1 – (Suite de: L'evenement) – fr ACRPP [073]

L'avenement du peuple voir a ce titre *see* L'evenment

L'avenir – London. n1-32. 5 oct-16 nov 1872 – 1 – (journal politique, litteraire) – fr ACRPP [073]

L'avenir – Leopoldville: [s.n.], jun 20, jul 7 1960 – us CRL [079]

L'avenir – Montreal, QC. 1900-01 – 1r – 1 – (Les vrai debats) – cn Library Assoc [071]

L'avenir – Montreal, QC. 1847-57 – 3r – 1 – cn Library Assoc [071]

L'avenir – Point-a-Pitre, Guadeloupe. 1907-1912 (1) – mf#67937 – us UMI ProQuest [079]

L'avenir – Port-au-Prince, Haiti: [s.n.] 1 annee n1-27. 30 dec 1899-7 juil 1900 – us CRL [079]

L'avenir – Paris. dec 1841-mars 1842 – 1 – (revue politique, litteraire et des modes) – fr ACRPP [073]

Avenir caledonien – dec 1954-23 dec 1987 – 213mf – 9 – mf#pmb doc393 – at Pacific Mss [980]

Avenir dans le passe : ou, les succes au paradis / Clairville, M – Paris, France. 1848? – 1r – us UF Libraries [440]

L'avenir de la bretagne : journal national breton et federaliste europeen – Brest. 1958-juin 1972 – 1 – fr ACRPP [073]

L'avenir de la france – [Paris]: E Briere, aug 1 1848 – us CRL [074]

Avenir de la guyane francaise / Chaton, Prosper – Cayenne, French Guiana. 1865 – 1r – us UF Libraries [972]

L'avenir de la vallee de l'orne : journal republicain – Joeuf. n412-488.1932-juin 1933 – 1 – fr ACRPP [073]

L'avenir de paris : toutes les informations en toute independance. – Paris. janv-juin 1919, janv-juin 1920, 1923, janv-juin 1926, juil 1927-juin 1928 – 1 – fr ACRPP [073]

L'avenir des minorites francaises au canada : discours prononce au 9eme congres general de l'association canadienne-francaise d'education d'ontario, a ottawa, le 12 octobre 1938 / Roy, Camille – Quebec: l'Action catholique, 1938 [mf ed 1992] – 1mf – 9 – mf#SEM105P1744 – cn Bibl Nat [305]

L'avenir du canada : discours prononce au parc sohmer a montreal, le 4 avril 1893 / Mercier, Honore – Montreal: Cie d'impr et de lithographie Gebhardt-Berthiaume, 1893 [mf ed 1974] – 1r – 5 – mf#SEM16P184 – cn Bibl Nat [971]

L'avenir du katanga – Elisabethville: V Nawezi – (Issues for jun 19-30 1961 filmed as pt of: Herbert C Weiss collection on the Belgian Congo jun 19-30 1960) – us CRL [074]

Avenir du pays el l'action nefaste de m foisset / Denis, Lorimer – Port-Au-Prince, Haiti. 1949 – 1r – us UF Libraries [972]
L'avenir du peuple canadien-francais / Nevers, Edmond de – Paris: H Jouve, 1896 [mf ed 1974] – 1r – 5 – mf#SEM16P189 – cn Bibl Nat [971]
L'avenir du tonkin – Hanoi, oct 1886-1907, 1910-15, 1922-juin 1941 [biwkly] – 1 – fr ACRPP [073]
L'avenir du tournais *see* Oui
L'avenir du travailleur : organe du parti ouvrier de la region nord – Tourcoing-Roubaix. fevr-juil 1887 – 1 – fr ACRPP [320]
L'avenir liberal – Ed. su Soir. Paris: Impr de l'Avenir liberal, mar 24 1871 – (Filmed as pt of: Commune de Paris newspapers. Newspapers on these reels are filmed chronologically, not alphabetically) – us CRL [074]
L'avenir national – Paris. 10 janv 1865-26 oct 1873 – 1 – fr ACRPP [073]
Avenir national – Paris. 4 juil-4 sept 1848 – 1 – fr ACRPP [073]
L'avenir politique du canada et des canadiens francais : questions de la federation imperiale, de l'independance, de l'annexion aux etats-unis / Gailly de Taurines, Charles – [Paris: s.n, 1891] [mf ed 1982] – 1mf – 9 – 0-665-17798-4 – mf#17798 – cn Canadiana [327]
L'avenir republicain *see* Memorial judiciaire de la loire
L'avenir socialiste – Paris. 1933-avr 1939 – 1 – fr ACRPP [335]
L'avenir syndical / Organe de la Confederation Nationale du Travail et des Unions et Federations des Syndicats Nationaux de France – Paris. n70-225. 29 sept 1917-1er dec 1924 – 1 – fr ACRPP [331]
L'avenire – New York NY, feb 16 1917 – 1r – 1 – (italian newspaper) – us IHRC [071]
L'avenire – Sleubenville OH, 1910* – 1r – 1 – (italian newspaper) – us IHRC [071]
L'avenire – Utica NJ, oct 5 1900-dec 9 1905 – 2r – 1 – (italian newspaper) – us IHRC [071]
Avenirs – Paris. 1947-72 – 5 – fr ACRPP [073]
Aventura 77 / Delegacion de la Juventud – Caceres: Imp. Sergio Dorado, 1977 – 1 – sp Bibl Santa Ana [946]
Aventura de outubro e a invasao de s paulo / Jardim, Renato – Rio de Janeiro, Brazil. 1932? – 1r – us UF Libraries [972]
Aventuras de diofanes / Orta, Teresa Margarida Da Silva E – Rio de Janeiro, Brazil. 1945 – 1r – us UF Libraries [972]
Aventuras del soldado desconocido cubano / Torriente Brau, Pablo De La – Havana, Cuba. 1962 – 1r – us UF Libraries [972]
Aventuras e aventureiros no brasil / Carvalho, Alfredo De – Rio de Janeiro, Brazil. 1929 – 1r – us UF Libraries [972]
Aventuras y desventuras del tercer diego garcia de paredes / Munoz de San Pedro, Miguel – Badajoz: Imp. de la Diputacion Prov., 1957. Rev. Est. Ex – sp Bibl Santa Ana [830]
L'aventure – Paris.n 5-9, fragm.juil 1927-mars 1929 – 1 – fr ACRPP [073]
Aventure de saint-foix : ou, le coup d'epee / Duval, Alexandre – Paris, France. 1802 – 1r – us UF Libraries [440]
Aventures de cow-boys / Vercheres, Paul [pseud] – Montreal: ed Police journal. n1 30 avril 1948-n726 24 janv 1962; ns: n1 21 fev 1962- [mf ed 1981] – 8r – 5 – (with ind; ceased with ns n13 23 janv 1963?) – mf#SEM16P319 – cn Bibl Nat [800]
Aventures de donald campbell : dans un voyage aux indes, par terre, et anecdotes piquantes sur l'originalite de son guide hassan artas / Campbell, Donald – Paris an 7 [1799] [mf ed Hildesheim 1995-98] – 2v on 4mf – 9 – €120.00 – 3-487-27434-5 – gw Olms [920]
Les aventures de pierre / Fortin, Charles-Henri – Quebec: le Centre pedagogique, [1960?] (mf ed 1991) – (= ser Coll petit jaseur) – 2mf – 9 – mf#SEM105P1421 – cn Bibl Nat [830]
Les aventures du prince romanic / Claudette – Montreal: les editions Varietes, 1943 [mf ed 1993] – (= ser Coll recits et legendes) – 1mf – 9 – (ill by francine) – mf#SEM105P1813 – cn Bibl Nat [830]
Aventures en guyane / Maufrais, Raymond – Paris, France. 1952 – 1r – us UF Libraries [972]
Les aventures etranges de l'agent 9e-13 : l'as des espions canadiens / Saurel, Pierre – Montreal: ed Police journal. n1 28 nov 1947-n970 28 sep 1966 (wkly) [mf ed 1976] – 8r – 5 – mf#SEM16P249 – cn Bibl Nat [360]
Les aventures extraordinaires de guy vercheres : l'arsene lupin canadien-francais – Montreal: ed Police journal. n1 8 oct 1948-n883 27 janv 1965 (wkly) [mf ed 1982] – 9r – 5 – mf#SEM16P326 – cn Bibl Nat [800]
Aventures lointaines : voyages aux iles sitka (ancienne amerique russe) / Frede, Pierre – Paris?: Libr de Firmin-Didot, 1890 – 2mf – 9 – mf#14995 – cn Canadiana [917]
Les aventures policieres d'albert brien : detective national des canadiens-francais / Valjean, Hercule – Montreal: ed Police journal. n1 9 avril 1948-n970 28 sep 1966 [mf ed 1981] – 11r – 5 – (with ind) – mf#SEM16P321 – cn Bibl Nat [800]
Avergal, N Vythinatha Aiyar et al *see* Aspects of the vedanta
Averill, A M *see* Baptist landmarkism tested by logic, history and scripture
Averkiev, D V *see* Ezhemesiachnoe izdanie
Averley, G *see* Bibliography of eighteenth-century legal literature
Averroes *see*
- Des averroes abhandlung
- Die hauptlehren des averroes nach seiner schrift, die widerlegung des gazali
- Philosophie und theologie
Averrois Cordubensis *see* De substantia orbis tractatus
Aversa, R *see*
- Logica institutionibus praeviis quaestionibus contexta
- Philosophia metaphysicam physicamque complectens questionibus contexta
Avertimenti et essamini intor a quelle cose che richiede a un bombardiero / Cataneo, G – Vinegia, 1582 – 1mf – 9 – mf#OA-256 – ne IDC [720]
Avertissement sur les disputes et le procede des missionnaires / [Drelincourt, C] – Charenton, 1654 – 3mf – 9 – (missing: title pg) – mf#CA-110 – ne IDC [240]
Avery, Asahel George *see* Our disrespect for law
Avery, David *see* The papers of david avery, 1746-1818
Avery, Diana Margaret *see* Upper pleistocene and holocene palaeoenvironments in the southern cape
Avery, E *see* Laws applicable to immigration and naturalization, 1953
Avery, Elroy Mckendree *see* Genesis of new port richey
Avery Institute of Afro-American History and Culture *see* Bulletin of the avery institute...
Avery, Ralph Emmett *see*
- America's triumph at panama
- Picturesque panama and the great canal
Aves de chile : en su clasificacion moderna / Housse, Emile – Santiago, Chile. 1945 – 1r – us UF Libraries [500]
Avesta – Falun, 1911 – 1r – 1 – sw Kungliga [078]
Avesta : die heiligen schriften der parsen...sammt der huzvaresch-uebersetzung / Spiegel, F – Wien. 2v. 1853-1858 – 13mf – 9 – mf#NE-20153 – ne IDC [956]
Avesta allehanda – Enkoeping, Sweden. 1877-81 – 2r – 1 – sw Kungliga [078]
Avesta eschatology compared with the books of daniel and revelations : being supplementary to zarathushtra, philo, the achaemenids and israel / Mills, Lawrence Heyworth – Chicago: Open Court 1908 [mf ed 1991] – 1mf [ill] – 9 – 0-524-00938-4 – mf#1990-2161 – us ATLA [230]
Avesta tidning – Sala, Sweden. 1882- – 1 – sw Kungliga [078]
Avesta tidning – Sala, Sweden. 1979- – 1 – sw Kungliga [078]
L'avesta, zoroastre et le mazdeisme / Hovelacque, Abel – Paris: Maisonneuve, 1878 [mf ed 1991] – 1mf – 9 – 0-524-01607-0 – (in french) – mf#1990-2546 – us ATLA [290]
Avestaposten – Hedemora, Sweden. 1897-1951 – 109r – 1 – sw Kungliga [078]
Avestaposten – Vasteras, Hedemora, Sweden. 1897-1900 – 2r – 1 – sw Kungliga [078]
L'aveugle – [Montreal]: impr Arbour & Dupont, 1912 [mf ed 1992] – 1mf – 9 – mf#SEM105P1575 – cn Bibl Nat [301]
L'aveugle de saint-eustache : grand roman canadien historique inedit / Feron, Jean – 2e ed. Montreal: Editions Edouard Garand, cop 1924 [mf ed 1987] – (= ser Le roman canadien) – 1mf – 9 – (ill by Albert Fournier) – mf#SEM105P853 – cn Bibl Nat [830]
Avezac-Macaya, Armand d' *see* Notice sur le pays et le peuple des yebous en afrique
Aviacion civil en el salvador / Gilbert, Glen Alexander – Montreal, Quebec. 1952 – 1r – us UF Libraries [380]
Avian communities in florida habitats / Engstrom,R Todd – Tallahassee, FL. 1993 – 1r – us UF Libraries [590]
Avian diseases – Kennett Square PA 1957+ – 1,5,9 – ISSN: 0005-2086 – mf#6684 – us UMI ProQuest [590]
Aviano Air Base [Italy] *see* Vigileer
Aviation / Harold, William G – s.l, s.l? . 1936 – 1r – us UF Libraries [629]
Aviation bases, flying fields / Braman, Sidney T – s.l, s.l? . 1936 – 1r – us UF Libraries [380]
Aviation law digest quarterly supplement – v1-7. jun 1941-sep 1947 (all publ) – 7mf – 9 – $10.50 – mf#LLMC 84-411 – us LLMC [340]
Aviation maintenance – Dec 1943-Nov 1948 – 1 – us L of C Photodup [629]
Aviation mechanics bulletin – Alexandria VA 1973 – 1,5,9 – ISSN: 0005-2140 – mf#8355 – us UMI ProQuest [629]
Aviation space and environmental medicine – Alexandria VA 1975+ – 1,5,9 – ISSN: 0095-6562 – mf#11403,01 – us UMI ProQuest [614]
Aviation week and space technology – New York NY 1916+ – 1,5,9 – ISSN: 0005-2175 – mf#364 – us UMI ProQuest [629]
Aviazione e marina – Genoa, Italy 1961-72 – 1 – mf#8337 – us UMI ProQuest [629]
Avicenna *see*
- Metahypsices compendium
- Metaphysica sive prima philosophia
Avicenna on theology / Alberry, A J – London, 1952 – 1mf – 8 – €3.00 – ne Slangenburg [180]
Avicenne / Carra de Vaux, Bernard, Baron – Paris: Felix Alcan 1900 [mf ed 1993] – (= ser Les grands philosophes) – 1mf – 9 – 0-524-07545-X – mf#1991-0127 – us ATLA [180]
Avicultural magazine – Bristol UK 1949+ – 1,5,9 – ISSN: 0005-2256 – mf#463 – us UMI ProQuest [590]
Die avignonesischen paepste, ihre machtfuelle und ihr untergang : vortrag / Hoefler, Karl Adolf Constantin, Ritter von – Wien: ...in Commission bei Karl Gerold, 1871 [mf ed 1990] – 1mf – 9 – 0-7905-5841-6 – (incl bibl ref) – mf#1988-1841 – us ATLA [241]
[Avila beach-] 5 cities times press recorder – CA. 1972-79 – 32r – 1 – $1920.00 – mf#B02037 – us Library Micro [071]
Avila Camacho, Manuel *see* Grandeza y afirmacion de mexico
Avila contra burgos y sus pueblos (anno 1670) – Avila – 1r – 5,6 – sp Cultura [946]
Avila, Francisco de *see* Arte de la lengua mexicana
Avila, G G d' *see* Teatro de las grandezas de la villa de madrid...
Avila, Jose de *see* Coleccion de noticias de muchas de las indulgencias plenarias y perpetuas que pueden ganar todos los fieles de christo
Avila, Julio Enrique *see* Vigia sin luz
Aviles Blonda, Maximo *see* Manos vacias
L'avion de france – Saigon: Avion de France [1937- [sep 1936-oct 1937; nov 24, 1937-feb 15, 1939; mar 1-29 1939] – 1r – 1 – mf#mf-11810 seam – us CRL [079]
Aviones sobre el pueblo / Montenegro, Carlos – La Habana: [s.n.] 1937 [mf ed 1977] – (= ser Blodgett coll) – 1mf – 9 – mf#w1055 – us Harvard [946]
Avis de m honey de la publication de son calendrier judiciaire et du tableau des honoraires ainsi que son prospectus de la publication d'un almanach des adresses, professionnelles, commerciales et litteraires du Canada pour l'annee 1857 – [Montreal?: s.n, 1857?] [mf ed 1986] – 1mf – 9 – 0-665-58847-X – mf#58847 – cn Canadiana [070]
Avis et rapports / France. Conseil Economique – juin 1947-86 – 1 – fr ACRPP [330]
Les avis et rapports du conseil economique et social – 1981- – €13.72y – (backfile: 1947-1980 €381.12) – fr Journal Officiel [350]
El Avisador *see* Por que no vas a la conferencia?
El avisador – Badajoz, 1888-91 – 5 – (numeros sueltos) – sp Bibl Santa Ana [073]
Avisador de badajoz – Badajoz.1862-66 – 9 – sp Bibl Santa Ana [074]
El avisador de badajoz – Badajoz, 1882 y 1885-1887. Numeros sueltos – 5 – sp Bibl Santa Ana [073]
Avi-Sha'ul, Mordekhai *see* Maharozet
Avisi particulari : ultimamente mandati dal magnifico m antonio egiptio maggior domo dell illustrissimo and eccellentissimo signor paulo giordano / Egiptio, A – n.p, [1572] – 1mf – 9 – mf#H-8337 – ne IDC [956]
Aviso – Wolfenbuettel DE, 1618-19 [single iss], 1620, 1621-22 [single iss], 1623 [many gaps], 1624 [single iss], 1625 [gaps] – 1r – 1 – gw Misc Inst [074]
Aviso de sanidad que trata de todos los generos de alimentos y del regimiento... / Nunez de Coria, F – Madrid, 1572 – 13mf – 9 – sp Cultura [610]
Avisos y documentos para la preservacin y cura de la peste / Diez Daza, A – Sevilla, 1599 – 1mf – 9 – sp Cultura [616]
Avity, Pierre d' *see* Description generale de l'afrique...
Avksentev et al *see* Ezhednevnaia bespartiinaia gazeta
Avksentev, N et al *see* Izdanie gruppy sotsialistov-revoliutsionerov
Avne bet ha-yotser / Weisz, Isaac – Paks, Hungary. 1900 – 1r – us UF Libraries [939]
Avner, Yehoshu'a Ze'ev *see* Tsir ne'eman
'Avni *see* The divan project
'Avnuellah al-Kazimi, 'Oerfi ve *see* The divan project
Avoca head-light – 1885 nov – 1 – mf#958913 – us WHS [071]
Avocado diseases / Stevens, H E – Gainesville, FL. 1922 – 1r – us UF Libraries [634]
Avocado production in florida / Wolfe, Herbert S – Gainesville, FL. 1934 – 1r – us UF Libraries [634]
Avocat / Roger, Francois – Paris, France. 1806 – 1r – us UF Libraries [440]
Avocat du beau sexe / Siraudin, Paul – Paris, France. 1862 – 1r – us UF Libraries [440]
Avocat patelin / Brueys – Paris, France. 1797 or 1798 – 1r – us UF Libraries [440]
Avocat patelin / Brueys – Paris, France. 1801 – 1r – us UF Libraries [440]
Avodat ha-kodesh / Gabbai, Meir Ben Ezekiel Ibn – Jerusalem, Israel. 1953/54 – 1r – us UF Libraries [939]
'Avodat Yisra'el – Philadelphia, PA. v1-2. 1934 – 1r – 1 – us UF Libraries [939]
Avolio, J *see* Six air columns
Avon advertiser – Andover, England 31 jan 2001-17 jul 2002 – 1 – (cont by: midweek andover advertiser [24 jul 2002-]) – uk British Libr Newspaper [072]
Avon baptist church – MA. 274p. 1780-1827 – 1 – $12.33 – (society book 1785-1844 (randolph, braintree, stoughton)) – us Southern Baptist [242]
Avon deanery magazine – [Windsor, NS?: s.n, 1893-189- or 19–] [mf ed jan 1893] – 9 – mf#P04783 – cn Canadiana [242]
Avon park / Darsey, Barbara Berry – s.l, s.l? . 1936 – 1r – us UF Libraries [978]
Avon park : points of interest / Darsey, Barbara Berry – s.l, s.l? . 1936 – 1r – us UF Libraries [978]
Avon Park, Florida / Darsey, Barbara Berry – s.l, s.l? . 1936-1938 – 1r – us UF Libraries [978]
L'avouerie de l'abbaye de saint-armand en pevele : memoires et travaux fac catholique de lille / Naz, R – fasc 32. Lille, 1927 – €3.00 – ne Slangenburg [241]
Avraham even-'ezra / Ben-Menachem, Naphtali – Jerusalem, Israel. 1943 – 1r – us UF Libraries [939]
Avraham garbov – Ramat Ha-Kovesh, Israel. 1940 – 1r – us UF Libraries [939]
Avramov, Vasil *see* Voinata mezhdu vizantiia i bulgariia v 986 godina i obsadata na sofiia ot imperatora vasilii 2 bulgaroubiets
Avrashov, G *see* Davosskii vestnik
Avrea et vere theologica commentatio d danielis hofmanni / Hoffmann, D – Magdebvrgi, 1600 – 1mf – 9 – mf#TH-1 mf 698 – ne IDC [242]
Avrekh, A I *see* Stolypin i tretia duma
Av-report – Berlin DE, 1978 9 jun-30 dec – 1r – 1 – gw Mikrofilm [380]
Avrigni, C J L d' *see* Jeanne d'arc a rouen
Avril, Adolphe d' *see*
- La chalde chretienne
- Documents relatifs aux eglises de l'orient et a leurs rapports avec rome
- St cyrille et st methode
Avril, Chantal *see* La cour d'appel de dijon – an huit – mille huit cent cinquante deux
Avril de Saint-Croix, Mme. Ghenia *see* Le feminisme
Avril, P *see* Travels into divers parts of europe and asia
Avrohom goldfaden un zigmunt mogulesko / Zylbercweig, Zalmen – Buenos Aires, Argentina. 1936 – 1r – us UF Libraries [939]
Avrom goldfaden / Mayzel, Nachman – Varshe, Poland. 1935 – 1r – us UF Libraries [939]
Avrora – St Petersburg, 1875-78 [bimthly] [1878 mthly] [mf ed Norman Ross Publ] – 34mf – 9 – us UMI ProQuest [640]
Avrupa bizi nasil taniyor / Hakki, Ismail [Alisan] – Dersaadet [Istanbul]: Kadir Matbaasi, 1329 [1910] – (= ser Ottoman literature, writers and the arts) – 1mf – 9 – $25.00 – us MEDOC [470]
Avtnomnaia sibir' : gaz oblastn, sotsialist i lit / ed by Pliiasunov, N et al – Tiumen' [Tobol gub]: Tobol otd Soiuza sibiriakov-avtonomistov 1918 [1918 27 marta-] – (= ser Asn 1-3) – n1 [1918] item 1, on reel n1 – 1 – mf#asn-1 001 – ne IDC [077]
Avto-emansipatsiia / Pinsker, Leon – St Petersburg, Russia. 1906 – 1r – us UF Libraries [939]
Avtomobil i vozdukhoplavanie – M., 1911-1912 – 44mf – 9 – (missing: 1912(5-24)) – mf#R-2325 – ne IDC [077]
Avtre veritable discovrs de la victoire des chrestiens contre les turcs : en la bataille nauale pres lepantho...1571 – Paris, 1571 – 1mf – 9 – mf#H-8173 – ne IDC [956]
Avtsinsky, Levi *see* Toldot yeshivat ha-yehudim be-kurland mi-shenat 321/1561 'ad shenat
Avulso – Rio Grande do Norte: Typ do Correio do Natal, 27 fev 1880 – (= ser Ps 19) – mf#P22B,04,189 – bl Biblioteca [079]
Avvakum Petrovich, Protopope *see* Zhitie protopopa avvakuma
L'avvenire de lavoratore : settimanale del partito socialista italiano, nella suizzera – Zuerich, Switzerland 4 jan-15 mar 1930 – 1 – (cont by: avanti: (l'avvenire de lavoratore) 29 mar 1930-12 may 1935) – uk British Libr Newspaper [335]

L'avvenire d'italia – Bologna, Italy 30 jan 1941-10 mar 1942, 26 may 1942-8 sep 1943 (imperfect) – 1 – uk British Libr Newspaper [074]

Awake thou that sleepest : and arise from the dead / Wesley, Charles – London, England. 1808 – 1r – us UF Libraries [240]

Awakened in a tavern – London, England. 18– – 1r – us UF Libraries [240]

Awakening in ireland – Edinburgh, Scotland. 1859 – 1r – us UF Libraries [240]

Awakening of afrikaner nationalism 1868-1881 / Van Jaarsveld, Floris Albertus – Cape Town, South Africa. 1961 – 1r – us UF Libraries [960]

The awakening of asian womanhood / Cousins, Margaret – Madras: Ganesh & Co, 1922 – (= ser Samp: indian books) – us CRL [305]

The awakening of scotland : a history from 1747 to 1797 / Mathieson, William Law – Glasgow: J Maclehose, 1910 – 1mf – 9 – 0-7905-6761-X – (incl bibl ref) – mf#1988-2761 – us ATLA [941]

The awakening of scotland, a history from 1747 to 1797 / Mathieson, William Law – Glasgow: J. Maclehose, 1910. xiv,303p – 1 – us Wisconsin U Libr [941]

The awakening of spring : a tragedy of childhood / Wedekind, Frank – 2nd ed. Philadelphia: Brown Bros 1910 [mf ed 1985] – 1r – 1 – (trans fr german by francis joseph ziegler. with: obras/lopez de ayala, adelardo) – mf#1248 – us Wisconsin U Libr [820]

Awarding attorneys' fees and managing fee litigation / Hirsch, Alan et al – 1994 – 2mf – 9 – $3.00 – mf#llmc99-037 – us LLMC [346]

Aware magazine – Carlisle UK 1991-95 – 1 – (cont: aware/harvester) – ISSN: 0017-8217 – mf#16627,03 – us UMI ProQuest [240]

Awareness – Toledo OH 1971-72 – 1 – ISSN: 0045-1231 – mf#7599 – us UMI ProQuest [639]

Awdah, Samih Ahmad *see* Jiyumurfulujiyat al-huwwat fi al-jabal al-akhdar

Awde, James et al *see* The minister at work

Awde, Robert *see* Canada

Awdry, Frances *see* In the isles of the sea

Awetik'ean, Gabriel *see* Kherakanouthiun hajkakan

Awful death of a drunkard – Runcorn, England. 18– – 1r – us UF Libraries [240]

Awgerian, Y [Aucher, P] *see* Dictionnaire francais-armenien-turc

Awin newsletter – 1968 nov-1971 oct – 1 – mf#1051492 – us WHS [071]

Awol press – v1 n9-1910 [1970 apr 1/15]; v1 n9,10 – 1 – mf#1051493 – us WHS [071]

Aworan ogun – London – 1r – 1 – (lacks: apr 1942 n22) – us CRL [079]

Awots = The spring – Riga. 144p. 1905-July 21, 1915. 6 – 1 – $245.76 – us Southern Baptist [242]

Awraq muassasat al-dirasat al-filastiniyah – Bayrut: Muassasat al-dirasat al-filastiniyah. n2,5-6,11-13,15-21. 1979-1983 – us CRL [079]

Awsa bulletin – 1979 oct-1982 may – 1 – mf#825190 – us WHS [071]

Awsby, Edith *see*
- Ruth seyton
- Three school friends

Awse reporter – 1945 mar 28-1975 sep – 1 – mf#1051494 – us WHS [071]

Axbridge and cheddar gazette – Axbridge, England 12 jan 1866-29 nov 1867 – 1r – 1 – uk British Libr Newspaper [072]

The axe : a journal of action against reaction / ed by Roberts, John H – Montreal. n1 jan 13th 1922- (wkly) [mf ed 1973] – 1r – 1 – (ceased v3 n36 sep 9th 1924?) – mf#SEM35P13 – cn Bibl Nat [073]

Axelson, Eric *see* Portugal and the scramble for africa, 1875-1891

Axenfeld, Karl et al *see* Missionswissenschaftliche studien

Axicon und ringpupille als bildformende elemente / Bickel, Gerhard – (mf ed 1995) – 2mf – 9 – €40.00 – 3-8267-2086-5 – mf#DHS 2086 – gw Frankfurter [530]

Axiomata ex commentariis ejus / Coke, Edward – 17th c – (= ser Holkham library family and political papers 253) – 1r – 1 – (comp by john brisco of lincoln's inn) – mf#96848 – uk Microform Academic [340]

Axiomatic basis and computational methods for optimal... / Padron, Mario – Gainesville, FL. 1969 – 1r – us UF Libraries [500]

The axioms of religion : a new interpretation of the baptist faith / Mullins, Edgar Young – Philadelphia: American Baptist Publ Society, 1908 – 1mf – 9 – 0-8370-8926-3 – (incl ind) – mf#1986-2926 – us ATLA [240]

Axon, William Edward Armytage *see* An architectural and general description of the town hall, manchester

Axt, Christian *see* Die natuerliche gestaltung der mesiodistalen frontzahnbreiten bei der totalprothetischen rekonstruktion

Axtell advertiser – Axtell, NE: John A Enochs, 1896 (wkly) – 2r – 1 – (cont by: axtell times) – us Bell [071]

Axtell, Daniel *see* Daniel axtell account book and letter entries, 1700-1711

The axtell republican – Axtell, NE: Jayne & Wilson. v1 n1. oct 10 1889- (wkly) – 2r – 1 – us Bell [071]

Axtell times – Axtell, Kearney Co, NE: Florence E Reynolds. v9 n42. feb 3 1905-v61 n24. sep 10 1953 (wkly) [mf ed feb 3 1905-oct 28 1937 (gaps) filmed 1985] – 2r – 1 – (cont: axtell advertiser. absorbed by: minden courier) – us NE Hist [071]

Axtell times – Axtell, NE: Florence E Reynolds. v61 n24. sep 10 1953 (wkly) – 5r – 1 – (cont: axtell advertiser; absorbed by: minden courier) – us Bell [071]

Axters, St *see* Geschiedenis van de vroomheid in de nederlanden

Ay qap – Troitsk, Russia 1911-15 [mf ed Norman Ross] – 3r – 1 – mf#nrp-1832 – us UMI ProQuest [077]

Ayala / Solsona y Baselga, Conrado – 1891 – 9 – sp Bibl Santa Ana [830]

Ayala, Balthazar *see* De jure et officiis bellicis et disciplina militari, libri 3

Ayala Duarte, Crispin *see* Ensayo critico y antologico acerca

Ayala, Juan Antonio *see* Lydia nogales

Ayala, Manuel Jose de *see* Diccionario de gobierno y legisalcion de indias...tomo 4, vol 1

Ayala Munoz, Ruben *see* Guia del inversionista

Ayalah sheluhah / Klatzkin, Naphtali Hirz – Warsaw, Poland. 1896 – 1r – us UF Libraries [939]

Ayalti, Hanan J *see* Tate un zun

Ayandigan – Tehran. shumarah-'i 3251-3419. 6 aban 1357-16 murdad 1358 [28 oct 1978-7 aug 1979] – 2r – 1 – $106.00 – (missing: n3256, 3279, 3287-3288, 3330, 3354-3355, 3397-3399, 3403) – us MEDOC [079]

Ayape, Eugenio *see* La calzada de oropesa, su santo cristo y sus monjas

Ayara / Moreno, Miguel – Panama, 1962 – 1r – us UF Libraries [972]

Ayarati motele / Chemerinsky, Hayim – Tel-Aviv, Israel. 1951 – 1r – us UF Libraries [939]

Aycinena Salazar, Luis *see* Procedimiento ex aequo et bono

Aycrigg, Benjamin *see* Memoirs of the reformed episcopal church and of the protestant episcopal church

Aydede – n1-90. 1338 [1922] – (= ser O & t journals) – 7mf – 9 – $110.00 – us MEDOC [956]

Aydin – (= ser Vilayet salnames) – 9 – (1300 [1883] 4mf $60; 1308 [1891] 18mf $290; 1312 [1894] 11mf $180; 1313 [1895] def'a 16 7mf $110; 1316 [1898], 1317 [1899] 8mf/yr $130 per 8mf; 1326 [1908] 12mf $195) – us MEDOC [956]

Aydin – Izmir: Vilayet Matbaasi, 1897-? Yayimliyan: Aydin Valiligi. n2111. 22 aug 1910 – (= ser O & t journals) – 1mf – 9 – $25.00 – us MEDOC [956]

Aydinlik : ictimai, terbiyevi, edebi aylik mecmuadir – Istanbul: Islamiye Mataasi, Amedi Matbaasi, Cihan Biraderler Matbaasi, 1921-25. n1-31. 1 haziran 1921-18 subat 1925 (special iss n1,2,5,8 1924-25) – (= ser O & t journals) – 15mf – 9 – $250.00 – us MEDOC [956]

Aye, Agnes *see* Chants et chansons en pays akye

Ayer 1871-1900 – Oxford, MA (mf ed 1995) – (= ser Massachusetts vital records) – 26mf – 9 – 0-87623-376-0 – (mf 1-15: town records 1871-1907. mf 16-18: births 1871-92. mf 19-20: marriages 1871-92. mf 21-22: deaths 1871-92. mf 23: deaths 1892-1901. mf 24: marriages 1892-1900. mf 25-26: births 1892-1900) – us Archive [978]

Ayer, Albert Azro *see* Historical sketch of the grande ligne mission

Ayer, Jacqueline *see* Voc emission reduction study at the hill air force base building 151 painting facility

Ayer, Joseph Cullen *see*
- The rise and development of christian architecture
- A source book for ancient church history
- Versuch einer darstellung der ethik joseph butlers

Ayer o el santo domingo de hace 50 anos / Gomez Alfau, Luis Emilio – Ciudad Trujillo, Dominican Republic. 1944 – 1r – us UF Libraries [972]

Ayine – sene 1-2 n1-72. 1337-39 [1921-23] [all publ] – (= ser O & t journals) – 5mf – 9 – $90.00 – us MEDOC [956]

A-ying *see*
- Hai kuo ying hsiung, i ming, cheng ch'eng-kung
- Hsiao shuo hsien t'an
- Hung hsuan-chia
- K'ang chan ch'i chien ti wen hsueh
- Pi hsueh hua
- Pu yeh ch'eng

Ayiti ap lite : se jounal dizyem depatman-manhattan – v1 n1-2 [1994 nov 2-1995 feb 1] – 1 – mf#4695434 – us WHS [071]

The aylesford union – Aylesford, NS: BYPU of the Upper Aylesford Baptist Church, [1897-1899?] – 9 – mf#P05055 – cn Canadiana [242]

Ayliff, John *see* History of the abambo

Aylik mecmua – Istanbul: Vatan Matbaasi, Cumhuriyet Matbaasi. Mueduerue: Kemal Salih. n1-12. nisan 1926-mart 1927 – (= ser O & t journals) – 7mf – 9 – $110.00 – us MEDOC [956]

Ayllon Laynez, Juan *see* Additiones ad varias resolutiones.

Aylmer, C *see* Nineteenth century books on china collection

Aylsworth, M B [comp] *see* Alumni souvenir

Aylsworth, Nicholas John *see* Moral and spiritual aspects of baptism

Aymon, J. *see* Tous les synodes nationaux des eglises reformees de france

Aymonier, Etienne *see*
- Le cambodge
- Geographie du cambodge

Aymonier, Etienne Francois *see* Histoire de l'ancien cambodge

Aymonier, M E *see* Les tchames et leurs religions

Ayn rand letter – New York NY 1973-76 – 1,5,9 – ISSN: 0045-124X – mf#10123 – us UMI ProQuest [100]

'Ayn shams – Cairo: Claudius Labib, 1900-04. yr 1 n1-yr 4 n12. tut 1617-nisi 1620 [coptic era][sep/oct 1900-sep 1904] – (= ser Arabic journals and popular press) – 1r – 1 – $425.00 – (in arabic & coptic) – us MEDOC [079]

Ayna – Samarkand, Uzbekistan 1913-15 [mf ed Norman Ross] – 2r – 1 – mf#nrp-1518 – us UMI ProQuest [077]

'Ayni *see* The divan project

Ayo, C *see* Las excelencias...y...propiedades del tabaco...

Ayr advertiser : or, west country journal – [Scotland] South Ayrshire, Air [i.e. Ayr]: Wilson & Paul jan 1844-dec 1878, jan 1893-dec 1950 (wkly) [mf ed 2004] – 98r – 1 – (cont by: ayr advertiser, or west country and galloway journal [feb 1853-nov 1968]; title varies slightly; imprint varies) – uk Newsplan [072]

Ayr advertiser (1995) : scotland's oldest weekly newspaper – Ayr: Clyde Weekly Press 1995-96 (wkly) – 1 – (cont: ayr advertiser & prestwick times; cont by: ayr advertiser and prestwick times (1996)) – ISSN: 1358-4286 – uk Scotland NatLib [072]

Ayr advertiser and prestwick times : scotland's oldest weekly newspaper – Ayr: Ayrshire Weekly Press 1996- (wkly) – 1 – (cont: ayr advertiser (1995)) – ISSN: 1358-4286 – uk Scotland NatLib [072]

Ayr and wigtownshire courier – [Scotland] South Ayrshire, Ayr: David Macarter 30 mar, 8 jun, 28 aug 1820 (wkly) [mf ed 2004] – 1r – 1 – uk Newsplan [072]

Ayr & district leader – Ayr: Ayrshire Leader Ltd 1991-2000 (wkly) [mf ed 6 jan 1995-] – 1 – (cont: ayrshire leader (kyle and carrick ed); cont by: ayr & south ayrshire leader) – ISSN: 1358-4332 – uk Scotland NatLib [072]

Ayr & south ayrshire leader – Ayr: Community Media Ltd 2000-03 (wkly) [mf ed 2001-] – 1 – (cont: ayr & district leader; cont by: leader. ayr & south ayrshire) – uk Scotland NatLib [072]

Ayr town crier – [Scotland] South Ayrshire, Kilmarnock: printed by Kilmarnock Herald Publ Co sep-dec 1938 & nov 1946 (mthly) [mf ed 2004] – 1r – 1 – (iss by ayr trades and labour council) – uk Newsplan [331]

Ayrault, Roger *see*
- Heinrich von kleist
- La legende de heinrich von kleist

Ayre, John *see*
- A compendious introduction to the study of the bible
- Grace of god that bringeth salvation hath appeared to all men

Ayre, John et al *see* An introduction to the critical study and knowledge of the holy scriptures

Ayrer, Jakob *see*
- Ayrers dramen

Ayrers dramen / ed by Keller, Adelbert von – Stuttgart: Litterarischer Verein, 1865 [mf ed 1993] – (= ser Blvs 76-80) – 5v – 1 – mf#8470 reels 15-16 – us Wisconsin U Libr [820]

Ayrers dramen / ed by Keller, Adelbert von – Stuttgart: Litterarischer Verein, 1865 [mf ed 1995] – (= ser Blvs 76-80) – 5v – 1 – mf#8470 reels 15-16 – us Wisconsin U Libr [820]

Ayres, Anne *see*
- Evangelical sisterhoods
- The life and work of william augustus muhlenberg

Ayres, Ebenezer *see* Cleveland, ohio, taxes, ms v.f. o

Ayres, Leonard Porter *see* Laggards in our schools

Ayres, Ph. *see* Emblemata amatoria

Ayres, Samuel Gardiner *see* Jesus christ our lord

Ayrshire agriculturist – [Scotland] South Ayrshire, Ayr: D Guthrie 29 sep 1843-dec 1848 (wkly) [mf ed 2004] – 4r – 1 – (cont: ayrshire and renfrewshire agriculturist and western counties' advertiser [3 oct 1846-25 sep 1847]; cont by: north british agriculturist and journal of horticulture) – uk Newsplan [630]

Ayrshire county news – [Scotland] South Ayrshire, Prestwick: Ayrshire County News 9 nov 1928-1 feb 1929 (wkly) [mf ed 2004] – 1r – 1 – uk Newsplan [072]

Ayrshire evening express and south-western counties observer – [Scotland] South Ayrshire, Ayr: J M Ferguson 12,14,16 jan 1878 (daily) [mf ed 2004] – 1r – 1 – uk Newsplan [072]

Ayrshire examiner – [Scotland] Kilmarnock: J Quigley 26 oct 1838 (wkly) [mf ed 2004] – 1r – 1 – (began in jul 1838; ceased in nov 1839) – uk Newsplan [072]

Ayrshire express – [Scotland] South Ayrshire, Ayr: Alex Grant 7 mar 1857-6 apr 1889 (wkly) [mf ed 2004] – 741v on 29r – 1 – (missing: 1863, 1877; 1860-61 at mf; absorbed: ayrshire courier; cont by: argus & express [10 jun 1871-28 jun 1873]; ayrshire argus & express & galloway review [5 jul 1873-11 mar 1876]; ayrshire argus & express [18 mar 1876-7 jul 1877]; argus & express [14 jul 1877-20 sep 1884]; ayrshire argus & express [26 sep 1884-6 apr 1889]) – uk Newsplan [072]

Ayrshire express and prestwick and troon chronicle – [Scotland] South Ayrshire, Ayr: Ferguson & Co 24 apr-11 sep 1908 (wkly) [mf ed 2004] – 1r – 1 – uk Newsplan [072]

Ayrshire monthly news letter and agricultural reporter – [Scotland] South Ayrshire, Irvine: M Dick 1 mar 1844-28 dec 1846 (mthly) [mf ed 2004] – 1r – 1 – uk Newsplan [630]

Ayrshire post (1987) – Ayr: Scottish & Universal Newspapers 1987- (wkly) [mf ed 1995-] – 1 – (cont: ayrshire post, and troon herald) – ISSN: 0963-8830 – uk Scotland NatLib [072]

Ayrshire times – [Scotland] South Ayrshire, Kilmarnock: publ...by J Brown 11 jul 1860-4 may 1861 (wkly) [mf ed 2004] – 1r – 1 – uk Newsplan [072]

Ayrshire weekly news and county advertiser – [Scotland] South Ayrshire, Saltcoats: James Hill Mearns 2 mar 1861-dec 1878, jan 1893-9 nov 1900 (wkly) [mf ed 2004] – 22r – 1 – (cont by: ayrshire weekly news and galloway press [apr 1866-nov 1900]; absorbed by: ayr observer and galloway chronicle [1853-1900]) – uk Newsplan [072]

Ayrshire world (central ed) – Irvine: Scottish & Universal Newspapers 1992-2003 (wkly) [mf ed 1999-] – 2r – 1 – (cont: ayrshire world (kilmarnock & irvine ed); merged with: ayrshire world (south ed) to form: ayrshire world (irvine, scotland)) – uk Scotland NatLib [072]

Ayrshire world (irvine, scotland) – 2003- (wkly) – 1 – (formed by union of: ayrshire world (central edition) and: ayrshire world (south edition)) – uk Scotland NatLib [072]

Ayrshire world (south ed) – Irvine: Scottish & Universal Newspapers 1992-2003 (wkly) [mf ed 13 jan 1995-] – 1 – (cont: ayrshire world (ayr ed); merged with: ayrshire world (central ed) to form: ayrshire world (irvine, scotland); not publ: n590-591) – uk Scotland NatLib [072]

Ayrton, Edward R *see*
- Pre-dynastic cemetary at el-mahasna
- Pre-dynastic cemetery at al mahasna

Ayrton, Edward R et al *see* Abydos

Aytoun, Robert Alexander *see* City centres of early christianity

Ayuntamiento colonial de la ciudad de guatemala / Chinchilla Aguilar, Ernesto – Guatemala, 1961 – 1r – us UF Libraries [972]

Ayurdharma : law of life – Bombay: Rajinder Mohan Sharma, Honorary Secretary of the Indian Ayurvedic Aid Society. v1 n1 jan 1965 – us CRL [615]

Ayurvedic Research Evaluation Committee *see* [Report]

Ayyildiz – Iskenderun: Ayyildiz matbaasi, [mar 28 1951-aug 29 1953] – 1r – us CRL [079]

The 'ayyub and mamluk sultans / Sell, Edward – London: Church Missionary Society, 1929.84p. Bibliog. footnotes – 1 – us Wisconsin U Libr [956]

Ayzinena, Jose de *see* Assertationes sacrae theologiae juxta mentem seraphici, subtilisque doctorum propugnandae

AZ – Wien: Neue AZ zeitungsverlagsgesellschaft mbH & Co KG, [1989-dec 12 1989-aug 1991 – us CRL [074]

AZ 1948 : i.e. ezerkilencszaznegyvennyolcadik evi amnesztia / Hungary. Laws, Statutes, etc – Hirlap-, Szaklap- es Konyvkiado 194-. 124p. LL-4008 – 1 – us L of C Photodup [348]

Az allgemeine zeitung fuer nordbaden und die pfalz – Mannheim, Germany 3 apr-30 nov 1951 – 1 – (cont: abend-zeitung fuer nordbaden und die pfalz [5 sep 1949-1 apr 1951]) – uk British Libr Newspaper [074]

Az allgemeine zeitung fuer wuerttemberg – Stuttgart, Germany 1 aug 1951-30 nov 1952 (imperfect) – 1 – (cont: wuerttembergische allgemeine zeitung [2 apr-31 jul 1951]) – uk British Libr Newspaper [074]

AZ am abend [arbeiterzeitung] : arbeiterzeitung – Vienna, Austria sep 1914-apr 1922 [mf ed Norman Ross] – 8r – 1 – (social democrat) – mf#nrp-1933 – us UMI ProQuest [074]

Az egyetemes europai jogtortenetnek rovid vazlata / Wenzel, Gusztav – Budapest, Pfeifer Ferdinand, 1877. 219 p. LL-4058 – 1 – us L of C Photodup [341]

Az est – Budapest, Hungary. Dec 1914-Feb 1917; Mar 1920-Apr 1921 – 4r – 1 – us L of C Photodup [079]

Az igazolo eljarasok zsebkonyve / Hungary. Laws, Statutes, etc – Budapest Franklin-Tarsulat 1945?. 90p. LL-4052 – 1 – us L of C Photodup [348]

Az iras = Hungarian news – Chicago: Az Iras, sep 1938-dec 22 1944; 1946-may 6 1949 – us CRL [071]

Az o-gyallal astrophysikai es meteorologiai observatoriumon vegzet megfigyelesek = Beobachtungen, angestellt am astrophysikalischen und meteorologischen observatorium in o-gyalla / O-Gyallal astrophysikai es meteorologiai observatorium – Budapest: Heisler J nyomsa [1896?]- (irreg) [mf ed 2006] – 1 – (cont: beobachtungen, angestellt am astrophysikalischen observatorium in o gyalla in ungarn) – mf#film mas 37493 – us Harvard [520]

Az uj nepbirosagi torveny (1947: 34. t.-c.) egyseges szerkezetben a hatalyos nepbirosagi rendeletekkel / Hungary. Laws, Statutes, etc – Budapest: Gergely, 1948 – 1 – mf#LL-4053 – us L of C Photodup [348]

Az ujsag – Budapest, Hungary 28 jun 1914-29 jun 1924 – 42r – 1 – uk British Libr Newspaper [077]

Az unitarius vallas : david ferenc koraban es azutan : irjak toebben / Szentmartoni, Kalman et al – Koloszvart: Nyomatott az Ellenzek Koenyvnyomdaban 1910 [mf ed 1993] – 1mf – 9 – 0-524-08668-0 – (in hungarian) – mf#1993-3193 – us ATLA [243]

Az ut – Budapest, Hungary 27 jul/aug-12/18 oct 1952 3/9 jan-19/25 dec 1954, 10/16 jul 1955-21/27 oct 1956 – 1r – 1 – uk British Libr Newspaper [077]

Az wuerttembergische abend-zeitung – Stuttgart, Germany 12 aug 1949-31 mar 1951 (imperfect) – 1 – (cont by: wuerttembergische allgemeine zeitung [2 apr-31 jul 1951]) – uk British Libr Newspaper [074]

Aza iz undzer land / Braslavsky, Mosheh – Pariz, France. 1948 – 1r – us UF Libraries [939]

Aza Monero, Alberto *see* Ritmos en la noche

Azad – Banaras, India. Jun 1950-Jul 1953 – 1r – 1 – us L of C Photodup [079]

Azad – Kazan, Russia 1906 [mf ed Norman Ross] – 1r – 1 – mf#nrp-661 – us UMI ProQuest [077]

Azad, 'Abd al-Rahman Sayf *see* Iran-i bastan

Azad buxara – Bukhara, Uzbekistan 1924 [mf ed Norman Ross] – 1r – 1 – mf#nrp-388 – us UMI ProQuest [077]

Azad hindustan – Rangoon, Burma. 1943-44 – 1r – 1 – us L of C Photodup [079]

Azadi – [Tehran]: Jibhah-'i Dimukratik-i Milli-i Iran. shumarah-'i 1-36. 8 farvardin 1358-15 bahman 1358 [29 mar 1979-4 feb 1980] – 1r – 1 – $53.00 – (missing: n22-23) – us MEDOC [079]

Azadlig – Baku, USSR. Aug 25 1990-June 27 1992 – 1r – 1 – us L of C Photodup [077]

Azais / Berr, Georges – Paris, France. 1948 – 1r – us UF Libraries [440]

Azais, Hyacinthe *see* Jugement impartial sur napoleon

Azais, Pierre H *see* Un mois de sejour dans les pyrenees

Azais, Pierre Hyacinthe *see* 12 sonates pour le violoncelle [et basse continue]

Azana, Manuel *see*
- Extract from pres azana's speech at valencia university, july 18, 1937
- Extracts from a speech delivered by the pres of the spanish republic, january 21, 1937
- Important decree issued by the president of the cabinet
- Texto integro del discurso pronunciado por don manuel azana diaz el dia 21 de enero de 1937 en el salon de sesiones del ayuntamiento de valencia

Azania news – Dar-es-Salaam: Dept of Publicity & Info, Pan Africanist Congress. [v1-13. 1966-78] – 1 – us CRL [070]

Azar de lecturas : critica / Feijoo, Samuel – Santa Clara, Cuba. 1961 – 1r – us UF Libraries [972]

Azar del jubilo / Sardinas Lleonart, Jose – Habana, Cuba. 1965 – 1r – us UF Libraries [972]

Azara, F de *see* Voyages dans l'amerique meridionale,...depuis 1781 jusqu'en 1801

Azara, Felix *see*
- Apuntaciones para la historia natural de las aves...tomo 1-2, asuncion
- Viajes por la america meridional

Azarakhsh – London. shumarah-'i 1,3,5-6. tir 1355-aban 2536 (1356) [jun/jul 1976-oct 1977] – 1r – 1 – $53.00 – (r also incl: 19 bahman danishju i, bisu-yi azadi, and sitiz) – us MEDOC [079]

Azarakhsh *see* Sitiz

Azaraksh *see*
- 19 bahman danishju'i
- Bisu-yi azadi

Azarbayjan / ed by Mudir, Mirza 'Aliquli – Tabriz. n4-8,13-15. 1 safar-jumada 1 1325 [16 mar-26 jun 1907] – 2r – 1 – $106.00 – (in persian and azeri) – us MEDOC [079]

Azarenko, A *see* Molodye oprichniki buzhuazii

Azariia, Monakh *see* Afonskii paterik ili zhizneopisaniia sviatykh, na sviatoi afonskoi gore prosiiavshchikh

Azaryahu, Joseph *see* Shihure histaklut

Azatian, V A [comp] *see* Sotsialisticheskoe stroitel'stvo sssr

Azb [arizona business] – Tempe AZ 1995+ – 1,5,9 – (cont: arizona business) – ISSN: 1079-4255 – mf#6383,01 – us UMI ProQuest [338]

'Azbi *see* The divan project

Azbuchnyi ukazatel russkoi povremennoi slovesnosti s 1735 po 1857 god / Vsevolodov – Spb., 1857 – 3mf – 9 – mf#R-7019 – ne IDC [077]

Azcarate, Carlos *see* El adulterio

Azcarate y Florez, Pablo de *see*
- The communist plot in spain
- Spain, past and future: an address to the cosmos society, oxford, on 15 may 1945

Azcuy Alon, Fanny *see* Jose joaquin palma

Azed *see* Dictionnaire de la langue francaise

Azedo de la Berrueza, Gabriel *see* Amenidades... de la vera alta y baxa

Azerbaidzhan : Ezhedn obshchestv -lit i polit gaz – Baku: [s n] 1919-20 [1918 [?]-1920 [27 apr] – (= ser Asn 1-3) – n1-284 [1919] n1-83 [1920] [gaps] item 2, on reel n1,2 – 1 – mf#asn-1 002 – ne IDC [077]

Azerbaycan – Baku, 1918-20. 27 muharrem, 31 ramazan 1337, 4 safer 1338 [1919] – (= ser O & t journals) – 1r – 1 – $75.00 – (in azeri; r also incl: maktab and ziya-yi kafkasiyah) – us MEDOC [956]

Azerbaycan *see*
- Maktab
- Ziya-yi kafkasiyah

Azerbaycan cumhuriyet keyfiyeti tesekkuelue ve simdiki vaziyeti / Emin, Resulzade Mehmet – Istanbul: Evkaf-i Islamiye Matbaasi, 1339M, 1341H [1923] – (= ser Ottoman histories and historical sources) – 3mf – 9 – $55.00 – us MEDOC [956]

A-zet pondelnik – V Praze: Melantrich [jul 4 1938-nov 27 1939] (wkly) – 11r – 1 – us CRL [077]

Azevedi, Thales De *see* Povoamento da cidade do salvador

Azevedo, Aluisio *see* Casa de pensao

Azevedo, Aroldo De *see*
- Geografia humana do brasil para o terceiro ano
- Regioes brasileiras
- Regiones e paisagens do brasil

Azevedo, Fay De *see* Democracia e parlamentarismo

Azevedo, Fernando De *see*
- Brazilian culture
- Canaviais e engenhos na vida politica do brasil
- Cultura brasileira
- Educacao publica em s paulo
- Trem corre para o oeste
- Universidades no mundo de amanha

Azevedo Filho, Leodegario A De *see* Introducao ao estudo da nova critica no brasil

Azevedo, J Lucio De *see* Novas epanaforas

Azevedo, Joao Lucio D' *see*
- Historia de antonio vieira
- Jesuitas no grao-para

Azevedo, Manuel Antonio Alvares De *see* Obras completas

Azevedo, Thales De *see*
- As elites de cor
- Gauchos

Azevedo, Vitor De *see* Feijo

Azg – Boston, MA: Azk Publ Co, [nov 1917-oct 15 1921] – 7r – us CRL [071]

Azg-pahak = Azk-bahag – Boston, MA: Azk-Bahag Publ Co, oct 19 1921-1922 – 4r – us CRL [071]

Azi et al *see* Tong wen guang hui quan shu

Aziatskii vestnik – Moscow. 1956-1963 (1) – 30mf – 9 – mf#1683 – ne IDC [077]

Azikiwe, Nnamdi *see* Liberia in world politics

L'azione coloniale : settimanale dell'istituto fascista dell'africa italiana – Rome, Italy 10 jan 1936-6 feb 1939, 8 may 1941-26 aug 1943 – 1 – (1941-43 very imperfect) – uk British Libr Newspaper [074]

Aziz, Abdul *see* The imperial treasury of the indian mughuls

Azizah de niamkoko / Crouzat, Henri – Paris, Presses de la cite [1959] – us CRL [944]

Azofeifa, Isaac Felipe *see* Vigilia en pie de muerte

Azopardi, Francesco *see* Le musicien pratique

Azoro / Alvarez Bravo, Armando – Habana, Cuba. 1964 – 1r – us UF Libraries [972]

Azoy lakh ikh / Gutman, Khaim – NYU York, NY. 1918 – 1r – us UF Libraries [939]

Azpiazu, J *see* Marquez, gabino. las enciclicas "rerum novarum", "quadragesimo anno" y "divini redemptorio" contra el comunismo, al alcance de todos. toledo, 1938

Aztecs / Davies, Nigel – London, England. 1973 – 1r – us UF Libraries [930]

Aztecs / Duran, Diego – New York, NY. 1964 – 1r – us UF Libraries [930]

The aztecs : their history, manners, and customs = Azteques / Biart, Lucien – Chicago: AC McClurg, 1892 – 1mf – 9 – 0-524-00691-1 – (in english) – mf#1990-2019 – us ATLA [930]

Aztlan – Los Angeles CA 1970+ – 1,5,9 – ISSN: 0005-2604 – mf#10593 – us UMI ProQuest [300]

Azuaga. Spain *see* Ordenanzas municipales

Azuar, Antonio *see* Medalla batida por la villa de alcantara, en honor del coronel maine

Azucar y poblacion en las antillas / Guerra, Ramiro – Habana, Cuba. 1935 – 1r – us UF Libraries [972]

Azucarero anuario de cuba / Ministerio Del Comercio Exterior – Vedado, Cuba. 1960 – 1r – us UF Libraries [025]

La azucena de quito : que broto en el florido campo de la iglesia en las indias occidentales... / Moran de Butron, Jacinto – en Lima: Por Joseph de Contreras...ano de 1702 – (= ser Books on religion...1543/44-c1800: biografias de religiosos) – 2mf – 9 – mf#crl-136 – ne IDC [241]

Azul / Dario, Ruben – Buenos Aires, Argentina. 1946 – 1r – us UF Libraries [972]

Azul / Dario, Ruben – Buenos Aires, Argentina. 1952 – 1r – us UF Libraries [972]

Azul / Dario, Ruben – San Salvador, El Salvador. 1961 – 1r – us UF Libraries [972]

Azul : pela arte – Curitiba, PR: Typ Beabacter, 04 mar-20 out 1900 – (= ser Ps 19) – mf#P17,02,128 – bl Biblioteca [440]

Azul cuarenta (cuentos del morenito damian) cuento / Rodriguez, Blanca Luz De – San Salvador, El Salvador. 1963 – 1r – us UF Libraries [972]

Azula Barrera, Rafael *see*
- De la revolucion al orden nuevo
- Poesia de la accion

Azulai, H J D *see* Ma'agal tov ha-shalem

Azulai, Hayyim Joseph David *see*
- Lev david
- Shem ha-gedolim

Azuni, Domenico A *see* Gemaelde von sardinien

Azurara, G E de *see* The chronicle of the discovery and conquest of guinea

[Azusa-] pomotropics – CA. jun 1900-1914 (incomplete) – 4r – 1 – $200.00 – mf#C02038 – us Library Micro [071]

Azyr, Felix Vicq d' et al *see* Encyclopedie methodique (ael3/12)

Azzo *see* Summa super codicem

Azzone Dei Porci *see* Summa super codicem

Die o b – Kaapstad: J A Smith [nov 12 1941-dec 17 1952] (wkly) – 4r – 1 – us CRL [079]

B a i c smoke signals – 1981 mar/apr – 1 – mf#634575 – us WHS [071]

B a n – 1983 sep/nov-1987 spring – 1 – mf#4868168 – us WHS [071]

B and c news – 1969 dec-1972; 1973 jan-1978 jun – 1 – mf#386271 – us WHS [071]

B b warfield's view of the authority of scripture / Trites, Allison Albert – 1962 – 1r – 1 – 0-8370-0720-8 – mf#1984-6103 – us ATLA [220]

B c and t news – 1978 sep-1985 sep; 1985 oct-1994 dec – 1 – mf#804194 – us WHS [071]

B c good templar – New Westminster, [BC: Grand Lodge of British Columbia, Independent Order of Good Templars, 1893-1894] [mf ed v1 n1 oct 16 1893-v1 n12 sep 15 1894] – 9 – ISSN: 1190-6707 – mf#P04510 – cn Canadiana [360]

B c guide – Vancouver: [s.n, 1899?-19–] [mf ed n15 jun 1900] – 9 – ISSN: 1190-6723 – mf#P04499 – cn Canadiana [917]

B c historical news – v12 n1-v13 n3 [1978 nov-1980 spring] – 1 – mf#641191 – us WHS [071]

B c lumber worker / International Woodworkers of America – 1940 aug 7-1949, 1950-51, v10 n2-v19 n24 [1941 jan 22-1949 sep 22] – 3r – 1 – (cont: b c lumber worker iwa bulletin; cont by: western canadian lumber worker) – mf#618436 – us WHS [634]

B c lumber worker iwa bulletin / International Woodworkers of America – v9 n7-1912 [1940 may 1-jul 24] – 1r – 1 – (cont: b c lumber worker union bulletin; cont by: b c lumber worker) – mf#1289902 – us WHS [634]

B c lumber worker union bulletin / International Woodworkers of America – 1939 jul 11-1940 apr 17 – 1r – 1 – (cont by: b c lumber worker iwa bulletin) – mf#1289903 – us WHS [634]

The b c mining exchange and investors' guide – Vancouver: [s.n, 1899] [mf ed v1 n5 may 1899-v1 n7 jul 1899] – 9 – mf#P04026 – cn Canadiana [622]

The b c mining exchange and investor's guide and mining tit-bits – Vancouver: [s.n, 1899-1906?] [mf ed v1 n10 oct 1889-v3 n12/1 dec/jan 1900/01] – 9 – mf#P04028 – cn Canadiana [622]

The a b c of taxation : with boston object lessons, private property in land, and other essays and addresses / Fillebrown, C B – Garden City, NY: Doubleday, Page & Co, 1916 – 3mf – 9 – $4.50 – (concerned with taxation on real property) – mf#LLMC 92-157 – us LLMC [336]

B c t f newsletter – v23 n8-v27 n7 [1984 feb 3-1988 mar 7] – 1 – mf#1494727 – us WHS [071]

B c teacher – v61 n3-v67 n1 [1982 jan/feb-1987 oct/nov] – 1 – mf#1494472 – us WHS [370]

B e f news – 1932 jun 25-1932 oct 1/nov – 1 – mf#1052748 – us WHS [071]

B e m news notes – Washington DC 1975-77 – 1,5,9 – mf#2975 – us UMI ProQuest [320]

B g beynon journal, 1813-1814 / Benyon, B G – [mf ed 1981] – 1r – 1 – (detailed daybook kept by lieutenant b g beynon of the british royal marines, who served on board hms menelaus, during the war of 1812 in the vicinity of baltimore, maryland, where his command fought american forces) – mf#ms1236 – us Western Res [355]

O b g yearbook [obstetrics and gynecology] – Nokomis FL 1960-72 – 1 – ISSN: 0473-6729 – mf#8792 – us UMI ProQuest [618]

B Greening Wire Co *see* Price list of lathing and reinforcing

B kahan-virgili / Yivo Institute For Jewish Research Research Training Division – Wilno, Lithuania. 1938 – 1r – us UF Libraries [939]

B M And T *see* Ibbuku lya syaa-zibwene

B m malabari : rambles with the pilgrim reformer / Singh, Jogendra – London: G Bell and Sons, 1914 – (= ser Samp: indian books) – us CRL [920]

B to b – Detroit MI 2001+ – 1,5,9 – (cont: advertising age's business marketing) – ISSN: 1530-2369 – mf#348.03 – us UMI ProQuest [650]

B z – Berlin DE. 1964 13 jul-1966 13 oct, 1967 6 may-1969 22 may, 1969 11 oct-1971 30 mar, 1971 3 aug- – 1 – (filmed by misc inst: 1968- [13r/yr, later 17r/yr]; filmed by loc 1962 [4r]) – gw Mikrofilm; gw Misc Inst; us L of C Photodup [074]

B z am mittag [berliner zeitung am mittag] – Berlin DE, 1904 22 oct-1906 31 mar, 1909 1 apr-30 jun, 1914 1 apr-30 jun, 1916 1 jul-30 sep, 1922 1 apr-30 jun, 1926 1 nov-31 dec, 1930 1 sep-30 sep, 1934 2 may-30 jun, 1939 1 jul-31 jul – 10r – 1 – (filmed by bnl: 1906 4 apr-1919 30 jul (gaps) [6r]) – gw Mikrofilm; uk British Libr Newspaper [074]

Ba, Amadou Hampate *see* L'empire peul du macina

Les ba de la kamtsha / Mertens, Joseph – Brussels. 1935-39. 3v – 1 – us CRL [960]

Ba hashiyah va bi hashiyah – [Tehran]: Nahzat-i Azadi-i Iran. shumarah-'i 1-4. aban 1341-day 1341 [oct 1962-dec 1962] – 1r – 1 – $53.00 – (title added on title pg: asnad-i nahat-i azadi-i iran) – us MEDOC [079]

Ba Hong *see* Vong quanh sai-gon

Ba Htwei, Thippan *see* Mui bhrai

Ba Kruin, U *see* Sak se kham upade a nhac khyup

Ba ma a myui sa mi / Saw Mon Nyin – Ran Kun: Si ha pum nhip tuik 1976 [mf ed 1990] – 1r with other items – 1 – (in burmese) – mf#mf-10289 seam reel 134/4 [§] – us CRL [305]

Ba ma' a tvan sa bho = Into hidden burma / Collis, Maurice – Ran kun Mrui': u Kui Kui Kri 1975 [mf ed 1995] – on pt of 1r – 1 – mf#11052 r1973 n3 – us Cornell [915]

Ba ma ca pe ba lai, bay lai / Cvam Rann, Mon – Ran kun: Rhve Aui ca pe tuik 1976 [mf ed 1990] – 1r with other items – 1 – (in burmese) – mf#mf-10289 seam reel 149/5 [§] – us CRL [480]

Ba ma' to lhan re samuin / Suta, Mon – Ran kun Mui Ca pay Ca pe [1975?] [mf ed 1994] – on pt of 1r – 1 – mf#11052 r1712 n5 – us Cornell [959]

Lo ba mhon khvan a ron lan : [a play] / Cin Aon Man, Vanna Kyo Than – Ran Kun: Ta ma mruin ca can 1970 [mf ed 1990] – 1r with other items – 1 – (in burmese) – mf#mf-10289 seam reel 149/3 [§] – us CRL [820]

Ba Mui *see* Pra jat nhan jat sa bhan

Ba San et al *see* Ca nay jan samuin ca tan mya

Ba shi nian dai xianggang bao zhang jian bao mu lu *see* Xianggang bao zhang jian bao

Ba Than, U *see* Rhe khet mran ma prannsa tan ca mya

Ba Than Win *see* Mran ma nuin nam lam ma kri nahn tam ta kri mya
Baader, Clemens Alois *see* Lexikon verstorbener baierischer schriftsteller des 18. und 19. jahrhunderts
Baader, Franz von *see*
- Blitzstrahl wider rom
- Fermenta cognitionis
- Ueber das durch die franzoesische revolution herbeigefuehrte beduerfniss einer neuern und innigern verbindung der religion mit der politik

Baader, Joseph *see*
- Nuernberger polizeiordnungen aus dem 13. bis 15. jahrhundert
- Verhandlungen ueber thomas von absberg
- Verhandlungen ueber thomas von absberg und seine fehden gegen den schwaebischen bund 1519-1530

Baal : drei fassungen / Brecht, Bertolt; ed by Schmidt, Dieter – 6. aufl. Frankfurt/Main: Suhrkamp 1971 [mf ed 19–] – 3mf – 9 – (comm by ed) – us OmniSys [820]
Baal : ein spiel / Avenarius, Ferdinand – Muenchen: G D W Callwey, c1920 [mf ed 1988] – 61p – 1 – mf#6970 – us Wisconsin U Libr [820]
Baal in the ras shamra texts / Kapelrud, Arvid S – G.E.C. Gad, 1952 – 9 – $10.00 – us IRC [290]
Ba'al Shem Tov *see* Des rabbi israel ben elieser, genannt baal-shem-tow
Baalen, Jan Karel van *see* De loochening der gemeene gratie
Baar, Carl *see* Judgeship creation in the federal courts
Ba-arets / Frischmann, David – Warsaw, Poland. 1913 – 1r – us UF Libraries [939]
Baars, A *see* Het proces sneevliet
Baart, Peter A *see*
- Deugden-spoor
- The roman court

Baartzes, Wesley Barry *see* 'N kritiese studie ten opsigte van die status van die plattelandse onderwyser in die ceres-tulbaghgebied
Baasch, Karen *see* Die crescentialegende in der deutschen dichtung des mittelalters
Baath, Albert Ulrik *see* Wagners sagor
Bab, Ali Muhammad Shirazi *see*
- Le beyaan arabe
- Le beyan persan

Bab, Julius *see*
- Durch das drama hauptmanns
- Gerhart hauptmann und seine besten buehnenwerke
- Goethe und die juden
- Goethes leben in seinen briefen
- Das leben goethes
- Richard dehmel
- Ueber den tag hinaus
- Wege zum drama
- Das wort friedrich hebbels

Baba, Mehmed 'Ali Hilmi Dede *see* The divan project
Baba padmanji : an autobiography / Padmanji, Baba; ed by Mitchell, John Murray – indian ed. Madras: Christian Literature Society, 1892 [mf ed 1995] – (= ser Yale coll) – 104p (ill) – 1 – 0-524-09911-1 – mf#1995-0911 – us ATLA [920]
Baba shamal – Tehran. dawrah-'i 2. shumarah-'i 124-148. 29 murdad 1326-6 isfand 1326 [20 aug 1947-23 mar 1947] – 1r – 1 – $53.00 – (missing: n139) – us MEDOC [079]
Babad diponagoro : serat babad dipa nagara karangani pun swargipijambah...ingkang angedalaken administratie djawi kando / Dipanagara Pangerannja – Doerakarta. 2v. 1908 – 3mf – 8 – mf#SE-1599 – ne IDC [959]
Babalik – Konya: Babalik Matbaasi, 1910-28. Sahib-i Imtiyaz: Yusuf Mazhar. n988. 11 eyluel 1922; 989,2021,2026,2432,2511,2558,2635,2851. 22 tesrinisani 1928 – (= ser O & t journals) – 1mf – 9 – $25.00 – us MEDOC [956]
Babalola, S A *see* Content and form of yoruba ijala
Babatunde Somade, H M *see* Chemical problems associated with the control of pests in stored groundniuts in west africa
Babbage, Charles *see*
- History of science and technology, series 3
- The ninth bridgewater treatise
- Observations addressed at the last anniversary

Babbage, Edward F *see*
- The "phat boy's" 16 years on the st lawrence
- The phat boy's delineations of the st lawrence river and its environs
- The phat boy's racy description of the st lawrence river and its environs

Babcock, Garth J *see* A single stage submaximal treadmill jog test to estimate vo2 max in subjects ages 30 to 39 years
Babcock, Maltbie Davenport *see* Letters from egypt and palestine
Babcock, Rufus *see*
- Forty years of pioneer life
- Memoirs of john mason peck, dd, 1864

Babcock, Willoughby Maynard 2 *see* Newspaper transcripts index
Bab-ed-din : the door of true religion: za-ti-et al-lah, el fi-da / Kheiralla, Ibrahim George – Chicago: C H Kerr 1897 [mf ed 1991] – (= ser Unity library 63) – 1mf – 9 – 0-524-01837-5 – mf#1990-2672 – us ATLA [290]
Babel : dat is verwarringhe der wederdooperen onder malkanderen... / Faukelius, H – Middelburgh, 1621 – 5mf – 9 – mf#PBA-179 – ne IDC [240]
Babel and bible : a lecture on the significance of assyriological research for religion: delivered before the german emperor / Delitzsch, Friedrich – Chicago: Open Court; London: Kegan Paul, Trench, Truebner [distributor] 1902 [mf ed 1986] – 1mf [ill] – 9 – 0-8370-9376-7 – (trans fr german by thomas j mccormack) – mf#1986-3376 – us ATLA [221]
Babel, Eugen *see* Graf adolf friedrich von schack
Babel und bibel : randglossen zu den beiden vortraegen friedrich delitzschs / Horovitz, Jakob – Frankfurt a M: J Kauffmann 1904 [mf ed 1993] – 1mf [ill] – 9 – 0-524-05676-5 – (incl bibl ref) – mf#1992-0526 – us ATLA [221]
Babel und das neue testament : ein vortrag / Fiebig, Paul – Tuebingen: Mohr 1905 [mf ed 1993] – (= ser [Sammlung gemeinverstaendlicher vortraege und schriften aus dem gebiet der theologie und religionsgeschichte] 42) – 1mf – 9 – 0-524-06129-7 – mf#1992-0796 – us ATLA [225]
Der babel-bibel-streit und die offenbarungsfrage : ein verzicht auf verstaendigung / Kittel, Rudolf – 2. unveraend aufl. Leipzig: A Deichert 1903 [mf ed 1992] – 1mf – 9 – 0-524-05223-9 – (incl bibl ref) – mf#1992-0356 – us ATLA [221]
Babell, Charles *see* [Recueil de pieces choisies a une et deux flutes]
Babelon, Ernest Charles Francois *see* Manual of oriental antiquities
Babelsberger stadtanzeiger *see* Stadtanzeiger fuer nowawes und neubabelsberg
The babes in the wood : a tragic comedy: a story of the italian revolution of 1848 / De Mille, James – Boston: W F Gill, 1875 – 2mf – 9 – (text in dble clms) – mf#06017 – cn Canadiana [830]
Babeuf, Emile *see* Proces de la conspiration, dite republicaine, de decembre 1830
Babiali – Siyasi, Mesleki, Mizah Mecmua. Matbaa Teknisyenlerinin Mesleki Mecmuasi. Sahibi: Aziz Uctay. Teknik Sekreter: Ibrahim Guezelce n1,3-9,13. 1 nisan 1949-mart 1950 – (= ser O & t journals) – 3mf – 9 – $55.00 – us MEDOC [956]
Babich, K E *see* Sibirskaia zemskaia derevnia
Le babillard du palais-royal – n1-139. Paris. juin-oct 1791 – 1 – fr ACRPP [073]
La babillarde : journal gaulois, extravagant, mondain, litteraire, anti-melancolique, artistique et joyeux – Paris. n3, 6-9. 11 oct-22 nov 1884 – 1 – fr ACRPP [073]
Babin, Basile Joseph *see* Bibliographie analytique des eveques et de quelques peres eudistes au canada
Babin, Maria Teresa *see*
- Critica literaria
- Fantasia boricua

Babinger, Franz et al *see* Die religionen der erde
Babington, C *see*
- Polychronicon ranulphi higden monachi cestrensis
- The repressor of overmuch blaming of the clergy

Babington, Churchill *see* The benefit of christ's death
Babington, Cynthia A *see* Traditional and non-traditional predictors of academic success
Babington, James P *see* Maturational pace and athletic potential
Babington, John Albert *see* The reformation
Babington, W P *see* Amasiah the son of zichri
Babington, William Dalton *see* Fallacies of race theories as applied to national characteristics
Babitonga : orgam imparcial, litterario e noticioso – Sao Francisco do Sul, SC: Typ Iniciadora, 14 mar-25 set 1885 – (= ser Ps 19) – mf#UFSC/BPESC – bl Biblioteca [079]
Babouk / Endore, S Guy – New York, NY. 1934 – 1r – us UF Libraries [972]
Babraham 1561-1950 – (= ser Cambridgeshire parish register transcript) – 4mf – 9 – £5.00 – uk CambsFHS [929]
Babson, Roger Ward *see* Central american journey
Babur : diarist and despot / Edwardes, Stephen Meredyth – London: A M Philpot, 1926 – (= ser Samp: indian books) – us CRL [954]
Babur, Emperor of Hindustan *see*
- The babur-nama in english
- Memoirs of hir-ed-din muhammed babur, emperor of hindustan

The babur-nama in english : memoirs of babur / Babur, Emperor of Hindustan – London: Luzac & Co, 1921- – (= ser Samp: indian books) – (trans fr original turki text by annette susannah beveridge) – us CRL [954]
Babushkin, K S *see* Nash put' [tiumen': 1919]
Babushkiny stariny / Krivopolena, Maria Dmitrievna; ed by Ozarovskaia, O E – Moskva: Gos izd-vo [1922] [mf ed Bloomington IN: Indiana Uni Lib, Preservation Dept 1984] – 1r – 1 – (russian folk literature) – us Indiana Preservation [390]
Babut, Ernest Ch *see*
- Le concil de turin
- Le concile de turin
- La plus ancienne decretale
- Priscillien et le pricillianisme
- Saint martin de tours

Babut, Ernest-Ch *see* Saint martin de tours
Baby fae collection / Loma Linda University Heritage Room – Riverside Co, CA. – 1r – 1 – $50.00 – (special compilation of newspapers) – mf#R40290 – us Library Micro [978]
Baby, Louis Francois Georges *see* Chateauguay
A baby of the frontier / Brady, Cyrus Townsend – New York, Toronto: F H Revell, c1915 – 4mf – 9 – 0-665-98897-4 – mf#98897 – cn Canadiana [830]
Baby talk – New York NY 2002+ – 1,5,9 – ISSN: 1529-5389 – mf#27065.04 – us UMI ProQuest [640]
Baby, William Lewis *see* Souvenirs of the past with illustrations
Babylon / Allen, Grant [Cecil Power] – London: Chatto & Windus, 1885 – 1mf – 9 – mf#56067 – cn Canadiana [830]
Babylon – London, England. 1851 – 1r – us UF Libraries [240]
Babylon and the old testament / Parrot, Andre – SCM Press, 1958 – 9 – $10.00 – us IRC [221]
Babylon, vol 1 / Allen, Grant [Cecil Power] – London: Chatto & Windus, 1885 – 4mf – 9 – (pt of cihm set; incl publ list) – mf#56068 – cn Canadiana [830]
Babylon, vol 2 / Allen, Grant [Cecil Power] – London: Chatto & Windus, 1885 – 4mf – 9 – (pt of cihm set) – mf#56069 – cn Canadiana [830]
Babylon, vol 3 / Allen, Grant [Cecil Power] – London: Chatto & Windus, 1885 – 4mf – 9 – (pt of cihm set) – mf#56070 – cn Canadiana [830]
Babylonian and assyrian laws, contracts and letters / Johns, Claude Hermann Walter – New York: Scribner, 1904 [mf ed 1992] – (= ser Library of ancient inscriptions 6) – 1mf – 9 – 0-524-04403-1 – (incl bibl ref) – mf#1992-0096 – us ATLA [340]
The babylonian and oriental record – London, 1(1886)-9(1901) – 36mf – 9 – €67.00 – (lacking: 7(1893)) – ne Slangenburg [930]
The babylonian and the hebrew genesis = Biblische und babylonische urgeschichte / Zimmern, Heinrich – London: David Nutt 1901 [mf ed 1985] – (= ser Der alte orient 3) – 1mf – 9 – 0-8370-5966-6 – (english by jane hutchison) – mf#1985-3966 – us ATLA [221]
The babylonian conception of heaven and hell = Die babylonisch-assyrischen vorstellungen vom leben nach dem tode / Jeremias, Alfred – London: David Nutt 1902 [mf ed 1989] – (= ser Der alte orient 4) – 1mf – 9 – 0-7905-1166-5 – (english trans by jane hutchison) – mf#1987-1166 – us ATLA [290]
Babylonian contract tablets in the metropolitan museum of art / ed by Moldenke, Alfred B – New York NY: Metropolitan Museum of Art 1893 [mf ed 1986] – 1mf – 9 – 0-8370-7569-6 – (incl ind) – mf#1986-1569 – us ATLA [470]
The babylonian genesis, the story of creation / Heidel, Alexander – U. of C. Press, 1942 – 9 – $10.00 – us IRC [290]
Babylonian influence on the bible and popular beliefs : tehom and tiamat, hades and satan: a comparative study of genesis 1.2 / Palmer, Abram Smythe – London: David Nutt 1897 [mf ed 1989] – (= ser Studies on biblical subjects 1) – 1mf [ill] – 9 – 0-7905-1726-4 – (incl bibl ref) – mf#1987-1726 – us ATLA [221]
Babylonian laws / Driver, G R – Oxford. 1954 – 9 – $15.00 – us IRC [900]
Babylonian life and history / Budge, Ernest Alfred Wallis – London: Religious Tract Soc 1884 [mf ed 1989] – (= ser By-paths of bible knowledge 5) – 1mf [ill] – 9 – 0-7905-3316-2 – (incl ind) – mf#1987-3316 – us ATLA [930]
Babylonian literature : lectures / Sayce, Archibald Henry – London: Samuel Bagster [1877] [mf ed 1986] – 1mf – 9 – 0-8370-8378-8 – (incl bibl ref) – mf#1986-2378 – us ATLA [470]
Babylonian liturgies : sumerian texts from the early period and from the library of ashurbanipal... / Langdon, S – Paris, 1913 – 6mf – 9 – (with int and ind) – mf#NE-423 – ne IDC [470]
Babylonian magic and sorcery : being "the prayers of the lifting of the hand"... / ed by King, Leonard William – London: Luzac 1896 [mf ed 1992] – 3mf [ill] – 9 – 0-524-02353-0 – mf#1990-2964 – us ATLA [130]
Babylonian magic and sorcery... / King, L W – London, 1896 – 4mf – 9 – mf#NE-380 – ne IDC [956]
Babylonian penitential psalms to which are added fragments of the epic of creation... / Langdon, S – Paris, 1927 – 4mf – 9 – mf#NE-4124 – ne IDC [470]
Babylonian records in the library of j pierpont morgan / ed by Clay, A T – New Haven, 1923 – 3mf – 9 – mf#NE-421 – ne IDC [956]
Babylonian religion and mythology / King, Leonard William – London: Kegan, Paul, Trench, Truebner 1903 [mf ed 1989] – (= ser Books on egypt and chaldaea 4) – 1mf – 9 – 0-7905-1338-2 – (incl bibl ref) – mf#1987-1338 – us ATLA [230]
Babylonian-assyrian birth-omens and their cultural significance / Jastrow, Morris – Giessen: Alfred Toepelmann 1914 [mf ed 1991] – (= ser Religionsgeschichtliche versuche und vorarbeiten 14/5) – 1mf – 9 – 0-524-00904-X – (incl bibl ref) – mf#1990-2127 – us ATLA [230]
Babylonians and assyrians : life and customs / Sayce, Archibald Henry – New York: Charles Scribner 1899 [mf ed 1988] – (= ser The semitic papers) – 1mf – 9 – 0-7905-0284-4 – mf#1987-0284 – us ATLA [930]
Babyloniens kultur und die weltgeschichte : ein briefwechsel / Koenig, Eduard – Berlin: E Runge [19–?] [mf ed 1989] – 1mf – 9 – 0-7905-3346-4 – mf#1987-3346 – us ATLA [930]
Babylonisch-assyrische geschichte / Tiele, Cornelis Petrus – Gotha: F A Perthes 1886-88 [mf ed 1992] – (= ser Handbuecher der alten geschichte 1//4) – 2v on 2mf – 9 – 0-524-03126-6 – (incl bibl ref) – mf#1990-3179 – us ATLA [930]
Babylonisch-assyrische grammatik / Ungnad, A – Muenchen, 1926 – 4mf – 8 – mf#H-356 – ne IDC [470]
Babylonisch-assyrische texte – Bonn: A Marcus & E Weber 1904 [mf ed 1986] – (= ser Kleine texte fuer theologische vorlesungen und uebungen 7) – 1mf – 9 – 0-8370-7526-2 – mf#1986-1526 – us ATLA [470]
Die babylonisch-assyrischen praesens- und praeteritalformen : im grundstamm der starken verba / Lindl, Ernest – Muenchen: Hermann Lukaschik 1896 [mf ed 1986] – 1mf – 9 – 0-8370-7715-X – (incl bibl ref) – mf#1986-1715 – us ATLA [470]
Die babylonisch-assyrischen vorstellungen vom leben nach dem tode : nach den quellen mit beruecksichtigung der alttestamentlichen parallelen / Jeremias, Alfred – Leipzig: J C Hinrichs 1887 [mf ed 1986] – 1mf – 9 – 0-8370-7224-7 – (in german & akkadian; incl bibl ref) – mf#1986-1224 – us ATLA [230]
Die babylonisch-assyrischen vorstellungen vom leben nach dem tode *see* The babylonian conception of heaven and hell
Babylonisch-astrales im weltbilde des thalmud und midrasch / Bischoff, Erich – Leipzig: J C Hinrichs 1907 [mf ed 1985] – 1mf [ill] – 9 – 0-8370-2348-3 – (incl ind) – mf#1985-0348 – us ATLA [270]
Babylonische beschwoerungreliefs : ein beitrag zur erklaerung der sog hadesreliefs / Frank, K – Leipzig, 1908 – (= ser Leipziger semitistische Studien) – 2mf – 9 – (leipziger semitistische studien, leipzig 1920 v3 pt3) – mf#NE-20113 – ne IDC [956]
Babylonische briefe aus der zeit der hammurapi-dynastie / Ungnad, Arthur – Leipzig: J C Hinrichs 1914 [mf ed 1986] – (= ser Vorderasiatische bibliothek 6) – 2mf – 9 – 0-8370-7747-8 – (incl bibl ref & ind; texts in german & akkadian; int & notes in german) – mf#1986-1747 – us ATLA [470]
Babylonische busspsalmen / Zimmern, Heinrich – Leipzig: J C Hinrichs 1885 [mf ed 1986] – 1mf – 9 – 0-8370-7118-6 – (incl bibl ref & ind) – mf#1986-1118 – us ATLA; ne IDC [470]
Die babylonische chronik : nebst einem anhang ueber die synchronistische geschichte p / Delitzsch, Friedrich – Leipzig: B G Teubner 1906 [mf ed 1986] – (= ser Abhandlungen der philologisch-historischen classe der koenigl saechsischen gesellschaft der wissenschaften 25/1) – 1mf – 9 – 0-8370-7691-9 – (incl bibl ref) – mf#1986-1691 – us ATLA [930]
Die babylonische fabel und ihre bedeutung fuer die literaturgeschichte / Ebeling, E – Leipzig, 1927 – (= ser Mitteilungen der altorientalischen Gesellschaft) – 1mf – 9 – (mitteilungen der altorientalischen gesellschaft. v2, pt 3) – mf#NE-20107 – ne IDC [956]
Die babylonische gebetsbeschwoerung / Kunstmann, W G – Leipzig, 1932 – (= ser Leipziger semitistische Studien) – 2mf – 9 – (leipziger semitistische studien, n.s. v2) – mf#NE-20035 – ne IDC [956]
Die babylonische gebetsbeschwoerung / Kunstmann, Walter G – Graefenhainichen, 1930 (mf ed 1993) – 2mf – 9 – €31.00 – 3-89349-322-0 – mf#DHS-AR 178 – gw Frankfurter [290]

Der babylonische gott tamuz / Zimmern, H – Leipzig, Teubner, 1909 – (= ser Abh koenigl saechs gesellschaft der wissenschaften 27) – 1mf – 9 – (abh koenigl saechs gesellschaft der wissenschaften v27) – mf#NE-20038 – ne IDC [956]
Babylonische hymnen und gebete in auswahl – Leipzig: JC Hinrichs, 1905 [mf ed 1989] – (= ser Der alte orient 7/3) – 1mf – 9 – 0-7905-2099-0 – (german trans fr akkadian by heinrich zimmern) – mf#1987-2099 – us ATLA; ne IDC [780]
Babylonische hymnen und gebete, zweite auswahl – Leipzig: J C Hinrichs 1911 [mf ed 1989] – (= ser Der alte orient 13/1) – 1mf – 9 – 0-7905-2039-7 – (german trans fr akkadian by heinrich zimmern) – mf#1987-2039 – us ATLA; ne IDC [780]
Die babylonische kosmogonie und der biblische schoepfungsbericht : ein beitrag zur apologie des biblischen gottesbegriffes / Kirchner, Aloys – Muenster i W: Aschendorff 1910 [mf ed 1989] – (= ser Alttestamentliche abhandlungen 3/1) – 1mf – 9 – 0-7905-2290-X – (incl bibl ref) – mf#1987-2290 – us ATLA [221]
Die babylonische kultur in ihren beziehungen zur unsrigen : ein vortrag / Winckler, Hugo – Leipzig: J C Hinrichs 1902 [mf ed 1990] – 1mf [ill] – 9 – 0-7905-3501-7 – (incl bibl ref) – mf#1987-3501 – us ATLA [930]
Das babylonische nimrodepos : keilschrifttext der bruchstuecke der sogenannten izdubarlegenden mit dem keilinschriftlichen sintfluthberichte / Gilgamesh; ed by Haupt, Paul – Leipzig: J C Hinrichs 1884-91 [mf ed 1986] – 1mf – 9 – 0-8370-9065-2 – (text in akkadian, notes in german; no more publ) – mf#1986-3065 – us ATLA [470]
Eine babylonische quelle fuer das buch job : eine literar-geschichtliche studie / Landersdorfer, Simon – Freiburg i B, St Louis MO: Herder 1911 [mf ed 1989] – (= ser Biblische studien 16/2) – 1mf – 9 – 0-7905-2978-5 – (in german & akkadian; incl bibl ref) – mf#1987-2978 – us ATLA [221]
Babylonische suehnriten : besonders mit ruecksicht auf priester und buesser / Schrank, Walther – Leipzig: J C Hinrichs 1908 – (= ser Leipziger semitistische studien 3/1) – 1mf – 9 – 0-524-02540-1 – (incl bibl ref) – mf#1990-3035 – us ATLA; ne IDC [290]
Babylonische talmud : hebraeisch und deutsch mit einschluss der vollstaendige misnah / ed by Goldschmidt, L – Berlin. v1-9. 1925 – €818.00 – (trans by ed) – ne Slangenburg [270]
Babylonische vertraege des berliner museums : in autographie, transscription und uebersetzung / ed by Peiser, Felix Ernst – Berlin: Wolf Peiser 1890 [mf ed 1986] – 2mf – 9 – 0-8370-9016-4 – (incl bibl ref & ind) – mf#1986-3016 – us ATLA [470]
Die babylonische weltschoepfung / Winckler, Hugo – Leipzig: J C Hinrichs 1906 [mf ed 1989] – (= ser Der alte orient 8/1) – 1mf – 9 – 0-7905-2096-6 – mf#1987-2096 – us ATLA [520]
Das babylonische weltschoepfungsepos / Delitzsch, Friedrich – Leipzig: S Hirzel 1896 [mf ed 1986] – (= ser Abhandlungen der philologisch-historischen classe der koenigl saechsischen gesellschaft der wissenschaften 17/2) – 1mf – 9 – 0-8370-7055-4 – (incl bibl ref) – mf#1986-1055 – us ATLA [930]
Die babylonischen busspsalmen und das alte testament / Bahr, Johannes – Berlin: Weidmann, 1903 [mf ed 1986] – (= ser Wissenschaftliche beilage zum jahresbericht des humboldt-gymnasiums zu berlin easter 1903) – 1mf – 9 – 0-8370-7361-8 – mf#1986-1361 – us ATLA [780]
Babylonisches im neuen testament / Jeremias, Alfred – Leipzig: J C Hinrichs 1905 [mf ed 1985] – 1mf – 9 – 0-8370-3781-6 – (incl bibl ref, general ind & ind of biblical & extra-biblical texts cited) – mf#1985-1781 – us ATLA [225]
Babylonisches im neuen testament / Karge, Paul – 1. & 2. aufl. Muenster i W: Aschendorff 1913 [mf ed 1993] – (= ser Biblische zeitfragen 6/9-10) – 2mf – 9 – 0-524-06142-4 – mf#1992-0809 – us ATLA [225]
Bac bi-monthly newsletter – 1975 jul/aug – 1 – mf#3945984 – us WHS [071]
Baca county banner *see* Baca county miscellaneous newspapers
Baca county miscellaneous newspapers – Springfield, CO (mf ed 1991) – 1r – 1 – (baca county banner (jun 25 1942-nov 12 1953), baca county republican (jan 15 1932-feb 12 1937), springfield beacon (apr 23 1887), springfield plainsman (mar 9 1937-mar 24 1939), stonington news (nov 21 1918), two buttes sentinel (mar 10 1910), walsh tab (jan 23-1930)' walsh topic (apr 15 1954-jun 27 1957)) – mf#MF Z99 B12 – us Colorado Hist [071]
Bacardi Y Moreau, Emilio *see* Cronicas de santiago de cuba
Baccalaureate sermons / Peabody, Andrew Preston – Boston: D Lothrop [c1885] [mf ed 1984] – 4mf – 9 – 0-8370-0869-7 – mf#1984-4198 – us ATLA [240]
Baccalaureate sermons and addresses / Terry, Milton Spenser – New York: Methodist Book Concern, c1914 [mf ed 1991] – 1mf – 9 – 0-7905-9710-1 – mf#1989-1435 – us ATLA [242]
Bacchae / Euripides – New York, NY. 19– – 1r – us UF Libraries [960]
The bacchae. / Euripides – Trans. into English rhyming verse with explan. notes by Gilbert Murray. 12th thousand.New York: Longman's Green, 1915. 94p – 1 – us Wisconsin U Libr [450]
Bacchus & ariadne : a grand ballet as performed at the kings theatre, haymarket 1797-8. composed [i.e. choreographed] by m gallet, the music [composed and] arranged for the piano forte by cesare bossi / Bossi, Cesare – London: Corri, Dussek & Co [1798] [mf ed 1989] – 1r – 1 – mf#pres. film 45 – us Sibley [790]
Bacchus, Francis *see* Essays
Bacchus marsh express – Australia 7 jan 1911-26 aug 1916 (imperfect) – 2r – 1 – uk British Libr Newspaper [079]
Baccus, Joseph H *see* Selected articles on minimum wages and maximum hours
Bach, Adolf *see*
– Geschichte der deutschen sprache
– Goethes leben im garten am stern
– Goethes rheinreise mit lavater und basedow im sommer 1774
– Lof der reinster vrowen
– Das rheinische marienlob
Bach, Carl Philipp Emanuel *see*
– Favourite solo for the violin & harpsichord by e bach
– Grosze passions cantate
– Heilig
– Herrn professor gellerts geistliche oden und lieder mit melodien
– Die israeliten in der wueste
– Kurze und leichte klavierstuecke
– Musikalisches vielerley
– Petites pieces pour le clavecin de c p e bach
– Sechs leichte clavier sonaten
– Sechs sonaten fuers clavier
– Six fuges pour le piano forte
– Versuch ueber sie wahre art das clavier zu spielen. 1. theil
Bach jahrbuch – Leipzig. 1904-50. 2 reels – 1 – us L of C Photodup [780]
Bach, Johann Christian *see*
– Four sonatas and two duetts for the harpsichord or piano forte
– Practical edition of four concertos for piano or harpsichord and orchestra...
– Quattro sonate notturne a due violini, 2 basso, a piuttosto viola
– Six concerts pour le clavecin, deux violons & un violoncelle
– Thirrd set of six concertos for the harpsichord or pianoforte
– Three favorite symphonies in eight parts
Bach, Johann Michael *see* Kurze und systematische anleitung zum general-bass und der tonkunst ueberhaupt
Bach, Johann Sebastian *see*
– 4 stimmiges choralbuch
– Le clavecin bien tempere
– Clavieruebung bestehend in einer aria
– Dritter theil der clavier uebung bestehend in verschiedenen vorspielen ueber die catechismus- und andere gesaenge, vor die orgel...
– Die kunst der fuge durch herrn johann sebastian bach ehemaligen capellmeister zu leipzig
– Litaney von martin luther und johann sebastian bach
– Musikalisches opfer sr koeniglichen majestaet in preussen & c allerunterthaenigst gewidmet...
– Practical organ accompaniment for the magnificat in d major and for the cantata no 106 of johann sebastian bach
– Six preludes a l'usage des commencants pour le clavecin
– Werke
Bach, Joseph *see*
– Die dogmengeschichte des mittelalters von christologische standpunkte
– Meister eckhart, der vater der deutschen speculation
Bach, Ludwig *see* Der glaube nach der anschauung des alten testamentes
Bach, Marcus *see* Strange altars
Bach, Robert *see* Die erwaehlung israels in der wueste
Bachajizade, 'Abd al-Rahman Bey *see*
– Dhail kitab al-fariq
– Kitab al-fariq al-makhluq wal-khaliq
Bachand, P *see* Speech on the budget by the hon p bachand, treasurer of the province of quebec
Bacharach, Charles K *see* Madrigals of claudio monteverdi
Bacharach, Siegfried *see* Nachrichtenblatt
Bachaumont, Louis Petit de *see*
– Memoires historiques, litteraires et critiques de bachaumont, depuis l'annee 1762 jusques 1788
– Memoires secrets de bachaumont
Die bach-drucke der hoboken-sammlung musikalischer erst- und fruehdruecke / Oesterreichische Nationalbibliothek Wien. Musiksammlung – [mf ed Hildesheim 1987] – (= ser Die europaeische musik) – 178mf – 9 – diazo €998.00 silver €1148.00 – gw Olms [780]
Bache, Kentish *see*
– Letter to the rev samuel davidson
– A letter to the rev samuel davidson...
Bache, Samuel *see*
– Examination of objections made to unitarianism by the rev j c miller, m a
– Lectures in exposition of unitarian views of christianity
Bacheler, Origen *see* Discussion on the existence of god and the authenticity of the bible
Bacheler, Otis Robinson *see* Hinduism and christianity in orissa
Bachelier de salamanque – Paris, France. 1815 – 1r – us UF Libraries [440]
Bacheller, Irving *see* The master
Bachelor of arts : a monthly magazine devoted to university interests and general literature – Killen TX 1895-98 – 1 – mf#2856 – us UMI ProQuest [378]
The bachelor of arts : a novel / Narayan, R K – London; New York: Thomas Nelson and Sons Ltd, 1937 – (= ser Samp: indian books) – us CRL [830]
Bachelor of divinity and master of theology theses / Pacific Theological College – Suva, Fiji. 1968-1993 – 33r – 1 – mf#PMB1084 – at Pacific Mss [200]
Bacher, Wilhelm *see*
– Die aelteste terminologie der juedischen schriftauslegung
– Die agada der babylonischen amoraeer
– Die agada der palaestinensischen amoraeer
– Die agada der tannaiten
– Die bibelexegese der juedischen religionsphilosophen des mittelalters vor maimuni
– Die bibelexegese moses maimunis
– Emendationes in plerosque sacrae scripturae veteris testamenti libros
– Die exegetische terminologie der juedischen traditionsliteratur
– Die hebraeische sprachwissenschaft vom 10. bis zum 16. jahrhundert
– Die juedische bibelexegese
– Leben und werke des abulwalid merwan ibn ganah (r jona)
– Die prooemien des alten juedischen homilie
– Tradition und tradenten in den schulen palaestinas und babyloniens
Bacheville, Barthelemy *see* Voyages des freres bacheville, capitaines de l'ex-garde, chevaliers de la legion d'honneur
Bachhofer, Ludwig *see* Early indian sculpture
Bachiller Cantaclaro *see* Hilvanes y zurzidos. lo que se llama perder el tiempo. ensayos poeticos
Bachiller Reganadientes *see* Demostractiones palmarias..
Bachiller Y Morales, Antonio *see*
– Cuba
– Historia de las medidas adoptadas por la administr...
Bachir, Ali *see* L'amour, le mariage, la justice selon le koran
Bachler, Levi R *see* An emg study of four elastic tubing closed kinetic chain exercises
Bachmair, Heinrich Franz S *see* Die neue zeit
Bachman, C *see* Emblematum sacra
Bachman, Catherine *see* Papers
Bachman, John *see*
– A defence of luther and the reformation
– Sermons
Bachmann, Adolf *see* Boehmen und seine nachbarlaender unter georg von podiebrad, 1458-61
Bachmann, Albert *see*
– Deutsche volksbuecher
– Die haimonskinder in deutscher uebersetzung des 16. jahrhunderts
– Mittelhochdeutsches lesebuch
– Morgant der riese in deutscher uebersetzung des 16. jahrhunderts
Bachmann, Beatrix *see* Wahn und wirklichkeit
Bachmann, Cordula *see* Ausbreitungsbiologische merkmalstypenspektrenanalyse von bryophytengesellschaften am beispiel eines ariden und eines tropischen standortes
Bachmann, Frederick William *see* Some german imitators of walter scot
Bachmann, Johannes *see*
– Alttestamentlichen untersuchungen, 1. buch
– Das buch der richter. erster band
– Dodekapropheton aethiopum
– Die klagelieder jeremiae in der aethiopischen bibeluebersetzung
– Praeparation und commentar zum jesaja
– Der prophet jesaia nach der aethiopischen bibeluebersetzung
Bachmann, Luise George *see* Bruckner
Bachmann, Philipp *see*
– Grundlinien der systematischen theologie
– J chr. k. v. hofmanns versoehnungslehre und der ueber sie gefuehrte streit
– The new message in the teaching of jesus
– Die persoenliche heilserfahrung des christen und ihre beduetung fuer den glauben nach dem zeugnisse der apostel
– Der zweite brief des paulus an die korinther
Bachofen, Charles *see* Essai sur l'ecclesiologie de zwingle
Bachofen, J C *see* Musikalisches halleluja
Bachor, Oskar-Wilhelm [comp] *see* Der kreis gerdauen
Bachrach, Jacob *see* Masa' la-arets ha-kedoshah
Bach's b-minor mass : a study in interpretation / Scharf, Warren A – U of Rochester 1961 [mf ed 1983] – 6mf – 9 – (with bibl) – mf#fiche 1088 – us Sibley [780]
Bach's canonic variations on von himmel hoch : an analysis and a practical edition / McClain, Charles S – U of Rochester 1967 [mf ed 19–] – 4mf – 9 – (with bibl) – mf#fiche 1148 – us Sibley [780]
Bach's matthaeuspassion / Leeuw, Gerardus van der – 7. druk. Amsterdam: Uitgeversmaatschappij Holland [1941?] [mf ed 1993] – 1mf – 9 – 0-524-08115-8 – mf#1993-9021 – us ATLA [780]
Die bach-sammlung – [mf ed 1998-2003] – (= ser Musikhandschriften der staatsbibliothek zu berlin – preussischer kulturbesitz 1) – 1619mf incl 267 col (1:24) – 9 – silver+col €6600.00 – ISBN-10: 3-598-34420-1 – ISBN-13: 978-3-598-34420-6 – (suppl 1 o/p; suppl 2 [294mf] €2965 [isbn: 978-3-598-33438-1]; with catalogue) – gw Saur [780]
Die bach-sammlung : supplement 2 / Fischer, Axel [comp]; ed by Sing-Akademie zu Berlin – [mf ed 2003] – (= ser Musikhandschriften der staatsbibliothek zu berlin – preussischer kulturbesitz) – 294mf (1:24) – 9 – silver €2965.00 – ISBN-10: 3-598-34440-6 – ISBN-13: 978-3-598-34440-4 – (int by ulrich leisinger) – gw Saur [780]
Bachye, Rabbi [Bahya ben Joseph ibn Pakuda] *see* The duties of the heart
Bacillus – [Scotland] Edinburgh: The Bacillus 22-25 nov 1905 (daily) – 4v on 1r – 1 – (a daily paper publ during the international fair in aid of the royal victoria hospital for consumption, in edinburgh, 1905) – uk Newsplan [072]
Bacilly, Benigne de *see* Remarques curieuses sur l'art de bien chanter
The back blocks of china : a narrative of experiences among the chinese, sifans, lolos, tibetans, shans and kachins, between shanghai and the irrawadi / Jack, Robert Logan – London: Edward Arnold 1904 [mf ed 1995] – (= ser Yale coll) – xxii/269p (ill) – 1 – 0-524-10007-1 – mf#1995-1007 – us ATLA [915]
Back, Claus *see* Der weg nach rom
The back handspring : comparison of kinematic variables of the center of gravity following three different hand placements / Yuen, Garry E – 1988 – 86p 1mf – 9 – $4.00 – us Kinesology [612]
Back home in kentucky – 1978 jan/feb-1985 may/jun – 1 – mf#1265626 – us WHS [071]
Back of the spanish rebellion / Fernsworth, Lawrence A – [s.l: s.n. 1936] – 9 – (repr fr "foreign affairs" 1936) – mf#w882, w883 – us Harvard [946]
Back porch pilot – 1976 oct-1979 apr; 1979 mar-1980 may – 1 – mf#657489 – us WHS [071]
Back porch radio pilot – 1980 jun-1981 jan – 1 – mf#674023 – us WHS [380]
Back, Samuel *see* Elischa ben abuja-acher
Back stage – New York NY 1977+ – 1,5,9 – ISSN: 0005-3635 – mf#11170.01 – us UMI ProQuest [790]
Back to bethlehem : modern problems in the light of the old faith / Willey, John H – New York: Eaton & Mains; Cincinnati: Jennings & Graham c1905 [mf ed 1989] – 1mf – 9 – 0-7905-2570-4 – (incl ind) – mf#1987-2570 – us ATLA [240]
Back to broward / 'Back To Broward' League – St Augustine, FL. 1916 – 1r – us UF Libraries [978]
'Back To Broward' League *see* Back to broward
Back to christ : some modern forms of religious thought / Spence, Walter – Chicago: A C McClurg 1900 [mf ed 1985] – 1mf – 9 – 0-8370-5505-9 – mf#1985-3505 – us ATLA [240]
Back to godhead – Philadelphia PA 1973-80 – 1,5,9 – ISSN: 0005-3643 – mf#7738 – us UMI ProQuest [240]
Back to holy church : experiences and knowledge acquired by a convert = Zurueck zur heiligen kirche / Ruville, Albert von; ed by Benson, Robert Hugh – London, New York: Longmans, Green 1910 [mf ed 1986] – 1mf – 9 – 0-8370-8220-X – (english trans by g schoetensack) – mf#1986-2220 – us ATLA [240]

Back to patmos : prophetic outlooks on present conditions / Simpson, Albert B – New York: Christian Alliance Pub Co, c1914 [mf ed 1992] – (= ser Christian & missionary alliance coll) – 2mf – 9 – 0-524-02266-6 – mf#1990-4273 – us ATLA [225]

Back to sunny seas / Bullen, Frank Thomas – London, England. 1905 – 1r – us UF Libraries [972]

Back to the old testament for the message of the new : an effort to connect more closely the testaments... / Curtis, Anson Bartie – Boston: Universalist Pub House c1894 [mf ed 1989] – 1mf – 9 – 0-7905-0640-8 – (incl bibl ref & ind) – mf#1987-0640 – us ATLA [221]

Back to the republic : the golden mean, the standard form of government / Atwood, Harry F – Chicago: Laird & Lee, 1918 – 2mf – 9 – $3.00 – mf#LLMC 92-182 – us LLMC [323]

Back to the trees / Culwick, Arthur Theodore – Cape Town, South Africa. 1965 – 1r – us UF Libraries [960]

Backes, I *see* S thomae de aquino quaestio de gratia capitis (s th 3, q 8) (fp40)

Backes, Nikolaus *see* Kardinal simon de brion (papst martin 4.)

Background and development of pedro menendez's contribution / Hoffman, Paul Everett – Gainesville, FL. no date – 1r – us UF Libraries [978]

Background and history of the general association of regular baptist churches / Stowell, Joseph M – 3rd ed. 1949 – 1 – $5.00 – us Southern Baptist [242]

Background materials relating to railroad retirement / U.S. Congress. Senate. Committee on Labor and Public Welfare. Subcommittee on Retirement – Washington, Govt. Print. Off., 1973. 99 p. LL-2240 – 1 – us L of C Photodup [343]

Background materials relating to the exercise of criminal jurisdiction over u.s. forces in japan under article 17 of the japan -united states administrative agreement : compiled by the office of the judge advocate, hq. affe78a (rear) apo 343 – Army-AG Admin Cen-Japan-150, 1962? – 1mf – 9 – $1.50 – mf#LLMC 96-074 – us LLMC [345]

The back-ground of assamese culture / Nath, R M – Shillong: AK Nath, 1948 – (= ser Samp: indian books) – us CRL [954]

The background of sacred story : life lessons from the less-known characters of the bible / Hastings, Frederick – London: Religious Tract Society, 1886 – 1mf – 9 – 0-8370-3519-8 – mf#1985-1519 – us ATLA [220]

The background of swedish immigration, 1840-1930 / Janson, Florence Edith Alfreda – Chicago, IL: The University of Chicago Press, [1931] [mf ed 1970] – (= ser Social service monographs 15; Library of american civilization 14878) – xi/517p on 1mf – 9 – us Chicago U Pr [304]

The background of the gospels : or judaism in the period between the old and new testaments / Fairweather, William – Edinburgh: T & T Clark, 1908 [mf ed 1989] – (= ser Cunningham lectures series 20) – 2mf – 9 – 0-7905-2766-9 – mf#1987-2766 – us ATLA [226]

Background to ballet / Melvin, Duncan – [London, 1947] – 1 – mf#*ZBD-*MGO pv16 – Located: NYPL – us Misc Inst [790]

Backgrounds of human fertility in puerto rico / Hatt, Paul K – Princeton, NJ. 1952 – 1r – us UF Libraries [304]

Backhoff, F I *see* Voyage as a moscovite envoy into china

Backhouse, Edmund *see* Annals and memoirs of the court of peking

Backhouse, Edward *see*
- Early church history
- Witnesses for christ and memorials of church life from the 4th to the 13th century

Backhouse, Edward et al *see* Biographical memoirs

Backhouse, J *see* A narrative of a visit to the mauritius and south africa

Backhouse, James *see*
- The life and correspondence of william and alice ellis, of airton
- The life and labours of george washington walker, of hobart town, tasmania
- A memoir of deborah backhouse of york
- A narrative of a visit to the mauritius and south africa

Backhouse, Sarah *see* Memoir of james backhouse

Backhus, Michaela et al *see* Public relations als ein bestandteil der unternehmenskommunikation im globalisierungsprozess

Backmann, Christine K *see* The effect of treadmill compliance and foot type electromyography of lower extremity muscles during running

Backnanger kreiszeitung : murrtal-bote – Backnang DE, 1987- – 6r/yr – 1 – gw Misc Inst [074]

Backoffner, Rudolph [comp] *see* Enthuellungen der geheimnisse der freimaurerei

Backpacker [1973] – Emmaus PA 1973-80 – 1,5,9 – ISSN: 0160-3329 – mf#9914.02 – us UMI ProQuest [790]

Backpacker [1981] – Emmaus PA 1981+ – 1,5,9 – (cont: backpacker including wilderness camping) – ISSN: 0277-867X – mf#9914.02 – us UMI ProQuest [790]

Backslider / Fuller, Andrew – Clipstone, England. 1802 – 1r – us UF Libraries [240]

Backslider – London, England. 18– – 1r – us UF Libraries [240]

Backslider : a memoir of dinah – London, England. 18– – 1r – us UF Libraries [240]

Backus, Edwin Burdette *see* The church and the social question

Backus, Isaac *see*
- A history of new england with particular reference to the denomination of christians called baptists
- Misc. materials
- Miscellaneous printed publications; also sermons by 38 other new england writers, 1746-1800
- The papers of isaac backus
- Ten diaries and other unpublished works in manuscript.

Backus, issac, papers, ms 71 – 1719-1805 – 1r – 1 – (primarily letters addressed to and copies of letters sent by rev issac backus, baptist minister and historian) – us Western Res [242]

Backus, James *see* James backus papers, ms 1548

Backus memorial baptist church – NORTH MIDDLEBORO, MA. 1756-1894 – 1 – $18.45 – us Southern Baptist [242]

The backwash of war : the human wreckage of the battlefield as witnessed by an american hospital nurse / La Motte, Ellen Newbold – New York, London: G P Putnam's Sons, 1916 – 1r – 1 – us Wisconsin U Libr [920]

Backwoods : cowcamp frolic / Huss, Veronica E – s.l, s.l? . 193? – 1r – us UF Libraries [978]

The backwoods of canada : being letters from the wife of an emigrant officer; illustrative of the domestic economy of british america / Traill, Catharine – London 1836 [mf ed Hildesheim 1995-98] – 1v on 3mf – 9 – €90.00 – 3-487-27121-4 – gw Olms [880]

The backwoods of canada : being letters from the wife of an emigrant officer, illustrative of the domestic economy of british america / Traill, Catherine Parr – new ed. London: C Knight, 1846 [mf ed 1983]] – 3mf – 9 – 0-665-41580-X – mf#41580 – cn Canadiana [304]

Bacmeister, Ernst *see*
- Der deutsche typus der tragoedie
- Kaiser konstantins taufe
- Der teure tanz

Bacmeister,Hartwig Ludw Christian *see* Russische bibliothek zur kenntnis des gegenwaertigen zustandes der literatur in russland

Bacon, Benjamin Wisner *see*
- The beginnings of gospel story
- Christianity old and new
- Commentary on the epistle of paul to the galatians
- The founding of the church
- The fourth gospel in research and debate
- The genesis of genesis
- An introduction to the new testament
- Jesus the son of god, or, primitive christology
- The making of the new testament
- The odore thornton munger
- The sermon on the mount
- The story of st paul
- The triple tradition of the exodus

Bacon, Catherine Jane *see* The effect of menstrual cycle phase on diffusing capacity of the lung

Bacon, Charles W *see* The american plan of government

Bacon, Dolores *see* Old new england churches and their children

Bacon, Eliza Ann *see* Memoir of rev. henry bacon

Bacon families association newsletter – 1981 oct 15-1984 jul – 1 – mf#842271 – us WHS [366]

Bacon, Francis *see*
- Bacon's essays
- The essays or counsels civil and moral of francis bacon
- The philosophical works of francis bacon, baron of verulam, viscount st albans
- The tvvoo bookes of francis bacon

Bacon, George Blagden *see* The sabbath question

Bacon, Grace Mabel *see* The personal and literary relations of heinrich heine to karl immermann

Bacon Hilda *see* Effects of aerobic exercise on the lipid profile levels of patients with moderate to severe burn injury

Bacon, James *see* The life and times of francis the first

Bacon, John *see*
- A letter to the right hon. sir robert peel, bart., m.p. on the appointment of a commission for promoting the cultivation and improvement of the fine arts
- Memoir of miss ann bacon

Bacon, Leonard *see*
- Four commemorative discourses
- The genesis of the new england churches
- Slavery discussed in occasional essays

Bacon, Leonard et al *see* Memorial of nathaniel w taylor

Bacon, Leonard Woolsey *see*
- God's wonderful work in france
- A history of american christianity
- How the rev. dr. stone bettered his situation
- An inside view of the vatican council
- Irenics and polemics
- The sabbath question

Bacon, Nath *see* Relation of the fearful estate of francis spira...

Bacon, R *see* For better relations with our latin american neighbors

Bacon, Reginald H *see* Benin, the city of blood

Bacon, Robert *see* Para el fomento de nuestras buenas

Bacon, Roger *see* Rogeri bacon opera quaedam hactenus inedita (rs15)

Bacon, Thomas *see* The township bonds of missouri, under the act of march 23rd, 1868

Bacon, Thomas Scott *see* It is written

Bacone chief – [Bacone OK]: publ by the Academic Students of Bacone College 1911-1920; Muskogee OK 1921-1951 [mf ed 2007] – (= ser Religious periodical literature of the hispanic and indigenous peoples of the americas, 1850-1985) – 1r – 1 – (lacks: 1916,1918,1926,1927[p13-20],1929-50; cont by: bacone college. warrior) – mf#2007i-s023 – us ATLA [378]

Bacons' complete preceptor for the clarinet, with a selection of airs, marches, &c – Philadelphia: A. Bacon, 181-. Includes: Hail to the Chief and Washington's March. Music-778, item 3 – 1 – us L of C Photodup [780]

Bacon's essays – London, England. 1920 – 1r – us UF Libraries [420]

Bacon's judicial repository *see* New york judicial repository

Bacot, Jacques *see*
- Dans les marches tibetaines
- Les mo-so

Bacquart, A *see* Une colonie de commerce francaise

Bacqueville de La Potherie, Claude-Charles *see* Voyage de l'amerique

Bacs-bodroger zeitung – Apatin, Yugoslavia. Apr 1923-May 1925 – 1r – 1 – us L of C Photodup [079]

Bacterial soft rot of potatoes in southern florida / Ruehle, George D – Gainesville, FL. 1940 – 1r – us UF Libraries [630]

Bacteriological reviews – Washington DC 1937-77 – 1,5,9 – (cont by: microbiological reviews) – ISSN: 0005-3678 – mf#92.02 – us UMI ProQuest [576]

Bacuez, Nicolas *see* The divine office

El baculo de san pedro de alcantara / Munoz de San Pedro, Miguel – Badajoz: dip prov, 1966 – 1 – sp Bibl Santa Ana [241]

Bacup advertiser – [NW England] Bacup Lib 9 dec 1967-12 feb 1970 – 1 – uk MLA; uk Newsplan [072]

Bacup and rossendale news – [NW England] Bacup Lib may 1863-dec 1890 – 1 – (title change: rossendale division gazette [1891-1901]) – uk MLA; uk Newsplan [072]

Bacup comet – [NW England] Bacup Lib nov-dec 1985 – 1 – uk MLA; uk Newsplan [072]

Bacup echo – [NW England] Bacup Lib jan 1972-apr 1977 – 1 – uk MLA; uk Newsplan [072]

Bacup times and rossendale advertiser – [NW England] Bacup Lib 8 apr 1865-4 apr 1891 – 1 – (title change: bacup & rawtenstall times [11 apr-24 dec 1891]; bacup times [jan 1892-nov 1965]) – uk MLA; uk Newsplan [072]

Baczko, Ferdinand von *see* Reise von posen durch das koenigreich polen und einen theil von russland

Bad and good guests – London, England. 1827 – 1r – us UF Libraries [240]

Bad awjuh al-ikhtilaf fi rasm ism al-makan al-wahid bi-huruf al-lughah al arabiyah fi al-mamlakah al-arabiyah al-saudiyah / Asad Sulayman Abduh – al-Kuwayt: Qism al-Jughrafiya bi-Jamiat al-Kuwayt wa-al-Jamiyah al-Jughrafiya al-Kuwaytiyah, [1985] – us CRL [956]

Bad british columbia blackout – n104-128 [1984 apr 6/20-1985 apr 12/25] – 1 – mf#963515 – us WHS [071]

Bad ems – struktur- und funktionswandel der baederstadt an der unterlahn : eine kulturgeographische untersuchung / Hapke, Ralf – 2000 – 4mf – 9 – 3-8267-2686-3 – mf#DHS 2686 – gw Frankfurter [943]

The bad lands cow boy – Little Missouri, Dakota [i.e. ND]; A T Packard. v1 n1 feb 7 1884-v3 n45 dec 23 1886 (wkly) – 1 – (publ in medora, nd nov 13 1885-dec 23 1886. missing: 1884 sep 11) – us North Dakota [071]

Bad lauterberger zeitung *see* Harz-kurier

Bad mergentheimer zeitung – Bad Mergentheim DE, 1955 3 jan-1960 – (= ser Bezirksausgabe von fraenkischen nachrichten, tauberbischofsheim) – 13r – 1 – gw Misc Inst [074]

Bad nauheim pudding – v1 n1-4; v2 n1; v3 n1 [1942 feb 7-mar 7; 1981 feb – 1 – mf#1701238 – us WHS [071]

Bad nauheimer anzeiger – Bad Nauheim DE, 1896 4 jan-1901, 1903-1914 [gaps] – 11r – 1 – (title varies: 26 mar 1910: oberhessische volksblaetter) – gw Mikrofilm [074]

Bad nauheimer zeitung und wetterauer zeitung *see* Wetterauer anzeiger und bad nauheimer zeitung

Bad news – v1 n1 [1970 sep 25] – 1 – mf#1584092 – us WHS [071]

Badajoz *see*
- Centanerio de colon y ferias
- Cuenta que el alcalde...de 1850
- Ejercicio de 1931. presupuesto ordinario formado para el referido ejercicio por la comision permanente y aprobado por el ayuntamiente pleno y e ilmo. sr.delegado de hacienda
- Ferias y fiestas de san juan. 1946
- Ferias y fiestas de san juan. guia de espectaculos 1953
- Fiestas de san jose. 1971
- Fiestas de san juan 1971
- Fiestas de san juan 1972
- Fiestas de san juan 1975
- Fiestas en honor de san roque 1945
- Fiestas y feria de la barriada de san roque... 1971
- Justas literarias de san juan organizadas por el excelentisimo ayuntamiento de badajoz
- Memoria presentada por la junta directiva a la asamblea general
- Memoria reglamentaria de secretaria general 1967
- Ordenanzas
- Ordenanzas de 1892
- Ordenanzas municipales
- Ordo divini officci...1828
- Presupuesto ordinario de gastos e ingresos ejercicio de 1961
- Presupuesto ordinario del interior correspondiente al ano 1951...badajoz
- Programa oficial de festjos que...con motivo de las fiestas de san juan 1969
- Programa oficial de las fiestas de san juan que...se celebraran...1968
- Programa oficial de las fiestas de san juan... 1973
- Propium festorum quae in...pacensis
- Reglamento de coro de la santa catedral de badajoz
- Reglamento para la escuela normal...badajoz
- Revista oficial de las fiestas de san juan 1979
- Tarifas y ordenanzas para la exaccion de arbitrios e impuestos municipales, aprobadas para el ejercicio economico de 1924-25

Badajoz apunte estructural y genetico / Rubio Recio, Jose Manuel – Badajoz: Imp. Diputacion Provincial, 1962 – sp Bibl Santa Ana [946]

Badajoz, Cruz Roja *see* Actuacion de la junta de senoras de la cruz roja de badajoz durante la campana de africa de 1921 y 1922

Badajoz. datos informativos / Subsecretaria de Turismo – sp Bibl Santa Ana [338]

El badajoz del siglo 16 / Guerra Guerra, Arcadio – Badajoz: imp. diputacion provincial, 1964 – sp Bibl Santa Ana [946]

Badajoz, Delegacion Provincial de Informacion y Turismo *see* Convocatoria y reglamento de la 1st asamblea provincial de cultura popular

Badajoz. Diocesis *see*
- La diocesis de badajoz. estadistica de 1970
- Ordo divini officci...proanno...1943

Badajoz. Diputacion Provincial *see* Reglamento general de funcinarios provinciales

Badajoz. Espana en Paz *see* Chronica de 25 anos

Badajoz. fiesta escolar. 1912 – Revista Semanal de la Escuela Nacional – 1 – sp Bibl Santa Ana [370]

Badajoz. Junta de Cofradias de Penitencia *see* Semana santa badajoz. 1971. programa oficial

Badajoz. Junta Diocesana *see* Obra de la propagacion de la fe

Badajoz. Santa Infancia *see* Obra pontificia de la santa infancia

Badajoz. Spain. Diputacion Provincial *see* Memoria..

Badajoz taurino / Cabanas Ventura, Felipe – Apuntes para la historia del toreo en Extremadura. 1896 – 9 – sp Bibl Santa Ana [790]

Badajoz. Tribunal Provincial de lo Contencioso-Administrativo de Badajoz *see* Pleito celebre. antecedentes. informes. sentencia dictada...en el pleito promovido por don saturnino sudon contra la resolucion del gobernador civil confirmando un acuerdo del ayuntamiento de alburquerque...

Badajoz-71 / Agenda Sindical – Madrid: Edita Ediciones Publicaciones Populares, 1971 – 1 – sp Bibl Santa Ana [946]
O badalo : orgao dedicado as pessoas que soffrem de hypocondria – Rio de Janeiro, RJ. 27 out-dez 1881 – (= ser Ps 19) – mf#P17,01,74 – bl Biblioteca [870]
Badan koordinasi perhimpunan- perhimpunan peladjar indonesia se-eropah, badan pekerdja / Madju terus – Praha, 1967(1-2) – 3mf – 9 – mf#SE-1801 – ne IDC [959]
Badan musjawarah angkatan 45 – Djakarta, 1964-1965 – 5mf – 9 – mf#SE-706 – ne IDC [959]
Badan musjawarah pengusaha nasional swasta dci djakarta raya / Aneka warta Bamunas Djaya – Djakarta, 1965 – 8mf – 9 – (missing: 1965 v1(1-5, 9, 10, 16)) – mf#SE-872 – ne IDC [959]
Badan musjawarat kebudajaan nasional : almanak seni – Djakarta, 1957 – 4mf – 9 – mf#SE-606 – ne IDC [959]
Badan musjawarat kebudajaan nasional – Djakarta, 1956-1957 nos 1-35 – 7mf – 9 – mf#SE-651 – ne IDC [959]
Badan pekerdja bulletin : badan koordinasi perhimpunan-perhimpunan peladjar indonesia se-eropah – Praha, 1966-1970(1) – 14mf – 9 – (missing: 1967(5-6); 1968(2-6)) – mf#SE-1324 – ne IDC [950]
Badan pemeriksa keuangan pemberitaan : indonesia – Djakarta, 1962-1963 – 9mf – 9 – mf#SE-1522 – ne IDC [959]
Badan penerangan persatuan mahasiswa krishnadwipajana / Dharma-budhi – Djakarta, 1953-1954 – 2mf – 9 – (missing: 1953-1954(1-3)) – mf#SE-358 – ne IDC [950]
Badan penerbit "antar nusa" – Makassar, 1958-1959 – 7mf – 9 – mf#SE-968 – ne IDC [950]
Badan penerbit karya martaco : karya; menudju ke kemerdekaan dan kemadjuan wanita – Djakarta, 1947-1950 – 2mf – 9 – (missing: 1947, v1; 1948, v2; 1949, v3; 1950, v4(1-5)) – mf#SE-899 – ne IDC [950]
Badan penerbit mangle / Mangle – Bogor, 1957-1971 – 165mf – 9 – (several issues missing) – mf#SE-921 – ne IDC [950]
Badan penerbit nasional : biwara – Jogjakarta, 1946-1947 – 3mf – 9 – (missing: 1946 v1(1-2, 4-7, 9)) – mf#SE-876 – ne IDC [959]
Badan penerbit nasional – Djakarta, April, 1950-1960 – 102mf – 9 – (several issues missing) – mf#SE-914 – ne IDC [950]
Badan penerbit pedoman / Siasat – Djakarta, 1947-1961. v1-15(1-707) – 414mf – 9 – (missing: 1947, v1(1-18, 40-42); 1948, v2(44-51, 54-78, 80-85, 88-91, 100-101, 106, 110); 1949, v3(117, 121-124); 1950, v4(164=, 189, 190); 1951, v5(193-207, 210, 213, 222, 223); 1952, v6(246-269); 1953, v7(294-295); 1954, v8(391); 1956, v10(457, 458, 463, 466, 484, 487); 1957, v11(503); 1958, v12(554); 1960, v14(690)) – mf#SE-962 – ne IDC [950]
Badan penerbit sikap / Sikap – Djakarta, 1948-1959 – 79mf – 9 – (missing: 1948, v1(2-3, 6-14, 16-21?); 1949, v2(8, 9, 18-52?); 1950, v3(1-9); 1954, v7(41(p8)-45(p7)?); 1956, v9(29-45, 48, 50); 1957, v10(24, 37); 1958, v11(13, 18, 27-38?); 1959, v12(13, 14, 16, 17)) – mf#SE-409 – ne IDC [950]
Badan penerbitan dewan nasional sobsi : bulletin sobsi – Djakarta, 1955-1956. v1-3(3/4) – 6mf – 9 – (missing: 1955, v1) – mf#SE-354 – ne IDC [959]
Badan pengawasan dan penjelenggaraan projek-projek industri laporan projek-projek dan home-office : indonesia – Djakarta, 1964 – 12mf – 9 – mf#SE-1523 – ne IDC [959]
Badan pengurusan kopra pedoman tataniaga departement perdagangan : indonesia – Djakarta, 1970 – 1mf – 9 – mf#SE-1524 – ne IDC [959]
Badan perdjoangan irian / Suara-Irian – Makassar, 1949-1954. v1-6(6) – 26mf – 9 – (missing: 1949, v1(1); 1950, v2(2, 7, 9-10, 12, 15-16, 19-24?); 1951, v3(1-2, 4-7, 10-11); 1952, v4(9-10, 13-16, 19-20); 1954, v6(2)) – mf#SE-1935 – ne IDC [959]
Badan perentjanaan pembangunan nasional bappenas = News bulletin secretariat national development planning agency / Indonesia – Djakarta 1964(1) – 1mf – 9 – mf#SE-1525 – ne IDC [959]
Badan perwakilan sementara rundingan – Makassar, 1949 – 11mf – 9 – mf#SE-214 – ne IDC [950]
Badan pimpinan perusahaan daerah laporan : indonesia – Djakarta, 1966-1967 – 2mf – 9 – mf#SE-1526 – ne IDC [959]
Badan pimpinan umum industri kimia laporan tahunan : indonesia – Djakarta, 1961 – 1mf – 9 – mf#SE-1527 – ne IDC [959]
Badan pimpinan umum perusahaan perkebunan dwi kora : kebun berdikari – Djakarta, 1965-1967 – 4mf – 9 – (missing: 1966 v1(9-11)) – mf#SE-164-1 – ne IDC [959]
Badan pimpinan umum perusahaan perkebunan karet negara laporan tahunan : indonesia – Djakarta, 1964 – 1mf – 9 – mf#SE-1528 – ne IDC [959]
Badan tenaga atom nasional madjalah journal badan tenaga atom nasional : indonesia – Djakarta, 1968-1971. v1-4(3/4) – 13mf – 9 – (missing: 1968, v1(4); 1969, v2(3-4); 1970, v3(2)) – mf#SE-1529 – ne IDC [959]
Badan urusan dagang / Madjalah niagawan negara – Djakarta, 1961. nos1-6 – 5mf – 9 – (missing: 1961(2-4)) – mf#SE-1816 – ne IDC [950]
Badan usaha penerbit almanak pertanian : almanak pertanian – Djakarta, 1953-1954 – 16mf – 9 – mf#SE-605 – ne IDC [959]
Badania nazw topograficznych starej wielkopolski / Kozierowski, Stanislaw – Poznan: Druk Uniw Poznanskiego, 1939 (mf ed 19–) – (= ser Harvard slavic humanities preservation microfilm project) – mf#ZQ-45 – us NY Public [914]
Badaro, F *see* L'eglise au bresil pendant l'empire et pendant la republique
Baddanta Singha *see* Pali pa ruit jo
Baddeley, Thomas *see* A sure way to find out the true religion
Baddiley, st michael – (= ser Cheshire monumental inscriptions) – 1mf – 9 – £2.50 – mf#368 – uk CheshireFHS [929]
Bade, Wilfrid *see*
– Gloria
– Gloria ueber der welt
– Tod und leben
Bade, Wilfried *see* Thiele findet seinen vater
Bade-anzeiger fuer sooden an der werra – Bad Sooden-Allendorf DE, 1883 25 may-1886, 1888-1915, 1920-21, 1924-1940 25 sep – 7r – 1 – (later: kuranzeiger bad sooden-werra, later: kuranzeiger bad sooden-allendorf) – gw Misc Inst [790]
Badeaux, Jean Baptiste *see* Journal des operations de l'armee lors de l'invasion du canada en 1775-76
Badekuren : novelle, kleines gefluegel: novelletten / Bluethgen, Viktor – 3. aufl. Anklam: H Wolter, [190-?] [mf ed 1993] – 93/106p – 1 – mf#8520 – us Wisconsin U Libr [830]
Baden, Hans Juergen *see* Der verschwiegene gott
Baden. Statistisches Landesamt *see* Statistisches jahrbuch fuer das land baden
Badenberg, Georg Robert *see* Sickness and healing
Badener tagblatt *see* Wochenblatt
Badener tagblatt [main edition] – Baden-Baden DE – 1 – (title varies: 24 dec 1948: badisches tagblatt. regional ed: Buehl 1946 9 jan-1981 [83r]; Gaggenau (ed murgtal) 1947 2 sep-1959, 1960 1 jul-1981 [71r]; Offenburg 1947 16 may-1979 30 nov [72r]; Rastatt 1947 2 sep-1981 [91r]) – gw Misc Inst [074]
Badener wochenblatt *see* Wochenblatt
Badenhorst, Dirk Cornelius *see* Ontmoeting en identiteit
Baden-Powell, Baden Henry *see*
– Hand-book of the manufactures and arts of the punjab
– A manual of the land revenue systems and land tenures of british india
– The origin and growth of village communities in india
– Short account of the land revenue and its administration in india
Baden-Powell, George Smyth *see*
– Protection and bad times
– The saving of ireland
Baden-Powell, R S S *see* Swaziland expedition, 1888-90; zululand campaign, 1888-90
Bader, Clarisse *see* La femme biblique
Bader, Gershom *see* R' yisrael ba'al shem tov
Bader, Josef *see* Meine fahrten und wanderungen im heimatlande
Bader, Karl Siegfried *see* Joseph von lassberg
Bader, Werner *see* Erweiterte b-bild-diagnostik in der mammasonographie mittels texturanalyse und speckle-muster-reduktion
Badgaesten – Laholm, 1911 – 1r – 1 – sw Kungliga [078]
Badgaesttidningen – Halmstad, Sweden. 1912-13 – 1r – 1 – sw Kungliga [078]
Badgaesttidningen – Halmstad, Sweden. 1913 – 1r – 1 – sw Kungliga [078]
Badgaesttidningen – Helsingborg, Sweden. 1914 – 1r – 1 – sw Kungliga [078]
The badge of conquest or the great sentimental grievance of the "irish people" – Dublin, [1870] – (= ser 19th c ireland) – 1mf – 9 – mf#1.1.1885 – uk Chadwyck [241]
Badger – v1 n1-8 (1944 jul 22-dec 25), v1 n9-11 (1945 jan 20-mar 29), v2 n1,2,3,4 (1945 apr 16, may/jun, sep 2, oct 27) [1]; 1894 oct 13-1897; 1898-01; 1902-1906 apr 7 [2]; v2 n11 (1904 dec 25) [3]; v1 n2,7 (1951 jul 15, dec 20), v1 n8-11 (1952 jan 25-may 1), v2 n11 (1956 dec) [4] – 1 – mf#1780011 [1]; 961951 [2]; 1780010 [3]; 1780085 [4] – us WHS [071]
Badger, Alfred G *see* Illustrated history of the flute
Badger american – 1923 apr-1924 oct – 1 – mf#1345277 – us WHS [071]
Badger bee – 1953 mar-1960 sep/oct – 1 – mf#469985 – us WHS [071]
Badger blade – 1902 mar 13-may 1, 1902 may 8-1903 dec 3, 1903 dec 10-1905 jun 15, 1905 jun 22-1907 jan 17, 1907 jan 24-1908 oct 15, 1908 oct 22-1910 may 19, 1910 may 26-1911 oct 5 [1]; 1920 aug 27-1921 may 26; 1921 jul 1-1924 sep 12; 1924 sep 19-1925 oct 23 [2] – 1 – mf#966767 [1]; 966779 [2] – us WHS [071]
Badger boy – 1899 mar-jun – 1 – mf#3582640 – us WHS [071]
Badger bulletin : official publication of the bow war – v1 n1-7, v2 n1-4 [1943 aug 30-1944;] + 2 undated & unnumbered bulletins; v1: special n(1943 nov 3] – 1 – mf#93761 – us WHS [071]
Badger bulletin – v17 n7-v27 n7 (1984 jan-1994 dec) [1]; 1947 mar 14-1953 oct 29; 1953 nov 5-1956 dec 21 [2] – 1 – mf#3319249 [1]; 1052783 [2] – us WHS [071]
Badger chess – Janesville WI: Badger Chess 1982-2000 [mf ed 1984-2003] – 5r – 1 – (bimthly jan/feb 1991- . jul/dec 1997-2000 have scattered pts with poor printing quality) – mf#8136 – us Wisconsin U Libr [790]
Badger common tater – v6 n7-v10 n12 [1953 jul-1958 dec] – 1 – mf#1110052 – us WHS [071]
Badger de molay – 1924 jan-1934 oct; 1934 dec-1960 dec; 1961 jan-1978 may – 1 – mf#1052785 – us WHS [071]
Badger entertainer – v1 n1-(v5 n2] [1983 dec-1988 feb/mar] – 1 – mf#1330349 – us WHS [071]
Badger express – 1940 may 25 – 1 – mf#916986 – us WHS [071]
Badger farm bureau news – 1930 nov-1949 dec; 1950-57; 1958-1966 oct 10; 1966 nov 5-1975 jun; 1975 jul-1985 – 1 – mf#1100332 – us WHS [630]
Badger folklore society newsletter – v1 n1-2 [1978 jun-aug] – 1 – mf#1053276 – us WHS [390]
Badger forty and eighter – 1972 nov-1979; 1980-1985 sep – 1 – mf#571864 – us WHS [071]
Badger, George Percy *see*
– The christians of assyria commonly called "nestorians"
– The nestorians and their rituals
Badger guardsman : intercom – v1 n1-v7 n1 [1956 oct-1963 jan] – 1 – mf#464008 – us WHS [071]
Badger herald / University of Wisconsin – Madison WI: Badger Herald Inc 1969- [mf ed 1985-] – 27r – 1 – (v25 n112 lacking some text on p4) – mf#6477 – us Wisconsin U Libr [071]
Badger jaycee / Wisconsin Junior Chamber of Commerce – 1953 jul-1962 may; 1963-1972 jan – 2r – 1 – mf#1052788 – us WHS [380]
Badger, Joseph *see*
– A memoir
– A memoir of rev joseph badger
Badger legionnaire – 1923-27; 1928-33; 1934-40; 1941-49; 1950-1956 jan; 1956-1963 aug; 1963 sep-1975 jun; 1975 jul-1982; 1983-1989 jun – 1 – mf#1110053 – us WHS [071]
Badger lutheran – 1949-54; 1955-58; 1959-64; 1965 jan 7-1969 may 8; 1969 may 22-1974 mar 8; 1974 apr 11-1978 dec 29; 1979-1981 dec 25 – 1 – mf#1278959 – us WHS [242]
Badger news – 1942 jun 18-23 – 1 – mf#3562293 – us WHS [071]
Badger ordnance news : the official publication of the badger ordnance works – v1 n1-v2 n19; v3 n1-v4 n21 [1942 apr 3-1943 aug 7; 1944 may 4-1945 sep 28] – 1 – mf#937627 – us WHS [350]
Badger ordnance world – v3 n25-v6 n11 [1955 sep 25-1958 jan 31] – 1 – mf#937629 – us WHS [071]
Badger prohibitionist – 1946 sep-1947 sep/oct – 1 – mf#931711 – us WHS [071]
Badger rails – v1 n1-2 [1980 aug/sep-oct/nov] – 1 – mf#903595 – us WHS [071]
Badger realtor : official publication of wisconsin realtors association – 1968 jun-1971 dec; 1976 jan-1979 sep – 1 – mf#967725 – us WHS [071]
Badger report : news from badger safe energy alliance – 1983 spring, fall-winter; 1984 spring-fall; 1985 summer-fall; 1986 winter/spring; 1987 spring; 1987/88 winter – 1 – mf#1672470 – us WHS [333]
Badger sportsman – 1944 aug-1952 dec; 1953-58; 1959-65; 1966-1972 aug; 1972 sep-1978; 1979-1984 jun; 1984 jul-1988 – 1 – mf#1052791 – us WHS [790]
Badger state – 1853 oct-1856; 1857 jan-1859 dec – 1 – mf#961044 – us WHS [071]
Badger state banner – 1868 apr 4-1869 dec 25; 1870-1905; 1906 jan 4-1907 may 30; 1907 jun 6-1909 mar 25; 1909 apr 1-1910 dec 8; 1910 dec 17-1912 sep 19; 1912 sep 26-1914 may 7; 1914 may 14-1916 mar 2; 1916 mar 9-1917 dec 20; 1917 dec 27-19 oct 23; 1919 oct 30-1921 aug 18-1923 may 24; 1921 aug 25-1923 may 24; 1923 may 31-1925 mar 5; 1925 mar 12-1926 mar 3 – 1 – mf#986198 – us WHS [071]
Badger state bulletin – 1956 nov-1976 aug 27 – 1 – mf#1052792 – us WHS [071]
Badger State Matchcover Club *see* Bulletin of the badger state...
Badger tales – 1946 jun 13-1961 jun 17 – 1 – mf#3562304 – us WHS [071]
Badger tidings – 1956 sep-1961; 1962 jan-1968 dec – 1 – mf#1110056 – us WHS [071]
Badgerland – 1973 dec 15-1977 nov 15; 1977 nov 30-1979 oct [1]; 1971 apr 30 [2] – 1 – mf#390659 [1]; 809004 [2] – us WHS [071]
Badgerland strider newsletter – 1983 feb-1986 jun; 1986 jul-1989 – 1 – mf#1664747 – us WHS [071]
Badgerland striders : [newsletter] – 1977 oct-1980; 1981-83 – 1 – mf#1670598 – us WHS [071]
Badham, Francis Pritchett *see* St mark's indebtedness to st matthew
Badia y Leblich, Domingo *see* Ali bey's el abassi reisen in afrika und asien in den jahren 1803 bis 1807
Badin, Adolphe *see* Jean-baptiste blanchard au dahomey.
Badin baptist church – NC. Apr 1836-Sep 1947 – 1 – $39.15 – us Southern Baptist [242]
Badin baptist church woman's missionary union – NC. Jul 1917-6 Oct 1939 – 1 – $11.25 – us Southern Baptist [242]
Badische abend-zeitung – Karlsruhe DE, 1949 23 jul-1967 30 jun – 1 – (title varies: 1 apr 1951: badische allgemeine zeitung, 1 jan 1960: allgemeine zeitung (main ed in mannheim), 1 apr 1966: suedwestdeutsche allgemeine zeitung (main ed in mannheim)) – gw Misc Inst [074]
Badische abendzeitung – Ettlingen DE, 1950, 6 feb-10 jun – 1 – (main ed in karlsruhe) – gw Misc Inst [074]
Badische abendzeitung [main edition] – Karlsruhe DE, 1949 23 jul-1967 30 jun – 1 – (title varies: 1 apr 1951: badische allgemeine zeitung, 1 jan 1960: allgemeine zeitung (ha in mannheim), 1 apr 1966: suedwestdeutsche allgemeine zeitung (ha in mannheim), regional ed: bretten 1950 6 feb-1951 30 jun, bruchsal 1950 28 feb-1951 30 jun, pforzheim 1949 2 sep-1953 31 mar) – gw Misc Inst [074]
Badische allgemeine zeitung / hardt – Karlsruhe DE, 1958 1 jul-1959 30 sep – 1 – gw Misc Inst [074]
Badische allgemeine zeitung / land – Karlsruhe DE, 1954 1 oct-1959 30 sep – 1 – gw Misc Inst [074]
Der badische gewerkschaftler : mitteilungsblatt fuer die gewerkschaften in der franzoesisch besetzten zone baden – Freiburg Br DE, 1946 jun-1947, 1949 [gaps] – 1r – 1 – mf#3868 – gw Mikropress [331]
Badische landesblaetter *see* Die biene
Badische Landes-Sternwarte zu Heidelberg [Koenigstuhl] *see*
– Veroeffentlichungen der badischen landes-sternwarte zu heidelberg [koenigstuhl]
– Veroeffentlichungen der landes-sternwarte und des planeteninstituts zu heidelberg [koenigstuhl]
badische landes-sternwarte zu heidelberg [koenigstuhl] *see* Veroeffentlichungen der badischen landessternwarte zu heidelberg [koenigstuhl]
Badische landeszeitung *see* Die biene
Badische landpost – Karlsruhe DE, 1890 18 jan-31 dec, 1898 [many gaps], 1901 [gaps] – 3r – 1 – gw Misc Inst [074]
Badische neueste nachrichten – Karlsruhe DE, 1950 4 jan-1952 10 jan, 1952 2 may-1953 30 may – 7r – 1 – (filmed by misc inst: 1954 jul-1977 jul; 1948 8 jan-5 aug [1r] 1969- [ca 9r/yr]) – gw Mikrofilm; gw Misc Inst [074]
Badische neueste nachrichten *see*
– Bruchsaler rundschau
– Woechentliches frag- und kundschaffts-blath
Badische presse – Karlsruhe DE, 1890, 1894-95, 1900-03, 1905-22, 1924-1929 30 sep, 1930-37, 1938 1 feb-30 apr, 1-31 jul, 1938 3 sep-1939 28 feb, 1939 1 apr-31 oct, 1939 1 dec-1944 31 aug – 95r – 1 – gw Misc Inst [074]
Badische rundschau – Karlsruhe DE, 1951 1 jul-1957 28 sep – 1r – 1 – gw Misc Inst [074]
Badische Sternwarte zu Heidelberg [Koenigstuhl] *see* Veroeffentlichungen der badischen sternwarte zu heidelberg [koenigstuhl]
Badische volks-zeitung *see* Woechentliches frag- und kundschaffts-blath

Badische volkszeitung / stadtausgabe [main edition] – Karlsruhe DE, 1953 1 jul-1968 31 may – 1 – (landesausgabe 1954 1 jul-1968 31 may; regional ed: mannheim 1956-1968 31 may; offenburg 1956-1968 31 may) – gw Misc Inst [074]
Badische warte – Karlsruhe DE, 1914 3 jun-1920 31 mar [gaps] – 2r – 1 – (with suppl) – gw Misc Inst [074]
Badische zeitung – Freiburg Br DE, 1947 8 jan-1954 31 mar – 1 – (covers ortenau & kinzigtal) – gw Misc Inst [074]
Badische zeitung – Freiburg im Breisgau, Germany. 1971-Oct 1973 – 17r – 1 – us L of C Photodup [074]
Badische zeitung *see*
– Karlsruher volksblatt
– Mannheimer morgenblatt
Badische zeitung 1841 – Karlsruhe DE, 1841 – 1r – 1 – gw Misc Inst [074]
Badische zeitung / laenderausgabe – Freiburg Br DE, 1949-57 – 1 – gw Misc Inst [074]
Badische zeitung [main ed] – Freiburg Br DE, 1947 8 jan-1950, 1951 2 jul-1952 30 jun, 1953 15 jan-1956 29 jun, 1960 29 jan-1963 – 19r until 1956 – 9 – (filmed by other misc inst: 1968- [ca 12r/yr]. regional & local ed: baden-baden, rastatt, buehl 1948 4 nov-1962, breisgauer nachrichten, emmendingen 1947-81, breisgauer nachrichten, emmendingen, nordlicher breisgau 1971 1 oct-1981 [45mf=90df], donaueschingen & konstanz 1947 8 jan-1948 31 oct, 1949 29 oct-1966 31, kaiserstuhl, emmendingen 1978 8 may-1981, singen 1947 19 aug-1950 1 oct, k=kenzingen [kenzinger wochenblatt] 1964 8 aug-1971 30 sep [25mf=49df], lahr, schwarzw 1949 15 oct-1954 31 mar, loerrach & das wiesental 1947 8 jan-1981, marnordbaden 1947 8 jan-23 dec [1r], markgraefler nachrichten 1949 15 oct-1981 [45r], rheinfelden, grenzach...1977 1 nov-1981, titisee-neustadt (rund um den hochfirst) later : schwarzwaelder anzeiger 1947 8 jan-1981 [89mf=177df], waldkirch 1956 2 jul-1960 31 jan [8mf=15df], 1949 15 oct-1971 30 sep, waldshut, st blasien, bonndorf...1972 1 apr-1981, weil, kandern...1973-1995 30 nov, wiesental- und hochrheinbote 1958-81, ueberlingen 1947 19 aug-1949, villingen-schwenningen: schwarzwald und baar 1947 8 jan-1948 31 oct, 1949 29 oct-1966 31 mar [29 oct 1949: villinger volksblatt], villingen-schwenningen: donau-post 1966 1 apr-1981 [1 dec 1967: ed donaueschingen/villingen, 1 apr 1972: badische zeitung / dv: bezirke donaueschingen/villingen], waldshut 1948 3 jan-1951 31 oct) – gw Misc Inst [074]
Badischer beobachter *see* Karlsruher anzeiger 1858
Badischer landesbote – Karlsruher DE, 1874 29 nov-1875 31 mar, 1876-1880 30 jun, 1881, 1883 1 jul-1885, 1887-1914 [gaps] – 1 – gw Misc Inst [074]
Badischer merkur *see* Mannheimer abendzeitung
Badisches gewerbeblatt *see* Mannheimer morgenblatt
Badisches landesprivatrecht / Dorner, Emil & Seng, Alfred – Halle (Saale): Waisenhaus, 1906 – (= ser Civil law 3 coll; Das buergerliche recht des deutschen reiches und preussens) – 8mf – 9 – (incl bibl ref and index) – mf#LLMC 96-576 – us LLMC [348]
Badisches magazin – Mannheim DE, 1811 1 aug-1812 30 may, 1813 1 jan-30 jun – 1r – 1 – gw Mikrofilm [074]
The badlands newsette – Medora, ND: Co. 2767, Camp SP8, Roosevelt State Park. v1 n1 sep 13 1935-sep 1937?// (wkly jan 2 1937-; semiwkly mar 20-dec 18 1936; wkly sep 13 1935-mar 6 1936 – 1 – (1st iss called ????. place of publ varies: medora, nd sep 13 1935-jul 10 1937; camp crook, sd jul 24 1927- . mimeographed nov 27 1936- . extra special ed publ jun 3 1937. not publ dec 4 1936, jul 17 1937) – mf#11461 – us North Dakota [071]
Badley, Brenton Hamline *see* Indian missionary directory and memorial volume
Badminton magazine of sports and pastimes – La Salle IL 1895-1923 – 1 – mf#4747 – us UMI ProQuest [790]
Badminton usa – Papillion NE 1972-76 – 1,5,9 – ISSN: 0045-1312 – mf#9072 – us UMI ProQuest [790]
Badre, Muhammad *see* The truth about islam
Badreux, Jean *see* Rectification du vocabulaire
Badruddin tyabji : a biography / Tyabji, Husain B – Bombay: Thacker & Co, 1952 – (= ser Samp: indian books) – us CRL [920]
Badstueber, Hubert *see* Friedrich von hagedorns jugendgedichte
Badstuebner, Frank *see* Netzwerkanalyse zur generierung der zustandsgleichungen
Badt, Benno Guilelmus *see* De oraculis sibyllinis a iudaeis compositis, pars 1
Badt-Strauss, Bertha *see* Annette von droste-huelshoff
Baechtold, J *see*
– Geschichte der deutschen literatur in der schweiz
– Hans salat
– Niklaus manuel
Baechtold, Jakob *see*
– Goethes iphigenie auf tauris
– Vier kritische gedichte
Baeck, Leo *see*
– Pharisees
– Wesen des judentums
Baeck, Louis *see* Etude socio-economique du centre extra-coutumier d'usumbura
Baeck, Samuel *see* Die geschichte des juedischen volkes und seiner literatur vom babylonischen exile bis auf die gegenwart mit einem anhange
Baedeker, K *see* Palestine et syrie
Baedeker, Karl *see* Rheinreise von strassburg bis rotterdam
Baedeker's handbooks for travellers – Greenwood Press – (= ser Research colls in travel and exploration) – 1891mf (20:1) – 9 – $7985.00 coll – (complete coll of 266 editions publ in english prior to ww2. with printed bibliography. subsets by publ series are also available as indicated below: ser n1: austria 26mf $220. ser n2: austria-hungary 20mf $175. ser n3: belgium & holland 95mf $570. ser n4: belgium & luxemburg 7mf $105. ser n5: the dominion of canada 20mf $175. ser n6: the eastern alps 86mf $510. ser n7: egypt: upper egypt; egypt: lower egypt; egypt & the sudan 66mf $500. ser n8: northern france 36mf $280. ser n9: southern france 56mf $365. ser n10: the riviera (&) south eastern france 9mf $115. ser n11: paris 136mf $720. ser n12: northern germany 98mf $560. ser n13: southern germany 87mf $510. ser n14: germany 9mf $115. ser n15: the rhine 119mf $605. ser n16: berlin & its environs 24mf $220. ser n17: great britain 85mf $510. ser n18: london 113mf $585. ser n19: greece 28mf $220. ser n20: northern italy 121mf $625. ser n21: central italy 101mf $560. ser n22: southern italy 98mf $550. ser n23: italy 22mf $200. ser n24: madeira 2mf $95. ser n25: norway & sweden 74mf $510. ser n26: palestine & syria 37mf $280. ser n27: russia 9mf $115. ser n28: spain & portugal 36mf $280. ser n29: switzerland 224mf $1000. ser n30: us 38mf $290. ser n31: mediterranean seaports & sea routes 9mf $115) – us UPA [910]
Die baeder am ostseestrande – Leipzig 1828 [mf ed Hildesheim 1995-98] – (= ser Fbc) – 1mf – 9 – €40.00 – 3-487-28909-1 – gw Olms [914]
Baegetr, Juan Jacobo *see* Noticias de la peninsula americana de california. introducion y notas por paul kirchhoff...
Baehler, Eduard *see*
– Erlebnisse eines schuldenbauers
– Leiden und freuden eines schulmeisters
– Nikolaus zurkinden von bern, 1506-1588
Baehnisch, Alfred *see* Die deutschen personennamen
Baehr, H A *see* Commercial precedents
Baehr, Karl Christian Wilhelm Felix *see* The books of the kings
Baehrens, Emil *see* Unedirte lateinische gedichte
Baehrens, W A *see* Ueberlieferung und textgeschichte der lateinischen origeneshomilien zum alten testament
Baehring, Bernhard *see*
– Johannes tauler und die gottesfreunde
– Thomas von kempen
Baelz, Walter *see* The gold fields of new ontario
Baenkelbuch : neue deutsche chansons / ed by Singer, Erich – Leipzig: E P Tal, 1920 [mf ed 1993] – 183p – 1 – mf#8361 – us Wisconsin U Libr [780]
Baentsch, Bruno *see*
– Altorientalischer und israelitischer monotheismus
– Exodus-leviticus-numeri
– Geschichtsconstruction oder wissenschaft?
– H st. chamberlains vorstellungen ueber die religion der semiten, spez. der israeliten
– Das heiligkeits-gesetz, lev 17-26
Baeprian khong krasuang su'ksathikan nangsuan phasa thai ru'ang thetsana su'apa / Vajiravudh, King of Siam – Phranakhon: Khurusapha 2501 [1958] [mf ed 1987] – 1r – 1 – (in thai. filmed with: ruam prakat khong khana.../ sathian wichailak) – mf#1856 – us Wisconsin U Libr [959]
Baer, Bernhard *see* Otsar sharshi leshon ha-kodesh
Baer, Heinz et al *see* Erbe und gegenwart
Baer, K E von *see* Kurzer bericht ueber wissenschaftliche arbeiten und reisen
Baer, K Z and Gr von Helmersen *see* Ueber die aelteren auslaendischen karten von russland bis 1700
Baer, Karl M *see* B'nai b'rith berlin
Baer newsletter – 1979 jan 9-1989 jan – 1 – mf#1574540 – us WHS [071]
Baerensprung, W et al *see* Monatschrift von und fuer mecklenburg
Baerle, K van *see*
– Blyde inkomst der allerdoorluchtighste koninginne
– Marie de medicis
– Marie de medicis, entrant dans amsterdam
– Medicae hospes
Baerwolf, Walther *see* Der graf von essex im deutschen drama
Baes, D *see* Scholastica commentaria in primam partem, summae theologicae s thomae aquinatis
Baesecke, Georg *see*
– Der deutsche abrogans
– Festgabe philipp strauch
– Fruehgeschichte des deutschen schrifttums
– Das glueckhafte schiff von zuerich
– Seelenwanderungen
– Die sprache der opitzischen gedichtsammlungen von 1624 und 1625
– Vorgeschichte des deutschen schrifttums
– Der wiener oswald
Baete, Ludwig *see*
– Die akte johannes schlaf
– Johann gottfried herder
Baete, Ludwig et al *see* Das johannes schlaf-buch
Baeteman, J *see* Dictionnaire amarigna-francais
Baethcke, Hermann *see* Des dodes danz
Baethgen, F *see* Cronica johannis vitodurani (mgh6:3.bd)
Baethgen, Friedrich *see*
– Evangelienfragmente
– Fragmente syrischer und arabischer historiker
– Mediaevalia (mgh schriften:17.bd)
– Die psalmen
– Tvrts mml svry
Baets, Maurice de *see* Mgr seghers, l'apotre de l'alaska
Baettbuch; ein christliche anleitung... / Wolf, H – Zuerych, Johann Wolff, 1594 – 8mf – 9 – mf#PBU-667 – ne IDC [240]
Baeume im wind : roman / Griese, Friedrich – Muenchen: A Langen/G Mueller 1937 [mf ed 2001] – 1r – 1 – (filmed with: das letzte gesicht & other titles) – mf#10489 – us Wisconsin U Libr [830]
Baeumer, Gertrud *see*
– Die deutsche frau in der sozialen kriegsfuersorge
– Die frau
– Goethe ueberzeitlich
– Goethes freundinnen
– Handbuch der frauenbewegung
– Ricarda huch
Baeumer, Suitbert *see* Geschichte des breviers
Baeumker, Alfarabi *see* Ueber den ursprung der wissenschaften
Baeumker, C *see*
– Avencebrolis (ibn gebirol fons vitae
– Die impossibilia des siger von brabant
– Witelo
Baeumker, Cl *see* Des alfred von sareshel (alfredus anglicus schrift de motu cordis
Baeumker, Cl Avencebrolis (Ibn Gebirol) *see* Fons vitae
Baeumker, Fr *see*
– Das inevitabele des honorius augustodunensis und dessen lehre
– Die lehre anselms von canterbury
Baeumker, Wilhelm *see* Das katholische deutsche kirchenlied in seinem singweisen
Baeyer, Johann Jacob *see* Ueber die strahlenbrechung in der atmosphaere
Baez, Cecilio *see* Bosquejo historico del brasil
Baez Finol, Vincencio *see* Venezuela
Baez, Paulino G *see* Poetas jovenes cubanos
Baeza, Flores *see* Cuatro poetas cubanos
Baffin, W *see* The voyages of...1612-1622
Bag Humas Universitas *see* Warta satyawatjana
Bagala, Yogesacandra *see* History of the indian association, 1876-1951
Bagatelles pour un massacre / Celine, Louis-Ferdinand – Paris: Denoel 1937 – us CRL [944]
Bagatta, J B *see* Admiranda orbis christiani
Bagce – Selanik: Asir Matbaasi, 1908-10. Mueduer ve Mueessisi: Abdurrahman Mehdi; Sahib-i Imtiyaz: Necib Necati. n41. 2 haziran 1325 [1909], 46. 10 temmuz 1325 [1909] – (= ser O & t journals) – 1mf – 9 – $25.00 – us MEDOC [956]
Bagchi, Prabodh Chandra *see*
– India and china
– Studies in the tantra, pt 1
Bagchi, Sitansusekhar *see* Inductive reasoning
Bagdat, Elise Constantinescu *see* La "querela pacis" d'erasme (1517)
Bagehot, Walter *see*
– English constitution
– Physics and politics
Bagenal, Philip Henry Dudley *see* The american irish
Der bagger – Koethen DE, 1966 jan-aug, 1967-1976 oct, 1977-1978 nov, 1979 jan-10 sep – 4r – 1 – (with gaps) – gw Misc Inst [074]
Baggs, C M *see* Letter addressed to the rev r burgess
The bagh caves in the gwalior state – London: India Society, 1927 – (= ser Samp: indian books) – (text by john marshall et al; forward [sic] by laurence binyon) – us CRL [750]
Baghdad – (= ser Vilayet salnames) – 9 – (1309 [1892] def'a 8, 1313-14k [1897] def'a 12 6mf/yr $90 per 6mf; 1316k [1898] def'a 14 5mf $75; 1317k [1899] def'a 15 5mf $75; 1318k [1900] 9mf $150; 1325 [1907] 6mf $90) – us MEDOC [956]
Baghdad news – Baghdad: Asia Print & Publ Co, jul 14 1964-dec 3 1967 – 10r – 1 – us CRL [079]
The baghdad observer – [Baghdad]: General Establishment for Press & Printing, dec 7 1967-jan/apr 1990; sep/dec 1990; may-aug-sep/dec 1992 – 1 – us CRL [079]
Baghdad press extracts – Baghdad: British Embassy Info Dept, [aug 10 1951-aug 28 1953]; [jan 25-sep 25 1954] – us CRL [070]
Baghdad times – 3 may 1921-26 feb 1923 (daily) – 3r – 1 – (imperfect) – mf#mf779 – uk British Libr Newspaper [079]
The baghela dynasty of rewah / Sastri, Hiranand – Calcutta: Govt of India, Central Publication Branch, 1925 – (= ser Samp: indian books) – us CRL [954]
Bagian bahasa, djawatan kebudajaan, kementerian po dan k – Djakarta, 1957-1959. v1-2(4) – 5mf – 9 – mf#SE-1784 – ne IDC [959]
Bagian bahasa, djawatan kebudajaan, kementerian pp dan k – Djakarta, 1951-1954. v1-4(1) – 2mf – 9 – (missing: 1951 v1; 1952 v2; 1953 v3(1-10, 12); 1954 v4(2-12)) – mf#SE-1783 – ne IDC [490]
Bagian bahasa, djawatan kebudajaan, kementerian pp dan k – Djakarta, 1951-1959 – 13mf – 9 – (missing: 1951-1953 v1-3(11); 1954 v4(1-2, 4, 7-12); 1955 v4(1-2)) – mf#SE-802 – ne IDC [959]
Bagian bahasa, djawatan kebudajaan, kementerian pp dan k – Djakarta, 1953-1959 – 22mf – 9 – (missing: 1954 v4(2-12); 1956 v1(1-2); 1958 v3(9); 1959 v4(1-2)) – mf#SE-801 – ne IDC [959]
Bagian dokumentasi, public relations, djawatan imigrasi / Warta imigrasi – Djakarta, 1950-1961 – 122mf – 9 – (missing: 1950-1951, v1-2; 1952, v3(1-11); 1953, v4(3-4, 6, 10-12); 1961, v12(6, 10)) – mf#SE-988 – ne IDC [959]
Bagian hubungan masjarakat, dpr-gr – Djakarta, May 1968-1972. v1-4(1-46) – 31mf – 9 – mf#SE-1866 – ne IDC [950]
Bagian madjalah, departemen perdagangan / Warta ekonomi untuk Indonesia – Djakarta, [1948]-1963. v1-16 – 231mf – 9 – (missing: [1948]-1951, v1-4; 1954, v7(14, 15, 25, 27, 28, 35); 1955, v8; 1959, v12(6-7); 1963, v16(41-44)) – mf#SE-306 – ne IDC [959]
Bagian penerangan kedutaan besar republik indonesia : aneka warta – London, 1951-1955 – 14mf – 9 – (missing: 1951, v1; 1952, v2; 1953, v3(1, 3, 8, 12, 15, 17-23, 26, 27); 1954, v4(29, 30, 40-end); 1955, v5(2, 3)) – mf#SE-1313 – ne IDC [959]
Bagian penjuluhan, djawatan bimbingan dan perawatan sosial, kementerian sosial ri – Jogjakarta, 1951-1963 – 3mf – 9 – (missing: 1951, v1; 1952, v2(1, 2, 5-end); 1953/1954, v3(1, 2, 4-end); 1958/1959, v6; 1959/1960, v7(3-end); 1961/1963, v8(5-8)) – mf#SE-865 – ne IDC [959]
Bagian psychologi, fakultas pedagogik, universitas gadjah mada / Warta-karya psychologi – Jogjakarta, 1961 V1(1) – 1mf – 9 – mf#SE-1991 – ne IDC [959]
Bagier, Guido Rudolf Georg *see* Das toenende licht
Baglan, glamorgan, parish church of st baglan : baptisms 1721-1925, burials 1721-1968, marriages 1721-1837 – [Glamorgan]: GFHS [mf ed c1999] – 1mf – 9 – £1.25 – uk Glamorgan FHS [929]
Baglan, st catherine's, monumental inscriptions – 2mf – 9 – £2.50 – uk Glamorgan FHS [929]
Baglione, G *see*
– Le vite de' pittori, scultori, architetti, ed intagliatori 1572-1642
– Le vite de' pittori, scultori et architetti 1572-1642
Baglivi, Georgii *see* Opera omnia medico-pratica et anatomica hac sexta editione accedit tractatus de vegetatione lapidum opus desideratum
Bagnani, Gilbert *see* Arbiter of elegance
Bagong aklat sa pilipino : aklat-sanayan / Del Rosario, Marissa E – [Manila?]: Philippine Book Co [c1969] [mf ed 1987] – 1r [ill] – 1 – mf#6774 – us Wisconsin U Libr [490]
Bagot, Daniel *see* Catechism
Bags and baggage – 1937 aug-1943 apr – 1 – mf#1052799 – us WHS [071]
Bagshawe, John B *see* The credentials of the catholic church
Baguenault de Puchesse, M Fernand *see* Histoire du concile de trente
Baguer, Mercedes *see* Mi madrastra
Baguer y Oliver, J *see* Floresta de disertaciones historico-medicas...
Baguidy, Joseph D *see*
– Considerations sur la conscience nationale
– Esquisse de sociologie haitienne
Bahadoor, Khan *see* Oudh
Baha'i magazine *see*
– Bahai news
– World unity
Baha'i news – n574-1957 8 [1979 jan-may]; n671-714 [1981 apr-1990 oct] – 1 – mf#798091 – us WHS [290]

Bahai news – Chicago IL, 1910-35 (bahai news 1910-11) [mf ed 2] – (= ser Christianity's encounter with world religions, 1850-1950) – 6r – 1 – (filmed with: star of the west 1911-22 and: baha'i magazine 1922-35. mainly in english with sects in persian) – mf#2-s196-198 – us ATLA [290]

Bahai teaching : quotations from the bahai sacred writings and several articles upon the history and aims of the teaching / Remey, Charles Mason – Washington DC: [s.n.] 1917 [mf ed 1992] – 1v on 1mf – 9 – 0-524-02038-8 – mf#1990-2813 – us ATLA [290]

Bahaism / Sell, Edward – London: Christian Literature Society for India 1912 [mf ed 1992] – (= ser The islam series) – 1mf – 9 – 0-524-02612-2 – mf#1990-3062 – us ATLA [290]

Bahaism and its claims : a study of the religion promulgated by baha ullah and abdul baha / Wilson, Samuel Graham – New York: Fleming H Revell c1915 [mf ed 1992] – 1mf – 9 – 0-524-02234-8 – (incl bibl ref) – mf#1990-2908 – us ATLA [290]

Bahaism, the modern social religion / Holley, Horace – New York: M Kennerley 1913 [mf ed 1991] – 1mf – 9 – 0-524-01832-4 – (incl bibl ref) – mf#1990-2667 – us ATLA [290]

Bahaism, the religion of brotherhood : and its place in the evolution of creeds / Skrine, Francis Henry – London, New York: Longmans, Green 1912 [mf ed 1991] – 1mf – 9 – 0-524-01382-9 – mf#1990-2394 – us ATLA [290]

Bahama argus – Nassau, Bahamas 16 jul 1831-14 jul 1832, 16 jul 1834-26 dec 1835 – 1 – uk British Libr Newspaper [079]

Bahama herald – New Providence Bahamas, 1864-29 dec 1866, 1867-24 aug 1877, 25 jan 1958, 31 dec 1960, 4 jan 1961-1 dec 1962, 1 feb, 27 jun 1964 – 15 1/2r – 1 – (aka: nassau herald) – uk British Libr Newspaper [079]

Bahama islands / Hassam, John Tyler – Cambridge, MA. 1899 – 1r – us UF Libraries [972]

Bahama islands – Nassau, Bahamas. 1926 – 1r – us UF Libraries [972]

Bahama islands / Rigg, J Linton – New York, NY. 1949 – 1r – us UF Libraries [972]

Bahama islands / Rigg, J Linton – Toronto, ON. 1951 – 1r – us UF Libraries [972]

Bahama islands / Shattuck, George Burbank – New York, NY. 1905 – 1r – us UF Libraries [972]

Bahama islands / Smith, Egbert T – Ft Myers, FL. 1950 – 1r – us UF Libraries [972]

Bahama songs and stories / Edwards, Charles Lincoln – New York, NY. 1942 – 1r – us UF Libraries [780]

Bahamas : isle of june / Bell, Hugh Maclachlan – New York, NY. 1934 – 1r – us UF Libraries [972]

Bahamas : session laws of bahamas – 1968-95 – 9 – $100.00 set – mf#402520 – us Hein [348]

Bahamas. Dept of Statistics *see* Statistical abstract 1969-1975

Bahamas guardian-news – Nassau, Bahamas 21 nov 1965-26 jun 1966 – 1/2r – 1 – uk British Libr Newspaper [079]

Bahamian folk lore / Fitz-James, James – Montreal: [s.n.] 1906 [mf ed 1999] – 1mf – 9 – 0-659-90475-6 – mf#9-90475 – cn Canadiana [390]

Bahamian interlude / Bruce, Peter Henry – London, England. 1949 – 1r – us UF Libraries [972]

Bahar – Tehran. sal-i 1, shumarah-i 1-12. 10 rabi al-sani 1328-25 zu'l qa'dah 1329 [21 apr 1970-17 nov 1911]. sal-i 2, shumarah-i 1-12. sh'aban 1339-jumadah al-avval 1341 [april 1921-dec 1922] – 1r – 1 – $325.00 – us MEDOC [079]

El bahar – Jakarta, Indonesia. 1966-1972 (1) – mf#67740 – us UMI ProQuest [079]

Bahar Allah *see*
- The book of ighan
- Hidden words

Baharistan-i-ghaybi : a history of the mughal wars in assam, cooch behar, bengal, bihar and orissa during the reigns of jahangir and shahjahan / Nathan, Mirza – Gauhati, Assam: Narayani Handiqui Historical Institute, 1936 – (= ser Samp: indian books) – (trans by m i borah) – us CRL [954]

Bahder, Karl von *see* Das lalebuch [1597]

Bahia / Tavares, Odorico – Rio de Janeiro, Brazil. 1961 – 1r – us UF Libraries [972]

Bahia (Brazil) Governor *see* Relatorios dos presidentes, 1a republica, 1892-1930

Bahia (Brazil) President *see* Relatorios dos presidentes, epoca do imperio, 1823-1889

Bahia de outrora / Querino, Manuel Raymundo – Salvador, Brazil. 1955 – 1r – us UF Libraries [972]

Bahia epigraphica e iconographica / Boccanera, Silio – Bahia, Brazil. 1928 – 1r – us UF Libraries [972]

Bahia, imagens da terra e do povo / Tavares, Odorico – Rio de Janeiro, Brazil. 1951 – 1r – us UF Libraries [972]

Bahia litteraria – Bahia: Typ do Diario da Bahia, 07 abr 1870 – (= ser Ps 19) – mf#P17,01,28 – bl Biblioteca [440]

Bahia no seculo 18 / Vilhena, Luiz Dos Santos – Salvador, Brazil. v1-3. 1969 – 1r – us UF Libraries [972]

Bahlcke, H *see* Die stellung der philanthropisten zum religionsunterricht

Bahler, Peter Benjamin *see* Electronic and computer music

Bahlke, William A *see* Chattel mortgage

Bahlmann, Paul *see* Die wiedertaeufer zu muenster

Bahlow, Ferdinand *see* Johann knipstro

The bahmanis of the deccan : an objective study / Sherwani, Haroon Khan – Hyderabad, Deccan: Manager of Publications, [1953] – (= ser Samp: indian books) – us CRL [954]

Bahmann, Reinhold [comp] *see* Matthias Claudius spricht zu uns

Bahn frei – Halle S DE, 1950, 25 feb-1995, 18 dec [gaps] – 4r – 1 – (title varies: waggonbau ammendorf) – gw Misc Inst [380]

Bahn, Karl *see* Marianne von willemer, goethes suleika

Die bahn swakopmund-windhoek / Gerding – Berlin: 1902. 406p. illus. maps – 1 – us Wisconsin U Libr [380]

Die bahn und der rechte weg des lao-tse : der chinesischen urschrift nachgedacht von alexander ular = Tao-te-king / Lao-tzu – Leipzig, 1921 (mf ed 1993) – 2mf – 9 – €31.00 – 3-89349-305-0 – mf#DHS-AR 161 – gw Frankfurter [180]

Der bahnbrecher *see* Mansfeld-echo

Bahnbrecher der deutsch reformierten kirche in der ver staaten von nord amerika *see* The pioneers of the reformed church in the united states of north america

Bahnmeester dod : en nedderduetsch drama in fief akten / Bossdorf, Hermann – Hamborg: R Hermes, 1919 [mf ed 1989] – (= ser Nedderduetsch boekerei 79) – 84p – 1 – mf#7053 – us Wisconsin U Libr [820]

Bahnwaerter thiel : novellistische studie / Hauptmann, Gerhart – Leipzig: P Reclam 1941 [mf ed 1990] – 1r – 1 – (aft by hans v huelsen. filmed with: die armseligen besenbinder / carl hauptmann) – mf#2700p – us Wisconsin U Libr [830]

Les baholoholo (congo belge) / Schmitz, Robert – Bruxelles, A DeWit [etc] 1912 – us CRL [960]

Bahosi – Mandalay, Burma. 1959-1964 – 24r – 1 – us L of C Photodup [079]

Bahr, Alice Harrison *see* College and undergraduate libraries

Bahr, Hermann *see*
- Der arme narr
- Drut
- Essays
- Fin de siecle
- Der franzl
- Die grosse suende
- Das hermann-bahr-buch
- Himmelfahrt
- Josephine
- Das konzert
- Kriegssegen
- The master
- Der meister
- "O mensch"
- O mensch!
- Die neuen menschen
- Renaissance
- Rezensionen
- Selbstbildnis
- Sendung des kuenstlers
- Spielerei
- Summula
- Die tante
- Theater
- Um goethe
- Zur kritik der moderne

Bahr, Johannes *see* Die babylonischen busspsalmen und das alte testament

Bahr-i sefid ve siyah bogazlarile marmara ve kara denizin ta'rifati – Muetercim: Faik Bey, 1294 [1878] – 3mf – 9 – $55.00 – us MEDOC [380]

Bahriye – (= ser Ministry and special interest salnames) – 9 – (1319 [1901] 4mf $60; 1326 [1908] 4mf $70; 1330 [1914] 5mf $75; 1340 [1924] 2mf $40) – us MEDOC [956]

Bahriye muezesi katalogu – [Istanbul]: Matbaa-yi Bahriye, 1917 – (= ser Ottoman histories and historical sources) – 3mf – 9 – $55.00 – us MEDOC [956]

Bahriyemiz tarihhcesi / Suekrue, Mehmet – [Istanbul]: Mertebin-i Osmaniye Matbaasi, 1328 [1912] – (= ser Ottoman histories and historical sources) – 1mf – 9 – $25.00 – us MEDOC [956]

Bahrs, Hans *see*
- Begegnung an der grenze
- Wir sind die glaeubigen

Bahtera ampera – Djakarta, 1963-1964 – 1mf – 9 – (missing: 1963 v1(1-11)) – mf#SE-340 – ne IDC [950]

Bahulikar, Balwant Narhar *see* Taraka-sangraha of annambhatta

Bai hoc ngay chua nhut : giai hia kinh – Ha Noi: [s.n. v5 n54-v7 n76 [aug 5 1928-jun 1 1930] – 1r – 1 – mf#mf-12548 seam – us CRL [079]

Bai tho cho ai / Minh Duc Hoai Trinh – [Saigon]: Thanh Truc 1974 [mf ed 1992] – on pt of 1r – 1 – mf#11052 r379 n2 – us Cornell [810]

Baian – Tambov, 1907-09 [mthly] – 16mf – 9 – us UMI ProQuest [780]

Baianul'khak – Kazan, Russia apr 1906-dec 1911 [mf ed Norman Ross] – 10r – 1 – us UMI ProQuest [077]

Baibele wa mushilo uwabamo icipingo ca kale ne cipingo cipya – London, England. 1957 – 1r – us UF Libraries [960]

Baiberi magwaro matsene amnari – London, England. 1949 – 1r – us UF Libraries [220]

Baiberi magwaro matsene amnari – London, England. 1961 – 1r – us UF Libraries [220]

Baiberi mazwi akacena amnari – London, England. 1957 – 1r – us UF Libraries [220]

Baie de samana / Justin, Joseph – Port-Au-Prince, Haiti. 1911 – 1r – us UF Libraries [972]

Baie des chaleurs railway : complete official record: official correspondence between his honor the lieutenant-governor and mr mercier, prime minister / Quebec (Province) – Montreal: Herald Co, 1891 – 1mf – 9 – mf#02212 – cn Canadiana [380]

Baie des chaleurs railway co : the facts relating to a portion of this road, viz the 60th to 80th mile, known as hogan's contract / Hogan, Michael J – Montreal?: s.n, 1893? – 1mf – 9 – mf#07236 – cn Canadiana [380]

La baie d'hudson : exploitation proposee de ses ressources de terre et de mer / Baillairge, Charles P Florent – S.l: s.n, 189-? – 1mf – 9 – mf#00088 – cn Canadiana [333]

Baie verte canal : notes respecting underground forest, etc: also synopsis of reports on baie verte canal from 1822 to 1874 – [S.l: s.n.], 1872 [mf ed 1980] – 1mf – 9 – 0-665-02213-1 – mf#02213 – cn Canadiana [627]

Baier, A H *see* Symbolik der roemisch-katholischen kirche

Baier, Adalbert *see* Das heidenroeslein

Baier, Daniel Marc *see* Ein verfahren zur etablierung von kulturen mit hohen kopienzahlen integrierter vektoren und seine anwendung in der modellierung von cml zellen

Baierische beytraege zur schoenen und nuetzlichen litteratur / ed by Westenrieder, Lorenz von – Muenchen 1779-81 – (= ser Dz) – 3 jge on 24mf – 9 – €240.00 – mf#k/n340 – gw Olms [430]

Der baierische landbot – Muenchen DE, 1790-91 – 2r – 1 – gw Misc Inst [074]

Der baierische landbote – Muenchen, Regensburg DE 1848-49 – 1 – (title varies: 25 oct 1825: der bayersche landbote, 24 apr 1827: der bayer'sche landbote, 3 jan 1832: der bayerische landbote, 1879: bayerischer landbote, fr 17 feb 1891 publ in regensburg) – gw Misc Inst [074]

Baierische national zeitung – Muenchen DE, 1807-19 – 14r – 1 – gw Misc Inst [074]

Baierisches seebuch : naturansichten und lebensbilder von den baierischen hochlandseen / Noe, Heinrich – Muenchen 1865 [mf ed Hildesheim: 1995-98] – (= ser Fbc) – viii/574p on 4mf – 9 – €120.00 – 3-487-29485-0 – gw Olms [914]

Baierland, Ortolf von *see* Aelterer deutscher 'macer' / ortolf von baierland: 'arzneibuch' / 'herbar' des bernhard von breidenbach / faerber- und maler-rezepte (cima13)

Baierlein, Eduard Raimund *see*
- The land of the tamulians and its missions
- Unter den palmen

Baiersche reise / Schrank, F von Paula von – Reston. 1978+ (1,5,9) – 4mf – 9 – mf#12791 – ne IDC [914]

The baiga / Elwin, Verrier – London: John Murray, 1940 – (= ser Samp: indian books) – (foreword by j h hutton) – us CRL [954]

Baigger, Hans-Christoph *see* Zur wissenschaftskritik bildungspolitischer analysen – unter besonderer beruecksichtigung des verhaeltnisses von kritisch-funktionalistischer und historisch-materialistischer theoriebildung

Baij Nath *see* Hinduism ancient and modern

Baiker, Armin *see* Konstruktion episomal replizierender vektoren fuer saeugetierzellen und untersuchung ihrer mitosischen stabilitaet

Baikie, James *see* Lands and peoples of the bible

Bail, P *see* Guide du cotillon

The bail reform act of 1984 / Golash, Deirdre – Washington: FJC, 1987 – 1mf – 9 – $1.50 – mf#LLMC 95-350 – us LLMC [340]

Bailados : breve noticia acerca da organizacao, da actividade e da obra da escola de bailado da prof. margarida de abreu / Lisbon. Portugal. Circulo de Iniciacao Coregrafica – Lisboa, 1948 – 1 – mf#*ZBD-*MGO pv29 – Located: NYPL – us Misc Inst [790]

Baildon, Samuel *see* The tea industry in india

Baildon, W P *see* Lincoln's inn

Baildon, W Paley *see* The coucher book of the cistercian abbey of kirkstall

Bailey, B *see* An exposition of the parables of our lord

Bailey branches – 1987 fall-1989 summer – 1 – mf#1581875 – us WHS [071]

Bailey, Cyril *see* The religion of ancient rome

Bailey, E B *see* Florida

Bailey, Edward L *see* History of the abington baptist association from 1807-1857

Bailey, Gilbert Stephen *see* History of the illinois river baptist association, and of its churches

Bailey, Harold Walter *see* The content of indian and iranian studies

Bailey, Henry *see* Analysis of blackstone's commentaries on the laws of england

Bailey, J D *see* Reverends philip mulkey and james fowler

Bailey, J W *see* Algae

Bailey, James *see*
- History of the seventh day baptists, 1802-1865
- History of the seventh-day baptist general conference

Bailey, James F *see*
- Federal rules of evidence
- Immigration and nationality acts

Bailey, Jeffery T *see* Free testosterone/cortisol responses to short term high-intensity resistance exercise overtraining

Bailey, John *see*
- Comfort for the feeble-minded
- Growth in grace

Bailey, John Read *see* Mackinac, formerly michilimackinac

Bailey, Joseph Whitman *see* The st john river in maine, quebec, and new brunswick

Bailey, L H *see* Florida plant immigrants

Bailey, Lawrence D *see* Collection

Bailey, Loring Woart *see*
- Notes on the geology and botany of digby neck
- On the acadian and st lawrence water-shed
- Report of explorations and surveys in portions of the counties of carleton, victoria, york and northumberland, new brunswick, 1885
- Report of explorations and surveys in portions of york and carleton counties, new brunswick
- Report on the pre-silurian (huronian) and cambrian
- Some nova scotian illustrations of dynamical geology
- The study of natural history and the use of natural science museums
- Triassic (?) rocks of digby basin

Bailey, Loring Woart et al *see* Report on the geology of southern new brunswick

Bailey, Mark L *see* Caffeine

Bailey on blackstone, 1822 *see* Analysis of blackstone's commentaries on the laws of england

Bailey, Robert E *see* The margaret cross norton working papers, 1924-1958

Bailey, Robert Wayne *see* Analysis of bartok's sixth string quartet

Bailey, Rufus W *see* Domestic duties of the family

Bailey, Rufus William *see* Scholar's companion

Bailey, Samuel Wordswoth *see* Homage of eminent persons to the book

Bailey, Silas *see* The american baptist preaching of the seventeenth and eighteenth centuries

Bailey, Solon Irving *see* The history and work of harvard observatory, 1839 to 1927

Bailey, T G *see*
- Kanauri vocabulary
- The languages of the northern himalayas

Bailey, Thomas Grahame *see* A history of urdu literature

Bailey, walter k, papers, ms 4665 – 1898-1970 – 1r – 1 – (family history, genealogy, and biographical information) – us Western Res [929]

Bailey, Wellesley Crosby *see*
- Glimpse at the indian mission-field and leper asylums in 1886-1887 [microform]
- The lepers of our indian empire

Bailey's cases in equity / South Carolina. Supreme Court – 1v. 1830-1831 (all publ) – (= ser Pre-nrs nominative equity reports) – 7mf – 9 – $10.50 – mf#LLMC 94-029 – us LLMC [342]

Bailey's law reports / South Carolina. Supreme Court – v1-2. 1828-1832 (all publ) – (= ser Pre-nrs nominative law reports) – 16mf – 9 – $24.00 – mf#LLMC 94-016 – us LLMC [340]

Bailie – [Scotland] Glasgow: printed by W Munro 23 oct 1872-28 apr 1926 (wkly) [mf ed 2003] – 2794v on 112r – 1 – uk Newsplan [072]

Bailin, Israel Ber *see* Alts in eyn lebn

Baillairge, Charles *see*
- Anticosti en 1900
- Rapport de m baillairge
- La vie, l'evolution, le materialisme

Baillairge, Charles P Florent *see*
- 20 ans apres, le club des 21 en 1879
- The abstract and concrete in education
- Adresse de bienvenue par m baillairge a la section de montreal des architectes du canada
- L'antiquite de la terre et de l'homme
- The aqueduct, quebec

– Asphalt block pavement
– La baie d'hudson
– Bribery and boodling, fraud, hypocrisy and humbug
– Clef du nouveau systeme de toiser tous les corps-segments, troncs et onglets de ces corps par une seule et meme regle...
– Clef du tableau stereometrique baillairge
– Clef synoptique
– Corporation de quebec, aux entrepreneurs d'acqueducs
– Dam construction
– Description et plan d'un nouveau calorifer a air chaud
– Dictionnaire d'homonymes, rimes, etc
– Divers
– Educational
– Etude ayant trait a la solution du probleme de determiner la hauteur atteinte par un projectile qui en retombant au niveau dont il a ete lance, a produit un effet connu
– Fires and fire-proof construction
– The free and liberal ventilation of sewers in its relation to the sanitation of our buildings
– Geometrie, toise et le tableau stereometrique
– Geometry, mensuration and the stereometrical tableau
– Le grec, le latin
– How best to learn to speak or teach a language
– Instructions to architects submitting competing designs for the new city hall, quebec
– Key to baillairge's stereometrical tableau
– Lettre microforme
– Masonry dams and retaining walls in general, concrete works, etc
– Memoire lu par l'auteur c baillairge devant la societe royale du canada
– Memoires lus devant la societe royale du canada 1882 & 1883
– Mr baillairge's address of welcome to the montreal section of canadian architects
– The navigation of the air
– New system of cubing any body by one and the same rule for elementary schools
– Nouveau dictionnaire d'homonymes, rimes, etc
– Nouveau dictionnaire francais, systeme "educationnel"
– Nouveau traite de geometrie et de trigonometrie rectiligne et spherique
– On the bearing and resisting strength of structures
– On the necessity of a school of arts for the dominion
– A paper read before the royal society of canada and before the can soc of civil engineers
– Papers read before the royal society of canada, 1882 and 1883
– A practical solution of the great social and humanitarian problem
– The progress of the nineteenth century
– The quebec land slide of 1889
– Radeau de sauvetage baillairge-hurly
– Rapport de la societe de geographie de quebec
– Rapport de l'ex-ingenieur de la cite, des travaux faits sous le maire...et durant le dernier tiers de siecle, 1866 a 1899
– Rapport de m baillairge
– Rapport du chevalier c baillairge
– Report of charles baillairge, engineer of the city of quebec
– Report of mr baillairge, engineer sic of the new quebec aqueduc
– Report of the city engineer, quebec
– Section de 10 milles
– Le stereometricon
– The stereometricon
– A summary of papers read at different times before the royal society of canada, the canadian association of civil engineers
– Supplementary report of the quebec corporation engineer of the north shore railway
– Technical education of the people in untechnical language
– Tentative de deduire des effets d'une explosion de chaudiere a vapeur, la pression par pouce carree sous laquelle la chaudiere a cede
– La ventilation libre des egouts en rapport avec l'hygiene de l'habitation
– Vocabulaire des homonymes simples de la langue francaise
– Vocabulary of english homonyms
– Why tidal energy

Baillairge, Frederic Alexandre *see* Ca et la
Baillairge, Frederic-Alexandre *see*
– A propos d'education
– Biografia del sir georges etienne cartier
– Le couvent
– Dictionnaire des verbes irreguliers et defectifs de la langue francaise
– Institutions de joliette
– La nature, la race, la sante
– La paroisse
– Son excellence mgr dom henri smeulders a joliette

Baillairge, Frederic-Alexandre [comp] *see* Philosophie
Baillairge, George Frederick *see*
– Alphabetical record
– Canada and newfoundland, etc- chronology
– Le canada de l'atlantique au pacifique et a la mer polaire, expeditions arctiques et voyages de decouverte au nord, etc, etc
– Esquisses biographiques
– Genealogie et notes historiques, etc
– Notices biographiques
– Tables donnant l'etendue et le progres de divers travaux publics, les distances, etc

Baillairge, Maurice *see* Derniers adieux de graziella
Baillairge's marine revolving steam express – S.l: s.n, 1897? – 1mf – 9 – mf#57832 – cn Canadiana [621]
Baillarge, George Frederick *see*
– Canada et terreneuve etc
– Canada from the atlantic to the pacific and arctic oceans, arctic voyages of discovery in the north and public works, etc, etc

Baillargeon, Cecile *see* Bibliographie analytique du rhumatisme articulaire aigu d'apres la documentation de la bibliotheque medicale de l'hopital du st sacrement et couvrant la periode de 1948-1952
Baillargeon, Charles-Francois *see* Circulaire
Baillaud, Emile *see* Sur les routes du sudan
Baillaud, Rene et al *see* Annales francaises de chronometrie
Baillet, Jules *see* Introduction a l'etude des idees morales dans l'egypte antique
Bailleul, Jacques C *see*
– Bibliomappe
– Histoire de napoleon

Bailleul, Louis *see*
– Les chasseurs de fourrures
– Les secrets de la maison blanche

Baillie, Joanna *see*
– Metrical legends of exalted characters
– Miscellaneous plays

Baillie, John *see*
– Rivers in the desert
– St augustine

Baillie, Laureen [comp] *see*
– American biographical archive (aba). supplement
– American biographical archive. series 2
– American biographical archive to 2
– Canadian biographical archive
– Indian biographical archive (india, pakistan, bangladesh, sri lanka)
– Skandinavisches biographisches archiv [sba]. neue folge

Baillie, Marianne *see*
– First impressions on a tour upon the continent in the summer of 1818
– Lisbon in the years 1821 [eighteen hundred and twenty-one], 1822 [eighteen hundred and twenty-two], and 1823

Baillie, Mrs., Marianne *see* Trifles in verse
Baillie-Grohman, William Adolph *see* Camps in the rockies
Baillieston, tollcross and shettleston express – [Scotland] Glasgow: G Hutcheson 31 mar 1896-13 mar 1900 (wkly) [mf ed 2003] – 207v on 2r – 1 – uk Newsplan [072]
Bailly, Jean Sylvain *see* Memoires de bailly
Bain, Alexander *see*
– The emotions and the will
– English composition and rhetoric
– John stuart mill
– Mental and moral science
– Moral science
– On the study of character
– Philosophical remains of george croom robertson

Bain, Andrew Geddes *see* Journals
Bain, Francis *see*
– Birds of prince edward island
– The natural history of prince edward island

Bain, James *see* Sketch of the history of st andrew's lodge of ancient, free and accepted masons
Bain, John A *see*
– The developments of roman catholicism
– The new reformation

Bain, John Wallace *see* God's songs and the singer
Bain, R Nisbet *see* Cossack fairy tales and folk-tales
Bainbridge, Harriette S *see* Life for soul and body
Bainbridge, William Folwell *see* Along the lines at the front
Baines, Edward *see* An address to the unemployed workmen of yorkshire and lancashire
Baines, KC *see* Fitness levels of children in north carolina
Baines, Peter Augustine *see*
– Inquiry into the nature, object and obligations of the religion of...
– Outlines of christianity
– Remonstrance in a third letter, addressed to charles abel moysey

Baines, T *see* Explorations in south-west africa
Baines, Thomas *see*
– Explorations in south-west africa
– History of the commerce and town of liverpool
– Yorkshire, past and present

Bainton, Roland Herbert *see* Early christianity
Bainvel, Jean Vincent *see*
– De magisterio vivo et traditione
– De scriptura sacra
– De vera religione et apologetica
– La foi et l'acte de foi

Baird, Annie Laurie Adams *see* Daybreak in korea
Baird, Charles Washington *see*
– A chapter on liturgies
– Eutaxia
– Histoire des refugies huguenots en amerique
– History of the huguenot emigration to america, vol 1
– History of the huguenot emigration to america, vol 2
– History of the huguenot emigration to america, vols 1 and 2

Baird, Frank *see* A short history of the presbyterian church in the parish of chipman
Baird, George W *see* Papers
Baird, Henry Martyn *see*
– History of the rise of the huguenots of france
– The huguenots and henry of navarre
– The huguenots and the revocation of the edict of nantes
– The life of the rev. robert baird, dd
– Modern greece
– The odore beza

Baird, John *see* Question of scientific torture
Baird, Julianne *see* Vocal serenata of the seventeenth and eighteenth centuries
Baird, Robert *see*
– Greek-english word-list
– Impressions and experiences of the west indies and north america in 1849
– The progress and prospects of christianity in the united states america
– Religion in america
– State and prospects of religion in america
– L'union de l'eglise et de l'etat dans la nouvelle-angleterre

Baird, Samuel John *see*
– A bible history of baptism
– The discussion on reunion
– The elohim revealed in the creation and redemption of man
– A history of the early policy of the presbyterian church in the training of her ministry, and of the first years of the board of education
– A history of the new school

Baird, Samuel John [comp] *see* A collection of the acts, deliverances, and testimonies of the supreme judicatory of the presbyterian church
Baird, Spencer Fullerton *see* The water birds of north america
Baird trust – Glasgow, Scotland. 1873 – 1r – us UF Libraries [240]
Baireuther politische zeitung *see* Bayreuther zeitung
Der bairische hiesel : volkserzaehlung aus bairen / Schmid, Herman – 2.aufl. Leipzig: Keil [18–?] – (= ser Gesammelte schriften. volks- und familien-ausgabe 9) – 1 – (bound with: der habermeister and sueden und norden) – mf#film mas c438 – us Harvard [390]
Bairn's bible / Stead, W T – London, England. 1900? – 1r – us UF Libraries [220]
O bairro de santana / Torres, Maria Celestina Teixeira Medndes – Sao Paulo, Brazil. 1970 – 1r – us UF Libraries [972]
Baiser au porteur / Scribe, Eugene – Paris, France. 1828 – 1r – us UF Libraries [440]
Baiser de l'aieul : piece en trois actes / Hippolyte, Dominique – Paris, France. 1924 – 1r – us UF Libraries [440]
Baisio, Guido de *see* Apparatus in librium 6 (siecle 14)
[Baja california-] la voz de la frontera – CA. nov 15 1886-nov 17 1888 – 1r – 1 – $60.00 – mf#C02039 – us Library Micro [071]
Baja California Sur. (Territory) *see* Boletin oficial
Baja-abc – 1974- – 26+ r – 1 – $1300.00 (subs $70/yr) – mf#R04159 – us Library Micro [079]
Bajareque / Celestrin, Heliodoro G – Habana, Cuba. 1951 – 1r – us UF Libraries [972]
Baji prabhou : a poem / Ghose, Aurobindo – Pondicherry: Arya Office, 1922 – (= ser Samp: indian books) – us CRL [810]
Bajo el chubasco / Izaguirre, Carlos – Tegucigalpa, Mexico. v1-2. 1945 – 1r – us UF Libraries [972]
Bajo el tiempo dificil : antologia de escritos de jose antonio / Primo de Rivera, Jose Antonio, Marques de Estella – [Barcelona]: Ediciones Arriba 1938 – 9 – mf#w722 – us Harvard [320]
Bajo la egida del generalisimo / Ducoudray, J H – Ciudad Trujillo, Dominican Republic. 1939 – 1r – us UF Libraries [972]
Bajo la garra / Abril Amores, Eduardo – Santiago, Cuba. 1922 – 1r – us UF Libraries [972]
Bajo la noche enferma / Meyreles Soler, Rafael – Ciudad Trujillo, Dominican Republic. 1946 – 1r – us UF Libraries [972]
Bajo las alas del aguila / Rodriguez Cerna, Jose – Guatemala, 1942 – 1r – us UF Libraries [972]
Bajo los austrias. la mujer espanola en la minerva literaria castellana / Perez de Guzman, Juan – Madrid: Escuela Tipografica Salesiana, 1923 – 1 – sp Bibl Santa Ana [440]
Bajo los cedros en flor / Ramirez Saizar, J – San Jose, Costa Rica. 1959 – 1r – us UF Libraries [972]
Bajo su mirada / Planchart, Enrique – Caracas, Venezuela. 1954 – 1r – us UF Libraries [972]
Bajracharya, M B *see* laswr buddhist sanskrit manuscripts (from nepal)
Bake, R W J C *see* Doorgraving der landengte van suez...
Baked and snack foods marketer – Minnetonka MN 1968 – 1,5,9 – ISSN: 0522-036X – mf#5920 – us UMI ProQuest [660]
Baker, Austin Hart *see* The american farmer's pictorial cyclopedia of live stock
Baker, Benjamin *see* The forth bridge
Baker city herald (baker city, or) – Baker City OR: [s.n.] -1911 [daily ex sun] – 1 – (cont: evening herald (baker city, or); cont by: baker herald (baker, or)) – us Oregon Lib [071]
Baker city herald (baker city, or: 1901) – Baker City OR: C W Hill, -1902 [daily ex sun] – 1 – (cont by: herald (baker city, or)) – us Oregon Lib [071]
Baker city herald (baker city, or: 1990) – Baker City OR: Baker City Herald Pub Co, 1990- [daily ex sat & sun] – 1 – (cont: democrat-herald (baker, or)) – us Oregon Lib [071]
Baker city weekly herald – Baker City OR: W S James, -1875 [wkly] – 1 – us Oregon Lib [071]
Baker county press – Macclenny, FL. 1931-1997 – 52r – (gaps) – us UF Libraries [071]
Baker county record – Baker OR: G L Jett, -1931 [semiwkly] – 1 – (absorbed: huntington news, pine valley herald; cont by: baker daily record) – us Oregon Lib [071]
Baker county record *see* Pine valley herald
Baker county reveille – Baker City OR: M H Abbott & Sons, -1887 [wkly] – 1 – (cont by: weekly reveille (1887-)) – us Oregon Lib [071]
Baker daily record – morning ed. Baker OR: Record Pub Co, 1931- [daily ex mon] – 1 – (cont: baker county record (1927-31)) – us Oregon Lib [071]
Baker democrat-herald – Baker OR: Baker Democrat-Herald Co, 1929-63 [daily ex sun] – 1 – (merger of morning democrat (-1929), evening baker herald (1928-29); cont by: democrat herald (1963-90)) – us Oregon Lib [071]
Baker, Frances J *see*
– First women physicians to the orient
– The story of the woman's foreign missionary society of the methodist episcopal church
– The story of the woman's foreign missionary society of the methodist episcopal church, 1869-1895

Baker, Franklin *see* Church establishment anti-christian
Baker, George Pierce *see* The principles of argumentation
Baker herald – Baker OR: Baker Herald Co, 1911-28 [daily ex sun] – 1 – (cont: baker city herald (-1911); cont by: evening baker herald (1928-29)) – us Oregon Lib [071]
Baker herald – Baker, OR : Baker Herald Co. v7 n195 (feb 20 1911)-v27 n304 (apr 7 1928). 1911-28 – 1 – (cont: baker city herald. cont by): evening baker herald) – us Oregon Hist [071]
Baker herald – Baker, ND: W A Kask, 1913; -v2 n29 jul 22 1915 (wkly) – 1 – (missing: 1914 aug 27; 1915 may 20) – mf#08260 – us North Dakota [071]
Baker, Herbert *see* Cecil rhodes
Baker, Ira Osborn *see* Engineers' surveying instruments
Baker, James *see* Evolution of mind
Baker, James Heaton *see* The sources of the mississippi
Baker, John Clapp *see* Baptist history of the north pacific coast
Baker, John Gilbert *see* Handbook of the irideae
Baker, John Kestell *see* Necessity of nature conservation legislation and the enforcement thereof in the gauteng province
Baker, Judith A *see* An evaluation of the effects of a smoking prevention program on middle school students' knowledge and attitudes concerning cigarette smoking
Baker, Kathy *see* Reimbursement of occupational therapy and physical therapy in hand rehabilitation
Baker, L Rebecca *see*
– Negro churches
– Negro education
– Negro ethnography
– Negro music

Baker, Lewis Carter *see* The fire of god's anger
Baker, Maude E *see* Pen symphonies
Baker, Naaman Rimnon *see* Constancy, and other poems
Baker newsletter – v1-v2 n2 [1976 aug-1977? sep?] – 1 – mf#379631 – us WHS [071]
Baker, Osmon Cleander *see* A guide-book in the administration of the discipline of the methodist episcopal church
Baker, Ray Stannard *see*
– Following the color line
– Papers

Baker, Richard St Barbe *see* Tambours africains
Baker, S W *see*
- The nile tributaries of abyssinia
- Die nilzufluesse in abessinien...

Baker, Samuel White *see*
- Exploration of the nile tributaries of abyssinia
- Ismailia

Baker, Shirley *see* Premier's (shirley baker's) letterbooks, 1873-74, 1880-90
Baker, Shirley W *see* English and tongan vocabulary
Baker, Shirley W et al *see* Tongan papers
Baker street journal – Bronx NY 1989+ – 1,5,9 – ISSN: 0005-4070 – mf#16875 – us UMI ProQuest [420]
Baker, Thomas E *see* A primer on the jurisdiction the u.s. courts of appeals
Baker, William *see* A plain exposition of the thirty-nine articles of the church of england
Baker, William King *see* A quaker warrior
Baker, William Mumford *see*
- The life and labours of the rev. daniel baker, d.d.
- The ten theophanies

Baker's and confectioner's journal – 1890 may 3-1895 apr; 1895 may 4-1897 jun 30; 1897 jul 15-1899; 1900-12; 1913-1915 aug 7; 1915 aug 14-1918; 1919-68; 1969 jan-oct; 1969 dec-1972 – 1 – mf#1388686 – us WHS [640]
Bakers' and confectioners' journal – 1888-1972 – (= ser Labor union periodicals, pt 3: food and agricultural industries) – 23r – 1 – $4340.00 – 1-55655-608-X – us UPA [660]
Bakers' journal – New York: [s.n.] v11 n1-4 may 4-25 1895 – us CRL [660]
Bakers' journal and deutsch-amerikanische backer-zeitung – Brooklyn, NY: [s.n.], 1895- [v21 n1-55] aug 13 1904-aug 26 1905. v29 n41-51 may 22-jul 31 1915 – us CRL [660]
Bakers review : the national monthly of bakery management – Minnetonka MN 1964-68 – 1 – mf#1667 – us UMI ProQuest [660]
Bakersfield – 1920-33; 1992- – (= ser California telephone directory coll) – 16r – 1 – $800.00 – mf#P00007 – us Library Micro [917]
Bakersfield news observer – 1992 dec 23-1993 jun 30; 1993 jul 7-dec 29; 1994 jan 5-jun 29; 1994 jul 6-dec 28; 1995 jan 4-jun 28; 1995 jul 5-dec 27; 1996 jan 3-dec 25; 1997 jan 1-jun 25; 1997 jul 2-dec 31; 1998 jan 7-jun 24; 1998 jul 1-dec 30 – 1 – mf#2622606 – us WHS [071]
[Bakersfield-] the bakersfield californian – CA. 1945-1950- – 78r – 1 – $4680.00 (subs $1500/y) – mf#C02040 – us Library Micro [071]
[Bakersfield-] the colored citizen – CA. apr-oct 1914 – 1r – 1 – $60.00 – mf#C02041 – us Library Micro [071]
[Bakersfield-] union labor journal – CA. oct 1943-1949 – 6r – 1 – $360.00 – mf#C02042 – us Library Micro [331]
Bakery and Confectionery Workers International Union of America *see* Local 37 news
Bakery and confectionery workers' international union of america. b and c news – Kensington MD 1973-78 – 1 – ISSN: 0001-043X – mf#8743 – us UMI ProQuest [331]
Bakery and Confectionery Workers' International Union of America et al *see* Local 3 bakery workers news
Bakery, Confectionery and Tobacco Workers International Union *see* Voice of local two
Bakery production and marketing – New York NY 1981-98 – 1,5,9 – ISSN: 0005-4127 – mf#12973 – us UMI ProQuest [660]
Bakewell, Robert *see* Travels
Bakh, Antonov et al *see* Sbornik statei
B'akharith hayamim = In the last days / Lesser, Abraham Jacob Gershon – Chicago, IL. 1897 – 1r – us UF Libraries [939]
Bakhtar – Isfahan. sal-i 1, shumarah-i 1-6. azar 1312-mihr va aban 1313 [nov 1933-sep,oct 1934]. sal-i 2, shumarah-i 1-11. azar 1313-mihr 1314 [nov 1934-sep 1935] – 1r – 1 – $395.00 – us MEDOC [079]
Bakhtar-i imruz – [Tehran:] Jabhah-'i Milli-i Iranyan-i muqim-i kharijah. dawrah-'i jadid, sal-i 1-2, shumarah-'i 2-3,5-30,32,34-35; dawrah-'i 3, shumarah-'i 1,3-5,7-17,20. 15 farvardin 1340-aban 1345 [4 apr 1961-nov 1966] – 1r – 1 – $53.00 – us MEDOC [079]
Bakhtar-i imruz, 1949-1953 – 7r – 1 – mf#NE-1805 – ne IDC [956]
Bakhtiarov, A A *see* Istoriia knigi na rusi
Baki *see* The divan project
Baking – Itasca IL 1989 – 1,5,9 – (cont: baking industry) – ISSN: 1041-3693 – mf#1583.01 – us UMI ProQuest [660]
Baking industries journal – Croydon UK 1968-80 – 1,5,9 – ISSN: 0005-4151 – mf#10263 – us UMI ProQuest [660]
Baking industry – Itasca IL 1963-87 – 1,5,9 – (cont by: baking) – ISSN: 0005-416X – mf#1583.01 – us UMI ProQuest [660]
Baking powders / Miller, H K – Lake City, FL. 1900 – 1r – us UF Libraries [660]
Bakinskaia stachka v dekabrie 1904 g – [s.l.]: Izd Iskry [1905?] – 1 – mf#80 – us Wisconsin U Libr [947]
Bakinski rabochii – Baku. 1955-1960. Incomplete – 1 – us NY Public [999]
Bakke, P Q *see* Development and evaluation of an interpretive nineteenth century american children's games program
Bakken, Angela J *see* A comparison of energy cost during forward and backward exercise on the precor c544 transport
Bakker, Debra L *see* Religious practices and high-risk behaviors of college student athletes and nonathletes
Bakker, F L *see* Jezus in de islam
Les bakongo dans leurs legendes / Struyf, Ivon – Bruxelles: G van Campenout, 1936 – 1 – us CRL [390]
Bakoulou : audience folklorique / Chevallier, Andre Fontanges Felicite – Port-Au-Prince, Haiti. 1950 – 1r – us UF Libraries [972]
Baksh, Ahmad *see* Correspondence 1914-1918
Bakst, N I *see* Letuchie listki
Bakst, V I *see* Letuchie listki
[Baku-] vyshka – Azerbaijan, USSR. 1971-1981 – 11r – 1 – $550.00 – mf#B63587 – us Library Micro [077]
Bakunin, Mikhail Aleksandrovich *see* God and the state
Die bakunisten an der arbeit / Engels, Friedrich – Leipzig – 1 – gw Mikropress [335]
[Bakusk-] bakiuskii rabochi – USSR. 1961-1981 – 21r – 1 – $1050.00 – mf#B63586 – us Library Micro [077]
Bal champetre au cinquieme etage : ou, rigolard che / Gregoire, Achille – Bruxelles, Belgium. 1830 – 1r – us UF Libraries [440]
Bal gangadhar tilak : his writings and speeches / Ghose, Aurobindo – Madras: Ganesh & Co, 1919 – (= ser Samp: indian books) – us CRL [954]
Balaban, Majer *see*
- Skizzen und studien zur geschichte der juden in polen
- Zur geschichte der juden in polen

Balabanov, M *see* Istoriia rabochei kooperatsii v rossii
Balabanov, M S *see* Obshchee uchenie o kooperatsii
Balabish / Wainright, G A – London, 1920 – (= ser Mees 37) – 6mf – 8 – €14.00 – ne Slangenburg [930]
Baladas espanolas / Barrantes Moreno, Vicente – 1865 – 9 – sp Bibl Santa Ana [946]
Baladas y canciones / Dario, Ruben – Madrid, Spain. 1923 – 1r – us UF Libraries [972]
Balade si idile / Cosbuc, George – Bucharest, Romania. 1964 – 1r – us UF Libraries [960]
Baladhuri, Ahmad Ibn Yahya' *see* Kitab futuh al-buldan
Balagh 'askari raqm : communiques of military operations / Fath (Organization) – [Jordan?]: Harakat al-Tahrir al-Watani al-Filastini "Fath", [85-424 (1967-1969)] (irreg) – 1r – 1 – us CRL [320]
Balaguer, G *see* Epidemia de tercianas...en varios pueblos de urgel...en 1785
Balaguer, Joaquin *see*
- Dominican reality
- Historia de la literatura dominicana
- Letras dominicanas
- Proceres escritores
- Realidad dominicana

Balai kursus tertulis ganaco / masa baru – Bandung, [196?] – 2mf – 9 – mf#SE-343 – ne IDC [950]
Balai pembangunan daerah – Djakarta, 1957-1958. v1-2(12) – 26mf – 9 – mf#SE-1956 – ne IDC [950]
Balai penelitian pendidikan bulletin bpp / Institut Keguruan dan Ilmu Pendidikan, Jogjakarta – Jogjakarta, 1968 – 1mf – 9 – mf#SE-164=9 – ne IDC [959]
Balai penerangan, markas tertinggi tentara repoeblik indonesia / Madjallah Tentara Repoeblik Indonesia – Jogjakarta, 1946 – 4mf – 9 – (missing: 1946 v1(2)) – mf#SE-594 – ne IDC [959]
Balai penjelidikan perusahaan2 gula / Madjalah perusahaan gula – Pasuruan, 1964-1968. v1-4(1/2) – 15mf – 9 – mf#SE-1808 – ne IDC [950]
Balai Pustaka *see* Kunang-kunang
Balai pustaka : bina pantjasila – Djakarta, 1966-1967 – 29mf – 9 – (missing: 1966 v1(12)) – mf#SE-984 – ne IDC [950]
Balai pustaka : indonesia; madjalah seni dan kebudajaan – Djakarta, 1949-1950 – 18mf – 9 – mf#SE-657 – ne IDC [959]
Balai teknologi makanan pewarta pusat djawatan pertanian rakjat – Djakarta, 1955-1956 – 6mf – 9 – mf#SE-836 – ne IDC [950]
Balaiada / Correa, Viriato – Sao Paulo, Brazil. 1927 – 1r – us UF Libraries [972]
Balakrishna, Ramachandra *see* Industrial development of mysore
Balan, Pietro *see* Clementis 7. epistolae per sadoletum scriptae
Balan, Robert *see* Materialeigenschaften von schutzhandschuhen und ihre beschaedigungen im verlaufe kieferorthopaedischer behandlungen
Balance – 1865 may 20 – 1 – mf#845944 – us WHS [071]
Balance / Camps, David – Habana, Cuba. 1964 – 1r – us UF Libraries [972]
Balance : the fundamental verity / Smith, Orlando Jay – Boston: Houghton, Mifflin 1904 [mf ed 1986] – 1mf – 9 – 0-8370-6385-X – (incl ind) – mf#1986-0385 – us ATLA [210]
Balance – [Northern Ireland] Belfast 3 dec 1850-mar 1851 [mf ed 2004] – 1r – 1 – uk Newsplan [072]
The balance : or, moral arguments for universalism / Mayo, Amory Dwight – Boston: BB Mussey and A Tompkins, 1847 – 1mf – 9 – 0-524-06431-8 – mf#1991-2553 – us ATLA [240]
Balance and state journal – La Salle IL 1802-11 – 1 – mf#3680 – us UMI ProQuest [320]
Balance control in bipedal animals : emg and kinematic analysis in chicks, implications for human balance control / Kato, Nobutaka – 1997 – 1mf – 9 – $4.00 – mf#PSY 2034 – us Kinesology [612]
Balance of external payments of puerto rico, 1942- / Sammons, Robert Lee – Rio Piedras, Puerto Rico. 1948 – 1r – us UF Libraries [336]
The balance of external payments of the gold coast for the fiscal years 1936/37 to 1938/39 – [Bristol, 1942] – us CRL [336]
Balance of payments statistics. yearbook – Washington DC 1946+ – 1,5,9 – (cont: balance of payments yearbook) – ISSN: 0252-3035 – mf#6527.01 – us UMI ProQuest [317]
Balance of payments. yearbook / International Monetary Fund – v1-22. 1938-69 – 1 – $240.00 – mf#0287 – us Brook [332]
Balance sheet – Mason OH 1919-93 – 1,5,9 – ISSN: 0005-4232 – mf#1774 – us UMI ProQuest [370]
Balancing the farm output / Spillman, William Jasper – New York, NY. 1927 – 1r – us UF Libraries [630]
Balancing the scales – 1980 winter-fall – 1 – mf#4867149 – us WHS [071]
Balanco da bossa / Campos, Augusto De – Sao Paulo, Brazil. 1968 – 1r – us UF Libraries [972]
Balandier, Georges *see*
- Daily life in the kingdom of the kongo
- Sociologie actuelle de l'afrique noire
- Vie quotidienne au royaume de kongo du 16e au 18e siecle

Balangero, Giovanni Battista *see* Australia e ceylan
Balans van het christendom / Leeuw, Gerardus van der – 3. druk. Amsterdam: H J Paris 1947 [mf ed 1993] – (= ser Bouwstenen voor een nieuwe samenleving 2) – 1mf – 9 – 0-524-08116-6 – mf#1993-9022 – us ATLA [240]
Balanza del comercio exterior de espana con las potencias extrangeras en 1792 – Madrid, 1803 – 7mf – 9 – sp Cultura [380]
Balanza del comercio exterior de espana con las potencias extrangeras en 1792 – Madrid, 1805 – 3mf – 9 – sp Cultura [380]
Balanza del comercio exterior de espana con las potencias extrangeras en 1826 – Madrid, 1828 – 3mf – 9 – sp Cultura [380]
Balapan / Almanak wanita – Djakarta, 1968 – 4mf – 9 – mf#SE-1310 – ne IDC [959]
Balaquer, Joaquin *see* Colon
Balazs, Bela *see* Der mantel der traeume
Balbaith, M J A *see* Confessions
Balbani, Niccolo *see* Life of galeazzo caracciolo
Balbas Capo, Vicente *see* Puerto rico a los diez anos de americanizacion
Balbi, Adriano *see*
- Atlas ethnographique du globe
- Essai statistique sur le royaume de portugal et d'algarve

Balbi, G *see*
- ...De ciuili and bellica fortitudine liber, ex myseriis poetae vergilii neune primum depromptus...
- ...De rebvs tvrcicis liber
- ...oratio habita cora clemete 7 de confoederatione nuper inita, paceque uniuersali, atque expeditione aduersus Turcas suspicienda

Balbi, Luigi (Aloysius) *see* Ecclesiastici concentus canendi 1, 2, 3 & 4 vocibus...
Balbian Verster, Jan Francois Leopold de *see* Ons mooi indie batavia
Balboa, descubridor del pacifico / Bayle, Constantino & Cabal, Juan – Madrid: Razon y Fe, 1944 – 1 – sp Bibl Santa Ana [946]
Balboa Troya y Quesada, Silvestre de *see* Espejo de paciencia
Balcania / L'Institut d'etudes et recherches balkaniques – Bucarest. 1938-45 – 1 – fr ACRPP [943]
Balcer, Georges *see* Rapport du secretaire
Balch, Elizabeth *see* Glimpses of old english homes
Balch, Emily Greene *see* The papers of emily greene balch, 1875-1961
Balch, George Thacher *see* Methods of teaching patriotism in the public schools
Balch, William Monroe *see* Christianity and the labor movement
Balch, William Ralston *see* The life of james abram garfield, late president of the united states
Balcony : the sydney review – Sydney, Australia 1965-66 – 1 – mf#2106 – us UMI ProQuest [400]
Baldaeus, P *see*
- Naauwkeurige beschryvinge van malabar en choromandel, derzelver aangrenzende rijcken, en het machtige eyland ceylon
- Wahrhaftige ausfuehrliche beschreibung der beruehmten ost-indischen kusten malabar und coromandel

Baldaeus, Philippus *see* Afgoderye der oost-indische heydenen
Balde, J *see* Iacobi balde s societate jesu urania victrix
Baldenecker, Udalrich *see* Six trios a un violon, taile et violoncello concertans, oeuvre 1
Baldensberger, Wilhelm *see* Der prolog des vierten evangeliums
Baldensperger, Fernand *see* Goethe en france
Baldensperger, Wilhelm *see* Das selbstbewusstsein jesu im lichte der messianischen hoffnungen seiner zeit
Balder, mythus und sage : nach ihren dichterischen und religioesen elementen untersucht / Kauffmann, Friedrich – Strassburg: K J Truebner 1902 [mf ed 1991] – (= ser Texte und untersuchungen zur altgermanischen religionsgeschichte 1) – 1mf – 9 – 0-524-01507-4 – (incl bibl ref) – mf#1990-2483 – us ATLA [290]
Balder the beautiful : the fire-festivals of europe and the doctrine of the external soul / Frazer, James George – London: Macmillan 1913 [mf ed 1993] – (= ser The golden bough 7) – 2v on 2mf – 9 – 0-524-05845-8 – (incl bibl ref & ind) – mf#1990-3509 – us ATLA [230]
Baldinger, E G *see* Magazin vor aerzte
Baldinger, Ernst Gotfried *see* Magazin vor aerzte
Baldinger, Ernst Gottfried *see*
- Medicinisches journal
- Neue arzneyen wider die medicinischen vorurtheile
- Neues magazin fuer aerzte

Baldinucci, F *see*
- Cominciamento e progresso dell'arte dell'intagliare in rame, colle vite di molti de'pi- eccellenti maestri della stessa professione
- Notizie de' professori del disegda cimabue in qua, per le quali si dimostra come...
- Tizie de' professori del diseg da cimabue in qua, per le quali si dimostra come, e per chi le bell' arti di pittura, scultura, e architettura lasciata la rozzezza delle maniere greca, e gottica, si sia in questi secoli ridotte all' antica loro perfezione
- Vita del cavaliere gio lorenzo bernino, scultore, architetto e pittore, scritta da f b fiorentino
- Vocabolario tosca dell'arte del diseg...

Baldivia Galdo, Jose Maria *see* La tradicion portuense de bolivia
Baldry, Alfred Lys *see*
- Albert moore
- Sir john everett millais

Balduino enrico / Geigel Sabat, Fernando Jose – Barcelona, Spain. 1934 – 1r – us UF Libraries [972]
Der baldur-mythos / Wesendonk, Mathilde – Dresden: Rumming 1875 – 1mf – 9 – mf#mw-9 – ne IDC [430]
Baldurs tod : ein maifestspiel / Pannwitz, Rudolf – Nuernberg: H Carl 1919 [mf ed 1991] – 1r – 1 – (filmed with: martin opitz / friedrich gundolf) – mf#2858p – us Wisconsin U Libr [820]
Baldus, Herbert *see* Ensaios de etnologia brasileira
Baldwin, Bernard *see*
- Biga boyowa
- Papuan notes and trobriand islands linguistic material
- Vocabulary of biga boyowa
- Vocabulary of bohilai

Baldwin, Brent T *see*
- The factors that division 1a football players at ball state university considered most important when deciding which university to attend during the recruiting process
- The factors that head football coaches at ncaa division 1a universities use to evaluate a potential athlete during the recruiting process

Baldwin bulletin – 1932 jan 1-1995 sep-dec [1]; 1873 oct 18, 1874 jan 7-1877 mar 30, 1877 apr 6-1879 jul 12, 1879 jul 19-nov 22 [2] – 1 – mf#1138140 [1]; 1138146 [2] – us WHS [071]
Baldwin, Charles Stealey *see* Expectations of lutheran military personnel as factors in shaping chaplain ministry
Baldwin City Cemetery Company *see* Oakwood cemetery records
Baldwin, Faith *see* Skyscraper souls
Baldwin, George Colfax *see* Representative women

Baldwin, H *see* Baldwin's reports of cases in the third circuit, 1828-1833
Baldwin, James Mark *see*
- Darwin and the humanities
- Development and evolution
- Fragments in philosophy and science
- Genetic theory of reality
- History of psychology
- Social and ethical interpretations in mental development

Baldwin, John Dennison *see*
- Ancient america

Baldwin, Joseph *see* The art of school management
Baldwin, Mark *see* Soil survey of the fort lauderdale area, florida
Baldwin, Maurice Scollard *see*
- Annual address of the bishop of huron to the synod of the diocese, june 16, 1891
- A break in the ocean cable
- Christian education
- Inaugural sermon preached in christ church cathedral, montreal
- Life in a look
- St paul
- The sunday school

[Baldwin park-] baldwin park and el monte herald press – CA. jan 1990- – 4r – 1 – $240.00 (subs $100/y) – mf#R02044 – us Library Micro [071]
[Baldwin park-] baldwin park bulletin – CA. jul 1982-dec 1990 – 17r – 1 – $1020.00 – mf#RC02043 – us Library Micro [071]
Baldwin, Robert *see* [Letter]
Baldwin, Roger Nash *see* Papers of roger nash baldwin (1885-1981)
Baldwin, Simeon Eben *see* The historic policy of the united states as to annexation
Baldwin, Stephen Livingstone *see* Foreign missions of the protestant churches
Baldwin, Susan *see* An evaluation of the physical fitness effects of a high school aerobic dance curriculum
Baldwin's reports of cases in the third circuit, 1828-1833 / Baldwin, H – Philadelphia: J Kay. 1v. 1837 (all publ) – (= ser Early federal nominative reports) – 7mf – 9 – $10.50 – mf#LLMC 81-429 – us LLMC [340]
Balen, Willem Julius Van *see*
- Antilia
- Ons gebiedsdeel curacao
- Venezuela

Balerma baptist church : church minutes – Hancock Co, GA 1878-98 – 1 – $10.00 – mf#6879 – us Southern Baptist [242]
Bales baptist church – Kansas City, MO – 1 – $149.99 – (. scrapbook, 1891-1958; minutes, dec 1936-61; letters, 1923-40; membership and account books, 1892-1936) – us Southern Baptist [242]
Balesteros Gaibrais, Manuel *see* Recuerdo y presencia de francisco pizarro
Balestrini C, Cesar *see* Economia minera y petrolera
Baletti a tre voci / Gastoldi, Giovanni Giacomo – 1594 – (= ser Mssa) – 1mf – 9 – €20.00 – mfchl 237 – gw Fischer [780]
Balfour, Alexander Hugh Bruce *see* An historical account of the rise and development of presbyterianism in scotland
Balfour, Andrew Jackson *see* The diocese of quebec
Balfour, Arthur J *see* Religion of humanity
Balfour, Arthur James, 1st Earl of *see* A defence of philosophic doubt
Balfour, Arthur James, 1st earl of *see* The foundations of belief
Balfour, Arthur James Balfour, Earl of *see* The religion of humanity
Balfour, Clara Lucas *see* Working women of this century
Balfour, Frederic Henry *see*
- The divine classic of nan-hua
- Taoist texts, ethical, political and speculative

Balfour, George *see* Trade and salt in india free
Balfour, Henry *see* The evolution of decorative art
Balfour, Robert G *see* Presbyterianism in the colonies
Balfour, Stewart *see* The unseen universe
Balfour, Thomas Alexander Goldie *see* God's two books
Balfour, Walter *see* An inquiry into the scriptural import of the words sheol, hades, tartarus, and gehenna
Balg, Gerhard Hubert *see* A comparative glossary of the gothic language
Balgarnie, Robert *see* Arkite worship
Balgy, Alexander *see* Historia doctrinae catholicae inter armenos unionisque eorum
Balham times *see* Clapham observer tooting
Bali post – Denpasar, Indonesia. Oct 1971-1993 – 65r – 1 – us L of C Photodup [079]
Balikesir – Balikesir. Sahib-i Imtiyaz ve Mueduer-i Mesul: Emin Vedad. n1. 8 eyluel 1338 [1922] – (= ser O & t journals) – 1mf – 9 – $25.00 – us MEDOC [956]
Balilla : azione coreografica in sei quadri. musica di carmine guarino / Adami, Giuseppe – Milano: G Ricordi, 1935 – 1 – mf#*ZBD-*MGTZ pv1-Res – Located: NYPL – us Misc Inst [790]
Balink, Albert *see* My paradise is hell
Baljon, Johannes Marinus Simon *see*
- Commentaar op de brieven van paulus aan de thessalonikers, efeziers, kolossers in aan filemon
- Commentaar op de katholieke brieven
- Commentaar op de openbaring van johannes
- Exegetisch-kritische verhandeling over den brief van paulus aan de galatiers
- De tekst der brieven van paulus aan de romeinen, de corinthiers en de galatiers

Balkan – Filibe, 1906-11. Sahib ve Muharriri: Ethem Ruhi. n526. 20 agustos 1324 [1908], 679,722,1134. 14 agustos 1326 [1910] – (= ser O & t journals) – 1mf – 9 – $25.00 – us MEDOC [956]
Balkan herald – Belgrade, Yugoslavia oct 1934-jun 1940 [mthly] – 1r – 1 – (wanting: v1 n2,5) – uk British Libr Newspaper [077]
Balkania – St Louis MO 1967-73 – 1,5,9 – ISSN: 0005-4321 – mf#7494 – us UMI ProQuest [949]
Balkansko zname – Gabrovo, Bulgaria. Jul 1955-87 – 23r – 1 – us L of C Photodup [077]
Balke, Martina *see* A multi-chase study of physical education resource teachers
Balkski svijet = Balkan world – Chicago: J R Palandech, sep 1917-mar 6 1919 – 1r – 1 – us CRL [071]
Balkwill, Francis Hancock *see* The testimony of the teeth to man's place in nature
Ball camp baptist church – KNOXVILLE, TN. 1818-Apr 1966 – 1 – $39.74 – us Southern Baptist [242]
Ball, Chad G *see* Exercise-induced muscle damage
Ball, Charles James *see* The ecclesiastical or deutero-canonical books of the old testament commonly called the apocrypha
Ball, E D *see* Some major celery insects in florida
Ball, Eli *see* The manual of the sacred choir
Ball, George Harvey *see* Christian baptism
Ball, Hugo *see* Flametti
Ball, James Dyer *see*
- The celestial and his religions
- Is buddhism a preparation or hindrance to christianity in china?
- Macao

Ball, John Thomas *see* The reformed church of ireland (1537-1886)
Ball, John Thomas. *see* The reformed church of ireland (1537-1886)
Ball, Kurt Herwarth *see*
- Guenter und christiane
- Im sommer danach

Ball, Louis, Jr *see* Music in the religious education of primary children
Ball, Mary C *see* A study of southern baptist vacation bible school music and its correlation with educational organization
Ball, Robert Stawell *see*
- An atlas of astronomy
- In the high heavens

Ball, Samuel *see* An account of the cultivation and manufacture of tea in china
Ball state university forum – Muncie IN 1960-89 – 1,5,9 – ISSN: 0888-188X – mf#6926 – us UMI ProQuest [378]
Ball, Thomas *see* The life of the renowned doctor preston
Ball, Thomas C *see* The effects of a carbohydrate-electrolyte replacement drink taken during high intensity exercise on sprint capacity at the end of exercise
Ball, Thomas Frederick *see* The london friends' meetings
Ball, Upendra Nath *see* Medieval india
Ball, William *see* The root of ritualism
Ball, William Edmund *see* St paul and the roman law
Ball, William Platt *see* Are the effects of use and disuse inherited?
Balla, Ignac *see* Romance of the rothschilds
A ballad book, or, popular and romantic ballads and songs.. – Edinburgh: Privately printed 1883 [mf ed Bloomington IN: Indiana Uni Lib, Preservation Dept 1984] – 2v on 1r – 1 – us Indiana Preservation [390]
A ballad of four religions *see* Si chiao ku ts'u (ccm222)
Ballad of gasparilla / Wayman, Herbert Edgar – Tampa, FL. 1933 – 1r – us UF Libraries [780]
The ballad of reading goal / Wilde, Oscar – repr by Boston: John W Luce & Co, n.d. – 1mf – 9 – $1.50 – mf#LLMC 92-229 – us LLMC [810]
Ballade am strom : roman / Betsch, Roland – Berlin: Grote, 1941, c1939 [mf ed 1989] – 651p – 1 – mf#7014 – us Wisconsin U Libr [830]
Ballade et danse des sylphes, op 5 : pour flute avec accompagnement de piano ou orchestre / Andersen, Joachim – Paris: C Joubert [188-] [mf ed 1994] – 1 – mf#pres. film 107 – us Sibley [780]
Die ballade in afrikaans / Vos, Willem Hermanus – 1955 [mf ed Bloomington IN: Indiana Uni Lib, Preservation Dept 1984] – 1r – 1 – us Indiana Preservation [390]
Balladen / Stucken, Eduard – 2. veraend aufl. Berlin: E Reiss 1920 [mf ed 1991] – 1r – 1 – (filmed with: totenhorn-sudwand / karl hans strobl) – mf#2907p – us Wisconsin U Libr [780]
Die balladen schillers im zusammenhang seiner lyrischen dichtung / Berger, Kurt – Berlin: Junker und Duennhaupt, 1939 [mf ed 1993] – (= ser Neue deutsche forschungen. abteilung neuere deutsche literaturgeschichte 23) – 84p – 1 – (incl bibl ref) – mf#8103 reel 2 – us Wisconsin U Libr [430]
Balladen und lieder / Edward, Georg – Grossenhain: Baumert & Ronge 1897 [mf ed 1989] – 1r – 1 – (filmed with: hallo welt! / kasimir edschmid) – mf#7204 – us Wisconsin U Libr [780]
Die balladen und ritterlichen lieder des freiherrn boerries von muenchhausen / Muenchhausen, Boerries, Freiherr von – 3. Aufl. Berlin: E Fleischel, 1908 – 1r – 1 – us Wisconsin U Libr [780]
Balladen vom geist / Hohlbaum, Robert – Berlin: K H Bischoff 1943 [mf ed 1990] – 1r – 1 – (filmed with: gestern / hugo von hofmannsthal) – mf#2729p – us Wisconsin U Libr [810]
Balladenbuch / Avenarius, Ferdinand – Stuttgart: Steingrueben, 1954, c1951 [mf ed 1993] – 552p – 1 – (incl ind) – mf#8354 – us Wisconsin U Libr [780]
The ballads and songs of scotland : in view of their influence on the character of the people / Murray, John Clark – Toronto: A Stevenson; London: Macmillan, 1874 – 3mf – 9 – mf#11213 – cn Canadiana [780]
Ballads and songs of the peasantry of england / Dixon, James Henry – London, England. 1864 – 1r – us UF Libraries [780]
Ballads en termes de la ley : originally written for the exclusive use of the trinity lawyers and other verses / Anson, Wiliam Reynell – Oxford: printed private circulation by Horace Hart, 1914? – 1mf – 9 – $1.50 – mf#LLMC 91-501 – us LLMC [810]
Ballads of acadia / Hannay, James – St John, NB: J A Bowes, 1909 – 1mf – 9 – 0-665-74478-1 – mf#74478 – cn Canadiana [810]
Ballads of lost haven : a book of the sea / Carman, Bliss – Boston: Lamson, Wolffe, 1897 [mf ed 1980] – 2mf – 9 – 0-665-00482-6 – mf#00482 – cn Canadiana [830]
Ballads of lost haven *see* Low tide on grand pre and ballads of lost haven
Ballagas, Emilio *see*
- Decimas por el jubilo martiano
- Nuestra senora del mar
- Obra poetica
- Orbita de emilio ballagas

Ballantine, James *see*
- Essay on ornamental art as applicable to trade and manufactures
- The life of david roberts, r a
- A treatise on painted glass

Ballantine, William *see*
- Letter to mr greville ewing
- A treatise on the statute of limitations

Ballantyne, James *see* Homes and homesteads in the land of plenty
Ballantyne, James Robert *see*
- Christianity contrasted with hindu philosophy
- Elements of hindi and braj bhakha grammar

Ballantyne, Robert Michael *see*
- An author's adventures
- The dog crusoe
- The dog crusoe and his master
- The iron horse
- The lighthouse
- The lonely island
- Personal reminiscences in book-making
- The prairie chief
- Silver lake
- Twice bought
- Ungava

Ballantyne, William *see* Professor john duncan, lld
Ballarat courier – Ballarat, Australia 20 jun 1887, 6 nov 1913-18 may 1920, 10 nov 1920-30 jun 1922 (wkly) (imperfect) – 1 – (cont by: courier [4 aug 1951-19 jul 1952]) – uk British Libr Newspaper [079]
Ballarat star *see* Star
Ballard, Addison *see* From talk to text
Ballard, B C [comp] *see* The british laws of the new hebrides
Ballard, Frank *see*
- Christian essentials
- Christian reality in modern light
- Haeckel's monism false
- The miracles of unbelief
- The omonism true
- The true god
- Why does not god intervene? and other questions

Ballard, Frank et al *see* Can we trust the bible?
Ballard, Stanley S *see* Phisics principles
Ballard, Susan *see*
- All the year round in japan
- Jottings from japan

Ballard, William Henry et al *see* Ontario high school arithmetic
Il ballarino di m fabritio caroso da sermoneta, diuiso in due trattati : nel primo de' quali si dimostra la diuersita de i nomi, che si danno a gli atti... / Caroso, Fabritio – In Venetia: Appresso Francesco Ziletti 1581 [mf ed 19–] – 2v in 1 on 1r – 1 – mf#film 1051 – us Sibley [780]
Ballat, Paul C *see* Effects of selected curriculum materials and teaching experience on the preactive planning of physical educators
Balleine, George Reginald *see* A history of the evangelical party in the church of england
Baller, Frederick William *see*
- Baller mandarin primer vocabularies
- The fortunate union
- Lessons in elementary wen-li
- Letters from an old missionary to his nephew
- Mandarin primer

Baller mandarin primer vocabularies : wade's romanization / Baller, Frederick William – [Peking: s.n, 1—] [mf ed 1995] – (= ser Yale coll) – 31p – 1 – 0-524-10120-5 – mf#1995-1120 – us ATLA [480]
Ballerini, Pietro *see* Petri ballerinii presbyteri veronensis de vi ac ratione primatus romanorum pontificum et de ipsorum infallibilitate in definiendis controversiis fidei liber singularis
Ballerini, Raffaele *see* Le prime pagine del pontificato di papa pio 9
Ballesteros Beretta, Antonio *see*
- Cristobal colon y el descubrimiento de america. 2 vol. barcelona, 1945
- Premio a la virtud

Ballesteros, Francisco A *see* Relacion del fallecimiento...molina
Ballesteros, Gaibrois *see* Manuel breviarios del pensamiento espanol...
Ballesteros Morientes, Alfredo *see* Reglamento para el regimen y administracion de la cofradia de la santisima virgen de navelonga, patron muy querido de cilleros de la diocesis de coria-caceres
Ballestros De Gaibrois, Mercedes *see* Vida de la avellaneda
Ballet, 1945-1950 / Haskell, Arnold Lionel – London [etc] Publ for the British Council, by Longmans, Green and Co [1951] – 1 – mf#*ZBD-*MGO pv 13 – Located: NYPL – us Misc Inst [790]
Ballet book : michel fokine and his ballet... – New York: E F Kalmus, c1935 – 1 – mf#*ZBD-*MGO pv 7 – Located: NYPL – us Misc Inst [790]
Ballet du temple de la paix : danse devant sa majestee a fontainebleu. mis en musique / Lully, Jean Baptiste – Paris: C Ballard 1685 [mf ed 19–] – 1r – 1 – (libretto by philippe quinault) – mf#film 511 – us Sibley [780]
Ballet in education; children's examinations : syllabus and rules: overseas / Great Britain. Royal Academy of Dancing – new ed. London, 1954 – 1 – mf#*ZBD-*MGO pv16 – Located: NYPL – us Misc Inst [790]
Ballet in red : music: don quichotte, pas de deux – minkus / Preston, David – [Brooklyn, NY: The Dance Mart, 1961] – 1 – mf#*ZBD-*MGO pv25 – Located: NYPL – us Misc Inst [790]
Ballet in the ussr / Lawson, Joan – [London]: SCR [1945] – 1 – mf#*ZBD-*MGO pv 10 – Located: NYPL – us Misc Inst [790]
Ballet news – New York NY 1979-86 – 1,5,9 – ISSN: 0191-2690 – mf#11849 – us UMI ProQuest [790]
Ballet stars – [n.p., 195?] – 1 – mf#*ZBD-*MGO pv21 – Located: NYPL – us Misc Inst [790]
Balletmuziek / Arntzenius, Louis Marie George – Bilthoven: H Nelissen [1958] – 1 – mf#*ZBD-*MGO pv19 – Located: NYPL – us Misc Inst [780]
[Ballets performed at the maryinsky theatre, leningrad, 1800-1903] – [New York? 194-?] – 1 – mf#*ZBD-*MGO pv25 – Located: NYPL – us Misc Inst [790]
Ballhorn, Friedrich *see* Grammatography
Balliere, Achille *see* Journal of a voyage from france to new caledonia
Ballila – Lynn MA, 1912* – 1r – 1 – (italian periodical) – us IHRC [073]
Ballin, Ada S *see* The science of dress in theory and practice
Ballina chronicle – Ballina, Ireland. may 1849-14 aug 1851 – 1 1/4r – 1 – (incorp with: connaught watchman) – uk British Libr Newspaper [072]
Ballina herald and mayo and sligo advertiser – Ballina, Ireland 22 oct 1891-10 nov 1892, 17 apr 1913-28 apr 1962 [mf jan-jul, sep-nov 1892] – 1 – (wanting 1925,1930; incorp with: western people) – uk British Libr Newspaper [072]
Ballina impartial : or trawly advertiser – Ballina, Ireland 13 jan 1823-26 dec 1825; 1 jan 1827-16 nov 1835 (wkly) – 1 – (discontinued) – uk British Libr Newspaper [072]

Ballina journal : and connaught advertiser – Ballina, Ireland 13 nov 1882-11 mar 1895 (wkly) – 5r – 1 – uk British Libr Newspaper [072]
Ballinasloe herald and east galway democrat – Ballinasloe, Ireland 12 oct 1985-30 jan 1988 [mf 1987-88] – 1 – (discontinued) – uk British Libr Newspaper [072]
Ballinger, J Kenneth *see* Miami millions
Ballinger, Margaret *see* From union to apartheid
Ballinrobe chronicle, and mayo advertiser – Ballinrobe, Ireland 22 sep 1866-7 dec 1867, 11 apr-1 oct 1903 [mf 1866-72, 1875-77, 1879, 1882-84, 1886-89, 1892-96] – 1 – (not publ 7 dec 1867-11 apr 1868; discontinued) – uk British Libr Newspaper [072]
Ballis, Oliviero *see* Canzonette amorose sprituali a tre voci
The ballot (london) – 2 jan 1831-25 dec 1831; 1 jan 1832-4 nov 1832 – (= ser 19th c british periodicals) – reel 1 – 1 – us Primary [073]
Ballou, Adin *see*
– Autobiography of adin ballou, 1803-1890
– Christian non-resistance
– Endless punishment rejected
– An exposition of views respecting the principal facts, causes, and peculiarities involved in spirit manifestations
– History of the hopedale community
– Practical christian socialism
– Primitive christianity and its corruptions
Ballou, Hosea *see*
– Ancient history of universalism
– Counsel and encouragement
– A voice to universalists
Ballou, Hosea et al *see* Series of letters in defence of divine revelation
Ballou, Howard Malcolm *see* Journal of a canoe voyage along the kauai palis, made in 1845
Ballou, Maturin Murray *see*
– Biography of rev. hosea ballou
– Equatorial america
Ballou, Moses *see* The divine character vindicated
Ballou's monthly magazine – La Salle IL 1855-93 – 1 – ISSN: 0730-9449 – mf#5249 – us UMI ProQuest [420]
Ballou's pictorial drawing-room companion – Boston. v. 1-17. may 1851-dec 1859. (incomplete) – 1 – us NY Public [073]
Ballou's pictorial drawing-room companion – La Salle IL 1851-59 – 1 – mf#5250 – us UMI ProQuest [640]
Ballow, Henry *see* A treatise of equity
Ballroom dancing in university education : developing a learning-theory based curricular model / Stubbs, Christopher R – 2000 – 186p on 2mf – 9 – $10.00 – mf#PE 4131 – us Kinesology [370]
Ballroom dancing to bronze medal standard / Great Britain. International Dance Teachers Association – 3rd ed. London [1962] – 1 – mf#*ZBD-*MGO pv20 – Located: NYPL – us Misc Inst [790]
Ball-room guide : a manual of dancing / Willock, H D – rev ed. Glasgow: J Cameron [186-?] – 1 – (contains the latest and most fashionable dances) – mf#*ZBD-*MGO pv11 – Located: NYPL – us Misc Inst [790]
Ballston Spa Area Historical Society et al *see* Grist mill
Balluseck, Lothar von *see* Dichter im dienst
Ballwin baptist church and mission – MO. 1952-61 – 1 – $22.40 – us Southern Baptist [242]
Ballyclare gazette and east antrim gazette – Ballyclared, Ireland 5 oct 1994- – 1 – uk British Libr Newspaper [072]
Ballymena advertiser – Ballymena, Ireland 7 jun 1873-16 jul 1892 (wkly) – 14r – 1 – (discontinued) – uk British Libr Newspaper [072]
Ballymena chronicle and antrim observer – Ballymena, Ireland 4 jan 1973- [mf 1986-] – 1 – uk British Libr Newspaper [072]
Ballymena guardian – Antrim, Ireland 1 nov 1973- [mf 1986-] – 1 – (cont: ballymena guardian & antrim standard [22 oct 1970-25 oct 1973]) – uk British Libr Newspaper [072]
Ballymena mail – Ballymena, Ireland [ns] 21 mar 1882, 9 feb-19 jul 1884 – 1 – (cont by: ballymena mail & larne [weekly] recorder [ns] 26 jul 1884-8 jul 1885) – uk British Libr Newspaper [072]
Ballymena mail and larne [weekly] recorder – Ballymena, Ireland 26 jul 1884-8 jul 1885 – 1r – 1 – (cont: ballymena mail [ns] 21 mar 1882, 9 feb-19 jul 1884) – uk British Libr Newspaper [072]
Ballymena observer – Ballymena, Ireland 22 aug 1857-18 apr 1985, 13 sep 1991-27 jan 1995 [mf 1991-95] – 1 – (wanting: nov, dec 1869; jan 1875-mar 1876; not publ fr 25 apr 1985-6 sep 1991(incorp in: ballymena times during this period)) – uk British Libr Newspaper [072]
Ballymena times – Ballymena, Ireland 15 apr 1985- [mf 1986-] – 1 – (cont: ballymena times [19 may 1966-20 aug 1970]) – uk British Libr Newspaper [072]
Ballymena weekly telegraph – Ballymena, Ireland 2 jun 1894-20 apr 1918 – 1 – (cont by: ballymena telegraph [27 apr 1918-11 dec 1920]) – uk British Libr Newspaper [072]
Ballymoney free press and northern counties advertiser – Ballymoney, Ireland 27 jan 1870, 15 may 1873-1 nov 1934 – 25r – 1 – (incorp with: coleraine chronicle) – uk British Libr Newspaper [072]
Ballymoney northern herald *see* Northern herald
Ballymoney times – Ballymoney, Ireland 11 sep 1991-29 mar 2000 – 1 – (cont: ballymoney times & ballymena observer [3 may 1989-4 sep 1991]; cont by: ballymoney & moyle times [5 apr 2000-]) – uk British Libr Newspaper [072]
Ballymoney times & ballymena observer – Ballymoney, Ireland 3 may 1989-4 sep 1991 – 1 – (cont by: ballymoney times [11 sep 1991-29 mar 2000]) – uk British Libr Newspaper [072]
Ballynahinch echo – Ballynahinch, Ireland 12 oct-23 nov 1988 – 1 – uk British Libr Newspaper [072]
Ballynahinch star – Ballynahinch, Ireland 22 nov 1974-28 mar 1975, 22 apr-2 jun 1989 – 5 1/2r – 1 – uk British Libr Newspaper [072]
Ballyshannon herald – Ballyshannon, Donegal. 10 jun 1831-?12 apr 1884 – 6r – 1 – (vols very incomplete; filmed with: county donegal general advertiser) – ie National [072]
Ballyshannon herald – Ballyshannon, Ireland 6 jan 1832-27 dec 1873 – 1 – (wanting: 1833, 1835, 1836, 1841, jan 1856, jan,feb 1864) – uk British Libr Newspaper [072]
Balmaceda / Nabuco, Joaquim – Sao Paulo, Brazil. 1949 – 1r – us UF Libraries [972]
Balmaceda, Julian C *see* Filipino language lexicon
Balmain observer – Balmain, aug 1884-dec1889, jan 1902-dec 1907, jan 1958-mar 1984 – 9r – at Pascoe [079]
Balmain observer and western suburbs advertiser – Sydney, Australia 10 sep 1892-11 nov 1893, 4 apr 1896-18 jun 1898, 14 jan 1899-21 nov 1903 (imperfect) – 1 – uk British Libr Newspaper [079]
Balme, Edward Balme Wheatley *see* Observations on the treatment of convicts in ireland
Balmer, Hans *see* Albert bitzius
Balmes en la encrucijada filosofica / Frutos Cortes, Eugenio – Zaragoza: Tip. La Academica, 1949 – sp Bibl Santa Ana [100]
Balmes, Jaime Luciano *see*
– Criterion
– Elements of logic
– European civilization
– Letters to a sceptic on religious matters
– Protestantism and catholicity compared in their effects on the civilization of europe
– Le protestantisme compare avec le catholicisme
Balneario de alange – Madrid: Imprenta Cosano – 1 – sp Bibl Santa Ana [946]
Balough, Elemer *see* The british empire
Balsa reports – 1972 feb, dec; 1973 mar; 1974 winter-1976 spring, winter; 1978 fall; 1980 winter; 1981 spring, winter; 1982 fall – 1 – mf#4877109 – us WHS [071]
Balsam lake ledger – 1898 nov 19-1899; 1900 jan-sep 13 – 1 – mf#1212225 – us WHS [071]
Balsan, Francois *see* Chez les femmes a crinieres du sud-angola
Balsbaugh, Christian Hervey *see* Glimpses of jesus
Balseiro, Jose Agustin *see*
– Palomas de eros
– Pureza cautiva
Balseiro, Jose Augustin *see* Vela mientras el mundo duerme
Balsells Rivera, Alfredo *see* Venadeado y otros cuentos
Balsham 1558-1950 – (= ser Cambridgeshire parish register transcript) – 9mf – 9 – £11.25 – uk CambsFHS [929]
Balsillie, David *see* An examination of professor bergson's philosophy
Baltacioglu *see* Kalbin goezu
Baltasar / Gomez De Avellaneda Y Arteaga, Gertrudis – Habana, Cuba. 1962 – 1r – us UF Libraries [972]
Baltasar / Gomez De Avellaneda Y Arteaga, Gertrudis – New York, NY. 1908 – 1r – us UF Libraries [972]
Baltasar cuartero y huerta y antonio vargas zuniga y montero de espinosa. marques de siete iglesias. indice... / Castro, Manuel – Madrid: Archivo Ibero-Americano, 1959 – 1 – sp Bibl Santa Ana [946]
Baltasar, Juan Antonio *see*
– Carta de edificacion
Baltazarini e il "balet comique de la royne" / Caula, Giacomo Alessandro – Firenze: Edizioni Sansoni Antiquariato, 1964 – 1 – mf#*ZBD-*MGO pv28 – Located: NYPL – us Misc Inst [790]
Balthasar huebmaier : the leader of the anabaptists / Vedder, Henry Clay – New York: G P Putnam 1905 [mf ed 1986] – (= ser Heroes of the reformation 8) – 1mf [ill] – 9 – 0-8370-9193-4 – (incl ind) – mf#1986-3193 – us ATLA [242]
Balthasar springers indienfahrt 1505/06 : wissenschaftliche wuerdigung der reiseberichte springers zur einfuehrung in den neudruck seiner 'meerfahrt' vom jahre 1509 / Schulze, Franz – Strassburg: J H E Heitz 1902 [mf ed 1993] – (= ser Drucke und holzschnitte des 15. und 16. jahrhunderts in getreuer nachbildung 8) – 1r [ill] – 1 – (incl bibl ref. filmed with: kleines deutsches sagenbuch / ed by will-erich peuckert) – mf#3367p – us Wisconsin U Libr [910]
The baltic and caucasian states. – Boston, New York: Houghton Mifflin Co., 1923. xx,269p. maps. Prepared under the care of Major-General Lord Edward Gleichen. Bibliographies – 1 – us Wisconsin U Libr [947]
Baltic biographical archive (baba) = Baltisches biographisches archiv (baba) / Frey, Axel [comp] – [mf ed 1995-98] – 436mf (1:24) – 9 – diazo €10,060.00 (silver €11,080 isbn: 978-3-598-33821-2) – ISBN-10: 3-598-33820-1 – ISBN-13: 978-3-598-33820-5 – (with printed ind) – gw Saur [947]
Baltic biographical archive. series 2 (baba2) = Baltisches biographisches archiv. neue folge (baba) / Frey, Axel [comp] – [mf ed 2003-06] – 319mf (1:24) – 9 – diazo €10,060.00 (silver €11,080 isbn: 978-3-598-35321-5) – ISBN-10: 3-598-35320-0 – ISBN-13: 978-3-598-35320-8 – (with printed ind) – gw Saur [947]
The baltic chronicle – Tilza, Latvia. Bi-monthly Organ of the Baltic Evangelical Mission. v1, 2, 19-39, incomplete. Publ. No. 6350b. One of two items on reel. 398p – 1 – us Southern Baptist [242]
The baltic independent – Estonia, 1999- – 1r per y – 1 – (backfile 1990-98 $85r) – us UMI ProQuest [077]
Baltic index = Baltikum index / ed by Bildarchiv Foto Marburg – Deutsches Dokumentationszentrum fuer Kunstgeschichte Philipps- Universitaet Marburg – [mf ed 2005-06] – (= ser Marburger index 4) – 415mf (1:24) in 4 installments – 9 – silver €3520.00 – ISBN-10: 3-598-35596-3 – ISBN-13: 978-3-598-35596-7 – (sold only as set) – gw Saur [700]
The baltic observer – Latvia, 1999- – 2r per y – 1 – (1992 1r $85. 1993-98 2r per y $170y) – us UMI ProQuest [077]
Baltijas baptisti un bunde, jeb baltijas baptistu tagadejs stahsoklis = The baltic baptists and the union, or the current condition of the baltic baptists / Frey, J A – Riga. 146p. 1887 – 1 – (publ n6347b. one of three items on reel) – us Southern Baptist [242]
Baltimore American Indian Center *see* Smoke signals
Baltimore baptist – 1883-97 – 5576p – 1 – $278.80 – (cont: the baptist, 1894. the evangel, 1895-97) – us Southern Baptist [242]
Baltimore business journal – Charlotte NC 1989-98 – 1,5,9 – ISSN: 0747-1823 – mf#16670 – us UMI ProQuest [650]
Baltimore city firefighter – 1980 jun, 1980 oct-1981 jun [1]; v1 n1-3 (1981 oct/nov-1982 feb/mar; 1982 jul-1983 apr) [2] – 1 – mf#665553 [1]; 1520156 [2] – us WHS [360]
Baltimore city reports / Maryland. Supreme Court – v1-4. 1888-1928 (all publ) – (= ser Maryland supreme court reports) – 31mf – 9 – $46.50 – mf#LLMC 84-152 – us LLMC [347]
Baltimore clipper – 1964 feb-1982 nov – 1 – mf#642630 – us WHS [071]
Baltimore commercial journal and lyford's price-current – Baltimore, Maryland. jan 23 1847-dec 8 1849 – 1r – 1 – us L of C Photodup [380]
Baltimore correspondent *see* Der deutsche correspondent
Baltimore County Historical Society *see* History trails
Baltimore daily clipper *see* American republican, and baltimore daily clipper
Baltimore daily commercial – Baltimore, Maryland. Oct 1965-Oct 1966 – 3r – 1 – (cont: baltimore clipper) – us L of C Photodup [071]
Baltimore daily gazette – Baltimore, Maryland. 1862-80 – 1 – us MD Archives [071]
Baltimore daily record – Baltimore, MD. v4-149, Jan 1890-Dec 1962. Incomplete. LL-02 – 1 – us L of C Photodup [340]
Baltimore evening news – Baltimore, Maryland. 1897-1918 – 1 – us MD Archives [071]
Baltimore firefighter – 1977 apr-1980 apr – 1 – mf#665552 – us WHS [360]
Baltimore first baptist church : scrapbook – BALTIMORE, MD. 1773-1968 – 1 – $28.53 – us Southern Baptist [242]
Baltimore GIs United *see* Open ranks
The baltimore jewish american – Baltimore. Md. 1908-09 – 1 – us AJPC [071]
Baltimore literary and religious magazine – La Salle IL 1835-41 – 1 – mf#4053 – us UMI ProQuest [073]
Baltimore literary monument – La Salle IL 1838-39 – 1 – mf#3941 – us UMI ProQuest [420]
Baltimore magazine for july 1807 : a literary magazine – La Salle IL 1807 – 1 – mf#3553 – us UMI ProQuest [420]
Baltimore medical and philosophical lyceum – La Salle IL 1811 – 1 – mf#3554 – us UMI ProQuest [610]
Baltimore medical and physical recorder – La Salle IL 1808-09 – 1 – mf#3555 – us UMI ProQuest [610]
Baltimore medical and surgical journal and review – La Salle IL 1833-34 – 1 – mf#3942 – us UMI ProQuest [610]
Baltimore monthly journal of medicine and surgery – La Salle IL 1830-31 – 1 – mf#4054 – us UMI ProQuest [610]
Baltimore monthly visitor – La Salle IL 1842 – 1 – mf#3943 – us UMI ProQuest [071]
Baltimore monument : a weekly journal, devoted to polite literature, science and the fine arts – La Salle IL 1836-38 – 1 – mf#3719 – us UMI ProQuest [073]
Baltimore morning sun *see* Sun
Baltimore Museum of Art *see*
– American prints, 1870-1950
– News-record of the baltimore museum of art
Baltimore museum of art record – Baltimore MD 1970-75 – 1,5,9 – ISSN: 0005-4518 – mf#7664 – us UMI ProQuest [060]
Baltimore philosophical journal and review – La Salle IL 1823 – 1 – mf#3556 – us UMI ProQuest [190]
Baltimore phoenix and budget – La Salle IL 1841-42 – 1 – mf#3747 – us UMI ProQuest [073]
Baltimore. Presbytery (Pres. Church in the USA) *see* Minutes, 1786-1906
Baltimore repertory of papers on literary and other topics – La Salle IL 1811 – 1 – mf#3557 – us UMI ProQuest [420]
Baltimore republican and daily argus – Baltimore, Maryland. 1842-61 – 1 – us MD Archives [071]
Baltimore review – 1958 sep 30 – 1 – mf#4780763 – us WHS [071]
Baltimore sun *see* Sun
Baltimore times – 1992 jul 27/aug 7-2001 jun 29/jul 5 [with gaps] – 1 – mf#2521561 – us WHS [071]
Baltimore weekly magazine – La Salle IL 1800-01 – 1 – mf#3558 – us UMI ProQuest [071]
Baltische bibliographie / Thomson, Erik – Wuerzburg: Holzner-Verlag 1957-62 [mf ed 1992] – (= ser Ostdeutsche beitraege aus dem goettinger arbeitskreis) – 2v on 10r – 1 – (incl ind. filmed with: ostdeutsche beitraege aus dem goettinger arbeitskreis) – mf#3180p – us Wisconsin U Libr [019]
Baltische blaetter – Koenigsberg (Kaliningrad RUS), 1848 – 1r – 1 – gw Misc Inst [077]
Baltische monatsschrift – Riga, 1859-1931. v1-62. – 593mf – 9 – mf#R-11289 – ne IDC [077]
Baltische post – Pernau (Paernu EW), 1927 22 jan-1928 30 aug – 1r – 1 – gw Misc Inst [077]
Baltische presse – Danzig (Gdansk PL), 1923 1 oct-1931 27 jun – 14r – 1 – gw Misc Inst [077]
Baltische rundschau – Wilna (Vilnius LT), 1998 nov-2002 jan – 1 – gw Misc Inst [077]
Baltische stimmen : wochenzeitung fuer stadt und land – Riga (LV), 1927 15 dec-1929 27 apr – 1r – 1 – gw Misc Inst [077]
Baltische studien – 15(1843)-40(1890) – 184mf – 9 – €351.00 – (nf: 6(1002)-14(1910). 46mf €88) – ne Slangenburg [947]
Die baltische tragoedie : eine romantrilogie / Vegesack, Siegfried von – Bremen: C Schuenemann 1935 [mf ed 1991] – 1r – 1 – (each work also publ separately. filmed with: vogt bartold / hans venatier & other titles) – mf#2923p – us Wisconsin U Libr [830]
Baltischer beobachter – Memel (Klaipeda LT), 1936 2 jan-30 sep, 1937 1 jan-31 mar & 1 jul-21 dec, 1938 2 jul-30 sep – 8r – 1 – gw Misc Inst [077]
Baltisch-litauische mitteilungen *see* Korrespondenz b
Baltz, Johanna *see* Fleurs des alpes
Baltzell, Amy *see* Psychological factors and resources related to rowers' coping in elite competition
Baltzer, Armin *see* Schweizerkunde
Baltzer, Florian *see* Diagnostischer und prognostischer wert neurologischer zusatzdiagnostik bei schwerem alkoholdelir

Baltzer, Otto *see*
- Beitraege zur geschichte der christologischen dogmas im 11ten und 12ten jahrhundert
- Judith in der deutschen literatur
- Praktische eschatologie
- Die sentenzen des petrus lombardus

O baluarte – Fortaleza, CE: Typ Apollo, 21 ago 1898; 06 jan 1899 – (= ser Ps 19) – mf#P18B,03,78 – bl Biblioteca [440]
Les baluba (congo belge) / Colle, R P – Bruxelles: Dewit 1913 – us CRL [960]
Baluba et lulua / Kalanda, Mabika – Bruxelles, Belgium. 1959 – 1r – us UF Libraries [960]
Baluchistan : 1901 (administrative report) – Delhi, 1941 – (= ser Census of india) – 1 – us CRL [315]
Baluzius, Stephanus *see*
- Concilia galliae narbonensis
- Vitae paparum avenionensium

Balwant vidyapeeth : journal of agriculture and scientific research – Bichpuri, India 1959-71 – 1 – ISSN: 0522-0718 – mf#6384.01 – us UMI ProQuest [630]
Balzac, B *see* Human law
Balzac, Honore De *see*
- Mercadet
- Romans et contes philosophiques

Balzak, Benjamin *see* Torat ha-adam
Balzani, Ugo *see*
- Italy
- The popes and the hohenstaufen

Balzli, Ernst *see* Zwuesche tuer u angle
Bam *see* Buenos aires musical
Bam, Nicola Rose-Anne *see* The role of inset in promoting multilingualism in western cape schools
Bama ca are asa sac nai ca tam mya / Ravhae Kauin Sa et al – Mantale: Athak Bama Nauin Nam Care Chara Asan 1966 [mf ed 1990] – 1r with other items – 1 – (in burmese) – mf#mf-10289 seam reel 158/2 [§] – us CRL [480]
Bama khit – Rangoon, Burma. Oct 1959-Mar 1960 – 3r – 1 – us L of C Photodup [079]
Bama ton su lay sama kri mya e khet pron chve nve pvai mhat tam : du ya chve nve pvai du yakyvan chvai hansatakha ruin – [s.l: Pran, nnvhan, rum 1963 [mf ed 1990] – [ill/pl] 1r with other items – 1 – (in burmese) – mf#mf-10289 seam reel 140/4 [§] – us CRL [320]
Bama tonsulaysama khet pron chvenve pvai e samuin : aca unto chvenve pvai unto kyerva caskuin kharuin – [s.l]: Pran, nnvhan, rum 1963 [mf ed 1990] – [ill/pl] 1r with other items – 1 – (in burmese) – mf#mf-10289 seam reel 142/2 [§] – us CRL [959]
Ba-matsor uva-kerav / Talmi, Efraim – Tel-Aviv, Israel. 1949 – 1r – us UF Libraries [939]
Bamberg, Elma L *see* "My home on the smoky"
Bamberg, Felix *see*
- Geschichte der orientalischen angelegenheit im zeitraume des pariser und des berliner friedens
- Ueber den einfluss der weltzustaende auf die richtungen der kunst

Bamberger, F *see* Die lehren des judentums nach den quellen
Bamberger hofkalender – 1764-1803 [mf ed 2002] – 80mf – 9 – €720.00 – 3-89131-386-1 – gw Fischer [390]
Bamberger, Johann Peter *see* Brittisches theologisches magazin
Bamberger, Selig *see* Maimonides' commentar zum tractat challah
Bamberger tagblatt – Bamberg DE, 1926-36 – 32r – 1 – mf#4598 – gw Mikropress [074]
Bamberger tagblatt *see* Taeglicher anzeiger
Bamberger volksblatt 1849 – Bamberg DE, 1849 1 feb-9 jun – 1r – 1 – gw Misc Inst [074]
Bamberger volksblatt 1871 – Bamberg DE, 1926-36 – 29r – 1 – (title varies: 1 may 1949: neues volksblatt / a, 1 apr 1953: bamberger volksblatt / b. with suppl: bamberger blaetter fuer fraenkische kunst und geschichte 1926 – 1936) – mf#4630 – gw Mikropress [074]
Bamberger zeitung 1795 – Bamberg DE, 1796-98, 1800-01, 1803 [gaps], 1804-05, 1806 [small gaps], 1807-1848 6 aug, 1848 1 jan-6 aug – 1r – 1 – (title varies: 1802: kurpfalzbaierische bamberger zeitung, 3 mar 1803: bamberger zeitung, 1810: fraenkischer merkur) – gw Misc Inst [074]
Bamberger zeitung 1848 – Bamberg DE, 1848 1 sep-1849 – 1r – 1 – gw Misc Inst [074]
Bambergischer schreib-calender / bamberger stadt- und landkalender – 1641-1957 [mf ed 2005] – 384mf – 9 – €1450.00 – 3-89131-460-4 – (publ with various titles; available: 1641-45, 1648-49, 1662-63, 1674, 1676, 1679, 1681-1942, 1946-47, 1949-57; not publ in 1943-45 & 1948; compiled fr holdings of the staatsbibliothek bamberg (mit historischem verein bamberg) & the staatsarchivs bamberg; filmed with: bambergischer schreib-calender – taschenausgabe 1687, 1689-92, 1694, 1695, 1698-1700, 1704-08, 1710-12, 1714, 1717, 1719-26, 1728-38, 1746 compiled fr holdings of the staatsbibliothek bamberg; bambergischer bauren-calender / gemeinnuetziger bauernkalender 1727 & 1805 compiled fr holdings of the staatsbibliothek bamberg) – gw Fischer [943]
Il bambino nell'arte attraverso i secoli / Maccone, Luigi – Bergamo. 1923 – 1 – us CRL [700]
Bamboo connection – 1985 sep 25-nov 4; 1986 feb 4,10, mar 11-dec 17; 1987 jan; 14-oct 30, nov/dec, 1988 mar 9-may 19, jun/jul, aug 19, sep/oct-nov/dec 15; 1989 jan/feb 2; special ed 1986 oct/nov, 1987 may/jun – 1 – mf#1840857 – us WHS [071]
Bamdad – Tehran, 1979-80. shumarah-'i 76-336. 15 murdad 1358-30 tir 1359 [6 aug 1979-21 jul 1980] – 2r – 1 – $150.00 – (missing: n145, 147, 150, 154, 257, 287, 293, 314, 325-330) – us MEDOC [079]
Ba-metsar – Tel-Aviv, Israel. 1941 or 1942 – 1r – us UF Libraries [939]
Bamileke de l'ouest cameroun / Tardits, Claude – Paris, France. 1960 – 1r – us UF Libraries [960]
Bamileke des fe'fe / Ngangoum, P F – Saint-Leger-Vauban, France. 1970 – 1r – us UF Libraries [960]
Ba-mivhan / Katznelson, Berl – Tel-Aviv, Israel. 1949 or 1950 – 1r – us UF Libraries [939]
Bamm, Peter *see*
- Der j punkt
- Die kleine weltlaterne

Bampfield, Francis *see* All in one
Bampflyde, C A *see* Report on the bird's nest caves of gormanton, british north borneo
Ban anh hung ca thang chap : tap tho – Ha-noi: Quan Doi Nhan Dan 1973 [mf ed 1992] – on pt of 1r – 1 – mf#11052 r71 n3 – us Cornell [810]
Ban Mo Tan Aon *see* Kui lui ni khte mran ma nuin nam samuin
Ban muang – Bangkok, Thailand. 1974-76; 1982-94 – 123r – 1 – us L of C Photodup [079]
The ban of the bori : demons and demon-dancing in west and north africa / Tremearne, Arthur John Newman – London: Heath, Cranton & Ouseley, [1914?] – 2mf – 9 – 0-524-03376-5 – (incl bibl ref) – mf#1990-3210 – us ATLA [390]
The ban of the bori : demons and demon-dancing in west and north africa / Tremearne, Arthur John Newman – London: Heath, Cranton & Ouseley, 1914. 504p.illus.plates – 1 – us Wisconsin U Libr [290]
Ban Ruan *see* Mogethin
Banaba : documents gathered by ken sigrah and papers of dr ron lampert on his archaeological excavation at te aka – 1965 – 1r – 1 – (available for ref) – mf#pmb1136 – at Pacific Mss [930]
Banaji, D R *see*
- Bombay and the sidis
- The gaikwads of baroda

Banana / Fawcett, William – London, England. 1921 – 1r – us UF Libraries [580]
Banani / Newman, Henry Stanley – New York, NY. 1969 – 1r – us UF Libraries [580]
Banaphula *see* Asabari
Banater bauernblatt – Temeschburg (Timisoara RO), 1921 1 jul-1922 13 oct – 1r – 1 – gw Misc Inst [630]
Banater beobachter – Grossbetschkerek (Veliki Beckerek YU), 1943 17-29 sep – 1r – 1 – (filmed by misc inst: 1942 3 jul-1943 31 aug (gaps) [3r]) – gw Misc Inst [077]
Banater beobachter – Zrenjanin, Yugoslavia. Jun 1942-Jun 1943 – 1r – 1 – us L of C Photodup [079]
Banater bote – Lugosch (Lugoj RO), 1929 21 apr-1930 15 may – 1r – 1 – gw Misc Inst [077]
Banater deutsche zeitung *see* Schwaebische volkspresse
Banater rundschau – Grossbetschkerek (Veliki Beckerek YU), 1938 2 oct-25 dec – 1r – 1 – gw Misc Inst [077]
Banater rundschau – Zrenjanin, Yugoslavia. Oct 1935-Apr 1939 – 2r – 1 – us L of C Photodup [079]
Banater schrifttum – Temeschburg (Timisoara RO), 1949-57 – 1 – (title varies: 1956: neue literatur; fr 1957 publ in bukarest) – gw Misc Inst [460]
Banater schrifttum – Temeschburg (Timisoara RO), 1949-57 – 1 – (title varies: 1956: neue literatur; publ in bukarest fr 1957) – gw Misc Inst [077]
Banater schulbote – Temeschburg (Timisoara RO), 1923-40 – 1 – gw Misc Inst [370]
Banater tagblatt – Temeschburg (Timisoara RO), 1920 1 aug-1938 14 jun – 7r – 1 – gw Misc Inst [077]
Banater volksblatt – Perjamosch (Perjamos/Lovrin RO), 1926 1 jul-1928 23 dec – 2r – 1 – (title varies: 1 jul 1928: illustriertes banater volksblatt) – gw Misc Inst [077]
Banbridge chronicle – Banbridge, Down. 1925-26, 2002- – 2r – 1 – (cont as: chronicle 12 sep 1985 -) – ie National [072]
Banbridge chronicle, gilford & rathfriland mail – Banbridge, Ireland 12 sep 1874-14 jun 1879 [mf 1874-96] – 1 – (cont by: banbridge chronicle, & downshire standard [18 jun 1879-8 mar 1968]; banbridge chronicle [15 mar 1968-5 sep 1985]; chronicle [19 sep 1985-27 oct 1988]; banbridge chronicle [3 nov 1988-]) – uk British Libr Newspaper [072]
Banbridge leader – Banbridge, Ireland 26 jan 1994- – 1 – uk British Libr Newspaper [072]
Banbury advertiser – Banbury, England 5 jul 1855-9 aug 1972 [mf 28 jan-2 sep 1897, 1912] – 1 – (lacking sep-dec 1897; discontinued) – uk Newsplan [072]
Banbury beacon – Banbury, England 23,30 may, 18-25 jul, 8 aug-19 sep, 3 oct-21 nov 1863; 5 jun 1868-30 sep 1905 – 1 – (lacking: jan-mar 1901, discontinued) – uk British Libr Newspaper [072]
Banbury cake – Banbury, England 15 aug 1973- [mf 1986-] – 1 – uk British Libr Newspaper [072]
Banbury citizen – Banbury, England 2 mar-17 aug 1989 [mf 1986-] – 1 – (cont: banbury focus [17 oct 1985-23 feb 1989; cont by: citizen (banbury) 24 aug 1989-10 jan 2003) – uk British Libr Newspaper [072]
Banbury & district citizen – Banbury, England 17 jan 2003-24 jun 2005 [mf 1986-] – 1 – (cont: citizen [24 aug 1989-10 jan 2003]; cont by: banbury & district review [1 jul 2005-]) – uk British Libr Newspaper [072]
Banbury focus – Banbury, England 5 nov 1970-19 nov 1971 – 1 – (cont by: focus (banbury) 26 nov 1971-10 oct 1985; very incomplete for 1970-71) – uk British Libr Newspaper [072]
Banbury guardian – Banbury, England 6 jul 1843-71, 1873-to date [mf 1984-] – 1 – (wanting 1872) – uk Newsplan [072]
Banbury herald, oxford times & midland counties advertiser – Banbury, England 27 jun 1867-27 feb 1869 – 1 – (1868 imperfect; cont: banbury herald, agricultural journal, & advertising chronicle [10 jan 1861-24 dec 1863]) – uk British Libr Newspaper [072]
Banbury herald & post – Banbury, England 10 jan-29 jun 1990 – 1 – uk British Libr Newspaper [072]
Banbury leader : the people's popular monthly journal – Banbury, England 6 jan 1909-28 oct 1912 – 1 – uk British Libr Newspaper [072]
Banc de la reine *see* Rapports judiciaires de quebec
Banca nazionale del lavoro quarterly review – Rome, Italy 1947+ – 1,5,9 – ISSN: 0005-4607 – mf#2187 – us UMI ProQuest [332]
Banca Sanchez SA *see* Estatutos
Banchieri, Adriano *see*
- Barca di venetia per padova...
- Direttorio monastico di canto fermo
- Eclesiastiche sinfonie dette canzoni in aria francese...
- Festino nella sera del giovedi grasso...
- Il metamorfosi musicale...
- La nobilissima, anzi asinissima compagnia delli briganti della bastina... compositione di camillo scaglieri dalla fratta
- Organo suonarino del p d adriano banchieri... opera 33
- La pazzia senile...
- Il zabaione musicale...

Banckwitz's illustrirte monatsblaetter – Leipzig DE, 1847 feb-jun – 1r – 1 – (title varies: mar 1847: banckwitz's illustrirtes wochenblatt) – gw Misc Inst [074]
Banckwitz's illustrirtes wochenblatt *see* Banckwitz's illustrirte monatsblaetter
Banco Comercial Israelita *see* Komertsyal bank
Banco De Angola Gabinete De Estudos Economicos *see* Economia do sisal
Banco De Angola, Lisbon Departamento De Estudos Economicos *see* Economic and financial survey of angola, 1960-1965
Banco de Guatemala *see* Boletin estadistico
Banco de la Nacion Argentina. Buenos Aires *see* Revista
Banco de la republica 1923-1948 / Otero Munoz, Gustavo – Bogota, Colombia. 1948 – 1r – us UF Libraries [332]
Banco De La Republica (Colombia) *see* Proceso historico del 20 de julio de 1810
El banco de la republica oriental del uruguay en el 25 : aniversario de su fundacion: 1896 -24 de agosto- 1921 / Uruguay. Banco de la Republica Oriental – Montevideo: A Barreiro y Ramos, 1921 (mf ed 19–) – 92p – mf#ZT-TB pv249 n4 – us Harvard [332]
Banco di Roma *see* Vademecum economico per l'aoi
Banco Do Nordeste Do Brasil *see* Sisal
Banco do Nordeste do Brasil Escritorio Tecnico de... *see* Electrificacao rural no nordeste
Banco Nacional de Mexico *see* Examen de la situacion economica de mexico
Bancos no brasil colonial / Aguiar, Pinto De – Salvador, Brazil. 1960 – 1r – us UF Libraries [332]
Bancroft blade – Bancroft, NE: J W Huntsberger. -v13 n1. jun 14 1901 (wkly) – 1r – 1 – (cont: bancroft independent. absorbed: bancroft enterprise. cont by: bancroft weekly blade) – us Bell [071]
Bancroft blade – Bancroft, NE: John G Neihardt. 53v. v15 n15. sep 25 1903-v67 n16. jun 17 1954 (wkly) [mf ed with gaps] – 14r – 1 – (cont: bancroft weekly blade. absorbed by: wisner news-chronicle) – us NE Hist [071]
Bancroft bugle – Walthill, NE: Burton Bargmann. 4v. v1 n1. apr 29 1971-v4 n3. may 9 1974 (wkly) [mf ed apr 29 1971-may 9 1974 filmed 1977] – 1r – 1 – (absorbed by: cuming county democrat, and: west point republican (1917). publ at walthill, apr 29-jul 29 1971; moved to bancroft aug 5 1971) – us NE Hist [071]
Bancroft, Charles *see* The footprints of time
Bancroft, E *see* Naturgeschichte von guiana in sued-amerika
Bancroft enterprise – Bancroft, NE: C C Sheaffer, 1895-v2 n14. may 8 1896 (wkly) – (= ser Bancroft blade) – 1r – 1 – (absorbed by: bancroft blade) – us Bell [071]
Bancroft, George *see*
- History of the formation of the constitution
- History of the united states of america
- History of the united states of america, vol 1
- History of the united states of america, vol 2
- History of the united states of america, vol 3
- History of the united states of america, vol 4
- History of the united states of america, vol 5
- History of the united states of america, vol 6
- History of the united states of america, vols 1-6
- Literary and historical miscellanies

Bancroft, Hubert Howe *see*
- Chronicles of the builders of the commonwealth
- History of british columbia, 1792-1887
- History of the northwest coast
- The native races of the pacific states of north america
- Reviews of hubert h bancroft's history of the pacific states
- The native races of the pacific states of north america
- The works of hubert howe bancroft

Bancroft independent – Bancroft, NE: J H Brayton (wkly) – 1r – 1 – (cont by: bancroft blade) – us Bell [071]
Bancroft, James R *see* The search for bargains
Bancroft, Jessie Hubbell *see* Games for the playground, home, school and gymnasium
Bancroft, Joseph Austen *see* On the amount of internal friction developed in rocks during deformation
Bancroft, R *see* A survey of the pretended holy discipline
Bancroft weekly blade – Bancroft, NE: Sinclair Bros. 3v. v13 n2. jun 21 1901-v15 n14. sep 18 1903 (wkly) – 1r – 1 – (cont: bancroft blade. cont by: bancroft blade (1903)) – us Bell [071]
Bancroftiana – n1-18,41-43; 52; 56-75; 78 [mar 1950-apr 1958, dec 1967-nov 1968, apr 1972, oct 1973-jun 1980, jul 1981] – 1 – mf#672574 – us WHS [071]
Band – 1987 spring-fall; 1989 mar/jun – 1 – mf#4775567 – us WHS [071]
Band of hope ritual : with responsive readings and temperance hymns / Cowie, J S, Mrs [comp] – Moncton, NB: Office of the Telephone, 1884 [mf ed 1980] – 1mf – 9 – mf#06036 – cn Canadiana [360]
De band tussen ambon en nederland / Graaf, H J de – Es-Gravenhage, 1969 – 1mf – 9 – mf#SE-160=0 – ne IDC [959]
Band wagon – n7,9 [1975 dec, 1976 sep] – 1 – mf#4780760 – us WHS [071]
Banda, H Kamuzu *see* The teaching of chinyanja
Banda, Hastings K *see* Speeches of dr hastings kamuzu banda, president of malawi
The bandahishn / ed by Anklesaria, E T D – Bombay, 1908 – 4mf – 9 – mf#NE-20152 – ne IDC [956]
Banda-linda de ippy : phonologie, derivation et composition / Cloarec-Heiss, France – Paris, France. 1969 – 1r – us UF Libraries [960]
Bandana baptist church – BANDANA, KY. 1900-May 1990 – 1 reel – 1 – $59.31 – (1,318p) – us Southern Baptist [242]
Bandaranaike, S W R D *see* Hand book of the ceylon national congress, 1919-1928
La bande joyeuse – Port-au-Prince: A gouraige & co. [1ere annee n10-n26]. (14 janv.-26 aout 1886) – 1 sheet – us CRL [079]
La bandeira italiana – Ouro Preto, MG: Typ Silva Cabral, nov 1884 – (= ser Ps 19) – bl Biblioteca [079]
Bandeira, Manuel *see*
- Antologia de poetas brasileiras bissextos contempo...
- Antologia dos poetas brasileiros da fase romantica
- Apresentacao da poesia brasileira
- Guia de ouro preto
- Guide d'ouro preto
- Nocoes de historia das literaturas

Bandeirantes and pioneers / Moog, Clodomir Vianna – New York, NY. 1964 – 1r – us UF Libraries [972]
Bandeirantes e pioneiros / Moog, Clodomir Vianna – Rio de Janeiro, Brazil. 1954 – 1r – us UF Libraries [972]
Bandeiras e sertanistas bahianos / Sousa Vianna, Urbino De – Sao Paulo, Brazil. 1935 – 1r – us UF Libraries [972]
Bandeirismo / Vasconcellos, Salomao De – Belo Horizonte, Brazil. 1944 – 1r – us UF Libraries [972]

Bandeirismo paulista e o recuo do meridiano / Ellis Junior, Alfredo – Sao Paulo, Brazil. 1938 – 1r – us UF Libraries [972]

Bandelow, Volker *see* Organisationsprobleme kommunaler kulturverwaltung

Bandera / Gay Calbo, Enrique – Habana, Cuba. 1945 – 1r – us UF Libraries [972]

Bandera : himno y escudo de cuba / Cuba Secretaria De Estado – Habana, Cuba. 1950 – 1r – us UF Libraries [972]

Bandera blanca / Rivera Natal, Facundo – Barcelona, Spain. 1962 – 1r – us UF Libraries [972]

La bandera regional – Plasencia, 1900. Various numeros – 5 – sp Bibl Santa Ana [073]

Bandera wolanda : dikaloewarkan saminggoe sekali = Netherlands' flag – 1901-03 – (= ser Vernacular press in the netherlands indies, c1855-1925, unit 1) – 18mf – 9 – €220.00 – mf#mmp134/6 – ne Moran [079]

Banderas oficiales y revolucionarias de cuba / Roig De Leuchsenring, Emilio – Habana, Cuba. 1950 – 1r – us UF Libraries [972]

Bandhu, Vishva *see* Siddha-bharati

Bandiera rossa : organ of the gruppi comunisti rivoluzionari, sezione italiana della 4 internazionale – Roma: [s. n.] 1950-2002 [mf ed 1988-2004] – 10r – 1 – (frequency varies; some pp missing) – mf#1188 – us Wisconsin U Libr [335]

Bandierantes e pioneiros / Moog, Clodomir Vianna – Rio de Janeiro, Brazil. 1961 – 1r – us UF Libraries [972]

Banditismo na bahia / Santos Maia, Eduardo – Belo Horizonte, Brazil. 1928 – 1r – us UF Libraries [972]

Bandlow, Heinrich *see*
- Koester hemp
- Naturdokter stremel

Bandoeng : the mountain city of the netherlands india / Reitsma, S A – Weltevreden, Batavia, [1932] – 1mf – 8 – mf#SE-1440 – ne IDC [959]

El bandolerismo / Zugasti, Julian de – v1 1876; v2 1876; v4 1877; v6 1877 – 9 – sp Bibl Santa Ana [800]

El bandolerismo...tomo 4 : parte primera. origenes del bandolerismo / Zugasti, Julian de – 2nd ed. Madrid: Fortanet v1. 1877 – 1 – sp Bibl Santa Ana [946]

O bandolim – Juiz de Fora, MG. 12 jan 1890 – (= ser Ps 19) – bl Biblioteca [079]

O bandolim : quarteto dedicado ao bello sexo do congresso do catete – Rio de Janeiro, RJ: Typ Lith Bittencourt, Vieira & C, 07 set-09 nov 1889 – (= ser Ps 19) – mf#P17,01,73 – bl Biblioteca [880]

Bandon recorder – Bandon OR: D E Stitt, -1910 [freq varies] – 1 – (cont by: semi-weekly bandon recorder) – us Oregon Hist [071]

Bandon recorder – Bandon OR: D E Stitt, -1910 [semiwkly] – 1 – (cont by: semi-weekly bandon recorder (1910-15)) – us Oregon Lib [071]

Bandon recorder – Bandon OR: Recorder Pub Co, Inc, [wkly] – 1 – (cont: semi-weekly bandon recorder) – us Oregon Hist [071]

Bandon recorder (bandon, or) – Bandon OR: Recorder Pub Co, Inc [semiwkly] – 1 – (began in 1915; cont: semi-weekly bandon recorder) – us Oregon Lib [071]

Bandon western world – Bandon OR: S Price & M Gillard-Juarez, 1983-86 [wkly] [mf ed 1983-88] – 4r – 1 – (cont: western world (bandon, or); cont by: western world (bandon, or: 1986)) – us Oregon Lib [071]

Bandtkie, Jerzy Samuel *see* Slownik dokladny jezyka polskiego i niemieckiego do podorecznego uzywania dla polakow i niemcow

Banduri, A *see*
- Imperium orientale sive antiquitates constantinopolitani

A bandurra – Maranhao: Typ Nacional, 15 jan-31 dez 1828 – (= ser Ps 19) – 1,5,6 – mf#P01,06,23 – bl Biblioteca [079]

Bandwagon – 1944 dec 15-1947 mar + suppls [1]; 1957 jun-1959 dec [2] – 1 – mf#1057975 [1]; 1052869 [2] – us WHS [071]

Bandwidth knowledge of results in motor skill performance and learning / Goodwin, Jeff E – Texas Woman's University, 1994 – 3mf – 9 – $12.00 – mf#PSY 1888 – us Kinesology [150]

Bandyopadhyay, Manik *see* Boatman of the padma

Bandyopadhyay, Pramathanath *see* A study of indian economics

Bandyopadhyaya, Shripada *see* The music of india

Bandyopadhyaya, Tarasankara *see* The eternal lotus

Bane, Anil Chandra *see* The eastern frontier of british india, 1784-1926

Banerjea, Jitendra Nath *see* The development of hindu iconography

Banerjea, Krishna Mohan *see* The arian witness

Banerjea, Pramathanath *see*
- A history of indian taxation
- Indian finance in the days of the company
- Provincial finance in india

Banerjea, Surendranath *see*
- A nation in making
- Speeches

Banerjee, Anil Chandra *see*
- Anglo-sikh relations
- Annexation of burma
- The cabinet mission in india
- Indian constitutional documents, 1757-1939
- Rajput studies

Banerjee, Brajendra Nath *see*
- Begam samru
- Begams of bengal
- Rajah rammohun roy's mission to england

Banerjee, Gooroodass *see* The education problem in india

Banerjee, Hiranmay *see* Genetic history of the problems of philosophy

Banerjee, Indubhusan *see* Evolution of the khalsa

Banerjee, Muraly Dhar *see* Genetic history of the problems of philosophy

Banerjee, Muralydhar *see* The desinamamala of hemacandra

Banerji, Albion Rajkumar *see* Through an indian camera

Banerji, G C [comp] *see* Brahmananda keshub chunder sen

Banerji, Nripendra Chandra *see* At the cross-roads, 1885-1946

Banerji, Projesh *see*
- Dance of india
- The folk-dance of india

Banerji, Rakhal Das *see*
- The age of the imperial guptas
- Bas reliefs of badami
- Eastern indian school of medieval sculpture
- The haihayas of tripuri and their monuments
- History of orissa
- The palaeography of the hathigumpha and the nanaghat inscriptions
- The temple of siva at bhumara

Banerji, S K *see* Humayun badshah

Banerji, Santi Kumar *see*
- A comparative study of the social background of adult education problems

Banes, Charles H *see* Benjamin griffith

Banes, dom *see*
- Commentarios ineditos a la tercera parte de santo tomae
- Scholastica commentaria in primam partem angelici doctoris s thomae
- Scholastica commentaria in primam partem angelici doctoris s thomae usque ad 63 quaestionem complectentia
- Scholastica commentaria in primam partem summae theologicae s thomae aquinatis
- Scholastica commentaria in secundam secundae angelici doctoris s thomae

Banes et molina : histoire, doctrines, critique metaphysique / Regnon, Theodore de – Paris: H Oudin 1883 [mf ed 1991] – 1mf – 9 – 0-7905-8563-4 – mf#1989-1788 – us ATLA [120]

Banet suad serhi / Esad – Hicaz: Hicaz Vilayeti Matbaasi, 1314 [1897] – (= ser Ottoman literature, writers and the arts) – 4mf – 9 – $60.00 – us MEDOC [470]

Banffshire advertiser (1881) : buckie and moray firth fishing and general gazette – Buckie: W F Johnston 1881-; Buckie: J & M Publ 1992- (wkly) [mf ed 3 jan 1995-] – 1 – (numbering ceased with: 97th yr n3974 (11th may 1977), resuming with: 105th yr n5463 (dec 9th 1986)) – uk Scotland NatLib [639]

Banffshire advertiser and buckie and moray firth fishing and general gazette – Buckie, Scotland 17 nov 1881-29 jun 1939 [mf 1923, 1981] – 1 – (cont by: banffshire advertiser & buckie & moray firth gazette [6 jul 1939-16 aug 1951]; banffshire advertiser [23 aug 1951-]) – uk British Libr Newspaper [072]

Banffshire herald – Keith: J Mitchell 1974- (wkly) [mf ed 1995-] – 1 – (cont: banffshire herald, strathisla advertiser & farming news; imprint varies) – uk Scotland NatLib [072]

Banffshire herald – Keith, Scotland 30 dec 1893- (wkly) [mf 1986-] – 1 – uk British Libr Newspaper [072]

Banffshire journal and northern farmer – Banff: Banffshire Journal Ltd 1951-96 (wkly) [mf ed 5 jan 1994-96] – 1 – (cont: banffshire journal, aberdeenshire mail, moray, nairn, and inverness review, and northern farmer [25 jul 1876-27 mar 1951]; cont by: banffshire journal [13 mar 1996-]) – ISSN: 1354-9634 – uk Scotland NatLib [072]

Banffshire journal & general advertiser – Banff, Scotland 30 sep,14 nov 1845, 6 jan, 17,24 mar, 12 may, 9 jun, 14,21 jul, 25 aug, 13,27 oct, 10 nov 1846-18 jul 1876 – 1 – (cont by: banffshire journal, aberdeenshire mail, moray, nairn & inverness review, & northern farmer [25 jul 1876-27 mar 1951]) – uk British Libr Newspaper [072]

Banffshire reporter – Portsoy, Scotland 16 jul 1869-1 dec 1920 – 1 – (wanting: 5 mar, 2 apr 1875, 5 jan-11 may, 17 nov-29 dec 1877) – uk British Libr Newspaper [072]

Bang, A C *see* Den norske kirkes geistlighed 1 reformations-aarhundredet (1536-1600)

Bang, A Chr *see*
- Hans nielsen hauge og hans samtid
- Den norske kirkes geistlighed i reformations-aarhundredet

Bang Giang *see* Manh vun van-hoc su

Bang, Herman *see* Gedanken zum sexualproblem

Bang, Hermann Joachim *see* Die taenzerin und andere erzaehlungen

Bang, W *see*
- Materialien zur kunde des aelteren englischen dramas
- Osttuerkische dialektstudien

Bang, Willy *see* Die altpersischen keilinschriften

Les bangala st (etat ind. du congo) / Overbergh, Cyr. van – Bruxelles, A DeWit [etc] 1907 – us CRL [960]

Bangalore spectator – Bangalore, India 6 jan 1877-27 mar 1895 (imperfect) – 1 – uk British Libr Newspaper [079]

Bange, W *see* Meister eckeharts lehre von goettlichen und geschoepflichen sein

Bangen, Johann Heinrich *see* Die roemische curie

Banghart, Peter D *see* Migrant labour in south west africa and its effects on ovambo tribal life

Bangkok chronicle – Bangkok. Thailand. 1939-41 – 9r – 1 – us L of C Photodup [079]

Bangkok post – Bangkok, Thailand. 1975- mthly updates – 1 – us Primary [079]

The bangkok post – Bangkok: Alexander MacDonald, 1953-88 – 1 – us CRL [079]

Bangkok review – Bangkok: Bangkok Review. v1-3 n3. nov 14 1934-feb 1 1936 – 2r – 1 – us CRL [079]

Bangkok times – Bangkok, Thailand. 1941-42 – 1r – 1 – us L of C Photodup [079]

Bangla Congress *see* Election manifesto and immediate programme

Bangladesh *see* Bangladesh gazette

Bangladesh gazette / Bangladesh – 1967- – 1 – us L of C Photodup [954]

Bangladesh observer – Dacca, Bangladesh. 1962-94 – 137r – 1 – us L of C Photodup [079]

Bangladesh observer *see* Pakistan observer

Bangor commercial – Bangor, ME. nov 27 1949-jan 16 1954 – 1 – us CRL [071]

Bangor daily commercial – Bangor, ME: Marcellus Emery, 1872-nov 25 1949 – 1 – us CRL [071]

The bangor daily news – Bangor, ME: [Bangor Pub Co], 1900-sep 1970 – 1 – us CRL [071]

Bangor independent – [1888 dec 1-1970 jul 16 [with gaps] – 1 – (with suppls) – mf#1161227 – us WHS [071]

Bangor observer & anglesea & carnarvonshire news & advertiser – [Wales] LLGC 30 mar 1883-12 sep 1884 [mf ed 2004] – 2r – 1 – uk Newsplan [072]

Bangs, John Kendrick *see* From pillar to post

Bangs, Nathan *see*
- An authentic history of the missions
- A discourse on occasion of the death of the rev wilbur fisk
- The errors of hopkinsianism detected and refuted
- An examination of the doctrine of predestination
- A history of the methodist episcopal church
- Letters to young ministers of the gospel
- The life of james arminius, d.d
- Methodist episcopacy
- The necessity, nature, and fruits, of sanctification
- The present state, prospects, and responsibilities of the methodist episcopal church
- The reformer reformed
- The reviewer answered
- A vindication of methodist episcopacy

Bangue nas alagoas / Diegues Junior, Manuel – Rio de Janeiro, Brazil. 1949 – 1r – us UF Libraries [972]

Banhidi, Zoltan *see* Learn hungarian

Die banier : [the only voice of 1 1/2 million coloureds in south africa] – Soutrivier [Stellenbosch : Die Banier, Jan 1962-feb 1966 [semimthly]; dec 1959-dec 1961 [mthly] – 2r – 1 – (began with dec 1959; ceased feb 1966; lacks: mar, aug 1961; p3-6 of feb 1964; jan 15(?) 1965; chiefly in afrikaans, some english) – mf#mf-10088 – us CRL [079]

Banier, Antoine *see* Explication historique des fables

Bank accounting & finance – London UK 1991-96 – 1,5,9 – ISSN: 0894-3958 – mf#17043 – us UMI ProQuest [650]

The bank act of 1844 : free trade in gold not incompatible with our standard of value... / Brookes, Henry – London: E Wilson, 1861 (mf ed 19–) – 48p – mf#ZT-TB pv4 n12 – us Harvard [332]

Bank automation quarterly – Indianapolis IN 1993 – 1,5,9 – (cont: banking software review) – mf#14951.04 – us UMI ProQuest [332]

Bank cases. 1878 / Moses, Raphel Jacob – New York: American Bankers' and Merchants' Agency, 1879. 148p. LL-2299 – 1 – us L of C Photodup [346]

Bank controllers report – New York NY 2002+ – 1,5,9 – ISSN: 1540-3912 – mf#26310 – us UMI ProQuest [332]

Bank Indonesia *see*
- Berita
- Bulletin
- Ekonomi, keuangan dan bank

Bank Industri Negara *see* Laporan

Bank investment consultant – New York NY 2003+ – 1,5,9 – mf#26669.02 – us UMI ProQuest [332]

Bank Koperasi Tani dan Nelajan *see* Gemah ripah

Bank loan report – New York NY 1992+ – 1,5,9 – ISSN: 1099-3398 – mf#18456 – us UMI ProQuest [332]

Bank marketing – Washington DC 1974+ – 1,5,9 – ISSN: 0888-3149 – mf#10211.02 – us UMI ProQuest [332]

Bank Negara Indonesia *see* Madjalah bank

Bank Negara Indonesia Unit 1 *see* Laporan tahun pembukuan

Bank Negara Indonesia Unit 3 *see*
- Laporan perkembangan
- Report

Bank negara indonesia unit 3 / Laporan – Djakarta, 1946-1970 – 22mf – 9 – mf#SE-264 – ne IDC [959]

Bank negara indonesia unit 3 – Djakarta, 1959-1972 – 60mf – 9 – (missing: 1959-1961, v1-3(1-3, 5-12); 1963, v5(1, 3-9, 11-12); 1964, v6(2-4, 7-12); 1965, v7(1-5, 7-12); 1966, v8(1-4, 7); 1968, v10(4, 8)) – mf#SE-1333 – ne IDC [959]

Bank note reporter – 1977-78; 1981-1983 jun; 1983 jul-1985; 1986-1988 jun – 1 – mf#519478 – us WHS [332]

Bank of america journal of applied corporate finance – New York NY 1995+ – 1,5,9 – (cont: continental bank journal of applied corporate finance) – ISSN: 1078-1196 – mf#17450.01 – us UMI ProQuest [332]

Bank of canada review = Revue de la banque du canada – Ottawa ON 1971+ – 1,5,9 – ISSN: 0045-1460 – mf#2783 – us UMI ProQuest [332]

Bank of england, monetary and financial statistics – London UK 2001+ – 1,5,9 – ISSN: 1365-7690 – mf#25897 – us UMI ProQuest [332]

Bank of england. quarterly bulletin – London UK 1960+ – 1,5,9 – ISSN: 0005-5166 – mf#7171 – us UMI ProQuest [332]

The bank of india : a proposal to establish a bank of issue for india on the model of the bank of england. three letters addressed to the hon edward stanhope... / Daniels, William H – London, [1879] – (= ser 19th c books on british colonization) – 1mf – 9 – mf#1.1.902 – uk Chadwyck [332]

Bank of Middleton [WI] *see* Downto business

Bank of Montreal *see*
- Annual general meeting
- Articles of association of the montreal bank
- By-laws for the management of the affairs of the bank of montreal
- List of stockholders...31st may, 1897
- Report of the directors to the shareholders at their...annual general meeting
- Rules and regulations adopted by president, directors and company of the bank of montreal
- Rules and regulations for the branches of the bank of montreal
- Statement of the result of the business of the bank for the year ending...

Bank of montreal : list of shareholders eligible for directors on the first monday in june, 1865 – S.l: s.n, 1865? – 1mf – 9 – mf#60958 – cn Canadiana [332]

Bank of Montreal. Annuity and Guarantee Funds Society *see* Charter and by-laws

Bank of montreal business review – Toronto ON 1980-92 – 9 – ISSN: 0005-531X – mf#35010 – us UMI ProQuest [332]

Bank of Montreal. Pension Fund Society *see* Act of incorporation and by-laws

Bank of New York and Trust Co *see* New home of the bank of new york and trust company

The bank of toronto : proceedings of the forty-first annual general meeting, wednesday, 16th june, 1897 – Toronto?: s.n, 1897? – 1mf – 9 – mf#60981 – cn Canadiana [332]

Bank operations & technology alert – New York NY 2001+ – 1,5,9 – ISSN: 1540-3947 – mf#28809 – us UMI ProQuest [332]

Bank Pembangunan Indonesia *see*
- Berita press release
- Laporan
- Report

Bank Pembangunan Sumatera Selatan *see* Laporan

Bank Pembangunan-Daerah Djawa-Timur *see* Laporan sekretariat dprdgr

Bank Potrebitel'skoi Kooperatsii *see* Kooperatsiia i finansy

Bank Rakjat Indonesia *see* Warta

Bank & s & l quarterly rating service – New York NY 2002+ – 1,5,9 – mf#22656 – us UMI ProQuest [332]

The bank screw : or, war and the gold discoveries in connexion with the money market / Malagrowther the less – London: Houlston & Stoneman, 1854 – (= ser 19th c economics) – 1mf – 9 – mf#1.1.434 – uk Chadwyck [332]
Bank systems & equipment – Manhasset NY 1972-88 – 1,5,9 – (cont by: bank systems & technology) – ISSN: 0146-0900 – mf#7233.01 – us UMI ProQuest [332]
Bank systems & technology – Manhasset NY 1990+ – 1,5,9 – (cont: bank systems and equipment) – ISSN: 1045-9472 – mf#7233.01 – us UMI ProQuest [332]
Bank Tabungan untuk Umum *see* Laporan tahunan
Bank Timur *see* Laporan direktur dan dewan komisaris
Bank Umum Nasional *see* Perkembangan bank umum nasional dalam tahun
Bank- und handels-zeitung – Berlin DE, 1862 30 jun-1865 30 jun – 9r – 1 – gw Misc Inst [332]
Banker and tradesman – Washington DC 1970-96 – 1 – ISSN: 0005-5409 – mf#8760 – us UMI ProQuest [333]
The bankers' journal and financial review – Toronto: F. Weir, [1889?-189- or 19–] – 9 – ISSN: 1190-6995 – mf#P04063 – cn Canadiana [332]
Bankers magazine – New York NY 1846-1997 – 1,5,9 – ISSN: 0005-545X – mf#3483 – us UMI ProQuest [332]
The bankers' magazine – v. 1-102. 1844-1916. V. 2-3, 9, 16, 19, 94 wanting – 1 – 848.00 – us L of C Photodup [332]
Bankers monthly – New York NY 1898-1993 – 1,5,9 – ISSN: 0005-5476 – mf#10484 – us UMI ProQuest [332]
Banker's safes / Goldie and McCulloch Co Ltd – Galt, Ont?: Jaffray Bros, 18–? – 1mf – 9 – mf#46493 – cn Canadiana [620]
Bankers' weekly circular and statistical record – New York NY 1845-46 – 1 – mf#2562 – us UMI ProQuest [332]
Banki i bankirskie kontory v rossii : spravochno-statisticheskie svedeniia o vsekh operiruiushchikh v rossii kreditnykh uchrezhdeniiakh / Evzlin, Z – Spb, 1904 – 7mf – 9 – mf#REF-145 – ne IDC [332]
Banki i kreditnye uchrezhdeniia sssr v 1924 g bankovskii spravochnik-ezhegodnik – M, 1924 – 2mf – 9 – (missing: title pg) – mf#REF-10 – ne IDC [332]
Banki i promyshlennost' v rossii : k voprosu o finansovom kapitale v rossii / Gindin, I F – M, L, 1927 – 4mf – 9 – mf#REF-175 – ne IDC [332]
Banki i razvitie sel'skogo khoziaistva v rossii v kontse 19-nachale 20 vv / Korelin, A P – M, 1986 – 1mf – 9 – mf#REF-505 – ne IDC [332]
Banki soiuza ssr / Livshits, F D – M, 1925 – 1mf – 9 – mf#REF-35 – ne IDC [332]
Der bankier reitet ueber das schlachtfeld : erzaehlung / Becher, Johannes Robert – Wien: Agis-Verlag, 1926 [mf ed 1995] – 91p – 1 – mf#8973 – us Wisconsin U Libr [880]
Bankim chandra : prophet of the indian renaissance, his life and art / Das, Matilal – Calcutta: DM Library, 1938 – (= ser Samp: indian books) – us CRL [920]
Bankim-tilak-dayananda / Ghose, Aurobindo – Calcutta: Arya Pub House, 1940 – (= ser Samp: indian books) – us CRL [920]
Banking / American Bankers Association – New York NY 1908-78 – 1,5,9 – (cont by: aba banking journal) – ISSN: 0005-5492 – mf#6992.01 – us UMI ProQuest [332]
Banking act of 1809 / Mississippi Legislature – 1v (all publ) – (= ser Mississippi supreme court reports) – 1mf – 9 – $1.50 – mf#LLMC 91-069 – us LLMC [346]
Banking and commerce : a practical treatise for bankers and men of business, together with the author's experience of banking life in england and canada during fifty years / Hague, George – New York: Bankers Pub Co, 1908 – 5mf – 9 – 0-665-76942-3 – mf#76942 – cn Canadiana [332]
Banking and commercial guide / Harcourt, Guy M – Houston: Dealy & Baker, 1891. 41 2, 21p. LL-417 – 1 – us L of C Photodup [346]
Banking and industrial finance in india / Das, Nabagopal – Calcutta: Modern Pub Syndicate, [1936] – (= ser Samp: indian books) – us CRL [332]
Banking and investment, 1789-1990 – 606mf – 9 – $5850.00 – 1-55655-434-6 – (p/g only $510) – us UPA [332]
Banking cases annotated / ed by Michie, Thomas J – Charlottesville: Michie Co.. v1-5. 1898-1903 (all publ) – 45mf – 9 – $67.00 – (a collection of all cases affecting banks decided by the courts of last resort in the us) – mf#LLMC 84-705 – us LLMC [336]
Banking growth in puerto rico / Di Venuti, Biagio – Baltimore, MD. 1955 – 1r – us UF Libraries [332]
Banking law journal – New York NY 1889+ – 1,5,9 – ISSN: 0005-5506 – mf#2381 – us UMI ProQuest [346]
Banking law journal – v1-25. 1889-1908 – 202mf – 9 – $303.00 – (updates planned) – mf#LLMC 84-412 – us LLMC [336]
Banking law journal – v1-77. 1889-1960 – 9 – $1705.00 set – ISSN: 0005-5506 – mf#100841 – us Hein [346]
Banking octopus and the silver question : an american financial history / Fogg, F M – Lansing, MI: Lansing Review Co, 1896 (mf ed 19–) – mf#ZT-560 – us Harvard [332]
Banking sector circular bnk / Commercial Advisory Foundation in Indonesia – Djakarta, 1970-1972. nos 1-5 – 3mf – 9 – mf#SE-1384 – ne IDC [659]
Banking software review – Indianapolis IN 1990-92 – 1,5,9 – (cont by: bank automation quarterly) – ISSN: 0892-6778 – mf#14951.04 – us UMI ProQuest [332]
Banking world – London UK 1992-95 – 1,5,9 – ISSN: 0737-6413 – mf#19609 – us UMI ProQuest [332]
Bankirskie kontory : rukovodstvo sluzhashchim v bankakh i bankirskikh kontorakh – Petrozavodsk, 1888 – 1mf – 9 – mf#REF-182 – ne IDC [332]
Banknotes – v1 n1-2 [1979 jun 1-sep 30] – 1 – mf#634157 – us WHS [332]
Bankovaia entsiklopediia / ed by Iasnopol'skii, L N – Kiev, 1914-1916. 2v – 16mf – 9 – mf#REF-149 – ne IDC [332]
Bankovskii spravochnik : vse kreditnye uchrezhdeniia soiuza s s r / ed by Varzar, VV – M, [1926] – 5mf – 9 – mf#REF-11 – ne IDC [332]
Bankovye komissionnye operatsii v selsko-khoziaistvennykh kreditnykh tovarishchestvakh / Vasys, I M – 1928 – 160p 2mf – 9 – mf#COR-373 – ne IDC [335]
Bankrotstvo antiproletarskikh partii v gruzii / Dzhangveladze, G A – Tbilisi, 1981 – 231p 3mf – 9 – mf#RPP-13 – ne IDC [325]
Bankrotstvo burzhuaznykh i melkoburzhuaznykh partii rossii v period podgotovki i pobedy velikoi oktiabrskoi sotsialisticheskoi revoliutsii / Komin, V V – 1965 – 644p 7mf – 9 – mf#RPP-22 – ne IDC [325]
Bankrotstvo melkoburzhuaznykh partii na donu / Sergeev, V N – Rostov n/D, 1979 – 152p 2mf – 9 – mf#RPP-41 – ne IDC [325]
Bankrupt and insolvent calendar – Dublin, Ireland 5 jan-28 dec 1846, 7 jan 1850-31 dec 1866 – 1 – (incorp with: stubbs' weekly gazette for ireland) – uk British Libr Newspaper [072]
Bankrupt register – New York: George T Deller. v1-2 + suppl. 1867-69 (all publ) – 23mf – 9 – $34.50 – (covers us district courts. cont by: national bankruptcy register reports) – mf#LLMC 89-201 – us LLMC [332]
Bankruptcy developments journal – Emory University. v1-17. 1984-2001 – 9 – $400.00 set – ISSN: 0890-7862 – mf#110091 – us Hein [346]
The bankruptcy magazine – v1 nos 1-12. jun 1897-may 1898 (all publ) – 11mf – 9 – $16.50 – (lacking: sep 1897) – mf#LLMC 84-413 – us LLMC [336]
The bankruptcy of india : an enquiry into the administration of india under the crown / Hyndman, Henry Mayers – London, 1886 – (= ser 19th c british colonization) – 3mf – 9 – mf#1.1.784 – uk Chadwyck [350]
Bankruptcy reform act of 1978 : a legislative history / ed by Resnick, Alan N & Wypyski, Eugene M – 1979 – 17v – 9 – $415.00 set – 0-89941-145-2 – (a collection of federal legislation documents concerned with the policies and legislation intent underlying the new act) – mf#301340 – us Hein [346]
Banks and banking : case law and index – New York: Case Law & Co. 1v + index vol. 1903 (all publ) – 18mf – 9 – $27.00 – (a complete series of condensed reports, federal, state and english, including canadian, australian, new zealand and hawaiian reports) – mf#LLMC 95-120 – us LLMC [346]
Banks, Charles Eugene *see* Authorized and authentic life and works of t de witt talmage
Banks, Charles W *see*
– Diaries
Banks, Edgar James *see* Jonah in fact and fancy
Banks, Florence Aiken *see* Coins of bible days
Banks herald – Banks, OR: H A Williams. v1 n1-v14 n2. nov 17 1910-nov 29 1923 – 1 – (ceased in 1923. cont by: beaverton review. aka: cornelius tribune (apr 23 1914-15)) – us Oregon Hist [071]
Banks herald (banks, or: 1910) – Banks OR: H A Williams, 1910- [wkly] – 1 – (absorbed by: beaverton review (-1941); ceased in 1923) – us Oregon Lib [071]
Banks herald (banks, or: 1926) – Banks OR: R Anderson, 1926- – 1 – (related to: beaverton review) – us Oregon Lib [071]
Banks, John Shaw *see*
– Christianity and the science of religion
– The development of doctrine from the early middle ages to the reformation
– The development of doctrine in the early church
– A manual of christian doctrine
– Martin luther, the prophet of germany
– Scripture and its witnesses
– The tendencies of modern theology
Banks, Joseph *see*
– Correspondence, 1766-1820
– History of science and technology, series 2
Banks, Louis Albert et al *see* T de witt talmage
Banks, Martha Burr *see* Heroes of the south seas
Banks news *see*
– Cornelius news
– West washington county news
Bankstown – canterbury express – Bankstown, oct 1984-jun 1995 – 19r – at Pascoe [079]
Bankura darpana – Bankura: [s.n., feb 1892-oct 1 1893; [jun 8 1903-mar 8 1943]; apr 16 1943-apr 8 1945; may 1 1956; [jun 8 1956-sep 16 1959] – us CRL [071]
Banna Dala *see* Rajadhiraj a re to pum
Banna dala, 880 – khan – 934 – Ran kun: Mran ma nuin nam Ca pe Pran pva re a san 1973 [mf ed 1993] – on pt of 1r – 1 – mf#11052 r588 n4 – us Cornell [959]
Bannatyne, Alexander M *see* Defence of the patronage act of 1874
Bannatyne, Andrew *see* Judicial statistics.
Banner – 1980 sep 3-1983; 1984; 1985 – 1 – mf#643779 – us WHS [071]
Banner / Ashtabula Co. Rock Creek – mar 1880-sep 1881 [wkly] – 1r – 1 – mf#B29232 – us Ohio Hist [071]
Banner : official organ of the sons of union veterans – 1929 feb-1944 jun; 1969-1982 jul – 1 – mf#615603 – us WHS [071]
Banner [1974] – Grand Rapids MI 1973+ – 1,5,9 – ISSN: 0005-5557 – mf#9842 – us UMI ProQuest [240]
Banner [aberdeen, scotland] – [Scotland] Aberdeen: George Cornwall jan 1844-dec 1850 (wkly) [mf ed 2004] – 4r – 1 – uk Newsplan [072]
Banner county news – Harrisburg, NE: C L Burgess. v2 n43. nov 16 1894-v62 n27. jul 7 1955 (wkly) [mf ed with gaps] – 17r – 1 – (formed by the union of: early day, and: labor wave. absorbed: banner county republican. absorbed by: gering courier (1900)) – us NE Hist [071]
Banner county republican – Harrisburg, NE: Republican Pub Co. v1 n18. nov 22 1895-v2 n17. nov 13 1896) (wkly) [mf ed with gaps filmed 1966?] – 1r – 1 – (absorbed by: banner county news) – us NE Hist [071]
The banner herald / ed by Progressive Primitive Baptists – Georgia. 23,516p. 1918-97 – 14r – 1 – mf#6886 – us Southern Baptist [071]
Banner of the constitution – La Salle IL 1829-32 – 1 – mf#4056 – us UMI ProQuest [323]
Banner of truth : combined with timothy young people's periodical – 1977 aug-1980; 1981-85; 1986-1989 jun – 1 – mf#554400 – us WHS [073]
Banner of truth – Grand Rapids MI 1989+ – 1 – mf#15295 – us UMI ProQuest [240]
Banner of ulster – Belfast, Ireland 10 jun 1842-31 aug 1869 (wkly) – 27r – 1 – (discontinued) – uk British Libr Newspaper [072]
Banner press / Medina Co. Wadsworth – oct 1907-sep 1942 (out of order) [wkly] – 15r – 1 – mf#B5997-6011 – us Ohio Hist [071]
Banner series / Meigs Co. Pomeroy – may 1867-dec 1868 [wkly] – 1r – 1 – mf#B123 – us Ohio Hist [071]
Banner und volksfreund – 1855 apr 10-sep 29; 1855 oct 1-1856 feb 29; 1856 mar 1-aug 30; 1856 sep 1-1857 feb 28; 1857 mar 2-aug 8; 1857 sep 1-1858 feb 27; 1858 mar 1-aug 31; 1858 sep 1-1859 feb 28; 1859 mar 1-aug 31; 1859 sep 1-1860 feb 28; 1860 mar 3-1861 aug 31; 1861 sep 1-1862 feb 28; 1862 mar 1-aug 30; 1862 sep 2-dec 31; 1865 apr 8; 1869 jan 24; 1879 feb 12-oct 24; 1879 oct 25-1880 may 11 – 1 – mf#1138604 – us WHS [071]
Banner-courier – Oregon City OR: Clackamas County Banner Pub Co Inc, 1919-50 [triwkly] – 1 – (began in 1919; merger of: oregon city courier (oregon city, or) and: clackamas county banner; merged with: oregon city enterprise (oregon city, or: daily), to form: enterprise-courier (oregon city, or)) – us Oregon Lib [071]
Bannerjea, Devendra Nath *see* India's nation builders
Banner-journal – 1926 mar 10-1939 dec 27; 1940-66; 1967-1979 dec 26; 1980-1997 dec – 1 – mf#1008022 – us WHS [071]
Bannerman, Charles *see* Charles bannerman papers
Bannerman, David Douglas *see*
– The church of christ
– Present position of the case of prof robertson smith
– The scripture doctrine of the church
– Worship, order, and polity of the presbyterian church
Bannerman, James *see*
– The church of christ
– Dialogues and detached sentences in the chinese language
– Inspiration
– Sermons
Banner-press – David City, NE: A H Morton & H E Hosch. v64 n17. dec 10 1953- (wkly) – 1 – (formed by the union of: people's banner and: butler county press) – us NE Hist [071]
Bannesianisme et molinisme / Regnon, Theodore de – Paris: Retaux-Bray 1890 [mf ed 1991] – 1mf – 9 – 0-7905-9452-8 – (no more publ? incl bibl ref) – mf#1989-2677 – us ATLA [120]
La banniere de marie immaculee – Ottawa: Peres oblats de Marie Immaculee, Juniorat du Sacrecoceur, 1893-1968 – 5r – 1 – us CRL [240]
Banning and arden's reports of patent cases in the u.s. circuit courts / Banning, H A & Arden, A – New York: L K Strouse. v1-5. 1874-81 (all publ) – 40mf – 9 – $60.00 – mf#LLMC 81-430 – us LLMC [346]
Banning, Emile Theodore *see* Africa and the brussels geographical conference
Banning, H A *see* Banning and arden's reports of patent cases in the u.s. circuit courts
Banning record – v2 n45,52-v3 n2,4-7,11 [1909 nov 25-1910 jan 13-27, feb 10-mar 3,31] – 1 – mf#918728 – us WHS [071]
[Banning-] record gazette – CA. sep 1972- – 70+ r – 1 – $4200.00 (subs $140/y) – mf#R02045 – us Library Micro [071]
Bannister, H *see* Missale gothicum, vol 1-2 (hbs52,54)
Bannister, Henry *see* Book of isaiah
Bannister, John William *see* Sketch of a plan for settling in upper canada
Bannister, Saxe [pseud] *see*
– British colonization and coloured tribes
– Humane policy
– Remarks on the indians of north america
Banos de banos / Diaz Perez, Nicolas – 1881 – 9 – sp Bibl Santa Ana [810]
Banos de montemayor. puerta de extremadura. apuntes 1971 / Hernandez Diaz, Erasmo – Caceres: Edit. Extremadura, 1971 – 1 – sp Bibl Santa Ana [946]
Banos de Velasco, J *see* L anneo seneca
Banque agricole and fonciere d'haiti / Deset, Enoch – Paris, France. 1882 – 1r – us UF Libraries [332]
Banque nationale d'haiti / Marcelin, Frederic – Paris, France. 1890 – 1r – us UF Libraries [332]
Banque nationale du Cambodge *see* Statuts
Banque nationale du cambodge – [Phnom-Penh 1955?] [mf ed 1989] – [ill] 1r with other items – 1 – (in cambodian & french) – mf#mf-10289 seam reel 028/03 [§] – us CRL [332]
La banque nationale du cambodge a quinze ans / Kim, Touch – [Phnom-Penh: Impr Sangkum Reastr Niyum] 1969 [mf ed 1989] – 1r with other items – 1 – mf#mf-10289 seam reel 027/01 [§] – us CRL [332]
Banquet addresses : annual banquet of the new york society of the order of the founders and patriots of america... – 8th-10th [1904-06], 12th [1908], 14th [1910] – 1 – mf#198189 – us WHS [080]
Banquet and ball in commemoration of the 50th anniversary of the reign of her most gracious majesty queen victoria – [s.l.: s.n. 1887?] [mf ed 1987] – 1mf – 9 – mf#27100 – cn Canadiana [390]
Le banquet donne a sir john a macdonald a quebec, le 15 octobre 1879 : discours / Chapleau, Joseph-Adolphe – S.l: s.n, 1879? – 1mf – 9 – mf#02570 – cn Canadiana [323]
Banquet offert a sir hector l langevin : ...ministre des travaux publics, par les citoyens de montreal a l'hotel windsor, le jeudi 8 octobre 1883 – Montreal: impr de "La Minerve", 1883 [mf ed 1980] – 1mf – 9 – 0-665-02505-X – mf#02505 – cn Canadiana [920]
Banquet to graduates of mcgill : friday, april 2nd, 1880 – [Montreal?: s.n, 1880?] [mf ed 1980] – 1mf – 9 – 0-665-00914-3 – mf#00914 – cn Canadiana [378]
Banquet to the hon f n blake, american consul : royal hotel, hamilton, 1st august, 1873 – [Hamilton, Ont: Hamilton Spectator, 1873?] [mf ed 1980] – 1mf – 9 – 0-665-02091-0 – mf#02091 – cn Canadiana [327]
Banquete / Santiago, Silviano – Rio de Janeiro, Brazil. 1970 – 1r – us UF Libraries [972]
Banquete de cavalleros y orden de vivir, ansi en tiempo de la sanidad como en... / Lobera de Avila, L – Alcala, 1542 – 9 – sp Cultura [615]

Banquete de nobles caballeros...e modo de bivir desde que se levantan hasta que se acuestan e traya del regimiento curativo y preservativo de las fiebres pestilenciales e de la pestilencia / Lobera de Avila, L – Augsburg, 1530 – 4mf – 9 – sp Cultura [615]
Bansavatar bisatr / Lak Sarum – Bhnam Ben: Panangar Tic Hun 2500 [1957] [mf ed 1990] – 1r with other items – 1 – (in khmer; also: pannagar tic hun, 2502 [1959] mf-10289 seam reel 118/1 [§]) – mf#mf-10289 seam reel 120/6 [§] – us CRL [959]
Bansavatar khmaer sankhep / Li Dham Ten – Bhnam Ben: Ron Bumb Sinh 2502 [1959] [mf ed 1990] – 1r with other items – 1 – (in khmer) – mf#mf-10289 seam reel 118/5 [§] – us CRL [959]
Bansavatar prades kambuja / Li Dham Ten – Bhnam Ben: Panangar Sen Nuan Huat [1964] [mf ed 1990] – 1r with other items – 1 – (in khmer) – mf#mf-10289 seam reel 118/3 [§] – us CRL [959]
Bansavatar prades kambuja sankhep / Jh H Men Jin – Kam Ban Cam: Pannagar Khmaer Samay 2507 [1964] [mf ed 1990] – 1r with other items – 1 – (in khmer) – mf#mf-10289 seam reel 117/12 [§] – us CRL [959]
Banta, Arthur Mangun *see* Studies on the physiology, genetics, and evolution
Banter – Halifax, NS: Morton's News Agency, [1874-18–] [mf ed v1 n7 oct 1 1874-v1 n8 oct 15 1874] – 9 – mf#P05139 – cn Canadiana [870]
Bantoe-filosofie / Tempels, Placide – Antwerpen, Belgium. 1946 – 1r – us UF Libraries [305]
Bantog na vida generoso asin ni flora na magna aquing hade – [Nueva Caceres: Libreria Mariana 190-?] [mf ed Bloomington IN: Indiana Uni Lib, Preservation Dept 1984] – (= ser Coll...in the bikol language) – 1r – 1 – us Indiana Preservation [400]
The bantu are coming : phases of south africa's race problem / Phillips, Ray Edmund – [2nd ed.] London: Student Christian Movt Press, [1930] – 1 – us CRL [960]
Bantu bayile nkuwa – Kasempa, Zambia. 19–? – 1r – us UF Libraries [305]
Bantu, boer, and briton / Macmillan, William Miller – Oxford, England. 1963 – 1r – us UF Libraries [305]
Bantu education – Pretoria, South Africa. 1957? – 1r – us UF Libraries [370]
Bantu folk lore : medical and general / Hewat, Matthew L – Cape Town, South Africa. 1906 – 1r – us UF Libraries [390]
Bantu folk tales : seven stories / Hertslet, Jessie – Cape Town, South Africa. 1946 – 1r – us UF Libraries [390]
The bantu in south african life / Brooks, Edgar Harry – Johannesburg: S A Institute of Race Relations, 1943 – 1 – us CRL [305]
Bantu law in south africa / Seymour, S M – Cape Town, South Africa. 1970 – 1r – us UF Libraries [305]
Bantu literature and life / Shepherd, Robert Henry Wishart – Lovedale, South Africa. 1955 – 1r – us UF Libraries [305]
Bantu mirror – Bulawayo, Rhodesia, jul 21 1956 – (= ser ST Clair Drake coll of Africana) – (issues filmed as pt of: st clair drake coll of africana) – us CRL [079]
Bantu philosophy / Tempels, Placide – Paris, France. 1959 – 1r – us UF Libraries [305]
Bantu tales / Price, Pattie – New York: E P Dutton, 1938 – us CRL [390]
Bantu word division / Guthrie, Malcolm – London, England. 1948 – 1r – us UF Libraries [470]
Bantu world – Johannesburg SA, 1932-46 – (title varies: world) – mf#MS05 – sa National [079]
Bantug nga historia ni bernardo carpio nga anac ni d a jimena sa espana *see* Historia sang dalagangan nga si bernardo carpio nga anac ni d sancho diaz cag ni d a jimena sa guinharian sa espana
Bantu-speaking tribes of south africa / Schapera, Isaac – London, England. 1937 – 1r – us UF Libraries [307]
Bantustan : a study in practical apartheid / Kruger, J D L – Queenstown, South Africa: Daily Representative, 1951 – 1 – us CRL [960]
Bantustans / Giniewski, Paul – Cape Town, South Africa. 1961 – 1r – us UF Libraries [305]
Bantysh-Kamenskii, D *see* Slovar dostopamiatnykh liudei russkoi zemli...
Bantysh-Kamenskii, D N *see* Deianiia znamenitykh polkovodtsev i ministrov sluzhivshikh v tsarstvovanie gosudaria imperatora petra velikogo
Bantysh-Kamenskii, N N *see* Obzor vneshnikh snoshenii rossii (po 1800 god)
Banvard, John *see* Family papers
Banville, Theodore Faullain De *see* Socrate et sa femme
Banyarwanda et barundi / Bourgeois, R – [Bruxelles, 1954-58] v3 – us CRL [960]
Banzay : diario en las trincheras coreanas / Caicedo Montua, Francisco A – Bogota, Colombia. 1961 – 1r – us UF Libraries [972]
Banzhaf, Johannes *see*
– Lachendes leben
– Lustiges volk
Bao dong phap – Grand journal quotidien d'information en langue annamite. Dir. Ngo-Van-Phu. no. 3-2144. Hanoi. 12 janv 1925-32 – 1 – fr ACRPP [079]
Bao dong-phap – Hanoi: Impr Bao dong-phap [jan 8 1926-jun 5 1932] – 7r – 1 – mf#mf-11119 seam – us CRL [079]
Bao to / Tran Trong Duc – [Saigon]: Tri-am 1974 [mf ed 1992] – on pt of 1r – 1 – mf#11052 r377 n7 – us Cornell [959]
Baoesastra indonesia-djawi : tjap-tjapan kaping tiga / Poerwadarminta, W J S – Djakarta: Gunseikanbu Kokumin Tosyokyoku (Bale Poestaka), 2605 (B P 1450) – 203p 3mf – 9 – mf#SE-2002 mf133-135 – ne IDC [490]
Baour-Lormian, Pierre Marie Francois Louis *see*
– Jerusalem delivree
– Mahomet 2
Bapat, Purushottam Vishvanath *see* Vimuttimagga and visuddhimagga
Bapinidu, Maganti *see* Indo-cambodia
Baptism / Ditzler, Jacob – Nashville TN: Southern Methodist Pub House 1884 [mf ed 1992] – 1mf – 9 – 0-524-05473-8 – (incl bibl ref) – mf#1990-5120 – us ATLA [242]
Baptism : how and for whom? / Colpitts, W W – Toronto: W Briggs; Montreal: C W Coates, 1896 [mf ed 1980] – 1mf – 9 – 0-665-00720-5 – mf#00720 – cn Canadiana [240]
Baptism : its subjects and mode viewed in connection with the heres... / Torry, James – Kirkwall, Scotland. 18– – 1r – us UF Libraries [242]
Baptism : not the means, but the symbol of the believer's union with... – Dublin, Ireland. 18– – 1r – us UF Libraries [242]
Baptism – Plymouth? England. 18– – 1r – us UF Libraries [242]
Baptism : sprinkling and pouring versus immersion: the question settled / King, David – Leicester: publ by...Churches of Christ in Great Britain & Ireland 1891 [mf ed 1992] – 1mf – 9 – 0-524-03321-8 – mf#1990-4681 – us ATLA [240]
Baptism : what saith the scripture? / Groves, Henry – London, England. 18– – 1r – us UF Libraries [242]
Baptism : [w]ho are the subjects and what is the mode?: being the substance of [t]wo discourses [pre]ached in the congregational chapel, london, c w, dec 9th, 1849 / Clarke, William Fletcher – [London, ON?: s.n.] – 0-665-91040-1 – (incl bibl ref) – mf#91040 – cn Canadiana [240]
Baptism : with reference to its import and modes / Beecher, Edward – New York: John Wiley 1849, c1848 [mf ed 1986] – 1mf – 9 – 0-8370-9363-5 – mf#1986-3363 – us ATLA [240]
Baptism : with reference to its import, modes, history, proper use, and the duty of parents to baptized children / Chapman, James L – Louisville KY: Morton & Griswold 1853 [mf ed 1992] – 1mf [ill] – 9 – 0-524-05472-X – mf#1990-5119 – us ATLA [242]
Baptism according to scripture / Hoare, Edward Hatch – London, England. 1850 – 1r – us UF Libraries [242]
Baptism according to st paul / Brown, Hugh Stowell – London, England. 18– – 1r – us UF Libraries [242]
Baptism and confirmation / Brooks, Phillips – New York: E P Dutton 1892, c1880 – 1mf – 9 – 0-7905-7501-9 – mf#1989-0726 – us ATLA [242]
Baptism and the baptists / Duncan, George – London: Baptist Tract & Book Soc 1885 [mf ed 1993] – 1mf – 9 – 0-524-07094-6 – mf#1991-2917 – us ATLA [242]
Baptism and the remission of sins / Mullins, E Y – 1906 – 1 – $5.00 – (a e dickenson. what baptist principles are worth to the world. 1889) – us Southern Baptist [242]
Baptism as taught in the scriptures / Lloyd, Rhys R – Boston: Congregational Sunday-School & Pub Soc c1895 [mf ed 1993] – 1mf – 9 – 0-524-07525-5 – mf#1991-3155 – us ATLA [220]
Baptism considered in its subjects and mode : in three letters to the reverend william elder, in which the nature of that ordnance is explained... / Ross, Duncan – Pictou, NS?: Weir Durham, 1825 – 1mf – 9 – mf#64125 – cn Canadiana [242]
Baptism discovered plainly and faithfully / Norcott, John – London, England. 1887 – 1r – us UF Libraries [242]
Baptism doth save / Smith, C A I – Plymouth, England. 1835 – 1r – us UF Libraries [242]
"Baptism, in a nutshell" examined / Ingels, Marion – St Louis MO: Christian Pub Co c1889 [mf ed 1992] – 1mf – 9 – 0-524-03319-6 – mf#1990-4679 – us ATLA [242]
The baptism in fire : the privilege and hope of the church in all ages / Smith, Charles Edward – Boston: D Lothrop, c1883 – 1mf – 9 – 0-524-06555-1 – mf#1991-2639 – us ATLA [242]
Baptism in plain english : or, an exposition of the divine command in relation to the initiatory rite of the christian church / Judd, Orrin Bishop – New York: Holman, Gray, 1853 [mf ed 1991] – 1mf – 9 – 0-7905-9007-7 – mf#1989-2232 – us ATLA [240]
Baptism in spirit and in fire / Challen, James – Philadelphia: James Challen 1859 [mf ed 1992] – 1mf – 9 – 0-524-03696-9 – mf#1990-4801 – us ATLA [242]
Baptism, jewish and christian / Hanauer, James Edward – London, New York: Longmans, Green 1906 [mf ed 1989] – (= ser Judaism and christianity: short studies 4) – 1mf – 9 – 0-7905-0950-4 – (incl bibl ref) – mf#1987-0950 – us ATLA [230]
Baptism misunderstood / Gatty, Alfred – London, England. 1849 – 1r – us UF Libraries [242]
The baptism of fire : and other sermons / Johnston, J Wesley – Toronto: William Briggs, 1888 – 1mf – 9 – 0-8370-7296-4 – mf#1986-1296 – us ATLA [242]
The baptism of roger williams : a review of rev. dr. w.h. whitsitt's inference / King, Henry Melville – Providence: Preston & Rounds, 1897 – 1mf – 9 – 0-7905-4984-0 – mf#1988-0984 – us ATLA [242]
The baptism of the ages and of the nations / Cathcart, William – Philadelphia: American Baptist Pub Soc, c1878 [mf ed 1989] – 1mf – 9 – 0-7905-4385-0 – (incl bibl ref) – mf#1988-0385 – us ATLA [240]
The baptism of the holy ghost / Mahan, Asa – New York: WC Palmer, Jr, 1870 – 1mf – 9 – 0-7905-8841-2 – mf#1989-2066 – us ATLA [242]
The baptism of the holy ghost / Mahan, Asa – New York: W.C. Palmer, jr., (c1870). viii,215p – 1 – us Wisconsin U Libr [242]
Baptism, the sacrament of regeneration : a sermon preached in the holy cross church, lockeport, ns, on evening of first sunday in lent / Gibbons, Simon – Yarmouth, NS?: Times Job Office, 188-? – 1mf – 9 – mf#64744 – cn Canadiana [240]
The baptismal controversy : its exceeding sinfulness / Hartzel, Jonas – Oskaloosa, Iowa: Central Book Concern, 1877 – 1mf – 9 – 0-524-02957-1 – mf#1990-4509 – us ATLA [242]
The baptismal question : a discussion of the baptismal question / Cooke, Parsons – Boston: Gould, Kendall & Lincoln, 1842 – 1mf – 9 – 0-524-07311-2 – mf#1991-3026 – us ATLA [242]
Baptismal regeneration : a doctrine of the church of england – London, England. 1842 – 1r – us UF Libraries [242]
Baptismal regeneration / Ferrar, William Hugh – London, England. 1865 – 1r – us UF Libraries [242]
Baptismal regeneration / Parker, Edward – Bristol, England. 1845? – 1r – us UF Libraries [242]
Baptismal regeneration as maintained by the church of england / Scholefield, James – Cambridge, England. 1850 – 1r – us UF Libraries [242]
Baptismal regeneration opposed to the doctrines and facts of the bi... / Fergusson, Archibald – London, England. 1854 – 1r – us UF Libraries [242]
Baptismal remission : or, the design of christian baptism / Hughey, George Washington – Cincinnati: Cranston & Stowe 1891 [mf ed 1992] – 1mf [ill] – 9 – 0-524-04543-7 – mf#1990-5050 – us ATLA [240]
The baptismal remission theory / Lofton, George A – 1 – 5.00 – us Southern Baptist [242]
Baptismal service according to the primitive mode – London, England. 18– – 1r – us UF Libraries [242]
Baptisms registers of the manea methodist circuit – 2mf – 9 – £2.50 – uk CambsFHS [242]
Baptist – (= ser Transcripts of nonconformist registers (pre 1837)) – 3mf – 9 – £5.00 – uk BedsFHS [242]
Baptist – 1965 apr, nov/dec; 1966 jan-may; 1969 jan/feb-mar/may; jun/aug; 1970 mar, oct; 1971 summer; 1972 summer; 1976 winter; 1977 jul/aug; 1985 sep/oct; 1986 jan/feb, jun/aug; 1987 autumn – 1 – mf#4828276 – us WHS [242]
Baptist adult union quarterly – 1930-61 – 1 – $269.22 – us Southern Baptist [242]
Baptist advocate : serving baptist homes and churches throughout louisiana – 1985 jul/aug – 1 – mf#3912614 – us WHS [242]
Baptist and commoner – Little Rock, AR. 23 Jul 1914-15 Feb 1939 – 1 – $432.33 – us Southern Baptist [242]
Baptist and congregational pioneers / Shakespeare, John Howard – 2nd ed. London: National Council of Evangelical Free Churches 1907 [mf ed 1990] – (= ser Eras of nonconformity 3) – 1mf – 9 – 0-7905-8209-0 – mf#1988-8092 – us ATLA [242]
The baptist and reflector – Brentwood, TN. 105,994p. 1835-1998 – 1 – mf#0230 – us Southern Baptist [242]
The baptist and slavery / Putnam, Mary B – 1840-45 – 1 – 5.00 – us Southern Baptist [242]
Baptist annals of oregon / Mattoon, C H – v1 and 2. 1844-1910 – 1 – $35.00 – us Southern Baptist [242]
Baptist annual register / Rippon, John – 4v. 1790-1802. (Complete) – 1 – $92.68 – us Southern Baptist [242]
Baptist annual register / ed by Rippon, John – London. UK. 1790-1802 – 1 – 54.15 – us ABHS [242]
Baptist argus – Louisville, KY. 1897-Apr 1908. and Baptist World. May 1908-19 – 1 – $966.73 – us Southern Baptist [242]
Baptist Association *see*
– Baptist messages from radio hour and television
– Confession of faith adopted by the baptist association met at philadelphia, a, 25 sep 1742
– Sketches of history of the baptist churches within the limits of the rappahannock associations in virginia
– Summary of church discipline; drawn up by the direction of the dover baptist association, va
Baptist Association. Charleston, South Carolina *see* An address from the charleston association...
Baptist Association. Connecticut *see* Second report of the connecticut society
Baptist Association. Louisiana *see*
– Luther rice
– Vernon
Baptist Association. Louisiana. *see* Papers
Baptist Association. Texas *see* Dallas
Baptist Associations *see*
– Annual
– Arkansas materials
– Index to regular and primitive baptist associations' annuals
– Misc. annuals
Baptist associations = Concord – TN. 1812-1908 – 1 – $32.48 – (1909-56) – us Southern Baptist [242]
Baptist Associations. Alabama *see* Annual
Baptist Associations. Arizona *see* Annual
Baptist Associations. Arkansas *see* Annual
Baptist Associations. California *see*
– Annual
– Annuals
Baptist Associations. Colorado, Southeastern, Arkansas Valley *see* Annual
Baptist Associations. Flint River. Alabama *see* Annuals
Baptist Associations. Florida *see* Annual
Baptist Associations. Georgia *see* Annual
Baptist Associations. Illinois *see*
– Alton industrial-williamson county
– Annual
– Olney
Baptist Associations. Illinois. Chicago Southern Baptist *see* Annual
Baptist Associations. Illinois. Saline *see* Annuals
Baptist Associations. Indiana, Brownstown *see* Annual
Baptist Associations. Jefferson County. Tennessee *see* Annuals
Baptist Associations. Kentucky *see*
– Annual
– Annuals
Baptist Associations. Kentucky. Blood River *see* Annuals
Baptist Associations. Kentucky. Greenup *see* Annuals
Baptist Associations. Kentucky. Laurel River *see* Annual
Baptist Associations. Louisiana *see*
– Annual
– Annuals
Baptist Associations, Maine. Androscroggin Free Will Baptist and United Baptist Annual Meetings *see* Minutes
Baptist Associations, Maine. Androscroggin Free Will Baptist Quarterly Meeting *see* Records
Baptist Associations, Maine. Anson Free Will Baptist Quarterly Meeting *see* Records
Baptist Associations, Maine. Bowdoinham Free Will Baptist Quarterly Meeting *see*
– Minutes
– Records
Baptist Associations, Maine. Bowdoinham Free Will Baptist Yearly Meeting *see* Records
Baptist Associations, Maine. Cumberland Baptist Association. *see*
– Records
Baptist Associations, Maine. Eastern Free Will Baptist Quarterly Conference *see* Records
Baptist Associations, Maine. Eastern Maine Baptist Association. Bible and Religious Tract Society *see* Minutes
Baptist Associations, Maine. Ellsworth Free Will Baptist Quarterly Meeting *see* Records
Baptist Associations, Maine. Exeter Free Will Baptist Quarterly Meeting *see* Records
Baptist Associations, Maine. Hancock Baptist Association *see* Records

Baptist Associations, Maine. Kennebec Baptist Association. Woman's Baptist Foreign Missionary Society *see* Records

Baptist Associations, Maine. Kennebec Free Will Baptist Yearly Meeting *see* Minutes

Baptist Associations, Maine. Lincoln Baptist Association *see* Minutes

Baptist Associations, Maine. Lincoln Baptist Association. *see* Records

Baptist Associations, Maine. Maine Central Free Will Baptist Yearly Meeting *see* Records

Baptist Associations, Maine. Maine Eastern Free Will Baptist Yearly Meeting *see* Records

Baptist Associations, Maine. Maine Free Baptist Association. Annual Meeting *see* Minutes

Baptist Associations, Maine. Oxford Free Will Baptist Quarterly Meeting *see* Records

Baptist Associations, Maine. Oxford Free Will Baptist Yearly Meeting *see* Records

Baptist Associations, Maine. Parsonfield Free Will Baptist Quarterly Meeting *see* Records

Baptist Associations, Maine. Penobscot Baptist Association. Female Missionary Society *see* Records

Baptist Associations, Maine. Penobscot Free Will Baptist Yearly Meeting *see*
– Records

Baptist Associations, Maine. South Aroostook Free Will Baptist Quarterly Meeting *see* Records

Baptist Associations, Maine. Waterville Free Will Baptist Quarterly Meeting *see* Records

Baptist Associations, Maine. Western Free Will Baptist Quarterly Conference *see* Records

Baptist Associations. Maryland *see* Annual

Baptist Associations. Michigan. Motor Cities *see* Annual

Baptist Associations. Mississippi *see* Annual

Baptist Associations. Missouri *see*
– Annual
– Eleven points river, shannon county

Baptist Associations. Missouri. Bethel Association. United Baptist *see* Annual

Baptist Associations. Missouri. Meramec Landmark Baptist Association *see* Annual

Baptist Associations. Missouri. Mineral Area *see* Annual

Baptist Associations, New Hampshire. Belknap Association of Free Will Baptist Churches *see* Records

Baptist Associations, New Hampshire. Belknap Association of Free Will Baptist Churches. Woman's Missionary Society *see* Records

Baptist Associations, New Hampshire. Belknap Free Will Baptist Quarterly Meeting *see* Records

Baptist Associations, New Hampshire. Portsmouth Baptist Association *see* Records

Baptist Associations, New Hampshire. Sandwich Free Will Baptist Quarterly Meeting. Ministers Conference *see* Records

Baptist Associations. New Mexico *see*
– Annual

Baptist Associations. New Mexico. Santa Fe *see* Annual

Baptist Associations. New York and Vermont. Miscellaneous Associations *see* Annual

Baptist Associations. North Carolina *see* Annual

Baptist Associations. North Carolina. Chowan Association *see* Annual

Baptist Associations. North Carolina. Eastern Association *see* Annual

Baptist Associations. North Carolina. Flat River Association *see* Annual

Baptist Associations. North Carolina. Tar River Association *see* Annual

Baptist Associations. Ohio *see* Annual

Baptist Associations. Oklahoma *see* Annual

Baptist Associations, Pennsylvania. Beaver Baptist Association *see* Manuscript minutes

Baptist Associations. Pennsylvania. Philadelphia Baptist Association *see* Annual

Baptist Associations, Pennsylvania. Redstone Baptist Association *see* Manuscript minutes

Baptist Associations. Primitive. Alabama *see* Annual

Baptist Associations. Primitive: Arkansas, California, Arizona, Delaware, 1844-1950 *see* Annual

Baptist Associations. Primitive. Florida *see* Annual

Baptist Associations. Primitive. Georgia *see* Annual

Baptist Associations. Primitive. Illinois, Iowa, Indiana, Kansas *see* Annual

Baptist Associations. Primitive. Kentucky *see* Annual

Baptist Associations. Primitive: Louisiana, Maine, Maryland, Michigan, Mississippi, Missouri, New York, Nebraska *see* Annual

Baptist Associations. Primitive. North Carolina *see* Annual

Baptist Associations. Primitive: Ohio, Oklahoma, Pennsylvania, South Carolina *see* Annual

Baptist Associations. South Carolina *see*
– Annual
– Annuals

Baptist Associations. South Carolina. Edgefield. Quarterly Meetings *see* Annual

Baptist Associations. Tennesse. Sweetwater/ Eastanalle *see* Annuals

Baptist Associations. Tennessee *see* Annual

Baptist Associations. Tennessee. Big Hatchie *see* Annuals

Baptist Associations. Tennessee. Concord *see* Annuals

Baptist Associations. Texas *see*
– Annual
– Annuals
– Red fork, salt fork, and wichita-archer-clay baptist associations. texas

Baptist Associations. Texas Bma *see* Annuals

Baptist Associations. Texas, Liberty *see* Annual

Baptist Associations. Utah-Idaho, Treasure Valley *see* Annuals

Baptist Associations. Virginia *see* Annual

Baptist Associations. Washington. Mount Pleasant *see* Annual

Baptist Associations. Washington. Puget Sound *see* Annual

Baptist banner – Atlanta/Augusta, GA. By James N. Ells. Dec 1862-Feb 1865. Scattered issues – 1 – us ABHS [242]

Baptist banner – Huntington, WV. Baptist General Association of West Virginia. 1891, 1893-97, 1898/99. Single reels available – 1 – us ABHS [242]

Baptist banner – IL. 1877-79, 1887-88 – 1 – $132.05 – us Southern Baptist [242]

Baptist basket – Louisville, KY. v9-15. 1881-94 – 1 – (v15 incomplete) – us Southern Baptist [242]

Baptist beacon – Southfield, MI. 2685p. 1986-99 – (= ser Michigan Baptist Advocate) – 1 – (formerly: michigan baptist advocate until may 1997) – mf#5865 – us Southern Baptist [242]

Baptist beacon – Vinita, OK. sep, apr 1949; jan, apr 1950 – 1 – $5.00 – us Southern Baptist [242]

The baptist beacon – Salem, Albany, McMinnville, Oregon. 1877-79. 33,710pp – 1 – us Southern Baptist [242]

Baptist beliefs / Mullins, Edgar Young – Louisville KY: Baptist World Publ Co 1912 [mf ed 1991] – 1mf – 9 – 0-7905-8528-6 – mf#1989-1753 – us ATLA [242]

Baptist biography, biblical recorder, obituary notices, 1835-1904 – 607p – 1 – $21.24 – us Southern Baptist [242]

Baptist boys and girls – 1904-22 – 1 – $136.92 – us Southern Baptist [242]

Baptist Brazilian Convention *see* Proceedings

Baptist builders in louisiana / ed by Durham, John Pinckney & Ramond, John S – Shreveport: Durham-Ramond, 1934 – 1 reel – 1 – $17.92 – (448p) – us Southern Baptist [242]

Baptist bulletin – Grand Rapids, MI. v1. 1933-May 1967 – 1 – $377.23 – us Southern Baptist [242]

A baptist century around the alamo / San Antonio Baptist Association – 212p. 1858-1958 – 1 – $7.42 – us Southern Baptist [242]

Baptist challenge – Little Rock, AR. v1-7. 1961-Mar 1967 – 1 – $7.49 – us Southern Baptist [242]

Baptist chorals – Lorenz Publication. 1888 – 1 – $5.00 – us Southern Baptist [242]

Baptist chronicle – Louisiana. 1888-1919; Baptist Message. 1920-82 – 1 – $1,975.08 – us Southern Baptist [242]

Baptist chronicle and literary register – KY. v1-3. 1830-32 – 1 – $33.60 – us Southern Baptist [242]

Baptist church and associations – ME. 10r. 9455p – 1 – us Southern Baptist [242]

The baptist church directory : a guide to the doctrines and discipline, officers and ordinances, principles and practices, of baptist churches / Hiscox, Edward Thurston – New York: Sheldon; Boston: Gould and Lincoln, 1864 – 1mf – 9 – 0-524-01331-4 – mf#1990-4080 – us ATLA [242]

Baptist church discipline, broadman press / Garrett, James L, Jr – NASHVILLE, TN – 1 – $5.00 – us Southern Baptist [242]

Baptist church in the great valley – VALLEY FORGE, PA. 1720-1942 – 1 – $102.96 – us Southern Baptist [242]

Baptist church of christ at berryville. clark county – BERRYVILLE, VA. 1841-60 – 1 – $9.72 – us Southern Baptist [242]

The baptist church of christ at woodbury – Woodbury, TN. oct 1844-feb 1888 – 1 – $11.52 – us Southern Baptist [242]

The baptist church of christ called providence on big river – Washington, MO. 11 sep 1831-feb 1894 – 1 – $13.23 – us Southern Baptist [242]

Baptist church of jesus chri louisville – KY. 24 Oct 1830-Mar 1840 – 1 – $5.00 – (later became walnut street baptist church) – us Southern Baptist [242]

Baptist church of nashville (the old school baptists) – NASHVILLE, TN. 3 May 1838, Jun 1839-22, Feb 1878 – 1 – $10.62 – us Southern Baptist [242]

Baptist church perpetuity : or, the continuous existence of baptist churches from the apostolic to the present day / Jarrel, Willis Anselm – Dallas TX: W A Jarrel 1894 [mf ed 1992] – 2mf – 9 – 0-524-03320-X – (incl bibl ref; int by w w everts, jr) – mf#1990-4680 – us ATLA [242]

Baptist church polity, doctrines, confession of faith – 1853-90 – 1 – $16.31 – (ten tracts) – us Southern Baptist [242]

Baptist church, polity, doctrines, confession of faith – 1845-75 – 1 – $16.24 – (nine authors) – us Southern Baptist [242]

Baptist churches of east tennessee : church genealogy, tabular form / Toomey, Glenn – 5r – 1 – $278.91 – us Southern Baptist [242]

The baptist commentator reviewed : two letters to the rev william jackson on christian baptism... / Taylor, Thomas – Halifax, NS?: J S Cunnabell, 1835 – 2mf – 9 – (incl bibl ref) – mf#64614 – cn Canadiana [240]

Baptist confessions of faith / McGlothlin, William Joseph – 1911 – 1 – $13.72 – us Southern Baptist [242]

Baptist confessions of faith / McGlothlin, William Joseph – Philadelphia: American Baptist Pub Soc 1911 [mf ed 1990] – 1mf – 9 – 0-7905-9509-5 – mf#1989-1214 – us ATLA [242]

Baptist Congress *see*
– Baptist congress
– Proceedings

Baptist congress : proceedings / Baptist Congress – 1882-1913 [complete] – 3r – 1 – mf#ATLA R0116 – us ATLA [242]

Baptist councils in america : a historical study of their origin and the principles of their development / Allison, William Henry – Chicago: G K Hazlitt 1906 [mf ed 1989] – 1mf – 9 – 0-7905-4368-0 – (incl bibl ref) – mf#1988-0368 – us ATLA [242]

Baptist courier – SC. 1869-1991 – 1 – $4,284.60 – (the working christian: 1869-77) – us Southern Baptist [242]

The baptist denomination : its history, doctrines, and ordinances / Haynes, Dudley C – New York: Sheldon, 1857 – 1mf – 9 – 0-524-05179-8 – mf#1990-5098 – us ATLA [242]

Baptist digest – Topeka, KA. 12,335p. dec 3 1945-1999 – 1 – mf#0725 – us Southern Baptist [242]

Baptist dishonesty : misquotations and other gross misrepresentations in the recent pamphlet on baptism of the baptist minister, rev a a cameron, of ottawa / Bethune, John – [Ottawa?: s.n, 1876?] [mf ed 1985] – 1mf – 9 – 0-665-37354-6 – mf#37354 – cn Canadiana [240]

Baptist doctrines : being an exposition, in a series of essays of representative baptist ministers, of the distinctive points of baptist faith and practice / ed by Jenkens, Charles Augustus – St Louis: Chancy R Barns 1880 [mf ed 1986] – 2mf – 9 – 0-8370-9001-6 – mf#1986-3001 – us ATLA [242]

Baptist Educational, Missionary and Sunday School Convention of South Carolina *see*
– Minutes of the baptist educational, missionary and sunday school convention of south carolina
– The...anniversary of the baptist educational, missionary and sunday-school convention of south carolina

Baptist Educational & Missionary Convention of South Carolina *see* Minutes of the... annual session of the baptist educational and missionary convention of south carolina

Baptist, Edward *see*
– Diary
– Letters to the pamphleteer

The baptist encyclopaedia : a dictionary of the doctrines, ordinances, usages...and of the general history of the baptist denomination in all lands / ed by Cathcart, William – rev ed. Philadelphia: Louis H Everts, 1883 – 15mf – 9 – 0-524-02456-1 – mf#1990-4315 – us ATLA [052]

Baptist faith and message *see* Southern baptist journal

Baptist Faith and Message Committee *see* Files

Baptist family in global village / ed by Langley, Thelma – 180p – 1 – $6.30 – us Southern Baptist [242]

Baptist fidelity – Publ by Conservative Baptists. jan, mar, may, jul 1948 – 1 – $5.00 – us Southern Baptist [242]

Baptist General Association. Kentucky. Executive Board *see* Minutes

Baptist General Association of Virginia *see* Religious herald

Baptist general association of west virginia, 1865-1915, woman's baptist missionary society of west virginia, ministers' fraternal union / [Hank, Arthur et al] [comp] – [s.l: s.n. 1915?] [mf ed 1993] – 5mf [ill] – 9 – 0-524-08773-3 – mf#1993-3278 – us ATLA [242]

Baptist General Conference of America *see*
– Standard

The baptist general convention and its work / Gambrell, James Bruton – 1918. 18p – 1 – 5.00 – us Southern Baptist [242]

Baptist General Missionary Convention. Board of Foreign Missions *see*
– Minutes of the triennial convention
– Minutes of triennial convention
– Proceedings of triennial convention

The baptist harmony / Burdett, Staunton S – 1834 – 1 – us Southern Baptist [242]

The baptist harmony / Burdett, Staunton S – 1842 – 1 – 5.00 – us Southern Baptist [242]

The baptist harp / American Baptist Publication Society – 1849 – 1 – us Southern Baptist [242]

Baptist headlight – Topeka, KS: [s.n.] v1 n1 sep 15 1893-v1 n32 aug 8 1894 (wkly, semimthly) [mf ed 1947] – (= ser Negro newspapers on microfilm) – 1r – 1 – (cont by: national baptist world) – us L of C Photodup [071]

Baptist herald – 1984 nov-1988 dec – 1 – mf#433100 – us WHS [242]

Baptist herald – Oakbrook Terrace IL 1973-95 – 1,5,9 – ISSN: 0005-5700 – mf#8793 – us UMI ProQuest [242]

The baptist herald – Cleveland, Ohio. 1923-72. 24,928p – 1 – 872.48 – us Southern Baptist [242]

Baptist heritage update : newsletter of the historical commission of the southern baptist convention and the southern baptist historical society – Tennessee. v1-15. 1985-99 – 1r – 1 – mf#7080 – us Southern Baptist [242]

Baptist history : from the foundation of the christian church to the close of the 18th century / Cramp, John Mockett – Philadelphia: American Baptist Publ Soc [186-?] [mf ed 1986] – 2mf – 9 – 0-8370-8977-8 – (incl bibl ref & ind) – mf#1986-2977 – us ATLA [242]

Baptist history / Wamble, Hugh – Research cards. 2740p – 5 – us Southern Baptist [242]

Baptist history and heritage – Brentwood TN 1985+ – 1,5,9 – ISSN: 0005-5719 – mf#15665 – us UMI ProQuest [242]

Baptist history of north dakota, 1879-1904 / Shanafelt, Thomas Miles – Huron SD: Huronite Printing [1904?] [mf ed 1992] – (= ser Baptist coll) – 1mf – 9 – 0-524-05004-X – mf#1990-5092 – us ATLA [242]

The baptist history of south dakota / Shanafelt, Thomas Miles – Sioux Falls: South Dakota Baptist Convention, c1899 [mf ed 1992] – (= ser Baptist coll) – 1mf – 9 – 0-524-03627-6 – (int by o a williams) – mf#1990-4787 – us ATLA [242]

Baptist history of the north pacific coast : with special reference to western washington, british columbia, and alaska / Baker, John Clapp – Philadelphia: American Baptist Pub Soc, c1912 [mf ed 1992] – (= ser Baptist coll) – 2mf – 9 – 0-524-03377-3 – mf#1990-4689 – us ATLA [242]

Baptist history vindicated / Christian, John Tyler – Louisville KY: Baptist Book Concern 1899 [mf ed 1993] – 1v on 1mf – 9 – 0-524-07153-5 – mf#1991-2942 – us ATLA [242]

The baptist home mission monthly – v1-31. 1878-1909 [complete] – 7r – 1 – mf#ATLA R0134 – us ATLA [242]

Baptist home missions in north america : including a full report of the proceedings and addresses of the jubilee meeting... – New York: Baptist Home Mission Rooms, 1883 [mf ed 1986] – 2mf – 9 – 0-8370-6153-9 – mf#1986-0153 – us ATLA [242]

Baptist home of philadelphia : a sketch of its origin and history / Nugent, George – Philadelphia: William Syckelmoore 1880 [mf ed 1993] – 1mf – 9 – 0-524-08491-2 – mf#1993-3136 – us ATLA [242]

Baptist horizon and bulletins on canadian southern baptists – Vancouver, BC. oct 1954-56; bulletins. 1957, 1959 – 1 – $5.00 – us Southern Baptist [242]

Baptist hour messages from radio and television – Fort Worth, TX. 3470p. 1941-53. – 1 – $254.38 – us Southern Baptist [242]

Baptist hymn and tune book – Plymouth collection, New York. 1868 – 1 – $22.47 – us Southern Baptist [780]

The baptist hymn book / Biddle, William P & Newborn, William – 694p – 1 – $24.29 – us Southern Baptist [780]

The baptist hymn book / Buck, William Calmes – Original and selected in two parts – 1 – $24.67 – us Southern Baptist [780]

Baptist hymn collection – From the New Orleans Baptist Theological Seminary. 6168p. 1790-1877 – 1 – $154.20 – (includes: broadus, andrew. collections of sacred ballads. 1790. fuller, richard and jeter, j.b. the psalmist with supplement. 1847; the psalmist, with music...1860) – us Southern Baptist [780]

Baptist hymn writers and their hymns / Burrage, Henry S – Portland ME: Brown Thurston c1888 [mf ed 1989] – 2mf [ill] – 9 – 0-7905-4496-2 – mf#1988-0496 – us ATLA [242]

Baptist hymnody in the us : selected titles from the hymnal collection of southern baptist theological seminary – 1616-1905 – 10r – 1 – $651.60 – (14,480p) – us Southern Baptist [242]

The baptist in history : five lectures / Mosher, Roswel Curtis – Albert Lea, Minn: Simonson and Whitcomb, 1900 – 1mf – 9 – 0-524-04882-7 – mf#1990-5080 – us ATLA [242]

Baptist informer : official organof the general baptist convention of north carolina – 1976 mar-may; 1984 jun-1995 dec – 1 – mf#1585894 – us WHS [242]

The baptist informer, bulletin / Fairmont. North Carolina. First Baptist Church – 1935-69.5034p – 1 – us Southern Baptist [242]

Baptist intermediate union quarterly 1 and 2 – 1922-61 – (= ser Intermediate BYPU Quarterly) – 1 – $299.81 – (title changes to: intermediate bypu quarterly. 1922-39) – us Southern Baptist [242]

The baptist irish society : its origin, history, and prospects / Belcher, Joseph et al – London: Printed for the Baptist Irish Society and sold by Houlston and Stoneman, 1845 – 1mf – 9 – 0-524-00503-6 – mf#1990-0003 – us ATLA [242]

The Baptist Joint Committee on Public Affairs *see* Report from the capital

Baptist Joint Committee on Public Affairs *see* Report from the capital

The baptist jubilee memorial / Winks, Joseph F – 1842 – 1 – 5.00 – us Southern Baptist [242]

Baptist landmarkism tested by logic, history and scripture / Averill, A M – 1880 – 1 – $5.00 – us Southern Baptist [242]

Baptist layman's book : a compend of baptist history, principles, practices, and institutions / Everts, William Wallace – Philadelphia: American Baptist Publ Soc c1887 [mf ed 1993] – 1mf – 9 – 0-524-07159-4 – mf#1991-2948 – us ATLA [242]

Baptist leader – 1990 1st qtr [cover only]; 1994 – 1 – mf#2756138 – us WHS [242]

Baptist leader – Philadelphia: American Baptist Publ Soc 1939- (1939-69) [mf ed 1970] – 62v on 30r – 1 – (formed by union of: adult leader, young people's leader, and children's leader; absorbed: judson journal [jan 1970]) – ISSN: 0005-5727 – mf0177 – us ATLA [242]

Baptist life – Columbia, MD. 23,847p. 1917-99 – 1 – (maryland baptist church life 1917-oct 1934. maryland baptist 1935-84. baptist true union 1985-93) – mf#0727 – us Southern Baptist [242]

A baptist manual : the polity of the baptist churches and of the denominational organisations / Soares, Theodore Gerald – Philadelphia: American Baptist Publ Soc, c1911 [mf ed 1993] – (= ser Baptist coll) – 1mf – 9 – 0-524-06883-6 – mf#1990-5302 – us ATLA [242]

Baptist manuscripts for the mennonite seminary library – Amsterdam. 156p – 1 – $5.46 – (20 letters and documents from the seventeenth century.) – us Southern Baptist [242]

Baptist married young people – oct 1956-61 – 1 – $57.40 – us Southern Baptist [242]

A baptist meeting-house : the staircase to the old faith, the open door to the new / Barrows, Samuel June – Boston: American Unitarian Assoc, 1885 [mf ed 1984] – 3mf – 9 – 0-8370-0974-X – (papers repr fr the christian register) – mf#1984-4342 – us ATLA [242]

Baptist memorial and monthly chronicle – Baptist Memorial and Monthly Record/ American Baptist Memorial. New York, 1842-56.Single reels available – 1 – us ABHS [242]

Baptist message – Louisiana, 1888-1999 – 66,813p – 1 – (cont: the baptist chronicle 1888-1919) – mf#0560 – us Southern Baptist [242]

Baptist messages from radio hour and television / Baptist Association – 1954-1977 – 1 – $132.93 – us Southern Baptist [242]

Baptist messenger – Oklahoma. may 1912-91 – 1 – $2,688.40 – us Southern Baptist [242]

Baptist Mission Press *see* Siwinowe kesibwi

Baptist Mission Union of America et al *see* Mission post

Baptist Missionary and Educational Convention. Indian Territory *see* Papers

Baptist Missionary and Educational Convention. Oklahoma and Indian Territories *see* Papers

Baptist missionary association of america directory and handbook – 1961-92 – 1 – $204.80 – us Southern Baptist [242]

Baptist missionary association of america yearbook – Jacksonville, TX. 1950-1991 – 1 – $529.11 – us Southern Baptist [242]

Baptist Missionary Association. Texas *see* Papers

Baptist missionary magazine – La Salle IL 1817-1909 – 1 – mf#4360 – us UMI ProQuest [242]

Baptist Missionary Society *see* Brief view of the baptist missions and translations

Baptist Missionary Society. Archives. London *see*
- Amalgamation of the general and particular baptist in england
- Hanserd knollys society materials
- London baptist association

Baptist Missionary Society. London *see*
- Archives, 1792-1914
- Index to archives
- Reports

Baptist monitor – 1969 mar 20-1980 dec 1 – 1 – mf#547440 – us WHS [242]

Baptist monitor and political compiler – KY. 1823-24 – 1 – $5.00 – us Southern Baptist [242]

The baptist movement in the continent of europe : a contribution to modern history / ed by Rushbrooke, James Henry – London: Carey Press: Kingsgate Press, 1915 – 1mf – 9 – 0-524-01750-6 – mf#1990-4142 – us ATLA [242]

The baptist movement of a hundred years ago, and its vindication : a discourse...jan 16 1868 / Weston, David – Boston: Gould & Lincoln, 1868 [mf ed 1991] – (= ser Baptist coll) – 1mf – 9 – 0-524-00978-3 – mf#1990-4036 – us ATLA [242]

Baptist new mexican – NM. 44,783p. 1909-1911, feb 15 1919-1999 – 1 – mf#0960 – us Southern Baptist [242]

Baptist news – IL. 1895-1902 – 1 – $96.91 – us Southern Baptist [242]

Baptist North America General Conference *see* Annual reports

Baptist observer – Indianapolis and Greensburg/ Seymour. Indiana Baptist Convention. Sept 1902-1968. Single reels available – 1 – 907.80 – us ABHS [242]

Baptist pamphlets / Arnold, Albert Nicholas – Philadelphia: American Baptist Publ Soc [1892?] [mf ed 1992] – 1v on 1mf – 9 – 0-524-04034-6 – mf#1990-4942 – us ATLA [242]

Baptist pamphlets – St Louis. MO. v1-9 – 1 – $158.96 – (mercantile library) – us Southern Baptist [242]

Baptist pamphlets. a / Hovey, Alvah – Philadelphia: American Baptist Publ Soc c[1892?] [mf ed 1992] – 1v on 1mf – 9 – 0-524-04047-8 – (incl bibl ref) – mf#1990-4955 – us ATLA [242]

Baptist pamphlets. c / Taylor, George Boardman et al – Philadelphia: American Baptist Publ Soc [1892?] [mf ed 1992] – (= ser Memorial series (philadelphia pa)) – 1v on 1mf – 9 – 0-524-04065-6 – mf#1990-4973 – us ATLA [242]

Baptist pamphlets. d / Norcott, John et al – Philadelphia: American Baptist Publ Soc [1892?] [mf ed 1992] – 1v on 1mf – 9 – 0-524-04230-6 – mf#1990-5021 – us ATLA [242]

Baptist peacemaker – Louisville, KY. 942p. v1-16. 1980-spring 1996 – 1 – mf#5838 – us Southern Baptist [242]

Baptist press news releases – 4958p. 1954-64 – 1 – us Southern Baptist [242]

The baptist principle in application to baptism and the lord's supper / Wilkinson, William Cleaver – new and enl ed. Philadelphia: American Baptist Publ Soc, c1897 – 1mf – 9 – 0-524-07773-8 – mf#1991-3341 – us ATLA [242]

Baptist principles reset : consisting of articles on distinctive baptist principles / Jeter, Jeremiah Bell et al – new enl ed. Dallas TX: Standard Pub 1902 [mf ed 1993] – 1mf – 9 – 0-524-08393-2 – mf#1993-3093 – us ATLA [242]

Baptist principles vindicated : in reply to the rev j w d gray's work on baptism / Tupper, Charles – Halifax, NS? : s.n, 1844 [mf ed 1984] – 3mf – 9 – 0-665-48687-1 – (incl bibl ref) – mf#48687 – cn Canadiana [242]

Baptist program – 1923-1970 – 1 – $317.17 – us Southern Baptist [242]

The baptist psalmody / Manly, Basil Jr. & Manly, Basil Sr. – 1850 – 1 – $27.09 – (1855 ed. $27.09. 1859 ed. $27.09) – us Southern Baptist [242]

Baptist quarterly *see* Baptist review

Baptist quarterly [1867] – La Salle IL 1867-77 – 1 – mf#4130 – us UMI ProQuest [242]

Baptist quarterly [1922] – Earls Barton UK 1922+ – 1,5,9 – ISSN: 0005-576X – mf#6266 – us UMI ProQuest [242]

Baptist quarterly review – La Salle IL 1879-92 – 1 – mf#3881 – us UMI ProQuest [242]

The baptist record – Mississippi. 71,142p. feb 1877-1999 – 1 – mf#0250 – us Southern Baptist [242]

Baptist reflector – Nashville, TN. 1877-82; Chattanooga, TN. 1885-89 – 4r – 1 – (american baptist reflector, chattanooga tn, 1882-85) – mf#6974 – us Southern Baptist [242]

Baptist reformation review – Nashville, TN. v2-5, 1973-76 – 1 – us ABHS [242]

The baptist reporter – London, Ont.: [s.n., 1893?-189- or 19–] – 9 – mf#P04799 – cn Canadiana [242]

Baptist review – 1963 apr 3-1978 apr; 1978 may-1986 nov/dec – 1 – mf#345731 – us WHS [242]

Baptist review – 1836-92 – 1 – $1194.20 – (christian review 1836-63; bibliotheca sacra 1863-66; baptist quarterly 1867-77; baptist review 1879-92) – us Southern Baptist [242]

Baptist sentinel – Dayton, WA; Dallas, OR 1894 (dec 27, only) 1899 – 1 – $83.72 – us Southern Baptist [242]

Baptist sentinel – Louisville. KY. v1-2. 1870-71 – 1 – $31.50 – us Southern Baptist [242]

The baptist short method with inquirers and opponents / Hiscox, Edward Thurston – Philadelphia: American Baptist Publication Society, c1868 – 1mf – 9 – 0-524-03935-6 – mf#1990-4929 – us ATLA [242]

Baptist songster or divine songs – Comp by R Winchell. 1829 – 1 – $8.05 – us Southern Baptist [242]

Baptist Southern Convention *see*
- Bulletins
- Index
- Minutes

Baptist standard – Nashville, TN. 1858-60 – 1 – $13.86 – us Southern Baptist [242]

Baptist standard – TX. 1892-1990 – 1 – $4100.40 – us Southern Baptist [242]

Baptist standard church directory and busy pastor's guide / Jordan, Lewis Garnett – Nashville: Sunday School Publ Board of the National Baptist Conventions, 1928 (mf ed 1976) – 1r – 1 – mf#ZZ-14246 – us NY Public [242]

Baptist State Conventions (American Baptist). Connecticut *see* Records

Baptist State Conventions. (American Baptist). Connecticut *see* Annual

Baptist State Conventions (American Baptist). Maine *see*
- Records
- Scrapbook

Baptist State Conventions (American Baptist). New Hampshire. Board of Promotion *see* Records

Baptist State Conventions (American Baptist). Oregon. Baptist Convention of the North Pacific Coast and Oregon Baptist Convention *see* Minutes, mission board

Baptist State Conventions. (American Baptist). Pennsylvania *see* Annual

Baptist State Conventions (American Baptist). South Dakota *see* Records

Baptist State Conventions (American Baptist). Vermont *see* Board of trustees, records

Baptist State Conventions. Kentucky *see* Annual

Baptist State Conventions. Louisiana *see* Annual

Baptist State Conventions. (Southern Baptist). Alabama *see* Annual

Baptist State Conventions. (Southern Baptist). Alaska *see* Annual

Baptist State Conventions. (Southern Baptist). Arizona *see* Annual

Baptist State Conventions. (Southern Baptist). Arkansas *see* Annual

Baptist State Conventions. (Southern Baptist). California *see* Annual

Baptist State Conventions. (Southern Baptist). District of Columbia *see* Annual

Baptist State Conventions. (Southern Baptist). Florida *see* Annual

Baptist State Conventions. (Southern Baptist). Georgia *see* Annual

Baptist State Conventions. (Southern Baptist). Illinois *see*
- Annual

Baptist State Conventions. (Southern Baptist). Kentucky *see* Annual

Baptist State Conventions. (Southern Baptist). Louisiana *see* Annual

Baptist State Conventions. (Southern Baptist). Maryland *see* Annual

Baptist State Conventions. (Southern Baptist). Mississippi *see* Annual

Baptist State Conventions. (Southern Baptist). Missouri *see* Annual

Baptist State Conventions. (Southern Baptist). New Mexico *see* Annual

Baptist State Conventions. (Southern Baptist). North Carolina *see* Annual

Baptist State Conventions. (Southern Baptist). Ohio *see* Annual

Baptist State Conventions. (Southern Baptist). Oklahoma *see* Annual

Baptist State Conventions. (Southern Baptist). South Carolina *see* Annual

Baptist State Conventions. (Southern Baptist). Tennessee *see*
- Annual
- Proceedings

Baptist State Conventions. (Southern Baptist). Texas *see* Annual

Baptist State Conventions. (Southern Baptist). Virginia *see*
- Annual
- Correspondence

Baptist student – 1922-61 – 1 – $487.06 – us Southern Baptist [242]

Baptist succession : a hand-book of baptist history / Ray, David Burcham – Cincinnati: Geo E Stevens 1873 [mf ed 1986] – 2mf – 9 – 0-8370-9104-7 – (incl bibl ref & ind) – mf#1986-3104 – us ATLA [242]

Baptist sunday school convention – Union County, SC. 114p. 1887-92 – 1 – $5.00 – us Southern Baptist [242]

The baptist system examined, the church vindicated, and sectarianism rebuked : a review of dr. fuller and others on baptism and the terms of communion / Seiss, Joseph Augustus – 3rd rev and enl ed. Baltimore: T Newton Kurtz, 1864 – 1mf – 9 – 0-524-08582-X – mf#1993-3167 – us ATLA [242]

Baptist times – London. 1857-1950 – 47.25r – 1 – mf#1. 2835.00 – us Southern Baptist [242]

The baptist times – 1871-1991+ – 96r – 1 – £4750.00 – (journal of the baptist movement) – mf#BPT – uk World [242]

Baptist Tract Society *see* Chairman's address at the annual meeting, 1870

Baptist training union magazine – 1926-61 – 1 – $689.08 – us Southern Baptist [242]

Baptist tribune – TX. 8 jan 1903-18 apr 1907 – 1 – $62.79 – us Southern Baptist [242]

Baptist Triennial Convention *see* Proceedings

Baptist Triennial Convention. Board of Foreign Missions of the Baptist General Missionary Convention *see* Minutes

Baptist true union – Columbia, MD. 1917-91 – 1 – $874.96 – (cont: maryland baptist church life, 1917-34; maryland baptist, 1935-84) – us Southern Baptist [242]

Baptist trumpet – 1895-1940 – 1 – $13.79 – (broken file) – us Southern Baptist [242]

Baptist trumpet – Killeen/Maud, TX. Old School or Primitive Baptist. 1937-71. Single reels available. Incomplete – 1 – us ABHS [240]

The baptist trumpet – Shipman VA: [s.n.] [mthly] [mf ed 2004] – v2 n12-v13 n11 (oct 1939-jul 1949) [gaps] on 1r – 1 – (publ monthly in the interest of the churches named in the directory and their communities; place of publ moved fr shipman to arrington, va between jun 1944-aug 1945; lacking several iss; some pgs damaged) – mf#2004-s031 – us ATLA [242]

Baptist union handbook – Publ by T H Crocott, Johannesburg, Africa. 1877-1963 – 1 – $218.89 – us Southern Baptist [242]

Baptist visitor – Newton and Dover, MD. 96p. 1866-77 – 1 – (incomplete) – us Southern Baptist [242]

Baptist waymarks : principles and usages of gospel churches, mainly from authentic sources, with notes and comments / Ford, Samuel Howard – Philadelphia: American Baptist Publ Soc c1903 [mf ed 1993] – 1mf – 9 – 0-524-08260-X – mf#1993-3015 – us ATLA [242]

Baptist, why and why not : twenty-five papers by twenty-five writers, and a declaration of faith / Dudley, Richard M et al – Nashville TN: Sunday School Board, Southern Baptist Convention c1900 [mf ed 1993] – 1mf – 9 – 0-524-07156-X – mf#1991-2945 – us ATLA [242]

Baptist Woman's Missionary Union. Auxiliary to Southern Baptist Convention *see*
- Annuals
- Yearbooks

Baptist Woman's Missionary Union. Kentucky *see* Annual

Baptist Woman's Missionary Union. Louisiana *see* Annual

Baptist Woman's Missionary Union. Mississippi *see* Annual

Baptist Woman's Missionary Union. South Carolina *see* Annual

Baptist Woman's Missionary Union. Spartan Woman's Missionary Union, Auxiliary to Spartan Baptist Association *see* Minutes

Baptist Woman's Missionary Union. Texas. Henderson County *see* Annual

Baptist world – McLean VA 1954+ – 1,5,9 – ISSN: 0005-5808 – mf#1940 – us UMI ProQuest [242]

Baptist world – v6 n6-v10 n4 [1972 jun-1976 oct/dec]; 1977; 1978 winter/spring; 1978 jun; 1979 spring/summer – 1 – mf#639513 – us WHS [242]

Baptist World Alliance *see*
- Baptist world alliance
- Relief ledgers
- Study papers

Baptist world alliance : proceedings / Baptist World Alliance – 1905-75 [complete] – 3r – 1 – ISSN: 0005-5805 – mf#ATLA R0117 – us ATLA [242]

Baptist World Alliance. Administrative and Executive Committees *see* Minutes

Baptist World Alliance. Advisory Committee *see* Minutes

Baptist World Alliance. Committee on Relief and Development *see* Minutes

Baptist World Alliance. Continental Committee *see* Minutes

Baptist World Alliance, Fifth Congress, Berlin *see* Papers

Baptist World Alliance. North American Baptist Fellowship *see* Minutes

The baptist world alliance second congress, philadelphia, june 19-25, 1911 : record of proceedings – Philadelphia, PA:...for the Philadelphia Cttee, 1911 [mf ed 1993] – (= ser Baptist coll) – 2mf – 9 – 0-524-07741-X – mf#1991-3309 – us ATLA [242]
Baptist World Congress (1905: London, England) *see* Authorized record of proceedings
Baptist young people – 1904-61 – 1 – $457.80 – (bypu quarterly 1904-28; senior bypu quarterly 1929-39; baptist young people's quarterly 1940-55; baptist young people 1956-) – us Southern Baptist [242]
Baptist young people's quarterly *see* Baptist young people
Baptista, Abel Dos Santos *see* Monografia etnografica sobre os macuas
Baptista, Jose Maria *see* Cronicas del bocono de ayer
Baptista, Juan *see*
– Advertencias
– Confessionario en lengua mexicana y castellana
Baptista paiarinus, chronica – Vicenza, 1556 – (= ser Holkham library manuscript books 466) – 1r – 1 – (written by bartholomaeus vellensis) – mf#2198 – uk Microform Academic [090]
Baptista Pereira, Antonio *see*
– Brasil e a raca
– Figuras do imperio e outros ensaios
– Pelo brasil maior
– Vultos e episodios do brasil
Baptisternes ugeblad – 1851-1961 – 1 – $1228.50 – (danish baptist) – us Southern Baptist [242]
Baptisti usuugingu (koguduse) pohikiri = Statutes of the baptist union (church) – Keilas, Estonia: "Kulwaja" trukk, 1926 – 1r – 1 – $5.00 – (one part of a two-part item) – us Southern Baptist [242]
The baptists / Rumble, L – 1 – 5.00 – us Southern Baptist [242]
The baptists : their origin, continuity, principles, spirit, polity, position, and influence / Jones, Tiberius Gracchus – Philadelphia: American Baptist Publ Society, [1860?] – 1mf – 9 – 0-7905-5098-9 – mf#1988-1098 – us ATLA [242]
The baptists / Vedder, Henry Clay – New York: Baker & Taylor, c1902 – 1mf – 9 – 0-8370-8954-9 – (incl ind) – mf#1986-2954 – us ATLA [242]
Baptists and liberty of conscience / Vedder, Henry Clay – Cincinnati: J R Baumes 1884 [mf ed 1993] – 1mf – 9 – 0-524-08602-8 – mf#1993-3187 – us ATLA [242]
The baptists and slavery, 1840-1845 / Putnam, Mary Burnham – Ann Arbor, Mich: G Wahr, 1913 – 2mf – 9 – 0-524-06592-6 – (incl bibl ref) – mf#1990-5258 – us ATLA [976]
The baptists and the american revolution / Cathcart, William – Philadelphia: SA George, 1876 – 1mf – 9 – 0-7905-4448-2 – (incl bibl ref) – mf#1988-0448 – us ATLA [975]
The baptists and the national centenary : a record of christian work, 1776-1876 / Weston, David et al; ed by Moss, Lemuel – Philadelphia: American Baptist Publ Society, 1876 – 1mf – 9 – 0-7905-5078-4 – mf#1988-1078 – us ATLA [242]
Baptists and their doctrines : sermons on distinctive baptist principles / Carroll, Benajah Harvey – New York: Fleming H Revell 1913 [mf ed 1993] – 1mf – 9 – 0-524-06808-9 – mf#1991-2795 – us ATLA [242]
The baptists in america / Cox, Francis Augustus & Hoby, J – 1836 – 1 – us Southern Baptist [242]
The baptists in america : a narrative of the deputation from the baptist union in england to the united states and canada / Cox, Francis Augustus & Hoby, James – 2nd rev ed. London: T Ward, 1836 – 2mf – 9 – 0-524-01935-5 – mf#1990-4159 – us ATLA [242]
The baptists in history : with an introduction on the parliament of religions / Lorimer, George Claude – Boston: Silver, Burdett, c1893 – 1mf – 9 – 0-8370-9008-3 – mf#1986-3008 – us ATLA [242]
Baptists in yorkshire, lancashire, cheshire and cumberland / Whitley, William Thomas et al – augm ed of 2 assoc vols for Baptist Hist Soc. London: Kingsgate Press 1913 [mf ed 1992] – 2mf [ill] – 9 – 0-524-04243-8 – (incl bibl ref) – mf#1990-5034 – us ATLA [242]
Baptists mobilized for missions / Vail, Albert Lenox – Philadelphia: American Baptist Publ Soc c1911 [mf ed 1993] – 1mf – 9 – 0-524-06560-8 – (incl bibl ref) – mf#1991-2644 – us ATLA [242]
The baptists of canada : a history of their progress and achievements / ed by Fitch, Ernest Robert – Toronto: Standard, c1911 – (= ser Mission Study Text Book) – 1mf – 9 – 0-7905-5036-9 – mf#1988-1036 – us ATLA [242]
The baptists of new hampshire / Hurlin, William et al – Manchester, NH: New Hampshire Baptist Convention, 1902 – 1mf – 9 – 0-524-03553-9 – mf#1990-4748 – us ATLA [242]
The baptists of yorkshire : being the centenary memorial volume of the yorkshire baptist association – Bradford: W Byles, 1912 – 4mf – 9 – 0-524-08826-8 – (incl bibl ref and ind) – mf#1993-3318 – us ATLA [242]
Baptists, the only thorough religious reformers / Adams, John Quincy – centennial ed, rev enl. New York: Sheldon 1876 [mf ed 1986] – 1mf – 9 – 0-8370-8880-1 – mf#1986-2880 – us ATLA [242]
Baptists, thorough religious reformers / Adams, John Quincy – 1857 – 1 – $5.88 – us Southern Baptist [242]
Baptists today – Oxford. 1960+ (1) 1960+ (5) 1960+ (9) – (= ser SBC Today) – 8r – 1 – (formerly: sbc today 1983-jul 1991) – mf#6097 – us Southern Baptist [242]
The baptists, who are they? and what do they believe? / Boggs, William Bambrick – [4th ed] Philadelphia: American Baptist Publ Soc, [1898?] [mf ed 1993] – 1mf – 9 – 0-524-07152-7 – (first printed in 1877) – mf#1991-2941 – us ATLA [242]
The baptists, who they are, and what they have done : a memorial series / Taylor, George Boardman – Philadelphia: American Baptist Publ Soc, [1873?] – 1mf – 9 – 0-524-07485-2 – mf#1990-5411 – us ATLA [242]
Baptists' why and why not / Frost, J M et al – 1900 – 1 – $16.03 – us Southern Baptist [242]
Baptistu zihnas lihdsekti : atbilde krimuldas luteranu mahzitajam j ehrmana kungam / Inkins, J – Riga: J A Freij, 1910 – 1 – (publ no. 6298 e. one of five items on a reel) – us Southern Baptist [242]
Baptisty (shtundisty) na kavkaze = Baptists (stundists) in the caucasus – Kiev, 1885 – 1 – $5.00 – us Southern Baptist [242]
Baptizing : biblical and classical / Day, Clinton D – Cincinnati: Jennings & Graham; New York: Eaton & Mains c1907 [mf ed 1989] – 1mf – 9 – 0-7905-0982-2 – (incl ind) – mf#1987-0982 – us ATLA [240]
Baptizing and teaching : evans' new exposition of ritual baptism and of baptism with the holy spirit, as seen when set back in their right places in the great system of revealed truth... / Evans, John Swanton – Toronto: W Briggs; Montreal: C Coates: Halifax NS: S Heustis, 1887 [mf ed 1980] – 8mf – 9 – 0-665-02925-X – mf#02925 – cn Canadiana [240]
Baptizo-dip-only : the world's pedobaptist greek scholarship, containing scores of answers to the author's questions... / Jarrel, Willis Anselm – Dallas: Texas Baptist Book House c1910 [mf ed 1993] – 1mf – 9 – 0-524-06740-6 – mf#1992-0943 – us ATLA [240]
Bapu gandhi / Piddington, Albert Bathurst – London: Williams & Norgate Ltd, 1930 – (= ser Samp: indian books) – us CRL [920]
Bapu ki prema prasadi : gandhi-yuga ki eka mahatvapurna patravali / Birala, Ghanasyamadasa – Bambai: Bharatiya Vidya Bhavana, 1977- (mf ed Bethlehem, PA: Mid-Atlantic Preservation Service, 1989) – 1mf – 9 – (in hindi) – mf#Sc Micro F-11318 – Located: NYPL – us Misc Inst [950]
Bapu's letters / by Kalelkar, Kaka – Ahmedabad: Navajivan Pub House, 1952- – (= ser Samp: indian books) – (transl from gujarati by arvindlal l mazmudar) – us CRL [860]
Bapu's letters to mira, 1924-1948 – Ahmedabad: Navajivan Pub House, 1949 – (= ser Samp: indian books) – us CRL [860]
Baquijano y Carrillo, Jose *see* Relectio extemporanea ad explanationem legis pamphilo 39
The bar *see*
– West virginia law review
The bar and legal world – no 12. 1903 (all publ) – 1mf – 9 – $1.50 – mf#LLMC 84-414 – us LLMC [340]
Bar briefs / North Dakota. State Bar Association – v1 dec 1924-v2 n12 1925 – 4mf – 9 – $6.00 – (in jan 1926, with v3 n2 title changed to: bar briefs and dakota law review; updates planned as add vols fall out of copyright) – mf#llmc97-583 – us LLMC [340]
Bar briefs *see* North dakota law review
Bar bulletin – Boston. no 1-167. 1924-40; v12-27. 1941-56 (all publ) – 9 – $165.00 set – mf#100881 – us Hein [340]
Bar bulletin : state bar of new mexico – v5-38. 1967-99 – 9 – $1561.00 set – (title varies: v1-18 as state bar of new mexico bulletin and advance opinions; v19-26 n10 as new mexico's news and views) – mf#400650 – us Hein [340]
The bar bulletin – Boston Bar Association. v1-27. 1924-56 (all publ) – 35mf – 9 – $58.00 – (lacking: v13) – mf#LLMC 84-415 – us LLMC [340]
Bar Eljokum, Schelomo *see* Jude spricht fur deutschland
Bar examination journal – 12v. 1871-94 (all publ) – 47mf – 9 – $70.00 – (cont as: the bar examination annual in 1893) – mf#LLMC 84-416 – us LLMC [340]
Bar examiner – Chicago IL 1931+ – 1,5,9 – ISSN: 0005-5824 – mf#3461 – us UMI ProQuest [340]
Bar examiner – v1-70. 1931-2001 – 9 – $581.00 set – ISSN: 0005-5824 – mf#100901 – us Hein [340]
The bar examiner – National Conference of Bar Examiners. v1-55. 1931-86 – 42mf (1:42) 2mf (1:24) – 9 – $192.00 – (lacking: v50. updates planned) – mf#LLMC 84-417 – us LLMC [340]
Bar harbor record – Bar Harbor, ME: Bar Harbor Press Co, feb 17 1887-feb 9 1916 – 1 – us CRL [071]
The bar harbor times – Bar Harbor, ME: Sherman Pub Co, apr 9 1924-jul 4 1968 – 1 – (issues for apr 9-dec 1924 filmed with: bar harbor times and bar harbor record, jan -apr 2 1924) – us CRL [071]
The bar harbor times – Bar Harbor, ME: W H Sherman, jul 11 1914-aug 19 1916 – (= ser The bar harbor times and bar harbor record) – (issues for jan-aug 19 1916 filmed with: bar harbor times and bar harbor record, aug 26-dec 1916) – us CRL [071]
The bar harbor times and bar harbor record – Bar Harbor, ME: W H Sherman, aug 26 1916-apr 2 1924 – (= ser The bar harbor times) – (issues for aug 26-dec 1916, and for jan-apr 2 1924 filmed with: bar harbor times) – us CRL [071]
Bar Hebraeus *see*
– Des gregorius abulfarag, gen bar-hebraeus, scholien zum buche daniel
– Gregorii abulfarag bar ebhraya in evangelium matthaei scholia
– Gregorii bar ebhraya in evangelium iohannis commentarius
– In duodecim prophetas minores scholia
Bar journal. state bar of new mexico – v1-5. 1995-99 – 9 – $70.00 set – ISSN: 0108-49793 – mf#116351 – us Hein [340]
Bar kochba – Berlin DE, 1919-21 – 1r – 1 – gw Misc Inst [074]
Bar kochba : blaetter fuer die heranwachsende juedische jugend – Berlin: Chaskel Zwi Kloetzel. v1-2. 1919/20, 1921 [complete] – (= ser German-jewish periodicals...1768-1945, pt 2) – 1r – 1 – $125.00 – mf#B25 – us UPA [939]
The bar leader – American Bar Assoc. v1-11 n3 1975-85 – 25mf – 9 – $37.50 – (updates available. lacking: v2) – mf#LLMC 84-418 – us LLMC [340]
Bar leader (aba) – v1-25. 1975-2001 – 9 – $289.00 set – ISSN: 0099-1031 – mf#401310 – us Hein [340]
Bar mitsvah deroshes – New York, NY. 1921 – 1r – us UF Libraries [939]
The bar reports : containing all the cases argued and determined in parliament, the house of lords, the privy council, the court of appeal in chancery, the rolls, v.c., queen's bench,.... / Great Britain. England – London: Horace Cox Co. v1-12. 1865-71 (all publ) – 114mf – 9 – $171.00 – mf#LLMC 95-225 – us LLMC [324]
Barabbas : a dream of the world's tragedy / Corelli, Marie – Montreal: Montreal News Co; Philadelphia: J B Lippincott, 1895, c1893 – 4mf – 9 – mf#26974 – cn Canadiana [880]
Barabbas the scapegoat : and other sermons and dissertations / Wratislaw, Albert Henry – London: J W Parker 1859 [mf ed 1986] – 1mf – 9 – 0-8370-9996-X – (in english & latin) – mf#1986-3996 – us ATLA [240]
Baraboo bulletin – 1882 aug 4-dec 22 – 1 – mf#953706 – us WHS [071]
Baraboo daily news – 1913 jan 4-jun 30 – 1 – mf#1138946 – us WHS [071]
Baraboo daily news and republic – 1929 feb 18-apr 8 – 1 – mf#1139105 – us WHS [071]
Baraboo daily republic – 1911 jan 3-1915 dec 31 [with gaps] – 1 – mf#959661 – us WHS [071]
Baraboo news – 1904 may 25-1906 feb 7; 1906 feb 7-1907 apr 3; 1907 apr 10-1908 apr 22; 1908 apr 29-1909 may 6; 1909 may 13-1910 apr 28; 1910 may 5-11 apr 27; 1911 may 4-dec 28 – 1 – mf#1002774 – us WHS [071]
Baraboo news republic – 1971 feb 8-1990 nov [with some gaps] – 1 – mf#1139137 – us WHS [071]
Baraboo news-republic – 1929 apr 9-1971 feb 6 [with gaps] – 1 – mf#1139133 – us WHS [071]
Baraboo republic – 1855 may 5-1923 sep 6 [with gaps] – 1 – mf#986197 – us WHS [071]
Baraboo sun – 1997 apr 10-aug 28; 1997 sep 4-dec 25; 1998 jan-jun; 1998 jul-nov 9 – 1 – mf#4258439 – us WHS [071]
Baraboo weekly bulletin – 1880 oct 15-1881; 1882 jan-may 26 – 1 – mf#953700 – us WHS [071]
Baraboo weekly news – 1912 jan 4-apr 18; 1912 apr 25-1913 apr 24; 1913 may 1-14 apr 2; 1914 apr 9-1915 mar 18; 1915 mar 25-1916 mar 9; 1916 mar 16-1917 feb 15; 1917 feb 22-1918 mar 7; 1918 mar 14-19 jun 12 – 1 – mf#1002775 – us WHS [071]
Barabudur : esquisse d'une histoire du bouddhisme, fondee sur la critique archeologique des textes / Mus, Paul – Hanoi: Impr d'Extreme-Orient, 1935 – 1 – us CRL [280]
Baracchi, P *see* Report of the board of visitors to the observatory
Barack, K A *see*
– Des teufels netz
– Gallus oheims chronik von reichenau
– Zimmerische chronik
Barado, Francisco *see*
– La elocuencia militar
– La historia militar de espana
– Literatura militar espanola
– Mis estudios historicos...
– Museo militar historia. indumentari..
– La pintura militar
– Sitio de amberes 1584-85
– El sitio de baler, de saturnino martin cerezo
– La vida militar en esdana
Bar-Adon, Dorothy Ruth (Kahn) *see* Twin villages of merhavia
Baraga bulletin : devoted to the cause of the apostle of the ottawas and chippewas – v11 n1-v12 n6; v16 n3-date [1956 jun-1958 may; 1962 jul-1982 apr] – 1 – mf#614803 – us WHS [305]
Baragua – Weehawken, NJ. 1970 mar-1981 jan – 1r – us UF Libraries [071]
Barahona Jimenez, Luis *see* Gran incognito
Barahona, Ruben *see* Breve historia de honduras
Baraita de-shemu'el yarhina'ah – Vilna, Lithuania. 1925 – 1r – us UF Libraries [939]
Baraita ma'aseh torah – Warsaw, Poland. 1884 – 1r – us UF Libraries [939]
Barajas, Gonzalo *see* A descriptive study of collegiate arena managers
Barajas Salas, Eduardo *see* Saqueo e incendio de valencia de mombuey en 1641 y un curioso documento de 1693 sobre este pueblo
Baralt, Rafael Maria *see* Resumen de la historia de venezuela desde el ano d...
Baranov, A A *see*
– Sbornik deistvuiushchikh zakonopolozhenii, postanovlenii, instruktsii i tsirkuliarov po potrebitelskoi kooperatsii
– Ustav potrebitelskogo obshchestva, upravliaemogo obshchim sobraniem ego chlenov
Baranov, P E *see* IAkutskoe zemstvo
Baranovitsher kuryer *see* Unzer weg
Baranowski, Richard M *see* Coreidae of florida
Barante, Amable G de *see*
– Histoire de la convention nationale
– Histoire des ducs de bourgogne de la maison de valois 1364-1477
– Histoire du directoire de la republique francaise
Barante, Amable-Guillaume-Prosper Brugiere *see* Etudes litteraires et historiques
Barantsov, M S *see* Kak rabotat revizionnoi komissii selsko-khoziaistvennogo kreditnogo tovarishchestva
Barao de macahubas : periodico scientifico, litterario e noticioso – Bahia: [s.n.] 21 abr 1886; 09 set 1888 – (= ser Ps 19) – mf#P18B,02,20 – bl Biblioteca [440]
Barao de rio branco / Paula Cidade, Francisco De – Rio de Janeiro, Brazil. 1941 – 1r – us UF Libraries [972]
Barash, Asher *see* Masa ve-harim
Barata, Agildo *see* Vida de um revolucionario
Barathvala, Pitambaradatta *see* The nirguna school of hindi poetry
Baratier, Albert Ernest Augustin *see* Souvenirs de la mission marchand
Baratta, Joseph Preston *see* World federalist movement
Barau, Justin *see* Memoire a la chambre des deputes
Baraza – [Dar es Salaam: s.n., aug 21/30-oct 18/24 1993; nov 1/7 1993-jan 17/23 1994; mar 7/13, jul 18/24-25/31, aug 15/21 1994; feb 21/27, feb 28/mar 6, jul 4/10-11/17, aug 8/14, 22/28, aug 30/sep 4, dec 15/18, 23/25 1994 – us CRL [079]
Baraza – Nairobi, Kenya: Baraza Ltd, 1979 – (= ser ST Clair Drake coll Of Africana) – (issues filmed as pt of: st clair drake coll of africana jun 9 1956) – us CRL [079]
Baraza and egyptian gazette *see* Egyptian mail
Barb fence regulator – v2 [1877 mar]; v6 [1881] – 1 – mf#1052922 – us WHS [071]
Barba, Alvaro Alonso *see* Traite de l'art metalique
Barba azul. opera / Hurtado, Antonio – 1869 – 9 – sp Bibl Santa Ana [890]
Barba Jacob, Porfirio *see*
– Poesias completas
– Terremoto de san salvador
Barba, P *see* Breve resumpta y tratado de la esencia, causas, pronostico..., y curacion de la peste
Barba Salinas, Manuel *see*
– Antologia del cuento salvadoreno, 1880-1955
– Memorias de un espectador
Barbadian – Bridgetown, Barbados 18 dec 1822-28 dec 1844, 3 jan 1846-29 dec 1849, 1 jan-31 dec 1853, 3 jan 1857-29 dec 1858, 4 jan 1860-9 dec 1861 – 1 – uk British Libr Newspaper [079]

Barbadian – Bridgetown, Barbados 6 feb 1839-8 apr 1840 (imperfect) – 1 – uk British Libr Newspaper [079]
Barbados / Barbados Development Board – Bridgetown, Barbados. 1962? – 1r – us UF Libraries [972]
Barbados / Makinson, David H – London, England. 1964 – 1r – us UF Libraries [972]
Barbados : our island home / Hoyos, F A – London, England. 1966 – 1r – us UF Libraries [972]
Barbados / Savage, Raymond – Philadelphia, PA. 1937 – 1r – us UF Libraries [972]
Barbados *see* Official gazette
Barbados advocate – Bridgetown, Barbados. 1988-1998 may – 105r – (gaps) – us UF Libraries [079]
Barbados advocate [daily ed] – Bridgetown, Barbados 21 jan 1926-21 may 1940, 15 feb 1941-19 nov 1961 – 1 – (cont by: advocate [home ed] 26 nov 1961-31 may 1963, 1 jan 1966-1 jun 1968) – uk British Libr Newspaper [079]
Barbados agricultural reporter – Bridgetown, Barbados 2 dec 1870-28 dec 1888, 31 dec 1895-30 jun 1922 (imperfect) – 1 – (cont: barbados agricultural reporter, & planter's scientific journal [15 jan, 7 jun, 10 sep, 13 nov, 17 dec 1845; 13 jul 1846]) – uk British Libr Newspaper [630]
Barbados and the confederation question, 1871-1885 / Hamilton, Bruce – London, England. 1956 – 1r – us UF Libraries [972]
Barbados book / Lynch, Louis – London, England. 1964 – 1r – us UF Libraries [972]
Barbados daily news – Bridgetown, Barbados jun 1960-12 dec 1963 – 1 – (cont by: daily news [13 dec 1963-26 sep 1965, 2 jan-31 jul 1966]; barbados daily news [27 jun-14 sep 1967]) – uk British Libr Newspaper [079]
Barbados Development Board *see* Barbados
Barbados diocesan history / Reece, James Ebenezer – London, England. 1928 – 1r – us UF Libraries [240]
Barbados globe and colonial advocate – Bridgetown Barbados 4 sep 1837, 8,15 feb 1838, 28 mar, 25 jul, 15 aug 1839, 2,9 sep 1839-9 apr 1840, 1 nov 1875-19 may 1926 (imperfect) – 47 1/2r – 1 – uk British Libr Newspaper [079]
Barbados herald – Bridgetown Barbados [ns] 3 apr 1879-27 apr 1885, 1 jun 1885-24 nov 1887, 9 jan 1888-30 jun 1892, 11 jul 1892-25 jul 1896 (imperfect) – 1 – uk British Libr Newspaper [079]
The barbados herald – Bridgetown. Barbados. Apr. 6, 13, 27; May 11, 1940 – 1 – us NY Public [079]
Barbados Legislature *see* Minutes of proceedings
Barbados observer – Bridgetown Barbados 4 jan 1958-28 nov 1964, 17,18 sep 1965, 8 jan-30 jul 1966 – 1 – uk British Libr Newspaper [079]
Barbados recorder – Bridgetown. Barbados 3 oct 1951-29 apr 1953, 4 jan 1958-30 dec 1959 – 7 1/2r – 1 – uk British Libr Newspaper [079]
Barbados standard – Bridgetown, Barbados 29 apr 1911-31 dec 1921 (daily) – 12r – 1 – (very imperfect, fr 17 feb 1912 onward consisting of weekend ed only) – uk British Libr Newspaper [079]
Barbados. Statistical Service *see* Abstracts of statistics 1956-1969
Barbados times – Bridgetown, Barbados 10 jan 1920-17 aug 1921 (imperfect) – 1r – 1 – uk British Libr Newspaper [079]
Barbara blomberg : historischer roman / Ebers, Georg – Stuttgart: Deutsche Verlags-Anstalt, [1893-97?] [mf ed 1993] – 2v on 5r – 1 – (filmed with: georg ebers gesammelte werke) – mf#3476p – us Wisconsin U Libr [830]
Barbara blomberg : ein stueck in drei akten mit vorspiel und epilog / Zuckmayer, Carl – Amsterdam: Bermann-Fischer 1949 [mf ed 1992] – 1r – 1 – (filmed with: eine selbstschau / heinrich zschokke) – mf#3073p – us Wisconsin U Libr [820]
Barbara heck : a tale of early methodism / Withrow, William Henry – Toronto: W Briggs; Montreal: C W Coates; Halifax: S F Huestis, 1895 [mf ed 1984] – 3mf – 9 – 0-665-48461-5 – (fr: the canadian methodist magazine for the year 1880) – mf#48461 – cn Canadiana [242]
The barbarian invasions of italy = Invasioni barbariche in italia / Villari, Pasquale – London: Unwin, 1913 – 2mf – 9 – 0-7905-7032-7 – (in english) – mf#1988-3032 – us ATLA [945]
Barbarie e trionfi : ossia le vittime illustri del san-si in cina nella persecuzione del 1900 / Ricci, P Giovanni – 2nd ed. Firenze: Tipografia Barbera, 1909 [mf ed 1995] – (= ser Yale coll) – viii/851p (ill) – 1 – 0-524-09077-7 – (in italian) – mf#1995-0077 – us ATLA [951]
Barbaro, D *see* La pratica della perspettiva...
Barbaro, J *see* Travels to tana and persia
Barbaroux, Charles Jean Marie *see* Memoires (inedits) de charles barbaroux
Barbarua, Srinath Duara *see* Tungkhungia buranji
Barbary coast / Bullard – New York, NY. 1913 – 1r – us UF Libraries [960]
Barbasco / Gonzalez Montalvo, Ramon – San Salvador, El Salvador. 1960 – 1r – us UF Libraries [972]
Barbauld, Anna Letitia *see* Eighteen hundred and eleven
Barbeau, Charles Marius *see* Huron and wyandot mythology
Barbeau, Victor *see* Cahiers de l'academie canadienne-francaise
Barbedo, Alceu *see* Fechamento do partido comunista do brasil
Barbeguiere, Jean Baptiste *see* La maconnerie mesmerienne
Barbella, Emanuele *see* Six duettos for the violin and violoncello
Barber, G L *see* Chicago law journal
Barber, Gershom Morse *see* A guide for notaries public and commissioners.
Barber, Heather *see* An examination of the sources and levels of perceived competence in male and female interscholastic coaches
Barber, Robert *see*
– Six sonatas for the piano forte or harpsichord
– Thomson's hymn to the seasons in score... opera 4
Barber, Robert Heberden *see* Supplementary bibliography of hawking
Barber, Thomas Walter *see* Scientific theology
Barber trade – 1961 jan – 1r – 1 – mf#4718414 – us WHS [640]
Barber, William Theodore Aquila *see*
– David hill
– Raymond lull
Barberena, Santiago Ignacio *see*
– Epoca antigua y de la conquista
– Epoca colonial
Barberey, Helene, Freifrau von *see*
– Elisabeth seton und das entstehen der katholischen kirche in den vereinigten staaten, pt 2
– Elisabeth seton und das entstehen der katholischen kirche in den vereinigten staaten, pt1
Barberi, A *see* Bullarii romani continuatio
Barberi, Giovanni *see* Compendio della vita e delle gesta di giuseppe balsamo
Barberino, F, da *see* Docvmenti d'amore di m. francesco barberino
Barberis, Robert *see* Ils sont fous ces liberaux
Barbers and Beauty Culturists Union of America *see* Beacon
Barber's shop – La Salle IL 1807-08 – 1 – mf#3559 – us UMI ProQuest [073]
The barber's trade union and other stories / Anand, Mulk Raj – London: Jonathan Cape, 1944 – (= ser Samp: indian books) – us CRL [830]
Barberton herald – Barberton SA, jul 1886-1964 [mf ed Pretoria: State Library] – 37r – 1 – mf#MS00067 – sa National [079]
Barberton, OH *see* City directory, 1928
Barbet DuBertrand, V R *see* Regne de louis 18
Barbet, I *see* Livre d'architecture d'autels et de cheminées...
Barbey d'Aurevilly, Jules *see* Les oeuvres completes
Barbey, W *see* Herborisations au levant...egypte, syrie et mediterranee...
Barbey-Boissier, C *see* Herborisations au levant... egypte, syrie et mediterranee...
Barbier, Antoine A *see* Dictionnaire des ouvrages anonymes
[Barbier d'Aucour, J] *see* Sentimens de cleante sur les entretiens d'ariste et d'eugene
Barbier de Montault, X *see* Traite d'iconographie chretienne
Le barbier de seville : opera comique en quatre actes, mis en musique sur la traduction italienne par le celebre sgr paisiello, et remis en francais d'apres la piece de mr de beaumarchais et parodie sous la musique, par mr framery... / Paisiello, Giovanni – Paris: chez Made Baillon...a la Muse Lyrique [1784?] [mf ed 1991] – 1r – 1r – mf#pres. film 109 – us Sibley [780]
Barbier, Emmanuel *see* Le devoir politique des catholiques
Barbier, Jules *see* Galathee
Barbier, Pierre *see* Roi cerf
The barbizon school of painters / Thomson, David Croal – London 1891 – (= ser 19th c art & architecture) – 5mf – 9 – mf#4.2.515 – uk Chadwyck [750]
The barbone parliament : first parliament of the commonwealth of england, 1653 / Glass, Henry Alexander – London: J. Clarke, 1899 – 1mf – 9 – 0-7905-5828-9 – mf#1988-1828 – us ATLA [941]
[Barbonius, J] *see* 57. morale sinne-beelden, aen sijne hoogheydt, den doorluchtigen ende hoogh-gheboren vorst Fredrick Hendrick prince van Orangien...
Barbosa De Oliveira, Albino Jose *see* Memorias de um magistrado do imperio
Barbosa, Jose *see* Relacoes luso-brasileiras
Barbosa, Jose Celso *see* Problema de razas
Barbosa, Martinho Da Rocha *see* Catecismo breve da doutrina crista em portugues e chichangane
Barbosa, Ruy *see*
– Directrizes de ruy barbosa
– Mocidade e exilio
– Oracao aos mocos
Barbot, J *see* A description of the coasts of north and south guinea
Barbour, A G *see*
– Account of the american church mission in shanghai and the lower yangtse valley
Barbour, Augustus *see* Making religion efficient
Barbour, Dorothy Dickinson *see*
– Chi-tu chiao hua ti chiao ting chiao yu [ccm2]
– Erh t'ung kuan li fa
Barbour, George Freeland *see*
– The ethical approach to theism
– A philosophical study of christian ethics
Barbour, George M *see* Florida for tourists
Barbour, John Humphrey *see* The beginnings of the historic episcopate as exhibited in the words of holy scripture and ancient authors
Barbour, Margaret Frazer *see*
– The child of the kingdom
– The soul-gatherer
Barbour, Ralph Henry *see* Cupid en route
Barbour's chancery appeals reports – New York. v1-3. 1845-48 (all publ) – 24mf – 9 – $36.00 – mf#LLMC 80-016 – us LLMC [340]
Barbour's supreme court cases / New York. (State). Supreme Court – v1-67. 1847-77 (all publ) – 536mf – 9 – $804.00 – mf#LLMC 80-009 – us LLMC [347]
Barbourville first baptist church – BARBOURVILLE, KY. 1804-77 – 1 – $10.53 – us Southern Baptist [242]
Barburger, bolstandiges / Luther, Martin – Printed by Christopher Saur, 1762 – 1 – $23.45 – (the hymn book for the practice of godliness in 649 christian and protestant psalms and hymns...) – us Southern Baptist [242]
Barbusse, Henri *see* Yeshu ha-notsri
Barca di venetia per padova... / Banchieri, Adriano – 1605 – (= ser Mssa) – 2mf – 9 – €35.00 – mfchl 175 – gw Fischer [780]
Barcarrota *see*
– Ferias y fiestas de septiembre 1970
– Ferias y fiestas en barcarrota. septiembre de 1946
– Ferias y fiestas patronales en honor de la virgen del soterrano. 1980
– Ordenanzas municipales
Barcelona, 1933 / Alvarez, Jose Ma. Leyendas & Japon, Cuentas del – Madrid: Razon y Fe, 1934 – 1 – sp Bibl Santa Ana [946]
Barcelona. archivo de la corona de aragon guia historica y descriptiva... / Udina Martorell, F – Madrid, 1986 – 9 – sp Cultura [946]
Barcelona bulletin – [Scotland] Glasgow: Guy A Aldred 15 may 1937 [mf ed 2003] – 1r – 1 – uk Newsplan [072]
Barcelona. (City). Ayuntamiento *see* Estadistica municipal
Barcena, Lucas *see*
– Antologia poetica, 1934-1954
– Tierra intima
Barchwitz-Krauser, O von *see* Six years with william taylor in south america
Barcia y Zambrana, Jose de *see* Epistola exhortatoria en orden a que los predicadores evangelicos no priven de la doctrina a las almas en los sermones de fiestas
Barckow, Klaus et al *see*
– Corvey – fuerstliche bibliothek corvey
– Corvey – fuerstliche bibliothek corvey-sachliteratur
– Corvey – fuerstliche bibliothek corvey-sachliteratur [deutschsprachige werke]
– Corvey – fuerstliche bibliothek corvey-sachliteratur [englischsprachige werke]
– Corvey – fuerstliche bibliothek corvey-sachliteratur [franzoesischsprachige werke]
Barclay, Alexander *see*
– A practical view of the present state of slavery in the west indies
Barclay, E D *see* A comparative view of the words bathe, wash, dip, sprinkle and pour of the english bible
Barclay, James Turner *see* The jerusalem mission
Barclay, John *see*
– An address delivered on saturday, the 16th march, 1878, in old st andrew's church, toronto
– A circular letter suggesting a petition to parliament for the protection of the temporalities fund
– A discourse in two parts, preached in st andrew's church, toronto
– A discourse preached in st andrew's church, toronto on the 24th of may, 1863
– Discourses preached in st andrew's church, toronto
– Extract from a sermon preached in st andrew's church, toronto, on the 30th april, 1865
– "He being dead yet speaketh"
– A sermon preached in st andrew's church, toronto
Barclay, Joseph *see* The talmud
Barclay, Lydia Ann *see* Selection from the letters of lydia ann barclay
Barclay, Rachel Mary *see* Centre broadsheet
Barclay, Robert *see*
– Apology for the true christian divinity
– Catechism wherein the christian principles and doctrines of the soc...
– On the communion or participation of the body and blood of christ
– Truth triumphant
Barclay, Wade Crawford *see* The worker and his bible
Barclays economic review – London UK 1993-96 – 1,5,9 – ISSN: 0956-5574 – mf#15718.03 – us UMI ProQuest [332]
Barco Centenera, Arcediano *see* Argentina y conquista del rio de la plata con otros acae cimientos de los reinos del peru. tucuman y estado de brasil. notas bibliograficas y biograficas de carlos navarro y lamarca
Barco de Centenera, M *see* Argentina...y conquista del rio de la plata con otros acaecimientos de los reynos...
O barco dos patoteiros – Recife, PE. 19 maio 1864 – (= ser Ps 19) – bl Biblioteca [079]
Barco Perez, Paulina *see* Decimas en honor...de la virgen
Bardach, Eugene *see*
– Implementation game
Bardaji y Buitrago, A *see* Contribuciones. impuestos, aranceles y gravamenes
Bardales B, Rafael *see* Neustro pueblo
Barddas : or, a collection of original documents illustrative of the theology, wisdom and usages of the bardo-druidic system of the isle of britain / Williams, John – Llandovery: DJ Roderic, 1862-74 [mf ed 1992] – 2mf – 9 – 0-524-04993-9 – (in welsh & english) – mf#1990-3451 – us ATLA [290]
Barde, Andre *see* Bon numero
Bardenhewer, Otto *see*
– Des h hippolytus von rom commentar zum buche daniel
– Des heiligen hippolytus von rom commentar zum buche daniel
– Geschichte der altkirchlichen litteratur
– Der name maria
– Patrology
– Polychronius
Bardesanes : der letzte gnostiker / Hilgenfeld, Adolf – Leipzig: T O Weigel 1864 [mf ed 1990] – 1mf – 9 – 0-7905-5951-X – (incl bibl ref) – mf#1988-1951 – us ATLA [290]
Bardesanes gnosticus syrorum primus hymnologus : commentatio historico-theologica / Hahn, August – Lipsiae: Sumtibus FCG Vogelii 1819 [mf ed 1990] – 1mf – 9 – 0-7905-6596-X – mf#1988-2596 – us ATLA [290]
Bardic studies of ireland – Dublin, Ireland. 1871 – 1r – us UF Libraries [490]
Bardill, Donald R *see* Journal of family social work
Bardin, A V *see* Fol'klor chkalovskoi oblasti
Bardin, Charles *see*
– On miracles
– On the importance of a religious education
– Render unto caesar the things which be caesar's, and unto god...
Bardos cubanos / Hills, Elijah Clarence – Boston, MA. 1901 – 1r – us UF Libraries [972]
Bardossy, Laszlo *see* A bardossy per. a magyar orszagos tudosito es a magyar tavirati iroda hivatalos kiadasaibol szerk
A bardossy per. a magyar orszagos tudosito es a magyar tavirati iroda hivatalos kiadasaibol szerk / Bardossy, Laszlo – Budapest, Hirado Konyvtar, 1945. 2 v. in 1. LL-4028 – 1 – us L of C Photodup [340]
Bardoux, Jacques *see*
– Le chaos espagnol
– Chaos in spain
– Staline contre l'europe
Bardsley, C W *see* A dictionary of english and welsh surnames
Bardsley, Joseph *see*
– Church of england right
– Greek church, her doctrines and principles contrasted with those of...
Bardwell, Horatio *see* Memoir of rev gordon hall
Bardwell, William *see*
– Healthy homes
– Westminster improvements
Bardy, Gustave *see* Didyme l'aveugle
Bare alberte : roman / Sandel, Cora – Oslo: Gyldendal 1939 – 1 – mf#1517 – us Wisconsin U Libr [830]
Barea, Arturo *see*
– La raiz rota
– Unamuno
Die barea-sprache / Reinisch, L – Wien, 1874 – 3mf – 9 – mf#NE-20257 – ne IDC [956]
Bareille, Georges *see* Le catechisme romain
Barellan leader – Barellan, jan 1928-dec 1929 – 1r – 9 – A$37.44 vesicular A$42.94 silver – at Pascoe [079]
Bareta dapit can volcan mayong sa albay can junio de 1897, pt1 : guinibong verso sa tataramon na bicol ni m p – Nueva Caceres: La Sagrada Familia 1897 [mf ed Bloomington IN: Indiana Uni Lib, Preservation Dept 1984] – (= ser Coll...in the bikol language) – 1r – 1 – us Indiana Preservation [490]

Barfuessele / Auerbach, Berthold – Stuttgart. J G Cotta, 1856 [mf ed 1993] – 255p – 1 – mf#8464 – us Wisconsin U Libr [830]
Barg aroyf / Charney, Daniel – Warsaw, Poland. 1935 – 1r – us UF Libraries [939]
[Bargagli, G] *see* Dialogo de' giuochi che nelle vegghie sanesi si usano di fare
Bargagli, S *see*
– Dell'imprese di scipion bargagli gentil'huomo sanese
– La prima parte dell'imprese
Bargas, Luis de *see* Peticion escrito de conclusiones al nuncio por el p. caceres con el p. juan de la serena y otros
Bargen, Melinda *see* The effect of a motor development program on preschool children's motor skills
Barger, Evert *see* Excavations in swat and explorations in the oxus territories of afghanistan
Barges, Jean Joseph Leandre *see* Vie du celebre marabout cidi abou-medien, autrement dit bou-medin
Barghoorn, Frederick Charles *see* Soviet image of the united states
Bargiel, Woldemar *see*
– 2tes trio fuer piano, violine und violoncello, op 20
– Erstes trio, in f dur, pianoforte, violine und violoncello
– Trio nr 3, b dur, op 37
Bargiel, Woldemar et al *see* Frederic chopin's (1810-1849) works
Bargslagsbladet – Koping, Sweden. 1890-1907 – 15r – 1 – sw Kungliga [078]
Bargslagsbladet – Koping, Sweden. 1979- – 1 – sw Kungliga [078]
Bargteheider zeitung – Bargteheide DE, 1932-37 – 16r – 1 – gw Misc Inst [074]
Bargy, Henry *see* La religion dans la societe aux etats-unis
Barham bridge *see* Koondrook / barham bridge
Barhebrus und seine scholien zur heiligen schrift / Goettsberger, Johann – Freiburg im Breisgau; St Louis, MO: Herder, 1900 [mf ed 1989] – (= ser Biblische studien 5/4-5) – 1mf – 9 – 0-7905-2224-1 – mf#1987-2224 – us ATLA [220]
Baric, Iva *see* Die entwicklung der praemaxillia und maxilla bei feten mit lippen-kiefer-gaumen-spalten
Barine, Arvede *see* Nevroses
Baring-Gould, Edith M E *see*
– In the year one in the far east
– With note-book and camera
Baring-Gould, S *see* Church in germany
Baring-Gould, Sabine *see*
– The book of were-wolves
– Church in germany
– The church revival
– English minstrelsie
– Legends of old testament characters
– Legends of the patriarchs and prophets
– The origin and development of religious belief
– Our inheritance
– The passion of jesus (first series)
– Post-mediaeval preachers
– A study of st paul
– Village conferences on the creed
Barisa – Addis Ababa: Dabata gazeta barisa, mar 12, apr 21, may 1,21 1976; mar 17-24, apr 7,21, may 12-jun 2 1977 – 1r – 1 – us CRL [079]
Barisal hitaishi – Bengal, India. Dec 1938-1939 – 1r – 1 – us L of C Photodup [079]
Barisan tani indonesia – Djakarta, 1960-1964 – 18mf – 9 – (missing: 1961, v2(apr, may, aug, sep, 21-26); 1962, v3(3, 5-26); 1963, v4(9/12-17/20)) – mf#SE-839 – ne IDC [959]
Barisan tani indonesia / Soeara tani – Jogjakarta, 1946-1947 – 1mf – 9 – (missing: 1946 v1(1)) – mf#SE-412 – ne IDC [959]
Barkai : the only hebrew-english-afrikaans fortnightly – Pretoria: State Library Corporate Communication, jan 1938-dec 1949 – 2r – 1 – mf#MS00391 – sa National [079]
Barkan, H *see* In shvere teg
Barker – 1976-84 – 1 – mf#962313 – us WHS [071]
Barker, Dudley *see* Swaziland
Barker, Edmund Henry *see* Literary anecdotes and contemporary reminiscences of professor porson and others
Barker, Ernest, Sir *see* The dominican order and convocation
Barker, George Frederick *see* Physics
Barker, George Stanley *see* An historical survey of black baptist hymnody in america
Barker, Henry *see* English bible versions
Barker, John Marshall *see* The saloon problem and social reform
Barker, Joseph *see*
– The cause of the distress at present prevailing in great britain and ireland
– Lectures on the church of england prayer book
Barker, Joseph Henry *see* True materialism
Barker, Lady [M A] *see* A year's housekeeping in south africa
Barker, Mary *see* Sir benjamin d'urban's administration of the eastern frontier
Barker, W *see* Modern atheism and the bible
Barker, Wharton *see*
– Memorandum on the commercial relations of the dominion of canada to the united states of america
– Our canadian relations
– Surplus revenues and canadian relations
Barker, William *see*
– Digging a little deeper
– Duration of future punishments
Barker, William B *see* A short historical account of the crimea
Barker's canadian monthly magazine – Kingston [Ont]: E J Barker, [1846]-1847 [mf ed v1 n2 jun 1846-v1 n12 apr 1847] – 9 – ISSN: 1190-7614 – mf#P04173 – cn Canadiana [073]
Barker's creek baptist church – ANDERSON COUNTY, SC. 272p. 1821-65, 1979-82 – 1 – $12.24 – us Southern Baptist [242]
Barkhoff, Harald *see* Handlungskontrolle und selbstkonzept(e) von hochleistungssportlern im roll- und eiskunstlauf in trainings- und wettkampfsituationen
Barking and dagenham citizen – Barking, England summer 1985- [mf 1986-] – 1 – (cont: london borough of barking & dagenham citizen [apr 1980-spring 1985]) – uk British Libr Newspaper [072]
Barking and dagenham express – Barking, England 1 dec 1984-10 sep 1994 [mf 1986-2002] – 1 – (cont by: barking express [17 sep 1994-19 aug 2000]) – uk British Libr Newspaper [072]
Barking and dagenham post – Barking, England 5 feb 1964-6 oct 1999 [mf 1977-] – 1 – (cont: dagenham post & barking & rainham guardian [20 jan 1928-26 dec 1930, 1 jan 1932-29 jan 1964]; cont by: post (barking & dagenham) 13 oct 1999-1 aug 2001) – uk British Libr Newspaper [072]
Barking and east ham advertiser, upton park, ilford and dagenham gazette – Barking, England 27 apr 1889-20 mar 1965 – 1 – (cont: barking & east ham advertiser, upton park, ilford & dagenham gazette [13 oct 1888-20 apr 1889]; cont by: barking & dagenham advertiser [27 mar 1965-25 jun 1971]) – uk Newsplan [072]
Barking & dagenham advertiser and barking recorder – Barking, London 2 jul 1971-14 aug 1981 [mf 1977-] – 1 – (cont: barking & dagenham advertiser [27 mar 1965-25 jun 1971]; cont by: dagenham & barking advertiser & barking recorder [21 aug 1981-7 apr 1989]) – uk British Libr Newspaper [072]
Barking & dagenham recorder – Barking, England 7 feb 1992- [mf 1977-] – 1 – (cont: dagenham & barking advertiser [14 apr 1989-31 jan 1992]) – uk British Libr Newspaper [072]
Barking & dagenham yellow advertiser – Barking, England 9 jan 1987-6 oct 1995 [mf 1986-] – 1 – (cont: yellow advertiser (barking & dagenham) 19 oct 1984-2 jan 1987; cont by: yellow advertiser (barking) 13 oct 1995-) – uk British Libr Newspaper [072]
Barking leader – Barking, England 7 sep 1994-17 nov 1995 – 1 – uk British Libr Newspaper [072]
Barkingside and hainault redbridge guardian *see* Redbridge guardian (gants hill barkingside and hainault ed)
The barkly and digger's news – Barkly West SA, 1898-1899 (wkly) [mf ed Cape Town: SA library 1982] – 1r – 1 – mf#MS00435 – sa National [079]
Barlaam und josaphat : franzoesisches gedicht des 13. jahrhunderts / Gui de Cambrai; ed by Zotenberg, Hermann & Meyer, Paul – Stuttgart: Litterarischer Verein 1864 [mf ed 1993] – (= ser Blvs 75) – 58r – 1 – (Preceded by: uebersicht ueber das 16. verwaltungsjahr des Litterarischen vereins [1863]. filmed with: bibliothek des literarischen vereins in stuttgart) – mf#3420p – us Wisconsin U Libr [440]
Barlaam und josaphat : franzoesisches gedicht des dreizehnten jahrhunderts / Cambrai, Gui de; ed by Zotenberg, Hermann & Meyer, Paul – Stuttgart: Litterarischer Verein, 1864 [mf ed 1993] – (= ser Blvs 75) – 419p – 1 – mf#8470 reel 15 – us Wisconsin U Libr [810]
Barlach, Ernst *see*
– Der gestohlene mond
– Der tote tag
Barletius, M *see*
– De vita moribvs ac rebvs praecipve adversvs tvrcas...
– Historia del magnanimo, et valoroso signor georgio castrioto, detto scanderbego, dignissimo principe de gli albani
Bar-Lewaw Mulstock, Itzhak *see* Placido
Barlicki, Norbert *see* Aleksander Debski
Barlow, A Ruffell *see* Studies in kikuyu grammar and idiom
Barlow, Alfred Ernest *see*
– On some dykes containing huronite
– On the origin and relations of the grenville and hastings series in the canadian laurentian
– The physical features and geology of the route of the proposed ottawa canal between the st lawrence river and lake huron
Barlow, J J *see* Our real danger, and how to meet it
Barlow, Joseph Lorenzo *see* Endless being
Barlow, T *see* Chronologia sacra
Barlowe, Raleigh *see* Land resource economics
Barmby, James *see* Gregory the great
Barmedman banner – Barmedman jan 1917-dec 1918 – 1mf – 9 – A$37.18 vesicular A$42.68 silver – at Pascoe [079]
Barmer zeitung 1833 – Wuppertal DE, 1847-1850 jun, 1855-60 – 10r – 1 – (title varies: 1852?: barmer buergerblatt, 4 nov 1860: barmer zeitung, 1 apr 1922: deutsches tageblatt, 1 jan 1924: westdeutsche allgemeine zeitung, 16 sep 1927: barmer zeitung) – gw Misc Inst [074]
Barmouth and county advertiser and district weekly news – [Wales] Gwynedd 1894, jan 1911-dec 1950 [mf ed 2003] – 24r – 1 – (cont as: barmouth advertiser and district weekly news [jan 1920-dec 1950]) – uk Newsplan [072]
[Barnabas] brief an die hebraeer : text mit angabe der rhythmen / ed by Blass, Friedrich – Halle (Saale): Max Niemeyer, 1903 [mf ed 1988] – 53p on 1mf – 9 – 0-7905-0252-6 – mf#1987-0252 – us ATLA [860]
Der barnabasbrief / Weiss, Johannes – Berlin: Wilhelm Hertz, 1888 – 1mf – 9 – 0-8370-1781-5 – mf#1987-6169 – us ATLA [227]
Barnabe d'Alsace, OFM *see*
– Questions de topographie palestinienne
– La ville de david
Barnabe, Michele *see* Bibliographie analytique de l'ile d'anticosti (reine du golfe)
Barnard, Cecil *see* Ivory trail
Barnard, Edouard Andre *see* L'agriculture au point de vue de l'emigration et de l'immigration
Barnard, Edouard-Andre *see*
– L'agriculture dans la province de quebec
– Beet sugar
– Cercles agricoles
– Du sucre de betteraves et de sa production economique dans la province de quebec
– Une lecon d'agriculture
– Manuel d'agriculture
– Nos ecoles d'agriculture
– Petit traite sur le dessechement et le drainage des terres, pouvant servir de texte aux conferences des cercles agricoles
Barnard, F A P *see* The higher education of women
Barnard, Frederick Augustus Porter *see* Should american colleges be open to women as well as to men?
Barnard, Howard Clive *see* The little schools of port-royal
Barnard, K H *see* Contributions to the knowledge of south african marine mollusca, pt 1
Barnard lines – v1 iss 1-v6 iss 12/14 [1981 spring-1986 winter/spring] – 1 – mf#1266285 – us WHS [071]
Barnard, P M *see*
– Biblical text of clement of alexandria
– Clement of alexandria, quis dives salvetur
Barnard, Percy Mordaunt *see* Quis dives salvetur
Barnard, Percy Mordaunt [comp] *see* The biblical text of clement of alexandria
Barnard, T H *see*
– Chipere
– Cipere
Barnard, William Theodore *see* The relations of railway managers and employees
Barnard's american journal of education *see* American journal of education [1855]
Barnaul'skii listok – Barnaul, Russia 1909-10 [mf ed Norman Ross] – 1 – mf#nrp-263 – us UMI ProQuest [077]
Barndopets vaelsignelse : eller, kan en kristen glaedja sig oefver det dop, hvilket han som barn mottagit? / Beyer, Johann Paul – Rock Island IL: Lutheran Augustana Book Concern c1893 [mf ed 1992] – 1mf – 9 – 0-524-05248-4 – mf#1991-2240 – us ATLA [240]
Barneby, William Henry *see*
– Life and labour in the far, far west
– Notes from a journal in north america in 1883
Barnedaaben : i lyset au guds ord og den kristne kirches historie / Petersen, W M H – Decorah IA: Lutheran Pub House 1899 [mf ed 1992] – 1mf – 9 – 0-524-04848-7 – mf#1990-1340 – us ATLA [242]
Barnes, Albert *see*
– The atonement in its relations to law and moral government
– Doeg the edomite
– Home missions
– An inquiry into the organization and government of the apostolic church
– An inquiry into the scriptural views of slavery
– Lectures on the evidences of christianity in the 19th century
– Life at threescore and ten
– Miscellaneous essays and reviews
– Notes, critical, illustrative, and practical, on the book of daniel
– Notes, critical, illustrative and practical on the book of job
– Notes, explanatory and practical
– Notes, explanatory and practical on the acts of the apostles
– Notes, explanatory and practical, on the book of revelation
– Notes, explanatory and practical, on the epistles of paul to the ephesians, philippians, and colossians
– Notes, explanatory and practical, on the first epistle of paul to the corinthians
– Notes, explanatory and practical, on the general epistles of james, peter, john and jude
– Notes, explanatory and practical, on the second epistle to the corinthians and the epistle to the galatians
– Presbyterianism
– The scriptural argument for episcopacy examined
– The throne of iniquity
– The way of salvation
Barnes, Arthur Stapylton *see*
– Blessed joan the maid
– The early church in the light of the monuments
– St peter in rome and his tomb on vatican hill
– The witness of the gospels
Barnes, Bertram *see* Goethe's knowledge of french literature
Barnes, Carolyn M *see* Effects of a creative dance program on the perceptual motor performance of trainable mentally retarded children
Barnes county citizen – Valley City, ND: E P Getchell. v1 n1 jul 2 1915- (wkly) – 1 – (issued with: pillsbury promoter. missing: 1921 feb 3; 1922 may 4, nov 9; 1924 jan 24; 1925 oct 8, nov 26, deec 3; 1926 feb 18-25, mar 4) – mf#11150-11152 – us North Dakota [071]
Barnes county citizen *see* Pillsbury promoter
Barnes county news – Valley City, ND: Don C Matchan. v29 n37 mar 4 1943-v31 n31 sep 28 1944 (wkly) – 1 – (cont: peoples opinion and good-will messenger; merged with: valley city times-record (valley city, nd: 1928) to form: valley city times-record and the barnes county news) – mf#10529 – us North Dakota [071]
Barnes county populist – Valley City, ND: Barnes Co Independent Central Committee. v1 n? oct 9 1894- (wkly) – 1 – mf#11453 – us North Dakota [071]
Barnes, Herbert *see* Nyanja-english vocabulary
Barnes, Irene H *see* Behind the great wall
Barnes, James *see*
– The hero of erie (oliver hazard perry)
– Naval actions of the war of 1812
Barnes, John *see* Complete triumph of moral good over evil
Barnes, John Arundel *see* Politics in a changing society
Barnes, Lela *see* Railroad collection
Barnes, Lemeul Call *see* John mason peck and one hundred years of home missions, 1817-1917
Barnes, Lemuel Call *see* Two thousand years of missions before carey
Barnes, Martha L *see* Using hypothesized measures of institutional readiness to predict success in receiving foundation grants among park and recreation agencies
Barnes, Mary Emelia Clark *see* New america
Barnes, mortlake and east sheen times – [London & SE] BLNL 1986- – 1 – (local ed of: richmond & twickenham times) – uk Newsplan [072]
Barnes, Ralph *see* Liber pontificalis of edmund lacy, bishop of exeter
Barnes & richmond gazette – [London & SE] Richmond 15 aug 1930-5 feb 1931 [mf ed 2003] – 1r – 1 – uk Newsplan [072]
Barnes, William Emery *see*
– An apparatus criticus to chronicles in the peshitta version
– The books of chronicles
– Canonical and uncanonical gospels
– A companion to biblical studies
– Haggai and zechariah
– Isaiah
– Lex in corde
– Malachi
– Pentateuchus syriace
– The peshitta psalter
– The two books of the kings
Barnes, William S *see* Maximal power output on the bicycle ergometer
Barnes, William W *see* The southern baptist convention, 1845-1953
Barnesborsky orol = The barnesboro eagle – Barnesboro, PA: A J Antos-ovi synovia, dec 1917-aug 5 1920 – 1r – 1 – us CRL [071]
Barnes's new brunswick almanack for the year of our lord 1875 : being third year after leap year... – Saint John NB: Barnes, [1875?] [mf ed 1984] – 2mf – 9 – 0-665-32010-8 – mf#32010 – cn Canadiana [030]
The barneston star – Barneston, NE: D R Mercer. v2 n22. aug 29 1890 (wkly) [mf ed 1990] – 1r – 1 – us NE Hist [071]
The barneston star – Barneston, NE: D R Mercer (wkly) – 1r – 1 – us Bell [071]

Barnet and finchley independent – Barnet, London 12 feb 1987-12 may 1988 [mf 1987-12 may 1988] – 1 – (cont: barnet independent [18 jun 1982-5 feb 1987]; cont by: barnet, finchley (... etc) independent [19 may 1988-29 mar 1990]) – uk British Libr Newspaper [072]
Barnet and whetstone express – Barnet, London 11 sep 1992-7 may 1993 – 1 – uk British Libr Newspaper [072]
Barnet borough times – Barnet, London 17 jan 1985-25 jan 1996 [mf 1986-] – 1 – (cont by: barnet & potters bar times [1 feb 1996-2 apr 1998]) – uk British Libr Newspaper [072]
Barnet echo – Barnet, England 30 jun 1988-1 jun 1989 – 1 – (cont by: barnet leader [8 jun 1989-26 apr 1990]) – uk British Libr Newspaper [072]
Barnet & finchley press – Barnet, England 30 apr 1987-19 nov 1992 [mf 1977-] – 1 – (cont: press [7 aug 1986-23 apr 1987]; cont by: barnet & whetstone press [26 nov 1992-14 sep 1994]; barnet & finchley press [21 sep 1994-13 mar 1996]; barnet & whetstone press [20 mar 1996-25 feb 1999]; press (barnet & whetstone) 4 mar 1999-31 may 2001) – uk British Libr Newspaper [072]
Barnet gazette, & tradesmen's courier for the parishes of chipping barnet, east barnet, etc – Barnet, England 6 dec 1856-28 dec 1861, 20 mar 1869 – 1 – uk British Libr Newspaper [072]
Barnet herald – Barnet, London feb-may 1890 – 1 – (discontinued) – uk British Libr Newspaper [072]
Barnet independent – Barnet, London 18 jun 1892-5 feb 1987 [mf 1986-93] – 1 – (cont by: barnet & finchley independent [12 feb 1987-12 may 1988]) – uk British Libr Newspaper [072]
Barnet leader – Barnet, London 8 jun 1989-26 apr 1990 – 1 – (cont: barnet echo [30 jun 1988-1 jun 1989]) – uk British Libr Newspaper [072]
Barnet local advertiser – [London & SE] Barnet 10 jan 1985-27 mar 1986 – 1 – (cont: barnet & district local advertiser [25 nov 1982-3 jan 1985]; cont by: barnet advertiser [3 apr 1986-18 apr 1996]) – uk Newsplan [072]
Barnet, Miguel *see*
– Isla de guijes
– Piedra fina y el pavorreal
Barnet post – Barnet, England 22 sep 1988-10 aug 1989 – 1 – uk British Libr Newspaper [072]
Barnet & potters bar advertiser – Barnet, England 30 jun 1999-30 may 2001 [mf 1986-2001] – 1 – (Microfilm: .) – uk British Libr Newspaper [072]
Barnet & potters bar times – Barnet, England 1 feb 1996-2 apr 1998 [mf 1986-] – 1 – (cont: barnet borough times [17 jan 1985-25 jan 1996]; cont by: barnet times [9 apr-17 dec 1998]) – uk British Libr Newspaper [072]
Barnet press – Barnet, London 28 dec 1861-29 mar, 10 may-21 jun 1862; 6 feb 1869-24 jun 1871; 4 jan 1873-30 dec 1876; 12 jan 1878-31 jul 1986 [mf 1977-] – 1 – (cont by: press [7 aug 1986-23 apr 1987]) – uk British Libr Newspaper [072]
Barnet press *see* Finchley free press
Barnet times – Barnet, London 9 apr-17 dec 1998 [mf 1986-] – 1 – (cont: barnet & potters bar times [1 feb 1996-2 apr 1998]; cont by: barnet & potters bar times [24 dec 1998-]) – uk British Libr Newspaper [072]
Barnet weekender – Barnet, England 25 sep 1987-24 jun 1988 – 1 – (cont: north london & herts weekender [21 feb 1986-18 sep 1987]) – uk British Libr Newspaper [072]
Barnett, Arthur Thomas *see* Why are betting and gambling wrong?
Barnett, Bion Hall *see* Reminiscences of fifty years in the barnett bank
Barnett, Claude A *see*
– The claude a barnett papers
Barnett, Don *see* Liberation support movement interview on angola
Barnett, Edith A *see* Common-sense clothing
Barnett, George [comp] *see* The british columbian and victoria guide and directory for 1863
Barnett, George E *see* Mediation, investigation and arbitration in industrial disputes
Barnett, Henrietta *see* Practicable socialism
Barnett, Lionel David *see*
– Antiquities of india
– Brahma-knowledge
– The heart of india
– Hinduism
– Some sayings from the upanishads
Barnett, Samuel *see*
– Practicable socialism
– Towards social reform
Barnett source – v1 n1-v3 n4 [1986 may-1989 feb] – 1 – mf#1548609 – us WHS [071]
Barnett, William T *see* Diary
Barnette, Henlee H *see* Clarence jordan: a prophet in blue jeans
Barnette, R M *see* Lysimeter studies with the decomposition of summer cover crops
Barnet-whetstone independent – Barnet, England 26 apr 1990-16 dec 1993 [mf 1986-93] – 1 – (cont: barnet, finchley (...etc) independent [19 may 1988-29 mar 1990]) – uk British Libr Newspaper [072]
Barneveld banner – 1897 jun 25-1899 oct 13 – 1 – mf#966272 – us WHS [071]
Barnewell and adolphus' reports : reports of cases argued and determined in the court of king's bench... / Barnewell, Richard V & Adolphus, John L – v1-5. 1831-34. London: Saunders & Benning, 1831-35 (all publ) – 56mf – 9 – $84.00 – mf#LLMC 95-246 – us LLMC [324]
Barnewell and alderson's reports : reports of cases argued and determined in the court of king's bench... / Barnewell, Richard V & Alderson, Edward H – v1-5. 1817-22. London: J Butterworth & Son, 1818-22 (all publ) 84-744 – 45mf – 9 – $67.00 – mf#LLMC 84-744 – us LLMC [324]
Barnewell and cresswell's reports : reports of cases argued and determined in the court of king's bench... / Barnewell, Richard V & Creswell, Creswell – v1-10. 1822-30. London: A Strahan, 1823- (all publ) – 102mf – 9 – $153.00 – mf#LLMC 84-745 – us LLMC [324]
Barnewell, Richard V *see*
– Barnewell and adolphus' reports
– Barnewell and alderson's reports
– Barnewell and cresswell's reports
Barney Cabrera, Eugenio *see* Geografia del arte en colombia, 1960
Barney, William L *see* Nineteenth century southern political leaders
Barni, Camille *see*
– Air varie pour violon et violoncelle
– Trois trios pour violon, alto et violoncelle, oeuvre 6
Barnick-Ben-Ezra, Barbara *see* Interpreting dance
Barnimer tageblatt – Berlin DE, 1929, 1930 1 apr-1932 30 sep, 1933 2 jan-25 feb – 10r – 1 – (cover: oranienburg, bernau, liebenwalde, altlandsberg, strausberg, werneuchen, kreise nieder- & oberbarnim) – gw Misc Inst [074]
Barnola, Pedro Pablo *see* Eduardo blanco
Barnoldswick and earby times – [NW England] Barnoldswick 1937-67, 1974, 1988 – 1 – uk MLA; uk Newsplan [072]
Barnoya Galvez, Francisco *see* Han de estar y estaran(cuentos y leyendas de gu...
Barns, Margarita *see*
– India, today and tomorrow
– The indian press
Barns, William Eddy *see* The labor problem
Barnsley chronicle – Barnsley, England 16 oct 1858-28 jan 1933 [mf 1864, 1865, feb-dec 1897, 1986-] – 1 – (cont by: barnsley chronicle & south yorkshire news (main ed) 4 feb 1933-14 may 1982) – uk British Libr Newspaper [072]
Barnsley echo – Barnsley, England 8 sep 1869-30 dec 1874 [mf sep-dec 1873] – 1 – (wanting jan-dec 1872, jan-aug 1873) – uk British Libr Newspaper [072]
Barnsley independent [1883] – Barnsley, England 6 jan 1883-2 sep 1939 [mf jan-dec 1888, jan-nov 1897, jan-dec 1912] – 1 – (incorp with: barnsley chronicle; wanting: jan-dec 1872, dec 1897; cont: barnsley times [7 apr 1855-30 dec 1882]) – uk British Libr Newspaper [072]
Barnsley independent [1996] – Barnsley, England 7 feb 1996- [mf 1986-] – 1 – (cont: independent [14 oct 1980-31 jan 1996]) – uk British Libr Newspaper [072]
Barnsley record – Barnsley, England 9 oct 1858-24 nov 1866 (wkly) – 4r – 1 – (discontinued) – uk British Libr Newspaper [072]
Barnsley sporting news – Barnsley, England 9 jul 1894-28 jun 1902 [mf 1896, 1898] – 1 – (wanting: dec 1895, jan-dec 1897; cont by: barnsley daily argus & sporting news [30 jun 1902-7 mar 1903]) – uk British Libr Newspaper [790]
Barnsley telephone – Barnsley, England 23 sep 1904-1 feb 1918, 2 jan-2 apr 1920 [mf jun 1911-dec 1912] – 1 – (wanting: jan-may 1911) – uk British Libr Newspaper [380]
Barnsley times – Barnsley, England 7 apr 1855-30 dec 1882 (wkly) – 15r – 1 – (wanting: jan-dec 1872, dec 1897; cont by: barnsley independent [6 jan 1883-2 sep 1939]) – uk British Libr Newspaper [072]
Barnstable 1643-1892 – Oxford, MA (mf ed 1987) – (= ser Massachusetts vital records) – 73mf – 9 – 0-931248-93-0 – (mf 1-6: births, marriages, deaths 1643-1714. mf 7-11: transcript of records 1643-1714. mf 12-13: index to transcripts 1643-1714. mf 14-17: town records 1713 -81. mf 18-22: transcript of records 1731-81. mf 23-26: index to transcripts 1713-81. mf 27-30: town records 1765-92. mf 31-34: town records 1793-1816. mf 35-39: transcript of records 1793-1816. mf 40-42: index to transcripts 1793-1816. mf 43-46: births, marriages, deaths 1816-75. mf 47-48: index to marriages 1816-75. mf 49-52: town records 1844-80. mf 53-56: family records 1861-90. mf 57-59: births, marriages, deaths 1844-64. mf 60-62: births 1865-91. mf 63-64: index to births 1865-91. mf 65: index to births 1844-91. mf 66-68: marriages 1853-92. mf 69-70: index to marriages 1853-92. mf 71-72: deaths 1865-91, vol 27. mf 73: index to deaths 1844-91) – us Archive [978]
Barnstead, New Hampshire. First Free Will Baptist Church *see* Records
Barnstein, Henry *see* The targum of onkelos to genesis
Barnston, George *see* The oregon treaty and the hudson's bay company
Barnston (wirral), christ church – (= ser Cheshire monumental inscriptions) – 2mf – 9 – £4.00 – mf#6 – uk CheshireFHS [929]
Barnton (northwich), christ church – (= ser Cheshire monumental inscriptions) – 2mf – 9 – £4.00 – mf#7 – uk CheshireFHS [929]
Barnum, Phineas Taylor *see* It has been my earnest endeavor for years to elevate the tone and character of public amusements
Barnum, Samuel Weed *see*
– Romanism as it is
Barnwell first baptist church – BARNWELL, SC. 1812-1912, 1923-72 – 1 – $48.15 – us Southern Baptist [242]
Baro, B *see* Fr joan duns scotus...per universam philosophiam,...contra adversantes defensus
Baro, P *see* De fide, eujusque ortu, et natura, plana ac dilucida explicatio
Der barocke geschichtsbegriff bei andreas gryphius / Kappler, Helmut – Frankfurt am Main: M Diesterweg 1936 [mf ed 1990] – (= ser Frankfurter quellen und forschungen zur germanischen und romanischen philologie 13) – 1r – 1 – (incl bibl ref. filmed with: monteur klinkhammer / erich grisar) – mf#2693p – us Wisconsin U Libr [430]
Das barocke geschichtsbild in lohensteins arminius / Wehrli, Max – Frauenfeld/Leipzig: Huber 1938 [mf ed 1991] – (= ser Wege zur dichtung 31) – 1r – 1 – (incl bibl ref. filmed with: der dichter siegfried lipiner (1856-1911) / hartmut von hartungen) – mf#2827p – us Wisconsin U Libr [430]
Baroda state, 1901 – Baroda. pt 4. 1921 (mf ed 1941) – (= ser Census of india) – 1 – (1911 administrative vol) – us CRL [315]
Barometer – v1 n10/11-v2 n1/2; v2 n4-v4 n1; [1943 jan/feb-mar/apr; 1943 fall-1944; fall/1945 winter] – 1 – mf#1131945 – us WHS [071]
Barometern – Kalmar, Sweden. 1979-. Oskarshamnstidningen – 1 – sw Kungliga [078]
Barometern – Kalmar, Sweden. 1841-1978 – 543r – 1 – (oskarshamnstidningen ed, 1868-78) – sw Kungliga [078]
Le baron americain / De Mille, James – Paris: C Levy, 1877 – 5mf – 9 – (trans by louis ulbach) – mf#32173 – cn Canadiana [830]
Baron, Auguste *see* Histoire de lyon pendant les journees des 21, 22 et 23 novembre 1831
Baron c c von der decken's reisen in ost-afrika in den jahren 1859 bis 1865 / Kersten, O – Leipzig, Heidelberg. 4pts. 1869-1879 – 37mf – 9 – mf#H-6129 – ne IDC [916]
Baron carl claus von der decken's reisen in ost-afrika : in den jahren 1859 bis 1865 – Leipzig [u a] 1869-70 [mf ed Hildesheim 1995-98] – 6v on 34mf – 9 – €340.00 – 3-487-26736-5 – gw Olms [916]
Baron Castro, Rodolfo *see*
– Espanolismo y antiespanolismo en la america hispana. la poblacion hispanoamericana a partir de la independencia
– Jose matias delgado y el movimiento insurgente de...
– Poblacion de el salvador
– Resena historica de la villa de san salvador
Baron, David *see*
– The ancient scriptures and the modern jew
– The history of the ten "lost" tribes
– The jewish problem, its solution
– The shepherd of israel and his scattered flock
Baron, L *see* L'expression du chant gregorien, vol 2
Baron report – 141-142, 144-161, 203, 208 [1982 jan 4-18; 1982 feb 15-oct 4; 1984 may 7, jul 16] – 1 – mf#783294 – us WHS [071]
Baron von haller / Bennett, J Risdon – London, England. 18– – 1r – us UF Libraries [240]
Baroni, Aldo *see* Cuba, pais de poca memoria
The baronial and ecclesiastical antiquities of scotland / Billings, Robert William – Edinburgh [1845-52] – (= ser 19th c art & architecture) – 11mf – 9 – mf#4.2.1305 – uk Chadwyck [720]
The baronial halls, and picturesque edifices of england / Hall, Samuel Carter – London 1848 – (= ser 19th c art & architecture) – 10mf – 9 – mf#4.2.1047 – uk Chadwyck [720]
Baronius, C *see* Generale kerckelycke historie van den gheboorte onzes h iesu christi tot het iaer 1624
Baronius, Caesar *see* Annales ecclesiastici
Barop-hombrucher volksblatt – Dortmund DE, apr 2-jun 29 1929, oct 1 1929-mar 31 1930, apr 4, jul 1 1930-mar 31 1931, jan 2-26, feb 7-jun 30, oct 2-dec 30 1933, apr 3-dec 31 1934, 1936-jun 30 1938, jan 2 1939-jun 29 1940 [gaps] – 11r – 1 – (aka: barop-hombrucher zeitung) – gw Misc Inst [074]
Barop-hombrucher zeitung *see* Barop-hombrucher volksblatt
Barotseland : eight years among the barotse / Stirke, Douglas Elliott Charles Romaine – New York, NY. 1969 – 1r – us UF Libraries [960]
Barou, Noah *see* Russian co-operative banking
Al-barq – Constantine. n22. 1927 – 1 – fr ACRPP [073]
Barr, Eleanor M *see* Records of the women's international league for peace and freedom, us section, 1919-59
Barr, J *see* Nineteenth century children's literature collection
Barr, James *see*
– Anglican church architecture
– Contending for the faith
Barr, John *see* Early religious history of
Barr, John O *see* Neural versus muscular responses to isometric strength training of the triceps brachii
Barr, L I *see* Course in lugbara
Barr, Lockwood *see* Will the railroads come back?
Barr, Robert *see*
– Cardillac
– A chicago princess
– The countess tekla
– The face and the mask
– From whose bourne
– The girl in the case
– In a steamer chair
– In the midst of alarms
– Jennie baxter, journalist
– Lord stranleigh abroad
– Lord stranleigh, philanthropist
– The mutable many
– One day's courtship
– The o'ruddy
– Over the border
– A prince of good fellows
– Revenge!
– Rhyme and reason
– A rock in the baltic
– Stranleigh's millions
– The strong arm
– The sword maker
– The tempestuous petticoat
– The unchanging east
– A woman in a thousand
– A woman intervenes
– Young lord stranleigh
Barr, Ruth B *see* Marianna
Barraba chronicle – Barraba 1923-40, 1941-42, 1942-49, 1944, 1946-52, 1952-59, 1961-69 – at Pascoe [079]
Barraba gazette – Barraba, jan 1969-dec 1996 – at Pascoe [079]
Barraba & manilla news – Barraba, jan 1898-dec 1907 – 2r – A$140.45 vesicular A$151.45 silver – at Pascoe [079]
Barrabases (cosas de mi pueblo) / Esteves, Luis Raul – Mexico City? Mexico. 1949 – 1r – us UF Libraries [972]
Barraca de feria / Fernandez De Castro, Jose Antonio – Habana, Cuba. 1933 – 1r – us UF Libraries [972]
Barradas, M *see* Emmanuelis barradas s i tractatus tres historico-geographici
Barradas, S *see* Commentaria in concordiam et historiam evangelicam
Barrado, Angel *see* La casa donde nacio san francisco de asis, patronato del estado espanol
Barrado Font, Francisco *see*
– Dominacion y guerras de espana en los paises bajos
– Introduccion a una obra historica
Barrado Manzano, Arcangel *see*
– Algunas actas capitulares de la provincia de san gabriel al principio del siglo 17 (anos 1601-1608)
– Antonio brasio...monumenta...
– Antonio da silva riego
– Arduino terzi
– As gavetas da torre fombo 4, lisboa 1964
– Basilio de sa. arturo
– Bibliografia missionaria anno 21 (1957) y anno 12 (1958)...
– Bibliografia missionaria. anno 23
– Bibliografia missionaria. anno 28
– La bula "inter graviores curas" de pio 7 en la orden franciscana y ulterior regimen general de la orden en espana
– Congreso de espiritualidad franciscana
– Constituciones de la provincia de san gabriel
– El consulado de buenos aires y sus proyecciones en la historia del rio de la plata
– Direccion general de archivos y bibliotecas
– Division bipartita de la provincia franciscana de san miguel de extremadura
– Doctoris subtilis et mariani joannis duris scoti (ofm). opera omnia...
– Doctoris...joannis dius scoti...opera omnia
– Domingo cresi...
– Dos misioneros franciscanos hermanos en el colegio de chillan
– Ephrem lougpre mystique franciscain
– Fundacion y fabrica del convento de san antonio de padua de almendralejo en la provincia franciscana de san gabriel
– Historia missionorum ordinis fratrum minoris 3. america septentrionalis. roma, 1968

BARROW

- Historia missionum ordinis fratum minorum 1. asia-centro orientalis et oceania. roma, 1967
- Jose lopez de toro y ramon paz remolar
- Joseph jortz
- Libros catolicos de espana...
- Manuscritos franciscanos de la biblioteca de vicente barrantes
- Maurice grafewski o.f.m. the supreme...
- Mayo documental, advertencia y prologo de r. caillet-bois. 12 tomos. buenos aires. 1961-1966
- Monumenta henricina, vol. 8. (1443-1445) y 9
- Necrologica
- Nuevas actas capitulares de la provincia des calza de san gabriel
- La provincia descalza de san gabriel y sus libros de patentes
- La provincia franciscana de san miguel infra tagum
- Soto, de iustitia et iure. 1. 1967 y suarez, tratado de leyes...2, 1967
- Tercer centenario de la canonizacion de san pedro de alcantara (1669-28 de abril-1969)

Barrado manzano, p arcangel ofm : las misiones franciscanas en bolivia conferencias... / Lejarza, Fidel de – Madrid: Missionalia Hispanica, 1949 – 1 – sp Bibl Santa Ana [240]

Barranquilla / Sojo Zambrano, Jose Raimundo – Barranquilla, Colombia. 1955 – 1r – us UF Libraries [972]

Barrantes Maldonado, Francisco *see* Relacion de la calificacion y milagros del santo crucifijo de galamea...dividida en dos libros

Barrantes Molina, Luis *see* Desde mi tonel

Barrantes Moreno, Vicente *see*
- Aparato bibliografico...extremadura
- Baladas espanolas
- Barros emeritenses
- Catalogo razonado y critico...extremadura
- Dias sin sol
- Discurso...academia de la historia
- Discursos leidos ante la real academia de la historia en la recepcion...el 14 de enero de 1872 con un biografia de este
- Discursos...academia espanola
- Discursos...academia...manuel de lo palacio
- Epistola religiosa...ceferino gonzalez...
- Estado de extremadura...isabel la catolica
- Estudios sobre los restos de ceramica romana
- Guerras piraticas de filipinas
- Historia general de filipinas
- Un historiador moderno de la tierra de la serena (d. nicolas perez jimenez)
- Indice de la biblioteca extremena..
- Informe sobre
- Informe (sobre) lettres intimes de j.m. alberoni...
- La instruccion primaria en filipinas. desde 1596 hasta 1868
- Juan de padilla
- Las jurdes y su leyendas
- Narraciones extremenas. la imprenta en extremadura
- Narraciones extremenas. la serrana de la vera..
- Noticia necrologica de don felipe-leon guerra
- Noticias sobre su obra. "el teatro tagalo"
- Las siete centurias de la ciudad de plasencia
- Sobre el derribo de una campana historica en badajoz
- El teatro tagalo
- Viaje electoral hecho con la bolsa a cuestas y el cuerpo molida a palos por a los infiernos del sufragio universal

Barras de Aragon, Francisco de las *see* Excursion a fregenal de la sierra y los jarales...

Barras, W *see* Proposed canonisation of mary queen of scots, cardinal beaton

Barras Y Prado, Antonio De Las *see* Habana a mediados del siglo 19

Barrasa, Joseph *see* Al padre miestro fray ioseph barrasa de la orden de nuestra senora de la merced

Bar'rasi – Qum: Mu'assasah-'i Ihya' va Nashr-i Miras-i Islami. shumarah-'i 11,15-18. 4 tir 1358-shahrivar 1358 [25 jun 1979-aug 1979] – 1r – 1 – $53.00 – us MEDOC [079]

Barrass, Edward *see*
- Class meetings
- Missionary scenes in many lands

Barratry. its origin, history and meaning in the maritime laws / New York. (City). Court of Common Pleas – New York: Baker & Godwin, 1872. 30p. LL-218 – 1 – us L of C Photodup [341]

Barraza, Melendez, Martin *see* Trayectoria del cuento salvadoreno

Barre 1718-1854 – Oxford, MA (mf ed 1995) – (= ser Massachusetts vital record transcripts to 1850) – 12mf – 9 – 0-87623-206-3 – (mf 1t-2t: vital records 1747-1809. mf 2t: marriages 1755-91. mf 2t-5t: births 1745-1854. mf 5t-6t: marriages 1781-1854. mf 6t-7t: deaths 1757-1845. mf 8t-10t: intentions 1792-1849. mf 10t: out-of-town marriages 1749-98. mf 10t-11t: births 1844-49. mf 11t-12t: marriages 1844-49. mf 12t: deaths 1844-49) – us Archive [978]

Barre 1718-1895 – Oxford, MA (mf ed 1987) – (= ser Massachusetts vital records) – 32mf – 9 – 0-87623-023-0 – (mf 1-3: births, marriages, deaths 1718-1810. mf 4-8: births, marriages, deaths 1752-1854. mf 9-16: index to b,m,d, prior to 1845. mf 17-19: births, marriages, deaths 1844-55. mf 20-28: births, marriages, deaths 1855-95. mf 29-32: index to b,m, d, 1844-95) – us Archive [978]

Barre, Louis-Francois-Joseph de la, vailly et al *see* Le grand dictionnaire historique (ael1/44.5)

Barre, M *see*
- Cassandre-agamemnon et colombine-cassandre, parodi
- Deux edmon
- Dugai-trouin
- Ecriteaux
- Gaspard l'avise
- Isle de la megalantropogenesie
- Lantara
- Mai des jeunes filles, ou, un passage de militaire
- Monet
- Peintre francais a londres

Barre, Stanislas Morrier *see*
- Aux meres canadiennes
- Cream raising by the centrifugal and other systems
- L'effect de la guerre sur nos methodes d'elevage et d'agriculture
- Essay on mr w h lynch's pamphlet entitled "scientific butter making"
- The farm pasteuriser
- Rapport sur la fabrication du beurre

Barred and disallowed case files of the southern claims commission, 1871-1880 / U.S. House of Representatives – (= ser Records of the united states house of representatives) – 4829mf – 9 – (with printed guide) – mf#M1407 – us Nat Archives [324]

Barreda, Ernesto Maria *see*
- Un camino en la selva
- Lucha de alas
- Una mujer

Barreintos, Alfonso Enrique *see* Gomez carrillo 30 anos despues

O barreirense : periodico hebdomadario – Sao Jose do Barreiro, SP: Typ do Barreirense, 23 dez 1876; jan, 22 abr 1877 – (= ser Ps 19) – mf#P18,01,80 – bl Biblioteca [079]

The barren ground of northern canada / Pike, W – London, New York, 1892 – 6mf – 9 – mf#N-347 – ne IDC [917]

Barren run baptist church – London. 1970-1981 (1) 1970-1981 (5) 1976-1981 (9) – 1r – 1 – $29.16 – mf#6561 – us Southern Baptist [242]

Barrenechea, Raul Porras *see* Cuadernos de historia del peru...

Barrer kantons-blatt – Barr, F. 1882-9 nov 1918, 1919, 1921-40, 1950-51, 1954 – 1 – (pts missing; cont by: journal de barr) – fr ACRPP [944]

Barrera, Claudio *see* Antologia de poetas jovenes de honduras desde 1935

Barrera, Isaac J *see* Lecturas para los grados superiores de la escuela

Barrera Moncada, Gabriel *see* Edad pre-escolar

Barreras Y Martinez Malo, Antonio *see* Prontuario de derecho constitucional cubano

Barrere, Bouguer *see* Neue reisen nach guiana, peru und durch das suedliche amerika

Barres, Maurice *see* The faith of france

Barret, George *see* The theory and practice of water colour painting

Barret, M l'Abbe *see* Cartulaire de marmoutier pour le perche

Barreto Rodriquez, Jesus *see* Jurisprudencia penal e casacion

Barrett, Alfred *see* Discourse on the modern mental philosophy viewed in its aspects on...

Barrett, Benjamin Fiske *see* Letters on the divine trinity

Barrett, Charlotte *see* Diary and letters of madame d'arblay ed by ger niece [charlotte barrett]

Barrett, Clifford *see* Ethics

Barrett, Douglas *see* Painting of the deccan

Barrett, Francis Thornton *see* On the selection of books for a reference library

Barrett, George Slatyer *see* The temptation of christ

Barrett, John Casebow *see* Bible

Barrett, John Pressley *see* The centennial of religious journalism

Barrett, Kate R *see* An interpretive inquiry of preservice teachers' reflections and development during a field-based elementary physical education methods course

Barrett, Maca *see* Caballo rojo

Barrett, S A *see* The ethno-geography of the pomo and neighboring indians

Barrett, Thomas Squire *see* Examination of gillespie

Barrette, J E T *see* Recit d'aventures dans le nord-ouest, etc

Barretto, Castro *see*
- Estudos brasileiros de populacao
- Povoamento e populacao

Barretto, Joseph *see* A dictionary of the persian and arabic languages

Las barriadas de...honrar al espiritu santo con los siguientes festejos...1976 / Parroquia del Espiritu Santo – Caceres: Tip. Rodriguez, 1976 – 1 – sp Bibl Santa Ana [946]

Las barriadas...honran al espiritu santo con los siguientes festejos, 1975 / Parroquia del Espiritu Santo – Caceres: Imp Rodriguez, 1975 – 1 – sp Bibl Santa Ana [240]

Barricada : official organ / Frente Sandinista de Liberacion Nacional. Nicaragua – Nicaragua: Frente Sandinista de Liberacion Nacional, Jul 25 1979-Dec 1991 – 56r – 1 – us L of C Photodup [079]

Barricada internacional : [english ed] – Mangue, Nicaragua 1986-97 – 1,5,9 – ISSN: 1013-9567 – mf#14917 – us UMI ProQuest [327]

Barrichina, P *see* Theses in cathedrae hippocratis concursu...

Barrie and the kailyard school / Blake, George – London, England. 1951 – 1r – us UF Libraries [960]

Barrientos, Alfonso Enrique *see*
- Cuento de amor
- Cuentos de belice
- Rafael heliodoro valle

Barrientos Casos, Luis Felipe *see* Los tres sindicalismos

Barrientos, Gaspar *see* Chorografia

Barrientos Lavin, Oscar *see* Estudio comparativo de la hipoteca minera y la hipoteca comun

Barrier collection of south asian political tracts / Barrier, Norman Gerald – 1 – us CRL [959]

Barrier daily truth – Broken Hill, jan-sep 1969, jan 1970-jun 1997 – at Pascoe [079]

Barrier miner – Broken Hill jan 1898-dec 1914, jan 1931-dec 1952 – 50r – 9 – A$3571.74 vesicular A$3846.74 silver – at Pascoe [079]

Barrier miner – Broken Hill jan 1969-nov 1974 – 20mf – 9 – (ceased publ nov 1974) – at Pascoe [079]

Barrier, N G *see* Political records filmed spring 1969

Barrier, Norman G *see* Tracts and miscellaneous printed material, university of missouri, summer of 1967, including india political pamphlets

Barrier, Norman Gerald *see* Barrier collection of south asian political tracts

Barriere, D *see*
- Villa aldobrandina tusculana sive varij illius hortorum et fontium prospectus

[Barriere, D] *see* Villa pamphilia eiusque palatium, cum suis prospectibus, statuae, fontes, vivaria, theatra, areolae, plantarum, viarumque ordines

Barriere, Jean Francois *see*
- La cour et la ville sous louis 14, louis 15 et louis 16
- Tableaux de genre et d'histoire, peints par differens maitres

Barriere, Theodore *see*
- Corneille qui abat des noix
- Faux bonshommes
- Malheur aux vaincus
- Monsieur qui suit les femmes
- Schoen, lieber joseph!
- Tete de linotte
- Vie de boheme

O barriga verde : semanrio illustrado – Florianopolis, SC. 11 set-18 nov 1928 – (= ser Ps 19) – bl Biblioteca [079]

Barriga, Victor *see* Los mercedarios en el peru en el siglo 16. documentos ineditos del archivo general de indias. tomo 1. roma, 1933

O barrigudo : periodico satyrico – Recife, PE: Typ Popular, 18 dez 1863 – (= ser Ps 19) – mf#P16,01,07 – bl Biblioteca [870]

Barrineau, Thomas Lorren *see* Teaching program in vocational agriculture for the gonzalez school

Barrington 1570-1950 – (= ser Cambridgeshire parish register transcript) – 7mf – 9 – £8.75 – uk CambsFHS [929]

Barrington, Arthur H *see* Anti-christian cults

Barrington, George *see*
- An account of a voyage to new south wales
- The history of new south wales
- A voyage to botany bay

Barrington, New Hampshire. First Baptist Church *see* Records

Barrio y Rufo, Jose *see* Apuntes para la historia...plasencia

Barrionuevo, G *see* Tratado sobre el laudano opiato...

Barrios ante la posteridad / Diaz, Victor Miguel – Guatemala, 1935 – 1r – us UF Libraries [972]

Barrios, Gilberto *see*
- Evocacion de rivas
- Tres almas

Barrios, Gonzalo *see* Dias y la politica

Barrios Y Carrion, Leopoldo *see* General calleja

The barrister : being anecdotes of the late tom nolan of the new york bar / Stansbury, Charles F – New York: Mab Press, 1902 – 3mf – 9 – $4.50 – mf#LLMC 95-166 – us LLMC [340]

The barrister – London: W Jenkinson. v1 nos 1,3,4. 1824 (all publ) – 1mf – 9 – $1.50 – (lacking: no 2) – mf#LLMC 84-419 – us LLMC [340]

The barrister – Toronto: Law Pub Co, [1894-1897] – 9 – mf#P05021 – cn Canadiana [340]

The barrister – Toronto. v1-3. 1894-97 (all publ) – 13mf – 9 – $19.50 – mf#LLMC 84-420 – us LLMC [340]

Barrister (aba) – v1-23. 1974-97 (all publ) – 9 – $249.00 set – ISSN: 0094-5277 – mf#100911 – us Hein [340]

Barristers' wives of new york, inc : annual news bulletin – 1980 may – 1 – mf#4881924 – us WHS [071]

Barro / Navas Miralda, Paca – Guatemala, 1951 – 1r – us UF Libraries [972]

Barro al acero en la roma de los chibchas / Camargo Perez, Gabriel – Cartagena? Colombia. 1961 – 1r – us UF Libraries [972]

Barro en la sangre / Silva, Fernando – Managua, Nicaragua. 1952 – 1r – us UF Libraries [972]

Barroco mineiro / Machado, Lourival Gomes – Sao Paulo, Brazil. 1969 – 1r – us UF Libraries [972]

Barroeta Schneidnagel, Santiago *see* Sucesos de cienfuegos

Barron, Alfred *see* Home talks. vol. 1

Barron county chronotype – 1874 sep 30-1877 dec 31; 1878-79; 1880-83; 1884-86; 1887-89; 1890-1890 jul 31 – 1 – mf#1012040 – us WHS [071]

Barron County Electric Cooperative *see* Annual report to the members...

Barron county independent – 1888 apr 26-1889 may 24 – 1 – mf#953716 – us WHS [071]

Barron county leader – 1951 jun 20-dec; 1952 jan-1956 jun 27 – 1 – mf#966485 – us WHS [071]

Barron county news-shield – 1918 nov 1-1998 dec [with gaps] – 1 – mf#983640 – us WHS [071]

Barron county shield – 1876 oct 6-1918 oct 31 [with gaps] – 1 – mf#1007832 – us WHS [071]

Barron county tribune – 1926 sep 15-1928 jul 5; 1928 jul 12-1929 nov 14 – 1 – mf#953715 – us WHS [071]

Barron g collier / Collin, Eric – s.l, s.l? . 193-? – 1r – us UF Libraries [978]

Barros, Adirson De *see* Ascensao e queda de miguel arraes

Barros Arana, Diego *see*
- Historia general de chile
- Proceso de pedro de valdivia y...

Barros, Domingos *see* Aeronautica brasileira

Barros emeritenses / Barrantes Moreno, Vicente – 1877 – 9 – sp Bibl Santa Ana [946]

Barros emeritenses / Barrantes Moreno, Vicente – S.L. s.i.s.a. – 1 – sp Bibl Santa Ana [946]

Barros, J de *see*
- L'asia...consigliero del christianissimo re di portogallo
- Da asia de joo de barros dos feitos, que os portuguezes fizeram no descubrimento

Barros, Jacy Rego *see* Senzala e macumba

Barros, Jayme De *see*
- Ocho anos de politica exterior del brasil
- Poetas do brasil
- Politica exterior do brasil

Barros, M Marques De *see* Litteratura dos negros

Barros, Silvia *see* Teatro infantil

Barroso, Antonio Emilio Vieira *see* Marajo

Barroso, Gustavo *see*
- Ao som da viola
- Bracelete de safiras
- Brasil
- Brasil em face do prata
- Brasil na lenda e na cartografia antiga
- Guerra do flores
- Guerra do rosas
- Heroes e bandidos
- Historia militar do brasil
- Mythes, contes et legendes des indiens
- Pero coelho de sousa
- Sertao e o mundo
- Sinagoga paulista
- Tamandare e nelson brasileiro
- Terra de sol

Barroso, Parsifal *see* Cearense

Barroux, R *see* Dagobert

Barrow, Bennet Hilliard *see* Plantation life in the florida parishes of louisiana, 1836-1846

Barrow, David *see* Involuntary, unmerited, perpetual, absolute, hereditary slavery examined, 1753-1819

Barrow guardian – [NE England] Cumbria 24 sep-dec 1910; jan 1911-28 jun 1947 [mf ed 2003] – 63r – 1 – uk Newsplan [072]

Barrow herald and furness advertiser – [NE England] Cumbria jan 1863-8 aug 1914 [mf ed 2003] – 88r – 1 – (cont as: barrow herald dalton advertiser and north lonsdale reflector [jan 1891-jun 1892]; barrow herald and dalton advertiser [jul 1892-dec 1894]; barrow herald [jan 1895-aug 1914]) – uk Newsplan [072]

Barrow, Isaac *see* Holy scripture and the pope's supremacy contrasted

Barrow, J *see*
- An account of travels into the interior of southern africa, in the years 1797 and 1798...
- The geography of hudson's bay
- Some account of the public life
- Travels in china
- A voyage to cochinchina, in the years 1792 and 1793

Barrow, John *see*
- Abrege chronologique
- A chronological history of voyages into the arctic regions undertaken chiefly for the purpose of discovering a north-east, north-west, or polar passage between the atlantic and pacific
- The eventful history of the mutiny and piratical seizure of h m s bounty
- Excursions in the north of europe
- A family tour through south holland
- Histoire chronologique des voyages vers le pole arctique
- The life of george lord anson
- Voyage a la cochinchine
- Voyage dans la partie meridionale de l'afrique

Barrow, K M *see* Three years in tristan da cunha

Barrow leader – [NE England] Cumbria 14 feb 1924-may 1927 [mf ed 2003] – 1r – 1 – uk Newsplan [072]

Barrow news and dalton chronicle – [NE England] Cumbria jan 1881-dec 1949 [mf ed 2003] – 150r – 1 – (cont as: barrow news [jul 1883-jun 1947]; barrow news and barrow guardian [jul 1947-dec 1949]) – uk Newsplan [072]

Barrow times – [NE England] Cumbria jan 1871-dec 1884 [mf ed 2003] – 25r – 1 – (missing: 1885; cont as: barrow, furness & north western daily times [1 jul 1871-apr 1873]; barrow daily times [may 1873-jun 1875]; barrow times [jul 1875-dec 1884]) – uk Newsplan [072]

Barrow-in-furness labour party, 1914-69 – (= ser Labour party in britain, origins and development at local level. series 2) – 7r – 1 – (with p/g. int by bryn trescartheric) – mf#97566 – uk Microform Academic [331]

Barrows, Charles Henry *see* The personality of jesus

Barrows, Clayton W *see* International journal of hospitality and tourism administration

Barrows, Comfort Edwin *see* The development of baptist principles in rhode island

Barrows, E C *see*
- Diary
- Diary of john comer

Barrows, Elijah Porter *see* Companion to the bible

Barrows, H D *see* A memorial and biographical history of the coast counties of central california

Barrows, Isabel C *see* Conference in the interest of physical training, boston, 1889

Barrows, John Henry *see*
- The christian conquest of asia
- Christianity, the world-religion
- Henry ward beecher
- I believe in god the father almighty

Barrows, Samuel June *see*
- A baptist meeting-house
- Jesus as a penologist

Barruel, Augustin *see* Abrege des memoires pour servir a l'histoire du jacobinisme

Barry, A *see* Christ in the midst of us

Barry advertiser (and gazette) – [Wales] Vale of Glamorgan 9 sep 1921-dec 1933 [mf ed 2003] – 11r – 1 – (cont as: barry advertiser [jan 1928-dec 1931]; barry advertiser and cinema gazette-barry advertiser and cinema review [1932]; barry advertiser and cinema review [1933]) – uk Newsplan [072]

Barry, Alfred *see*
- The atonement of christ
- England's mission to india
- First words in australia
- Introduction to the study of the old testament
- Lectures on architecture delivered at the royal academy by the late edward m barry...
- The manifold witness, for christ
- Masters in english theology
- On some of the present needs of the church of england
- The position of the laity in the church
- Six sermons on the bible
- Some lights of science on the faith
- What is natural theology?

Barry, Alfred, Bishop of Sydney *see*
- The architect of the new palace at westminster
- The life and works of sir charles barry

Barry, Alfred et al *see* Masters in english theology

Barry, Charles *see*
- Illustrations of the new palace of westminster. first series
- Illustrations of the new palace of westminster. second series
- Studies and examples of the modern school of english architecture
- The travellers' club house
- The travellers' club house, by charles barry, architect

Barry, Dawn M *see* Energy expenditure of step training vs low impact aerobics using three common movement patterns

Barry & district news – Barry, Wales 13 nov 1925-28 jun 1962 [mf 1889-1950, 1986-] – 1 – (cont: barry dock news [5 jul 1889-6 nov 1925]; cont by: barry & district news & barry herald [5 jul 1962-31 dec 1964]; barry & district news [7 jan 1965-]) – uk British Libr Newspaper [072]

Barry dock news – [Wales] Vale of Glamorgan 5 jul 1889-dec 1950 [mf ed 2002] – 58r – 1 – (cont: barry dock news,penarth express and vale of glamorgan reporter [jul 1889-dec 1893]; cont by; barry & district news [jan 1926-dec 1932]; barry & district news and vale of glamorgan chronicle [1934-dec 1950]) – uk Newsplan [072]

Barry, E *see* Barrymore records of the barrys of county cork

Barry, Edward Middleton *see* Lectures on architecture delivered at the royal academy by the late edward m barry...

Barry, George Duncan *see* The transfiguration of our lord

Barry, glamorgan, parish church of porthkerry road, wesleyan : baptisms 1889-1930, marriages 1893-1934; & cadoxton wesleyan methodist church, marriages 1877-1925 – 1mf – 9 – 155=1mf – uk Glamorgan FHS [929]

Barry, glamorgan, parish church of st nicholas : baptisms 1724-1900, burials 1724-1900, marriages 1724-1837 – 1mf – 9 – £1.25 – uk Glamorgan FHS [929]

Barry herald – Barry, Wales 21 feb 1896-21 jun 1962 [mf 1912] – 1 – (wanting 1897; incorp with: barry and district news) – uk British Libr Newspaper [072]

Barry, James *see*
- A series of etchings
- The works of james barry, esq

Barry, John R *see* Client motivation for rehabilitation

Barry, Phillips *see* Folk music in america

Barry railway & docks accident books – 1916-27 – 1mf – 9 – £1.25 – uk Glamorgan FHS [380]

Barry, st nicholas, monumental inscriptions : and merthyr dyfan ss dyfan & teilo; porthkerry st curig – 1mf – 9 – £1.25 – uk Glamorgan FHS [929]

Barry sullivan and his contemporaries : a histrionic record / Sillard, Robert M – London: T F Unwin, 1901 (mf ed 19–) – 1 – mf#*ZC-30 – us NY Public [790]

Barry the soldier – London, England. 18– – 1r – us UF Libraries [240]

Barry trades directories 1897 : directory transcript – 2mf – 9 – £2.50 – uk Glamorgan FHS [380]

Barry trades directories 1906 : directory transcript – 2mf – 9 – £2.50 – uk Glamorgan FHS [380]

Barry trades directories 1914 : directory transcript – 3mf – 9 – £3.75 – uk Glamorgan FHS [380]

Barry, William *see* Outline of the life of john henry, cardinal newman

Barry, William Francis *see*
- Ernest renan
- Heralds of revolt
- Newman
- The papacy and modern times
- The papal monarchy
- The tradition of scripture

Barrymore records of the barrys of county cork : from the earliest to present times with pedigrees / Barry, E – Cork: Guy & Co, 1902 – 1r – 1 – us Western Res [920]

Barrymore, William *see* Davy jones

Barsanti, Alfred *see* Missionnaires d'asie

Barsanti, Francesco *see*
- [1-4] sonatas of three parts for two violins, a violoncello and thorough bass for the harpsichord made out of geminianis solos
- Six sonatas for two violins and a bass, opera sesta

[Barstow-] barstow college student newspaper – CA. 1961-79 – 1r – 1 – $60.00 – mf#R02046 – us Library Micro [071]

[Barstow-] barstow printers review – CA. jun 1943-oct 1958 – 17r – 1 – $1020.00 – mf#C02047 – us Library Micro [680]

[Barstow-] desert dispatch – CA. 1971- – 92+ r – 1 – $5520.00 (subs $170/y) – mf#RC02048 – us Library Micro [071]

Barsukov, N *see* Istochniki russkoi agiografii

Barsukov, N P *see* Istochniki russkoi agiografii

Bartas: his devine weekes and workes / Du Bartas, Guillaume de Salluste – 1605 – 9 – us Scholars Facs [810]

Bartel, Kirsten *see* Sprache und erfahrung

Bartell, Edmund *see* Hints for picturesque improvements in ornamented cottages

Bartell, Laura B *see* Law clerk handbook

Bartels, Adolf *see*
- Bauernspiegel
- Deutschvoelkische gedichte
- Dietrich sebrandt
- Die dithmarscher
- Einfuehrung in die weltliteratur
- Friedrich hebbel und otto ludwig
- Geschichte der deutschen literatur
- Goethe der deutsche
- Hebbels herkunft und andere hebbel-fragen
- Jeremias gotthelf
- Klaus groth
- Wilde zeiten: rolves karsten

Bartels, Petrus *see* Johannes a lasco

Bartels, Rudolf *see* Zu schillers "das ideal und das leben"

Barter bulletin / Wisconsin Association of Manufacturers and Commerce – 1976 mar-1977, 1978 mar-1979 win – 2r – 1 – (cont: barter bulletin [madison, wi]) – mf#645680 – us WHS [380]

Barter bulletin / Wisconsin State Chamber of Commerce – 1947 jul-1959, 1960-1969 may, 1969 dec-1975 oct – 3r – 1 – (cont by: barter bulletin [milwaukee wi]) – mf#645677 – us WHS [380]

Barter, William Brudenell *see* Lord morpeth's remarks on "the tracts for the times" considered

Bartgis's virginia gazette : and the winchester advertiser – 1790 jan 6-1791 nov 26 – 1 – mf#882602 – us WHS [071]

Barth, Auguste *see*
- L'inscription sanscrite de han chey
- Inscriptions sanscrites du cambodge
- The religions of india

Barth, Auguste et al *see* [Inscriptions sanscrites du cambodge et de campa]

Barth, C *see*
- Die interpretation des neuen testaments in der valentinianischen gnosis

Barth, Christian Gottlob *see* A general history of the world

Barth, Christian Gottlob, 1799-1862 *see* History of the christian church

Barth Family *see* Papers

Barth, Fritz *see*
- Calvin und servet
- Calvins persoenlichkeit und ihre wirkungen auf das geistige leben der neuzeit
- Einleitung in das neue testament
- The gospel of st john and the synoptic gospels
- Die hauptprobleme des lebens jesu

Barth, H *see* Reise von trapezunt durch die noerdliche haelfte kleinasiens nach scutari im herbst 1858

Barth, Hans *see* Roemische asche

Barth, Heinrich *see* Reisen und entdeckungen in nord- und central-afrika in den jahren 1849 bis 1855

Barth, Hermann von *see* Aus den noerdlichen kalkalpen

Barth, J *see*
- Etymologische studien zum semitischen insbesondere zum hebraeischen lexicon
- Sprachwissenschaftliche untersuchungen zum semitischen
- Wurzeluntersuchungen zum hebraeischen und aramaeischen lexicon

Barth, Jakob *see*
- Beitraege zur erklaerung des jesaia
- Die nominalbildung in den semitischen sprachen
- Die pronominalbildung in den semitischen sprachen

Barth, Joseph *see* The ethics of felix adler

Barth, Karl *see*
- The epistle to the romans
- The german church struggle
- Der roemerbrief

Barthe, Edouard *see* Monument a la gloire de marie

Barthe, Georges Isidore *see* Drames de la vie reelle

Barthe, Joseph Guillaume *see*
- Le canada reconquis par la france
- Lettre sur le canada

Barthe, Ulric *see* Wilfred laurier a la tribune

Barthel, Emil G *see* Nikolaus lenau's saemmtliche werke in einem bande

Barthel, Ernst *see*
- Goethe, das sinnbild deutscher kultur
- Goethes relativitaetstheorie der farbe
- Goethes wissenschaftslehre in ihrer modernen tragweite

Barthel, Helene *see* Der emmentaler bauer bei jeremias gotthelf

Barthel, Ludwig Friedrich *see*
- Das maedchen phoebe
- Zwischen krieg und frieden

Barthel, Max *see*
- Revolutionaere gedichte
- Ueberfluss des herzens
- Das unsterbliche volk

Barthelemon, Cecilia Maria *see* Six sonatas for the harpsichord or piano forte

Barthelemy, Edouard de *see* Catalogue des gentilshommes de normandie...

Barthelemy, Hippolyte *see* L'alsace et la lorraine comment elles redeviendront francaises

Barthelemy Saint-Hilaire, Jules *see*
- The buddha and his religion
- De l'ecole d'alexandrie
- Des vedas
- Mahomet et le coran

Barthelemy-Hadot, Marie-Adele *see* Maclovie, comtesse de warberg

Barthelmy, M P *see* Biographie et galerie historique des contemporains...

Barthold, W *see*
- Nachrichten ueber den aral-see und den unteren lauf des amu-darja von den aeltesten zeiten bis zum 17. jahrhundert
- Turkestan down to the mongol invasion

Bartholdy, Georg Wilhelm et al *see* Woechentliche unterhaltungen ueber die charakteristik der menschheit

Bartholdy, Jakob L *see* Voyage en grece, fait dans les annees 1803 et 1804

Bartholino, Tomas *see* El incendio de la biblioteca

Bartholm ess, Christian *see* Histoire critique des doctrines religieuses de la philosophie moderne

Bartholmess, Christian *see* Huet, evaeque d'avranches, ou, le scepticisme theologique

Bartholomaei, Joh Christian *see* Nova acta historico-ecclesiastica

Bartholomaeus, Anglicus *see* Medieval lore

Bartholomaeus Exoniensis *see* Contra fatalitatis errorem

Bartholomaeus, G *see*
- De afflictione tam captivorvm quam etiam sub turcae tributo viuentium christianorum...
- De turcarum ritv et caeremoniis...
- De tvrcarvm moribvs epitome
- Libellvs vere christiana...
- Profetia de i tvrchi, della loro rouina, o la conuersione alla fede di christo per forza della spada chrstiana
- Prophetia de maometani, et altre cose turchesche...

Bartholomaeus ziegenbalg : oder, die ersten anfaenge der lutherischen mission unter den tamulen in ostindien / Frey, August Emil – Allentown, PA: Brobst, Diehl, 1883 [mf ed 1986] – (= ser Missions-bibliothek fuer jung und alt 2) – 1mf – 9 – 0-8370-7138-0 – mf#1986-1138 – us ATLA [242]

Bartholomaeus ziegenbalg : der vater der evangelischen tamulenmission, eine jubilaeumsgabe / Gehring, Alwin – 2. erw aufl. Leipzig: Verlag der Ev-luth Mission, 1907 [mf ed 1995] – (= ser Yale coll) – 104p (ill) – 1 – 0-524-09832-8 – (in german) – mf#1995-0832 – us ATLA [240]

Bartholomeus de Cotton *see* Historia anglicana (ad 449-1298) (rs16)

Bartholomew, Alfred *see* Hints relative to the construction of fire-proof buildings

Bartholomew, Allen R *see* Won by prayer

Bartholomew, Charles Charles *see* Connection of the holy sacraments with the spiritual life and their...

Bartholomew, Elam *see* Diaries

Bartholomew, John Glass *see*
- The altar
- The comforter

Bartholomin, Mr *see* Pizarro o la conquista del peru

[Bartle-] mccloud river pioneer – CA. aug 24 1889-dec 31 1893 (wkly) – 1r – 1 – $60.00 – mf#B02049 – us Library Micro [071]

Bartlesville first baptist church – BARTLESVILLE, OK. 1904-54 – 1 – $63.63 – us Southern Baptist [242]

Bartlet, James Vernon *see*
- The apostolic age
- Christianity in history
- The earlier pauline epistles
- St mark

Bartlett, Caroline J *see* Woman's call to the ministry

Bartlett, Charles Lafayette *see* Trial by jury in contempt proceedings

Bartlett, D W *see* Bartlett's digest, election cases in the house of representatives

Bartlett, Ellis Ashmead *see* British, natives and boers in the transvaal...

Bartlett, Frederic C *see* Remembering

Bartlett, Frederic P *see* Puerto rico y su problema de poblacion

Bartlett, James Herbert *see*
- District steam supply
- The manufacture, consumption and production of iron, steel, and coal in the dominion of canada
- The manufacture of iron in canada

Bartlett, John Russell *see* Letters of roger williams to winthrop

Bartlett, Josiah *see*
- Family papers
- Microfilm edition of papers of josiah bartlett in the years 1774-1794

Bartlett, Robert E *see* African political ephemera, 1958-1966

Bartlett, S T [comp] *see* The junior league hand-book

Bartlett, Samuel Colcord *see*
- Christian relations of the east and the west
- The duty and the limitations of civil disobedience
- From egypt to palestine through sinai, the wilderness and the south country
- Historical sketch of the hawaiian mission
- Historical sketch of the missions of the american board among the north american indians
- Historical sketch of the missions of the american board in the sandwich islands, micronesia, and marquesas
- Lectures on modern universalism
- Sketches of the missions of the american board
- Sources of history in the pentateuch

Bartlett, Vernon *see* I accuse

Bartlett, W H *see* Pilgrimage through the holy land

Bartlett's digest, election cases in the house of representatives : 1865-1871 / Bartlett, D W – Washington: GPO. 41st Congress 2nd session. 1871? – 10mf – 9 – $15.00 – mf#LLMC 95-123 – us LLMC [340]

Bartley, E T *see* Lilian's retrospect

Bartley inter-ocean – Bartley, NE: Professor Wm Smith. -v61 n12. mar 7 1947 (wkly) – 16r – 1 – (merged with: indianola reporter to form: red willow county reporter; v2 n38-v15 n11 called also whole n90-740) – us Bell [071]

Bartlow 1573-1950 – (= ser Cambridgeshire parish register transcript) – 4mf – 9 – £5.00 – uk CambsFHS [929]

Bartok, Bela *see*
- Magyar nepdal
- Serbo-croatian folk songs
- Das ungarische volkslied

Bartok's mikrokosmos : an analysis of its technical difficulties / Sielska, Marya – U of Rochester 1947 [mf ed 19–] – 2mf – 9 – (with bibl) – mf#fiche 179 – us Sibley [780]

Bartol, Cyrus Augustus *see*
- Discourses on the christian spirit and life
- The five ministers
- In remembrance-address on occasion of the death of charles greely loring
- Radical problems
- The rising faith
- The word of the spirit to the church

Bartolache, Jose Ignacio *see* Manifiesto satisfactorio anunciado en la gazeta de mexico [tom 1 n53]

Bartol'd, Vasilii Vladimirovich *see* Mussulman culture

Bartold, Vasilii Vladimirovich *see* Zur geschichte des christentums in mittel-asien bis zur mongolischen eroberung

Bartoli, D *see* Dell'istoria della compagnia di ges- l'asia

Bartoli, Daniello *see* Istoria della compagnia di gesu il giappone

Bartoli, Giorgio *see* The primitive church and the primacy of rome

Bartoli, P *see* Admiranda romanarum antiquitatum veteris sculpturae vestigia...notis i p bellorii illustrata

Bartolini, Domenico *see* Di s. zaccaria papa e degli anni del suo pontificato

Bartolini e la cerrito; ossia, dell'onorare e premiare gli artisti : ragionamento del dottor quirico filopanti [pseud] / Filopanti, Quirico [pseud] – Bologna: Pei Tipi delle Muse, 1845 – 1 – mf#*ZBD-*MGO pv27 – Located: NYPL – us Misc Inst [790]

Bartolo, P S *see* Columna antoniniana marci aurelii antonini augusti...

Bartolo, Salvatore di *see* Les criteres theologiques

Bartolome deya – Minorca, Spain. v873. 1765-1779? – 1r – us UF Libraries [324]

Bartolome jose gallardo... / Blanco Garcia, Francisco – Madrid: Saenz de Jubera, 1909 – 1 – sp Bibl Santa Ana [440]

Bartolome pons – Minorca, Spain. v269-272. 1759-1774 – 1r – (gaps) – us UF Libraries [324]

Bartolozzi and his works / Tuer, Andrew W – London [1881] – (= ser 19th c art & architecture) – 6mf – 9 – mf#4.2.1217 – uk Chadwyck [760]

Barton 1600-1950 – (= ser Cambridgeshire parish register transcript) – 5mf – 9 – £6.25 – uk CambsFHS [929]

Barton, Andrew R *see* The effects of a crosstraining program on strength development

Barton, Benjamin Smith *see* New views of the origin of the tribes and nations of america

Barton, Bruce *see* Man nobody knows

Barton, Charles *see* Modern precedents in conveyancing.

Barton, Dunbar P *see* The story of the inns of court

Barton, E *see* Apologetic postscript to the rhapsody

Barton, G A *see* Miscellaneous babylonian inscriptions

Barton, G B *see* The draft bill to constitute the commonwealth of australia

Barton, George *see* Angels of the battlefield

Barton, George A *see*
- Hilprecht's fragment of the babylonian deluge story
- The origin and development of babylonian writing
- The religions of the world

Barton, George Aaron *see*
- A critical and exegetical commentary on the book of ecclesiastes
- The heart of the christian message
- A sketch of semitic origins

Barton in the clay – (= ser Bedfordshire parish register series) – 2mf – 9 – £5.00 – uk BedsFHS [929]

Barton, James L *see*
- Educational missions
- Human progress through missions

Barton, James Levi *see*
- The missionary and his critics
- The unfinished task of the christian church

Barton, L E *see* Autobiographical material

Barton, William *see* The princes of india

Barton, William E *see* The book of enlightenment for the instruction of the inquirer

Barton, William Eleazar *see*
- Congregational creeds and covenants
- Day by day with jesus
- The law of congregational usage
- Old plantation hymns
- A pocket congregational manual

Barton, William Eleazar et al *see* His life

Bartosch, Alexander *see* Theodyrene

Bartoszewicz, Kazimierz *see* Antysemityzm w literaturze polskiej 15-17 w...

Bartow Board of Trade *see* Bartow, polk county, florida

Bartow, polk county, florida / Bartow Board of Trade – Bartow, FL. 1915? – 1r – us UF Libraries [978]

Bartram, Ulrike *see* Untersuchungen ueber die effektivitaet von strategien zur vermeidung der roetelnembryopathie

Bartram, W *see* Travels through north and south carolina, georgia, east and west florida, the cherokee country...

Bartram, William *see*
- Recollections of seven years residence at the mauritius
- Voyage dans les parties sud de l'amerique septentrionale savoir

Bart's journal – London UK 1975-80 – 1,5,9 – mf#12841 – us UMI ProQuest [360]

Bartsch, Adam von *see* Le peintre graveur

Bartsch, Heinrich *see* Noch weht preussens fahne

Bartsch, J J *see* Des roemischen kaisers lieb-lob- und gluecks-werther davidischer achilles, der baeyrische mars

Bartsch, Karl *see*
- Albrecht von halberstadt und ovid im mittelalter
- Anmerkungen zu konrads trojanerkrieg
- Demantin
- Denkmaeler der provenzalischen litteratur
- Deutsche dichtungen des mittelalters
- Deutsche liederdichter des 12. bis 14. jahrhunderts
- Die deutsche treue in sage und poesie
- Die erloesung
- Ernst herzog
- Hugo von montfort
- Karl der grosse von dem stricker
- Konrad, der pfaffe
- Kudrun
- Meisterlieder der kolmarer handschrift
- Meleranz
- Mitteldeutsche gedichte
- Das nibelungenlied
- Reinfrid von braunschweig
- Untersuchungen ueber das nibelungenlied
- Walther von der vogelweide
- Wolfram's von eschenbach parzival und titurel

Bartsch, Rudolf Hans *see*
- Als oesterreich zerfiel
- Brueder im sturm
- Ewiges arkadien!
- Der falke vom mons regius
- Der steirische weinfuhrmann
- Unerfuellte geschichten
- Zwoelf aus der steiermark

Bartscherer, Agnes *see*
- Paracelsus, paracelsisten und goethes faust
- Zur kenntnis des jungen goethe

Bartz, Karl *see* Lilienbanner und preussenaar

Barua, Arabinda *see* The petakopadesa

Barua, Beni Madhab *see*
- A history of pre-buddhistic indian philosophy
- Prolegomena to a history of buddhist philosophy

Barua, Benimadhab *see*
- Old brahmi inscriptions in the udayagiri and khandagiri caves
- Prakrit dhammapada

Barua, Birinchi Kumar *see*
- Assamese literature
- A cultural history of assam

Barua, Kanaklal *see* Early history of kamarupa

Baruch, Amy R *see* Effects of caffeine on central on peripheral hemodynamics at rest and during exercise

Baruch, Isaac Loeb *see* Roshe perakim ba-talmud

Baruch, Jacques *see* La democratie au cambodge

Baruch the scribe / Mountain, Jacob Henry Brooke – London, England. 1841 – 1r – us UF Libraries [240]

Baruffaldi, G *see* Vite de' pittori e scultori ferrarese...

Barukh shpinozah / Zeitlin, Hillel – Varsha, Poland. 1900 – 1r – us UF Libraries [939]

Barvo y Bravo Fernando *see* Bimilenario de la fundacion de la colonia norba caeserina

Barwick, Steven *see* Madrigals of matheo flecha

Barwin, F L *see*
- Pontiac
- Der verrat von detroit

Barwin, Victor *see* Millionaires and tatterdemalions

Barwon express – Walgett, jun 29 1901-jun 3 1905 – 1r – 9 – A$60.76 vesicular A$66.26 silver – at Pascoe [079]

Bary, Erwin de *see* Le dernier rapport d'un europeen sur ghat et les touareg de l'air

Bary, Maxime De *see* Grand gibier et terres inconnues

Barykada – Warsaw, [Poland] 1942-44 [mf ed Norman Ross] – 1r – 1 – (filmed with: miecz i plug [1940-44]) – mf#nrp-2115 – us UMI ProQuest [077]

Barykada *see* Miecz i plug

Bar-Yossef, Yehoshua *see* Kol ha-yetsarim

Barz, Andre *see*
- Literaturunterricht und massenmedien
- Psychologische aspekte des darstellenden spiels

Barzigar – Nashriyah-'i Haftigi-i Kishavarzi va Damdari. n18-409/410. 20 urdibihisht 1359-22 isfand 1366 – 3r – 1 – $159.00 – (missing: n19, 21, 33, 43, 45, 85, 87, 91, 111, 115, 130, 144, 190, 218, 220, 258, 306, 352, 358) – us MEDOC [079]

Barzilay, Jacques *see* Dictionnaire geographique et descriptif de l'italie

Bas reliefs of badami / Banerji, Rakhal Das – Calcutta: Govt of India, Central Publ Branch, 1928 – (= ser Samp: indian books) – us CRL [730]

Bas Y Cortes, Vincente *see* Cartas al rey acerca de la isla de cuba

Basa field notes / Schwab, George – [Evanston, 1965] – us CRL [500]

Basak, Radhagovinda *see* The history of north-eastern india

Basakabaka be buganda / Kagwa, Apolo – [s.l: s.n, 19–?] – 1 – us CRL [960]

Basalenque, Diego *see* Historia de la provincia de san nicolas de tolenti

Basalt High School [CO] *see* Montane

Basar, Suekuefe Nihal *see* Renksiz istirab

Basari – Manisa: Basari Murettiphanesi, [jul 9 1951-feb 22 1952] – 1r – 1 – us CRL [079]

Basaroff, Archpriest *see* The russian orthodox church

Basar-zeitung – 1916 mar 2,7 – 1 – mf#1165547 – us WHS [071]

Bas-Canada *see*
- An abstract of the most material parts of an act...
- Act to incorporate the quebec fire-assurance company
- Acte pour la decision sommaire des petites causes
- Bill
- Copy of the charter of the corporation of saint nicolet in lower canada
- [Ordonnance pour pourvoir a l'amelioration des chemins dans le voisinage de la cite de Montreal et y conduisant et pour etablir un fonds pour cet objet 3 victoria, cap 31]
- Public accounts for...
- Rapport des commissaires nomme [sic] pour l'exploration du pays, borne par les rivieres saguenay, saint-maurice et saint-laurent
- Rapport des commissaires nommes pour l'exploration du pays entre les rivieres st maurice et outaouais
- Rapport des commissaires pour explorer le saguenay
- Regles et reglements de police pour les fauxbourgs et la cite de montreal
- Report of the commissioners apointed to explore the country between the st maurice and the ottawa
- Report of the commissioners appointed under the lower canada act
- Report of the commissioners for exploring the country lying between the rivers saguenay, saint maurice and saint lawrence
- Report of the commissioners for exploring the saguenay
- Reports of commissioners for roads and other internal communications
- Statement of the public revenue and expenditure of lower canada for the year...

Bas-Canada. Commissaires nommes pour l'exploration du pays entre les rivieres Saint-Maurice et Outaouais Canada *see* Rapport des commissaires nommes en vertu de l'acte de la 9e. geo 4, chap 29

Bas-Canada. Cour du Banc du Roi *see* Reports of cases argued and determined in the courts of king's bench and in the provincial court of appeals of lower canada

Bas-Canada. Cour du Banc du Roi (District de Montreal) *see* Rules and orders of practice, for the court of king's bench, district of montreal

Bas-Canada. Cour du Banc du Roi (District de Quebec) *see* Table of fees

Le bas-canada entre le moyen-age et l'age moderne / Gingras, Apollinaire – Quebec?: "Canadien", 1880 – 1mf – 9 – mf#53330 – cn Canadiana [241]

Bas-Canada. Gouverneur *see*
- Message de son excellence le gouverneur en chef
- Message from his excellency the governor in chief

Bas-Canada. Laws, Statutes etc *see* Articles enacted in the act intituled "an act to repeal a certain act therein-mentioned

Bas-Canada. Milice. Volunteer Corps *see* Rules and regulations, schedules of pay, etc

Bas-Canada. Parlement *see* Provincial parliament of lower-canada

Bas-Canada. Parlement. Chambre d'Assemblee *see*
- Copy of the fourth report of the standing committee of grievances made to the assembly of lower canada
- Fourth report of the standing committee of grievances
- Rapport...nomme pour s'enquerir de l'etat actuel de l'education dans la province du bas-canada
- Second rapport...sur diverses communications de son excellence le gouverneur en chef lord aylmer, sur le sujet des finances de la province du bas-canada

Bas-Canada. Parlement. Chambre d'assemblee *see*
- Bill introduit dans la chambre d'assemblee...pour mieux regler la milice de cette province
- Enquete...sur les evenements du 21 mai 1832, a montreal
- Extrait des instructions royales a son excellence le tres-honorable george, comte de dalhousie...
- Fifth report of the standing committee on roads and public improvements
- First report of the committee...on that part of the speech of his excellency the governor in chief
- First report of the special committee of the house of assembly on the engrossed bill from the legislative council to repeal certain parts of the judicature act
- First report of the standing committee on roads and public improvements
- House of assembly, wednesday, 21st february 1827
- Memoire accompagnant la requete presentee a la chambre d'assemblee
- Minutes des temoignages et rapport du comite special de la chambre d'assemblee du bas-canada
- Minutes of evidence and report of the special committee...
- Premier rapport du comite de la chambre d'assemblee
- Premier rapport du comite special sur les communications interieures
- Proceedings in the assembly of lower-canada
- Proceedings of the house of assembly upon the petition of the electors of the county of bedford against the legality of the election for the said county
- Questions submitted by a special committee of the house of assembly of lower canada to the curates of the diocese of quebec
- Rapport du comite special auquel a ete refere cette partie de la harangue de son excellence relative a l'organization [sic] de la milice
- Rapports du comite special sur les chemins et autres communications interieures
- Rapports et temoignages...
- Rapport...sur le departement du bureau de la poste dans la province du bas-canada
- Regles et reglements permanents...
- Report of a special committee of the house of assembly appointed to enquire into the state of education in this province
- Report of a special committee...appointed to enquire into the state of education in this province
- Report of the select committee, on the public accounts of the province
- Report of the special committee of the house of assembly on the post office department in the province of lower canada
- Report of the special committee to whom was referred that part of his excellency's speech which referred to the organization of the militia
- Report of the standing committee of education and schools
- Report...on education
- Report...on the petition of certains inhabitants of the district of gaspe
- Report...on the petitions against the road laws and the office of grand-voyer
- Reports and evidence of the special committee...to whom were referred the petition of the inhabitants of the county of york, that of the inhabitants of the city of montreal
- Reports from the special committee on roads and other internal communications
- Second et troiseme rapports du comite special
- Second rapport du comite permanent de griefs

– Second report from the special committee on various communications from his excellency the governor in chief lord aylmer
– Second report of the standing committee on roads and public improvements
– Standing committee on roads and public improvements, 1831-1832
– Third report of the standing committee of grievances
– Third report...on that part of the speech of his excellency the governor in chief
– Troisieme rapport du comite permanent des griefs

Bas-Canada. Parlement. Chambre d'Assemblee. Bibliotheque *see*
– Catalogue des livres appartenant a la bibliotheque de la chambre d'assemblee
– Catalogue of books in the library of the house of assembly

Bas-Canada. Parlement. Chambre d'assemblee. Bibliotheque *see* Library of the house of assembly

Bas-Canada. Parlement. Conseil legislatif *see*
– Extract from the journals of the legislative council of the 2d of march, 1814
– Extracts from the journals of the legislative council of the province of lower-canada from the year 1795 to 1813 inclusive
– Rapport, etc
– Remembrances
– A report from the special committee...to whom the petition from several merchants and ship-owners of the port of quebec, was referred

Basch, Carl et al *see* Der juedische handwerker

Basch, Johannes *see* Gesammt-verlags-katalog des deutschen buchhandels und des mit ihm im direkten verkehr stehenden auslandes

Baschiri, Ayse *see* Das vorkommen einiger metalle in teedrogen und heiltees

Baschkaer zeitung – Apatin (YU), 1930, 1932, 1935, 1938-40 – 2r – 1 – gw Misc Inst [077]

Bascom, Flavel *see* Autobiography, 1881-1891

Bascom, Henry Bidleman *see* The little iron wheel

Bascom, John *see*
– Comparative psychology
– Ethics
– Evolution and religion
– The goodness of god
– Growth of nationality in the united states
– Natural theology
– A philosophy of religion
– Philosophy of rhetoric
– Problems in philosophy
– Science, philosophy and religion
– Sermons and addresses
– Social theory
– Things learned by living
– The words of christ as principles of personal and social growth

Bascom, William Russell *see* "Secret societies"

Le bas-congo, etat religieux et social / Philippart, Leon – Louvain: Saint-Alphonse, 1929 – 1 – us CRL [306]

Basdan (1948-1949) and yeni basdan (1950) – Istanbul: Aziz Nesin, Rifat Ilgaz [all publ] – (= ser O & t journals) – 4mf – 9 – $60.00 – (publ in istanbul by aziz nesin, then rifat ilgaz after sabahattin ali's death) – us MEDOC [956]

Base line : a newsletter of the map & geography round... – Chicago IL 2002+ – 1,5,9 – ISSN: 0272-8532 – mf#35655 – us UMI ProQuest [912]

Baseball and italian-americans : how baseball helped italian-americans assimilate into mainstream america / Iaia, Jim – 1998 – 1mf – 9 – $4.00 – mf#PE 4025 – us Kinesology [306]

Base-ball ballads / Rice, Grantland – Nashville, TN. 1910 – 1r – us UF Libraries [780]

Baseball card news – v8 n1-13 [1989 jan 6-jun 23]; 1989 jul-dec; 1990 jan-dec; 1991 jan-dec; 1992 jan-mar – 1 – mf#1667830 – us WHS [790]

Baseball card price guide monthly – 1988 apr-1989 jun – 1 – mf#1545491 – us WHS [790]

Baseball cards – v1 n1=1 [1981 spring]; v3 n1=6 [1983 fall]; v4 n1-3=7-9 [1984 apr-aug]; v5 n3=13 [1985 aug]; v6 n1,3,5=15,17,19 [1986 apr, aug, dec]; v7 n1,3-6; 8-9=20,21-24, 26-17 [1987 feb, jun-sep, nov-dec]; v8 n1-3=n28-30 [1988 jan-mar]; v9 n2-11=n42-52 [1989 feb-dec]; v10 n1-6=n53-62 [1990 jan-jun]; v10 n7-10=n59-62 [1990 jul-oct] – 1 – mf#1549293 – us WHS [790]

Baseball digest – Evanston IL 1971+ – 1,5,9 – ISSN: 0005-609X – mf#6176 – us UMI ProQuest [790]

Baseball follows the flag : america and ist national pastime during the world wars / Pustz, Matthew J – 1990 – 3mf – 9 – $12.00 – mf#PE 4030 – us Kinesology [790]

Baseball players chronicle. american chronicle of sports and pastimes – New York, NY. 1867, 1868 – 1r – 1 – $40.00r – us Notre Dame [790]

Baseball scrapbooks – ca 1853-82(-1912) – 1r – 1 – $40.00r – us Notre Dame [790]

Basecke, M *see* The relationships among exercise blood lactate response, muscle blood flow, and oxidative adaptation to endurance training in the rat

Based on byzantinische zeitschrift : byzantine studies. author index. / ed by Allen, J S – Dumbarton Oaks – 267mf – 9 – mf#O-2079 – ne IDC [720]

Basedow, A *see* Die inclusen in deutschland

Basedow, J B et al *see* Paedagogische unterhandlungen

Basedow, Joh Bernh *see* Philantropisches archiv

Basedow, Johann Bernhard *see* Philalethie

Basel. kunsthalle catalogue – (= ser Art exhibition catalogues on microfiche) – (individual titles not listed separately) – uk Chadwyck [700]

Baseless fears : professional baseball's wary relationship with radio, 1921-1934 / Smith, Lowell D – 1995 – 1mf – 9 – $4.00 – mf#PE 4041 – us Kinesology [650]

Baselt, Randall *see* Journal of analytical toxicology

Bases de la 1st exposicion de fotografias... / Obra Sindical de Educacion y Descanso – Caceres: Tip. El Noticiero, s.a. – 1 – sp Bibl Santa Ana [770]

Bases essentielles d'un redressement economique / Nicolas, Schiller – Port-Au-Prince, Haiti. 194- – 1r – us UF Libraries [972]

Bases of belief : an examination of christianity as a divine revelation by the light of recognised facts and principles in four parts / Miall, Edward – London: Arthur Hall, Virtue 1853 [mf ed 1985] – 2mf – 9 – 0-8370-4424-3 – mf#1985-2424 – us ATLA [240]

The bases of design / Crane, Walter – London 1898 – (= ser 19th c art & architecture) – 4mf – 9 – mf#4.2.715 – uk Chadwyck [740]

Bases of religious belief : historic and ideal: an outline of religious study / Tyler, Charles Mellen – New York: G P Putnam 1897 [mf ed 1985] – 1mf – 9 – 0-8370-5591-1 – (incl ind) – mf#1985-3591 – us ATLA [210]

Bases of yoga / Ghose, Aurobindo – Calcutta: Arya Pub House, 1936 – (= ser Samp: indian books) – us CRL [280]

Bases orientadoras para la fijacion de honorarios aprobados por la junta general de este ilustre colegio, el 31 de marzo de 1961 / Colegio Provincial de Abogados de Badajoz – Badajoz: Tip. Clasica, 1965 – sp Bibl Santa Ana [946]

Bases y reglamentos...circulo concordia / Belmonte, Francisco – 1896 – 9 – sp Bibl Santa Ana [240]

Basetti-Sani, Julio *see* Mohammed et saint francois...

Basevi, Abramo *see*
– Introduction a un nouveau systeme d'harmonie
– Introduzione ad un nuovo sistema d'armonia
– Studj sull'armonia

Basham, Arthur Llewellyn *see* History and doctrines of the ajivikas

Bashford, James Whitford *see*
– China
– China and methodism
– God's missionary plan for the world
– Wesley and goethe

Bashkimi – Tirana, Albania. Nov 1943-Dec 1976 – 45r – 1 – us L of C Photodup [079]

Bashkimi – Tirane Albania, 1976-1991 – 16r – 1 – gw Mikropress [077]

Bashkimi – Tirane, Albania 28 nov 1944-30 dec 1991 – 1 – (wanting 1982; cont: bashkimi i kombit [28 nov 1943-29 oct 1944]) – uk British Libr Newspaper [077]

Bashor, Stephen Henri *see* The waynesboro' discussion on baptism, the lord's supper, and feet-washing

Bashor, Stephen Henry *see*
– The gospel hammer and highway grader
– A sermon on baptism
– Where is holsinger?

Basic and applied social psychology – Mahwah NJ 1989+ – 1,5,9 – ISSN: 0197-3533 – mf#17602 – us UMI ProQuest [150]

Basic and the teaching of english in india / Myers, Adolph – Cambridge: Publ for the Orthological Institute by the Times of India Press, Bombay, 1938 – (= ser Samp: indian books) – us CRL [420]

Basic bantu / Hopkin-Jenkins, K – Pietermaritzburg, South Africa. 1947 – 1r – us UF Libraries [470]

The basic conception of buddhism / Bhattacharya, Vidhushekhara – Calcutta: University of Calcutta, 1934 – (= ser Samp: indian books) – us CRL [280]

Basic course in arranging for school orchestras / Whybrew, William Ernest – U of Rochester 1953 [mf ed 19–] – 2v on 5mf – 9 – (with app & bibl) – mf#fiche 1061 – us Sibley [780]

Basic data on the other american republics / United States Office Of Inter-American Affairs – Washington, DC. 1945 – 1r – us UF Libraries [972]

Basic documents of the league of arab states – New York: The Arab Info Center, 1955 – us CRL [327]

Basic education / Gandhi, Mahatma – Ahmedabad: Navajivan Pub House, 1951 – (= ser Samp: indian books) – us CRL [370]

Basic elements of the christian faith *see* Chi-tu chiao te chi pen hsin yang [ccm321]

Basic english / Ogden, C K – London, England. 1932 – 1r – us UF Libraries [420]

Basic ideas in religion : or, apologetic theism / Micou, Richard Wilde; ed by Micou, Paul – New York: Association Press 1916 [mf ed 1991] – 2mf [ill] – 9 – 0-7905-9815-9 – (incl bibl ref) – mf#1989-1540 – us ATLA [210]

Basic melody of the togaku pieces of the gagaku repertoire / Garfias, Robert – Los Angeles: U of California 1958 [mf ed 19–] – 8mf – 9 – mf#fiche 997 – us Sibley [780]

Basic principles of revisionism – London, 1929 – 1mf – 9 – mf#J-28-20 – ne IDC [956]

Basic swimming analyzed / Harris, Marjorie M – Boston: Allyn & Bacon [1969] [mf ed 1985] – 1r [ill] – 1 – (filmed with: history of the corporation of birmingham / bunce, j t) – mf#6719 – us Wisconsin U Libr [790]

Basic truths of the christian faith / Willett, Herbert Lockwood – Chicago: Christian Century Co 1903 [mf ed 1993] – 1mf [ill] – 9 – 0-524-07331-7 – mf#1991-3046 – us ATLA [240]

Basier, Isaac *see* The correspondence of isaac basire

Basiese elemente van 'n effektiewe jeugbediening / Engelbrecht, Gert Hermias – Uni of South Africa 2001 [mf ed Johannesburg 2001] – 3mf – 9 – (incl bibl ref) – mf#mfm14687 – sa Unisa [240]

Basil, Fritz *see* Wacholder

Basil, Saint, Bishop of Caesarea *see* The book of saint basil the great, bishop of caesarea in cappadocia, on the holy spirit

Basildes am ausgange des apostolischen zeitalters : als erster zeuge fuer alter und autoritaet neutestamentlicher schriften: insbesondere des johannesevangeliums / Hofstede de Groot, Petrus – verm ausg. Leipzig: J C Hinrichs, 1868 [mf ed 1990] – 1mf – 9 – 0-7905-7003-3 – (incl bibl ref) – mf#1988-3003 – us ATLA [225]

Basili, Andrea *see* Musica universale armonico pratica dettata dall'istinto

Basilica del siglo 7 en burguillos / Martinez Martinez, Matias Ramon – Madrid: Tip. de Fortanet, 1898 – sp Bibl Santa Ana [946]

Basilica, monasterium et le culte de st martin de tours / Bosch, J van den – Nijmegen, 1959 – 4mf – 8 – €11.00 – ne Slangenburg [241]

Basilicae veteris vaticanae descriptio... / Angelis, Pedro de – Roma, 1646 – 7mf – 9 – mf#O-1045 – ne IDC [700]

Das basilidianische system : mit besonderer ruecksicht auf die angaben des hippolytus / Uhlhorn, Gerhard – Goettingen: Dieterich, 1855 – 1mf – 9 – 0-7905-6378-9 – (incl bibl ref) – mf#1988-2378 – us ATLA [180]

Basilio de sa. arturo : documentacao... / Barrado Manzano, Arcangel – Madrid: Archivo Ibero-Americano, 1959 – 1 – sp Bibl Santa Ana [240]

Basilique de st-pierre de rome – Montreal: s.n, 1870 – 1mf – 9 – mf#03903 – cn Canadiana [720]

Basilius des Grossen (Basil The Great, Saint (Basil of Caesarea)) *see*
– Ausgewaehlte briefe, 1. bd (bdk46 1.reihe)
– Ausgewaehlte homilien und predigten, 2. bd (bdk47 1.reihe)

Basin research – Oxford UK 1988+ – 1,5,9 – ISSN: 0950-091X – mf#15614 – us UMI ProQuest [333]

Basiner, T F J *see* Naturwissenschaftliche reise durch die kirgisensteppe nach chiwa

Basingstoke extra – Basingstoke, England 12 sep 2001- – 1 – (cont: gazette extra [16 feb 2000-5 sep 2001]) – uk British Libr Newspaper [072]

Basingstoke extra – Basingstoke, England 12 sep 2001- – 1 – (cont: gazette extra [16 feb 2000-5 sep 2001]) – uk British Libr Newspaper [072]

Basingstoke gazette – Basingstoke, England 3 oct 1969-9 may 1975 – 1 – (iss twice wkly fr 1981 to aug; cont: hants & berks gazette & basingstoke journal [7 jan 1966-26 sep 1969]; cont by: gazette (basingstoke & district ed) 16 may 1975-9 jan 1981) – uk British Libr Newspaper [072]

Basingstoke gazette extra – Basingstoke, England 13 may 1981-17 nov 1999 – 1 – (cont: gazette extra [20 aug 1980-6 may 1981]; cont by: extra: for basingstoke & north hampshire [24 nov 1999-9 feb 2000]) – uk British Libr Newspaper [072]

Basingstoke midweek gazette – Basingstoke, England 27 feb 1973-13 may 1975 – 1 – (cont by: gazette (midweek ed) 20 may 1975-18 dec 1979) – uk British Libr Newspaper [072]

Basingstoke & north hampshire business gazette – oct 1984-jul 1985 – 1 – (cont by: basingstoke & hampshire business gazette [aug 1985-feb 1987]) – uk British Libr Newspaper [650]

Basingstoke & north hampshire gazette – Basingstoke, England 24 feb 1986-9 jan 1998 [mf 1979-] – 1 – (cont: basingstoke & north hants gazette [16 jan 1981-21 feb 1986]; cont by: gazette (basingstoke & north hampshire) 12 jan 1998-31 aug 2001]) – uk British Libr Newspaper [072]

Basingstoke & north hampshire gazette – Basingstoke, England 7 sep 2001-5 jun 2005 [mf 1979-] – 1 – (cont: gazette [12 jan 1998-31 aug 2001]; cont by: gazette (basingstoke, tadley) 9 jun 2005-) – uk British Libr Newspaper [072]

Basingstoke & north hampshire gazette [northern ed] – Basingstoke, England 7 sep 2001-22 feb 2002 – 1 – (discontinued; replaced (in pt?) by: tadley gazette; cont: gazette (northern ed) 7 jan 2000-31 aug 2001) – uk British Libr Newspaper [072]

Basingstoke & north hants gazette – Basingstoke, England 16 jan 1981-21 feb 1986 [mf 1979-] – 1 – (iss twice wkly fr 1981 to aug; cont: gazette [16 may 1975-9 jan 1981]; cont by: basingstoke & north hampshire gazette [24 feb 1986-9 jan 1998]) – uk British Libr Newspaper [072]

Basis *see* Madjallah bulanan basis untuk soal-soal kebudajaan umum

Basis calmly considered – Edinburgh, Scotland. 1820 – 1r – us UF Libraries [240]

A basis for community development planning on ponape *see* Land tenure and power / a basis for community development planning on ponape

Basis of a development program for colombia : appendix / World Bank – Washington? DC. 1950? – 1r – us UF Libraries [307]

Basis of a development program for colombia / World Bank – Washington, DC. 1950 – 1r – us UF Libraries [307]

Basis of a development program for columbia : report / World Bank – Washington, DC. 1950 – 1r – us UF Libraries [307]

The basis of an indo-british treaty / Panikkar, Kavalam Madhava – New Delhi: Indian Council of World Affairs; Bombay: Oxford University Press, 1946 – (= ser Samp: indian books) – us CRL [954]

The basis of anglican fellowship in faith and organization : an open letter to the clergy of the diocese of oxford / Gore, Charles – new ed being the 5th impression, with a preface. London: AR Mowbray; Milwaukee, USA: Young Churchman Co, 1914 – 1mf – 9 – 0-7905-9938-4 – mf#1989-1663 – us ATLA [241]

The basis of assurance in recent protestant theologies / Robins, Henry Burke – 1912 – 1mf – 9 – 0-7905-9089-1 – mf#1989-2314 – us ATLA [242]

The basis of morality / Besant, Annie Wood – Adyar: India, Theosophical pub house, 1915 [mf ed 1981] – 40p – 1 – mf#8547 – us Wisconsin U Libr [290]

The basis of morality = Ueber das fundament der moral / Schopenhauer, Arthur – London: Swan Sonnenschein, 1903 – 1mf – 9 – 0-7905-9875-2 – (in english) – mf#1989-1600 – us ATLA [170]

Basis of union : agreed upon by the associate and general associate – Edinburgh, Scotland. 1820 – 1r – us UF Libraries [240]

Ein basismodell zur immunantwort : ist die komplexitaet des immunsystems auf basale wechselwirkungen reduzierbar? / Mayer, Herbert – (mf ed 1997) – 2mf – 9 – €40.00 – 3-8267-2427-5 – mf#DHS 2427 – gw Frankfurter [574]

Basivava (ny) – Tananarive. 1907-2 avr 1909. (no.17-133). – 1 – (lacking: n98-99) – fr ACRPP [073]

Baskerville, LF *see* Attitudes, perceptions and coping skills of long-term breast cancer survivors

Basketball / Naismith, James – 1894 – 1mf – 9 – $3.00 – us Kinesology [790]

Basketball digest – Evanston IL 1977+ – 1,5,9 – ISSN: 0098-5988 – mf#11614 – us UMI ProQuest [790]

Basketball tournament plan – 1933-65 – 1 – mf#683533 – us WHS [790]

Die basler bearbeitung von lambrechts alexander / ed by Werner, Richard Maria – Stuttgart: Litterarischer Verein, 1881 (Tuebingen: L F Fues) [mf ed 1993] – (= ser Blvs 154) – 230p – 1 – (incl bibl ref and ind) – mf#8470 reel 32 – us Wisconsin U Libr [430]

Die basler bearbeitung von lambrechts alexander / ed by Werner, Richard Maria – Stuttgart: Litterarischer Verein, 1881 (Tuebingen: L F Fuess) – 1 – (incl bibl ref and ind) – us Wisconsin U Libr [430]

Die basler hexenprozesse in dem 16ten und 17ten jahrhundert / Fischer, Friedrich – Basel: Schweighauser, 1840 – 1mf – 9 – 0-524-05312-X – mf#1990-1430 – us ATLA [949]

Das basler konzil : seine berufung und leitung, seine gliederung und seine behoerdenorganisation / Lazarus, Paul – Berlin: E Ebering, 1912 – (= ser Historische Studien) – 1mf – 9 – 0-7905-6113-1 – (incl bibl ref) – mf#1988-2113 – us ATLA [940]
Die basler mission in indien : zugleich als festschrift zum 50 jaehrigen jubilaeum der kanara-mission / ed by Stolz, C – Basel: Verlag der Missionsbuchhandlung, 1884 [mf ed 1995] – (= ser Yale coll) – 108p – 1 – 0-524-10061-6 – (in german) – mf#1995-1061 – us ATLA [240]
Basler nachrichten – Basel CH, 1914 18 oct-1919 10 aug, 1939 31 aug-1940 16 apr, 1941-1942 23 dec, 1943 11 feb-1946, 1956 2 feb-dec – 87r – 1 – (title varies: 31 jan 1977: basler zeitung. filmed by misc inst: 1931 nov-dec [1r]; 1950-1977 jun. with suppl: basler magazin 1977-; 3 [drei] 1991 28 feb-) – uk British Libr Newspaper; gw Misc Inst [074]
Basler zeitschrift : fuer geschichte und altertumskunde – 1(1902)-44(1945) – 303mf – 9 – €548.00 – ne Slangenburg [900]
Basnage, J *see*
– Histoire de la religion des eglises reformees
– Traite de la conscience...
Basoche / Messager, Andre – Paris, France. 1900 – 1r – us UF Libraries [440]
Les basonge (etat ind. du congo) / Overbergh, Cyr. van – Bruxelles, A DeWit [etc] 1908 – us CRL [960]
The basque priests persecuted by the fascist forces – n.p. 193? Fiche W 749. (Blodgett Collection of Spanish Civil War Pamphlets) – 9 – us Harvard [946]
Basque-americans and sequential theory of migration and adaptation / McCall, Grant Erwin – 1r – 1 – $50.00 – mf#C603005 – us Library Micro [304]
Basra – (= ser Vilayet salnames) – 9 – (1308 [1891] def'a 1 3mf $55; 1318 [1900] 5mf $75) – us MEDOC [956]
Basra, Amarjit S *see* Journal of crop production
Bass, E *see* Die merkmale der israelitischen prophetie nach der traditionellen auffassung des talmud
[Bass lake-] mountain press – CA. jan 5 1977-1978 – 2r – 1 – $120.00 – mf#B02050 – us Library Micro [071]
Bass, Samuel *see* Hofim
Bass, Scott A *see* Journal of aging and social policy
El bassair – Alger. mars 1936-39, 1947-mars 1956 – 1 – fr ACRPP [073]
Bassano, F da *see* Vocabolario tigray-italiano e repertorio italiano-tigray
Bassas dansas & branlos practicantes *see* Antonius de arena provencalis, de bragardissima villa de soleriis
...Bassas dansas...augmentatus... / Arena, Antonius – nova novorum novissima. Bartholomeum Bollam...1670 – 9 – (contains also: poema macaronicum huguenotico) – us Sibley [780]
Bassermann, Heinrich *see* Der glaube an jesus christus
Basset, Andre *see*
– Le berbere a l'ecole nationale des langues orientales vivantes
– Langue berbere
– Quatre etudes de linguistique berbere
Basset, Bernard *see* Chorokodza
Basset clarinet of anton stadler and its music / Poulin, Pamela Lee – U of Rochester 1976 [mf ed 19–] – 5mf – 9 – (with bibl) – mf#fiche 830, 644 – us Sibley [780]
Basset, Fletcher S *see* Sea phantoms
Basset, Joshua *see* Essay towards a proposal for catholic communion
Basset, R *see* Histoire de la conquete de l'abyssinie (16e siecle)
Bassett, Ancel Henry *see* A concise history of the methodist protestant church
The bassett bulletin – Bassett, NE: B S Garretson. v1 n1. aug 20 1908- (wkly) [mf ed filmed 1973] – 1r – 1 – us NE Hist [071]
Bassett eagle – Bassett, NE: W T Phillips, -1902// (wkly) [mf ed v1 n32. aug 22 1895-feb 22 1902 (gaps)] – 2r – 1 – (cont by: newport eagle) – us NE Hist [071]
Bassett, Henry Lawrence *see* Adventures in samoa
Bassett herald – Bassett, NE: R A Sanders, -oct 3 1889// (wkly) – 1r – 1 – (merged with: rock county republican to form: republican-herald (bassett ne)) – us Bell [071]
Bassett, James *see*
– Criminal pleading and practice
– Persia, eastern mission
Bassetti-Sani, Julio *see* La vierge inmaculee et le probleme de l'apostolat musulman...
Bassetts daily chronicle *see* Southern chronicle
Der bassgeiger / das verhexte buch : zwei berliner geschichten / Kretzer, Max – Leipzig: P Reclam [1910?] [mf ed 1990] – 1r – 1 – (filmed with: kotzebue in england / walter sellier) – mf#2776p – us Wisconsin U Libr [830]
Bassi, M *see* Dispareri in materia d'architettura et perspettiva...
Bassieres, Eugene *see* Notice sur la guyane
Bassingbourn 1558-1950 – (= ser Cambridgeshire parish register transcript) – 15mf – 9 – £18.75 – uk CambsFHS [929]
Les bassins du niger : etude de geographie physique et de paleographie... / Urvoy, Y – Paris: Larose, 1942 – us CRL [960]
Bassnett, Thomas *see* The true theory of the sun
Basso, Hamilton *see* Quota of seaweed
Bassompierre, Francois de *see*
– Ambassade du mareschal de bassompierre en espagne l'an 1621
– Ambassade du mareschal de bassompierre en suisse l'an 1625
– Memoires du mareschal de bassompierre
– Nouveaux memoires du marechal de bassompierre...
Basson, P A et al *see* "Grootlamsiekte"
Bassoppo, Paul *see* Manana a maria musante
Bassus triciniorum, in secundo tomo contentorum – 1560 – (= ser Mssa) – 2mf – 9 – €35.00 – mfchl 463 – gw Fischer [780]
Basta, Susan M *see* Pressure sore prevention self-efficacy and outcome expectations in the spinal cord-injured
Basta ya! – 1970 apr, oct-dec; 1971 aug – 1 – mf#788515 – us WHS [071]
Bastads tidning – 1895-97 – 9 – sw Kungliga [078]
Bastads tidning – Laholm, Sweden. 1883-86 – 2r – 1 – sw Kungliga [078]
Bastard-D'Estang, Henri de *see* Les parlements de france
Bastasch, Jeanne D *see* The effects of integrating geometry into physical education
Basterra, Ramon de *see* Una empresa el siglo 18
Bastgen, Hubert *see*
– Bayern und der heilige stuhl in der ersten haelfte des 19. jahrhunderts
– Die besetzung der bischofssitze in preussen in der ersten haelfte des 19. jahrhunderts
– Der heilige stuhl und die heirat der prinzessin elisabeth von bayern mit dem kronprinzen friedrich wilhelm von preussen
– Das herzogspaar ferdinand und julie von anhalt-koethen, die anfaenge der katholischen pfarrei koethen und der heilige stuhl
– Saemtliche aufsaetze und miszellen
Bastian, A *see* Die deutsche expedition der loango-kueste
Bastian, Adolf *see*
– Der buddhismus in seiner psychologie
– Deutsche revue
– Die samoanische schoepfungs-sage und anschliessendes aus der suedsee
– Die seele indischer und hellenischer philosophie in den gespenstern modernen geisterseherei
Bastian, Becca *see* Sacred ground
Bastian, C Don *see* The de soysa charitaya
Bastian, Henry *see* Brain as organ of the mind
Bastian, Henry Charlton *see* The beginnings of life
Bastide, Jean Francois de *see* Lettre a m rousseau
[Bastide, M-A de la] *see* Reponse au livre de m l'eveque de condom
Bastide, Roger *see* A poesia afro-brasileira
Bastien, Gilles *see* L'abc du hatha-yoga pour enfants de 6 a 12 ans
Bastien, Remy *see* Anthologie du folklore haitien
Bastien, Rene *see* Cheveu sur la langue
Bastier, Paul *see* L'esoterisme de hebbel
La bastille : memoires pour servir a l'histoire secrete du gouvernement francais; depuis le 14e siecle jusqu'en 1789 / Dufey, Pierre Joseph Spiridion – Paris 1833 [mf ed Hildesheim 1995-98] – 1v on 3mf – 9 – €90.00 – ISBN-10: 3-487-25921-4 – ISBN-13: 978-3-487-25921-5 – gw Olms [944]
The bastille / Bingham, D – New York: Scribner & Welford. 2v. 1888 – 12mf – 9 – $18.00 – mf#LLMC 92-124 – us LLMC [340]
La bastille devoilee : ou recueil de pieces authentiques pour servir a son histoire / Charpentier, Toussaint von – Paris 1789-1790 [mf ed Hildesheim 1995-98] – 2v on 1mf – 9 – €40.00 – ISBN-10: 3-487-25924-9 – ISBN-13: 978-3-487-25924-6 – gw Olms [933]
Bastin, Georges *see* La notion d'adaptation en traduction
Bastingius, J *see* In catechesin religionis christianae...
Bastomski, Solomon *see*
– Baym kval
– Far undzer shul
Le baston de la foy chrestienne : liure tresutile a tous chrestiens... / Bres, G de – Lyon, 1555 – 5mf – 9 – mf#PBA-431 – ne IDC [240]
Le baston de la foy chrestienne : pour s'armer contre les ennemis... / Bres, G de – [Geneve], Nicolas Barbier & Thomas Courteau, 1558+index – 6mf – 9 – mf#PBA-447 – ne IDC [240]
Le baston de la foy chrestienne : propre pour rembarrer les ennemis... / Bres, G de – [Geneve], Nicolas Barbier & Thomas Courteau, 1558+index – 9mf – 9 – mf#PBA-444 – ne IDC [240]
Le baston de la foy chrestienne... / Bres, G de – Caen, Pierre le Chandelier, 1564 – 7mf – 9 – mf#PBA-441 – ne IDC [240]
Le baston de la foy chrestienne... / Bres, G de – [Caen, Pierre Philippe, 1558] – 7mf – 9 – mf#PBA-440 – ne IDC [240]
Le baston de la foy chrestienne... / Bres, G de – [Caen, Pierre Philippe], 1560 – 9mf – 9 – mf#PBA-446 – ne IDC [240]
Le baston de la foy chrestienne... / Bres, G de – [Geneve], Clavde Dehvchin, 1563+index – 7mf – 9 – mf#PBA-443 – ne IDC [240]
Le baston de la foy chrestienne... / Bres, G de – Geneve: Guillaume Regnoult, 1562 – 7mf – 9 – mf#PBA-435 – ne IDC [240]
Le baston de la foy chrestienne... / Bres, G de – Geneve, lean Bonnefoy, 1565 – 6mf – 9 – mf#PBA-432 – ne IDC [240]
Le baston de la foy chrestienne... / Bres, G de – [Geneve], Nicolas Barbier & Thomas Courteau, 1559+index – 7mf – 9 – mf#PBA-448 – ne IDC [240]
Le baston de la foy chrestienne... / Bres, G de – [Geneve], Nicolas Barbier & Thomas Courteau, 1561 – 5mfmf – 9 – mf#PBA-427 – ne IDC [240]
Le baston de la foy chrestienne... / Bres, G de – [Geneve], Thomas Courteau, 1565 – 6mf – 9 – mf#PBA-428 – ne IDC [240]
Le baston de la foy chrestienne... / Bres, G de – [Lyon], Ian Martin, 1562 – 7mf – 9 – mf#PBA-429 – ne IDC [240]
Le baston de la foy chrestienne... / Bres, G de – Lyon, [J. Saugrain], 1562 – 7mf – 9 – mf#PBA-430 – ne IDC [240]
Le baston de la foy chrestienne... / Bres, G de – n.p, 1562+index – 9mf – 9 – mf#PBA-442 – ne IDC [240]
Bastos De Avila, Jose *see* Questoes de anthropologia brasileira
Bastos, Humberto *see*
– Economia brasileira e o mundo moderno
– Marcha do capitalismo no brasil
– Pensamento industrial no brasil
– Rui barbosa
Bastos, Joaquim Justino Alves *see* Encontro com o tempo
Bastos, Tacary Assis *see* Positivismo e a realidade brasileira
Bastos tigre e 'la belle epoque' / Menezes, Raimundo De – Sao Paulo, Brazil. 1966 – 1r – us UF Libraries [972]
Basu, Anathnath *see* Education in modern india
Basu, B D *see*
– The sacred books of the hindus
– The sukraniti
Basu, Baman Das *see*
– The consolidation of the christian power in india
– History of education in india under the rule of the east india company
– India under the british crown
– My sojourn in england
– Rise of the christian power in india
– Ruin of indian trade and industries
– Story of satara
Basu, Durga Das *see* Annotated constitution of india
Basu, Lotika *see* Indian writers of english verse
Basu, Nirmal Kumar *see* Studies in gandhism
Basu, Praphullachandra *see* Indo-aryan polity
Basu, Saroj Kumar *see* Recent banking developments
Basulto De Montoya, Flora *see* Tierra procer
Basuto / Ashton, Edmund Hugh – London, England. 1952 – 1r – us UF Libraries [960]
Basuto / Ashton, Edmund Hugh – London, England. 1967 – 1r – us UF Libraries [960]
Basuto fireside tales / Savory, Phyllis – Cape Town, South Africa. 1962 – 1r – us UF Libraries [390]
Basutoland / Coates, Austin – London, England. 1966 – 1r – us UF Libraries [960]
Basutoland / Martin, Minnie – New York, NY. 1969 – 1r – us UF Libraries [960]
Basutoland *see* Memorandum of development plans
Basutoland. Administrative Reforms Committee *see* Report of the administrative reforms committee, april-jul 1954
Basutoland, bechuanaland protectorate and swaziland / Great Britain Colonial Office – London, England. 1960 – 1r – us UF Libraries [960]
Basutoland Delimitation Commission *see* Report of the delimitation commission, 1965
Basutoland. Dept of Agriculture *see* Agricultural survey, 1949-50
The basutoland general election 1965 : report / Sanders, P B – [S.l: s.n. 1965?] – us CRL [960]
Basutoland. National Council. Maseru *see*
– Legislative council debates
– Proceedings
Basutoland National Council. Select Committee on Discriminatory Legislation *see* Report of the select committee on discriminatory legislation
Basutoland. National Council Select Committee on Labour Organisation *see* Report of the select committee on labour organisation
Basutoland. National Council Select Committee on the Liquor Proclamation *see* Report
Basutoland. National Council Select Committee on Wills, Estates and Marriages *see* Report of the select committee on wills, estates, and marriages
Basutoland Parliament Senate *see*
– Minutes of proceedings of the senate
– Parliamentary debates (hansard) official report
Basutoland reports – Cape Town, South Africa. v1-3b. 1964 – 1r – us UF Libraries [960]
Basutos : or, twenty-three years in south africa / Casalis, Eugene Arnaud – Cape Town, South Africa. 1965 – 1r – us UF Libraries [960]
The basutos : or, twenty-three years in south africa / Casalis, E – London, 1861 – 5mf – 9 – mf#HT-27 – ne IDC [916]
Bat, A G *see*
– Sbornik deistvuiushchikh zakonopolozhenii, postanovlenii, instruktsii i tsirkuliarov po potrebitelskoi kooperatsii
– Ustav potrebitelskogo obshchestva, upravliaemogo obshchim sobraniem ego chlenov
Bat he-'ashir / Rabinovitz, Alexander Siskind – Warsaw, Poland. 1898 – 1r – us UF Libraries [939]
Bat scan – v22 n3-1910 [1983 mar-oct]; 1984: mar, sep 1986 feb – 1 – mf#1044272 – us WHS [071]
Ba-ta-clam : chinoiserie franco-bresilienne – Rio de Janeiro, RJ: Imp et Lith de Ba-ta-clan, 01 jun 1867-out 1870; jul-30 set 1871 – (= ser Ps 19) – mf#P03,04,02-05 – bl Biblioteca [321]
La bataille – Paris, France 13 jan 1889-23 apr 1892 (wkly) – 1 – (cont by: marseillaise (ns) 24 apr 1892-30 jun 1893) – uk British Libr Newspaper [074]
La bataille – Paris. 10 mai-15 oct 1882, 28 mai 1883-25 janv 1886, 13 janv 1889-1906, 1897-1906 – 1 – (contient: le courrier quotidien de l'exposition de 1889 [n1-95 6 mai-7 aout 1889]) – fr ACRPP [073]
La bataille / Federation du Nord du Parti Socialiste – Lille. S.F.I.O. 1936-mai 1940 – 1 – fr ACRPP [325]
La bataille – Paris, France (ns) 7 dec 1944-17 jan 1950 (wkly) – 1 – ("la bataille" was founded as paris ed of the algiers newspaper "la marseillaise", wh was publ in london fr 14 jun 1942-3 jul 1943 & in algiers fr 4 dec 1943-12 feb 1945; cont by: rouge et le noir. la bataille (ns) 31 jan-22 dec 1950) – uk British Libr Newspaper [074]
La bataille – Paris. 30 nov 1944-1950 [wkly] – 1 – (l'hebdomadaire independant, politique et litteraire. devenu: le rouge et le noir. la bataille) – fr ACRPP [073]
La bataille : organe quotidien syndicaliste – Paris, France 5 nov 1915-12 aug 1919 (very imperfect) – 1 – uk British Libr Newspaper [074]
La bataille – Port-au-Prince, Haiti: [s.n.]. [1ere annee n4-n23]. (14 juin-13 dec 1902) – 2 sheets – us CRL [079]
La bataille de chateauguay / Sulte, Benjamin – Quebec: R Renault, 1899 [mf ed 1981] – 2mf – 9 – mf#24413 – cn Canadiana [355]
Bataille de denain / Theaulon, M – Paris, France. 1816 – 1r – us UF Libraries [025]
La bataille economique, sociale ouvriere – Paris. fevr 1921-mars 1938 – 1 – (puis: syndicale et sociale.) – fr ACRPP [073]
Bataille, Henry *see* Masque
La bataille socialiste – n21-78. Paris. juil 1929-avr 1934. mq no. 22-27, 29, 47, 50, 52, 55, 62, 65 – 1 – fr ACRPP [325]
La bataille syndicaliste / C G T – n1-45. Paris. 1er mai 1922-25 oct 1925 – 1 – (mq n41) – fr ACRPP [320]
La bataille syndicaliste – Paris. 27 avr 1911-15 dec 1920 – 1 – fr ACRPP [320]
Batalha do petroleo brasileiro / Victor, Mario – Rio de Janeiro, Brazil. 1970 – 1r – us UF Libraries [972]
La batalla – Barcelona: Impr Myria, may 23-sep 6. sep 19, oct 17-dec 11 1930; feb 12, mar 5, apr 2-may 28. jun 11-18. jul 9, 23, aug 13-20, sep 3 1931 – 1r – 1 – us CRL [074]
Batalla contra el comunismo en colombia / Nieto Rojas, Jose Maria – Bogota, Colombia. 1956 – 1r – us UF Libraries [972]
Batalla da opiniao publica / Rabelo, Genival – Rio de Janeiro, Brazil. 1970 – 1r – us UF Libraries [972]
Batalla de guatemala / Toriello Garrido, Guillermo – Buenos Aires, Argentina. 1956 – 1r – us UF Libraries [972]
Batalla de guatemala / Toriello Garrido, Guillermo – Mexico City? Mexico. 1955 – 1r – us UF Libraries [972]

La batalla de huamachuco / Valenzuela, Raimundo del Rio – Santiago [Chile]: Impr Gutenberg, 1885 [mf ed 1998, 2005] – 1r – 1 – mf#28337; 36462 – us Harvard [972]
La batalla de la albuera / Blake, Joaquin – 1811 – 9 – sp Bibl Santa Ana [946]
Batalla de las carreras / Herrera, Cesar A – Santo Domingo, Dominican Republic. 1971 – 1r – us UF Libraries [972]
Batalla de san pedro perulapan : (25 de septiembre...) / Asensio Menendez, Jose – Mexico City? Mexico. 1941 – 1r – us UF Libraries [972]
La batalla de zalaca. episodio historico-extremeno / Hurtado de Mendoza, Publio – Caceres: Tip. Enc. y Lib. de L. Jimenez, 1909 – 1 – sp Bibl Santa Ana [946]
Batalla del santuario / Posada Gutierrez, Joaquin – Bogota, Colombia. 1936 – 1r – us UF Libraries [972]
Batalla por la produccion / Berrios Rodriguez, Brigido – Ponce, Puerto Rico. 1947 – 1r – us UF Libraries [972]
Das batanaeische giebelgebirge : excurs ueber ps 68, 16 zu delitzsch' psalmencommentar (aufl 4 1883) / Wetzstein, Johann Gottfried – Leipzig: Doerffling & Franke, 1884 [mf ed 1988] – 1mf – 9 – 0-7905-0298-4 – mf#1987-0298 – us ATLA [221]
Le batave : ou le nouvelliste etranger – Paris. n1-500. fevr 1793-juin 1794 – 1 – (puis ou le sansculotte. devenu: le sans-culotte) – fr ACRPP [073]
Die bataver : historischer roman aus der voelkerwanderung (a 69 n chr) / Dahn, Felix – Leipzig: Breitkopf & Haertel, 1898 – 6r – 1 – us Wisconsin U Libr [830]
Batavia in post-war days / Ellis – Batavia, 1948 – 1mf – 9 – mf#SE-1444 – ne IDC [959]
Batavia sacra of kerkelijke historie en oudheden van batavia – Antwerpen v1-3 1715-16; Aanhangsel op Utrecht, 1744 – €86.00 – ne Slangenburg [242]
Bataviaasch nieuwsblad – 1885-1926 – 2746mf – 9 – €10,700.00 set – (also available in subsets: 1885-89 [175mf] €760 [m191]; 1890-94 [250mf] €1085 [m192]; 1895-99 [313mf] €1360 [m193]; 1900-04 [455mf] €1975 [m194]; 1905-09 [450mf] €1950 m195]; 1915-19 [343mf] €1485 [m197]; 1920-24 [242mf] €1050 [m198]; 1925-26 [117mf] €505 [m199]) – mf#m190 – ne MMF Publ [079]
Bataviaasch nieuwsblad : vijftig jaren, 1885-1935 – Batavia, 1935 – 3mf – 8 – mf#SE-1455 – ne IDC [074]
Bataviasch koloniale courant, 1810-11 / Netherlands. Royal Library. The Hague. Newspaper Dept – 1810-27 – 145mf – 9 – €535.00 – (cont by: java government gazette from 1812-16; cont by: bataviasche courant 1810-27; 1810-11 8mf €32.50 m141, 1812-16 35mf €137.50 m142, 1816-27 102 mf €405 m143) – mf#M140 – ne MMF Publ [324]
Bataviasche historische, geographische, huishoudelyke en reis-almanach : of, nuttig en noodzakelyk handboek, voor hun die naar oost-indien varen, of kundigheid van de gewesten begeeren / Hofhout, Johannes – Rotterdam: J Hofhout 1789 [mf ed Rochester NY: Archival Microfilming Services for Yale Uni Lib 1996 – 1r – 1 – (incl: nederduitsch en maieidisch woordenboek) – us CRL [915]
Bataviasche koloniale courant – Batavia: 's Lands Drukkerij (wkly) [mf ed Jakarta: National Library of Indonesia] – jan 4-aug 2 1811 on reel 1 – 1 – (began in 1810; ceased in 1811; cont by: java courant) – mf#mf-7738 seam – us CRL [079]
Bataviasche studenten almanak voor het jaar 1931 : eerste jaargang – Batavia, 1931 – 4mf – 8 – mf#SE-1443 – ne IDC [378]
Batayang aklat sa wika : ikalimang baitang / Manalastas Laraya, Consorcia – Quezon City, Philippines: Bustamente Press 1960 [mf ed 1987] – 1r [ill] – 1 – mf#6774 – us Wisconsin U Libr [490]
Batchelor, Henry *see*
– Advisory councils
– The christian fulfilments and uses of levitical sin-offering
Batchelor, J *see* The ainu of japan
Batchelor, John *see* The ainu and their folk-lore
Bate, Fanny *see* Gems of hope in memory of the faithful departed
Bate, Francis *see* The naturalistic school of painting
Bate, H Maclear *see* South africa without prejudice
Bate, John *see* Emigration
Bate, John Drew *see* Examination of the claims of ishmael
Bate, Percy H *see* The english pre-raphaelite painters
Bateau de blanchisseuses / Villeneuve, Ferdinand De – Paris, France. 1832 – 1r – us UF Libraries [440]
O batel – Granja, CE: [s.n.] 24 jan 1892 – (= ser Ps 19) – mf#P18B,03,09 – bl Biblioteca [079]
O batel – Sobral, CE: Typ da Gazeta do Sobral, 21 nov 1886 – (= ser Ps 19) – mf#P17,03,60 – bl Biblioteca [079]
Bateman, F Foster *see* Reply to the montreal harbour engineer's report on the st lawrence bridge and manufacturing scheme
Bateman, Frederic *see* Darwinism tested by language
Bateman, Josiah *see* Martiniere
Bateman, Lee La Trobe *see* Florida trucking for beginners
Bateman, Warner Mifflin *see* Warner m bateman papers, 1849-1897
Bateman, William *see* Lord bateman's plea for limited protection or for reciprocity in free trade
Batemann, Joylyn *see* Choreographing as teaching/teaching as choreographing
Bates, Chrisenberry Lee *see* Federal procedure at law.
Bates, Darrell *see* Fly-switch from the sultan
Bates, Elisha *see*
– The doctrines of friends
– An examination of certain proceedings and principles of the society of friends, called quakers
Bates, Henry W *see* The naturalist on the river amazon
Bates, Henry Walter *see*
– The naturalist on the river amazons
– Naturalista no rio amazonas
Bates, J H *see* Christian science and its problems
Bates, John *see* Imputation
Bates, Lindon Wallace *see* Path of the conquistadores
Bates, Miner Searle *see* Chi-tu chiao yu kung ch'an chu i [ccm4]
Bates, MK *see* Effect of peer group presence on the gross motor performance of young children
Bates, O *see* The eastern libyans
Bates, Ralph *see* En la espana leal ha nacido un ejercito
Bates, Robert H et al *see* Fve miles high
Bates, Samuel Penniman *see* A brief history of the one hundredth regiment (roundheads)
Bates, W C *see* Exponential outline with definitions of blackstone's commentaries.
Bates, Walter *see* Kingston and the loyalists of the spring fleet of a d 1783
Bates, William *see* George cruikshank
Bateson, Thomas *see* Second set of bateson's madrigals
Bateson, W *see* Journal of genetics
Bateson, William *see* Mendel's principles of heredity
Batesville first baptist church : church records – BATESVILLE, AR. jun 1883-1930, 1937-aug 1942, mar 1946-1977 – 1 – $87.93 – us Southern Baptist [242]
Bath and wilts chronicle – Bath, England 16 sep 1911-11 apr 1925 [mf jul-dec 1896, 1911, sep-dec 1912, 1986-] – 1 – (cont: bath daily chronicle & argus [8 oct 1903-15 sep 1911]; cont by: bath & wilts chronicle & herald [14 apr 1925-10 jun 1961]) – uk British Libr Newspaper [072]
Bath argus and west of england advertising register [daily ed] – Bath, England 31 oct 1877-29 apr 1892 – 1 – (cont: evening argus: publ as a daily ed of: bath argus (saturday) weekly paper [20 jan 1876-30 oct 1877]; cont by: bath daily argus [2 may 1892-28 jul 1893]) – uk British Libr Newspaper [072]
Bath daily chronicle – Bath, England 7 aug 1883-14 feb 1900 [mf jul-dec 1896, 1911, sep-dec 1912, 1986-] – 1 – (wanting: jan-jun 1897; cont: bath evening chronicle [12 jun 1877-4 aug 1883]; cont by: daily chronicle & argus [15 feb 1900-7 oct 1903]; bath daily chronicle and argus [8 oct 1903-15 sep 1911]; bath & wilts chronicle [16 sep 1911-11 apr 1925]) – uk British Libr Newspaper [072]
Bath guardian – Bath, England 1 feb 1834-15 aug 1835 (wkly) – 1 – (cont by: bath & devizes guardian, & general advertiser for somerset, wilts, & gloucester [22 aug 1835-29 apr 1837]; bath guardian, & general advertiser for somerset, wilts, & gloucester [6 may 1837-27 jul 1839]) – uk British Libr Newspaper [072]
Bath house regulations / Youngblood, Alice – s.l, s.l? . 193-? – 1r – us UF Libraries [978]
Bath journal – Bath, England 19 jan 1744-13 feb 1757 – 1 – (cont by: boddely's bath journal [14 feb 1757-29 dec 1766, 4 jan 1768-31 dec 1770, 6 jan 1772- 1 mar 1773]; bath journal: (bath journal & general advertiser) 8 mar-27 dec 1773, 2 jan-25 dec 1775, 19 aug 1776, 15 feb 1779-29 dec 1783, 2 jan 1786-28 dec 1789, 3 jan 1791-31 dec 1798, 20 jan 1800-25 dec 1809, 4 jan-27 dec 1813, 2 jan 1815-30 dec 1816, 3 jan 1820-31 dec 1821; keenes' bath journal & general advertiser [7 jan 1822-27 dec 1824, 5 jan-28 dec 1878]) – uk British Libr Newspaper [072]
Bath. Presbytery (Pres. Church in the USA) *see* Minutes, 1817-1862
Bath weekly argus and west of england advertising register – Bath, England 17 apr 1897-21 oct 1911 – 1 – (incorp with: bath chronicle; wanting: 1872, 1896, 1898; cont: bath argus & west of england advertising register (weekly ed) 23 jul 1870-10 apr 1897) – uk British Libr Newspaper [072]
Ba'that al-'ilmiyah fi 'ahd muhammad 'ali / Tusun, 'Umar – Al-Iskandariyah, Egypt. 1934 – 1r – us UF Libraries [956]
Bathe, Johannes Clemens *see* Die bewegungen und haltungen des menschlichen koerpers in heinrich von kleists erzaehlungen
Batheos thlibetai *see* Answer of the great church of constantinople to the papal encyclical on union
Bathini na abantu? : Ucwangiciso-ntsapho / Lempilo, Isebe – [Pretoria: Dept of Health 1978?] [mf ed Pretoria, RSA: State Library [199-]] – 8p on 1r with other items – 5 – (in xhosa) – mf#op 06758 r24 – us CRL [360]
Batho, Cyril *see* The distribution of stress in certain tension members
Bathory et possevino : documents inedits sur les rapports du saint-siege avec les slaves / ed by Pierling, Paul – Paris: Ernest Leroux 1887 [mf ed 1991] – 1mf – 9 – 0-524-01237-7 – mf#1990-0376 – us ATLA [241]
Bathurst advocate – Bathurst, feb 1848-dec 1849 – 1r – A$38.28 vesicular A$43.78 silver – at Pascoe [079]
Bathurst argus – Bathurst, apr 1904-dec 1907 – 4r – A$262.02 vesicular A$284.02 silver – at Pascoe [079]
Bathurst courier *see* Perth courier
Bathurst free press – Bathurst. 1872, 1882, jan 1885-dec 1897 – 12r – A$462.00 vesicular A$528.00 silver – at Pascoe [079]
Bathurst free press – Bathurst, jan 1850-dec 1862 – 6r – A$303.64 vesicular A$336.64 silver – at Pascoe [079]
Bathurst free press – Bathurst, jan 1899-mar 1904 – 5r – A$338.01 vesicular A$365.51 silver – at Pascoe [079]
Bathurst, Henry *see* Charge delivered to the clergy of the diocese of norwich
Bathurst (Ont: District). Council *see* Rules and regulations and part of the bye-laws of the municipal council of the bathurst district
Bathurst post – Bathurst, aug 1881-mar 1922 (misc iss) – 1r – A$38.50 vesicular A$44.00 silver – at Pascoe [079]
Bathurst schools discussed in the legislature : sifting the evidence: the investigation at bathurst and the finding of judge fraser criticised by mr pitts – Fredericton, NB: Reporter Office, 1893? – 1mf – 9 – mf#06029 – cn Canadiana [377]
Bathurst times – Bathurst, jan 1867-mar 1904 – 28r – A$1078.00 vesicular A$1232.00 silver – at Pascoe [079]
Bathurst times – Bathurst, jan 1908-dec 1939 – 28r – A$2038.48 vesicular A$2192.48 silver – at Pascoe [079]
Bathurst times – Bathurst, nov 1860-apr 1862 – 1r – 9 – A$54.74 vesicular A$60.24 silver – at Pascoe [079]
Bathurst, W A *see* Chruch question and the approaching election
Batiendo la esperanza / Rodriguez Rubio, Inocencia – Madrid: Graf. Saldana, 1973 – 1 – sp Bibl Santa Ana [946]
Batiffol, Pierre *see*
– L'abbaye de rossano
– Catholicisme et papaute
– L'eglise naissante et le catholicisme
– L'enseignement de jesus
– Histoire du breviaire romain
– La litterature grecque
– Les odes de salomon
– Orpheus et l'vangile
– La paix constantinienne et le catholicisme
– Primitive catholicism
Batilo. egloga / Melendez Valdes, Juan – 1780 – 9 – sp Bibl Santa Ana [810]
Le batiment – Paris, France jan 1961-dec 1964 (mthly) – 1 – uk British Libr Newspaper [074]
Batiment et travaux publics *see* Moniteur de l'entreprise et de l'industrie
Le batiment: travaux publics et particuliers – Paris, France 2 jan 1960-19 dec 1964 (wkly) – 1 – uk British Libr Newspaper [690]
Batissier, L *see*
– Elements d'archéologie nationale, procedes d'une histoire de l'art monumental chez les anciens
– Histoire de l'art monumentale dans l'antiquite et au moyen age
Batista Y Cuba, Juan Wilfredo *see* Effect of mulch and chemical treatments on microbiological action i...
Batista Y Zaldivar, Fulgencio *see*
– Ideario de batista
– Revolucion social o politica reformista...
Bat'kivshchyna – Toronto, Canada 8 oct 1955-feb/mar 1986 [mf 1955-75] – 1 – (in cyrillic) – uk British Libr Newspaper [071]
Bat'kivshchyna: organ ukrains'koho narodnoho frontu – Paris, France jan [1944] – 1 – (in cyrillic) – uk British Libr Newspaper [074]
Batley examiner – Batley, England 21 jul 1893-26 oct 1895 (wkly) – 2r – 1 – (discontinued) – uk British Libr Newspaper [072]
Batley, H W *see* A series of studies for domestic furniture decorating
Batley news – Batley, England 6 jan 1883-27 jun 1959 (wkly) – 1 – (cont by: batley news & reporter [4 jul 1959-20 may 1971]; Batley News [27 may 1971- (mf 1986-)]) – uk British Libr Newspaper [072]
Batley reporter and guardian – Batley, England 17 Jul 1869-20 jun 1959 (wkly) [mf 1897,1898,1910] – 1 – (incorp with: batley news) – uk British Libr Newspaper [072]
Batliwala, S S *see* Makers of new china
Baton Rouge Bicentennial Commission *see* Bicentennial banner
Baton rouge chronicle – 1993 oct-dec; 1994 feb-mar, may-jun, sep – 1 – mf#3056276 – us WHS [071]
Baton rouge post – 1991 jul 4; 1992 jul 2; 1996 dec 26; 1997 jan 23, feb 6 – 1 – mf#3056265 – us WHS [071]
Baton rouge tribune – 1993 feb – 1 – mf#3912408 – us WHS [071]
Batra, Ram Lal *see* Science and art of indian music
Batrakov, N I *see* Kur'er syzrani
Batrell, Ricardo *see* Para la historia
Batres Jauregui, Antonio *see* Estudios historicos y literarios
Batres Montufar, Jose *see*
– Poesias
– Poesias de jose batres montufar
– Poesias de jose batres montufar (natural de guatem...
Batschkaer zeitung – Apatin, Yugoslavia. 1931; 1933-38; 1941-43 – 7r – 1 – us L of C Photodup [079]
Bats'kaushchyna = Fatherland – Munich, Germany 2 nov 1952, 25 jan, 22 feb 1953-dec 1966 – 1 – (white ruthenian newspaper; in cyrillic) – uk British Libr Newspaper [074]
Batson, Alfred *see* Vagabond's paradise
Batsto citizens gazette – 1966 mar-1982 fall – 1 – mf#614656 – us WHS [071]
Batt, Kurt *see* Revolte intern
La battaglia taliana a quatro voci : composta da m mathias fiamengo, maestro di capella del domo di milano... / Werrecore, Mathias – Venetiis: Hieronymum Scotum 1550 [mf ed 19–] – 1mf – 9 – mf#fiche 219 – us Sibley [780]
Battaglione garibaldi, ottobre 1936-aprile 1937 / Vita, A de – Parigi: Ed di coltura sociale 1937 – 9 – mf#w1248 – us Harvard [946]
Battaille, Louis Nicolas *see* Lapin
La batte : gazette satirique – n1-15. Paris. mars-oct 1888 [bimnthly] – 1 – fr ACRPP [870]
Batteau and banaboo / White, Walter Grainge – Brigg, England. no date – 1r – us UF Libraries [972]
Battelle research outlook – Columbus OH 1969-72 – 5,9 – (cont by: research outlook) – ISSN: 0522-4810 – mf#3486.01 – us UMI ProQuest [660]
Battelle technical review – Columbus OH 1952-68 – 1 – ISSN: 0522-4829 – mf#1563 – us UMI ProQuest [020]
Batten, Loring Woart *see* The old testament
Batten, Samuel Zane *see*
– The christian state
– The social task of christianity
Battered women : issues of public policy / U.S. Commission on Civil Rights – Washington: GPO, jan 30-31 1978 – 8mf – 9 – $12.00 – mf#LLMC 94-330 – us LLMC [322]
Battersby, Hannah S *see* Home lyrics
Battersby, Thomas Stephenson Francis *see* The secret policy of the land act
Battersby, W J *see*
– The complete catholic directory, almanack and registry...
– A complete catholic registry, directory and almanack...
Battersby's dominion pocket railway and travellers guide (with map), n107, jan, 1885 – Montreal: D Battersby, [1885?] [mf ed 1980] – 2mf – 9 – 0-665-02999-3 – (incl ind) – mf#02999 – cn Canadiana [917]
Battersea and wandsworth observer *see* South lambeth battersea and wandsworth times
Battersea council of action : strike bulletin – W&sworth, England 11 may 1926 – 1 – uk British Libr Newspaper [331]
Battersea guardian – W&sworth, England 24 may 1990-20 jul 1995 – 1 – uk British Libr Newspaper [072]
Battersea news – Wandsworth, London 9 apr 1965-13 jul 1990 [mf 1986-91, 1998-2000] – 1 – (cont: battersea boro' news [5 jan 1906-2 apr 1965]; cont by: wandsworth borough news (battersea & clapham ed) [20 jul 1990-7 jun 1991, 3 jul 1998-30 jun 2000]) – uk British Libr Newspaper [072]
Battersea news – W&sworth, England 3 jul 1998-26 oct 2001 – 1 – uk British Libr Newspaper [072]

Battersea vanguard – [London & SE] Wandsworth 14 jul 1907-jul 1908 [mf ed 2003] – 1r – 1 – uk Newsplan [072]
Batteux, Charles *see* Les beaux arts reduits a un meme principe
Battey, Thomas C *see* The life and adventures of a quaker among the indians
Batthyany, Vince *see*
– Reise durch einen theil ungarns, siebenbuergens, der moldau und buccovina im jahr 1805
– Reise nach constantinopel
Battie, William *see* A treatise on madness
Battier, Alcibiade Fleury *see* Sous les bambous
Battishill, Jonathan *see* Six anthems and ten chants
Battista : or, the labourer called at the eleventh hour – London, England. 18-- – 1r – us UF Libraries [240]
Battistella, Antonio *see* Il s. officio e la riforma religiosa in friuli
Battle chart of the united states : containing an account of the principal battles fought by the american since the commencement of the revolution – [New York?: s.n, 1848?] [mf ed 1986] – 1mf – 9 – 0-665-60328-2 – mf#60328 – cn Canadiana [355]
The battle creek blade – [Burnett, NE]: A E Shelton. v1 n1. apr 2 1885- (wkly) – 1r – 1 – (publ also in battle creek may 1 1885-) – us Bell [071]
Battle creek enterprise – Battle Creek, NE: D W Bryan. v1 n1. apr 20 1887- (wkly) [mf ed 1969] – 1 – us NE Hist [071]
Battle creek enterprise – Battle Creek, NE: D W Bryan. v1 n1. apr 20 1887- (wkly) – 22r – 1 – (date dropped with 36th yr n48; resumed 38th yr n22 nov 15 1923) – us Bell [071]
Battle creek republican – Battle Creek, NE: C F Montross. oct 24 1895-sep 20 1901 (wkly) [mf ed with gaps filmed 2000] – 6r – 1 – us NE Hist [071]
Battle creek republican – Battle Creek, NE: C F Montross (wkly) – 3r – 1 – us Bell [071]
Battle cry – 1983 jun-1986 dec; 1987 jan-1988 mar – 1 – mf#1508987 – us WHS [071]
Battle for rhodesia / Reed, Douglas – Cape Town, South Africa. 1966 – 1r – us UF Libraries [960]
[Battle mountain-] battle mountain bugle – NV. 1957-1962; 1976- – 27r – 1 – $1620.00 (subs $90y) – mf#N04406 – us Library Micro [071]
[Battle mountain-] central nevadan – NV. 1885-1907 [wkly] – 9r – 1 – $540.00 – mf#U04407 – us Library Micro [071]
[Battle mountain-] herald and central nevadan – NV. 1907-11 [wkly] – 1r – 1 – $60.00 – mf#U04408 – us Library Micro [071]
[Battle mountain-] landess free press – NV. 1881-82 [wkly] – 1r – 1 – $60.00 – mf#U04409 – us Library Micro [071]
[Battle mountain-] measure for measure – NV. mar-oct 1875 [wkly] – 1r – 1 – $60.00 – mf#U04410 – us Library Micro [071]
[Battle mountain-] messenger – NV. 1881-84 [wkly] – 1r – 1 – $60.00 – mf#U04411 – us Library Micro [071]
[Battle mountain-] reese river valley times – NV. 1980 – 1r – 1 – $60.00 – mf#N03699 – us Library Micro [071]
[Battle mountain-] scout – NV. 1913-57 (incomplete) [wkly] – 14r – 1 – $840.00 – mf#U04412 – us Library Micro [071]
[Battle mountain-] valley times – NV. 1981 – 1r – 1 – $60.00 – mf#N03700 – us Library Micro [071]
Battle of fort george / Cruikshank, Ernest Alexander – Niagara, Ont?: s.n, 1896 – 1mf – mf#06038 – cn Canadiana [355]
The battle of fort george / Cruikshank, Ernest Alexander – Welland [Ont]: Tribune Print, 1904 – 1mf – 9 – 0-665-76091-4 – mf#76091 – cn Canadiana [355]
The battle of lundy's lane, 25th july, 1814 : a historical study / Cruikshank, Ernest Alexander – Welland, Ont?: Lundy's Lane Historical Society, 1893? – 1mf – 9 – mf#27004 – cn Canadiana [355]
Battle of majuba hill / Ransford, Oliver – New York, NY. 1968, c1967 – 1r – us UF Libraries [960]
Battle of natural bridge – s.l, s.l? . 193-? – 1r – us UF Libraries [978]
The battle of queenston heights, october 13th, 1812 / Curzon, Sarah Anne – Toronto?: s.n, 1899 – 1mf – 9 – (with a sketch of her life and works by lady edgar) – mf#29313 – cn Canadiana [971]
The battle of standpoints : the old testament and the higher criticism / Cave, Alfred – London: Eyre and Spottiswoode, 1890. Beltsville, Md: NCR Corp, 1978 (1mf); Evanston: American Theol Lib Assoc, 1984 (1mf) – (= ser Biblical crit – us & gb) – 9 – 0-8370-0739-9 – (incl bibl ref) – mf#1984-1063 – us ATLA [220]
The battle of the plains : the greatest event in canadian history / Harper, John Murdoch – Toronto: Musson Book Co, c1909 – (= ser Studies in verse and prose) – 4mf – 9 – 0-665-73152-3 – (incl biogr of major leaders in the battle) – mf#73152 – cn Canadiana [971]
The battle of the plains / Harper, John Murdoch – Quebec: publ for aut, 1895? – 1mf – 9 – mf#53973 – cn Canadiana [971]
Battle of the standpoints / Cave, Alfred – London, England. 1890 – 1r – us UF Libraries [240]
The battle of trenton / Hewitt, James – A sonata for the piano-forte dedicated to General George Washington. New York: James Hewitt 1797. MUSIC 282 – 1 – us L of C Photodup [780]
The battle with tuberculosis and how to win it : a book for the patient and his friends / King, Dougall Macdougall – Philadelphia: J B Lippincott, c1917 [mf ed 1995] – 3mf – 9 – 0-665-77295-5 – (incl app) – mf#77295 – cn Canadiana [616]
The battleaxe, or gazette of the church army – London. 2 Apr 1883-30 Jan 1886. -f. – 47 feet – 1 – uk British Libr Newspaper [074]
Battle-pieces and aspects of the war / Melville, William – 1866 – 9 – us Scholars Facs [890]
Battles of the american revolution, 1775-1781 : historical and military criticism with topographical illustration / Carrington, Henry Beebee – New York: A S Barnes, c1876 [mf ed 1980] – 9mf – 9 – 0-665-02553-X – mf#02553 – cn Canadiana [975]
Battles of the boer war / Pemberton, William Baring – London, England. 1969, c1964 – 1r – us UF Libraries [960]
The battles of the world : or, cyclopedia of battles, sieges, and important military events / Borthwick, John Douglas – Montreal: J Muir, 1866 – 6mf – 9 – mf#48000 – cn Canadiana [355]
Battlesden – (= ser Bedfordshire parish register series) – 1mf – 9 – £3.00 – uk BedsFHS [929]
Battleship New Jersey Historical Museum Society *see* Ship's newsletter bb-62
Batts, Michael S *see* Bruder hansens marienlieder
Batty, Beatrice *see* Forty-two years amongst the indians and eskimo
Batty, J A Staunton *see* Our opportunity in china
Batuah, A D *see* Tambo minangkabau dan adatnja
Les batuecas y las jurdes / Beauchet, Ludovico – 1895 – 9 – sp Bibl Santa Ana [946]
Baturinskii, D A *see* Agrarnaia politika tsarskogo pravitel'stva i krest'ianskii pozemel'nyi bank
O baturite : neutro entre partidos politicos – Baturite, CE: Typ do Baturite, 28 jul 1878 – (= ser Ps 19) – bl Biblioteca [079]
Der bau : der kampf um ein werk: roman / Seidl, Florian – Braunschweig: G Westermann c1937 [mf ed 1991] – 1r – 1 – (filmed with: heinrich seidel und der deutsche humor / alfred biese) – mf#2942p – us Wisconsin U Libr [830]
Der bau des tempels salomo's nach der koptischen bibelversion / Brugsch, Heinrich Karl – Leipzig: J C Hinrichs, 1877 – 1mf – 9 – 0-7905-3308-1 – mf#1987-3308 – us ATLA [220]
Bau einer kalibrierstation fuer thermokoppel und waermeleitfaehigkeit von kunststoffen / Wolfrum, Renate – (mf ed 1995) – 1mf – 9 – €30.00 – 3-8267-2214-0 – mf#DHS 2214 – gw Frankfurter [660]
Bau eines rundlaufthermostaten fuer das festkoerperdilatometer und messung des ausdehnungsverhaltens und der volumenrelaxation von polymeren / Kreuzer, Wolfgang – (mf ed 1995) – 1mf – 9 – €30.00 – 3-8267-2213-2 – mf#DHS 2213 – gw Frankfurter [660]
Bau, Mingchien Joshua *see* Tsai hua wai ch'iao chih ti wei
Bau und test eines wirbelstrom-septums fuer delta / Albers, Jan – (mf ed 1996) – 1mf – 9 – €30.00 – 3-8267-2286-8 – mf#DHS 2286 – gw Frankfurter [530]
Bau-anschlag : oder richtige anweisung...alle... bau-kosten ausfuendig zu machen / Penther, J F – Augsburg, 1743 – 5mf – 9 – mf#OA-107 – ne IDC [720]
Der bauarbeiter – Rosslau DE, 1959 19 dec-1960 8 apr – 1r – 1 – (bau- und montagebetrieb) – gw Misc Inst [690]
Bauch, Bruno *see* Goethe und die philosophie
Bauch, Gustav *see* Ueber die historia romana des paulus diaconus
Baucke, Florian *see* Memorias del p florian paucke
Baud, Benjamin *see* Architectural illustrations of windsor castle
Baud, C *see* Observations sur les cordes a instrumens de musique, tant de boyau que soie
Baud-Bovy, Samuel *see* Chansons du dodecanese
Baudelaire, Charles *see* Strandgut
Baudert, Samuel *see* Die evangelische mission
Baudesson, Henry *see* Indo-china and its primitive people
Baudier, M *see* The history of the court of the king of china
Baudin Hesmo, Manuel *see* El obispo de quito don alonso de la pena montenegro
Baudin, Noel *see* Fetichism and fetich worshipers
Baudin, P *see* Fetichism and fetich worshipers
Baudissin, Eva Fanny Bernhardine Tuerk, graefin von *see* Die grosse woge
Baudissin, Ida, Graefin *see* Durch sturm und not
Baudissin, W W *see* Adonis und esmun
Baudissin, Wolf Ernst Hugo Emil, Graf von *see* Der kleine gerd
Baudissin, Wolf Wilhelm *see*
– Die alttestamentliche wissenschaft und die religionsgeschichte
– August dillman
– Jahve et moloch, sive, de ratione inter deum israelitarum et molochum intercedente
Baudissin, Wolf Wilhelm, Graf *see* Einleitung in die buecher des alten testamentes
Baudissin, Wolf Wilhelm, Graf von *see*
– Adonis und esmun
– Die alttestamentliche spruchdichtung
– Eulogius und alvar
– Studien zur semitischen religionsgeschichte
– Zur geschichte der alttestamentlichen religion in ihrer universalen bedeutung
Bauditz, Sophus *see* Fra pol til pol
Baudoin, J *see*
– Emblemes divers
– Iconologie
– Recueil d'emblemes divers
Baudoncourt, Jacques de *see* Histoire populaire du canada
Baudot de Juilly, Nicolas *see* Histoire et regne de louis 11
Baudot, Jules *see*
– Hymnes latines et hymnaires
– The lectionary
Baudouin, Charles-Gabriel *see* Travaux franc-maconniques
Baudouin, Philibert *see*
– Index to incorporated bodies and to private and local law...
– Supplement no 1 to the index to incorporated bodies and to private and local law
– Table de concordance du code de procedure civile...
Baudri *see* Der erzbischof von koeln johannes cardinal von geissel und seine zeit
Baudrillart, Alfred *see*
– The catholic church, the renaissance and protestantism
– L'enseignement catholique dans la france contemporaine
– Vie de mgr. d'hulst
Baudu, Paul *see* Vieil empire, jeune eglise
Bau-echo – Halberstadt DE, 1960 5 feb-1963 27 jun – 1r – 1 – gw Misc Inst [074]
Der bauer – Herbestahl (B), 1933 7 jan-23 dec, 1934 6 jan-29 dec, 1937 3 jan-1939 – 1 – gw Misc Inst [630]
Bauer, Adolf *see*
– Die chronik des hippolytos im matritensis graecus 121
– Die chronik des hippolytos in matritensis graecus 121
Bauer, Albert *see*
– Das feld unserer ehre
– Folkert der schoeffe
Bauer, B *see* Philo, strausz und renan und das urchristentum
Bauer, Bruno *see*
– Die apostelgeschichte
– Kritik der evangelien und geschichte ihres uhrsprungs
– Kritik der evangelischen geschichte
– Kritik der evangelischen geschichte des johannes
– Kritik der paulinischen briefe
– Philo, strauss und renan und das urchristenthum
– Die religion des alten testaments
Bauer, F C *see*
– Die christliche lehre von der versoehnung in ihrer geschichtlichen entwicklung
– Paulus der apostel jesus christi
Bauer, Franz *see* Georg buechner
Bauer, Friedrich *see* Sterne'scher humor in immermanns "muenchhausen"
Bauer, H *see* Die psychologie alhazens
Bauer, Hans *see* Zur entzifferung der neuentdeckten sinaischrift und zur entstehung des semitischen alphabets
Bauer, Heinrich *see* Florian geyer
Bauer, Hugo *see* Der burggraf von nuernberg
Der bauer im deutschen liede – Berlin: Mayer und Mueller 1890 [mf ed Bloomington IN: Indiana Uni Lib, Preservation Dept 1984] – 1r – 1 – us Indiana Preservation [390]
Bauer, Ina *see* Frauen im management
Bauer, Jeremy *see* Kinetics and kinematics of prepubertal children
Bauer, Johannes *see*
– Ungedruckte predigten schleiermachers aus den jahren 1820-1828
– Werke
Bauer, Josef Martin *see*
– Achtsiedel
– Am anderen morgen
– Das maedchen auf stachet
– Der sonntagsluegner
Bauer, Karl *see* Goethes kopf und gestalt
Bauer, Louis *see* Lectures on causes, pathology and treatment of joint diseases
Bauer, Ludwig Caesar *see* Und der liebe sonnenschein
Bauer, Martin *see*
– Herstellung und spektroskopische charakterisierung matrix-isolierter silbercluster
– Photothermische untersuchungen an kleinen silberpartikeln
Bauer Paiz, Alfonso *see* Catalogacion de leyes y disposiciones de trabajo d...
Bauer, Shari R *see* Comparison of hydrostatic weighing to bioelectrical impedance analysis in women greater than thirty percent body fat
Bauer, Walter *see* Lehrbuch der neutestamentlichen theologie
Bauer, Wilhelm *see* Die ethik soeren kierkegaards
Bauerhorst, Kurt *see* Bibliographie der stoff- und motivgeschichte der deutschen literatur
Bauerman, Hilary *see* Report on the geology of the country near the forty-ninth parallel of north latitude west of the rocky mountains
Bauern, bonzen und bomben : roman / Fallada, Hans – Berlin: Vier Falken c1931 [mf ed 1989] – 1r – 1 – (filmed with: ein kleiner deutscher / ernst dittmer) – mf#7178 – us Wisconsin U Libr [830]
Bauernaufstand vom jahre 1381 in der englischen poesie / Eberhard, Oscar – Heidelberg, Germany. 1917 – 1r – us UF Libraries [420]
Bauern-echo – Berlin, Germany. 1962-76 – 27r – 1 – us L of C Photodup [074]
Bauern-echo – Berlin DE, 1948 18 jul-1954, 1991 – 13r – 1 – (filmed by misc inst: 1992 2 jan-31 jul, 1954 1 aug-1990 [73r]; title varies: 1 aug 1990: deutsches landblatt, ed a for potsdam, berlin, cottbus, frankfurt/o) – mf#6060 – gw Mikropress; gw Misc Inst [630]
Bauernerbe : erzaehlungen und schildereien / Huggenberger, Alfred – Berlin: Aehrenlese Verlag 1943 [mf ed 1995] – 1r – 1 – (filmed with: peter michel / friedrich huch) – mf#3883p – us Wisconsin U Libr [390]
Bauernfeind, O *see* Der roemerbrieftext des origenes
Bauernfeld, Eduard von *see*
– Alkibiades
– Bauernfelds ausgewaehlte werke in vier baenden
– Eduard von bauernfelds gesammelte aufsaetze
– Fortunat
– Die republik der thiere
– Der selbstquaeler
– Der vater
– Die verlassenen
– Zwei familien
Bauernfelds ausgewaehlte werke in vier baenden – Leipzig: Hesse, [1905?] [mf ed 1998] – 4v in 1 – 1 – (crit biogr int ed by emil horner) – mf#9961 – us Wisconsin U Libr [802]
Bauernfeldt, Eduard von *see* Der vater
Der bauernfreund – Mangelsdorf DE, 1957-1960 mar [gaps] – 1r – 1 – gw Misc Inst [630]
Der bauernfuerst : roman / Schuecking, Levin – Leipzig: F A Brockhaus 1851 [mf ed 1995] – 2v on 1r – 1 – (filmed with: sueden und norden / hermann schmid) – mf#3738p – us Wisconsin U Libr [830]
Der bauerngeneral : ein mitkaempfer erzaehlt von der tragik deutschen kaempfertums auf amerikanischer erde: roman / Weyland, Hans – Muenchen: F Eher 1939 [mf ed 1993] – 1r – 1 – (filmed with: 1000 tage westfront / franz wallenborn & other titles) – mf#7830 – us Wisconsin U Libr [830]
Das bauernhaus in palaestina : mit ruecksicht auf das biblische wohnhaus / Jaeger, Karl – Goettingen: Vandenhoeck & Ruprecht, 1912 – 1mf – 9 – 0-8370-3756-5 – mf#1985-1756 – us ATLA [230]
Der bauernhochzeitsschwank : meier betz und metzen hochzit / ed by Wiessner, Edmund – Tuebingen: M Niemeyer, 1956 [mf ed 1993] – (= ser Altdeutsche textbibliothek n48) – 64p – 1 – (incl bibl ref) – mf#8193 reel 4 – us Wisconsin U Libr [390]
Der bauernkanzler / Kath, Lydia – Feldpostausg. Berlin: Junge Generation Verlag [1944] [mf ed 1990] – 1r [ill] – 1 – (filmed with: on the eve / leopold kampf) – mf#2752p – us Wisconsin U Libr [830]
Der bauernrebell : roman aus der tirolergeschichte / Schmid, Herman – 2.aufl. Leipzig: Keil [18-?] – (= ser Gesammelte schriften. volks- und familien-ausgabe 33-34 nf1-2) – 2v in 1 – 1 – mf#film mas c438 – us Harvard [830]

Bauernruf – Duesseldorf DE, 1957 nov-1969 – 2r – 1 – gw Misc Inst [630]
Bauernschritt : erzaehlung / Zierer, Maria Steinmueller – Stuttgart: J G Cotta 1943 [mf ed 1992] – 1r – 1 – (filmed with: alle wipfel rauschen heimat / wolfgang zenker) – mf#3067p – us Wisconsin U Libr [830]
Bauernspiegel : oder, lebensgeschichte des jeremias gotthelf, von ihm selbst beschrieben / ed by Bartels, Adolf – Leipzig: M Hesses, [190-?] – 432p – 1 – mf#8518 – us Wisconsin U Libr [920]
Der bauern-spiegel : oder, lebensgeschichte des jeremias gotthelf, von ihm selbst geschrieben / ed by Mueller, Ernst – Erlenbach, Zuerich: E Rentsch, 1921 [mf ed 1993] – (= ser Saemtliche werke in 24 baenden 1) – 386p – 1 – (incl bibl ref) – mf#8522 reel 1 – us Wisconsin U Libr [920]
Bauernstolz : dorfgeschichten aus dem weserlande / Strauss und Torney, Lulu von – Berlin: E Fleischel 1921 [mf ed 1991] – 1r – 1 – (filmed with: ekkehard / j von scheffel) – mf#2846p – us Wisconsin U Libr [830]
Das bauerntum im grenz- und volksdeutschen roman der gegenwart / Luis, Werner – Berlin: Junker & Duennhaupt 1940 [mf ed 1992] – (= ser Neue deutsche forschungen. abteilung neuere deutsche literaturgeschichte 268) – 2r – 1 – (incl bibl ref) – mf#3185p – us Wisconsin U Libr [430]
Der bauernzorn : eine erzaehlung aus dem grossen bauernkrieg / Zacharias, Alfred – Dresden: W Heyne [194-?] [mf ed 1993] – 1r [ill] – 1 – (filmed with: die dichterische entwicklung j f w zachariaes / otto hermann kirchgeorg) – mf#7969 – us Wisconsin U Libr [830]
Das baugewerbe – Karlsruhe DE, 1885 2 may-1886 2 apr – 1 – gw Misc Inst [690]
Baugey, Georges *see* De la condition legale du culte israelite en france et en algerie
Baugh, Folliott *see* Almsgiving
Baugher, Henry Louis *see* Annotations on the gospel according to st luke
Baughman, Robert Williamson *see* Post offices
Baughman-bachman quaterly – 1987 jan-1989 oct – 1 – mf#1616261 – us WHS [071]
Baugy, Louis Henri *see* Journal d'une expedition contre les iroquois en 1687
Bauhofer, Janos Gyoergy *see* History of the protestant church in hungary from the beginning of the reformation to 1850
Die bauhuette : illustrirte freimaurerzeitung / ed by Findel, J G – Leipzig 1858-1932 [mf ed 2002] – 74v on 383mf [ill] – 9 – €1780 diazo €2136 silver – 3-89131-389-6 – (pref by ed; zeitschrift für deutsche freimaurerei) – gw Fischer [360]
Bauingenieur – Heidelberg, Germany 1982 – 1,5,9 – ISSN: 0005-6650 – mf#13142 – us UMI ProQuest [690]
Die bauinschriften sanheribs / Sennacherib, King of Assyria; ed by Meissner, Bruno & Rost, Paul – Leipzig: Eduard Pfeiffer, 1893 – 1mf – 9 – 0-8370-7739-7 – (incl indes. texts in german and akkadian; commentary in german) – mf#1986-1739 – us ATLA [470]
Ein baukran stuerzt um : berichte aus der arbeitswelt / ed by Bredthauer, Karl D et al – Muenchen: Piper, c1970 – 1r – 1 – us Wisconsin U Libr [430]
Die baukunst betreffend : sammlung nuetzlicher aufsaetze und nachrichten – Berlin, 1797-1806. 6v – 25mf – 9 – mf#0A-109 – ne IDC [720]
Baulaender bote und boxberger anzeiger – Adelsheim DE, 1914 27 jul-1918 21 sep – 2r – 1 – gw Misc Inst [074]
Baulin, A G *see* Partiia o promyslovoi kooperatsii i kustarnoi promyshlennosti
Bauls – v1-2 n1 [1969 jan 6-sep 24] – 1 – mf#764799 – us WHS [071]
Baum, A *see* Magistrat und reformation in strassburg bis 1529
Baum, Adolf *see* Magistrat und reformation in strassburg bis 1529
Baum, Ernst *see* Philipp hafners gesammelte werke
Ein baum im odenwald : novelle / Roquette, Otto – Breslau: S Schottlaender, 1884 – 1r – 1 – us Wisconsin U Libr [830]
Baum, J C *see* Premiere liturgie des eglises reformees de france
Baum, Johann Wilhelm *see*
– Capito und butzer, strassburgs reformatoren
– Franz lambert von avignon
– The odor beza
Baum, Kurt *see* Am leben entlang
Baum, Oskar *see* Das volk des harten schlafs
Baum, William M *see* The essays, debates, and proceedings
Bauman, Irwin Wiegner *see* Der kampf der giessener theologischen fakultaet gegen zinzendorf und die bruedergemeine, 1740-1750
Bauman, K I *see* Khlebnaia selskokhoziaistvennaia kooperatsiia i sotsialisticheskoe pereustroistvo krestianskogo khoziaistva
Bauman, Mara J *see* The effect of exercise training on fasting blood glucose levels in adolescents
Baumann, Eberhard *see* Der aufbau der amosreden
Baumann, Franz Ludwig *see*
– Quellen zur geschichte des bauernkrieges aus rotenburg an der tauber
– Quellen zur geschichte des bauernkriegs aus rotenburg an der tauber
– Quellen zur geschichte des bauernkriegs in oberschwaben
Baumann, Hans *see*
– Alexander
– Atem einer floete
– Das heimliche haus
– Der kreterkoenig
– Der wandler krieg
– Wir zuenden das feuer
Baumann, Johannes *see* Mathematische verfahren zur bewertung von standort- und auslieferungsalternativen im handel
Baumann, Julius *see*
– Die gemutsart jesu
– Ueber religionen und religion
Baumann, Klaus-Dieter *see*
– Integrative fachtextsortenstilistik
– Leipziger arbeiten zur fachsprachenforschung
– Sozio-semantische besonderheiten historiographischer fachtexte des englischen
Baumann, Markus *see* Bestimmung der molekularen hyperpolisierbarkeit organischer materialien mit der methode des selbstbeugungseffektes
Baumann, Michael A *see* Veraenderung der entladungsmuster der verschiedenen phasentypen medullaerer respiratorischer neurone durch acetylcholin sowie spezifische agonisten und antagonisten
Baumann, Simone *see*
– Die entwicklung der dermato-venerologie an der fakultaet/dem bereich medizin der karl-marx-universitaet von 1945 bis 1975
– Die wirksamkeit walter kruses als direktor des hygieninstitutes an der leipziger universitaet von 1913-1933/34
Baumannn, Frank *see* Roentgen- und photoelektronenspektroskopische untersuchungen von verbindungen der 3d-uebergangsmetalle und dem 5d-uebergangsmetall platin
Baumbach, Friedrich August *see* Douze romances
Baumbach, Johann Carl [comp] *see* Sammlung von clavier und singe-stuecken verschiedener verfertiger
Baumbach, Rudolf *see*
– Abenteuer und schwaenke
– Ausgewaehlte maerchen und gedichte
– Es war einmal
– Frau holde
– Horand und hilde
– Kaiser max und seine jaeger
– Krug und tintenfass
– Lieder eines fahrenden gesellen
– Mein fruehjahr
– Der pathe des todes
– Sommermaerchen
– Von der landstrasse
Baumgaertel, Friedrich *see* Elohim ausserhalb des pentateuch
Baumgaertner, Karl Heinrich *see* Kranken-physiognomik
Baumgard, Otto Wilhelm Gustav *see* Gutzkows dramaturgische taetigkeit am dresdener hoftheater
Baumgart, Gertrud *see* Goethes lyrische dichtung in ihrer entwicklung und bedeutung
Baumgart, Hermann *see*
– Goethes faust als einheitliche dichtung
– Goethes "geheimnisse" und seine "indischen legenden"
– Goethes lyrische dichtung in ihrer entwicklung und bedeutung
– Goethe's maerchen
Baumgart, Wolfgang *see* Der wald in der deutschen dichtung
Baumgartel, Elise J *see* The cultures of prehistoric egypt
Baumgarten, Alexander Gottlieb *see* Initia philosophiae practicae primae acroamatice
Baumgarten, Dietrich *see* Deutsche finanzpolitik
Baumgarten, Franz Ferdinand *see* Das werk conrad ferdinand meyers
Baumgarten, Fritz *see* Der wilde graf (wilhelm von fuerstenberg) und die reformation im kinzigthal
Baumgarten, Harald *see* Der grillenpfiff
Baumgarten, Hermann *see*
– Jacob sturm
– Karl 5. und die deutsche reformation
– Vor der bartholomaeusnacht
Baumgarten, Johann *see* Die komischen mysterien des franzoesischen volkslebens in der provinz
Baumgarten, M *see* Die apostelgeschichte
Baumgarten, Michael *see*
– The acts of the apostles
– Die aechtheit der pastoralbriefe
– Doctrina iesu christi de lege mosaica ex oratione montana
– Die nachtgesichte sacharias
Baumgarten, Otto *see*
– Neue bahnen
– Volksschule und kirche
Baumgarten, Otto et al *see* Unsere religioesen erzieher
Baumgarten, Paul Maria *see* Die vulgata sixtina von 1590 und ihre einfuehrungsbulle
Baumgartner, Alexander *see*
– Calderon
– Goethe
– Lessing's religioeser entwicklungsgang
– Longfellow's dichtungen
Baumgartner, Amy *see* Factors that influence division 2 recruited female intercollegiate soccer student-athletes in selecting their university of their choice
Baumgartner, Antoine Jean *see* Introduction a l'etude de la langue hebraique
Baumgartner, Anton Jean *see* Calvin hebraisant et interprete de l'ancien testament
Baumgartner, Gabriele [comp] *see*
– Polskie archiwum biograficzne
– Polskie archiwum biograficzne (pab1). supplement
– Polskie archiwum biograficzne. seria nowa
– Polskie archiwum biograficzne. seria nowa (pab2) supplement
Baumgartner, Matthias *see*
– Die erkenntnislehre des wilhelm von auvergne
– Die philosophie des alanus de insulis
Baumgartner, Renee M *see* Intercollegiate athletics and organizational culture
Baumgartner, Ted A *see*
– Descriptive and predictive discriminant analysis of the golf ability of college males
– Physical activity patterns and characteristics of high school students in a governor's honors program
Baumholder community news – 1980 feb 1-1981 aug 24 – 1 – mf#1476961 – us WHS [071]
Baumholder gig sheet – v1 n1 – 1 – mf#720856 – us WHS [071]
Baumholder Military Community *see*
– Champion times
– Community news
Baumkoller, A *see* Le mandat sur la palestine
Baumstark, A *see*
– Abendlaendische palaestinapilger des ersten jahrtausends und ihre berichte
– Die christlichen literaturen des orients
– Das palimpsestsakramentar im cod aug 112
Baumstark, Anton *see*
– Abendlaendische palaestinapilger des ersten jahrtausends und ihre berichte
– Die christlichen literaturen des orients
– Denkmaeler der entstehungsgeschichte des byzantinischen ritus
– Festbrevier und kirchenjahr der syrischen jakobiten
– Die konstantinopolitanische messliturgie vor dem 9 jahrhundert
– Liturgia romana e liturgia dell'esarcato
– Die petrus- und paulusacten in der litterarischen ueberlieferung der syrischen kirche
Baunard *see* The odulfe
Baunard, Louis *see*
– Histoire de la venerable mere madeleine-sophie barat
– Histoire de mme duchesne
– Theodulfe
Baunard, Louis, Abbe *see* The life of mother duchesne
Baur, A *see* Zwinglis theologie, ihr werden und ihr system
Baur, Albrecht [comp] *see* Schleiermacher's christliche lebensanschauungen
Baur, August *see*
– Johann calvin
– Zwinglis theologie
Baur, F C *see* Die tuebinger schule und ihre stellung zur gegenwart
Baur, Ferdinand Christian *see*
– Apollonius von tyana und christus
– Das christenthum und die christliche kirche der drei ersten jahrhunderte
– Das christliche des platonismus, oder, sokrates und christus
– Die christliche gnosis, oder, die christliche religions-philosophie in ihrer geschichtlichen entwiklung
– Die christliche kirche des mittelalters in den hauptmomenten ihrer entwicklung
– Die christliche kirche vom anfang des vierten bis zum ende des sechsten jahrhunderts in den hauptmomenten ihrer entwicklung
– Die christliche lehre von der dreieinigkeit und menschwerdung gottes in ihrer geschichtlichen entwicklung
– Die christliche lehre von der versoehnung
– The church history of the first three centuries
– De ebionitarum origine et doctrina
– Das dogma der alten kirche
– Das dogma der neueren zeit
– Das dogma des mittelalters
– Drei abhandlungen zur geschichte der alten philosophie und ihres verhaeltnisses zum christenthum
– Die epochen der kirchlichen geschichtschreibung
– Der gegensatz des katholicismus und protestantismus
– Die ignatianischen briefe und ihr neuester kritiker
– Kirchengeschichte der neueren zeit
– Kritische untersuchungen ueber die kanonischen evangelien
– Lehrbuch der christlichen dogmengeschichte
– Das manichaeische religionssystem
– Das markusevangelium nach seinem ursprung und charakter
– Paul
– Paul, the apostle of jesus christ
– Paulus, der apostel jesu christi
– Die sogenannten pastoralbriefe des apostels paulus
– Die tuebinger schule und ihre stellung zur gegenwart
– Ueber den ursprung des episcopats in der christlichen kirche
Baur, Ferdinand Friedrich *see*
– Die christliche kirche des mittelalters in den hauptmomenten ihrer entwicklung
– Kirchengeschichte der neueren zeit
Baur, Gustav *see*
– Die berechtigung der theologie als eines nothwendigen gliedes im gesamtorganismus der wissenschaft
– Geschichte der alttestamentlichen weissagung, theile 1
Baur, Johannes *see* Giovanni gentiles philosophie und paedagogik
Baur, L *see*
– Gundissalinus' de divisione philosophiae
– Die philosophie des robert grosseteste bischofs von lincoln
– Die philosophische werke des robert grosseteste, bischofs von lincoln
Baur, Ludwig *see* De divisione philosophiae
Baur, Ludwig et al *see* Roger bacon
Baur, Thorsten *see* Cost benchmarking als instrument des kostenmanagements
Baur, Tobias *see*
– Dimensionen der alterung im sozialstaat
Baur, Wilhelm *see*
– Religious life in germany during the wars of independence
– Der sonntag und das familienleben
Bauschinger, Julius *see* Annalen der kaiserlichen universitaets-sternwarte in strassburg
Bauschinger, Julius et al *see* Veroeffentlichungen der universitaetssternwarte zu leipzig
Baussern, Waldemar von *see* Peter cornelius (1824-1874) musical works
Bausset, Louis F de *see* Memoires anecdotiques sur l'interieur du palais et sur quelques evenemens de l'empire
Bausteine aus dem or[ient] naumburg a/s / ed by Haage, F W – Naumburg: H Sieling [1855?] – 3mf – 9 – mf#vrl-41 – ne IDC [366]
Bausteine zur deutschen literaturgeschichte : aeltere deutsche dichtung / Becker, Hendrik – Halle (Saale): M Niemeyer, 1957 [mf ed 1993] – ix/401p/[160]pl – 1 – mf#8072 – us Wisconsin U Libr [430]
Bautain, Louis *see* The art of extempore speaking
El bautismo : renovacion del rito... / Bautismo, El – Caceres: Tip. Extremadura, 1978 – sp Bibl Santa Ana [920]
Bautismo de los fetos abortivos...cesarea / Riva, Juan Antonio de la – 1817 – 9 – sp Bibl Santa Ana [240]
Bautismo, El *see* El bautismo
Bautz, Joseph *see* Weltgericht und weltende
Bautzener geschichtsblaetter – Bautzen DE, 1909-13; 1925-30 – 1r – 1 – gw Misc Inst [074]
Bautzener nachrichten *see* Budissinische woechentliche nachrichten
Bautzener tageblatt – Bautzen DE, 1898 3 jan-1945 18 apr – 90r – 1 – gw Misc Inst [074]
Bauxite bulletin – Weipa, jun 1966-dec 1985 – 10r – at Pascoe [079]
Bauza, Guillermo *see*
– Con los brazos abiertos
– Don cristobal
– Filo del ensueno
Bauza, Obdulio *see*
– Casa solariega
– Hogueras de cal
– Voces esperadas
Les bavards – Port-au-Prince: Paulin Andreoli. [1ere annee n1-2e annee n14]. (21 dec 1876-22 mars 1877) – 1 sheet – us CRL [079]
Bavaria (Germany). Bayerische Staatskanzlei *see* Dokumente zum aufbau des bayerischen staates
Bavaria. Laws, Statutes, etc *see* Gesetz uber die verwaltungsgerichtsbarkeit in bayern, wurttemberg-baden und hessen, mit kommentar von paulus van husen
Bavaria. Statistisches Landesamt *see*
– Beitraege zur statistik bayerns
– Statistisches jahrbuch fur das koenigreich bayern
Bavinck, H *see* De ethik van ulrich zwingli
Bavinck, Herman *see*
– De ethiek van ulrich zwingli
– Johannes calvijn
– The philosophy of revelation
– De welsprekendheid

The bawan akhari : or, guru arjan's alphabet / Macauliffe, Max – [S.l.: s.n., 19–?] – 1mf – 9 – 0-524-07490-9 – mf#1991-0111 – us ATLA [280]

Bawenda of the spelonken (transvaal) / Wessmann, R – New York, NY. 1969 – 1r – us UF Libraries [960]

Bawl street journal – v45 n1 [1972 jun 2]; v9-11 [1933-35] – 1 – mf#1017597 – us WHS [071]

Bawr, Alexandrine Sophie Goury De Champgrand *see* Suite d'un bal masque, comedie en un acte et en pr...

Bax, Ernest Belfort *see*
- Jean paul marat
- The peasants war in germany, 1525-1526
- Rise and fall of the anabaptists

Baxmann, Rudolf *see* Die politik der paepste von gregor 1. bis auf gregor 7.

Baxmann, Rudolf et al *see* Vortraege fuer das gebildete publikum. dritte sammlung

Baxter, Elizabeth *see* The story of the kurku mission

Baxter, Garrett *see* Political economy

Baxter, James Phinney *see*
- Early voyages to america
- Introduction of the ironclad warship
- The pioneers of new france in new england
- What caused the deportation of the acadians?

Baxter, John *see* Protestant assertions examined and refuted

Baxter, John Babington Macaulay [comp] *see* Historical records of the new brunswick regiment, canadian artillery

Baxter, Lucy E (Barnes) [pseud: Leader Scott] *see*
- The cathedral builders
- The renaissance of art in italy

Baxter, Matthew *see* Methodism

Baxter, Michael Paget *see* Coming wars

Baxter, Richard *see*
- The life of rev. richard baxter, a.d. 1615-a.d. 1691
- The reformed pastor
- The saints' everlasting rest

Baxter, Robert *see*
- The irish tenant-right question examined by a comparison of the law and practice of england with...ireland
- Irvingism

Baxter, Samuel John *see* The development of a ministry at the immanuel baptist church, toronto, canada, to integrate multicultural peoples

The baxter treatises : from the papers of richard baxter (1615-91) at dr williams' library london – London (mf ed 2000) – 6r – 1 – £310.00 – (catalogue by roger thomas incl) – mf#DWB – Dist. US: us UMI ProQuest – uk World [941]

Baxter, William *see*
- Life of elder walter scott
- Life of knowles shaw, the singing evangelist

Baxter, William Edward *see* America and the americans

Baxter, Wl *see* Sabbath not for the jew but for man

Baxter's contested election cases / New York. (State) – 1v. 1777-1899 (all publ) – 9mf – 9 – $13.50 – mf#LLMC 80-017 – us LLMC [340]

Bay area free press – v1 n1-8 [1969 nov 11-1970 mar 10] – 1 – mf#1053013 – us WHS [071]

Bay area health liberation news / Medical Committee for Human Rights (US) – 1972 mar/apr-1975 feb – 1r – 1 – (cont: health liberation news bay area) – mf#937290 – us WHS [360]

Bay area painters news – v2 n9-v16 n11 [1967 mar 31-1984 nov] – 1 – mf#347415 – us WHS [071]

Bay Area Radical Teachers' Organizing Collective *see* No more teachers dirty looks

Bay area report – 1982 jun – 1 – mf#4868237 – us WHS [071]

Bay area socialist – n16; 1968 feb-1974 sep – 1 – mf#347413 – us WHS [071]

Bay Area Typographical Union No 21 *see* Northern california labor with typographical bulletin official bulletin of bay area typographical union no 21

Bay area women's news and community calendar – v1 n2-v2 n1 [1987 may/jun-1988 mar/apr] – 1 – mf#1546317 – us WHS [071]

Bay area worker – v2 n9-v4 n2 [1972 apr-1975 jun] – 1 – mf#228976 – us WHS [071]

Bay banner – v2 n3-v9 n22 [1953 feb 4-1961 oct] – 1 – mf#599779 – us WHS [071]

Bay breeze – 1987 oct-1994 dec; 1989 oct-1990 sep – 1 – mf#1110631; 1653931 – us WHS [071]

Bay bugle – Paihia NZ. nov 1981-dec 1983 [fortnightly] – 1r – 1 – mf#12.18 – nz Nat Libr [079]

Bay Cities Metal Trades and Industrial Union Council *see* Labor review

Bay city chronicle – Bay City OR: A N Merrill, [wkly] [mf ed 1968] – 1r – 1 – us Oregon Lib [071]

Bay city press – 1860 jun 30-1862 apr 19 – 1 – mf#918759 – us WHS [071]

Bay city tribune – Bay City OR: J S Dellinger, 1891- [wkly] – 1 – us Oregon Lib [071]

Bay county beacon tribune – Pensacola, FL. 1918 nov 15-1926 dec – 2r – (gaps) – us UF Libraries [071]

Bay County Genealogical Society [MI] *see* Chips and ships

Bay county herald – Pensacola, FL. 1931 aug 27-1934 – 3r – (gaps) – us UF Libraries [071]

Bay di andn / Blumstein, Isaac – Mendosa, Argentina. 1935 – 1r – us UF Libraries [939]

Bay di bregen fun dnyester / Feigenberg, Rachel – Varsha, Poland. 1925 – 1r – us UF Libraries [939]

Bay guardian – v1 n1-v2 n11 [1966 oct 27-1968 jun 18] – 1 – mf#1097747 – us WHS [071]

Bay guild news : official publication of the san francisco-oakland newspaper guild – v34 n4-v42 n10 [1977 jun-1985 oct] – 1 – mf#946528 – us WHS [071]

Bay guildsman : official publication of the san francisco-oakland newspaper guild – v32 n1-v34 n3 [1975 jan-1977 apr/may] – 1 – mf#946517 – us WHS [071]

Bay leaves – 1939 sep 14-1941 jun 12, (1941 jun 19-1942 oct) [1]; (1934 aug 23-1938 jan) [2]; 1947 jan 23-1949 mar 10; 1949 mar 17-1950 aug 24; 1950 aug 31-1951 dec 13; 1952 jan 7-1954 may 20 [3] – 1 – mf#943976 [1]; 943916 [2]; 943978 [3] – us WHS [071]

Bay leaves and lake geneva observer – [1938 jan-1939 jun 15] – 1 – mf#943918 – us WHS [071]

Bay nakht oyfn alten mark / Peretz, Isaac Leib – Varshe, Poland. 1909 – 1r – us UF Libraries [939]

Bay news – Coos Bay OR: W C Swanson, 1985- [wkly] – 1 – us Oregon Lib [071]

Bay of plenty beacon – Whakatane, NZ. 5 aug-20 nov 1964, 25 nov 1964-31 mar 1965, 19 nov 1965-30 apr 1966, may-jun 1973, aug-nov 1973, jan 1974-apr 1975, jul 1975-jun 1981, jan 1982-dec 1989 – 94r – 1 – (title changes to: whakatane beacon fr may 1973) – mf#16.8 – nz Nat Libr [079]

Bay of plenty farmer – Tauranga, NZ. 1982-84 – 1r – 1 – mf#16.31 – nz Nat Libr [079]

Bay of plenty hometalk – oct 1980-mar 1981 – 1r – 1 – mf#16.32 – nz Nat Libr [079]

Bay of plenty mirror – Whakatane, NZ. oct 1972-may 1973; jul-dec 1973; jan-apr 1974; jul-aug 1974; 1-11 oct 1974 – 1 – mf#16.12 – nz Nat Libr [079]

Bay of plenty times – Tauranga, NZ. nov 1875-1886; 2 jan 1963-10 apr 1963; 30 nov 1963-4 mar 1964; 8 dec 1965-29 jan 1966; 16 feb-15 may 1971; 2 jan-31 mar 1973; may 1974-feb 2002 – 1 – mf#16.5 – nz Nat Libr [079]

Bay pilot – St Andrews, NB. 1878-89 – 4r – 1 – cn Library Assoc [971]

Bay reporter – Coos Bay OR: Empire Pub Co, [wkly] – 1 – (cont: empire builder (1977-78)) – us Oregon Lib [071]

Bay state banner – 1968 aug 15-29, oct 10-1969 dec 11; 1969 dec 18-1971 feb 25; 1971 mar 4-1972 dec 28; 1973 dec 6-1974 oct 31; 1973 jan 4-1973 nov 29; 1974 nov 7-1975 jun 26 – 1 – mf#501653 – us WHS [071]

Bay state employee – 1977 may-1979 jun/jul; 1979 aug-1984 jun/jul; 1985 jan-nov/dec – 1 – mf#499189 – us WHS [071]

Bay state librarian – Wakefield MA 1970-94 – 1,5,9 – ISSN: 0005-6944 – mf#5956 – us UMI ProQuest [020]

Bay sun – jul 1983-apr 1988 – 1 – (incl mount extra) – mf#16.21 – nz Nat Libr [079]

Bay sun – Tauranga, NZ. nov 1975-jun 1983 – 43r – 1 – (incl mount extra fr jul 1983; aka: bay sun spree) – mf#16.21 – nz Nat Libr [079]

Bay sun *see* Bay sun spree

Bay sun spree – Tauranga, NZ. jan-jun 1986 – 1r – 1 – (aka: bay sun) – mf#16.38 – nz Nat Libr [079]

Bay view observer – 1937 jan 14-1939 dec 21 – 1 – mf#1166065 – us WHS [071]

Bay viewer – 1987 aug 6-1997 oct-dec – 1 – mf#1212416 – us WHS [071]

Ba-yamim ha-hem : sipuro shel zaken / Steinberg, Judah – Berlin, Germany. 1923 – 1r – us UF Libraries [939]

Bayan 'amaliyat raqm : communiques of military operations / Jabhah al-Sha'biyah li-Tahrir Filastin – [Amman, Jordan?]: Jabhah al-Sha'biyah li-Tahrir Filastin, [50-270 (1968-1969)] (irreg) – 1r – 1 – us CRL [320]

Bayan ul-haq – Kazan, Russia 1912-14 [mf ed Norman Ross] – 14r – 1 – ISSN: nrp-663 – us UMI ProQuest [077]

Bayard de LaVingtrie, Ferdinand M *see*
- Voyage dans l'interieur des etats-unis a bath, winchester, dans la vallee de shenandoha, etc
- Voyage de terracine a naples

Bayard, James A *see* Papers

Bayard, Jean-Francois-Alfred *see*
- Capitaine charlotte
- Changement de main
- Couleurs de marguerite
- Deux couronnes
- Deux font la paire
- Etourneau
- Fils de famillle
- Gamin de paris
- Gants jaunes
- Mademoisellemimi pinson
- Mari a la campagne
- Marie mignot
- Marquise de carabas
- Mathilde, ou, la jalousie

Bayard, M *see*
- Reine de seize ans
- Reveil du lion

Bayard, M (Jean-Francois-Alfred) *see*
- Petit-fils
- Phoebus
- Premieres armes de richelieu

Bayard, Richard H *see* Papers

The bayard rustin papers / ed by Bracey, John H Jr & Meier, August – 22r – 1 – $3125.00 – 1-55655-064-2 – (with p/g) – us UPA [320]

Bayard transcript – Bayard, NE: Transcript Pub Co. v4 n23. may 27 1892 (wkly) [mf ed feb 17 1893- (gaps) filmed 1969] – 1 – (cont: chimney rock transcript) – us NE Hist [071]

Bayard transcript – Bayard, NE: Transcript Pub Co (wkly) – 28r – 1 – (cont: chimney rock transcript) – us Bell [071]

Bayard, William *see*
- An address delivered at the opening of the training school for nurses
- An address on water in relation to disease
- Address upon the progress of medical science
- An address upon the progress of medicine, surgery and hygiene, during the last 100 years
- An address upon the use and abuse of alcoholic drinks
- History of the general public hospital in the city of saint john, nb
- Presidential address at the meeting of the maritime medical association

Bayer, Edmund *see*
- Danielstudien
- Das dritte buch esdras

Bayer, Emily C *see* The compilation of items and calibration for a survery of infant motor behavior

Bayer, Josef *see* Studien und charakteristiken

Bayer, Maximilian *see* Die helden der naukluft

Bayer, Robert von *see*
- Auf abschuessiger bahn
- Auf der station
- Der kampf um's dasein
- Nomaden
- Sesam
- Sphinx
- Der weg zum glueck

Bayerisch land und bayerisch volk / Schlicht, Josef – Muenchen: M Hullter 1875 [mf ed 1997] – 1r – 1 – mf#film mas 26815 – us Harvard [914]

Bayerisch land und volk : offizielles organ des landesverbandes zur hebung des fremdenverkehrs in bayern / Landesverband zur Hebung des Fremdenverkehrs in Bayern – Muenchen: W Braun 1890-92 [mf ed 1978] – 3v on 1r [ill] – 1 – (imprint varies) – mf#film mas c331 – us Harvard [914]

Die bayerische abendmahlsgemeinschaftsfrage : ein anfang eingehenderer eroerterung / Delitzsch, Franz – Erlangen: Theodor Blaesing, 1852 – 1mf – 9 – 0-8370-9611-1 – (incl bibl ref) – mf#1986-3611 – us ATLA [240]

Bayerische arbeiter-zeitung *see* Die perspektive

Bayerische bibliothek – Bamberg: Buchnersche Verlagsbuchhandlung 1890-92 [mf ed 1978] – 29v on 2r [ill] – 1 – mf#film mas c280 – us Harvard [914]

Bayerische blaetter fuer das bayerische gymnasialschulwesen – Muenchen: Lindauersche Universitaetsbuchhandlung 1917-35 (qrtly) [mf ed 1978] – 19v on 17r [ill] – 1 – (cont: blaetter fuer das gymnasial-schulwesen; iss by: bayerischer gymnasiallehrerverein 1917-24; and: verein bayerischer philologen 1925-35; imprint varies) – mf#film mas c396 – us Harvard [373]

Der bayerische eilbote – Muenchen DE, 1848 jul-1851 – 3r – 1 – gw Misc Inst [074]

Bayerische israelitische gemeinde zeitung – Muenchen DE, 1926-37 [gaps] – 1 – gw Misc Inst [270]

Bayerische israelitische gemeindezeitung : nachrichtenblatt der israelitischen kultusgemeinden muenchen, augsburg, bamberg, und des verbandes bayerischer israelitischer gemeinden – Munich: B Heller. v1-14. 1925-38 – (= ser German-jewish periodicals...1768-1945, pt 3) – 5r – 1 – $770.00 – mf#B426 – us UPA [939]

Bayerische landboetin – Muenchen DE, 1848-49 – 1r – 1 – gw Misc Inst [074]

Bayerische landeszeitung 1949 – Muenchen DE, 1949 21 jan-1951 jun – 1 – (title varies: 28 jan 1951: muenchner allgemeine) – gw Misc Inst [074]

Bayerische ostmark : bayerische grenzmarkzeitung marktredwitz – Marktredwitz DE, 1936 25 may-1938 2 jan – 3r – 1 – (subtitle varies: 1 oct 1936: tageszeitung fuer marktredwitz und wunsiedel; 16 nov 1936: tageszeitung fuer marktredwitz, wunsiedel, arzberg; 1 nov 1937: marktredwitzer tagblatt, bayerische ostmark. fichtelgebirgs-kurier, tageszeitung fuer marktredwitz, wunsiedel, arzberg, fuer das fichtelgebirge, den steinwald und die angrenzende oberpfalz) – gw Misc Inst [074]

Bayerische ostmark – Selb DE, 1978 2 oct- – ca 8r/yr – 1 – (filmed by other misc inst: 1934 1 oct-1935 30 mar, 1935 1 oct-31 dec, 1936 1 feb-1939 31 may, 1939 1 jul-1941 30 nov, 1948 4 dec-1972 (local pp), 1974-83 (local pp), 1996, 1998-1999 19 feb; until 1 nov 1935 incl: selb, rehau, wunsiedel, marktredwitz, subtitle: 2 nov 1935-3 feb 1936: selber tagblatt, between 4 feb 1936-30 nov 1941 incl: kreis selb, title varies: 4 dec 1948: frankenpost, 2 jan 1984: selber tagblatt, regional ed of frankenpost, hof) – gw Misc Inst [074]

Bayerische ostmark *see*
- Fraenkisches volk
- Fraenkisches volk [main edition]

Bayerische ostwacht – Landshut DE, 1933 2 oct-1944 18 dec – 1 – (main ed in bayreuth; title varies: 1 oct 1934: bayerische ostmark, 1 aug 1942: landshuter rundschau, 1 mar 1943: landshuter kurier) – gw Misc Inst [074]

Bayerische ostwacht *see* Fraenkisches volk [main edition]

Die bayerische presse – Wuerzburg DE, 1849 1 dec-1851 8 feb – 2r – 1 – gw Misc Inst [074]

Bayerische radio-zeitung und bayernfunk – Muenchen DE, 1924 3 aug-1940 22 dec – 24r – 1 – (title varies: 1 jan 1925: sueddeutscher rundfunk / a, 1 may 1928: bayerische radiozeitung; with suppls) – gw Mikrofilm [380]

Bayerische rundschau – Kulmbach DE, 1977- – ca 7r/yr – 1 – gw Misc Inst [074]

Bayerische staatszeitung *see* Muenchener politische zeitung

Bayerische staatszeitung 1913 – Muenchen DE, 1913-1934 30 jun, 1946 1 jun-1999 – 91r – 1 – gw Mikrofilm [350]

Das bayerische vaterland *see* Das bayrische vaterland

Der bayerische volksfreund – Muenchen DE, 1848-49 – 1r – 1 – (with suppl) – gw Misc Inst [074]

Bayerische volkspartei-korrespondenz – Muenchen DE, 1921 30 may-1923 18 nov – 5r – 1 – mf#5375 – gw Mikropress [325]

Bayerische wochenschrift – Muenchen DE, 1859 2 apr-24 sep – 1r – 1 – gw Misc Inst [074]

Bayerischer kurier – Muenchen DE, 1918-21 – 8r – 1 – (filmed by bnl: 1916 3 sep-1922) – mf#5391 – gw Mikropress; uk British Libr Newspaper [074]

Bayerisches jahrbuch fuer volkskunde – Regensburg : Josef Habbel 1950- [mf ed Bloomington IN: Indiana Uni Lib, Preservation Dept 1984] – (= ser Beitraege zur muenchener volkskunde) – 1 – (aka: beitraege zur muenchener volkskunde 1958) – us Indiana Preservation [300]

Bayerisches landesprivatrecht / Oertman, Paul – Halle (Saale): Waisenhaus, 1903 – (= ser Civil law 3 coll; Archiv fuer theorie und praxis des allgemeinen deutschen handels- und wechselrecht) – 8mf – 9 – mf#LLMC 96-573 – us LLMC [346]

Bayerisches volksblatt – (Regensburg-) Stadtamhof DE, 1849 1 mar-30 dec – 1 – (with suppl) – gw Misc Inst [074]

Das bayerland – Muenchen: R Oldenbourg. v1-37 [1890]-1990 [mf ed 1978] – 37v on 18r – 1 – (lacks: v30 n24.) – mf#film mas c333 – us Harvard [943]

Bayerle, Bernard Gustav *see*
- Das christliche alterthum
- Ein vollstaendiges leben jesu, seiner heiligen mutter maria, und der uebrigen heiligen seiner zeit

Bayern und der heilige stuhl in der ersten haelfte des 19. jahrhunderts : nach den akten des wiener nuntius severoli und der muenchener nuntien serra-cassano, mercy d'argenteau und viale-prela... / Bastgen, Hubert – Muenchen, 1940 (mf ed 1993) – 7mf – 9 – 3-89349-253-4 – mf#DHS-AR 107 – gw Frankfurter [240]

Bayern-2. parteitag / Sozialdemokratische Partei Deutschlands – Muenchen, 30 Sept-1 Oct 1894 – 1 – gw Mikropress [335]

Bayern-3. parteitag / Sozialdemokratische Partei Deutschlands – Nuernberg, 12-13 Jul 1896 – 1 – gw Mikropress [335]

Bayern-4. parteitag / Sozialdemokratische Partei Deutschlands – Wuerzburg, 30-31 Oct 1898 – 1 – gw Mikropress [335]

Bayern-7. parteitag / Sozialdemokratische Partei Deutschlands – Augsburg, 26-27 Jun 1904 – 1 – gw Mikropress [335]

BAYERN-8

Bayern-8. parteitag / Sozialdemokratische Partei Deutschlands – Schweinfurt, 4-5 Mar 1906 – 1 – gw Mikropress [335]
Bayern-13. parteitag / Sozialdemokratische Partei Deutschlands – Neustadt a.H., 18-20 Jul 1914 – 1 – gw Mikropress [335]
Bayern-kurier – Muenchen DE, 1950 3 jun-1965 – 7r – 1 – (filmed by other misc inst: 1969 5 jul-20 dec, 1975 4 jan-29 nov) – mf#12037 – gw Mikropress; gw Misc Inst [074]
Der bayersche landbote *see* Der baierische landbote
Bayersches volksblatt – Wuerzburg DE, 1829-32 – 2r – 1 – (title varies: 1830: bayerisches volksblatt) – gw Misc Inst [074]
Der bayerwald-bote – Regen DE, 1983 1 jun – ca 9r/yr – 1 – (bezirksausgabe von passauer neue presse, passau) – gw Misc Inst [074]
Bayerwald-echo – Cham DE, 1977- – ca 12r/yr – 1 – (bezirksausgabe von mittelbayerische zeitung, regensburg) – gw Misc Inst [074]
Bayet, Albert *see* La morale laique et ses adversaires
Bayete : cronicas africanas do atlantico ao indico / Rocha, Hugo – [O Porto?: Oficinas Graficas do Comercio do Porto] 1933 – 1 – (filmed with: barahona e costa, henrique cesar da silva: apontamentos para a historia de guerra de zambezia 1871-75 et al) – us CRL [960]
Bayeux, Adolphe Auguste *see* Nos aieux
Bayfield county journal – 1978 feb 9-jul 20 – 1 – mf#955080 – us WHS [071]
Bayfield county press – 1980 jun-dec-1989 [1]; 1882 nov 25-1936 aug 13 (with gaps) [2] – 1 – mf#1005098 [1]; 1005095 [2] – us WHS [071]
Bayfield, Henry Wolsey *see* Directions de navigation pour l'ile de terreneuve et la cote du labrador et pour le golfe et le fleuve st-laurent
Bayfield mercury – 1857 apr 18-sep 5; 1857 aug 22 – 1 – mf#1214410 – us WHS [071]
Bayfield press – 1870 oct 13-1872 jun 15 [1]; 1877 jun 13-1881 mar; 1878 oct 23-1879 may 21; 1881 jul-1882 nov 18 [2] – 1 – mf#1214379 [1]; 1005093 [2] – us WHS [071]
Bayfield progress – 1909 jun 3-11; 1912-1927 mar 30 – 1 – mf#1214409 – us WHS [071]
Bayford, Augustus Frederick *see* The judgment of the right hon. stephen lushington, d.c.l...
Baygram – v9 n6-8,10-12,14,17-18,23 [1983 mar 15-apr 15, may 15-jun 15, jul 15, sep 1-15, dec 1]; v10 n1,15 [1984 jan 1, aug 1]; v11 n1-3,5,7-10,12,14-16,19-1924 [1985 jan 1-feb 1, mar 1, apr 1-may 15, jun 15, jul 15-aug 15, oct 1-dec 15] – 1 – mf#1728122 – us WHS [071]
Baykara, Sultan Hueseyin *see* The divan project
Bayle, Constantino *see*
– 2nd congreso de geografia e historia hispano-americanas
– El 4 centenario del descubrimiento de california
– El 4th centenario de la fundacion de lima
– El 26th congreso internacional de americanistas
– A caza de testamentos. una pieza mayor
– A los lectores antiguos y nuevos de "razon y fe"
– A orillas del orinoco y a orillas del tamesis
– Abolicion oficial del laicismo en las escuelas
– Actas del cabildo de caracas
– Acuerdos del extinguido cabildo de buenos aires. serie 2, tomo 4 (1719 a 1722). buenos aires, 1927
– Acuerdos del extinguido cabildo de buenos aires. serie 2, tomo 9 y serie 3, tomo 9
– Adicion a la relacion descriptiva de los mapas planos...archivo general de indias
– Algo mas sobre las bulas alejandrinas
– Algunos puntos de historia acerca de la historia del coloniaje en el ecuador
– El alma cristiana de cortes. en el centenario de su muerte
– America en tiempo de felipe 2 segun el cosmografo cronista juan lopez de velasco. el territorio espanol de ifni
– El amor al libro...
– El ano pedagogico hispanoamericano
– Un apostol con temple de martin. la beata filipina duchense
– D'apres les paraboles histoires vraies...
– Apuntes de recuerdos
– Archivo de la nacion
– Archivo del general miranda
– Archivo general de la nacion
– Los argonautas ingleses de ultima hora
– Armaes do museu paulista, tomos 1 y 2
– Armes for red spain
– Arte en america y filipinas
– Balboa, descubridor del pacifico
– Belisario pena
– Bibliografia. historias del imperio lusitano
– Bibliografia jesuistica de mainas
– Bibliografia sobre las misiones de mainas. un misionero misionologo
– Biblioteca goathemala de la sociedad de geografia e historia...
– Biografia documentada
– Biografias. antonio maura, de taxonera, la emperatriz eugenia, de cabal y pedro de alvarado, de baron castro
– Boletin de historia americana
– Boletin de historias americanas
– Las borracheras y el problema de las conversaciones en indias
– Breve historia de mexico
– Browne, knight and a day. leipzig 1928
– Las bulas alejandrinas de 1493 referentes a las indias
– El caballo de batalla de los nuevos cruzados en la america espanola
– Cabildos de indios en la america espanola
– Cabrera, pablo...
– Campana del brasil. antecedentes coloniales. tomo 1
– La campana protestante en america
– Campana reaundada
– Campanas en el rif y gebala, del general berenguer
– El campo propio del sacerdote secular en la evangelizacion americana
– El caos de religiones nuevas
– Carabelas de espana...pinzon, juan de la cosa, diaz de solis juan sebastian elcano
– El caracter
– La carcel de mujeres de madrid
– Cardenal antonio caggiano, obispo de rosario. la figura de san francisco solano y su actuacion en el locuman...
– Un caso curioso de derecho y de anatomia mineral
– Las castas del mexico colonial...1924
– Los catalanes en grecia...
– Catalogo de documentos relativos a las islas filipinas existentes en el archivo de indias de sevilla
– Catalogo de los documentos relativo a las islas filipinas existentes en el archivo de indias de sevilla. tomo 7
– Catalogo de los documentos relativos a las islas filipinas...
– Catalogo de los fondos americanos del archivo de protocolos de sevilla...
– Catalogo de pasajeros a indias durante los siglos 16, 17 y 18. vol 2
– Centenario de don fray juan de zumarraga, el 4th
– El centenario de magallanes
– The church in spain
– Ciriaco perez bustamante, la fundacion de un imperio...
– Los clerigos y la extirpacion de la idolatria entre los neofitos americanos
– El clero secular y la evangelizacion de america
– Coleccion de diarios y relaciones para la historia de los viajes y descubrimientos. vol 1...vol 2...madrid, 1943
– Coleccion de documentos ineditos para la historia de iberoamerica
– Coleccion de documentos relativos al adelantado capitan don sebastian de belalcazar 1535-1560
– Coleccion general de documentos relativos a las islas filipinas
– Colon italiano? colon espanol?
– El coloniaje y sus detractores
– Communist attack in great britain, london 1938
– Como se alhajaban casas e iglesias en maynas
– Compania fundidora de fierro y acero de monterrey, s.a. ...
– Un complot terrorista en el siglo 15th. madrid, 1927
– La comunion entre los indios americanos
– Las comuniones en la espana roja
– El comunismo en espana. cinco anos en el partido
– La conferencia de lausana sobre fe y disciplina
– Congreso eucaristico de dublin
– El congreso mariano de sevilla
– El congreso y exposicion
– El convento de tepotzotlan...1924
– Corona funebre
– La coronacion de la virgen de guadalupe
– El corpus de los neofitos americanos
– Un corresponsal extranjero y unas teorias
– Cortes y la evangelizacion de nueva espana
– Cristobal colon
– Cronica del congreso eucaristico nacional
– La cruz en la conquista de america
– Cuando y donde se ordeno bartolome de las casas
– La cuestion religiosa en america
– La cuestion romana y el marques de comillas
– El culto a la eucaristia en la espana roja
– El cura de santa cruz
– El cura santa cruz
– Cura y mil veces cura. barcelona, 1928
– Damaso de la presentacion. vida del p. jose ma del montecarmelo
– De historias americanas
– Los demanes en la conquista de america, buenos aires, 1943
– Descubridores jesuitas del amazonas
– Descubridores jesuitas del amazonas, breve descripcion
– Documentos historicos coleccionados po...seccion geografia...
– Documentos para la historia argentina
– Documentos para la historia argentina. tomo 20, iglesia. cartas antiguas de la provincia del paraguay, avila y tucuman de la compania de jesus (1609-1614). buenos aires, 1927
– Don bosco
– Don constituyentes del ano 1824. biografias de don miguel ramos arizpe y d. lorenzo zavala, mexico. museo nacional de arqueologia, 1925
– Don pedro de alvarado, conquistador del reino de guatemala. madrid, 1927
– Don pedro de valdivia...badajoz, 1928
– Los dorados ingleses
– Dos testigos abrumadores contra los nacionales
– Duchaussois
– Eduardo posada
– Educacion de la mujer en america
– El dorado fantasma
– Elementos griegos y latinos que entran en la composicion de numerosos tecnicismos espanoles, franceses e ingleses
– El emmo cardenal goma
– En el pais de los eternos hielos
– En la catedral de toledo
– En plein confict
– Ensayo politico sobre el reino de la nueva espana
– Ensayos historicos
– La ensenanza de lenguas civilizadas a los barbaros. un caso de teologia pastoral misionera
– Entre dos filos
– Ernesto pinto. el santo del siglo
– Espana
– Espana ante la independencia de los estados unidos por el dr juan f...
– Espana en indias
– Espana en trento. 2. el concilio de trento en las indias espanolas
– Espana y el clero indigena de america
– Los espanoles y magallanes en la expedicion del estrecho
– El espiritu de santa teresa y el de san ignacio
– El espiritu genuino de falange espanola. es catolico?
– Espistolario de nueva espana 1505-1812 recopilado por francisco del paso y troncoso, tomo 16, mejico, 1942
– Estadistica de sangre y de gloria
– Estadisticas sangrientas
– Estado general de las fundaciones hechas por don jose escandon...tomos 1, 2 y 3
– Estados unidos mexicanos...
– La estela de un campesino
– El estrecho de magallanes
– Estudios hispanoamericanos
– El evangelio comentado conferencias por radio
– Excursion por el campo teosofico
– Excursion por el campo teosofico 2
– La exposicion internacional de barcelona
– La exposicion misional del vaticano
– La extirpacion de la idolatria en el peru del p. pablo joseph de arriaga
– Facultad de filosofia y letras. universidad de buenos aires. publicaciones historicas
– Felipe 2nd y la evangelizacion de america
– Figueiredo de estudios de historia americana
– Filantropia sospechosa
– La fin de l'empire espagnol d'amerique par marins andre
– First expedition of vargas into new mexico, 1692
– Fto vida y virtudes del venerable siervo de dios marcelino champagnat. barcelona, fto. 1929
– Los fundadores de bogota
– Gallions reach. leipzig, 1928
– Garcia moreno
– Garcia moreno y la instruccion publica por julio tobar donoso
– Geografia de espana...
– Geografia economica
– Gesnalda dello spirito sancto
– Guadalupe de extremadura en indias
– Hacia alla y para aca...
– Hernandez marcial
– Os herois de coaro e pirapo
– Le herpeur, m.l'oratoir de france...
– Los hijos de lutero entre los hijos del sol
– Histoire de notre dame de lourdes
– Histoire du cure santa cruz.
– Historia de america espanola 1920-1925
– Historia de la republica de el salvador...
– Historia de mexico, de francisco benegas galvan
– Historia de nuestra senora de guadalupe...
– Historia peregrina de un inca andaluz
– Historia verdadera de la conquista de la nueva espana
– Las historias del origen de las indias de esta provincia de guatemala
– Historical records and studies thomas f mechan
– Historical records and studies...new york, catholic historical society. 1929
– Ideales misioneros de los reyes catolicos
– La iglesia y la educacion popular en indias
– La iglesia y la masoneria en venezuela...
– El ilustrisimo fray hipolito sanchez rangel, primer obispo de maynas
– Impedimientos de misioneros
– Los imperialismos de juan gines de sepulveda en su
– In spain with the international brigade
– Instituto historico e geographico brasileiro...
– Interferencias
– Isagoge historica apologetica de las indias occidentales y especial de la provincia de san vicente de chiapa y guatemala, de la orden de predicadores. guatemala, 1935
– Un jesuita "a palos", jeronimo del portillo
– Los jesuitas en la provincia de quito de 1570 a 1774
– Jorge ricardo bejarano narino, su vida sus infortunios, su talla historica
– Juegos antiguos en america
– La junta para ampliacion de estudios. rectificacion y comentarios
– Justificaciones historicas
– Lance curioso en visita pastoral, missionalia hispanica, 1948
– Lecturas de historia nacional relacionadas con el santisimo sacramento. santiago de chile, 1928
– Legislacion sobre indios del rio de la plata en el siglo 16th. madrid. 1928
– Letrados y politicos
– Letture di filosofia...
– Libro primero de cabildos de la villa de san miguel de ibarra 1606-1617
– Los libros del cabildo de quito
– Literatura eclesiastica, cronica de la semana pro seminario celebrada en toledo los d'ias 4-10 de noviembre de 1935...
– La loca del sacramento
– Lucia in london. leipzig 1928
– Magallanes en 1925...por manuel zorrilla
– Manuel jimenez fernandez...
– Manuel vicente villaran, ex rector de la universidad. la universidad de san marcos, de lima...
– El mariscal de berwick
– La medicina y las misiones
– Meditaciones de la vida de cristo...
– Mejico.-la era de los martires
– Ministerio de trabajo y prevision. aportacion de los colonizadores espanoles a la prosperidad de america
– Ministerio de trabajo y prevision. catalogo pasajeros a indias...vol 1
– Ministerio de trabajo y prevision. disposiciones complementarias de las leyes de indias. tres tomos
– Ministerio de trabajo y prevision. seleccion de las leyes de indias...
– Un misionero y misionologo desconocido
– Misiones de mainas hacia la mitad del siglo 16
– Las misiones, defensa de las fronteras mainas
– Los modernos evangelizadores en la tierra de la santa cruz
– Monumenta histotica societatis jesu...romae, 1946
– Motolinia's history of the indians of nem spain
– El mundo catolico y la carta colectiva del episcopado espanol
– Los municipios y los indios
– Museo de america. sus precursores en el siglo 16
– Museo intelectual. vanguardia. mexico, 1928
– Museo nacional de arqueologia
– Naciones de lengua espanola
– Naturismo y afines
– The nigger of the "narcissus"
– Los ninos indigenas en la cristianizacion de america una pagina conmovedora en la historia
– Notas acerca del teatro religioso en la america colonial
– Notas y textos
– Notas y textos. planes antiguos de seminarios de misiones y de reclutar clero secular para la evangelizacion de america
– Noticia de un libro viejo y de una gloria olvidada
– Noticias bibliograficas
– Noticias generales
– Nuestra senora de lourdes
– La nueva conquista de la america espanola
– Nueva fase de la compana
– Nueva fase de una campana
– Los nuevos beatos martires del paraguay, simiente de las redenciones
– Los nuevos santos ingleses
– Ordenes religiosas no misioneras en indias
– P antonio maria de barcelona. o.m. martiri della rivolutione maxista nella spana...
– P fr tomas de san rafael, carmelita descalzo
– P silverio de santa teresa, c.d. obras de santa teresa de jesus
– Una pagina de geografia aneja
– Papeles del archivo. publicaciones del archivo general de la nacion
– Pastelerias en lima primitiva
– La patria del almirante...
– El patronato espanol en el virreyno del peru durante el siglo 16
– Patzcuaro. texto de manuel toussaint. mexico, 1942
– El petroleo de colombia
– Un pintor quiteno y un cuadro admirable del siglo 16 en el museo arqueologico nacional, 1929
– La plaga social de la blasfemia
– El poeta verdaguer y el marques de comillas
– Os portugueses no mundo
– Postrera voluntad y testamento de hernando cortes
– Pour le pape...
– Los precursores del congreso eucaristico de buenos aires. la devocion al santisimo de los conquistadores y pobladores de america
– Los precursores ideologicos de la guerra de la independencia 1789-1794

- La predicacion sagrada segun los documentos pontificios y doctrina de los santos padres
- La primavera de la vida
- Los principes de la literatura...
- El problema religioso en america
- La propaganda protestante en la america espanola
- El protector de indios
- El protestantismo en la america espanola
- El proximo congreso de las juventudes hispano-americanas
- Que pasa en espana?
- Que pasa en espana? a los catolicos de mundo
- Que pasa en mejico?
- Que pasa en mejico
- Lo que quieren los espanoles de buena voluntad...
- Quien a sido alguien en venezuela?
- Raimundo riva. lecturas historicas
- Ramon l. gomez. los privilegios de la america latina en su parte historico-cronologica
- La reciente enciclica del papa sobre la unidad religiosa
- Relacion de la visita general que en la diocesis de caracas y venezuela hizo el ilmo. sr. d. mariano marti, del consejo de s.m. 1771-1784
- Rene bazin
- La represion de la blasfemia
- Respuesta a una pregunta interesante
- El restablecimiento del culto en la espana roja
- Retiro espiritual para comunidades religiosas...
- La ruta de lo desconocido...
- Un sacerdote cautivo de los araucanos
- Sagrada biblia
- La salud de la america espanola. paris, 1926
- San eulogio de cordoba
- San eulogio de cordoba. madrid, 1928
- San francisco javier. el hombre y el santo
- Sangre en los tapehuanes
- Santa maria en indias
- La s.c. de propaganda fide e le missioni del ginppane
- Semblanza de santa teresa de jesus
- Septembre 1792. histoire politique des massacres...
- El siglo jeroglifico azteca...1897
- Los sin-dios. la campana de nuestros dias
- Sobre la "escuela nueva"
- La supresion del juego
- El taumaturgo catalan beato salvador de horta
- El teatro indigena de america
- El teosofismo
- La tormenta que viene de oriente
- Tres conquistadores y pobladores de la nueva espana
- Une. consignas aprobadas por el primer congreso nacional uncista reunido en caracas...
- United nations estado misionero
- Vallee-pousin, louis de la nirvana...
- Variedades
- Una victima de amor divino
- Vida de la sierva de dios sor maria de los dolores y patrocinio, por la r.m. sor maria isabel de jesus...
- Vida de san bernardo abad de claraval
- Vida del beato juan de avila, por un sacerdote. madrid, 1928
- Vida del segoviano rodrigo de contreras, del marques de lozoya y coleccion de las memorias...de los virreyes del peru de ricardo beltran rozpide
- La vida social en la colonia segun los "libros del cabildo de quito"
- Vida y obras de santa maria margarita de alocoque, 3 tomos. examen de libros
- La virgen del pilar y america
- The woman who stole everything
- Y grosso
- Yela utrilla, juan f. catedratico y doctor en ciencias historicas...

Bayle, constantino. el protector de indios. sevilla, 1944 / Saenz de Santa Maria, C – Madrid: Razon y Fe, 1946 – 1 – sp Bibl Santa Ana [946]

Bayle, P *see* Oeuvres diverses

Bayle, Pierre *see* Dictionaire historique et critique (ael1/45)

Bayle, Pierre Georges *see* La saisie et la vente judiciaire des fonds de commerce

Baylee, Joseph *see*
- God, man, and the bible
- Thoughts upon questions of existing controversy, in letters to a friend

Bayley, Arthur Rutter *see* The great civil war in dorset, 1642-1660

Bayley, Frederick W *see* Four years' residence in the west indies

Bayley, James Roosevelt *see* A brief sketch of the early history of the catholic church on the island of new york

Baylis, Samuel Mathewson *see* Camp and lamp

Baylor business review – Waco TX 1988+ – 1,5,9 – (cont: baylor business studies) – ISSN: 0739-1072 – mf#15719 – us UMI ProQuest [338]

Baylor business studies – Waco TX 1949-83 – 1,5,9 – (cont by: baylor business review) – ISSN: 0005-724X – mf#5140 – us UMI ProQuest [650]

Baylor law review – v1-52. 1948-2000 – 5,6,9 – $908.00 set – (v1-36 1948-84 in reel $422.00. v37-52 1985-2000 in mf $486.00) – ISSN: 0005-7274 – mf#100931 – us Hein [340]

Baylor University College of Medicine. Houston, Texas *see* Catalogs and college records

Bayly, Anselm *see* Practical treatise on singing and playing with just expression and real elegance

Bayly, Joseph *see* Congo crisis

Bayly, Mary *see* The story of our english bible and what it cost

Baym kval / Bastomski, Solomon – Vilne, Lithuania. 1920 – 1r – us UF Libraries [939]

Baym kval / Bastomski, Solomon – Vilne, Lithuania. 1923 – 1r – us UF Libraries [939]

Bayne, Peter *see*
- The chief actors in the puritan revolution
- The free church of scotland
- Martin luther
- The testimony of christ to christianity

Bayne, R *see* Converted dealer

Baynes, Arthur Hamilton *see* My diocese during the war

Baynes, Herbert *see*
- Ideals of the east
- The way of the buddha

Baynes, Paul *see* Entire commentary upon the whole epistle of the apostle paul to the ephesians

Baynes, Thomas Spencer *see* An essay on the new analytic of logical forms

Baynton, wharton, and morgan papers, 1757-1787 – (mf ed 1967) – 10r – 1 – silver $300 diazo $200 – (with guide comp by donald h kent et al (1967) $4.25 isbn: 0-911124-20-9) – us Penn Hist [380]

Bayo, Alberto *see* Versos revolucionarios

Bayo, Armando *see*
- Africa contra el colonialismo
- Gran revolucion africana

Bayol, Jean *see* Voyage en senegambie

Bayonet – 1988 jan 8-dec; 1989 jan 6-dec 22; 1993 feb 12-sep 24 – 1 – mf#639575 – us WHS [071]

Bayony Posada, Nicolas *see* Escritas literarios de rufino jose cuervo

Bayou plaquemine baptist church : church minutes – Plaquemine, LA. 1289p. 1946-98 – 1 – $58.01 – mf#7027 – us Southern Baptist [242]

Bayou rouge baptist church : church minutes – Evergreen, LA. 1918p. jul 25 1841-jan 10 1999 – 1 – $86.31 – mf#7028 – us Southern Baptist [242]

Bayreuth : 1876-1896 / Weingartner, Felix – Berlin: S Fischer 1897 – 1mf – 9 – mf#wa-125 – ne IDC [780]

Bayreuth 1897 : manuel pour les visiteurs de bayreuth / Wild, Friedrich – Leipzig, Baden-Baden: C Wild 1897 – 2mf – 9 – mf#wa-114 – ne IDC [780]

Bayreuth-album 1896 – Elberfeld: S Lucas [1896] – 2mf – 9 – mf#wa-2 – ne IDC [780]

Bayreuther blaetter. deutsche zeitschrift im geiste richard wagners – v. 1-61. 1878-1938 – 1 – us L of C Photodup [780]

Bayreuther briefe vom reinen thoren : "parsifal" von richard wagner / Lindau, Paul – 5. aufl. Breslau: S Schottlaender 1883 – 1mf – 9 – mf#wa-62 – ne IDC [780]

Bayreuther erinnerungen: freundschaftliche briefe / Pohl, Richard – Leipzig: Kahnt [1877] – 1mf – 9 – mf#wa-83 – ne IDC [780]

Bayreuther intelligenz-zeitung – Bayreuth DE, 1769, 1773-1781 sep, 1782-90, 1793 jul-dec, 1795, 1797, 1798-1803, 1804-06, 1807-jul 1808 – 10r – 1 – (with gaps) – gw Misc Inst [074]

Bayreuther kurier *see* Fraenkisches volk [main edition]

Bayreuther zeitung – Bayreuth DE, 1752-55, 1758-78, 1779 [gaps], 1780-1800, 1802, 1804-05, 1807, 1813-19, 1848-49 – 46r – 1 – (other titles: bayreuther zeitungen, baireuther politische zeitung; title varies: 1863: neue bayreuther zeitung; filmed by other misc inst: 1848-49 [3r]) – gw Misc Inst [074]

Das bayrische vaterland – Muenchen DE, 1869 23 mar-1934 26 sep – 1 – (title varies: 3 may 1872: das bayerische vaterland) – gw Misc Inst [943]

Bayrischer Philologenverband *see* Fachgruppe deutsch-geschichte. interpretationen moderner lyrik

Bayrisches schnurrenbuch / Poddel, Peter – Stuttgart: W Kohlhammer 1942 [mf ed 1993] – 1r [ill] – 1 – (filmed with: dreissig neue erzähler des neuen deutschland / ed & int by wieland herzfelde) – mf#3364p – us Wisconsin U Libr [390]

Bayro, P *see* ...De Medendis Humani Corporis. Enchiridion Vulgo Veni Mecum...

Bays, Davis H *see* The doctrines and dogmas of mormonism

Bay's law reports / South Carolina. Supreme Court – v1-2. 1783-1804 (all publ) – (= ser Pre-nrs nominative law reports) – 13mf – 9 – $19.50 – mf#LLMC 94-009 – us LLMC [340]

Bays news – Auckland, NZ. 1981-83 – 1r – 1 – mf#11.49 – nz Nat Libr [079]

Bayswater chronicle – [London & SE] Westminster Archives 20 jun 1860-25 dec 1926, 1 jan 1938-8 sep 1939 – 1 – (lacking: 1874, feb 1875; cont as: west london chronicle [7 jan 1944-31 dec 1949]; indicator [7 jan 1950-26 dec 1953]; both titles to be filmed) – uk Newsplan [072]

Bayt-i laylah : or, persian distichs, from various authors, in which the beauties of the language are exhibited in a small compass, and may be easily remembered / Weston, Stephen – London: printed for aut by S Rousseau, 1814 – (= ser 19th c books on linguistics) – 2mf – 9 – mf#2.1.20 – uk Chadwyck [470]

Bazaar! : the public will please to take notice that the ladies' bazaar...will be held on tuesday, the 22nd day of february, 1848 – Newmarket, Ont?: Porter, 1848? – 1mf – 9 – mf#43670 – cn Canadiana [360]

The bazaar – Rodney, Ont: Bazaar Print Co, [1887-18– or 19–] – 9 – mf#P04857 – cn Canadiana [073]

Bazaar gazet – New York, NY. The Daily Gazette of the Bazaar for the Jewish War Sufferers. 1916 – 1 – us AJPC [071]

Bazaar paintings of calcutta : the style of kalighat / Archer, William George – London: Her Majesty's Stationery Office, 1953 – (= ser Samp: indian books) – us CRL [750]

Bazaine, Achille Francois *see*
- Pieces annexes au rapport sur l'affaire du marechal bazaine.
- Proces bazaine.

Der bazar : illustrirte damen-zeitung – Berlin 1863-65, 1867, 1869-71, 1873-75, 1881-82, 1884, 1886 [single iss], 1887 [gaps], 1914 1 jan-11 mai [gaps], 1914 17 aug-1915 [gaps], 1917 & 1931 [mf ed 2004] – 10r – 1 – gw Mikrofilm [640]

Le bazar – [Sorel, Quebec?]: J A Chenevert, v1 n1 (1er dec 1888)- – 9 – mf#P04089 – cn Canadiana [360]

The bazar book of decorum: the care of the person, manners, etiquette, and ceremonials – New York: Harper & Brothers, 1871. 278p. Includes index – 1 – us Wisconsin U Libr [390]

Bazar de las sorpresas / Mazas Garbayo, Gonzalo – Habana, Cuba. 1957 – 1r – us UF Libraries [972]

Bazar kampen – v3 [1902 feb 20-22] – 1 – mf#350624 – us WHS [071]

Bazar litterario de educacao e de recreacao – Rio de Janeiro, RJ: Typ de S Vicente de Paulo, 01 out 1878-15 jun 1879 – (= ser Ps 19) – mf#P17,01,83 – bl Biblioteca [079]

Bazar nisse – v6-ns v6 [1901 nov23-1907 feb 21-23] – 1 – mf#350633 – us WHS [071]

Bazar pour la cathedrale de montreal : sous la haute direction de mgr edouard chs fabre, eveque de montreal – Montreal: s.n, 1886 – 1mf – 9 – mf#08907 – cn Canadiana [360]

Bazar volante – Rio de Janeiro, RJ: Typ do Bazar Volante, 27 set 1863-set 1864; out 1865-30 dez 1866 – (= ser Ps 19) – mf#P3,1,10-13 – bl Biblioteca [079]

Bazar volante *see* O arlequim

Bazard, Palmyre *see* La religion saint simonienne

Bazaz, Prem Nath *see* Inside kashmir

Bazazzaga – Zaria: Gaskiya Corp, [dec 1956-sep 4 1958] – 1r – 1 – (filmed with: zaruma and other hausa newspapers) – us CRL [960]

Bazett, L Margery *see* After-death communications

Bazhenov, A *see*
- Irtysh
- Den' kazaka: ustraevaemyi sluzhashchimi sluzhby sborov omsk zh d v pol'zu mobilizovannykh kazakov i dobrovol'tsev pri uchastii zhurn "irtysh"

Bazhov, Pavel Petrovich *see* Malakhitovaia shkatulka

Bazile, Corneille *see* Terreur noite a la guadeloupe

Bazin, Jacques Rigomer *see* Jacqueline d'olzebourg

Bazin, Jean-Marie *see* 'Notes sur la mission' by father jean-marie bazin

Bazin, Robert *see*
- Histoire de la litterature americaine de langue es...
- Historia de la literatura americana en lengua espa...

Baznicas zinas – 1952 mar 9-1967; 1968-78; 1979-1987 apr/may – 1 – mf#1363328 – us WHS [071]

Bazot, Etienne-Francois *see*
- Contes maconiques dedies aux soeurs et aux freres
- Manuel du franc-macon

Bazu-yi inqilab – Tehran: Junbish-i Kargaran-i Musalman, 1980. sal-i 1, shumarah-'i 1-32. 13 murdad 1359-5 aban 1359 [4 aug-28 oct 1980] – 1r – 1 – $53.00 – us MEDOC [079]

Bazzocchini, Benvenuto *see* L'emmaus di s luca

Bb news – 45th [1992 aug 1/4] – 1 – mf#4027793 – us WHS [071]

Bbb report – 1967 feb 28-1987 jun/jul – 1 – mf#630790 – us WHS [071]

The bbc and the general strike of 1926 – 4r – 1 – (with p/g) – mf#97608 – uk Microform Academic [380]

BBC (British Broadcasting Corporation) Home Service *see* Nine o'clock news, 1939-1945

BBC (British Broadcasting Corporation) Radio *see* Author and title catalogue of transmitted drama, poetry and features 1929-1975

BBC (British Broadcasting Corporation) Television *see* Author and title catalogue of transmitted drama and features, 1936-1975

Bbc handbooks, annual reports and accounts, 1927-2002 / British Broadcasting Corporation – 20r – 1 – (with p/g) – mf#97602 – uk Microform Academic [380]

Bc studies – Vancouver BC 1972+ – 1,5,9 – ISSN: 0005-2949 – mf#8209 – us UMI ProQuest [300]

Bcnu pdate : the magazine of the british columbia nurses' union – 1991 win/spring-1992 oct – 1 – mf#2989193 – us WHS [610]

Bcnu reports – 1982 jan/feb-1990 nov/dec – 1 – mf#1052742 – us WHS [071]

Bd 4 abt 1a und 1b der auf anordnung der schweizerischen bundesbehoerden veranstalteten sammlung der aeltern eidgenoessischen abschiede : abschiede, die eidgenoessischen, aus dem zeitraume von 1521-1532 / ed by Strickler, J – Brugg, 1873-1876 – 37mf – 9 – mf#ZWI-13 – ne IDC [240]

Bdard, Pierre-Stanislas *see* A tous les electeurs du bas canada

Be dan panna nhan loki padesa kyam / Kam Man, U' – Ran kun: Canda Van Ca pe 1976 [mf ed 1995] – on pt of 1r – 1 – mf#11052 r1951 n8 – us Cornell [480]

Be not many masters / Woodford, James Russell – London, England. 1848 – 1r – us UF Libraries [240]

Be not schismatics, be not martyrs, by mistake / Hamilton, William – Edinburgh, Scotland. 1843 – 1r – us UF Libraries [240]

Be not weary in well-doing – London, England. 18– – 1r – us UF Libraries [240]

B/E Productions [Oconomowoc, WI] *see* From one business to another

BE radio – Shawnee Mission KS 2001 – 1,5,9 – ISSN: 1081-3357 – mf#21289.01 – us UMI ProQuest [380]

Be ready! / Watts, H – London, England. 18– – 1r – us UF Libraries [240]

Be true! : a few words to the confirmed youth of the evangelical lutheran church / Cooperrider, George Trout – Columbus OH: Lutheran Book Concern [19–?] [mf ed 1992] – 1mf – 9 – 0-524-04767-7 – mf#1991-2153 – us ATLA [242]

Be yeu : truyen dai / Nha Ca – Saigon, Thuong Yeu 1974 [mf ed 1992] – on pt of 1r – 1 – mf#11052 r232 n3 – us Cornell [830]

Be your own lawyer : a business manual containing a synopsis of the mercantile, or business laws of ontario / Anger, William Henry – Toronto: printed for the aut, 1895 [mf ed 1980] – 2mf – 9 – 0-665-02407-X – (incl ind) – mf#02407 – cn Canadiana [346]

Be your own lawyer, or, secrets of the law office : giving in concise form the mercantile or business laws of canada, the technical points and main features of the law / Anger, William Henry – Toronto: W H Anger, 1896 [mf ed 1983] – 2mf – 9 – (incl ind) – mf#10517 – cn Canadiana [346]

Beach, Abijah Ives *see* Papers

Beach, Charles Fisk *see*
- Commentaries on the law of receivers.
- Individual evangelism
- A treatise on the modern law of contracts.
- The trust, an economic evolution

Beach, David Nelson *see* The newer religious thinking

Beach, Douglas Martyn *see* Phonetics of the hottentot language

Beach family magazine – 1926 jan 1-32 aug 1 – 1 – mf#1053025 – us WHS [640]

Beach, Harlan P *see* A geography and atlas of protestant missions

Beach, Harlan Page *see*
- The cross in the land of the trident
- Dawn on the hills of t'ang
- India and christian opportunity
- Knights of the labarum
- Princely men in the heavenly kingdom

Beach, Harlan Page et al *see* Protestant missions in south america

Beach, James Mark *see* Christ and the covenant

Beach resort news – Delake OR: DeLake Pub Co, [wkly] – 1 – (cont by: lincoln coast news (-1930)) – us Oregon Lib [071]

Beach resort news (lincoln city, or) – Delake OR: G Garland Sittser, -1939 [wkly] – 1 – (began in 1930; cont: lincoln coast news; absorbed: lincoln county press. merged with: lincoln coast guard, to form: north lincoln coast guard in combination with the beach resort news; iss for aug 21 1931-dec 18 1936 called: beach resort news combined with lincoln county press) – us Oregon Lib [071]

Beach resort news (lincoln city, or) *see* Lincoln county press

Beach, Rex *see* Winds of chance

Beach, Rex Ellingwood *see* Miracle of coral gables

Beach, Thelma *see* Form of the brahms double concerto

A beachcomber in the orient / Foster, Harry La Tourette – New York: Blue Ribbon Books [1923] [mf ed 1987] – 395p – 1 – mf#6833 – us Wisconsin U Libr [915]

Beachcomber's island sun – Holmes Beach, FL. v3 n49-v4 n23. 1992 jan-jun – 1r – (1992 feb 6) – us UF Libraries [071]

Beachhead Collective *see* Free venice beachhead

Beacock, D V *see* Heredity and environment beginning with the primordial cell

Beacon / Barbers and Beauty Culturists Union of America – 1946 sep 15-1955 jun – 1r – 1 – mf#1110168 – us WHS [640]

Beacon – Cumberland WI. 1984 oct-1985 oct, 1985 nov-1986 jun, 1986 jul-1987 feb, 1987 mar-may – 4r – 1 – mf#1222787 – us WHS [071]

Beacon : the fearless weekly – Ottawa, Ontario; Hull, Quebec. v1-3. dec 12 1929-jul 2 1932// ? – 1r – 1 – Can$110.00 – cn McLaren [071]

Beacon / Lake Co. Fairport Harb – feb 1959-oct 1960 [wkly] – 1r – 1 – mf#B6108 – us Ohio Hist [071]

Beacon / Lake Co. Fairport Harb – (sep-dec 1935), jan 1936-feb 1959 [wkly] – 16r – 1 – mf#B33282-33297 – us Ohio Hist [071]

Beacon – [Scotland] Edinburgh: printed by D Stevenson 6 jan-22 sep 1821 (wkly) [mf ed 2003] – 1r – 1 – uk Newsplan [072]

The beacon – Boston MA, 1933-73 – 6r – 1 – us IHRC [073]

The beacon – Cleveland OH, 1933 – 1r – 1 – (italian newspaper) – us IHRC [071]

The beacon *see* Miscellaneous newspapers of larimer county

Beacon [edinburgh] : published by church of Scotland – Edinburgh, Scotland [ns] 29 oct 1892, 15 nov 1892-16 oct 1893 – 1 – uk British Libr Newspaper [242]

Beacon hill baptist church – Somerset KY nov 1967-5 jan 1983 – 1 – $16.11 – us Southern Baptist [242]

Beacon journal / Summit Co. Akron – (1840-48), 1849-68, 1871-1939 – 333r – 1 – mf#B4654-4987 – us Ohio Hist [071]

Beacon journal index / Summit Co. Akron – 1841-45, 1849-1939 – 15r – 1 – mf#B4639-4653 – us Ohio Hist [071]

Beacon light – Vale OR: Hurley & Kautzman, -1912 [wkly] – 1 – (absorbed: malheur booster; oregon oriano (1905-12)) – us Oregon Lib [071]

Beacon light *see*
– Malheur booster
– Oregon oriano

The beacon light – Oakdale, NE: Bert Kautzman (wkly) [mf ed v8 n23. mar 24-apr 7 1893] – 1r – 1 – us NE Hist [071]

Beacon light and holt county independent, consolidated – O'Neill, NE: H Kautzman, -jun 11 1897// (wkly) [mf ed nov 9 1894-jun 11 1897 (gaps) filmed 1998] – 2r – 1 – (formed by the union of: beacon light and: holt county independent. cont by: holt county independent (1897)) – us NE Hist [071]

Beacon news – 1956 feb-1959 nov; 1960-62; 1963-68; 1969 jan-1970 dec – 1 – mf#1053030 – us WHS [071]

The beacon of the truth : or testimony of the coran of the christian religion – London, 1894 – (trans fr arabic by sir william muir) – ne Slangenburg [230]

The beacon of truth : or, testimony of the coran to the truth of the christian religion / Minaar ul Hakk – London: Religious Tract Society, 1894 – 1mf – 9 – 0-524-01848-0 – mf#1990-2683 – us ATLA [230]

Beacon [st &rews, canada] – St &rews, Canada 7 jan 1915-1 feb, 28 jun-9 aug 1919 (imperfect) – 1 – (not publ between 1 feb-28 jun 1919; cont: st &rews beacon [27 aug-31 dec 1914]) – uk British Libr Newspaper [072]

Beacon: the paper of the barbados labour party and of the barbados workers' union – Bridgetown Barbados 4 jan 1958-5 dec 1964, 18 sep 1965, 8 jan-28 may 1966 – 1 – uk British Libr Newspaper [331]

A beacon to the society of friends / Crewdson, Isaac – London: Hamilton, Adams, 1835 [mf ed 1993] – (= ser Society of friends (quakers) coll) – 1mf – 9 – 0-524-07560-3 – mf#1991-3180 – us ATLA [243]

Beacon-observer – Overton, NE: Taylor Print Service. 76th yr n28. oct 4 1973- (wkly) [mf ed 1977] – 1 – (lacks: jan 9 1975. formed by the merger of: elmcreek beacon and: overton observer. publ in elm creek ne nov 8 1973- . cont the numbering of: elmcreek beacon) – us NE Hist [071]

Beadle, Delos White *see*
– The american lawyer and businessman's formbook...
– Canadian fruit, flower, and kitchen gardener

Beadle, John Hanson *see* Proposals for publishing the history of mormonism

Beadle's dime library – 1878-98. 967 issues – 1 – us L of C Photodup [830]

Beadle's half dime singer's library. – .New York. 2nd ed. no. 12. 1878 – 1 – us NY Public [780]

Beadle's half-dime library – 1877-1905. 826 issues – 1 – us L of C Photodup [830]

Beadle's monthly : a magazine of to-day – La Salle IL 1866-67 – 1 – mf#3891 – us UMI ProQuest [071]

The beads from taxila / Beck, Horace C; ed by Marshall, John – Delhi: Manager of Publications, 1941 – (= ser Samp: indian books) – us CRL [930]

Beaglehole, Ernest *see* Notes on hopi economic life

Beal, Edward *see* The law of bailments

Beal, Samuel *see*
– Abstract of four lectures on buddhist literature in china
– Buddhism in china
– A catena of buddhist scriptures from the chinese
– The romantic legend of saakya buddha
– The romantic legend of sakya buddha

Beal, William James *see*
– Grasses of north america, v1
– Grasses of north america, v2
– Grasses of north america, vols 1 and 2

Beale, J F *see* Lives and labors of eminent divines

Beale, Joseph H, Jr *see* Beale's cases on conflict of laws

Beale, Joseph H, Jr. *see*
– A selection of cases on the law of carriers
– Selections from a treatise on the conflict of laws

Beale's cases on conflict of laws / Beale, Joseph H, Jr – Cambridge, MA: Harvard University Press. 2v. 1907 (all publ) – 18mf – 9 – $27.00 – mf#LLMC 95-047 – us LLMC [340]

Beals, Carleton *see*
– Crime of cuba
– Rifle rule in cuba

Beals, Zephaniah Charles *see* China and the boxers

Beam – 1992 jan-dec 18, 1993 jan 9-dec 17 [1]; 1986 mar 14-1987 may 1, 1992, 1993 [2] – 1 – mf#639569 [1]; 1212128 [2] – us WHS [071]

Beam – Fort Worth, TX. 1954-61 – 1 – $116.76 – us Southern Baptist [242]

Beam interactions with materials and atoms *see* Nuclear instruments and methods in physics research sect b

Beaman, Middleton G *see* Index-analysis of the federal statutes, 1789-1907

Beamish, North Ludlow *see* The discovery of america by the northmen

Beamrider – v40 n12 [1986 oct 14]; v41 n7-8 [1987 jul 24-aug 11]; v42 n16, 25-27 [1988 aug 11, dec 1-29]; v44 n4-6,17,20 – 1 – mf#1426910 – us WHS [071]

Beams of light – v4-45. 1909-50 [gaps] – (= ser Mennonite serials coll) – 5r – 1 – mf#ATLA 1994-S010 – us ATLA [242]

Beamten-blatt *see* Rheinisch-westfaelische beamten-zeitung

Der beamtenbund – Bonn-Bad Godesberg DE, 1961-88 – 1 – gw Misc Inst [350]

Der beamtenbund *see* Die gemeinschaft

Bean, Edwin F *see* Beans history and directory of nevada county

Bean, J V et al *see* Robb's family physician

Bean leaf-hopper and hopperburn with methods of control / Beyer, A H – Gainesville, FL. 1922 – 1r – us UF Libraries [630]

Beans history and directory of nevada county / Bean, Edwin F – Nevada Co, CA. 1867 – 1r – 1 – $50.00 – mf#B40246 – us Library Micro [978]

Bear creek baptist church. decatur county – PARSONS, TN. 1842-feb 1929 – 1 – $15.57 – us Southern Baptist [242]

Bear facts – 1984 oct-1993 mar – 1 – mf#1110170 – us WHS [071]

Bear hills native voice – 1982 nov 18-1983; 1984-85; 1986-87; 1988 jan-jun 16 – 1 – mf#916477 – us WHS [071]

Bear, James Edwin *see* The mission work of the presbyterian church in the united states in china, 1867-1952

Bear lake record – Bear Lake, PA. -w 1895-1896 – 13 – $25.00r – us IMR [071]

Bear river news *see* Wheatland newspapers

Bear springs baptist church. stewart county – DOVER, TN. 1960-jul 1968 – 1 – $7.36 – us Southern Baptist [242]

Bear talk – 1981 nov-1986 oct – 1 – mf#1288344 – us WHS [071]

Bear Tribe Medicine Society *see* Many smokes

Bearbeitung der waehrend der totalen mondfinsterniss 1884 oct 4 und 1888 jan 28 beobachteten sternbedeckungen / Struve, Ludwig – Dorpat: C Mattiesen 1893 [mf ed 1998] – 1r – 1 – mf#film mas 28404 – us Harvard [520]

Die bearbeitungen des "verbrechers aus verlorener ehre" : mit benutzung ungedruckter briefe von und an herrn kurz / Stoess, Willi – Stuttgart: Metzler, 1913 [mf ed 1992] – (= ser Breslauer beitraege zur literaturgeschichte. neue folge 37) – viii/74p – 1 – (incl bibl ref) – mf#8014 reel 4 – us Wisconsin U Libr [430]

Beard, Augustus Field *see*
– A crusade of brotherhood
– The story of john frederic oberlin

Beard, Charles *see*
– Martin luther and the reformation in germany
– Outlines of christian doctrine
– Port royal
– The reformation of the 16th century

Beard, Charles A *see*
– American government and politics
– Economic origins of jeffersonian democracy

Beard, Glenn C *see* The effect of carbonated solutions on gastric emptying during prolonged cycling

Beard, J R *see* Life of christ the source and pattern of christian influence

Beard, John R *see* How did we come by the reformation?

Beard, John Reilly *see* Divinity and atonement of jesus christ scripturally expounded

Beard, John Relly *see*
– Christ the interpreter of scripture
– Letters on the grounds and objects of religious knowledge
– A manual of christian evidence
– A revised english bible the want of the church and the demand of the age
– Unitarianism exhibited in its actual condition

Beard, Mary Ritter *see* Short history of the american labor movement

Beard, Richard *see*
– Lectures on theology
– Why am i a cumberland presbyterian?

Beardslee, John Walter *see*
– The bible among the nations
– Outlines of an introduction to the old testament

Beardsley, Aubrey Vincent *see*
– A book of fifty drawings by aubrey beardsley
– The early work of aubrey beardsley
– A second book of fifty drawings by aubrey beardsley

Beardsley, Eben Edwards *see*
– Addresses and discourses
– Life and correspondence of samuel johnson, d.d
– Life and correspondence of the right reverend samuel seabury, d.d

Beardsley, Frank Grenville *see*
– Christian achievement in america
– A history of american revivals

The bear-hunters of the rocky mountains / Bowman, Anne – Boston: Crosby and Nichols, 1862 – 6mf – 9 – mf#42949 – cn Canadiana [830]

Bearing of morals on religion / Clifford, W K – London, England. 1877 – 1r – us UF Libraries [240]

The bearing of recent discovery on the trustworthiness of the new testament / Ramsay, William Mitchell – 2nd ed. London: Hodder and Stoughton, 1915 – (= ser James Sprunt Lectures) – 2mf – 9 – 0-524-05627-7 – mf#1992-0482 – us ATLA [225]

Bearing of the american revival on the duties and hopes of british... / James, John Angell – Glasgow? Scotland. 18– – 1r – us UF Libraries [240]

The bearing of the evolutionary theory on the conception of god : a study in contemporary interpretations of god in terms of the doctrine of evolution / Kawaguchi, Ukichi – 1916 – 2mf – 9 – 0-524-07939-0 – mf#1991-0189 – us ATLA [210]

The bearing of the theory of evolution on christian doctrine / Betts, John Arthur – London: SPCK, 1897 – 1mf – 9 – 0-524-07226-4 – mf#1991-2967 – us ATLA [240]

Bearing stresses on surfaces inclined to the direction of the grain / Sawyer, William L – s.l, s.l, s.l? 1937 – 1r – us UF Libraries [630]

The bearings of modern commerce on the proress of modern missions : the annual sermon before the bishops, clergy and laity constituting the board of missions of the protestant episcopal church in the united states / Stone, John Seely – New York: William Osborn, 1839 – 1mf – 9 – 0-7905-6628-1 – mf#1988-2628 – us ATLA [242]

Bearings of popery on the priesthood of christ / Dods, Marcus – Edinburgh, Scotland. 1837 – 1r – us UF Libraries [240]

The bearings of the darwinian theory of evolution on moral and religious progress / Weiss, Frederick Ernest – London: Philip Green, 1909 – (= ser Essex Hall Lecture) – 1mf – 9 – 0-524-07840-8 – mf#1991-3387 – us ATLA [210]

Bearliner, A *see* Gedenkblatt an professor a bearliner

Be'arvot argentinah / Maidanik, Marcos – Buenos Aires, Argentina. 1948 – 1r – us UF Libraries [939]

Beata virgo maria in suo conceptu immaculata ex monumentis omnium seculorum demonstrata : accedit amplissima literatura / Roskovany, Augustino de – Budapestini: Typis Athenaei 1873-81 [mf ed 1986] – 9v on 20mf – 9 – 0-8370-9108-X – (incl ind) – mf#1986-3108 – us ATLA [241]

Beater, Jack *see* Sea avenger

La beatificacion del venerable sebastian de aparicio. mexico, 1934 / Ocaranza, Fernando – Madrid: Razon y Fe, 1935 – 1 – sp Bibl Santa Ana [972]

The beatitudes : or, some christian fundamentals / McCann, Samuel Napoleon – Elgin, Ill: Brethren Pub House, 1913 – 1mf – 9 – 0-524-04058-3 – mf#1990-4966 – us ATLA [240]

The beatitudes and other sermons / Maclaren, Alexander – London: Alexander and Shepheard, 1896 – 1mf – 9 – 0-7905-1353-6 – mf#1987-1353 – us ATLA [240]

The beatitudes of christ : a study of the way of the blessed life / Johnston, Howard Agnew – Chicago:Winona, 1905 – 1mf – 9 – 0-8370-3789-1 – mf#1985-1789 – us ATLA [220]

El beato sanz y companeros martires del orden de predicadores / Fernandez Arias, Evaristo – Manila: Establecimiento tipografico del Colegio de Santo Tomas 1893 [mf ed 1995] – (= ser Yale coll) – 1r [ill] – 1 – 0-524-09712-7 – (in spanish. filmed with other works) – mf#1995-0712 – us ATLA [241]

Beatrice courier – Beatrice, NE: Ritchey & Conlee. v1 n17. may 4 1875-1881// (wkly) [mf ed may 4 1875-nov 12 1879 (gaps)] – 2r – 1 – (cont by: gage county independent) – us NE Hist [071]

The beatrice daily express – Beatrice, NE: M A Brown. 1st yr n253. nov 11 1884-v39 n23. apr 26 1924 (daily ex sun) [mf ed with gaps] – 69r – 1 – (absorbed: beatrice weekly express (1911). absorbed by: beatrice daily sun) – us NE Hist [071]

Beatrice daily sun – Beatrice, NE: G P Marvin, jul 8 1902 (daily ex sun) – 182r – 1 – (absorbed: beatrice daily express apr 27 1924 and: beatrice times nov 25 1952) – us Bell [071]

Beatrice daily sun – Beatrice, NE: G P Marvin. v1 n2. jul 9 1902- (daily ex sun) [mf ed aug 23-dec 31 1940] – 2r – 1 – (absorbed: beatrice daily express apr 27 1924 and: beatrice times nov 25 1952) – us NE Hist [071]

Beatrice daily sun – Beatrice, NE: G P Marvin, jul 8 1902- (daily ex sun) – 28r – 1 – (absorbed: beatrice daily express, apr 27 1924 and: beatrice times, nov 25 1952 (daily ex sun)) – Located: u nebraska-lincoln libraries – us Misc Inst [071]

Beatrice daily sun – Beatrice, NE: G P Marvin, jul 8 1902 (daily ex sun) – 20r – 1 – (absorbed: beatrice daily express [apr 27 1924] and: beatrice times [nov 25 1952]; vol & iss numbering dropped with may 15-16 1976) – us Microfilm Corp [071]

Beatrice daily times – Beatrice, NE: Times Pub Co. 12v. v1 n1. jul 1 1892-v12 n34. feb 15 1898 (daily ex sun) [mf ed with gaps] – 6r – 1 – (formed by the union of: daily democrat and: beatrice republican. cont by: beatrice evening times. issues for jul 19 1892-feb 15 1898 called 6th yr n212-v12 n34) – us NE Hist [071]

Beatrice evening times – Beatrice, NE: W S Tilton. v12 n35. feb 16 1898-oct 14 1902 (daily ex sun) [mf ed with gaps] – 7r – 1 – (ceased nov 1902. cont: beatrice daily times. other ed: beatrice weekly times) – us NE Hist [071]

Beatrice express – Beatrice, NE: Theodore Coleman & Co (wkly) [mf ed v2 n1. apr 15 1871-feb 25 1878] – 4r – 1 – (cont: beatrice clarion. cont by: beatrice weekly express) – us NE Hist [071]

The beatrice news – Beatrice, NE: Charles W Clarke. v1 n1. jun 24 1926- (wkly) [mf ed -aug 19 1949 with gaps] – 6r – 1 – us NE Hist [071]

Beatrice post – Beatrice, NE: Chr Kiefer und H H Fast. 6v. 1892-6 jahrg n3. 25 mar 1897 (wkly) [mf ed 4 jahrg n26. 5 sep 1895-97 (gaps) filmed 1975?] – 1r – 1 – (in german. cont by: nebraska post) – us NE Hist [071]

Beatrice presse – Beatrice, NE: Paul Springer, 1888 (wkly) [mf ed jahrg 2 n40. 28 aug 1890 filmed [1990]] – 1r – 1 – (in german) – us NE Hist [071]

Beatrice record / Taxpayers Protective League (Beatrice, Nebraska) – Beatrice, NE: [Taxpayers Protective League] v1 n1. sep 17 1925- (wkly) [mf ed [1993]] – 1r – 1 – us NE Hist [071]

The beatrice republican – Beatrice, NE: J W Hill. v5 n5. jan 16 1886-v11 n28. jun 25 1892 (wkly) [mf ed with gaps] – 2r – 1 – (cont: gage county independent. merged with: daily democrat, to form: beatrice daily times) – us NE Hist [071]

The beatrice times – Beatrice, NE: Beatrice Times Co. 11v. v1 n142. aug 27 1942-v11 n189. nov 23 1952 (daily ex mon) [mf ed with gaps] – 46r – 1 – (cont: times. absorbed by: beatrice daily sun) – us NE Hist [071]

Beatrice weekly express – Beatrice, NE: M A Brown. -v32 n25. sep 26 1901 (wkly) [mf ed v9 n43. feb 3 1879-sep 26 1901 (gaps)] – 5r – 1 – (cont: beatrice express. cont by: beatrice semi-weekly express) – us NE Hist [071]

Beatrice weekly express – Beatrice, NE: Beatrice Express Pub Co. 1v. 39th yr n78. oct 12 1911-39th yr n97. mar 7 1912 (wkly) – 1r – 1 – (cont: semi-weekly express (1909). absorbed by: beatrice daily express) – us NE Hist [071]

Beatrice weekly express – Beatrice, NE: Beatrice Express Pub Co. 1v. 39th yr n78. oct 12 1911-39th yr n97. mar 7 1912 (wkly) – 1r – 1 – (cont: semi-weekly express (1909); absorbed by: beatrice daily express; other ed: beatrice daily express) – us NE Hist [071]

Beatrice weekly times – Beatrice, NE: Times Pub Co, dec 3 1892-v31 n10. feb 4 1909 (wkly) [mf ed with gaps] – 5r – 1 – (absorbed by: semi-weekly express (1909); iss for oct 4-oct 18 1895 called v9 n42-v9 n44; iss for nov 1 1895-feb 4 1909 called v14 n46-v31 n10; other ed: beatrice daily times 1892-98 and: beatrice evening times 1898-1902) – us NE Hist [071]

Beatrice-du-Saint-Sacrement, soeur *see* Bio-bibliographie analytique 1917-1941 de monsieur l'abbe pierre gravel cure de boischatel

Beatriche / Korvin-Piotrovskii, Vladimir L'vovich – Berlin: Knigoizd-vo "Slovo", 1929 [mf ed 2002] – 1r – 1 – (filmed with: rafael' / boris zaitsev (1924)) – mf#5238 – us Wisconsin U Libr [820]

Beatson, Robert *see* A political index to the histories of great britain and ireland

Beattie, Francis Robert *see*
- Apologetics
- An examination of the utilitarian theory of morals
- The methods of theism
- The presbyterian standards
- Radical criticism

Beattie, James *see*
- Essays
- History of the church of scotland during the commonwealth

Beattie, Malcolm Hamilton *see* On the hooghly

Beattie, William *see* Waldenses, or protestant valleys of piedmont and dauphiny

Beatty, Alfred Chester *see* Library of a chester beatty

[Beatty-] bullfrog miner – NV. 1905-08 [wkly] – 3r – 1 – $180.00 – mf#U04413 – us Library Micro [071]

Beatty, Charles C *see* Record of the family of charles beatty who emigrated from ireland to america in 1792

Beatty family scrapbook, 1790-1850 (1849-1850) – [mf ed 1983] – 1r – 1 – mf#ms4751 – us Western Res [978]

Beatty, Paul B *see* A history of the lutheran church in guyana

Beatty, Samuel G *see*
- Book-keeping, by single and double entry
- Book-keeping by single and double entry
- The canadian accountant

[Beatty-] transvaal miner – NV. 1906 – 1r – 1 – $60.00 – mf#U04414 – us Library Micro [622]

Beatty, William Henry *see* The boards of trade general arbitrations act (1894)

[Beatty-beatty-] amargosa times – NV. 1982-1983 – 1r – 1 – $60.00 – mf#N03701 – us Library Micro [071]

Beau, Claude le *see*
- Avantures du sr c lebeau, avocat en parlement
- Geschichte des herrn c le beau, advocat im parlament

Beau monde : or, literary and fashionable magazine – La Salle IL 1806-09 – 1 – mf#5252 – us UMI ProQuest [740]

The beau monde 1806-08 *see* Ackermann's 'repository of arts'

Beau monde and monthly register – La Salle IL 1809-10 – 1 – mf#3003 – us UMI ProQuest [071]

Le beau navire – Revue de la poesie. Dir. Maurice Chapelan. no. 1-9. Paris. nov 1934-juin 1939 – 1 – fr ACRPP [410]

Le beau roman d'amour de rolande desormeaux : sa jeunesse...sa carriere artistique...sa vie sentimentale... / Brousseau, Serge – Montreal: editions des Succes populaires, 1964 [mf ed 1987] – 9 – (pref by lucille dumont) – mf#SEM105P801 – cn Bibl Nat [920]

Beaubien : chroniques judicaires / Canada. Quebec. (Province) – 1v. 1905-06 – 4mf – 9 – $6.00 – (correspondances judicaires publiees sous forme de "chroniques" dans le journal "le soleil") – mf#LLMC 81-068 – us LLMC [340]

Beaubien, Charles Philippe *see*
- Les amen de monsabre
- Ecrin d'amour familial

Beaubien, Louis *see*
- Le chemin de fer
- Direction pour la culture en vert du ble-d'inde et son ensilage
- Discours prononce a la seance du 3 juin 1892 de l'assemblee legislative de la province de quebec
- Etude sur l'education agricole
- Silos and pasture lands
- A speech delivered at the session of june 3rd, 1892

Beauchamp, Alph. de *see* The life of ali pacha

Beauchamp, Alphonse de *see*
- Memoires secrets et inedits
- Vie de louis 18

Beauchamp, Henry King *see* Hindu manners, customs and ceremonies

Beauchamp, Joseph *see*
- Guide indispensable au peuple
- Repertoire general de jurisprudence canadienne... 1770-1913 and supplement 1913-23

Beauchamp, William M *see* The iroquois trail

Beauchemin, Neree *see* Les floraisons matutinales

Beauchesne, Alcide H de *see* Louis 17

Beauchet, Ludovico *see* Les batuecas y las jurdes

Beauclerk, Charles, Lord *see* Lithographic views of military operations in canada under his excellency sir john colborne...

Beauclerk, George R *see* A journey to marocco

Beau-cocoa – New York NY 1968-73 – 1 – ISSN: 0067-4737 – mf#7419 – us UMI ProQuest [400]

Beaucoup baptist church – PINCKNEYVILLE, IL. 15 apr 1915-72 – 1 – $50.31 – us Southern Baptist [242]

Beaucoup de bruit pour rien / Legendre, Louis – Paris, France. 1887 – 1r – us UF Libraries [440]

Beaudoin, Gilles *see* Bio-bibliographie du reverend pere gonzalve poulin

Beaudoin, Jean *see* Journal d'une expedition de d'iberville

Beaudry, David-Hercule *see*
- Le conseiller du peuple
- Precis historique de l'execution de jean-bapt desforges et de marie-anne crispin, veuve jean-baptiste gobier dit belisle

Beaudry, Edouard Alexis *see* Le questionnaire annote du code civil du bas-canada

Beaudry, Francois-Xavier *see* Testament solennel de mr f x beaudry

Beaudry, Joseph Alphonse Ubalde *see*
- Code des cures, marguilliers et paroissiens
- Des puits et des aqueducs
- Rapport de l'aqueduc de quebec
- Report on the quebec water works

Beaudry, Louis Napoleon *see*
- Face a face
- Historic records of the fifth new york cavalry, first ira harris guard
- Spiritual struggles of a roman catholic

Beaudry, Pauline *see* Bibliographie analytique de baie comeau sur la cote-nord du saint-laurent

Beaufils, G *see* La vie de la venerable mere jeanne de lestonac

Beaufort baptist church – Beaufort Co, SC. 1795p. 1840-1965, 1969-84 – (= ser Deacons' Meetings) – 1 – $80.78 – (deacons' meetings 1962-84) – mf#5003-31 – us Southern Baptist [242]

Beaufort county historical society papers – [mf ed [S.I.]: Association for Information & Image Management] – 1r – 1 – mf#45-351 – us South Carolina Historical [978]

Beaufort County. North Carolina. First Baptist Church *see* Washington church book

Beaufort courier – Beaufort West SA, 1869- – 1 – sa National [079]

Beaufort, Francis *see*
- Karamania
- Karamanien

Beaufort Marine Corps Air Station *see* Jet stream

Beaufort merchant's account book, 1785-1791 / Verdier, John Mark – 9 – mf#34/326 – us South Carolina Historical [380]

Beaufort, W L *see* Bible

Beaufoy, Mark *see*
- Mexican illustrations
- Tour through parts of the united states and canada

Beaugrand, Honore *see*
- Across the continent via the canadian pacific railway
- Anita
- La chasse galerie
- De montreal a victoria par le transcontinental canadien
- Les feux-follets
- Jeanne la fileuse
- Lettres de voyage
- Melanges
- New studies of canadian folk lore
- Six mois dans les montagnes-rocheuses

The beauharnois canal question / Girouard, Desire – [Montreal?: Herald], 1873 – 1mf – 9 – 0-665-91571-3 – mf#91571 – cn Canadiana [380]

Beaujeu, Monongahela de *see* The hero of the monongahela

Beaujour, Felix de *see* Voyage militaire dans l'empire othoman

Beaulieu Roy, Therese et al *see* "L'ordinaire"

Beaumarchais, Pierre Augustin Caron De *see*
- Sevilla berberi
- Tarare

Beaumarchais, Pierre Augustin Caron de *see* Le barbier de seville

Beaumont, Francis *see* Arbaces und panthea

Beaumont, J A *see* Travels in buenos ayres, and the adjacent provinces of the rio de la plata

Beaumont, Joseph *see* City of refuge

The beaumont library catalogue : reflecting the social and literary world of the late 1700s – late 1700s [mf ed Microforms International Marketing Corp] – 3mf – 9 – (with p/g ed by william paton & anthony davis. contains handwritten records of entries from the personal libraries of sir george howland beaumont and lady margaret. reveals one aspect of the intellectual, literary, and social interests in england during the time of the french revolution) – us UMI ProQuest [941]

Beaumont, Pierre de *see* Contes africains

Beaumont, R C de *see* Souvenir du banquet laurier, boston, mass, hotel vendome, mardi, 17 novembre 1891

Beaunier, dom *see*
- Abbayes et prieures de l'ancienne france
- Recueil historique des archeveches, eveches, abbayes et prieures de france
- Recueil historique des archeveches, eveches et prieures de france

Beaunoir, M De *see* Fanfan et colas

Beauplan, Amedee De *see* Dame du second

Beauregard, Cherry N *see* Tuba

Beausobre, M de *see* Histoire critique de manichee et du manicheisme

Beausoleil, Cleophas *see*
- Adresse de m c beausoleil
- Discours de m beausoleil, mp sur la reciprocite avec les etats-unis
- La reciprocite

Beausoleil, Joseph Maxime *see*
- Le dernier chant des serins de laval
- "Entre nous"
- La trompette de la metempsycose universitaire

La beaute des femmes : dans les poetes provencaux et dans la tradition populaire... / Prato, Stanislau – Paris: A Dupret [1888?] [mf ed Bloomington IN: Indiana Uni Lib, Preservation Dept 1984] – 1r – 1 – us Indiana Preservation [390]

Les beautes de la cantate du prince de galles – [S.l: s.n, 18–?] [mf ed 1984] – 1mf – 9 – 0-665-45458-9 – mf#45458 – cn Canadiana [780]

Beautes de la marine : ou recueil des traits les plus curieux, concernant les marin voyageurs, et les marins militaires des temps modernes / Caillot, Antoine – Paris 1823 [mf ed Hildesheim 1995-98] – 2v on 6mf – 9 – €120.00 – 3-487-29971-2 – gw Olms [910]

Beautes de l'histoire des etats-unis de l'amerique septentrionale : ou precis des evenemens les plus remarquables concernant ces differens etats, jusques et compris les deux dernieres guerres, et la paix de 1815... / Nougaret, Pierre Jean Baptiste – Paris: Chez Brunot-Labbe...1817 [mf ed 1984] – 6mf – 9 – 0-665-14226-9 – mf#14226 – cn Canadiana [917]

Beautes de l'histoire du canada : ou epoques remarquables, traits interessans, moeurs, usages, coutumes des habitans du canada... / Bossange, Gustave – Paris: Bossange freres, libraires...1821 [mf ed 1983] – 6mf – 9 – 0-665-44361-7 – mf#44361 – cn Canadiana [971]

Beautes du bresil / Henriot, Emile – Paris, France. 1946 – 1r – us UF Libraries [972]

Beauticians journal and guide – 1949 sep – 1 – mf#4718441 – us WHS [640]

Beauties of antiquity : or, remnants of feudal splendor and monastic times engraved in aquatinta / Hassell, John – London 1807 [mf ed Hildesheim 1995-98] – 2mf – 9 – €60.00 – 3-487-28824-9 – gw Olms [930]

Beauties of german literature : as exemplified by the works of pichler, richter, zschoekke, and tieck – London, New York: F Warne, [1868?] – (= ser Chandos classics) – 1 – (incl biogr notices) – mf#8363 – us Wisconsin U Libr [430]

The beauties of ireland : being original delineations, topographical, historical, and biographical, of each county / Brewer, James – London 1825-26 [mf ed Hildesheim 1995-98] – 2v on 8mf – 9 – €160.00 – 3-487-27849-9 – gw Olms [914]

Beauties of priestcraft : or, a short history of shakerism / Whitbey, John – New Harmony, IN: New Harmony Gazette, 1826 – 1r – 1 – (printed for the author at the office of the gazette; from historical & philosophical society of ohio) – us Western Res [243]

Beauties of samuel rutherford – Edinburgh, Scotland. 18– – 1r – us UF Libraries [240]

Beauties of the evangelical magazine – La Salle IL 1802-03 – 1,5,9 – mf#3560 – us UMI ProQuest [240]

Beauties of the st lawrence : the tourist's ideal trip via the richelieu and ontario navigation company's steamers / Foran, Joseph Kearney – [Quebec?: s.n, 1893?] – 1mf – 9 – 0-665-90959-4 – mf#90959 – cn Canadiana [917]

Beauties of wiltshire : displayed in statistical, historical, and descriptive sketches; interspersed with anecdotes of the arts / Britton, John – London 1801-25 [mf ed Hildesheim 1995-98] – 3v on 9mf – 9 – €180.00 – 3-487-27960-6 – gw Olms [880]

Beautifier – 1986 sep-nov – 1 – mf#4865645 – us WHS [640]

Beautiful / Lee, Vernon – Cambridge, England. 1913 – 1r – us UF Libraries [720]

The beautiful gleaner : a hebrew pastoral story: being familiar expositions of the book of ruth / Braden, William – 2nd ed. London: James Clarke, 1872 – 1mf – 9 – 0-8370-2432-3 – mf#1985-0432 – us ATLA [220]

Beautiful houses : being a description of...artistic homes / Haweis, Mary Eliza (Joy) – London 1882 – (= ser 19th c art & architecture) – 2mf – 9 – mf#4.2.37 – uk Chadwyck [720]

Beautiful joe : an autobiography / Saunders, Marshall – Toronto: Standard Pub Co, 1898 – (= ser Phoenix series) – 4mf – 9 – (int by hezekiah butterworth) – mf#32902 – cn Canadiana [920]

The beautiful life of francis e willard : a memorial volume / Gordon, Anna Adams – Chicago: Woman's Temperance Pub Assoc [c1898] [mf ed 1984] – (= ser Women & the church in america 108) – 5mf – 9 – 0-8370-1249-X – (int by lady henry somerset) – mf#1984-2108 – us ATLA [920]

Beautiful melrose, florida – Melrose, FL. 19–? – 1r – us UF Libraries [978]

A beautiful rebel : a romance of upper canada in eighteen hundred and twelve / Campbell, Wilfred – Toronto: Westminster, c1909 – 4mf – 9 – 0-665-73992-3 – mf#73992 – cn Canadiana [830]

Beautiful santa cruz county / Francis, Phil – Santa Cruz Co, CA. 1896 – 1r – 1 – $50.00 – mf#B40265 – us Library Micro [978]

Beautifying india / Randhawa, Mohindar Singh – Delhi: Rajkamal Publ, 1950 – (= ser Samp: indian books) – us CRL [710]

Beauty and art / Heaton, John Aldam – London 1897 – (= ser 19th c art & architecture) – 3mf – 9 – mf#4.2.127 – uk Chadwyck [700]

Beauty and power of holiness in heart and life – New York: M French [mthly] [mf ed 2005] – 2v on 1r – 1 – (latest iss consulted: v14 n11 (nov 1863) ed by mr & mrs austa malinda french; mf: v14 n1-11 (jan-nov 1863) [lacks: 1863 ind?]; cont: beauty of holiness in heart and life; absorbed by: guide to and beauty of holiness) – mf1064 – us ATLA [210]

Beauty classic magazine – v2 n1-3 [1985 winter-[summer]]; v3 n2-4 [1986-1987]; v4 n1 [1987] – 1 – mf#4717747 – us WHS [640]

The beauty, history, romance and mystery of the canadian lake region / Campbell, Wilfred – Toronto: Musson, c1914 – 4mf – 9 – 0-665-74833-7 – mf#74833 – cn Canadiana [917]

Beauty of holiness – Xenia OH, New York: M French. v7-11 1856-60 [mf ed 2005] – 5v on 2r – 1 – (cont: beauty of holiness and sabbath miscellany; cont by: beauty of holiness in heart and life; lacks: v8 (1957) [ind only], v9 n5 (may 1858) [covers only]) – mf#S1062 – us ATLA [240]

Beauty of holiness / Mant, Richard – Belfast, Northern Ireland. 1843 – 1r – us UF Libraries [240]

The beauty of holiness : ten lectures on external religious observances / Lee, Frederick George – 3rd ed. London: GJ Palmer, 1869 – 1mf – 9 – 0-7905-8500-6 – mf#1989-1725 – us ATLA [240]

The beauty of holiness and sabbath miscellany – Columbus OH: Scott & Bascom. v1-6 1853-55 [mf ed 2005] – 6v on 1r – 1 – (lacks: v3 n4; ind for v5?]; cont by: beauty of holiness) – mf#S1061 – us ATLA [242]

Beauty of holiness in heart and life – New-York: M French. v12-13 1861-62 [mf ed 2005] – 2v on 1r [ill] – 1 – (cont: beauty of holiness; cont by: beauty and power of holiness in heart and life) – mf1063 – us ATLA [240]

The beauty of immanuel : his name shall be called wonderful / Halsey, Leroy J – Philadelphia: Presbyterian Board of Publ, c1860 – 1mf – 9 – 0-7905-1528-8 – mf#1987-1528 – us ATLA [240]

Beauty of the liturgy of the church of england / Pratt, William Henry – Belfast, Northern Ireland. 1822 – 1r – us UF Libraries [241]

Beauty product marketing – San Diego CA 1989 – 1,5,9 – (cont: product marketing) – ISSN: 1040-5526 – mf#7506.06 – us UMI ProQuest [650]
Beauty spots of florida – s.l, s.l? . 193-? – 1r – us UF Libraries [978]
Beauty talk – v3 n10 [1994 oct], v4 n11 [1996 nov], v5 n1-v6 n2 [1996-97 apr/may], v7 n6 [1997 dec/1998 jan], v8 n1-3 [1998 feb/mar-jun/jul] – 1r – 1 – mf#3149248 – us WHS [640]
Beauty trade – 1954 oct-1961 nov; 1962 apr-1966 nov; 1967 jan-1971 dec; 1972 mar-1978 oct – 1 – mf#4723551 – us WHS [640]
Beauvoir, Vilfort *see* Controle financier du gouvernement des etats-unis
Beauvois, Eugene *see*
– La decouverte du nouveau monde par les irlandais et les premieres traces du christianisme en amerique avant l'an 1000
– Les derniers vestiges du christianisme preche du 10e au 14e siecle dans le markland et la grande irlande
– Les gallois en amerique au 12e siecle
– Origines et fondation du plus ancien eveche du nouveau monde
– Les papas du nouveau-monde rattaches a ceux des iles britanniques et nordatlantiques
Les beaux arts reduits a un meme principe / Batteux, Charles – Paris: Durand 1746 [mf ed 19–] – 1r – 1 – mf#film 1026 – us Sibley [780]
Beaux yeux and jeunes coeurs soyez fidelles : the favorite french air & gavotte, sung by sigra storace...in the opera of la cameriera astuta / Storace, Stephen – London: Birchall & Andrews [1788] [mf ed 19–] – 1r – 1 – mf#pres. film 73 – us Sibley [780]
Beaux-arts : Chronique des arts et de la curiosite – Dir. G. Wildenstein. Paris. 1923-juin 1940 – 1 – fr ACRPP [700]
Beavan, Charles *see* Beavan's reports
Beavan's reports : reports of cases argued and determined in rolls court / Beavan, Charles – v1-36. 1838-66. London: Saunders & Benning/Stevens & Norton, 1840-69 (all publ) – 276mf – 9 – $414.00 – mf#LLMC 95-283 – us LLMC [324]
Beaven, E W *see* Remnancy
Beaven, James *see* An account of the life and writings of s irenaeus, bishop of lyons and martyr
Beaver – 1916 aug; 1920 jul; 1922 jul; 1922 jan-1931 aug – 1 – mf#2473124 – us WHS [071]
The beaver : kanadische armee – Soest DE, 1957 3 may-1970 9 oct – 1 – gw Misc Inst [355]
Beaver and Toronto Mutual Fire Insurance Company. Annual meeting (2e: 1871 : Toronto, Ontario) *see* Proceedings at the second annual meeting...held march 21-23, 1871
Beaver argus and radical – Beaver, PA. -w 1896-1912 – 13 – $25.00r – us IMR [071]
Beaver baptist church – Cynthiana, KY. may 1809-nov 1896; 1905-jul 1910; nov 1919-1920 – (= ser WMU Records) – 1 – $47.25 – (wmu records, 1919-55) – us Southern Baptist [242]
Beaver briefs – v8 n4-v19 n4 [1976 oct-1987 fall] – 1 – mf#1573042 – us WHS [071]
Beaver city times – Beaver City, NE: Times Pub Co. -v28 n3. jan 10 1902 (wkly) [mf ed v20 n1. jan 11 1894-jan 10 1902 (gaps)] – 2r – 1 – (cont: beaver city weekly times. absorbed: hendly hustler. merged with: beaver valley tribune to form: beaver city times-tribune) – us NE Hist [071]
Beaver city times-tribune – Beaver City, NE: Frankie John. v117 n23. jun 7 1990- (wkly) [mf ed 1991-] – 1 – (cont: times-tribune (1989)) – us NE Hist [071]
Beaver city times-tribune – Beaver City, NE: Daryl D and Faye C Killough. 15v. v102 n18. may 1 1975-[v116] n17. apr 27 1989 (wkly) [mf ed filmed 1979-91] – 6r – 1 – (cont: times-tribune. cont by: times-tribune (1989)) – us NE Hist [071]
Beaver city times-tribune – Beaver City, NE: Merwin Pub Co. v28 n4. jan 17 1902-06// (wkly) [mf ed -mar 30 1906 (gaps)] – 3r – 1 – (formed by the union of: beaver city times and: beaver valley tribune. cont by: times-tribune. cont numbering of: beaver city times) – us NE Hist [071]
Beaver creek baptist church – Kershaw County, SC. 1868-83 – 1 – $5.00 – us Southern Baptist [242]
Beaver Creek Baptist Church. Henry County, Virginia *see* History of beaver creek baptist church
The beaver crossing bugle – Beaver Crossing, NE: H C Hensel. 3v. v1 n1. apr 24 1887-v3 n14. jul 24 1889 (wkly) [mf ed with gaps] – 1r – 1 – us NE Hist [071]
Beaver crossing newsletter – [Beaver Crossing, NE: Dorothy Christian] 1v. v1 n. jan 4 1977-v2 n1. jan 4 1978 (wkly) [mf ed filmed 1981] – 1r – 1 – (cont by: life at beaver crossing) – us NE Hist [071]
Beaver crossing times – Beaver Crossing, NE: F C Diers. 2nd yr n27. jul 5 1906- (wkly) [mf ed -may 23 1968 with gaps] – 23r – 1 – (cont: pride of beaver crossing. vol numbering dropped with feb 23 1956 issue) – us NE Hist [071]
Beaver dam argus – 1860 dec 7-1956 jul 12 [with gaps] – 1 – mf#986195 – us WHS [071]
Beaver dam baptist church – Beaufort County, SC. 1834-oct 1968 – 1 – $43.56 – us Southern Baptist [242]
Beaver dam baptist church – Fountain City, TN. By Mary Sue Beggs. 1959 – 1 – $5.00 – us Southern Baptist [242]
Beaver dam baptist church – Shelby, NC. 1850-1950 – 1 – $5.00 – us Southern Baptist [242]
Beaver dam baptist church – Fountain City, TN. 1802-1959 – 1 – $63.90 – (sunday school records, jan 1900-12) – us Southern Baptist [242]
Beaver dam daily citizen – Beaver Dam WI, 1915 jun 10-12, 1914 dec 10-1915 jun 9, 1914 jun 22-dec 9, 1913 dec 20-1914 jun 20, 1913 jul 1-dec 19, 1913 jan 2-jun 30, 1912 aug 1-dec 31, 1912 feb 14-jul 31, 1911 aug 24-1912 feb 13, 1911 feb 20-aug 23 – 10r – 1 – (cont by: daily citizen (beaver dam wi: 1911)) – mf#1139732 – us WHS [074]
Beaver dam daily citizen – Beaver Dam WI, 1930 dec 12-1953 jul-dec [with gaps] – 37r – 1 – (cont: daily citizen (beaver dam, wi: 1915); cont by: daily citizen (beaver dam, wi: 1971)) – mf#1139734 – us WHS [074]
Beaver dam democrat – 1859 aug 13; 1860 jan 7-1861 dec 28 – 1 – mf#926936 – us WHS [071]
Beaver dam republican – 1853 feb 10-1855 apr 18 – 1 – mf#955257 – us WHS [071]
Beaver dam sentinel – [1854 oct 26-1855 apr 12] – 1 – mf#1093920 – us WHS [071]
Beaver, Kathryn L *see* Impact of the acquired immunodeficiency syndrome (aids)
The beaver lake tragedy : a full and particular account of the whole proceedings in the above extraordinary crimes... / Slavin, Patrick, Sr – NY: publ for B O'Brien, 1857 [mf ed 1983] – 1mf – 9 – 0-665-43305-0 – mf#43305 – cn Canadiana [345]
Beaver, P *see* African memoranda
Beaver, Philip *see* African memoranda
Beaver. Presbytery (Pres. Church in the USA) *see* Minutes, 1833-1870
The beaver radical – Beaver, PA. Dec 11 1868-Dec 24 1869; Jan 7 1870; Dec 15 1871; March 15 1872 – 2r – 1 – us L of C Photodup [071]
Beaver state herald – Gresham, Montavilla OR: Beaver State Pub Co, -1914 [wkly] – 1 – (merged with: mount scott news (1906-14) to form: mt scott herald (1914-23)) – us Oregon Lib [071]
Beaver state herald *see*
– Mt scott herald
– Mount scott news
Beaver state news – Hubbard OR: R B Conover, [wkly] – 1 – us Oregon Lib [071]
Beaver valley labor history journal – 6 1 v1 n1-v3 n1 [1979 mar-1981 jul] – 1 – mf#669174 – us WHS [331]
Beaver valley mercury – Danbury, NE: Illustrative Publ. v1 n23. nov 19 1936-v4 n34. dec 21 1939 (wkly) [mf ed with gaps filmed 1978] – 2r – 1 – (absorbed by: mccook republican) – us NE Hist [071]
The beaver valley news – St Edward, NE: P A BArrows, 1891 (wkly) [mf ed v2 n5. jul 29 1892 filmed [1983] – 1r – 1 – us NE Hist [071]
Beaver valley tribune – Beaver City, NE: F N Merwin. v5 n1. apr 17 1890-v16 n41. jan 10 1902 (wkly) [mf ed with gaps filmed -1992] – 4r – 1 – (merged with: beaver city times to form: beaver city times-tribune) – us NE Hist [071]
Beaverbrook, Max Aitken, Baron *see* Canada in flanders
Beavercreek daily news / Montgomery Co. Dayton – apr 1977-jul 1979 [daily] – 18r – 1 – mf#B25612-25629 – us Ohio Hist [071]
Beaverdam baptist church – Asheville, NC. 1952-75 – 1 – $53.37 – us Southern Baptist [242]
Beaverdam baptist church – Fair Play, SC. 1868-1963 – 1 – $38.07 – us Southern Baptist [242]
Beaverdam baptist church – Wilkes County, GA. 1836-1923 – 1 – $32.99 – us Southern Baptist [242]
Beaverton enterprise – Beaverton OR: S M Brown, 1927-51 [wkly] [mf ed 1961] – 6r – 1 – (merged with: aloha news (1927-51) and: tigard sentinel (1924-51) and: multnomah press (1926-51), to form: valley news (beaverton, or)) – us Oregon Lib [071]
Beaverton enterprise *see*
– Aloha news
– Tigard sentinel
Beaverton express – Ontario, CN. jan 1940-dec 1979 – 18r – 1 – cn Commonwealth Imaging [071]
Beaverton review – Beaverton OR: J H Hulett, -1941 [wkly] – 1 – (absorbed: banks herald (1910); cont by: banks herald (1926); 1925-26 incl newspaper publ during school terms by beaverton high school; beginning with jan 21 1926 iss, the banks herald resumes independent publ) – us Oregon Lib [071]
Beaverton review *see*
– Banks herald (banks, or: 1910)
– Banks herald (banks, or: 1926)
Beaverton times – Beaverton OR: A J Hicks, [wkly] [mf ed 1974] – 1r – 1 – (cont: owl (1912-)) – us Oregon Lib [071]
Beaverton valley times – Beaverton OR: Times Pub, 1989- [wkly] [mf ed 1991-] – 14r – 1 – (cont: valley times (1962-89)) – us Oregon Lib [071]
Beawes, Wyndham *see* Lex mercatoria rediviva
Beazley, C R *see* The text and versions of john de plano carpini and william de rubruquis
Beazley, Charles Raymond *see* John and sebastian cabot
Bebashi news – 1989 oct-1989 feb; 1992 summer – 1 – mf#4848520 – us WHS [071]
Bebbington news – [NW England] Bebbington 1936-30 jun 1956 – 1 – (title change: bebbington news & advertiser 7 jul 1956-1969; bebbington news 1970-) – uk MLA; uk Newsplan [072]
O bebe – Tabuleiro Grande, MG. 05 jun 1897 – (= ser Ps 19) – bl Biblioteca [079]
Bebel, A *see* Gegen den militarismus und gegen die neuen steuern
Bebel, Auguste *see*
– La femme et le socialisme
– Massovaia politicheskaia stachka i sotsialdemokratiia
Bebel, Heinrich *see* Heinrich bebels facetien drei buecher
Beberapa fasal ekonomi : djalan ke ekonomi dan kooperasi / Hatta, M – Djakarta: Oesaha Baroe "Penjiar" (2602) – 133p 1mf – 9 – mf#SE-2002 mf189-190 – ne IDC [330]
Beberapa penggalan dari sedjarah perdjoeangan oemmat islam : nomor-peringatan setahoen "asj-sjoe'lah" / Asj-Sju'llah – Djakarta: Gunseikanbu Syumubu, 2605 – 93p 1mf – 9 – mf#SE-2002 mf20 – ne IDC [260]
Bebermeyer, Gustav *see*
– Heinrich bebels facetien drei buecher
– Hermann flayders ausgewaehlte werke
Bebraer nachrichten : sontraer anzeiger mit den mitteilungen des kreises rotenburg a f – obersuhler zeitung – Bebra DE, 1933-1935 29 jun – 6r – 1 – gw Misc Inst [074]
Bebraer tageblatt – Bebra DE, 1903-08, 1909 apr-1944 – 58r – 1 – (filmed with suppl; title varies: 1 jan 1914: bebraer tageszeitung, later: ns-tageblatt fuer den kreis rotenburg a f und nachbargebiete, 15 aug 1942: ns-tageblatt fuer ostkurhessen und nachbargebiete) – gw Misc Inst [074]
Bebraer zeitung – Bebra DE, 1902-1912 29 sep – 14r – 1 – (filmed with suppl) – gw Misc Inst [074]
Bebraer zeitung und eisenbahn-anzeiger – Bebra, Fulda DE, 1896 4 jan-28 apr – 1r – 1 – (filmed with suppl) – gw Misc Inst [380]
The bec missal (hbs94) / Hughes, A – 1963 – (= ser Henry bradshaw society (hbs)) – 6mf – 8 – €14.00 – ne Slangenburg [241]
Becados / Miranda, Anisia – Habana, Cuba. 1965 – 1r – us UF Libraries [972]
Beccadelli, Antonius [Panormita] *see* Libellus hermaphroditi
Beccari C *see* Expedionis aethiopicae
Beccari, C *see* Rerum aethiopicarum scriptores occidentales inediti a saeculo 16 ad 19
Beccles and bungay journal – Beccles 4 feb 1933-to date – 1 – (n14825 4 feb 1933-to date previously norwich mercury, east suffolk ed) – uk Newsplan [072]
Beccles weekly news – [East Midlands] Suffolk 29 jun 1858-dec 1893, 1898-1 mar 1926 [mf ed 2004] – 58r – 1 – (missing: 1896-97, 1912; cont by: beccles and bungay weekly news [jan 1861-dec 1867]; east suffolk gazette [jan 1868-mar 1926]) – uk Newsplan [072]
Becerra Gonzalez, Maria *see* Derecho minero de mexico y vocabulario con definic...
Becerra, Longino *see* Problema agrario en honduras
Becerra Tanco, Luis *see* Felicidad de mexico en la admirable aparicion de la virgen maria nra. senora de guadalupe, y origen de su milagrosa imagen
Becerra y Valcarzel, Diego *see*
– Carta pastoral
– De iure sacrorum
Becerro de Bengoa, Ricardo *see*
– Cartilla politica donosiana
– Ensayo para una teoria de extremadura
– Hacia la union de los pueblos latinos
– La hermandad de alfereces y el destino de espana
– La idea tradicional. destino de espana por
– El movimiento de union latina en extremadura
– El muno hispanico y su reitegracion historica bajo el signo de guadalupe
– Nacional-integrismo
– Programacion ideologica del bimilenario de merida
– Reorganizacion del frente nacional de excombatientes
Bech, Birger *see*
– Five years in a sailor's life
– The unknown
Bech, Fedor *see* Hartmann von aue
Bechard, Auguste *see*
– L'ancien quebec, descriptions, nos archives, etc
– Biographie de m francois vezina
– La gaspesie en 1888
– Histoire de la paroisse de saint-augustin
– Histoire de l'ile-aux-grues et des iles voisines
– L'hon pierre garneau
– L'honorable a-n morin
– L'honorable joseph-g blanchet
– M l'abbe francois pilote
Becher, Heinrich *see* Die ausfuehrungsgesetze zum buergerlichen gesetzbuche
Becher, Hubert *see*
– Der deutsche primas
– Ernst juenger
Becher, Johannes Robert *see*
– Abschied
– Anders ist der neue tag
– Der bankier reitet ueber das schlachtfeld
– Der befreier
– Dank an stalingrad
– Dichtung
– Es wird zeit
– Gedichte fuer ein Volk
– Der gestorbene
– Gewissheit des siegs und sicht auf grosse tage
– Der gluecksucher und die sieben lasten
– Die hohe warte
– Hymnen
– Das neue gedicht
– Roter marsch / der leichnam auf dem thron / die bombenflieger
– Ein staat wie unser staat
– Um gott
– Verfall und triumph
– Vom anderswerden
– Wiedergeburt
– Zion
Becherwahrsagung bei den babyloniern / Hunger, J – Leipzig, 1903 – (= ser Leipziger semitistische Studien) – 1mf – 9 – (leipziger semitistische studien, 1904 v1 pt1) – mf#NE-20109 – ne IDC [956]
Becherwahrsagung bei den babyloniern : nach zwei keilschrifttexten aus der hammurabi-zeit / Hunger, Johannes – Leipzig: J C Hinrichs 1903 [mf ed 1986] – 1mf – 9 – 0-8370-7068-6 – (incl bibl ref) – mf#1986-1068 – us ATLA [290]
Bechet, Eugene *see* Cinq ans de sejour au soudan francais
Bechmann, Trude *see* Jeannette
Bechstein, Johann Matthaeus *see* Diana
Bechstein, Ludwig *see*
– Deutsches sagenbuch
– Ein dunkles loos
– Faustus
– Luther
– Neues deutsches maerchenbuch
– Der ring
– Thueringens koenigshaus
Bechstein, Ludwig [comp] *see* Deutsche maerchen und sagen
Bechstein, Reinhold *see*
– Altdeutsche maerchen, sagen und legenden
– Gottfried's von strassburg tristan
– Heinrich und kunigunde
– Heinrich's von freiberg tristan
– Ulrich's von lichtenstein frauendienst
Bechtel, Christine *see* Musikdidaktische aufgaben im kinderchor (1984)
Bechter, Barbara *see* Der garten von vaux-le-vicomte
Bechtold, Fritz *see* Nanga parbat adventure
Bechtold, O *see* Der "ruf nach synoden" als kirchenpolitische erscheinung im jungen erzbistum freiburg (1827-1860)
Bechuana fireside tales / Savory, Phyllis – Cape Town, South Africa. 1965 – 1r – us UF Libraries [390]
The bechuana of south africa / Crisp, William – London: Society for Promoting Christian Knowledge, 1896 – 1 – us CRL [960]
Bechuana spelling-book : buka ea likaelo tsa eintla... = Lichuanelo tsa molemo / Moffat, Robert – Cape Town: SA Library 1980 – 1r – 1 – (text in english and tlhaping dialect of tswana) – sa National [470]
Bechuanaland / Munger, Edwin S – London, England. 1965 – 1r – us UF Libraries [960]
Bechuanaland daily news – Gaborone: Bechuanaland Govt Info Service, sep 1 1965-sep 16 1966 – us CRL [079]
Bechuanaland Protectorate. African Advisory Council *see* Minutes of the...session of the african advisory council
The bechunanas, the cape colony, and the transvaal : proceedings of the public meeting held at the mansion house, london, on tuesday, november 27th, 1883 / Aborigines Protection Society, London – London, 1884 – (= ser 19th c books on british colonization) – 1mf – 9 – mf#1.1.3710 – uk Chadwyck [960]

Beck, Aaron N *see* Civil war diary
Beck, Belinda R *see* An investigation of anatomical structures associated with the site of medial tibial stress syndrome, often referred to as "shin splints"
Beck, Carl *see*
– Flores musice omnis cantus gregoriani
– Schleiermacher als mann der kirche
– Schleiermacher, ein deutscher mann
Beck, Dietrich *see* Die kirchlichen simultanverhaeltnisse der rheinprovinz unter besonderer beruecksichtigung des ryswicker friedens
Beck, Friedrich *see*
– Einkehr
– Singgedichte
Beck, George Fairley *see* Daybreak
Beck, Henry *see* Flute book
Beck, Henry Houghton *see* Cuba's fight for freedom and the war with spain
Beck, Herbert *see* Mittelalterliche skulpturen in barockaltaeren
Beck, Hermann *see*
– Kaspar klee von gerolzhofen
– Die religioese volkslitteratur der evangelischen kirche deutschlands in einem abriss ihrer geschichte
Beck, Horace C *see* The beads from taxila
Beck, James M *see* The evidence in the case
Beck, Jean-B *see*
– Le chansonnier cange (bibl nat paris fonds fr n846). les chansonniers des troubadours et des trouveres, no 1 facsimile-edition par jean beck
– Le manuscrit du roi (bibl nat paris fonds fr n844). les chansonniers des troubadours et des trouveres, no 2 facsimile-edition par jean beck
– Die melodien der troubadours. troubadours und trouveres
Beck, Johann Tobias *see*
– Einleitung in das system der christlichen lehre, oder, propaedeutische entwicklung der christlichen lehrwissenschaft
– Erklaerung der briefe petri
– Erklaerung der offenbarung johannes, cap 1-12
– Erklaerung der propheten micha und joel
– Erklarung der propheten nahum und zephanja
– Die ethische erscheinung des christlichen lebens
– Die genetische anlage des christlichen lebens
– Leitfaden der christlichen glaubenslehre
– Outlines of biblical psychology
– Die paedagogische entwicklung des christlichen lebens
– Vorlesungen ueber christliche glaubenslehre
Beck, John S *see* Service of the church of the redeemer, brighton, mass
Beck, Karl *see*
– Lieder vom armen mann
– Naechte
– Taeubchen im nest, 1860
Beck, Karl Isidor *see* Gedichte
Beck, W C A *see* Der panamakanal...
Beck, William *see*
– The friends
– The london friends' meetings
Beckenbach, J R *see*
– Fertility program for celery production on everglades organic soils
– Functional relationships between boron and various anions in the nutrition of the tomato
Beckenham advertiser – Bromley, England 9 sep 1982-30 jan 1986 [jan 1986] – 1 – (cont, in pt, of: beckenham & penge advertiser; fr feb 1986 onward amalg with: penge advertiser & publ as: beckenham & penge advertiser) – uk British Libr Newspaper [072]
Beckenham and penge record – Bromley, England 31 jul 1972-26 sep 1990 [1986-] – 1 – (cont by: beckenham & penge news shopper [3 oct 1990-27 feb 1991]) – uk British Libr Newspaper [072]
Beckenham & district times – Bromley, England 6 jan 1905-25 may 1928 – 1 – (wanting: 1914; cont by: beckenham & kentish times [1 jun 1928-29 aug 1941]) – uk British Libr Newspaper [072]
Beckenham journal, and penge and sydenham advertiser – Bromley, England 1 sep 1876, 20 jan 1883-5 dec 1885 [mf jan-dec 1912, 1979-82] – 1 – (replaced by: beckenham times [1985-9]) – uk British Libr Newspaper [072]
Beckenham & penge advertiser – [London & SE] Bromley 12 jul 1888-20 may 1909, 1 jan 1914-2 sep 1982, 7 feb 1986-10 nov 1989 [mf 7 feb 1986-10 nov 1989] – 1 – (wanting 1897; publ as 2 ed between 9 sep 1982 & 30 jan 1986 : beckenham advertiser & penge advertiser; cont as 2 ed fr 17 nov 1989 onward: bromley & beckenham advertiser & penge and annerley advertiser) – uk Newsplan; uk British Libr Newspaper [072]
Beckenham & penge news shopper – Bromley, England 3 oct 1990-27 feb 1991 [mf 1986-] – 1 – (cont: beckenham & penge record [31 jul 1972-26 sep 1990]; cont by: news shopper (beckenham & penge) 6 mar 1991-) – uk British Libr Newspaper [072]
Beckenham times – Bromley, England 12 dec 1985-26 oct 1989 [mf 1986-9] – 1 – (amalg with: bromley times & the chislehurst times & subsequently publ as: bromley beckenham chislehurst times [2 nov 1989-]) – uk British Libr Newspaper [072]
Becker, A *see* Papst urban 2 (1088-1099) (mgh schriften:19.bd 1.teil)
Becker, Aaron *see* Ha-mediniyut ha-miktso'it veha-kalkalit shel ha-histadrut
Becker, B *see* Zedekunst
Becker, Bernard Henry *see* Disturbed ireland
Becker, Bernhard *see*
– Bronnen tot de kennis van hret leven en de werken van d van coornhert
– Zinzendorf und sein christentum
Becker, C F *see* Lieder und weisen vergangener jahrhunderte
Becker, Carl *see* Die braut des spaniers
Becker, Carl Heinrich *see* Christianity and islam
Becker, Christoph *see* Lichtenstein
Becker, Emile *see* Le reverend pere joseph gonnet de la compagnie de jesus
Becker, Ernest *see* Birth and death of meaning
Becker, Ernst *see* Annalen der kaiserlichen universitaets-sternwarte in strassburg
Becker, Frank Silvester *see* The excise and hotel laws of the state of new york
Becker, H F et al *see* Neue monatsschrift von und fuer mecklenburg
Becker, Hendrik *see* Bausteine zur deutschen literaturgeschichte
Becker, Henrietta K *see* Kleist and hebbel
Becker, Jeronimo *see* Carta y otros documentos de hernando cortes
Becker, Johann Philip *see* Neue stunden der andacht
Becker, Karl Friedrich *see* Karl friedrich becker's weltgeschichte
Becker, Monika *see* Kreativitaet und historismus
Becker, Nicolaus *see* Gedichte
Becker, Otto Eugen Hasso *see* Das australische abenteuer
Becker, Peter *see*
– Hill of destiny
– Path of blood
– Rule of fear
– Sandy tracks to the kraals
Becker, Philipp August *see* Der suedfranzoesische sagenkreis und seine probleme
Becker, R B *see*
– Circulatory system of the cow's udder
– Effect of calcium-deficient roughages upon mild production and welfare of dairy cows
– Salt sick
– Stiffs or sweeny (phosphorus deficiency) in cattle
Becker, Rud Z *see* Nationalzeitung der teutschen
Becker, Rudolf *see* Christian weises romane und ihre nachwirkung
Becker, Susan L *see* An examination of the relationship among target structures, team motivational climate, and achievement goal orientation
Becker, W A *see* Charikles
Becker, Werner *see*
– Sportunterricht und hochschulsport in frankreich
– Untersuchungen zur veraenderung der informationsverarbeitungsfaehigkeit von 9 bis 12jaehrigen kindern im verlauf von lernprozessen unterschiedlicher modalitaetspraeferenzen
Becker, Wilhelm *see*
– Die geheimen gesellschaften mit vollem rechte verurtheilt von der katholischen kirche
– Sterne und sternsysteme
Becker, Wilhelm Adolph *see* Gallus
Becket 1765-1900 – Oxford, MA (mf ed 1988) – (= ser Massachusetts vital records) – 35mf – 9 – 0-87623-068-0 – (mf 1-9: town records 1765-1835. mf 10-14: proprietors records 1737-71. mf 15-23: towen records 1836-56. mf 24-27: index to births, marriages, deaths 1850-58. mf 28-29: births, marriages, deaths 1844-58. mf 30-35: births, marriages, deaths 1859-1900) – us Archive [978]
Becket, archbishop of canterbury : a biography / Robertson, James Craigie – London: J Murray 1859 [mf ed 1990] – 1mf – 9 – 0-7905-6674-5 – (incl bibl ref) – mf#1988-2674 – us ATLA [241]
Becket, Hugh W *see* Record of winter sports, 1883-84
Beckett 1753-1849 – Oxford, MA (mf ed 1995) – (= ser Massachusetts vital record transcripts to 1850) – 4mf – 9 – 0-87623-207-1 – (mf 1t-3t: marriages & intentions 1765-1849; births & deaths 1753-1841. mf 4t: b,m,d 1844-49) – us Archive [978]
Beckett, John Edgar *see* Hints on agriculture
Beckett, William *see* Christian's desire to depart
Beckford, William *see* Italy
Beckh, Hermann *see* Buddhismus (buddha und seine lehre)
Beckherrn, Richard *see* M opitz, p ronsard und d heinsius
Beckley heights baptist church – Dallas, TX. 22 Apr 1953-12 Feb 1964 – 1 – $17.82 – us Southern Baptist [242]
Beckman, Erik Richard *see* The massacre at sianfu
Beckman, J *see*
– Die kirchliche ordnung der taufe
– Quellen zur geschichte des christlichen gottesdienstes
Beckman, S *see* Die gottesanrede im ante-sanctus
Beckmann, Christoph *see* Tagesprofil von leukozytensubpopulationen im peripheren blut gesunder probanden
Beckmann, J *see* Litteratur der aelteren reisebeschreibungen...
Beckmann, Joachim *see* Kirchliches jahrbuch fuer die evangelische kirche in deutschland, 1933-1944
Beckmann, Joh *see* Beytraege zur oekonomie, technologie, polizey- und cameralwissenschaften
Beckmann, Joh [v1-18] *see* Physikalisch-oekonomische bibliothek
Beckmann, Johann *see*
– Physikalisch-oekonomische bibliothek
– Vorbereitung zur waarenkunde oder zur kenntnis de vornehmsten auslaendischen waaren
The beckoning hand : and other stories / Allen, Grant – London: Chatto and Windus, 1887 – 5mf – 9 – (with a frontispiece by townley green) – mf#05009 – cn Canadiana [830]
Beck's journal of decorative art – 1886-88 [mf ed Chadwyck-Healey] – (= ser Rare 19th century american art journals) – 18mf – 9 – uk Chadwyck [740]
Beckstrat, Bernd *see* Rund um den kohlenpott
Beckumer kreisanzeiger *see* Westfaelische nachrichten [main edition]
Beckumer zeitung – Beckum 1847 12 jun, 1932 1 jul-30 sep, 1933 1 jul-30 sep [mf ed 2004] – 2r – 1 – gw Mikrofilm [074]
Beckwith, Clarence Augustine *see* Realities of christian theology
Beckwourth, James P *see* The life and adventures of james p beckwourth
Become as little children – London, England. 1827 – 1r – us UF Libraries [240]
Becon, Thomas *see* Writings of the rev. thomas becon
Becontree guardian and chadwell heath news – Barking, London 23 mar-21 dec 1923, 7 mar-26 dec 1924, 14 jan-23 dec 1927 (wkly) – 2 1/2r – 1 – (incorp with: dagenham post & barking & rainham guardian) – uk British Libr Newspaper [072]
Becot, Joseph *see* De l'organisation de la justice repressive aux principales epoques historiques
Becvarovsky, Antonin Frantisek *see* Trois sonatas pour piano forte avec violon et violoncelle obliges, oeuvre 3me
Beda lam kabya mya : [poems] / Jo Gyi – Yan Kun: Kan Ko Mre Cape 1970 [mf ed 1990] – 1r with other items – 1 – (in burmese) – mf#mf-10289 seam reel 150/5 [§] – us CRL [810]
Beda Venerabilis *see*
– Expositio actuum apostolorum. retractatio in actus apostolorum. nomina regionum atque locorum de actibus apostolorum. in epistulas 7 catholicas
– In tobiam. in proverbia. in cantica canticorum. in habacuc
Bedale, Charles Lees *see* The social teaching of the bible
Bedan lakkhana / Kyaw, U, Hsu htu pan – Ran kun: Chu Thu Pan Ca pe 1975 [mf ed 1995] – on pt of 1r – 1 – mf#11052 r1966 n2 – us Cornell [959]
Bedan pon khyup kyam / Cin Chan, Cha ra – Ran kun Mrui': u the on, Mrac man Eravati Ca pe 1975 [mf ed 1995] – on pt of 1r – 1 – mf#11052 r1980 n6 – us Cornell [959]
Bedard, Denyse *see* Bio-bibliographie de jules-s lesage
Bedard, Elzear *see* Report
Bedard, J Alphonse *see* Le guide commercial pour la ville de detroit et ses environs
Bedard, Jean Baptiste Charles *see* Declaration et observations presentees
Bedard, Joseph-Edouard *see*
– Code municipal de la province de quebec annote 1898-1902
– Code municipal de la province de quebec annote mis au courant de la legislation et de la jurisprudence
– L'ivrognerie et la loi des licences
Bedard, M H *see* Le jeune homme et la litterature
Bedard, Pierre-Stanislas *see* To the commons' house of assembly of lower canada
Bedard, Suzanne *see* Bibliographie analytique de l'oeuvre de charlotte savary
Bedarf es einer besonderen inspirationslehre? : vortrag gehalten auf der theologischen konferenz in kiel am 7. juli 1891 / Kier, P O – Kiel: Ernst Homann 1891 [mf ed 1985] – 1mf – 9 – 0-8370-3895-2 – mf#1985-1895 – us ATLA [220]
Bedatthadipani : nnon kan brui haj kyam ran kri path nisya / Kavindabhi Saddhammadharadhaja – Ran kun: Jambu Sangaha Pitakat Ca pe: Phran khyi ra thana, Sutavati Ca aup tuik 1981- [mf ed 1990] – 1r with other items – 1 – (at head of title: pu sim kyon cha ra to u silacara thap mam phrann cvak re sa pru cu so; in burmese & pali) – mf#mf-10289 seam reel 134/3 [§] – us CRL [130]
Bedava gazete [akbaba' nin ilavesi] – Istanbul: Sabah Matbaasi, 1925. Mueduer-i Mesul: Yusuf Ziya [Ortac] n1-5. 24 mart-9 nisan 1341 [1925] – (= ser O & t journals) – 1mf – 9 – $25.00 – us MEDOC [956]
Beddoes, Thomas *see* Authentic and useful history of issac jenkins, his wife...
Beddome, R H *see* Handbook of the ferns of british india, ceylon and malaya peninsular
Beddow, J J *see* Evening communions
Beddy, Joseph Fawcett *see* Faithful minister
Bede, Saint *see* Bede's ecclesiastical history of the english people
Bedel, J *see* La vie du r p piere fourier
Bedell, Kenneth B *see* Swaziland methodist experiment with racial blending
Bedell, Ralph Clairon *see* Relationship between the ability to recall...
Bedencken ob der verraehter judas auch an dem tisch desz herren gesessen... / Bullinger, Heinrich – N.p., Nicolaus Erbenius, 1596 – 1mf – 9 – mf#PBU-267 – ne IDC [240]
Bedenken der theologischen facultaeten der landesuniversitaet jena und der universitaeten zu berlin, goettingen und heidelberg : ueber das rescript des herzoglichen consistoriums zu altenburg vom 13. nov. 1838 (den kirchlichen separatismus in der ephorie ronneburg betreffend) und ueber zwei verwandte fragen – Altenburg: In Commission der Schnuphase'schen Buchhandlung 1839 [mf ed 1992] – 1mf – 9 – 0-524-02607-6 – mf#1990-0659 – us ATLA [230]
Bede's ecclesiastical history of the english people : a revised translation = Historia ecclesiastica gentis anglorum / Bede, Saint – London: G Bell 1917 [mf ed 1992] – (= ser Bohn's antiquarian library) – 2mf – 9 – 0-524-03754-X – (int, life & notes by a m sellar) – mf#1990-1101 – us ATLA [240]
Die bedeutendsten romane philipps von zesen und ihre literargeschichtliche stellung / Gartenhof, Kaspar – Nuernberg: G P J Bieling-Dietz, 1912 – 1r – 1 – (incl bibl ref) – us Wisconsin U Libr [430]
Die bedeutung buergerlicher und kuenstlerischer lebensform fuer goethes leben und werk : dargestellt am faust 1. teil / Kurzweil, Benedikt – Limburg: Limburger Vereinsdruckerei, 1933 – 1 – (incl bibl ref) – us Wisconsin U Libr [430]
Die bedeutung calvins und des calvinismus fuer die protestantische welt : im lichte der neueren und neuesten forschung / Knodt, Emil – Giessen: A Toepelmann, 1910 – (= ser Vortraege der theologischen Konferenz zu Giessen) – 1mf – 9 – 0-524-04262-4 – (incl bibl ref) – mf#1991-2046 – us ATLA [242]
Die bedeutung der allgemeinen sittenlehre des buddhismus / Yasuda, Minori – Jena: B Engau, 1893 – 1mf – 9 – 0-524-02728-5 – (incl bibl ref) – mf#1990-3131 – us ATLA [280]
Die bedeutung der antiochenischen schule auf dem exegetischen gebiete : nebst einer abhandlung ueber die aeltesten christlichen schulen / Kihn, Heinrich – Weissenburg: C F Meyer, 1866 – 2mf – 9 – 0-524-00049-2 – mf#1989-2749 – us ATLA [225]
Die bedeutung der atlantik-pacifik-eisenbahn fuer das reich gottes : eine festschrift / Plath, Karl Heinrich Christian – Berlin: W Schultze, 1871 – 1mf – 9 – 0-524-02863-X – (incl bibl ref) – mf#1990-0720 – us ATLA [240]
Die bedeutung der im art. 119 i 2 der reichsverfassung ausgesprochenen gleichberechtigung der geschlechter de lege ferenda fuer die personenrechtliche stellung der ehefrau / Keßler, Aloys Wilhelm – Erlangen, 1934 (mf ed 1995) – 1mf – 9 – €24.00 – mf#DHS-AR 3162 – gw Frankfurter [342]
Die bedeutung der juden fuer erhaltung und wiederbelebung der wissenschaften im mittelalter *see* The importance of the jews for the preservation and revival of learning during the middle ages
Bedeutung der nagelhistologie fuer die diagnostik der onychomykose : direkter vergleich zwischen der histologischen und mykologisch-kulturellen untersuchung / Lacour, Till – 1996 – 1mf – 9 – 3-8267-2374-0 – mf#DHS 2374 – gw Frankfurter [616]
Die bedeutung der reformierten theologie fuer die religioese lage der gegenwart : vortrag auf der hauptversammlung des reformierten bundes, herford, 30. august 1905 / Lang, August – Neukirchen, Kreis Moers: Verlag der Buchh des Erziehungsvereins, [1905?] – 1mf – 9 – 0-7905-6758-X – mf#1988-2758 – us ATLA [242]
Die bedeutung der somatotropen achse fuer muskelwachstum und reproduktion bei landwirtschaftlichen nutztieren / Sauerwein, Helga – (mf ed 1994) – 3mf – 9 – €49.00 – 3-8267-2014-8 – mf#DHS 2014 – gw Frankfurter [630]
Die bedeutung der werturteile fuer das religioese erkennen / Scheibe, Max – Halle a. S:Max Niemeyer, 1893 – 1mf – 9 – 0-8370-5080-4 – (incl bibl ref) – mf#1985-3080 – us ATLA [200]

Die bedeutung der wirtschaftlichen kooperation fuer die wirtschaftsentwicklung chinas am beispiel joint ventures / Chen, Xinhua – (mf ed 1995) – 2mf – 9 – €40.00 – 3-8267-2189-6 – mf#DHS 2189 – gw Frankfurter [337]

Die bedeutung der wortsippe kvd im hebraeischen / Caspari, Wilhelm – Leipzig: A Deichert, 1908 – 1mf – 9 – 0-8370-9213-2 – (incl bibl ref and index) – mf#1986-3213 – us ATLA [470]

Die bedeutung des aesthetischen in der evangelischen religion – noch ein wort ueber den christlichen dienst / Gross, G & Schlatter, Adolf von – Guetersloh: C Bertelsmann, 1905 – (= ser Beitraege zur foerderung christlicher theologie) – 1mf – 9 – 0-7905-9212-6 – mf#1989-2437 – us ATLA [242]

Die bedeutung des bergbaus bei goethe und in der deutschen romantik / ed by Duerler, Josef – Frauenfeld: Huber & Co Aktiengesellschaft, 1936 [mf ed 1999] – (= ser Wege zur dichtung 24) – 89p (ill) – 1 – (incl bibl ref) – mf#10194 – us Wisconsin U Libr [430]

Die bedeutung des geschichtlichen in der religion / Hase, Karl Alfred von – Leipzig: Breitkopf und Haertel, 1874 – 1mf – 9 – 0-7905-4808-9 – mf#1988-0808 – us ATLA [210]

Die bedeutung des hieronymus fuer die alttestamentliche textkritik / Nowack, Wilhelm – Goettingen: Vandenhoeck & Ruprecht, 1875 – 1mf – 9 – 0-524-05050-3 – (incl bibl ref) – mf#1992-0303 – us ATLA [221]

Die bedeutung des musikalischen und akustischen in e.t.a. hoffmanns literarischem schaffen / Schaeffer, Carl – Marburg a.L.: N G Elwert, 1909 – 1r – 1 – (incl bibl ref (3rd prelim. page)) – us Wisconsin U Libr [430]

Die bedeutung des niblungenlieder fuer die deutsche nation / Bergmann, Anton – Karlsruhe i. Baden, 1924 [mf ed 1987] – 24p – 1 – mf#8692 – us Wisconsin U Libr [390]

Die bedeutung des todes jesu : nach seinen eigenen aussagen auf grund der synoptischen evangelien / Hollmann, Georg – Tuebingen: J C B Mohr, 1901 – 1mf – 9 – 0-7905-1108-8 – mf#1987-1108 – us ATLA [240]

Die bedeutung des vergeltungsgedankens fuer die ethik jesu, dargestellt im anschluss an die synoptischen evangelien / Karner, Friedrich Karl – Oedenburg-Sopron, 1927 (mf ed 1993) – 2mf – 9 – €31.00 – 3-89349-325-5 – mf#DHS-AR 179 – gw Frankfurter [240]

Die bedeutung funktionaler probleme fuer die medizinische versorgung aelterer menschen : das versorgungskonzept des geriatrischen assessment / Pientka, Ludger – (mf ed 1995) – 6mf – 9 – €62.50 – 3-8267-2168-3 – mf#DHS 2168 – gw Frankfurter [362]

Die bedeutung richard simons fuer die pentateuchkritik / Stummer, Friedrich – Muenster in Westf: Aschendorff, 1912 – 1mf – 9 – 0-7905-3237-9 – (incl bibl ref) – mf#1987-3237 – us ATLA [221]

Die bedeutung von gruppen- und minderheitenrechten fuer die suedafrikanische verfassungsentwicklung / Weber, Rolf – (mf ed 1993) – 3mf – 9 – €49.00 – 3-89349-738-2 – mf#DHS 738 – gw Frankfurter [322]

Bedeutung von mangelernaehrung im alter unter besonderer beruecksichtigung der therapie mittels perkutaner endoskopischer gastrostomie : eine laengsschnittanalyse bei 252 multimorbiden alterspatienten / Krys, Ute – (mf ed 1998) – 2mf – 9 – €40.00 – 3-8267-2552-2 – mf#DHS 2552 – gw Frankfurter [618]

Die bedeutungen der wortsippe "kbd" im hebraeischen / Caspari, Wilhelm – Leipzig, 1908 – 2mf – 9 – mf#NE-485 – ne IDC [470]

Bedford – (= ser Bedfordshire parish register series) – 2mf – 9 – £5.00 – uk BedsFHS [929]

Bedford 1693-1849 – Oxford, MA (mf ed 1995) – (= ser Massachusetts vital record transcripts to 1850) – 5mf – 9 – 0-87623-208-X – (mf 1t-4t: vital records 1693-1887. mf 3t-4t: marriages 1810-43. mf 4t: out-of-town marriages 1731-99; births 1840-42. mf 5t: births 1836-49; marriages & deaths 1843-49) – us Archive [978]

Bedford advertiser – Bedford SA, 12 july 1878-25 dec 1885 – 4r – 1 (diazo also available at reduced price) – sa National [079]

Bedford county press – Bedford, PA., 1868-1984 – 13 – $25.00r – us IMR [071]

Bedford county press – Everett, PA., 1875-1959 – 13 – $25.00r – us IMR [071]

Bedford enterprise – Bedford SA, 1885-99 – 1 – mf#MS00180 – sa National [079]

Bedford, F *see*
- The holy land, egypt, constantinople, athens, etc, etc
- The treasury of ornamental art

Bedford gazette – Bedford, PA. -w 1832-1847 – 13 – $25.00r – us IMR [071]

Bedford guardian – Bedford SA, 1881-84 – 2r – 1, 16 diazo available – sa National [960]

Bedford inquirer – Bedford, PA. -w 1902-1930 – 13 – $25.00r – us IMR [071]

Bedford, New Hampshire. Bedford Baptist Church *see* Records

Bedford – st john – (= ser Bedfordshire parish register series) – 2mf – 9 – £5.00 – uk BedsFHS [929]

Bedford – st mary monumental inscriptions – (= ser Bedfordshire parish register series) – 2mf – 9 – £5.00 – uk BedsFHS [929]

Bedford – st pauls – (= ser Bedfordshire parish register series) – 3pt on 9mf – 9 – £18.50 – uk BedsFHS [929]

Bedford – st peter – (= ser Bedfordshire parish register series) – 1mf – 9 – £3.00 – uk BedsFHS [929]

Bedford, Thomas *see* A treatise of the sacraments

Bedford times-register – Bedford, OH, apr 12 1973-nov 24 1988 – 15r – 1 – (weekly cleveland suburban newspaper) – us Western Res [071]

Bedford-Jones, T *see*
- The apostolic rite of confirmation
- The charity that covers a multitude of sins
- Congregational music and some of its hindrances
- Day of supplication, war in south africa
- The dead queen
- Edification
- The goodness of god
- Gregorian chants for canticles and psalter
- A letter to the rev henry wilson
- Love of the world
- New year's sermon, lord, what wilt thou have me to do?
- Paper containing a proposition on the education of men for the ministry in the diocese of ontario
- Perfecting holiness
- A sermon preached in the church of st alban the martyr, ottawa, on trinity sunday evening, may 23rd, 1875

Bedfordshire advertiser *see* Luton times etc

Bedfordshire historical record society – v1-46. 1913-67 – (= ser Publications of the english record societies, 1835-1972) – 133mf – 9 – uk Chadwyck [941]

Bedfordshire parish poor law papers 1622-1834 – (= ser Bedfordshire parish register series) – 9 – £2.75 – (incl ind) – uk BedsFHS [929]

Bedfordshire strays index, vol 1 + 2 – 4mf – 9 – £5.00 – uk BedsFHS [941]

Bedford-Stuyvesant Youth in Action Community Corporation *see* Comet

Bedi, Baba Pyare Lal *see* Harvest from the desert

Bedier, Joseph *see* Les chansons de colin muset

Bedingfield, Lady *see* Aristocratic women

Bedjan, Paul *see*
- Acta martyrum et sanctorum
- Le livre d'heraclide de damas

Bednarczyk, Janet H *see* The effect of mass on the kinematics of steady state wheelchair propulsion in adults and children with spinal cord injury

Bednyi, Dem'ian *see* Vpered i vyshe!

Be-doro shel bialik / Keshet, Yeshurun – Tel-Aviv, Israel. 1942/1943 – 1r – us UF Libraries [939]

Bedoya Cardona, Ernesto *see* De desterrado a presidente

Bedoya, Juan Manuel *see* Oracion funebre

Bedoya, Victor A *see* Etnologia y conquistas del tolima y la hoya del qu...

Bedrifts-okonomen – Oslo, Norway 1977-80 – 1,5,9 – ISSN: 0005-7606 – mf#10197 – us UMI ProQuest [338]

Bedrinana, Francisco C *see* Vida y aventura de rodrigo de xerez

Bedrock democrat – Baker City OR: Abbot & M'arteur, [wkly] – 1 – (began with may 11 1870 iss; related to daily ed: morning democrat, baker democrat; cont by: weekly bedrock democrat) – us Oregon Lib [071]

Bedrock democrat (baker city, or) – Baker City OR: [s.n.] [wkly] – 1 – (related to daily ed: morning democrat (baker city, or)) – us Oregon Lib [071]

Bedrock democrat (baker city, or) *see* Morning democrat (baker city, or)

The bedroom and boudoir / Broome, Mary Anne (Stewart) Barker, lady – London 1878 – (= ser 19th c art & architecture) – 2mf – 9 – mf#4.2.34 – uk Chadwyck [740]

Bedside nurse : nursing care – Lancaster PA 1968-73 – 1 – ISSN: 0005-7665 – mf#2564.02 – us UMI ProQuest [610]

Bedsole, Malcolm R *see* Effect of time of turning and method of supplementing green manures

Die bedudinghe naden sinne van sunte augustijns regule / ed by Flou, Karel de – Gent, 1901 – 6mf – 8 – €14.00 – ne Slangenburg [241]

Beduerfen wir fuer unser christentum einer aeussern autoritaet im wort gottes? / Oehler, Theodor – Basel: Missionsbuchhandlung 1906 [mf ed 1992] – 1mf – 9 – 0-524-03244-0 – mf#1990-0872 – us ATLA [220]

Beduerfnisse alter menschen in psychiatrischer behandlung : untersuchung der aeusserungen ueber beduerfnisse von patienten einer geronto-psychiatrischen tagesklinik und ihre abbildung in diagnose und therapie / Bramesfeld, Anke – (mf ed 1996) – 2mf – 9 – €40.00 – 3-8267-2355-4 – mf#DHS 2355 – gw Frankfurter [618]

Beduschi, Nilton *see* Organizacao-padrao dos departamentos municipais

Bedwas, (mon), hephzibah welsh baptist, monumental inscriptions – 2mf – 9 – £2.50 – uk Glamorgan FHS [929]

Bedwas, (mon), st barrwg, monumental inscriptions – 2mf – 9 – £2.50 – uk Glamorgan FHS [929]

Bedwell, Cyril E A *see* A brief history of the middle temple

Bedwellty, st sennan, monumental inscriptions – 2mf – 9 – £2.50 – uk Glamorgan FHS [929]

Bedworth citizen – Bedworth, England 14 may 1987-28 mar 1991 – 1 – (variant ed of: coventry citizen) – uk British Libr Newspaper [072]

Bedworth echo – Bedworth, England 13 sep 1979- – 1 – uk British Libr Newspaper [072]

Bedworth & foleshill news and coventry chronicle – [West Midlands] Bedworth 26 may 1900-15 nov 1924 [mf ed 2003] – 1 – (missing: 1904; cont by: bedworth & foleshill news, coventry chronicle and warwickshire county graphic (+ suppl) [22 nov 1924-3 jan 1925]; coventry chronicle, bedworth & foleshill news and warwickshire county graphic [jan 1926-dec 1930]; coventry chronicle, bedworth & foleshill news and tribune [9 may 1930-14 may 1948]) – uk Newsplan; uk British Libr Newspaper [072]

Bedworth guardian – Bedworth, England 4 jan 1873-14 mar 1908, 26 sep 1919-9 may 1941 [mf 1896,1898] – 1 – (wanting: 1897; incorp with: nuneaton chronicle) – uk British Libr Newspaper [072]

Bedworth guardian and warwickshire miners' saturday evening journal – [West Midlands] Warwickshire jan 1873-14 mar 1908; 26 sep 1919-9 may 1941 [mf ed 2003] – 49r – 1 – (missing: 1878-79, 1896-98; cont as: bedworth guardian and foleshill weekly post [jan 1881-dec 1919]; bedworth guardian [jan 1920-dec 1922]; bedworth guardian and foleshill courier [jan 1923-may 1941]) – uk Newsplan [072]

Bedworth, Thomas *see* The power of conscience exemplified in the genuine and extraordinary confession of thomas bedworth.

Bedworth times – Bedworth, England 16 jan 1875-5 feb 1876 (wkly) – 1 – (discontinued) – uk British Libr Newspaper [072]

Bee – 1884 nov 25-1932 apr 21 [1]; v74 n11796 (1893 aug 12) [2]; 1962 dec 1-1998 sep-dec [3] – 1 – mf#1005379 [1]; 880919 [2]; 1005382 [3] – us WHS [071]

Bee / Brown Co. Ripley – 1906-9/18,19-20,27-1938 [wkly] – 13r – 1 – mf#B7684-7696 – us Ohio Hist [071]

Bee / Brown Co. Ripley – apr 1878-apr 1880 [wkly] – 1r – 1 – mf#B9665 – us Ohio Hist [071]

Bee / Brown Co. Ripley – jul 1947-dec 1983 [wkly] – 24r – 1 – mf#B22883-22906 – us Ohio Hist [071]

Bee / Brown Co. Ripley – sep 1850-may 1852 (some damaged) [wkly] – 1r – 1 – mf#B5538 – us Ohio Hist [071]

Bee – [Pictou, N.S.?]: J. Dawson, 1836-1838 – 9 – (cont by: mechanic and farmer) – mf#P04553 – cn Canadiana [630]

Bee – Killen TX 1759-1759 – 1 – mf#3099 – us UMI ProQuest [073]

Bee / Lucas Co. Toledo – jul-dec 1889 [daily] – 2r – 1 – mf#B34538-34539 – us Ohio Hist [071]

Bee / Lucas Co. Toledo – jul-dec 1890 [daily] – 2r – 1 – mf#B34666-34667 – us Ohio Hist [071]

The bee – Hillsburg [Ont]: G A Lacey, [1881-18–?] – 9 – mf#P04985 – cn Canadiana [071]

Bee and the phillips times – Phillips WI. 1932 apr 28/may 26-1962 jan/nov 29 – 21r – 1 – (with gaps; cont: bee (phillips wi: 1884); prentice news; prentice news; cont by: bee (phillips wi: 1962)) – mf#1005380 – us WHS [071]

Bee, bulletin of environmental education – Brighton UK 1975-80 – 1,5,9 – ISSN: 0045-1266 – mf#10939 – us UMI ProQuest [333]

Bee culture – Medina OH 1993+ – 1,5,9 – (cont: gleanings in bee culture) – ISSN: 1071-3190 – mf#19637 – us UMI ProQuest [630]

Bee line – New York (State). 1984 apr-1993 feb – 1r – 1 – mf#1053584 – us WHS [071]

Bee / news / Preble Co. Eldorado – 1908-1927, 1930-jan 1931 [wkly] – (= ser News) – 7r – 1 – (title changes fr: bee to news) – mf#B11327-11333 – us Ohio Hist [071]

Bee, or literary weekly intelligencer – La Salle IL 1790-94 – 1 – mf#3906 – us UMI ProQuest [071]

Bee (portland, or) – Portland OR: John F Dillin Jr, 1906- [wkly] – 1 – us Oregon Lib [071]

Bee reviv'd *see* Eighteenth century journals

Bee revived : or, the universal weekly pamphlet – La Salle IL 1733-35 – 1 – mf#4205 – us UMI ProQuest [071]

Bee, T *see* Bee's reports of cases in the district court of south carolina, 1792-1805

Bee (tygh valley, or) – Tygh Valley OR: Elmer O Shepherd, [wkly] – 1 – (began apr 20 1905) – us Oregon Lib [071]

Bee-argus / Paulding Co. Antwerp – dec 29, 1970-dec 1982 [wkly] – 6r – 1 – mf#B13014-13019 – us Ohio Hist [071]

Bee-argus / Paulding Co. Antwerp – mar 1918-apr 1919 [wkly] – 1r – 1 – mf#B921 – us Ohio Hist [071]

Bee-argus / Paulding Co. Antwerp – nov 1931-sep 1947, oct 1953-sep 1971 [wkly] – 12r – 1 – mf#B32647-32658 – us Ohio Hist [071]

Beebe, Charles William *see*
- Beneath tropic seas
- High jungle
- Jungle days
- Jungle peace
- Nonsuch

Beebe, William *see*
- Edge of the jungle
- Jungle peace

Beebeejaun, Bibi Mehtaab Parveen *see* The synthesis of aromatic carboxyl functionalized polymers by atom transfer radical polymerization

Beech branch baptist church – Luray, SC. 1918-75 – 1 – $5.31 – us Southern Baptist [242]

Beech island baptist church – Aiken County, SC. 1953-67 – 1 – $10.62 – us Southern Baptist [242]

Beech, Mervyn Worcester Howard *see* The suk

Beecher, Charles *see*
- Autobiography, correspondence etc of lyman beecher
- The bible, a sufficient creed
- The eden tableau, or, bible object-teaching
- The metronome
- Redeemer and redeemed
- A review of the "spiritual manifestations"

Beecher, Edward *see*
- Baptism
- The concord of ages
- The conflict of ages
- The papal conspiracy exposed, and protestantism defended, in the light of reason, history, and scripture

Beecher Henry Ward *see* Essence of religion

Beecher, Henry Ward *see*
- Bible studies
- Christian self-denial
- Divine compassion
- Evolution and religion
- Fruits of the spirit
- The life of jesus, the christ
- The life of jesus the christ
- Nature's warning
- New star papers, or, views and experiences of religious subjects
- Nicodemus and the re-birth
- The overture of angels
- Patriotic addresses
- Plymouth pulpit
- Report of a sermon...charlottetown, sabbath morning, aug 10th, 1879
- Selected sermons as delivered by henry ward beecher, in plymouth church, brooklyn
- A summer parish
- Twelve lectures to young men, on various important subjects
- Woman's influence in politics

Beecher, Henry Ward (Mrs) *see* All around the house

Beecher, Henry Ward, Mrs *see* Letters from florida

Beecher, Leonard James *see* Language teaching in kikuyu schools

Beecher, Lyman *see*
- Autobiography, correspondence etc of lyman beecher
- The bible a code of laws
- The faith once delivered to the saints
- The gospel according to paul
- Lectures on political atheism and kindred subjects
- Lectures on scepticism
- Letters of the rev. dr. beecher and rev. mr. nettleton on the "new measures" in conducting revivals of religion
- Sermons
- Something has been done during the last forty years
- Views of theology

Beecher, Mary A *see* Index of presbyterian ministers

Beecher, Thomas Kinnicut *see* Our seven churches

Beecher, William Constantine et al see A biography of rev henry ward beecher
Beecher, Willis Judson see
– The dated events of the old testament
– Farmer tomkins and his bibles
– Index of presbyterian ministers
– The prophets and the promise
– The teaching of jesus concerning the future life
Beechey, F W see Narrative of a voyage to the pacific and beering's strait to co-operate with the polar expeditions
Beechey, Frederick W see
– Proceedings of the expedition to explore the northern coast of africa
– Reise nach dem stillen ocean und der beeringsstrasse
Beeching, H C see Church and state
Beeching, Henry Charles et al see The bible doctrine of atonement
Beechmont baptist church – New York. 1804-1807 (1) – 1 – $156.69 – mf#3592 – us Southern Baptist [242]
Beechridge baptist church – Hatton, KY. Feb 1914-Oct 1940 – 1 – $9.72 – us Southern Baptist [242]
Beef – Shawnee Mission KS 1985+ – 1,5,9 – ISSN: 0005-7738 – mf#15036 – us UMI ProQuest [636]
Beef cattle improvement in florida / Knapp, Bradford – Gainesville, FL. 1935 – 1r – us UF Libraries [636]
Beef production in florida / Shealy, A L – Gainesville, FL. 1933 – 1r – us UF Libraries [636]
The bee-hive – London: Publ by George Potter for the Trades' Newspaper Co, jul 4-dec 1868; 1869-76 – us CRL [072]
The bee-hive – (The Penny Bee-Hive The Industrial Review, Social and Political). London. -w. Oct 1862-Dec 1878. (16 reels) – 1 – uk British Libr Newspaper [330]
The beehive – Toronto: Pub...by Rogers & Larminie, [1874-18– or 19–] – 9 – mf#P04358 – cn Canadiana [917]
Bee-hive cottage / Cameron, Mrs – London, England. 18– – 1r – us UF Libraries [240]
The bee-hive newspaper – London: Publ by George Potter for the Trades' Newspaper Co, jan 1863-jun 1868 – us CRL [072]
Beehler, William Henry see The cruise of the brooklyn
Beeinflussung von sensorischer reizschwelle und urodynamischen parametern durch lidocain-haltiges gleitgel fuer topische anwendung in der urethra / Eggersmann, Christian – (mf ed 1998) – 1mf – 9 – €30.00 – 3-8267-2563-8 – mf#DHS 2563 – gw Frankfurter [616]
Beek, C I M I van see Passio sanctarum perpetuae et felicitatis latine et graece (fp43)
[Beek, K] see Le triomphe royal
Beekman, A W H see Het kasteel"de slangenburg" en zijn kunstschatten
Beeld-snyders kunstkabinet... / Bossuit, F van – Amsterdam, 1727 – 6mf – 9 – mf#0-1148 – ne IDC [700]
Beelen, J Th see Epsitolae binae de virginitate, syriace
Beelen, Jan Theodor see Grammatica graecitatis novi testamenti
Beeman family newsletter – 1977 aug-1979 dec – 1r – 1 – (cont by: b'man newsletter (1980)) – mf#403353 – us WHS [640]
Beeman, Thomas O see Ritualism, doctrine not dress
Beemelmans, Fr see Zeit und ewigkeit nach thomas von aquino
Beemer, J J see Circular letter from the president, pontiac pacific junction railway co...
Beemer times – Beemer, NE: A D Beemer. 61v. mar 11 1886-v61 n3. jan 2 1947 (wkly) – (= ser Wisner News-Chronicle) – 14r – 1 – (absorbed by: wisner news-chronicle) – us Bell [071]
Beer, B see Das buch der jubiaeen und sein verhaeltniss zu den midraschim
Beer, Bernhard see
– Leben abraham's nach auffassung der juedischen sage
– Leben moses
Beer drinking and business – [Toronto?: Dominion Alliance for the Suppression of the Liquor Traffic, 189-?] [mf ed 1992] – 1mf – 9 – 0-665-90910-1 – (original iss in ser: campaign leaflets) – mf#90910 – cn Canadiana [360]
Beer, Georg see
– Mose und sein werk
– Saul, david, salomo
– Der text des buches hiob
Beer, Gottfr Ludw see Magazin fuer die brandenburgisch-bayreuthische geschichte
Beer, Joseph W see
– The jewish passover and the lord's supper
– A summary of religious faith and practice
Beer, Marie de see A supply chain of consumer goods
Beer, Michael see Saemmtliche werke
Beer, Oskar see Hebbels judith und maria magdalena im urteil seiner zeitgenossen
Beer Pastor, Oscar see Practicas fiscales
Beer, Paul see Philosophische aufsaetze
Beer, Rudolf see
– De compositione hominis
– De ente praedicamentali: from the unique vienna ms.; quaestiones 13 logicae et philosophicae: from the unique prague ms
Beer shipped into wisconsin : from out-of-state breweries and wholesalers – Wisconsin. 1980 aug-1985 jun – 1r – 1 – (cont: beer imported from out-of-state breweries and wholesalers; cont by: beer shipments into wisconsin (in barrels)) – mf#589525 – us WHS [660]
Be'er yitshak / Levinson, Isaac Baer – Warsaw, Poland. 1899 – 1r – us UF Libraries [939]
Beerbohm tree collection : from the university of bristol – 9 – (pt1: prompt copies [439mf] £2100; pt2:... [443mf] £2100; pt3: agreements & press cuttings [458mf] £2100; pt4: correspondence, business & production records [111mf] £650; with d/g) – uk Matthew [790]
Beer-Hofmann, Richard see
– Der graf von charolais
– Jaakobs traum
– Jacob's dream
– Vorspiel auf dem theater zu koenig david
Beermann, Gustav see Die koridethi evangelien th038
Beermann, Kerstin see Erzeugung von phototaktischen verhaltensvarianten aus den mutanten d1, km1 und flx15 des archaebakteriums halobakterium salinarium unter verwendung der mutagene ethylmethansulfonat und n-methyl-n'-nitro-n-nitrosoguanidin
Be-erot avraham / Klein, Albert – Tirnoya, Bulgaria? v1-2. 1928-1931 or 1932 – 1r – us UF Libraries [939]
Beers, William George see
– Annual meeting of the royal college of dental surgeons, ontario
– The canadian mecca
– The discriminate use of amalgam for filling teeth
– Lacrosse
– Observations in the mouth during pregnancy and the catamenia
– Over the snow
– Patriotic speech, in reply to the toast of professional annexation
– Young canada's reply to "annexation"
Bees, Nikos A see Hippolyts schrift ueber die segnungen jakobs – hippolyts danielcommentar in handschrift no 573 des meteoronklosters
Bee's reports of cases in the district court of south carolina, 1792-1805 / Bee, T – Philadelphia: Farrand. 1v. 1810 (all publ) – (= ser Early federal nominative reports) – 6mf – 9 – $9.00 – mf#LLMC 81-431 – us LLMC [347]
Beesly, Augustus Henry see Sir john franklin
Beeson, C H see Hegemonius acta archelai (gcsej7)
Beeson, Luana J see Health knowledge competencies and essential health skills of entry level college freshmen enrolled in oregon's research universities
Beeston and west notts gazette echo see Gazette and echo (beeston and west notts gazette echo)
Beet, Joseph Agar see
– The church, the churches, and the sacraments
– A commentary on st paul's epistle to the galatians
– A commentary on st paul's epistle to the romans
– A commentary on st paul's epistles to the corinthians
– A commentary on st paul's epistles to the ephesians, philippians, colossians, and to philemon
– The credentials of the gospel
– The firm foundation of the christian faith
– Holiness symbolic and real
– The immortality of the soul
– The last things
– A manual of theology
– Nature and christ
– The new life in christ
– The new testament
– The old testament
– A shorter manual of theology
– A theologian's workshop, tools, and methods
– Through christ to god
– A treatise on christian baptism
Beet sugar : its economical production in the province of quebec / Barnard, Edouard-Andre – Quebec?: s.n, 1877? – 1mf – 9 – (also available in french) – mf#62734 – cn Canadiana [635]
Beet sugar : its economical production in the province of quebec: a paper read before the district of bedford agricultural association, on the 9th march, 1877 / Barnard, Edouard-Andre – Quebec?: s.n, 1877? – 1mf – 9 – mf#03369 – cn Canadiana [635]
Beet sugar enterprise – Grand Island, NE: M A Lunn. v1 n1. may 1890- (mthly) [mf ed 1996] – 1r – 1 – us NE Hist [338]
Beet sugar industry : its adaptability to canada, favorable prospects of success... / Lawder, Robert H – Montreal?: s.n, 1895 – 1mf – 9 – mf#08879 – cn Canadiana [635]
The beet sugar industry : it can be successfully developed in canada under a reasonably liberal government policy / Lawder, Robert H – Ottawa?: s.n, 1895 – 1mf – 9 – mf#08946 – cn Canadiana [635]
Beet, William Ernest see
– The early roman episcopate to a.d. 384
– The medieval papacy
Beethoven : drame en trois actes / Fauchois, Rene – Tourville-la-Riviere (Seine-Inferieure), France. 1938 – 1r – us UF Libraries [440]
Beethoven : impressions of contemporaries – New York, NY. 1926 – 1r – us UF Libraries [025]
Beethoven / Turner, W J – London, England. 1927 – 1r – us UF Libraries [780]
Beethoven and the sonata-allegro form as revealed in the pianoforte sonatas / Waters, Edward Neighbor – U of Rochester 1928 [mf ed 1974] – 3mf – 9 – mf#fiche 84 – us Sibley [780]
Beethoven et wagner : essais d'histoire et de critique musicales / Wyzewa, Teodor de – Paris: Perrin 1898 – 3mf – 9 – mf#wa-124 – ne IDC [780]
Beethoven, Ludwig van see
– [Sketches for portions of "quoniam tu solus sanctus"]
– Werke
Beethoven. sugestiones / Hindos, Jose de – Ensayo intimo.Caceres.Imp.Mod. 1927 – 1 – sp Bibl Santa Ana [780]
Beethoven, teosofo, un capitulo de la obra el... drama lirico de... / Roso de Luna, Mario – Pontevedra: Tip. Vda. e Hijos de Antunez, 1915 – 1 – sp Bibl Santa Ana [780]
Beethoven und wagner / Sternfeld, Richard – Charlottenburg [-Berlin]: Verlag der Allgemeinen Musikzeitung 1886 – 1mf – 9 – mf#wa-106 – ne IDC [780]
Die beethoven-sammlung / ed by Staatsbibliothek zu Berlin – Preussischer Kulturbesitz – [mf ed 2002-05] – (= ser Musikhandschriften der staatsbibliothek zu berlin – preussischer kulturbesitz 3) – 462mf incl 321 col mf (1:24) in 4 installments – 9,15 – silver+col mf €8372.00 – ISBN-10: 3-598-34427-9 – ISBN-13: 978-3-598-34427-5 – gw Saur [780]
Beet-root and beet-root sugar : the description of all the processes of manufacture... / Cull, Edward Lefrey – Toronto?: s.n, 1874 – 1mf – 9 – (1st ed publ under the title: the whole history and mystery of beet-root and beet-root sugar) – mf#23943 – cn Canadiana [635]
Beets, Henry see Ds. willem hendrik frieling
Beets, Nicolaas see Life and character of j. h. van der palm, d.d
Der befehl des gewissens : ein roman von den wirren der nachkriegszeit und der ersten erhebung / Zoeberlein, Hans – 17. aufl. Muenchen: Zentralverlag der NSDAP, F Eher c1937 [mf ed 1992] – 1r – 1 – (filmed with: auf biegen und brechen/ erwin zindler) – mf#3069p – us Wisconsin U Libr [830]
Befehl ist befehl : erzaehlungen / Klucke, Walther Gottfried – Berlin: Bong 1941 [mf ed 1990] – 1r – 1 – (filmed with: heinrich von kleist / hermann graef) – mf#2772p – us Wisconsin U Libr [830]
Die befestigungen von herakleia am latmos / Krischen, F – Berlin, 1912 – 2mf – 9 – mf#NE-106 – ne IDC [956]
Le beffroi – Art et litterature modernes Dir. Leon Bocquet. no. 64-104. Roubaix. avr mai 1906-sept oct 1913 – 1 – fr ACRPP [800]
Befindlichkeitsveraenderungen durch musik : ein vergleich zwischen psychotisch kranken und psychisch gesunden / Leuwer, Martin – (mf ed 1995) – 2mf – 9 – €40.00 – 3-8267-2190-X – mf#DHS 2190 – gw Frankfurter [616]
Befolkning i oldtiden – Stockholm: A Bonnier 1936 [mf ed Bloomington IN: Indiana Uni Lib, Preservation Dept 1984] – 1r – 1 – us Indiana Preservation [390]
Befolkning under medeltiden – Stockholm: A Bonnier 1938 [mf ed Bloomington IN: Indiana Uni Lib, Preservation Dept 1984] – 1r – 1 – us Indiana Preservation [390]
Before and after independence : a collection of the most important and soul-stirring speeches delivered by jawaharlal nehru, during the most important and soul-stirring years in india's history, 1922-1950 / ed by Bright, J S – New Delhi: Indian Print Works, [195-] – (= ser Samp: indian books) – us CRL [850]
Before the altar : or, a series of annotated propositions on liturgics / Schuette, Conrad Herman Louis – Columbus OH: Lutheran Book Concern 1894 [mf ed 1991] – 1mf – 9 – 0-524-01092-7 – mf#1990-4057 – us ATLA [242]
Before the coming of the loyalists / Haight, Canniff – Toronto: Haight, 1897 [mf ed 1980] – 1mf – 9 – 0-665-05145-X – mf#05145 – cn Canadiana [975]
Before the great pillage : with other miscellanies / Jessopp, Augustus – London: T F Unwin 1901 [mf ed 1990] – 1mf – 9 – 0-7905-5285-X – mf#1988-1285 – us ATLA [941]
Before the mast and behind the pulpit / Allan, Alexander M – [19–?] – 1r – 1 – 0-8370-1134-5 – mf#1984-B115 – us ATLA [240]
Before the military commission convened by the commanding general, united states army forces, western pacific. proceedings / Yamashita, Tomoyuki – Manila. On film: v1-34, 1945. LL-030 – 1 – (exhibits. manila. on film: v1-4, 1945. II-030. 1) – us L of C Photodup [355]
Before the table : an inquiry, historical and theological, into the true meaning of the consecration rubric in the communion service of the church of england... / Howson, John Saul – London: Macmillan 1875 [mf ed 1992] – 1mf – 9 – 0-524-03160-6 – mf#1990-4609 – us ATLA [242]
Der befreier – Erfurt DE, 1921 – 1 – gw Misc Inst [074]
Der befreier : rede, gehalten am 28. august 1949 im nationaltheater weimar zur zweihundertsten wiederkehr des geburtstages von johann wolfgang von goethe / Becher, Johannes Robert – Berlin: Aufbau-Verlag, 1949 [mf ed 1993] – 55p – 1 – mf#8655 – us Wisconsin U Libr [320]
Die befreite deutsche wortkunst / Holz, Arno – Wien: Avalun-Verlag, 1921 – 80p – 1 – us Wisconsin U Libr [430]
Beg, Abdulla Anwar see
– The life and odes of ghalib
– The poet of the east
Beg, Aribozli Nu'man Mahir see The divan project
Beg, Hersekkli 'Arif Hikmet see The divan project
Beg, 'Izzet see The divan project
Beg, Leskofceli Galib see The divan project
Beg, Munse'at-i 'Izzet see The divan project
Beg, Munse'at-i Nu'man Mahir see The divan project
Bega district news – Bega. jan 1969-apr 1994, sep 1994-jun 1995 – at Pascoe [079]
Bega district news – Bega, oct 1923-dec 1933 – 5r – A$192.50 vesicular A$220.00 silver – at Pascoe [079]
Bega gazette – Bega, feb 1865-dec 1899 (misc. yrs) – 5r – A$209.48 vesicular A$231.48 silver – at Pascoe [079]
Bega standard – Bega, jan 1876-sep 1923 – 18r – A$1115.40 vesicular A$1214.40 silver – at Pascoe [079]
Begam samru / Banerjee, Brajendra Nath – Calcutta: M C Sarkar & Sons, 1925 – (= ser Samp: indian books) – us CRL [920]
Begams of bengal : mainly based on state records / Banerjee, Brajendra Nath – Calcutta: SK Mitra & Bros, 1942 – (= ser Samp: indian books) – us CRL [305]
Begebenheiten des capitains von der russisch-kaiserlichen marine golownin : in der gefangenschaft bei den japanern in den jahren 1811, 1812 und 1813, nebst seinen bemerkungen ueber d japan reich und volk... – Leipzig 1817-1818 [mf ed Hildesheim 1995-98] – 2v on 5mf – 9 – €100.00 – 3-487-27546-5 – (trans fr russian) – gw Olms [920]
Begegnung an der grenze / Bahrs, Hans – Berlin: Nordland Verlag, [c1942] [mf ed 1989] – (= ser Nordland-buecherei 27) – 67p – 1 – mf#6972 – us Wisconsin U Libr [830]
Die begegnung auf dem riesengebirge : novelle / Kolbenheyer, Erwin Guido – Muenchen: A Langen/G Mueller, 1934, c1932 – 1r – 1 – us Wisconsin U Libr [830]
Begegnungen mit menschen, buechern, staedten / Zweig, Stefan – Wien: H Reichner, c1937 [mf ed 1992] – 478p – 1 – mf#7982 – us Wisconsin U Libr [430]
Begegnungen und wuerdigungen : literarische portraets von carl spitteler bis klaus mann / ed by Goldammer, Peter – 1. aufl. Rostock: Hinstorff 1984 [mf ed 1992] – (= ser Die sammlung deutschsprachige literatur in laengsschnitten) – 1r – 1 – (incl bibl ref. filmed with: historical chart of english literature for use in schools and colleges / nelson lewis greene) – mf#3160p – us Wisconsin U Libr [430]
Begemann, Wilhelm see
– Der alte und angenommene schottische ritus und friedrich der grosse
– Die fruchtbringende gesellschaft und johann valentin andreae
– Schwache prateritum der germanischen sprachen
– Vorgeschichte und anfaenge der freimaurerei in schottland
Beger, Kai-Uwe see Zur gesellschaftlichen produktion von armut
Begg, Alexander see
– The directory of mines (corrected and published quarterly)
– Enquire within for information about manitoba and the boundless wheat fields of the new northwest
– The great north-west of canada
– Notes on vancouver island
– Review of the alaskan boundary question
– Seventeen years in the canadian north-west

Begg and lynch's hand-book and general guide to british columbia – Victoria, BC: B C Guide Pub Co, [1893] [mf ed v1 n1 apr 1893-v1 n3 jun 1893] – 9 – mf#P04494 – cn Canadiana [917]

Begg, James *see*
- Anarchy in worship
- Covenanting struggle
- Creeds and consistency
- Hand of god in the disruption and the vital importance of free chur...
- Hints on health
- History of the act of queen anne, 1711
- Protestant classes necessary, or, the importance of studying the...
- Purity of worship in the presbyterian church
- Reply to sir james graham's letter
- Scottish public affairs, civil and ecclesiastical
- Seat rents brought to the test of scripture, law, reason, and experience...

Begg, James A *see* Scriptural argument for the coming of the world at the commencement of the millennium

The beggars of holland and the grandees of spain : a history of the reformation in the netherlands, from a.d. 1200 to 1578 / Mears, John William – Philadelphia: Presbyterian Publ Comm, c1867 – 2mf – 9 – 0-7905-5307-4 – mf#1988-1307 – us ATLA [949]

Beggar's opera... : with the additional alterations, by dr arne, for the voice, harpsichord & violin, the basses entirely new / [Pepusch, John Christopher] – London: Preston [181-?] [mf ed 19–] – 1mf – 9 – mf#fiche 21 – us Sibley [780]

The beggers ape / Niccols, Richard – 1627 – 9 – 5.00 – us Scholars Facs [810]

Beggi, Francesco Orzzio *see* The incubi of rome and venice

Beggs, Charles *see* The first steps to irish liberty

Begg's monthly and general guide to british columbia – Victoria, BC: A Begg, [1893-189-] [mf ed v1 n4 jul 1893] – 9 – mf#P04495 – cn Canadiana [917]

Begin, Emile A *see* Histoire des duches de lorraine et de bar

Begin, Louis Nazaire *see*
- Chronologie de l'histoire des etats-unis d'amerique
- Chronologie de l'histoire du canada

Begin, Louis-Nazaire *see*
- Catechisme de controverse premiere partie
- Chronologie de l'histoire des etats-unis d'amerique
- Chronologie de l'histoire du canada
- Le culte catholique
- La primaute et l'infaillibilite des souverains pontifes
- La sainte ecriture et la regle de foi

Beginner teacher – Nashville, TN. 1931-62 – 1 – $254.73 – us Southern Baptist [242]

Beginner teacher and pupil book : years 1-2 – 1928, 1928-33 – 1 – $57.68 – (better home 1935-46) – us Southern Baptist [242]

The beginners' department of the church school / Athearn, Walter Scott – [Des Moines, Iowa: Dept of Religious Education, Drake University], c1913 – 1mf – 9 – 0-524-06702-3 – mf#1991-2732 – us ATLA [240]

The beginners of a nation : a history of the source and rise of the earliest english settlements in america, with special reference to the life and character of the people / Eggleston, Edward – New York: D Appleton, 1896 [mf ed 1990] – 1mf – 9 – 0-7905-5936-6 – mf#1988-1936 – us ATLA [975]

The beginning : its when and its how / Ponton, Mungo – London, 1871 – (= ser 19th c evolution & creation) – 7mf – 9 – mf#1.1.1520 – uk Chadwyck [577]

Beginning and progress of our sunday schools in canada / Thomas, C A – 1875 – 1r – 1 – $35.00 – (german & english trans) – mf[s]-111 – us Commission [242]

Beginning at jerusalem : studies in historic communions of christendom / Lacey, Thomas James – New York: Edwin S Gorham c1909 [mf ed 1986] – 1mf – 9 – 0-8370-8687-6 – mf#1986-2687 – us ATLA [240]

The beginning of the end / Mahtab, Harekrushna – Calcutta: Book Co, [1949] – (= ser Samp: indian books) – us CRL [954]

The beginning of things in nature and in grace : or, a brief commentary on genesis / Wight, Joseph K – Boston: Sherman, French, 1911 [mf ed 1993] – 1mf – 9 – 0-524-06063-0 – mf#1992-0776 – us ATLA [221]

Beginning somali history / Gilbert, Paul S ET A L – Afgoi: National Teacher Educ Center Press, 1967 – us CRL [960]

Beginning vai / Terplan, Elizabeth Solinsky – San Francisco, CA. 1965 – 1r – us UF Libraries [025]

Beginnings in india / Stock, Eugene – London: Central Board of Missions and SPCK, 1917 [mf ed 1995] – (= ser Yale coll; Romance of missions) – 1 – 0-524-09043-2 – mf#1995-0043 – us ATLA [954]

The beginnings of christianity = Anfaenge unserer religion / Wernle, Paul; ed by Morrison, William Douglas – London: Williams and Norgate; New York: GP Putnam, 1903-1904 – (= ser Theological Translation Library) – 2mf – 9 – 0-8370-9596-4 – (in english) – mf#1986-3596 – us ATLA [240]

The beginnings of christianity : with a view of the state of the roman world at the birth of christ / Fisher, George Park – New York: Charles Scribner's Sons, 1901. Beltsville, Md: NCR Corp, 1977 (7mf); Evanston: American Theol Lib Assoc, 1984 (7mf) – 9 – 0-8370-0133-1 – (incl bibl ref and index) – mf#1984-0019 – us ATLA [240]

The beginnings of english christianity : with special reference to the coming of st. augustine / Collins, William Edward – London: Methuen, 1898 – (= ser The Churchman's Library) – 1mf – 9 – 0-7905-4176-9 – (incl bibl ref) – mf#1988-0176 – us ATLA [240]

The beginnings of english utilitarianism / Albee, Ernest – Boston: Ginn, 1897 [mf ed 1987] – iv/101p – 1 – mf#1935 – us Wisconsin U Libr [170]

The beginnings of gnostic christianity / Rylands, Louis Gordon – London: Watts, 1940 – 1mf – 9 – 0-524-08136-0 – mf#1993-9042 – us ATLA [220]

The beginnings of gnostic christianity / Rylands, Louis Gordon – London: Watts, 1940. viii,300p – 1 – us Wisconsin U Libr [240]

The beginnings of gospel story : a historico-critical inquiry into the sources and structure of the gospel according to mark / Bacon, Benjamin Wisner – New Haven, Conn.: Yale University Press; London: Henry Frowde, 1909 – 1mf – 9 – 0-8370-9362-7 – (incl bibl ref) – mf#1986-3362 – us ATLA [225]

The beginnings of hindu pantheism / Lanman, Charles Rockwell – Cambridge, Mass, USA: CW Sever, 1890 – 1mf – 9 – 0-524-01780-8 – mf#1990-2628 – us ATLA [280]

The beginnings of history according to the bible and the traditions of oriental peoples : from the creation of man to the deluge = Origines de l'histoire d'apres la bible et les traditions des peuples orientaux / Lenormant, Francois – New York: Charles Scribner, 1883, c1881 – 2mf – 9 – 0-8370-6995-5 – (in english. incl bibl ref) – mf#1986-0995 – us ATLA [930]

The beginnings of indian historiography and other essays / Ghoshal, Upendra Nath – Calcutta: Ramesh Ghoshal, 1944 – (= ser Samp: indian books) – us CRL [954]

The beginnings of libraries / Richardson, Ernest Cushing – Princeton: Princeton University Press, 1914 – 1mf – 9 – 0-7905-7137-4 – (incl bibl ref) – mf#1988-3137 – us ATLA [020]

The beginnings of life : being some account of the nature, modes of origin and transformations of lower organisms / Bastian, Henry Charlton – London, 1872 – (= ser 19th c evolution & creation) – 2v on 15mf – 9 – mf#1.1.1584 – uk Chadwyck [575]

The beginnings of methodism in colorado *see*
- Theses on methodism

The beginnings of new england : or, the puritan theocracy in its relations to civil and religious liberty / Fiske, John – Boston: Houghton, Mifflin, 1902 – 1mf – 9 – 0-524-02738-2 – (incl bibl ref) – mf#1990-4413 – us ATLA [975]

Beginnings of nyasaland and north-eastern rhodesia, 1859-95 / Hanna, Alexander John – Oxford, England. 1969 – 1r – us UF Libraries [960]

The beginnings of quakerism / Braithwaite, William Charles – London: Macmillan, 1912 – 2mf – 9 – 0-7905-4149-1 – (incl bibl ref) – mf#1988-0149 – us ATLA [243]

The beginnings of the art in eastern india : with special reference to sculptures in the indian museum, calcutta / Chanda, Ramaprasad – Calcutta: Govt of India Central Publication Branch, 1927 – (= ser Samp: indian books) – us CRL [730]

The beginnings of the church / Scott, Ernest Findlay – New York: Scribner, 1914 – (= ser Ely Lectures) – 1mf – 9 – 0-524-04851-7 – mf#1990-1343 – us ATLA [240]

The beginnings of the historic episcopate as exhibited in the words of holy scripture and ancient authors / Barbour, John Humphrey – New York: E & JB Young, 1887 – 1mf – 9 – 0-524-06113-0 – mf#1992-0780 – us ATLA [240]

The beginnings of the moravian mission in alaska / Hamilton, John Taylor – [Bethlehem, PA: The Comenius press, 1890] (mf ed 19–) – (= ser History of the Pacific Northwest, PNW) – 23p – mf#ZH-396 – us NY Public [243]

The beginnings of the temporal sovereignty of the popes, a d 754-1073 = Premiers temps de l'etat pontifical (754-1073) / Duchesne, Louis – London: Kegan Paul, Trench, Truebner, 1908 [mf ed 1989] – (= ser The international catholic library 11) – 1mf – 9 – 0-7905-4563-2 – (english trans by arnold harris mathew. incl bibl ref) – mf#1988-0563 – us ATLA [241]

The beginnings of the theological seminary of the reformed church in the united states, from 1817 to 1832 / Appel, Theodore – Philadelphia: Reformed Church Publ Board, 1886 – 1mf – 9 – 0-524-03606-3 – mf#1990-4766 – us ATLA [242]

The beginnings of the wesleyan movement in america and the establishment therein of methodism / Atkinson, John – New York: Hunt & Eaton; Cincinnati: Cranston & Curts, 1896 – 2mf – 9 – 0-7905-5561-1 – mf#1988-1561 – us ATLA [242]

Beginnings of vijayanagara history / Heras, H – Bombay, 1929; Madrid: Razon y Fe, 1931 – 1 – sp Bibl Santa Ana [954]

The beginnings of yale : 1701-1726 / Oviatt, Edwin – New Haven: Yale University Press, 1916 – 2mf – 9 – 0-524-07530-1 – mf#1991-3160 – us ATLA [378]

De beginselen van gods koninkryk in den mensch : uitgedrukt in zinne-beelden / [Huygen, Pieter] – t'Amsterdam: Wed. Pieter Arents, 1690 – 2mf – 9 – mf#0-644 – ne IDC [090]

De beginselen van gods koninkryk in den mensch : uytgedrukt in verscheide zinne-beelden... / Huygen, Pieter & Huygen, Jan – ed 7. t'Amsterdam: Jacob ter Beek, 1738 – 5mf – 9 – mf#O-3233 – ne IDC [090]

Beginselen van separatie : critisch en historisch onderzocht / Hospers, Gerrit Hendrik – Cleveland OH: Pub House of the Reformed Church 1897 [mf ed 1993] – 1mf – 9 – 0-524-06625-6 – mf#1991-2680 – us ATLA [242]

Beginzelen der vesting-bouw / Bruist, B – Amsterdam, 1705 – 3mf – 9 – mf#OA-141 – ne IDC [720]

Der begleiter auf dem weser-dampfschiffe von muenden nach bremen / Boclo, Ludwig – Goettingen 1844 [mf ed Hildesheim 1995-98] – 2mf – 9 – €60.00 – 3-487-29522-9 – gw Olms [914]

Begley, Walter *see*
- Biblia anagrammatica
- Biblia cabalistica

Begraebnissplatz der opler – s.l, s.l? . 1755 – 1r – us UF Libraries [920]

Begrebet angest : en simpel psychologisk-paapegende cverveielse i retning af det dogmatiske problem om arvesynden / Kierkegaard, Soeren – Kobenhavn: C A Reitzel 1844 [mf ed 1990] – 1mf – 9 – 0-7905-7953-7 – mf#1989-1178 – us ATLA [210]

Begreppet herrens tjaenare hos andre-esaias : kritisk-exegetisk undersoekning / Lundborg, Matheus – Lund: Gleerupska Universitets-Bokhandeln [189-?] [mf ed 1986] – 1mf – 9 – 0-8370-9559-X – (in swedish; incl bibl ref) – mf#1986-3559 – us ATLA [221]

Der begriff der bekehrung : im lichte der heiligen schrift, der kirchengeschichte und der forderungen des heutigen lebens: eine untersuchung / Herzog, Johannes – Giessen: J Ricker (Alfred Toepelmann) 1903 [mf ed 1985] – 1mf – 9 – 0-8370-4883-4 – (incl bibl ref) – mf#1985-2883 – us ATLA [200]

Der begriff der diatheke im hebraeerbrief / Riggenbach, Eduard – Leipzig: A Deichert 1908 [mf ed 1989] – 1mf – 9 – 0-7905-3212-3 – (in german, latin & greek) – mf#1987-3212 – us ATLA [200]

Der begriff der gnade im neuen testament : eine biblisch-theologische untersuchung / Voemel, Rud – Guetersloh: C Bertelsmann 1903 [mf ed 1991] – (= ser Beitraege zur foerderung christlicher theologie 7/1903/5) – 1mf – 9 – 0-524-00197-9 – (filmed with: tertullians dogmatische und ethische grundanschauungen by wilhelm vollert) – mf#1989-2897 – us ATLA [225]

Der begriff der heiligkeit im neuen testament : eine von der haager gesellschaft zur verteidigung der christlichen religion gekroente preisschrift / Issel, Ernst – Leiden: E J Brill 1887 [mf ed 1985] – 1mf – 9 – 0-8370-3733-6 – (incl bibl ref) – mf#1985-1733 – us ATLA [225]

Der begriff der katholicitaet der kirche und des glaubens : nach seiner geschichtlichen entwicklung / Soeder, Rudolf – Wuerzburg: Leo Woerl 1881 [mf ed 1990] – 1mf – 9 – 0-524-00347-5 – (incl bibl ref) – mf#1989-3047 – us ATLA [241]

Der begriff der offenbarung / Herrmann, Wilhelm – Giessen: J Ricker 1887 [mf ed 1992] – (= ser Vortraege der theologischen konferenz zu giessen 3. folge) – 1mf – 9 – 0-524-02642-4 – (filmed with: bericht ueber den gegenwaertigen stand der forschung auf dem gebiet der vorreformatorischen zeit by karl mueller) – mf#1990-0666 – us ATLA [240]

Der begriff der wahrheit in dem evangelium und den briefen des johannes / Buechsel, Friedrich – Guetersloh: C Bertelsmann 1911 [mf ed 1993] – (= ser Beitraege zur foerderung christlicher theologie 15/3) – 1mf – 9 – 0-524-05905-5 – (incl bibl ref) – mf#1992-0662 – us ATLA [227]

Der begriff des betriebes / Lostorf, Edmund Willi – Bern, 1938 – 1 – gw Mikropress [650]

Der begriff des kunstwerks in goethes aufsatz von deutscher baukunst (1772) und in schillers aesthetik : vortrag gehalten auf der 46. versammlung deutscher philologen und schulmaenner zu strassburg i.e. / Gneisse, Karl – Strassburg: Heitz 1901 [mf ed 1990] – 1r – 1 – (filmed with: goethe und weimar / ernst schrumpf) – mf#7378 – us Wisconsin U Libr [430]

Der begriff des uebernatuerlichen : sein dialektischer charakter und das princip der identitaet / Tillich, Paul – Koenisberg Nm: H Madrasch 1915 [mf ed 1991] – 1mf – 9 – 0-524-00400-5 – (incl bibl ref) – mf#1989-3100 – us ATLA [230]

Der begriff des volksgeistes in ernst moritz arndts geschichtsanschauung : ein beitrag zur geschichte der geschichtswissenschaft vorgelegt von rudolf kruegel / Kruegel, Rudolf – Langensalza: H Beyer 1914 [mf ed 1988] – 1r – 1 – (filmed with: ernst mortiz arndt / ernst muesebeck) – mf#6955 – us Wisconsin U Libr [100]

Der begriff dogma entwickelt aus der entscheidung ueber die unbefleckte empfaengniss mariae / Enders, Barthol – Amberg: Hermann v Train 1857 [mf ed 1986] – 1mf – 9 – 0-8370-7790-7 – (incl bibl ref) – mf#1986-1790 – us ATLA [241]

Der begriff "edel" bei goethe / Liederwald, Carl – Greifswald, 1913 (mf ed 1994) – 1mf – 9 – €24.00 – 3-8267-3082-8 – mf#DHS-AR 3082 – gw Frankfurt [430]

Der begriff klassisch bei herder / Oelsner, Werner – Wuerzburg: K Triltsch 1939 [mf ed 1991] – 1r – 1 – (incl bibl ref. filmed with: goethe und seine eltern / herman kruger-westend) – mf#7473 – us Wisconsin U Libr [430]

Begriff und aufgabe der dogmengeschichte / Haase, Felix – Breslau: Goerlich & Coch 1911 [mf ed 1990] – 1mf – 9 – 0-7905-6956-5 – (incl bibl ref) – mf#1988-2996 – us ATLA [240]

Die begriffe der zeit und ewigkeit im spaeteren platonismus / Leisegang, H – 1913 – (= ser Bgphma 13/4) – €5.00 – ne Slangenburg [180]

Die begriffe fleisch und geist im biblischen sprachgebrauch / Wendt, Hans Hinrich – Gotha: Friedr Andr Perthes, 1878 – 1mf – 9 – 0-8370-9344-9 – (incl bibl ref) – mf#1986-3344 – us ATLA [220]

Die begriffe geist und leben bei paulus : in ihren beziehungen zu einander: eine exegetisch-religionsgeschichtliche untersuchung / Sokolowski, Emil – Goettingen: Vandenhoeck & Ruprecht, 1903 – 1mf – 9 – 0-8370-6410-4 – (incl bibl ref and index) – mf#1986-0410 – us ATLA [220]

Die begriffe pflicht und tugend in der sittenlehre kant's und schleiermacher's : eine vergleichende studie / Ewh, Paul – Witten: CL Krueger, 1891 – 1mf – 9 – 0-524-00024-7 – mf#1989-2724 – us ATLA [170]

Die begruendung der erkenntnis nach dem hl augustinus / Hessen, J – 1916 – (= ser Bgphma 19/2) – €7.00 – ne Slangenburg [100]

Begruendung des entwurfes eines rechtes der erbfolge entwurfes eines einfuehrungsgesetzes *see* Entwurf eines rechtes der erbfolge fuer das deutsche reich

Die begruendung unserer sittlich-religioesen ueberzeugung / Koestlin, Julius – Berlin: Reuther & Reichard, 1893 [mf ed 1985] – 1mf – 9 – 0-8370-3981-9 – mf#1985-1981 – us ATLA [240]

Begue-Clavel, F-T *see* Histoire pittoresque de la francmaconnerie et des societes secretes anciennes et modernes

Begueule : ou, la princesse et le charbonnier / Brazier, Nicholas – Paris, France. 1826 – 1r – us UF Libraries [440]

Beguillet, Edme *see* Description historique de paris

Beguin, P *see*
- The saurus fratris salimbene de adam
- Thesaurus fontium franciscanorum

Beguinot, Francesco *see* Il berbero nefusi di fassato

Begutex *see* Der gummiwerker

Beha ullah (the glory of god) / Kheiralla, Ibrahim George – Chicago: I G Kheiralla 1900 [mf ed 1992] – 2mf [ill] – 9 – 0-524-02215-1 – mf#1990-2889 – us ATLA [290]

Behaghel, Otto *see*
- Heinrichs von veldeke eneide
- Heliand
- Der heliand und die altsaechsische genesis

Behaghel, Wilhelm *see* Die gewerbliche stellung der frau im mittelalterlichen koeln

Behagle, Ferdinand de *see* Au pays de l'esclavage

Behan, John *see* On dr maguire's pamphlet

Die behandlung der einzelnen stoffelemente in den epen veldekes und hartmans / Roetteken, Hubert – Halle: E Karras, 1887 – 1r – 1 – us Wisconsin U Libr [430]

Behandlung der instabilen angina pectoris : unter beruecksichtigung der braunwaldklassifikation mit ballonangioplastie (ptca). akutdaten und langzeitbeobachtung von 231 patienten / Stolzenburg, Thomas Heiko – (mf ed 1998) – 2mf – 9 – €40.00 – 3-8267-2528-X – mf#DHS 2528 – gw Frankfurter [617]

Die behandlung des sittlichen problems in schillers "kampf mit dem drachen", der erzaehlung bei livius 8, 7, kleists "prinz von homburg" und sophokles' "antigone" / Seiler, Friedrich – Eisenberg: P Kaltenbach, 1890 – 1r – 1 – (incl bibl ref) – us Wisconsin U Libr [430]

Die behandlung von goethes "faust" in den oberen klassen hoeherer schulen / Haehnel, K – 2., verb u verm Aufl. Gera: T Hoffmann, 1896 – 1r – 1 – (incl bibl ref) – us Wisconsin U Libr [430]

Behar proverbs / Christian, John – London: K Paul, Trench, Truebner 1891 [mf ed Bloomington IN: Indiana Uni Lib, Preservation Dept 1984] – 1r – 1 – us Indiana Preservation [390]

Behauptung der himmlischen musik aus den gruenden der vernunft, kirchen-lehre und heiligen schrift... / Mattheson, Johann – Hamburg: C Herold 1747 [mf ed 19–] – 3mf – 9 – mf#fiche 436 – us Sibley [780]

Behave kindly – London, England. 18– – 11r – us UF Libraries [240]

Behavior and philosophy – Concord MA 1990+ – 1,5,9 – (cont: behaviorism) – ISSN: 1053-8348 – mf#12833.01 – us UMI ProQuest [150]

Behavior genetics – Dordrecht, Netherlands 1970+ – 1,5,9 – ISSN: 0001-8244 – mf#10849 – us UMI ProQuest [150]

Behavior modification – Newbury Park CA 1983+ – 1,5,9 – ISSN: 0145-4455 – mf#14005 – us UMI ProQuest [150]

Behavior research methods – Austin TX 2005+ – 1,5,9 – ISSN: 1554-351X – mf#5868.02 – us UMI ProQuest [150]

Behavior research methods and instrumentation – Austin TX 1968-83 – 1,5,9 – (cont by: behavior research methods, instruments, and computers: a journal of the psychonomic society, inc) – ISSN: 0005-7878 – mf#5868.02 – us UMI ProQuest [150]

Behavior research methods, instruments, & computers : a journal of the psychonomic society, inc / Psychonomic Society, Inc – Austin TX 1984+ – 1,5,9 – (cont: behavior research methods and instrumentation) – ISSN: 0743-3808 – mf#5868.02 – us UMI ProQuest [150]

Behavior science notes – Newbury Park CA 1966-73 – 1,5 – (cont by: behavior science research) – ISSN: 0005-7886 – mf#2084.02 – us UMI ProQuest [150]

Behavior science research – Newbury Park CA 1974-92 – 1,5,9 – (cont: behavior science notes; cont by: cross-cultural research) – ISSN: 0094-3673 – mf#2084.02 – us UMI ProQuest [300]

Behavior today – Providence RI 1970-92 – 1,5,9 – ISSN: 0005-7924 – mf#9802 – us UMI ProQuest [301]

Behavioral and brain sciences – Cambridge UK 1978+ – 1,5,9 – ISSN: 0140-525X – mf#13015 – us UMI ProQuest [150]

Behavioral and social sciences librarian / ed by Stover, Mark – mf#0163-9269 – us Haworth [300]

Behavioral assessment – Oxford UK 1979-92 – 1,5,9 – ISSN: 0191-5401 – mf#49285 – us UMI ProQuest [150]

Behavioral counseling and community interventions – Dordrecht, Netherlands 1983 – 1,5,9 – (cont: behavioral counseling quarterly) – ISSN: 0749-1301 – mf#12180.01 – us UMI ProQuest [150]

Behavioral counseling quarterly – Dordrecht, Netherlands 1981-82 – 1,5,9 – (cont by: behavioral counseling and community interventions) – ISSN: 0190-1028 – mf#12180.01 – us UMI ProQuest [150]

Behavioral determinants of insulin resistance : in non-diabetic patients with coronary disease / Littrell, Tanya R – 2000 – 83p on 1mf – 9 – $5.00 – mf#PH 1702 – us Kinesology [616]

Behavioral disorders – Olathe KS 1975+ – 1,5,9 – ISSN: 0198-7429 – mf#12891 – us UMI ProQuest [370]

Behavioral ecology – Oxford UK 1990+ – 1,5,9 – ISSN: 1045-2249 – mf#17993 – us UMI ProQuest [574]

Behavioral ecology and sociobiology – Dordrecht, Netherlands 1981+ – 1,5,9 – ISSN: 0340-5443 – mf#13143 – us UMI ProQuest [304]

Behavioral health management – 1994+ – 1,5,9 – (cont: addiction and recovery) – ISSN: 1075-6701 – mf#16861.05 – us UMI ProQuest [362]

Behavioral interventions – Hoboken NJ 1994+ – 1,5,9 – (cont: behavioral residential treatment) – ISSN: 1072-0847 – mf#13456.01 – us UMI ProQuest [362]

Behavioral medicine – Washington DC 1988+ – 1,5,9 – (cont: journal of human stress) – ISSN: 0896-4289 – mf#12489,01 – us UMI ProQuest [150]

Behavioral neuroscience – Washington DC 1983+ – 1,5,9 – ISSN: 0735-7044 – mf#13352 – us UMI ProQuest [150]

Behavioral research in accounting – Sarasota FL 1992+ – 1,5,9 – ISSN: 1050-4753 – mf#19423 – us UMI ProQuest [150]

Behavioral residential treatment – Hoboken NJ 1992-93 – 1,5,9 – (cont by: behavioral interventions) – ISSN: 0884-5581 – mf#13456.01 – us UMI ProQuest [150]

Behavioral science – Hoboken NJ 1956-96 – 1,5,9 – ISSN: 0005-7940 – mf#1451 – us UMI ProQuest [150]

Behavioral sciences & the law – Hoboken NJ 1983+ – 1,5,9 – ISSN: 0735-3936 – mf#13077 – us UMI ProQuest [344]

Behaviorism – Concord MA 1980-89 – 1,5,9 – (cont by: behavior and philosophy) – ISSN: 0090-4155 – mf#12833.01 – us UMI ProQuest [150]

Behaviorism / Watson, John B – New York, NY. no date – 1r – us UF Libraries [150]

Behaviour problems in the classroom : a model for teachers to assist learners with unmet emotional needs / Weeks, Franscina Hester – Uni of South Africa 2001 [mf ed Johannesburg 2001] – 10mf – 9 – (incl bibl ref) – mf#mfm15033 – sa Unisa [370]

Behaviour research and therapy – Oxford UK 1963+ – 1,5,9 – ISSN: 0005-7967 – mf#49019 – us UMI ProQuest [150]

Behavioural brain research – Oxford UK 1980+ – 1,5,9 – ISSN: 0166-4328 – mf#42067 – us UMI ProQuest [150]

Behavioural processes – Oxford UK 1976+ – 1,5,9 – ISSN: 0376-6357 – mf#42216 – us UMI ProQuest [150]

Beheim, Michel *see* Die gedichte des michel beheim

Beheim-Schwarzbach, Eberhard *see* Dramenformen des barock

Beheim-Schwarzbach, Martin *see*
– Die herren der erde
– Novalis (friedrich von hardenberg)

Be-hilahem yisrael / Ben-Gurion, David – Tel-Aviv, Israel. 1952 – 1r – us UF Libraries [939]

Behind the arras : a book of the unseen / Carman, Bliss – Boston, New York: Lamson, Wolffe, 1895 – 2mf – 9 – (with designs by t b meteyard) – mf#03841 – cn Canadiana [890]

Behind the cotton curtain : newsletter of southern democratic socialists – v1 n2-6 [1981 jul/aug-1982 may/jun] – 1r – 1 – mf#655231 – us WHS [325]

Behind the creek – Wausau WI. 1901 jun 4-1902 jun – 1r – 1 – (cont by: north wausau news) – mf#946947 – us WHS [071]

Behind the great wall : the story of the c e z m s work and workers in china, with numerous illustrations / Barnes, Irene H – London: Marshall Brothers; Church of England Zenana Missionary Society, [1896] [mf ed 1995] – (= ser Yale coll) – viii/184p (ill) – 1 – 0-524-09317-2 – (pref by handley c g moule) – mf#1995-0317 – us ATLA [242]

Behind the headlines – Toronto ON 1970+ – 1,5,9 – ISSN: 0005-7983 – mf#6152 – us UMI ProQuest [321]

Behind the lianas / Larsen, Henry A – Edinburgh, Scotland. 1958 – 1r – us UF Libraries [972]

Behind the racial tensions in south africa / Whyte, Quintin – Johannesburg, South Africa. 1953 – 1r – us UF Libraries [321]

Behind the scenes / Freeman Institute – v1 n1 [1978 dec] – 1r – 1 – (cont; behind the scenes in washington) – mf#638851 – us WHS [071]

Behind the scenes at the front / Adam, George Jefferys – London: Chatto & Windus, 1915 [mf ed 1989] – viii/239/1p – 1 – mf#2763 – us Wisconsin U Libr [933]

Behind the scenes in washington – iss5 v1 n1-iss6 n7 [1977 jun-1978 jul] – 1r – 1 – (cont by: behind the scenes) – mf#638850 – us WHS [071]

Behind the veil : a poem / De Mille, James – Halifax, NS: T Allen, 1893 [mf ed 1980] – 1mf – 9 – 0-665-02706-0 – mf#02706 – cn Canadiana [810]

Behind the veil in persia and turkish arabia / Griffith, M E Hume- – London, 1909 – 5mf – 9 – mf#HTM-72 – ne IDC [915]

The behistan inscription of king darius / Tolman, Herbert Cushing – Nashville, Tenn.: Vanderbilt University, 1908 – (= ser Vanderbilt University Studies) – 1mf – 9 – 0-7905-3189-5 – (incl bibl ref) – mf#1987-3189 – us ATLA [470]

Behl, Carl Friedrich Wilhelm *see*
– Gerhart hauptmann
– Gerhart hauptmanns leben chronik und bild
– Zwiesprache mit gerhart hauptmann

Behlen, Stephan *see*
– Archiv der forst und- jagd-gesetzgebung der deutschen bundesstaaten
– Der spessart

Behm, Heinrich M Th *see* Ueber den verfasser der schrift, welche den titel "hirt" fuehrt

Behn, Aphra *see* Oronoko

Behn, Wolfgang *see* The dissident press of revolutionary iran

Behold! – 1986 sep-1989 jun – 1r – 1 – mf#1573043 – us WHS [071]

Behold he cometh with clouds – Kelso, Scotland. 1842 – 1r – us UF Libraries [240]

Behold the man! = Sehet welch ein mensch! / Delitzsch, Franz – New York: T Whittaker [1889?] [mf ed 1989] – 1mf – 9 – 0-7905-1871-6 – (trans fr german by elizabeth c vincent) – mf#1987-1871 – us ATLA [240]

Behold the morning! : the imminent and premillennial coming of jesus christ / Wimberly, Charles Franklin – New York: FH Revell, c1916 [mf ed 1991] – 1mf – 9 – 0-7905-8747-5 – mf#1989-1972 – us ATLA [240]

Behold the west indies / Oakley, Amy Ewing – New York, NY. 1941 – 1r – us UF Libraries [972]

Behr, August von *see* Meine reise durch schlesien, galicien, podolien

Behr, Thomas *see* Perspektiven einer koerperorientierten erwachsenenbildung

Behramjee, P *see* Dinkard

Behrend, Christian *see* Pikrinsaeuremetabolismus in nocardioides sp. cb-22

Behrend, Erich *see* Theodor fontanes roman "der stechlin"

Behrend, Fritz *see*
– Anbind- oder fangbriefe

Behrends, Adolphus Julius Frederick *see*
– The christ of nineteen centuries
– The old testament under fire
– Socialism and christianity
– The world for christ

Behrendt, Mike *see* Die verstaendlichkeit von fachtexteni in abhaengigkeit vom gegenstandsbereich

Behrendt, Richard Fritz Walter *see* Modern latin america in social science literature

Behrens, Christian *see* Die rolle des cd95-liganden bei der abstossung von transplantattumoren

Behrens, Christoph [comp] *see* Fuersten-postkarten

Behrens, Emil *see* Assyrisch-babylonische briefe

Behrens, G *see* Das fruehchristliche und merowingische mains

Behring sea arbitration : papers relating to the proceedings...constituted under article 1 of the treaty concluded at washington on the 29th february, 1892... – London: Printed for HMSO by Harrison & Sons, 1893 – 2mf – 9 – mf#14453 – cn Canadiana [341]

The behring sea question / Lash, Zebulon Aiton – [Toronto?: s.n, 1893?] – 1mf – 9 – 0-665-15424-0 – mf#15424 – cn Canadiana [639]

Behrmann, Georg *see* Das buch daniel

Behrmann, Willi *see* Feuer der nacht

Behr-Pinnow, C von *see* Die vererbung bei den dichtern a bitzius, c f meyer und g keller

Bei den patagoniern : ein damenritt durch unerforschte jagdgruende / Dixie, Florence C – Leipzig 1882 [mf ed Hildesheim 1995-98] – 1v on 2mf – 9 – €60.00 – 3-487-26837-X – gw Olms [918]

Bei goethe zu gaste : neues von goethe, aus seinem freundes- und gesellschaftskreise / Gaedertz, Karl Theodor – Leipzig: G Wigand 1900 [mf ed 1990] – 1r [ill] – 1 – (filmed with: goethe und karl august / heinrich duntzer) – mf#2806p – us Wisconsin U Libr [430]

Die beicht nach caesarius von heisterbach / Koeniger, Albert Michael – Muenchen: Lentner 1906 [mf ed 1990] – (= ser Veroeffentlichungen aus dem kirchenhistorischen seminar muenchen 2/10) – 1mf – 9 – 0-7905-6301-0 – (incl bibl ref) – mf#1988-2301 – us ATLA [240]

Beicht- und communionbuch fuer evangelische christen : zum gebrauch sowol in, als ausserhalb des gotteshauses / Loehe, Wilhelm – 5. verm verb aufl. Nuernberg: G Loehe 1871 [mf ed 1992] – 1mf [ill] – 9 – 0-524-05380-4 – mf#1991-2286 – us ATLA [242]

Beicht- und Suendenspiegel *see* Die zehn gebote (mxt3)

Die beichte und absolution / Kliefoth, Theodor – Schwerin: Stiller, 1856 – 2mf – 9 – 0-524-04215-2 – mf#1990-5006 – us ATLA [240]

Das beichtsiegel : novelle / Bergengruen, Werner – Freiburg/Br: Christophorus-Verlag, 1948 [mf ed 1995] – 106p – 1 – mf#8976 – us Wisconsin U Libr [830]

Die beiden aeltesten lateinischen fabelbuecher des mittelalters : des bischofs cyrillus speculum sapientiae und des nicolaus pergamenus dialogus creaturarum / ed by Graesse, J G Th – Stuttgart: Litterarischer Verein, 1880 (Tuebingen: H Laupp) [mf ed 1993] – (= ser Blvs 148) – 309p – 1 – mf#8470 reel 31 – us Wisconsin U Libr [390]

Die beiden auswanderer : schauspiel in zwei abtheilungen und fuenf aufzuegen / Gutzkow, Karl – [S.l: s.n, 1857?] [mf ed 1993] – 70p – 1 – mf#8668 – us Wisconsin U Libr [820]

Die beiden briefe pauli an die thessalonicher *see* The two epistles of paul to the thessalonians

Die beiden ersten erasmus-ausgaben des neuen testaments und ihre gegner / Bludau, August – Freiburg i B, St Louis MO: Herder, 1902 – (= ser Biblische studien) – 1mf – 9 – 0-7905-1923-2 – (incl bibl ref) – mf#1987-1923 – us ATLA [225]

Die beiden fassungen von goethes die leiden des jungen werthers : eine stilpsychologische untersuchung / Riess, Gertrud – Breslau: Trewendt & Granier, 1924 – 1r – 1 – (incl bibl ref) – us Wisconsin U Libr [430]

Die beiden fassungen von schillers abhandlung "ueber naive und sentimentalische dichtung" : ein beitrag zur aesthetik schillers / Krancke, Adolf – Goettingen: L Hofer, 1911 – 1r – 1 – (incl bibl ref) – us Wisconsin U Libr [430]

Die beiden friesen *see* Nordnordwest / die beiden friesen

Die beiden genossen : sozialer roman / Kretzer, Max – 5. aufl. Leipzig: P List c1919 [mf ed 1995] – 1r – 1 – (filmed with: die tuerken vor wien / richard kralik) – mf#3909p – us Wisconsin U Libr [830]

Die beiden gewoehnlichen aethiopischen gregorius-anaphoren / Loefgren, O – Roma, 1933 – 2mf – 8 – €5.00 – (trans with ann by s euringer) – ne Slangenburg [243]

Die beiden griechischen klementinen-epitomen und ihre anhaenge / Paschke, F – Berlin, 1966 – (= ser Tugal 5-90) – 6mf – 9 – €14.00 – ne Slangenburg [240]

Die beiden letzten lebensjahre von johannes calvin / Zahn, Adolf – revidierter Neudruck. Stuttgart: Union deutsche Verlagsgesellschaft, 1898 – 1mf – 9 – 0-7905-6859-4 – mf#1988-2859 – us ATLA [242]

Die beiden schwerter, lukas 22, 35-38 : ein stueck aus der besonderen quelle des lukas / Schlatter, Adolf von – Guetersloh: C Bertelsmann, 1916 – (= ser Beitraege zur foerderung christlicher theologie) – 1mf – 9 – 0-524-06158-0 – mf#1992-0825 – us ATLA [220]

Die beiden tubus / Kurz, Hermann – Stuttgart: K Mayer, 1946 – 1r – 1 – us Wisconsin U Libr [830]

Beidleman, B A *see* Energy balance and the components of total daily energy expenditure in endurance trained and untrained women

Beielstein, Felix Wilhelm *see*
– Der grosse imhoff
– Rauch an der ruhr

Beier, Ulli *see* Mbari notebooks

Beierwaltes, Andreas *see* Die "communitarians"

Beij, B de *see* Getuigenis der 25jarige evangelie-bediening

O beija flor : jornal de instruccao e recreio – Rio de Janeiro, RJ. 07 arb 1849-1852 – (= ser Ps 19) – mf#DIPER – bl Biblioteca [073]

O beija flor – Para: Typ de Mendonca e Baena, 14 jul 1850-23 mar 1851 – (= ser Ps 19) – bl Biblioteca [079]

[Beijing-] beijing review – CC. 1979- – 21r – 1 – $1050.00 (subs $70y) – mf#R04158 – us Library Micro [079]

O beijo : publicacao semanal de modinhas recitativos, lundus e poesias... – Rio de Janeiro, RJ: Typ Economica, 11 mar 1881 – (= ser Ps 19) – mf#P17,01,121 – bl Biblioteca [440]

Beik, Kazimir *see* Zur entstehungsgeschichte von goethes torquato tasso

Beilage – Filmed on separate reels from 1891-1907 – 1 – us NY Public [074]

Beilage zum niederrheinischen kurier fuer das konstitutionelle deutschland – Strassburg (Strasbourg F), 1830 9 dec-1831 – 1 – (title varies: n23 1831: das konstitutionelle deutschland) – gw Misc Inst [323]

Beilagen zu den stenographischen berichten ueber die offentlichen... – Berlin, Germany. 1920 – 1r – us UF Libraries [943]

Beilby, Kristine M *see* Predictors of falls in elderly home care clients residing in assisted living facilities

Beim lampenlicht : erzaehlungen / Wildermuth, Ottilie; ed by Willms, Agnes – Stuttgart: Gebrueder Kroener 1878 [mf ed 1995] – 1r – 1 – (aus ihrem nachlasse gesammelt und ergaenzt von ihrer tochter agnes willms) – mf#3764p – us Wisconsin U Libr [830]

Beiner, Marcus *see* Kommunikationsgemeinschaft und kontraktualismus

Being an asperger vs being real : the story of anneli and helouise: a case study and a testimonio / Froneman, Helouise – Pretoria: Vista University 2003 [mf ed 2003] – 2mf – 9 – (incl bibl ref) – mf#mfm15239 – sa Unisa [616]

The being and attributes of god / Hall, Francis Joseph – New York: Longmans, Green 1909 [mf ed 1990] – (= ser Dogmatic theology 3) – 1mf – 9 – 0-7905-3889-X – (incl bibl ref) – mf#1989-0382 – us ATLA [210]

Being and glory of god / Shepard, Thomas – Aberdeen, Scotland. 1848 – 1r – us UF Libraries [240]

The being of god : moral government and theses in theology / Squier, Miles Powell; ed by Boyd, James Robert – Rochester, NY: E Darrow & Kempshall, 1868 – 1mf – 9 – 0-8370-5511-3 – (incl bibl ref) – mf#1985-3511 – us ATLA [210]

The being of god as unity and trinity / Steenstra, Peter Henry – Boston: Houghton, Mifflin; Cambridge: Riverside Press, 1891 – 1mf – 9 – 0-8370-2369-6 – (incl bibl ref) – mf#1985-0369 – us ATLA [210]

Being respectable / Flandrau, Grace – New York, NY. 1923 – 1r – us UF Libraries [025]

Being single – 1984 jul/aug, 1988 mar/apr, 1993 sep/oct-1995 jul/aug-nov/dec – 1r – 1 – mf#2844035 – us WHS [306]

Being-black-in-the-world / Manganyi, N C – [Johannesburg]: SPRO-CAS/Raven, [1973] – us CRL [305]

Beinker, Nele Karen *see* Die entzuendliche aktivitaet und der eisengehalt der leber bei chronischer hepatitis b und c

Beira news and east coast chronicle : noticias da beira – Beira, Mozambique 4 sep 1917-30 dec 1921, 2 apr 1930-28 dec 1939, 20 jan 1941-30 apr 1953 (semiwkly) – 1 – (in english & portugese) – uk British Libr Newspaper [079]

Beira post = Correio da beira – Beira, Mozambique 23 mar 1898-25 aug 1917 (wkly) (imperfect) – 8r – 1 – (in english & portugese) – uk British Libr Newspaper [079]

Beirut – (= ser Vilayet salnames) – 9 – (1318 [1900] def'a 2 5mf $75; 1332-33 [1917] 11mf $180; 1333-35 [1917] 9mf $150) – us MEDOC [956]

Beis, N *see* Hippolyts schrift ueber die segnungen jakobs

Beissel, Stephan *see*
- Bilder aus der geschichte der altchristlichen kunst und liturgie in italien
- Entstehung der perikopen der roemischen meszbuecher
- Die verehrung der heiligen und ihrer reliquien in deutschland bis zum beginne des 13. jahrhunderts
- Die verehrung der heiligen und ihrer reliquien in deutschland waehrend der zweiten haelfte des mittelalters
- Wallfahrten zu unserer lieben frau in legende und geschichte

Beissier, Fernand *see* Roman d'un notaire

Beissner, Friedrich *see*
- Geschichte der deutschen elegie
- Hoelderlin

Der beitch – New York. 1908 – 1 – us AJPC [240]

Beith, Alexander *see*
- Compulsion of the gospel
- Letter to patrick arkley, esq, advocate

Beith echo and northern district advertiser – [Scotland] North Ayrshire, Beith: J Ferguson jan 1904-jan 1908; sep 1910-14 mar 1913 (wkly) [mf ed 2004] – 4r – 1 – uk Newsplan [072]

Beith, James *see* Powers that be

Beitia, eugenio. apostolado de los seglares 2nd edicion. madrid, 1939 / Guerrero, E – Madrid: Razon y Fe, 1943 – 1 – sp Bibl Santa Ana [946]

Beitraege zu den mitteln der volkserziehung im geiste der menschenbildung – Trogen (CH), 1832-33 – 1r – 1 – gw Misc Inst [170]

Beitraege zu der lehre von den griechischen praepositionen / Mommsen, Tycho – Berlin: Weidmann 1895 [mf ed 1993] – 2mf – 9 – 0-524-05927-6 – mf#1992-0684 – us ATLA [450]

Beitraege zu j a bengel's schriftenklaerung und bemerkungen desselben : zu dem gnomon novi testamenti aus handschriftlichen aufzeichnungen – Leipzig: Fues 1865 [mf ed 1988] – 1mf – 9 – 0-7905-0120-1 – (in german, greek & latin) – mf#1987-0120 – us ATLA [225]

Beitraege zu klopstocks zeitmessung : [tonausdruck, zeitausdruck, tonverhalt] / Pawel, Jaro – s.l: s.n. 1903 [mf ed 1990] – 1r – 1 – (Filmed with: klopstocks leben und werke / karl heinemann) – mf#2770p – us Wisconsin U Libr [430]

Beitraege zu luthers liturgischen reformen : 1. luthers lateinische und deutsche litanei von 1529, 2. luthers deutsche versikel und kollekten / Drews, Paul – Tuebingen: J C B Mohr 1910 [mf ed 1990] – (= ser Studien zur geschichte des gottesdienstes und des gottesdienstlichen lebens) – 1mf – 9 – 0-7905-6166-2 – (incl bibl ref) – mf#1988-2166 – us ATLA; ne Slangenburg [242]

Beitraege zu paul heyses novellentechnik / Klein, John Frederick – s.l: s.n. 1920 [mf ed 1990] – 1r – 1 – (incl bibl ref. filmed with: der wollmarkt / h clauren) – mf#2725p – us Wisconsin U Libr [430]

Beitraege zu uhland : uhlands jugenddichtung / Naegele, Eugen – Tuebingen: W Armbruster & O Riecker 1893 [mf ed 1991] – (= ser Nachrichten ueber das koenigliche gymnasium zu tuebingen 1892-93) – 1r – 1 – (incl number of unpubl early poems; incl bibl ref. filmed with: ludwig uhland als dichter und patriot / hermann dederich) – mf#2930p – us Wisconsin U Libr [430]

Beitraege zum studium der gedichte von j.m.r. lenz / Anwand, Oscar – Muenchen, 1897 (mf ed 1995) – 1mf – 9 – €24.00 – 3-8267-3129-8 – mf#DHS-AR 3129 – gw Frankfurter [430]

Beitraege zum teutschen privat-rechte / ed by Arnold, Friedrich Christian von – Ansbach: C Bruegel. v1-2. 1840-42 – (= ser Civil law 3 coll) – 18mf – 9 – (incl bibl ref and index) – mf#LLMC 96-533 – us LLMC [346]

Beitraege zur abendmahlslehre tertullians / Leimbach, Carl Ludwig – Gotha: F A Perthes 1874 [mf ed 1990] – 1mf – 9 – 0-7905-5171-3 – (incl bibl ref) – mf#1988-1171 – us ATLA [240]

Beitraege zur allgemeinen vergleichenden sprachkunde : erstes heft. die praepositionen / Lisch, Georg Christian Friedrich – Berlin: Nauck 1826 – (= ser Whsb) – 1mf – 9 – €20.00 – mf#Hu 011 – gw Fischer [410]

Beitraege zur alten geschichte / Klio – Leipzig. v1-16. 1901-1920 – 220mf – 8 – mf#H-682 – ne IDC [930]

Beitraege zur altertumskunde des orients / Landau, Wilh, Freiherr von – Leipzig: Eduard Pfeiffer 1893-1906 [mf ed 1986] – 1mf – 9 – 0-8370-7301-4 – (incl bibl ref) – mf#1986-1301 – us ATLA [930]

Beitraege zur askanischen volkskunde / Stephan, Oskar – 1925 [mf ed Bloomington IN: Indiana Uni Lib, Preservation Dept 1984] – 1r – 1 – us Indiana Preservation [390]

Beitraege zur assyriologie und vergleichenden semitischen sprachwissenschaft / ed by Delitzsch, Franz & Haupt, P – Leipzig. v1-10. 1890-1927 – 67mf – 9 – mf#NE-20060 – ne IDC [470]

Beitraege zur assyriologie und vergleichenden semitischen sprachwissenschaft / Beitraege zur assyriologie und semitischen sprachwissenschaft = Contributions to assyriology, semitic languages and philology – Leipzig, 1889-1927 [mf ed 2] – (= ser Christianity's encounter with world religions, 1850-1950) – 3r – 1 – (in german) – mf#2-s139-140 – us ATLA [470]

Beitraege zur athanasius *see* Athanasiana

Beitraege zur auslegung von richard wagners "ring des nibelungen" / Grisson, Rudolf Hermann Rulemann – Leipzig: A Klein 1934 [mf ed 1991] – 1r – 1 – (incl bibl ref; also publ under title: herrscherdaemmerung und deutschlands erwachen in wagners "ring des nibelungen"; filmed with: richard wagner's tondrama) – mf#3023p – us Wisconsin U Libr [780]

Beitraege zur bayerischen kirchengeschichte – 1(1894)-26(1919) – 132mf – 9 – €252.00 – ne Slangenburg [241]

Beitraege zur bedeutung von lipoproteinen und thromboxan fuer klinische und experimentelle artherosklerose und zur antiatherosklerotischen wirkung von hdl / Beitz, Agathe – (mf ed 1997) – 2mf – 9 – €40.00 – 3-8267-2406-2 – mf#DHS 2406 – gw Frankfurter [574]

Beitraege zur befoerderung der menschenkenntniss : besonders in ruecksicht unserer moralischen natur / ed by Pockels, Karl Friedrich – Berlin 1788-89 – (= ser Dz) – 2st on 2mf – 9 – €60.00 – mf#k/n5146 – gw Olms [170]

Beitraege zur belehrung und erholung – Osnabrueck DE, 1848 20 may-1849 29 dec – 1r – 1 – gw Misc Inst [080]

Beitraege zur beruhigung und aufklaerung ueber diejenigen dinge, die dem menschen unangenehm sind oder sein koennen : und zur naehern erkenntnis der leidenden menschheit / ed by Fest, Joh Samuel – Leipzig 1789-97 – (= ser Dz. abt philosophie) – 5v[zu je 3st] on 26mf – 9 – €260.00 – (bd 5 st 2f ed by christian victor kindervater) – mf#k/n557 – gw Olms [100]

Beitraege zur beschreibung von schlesien – Brieg (Brzeg PL), 1783 n1-4, 1784 n3, 1785 n4, 1786 n6, 1789 n9, 1791 n10, 1796 n13 – 1 – gw Misc Inst [943]

Beitraege zur beurtheilung der septuaginta : eine wuerdigung wellhausenscher textkritik / Jahn, Gustav – Kirchhain N-L: Max Schmersow, [1902] [mf ed 1985] – 1mf – 9 – 0-8370-3758-1 – mf#1985-1758 – us ATLA [221]

Beitraege zur beurtheilung g e lessing's / Mayr, Richard – Wien: A Hoelder 1880 [mf ed 1995] – 1r – 1 – (incl bibl ref. filmed with: gotthold ephraim lessing, 1729-1781 / siegfried seidel & other titles) – mf#3690p – us Wisconsin U Libr [840]

Beitraege zur biblischen landes- und altertumskunde – v68. 1949-51 [complete] – Inquire – 1 – mf#ATLA 1994-S505 – us ATLA [930]

Beitraege zur bretonischen und celtisch-germanischen heldensage / Schulz, Albert – Quedlinburg; Leipzig: G Basse 1847 [mf ed 1993] – (= ser Bibliothek der gesammten deutschen national-literatur [von der aeltesten bis auf die neuere zeit] 2/3) – 10r – 1 – (incl bibl ref) – mf#3394p – us Wisconsin U Libr [430]

Beitraege zur caritativen taetigkeit des gallitzinkreises / Plugge, Heinrich – [Muenster, 1934] (mf ed 1996) – 2mf – 9 – €31.00 – 3-8267-3206-5 – mf#DHS-AR 21 – gw Frankfurter [943]

Beitraege zur deskriptiven poetik in den mittelhochdeutschen volksepen und in der thidreksaga / Huennerkopf, Richard – Heidelberg, 1914 (mf ed 1994) – 1mf – 9 – €24.00 – 3-8267-3104-2 – mf#DHS-AR 3104 – gw Frankfurter [430]

Beitraege zur deutschen literaturwissenschaft : nr 1 (1907)-nr 40 (1931) / ed by Elster, Ernst – Marburg: N G Elwert, 1907-31 [mf ed 1992] – 40v – 1 – mf#8004 – us Wisconsin U Libr [430]

Beitraege zur entwicklung religionssystematischen denkens im judentum des 19. jahrhunderts / Schoeps, Hans-Joachim – Dresden, 1934 – 9 – 3-89349-258-5 – gw Frankfurter [270]

Beitraege zur entwicklungsgeschichte des judentums con ca 400 v chr bis ca 1000 chr / Khvolson, Daniil Avraamovich – Leipzig: H Haessel 1910 [mf ed 1991] – 1mf – 9 – 0-524-01774-3 – mf#1990-2622 – us ATLA [270]

Beitraege zur erklaerung der apostelgeschichte : auf grund der lesarten des codex d und seiner genossen / Belser, Johannes Evangelist – Freiburg i B, St Louis MO: Herder 1897 [mf ed 1986] – 1mf – 9 – 0-8370-9527-1 – (incl bibl ref) – mf#1986-3527 – us ATLA [226]

Beitraege zur erklaerung der persischen keilinschriften : erstes heft / Holtzmann, Adolf – Carlsruhe: G Holtzmann 1845 [mf ed 1986] – 1mf – 9 – 0-8370-7639-0 – (no more publ) – mf#1986-1639 – us ATLA [490]

Beitraege zur erklaerung des buches daniel / Meinhold, Johannes – Leipzig: Doerffling & Franke 1888 [mf ed 1985] – 1mf – 9 – 0-8370-4376-X – (no more publ? incl bibl ref) – mf#1985-2376 – us ATLA [221]

Beitraege zur erklaerung des jesaia / Barth, Jakob – Karlsruhe: H Reuther 1885 – 1mf – 9 – 0-8370-2195-2 – mf#1985-0195 – us ATLA [221]

Beitraege zur erklaerung des koraan / Hirschfeld, Hartwig – Leipzig: O Schulze 1886 [mf ed 1991] – (= ser [Morgenlaendische forschungen] 9) – 1mf – 9 – 0-524-01372-1 – (incl bibl ref) – mf#1990-2384 – us ATLA [260]

Beitraege zur erklaerung und kritik des buches tobit / Mueller, Johannes – Giessen: Alfred Toepelmann 1908 [mf ed 1989] – (= ser Beihefte zur zeitschrift fuer die alttestamentliche wissenschaft 13) – 1mf – 9 – 0-7905-1537-7 – (in german & greek; incl bibl ref; filmed with: alter und herkunft des achikar-romans und sein verhaeltnis zu aesop by rudolf smend) – mf#1987-1537 – us ATLA [221]

Beitraege zur erklaerung und textkritik des buches tobias / Schulte, Adalbert – Freiburg i. Breisgau; St Louis MO: Herder 1914 [mf ed 1989] – (= ser Biblische studien 19/2) – 1mf – 9 – 0-7905-1915-1 – mf#1987-1915 – us ATLA [221]

Beitraege zur eschatologie des islams / Rueling, Josef Bernhard – Leipzig: O Harrassowitz 1895 [mf ed 1992] – 1mf – 9 – 0-524-02049-3 – mf#1990-2824 – us ATLA [260]

Beitraege zur evangelien-kritik / Bleek, Friedrich – Berlin: G Reimer 1846 [mf ed 1986] – 1mf – 9 – 0-8370-9531-X – (incl ind) – mf#1986-3531 – us ATLA [225]

Beitraege zur flora von aegypten und arabien / Fresenius, J B G W – [Frankfurt a. Main, 1834] – (= ser Museum Senckenbergianum) – 4mf – 8 – (museum senckenbergianum, 1833-1834) – mf#5127 – ne IDC [580]

Beitraege zur frauenbiologie : die juedischen rituellen sexualvorschriften / Weissenberg, S – Berlin, Koeln, 1927 (mf ed 1993) – 1mf – 9 – €24.00 – 3-89349-357-3 – mf#DHS-AR 357 – gw Frankfurter [270]

Beitraege zur genaueren kenntnis der attischen gerichtssprache aus den zehn rednern / Schodorf, Konrad – Wuerzburg: A Stuber 1904 [mf ed 1990] – (= ser Beitraege zur historischen syntax der griechischen sprache 17) – 1mf – 9 – 0-8370-1598-7 – mf#1987-6080 – us ATLA [450]

Beitraege zur geographie, geschichte und statenkunde / ed by Fabri, Joh Ernst – Nuernberg 1796 – (= ser Dz. historisch-geographische abt) – 6st on 7mf – 9 – €140.00 – mf#k/n1273 – gw Olms [900]

Beitraege zur geographie und ethnographie babyloniens im talmud und midrasch / Berliner, Abraham – Berlin: J Gorzelanczyk 1884 [mf ed 1985] – 1mf – 9 – 0-8370-2290-8 – (incl ind of geographical names in hebrew) – mf#1985-0290 – us ATLA [910]

Beitraege zur geschichte august hermann francke's : enthaltend den briefwechsel francke's und spener's / ed by Kramer, Gustav – Halle: Verlag der Buchhandlung des Waisenhauses 1861 [mf ed 1991] – 2mf – 9 – 0-524-00565-6 – (incl bibl ref) – mf#1990-0065 – us ATLA [242]

Beitraege zur geschichte der arbeiterbewegung – 1959-1978, 1981-1989 – 690mf – 1 – gw Mikropress [335]

Beitraege zur geschichte der arbeiterbewegung des niederrheinisch-westfaelischen bergbaues / Velsen, Wilhelm von – Essen 1940 – 1 – gw Mikropress [943]

Beitraege zur geschichte der arbeiterbewegung im rheinisch-westfaelischen industriegebiet / Umbreit, Robert – Dortmund 1932 – 1 – gw Mikropress [331]

Beitraege zur geschichte der christologischen dogmas im 11ten und 12ten jahrhundert / Baltzer, Otto – Leipzig: A Deichert, 1898 [mf ed 1990] – (= ser Studien zur geschichte der theologie und der kirche 3/1) – 1mf – 9 – 0-7905-3570-X – (incl bibl ref) – mf#1989-0063 – us ATLA [240]

Beitraege zur geschichte der deutschen sprache und literatur – Halle. v1-44. 1874-1920 – 291mf – 9 – mf#H-10025 – ne IDC [430]

Beitraege zur geschichte der evangelischen kirche in russland / Dalton, Hermann – Berlin: Reuther & Reichard, 1887-1905. Chicago: Dep of Photodup, U of Chicago Lib, 1978 (1r); Evanston: American Theol Lib Assoc, 1984 (1r) – 1 – 0-8370-0603-1 – (incl bibl ref and index) – mf#1984-T078 – us ATLA [242]

Beitraege zur geschichte der griechischen philosophie und religion / Wendland, Paul & Kern, Otto – Berlin: G Reimer, 1895 [mf ed 1989] – 1mf – 9 – 0-7905-3115-1 – (incl bibl ref) – mf#1987-3115 – us ATLA [180]

Beitraege zur geschichte der hebraeischen und aramaeischen studien / Perles, Joseph – Muenchen: Theodor Ackermann 1884 [mf ed 1985] – 1mf – 9 – 0-8370-4704-8 – (in german, aramaic, hebrew, italian & latin; incl bibl ref & ind) – mf#1985-2704 – us ATLA [470]

Beitraege zur geschichte der kreise neuss-grevenbroich *see* Neuss-grevenbroicher zeitung

Beitraege zur geschichte der kreuzzuege / Roehricht, Reinhold – Berlin: Weidmannsche Buchhandlung, 1874-78 [mf ed 2003] – 2v in 1 on 1r – 1 – (incl bibl ref & ind) – mf#b00653 – us ATLA [931]

Beitraege zur geschichte der kunst und der kunsttechnik aus mittelhochdeutschen dichtungen / Ilg, A – Wien, 1892. v5 – 3mf – 9 – mf#O-517 – ne IDC [700]

Beitraege zur geschichte der lehre vom parallelismus der individual- und der gesamtentwicklung / Kleinsorge, John Arnold – Jena, 1899 (mf ed 1994) – 1mf – 9 – €19.00 – 3-8267-3078-X – mf#DHS-AR 3078 – gw Frankfurter [120]

Beitraege zur geschichte der mystik in der reformationszeit / Hegler, Alfred; ed by Koehler, Walther – Berlin: C A Schwetschke 1906 – (= ser Archiv fuer reformationsgeschichte) – 1mf – 9 – 0-7905-5941-2 – (incl bibl ref) – mf#1988-1941 – us ATLA [242]

Beitraege zur geschichte der schweizerisch-reformierten kirche : zunaechst derjenigen des kantons bern / ed by Trechsel, F] – Bern, Jenni. 4pts. 1841-1842 – 8mf – 9 – mf#PBU-454 – ne IDC [242]

Beitraege zur geschichte der theologischen facultaet in freiburg / Koenig, Joseph – Freiburg i. B.: H M Poppen 1884 [mf ed 1990] – 1mf – 9 – 0-7905-6238-3 – (incl bibl ref) – mf#1988-2238 – us ATLA [378]

Beitraege zur geschichte der westlichen araber / Mueller, M J – Muenchen, 1866-1878. 2pts – 4mf – 9 – mf#NE-334 – ne IDC [956]

Beitraege zur geschichte des hexenglaubens und des hexenprocesses in siebenbuergen / Mueller, Friedrich – Braunschweig: C A Schwetschke 1854 [mf ed 1991] – 1mf – 9 – 0-524-01855-3 – (incl bibl ref) – mf#1990-2690 – us ATLA [230]

Beitraege zur geschichte des jesuiten-ordens / Friedrich, Johann – Muenchen:...Akademische Buchdruckerei von F Straub 1881 [mf ed 1990] – 1mf – 9 – 0-7905-5875-0 – (in german & latin; incl bibl ref) – mf#1988-1875 – us ATLA [241]

Beitraege zur geschichte des jesuitenordens / Reusch, Franz Heinrich – Muenchen: C H Beck 1894 [mf ed 1990] – 1mf – 9 – 0-7905-7133-1 – (incl bibl ref) – mf#1988-3133 – us ATLA [241]

Beitraege zur geschichte des rundfunks – Berlin DE, n2 1967-72 – 1r (doubled) – 1 – (filmed by misc inst: 1973-1986 n1, 1987-1989 n3; incl ind 1967-73) – mf#12960 – gw Mikropress; gw Misc Inst [380]

Beitraege zur geschichte des spanischen protestantismus und der inquisition : im sechzehnten jahrhundert / Schaefer, Ernst – Guetersloh: C Bertelsmann 1902 [mf ed 1990] – 3v on 5mf – 9 – 0-7905-8147-7 – (incl bibl ref) – mf#1988-6094 – us ATLA [242]

Beitraege zur geschichte dortmunds und der grafschaft mark / ed by Historischer Verein fuer Dortmund und die Grafschaft Mark – Dortmund : Historischer Verein für Dortmund und die Grafschaft Mark 1875- [v1-32 1875-1925] [mf ed 1978] – 4r – 1 – mf#film mas c358 – us Harvard [943]

Beitraege zur geschichte und frage nach den mitarbeitern der "frankfurter gelehrten anzeigen" vom jahre 1772 : auch ein kapitel zur goethe-philologie / Braeuning-Oktavio, Hermann – Darmstadt: L Vogelsberger 1912 [mf ed 1990] – 1r – 1 – (incl bibl ref. filmed with: goethe und pestalozzi / gottfried bohnenblust) – mf#7384 – us Wisconsin U Libr [430]

Beitraege zur geschichte und zum verstaendnis des massbegriffs / Winterscheidt, Heinrich – Bonn, 1909 (mf ed 1994) – 2mf – 9 – €31.00 – 3-8267-3056-9 – mf#DHS-AR 3056 – gw Frankfurter [100]

Beitraege zur jesaiakritik : nebst einer studie ueber prophetische schriftstellerei / Giesebrecht, Friedrich – Goettingen: Vandenhoeck & Ruprecht 1890 [mf ed 1985] – 1mf – 9 – 0-8370-3269-5 – (incl app) – mf#1985-1269 – us ATLA [221]

Beitraege zur kenntnis der assyrisch-babylonischen medizin : texte mit umschrift, uebersetzung und kommentar / Kuechler, Friedrich – Leipzig: JC Hinrichs, 1904 – (= ser Assyriologische bibliothek) – 1mf – 9 – 0-8370-7556-4 – (incl bibl ref and ind) – mf#1986-1556 – us ATLA [930]

Beitraege zur kenntnis der babylonischen religion / Zimmern, H – Leipzig, 1896-1901. 3 pts – 7mf – 9 – mf#NE-471 – ne IDC [956]

Beitraege zur kenntnis der babylonischen religion / Zimmern, Heinrich – Leipzig: J C Hinrichs 1896-1901 [mf ed 1986] – 1mf – 9 – 0-8370-7119-4 – (incl bibl ref & glos) – mf#1986-1119 – us ATLA [290]

Beitraege zur kenntnis der byzantinischen liturgie : texte und studien / ed by Engdahl, Richard – Berlin: Trowitzsch, 1908 [mf ed 1990] – (= ser Neue studien zur geschichte der theologie und der kirche 5) – 1mf – 9 – 0-7905-7227-3 – (text in greek & latin. discussion in german & greek. incl bibl ref) – mf#1988-3227 – us ATLA [243]

Beitraege zur kenntnis der religionsphilosophischen anschauungen des flavius josephus / Lewinsky, Abraham – Breslau: Preuss & Juenger 1887 [mf ed 1985] – 1mf – 9 – 0-8370-4099-X – (incl bibl ref) – mf#1985-2099 – us ATLA [270]

Beitraege zur kenntnis des forstwesens in deutschland – Leipzig: Baumgartnersche Buchh 1819-21 [1-4 heft (1819-1821)] [mf ed 2005] – 4v on 1r – 1 – (frequency varies) – mf#film mas 36912 – us Harvard [634]

Beitraege zur kenntnis des islamischen vereinswesens auf grund von bast madad et-taufiq / Thorning, Hermann – Berlin: Mayer & Mueller 1913 [mf ed 1992] – (= ser Tuerkische bibliothek 16) – 1mf – 9 – 0-524-02667-X – (in german & arabic; incl bibl ref) – mf#1990-3097 – us ATLA [260]

Beitraege zur kenntnis von klingers sprache und stil in seinen jugend-dramen / Philipp, Richard – Freiburg i.Br: U Hochreuther 1909 [mf ed 1990] – 1r – 1 – (incl bibl ref. filmed with: neue studien uber heinrich von kleist / berthold schulze) – mf#2767p – us Wisconsin U Libr [430]

Beitraege zur kenntniss der literatur, kunst, mythologie und geschichte des alten aegypten : erstes heft. mit vier lithographischen tafeln / Seyffarth, Gustav – Leipzig: Barth 1826 – (= ser Whsb) – 1mf – 9 – €20.00 – mf#Hu 492 – gw Fischer [470]

Beitraege zur kenntniss des innern von russland / Erdmann, Johann F – Riga [u. a.] 1822-26 [mf ed Hildesheim 1995-98] – (= ser Fbc) – 3v on 9mf – 9 – €180.00 – 3-487-29021-9 – gw Olms [947]

Beitraege zur kenntniss des russischen reiches und der angraenzenden laender asiens – Spb, 1839-1896 – 324mf – 9 – mf#R-1667 – ne IDC [915]

Beitraege zur kirchengeschichte, archaeologie und liturgik / Hefele, Karl Joseph von – Tuebingen: H. Laupp 1864 [mf ed 1990] – 2v on 3mf [ill] – 9 – 0-7905-4749-X – (incl bibl ref) – mf#1988-0749 – us ATLA [240]

Beitraege zur kirchenverfassungsgeschichte und kirchenpolitik : insbesondere des protestantismus / Hundeshagen, K B – Wiesbaden, 1864 – 6mf – 9 – mf#ZWI-39 – ne IDC [242]

Beitraege zur kritischen wuerdigung der dramatischen dichtungen theodor koerners / Struker, Johannes [comp] – [s.l: s.n.] 1910 [mf ed 1990] – 1r – 1 – (incl bibl ref. filmed with: die begegnung auf dem riesengebirge / e g kolbenheyer) – mf#2775p – us Wisconsin U Libr [430]

Beitraege zur kunde der indogermanischen sprachen – Goettingen. v1-30. 1877-1907+ind – 210mf – 8 – mf#H-1379 – ne IDC [400]

Beitraege zur kunde ehst-, liv- und kurlands – Reval, 1868-1915. v1-8 – 68mf – 9 – mf#R-1668 – ne IDC [077]

Beitraege zur laender- und staatenkunde der tartarei : aus russischen berichten / ed by Ehrmann, Theophil F – Weimar 1804 [mf ed Hildesheim 1995-98] – 1v on 1mf – 9 – €40.00 – 3-487-26586-9 – gw Olms [947]

Beitraege zur landeskunde von suedwestafrika / Jaeger, Fritz & Waibel, Leo – Berlin: E S Mittler, 1920-21 – 1 – us CRL [960]

Beitraege zur literaturgeschichte und – methodologie : gemeinsamer studienband von germanisten der friedrich-schiller-universitaet jena und der alexandru-ion-cuza-universitaet iasi / ed by Fassel, Hort & Hammer, Klaus – Jena: Die Universitaet 1982 [mf ed 1992] – (= ser Wissenschaftliche beitraege der friedrich-schiller-universitaet jena) – 1r – 1 – (incl bibl ref. filmed with: historical chart of english literature for use in schools and colleges) – mf#3160p – us Wisconsin U Libr [430]

Beitraege zur muhammedanischen dogmatik / Krehl, Ludolf – Leipzig: Breitkopf & Haertel 1885 [mf ed 1991] – 1mf – 9 – 0-524-01568-6 – (no more publ?) – mf#1990-2522 – us ATLA [260]

Beitraege zur naturgeschichte der maskarenischen insel in die beiden organischen naturreiche und mehrere neue entdeckungen in denselben betreffend : ein anhang zu der teutschen uebersetzung dieser reise / Bory de Saint-Vincent, Jean B – Weimar 1805 [mf ed Hildesheim 1995-98] – 1v on 2mf – 9 – €60.00 – 3-487-26567-2 – gw Olms [574]

Beitraege zur palaeontologie und geologie oesterreich-ungarns und des orients – Killen TX 1882-1915 (1) – 1 – mf#3301 – us UMI ProQuest [560]

Beitraege zur philosophischen anthropologie, psychologie und den damit verwandten wissenschaften / ed by Wagner, Joh Mich – Wien 1794, 1796 – (= ser Dz. abteilung naturwissenschaft) – 2v on 5mf – 9 – €100.00 – mf#k/n3354 – gw Olms [120]

Beitraege zur psychologie alberts des grossen / Schneider, A – Muenster, 1903/1906 – (= ser Bgphma) – 11mf – 8 – €21.00 – ne Slangenburg [140]

Beitraege zur rabbinischen sprach- und alterthumskunde / Eisler, Leopold – Wien: Herzfeld & Bauer 1872 [mf ed 1986] – 1mf – 9 – 0-8370-9141-1 – (incl ind) – mf#1986-3141 – us ATLA [470]

Beitraege zur reformationsgeschichte / ed by Friedlaender, G – Berlin, Enslinesche Buchhandlung [Ferdinand Mueller], 1837 – 4mf – 9 – mf#PBU-427 – ne IDC [242]

Beitraege zur richtigen wuerdigung der evangelien und der evangelischen geschichte / Wieseler, Karl – Gotha: Friedrich Andreas Perthes 1869 [mf ed 1986] – 1mf – 9 – 0-8370-9278-7 – (incl bibl ref & ind) – mf#1986-3278 – us ATLA [226]

Beitraege zur sektengeschichte des mittelalters / Doellinger, Ignaz von – (1. theil: geschichte der gnostisch-manichaeischen sekten muenchen, 1890 €12. 2. theil: dokumente vornehmlich zur geschichte der valdesier und katharer, muenchen, 1890 €25) – ne Slangenburg [290]

Beitraege zur semitischen sprachwissenschaft / Noeldeke, T – Strassburg, 1904 – 3mf – 9 – mf#NE-480 – ne IDC [470]

Beitraege zur semitischen sprachwissenschaft / Noeldeke, Theodor – Strassburg: Karl J Truebner 1904 [mf ed 1986] – 1mf – 9 – 0-8370-8365-6 – (incl bibl ref) – mf#1986-2365 – us ATLA [470]

Beitraege zur semitischen sprachwissenschaft / Noldeke, Th – Strassburg, 1904 – 3mf – 8 – €7.00 – ne Slangenburg [470]

Beitraege zur spracherklaerung des neuen testaments : zugleich eine wuerdigung der recension meines commentars zum briefe an die roemer von d. fritzsche / Tholuck, August – Halle: Eduard Anton 1832 [mf ed 1988] – 1mf – 9 – (incl bibl ref & ind) – mf#1987-0171 – us ATLA [225]

Beitraege zur statistik bayerns / Bavaria. Statistisches Landesamt – v. 1-207. 1853-1958. Scattered volumes wanting – 1 – 616.00 – us L of C Photodup [943]

Beitraege zur statistik der stadt frankfurt – v. 1-5. 1858-90 – 1 – 82.00 – us L of C Photodup [943]

Beitraege zur statistik mecklenburgs / Mecklenburg-Schwerin. Statistisches Landesamt – v. 1-16. 1856-1910. V. 16 Wanting – 1 – us L of C Photodup [943]

Beitraege zur stilistik von hoelderlin "tod des empedokles" / Schmidt, Wolfgang – Marburg a.L: N G Elwert 1927 [mf ed 1992] – (= ser Beitraege zur deutschen literaturwissenschaft 28) – 1r – 1 – (incl bibl ref. filmed with: johann rist als weltlicher lyricker / oskar kern) – mf#3098p – us Wisconsin U Libr [430]

Beitraege zur technik in hebbels tagebuch / Hoestermann, Emilie – Bonn: H Ludwig 1917 [mf ed 1990] – 1r – 1 – (incl bibl ref. filmed with: hebbels dithmarschenfragment / heinrich bender) – mf#2704p – us Wisconsin U Libr [430]

Beitraege zur textkritik von origenes' johannescommentar / Koetschau, Paul – Leipzig: J C Hinrichs, 1905 [mf ed 1989] – (= ser Tugal 28/2) – 1mf – 9 – 0-7905-1723-X – (in german, greek & latin. incl ind) – mf#1987-1723 – us ATLA [240]

Beitraege zur textkritik von origenes' johannescommentar / Koetschau, Paul – Leipzig, 1905 – (= ser Tugal 2-28/2a) – 2mf – 9 – €5.00 – ne Slangenburg [240]

Beitraege zur typologie und symptomatologie der arbeitskurve / Remplein, Heinz – Leipzig: J A Barth, 1942 [mf ed 2002] – (= ser Beihefte...zur zeitschrift fuer angewandte psychologie und charakterkunde 91) – 1r – 1 – (filmed with (reel 12): die sprache der menschlichen leibeserscheinung / von ludwig eckstein (v92 1943) & other title. incl bibl ref) – mf#5153 reel 12 – us Wisconsin U Libr [150]

Beitraege zur vaterlandskunde fuer inneroesterreichs einwohner / ed by Kindermann, Joseph Karl – Graez 1790 – (= ser Dz. historisch-geographische abt) – 2v on 6mf – 9 – €120.00 – mf#k/n1233 – gw Olms [943]

Beitraege zur verstaendigung ueber begriff und wesen der sittlich-religioesen erfahrung / Petran, Ernst – Guetersloh:C Bertelsmann 1898 [mf ed 1985] – 1mf – 9 – 0-8370-4718-8 – (incl bibl ref) – mf#1985-2718 – us ATLA [240]

Beitraege zur voelker- und laenderkunde / ed by Sprengel, Matthias Christian et al – Leipzig 1781-90 – (= ser Dz. historisch-geographische abt) – 14pt on 29mf – 9 – €290.00 – mf#k/n1113 – gw Olms [305]

Beitraege zur weiterentwicklung der christlichen religion / ed by Deissmann, Gustav Adolf et al – Muenchen: J F Lehmann 1905 [mf ed 1985] – 1mf – 9 – 0-8370-2255-X – mf#1985-0255 – us ATLA [240]

Beitraege zur weitern ausbildung der deutschen sprache / ed by Campe, Joachim Heinrich – Braunschweig 1795-97 – (= ser Dz) – 3v[=8st] on 11mf – 9 – €110.00 – mf#k/n889 – gw Olms [430]

Beitraege zur wieland-biographie : aus ungedruckten papieren / Wieland, Christoph Martin; ed by Funck, Heinrich – Freiburg i. B: J C B Mohr 1882 [mf ed 1991] – 1r – 1 – (filmed with: die wahre geschichte vom wiederhergestellten kreuz / franz werfel) – mf#2958p – us Wisconsin U Libr [430]

Beitraege zur wuerdigung von karl gutzkow als lustspieldichter : mit einem einleitenden teil ueber ein unbekanntes tagebuch / Mueller, Peter – Marburg a.L: N G Elwert 1910 [mf ed 1992] – (= ser Beitraege zur deutschen literaturwissenschaft 16) – 1r – 1 – (incl bibl ref. filmed with: johann rist als weltlicher lyricker / oskar kern) – mf#3098p – us Wisconsin U Libr [430]

Beitrag der markophyten zu den schwebstoffen der tide-llbe / Hoberg, Marcus – (mf ed 1997) – 2mf – 9 – €40.00 – 3-8267-2482-8 – mf#DHS 2482 – gw Frankfurter [574]

Ein beitrag zu theodor storm's stimmungskunst / Stamm, Hermann – Eckernfoerde: C Heldt, 1914 – 1 – (includes bibliographical references) – us Wisconsin U Libr [430]

Ein beitrag zum entwurf von zeitreihenreglern / Fux, Manfred – (mf ed 1995) – 2mf – 9 – €40.00 – 3-8267-2140-3 – mf#DHS 2140 – gw Frankfurter [621]

Beitrag zur bestimmung der parameter und zur detektion der struktur von zweiphasenstroemungen mittels ultraschall / Hofmann, Bernd & Rockstroh, Manfred – (mf ed 1993) – 1mf – 9 – €30.00 – 3-89349-746-3 – mf#DHS 746 – gw Frankfurter [621]

Beitrag zur charakterisierung der permeabilitaet flaechiger verstaerkungsmaterialien / Shafi, Vahid – (mf ed 1996) – 2mf – 9 – €40.00 – 3-8267-2339-2 – mf#DHS 2339 – gw Frankfurter [621]

Ein beitrag zur christologie des alten testamentes : mit beruecksichtigung / Schaffnit, K – Duesseldorf: C Schaffnit, [1892?] – 1mf – 9 – 0-8370-5073-1 – (incl bibl ref) – mf#1985-3073 – us ATLA [221]

Ein beitrag zur dreidimensionalen finite-elemente-berechnung von geschichteten verbundwerkstoffen / Meyer, Ralf I – 1995 – v/160p – 9 – 3-8267-1058-4 – gw Frankfurter [620]

Beitrag zur flora aethiopiens / Schweinfurth, G A – Berlin, 1867 – 8mf – 8 – mf#320 – ne IDC [580]

Ein beitrag zur frage ueber die fremdwoerter im koraan / Dvorak, Rudolf – Muenchen: F Straub, 1884 – 1mf – 9 – 0-524-01542-2 – (incl bibl ref) – mf#1990-2496 – us ATLA [260]

Beitrag zur geologie des "massif de ceze", oestlicher teil, gard (frankreich) : stratigraphie und mikrofazielle untersuchungen / Breyer, Ralf – (mf ed 1992) – 3mf – 9 – €49.00 – 3-89349-476-6 – mf#DHS 476 – gw Frankfurter [550]

Ein beitrag zur geschichte der assyriologie in deutschland / Winckler, Hugo – Leipzig: Eduard Pfeiffer, 1894 [mf ed 1986] – 1mf – 9 – 0-8370-7756-7 – (in german) – mf#1986-1756 – us ATLA [470]

Beitrag zur geschichte des rundfunks – Berlin, 1967-1972 – 1r – 1 – gw Mikropress [380]

Beitrag zur geschichte und soziologie des ruhraufstandes vom marz-april 1920 / Colm, Gerhard – Essen a. d. R., 1921 – 1 – gw Mikropress [300]

Ein beitrag zur kenntnis der genstruktur der phosphoenolpyruvat-carboxylase hoeherer pflanzen : psilotum nudum, welwitschia mrabilis und tillandsia usneoides / Glasow, Catharina von – (mf ed 1995) – 1mf – 9 – €30.00 – 3-8267-2186-1 – gw Frankfurter [574]

Ein beitrag zur kenntnis des sprachgebrauchs klopstocks / Wuerfl, Christoph – Bruenn: C Winiker, 1883 – 1r – 1 – (incl bibl ref) – us Wisconsin U Libr [430]

Ein beitrag zur kritik von lessings laokoon / Brill, Bernhard – s.l: s.n. 18–? [mf ed 1990] – 1r – 1 – (filmed with: die freunde machen den philosophen, der englander, der waldbruder von jakob michael reinhold lenz / ilse kaiser) – mf#2823p – us Wisconsin U Libr [430]

Beitrag zur vorgeschichte der aufloesung der kloester in england und wales speciell unter der regierung heinrichs 8 / Wilson, Gilbert B – Halle, 1900 (mf ed 1993) – 1mf – 9 – €24.00 – 3-89349-306-9 – mf#DHS-AR 162 – gw Frankfurter [941]

Beitrage zu einfach-praktischen prufungen verschiedener handelswaren / Suepke, H F W – Braunschweig, 1842 – 1 – gw Mikropress [380]

Beitrarge zur geographie palastinas / Hildesheimer, Hirsch – Berlin, Germany. 1885 – 1r – us UF Libraries [956]

Beitz, Agathe *see*
- Beitraege zur bedeutung von lipoproteinen und thromboxan fuer klinische und experimentelle artherosklerose und zur antiatherosklerotischen wirkung von hdl
- Ueber einige aminopeptidasen aus euglena gracilis und ihre wechselwirkung mit einer zelleigenen inhibitorfraktion

Bejaardesorg / South Africa. Department of National Health and Population Development [Departement van Nasionale Gesondheid en Bevolkingsontwikkeling – Pretoria: Dept van Nasionale Gesondheid en Bevolkingsontwikkeling 1986 [mf ed Pretoria, RSA: State Library [199-]] – 106p [ill] on 1r with other items – 5 – (incl bibl ref) – mf#op 08222 r24 – us CRL [362]

Bejar, Duque de *see* Nueva disposicion y regimen que su excelencia ha dado para los colegios de ninas pobres. huerfanas de su villa y tierra de bejar en este ano del senor de mil setecientos y veinticinco

Bejarano, Jorge *see* Alimentacion y nutricion en colombia

Bejarano y Sanchez, Eloy *see*
- Aguas azoadas..
- La educacion integral
- La educacion medica integral
- Memoria presentada en el 14th congreso internacional de medicina por...

Bejr Dum Kravil *see* Mak thyn

Bejr Sal *see* Bidhi pracam tap bir khae

Die bekämpfung des christentums durch den römischen staat : bis zum tode des kaisers julian, 363 / Linsenmeyer, Anton – Muenchen: J.J. Lentner, 1905 – 1mf – 9 – 0-7905-5174-8 – (incl bibl ref) – mf#1988-1174 – us ATLA [240]

Bekannte und unbekannte grossen : skizzen und novelletten aus der kunst-und theaterwelt / Haffner, Karl – Wien: Selbstverlag der Witwe des Verfassers Fran Elise Haffner...1884 [mf ed 1981] – 1r – 1 – mf#228 – us Wisconsin U Libr [880]

Bekanntmachungen fuer die stadt dortmund *see* Bekanntmachungen fuer gross-dortmund

Bekanntmachungen fuer gross-dortmund – Dortmund DE, aug 25 1945-apr 23 1976, may 14 1976-86 – 1 – (title varies: 12 dec 1947: bekanntmachungen fuer die stadt dortmund, 17 may 1950: bekanntmachungen) – gw Misc Inst [350]

Bekanntnusz desz waaren gloubens... / Bullinger, Heinrich – Zuerych, Christoffel Froschower, 1566 – 2mf – 9 – mf#PBU-227 – ne IDC [240]

Bekantnuss vom heyligen abendmal in sechtzehen predigt getheylet / Mathesius, J – Nuernberg, 1567 – 5mf – 9 – mf#TH-1 mf 958-962 – ne IDC [242]

Bekarevich, A *see* Golos fronta

Beke, C T *see*
- The british captives in abyssinia
- Letters on the commerce and politics of abessinia and other parts of eastern africa
- On the geographical distribution of the languages of abyssinia and the neighbouring countries

Beke, Charles Tilstone *see* A few words with bishop colenso on the subject of the exodus of the israelites and the position of mount sinai

Bekehirnok : a baptista agyhaz lapja – v10-35. 1966-91 (incomplete) – Inquire – 1 – mf#ATLA S0335 – us ATLA [240]

Bekehrung armeniens durch den heiligen gregor illuminator : nach national-historischen quellen bearbeitet / Samueljan, M – Wien: Mechitharisten-Congregations-Buchh 1844 [mf ed 1992] – 1mf – 9 – 0-524-02990-3 – mf#1990-0777 – us ATLA [240]

Die bekehrung johannes calvins / Lang, August – Leipzig: A Deichert 1897 [mf ed 1990] – (= ser Studien zur geschichte der theologie und der kirche 2/1) – 1mf – 9 – 0-7905-7008-4 – (incl bibl ref) – mf#1988-3008 – us ATLA [242]

Die bekehrung menno simons' und sein ausgang aus der roemischen kirche / Menno Simons – Elkhart IN: Mennonitische Verlagshandlung 1883 [mf ed 1992] – 1mf – 9 – 0-524-05441-X – (in german) – mf#1990-1473 – us ATLA [242]

Bekehrung und gnadenwahl : fuer jeden christen / Zorn, Carl Manthey – St Louis MO: Concordia Pub House 1902 [mf ed 1993] – 2v on 1mf – 9 – 0-524-05781-8 – mf#1991-2337 – us ATLA [240]

Bekendtnis doctoris tilemanni heshvsii von der persoenlichen vnd in alle ewigkeit vnzertrenlichen vereinigung beyder naturen in jhesu christo / Hesshusen, T – [Eisleben], 1585 – 1mf – 9 – mf#TH-1 mf 612 – ne IDC [242]

Ein bekenntnis : novelle / Storm, Theodor – Berlin: Gebrueder Paetel, 1888 – 1r – 1 – us Wisconsin U Libr [830]

Das bekenntnis der evangelisch-lutherischen kirche in der konsequenz seines prinzips / Thomasius, Gottfried – Nuernberg: A Recknagel, 1848 – 1mf – 9 – 0-7905-7546-9 – (incl bibl ref) – mf#1989-0771 – us ATLA [242]

Das bekenntniss der lutherischen kirche von der versoehnung und die versoehnungslehre d. chr. k. v. hofmann's / Thomasius, Gottfried – Erlangen: T Blaesing, 1857 – 1mf – 9 – 0-7905-7480-2 – (incl bibl ref) – mf#1989-0705 – us ATLA [242]

Die bekenntnisschriften der altprotestantischen kirche deutschlands / ed by Heppe, Heinrich – Cassel: T. Fischer, 1855 – 2mf – 9 – 0-7905-8058-6 – mf#1988-6039 – us ATLA [242]

Die bekenntnisschriften der reformierten kirche : in authentischen texten mit geschichtlicher einleitung und register / ed by Mueller, Ernst Friedrich Karl – Leipzig: A. Deichert, 1903 – 3mf – 9 – 0-7905-8061-6 – (incl bibl ref) – mf#1988-6042 – us ATLA [242]

Die bekenntnisschriften der reformirten kirche deutschlands / ed by Heppe, Heinrich – Elberfeld: RL Friderichs, 1860 – 1mf – 9 – 0-7905-8218-X – mf#1988-6118 – us ATLA [240]

Bekenntnisse / Dehmel, Richard – 3. und 4. aufl. Berlin: S Fischer, 1926 [mf ed 1989] – (= ser Gesammelte werke in einzelausgaben) – 204p – 1 – mf#7170 – us Wisconsin U Libr [860]

Bekenntnisse, 7. bd (bdk18 1.reihe) / Augustinus (Augustine, Saint, Bishop of Hippo) – (= ser Bibliothek der kirchenvaeter. 1. reihe (bdk 1.reihe)) – €15.00 – ne Slangenburg [241]

Die bekenntnisse und die wichtigsten glaubenszeugnisse der griechisch-orientalischen kirche : im originaltext, nebst einleitenden bemerkungen – Leipzig: J C Hinrichs, 1904 [mf ed 1986] – (= ser Thesauros tes orthodoxias) – 1mf – 9 – 0-8370-7489-4 – (in greek & german. incl bibl) – mf#1986-1489 – us ATLA [243]

Bekentnis d georgij maioris von dem artickel der iustification / Major G – Wittemberg, 1558 – 1mf – 9 – mf#TH-1 mf 918 – ne IDC [242]

Bekentnis vnterricht vnd vermanung der pfarrhern vnd prediger der christlichen kirchen zu magdeburgk / [Amsdorff, N von] – [Magdeburg, 1550] – 2mf – 9 – mf#TH-1 mf 17-18 – ne IDC [242]

Bekentnis von seinem glauben vnd lere/ geschrieben an eynen widderteuffer / Bugenhagen, J – Wittenberg, 1529 – 1mf – 9 – mf#TH-1 mf 134 – ne IDC [242]

Bekentniss vnnd erklerung auffs interim durch der erbarn stedte / [Aepinus, J] – Magdeburg, [1548] – 3mf – 9 – mf#TH-1 mf 2-4 – ne IDC [242]

Bekentnus von etlichen irthumen maioris / Menius, J [and Flacius Illyricus d A, M] – np, [1557] – 1mf – 9 – mf#TH-1 mf 1164 – ne IDC [242]

Beker, Gabriele *see* Zumutbarkeit von kernenergierisiken

Beker, Jerome *see* Child and youth services

Bekes megyei nepujsag – Bekescsaba, Hungary. 1962-90 – 58r – 1 – us L of C Photodup [079]

Bekker, Ernst Immanuel *see* System und sprache des entwurfes eines buergerlichen gesetzbuches fuer das deutsche reich

Bekkeri, Imm *see*
- Annales
- Atheniensis historiarum libri 10
- Chronographia
- De officialibus palatii constantinopolitani
- Ephraemius
- Excerpta de antiquitatibus constantinopolitani
- Georgius phrantzes, ioannes cananus, ioannes anagnostes
- Historia
- Historiarum quae supersunt
- Ioannes lydus
- Theophanes continuatus, ioannes cameniata, symeon magister, georgius monachus
- Zosimus

Bekkerus, Imm *see*
- Breviarium historiae metricum
- De michael et androico palaeologis libri 13
- Descriptio templi sanctae sophiae [cshb32]
- Historia
- Historia politica et patriarchica
- Historiarum libri 8
- Merobaudes et corippus

Bekki, A *see* Geopolitiek asia timoer raja

Beknopte geschiedenis der katholieke missie in suriname / Coll, Cornelius van – Gulpen: M Albert 1884 [mf ed 1992] – 1mf [ill] – 9 – 0-524-02520-7 – (incl bibl ref) – mf#1990-0620 – us ATLA [241]

Beknopte geschiedenis van de vereeniging "ambonsch studiefonds" (1909-1917) met een korte toespraak tot het ambonsche volk / Manusama, A T – Weltevreden, 1917 – 1mf – 8 – mf#SE-1430 – ne IDC [959]

Beknopte geschiednis van de kolonie suriname / Bibaz, R Bueno – Paramaribo, Surinam. 1928 – 1r – us UF Libraries [972]

Bekoropoka / Lavondes, Henri – Paris, France. 1967 – 1r – us UF Libraries [960]

Der bekraentze weiher : erzaehlungen / Britting, Georg – Muenchen: A Langen/G Mueller, 1937 [mf ed 1989] – 106p – 1 – mf#7088 – us Wisconsin U Libr [830]

Bek's daily : bulletin ekonomi keuangan – Jakarta: P T Biro Penerangan Ekonomi nov 1974- (daily ex sun (& sat)) [mf ed Ithaca NY: [John M Echols Collection] Cornell University 2002] – 16r [1974-jul 1976 [gaps], some iss wanting] – 1 – (in english; saturday iss have title: bek's daily weekender nov 1974- bulletin ekonomi keuangan; cont: bulletin ekonomi keuangan (bek's english ed)) – mf#mf-12821 seam – us CRL [330]

Bek's djakarta digest – Djakarta, 1970-1971 nos 1-367 – 91mf – 9 – (missing: 1970(1, 183); 1971(347, 366)) – mf#SE-1321 – ne IDC [959]

Bel : the christ of ancient times / Radau, Hugo – Chicago: Open Court, 1908 [mf ed 1992] – 1mf – 9 – 0-524-03372-2 – (incl bibl ref) – mf#1990-3206 – us ATLA [290]

Bela bartok's music for string instruments, percussion and celesta / Bigelow, Ralph Emerson – U of Rochester 1953 [mf ed 19–] – 3mf – 9 – mf#fiche697 – us Sibley [780]

Bela crkvaer volksblatt – Weisskirchen (Bela Crkva YU), 1940 7 jan-1941 30 mar – 1r – 1 – (filmed by loc: oct 1923-39 [4r]) – gw Misc Inst; us L of C Photodup [077]

Die belagerung von neuss *see* Der feigling, die belagerung von neuss

Belain d'esnambuc / Joyau, Auguste – Paris, France. 1950 – 1r – us UF Libraries [972]

Belamy, Theodore *see* Rome

Belaney, R *see* The bible and the papacy

Belaney, Robert *see*
- Formation and growth of society out of christian marriage and its c...
- The kingdom of god on earth

Belanger, C P *see* Voyage aux indes orientales par le nord de l'europe, les provinces du caucase, la georgie, l'armenie et la perse, suivi de details...

Belanger, Charles *see* Voyage aux indes-orientales

Belanger, Claudine *see* A st-fabien, en s'amusant, cuisinons

Belanger, Jules [comp] *see* Almanach judiciaire et commercial pour l'annee 1871

Belanger, Louis Charles *see* Manual of the duties of road and rural inspectors

Belanger, Pauline *see* Les ministres de la couronne du quebec, 1867-1964

Belanger, Pierrette *see* Bibliographie analytique de l'oeuvre de monsieur auguste viatte

Belanger, Rene *see* Origine et histoire de la dime ecclesiastique premiere partie

Belarusian parliamentary papers 1990-1995 : minutes of the 12th congressional session of the supreme soviet of the republic of belarus = Parlamentskie dakumenti belarusi – 937mf coll – 9 – $5,600.00 coll – us UMI ProQuest [324]

Belaruski holas = Byelorussian voice – Toronto, Canada 1 apr, 1 jun 1955; mar 1956-jan 1992 [mf 1955-75] – 1 – (in cyrillic) – uk British Libr Newspaper [071]

Belaunde, Victor Andres *see*
- El debate constitucional. discursos en la asamblea 1931-1933. lima 1933
- Meditaciones peruanas

"Belauscht!" : gedichte und sprueche / Hubel, Henni – 2.aufl. New York: A Baeumer, 1908 [mf ed 1991] – 90p – 1 – mf#7489 – us Wisconsin U Libr [800]

Belausteguidoitia, Ramon De *see* Con sandino en nicaragua

Belaval, Emilio S *see*
- Areyto
- Circe o el amor
- Cuentos de la universidad
- Cuentos para fomentar el turismo
- Literatura de transicion
- Vida

Die belchenstimme – Gebweiler, Elsass (Guebwiller F), 1896 10 dec-1908 20 mar [gaps] – 1 – (title varies: 1902: gebweiler anzeiger, also: gebweiler volksblatt) – fr ACRPP [074]

Belcher, Alexander Emerson *see*
- Poems and patriotic verses
- What i know about commercial travelling

Belcher, E *see* Narrative of the voyage of hms samarang, during the years 1843-1846

The belcher islands of hudson bay / Flaherty, R J – New York, 1918. v5 – 1mf – 9 – mf#N-210 – ne IDC [917]

Belcher, Jonathan *see* Governor jonathan belcher letter books, 1723-1754

Belcher, Joseph *see*
- The clergy of america
- George whitefield
- Historical sketches of hymns, their writers, and their influence
- The religious denominations in the united states
- Robert raikes
- William carey

Belcher, Joseph et al *see* The baptist irish society

Belcher's farmer's almanack for the year of our lord 1853 : being the first after bissextile, or leap-year...calculated for halifax – Halifax, NS: C H Belcher, [1853?] [mf ed 1983] – 2mf – 9 – 0-665-32011-6 – mf#32011 – cn Canadiana [030]

Belchertown 1765-1893 – Oxford, MA (mf ed 1986) – (= ser Massachusetts vital records) – 35mf – 9 – 0-87623-036-2 – (mf 1-6: b,m,d & i 1773-1841. mf 7: index: births by families 1765-1849. mf 8-15: births by families 1765-1849. mf 16-17: index: b,m,d 1843-57 bk a. mf 18-19: b,m,d 1844-57. mf 20-22: index to births 1857-92 vol b. mf 23-24: births 1857-92 vol b. mf 25-27: index to marriages 1857-93 vol b. mf 28-29: marriages 1857-93 vol b. mf 30-32: index to deaths 1857-92 vol b. mf 33-35: deaths 1857-92 vol b) – us Archive [978]

Belck, W *see* Die kelischin-stele und ihre chaldisch-assyrischen keilinschriften

Belden, A Russell *see* History of the cayuga baptist association

Belden progress – Belden, NE: R B Crellin. v18 n48. dec 28 1911-v48 n26. aug 27 1942) (wkly) [mf ed with gaps] – 7r – 1 – (absorbed by: laurel advocate) – us NE Hist [071]

Belding, Albert Martin *see*
- A heart-broken coroner and other wonders
- Sir john thompson
- Transvaal souvenir

O belecho : orgao dos filhos da candinha – Fortaleza, CE: Typ Guttengerg, 01 nov 1899 – (= ser Ps 19) – mf#P18B,03,75 – bl Biblioteca [870]

Belediye buetceleri : (1930 senesi mahsus olarak belediye meclisleri tarafindan tanzim olunup tasdika iktiran eden) – Istanbul: Matbaacilik ve Nesriyat Tuerk Anomim Sirketi, 1932 – 86mf – 9 – $1340.00 set – (t.c. dahiliye vekaleti, mahalli idareler umum mueduerluegue) – us MEDOC [350]

Die belegschaft der bergwerke und salinen im oberamtsbezirk dortmund nach der zahlung vom 16. 12. 1893 / Taeglichsbeck, D – Dortmund 1895 – 1 – gw Mikropress [331]

Beleno C, Joaqin *see* Luna verde

Beleno C, Joaquin *see* Gamboa road gang

Belfast advertiser – Belfast, Ireland 24 oct 1879-15 dec 1880 – 1 – (cont by: belfast weekly advertiser (ns) 23 dec 1880-24 feb 1886) – uk British Libr Newspaper [072]

Belfast advertiser and literary gazette – Belfast, Ireland 5 nov-31 dec 1847 (wkly) – 1 – uk British Libr Newspaper [072]

Belfast and newry standard – Belfast, Ireland 3 may-24 may, 19 jul-23 aug, 22 nov 1889-27 nov 1891 – 1 – (cont by: newry & belfast standard [4 dec 1891-9 jun 1899 (mf 1889-96)]) – uk British Libr Newspaper [072]

Belfast commercial chronicle – Belfast, Ireland 18 feb 1805-23 dec 1812, 1 jan 1816-31 dec 1817, 19 aug 1820-30 dec 1822, 1 jan 1828-30 dec 1829, 1 jan-31 dec 1831 (imperfect) – 1 – uk British Libr Newspaper [072]

Belfast daily mercury – Belfast, Ireland 19 apr 1854-2 nov 1861 – 1 – (discontinued; cont: belfast mercury [29 mar 1851-17 apr 1854]) – uk British Libr Newspaper [072]

Belfast daily post – Belfast, Ireland 20 mar-21 apr 1882 – 1 – uk British Libr Newspaper [072]

Belfast election – Belfast, Ireland 30 sep-18 nov 1868 (wkly) – 1 – uk British Libr Newspaper [325]

Belfast evening star – Belfast, Ireland 29 jan-30 may 1890 – 1 – (discontinued) – uk British Libr Newspaper [072]

Belfast evening telegraph – Belfast, Ireland 20 mar 1871-18 apr 1918 (wkly) – 1 – (cont by: belfast telegraph [19 apr 1918-]) – uk British Libr Newspaper [072]

Belfast gazette / Northern Ireland – 1921-73. 15 reels – 1 – $500.00 – us Trans-Media [324]

The belfast gazette / Northern Ireland – 1957-1968- – 1 – us NY Public [941]

Belfast gazette, portland & warnambool advertiser – Port Fairy, aug 1842-dec 1849 – 1r – A$62.26 vesicular A$67.76 silver – at Pascoe [079]

Belfast: irish news *see* Irish weekly and ulster examiner

Belfast labour chronicle : the organ of the belfast trades council & labour representation committee – Belfast, Ireland nov 1904-19 may 1906 – 1 – (wanting: dec 1904, aug,4 nov,2-23 dec 1905, 13 jan,10 mar,7 apr-19 may 1906) – uk British Libr Newspaper [331]

Belfast linen trade circular – Belfast, Ireland 20 feb 1852-24 dec 1885 [mf 1885] – 1 – (incorp with: irish textile journal) – uk British Libr Newspaper [670]

Belfast mercantile register and weekly advertiser – Belfast, Ireland 7 jan 1840-23 mar 1852 – 1 – (cont by: mercantile journal & statistical register (ns) 30 mar 1852-25 apr 1893) – uk British Libr Newspaper [330]

Belfast mercury – Belfast, Ireland 29 mar 1851-17 apr 1854 – 1 – (cont by: belfast daily mercury [19 apr 1854-2 nov 1861]) – uk British Libr Newspaper [072]

Belfast monthly magazine – La Salle IL 1808-14 – 1 – mf#4206 – us UMI ProQuest [073]

Belfast morning news – Belfast, Ireland 23 nov 1857-27 apr 1882 (daily) – 1 – (cont by: morning news [1 may-22 jul 1882]) – uk British Libr Newspaper [072]

Belfast news-letter – Belfast, Ireland 18 jul 1769-1 sep 1962 – 1 – (cont: belfast news-letter & general advertiser [3 oct 1738-14 jul 1769], index [1738-1800] mf; cont by: news letter [3 sep 1962-]) – uk British Libr Newspaper [072]

Belfast prices current – Belfast, Ireland 3 jan-19 dec 1850 – 1 – uk British Libr Newspaper [338]

Belfast protestant journal – Belfast, Ireland 4 may 1844-27 jul 1850 – 3r – 1 – uk British Libr Newspaper [242]

Belfast shipping & commercial list – Belfast, Ireland 7 jan 1808-21 apr 1810 (imperfect) – 1 – uk British Libr Newspaper [380]

Belfast strike bulletin – Belfast, Ireland 25 jan-19 feb 1919 – 1 – uk British Libr Newspaper [331]

Belfast telegraph – Belfast, Ireland 19 apr 1918- [mf 1986-] – 1 – (cont: belfast evening telegraph [20 mar 1871-18 apr 1918]) – uk British Libr Newspaper; us CRL [072]

Belfast telegraphic circular – Belfast, Ireland 13 mar 1854-25 aug 1855 – 1 – uk British Libr Newspaper [072]

Belfast times – Belfast, Ireland 1 jan-31 may 1872 – 1 – (cont by: belfast daily times [1 jun-10 aug 1872] (discontinued)) – uk British Libr Newspaper [072]

Belfast trades council, 1881-1951 – (= ser Labour party in britain, origins and development at local level. series 1) – 5r – 1 – (int by john w boyle) – mf#97278 – uk Microform Academic [331]

Belfast weekly mail – Belfast, Ireland 19 nov 1852-15 sep 1854 – 1 – uk British Libr Newspaper [072]

Belfast weekly news – Belfast, Ireland 6 jul 1855-18 jun 1942 [mf 1857-59, 1861-90, 1892-96] – 1 – (amalg with: belfast news-letter) – uk British Libr Newspaper [072]

Belfast weekly post – Belfast, Ireland 1 apr 1882-28 jun 1884 – 1 – (discontinued) – uk British Libr Newspaper [072]
Belfast weekly star – Belfast, Ireland 7 jun 1890-25 jul 1891 – 1 – (discontinued) – uk British Libr Newspaper [072]
Belfast weekly telegraph – Belfast, Ireland 28 feb 1873-21 feb, 15 aug 1874-30 oct 1964 [mf 28 feb 1873-21 feb 1874, 1931-33, 1935-49 – 1 – (wanting 1930; discontinued) – uk British Libr Newspaper; uk Newsplan [072]
Belford, Michele L *see* The effect of arm and leg versus legs alone exercise on the stairmaster 4000pt in females
Belford's monthly – La Salle IL 1888-93 – 1 – mf#5253 – us UMI ProQuest [073]
Belfrage, Henry *see*
– Examples and counsels for the moral guidance of youth
– Guide to the lord's table
La belge aux gants noirs : drame en trois actes / Lacerte, Adele Bourgeois – [Ottawa?: s.n.] 1920 [mf ed 1995] – 1mf – 9 – 0-665-74772-1 – mf#74772 – cn Canadiana [820]
Les belges au guatemala, 1840-1845 / Fabri, Joseph – Bruxelles, 1955 – 1 – us CRL [972]
Belges dan l'afrique centrale – Bruxelles, Belgium. v1-3. 1886 – 1r – us UF Libraries [960]
Belgian congo / Great Britain Naval Intelligence Division – London, England. 1944 – 1r – us UF Libraries [960]
Belgian congo : some recent changes / Slade, Ruth – London, England. 1960 – 1r – us UF Libraries [960]
Belgian Congo Archives *see* Plan de classification a l'usage de l'administration d'afrique
Belgian Congo. Secretariat general *see* Bulletin administratif et commercial
Belgian Congo Service De L'information *see* Congo belge, 1944
Belgian news and continental advertiser – Brussels, Belgium 30 dec 1893-25 sep 1896 (wkly) – 2 1/2r – 1 – uk British Libr Newspaper [074]
Belgian times & news & european express – Brussels, Belgium 12 jan 1907-26 sep 1908 – 1 – (cont: european express & belgian times & news [9 nov 1901-5 jan 1907]; cont by: continental review [3 oct 1908-21 oct 1909]) – uk British Libr Newspaper [072]
The belgian traveller : or, a tour through holland, france, and switzerland, during the years 1804 and 1805 in a series of letters from a nobleman to a minister of state / Stewarton – London 1806 [mf ed Hildesheim 1995-98] – 4v on 8mf – 9 – €160.00 – 3-487-27843-X – gw Olms [860]
Belgian underground press in world war 2 – (= ser World war 2 research colls) – 578mf (18:1) – 9 – $4965.00 – (with p/g) – us UPA [070]
Belgique – Paris. devenu: Paris-Bruxelles puis Paris-mondial. 29 aout 1944-45 – 1 – fr ACRPP [074]
La belgique coloniale – Brussels, Belgium. La Belgique maritime et coloniale. -w. 1895-1932. 27 reels – 1 – uk British Libr Newspaper [949]
La belgique coloniale – Brussels, Belgium 10 nov 1895-25 jun 1905 – 1 – (cont by: belgique maritime et coloniale [2 jul 1905-15/16 aug 1914, 30 jan 1919-25 dec 1921]; belgique maritime, coloniale et economique [1 jan 1922-25 dec 1932]) – uk British Libr Newspaper [074]
La belgique maconnique / Freemasons [Belgium] – Bruxelles: Tillot 1887 – 3mf – 9 – mf#vrl-14 – ne IDC [366]
La belgique maritime, coloniale et economique – Brussels, Belgium 1 jan 1922-25 dec 1932 – 1 – (discontinued; cont: la belgique maritime et coloniale [2 jul 1905-15/16 aug 1914, 30 jan 1919-25 dec 1921]) – uk British Libr Newspaper [380]
La belgique sous la domination etrangere *see* Arthur young's tour in ireland [1776-1779]
Belgischer kurier – Brussels, Belgium 2-27 mar, 18 apr-30 jun 1916; 3,12,13,15 apr, 20 may, 1,2,8,18 jun 1917 – 1 – uk British Libr Newspaper [074]
Belgischer-bilder kurier – Brussels, Belgium [mar-jun 1916] – 1 – uk British Libr Newspaper [072]
Belgium *see* Moniteur belge
Belgium and western germany in 1833 : including visits to baden-baden, wiesbaden, cassel, hanover, the harz mountains etc / Trollope, Frances – London 1834 [mf ed Hildesheim 1995-98] – (= ser Fbc) – 2v on 4mf – 9 – €120.00 – 3-487-27793-X – gw Olms [914]
Belgium (before 1830) *see* Le parti liberal et le gros bon sens
Belgium. Commission Centrale de Statistique *see* Statistique generale de la belgique
Belgium. Conseil Colonial *see* Compte rendu analytique des seances
Belgium. Ministere de l'Interieur *see* Bulletin des commissiones royales d'art et d'archeologie
Belgium. Ministere des Affaires Economiques et des Classes Moyennesinstitut National de Statistique *see* Annuaire statistique de la belge 1870-1962
Belgium Ministere Des Affaires Estrangeres Service *see* Rwanda and burundi in 1962
Belgium. Ministere des Colonies *see* Aus den archiven des belgischen kolonialministeriums berlin, 1916
Belgium. Ministry of Foreign Affairs *see*
– Gift from the ministry of foreign affairs and external commerce of belgium
– Records relating to north america, 1834-1899
Belgium Office Du Tourisme Du Congo Belge Et Du Ruanda-Urundi *see*
– Guide du voyageur au congo belge et au ruanda-urundi
– Visitez le congo belge
Belgium.Office du Tourisme du Congo belge et du Ruanda-Urundi *see* Traveller's guide to the belgian congo and ruanda-urundi
Belgorodskaia pravda – Belgorod, Russia 1973-88 [mf ed Norman Ross] – 5r – 1 – mf#nrp-273 – us UMI ProQuest [077]
The belgrade herald – Belgrade, NE: O M Mayfield, sep 29 1900 (wkly) [mf ed v1 n2 oct 6 1900)-aug 30 1945 (gaps)] – 27r – 1 – us NE Hist [071]
Belgrader nachrichten – Belgrad (YU), 1915 15 dec-1917 30 jun, 1918 1 jan-27 oct – 1 – gw Misc Inst [077]
Belgrader zeitung – Belgrade, Yugoslavia. Nov 1924-Mar 1926 – 4r – 1 – us L of C Photodup [079]
Belgrano, Mario *see*
– La francia y la monarquia en el plata (1818-1820). la politica del duque de richelieu. misiones...buenos aires, 1933
– Rivodavia y sus gestiones diplomaticas con espana. (1815-1820). 2nd ed. buenos aires, 1934
Belgravia – La Salle IL 1866-99 – 1 – mf#3910 – us UMI ProQuest [790]
Beliaev, V I *see* Arabskie rukopisi sobraniia leningradskogo gosudarstvennogo universiteta
Beliaev, V N *see* Kak organizovat selskokhoziaistvennoe kreditnoe tovarishchestvo
Belic, Aleksandar *see* La macedoine
Belice / Asturias, Francisco – Guatemala, 1941 – 1r – us UF Libraries [972]
Belice : defensa de los derechos de mexico / Fabela, Isidro – Mexico City? Mexico. 1944 – 1r – us UF Libraries [972]
Belice / Estrada De La Hoz, Julio – Guatemala, 1949 – 1r – us UF Libraries [972]
Belice : estudio historico, politico y legal sobre / Martinez Alomia, Santiago – Campeche, Mexico. 1945 – 1r – us UF Libraries [972]
Belice, 1663 (?)-1821 / Calderon Quijano, Jose Antonio – Sevilla, Spain. 1944 – 1r – us UF Libraries [972]
Belice es de guatemala / Hurtado Aguilar, Luis A – Guatemala, 1958 – 1r – us UF Libraries [972]
Belice mexicano / Gallegos, Anibal – Mexico City? Mexico. 1951 – 1r – us UF Libraries [972]
Belice pertenece a guatemala – Guatemala, 1947 – 1r – us UF Libraries [972]
Belidor, [B F] *see*
– Architecture hydraulique...
– Uveau cours de mathematique...l'usage de l'artillerie et du genie
Belief / Chaney, George Leonhard – Boston: Roberts 1889 [mf ed 1985] – 1mf – 9 – 0-8370-3213-X – mf#1985-1213 – us ATLA [210]
Belief and life : studies in the thought of the fourth gospel / Selbie, William Boothby – New York: Charles Scribner's 1917 [mf ed 1992] – (= ser The short course series) – 1mf – 9 – 0-524-05290-5 – mf#1992-0391 – us ATLA [225]
The belief and worship of the anglican church / Knowles, Archibald Campbell – Philadelphia: George W Jacobs, 1894 – 1mf – 9 – 0-524-03166-5 – mf#1990-4615 – us ATLA [241]
Belief in a personal god / Huizinga, Arnold van Couthen Piccardt – Boston: Sherman, French 1910 [mf ed 1990] – 1mf – 9 – 0-7905-3971-3 – mf#1989-0464 – us ATLA [210]
Belief in god : an examination of some fundamental theistic problems / Savage, Minot Judson – Boston: George H Ellis 1881 [mf ed 1985] – 1mf – 9 – 0-8370-5053-7 – mf#1985-3053 – us ATLA [210]
Belief in god : its origin, nature, and basis / Schurman, Jacob Gould – New York: Charles Scribner 1907 [mf ed 1985] – 1mf – 9 – 0-8370-5170-3 – mf#1985-3170 – us ATLA [210]
The belief in god and immortality : a psychological, anthropological and statistical study / Leuba, James Henry – Boston: Sherman, French, 1916 – 1mf – 9 – 0-7905-7898-0 – (incl bibl ref) – mf#1989-1123 – us ATLA [210]
Belief in mental imagery in free throw performance / Nordeen, Lisa M – Springfield College, 1994 – 1mf – 9 – $4.00 – mf#PSY1857 – us Kinesology [611]
The belief in personal immortality / Haynes, Edmund Sidney Pollock – New York: Putnam, 1913 – (= ser Science Series) – 1mf – 9 – 0-7905-3900-4 – (incl bibl ref) – mf#1989-0393 – us ATLA [210]
Belief in the divinity of jesus christ = Foi en la divinite de jesus-christ / Didon, Henri – London: K Paul, Trench, Truebner; New York: Benziger 1894 [mf ed 1990] – 1mf – 9 – 0-7905-3721-4 – (trans fr french) – mf#1989-0214 – us ATLA [210]
The belief of the first three centuries concerning christ's mission to the underworld / Huidekoper, Frederic – New York: James Miller, 1876, c1854 – 1mf – 9 – 0-7905-1898-8 – (incl ind) – mf#1987-1898 – us ATLA [240]
Beliefs about the bible / Savage, Minot Judson – Boston: Geo H Ellis, 1900, c1883 [mf ed 1985] – 1mf – 9 – 0-8370-5054-5 – mf#1985-3054 – us ATLA [220]
The beliefs of unbelief : studies in the alternatives to faith / Fitchett, William Henry – New York: Eaton & Mains, c1907 [mf ed 1985] – 1mf – 9 – 0-8370-3152-4 – mf#1985-1152 – us ATLA [210]
Believe and be saved – Kilmarnock? Scotland. 18– – 1r – us UF Libraries [240]
Believe and live – Kelso, Scotland. 1842 – 1r – us UF Libraries [240]
Believer / Wilde, Lady – Dublin, Ireland. 1885 – 1r – us UF Libraries [240]
The believer born of almighty grace / Dabney, Robert Lewis – 9 – $50.00 – us Presbyterian [240]
Believer immersion as opposed to unbeliever sprinkling : in two essays, first on the abrahamic covenant, second on christian baptism... / Crawford, Alexander – Charlottetown, PEI?: s.n, 1827 – 2mf – 9 – mf#64657 – cn Canadiana [240]
Believer's comfort in temptation and affliction – London, England. 1792 – 1r – us UF Libraries [240]
The believers' hope : or, christ coming for his people / Marsh, Frederick Edward – New York: Gospel Pub House, [19–?] [mf ed 1992] – (= ser Christian & missionary alliance coll) – 1mf – 9 – 0-524-03724-8 – mf#1990-4829 – us ATLA [240]
Believer's joy in god / M'cheyne, Robert Murray – Edinburgh, Scotland. 1858 – 1r – us UF Libraries [240]
Believer's life in heaven / Winslow, Octavius – London, England. 18– – 1r – us UF Libraries [240]
Believers' Meeting for Bible Study (12th: 1888: Niagara-on-the-Lake) *see* Report on the believers' meeting for bible study
Believers' meeting for Bible Study (14th : 1890 : Niagara-on-the-Lake, Ont) *see* A week of blessing
Believer's triumph over sin and death / Room, Charles – London, England. 1835 – 1r – us UF Libraries [240]
Belig *see* The divan project
Belin, Jean Paul *see* Learn bemba by speaking it
Belin, M A *see* Histoire de la latinite de constantinople
Belinde : ein liebesstueck in fuenf aufzugen / Eulenberg, Herbert – 6. aufl. Leipzig: E Rowohlt 1913, c1912 [mf ed 1989] – 1r – 1 – (filmed with: bozena / marie von ebner-eschenbach) – mf#7268 – us Wisconsin U Libr [820]
Belinskii, V E *see* Russkii geraldicheskii slovar
Belinsky, Vissarion Grigoryevich *see* Sobranie sochinenii v g bielinskago
Belisario pena / Bayle, Constantino – Buenos Aires: R. Herrando y Cia, Impresores, 1915. Sep. Rev. Estudios – 1 – sp Bibl Santa Ana [240]
Belisle, Louis-Alexandre *see*
– Histoire de blondine
– La petite souris grise
Belize / Bianchi, William J – New York, NY. 1959 – 1r – us UF Libraries [972]
Belize advertiser – Belize 21 may 1881-25 oct 1884, 25 dec 1886, jan 1887 (imperfect) – 1 – (cont by: belize advertiser & british honduras gazette (ns) 26 feb 1887-19 may 1888 (imperfect); belize advertiser [28 jul 1888-9 nov 1889 (imperfect)]) – uk British Libr Newspaper [079]
Belize billboard – Belize 1 jan 1957-16 may 1971 (daily) (imperfect) – 1 – (fr 7 sep 1969 onward sunday iss entitled "sunday billboard" & fr 13 nov 1969 onward thurs iss entitled "thursday billboard") – uk British Libr Newspaper [079]
Belize (british honduras) : an anglo-guatemalan con / Mendoza, Jose Luis – London, England. 1948 – 1r – us UF Libraries [972]
Belize. Central Planning Unit *see* Abstract of statistics 1961-1970/1972
Belize independent – Belize 11 oct -27 dec 1888, 16 apr 1894-17 apr 1896 – 1 – uk British Libr Newspaper [079]
Belize independent – Belize City, Belize 1 jan 1930-27 jul 1938, 4 jan 1939-8 dec 1943, 14 jun-20 sep 1944, 7 feb [28 mar 1945] – 1 – (not publ between 8 dec 1943-14 jun 1944, or between 20 sep 1944-7 feb 1945) – uk British Libr Newspaper [079]
Belize times – Belize 1 jan 1959-22 nov 1972 (daily) (imperfect) – 30r – 1 – uk British Libr Newspaper [079]
Bell / Brotherhood of Marine Officers – v2 n2-v17 [1971 apr-1978 apr] – 1r – 1 – mf#499187 – us WHS [355]
Bell, Alexander Graham *see* Growth of the oral method of instructing the deaf
Bell, Alexander Melville *see*
– Explanatory lecture on visible speech, the science of universal alphabetics
– The faults of speech
– On teaching reading in public schools
– The principles of elocution
– Sounds and their relations
Bell, Andrew *see*
– Dr bell's system of instruction
– General james wolfe, his life and death
Bell, Archie *see* Spell of the caribbean islands
Bell, Arthur John *see*
– Whence comes man
– Why does man exist?
Bell, Aubrey F G *see*
– Francisco sanchez el brocense
– Studies in portuguese literature
Bell, Charles *see*
– Essays on the anatomy of expression in painting
– The hand
Bell, Charles Dent *see* Henry martyn
Bell, Charles Napier *see*
– Continuation of henry's journal
– Henry's journal, covering adventures and experiences in the fur trade on the red river 1799-1801
– The historical and scientific society of manitoba
– Navigation of hudson bay and straits
– Old time milling
– The olden time
– Original letters and other documents relating to the selkirk settlement
– Our northern waters
– The selkirk settlement and the settlers
– Some historical names and places of the canadian north-west
– Some red river settlement history
– Tangweers
– Winnipeg
Bell, Clark, 1832-1918 *see* Criminal abortion and the new english criminal evidence act
Bell, Earl S *see* Evangelical beginnings in the arizona territory
Bell, Edward *see*
– The lay of the nibelungs
– Selected prose works
Bell, Edwin *see* A treatise on the law of landlord and tenant in canada
Bell, Emily Lagow *see* My pioneer days in florida, 1876-1898
Bell, Evans *see*
– The empire in india
– The great parliamentary bore
– Retrospects and prospects of indian policy
Bell, F A *see* Ungarn in wort und bild
Bell, G M *see* Airdrie literary album
[Bell gardens-] bell gardens review – CA. 1974-1981 – 11r – 1 – $660.00 – mf#H03152 – us Library Micro [071]
Bell, George *see*
– The consolidated municipal act, 1883
– Religious teaching in secondary schools
– Rough notes by an old soldier
– Sunday-school conventions
Bell, George Kennedy Allen *see* The meaning of the creed
Bell, George T *see* The passenger department of canadian steam railways
Bell, Henry *see* Development of christ's humanity
Bell, Herbert Clifford Francis *see* Guide to british west indian archive materials
Bell, Hermann *see*
– Nubian-english-arabic dictionary
– Survey of nubian place names
Bell, Hugh Maclachlan *see* Bahamas
Bell, J Munro [comp] *see* Chippendale, sheraton and hepplewhite furniture designs
Bell, James *see* Critical researches in philology and geography
Bell, John *see*
– Critical researches in philology and geography
– List of plants of the manitoulin islands, lake huron
– A miracle of modern missions
– Observations on italy
– Papers
– Voyages depuis st petersbourg en russie

Bell, John Allison *see* Chebucto and other poems
Bell, Josiah Jones *see* In canada's national park
Bell journal of economics – Santa Monica CA 1975-83 – 1,5,9 – (cont: bell journal of economics and management science; cont by: rand journal of economics) – ISSN: 0361-915X – mf#10122.02 – us UMI ProQuest [338]
Bell journal of economics and management science – Santa Monica CA 1970-74 – 1,5,9 – (cont by: bell journal of economics) – ISSN: 0005-8556 – mf#10122.02 – us UMI ProQuest [330]
Bell, Kenneth Ray *see* The development of a program of student financial assistance for east coast bible college
Bell labs technical journal – Hoboken NJ 1996+ – 1,5,9 – ISSN: 1089-7089 – mf#25433 – us UMI ProQuest [380]
Bell, Lallah *see* Princess and poet
Bell, Leon G *see* Report...on the exploration made of the route of the huron and ottawa railway from ottawa city to parry sound
Bell, M M S *see* The politics of administration
Bell, Malcolm *see* Edward burne-jones
Bell, Nancy R E (Meugens) *see* Representative painters of the 19th century
Bell of a florida spanish mission / Williams, Emma Rochelle – s.l, s.l? . 193-? – 1r – us UF Libraries [978]
Bell, R *see* The origin of islam in its christian environment
Bell, Ralph Graham *see* Rhodesia
Bell, Richard *see* Origin of islam in its christian environment
Bell ringer : albany battalion newsletter – 1989 oct-1993 jul – 1r – 1 – mf#1757327 – us WHS [071]
Bell, Robert *see*
– Alexander Murray
– Ancient poems, ballads and songs of the peasantry of england
– Forest fires in northern canada
– The forests of canada
– The geological history of lake superior
– The laurentian and huronian systems in the region north of lake huron
– Marble island and the north-west coast of hudson's bay
– The mineral resources of the hudson's bay territories
– Observations on the conference of the rev thomas chalmers
– On the commercial importance of hudson's bay
– On the occurrence of mammoth and mastodon remains around hudson's bay
– The origin of gneiss and some other primitive rocks
– A plea for pioneers
– Pre-paleozoic decay of crystalline rocks north of lake huron
– Recent explorations to the south of hudson bay
– Reports on the geology of the basin of moose river and of lake of the woods and adjacent country, 1881
– Rising of the land around hudson bay
Bell, Solomon *see* The polar regions of the western continent explored
Bell street chapel discourses : containing selections from the writings of james eddy, providence, rhode island, 1889-1899 / Spencer, Anna Garlin – Providence, RI: Journal of Commerce, [1899?] [mf ed 1985] – 1mf – 9 – 0-8370-5508-3 – mf#1985-3508 – us ATLA [240]
Bell system technical journal – New Providence NJ 1922-83 – 1,5,9 – (cont by: at and t bell laboratories technical journal : a journal of the at&t companies) – ISSN: 0005-8580 – mf#58.02 – us UMI ProQuest [380]
Bell, T *see* The zoology of the voyage of hms beagle...during the years 1832-1836
Bell, T P *see* Correspondence files of corresponding secretaries, baptist sunday school board
Bell telephone laboratories, inc. bell laboratories record – Short Hills NJ 1925-82 – 1,5,9 – ISSN: 0005-8564 – mf#787.01 – us UMI ProQuest [380]
Bell telephone magazine – Berkeley Heights NJ 1922-83 – 1,5,9 – ISSN: 0096-8692 – mf#444 – us UMI ProQuest [380]
Bell, Thomas Evans *see*
– "Our great vassal empire"
– Retrospects and prospects of indian policy
Bell, William Melvin *see* The love of god
Bella fluminense : jornal variado – Rio de Janeiro, RJ: Typ Popular de Azevedo Leite, 10 jul 1864 – (= ser Ps 19) – mf#DIPER – bl Biblioteca [079]
Bella, S della *see* Six views of pratolino
Bellaire baptist church – Bermott, AR. 1940-75 – 1 – $27.45 – us Southern Baptist [242]
Bellaire, J P *see* J p bellaire's infanterie-hauptmann's...
Bellamy, Edward *see* Equality
Bellamy, Joseph *see*
– Letters and other papers of joseph bellamy
– The works of joseph bellamy, d.d., first pastor of the church in bethlem, conn
Bellamy, Julien *see* La theologie catholique au xix siecle
Bellarmin, R *see* Disputationes de controversiis christianae fidei, adversus huius temporis haereticos...
Bellarmino, Roberto Francesco Romolo, Saint *see* Declaracion copiosa de las quatro partes mas essenciales, y necessarias de la doctrina christiana
Bellas artes en guatemala / Diaz, Victor Miguel – Guatemala, 19344 – 1r – us UF Libraries [972]
Las bellas artes plasticas en sevilla, tomo 1 / Cascales Munoz, Jose – Toledo: Tip Colegio de Huerfanos, 1928 – 1 – sp Bibl Santa Ana [730]
Las bellas artes plasticas en sevilla...desde el siglo 13 hasta nuestros dias...tomo 2 / Cascales Munoz, Jose – Toledo: Imp Colegio de Huerfanos, 1929 – 1 – sp Bibl Santa Ana [730]
Bellay, A *see* L'enseignement des jesuites au canada
Bellcore exchange – Piscataway NJ 1989-92 – 1 – ISSN: 1040-2020 – mf#16438.02 – us UMI ProQuest [380]
Belle amour / Marchand, Leopold – Paris, France. 1931 – 1r – us UF Libraries [440]
Belle eveillee / Franc-Nohain – Paris, France. 1931, c1927 – 1r – us UF Libraries [440]
Belle glade herald – Belle Glade, FL. 1985-1989 – 5r – us UF Libraries [071]
Belle glade record – Belle Glade, FL. 1929 jul 12-nov 1 – 1r – us UF Libraries [071]
Belle, Manfred *see* Der entwicklungspolitische runde tisch in der ddr und im vereinigten deutschland
Belle mirror – Oconomowoc WI. [1869 jun 30-1870 aug 6] – 1r – 1 – (cont by: oconomowoc times) – mf#958218 – us WHS [071]
Belle Springs Creamery. Abilene, Kansas *see* Journal and inventories
Bellefeuille, Edouard Lefebvre de *see* Les edits et ordonnances royaux et le conseil superieur de quebec
Bellefeuille, Edouard Lefebvre de [comp] *see* Le code civil du bas-canada (en force depuis le 1er aout 1866)
Bellefonte gazette – Bellefonte, PA. -w 1889-1912 – 13 – $25.00r – us IMR [071]
Bellefonte republican – Bellefonte, PA. -w 1889-1931 – 13 – $25.00r – us IMR [071]
Bellegarde, Dantes *see*
– Dessalines a parle
– Haiti et son peuple
– Haitien parle
– Histoire du peuple haitien, 1492-1952
– Nation haitienne
– Occupation americaine d'haiti
– Pages d'histoire
– Pour une haiti heureuse
– Republique d'haiti et les etats-unis
– Resistance haitienne
Bellegarde ou l'enfant indien adopte : histoire canadienne – [Paris?: s.n.] 1833 [mf ed 1985] – 2v on 1mf – 9 – 0-665-48257-4 – (trans fr english. int by ph chasles) – mf#48257 – cn Canadiana [830]
Belleli, Lazare *see*
– An independent examination of the assuan and elephantine aramaic papyri
– Interpretations erronees et faux monuments
Bellemare, Alphonse *see* La croisade canadienne
Bellemare, Bertrand *see* Sentence arbitrale et rapport ecrit a l'occasion de l'arbitrage entre la canadian carborundum company ltd de shawinigan falls
Belle-mere / Scribe, Eugene – Paris, France. 1826 – 1r – us UF Libraries [440]
La belle-nivernaise : from selected stories / Daudet, Alphonse – Toronto: Morang Educational Co, 1913, c1901 – 2mf – 9 – 0-665-73898-6 – (selected stories in french. notes in english. incl biogr sketch of daudet) – mf#73898 – cn Canadiana [830]
Bellermann, Christian F *see* Erinnerungen aus suedeuropa
Bellermann, Heinrich *see* Die groesse der musikalische intervalle als grundlage der harmonie
Bellermann, Johann Joachim *see* Phoeniciae linguae vestigia in melitensi
Bellermann, Ludwig *see* Schillers werke
Bellerophon : of lust tot wijsheyd – Amsterdam: Dirck Pietersz, 1638 – 4mf – 9 – mf#O-3143 – ne IDC [090]
Bellerophon : of lust tot wysheyt / [Pers, D P] – [Amsterdam: Dirck Pietersz, 1648] – 8mf – 9 – mf#O-3250 – ne IDC [090]
Bellerophon : tragedie mise en musique / Lully, Jean Baptiste – 2e ed, Paris: C Ballard 1714 [mf ed 19–] – 1r – 1 – mf#film 554 – us Sibley [780]
Bellerophon... / [Pers, D P] – t'Amstelredam: Dirk Pietersz., 1614 – 2mf – 9 – mf#O-712 – ne IDC [090]
Bellerophon... / Pers, D P – t'Amstelredam: Willem van Beaumont, 1656-62 – 8mf – 9 – mf#O-3251 – ne IDC [090]
Bellerose, Joseph Hyacinthe *see* L'orangisme et le catholicisme
Bellerose, Joseph-Hyacinthe *see*
– Assemblee a saint-hyacinthe le 8 decembre 1885 pour protester contre l'execution de riel
– Discours de l'hon m bellerose
Bellerose, L H *see*
– Petit manuel d'agriculture a l'usage des ecoles
– Traite elementaire d'arithmetique
Belles et fieres antilles / Leblond, Marius – Paris, France. 1937 – 1r – us UF Libraries [972]
Bellesheim, Alphons *see*
– Geschichte der katholischen kirche in irland von der einfuehrung des christenthums bis auf die gegenwart
– Geschichte der katholischen kirche in schottland
– History of the catholic church in scotland
– Wilhelm cardinal allen (1532-1594) und die englischen seminare auf dem festlande
Bellessort, Andre *see* L'apotre des indes et du japon
Bellett, G *see* Benefits of affliction
Bellett, John Crosthwaite *see* God's witness in prophecy and history
Bellett, John Gifford *see*
– The epistle to the ephesians
– The evangelists
Belleview baptist church – Boone County, KY. 932p. 1803-1914 – 1 – $41.94 – (formerly: cedar creek baptist church; name changed sep 12 1885) – us Southern Baptist [242]
Belleville advocate – Belleville IL. 1854 nov 8 – 1r – 1 – (cont: representative and gazette; illinois independent) – mf#845856 – us WHS [071]
Belleville intelligencer – Belleville, ON. 1862-73 – 11r – 1 – cn Library Assoc [071]
Belleville labor news / Belleville Trades and Labor Assembly – v1 n11-v13 n10 [1949 jan-1960 oct] – 1r – 1 – (cont by: southwestern illinois labor news) – mf#1053147 – us WHS [331]
Belleville news – Belleville WI. 1895 jan 24-oct 18 – 1r – 1 – (cont by: albany journal (albany wi: 1895)) – mf#936857 – us WHS [071]
Belleville recorder – Belleville WI. [1926 jun 10/1927 sep 1]-1999 [few gaps] – 44r – 1 – (cont: new glarus post and belleville recorder) – mf#937076 – us WHS [071]
Belleville recorder – Belleville WI. [1902 oct 3/1903 dec 25]-[1923 jul 6/1924 jul 4] [small gaps] – 15r – 1 – (cont: sugar river recorder; cont by: new glarus post; new glarus post and belleville recorder) – mf#937073 – us WHS [071]
Belleville Trades and Labor Assembly *see* Belleville labor news
Belleviller post und zeitung : wochenausgabe [weekly edition] – Belleville IL (USA), mar 2 1922-jan 4 1923 – 1r – 1 – Dist. gw Mikrofilm – gw Misc Inst [071]
Bellevue baptist church – Owensboro, KY. dec 1958-67 – 1 – $14.22 – us Southern Baptist [242]
The bellevue broadcaster – Bellevue, NE: Charles Clarey and Thelma Meyers, 1930 (wkly) [mf ed v2 n32 aug 11 1932-may 12 1933 with gaps] – 1r – 1 – (issues for mar 31-may 12 1933 called vol no 1 n1-vol no 1 n7) – us NE Hist [071]
Bellevue central high school publications / Huron Co. Bellevue – (1926-35, 1938-71) [irreg] – 1r – 1 – mf#B2452 – us Ohio Hist [071]
The bellevue enterprise – Bellevue, NE: William H Toy. v1 n1. apr 12 1888- (semiwkly) [mf ed -may 12 1888 (gaps) filmed 1973] – 1r – 1 – us NE Hist [071]
Bellevue gazette – Bellevue, NE: University Printing Co, 1904 (wkly) [mf ed v2 n46. mar 2 1906-v7 n1. apr 15 1910 with gaps] – 1r – 1 – us NE Hist [071]
The bellevue gazette – Bellevue, NE: S A Strickland & Co. v1 n1. oct 23 1856-oct 1858// (wkly) [mf ed -sep 2 1858 with gaps] – 1r – 1 – us NE Hist [071]
Bellevue, Jean *see* Une visite chez le capitaine b...
Bellevue leader – Bellevue, NE: [s.n.] (wkly) [mf ed v10 n13. feb 3 1982-] – 1 – (absorbed: bellevue press. vol and numbering dropped nov 21 1984) – us NE Hist [071]
Bellevue newspaper – Bellevue, NE: Larry W Davis Sr. (wkly) [mf ed v2 n32-B. aug 5-sep 16 1981 filmed 1984] – 1r – 1 – (cont: bellevue-guide) – us NE Hist [071]
Bellevue. Ohio. Methodist Episcopal Church *see* Church records, ms 2041
Bellevue press – Bellevue, NE: Eloine Gebbie. v1 n1. dec 7 1945-jan 20 1982// (wkly) [mf ed -sep 30 1981 with gaps] – 23r – 1 – (absorbed: bellevue press tuesday morning sarpy county gazette; absorbed by: bellevue leader) – us NE Hist [071]
Bellevue press tuesday morning sarpy county gazette – Bellevue, NE: Eloine Gebbie & J B Gebbie Jr. 1v. v8 n21. nov 15 1960-v9 n18. oct 31 1961 (wkly) [mf ed 1980] – 1r – 1 – (cont: sarpy county gazette. absorbed by: bellevue press) – us NE Hist [071]
The bellevue record – Bellevue, NE: Fred L Wertz, 1898 (wkly) [mf ed v1 n3. nov 9 1898-jan 11 1899 (gaps) filmed 1973] – 1r – 1 – us NE Hist [071]
Bellevue-guide – Bellevue, NE: Bellevue Guide Inc. v5 n10. mar 4 1970-1980// (wkly) [mf ed -sep 9 1970 filmed in 1979] – 1r – 1 – (cont by: bellevue newspaper) – us NE Hist [071]
Bellew, Henry Walter *see* The races of afghanistan
Bellezas del alma: la caridad / Rodriguez Marcos, Antonio – 1885 – 9 – sp Bibl Santa Ana [240]
Le bellezze della citta di firenze / Bocchi, F; ed by Cinelli, G – Firenze, 1677 – 8mf – 9 – mf#O-960 – ne IDC [720]
[Bellflower-] herald enterprise – CA. 1926-1959 (scats) – 42r – 1 – $2520.00 – (aka: news-bulletin; shipping news) – mf#H03151 – us Library Micro [071]
Belli, Onorio *see* A description of some important theatres
Belli, Piero *see* De re militari et bello tractatus
Bellido, Jose *see* Vida de la v m r m maria anna agueda de s ignacio
Bellifortis / feuerwerkbuch (cf-lp3) : farbmikrofiche-edition der bilderhandschriften goettingen, niedersaechsische staats- und universitaetsbibliothek, 2°cod ms philos 64 cim und 64a cim / Kyeser, Konrad – (mf ed 1995) – (= ser Codices figurati – libri picturati (cf-lp) 3) – 9 color mf – 15 – €395.00 – 3-89219-303-7 – (int & description by udo friedrich; ann by fidel raedle) – gw Lengenfelder [090]
Bellin, Jacques-Nicolas *see* Le petit atlas maritime
Belling, Detlev W et al *see* Das selbstbestimmungsrecht minderjaehriger bei medizinischen eingriffen
Belling, Laura R *see* The relationship between social physique anxiety and physical activity
Bellingen courier sun – Bellingen, jan-dec 1965 – at Pascoe [079]
Bellingham 1698-1849 – Oxford, MA (mf ed 1995) – (= ser Massachusetts vital record transcripts to 1850) – 7mf – 9 – 0-87623-209-8 – (mf 1t-3t: births & deaths 1698-1844. mf 2t-3t: marriages 1752-1827. mf 3t-4t: intentions 1739-1827. mf 4t: marriages 1739-70. mf 5t: intentions & marriages 1827-49. mf 6t: births & deaths 1814-46. mf 7t: b,m,d 1844-49) – us Archive [978]
[Bellingham-] journal of ethnic studies – WA. 1973-1975 – 1r – 1 – $110.00 – mf#R05406 – us Library Micro [305]
Bellingshausen, F von *see* Forschungsfahrten im suedlichen eismeer 1819-1821
Bellman – Hull, England 7 jan 1880-3 feb 1883 [mf 1882] – 1 – (discontinued) – uk British Libr Newspaper [072]
Bellman, Carl Michael *see* Fredmans epistlar
Bellman [richmond] : a recreation journal for richmond twickenham kingston – Richmond upon Thames, London 8 jan-aug 1887 – 1 – (cont by: bellman & commentator for richmond [11 oct-15 nov 1887]) – uk British Libr Newspaper [072]
Bello, A *see* Itinerario y pensamiento de los jesuitas expulsos de chile (1767-1815)
Bello, Andres *see*
– Odes of bello, olmedo and heredia
– Opusculos gramaticales
– Pensamiento vivo de andres bello
Bello en colombia / Instituto Caro Y Cuervo – Bogota, Colombia. 1952 – 1r – us UF Libraries [972]
Bello, Julio *see* Memorias de um senhor de engenho
Bello Lozano, Humberto *see* Cronica procesal
Belloc, Hilaire *see* Four men
Bellogin Garcia, Andres *see* Vida y hazanas de alvar nunez cabeza de vaca. madrid, 1928
Bellona : ein militaerisches journal / ed by Seidel, Karl von – Dresden 1781-87 – (= ser Dz) – 20st on 20mf – 9 – €200.00 – mf#k/n3967 – gw Olms [355]
Bellori, G P *see*
– Ritratti di alcuni celebri pittori del secolo 17...con le vite de'medesimi...
– Le vite de' pittori, scultori, et architetti moderni
– Le vite de' pittori, scultori et architetti moderni...
– Le vite de pittori, scvltori et architetti moderni, scritte da gio
Bellorius, i p veteres arcus augustorum triumphis insignes... – Romae, 1690 – 3mf – 9 – mf#O-144 – ne IDC [700]
Belloso Rodriguez, Pedro *see*
– Con lo que tengo a bordo
– Millonario de pobreza
– El nombre nuestro de cada dia
– Los otros, el paisaje y yo
– Poemas
– Poemas de campo y pueblo
– Salterio de mis horas
Bellot, Hugh Hale L *see*
– Gray's inn and lincoln's inn
– The inner and middle temple
– Ireland and canada

Bellot, Joseph Rene *see* Voyage aux mers polaires a la recherche de sir john franklin
Bellow, Saul *see* Great jewish short stories
Bellows, Elizabeth *see* John bellows
Bellows, Henry Whitney *see*
– An appeal in behalf of the further endowment of the divinity school of harvard university
– The old theology and the new
– Twenty-four sermons
Bellows, John *see*
– John bellows
– The truth about the transvaal war and the truth about war
Bellows, Russell Nevins *see*
– Twenty-four sermons
– Unitarian church directory and missionary handbook, 1884-1885
Belloy, Pierre-Laurent Buyrette De *see*
– Gaston et bayard
– Zelmire
Bell's dictionary : a dictionary of the law of scotland – Edinburgh. 2v. 1816 – 11mf – 9 – $16.50 – mf#LLMC 95-409 – us LLMC [340]
Bell's life in london and sporting chronicle – [London & SE] Westminster Archives 3 mar 1822-29 may 1886 [wkly] – 67r – 1 – uk Newsplan [072]
Bells life in sydney – Sydney, 1845-70 – 5r – 1 – A$192.50 vesicular A$220.00 silver – at Pascoe [073]
Bell's literary intelligencer and new national omnibus – La Salle IL 1834 – 1 – mf#4207 – us UMI ProQuest [073]
Bell's new weekly messenger – London, England 1 jan 1832-25 mar 1855 (wkly) – 23 1/4r – 1 – (incorp with: news of the world) – uk British Libr Newspaper [072]
The bells of christmas / Young, Egerton Ryerson – Toronto: W Briggs, 18–? – 1mf – 9 – mf#50402 – cn Canadiana [830]
The bells of england / Raven, J J – London, 1906 – 7mf – 8 – mf#H-1351 – ne IDC [700]
The bells of is : or, voices of human need and sorrow: echoes from my early pastorate / Meyer, Frederick Brotherton – New York: Fleming H Revell, c1894 – 1mf – 9 – 0-8370-6335-3 – mf#1986-0335 – us ATLA [920]
Bellshill speaker – Bellshill: D MacLeod 1984- (wkly) [mf ed 7 jul 1994-] – 1 – (cont: bellshill speaker & mid-lanarkshire gazette) – ISSN: 1355-0861 – uk Scotland NatLib [072]
Bell-Smith, Frederick Marlett *see* A full description of the two historical paintings of the funeral of the late sir john s d thompson...
Belluco, Bartolome *see*
– De sacra praedicationes in o.f.m...
– Legislatio (ofm) de musica sacra. studium historico-iuridicum
Bellue, Pierre *see* L'ermite toulonnais faisant suite a l'ermite en province de m de jouy
Bellum judaicum / Josephus, Flavius [Joseph Ben Matthias] – 16th c – (= ser Holkham library manuscript books 454) – 1r – 1 – (gelenius's revision of rufinus's latin trans) – mf#96531 – uk Microform Academic [939]
Bellum punicum 1... / Bruni, Leonardo [Leonardo Aretino] – 15th c – (= ser Holkham library manuscript books 478,488) – 1r – 1 – (filmed with: lapus castelliunculus: opuscula 12) – mf#96536 – uk Microform Academic [450]
Bellum punicum, cathaginese, gallicum *see* Livius, books 31-40/dictys...
Bellum trojanum *see* Livius, books 31-40/dictys...
Bellview baptist church – Spartanburg County, SC. 1891-1973 – 1 – $39.11 – us Southern Baptist [242]
Bellwood gazette – Bellwood, NE: Wm H McGaffin. v7 n23. jun 24 1892-v53 n47. dec 15 1939 (wkly) [mf ed with gaps] – 3r – 1 – (absorbed by: people's banner (david city, ne). suspended temporarily foll jul 7 1939 issue; resumed with jul 28 1939) – us NE Hist [071]
Bellwood gazette – Bellwood, NE: Wm H McGaffin. -v53 n47. dec 15 1939 (wkly) [mf ed v7 n23. jun 24 1892-dec 15 1939 (gaps)] – 3r – 1 – (absorbed by: people's banner. suspended temporarily foll jul 7 1939; resumed jul 28 1939) – us NE Hist [071]
Bellwood, W A *see*
– Whither the transkei
Belly, Felix *see*
– Travers l'amerique centrale
– A travers l'amerique centralele nicaragua et le canal interoceanique
Belmont, August *see* The papers of august belmont, jr, 1827-1965
Belmont bee – Belmont WI. 1894 mar 1, may 17-jun 7, aug 30, 1898 may 26-1901 aug 22 – 1r – 1 – mf#955206 – us WHS [071]
[Belmont-] belmont courier – CA. jul 18 1936-jun 2 1944 – 3r – 1 – $180.00 – mf#C03154 – us Library Micro [071]
[Belmont-] belmont courier bulletin – CA. jan 6 1953-sep 2 1954; feb 1961-81 – 21r – 1 – $1260.00 – mf#B02051 – us Library Micro [071]
[Belmont-] belmont enterprise – CA. aug-dec 1960 – 1r – 1 – $60.00 – mf#H04002 – us Library Micro [071]
Belmont chronicle / Belmont Co. Saint Clairsv – 1931-32,35-44,7/45-4/61,62-1965 [wkly] – 12r – 1 – mf#B23283-23294 – us Ohio Hist [071]
Belmont chronicle / Belmont Co. Saint Clairsv – jan 1873-dec 1906 [wkly] – 15r – 1 – mf#B9967-9981 – us Ohio Hist [071]
Belmont chronicle / Belmont Co. Saint Clairsv – jan 1966-dec 1973 [wkly] – 4r – 1 – mf#B32116-32119 – us Ohio Hist [071]
Belmont chronicle / Belmont Co. Saint Clairsv – v1 n1. (7/1836-4/1854), 8/1854-12/1872 [wkly] – 9r – 1 – mf#B188-196 – us Ohio Hist [071]
The belmont circle – Tokyo. 1970-1987 (1) 1971-1987 (5) 1976-1987 (9) – 2r – 1 – (the circle 1991-1999) – mf#5826 – us Southern Baptist [071]
Belmont Co. Barnesville *see*
– Enterprise
– Miscellaneous newspapers
– Saturday whetstone
– Whetstone
Belmont Co. Bellaire *see*
– Daily independent
– Daily leader
– Democrat
– Herald
Belmont Co. Flushing *see* News
Belmont Co. Martins Ferry *see*
– Daily times
– Evening times
– Times-leader
Belmont Co. Saint Clairsv *see*
– Belmont chronicle
– Gazette
– Gazette-chronicle
– Independent republican
– National historian
[Belmont-] courier – NV. 1874-1901 [wkly] – 10r – 1 – $600.00 – mf#U04415 – us Library Micro [071]
Belmont, Francois Vachon de *see* Histoire de l'eau-de-vie en canada
Belmont gazette – Belmont WI. 1836 oct 25-1837 apr 12 – 1r – 1 – mf#955205 – us WHS [071]
Belmont heights baptist church – Nashville, TN. 1909-79 – 1 – $356.27 – us Southern Baptist [242]
[Belmont-] mountain champion – NV. 1868-69 – 1r – 1 – $60.00 – mf#U04416 – us Library Micro [071]
[Belmont shore-] enterprise – CA. jan-jun 1967 – 1r – 1 – $60.00 – mf#H04003 – us Library Micro [071]
[Belmont shore-] local enterprise – CA, jul-dec 1960 – 1r – 1 – $60.00 – mf#H04004 – us Library Micro [071]
[Belmont-] silver bend reporter – NV. 1867-68 – 1r – 1 – $60.00 – mf#U04417 – us Library Micro [071]
[Belmont-] silver bend weekly reporter – NV. 1867 – 1r – 1 – $60.00 – mf#U04418 – us Library Micro [071]
Belmont success – Belmont WI. 1903 jun 18/1909-1958/61 – 9r – 1 – (with small gaps; cont by: republican-journal (darlington wi); republican journal and the belmont success) – mf#955250 – us WHS [071]
Belmont University *see* The belmont vision
The belmont vision / Belmont University – Nashville, TN. jan 1953-may 1985; sep 1985-apr 1993 (and scattered iss 1952-74); sept 1993-may 2000 – 3r – 1 – mf#5771 – us Southern Baptist [378]
[Belmont-carlmont-] enquirer bulletin – CA. 1982-1985 – 4r – 1 – $240.00 – (cont: courier bulletin. cont by: san carlos enquirer bulletin) – mf#B03153 – us Library Micro [071]
Belmonte, Francisco *see*
– Bases y reglamentos...circulo concordia
– Memoria del jurado...dona isabel 2
Belmore y enriqueta, o, la medalla de oro : pt 1: belmore y enriqueta – Manila: Fajardo 1904 [mf ed Bloomington IN: Indiana Uni Lib, Preservation Dept 1984] – (= ser Coll...in the tagalog language 2) – 1r – 1 – us Indiana Preservation [490]
[Belo horizonte-] revista brasileira de ciencias socicais – BL. 1961-63 – 1r – $50.00 – mf#R60217 – us Library Micro [300]
[Belo horizonte-] revista brasileira do estudos politicos – BL. 1956-1979 – 8r – $400.00 – mf#R04163 – us Library Micro [079]
Beloe, William *see* Sermon preached at the parish church of allhallows, london wall,
Beloit alumnus – 1913 dec, 1914 oct-1918 oct, 1919 mar, 1920 apr-jun, 1921 feb-1923 nov, 1924 jan-1929 jun – 2r – 1 – (cont by: beloit college bulletin. alumni issue) – mf#3262262 – us WHS [378]
Beloit chronicle – Beloit WI. 1981 mar 26-1982 dec 10 – 1r – 1 – (cont by: chronicle (beloit wi)) – mf#955602 – us WHS [071]
Beloit College *see*
– Bulletin of beloit college
– Round table
Beloit college bulletin – 1933 oct-1938 jun – 1r – 1 – (cont: beloit alumnus; cont by: beloit college bulletin; the alumnus) – mf#3262315 – us WHS [378]
Beloit college bulletin – 1938 oct-1945 oct – 1r – 1 – (cont; beloit college bulletin. alumni issue; cont by: bulletin of beloit college. the alumnus) – mf#3262284 – us WHS [378]
Beloit college monthly – v7 [1859 oct-1860 jul] – 1r – 1 – (cont by: beloit monthly) – mf#688371 – us WHS [378]
Beloit crescent – Beloit WI. 1872 jul 10 – 1r – 1 – mf#955851 – us WHS [071]
Beloit daily independent – Beloit WI. 1928 sep 4-29 mar 30, 1929 apr 1-1931 may 29, 1931 jun 5-1934 aug 17 – 3r – 1 – mf#3475013 – us WHS [071]
Beloit daily news – 1945 aug 14 – 1r – 1 – (cont: beloit daily grit) – mf#1124535 – us WHS [071]
Beloit independent – Beloit WI. 1923 jul 13-1925 sep 25, 1925 oct 2-1927 jun 3, 1927 jun 7-1928 sep 7 – 3r – 1 – mf#955847 – us WHS [071]
Beloit journal of politics, literature, and general – Beloit WI. 1848 jun 20-1848 jul 20 – 1r – 1 – (cont by: beloit journal (beloit wi: 1855)) – mf#956825 – us WHS [071]
Beloit outlook – Beloit WI. 1880 jan 3 [prospectus], feb 7-jun – 1r – 1 – (cont by: beloit weekly outlook) – mf#955493 – us WHS [071]
Beloit poetry journal – v1-21. 1950-71 – 1 – us AMS Press [810]
Beloiter – v1 n1 [1976 aug] – 1r – 1 – (cont by: beloit quarterly) – mf#351002 – us WHS [071]
Belokonskii, I P *see*
– Derevenskie vpechatleniia (iz zapisok zemskogo statistika)
– Zemskoe dvizhenie
Belokurov, N G *see* Instruktsiia o poriadke kratko-srochnogo kreditovaniia kustarno-promyslovoi kooperatsii i operativnyi uchet
Belokurov, S A *see* Ukazatel ko vsem periodicheskim izdaniiam imperatorskogo obshchestva istorii i drevnostei rossiiskikh ... po 1915
Belolikov, V Z *see* Inok nikodim starodubskii
Belon, P *see* Les observations de plvsievrs singvlaritez and choses memorables, trouuees en grece, asie, iudee, egypte, arabie...
Belorusskaia ssr v tsifrakh : k 10-letiiu sushchestvovaniia bssr, 1919-1929 – Minsk, 1929 – 534p 6mf – 9 – mf#RHS-11 – ne IDC [314]
Belorusskii kooperator – Minsk, 1922(1) – 1mf – 9 – mf#COR-549 – ne IDC [335]
Belot, Adolphe *see*
– Testament de cesar girodot
[Belou(e)tte, N] *see* Societas humana in nativis seminibus sita, artibus roborata, adversitatibus confirmata, prosperitate ac otio caduca
Belousov, I M *see* Rossiiskii soiuz obshchestv vzaimnogo ot ognia strakhovaniia
Belov, S I *see* Atbasarskaia zhizn'
The beloved : an iowa boy in the jungles of africa: charles warner mccleary, his life, letters and work / Halsey, A W et al; ed by Hinkhouse, John Frederick – Fairfield, IA: Publ by Friends, 1909 – 1mf – 9 – 0-8370-6586-0 – (incl ind) – mf#1986-0586 – us ATLA [920]
The beloved physician : or, the life and travels of luke the evangelist / Alcott, William Andrus; ed by Kidder, Daniel Parish – New-York: G Lane & CB Tippett, 1845 – 1mf – 9 – 0-524-05896-2 – mf#1992-0653 – us ATLA [240]
The beloved physician of tsang chou : life-work and letters of dr arthur d peill, frcse / ed by Peill, J – London: Headley Bros, [19–] – 1mf – 9 – 0-8370-6300-0 – mf#1985-0300 – us ATLA [920]
Belozerov, A *see* Ukazatel knig, izdannykh na russkom jazyke, po predmetam otnosiashchimsia do gornoi chasti
Belper & alfreton chronicle & mid-derbyshire post – Belper, England 14 feb 1885-8 feb 1901 [mf 1898] – 1 – (incorp with: derby mercury) – uk British Libr Newspaper [072]
Belper allestree and duffield news – Belper, England 21 may 1992-28 dec 1995 [mf 1986-] – 1 – (cont: belper news, allestree & duffield news [10 oct 1985-14 may 1992]; cont by: belper news [4 jan 1996-]) – uk British Libr Newspaper [072]
Belper express – Belper, England 22 jun 1989- – 1 – uk British Libr Newspaper [072]
Belper journal and times – [East Midlands] Belper 18 jun 1870-23 jan 1875 [mf ed 2004] – 5r – 1 – (discontinued) – uk Newsplan; uk British Libr Newspaper [072]
Belper news & derbyshire telephone & mid-derbyshire mail, & peoples advocate – Belper, England 16 aug 1901-28 sep 1917 – 1 – (wanting: 1908; cont: the belper news & mid-derbyshire mail [6 jan 1899-19 aug 1901]; cont by: belper news [ns] 5 oct 1917-3 oct 1985) – uk British Libr Newspaper [072]
Belper weekly times, and derbyshire county herald – [East Midlands] Belper 29 jun 1861-8 aug 1868 [mf ed 2004] – 4r – 1 – (discontinued) – uk Newsplan; uk British Libr Newspaper [072]
Belsare, Malhar Bhikaji *see* An etymological gujarati-english dictionary
Belser, Johannes Evangelist *see*
– Die apostelgeschichte
– Beitraege zur erklaerung der apostelgeschichte
– Die briefe des apostels paulus an timotheus und titus
– Die briefe des heiligen johannes
– Einleitung in das neue testament
– Der epheserbrief des apostels paulus
– Die epistel des heiligen jakobus
– Die geschichte des leidens und sterbens, der auferstehung und himmelfahrt des herrn
– Die selbstvertheidigung des heiligen paulus im galaterbriefe (1,11 bis 2,21)
– Der zweite brief des apostels paulus an die korinther
Belsham, Jacobus *see* Canadia, ode
Belsham, Thomas *see*
– Importance of right sentiments concerning the person of christ
– Sermon occasioned by the death of the rev theophilus lindsey
Belsheim, J *see* Codex vercellensis
Bel'skij sovet r ki arm deputatov *see* Izvestiia bel'skogo soveta rabochikh, krest'ianskikh i armejskikh deputatov
Belsterling, Charles S *see* Belsterling's digest of decisions of the federal courts and the i.c.c. in the matter of transit privileges
Belsterling's digest of decisions of the federal courts and the i.c.c. in the matter of transit privileges / Belsterling, Charles S – Pittsburgh: C S Belsterling. 1v. 1913 – 4mf – 9 – $6.00 – mf#LLMC 95-119 – us LLMC [347]
Belt, William *see* Conversations on the office of sponsors for infants
Beltaine: the organ of the irish literary theatre. london. v. 1 no. 1-3. may 1899-apr 1900 – 1 – us NY Public [420]
Beltane the smith : a romance / Farnol, Jeffery – Toronto: Musson [1915?] [mf ed 1999] – 6mf – 9 – 0-659-90262-1 – mf#9-90262 – cn Canadiana [830]
Belton first baptist church – Belton, SC. 11 oct 1861-16 apr 1911 – 1 – $27.81 – us Southern Baptist [242]
Belton, Francis George *see* A manual for confessors
Beltrami, Giacomo C *see*
– Le mexique
– A pilgrimage in europe and america
Beltran, Juan Gregorio *see* Historia del brasil
Beltran y Rozpide, Ricardo *see*
– America en tiempo de felipe 2 segun el cosmografo cronista juan lopez de velasco. el territorio espanol de ifni
– Apuntamientos sobre el adelantamiento de yucatan, de amalio huarte y echenique
– Geografia
Belustigungen des verstandes und des witzes / ed by Schwabe, Johann Joachim – Leipzig 1741-45 [mf ed Hildesheim 1977] – (= ser Allgemeinwissenschaftliche und literarische zeitschriften des 17. und 18. jahrhunderts) – 55mf – 9 – diazo €218.00 silver €268.00 – gw Olms [400]
Belvacensis, Vincentius *see* De morali principis institutione
Belvalkar, Shripad Krishna *see*
– An account of the different existing systems of sanskrit grammar
– History of indian philosophy
Belvedere first baptist church. aiken county – Belvedere, SC. 812p. 1955-83. – 1 – $36.54 – us Southern Baptist [242]
Belvidere news – Belvidere, NE: Miller & Ross, 1890-98// (wkly) – 1r – 1 – (absorbed by: people's champion (hebron ne); cont as a separately numbered sect in: people's champion (hebron ne) mar 1898-feb 1901) – us Bell [071]
Belvior eagle – 1992 mar 17-dec 11, 1993 jan-jun 24, 1993 jul 1-dec 16 – 3r – 1 – mf#5486796 – us WHS [071]
Belvis Trejo, A *see* Suplica general y ultima que insinua a la...villa de madrid...(sobre la limona)
Belvoir Literary Society *see* Records of the meetings
Belwe, Andreas *see* Der mensch ist im gegenteil
Belydenisse : ofte verklaringhe van 't gevoelen der leeraren, die in de gheunieerde nederlanden remonstranten worden ghenaemt... / Wtenbogaert, J – Ed 3. n.p, 1630 – 3mf – 9 – mf#PBA-364 – ne IDC [240]
Belyea, Harold Cahill *see* Forest measurement
Belzer, Edwin G *see* The nature and status of health-promotion programs in nova scotian goods-producing industries
Belzer, Edwin G Jr. *see* Palliative care
Belzig-reetz-wiesenburger zeitung – Wiesenburg (Mark) DE, 1924 1 jan-28 jun, 1925-26, 1928, 1930-42 – 1 – gw Misc Inst [074]
Belzile, Marie-Paule *see* Bio-bibliographie du docteur philippe hamel

Belzoni, Giovanni B *see*
- Le jeune voyageur
- Narrative of the operations and recent discoveries within the pyramids, temples, tombs, and excavations, in egypt and nubia

Belzoni, Giovanni Battista *see* Catalogue of the various articles of antiquity

O bem da ordem – Rio de Janeiro, RJ: Typ Real, 1821 – (= ser Ps 19) – mf#P01,03,01 – bl Biblioteca [320]

Bem, I I *see* Den' kooperatsii – 21 dekabria 1844 [lxxv] – 21 dekabria 1919 g

O bem publico : folha imparcial – Pindamonhangaba, SP: Typ do Bem Publico, 02, 11 nov 1877; fev 1878; maio-jun, ago, nov 1879; 13, 27 jun 1880 – (= ser Ps 19) – mf#P18,01,121 – bl Biblioteca [321]

Bema : the official publication of the diocese / Armenian Church of America – 1980 apr-1986 dec – 1r – 1 – (cont: armenian church, hayastanyaitz yegeghetzy; cont by: armenian church (1987)) – mf#1218228 – us WHS [240]

The bema – Saint Martins, NB: Union Baptist Seminary, [189–1893?] – 9 – mf#P04530 – cn Canadiana [378]

Beman, Nathan S S *see* Episcopacy exclusive

Bemba grammar / Van Sambeek, J – Cape Town, South Africa. 1955 – 1r – us UF Libraries [470]

Bemba grammar notes for beginners / Hoch, E – s.l, s.l? . 19–? – 1r – us UF Libraries [470]

Bemba marriage and present economic conditions / Richards, A – (= ser Institute for social research, university of zambia. papers 4) – 3mf – 7 – mf#363/4 – uk Microform Academic [960]

Bement gazette – Bement IL. 1881 dec 10 – 1r – 1 – mf#1159477 – us WHS [071]

Bemerkungen auf einer alpen-reise ueber den bruenig, bragel, kirenzenberg : und ueber die flueela, den maloya und spluegen / Kasthofer, Carl – Bern 1825 [mf ed Hildesheim 1995-98] – 2mf – 9 – €60.00 – 3-487-29346-3 – gw Olms [914]

Bemerkungen auf einer alpen-reise ueber den susten, gotthard, bernardin : und ueber die oberalp, furka und grimsel; mit erfahrungen ueber die kultur der alpen und einer vergleichung des wirthschaftlichen ertrags der buendenschen und bernischen alpen / Kasthofer, Carl – Aarau 1822 [mf ed Hildesheim 1995-98] – 2v on 3mf – 9 – €90.00 – 3-487-29347-1 – gw Olms [914]

Bemerkungen auf einer reise aus norddeutschland ueber frankfurt nach dem suedlichen frankreich im jahr 1819 / Mutzenbecher, Johann D – Leipzig 1822 [mf ed Hildesheim 1995-98] – 2mf – 9 – €60.00 – 3-487-29746-9 – gw Olms [914]

Bemerkungen auf einer reise aus thueringen nach wien im winter 1805 bis 1806 / Bertuch, Carl – Weimar 1808 [mf ed Hildesheim 1995-98] – 1v on 2mf – 9 – €60.00 – 3-487-29408-7 – gw Olms [914]

Bemerkungen auf einer reise durch das innere der vereinigten staaten von nord-amerika im jahre 1819 : besonders in beziehung auf die an den fluessen sangoemo und onapischquasippi im norden des illinois-staats belegenen... / Ernst, Ferdinand – Hildesheim 1820 [mf ed Hildesheim 1995-98] – 1v on 2mf – 9 – €60.00 – 3-487-27151-6 – gw Olms [917]

Bemerkungen auf einer reise durch die niederlande nach paris : im eilften jahre der grossen republik / [Sierstorpff, C H von] – [Hamburg], 1804. 2v – 13mf – 9 – mf#HT-269 – ne IDC [914]

Bemerkungen auf einer reise durch die vereinten staaten von nord-amerika / in den jahren 1817, 1818 und 1819 / Harris, William T – Weimar 1822 [mf ed Hildesheim 1995-98] – 1v on 2mf – 9 – €60.00 – 3-487-26496-X – gw Olms [917]

Bemerkungen auf einer reise durch einen theil der schweiz und einige ihrer naechsten umgebungengeschrieben im bluethen-monath / Erbach, Albrecht – Heidelberg 1809 [mf ed Hildesheim 1995-98] – (= ser Fbc) – 280p on 2mf – 9 – €60.00 – 3-487-29370-6 – gw Olms [914]

Bemerkungen auf einer reise durch einen theil von teutschland, der schweiz, italien und frankreich : im jahre 1806 – Koenigsberg 1809 [mf ed Hildesheim 1995-98] – 1v on 2mf – 9 – €60.00 – 3-487-27782-4 – gw Olms [914]

Bemerkungen auf einer reise durch england / Broling, Gustav – Giessen 1828 [mf ed Hildesheim 1995-98] – (= ser Fbc) – 2v on 4mf – 9 – €120.00 – 3-487-27973-8 – gw Olms [914]

Bemerkungen auf einer reise durch frankreich, spanien, und vorzueglich portugal / Link, Heinrich F – Kiel 1801-04 [mf ed Hildesheim 1995-98] – 3v on 7mf – 9 – €140.00 – 3-487-29806-6 – gw Olms [914]

Bemerkungen auf einer reise durch schlesien, boehmen und einen theil von oestreich nach salzburg / Haak – Koenigsberg [u.a.] 1829 [mf ed Hildesheim: 1995-98] – (= ser Fbc) – xiv/328p on 2mf – 9 – €60.00 – 3-487-29418-4 – gw Olms [914]

Bemerkungen auf einer reise im jahre 1827 durch die beskiden ueber krakau und wieliczka nach den central-karpathenals : beitrag zur characteristik dieser gebirgsgegenden und ihrer bewohner / Sydow, Albrecht von – Berlin 1830 [mf ed Hildesheim 1995-98] – 3mf – 9 – €90.00 – 3-487-29158-4 – gw Olms [914]

Bemerkungen auf einer reise um die welt in den jahren 1803 bis 1807 / Langsdorff, G H von – Frankfurt am Mayn, 1812. 2v – 8mf – 9 – mf#H-6111 – ne IDC [910]

Bemerkungen auf einer reise um die welt in den jahren 1803 bis 1807 / Langsdorff, Georg H von – Frankfurt am Mayn 1812 [mf ed Hildesheim 1995-98] – 2v on 9mf – 9 – €180.00 – 3-487-26629-6 – gw Olms [910]

Bemerkungen auf einer reise von breslau ueber salzburg, durch tyrol, die suedliche schweiz nach rom, neapel und paestum : im jahre 1818 / Charpentier, Toussaint von – Leipzig 1820 [mf ed Hildesheim 1995-98] – (= ser Fbc) – 2v on 4mf – 9 – €120.00 – 3-487-27767-0 – gw Olms [914]

Bemerkungen einer reise im russischen reich im jahre 1772 / Georgi, J G – New York. 1873-1930 (1) – 16mf – 9 – mf#5551 – ne IDC [914]

Bemerkungen ueber den evangelischen religionsunterricht an hoeheren lehranstalten / Ubbelohde, Wilh. – Oldenburg: Gerhard Stalling 1877 [mf ed 1986] – 1mf – 9 – 0-8370-7594-7 – mf#1986-1594 – us ATLA [377]

Bemerkungen ueber die beduinen und wahaby / Burckhardt, John L – Weimar 1831 [mf ed Hildesheim 1995-98] – 1v on 4mf – 9 – €120.00 – 3-487-26469-2 – gw Olms [916]

Bemerkungen ueber natur, kunst und wissenschaft : auf einer reise ueber berlin und den harz nach hamburg zu der versammlung der naturforscher und aerzte im jahre 1830, nebst der rueckreise ueber copenhagen / Pontin, Magnus M – Hamburg 1832 [mf ed Hildesheim 1995-98] – 2mf – 9 – €60.00 – 3-487-29527-X – gw Olms [914]

Bemerkungen ueber rio de janeiro und brasilien : waehrend eines zehnjaehrigen aufenthalts daselbst, vom jahre 1808 bis 1818 / Luccock, John – Weimar 1821 [mf ed Hildesheim 1995-98] – 2v on 7mf – 9 – €140.00 – 3-487-26498-6 – gw Olms [918]

Bemerkungen ueber russland : auf einer reise gemacht im jahre 1792 und 93; mit statistischen und meteorologischen tabellen / Sternberg, Joachim von – [s. l.] 1794 [mf ed Hildesheim 1995-98] – (= ser Fbc) – 2mf – 9 – €60.00 – 3-487-29012-X – gw Olms [914]

Bemetzrieder, Anton *see*
- Art of modulating illustrated in one grand lesson and two preludes
- General instructions in music
- Lecons de clavecin
- Methode et reflexions sur les lecons de musique
- Music made easy to every capacity
- Nouvelles lecons de clavecin
- Reflexions sur les lecons de musique
- Traite de musique

Bemidji State University *see* Oshkaabewis native journal

Bemies, Charles Otto *see* Church in the country town

Bemister, George *see*
- Mr bemister's report of the flood at berthier
- Railway routes from montreal

Be-mizreh ha-zeman / Steinman, Eliezer – Tel-Aviv, Israel. 1930/31 – 1r – us UF Libraries [939]

Bemmann, Helga *see* Fuers publikum gewaehlt-erzaehlt

Bemrose, William *see*
- Bow, chelsea, and derby porcelain
- The life and works of joseph wright a r a
- Manual of wood carving

O bem-te-vi : jornal joco-serio – Rio de Janeiro, RJ: Typ S C A Quintanilha, 31 ago 1867 – (= ser Ps 19) – mf#P17,01,116 – bl Biblioteca [880]

O bemtivi : orgao de chicana – Fortaleza, CE. 03 abr 1892 – (= ser Ps 19) – mf#P17,01,35 – bl Biblioteca [079]

Ben gevuloth / Levin, Emma – Merhavya, Israel. 1943/44 – 1r – us UF Libraries [939]

Ben jonsons volpone : eine lieblose komoedie in drei akten / Zweig, Stefan – Potsdam: G Kiepenheuer, 1926, c1925 [mf ed 1992] – (= ser Die liebhaberbibliothek) – 148p (ill) – 1 – mf#7801 – us Wisconsin U Libr [420]

Ben khung cua so : tap truyen ngan / Son Tung – Ha-noi: Lao Dong 1974 [mf ed 1992] – on pt of 1r – 1 – mf#11052 r15 n1 – us Cornell [959]

Ben milhamah ve-shalom / Tenenbaum, Joseph – Jerusalem, Israel. 1960 – 1r – us UF Libraries [939]

Ben owen : a lancashire story / Perrett, Jennie – Toronto: W Briggs, 1882 – 2mf – 9 – 0-665-53744-1 – (incl publ ist) – mf#53744 – cn Canadiana [830]

Ben owen : a lancashire story / Perrett, Jennie – Toronto: W Briggs, 1881 [mf ed 1980] – 2mf – 9 – 0-665-00763-9 – mf#00763 – cn Canadiana [830]

Benabides Checa, Jose *see* Notas para sus biografias y para la historia documental de la santa iglesia catedral y ciudad de plasencia

Benacci, V *see* Descrittione degli apparati fatti in bologna per la venuta di n s papa clemente 8...

Il benadir / Mantegazza, Vico – Milan: Fratelli Treves, 1908 – 1 – us CRL [945]

Be-nahal-perat / Zuta, Haim Arieh – Jerusalem, Israel. 1924 or 1925 – 1r – us UF Libraries [939]

Benalla standard & north eastern ensign – Benalla, may 1895-dec 1909 – 2r – at Pascoe [079]

Ben-Ami, Mordecai *see* Kitve ben-ami

Benaming : getal en traktementen der inlandse hoofden op java, 1839 – 1mf – 8 – mf#SD-111 – ne IDC [959]

Der benanbrief : eine moderne leben-jesu-faelschung des herrn ernst edler von der planitz / Schmidt, Carl – Leipzig, 1921 – (= ser Tugal 3-44/1) – 2mf – 9 – €5.00 – ne Slangenburg [227]

Benard de LaHarpe, Jean B *see* Journal historique de l'etablissement des francais a la louisiane

Benares, the sacred city : sketches of hindu life and religion / Havell, Ernest Binfield – 2nd ed. London: W Thacker, [1912?] [mf ed 1995] – (= ser Yale coll) – xiii/226p (ill) – 1 – 0-524-09879-4 – mf#1995-0879 – us ATLA [280]

Benari, Nahum *see* Erkhe ruah ve-sifrut

Benary, Ferdinand *see* De hebraeorum leviratu

Benattar, S C *see* Le bled en lumiere, folklore tunisien

Benavente, Juan Alfonso de *see* Tractatus de penitentiis (i. 1500)

Ben-Avi, Itzhak *see* Ukhlesse artsenu

Benavides Checa, Jose *see* El fuero de plasencia

Benavides Llorente, Daniel *see* Realizacion del fuero del trabajo 1 conferencia...18 de enero de 1950. precede al titulo jefatura provincial de fet y de las jons

Ben-avigdor le-hag-yovlo – Warschau, Poland. 1916 – 1r – us UF Libraries [939]

Bench and bar – Chicago. v1-2. 1869-71. ns: v1-3. 1871-74 5v – (= ser Historical legal periodical series) – 1 – $55.00 – mf#408840 – us Hein [340]

Bench and bar : a complete digest of the wit, humor, asperities, and amenities of the law / Bigelow, Lee Eugene – New York: Harper & Bros, 1871 – 6mf – 9 – $9.00 – mf#LLMC 91-062 – us LLMC [870]

Bench and bar – Frankfort KY 1995+ – 1,5,9 – (cont: kentucky bench and bar) – mf#3271,02 – us UMI ProQuest [340]

Bench and bar / Lawyers Club of Detroit – Lawyers Club of Detroit. v1-6. 1921-26 (all publ) – 11mf – 9 – $16.50 – (lacking: v1 n3. v2 n7-8) – mf#LLMC 84-424 – us LLMC [340]

Bench and bar : a monthly magazine for lawyers – New York: Bench & Bar Co. os: v1-28. 1906-12; ns: v1-15. 1912-20 (all publ) – 69mf – 9 – $103.00 – (lacking: os v25-27. ns v5-8) – mf#LLMC 84-422 – us LLMC [340]

The bench and bar / Lawyers' Association of the 8th Judicial Circuit of Missouri – v1-7. 1935-42 – 4mf – 9 – $6.00 – mf#LLMC 84-423 – us LLMC [340]

The bench and bar of cleveland / Kennedy, James Henry – Cleveland, Cleveland Printing and Publishing Co., 1889. 358 p. LL-636 – 1 – us L of C Photodup [340]

The bench and bar of georgia: memoirs and sketches / Miller, Stephen Franks – Philadelphia: Lippincott, 1858. 2v. LL-316 – 1 – us L of C Photodup [340]

Bench and bar of minnesota – v1-58. 1934-2001 – 9 – $1347.00 set – ISSN: 0276-1505 – mf#401291 – us Hein [340]

The bench and bar of minnesota – v1-23. 1943-66 – 205mf – 9 – $307.00 – mf#LLMC 84-421 – us LLMC [340]

The bench and bar of new york / Proctor, Lucien Brock – New York: Diossy & Co, 1870 – 9mf – 9 – $13.50 – mf#LLMC 91-009 – us LLMC [340]

The bench and bar of saratoga county. / Mann, Enos R – Ballston, N.Y.: Waterbury & Inman, 1876. 391p. LL-628 – 1 – us L of C Photodup [340]

Bench mark – 1990 feb-1993 dec – 1r – 1 – mf#1759151 – us WHS [071]

Ben-chananja : monatsschrift fuer juedische theologie / ed by Loew, Leopold – Szegedin 1858-67 [mf ed Hildesheim 1993] – (= ser Bibliothek des deutschen judentums) – 10v on 86mf – 9 – diazo €318.00 silver €368.00 – gw Olms [270]

Ben-chananja : monatsschrift fuer juedische theologie und fuer juedisches leben in gemeinde, synagoge und schule – Szegedin 1858-67 [complete] – (= ser German-jewish periodicals...1768-1945, pt 1) – 10v on 3r – 1 – $325.00 – mf#B27 – us UPA [270]

Benchbook for us district court judges – 1996 – 3mf – 9 – $4.50 – mf#llmc99-021 – us LLMC [347]

Benchmark : quarterly review of the constitution and the courts – v1-5. 1985-93 all publ – 9 – $67.00 set – (none publ 1989, 1992. ceased with v5 n2) – ISSN: 0743-0310 – mf#110241 – us Hein [342]

Benchmarking – Bradford UK 2001+ – 1,5,9 – ISSN: 1463-5771 – mf#31604,01 – us UMI ProQuest [650]

Bend bulletin (bend, or: 1903) – Bend OR: M Lueddemann, [wkly] [mf ed 1966] – 1r – 1 – (began in 1903; ceased in 1931; related to daily ed: daily bulletin (bend or 1916-17) and: bend bulletin (bend, or: 1917)) – us Oregon Lib [071]

Bend bulletin (bend, or: 1903) *see*
- Bend bulletin (bend, or: 1917)
- Daily bulletin [bend, or]

Bend bulletin (bend, or: 1917) – Bend OR: G P Putnam, 1917-63 [daily ex sun] – 1 – (related to: bend bulletin (bend, or: 1903); cont: daily bulletin (bend, or); absorbed: central oregon press; cont by: bulletin (bend, or)) – us Oregon Lib [071]

Bend bulletin (bend, or: 1917) *see* Bend bulletin (bend, or: 1903)

Bend daily press – Bend OR: Bend Press Pub Co, [daily] [mf ed 1966] – 1r – 1 – (related to weekly ed: bend press, (1922)) – us Oregon Lib [071]

Bend daily press *see* Bend press

Bend free press – Bend OR: S D Pierce, 1936 [wkly] [mf ed 1965] – 1r – 1 – (absorbed by: free press (bend or, 1932); ceased in aug 1936) – us Oregon Lib [071]

Bend of the river – 1972 dec-1976, 1977-80, 1981-1984 jul, 1984 aug-1986, 1987 – 5r – 1 – mf#976378 – us WHS [071]

Bend pilot – Bend OR: T H Mark, 1940-50 [wkly] [mf ed 1966] – 5r – 1 – (cont: free press (bend or: 1932)) – us Oregon Lib [071]

Bend press – Bend OR: Bend Press Pub Co, [wkly] [mf ed 1968] – 1r – 1 – (related to: bend daily press; cont by: central oregon press (-1926)) – us Oregon Lib [071]

Bend press *see* Bend daily press

Benda, Georg *see*
- [Ariadne auf naxos]
- Concertino per il cembalo

Benda, Julien *see*
- Une philosophie pathetique
- Trahison des clercs

Bendahman, Jadwiga *see* Der reduplizierte aorist in den indogermanischen sprachen

Bendavid, Isaac Besht *see* Goldwin smith and the jews

Bender, E P *see* Report of survey of french river, georgian bay, lake huron

Bender, Heinrich *see* Hebbels dithmarschenfragment

Bender, Henry Richard *see*
- The problem of consolation
- Twentieth century interpretation of paul's epistle to the ephesians

Bender, Prosper *see*
- Canada's actual condition
- A canadian view of annexation
- The disintegration of canada
- The french canadians in new england
- The french-canadian conteur of the olden days
- The french-canadian peasantry
- A new france in new england
- A rejoinder to dr hughes
- Visions of old quebec

Bender, Wilhelm *see*
- Johann konrad dippel
- Schleiermachers philosophische gotteslehre
- Schleiermachers theologie mit ihren philosophischen grundlagen

Bender's corporation manual *see* 1924 supplement to rosbrook on new york corporations (2d ed) and bender's corporation manual

Bender's corporation manual, state of new york. / Rosbrook, Alden Ivan – Albany: Bender, 1923. 463p. LL-1428 – 1 – us L of C Photodup [348]

Bendicenos senor / Aradillas Agudo, Antonio – Madrid: Editado por PPC y el Secretariado del Apostolado Rural de los HH de A.C., 1960 – sp Bibl Santa Ana [240]

Bendig, Helmut *see* Die rolle italiens bei der entstehung der europaeischen wirtschaftsgemeinschaft bis 1958

Bendigo advertiser – Bendigo, Australia 18 aug 1854-1859, 20 jun 1861-1918, 8 feb 1919-30 jun 1922, 1 jan 1952-31 jan 1953 – 235r – 1 – (wanting 1860; 1856,1857,1862 very imperfect; fr jul 1917-dec 1918, sat iss only) – uk British Libr Newspaper [079]
Bendigo mercury – Bendigo, Australia 9,25 jun, 28 aug 1858; 12,22 jan, 7,14 apr, 20 may, 1 jul, 13,22 aug 1859 (wkly) – 1 – uk British Libr Newspaper [079]
Bendix Local No 9, UAW *see* Plain facts
Bendix technical journal – Southfield MI 1968-73 – 1,5,9 – ISSN: 0005-8718 – mf#5053 – us UMI ProQuest [600]
Bendrey, Vasudeo Sitaram *see*
– A study of muslim inscriptions
– Tarikh-i-ilahi
Bene bilu / Ben-Zion, S – Tel-Aviv, Israel. 1929 or 1930 – 1r – us UF Libraries [939]
Bene ha-yoreh / Smoly, Eliezer – Tel-Aviv, Israel. 1937 – 1r – us UF Libraries [939]
"Bene mosheh" u-tekufatam / Tchernowitz, Samuel – Warsaw, Poland 1914 – 1r – 1 – us UF Libraries [939]
Bene yisakhar / Dynow, Zevi Elimelech – Lemberg, Ukraine. v1-2. 1909 – 1r – us UF Libraries [939]
Beneath tropic seas / Beebe, Charles William – New York, NY. 1928 – 1r – us UF Libraries [972]
Benecke, Heinrich *see* Wilhelm vatke in seinem leben und seinen schriften
Benecken, Friedrich Burch *see* Jahrbuch fuer die menschheit
Benedetti, F *see* L'imprese della m c di d filippo d'austria 2 re di spagna
Benedetti, Mario *see* El pais de la cola de paja
Benedetti, Vincent *see* Ma mission en prusse
Benedetto da Mantova *see*
– Aonio paleario and his friends
– The benefit of christ's death
Benedetto, L F *see* Marco polo/il milione
Benedict 14, Pope *see* Para la venidera memoria del negocio
Benedict broadcaster – Benedict, NE: LeRoy Overstreet. -v9 n31. jun 11 1941 (wkly) [mf ed v4 n40. aug 19 1936-jun 11 1941] – 1r – 1 – (absorbed by: york daily news-times) – us NE Hist [071]
Benedict, David *see*
– Fifty years among the baptists
– A general history of the baptist denomination in america and other parts of the world
– A history of all religions
– History of the donatists
Benedict de spinoza : his life, correspondence, and ethics / Willis, Robert – London: Truebner 1870 [mf ed 1991] – 2mf – 9 – 0-524-00216-9 – mf#1989-2916 – us ATLA [170]
Benedict gazette – Benedict, NE: Vanzandt Crownover, 1891- (wkly) [mf ed v2 n17. apr 23 1892 filmed 1973] – 1r – 1 – us NE Hist [071]
Benedict herald – Benedict, NE: W E Muth. -v3 n16. jun 27 1902 (wkly) [mf ed v2 n7. apr 26 1901-jun 271902 (gaps)] – 1r – 1 – (cont: benedict weekly herald. absorbed by: teller) – us NE Hist [071]
Benedict, Laura Watson *see* Study of bagobo ceremonial, magic and myth
Benedict, Michael A *see* Reliability in the measurement of muscle fiber composition and the histochemical staining for glycogen
Benedict news-herald – Benedict, NE: G R Douglas. 6v. v1 n1. jul 11 1902-v6 n39. jun 15 1910 (wkly) [mf ed with gaps] – 1r – 1 – us NE Hist [071]
Benedict, R D *see* Benedict's reports of cases in the district courts of the u.s. (2nd circuit), 1865-1879
Benedict, Ruth *see* Chrysanthemum and the sword
Benedict, Saint, Abbot of Monte Cassino *see* Die althochdeutsche benediktinerregel des cod sang 916
Benedict weekly herald – Benedict, NE: W E Muth, 1900 (wkly) [mf ed v1 n18. jul 13-dec 14 1900 (gaps)] – 1r – 1 – (cont by: benedict herald) – us NE Hist [071]
Benedicti regula monachorum / ed by Woelffin, E – Lipsiae, 1795 – 3mf – 8 – €7.00 – ne Slangenburg [241]
Benedicti regula monasteriorum, sancti / ed by Cuthbertus Butler, D – ed 3a. Friburgi Brisg, 1935 – €12.00 – ne Slangenburg [241]
Benedictine Monk Of Termonde (Belgium) *see* Katekisima thukhu ya pfhundzo ya vakreste
Benedictine pioneers in australia / Birt, Henry Norbert – St Louis MO: B Herder 1911 [mf ed 1992] – 2v on 3mf [ill] – 9 – 0-524-03928-3 – mf#1990-4922 – us ATLA [241]
Benedictines – Kansas City KS 1946+ – 1,5,9 – ISSN: 0005-8726 – mf#10678 – us UMI ProQuest [241]
La benediction abbatiale : allocution prononcee a la benediction de dom pacome gaboury a la trappe de notre-dame, a oka, le 13 novembre 1913 / Emard, Joseph-Medard – Valleyfield [Quebec: s.n.] 1913 [mf ed 1994] – 1mf – 9 – 0-665-73188-4 – mf#73188 – cn Canadiana [241]
Benediction du nouveau seminaire de st germain de rimouski – [Rimouski?]: [s.n.], [1878?] [mf ed 1980] – 1mf – 9 – 0-665-02319-7 – mf#02319 – cn Canadiana [241]
Benediction du nouveau seminaire de ste-therese, le 26 juin 1883 – Montreal: Beauchemin & Valois, libraires-impr, 1883 [mf ed 1980] – 1mf – 9 – 0-665-02318-9 – mf#02318 – cn Canadiana [241]
Benediction of a church; litany of the saints; blessing of a bell; litany of the b virgin – Kamloops, BC?: s.n, 1893? – 1mf – 9 – mf#14616 – cn Canadiana [241]
Benediction solennelle du t r p dom m antoine : abbe de n d du lac des deux montagnes d'oka: a l'eglise de notre-dame de montreal, le 29 juin 1892 – Montreal: s.n, 1892 – 1mf – 9 – mf#04043 – cn Canadiana [241]
The benedictional of archbishop robert (hbs24) / Wilson, H Austin – 1903 – (= ser Henry bradshaw society (hbs)) – 4mf – 8 – €11.00 – ne Slangenburg [241]
The benedictional of john longlonde (hbs64) / Woolley, R M – 1927 – (= ser Henry bradshaw society (hbs)) – 2mf – 8 – €5.00 – ne Slangenburg [241]
Benedictionale des diozese meissen von 1512 / ed by Schoenfelder, A – Paderborn, 1904 – €5.00 – ne Slangenburg [241]
Das benedictionale des diozese meissen von 1512 / ed by Schoenfelder, A – Paderborn, 1904 – 2mf – 8 – €5.00 – ne Slangenburg [241]
Benedictiones – Chishawasha, Zimbabwe. 1932 – 1r – us UF Libraries [960]
Benedictions : or, the blessed life / Cumming, John – Boston: John P Jewett, 1858 [mf ed 1984] – 4mf – 9 – 0-8370-0962-6 – mf#1984-4313 – us ATLA [240]
Benedictis, J B *see* Philosophia peripatetica
The benedictonals of freising (hbs88) / Amiet, R – 1974 – (= ser Henry bradshaw society (hbs)) – 3mf – 8 – €7.00 – ne Slangenburg [241]
Benedict's reports of cases in the district courts of the u.s. (2nd circuit), 1865-1879 / Benedict, R D – New York: Baker. v1-10. 1869-82 (all publ) – (= ser Early federal nominative reports) – 71mf – 9 – $106.00 – mf#LLMC 81-432 – us LLMC [324]
Benedictus illustratus sive disquisitionum monasticarum libri 12 / Haeften, Benedictus van – Antverpiae. v1-2. 1644 – 2v on 36mf – 8 – €117.00 – ne Slangenburg [241]
Benedictus van Canfield *see* Den regel der volmaectheyt
Benedikt von aniane : werk und persoenlichkeit / Narberhaus, Jozef – Muenster, 1930 – 2mf – 8 – €14.00 – ne Slangenburg [241]
Das benediktbeurer passionsspiel; das st galler passionsspiel : nach den handschriften / ed by Hartl, Eduard – Halle/Saale: M Niemeyer, 1952 [mf ed 1993] – (= ser Altdeutsche textbibliothek n41) – 131p – 1 – mf#8193 reel 4 – us Wisconsin U Libr [241]
Die benediktiner in alabama : und geschichte der gruendung von st. bernard / Reger, Ambrose – Baltimore: Kreuzer, 1898 – 1mf – 9 – 0-524-05589-0 – mf#1991-2313 – us ATLA [241]
Benediktiner monnik der abtdye te egmont : met taal- n historikundige aanteekeningen opgehelderd door mr gerard van loon / Rymchronyk van Klaas Kolyn – 's-Graavenhaage, 1745 – €40.00 – ne Slangenburg [241]
Der benediktinische abt / Hegglin, P – St Ottilien, 1961 – 3mf – 8 – €7.00 – ne Slangenburg [241]
Benediktinisches ordensrecht / Mayer, P H S – Beuron. v1-4. 1929 – 17mf – 8 – €32.00 – ne Slangenburg [241]
Benedix, Roderich *see*
– Haustheater
– Nein
– Volkstheater
Beneficiaire / Theaulon, M – Paris, France. 1825 – 1r – us UF Libraries [440]
Beneficial effects of christianity on the temporal concerns of mank... / Porteus, Beilby – London, England. 1806 – 1r – us UF Libraries [240]
The beneficial influence of a well regulated nationality : a sermon delivered before the st andrew's society, of montreal, on st andrew's day, nov 30th, 1857, in saint gabriel street scotch church / Kemp, Alexander Ferrie – Montreal?: s.n, 1857 [mf ed 1984] – 1mf – 9 – 0-665-45232-2 – mf#45232 – cn Canadiana [242]
Beneficio di cristo *see* The benefit of christ's death
Benefit news / National Mutual Benefit (Madison WI) – 1931 sep-1944 aug, 1944 sep-1956, 1957-61, 1962-80 – 4r – 1 – (cont: beaver (madison wi)) – mf#1053162 – us WHS [360]
The benefit of christ's death = Beneficio di cristo / Benedetto da Mantova; ed by Babington, Churchill – London: Bell & Daldy; Cambridge: Deighton, Bell, 1855 [mf ed 1990] – 1v on 1mf – 9 – 0-7905-8138-8 – (int by churchill babington) – mf#1988-6085 – us ATLA [240]
Benefit Trust Life Insurance Co et al *see* Railway employees' journal
Benefits law journal – v1-12. 1988-99 – 9 – $597.00 set – ISSN: 0897-7992 – mf#112161 – us Hein [340]
Benefits of affliction / Bellett, G – Bridgnorth, England. 1842? – 1r – us UF Libraries [240]
The benefits of audio-visual technology in addressing racial profiling : hearing before the committee on government reform, house of representatives, 107th congress, 1st session, july 19 2002 / United States. Congress. House. Committee on Government Reform – Washington: US GPO 2002 [mf ed 2002] – 2mf – 9 – 0-16-068523-0 – (incl bibl ref) – us GPO [350]
Benefits of baptism – London, England. 18– – 1r – us UF Libraries [242]
Benefits of the reformation – London, England. 18– – 1r – us UF Libraries [242]
Benefits quarterly – Brookfield WI 1987+ – 1,5,9 – ISSN: 8756-1263 – mf#15720 – us UMI ProQuest [650]
Beneke, Walter *see* Paraiso de los imprudentes
Benekendorf, Karl Friedrich von *see* Berliner beytraege zur landwirtschaftswissenschaft
Benelux-kunstindex : bilddokumentation zur kunst in belgien, in den niederlanden und luxemburg / ed by Bildarchiv Foto Marburg – Deutsches Dokumentationszentrum fuer Kunstgeschichte Philipps- Universitaet Marburg – [mf ed 1999] – 251mf (1:24) – 9 – silver €3560.00 – ISBN-10: 3-598-34400-7 – ISBN-13: 978-3-598-34400-8 – gw Saur [700]
Benemeritos de la patria y ciudadanos de honor de... / Solera Rodriguez, Guillermo – San Jose, Costa Rica. 1964 – 1r – us UF Libraries [972]
Benes, Edvard *see* My war memoirs
Beneshevich, V N *see* Opisanie grecheskikh rukopisei monastyria sviatoi ekateriny na sinae 1, 3:1
Benet, Stephen Vincent *see* A treatise on military law and the practice of courts-martial
Benet Y Castellon, Eduardo *see*
– Birin
– Ensayo de haikai antillano
– Jabuquito de haikais
– Triptico
– Vida y yo
– Yo, pecador
Benevolence baptist church – Washington. 1961-1973 (1) – 1r – 1 – $35.55 – mf#6526 – us Southern Baptist [242]
The benevolence of the gospel toward the poor : a discourse. delivered at madison university, hamilton, new-york... / Fuller, Richard – Baltimore: GF Adams, 1848 – 1mf – 9 – 0-524-08361-4 – mf#1993-3061 – us ATLA [226]
Benevolent and Protective Order of Elks *see* Elk's gulch gazette
The benevolent banner – North Topeka, KS. v1 n1-21 may 21-oct 22 1887 (mf ed 1947) – (= ser Negro Newspapers on Microfilm) – 1r – 1 – us L of C Photodup [071]
Beneze, Emil *see* Das traummotiv in altdeutscher dichtung
Ben-Ezra, Victor *see* The effect of swim training on plasma somatomedin-c levels in 8- to 10-year-old children
Benfey, Theodor *see* Die persischen keilinschriften
Los benficios del telefono / Garcia Garcia, Juan – Caceres: Tall. El Noticiero, 1953 – 1 – sp Bibl Santa Ana [820]
Bengal and assam, behar and orissa : their history, people, commerce, and industrial resources / Playne, Somerset [comp]; ed by Wright, Arnold – London: Foreign and Colonial Compiling and Pub Co, 1917 – (= ser Samp: indian books) – us CRL [954]
Bengal as a field of missions / Wylie, Macleod – London: W H Dalton; Calcutta: Thacker, Spink 1854 [mf ed 1994] – 5mf – 9 – 0-524-08825-X – mf#1993-3317 – us ATLA; ne IDC [240]
Bengal, bihar, and orissa, sikkim / O'Malley, Lewis Sydney Steward – Cambridge: University Press, 1917 – (= ser Samp: indian books) – us CRL [954]
The bengal dispensatory: chiefly compiled from the works of roxburgh, wallich, ainslie, wight, arnot, royle, pereira, lindley, richard and fee / O'Shaughnessy, William Brooke – Calcutta: W. Thacker, St. Andrew's Library, 1842. xxiii,794p. plates – 1 – us Wisconsin U Libr [954]
Bengal Govt. Dept of Agriculture, Forest and Fisheries *see* Agricultural statistics by plot to plot enumeration in bengal, 1944-45
Bengal haggis : the lighter side of indian life / Archbold, William Arthur Jobson – London: Scholartis Press, 1928 – (= ser Samp: indian books) – us CRL [954]
Bengal hurkaru, 1822-66 – 100r – 1 – mf#4897 – uk Microform Academic [079]
Bengal in 1756-1757 / ed by Hill, S C – London: John Murray, 1905 – (= ser Samp: indian books) – us CRL [954]
Bengal in the sixteenth century ad / Das Gupta, J N – [Calcutta]: University of Calcutta, 1914 – (= ser Samp: indian books) – us CRL [301]
Bengal (India) *see* A digest of the law of landlord and tenant
Bengal journey : a story of the part played by women in the province, 1939-1945 / Godden, Rumer – London, New York: Longsmans, Green & Co, 1945 – (= ser Samp: indian books) – us CRL [954]
Bengal lancer / Yeats-Brown, Francis – London: Victor Gollancz, 1930 – (= ser Samp: indian books) – us CRL [915]
The bengal methodist – [s.l: s.n.] [mthly] [mf ed 2006] – n2-14 (jan-dec 1882) on 1r – 1 – (n3 omitted in numbering; damaged: n2 (jan 1882) p35-38) – mf#2006c-s024 – us ATLA [242]
The bengal mission – London: Church Missionary Society, 1903 – (= ser The Missions of the Church Missionary Society) – 1mf – 9 – 0-524-06078-9 – mf#1991-2391 – us ATLA [240]
Bengal nawabs : containing azad-al-husaini's naubahar-i-murshid quli khani, karam 'ali's muzaffarnamah, and yusuf 'ali's ahwal-i-mahabat jang – Calcutta: Asiatic Society, 1952 – (= ser Samp: indian books) – (transl into english by jadu nath sarkar) – us CRL [954]
Bengal Relief Committee *see* Correspondence, 1943-1947
Bengal tenancy bill : a speech delivered on moving for leave to introduce the bill into the legislative council of the governor general / Ilbert, Peregrine – Simla: Govt Central Branch Press, 1883 – (filmed with: a historical sketch of fyzabad tehsil by p carnegy, lucknow, 1870) – us CRL [323]
The bengal tragedy / Ghosh, Tushar Kanti – Lahore: Hero Publications, 1944 – (= ser Samp: indian books) – us CRL [360]
Bengal under the lieutenant-governors : being a narrative of the principal events and public measures during their periods of office, from 1854 to 1898 / Buckland, C E – Calcutta: Kedarnath Bose, 1902 – (= ser Samp: indian books) – us CRL [954]
Bengalee – Calcutta, India. 1907-Aug 1932 – 71r – 1 – us L of C Photodup [079]
The bengalee – Calcutta, India. -w. 1886-89; 1900-06, 1917. 34 reels – 1 – uk British Libr Newspaper [079]
Bengalees of tomorrow / Oyajeda Ali, Esa – Calcutta: Das Gupta & Co, 1945 – (= ser Samp: indian books) – us CRL [954]
Bengalese : official magazine of the american provinces of the congregation of holy cross – Notre Dame IN: [Bengal Foreign Mission Society], 1919-56 [mf ed 2005] – 38v on 5r – 1 – (cont by: holy cross missions) – mf#2005C-s007 – us ATLA [241]
The bengali drama : its origin and development / Guha-Thakurta, Prabhucharan – London: Kegan Paul, Trench, Trubner & Co, 1930 – (= ser Samp: indian books) – us CRL [790]
Bengali grammar / Yates, William & Wenger, John – rev ed. Calcutta: J W Thomas, Baptist Mission Press; London: Luzac, 1885 [mf ed 1995] – (= ser Yale coll) – vii/136p – 1 – 0-524-09342-3 – mf#1995-0342 – us ATLA [490]
Bengali literature / Ghosh, Jyotish Chandra – London, New York: Oxford University Press, 1948 – (= ser Samp: indian books) – us CRL [490]
Bengali literature / Ray, Annadasankar & Ray, Lila – Bombay: Publ for PEN All-India Centre by International Book House, 1942 – (= ser Samp: indian books) – us CRL [490]
The bengali ramayanas : being lectures delivered to the calcutta university in 1916, as ramtanu lahiri research fellow in the history of bengali language and literature / Sen, Dineshchandra – [Calcutta]: The University, 1920 – (= ser Samp: indian books) – us CRL [490]
Bengali selections : with translations and a vocabulary / Haughton, Graves Chamney – London: Kingsbury, Parbury, and Allen 1822 – (= ser Whsb) – 3mf – 9 – €40.00 – mf#Hu 282 – gw Fischer [490]
Bengali self-taught : by the natural method with phonetic pronunciation / Chatterji, Suniti Kumar – London: E Marlborough & Co, [1927] – (= ser Samp: indian books) – us CRL [490]
Bengel, Johann Albrecht *see* Dr johann albrecht bengels auslegung des neuen testaments
Bengesco, G *see* Essai d'une bibliographie sur la question d'orient
Bengoechea, I *see* La virgen maria en la vida y en la obra de benito arias montano

BENGOUGH

Bengough, John Wilson *see*
- Bengough's popular readings
- The breach of promise trial
- Bunthorne abroad
- A caricature history of canadian politics
- Cartoons of the campaign
- The gin mill primer
- Grip
- In many keys
- Motley
- The prohibition aesop
Bengough, Thomas *see* Bengough's cosmopolitan shorthand writer
Bengough's cosmopolitan shorthand writer – Toronto: T Bengough, [1881?-1883] [mf ed v3 n1/2 may/jun 1882-v3 n12 apr 1883] – 9 – mf#P04189 – cn Canadiana [650]
Bengough's illustrated monthly – Toronto: Bengough, [1885-18– or 19–] [mf ed v1 n1 feb 1885] – 9 – mf#P04357 – cn Canadiana [073]
Bengough's popular readings / ed by Bengough, John Wilson – Toronto: Bengough, Moore & Bengough, 1882-[18– or 19–] [mf ed n1 [1882]] – 9 – mf#P05158 – cn Canadiana [410]
Bengtsforstidningen dalslanningen – Bengtsfors, Sweden. 1979- – 1 – sw Kungliga [078]
Benguela railway / Companhia Do Caminho De Ferro De Benguela – Lisbon, Portugal. 1960 – 1r – us UF Libraries [380]
O benguella – Benguella: Tavares & Co [feb 5 1910] (wkly) – (= ser Coll of lusophone african newspapers and serials) – 1r – 1 – us CRL [079]
Ben-Gurion, David *see* Be-hilahem yisrael
Benham and Froud *see* Illustrations of art metal and woodwork
Benham, Daniel *see* Sketch of the life of jan august miertsching
Benham, William *see*
- How to teach the old testament
- The johannine books
- The lives of the popes
- St john and his work
- A short history of the episcopal church in the united states
Ben-hur / Wallace, Lew – London, England. 1954 – 1r – us UF Libraries [790]
Beni hasan: part 2 / Newberry, Percy E – 1893 – 9 – $10.00 – us IRC [930]
[Benicia-] solano county herald – CA. nov 5 1855-oct 30 1858 – 1r – 1 – $60.00 – mf#B02052 – us Library Micro [071]
Benicia-vallejo – 1921-31, 1933-45, 1992- – (= ser California telephone directory coll) – 13r – 1 – $650.00 – mf#P39 – us Library Micro [917]
Benignus, Wilhelm *see*
- Gedichte und aufsaetze
- Weltstromlieder
Beni-hasan, 4: zoological and other details / Griffith, FL – 1900 – 9 – $10.00 – us IRC [930]
Benin games and sports / Egharevba, Jacob U – Special ed. Sapele: Central Press, 1951 – 1 – us CRL [790]
Benin. Institut nationale de l'Analyse Economique *see* Annuaire statistique 1965-1975
Benin law and custom / Egharevba, Jacob U – 3rd ed. [Benin City: J U Egharevba, 1949] – 1 – us CRL [390]
The benin massacre / Boisragon, Alan – 2nd ed. London: Metheun, 1898 – 1 – us CRL [960]
Benin, the city of blood / Bacon, Reginald H – London, New York: E Arnold 189[7] [mf ed [Sl. s.n. 19–?]] – 1r – 1 – (filmed with: notes africaines, apr 1942-oct 1946) – us CRL [960]
Beninati, Giuseppe *see* L'immacolata davanti al razionalismo
[Beniowksy] : opera en trois actes en prose / Boieldieu, Francois Adrien – [Paris: Erard c1800] [mf ed 1992] – 1mf – 1 – (libretto by alexandre duval, based on a von kotzebue's graf benjowski; french words) – mf#pres. film 113 – us Sibley [780]
Beniowski, M A *see* Voyages et memoires de mauric-auguste, comte de benyowsky...
Beniowski, Moritz A von *see* Voyages et memoires de maurice-auguste, comte de benyowsky
Beniprasada *see* A few suggestions on the problems of the indian constitution
Benisch, Abraham *see* Bishop colenso's objections to the historical character of the pentateuch and the book of joshua (contained in pt 1)
Benisch-Darlang, Eugenie *see* Mit goethe durch die schweiz
Benitez, Adigio *see*
- Dias como llamas
- Manuel ascunce elegia
Benitez, Fernando *see* In the footsteps of cortes
Benitez, Jose Antonio *see* Puerto rico and the political destiny of america
Benito arias montano (1527-1598) / Arias Montano, Benito & Rekers, Bernard – Amsterdam, 1961 – 1 – sp Bibl Santa Ana [920]
Benito arias montano (1527-1598) / Rekers, Bernard – London: Warburg Institute, 1972 – 1 – sp Bibl Santa Ana [240]
Benito arias montano. datos biograficos / Lujan Garcia, Jose – Malaga: Revista Espanola de Estudios Biblicos, 1928 – 1 – sp Bibl Santa Ana [780]
Benito arias montano. extractos de su vida / Doetsch, Carlos – Madrid: Imp. de Blass y Cia, 1920 – 1 – sp Bibl Santa Ana [780]
Benito arias montano, padre de la arqueologia biblica / Santos Olivera, Balbino – Malaga: Revista Espanola de Estudios Biblicos, 1928 – 1 – sp Bibl Santa Ana [240]
Benito arias montano, "poeta laureatus" / Lopez de Toro, Jose – Madrid: Revista de Archivos, Bibliotecas y Museos, 1954. pp. 167-188 – 1 – sp Bibl Santa Ana [810]
Benito arias montano y aubrey f.g. beel / Arconada, Mariano – Malaga: Revista Espanola de Estudios Biblicos, 1928 – 1 – sp Bibl Santa Ana [240]
Los beniverman en merida y badajoz / Codera, Francisco – Zaragoza: Mariano Escar, 1904 – 1 – sp Bibl Santa Ana [946]
Benjamin britten : die kammermusik fuer streicher / Eschment, Ulrich-Alexander – (mf ed 1996) – 1mf – 9 – €30.00 – 3-8267-2281-7 – mf#DHS 2281 – gw Frankfurter [780]
Benjamin c yancey papers / Yancey, Benjamin C – 1800-1931. University of North Carolina Library. Guide – 1 – $288.00 – us CIS [920]
Benjamin constant : esboco de uma apreciacao sintetica da vida e da obra do fundador da republica brazileira / Mendes, Raymundo Teixeira – Rio De janeiro, Brazil: Sede central da Igreja pozitivista do Brazil 1913 – (= ser Apostolado pozitivista do brazil. [publicacoes] 120) – 1r – 1 – us UF Libraries [972]
Benjamin constant / Neiva, Venancio De Figueiredo – Rio de Janeiro, Brazil. 1952 – 1r – us UF Libraries [440]
Benjamin f perry papers / Perry, Benjamin F – 1822-1933. University of North Carolina Library. Guide – 1 – $36.00 – us CIS [976]
Benjamin f stickney papers *see* Stickney, benjamin f, papers, ms 3450
Benjamin franklin's account books : 1713-1874 / Franklin, Benjamin – 1977 – 3r – 1 – $390.00 – (incl printed guide) – mf#S1851 – us Scholarly Res [920]
Benjamin, George J *see* Contribution a l'histoire diplomatique et contempo...
Benjamin griffith : biographical sketches contributed by friends / ed by Banes, Charles H – Philadelphia: American Baptist Publ Soc [1894?] [mf ed 1993] – 1mf [ill] – 9 – 0-524-06703-1 – mf#1991-2733 – us ATLA [242]
Benjamin harrison papers – 151r – 1 – $5,285.00 – (with guide) – Dist. us Scholarly Res – us L of C Photodup [975]
Benjamin hederichs lateinisch-deutsche, deutsch-lateinische und griechisch-lateinische, lateinisch-griechische woerterbuecher – 1675-1748 [mf ed 1988] – 66mf – 9 – €340.00 – 3-89131-032-3 – (int vol by franz josef hausmann: "altsprachliche lexikographie im zeitalter des barock. die woerterbuecher des benjamin hederich (1675-1748)") – gw Fischer [040]
Benjamin hellier : his life and teaching: a biographical sketch, with extracts from his letters, sermons, and addresses / ed by Hellier, Anna M & Hellier, John Benjamin – London: Hodder & Stoughton 1889 [mf ed 1991] – 2mf – 9 – 0-7905-8996-6 – mf#1989-2221 – us ATLA [240]
Benjamin jowett : master of balliol / Tollemache, Lionel Arthur – London: E Arnold [1895?] [mf ed 1992] – 1mf – 9 – 0-524-04627-1 – (incl bibl ref) – mf#1990-1287 – us ATLA [378]
The benjamin lincoln papers, 1635-1964 – [mf ed 1967] – 13r – 1 – (with p/g) – us MA Hist [355]
Benjamin morse papers, ms v.f. m / Morse, Benjamin – 1805-09 – 1r – 1 – (morse's justice of the peace docket book) – us Western Res [920]
Benjamin, Of Tudela *see* Masa'ot
Benjamin robert haydon : correspondence and table-talk. with a memoir by his son – London 1876 – (= ser 19th c art & architecture) – 12mf – 9 – mf#4.2.1707 – uk Chadwyck [750]
Benjamin, Samuel Greene Wheeler *see*
- The atlantic islands as resorts of health and pleasure
- The cruise of the alice may in the gulf of st lawrence and adjacent waters
- The turk and the greek
Benjamin, Walter *see* Ursprung des deutschen trauerspiels
Benkard, Christian *see* Unter deutschen palmen
Benkartek, Dietmar *see*
- Ein interpretierendes worterbuch der nominalabstrakta im "narrenschiff" sebastians brants von abenteuer bis zwietracht
- Zur interpretation dostojevskijs aus der sicht sowjetischer literaturkritiker
Benke, Bertha *see* Diaries and papers of bertha and herman benke
Benke, Herman C *see* Diaries and papers of bertha and herman benke
Benkelman chronicle – Benkelman, NE: J P Israel, aug 1897-v9 n38. may 11 1906 (wkly) [mf ed v1 n2. sep 2 1897)-may 11 1906 (gaps)] – 3r – 1 – (absorbed: dundy county journal. merged with: benkelman news to form: news-chronicle) – us NE Hist [071]
Benkelman news – Benkelman, NE: J F Haskin. -v13 n51. may 11 1906 (wkly) [mf ed v6 n50. mar 24 1899-oct 10 1902 (gaps)] – 1r – 1 – (merged with: benkelman chronicle to form: news-chronicle) – us NE Hist [071]
Benkelman post – Benkelman, NE: C L Ketler. -v7 n17. apr 28 1922 (wkly) [mf ed v5 n1. jan 2 1920-apr 28 1922] – 1r – 1 – (merged with: news-chronicle to form: benkelman post and news-chronicle) – us NE Hist [071]
Benkelman post and news-chronicle – Benkelman, NE: C L Ketler. vol n29 n[50] may 5 1922- (wkly) – 1 – (formed by the union of: benkelman post and: news-chronicle) – us NE Hist [071]
Benkowitz, Carl F *see*
- Helios der titan oder rom und neapel
- Das italienische kabinet
- Reise von glogau nach sorrent
- Reisen von neapel in die umliegenden gegenden
Ben-Menachem, Naphtali *see* Avraham even-'ezra
Benn, Alfred William *see*
- Greek philosophers
- History of ancient philosophy
- The history of english rationalism in the nineteenth century
- Revaluations
Benn, George *see* A history of the town of belfast
Benn, Gottfried *see*
- Ausdruckswelt
- Fragmente
- Gehirne
- Goethe und die naturwissenschaften
- Der neue staat und die intellektuellen
- Soehne
- Die stimme hinter dem vorhang und andere szenen
- Trunkene flut
Bennasar, P Guillermo *see* Diccionario tiruray-espanol
Benndorf, Kurt *see* Der musicalische quack-salber
Benner, Enos *see* Abhandlung ueber die rechenkunst oder practische arithmetik zum gebrauch fuer schulen
Benner's prophecies of future ups and downs in prices : what years to make money on pig-iron, hogs, corn, and provisions – Toronto: Belford Bros, 1877 [mf ed 1979] – 2mf – 9 – 0-665-00906-2 – (incl ind) – mf#00906 – cn Canadiana [332]
Bennet, G *see* Journal of voyages and travels... to...the south sea islands, china, india h and c
Bennet, Grace *see* Memoirs 1715-1803
Bennet sun – Bennet, NE: Sun Pub Co. 51v. v1 n1. jan 6 1911-v51 n37. oct 5 1961 (wkly) [mf ed with gaps filmed -1977] – 15r – 1 – (cont by: lincolnland sun) – us NE Hist [071]
The bennet union – Bennet, NE: C F Collins. v4 n48. nov 24 1892-jun 28 1900 (wkly) [mf ed with gaps] – 2r – 1 – (issues for feb 28-dec 26 1895 called v7 n6-v8 n44 and also full n578-full n623. issues for jan 2 1896-jun 28 1900-called v9 n1-v13 n36 and also full n469-full n709) – us NE Hist [071]
Bennett, Almon *see* Almon bennett's platform bee house
Bennett, Arnold *see*
- Milestones
- The woman who stole everything
Bennett, Charles Edwin *see* Syntax of early latin
Bennett, Charles W *see* History of the philosophy of pedagogics
Bennett, Charles Wesley *see* Christian archaeology
Bennett, Edmund Hatch *see*
- Bennett's fire insurance cases, 1729-1875
- The four gospels from a lawyer's standpoint
- A selection of leading cases in criminal law
Bennett, F D *see* Narrative of a whaling voyage round the globe from the year 1833 to 1836
Bennett, Frank *see* Forty years in brazil
Bennett, Frank David *see* Southern character as presented by american playwrights
Bennett, Frederick *see* The story of w. j. e. bennett
Bennett, George Hanneman *see*
- Illustrated history of british guiana
Bennett, Hugh H *see* Soils and agriculture of the southern states
Bennett, J L *see* Outlines of trial procedure
Bennett, J Risdon *see* Baron von haller
Bennett, James *see*
- The history of dissenters
- The history of dissenters during the last thirty years (from 1808 to 1838)
Bennett, James Risdon *see* The diseases of the bible
Bennett, James William *see* Breed of barren metal, or, currency and interest
Bennett, John *see* The treasure of peyre gaillard
Bennett, John J *see* Miscellaneous botanical works, 1886-88
Bennett, John R *see* The bible in the schools
Bennett, John Whitchurch *see* Ceylon and its capabilities
Bennett, Susan B *see* Credentials of cardiac rehabilitation personnel
Bennett, W H *see*
- The bible story retold for young people
- The mishna as illustrating the gospels
Bennett, W J E *see* S john the baptist's day, 1854
Bennett, Wendell Clark *see* Excavations in the cuenca region, ecuador
Bennett, William Henry *see*
- Biblical introduction
- The book of jeremiah
- The books of chronicles
- The general epistles
- Joshua and the conquest of palestine
- The mishna as illustrating the gospels
- A primer of the bible
- The religion of the post-exilic prophets
Bennett, William Henry et al *see*
- Christ and civilization
- Faith and criticism
Bennett, William J E *see*
- Distinctive errors of romanism
- First letter to the right honourable lord john russell
Bennett, William J. E *see*
- On methodism and the swedenborgians
- On presbyterianism and irvingism
- On romanism
Bennett, William James Early *see* On anabaptism, the independents, and quakerism
Bennett, William Wallace *see* Memorials of methodism in virginia
Bennett's fire insurance cases, 1729-1875 : u.s. and great britain / Bennett, Edmund Hatch – New York: Hurd & Houghton. v1-5. 1872-1877 (all publ) – 47mf – 9 – $70.00 – mf#LLMC 95-131 – us LLMC [346]
Bennetts, Robert E *see* Snail kite
Benni, Cyril Benham *see* The tradition of the syriac church of antioch
Bennie, W G *see*
- Imibengo
Benning leader – Columbus, Fort Benning GA. 1992 oct 9-1993 jul 2, 1993 aug 8-dec 31 – 2r – 1 – (cont: benning patriot) – mf#2578237 – us WHS [071]
Benninghoff, Ludwig *see* Der kreis (mme4)
Bennington herald – Bennington, NE: Cortes J Wilcox, 1904 (wkly) [mf ed v14 n5. jan 3 1908-jul 22 1927 (gaps)] – 6r – 1 – (merged with: waterloo gazette; millard courier and: elkhorn exchange to form: douglas county gazette; iss for dec 2 1910-jul 22 1927 called v7 n[1]-v23 n34) – us NE Hist [071]
Benno papentrigk's schuettelreime : wie er sie seiner freundschaft auf den ostertisch zu legen pflegte / Papentrigk, Benno – [s.l: s.n.] 1946 (Ansbach: Druck der Fraenkischen Landeszeitung) [mf ed 1991] – 1r – 1 – (filmed with: waiblinger / wolfgang kirchbach & other titles) – mf#2758p – us Wisconsin U Libr [810]
Benois, A *see* Khudozhestvennye sokrovishcha rossii
Benoist, Charles *see* Les ouvriers de l'aiguille a paris
[Benoist, E] *see*
- Histoire de l'edit de nantes
- Histoire et apologie de la retraite des pasteurs...
Benoit 14, Pope *see* Encyclique vix pervenit sur les contrats, 1er novembre 1745
Benoit, A *see* Saint gregoire de nazianze
Benoit, Camille *see* Les motifs typiques des maitres-chanteurs de nuremberg, comedie musicale par richard wagner
Benoit de canfield (1562-1610) : sa vie, sa doctrine et son influence / Veghel, Optatus de – Romae, 1949 – 14mf – 8 – €27.00 – ne Slangenburg [241]
Benoit de Sainte-Maure *see* Roman de troie (cima10)
Benoit, F *see* A l'abbaye de montmajour
Benoit, Henry E *see* L'eglise anglicane avant la reforme, abrege d'histoire ecclesiastique
Benoit, Julien *see* General andre fontanges chevallier
Benoit, Renee de van Berchem *see* Souvenirs et lettres
Benoit-Jean *see* Jacques bonhomme
Benoni : son of my sorrow / Humphriss, Deryck – Benoni, South Africa. 1968 – 1r – us UF Libraries [960]
Benony, D S *see* Evenements de fevrier, mai, juillet 1911
Benrath, Karl *see*
- An den christlichen adel deutscher nation
- Geschichte der reformation in venedig
- Julia gonzaga
- Luther im kloster, 1505-1525
- Zur geschichte der marienverehrung
Benrath, Paul *see* Goethe und luther

Benrather tageblatt – (Duesseldorf-) Benrath DE, 1953 11 apr- 955 21 feb [gaps] – 3mf=5df – 1 – (with suppl: am rheinesstrand 1919 13 jul-1921 27 mar [1r], gute geister 1913-17 [1r publ in berlin-charlottenburg], die illustrierte 1928 n38-1933 [2r publ in berlin], illustriertes sonntags-blatt 1921-22 [1r publ in stuttgart], das leben im bild 1934-40 [3r], sport 1913-16 [1r publ in muenchen], rundblick der woche 1950 7 may-1951 [1r]; with gaps) – gw Mikrofilm; gw Misc Inst [074]
Benrather zeitung – (Duesseldorf-) Benrath DE, jul-sep 1911, 1912-14, apr 1915-mar 1922 – 14r – 1 – (with suppl: illustrierte sonntags-zeitung (gelsenkirchen) 1915 n48, 1916 [1r], illustrierte zeitung (benrath) 1912 n2-1914 n47 [1r], illustriertes sonntagsblatt (stuttgart, gelsenkirchen) 1911 n27-39, 1917 n1-1921 n2 [1r], der rheinlaender jul 6 1911-dec 10 1912 [1r], sport anzeiger im bild nov 3-dec 29 1924 [1r], wochenbild 1921 n13-1922 n13 [1r]; with gaps) – gw Misc Inst [074]
Ben's literary advertiser, and register of engravingy, works on the fine arts, etc (london) – jan 1832-dec 1840 – (= ser 19th c british periodicals) – r56 – 1 – us Primary [700]
Bensalem, or, the new economy : a dialogue for the industrial classes on the financial question / Galbraith, Thomas – New York: T Galbraith, Jr, 1874 – 1mf – 9 – mf#23902 – cn Canadiana [332]
Bensasson, Maurice Jacques *see* Israelites espanoles
Bensberger volkszeitung *see* Bensberg-gladbacher anzeiger
Bensberg-gladbacher anzeiger – Bergisch Gladbach DE, 1870 6 jul-31 dec, 1871 18 jan-1872, 1873 1 oct-1874 26 sep, 1878-88, 1889 19 jan-1894, 1895 19 jan-1923 3 nov, 1924 1 feb-1929 – 33r – 1 – (title varies: 4 apr 1907: bensberger volkszeitung) – gw Misc Inst [074]
Ben-Shalom, Avraham *see* Deep furrows
Bension, Ariel *see* Hilula
Bensley, A A *see* Solomon islands diary
Bensly, Robert L *see*
- Epistle to the hebrews
- The fourth book of ezra
- The missing fragment of the latin translation of the fourth book of ezra

Bensly, Robert Lubbock *see*
- The fourth book of ezra
- The fourth book of maccabees and kindred documents in syriac
- The harklean version of the epistle to the hebrews, chap. 11. 28-13. 25

Bensly, Robert Lubbock et al *see* The four gospels in syriac
Benson, Arthur Christopher *see*
- The leaves of the tree
- The life of edward white benson
- William laud

Benson, Christopher *see*
- Certain and sufficient maintenance the right of christ's ministers
- Christian preaching considered
- Discourses upon tradition and episcopacy
- Protestant zeal recommended

Benson county courier – Leeds, ND: Leslie Strand. v1 n1 oct 6 1949-v5 n22 feb 18 1954 (wkly) – 1 – (absorbed by: benson county farmers press) – mf#03815-03816 – us North Dakota [071]
Benson county farmers press : [official paper benson county and minnewaukan] – Minnewaukan, ND: [Benson Co Farmers Press, Inc] v36 n27 jun 26 1919- (wkly) – 1 – (cont: north dakota siftings; absorbed: esmond bee (esmond, nd: 1902), and benson county courier; missing: 1970 jun 4; currently publ) – mf#03701-08373 etc – us North Dakota [071]
Benson, David P *see* Church music in theory and practice in selected baptist churches, an exploratory study
Benson, E F *see* Lucia in london. leipzig 1928
Benson, Edward White *see*
- Addresses on the acts of the apostles
- The apocalypse
- The cathedral
- Christ and his times
- Cyprian
- Living theology
- The seven gifts

Benson, Elizabeth P *see*
- Pre-columbian art

Benson, Henry Clark *see* Life among the choctaw indians
Benson, Jane *see* Quaker pioneers in russia
Benson, Joseph *see*
- Farther defence of the methodists
- Works

Benson, L *see* The book of remarkable trials and notorious characters
Benson, Louis F *see* The english hymn
Benson, Louis FitzGerald *see*
- The chapel hymnal
- The hymnal
- Studies of familiar hymns

Benson magazine of research – v1 n1-v4 n2 [1980 may-1983 nov] – 1r – 1 – mf#1265703 – us WHS [071]
Benson, Margaret *see* The temple of mut in asher
Benson, Mary *see*
- South africa

Benson, Nels *see* Leaching of nitrogen from certain florida soils after the applicati...
Benson, Richard Meux *see* Life of father goreh
Benson, Robert *see* Sketches of corsica
Benson, Robert Hugh *see*
- Back to holy church
- The light invisible
- A mystery play
- Non-catholic denominations

Benson sun – Omaha, NE: Stanford Lipsey. 4v. v80 n18. mar 10 1977)-v83 n135. aug 31 1983 (wkly) [mf ed with gaps filmed 1977-83] – 20r – 1 – (merged with: south omaha sun; north omaha sun; dundee sun; northwest sun, and west omaha sun to form: omaha sun) – us NE Hist [071]
Benson trace – v1 n1-v4 n1 [1980 apr/jun-1983 apr/jun] – 1r – 1 – mf#1265696 – us WHS [071]
Benson, W J P *see* West indies and british guiana
Benson, William Arthur Smith *see* Elements of handicraft and design
Bensow, Oscar *see*
- Die bibel, das wort gottes
- Glaube, liebe und gute werke
- Die lehre von der kenose
- Die lehre von der versoehnung

Bent, J T *see*
- The ruined cities of mashonaland
- The sacred city of the ethiopians

Bent, James Theodore *see* The ruined cities of mashonaland
Bent, Samuel Arthur *see* Why was louisburg twice beseiged?
Bent, Theodore *see* Ruined cities of mashonaland
The bent twig / Fisher, Dorothy Canfield – Toronto: Copp, Clark, 1915 [mf ed 1998] – 6mf – 9 – 0-665-65475-8 – mf#65475 – cn Canadiana [830]
Bentancourt, Luis Victoriano *see* Articulos de costumbres
Bentara – Ende, [1949]-1957 – 7mf – 9 – (missing: [1949]-1956, v1-8(1-23, 25, 26); 1956, v9(2-30); 1957, v10(2, 3, 5-12, 14, 15, 17, 18)) – mf#SE-983 – ne IDC [950]
Bente, Friedrich *see* Was steht der vereinigung der lutherischen synoden amerikas im wege?
Bentes, Paulo *see* Porongo
Benthall, John *see* Songs of the hebrew poets in english verse
Bentham, G *see* Botany of the voyage of hms sulphur [the]
Bentham guardian – [NW England] Bentham 1976 – 1 – uk MLA; uk Newsplan [072]
Bentley, Eliza *see* Precious stones for zion's walls
Bentley Historical Library *see* Annual report of the..., michigan historical collections
Bentley, John *see*
- Essays relative to the habits, character, and moral improvement of the hindoos
- A historical view of the hindu astronomy

Bentley library annual : annual report of the bentley historical library, michigan historical collections – 1973/74, 1975/76 – 1r – 1 – (cont: report of the michigan historical collections, michigan historical collections) – mf#781650 – us WHS [020]
Bentley, W H *see* Pioneering on the congo
Bentley, William Preston *see* Illustrious chinese christians
Bentley's miscellany – La Salle IL 1837-68 – 1 – mf#5256 – us UMI ProQuest [073]
Bentley's quarterly review – La Salle IL 1859-60 – 1 – mf#4186 – us UMI ProQuest [073]
Benton advocate – Benton WI. 1901 sep 12/1903 may 7-1956/1959 mar 27 – 32r – 1 – (with gaps) – mf#961323 – us WHS [071]
Benton, Alexander Hay *see* Indian moral instruction and caste problems
Benton bulletin (philomath, or) – Philomath OR: Jim & Carolyn Gill, 1976- [wkly] – 1 – us Oregon Lib [071]
Benton county courier – Corvallis, OR: A E Frost and M J Brown. v9 n21-v11 n2. mar 18 1915-dec 27 1917 – 1 – (cont: benton county republican (1906) and daily republican; cont by: semi-weekly benton county courier) – us Oregon Hist [071]
Benton county courier *see* Daily republican [corvallis, or]
Benton county courier (corvallis or: 1915) *see* Benton county republican (corvallis, or: 1906)
Benton county courier (corvallis, or: 1915) – Corvallis OR: A E Frost & M J Brown, 1915-17 [wkly] – 1 – (merger of: benton county republican (corvallis, or: 1906), daily republican (corvallis, or); cont by: semi-weekly benton county courier) – us Oregon Lib [071]
Benton county courier (corvallis, or: 1919) – Corvallis OR: A E Frost, 1919- [semiwkly] – 1 – (ceased in 1925; cont: semi-weekly benton county courier; cont by: benton independent (corvallis, or); iss for feb 6 1923-apr 22 1923 called: morning courier) – us Oregon Lib [071]
Benton county herald (corvallis, or) – Corvallis OR: G E Hamilton, 1932-78 [wkly] [mf ed 1962-78] – 16r – 1 – (cont: benton independent (corvallis, or); merged with: greater oregon to form: weekly oregon herald (albany, or); iss for aug 12 1960-sep 14 1961 incl suppl with title: social security news) – us Oregon Lib [071]
Benton county republican *see*
- Daily republican [corvallis, or]

Benton county republican (corvallis, or: 1906) – Corvallis OR: Smith & Morgan, 1906-15 [wkly] [mf ed 1970] – 2r – 1 – (merged with: daily republican (corvallis, or), to form: benton county courier (corvallis or: 1915); related to: benton county republican (corvallis, or: 1909), tri-weekly republican (corvallis, or), daily republican (corvallis, or)) – us Oregon Lib [071]
Benton county republican (corvallis, or: 1906) *see* Tri-weekly republican (corvallis, or)
Benton county republican (corvallis, or: 1909) – Corvallis OR: Republican Pub Co, 1909 [semiwkly] [mf ed 1973] – 1r – 1 – (related to: benton county republican (corvallis, or: 1906); cont by: tri-weekly republican (corvallis, or)) – us Oregon Lib [071]
Benton county republican (corvallis, or: 1909) *see* Benton county republican (corvallis, or: 1906)
Benton county review – Philomath OR: F S Minshall, -1964 [wkly] [mf ed 1960-67] – 12r – 1 – (began in 1904. 1925-38 incl newspapers pub by local high schools and philomath college) – us Oregon Lib [071]
Benton democrat (corvallis, or) – Corvallis OR: G W Quivey & J A Miller [wkly] – 1 – (began in 1871) – us Oregon Lib [071]
Benton first baptist church – Benton, TN. 1836-aug 1946 – 1 – $40.95 – us Southern Baptist [242]
Benton independent – Corvallis, Benton County, OR : A W Lawrence. v1 n2-v27 n52. apr 18 1924-mar 24 1932 – 1 – (cont: benton county courier (1919); cont by: benton county herald; aka: benton independent and benton county courier) – us Oregon Hist [071]
Benton independent (corvallis, or) – Corvallis OR: A W Lawrence [wkly] – 1 – (began in 1924; absorbed: benton county courier (corvallis, or: 1919); cont by: benton county herald (corvallis, or)) – us Oregon Lib [071]
Benton, Josiah Henry *see* The book of common prayer
Benton leader – Corvallis OR: M L Pipes [wkly] [mf ed 1973] – 1r – 1 – (began in 1882) – us Oregon Lib [071]
Benton news – Benton, North Chicago etc...IL. v6 n12 1930 aug 7 – 1r – 1 – mf#1010923 – us WHS [071]
Benton, P A *see* The languages and people of bornu
Benton, Thomas Hart *see* Thirty years' view
Bents news – Sydney, apr 1839-jun 1839 – 1r – A$29.52 vesicular A$35.02 silver – at Pascoe [079]
Bentwich, Herbert *see* Future of our schools
Bentwich, Norman De Mattos *see*
- Jews in our time
- Necessity of ceremonial in judaism

Bentz, Paul A *see* Canal zone code, 1934
Die benuetzung der antike in wielands moralischen briefen : beitrag zur entwicklungs-geschichte der deutschen literatur im 18. jahrhundert / Doell, M – Eichstaett: Ph Broenner, 1903 – 1 – (incl bibl ref) – us Wisconsin U Libr [430]
Die benutting van ervaringsleer van jeugdiges in die begeleiding tot geestelike volwassenheid / Nieuwenhuis, Frederick Johannes – Uni of South Africa 2000 [mf ed Johannesburg 2000] – 3mf [ill] – 9 – (text in afrikaans; abstract in afrikaans and english; incl bibl ref) – mf#mfm14853 – sa Unisa [305]
Die benutzung der pflanzenwelt in der alttestamentlichen religion : eine studie / Lundgreen, Friedrich – Giessen: Alfred Toepelmann, 1908 – (= ser Beihefte zur zeitschrift fuer die alttestamentliche wissenschaft) – 1mf – 9 – 0-8370-4202-X – (incl ind of biblical citations) – mf#1985-2202 – us ATLA [221]
Benvenuto cellini had no prejudice against bronze / Graves, Anna Melissa – Baltimore, MD. 1943 – 1r – us UF Libraries [730]
Benvenutus van Venraai *see* Handleiding der patrologie
Benwick 1851-1950 – (= ser Cambridgeshire parish register transcript) – 4mf – 9 – £5.00 – uk CambsFHS [929]
Ben-Yami, M'Nakhem *see* Report on the fisheries in ethiopia
Ben-Yehuda, Eliezer *see* Erets yisrael
Ben-Yehuda, Hemda *see*
- Nose ha-degel
- Sipurim me'hayye ha-haluzim

Ben-Yehudah, Barukh *see*
- Kol hahinukh ha-tsiyoni
- Tenu'at-morim le-ma'an tsiyou u-ge'ulatah

Benyon, B G *see* B g beynon journal, 1813-1814
Benz, E *see*
- Marius victorinus und die entwicklung der abendlaendischen willensmetaphysik
- Zur ueberlieferung der matthaeuserklaerung des origenes

Benz, Karl *see*
- Die ethik des apostels paulus
- Die stellung jesu zum alttestamentlichen gesetz

Benz, Richard *see*
- Genius im wort
- Der gestiefelte kater / das rotkaeppchen
- Goethes goetz von berlichingen in zeichnungen von franz pforr
- Jean paul
- Romantik aus schriften, briefen, tagebuechern

Benz, Wolfgang *see* Die "judenfrage"
Benze, Charles Theodore *see* The confessional principle and the confessions of the lutheran church
Benzeev, Israel *see* Yehudim ba-'arav
Benzenberg, Johann F *see* Die verwaltung des staatskanzlers fuersten von hardenberg
Benzenberg, Johann Friedrich et al *see* Versuche die entfernung, die geschwindigkeit und die bahnen der sternschnuppen zu bestimmen
Benzinger, I *see* Hebraeische archaeologie
Benzinger, Immanuel *see*
- Bilderatlas zur bibelkunde
- Die buecher der koenige
- Geschichte israels bis auf die griechische zeit
- Wie wurden die juden das volk des gesetzes?

Ben-Zion, S *see*
- Bene bilu
- Kotel ha-maharavi be-divre yeme yisra'el uve-masorto ve-sifruto

Benzmann, Hans *see*
- Die deutsche ballade
- Meine heide

Benzoni, Girolamo *see* Das sechste theil der neuwen welt
Ben-Zvi, I *see* Massa'ot erez-yisra'el le-rabbi mosheh bosola
Beobachter – West Bend WI. 1888 jan 6/1889 may 10-1910 mar 11/aug 26 – 21r – 1 – (with gaps; cont by: west bend beobachter) – mf#1097646 – us WHS [071]
Der beobachter – Kassel DE, 1836-38 – 2r – 1 – gw Misc Inst [074]
Der beobachter – Zary, Poland. Dec 1934-Oct 1935 – 4r – 1 – us L of C Photodup [077]
Der beobachter *see*
- Arnstaedtische woechentliche anzeigen und nachrichten
- Der beobachter am eulenthal
- Duesseldorfer beobachter
- Der hochwaechter

Der beobachter am eulenthal – Waldenburg [Walbrzych PL], Schweidnitz [Swidnica PL], 1837-46 – 3r – 1 – (title varies: 1843: der beobachter) – gw Misc Inst [074]
Der beobachter an der bergisch-maerkischen eisenbahn *see* Hermann
Der beobachter an der elbe – Hamburg DE, 1865 1 apr 1-30 sep – 1 – gw Misc Inst [914]
Der beobachter an der enz und in der pfalz – Pforzheim DE, 1832 1 mar-1832 29 dec – 1r – 1 – gw Misc Inst [074]
Beobachter an der haar – Hamm (Westf) DE, 1957 22 nov-1959, 1965 mar [gaps] – 1 – (filmed by other misc inst: 1954-69) – gw Misc Inst [074]
Beobachter an der losse – Hessisch-Lichtenau DE, 1898 25 jun-1910 – 5r – 1 – (with suppl) – gw Misc Inst [074]
Der beobachter an der spree – Berlin DE, 1819-23, 1825-1827 25 jun, 1828-1830 28 jun, 1831-35, 1837-48, 1850, 1852-54 – 1 – (incl suppl: 1869 26 apr) – gw Misc Inst [074]
Beobachter im iser- und riesengebirge *see* Der bote aus dem riesengebuerge
Beobachter im saaletal – Halle S DE, 1931, 1932 [single iss], 1934-40 – 1 – gw Misc Inst [074]
Der beobachter vom donnersberg – Mainz DE, 1849-49 – 3r – 1 – (title varies: 30 dec 1801: mainzer zeitung, 1 jan 1806: neue mainzer zeitung, 20 dec 1807: neue mainzer zeitung, 5 oct 1809: gazette de mayence. mainzer zeitung, 1 feb 1812: journal du mont-tonnere. der donnersberger, 5 may 1814: mainzer zeitung, 12 nov 1822: anzeigeblatt der mainzer zeitung, 5 dec 1822: rhenus. neue mainzer zeitung, 8 dec 1822: neue mainzer zeitung, 1 jan 1835: mainzer zeitung, 19 nov 1850: neue mainzer zeitung, 20 nov 1850: mainzer abendpost; with suppls) – gw Misc Inst [074]
Die beobachterin an der spree und havel – Berlin DE, 1819 18 jan-28 jun – 1r – 1 – gw Misc Inst [074]

Beobachtungen, angestellt am astrophysikalischen observatorium in o gyalla in ungarn – Halle: Druck & Verlag von H W Schmidt 1879-94 [mf ed 2006] – 16v on 2r – 1 – (frequency varies 1872/1878-1888/1889; biennial 1890/1891-1892/1893; cont by: az o-gyallal astrophysikai es meteorologiai observatoriumon vegzet megfigyelesek; numbering begins with 2. bd (jahre 1890); v8 iss in 2pt; title varies slightly; vols for 1879-83 have title: beobachtungen, angestellt am astrophysikalischen observatorium in o gyalla; vols for 1884-1892/1893 have title: beobachtungen, angestellt am astrophysikalischen observatorium in o gyalla (ungarn)) – mf#film mas c6086 – us Harvard [520]
Beobachtungen auf reisen in und ausser deutschland : [fortsetzung und beschluss] / Niemeyer, August H – Halle [u a] 1820-26 [mf ed Hildesheim 1995-98] – 6v on 16mf – 9 – €160.00 – 3-487-27809-X – gw Olms [910]
Beobachtungen des grossen cometen von 1807 : sammt einem nachtrage zu dem aphroditographischen fragmenten / Schroeter, Johann Hieronymus – Goettingen: Vandenhoek-Ruprecht 1811 [mf ed 1998] – 1r [pl/ill] – 1 – mf#film mas 28300 – us Harvard [520]
Beobachtungen von cometen angestellt auf der sternwarte zu helsingfors im winter und fruehjahr 1885-1886 / Donner, Anders Severin – Helsingfors: Druckerei der finnischen Litteratur-Gesellschaft 1888 [mf ed 1998] – 1r [pl/ill] – 1 – mf#film mas 28409 – us Harvard [520]
Beobachtungs-ergebnisse der koeniglichen sternwarte zu berlin / ed by Struve, Hermann – Berlin: F Duemmler 1881-1914 [mf ed 1999] – 16v on 1r [ill] – 1 – (companion publ to: astronomische beobachtungen auf der koeniglichen sternwarte zu berlin; absorbed by: veroeffentlichungen der koeniglichen sternwarte zu berlin-babelsberg) – mf#film mas c4284 – us Harvard [520]
Beogradska nedelja – Belgrade, Yugoslavia. Sept 1961-Nov 1962 – 1r – 1 – us L of C Photodup [949]
Be-ohole torah bi-yeme ha-milhamah – Jerusalem, Israel. 1943 – 1r – us UF Libraries [939]
Beowulf / ed by Heyne, Moritz – 10.aufl. Paderborn: F Schoeningh, 1913 [mf ed 1990] – (= ser Bibliothek der aeltesten deutschen litteratur-denkmaeler v3; Angelsaechsische denkmaeler v1) – 328p – 1 – (text in anglo-saxon; prefatory material in german) – mf#7453 – us Wisconsin U Libr [420]
Bepswa, Kenneth S *see* Ndakamuda dakara afa
Ber, Of Bolechow *see* Zikhronot r' dov
Beraber – n1-9. 1952-53 [all publ] – (= ser O & t journals) – 1mf – 9 – $25.00 – us MEDOC [956]
Beradt, Martin *see* Die verfolgten
Beranger, Pierre Jean de *see*
– Musique des chansons de beranger
– Songs from beranger
Berar. India (State). Superintendent of Census Operations *see* Miscellany
Berard, Jean-Antoine *see* L'art du chant
Berard, Victor *see* De l'origine des cultes arcadiens
Berardi, Angelo *see*
– Arcani musicali suelati dalla vera amicitia
– Docvmenti armonici di d angelo berardi...
– Miscellanea mviscale di d angelo berardi...
– Il perche musicale
Berardi, Maria Helena Petrillo *see* Santo amaro
Berardini, Lorenzo *see* Frate angelo da chiarino... osimo 1964
Berardo Garcia, Jose *see* Explosion de mayo
Beratungslehrer : eine neue rolle im system / Grewe, Norbert – Neuwied, Frankfurt a.M.: Luchterhand 1990 (mf ed 1996) – (= ser Praxishilfen schule luchterhand) – 4mf – 9 – €45.00 – 3-8267-9694-2 – mf#DHS 9694 – gw Frankfurter [370]
Berault-Bercastel, Antoine Henri de *see*
– Histoire de l'eglise
– Histoire generale de l'eglise
Berben, Abdon *see* Aguas bicarbonatadas calcicas de alange
Le berbere a l'ecole nationale des langues orientales vivantes / Basset, Andre – Paris: Impr Nationale de France, 1948 – 1 – us CRL [470]
Il berbero nefusi di fassato : grammatica, testi raccolti dalla viva voce, vocabolarietti / Beguinot, Francesco – 2. rev miglio. ed. Roma: Instituto per l'Oriente, 1942 – us CRL [470]
Berbig, Johannes *see* Revolte in ochsenfurt
Le berceau de l'islam : l'arabie occidentale a la veille de l'hegire / Lammens, Henri – Romae: Sumptibus Pontificii Instituti Biblici, 1914 [mf ed 1991] – (= ser Scripta pontificii instituti biblici) – 1mf – 9 – 0-524-01907-X – (in french. incl bibl ref) – mf#1990-2720 – us ATLA [260]
Le berceau historique des mysteres de la franc-maconnerie : ou, tableau de l'histoire ancienne et moderne de l'ordre / Kiener, Francois-Joseph – 3e ed. Paris: Pommeret et Moreau 1860 – 1mf – 9 – (with portrait of aut) – mf#vrl-69 – ne IDC [366]
Berchem, M van *see* Amida
Berchmans-Boes, Johannes *see* An der pforte des todes
Berchtenbreiter, Maria *see*
– Die hexen von spoek
– Die stadt wundert sich ueber orlian
Berchtesgadener anzeiger – Berchtesgaden DE, 1 sep 1952-30 nov 1961; 1962-30 apr 1976 – 52r – 1 – (filmed by misc inst: may 1976- [ca 4r/yr]) – gw Mikrofilm; gw Misc Inst [074]
Berchtold, Joseph *see* Die unvereinbarkeit der neuen paepstlichen glaubensdekrete mit der bayerischen staatsverfassung
Bercy, Beauge *see* Peripeties d'une democratie
Berdiaev, N A *see*
– Khristianstvo i aktivnost' cheloveka
– O samoubiistvie
Berdiaev, Nikolai *see*
– Konstantin leon'tev
– O samoubiistvie
Berdiansk Sovet rk i kd *see* Izvestiia berdianskogo soveta rabochikh, soldatskikh i krest'ianskikh deputatov
Berdianskaia zhizn' : gaz obshchestv i radik -progres / ed by El'shtein, N S – Berdiansk [Tavr gub]: Berdian koop t-vo "Pechatnoe delo" 1918 [1917 16 iiulia-] – (= ser Asn 1-3) – n100-158 [1918] [gaps] item 1, on reel n1 – 1 – mf#asn-2 001 – ne IDC [077]
Berdichevskii, N G *see*
– Deistvuiushchee kooperativnoe zakonodatelstvo
– Deistvuiushchee zakonodatelstvo o potrebitelskoi kooperatsii
– Dekret o potrebitelskoi kooperatsii, 20 maia 1924 g
– Zakon o zhilishchnoi kooperatsii:
Berdichevsky, Micah Joseph *see* Peri sefer
Berdrow, Otto *see* Rahel varnhagen
Berdu, Gabriel *see* Tratado de el tercer orden del glorioso patriarcha santo domingo de guzman
Berea *see* Judge
Berea baptist church – Edgefield County, SC. 36p. 1954-76 – 1 – $5.00 – us Southern Baptist [242]
Berea baptist church – Berea, KY. feb 1896-sept 1985 – 1 – $96.21 – (lacking: 1902-07, 1909, aug 1911-aug 1912, oct 1912-mar 1913) – us Southern Baptist [242]
The berean : a manual for the help of those who seek the faith of the primitive church / Noyes, John Humphrey – Putney, Vt: Office of the Spiritual magazine, 1847 – 6mf – 9 – 0-524-03656-X – mf#1990-1084 – us ATLA [240]
Berean, a religious publication – La Salle IL 1824-28 – 1 – mf#4359 – us UMI ProQuest [240]
Berean, or, scripture-searcher – La Salle IL 1802-10 – 1 – mf#4421 – us UMI ProQuest [220]
The bereans : a discourse on the subject of our public schools / Irvine, R – Augusta, GA: Sainsimon & Morrison, 1878 – 1mf – 9 – 0-8370-7549-1 – mf#1986-1549 – us ATLA [220]
Berechiah Ben Natronai *see* Mishle shu'alim
Berechnung und dimensionierung permanenterregter gleichstromlinearantriebe der feinwerktechnik / Roemer, Olaf – (mf ed 1994) – 2mf – 9 – €40.00 – 3-89349-883-4 – mf#DHS 883 – gw Frankfurter [620]
Die berechtigung der theologie als eines nothwendigen gliedes im gesamtorganismus der wissenschaft : vortrag auf der conferenz zu meissen am 10. juni 1874 / Baur, Gustav – Gotha: Friedrich Andreas Perthes 1875 [mf ed 1985] – 1mf – 9 – 0-8370-2215-0 – mf#1985-0215 – us ATLA [210]
Berechtigung und praktische anwendung des verbots der werbung mit sonderangeboten bei mengenmaessig beschraenkter abgabe : eine kritische betrachtung des paragraphen 6 d abs. 1 nr. 2 uwg / Conrad, Christine – (mf ed 1993) – 3mf – 9 – €49.00 – 3-89349-820-6 – mf#DHS 820 – gw Frankfurter [346]
Berechtigung und zuversichtlichkeit des bittgebets : vortrag auf der saechsischen pastoralkonferenz zu halle a.d. s / Kaehler, Martin – Halle: J Fricke 1888 [mf ed 1990] – 1mf – 9 – 0-7905-7409-8 – mf#1989-0634 – us ATLA [240]
Die beredsamkeit eine tugend *see* Eloquence a virtue
Die beredsamkeit j enoch powells / Lang, Hartmut – Frankfurt a.M., 1972 – 3mf – 9 – 3-89349-749-8 – gw Frankfurter [943]
Beremant, Gordon et al *see* The cases of the u.s. court of appeals for the d.c. circuit
Berend, Alice *see* Der glueckspilz
Berend, Julia Z *see* Influence of diet and the menstrual cycle on lactate concentration during increasing exercise intensities
Berendsohn, Walter Arthur *see*
– Grundformen volkstuemlicher erzaehlerkunst in den kinder- und hausmaerchen der brueder grimm
– Der impressionismus hofmannsthals als zeiterscheinung
– Noch ein stueck knabendichtung goethes
– Zur methode der reimuntersuchung im streit um goethes "joseph"
Berendts Alexander *see* Vom juedischen kriege
Berendts, Alexander *see*
– Die handschriftliche ueberlieferung der zacharias- und johannes-apokryphen
– Studien ueber zacharias-apokryphen und zacharias-legenden
– Ueber die bibliotheken der meteorischen und ossa-olympischen kloester
– Das verhaeltnis der roemischen kirche zu den kleinasiatischen vor dem nicaenischen konzil
– Die zeugnisse vom christentum im slavischen "de bello judaico" des josephus
Berengarius turonensis : oder, eine sammlung ihn betreffender briefe / ed by Sudendorf, Hans – Hamburg: F & A Perthes 1850 [mf ed 1990] – 1mf – 9 – 0-7905-6800-4 – (incl bibl ref) – mf#1988-2800 – us ATLA [090]
Berenger, Laurent P *see* Les soirees provencales
Berenger-Feraud, Laurent-Jean-Baptiste *see* Superstitions et survivances
Berenguer Carisomo, Arturo *see* Medio siglo de literatura americana
Berens, August *see*
– Fruehlingsboten
– Gnade und wahrheit
Berens, Edward *see*
– The history of the prayer book of the church of england
– Pastoral advice to married persons
– Pastoral advice to servants
– Pastoral watchfulness and zeal
Berens, Josefa *see*
– Der femhof
– Frau Magdlene
– Einer sippe gesicht
Berens, Lewis Henry *see* The digger movement in the days of the commonwealth
Berens, S L *see* Nansen in the frozen world
Berenson, Bernard *see*
– Florentine painters of the renaissance
– Italian painters of the renaissance
Berenson, Bernhard *see* Venetian painting
Berenson, Senda *see* Line basketball, or basketball for women
Beresford Hope, Alexander James Beresford *see*
– The social influence of the prayer book
– Worship in the church of england
Beresford, John *see* The correspondence of the right hon john beresford
Beresford, John Davys *see* All or nothing
Beresford, William *see*
– The correspondence of the right hon john beresford
– Der kapitaine portlock's und dixon's reise um die welt
– Voyage autour du monde, et principalement a la cote nord-ouest de l'amerique
– A voyage round the world, but more particularly to the north-west coast of america
Beresford-Hope, Alexander James Beresford *see*
– The art-workman's position
– The condition and prospectus of architectural art
– The english cathedral of the nineteenth century
– Public offices
El beresh 1 : the tomb of tehuti-hetep / Newberry, Percy E – 9 – $10.00 – us IRC [930]
El beresh, pt 2 / Griffith, Fl & Newberry, Percy E – 9 – $10.00 – us IRC [930]
Berestov, P V *see* Kustanaiskii listok
Beretning om hauges synodes...kinamissions-aarsmode / Hauge's Norwegian Evangelical Lutheran Synod – Red Wing MN: Hauges synodes trykkeri. 3rd-7th (1896-1900) [mf ed 2005] – 1r – 1 – mf#2005c-s025 – us ATLA [242]
Beretty, D W *see* Van 13 momenten uit een 13-jarig bestaan
Berezin, M M *see* Planirovanie khoziaistvennoi deiatelnosti soiuzov i nizovoi seti promyslovoi kooperatsii
Berezin-Shiriaev, I *see*
– Dopolnitelnye materialy dlia bibliografii, ili opisanie russkikh i inostrannykh knig, graviur i portretov, nakhodiashchikhsia v biblioteke liubitelia n n
– Materialy dlia bibliografii ili obozrenie russkikh i inostrannykh knig nakhodiashchikhsia v biblioteke liubitelia istoricheskikh nauk i slovestnosti n n
– Obzor knig, broshiur, khudozhestvennykh izdanii, graviur i portretov, russkikh i nekotorykh inostrannykh, nakhodiashchikhsia v moei biblioteke, ili okonchatelnye materialy dlia bibliografii...
– Opisanie nakhodiashchikhsia u s-peterburgskogo pervoi gildii kuptsa i otomstvennogo pochetnogo grazhdanina fedora egorovicha sokurova knig, broshiur, ukazov i estampov...
– Opisanie russkikh i inostrannykh knig, nakhodiashchiksia v biblioteke liubitelia istoricheskikh nauk,
– Poslednie materialy dlia bibliografii, ili opisanie knig, broshiur, khudozhestvennykh izdanii...
Berg, Christina M *see* Myopia education 101
Berg, Dagmar *see* Das phaenomen culture shock
Berg, Elliot *see* Recruitment of a labor force in sub-saharan africa
Berg, Elliot J *see* Trade unions in french west africa
Berg, Emil P *see* The conversion of india
Berg frei! : organ des touristenvereins "die naturfreunde" – Aussig (Usti nad Labem CZ), 1930 n3-1936 n4 – 1r – 1 – gw Misc Inst [790]
Berg, Guenther Heinrich von *see* Teutsches staats-magazin
Berg, J van den *see* Constrained by jesus' love
Berg, Johannes van den *see* Constrained by jesus' love
Berg, Joseph Frederic *see* The influence of the septuagint upon the pesittaa psalter
Berg, Joseph Frederick *see*
– The bible vindicated against the aspersions of joseph barker
– Farewell words to the first german reformed church, race street, philadelphia
– Lectures on romanism
– The second advent of jesus christ not premillennial
– The stone and the image
Berg, Leo *see*
– Aus der zeit, gegen die zeit
– Heine, nietzsche, ibsen
– The superman in modern literature
– Der uebermensch in der modernen litteratur
– Zwischen zwei jahrhunderten
Berg, Richard K *see* Government in the sunshine act
Berg, Robert et al *see* Vortraege fuer das gebildete publikum
Berg, Rose Monica [comp] *see* Bibliography of management literature
Berg, Theresa A *see* Addressing, analyzing, and challenging social issues and problems in the coaching profession
Berg- und gletscher-fahrten in den hochalpen der schweiz : zweite sammlung / Studer, Gottlieb L – Zuerich 1863 [mf ed Hildesheim 1995-98] – (= ser Fbc) – 3mf – 9 – €90.00 – 3-487-29353-6 – gw Olms [914]
Berg und gletscher-reisen in den oesterreichischen hochalpen / Ruthner, Anton von – Wien 1864 [mf ed Hildesheim: 1995-98] – (= ser Fbc) – xvii/414p on 3mf – 9 – €90.00 – 3-487-29428-1 – gw Olms [914]
Berg- und huettenmaennische rundschau : organ fuer die interessen des bergbaues, huettenbetriebes und verwandter industrieen – Katowice, Poland 5 oct 1904-20 mar 1921 – 5r – 1 – uk British Libr Newspaper [622]
Berg- und huettenmaennische zeitung – Leipzig, Germany 3 jan 1842-25 dec 1844, 7 jan 1846-29 dec 1847, 3 jan-26 dec 1849, 1 jan 1851-30 dec 1904 – 1 – (in 1842 publ at nordhausen; fr 1843-63 at freiberg) – uk British Libr Newspaper [622]
Bergaigne, Abel *see* La religion vedique d'apres les hymnes du rig-veda
Bergamin, Jose *see* Maranon's betrayal
Bergano, Diego *see* Arte de la lengua pampanga
Bergano Y Villegas, Simon *see* Poemas
Berganza, F *see* Antiguedades de espana propugnadas en las noticias de sus reyes y condes...
Bergarbeiter : schauspiel in einem akt / Maerten, Lu – Stuttgart: J H W Dietz 1909 [mf ed 1990] – 1r – 1 – (filmed with: nikolaus lenau / eduard castle) – mf#2821p – us Wisconsin U Libr [820]
Der bergarbeiterausstand und die techn. grubenbeamten gelsenkirchen 1905 / Bertenburg, Carl – 1 – gw Mikropress [330]
Bergarbeiter-mitteilungen / ed by Arbeitsausschuss Freigewerkschaftlicher Bergarbeiter – London (GB), 1936 jul-dec, 1937-1938 nov [gaps], 1939 jan-mar – 1r – 1 – gw Misc Inst [331]
Bergarbeiter-streik im ruhrgebiet im fruehjahr 1912 – Coeln o. J – 1 – gw Mikropress [331]
Der bergarbeiterstreik und die untersuchungskommissionen – Bochum 1905 – 1 – gw Mikropress [331]
Der bergarbeiterstreik vom mai 1889 im rheinisch-westfalischen industriegebiet unter besonderer beruecksichtigung der stellung kaiser wilhelm 2. und furst / Hahn, Wilhelm – 1 – gw Mikropress [943]
Die bergarbeiter-verhaeltnisse in grossbritannien / Hasse, R & Kruemmer, G – Saarbruecken, 1891 – 1 – gw Mikropress [331]
Bergarbeiter-zeitung – London (GB), 1936 jul-39 [gaps] – 1r – 1 – gw Misc Inst [331]
Bergarbeiter-zeitung *see*
– Deutsche berg- und huetten-arbeiter-zeitung
– Glueckauf!
Die bergbau- industrie *see* Deutsche berg- und huetten-arbeiter-zeitung
Bergbau- und huettenkombinat, bt huette *see* Unser eisen
Die bergbau-industrie *see* Glueckauf!
Bergcultures – Djakarta, 1926-1942 v1-26 – 667mf – 9 – (cont as: menara perkebunan djakarta, 1958-1970 v27-39. missing: 1965, v34(9); 1969, v38(1-2, 11-12); 1970, v39(3-end)) – mf#SE-833 – ne IDC [959]

Berge in flammen : ein roman aus den schicksalstagen suedtirols / Trenker, Luis – Berlin: Neufeld & Henius c1931 [mf ed 1991] – 1r – 1 – (filmed with: paul de lagarde / edward schroder) – mf#2916p – us Wisconsin U Libr [830]
Berge und menschen : roman / Federer, Heinrich – Berlin: G Grote 1919, c1911 [mf ed 1996] – (= ser Grote'sche sammlung von werken zeitgenoessischer schriftsteller 103) – 1r – 1 – (filmed with: vae victis / nataly von eschstruth) – mf#3927p – us Wisconsin U Libr [830]
Bergedorfer nachrichten – Hamburg DE, 1881-86 [single iss] – 1r – 1 – gw Misc Inst [074]
Bergedorfer zeitung – Hamburg DE, 1976- – ca 6r/yr – 1 – gw Misc Inst [074]
Bergel, Joseph *see*
– Der himmel und seine wunder
– Studien ueber die naturwissenschaftlichen kenntnisse der talmudisten
Bergel, Sigmund *see* Moses montefiore und der orden
Bergemann, G Ph *see* Stettinischer schauplatz der vernunft und des geschmaks
Bergen, John Tallmadge *see* Evidences of christianity
Bergengruen, Werner *see*
– Das beichtsiegel
– Des knaben plunderhorn
– E T A hoffmann
– Das feuerzeichen
– Das kaiserrech in truemmern
Bergenthal, Ferdinand *see* Das werk georges
Berger, Alfred, Freiherr von *see* Gesammelte schriften
Berger, Andreas *see* Harmoniae seu cantiones sacrae...
Berger, Arnold Erich *see*
– Der junge herder und winckelmann
– Klopstocks sendung
– Die kulturaufgaben der reformation
Berger, Berta *see* Der moderne deutsche bildungsroman
Berger, Christoph Heinrich von *see* Commentatio de personis vvlgo larvis sev mascheris, von der carnavals-lvst
Berger, Daniel *see* History of the church of the united brethren in christ
Berger, E W *see*
– Whitefly conditions in 1906
– Whitefly control
– Whitefly studies in 1908
Berger, Ernst Hugo *see* Mythische kosmographie der griechen
Berger, Heinrich *see*
– R benjamin b jehuda und sein commentar zu esra und nehemia
Berger, Johann *see*
– Leben und wirken des hochseligen johannes nep. neumann
– Life of right rev. john n. neumann, d.d.
Berger, Karl *see*
– Schiller
– Theodor koerner
Berger, Karl Heinz *see*
– Die affenschande
– Deutsche balladen
– Nettesheim
Berger, Karl-Heinz *see* Westdeutsche prosa
Berger, Kurt *see*
– Die balladen schillers im zusammenhang seiner lyrischen dichtung
– Menschenbild und heldenmythos
– Rainer maria rilkes fruehe lyrik
Berger, Ludwig *see* Griseldis
Berger, Philippe *see*
– L'ange d'astarte
– M ernest renan et la chaire d'hebreu au college de france
Berger, Richard A *see*
– An evaluation of a home-based exercise program involving non-exertional hypoxemic and exertional hypoxemic chronic obstructive pulmonary diseased patients
– Physiological response of trained cyclists to various cycling handlebar postures
Berger, Rutherford C *see* Design criteria for lateral dikes in estuaries
Berger, Samuel *see*
– La bible francaise au moyen age
– De l'histoire de la vulgate en france
– Histoire de la vulgate pendant les premiers siecles du moyen age
Berger, Siegfried *see*
– Die goettin laechelt
– Regine und die ahnherren
Berger, Uwe *see*
– Anders ist der neue tag
– Deutsches gedichtbuch
Bergerak? / Tan, Boen Soan – Soerabaia: Tan's Drukkerij, 1935 [mf ed 1998] – (= ser Penghidoepan 124) – 1r – 1 – (coll as pt of the colloquial malay collection. filmed with: multi-millionair / ong khing han) – mf#10002 – us Wisconsin U Libr [830]
Bergeron, Helene *see* Bio-bibliographie analytique de rene ouvrard
Bergeron, Juliana *see* Chanoine jean bergeron, 1868-1956
Bergeron, P *see*
– Relation des voyages en tartarie de fr guillaume de rubriquis, fr iean du pian carpin...
– Traicte des tartares
– Traite des tartares
– Voyages faits principalement en asie dans les 12, 13, 14, et 15 siecles...
Bergerson, Mark *see* A comparison of the effects of a wrestling practice and a weightlifting workout on the body fat percent of wrestlers
Berges, W *see* Die fuerstenspiegel des hohen und spaeten mittelalters (mgh schriften:2.bd)
Der berggeist – Koeln DE, 1856 jul-1860 28 dec, 1866 2 jan-1885 29 dec – 14r – 1 – uk British Libr Newspaper [074]
Berggeschichten / Achleitner, Arthur – Stuttgart: Adolf Bonz, 1905 [mf ed 1995] – 309p – 1 – mf#8917 – us Wisconsin U Libr [830]
Berggren, Jakob *see*
– Bibel und josephus ueber jerusalem und das heilige grab
– Reisen in europa und im morgenlande
Berggren, S *see* Musci et hepaticae spetsbergenses
Bergh, Alfred von *see* Letzte reisebriefe von alfred von bergh ueber portugal und spanien
Bergh, Christine E *see* The relationship between sex-role classification and activity, and trait and state anxiety
Bergh, Johan Arndt *see*
– History of the norwegian lutheran church in america
– Den norsk lutherske kirkes historie i amerika
– Stenografisk referat af forhandlingerne ved frikonferensen
Bergh, Johan Arndt et al *see* Fra ungdomsaar
Bergh, L Ph van den *see* Handboek der middelnederlandse geographie
Bergh, Suzette *see* Die invloed van bekering op pastorale psigoterapie en voorligting
Bergh van Eysinga, Gustaaf Adolf van den *see*
– Indische einfluesse auf evangelische erzaehlungen
– Onderzoek naar de echtheid van clemens' eersten brief aan de corinthiers
– Radical views about the new testament
Berghaeuser, Wilhelm *see* Die darstellung des wahnsinns im englischen drama bis zum ende des 18. jahrhunderts
Berghaus, Heinrich K *see* Allgemeine laender- und voelkerkunde
Bergholtz, Gustav Fredric *see* The lord's prayer in the principal languages, dialects and versions of the world
Bergholz, Harry *see* Ueber den tag hinaus
Bergisch gladbacher zeitung – Bergisch Gladbach DE, 1891-92; 1893 11 jan-1902; 1904-1905 29 dec; 1906 8 jan-1908 31 mar – 9r – 1 – (with gaps) – gw Misc Inst [074]
Bergische arbeiterstimme – Solingen, Duesseldorf DE, 1901-1920 jun, 1921-1933 20 feb – 67r – 1 – (with suppl: die rote streikfront 1931 jan) – mf#3980 – gw Mikropress [331]
Bergische heimat – Wermelskirchen DE, 1926 n1-15 – 1r – 1 – (suppl to: wermelskircher zeitung) – gw Misc Inst [074]
Bergische heimatblaetter – Solingen DE, 1928-34 – 1r – 1 – (suppl to: bergische zeitung 1868) – gw Misc Inst [074]
Bergische landeszeitung *see* Koelnische rundschau
Bergische landeszeitung 1949 – Bergisch Gladbach DE, 1949 28 sep-1952 20 aug – 8r – 1 – gw Mikrofilm [074]
Bergische morgenpost *see* Rheinische post / interzonenausgabe
Bergische morgenpost, bmiii, bm-le – Remscheid-Lennep DE, 1950-1951 30 jun, 1951 1 oct-1953 27 may – 9mf – 9 – (bezirksausgabe von rheinische post, duesseldorf) – gw Mikrofilm [074]
Bergische rundschau *see* Koelnische rundschau
Bergische tageszeitung – Wuppertal 1921 24 dec [jub-ausg], 1953 1 okt-1954 31 mar [mf ed 2004] – 1 – (vbg: wuppertal & bergisches land; vlg in gelsenkirchen-buer) – gw Mikrofilm [074]
Bergische volksstimme – Wuppertal DE, 1877-1878 1 nov – 4r – 1 – gw Misc Inst [074]
Bergische wochenpost – Wuppertal DE, 1953 4 jul-1954 25 sep – 1 – (district ed of wuppertaler stadt-anzeiger) – gw Misc Inst [074]
Bergische zeitung – Velbert DE, 1950-1975 31 jan [gaps] – 114r – 1 – (filmed by misc inst: 1998- [ca 11r/yr], 1958-60; title varies: 21 jan 1888: velberter zeitung, 15 oct 1949: velberter zeitung, niederbergische heimat; incl suppl: niederbergische heimat 1930 9 nov-1931 18 oct [1r missing n6-8]. publ in velbert between 7 jan 1882-nov 1974, afterwhich sold to: westdeutsche allgemeine, essen) – gw Mikrofilm; gw Misc Inst [074]
Bergischer volksbote *see* Rhein-wupper-zeitung
Bergisches archiv – Wuppertal DE, 1809-1811 jun – 2r – 1 – (title varies: 1810: grossherzoglich bergisches archiv) – gw Misc Inst [025]
Bergisches volksblatt – Solingen DE, 1849 29 jun-28 dec – 1r – 1 – (title varies: 1 jul 1868: solinger zeitung) – gw Misc Inst [074]
Bergisch-gladbacher volkszeitung (heiderscheid zeitung) *see* Volksblatt fuer bergisch gladbach und umgebung
Bergisch-maerkische zeitung – Hagen, Westf DE, 1934 28 jun-30 sep, 1936 jan-jun, 1936 oct-1937 mar, 1937 jul-sep, 1938 jan-30 apr – 1 – (main ed in wuppertal) – gw Misc Inst [074]
Bergisch-maerkische zeitung *see* Provinzial-zeitung
Bergk, Johann A *see* Reise in holland im jahre 1806 [achtzehnhundertsechs]
Der bergknappe : zeitschrift fuer christliche bergleute – Essen DE, 1895 23 nov [trial no], 1896-1932 – 9r – 1 – mf#3248 – gw Mikropress [074]
Bergkreyen, auff zwo stimmen componirt – 1551 – (= ser Mssa) – 1mf – 9 – €20.00 – mfchl 397 – gw Fischer [780]
Bergkristalle : novellen und erzaehlungen aus der schweiz – Bern: B F Haller 1876-79 [mf ed 1979] – 20v in 7 on 1r – 1 – (v1-15: novellen und erzaehlungen von arthur bitter; v16-20: novellen und erzaehlungen von j j romang) – mf#film mas 8366 – us Harvard [830]
Berglar-Schroeer, Paul *see*
– Eid bleibt eid
– Der feuerspeiende berg
Bergliederbuechlein : historisch-kritische ausgabe / ed by Mincoff-Marriage, Elizabeth – Leipzig: K W Hiersemann, 1936 [mf ed 1993] – (= ser Blvs 285) – xviii/313p – 1 – (incl bibl ref) – mf#8470 reel 56 – us Wisconsin U Libr [780]
Bergmaennisches journal – Freiburg, Germany 1788-93 – 1 – mf#2273 – us UMI ProQuest [622]
Bergman, Richard *see* Collected field reports on the phonology of tampulma
Bergmann, Anton *see* Die bedeutung des niblungenlieder fuer die deutsche nation
Bergmann, E von *see* Hieratische und hieratisch-demotische texte der sammlung aegyptischer alterthuemer des allerhoechsten kaiserhauses
Bergmann, Gustav *see* Umweltgerechtes produkt-design
Bergmann, Herta *see* Ich moechte nach hause
Bergmann, Joseph *see* Das ambraser liederbuch vom jahre 1582
Bergmann, Julius *see* Grundlinien einer theorie des bewusstseins
Bergmann, Wolfgang *see* Goethes morphologie
Bergna, Costanzo *see* Tripoli dal 1510 al 1850
Bergner, H *see* Handbuch der buerglichen kunstaltertuemer
Bergnubische sprache : (dialekt von gebel delen) / Kauczor, Daniel – Wien, Austria. 1920 – 1r – us UF Libraries [470]
Bergobzoomer, Johann Baptist *see* In der noth lernet man die freunde kennen
Bergpostilla : oder sarepta darinn von allerley bergkwerck vnd metallen, guter bericht gegeben wird mit troestlicher vnd lehrhaffter erklerung aller sprüech, so in heiliger schrifft von metall reden / Mathesius, J – Nuernberg, 1578 – 6mf – 9 – mf#TH-1 mf 1143-1148 – ne IDC [242]
Die bergpredigt : ihr aufbau, ihr urspruenglicher sinn und ihre echtheit, ihre stellung in der religionsgeschichte und ihre bedeutung fuer die gegenwart / Weinel, Heinrich – Leipzig: B G Teubner, 1920. Chicago: Dep of Photodup, U of Chicago Lib, 1970 (1r); Evanston: American Theol Lib Assoc, 1984 (1r) – (= ser Aus natur und geisteswelt) – 1 – 0-8370-0543-4 – mf#1984-B229 – us ATLA [220]
Die bergpredigt : verdeutscht und vergegenwaertigt / Mueller, Johannes – 3. aufl. Muenchen: Beck 1911 [mf ed 1992] – 1mf – 9 – 0-524-03984-4 – mf#1992-0027 – us ATLA [220]
Die bergpredigt *see* Commentary on the sermon of the mount
Die bergpredigt (matth. 5-7, luk. 6, 20-49) / Heinrici, Carl Friedrich Georg – Leipzig: Alexander Edelmann 1905 [mf ed 1985] – 1mf – 9 – 0-8370-3549-X – (incl bibl ref) – mf#1985-1549 – us ATLA [225]
Die bergpredigt nach matthaeus und lucas : exegetisch und kritisch untersucht / Achelis, Ernst Christian – Bielefeld: Velhagen & Klasing 1875 [mf ed 1989] – 2mf – 9 – 0-7905-1620-9 – (incl bibl ref) – mf#1987-1620 – us ATLA [225]
Bergpredigten : gehalten auf der hoehe der zeit unter freiem himmel und zu schimpf und spott unseren feinden den schwaechen, lastern und irrthuemern der cultur gewidmet / Rosegger, Peter – Wien: A Hartleben 1885 [mf ed 1995] – (= ser Bibliothek der deutschen literatur) – 1r – 1 – (filmed with: als ich jung noch war) – mf#3721p – us Wisconsin U Libr [840]
Bergpsalmen : dichtung / Scheffel, Joseph Viktor von – 6. aufl. Stuttgart. A Bonz 1895 [mf ed 1991] – 1r [ill] – 1 – (filmed with: der bluhende stab / ruth schaumann) – mf#2868p – us Wisconsin U Libr [810]
De bergrede / Oort, Henricus – Assen: L Hansma, 1905 [mf ed 1985] – 100p on 1mf – 9 – 0-8370-4619-X – mf#1985-2619 – us ATLA [220]
De bergrede en andere synoptische fragmenten : een historisch-kritisch onderzoek met een inleiding over enkele leemten in de methode van de kritiek der evangelien / Pierson, A – Amsterdam: P N van Kampen & Zoon, 1878 [mf ed 1989] – 260p on 1mf – 9 – 0-7905-3100-3 – mf#1987-3100 – us ATLA [225]
Bergreihen : ein liederbuch des 16. jahrhunderts / ed by Meier, John – Halle: Max Niemeyer, 1892 – (= ser Neudrucke deutscher literaturwerke des 16. und 17. jahrhunderts n99-100) – xvi/122p – 1 – (incl bibl ref and ind) – mf#8413 reel 5 – us Wisconsin U Libr [780]
Bergreisen / Fischer, Christian A – Leipzig 1804/05 [mf ed Hildesheim 1995-98] – (= ser Fbc) – 2v on 4mf – 9 – €120.00 – 3-487-27753-0 – gw Olms [910]
Bergslagskuriren *see* Orebrokuriren
Bergslagsposten – Lindesberg, Sweden. 1882-85 – 2r – 1 – sw Kungliga [078]
Bergslagsposten – Lindesberg, Sweden. 1979- – 1 – sw Kungliga [078]
Bergson : an exposition and criticism from the point of view of st thomas aquinas / Gerrard, Thomas John – London: Sands; St Louis MO: B Herder [1913?] [mf ed 1990] – 1mf – 9 – 0-7905-3873-3 – mf#1989-0366 – us ATLA [140]
Bergson and religion / Miller, Lucius Hopkins – New York: H Holt 1916 [mf ed 1991] – 1mf – 9 – 0-7905-9519-2 – (incl bibl ref) – mf#1989-1224 – us ATLA [200]
Bergson and the modern spirit : an essay in constructive thought / Dodson, George Rowland – Boston: American Unitarian Assoc 1913 [mf ed 1990] – 1mf – 9 – 0-7905-7337-7 – mf#1989-0562 – us ATLA [140]
Bergson, H *see* L'intuition philosophique
Bergson, Henri *see*
– Dreams
– L'evolution creatrice
– The introduction to a new philosophy
– An introduction to metaphysics
– Laughter
– Matiere et memoire
– Matter and memory
– The meaning of the war
– La perception du changement
– Time and free will
Der bergsteiger *see* Mitteilungen des saechsischen bergsteigerbundes
Bergstraesser anzeigeblatt *see* Bergstraesser anzeiger
Bergstraesser anzeiger – Bensheim DE, 1976- – 10r/yr – 1 – (title varies: 1969: bergstraesser anzeigeblatt) – gw Misc Inst [074]
Bergstraesser, G *see* Nichtkanonische koranlesarten im muhtasab des ibn ginni
Bergstresser, Peter *see*
– Vain excuses answered
– The waynesboro' discussion on baptism, the lord's supper, and feet-washing
Berguer, Henry *see* Calvin aujourd'hui
Berguer, Henry et al *see* Jubile de calvin a geneve
Berguizas, Francisco P *see* Obras poeticaside pindaro..
Die bergung : eine erzaehlung / Leip, Hans – Stuttgart: J G Cotta, 1944, c1939 – 1r – 1 – us Wisconsin U Libr [830]
Der bergwerksfreund – Eisleben DE, 1839 20 jan-1847 22 sep – 5r – 1 – uk British Libr Newspaper [622]
Bergwerks-und industrie-anzeiger – Berlin, Germany 4 jan 1859-9 nov 1865 – 1 – uk British Libr Newspaper [338]
Der bergwirth : geschichte aus den bayrischen bergen / Schmid, Herman – 2.aufl. Leipzig: Keil [18–?] – (= ser Gesammelte schriften. volks- und familien-ausgabe 30) – 1 – (bound with: die zuwider-wurzen) – mf#film mas c438 – us Harvard [830]
Beri, S G *see* Indian economics
Beria, Giovanni Battista *see*
– Concerti musicali a dve, tre e qvattro voci
– Motteti a due, tre, e quattro voci co'l te deum...opera seconda
– Il primo libro delli motetti concertati a vna, dve, tre, qvattro, e cinqve voci
Beriault, Raymond *see* Khmers
Bericht / Deutscher Landarbeiter-Verband – Berlin [mf ed 1985] – 1r – 1 – (sometimes incl niederschrift der generalversammlung) – mf#6613 – us Wisconsin U Libr [331]
Bericht / Historischer Verein Bamberg – Bamberg: C C Buchners Verlag 1941 (annual) [mf ed 198-] – 1v on 6r – 1 – (cont: bericht des historischen vereins fuer die pflege der geschichte des ehemaligen fuerstbistums bamberg zu bamberg; cont by : bericht [historischer verein fuer die pflege der geschichte des ehemaligen fuerstbistums zu bamberg]) – mf#film mas c362 – us Harvard [943]

Bericht der krancken / Bullinger, Heinrich – [Zuerich, Christoffel Froschouer], 1535 – 2mf – 9 – mf#PBU-126 – ne IDC [240]

Bericht der naturforschenden gesellschaft zu bamberg – Bamberg, Germany: [J M Reindl] – 1r – 1 – (cont: ueber das bestehen und wirken der naturforschenden gesellschaft) – mf#1372 – us Wisconsin U Libr [500]

Bericht der thora-lehranstalt (jeschiwa) – Frankfurt am Main, Germany. 19–? – 1r – us UF Libraries [939]

Bericht des central-ausschuss fuer die innere mission der deutschen evangelischen kirche (fw4) – Berlin-Dahlem 1849/52(1853)-1927/30(1931) [mf ed 2004] – (= ser Freie wohlfahrtspflege (fw) 4) – 65v on 62mf – 9 – €390.00 – 3-89131-458-2 – (v1 1849/52 (1853) with title: bericht ueber die wirksamkeit des central-ausschusses fuer die innere mission der deutschen evangelischen kirche, v61 1919: bericht ueber die wirksamkeit des central-ausschusses fuer die innere mission der deutschen evangelischen kirche) – gw Fischer [242]

Bericht des hauptburos an den 17 / Jewish National Fund – Jerusalem, Israel. 1931 – 1r – us UF Libraries [939]

Bericht des obergerichts von schaffhausen an den grossen rath des kantons schaffhausen ueber die geschaftsfuhrung sammtlicher gerichtsstellen und den zustand des gerichtswesens in dem amtsjahre vom 1. juni 1871 bis und mit dem 31. mai 1872 / Schaffhausen. (Canton). Obergericht – Schaffhausen: Gelzer, 1873. 24p. LL-4021 – 1 – us L of C Photodup [340]

Bericht des obergerichts von schaffhausen an den grossen rath des kantons schaffhausen ueber die geschaftsfuhrung sammtlicher gerichtsstellen und den zustand des gerichtswesens vom 1. juni 1873 bis und mit dem 31. mai 1874 / Schaffhausen. (Canton). Obergericht – Schaffhausen: Gelzer, 1875. 27p. LL-4020 – 1 – us L of C Photodup [340]

Bericht des reichskohlenrates ueber die kohlenwirtschaft – Berlin DE, 1921-38 – 2r – 1 – mf#6357 – gw Mikropress [622]

Bericht eines forschers im tropischen suedafrika : aus dem englischen / Galton, Francis – Leipzig 1854 [mf ed Hildesheim 1995-98] – 1v on 2mf – 9 – €60.00 – 3-487-27192-3 – gw Olms [916]

Bericht etlicher fuernemesten stuecke, den juengstentage, vnd was darauff folgen wirdt, betreffend / Waldner W – Regenspurg, [1564] – 2mf – 9 – mf#TH-1 mf 1465-1466 – ne IDC [242]

Bericht fuer der periode...dem comite der nicolai-hauptsternwarte ueber deren thaetigkeit abgestattet von director der sternwarte / Nikolaevskaia glavnaia astronomicheskaia observatoriia – St Petersburg: Buchdr der Kaiserlichen Akademie der Wissenschaften 1890- [mf ed 2006] – 1r – 1 – (trans fr russian; cont: nikolaevskaia glavnaia astronomicheskaia observatoriia. jahresbericht am...dem comite der nicolai-hauptsternwarte abgestattet vom director der sternwarte; cont by: nikolaevskaia glavnaia astronomicheskaia observatoriia jahresbericht ueber die thatigkeit der kaiserlichen nicolai-hauptsternwarte) – mf#film mas 37493 – us Harvard [520]

Bericht ob man on die tauffe vnd empfahungen des leibs vnd bluts christi allein durch den glauben kuenne selig werden / Corvinus, A – [Magdeburgk], 1538 – 1mf – 9 – mf#TH-1 mf 351 – ne IDC [242]

Bericht over de padi-gewassen, 1694 / Sura, W – 1mf – 8 – mf#SD-102 mf 9 – ne IDC [630]

Bericht uber eine reise ins gebiet / Martin, Karl – s.l, s.l? . 1885? – 1r – us UF Libraries [910]

Bericht uber eine reise nach niederlandisch west-indien / Martin, Karl – Leiden, Netherlands. v1-2. 1888 – 2r – us UF Libraries [918]

Bericht uber entstechung des vereins und kassenlegung bis... / Judisches Altersheim Und Siechenhaus In Oxtpreussen – Allenstein, Poland. 1907 – 1r – us UF Libraries [943]

Bericht ueber anlage des herbariums waehrend der reisen nebst erlaeuterung der topographischen angaben / Schlagintweit-Sak_nl_nski, H A R von – Muenchen: Verlag der k Akademie, 1876 – 2mf – 9 – mf#BT-315 – ne IDC [910]

Bericht ueber das...geschaeftsjahr...des verbandes der landwirtschaftlichen genossenschaften im koenigreiche sachsen / Verband der landwirtschaftlichen Genossenschaften im Koenigreiche Sachsen – Dresden: [s.n.] [mf ed 1981] – 1r – 1 – (began in 1891; ceased with 27. geschaeftsjahr (1917/18). 1914/15 lacking p32-33. filmed with: jahres-bericht des vorstandes fuer... / verband der maler, lackierer,...) – mf#7703 reel 91 n1 – us Wisconsin U Libr [630]

Bericht ueber den ersten kongress : rom, 5.-10. september 1955 / ed by Internationale Vereinigung fuer Germanische Sprach- und Literaturwissenschaft – [S.l.] : Die Vereinigung, 1958 (Verona : Stamperia Valdonega) [mf ed 1993] – 49p – 1 – mf#7845 – us Wisconsin U Libr [430]

Bericht ueber den gegenwaertigen stand der forschung auf dem gebiet der vorreformatorischen zeit *see* Der begriff der offenbarung

Bericht ueber den parteitag / Communist Party. Germany – Berlin. no. 1-12, 15. 1918-1929, 1946 – 1 – us NY Public [335]

Bericht ueber die hochschule fuer die wissenschaft des judenthums in berlin – Berlin DE, 1874-1938 – 3r – 1 – us UMI ProQuest [939]

Bericht ueber die lage in deutschland : auslandsbuero "neu beginnen" – Prag (CZ), 1933 oct-1936 sep – 1r – 1 – gw Misc Inst [943]

Bericht ueber die wirksamkeit des central-ausschusses fuer die innere mission der deutschen evangelischen kirche *see* Bericht des central-ausschuss fuer die innere mission der deutschen evangelischen kirche (fw4)

Bericht ueber die...jahre-versammlung des forstverein fuer das grossherzogthum hessen zu... / Forstverein fuer das Grossherzogthum Hessen. Jahres-Versammlung – Darmstadt: J C Herbert. 2.jahres-versammlung 1877 [mf ed 2005] – 1v on 1r – 1 – (cont: forstverein fuer das grossherzogthum hessen. bericht ueber die ... versammlung des forstverein fuer das grossherzogthum hessen zu...; cont by: forstverein fuer das grossherzogthum hessen. jahre-versammlung des forstverein fuer das grossherzogthum hessen zu...) – mf#film mas 99999 – us Harvard [634]

Bericht ueber die...versammlung des forstverein fuer das grossherzogthum hessen zu... / Forstverein fuer das Grossherzogthum Hessen. Versammlung – Darmstadt: J C Herbert. 1. versammlung 1876] [mf ed 2005] – 1v on 1r – 1 – (cont by: forstverein fuer das grossherzogthum hessen. bericht ueber die... jahre-versammlung des forstverein fuer das grossherzogthum hessen zu...) – mf#film mas 99999 – us Harvard [634]

Bericht ueber die...versammlung des forstverein fuer das grossherzogthum hessen zu... / Forstverein fuer das Grossherzogthum Hessen. Versammlung – Gruenberg: H Robert [1883]-[5-17 versammlung (1882-1910)] [mf ed 2005] – 1r [ill] – 1 – (cont: forstverein fuer das grossherzogthum hessen. jahre-versammlung des forstverein fuer das grossherzogthum hessen zu...; imprint varies) – mf#film mas 99999 – us Harvard [634]

Bericht ueber die...versammlung deutscher forstmaenner zu... / Versammlung Deutscher Forstmaenner – Berlin: Verlag von Wiegandt & Hempel 1873-1900 (annual) [1.-27.(1872-1899)] [mf ed 2005] – 27v on 3r [ill] – 1 – (cont by: deutscher forstverein. hauptversammlung. bericht ueber die...hauptversammlung des deutschen forstvereins; imprint varies) – mf#film mas c6060 – us Harvard [634]

Bericht ueber drei reisen in lydien und der suedlichen aiolis : ausgefuhrt 1906, 1908, 1911 / Keil, J & Premerstein, A von – Wien, 1910-1915. v53, 54, 57 – 14mf – 9 – mf#H-631 – ne IDC [915]

Bericht ueber eine, im jahre 1840, in die oestliche dsungarische kirgisensteppe unternommene reise / Schrenk, A – Spb, 1845. v7 – 3mf – 9 – mf#R-1667 – ne IDC [915]

Bericht ueber meine reise nach texas im jahre 1846 : die verhaeltnisse und den zustand dieses landes betreffend / Sommer, Karl von – Bremen 1847 [mf ed Hildesheim 1995-98] – 1v on 1mf – 9 – €40.00 – 3-487-27133-8 – gw Olms [917]

Bericht von dem exorcismo bey der tauffe / Coler, J – [Magdeburgk], 1590 – 1mf – 9 – mf#TH-1 mf 338 – ne IDC [242]

Bericht von der vbiquitet / Andreae d A, J – Tuebingen, 1589 – 1mf – 9 – mf#TH-1 mf 19 – ne IDC [242]

Bericht wie die, so...mit...fragen versuocht werdend, antworten...moegind / Bullinger, Heinrich – Zuerych, Christoffel Froschower, 1559 – 3mf – 9 – mf#PBU-209 – ne IDC [240]

Berichte de rheinischen mission gesellschaft – Barmen, Germany. 1830-1964 – 37r – 1 – mf#MS00035 – sa National [943]

Berichte der deutschen chemischen gesellschaft / Deutsche Chemische Gesellschaft – Berlin, Germany. v33 sonderheft-v37 pt2. 1900-1904 – 13r – us UF Libraries [540]

Berichte des forst-vereines fuer oesterreich ob der enns – [Linz?]: Forst-Verein fuer Oesterreich ob der Enns, 1856-[1896] (Linz: J Feichtinger's Erben) [1.heft-36.bd(1856-96)] [mf ed 2004] – 35v on 4r [ill] – 1 – (irreg 1856-69, annual 1871-92, 4nos/yr 1896; cont by: berichte des forst-vereines fuer oesterreich und salzburg; v35 never publ; some vols accompanied by suppl; publ gmunden 1880-96) – mf#film mas c6004 – us Harvard [634]

Berichte des forst-vereines fuer oesterreich und salzburg / Forst-Verein fuer Oesterreich und Salzburg – Gmunden: Im Verlage des Vereines [1897-1920] ([s.l.]: Joh Habacher) [37.bd-59.jahrg(1897-1920)] [mf ed 2004] – 2r [ill] – 1 – (frequency varies, 1897-1919, annual, 1920; cont: berichte des forst-vereines fuer oesterreich ob der enns; some iss publ in combined form; v44 complete in 1 no) – mf#film mas c6005 – us Harvard [634]

Berichte ueber die biologisch-geographischen untersuchungen in den kaukasuslaendern / Radde, G – Tiflis, 1866 – 5mf – 9 – mf#AR-1622 – ne IDC [914]

Berichte ueber die verhandlungen der naturforschenden gesellschaft zu freiburg i.b. – Freiburg i.B., 1858-1882. v1-8 – 155mf – 9 – mf#8703c – ne IDC [590]

Berichte ueber die zur bekanntmachung geeigneten verhandlungen der koenigl preuss akademie der wissenschaften zu berlin – Berlin, 1836-1855. v1-20 – 2078mf – 8 – (cont as: monatsberichte der koeniglich preussischen akademie der wissenschaften zu berlin, berlin 1856-1881 v1-46 register 1836-1873; sitzungsberichte der koeniglich preussischen akademie der wissenschaften zu berlin, berlin 1882-1926) – mf#H-411c – ne IDC [956]

Berichte ueber reisen im sueden von ost-sibierien im auftrage der kaiserlichen russischen geographischen gesellschaft : ausgefuehrt in den jahren 1855 bis incl 1859 / Radde, G – Spb, 1861. v23 – 13mf – 9 – mf#R-1667 – ne IDC [915]

Berichte und protokoll vom verbandstag... / Verband der Lithographen, Steindrucker und verwandten Berufe – Berlin: Hass 1925- [mf ed 1981] – 1r – 1 – (cont: verband der lithographen, steindrucker und verwandten berufe (deutscher senefelder-bund) rechenschaftsberichte und protokoll des verbandstages; filmed with: jahrbuch der innung bund der bau-, maurer–und zimmermeister zu berlin...) – mf#7703 reel 54 n2 – us Wisconsin U Libr [680]

Berichte westpreussischen botanisch-zoologischen vereins zu danzig – Washington. 1973-1980 (1) 1973-1980 (5) 1977-1980 (9) – 90mf – 9 – mf#8629 – ne IDC [590]

Berichten uitgegeven door het utrechtsche : studenten sendinggezelschap – 1896-1908 [complete] – 3r – 1 – mf#ATLA S0741 – us ATLA [378]

Berichten van de dienst van economische zaken – Hollandia, 1961-1962(1-34) – 10mf – 9 – (missing: 1961(1, 4-5, 9); 1962(13-15, 18-19, 24-25, 28-31)) – mf#SE-1843 – ne IDC [959]

Die berichtigte lutherbibel : rektoratsrede mit anmerkungen / Kamphausen, Adolf – Berlin: Reuther & Reichard, 1894 – 1mf – 9 – 0-8370-3842-1 – (incl bibl ref) – mf#1985-1842 – us ATLA [220]

Bericht...ortsausschuss halle a.s., sowie des arbeitersekretariats halle a.s., fuer das jahr... / Allgemeiner deutscher Gewerkschaftsbund. Ortsausschuss Halle – Halle: Hallesche Druckerie-Gesellschaft 1929 [mf ed 1981] – (= ser Serial publications of german trade unions) – 1r – 1 – (1929 contains summary of 1922-28. filmed with: bericht / verband der deutschen buchdrucker; jahresbericht...; protokoll der...konferenz des reichsbeirats der betriebsräte und konzernvertreter der metallindustrie) – mf#7703 reel 88 n1 – us Wisconsin U Libr [331]

Berieko press / Mingguan berita ekonomi – Surakarta, 1962-1967. v1-5(14) – 21mf – 9 – (missing: 1962-1965 v1-4(1-35, 38, 40, 52, 53, 55, 82-87, 90, 93-96, 123)) – mf#SE-275 – ne IDC [959]

Berigten nopens den toestand en de vorderingen van het werk der inlandsche evangelisten in china / Vereeniging ter Bevordering van het Christendom in China – Amsterdam: J C Loman Jr, 1850 [mf ed 1995] – (= ser Yale coll; China en het evangelie 3) – 18p – 1 – 0-524-09484-5 – (in dutch) – mf#1995-0484 – us ATLA [240]

Bering straits agluktuk / Bering Straits Native Corporation – 1975 nov-1976 feb, 1982 aug, 1982 oct/nov-1984 apr, 1984 sep/oct-1986 jan – 2r – 1 – mf#1048025 – us WHS [071]

Bering Straits Native Corporation *see*
- Bering straits agluktuk
- Kawerak news
- Kawerak nipliksuk

Beringen, Heinrich von *see*
- Das schachgedicht

Beringer, fr *see* Die ablaesse, ihr wesen und gebrauch

Bering's voyages / ed by Golder, F A – New York, 1922-1925. v1-2 – 14mf – 9 – mf#N-224 – ne IDC [910]

Berington, Joseph *see* The literary history of the middle ages

Berit ha-levi / Grajevsky Jacob Osher – Jerusalem, Israel. 1902-03 – 1r – us UF Libraries [939]

Berita / Bank Indonesia – Djakarta, 1953-1960(20) – 30mf – 9 – (missing: 1953(1)) – mf#SE-259 – ne IDC [959]

Berita / Ikatan Dokter Indonesia – Djakarta, 1952-1959 – 13mf – 9 – (missing: 1952(jan-aug, oct-dec); 1953(1); 1954(jan-feb); 1955; 1956; 1957; 1958; 1959(jan, feb, apr-jun)) – mf#SE-994 – ne IDC [959]

Berita baperki – Djakarta, 1954 – 4mf – 9 – mf#SE-339 – ne IDC [950]

Berita baperki – Jogjakarta, 1956 – 5mf – 9 – mf#SE-344 – ne IDC [959]

Berita bibliografi – Djakarta, 1956-1961 – 21mf – 9 – mf#SE-627 – ne IDC [959]

Berita bibliografi – Djakarta, 1963-1966 – 14mf – 9 – mf#SE-628 – ne IDC [959]

Berita buana – Djakarta, Indonesia. Feb 1976-Dec 1993 – 77r – 1 – us L of C Photodup [079]

Berita DKA *see* Perusahaan negara kereta api

Berita ekonomi – Djakarta, 1953-1958/1959 v1-7 – 49mf – 9 – (missing: 1953, v1(1-6); 1955, v2-3(18-35); 1956, v4(43); 1957, v6(59-60)) – mf#SE-274 – ne IDC [959]

Berita ekonomi indonesia = Indonesia economic news / Indonesia. Kementerian perekonomian – Djakarta, 1948-1955 – 47mf – 9 – (missing: 1948-1950; 1951(109, 111, 114, 122); 1954(161)) – mf#SE-291 – ne IDC [330]

Berita IKIP *see* Institut keguruan dan ilmu pengetahuan

Berita Indonesia *see* Kedutaan besar republik indonesia

Berita indonesia – Bangkok [1958]1960-1963. v1-6 – 14mf – 9 – (missing: [1958], v1; 1959, v2; 1960, v3(1-6, 10, 11, 15-17, 19-[24]); 1961, v4(1-11, [24]); 1962, v5(21-[24])) – mf#SE-995 – ne IDC [959]

Berita LIPI *see* Lembaga ilmu pengetahuan indonesia

Berita mapie / Madjelis Perniagaan Indonesia di Eropa – Amsterdam, 1951/1952 – 50mf – 9 – mf#SE-1798 – ne IDC [959]

Berita negara ri suppl 5 pertjetakan negara ri / Indonesia Neratja ringkas Bank Indonesia – Djakarta, 1950-1961 – 31mf – 9 – (missing: 1950(84-88); 1951(90-92); 1957(26-30, 35); 1958) – mf#SE-223 – ne IDC [959]

Berita organisasi / Partai Nasional Indonesia – Djakarta, 1966-1972 – 10mf – 9 – (missing: 1966, v1; 1967, v2; 1968, v3(1-4, 6-8); 1969, v4(12)) – mf#SE-1868 – ne IDC [959]

Berita panitia hadji indonesia – Djakarta, 1953-1957 – 3mf – 9 – (missing: 1953, v1(1-6, 8-19, 21-end); 1954-1957, v2-4(1)) – mf#SE-1352 – ne IDC [959]

Berita partai / Partai Sjarikat Islam Indonesia – Djakarta, 1963-1965. v1-4(1) – 1mf – 9 – (missing: 1963-1964 v1-3) – mf#SE-1869 – ne IDC [959]

Berita pon ke-2 panitija besar / Pekan Olahraga Nasional – Djakarta, 1951 – 2mf – 9 – (missing: 1951(1)) – mf#SE-1882 – ne IDC [959]

Berita press release / Bank Pembangunan Indonesia – Djakarta, 1962 – 1mf – 9 – mf#SE-273 – ne IDC [959]

Berita Repoeblik Indonesia Departemen Penerangan *see* Indonesia (republic, 1945-1949)

Berita resmi indonesia timur lampiran-tambahan : east indonesia – Makassar, 1949-1950 – 5mf – 9 – mf#SE-212 – ne IDC [959]

Berita resmi indonesia timur staatscourant van oost-indonesie – Makassar, [1949]-1950 – 6mf – 9 – (missing: [1949]; 1950(2, 4-11, 16-22, 25-26)) – mf#SE-211 – ne IDC [959]

Berita tuberculosea indonesiensis / Departemen Kesehatan – Jogjakarta, [1954]-1963 – 4mf – 9 – (missing: [1954]-1958, v1-5(1-3); 1959/60, v6-7) – mf#SE-852 – ne IDC [959]

Berita unsrat biro publikasi : penerbitan unsrat / Universitas Sam Ratulangi – Manado, May, 1969 – 1mf – 9 – mf#SE-1978 – ne IDC [959]

Berita yudha – Djakarta, Indonesia: Edisi Pusat, 1965-1993 – 87r – 1 – us L of C Photodup [079]

Berita-negara republik indonesia pertjetakan negara ri : indonesia – Djakarta, 1950-1972 – 579mf – 9 – (missing: 1958; 1968-1969; 1971(35, 47)) – mf#SE-219 – ne IDC [959]

Berjano, Daniel Escobar *see*
- Antigua certa de hermandad entre plasencia y talavera
- Notas epigraficas. caceres

Berjano Escobar, Daniel *see* Extremadura en las obras de cervantes

Berk – Istanbul: A K Tuzluyan-Idare-i Sirket-i Muerettibiyye Matbaasi, 1302-03 [1885-86]. Sahib-i Imtiyaz: Mehmed Remzi. n1-12. 1302-03 [1885-86] – (= ser O & T journals) – 5mf – 9 – $75.00 – us MEDOC [956]

Berkala berita / Corps Bukit Barisan – Medan, 1968. v1(1) – 2mf – 9 – mf#SE-1397 – ne IDC [950]

Berkala "pembangunan" – Medan, 1959-1963. 4 v – 42mf – 9 – (missing: 1961, v3(1, 3-10); 1962-1963, v4(1-5)) – mf#SE-875 – ne IDC [950]

Berkala sarbumusi bagian penerangan, dewan pimpinan pusat, sarekat buruh muslimin indonesia / Sarekat Buruh Muslimin Indonesia – Djakarta, 1968-1969(1-12) – 7mf – 9 – (missing: 1968(1)) – mf#SE-1919 – ne IDC [959]

Berkeley / Fraser, Alexander Campbell – Edinburgh: Blackwood, 1881 [mf ed 1990] – (= ser Philosophical classics for english readers 3) – 1mf – 9 – 0-7905-7296-6 – mf#1989-0521 – us ATLA [140]

Berkeley and spiritual realism / Fraser, Alexander Campbell – London: A Constable, 1908 [mf ed 1990] – (= ser Philosophies ancient and modern (london, england)) – 1mf – 9 – 0-7905-3745-1 – (incl bibl ref) – mf#1989-0238 – us ATLA [140]

Berkeley barb – 1966 dec 5, 1967 jan 13/1968 sep 5-1977 sep 2/1978 jun 1 – 12r – 1 – (with gaps) – mf#3155418 – us WHS [071]

[Berkeley-] berkeley citizen – CA. apr 1966-mar 1967 – 1r – 1 – $60.00 – mf#C02055 – us Library Micro [071]

[Berkeley-] berkeley daily gazette – CA. nov 1894 – 372r – 1 – $22,320.00 – (aka: north east bay independent and gazette) – mf#BC02057 – us Library Micro [071]

[Berkeley-] berkeley post – CA. 1979-aug 1980; 1981 (incomplete), an-jul 14 1985 – 9r – 1 – $540.00 – (cont by: tri city post) – mf#B02058 – us Library Micro [071]

[Berkeley-] berkeley times – CA. sep 1 1920-mar 25 1922; aug 1984-aug 1985 – 4r – 1 – $240.00 – (aka: berkeley daily times) – mf#C03158 – us Library Micro [071]

[Berkeley-] berkeley tribe – CA. jul 18 1969-may 19 1972 – 4r – 1 – $240.00 – mf#B02059 – us Library Micro [071]

[Berkeley-] california legionnaire – CA. aug 1940-1967 – 5r – 1 – $300.00 – mf#C02053 – us Library Micro [071]

[Berkeley-] california monthly – CA. oct 1981-1982 – 2r – 1 – $120.00 – mf#R02054 – us Library Micro [073]

[Berkeley-] daily californian – CA: UC of Berkeley, 1897-1908; 1910-18; 1920-58 – 61r – 1 – $3660.00 – mf#C02056 – us Library Micro [071]

Berkeley daily times *see* [Berkeley-] berkeley times

[Berkeley-] el informador – CA. 1967- – 1r – 1 – $60.00 – mf#R03156 – us Library Micro [071]

Berkeley, G F H *see* The campaign of adowa and the rise of menelik

Berkeley, G H F *see* Letter books, 1847-8

[Berkeley-] gay sunshine – CA. n1-46. 1970-1980 – 1r – 1 – $60.00 – mf#R04006 – us Library Micro [071]

Berkeley, George *see* The works of george berkeley

Berkeley High School *see* Pack rat

[Berkeley-] hilltop mirror – CA. oct 1944-sep 1946; oct 1948-sep 1949; sep 1956-sep 1957 – 3r – 1 – $180.00 – mf#B06013 – us Library Micro [071]

Berkeley journal of employment and labor law – Berkeley CA 1993+ – 1,5,9 – (cont: industrial relations law journal) – ISSN: 1067-7666 – mf#11959,01 – us UMI ProQuest [344]

Berkeley journal of employment and labor law – University of California at Berkeley. v1-22. 1976-2 – 9 – $404.00 set – (v1-16 1976-95 in reel $247.00; v17-22 1996-2 in mf $157.00; title varies: v1-13 1976-92 as industrial relations law journal) – ISSN: 1067-7666 – mf#103371 – us Hein [344]

Berkeley journal of international law – v1-19. 1983-2 – 1 – $439.00 set – (title varies: v1-13 1983-96 as international tax and business lawyer) – ISSN: 0741-4269 – mf#109281 – us Hein [343]

Berkeley journal of sociology – [Berkeley CA]: University of California, Berkeley [1959-] [mf ed 1988-] – 1 – (cont: berkeley publications in society and institutions) – mf#1497 – us Wisconsin U Libr [301]

[Berkeley-] la prensa libre – CA. 1969 – 1r – 1 – $60.00 – mf#R03157 – us Library Micro [071]

Berkeley, Lowry E *see* Measure and method of christian liberality

Berkeley, M J *see* Journal of the linnean society

Berkeley news – v1 n2-1915 [1975 mar 20-oct 3] – 1r – 1 – mf#351003 – us WHS [071]

[Berkeley-] people's world – CA. oct 1943-1986 – 67r – 1 – $4020.00 – (cont: see san francisco) – mf#C02060 – us Library Micro [071]

[Berkeley-] plexus – CA. mar 15 1974-dec 1986 – 3r – 1 – $180.00 – mf#C03583 – us Library Micro [071]

Berkeley prout weekly – v1 n3-5 [1984 nov 2/19, dec 17] – 1r – 1 – mf#1304013 – us WHS [071]

Berkeley technology law journal – Berkeley CA 1996+ – 1,5,9 – (cont: high technology law journal) – ISSN: 1086-3818 – mf#15677,01 – us UMI ProQuest [346]

Berkeley technology law journal – University of California at Berkeley. v1-16. 1986-2 – 9 – $515.00 set – (title varies: 1-10 1986-95 as high technology law review) – mf#109781 – us Hein [346]

[Berkeley-] the berkeley monthly – CA. v9 n8 may 1979-v18 n2 nov 1987 – 8r – 1 – $480.00 – mf#B06012 – us Library Micro [071]

[Berkeley-] the berkeley/tri-city post – CA. jul 17 1985- – 19r – 1 – $1140.00 (subs $90/y) – (cont: berkeley post) – mf#B03159 – us Library Micro [071]

[Berkeley-] the express – CA. 1979- – 34r – 1 – $2040.00 (subs $120/y) – mf#B05027 – us Library Micro [071]

Berkeley tribe / Red Mountain Tribe – Berkeley CA. 1969 jul 10-1970 sep 4, 1970 aug 21-1972 may – 2r – 1 – (cont: barb on strike) – mf#764811 – us WHS [071]

[Berkeley-] voice – CA. aug 1984-dec 1987, jan 1989-jun 1991, 1993 – 4r – 1 – $240.00 – (aka: california voice) – mf#B06014 – us Library Micro [071]

[Berkeley-] weekly californian – CA. oct 6 1926-apr 13 1927 – 1r – 1 – $60.00 – mf#C03160 – us Library Micro [071]

Berkeley women's law journal – Berkeley CA 1985+ – 1,5,9 – ISSN: 0882-4312 – mf#17803.01 – us UMI ProQuest [340]

Berkeley women's law journal – v1-16. 1985-2001 – 9 – $278.00 set – ISSN: 0882-4312 – mf#109741 – us Hein [340]

[Berkely-] information systems on latin america – CA. jan 6 1973-jun 1980 – 16r – 1 – $960.00 – mf#R03155 – us Library Micro [000]

Berkemeier, Gottlieb E *see* Wartburg-klaenge und gesaenge

Berkemeyer, F *see* Pastor and people

Berkenkamp, F *see* Vermaaklyk lusthof van zede-en zinnebeelden waar in voorkomen leerzaame, nuttige en aangenaame opwekkingen tot deugd en godtsvrucht

[Berkenkamp, F] *see* Vermaaklyk lusthof van zede- en zinnebeelden waar in voorkomen leerzame, nuttige en aangenaame opwekkingen tot deugd en godsvrucht

Berkenmeyer, Paul *see*
- Le curieux antiquaire, ou recueil geographique et historique
- P I berkenmeyers geographische fragen
- P I berkenmeyers poetische anleitung zur universal-histoie

Berkhamsted & district gazette – Berkhamsted, England 16 nov 1984-16 may 1991 [mf 1980-91] – 1 – (cont: berkhamsted gazette & tring & district news [1 oct 1904-9 nov 1984 (wanting 1912)]; cont by: berkhamsted gazette [23 may-26 sep 1991]) – uk British Libr Newspaper [072]

Berkhamsted times & advertiser for herts, bucks, & beds – Berkhamsted, England 17 apr 1875-10 may 1884 – 1 – (cont by: berkhamsted times, tring telegraph, chesham news & advertiser for herts, bucks & beds [16 may 1884-28 sep 1900]) – uk British Libr Newspaper [072]

Berkhamsted & tring herald & post – Berkhamsted, England 10 jun 1993-1 sep 1994 – 1 – (cont: berkhamsted herald & post [24 aug 1989-3 oct 1991]; cont by: herald & post (berkhamsted & tring) 15 sep 1994-28 sep 1995) – uk British Libr Newspaper [072]

Berkhof, Louis *see*
- Christendom en leven
- De drie punten in alle deelen gereformeerd
- Premillennialisme

Berkhof, Louis et al *see* Waar het in de zaak janssen om gaat

Berkley 1670-1900 – Oxford, MA (mf ed 1991) – (= ser Massachusetts vital records) – 32mf – 9 – 0-87623-139-3 – (mf 1-5: town records 1670-1824. mf 6-12: town records 1735-1834. mf 13-22: town records 1748-1838. mf 23-25: vital records 1737-1887. mf 26-27: vital records 1844-59. mf 28: marriages 1860-1900. mf 29-30: births 1860-1900. mf 31-32: deaths 1860-1902) – us Archive [978]

Berkoff, David C *see* Chloroform exposure and dose determination associated with competitive swimmers during a two-hour swim practice

Berkov, P N *see*
- Istoriia russkoi zhurnalistiki 18 veka
- Russkaia narodnaia drama 17-20 vekov
- Satiricheskie zhurnaly n i novikova

Berks and oxon advertiser – Wallingford, England 7 jun 1889-15 may 1942, 7 jan 1949-25 dec 1959 (wkly) [mf jan-jul 1911] – 1 – (not publ between 15 may 1942 & 7 jan 1949; wanting jan 1896-dec 1898, aug-dec 1911; cont by: wallingford & district news and berks & oxon advertiser [1 jan 1960-24 nov 1962]; wallingford & didcot news & berks and oxon advertiser [1 dec 1962-9 feb 1963]) – uk British Libr Newspaper [072]

Berks county democrat – Boyertown, PA. 1902-29. 10 rolls – 13 – $25.00r – us IMR [071]

Berks county democrat – Reading, PA., 1858-1859 – 13 – $25.00r – us IMR [071]

Berks of old – v1 n1-v5 n5 [1983 may-1989 feb] – 1r – 1 – mf#1596623 – us WHS [071]

Berkshire County W T A Society *see* Voice of truth

Berkshire review – Williamstown MA 1965-86 – 1,5,9 – ISSN: 0005-920X – mf#7012 – us UMI ProQuest [073]

Berkun, Arthur *see* Kamerad bursche

Berla, Alois *see* Frater thomerl

Berlage, Hendrik Petrus *see* Disquisitio exegetico-theologica de formulae paulinae pistis iesou christou significatione

Berle, Adolf Augustus *see* Christianity and the social rage

Berleant, Arnold *see* Fugue in the orchestral works of bartok

Berlepsch, Emilie von *see* Caledonia

Berlepsch, Hermann A *see* Die alpen in natur- und lebensbildern

Berlepsch, Hermann Alexander von *see* Concordanz der deutschen national-literatur

Berliere, Ursmer *see* Documents inedits pour servir a l'histoire ecclesiastique de la belgique

Berlin : ein buch fuer junge und alte preussen – Berlin 1852 [mf ed Hildesheim 1995-98] – (= ser Fbc) – 2mf – 9 – €60.00 – 3-487-29562-8 – gw Olms [914]

Berlin : illustrirte montags-zeitung – Berlin DE, 1857 – 1r – 1 – gw Misc Inst [074]

Berlin am mittag – Berlin DE, 1947 4 feb-1948 1 feb – 1r – 1 – gw Misc Inst [074]

Berlin am morgen – Berlin DE, 1929 15 mar-1933 12 feb [gaps] – 14r – 1 – (title varies: 15 mar-19 apr 1929: die welt am morgen) – gw Misc Inst [074]

Berlin and its environs / Karl Baedeker (Firm) – Leipzig, Germany. 1923 – 1r – us UF Libraries [943]

Berlin and its treasures : being a series of views of the principal buildings, churches, monuments etc – Leipzig 1853-58 – (= ser 19th c art & architecture) – 7mf – 9 – mf#4.2.1762 – uk Chadwyck [720]

Berlin baptist church documents 1837-41 / Oncken Archives – Hamburg, Germany. 60p – 1 – us Southern Baptist [242]

Berlin bei nacht : culturbilder / Rasch, Gustav – Berlin [1871] [mf ed Hildesheim: 1995-98] – (= ser Fbc) – 246p on 2mf – 9 – €60.00 – 3-487-29566-0 – gw Olms [914]

Berlin, Charles *see* Hebrew books from the harvard college library

Berlin city courant – Berlin WI. 1859 aug 25-1863, 1861 jul 9-oct 24, 1863 aug 6-1864 may 26 – 3r – 1 – (cont by: berlin courant) – mf#966167 – us WHS [071]

Berlin courant – Berlin WI 1864 jun 3/1865 dec 28, 1866-68, 1869-71, 1872 jan 4/1875 apr 24-1912 jan 5/1915 dec 23 – 14r – 1 – (with gaps; cont: berlin city courant) – mf#927911 – us WHS [071]

Berlin crisis *see* Foreign office files: united states of america, series 3

The berlin crisis, 1958-1962 – [mf ed Chadwyck-Healey] – (= ser National security archive, washington dc: the making of us policy) – 2900+ docs on 460mf – 9 – (with 2v p/g & ind) – uk Chadwyck [327]

Berlin daily courant – Berlin WI. 1885 jul 2, 1885 jul 2-dec 31, 1886 apr 23 – 3r – 1 – mf#961324 – us WHS [071]

Berlin, Dorothea *see* Erinnerungen an gustav nachtigal

Berlin evening journal – Berlin WI. 1881 oct 7/dec, 1882 jan/aug 8-1942 jan-mar 28 – 122r – 1 – (with gaps; cont: evening journal, tri-county news, tri-county news, redgranite times; cont by: berlin journal) – mf#961582 – us WHS [071]

Berlin. Germany. Deutsches Institut fur Wirtschaftsforschung *see* Statistisches kompendium uber die sowjetische besatzungszone

Berlin hoert und sieht – Berlin DE, 1932 30 oct-1941 26 may – 11r – 1 – gw Mikrofilm [074]

Berlin. Institut fuer Marxismus-Leninismus *see* Geschichte der deutschen arbeitsbewegung in 15 kapiteln

Berlin journal – Berlin WI. 1870 aug 30/1874 dec 29, 1875-78, 1879-81, 1882 jan 1-20 – 4r – 1 – (cont by: berlin weekly journal) – mf#930438 – us WHS [071]

Berlin journal – Berlin WI. 1942 apr 2/dec 31-2000 sep-dec – 87r – 1 – (with gaps; cont: berlin evening journal) – mf#984304 – us WHS [071]

Berlin journal-courant – Berlin WI. 1916 feb 3-nov 30, 1916 dec 7-1918 jun 20, 1918 jun 27-1920 jan 1, 1920 jan 8-1921 jul 14, 1922 jul 21-1922 jun 29 – 5r – 1 – (cont: berlin weekly journal) – mf#961575 – us WHS [071]

Berlin, Knud Kugleberg *see* Danemarks recht auf gronland

Berlin. Literaturarchiv-Gesellschaft *see* Jahresbericht ueber die wissenschaftlichen erscheinungen auf dem gebiete der neuren deutschen literatur

Berlin, Naphtali Zevi Judah *see* Sh u-t meshiv davar

Berlin observer – Berlin. 1983 jul 15-1986 aug 29, 1992 jan 10-1994 jul 15 – 2r – 1 – (cont: grooper; berlin tabulator) – mf#1125201 – us WHS [071]

Berlin record – Ontario, CN. jan 1907-dec 1916 – 21r – 1 – (cont: kitchener-waterloo record; cont by: the record) – cn Commonwealth Imaging [071]

The berlin record *see* The record

Berlin, rom, tokio – Berlin DE, 1939-42, 1944 – 4r – 1 – gw Misc Inst [074]

Berlin. Staatliche Museen *see* Herwarth walden und die europaeische avantgarde

Berlin. Staatliche Museen. Museum fuer Voelkerkunde *see* Fuehrer durch die sammlungen des museums fuer voelkerkunde

Berlin. Statistisches amt *see* Statistiches jahrbuch der stadt berlin

Berlin tabulator – Berlin. 1981 apr 30-1985 feb 9, 1985 feb 15-1987 dec 18, 1988-1990 apr – 3r – 1 – (cont: tabulator (berlin, germany); cont by: berlin observer) – mf#1013369 – us WHS [071]

Berlin und die berliner : in wort und Bild / Loeffler, Ludwig – Leipzig 1856 [mf ed Hildesheim: 1995-98] – (= ser Fbc) – 138p on 1mf – 9 – €40.00 – 3-487-29563-6 – gw Olms [914]

Berlin und wien : ein skizzenbuch / Proehle, Heinrich – Berlin 1850 [mf ed Hildesheim: 1995-98] – (= ser Fbc) – x/186p on 2mf – 9 – €60.00 – 3-487-29560-1 – gw Olms [914]

Berlin weekly journal – Berlin WI. 1882 jan 27/1884 jun-1915 jun 10/1916 jan 27 – 1 – (with gaps; cont: berlin journal; cont by: berlin journal-courant) – mf#961577 – us WHS [071]

Berlin weekly times – Berlin, NE: I N Hunter & Son, 1915-v3 n45. sep 3 1918 (wkly) [mf ed v3 n32. jan 15-sep 3 1918 (gaps) filmed 1979] – 1r – 1 – (cont by: otoe weekly times) – us NE Hist [071]

Berlin wie es ist : ein gemaelde des lebens dieser residenzstadt und ihrer bewohner, dargestellt in genauer verbindung mit geschichte und topographie / Kertbeny, Karoly M von – Berlin 1831 [mf ed Hildesheim: 1995-98] – (= ser Fbc) – viii/337p on 3mf – 9 – €90.00 – 3-487-29565-2 – gw Olms [914]

Berlin-Brandenburgischen Akademie der Wissenschaften *see* Acta borussica neue folge

Berlinck, Eodoro Lincoln *see* Fatores adversos na formacao brasileira

Der berliner – Berlin DE, 1945 2 aug-1946 30 apr, 1969-1972 30 jun – 26r – 1 – (filmed by bnl: 1945 2 aug-1954 [46r]; filmed by misc inst: 1958-1962 31 mar, 1962 13 may-17 jun, 26 jun-16 sep, 1962 6 nov-1966, 1945 2 aug-1972 30 jun [gaps]; merged with telegraf 1 may 1946 [der berliner: 2 aug 1945-30 apr 1946, telegraf: 22 mar 1946-30 jun 1972]; incl suppl: bild der zeit / berlin im bild 1950 9 jul-1951 28 oct [1r]) – mf#7138 – gw Mikropress; uk British Libr Newspaper; gw Misc Inst [074]

Die berliner abendblaetter heinrich von kleists : ihre quellen und ihre redaktion / Sembdner, Helmut – Berlin: Weidmann, 1939 [mf ed 1994] – (= ser Schriften der kleist-gesellschaft 19) – 16/402/7p – 1 – (incl bibl ref and ind) – mf#8707 – us Wisconsin U Libr [070]

Berliner, Abraham *see*
- Aus dem leben der juden deutschlands im mittelalter
- Beitraege zur geographie und ethnographie babyloniens im talmud und midrasch
- Censur und confiscation hebraischer bucher im kircenstaate
- Magazin fuer die wissenschaft des judentums
- Synagogal-poesieen
- Targum onkelos

Berliner adressbuch – 1891-1941. Scattered volumes wanting – 1 – us L of C Photodup [943]

Berliner adressbuch : adressbuch fuer berlin und seine vororte = Berlin directory : directory for berlin and its suburbs 1919-1932 / ed by Umlauf, Konrad – 1919-32 [mf ed 1984] – 981mf (1:24) – 9 – silver €2980.00 – ISBN-10: 3-598-30284-3 – ISBN-13: 978-3-598-30284-8 – gw Saur [914]

Berliner allgemeine musikalische zeitung / ed by Marx, Adolf Bernhard – Berlin: Schlesingersche Buch- Musikhandlung 1824-[1830] [mf ed 19–] – 7v on 66mf – 9 – ("musikbeilagen" iss with v1-6) – mf#fiche 694 – us Sibley [780]

Berliner allgemeine musikalische zeitung 1824-1830 – [mf ed 1994] – 37mf (1:24) + suppl – 9 – diazo €1260.00 (silver €1590 isbn: 978-3-598-33805-2) – ISBN-10: 3-598-33804-X – ISBN-13: 978-3-598-33804-5 – gw Saur [780]

Berliner allgemeine zeitung – Berlin DE, 1861 14 dec-1863 – 6r – 1 – gw Misc Inst [074]

Berliner allgemeine zeitung *see* Das deutsche blatt

Berliner arbeiterzeitung – Berlin DE, 1927 9 jan-1928 25 nov, 1929 3 feb-1930 30 dec – 2r – 1 – mf#5867 – gw Mikropress [331]
Berliner astronomisches jahrbuch fuer... *see* J f encke's astronomische abhandlungen
Der berliner baer – Berlin DE, 1930 jan-jun – 2r – 1 – gw Misc Inst [074]
Berliner beytraege zur landwirtschaftswissenschaft / ed by Benekendorf, Karl Friedrich von – Berlin 1771-91 – (= ser Dz) – 8v+ind on 39mf – 9 – €390.00 – (v8 ed by gottfried ludolf grassmann) – mf#k/n2904 – gw Olms [630]
Berliner blatt – Berlin DE, 1907 jan-mar – 1r – 1 – gw Misc Inst [074]
Berliner boersen-courier – Berlin DE, 1895 1 apr-1933 [gaps] – 253r – 1 – (filmed by other misc inst: 1878 jul-dec, 1879 apr-dec, 1880 apr, may, aug-dec, 1885 may, 1921 sep-oct, 1924 jun, jul, oct, 1925 jan, may, jun, aug, 1926 aug [13r]. incl suppl: bilder-courier 1924 28 mar-31 dec [1r]) – gw Misc Inst [332]
Berliner boersen-zeitung – Berlin DE, 1926-27, 1939 16 jan-1941 28 feb, 1941 1 jul-31 oct, 1942 1 jan-30 apr, 1 jul-31 aug [gaps] – 30r – 1 – (filmed by misc inst: 1856-63, 1864 apr-1884 aug, 1884 oct-1944 aug [796r]; 1932 jul-1933 jan, 1933 mar-1939 mar, 1940-1941 nov, 1942-43 [69r]. with suppl: boerse des lebens 1871 [single iss]) – gw Mikrofilm; gw Misc Inst [074]
Berliner borsen-zeitung – Berlin, Germany. Jun 1934-Aug 1944 – 83r – 1 – us L of C Photodup [074]
Berliner caricaturen und silhouetten : die nichtdemokratische presse berlins – Berlin, Bremen DE, 1850 – 1 – gw Misc Inst [740]
Berliner correspondenz – Berlin DE, 1894 10 dec-1895, 1897-1907, 1910 19 jan-1914 2 sep – 3r – 1 – gw Misc Inst [074]
Berliner, Emile *see* Conclusions
Die berliner estafette – Berlin DE, 1827 10 jul-31 dec – 1r – 1 – gw Misc Inst [074]
Die berliner estaffette – Berlin DE, 1827 10 jul-31 dec – 1r – 1 – gw Misc Inst [074]
Der berliner figaro – Berlin DE, 1832-1833 9 may, 1834, 1836-42, 1844-47, 1875 2 nov-1876 22 mar – 16r – 1 – gw Misc Inst [074]
Berliner fliegende blaetter : beilage zum neuen berliner tageblatt – Berlin DE, sep 19 1875-jun 22 1876 – 1r – 1 – mf#6012 – gw Mikropress [074]
Berliner fremdenblatt – Berlin DE, 1837 3 jan-30 jun, 1838, 1839 1 jul-1840 30 jun, 1841 – 5r – 1 – gw Misc Inst [074]
Die berliner handschrift der sahidischen apostelgeschichte / Hintze, F & Schenke, H M – Berlin, 1970 – (= ser Tugal 5-109) – 3mf – 9 – €17.00 – ne Slangenburg [240]
Die berliner handschrift des sahidischen psalters / ed by Rahlfs, Alfred – Berlin: Weidmann, 1901 – (= ser Abhandlungen der koeniglichen gesellschaft der wissenschaften zu goettingen) – 1mf – 9 – 0-524-02770-6 – mf#1987-6464 – us ATLA [090]
Berliner illustrierte nachrichten *see* Berliner nachrichten
Berliner illustrierte nachtausgabe *see* Nachtausgabe-der tag
Berliner illustrierte zeitung – Berlin DE, 1894 18 aug-30 dec, 1905, 1909, 1911-14, 1920-23, 1924 6 jul-28 dec, 1925 5 jul-1926, 1927 31 dec-1928, 1938, 1944 6 jul-1945 29 apr – 15r – 1 – (filmed by misc inst: 1895-1904, 1906-08, 1910, 1914 2 aug-1919 [gaps], 1924 6 jan-29 jun, 1925 1 jan-28 jun, 1927, 1929-37, 1938 17 mar-1941 [gaps], 1943-1945 15 feb (gaps) [54r]; 1891 14 dec (probe-nr), 1892-1901, 1927. aka: berliner illustrirte zeitung) – gw Mikrofilm; gw Misc Inst [074]
Berliner Industrie Und Handeskammer *see* Wirtschaftblatt
Berliner intelligenzblatt – Berlin DE, 1788 oct-dec, 1792 jul-dec, 1807-09, 1811-12, 1816-1910, 1912-1922 sep – 498r – 1 – (title varies: 1783: neues berliner intelligentz-blatt; incl suppl: besondere beylage 1802-06 [2r], gemeinnuetziger anzeiger 1810 jul-dec, 1811 jul-dec, 1816, 1818, 1819, 1821 [3r]) – gw Misc Inst [074]
Berliner journal – Kitchener, ON: Rittinger & Motz, 1859-79 – 6r – 1 – ISSN: 1181-3960 – cn Library Assoc [071]
Berliner journal 1924 – Berlin DE, 1924-1933 7 jan – 1r – 1 – gw Misc Inst [074]
Berliner kirchen briefe – Berlin-Grunewald: Lutherischen Verlagshaus. n1-75/76; ns: n1-20. 1961-85 [irreg] [mf ed 1988-89] – 3r – 1 – (lacks: n65 p4-5) – mf0842 – us ATLA [242]
Berliner kurier am abend *see* Bz am abend
Berliner kurier am morgen – Berlin DE, 1992 2 jan-31 jul – 2r – 1 – gw Misc Inst [074]
Berliner lokal-anzeiger – Berlin: A Scherl, 1928 (r1-12); jan-mar 1933 (r13-15) – 15r – 1 – us CRL [072]
Berliner lokal-anzeiger – Berlin DE, 1900 1 aug-1919 30 nov, 1920-26, 1927 feb, mar, 1927 1 may-1930 31 jul, 1930 1 sep-1932 31 may, 1932 1 aug-1933 25 apr, 1933 1 jun-1937, 1938 1 apr-1944 31 aug – 334r – 1 – (filmed by misc inst: 1887 aug-1911, 1915 sep-18 oct, 1916 24 sep-dec, 1918 28 apr-13 sep, 1922 may, jun, sep-dec, 1923 mar-jun, 1924 oct-nov, 1939 aug-dec, 1940 mar-aug [119r]; incl suppls: bilder vom tage 1910 1 jul-31 aug, 1912 10 jan, 7 mar, 1914 16 mar-1917 [gaps], 1919 20 apr, sport-echo 1923 19 oct-1931 mar, unterhaltungsbeilage 1929-1931 7 apr; die weite welt 1925-26, 1931 22 nov-1933 24 sep [1r]) – gw Misc Inst [074]
Berliner missionsberichte aus sud-afrika : 1836-1939 / ed by Berliner Missionsgesellschaft – Berlin: Verl d Berliner Missionsgesellschaft. v for 1939-40 called 116-117 jahrg (mthly) [mf ed Pretoria: State Library Corporate Communication, 1994] – 544mf – 9 – 0-7989-1198-0 – (suspended jun 1940-1946. ceased with 1948 n2) – mf#MFM05609 – sa National [943]
Berliner Missionsgesellschaft *see* Berliner missionsberichte aus sud-afrika
Der berliner mittag – Berlin DE, 1928 jan-18 jun – 2r – 1 – Dist. gw Mikrofilm – gw Misc Inst [074]
Berliner mittagszeitung *see* Das kleine journal 1883
Berliner modenspiegel – Berlin DE, 1838-46, 1848 – 6r – 1 – (title varies: 1835: berliner modenspiegel in- und auslaendischer originale. filmed with suppls) – gw Misc Inst [640]
Berliner montag – Berlin DE, 1948 26 jul-1952 – 2r – 1 – gw Misc Inst [074]
Berliner montags-echo – Berlin DE, 1947 22 dec-1963 18 feb – 7r – 1 – (until 4 jan 1948: montags-echo) – gw Mikrofilm [074]
Berliner montags-post – Berlin DE, 1854 25 dec-1864 – 3r – 1 – Dist. gw Mikrofilm – gw Misc Inst [074]
Berliner montagspost – Berlin DE, 1920 10 may-1938, 1940-1942 28 dec – 14r – 1 – (with gaps) – gw Misc Inst [074]
Berliner montags-zeitung – Berlin DE, 1861 7 jan-1876 – 1 – gw Misc Inst [074]
Berliner morgenpost – Berlin DE, 1898 20 sep, 1 nov-1905 mar, 9 may, 1905 jun-27 aug, 1905 oct-1906, 1908 apr-jun, 1909 oct, 1920 aug-sep, 1924 mar-apr, 1926 sep-oct, 1929 feb, 1934 jul-aug, 1939 mar, 1941 mar-apr, 1942 jul-sep, 1944 oct-1945 25 apr, 1952 26 sep- – 375r until 12 may 2003 [excl 1998] – 1 – (filmed by misc inst: 1907-1908 mar, 1908 jul-1909 aug, 1909 nov-1920 jun, 1920 oct-1923, 1924 apr-1926 aug, 1926 nov-1929 jan, 1929 mar-1934 jun, 1934 sep-1939 feb, 1939 apr-1941 feb, 1941 may-1942 jun, 1942 oct-1944 sep [239r]; 1973 1 jun-1977 sep; 1976- [ca 13r/yr]. with suppl: berliner illustrierte zeitung 1997 4/5 jan) – gw Mikrofilm; gw Misc Inst [074]
Berliner morgen-zeitung – Berlin DE, 1891, 1892 jul-sep, 1893 jul-1903, 1904 apr-1907 mar, 1907 jul-1919, 1920 may-1923, 1924 apr-1928 mar, 1928 jul-1939 15 feb – 124r – 1 – gw Misc Inst [074]
Berliner nachrichten – Berlin DE, 1922-23, 1927-28, 1930-1931 27 nov, 1932-1933 7 mar, 1933 10 jul-1934 13 oct – 3r – 1 – (with gaps; title varies: n42 1927: berliner illustrierte nachrichten, 10 nov 1928: berliner tribuene) – gw Misc Inst [074]
Berliner nachtausgabe *see* Nachtausgabe-der tag
Berliner neueste nachrichten *see* Coepenicker dampfboot
Berliner neueste nachrichten 1881 – Berlin DE, 1914 25 jul-1915 jun, 1915 30 sep-dec, 1916 apr-1917 mar, 1917 jul-dec, 1918 apr-nov – 10r – 1 – (filmed by misc inst: 1889-30 jun 1919 (gaps) [130r]. with suppls: deutscher hausfreund 1889-1918 [gaps]; mode und handarbeit 1897-1914 [gaps]) – mf#5116 – gw Mikropress; gw Misc Inst [074]
Berliner neuigkeitsbote fuer gebildete staende *see* Der neuigkeitsbote
Berliner ostend-zeitung – Berlin DE, 1880 2 oct-1881 – 2r – 1 – (earlier: der ost-district) – gw Misc Inst [074]
Berliner paedagogische zeitung – Berlin DE, 1877-87 – 1 – (title varies: 2 mar 1876: paedagogische zeitung; 1919: allgemeine deutsche lehrerzeitung) – gw Misc Inst [370]
Berliner pfennig-blaetter – Berlin DE, 1847 – 1r – 1 – (filmed by other misc inst: 1844 Apr, 1844 1 jul-1849, 1853, 1855-57, 1860, 1862, 1864-68) – gw Misc Inst [074]
Berliner politisches wochenblatt – Berlin DE, 1831 8 oct-1837, 1841 – 3r – 1 – gw Misc Inst [320]
Berliner reform – Berlin DE, 1861 2 oct-1862; 1864-1866 30 jun [gaps] – 5r – 1 – gw Misc Inst [074]
Berliner skizzen / Kretzer, Max – Berlin: C Duncker 1898 [mf ed 1995] – 1r – 1 – (filmed with: die buchhalterin / max kretzer & other titles) – mf#3910p – us Wisconsin U Libr [880]
Berliner skizzen : neve vorstadtgeschichten / Seidel, Heinrich – Leipzig: A G Liebeskind 1894 – 1 – mf#film mas c419 – us Harvard [880]
Berliner stadtblatt *see* Der sozialdemokrat 1946
Berliner stimme – Berlin, 20 Oct 1951-1967 – 8r – 1 – gw Mikropress [074]
Berliner stimme : sozialdemokratische wochenzeitung – Berlin DE, oct 20 1951-67 – 8r – 1 – (cont: der sozialdemokrat [west-berlin]) – mf#2090 – gw Mikrofilm [074]
Berliner stimmen – Berlin DE, 1926-31 – 1 – gw Misc Inst [074]
Berliner sueden *see*
– Neue tempelhofer zeitung
– Tempelhof-mariendorfer zeitung
Berliner tageblatt – Berlin, Germany. Dec 1899-Jan 1939 – 215r – 1 – us L of C Photodup [074]
Berliner tageblatt – Berlin DE, 1914-1939 30 jan [gaps] – 153r – 1 – (filmed by misc inst: 1895 1 jan-28 feb, 1895 1 apr-1900 31 mar, 1900 1 may-1913 [gaps], 1935 feb, 1872 1 jan-1939 31 jan (gaps) [574r]; incl suppls: berliner sonntagsblatt 1875-79 [gaps], literarische rundschau 1898 11 dec-1908 (gaps) [4r], technische rundschau 1897-1935 [gaps]) – mf#354 – gw Mikropress; gw Misc Inst [074]
Berliner tageblatt : wochenendausgabe fuer ausland und uebersee – Berlin DE, 1914-18, 1922-30 – 11r – 1 – (1927: berliner tageblatt / monatsausgabe fuer ausland und uebersee) – gw Misc Inst [074]
Berliner theater-zeitung [...] – Berlin DE, 1837-40 [gaps] – 1r – 1 – gw Misc Inst [790]
Berliner tribuene *see* Berliner nachrichten
Berliner vereinsbote – Berlin DE, 1912-18, 1921-38 – 1 – (title varies: 1901 n20: israelitische rundschau, 1902 n40: juedische rundschau; with suppl: literaturblatt 1905-08) – gw Misc Inst [939]
Berliner vereins-zeitung – Berlin DE, 1904-1911 26 jan [gaps] – 3r – 1 – gw Misc Inst [074]
Berliner volksblatt – Berlin DE, 1884 1 apr-1933 28 feb [gaps] – 168r – 1 – (filmed by misc inst: 1914 25 jun-30 sep, 1916 1 jan, 31 mar, 1 oct-31 dec, 1919 30 sep-31 dec, 1922 31 mar-1 jul, 1923 1 jul-1924 30 sep [gaps], 1930 1 jul-1931 28 apr, 1932 1 jul-1933 28 feb. title varies: 1 jan 1891: vorwaerts, cont as: neuer vorwaerts in karlsbad; incl suppls: die neue welt 1914-19, volk und zeit / bilder zum vorwaerts 1919 jul-1933 mar) – mf#195 – gw Mikropress; gw Misc Inst [074]
Berliner volks-zeitung *see* Urwaehler-zeitung
Berliner vororts-zeitung – Berlin DE, 1924-33, 1934 28 jan-25 feb – 2r – 1 – gw Misc Inst [074]
Berliner wespen – Berlin DE, 1874 23 jan-1875, 1880 jan-26 nov, 1881-82 – 3r – 1 – gw Misc Inst [074]
Der berliner westen *see* Berlin-wilmersdorfer zeitung
Berliner wochenblatt zur belehrung und unterhaltung *see* Nuetzliches und unterhaltendes berlinisches wochenblatt fuer den gebildeten buerger und denkenden landmann
Berliner zeitung – Berlin, Germany. Jul 1946-sept 1947; 1962-apr 1973 – 39r – 1 – us L of C Photodup [074]
Berliner zeitung 1877 – Berlin DE, 1883-1886 aug, 1887 jan-aug, 1888, 1889 may-dec, 1890 may-1892 mar, 1892 jul-1898 jun, 1898 oct-1899 mar, 1899 jul-dec, 1900 apr-jun, 1900 oct-1901 mar, 1901 jul-sep, 1902 jul-1903, 1904 jul-sep, 1905 jan-mar – 1 – (with suppl: gerichtslaube 1883-1903 [gaps]) – gw Misc Inst [074]
Berliner zeitung 1945 – Berlin DE, 22 may 1945-48; 1950-30 jun 1957 – 18r – 1 – (filmed by mikropress: 1945 21 may-1993 [157r], 1995- order#1065; filmed by misc inst: 1945 21 may-1947 [5r]; 1945 21 may-1991 (gaps) [120r]; 1992- [10r/yr]; 1958 2 jan-31 mar, 1968 17 aug-1969) – gw Mikrofilm; gw Mikropress; gw Misc Inst [074]
Berliner zeitungs-halle – Berlin DE, 1846 1 oct-1848 12 dec – 5r – 1 – gw Misc Inst [074]
Berlinerinnen : zwei frauenschicksale / Fontane, Theodor; ed by Langenbucher, Hellmuth – Bayreuth: Gauverlag, 1944 [mf ed 1989] – (= ser Bayreuther feldpostausgaben) – 152p – 1 – mf#7248 – us Wisconsin U Libr [880]
Berlin-friedenauer tageblatt – Berlin DE, 1924 19 dec-1925 31 jan, 1925 1 jul-1933 29 sep – 1 – gw Misc Inst [074]
Berlingske politiske og avertissements-tidende – Copenhagen, Denmark 2 jan 1883-31 dec 1935 – 1 – (cont: danske statstidende [3 oct 1808-31 dec 1882]; cont by: berlingske tidende [1 jan 1936-30 jan 1977, 1 may 1977-]) – uk British Libr Newspaper [078]
Berlingske tidende – Kobenhavn: Interessentskabet Berlingske Tidende, jul 1938-55 – 204r – 1 – us CRL [079]
Berlingske tidende – Copenhagen, Denmark 1 jan 1936-30 jan 1977, 1 may 1977- – 1 – (wanting: 16-31 dec 1969; not publ between 30 jan-1 may 1977; cont: berlingske politiske og avertissements-tidende [2 jan 1883-31 dec 1935]) – uk British Libr Newspaper [074]
Berlinguet, Francois Xavier *see* Rapport et plans sur des ameliorations generales dans le havre de quebec
Berlinguet, François-Xavier *see* Report and plans on general improvements in the quebec harbour
Berlinische blaetter / ed by Biester, Johann Erich – Berlin 1797-98 – (= ser Dz. abt literatur) – 2v on 20mf – 9 – €200.00 – mf#k/n4613 – gw Olms [430]
Berlinische blaetter *see* Berlinische monatsschrift 1783-96 / berlinische blaetter 1797-98 / neue berlinische monatsschrift 1799-1811
Berlinische monatschrift – Berlin, 1783-1790. v1-16 – 134mf – 8 – mf#H-574 – ne IDC [700]
Berlinische monatsschrift / ed by Gedike, Friedrich et al – Berlin 1783-96 – (= ser Dz) – 28v on 197mf – 9 – €1182.00 – mf#k/n369 – gw Olms [074]
Berlinische monatsschrift 1783-96 / berlinische blaetter 1797-98 / neue berlinische monatsschrift 1799-1811 / ed by Gedike, Friedrich & Biester, Johann Erich – Berlin, Stettin 1783-1811 [mf ed Hildesheim 1992] – 58v on 363mf – 9 – diazo €1498.00 silver €1890.00 – gw Olms [074]
Berlinische nachrichten von staats- und gelehrten sachen – Berlin DE, 1740 30 jun-1811 29 jun, 1812-1874 31 oct [gaps] – 150r – 1 – (filmed by misc inst: 1849 jan-mar, 1872 jan-mar [2r]; title varies: 4 jun 1872: spenersche zeitung) – gw Misc Inst [943]
Berlinische privilegirte zeitung – Berlin DE, 1844 n122-129, 1903 15-31 dec – 2r – 1 – (filmed by bnl: 1816-1919 [1318r], filmed by mikropress: 1812-15 [3r], 1918-1933 [86r], filmed by misc inst: 1725, 1729, 1749 8 may-9 oct, 1761-1800 [gaps], 1804-11 [gaps], 1816 2 jul-31 dec, 1818 2 jul-1827, 1828 1 apr-1830, 1843 30 jan-1846 24 dec [gaps], 1847-1917, 1934 jan-mar, 1909-23 [647r]; title varies: 1779: koeniglich berlinische privilegirte staats- und gelehrte zeitung, 1785: koeniglich privilegirte berlinische zeitung von staats- und gelehrten sachen, 24 dec 1911: vossische zeitung, (prior 1721 numerous earlier titles); sonntagsbeilage: 1875-88, 9 mar 1890 [gaps], 1899-1916 [gaps] [10r]; filmed with: bibliographisches repertorium 1858-1903, zeitbilder: 1914-19, 1930-25 mar 1934) – gw Mikrofilm; uk British Libr Newspaper; gw Mikropress; gw Misc Inst [074]
Berlinische sammlungen zur befoerderung der arzneywissenschaft, der naturgeschichte der haushaltungskunst, cameralwissenschaft und der dahin einschlagenden litteratur / ed by Martini, Friedrich Heinrich Wilhelm – Berlin 1768-79 – (= ser Dz. abteilung naturwissenschaft) – 10v[zu je 6 h] on 45mf – 9 – €450.00 – mf#k/n3245 – gw Olms [500]
Berlinisches archiv der zeit und ihres geschmacks / ed by Meyer, Fr Ludw Wilh – Berlin 1795-1800 – (= ser Dz. abt literatur) – 49mf – 9 – €490.00 – (1799f ed by fr eberh rambach & ignatz aurelius fessler) – mf#k/n4603 – gw Olms [943]
Berlinisches journal fuer aufklaerung / ed by Fischer, Gottlob Nathanael et al – Berlin 1788-90 – (= ser Dz) – 8v[=25st] on 16mf – 9 – €160.00 – mf#k/n706 – gw Olms [370]
Berlinisches magazin der wissenschaften und kuenste / ed by Wippel, Wilh Jakob – Berlin 1782-84 – (= ser Dz) – jg 1 [st 1-4]; jg 2 [st 1] on 7mf – 9 – €140.00 – mf#k/n364 – gw Olms [074]
Berlin-lichtenberger tageblatt – Berlin DE, 1919, 1922 2 jan-31 mar, 1930 9 aug-1933, 1936-1939 30 jun, 1939 2 oct-1943 13 mar – 1 – (title varies: 2 jan 1922: lichtenberger tageblatt, 2 feb 1936: lichtenberger anzeiger und tageblatt, 1 jan 1939: anzeiger und tageblatt, 1 apr 1941: karlshorst-lichtenberger nachrichten) – gw Misc Inst [074]
Berlin-Neubart, Heinrich *see* Historia de la imagineria colonial en guatemala
Berlin-schoeneberger tageblatt – Berlin DE, 1920 1 apr-1921 31 mar, 1924 2 jan-28 jun, 1924 1 oct-1926 30 sep [gaps] – 1 – gw Misc Inst [074]
Berlin-tegeler-anzeiger – Berlin DE, 1919-1920 30 jun, 1922 4 jan-28 jun – 1 – (with gaps) – gw Misc Inst [074]
Berlin-wilmersdorfer zeitung – Berlin DE, 1908 may 9-18, 1919 jul-1925 jun, 1925 sep-1926 mar, 1926 jul-1931 mar, 1931 jul-1932, 1934 jan-apr, 1934 jul-1938 apr, 1939 jul-1944 jun – 87r – 1 – (title varies: 3 may 1920: der berliner westen, 1 jan 1933: der westen, beginning: wilmersdorfer zeitung) – Dist. gw Mikrofilm – gw Misc Inst [074]
Berlioz, Hector *see*
– Damnation de faust
– [L'imperiale; cantata for the paris exhibition, op 26]
– Werke

Berlyn, Graeme P *see* Journal of sustainable forestry
Berman, L *see* S-peterburgskiia evreiskiia uchilishcha
Bermann, Moriz *see* Alt-wien in geschichten und sagen fuer die reifere jugend
Bermant, George et al *see* Protracted civil trials
Bermant, Gordon *see*
– Alternative dispute resolution in a bankruptcy court
– Conduct of the voir dire examination
– Jury selection procedures in u.s. district courts
– Preparing a u.s. court for automation
– The quality of advocacy in the federal courts
– The voir dire examination, juror challenges, and adversary advocacy
Bermant, Gordon et al *see* Chapter 11 venue change by large public companies
Bermejo, Fernando *see* Campanulas
Bermejo, Vladimiro *see* Vida y hechos del conquistador del peru
Bermingham, Joseph Aldrich *see* The rise and decline of irish industries
Bermondsey, Eng St Mary Magdalene (Parish) *see* Parish registers of st mary magdalene, bermondsey, 1548-1609
Bermondsey news – Southwark, England 6 oct 1987-21 jul 1988 – 1 – (cont by: southwark & bermondsey news'[4 aug 1988-31 dec 1992]; southwark news [14 jan 1993-]) – uk British Libr Newspaper [072]
Bermondsey & rotherhithe advertiser : a local journal for the south-east of london – Southwark, London 25 may 1868-15 aug 1874 [wkly] – 1 – (cont by: bermondsey & rotherhithe advertiser & southwark journal [12 sep 1874-21 jan 1882]) – uk British Libr Newspaper [072]
Bermuda and the american revolution : 1760-1783 / Kerr, Wilfred Brenton – Princeton, NJ. 1936 – 1r – us UF Libraries [972]
Bermuda baptist church – Dillon County, SC. 1944-45, 1957-72 – 1 – $5.04 – us Southern Baptist [242]
Bermuda colonist – Hamilton, Bermuda 9 nov 1870-3 may 1871, 7 jan 1874-27 aug 1879, 29 jun 1881-12 apr 1882, 3 jan 1883-26 dec 1888, 2 jan 1895-28 dec 1914 (daily) (imperfect) – 1 – (fr 1870-1884 publ at st george; cont by: bermuda colonist & daily news [1 jan 1915-27 mar, 2 aug-21 dec 1920, incorp with: royal gazette & colonist daily]) – uk British Libr Newspaper [079]
Bermuda days / March, Bertha – New York, NY. 1929 – 1r – us UF Libraries [972]
Bermuda digest of statistics 1973-1974 / Bermuda. Statistical Office – (= ser Latin american & caribbean...1821-1982) – 2mf – 9 – uk Chadwyck [318]
Bermuda gazette – Hamilton, Bermuda 17 jan-14 feb 1784 (imperfect) – 1 – (cont by: bermuda gazette & weekly advertiser [21 feb 1784-1 jan, 23 jul 1785, 7 jan 1786-29 dec 1787, 10 jan 1789-17 feb 1798]; bermuda gazette [24 feb 1798-16 mar 1804]; bermuda gazette & weekly advertiser [23 apr 1804-28 dec 1805, 12 oct 1811-25 dec 1813, 7 jan 1815-5 oct 1816]; bermuda gazette, & hamilton & st george's weekly advertiser [12 oct 1816-8 aug 1818, 2 jan 1819-29 sep 1821]; bermuda gazette [6 oct 1821-20 nov 1824, 1 jan,19 nov 1825, 28 jan-9 dec 1826, 7 jul, 1 sep 1827]; royal gazette, bermuda commercial & general advertiser & recorder [29 jan 1828-29 dec 1903, 23 feb 1904-30 jul 1918, 1 oct 1918-30 dec 1920]) – uk British Libr Newspaper [079]
Bermuda in the old empire / Wilkinson, Henry Campbell – London, England. 1950 – 1r – us UF Libraries [972]
Bermuda in three colors / Wells, Carveth – New York, NY. 1935 – 1r – us UF Libraries [972]
Bermuda journey / Zuill, W S – Hamilton, Bermuda. 1946 (1965 printing) – 1r – us UF Libraries [919]
Bermuda mid-ocean news and colonial government gazette – Hamilton, Bermuda 2 jan 1958-8 aug 1970 (daily) (imperfect) – 1 – uk British Libr Newspaper [079]
Bermuda news pictorial – Hamilton, Bermuda 31 jan 1960-18 nov 1961 (imperfect) – 1 – uk British Libr Newspaper [079]
Bermuda past and present / Hayward, Walter Brownell – New York, NY. 1910 – 1r – us UF Libraries [972]
Bermuda recorder – Hamilton, Bermuda 3 jan 1959-5 may, 6 oct, 10 nov 1967; 13 apr 1968-3 jan,14 mar 1969-25 jul 1970 – 1 – uk British Libr Newspaper [079]
Bermuda sampler, 1815-1850 / Zuill, William – Hamilton, Bermuda. 1937? – 1r – us UF Libraries [972]
Bermuda. Statistical Office *see* Bermuda digest of statistics 1973-1974
Bermuda sun – Hamilton, Bermuda. Oct 1 1993-June 24 1994; July 1-Dec 30 1994; 1995 – 5r – 1 – us L of C Photodup [079]
Bermuda sun and official government gazette – Hamilton, Bermuda 30 may-12 oct 1964, 7 jan-30 dec 1967, 13 apr-21 dec 1968, 1 feb 1969-11 jul 1970 – 1 – uk British Libr Newspaper [079]
Bermuda sun weekly – Hamilton Bermuda 24 oct 1964-31 dec 1966 – 1 – uk British Libr Newspaper [079]
The bermuda times – Hamilton, Bermuda. Mar 4 1987-Dec 15 1989; 1990-June 1994; Aug 5 1994-1995 – 9r – 1 – us L of C Photodup [079]
Bermuda's 'oldest inhabitants' / Smith, Louisa Hutchings – Sevenoaks, England. 1934 – 1r – us UF Libraries [972]
Bermuda's story / Tucker, Terry – Hamilton, Bermuda. 1959 – 1r – us UF Libraries [972]
Bermudes, Felix *see* Sem armas no meio das feras
Bermudez de la Torre y Solier, Pedro Jose *see*
– Censura del senor doctor d pedro joseph bermudez de la torre y solier, alguacil mayor de corte desta real audiencia de lima
– Triunfos del santo oficio peruano
Bermudez de Pedraza, F *see* Arte legal para estudiar jurisprudencia con la paratitla y exposicion...
Bermudez, Jose Alejandro *see* Compendio de la historia de colombia
Bermudez M, Antonio *see* Ester
Bermudez Meza, Antonio *see* Prismas
Bermudez, Nestor *see* Mensajeros del ideal
Bermudez Plata, C *see* Sevilla. archivo general de indias. seccion de contratacion. catalogo de pasajeros de indias...(1509-1599)
Bermudez Plata, Cristobal *see* Catalogo de pasajeros a indias durante los siglos 16, 17 y 18. vol 2 (1535-1538)
Bermudez, Ricardo J *see* Cuando la isla era doncella
Bermudian : a commercial, literary, and political weekly journal – Hamilton, Bermuda 9 feb-2 mar, 27 apr, 4,25 may-15 jun, 20 jul-17 aug, 14 sep-19 oct, 9 nov 1839-15 feb 1840; 13 jan, 3-17 mar 1858; 5 jul 1871- 31 dec 1873; 12 oct-14 dec 1875 – 1 – uk British Libr Newspaper [079]
Bermunkas = Wage worker – Chicago: General Exec Board of Industrial Workers of the World, 1923-; 1948-feb 1952 – 2r – 1 – us CRL [331]
Bern : black education resource newsletter / Education Resource Associates – 1985 jan/feb – 1r – 1 – mf#5132263 – us WHS [370]
Bern. Sektion des Schweizerischen Verbandes fuer Frauenstimmrecht *see* Jahrbuch der schweizerfrauen
Bernab y Thomas, Alfred *see* After coronado. spanish exploration northeast of new mexico, 1696-1727...
Bernadau, Pierre *see* Tableau de bordeaux
Bernadi Mas, Jose *see* Momentos
Bernadine a Piconio *see* An exposition of the epistles of st paul
Bernadskogo, V N et al *see* Ocherki istorii karelii
Bernal, Calixto *see* Vindicacion
Bernal diaz del castillo / Cunninghame Graham, Robert Bontine – London, England. 1915 – 1r – us UF Libraries [972]
Bernal Escobar, Alejandro *see* Educacion en colombia
Bernaldez, Andres *see*
– Historia de los reyes catolicos
– Historia de los...don fernando y..
Bernaldez, Fernando *see* Resena sobre la traida de aguas a badajoz
Bernaldo De Quiros, Constancio *see*
– Cursillo de criminologia y derecho penal
– Lecciones de legislacion penal comparada
Bernaldo, Ruben *see* Puerta inicial
Bernanos no brasil / Sarrazin, Hubert – Petropolis, Brazil. 1968 – 1r – us UF Libraries [972]
Bernard, A *see* Recueil des chartes de l'abbaye de cluny
Bernard, Auguste et al *see* Apologetische vortraege
Bernard, Bayle (Mrs) *see* Retrospections of america, 1797-1911
Bernard, David *see* Light on masonry
Bernard de Clairvaux, Saint *see*
– Life and works of saint bernard, abbot of clairvaux
– Predigten des heiligen bernhard in altfranzoesischer uebertragung
Bernard de montfaucon et les bernardins 1715-1750 / Broglie, E de – Paris. v1-2. 1891 – €25.00 – ne Slangenburg [241]
Bernard delicieux et l'inquisition albigeoise (1300-1320) / Haureau, Barthelemy – Paris: Hachette 1877 [mf ed 1990] – 1mf – 9 – 0-7905-6292-8 – (in french & latin) – mf#1988-2292 – us ATLA [944]
Bernard, Dominique F *see* Exercice du pouvoir actuel et les desiderata du pe...
Bernard, Esther *see*
– Briefe waehrend meines aufenthalts in england und portugal an einen freund
– Neue briefe durch england und portugal
Bernard, Henry Norris *see* The mental characteristics of the lord jesus christ
Bernard, J *see* The irish liber hymnorum (hbs13-14)
Bernard, J H *see* The odes of solomon (ts8/3)
Bernard jean bettelheim : first missionary to okinawa / Bull, Earl – 1r – 1 – $35.00 – mf#um-233 pt1-2 – us Commission [242]
Bernard, John *see* Retrospections of america, 1797-1911
Bernard, John Henry *see*
– From faith to faith
– The present position of the irish church
– Studia sacra
Bernard, John Hernry *see* Kant's critical philosophy for english readers
Bernard, Joseph Carl *see* Faust, opera [in three acts by j c bernard music by] spohr
Bernard, Louis *see* Resume de l'histoire de bretagne
Bernard, M *see* La philosophie religieuse de gabriel marcel
Bernard, Marc *see* En hydravion au-dessus du continent noir
Bernard, Mathieu Adolphe *see* Manuel de droit constitutionnel et administratif
Bernard of clairvaux : the times, the man, and his work / Storrs, Richard Salter – New York: Scribner 1892 [mf ed 1990] – 2mf – 9 – 0-7905-6323-1 – (english text, notes in latin & french) – mf#1988-2323 – us ATLA [241]
Bernard of Clairvaux, Saint *see* Cantica canticorum
Bernard, P *see* Explication de l'edict de nantes par les autres edicts de pacification, declarations et arrests de reglement
Bernard Quaritch, Ltd *see* The numbered catalogues numbers 45-888
Bernard, Richard *see* Ruth's recompence
Bernard, Richard B *see* A tour through some parts of france, switzerland, savoy, germany and belgium
Bernard shaw's phonetics / Saxe, Joseph – Copenhagen, Denmark. 1936 – 1r – us UF Libraries [420]
Bernard, Theos *see*
– Hindu philosophy
– Philosophical foundations of india
Bernard, Thomas Dehany *see*
– The central teaching of jesus christ
– The progress of doctrine in the new testament
– The songs of the holy nativity
– The witness of god
Bernardes / Amora, Paulo – Sao Paulo, Brazil. 1964 – 1r – us UF Libraries [972]
Bernardez, Manuel *see* Brasil
Bernardi Abbatis Casinensis *see* In regulam s benedicti expositio
Bernardi, Bernardo *see* Mugwe
Bernardino, S *see* Tractatus de septem donis spiritus sancti
Bernardo de palissy / Tapia Y Rivera, Alejandro – San Juan, Puerto Rico. 1944 – 1r – us UF Libraries [730]
Bernardo pereira de vasconcellos e seu temp / Sousa, Octavio Tarquinio de – Rio de Janeiro. 1937 – 1 – us CRL [920]
Bernardo pereira de vasconcellos e seu tempo / Sousa, Octavio Tarquinio De – Rio de Janeiro, Brazil. 1937 – 1r – us UF Libraries [972]
Bernardston 1735-1897 – Oxford, MA (mf ed 1987) – (= ser Massachusetts vital records) – 25mf – 9 – 0-87623-055-9 – (mf 1-4: town & vital records 1762-86. mf 5-9: town & vital records 1786-1815. mf 10-11: vital records 1747-1815. mf 12-18: proprietors records 1735-1819. mf 19: proprietors records 1819-55. mf 20-21: b,m,d 1844-60. mf 22-25: b,m,d 1861-97) – us Archive [978]
Bernardus de Parentinis *see* Lilium sive elucidarius difficultatum cira officium missae
Bernardus gutolfi monachi : seu vita sanctissimi p n bernardi per monachum gutolphi / ed by Heimb, Theop – Norimbergae. v1-2. 1743-46 – v1 22mf v2 27mf – 8 – €94.00 – ne Slangenburg [241]
Bernatskii, M *see* Denezhnaia reforma v sovetskoi rossii
Bernatz, Johann Martin *see* Scenes in ethiopia
Bernatzik, Hugo Adolf *see* Geheimnisvolle inseln tropen-afrikas
Bernau, J H *see* Missionary labours in british guiana
Bernauer generalanzeiger – Berlin DE, 1925 4 jan-29 oct, 1926 1 apr-29 jun, 1927 4 jan-29 mar, 1932-33, 1934 3 apr-1935 30 mar, 1935 1 jul-1936 31 mar – 1 – (publ in [berlin-]pankow) – gw Misc Inst [074]
Bernay, Alexandre de *see* Li romans d'alixandre
Bernays, Edward L *see* Propaganda
Bernays, Jacob *see*
– The ophrastos' schrift ueber frommigkeit
– Ueber das phokylideische gedicht
Bernays, Michael *see*
– Briefe von und an michael bernays
– Schriften zur kritik und litteraturgeschichte
– Ueber kritik und geschichte des goetheschen textes
Berncastle, J *see* A voyage to china
Berndorff, Hans Rudolf *see* Shiva und die galgenblume
Berndt, Christina *see* Cd4- und cxcr4-vermittelte apoptose als moeglicher mechanismus der t-zell-depletion bei aids
Berndt, Inge *see* Was haltet ihr von jesus?
Berneisen, Ewald *see* Hoffmann von fallersleben als vorkaempfer und erforscher der niederlaendisch-vlaemischen litteratur
Berner beitraege zur geschichte der schweizerischen reformationskirchen / Billeter, M et al; ed by Nippold, F – Bern, 1884 – 5mf – 9 – mf#ZWI-18 – ne IDC [242]
Berner, Lewis *see* Mayflies of florida
Die berner taeufer bis 1532 / McGlothlin, William Joseph – Berlin: E Ebering, 1902 – 1mf – 9 – 0-7905-6764-4 – (incl bibl ref) – mf#1988-2764 – us ATLA [240]
Die berner-chronik des valerius a / Anshelm, V – Bern, Wyss, 1884-1901. 6 v – 32mf – 9 – mf#PBU-459 – ne IDC [240]
Bernes, J *see* Papel que responde a...damian de mayorga y guzman, medico...sobre...la calentura maligna
Bernet Kempers, August Johan *see* The bronzes of nalanda and hindu-javanese art
Bernewitz, Elsa *see*
– Die entrueckten
– Die zeitalter
Berney, Michael *see* Transition guide
Berney, Saffold *see* Hand-book of alabama
Bernfeld, Siegfried *see* Jerubbaal
Bernfeld, Simon *see*
– Dahat elohim
– Der talmud
Bernfield, Simon *see*
– Die lehren des judentums nach den quellen
– Muhamed
Bernhagen, Joerg *see* Die ernaehrungsbedingten mangelkrankheiten der erwachsenen feldarbeitersklaven im antebellum sueden der usa. 1810-1860
Bernhard felsenthal : teacher in israel / Felsenthal, Emma – New York, NY. 1924 – 1r – us UF Libraries [920]
Bernhard pankok : das gebrauchsgraphische werk / Heinen, Mechthild – (mf ed 1993) – 6mf – 9 – €62.50 – 3-89349-706-4 – mf#DHS 706 – gw Frankfurter [700]
Bernhard von breydenbach and his journey to the holy land, 1843-1844 : a bibliography / Davies, H W – 4mf – 9 – mf#HT-278 – ne IDC [915]
Bernhard, W *see* Allgemeines deutsches lieder-lexikon
Bernhardi, Theodor von *see* Volksmaehrchen und epische dichtung
Bernhardische und eckhartische mystik in ihren beziehungen und gegensaetzen : eine dogmengeschichtliche untersuchung / Bernhart, Joseph – Kempten: J Koesel 1912 [mf ed 1992] – 1mf – 9 – 0-524-03215-7 – (incl bibl ref) – mf#1990-0843 – us ATLA [230]
Bernhardt, August *see* Forstliche zeitschrift
Bernhardt, Ernst *see* Vulfila
Bernhardt, G de *see* Handbook of commercial treaties etc, between great britain and foreign powers
Bernhardt, Wilhelm *see*
– Einführung in goethe's meisterwerke; selections from goethe's poetical and prose works
– Ludwig uhlands politische betaetigungen und anschauungen
Bernhardy, Gottfried *see* Grundriss der griechischen litteratur
Bernhart, Joseph *see*
– Bernhardische und eckhartische mystik in ihren beziehungen und gegensaetzen
– Vom papste
Bernheim, Ernst *see*
– Lehrbuch der historischen methode und der geschichtsphilosophie
– Quellen zur geschichte des investiturstreites
– Das wormser konkordat und sein vorurkunden
Bernheim, Gotthardt Dellmann *see* The history of the evangelical lutheran synod and ministerium of north carolina
Bernheim, Roger *see* Die terzine in der deutschen dichtung von goethe bis hofmannsthal
Bernheim-jeune et cie catalogue – Paris – (= ser Art exhibition catalogues on microfiche) – (individual titles not listed separately) – uk Chadwyck [700]
Bernhoeft, Franz *see* Kauf, miethe und verwandte vertraege in dem entwurfe eines buergerlichen gesetzbuches fuer das deutsche reich
Bernice first baptist church – Bernice, LA. 1900-17 – 1 – $7.56 – us Southern Baptist [242]
Bernier, Bernard *see*
– Memorials respecting the working of the laws governing reformatory and industrial schools
– Reponse a quelques observations formulees...29 mars 1893
Bernier, Francois *see*
– Travels in the mogul empire
– Travels in the mogul empire, a.d. 1656-1668
Bernieres de Louvigny, Jean de *see* Den inwendighen christenen

Bernier-Lesieur, Raymond see Bibliographie analytique des etudes pedologiques des sols des comtes dans la province de quebec

Bernikov, Ilya Stepanovich see Kratkii kurs tserkovnago prava pravoslavnoi tserkvi

Berninger, Gertrud see Adern in marmor

Bernis, Francois J de see Correspondance du cardinal de bernis, ministre d'etat

Bernisches magazin der natur, kunst und wissenschaften / ed by Wyttenbach, Jakob Samuel – Bern 1776-80 – (= ser Dz) – 10mf – 9 – €100.00 – mf#k/n315 – gw Olms [074]

Bernisches mausoleum : oder vorderst gott zur ehr, lob und dank... – Bern, Bondelin. 6pts. 1740-1742 – 13mf – 9 – mf#ZWI-43 – ne IDC [240]

Bernkasteler zeitung see Gemeinnuetziges wochenblatt

Bernkast'ler tageblatt see Gemeinnuetziges wochenblatt

Bernkast'ler wochenblatt see Gemeinnuetziges wochenblatt

Bernkast'ler zeitung see Gemeinnuetziges wochenblatt

Berno, K A see An examination of rider arousal in the three phases of an equestrian combined training event

Bernoulli, A Basler see Die chroniken des karthaeuser klosters in klein-basel

Bernoulli, C A see Christentum und kultur

Bernoulli, Carl Albrecht see
- Die heiligen der merowinger
- Das konzil von nicaea
- Der schriftstellerkatalog des hieronymus
- Die wissenschaftliche und die kirchliche methode in der theologie

Bernoville, Gactan La Croix de Sang see Histoire du cure santa cruz.

Bernstein, A see Some jewish witnesses for christ

Bernstein, Aaron David see
- Die jahre der reaktion
- Ursprung der sagen von abraham, isaac und jacob

Bernstein, Bela see Negyvennyolcas magyar szabadsagharc es a zsidok

Bernstein, Eduard see Dokumente zum weltkrieg 1914

Bernstein, Herman see Truth about "the protocols of zion"

Bernstein, Ignatz see Katalog dziel tresci przyslowiowej skladajacych bibljoteke ignacego bernsteina

Bernstein, P see Der buddhismus und das christentum vor dem forum des philosophischen and ethischen denkens

Bernstein, S see Shomre ha-homot

Bernstein, Simon see Der zionismus

Bernt, Alois see Handbuch der deutschen literaturgeschichte

Bernus, Alexander von see Das reich (klp10)

Bernus, Auguste see Richard simon et son histoire critique du vieux testament

Berolzheimer, Fritz see The world's legal philosophies

Berquin, M (Arnaud) see Jeunes officiers

Berquin-Duvallon see Schilderung von louisiana

Berr, Georges see
- Azais
- Maitre bolbec et son mari
- Monsieur beverley

Berres, Frauke Rita see Nursing bottle caries

Berrien, John M. see berrien papers

Berrigan advocate – Berrigan jan 1899-dec 1904 – 1mf – 9 – A$68.99 vesicular A$74.49 silver – at Pascoe [079]

Berrima district leader – Berrima jun 1898-nov 1901 – 1mf – 9 – A$71.85 vesicular A$77.35 silver – at Pascoe [079]

Berrima district post – Berrima, jan 1960-dec 1968 – 9mf – 9 – A$634.88 vesicular A$684.38 silver – at Pascoe [079]

Berrima district post – Berrima, jan 1969-dec 1983 – 9 – at Pascoe [079]

Berrima district post – Moss Vale – 9 – at Pascoe [079]

Berring, Robert C see Legal reference services quarterly

Berrios, Jose David see Elementos de gramatica de la lengua keshua

Berrios Rodriguez, Brigido see Batalla por la produccion

Berrows worcester journal, 1712-1850 : from worcester public library – 30r – 1 – (formerly known as: the worcester postman) – mf#96913 – uk Microform Academic [074]

Berrow's worcester journal (a) – Worcester, England 11 oct 1753-9 mar 1820 – 1 – (cont: worcester journal [30 jun 1748-4 oct 1753]; some iss are mutilated) – uk British Libr Newspaper [072]

Berrow's worcestershire journal – Worcester, England 8 jun 1985-19 jun 1987 [mf 1986-] – 1 – (cont: berrow's journal [26 jan 1984-1 jun 1985]; cont by: berrow's worcester journal [26 jun 1987-]) – uk British Libr Newspaper [072]

Berry, Charles A see Mischievous goodness

Berry, Charles Treat see An historical survey of the first presbyterian church, caldwell, nj

Berry, D M see The sister martyrs of ku cheng

Berry, Edward Wilber see Tree ancestors

Berry, George Keys see The eight leading churches

Berry, George Ricker see
- Book of proverbs
- The interlinear literal translation of the hebrew old testament
- A new greek-english lexicon to the new testament
- The old testament among the semitic religions

Berry, Jack see
- Proceedings...conference on african languages and literatures (1966: northwestern university)
- Pronunciation of ewe
- Pronunciation of ga
- Spoken art in west africa

Berry, James R see Foreign student-athletes and their motives for attending north carolina ncaa division 1 institutions

Berry, John Cutting see Points of etiquette which we should know and observe in our social relations with the japanese

Berry pickin – 1974 sep-1975 sep – 1r – 1 – mf#351005 – us WHS [634]

Berry register – Berry, jan-dec 1894; jan 1898-dec 1905 – 3r – A$148.46 vesicular A$164.96 silver – at Pascoe [079]

Berry, Ronald see Journal of internet commerce

Berry, Thomas Sterling see Christianity and buddhism

Berry, William Grinton see Bishop hannington

Berryer see Revolutionary justice in spain

Berryhill baptist church – Charlotte, NC. 1895-1960. Bulletins. 1954-60 – 1 – $48.96 – us Southern Baptist [242]

Berryman, John R see Berryman's digest of the law of insurance

Berryman's digest of the law of insurance : being an analysis of fire, marine, life and accident insurance cases / Berryman, John R – Chicago: Callaghan. 1v. 1888 (all publ) – 10mf – 9 – $15.00 – (cont: sansum's digest of insurance cases) – mf#LLMC 95-130 – us LLMC [346]

Berryville first baptist church – Berryville, AR. 1891-1982 – 1 – $171.09 – us Southern Baptist [242]

Bers, E see Voprosy nashego vremeni

Bers, G see Lavenir

Bersarabie in nayntsen akhtsen / Gutman, Golde – Buenos Aires, Argentina. 1940 – 1r – us UF Libraries [939]

Bersaucourt, Albert De see Etudes et recherches

Bershadski, Isaiah see Neged ga-zerem

Bersier, E see Projet de revision de la liturgie des eglises reformees de france prepare sur l'invitation du synode general officieux

Bersier, Eugene see
- Histoire du synode general de l'eglise reformee de france, paris, juin-juillet 1872
- Liturgie a l'usage des eglises reformees

Berstl, Julius see The tentmaker

Bert, Georg see
- Aphrahat's des persischen weisen homilien

Bert, Paul see
- Les colonies francaises
- La morale des jesuites
- Preface to la morale des jesuites

Bert, Pierre Nicolas see Esprit de parti

Bertachinus, Ioan Firm see Repertorium...iuris

Bertanam kapas di djawa : diterbitkan dengan izin hodohan / Ishikawa, T – Djakarta: Djawa Gunseikanboe (Balai Poestaka), 2603 (B P n1526) – 27p 1mf – 9 – mf#SE-2002 mf47 – ne IDC [959]

Bertaux, Felix see Panorama de la litterature allemande contemporaine

Bertelli, P see Vite degl' imperatori de tvrchi...

Bertenburg, Carl see Der bergarbeiterausstand und die techn. grubenbeamten gelsenkirchen 1905

Berthe, Augustine see Garcia moreno

Bertheau, Ernst see
- Die buecher esra, nechemia und ester
- Die sieben gruppen mosaischer gesetze in den drei mittleren buechern des pentateuchs
- Die sprueche salamo's – der prediger salomo's

Berthelon, Christiane see Expression du haut degre en francais contemporain

Berthelot, Amable see Dissertation sur le canon de bronze que l'on voit dans le musee de m. chasseur a quebec

Berthelot, Hector see
- Les mysteres de montreal

Berthelot, Marcellin see Science et morale

Berthelot, Rene see La sagesse de shakespeare et de goethe

Berthevin, Jules see Recherches historiques sur les derniers jours des rois de france, leurs funerailles, leurs tombeaux

Berthier, Hugues Jean see Manuel de la langue malgache (dialecte merina)

Berthier, Jean-Baptiste see Compendium theologiae dogmaticae et moralis

Berthier, Joachim Joseph see L'etude de la somme theologique de saint thomas d'aquin

Berthier, Louis see Memoires du marechal berthier

Bertholdt, Leonhard see
- Disseritur de praecipuis ad primas causas christianismi formaliter spectati penetrandi subsidiis
- Handbuch der dogmengeschichte

Bertholet, Alfred see
- Buddhismus und christentum
- Daniel und die griechische gefahr
- Die eigenart der alttestamentlichen religion
- Die israelitischen vorstellungen vom zustand nach dem tode
- Die juedische religion von der zeit esras bis zum zeitalter christi
- Religionsgeschichtliches lesebuch
- Seelen-wanderung
- Die stellung der israeliten und der juden zu den fremden
- The transmigration of souls

Berthoud, Aloys see Le calvinisme de l'avenir

Berthoud van Berchem, Jacob see Itineraire de la vallee de chamonix, d'une partie du bas-vallais et des montagnes avoisinantes

Berthre de Bourniseaux, Pierre see Histoire de louis 16

Berti, Jose see Espejismo de la selva

Bertie township herald – Ontario, CN. jan 1930-dec 1931 – 2r – 1 – cn Commonwealth Imaging [071]

Bertillon 166 / Soler Puig, Jose – Habana, Cuba. 1960 – 1r – us UF Libraries [972]

Bertillon, Alphonse see Identification anthropometrique; instructions signaletiques

Bertin, George see Abridged grammars of the languages of the cuneiform inscriptions

Der bertin-altar aus st-omer im kaiser-friedrich-museum zu berlin / Klemm, Wilhelm Bernhard – Leipzig, 1913 (mf ed 1993) – 2mf – 9 – €31.00 – 3-89349-326-3 – mf#DHS-AR 180 – gw Frankfurter [720]

Bertini, Giovanni see La rivoluzione spagnola

Bertolacci, Anthony see A view of the agricultural, commercial, and financial interests of ceylon

Bertol-Braivil see Main droite et main gauche

Bertold haller / Pestalozzi, C – Elberfeld: R L Friderichs, 1861 – 1mf – 9 – mf#PBU-457 – ne IDC [240]

Bertold haller oder die reformation von bern / Kirchhofer, M – Zuerich: Orell Fuessli, 1828 – 3mf – 9 – mf#PBU-452 – ne IDC [242]

Bertoldo : ein beitrag zur jugendentwicklung michelangelos / Rohwaldt, Karl – Berlin, 1896 (mf ed 1995) – 1mf – 9 – €24.00 – 3-8267-3142-5 – mf#DHS-AR 3142 – gw Frankfurter [700]

Bertolino-Green, Dianne Lyn see Letty m. russell as pastoral theologian

Bertololy, Paul see
- Dora holdenrieth
- Liebe

Bertolt brecht und die geisteswelt des fernen ostens / Kim, Dae Tschong – Heidelberg, 1969 (mf ed 1994) – 2mf – 9 – €31.00 – 3-89349-996-2 – mf#DHS-AR 996 – gw Frankfurter [430]

Berton, Eugene see L'eglise de calvin a strasbourg (1538-1541)

Berton, Henri see
- Le concert interrompu
- Les deux mousquetaires
- Francois de foix
- Montano et stephanie
- Ninon chez mme de sevigne
- Valentin ou le paysan romanesque

Berton, Pierre see La rencontre

Bertoni, F see [Olimpiade]

Bertoni, G see Archivum romanicum

Bertos, Rigas see Jacopo torriti

Bertram, Adolf see The odoreti, episcopi cyrensis, doctrina christologica

Bertram, Chr Aug von see Litteratur- und theaterzeitung

Bertram, Christ August von see Ephemeriden der litteratur und des theaters

Bertram, Christian August von see Annalen des theaters

Bertram, Ernst see
- Deutsche arbeiten der universitaet koeln
- Heinrich von kleist
- Der rhein
- Strassburg
- Theodor fontanes briefe
- Von deutschem schicksal

Bertram, Joachim Christoph see Litterarische abhandlungen

Bertram, Johannes see Goethes faust im blickfeld des 20. jahrhunderts

Bertram, Kate see Caribbean cruise

Bertram, Kurt see Nachlass hans von seeckt (bestand n 247) bd 19

Bertram, Oswald see Geschichte der cansteinschen bibelanstalt in halle

Bertrand Agramonte, Lourdes see Tornasol

Bertrand de Molleville, Antoine see
- Memoires particuliers
- Memoires secrets pour servir a l'histoire de la derniere annee du regne de louis 16, roi de france

Bertrand herald – Bertrand, NE: H E Waters. 39th yr n8. oct 5 1928- [mf ed with gaps] – 1 – (cont: independent herald) – us NE Hist [071]

Bertrand, Isidore see
- Aujourd'hui et demain
- La trinite d'en bas

Bertrand, Jean see Congo belge

Bertrand, Jean Jacques Achille see
- Cervantes en el pais de fausto
- L tieck et le theatre espagnol

Bertrand, Joseph see
- Lettres edifiantes et curieuses de la nouvelle mission du madure
- Memoires historiques sur les missions des ordres religieux
- La mission du madure

The bertrand leader – Bertrand, NE: J L Witters. v1 n[1] sep 19 1896)-apr 7 1898// (wkly) [mf ed -mar 31 1898 (gaps)] – 1r – 1 – (absorbed by: phelps county journal) – us NE Hist [071]

Bertrand l'horloger / Premaray, Jules De – Paris, France. 1843 – 1r – us UF Libraries [440]

Bertrand, Lionel see
- Bibliotheque sulpicienne, vol 1
- Bibliotheque sulpicienne, vol 2
- Bibliotheque sulpicienne, vol 3
- Bibliotheque sulpicienne, vols 1-3

Bertrand, Louis see Saint augustin

The bertrand times – Bertrand, NE: C Clinton Page. v1 n44. oct 18 1895-may 1 1896 (wkly) – 1r – 1 – us NE Hist [071]

Bertrin, Georges see Lourdes

Bertsch, Hugo see
- Bob, der sonderling
- Die geschwister

Bertsche, Karl see
- Grillen und pillen aus abraham a sancta clara
- Neue predigten
- Neun neue predigten

Bertuch, Carl see Bemerkungen auf einer reise aus thueringen nach wien im winter 1805 bis 1806

Bertuch, Friedrich J see Peru

Bertuch, Friedrich Johann Justin et al see Allgemeine literatur-zeitung

Bertuch, Friedrich Justin see London und paris

Bertuch, Friedrich Justin et al see Journal des luxus und der moden

Berube, Joseph-Francois see Memoire au cardinal barnabo

Berufsauslese und anpassung einer fabrikarbeiterinnengruppe : (auf grund von untersuchungen in einer hamburger hartgummiwarenfabrik) / Mais, Clara Maria – Hamburg, 1928 (mf ed 1995) – 1mf – 9 – €24.00 – 3-8267-3171-9 – mf#DHS 3171 – gw Frankfurter [331]

Der berufsbeamte – Duesseldorf DE, 1927-33 [gaps] – 2r – 1 – gw Misc Inst [350]

Die berufsbegabung der alttestamentlichen propheten / Giesebrecht, Friedrich – Goettingen: Vandenhoeck & Ruprecht, 1897 – 1mf – 9 – 0-8370-3270-9 – mf#1985-1270 – us ATLA [221]

Das berufsbewusstsein jesu mit beruecksichtigung geschichtlicher analogien untersucht / Fritzsche, Volkmar – Zittau, 1905 (mf ed 1993) – 1mf – 9 – €24.00 – 3-89349-346-8 – mf#DHS-AR 199 – gw Frankfurter [240]

Berufsbezogene possible selves in der betrieblichen weiterbildung / Fliegen, Ina – 2000 – 3mf – 9 – 3-8267-2687-1 – mf#DHS 2687 – gw Frankfurter [300]

Das berufsideal der volksschullehrerin : unter besonderer beruecksichtigung des berufsmotives und des berufsvorbildes / Weinand, Maria – Koeln, 1931 (mf ed 1995) – 1mf – 9 – €24.00 – 3-8267-3169-7 – mf#DHS-AR 3169 – gw Frankfurter [370]

Berufsorientierung und selbstorientierung : praktischer versuch in der gymnasialen oberstufe. verlauf – analyse – kritik / Liliensiek, Peter – (mf ed 1998) – 5mf – 9 – €59.00 – 3-8267-2537-9 – mf#DHS 2537 – gw Frankfurter [373]

Die berufs-reise nach america : briefe der generalin von riedesel auf dieser reise und waehrend ihres sechsjaehrigen aufenthalts in america zur zeit des dortigen krieges in den jahren 1776 bis 1783 nach deutschland geschrieben / Riedesel, Friederike C von – Berlin 1801 [mf ed Hildesheim 1995-98] – 1v on 3mf – 9 – €90.00 – 3-487-26909-0 – gw Olms [860]

Die berufung zur erbschaft und die letztwilligen verfuegungen ueberhaupt nach dem entwurfe eines buergerlichen gesetzbuches fuer das deutsche reich / Petersen, Julius – Berlin: J Guttentag, 1889 – (= ser Civil law 3 coll; Beitraege zur erlaeuterung und beurtheilung des entwurfes eines buergerlichen gesetzbuches fuer das deutsche reich) – 2mf – 9 – (incl bibl ref) – mf#LLMC 96-604 – us LLMC [346]

Be-rumo shel 'olam / Wichniansky, Solomon Jacob – Odessa, Ukraine. 1894 – 1r – us UF Libraries [939]

Berurim / Yashar, Baruch – Tel-Aviv, Israel. 1953 – 1r – us UF Libraries [939]
Bervin, Antoine *see*
– Louis-edouard pouget
– Mission a la havane
– Vie etourdissante de jean lucksa
Berwick advertiser – Berwick-upon-Tweed, England 22 nov-27 dec 1823, 1 jan 1825-23 feb 1983 [mf 1808-1820, 1823, 1897-98, mar-dec 1912] – 1 – (wanting: jan-feb 1912; cont: british gazette, & berwick advertiser [2 jan 1808-16 dec 1820, 4 jan-15 nov 1823]; cont by: berwickshire & berwick advertiser [3 mar 1983-30 aug 1984]) – uk British Libr Newspaper [072]
The berwick advertiser – England, 1830; 1834; 1838; 1840; 1862; 1865; 1870-71; 1873-78; 1880-92; 1897; 1904-05; 1908-11; 1916-17; 1919-50; 1980-86; 1988; 1993- – 83+ r – 1 – uk British Libr Newspaper [072]
Berwick and kelso warder – [Scotland] Berwick-upon-Tweed: J Ramsay 14 nov 1835-dec 1898 (wkly) [mf ed 2004] – 47r – 1 – (cont by: berwick warder [jan 1859-dec 1884]; border counties gazette & agriculturist [jan 1884-dec 1885]; border counties gazette & berwick warder [jan 1886-dec 1898]) – uk Newsplan [072]
Berwick gazette – Berwick-upon-Tweed, England 20 oct 1988- [mf 1986-] – 1 – (cont: berwick leader [13 sep 1984-13 oct 1988]) – uk British Libr Newspaper [072]
Berwick, George *see* The forces of the universe
Berwick journal – Berwick-upon-Tweed, England 29 dec 1855-9 feb 1928 [mf Feb-Dec 1874, 1897, 1912] – 1 – (wanting: jan 1874, 1911; cont: illustrated berwick journal [16 jun-22 dec 1855]; cont by: berwick journal & north northumberland news [16 feb 1928-28 mar 1957]) – uk British Libr Newspaper [072]
Berwick warder – Berwick-upon-Tweed, England 31 dec 1858-12 sep 1884) – 1 – (cont: berwick & kelso warder [14 nov 1835-24 dec 1858]; cont by: border counties gazette & agriculturist [19 sep 1884-9 oct 1885]) – uk British Libr Newspaper [072]
Berwick y de Alba, Duque de *see* El mariscal de berwick
Berwickshire news (1880) : the county newspaper – Berwick: G F Steven n545 [jan 6 1880]-88th yr n4449 [apr 1 1957] (wkly) – 1 – (cont: berwickshire news and general advertiser; merged with: berwickshire advertiser to form: berwickshire news & the berwickshire advertiser; publ varies: a steven 1911-19; northumberland & berwickshire newspapers ltd 1924-mar 25 1957; tweedale press apr 1 1957) – uk Scotland NatLib [072]
Berwickshire news and east lothian herald : incorporating the berwickshire advertiser : established 1893-1957 – Berwick-upon-Tweed: Tweedale Press Group (wkly) [mf ed 1995-] – 1 – (cont: berwickshire and berwick advertiser (berwickshire ed)) – ISSN: 0964-2846 – uk Scotland NatLib [072]
Berwickshire news and general advertiser – Berwick: G F Steven jul 6 1869- (wkly) – 1 – (cont by: berwickshire news (1880)) – uk Scotland NatLib [072]
Berzelius, Jons Jakob *see* Leerboek der scheikunde
Berzinskiya, Efrayim Duber *see* Pelite efrayim
La besace d'amour : grand roman canadien historique inedit / Feron, Jean – Montreal: editions Edouard Garand, 1925 [mf ed 1987] – 1mf – 9 – (ill by albert fournier) – mf#SEM105P855 – cn Bibl Nat [830]
Besant, Anne Wood *see* Our corner
Besant, Annie *see* Bhagavad-gita
Besant, Annie (Wood) *see* Force no remedy
Besant, Annie Wood *see*
– Ancient ideals in modern life
– The ancient wisdom
– Annie besant
– Avataras
– The basis of morality
– Birth of new india
– Buddhist popular lectures
– The building of the kosmos and other lectures
– The changing world
– Death- and after?
– Esoteric christianity
– For india's uplift
– Four great religions
– Fruits of christianity
– How india wrought for freedom
– The immediate future and other lectures
– India
– India, bond or free
– Indian ideals in education, philosophy and religion, and art
– Initiation
– Introduction to theosophy
– Is christianity a success?
– Is theosophy anti-christian?
– Karma
– Lectures on political science
– Man, whence, how and whither
– My path to atheism
– Myth of the resurrection
– Natural history of the christian devil
– Natural religion versus revealed religion
– On the atonement
– The path of discipleship
– Popular lectures on theosophy
– Reincarnation
– The seven principles of man
– The story of the great war
– Theosophy
– Theosophy and the new psychology
– Theosophy in relation to human life
– Thought power, its control and culture
– The true basis of morality
– Vegetarianism in the light of theosophy
– The wisdom of the upanishats
– World problems of to-day
Besant, Walter *see*
– Constantinople
– The rise of the empire
– Studies in early french poetry
Besarabiyah – Tel-Aviv, Israel. 1941 – 1r – us UF Libraries [939]
Besault, Lawrence De *see*
– President trujillo
– President trujillo, his work and the dominican rep
Beschaeftigungen der berlinischen gesellschaft naturforschender freunde – Berlin, 1775-1779. v1-4 – 350mf – 9 – mf#8608c – ne IDC [590]
Beschaeftigungen der berlinischen gesellschaft naturforschender freunde / ed by Martini, Friedrich Heinrich Wilhelm – Berlin 1775-79 – (= ser Dz. abteilung naturwissenschaft) – 4v on 17mf – 9 – €170.00 – (1779 ed by friedrich wilhelm otto) – mf#k/n3259 – gw Olms [500]
Beschaeftigungen des geistes und des herzens / ed by Muechler, Joh Georg – Berlin 1755-57 – (= ser Dz) – 2v on 6mf – 9 – €120.00 – mf#k/n4964 – gw Olms [870]
Bescheiden omtrent zijn bedrijf in indie / Coen, Jan Pieterszoon – 's Gravenhage 1919-53 [mf ed 2004] – (= ser Rare printed sources and reference works for the history of dutch colonialism) – 7v in 8 on 77mf – 9 – €750.00 – mf#mmp115 – ne Moran [959]
Beschi, C J *see* Instructions to catechists
Beschi, Constantino Giuseppe *see* A grammar of high tamil
Beschi, Costantino Giuseppe *see* The adventures of the gooroo noodle
Beschi, Costanzo Giuseppe *see*
– Adventures of the gooroo paramartan
– Grammar of the high dialect of the tamil language, termed shen-tamil
Die beschneidung in ihrer geschichtlichen, ethnographischen, religioesen und medicinischen bedeutung / Steinschneider, Moritz et al; ed by Glassberg, Abraham – Berlin C: C Boas, 1896 – 1mf – 9 – 0-7905-3078-3 – (incl bibl ref) – mf#1987-3078 – us ATLA [390]
Beschouwende en praktikale godgeleerdheit... / Mastricht, P van – Rotterdam, 1749-53. 4v – 39mf – 9 – mf#PBA-263 – ne IDC [240]
Beschouwing der wereld : bestaande in hondert konstige figuuren... / Luyken, Jan – Amsteldam: Wed P Arentz, en K vander Sys, 1708 – 8mf – 9 – mf#0-350 – ne IDC [090]
Beschouwinge van zion... / Lodensteyn, J van – Utrecht, 1674-77. 4pts – 3mf – 9 – mf#PBA-232 – ne IDC [240]
Beschouwinge van zion... / Lodensteyn, J van – Utrecht, 1674-78. 5pts – 3mf – 9 – mf#PBA-233 – ne IDC [240]
Beschreibung der aegyptischen sammlung des niederlaendischen reichsmuseums der altertuemer in leiden / Boeser, P A A – Haag, 1909-1910 – 10mf – 8 – mf#NE-20068 – ne IDC [956]
Beschreibung der aegyptischen sammlung des niederlaendischen reichsmuseums der altertuemer in leiden / Boeser, P A A – Haag, 1911-1932 – 27mf – 9 – mf#NE-393 – ne IDC [930]
Beschreibung der ehemaligen venetianischen besitzungen auf dem festen lande und an den kuesten von griechenland / Grasset-Saint-Sauveur, Andre – Weimar 1801 [mf ed Hildesheim 1995-98] – 1v on 2mf – 9 – €60.00 – 3-487-26610-5 – gw Olms [914]
Beschreibung der insel st helena : nach ihrer geognostischen beschaffenheit und bildung, nebst nachrichten von dem klima, der naturgeschichte und den bewohnern derselben; aus dem englischen / Duncan, Francis – Weimar 1807 [mf ed Hildesheim 1995-98] – 1v on 2mf – 9 – €60.00 – 3-487-26556-7 – gw Olms [914]
Beschreibung der kaiserstadt constantinopel : ihrer umgebungen, der sitten und gebraeuche daselbstaus zuverlaessigen quellen / Zrecin, J – [Darmstadt] 1828 [mf ed Hildesheim 1995-98] – 1mf – 9 – €40.00 – 3-487-29081-2 – gw Olms [915]
Beschreibung der laender zwischen den fluessen terek und kur am caspischen meere / Marschall von Bieberstein, F A – Frankfurt am Main, 1800 – 7mf – 9 – mf#793 – ne IDC [914]
Beschreibung der raysz leonardt rauwolffen... : so er vor diser zeit gegen auffgang inn die morgenlender, fuernemlich syriam... / Rauwolff, L – New York. 1968-1988 (1) 1984-1988 (5) 1984-1988 (9) – 6mf – 9 – mf#9234 – ne IDC [915]
Beschreibung der reisen und entdeckungen im noerdlichen und mittlern africa in den jahren 1822 bis 1824 / Denham, Dixon – Weimar 1827 [mf ed Hildesheim 1995-98] – 1v on 5mf – 9 – €100.00 – 3-487-26482-X – gw Olms [916]
Beschreibung der russischen provinzen zwischen dem kaspischen und schwarzen meere / Klaproth, J [H] von – Berlin, 1814 – 3mf – 9 – mf#AR-1602 – ne IDC [914]
Beschreibung der stadt rom : mit beitraegen von barthold georg niebuhr und einer geognostischen abhandlung von f hoffmann; erlaeutert durch plaene, aufrisse und ansichten von den architekten knapp und stier... / Platner, Ernst – Stuttgart [u. a.] 1830-42 [mf ed Hildesheim 1995-98] – (= ser Fbc) – 5v on 24mf – 9 – €240.00 – 3-487-29245-9 – gw Olms [914]
Beschreibung des allegorischen feuerwerkes : welches an dem gedaechtnis-feste der... thron-besteigung...katharina der zweiten...zu st petersburg...den 28 junii 1763... – Spb, 1763 – 1mf – 9 – mf#0-1119 – ne IDC [700]
Beschreibung des koenigreichs hannover / Sonne, Heinrich D – Muenchen 1829-34 [mf ed Hildesheim 1995-98] – 5v on 14mf – 9 – €140.00 – 3-487-29530-X – gw Olms [914]
Beschreibung des kurfuerstenthums hessen / Landau, Georg – Kassel 1842 [mf ed Hildesheim: 1995-98] – (= ser Fbc) – x/649p on 4mf – 9 – €120.00 – 3-487-29500-8 – gw Olms [914]
Beschreibung des toedtlichen kriegs... / Langhans, J0 – Basel, 1619 – 1mf – 9 – mf#PBU-699 – ne IDC [240]
Beschreibung einer englischen missions-reise nach dem suedlichen stillen ocean : in den jahren 1796, 1797 und 1798 im schiffe duff, unter commando des capitains james wilson / Wilson, James – Weimar 1800 [mf ed Hildesheim 1995-98] – 1v on 3mf – 9 – €90.00 – 3-487-26611-3 – gw Olms [910]
Beschreibung einer reise durch deutschland und die schweiz im jahre 1781 : nebst bemerkungen ueber gelehrsamkeit, industrie, religion und sitten / Nicolai, Friedrich – Berlin [u a] 1783-95 [mf ed Hildesheim 1995-98] – 10v on 36mf – 9 – €360.00 – 3-487-29622-5 – gw Olms [914]
Beschreibung einer reise in das achenthal : und durch einige gegenden des isarkreises – Muenchen 1830 [mf ed Hildesheim 1995-98] – (= ser Fbc) – 2mf – 9 – €60.00 – 3-487-29484-2 – gw Olms [914]
Beschreibung einer reise in das indische meer in der fregatte risus : nach dem cap der guten hoffnung, den inseln bourbon, frankreich und den seschellen; nach madras und den inseln java, st paul und amsterdam; waehrend der jahre 1810 und 1811 / Prior, James – Weimar 1819 [mf ed Hildesheim 1995-98] – 1v on 2mf – 9 – €60.00 – 3-487-26508-7 – gw Olms [910]
Beschreibung einer reise nach st petersburg, stockholm und kopenhagen / Woltmann, J F – Hamburg 1833 [mf ed Hildesheim 1995-98] – (= ser Fbc) – 1v on 2mf – 9 – €60.00 – 3-487-27798-0 – gw Olms [914]
Beschreibung einer reise nach stuttgart und strasburg im herbste 1801 : nebst einer kurzen geschichte der stadt strasburg waehrend der schreckenszeit / Meiners, Christoph – Goettingen 1803 [mf ed Hildesheim: 1995-98] – (= ser Fbc) – 534p on 4mf – 9 – €120.00 – 3-487-29491-5 – gw Olms [914]
Beschreibung einer reise nach surinam und des aufenthaltes daselbst... : so wie von des verfassers rueckkehr nach europa ueber nord-amerika / Sack, Albert von – Berlin 1821 [mf ed Hildesheim 1995-98] – 2v on 6mf – 9 – €120.00 – 3-487-26882-5 – gw Olms [910]
Beschreibung meiner reise von hamburg nach brasilien im juni 1824 : nebst nachrichten ueber brasilien bis zum sommer 1825 und ueber die auswanderer dahin / Schumacher, P H – Braunschweig 1826 [mf ed Hildesheim 1995-98] – 1v on 1mf – 9 – €40.00 – 3-487-26863-9 – gw Olms [910]
Beschrijvende catalogus der pamfletten-verzameling : van de boekerij der remonstrantsche kerk te amsterdam / Rogge, Hendrik Cornelis – [Amsterdam: J. H. Scheltema, 1862-1865]. Chicago: Dep of Photodup, U of Chicago Lib, 1974 (1r); Evanston: American Theol Lib Assoc, 1984 (1r) – 1 – 0-8370-0581-7 – mf#1984-B396 – us ATLA [240]
Beschrijving der nederlandsche bezittingen in oost-indie / Aa, Abraham Jacob von der – Amsterdam: J F Schleijer, 1846 [mf ed 1987] – (ill) – 1 – (incl bibl footnotes) – mf#1852 – us Wisconsin U Libr [959]
Beschrijving van de egyptische verzameling in het rijksmuseum van oudheden te leiden / Boeser, P A A – Leiden, 1905 – 6mf – 8 – mf#NE-20069 – ne IDC [956]
Beschrijving van den kraton van Groot-Atjeh zijne verdedigingskracht en bewapening *see* Merkwaardigheden er in aangetroffen
Beschrijving van guiana, of de wilde kust, in zuid-america,... / Hartsinck, J J – Amsterdam, 1770. 2v – 11mf – 9 – mf#H-6180 – ne IDC [590]
Beschrijvinghe van alle de nederlanden... / Guicciardijn, L – Amsterdam, 1612 – 14mf – 9 – mf#0A-149 – ne IDC [720]
Das beschuetzte orchestre : oder desselben zweyte eroeffnung... / Mattheson, Johann – Hamburg: Schillerische Buchladen 1717 [mf ed 19–] – 3mf – 9 – (greater pt of work was written in refutation of j h buttstett's ut, re, mi, fa, sol, la, tota, musica) – mf#fiche 437 – us Sibley [780]
Beseda – Canberra City, ACT: The Czecholovak Australian Association of Canberra: cis1 roc1. srp 1964- (mthly) – 2r – 1 – us NE Hist [071]
Beseda – no. 1-7. 1923-25 – 1 – us L of C Photodup [460]
Beseda *see* Illiustrirovannyi literaturnyi ezhemesiachnyi zhurnal
Beseda venkovske rodiny – Prague, Czech Republic 27 oct 1950-12 jan 1951, 4 jan 1952-27 feb 1953 – 1 – uk British Libr Newspaper [077]
Beseduiushchii grazhdanin *see* Ezhemesiachnoe izdanie...
Beseduyushchii grazhdanin – St. Petersburg. 1789. v. 1-3 – 1 – us NY Public [999]
Besedy po russkomu raskolu i sektantstvu = Discussions on the russina schism and sectarianism / Flegmatov, Andrei – Tsaritsyn, 1892 – 1r – 1 – $52.84 set – (one of three items on reel by same author: poucheniya i besedy po russkomu i sektantstvu. – exhortations and idscussions on the russian schism and sectarianism. – 3rd ed 1899; poucheniya i besedy po russkomu raskolu i sektantstvu. – exhortations and discussions on the russian schism and sectariansim. – 2nd ed 1897) – us Southern Baptist [242]
Besedy v obshchestve liubitelei rossiiskoi slovesnosti pri imperatorskom moskovskom universitete – M., 1867-1871. v1-3 – 13mf – 9 – mf#1686 – ne IDC [077]
Beseler, Horst et al *see* Proben junger erzaehler
Besemeres, John Daly *see* Success in india
Besenval, Pierre V de *see*
– Memoires de m le baron de besenval
– Memoires du baron de besenval
Beser – Sahibi ve Nesriyat Mueduerue: Abidin Nesimi. n1. 1 ocak 1949 – (= ser O & t journals) – 1mf – 9 – $25.00 – us MEDOC [956]
Die besessenheit : mit besonderer beruecksichtigung der lehre der hl. vaeter / Leistle, David – Dillingen: L Keller's Wwe, [1886?] – 1mf – 9 – 0-524-01968-1 – (incl bibl ref) – mf#1990-2759 – us ATLA [210]
Die besetzung der bischofssitze in preussen in der ersten haelfte des 19. jahrhunderts : mikroedition von der ersten, durch kriegseinwirkung fast vollstaendig vernichteten auflage / Bastgen, Hubert – Paderborn, 1941 (mf ed 1993) – 4mf – 9 – €74.00 – 3-89349-254-2 – mf#DHS-AR 108 – gw Frankfurter [240]
Die besetzung des paepstlichen stuhls : unter den kaisern heinrich 3. und heinrich 4. / Martens, Wilhelm – Freiburg i.B: J C B Mohr, 1887 [mf ed 1986] – 1mf – 9 – 0-8370-7648-X – mf#1986-1648 – us ATLA [241]
Be-sha'ah zu – Yafo, Israel. n1-3. 1916 – 1r – us UF Libraries [939]
Besichtigung dess neuen zu marpurg aussgesteckten trophie der calvinischen warheit : darinnen vornemlich von der sacramentirischen analogia dess brotbrechens gehandelt wirdt / Mentzer, B – Giessen, Hampel, 1609 – 1mf – 9 – mf#TH-1 mf 1168 – ne IDC [242]
Beside the bamboo / Macgowan, John – London: London Missionary Society, 1914 [mf ed 1995] – (= ser Yale coll) – 191p (ill) – 1 – 0-524-09138-2 – mf#1995-0138 – us ATLA [915]
Beside the bowery / Denison, John Hopkins – New York: Dodd, Mead 1914 [mf ed 1990] – 1mf [ill] – 9 – 0-7905-4553-5 – mf#1988-0553 – us ATLA [360]
The besiegers' prayer : or, a christian nation's appeal to the god of battles for success in the righteous war: a sermon / Carroll, John – Toronto?: s.n, 1855 (Toronto: Guardian) – 1mf – mf#47160 – cn Canadiana [240]
Besier, Rudolf *see* Miss ba...

Beskerm u kind = Protect your child / South Africa. Department of National Health and Population Development [Departement van Nasionale Gesondheid en Bevolkingsontwikkeling – Pretoria: Dept van Nasionale Gesondheid en Bevolkingsontwikkeling 1986 [mf ed Pretoria, RSA: State Library [199-]] – 46p [ill] on 1r with other items – 5 – (also available in english, northern sotho, sotho, tsonga, tswana, venda, xhosa & zulu) – mf#op 08446 r25 – us CRL [362]
Beskidenlaendische deutsche zeitung *see* Bielitz-bialer deutsche zeitung
Beskow, Gustaf Emanuel *see*
– Herren aer oefwersteprest
– Nagra betraktelser oefwer p. waldenstroems skrift, om foersoningens betydelse
– Den svenska missionen
Beskrovnyi, L T *see* Russkaia armiia i flot v 18 veke
Besler, Samuel *see*
– Gaudij paschales iesv christi redivivi
– Hymnor[um] et threnodiarvm sanctae crvcis in devotam passionis iesu christi dei et hominis commemorationem fascicvlvs ad hebdomadam magnam...
– Hymnor[um] et threnodiarvm sanctae crvcis in salvtarem passionis iesu christ, dei et hominis, memoriam pars tertia...
– Threnodiarvm sanctae crvcis in salutiferam passionis d n i c recordationem, continvatio historica
El beso? / Aradillas Agudo, Antonio – Madrid: Ediciones Studium, 1960 – sp Bibl Santa Ana [946]
Besof maarav – Los Angeles, CA. Spring 1981-Fall/Winter 1982 – 1 – us AJPC [071]
Besonders meublirt- und gezierte todten-capelle / Abraham...Sancta Clara – Wuertzburg, Nuernberg: Druck Martin Frantz Hertz, 1729 – 6mf – 9 – mf#O-1529 – ne IDC [090]
Besonders meublirt- und gezierte todten-capelle : oder allgemeiner todten-spiegel... / Abraham...Sancta Clara – Wuertzburg, Nuernberg: Dructo Martin Franz Herz, 1710 – 6mf – 9 – mf#O-1528 – ne IDC [090]
Los besos bajo tierra (poemas del amor y del desamos) / Camino Burgos, Luis G – Caceres: Talleres Extremadura – 1 – sp Bibl Santa Ana [946]
Besouchet, Lidia *see*
– Jose ma paranhos, visconde do rio branco
– Literatura del brasil
O besouro : orgao prosaico – Fortaleza, CE. 20 abr 1892 – (= ser Ps 19) – mf#P17,01,36 – bl Biblioteca [079]
O besouro : periodico critico, noticioso e litterario – Maceio, AL. 02 mar 1878-08 mar 1879 – (= ser Ps 19) – mf#P17,01,02 – bl Biblioteca [410]
Bespartiinyi, dvukhnedelnyi zhurnal : pod rukovodstvom m kolbert – LiFge, 1911 – 1mf – 9 – mf#R-18100 – ne IDC [077]
Bespartiinyi, nezavisimyi zhurnal – Spb., 1911. v1-6 – 4mf – 9 – mf#R-3764 – ne IDC [077]
Bespartiinyi zhurnal dlia kresteian i seleskoi intelligentsii : kresteianskoe delo – M., 1909-1912 – 31mf – 9 – (missing: 1910(3, 6, 9-12, 16, 19); 1911(3, 6); 1912) – mf#R-2350 – ne IDC [077]
Bespartiinyi zhurnal literatury, nauki, iskusstva i obshchestvennoi zhizni – Brookfield. 1937+ (1) 1971+ (5) 1977+ (9) – 256mf – 9 – (missing: 1913(3-11); 1916(6-12); 1917(2-12); 1918(1-12)) – mf#1809 – ne IDC [077]
Bespiegelingen over gods wysheid in't bestier der schepselen, en eerkroon voor de caab de goede hoop / Marre, Jan de – te Amsterdam: By Adriaan Wor en de Erve G onder de Linden 1746 – (= ser [Travel descriptions from south africa, 1711-1938]) – 3mf – 9 – mf#zah-15 – ne IDC [916]
Bess, Bernhard *see* Unsere religioesen erzieher
Bess, Henry Alver *see* Biology, life history, and control of the diamond-back moth
Bessa, Manuel Negreiros *see* Macambira (bromelia forrageira)
Bessarabischer beobachter – Sarata (UA), 1932 1 jul-1934 26 jul – 1r – 1 – gw Misc Inst [077]
Bessarabskie gubernskie vedomosti – Chisinau, Moldova 1854-1917 [mf ed Norman Ross] – 46r – 1 – mf#nrp-439 – us UMI ProQuest [077]
Bessarion : studie zur geschichte der renaissance / Rocholl, Rudolf – Leipzig: A Deichert 1904 [mf ed 1991] – 1mf – 9 – 0-524-01530-9 – (incl bibl ref) – mf#1990-0436 – us ATLA [931]
Bessarion, Cardinal *see* Lettere, and orazioni... scritte a principi d'italia intorno al collegarsi, et imprender guerra contro al turco
Bessarione : pubblicazione periodica di studi orientali – Roma, 1896-1922. 26v – 200mf – 9 – mf#AR-1775 – ne IDC [956]
Bessarione : pubblicazione periodica di studi orientali – 1(1896)-9(1901) – 107mf – 9 – €204.00 – (serie 3: v9(1912) 7mf €15. v34(1918) 6mf €14) – ne Slangenburg [950]
Besse, Henry True *see* Church history
Besse, J et al *see* Abbayes et prieures de l'ancienne france (afm36)
Besse, J M *see*
– Abbayes et prieures de l'ancienne france
– Les moines de l'afrique romaine (4e et 5e siecle)
– Les moines de l'ancienne france
– Les moines d'orient anterieurs au concile de chalcedoine
Besse, Leon *see*
– Father beschi of the society of jesus
– La mission du madure
Bessel, Friedrich Wilhelm *see* Untersuchungen ueber die scheinbare und wahre bahn des im jahre 1807 erschienenen grossen kometen
Bessel, Friedrich Wilhelm et al *see* J f encke's astronomische abhandlungen
Bessell, A *see* Tristan und isolde
Bessels, Emil *see*
– Die amerikanische nordpol-expedition
– Einige worte uber die inuit (eskimo) des smith-sundes
– The northern most inhabitants of the earth
– Smith sound and its exploration
Bessemer – Bessemer AL. n63 [1888 aug 11] – 1r – 1 – mf#853401 – us WHS [071]
Bessemer [city directory] : listing – 1893/94 – 1r – 1 – mf#2877283 – us WHS [071]
Besser, Wilhelm Friedrich *see*
– Apostlernes gjerninger forklaret i bibellaesninger
– Paulus
Bessey, William E *see* Evidences of ancient civilization in america
Bessieres, A *see* Pour le pape...
Bessler, Johannes Ferdinand *see* Unterricht und uebung in der religion
Bessodes, Maurice *see* Saint roch. histoire et legendes
Besson *see* Opinion d'un chiffonier de paris sur monsieur lamartine
Besson, Antonin *see* Le projet de reforme de la procedure penale, rapport
Besson, Louis Francois Nicolas, monseigneur *see* L'homme-dieu
Besson, Maurice *see* Histoire des colonies francaises
Besson, Paul *see*
– Etudes sur le theatre contemporain en allemagne
– Michel servet
Bessonet, George *see* The katharist book of perfection
Bessonnet, Rene *see* Essai sur les hallucinations conscientes
Bessonov, S V *see* Nadzor nad knigoi. opyt sistematizatsii materialov o tsenzure v dopetrovskuiu epokhu
Bessoth, Richard *see* Organisationsklima an schulen
Bessy conway : or, the irish girl in america / Sadlier, James, Mrs – New York: D & J Sadlier, 1861 [mf ed 1995] – 4mf – 9 – 0-665-94767-4 – (original issued in ser: parlor and cottage library) – mf#94767 – cn Canadiana [830]
Bessy lesley : the reeler – London, England. 18-- – 1r – us UF Libraries [240]
Best and smith's reports : reports of cases argued and determined in the court of queen's bench and the court of exchequer chamber on appeal... / Best, William M & Smith, George J P – v1-20. 1863-70. London: H Sweet, 1863-71 (all publ) – 110mf – 9 – $165.00 – mf#LLMC 84-748 – us LLMC [324]
Best and the cheapest – London, England. 18-- – 1r – us UF Libraries [240]
BEST (Baptist Education Study Task) *see* Correspondence, 1964-67
Best, Henry *see* The farming and account books of 1641
Best, John *see* Transalpine memoirs
Best known works of voltaire – New York, NY. 1927 – 1r – us UF Libraries [440]
Best known works of w s gilbert / Gilbert, W S – Garden City, NY. no date – 1r – us UF Libraries [780]
Best, Lisa A *see* Sport psychological manual for coaches of high school soccer players
The best methods of counteracting modern infidelity : a paper read...new york, oct 6 1873 = Die besten methoden der bekaempfung des modernen unglaubens / Christlieb, Theodor – New York: Harper, 1874, c1873 [mf ed 1985] – 1mf – 9 – 0-8370-2660-1 – (in english) – mf#1985-0660 – us ATLA [210]
Best, Nolan Rice *see*
– Beyond the natural order
– Yes, "it's the law" and it's a good law
Best of health – 1987 apr/may-1991 winter – 1r – 1 – mf#1839647 – us WHS [613]
Best practices and benchmarking in healthcare – Oxford UK 1996-97 – 1,5,9 – ISSN: 1085-0635 – mf#22295 – us UMI ProQuest [360]
Best purchase – Dublin? Ireland. 18-- – 1r – us UF Libraries [240]
Best, R *see* The martyrology of tallaght (hbs12)
Best, Randolph B *see* Papers of the maryland state colonization society, 1817-1902
Best, Ron *see* Teaching skills for learning
Best sellers – Scranton PA 1941-87 – 1,5,9 – ISSN: 0005-9625 – mf#1916 – us UMI ProQuest [070]
Best stories of modern bengal / ed by Gupta, Dilip K – Calcutta: Signet Press, 194- – (= ser Samp: indian books) – (trans by nilima devi) – us CRL [830]
Best, Walter *see* Die generalin und andere geschichten
Best, William M *see* Best and smith's reports
Der bestand preussische akademie der kuenste – kaiserreich, weimarer republik, nationalsozialismus, nachkriegszeit (1871-1955) / ed by Stiftung Archiv der Akademie der Kuenste Berlin – [mf ed 1994-98] – 3337mf (1:24) – 9 – diazo €16,950.00 (silver €21,750 isbn: 978-3-598-33759-8) – ISBN-10: 3-598-33758-2 – ISBN-13: 978-3-598-33758-1 – gw Saur [700]
Der bestand preussische akademie der kuenste – kaiserreich, weimarer republik, nationalsozialismus, nachkriegszeit (1871-1955) teil 1 : die sektionen fuer die bildenden kuenste, fuer musik und fuer dichtkunst / ed by Stiftung Archiv der Akademie der Kuenste Berlin – [mf ed 1994] – 1218mf (1:24) – 9 – diazo €5650.00 (silver €7250 isbn: 978-3-598-33705-5) – ISBN-10: 3-598-33704-3 – ISBN-13: 978-3-598-33704-8 – (with guide) – gw Saur [700]
Der bestand preussische akademie der kuenste – kaiserreich, weimarer republik, nationalsozialismus, nachkriegszeit (1871-1955) teil 2 : praesident, mitglieder, staendige sekretaere, statuten und senatsprotokolle – [mf ed 1995] – 969mf (1:24) – 9 – diazo €5650.00 (silver €7250 isbn: 978-3-598-33715-4) – ISBN-10: 3-598-33714-0 – ISBN-13: 978-3-598-33714-7 – (with guide) – gw Saur [700]
Der bestand preussische akademie der kuenste – kaiserreich, weimarer republik, nationalsozialismus, nachkriegszeit (1871-1955) teil 3 : austellungen und kunstpreise – [mf ed 1998] – 1150mf (1:24) – 9 – diazo €5650.00 (silver €7250 isbn: 978-3-598-33725-3) – ISBN-10: 3-598-33724-8 – ISBN-13: 978-3-598-33724-6 – (with guide) – gw Saur [700]
Beste, Henry *see*
– Four years in france
– Italy as it is
Beste, Konrad *see*
– Die drei esel der doktorin loehnefink
– Grillparzers verhaeltnis zur politischen tendenzliteratur seiner zeit
– Das heidnische dorf
Der beste und kuerzeste weg zur vollkommenheit / Nieremberg, Juan Eusebio – 2., verb. Aufl. Freiburg im Breisgau; St. Louis, Mo.: Herder, 1906 – (= ser Aszetische bibliothek) – 2mf – 9 – 0-8370-7317-0 – (incl bibl ref) – mf#1986-1317 – us ATLA [240]
Die besten methoden der bekaempfung des modernen unglaubens *see* The best methods of counteracting modern infidelity
Bestendige antwort etlicher fragstueck / Dathenus, P – Heidelberg, 1572 – 1mf – 9 – mf#PBA-166 – ne IDC [240]
Bestendige antwort etlicher fragstueck : so die predicanten zu frankfurt am mayn...in truck zu warnung haben aussgeben lassen... / Dathenus, P – Heidelberg, 1572 – 1mf – 9 – mf#PBA-158 – ne IDC [240]
Bestendige bekantnus d samuel hubers ob gott durch seinen lieben son jesum christum nur allein etlich wenig menschen order zumal alle menschen vom tode allesampt erloest habe / Huber, S – np, 1598 – 1mf – 9 – mf#TH-1 mf 742 – ne IDC [242]
Bestendige entdeckung des caluinischen geists welcher sich vnterstehe das leiden jhesu christi fuer vnsere suende zu verlaeugnen vnd auffzuheben / Huber, S – Wittemberg, 1593 – 3mf – 9 – mf#TH-1 mf 711-713 – ne IDC [242]
Bestimmung der molekularen hyperpolisierbarkeit organischer materialien mit der methode des selbstbeugungseffektes / Baumann, Markus – (mf ed 1997) – 2mf – 9 – €40.00 – 3-8267-2400-3 – mf#DHS 2400 – gw Frankfurter [540]
Bestimmung der renalen clearance im rahmen der nuklearmedizinischen nierendiagnostik / Mende, Traute – 1993 – 9 – 3-89349-453-7 – gw Frankfurter [616]
Die bestimmung des fachlichkeitsgrades von texten der industriesoziologie des englischen und deutschen / Dohms, Evelyn – (mf ed 1995) – (= ser Leipziger arbeiten zur fachsprachenforschung) – 1mf – 9 – €30.00 – 3-8267-2177-2 – mf#DHS 2177 – gw Frankfurter [410]
Bestimmung des mondhalbmessers : aus den waehrend der totalen mondfinsterniss 1884 october 4 beobachteten sternbedeckungen / Struve, Ludwig – Dorpat: C Mattiesen 1889 [mf ed 1998] – 1r – 1 – (incl bibl ref) – mf#film mas 28404 – us Harvard [520]
Bestimmung eines elementrasters in blutplasma und vollblut bei schwangeren frauen und deren neugeborenen kindern / Pfeiffer, Melanie – (mf ed 1997) – 2mf – 9 – €40.00 – 3-8267-2499-2 – mf#DHS 2499 – gw Frankfurter [615]
Bestimmung von hla klasse 1 an zur autovaccination bestimmten, epithelmarker-charakterisierten tumorzellen / Kleef, Rald – (mf ed 1994) – 1mf – 9 – €24.00 – 3-89349-890-7 – gw Frankfurter [611]
Bestimmungsgruende und probleme staatlicher und institutioneller familienpolitik am beispiel der bundesrepublik deutschland / Schluchter, Wolfgang – Heidelberg, 1973 – 2mf – 9 – 3-89349-723-4 – gw Frankfurter [321]
Bestimmungstabellen zur flora von aegypten / Ramis, A I – Jena, 1929 – 3mf – 9 – mf#11526 – ne IDC [580]
Der bestirnte himmel ueber mir : ein kant-roman / Treptow, Alfred – Berlin: Steuben Verlag P G Esser c1939 [mf ed 1991] – 1r – 1 – (filmed with: leuchtendes land / luis trenker) – mf#2917p – us Wisconsin U Libr [830]
Bestmann, Hugo Johannes *see*
– Die anfaenge des katholischen christentums und des islams
– Encyclopaedie der theologie
– Die katholische sitte der alten kirche in ihrer geschichtlichen entwicklung
– Die sittlichen stadien in ihrer geschichtlichen entwicklung
– Die theologische wissenschaft und die ritschl'sche schule
Beston, Henry *see* The book of gallant vagabonds
Der bestrafte betrueger : tragisch-pantomimisches ballet in fuenf aufzuegen. aufzufuehren auf den kaiserl. koenigl. schaubuehnen / Muzzarelli, Antonio – Wien, 1793 – 1 – mf#*ZBD-*MGTZ pv5-Res – Located: NYPL – us Misc Inst [790]
Best's review – Oldwick NJ 2000+ – 1,5,9 – ISSN: 1527-5914 – mf#29389 – us UMI ProQuest [360]
Best's review life/health insurance edition – Oldwick NJ 1906-1999 – 1,5,9 – ISSN: 0005-9706 – mf#1762 – us UMI ProQuest [368]
Best's review property/casualty insurance edition – Oldwick NJ 1977-99 – 1,5,9 – (cont: best's review property/liability insurance ed) – ISSN: 0161-7745 – mf#379,01 – us UMI ProQuest [368]
Best's review property/liability insurance ed – Oldwick NJ 1938-76 – 1,5,9 – (cont by: best's review property/casualty insurance edition) – ISSN: 0005-9714 – mf#379.01 – us UMI ProQuest [368]
Ein besuch in tobruk an der kueste von marmarica... / Schweinfurth, G – [Berlin, 1883] – 1mf – 9 – mf#13063 – ne IDC [956]
Beswick, Delbert Meacham *see* Organ works of dietrich buxtehude
Bet avot / Hershman, Shelomoh Zalman – Berlin, Germany. 1888 – 1r – us UF Libraries [939]
Bet ha-behirah 'al masekhet 'avodah zarah / Meiri, Menahem Ben Solomon – Jerusalem, Israel. 1944 – 1r – us UF Libraries [939]
Bet ha-levi / Soloveichik, Joseph Baer – Vilna, Lithuania. 1910 – 1r – us UF Libraries [939]
Bet hilel / Ring, Max – Tel-Aviv, Israel. 19-- – 1r – us UF Libraries [939]
Bet midrash shemu'el / Hirschowitz, Abraham Eber – Jerusalem, Israel. v1-2. 190-? – 1r – us UF Libraries [939]
Bet page / Sivitz, Moses Simon – Jerusalem, Israel. 1904 – 1r – us UF Libraries [939]
Bet rabi / Helmann, Chayim Meir – Berdichev, Ukraine. 1903 – 1r – us UF Libraries [939]
Bet tzedek news – Los Angeles, CA.1980-84 – 1 – us AJPC [071]
Bet ya'akov – Ungvar, Ukraine 1868 [mf ed Norman Ross] – 1r – 1 – (in hebrew; with: mitspeh [st petersburg] undated, with: ha-yekov [st petersburg] undated, with: degel ha-torah [warsaw] 1921-23, with: mesilot [warsaw] 1935-37) – mf#nrp-1868 – us UMI ProQuest [939]
Beta, Ottomar *see* 'Old-iniquity'
Le betail canadien / Couture, Joseph-Alphonse – Quebec?: La Semaine Commerciale, 1900? – 1mf – 9 – mf#05399 – cn Canadiana [636]
Betancourt Agramonte, Eugenio *see*
– Ignacio agramonte y la revolucion cubana
Betancourt Agramonte, Oscar *see* Cartas a fidel castro
Betancourt, Gaspar *see* Aristas
Betancourt, Romulo *see*
– Romulo betancourt
– Venezuela
Betancourt y el comunismo / Conte Aguero, Luis – Miami, FL. 1962 – 1r – us UF Libraries [972]
Betancourt Y Miranda, Angel C *see* Legislacion hipotecaria vigente en la republica de...
Betancur, Cayetano *see*
– Introduccion a la ciencia del derecho
– Sociologia de la autenticidad y la simulacion

BETTELHEIM

Betancur Jaramillo, Carlos *see* Regimen legal de los concubinos en colombia
Betbuechlin gestellet durch andream musculum doctor gemehret vnd gebessert / Musculus, A – Frankfurt an der Oder, [1560] – 2mf – 9 – mf#TH-1 mf 1211-1212 – ne IDC [242]
Bete de musseau / Thoby-Marcelin, Philippe – New York, NY. 1946 – 1r – us UF Libraries [972]
La bete noire – Paris. n1-4. mai-juil 1935 – 1 – fr ACRPP [073]
De beteekenis der historische studie van het oostersch-grieksch christendom : rede uitgesproken bij de aanvaarding van het ambt van buitengewoon hoogleeraar aan de rijks-universiteit te leiden... / Zwaan, Johannes – Haarlem: F Bohn, 1912 [mf ed 1992] – 24p on 1mf – 9 – 0-524-02998-9 – (incl bibl ref) – mf#1990-0785 – us ATLA [243]
De beteekenis die de oudere waarnemingen nog heden voor de sterrenkunde hebben : rede / Sande Bakhuyzen, Ernst Frederik van de – Leiden: E J Brill, 1909 [mf ed 1993] – 40p on 1mf – 9 – 0-524-05998-5 – mf#1992-0735 – us ATLA [520]
Die beteiligung der christen am oeffentlichen leben in vorconstantinischer zeit : ein beitrag zur aeltesten kirchengeschichte / Bigelmair, Andreas – Muenchen: JJ Lentner, 1902 [mf ed 1989] – (= ser Veroeffentlichungen aus dem kirchenhistorischen seminar muenchen 8) – 1mf – 9 – 0-7905-4088-6 – (incl bibl ref) – mf#1988-0088 – us ATLA [230]
Die beteiligung der religionsgesellschaften an staatlichen aufgaben : gegenueberstellung der vehaeltnisse vor und nach dem jahre 1918 / Jung, Hermann – Erlangen, 1931 (mf ed 1995) – 1mf – 9 – €24.00 – 3-8267-3183-2 – mf#DHS-AR 3183 – gw Frankfurter [340]
De betekenis van het woord solitudo in de collationes van cassianus / Kabel, Michael – 1959 – 4mf – 8 – €11.00 – ne Slangenburg [241]
Die betekenis van walvisbaai as hawe vir suidwes-africa / Smit, Phillipus – [Stellenbosch] Universiteit van Stellenbosch 1962 – us CRL [960]
Der betende gerechte der psalmen : historischkritische untersuchung als beitrag zu einer einleitung in den psalter / Engert, Thaddaeus – Wuerzburg:Goebel & Scherer, 1902 – 1mf – 9 – 0-8370-3062-5 – mf#1985-1062 – us ATLA [220]
Betes rares de la jungle africaine / Sanderson, Ivan Terrance – Paris, France. 1938 – 1r – us UF Libraries [960]
Beth, Karl *see*
– Die entwicklung des christentums zur universal-religion
– Der entwicklungsgedanke und das christentum
– Die moderne und die prinzipien der theologie
– Die orientalische christenheit der mittelmeerlaender
– Die orientalische christenheit der mittelmeerlaender, reisestudien zur statistik und symbolik der griechischen, armenischen und koptischen kirche
– Religion und magie bei den naturvoelkern
– Das wunder
– Wunder jesu
– Die wunder jesu
Beth sifrenu – (New York). 1920 – 1 – us AJPC [073]
Beth vaad la'chachomim – (New York). 1903 – 1 – us AJPC [073]
Bethabara, south carolina – Bethabara. Laurens County, SC – 1 – $16.65 – us Southern Baptist [242]
Bethada naem nrenn = Lives of irish saints / Plummer, Ch – Oxford, 1922 – 2v – €32.00 – ne Slangenburg [241]
The bethal trial: state vs. z. mothopeng and 17 others, in the supreme court of south africa, southeastern local division / Mothopeng, Zephania – 1978 – 1 – us CRL [960]
Bethanien : bibelstunden ueber den philipperbrief zum gebrauche insbesondere fuer diakonissenanstalten, kirchliche gemeinschaften und das christliche haus / Borrmann, A – Guetersloh: C Bertelsmann 1914 [mf ed 1992] – 1mf – 9 – 0-524-05208-5 – mf#1992-0341 – us ATLA [225]
Bethany advertiser – Bethany, PA., 1841 – 13 – $25.00r – us IMR [071]
Bethany baptist church – Burnt Corn, AL. 1821-1956 – 1 – $20.70 – us Southern Baptist [242]
Bethany baptist church – Jefferson Davis County, MS. 1820-1949 – 1 – $36.90 – us Southern Baptist [242]
Bethany baptist church – Milam, TX. 718p. 1892-1978 – 1 – $32.31 – (lacks: apr 1918-jan 1921) – us Southern Baptist [242]
Bethany baptist church – Mccormick County, SC. 142p. 1956-83 – 1 – $6.39 – us Southern Baptist [242]
Bethany baptist church – Neshoba County, MS. aug 1884-91 – 1 – $5.00 – us Southern Baptist [242]
Bethany baptist church – Williamsburg County, SC. 1890-1921 – 1 – $5.00 – us Southern Baptist [242]
Bethany baptist church. flat river association – NC. 1883-98 – 1 – $5.04 – us Southern Baptist [242]
Bethany Baptist Church, Hunter, KS *see* Universal church record and clerk's books
Bethany baptist church. kings mountain association – Grover, NC. 1947-63 – 1 – $6.93 – us Southern Baptist [242]
Bethany Lutheran College et al *see* Lutheran synod quarterly
Bethany Mission for Colored Children *see*
– Annual report of bethany mission for colored children
– Annual report of the bethany mission for colored people
Bethcar baptist church – AikenCounty, SC. 380p. 1863-76, 1894-34 – 1 – $17.10 – us Southern Baptist [978]
Bethel baptist church – New Design, IL. 1806-1951 – 1 – $6.57 – (church history on front of film) – us Southern Baptist [242]
Bethel baptist church : church membership – Franklinton, LA. 1907-45 – 1 – $13.14 – mf#6850 – us Southern Baptist [242]
Bethel baptist church – Morristown, TN. 1422p. feb 1874-aug 1889, apr 1893-apr 1896, jan 1938-sept 1975 – 1 – $63.99 – (financial records, 1894-1906, 1917-19, 1960-64) – us Southern Baptist [242]
Bethel baptist church – Butler, GA. 1838-1956. 734p – (= ser Hopeful Baptist Church) – 1 – $33.03 – (formerly called: hopeful baptist church 1838-54) – mf#6808 – us Southern Baptist [242]
Bethel baptist church – Heard County, GA. 1828-1901 – 1 – $14.40 – us Southern Baptist [242]
Bethel baptist church – Jasper County, GA. 1853-1900 – 1 – $14.40 – us Southern Baptist [242]
Bethel baptist church – Oconee Co, SC. 1799p. 1831-96, 1904-11, 1931-76, may 1976-oct 1989 – 1 – $80.96 – mf#5003-42c – us Southern Baptist [242]
Bethel baptist church – Saluda County, SC. 1853-69 – 1 – $13.37 – us Southern Baptist [242]
Bethel baptist church – Shelby, NC. 1940-63 – 1 – $11.48 – us Southern Baptist [242]
Bethel baptist church – Tampa, FL. 1902-mar 1971 – 1 – $115.65 – us Southern Baptist [242]
Bethel baptist church / Taylor, W W – Pitt County, NC. 1887-1953 – 1 – $5.00 – us Southern Baptist [242]
Bethel baptist church – Townsend, TN. 1840-1924, 24 aug 1947-nov 1963 – 1 – $31.50 – us Southern Baptist [242]
El bethel baptist church. cherokee county. south carolina : church records – 1803-1950 – 1 – us Southern Baptist [242]
Bethel, Maine. Free Will Baptist Church *see* Records
Bethel. Miami County. Ohio. Honey Creek Baptist Church *see* Honey creek baptist church records, 1811-1844
Bethel Park Federation of Teachers *see*
– Bft, aft news
– Bpft, aft action
– Bpft news
– Bpft united teachers
Bethell, Arnold Talbot *see*
– Early settlers of the bahama islands
– Early settlers of the bahamas and colonists of nor...
Bethell, Christopher *see* Charge delivered to the clergy of the diocese of bangor
Bethell, John A *see* Pinellas
Bethel's voice – 1984 apr 15, 1986 jun 29, 1989 mar 12, 1990 mar 13 – 1r – 1 – mf#4023725 – us WHS [242]
Bethel-United Apostolic Church *see* Voice of holiness
Bethesda baptist church – Hinds County, MS. 1846-1937 – 1 – $43.02 – us Southern Baptist [242]
Bethge, Friedrich *see* Marsch der veteranen
Bethisy, Jean Laurent de *see* Exposition de la theorie et de la pratique de la musique
Bethke, T *see* Effects of relationship status, setting and sex of perpetrator on college student evaluations of dating violence
Bethleem : le sanctuaire de la nativite / Vincent, Hughes & Abel, Felix-Marie – Paris: Victor Lecoffre 1914 [mf ed 1989] – 3mf [ill] – 9 – 0-7905-2440-6 – (incl bibl ref & ind) – mf#1987-2440 – us ATLA [240]
Bethlehem / Faber, Frederick William – new ed. London: Burns & Oates [1860?] [mf ed 1992] – 2mf – 9 – 0-524-04452-X – mf#1992-0121 – us ATLA [240]
Bethlehem baptist church – Colleton County, SC. 1911-1924, 1954-1989 – (= ser BYPU) – 1 reel – 1 – $22.59 – (bypu 1925-1929) – us Southern Baptist [242]
Bethlehem baptist church – Winona, MS. 116p. 1849-1866 – 1 – $5.22 – (historical sketches) – us Southern Baptist [242]
Bethlehem baptist church : irwin county – Reston. 1948+ (1) 1972+ (5) 1975+ (9) – 1r – 1 – $28.58 – mf#6520 – us Southern Baptist [242]
Bethlehem baptist church – Kings Mountain, NC. 1854-1963 – 1 – $65.88 – us Southern Baptist [242]
Bethlehem baptist church – Louisville, KY. 1960-65 – 1 – $25.74 – us Southern Baptist [242]
Bethlehem baptist church – Round-o, SC. 1913-80 – 1 – $13.68 – (minutes scattered 1942-53, 1954-79, only may of 1980) – us Southern Baptist [242]
Bethlehem baptist church : warthen, georgia – Washington Co, GA. 1044p – 1 – $46.98 – (church minutes 1791-1912; church minutes 1911-99; cemetery list of people buried (compiled in 1961), history 1790-1990) – mf#2017-1 – us Southern Baptist [242]
Bethlehem baptist church. sandy creek association – NC. 200p. 1834-1919 – 1 – $9.00 – us Southern Baptist [242]
Bethlehem church. sandy creek association – NC. 1834-1914 – 1 – $9.00 – us Southern Baptist [242]
Bethlehem plant news / United Steelworkers of America. (Bethlehem Plant PA) – 1985 nov-1987 apr, jun-1988 apr, jun-aug, oct – 1r – 1 – (cont by: news and views of the tri-local) – mf#1290026 – us WHS [670]
Bethnal green chronicle and east end weekly news – Tower Hamlets, London 6 dec 1873-7 feb 1874 – 1 – (cont by: east london standard & bethnal green chronicle [14 feb 1874]) – uk British Libr Newspaper [072]
Bethnal green news and the shoreditch guardian and the hackney argus – [London & SE] Tower Hamlets 30 jan 1909; 7 dec 1912; 19 apr, 9, 23 aug, 11 oct 1913; 5 feb 1921; 24 nov 1923 [mf ed 2003] – 1r – 1 – (iss dated 30 jan 1909 incomplete) – uk Newsplan [072]
Bethnal green times – Tower Hamlets, London 4 jan 1862-10 jul 1869 – 1 – uk British Libr Newspaper [072]
Bethsaida baptist church – Fayette County, GA. 1830-1913 – 1 – $29.16 – us Southern Baptist [242]
"Bethshean" / Fisher, Clarence S – University of Pennsylvania Museum Journal: Dec 1923 – 9 – $10.00 – us IRC [930]
Bethuel baptist church. greenville county – Greenville, SC. 1949-1977 – 1 – $18.09 – us Southern Baptist [242]
Bethune, Alexander Neil *see*
– Memoir of the right reverend john strachan...
– Memoir of the right reverend john strachan, d.d., ll. d
– A sermon preached in the church of st george the martyr, toronto
Bethune, George Washington *see*
– Expository lectures on the Heidelberg catechism
– A word to the afflicted
Bethune, Joanne *see* Biographical sketch of joanne bethune
Bethune, John *see* Baptist dishonesty
Bethune, Mary McLeod *see*
– Mary mcleod bethune papers
– The papers of mary mcleod bethune, 1875-1955
Bethune, Norman *see*
– Crime on the road malaga-almeria
– El crimen del camino malaga-almeria
Bethune, Thomas Greene *see* A collection of published music including 17 piano pieces and 3 vocal selections
Bethune-Baker, James Franklin *see*
– The influence of christianity on war
– An introduction to the early history of christian doctrine
– The meaning of homoousios in the "constantinopolitan" creed
– The meaning of homoousios in the 'constantinopolitan creed'
– Nestorius and his teaching
– The old faith and the new learning
Bethune-cookman college continuing education newsletter – Daytona Beach FL. 1990 apr – 1r – 1 – mf#4025966 – us WHS [374]
Bethune-Cookman College [Daytona Beach FL] *see* Clarion
Bethusy-Huc, Valeska von Reiswitz-Kaderzin, Graefin von [pseud: Moritz von Reichenbach] *see* Der aelteste sohn
Beti, Mongo *see* King lazarus
Der betler = The beggar – Philadelphia, PA. 1908 – 1 – us AJPC [830]
Betley & balterley (staffs) poor rate surveys 1824 – [South Cheshire FHS] – (= ser Cheshire church registers) – 2mf – 9 – £4.00 – mf#369 – uk CheshireFHS [929]
Betley (staffs), st margaret – (= ser Cheshire monumental inscriptions) – 3mf – 9 – £4.50 – mf#109 – uk CheshireFHS [929]
Be-tokh ha-homot / Porush, Menahem Mendel – Jerusalem, Israel. 1948 – 1r – us UF Libraries [939]
Beton arme : le journal haitien – 1994 may 17/31-jun7/21, jul 19/aug 2 – 1 – mf#3055722 – us WHS [071]
Beton arme – n1-71. Paris. 1957-mai 1967 – 5 – fr ACRPP [073]
Betrachtung der littauischen sprache : in ihrem ursprunge, wesen und eigenschaften / Ruhig, Philipp – Koenigsberg: Hartung 1745 – (= ser Whsb) – 1mf – 9 – €20.00 – mf#Hu 199b – gw Fischer [460]
Betrachtungen ueber die jetzige crise des ottomanischen reichs : ihre wirkenden ursachen und wahrscheinlichen folgen / Paris, Jean J – Leipzig 1822 [mf ed Hildesheim 1995-98] – (= ser Fbc) – 2mf – 9 – €60.00 – 3-487-29134-7 – gw Olms [932]
Betrachtungen ueber die mystik in goethe's "faust" / Hartmann, Franz – Leipzig: W Friedrich [1900] [mf ed 1990] – 1r – 1 – (filmed with: goethe's faust / ed by albert gruen) – mf#7347 – us Wisconsin U Libr [430]
Betrachtungen ueber die neuesten historischen schriften – Altenburg, Halle S DE, 1769-78 – 7r – 1 – (title varies: 1774: fortgesetzte betrachtungen ueber die neuesten historischen schriften) – gw Misc Inst [900]
Betrachtungen ueber die verhaeltnisse der katholischen kirche im umfange des deutschen bundes / Wessenberg, Ignaz Heinrich, Freiherr von – O.O., 1818 (mf ed 1992) – 1mf – 9 – €24.00 – 3-89349-080-9 – mf#DHS-AR 53 – gw Frankfurter [241]
Betrachtungen zum leben jesu (cima22) : farbmikrofiche-edition der handschrift liege, bibliotheque generale de l'universite, ms wittert 71 – (mf ed 1991) – (= ser Codices illuminati medii aevi (cima) 22) – 41p on 8 color mf – 15 – €335.00 – 3-89219-022-4 – (int & description by hans-walter stork) – gw Lengenfelder [090]
Betrayal in india / Karaka, Dosoo Framjee – London: Victor Gollancz; Bombay: Distributors in India [etc], Allied Publ, 1950 – (= ser Samp: indian books) – us CRL [954]
The betrayal of freedom : a study in nehru's political ideas / Krishnamurti, Y G – Bombay: Popular Book Depot, 1944 – (= ser Samp: indian books) – (foreword by bhulabhai j desai) – us CRL [320]
Die betrekkinge tussen nederland en suid-afrika, 1946-1961 / Hendricks, Wayne Graham – U of the Western Cape 1984 [mf ed S.l: s.n. 1984] – 3mf – 9 – (incl bibl) – sa Misc Inst [327]
Der betrieb – Duesseldorf DE, 1848 jan-jul – 1r – 1 – (suppl of handelsblatt duesseldorf) – gw Misc Inst [380]
Der betriebs-aktivist – Zeitz, Profen DE, 1949-1968 25 mar [gaps] – 2r – 1 – gw Misc Inst [331]
Betriebsecho – Halberstadt DE, 1955 14 jan-1961 8 dec [gaps] – 1r – 1 – (title varies: press- und stanzwerk raguhn/halberstadt) – gw Misc Inst [074]
Betriebsraete : zeitschrift fuer funktionaere der metallindustrie – 15 Apr 1920-22 Dec 1923 – 2r – 1 – gw Mikropress [331]
Betriebsraetezeitschrift der funktionaere der metallindustrie – Stuttgart DE, 1920 15 apr-1923 22 dec – 2r – 1 – mf#11264 – gw Mikropress [331]
Die betriebsrechnung / Kaefer, Karl – Zurich, 1943 – 1 – gw Mikropress [943]
Die betriebssteuer : eine alternative zur derzeitigen unternehmensbesteuerung / Boeer, Bjoern – (mf ed 1994) – 1mf – 9 – €30.00 – 3-8267-2025-3 – mf#DHS 2025 – gw Frankfurter [336]
Betriebsstimme – Meuselwitz, Zeitz DE, 1950 17 sep-1959 27 jun [gaps] – 2r – 1 – gw Misc Inst [331]
Betriebsstimme – Meuselwitz, Zeitz DE, 1950 17 sep-1959 27 jun [gaps] – 2r – 1 – gw Misc Inst [331]
Betriebszeitung der bremer strassenbahn ag und bremer vorortbahnen gmbh – Bremen DE, 1936-39 – 1r – 1 – gw Misc Inst [380]
Die betruebte pegnesis : den leben, kunst- und tugend-wandel des seelig-edlen floridans, h. sigm. von birken com. pal. caes / Blumen-Gesellschaft an der Pegnitz – Nuernberg: Christian Sigm. Froberg, 1684 – 5mf – 9 – mf#0-26 – ne IDC [090]
Betsch, Roland *see*
– Ballade am strom
– Regie-express d 21
Bett, A *see* Changing attitudes toward physically disabled persons using a videotape sport intervention
Bettany, George Thomas *see* The world's religions
Bettelheim, Anton *see*
– Anzengrubers werke in vierzehn teilen
– Briefe von ludwig anzengruber
– Gesammelte schriften
– Letzte dorfgaenge
– Ludwig anzengruber
– Marie von ebner-eschenbach
Bettelheim, Anton et al *see*
– Gesammelte werke
– Ludwig anzengrubers gesammelte werke
Bettelheim, Bernard Jean *see* Letter from b j bettelheim, md, missionary in lewchew

Bettelheim, Samuel *see* Zuruck zur bibel!

Bettelheim-Gabillon, Helene *see* Betty paolis gesammelte aufsaetze

Betten, Francis Sales *see* The roman index of forbidden book

Bettencourt, Gastao De *see* Folclore no brasil

Bettencourt Vasconcellos Corte Real do Canto, Vital de *see* Descripcao historica, topographica e ethnographica do districto de s. joao baptista d'ajudia e do reino de dahome na costa da mina

Better Boys Foundation *see* Black theater bulletin

The better covenant practically considered : from hebrews 8, 6, 10-12, with a supplement on philippians 2, 12, 13 / Goode, Francis – Philadelphia: Smith & English [distributor] 1855 [mf ed 1993] – 1mf – 9 – 0-524-06919-0 – mf#1992-1012 – us ATLA [240]

Better crops with plant food – Norcross GA 1923+ – 1,5,9 – ISSN: 0006-0089 – mf#2113 – us UMI ProQuest [635]

Better days for working people / Blaikie, William Garden – London: Alexander Strahan 1865 [mf ed 1991] – 1mf – 9 – 0-524-00622-9 – mf#1990-0122 – us ATLA [360]

Better gifts and the more excellent way / Wordsworth, Charles – London, England. 1846 – 1r – us UF Libraries [240]

Better health / Methodist Hospital (Madison WI) – v52 n1-v57 n4 [1979 fall-1984 fall/ winter] – 1r – 1 – (cont: philanthropies) – mf#912577 – us WHS [360]

Better home *see* Beginner teacher and pupil book

Better investing – Royal Oak MI 1967+ – 1,5,9 – ISSN: 0006-016X – mf#3136 – us UMI ProQuest [332]

Better living today : a publication of the universal foundation for better living, inc – 1985 aug 18 – 1r – 1 – mf#4717868 – us WHS [640]

Better nutrition – El Segundo CA 1996+ – 1,5,9 – (cont: better nutrition for today's living) – mf#18407,01 – us UMI ProQuest [613]

Better nutrition for today's living – El Segundo CA 1994-95 – 1 – (cont by: better nutrition) – mf#18407.01 – us UMI ProQuest [613]

Better prospects of the church / Hare, Julius Charles – London, England. 1840 – 1r – us UF Libraries [240]

Better roads – Park Ridge IL 1990-96 – 1,5,9 – ISSN: 0006-0208 – mf#18358 – us UMI ProQuest [625]

Better support of the ministry and its relations to the spiritual l... / M'candlish, John M – Edinburgh, Scotland. 1889? – 1r – us UF Libraries [240]

The better worker – Booker Washington Birthplace VA: Better Workers' Institute 1948-58 [v9 n5-v16 n2 (nov 1948-feb 1952) [mthly] [mf ed 2005] – 21v on 1r – 1 – (cont: negro worker (tuskegee, al); lacks: v9 n6-, v13 n1 (dec 1948-jul 1950), v13 n3,5 (sep, nov 1950), v14 n2-6,8-12, v15 n1-2,4-12, v16 n1 (jul 1944); some iss damaged) – mf#2005-s109 – us ATLA [331]

Better World Educational Corporation *see* Free for all

Betteridge, Harold T *see* Die alpen

Betteridge, Walter Robert *see*
- Book of deuteronomy
- Exodus

Bettermann, Henrik *see* Komplexitaetsanalyse der rr-dynamik im 24-stunden-ekg

Bettex, Frederic *see*
- The bible the word of god
- The first page of the bible
- Naturstudium und christentum
- La religion et les sciences de la nature
- What think ye of christ?
- The word of truth
- Wunder
- Zweifel?

Betti, Emilio *see* Hermeneutik als allgemeine methodik der geisteswissenschaften

Bettina / Seidel, Ina – Stuttgart: J G Cotta 1944 [mf ed 1988] – (= ser Die dichter der deutschen) – 1r – 1 – (filmed with: bettina von arnims aufruf zur revolution und zum volkerbunde / ed & int by curt moreck) – mf#6958 – us Wisconsin U Libr [830]

Bettina von arnim (1785-1859) : ein erinnerungsblatt zu ihrem hundertsten geburtstage / Alberti, Conrad – Leipzig: O Wigand, 1885 [mf ed 1988] – 135/1p – 1 – mf#6958 – us Wisconsin U Libr [430]

Bettina von arnims aufruf zur revolution und zum voelkerbunde : gespraeche mit daemonen – Muenchen: H Schmidt, [1919] [mf ed 1988] – (= ser Arnim, Bettina von) – xi/252p – 1 – (int by curt moreck) – mf#6958 – us Wisconsin U Libr [943]

Bettina von arnims polenbroschuere / ed by Pueschel, Ursula – Berlin: Henschelverlag, 1954 [mf ed 1993] – 175p – 1 – (originally publ in 1848 under title: an die aufgeloes'te preussische national-versammlung; incl bibl ref) – mf#8459 – us Wisconsin U Libr [430]

Bettison, David G *see* Demographic structure of seventeen villages in the peri-urban area of blantyre-limbe, nyasaland

Der bettler in der schottischen dichtung / Raske, Karl – Berlin, 1908 (mf ed 1994) – 2mf – 9 – €31.00 – 3-8267-3059-3 – mf#DHS-AR 3059 – gw Frankfurter [420]

Betts, John Arthur *see* The bearing of the theory of evolution on christian doctrine

Betts, John Thomas *see*
- 17 opuscules
- Juan de valdes' commentary upon the gospel of st matthew
- Spiritual milk, or, christian instruction for children

Betts, Peter John et al *see* Eigene wege 1960

Bettws, glamorgan, parish church of st david : baptisms 1721-1926, burials 1723-1950, marriages 1722-1836 – [Glamorgan]: GFHS [mf ed c2000] – 1mf – 9 – £1.25 – uk Glamorgan FHS [929]

Bettws, sardis baptist, monumental inscriptions – 1mf – 9 – £1.25 – uk Glamorgan FHS [929]

Bettws, st david, monumental inscriptions – 2mf – 9 – £2.50 – uk Glamorgan FHS [929]

Betty o'neal concentrator – NV. jun 1924 [wkly] – 1r – 1 – $60.00 – mf#U04420 – us Library Micro [071]

Betty paolis gesammelte aufsaetze / ed by Bettelheim-Gabillon, Helene – Wien: Literarischer Verein 1908 [mf ed 1993] – (= ser Schriften des literarischen vereins in wien 9) – 5r – 1 – mf#3333p – us Wisconsin U Libr [802]

Between caesar and jesus / Herron, George Davis – New York: Thomas Y Crowell c1899 [mf ed 1989] – 1mf – 9 – 0-7905-1894-5 – mf#1987-1894 – us ATLA [230]

Between hearts : the letters, diaries and manuscripts of vita sackville-west and harold nicholson – (= ser Vita Sackville-West and Harold Nicholson Manuscripts) – 15r – 1 – (previous title: vita sackville-west and harold nicholson manuscripts. with printed guide. from sissinghurst castle, kent: the huntington library, california; and other libraries) – mf#C35-28020 – us Primary [420]

Between heathenism and christianity : being a translation of seneca's de providentia, and plutarch's de sera numinis vindicta / Super, Charles William – Chicago: Fleming H Revell 1899 [mf ed 1986] – 1mf – 9 – 0-8370-6418-X – (in english; incl bibl) – mf#1986-0418 – us ATLA [230]

Between our selves – v1 n1-3 [1985 winter-fall/ winter] – 1r – 1 – mf#1133945 – us WHS [071]

Between the bradlaughs – Twickenham, England. 1879 – 1r – us UF Libraries [240]

Between the bridges / Woodworkers' Industrial Union of Canada – n1-2 [1949 feb 21-apr 11] – 1r – 1 – mf#681691 – us WHS [331]

Between the lines – Washington Crossing PA 1942-76 – 1,5,9 – ISSN: 0006-0305 – mf#2307 – us UMI ProQuest [240]

Between the planets / Watson, Fletcher Guard – 1st ed. Philadelphia: Blakiston Co c1941 [mf ed 1998] – 1r [ill] – 1 – (incl ind) – mf#film mas 28218 – us Harvard [520]

Between the testaments : or, interbiblical history / Gregg, David – New York: Funk & Wagnalls 1907 [mf ed 1992] – 1mf – 9 – 0-524-05037-6 – mf#1992-0290 – us ATLA [221]

Between two continents / Wilhelm, Prince Of Sweden – London, England. 1922 – 1r – us UF Libraries [972]

Betz, H D *see* Lukian von samosata und das neue testament

Betzendoerfer, Walter *see*
- Glauben und wissen bei den grossen denkern des mittelalters
- Hoelderlins studienjahre im tuebinger stift

Betzi ou l'amour comme il est : roman qui n'en est pas un – Paris: Remouard, 1801 – (= ser Les femmes [coll]) – 4mf – 9 – mf#9759 – fr Bibl Nationale [830]

Beuker, Henricus *see* Leerredenen

Beukes, Catharina F *see* Kwagandaganda

Beukes, Wiets Taylor Heyman *see* Hauptling in der gesellschaft der sud...

Beulah / Evans, Augusta Jane – [Toronto?: s.n, 188-?] [mf ed 1993] – 3mf – 9 – 0-665-92083-0 – (in dble clms. original iss in ser: [robertson's cheap series]) – mf#92083 – cn Canadiana [830]

Beulah Baptist Association *see* Minutes of the...annual session of the beulah baptist association

Beulah baptist church – Abbeville County, SC. 1960-72 – 1 – $5.00 – us Southern Baptist [242]

Beulah baptist church – Florence County, SC. 1872-1924 – (= ser History of Church) – 1 – $9.36 – (history of church. 1872-1966) – us Southern Baptist [242]

Beulah baptist church – Richland County, SC. 1806-83.208p – 1 – $9.36 – us Southern Baptist [242]

Beulah baptist church – Tippah County, MS. 1848-jun 1885 – 1 – $11.88 – us Southern Baptist [242]

Beulah baptist church – Union County, SC. 1881-Feb 1973 – 1 – $20.70 – us Southern Baptist [242]

Beulah christian – Providence RI: F A Hillery 1892-1911 [mf ed 1982] – 20v on 5r – 1 – (lacks several iss; some pp damaged; some iss accomp by suppls; formed by union of: beulah items and: bible christian; merged with: pentecostal era to form: pentecostal christian) – mf0420b – us ATLA [240]

Beulah items – Providence RI: F A Hillery 1888-92 [mf ed 1982] – 5v on 1r – 1 – (lacks: v2 n8-9,11-12; v3 n1,5; v4 n4; v5 n1; damaged: v3 n7-12, v4 n1-12; merged with: bible christian to form: beulah christian) – mf0420a – us ATLA [240]

Beulah-land : or, words of cheer for christian pilgrims / Cuyler, Theodore Ledyard – New York: American Tract Society, c1896. Chicago: Dep of Photodup, U of Chicago Lib, 1977 (1r); Evanston: American Theol Lib Assoc, 1984 (1r) – 1 – 0-8370-0163-3 – mf#1984-T052 – us ATLA [240]

Beumelburg, Werner *see*
- Das eherne gesetz
- Der feigling, die belagerung von neuss
- Gruppe bosemueller
- Die hengstwiese
- Joerg
- Kaiser und herzog
- Kampf um spanien
- Der kuckuck und die zwoelf apostel
- Mont royal
- Preussische novelle
- Der strom
- Wen die goetter lieben

Beumer, Rebecca L *see* Determining the presence of lifeguards during competitive swimming events at mid-american conference universities

Beuordering van wetenskap van die suid-afrikanse : verenigung – v4-16. 1966-68 [complete] – 1r – 1 – mf#ATLA S0676A – us ATLA [073]

Beurden, A F van *see* Het missale van de kerk te wijk bij heusden

Beurhaus, Friedrich *see* Erotematum musicae libri duo

Beurhusius, Friedrich *see* Erotematum musicae libri duo

Beurmann, Eduard *see* Deutschland und die deutschen

Beurmann, M von *see* Vocabulary of the tigre language

Das beurteilen in der fachsprachlichen kommunikation / Fiss, Sabine – Chemnitz 1983 (mf ed 1995) – 2mf – 9 – €40.00 – 3-8267-2118-7 – mf#DHS-AR 2118 – gw Frankfurter [400]

Beurteilung des effektes organischer loesungsmittel auf das hoervermoegen / Heitmann, P & Bolt, H M – (mf ed 1996) – 1mf – 9 – €30.00 – 3-8267-2310-4 – mf#DHS 2310 – gw Frankfurter [616]

Beusechem, J M *see* Statistiek van java en madura door j m beusechem, 1836

Beuthener zeitung *see* Ostdeutsche morgenpost

Beutler, Ernst *see* Goethe in briefen und gespraechen

Bevan, Anthony Ashley *see*
- The hymn of the soul
- A short commentary on the book of daniel for the use of students

Bevan, D Barclay *see* Minister's letter, n8

Bevan, Edwyn Robert *see*
- Indian nationalism
- Jerusalem under the high-priests
- Stoics and sceptics

Bevan, Frances *see* Three friends of god

Bevan, Joseph Gurney *see* Extracts form the letters and other writings of the late joseph gurney bevan

Bevan, Llewelyn D *see* "The blessedness of giving" and "perilous times"

Bevan, W L *see*
- Case of the church in wales
- Is the church in wales an alien institution?

Bevan's treaties : treaties and other international agreements of the united states, 1776-1949 / U.S. Dept of State – v1-13 [all publ] – (= ser Department of state treaty publications) – 160mf – 9 – $240.00 – (incl ind) – mf#llmc 79-440 – us LLMC [341]

Beverage industry – Deerfield IL 1974+ – 1,5,9 – ISSN: 0148-6187 – mf#9680,02 – us UMI ProQuest [660]

Beverage world – New York NY 1975+ – 1,5,9 – (cont: soft drinks) – ISSN: 0098-2318 – mf#289,01 – us UMI ProQuest [660]

Beveregio, G *see* Synodicon sive pandectae canonum apostolorum et conciliorum ab ecclesia graeca receptorum

Beveridge, Albert J *see* The life of john marshall

Beveridge committee report on the welfare state *see* State provision for social need

Beveridge, Henry *see* The trial of maharaja nanda kumar

Beveridge, John *see* Covenanters

Beveridge, M P *see* Sanganai namai chamunorwa

Beveridge papers from the british library of political and economic science *see* State provision for social need, series 2

Beveridge, Thomas Hanna et al *see* The church memorial

Beveridge, William *see* Resolutions respecting religion

Beverland, Adriaan *see*
- De fornicatione cavenda admonitio
- De stolatae virginitatis jure lucubratio academica
- Hadriani beverlandi de peccato originali... dissertatio

Beverley advertiser – Beverley, England 4 sep 1992-5 may 1995 – 1 – (cont by: beverley & east yorkshire advertiser [12 may 1995-]) – uk British Libr Newspaper [072]

Beverley & district star – Beverley, England 3 jan 1986-12 may 1988 – 1 – (replaced by: beverley weekly journal) – uk British Libr Newspaper [072]

Beverley & east riding telegraph – Beverley, England 4 may 1895-26 dec 1896 [mf may-dec 1895, 1898, 1900] – 1 – (cont by: east riding telegraph [1 jan 1898-7 nov 1903]) – uk British Libr Newspaper [072]

Beverley echo – Beverley, England 20 jan 1885-8 jul 1903 [mf 1897] – 1 – (wanting 1896; discontinued) – uk British Libr Newspaper [072]

Beverley guardian – Beverley, England 5 jan 1856- [mf 1986-] – 1 – uk British Libr Newspaper [072]

Beverley independent – Beverley, England 14 Apr 1888-25 Mar 1911 (wkly) [mf 1894, 1897] – 1 – (discontinued) – uk British Libr Newspaper [072]

Beverley recorder and general advertiser – Beverley, England 21 mar 1857-5 oct 1867 – 1 – (cont: beverley weekly recorder [7 jul 1855-15 mar 1857]; cont by: beverley weekly recorder & general advertiser [12 oct 1867-29 dec 1883]; beverley recorder [5 jan 1884-30 dec 1916 (mf 1912)]) – uk British Libr Newspaper [072]

Beverley target – Beverley, England 28 sep 1989- – 1 – uk British Libr Newspaper [072]

Beverley weekly journal – Beverley, England 19 may 1988-29 jun 1989 – 1 – uk British Libr Newspaper [072]

Beverly 1650-1849 – Oxford, MA (mf ed 1995) – (= ser Massachusetts vital record transcripts to 1850) – 38mf – 9 – 0-87623-210-1 – (mf 1t-2t: births 1650-1704. mf 1t: deaths 1660-1704. mf 2t: marriages: 1666-1704. mf 2t-3t: intentions 1695-1722. mf 3t-4t: intentions 1771-72. mf 3t-10t: births 1670-1779. mf 7t-9t: marriages 1689-1766. mf 9t-11t: intentions 1716-67. mf 9t-10t: deaths 1686-1795. mf 11t-15t: births 1750-1854. mf 12t-17t: deaths 1772-1872. mf 18t-21t. intentions 1772-1813. mf 21t-23t: marriages 1771-1833. mf 23t-25t: intentions 1813-33. mf 25t-27t: deaths 1770-1857. mf 27t-29t: intentions 1834-49. mf 27t-30t: marriages 1834-43. mf 30t: births & deaths 1811-49. mf 31t: vital records 1813-49. mf 31t-34t: births 1844-49. mf 34t-35t: marriages 1843-49. mf 35t-37t: deaths 1843-49. mf 37t: births 1818-45. mf 38t: out-of-town marrs 1683-1799) – us Archive [978]

Beverly 1653-1907 – Oxford, MA (mf ed 1989) – (= ser Massachusetts vital records) – 151mf – 9 – 0-87623-102-4 – (mf 1-35: vital records 1653-1850. mf 36-41: town records 1685-1711. mf 42-45: proprietors 1698-1817. mf46-56: births & deaths 1750-1845. mf 57: out-of-town marriages 1683-1798. mf 58-61: index: marriages 1723-1813. mf 62-74: marriages & intentions 1770-1853. mf 75-77: index: vital records 1843-50. mf 78-81: vital records 1843-50. mf 82-83: index: deaths 1851-82. mf 84-90: deaths 1851-82. mf 91-92: index: deaths 1883-92. mf 93-94: deaths 1883-92. mf 95-97: index: deaths 1893-1907. mf 98-102: deaths 1893- 1907. mf 103-105: index: marriages 1851-86. mf 106-112: marriages 1837-86. mf 113-114: index: marriages 1887-92. mf 115-116: marriages 1887- 92. mf 117-119: index: marriages 1893-1907. mf 120-124: marriages 1893-1907. mf 125-134: marriage intentions 1873-97. mf 135-136: index: births 1851-79. mf 137- 143: births 1851-79. mf 144-145: index: births 1880-92. mf 146-48: births 1880-92. mf 149-151: index: births 1893-1907. mf 152-56: births 1893-1907) – us Archive [978]

Beverly hills – 1945-50; 1977-87; 1989-92 – (= ser California telephone directory coll) – 26r – 1 – $1300.00 – mf#P00008 – us Library Micro [917]

Beverly hills bar association journal – v1-28, 31-32. 1967-98 – 9 – $371.00 set – (title varies: v1-11 1967-80 as journal of the beverly hills bar association) – ISSN: 1051-628X – mf#110071 – us Hein [340]

Beverly hills / santa monica – 1993- – (= ser California telephone directory coll) – 3r – 1 – $150.00 – mf#P00009 – us Library Micro [917]

Beverovicius, J *see* Epistolica quaestio de vitae termino...fatali, an mobili?

Bevolkingsensus, 1960 : steekproeftabellasie / South Africa Bureau Of Census And Statistics – Pretoria? South Africa. n1-8. 1962-1965 – 1r – us UF Libraries [960]
Bewaehrung : gedichte / Marteau, Eugen Henrik – [Bayreuth]: Gauverlag Bayreuth 1943 [mf ed 1993] – 1r – 1 – (filmed with: die fahrt nach letztesand / martin luserke) – mf#7604 – us Wisconsin U Libr [810]
Bewaehrung der herzen : novelle / Wittek, Erhard – feldausg. Guetersloh: C Bertelsmann [1943] [mf ed 1992] – 1r – 1 – (filmed with: miniaturen / georg witkowski) – mf#3059p – us Wisconsin U Libr [830]
Beware of idolatry / Irons, Joseph – London, England. 1845? – 1r – us UF Libraries [240]
Bewegingen preanger regentschappen (tjiandjur) – Mailrapport. n642a. 1885 – 1mf – 8 – mf#SD-101 mf 1 – ne IDC [959]
Die bewegung – Muenchen DE, 1936-42 [gaps] – 1 – (filmed by other misc inst: 1930-1931 n17) – gw Misc Inst [074]
Die bewegungen und haltungen des menschlichen koerpers in heinrich von kleists erzaehlungen / ed by Bathe, Johannes Clemens – Tuebingen: H Laupp, 1917 [mf ed 1991] – 80p – 1 – (incl bibl ref) – mf#7514 – us Wisconsin U Libr [430]
Der beweis des christenthums *see* Natural religion
Der beweis fuer das dasein gottes und seine persoenlichkeit : mit ruecksicht auf die herkoemmlichen gottesbeweise / Melzer, Ernst – Neisse: Josef Graveur, [1910?] – 1mf – 9 – 0-7905-9514-1 – mf#1989-1219 – us ATLA [210]
Der beweis fuer die wahrheit des christentums : ein beitrag zur apologetik / Steude, E Gustav – Guetersloh: C Bertelsmann, 1899 [mf ed 1991] – (= ser Beitraege zur foerderung christlicher theologie 3/5) – 1mf – 9 – 0-524-00149-9 – mf#1989-2849 – us ATLA [240]
Bewer, Julius August *see*
– Die anfaenge des nationalen jahweglaubens
– The history of the new testament canon in the syrian church...
– The story of hosea's marriage
Bewer, Max *see* Ein goethepreis
Bewertung der tv-serie auf empirischer grundlage / Boll, Uwe – (mf ed 1994) – 2mf – 9 – €40.00 – 3-8267-2043-1 – mf#DHS 2043 – gw Frankfurter [790]
Bewick gleanings : being impressions from copperplates and wood blocks / Boyd, Julia – Newcastle-upon-Tyne 1886 – (= ser 19th c art & architecture) – 4mf – 9 – mf#4.1.350 – uk Chadwyck [760]
Bewick memento : catalogue with purchasers' names and prices realised of the scarce and curious collection of books...sold by auction at newcastle-upon-tyne on february 5th, 6th, 7th, and august 26th 1884 – London, New York. 2pt. 1884 – (= ser 19th c art & architecture) – 2mf – 9 – mf#4.1.4 – uk Chadwyck [720]
Bewick, Thomas *see* A memoir of thomas bewick
Bewick, William *see* Life and letters of william bewick
The bewildered querists and other nonsense / Crofton, Francis Blake – New York: G P Putnam's Sons, 1875 – 2mf – 9 – mf#06825 – cn Canadiana [880]
Bewilligung vnd confirmation eines burgermeisters... / Bullinger, Heinrich & Jud, L – [Zuerich, Christoph Froschouer], 1532 – 1mf – 9 – mf#PBU-536 – ne IDC [240]
Bewilligung vnd confirmation...ueber die restitution vnd verbesserung ettlicher maenglen vnd miszbruechen... / Bullinger, Heinrich – Zuerich, 1532 – 1mf – 9 – mf#PBU-259 – ne IDC [240]
Bewley family roots – 1982 mar-jun – 1r – 1 – (cont: bewley family roots newsletter; cont by: bewley roots) – mf#1758471 – us WHS [929]
Bewley family roots newsletter – 1981 mar-dec – 1r – 1 – (cont by: bewley family roots) – mf#1053278 – us WHS [929]
Bewley roots – 1982 sep/dec-1983 mar – 1r – 1 – (cont: bewley family roots; cont by: bewley roots newsletter; bewley roots newsletter) – mf#1759039 – us WHS [929]
Bewley roots family newsletter – 1985 mar, sep-1987 mar – 1r – 1 – (cont: bewley roots newsletter, bewley roots newsletter; cont by: bewley roots newsletter (rineyville ky: 1987)) – mf#1759074 – us WHS [929]
Bewley roots newsletter – 1983 jul-1984 sep – 1r – 1 – (cont: bewley roots; cont by: bewley roots family newsletter) – mf#1759051 – us WHS [929]
Bewley roots newsletter – 1987 jun/dec – 1r – 1 – (cont: bewley roots family newsletter) – mf#1759113 – us WHS [929]
Bewley, William Fleming *see* Diseases of glasshouse plants
Das bewusstsein der gnade : die erloesungslehre bis zur lehre von der kirche / Rothe, Richard; ed by Schenkel, Daniel – Heidelberg: JCB Mohr, 1870 – 1mf – 9 – 0-8370-9734-7 – (incl bibl ref) – mf#1986-3734 – us ATLA [240]
Das bewusstsein der gnade : die lehre von der kirche bis zum schlusse / Rothe, Richard; ed by Schenkel, Daniel – Heidelberg: JCB Mohr, 1870 – 1mf – 9 – 0-8370-9735-5 – (incl bibl ref and index) – mf#1986-3735 – us ATLA [240]
Das bewusstsein der suende / Rothe, Richard; ed by Schenkel, Daniel – Heidelberg: JCB Mohr, 1870 – 1mf – 9 – 0-8370-9736-3 – (incl bibl ref) – mf#1986-3736 – us ATLA [240]
The bexar archives, 1717-1836 : colonial archives of texas during the spanish and mexican periods – 1 – $29,420.00 coll – (the bexar archives: 1717-1803 31r isbn 1-55655-042-1 $4850. 1804-21 38r isbn 1-55655-043-x $5945. 1822-36 103r isbn 1-55655-044-8 $16,105. bexar archives translations 26r isbn 1-55655-046-4 $4080. with p/g) – us UPA [975]
Bexhill quarterly – [London & SE] East Sussex, Bexhill Ref Lib may-nov 1914 [mf ed 2003] – 1r – 1 – uk Newsplan [072]
Bexhill-on-sea observer – [London & SE] East Sussex, Bexhill Ref Lib may 1896-1897, 1904, 1911, BLNL 1986- [mf ed 2003] – 1 – uk Newsplan [072]
Bexley & eltham leader – Bexley, England 13 dec 1985-7 jul 1989 [mf 1986-9] – 1 – (replaced by: bexley & eltham comet leader; cont: leader (bexley borough & eltham ed) [16 feb 1984-5 dec 1985]; bexley & eltham times leader [11 jun 1993-24 feb 1995]; leader (bexley & eltham ed) 3 mar 1995-7 jun 1996; bexley & eltham leader [14 jun 1996-22 dec 2000]) – uk British Libr Newspaper [072]
Bexley heath, bexley & district times, and dartford chronicle – Bexley, England 3 jan 1905-27 sep 1918 [mf jan-jul 1912] – 1 – (wanting: aug-dec 1912) – uk British Libr Newspaper [072]
Bexley heath & erith observer – Bexley, England 5 jan 1878-16 may 1902 – 1 – (wanting: 1897; cont: bexley heath, dartford & erith observer [1 jan-17 sep 1870, 15 may 1875-29 dec 1877]; cont by: bexley heath & bexley observer [23 may 1902-26 sep 1918]) – uk British Libr Newspaper [072]
Bexley heath, welling and district free press, and entertainment guide – Bexley, England 17 jun 1933-30 oct 1937 – 1 – uk British Libr Newspaper [072]
Bexleyheath, welling & district free press – [London & SE] Bexley 17 jun 1933-30 oct 1937 [mf ed 2003] – 5r – 1 – uk Newsplan [072]
Bexleyheath & welling news shopper – Bexley, England 10 feb 1983-27 feb 1991 [mf 1986-] – 1 – (cont: news shopper [21 sep 1967-3 feb 1983]; cont by: news shopper (bexleyheath & welling) 6 mar 1991-) – uk British Libr Newspaper [072]
Bexleyheath & welling times – Bexley, England 12 dec 1985-28 sep 2000 [mf 1986-2000] – 1 – (cont: bexleyheath & welling observer and kentish times [3 jan 1958-5 dec 1985]; cont by: kentish times (bexleyheath & welling ed) 5 oct 2000-27 sep 2002) – uk British Libr Newspaper [072]
Bey, Alican Serif *see* Harflerimizin muedafaasi
Bey, Feridun *see* Mecmu'a-i muenseat-i selatin
Bey, Mehmet Ata *see* Gueft ue senid
Bey, Mustafa Koci *see* Koeci bey risalesi
Bey, Sait *see* Hiziragazade arif aga'nin mahdumu sait bey divani
Le beyaan arabe : le livre sacre de baabysme de seyy ed ali mohammed, dit le baab = Bayan / Bab, Ali Muhammad Shirazi – Paris: Ernest Leroux, 1905 – (= ser Bibliotheque orientale elzevirienne) – 1mf – 9 – 0-524-00683-0 – (in french) – mf#1990-2011 – us ATLA [290]
Le beyan persan / Bab, Ali Muhammad Shirazi – Paris: P Geuthner, 1911-1914 – 2mf – 9 – 0-524-07067-9 – mf#1991-0049 – us ATLA [290]
Beyan uel-hak – Istanbul: Yeni Ikdam Matbaasi, Mahmud Bey Matbaasi. n1-182. 22 eyluel 1324-22 tesrinievvel 1328 [1906-10] – (= ser O & t journals) – 22mf – 9 – $350.00 – (publ by: cemiyet-i ilmiye-i osmaniye. sahib-i imtiyaz: ahmed efendi) – us MEDOC [956]
Beyen, Petrus *see* Korte verhandeling over het zingen en speelen in de hervormde kerk van nederland
Beyer, A H *see* Bean leaf-hopper and hopperburn with methods of control
Beyer, Henry Otley *see* Population of the philippine islands in 1916
Beyer, Johann Paul *see* Barndopets vaelsignelse
Beyer, Konrad *see* Friedrich rueckert
Beyer, Roberta *see* Motor proficiency of males with attention deficit hyperactive disorder and males with learning disabilities
Beyer, Rudolph von *see* Meine begegnung mit goethe und anderen grossen zeitgenossen
Beyerhaus, Gisbert *see*
– Studien zur staatsanschauung calvins
Beyerlein, Franz Adam *see*
– Jena oder sedan?
– 'Jena' or 'sedan'?
– Taps
Beyn hashmoshes / Palepade, Benzion – Buenos Ayres, Argentina. 1951 – 1r – us UF Libraries [939]
Beyne, Pierre *see* Manual de l'emprunteur sur warrants commerciaux et sur warrants agricoles
Beyoglu alemi – Sahibi ve Mueduerue: M Nizamettin. n3. 3 temmuz 1930 – (= ser O & t journals) – 1mf – 9 – $25.00 – us MEDOC [956]
Beyond architecture / Porter, Arthur Kingsley – Boston, MA. 1918 – 1r – us UF Libraries [720]
Beyond black magazine [bbm] – 1989 fall – 1r – 1 – mf#4852767 – us WHS [071]
Beyond cybernetics : connecting the professional and personal selves of the therapist / Marovic, Snezana – Uni of South Africa 2000 [mf ed Johannesburg 2000] – 5mf [ill] – 9 – (incl bibl ref) – mf#mfm15040 – sa Unisa [150]
Beyond good and evil / Nietzsche, Friedrich Wilhelm – New York, NY. 1917 – 1r – us UF Libraries [190]
The beyond that is within : and other addresses / Boutroux, Emile – London: Duckworth 1912 [mf ed 1990] – 1mf – 9 – 0-7905-3688-9 – mf#1989-0181 – us ATLA [100]
Beyond the city lights / Green, Lawrence George – Cape Town, South Africa. 1957 – 1r – us UF Libraries [960]
Beyond the grave = Ueber den zustand nach dem tode / Cremer, Hermann – New York: Harper 1886, c1885 [mf ed 1989] – 1mf – 9 – 0-7905-0934-2 – (incl ind, trans fr german by samuel t lowrie, int by a a hodge) – mf#1987-0934 – us ATLA [220]
Beyond the hills of dream / Campbell, Wilfred – Boston; New York: Houghton, Mifflin, 1899 [mf ed 1980] – 2mf – 9 – 0-665-00405-2 – mf#00405 – cn Canadiana [810]
Beyond the hills of dream / Campbell, Wilfred – Toronto: G N Morang, 1900 [mf ed 1981] – 2mf – 9 – mf#26764 – cn Canadiana [810]
Beyond the mexique bay / Huxley, Aldous – New York, NY. 1934 – 1r – us UF Libraries [420]
Beyond the natural order : essays on prayer, miracles and the incarnation / Best, Nolan Rice – New York: Fleming H Revell c1908 [mf ed 1985] – 1mf – 9 – 0-8370-2683-0 – mf#1985-0683 – us ATLA [210]
Beyond the pir panjal : life among the mountains and valleys of kashmir / Neve, Ernest Frederic – London; Leipsic: T Fisher Unwin, 1912 [mf ed 1995] – (= ser Yale coll) – xvi/320p (ill) – 1 – 0-524-09029-7 – mf#1995-0029 – us ATLA [920]
Beyond the rhine : memories of art and life in germany before the war / Henry, Marc – London: Constable & Co Ltd 1918 [mf ed 1985] – 1r – 1 – (filmed with: zapiski o moei zhizni / grech, n i) – mf#1491 – us Wisconsin U Libr [943]
Beyond the shadow : or, the resurrection of life / Whiton, James Morris – New York: Thos Whittaker 1898 [mf ed 1991] – (= ser Freedom of faith series) – 1mf – 9 – 0-7905-8627-4 – (originally iss in the us in 1881 under title: the gospel of the resurrection) – mf#1989-1852 – us ATLA [240]
Beyond the smoke that thunders / Cullen, Lucy Pope – New York, NY. 1940 – 1r – us UF Libraries [960]
Beyond the utmost purple rim : abyssinia, somaliland, kenya colony, zanzibar, the comoros, madagascar / Powell, E A – London, 1925 – 6mf – 9 – mf#NE-20228 – ne IDC [916]
Beys, Martina *see* Evaluation des asthma-verhaltenstrainings (avt)
Beyschlag, Willibald *see*
– Der altkatholicismus
– Die auferstehung und ihre neueste bestreitung
– Die christologie des neuen testaments
– Das leben jesu
– New testament theology
– Ueber das "leben jesu" von renan
– Ueber die bedeutung des wunders im christenthum
– Welchen gewinn hat die evangelische kirche aus den neuesten verhandlungen ueber das leben jesu zu ziehen?
– Woran fehlt's uns glaeubigen predigern, um in weiterem umfange geistliches leben zu wecken?
– Zur deutsch-christlichen bildung
– Zur johanneischen frage
– Zur verstaendigung ueber den christlichen vorsehungsglauben
Beyschlag, Willibald et al *see* Vortraege fuer das gebildete publikum. zweite sammlung
Beyssac, J *see* Abbayes et prieures de l'ancienne france (afm37)
Beytraege zu einer bibliothek fuers volk – Hannover DE, 1783-86 – 2r – 1 – gw Misc Inst [074]
Beytraege zum teutschen recht / ed by Siebenkees, Johann Christian – Nuernberg, Altdorf 1786-90 – (= ser Dz) – 6pt on 12mf – 9 – €120.00 – mf#k/n2581 – gw Olms [342]
Beytraege zur aufnahme des bluehenden wohlstands der staaten / ed by Griesheim, Christ Ludwig von – Leipzig, Zittau 1766 – (= ser Dz) – 1v on 5mf – 9 – €100.00 – mf#k/n2885 – gw Olms [943]
Beytraege zur befoerderung theologischer und andrer wichtigen kenntnisse von kielischen und auswaertigen gelehrten / ed by Cramer, Johann Andreas – Kiel, Hamburg 1777-83 – (= ser Dz. abt theologie) – 4v on 12mf – 9 – €120.00 – mf#k/n2134 – gw Olms [200]
Beytraege zur befoerderung vernuenftigen denkens in der religion / ed by Corrodi, Heinrich – Frankfurt, Leipzig 1780-88, Winterthur 1789-94 – (= ser Dz. abt theologie) – 18iss on 35mf – 9 – €350.00 – mf#k/n2146 – gw Olms [200]
Beytraege zur critischen historie der deutschen sprache, poesie und beredsamkeit / ed by Deutsche Gesellschaft, Leipzig – Leipzig 1732-44 [mf ed Hildesheim 1977] – (= ser Allgemeinwissenschaftliche und literarische zeitschriften des 17. und 18. jahrhunderts) – 74mf – 9 – diazo €328.00 silver €388.00 – gw Olms [430]
Beytraege zur erlaeuterung der kirchen-reformations-geschichten / ed by Fuesslin, J C – Zuerich, Conrad Orell, Heidegger 1741-1753. 5 v – 27mf – 9 – mf#PBU-421 – ne IDC [242]
Beytraege zur erweiterung der geschichtskunde / ed by Meusel, Johann Georg – Augsburg 1780-82 – (= ser Dz. historisch-geographische abt) – 2pt on 4mf – 9 – €120.00 – mf#k/n1105 – gw Olms [900]
Beytraege zur europaeischen laenderkunde, die moldau, wallachey, bessarabien und bukowina : neueste darstellung dieser laender... / Karacsay, Fedor – Wien [1818] [mf ed Hildesheim 1995-98] – 1v on 1mf – 9 – €40.00 – 3-487-26834-5 – gw Olms [914]
Beytraege zur genauern kenntniss der spanischen besitzungen in amerika / Fischer, Christian A – Dresden 1802 [mf ed Hildesheim 1995-98] – 1v on 2mf – 9 – €60.00 – 3-487-26910-4 – gw Olms [970]
Beytraege zur geschichte der philosophie / ed by Fuelleborn, Georg Gustav – Jena 1791-99 – (= ser Dz. abt philosophie) – 3v[=st1-12] on 15mf – 9 – €150.00 – mf#k/n566 – gw Olms [100]
Beytraege zur geschichte der teutschen justizpflege im 18. jahrhundert / ed by Staudner, Johann Leopold – Nuernberg 1789 – (= ser Dz) – 8mf – 9 – €160.00 – mf#k/n2584 – gw Olms [340]
Beytraege zur geschichte und statistik der evangelischen kirche / Augusti, Johann Christian Wilhelm – Leipzig: Dyk 1837-38 [mf ed 1993] – 3v on 10mf – 9 – 0-524-08726-1 – mf#1993-3231 – us ATLA [242]
Beytraege zur geschichte von baiern / ed by Mederer, Johann Nepomuk – Regensburg, Ingolstadt 1777-1793 – (= ser Dz. historisch-geographische abt) – 5st on 6mf – 9 – €120.00 – mf#k/n1090 – gw Olms [943]
Beytraege zur historie und aufnahme des theaters / ed by Lessing, Gotthold Ephraim et al – Stuttgart 1750 – (= ser Dz) – 4st on 4mf – 9 – mf#k/n4125 – gw Olms [790]
Beytraege zur kenntniss norwegens : gesammelt auf wanderungen waehrend der sommermonate der jahre 1821 und 1822 / Naumann, Carl F – Leipzig 1824 [mf ed Hildesheim 1995-98] – (= ser Fbc) – 2v on 5mf – 9 – €100.00 – 3-487-28919-9 – gw Olms [914]
Beytraege zur litteratur besonders des sechszehnten jahrhunderts / ed by Strobel, Georg Theodor – Nuernberg, Altdorf 1784-86 – (= ser Dz. abt literatur) – 2v[zu je 2st] on 6mf – 9 – €120.00 – mf#k/n4539 – gw Olms [410]
Beytraege zur mainzer geschichte mit urkunden / ed by Schunk, Johann Peter – Mainz 1788-90 – (= ser Dz) – 3v on 9mf – 9 – €180.00 – mf#k/n1273 – gw Olms [943]
Beytraege zur naehern kenntnis des schweizerlandes / ed by Schinz, H Rudolph – Zuerich 1783-87 – (= ser Dz. historisch-geographische abt) – 5iss on 5mf – 9 – €100.00 – mf#k/n1139 – gw Olms [949]
Beytraege zur oekonomie, technologie, polizey- und cameralwissenschaften / ed by Beckmann, Joh – Goettingen 1779-91 – (= ser Dz) – 12pt on 12mf – 9 – €120.00 – mf#k/n2923 – gw Olms [330]

Beytraege zur statistik und geographie vorzueglich von deutschland aus der neuesten literatur / ed by Roesler, Chr Friedr – Tuebingen 1780-82 – (= ser Dz. historisch-geographische abt) – 3st on 6mf – 9 – €120.00 – mf#k/n1106 – gw Olms [914]

Beytraege zur vaterlaendischen historie, geographie, statistik und landwirtschaft sammt einer uebersicht der schoenen litteratur / ed by Westenrieder, Lorenz von – Muenchen 1788-1817 – (= ser Dz. historisch-geographische abt) – 30mf – 9 – €300.00 – mf#k/n1163 – gw Olms [943]

beytrag zu einer historischen, politischen und statistischen entwicklung der von napoleon bonaparte waehrend seines obercommando und seiner regierung befolgten : maassregeln und entwuerfe, mit einer sammlung dahin gehoeriger staatsschriften – Hamburg 1814 [mf ed Hildesheim 1995-98] – 1v on 3mf – 9 – €90.00 – 3-487-26236-3 – gw Olms [940]

Beza, Theodor de *see*
- Ad acta colloquii montisbelgardensis tubingae edita ia
- Ad danielis hofmanni demonstrationes ad oculum...
- Ad gilberti genebrardi accusationem
- Ad tractationem de ministrorum evangelii gradibus ab saravia editam responsio
- Apologia pro justificatione adversus lescalium
- Chrestiennes meditations...
- Confession de la foy chrestienne...
- Confession de la foy chrestienne
- De controversiis in coena
- Epistolae theologicae
- Icones
- Icones, id est verae imagines virorum doctrina simul et pietate illustrium...quibus adiectae sunt nonnullae picturae quas emblemata vocant
- Icones id est verae imagines virorvm doctrina simvl et pietate illvstrivm..., partim vera religio in variis orbis...
- Iobvs... partim commentarijs partim paraphrasi illustratus, cui etiam additus est ecclesiastes...
- The iudgment of a most reverend and learned man, from beyond the seas, concerning a threefolde order of bishops...
- Jobus...eccleciastes
- Lex dei, moralis, ceremonialis, et politica...
- The life of john calvin
- Novum testamentum domini nostri jesu christi
- Poemata varia
- Response aux cinq premieres et principales demandes de jean hay
- Responsio ad qvaestionvm et responsionvm danielis hofmanni...
- Sermons sur l'histoire de la passion
- Sermons sur l'histoire de la resurrection
- Sermons sur...le cantique des cantiques...
- Tractationes theologicae
- Tractatus de vera excommunicatione et christiano presbyterio
- Les vrais pourtraits des hommes illustres en piete et doctrine, du travail desquels dieu s'est servi en ces derniers temps...

Bezae codex cantabrigiensis : being an exact copy, in ordinary type, of the celebrated uncial graeco-latin manuscript of the four gospels and acts of the apostles... / ed by Scrivener, Frederick Henry Ambrose – Cambridge: Deighton, Bell, 1864. Chicago: Dep of Photodup, U of Chicago Lib, 1978 (1r); Evanston: American Theol Lib Assoc, 1984 (1r) – 1 – 0-8370-1123-X – (incl bibl ref) – mf#1984-T135 – us ATLA [220]

Beza's icones : contemporary portraits of reformers of religion and letters / McCrie, Charles Greig – London: Religious Tract Soc 1906 [mf ed 1990] – 1mf – 9 – 0-7905-8121-3 – mf#1988-8038 – us ATLA [242]

Bezbozhie pobedit = Atheism triumphs / Tikhomirov, P – Simferopol, 1930 – 1 – $74.32 – (one of 13 titles by soviet authors on reel) – us Southern Baptist [242]

[Beze, T] *see* Dv droit des magistrats svr leurs svbiets...

Bezerra De Freitas, Jose *see*
- Forma e expressao no romance brasileiro
- Historia da literatura brasileira

Bezerra, Felte *see* Etnias sergipanas

Bezerra, Joao Climaco *see* Nao ha estrelas no ceu...

Bezerra Tanco, Luis *see* Felicidad de mexico... guadalupe extremuros

Bezgin, I G *see* Opisanie vsekh russkikh knig i povremennykh izdanii, vyshedshikh s 1708 goda

Bezhenskaia pravda : organ vserossijskogo soiuza bezhentsev – 1917 [mf ed Norman Ross] – 1r – 1 – mf#nrp-86 – us UMI ProQuest [077]

Bezhetsk Sovet rk i kd *see* Izvestiia bezhetskogo soveta krest'ian, rabochikh i krasno-armejskikh deputatov. tversk-gub

Bezhetskii vestnik : sotsialisticheskaia gazeta – Bezhetsk, Russia 1917 [mf ed Norman Ross] – 1r – 1 – mf#nrp-332 – us UMI ProQuest [077]

Die beziehung des christentums zum griechischen heidentum : im urteil der vergangenheit und gegenwart / Glawe, Walther – Berlin: E Runge 1913 – (= ser Biblische zeit- und streitfragen 8/8) – 1mf – 9 – 0-7905-2957-2 – mf#1987-2957 – us ATLA [230]

Die beziehung zwischen elektromyographischen untersuchungen bei sportlicher betaetigung von zerebralparetikern : unter besonderer beruecksichtigung des krafttrainings mit erkenntnissen aus der literatur / Kracht, Arnim – (mf ed 1997) – 2mf – 9 – €40.00 – 3-8267-2445-3 – mf#DHS 2445 – gw Frankfurter [612]

Die beziehungen des dramatikers achim von arnim zur altdeutschen literatur / Bottermann, Walther – Goettingen 1895 (mf ed 1995) – 1mf – 9 – €24.00 – 3-8267-3121-2 – mf#DHS-AR 3121 – gw Frankfurter [430]

Die beziehungen des dramatikers achim von arnim zur altdeutschen litteratur / Bottermann, Walther – Goettingen: Druck der Dieterich'schen Univ.-Buchdruckerei 1895 [mf ed 1988] – 1r – 1 – (incl bibl ref; filmed with: ariel's offenbarungen / ludwig achim von arnim) – mf#6956 – us Wisconsin U Libr [430]

Die beziehungen des sumerischen zum baskischen, westkaukasischen und tibetischen / Bouda, K – Leipzig, 1938 – (= ser Mitteilungen der altorientalischen Gesellschaft) – 1mf – 9 – (mitteilungen der altorientalischen gesellschaft. v12, pt 3)) – mf#NE-20106 – ne IDC [956]

Die beziehungen von roem. 1-3 zur missionspraxis des paulus / Weber, Emil – Guetersloh: C Bertelsmann, 1905 – (= ser Beitraege zur foerderung christlicher theologie) – 1mf – 9 – 0-524-06221-8 – mf#1992-0859 – us ATLA [220]

Beziehungen zwischen den symbiosepartnern in der orchideen-mycorrhiza / Rohm, Eva Maria – (mf ed 1993) – 5mf – 9 – €62.50 – 3-89349-718-8 – mf#DHS 718 – gw Frankfurter [574]

Bezirksbote : organ aller demokratischen parteien – Neunkirchen, Austria 6 jul 1946-7 feb 1948 – 1r – 1 – uk British Libr Newspaper [320]

Bezoek aan een nederlandsche stad in de 16e eeuw / Telting, A – (Deventer). Den Haag, 1906 – €7.00 – ne Slangenburg [949]

Bezold, Carl *see*
- Kurzgefasster ueberblick ueber die babylonischassyrische literatur
- Oriental diplomacy
- Orientalische studien
- Die schatzhoehle "me'arath gazze"
- The tell el-amarna tablets in the british museum
- Ueber keilinschriften

Bezold, Friedrich von *see*
- Geschichte der deutschen reformation
- Zur geschichte des husitentums

Bezold, Friedrich von et al *see* Der protestantismus am ende des 19. jahrhunderts in wort und bild

Bezold, G von *see* Die kirchliche baukunst des abendlandes, historisch und systematisch dargestellt

Bezuidenhout, Cynthia Anne *see* Evaluation of the first year of the training of english-speaking primary school mathematics teachers in the transvaal

Bezuidenhout, Elsie Petronella *see* The educational psychological effect of the cochlear implant on the hearing-impaired child's family

Bezzenberger, H E *see* Maere von sente annen

Bf and sc news / Black Faculty and Staff Caucus (Baton Rouge LA) – 1996 aug 28 – 1r – 1 – mf#3912607 – us WHS [378]

Bft, aft news / Bethel Park Federation of Teachers – v4 n1 [1977 dec] – 1r – 1 – (cont: bpft united teachers; cont by: bpft, aft action) – mf#647052 – us WHS [370]

Bgr / University of Rhode Island – 1975 jul/aug-1980 may/jun – 1r – 1 – (cont: newsletter (university of rhode island. bureau of government research); cont by: bgr newsletter) – mf#615539 – us WHS [378]

Bgr [business and government review] – Columbia MO 1960-70 – 1 – ISSN: 0521-9574 – mf#5741 – us UMI ProQuest [350]

Bgr newsletter / University of Rhode Island – 1980 summer-1982 spring – 1r – 1 – (cont; bgr (kingston ri)) – mf#615600 – us WHS [378]

Bha kri rvhan : [a novel] / Buil Ba Kui – Ran kun: Gun ron Pum nhip tuik 1959 [mf ed 1990] – 1r with other items – 1 – (in burmese) – mf#mf-10289 seam reel 192/9 [§] – us CRL [830]

Bha lai 'hai se dan : [a novel] / Sin, Da gun – Ran kun: U Bha Sin 1954 [mf ed 1990] – (= ser Kambha a lan ca pe can 1) – 1r with other items – 1 – (in burmese) – mf#mf-10289 seam reel 191/3 [§] – us CRL [830]

Bha Mon, Sa khan *see* Vamsanu a re to pum samuin

Bha Ran, Hamsavati U *see* Mran ma ca a khre pra

Bha Rhan Bhuil mhu *see* Buil mhu bha rhan e a no ra tha a ran ka mran ma nuin nam

Bha Rup, U *see* A kkhara sankhya sac a thak tan mya a phui

Bha sa pran cape : papers presented at a conference on translation, rangoon, 23-27 sep 1969 / Van, U et al – Yan Kun: Cape Biman 1970 [mf ed 1990] – 1r with other items – 1 – (in burmese or english) – mf#mf-10289 seam reel 162/5 [§] – us CRL [410]

Bha Son, Sa khan *see* Tuin prann ran su

Bha va kha yi / Mon Mon Kri, U – [Yan Kun]: Capay U Cape 1976 [mf ed 1990] – 1r with other items – 1 – (in burmese) – mf#mf-10289 seam reel 155/6 [§] – us CRL [959]

Bha Van *see* Sai ta pvan ut ta khyap

Bhaca society / Hammond-Tooke, W D – Cape Town, South Africa. 1962 – 1r – us UF Libraries [960]

Bhaduri, Sadananda *see* Studies in nyaya-vaisesika metaphysics

Bhadury, Manjulika *see* The art of hindu dance

Bhaganagar struggle : a brief history of the movement led by hindu maha sabha in hyderabad in 1938-39 / Date, S R – Pune: Date, [1940?] – 1 – us CRL [954]

Bhagavad gita : "the songs of the master" / Bhagavadgita – Flushing, New York: Charles Johnston, 1908 [mf ed 1995] – (= ser Yale coll) – lxii/60p – 1 – 0-524-09711-9 – (trans, int and commentary by charles johnston) – mf#1995-0711 – us ATLA [490]

The bhagavad gita : or, the lord's lay, with commentary and notes, as well as references to the christian scriptures – London: Treubner, [1887] [mf ed 1995] – (= ser Yale coll) – ix/283p – 1 – 0-524-10197-3 – (trans fr sanskrit by mohini m chatterji) – mf#1995-1197 – us ATLA [280]

The bhagavad gita – Cambridge, Mass: Harvard University Press, 1946-1952 – (= ser Samp: indian books) – (trans and interpreted by franklin edgerton) – us CRL [490]

Bhagavad-gita : or, the lord's song – London: J M Dent 1905 [mf ed 1993] – (= ser The temple classics) – 1mf [ill] – 9 – 0-524-07068-7 – (trans by lionel d barnett) – mf#1991-0050 – us ATLA [280]

Bhagavad-gita : the song of god – Madras, India: Sri Ramakrishna Math, 1945 – (= ser Samp: indian books) – (trans by swami prabhavananda and christopher isherwood) – us CRL [280]

Bhagavad-gita : with samskrt text, free translation into english, an introduction to samskrt grammar, and a complete word-index / ed by Besant, Annie & Das, Bhagavan – Madras, India: Theosophical Pub House, 1950 – (= ser Samp: indian books) – us CRL [280]

Bhagavadgita : the song of the lord – London: J Murray, 1931 – (= ser Samp: indian books) – (transl with introduction and notes by edward j thomas) – us CRL [280]

Bhagavadgita *see* Bhagavad gita

The bhagavad-gita : or, a discourse between krishna and arjuna on divine matters. a sanskrit philosophical poem – Hertford: Stephen Austin, 1855 [mf ed 1995] – (= ser Yale coll) – cxix/155p – 1 – 0-524-09668-6 – (trans with copious notes, int on sanskrit philosophy, and other matter by j cockburn thomson) – mf#1995-0668 – us ATLA [280]

The bhagavadgita : with an introductory essay, sanskrit text, english translation, and notes – London: George Allen and Unwin, 1948 – (= ser Samp: indian books) – us CRL [490]

The bhagavad-gita and modern scholarship / Roy, Satis Chandra – London: Luzac & Co, 1941 – (= ser Samp: indian books) – us CRL [280]

Bhagavad-gita, des erhabenen sang – Jena: E Diederichs 1912 [mf ed 1993] – (= ser Religioese stimmen der voelker. die religion des alten indien 2) – 1mf – 9 – 0-524-07493-3 – mf#1991-0114 – us ATLA [280]

The bhagavad-gita interpreted in the light of christian tradition / Sampson, Holden Edward – London: W Rider 1918 [mf ed 1993] – 1mf – 9 – 0-524-07501-8 – mf#1991-0122 – us ATLA [230]

The bhagavadgita (stbe8) : with the sanatsugatiya and the anugita – 1882 – (= ser Sacred book of the east (sbte)) – 8mf – 8 – €17.00 – (trans by kashinath trimbak telang) – ne Slangenburg [280]

The bhagavadgita with the sanatsugatiya and the anugita – Oxford: Clarendon Press, 1908 – (= ser Samp: indian books) – (trans by kashinath trimbak telang) – us CRL [490]

Bhagwat, Durga *see* Romance in sacred lore

Bhakti djaya : tantangan nasional – Djakarta, 1969-1970 – 15mf – 9 – mf#SE-1957 – ne IDC [950]

Bhakti Hridaya Bon, swami *see* The geeta

Bhakti Pradipa Tirtha, Tridandibhiksu *see* Sri caitanya mahaprabhu

The bhakti sutras of narada : with explanatory notes and an introduction by the translator – Allahabad: Panini Office, 1917 – (= ser Samp: indian books) – (trans by nandalal sinha) – us CRL [280]

The bhakti-ratnavali : with the commentary of visnu puri – Bhakti: Panini Office, 1918 – (= ser Samp: indian books) – (trans by a b allahabad) – us CRL [490]

Bhakti-yoga : a course of lectures by swami vivekananda / Vivekananda, Swami – Vidya Vinodini Press, 1904 – (= ser Samp: indian books) – (trans by k m trichur) – us CRL [280]

Bhakti-yoga / Vivekananda, Swami – Almora: Advaita Ashrama, 1938 – (= ser Samp: indian books) – us CRL [280]

Bhama khvan pranna / Bhui Kri, U – Mantale: Khic nnui pum nhip tuik 1969 [mf ed 1990] – [pl/ill] 1r with other items – 1 – mf#mf-10289 seam reel 143/5 [§] – us CRL [790]

Bhaminivilasa of panditaraja jagannatha : critically edited with his own commentary called "casaka" in sanskrit and translation and notes in english / Sharma, Har Dutt – Poona: Oriental Book Agency, 1935 – (= ser Samp: indian books) – us CRL [920]

Bhamo expedition : report on the practicability of re-opening the trade route, between burma and western china / Bowers, Alexander – Rangoon: American Mission Press, 1869 – (= ser 19th c books on china) – 3mf – 9 – (with app) – mf#7.1.54 – uk Chadwyck [915]

Bhandare, M S *see* The bharatamanjari of kshemendra

Bhandarkar, D R *see* India

Bhandarkar, Devadatta Ramakrishna *see*
- Ancient indian numismatics
- Asoka
- Some aspects of ancient hindu polity
- Some aspects of ancient indian culture

Bhandarkar, Ramkrishna Gopal *see*
- Early history of the dekkan
- A peep into the early history of india
- Vaisnavism, saivism and minor religious systems
- Vaisnavism, saivism, and minor religious systems
- Wilson philological lectures on sanskrit and the derived languages delivered in 1877

Bhaneman, Carl P *see* A survey of division 2 athletic and physical education fiscal trends

Bhanja, K C *see*
- Lure of the himalaya
- Mystic tibet and the himalaya

Bharat – Allahabad, India. 3 Sept 1946-Aug 1966 – 58r – 1 – us L of C Photodup [079]

The bharat jyoti – Bombay, India. 1962-64 – 3r – 1 – us L of C Photodup [079]

Bharata ka rajapatra / India – 1967- – 1 – 600.00y – us L of C Photodup [954]

Bharata Muni et al *see* Tandava laksanam

Bharata shakti : collection of addresses on indian culture / Woodroffe, John George – Madras: Ganesh & Co, 1921 – (= ser Samp: indian books) – us CRL [954]

Bharata-kaumudi : studies in indology in honour of dr radha kumud mookerji – Allahabad: Indian Press, 1945- – (= ser Samp: indian books) – us CRL [954]

The bharatamanjari of kshemendra : aranya parva / ed by Bhandare, M S – Bombay: Standard Pub Co 1919 [mf ed 1995] – (= ser Samp: indian books) – 2v on 1r – 1 – (filmed with other items; int, trans, app & various readings by ed) – mf#mf-11323 reel 027 – us CRL [810]

Bharatan Kumarappa *see* The indian literatures of today

Bharati, Shuddhananda *see* Alvar saints

Bharatiya samachar – New Delhi, India. 8 Mar 1941-15 Dec 1943; 1944-May 1945 – 2r – 1 – us L of C Photodup [079]

Bharavi *see* Kiratarjuniyam cantos 1-3

Bhartrhari *see*
- Bhartriharis sententiae et carmen quod chauri nomine circumfertur eroticum
- The century of life
- The nitisataka, sringarasataka and vairagyasataka of bhartrihari
- The vairagya-satkam

Bhartriharis sententiae et carmen quod chauri nomine circumfertur eroticum : ad codicum mstt fidem edidit latine vertit et commentariis instruxit petrus a bohlen / Bhartrhari – Berolini: Duemmler 1833 – (= ser Whsb) – 3mf – 9 – €40.00 – (incl: variae lectiones ad bohlenii editionem bhartriharis sententiae pertinentes e codicibus extr / anton schiefner & a weber) – mf#Hu 251 – gw Fischer [490]

Bharucha, Sheriarji Dadabhai *see* A brief sketch of the zoroastrian religion and customs

Bhasa : a study / Pusalker, Achut Dattatraya – Lahore: Mehar Chand Lachhman Das, [1940] – (= ser Samp: indian books) – us CRL [490]

Bhasa *see*
- Pratima
- Pratimanatakam / bhasapranitam = pratima nataka of bhasa

Bhasa and the authorship of the thirteen trivandrum plays / Sastri, Hiranand – Calcutta: Govt of India, Central Publication Branch, 1926 – (= ser Samp: indian books) – us CRL [490]
Bhasa pali jan tampun : paep rian thmi / Nuk, Thaem – Bhnam Ben: Maha vilalay qaksar sastr nyn manuss sastr [1962?] [mf ed 1990] – 1r with other items – 1 – (in khmer & pali) – mf#mf-10289 seam reel 123/4 [§] – us CRL [490]
Bhasa pran Ca pe Nhi nho Pha lhay Pvai [1969: Rangoon, Burma] *see* Bha sa pran cape
Bhasa-pariccheda with siddhanta-muktavali / Bhattacarya, Visvanatha Nyayapancanana – Mayavati, Almora: Advaita Ashrama, 1940 – (= ser Samp: indian books) – (trans by swami madhavananda) – us CRL [490]
Bhasit samray / Tau Hin – Bhnam Ben: Kmemarah Pannagar 2497 [1955] [mf ed 1990] – 1r with other items – 1 – (in khmer) – mf#mf-10289 seam reel 113/6 [§] – us CRL [170]
Bhaskara menon / Appantampuran; ed by Varma, A R Rajaraja – [Trivandrum]: Kulakunnathu Gopala Menon, [1909] – (= ser Samp: indian books) – us CRL [954]
Bhatanagara, Ramaratana *see* The rise and growth of hindi journalism, 1826-1945
Bhate, Govinda Cimanaji *see* History of modern marathi literature, 1800-1938
Bhatt, Gajanan Umashankar *see* The system of education in germany since the war
Bhatta narayana's venisamharam : edited with an introduction, a literal english translation, exhaustive grammatical, critical and exegstical notes, and useful appendices / ed by Devasthali, G V – Bombay: Book-sellers Pub Cop, 1953 – (= ser Samp: indian books) – us CRL [280]
Bhattacarya, Visvanatha Nyayapancanana *see* Bhasa-pariccheda with siddhanta-muktavali
Bhattacaryya, Haridasa *see* The foundations of living faiths
Bhattacharya, Asutosh *see*
– An introduction to the study of the medieval bengali epics
– Studies in post-samkara dialectics
Bhattacharya, Batuknath *see* The "kalivarjyas"
Bhattacharya, Bhabani *see*
– India cavalcade
– So many hungers!
Bhattacharya, Jogendra Nath *see* Hindu castes and sects
Bhattacharya, Shiva Chandra Vidyarnava *see* Principles of tantra
Bhattacharya, Vidhushekhara *see*
– The agamasastra of gaudapada
– The basic conception of buddhism
Bhattacharyya, Benoytosh *see*
– The indian buddhist iconography
– An introduction to buddhist esoterism
– Sanskrit culture in a changing world
Bhattacharyya, Hari Mohan *see* The principles of philosophy
Bhattacharyya, Kokileswar, Pandit *see* An introduction to adwaita philosophy
Bhattacharyya, Krishnachandra *see* Studies in vedantism
Bhattacharyya, N C *see* Some bengal villages
Bhattacharyya, P N *see* A hoard of silver punch marked coins from purnea
Bhattacharyya, Sudhindra Nath *see* A history of mughal north-east frontier policy
Bhattagopinathadiksitaviracita samskararatnamala... / Gopinatha, Diksita, Bhatta – Punyakhyapattane: Anandasramamudrapalaye 1899 [mf ed 1982] – (= ser Ananda' sramasamskrtagranthavalih 39) – 1r – 1 – mf#241 – us Wisconsin U Libr [280]
Bhattoji Diksita *see* The siddhanta kaumudi
Bhava adhippay nhan bhava sacca / Nanda Sin Jam – Ran kun Mrui: Lhuin Sac Ca pe 1981 [mf ed 1987] – 1r with other items – 1 – us Univ Ill Libr [480]
Bhava adhippay nhan bhava sacca / Nanda Sin Jam – Ran kun: Yogi Pum nhip tuik 1975 [mf ed 1994] – on pt of 1r – 1 – mf#11052 r1835 n4 – us Cornell [480]
Bhava nhan ca pe / Tak Tui – Ran kun: Nu Yan Ca pe phran khyi re Thana, San Bhava Ca up tuik 1966 [mf ed 1990] – 1r with other items – 1 – (in burmese; with bibl) – mf#mf-10289 seam reel 196/7 [§] – us CRL [830]
Bhava ra nam nhan pai tan sam kabya mya : [poems] / Thi la Cac su – Ran kun: Mran ma Rvhe Prann Ca up Tuik: Phran khyi re, Ca pe Loka Ca up Tuik 1986 [mf ed 1990] – 1r with other items – 1 – (in burmese) – mf#mf-10289 seam reel 155/1 [§] – us CRL [810]
Bhava thui thui : [short stories] / So ta Chve – Ran kun: U San Kray, Mra Ca pe 1962 [mf ed 1990] – (= ser Mra ca can 33) – 1r with other items – 1 – (in burmese; At head of title: 1961 – khu nhac Ca pe Biman ta thon chu ra) – mf#mf-10289 seam reel 201/6 [§] – us CRL [830]
Bhava ton tan nhan a khra vatthu tui mya : [short stories] / Khyac Sak U – Ran kun: Ca pe bi man a phvai 1983 [mf ed 1990] – 1r with other items – 1 – (in burmese) – mf#mf-10289 seam reel 149/8 [§] – us CRL [830]
Bhavabhuti *see*
– Bhavabhuti's malatimadhava
– Bhavabhuti's uttaracharitam
– Uttararamacaritam
Bhavabhuti's malatimadhava : with the commentary of jagaddhara / ed by Kale, M R – Bombay: Oriental Pub Co, 1908 – (= ser Samp: indian books) – (with a literal english trans, notes, & int) – us CRL [180]
Bhavabhuti's uttaracharitam : with sanskrit commentary, english translation, critical and explanatory notes, etc and introduction / Ray, Saradaranjan – Calcutta: Kohinur Print Works, 1924 – (= ser Samp: indian books) – (rev and enl by kumudranjan ray) – us CRL [180]
Bhave, S S *see* Vikramorvasiya
Bhavnagar state census – Bhavnagar. pts1-2. 1931 – (= ser Census reports of india) – 1 – us CRL [315]
Bhe, pu, rhin : sui 'ma hut, bama 'nuin nam re kui kuin lhup khai' so tho sum u / Lha Kvan – Ran kun: Van Sui Ca up tuik 1976 [mf ed 1994] – on pt of 1r – 1 – mf#11052 r1739 n1 – us Cornell [959]
Bheme, Herman *see* Ueber das verhaeltnis heinrich von kleists zu c.m. wieland
Bhi Li Thun *see*
– Puppha tael khnum sralan
– Tamnak dyk bhlian kroy pangas!
Bhi newsletter / Black and Hispanic Images, Inc – 1988 jul/aug – 1r – 1 – mf#5132369 – us WHS [305]
Bhide, A S *see* Veer savarkar's "whirl-wind propaganda"
Bhishma parva – Calcutta: Bharata Press 1887 [mf ed 1993] – 2mf – 9 – 0-524-08009-7 – (trans chiefly by kesari mohan ganguli) – mf#1991-0231 – us ATLA [490]
Bhloen maranah : pralom lok jivit bit samay paccuppann / Hak Chai Huk – Bhnam Ben: Ron Bumb Viriyah 2058 [1966] [mf ed 1990] – 1r with other items – 1 – (in khmer) – mf#mf-10289 seam reel 097/10 [§] – us CRL [959]
Bhoja raja / Srinivasa Iyengar, P T – Madras: Methodist Pub House, 1931 – (= ser Samp: indian books) – us CRL [920]
Bhoja's srngara prakasa / Raghavan, Venkatarama – Bombay: Karnatak Pub House, 1940- – (= ser Samp: indian books) – us CRL [810]
Bhoodan yajna = Land-gifts mission / Vinoba, 1895-1982 – Ahmedabad: Navajivan Pub House, 1953 – (= ser Samp: indian books) – us CRL [954]
Bhu ran kham ma lui khyan pon – Ran kun Mrui': Mui Pvan Ca pe tuik 1975 [mf ed 1993] – on pt of 1r – 1 – mf#11052 r640 n9 – us Cornell [959]
Bhui Kri, U *see* Bhama khvan pranna
Bhui San, Cac kuin U *see*
– Cac kuin u bhui san / aphum pan khyi u cam tui
– Cit tim mon
– Kra rann
– Kra vat rann
– Loka gun ron
– Loka kre mhum
– Loka vat rann
– Lu vat rann
– Mon mon
– Na mai kyvan
– Na mai kyvan nan tvan sum vatthu to kri
– Pa lum pay
Bhui San, Cac kuin U Bhui San re *see* Ta kay yokkya
Bhui Van, U *see*
– Ci pva re samavayama
– Samavayama sa bho ta ra nhan lak tve panna
Bhum kyvan : [a novel] / Lan Yun Sac Lvan – Ran kun: Sac Lvan Ca pe phran khyi re 1958 [mf ed 1990] – 1r with other items – 1 – (in burmese) – mf#mf-10289 seam reel 195/1 [§] – us CRL [830]
Bhumi vilya : thnak di 6 – [Bhnam Ben]: Krasuan Qaparam 1984 [mf ed 1990] – 1r with other items – 1 – (in khmer) – mf#mf-10289 seam reel 122/6 [§] – us CRL [910]
Bhun Kyo, Nat mok *see* Angalip mran ma cac samuin
Bhun Mran, U *see* Budha bhasa pui karan pe ca samuin [1851-1970]
Bhun Nuin, Takkasuil *see*
– Cim ne u mann kyvan to mre
– Su nay khyan lui bhai chak r kho mrann khuin
Bhushan, V N *see*
– The hawk over heron
– The moving finger
– The peacock lute
Bhuyan, S K *see* Tungkhungia buranji
Bhuyan, Suryya Kumar *see*
– Anglo-assamese relations, 1771-1826
– Annals of the delhi badshahate
– Early british relations with assam
– Lachit barphukan and his times
Bi an duoi mai truong / Vu, Hanh – [Saigon] Van Chau [1974] [mf ed 1992] – on pt of 1r – 1 – mf#11052 r87 n7 – us Cornell [480]
Bi koh tralac mak guk brai sa / Samn Hael – Bhnam Ben: Ron Bumb Camroen Ratth 2515 [1971] [mf ed 1990] – 1r with other items – 1 – mf#mf-10289 seam reel 108/1 [§] – us CRL [959]
Bi mat phim truong / Hoang Ha – Saigon: Ve Tinh 1975 [mf ed 1992] – on pt of 1r – 1 – mf#11052 r131 n1 – us Cornell [959]
Bi sangram nau viat nam rahut tal sangram tael cheh loe ti yoen / S'un S'ilun – Bhnam Ben: Ron Bu mb Cam Roen Ratth 2514 [1970] [mf ed 1990] – 1r with other items – 1 – (in khmer; incl bibl ref) – mf#mf-10289 seam reel 118/8 [§] – us CRL [959]
Biach, Adolf *see* Biblische sprache und biblische motive in wielands oberon
Bialik, Hayyim Nahman *see* Law and legend
Bialluch, Max *see* Das lachende dorf
Bialystoker zeitung – Bialystok, Poland. Feb-Sept 1916 – 1r – 1 – us L of C Photodup [077]
Biana, J *see* Tratado de peste, sus causas y curacion...
Bianchi, A de *see* Viaggi in armenia, kurdistan e lazistan
Bianchi, Francesco *see*
– Dictionarivm latino epiroticvm
– Favorite recitative and air tu gran dio...
– Misera a che risolvo
– Non vi chiedo eterni dei
Bianchi, G *see* Alla terra dei galla
Bianchi, Lorenzo *see*
– Italien in eichendorffs dichtung
– Der junge josef goerres und friedrich hoelderlins hyperion
– Studien zur beurteilung des abraham a santa clara
Bianchi, Nerino *see* Della vita e delle opere di terenzio mamiani
Bianchi, Pietro Antonio *see* Sacri concentus octonis vocibus...
Bianchi, William J *see* Belize
Bianchini, Francesco *see* Francisci blanchini veronensis...de tribus generibus instrumentorum musicae veterum organicae dissertatio
Biancolelli, Pierre-Francois *see* Agnes de chaillot
Bianconi, Alfonso M *see* Vita del b francesco de capillas dell'ordine dei predicatori
Biang lala – Batawi: J M Arnold [1867-72] (wkly) [mf ed Jakarta: National Library of Indonesia 1984, 1989] – 3r – 1 – (iss for jan 14 1870-71 have title: biang-lala; publ by: ogilvie & co, apr 1868-mar 1869; h m van dorp apr-dec 1869; j m arnold jan 14-nov 1870; rehoboth-zending-pers dec 3 1870-72; organ of: genootschap van in- en uitwendinge zending dec 3 1870-72) – mf#mf-7741 seam, mf-11283 seam – us CRL [079]
Bi-annual digest of statistics 1966-1976 / Mauritius. Central Statistics Office – (= ser African official statistical serials, 1867-1982) – 28mf – 9 – uk Chadwyck [316]
Bi-annual publication of napa valley genealogical and biographical society – v3 n1-2 [1985 spring-fall] – 1r – 1 – mf#1219560 – us WHS [929]
Bianquis, Genevieve *see* Henri heine
Bianquis, Jean *see* L'oeuvre des missions protestantes a madagascar
Biard, Francois Auguste *see* Dois anos no brasil
Biard, Pierre *see* Relation de la nouvelle france, de ses terres, naturel du pais, et de ses habitants
Biart, Lucien *see*
– A travers l'amerique
– The aztecs
– My rambles in the new world
Bibaud, Francois Marie Uncas Maximilien *see*
– Memorial des honneurs etrangers conferes a des canadiens ou domicilies de la puissance du canada
– Le memorial des vicissitudes et des progres de la langue francaise en canada redige dans un hameau de la seigneurie dequire en 1870
Bibaud, jeune [comp] *see* Bibliotheque canadienne
Bibaud, Maximilien *see*
– Commentaires sur les lois du bas-canada
– L'honorable l a dessaules
– Napoleon 1 et Napoleon 3
Bibaz, R Bueno *see* Beknopte geschiednis van de kolonie suriname
Die bibel : oder, die schriften des alten and neuen bundes – Leipzig: FA Brockhaus, 1858-1868 – 8mf – 9 – 0-524-03870-8 – mf#1987-6483 – us ATLA [220]
Die bibel als kanon : drei vortraege / Volck, Wilhelm – Dorpat [Tartu, Estonia]: EJ Karow, 1885 – 1mf – 9 – 0-8370-5653-5 – mf#1985-3653 – us ATLA [220]
Die bibel bernhard overbergs / Kruchen, Gottfried – Muenster, 1956 – 386p – 9 – 3-89349-242-9 – gw Frankfurter [220]
Die bibel, das buch der menschheit / Kaehler, Martin – Berlin: M Warneck, 1904 – 1mf – 9 – 0-524-08082-8 – (incl bibl ref) – mf#1992-1142 – us ATLA [220]
Die bibel, das wort gottes : eine darstellung und verteidigung der bleibenden wahrheit der lutherischen lehre von der inspiration der heiligen schrift / Bensow, Oscar – Guetersloh: C Bertelsmann, 1909 – (= ser Biblische volksbuecher) – 1mf – 9 – 0-7905-0486-3 – mf#1987-0486 – us ATLA [220]
Die bibel des josephus : untersucht fuer buch 5-7 der archaeologie / Mez, Adam – Basel: In Kommission bei Jaeger & Kober, 1895 [mf ed 1989] – 1mf – 9 – 0-7905-3150-X – (in german, greek & hebrew) – mf#1987-3150 – us ATLA [221]
Bibel oder heilige geschrifft : gsangsweyss in drue lieder uffs kuertzest zusamen verfasset und gestallt / Aberlin, Joachim – Zuerich: Christoffel Froschauer 1551 – (= ser Hqab. literatur des 16. jahrh.) – 2mf – 9 – €30.00 – mf#1551 – gw Fischer [780]
Bibel und babel / Hornburg-Stralsund, Johannes – Potsdam, Germany. 1903 – 1r – us UF Libraries [939]
Die bibel und die suedarabische altertumsforschung / Landersdorfer, Simon – 1. & 2. aufl. Muenster i W: Aschendorff 1910 [mf ed 1992] – (= ser Biblische zeitfragen) – 1mf – 9 – 0-524-04101-6 – (incl bibl ref) – mf#1992-0059 – us ATLA [220]
Bibel und gemeinde – Wuppertal-Vohwinkel: R Brockhaus. v56- [1956-] [mf ed 1985-] – 12r – 1 – (cont: nach dem gesetz und zeugnis; lacks: v60-61 [1960-61], v79 [1979] n4) – ISSN: 0006-5061 – mf#S0676 – us ATLA [220]
Bibel und josephus ueber jerusalem und das heilige grab : wider robinson und neuere sionspilger als anhang zu reisen im morgenlande / Berggren, Jakob – Lund: J Berggren 1862 [mf ed 1993] – 2mf – 9 – 0-524-06510-1 – mf#1992-0894 – us ATLA [915]
Bibel und kirche – Stuttgart: Katholisches Bibelwerk. v2- [1947-] [mf ed 198-] – 1 – (lacks: 1952 n4; organ of: katholisches bibelwerk stuttgart, 1947-; and of: schweizerisches katholisches bibelwerk and oesterreichisches katholisches bibelwerk, 3. quartal 1971-) – ISSN: 0006-0623 – mf#S0674 – us ATLA [240]
Bibel und natur : vorlesungen ueber die mosaische urgeschichte und ihr verhaeltniss zu den ergebnissen der naturforschung / Reusch, Franz Heinrich – 4. bed. vermehrte und theilweise umgearb. aufl. Bonn: Eduard Weber's Verlags-Buckhandlung, 1876. Chicago: Dep of Photodup, U of Chicago Lib, 1975 (1r); Evanston: American Theol Lib Assoc, 1984 (1r) – 1 – 0-8370-0540-X – (incl bibl ref and ind) – mf#1984-B477 – us ATLA [220]
Bibel und natur in der harmonie ihrer offenbarungen / Zollmann, Theodor – Hamburg: Agentur des Rauhen Hauses 1869 [mf ed 1986] – 1mf – 9 – 0-8370-9358-9 – mf#1986-3358 – us ATLA [210]
Bibel und naturwissenschaft / Schmitt, Alois – 1. & 2. aufl. Muenster i W: Aschendorff 1910 [mf ed 1992] – (= ser Biblische zeitfragen) – 1mf – 9 – 0-524-04112-1 – (incl bibl ref) – mf#1992-0070 – us ATLA [210]
Bibel und talmud in ihrer bedeutung fuer philosophie und kultur : text, uebersetzung und erklaerung auserlesener stuecke / Fischer, Bernard – 2. ausg. Leipzig: Johann Ambrosius Barth 1881 [mf ed 1985] – 1mf – 9 – 0-8370-3137-0 – (hebrew text with parallel german trans) – mf#1985-1137 – us ATLA [221]
Bibel und wissenschaft : grundsaetze und deren anwendung auf die probleme der biblischen urgeschichte, hexaemeron, sintflut, voelkertafel, sprachverwirrung... / Schoepfer, Aemilian – Brixen: Buchh. des Katholisch-Politischen Pressvereins 1896 [mf ed 1985] – 1mf – 9 – 0-8370-5133-9 – (incl bibl ref) – mf#1985-3133 – us ATLA [210]
Bibelatlas : zehn karten zu bunsens bibelwerk / Lange, Henry – Leipzig: F A Brockhaus 1860 [mf ed 1992] – (= ser Vollstaendiges bibelwerk fuer die gemeinde 10) – 1mf [ill] – 9 – 0-524-02784-6 – mf#1987-6478 – us ATLA [220]
Bibelblaetter – Basel: Bibelgesellschaft in Basel [1853-1914] [mf ed 1996] – 62v on 22r – 1 – (organ of: bibelgesellschaft in basel; vols for 1853-56 iss as suppl to: magazin fuer die neueste geschichte der evangelischen missions- und bibelgesellschaften; vol for 1857-1914 as suppl to: evangelisches missions-magazin) – mf#1996-S523 – us ATLA [220]
Die bibelexegese der juedischen religionsphilosophen des mittelalters vor maimuni / Bacher, Wilhelm – Budapest, 1892 – 4mf – 8 – €10.00 – ne Slangenburg [221]
Die bibelexegese moses maimunis / Bacher, Wilhelm – Budapest, 1896 – 4mf – 8 – €10.00 – ne Slangenburg [221]
Bibelforskaren – Stockholm: Z Haeggstroem. v1-40 [1884-1923] [mf ed 1997] – 40v on 8r – 1 – mf#1997-S003 – us ATLA [220]

Die bibelfrage in der gegenwart : fuenf vortraege / Klostermann, D et al – Berlin: Fr. Zillesen, 1905 – 1mf – 9 – 0-8370-2332-7 – mf#1985-0332 – us ATLA [220]

Bibelgeschichte : das ewige reich gottes und das leben jesu / Bunsen, Christian Karl Josias, Freiherr von; ed by Holtzmann, Heinrich Julius – Leipzig: F A Brockhaus 1865 [mf ed 1992] – (= ser Vollstaendiges bibelwerk fuer die gemeinde 9) – 2mf – 9 – 0-524-02771-4 – mf#1987-6465 – us ATLA [225]

Die bibelkritik im religionsunterricht / Hahn, Traugott – Berlin: Edwin Runge 1910 [mf ed 1989] – (= ser Biblische zeit- und streitfragen 6/2) – 1mf – 9 – 0-7905-2592-5 – mf#1987-2592 – us ATLA [220]

Bibel-lexikon : realwoerterbuch zum handgebrauch fuer geistliche und gemeindeglieder / ed by Schenkel, Daniel – Leipzig: F A Brockhaus 1869-75 [mf ed 1990] – 5v on 8mf [ill] – 9 – 0-8370-1720-3 – mf#1987-6118 – us ATLA [052]

Bibelot : a reprint of poetry and prose for book lovers, chosen in part from scarce editions and sources not generally known – La Salle IL 1895-1925 – 1 – mf#5257 – us UMI ProQuest [800]

Der bibel'sche orient : eine zeitschrift in zwanglosen heften – Munich: Fleischmann. n1-2. 1821 [complete] – (= ser German-jewish periodicals...1768-1945, pt 3) – 1r – 1 – $165.00 – mf#B31 – us UPA [939]

Bibelske og kirkehistoriske skisser og afhandlinger / Sverdrup, Georg; ed by Helland, Andreas – Minneapolis, MN: Frikirkens Boghandels Forlag, 1909 [mf ed 1993] – (= ser Professor georg sverdrups samlede skrifter i udvalg 1; Lutheran coll) – 1mf – 9 – 0-524-06323-0 – mf#1991-2496 – us ATLA [242]

Bibelstudien : beitraege, zumeist aus den papyri und inschriften, zur geschichte der sprache, des schrifttums und der religion des hellenistischen judentums und des urchristentums / Deissmann, Gustav Adolf – Marburg: N G Elwert 1895 [mf ed 1986] – 1mf [ill] – 9 – 0-8370-9373-2 – (incl bibl ref & ind) – mf#1986-3373 – us ATLA [220]

Bibelstudien, 1. abtheilung / Hoelemann, Hermann Gustav – Leipzig: E Haynel 1859 [mf ed 1993] – 3mf – 9 – 0-524-07964-1 – (no more publ) – mf#1992-1119 – us ATLA [220]

Bibelstunden ueber das evangelium st matthaei / Ryle, John Charles – Berlin: Wilhelm Schultze 1857 [mf ed 1993] – 1mf – 9 – 0-524-06800-3 – mf#1992-0963 – us ATLA [225]

Bibelurkunden : geschichte der buecher und herstellung der urkundlichen bibeltexte / Bunsen, Christian Karl Josias, Freiherr von – Leipzig: F A Brockhaus 1860-70 [mf ed 1992] – (= ser Vollstaendiges bibelwerk fuer die gemeinde 5-8) – 4v on 7mf – 9 – 0-524-03877-5 – (pt2-4 ed by heinrich julius holtzmann) – mf#1987-6490 – us ATLA [220]

Bibelwissenschaft und religionsunterricht : sechs thesen / Kautzsch, Emil – Halle (Saale): Eugen Strien 1900 [mf ed 1986] – 1mf – 9 – 0-8370-7877-6 – (incl bibl ref) – mf#1986-1877 – us ATLA [377]

Biber, George Edward *see* Fiat justitia

Biber, Menachem Mendel *see* Mazkeret li-gedole ostrha

Biberacher tagblatt *see* Nuetzliches unterhaltungs- und wochenblatt fuer verschiedene leser

Biberacher unterhaltungsblatt – Biberach a.d. Riss DE, 1863-1921 – 5r – 1 – gw Misc Inst [074]

Biberfeld, E *see* Der sabbath

Biberstein-Kasimirski, Albert de *see* Le koran

Bibesco, Antoine *see* Mon heritier

Bibiena, F G da *see* L'architettura civile preparata su la geometria, e ridotta alle prospettive

Bibijaguas : cuentos / Agostini, Victor – Habana, Cuba. 1963 – 1r – us UF Libraries [972]

"The bible" : an examination of the article under that title in the "encyclopædia britannica" / Wood, Walter – Edinburgh, Scotland: MacNiven & Wallace 1880 – (= ser Scottish tracts for the times 1) – 1r – 1 – us UF Libraries [221]

Bible : and the bible only, the religion of protestants / Neale, J M – London, England. 1852 – 1r – us UF Libraries [220]

Bible : the best story book – London, England. 18– – 1r – us UF Libraries [220]

Bible : its form and its substance / Stanley, Arthur Penrhyn – Oxford, England. 1863 – 1r – us UF Libraries [220]

Bible : a nation's safeguard and glory / Stock, John – London, England. 18– – 1r – us UF Libraries [220]

Bible : the only safe basis of national education / Barrett, John Casebow – London, England. 1838 – 1r – us UF Libraries [220]

Bible / Smith, William Robertson – 9th ed [New York: Samuel L Hall, 1878] [mf ed 1984] – (= ser Biblical crit – us & gb 87) – 1mf – 9 – 0-8370-1583-9 – mf#1984-1087 – us ATLA [220]

Bible : the sure word of god, the hope and consolation of man / Beaufort, W L – Dublin, Ireland. 1825 – 1r – us UF Libraries [240]

Bible : the test of truth – London, England. 18– – 1r – us UF Libraries [220]

Bible : what is it? whence it came? how came it? wherefore came it? / Morris, Alfred John – London, England. 18– – 1r – us UF Libraries [220]

Bible : what it is, and what it is not / Martineau, James – Liverpool, England. 1839 – 1r – us UF Libraries [220]

Bible : why we should value it / Stock, John – London, England. 18– – 1r – us UF Libraries [220]

Bible *see* Sheng ching (ccm289)

The bible : a general introduction / Alleman, Herbert Christian – Philadelphia: Lutheran Publ Soc, c1914 – (= ser Lutheran Teacher-Training Series for the Sunday School) – 1mf – 9 – 0-524-05651-X – mf#1992-0501 – us ATLA [220]

The bible : its origin, its significance, and its abiding worth / Peake, Arthur Samuel – London, New York: Hodder and Stoughton, 1913 – 2mf – 9 – 0-7905-1728-0 – (incl ind) – mf#1987-1728 – us ATLA [220]

The bible : a missionary book / Horton, Robert Forman – Edinburgh: Oliphant Anderson & Ferrier; New York: Pilgrim Press [distributor, 1908?] – 1mf – 9 – 0-8370-6066-4 – mf#1986-0066 – us ATLA [220]

The bible : a scientific revelation / Adams, Charles Coffin – New York: James Pott, 1882 – 1mf – 9 – 0-8370-2040-9 – (spine title: the bible scientific. incl ind) – mf#1985-0040 – us ATLA [220]

The bible, a book for mankind : a discourse before the american bible society / Storrs, Richard Salter – [S.l.: s.n.], 1896 (Brooklyn, NY: Eagle Press) – 1mf – 9 – 0-524-05824-5 – mf#1992-0651 – us ATLA [220]

The bible a code of laws : a sermon delivered in park street church, boston, sep 3 1817... / Beecher, Lyman – Andover: Printed by Flagg and Gould, 1818 – 1mf – 9 – 0-7905-3241-7 – mf#1987-3241 – us ATLA [240]

The bible a miracle : or, the word of god its own witness: the supernatural inspiration of the scriptures shown from their literary, theological, moral, and political excellence / MacDill, David – Philadelphia: Wm S Rentoul, 1872, c1871 [mf ed 1991] – 2mf – 9 – 0-524-00058-1 – mf#1989-2758 – us ATLA [220]

The bible, a sufficient creed : being two discourses / Beecher, Charles – Fort Wayne: Times & Press Job Office, 1846 – 1mf – 9 – 0-7905-0550-9 – mf#1987-0550 – us ATLA [220]

The bible advocate – Montreal: Montreal Auxiliary Bible Society, 1837-1838 – 9 – ISSN: 1190-6944 – mf#P04111 – cn Canadiana [220]

Bible alive / Deliverance Evangelistic Church (Philadelphia PA) – 1979 oct – 1r – 1 – mf#4023710 – us WHS [242]

The bible among the nations : a study of the great translations / Beardslee, John Walter – Chicago: Fleming H Revell, 1899 [mf ed 1985] – 1mf – 9 – 0-8370-2221-5 – mf#1985-0221 – us ATLA [220]

The bible analyzed in twenty lectures / Kelso, John Russell – New York: Truth Seeker Office, c1884 – 2mf – 9 – 0-524-05677-3 – mf#1992-0527 – us ATLA [220]

The bible and astronomy : an exposition of the biblical cosmology and its relations to natural science = Bibel und astronomie / Kurtz, Johann Heinrich – Philadelphia: Lindsay & Blakiston 1857 [mf ed 1992] – 2mf – 9 – 0-524-03978-X – (trans fr 3rd improved german ed by t d simonton) – mf#1992-0021 – us ATLA [230]

The bible and babylon : a brief study in the history of ancient civilization / Koenig, Eduard – 9th rev enl ed. Burlington IA: German Literary Board 1903 [mf ed 1986] – 1mf – 9 – 0-8370-9396-1 – (trans fr german by charles ebert hay; incl bibl ref) – mf#1986-3396 – us ATLA [221]

Bible and christianity – London, England. 18– – 1r – us UF Libraries [220]

The bible and criticism : four lectures / Rainy, Robert – London: Hodder & Stoughton, 1878 – 1mf – 9 – 0-8370-4828-1 – mf#1985-2828 – us ATLA [220]

The bible and english prose style : selections and comments – Boston, MA: D C Heath, 1892 – 1mf – 9 – 0-7905-1508-3 – mf#1987-1508 – us ATLA [220]

Bible and filial piety *see* Sheng ching yu chung-kuo hsiao tao (ccm194)

Bible and history studies : outlines, papers and lectures / Sanderson, Eugene Claremont – Eugene OR. Church & School Pub Co 1912 [mf ed 1993] – 1mf – 9 – 0-524-05999-3 – (int by i n mccash) – mf#1992-0736 – us ATLA [240]

The bible and its books / Hamill, Howard Melancthon – Nashville, TN: Pub House of the ME Church, South, 1903 – 1mf – 9 – 0-524-06206-4 – mf#1992-0844 – us ATLA [220]

The bible and its critics : an enquiry into the objective reality of revealed truths: being the boyle lectures for 1861 / Garbett, Edward – London: Seeley and Griffiths; B Seeley, 1861 – 1mf – 9 – 0-7905-0892-3 – (incl bibl ref) – mf#1987-0892 – us ATLA [220]

The bible and its interpreter / Casey, Patrick H – 2nd ed. Philadelphia: John Jos McVey, 1900 – 2mf – 9 – 0-524-06676-0 – mf#1992-0929 – us ATLA [220]

The bible and its interpreters : the popular theory, the roman theory, the literary theory, the truth / Irons, William Josiah – London: J T Hayes, 1865 – 1mf – 9 – 0-524-07539-5 – mf#1992-1082 – us ATLA [220]

The bible and its literature : an inaugural address...jan 20 1841 / Robinson, Edward – New York: Office of the American Biblical Repository, and the American Eclectic, 1841 [mf ed 1984] – (= ser Biblical crit – us & gb 35) – 1mf – 9 – 0-8370-0711-9 – mf#1984-1035 – us ATLA [220]

The bible and its study : promptings and helps to an intelligent use of the bible / Sears, Barnas et al – Philadelphia: John D Wattles, [19–?] – 1mf – 9 – 0-524-03990-9 – mf#1992-0033 – us ATLA [220]

The bible and its theology : a review, comparison, and re-statement / Smith, George Vance – 5th rev ed. London, New York Longmans, Green 1901 [mf ed 1985] – 1mf – 9 – 0-8370-5282-3 – (incl bibl ref) – mf#1985-3282 – us ATLA [220]

The bible and its transmission : being an historical and bibliographical view of the hebrew and greek texts, and the greek, latin and other versions of the bible (both ms. and printed) prior to the reformation / Copinger, Walter Arthur – London: Henry Sotheran, 1897 – 8mf – 9 – 0-524-07478-X – mf#1992-1110 – us ATLA [012]

Bible and lord shaftesbury / Burgess, Henry – Oxford, England. 1856 – 1r – us UF Libraries [220]

The bible and men of learning : in a course of lectures / Mathews, James McFarlane – New York: Daniel Fanshaw, 1855 [mf ed 1985] – 1mf – 9 – 0-8370-4319-0 – (incl ind) – mf#1985-2319 – us ATLA [210]

The bible and modern criticism / Anderson, Robert – 5th ed. London: Hodder and Stoughton, 1905. Beltsville, Md: NCR Corp, 1977 (4mf); Evanston: American Theol Lib Assoc, 1984 (4mf) – 9 – (incl bibl ref and ind) – mf#1984-0043 – us ATLA [220]

The bible and modern investigation : three lectures delivered to clergy at norwich at the request of the bishop, with an address on the authority of holy scripture / Wace, Henry – London, New York: Society for Promoting Christian Knowledge, 1903. Beltsville, Md: NCR Corp, 1978 (2mf); Evanston: American Theol Lib Assoc, 1984 (2mf) – (= ser Biblical crit – us & gb) – 9 – 0-8370-0224-9 – mf#1984-1054 – us ATLA [220]

The bible and modern life – bible words and phrases = Bible and modern life / Auerbach, Joseph Smith – New York: Harper, 1914 – 1mf – 9 – 0-524-05785-0 – mf#1992-0612 – us ATLA [220]

Bible and modern thought / Marshall, Thomas L – London, England. 1863 – 1r – us UF Libraries [230]

The bible and modern thought / Birks, Thomas Rawson – Cincinnati: Poe & Hitchcock, 1867 [mf ed 1985] – 1mf – 9 – 0-8370-2345-9 – (ed's pref signed by isaac william wiley) – mf#1985-0345 – us ATLA [220]

The bible and modern thought / Emerson, George Homer – Boston: Universalist Pub House, 1890 – 1mf – 9 – 0-524-04571-2 – mf#1992-0159 – us ATLA [220]

The bible and other sacred books : a contribution to the study of apologetics and comparative theology / Terry, Milton Spenser – New York: Hunt & Eaton; Cincinnati: Cranston & Stowe, 1890 – 1mf – 9 – 0-7905-8600-2 – (incl bibl ref) – mf#1989-1825 – us ATLA [220]

The bible and reason against atheism : in a series of letters to a friend / [Edwards, Martin Luther] – Chicago: ML Edwards, 1881 [mf ed 1993] – 1mf – 9 – 0-524-05802-4 – mf#1992-0629 – us ATLA [220]

Bible and science : the bible and other ancient literature in the nineteenth century / Townsend, Luther Tracy – New York: Chautauqua Press, CLSC Dept, 1889, c1884 – 1mf – 9 – $12.50 – 0-8370-5657-8 – (originally iss in 1874 under title: the bible and science, and in 1883 under title: the bible in the light of modern science) – mf#1985-3657 – us ATLA [210]

The bible and science / Brunton, Thomas Lauder – London: Macmillan, 1881 – 2mf – 9 – 0-8370-2496-X – mf#1985-0496 – us ATLA [210]

The bible and science : or, the world-problem / Lewis, Tayler – Schenectady: G Y Van Debogert 1856 [mf ed 1989] – 1mf – 9 – 0-7905-2919-X – mf#1987-2919 – us ATLA [210]

The bible and slavery : in which the abrahamic and mosaic discipline is considered in connection with the most ancient forms of slavery, and the pauline code on slavery as related to roman slavery and the discipline of the apostolic churches / Elliott, Charles – Cincinnati: L Swormstedt & A Poe for the Methodist Episcopal Church, 1857 – 1mf – 9 – 0-7905-4357-5 – mf#1988-0357 – us ATLA [230]

The bible and social reform : or, the scriptures as a means of civilization / Tyler, Ransom Hebbard – Philadelphia: James Challen, 1860 – 1mf – 9 – 0-524-06060-6 – mf#1992-0773 – us ATLA [230]

The bible and spiritual criticism / Pierson, Arthur Tappan – New York: Baker & Taylor, 1905 – (= ser Exeter Hall Lectures) – 1mf – 9 – 0-7905-1834-1 – mf#1987-1834 – us ATLA [220]

The bible and spiritual life / Pierson, Arthur Tappan – New York: Gospel Publ House, [1908?] [mf ed 1989] – 2mf – 9 – 0-7905-1783-3 – mf#1987-1783 – us ATLA [220]

The bible and the british museum / Habershon, Ada Ruth – London: Morgan & Scott, 1909 [mf ed 1989] – 1mf – 9 – 0-7905-1604-7 – (incl ind) – mf#1987-1604 – us ATLA [060]

The bible and the catholic church : a lecture. delivered in the church of the immaculate conception, new lebanon, n.y / Moriarty, James Joseph – Albany: Van Benthuysen, 1871 – 1mf – 9 – 0-7905-0144-9 – mf#1987-0144 – us ATLA [220]

Bible and the child / Martineau, James – London, England. 1845 – 1r – us UF Libraries [220]

The bible and the critic / Brodie-Brockwell, Charles Alexander – [Toronto?: s.n, 1909?] – 1mf – 9 – 0-665-87873-7 – mf#87873 – cn Canadiana [220]

The bible and the critics : a reply to modern criticism etc by prof george adam smith, glasgow / M'Ewan, John – Edinburgh: R W Hunter, 1902 [mf ed 1989] – 1mf – 9 – 0-7905-1468-0 – mf#1987-1468 – us ATLA [220]

The bible and the east / Conder, C R – Edinburgh: William Blackwood, 1896 – 1mf – 9 – 0-7905-1639-X – (incl ind) – mf#1987-1639 – us ATLA [220]

The bible and the monuments / Hatcher, Eldridge Burwell – Richmond, VA: Whittet & Shepperson, 1895 – 1mf – 9 – 0-524-05613-7 – mf#1992-0468 – us ATLA [220]

The bible and the monuments : the primitive hebrew records in the light of modern research / Boscawen, William Saint Chad – 2nd ed. London; New York: Eyre and Spottiswoode, 1895 – 1mf – 9 – 0-7905-0307-7 – (incl bibl ref) – mf#1987-0307 – us ATLA [220]

The bible and the papacy / Belaney, R – London: Kegan Paul, Trench, 1889 – 1mf – 9 – 0-8370-6885-1 – mf#1986-0885 – us ATLA [220]

The bible and the prayer book : mistranslations, mutilations and errors, with references to paganism / Dixon, Benjamin Homer – Toronto: Toronto Willard Tract Depository, 1895? – 3mf – 9 – mf#27034 – cn Canadiana [242]

Bible and the school board / Davidson, James – Edinburgh, Scotland. 18– – 1r – us UF Libraries [220]

"The bible and the square" : being a masonic mirror and guide, containing scriptual and masonic teachings... / Akerman, William – Montreal: W Akerman, 1875 [mf ed 1980] – 1mf – 9 – 0-665-03243-9 – mf#03243 – cn Canadiana [360]

The bible and the sunday school / Newton, Richard et al; ed by Crafts, Wilbur Fisk – 2nd ed. Boston: Lee & Shepard, [1878?] – 1mf – 9 – 0-7905-1646-2 – mf#1987-1646 – us ATLA [220]

The bible and wine : respectfully addressed to all believers in the bible who make, sell or drink as a beverage intoxicating liquors / Campbell, Alexander – Montreal?: J C Beckett, 1870 – 1mf – 9 – mf#10277 – cn Canadiana [230]

The bible and woman : a critical and comprehensive examination of the teaching of the scriptures concerning the position and sphere of woman / Hayden, M P – Cincinnati: Standard Pub. Co., 1902. El Segundo, Ca: Micro Publication Systems, 1981 (1mf); Evanston: American Theol Lib Assoc, 1984 (1mf) – (= ser Women & the church in america) – 9 – 0-8370-1440-9 – mf#1984-2160 – us ATLA [230]

The bible and woman suffrace / Hooker, John – Hartford: Printed by Case, Lockwood & Brainard, 1874. Beltsville, Md: NCR Corp, 1978 (1mf); Evanston: American Theol Lib Assoc, 1984 (1mf) – (= ser Women & the church in america; Tracts of connecticut woman suffrage Association) – 9 – 0-8370-0729-1 – mf#1984-2022 – us ATLA [230]

The bible argument for socialism / Wilson, Jackson Stitt – Berkeley, CA: JS Wilson, 1911 – (= ser Social Crusade Series) – 1mf – 9 – 0-524-03597-0 – (incl bibl ref) – mf#1990-1057 – us ATLA [335]

The bible as an educator / Northrup, Birdsey Grant – [S.l.]: Yokohama Seishi Bunsha, 1895 – 1mf – 9 – 0-524-05928-4 – mf#1992-0685 – us ATLA [220]

The bible as english literature / Gardiner, John Hays – New York: Charles Scribner, 1906 – 1mf – 9 – 0-8370-9865-3 – (incl ind) – mf#1986-3865 – us ATLA [220]

The bible as it is : genesis to judges. a simple method of mastering and understanding the bible / Patterson, Alexander – Chicago: Winona, 1906 – 1mf – 9 – 0-7905-1557-1 – mf#1987-1557 – us ATLA [220]

The bible as literature : an introduction / Wood, Irving Francis – New York: Abingdon, c1914 – 1mf – 9 – 0-7905-0537-1 – mf#1987-0537 – us ATLA [220]

The bible as literature / Moulton, Richard Green et al – New York:Thomas Y. Crowell, c1896 – 1mf – 9 – 0-8370-3445-0 – (incl bibl ref) – mf#1985-1445 – us ATLA [220]

The bible as literature / Wilson, John Mills – Boston: Unitarian Sunday-School Society, c1909 – (= ser Beacon series (boston, mass.)) – 1mf – 9 – 0-524-04291-8 – mf#1992-0083 – us ATLA [220]

Bible atlas : (non-sectarian) / Maccoun, Townsend – New York, NY. 1912 – 1r – us UF Libraries [220]

The bible atlas of maps and plans : to illustrate the geography and topography of the old and new testaments and the apocrypha / Clark, Samuel – 6th ed. London: SPCK; New York: E & J B Young 1900 [mf ed 1990] – 2mf – 9 – 0-8370-1904-4 – mf#1987-6291 – us ATLA [220]

Bible band topics for weekly meetings / Church of God in Christ – 1990 fall – 1r – 1 – mf#4023719 – us WHS [242]

Bible banner – Fort Worth TX: Foy E Wallace. v1-12 [1938-49] [mf ed 1996] – 12v on 2r – 1 – (imprint varies; lacks:suppl; publ suspended: nov 1944-may 1945; v12 n1 called n3; suppl publ in jul-aug 1944; cont: gospel guardian (oklahoma city ok); cont by: gospel guardian) – ISSN: 0161-9888 – mf#1996-S531 – us ATLA [220]

Bible biography : a portrayal of the characters in holy writ / Whitteker, John Edwin – Philadelphia: United Lutheran Publication House, c1901 – 1mf – 9 – 0-524-06584-5 – mf#1992-0927 – us ATLA [220]

The bible by coverdale 1805 : remarks on the titles, the year of publication, the preliminary, the water-marks, etc with facsimiles / Fry, Francis – London: Willis & Sotheran; Bristol: Lasbury, 1867 – 1mf – 9 – 0-8370-9469-0 – (incl bibl ref) – mf#1986-3469 – us ATLA [220]

Bible characters : ahithophel to nehemiah / Whyte, Alexander – New York: Fleming H Revell, 1899 – 1mf – 9 – 0-8370-9670-7 – mf#1986-3670 – us ATLA [221]

Bible characters : being selections from sermons of alexander gardiner mercer... / Mercer, Alexander Gardiner – New York: G P Putnam's Sons, 1885 [mf ed 2004] – 1r – 1 – 0-524-10488-3 – (with brief memoir of aut by manton marble) – mf#b00703 – us ATLA [220]

Bible characters / Moody, Dwight Lyman – Chicago: Fleming H Revell, c1888 – 1mf – 9 – 0-7905-3152-6 – mf#1987-3152 – us ATLA [220]

Bible characters / Reade, Charles – New York: Harper, 1889 – 1mf – 9 – 0-8370-4848-6 – mf#1985-2848 – us ATLA [220]

The bible christians : their origin and history (1815-1900) / Bourne, Frederick William – [London?]: Bible Christian Book Room, 1905 – 2mf – 9 – 0-524-06363-X – mf#1990-5233 – us ATLA [240]

Bible chronology carefully unfolded / Goodenow, Smith Bartlett – New York: Fleming H Revell 1896 [mf ed 1985] – 1mf – 9 – 0-8370-3336-5 – mf#1985-1336 – us ATLA [220]

Bible chronology from abraham to the christian era / Auchincloss, William Stuart – New York: D van Nostrand, 1905 [mf ed 1985] – 1mf – 9 – 0-8370-2126-X – mf#1985-0126 – us ATLA [221]

Bible chronology vindicated by its own internal evidence : dates tabulated / Brown, Oliver May – Cleveland: Christian Messenger Publ Co, c1901 – 1mf – 9 – 0-8370-2479-X – (includes chronological tables) – mf#1985-0479 – us ATLA [220]

Bible Churchmen's Missionary Society *see* Record of a...year

The bible collector : official organ of: international society of bible collectors – [La Mirada CA: International Society of Bible Collectors [1965-84] [mf ed 1982-85] – 2r – 1 – (iss for jan/mar 1965-oct/dec 1965 lack vol numbering but constitute "1st year"; some nos iss in combined form; incl ind) – mf#S0385 – us ATLA [220]

Bible collectors' world – Oak Creek WI 1989 – 1 – ISSN: 0883-9204 – mf#16078 – us UMI ProQuest [220]

Bible criticism and the average man / Johnston, Howard Agnew – New York: Fleming H Revell, 1902 – 1mf – 9 – 0-8370-3790-5 – (incl ind) – mf#1985-1790 – us ATLA [220]

Bible dictionary – Madras: Christian Vernacular Education Society 1862 [mf ed 1995] – (= ser Yale coll) – 1r [ill] – 1 – 0-524-10002-0 – (in tamil) – mf#1995-1002 – us ATLA [052]

Bible difficulties and how to meet them : a symposium / ed by Atkins, Frederick Anthony – New York: Fleming H Revell c1891 [mf ed 1985] – 1mf – 9 – 0-8370-2121-9 – mf#1985-0121 – us ATLA [220]

Bible doctrine for young disciples / Power, Frederick Dunglison – Chicago: Fleming H Revell, c1899 – (= ser Bethany c e reading courses) – 2mf – 9 – 0-524-07909-9 – mf#1991-3454 – us ATLA [240]

The bible doctrine of atonement : six lectures given in westminster abbey / Beeching, Henry Charles et al – New York: Dutton 1907 [mf ed 1992] – 1mf – 9 – 0-524-05275-1 – (incl bibl ref) – mf#1992-0376 – us ATLA [240]

The bible doctrine of future punishment as taught in the epistles of paul – Boston: Berean Pub Assoc, 1868 – 2mf – 9 – 0-524-04725-1 – mf#1991-2130 – us ATLA [227]

The bible doctrine of man : or, the anthropology and psychology of scripture / Laidlaw, John – new rev ed. Edinburgh: T & T Clark, 1895 [mf ed 1985] – (= ser Cunningham lectures 7th ser) – 1mf – 9 – 0-8370-4037-X – (incl bibl & ind) – mf#1985-2037 – us ATLA [220]

The bible doctrine of the future / Lowber, James William – St Louis, MO: Christian Pub Co, 1906 – 1mf – 9 – 0-524-06550-0 – mf#1991-2634 – us ATLA [220]

The bible doctrine of the middle life as opposed to swedenborgianism and spiritism / Graves, James Robinson – Memphis, Tenn: Baptist Book House, c1873 – 1mf – 9 – 0-524-03843-0 – mf#1990-4890 – us ATLA [130]

The bible doctrine of the soul : an answer to the question, is the popular conception of the soul that of holy scripture? / Ives, Charles Linnaeus – New Haven, CT: Judd & White, 1873 – 1mf – 9 – 0-8370-3738-7 – (incl ind and appendix containing biblical citations) – mf#1985-1738 – us ATLA [220]

The bible doctrine of the soul : or, man's nature and destiny as revealed / Ives, Charles Linnaeus – [rev ed]. Philadelphia: Claxton, Remsen, & Haffelfinger, 1878, c1877 – 1mf – 9 – 0-8370-3739-5 – (includes appendixes of biblical references) – mf#1985-1739 – us ATLA [220]

Bible doctrines, alphabetically arranged : being hints, helps, and illustrations of scripture truths, for the use of christian workers, and the instruction and edification of christian readers / Ritchie, Andrew – Chicago: F H Revell c1886 [mf ed 1991] – 1mf – 9 – 0-8370-1914-1 – mf#1987-6301 – us ATLA [240]

Bible echoes in ancient classics / Ramage, Craufurd Tait – Edinburgh: Adam & Charles Black, 1878 [mf ed 1989] – 1mf – 9 – 0-7905-3210-7 – (in english, greek & latin. incl bibl ref) – mf#1987-3210 – us ATLA [220]

The bible educator / ed by Plumptre, Edward Hayes – London, New York: Cassell, Petter & Galpin [18–] – 4v in 2 on 4mf [ill] – 9 – 0-8370-1200-7 – mf#1987-6030 – us ATLA [220]

La bible en france : ou, les traductions francaises des saintes ecritures / Petavel, Emmanuel – Paris: Librairie Francais et Etrangere, 1864 – 1mf – 9 – 0-8370-4714-5 – (incl bibl ref & index) – mf#1985-2714 – us ATLA [220]

Bible english : chapters on old and disused expressions in the authorized version of the scriptures and the book of common prayer / Davies, Thomas Lewis Owen – London: G Bell, 1875. Chicago: Dep of Photodup, U of Chicago Lib, 1972 (1r); Evanston: American Theol Lib Assoc, 1984 (1r) – 1 – 0-8370-0096-3 – (incl ind) – mf#1984-B305 – us ATLA [220]

La bible et les decouvertes modernes en palestine, en egypte et en assyrie / Vigouroux, Fulcran – 5e rev augm ed. Paris: Berche et Tralin, 1889 – 6mf – 9 – 0-8370-1974-5 – (incl bibl ref) – mf#1987-6361 – us ATLA [220]

Bible et terre sainte – 1(1957)-166(1974) – 221mf – 9 – €422.00 – ne Slangenburg [220]

Bible et vie chretienne – 1(1953)-83(1968) – 170mf – 9 – €351.00 – ne Slangenburg [220]

The bible for children – New York: Century, 1902 – 2mf – 9 – 0-524-08170-0 – mf#1992-1156 – us ATLA [220]

The bible for home and school *see* Commentary on the book of deuteronomy

The bible for learners = De bijbel voor jongelieden / Oort, Henricus & Hooykaas, Isaac – Boston: Roberts, 1881-96 [mf ed 1990] – 5mf – 9 – 0-8370-1676-2 – (in english) – mf#1987-6104 – us ATLA [220]

Bible for man, not man for the bible – Ramsgate, England. 1870 – 1r – us UF Libraries [220]

Bible for the world / Somervile, Alexnader Niel – London, England. 18– – 1r – us UF Libraries [220]

La bible francaise au moyen age : etude sur les plus anciennes versions de la bible ecrites en prose de langue d'oil / Berger, Samuel – Paris: Imprimerie Nationale, 1884 [mf ed 1989] – 2mf – 9 – 0-7905-1261-0 – (incl ind) – mf#1987-1261 – us ATLA [220]

The bible hand-book : an introduction to the study of sacred scripture / Angus, Joseph – 2nd rev ed. New York: Carlton & Lanahan, c1868 – 2mf – 9 – 0-524-03961-5 – mf#1992-0004 – us ATLA [220]

A bible hand-book, theologically arranged : designed to facilitate the finding of proof-texts on the leading doctrines of the bible / Holliday, Fernandez C – Cincinnati: Hitchcock & Walden, 1869 [mf ed 1984] – 4mf – 9 – 0-8370-0185-4 – mf#1984-0038 – us ATLA [220]

Bible harmony : a study of the bible as a whole, showing that from genesis to the revelation it is a perfectly harmonious history of the progressive creation of man / Adams, Arthur Prince – 2nd ed. Beverly MA: Droweht Pub Co 1890 [mf ed 1993] – 1mf – 9 – 0-524-05782-6 – (rev ed of: bible theology and endless torments not scriptural) – mf#1992-0609 – us ATLA [220]

The bible hell : the words rendered hell in the bible / Hanson, John Wesley – 4th ed. Boston: Universalist Pub House, 1888 – 1mf – 9 – 0-524-04434-1 – mf#1991-2099 – us ATLA [220]

Bible history / Spalding, B J – New York: Schwartz, Kirwin & Fauss, c1883 – 1mf – 9 – 0-8370-5334-X – (incl ind) – mf#1985-3334 – us ATLA [220]

Bible history for schools and the home – Rock Island, IL: Augustana Book Concern, 1911 – 1mf – 9 – 0-524-07118-7 – mf#1992-1034 – us ATLA [220]

A bible history of baptism / Baird, Samuel John – Philadelphia: James H Baird, 1882 [mf ed 1989] – 2mf – 9 – 0-7905-0786-2 – (incl bibl ref & ind) – mf#1987-0786 – us ATLA [220]

The bible – how to teach the bible / Hovey, Alvah & Gregory, John Milton – Philadelphia: Griffith and Rowland Press, [19–?] – 1mf – 9 – 0-524-06842-9 – mf#1992-0984 – us ATLA [220]

Bible illustrations from the new hebrides : with notices of the progress of the mission / Inglis, John – London, New York: Thomas Nelson, 1890 – (= ser Yale coll) – xi/356p – 1 – 0-524-10164-7 – mf#1995-1164 – us ATLA [220]

The bible in brazil : colporter experiences / Tucker, Hugh Clarence – New York: Young People's Missionary Movt of the US & Canada c1902 [mf ed 1986] – 1mf – 9 – 0-8370-6708-1 – (incl ind) – mf#1986-0708 – us ATLA [220]

The bible in browning : with particular reference to the ring and the book / Machen, Minnie Gresham – New York: Macmillan, 1903 – 1mf – 9 – 0-7905-1227-0 – (incl ind) – mf#1987-1227 – us ATLA [420]

The bible in english literature / Work, Edgar Whitaker – New York: Fleming H Revell, c1917 – 1mf – 9 – 0-524-04419-8 – (incl bibl ref) – mf#1992-0112 – us ATLA [420]

The bible in ethiopic – Asmara: Franciscan Press, 1919-1926 – 1r – 1 – 0-8370-1011-X – mf#1984-B517 – us ATLA [220]

The bible in ethiopic (ge'ez) : beluy kidan (old testament), haddis kidan (new testament) – Asmarae. 5v. 1934 – 37mf – 9 – mf#J-411-1 – ne IDC [220]

The bible in modern light : a course of lectures before the bible department of the woman's club, omaha / Conley, John Wesley – Philadelphia: Griffith and Rowland Press, 1904 – 1mf – 9 – 0-524-05800-8 – mf#1992-0627 – us ATLA [220]

The bible in our modern world / Sheldon, Frank Milton – Boston: Pilgrim Press, c1917 – 1mf – 9 – 0-524-05822-9 – mf#1992-0649 – us ATLA [220]

The bible in our public schools : a sermon preached before the presbytery of lyons, n.y. sep 13, 1870 in which is found a brief reply to dr. spear's argument for excluding the bible from our public schools / Rudd, George R – Lyons, NY: Office of the Republican, 1870 – 1mf – 9 – 0-8370-7579-3 – mf#1986-1579 – us ATLA [377]

The bible in public schools : address upon a resolution to petition the board of education to exclude the bible from public schools / Kilgore, Damon Y – 2nd ed. Philadelphia: The League, [ca 1875] – 1mf – 9 – 0-8370-7552-1 – mf#1986-1552 – us ATLA [377]

The bible in public schools : a sermon / Spear, Samuel Thayer – New York: Wm C Martin, 1870 – 1mf – 9 – 0-8370-7589-0 – mf#1986-1589 – us ATLA [377]

Bible in schools plans of many lands : documents gathered and compiled for council of church boards of education, 1914 – superintendents' ed. Washington, DC: Illustrated Bible Selections Commission, c1914 – 1mf – 9 – 0-7905-1509-1 – mf#1987-1509 – us ATLA [220]

The bible in shakspeare [sic] : a study of the relation of the works of william shakspeare [sic] to the bible... / Burgess, William – New York, Toronto: F H Revell, c1903 – 4mf – 9 – 0-665-66577-6 – mf#66577 – cn Canadiana [410]

The bible in spain : or, the journeys, adventures, and imprisonments of an englishman in an attempt to circulate the scriptures in the peninsula / Borrow, George Henry – New York: G P Putnam, 1899 [mf ed 1986] – 2mf – 9 – 0-8370-6164-4 – (incl bibl ref, app, glos & ind) – mf#1986-0164 – us ATLA [914]

The bible in the common schools : superior court of cincinnati in general term, february, 1870: john d. minor et als. versus the board of education of the city of cincinnati et als. / Storer, Bellamy – Cincinnati: Robert Clarke, 1870 – 1mf – 9 – 0-8370-8791-0 – mf#1986-2791 – us ATLA [377]

Bible in the furnace : a review of prof w r smith's article "bible" in the "encyclopaedia britannica" / Whitmore, Charles John – Edinburgh: Maclaren & Macniven, 1877. Chicago: Dep of Photodup, U of Chicago Lib,1978 (1r); Evanston: American Theol Lib Assoc, 1984 (1r) – (= ser Case of william robertson smith in the free church of scotland) – 1 – 0-8370-0644-9 – mf#1984-6268 – us ATLA [220]

The bible in the light of nature, of man, and of god : vol 1: to the call of abraham also in its essential relations to the religions of the world / Chisholm, Alexander – Inverness: A Chisholm, 1891 – 1mf – 9 – 0-8370-2654-7 – (no more published) – mf#1985-0654 – us ATLA [220]

The bible in the nineteenth century : eight lectures / Carpenter, Joseph Estlin – London, New York: Longmans, Green, 1903. Beltsville, Md: NCR Corp, 1978 (6mf); Evanston: American Theol Lib Assoc, 1984 (6mf) – (= ser Biblical crit – us & gb) – 9 – 0-8370-0671-6 – (incl bibl ref) – mf#1984-1009 – us ATLA [220]

The bible in the public schools : opinions of individuals and of the press, and judicial decisions – New York: J W Schermerhorn, 1870 – 1mf – 9 – 0-8370-8543-8 – mf#1986-2543 – us ATLA [377]

The bible in the public schools : proceedings and addresses at the mass meeting, pike's music hall, cincinnati, tuesday evening, september 28, 1869: with a sketch of the anti-bible movement – Cincinnati: Gazette Steam Book and Job Printing House, 1869 – 1mf – 9 – 0-8370-7554-8 – mf#1986-1554 – us ATLA [377]

Bible in the school – Glasgow, Scotland. 1870 – 1r – us UF Libraries [220]

The bible in the schools / Bennett, John R – Edgerton, WI: F W Coon, 1889 – 1mf – 9 – 0-8370-9604-9 – mf#1986-3604 – us ATLA [377]

The bible in the workshop, or, christianity the friend of labor / Mears, John William – New York: Scribner, 1857, c1856 – 1mf – 9 – 0-524-00769-1 – mf#1990-0201 – us ATLA [240]

The bible in theology : an address. delivered before the national conference of unitarian and other christian churches / Fenn, William Wallace – Boston: American Unitarian Association, 1892 – 1mf – 9 – 0-524-05803-2 – mf#1992-0630 – us ATLA [220]

Bible index and christian sentinel – Toronto: [s.n., 1881?-188- or 19–] [mf ed v2 n4 apr 1882; v2 n10 oct 1882] – 9 – mf#P06057 – cn Canadiana [240]

The bible indicator – Owen Sound, Ont: [s.n, 1868?-18– or 19–] – 9 – mf#P06060 – cn Canadiana [220]

Bible institute series, no 2 / Cobb, Edward M – Manchester, IN: Manchester College Print, [1899?] – 1mf – 9 – 0-524-03702-7 – mf#1990-4807 – us ATLA [220]

Bible interpretation : or, the bible its own interpreter. word studies / Lansing, John A – Cambridge MA: University Press c1916 [mf ed 1993] – 1mf – 9 – 0-524-05681-1 – mf#1992-0531 – us ATLA [220]

The bible, is it the word of god? / Reed, James et al – Boston: Massachusetts New-Church Union, 1899 – (= ser Bennett lectures) – 1mf – 9 – 0-524-05689-7 – mf#1992-0539 – us ATLA [220]

The bible, its origin and nature : seven lectures / Dods, Marcus – New York: Charles Scribner, 1905 – 1mf – 9 – 0-8370-9463-1 – mf#1986-3463 – us ATLA [220]

Bible league essays in bible defence and exposition / Leavitt, John McDowell – New York: Bible League Book Co, 1909, c1908 [mf ed 1985] – 1mf – 9 – 0-8370-4369-7 – mf#1985-2369 – us ATLA [220]

Bible lessons / Abbott, Edwin Abbott – 3rd ed. London, New York: Macmillan, 1871 – 1mf – 9 – 0-7905-3120-8 – mf#1987-3120 – us ATLA [220]

Bible literature : an introductory view of the bible and its books for the general reader and sixth grade text-book for schools and colleges / Haas, John Augustus William – Philadelphia: General Council Lutheran Publ House 1906 [mf ed 1993] – 1mf – 9 – 0-524-05674-9 – mf#1992-0524 – us ATLA [220]

A bible manual : intended to furnish a general view of the holy scriptures, as introductory to their study / Crosby, Howard – New York: University Pub Co, 1869 [mf ed 1993] – 2mf – 9 – 0-524-06197-1 – mf#1992-0835 – us ATLA [220]

The bible message for modern manhood / Thoms, Craig Sharp – Philadelphia: Griffith & Rowland Press, 1912 – 1mf – 9 – 0-524-05424-X – (incl bibl ref) – mf#1992-0434 – us ATLA [220]

Bible miniatures : character sketches of 150 heroes and heroines of holy writ / Wellis, Amos Russell – New York: Fleming H Revell, c 1909 [mf ed 1985] – 1mf – 9 – 0-8370-5777-9 – (incl ind) – mf#1985-3777 – us ATLA [220]

The bible mode of baptism : new light on the subject / Mahaffey, James Ervin – 4th ed. Columbia, SC: State Co, [1910?] – 1mf – 9 – 0-524-08414-9 – mf#1993-0029 – us ATLA [220]

Bible moralisee : bodleian library mss. 270b, sc. 2937. folios iv-224v – c1250 – 7r – 14 – mf#C536-542 – uk Microform Academic [220]

Bible myths : and their parallels in other religions / Doane, Thomas William – New York, NY. 1882 – 1r – us UF Libraries [230]

Bible myths and their parallels in other religions : being a comparison of the old and new testament myths and miracles with those of heathen nations of antiquity... / Doane, Thomas William – 4th ed. New York: Commonwealth Co, c1882 – 2mf – 9 – 0-524-03114-2 – (incl bibl ref) – mf#1990-3167 – us ATLA [220]

The bible narrative and heathen traditions : the traces of the facts mentioned in genesis in the traditions of all nations / Peet, Stephen Denison – [S.l: s.n, 187-?] – 1mf – 9 – 0-524-01454-X – mf#1990-2449 – us ATLA [220]

La bible ne suffit pas pour enseigner les verites necessaires au salut / Damen, Arnold – Ottawa?: Impr du Canada, 1880 – 1mf – 9 – mf#06196 – cn Canadiana [230]

Bible, new testament = Astuatsashunch nor ktakaran – Amsterdam, 1698[-1700] – 11mf – 9 – mf#AR-436 – ne IDC [243]

The bible not of man : or, the argument for the divine origin of the sacred scriptures: drawn from the scriptures themselves / Spring, Gardiner – New-York: American Tract Society, c1847 – 1mf – 9 – 0-7905-0386-7 – mf#1987-0386 – us ATLA [220]

Bible NT Mark *see*
– Mak

Bible NT Shona *see* Chitenderano chitsva

Bible. o t psalms. bisaya. 1912 : ang basahon sa mga salmos: bahin sa biblia kun balaang kasulatan – Manila, P I: Philippine Agency, American Bible Society 1912 [mf ed Bloomington IN: Indiana Uni Lib, Preservation Dept 1984] – (= ser Coll...in the bisaya language 1) – 1r – 1 – (aka: ang mga salmos) – us Indiana Preservation [221]

The bible of our lord and his apostles : the septuagint considered in its relation to the gospel in its history and as an interpreter of the old testament / Carleton, James G – Dublin: Hodges, Figgis; London: Simpkin, Marshall, 1888 – 1mf – 9 – 0-7905-3318-9 – (incl bibl ref) – mf#1987-3318 – us ATLA [220]

The bible of st. mark : st. mark's church, the altar & throne of venice / Robertson, Alexander – London: G Allen, 1898 – 2mf – 9 – 0-7905-6724-5 – mf#1988-2724 – us ATLA [720]

The bible of the reformation : its translators and their work / Heaton, William James – London: Francis Griffiths, 1910 – 1mf – 9 – 0-8370-9154-3 – (incl bibl ref) – mf#1986-3154 – us ATLA [220]

Bible on the present crisis : the republic of the united states, and its counterfeit presentment, the slave power and the southern confederacy... – New-York: S Tousey c1863 [mf ed 1998] – 1r – 1 – mf#27868 – us Harvard [220]

Bible on the rock – Edinburgh, Scotland. 1877 – 1r – us UF Libraries [220]

Bible on the rock : a letter to principal rainy, on his speech in the free church commission, and on professor w r smith's articles in the 'encyclopaedia britannica' / Wilson, Robert – 2d ed. Edinburgh: James Gemmell, 1877. Chicago: Dep of Photodup, U of Chicago Lib, 1978 (1r); Evanston: American Theol Lib Assoc, 1984 (1r) – (= ser Case of william robertson smith in the free church of scotland) – 1 – 0-8370-0639-2 – mf#1984-6276 – us ATLA [220]

The bible on women's public speaking / Eaton, Thomas Treadwell – 1895. 50p – 1 – $5.00 – us Southern Baptist [400]

The bible on women's public speaking / Eaton, Thomas Treadwell – Louisville, Ky.: Baptist Book Concern, c1895. El Segundo, Ca: Micro Publication Systems, 1981 (1mf); Evanston: American Theol Lib Assoc, 1984 (1mf) – (= ser Women & the church in america) – 9 – 0-8370-1417-4 – mf#1984-2152 – us ATLA [240]

Bible outline / Young, Emanuel Sprankel – 3rd ed. Elgin, IL: Brethren Pub House, 1900 [mf ed 1992] – (= ser Series of bible studies; Church of the brethren coll) – 1mf – 9 – 0-524-03869-4 – mf#1990-4916 – us ATLA [220]

Bible outlines : comprehensive epitomes of the leading features of the books of the old and new testaments / Scott, W – London: Alfred Holness; Glasgow: R L Allan, [1879] – 1mf – 9 – 0-8370-5196-7 – mf#1985-3196 – us ATLA [220]

The bible, prayer book, and terms in our china missions : addressed to the house of bishops / Schereschewsky, Samuel Isaac Joseph, Bishop – [Geneva NY: s.n, 1888?] [mf ed 1995] – (= ser Yale coll) – 16p – 1 – 0-524-10025-X – mf#1995-1025 – us ATLA [220]

Bible. Presbyterian Church. General Synod *see* Minutes, 1938-1955

Bible. Presbyterian Church. General Synod. Collingswood *see* Minutes, 1956-1985

Bible. Presbyterian Church. General Synod. Columbus *see* Minutes, 1956-1960

Bible problems and the new material for their solution : a plea for thoroughness of investigation addressed to churchmen and scholars / Cheyne, Thomas Kelly – New York: Putnam's, 1904 – 1mf – 9 – 0-8370-2641-5 – mf#1985-0641 – us ATLA [220]

Bible proofs of universal salvation : containing the principal passages of scripture that teach the final holiness and happiness of all mankind / Hanson, John Wesley – 10th ed. Boston: Universalist Pub House, 1903 – 1mf – 9 – 0-524-06419-9 – mf#1991-2541 – us ATLA [220]

The bible reader's encyclopaedia and concordance : based upon the bible reader's manual by c.h. wright / ed by Clow, William Maccallum – London: Collin's Clear-Type Press, [19–?] – 2mf – 9 – 0-524-07334-1 – (incl ind) – mf#1992-1065 – us ATLA [052]

Bible readers guide – 1955-61 – 1 – us Southern Baptist [242]

The bible readers' manual : or, aids to biblical study for students of the holy scriptures / ed by Wright, Charles Henry Hamilton – London: William Collins; New York: International Bible Agency [1892?] [mf ed 1986] – (= ser Aids to biblical study for students of the holy scripture) – 1mf – 9 – 0-8370-9755-X – (incl ind) – mf#1986-3755 – us ATLA [220]

Bible reading in the early church = Ueber den privaten gebrauch der heiligen schriften in der alten kirche / Harnack, Adolf von – New York: G P Putnam; London: Williams & Norgate, 1912 – 1mf – 9 – 0-7905-1411-7 – (incl bibl ref and indexes) – mf#1987-1411 – us ATLA [220]

Bible readings : precepts and outlines. second grade text book in the lutheran graded system for intermediate schools / Schmauk, Theodore Emanuel – Philadelphia: United Lutheran Publication House, [1905?] – 1mf – 9 – 0-524-04082-6 – mf#1991-2027 – us ATLA [220]

Bible readings and bible studies / Rosenberger, Isaac J – Elgin, IL: Brethren Pub House, 1909 – 1mf – 9 – 0-524-03857-0 – mf#1990-4904 – us ATLA [220]

The bible references of john ruskin / Gibbs, Mary & Gibbs, Ellen – New York: Oxford University Press, American Branch; London: George Allen, 1898 – 1mf – 9 – 0-7905-0050-7 – (incl ind) – mf#1987-0050 – us ATLA [420]

Bible revision / Porter, J Scott – London, England. 1857 – 1r – us UF Libraries [220]

The bible rule of temperance : total abstinence from all intoxicating drink / Duffield, George – New York: National Temperance Society and Publication House, 1868 – 1mf – 9 – 0-7905-1652-7 – mf#1987-1652 – us ATLA [220]

Bible Sabbath Association *see* Sabbath sentinel

The bible school to-day / Hardin, John Huffman – St Louis, MO: Christian Pub Co, c1907 – 1mf – 9 – 0-524-04261-6 – mf#1991-2045 – us ATLA [220]

Bible, science, and faith / Zahm, John Augustine – Baltimore: John Murphy 1894 [mf ed 1985] – 1mf – 9 – 0-8370-5940-2 – (incl bibl ref) – mf#1985-3940 – us ATLA [210]

Bible seller – London, England. 18– – 1r – us UF Libraries [220]

Bible side-lights from the mound of gezer : a record of excavation and discovery in palestine / Macalister, Robert Alexander Stewart – London: Hodder and Stoughton, 1906 – 1mf – 9 – 0-7905-1226-2 – (incl ind) – mf#1987-1226 – us ATLA [930]

Bible societies and the baptists / Bitting, C C – 1897 – 1 – 5.00 – us Southern Baptist [242]

Bible societies and the baptists / Bitting, Charles Carroll – Philadelphia: American Baptist Publication Society, 1897 – 1mf – 9 – 0-524-04066-4 – mf#1991-2011 – us ATLA [220]

Bible society of india and ceylon. annual reports – India, 1944-67 [mf ed 2001] – (= ser Christianity's encounter with world religions, 1850-1950) – 6r – 1 – mf#2000-s009-013 – us ATLA [220]

Bible society recorder – Toronto: Upper Canada Bible Society, [1870?-1892] – 9 – mf#P04833 – cn Canadiana [220]

Bible songs : a collection of psalms set to music for use in church and evangelistic services, prayer meetings, sabbath schools, young people's societies, and family worship – rev enl ed. Pittsburgh: United Presbyterian Board of Publication, 1907 – 3mf – 9 – 0-524-07270-1 – mf#1991-3011 – us ATLA [780]

Bible stories – Mariannhill, South Africa. 1901 – 1r – us UF Libraries [220]

Bible stories – London, England. v1-2. 1802 – 1r – (missing: p5-8) – us UF Libraries [220]

Bible stories and poems : from creation to the captivity – superintendents' ed. Washington, DC: Illustrated Bible Selections Commission, c1914 – 1mf – 9 – 0-7905-1502-4 – mf#1987-1502 – us ATLA [220]

Bible stories for the village congregation / MacDonald, Margaret J R – Madras: Christian Literature Society for India, 1919 [mf ed 1995] – (= ser Yale coll) – iv/82p – 1 – 0-524-09544-2 – mf#1995-0544 – us ATLA [220]

Bible story : first grade text-book in lutheran lesson series for intermediate sunday-schools – Philadelphia: United Lutheran Publication House, c1897 – 1mf – 9 – 0-524-07622-7 – mf#1991-3229 – us ATLA [220]

The bible story retold for young people / Bennett, W H – New York: Macmillan, 1914, c1898 – 1mf – 9 – 0-7905-3124-0 – mf#1987-3124 – us ATLA [220]

Bible student – Beamsville, CN. 1904-13 – 1r – 1 – cn Commonwealth Imaging [220]

The bible students' cyclopaedia : or, bible marking and reading, rapid system of memorizing biblical facts, treasury for the home circle in prose and verse / Snead, Littleton Upshur – Brooklyn, NY: Christian Alliance Pub, c1900 – 4mf – 9 – 0-524-08814-4 – mf#1993-3306 – us ATLA [220]

Bible studies : contributions chiefly from papyri and inscriptions to the history of the language, the literature, and the religion of hellenistic judaism and primitive christianity = Bibelstudien / Deissmann, Gustav Adolf – Edinburgh: T & T Clark 1901 [mf ed 1986] – 1mf – 9 – 0-8370-9372-4 – (incl bibl ref & ind) – mf#1986-3372 – us ATLA [220]

Bible studies : readings in the early books of the old testament, with a familiar comment / Beecher, Henry Ward; ed by Howard, John R – New York: Fords, Howard, & Hulbert, 1893 – 1mf – 9 – 0-8370-2234-7 – mf#1985-0234 – us ATLA [220]

Bible studies from the new testament : covering the international sunday school lessons for 1890 / Pentecost, George Frederick – New York: A S Barnes, c1889 – 1mf – 9 – 0-524-05931-4 – mf#1992-0688 – us ATLA [225]

Bible studies in missions / Ober, Charles Kellogg – New York: International Committee of YMCA, c1899 – 1mf – 9 – 0-524-06851-8 – mf#1992-0993 – us ATLA [220]

Bible studies in the life of paul : historical and constructive / Sell, Henry Thorne – Chicago: Fleming H Revell, c1904 – 1mf – 9 – 0-524-03991-7 – mf#1992-0034 – us ATLA [225]

Bible studies in vital questions / Sell, Henry Thorne – New York: Fleming H Revell, c1916 – 1mf – 9 – 0-524-06580-2 – mf#1992-0923 – us ATLA [220]

Bible studies on santification and holiness / MacGillivray, J D – Chicago: F H Revell, c1899 – 1mf – 9 – 0-7905-3387-1 – mf#1987-3387 – us ATLA [220]

Bible studies on the sabbath question : for the use of pastors... / Main, Arthur Elwin – Plainfield, NJ: Sabbath School Board of the 7th Day Baptist General Conference, 1909, c1910 – 1mf – 9 – 0-7905-1229-7 – mf#1987-1229 – us ATLA [220]

Bible study by doctrines : twenty-four studies of great doctrines / Sell, Henry T – Chicago: Fleming H Revell, c1897 – 1mf – 9 – 0-8370-5345-5 – mf#1985-3345 – us ATLA [220]

A bible study on prayer / Gamertsfelder, Solomon Jacob – Cleveland, O[hio]: Pub House of the Evangelical Assoc, c1907 [mf ed 1991] – (= ser Albright series) – 1mf – 9 – 0-524-00026-3 – mf#1989-2726 – us ATLA [220]

Bible study popularized / Lee, Frank Theodosius – Chicago: Winona, 1904 – 1mf – 9 – 0-7905-1965-8 – mf#1987-1965 – us ATLA [220]

Bible teachings : a summary view of christian doctrine and christian character / Stump, Joseph – Philadelphia: United Lutheran Publ House, c1902 – 1mf – 9 – 0-524-04779-0 – mf#1991-2165 – us ATLA [240]

Bible teachings in nature / Macmillan, Hugh – London, New York: Macmillan 1867 [mf ed 1986] – 1mf – 9 – 0-8370-9295-7 – mf#1986-3295 – us ATLA [210]

The bible text cyclopedia : a complete classification of scripture texts in the form of an alphabetical list of subjects / Inglis, James – 1st american from 7th English ed. Philadelphia: J B Lippincott, [187-?] – 2mf – 9 – 0-524-08412-2 – mf#1993-0027 – us ATLA [220]

The bible, the baptist and the board system / Scarboro, J A – 1904 – 1 – us Southern Baptist [242]

The bible, the best book in the world : an address / Stucker, Edwin S – Chicago: Fleming H Revell, 1902 – 1mf – 9 – 0-524-06161-0 – mf#1992-0828 – us ATLA [220]

The bible, the church, and the reason : the three great fountains of divine authority / Briggs, Charles Augustus – New York: Scribner's, 1892 – 1mf – 9 – 0-8370-2446-3 – (incl ind of subjects and biblical passages cited) – mf#1985-0446 – us ATLA [220]

The bible, the koran, and the talmud : or, biblical legends of the mussulmans = Biblische legenden der muselmaenner / Weil, Gustav – London: Longman, Brown, Green and Longmans, 1846 – 1mf – 9 – 0-524-01388-8 – (in english) – mf#1990-2400 – us ATLA [230]

The bible, the missal, and the breviary : or, ritualism self-illustrated in the liturgical books of rome. containing the text of the entire roman missal, rubrics, and prefaces / Lewis, George – Edinburgh: T & T Clark, 1853 – 3mf – 9 – 0-8370-6073-7 – (incl bibl ref) – mf#1986-0073 – us ATLA [240]

The bible, the rod, and religion, in common schools : the ark of god on a new cart: a sermon / Smith, Matthew Hale et al – Boston: Redding, 1847 – (= ser Congregational coll) – 1mf – 9 – 0-524-05573-4 – (together with: a review of the sermon by wm b fowle and: strictures on the sectarian character of the common school journal) – mf#1991-2307 – us ATLA [377]

Bible, the teachers, the children – London, England. 1896 – 1r – us UF Libraries [220]

The bible the word of god = Bibel gottes wort / Bettex, Frederic – New York: Hodder & Stoughton; George H Doran, c1904 – 1mf – 9 – 0-7905-0802-8 – mf#1987-0802 – us ATLA [220]

Bible theology and modern thought / Townsend, Luther Tracy – Boston: Lee & Shephard; New York: Charles T Dillingham, 1883, c1882 – 1mf – 9 – 0-8370-5604-7 – mf#1985-3604 – us ATLA [210]

Bible threatenings explained : or, passages of scripture sometimes quoted to prove endless punishment shown to teach consequences of limited duration / Hanson, John Wesley – Boston: Universalist Pub House, 1893 – 1mf – 9 – 0-524-06420-2 – mf#1991-2542 – us ATLA [220]

Bible today – Collegeville MN 1962+ – 1,5,9 – ISSN: 0006-0836 – mf#6584 – us UMI ProQuest [220]

Bible training / Stow, David – Glasgow, Scotland. 1837 – 1r – us UF Libraries [220]

Bible translating / Nida, E – London, 1961 – 8mf – 8 – €17.00 – ne Slangenburg [220]

Bible translator – New York NY 1950+ – 1,5,9 – ISSN: 0006-0844 – mf#2139 – us UMI ProQuest [220]

The bible true : or, the cosmogony of moses compared with the facts of science / Fly, Elijah M – Philadelphia: Claxton, Remsen & Haffelfinger, 1871 – 2mf – 9 – 0-524-05724-9 – mf#1992-0567 – us ATLA [220]

The bible true to itself : a treatise on the historical truth of the old testament / Moody Stuart, A – 2nd ed. London: J Nisbet, 1885 – 2mf – 9 – 0-7905-3272-7 – mf#1987-3272 – us ATLA [220]

BIBLICAL

Bible truths, with shak[e]spearian parallels / Selkirk, James Brown – 3rd ed. London: Hodder and Stoughton, 1872 – 1mf – 9 – 0-524-08511-0 – mf#1993-0036 – us ATLA [220]

The bible under higher criticism : a review of current evolution theories about the old testament / Dewart, Edward Hartley – Toronto: W Briggs, 1900 – 3mf – 9 – mf#05192 – cn Canadiana [220]

The bible under trial : in view of present-day assaults on holy scripture / Orr, James – New York: AC Armstrong, 1907 [mf ed 1985] – 1mf – 9 – 0-8370-4636-X – (incl bibl ref, ind & app) – mf#1985-2636 – us ATLA [220]

Bible union of china. bulletin : organ of the bible union of china – Shanghai, 1921-38 [mf ed 2001] – (= ser Christianity's encounter with world religions, 1850-1950) – 3r – 1 – mf#2001-s091-092 – us ATLA [220]

The bible verified / Archibald, Andrew Webster – Philadelphia: Presbyterian Bd of Publ & Sabbath-School Work, c1890 – 1mf – 9 – 0-8370-2111-1 – mf#1985-0111 – us ATLA [220]

The bible versus infidelity / Adams, Henry – St John, NB: E J Armstrong, 1895 – 1mf – 9 – mf#06152 – cn Canadiana [240]

The bible versus the secretary / Sprague, Franklin Monroe – Boston: Stratford, 1923 – 1mf – 9 – 0-524-08144-1 – mf#1993-9050 – us ATLA [220]

The bible view of the jewish church : in thirteen lectures delivered during january-april, 1888, in the fourth avenue presbyterian church, n.y. / Crosby, Howard – New York: Funk & Wagnalls, 1888 – 1mf – 9 – 0-8370-9929-3 – mf#1986-3929 – us ATLA [270]

The bible vindicated against the aspersions of joseph barker / Berg, Joseph Frederick – Philadelphia: W S Young, 1854 – 1mf – 9 – 0-524-05791-5 – mf#1992-0618 – us ATLA [220]

Bible vs tradition : in which the true teaching of the bible is manifested, the corruptions of theologians detected, and the traditions of men exposed / Ellis, Aaron – 3rd ed. New-York: Pub at the Office of Bible examiner, c1853 – 1mf – 9 – 0-524-06737-6 – mf#1992-0940 – us ATLA [220]

The bible way : an antidote to campbellism / Black, J F – Cincinnati: Jennings and Graham, c1906 – 1mf – 9 – 0-524-06981-6 – mf#1991-2834 – us ATLA [220]

Bible Way Churches of Our Lord Jesus Christ World Wide *see* Bible way news voice

Bible way news voice / Bible Way Churches of Our Lord Jesus Christ World Wide – v30 n19-v30 n24 [1977 jul/aug-1980 oct/dec], 1993 aug/sep – 2r – 1 – mf#679640 – us WHS [220]

The bible, who wrote it? / Pendleton, CS – v1; The English Bible, how did we get it?, v1; The English Bible; how did we get it?, v2. 308p – 1 – $10.78 – us Southern Baptist [242]

Bible word-book : a glossary of scripture terms which have changed their popular meaning, or are no longer in general use / Swinton, William; ed by Conant, Thomas Jefferson – New York: Harper, 1876 – 1mf – 9 – 0-7905-0117-1 – mf#1987-0117 – us ATLA [220]

Bible work in bible lands : or, events in the history of the syria mission / Bird, Isaac – Philadelphia: Presbyterian Board of Publ, c1872 – 1mf – 9 – 0-8370-6019-2 – mf#1986-0019 – us ATLA [240]

Bible Zulu *see* Ibaible eli ingcwele

Bible zulu 1946 : ibaible eli ingcwele – London, England. 1946 – 1r – us UF Libraries [960]

The bible's authority supported by the bible's history : in four chapters / Gauss, J H – St Louis: Buxton & Skinner 1896 [mf ed 1985] – 1mf – 9 – 0-8370-3234-2 – mf#1985-1234 – us ATLA [220]

Les bibles et les initiateurs religieux de l'humanite / Leblois, Louis – Paris: Fischbacher, 1883-1888 – 7mf – 9 – 0-524-08052-6 – (incl bibl ref) – mf#1991-0268 – us ATLA [220]

The bibles in the caxton exhibition 1877 : or, a bibliographical description... / Stevens, Henry – rev corr ed. London: Henry Stevens IV; New-York: Scribner Welford & Armstrong, 1878 [mf ed 1988] – 1mf – 9 – 0-7905-0392-1 – mf#1987-0392 – us ATLA [220]

Bibles of england / Edgar, Andrew – London, England. 1889 – 1r – us UF Libraries [220]

The bibles of england : a plain account for plain people of the principal versions of the bible in english / Edgar, Andrew – London: Alexander Gardner, 1889 – 1mf – 9 – 0-8370-3027-7 – (includes appendixes on scottish versions, theocracy and the word mass) – mf#1985-1027 – us ATLA [220]

The bibles of other nations : being selections from the scriptures of the chinese, hindoos, persians, buddhists, egyptians, and mohammedans. to which is added, the teaching of the twelve apostles and selections from the talmud and apocryphal gospels – Manchester: Brook and Chrystal, 1885 – 1mf – 9 – 0-524-07608-1 – mf#1991-0134 – us ATLA [200]

A bible-school vision / Welshimer, Pearl Howard – Cincinnati, O[hio]: Standard Pub Co, c1909 [mf ed 1992] – 1mf – 9 – 0-524-04283-7 – mf#1991-2067 – us ATLA [240]

The bible-work : or bible reader's commentary. the new testament in two volumes – 2nd ed. New York: Funk & Wagnalls, 1889, c1883 – 15mf – 9 – 0-524-08072-0 – mf#1992-1132 – us ATLA [220]

Biblia – Killen TX 1889-1905 – 1 – mf#3289 – us UMI ProQuest [020]

Biblia anagrammatica : or, the anagrammatic bible: a literary curiosity / Begley, Walter – [S.l.]: Privately printed for the author, 1904 (London: Hazell, Watson, and Viney) – 1mf – 9 – 0-7905-0907-5 – (text in latin and english) – mf#1987-0907 – us ATLA [220]

Biblia armenica / ed by Zohrabean, J – Venezia, 1805 – €121.00 – ne Slangenburg [221]

Biblia. a.t. (siecle 13) – Calahorra – 1r – 5,6 – sp Cultura [220]

Biblia cabalistica : or, the cabalistic bible. showing how the various numerical cabalas have been curiously applied to the holy scriptures / Begley, Walter – London: David Nutt, 1903 – 1mf – 9 – 0-7905-1862-7 – mf#1987-1862 – us ATLA [220]

La biblia de guadalupe : un interesante codice desconocido (1) / Zamora, Hermenegildo – Madrid: CSIC, 1967 – 1 – sp Bibl Santa Ana [020]

Biblia filipina : primera piedra para un genesis cientifico expuesto segun las rectificaciones de jesus / Aglipay y Labayan, Gregorio – Barcelona: Antonio Virgili, 1908 – 1mf – 9 – 0-8370-9361-9 – (no more published?) – mf#1986-3361 – us ATLA [210]

Biblia hebraica / ed by Kittel, Rudolf – Lipsiae: J C Hinrichs, 1905-1906 – 13mf – 9 – 0-8370-1851-X – mf#1987-6238 – us ATLA [221]

Biblia hoje – Rio de Janeiro: Publicacao de tempo e presenca editora LTDA, n4 (1971); n32,36 (feb, nov 1975); n37 (feb 1976); n40-46 (jul 1976-may 1978); n48-52 (sep 1978-mar 1979) – 1r – us CRL [079]

Biblia ia ana / Alves, P A – Chipanga, Zimbabwe. 1939 – 1r – us UF Libraries [960]

Biblia latina cum glossa ordinaria – Strassburg: Adolph Rusch, c1480 – 137mf – 8 – €263.00 – ne Slangenburg [220]

Biblia latina de gutenberg : (i maguntiae, j gutenberg anno 1455) – Burgos – 1r – 5,6 – sp Cultura [220]

Biblia pauperum : nach dem original in der lyceumsbibliothek zu constanz / ed by Laib & Schwarz, Franz Joseph – 2. unveraend aufl, neue ausg. Freiburg i B: Herder, 1892 – 1mf – 9 – 0-524-04830-4 – mf#1990-1322 – us ATLA [220]

Biblia pauperum *see*
– Apokalypse / ars moriendi / biblia pauperum / antichrist / fabel vom kranken loewen / kalendarium und planetenbuecher / historia david
– Die zehn gebote

Biblia rabbinica : cum targum et commentaris rabbinorum – Amsterdam. v1-4. s.a. – 127mf – 8 – €242.00 – ne Slangenburg [270]

Biblia romanica de san pedro de cardena (siecle 12) – Burgos – 1r – 5,6 – sp Cultura [220]

Biblia sacra : cum universis fr vatabli et variorum interpretum annotationibus – Parisiss. v1-2. 1729-1745 – 92mf – 8 – €176.00 – ne Slangenburg [220]

Biblia sacra cum glossa, interlineari et ordinari / Nicolaus de Lyra – Lugduni. v1-6. 1545 – 8 – €493.00 – ne Slangenburg [220]

Biblia sacra cum glossis, interlineari et ordiniaria / Lyra, Nicolaus de – Lugduni. v1-6. 1545 – 8 – €493.00 – (v1 pentateuch (42mf). v2 et expositionibus: iosue-esther (36mf)) – ne Slangenburg [221]

Biblia sacra hebraice, chaldaice, graece, latine / ed by Montanus, A – Antverpiae. v1-8. 1572 – 99mf – 8 – €189.00 – ne Slangenburg [220]

Biblia sacra polyglotta / Brian Walton – Londini. v1-6. 1653-1657 – 6v on 328mf – 8 – €625.00 – ne Slangenburg [220]

Biblia sacra regia... / Arias Montano, Benito – Amberes: Christof Plantinus, 1569 – 1 – sp Bibl Santa Ana [240]

Biblia sacra vulg ed : ...cum scholiis...j marianae, et notationibus e sa – Antverpiae. v1-2. 1624 – 2v on 91mf – 8 – €174.00 – ne Slangenburg [220]

Biblia sacra vulgatae editionis sixti 5 pont max iussu recognita et clementis 8 auctoritate edita – Ratisbonae; Neo Eboraci [New York]: F Pustet, 1914 – 3mf – 9 – 0-8370-1951-6 – mf#1987-6338 – us ATLA [220]

[Biblia sancti petri rodensis] – 10th-11th c – us CRL [999]

Biblia (siecle 14) – Calahorra – 1r – 5,6 – sp Cultura [220]

Biblia (siecle 15) – Barcelona – 1r – 5,6 – sp Cultura [220]

Biblia, tai esti : wissas szwentas rasstas, seno ir naujo testamento, pagal wokisska perstattima d mertino luteraus, su kickwieno perskyrimo trumpu pranossimmu...nu kellue mokytoju lietuwj'lietuwisskay perstattytas – Karalauczuje: Kanter 1755 – (= ser Whsb) – 9mf – 9 – €135.00 – (foreword by johann jacob quandt) – mf#Hu 202 – gw Fischer [225]

Biblia, tai esti : wissas szwentas rasstas, seno ir naujo testamento, su kickwieno perskyrimo trumpu pranossimmu...nu kellue mokytoju lietuwj'lietuwisskay perstattytas – Karalauczuje: Degen 1816 – (= ser Whsb) – 21mf – 9 – €145.00 – (foreword by ludwig jedemin rhesa) – mf#Hu 204 – gw Fischer [225]

Biblia vulgata y vetus latina (siecle 9-10). despues de su restauracion – Madrid – 1r – 5,6 – sp Cultura [220]

Bibliander, T *see*
– Ad illustrissimos germaniae principes et optimates liberarum at imperialium ciuitatum oratio...de restituenda pace in germanico imperio caeterisque politijs...
– Ad nominis christiani socios consultatio qu nam ratione turcarum dira potentia reppelli possit ac debeat...populo christiano...
– Ad omniu ordinum reip
– Amplior consideratio decreti synodalis tridentini
– Christiana et catholica doctrina, fides, opera, ecclesia diui petri apostoli...per theodorvm bibliandrvm collecta
– Christianismvs sempiternvs, vervs, certvs et immvtabilis...
– Concilium sacrosanctvm domini nostri iesu christi, angelorum, apostolorum...decreta sacrosancti concilij... regis sapientissime solomonis sermo de sapientia uera...
– De fatis monarchiae romanae somnium vaticanum esdrae prophetae, quod theodorus bibliander interpretatus est...
– De legitima uindicatione christianismi ueri et sempiterni...libri antisophistici tres scripti...
– De mysterijs salutiferae passionis et mortis iesv messiae
– De ratione communi omnium linguarum literaru commentarius...explicatio doctrinae recte...vivendi
– De ratione temporvm, christianis rebus et cognoscendis et explicandis accomodata, liber unus
– De summa trinitate et fide catholica...
– Institvtionvm grammaticarvm de lingva hebraea liber unus...
– Machvmetis saracenorvm principis eivsque svccessorvm vitae ac doctrina ipseque alcoran... quae...petrus abbas cluniacensis...ex arabica lingua in latinam transferri curauit...
– Oratio theodori bibliandri ad enarrationem esaiae prophetarum principis dicta tiguri 3 idus ianuarij...natali christi domini anno 1802
– Quomodo legere oporteat sacras scripturas, praescriptiones propheticae...compendium quoque doctrinae christianae ex diui augustini libris collectum, additum est per theodorvm bibliandrum
– Sermo divinae maiestatis voce pronunciatus in mote sinai...
– Temporum...condito mundo usque ad ultimam ipsius aetatem supputatio...
– Theodori bibliandri de optimo genere grammaticorum hebraicorum commentarius

Biblica – 1(1920)-27(1946) – 214mf – 9 – €408.00 – ne Slangenburg [220]

Biblical and literary essays / Davidson, Andrew Bruce; ed by Paterson, James Alexander – 2nd ed. London: Hodder & Stoughton, 1903 [mf ed 1984] – (= ser Biblical crit – us & gb 13) – 5mf – 9 – 0-8370-0257-5 – mf#1984-1013 – us ATLA [080]

Biblical and patristic relics of the palestinian syriac literature : from mss in the bodleian library and in the library of saint catherine on mount sinai / ed by Gwilliam, George Henry et al – Oxford: Clarendon Press, 1896 [mf ed 1990] – (= ser Anecdota oxoniensia) – 1mf – 9 – 0-8370-1668-1 – (texts in syriac & english) – mf#1987-6098 – us ATLA [470]

Biblical and patristic relics of the palestinian syriac literature – Oxford. v1-pt9. 1896 – (= ser Anecdota oxoniensia. semitic series) – 4mf – 8 – €11.00 – ne Slangenburg [221]

Biblical and practical theology / Chapell, Frederic Leonard – Philadelphia: H Chapell, 1901 – 1mf – 9 – 0-7905-7705-4 – mf#1989-0930 – us ATLA [240]

Biblical and semitic studies : critical and historical essays – New York: Scribner's 1901 [mf ed 1985] – 1mf – 9 – $12.50 – 0-8370-2334-3 – mf#1985-0334 – us ATLA [220]

Biblical and theological studies – New York: Charles Scribner, 1912 – 2mf – 9 – 0-7905-1774-4 – (incl bibl ref) – mf#1987-1774 – us ATLA [220]

Biblical anthropology compared with and illustrated / Astley, Hugh John Dunkinfield – London, England. 1929 – 1r – us UF Libraries [220]

Biblical antiquities : a hand-book for use in seminaries, sabbath-schools, families and by students of the bible / Bissell, Edwin Cone – Philadelphia: American Sunday-School Union, 1888 – (= ser Green fund book) – 1mf – 9 – 0-8370-9364-3 – (incl ind) – mf#1986-3364 – us ATLA [220]

The biblical antiquities of philo = Liber antiquitatum biblicarum / Pseudo-Philo – London: SPCK, 1917 [mf ed 1992] – (= ser Translations of early documents. series 1, palestinian jewish texts (pre-rabbinic) 12) – 1mf – 9 – 0-524-04586-0 – (now 1st trans fr old latin version by montague rhodes james) – mf#1992-0174 – us ATLA [221]

Biblical apocalyptics : a study of the most notable revelations of god and of christ in the canonical scriptures / Terry, Milton Spenser – New York: Eaton & Mains; Cincinnati: Curts & Jennings, 1898 [mf ed 1986] – 2mf – 9 – 0-8370-9509-3 – (incl bibl ref & ind) – mf#1986-3509 – us ATLA [220]

Biblical archaeologist – Boston MA 1938-97 – 1,5,9 – (cont by: near eastern archaeology) – ISSN: 0006-0895 – mf#3127.01 – us UMI ProQuest [930]

Biblical archaeology review – Washington DC 1998-2000 – 1,5,9 – ISSN: 0098-9444 – mf#25207 – us UMI ProQuest [930]

Biblical atlas and scripture gazetteer : with geographical descriptions and copious bible references – [4th ed] London: Religious Tract Soc [1890?] [mf ed 1992] – 1mf [ill] – 9 – 0-524-05102-X – (rev ed of: new biblical atlas and scripture gazetteer) – mf#1992-0323 – us ATLA [220]

Biblical commentary on st paul's epistles to the galatians, ephesians, colossians, and thessalonians / Olshausen, Hermann – Edinburgh: T & T Clark, 1851. Chicago: Dep of Photodup, U of Chicago Lib, 1973 (1r); Evanston: American Theol Lib Assoc, 1984 (1r) – (= ser Clark's foreign theological library) – 1 – 0-8370-0415-2 – (incl bibl) – mf#1984-B356 – us ATLA [227]

Biblical commentary on st paul's epistles to the philippians, to titus, and the first tp timothy : in continuation of the work of olshausen / Wiesinger, August – Edinburgh: T & T Clark 1851 [mf ed 2004] – (= ser Clark's foreign theological library 23) – 1r – 1 – 0-524-10513-8 – (trans fr german by john fulton) – mf#b00726 – us ATLA [227]

Biblical commentary on st paul's first and second epistles to the corinthians / Olshausen, Hermann – Edinburgh: T & T Clark, 1855 [mf ed 2002] – (= ser Clark's foreign theological library 20) – 1r – 1 – (trans fr german with notes by john edmund cox) – mf#b00641 – us ATLA [227]

Biblical commentary on the epistle to the hebrews : in continuation of the work of olshausen / Ebrard, Johannes Heinrich August – Edinburgh: T & T Clark 1853 [mf ed 1986] – (= ser Clarke's foreign theological library 32) – 1mf – 9 – 0-8370-9617-0 – (trans fr german by john fulton) – mf#1986-3617 – us ATLA [227]

Biblical commentary on the new testament = Biblischer commentar ueber saemmtliche schriften des neuen testaments / Olshausen, Hermann – 1st american ed. New York: Sheldon, Blakeman, 1858, c1856 [mf ed 1989] – 6v on 9mf – 9 – 0-8370-1199-X – (rev after 4th german ed by a c kendrick. english trans by david fosdick, jr) – mf#1987-6029 – us ATLA [225]

Biblical commentary on the prophecies of isaiah = Biblischer commentar ueber den propheten iesaia / Delitzsch, Franz – New York: Funk & Wagnalls, [ca 1890] – (= ser The Foreign Biblical Library) – 3mf – 9 – 0-7905-1588-1 – (in english) – mf#1987-1588 – us ATLA [221]

Biblical commentary on the prophecies of isaiah = Biblischer commentar ueber den propheten iesaia / Delitzsch, Franz – Edinburgh: T & T Clark, New York: Scribner & Welford [dist] 1890 [mf ed 1989] – (= ser Clark's foreign theological library 42,44) – 3mf – 9 – 0-8370-1315-1 – (trans fr 4th ed, int by s r driver) – mf#1987-6048 – us ATLA [221]

Biblical commentary on the proverbs of solomon = Salomonische spruechbuch / Delitzsch, Franz – Edinburgh: T & T Clark 1874-75 [mf ed 1993] – (= ser Clark's foreign theological library 43,47) – 2v on 2mf – 9 – 0-524-08309-6 – (trans fr german by m g easton) – mf#1993-4 – us ATLA [221]

Biblical criticism : a brief discussion of its history, principles and methods / Haas, John Augustus William – Philadelphia: General Council Lutheran Publ House 1903 [mf ed 1993] – 1mf – 9 – 0-524-07653-7 – mf#1992-1094 – us ATLA [220]

Biblical criticism / Stubbs, William – London: SPCK; New York: E S Gorham, 1905 [mf ed 1988] – 1mf – 9 – 0-7905-0352-2 – (pref by montagu burrows) – mf#1987-0352 – us ATLA [220]

Biblical criticism and modern thought : or, the place of the old testament documents in the life of today / Jordan, William George – Edinburgh: T & T Clark, 1909 – 1mf – 9 – 0-8370-3800-6 – (based on the chancellor's lectureship of queen's university for 1906-1907. incl bibl ref and index) – mf#1985-1800 – us ATLA [220]

Biblical dogmatics : an exposition of the principal doctrines of the holy scriptures / Terry, Milton S – New York: Eaton & Mains; Cincinnati: Jennings & Graham, c1907 – 2mf – 9 – 0-7905-2435-X – (incl ind) – mf#1987-2435 – us ATLA [220]

Biblical dogmatics / Voigt, Andrew George – [s.l]: LBP c1917 [mf ed 1993] – 1mf – 9 – 0-524-08661-3 – mf#1993-2121 – us ATLA [240]

Biblical epochs / Hart, Burdett – Philadelphia: Presbyterian Board of Publ, 1896 – 1mf – 9 – 0-8370-3503-1 – mf#1985-1503 – us ATLA [220]

Biblical eschatology / Hovey, Alvah – Philadelphia: American Baptist Publ Soc, c1888 – 1mf – 9 – 0-8370-3677-1 – (includes appendix & indexes) – mf#1985-1677 – us ATLA [220]

The biblical eschatology : its relation to the current presbyterian standards and the basal principles that must underlie revision / Cheever, Henry Theodore – Worcester: F S Blanchard, 1893 – 1mf – 9 – 0-7905-0924-5 – (incl bibl ref) – mf#1987-0924 – us ATLA [220]

Biblical essays / Kenrick, John – London: Longman, Green, Longman, Roberts, & Green, 1864 – 1mf – 9 – 0-524-06046-0 – mf#1992-0759 – us ATLA [220]

Biblical essays / Lightfoot, Joseph Barber – 2nd ed. London, New York: Macmillan, 1904 – 2mf – 9 – 0-8370-9556-5 – (incl bibl ref and indexes) – mf#1986-3556 – us ATLA [220]

Biblical essays / Lightfoot, Joseph Barber – London, New York: Macmillan, 1893 – 2mf – 9 – 0-8370-9880-7 – (incl bibl ref and indexes) – mf#1986-3880 – us ATLA [220]

Biblical essays / Lightfoot, Joseph Barber – London; New York: Publ by the Trustees of the Lightfoot Fund: Macmillan, 1893 – 2mf – 9 – 0-8370-9400-3 – (incl bibl ref and indexes) – mf#1986-3400 – us ATLA [220]

Biblical essays : or, exegetical studies on the books of job and jonah, ezekiel's prophecy of gog and magog... / Wright, Charles H H – Edinburgh: T & T Clark, 1886 – 1mf – 9 – 0-8370-5919-4 – (incl ind of biblical passages cited subject index) – mf#1985-3919 – us ATLA [220]

Biblical exegesis, federal theology, and johannes cocceius : developments in the interpretatioon of hebrews 7:1-10:18 / Lee, Brian J – [mf ed 2003] – 1r – 1 – mf#d00006 – us ATLA [225]

Biblical fragments from mount sinai / ed by Harris, James Rendel – London: C J Clay, 1890 [mf ed 1986] – 1mf – 9 – 0-8370-9153-5 – (text in greek & syriac; int in english) – mf#1986-3153 – us ATLA [220]

Biblical geography and history / Kent, Charles Foster – New York: Scribner, 1911 – 1mf – 9 – 0-524-06464-4 – (incl bibl ref) – mf#1992-0892 – us ATLA [220]

The biblical geography of asia minor, phoenicia, and arabia / Rosenmueller, Ern Frid Car – Edinburgh: Thomas Clark, 1841 – (= ser The biblical cabinet) – 1mf – 9 – 0-524-05110-0 – (incl bibl ref) – mf#1992-0331 – us ATLA [220]

Biblical gleanings : or, a collection of passages of scripture that have been generally considered to be mistranslated in the received english version / Wemyss, Thomas – York: printed by Thomas Wilson; London: sold by Ogle & Baynes [1815?] [mf ed 1986] – 1mf – 9 – 0-8370-9995-1 – mf#1986-3995 – us ATLA [220]

Biblical hermeneutics : or, the art of scripture interpretation / Seiler, Georg Friedrich – London: Frederick Westley and A H Davis, 1835 – 2mf – 9 – 0-524-06744-9 – mf#1992-0947 – us ATLA [220]

Biblical hermeneutics and hebraism in the early 17th century : as reflected in the work of john weemse (1579-1636) / Shim, Jai-Sung – Grand Rapids MI: Calvin Theological Seminary, 1998 [mf ed 1999] – 1r – 1 – $130.00 – mf#1999-B003 – us ATLA [220]

Biblical history : a lecture...new york, sep 19 1889 / Briggs, Charles Augustus – New York: Scribner 1889 [mf ed 1989] – 1mf – 9 – 0-7905-1574-1 – (with app) – mf#1987-1574 – us ATLA [220]

The biblical history of the hebrews / Foakes-Jackson, Frederick John – 3rd enl ed. Cambridge: W Heffer, 1909 – 2mf – 9 – 0-524-04456-2 – (incl bibl ref) – mf#1992-0125 – us ATLA [220]

The biblical history of the hebrews to the christian era / Foakes-Jackson, Frederick John – New York: G H Doran Co, [1920] (mf ed 1995) – 1r – 1 – (incl bibl ref and ind) – mf#*ZP-1498 – us NY Public [221]

Biblical ideas of atonement : their history and significance / Burton, Ernest DeWitt et al – Chicago: University of Chicago Press 1909 [mf ed 1989] – 1mf – 9 – 0-7905-2761-8 – mf#1987-2761 – us ATLA [240]

Biblical inspiration and christ / Vincent, Marvin Richardson – New York: Anson D F Randolph c1894 [mf ed 1985] – 1mf – 9 – 0-8370-5638-1 – mf#1985-3638 – us ATLA [240]

Biblical introduction / Bennett, William Henry – New York: Thomas Whittaker, 1899 – 2mf – 9 – 0-524-00508-7 – mf#1990-0008 – us ATLA [220]

Biblical lectures : before the young men's christian association, portland, oregon / Driver, I D – Portland, OR: Himes, 1888 – 1mf – 9 – 0-524-05667-6 – mf#1992-0517 – us ATLA [220]

Biblical lectures : ten popular essays on general aspects of the sacred scriptures / Gigot, Francis Ernest – Baltimore, MD: John Murphy, c1901 – 1mf – 9 – 0-7905-8795-5 – mf#1989-2020 – us ATLA [220]

Biblical libraries : a sketch of library history from 3400 b.c. to a.d. 150 / Richardson, Ernest Cushing – Princeton: Princeton University Press; London: Oxford University Press, 1914 – 1mf – 9 – 0-7905-8049-7 – (incl bibl ref) – mf#1988-6030 – us ATLA [020]

Biblical literature and its backgrounds : being a gathering together from far and near of divers and sundry facts and opinions... / Macarthur, John Robertson – New York, London: D Appleton-Century Co [c1936] [mf ed 1986] – 1r – 1 – (filmed with: guerino il meschino / barberino, a) – mf#7240 – us Wisconsin U Libr [220]

Biblical manuscripts and books in the library of the jewish theological seminary (mostly from the sulzberger collection) : exhibited at the annual meeting of the society of biblical literature and exegesis held at the seminary, december 29-30, 1913, new york / Jewish Theological Seminary of America. Library – [New York?: s.n., 1913?] – 1mf – 9 – 0-524-05403-7 – mf#1992-0413 – us ATLA [012]

Biblical philistines : origins and identity / Fugitt, Stephen Mark – Uni of South Africa 2000 [mf ed Pretoria: UNISA 2000] – 3mf – 9 – (incl bibl ref) – mf#mfm14716 – sa Unisa [221]

Biblical psychology : in four parts / Forster, Jonathan Langstaff; ed by Forster, Henry L – London: Longmans, Green 1873 [mf ed 1985] – 1mf – 9 – 0-8370-3165-6 – (incl ind) – mf#1985-1165 – us ATLA [220]

Biblical quotations in old english prose writers / ed by Cook, Albert Stanburrough – London; New York: Macmillan, 1898 – 1mf – 9 – 0-8370-1788-2 – mf#1987-6176 – us ATLA [420]

Biblical quotations in old english prose writers : second series / ed by Cook, Albert Stanburrough – New York: Charles Scribner; London: Edward Arnold, 1903 – 1mf – 9 – 0-7905-0006-X – (incl ind) – mf#1987-0006 – us ATLA [420]

The biblical recorder – Raleigh, NC. 99,516p. 1834-1999 – 1 – mf#0444 – us Southern Baptist [242]

Biblical recorder and southern watchman – South Carolina. 3 Mar 1838-26 Dec 1840 – 1 – us Southern Baptist [242]

Biblical repertory : a collection of tracts in biblical literature – v1-4. 1825-28 [complete] – 2r – 1 – mf#ATLA S0205 – us ATLA [220]

The biblical repertory and princeton review : index – 1825-68 [complete] – 1r – 1 – mf#ATLA S0209 – us ATLA [240]

Biblical repository and classical review – La Salle IL 1831-50 – 1 – mf#4057 – us UMI ProQuest [220]

Biblical research – Notre Dame IN 1956+ – 1,5,9 – ISSN: 0067-6535 – mf#8940 – us UMI ProQuest [220]

Biblical researches and travels in russia : including a tour in the crimea, and the passage of the caucasus; with observations on the state of the rabbinical and karaite jews, and the mohammedan and pagan tribes... / Henderson, Ebenezer – London 1826 [mf ed Hildesheim 1995-98] – (= ser Fbc) – 4mf – 9 – €120.00 – 3-487-29016-2 – gw Olms [914]

Biblical researches in palestine : and in the adjacent regions: a journal of travels in the year 1838 / Robinson, Edward & Smith Eli – Boston: Crocker & Brewster, 1856 [mf ed 1984] – (= ser Biblical crit – us & gb 36) – 3v on 22mf – 9 – 0-8370-0251-6 – (incl bibl ref & ind) – mf#1984-1036 – us ATLA [915]

Biblical review – Killen TX 1846-50 – 1 – mf#5258 – us UMI ProQuest [220]

The biblical review – v1-17. 1916-32 [complete] – Inquire – 1 – mf#ATLA 1993-S511 – us ATLA [220]

The biblical scheme of nature and man : four lectures. delivered in the bowdon downs congregational church / Mackennal, Alexander – Manchester: Brook & Chrystal, 1886 – 1mf – 9 – 0-524-06575-6 – mf#1992-0918 – us ATLA [242]

Biblical scholarship and inspiration : two papers / Evans, Llewelyn Joan – Cincinnati: Robert Clarke, 1891 – 1mf – 9 – 0-8370-3080-3 – (incl bibl ref) – mf#1985-1080 – us ATLA [220]

Biblical standpoint : views of the sonship of christ, the comforter, and trinity / Wilbur, Asa – 2nd rev enl ed. Boston: A Williams 1875 [mf ed 1985] – 1mf – 9 – 0-8370-2972-4 – (incl app) – mf#1985-0972 – us ATLA [240]

Biblical study : its principles, methods, and history / Briggs, Charles Augustus – 4th ed. New York: Charles Scribner 1894, c1883 [mf ed 1986] – 2mf – 9 – 0-8370-6089-3 – (incl ind) – mf#1986-0089 – us ATLA [220]

Biblical teachings concerning the sabbath and the sunday / Lewis, Abram Herbert – 2nd rev ed. Alfred Centre NY: American Sabbath Tract Soc 1888 [mf ed 1989] – 1mf – 9 – 0-7905-2979-3 – mf#1987-2979 – us ATLA [220]

The biblical text of clement of alexandria : in the four gospels and the acts of the apostles / Barnard, Percy Mordaunt [comp] – Cambridge: University Press 1899 [mf ed 1989] – (= ser Texts and studies (cambridge, england) 5/5) – 1mf – 9 – 0-7905-1860-0 – (incl bibl ref; in greek & english) – mf#1987-1860 – us ATLA [226]

Biblical text of clement of alexandria (ts5/5) : in the four gospels and the acts of the apostles / ed by Barnard, P M – 1899 – (= ser Texts and studies (ts)) – 2mf – 9 – €5.00 – (int by f c burkitt) – ne Slangenburg [226]

Biblical theology bulletin – South Orange NJ 1971+ – 1,5,9 – ISSN: 0146-1079 – mf#11231 – us UMI ProQuest [220]

Biblical theology of the new testament = Biblische theologie des neuen testamentes / Schmid, Christian Friedrich – Edinburgh: T & T Clark 1870 [mf ed 1986] – (= ser Clark's foreign theological library 27) – 2mf – 9 – (incl ind, trans fr 4th german ed by g h venables) – mf#1986-3900 – us ATLA [225]

Biblical theology of the new testament / Weidner, Revere Franklin – 2nd rev ed. Chicago: Fleming H Revell. 2v. c1891 – 2mf – 9 – 0-7905-0450-2 – (incl ind) – mf#1987-0450 – us ATLA [225]

Biblical theology of the old testament / Weidner, Revere Franklin – 2nd rev enl ed. New York: Fleming H Revell, c1896 – 1mf – 9 – 0-8370-9838-6 – (incl ind) – mf#1986-3838 – us ATLA [221]

Biblical things not generally known : first series – London: Elliot Stock, 1879 – 1mf – 9 – 0-8370-2335-1 – (incl ind of subjects and biblical passages cited) – mf#1985-0335 – us ATLA [220]

The biblical view of marriage *see* Sheng ching chih hun yin kuan (ccm88)

The biblical view of the soul / Waller, G – London, New York: Longmans, Green, 1904 – 1mf – 9 – 0-7905-2202-0 – (incl ind) – mf#1987-2202 – us ATLA [220]

Biblical viewpoint – Greenville SC 1985+ – 1,5,9 – ISSN: 0006-0925 – mf#15162 – us UMI ProQuest [220]

Biblical world – La Salle IL 1882-1920 – 1 – ISSN: 0190-3578 – mf#5259 – us UMI ProQuest [220]

Biblicarum quaestionum decas / Patrizi, Francesco Saverio – Romae: Ex typographia polyglotta, 1877 – 1mf – 9 – 0-524-05930-6 – mf#1992-0687 – us ATLA [220]

Biblico-theological lexicon of new testament greek = Biblisch-theologisches woerterbuch der neutestamentlichen graecitaet / Cremer, Hermann – 3rd English ed. Edinburgh: T & T Clark, 1880 – 7mf – 9 – 0-524-07961-7 – (in english) – mf#1992-1116 – us ATLA [052]

Bibliia...sirech'knigi vetkhago i novago zaveta po iazyku slavensku... – Ostrog: Ivan Fedorov, 1581 – 23mf – 9 – mf#RHB-5 – ne IDC [460]

Bibliolatry : false and true / Heard, J B – Belfast, Northern Ireland. 1854 – 1r – us UF Libraries [240]

Bibliofilia sentimental / Castaneda, Vicente – Valencia: Editorial Castalia, 1949 – sp Bibl Santa Ana [020]

Bibliografi nasional indonesia kumulasi 1945-1963 – [Djakarta], 1965 – 12mf – 9 – mf#SE-629 – ne IDC [959]

Bibliografia brasileira de administracao publica e / Richardson, Ivan L – Rio de Janeiro, Brazil. 1964 – 1r – us UF Libraries [350]

Bibliografia de d. vicente barrantes. (1829-1898) / Rodriguez Monino, Antonio – Badajoz, 1946 – 1 – sp Bibl Santa Ana [920]

Bibliografia de don aristides rojas, 1826-1894 / Biblioteca Nacional (Venezuela) – Caracas, Venezuela. 1944 – 1r – us UF Libraries [972]

Bibliografia de extremadura / Sanchez Loro, Domingo – Caceres: Departamento de Seminarios, 1951 – 1 – sp Bibl Santa Ana [010]

Bibliografia de la lengua guarani. buenos aires, 1930 / Medina, Jose Toribio – Madrid: Razon y Fe, 1931 – 1 – sp Bibl Santa Ana [440]

Bibliografia de la lengua valenciana / Ribelles Comin, Jose – 3v. 1915-31 – 1,9 – us AMS Press [440]

Bibliografia de la provincia dominicana de colombia / Mesanza, Andres – Caracas, 1929; Madrid: Razon y Fe, 1931 – 1 – sp Bibl Santa Ana [972]

Bibliografia de las publicaciones / Munoz de San Pedro, Miguel – Valencia: Artes Graficas Solers, 1966 – 1 – sp Bibl Santa Ana [070]

Bibliografia de r. foulche-delbosc new york / Foulche Delbosc, Isabel & Puyol, Julio – 1933. Extrait de la revue Hispanique. Tome 91 – 1 – sp Bibl Santa Ana [010]

Bibliografia de rufino jose cuervo / Torres Quintero, Rafael – Bogota, Colombia. 1951 – 1r – us UF Libraries [972]

Bibliografia e indice da geologia da amazonia lega – Belem, Brazil. 1969 – 1r – us UF Libraries [550]

Bibliografia etiopica : catalogo descrittivo e ragionato degli scritti pubblicati dalla invenzione della stampa fino a tutto il 1891, intorno alla etiopia e regioni limitrofe / Fumagalli, Guiseppe – Milan: U Hoepli, 1893 – 1 – us CRL [960]

Bibliografia etiopica : in continuazione alla "bibliografia etiopica" di g. fumagalli / Zanutto, Silvio – Roma: a cura del Ministero delle colonie, Sindicato italiano arti grafiche, 1932-36. 2v – 1 – us CRL [960]

Bibliografia general espanola e hispanoamericana – 16v. 1925-42 – 1,9 – us AMS Press [010]

Bibliografia giuridica coloniale (italia) – Rome: Edizione de 'Il Codice tributario dell' Africa italiana", 1943 – 1 – us CRL [340]

Bibliografia hispanica – 13v. 1942-54 – 1,9 – us AMS Press [010]

Bibliografia. historias del imperio lusitano / Bayle, Constantino – Madrid: Razon y Fe, 1946 – 1 – sp Bibl Santa Ana [019]

Bibliografia jesuistica de mainas / Bayle, Constantino – Madrid: Missionalia Hispanica, 1949 – 1 – sp Bibl Santa Ana [240]

Bibliografia manchega. bibliografia de las provincias de albacete, ciudad real, cuenca y toledo / Cotta y Marquez de Prado, Fernando de – Madrid, 1961. Sep. Rev. Estudios Regionales La Mancha – sp Bibl Santa Ana [946]

Bibliografia missionaria – Vatican City, Italy: Pontifical Missionary Library...1934-98 [mf ed 2000] – (= ser Christianity's encounter with world religions, 1850-1950) – 8r – 1 – (in italian) – mf#2000-s000-001 – us ATLA [241]

Bibliografia missionaria anno 21 (1957) y anno 12 (1958)... / Barrado Manzano, Arcangel – Madrid: Archivo Ibero Americano, 1959 – 1 – sp Bibl Santa Ana [012]

Bibliografia missionaria. anno 23 / Barrado Manzano, Arcangel – Madrid: Arch. Ibero Americano, 1961 – 1 – sp Bibl Santa Ana [240]

Bibliografia missionaria. anno 28 / Barrado Manzano, Arcangel – Madrid: Arch. Ibero Americano, 1964 – 1 – sp Bibl Santa Ana [240]

Bibliografia missionaria anno 30, 1966. roma, 1967 / Rommeiskichen, Giovanni – Madrid: Graf. Calleja, 1967 – 1 – sp Bibl Santa Ana [240]

Bibliografia missionaria anno 39, 1965. roma, 1966 / Rommerskichen, Giovanni – Madrid: Graf. Calleja, 1966 – 1 – sp Bibl Santa Ana [240]

Bibliografia puertorriquena (1493-1930). madrid, 1932 / Pedreira, Antonio S – Madrid: Razon y Fe, 1935 – 1 – sp Bibl Santa Ana [920]

Bibliografia sobre las misiones de mainas. un misionero misionologo / Bayle, Constantino – Madrid: Ediciones Jura, 1949 – 1 – sp Bibl Santa Ana [240]

Bibliografia sobre los mixtecas (1719-1991) : antropologia fisica, antropologia social y aplicada, arqueologia, historia, codieces y linguistica / Dittmar, Manuela – (mf ed 1994) – 1mf – 9 – €30.00 – 3-8267-2057-1 – mf#DHS 2057 – gw Frankfurter [574]

Bibliografia...anno 28, 1964 / Rommerskichen, Giovanni – Roma, 1965; Madrid: Graf Calleja, 1966 – 1 – sp Bibl Santa Ana [946]

Bibliografias cubanas / Peraza Sarausa, Fermin – Washington, DC. 1945 – 1r – us UF Libraries [972]

Bibliograficeskij obzor apokrifov v juznoslavjanskoj i russkoj pismennosti / Jacimirskij, A J – 1921 – 6mf – 8 – mf#R-175 – ne IDC [243]

Bibliograficheskaia letopis – Spb., 1914-1917. v1-3 – 12mf – 9 – mf#R-3501 – ne IDC [077]

Bibliograficheski yezhegodnik – Moscow. v. 1-8. 1911-1914, 1921 22-1924 – 1 – us NY Public [010]

Bibliograficheskie listy russkogo bibliologicheskogo obshchestva – Pg., 1922. v1-3 – 2mf – 9 – (missing: 1922, v1) – mf#R-4298 – ne IDC [077]

Bibliograficheskie zapiski – Baltimore. 1953-1991 (1) 1971-1991 (5) 1975-1991 (9) – 37mf – 9 – mf#1689 – ne IDC [077]

Bibliograficheskii desiatigodnik po kooperatsii : kommunisticheskaia kooperativnaia literatura 1917-1927 – 1928 – 119p 2mf – 9 – mf#COR-527 – ne IDC [335]

Bibliograficheskii ezhegodnik – M., 1911-1924. v1-8 – 53mf – 9 – mf#R-3487 – ne IDC [077]

Bibliograficheskii listok – Spb., 1902-1903. v1-12 – 12mf – 9 – mf#R-4297 – ne IDC [077]

Bibliograficheskii obzor izdanii tsentralnogo statisticheskogo komiteta, vyshedshikh po 1-e avgusta 1895 goda – Spb., 1895 – 3mf – 9 – mf#R-5628 – ne IDC [077]

Bibliograficheskii obzor zemskoi statisticheskoi i otsenochnoi literatury so vremeni uchrezhdeniia zemstv 1864-1903 g / ed by Karavaev, V F – 8mf – 8 – mf#RZ-162 – ne IDC [314]

Bibliograficheskii spisok literaturnykh trudov kievskago mitropolita evgeniia bolkhovitinova / Shmurlo, E F – 1888 – 76p 3mf – 8 – mf#R-7270 – ne IDC [243]

Bibliograficheskii ukazatel / Istoriia russkoi literatury 18 veka; ed by Stepanov, V P & Stennik, I V – 1968 – 14mf – 8 – mf#R-6185 – ne IDC [947]

Bibliograficheskii ukazatel knig i statei o slavianskikh pervouchiteliakh sv kirille i mefodie – 1885 – 22p 1mf – 8 – mf#R-5878 – ne IDC [243]

Bibliograficheskii ukazatel knig i statei, otnosiashchikhsia do obshchestv, osnovannykh na nachalakh vzaimnosti, artelei, polozheniia rabochego sosloviia i melkoi kustarnoi promyshlennosti v rossii / Mezhov, V I – 1872-1889 – 5v 5mf – 9 – (missing:v2 suppl 1 p1-72) – mf#COR-535 – ne IDC [335]

Bibliograficheskii ukazatel' knig i zakliuchaiushchikhsia v nikh statei obnimaiushchii deiatel'nost' statisticheskikh komitetov s samogo nachala ikh uchrezhdeniia vplot' do 1873 g / Mezhov, V I – Spb, 1873. iv/128p – 3mf – 8 – mf#R-7152 – ne IDC [314]

Bibliograficheskii ukazatel kooperativnoi literatury za 1927 g / Brovkin, T M – 1929 – 171p 2mf – 9 – mf#COR-529 – ne IDC [335]

Bibliograficheskii ukazatel perevodnoi belletristiki v russkikh zhurnalakh za piat let 1897-1901 g / Braginskii, D – Spb., 1902 – 2mf – 9 – mf#R-7084 – ne IDC [077]

Bibliograficheskii ukazatel rabot nauchnykh sotrudnikov instituta vysokomolekuliarnykh soedinenii an sssr / Akademiia nauk SSSR, Institut vysokomolekuliarnykh soedinenii i Biblioteka Akademii nauk SSSR – Leningrad: Biblioteka 1961-1977/79 – us CRL [947]

Bibliograficheskii ukazatel' sochinenii otdel'no izdannykh i statei pomeshchennykh v povremennykh izdaniiakh : literatura dolgosrochnogo kredita – Spb, 1901 – 3mf – 8 – mf#R-7136 – ne IDC [332]

Bibliograficheskii ukazatele za 1858 i 1859 gg zhurnal ministerstva vnutrennykh del / Mezhov, V I – M., 1860-1861 – 11mf – 9 – mf#R-4358 – ne IDC [077]

Bibliograficheskii vestnik – Spb., 1902-1905; 1908-1911 – 313mf – 9 – (missing: 1904(4, 50, 52); 1909(23); 1910(1, p 1-4)) – mf#R-4318 – ne IDC [077]

Bibliograficheskoe obozrenie drevneslavianskoi i russkoi pismennosti i drugikh pamiatnikov ot 16 do nachala 20 v / Burtsev, A E – 1904. 5v – 48mf – 8 – mf#R-4669 – ne IDC [947]

Bibliografiia finansov, promyshlennosti i torgovli : so vremen petra velikogo po nastoiashchee vremia (s 1714 po 1879 god vkliuchitel'no) / Karataev, S I – Spb, 1880 – 5mf – 9 – mf#COR-532 – ne IDC [332]

Bibliografiia po istorii rimskoi literatury v rossii s 1709 po 1889 god / Naguevskii, D I – Kazan, 1889 – 2mf – 8 – mf#R-7173 – ne IDC [947]

Bibliografiia russkoi periodicheskoi pechati 1703-1900 / Lisovskii, N M – Spb., 1895-1913 – 12mf – 9 – mf#151 – ne IDC [077]

Bibliografiia russkoi periodicheskoi pechati 1703-1900 gg : materialy dlia istorii russkoi zhurnalistiki / Lisovskii, N M – Washington. 1956+ (1) 1965+ (5) 1970+ (9) – 20mf – 9 – mf#1151 – ne IDC [077]

Bibliografiia ukrainskoi presi 1816-1916 / Ignatienko, V – New York. 1947+ (1) 1971+ (5) 1977+ (9) – 5mf – 9 – mf#957 – ne IDC [077]

Bibliografija edvarda kardelja / Bulovec, Stefka – Ljubljana: Komunist 1980 [mf ed 1986] – 1r [ill] – 1 – (int also in english, albanian, hungarian, macedonian, russian, and serbo-croatian (roman); incl ind) – mf#6616 – us Wisconsin U Libr [010]

Bibliografiya periodiki – Moscow. v. 1-4. 1923 – 1 – us NY Public [010]

Bibliografhie de l'oeuvre de madame francoise gaudet-smet de la societe des ecrivains canadiens : et du cercle de femmes journalistes, ecrivain de presse, radio et tv, specialiste en arts domestiques / Marie de St-Gilles, soeur – 1963 [mf ed 1979] – (= ser Bibliographies du cours...1947-66) – 3mf – 9 – mf#SEM105P4 – cn Bibl Nat [070]

Bibliographer : a journal of book-lore – Killen TX 1881-84 – 1 – mf#2858 – us UMI ProQuest [070]

Bibliographer's manual of american history / Bradford, Thomas Lindsley – 5v. 1907-10 – 1,9 – us AMS Press [019]

The bibliographer's manual of english literature / Lowndes, William Thomas – With Appendix. 8 v. 1864-65 – 1,9 – us AMS Press [018]

Bibliographia brentiana : bibliographisches verzeichnis der gedruckten und ungedruckten schriften und briefe des reformators johannes brenz / Koehler, Walther – Berlin: C.A. Schwetschke, 1904 – 1mf – 9 – 0-7905-8060-8 – mf#1988-6041 – us ATLA [012]

Bibliographia calviniana : catalogus chronologicus operum calvini / ed by Erichson, Alfred – Berolini: C.A. Schwetschke, 1900 – 1mf – 9 – 0-7905-8028-4 – mf#1988-6009 – us ATLA [012]

Bibliographia catholica americana, pt 1 : a list of works written by catholic authors and published in the united states / Finotti, Joseph Maria – New York: Catholic Pub House, 1872 [mf ed 1990] – 1mf – 9 – 0-7905-8140-X – (no more publ) – mf#1988-6087 – us ATLA [241]

Bibliographia geographica palaestinae / Tobler, T – [Leipzig, 1869] – 3mf – 9 – mf#HT-289 – ne IDC [915]

Bibliographia geographica palaestinae / Tobler, T – Leipzig: S Hirzel, 1867 – 3mf – 9 – mf#H-2915 – ne IDC [915]

Bibliographia geographica palaestinae : zunaechst kritische uebersicht gedruckter und ungedruckter beschreibungen der reisen ins heilige land / Tobler, Titus – Leipzig: S Hirzel, 1867 – 1mf – 9 – 0-7905-3173-9 – mf#1987-3173 – us ATLA [915]

Bibliographia zoologiae et geologiae / Agassiz, Louis – v1-4. 1848-54 – (= ser Ray society publications 13/18/22/26) – 4r 52mf – 1,7 – mf#9/85948 – uk Microform Academic [550]

Bibliographia zoologiae et geologiae : a general catalogue of all books, tracts, and memoirs on zoology and geology / Agassiz, Louis; ed by Strickland, H E – corr enl ed. London, 1848-54 – (= ser 19th c evolution & creation) – 26mf – 9 – mf#1.1.10933 – uk Chadwyck [015]

Bibliographic annual in speech communication – Washington DC 1970-75 – 1,5,9 – ISSN: 0067-6837 – mf#6274 – us UMI ProQuest [370]

Bibliographic survey : the negro in print – Washington DC 1965-71 – 1 – ISSN: 0006-1263 – mf#3226 – us UMI ProQuest [305]

A bibliographical account of english theatrical literature : from the earliest times to the present day / Lowe, Robert William – London: John C Nimmo, 1888 – (= ser 19th c publishing...) – 5mf – 9 – mf#3.1.109 – uk Chadwyck [790]

A bibliographical description of the editions of the new testament, tyndale's version in english : with numerous readings, comparisons of texts and historical notices / Fry, Francis – London: Henry Sotheran, 1878 [mf ed 1988] – 4mf – 9 – 0-7905-0127-9 – (incl ind) – mf#1987-0127 – us ATLA [225]

Bibliographical notes on histories of inventions and books of secrets : six papers read to the archaeological society of glasgow april 1882-jan 1888 / Ferguson, John – Glasgow, 1885-98 – (= ser 19th c publishing...) – 3mf – 9 – (iss in 8pts: pts1 & 2 dated 1896 are of an ed of 150 reprd copies, with the imprint: glasgow: printed at the university press by robert maclehose and co. ind dated 1898 also has this imprint. pts3,4,5 & 6 dated respectively 1885, 1888, 1889 and 1890 are of an ed of 100 reprd copies, publ by strathern & freeman. incl the suppl to notes on books of secrets, pt2) – mf#3.1.42 – uk Chadwyck [070]

Bibliographical society. london. transactions, series 1 – v1-15. 1892-1919 – 9 – $108.00 – mf#0100 – us Brook [010]

Bibliographie : cinquante ouvrages traitant de l'adolescence / Houle, Alphonse – 1964 [mf ed 1979] – (= ser Bibliographies du cours...1947-66) – 3mf – 9 – (with ind) – mf#SEM105P4 – cn Bibl Nat [305]

Bibliographie : les monuments historiques de quebec, 1854-1901 / St-Roger, soeur – 1963 [mf ed 1979] – (= ser Bibliographies du cours...1947-66) – 1mf – 9 – (with ind) – mf#SEM105P4 – cn Bibl Nat [720]

Bibliographie : les monuments historiques de quebec, 1901-1920 / Ste-Martine, soeur – 1963 [mf ed 1978] – (= ser Bibliographies du cours...1947-66) – 2mf – 9 – (with ind) – mf#SEM105P4 – cn Bibl Nat [720]

Bibliographie (1915 a 1940) du dr georges maheux : entomologiste, membre de la societe royale du canada / Naud, Denise – 1963 [mf ed 1979] – (= ser Bibliographies du cours...1947-66) – 2mf – 9 – (with ind) – mf#SEM105P4 – cn Bibl Nat [616]

Bibliographie 1950-1958 de la bienheureuse marguerite d'youville : fondatrice des soeurs de la charite (soeurs grises) de montreal / Sainte-Fernande, soeur – 1963 [mf ed 1979] – (= ser Bibliographies du cours...1947-66) – 1mf – 9 – mf#SEM105P4 – cn Bibl Nat [241]

Bibliographie analytique : ecrits des peres oblats de marie-immaculee sur la sainte-vierge dans les annales de notre-dame du cap, 1902-1962 / Marie-des-Anges, soeur – 1963 [mf ed 1979] – (= ser Bibliographies du cours...1947-66) – 4mf – 9 – (with ind; pref by paul-henri barabe) – mf#SEM105P4 – cn Bibl Nat [241]

Bibliographie analytique de alain grandbois / Gagnon, Huguette – 1964 [mf ed 1979] – (= ser Bibliographies du cours...1947-66) – 2mf – 9 – (with ind; pref by charles-marie boissonneault) – mf#SEM105P4 – cn Bibl Nat [010]

Bibliographie analytique de baie comeau sur la cote-nord du saint-laurent / Beaudry, Pauline – 1959 [mf ed 1978] – (= ser Bibliographies du cours...1947-66) – 1mf – 9 – (with ind; incl english text) – mf#SEM105P4 – cn Bibl Nat [917]

Bibliographie analytique de c-j magnan / Laberge, Raymond-Marie – 1953 [mf ed 1978] – (= ser Bibliographies du cours...1947-66) – 1mf – 9 – (with ind; pref by j-chs magnan) – mf#SEM105P4 – cn Bibl Nat [010]

Bibliographie analytique de ernest pallascio-morin : journaliste, auteur dramatique, conferencier / Marie-des-Lys, soeur – 1961 [mf ed 1978] – (= ser Bibliographies du cours...1947-66) – 2mf – 9 – (with ind; pref by roger brien) – mf#SEM105P4 – cn Bibl Nat [070]

Bibliographie analytique de francoise l-roy : journaliste, vice-presidente du cercle des femmes journalistes de montreal, membre de la societe des ecrivains canadiens / Saint-Jean-Marie, soeur – 1964 [mf ed 1979] – (= ser Bibliographies du cours...1947-66) – 2mf – 9 – (with ind; pref by paul-e gosselin) – mf#SEM105P4 – cn Bibl Nat [070]

Bibliographie analytique de jacques de monleon / Vachon, Madeleine E – 1963 [mf ed 1979] – 1mf – 9 – (with ind; pref by emmanuel trepanier) – mf#SEM105P4 – cn Bibl Nat [010]

Bibliographie analytique de joseph-thomas leblanc / Blouin, Gisele – 1950 [mf ed 1978] – (= ser Bibliographies du cours...1947-66) – 2mf – 9 – (with ind; pref by luc lacourciere) – mf#SEM105P4 – cn Bibl Nat [010]

Bibliographie analytique de la chanson de folklore : ouvrages parus au canada de 1950 a 1962 / Dupuis, Monique – 1964 [mf ed 1979] – (= ser Bibliographies du cours...1947-66) – 1mf – 9 – (with ind; pref by aut) – mf#SEM105P4 – cn Bibl Nat [780]

Bibliographie analytique de la delinquance juvenile : ou de l'inadaptation familiale et sociale des mal-aimes, evolution des opinions, des idees emises dans les causes de la prevention et la rehabilitation des adultes de demain, 1955-1965 / Chapdelaine, Cecile – 1966 [mf ed 1979] – (= ser Bibliographies du cours...1947-66) – 8mf – 9 – (with ind) – mf#SEM105P4 – cn Bibl Nat [360]

Bibliographie analytique de la federation des guides catholiques de la province de quebec / Drolet, Bernadette – 1963 [mf ed 1979] – (= ser Bibliographies du cours...1947-66) – 1mf – 9 – (with ind; pref by simone pare) – mf#SEM105P4 – cn Bibl Nat [241]

Bibliographie analytique de la genealogie dans les comtes de saint-maurice, maskinonge, champlain / Lemay, Leona – 1964 [mf ed 1979] – (= ser Bibliographies du cours...1947-66) – 2mf – 9 – (with ind; pref by pierre matte) – mf#SEM105P4 – cn Bibl Nat [929]

Bibliographie analytique de la litterature musicale canadienne francaise / Laflamme, Claire – 1950 [mf ed 1978] – (= ser Bibliographies du cours...1947-66) – 1mf – 9 – (with ind; pref by henri gagnon) – mf#SEM105P4 – cn Bibl Nat [780]

Bibliographie analytique de la litterature pedagogique canadienne francaise de 1790 a 1900 / Gagnon, Gilberte – 1951 [mf ed 1979] – (= ser Bibliographies du cours...1947-66) – 1mf – 9 – (with ind; pref by clement lockquell) – mf#SEM105P4 – cn Bibl Nat [440]

Bibliographie analytique de la litterature pedagogique canadienne-francaise depuis 1900 / Ratte, Alice – 1951 [mf ed 1979] – (= ser Bibliographies du cours...1947-66) – 2mf – 9 – (with ind; pref by clement lockquell) – mf#SEM105P4 – cn Bibl Nat [440]

Bibliographie analytique de la psychologie infantile, 1948 a 1952 / Dumas, Rollande – 1952 [mf ed 1978] – (= ser Bibliographies du cours...1947-66) – 1mf – 9 – (with ind) – mf#SEM105P4 – cn Bibl Nat [150]

Bibliographie analytique de la reliure au canada francais / Desrochers-Leduc, Lucienne – 1953 [mf ed 1978] – (= ser Bibliographies du cours...1947-66) – 1mf – 9 – (with ind; pref by aut) – mf#SEM105P4 – cn Bibl Nat [680]

Bibliographie analytique de la vallee de la matapedia / Marie de Saint-Joseph-Jean, soeur – 1964 [mf ed 1979] – (= ser Bibliographies du cours...1947-66) – 2mf – 9 – (with ind) – mf#SEM105P4 – cn Bibl Nat [917]

Bibliographie analytique de la vie personnelle de l'infirmiere d'apres la documentation des revues d'infirmieres en langue francaise de la province de quebec : couvrant la periode de 1951 a 1961 exclusivement / Munger, Angele – 1962 [mf ed 1979] – (= ser Bibliographies du cours...1947-66) – 1mf – 9 – (with ind) – mf#SEM105P4 – cn Bibl Nat [360]

Bibliographie analytique de l'abbe anselme longpre...du diocese de saint-hyacinthe, premiere partie (1927-1947) / Joseph-Marie, soeur – 1963 [mf ed 1979] – (= ser Bibliographies du cours...1947-66) – 2mf – 9 – (with ind) – mf#SEM105P4 – cn Bibl Nat [241]

Bibliographie analytique de l'abbe henri-raymond casgrain / Sainte-Thecle, soeur – 1955 [mf ed 1978] – (= ser Bibliographies du cours...1947-66) – 2mf – 9 – (with ind; pref by albert tessier) – mf#SEM105P4 – cn Bibl Nat [241]

Bibliographie analytique de l'amiante du canada / Marcoux, Lucile – 1953 [mf ed 1979] – (= ser Bibliographies du cours...1947-66) – 1mf – 9 – (with ind; pref by lucien lavigne) – mf#SEM105P4 – cn Bibl Nat [622]

Bibliographie analytique de l'histoire d'acadie / Corrivault, Blaise – 1950 [mf ed 1978] – (= ser Bibliographies du cours...1947-66) – 1mf – 9 – (with ind; pref by francis bourque) – mf#SEM105P4 – cn Bibl Nat [971]

Bibliographie analytique de l'honorable juge sir adolphe-basile routhier, homme de lettres / Sainte-Janviere, soeur – 1952 [mf ed 1979] – (= ser Bibliographies du cours...1947-66) – 1mf – 9 – (pref by paul-emile gosselin) – mf#SEM105P4 – cn Bibl Nat [340]

Bibliographie analytique de l'ile d'anticosti (reine du golfe) / Barnabe, Michele – 1959 [mf ed 1978] – (= ser Bibliographies du cours...1947-66) – 1mf – 9 – (with ind) – mf#SEM105P4 – cn Bibl Nat [910]

Bibliographie analytique de l'ileaux-coudres / Amyot, Michel – 1952 [mf ed 1979] – (= ser Bibliographies du cours...1947-66) – 1mf – 9 – (with ind; pref by michel amyot) – mf#SEM105P4 – cn Bibl Nat [917]

Bibliographie analytique de l'oeuvre, 1941-1960, de monsieur pierre-paul turgeon, notaire / Pelletier, Carmen – 1964 [mf ed 1979] – (= ser Bibliographies du cours...1947-66) – 1mf – 9 – (with ind; pref by henri turgeon) – mf#SEM105P4 – cn Bibl Nat [340]

Bibliographie analytique de l'oeuvre de beraud de saint maurice / Sainte-Hildegarde, mere – 1963 [mf ed 1979] – (= ser Bibliographies du cours...1947-66) – 1mf – 9 – (with ind, pref by herve biron) – mf#SEM105P4 – cn Bibl Nat [241]

Bibliographie analytique de l'oeuvre de bertrand vac / Guilbault, Renee – 1963 [mf ed 1979] – (= ser Bibliographies du cours...1947-66) – 1mf – 9 – (with ind; pref by jean-charles bonenfant) – mf#SEM105P4 – cn Bibl Nat [010]

Bibliographie analytique de l'oeuvre de bruno lafleur / Laki, Georgette – 1964 [mf ed 1979] – (= ser Bibliographies du cours...1947-66) – 1mf – 9 – (with ind; pref by jean-charles bonenfant) – mf#SEM105P4 – cn Bibl Nat [010]

Bibliographie analytique de l'oeuvre de charlotte savary / Bedard, Suzanne – 1951 [mf ed 1979] – (= ser Bibliographies du cours...1947-66) – 1mf – 9 – (with ind; pref by georges letourneau) – mf#SEM105P4 – cn Bibl Nat [920]

Bibliographie analytique de l'oeuvre de du docteur jean-baptiste jobin : president du college des medecins et chirurgiens de la province de quebec / Desjardins, Jeannette – 1963 [mf ed 1979] – (= ser Bibliographies du cours...1947-66) – 2mf – 9 – (with ind; pref by charles-marie boissonnault) – mf#SEM105P4 – cn Bibl Nat [610]

Bibliographie analytique de l'oeuvre de i-w jones : geologue, directeur des services geologiques, ministere des richesses naturelles / Fortier, Marie-Marthe – 1962 [mf ed 1978] – (= ser Bibliographies du cours...1947-66) – 1mf – 9 – (with ind; pref by p e grenier) – mf#SEM105P4 – cn Bibl Nat [550]

Bibliographie analytique de l'oeuvre de jean-paul legare : journaliste de rimouski, president du journal l'echo du bas st-laurent, 1945-1961 / Sainte-Madeleine-du-Calvaire, soeur – 1964 [mf ed 1979] – (= ser Bibliographies du cours...1947-66) – 4mf – 9 – (with ind; pref by adrien begin) – mf#SEM105P4 – cn Bibl Nat [070]

Bibliographie analytique de l'oeuvre de l'abbe arthur maheux pour les annee / Nadeau, Marie-Marthe – 1947 [mf ed 1978] – (= ser Bibliographies du cours...1947-66) – 1mf – 9 – (with ind) – mf#SEM105P4 – cn Bibl Nat [241]

Bibliographie analytique de l'oeuvre de l'abbe jean holmes : un des fondateurs de l'universite laval / Lefaivre, Louise – 1964 [mf ed 1978] – (= ser Bibliographies du cours...1947-66) – 1mf – 9 – (with ind; pref by don guay) – mf#SEM105P4 – cn Bibl Nat [378]

Bibliographie analytique de l'oeuvre de l'abbe roch duval : licencie en orientation, debut a septembre 1964 / Pouliot, Aline – 1964 [mf ed 1979] – (= ser Bibliographies du cours...1947-66) – 1mf – 9 – (with ind; pref by gerard dion) – mf#SEM105P4 – cn Bibl Nat [241]

Bibliographie analytique de l'oeuvre de leopold lamontagne de la societe royale du canada : doyen de la faculte des lettres a l'universite laval / Roy, Reina L – 1963 [mf ed 1979] – (= ser Bibliographies du cours...1947-66) – 2mf – 9 – (with ind; pref by luc lacourciere) – mf#SEM105P4 – cn Bibl Nat [378]

Bibliographie analytique de l'oeuvre de l'honorable senateur cyrille vaillancourt : president-gerant de l'union regionale des caisses populaires desjardins, district de quebec... / Boutin, Rose-Anne – 1961 [mf ed 1978] – (= ser Bibliographies du cours...1947-66) – 3mf – 9 – (with ind; pref by ferdinand ouellet) – mf#SEM105P4 – cn Bibl Nat [010]

Bibliographie analytique de l'oeuvre de l'honorable senateur cyrille vaillancourt... / Boutin, Rose-Anne – 1961 [mf ed 1978] – (= ser Bibliographies du cours...1947-66) – 2mf – 9 – (with ind; pref by ferdinand ouellet) – mf#SEM105P4 – cn Bibl Nat [410]

Bibliographie analytique de l'oeuvre de louis-philippe roy : commandeur de l'ordre de saint-gregoire le grand... / Sainte-Marie-Cleophas, soeur – 1964 [mf 1979] – (= ser Bibliographies du cours...1947-66) – 10mf – 9 – (pref by omer-jules desaulniers) – mf#SEM105P4 – cn Bibl Nat [920]

Bibliographie analytique de l'oeuvre de louis-philippe roy... : commandeur de l'ordre de saint-gregoire le grand...redacteur en chef de l'action catholique, 1949-1960 / Sainte-Marie-Cleophas, soeur – 1964 [mf ed 1979] – (= ser Bibliographies du cours...1947-66) – 10mf – 9 – (with ind; pref by omer-jules desaulniers) – mf#SEM105P4 – cn Bibl Nat [070]

Bibliographie analytique de l'oeuvre de louis-philippe roy... : directeur de l'action, 1920-1948 / Saint-Majella, soeur – 1964 [mf ed 1979] – (= ser Bibliographies du cours...1947-66) – 6mf – 9 – (with ind; pref by wheeler dupont) – mf#SEM105P4 – cn Bibl Nat [070]

Bibliographie analytique de l'oeuvre de m albert rioux : agronome, maitre es-sciences agricoles, docteur es-sciences sociales, economiques et politiques / Carbonneau, Leopold – 1952 [mf ed 1978] – (= ser Bibliographies du cours...1947-66) – 2mf – 9 – (with ind; pref by jean-charles bonenfant) – mf#SEM105P4 – cn Bibl Nat [630]

Bibliographie analytique de l'oeuvre de m avila bedard... : sous-ministre des terres et forets dans le cabinet provincial / Houde, Marguerite A – 1958 [mf ed 1978] – (= ser Bibliographies du cours...1947-66) – 3mf – 9 – (with ind; pref by j a breton) – mf#SEM105P4 – cn Bibl Nat [630]

Bibliographie analytique de l'oeuvre de m emile castonguay / Daigle, Louise – 1964 [mf ed 1979] – (= ser Bibliographies du cours...1947-66) – 2mf – 9 – (with ind; pref by charles-marie boissonnault) – mf#SEM105P4 – cn Bibl Nat [010]

Bibliographie analytique de l'oeuvre de m eugene l'heureux : journaliste, membre de la societe royale du canada, 1918-1929 / Sainte-Daniella, soeur – 1962 [mf ed 1979] – (= ser Bibliographies du cours...1947-66) – 2mf – 9 – (with ind; pref by joseph dandurand) – mf#SEM105P4 – cn Bibl Nat [070]

Bibliographie analytique de l'oeuvre de m eugene l'heureux : journaliste, membre de la societe royale du canada et de l'academie canadienne, 1940-1949 / Saint-Louis-Daniel, soeur – 1964 [mf ed 1979] – (= ser Bibliographies du cours...1947-66) – 6mf – 9 – (with ind; pref by lorenzo pare) – mf#SEM105P4 – cn Bibl Nat [070]

Bibliographie analytique de l'oeuvre de m l'abbe arthur maheux de la societe royale du canada : archiviste du seminaire de quebec / Pelletier, Jacqueline – [1956] (mf ed 1978) – (= ser Bibliographies du cours...1947-66) – 1mf – 9 – (with ind; pref by alphonse-marie parent) – mf#SEM105P4 – cn Bibl Nat [241]

Bibliographie analytique de l'oeuvre de m l'abbe louis o'neill : aumonier a l'academie de quebec, professeur de morale sociale / Morin, Romuald – 1962 [mf ed 1979] – (= ser Bibliographies du cours...1947-66) – 1mf – 9 – (with ind; pref by gustave tardif) – mf#SEM105P4 – cn Bibl Nat [300]

Bibliographie analytique de l'oeuvre de m louis-philippe robidoux de la societe royale du canada : redacteur en chef de la tribune de sherbrooke / Saint-Fidele, soeur – 1964 [mf ed 1979] – (= ser Bibliographies du cours...1947-66) – 4mf – 9 – (with ind; pref by conrad groleau) – mf#SEM105P4 – cn Bibl Nat [070]

Bibliographie analytique de l'oeuvre de madame marcelle lepage-thibaudeau : licenciee en sciences naturelles, licenciee en phychologie et psychotherapie / Cantin, Louise – 1964 [mf ed 1979] – (= ser Bibliographies du cours...1947-66) – 2mf – 9 – (with ind; pref by zephirin rousseau) – mf#SEM105P4 – cn Bibl Nat [500]

Bibliographie analytique de l'oeuvre de mademoiselle gertie kathleen hart : diplomee de l'universite laval, de l'universite d'oxford, de la faculte des lettres de la sorbonne / Marie-de-la-Charite, soeur – 1962 [mf ed 1978] – (= ser Bibliographies du cours...1947-66) – 1mf – 9 – (with ind; pref by mademoiselle hart) – mf#SEM105P4 – cn Bibl Nat [378]

Bibliographie analytique de l'oeuvre de maitre eugene l'heureux : journaliste, 1950-1960, bibliothecaire-adjoint au parlement / Marie-de-la-Salette, soeur – 1964 [mf ed 1978] – (= ser Bibliographies du cours...1947-66) – 6mf – 9 – (with ind; pref by raymond dube) – mf#SEM105P4 – cn Bibl Nat [070]

Bibliographie analytique de l'oeuvre de maitre eugene l'heureux : membre de la societe royale du canada et de l'academie canadienne, redacteur en chef du progres du saguenay et de l'action catholique, 1929-1940 / Marie-des-Miracles, soeur – 1964 [mf ed 1979] – (= ser Bibliographies du cours...1947-66) – 3mf – 9 – (with ind; pref by arthur maheux) – mf#SEM105P4 – cn Bibl Nat [241]

Bibliographie analytique de l'oeuvre de marcel clement : licencie es lettres, diplome d'etudes superieures de philosophie (sorbonne)... / Joly, Monique – 1950 [mf ed 1978] – (= ser Bibliographies du cours...1947-66) – 1mf – 9 – (with ind) – mf#SEM105P4 – cn Bibl Nat [378]

Bibliographie analytique de l'oeuvre de marcel dube / Laforest, Marthe – 1964 [mf ed 1978] – (= ser Bibliographies du cours...1947-66) – 2mf – 9 – (with ind; pref by louis-georges carrier) – mf#SEM105P4 – cn Bibl Nat [010]

Bibliographie analytique de l'oeuvre de mgr arthur maheux... : archiviste du seminaire de quebec, membre de la societe royale du canada / Thibault, Priscille – 1964 [mf ed 1978] – (= ser Bibliographies du cours...1947-66) – 2mf – 9 – (with ind; pref by maurice lebel) – mf#SEM105P4 – cn Bibl Nat [241]

Bibliographie analytique de l'oeuvre de monseigneur albert tessier... : de la societe royale du canada, de la societe des dix, visiteur en chef des instituts familiaux / Elisabeth-de-la-Trinite, soeur – 1962 [mf ed 1978] – (= ser Bibliographies du cours...1947-66) – 2mf – 9 – (with ind; pref by raymond douville) – mf#SEM105P4 – cn Bibl Nat [971]

Bibliographie analytique de l'oeuvre de monseigneur arthur maheux de la societe royale du canada : archiviste au seminaire de quebec / Bois, Jacqueline – 1964 [mf ed 1979] – (= ser Bibliographies du cours...1947-66) – 1mf – 9 – (with ind; pref by benoit garneau) – mf#SEM105P4 – cn Bibl Nat [241]

Bibliographie analytique de l'oeuvre de monseigneur victorin germain... : directeur-administrateur de la sauvegarde de l'enfance / Marie-Liberatrice, soeur – 1962 [i.e. 1963] (mf ed 1978) – (= ser Bibliographies du cours...1947-66) – 3mf – 9 – (with ind; pref by soeur saint ferdinand) – mf#SEM105P4 – cn Bibl Nat [360]

Bibliographie analytique de l'oeuvre de monsieur auguste viatte : docteur es lettres, professeur titulaire de litterature francaise a l'universite laval / Belanger, Pierrette – 1948 [mf ed 1978] – (= ser Bibliographies du cours...1947-66) – 1mf – 9 – (with ind) – mf#SEM105P4 – cn Bibl Nat [440]

Bibliographie analytique de l'oeuvre de monsieur charles-marie boissonnault : historien, poete, critique et publiciste / Lapointe, Madeleine – 1959 [mf ed 1978] – (= ser Bibliographies du cours...1947-66) – 1mf – 9 – (with ind; pref by francis desroches) – mf#SEM105P4 – cn Bibl Nat [920]

Bibliographie analytique de l'oeuvre de monsieur charles-marie boissonnault : historien, poete et critique / Guire, Juliette de – 1950 [mf ed 1979] – (= ser Bibliographies du cours...1947-66) – 1mf – 9 – (with ind; pref by marcel trudel) – mf#SEM105P4 – cn Bibl Nat [920]

Bibliographie analytique de l'oeuvre de monsieur l'abbe jean-baptiste gauvin... : principal de l'ecole normale de mont-joli / Marie des Cherubins, soeur – 1963 [mf ed 1979] – (= ser Bibliographies du cours...1947-66) – 2mf – 9 – (with ind; pref by richard joly) – mf#SEM105P4 – cn Bibl Nat [378]

Bibliographie analytique de l'oeuvre de monsieur le chanoine georges panneton, trois-rivieres / Marie de Saint-Alphonse-de-Jesus, soeur – 1963 [i.e. 1964] (mf ed 1979) – (= ser Bibliographies du cours...1947-66) – 2mf – 9 – (with ind; pref by antonio magnan) – mf#SEM105P4 – cn Bibl Nat [241]

Bibliographie analytique de l'oeuvre de monsieur le chanoine paul-emile crepeault, 1944-1964 / Gingras, Jean-Jules – 1965 [mf ed 1979] – (= ser Bibliographies du cours...1947-66) – 1mf – 9 – (with ind; pref by georges bherer) – mf#SEM105P4 – cn Bibl Nat [241]

Bibliographie analytique de l'oeuvre de monsieur pierre-h ruel : doyen, faculte des sciences de l'education, universite de sherbrooke / Saint-Raymond-du-Sauveur, soeur – 1964 [mf ed 1979] – (= ser Bibliographies du cours...1947-66) – 2mf – 9 – (with ind. pref by maurice o'bready) – mf#SEM105P4 – cn Bibl Nat [370]

Bibliographie analytique de l'oeuvre de monsieur rene pomerleau : docteur es sciences de la societe royale du canada, pathologiste forestier, laboratoire des recherches forestieres, gouvernement du canada, quebec / Maheux, Laura – 1964 [mf ed 1979] – (= ser Bibliographies du cours...1947-66) – 2mf – 9 – (with ind; pref by georges maheux) – mf#SEM105P4 – cn Bibl Nat [634]

Bibliographie analytique de l'oeuvre de odilon arteau : redacteur a l'action catholique / Sainte-Anne-de-Marie, soeur – 1964 [mf ed 1979] – (= ser Bibliographies du cours...1947-66) – 6mf – 9 – (with ind; pref by louis-philippe roy) – mf#SEM105P4 – cn Bibl Nat [241]

Bibliographie analytique de l'oeuvre de olivette lamontagne / Gagnon, Jacques Etiennette – 1957 [mf ed 1978] – (= ser Bibliographies du cours...1947-66) – 1mf – 9 – (with ind; pref by aime plamondon) – mf#SEM105P4 – cn Bibl Nat [010]

Bibliographie analytique de l'oeuvre de reverende soeur sainte-claire-de-rimini (1887-) : soeur de la charite de quebec / Sainte-Apollonie, soeur – 1962 [mf ed 1979] – (= ser Bibliographies du cours...1947-66) – 1mf – 9 – (with ind; pref by edgar dion) – mf#SEM105P4 – cn Bibl Nat [241]

Bibliographie analytique de l'oeuvre de soeur marie-emmanuel : religieuse ursuline, professeur de litterature francaise au college des ursulines de quebec / Sainte-Madeleine, soeur – 1955 [mf ed 1978] – (= ser Bibliographies du cours...1947-66) – 1mf – 9 – (with ind; pref by mere saint francois de sales) – mf#SEM105P4 – cn Bibl Nat [241]

Bibliographie analytique de l'oeuvre de son excellence rev mgr napoleon-alexandre labrie... : eveque titulaire de hilta / Guilbault, Germaine – 1964 [mf ed 1979] – (= ser Bibliographies du cours...1947-66) – 2mf – 9 – (with ind; pref by jean-paul pelletier) – mf#SEM105P4 – cn Bibl Nat [241]

Bibliographie analytique de l'oeuvre dom albert jamet de l'abbaye benedictine de solesmes / Sainte Therese-de-l'Enfant-Jesus, mere – 1960 [mf ed 1978] – (= ser Bibliographies du cours...1947-66) – 1mf – 9 – (with ind; pref by marie-emmanuel) – mf#SEM105P4 – cn Bibl Nat [241]

Bibliographie analytique de l'oeuvre du docteur de la broquerie fortier... : professeur titulaire de clinique pediatrique... / Paul de Rome, soeur – 1962 [mf ed 1979] – (= ser Bibliographies du cours...1947-66) – 2mf – 9 – (with ind; pref by charles-auguste gauthier) – mf#SEM105P4 – cn Bibl Nat [618]

Bibliographie analytique de l'oeuvre du docteur jean-charles miller : medecin-psychiatre a l'ecole la jemmerais, 1898-1952 / Montciel, Marie-de – 1963 [mf ed 1979] – (= ser Bibliographies du cours...1947-66) – 3mf – 9 – (with ind; pref by mere marie-de-graces) – mf#SEM105P4 – cn Bibl Nat [616]

Bibliographie analytique de l'oeuvre du docteur louis-edmond hamelin : professeur titulaire de geographie a l'universite laval... / Saint-Alphonse-de-Liguori, Marie de – 1962 [i.e. 1963] [mf ed 1978] – (= ser Bibliographies du cours...1947-66) – 2mf – 9 – (with ind; pref by fernand grenier) – mf#SEM105P4 – cn Bibl Nat [900]

Bibliographie analytique de l'oeuvre du docteur marcel langlois : professeur titulaire a l'universite laval, certifie du college royal en pediatrie / Marie Gemma, soeur – 1961 [mf ed 1978] – (= ser Bibliographies du cours...1947-66) – 1mf – 9 – (with ind) – mf#SEM105P4 – cn Bibl Nat [618]

Bibliographie analytique de l'oeuvre du docteur marcel langlois : professeur titulaire a l'universite laval, certifie du college royal en pediatrie, membre emerite de l'association des administrateurs d'hopitaux de la province de quebec / Saint-Leonce, soeur – 1962 [mf ed 1978] – (= ser Bibliographies du cours...1947-66) – 1mf – 9 – (with ind) – mf#SEM105P4 – cn Bibl Nat [610]

Bibliographie analytique de l'oeuvre du docteur pierre jobin... : professeur titulaire et directeur du departement d'anatomie de la faculte de medecine de l'universite laval, 1927-1964 / Christine-Marie, soeur – 1964 [mf ed 1979] – (= ser Bibliographies du cours...1947-66) – 2mf – 9 – (with ind; pref by andre jobin) – mf#SEM105P4 – cn Bibl Nat [611]

Bibliographie analytique de l'oeuvre du docteur roland desmeules... : professeur titulaire a l'universite laval, chef du departement de medecine a l'hopital laval... / Sainte-Veronique, soeur – 1964 [mf ed 1979] – (= ser Bibliographies du cours...1947-66) – 2mf – 9 – (with ind; pref by soeur saint-ferdinand) – mf#SEM105P4 – cn Bibl Nat [616]

Bibliographie analytique de l'oeuvre du r p j-hermann poisson / Saint-Damase-Marie, soeur – 1962 [mf ed 1979] – (= ser Bibliographies du cours...1947-66) – 1mf – 9 – (with ind; pref by damase laberge) – mf#SEM105P4 – cn Bibl Nat [010]

Bibliographie analytique de l'oeuvre du r p marcel dubois... / Marcotte, Beatrice – 1957 [mf ed 1978] – (= ser Bibliographies du cours...1947-66) – 1mf – 9 – (with ind; pref by camille pacreau) – mf#SEM105P4 – cn Bibl Nat [241]

Bibliographie analytique de l'oeuvre du r pere philias f bourgeois de la congregation de sainte-croix : professeur de litterature et d'histoire a l'universite saint-joseph / Saint-Marc, mere – 1964 [mf ed 1979] – (= ser Bibliographies du cours...1947-66) – 2mf – 9 – (with ind) – mf#SEM105P4 – cn Bibl Nat [400]

Bibliographie analytique de l'oeuvre du reverend pere alexis de barbezieux, capucin / Dumas, Gabriel-Marie – 1956 [mf ed 1978] – (= ser Bibliographies du cours...1947-66) – 1mf – 9 – (with ind) – mf#SEM105P4 – cn Bibl Nat [241]

Bibliographie analytique de l'oeuvre du reverend pere edmond gaudron... : professeur d'histoire de la philosophie ancienne a la faculte de la philosophie de l'universite laval de quebec / Saint-Jean-de-Brebeuf-Hudon, mere – 1956 [mf ed 1978] – (= ser Bibliographies du cours...1947-66) – 1mf – 9 – (with ind; pref by marie-de-l'annonciation) – mf#SEM105P4 – cn Bibl Nat [180]

Bibliographie analytique de l'oeuvre du reverend pere eugene lefebvre... : directeur de l'oeuvre du pelerinage au sanctuaire de sainte-anne-de-beaupre / Roy, Jean – 1960 [mf ed 1978] – (= ser Bibliographies du cours...1947-66) – 1mf – 9 – (with ind; pref by lucien gagne) – mf#SEM105P4 – cn Bibl Nat [241]

Bibliographie analytique de l'oeuvre du reverend pere jean bousquet : lecteur en theologie / Marie-Helene-de Rome, soeur – 1964 [mf ed 1979] – (= ser Bibliographies du cours...1947-66) – 1mf – 9 – (with ind; pref by antonin lamarche) – mf#SEM105P4 – cn Bibl Nat [241]

Bibliographie analytique de l'oeuvre d'un grand chroniqueur louis-philippe audet...1953-1962 / Louis-Ernest, soeur – 1963 [mf ed 1979] – (= ser Bibliographies du cours...1947-66) – 3mf – 9 – (with ind; pref by albert tessier) – mf#SEM105P4 – cn Bibl Nat [410]

Bibliographie analytique de louis hemon / Carpentier, Denyse – 1962 [i.e. 1963] (mf ed 1978) – (= ser Bibliographies du cours...1947-66) – 1mf – 9 – (with ind; pref by paul carpentier) – mf#SEM105P4 – cn Bibl Nat [010]

Bibliographie analytique de louis-alexandre belisle : auteur et editeur / Blouin, Gervaise – 1953 [mf ed 1978] – (= ser Bibliographies du cours...1947-66) – 1mf – 9 – (with ind; pref by roland morin) – mf#SEM105P4 – cn Bibl Nat [070]

Bibliographie analytique de luc lacourciere : titulaire de la chaire de folklore, universite laval, quebec / Couture, Marguerite – 1950 [mf ed 1978] – (= ser Bibliographies du cours...1947-66) – 1mf – 9 – (with ind; pref by luc lacourciere) – mf#SEM105P4 – cn Bibl Nat [390]

Bibliographie analytique de m jean-charles falardeau : sociologue, directeur du departement de sociologie a la faculte des sciences sociales de l'universite laval / Legare, Claire – 1953 [mf ed 1979] – (= ser Bibliographies du cours...1947-66) – 1mf – 9 – (with ind; pref by gonzalve poulin) – mf#SEM105P4 – cn Bibl Nat [301]

Bibliographie analytique de madame helene b beausejour / Gagnon, Marcelle – 1964 [mf ed 1978] – (= ser Bibliographies du cours...1947-66) – 1mf – 9 – (with ind; pref by marguerite a hebert) – mf#SEM105P4 – cn Bibl Nat [010]

Bibliographie analytique de madame yolande chene / Morissette, Rachel – 1963 [mf ed 1979] – (= ser Bibliographies du cours...1947-66) – 1mf – 9 – (pref by guy laviolette) – mf#SEM105P4 – cn Bibl Nat [010]

Bibliographie analytique de mademoiselle simone pare : professeur a l'ecole de service social, faculte des sciences sociales, universite laval / Gignac, Francoise – 1958 [mf ed 1978] – (= ser Bibliographies du cours...1947-66) – 2mf – 9 – (with ind; pref by gilles-marie belanger) – mf#SEM105P4 – cn Bibl Nat [300]

Bibliographie analytique de marie-therese 1944-1961 (mlle marie-therese chevalier) : responsable du service marial de montreal et redactrice du digeste marial de 1948 a 1961 / Marie de la Recouvrance, soeur – 1962 [mf ed 1979] – (= ser Bibliographies du cours...1947-66) – 2mf – 9 – (with ind; pref by henri-marie guindon) – mf#SEM105P4 – cn Bibl Nat [360]

Bibliographie analytique de maxine : membre de la societe des ecrivains canadiens et de la societe des ecrivains pour la jeunesse / Lemay, Clotilde – 1950 [mf ed 1978] – (= ser Bibliographies du cours...1947-66) – 1mf – 9 – (with ind) – mf#SEM105P4 – cn Bibl Nat [410]

Bibliographie analytique de mgr louis-joseph aubin : principal de l'ecole normale n-d du bon-conseil, chicoutimi / Madeleine-de-la-Croix, soeur – 1961 [mf ed 1978] – (= ser Bibliographies du cours...1947-66) – 2mf – 9 – (with ind; pref by jacques tremblay) – mf#SEM105P4 – cn Bibl Nat [378]

Bibliographie analytique de mme paule develuy : membre de la societe des ecrivains canadiens... / Brulotte, Marie-Berthe – 1964 [mf ed 1979] – (= ser Bibliographies du cours...1947-66) – 2mf – 9 – (with ind; pref by jean-paul labelle) – mf#SEM105P4 – cn Bibl Nat [410]

Bibliographie analytique de monseigneur felix-antoine savard : professeur de poesie francaise a l'universite laval / Therese du Carmel, soeur – 1964 [mf ed 1979] – (= ser Bibliographies du cours...1947-66) – 3mf – 9 – (with ind; pref by luc lacourciere) – mf#SEM105P4 – cn Bibl Nat [440]

Bibliographie analytique de monsieur alphonse desilets / Gresley, Joseph-Edouard le – 1950 [mf ed 1978] – (= ser Bibliographies du cours...1947-66) – 1mf – 9 – (with ind; pref by a hubert) – mf#SEM105P4 – cn Bibl Nat [010]

Bibliographie analytique de monsieur andre giroux : publiciste au ministere de l'industrie et du commerce / Giroux, Yvette – 1948 [mf ed 1979] – (= ser Bibliographies du cours...1947-66) – 1mf – 9 – (with ind; pref by jean marchand) – mf#SEM105P4 – cn Bibl Nat [380]

Bibliographie analytique de monsieur francis desroches / Pouliot, Ghislaine – 1959 [mf ed 1978] – (= ser Bibliographies du cours...1947-66) – 1mf – 9 – (with ind; pref by michel levasseur) – mf#SEM105P4 – cn Bibl Nat [010]

Bibliographie analytique de monsieur l'abbe armand dube : cure de kamouraska / Perusse, Claire – 1957 [mf ed 1978] – (= ser Bibliographies du cours...1947-66) – 1mf – 9 – (with ind; pref by raymond boucher) – mf#SEM105P4 – cn Bibl Nat [241]

Bibliographie analytique de monsieur l'abbe roland dufour : directeur de la revue temoignages, aumonier a l'institut familial n-d du bon-conseil, chicoutimi / Sainte-Candide, soeur – 1962 [mf ed 1978] – (= ser Bibliographies du cours...1947-66) – 2mf – 9 – (with ind; pref by gerard desgagne) – mf#SEM105P4 – cn Bibl Nat [360]

Bibliographie analytique de monsieur yvon theriault : conseiller en relations publiques / Defoy, Louisa – 1960 [mf ed 1978] – (= ser Bibliographies du cours...1947-66) – 1mf – 9 – (with ind; pref by marcel panneton) – mf#SEM105P4 – cn Bibl Nat [650]

Bibliographie analytique de paul legendre : realisateur a radio-canada, membre de la "societe des ecrivains canadiens" / Roland, Jacqueline – 1953 [mf ed 1978] – (= ser Bibliographies du cours...1947-66) – 1mf – 9 – (with ind; pref by charles de koninck) – mf#SEM105P4 – cn Bibl Nat [380]

Bibliographie analytique de reine malouin : membre de la societe des ecrivains canadiens, directrice de la societe des poetes du canada et societaire de l'academie de la ballade francaise / Levasseur, Carmelle – 1950 [mf ed 1978] – (= ser Bibliographies du cours...1947-66) – 1mf – 9 – (with ind; pref by gerard martin) – mf#SEM105P4 – cn Bibl Nat [440]

Bibliographie analytique de renee des ormes / Talbot, Bernadette – 1947 [mf ed 1978] – (= ser Bibliographies du cours...1947-66) – 1mf – 9 – (with ind; pref by victorin germain) – mf#SEM105P4 – cn Bibl Nat [010]

Bibliographie analytique de simone bussieres / Gagnon, Francoise – 1963 [mf ed 1979] – (= ser Bibliographies du cours...1947-66) – 1mf – 9 – (with ind) – mf#SEM105P4 – cn Bibl Nat [010]

Bibliographie analytique de son excellence monseigneur jean-marie fortier : eveque de gaspe / Madeleine-de-Galilee, soeur – 1965 [mf ed 1979] – (= ser Bibliographies du cours...1947-66) – 3mf – 9 – (with ind; pref by marcel-jacques drouin) – mf#SEM105P4 – cn Bibl Nat [241]

Bibliographie analytique des ecrits canadiens sur l'oeuvre et la personnalite de cornelius krieghoff / Cimon, Constance – 1964 [mf ed 1979] – (= ser Bibliographies du cours...1947-66) – 1mf – 9 – (with ind; pref by marius barbeau) – mf#SEM105P4 – cn Bibl Nat [750]

Bibliographie analytique des ecrits de jean hubert : editorialiste a l'action / Labrie, Jean-Marc – 1964 [mf ed 1979] – (= ser Bibliographies du cours...1947-66) – 4mf – 9 – (with ind; pref by alfred rouleau) – mf#SEM105P4 – cn Bibl Nat [070]

Bibliographie analytique des ecrits publies au canada francais de 1930 a 1960 : sur la peinture religieuse / Marie de Sainte-Jeanne-de-Domremy, soeur – 1964 [mf ed 1979] – (= ser Bibliographies du cours...1947-66) – 3mf – 9 – (with ind; pref by raymonde gravel) – mf#SEM105P4 – cn Bibl Nat [750]

Bibliographie analytique des ecrits publies par les soeurs de la charite de quebec, 1942-1961 / Marguerite-des-Varennes, soeur – 1964 [mf ed 1979] – (= ser Bibliographies du cours...1947-66) – 2mf – 9 – (with ind; pref by mere marie-de-graces) – mf#SEM105P4 – cn Bibl Nat [241]

Bibliographie analytique des etudes pedologiques des sols des comtes dans la province de quebec / Bernier-Lesieur, Raymond – 1956 [mf ed 1978] – (= ser Bibliographies du cours...1947-66) – 3mf – 9 – (with ind; pref by aut) – mf#SEM105P4 – cn Bibl Nat [550]

Bibliographie analytique des eveques et de quelques peres eudistes au canada / Babin, Basile Joseph – 1949 [mf ed 1979] – (= ser Bibliographies du cours...1947-66) – 2mf – 9 – (with ind; pref by r bernier) – mf#SEM105P4 – cn Bibl Nat [241]

Bibliographie analytique des iles-de-la-madeleine / Harvey, Leonise – 1964 [mf ed 1979] – (= ser Bibliographies du cours...1947-66) – 3mf – 9 – (with ind; pref by ovide hubert) – mf#SEM105P4 – cn Bibl Nat [917]

Bibliographie analytique des notices biographiques et des ecrits des soeurs de la charite de quebec decedees, 1851-1917 / Sainte-Heliene-de-Marie, soeur – 1963 [mf ed 1979] – (= ser Bibliographies du cours...1947-66) – 2mf – 9 – (with ind; pref by edgar larochelle) – mf#SEM105P4 – cn Bibl Nat [241]

Bibliographie analytique des notices biographiques et des ecrits des soeurs de la charite de quebec decedees, 1918-1938 / Sainte-Mariette, soeur – 1963 [mf ed 1979] – (= ser Bibliographies du cours...1947-66) – 2mf – 9 – (with ind; pref by emile turgeon) – mf#SEM105P4 – cn Bibl Nat [241]

Bibliographie analytique des notices biographiques et des ecrits des soeurs de la charite de quebec decedees, 1939-1959 / Sainte-Therese-de-l'Enfant-Jesus, soeur – 1964 [mf ed 1979] – (= ser Bibliographies du cours...1947-66) – 3mf – 9 – (with ind; pref by joseph gingras) – mf#SEM105P4 – cn Bibl Nat [241]

Bibliographie analytique des oeuvres de m jean-marie laurence : directeur linguistique et chef des services d'annonceurs des reseaux francais de radio-canada / Clarence, frere – 1964 [mf ed 1979] – (= ser Bibliographies du cours...1947-66) – 2mf – 9 – (with ind; pref by frere marcel) – mf#SEM105P4 – cn Bibl Nat [440]

Bibliographie analytique des ouvrages de langue francaise sur l'histoire de la ville de quebec au 19e siecle / Cote, Athanase – 1964 [mf ed 1979] – (= ser Bibliographies du cours...1947-66) – 2mf – 9 – (pref by robert sylvain) – mf#SEM105P4 – cn Bibl Nat [971]

Bibliographie analytique des ouvrages edites par les presses universitaires laval, 1950 a 1957 / Pare, Richard – 1959 [mf ed 1978] – (= ser Bibliographies du cours...1947-66) – 1mf – 9 – (with ind; pref by marcel hudon) – mf#SEM105P4 – cn Bibl Nat [070]

Bibliographie analytique des recueils biographiques canadiens : francais et anglais / Rainville, Lucie – 1964 [mf ed 1979] – (= ser Bibliographies du cours...1947-66) – 2mf – 9 – (incl english text; with ind; pref by emilia b allaire) – mf#SEM105P4 – cn Bibl Nat [920]

Bibliographie analytique des travaux de joseph risi... / Lacroix, Celine – 1959 [mf ed 1978] – (= ser Bibliographies du cours...1947-66) – 1mf – 9 – (with ind; pref by lucien montreuil) – mf#SEM105P4 – cn Bibl Nat [010]

Bibliographie analytique des travaux de paul-antoine giguere : directeur du departement de chimie, faculte des sciences, universite laval / Tanguay, Marthe – 1959 [mf ed 1978] – (= ser Bibliographies du cours...1947-66) – 1mf – 9 – (pref by henri demers) – mf#SEM105P4 – cn Bibl Nat [540]

Bibliographie analytique des travaux de paul-edouard gagnon / Fortin, Isabelle – 1958 [mf ed 1978] – 1mf – 9 – (pref by j-b parent) – mf#SEM105P4 – cn Bibl Nat [010]

Bibliographie analytique du basson / Laberge, Gaetan – 1954 [mf ed 1978] – (= ser Bibliographies du cours...1947-66) – 1mf – 9 – (with ind; pref by omer letourneau) – mf#SEM105P4 – cn Bibl Nat [780]

Bibliographie analytique du docteur emile gaumond chef du service de dermato-syphiligraphie, hotel-dieu, quebec : professeur titulaire de dermatologie et syphiligraphie theorique, universite laval / Audet, M-R – 1953 [mf ed 1978] – (= ser Bibliographies du cours...1947-66) – 1mf – 9 – (with ind; pref by sylvio leblond) – mf#SEM105P4 – cn Bibl Nat [616]

Bibliographie analytique du docteur louis-georges godin (1897-1932) / Giroux, Pauline – 1964 [mf ed 1979] – (= ser Bibliographies du cours...1947-66) – 2mf – 9 – (with ind; pref by albert tessier) – mf#SEM105P4 – cn Bibl Nat [610]

Bibliographie analytique du major louis-alexandre plante / Plante, Therese – 1953 [mf ed 1979] – (= ser Bibliographies du cours...1947-66) – 1mf – 9 – (with ind; alphonse desilets) – mf#SEM105P4 – cn Bibl Nat [355]

Bibliographie analytique du reverend pere philippe deschamps : professeur agrege a la faculte de pedagogie et d'orientation de l'universite laval / Charlotin, Marie-Joseph – 1950 [mf ed 1978] – (= ser Bibliographies du cours...1947-66) – 1mf – 9 – (with ind) – mf#SEM105P4 – cn Bibl Nat [378]

Bibliographie analytique du rhumatisme articulaire aigu d'apres la documentation de la bibliotheque medicale de l'hopital du st sacrement et couvrant la periode de 1948-1952 / Baillargeon, Cecile – 1954 [mf ed 1978] – (= ser Bibliographies du cours...1947-66) – 1mf – 9 – (with ind; pref by antonio martel) – mf#SEM105P4 – cn Bibl Nat [616]

Bibliographie analytique du sujet pedagogique : l'ecole active en general a l'ecole primaire elementaire / Roy, Marguerite – 1960 [mf ed 1978] – (= ser Bibliographies du cours...1947-66) – 1mf – 9 – (with ind) – mf#SEM105P4 – cn Bibl Nat [370]

Bibliographie analytique partielle de la cote-nord / Coulombe, Marie-Anne – 1950 [mf ed 1978] – (= ser Bibliographies du cours...1947-66) – 2mf – 9 – (with ind; pref by damase potvin) – mf#SEM105P4 – cn Bibl Nat [917]

Bibliographie analytique "pour mieux servir", de 1956 a 1960 / Monique-Madeleine, soeur – 1962 [mf ed 1979] – (= ser Bibliographies du cours...1947-66) – 1mf – 9 – (with ind) – mf#SEM105P4 – cn Bibl Nat [010]

Bibliographie analytique, precedee d'une biographie, de monseigneur alphonse-marie parent... : vice-recteur de l'universite laval / Marie-de-Saint-Denis l'Aeropagite, soeur – 1964 [mf ed 1979] – (= ser Bibliographies du cours...1947-66) – 2mf – 9 – (with ind; pref by jean-baptiste parent) – mf#SEM105P4 – cn Bibl Nat [378]

Bibliographie analytique precedee d'une biographie, de monsieur l'abbe alexandre paradis... : aumonier au foyer villa-maria, saint-alexandre, kamouraska / Sainte-Rachel-Therese, soeur – 1964 [mf ed 1979] – (= ser Bibliographies du cours...1947-66) – 3mf – 9 – (with ind; pref by edgar larochelle) – mf#SEM105P4 – cn Bibl Nat [241]

Bibliographie analytique, precedee d'une biographie, du henri-marie guindon... : docteur en theologie de la societe canadienne d'etudes mariales / Saint-Antoine-de-Padoue, soeur – 1962 [mf ed 1978] – (= ser Bibliographies du cours...1947-66) – 2mf – 9 – (with ind; pref by herve gagne) – mf#SEM105P4 – cn Bibl Nat [241]

Bibliographie analytique, precedee d'une biographie, du rev pere florian lariviere : recteur du college des jesuites, quebec, 1943-1962 / Sainte-Agilberte, Soeur – 1963 [mf ed 1979] – (= ser Bibliographies du cours...1947-66) – 1mf – 9 – (with ind; pref by maurice ruest) – mf#SEM105P4 – cn Bibl Nat [378]

Bibliographie analytique precedee d'une biographie du rev pere jean-paul dallaire : college des jesuites, quebec / Marie-de-Sion, soeur – 1960 [mf ed 1978] – (= ser Bibliographies du cours...1947-66) – 2mf – 9 – mf#SEM105P4 – cn Bibl Nat [241]

Bibliographie analytique, precedee d'une biographie, du reverend pere francis goyer : congregation des peres du tres saint-sacrement, quebec, 1948-1962 / Saint-Donat-Joseph, soeur – 1963 [mf ed 1979] – (= ser Bibliographies du cours...1947-66) – 1mf – 9 – (with ind; pref by marcel langlois) – mf#SEM105P4 – cn Bibl Nat [241]

Bibliographie analytique sur la methodologie de l'histoire du canada (1950-1962) / Gabriel-de-l'Annonciation, soeur – 1963 [mf ed 1978] – (= ser Bibliographies du cours...1947-66) – 1mf – 9 – (with ind; pref by soeur marie-gemma) – mf#SEM105P4 – cn Bibl Nat [971]

Bibliographie analytique sur l'artisanat canadien / Saint-Achillas, soeur – 1962 [mf ed 1978] – (= ser Bibliographies du cours...1947-66) – 2mf – 9 – (with ind; pref by albert tessier) – mf#SEM105P4 – cn Bibl Nat [740]

Bibliographie analytique sur le forum catholique de montreal (catholic inquiry forum) 1952-1962 / Florence-du-Sacre-Coeur, soeur – 1963 [mf ed 1979] – (= ser Bibliographies du cours...1947-66) – 2mf – 9 – (with ind; pref by irenee beaubien) – mf#SEM105P4 – cn Bibl Nat [241]

Bibliographie analytique sur l'hygiene mentale preventive des jeunes de la province de quebec, 1925-1955 / Marie-Patricia, soeur – 1963 [mf ed 1979] – (= ser Bibliographies du cours...1947-66) – 1mf – 9 – (with ind) – mf#SEM105P4 – cn Bibl Nat [616]

Bibliographie annotee d'ouvrages genealogiques a la bibliotheque du parlement : indiquant d'autres bibliotheques canadiennes possedant les memes ouvrages = Annotated bibliography of genealogical works in the library of parliament (with location in other libraries in Canada) / Mennie-de Varennes, Kathleen – Ottawa: Bibliotheque du Parlement, 1963 [mf ed 1976] – 1r – 5 – mf#SEM16P270 – cn Bibl Nat [929]

Bibliographie antonienne : ou, nomenclature des ouvrages livres, revues, brochures, feuilles, etc, sur la devotion a s antoine de padoue, publies dans la province de quebec de 1777 a 1909 / Hugolin, pere – [Quebec?: s.n.] 1910 [mf ed 1995] – 1mf – 9 – 0-665-74642-3 – mf#74642 – cn Canadiana [012]

Bibliographie canadienne de l'accreditation des hopitaux, 1955-1962 / Sainte-Anne-du-Sauveur, soeur – 1962 [mf ed 1978] – (= ser Bibliographies du cours...1947-66) – 2mf – 9 – (with ind; pref by jean morisset) – mf#SEM105P4 – cn Bibl Nat [360]

Bibliographie canadienne des archives medicales d'un hopital, 1944-1950 / Ste-Antonie, soeur – 1960 [mf ed 1978] – (= ser Bibliographies du cours...1947-66) – 1mf – 9 – (with ind; pref by mathieu samson) – mf#SEM105P4 – cn Bibl Nat [610]

Bibliographie canadienne des archives medicales d'un hopital, 1951-1956 / Saint-Pierre, Jeanne-M – 1961 [mf ed 1978] – (= ser Bibliographies du cours...1947-66) – 2mf – 9 – (with ind; pref by jean caron) – mf#SEM105P4 – cn Bibl Nat [610]

Bibliographie collective des auteurs de la region du saguenay / St-Pierre, Jacqueline – 1964 [mf ed 1979] – (= ser Bibliographies du cours...1947-66) – 2mf – 9 – (with ind) – mf#SEM105P4 – cn Bibl Nat [440]

Bibliographie de certains maristes (peres, eveques et missionnaires) americains et neo-zelandais / Bisson, Fernand – 1953 [mf ed 1978] – (= ser Bibliographies du cours...1947-66) – 3mf – 9 – mf#SEM105P4 – cn Bibl Nat [241]

Bibliographie de joseph belleau / Vallee, Camille – 1951 [mf ed 1979] – (= ser Bibliographies du cours...1947-66) – 1mf – 9 – (with ind; pref by jean-c falardeau) – mf#SEM105P4 – cn Bibl Nat [010]

Bibliographie de joseph-edmond roy / Verret, Madeleine – 1947 [mf ed 1979] – (= ser Bibliographies du cours...1947-66) – 1mf – 9 – (with ind; pref by gerard martin) – mf#SEM105P4 – cn Bibl Nat [010]

Bibliographie de la croisade eucharistique : mouvement d'"action catholique des enfants" (pie 12) / Helene de la Presentation, soeur – 1964 [mf ed 1979] – (= ser Bibliographies du cours...1947-66) – 2mf – 9 – (with ind; pref by blondin dube) – mf#SEM105P4 – cn Bibl Nat [241]

Bibliographie de la litterature francaise de 1800 a 1930 / Thieme, Hugo P – 3v. 1933 – 1,9 – us AMS Press [018]

Bibliographie de la peinture au canada / Charbonneau, Jeannine – 1952 [mf ed 1978] – (= ser Bibliographies du cours...1947-66) – 1mf – 9 – (with ind) – mf#SEM105P4 – cn Bibl Nat [750]

Bibliographie de la poesie canadienne-francaise 1935-1958 / Fortier, Suzanne – 1960 [mf ed 1978] – (= ser Bibliographies du cours...1947-66) – 1mf – 9 – (with ind) – mf#SEM105P4 – cn Bibl Nat [018]

Bibliographie de la psychologie rationnelle au canada francais, 1945-1963 / Germaine-Marie, soeur – 1964 [mf ed 1979] – (= ser Bibliographies du cours...1947-66) – 1mf – 9 – (with ind; pref by mere marie-lucienne) – mf#SEM105P4 – cn Bibl Nat [150]

Bibliographie de l'afrique equatoriale francaise / Bruel, Georges – Paris: E Larose, 1914 – 1 – us CRL [010]

Bibliographie de l'hotel-dieu d'alma / Marie-des-Sept-Douleurs, soeur – 1960 [i.e. 1961] (mf ed 1978) – (= ser Bibliographies du cours...1947-66) – 1mf – 9 – (with ind; pref by victor tremblay) – mf#SEM105P4 – cn Bibl Nat [360]

Bibliographie de l'hotel-dieu st-vallier de chicoutimi, 1879-1889 / Marie de la Grace, soeur – 1962 [mf ed 1978] – (= ser Bibliographies du cours...1947-66) – 1mf – 9 – (with ind; pref by pere gerard plourde) – mf#SEM105P4 – cn Bibl Nat [360]

Bibliographie de l'ile d'orleans / Leclerc, Marcel – 1950 [mf ed 1978] – (= ser Bibliographies du cours...1947-66) – 1mf – 9 – (with ind) – mf#SEM105P4 – cn Bibl Nat [917]

Bibliographie de l'oeuvre de georges duhamel de l'academie francaise : president de l'alliance francaise universelle, membre de l'academie de medecine et membre de l'academie des sciences morales / Lizotte, Marguerite – 1947 [mf ed 1978] – (= ser Bibliographies du cours...1947-66) – 1mf – 9 – (with ind; pref by jean delage) – mf#SEM105P4 – cn Bibl Nat [378]

Bibliographie de l'oeuvre de gerard tremblay : s-min, ministere du travail, gouv prov, ancien professeur a l'universite laval / Bourgoing, Andre – 1963 [mf ed 1979] – (= ser Bibliographies du cours...1947-66) – 1mf – 9 – (with ind) – mf#SEM105P4 – cn Bibl Nat [378]

Bibliographie de l'oeuvre de jeanne d'aigle (daigle) : membre de la societe des ecrivains canadiens, membre de la societe des ecrivains pour la jeunesse... / Marie Aimee-des-Anges, soeur – 1961 [mf ed 1978] – (= ser Bibliographies du cours...1947-66) – 2mf – 9 – (with ind; pref by madame reine malouin) – mf#SEM105P4 – cn Bibl Nat [400]

Bibliographie de l'oeuvre de la venerable anne-marie rivier : une femme-apotre, 1768-1838 / Marie Saint-Jean-Eudes, soeur – 1964 [mf ed 1979] – (= ser Bibliographies du cours...1947-66) – 1mf – 9 – (with ind) – mf#SEM105P4 – cn Bibl Nat [920]

Bibliographie de l'oeuvre de l'abbe anselme longpre : licencie en philosophie, licencie en theologie, bachelier en droit canon, 1947-1964 (deuxieme partie) / Sainte-Angele, soeur – 1964 [mf ed 1979] – (= ser Bibliographies du cours...1947-66) – 2mf – 9 – (with ind) – mf#SEM105P4 – cn Bibl Nat [920]

Bibliographie de l'oeuvre de louis berube / Levesque, Ginette – 1962 [i.e. 1963] (mf ed 1978) – (= ser Bibliographies du cours...1947-66) – 1mf – 9 – (with ind; pref by raymond boucher) – mf#SEM105P4 – cn Bibl Nat [010]

Bibliographie de l'oeuvre de louis-philippe audet / Chandonnet, Gemma – 1954 [mf ed 1978] – (= ser Bibliographies du cours...1947-66) – 2mf – 9 – (with ind) – mf#SEM105P4 – cn Bibl Nat [010]

Bibliographie de l'oeuvre de louis-philippe audet : l sc de la societe des ecrivains canadiens, professeur a l'universite laval / Cossette, Angele – 1948 [mf ed 1979] – (= ser Bibliographies du cours...1947-66) – 1mf – 9 – (with ind; pref by georges maheux) – mf#SEM105P4 – cn Bibl Nat [400]

Bibliographie de l'oeuvre de m jean-baptiste caouette : poete canadien / Legare, Helene – 1964 [mf ed 1979] – (= ser Bibliographies du cours...1947-66) – 1mf – 9 – (with ind; pref by mario bussanga) – mf#SEM105P4 – cn Bibl Nat [440]

Bibliographie de l'oeuvre de maurice lebel : secretaire a la faculte des lettres et professeur de litterature grecque a l'universite laval / Pare, Rosario – 1947 [mf ed 1978] – (= ser Bibliographies du cours...1947-66) – 1mf – 9 – (with ind; pref by j-m blanchet) – mf#SEM105P4 – cn Bibl Nat [450]

Bibliographie de l'oeuvre de monsieur gerard morisset : membre de la societe royale du canada, 1950-1962 / Garneau, Marthe – 1947 [mf ed 1979] – (= ser Bibliographies du cours...1947-66) – 1mf – 9 – (wit ind; pref by paul-e plamondon) – mf#SEM105P4 – cn Bibl Nat [010]

Bibliographie de l'oeuvre de monsieur l'abbe adrien bouffard : ptre, secretaire national de l'union pontificale missionnaire du clerge... / Pierre-de-la-Croix, soeur – 1962 [mf ed 1979] – (= ser Bibliographies du cours...1947-66) – 2mf – 9 – (with ind) – mf#SEM105P4 – cn Bibl Nat [241]

Bibliographie de l'oeuvre de monsieur l'abbe andre jobin : principal de l'ecole normale notre-dame, st-roch, quebec... / Celine-de-la-Presentation, Soeur – 1964 [mf ed 1979] – (= ser Bibliographies du cours...1947-66) – 2mf – 9 – (with ind; pref by j c racine) – mf#SEM105P4 – cn Bibl Nat [241]

Bibliographie de l'oeuvre de monsieur rolland dumais : conseiller pedagogique en sciences naturelles [a] la commission des ecoles catholiques de quebec / Garneau, Robert & Matte, Paul-Henri – 1964 [mf ed 1979] – (= ser Bibliographies du cours...1947-66) – 2mf – 9 – (with ind) – mf#SEM105P4 – cn Bibl Nat [500]

Bibliographie de l'oeuvre de monsieur rolland dumais : conseiller pedagogique en sciences naturelles a la commission des ecoles catholiques de quebec, directeur-fondateur du camp marie-victorin / Morin, Jacques & Matte, Paul-Henri – 1964 [mf ed 1979] – (= ser Bibliographies du cours...1947-66) – 2mf – 9 – (with ind) – mf#SEM105P4 – cn Bibl Nat [500]

Bibliographie de l'oeuvre de sa grandeur monseigneur paul-eugene roy, 1859-1926 : dix-huitieme eveque et huitieme archeveque de quebec / Marie-du-Perpetuel-Secours, soeur – 1964 [mf ed 1979] – (= ser Bibliographies du cours...1947-66) – 3mf – 9 – (with ind; pref by arthur maheux) – mf#SEM105P4 – cn Bibl Nat [241]

Bibliographie de l'oeuvre du docteur albert jobin, 1867 a 1952 / Raymond-Marie, soeur – 1964 [mf ed 1979] – (= ser Bibliographies du cours...1947-66) – 2mf – 9 – (with ind; pref by louis-philippe roy) – mf#SEM105P4 – cn Bibl Nat [610]

Bibliographie de l'oeuvre du reverend pere gaston carriere : professeur de philosophie a l'universite d'ottawa et historien de la congregation / Louis-Bernard, soeur – 1963 [mf ed 1979] – (= ser Bibliographies du cours...1947-66) – 2mf – 9 – (with ind) – mf#SEM105P4 – cn Bibl Nat [100]

Bibliographie de l'oeuvre du reverend pere hector-l bertrand : fondateur du comite des hopitaux du quebec... / Deschenes, Jean-Claude – 1963 [mf ed 1979] – (= ser Bibliographies du cours...1947-66) – 2mf – 9 – (with ind) – mf#SEM105P4 – cn Bibl Nat [360]

Bibliographie de l'oeuvre musicale du frere barnabe : docteur en musique, membre de la societe des compositeurs canadiens / Traversy, Paul – 1963 [mf ed 1979] – (= ser Bibliographies du cours...1947-66) – 1mf – 9 – (with ind; pref by frere charles) – mf#SEM105P4 – cn Bibl Nat [780]

Bibliographie de lotbiniere / Couture, Suzanne – 1959 [mf ed 1978] – (= ser Bibliographies du cours...1947-66) – 1mf – 9 – (with ind) – mf#SEM105P4 – cn Bibl Nat [010]

Bibliographie de m jean-paul gelinas / Marie Simeon, soeur – 1962 [mf ed 1978] – (= ser Bibliographies du cours...1947-66) – 1mf – 9 – (with ind; pref by raymond lecours) – mf#SEM105P4 – cn Bibl Nat [241]

Bibliographie de madagascar / Grandidier, Guillaume – Paris: Comite de Madagascar, 1906 [i.e. 1905]-57 – 1 – us CRL [960]

Bibliographie de madame ella charland-ostiguy / Ostiguy, Flore-Ella – 1950 [mf ed 1979] – (= ser Bibliographies du cours...1947-66) – 1mf – 9 – (with ind; pref by mere gabriel-marie) – mf#SEM105P4 – cn Bibl Nat [010]

Bibliographie de madame emma boivin-vaillancourt / Lockwell, Huguette – [1954?] (mf ed 1978) – (= ser Bibliographies du cours...1947-66) – 1mf – 9 – (with ind; pref by marie-elzear) – mf#SEM105P4 – cn Bibl Nat [010]

Bibliographie de madame gabrielle roy / Gauthier, Georges – 1960 [mf ed 1978] – (= ser Bibliographies du cours...1947-66) – 2mf – 9 – (with ind; pref by therese m miller) – mf#SEM105P4 – cn Bibl Nat [010]

Bibliographie de mere isabelle sormany : dite ladauversiere, religieuse hospitaliere de saint-joseph, premiere superieure generale 1897-1957 / Albert, Maria – 1962 [mf ed 1979] – (= ser Bibliographies du cours...1947-66) – 3mf – 9 – (with ind; pref by livain chiasson) – mf#SEM105P4 – cn Bibl Nat [010]

Bibliographie de mme marthe lemaire-duguay / Dery, Marie-Claire – 1962 [mf ed 1978] – (= ser Bibliographies du cours...1947-66) – 2mf – 9 – (with ind; pref by alphonse roux) – mf#SEM105P4 – cn Bibl Nat [010]

Bibliographie de monsieur l'abbe emile begin : professeur a l'universite laval, ouvrage imprime et articles de revues [1933-1963] / Marie-de-la-Sainte-Enfance, soeur – 1963 [mf ed 1979] – (= ser Bibliographies du cours...1947-66) – 2mf – 9 – (with ind; pref by pierre-paul turgeon) – mf#SEM105P4 – cn Bibl Nat [378]

Bibliographie de monsieur l'abbe henri grenier : docteur en philosophie, docteur en theologie, docteur en droit canon / Marie de St-Didier, soeur – 1948 [mf ed 1978] – (= ser Bibliographies du cours...1947-66) – 1mf – 9 – (with ind; pref by paul-emile gosselin) – mf#SEM105P4 – cn Bibl Nat [241]

Bibliographie de monsieur l'abbe honorius provost : du seminaire de quebec, archiviste a l'universite laval / Labbe, Edith – 1962 [mf ed 1978] – (= ser Bibliographies du cours...1947-66) – 1mf – 9 – (with ind; pref by paul emile gosselin) – mf#SEM105P4 – cn Bibl Nat [241]

Bibliographie de monsieur l'abbe j w laverdiere : geologue-pretre / Chaperon, Elisee – 1957 [mf ed 1978] – (= ser Bibliographies du cours...1947-66) – 2mf – 9 – (with ind) – mf#SEM105P4 – cn Bibl Nat [550]

Bibliographie de monsieur l'abbe j-alfred tremblay / Marie-de-la-Croix, soeur – 1961 [mf ed 1978] – (= ser Bibliographies du cours...1947-66) – 1mf – 9 – (with ind; pref by raymond desgagne) – mf#SEM105P4 – cn Bibl Nat [241]

Bibliographie de monsieur l'abbe paul-emile gosselin : professeur de philosophie au seminaire de quebec, secretaire general du comite de la survivance francaise en amerique / Marie-des-Anges, soeur – 1948 [mf ed 1979] – (= ser Bibliographies du cours...1947-66) – 1mf – 9 – (with ind; pref by honorius provost) – mf#SEM105P4 – cn Bibl Nat [190]

Bibliographie de monsieur le cure l boisseau : cure de new carlisle / Alberta-Marie, soeur – 1961 [mf ed 1978] – (= ser Bibliographies du cours...1947-66) – 1mf – 9 – (with ind; pref by j b carignan; ill by soeur alina-marie) – mf#SEM105P4 – cn Bibl Nat [241]

Bibliographie de monsieur richard joly : debut a 1954 / Larrivee, Micheline – 1964 [mf ed 1979] – (= ser Bibliographies du cours...1947-66) – 1mf – 9 – (with ind; pref by eddy slater) – mf#SEM105P4 – cn Bibl Nat [010]

Bibliographie de monsieur richard joly, 1955 a 1963 / Larrivee, Pierrette – 1964 [mf ed 1979] – (= ser Bibliographies du cours...1947-66) – 1mf – 9 – (with ind; pref by roch duval) – mf#SEM105P4 – cn Bibl Nat [010]

Bibliographie de nos grands prix de la province, 1944-1954 / Saint-Hilaire, Alphonse – 1955 [mf ed 1978] – (= ser Bibliographies du cours...1947-66) – 3mf – 9 – (with ind; pref by jean-baptiste soucy) – mf#SEM105P4 – cn Bibl Nat [700]

Bibliographie de paul-andre lamontagne / Guay, Marcel – 1956 [mf ed 1978] – (= ser Bibliographies du cours...1947-66) – 1mf – 9 – (with ind; pref by paul-andre lamontagne) – mf#SEM105P4 – cn Bibl Nat [010]

Bibliographie de soeur marie de saint-paul-de-la-croix, (georgianna juneau) (1873-1940) / Marie de Saint-Joseph-du-Redempteur, soeur – 1962 [mf ed 1978] – (= ser Bibliographies du cours...1947-66) – 2mf – 9 – (with ind; pref by victorin germain) – mf#SEM105P4 – cn Bibl Nat [241]

Bibliographie de stanislas vachon : ses oeuvres, ses ecrits / Legare, Denise – 1955 [mf ed 1978] – (= ser Bibliographies du cours...1947-66) – 1mf – 9 – (with ind; pref by jacques legare) – mf#SEM105P4 – cn Bibl Nat [010]

Bibliographie der arbeiten von prof dr helmut breuer und dr maria weuffen : sowie anderer mitglieder der forschungsgruppe "prophylaktische diagnostik" an der ernst-moritz-arndt universitaet greifswald / Amse, Corina – (mf ed 1995) – 1mf – 9 – €30.00 – 3-8267-2156-X – mf#DHS 2156 – gw Frankfurter [378]

Bibliographie der deutschen universitaeten : systematisch geordnetes verzeichnis der bis ende 1899 gedruckten buecher und aufsaetze ueber das deutsche universitaetswesen. im auftrag des preussischen unterrichts-ministeriums bearbeitet. pt1 / ed by Erman, Wilhelm & Horn, Ewald – Leipzig, Berlin 1904 (mf ed 1993) – 9mf – 9 – €110.00 – 3-89349-223-2 – mf#DHS-AR 136 – gw Frankfurter [378]

Bibliographie der deutschen universitaeten : systematisch geordnetes verzeichnis der bis ende 1899 gedruckten buecher und aufsaetze ueber das deutsche universitaetswesen. im auftrag des preussischen unterrichts-ministeriums bearbeitet. pt2 / ed by Erman, Wilhelm & Horn, Ewald – Leipzig, Berlin, 1904 (mf ed 1993) – 13mf – 9 – €140.00 – 3-89349-323-9 – mf#DHS-AR 137 – gw Frankfurter [378]

Bibliographie der deutschen universitaeten : systematisch geordnetes verzeichnis der bis ende 1899 gedruckten buecher und aufsaetze ueber das deutsche universitaetswesen. im auftrag des preussischen unterrichts-ministeriums bearbeitet. pt3 / ed by Erman, Wilhelm & Horn, Ewald – Leipzig, Berlin, 1905 (mf ed 1993) – 2mf – 9 – €50.00 – 3-89349-324-7 – (with ind) – mf#DHS-AR 142 – gw Frankfurter [378]

Bibliographie der sozialwissenschaften – v1-39. 1905-43 – 9 – $804.00 – (in german) – mf#0101 – us Brook [013]

Bibliographie der stoff- und motivgeschichte der deutschen literatur / Bauerhorst, Kurt – Berlin: W de Gruyter & Co, 1932 [mf ed 1993] – (= ser Stoff- und motivgeschichte der deutschen literatur) – xvi/118p – 1 – mf#7840 – us Wisconsin U Libr [430]

Bibliographie des biographies des religieuses augustines hospitalieres de la misericorde de jesus : decedees au monastere de l'hotel-dieu de levis, 1892-1962 / Marie-de-la-Portectin, soeur – 1962 [mf ed 1978] – (= ser Bibliographies du cours...1947-66) – 3mf – 9 – (with ind; pref by joseph lehoux) – mf#SEM105P4 – cn Bibl Nat [360]

Bibliographie des biographies des religieuses decedees a l'hotel-dieu du sacre-coeur de jesus de quebec : 1825-1960 / Catherine-de-Saint-Augustin, soeur – 1961 [mf ed 1978] – (= ser Bibliographies du cours...1947-66) – 1mf – 9 – (with ind; pref by mere saint-zephirin) – mf#SEM105P4 – cn Bibl Nat [360]

Bibliographie des biographies des religieuses decedees a l'hotel-dieu du sacre-coeur de jesus de quebec, 1879-1925 / Sainte-Monique, soeur – 1961 [mf ed 1978] – (= ser Bibliographies du cours...1947-66) – 1mf – 9 – (with ind; pref by marie-de-l'eucharistie) – mf#SEM105P4 – cn Bibl Nat [360]

Bibliographie des biographies des religieuses decedees a l'hotel-dieu saint-vallier de chicoutimi (1884-1959) / Sainte Marie-Madeleine, soeur – 1961 [mf ed 1978] – (= ser Bibliographies du cours...1947-66) – 1mf – 9 – (with ind; pref by joseph lalancette) – mf#SEM105P4 – cn Bibl Nat [360]

Bibliographie des ecrits de freud : en francais, allemand et anglais / Dufresne, Roger – Paris: Payot, 1973 [mf ed 1998] – (= ser Coll science de l'homme) – 4mf – 9 – (freud, sigmund) – mf#SEM105P2865 – cn Bibl Nat [150]

Bibliographie des ecrits du tres rev pere pascal d'ottawa, ex-prov de l'ordre des freres mineurs capucins, ex-prof de patrologie et d'histoire sainte / Bourdages, Jeanne – 1962 [mf ed 1978] – (= ser Bibliographies du cours...1947-66) – 2mf – 9 – (pref by edgar godin) – mf#SEM105P4 – cn Bibl Nat [920]

Bibliographie des journaux de quebec : 1889-1940 / Pettigrew, Renee – 1952 [mf ed 1979] – (= ser Bibliographies du cours...1947-66) – 1mf – 9 – (with ind; pref by bruno lafleur) – mf#SEM105P4 – cn Bibl Nat [073]

Bibliographie des memellandes / Szameitat, Max – Wuerzburg: Holzner-Verlag 1957 [mf ed 1993] – (= ser Ostdeutsche beitraege aus dem goettinger arbeitskreis 7) – 10r – 1 – (incl ind) – mf#3180p – us Wisconsin U Libr [014]

Bibliographie des oeuvres du tres reverend pere frederic janssoone de ghyvelde : apotre de palestine et de notre-dame du cap...1838-1916 / Marie Saint Salvy, soeur – 1963 [mf ed 1979] – (= ser Bibliographies du cours...1947-66) – 2mf – 9 – (pref by soeur saint olivier) – mf#SEM105P4 – cn Bibl Nat [241]

Bibliographie des oeuvres litteraires publiees au canada de robert rumilly : ecrivain, membre de l'academie canadienne-francaise, traducteur au senat canadien a ottawa / Desautels, Adrien – 1947 [mf ed 1978] – (= ser Bibliographies du cours...1947-66) – 1mf – 9 – mf#SEM105P4 – cn Bibl Nat [440]

Bibliographie des ouvrages concernant la temperance : livres, brochures, journaux... etc, imprimes a quebec et a levis depuis l'etablissement de l'imprimerie (1764) jusqu'a 1910 / Hugolin, pere – [Quebec?: s.n.] 1910 [mf ed 1995] – 2mf – 9 – 0-665-73998-2 – mf#73998 – cn Canadiana [013]

Bibliographie des ouvrages relatifs a la senegambie et au soudan occidental / Colzel, M – Paris, C Delagrave 1891 – us CRL [960]

Bibliographie des principales editions originals d'ecrivains francais du 15 au 18 siecle / LePetit, Jules – 1v. 1888 – 1,9 – us AMS Press [014]

Bibliographie des quotidiens de langue francaise parus dans la province de quebec depuis 1867 / Boucher, Louis – 1957 [mf ed 1978] – (= ser Bibliographies du cours...1947-66) – 1mf – 9 – (with ind; pref by me jean-charles bonenfant) – mf#SEM105P4 – cn Bibl Nat [440]

Bibliographie des revues et journaux litteraires des 19e et 20e siecles, vol 1 / Place, Jean-Michel & Vasseur, Andre – 1 – (l'academie francaise: revue d'art et de litterature, ed. saint-georges de bouhelier. paris, feb-mar 1893 (nos 1-2); l'assomption: essais d'art catholique, ed. saint-georges de bouhelier. paris, 10 mar-apr 1893 (nos 1-2); l'annonciation: livet de reve et d'amour, ed. saint-georges de bouhelier. paris, 10 jun-1 oct 1894 (nos 1-5); les agapes, ed. paul d'orsay. paris, 10 nov-14 dec 1840 (nos 1-2); la coupe: recueil mensuel d'art et d'ethique, ed. joseph loubet, richard wemau. montpellier, may 1895-jun 1898 (nos 1-15); le courrier social illustre: philosophie, art, science, ed. andre ibels. paris, 1 nov-31 dec 1894 (nos 1-4); le fou: journal litteraire, ed. edouard guillaumet. paris, 26 feb-4 jun 1883 (nos 1-12); les grimaces: pamphlet hebdomadaire, ed. octave mirbeau. paris, 21 jul 1883-12 jan 1884 (nos 1-26); les ibis: revue litteraire et artistique, ed. henri degron, tristan klingsor. paris, apr-dec 1894 (nos 1-4); la renaissance litteraire et artistique, ed. jean aicard, emile blemont. paris, 27 apr 1872-3 may 1874 (nos 1-99); le reveil, ed. adolphe granier de cassagnac. paris, 2 jan 1858-16 apr 1859 (nos 1-68); le saint-graal: cahiers d'art et d'esthetique, ed. emmanuel signoret. paris, etc., 25 jan 1892-feb 1899 (nos 1-20); la syrinx, ed. joachim gasquet, paul souchon. aix-en-provence, jan 1892-feb 1894 (nos 1-13); la variete: revue litteraire, ed. leconte de lisle. rennes, 1 apr 1840-1 mar 1841 (nos 1-12)) – fr ACRPP [014]

Bibliographie des revues et journaux litteraires des 19e et 20e siecles, vol 2 / Place, Jean-Michel & Vasseur, Andre – 1 – (le banquet: publication mensuelle ed fernand gregh, paris mar 1892-mar 1893 (n1-8); les cahiers occitans ed comite d'action de la ligue occitane, paris feb-aug 1899 (n1-2); les chroniques: revue litteraire et artistique ed charles le goffic, maurice barres, paris 1 nov 1886-oct; nov 1887 (n1-12); la conque ed pierre louys, paris 15 mar 1891-jan 1892 (n1-11); les contemporains: journal hebdomadaire ed alfred le petit, felicien champsaur, paris dec 1880-7 dec 1881 (n1-43); l'enclos: arts, dits et faits, pour le mieux ed louis lumet, paris apr 1894-jan, feb 1899 (n1-36; 37); les femmes du jour, paris apr 1886-1892 (n1-11); les hommes d'aujourd'hui ed cinqualbre, later vanier, paris 13 sep 1878-jan 1899 (n1-469); l'idee libre: revue mensuelle de litterature et d'art ed emile besnus, paris, apr 1892-95 (n1-36) (combines with le reve et l'idee: documents sur le naturisme, to form: documents sur le naturisme, q.v.); la reve et l'idee ed maurice le blond, paris may 1894-may 1895 (n1-5), cont as: documents sur le naturisme ed maurice le blond, paris nov 1895-11 sep 1896 (n1-11), cont as: la revue naturiste ed maurice le blond, paris mar 1897-nov 1901 (n1-35); la revue rouge de litterature et d'art ed gustave langlet, later f hache, paris 1896-apr 1898 (n1-8); les taches d'encre: gazette mensuelle ed maurice barres, paris 5 nov 1884-feb 1885 (n1-4); la treve-dieu: revue d'art et de litterature ed yves berthou, le havre 1897 (n1-12); la vie litteraire ed albert collignon, paris 28 oct 1875-26 sep 1878 (n1-153)) – fr ACRPP [014]

Bibliographie des revues et journaux litteraires des 19e et 20e siecles, vol 3 / Place, Jean-Michel & Vasseur, Andre – 1 – (accords. cahiers mensuels de litterature. dir. andre desson, andre harlaire. paris. mai-oct; nov 1924 (no. 1-3; 4) 1. action. cahiers individualistes de philosophie et d'art. dir. florent fels. paris. fevr 1920-mars; avr 1922 (no. 1-12) 1. aventure. dir. marcel arland, rene crevel, georges limbour, roger vitrac. paris. nov 1921-janv 1922 (no. 1-3) suivi de: des. dir. marcel arland. paris. avr 1922 (no. 1) 1. les cahiers idealistes francais. dir. edouard dujardin. paris. fevr 1917-juin 1920 (no. 1-37. fevr 1921-fevr 1928 (n.s. no. 1-16) 1. intentions. revue mensuelle. dir. pierre-andre may. paris. 1922-24 (no. 1-28; 30) 1. interventions. gazette internationale des lettres et des arts modernes. dir. paul dermee. paris. dec 1923-janv 1924 (no. 1-2) suivi de: le mouvement accelere. organe accelerateur de la revolution artistique et litteraire. dir. paul dermee. paris. nov 1924 (no. 1) 1. inversions. paris. 15 nov 1924-1er mars 1925 (no. 1-4) suivi de: l'amitie. paris. avr 1925 (no. 1) 1. le mouton blanc. revue mensuelle. dir. jean hytier. lyon. sept 1922-nov 1924 (no. 1-7) 1. l'oeuf dur. dir. gerard rosenthal. paris. mars 1921-ete 1924 (no. 1-16) 1. la revue europeenne. dir. edmond jaloux, valery larbaud, andre germain, philippe soupault. paris. 1er mars 1923-juil 1931 (no. 1-101) 1. sic. sons, idees, couleurs, formes. dir. pierre albert-birot. paris. 1916-19 (no. 1-54) suivi de: paris. dir. pierre albert-birot. paris. nov 1924 (no. 1) 1. surrealisme. revue mensuelle. dir. ivan goll. paris. oct 1924 (no. 1) 1) – fr ACRPP [014]

Bibliographie des revues et journaux litteraires des 19e et 20e siecles, vol 4 / Place, Jean-Michel & Vasseur, Andre – 1 – (cabaret voltaire. dir. hugo ball. zurich. juin 1916 (no. 1) 1. cannibale. dir. francis picabia. paris. 25 avr-25 mai 1920 (no. 1-2) 1. le coeur a barbe. journal transparent. dir. paul eluard, georges ribemont-dessaignes, tristan tzara. paris. avr 1922 (no. 1) 1. dada. dir. tristan tzara. zurich, paris. juil 1917-sept 1921 (no. 1-8) 1. der zeltweg. dir. otto flake, walter serner, tristan tzara. zurich. nov 1919 (no. 1) 1. litterature. dir. louis aragon, andre breton, philippe soupault. paris. mars 1919-aout 1921 (no. 1-20). 1er mars 1922-juin 1924 (n.s. no. 1-13) 1. manometre. dir. emile malespine. lyon. juil 1922-janv 1928 (no. 1-9) 1. la pomme de pins. dir. francis picabia. saint-raphael. 25 fevr 1922 (no. 1) 1. projecteur. dir. celine arnuald. paris. 21 mai 1920 (no. 1) 1. 391. dir. francis picabia. barcelone, new york, zurich, paris. 25 janv 1917-oct 1924 (no. 1-19) 1. z. dir. paul dermee. paris. mars 1920 (no. 1-2) 1) – fr ACRPP [014]

Bibliographie des revues et journaux litteraires des 19e et 20e siecles, vol 1-4 / Place, Jean-Michel & Vasseur, Andre – 1 – fr ACRPP [014]

Bibliographie d'ouvrages anciens de medecine gardes a l'hopital general (1669-1874) / Marie de l'Assomption, soeur – 1961 [mf ed 1978] – (= ser Bibliographies du cours...1947-66) – 1mf – 9 – (with ind; pref by louis-napoleon larochelle) – mf#SEM105P4 – cn Bibl Nat [610]

Bibliographie d'ouvrages ayant trait a l'afrique en general dans ses rapports avec l'exploration and la civilisation de ses contrees... / Kayser, Gabriel – Bruxelles, 1887 – 1 – us CRL [960]

Bibliographie d'ouvrages ayant trait...l'afrique en general dans ses rapports avec l'exploration, et la civilisation de ces contrees / Kayser, G – Bruxelles, 1887 – 4mf – 9 – mf#A-251 – ne IDC [916]

Bibliographie du comte de lotbiniere / Tousignant, J Laureat – 1961 [mf ed 1978] – (= ser Bibliographies du cours...1947-66) – 1mf – 9 – mf#SEM105P4 – cn Bibl Nat [010]

Bibliographie du congo, 1880-1895.. / Wauters, Alphonse Jules – Brussels. 1895 – 1 – us CRL [960]

Bibliographie du congo belge et du ruanda-urundi : regime foncier / Heyse, Theodore – Bruxelles: G van Campenhout, 1947 – 1 – us CRL [960]

La bibliographie du droit naturel dans la philosophie chretienne du commencement de l'epoque jusqu'au dix-septieme siecle / Laupacis, Benedict – 1952 [mf ed 1979] – (= ser Bibliographies du cours...1947-66) – 1mf – 9 – (with ind) – mf#SEM105P4 – cn Bibl Nat [240]

Bibliographie du quebec : index... / Bibliotheque nationale du Quebec. Ministere des affaires culturelles – Montreal: La Bibliotheque, 1968/1973- (irreg) [mf ed 1984-] – 9 – mf#SEM105P415 – cn Bibl Nat [971]

Bibliographie du quebec : liste des publications quebecoises ou relatives au quebec etablies / Bibliotheque nationale du Quebec. Ministere des affaires culturelles – Montreal: la Bibliotheque. v[1] 1968- (varies) [mf ed 1984-] – 9 – mf#SEM105P414 – cn Bibl Nat [971]

Bibliographie du r p arcade-m monette / Paradis, Andre – 1964 [mf ed 1979] – (= ser Bibliographies du cours...1947-66) – 1mf – 9 – mf#SEM105P4 – cn Bibl Nat [010]

Bibliographie du rev f eloi-gerard : mariste, genealogiste, historien, pedagogue / Savard, Leo – 1955 [mf ed 1978] – (= ser Bibliographies du cours...1947-66) – 2mf – 9 – (with ind) – mf#SEM105P4 – cn Bibl Nat [920]

Bibliographie du reverend frere m-cyrille : cote des ecoles chretiennes / Lefebvre, Gerard – 1965 [mf ed 1979] – (= ser Bibliographies du cours...1947-66) – 2mf – 9 – (with ind) – mf#SEM105P4 – cn Bibl Nat [241]

Bibliographie du reverend frere robert / Marie-Elie, frere – 1962 [mf ed 1979] – (= ser Bibliographies du cours...1947-66) – 1mf – 9 – (with ind) – mf#SEM105P4 – cn Bibl Nat [241]

Bibliographie du rocher de grand-mere (monument sis en la cite du meme nom) : presentee a la faculte des lettres de l'universite laval / Lord, Marcel A – 1964 [mf ed 1979] – (= ser Bibliographies du cours...1947-66) – 1mf – 9 – (with ind) – mf#SEM105P4 – cn Bibl Nat [917]

Bibliographie du theatre canadien-francais avant 1900 / Ouellet, Therese – 1949 [mf ed 1979] – (= ser Bibliographies du cours...1947-66) – 1mf – 9 – (with ind) – mf#SEM105P4 – cn Bibl Nat [790]

Bibliographie du theatre canadien-francais de 1900-1955 / Bilodeau, Francoise – 1956 [mf ed 1978] – (= ser Bibliographies du cours...1947-66) – 1mf – 9 – (with ind) – mf#SEM105P4 – cn Bibl Nat [790]

Bibliographie ethnographique de l'afrique equatoriale francaise, 1914-48 – Paris: Impr nationale, 1949 – 1 – us CRL [305]

Bibliographie franciscaine : inventaire des revues, livres, brochures et autres ecrits publies par les franciscains du canada de 1890 a 1915 / Hugolin, pere – [Quebec?: s.n.] 1916 [mf ed 1995] – 2mf – 9 – 0-665-74641-5 – mf#74641 – cn Canadiana [012]

Bibliographie geographique de l'egypte / Lorin, Henri – Cairo: Impr de l'Institut francais d'archeologie orientale du Caire. 2v. 1928-29 – 1 – us CRL [916]

Bibliographie necrologique des religieuses hospitalieres de saint-joseph : province de notre-dame de l'assomption, vallee-lourdes, nb / Duplessis, T – 1962 [mf ed 1979] – (= ser Bibliographies du cours...1947-66) – 2mf – 9 – (with ind; pref by albert guyot) – mf#SEM105P4 – cn Bibl Nat [360]

Bibliographie neuer erscheinungen aller laender auf dem gebiete der naturgeschichte und der exakten wissenschaften / Naturae Novitates – Berlin, 1879-1943. v1-65 – 319mf – 9 – mf#8720 – ne IDC [500]

Bibliographie pratique de la litterature grecque : des origines a la fin de la periode romaine / Masqueray, Paul – Paris: C Klincksieck, 1914 – 1mf – 9 – 0-8370-1908-7 – mf#1987-6295 – us ATLA [014]

Bibliographie raisonee des ouvrages concernant le dahomey / Pawlowski, Auguste – Paris: L Baudoin, 1895 – 1 – us CRL [010]

Bibliographie relative a l'histoire de la nouvelle-france, 1516-1700 : avec notes analytiques et critiques / Lavoie, Amedee – 1949 [mf ed 1978] – (= ser Bibliographies du cours...1947-66) – 1mf – 9 – (with ind) – mf#SEM105P4 – cn Bibl Nat [971]

Bibliographie sur la devotion au sacre-coeur chez les missionnaires du sacre-coeur francais et canadiens / Robert, Lionel – 1962 [i.e.] 1963 [mf ed 1978] – (= ser Bibliographies du cours...1947-66) – 1mf – 9 – (pref by paul-emile drouin) – mf#SEM105P4 – cn Bibl Nat [241]

Bibliographie sur la methodologie de la lecture / Marie-Paul, frere – 1947 [mf ed 1978] – (= ser Bibliographies du cours...1947-66) – 1mf – 9 – (with ind) – mf#SEM105P4 – cn Bibl Nat [370]

Bibliographie sur la methodologie du catechisme / Pierre, frere – 1948 [mf ed 1979] – (= ser Bibliographies du cours...1947-66) – 1mf – 9 – (with ind) – mf#SEM105P4 – cn Bibl Nat [241]

Bibliographie sur le centre de service social du diocese des trois-rivieres / Rouyn, Solange de – 1962 [i.e. 1963] (mf ed 1979) – (= ser Bibliographies du cours...1947-66) – 2mf – 9 – (with ind; pref by jules perron) – mf#SEM105P4 – cn Bibl Nat [360]

Bibliographie sur le cinema : essai de bibliographie des publications canadiennes-francaises sur le cinema de 1940 a 1960 / Roy, Jean-Luc – 1963 [mf ed 1978] – (= ser Bibliographies du cours...1947-66) – 2mf – 9 – (with ind) – mf#SEM105P4 – cn Bibl Nat [790]

Bibliographie sur le scoutisme catholique dans la province de quebec / Hamel, Andre – 1948 [mf ed 1979] – (= ser Bibliographies du cours...1947-66) – 1mf – 9 – (with ind) – mf#SEM105P4 – cn Bibl Nat [360]

Bibliographie sur l'ophtalmo-oto-rhino-laryngologie : travaux publies de 1940 a 1950... / Ste-Therese de Lisieux, soeur – 1950 [mf ed 1979] – (= ser Bibliographies du cours...1947-66) – 1mf – 9 – (with ind; pref by g-t gauthier) – mf#SEM105P4 – cn Bibl Nat [617]

Bibliographie wagnerienne francaise / Silege, Henri [comp] – Paris: Fischbacher 1902 – 1mf – 9 – mf#wa-104 – ne IDC [780]

Bibliographie zur juedisch-hellenistischen und intertestamentarischen literatur 1900-1965 / Delling, G – 1969 – (= ser Tugal 5-106) – 3mf – 9 – €7.00 – ne Slangenburg [230]

Bibliographie zur tristansage / Kuepper, Heinz – Jena: E Diederich, 1941 (Leipzig: Radelli und Hille) [mf ed 1993] – (= ser Deutsche arbeiten der universitaet koeln 17) – 127p – 1 – mf#8215 reel 2 – us Wisconsin U Libr [390]

Bibliographie zur volkskunde der donauschwaben / Rez, Heinrich – Muenchen: E Reinhardt 1935 [mf ed Bloomington IN: Indiana Uni Lib, Preservation Dept 1984] – 1r – 1 – us Indiana Preservation [390]

Les bibliographies du cours de bibliotheconomie de l'universite laval, 1947-1966 : [index] / Thouin, Richard [comp] – 1mf – 9 – (with ind) – mf#SEM105P4 – cn Bibl Nat [020]

Bibliographische mitteilungen ueber die rechtsstellung der frau im deutschen reich und in oesterreich (hq2) : gesetzgebung – rechtsprechung – schrifttum – 1935-41 [mf ed 1991] – (= ser Hq) – 21v on 3mf – 9 – €30.00 – 3-89131-043-9 – gw Fischer [305]

Bibliographischer vierteljahresbericht fuer die juedische literatur / ed by Lewin, Reinhold – Leipzig: M W Kaufmann. v1 n1. 1914 – (= ser German-jewish periodicals...1768-1945, pt 3) – 1r – 1 – $165.00 – mf#B32 – us UPA [470]

Bibliographischer zugang zur griechischen philosophie : zugangsbibliographie/philosophie 1 / Aul, Joachim – (mf ed 1997) – 5mf – 9 – €59.00 – 3-8267-2504-2 – mf#DHS 2504 – gw Frankfurter [180]

Bibliography – s.l, s.l? . 193-? – 1r – us UF Libraries [978]

Bibliography and index to the works of theodore parker / ed by Wendte, Charles William – centenary ed. Boston: American Unitarian Association, [1913?] – 2mf – 9 – 0-524-07473-9 – mf#1991-3133 – us ATLA [012]

Bibliography and reference list of the history and literature : relating to the adoption of the constitution of the united states / Ford, Paul L – Brooklyn, 1896 – 1mf – 9 – $1.50 – mf#LLMC 84-808 – us LLMC [342]

Bibliography for conductors of college and community orchestras : a selective annotated list of written materials on organization, conducting, and general interpretive background / Johnson, Merton Bainer – U of Rochester 1966 [mf ed 19–] – 2mf – 9 – mf#fiche 496 – us Sibley [780]

Bibliography, history of duval county / Collar, Jimmie Oliver – s.l, s.l? . 1936 – 1r – us UF Libraries [978]

Bibliography in christian education for seminary and college libraries / Wyckoff, D Campbell – Dayton OH: Commission on Christian Educ, American Assoc of Theological Schools; Philadelphia PA: Board of Christian Educ, United Presbyterian Church in the USA 1968-88 [mf ed 1990] – 5mf – 9 – (iss as add to: bibliography in christian education for presyterian college libraries; cont: suggested bibliography in christian education for seminary and college libraries) – mf#1987-Z001 – us ATLA [240]

Bibliography of african christian literature / Conference of Missionary Societies in Great Britain and Ireland – London, 1923 – 1 – us CRL [240]

Bibliography of african christian literature / Rowling, F & Wilson, C E – London: The Conference, 1923 – 1r – 1 – 0-8370-0544-2 – mf#1984-B254 – us ATLA [012]

Bibliography of agriculture [1964] – Phoenix AZ 1964-99 – 1,5,9 – ISSN: 0006-1530 – mf#1693 – us UMI ProQuest [016]

Bibliography of agriculture. annual cumulative indexes – Phoenix AZ 1975-96 – 1,5,9 – ISSN: 1082-6408 – mf#11091 – us UMI ProQuest [016]

Bibliography of american church history *see* History of the disciples of christ, the society of friends, the united brethren in christ and the evangelical association

Bibliography of american hymnals / ed by Ellinwood, Leonard & Lockwood, Elizabeth – 9 – $95.00 – (a listing of 7,500 citations with imprint, yr of publ, compiler, pagination, location of copy indexed, intended religious denomination and name of indexer. companion to the dictionary of american hymnology) – us Univ Music [012]

A bibliography of american natural history / Meisel, M – Washington. 1937-1973 (9) – 19mf – 9 – mf#9273 – ne IDC [500]

Bibliography of american women – 47r – 1 – (bibliography contains ca 50,000 titles and covers monographs written by women in all major fields of study from 1500-1904. arranged in chronological, alphabetical and subject order) – mf#C36-13300 – us Primary [305]

Bibliography of asian studies – Ann Arbor MI 1970-91 – 1,5,9 – ISSN: 0067-7159 – mf#6151 – us UMI ProQuest [019]

A bibliography of british somaliland / Viney, N M – London, 1937 – 1mf – 8 – mf#A-262 – ne IDC [960]

A bibliography of congo languages / Starr, Frederick – (University of Chicago, Dept. of Anthropology. Bulletin V). Chicago. 1908 – 1 – us CRL [490]

Bibliography of eighteenth-century legal literature : a subject and author catalog of law treatises and all law-related literature held in the main legal collections in england / Adams, J N & Averley, G – [mf ed Chadwyck-Healey, 1982] – 6mf – 9 – uk Chadwyck [340]

Bibliography of electrical recordings in the cns and related literature – Los Angeles CA 1972-1973 (1) – ISSN: 0084-7879 – mf#7777 – us UMI ProQuest [013]

Bibliography of indian coins / Singhal, C R; ed by Altekar, A S – Bombay: Numismatic Society of India, 1950-1952 – (= ser Samp: indian books) – us CRL [730]
Bibliography of management literature : up to february, 1927 / Berg, Rose Monica [comp] – New York: American Soc of Mechanical Engineers, c1927 (mf ed 19–) – 67p – mf#ZT-TB+ pv467 n2 – us Harvard [650]
Bibliography of massachusetts vital records 1620-1905 / Holbrook, Jay Mack – 8th ed. Oxford MA 2000 – 925p on 3mf – 9 – 0-87623-413-9 – (mf1: abington to groton. mf2: hailfax to norton. mf3: oakham to yarmouth. this ed annotates over 400 microfiche colls that provide vital & other public records for 313 towns. also catalogues 194 towns (290mf) incl in the old printed series of massachusetts vital records to 1850) – us Archive [019]
A bibliography of mughal india, 1526-1707 a.d. / Sharma, Sri Ram – Bombay: Karnatak Pub House, [194-] – (= ser Samp: indian books) – (with a foreword by jadunath sarkar) – us CRL [954]
A bibliography of nigerian history / Jenkins, George – [s.l.]: Microsystems Inc, 1962 – us CRL [960]
A bibliography of nigerian history : preliminary draft / Oni-Orisan, B A – 1968 – us CRL [960]
A bibliography of nineteenth-century legal literature / Adams, J N & Davies, M J – 3v. 1992-95 – 9 – £1,200.00 – 0-907977-42-1 – (subject catalogue on diazo mf and cd-rom edition) – uk Chadwyck [340]
A bibliography of ralph waldo emerson / Cooke, George Willis – Boston: Houghton, Mifflin, 1908 [mf ed 1991] – (= ser Unitarian/universalist coll) – 1mf – 9 – 0-524-01079-X – mf#1990-4044 – us ATLA [014]
A bibliography of ramayana / Gore, N A – Poona: The Author, 1943 – (= ser Samp: indian books) – us CRL [490]
Bibliography of recorded music for dance / Acker, Doris M – [New York, 1947] – 1r – 1 – mf#*ZBD-*MGO pv14 – Located: NYPL – us Misc Inst [780]
Bibliography of the members of the royal society of canada : from the volume of "transactions" for 1894 – [S.l: s.n, 1894?] – 1mf – 9 – (incl some french text) – mf#00885 – cn Canadiana [019]
Bibliography of the new hebrides islands, 1610-1942 / Ferguson, J A – 1r – 1 – mf#PMB1126 – at Pacific Mss [019]
Bibliography of the semitic languages of ethiopia / Leslau, W – New York, 1946 – 2mf – 8 – mf#A-253 – ne IDC [470]
Bibliography of the spanish civil war 1936-1939 / Institute of History of the Spanish Civil War – 1r – 1 – $40.00 – mf#B50562 – us Library Micro [946]
Bibliography of the status of south-west africa up to june 30th / Loening, Luise Susanne Ernestine – Rondenbosch, South Africa. 1951 – 1r – us UF Libraries [960]
Bibliography of the utes – Denver, CO, 1951 (mf ed) – 1r – 1 – mf#MF Ut2a – us Colorado Hist [305]
A bibliography of unfinished books in the english language, with annotations / Corns, Albert Reginald & Sparke, Archibald – 1v. 1915 – 1,9 – us AMS Press [010]
A bibliography on development planning in nigeria, 1955-1968 / Akinyotu, Adetunji – Ibadan: Nigerian ISER, 1968 – us CRL [013]
Bibliography on the hypothalamic- pituitary-gonadal system – Los Angeles CA 1971-73 – 1 – ISSN: 0084-7887 – mf#7776 – us UMI ProQuest [611]
Bibliography on the nigerian civil war / Cervenka, Zdenek – Chicago, U of Chicago, Photodup Dep, [19–?] (mf ed) – us CRL [960]
Bibliography on the use of hydrocyanic acid gas as a fumigant / University Of Florida Agricultural Experiment Station – Gainesville, FL. 1935 – 1r – us UF Libraries [630]
Bibliologisher zamlbukh / Akademiia Nauk Ursr, Kiev Instytut Ievreis'koi Proletars'koi – Charkow, Ukraine. 1930 – 1r – us UF Libraries [939]
Bibliomappe : ou livre-cartes. lecons methodiques de chronologie et de geographie / Bailleul, Jacques C – Paris 1824-26 [mf ed Hildesheim 1995-98] – 2v on 10mf – 9 – €100.00 – 3-487-29973-9 – gw Olms [912]
Bibliopegia : or, the art of bookbinding in all its branches / Arnett, John Andrews – London: R Groombridge; New York: W Jackson, 1835 – us CRL [680]
Le bibliophile canadien : bulletin d'ouvrages canadiens etc etc – Quebec: J O Filteau, 1892 – 1mf – 9 – mf#01066 – cn Canadiana [010]
Biblorum sacrorum iuxta vulgatam clementinam – Roma, Italy. 1946 – 1r – us UF Libraries [025]
Biblorum sacrorum latinae versiones antiquae, seu vetus italica / Sabatier, Paul – Remis. tom1-3. 1743 – 139mf – 8 – €266.00 – ne Slangenburg [221]
Biblioteca argentina de libros raros americanos, tomo 4 : fr. joseph antonio de san alberto... – Madrid: Razon y Fe, 1927 – 1 – sp Bibl Santa Ana [972]
Biblioteca de catalunya / Institute d'estudis Catalans. Barcelona – 1921-36. 13v – 1 – 69.00 – us L of C Photodup [800]
Biblioteca de historia nacional – Bogota. n. 1-83. (wanting 68 and 70) – 1 – 881.00 – us L of C Photodup [972]
La biblioteca de jules janin / Lacroix, Pablo – Valencia: Editorial Castalia, 1950 – sp Bibl Santa Ana [020]
BIBLIOTECA de las tradiciones Espanolas. Sevilla, Guichot y Compania *see*
– De los maleficios y los demonios
– Juegos infantiles de extremadura
La biblioteca de tradiciones populares / Machado Alvarez, Antonio – Tomo II. 1884 – 9 – (tomo 3 1884) – sp Bibl Santa Ana [390]
La biblioteca del camarista de castilla d fernando jose de velasco y ceballos / Escagedo Salmon, Mateo – Santander. Lib. Moderna, 1932 – 1 – sp Bibl Santa Ana [020]
Biblioteca dell' economista – Turin. ser. 1 v. 1-ser. 5 v. 20. 1850-1923. Incomplete – 1 – us NY Public [330]
Biblioteca domestica – Rio de Janeiro, RJ: Typ a Vapor de Adolpho de Castro Silva & C, 13 jun 1885 – (= ser Ps 19) – mf#P17,01,110 – bl Biblioteca [640]
Biblioteca encantada / Diaz Montero, Anibal – San Juan, Puerto Rico. 1957 – 1r – us UF Libraries [972]
Biblioteca erasmista de diego mendez / Almoina, Jose – Ciudad Trujillo, Dominican Republic. 1945 – 1r – us UF Libraries [972]
Biblioteca goathemala de la sociedad de geografia e historia... / Bayle, Constantino – Madrid: Razon y Fe, 1939 – 1 – sp Bibl Santa Ana [972]
Biblioteca historica de puerto-rico / Tapia Y Rivera, Alejandro – San Juan, Puerto Rico. 1945 – 1r – us UF Libraries [972]
Biblioteca historico-genealogica asturiana... / Alvarez de la Rivera, Senen – Madrid, Razon y Fe, 1926 – 1 – sp Bibl Santa Ana [920]
Biblioteca Luis-Angel Arango *see* Incunables bogotanos, sigle 18
Biblioteca manual medico-practica... / Carrion, C – Barcelona, 1745 – 7mf – 9 – sp Cultura [610]
Biblioteca militar espanola / Garcia de la Huerta, Vicente – 1760 – 9 – sp Bibl Santa Ana [355]
Biblioteca Nacional (Brazil) *see* Amazonia brasileira
Biblioteca Nacional De Guatemala *see* Algunos juicios de escritos guatemaltecos...
Biblioteca Nacional, Madrid *see*
– Author catalogues
– Catalogo de la coleccion gomez-imaz
– Catalogo de publicaciones periodicas
– Catalogo de varios especiales
Biblioteca Nacional (Venezuela) *see* Bibliografia de don aristides rojas, 1826-1894
Biblioteca Publica do Para *see* O 31 de agosto
Biblioteca Publica Municipal *see* Catalogo de obras que existen en esta biblioteca en 30 de marzo de 1952
Las bibliotecas de espana...publicas / Diaz Perez, Nicolas – 1885 – 9 – sp Bibl Santa Ana [020]
Las bibliotecas de la antiguedad estudio / Lipsio, Justo – Valencia: Editorial Castalia, 1948 – (pref and notes by jose lopez de toro) – sp Bibl Santa Ana [946]
Biblioteconomia brasileira / Russo, Laura Garcia Moreno – Rio de Janeiro, Brazil. 1966 – 1r – us UF Libraries [972]
Biblioteka Akademii Nauk (BAN) *see* Nep rare editions
Biblioteka Akademii Nauk (BAN), St Petersburg *see* Asian books in the russian language
Biblioteka deshevaia obshchestvennaia *see* Zhurnal bibliografii i belletristiki
Biblioteka Ossolinskich *see* Pismo historii, literaturze, umiejetnosciom i rzeczom narodowym poswiecone
Biblioteka soiuza selsko-khoziaistvennoi, kreditnoi i kustarno-promyslovoi kooperatsii : rekomendatelnyi spisok knig dlia komplektovaniia i popolneniia bibliotek selskokhoziaistvennoi... – 1928 – 2 – 9 – mf#COR-528 – ne IDC [335]
Biblioteka stefana iavorskogo / Maslov, S I – Kiev, 1914 – 3mf – 8 – mf#R-4457 – ne IDC [947]
Biblioteka vestnika vinodeliia – Odessa, 1909-1913 – 25mf – 9 – (missing: 1909, v1-4; 1912, v15-16) – mf#R-1538 – ne IDC [077]
Biblioteka vracha – M., 1894-1899 – 365mf – 9 – (missing: 1894(9); 1898(5, 10-11); 1899(3-10)) – mf#R-1539 – ne IDC [077]
Bibliotekar – M., 1910-1915 – 68mf – 9 – mf#R-3498 – ne IDC [077]
Bibliotekar – Sofia, Bulgaria 1969-90 – 1,5,9 – ISSN: 0204-7438 – mf#5025 – us UMI ProQuest [020]
Biblioteki moskovskogo glavnogo arkhiva ministerstva inostrannykh del, katalog rukopisiam, otnosiaschimsia do moskvy, moskovskoi gubernii, ikh tserkvei i monastyrei / Tokmakov, I F – 1879 – 19p 1mf – 9 – mf#R-11,169 – ne IDC [243]
Biblioteki moskovskogo glavnogo arkhiva ministerstva inostrannykh del, katalog rukopisiam, otnosiashchimsia do tserkovnoi istorii / Tokmakov, I F – 1880 – 25p 1mf – 9 – mf#R-11,151 – ne IDC [243]
Biblioteksbladet – Stockholm, Lund, Sweden 1965-80 – 1,5,9 – ISSN: 0006-1867 – mf#2039 – us UMI ProQuest [020]
Bibliotheca... / ed by Gessner, K et al – Tigvri: Christoph Froschover, 1583 – 10mf – 9 – mf#PBU-475 – ne IDC [240]
Bibliotheca aegyptiaca : repertorium ueber die bis zum jahre 1857 in bezug auf aegypten, seine geographie, landeskunde, naturgeschichte... / Jolowitz, Heimann – Leipzig: W Engelmann, 1858 – 1 – us CRL [956]
Bibliotheca aethiopica / Goldschmidt, L – Leipzig, 1893 – 1mf – 9 – mf#NE-20270 – ne IDC [960]
Bibliotheca africana – Innsbruck, 1924-34 – 1 – us CRL [960]
Bibliotheca americana / Roorbach, Orville A – 1852. Supplement to the Bibliotheca americana. 1855. Addenda to the Bibliotheca americana. 1858. Volume 4 of the Bibliotheca americana – (1861. 4 v. 1,9) – us AMS Press [010]
Bibliotheca annua – 4v. 1700-03 – 1,9 – us AMS Press [020]
Bibliotheca anti-quakeriana : or, a catalogue of books adverse to the society of friends... / Smith, Joseph – London: Joseph Smith, 1873 [mf ed 1990] – 2mf – 9 – 0-7905-8228-7 – mf#1988-6128 – us ATLA [243]
Bibliotheca anti-trinitariorum : compendium historiae ecclesiasticae unitariorum / Sandius, C – Freistadii, 1684 – 4mf – 8 – €12.00 – ne Slangenburg [240]
Bibliotheca ascetica antiquo-nova / Pez, Bernhard (Leopold) – Ratisbonae. v1-10. 1723-1733 – 65mf – 8 – €124.00 – ne Slangenburg [241]
Bibliotheca augustiniana / Ossinger, J F – Ingolstadii, 1768 – €82.00 – ne Slangenburg [241]
Bibliotheca benedictino-mauriana / Pez, Bern – Augustae-Vindelicorum, 1716 – 6mf – 8 – €14.00 – ne Slangenburg [241]
Bibliotheca biblica : a select list of books on sacred literature / Orme, William – Edinburgh: Adam Black; London: Longman, Hurst, Rees, Orme, Brown and Green, 1824 – 2mf – 9 – 0-7905-2193-8 – (incl ind) – mf#1987-2193 – us ATLA [220]
Bibliotheca bio-bibliografica della terra santa e dell' oriente francescano / Golubovich, G – Firenze. 5v. 1906-1927 – 72mf – 8 – mf#U-673 – ne IDC [700]
Bibliotheca brasileira – Rio de Janeiro, RJ: Typ Preserveranca, jul-set 1863 – (= ser Ps 19) – mf#P02A,03,17 – bl Biblioteca [972]
Bibliotheca britannica / Watt, Robert – 4v. 1824 – 1,9 – us AMS Press [010]
Bibliotheca canadensis : a catalogue of a very large collection of books and pamphlets relating to the history, the topography, the manners and customs of the indians, the trade and government of north america... / R W Douglas and Co – Toronto: R W Douglas, [1887?] [mf ed 1987] – 1mf – 9 – 0-665-68318-9 – mf#68318 – cn Canadiana [020]
Bibliotheca canadensis : or, a manual of canadian literature / Morgan, Henry James – [Ottawa?: G E Desbarats], 1867 – 5mf – 9 – mf#11068 – cn Canadiana [410]
Bibliotheca canadensis – [Quebec] : O Frechette, [1881] – 9 – ISSN: 1190-6901 – mf#P04151 – cn Canadiana [070]
Bibliotheca cluniacensis / ed by Marrier, Martin – Bruxelles-Paris, 1915 – 45mf – 8 – €86.00 – ne Slangenburg [241]
Bibliotheca cooperiana : catalogue of a further portion of the library of charles purton cooper... / Sotheby and Wilkinson – [London], 1856 – (= ser 19th c publishing...) – 2mf – 9 – mf#3.1.12 – uk Chadwyck [020]
Bibliotheca diabolica : being a choice selection of the most valuable books relating to the devil, his origin, greatness, and influence / Kernot, Henry – [New York]: Scribner, Welford & Armstrong [distributor], 1874 – 1mf – 9 – 0-524-00564-8 – mf#1990-0064 – us ATLA [012]
Bibliotheca doellingeriana : katalog der bibliothek des verstorbenen kgl universitaets-professors j j j von doellinger... – Muenchen: J. Lindauer, 1893 [mf ed 1990] – 2mf – 9 – 0-7905-8026-8 – mf#1988-6007 – us ATLA [012]
Bibliotheca eliotae: eliotis librarie / Elyot, Thomas – Augmented by Thomas Cooper, London, 1548. Reprint Delmar, 1975. Introd. by Lillian Gottesman, Bronx Community College.Latin-English dictionary widely used in Tudor times – 9 – 60.00 – us Scholars Facs [450]
Bibliotheca fratrum polonorum – 8 – (fausti socini senensis: opera omnia v1-2, irenopoli 1656 59mf €136. joh crellii franci: opera omnia v1-4, eleutheropoli 1656 86mf €164. slichtingii de bukowiec, j: commentaria posthuma in plerosque novi testamenti libros v1-2, irenopoli 1656 v1 12mf v2 18mf €57. joh lud wolzogenii: opera omnia v1-3, irenopoli 1656 54mf €103. sam przipcovii: omnia opera, eleutheropoli 1692 31mf €60) – ne Slangenburg [240]
Bibliotheca hispana vetus / Antonio, Nicolas – 2v. 1788 – 1,9 – us AMS Press [010]
Bibliotheca historica : oder, systematisch geordnete uebersicht der in deutschland und dem auslande auf gebiete der gesammten geschichte neu erschienen buecher – Goettingen: Vandenhoeck & Ruprecht's Verlag. 20v 1863-82 [v1-9] [mf ed 1979] – 3r – 1 – (fr 1853-61 this formed pt of the bibliotheca historico-geographica [1.-9. jahrg] wh after 1861 was iss in 2 parallel ser: 1. bibliotheca historica. 2. bibliotheca geographico-statistica et oeconomico-politica; publ discontd in 1882; resumed in 1887 under title: bibliotheca historica. vierteljaehrliche systematisch geordnete uebersicht der auf dem gebiete der gesammten geschichte in deutschland und dem auslande neu erschienenen schriften und zeitschriften-aufsaetze ed by oscar masslow. neue folge: 1. jahrg 1887 [fortsetzung der ersten 30. jahrgaenge, 1853-82] goettingen 1888) – mf#film mas c607 – us Harvard [900]
Bibliotheca historica medii aevi : wegweiser durch die geschichtswerke des europ mittelalters bis 1500 / Potthast, A – Berlin. v1-2. 1896 – €76.00 – ne Slangenburg [931]
Bibliotheca historica medii aevi : wegweiser durch die geschichtswerke des europaeischen mittelalters bis 1500 / Potthast, August – 2. verb verm aufl. Berlin: W Weber, 1896 [mf ed 1990] – 2v on 5mf – 9 – 0-7905-8046-2 – (incl bibl ref) – mf#1988-6027 – us ATLA [931]
Bibliotheca historico-naturalis physico-chemica et mathematica : oder, systematisch geordnete uebersicht der in deutschland und dem auslande auf dem gebiete der gesammten naturwissenschaften und der mathematik neu erschienen buecher – Goettingen: Vandenhoeck & Ruprecht. jahrg 2-37 (1852-87) [1852]-1888 [mf ed 1979 v1-9] – 4r – 1 – (semiannual 1852-85; quarterly 1886-87; cont: bibliotheca historico-naturalis et physico-chemica, oder, systematisch geordnete uebersicht der in deutschland und dem auslande auf dem gebiete der gesammten naturwissenschaften neu erschienen buecher; title varies) – mf#film mas c606 – us Harvard [500]
Bibliotheca historico-philologico theologica bremensis – 1(1718)-8(1725). Amstelodami, 1720-1725 – 96mf – 9 – €178.00 – ne Slangenburg [200]
Bibliotheca instituta et collecta primum a conrado gesnero / Simler, J – Zuerich, Christoph Froschauer, 1574 – 8mf – 9 – mf#PBU-410 – ne IDC [240]
Bibliotheca lindesiana : list of manuscripts and examples of metal and ivory bindings exhibited to the bibliographical society at the grafton galleries 13th june 1898 by the president / Lindsay, James Ludovic, 26th Earl of Crawford – [Aberdeen], 1898 – (= ser 19th c publishing...) – 1mf – 9 – mf#3.1.54 – uk Chadwyck [090]
Bibliotheca lutherana : a complete list of the publications of all the lutheran ministers in the united states / Morris, John Gottlieb – Philadelphia: Lutheran Board of Publication: Lutheran Book Store, 1876, 1875 – 1mf – 9 – 0-7905-5722-3 – mf#1988-1722 – us ATLA [012]
Bibliotheca mathematica – Killen TX 1887-1913 – 1 – mf#6105 – us UMI ProQuest [510]
Bibliotheca mathematica : zeitschrift fuer geschichte der mathematischen wissenschaften – v1-14. 1900-1914 – 1,5 – $216.00 – (in german) – mf#0102 – us Brook [510]
Bibliotheca medievalis graeca / Sathas – 1872-1894. 7v – 56mf – 9 – mf#OA-246 – ne IDC [720]
Bibliotheca missionum / ed by Streit, R – Muenster, 1916-1924.v1-2 – 22mf – 9 – mf#H-3050c – ne IDC [240]
Bibliotheca mysticorum selecta : tribus constans partibus... / Poiret, P – Amsterdam, 1708 – 5mf – 9 – mf#PPE-221 – ne IDC [240]
Bibliotheca novi testamenti graeci : cuius editiones ab initio typographiae ad nostram aetatem impressas quotquot reperiri potuerunt / Reuss, Eduard – Brunsvigae: apud C A Schwetschke, 1872 – 1mf – 9 – 0-7905-3166-6 – mf#1987-3166 – us ATLA [225]
Bibliotheca orientalis : or a complete list of books...on the history...of the east / ed by Friederici, C – London, etc, 1876-1883 – 10mf – 9 – mf#AR-2140 – ne IDC [956]

Bibliotheca orientalis clementino-vaticana / ed by Assemanus, J S – Romae. v1-3. 1719-1728 – 3v on 122mf – 8 – €254.00 – (pt1: de scriptoribus syris orthodoxis, romae 1719 €56. pt2: de scriptoribus syris monophysitis, romae 1721 €60. pt3/1a: de scriptoribus syris nestorianis, romae 1725 €60. pt3/2a: de syris nestorianis, romae 1728 €79) – ne Slangenburg [240]

Bibliotheca palatina : stampati palatini = Printed books / ed by Boyle, Leonard – [mf ed 1989-95] – 21,103mf (1:24) – 9 – silver €35,600.00 – ISBN-10: 3-598-32880-X – ISBN-13: 978-3-598-32880-0 – (ind sold separately) – gw Saur [070]

Bibliotheca parva theologica : a catalogue of books recommended to students in divinity, with a selection of the best editions of the fathers of the church – Oxford: John Henry Parker, 1851 – 1mf – 9 – 0-524-00298-3 – mf#1989-2998 – us ATLA [012]

Bibliotheca patrum cisterciensium / Tissier, Bertrandus – Bonofonte. v1-8. 1660-69 – 106mf – 8 – €371.00 – (v1: 1660 12mf. v2: 1662 16mf. v3: 1660 12mf. v4: 1662 14mf. v5: 1662 17mf. v6: 1664 6mf. v7: 1669 16mf. v8: 1669 13mf) – ne Slangenburg [241]

Bibliotheca philologica classica – Berlin: S Calvary 1875-1938 [mf ed 1979] – 65v on 9r – 1 – (also iss as: beiblatt zum jahresbericht ueber dei fortschritte der klassischen altertumswissenschaft) – mf#film mas c562 – us Harvard [450]

Bibliotheca premonstratensis ordinis / Paige, Johannis, Le – Parisiis, 1633 – 48mf – 8 – €92.00 – ne Slangenburg [240]

Bibliotheca rabbinica : eine sammlung alter midraschim. zum ersten male ins deutsche uebertragen / ed by Wuensche, August – Leipzig. v1-11. 1880-1885 – 9 – €118.00 – ne Slangenburg [270]

Bibliotheca reformatoria neerlandica / ed by Cramer, J & Pijper, F – 's-Gravenhage. v1-10. 1903-14 – €328.00 – ne Slangenburg [242]

Bibliotheca rerum germanicarum / ed by Jaffe, Ph – Berolini. v1-6. 1864-73 – 81mf – 8 – €502.00 – ne Slangenburg [240]

Bibliotheca rhetorum praecepta et exempla complectens quae ad poeticam facultatem pertinent... / Jay, G F le – Venetia: Typographia Balleoniana, 1747 – 9mf – 9 – mf#0-00 – ne IDC [090]

Bibliotheca sacra *see* Baptist review

Bibliotheca sacra [1843] – La Salle IL 1843 – 1 – mf#3944 – us UMI ProQuest [240]

Bibliotheca sacra [1844] – Dallas TX 1844+ – 1,5,9 – ISSN: 0006-1921 – mf#850 – us UMI ProQuest [240]

Bibliotheca sacra and american biblical repository – 8(1851)-20(1863) – 204mf – 9 – €389.00 – ne Slangenburg [220]

Bibliotheca sacra and theological review – 1(1844)-7(1850) – 98mf – 9 – €187.00 – ne Slangenburg [240]

Bibliotheca sacra dallas theological seminary – 21(1864)-39(1882) – 267mf – 9 – €509.00 – (with ind: 1(1844)-30(1873) 5mf €12) – ne Slangenburg [240]

Bibliotheca sacri ordinis cisterciensis / Visch, C de – Koeln, 1656 – 10mf – 8 – €35.00 – ne Slangenburg [241]

Bibliotheca sancta / Sixtus of Siena – Lyon, 1575. 2v – 15mf – 9 – mf#CA-60 – ne IDC [240]

Bibliotheca scriptorum historiae naturalis / Boehmer, G R – Leipzig, 1785-1789. 5v – 66mf – 9 – mf#8438 – ne IDC [590]

Bibliotheca universalis / Gessner, K – Zuerich, Christoph Froschauer, 1545 – 23mf – 9 – mf#PBU-407 – ne IDC [240]

Bibliothecae apostolicae vaticanae codicum manuscriptorum catalogus, vol 1 : codics ebraicos et samaritanos / Assemanus, J S & Assemanus, S E – Romae, 1756 – 75mf – 8 – €48.00 – ne Slangenburg [240]

Die "bibliothecae" des jean jacques manget (ael3/20) – [mf ed 1996] – (= ser Archiv der europaeischen lexikographie: fach-enzyklopaedien) – 174mf – 9 – €1190.00 set – 3-89131-216-4 – (contains: bibliotheca medico-practica, genf 1695-98 [4v on 46mf] €380; bibliotheca anatomica, 2nd ed genf 1699 [2v on 26mf] €210; bibliotheca chemica (curiosa) genf 1702 [2v on 20mf] €170; bibliotheca pharmaceutico-medica, koeln 1703 [2v on 24mf] €190; bibliotheca chirurgica, genf 1721 [4v on 28mf] €220; bibliotheca scriptorum medicorum veterum et recentiorum, genf 1731 [4pts in 2v on 29mf] €230) – gw Fischer [610]

Bibliothecae mediceae laurentianae et palatinae codicum mms : orientalium catalogus / Assemani, E – Florentiae, 1742 – 38mf – 8 – €73.00 – ne Slangenburg [240]

Bibliothecae syriacae – Gottingae: L Horstmann, 1892 – 1mf – 9 – 0-8370-1789-0 – mf#1987-6177 – us ATLA [221]

Bibliotheek van de Koninklijke Vereeniging ter bevordering van de belangen des Boekhandels *see* The correspondence of marc-michel rey, 1747-1778

Bibliotheek van nederlandsche kerkgeschiedschrijvers : opgave van hetgeen nederlanders over de geschiedenis der christelijke kerk geschreven hebben / Sepp, Christiaan – Leiden: EJ Brill, 1886 – 2mf – 9 – 0-524-01894-4 – mf#1990-0521 – us ATLA [240]

Bibliothek der aeltesten deutschen litteratur-denkmaeler – Paderborn: F Schoeningh, 1874-1923 [mf ed 1993] – 13v – 1 – mf#8437 – us Wisconsin U Libr [430]

Bibliothek der alten litteratur und kunst mit ungedruckten stuecken aus der escurialbibliothek und andern / ed by Tychsen, Thomas Christian et al – Goettingen 1786-94 – (= ser Dz. abt literatur) – 10st on 16mf – 9 – €160.00 – mf#k/n4555 – gw Olms [430]

Bibliothek der deutschen literatur : mikrofiche-gesamtausgabe nach angaben des taschengoedeke – [mf ed 1990-2005] – 22,810mf (1:24) – 9 – diazo €19,800.00 (silver €34,000 isbn: 978-3-598-50001-5) – ISBN-10: 3-598-50000-9 – ISBN-13: 978-3-598-50000-8 – (incl bibl, ind and suppl 1) – gw Saur [430]

Bibliothek der deutschen literatur. zweites supplement / Frey, Axel [comp] – [mf ed 2002-05] – 2131mf (1:24) in 7 installments – 9 – diazo €13,800.00 (silver €16,800 isbn: 978-3-598-53308-2) – ISBN-10: 3-598-53307-1 – ISBN-13: 978-3-598-53307-5 – gw Saur [430]

Bibliothek der frauenfrage in deutschland nach sveistrup – [mf ed 2001] – (= ser Hq 40) – ca 1000mf in 12-14 installments – 9 – €4560.00 per installment – 3-89131-300-4 – (lfg1: isbn 3-89131-301-2 [1998]; lfg2: isbn 3-89131-302-0 [1999]; lfg3: isbn 3-89131-303-9 [1999]; lfg4: isbn 3-89131-304-7 [2000]; lfg5: isbn 3-89131-305-5 [2000]; lfg6: isbn 3-89131-306-3 [2001]; lfg7: isbn 3-89131-307-1 [2002]; lfg8: isbn 3-89131-308-x [2002]; lfg9: isbn 3-89131-309-8 [2002]; lfg10: isbn 3-89131-310-1 [2003]; lfg11: isbn 3-89131-311-x [2004]; lfg 12: isbn 3-89131-312-8 [2005]) – gw Fischer [305]

Bibliothek der gesammten deutschen national-literatur : von der aeltesten bis auf die neuere zeit – Quedlinburg, Leipzig: G Basse, 1835-1872 [mf ed 1993] – 47v in 51/pl – 1 – (work projected for [1. abt] 13. bd, 1.-2. t, konrad von wuerzburg's der trojanische krieg, was never publ in the series. some vols publ out of sequence) – mf#8438 – us Wisconsin U Libr [430]

Bibliothek der grazien : eine monatschrift fuer liebhaberinnen und freunde des gesangs und klaviers / ed by Bossler, Heinrich Philipp – Speier: H P Bossler [1789?] [mthly] [mf ed 19–] – 1r – 1 – mf#film 1028, 1676 – us Sibley [780]

Bibliothek der kirchenvaeter. 1. reihe (bdk 1.reihe) – Muenchen. 63v. 1911-1933 – 63v on 472mf – 8 – €882.00 – (individual titles also listed separately) – ne Slangenburg [240]

Bibliothek der kirchenvaeter. 2. reihe (bdk 2.reihe) – Muenchen. v1-20. 1932-1938 – 9 – €250.00 – (vols also listed separately) – ne Slangenburg [240]

Bibliothek der maehrischen staatskunde / ed by Hankenstein, Johann Aloys Hanke von – Wien 1786 – (= ser Dz. historisch-geographische abt) – 1v on 3mf – 9 – €90.00 – mf#k/n1173 – gw Olms [320]

Bibliothek der neuesten und wichtigsten reisebeschreibungen zur erweiterung der erdkunde / Sprengel, Matthias C – Weimar 1806-14 [mf ed Hildesheim 1995-98] – 3v on 5mf – 9 – €100.00 – 3-487-26569-9 – gw Olms [910]

Bibliothek der paedagogischen literatur – Leipzig DE, 1803 sep-dec – 1r – 1 – gw Mikrofilm [370]

Bibliothek der redenden und bildenden kuenste – Leipzig 1806-11 – (= ser Dz. abt literatur) – v1-8[zu je 2st] on 24mf – 9 – €240.00 – mf#k/n4683 – gw Olms [700]

Bibliothek der schoenen wissenschaften und der freyen kuenste / ed by Mendelssohn, Friedrich Nicolai et al – Leipzig 1757-65 – (= ser Dz. abt literatur) – 12v+ind on 39mf – 9 – €390.00 – (fr v5 ed by christian felix weisse) – mf#k/n4430 – gw Olms [700]

Bibliothek der symbole und glaubensregeln der alten kirche / ed by Hahn, August – 3rd rev and enl ed. Breslau: E Morgenstern, 1897 [mf ed 1990] – 1mf – 9 – 0-7905-8104-3 – (in greek & latin. 1st publ 1842. incl bibl ref) – mf#1988-6066 – us ATLA [240]

Eine bibliothek der symbole und theologischer tractate zur bekaempfung des priscillianismus und westgothischen arianismus aus dem 6. jahrhundert : ein beitrag zur geschichte der theologischen litteratur in spanien / Kuenstle, Karl – Mainz: F Kirchheim 1900 [mf ed 1991] – (= ser Forschungen zur christlichen litteratur- und dogmengeschichte 1/4) – 1mf – 9 – 0-7905-9406-4 – mf#1989-2631 – us ATLA [240]

Bibliothek der unterhaltung und des wissens – Stuttgart: Union Deutsche Verlagsgesellschaft, 1876-[1944?] (irregular) [mf ed 1994] – 1 – (began with 1876. ceased with 1944?) – mf#8675 – us Wisconsin U Libr [430]

Bibliothek des deutschen museums : alphabetischer katalog und schlagwortkatalog = Library catalogue of the deutsche museum / ed by Bibliothek des deutschen Museums. Muenchen – [mf ed 1981-82] – 612mf (1.42) – 9 – silver €3980.00 – ISBN-10: 3-598-30397-1 – ISBN-13: 978-3-598-30397-5 – (alphabetischer katalog: 313mf; schlagwortkatalog 299mf (1:42)) – gw Saur [060]

Bibliothek des deutschen Museums. Muenchen *see* Bibliothek des deutschen museums

Bibliothek des deutschen patentamtes muenchen : kreuzkatalog = Library of the german patent office munich – 1945-74 [mf ed 1983] – 167mf (1:42) – 9 – diazo €1790.00 (silver €1980.00 isbn: 978-3-598-30453-8) – ISBN-10: 3-598-30454-4 – ISBN-13: 978-3-598-30454-5 – gw Saur [346]

Bibliothek des literarischen vereins in stuttgart – Stuttgart: Literarischer Verein, 1842- (bimthly) [mf ed 1993] – 1 – (each iss has also a distinctive title) – mf#8470 – us Wisconsin U Libr [430]

Bibliothek fuer denker und maenner von geschmack / ed by Winkopp, Peter Adolf et al – Gera 1783-91 – (= ser Dz. abt literatur) – 4v on 15mf – 9 – €150.00 – mf#k/n4534 – gw Olms [430]

Bibliothek fuer officiere / ed by Scharnhorst, Georg von – Goettingen 1785 – (= ser Dz) – 4st on 6mf – 9 – €120.00 – mf#k/n3972 – gw Olms [355]

Bibliothek fuer Zeitgeschichte, Stuttgart *see* Systematischer katalog der bibliothek fuer zeitgeschichte, stuttgart

Bibliothek knaake : katalog der sammlung von reformationsschriften des begruenders der weimarer lutherausgabe / Knaake, Joachim Karl Friedrich – Leipzig: O Weigel, 1908 – 10mf – 9 – 0-524-08782-2 – mf#1993-1090 – us ATLA [012]

Bibliothek russischer denkwuerdigkeiten – Stuttgart. v1-7. 1894-1895 – 39mf – 8 – mf#R-1680 – ne IDC [460]

Bibliothekar – Leipzig, Germany 1962-66 – 1 – ISSN: 0006-1964 – mf#2690 – us UMI ProQuest [020]

Bibliothekstechnisches aus der vatikana / Ehrle, Franz – [s.l: s.n. 1916?] [mf ed 1992] – 1mf – 9 – 0-524-04129-6 – mf#1990-1199 – us ATLA [240]

La bibliotheque a cinq cents – Montreal: Poirier, Bessette, [1886-189-?] – 9 – mf#P04017 – cn Canadiana [440]

Bibliotheque ancienne et moderne – Amsterdam. 1714-27 (1-27) – 1 – fr ACRPP [073]

Bibliotheque angloise – ou Histoire litteraire de la Grande-Bretagne. Amsterdam. 1717-28 (I-XV) – 1 – fr ACRPP [420]

Bibliotheque britannique – ou Histoire des ouvrages des savants de la Grande-Bretagne. La Haye. avr 1733-mars 1747 (I-XXII) – 1 – fr ACRPP [420]

Bibliotheque canadienne – Levis: Pierre-Georges Roy, 1898-[189- ou 19–] – 9 – (ceased 189-?) – mf#P04153 – cn Canadiana [020]

Bibliotheque canadienne : ou annales bibliographiques / Bibaud, jeune [comp] – Montreal: impr par Cerat et Bourguignon, [1858] [mf ed 1982] – 1mf – 9 – 0-665-32619-X – mf#32619 – cn Canadiana [019]

La bibliotheque canadienne : ou miscellanees historiques, scientifiques et litteraires, vols 1-9 – Montreal: M Bibaud. 9v. 1825-1830 – 0-665-47584-5 – mf#47584 – cn Canadiana [880]

La bibliotheque canadienne-francaise – Quebec: Societe Saint-Vincent de Paul, 1896-1897 [mf ed v1 n1 sep 1896-v1 n12 dec 1897] – 9 – mf#P04014 – cn Canadiana [440]

Bibliotheque curieuse historique et critique : ou catalogue raisonne de livres difficiles a trouver / ed by Clement, David – Goettingen 1750-60 [mf ed 1998] – 9v on 47mf – 9 – €330.00 – 3-89131-245-8 – gw Fischer [900]

Bibliotheque de feu jos adolp gariepy, de longueuil : pour etre vendue par encan dans la batisse de l'aqueduc, a longueuil, p q, mardi, le 7 avril 1896 – [Longueuil, Quebec?: s.n, 1896?] – 1mf – 9 – 0-665-94418-7 – mf#94418 – cn Canadiana [020]

Bibliotheque de la Societe Psychanalytique de Paris *see* Almanach fuer das jahr [...]

Bibliotheque de l'ecole des chartes – 1(1839)-116(1958) – 1778mf – 9 – €3389.00 – ne Slangenburg [020]

Bibliotheque de l'ecole des chartes – Paris. v1-96. 1839-1934 – 1 – $1350.00 – (in french. v96-139 1935-81 $652 [0103]) – mf#0104 – us Brook [020]

Bibliotheque des benedictins de la congregation de saint-vanne et saint-hydulphe (afm29) / Godefroy, J – 1925 – (= ser Archives de la france monastique (afm)) – €14.00 – ne Slangenburg [241]

La bibliotheque des bonnes lectures illustrees – Trois-Rivieres, Quebec: Societe de publ des bonnes lectures illustrees, [180-?-180-?] – 9 – ISSN: 1190-7851 – mf#P04067 – cn Canadiana [440]

Bibliotheque des ecoles francaises d'athens et de rome – v1-102 – 9 – $768.00 – (in french) – mf#0105 – us Brook [450]

Bibliotheque des sciences et des beaux-arts – La Haye. 1754-80 (I-L) – 1 – fr ACRPP [500]

La bibliotheque francaise – Montreal: Societe des Publications Francaises, 1887-[189- ou 19–] – 9 – mf#P04029 – cn Canadiana [440]

Bibliotheque francoise – ou Histoire litteraire de la France. Amsterdam. 1723-46 – 1 – fr ACRPP [440]

Bibliotheque geographique et instructive des jeunes gens : ou recueil de voyages interessantsdans toutes les parties du monde; pour l'instruction et l'amusement de la jeunesse; traduit de l'allemand et orne de cartes et figures / Campe, Joachim H – Paris [u.a.] 1802-1807 [mf ed Hildesheim 1995-98] – 72v on 108mf – 9 – €648.00 – 3-487-29948-8 – gw Olms [910]

Bibliotheque germanique – ou Histoire litteraire de l'Allemagne, de la Suisse et des pays du Nord. Amsterdam. 1720-41 – 1 – fr ACRPP [430]

Bibliotheque historique de la france : contenant le catalogue de tous les ouvrages, tant imprimez que manuscripts, qui traitent de l'histoire de ce roiaume, ou qui y ont rapport; avec des notes critiques et historiques / Long, J le – Paris, 1719 – 52mf – 9 – mf#H-124 – ne IDC [700]

La bibliotheque illustree – Montreal: Patry & Peeters, [1898-19–] – 9 – mf#P04090 – cn Canadiana [440]

Bibliotheque internationale de l'alliance scientifique universelle, tome 1, fascicule 3 – Quebec: L Brousseau, 1892 – 1mf – 9 – mf#00784 – cn Canadiana [440]

Bibliotheque linguistique americaine – Microcard Editions – 70mf (24:1) – 9 – $480.00 – us UPA [490]

La bibliotheque medicale de l'hotel-dieu de quebec, 17-18-19e siecles / Langlois, Marguerite – 1961 [mf ed 1978] – (= ser Bibliographies du cours...1947-66) – 4mf – 9 – (pref by j-l petitclerc) – mf#SEM105P4 – cn Bibl Nat [610]

Bibliotheque National. France. Dept des Imprimes *see* Catalogue de l'histoire de l'afrique

Bibliotheque nationale : catalogues du departement de la musique – 1836mf – 9 – £7,995.00 – uk Chadwyck [780]

Bibliotheque nationale : l'inventaire des catalogues des manuscrits – 9 – £14,100.00 – uk Chadwyck [090]

Bibliotheque nationale departement des manuscrits. Departement des Manuscrits *see* Inventaire des instruments de recherche

Bibliotheque nationale du Quebec *see* Politique de conservation du patrimoine documentaire

Bibliotheque nationale du Quebec. Bureau de la bibliographie retrospective *see* Bqr 1821-1967 t 1-22

Bibliotheque nationale du Quebec. Departement des manuscrits *see* Collection de musique canadienne

Bibliotheque nationale du Quebec. Ministere des affaires culturelles *see*
- Bibliographie du quebec

Bibliotheque Nationale. France *see*
- The author catalogues of printed books
- Catalogue de l'histoire de france
- Catalogue general des periodiques des origines a 1959
- Catalogues du departement des arts du spectacle

Bibliotheque Nationale. Quebec *see*
- Fichier d'autorite
- Plans d'assurances de villes du quebec

Bibliotheque orientale : ou dictionnaire universel supplement / Visdelou, C de & Galand, C – Maestricht, 1780 – 1mf – 9 – €42.00 – mf#HT-567 – ne IDC [915]

Bibliotheque orientale : ou dictionnaire universel supplement / ed by Visdelou, C & Galand, A – Maestricht, 1780 – €42.00 – (contenant...tout cequi regarde la connaissance des peuples de l'orient...par m d'herbelot de molainville) – ne Slangenburg [050]

Bibliotheque paroissiale de Notre-Dame *see* Catalogue de la bibliotheque de l'oeuvre des bons livres, erigee a montreal

Bibliotheque raisonnee des ouvrages des savans de l'europe – Amsterdam. 1752 – 1 – fr ACRPP [073]

Bibliotheque sulpicienne, vol 1 : ou histoire litteraire de la compagnie de saint-sulpice / Bertrand, Lionel – Paris: A Picard, 1900 [mf ed 1980] – 7mf – 9 – (incl ind and bibl ref) – mf#02061 – cn Canadiana [241]

Bibliotheque sulpicienne, vol 2 : ou histoire litteraire de la compagnie de saint-sulpice / Bertrand, Lionel – Paris: A Picard, 1900 [mf ed 1980] – 7mf – 9 – (incl ind and bibl ref) – mf#02062 – cn Canadiana [241]

Bibliotheque sulpicienne, vol 3 : ou histoire litteraire de la compagnie de saint-sulpice / Bertrand, Lionel – Paris: A Picard, 1900 [mf ed 1980] – 6mf – 9 – 0-665-02063-5 – (incl ind and bibl ref) – mf#02063 – cn Canadiana [241]

Bibliotheque sulpicienne, vols 1-3 : ou histoire litteraire de la compagnie de saint-sulpice / Bertrand, Lionel – Paris: A Picard, 1900 – 1mf – 9 – mf#02060 – cn Canadiana [241]

Bibliotheque universelle des voyages : on notice complete et raisonnee de tous les voyages anciens et modernes dans les differentes parties du monde... / Boucher de la Richarderie, G – Paris. 6v. 1808 – 36mf – 9 – mf#HT-274 – ne IDC [910]

Bibliotheque universelle des voyages effectues par mer : ou par terre dans les diverses parties du monde depuis les premieres decouvertes jusqu'a nos jours... / Montemont, Albert – Paris: Armand-Aubree. 46v. 1833-36 [mf ed 1984] – 46v on 1mf – 9 – mf#46428 – cn Canadiana [910]

Bibliotheque universelle des voyages effectues par mer ou par terre dans les diverses parties du monde, depuis les premieres decouvertes jusqu'a nos jours : contenant la description des moeurs, coutumes, gouvernemens, cultes, sciences et arts... / ed by Montemont, Albert – Paris 1833-36 [mf ed Hildesheim 1995-98] – 46v on 138mf – 9 – €828.00 – 3-487-26466-8 – gw Olms [910]

Les bibliotheques paroissiales : bibliographie analytique de la litterature francaise parue sur le sujet dans la province de quebec / Turcotte, Fernande – 1952 [mf ed 1979] – (= ser Bibliographies du cours...1947-66) – 1mf – 9 – (with ind) – mf#SEM105P4 – cn Bibl Nat [440]

Bibliotheques publiques : les breviaires manuscrits des bibliotheques publiques de france / ed by Leroquais, V – Paris, 1934. 5v – 54 – 9 – mf#O-477 – ne IDC [700]

Bibliotheques publiques : les pontificaux manuscrits... / ed by Leroquais, V – Paris, 1937. 4v – 28mf – 9 – mf#O-478 – ne IDC [700]

Biblische archaeologie / Schegg, Peter; ed by Wirthmueller, Johann Baptist – Freiburg i B; St Louis MO: Herder 1887 [mf ed 1989] – (= ser Theologische bibliothek) – 2v on 2mf – 9 – 0-7905-2059-1 – mf#1987-2059 – us ATLA [220]

Biblische dogmatik alten und neuen testaments : oder kritische darstellung der religionslehre des hebraismus, des judenthums und urchristenthums, zum gebrauch akademischer vorlesungen / Wette, Wilhelm Martin Leberecht de – 3rd impr ed. Berlin: G Reimer, 1831 [mf ed 1984] – 4mf – 9 – 0-8370-1094-2 – mf#1984-4492 – us ATLA [242]

Biblische exegese in ihren beziehungen zur semitischen philologie / Yahuda, Abraham Shalom – Berlin, Germany. 1906 – 1r – us UF Libraries [470]

Biblische geschichte : der heiligen schrift / Kurtz, Johann Heinrich – 12. Aufl. Berlin: Justus Albert Wohlgemuth, 1865 – 1mf – 9 – 0-8370-4020-5 – mf#1985-2020 – us ATLA [220]

Die biblische geschichte des alten testaments : kurze auslegung der alttestamentlichen geschichtsbuecher / Stoeckhardt, George – St Louis, MO: Concordia Pub House, 1906 – 1mf – 9 – 0-524-03996-8 – mf#1992-0039 – us ATLA [221]

Die biblische geschichte des neuen testaments : kurze auslegung der evangelien und apostelgeschichte / Stoeckhardt, George – St Louis, MO: Concordia Pub House, 1906 – 1mf – 9 – 0-524-05422-3 – mf#1992-0432 – us ATLA [225]

Biblische hermeneutik / Hofmann, Johann Christian Konrad von; ed by Volck, Wilhelm – Noerdlingen: C H Beck, 1880 – 1mf – 9 – 0-524-05103-8 – mf#1992-0324 – us ATLA [220]

Biblische koenigsdramen in der franzoesischen tragoedie des 16. und 17. jahrhunderts / Carlebach, David – Halberstadt, 1912 (mf ed 1993) – 1mf – 9 – €24.00 – 3-89349-331-X – mf#DHS-AR 185 – gw Frankfurter [440]

Biblische legenden der muselmaenner / Weil, Gustav – Frankfurt a M: Literarische Anstalt, 1845 – 1mf – 9 – 0-524-02103-1 – mf#1990-2867 – us ATLA [260]

Biblische liebeslieder : das sogenannte hohelied salomos: unter steter beruecksichtigung der uebersetzungen goethes und herders im versmasse der urschrift / Haupt, Paul – Leipzig: J C Hinrichs; Baltimore: John Hopkins, 1907 – 1mf – 9 – 0-8370-3530-9 – (incl ind comapring the version of the song of solomon in the bk with the traditional text) – mf#1985-1530 – us ATLA [220]

Biblische mythologie des alten und neuen testamentes : versuch einer neuen theorie zur aufhellung der dunkelheiten und scheinbaren widersprueche in den canonischen buechern der juden und christen / Nork, F – Stuttgart: JF Cast, 1842-1843 – 3mf – 9 – 0-524-06522-5 – mf#1992-0906 – us ATLA [220]

Die biblische poesie : besonders die alttestamentliche, und ihre behandlung in der schule / Traeckner, Chr – Gotha: C F Thienemann, 1902 – 1mf – 9 – 0-8370-8554-3 – (incl bibl ref) – mf#1986-2554 – us ATLA [270]

Der biblische samson / Zapletal, Vincenz – Freiburg (Schweiz): O Gschwend, 1906 – 1mf – 9 – 0-8370-9676-6 – mf#1986-3676 – us ATLA [221]

Der biblische schoepfungsbericht : ein exegetischer versuch / Hummelauer, Franz von – Freiburg i B: Herder, 1877 – (= ser Ergaenzungshefte zu den "Stimmen aus Maria-Laach") – 1mf – 9 – 0-524-08607-9 – mf#1993-0042 – us ATLA [220]

Der biblische schoepfungsbericht (gen 1, 1 bis 2, 3) / Kaulen, Franz – Freiburg i B: Herder, 1902 – 1mf – 9 – 0-524-05914-4 – mf#1992-0671 – us ATLA [221]

Biblische sprache und biblische motive in wielands oberon / Biach, Adolf – Bruex: M Herzum, 1897 [mf ed 1992] – 31p – 1 – (incl bibl ref) – mf#7763 – us Wisconsin U Libr [430]

Die biblische theologie : einleitung in's alte und neue testament und darstellung des lehrgehaltes der biblischen buecher nach ihrer entstehung und ihrem geschichtlichen verhaeltniss: ein handbuch zum selbstunterricht / Noack, Ludwig – Halle: C E M Pfeffer, 1853 – 1mf – 9 – 0-8370-9722-3 – (incl bibl ref) – mf#1986-3722 – us ATLA [220]

Biblische theologie des alten testaments / Kautzsch, E – Tuebingen: J C B Mohr, 1911 – 1mf – 9 – 0-7905-1121-5 – (incl bibliographies and ind) – mf#1987-1121 – us ATLA [221]

Biblische theologie des neuen testamentes *see* Biblical theology of the new testament

Die biblische theologie und ihre gegner / Myrberg, Otto Ferdinand – Guetersloh: C Bertelsmann, 1892 – 1mf – 9 – 0-8370-8847-X – mf#1986-2847 – us ATLA [220]

Biblische und babylonische urgeschichte *see* The babylonian and the hebrew genesis

Die biblische und die babylonische gottesidee : die israelitische gottesaufassung im lichte der altorientalischen religionsgeschichte / Hehn, Johannes – Leipzig: J C Hinrichs 1913 [mf ed 1989] – 2mf [ill] – 9 – 0-7905-1151-7 – (incl bibl ref & ind) – mf#1987-1151 – us ATLA [221]

Die biblische und kirchliche lehre vom antichrist / Philippi, Ferdinand – Guetersloh: Bertelsmann, 1877 [mf ed 1991] – 1mf – 9 – 0-7905-9579-6 – mf#1989-1304 – us ATLA [220]

Der biblische urgeschichte / Nikel, Johannes – 3. aufl. Muenster in Westf: Aschendorff 1910 [mf ed 1989] – (= ser Biblische zeitfragen 2/3) – 1mf – 9 – 0-7905-3206-9 – (incl bibl ref) – mf#1987-3206 – us ATLA [221]

Die biblische urgeschichte : in ihrem verhaeltnis zu den urzeitsagen anderer voelker, zu den israelitischen volkserzaehlungen und zum ganzen der heiligen schrift / Lotz, Wilhelm – Leipzig: A Deichert (Georg Boehme), 1907 – 1mf – 9 – 0-8370-4182-1 – mf#1985-2182 – us ATLA [220]

Die biblische wahrheit in ihrer harmonie mit natur und geschichte : ein lehr- und lesebuch zur orientirung in den religioesen wirren unserer zeit / Hamberger, Julius – Muenchen: Carl Merhoff, 1877 – 1mf – 9 – 0-8370-5116-9 – mf#1985-3116 – us ATLA [220]

Biblische zeitfragen – Muenster in Westfalen: Aschendorff [v9 (1919-21)] [mf ed 2005] – 1r with other items – 1 – (gemeinverstaendlich eroertert; ein broschuerenzyklus) – mf1074 – us ATLA [220]

Die biblische zeitrechnung : vom auszuge aus aegypten bis zum beginne der babylonischen gefangenschaft / Lederer, Carl – Speier: In Kommission der Ferd. Kleeberger'schen Buchhandlung, [1888?] – 1mf – 9 – 0-8370-4077-9 – (incl bibl ref) – mf#1985-2077 – us ATLA [220]

Biblische zeitschrift – 1(1903)-24(1938/39) – 191mf – 9 – €364.00 – ne Slangenburg [220]

Biblische zeitschrift – Paderborn, Germany 1977+ – 1,5,9 – ISSN: 0006-2014 – mf#11395 – us UMI ProQuest [220]

Die biblischen frauen des alten testamentes / Zschokke, Hermann – Freiburg im Breisgau; St Louis: Herder, 1882 – 2mf – 9 – 0-8370-5972-0 – (incl bibl ref and index) – mf#1985-3972 – us ATLA [221]

Die biblischen vorstellungen vom teufel und ihr religioeser werth : ein beitrag zu der frage, giebt es einen teufel? ist der teufel ein gegenstand des christlichen glaubens? / Laengin, Georg – Leipzig: Wigand, 1890 – 1mf – 9 – 0-524-01779-4 – mf#1990-2627 – us ATLA [220]

Die biblischen wundergeschichten / Wimmer, Richard – 3. u 4. Aufl. Freiburg i.B.: J C B Mohr (Paul Siebeck), 1890 – 1mf – 9 – 0-8370-5867-8 – mf#1985-3867 – us ATLA [240]

Biblischer commentar ueber den propheten iesaia *see* Biblical commentary on the prophecies of isaiah

Biblischer commentar ueber die psalmen *see* Commentary on the book of psalms

Biblisches realwoerterbuch : zum handgebrauch fuer studirende, candidaten, gymnasiallehrer und prediger / Winer, Georg Benedikt – 3. sehr verb und verm Aufl. Leipzig: CH Reclam, 1847-1848 – 4mf – 9 – 0-8370-1846-3 – mf#1987-6234 – us ATLA [052]

Die biblisch-prophetische theologie : ihre fortbildung durch chr. a. crusius und ihre neueste entwickelung seit der christologie hengstenbergs / Delitzsch, Franz – Leipzig: Gebauer, 1845 – (= ser Biblisch-theologische und apologetisch-kritische studien) – 1mf – 9 – 0-7905-3328-6 – (incl bibl ref) – mf#1987-3328 – us ATLA [240]

Biblishe mayselakh / Pat, Jacob – Byalistok, Poland. 191-? – 1r – us UF Libraries [939]

Biblisk historia for hemmet och skolan / ed by Scandinavian Evangelical Lutheran Augustana Synod of North America – Rock Island, IL: Augustana-synodens, 1887 – 1mf – 9 – 0-524-05197-6 – mf#1991-2233 – us ATLA [220]

Bibliska beraettelser ur nya och gamla testamentet : barnens foersta laerobak i bibliska historien / Zetterstrand, Ernst Adrian – Rock Island IL: Augustana Book Concern c1904 [mf ed 1992] – 1mf – 9 – 0-524-06455-5 – (in swedish) – mf#1991-2577 – us ATLA [220]

Bibliska studier [1st series] = Etudes bibliques. premiere serie / Godet, Frederic Louis – Upsala: W Schultz 1879 [mf ed 1986] – 1mf – 9 – 0-8370-9472-0 – (incl & bibl ref) – mf#1986-3472 – us ATLA [220]

Bibliska studier [2nd series] = Etudes bibliques. deuxieme serie / Godet, Frederic Louis – Upsala: W Schultz 1878 [mf ed 1986] – 1mf – 9 – 0-8370-9471-2 – (incl & bibl ref; in swedish) – mf#1986-3471 – us ATLA [225]

Bibliusoegur og agrip af kirkjusoegunni handa boernum / Klaveness, Th & Jonsson, Sigurur – 4. utgafa Reykjavik: Bokaverzlun Sigfusar Eymundssonar, 1910 [mf ed 1989] – 1mf – 9 – 0-7905-3081-3 – (in icelandic) – mf#1987-3081 – us ATLA [220]

Bican, Ahmed *see* The divan project

Bicen iowa / Iowa American Revolution Bicentennial Commission – 1971 may-1976 aug – 1r – 1 – mf#354461 – us WHS [978]

Bi-centenario do nascimento do patriarca da indepe – Brasilia, Brazil. 1964 – 1r – us UF Libraries [972]

Bicentenary of the assembly of divines at westminster – Edinburgh, Scotland. 1843 – 1r – us UF Libraries [240]

Bicentenary of the founding of the colony of sierra leone, 1787-1987 : international symposium on sierra leone, may 19-21 1987: miatta conference centre, brookfields, freetown / International Symposium on Sierra Leone (1987: Freetown, Sierra Leone) – [Freetown: s.n, 1987?] – 1r – 1 – us CRL [960]

Bicentennial / Tussekiah Baptist Church – 1776-1976 – 1 – 5.00 – us Southern Baptist [242]

Bicentennial banner / Baton Rouge Bicentennial Commission – v1 n1-2 [1975 jun-oct] – 1r – 1 – mf#352191 – us WHS [975]

Bicentennial celebration commission newsletter / New Jersey Bicentennial Commission – v1-3 n2 [1974 sep-1976 dec] – 1r – 1 – mf#354460 – us WHS [975]

Bicentennial chronicle – v1 n1-v3 n4 [1970 oct-1972 summer] – 1r – 1 – (cont by: colonial heritage) – mf#1053343 – us WHS [071]

Bicentennial Commission of Pennsylvania *see* Pennsylvania bicentennial news

Bicentennial Council of the Thirteen Original States *see* Newsletter of the great american achievements program

Bicentennial Council of the Thirteen Original States Fund *see* Historical quarterly of the bicentennial council of the thirteen original states fund, inc

Bicentennial in texas / American Revolution Bicentennial Commission of Texas – v1 n1-v1 n4 [1972 fall-1973 spring, 1973 jul 4] – 1r – 1 – (cont by: emergence '76 (arlington tx)) – mf#811593 – us WHS [975]

Bicentennial in texas / American Revolution Bicentennial Commission of Texas – 1976 jan/feb-summer – 1r – 1 – (cont: emergence '76 (arlington tx)) – mf#366598 – us WHS [975]

Bicentennial news / American Revolution Bicentennial Commission – 1973 apr 25-1976 jul 5 – 1r – 1 – mf#354477 – us WHS [975]

Bicentennial times / American Revolution Bicentennial Administration – v1 n1-v3 n12 [1973 dec-1976 dec] – 1r – 1 – (cont: bicentennial newsletter) – mf#354472 – us WHS [975]

Bicentennial times / Revolutionary War Bicentennial Commission – v1 n1-14 [1973 dec-1976 jul] – 1r – 1 – mf#599494 – us WHS [975]

Bicester advertiser (a) – Bicester, England 7 july 1855-26 jan 1866 (wkly) – 5r – 1 – (discontinued) – uk British Libr Newspaper [072]

Bicester advertiser (b) – Bicester, England 10 jan 1879- [mf 1986-] – 1 – (wanting: 1896, 1897) – uk British Libr Newspaper [072]

Bicester herald – Bicester, England 9 jun 1855-6 jul 1917 – 1 – (wanting: 1897) – uk British Libr Newspaper [072]

Bicester review – Bicester, England 4 jul 1986-4 may 1990 – 1 – (cont by: review (bicester) 11 may 1990-) – uk British Libr Newspaper [072]

Biche au bois : ou, le royaume des fees / Cogniard, Theodore – Paris, France. 1845? – 1r – us UF Libraries [440]

Bicinia gallica, latina et germanica – 1545 – (= ser Mssa) – 4mf – 9 – €60.00 – mfchl 90 – gw Fischer [780]

Bicinia gallica, latina, germanica. secundus tomus – 1545 – (= ser Mssa) – 6mf – 9 – €80.00 – mfchl 91 – gw Fischer [780]

Bicinia sacra... / Friderici, Daniel – 1642 – (= ser Mssa) – 2mf – 9 – €35.00 – mfchl 225 – gw Fischer [780]

Bicinia, sive cantiones suavissimae duarum vocum – 1590 / 1609 – (= ser Mssa) – 2mf/2mf – 9 – €35.00/€35.00 – mfchl 143 / 144 – gw Fischer [780]

Bick, Ch *see* Ugwalo olutsha lwokufunda isi-ngisi

Bickel, Gerhard *see* Axicon und ringpupille als bildformende elemente

Bickell, Gustav *see*
- The lord's supper and the passover ritual
- Metrices biblicae regulae exemplis illustratae
- Outlines of hebrew grammar
- Der prediger ueber den wert des daseins

Bickelmann, Ingeborg *see* Goethes "werther" im urteil des 19. jahrhunderts

Bickermann, Joseph *see* K samopoznaniiu evreia

Bickerstaff, Isaac *see* Padlock

Bickersteth, Edward Henry *see*
- Christian duty of feeding the poor of the flock
- Convictions of balaam
- Discourse on justification by faith
- Divine warning to the church, at this time
- National humiliation and prayer
- Practical address to british christians
- A practical and explanatory commentary on the new testament
- The rock of ages
- Sacred chronology, and the arrangement of the apocalypse
- Scriptural guide for ministers in these days
- The works of rev. e. bickersteth

Bickersteth, Emily *see* Extracts from woman's service on the lord's day

Bickersteth, H V *see* Confirmation

Bickersteth, Marion (forsyth) *see* Edward bickersteth

Bickersteth, Montagu Cyril *see* Letters to a godson

Bickersteth, Robert *see* Convictions of agrippa

Bickerton, Derek *see* Murders of boysie singh

Bickley, Augustus Charles *see* George fox and the early quakers

Bickmore, A S *see* Travels in the east indian archipelago

Bickum, Bonnie D *see* The history of graded exercise testing in cardiac rehabilitation

Bi-county argus – De Soto, Ferryville etc WI. 1917 mar 22, 1919 jul 17 – 1r – 1 – mf#1220917 – us WHS [071]

The bicycle – Hamilton [Ont.: s.n., 1882-18– or 19–] – 9 – mf#P04858 – cn Canadiana [790]

Bicycling – Emmaus PA 1976+ – 1,5,9 – ISSN: 0006-2073 – mf#10977 – us UMI ProQuest [790]

Bicz bozy = God's whip – Chicago IL, 1912, 1915-17, 1934 – 1r – 1 – (polish newspaper) – us IHRC [071]

Bidar : its history and monuments / Yazdani, Ghulam – London: Oxford University Press, 1947 – (= ser Samp: indian books) – us CRL [954]

Biddenham – (= ser Bedfordshire parish register series) – 1mf – 9 – £3.00 – uk BedsFHS [929]

Biddenham, st james monumental inscriptions – Bedfordshire Family HS 1977 – (= ser Bedfordshire parish register series) – 1mf – 9 – £1.25 – uk BedsFHS [929]

Biddle, George Washington *see* A sketch of the professional and judicial character of the late george sharswood

Biddle, Nicholas *see* Papers

Biddle, William P *see* The baptist hymn book

Biddulph, John *see* The pirates of malabar

Biddulph manuscript, extracts from the... : from hereford city library – 1r – 1 – mf#3638 – uk Microform Academic [980]

Biddulph, Thomas Tregenna *see*
- Christian charity, exerting itself by means of missionary incitement for the correction of hindoo immorality
- Septuagenarian confession of faith

Bidegain, Jean *see*
- Le grand orient de france
- Magistrature & justice maconniques

Bidermann, E *see* Ehren-gebu oesterreichischer helden-tugenden

Bidez, J *see*
- Kirchengeschichte
- Philostorgius kirchengeschichte

Bidez, Joseph *see*
- La tradition manuscrite de sozom ene et la tripartite de theodore le lecteur
- La tradition manuscrite de sozomene

Bidhi pracam tap bir khae / Bejr Sal – Bhnam Ben: Butdh Sasanapandity 2508 [1966] [mf ed 1990] – 1r with other items – 1 – (in khmer) – mf#mf-10289 seam reel 117/1 [§] – us CRL [520]

[Bidloo, G] *see*
- De publijke intrede van william de 3...gedaen in 's gravenhage op den 5 februarij 1691...
- Relation du voyage de sa majest, britannique en hollande
- Relation du voyage de sa majeste britannique en hollande...le 31 de janvier, jusqu'...son retour ...au mois d'avril 1691...

Bidpai *see* The earliest english version of the fables of bidpai / "the morall philosophie of doni" by sir thomas north

Bidrag til en skildring af guinea-kysten og dens indbyggere : og til en beskrivelse over de danske colonier paa denne kyst, samlede under mit ophold i afrika i aarene 1805 til 1809 / Monrad, Hans Christian – Kobenhavn: A Seidelin, 1822 – 1 – us CRL [960]

Bidrag till finlands officiela statistik 1885-1914 / Finland. Statisticka Centralbyran – (= ser European official statistical serials, 1841-1984) – 39mf – 9 – uk Chadwyck [314]

Bidston, st oswald – (= ser Cheshire monumental inscriptions) – 2mf – 9 – £4.00 – mf#8 – uk CheshireFHS [929]

Bidston, st oswald 1581-1700 – (= ser Cheshire church registers) – 3mf – 9 – £4.50 – mf#375 – uk CheshireFHS [929]

Bidwell, Charles Toll *see* Isthmus of panama

Bidwell, George *see* Forging his chains

Bidwell, John *see* John bidwell, pioneer

Bidwell news *see* [Fort bidwell-] bidwell gold nugget (bidwell news)

Bidyabinod, B B *see* Fragment of a prajnaparamita manuscript from central asia

Bie, C de *see*
- Echos weder-klanck passende op den gheestelycken wecker tot godtvruchtighe oeffeningen...
- Faems weer-galm der neder-duytsche poesie van cornelio de bie tot lyer...
- Het gulden cabinet van de edele vrye schilderconst ontsloten door den lanck ghewenschten vrede tusschen de twee machtighe croonen van spaignien en vranckryk
- Den sedighen toet-steen vande onverdraeghelycke welde verthoont in 't leven van den verloren sone

Bie, L Th *see* Koey-tjoe say ma-tiauw

[Bieber-] argus gazette – CA. jun 1948-nov 1956 – 4r – 1 – $240.00 – mf#C02061b – us Library Micro [071]

[Bieber-] big valley gazette – CA. jan 29 1893-apr 9 1948 – 22r – 1 – $1320.00 – mf#C02062 – us Library Micro [071]

Bieber, Hugo *see* Der kampf um die tradition

[Bieber-] mountain tribune – CA. may 6 1881-dec 1892 – 3r – 1 – $180.00 – mf#C03584 – us Library Micro [071]

Biebuyck, Daniel P *see* Mitamba

Biechteler, Benedikt *see* Vox suprema oloris parthenii

Biedenkapp, Georg *see* Brennende lieder und strophen

Biedermann, Alois Emanuel *see*
- Christliche dogmatik
- Die freie theologie, oder, philosophie und christenthum in streit und frieden
- Die pharisaeer und sadducaeer

Biedermann, Flodoard, Freiherr von *see*
- Goethe als raetseldichter
- Goethe und dresden

Biedermann, Flodoard, Freiherr von [comp] *see* Chronik von goethes leben

Biedermann, Karl *see* Vorlesungen ueber socialismus und socialpolitik

Biedermann, Michael C *see* Correlation between muscle relaxation and sarcoplasmic reticulum ca2+-atpase during fatigue

Biedermann, Woldemar, Freiherr von *see*
- Goethes briefwechsel mit friedrich rochlitz
- Goethes gespraeche
- Zu goethe's gedichten

Das biedermeier im spiegel seiner zeit : briefe, tagebuecher, memoiren, volksszenen und aehnliche dokumente / Hermann, Georg [comp] – Berlin: Deutsches Verlagshaus, 1913 [mf ed 1989] – 415p (ill) – 1 – mf#7051 – us Wisconsin U Libr [430]

Biederstaedt, Birgit *see* Aspekte des funktional-semantischen feldes der art und weise im modernen englisch

Biednota : izd. tsentralnogo komiteta rossiiskoi kommunisticheskoi partii (bolshevikov) – Moskva: Komitet, [sep 8 1921-apr 5 1923] – 1r – us CRL [320]

Bieger, Juergen *see* Mein freund johannes

Biehler, E *see* Four methods of teaching english to maswina

Biel, G *see*
- Canonis misse expositio ediderunt
- Sermones dominicales de tempore...

Biel, Gabriel *see*
- Collectorium in 4 libros sententiarum
- Sacri canonis missae tam mystica quam litteralis expositio
- Sermones de festivitatibus christi – passionis doninicae sermo historialis
- Sermones de festivitatibus gloriosae virginis mariae
- Sermones de sanctis. in officium industrii henrici gran
- Sermones dominicales de tempore

Bieleck, Rudolph *see*
- Des vaters fluch
- Der menschenfeind

Bielefeld, Charles Frederick *see* On the use of improved papier-mache in furniture...interior decoration...art

Bielefelder kreisblatt *see* Oeffentliche anzeigen des districts...

Bieler, L *see* The life and legend of st patrick

Bielfeld, H *see* A guide to painting on glass

Bielfeld, H A *see* Gedichte

Bielitz-bialer deutsche zeitung – Bielitz-Biala (Bielsko-Biala PL), 1924 16 apr-1934 okt – 1 – (title varies: sep 1930: beskidenlaendische deutsche zeitung) – gw Misc Inst [077]

Bielitzer volksstimme – Bielitz-Biala (Bielsko-Biala PL), 1925 aug-dec, 1926 jan-oct, 1927-30 – 1 – (title varies: 4 sep 1920: volksstimme; with suppl) – gw Misc Inst [077]

Bielschowsky, Albert *see*
- Goethe, sein leben und seine werke
- The life of goethe

Biemond, Catharina Elizabeth *see* Simulasie in geskiedenisonderrig in die primere skool

Bien = The bee – San Francisco CA. 1941 jun 5-1942 dec 31, 1943 jan 7-1944 dec 31, 1954 jan 7-1955 jun 16, 1955 jun 23-1956 nov 8, 1956 nov 15-1958 apr 3, 1958 apr 10-1958 dec 25 – 6r – 1 – mf#770042 – us WHS [071]

Der bien boeck / Thomas of Cantimpre – Peter van Os premter tot Swolle, 1488 – €19.00 – ne Slangenburg [241]

Le bien public – Paris: Impr Dubuisson et Ce, mar 28-29,31, apr 1-3,6-8,11-14,18-19,21 1871 – (filmed as pt of: commune de paris newspapers. on these reels, newspapers are filmed chronologically) – us CRL [074]

Le bien public – n1-193. Paris. 5 mars 1871-30 juin 1878, 8 juin 1882-21 janv 1883; no. 99-144. 7 janv 1883 sic-17 janv 1884; no. 1-7. 15 nov-24 dec 1891; no. 1-32. janv-9 oct 1892 [daily] – 1 – (mq n8, 24, 29; no. 1. 3 dec 1893; no. 1-2. 4 juil., 14 dec 1894) – fr ACRPP [320]

Le bien public – Paris. 24 mai-12 dec 1848 – 1 – fr ACRPP [073]

Le bien public *see* Le pour et le contre

El bien publico – Montevideo, Uruguay. sep 1955-dec 1962 [daily] – 49r – 1 – (imperfect) – uk British Libr Newspaper [079]

Bienang queralandalanan ning g'inutang jesuchristo, cabang quetiya quing yatu, angga quing pangalasanat pangamate quing cruz ulin ing pamanaclungna quing casalanan tamu – Manila: Fajardo 1905 [mf ed Bloomington IN: Indiana Uni Lib, Preservation Dept 1984] – 1r – 1 – us Indiana Preservation [241]

Biencourt de Poutrincourt et de Saint-Just, Jean de, Baron *see* Factum du proces entre jean de biencourt, sr de poutrincourt et les peres biard et masse, jesuites

Die biene – 1808-10 [mf ed 1997] – (= ser Die zeitschriften des august von kotzebue) – 19mf – 9 – €200.00 – 3-89131-232-6 – gw Fischer [430]

Die biene – Karlsruhe DE, 1849 25 jul-1922 31 may – 1 – (filmed by other misc inst: 1849 25 jul-30 dec. title varies: 1 mar 1850: badische landesblaetter; 1 jun 1850: badische landeszeitung. incl suppl: karlsruher unterhaltungsblatt 1850 [1r]) – gw Misc Inst [074]

Die biene – Cleveland, OH. aug 9 1873-dec 30 1882 – (= ser Ethnic newspapers) – 4r – 1 – (german language labor newspaper, publ at varying frequencies) – us Western Res [071]

'Die biene auf dem missionsfelde' / Gossner Mission, Berlin – 1834-1857 – 1r – 1 – mf#pmb552 – at Pacific Mss [240]

Die biene oder neue kleine schriften / ed by Kotzebue, August von – Koenigsberg 1808-10 – (= ser Dz) – 7v on 21mf – 9 – €210.00 – mf#k/n6123 – gw Olms [500]

Biener, Christian Gottlob *see* Commentarii de origine et progressu legum iuriumque germanicorum

Bienert, T *see* Lepidopterologische ergebnisse einer reise in persien in den jahren 1858 und 1859

Le bien-etre social – Journal politique hebdomadaire. Bruxelles. 1858-aout 1860. I.F.H.S – 1 – fr ACRPP [949]

Le bienfaiteur – Joliette [Quebec]: Comite du Monument Joliette, [1892-189- ou 19–] – 9 – mf#P04091 – cn Canadiana [971]

Le bienhereux martin de porres...paris / Fumet, Stanislas – Madrid: Razon y Fe, 1935 – 1 – sp Bibl Santa Ana [920]

Bienheureuse marguerite d'youville : fondatrice des soeurs de la charite (soeurs grises) de montreal: bibliographie canadienne, 1938-1949 / Saint-Hyacinthe, soeur – 1963 [mf ed 1979] – (= ser Bibliographies du cours...1947-66) – 2mf – 9 – (with ind; pref by abbe jean mercier) – mf#SEM105P4 – cn Bibl Nat [241]

Le bienheureux fra giovanni angelico de fiesole (1387-1455) / Cochin, Henry – 3. ed. Paris: V Lecoffre, 1906 – (= ser Les Saints) – 1mf – 9 – 0-524-03277-7 – (incl bibl ref) – mf#1990-0888 – us ATLA [750]

Bienheureux martyrs de l'ouganda – Namur, Belgium. 1934 – 1r – us UF Libraries [960]

Biennial – 1980 jun-1981 dec – 1 – mf#713945 – us WHS [071]

Biennial and report of the president, secretary and official auditor / Bricklayers, Masons, and Plasterers International Union of America – 1918/20 – 1r – 1 – (cont; annual report of president and secretary of the bricklayers, masons and plasterers'international union of america, bricklayers, masons, and plasterers international union of america) – mf#1427152 – us WHS [690]

Biennial report / Arizona. Territory. Prisons – Phoenix. 1891/92-97/98 – 1 – $23.00r – us L of C Photodup [360]

Biennial report / Florida Geological Survey – Tallahassee, FL. 15th-16th. 1987-1990 – 1r – us UF Libraries [550]

Biennial report / Florida Geological Survey – Tallahassee, FL. 4th-14th. 1939-1960 – 11r – us UF Libraries [550]

Biennial report of the department of local affairs and development – 1967/69-1977/79 – 1r – 1 – mf#277291 – us WHS [350]

Biennial report of the free library commission of wisconsin / Wisconsin Free Library Commission – 1895/96-1910/12 – 1r – 1 – (cont by: biennial report of the wisconsin free library commission) – mf#569754 – us WHS [020]

Biennial report of the industrial school for colored girls of delaware / Industrial School for Colored Girls of Delaware (Marshallton DE) – Wilmington DE: C H Gray, 1930-32 [mf ed 2004] – 2v on 1r – 1 – mf#2004-s028 – us ATLA [365]

Biennial report of the secretary of state / Nevada. State Dept – Carson City. 1883-1884, 1889-1890 – 1 – us NY Public [978]

Biennial report of the state board of control of wisconsin – 1903/04-1905/06, 1925/26-1927/28 – 1r – 1 – mf#162933 – us WHS [350]

Biennial report of the state historical and natural history society of colorado / Colorado. State Historical and Natural History Society – Denver, CO: The Collier & Cleveland Lithographing Co, 1889 (mf ed 1976) – 1r – 1 – mf#MF C714hnhb – us Colorado Hist [978]

Biennial report of the wisconsin free library commission / Wisconsin Free Library Commission – 1912/14-1938/40 – 1r – 1 – (cont; biennial report of the free library commission of wisconsin) – mf#569758 – us WHS [020]

Biennial report to state board of conservation / Florida Geological Survey – Tallahassee, FL. 1st-3rd. 1933-1938 – 1r – us UF Libraries [550]

Biennial reports (a-g), 1882-1936 / Alabama – Quarterly. 1935-80. 16 reels – 1 – $35.00r – us Trans-Media [340]

Biennial survey of education – Washington DC 1916-52 – 1 – mf#5790 – us UMI ProQuest [370]

[Biens, C P] *see*
- Handt-boecxken der christelijcke gedichten, sinne-beelden ende liedekens, tot troost ende vermaeck der gelooviger zielen
- Profytelyck cabinet, voor den christelijcken jongelingh

Bienvenida. Ayuntamiento *see*
- Fiestas patronales en honor de ntra sra de los milagros, 1979
- Fiestas patronales en honor de nuestra senora de los milagros. septiembre 1970

Bienvenue a son altesse royale le duc d'york et de cornwall, sept 1901 / Frechette, Louis – Montreal: Granger Freres, 1901 – 1mf – 9 – 0-665-74258-4 – mf#74258 – cn Canadiana [810]

Bienville first baptist church. bienville, louisiana : church records – 1894-Apr 1969 – 1 – us Southern Baptist [242]

Bierbau, Otto Julius *see* Stilpe

Bierbaum, Otto Julius *see*
- Gesammelte werke
- Gugeline
- Irrgarten der liebe
- Moderner musen-almanach
- Die schatulle des grafen thruemmel und andere nachgelassene gedichte

Bierbaum, Otto Julius et al *see* Deutsche chansons

Bierbower, Austin *see* The virtues and their reasons

Bierck, Harold Alfred *see* Vida publica de don pedro gual...

La biere et les boissons fermentees – Paris, France 15 jan 1899-may/jul 1913 – 1 – uk British Libr Newspaper [074]

Bierer, Everard *see* The evolution of religions

Biergans, F Th *see* Brutus

Biermann, Johannes *see* Sachenrecht

Bierwirth, Gerhard *see* Die problematik des englischen schauerromans

El bierzo : nuevas lapidas romanas / Roso de Luna, Mario – Madrid: Fortanet, 1912 – sp Bibl Santa Ana [946]

Biesanz, John Berry *see* Costa rican life

Biese, Alfred *see*
- Deutsche literaturgeschichte
- Fritz reuter, heinrich seidel und der humor in der neueren deutschen dichtung
- Goethes bedeutung fuer die gegenwart
- Heinrich seidel und der deutsche humor
- Theodor storm und der moderne realismus

Biesenbach, G B *see* Organisasieklimaatskepping as bestuurstaak van die skoolhoof in skole onder die kaaplandse departement van onderwys

Biesenthal, Johannes Heinrich Raphael *see* Das trostschreiben des apostels paulus an die hebraeer

Biester, Johann Erich *see*
- Berlinische blaetter
- Berlinische monatsschrift 1783-96 / berlinische blaetter 1797-98 / neue berlinische monatsschrift 1799-1811
- Neue berlinische monatsschrift

Biesterveld, Petrus *see* Calvijn als bedienaar des woords

Die bif : blaetter idealer frauenfreundschaften – Berlin DE, 1925 n2, 3 – 1r – 1 – gw Misc Inst [305]

Biffart, Max *see* Deutschland

Bifur – n1-8. Paris. mai 1929-juin 1931 – 1 – fr ACRPP [073]

The big apple : "the latest modern dance" / Goldman, Norma – Newark, NJ: Dancers Art Guild, c1938 – 1 – mf#*ZBD-*MGO pv16 – Located: NYPL – us Misc Inst [790]

Big apple dyke news – v1 n1-v4 n2 [1981 mar-1984 feb/mar], v5 n1-v6 n1 [1985 spring-1986 spring], v8 n1 [1988 summer] – 1r – 1 – mf#1330856 – us WHS [305]

[Big bear-] big bear lake limelight – CA. aug 1848-1958 – 3r – 1 – $180.00 – mf#C02066 – us Library Micro [071]

[Big bear-] big bear life – CA. apr 1929-1959 – 1r – 1 – $60.00 – mf#R02064 – us Library Micro [071]

[Big bear lake-] the grizzly – CA. may 1941-1961 – 13r – 1 – $780.00 – mf#R02065 – us Library Micro [071]

[Big bear-] mountaineer – CA. apr 1933-nov 1938 – 1r – 1 – $60.00 – mf#R02063 – us Library Micro [071]

Big bend-vernon bulletin – Mukwonago WI. 1982 may 18-1983 jun 28, 1983 jul-1984 dec, 1985 jan-1986 apr 29, 1986 may 6-1987 feb 24, 1987 mar-dec, 1988, 1989, 1990, 1991, 1992, 1993 jan 4-dec 27 – 11r – 1 – (cont by: Muskego times-record; Chief (Mukwonago WI)) – mf#2907244 – us WHS [071]

Big bone baptist church. union city, kentucky : church records – 1823-1948; Ladies Missionary Society and Aid Society, 1913-27. 1744p – 1 – 69.76 – us Southern Baptist [242]

Big boulevard – Long Beach CA 1973-75 – 1,5,9 – mf#8641 – us UMI ProQuest [071]

Big business and the public / Brookings, Robert Somers – Garden City, NY: Country Life Press, 1926 (mf ed 19–) – 14p – mf#ZT-TN pv89 n8 – us Harvard [338]

Big country news – Alberta, CN. jan 1975-dec 1978 – 2r – 1 – cn Commonwealth Imaging [071]

Big Creek Association (TN) *see* Annual session

Big creek baptist church : church records – Coolidge, GA. 1440p. aug 1882-sep 1982 – 1 – $64.80 – us Southern Baptist [242]

Big creek baptist church. anna, illinois : church records – 1852-1983 – 1 – 5.00 – us Southern Baptist [242]

Big cypress / Munroe, Kirk – Boston, MA. 1894 – 1r – us UF Libraries [978]

Big e magazine : familygram of the uss enterprise – 1990 fall, 1991 fall – 1r – 1 – mf#2341567 – us WHS [355]

Big farmer – Frankfort IL 1950-79 – 1,5,9 – (cont by: big farmer entrepreneur) – ISSN: 0006-2189 – mf#302.01 – us UMI ProQuest [630]

Big farmer entrepreneur – Frankfort IL 1980-83 – 1,5,9 – (cont: big farmer) – ISSN: 0274-6050 – mf#302.01 – us UMI ProQuest [630]

Big fat – Ann Arbor MI 1970 – 1 – mf#6106 – us UMI ProQuest [780]

Big five era – 1902 may – 1r – 1 – (cont: gold nugget) – mf#1053348 – us WHS [622]

Big game fishermen's paradise / Kaplan, Moise N – Tallahassee, FL. 1936 – 1r – us UF Libraries [978]

Big Hatchie. Tennessee. Baptist Associations *see* Annuals

The big heart / Anand, Mulk Raj; ed by Ramanathan, P – Madras: C Subbiah Chetty & Co, [between 1945 and 1953] – (= ser Samp: indian books) – us CRL [490]

Big house, mister? / Richardson, Martin D – s.l, s.l? . 193? – 1r – 1 – us UF Libraries [978]

Big Little Book Collectors Club of America *see* Big little times

Big little times / Big Little Book Collectors Club of America – 1982-1984 sep/oct, 1984 nov-1988 dec – 2r – 1 – mf#869599 – us WHS [071]

Big mama rag – Denver CO. 1972 nov-1978 may, 1978 jun-1981 dec, 1982 jan-1984 apr, v1 n1-v1 n3, v1 n5-v7 n6 [1973 oct/nov-1979 jul] – 4r – 1 – mf#593943 – us WHS [071]

Big muddy gazette – v3 n1-v4 n9 [1970 sep-1972 apr 21/may 3] – 1r – 1 – (cont by: all american rag) – mf#968098 – us WHS [071]

[Big pine-] the big pine citizen – CA. jan 8 1914-jun 14 1933 – 5r – 1 – $300.00 – mf#B02069 – us Library Micro [071]

Big red news : the newsletter of the democratic socialist alliance – 1980 sep-1983 aug – 1r – 1 – (cont by: creeping socialist) – mf#688761 – us WHS [325]

Big river express – Grafton, mar 1973-dec 1978 – 6r – at Pascoe [079]

Big river news – v1 n1-v13 n5=188 [1973 jun 10-1984 jun] – 1r – 1 – mf#708645 – us WHS [071]

Big rock baptist church. stewart county. big rock, tennessee : church records – 1938-67 – 1 – us Southern Baptist [242]

Big rock candy mountain – Menlo Park CA 1970-71 – 1 – ISSN: 0006-2197 – mf#7741 – us UMI ProQuest [333]

Big sky flyer – 1981 may-1993 dec, 1982 jan-1989 dec, 1990 jan-1993 dec – 3r – 1 – mf#3283378 – us WHS [071]

Big spring / Comstock, Bertha A – s.l, s.l? . 1939 – 1r – us UF Libraries [978]

Big spring baptist church. severns valley association. kentucky : church records – 1884-1913, 1957-68 – 1 – us Southern Baptist [242]

Big springs baptist church. hardin county. kentucky : church records – 1816-1940 – 1 – us Southern Baptist [242]

Big springs enterprise – Big Springs, NE: Ray A Evans. 1v. v1 n1. feb 21 1952-v1 n31. sep 18 1952 (wkly) – 1r – 1 – (cont by: enterprise) – us NE Hist [071]

Big springs enterprise – Big Springs, NE: Herbert M Fisbeck. 1v. v10-11. n1-16. feb 15-may 31 1963 (wkly) – 2r – 1 – (split from: keith county news (1897). absorbed by: julesburg grit-advocate) – us NE Hist [071]

Big springs journal – Big Springs, NE: Frank B Hartman. 2v. v1 n1- apr 6 1911-v2 n35. nov 16 1912 (wkly) [mf ed with gaps filmed 1984] – 1r – 1 – (absorbed by: chappell register) – us NE Hist [071]

Big springs news – Big Springs, NE: Big Springs News Printing (wkly) [mf ed v1 n44. oct 24 1918 filmed 1999] – 1r – 1 – us NE Hist [071]

Big springs news – Big Springs, NE: Wm L Wolfe, 1930-v5 n39. apr 4 1935 (wkly) [mf ed v1 n38. mar 12 1931-apr 4 1935] – 2r – 1 – (cont: deuel county herald) – us NE Hist [071]

Big springs news – Big Springs, NE: W L Wolfe. v6 n13. oct 3 1935- (wkly) [mf ed -oct 23 1936 (gaps)] – 1r – 1 – (split from: deuel county herald. claims to be the cont of the former deuel county herald but the 2 titles were publ concurrently) – us NE Hist [071]

The big springs news – Big Springs, NE: C E Grisham, 1923 (wkly) [mf ed v1 n19. aug 10 1923-oct 29 1926 (gaps) filmed 1999] – 1r – 1 – us NE Hist [071]

Big stevens creek baptist church. edgefield district. south carolina : church records – Jun 1803-1901 – 1 – us Southern Baptist [242]

The big stick : illustrated yiddish journal of humor and satire – (New York), 1909-27 – (= ser Der groyser kundes) – 1 – us AJPC [870]

The big stick – London, England. -f. 1.15 Oct 1920. 3 ft – 1 – uk British Libr Newspaper [072]

Big thompson valley news *see* Miscellaneous newspapers of larimer county

Big us – Cleveland OH. v1 n6-v3 n1 [1968 dec 6-1969 oct 14] – 1r – 1 – (cont by: burning river news) – mf#1056241 – us WHS [071]

Biga boyowa : a notional study of the trobriand islands language / Baldwin, Bernard – n.d. – 1r – 1 – mf#pmb41 – at Pacific Mss [490]

Les bigailles – Port-au-Prince, (H): Impr de l'Abeille, [1901- – us CRL [079]

Les bigailles – [Port-au-Prince]: Impr Bouchereau & Co, 1876- – us CRL [079]

Bigandet, Paul *see* Voyage en birmanie

Bigandet, Paul Ambrose, Bishop *see* The life or legend of gaudama, the buddha of the burmese

Les bigarrures et touches du seigneur des accords / Tabourot, E – Rouen: Loys du Mesnil, 1640 – 9mf – 9 – mf#O-863 – ne IDC [090]

Bigelmair, Andreas *see* Die beteiligung der christen am oeffentlichen leben in vorconstantinischer zeit

Bigelow, Andrew *see* Leaves from a journal

Bigelow, Harry Augustus *see* Introduction to the law of real property

Bigelow, John *see*
- Molinos the quietist
- Wit and wisdom of the haytians

Bigelow, Lee Eugene *see*
- Bench and bar
- Legendary
- Story of the jacksonville ferry services

Bigelow, Melville M *see*
- Bigelow's reports of all the public life and accident insurance cases
- An index of the cases overruled, reversed, denied, doubted, modified, limited, explained, and distinguished

Bigelow, Poultney *see* White man's africa

Bigelow, Ralph Emerson *see* Bela bartok's music for string instruments, percussion and celesta

Bigelow Society *see* Forge

Bigelow, William Sturgis *see* Buddhism and immortality

Bigelow's reports of all the public life and accident insurance cases : determined in the american courts prior to january 1871, with notes on the english cases / Bigelow, Melville M – New York: Hurd & Houghton. v1-5. 1874-77 (all publ) – 45mf – 9 – $67.00 – mf#LLMC 95-129 – us LLMC [347]

Bigg, Charles *see*
- The christian platonists of alexandria
- The church's task under the roman empire
- A critical and exegetical commentary on the epistles of st peter and st jude
- Neoplatonism
- The origins of christianity
- The spirit of christ in common life
- Unity in diversity
- Wayside sketches in ecclesiastical history

Biggar, Emerson Bristol *see*
- Anecdotal life of sir john macdonald
- The boer war
- Canada's approaching peril
- Canada's crisis
- The canadian farmer, the general consumer and the wool tariff
- The canadian railway problem
- Hydro-electric development in ontario
- Reciprocity
- Sauvons nos forets

Biggar, Henry Percival *see* The early trading companies of new france

Biggers, Earl Derr *see* Love insurance

Biggin hill news – Bromley, England 12 dec 1966-21 feb 1981, 11 april 1991- [mf 1991-] – 1 – (wanting: 1973-75, 1977-78) – uk British Libr Newspaper [072]

Biggleswade – (= ser Bedfordshire parish register series) – 2mf – 9 – £5.00 – uk BedsFHS [929]

Biggleswade, st andrew monumental inscriptions – Bedfordshire Family HS 2000 – (= ser Bedfordshire parish register series) – 1mf – 9 – £1.25 – uk BedsFHS [929]

Biggs, Asa *see* Presentation of portrait of honorable asa biggs to united states district court

[Biggs-] biggs news – CA. apr-dec 1928; 1930; 1932; jan 1934-oct 1935; jan 1936-jan 1938; 1939-1941; 1945-1954; 1955-1990 – 35r – 1 – $2100.00 – mf#B03162 – us Library Micro [071]

[Biggs-] biggs weekly news – CA. nov 1924-dec 1930 – 2r – 1 – $120.00 – mf#C03163 – us Library Micro [071]

Biggs, James *see* Hints for finding out truth

Biggs, Johanna Wilhelmina Margareta *see* Die rol van 'n pastorale berader in 'n gemeentediensjaarspan

Biggs, Joseph et al *see* A concise history of kehukee baptist association, nc, pts 1 and 2

Biggs, Mary E *see* Study of modern foreign languages in denver, 1874-1934

[Biggs-] sunshine valley news – CA. mar 18 1910-may 12 1916 (wkly) – 2r – 1 – $120.00 – mf#B02068 – us Library Micro [071]

[Biggs-] the biggs argus – CA. feb 25 1888-dec 1892; 1894; 1897; 1901-02; 1905; jun-dec 1906; 1908; 1910-11 – 4r – 1 – $240.00 – mf#B02067 – us Library Micro [071]

Big-head : (osteoj-porosis) / Bitting, A W – Lake City, FL. 1894 – 1r – us UF Libraries [630]

Bigland, Eileen *see* Lake of the royal crocodiles

Bigmore, Edward Clements *see* The printed book, its history, illustration, and adornment

Bignami, Enrico *see* La plebe

Bignell, John *see* Return to an address of the house of assembly to his excellency the governor general, dated the 12th june, 1851

Bigney, Laura *see* Prize essays on tobacco

Bignold, H B *see* Imperial statutes in force in new south wales...

Bignon, Louis *see*
- Histoire de france
- Histoire de france sous napoleon
- Lettre a un ancien ministre d'un etat d'allemagne sur les differends de la maison d'anhalt avec la prusse

Bigot, Francois *see* Lettres de l'intendant bigot au chevalier de levis

Bigot, Jacques *see* Copie d'une lettre escrite par le pere jacques bigot de la compagnie de jesus, l'an 1684

Big-r trades directories – 1997 – 2mf – 9 – £2.50 – (all-wales – incl glamorgan families) – uk Glamorgan FHS [380]

Bihar : the heart of india / Houlton, John – Bombay: Orient Longmans, 1949 – (= ser Samp: indian books) – us CRL [954]

Bihar and orissa – Patna: Supt Govt Print, Bihar and Orissa. pt 3. 1921 – (= ser Census of india) – us CRL [315]

Bihar and orissa during the fall of the mughal empire : with a detailed study of the marathas in bengal and orissa / Sarkar, Jadunath – Patna: Patna University, 1932 – (= ser Samp: indian books) – us CRL [954]

The bihar gazette / Bihar. India. (State) – Patna. 1945-1966 – 1 – us NY Public [324]

Bihar (India). Political Dept Patna, Political Dept, 1941 *see* Report on the press in bihar

Bihar. India. (State) *see* The bihar gazette

Bihe and garenganze : or, four years' further work and travels in central africa / Arnot, Frederick Stanley – London, [1893] – 2mf – 9 – mf#HTM-5 – ne IDC [916]

Bihl, Josef *see* Die gestalt der wortform und des satzes unter einwirkung des rhythmus bei chaucer und gower

Bihlmeyer, K *see* Deutsche schriften

Bihtebuoch dabey die bezeichenunge der heil. messe : beichtbuch aus dem 14. jahrhundert / ed by Oberlin, Jeremias Jacob – Strasburg 1784 – (= ser Whsb) – 2mf – 9 – €30.00 – (glosses by ed) – mf#Hu 150 – gw Fischer [240]

Bij de reuzen and dwergen van ruanda / Overschelde, Gerard Van – Tielt, Belgium. 1947 – 1r – us UF Libraries [960]

The bijak of kabir – Hamirpur: the author, 1917 [mf ed 1995] – (= ser Yale coll) – iv/236p (ill) – 1 – 0-524-09440-3 – (trans into english by ahmad shah) – mf#1995-0440 – us ATLA [810]

Bijapur and its architectural remains : with an historical outline of the 'adil-shahi dynasty / Cousens, Henry – Bombay: Printed at the Govt Central Press, 1916 – (= ser Samp: indian books) – us CRL [720]

Bijapur inscriptions / Nazim, Muhammad – Delhi: Manager of Publ, 1936 – (= ser Samp: indian books) – us CRL [730]

Bijapur, old capital of the adil shahi kings : guide to its ruins with historical outline / Cousens, Henry – Poona: Printed at the Orphanege Press, 1889 – (= ser Samp: indian books) – us CRL [930]

De bijbel voor jongelieden *see* The bible for learners

Bijblad op het staatsblad van nederlandsch indie – Batavia, 1857-1949. v1-78 – 1081mf – 9,8 – (cont as: tambahan lembaran-negara ri djakarta, 1950-1969. v1-20 nos 1-2979. several iss missing) – mf#SE-32; SE-226 – ne IDC [959]

Bijdrage tot de kennis van hat vedische ritueel, jaiminiyasrautasutra / Gaastra, Dieuke – Leiden: EJ Brill, 1906 – 1mf – 9 – 0-524-01960-6 – mf#1990-2751 – us ATLA [280]

Bijdrage tot de kennis van het gereformeerd protestantisme / Gooszen, M A – 1887. v21 (p 505-554) – (= ser Geloof en Vrijheid) – 1mf – 9 – mf#PBU-438 – ne IDC [242]

Bijdrage tot de tekstkritiek van richteren 1-16 / Doornink, Adam van – Leiden: E J Brill, 1879 [mf ed 1985] – 1mf – 9 – 0-8370-2950-3 – mf#1985-0950 – us ATLA [221]

Bijdragen en mededelingen van het historisch genootschap – Utrecht, 1(1878)-66(1948) – 522mf – 9 – €995.00 – ne Slangenburg [900]

Bijdragen tot de kennis van het hindoeisme op java / Brumund, Jan Frederik Gerrit – Batavia: Lang, 1868 – 1r – 1 – 0-8370-1511-1 – mf#1984-B233 – us ATLA [280]

Bijdragen voor de geschiedenis van het bisdom haarlem – 1(1873)-65(1958) – 405mf – 9 – €701.00 – (incl ind) – ne Slangenburg [242]

Bijl, Pieter van der *see* Studies on selected therapeutic and toxic agents in stomatology (1981-1998)

Bijou – 1828-30 – (= ser English gift books and literary annuals, 1823-1857) – 13mf – 9 – uk Chadwyck [800]

Bijou, Cajuste *see* Campagne contre le papier-monnaie

Bijouterie / Roger-Miles, Leon Octave Jean Roger – Paris, France. 1895 – 1r – us UF Libraries [025]

Bijskaia pravda : organ bijskogo soveta rabochikh, krest'ianskikh i krasno-armejskikh deputatov – Bijsk, Russia 1918 [mf ed Norman Ross] – 1r – 1 – mf#nrp-337 – us UMI ProQuest [077]

Bijskii rabochii – Bijsk, Russia 1973-85 [mf ed Norman Ross] – 4r – 1 – mf#nrp-338 – us UMI ProQuest [077]

Bijutsu no nihon – v1 n1-v15 n8. 1909-23 – 15r – 1 – Y225.000 – (with 68p guide; in japanese) – ja Yushodo [700]

Bijvoegsel van het officieel Nieuwsblad *see* Nationale merken

Bike world – Emmaus PA 1974-80 – 1,5,9 – ISSN: 0098-8650 – mf#9871 – us UMI ProQuest [790]

Bikerman, I *see* Rossiiskaia revoliutsiia i gosudarstvennaia duma

Biko, B S *see* Black viewpoint

Bikupan – Malmo, Sweden. 1868-69 – 1 – sw Kungliga [078]

Bikure nisan / Jaffe, Abraham Nissan – Vilna, Lithuania. 1919 – 1r – us UF Libraries [939]

Bikure ribal / Levinsohn, Isaac Baer – Warsaw, Poland. 1900 – 1r – us UF Libraries [939]

Bikure tsiyon / Zaks, Bencion Lejb – Nemaksciai, Lithuania. 19– – 1r – us UF Libraries [939]

Bikure ya'akov / Rabinowitz, Jacob – London, England. 1899 – 1r – us UF Libraries [939]

Bilac, Olavo *see* Poesias

Bilaga til elfsborgs lans annonsblad – Vanersborg, Sweden. 1891 – 1 – sw Kungliga [078]

Bilaga till "om existens, tid och localitet i svenskan" : testbatteriet / Rahkonen, Matti – Jyvaeskylae: Institutionen foer nordiska sprak vid Jyvaeskylae universitet, 1982 (mf ed 1989) – 1mf – 9 – mf#*XM-18,664 – us NY Public [430]

Bilan de l'uvre de norodom sihanouk pendant le mandat royal de 1952 a 1955 / Sam Sary – [Phnom-Penh: Impr Albert Portail 1955] [mf ed 1989] – [facs] 1r with other items – 1 – (pref by samdach preah upayuvareach) – mf#mf-10289 seam reel 005/03 [§] – us CRL [959]

Bilan de m baillairge : comme architecte, ingenieur, arpenteur-geometre... – S.l: s.n, 18–? – 1mf – 9 – mf#00090 – cn Canadiana [624]

Le bilan dogmatique de l'orthodoxie regante / Lobstein, Paul – Paris: Librairie Fischbacher, 1891. Chicago: Dep of Photodup, U of Chicago Lib, 1975 (1r); Evanston: American Theol Lib Assoc, 1984 (1r) – (= ser HIS etudes christologiques) – 1 – 0-8370-0563-9 – (incl bibl ref) – mf#1984-6053 – us ATLA [240]

Bilan dogmatique de l'orthodoxie regnante *see* Collected works

Bilan, John *see* Science and common sense

Die bilanz der revolution : ein rueckblick und ein ausblick / Stroebel, Heinrich – Berlin: Verlag Neues Vaterland, E Berger & Co, [1919] [mf ed 1987] – (= ser Flugschriften des bundes neues vaterland. [n f] 17) – 24p – 1 – mf#6929 – us Wisconsin U Libr [943]

Die bilanzrechtlichen beschluesse der grossen senate von rfh und bfh : kritische darstellung / Hildebrand, Ulrich – (mf ed 1992) – 2mf – 9 – €49.00 – 3-89349-596-7 – mf#DHS 596 – gw Frankfurter [330]

Bilateral lower extremity function during the support phase of running / McCaw, Steven T – 1989 – 151p 2mf – 9 – $4.00 – us Kinesology [612]

Bilateral staff conversations with latin american republics / Munden, Cecil L – New Orleans. 1947. 166p – 1 reel – 1 – $16.00 – us L of C Photodup [977]

Bilby, J W *see* Among unknown eskimo

Bilby, Thomas *see* Young folk's illustrated book of birds

Bild *see* Bild-zeitung

Bild am sonntag – Essen DE, 2002 3 feb-29 dec – 2mf=4df – 1 – (filmed by misc inst: 1956 29 apr-1964 7 jun [gaps], 1964 13 oct-1980 31 jan, 1980 20 apr-1997 12 oct, 1998 8 mar-2002 27 jan. ha in hamburg) – gw Mikrofilm; gw Misc Inst [074]

Bild am sonntag – Hamburg DE, 1954 15 aug-1999 27 jun [gaps] – 144r – 1 – gw Mikrofilm [074]

Bild der zeit / berlin im bild *see* Der berliner

Das bild des christentums bei den grossen deutschen idealisten : ein beitrag zur geschichte des christentums / Luelmann, Christian – Berlin: C A Schwetschke, 1901 – 1mf – 9 – 0-8370-8693-0 – (incl bibl ref) – mf#1986-2693 – us ATLA [240]

Bild des fuehrers : gedichte / Vesper, Will; ed by Schmidkunz, Walter – [Muenchen]: Muenchner Buchverlag [194-?] [mf ed 1991] – (= ser Muenchner lesebogen 48) – 1r – 1 – (filmed with: blumbergshof / siegfried von vegesack) – mf#2944p – us Wisconsin U Libr [810]

Das bild des zahnarztes in der oeffentlichkeit / Tonn, Anke – (mf ed 1999) – 2mf – 9 – €40.00 – 3-8267-2658-8 – mf#DHS 2658 – gw Frankfurter [617]

Das bild in der dramatischen sprache grillparzers / Cafasso, Arthur – Leoben: Im Verlage des Landes-Obergymnasiums 1884 [mf ed 1990] – (= ser Jahresbericht des landes-obergymnasiums zu leoben 1884) – 1r – 1 – (incl bibl ref. filmed with: grillparzers verhaltnis zur politischen tendenzliteratur seiner zeit / konrad beste) – mf#2690p – us Wisconsin U Libr [430]

Das bild in dir : roman / Wilhelm, Wolfgang – Berlin: W Limpert 1942 [mf ed 1991] – 1r – 1 – (filmed with: armut / anton wildgans) – mf#3052p – us Wisconsin U Libr [830]

Bild [main edition] – Hamburg DE – 1 – (regional ed: duesseldorf 1998- [4r/yr], printed in essen-kettwig; s [= suedwestfalen] 1982-1995 23 may) – gw Misc Inst [074]

Das bild meines lebens / Gerhard, Adele – Wuppertal: Abendland-Verlag 1948 [mf ed 1990] – 1r – 1 – (filmed with: zeitgenoessische dichter) – mf#7297 – us Wisconsin U Libr [430]

Bild / r *see* Bild-zeitung

Bild / ro [=ruhr-ost] *see* Bild-zeitung

Bild / s [suedwestfalen] – Essen-Kettwig DE, 1982-1995 23 may – 1 – (ha in hamburg) – gw Misc Inst [074]

Bild und begriff : studien ueber die beziehungen zwischen kunst und wissenschaft / Kuczynski, Juergen & Heise, Wolfgang – 1. aufl. Berlin: Aufbau-Verlag, c1975 [mf ed 1993] – 463p – 1 – (incl bibl ref) – mf#8236 – us Wisconsin U Libr [430]

Bild und film – Moenchengladbach, 1912/13 – 1 – gw Misc Inst [790]

Bild und schule *see* Bild-archiv / bild und schule

Das bild von richard wagners tristan und isolde in der deutschen literatur / Park, Rosemary – Jena: E Diederich, [1935?] [mf ed 1993] – (= ser Deutsche arbeiten der universitaet koeln 9) – 141p – 1 – (incl bibl ref) – mf#8215 reel 2 – us Wisconsin U Libr [390]

Bild-archiv / bild und schule – Muenchen DE, 1920-21 – 1 – gw Mikrofilm [370]

Bildarchiv Foto Marburg – Deutsches Dokumentationszentrum fuer Kunstgeschichte Philipps- Universitaet Marburg *see*
- Armenien-index
- Baltic index
- Benelux-kunstindex
- British art index
- Griechenland-index
- Italien index
- Italien-index. neue folge
- Marburger index
- Schweiz-index
- Spanien-und portugal-index

Bildarchiv Foto Marburg – Deutsches Dokumentationszentrum fuer Kunstgeschichte Philipps-Universitaet Marburg *see*
- Aegypten-index
- Index photographique de l'art en france
- Oesterreich-index
- Tschechoslowakei-index

Bilden ungeloeste fragen ein hindernis fuer den glauben? : vortrag / Heim, Karl – 3. Aufl. Ascona: C v Schmidtz, 1906 – 1mf – 9 – 0-7905-3902-0 – mf#1989-0395 – us ATLA [240]

Bilder aus china / Faber, Ernst – Barmen: Im Verlage des Missionshauses, 1877 [mf ed 1995] – (= ser Yale coll; Rheinische missions-traktate 12-13) – 2v in 1 (ill) – 1 – (in german) – mf#1995-1179 – us ATLA [951]

Bilder aus constantinopel : eine schilderung des lebens, der sitten und gebraeuche in dieser hauptstadt / Fliegner, Ferdinand – Breslau 1853 [mf ed Hildesheim 1995-98] – (= ser Fbc) – 3mf – 9 – €90.00 – 3-487-29123-1 – gw Olms [915]

Bilder aus dem kaukasus / Hahn, C von – Leipzig, 1900 – 4mf – 9 – mf#AR-1595 – ne IDC [914]

Bilder aus dem leben jesu : biblische vortraege / Lehmann, Ernst Gottlob – Leipzig: J C Hinrichs 1875 [mf ed 1985] – 1mf – 9 – 0-8370-4083-3 – mf#1985-2083 – us ATLA [240]

Bilder aus dem schwarzwald / Buehrlen, Friedrich L – Stuttgart 1828/31 [mf ed Hildesheim: 1995-98] – (= ser Fbc) – 2v on 4mf – 9 – €120.00 – 3-487-29587-3 – gw Olms [914]

Bilder aus der berliner mission in lukhang-suedchina : nach den berichten des missionaers rhein / Gurr, Paul – Berlin: Berliner evang Missionsgesellschaft, [1909] [mf ed 1995] – (= ser Yale coll) – 85p (ill) – 1 – 0-524-09036-X – (in german) – mf#1995-0036 – us ATLA [242]

Bilder aus der deutschen jesuitenmission puna / Doering, Heinrich – Aachen: Xaverius-Verlag, 1918 [mf ed 1995] – (= ser Yale coll; Abhandlungen aus missionskunde und missionsgeschichte 7) – 81p (ill) – 1 – 0-524-09852-2 – (in german) – mf#1995-0852 – us ATLA [241]

Bilder aus der geschichte der altchristlichen kunst und liturgie in italien / Beissel, Stephan – Freiburg i B, St Louis MO: Herder 1899 [mf ed 1990] – 1mf [ill] – 9 – 0-7905-5860-2 – (text in german, notes in latin; incl bibl ref) – mf#1988-1860 – us ATLA [700]

Bilder aus der schweiz : zwei erzaehlungen / Gotthelf, Jeremias [Albert Bitzius]; ed by Braasch, Theodor – New York: F S Crofts, c1937 [mf ed 1993] – xiii133p – 1 – (incl bibl ref) – mf#8518 – us Wisconsin U Libr [880]

Bilder aus einem leben : erinnerungen eines ostpreussischen juden / Rosenberg, Curt – Wuerzburg: Holzner Verlag 1962 [mf ed 1992] – (= ser Ostdeutsche beitraege aus dem goettinger arbeitskreis) – 10r – 1 – mf#3180p – us Wisconsin U Libr [939]

Bilder aus griechenland und der levante : mit einem vorworte vom professor zeune / Byern, E von – Berlin 1833 [mf ed Hildesheim 1995-98] – (= ser Fbc) – 3mf – 9 – €90.00 – 3-487-29129-0 – gw Olms [914]

Bilder aus italien / Oefele, Aloys von – Frankfurt am Main 1833 [mf ed Hildesheim 1995-98] – (= ser Fbc) – 2v on 5mf – 9 – €100.00 – 3-487-29266-1 – gw Olms [914]

Bilder aus japan / Fischer, Adolf – Berlin: Georg Bondi, 1897 [mf ed 1995] – (= ser Yale coll) – 412p (ill) – 1 – 0-524-09919-7 – (in german) – mf#1995-0919 – us ATLA [950]

Bilder aus japan : schilderung des japanischen volkslebens / Kleist, Hugo – Leipzig [1890] [mf ed Hildesheim 1995-98] – 1v on 2mf – 9 – €60.00 – 3-487-27539-2 – gw Olms [915]

Bilder aus sued-tirol : und von den ufern des gardasees / Noe, Heinrich – Muenchen 1871 [mf ed Hildesheim: 1995-98] – (= ser Fbc) – 414p on 3mf – 9 – €90.00 – 3-487-29438-9 – gw Olms [914]

Bilder der zeit – Leipzig DE, 1855 – 1r – 1 – gw Misc Inst [074]

Bilder des orients von h steiglitz : fuer eine singstimme mit begleitung des pianoforte...op 140 heft 3 / Marschner, Heinrich – Leipzig: Fr Keistner [1849] [mf ed 1991] – 1r – 1 – (song cycle for voice & piano; german words) – mf#pres. film 97 – us Sibley [780]

Bilder im sinnspruch und gleichniss / Weninger, Francis Xavier – Cincinnati: Jos A Hemann. 7v. 1855-57 – 7mf – 9 – 0-8370-6854-1 – mf#1986-0854 – us ATLA [240]

Bilder nach skulpturen und gemaelden der sammlung / Zurich. Kunsthaus – 1936, 1959 – 2mf – 9 – uk Chadwyck [730]

Bilder und symbole babylonisch-assyrischer goetter / Frank, K – Leipzig, 1906 – (= ser Leipziger semitistische Studien) – 1mf – 9 – (leipziger semitistische studien, leipzig 1908 v2 pt2) – mf#NE-20110 – ne IDC [956]

Bilder ur goethes faust / Rydberg, Viktor – Stockholm: A Bonnier, 1897 [mf ed 1993] – 62p – 1 – mf#8605 – us Wisconsin U Libr [430]

Bilder vom tage *see* Berliner lokal-anzeiger

Bilderatlas zur bibelkunde : ein handbuch fuer den religionslehrer u. bibelfreund / Frohnmeyer, Ludwig Johannes & Benzinger, Immanuel – Stuttgart: T Benzinger, 1905 – 1mf – 9 – 0-524-02776-5 – mf#1987-6470 – us ATLA [220]

Bilderbogen *see* Zeitung der 10. armee

Bilderbuch aus england / Fontane, Theodor; ed by Fontane, Friedrich – Berlin: G Grote, 1938 [mf ed 1989] – xx/250p – 1 – (int by hanns martin elster) – mf#7248 – us Wisconsin U Libr [914]

Bilderbuch der letzten 10 jahre 1945-1955 *see* Quick

Der bildercatechismus des fuenfzehnten jahrhunderts : und die catechetischen hauptstuecke in dieser zeit bis auf luther / Geffcken, Johannes – Leipzig: TO Weigel, 1855 – 1mf – 9 – 0-524-04835-5 – mf#1990-1327 – us ATLA [240]

Bilder-conversations-lexikon fuer das deutsche volk (ael1/28) – Leipzig 1837-41 [mf ed 1995] – (= ser Archiv der europaeischen lexikographie, abt 1: enzyklopaedien) – 4v on 32mf – 9 – €200.00 – 3-89131-206-7 – gw Fischer [030]

Bilder-magazin – Leipzig DE, 1842 – 1 – gw Misc Inst [074]

Bilder-magazin fuer allgemeine weltkunde – Leipzig DE, 1834-35 – 1r – 1 – gw Misc Inst [910]

Bildermappe : mit 273 abbildungen samt erklaerungen zur religion babyloniens und assyriens / Jastrow, Morris – Giessen: Alfred Toepelmann 1912 [mf ed 1993] – 1mf [ill] – 9 – 0-524-06927-1 – (incl bibl ref) – mf#1990-3553 – us ATLA [930]

Bildersaal fuer geschichte, natur und kunst – Karlsruhe DE, 1833-36 – 1r – 1 – gw Misc Inst [074]

Bilderschriften der renaissance : hieroglyphik und emblematik in ihren beziehungen und fortwirkungen / Volkmann, L – Leipzig: Hiersemann, 1923 – 2mf – 9 – mf#O-1969 – ne IDC [090]

Die bildersprache des alten testaments : ein beitrag zur aesthetischen wuerdigung des poetischen schrifttums im alten testament / Wuensche, August – Leipzig: Eduard Pfeiffer, 1906 – 1mf – 9 – (incl bibl ref and index) – mf#1986-3279 – us ATLA [221]

Bildgesteuerte zugangstechniken in der minimal invasiven chirurgie : stand der technik und entwicklung neuer bildgesteuerter verfahren / Melzer, Andreas – 1998 – 3mf – 9 – 3-8267-2567-0 – mf#DHS 2567 – gw Frankfurter [617]

Bildhaftigkeit im franzosischen aroot / Schultz, Irmgard – Giessen, Germany. 1936 – 1r – us UF Libraries [960]

Der bildhauer : ein roman / Zobeltitz, Hanns von – Stuttgart: Deutsche Verlags-Anstalt 1906 [mf ed 1992] – 1r – 1 – (filmed with: auf biegen und brechen/ erwin zindler) – mf#3069p – us Wisconsin U Libr [830]

Der bildhauer kurt lehmann : das plastische werk. ein beitrag zur bildhauerkunst des 20. jahrhunderts / Bury, Karin – (mf ed 1995) – 6mf – 9 – €62.50 – 3-8267-2171-3 – mf#DHS 2171 – gw Frankfurter [730]

Bildjournalen – Stockholm, 1954-80 – 32r – 1 – sw Kungliga [078]

Der bildliche ausdruck in der prosa eduard moerikes / Kappenberg, Hans – Greifswald: H Adler 1914 [mf ed 1990] – 1r – 1 – (incl bibl ref. filmed with: ueber die galgenlieder / christian morgenstern) – mf#2838p – us Wisconsin U Libr [430]

Die bildnisse carl augusts von weimar / Wahl, Hans; ed by Wahl, Hans – Weimar: Goethe-Gesellschaft, 1925 [mf ed 1993] – (= ser Schriften der goethe-gesellschaft 38) – 62p/48pl (ill) – 1 – mf#8657 reel 9 – us Wisconsin U Libr [750]

Bildnisse der beruehmtesten menschen aller voelker und zeiten (ael1/25) : supplementband zu jedem biographischen woerterbuch, besonders zu dem conversations-lexikon / ed by Seemann, Otmar – Zwickau, 1818-32 [mf ed 1994] – (= ser Archiv der europaeischen lexikographie, abt 1: enzyklopaedien) – 35iss on 5mf – 9 – €130.00 – 3-89131-192-3 – gw Fischer [030]

Die bildnisse goethes / ed by Schulte-Strathaus, Ernst – Muenchen: G Mueller, 1910 [mf ed 1989] – (= ser Propylaeen-ausgabe von goethes saemtlichen werken suppl 1) – 100p/167pl (ill) – 1 – mf#6975 – us Wisconsin U Libr [430]

Die bildnisse wielands / Weizsaecker, Paul – Stuttgart: W Kohlhammer, 1893 – 1r – 1 – (incl bibl ref and indexes) – us Wisconsin U Libr [430]

Die bildschnitzer von weilburg / Eckstein, Ernst – Stuttgart: J G Cotta, [between 1910 and 1929] – 1r – 1 – us Wisconsin U Libr [430]

Die bildspielkunst – Berlin DE, 1913 15 aug – 1 – gw Mikrofilm [790]

Bildt, Carl Nils Daniel Bildt, Freiherr von *see* The conclave of clement 10 (1670)

Die bildung der evang. theologen fuer den praktischen kirchendienst : eine denkschrift zur fuenfundzwanzigjaehrigen stiftungsfeier des evang.-protestantischen predigerseminars in heidelberg / Schenkel, Daniel – Heidelberg: JCB Mohr, 1863 – 1mf – 9 – 0-7905-6828-4 – mf#1988-2828 – us ATLA [242]

Die bildung von rueckstellungen fuer rekultivierung, sanierung und nachsorge bei oberirdischen deponien nach handels- und steuerrecht / Ossendot, Ralf – (mf ed 1996) – 2mf – 9 – €40.00 – 3-8267-2367-8 – mf#DHS 2367 – gw Frankfurter [336]

Das bildungsideal der deutschen klassik und die moderne arbeitswelt / Litt, Theodor – 7. aufl. Bochum: F Kamp, 1967 [mf ed 1992] – (= ser Kamps paedagogische taschenbuecher. blaue reihe, allgemeine paedagogik v3) – 152p – 1 – mf#8210 – us Wisconsin U Libr [190]

Bildungstheorie – literaturdidaktik – musikdidaktik : fuenf beitraege zum zusammenhang bildungstheoretischer grundlagen und didaktik-konzeptionen der literatur und musik / Pongratz, Gregor – (mf ed 1998) – 5mf – 9 – €59.00 – 3-8267-2540-9 – mf#DHS 2540 – gw Frankfurter [370]

Der bildungsverein / volksbildung – Berlin DE, 1896, 1898-1920 – 1 – gw Mikropress [370]

Der bildwart – Berlin DE, 1925-30, 1931 may-1932 oct, 1935 feb – 3r – 1 – gw Mikrofilm [074]

Bild-zeitung – Essen DE, 1964 1 jul-1967, 1995 24 may-1998 – 12r [fr 1995] – 1 – (covers rhein-ruhr; title varies: 13 sep 1971: bild / r, 12 sep 1975: bild / ro [=ruhr-ost], printed in essen-kettwig; filmed by misc inst: 1968-81, 1995 24 may- [ca 2r/yr]) – gw Mikrofilm; gw Misc Inst [074]

Bild-zeitung – Esslingen a. Neckar DE, 1968-74, 1981 15 jul-31 dec – 1 – (title varies: 13 sep 1971: bild. covers: region stuttgart; main ed in hamburg) – gw Misc Inst [074]

Bild-zeitung – Hamburg DE, 1952 24 jun-1999 30 jun – 238r – 1 – (title varies: 13 sep 1971: bild; filmed by misc inst: 1975 – 1981, 14 jul, 1952 24 jun- [until 2002 249r]) – gw Mikrofilm; gw Misc Inst [074]

Biley, Edward *see*
- The elohistic and jehovistic theory minutely examined
- A supplement to the horae paulinae of archdeacon paley

Bilgi mecmuasi – Istanbul: Matbaa-i Amire, 1913-14. Yayimliyan: Tuerk Bilgi Dernegi; Mueduer: Celal Sahir. n1-7. tesrinisani 1329 [1913]-nisan 1330 [1914] – (= ser O & t journals) – 12mf – 9 – $195.00 – us MEDOC [956]

Bilgrami, Syed Husain *see* Address

Bilguer, Dr von *see* Gregor der grosse

Bilhana : an indian romance / Seshadri, P – Madras: Srinivasa Varadachari & Co, 1914 – (= ser Samp: indian books) – (adapted from sanskrit by p seshadri) – us CRL [830]

Bilhana *see* Sasikala ane caurapancasika

Bilharzia: beware! = Bilharzia, pasop! / Pitchford, R J – Pretoria: Dept of Health 1979 [mf ed Pretoria, RSA: State Library [199-]] – 24p [ill] on 1r with other items – 5 – mf#op 06906 r23 – us CRL [616]

Bilharzia (human redwater) disease = Bilharzia-(rooiwater) siekte onder mense / South Africa. Department of Public Health] – [Pretoria: Govt Printer] 1929 [mf ed Pretoria, RSA: State Library [199-]] – 1r with other items – 5 – mf#r23 – us CRL [616]

Bilharzia in south Africa = Bilharzia in suid-afrika / Gear, James Henderson Sutherland et al – 2nd ed. Pretoria: Dept of National Health & Population Development 1988 [mf ed Pretoria, RSA: State Library [199-]] – 1r with other items – 5 – (in english & afrikaans) – mf#op 08722 r23 – us CRL [616]

Bilharzia in south africa = Bilharzia in suid-afrika / Gear, James Henderson Sutherland et al – Pretoria: Dept of Health [1977?] [mf ed Pretoria, RSA: State Library [199-]] – 39p [ill] on 1r with other items – 5 – (in english & afrikaans) – mf#op 06642 r23 – us CRL [616]

Bilhaud, Paul *see* Esperances

Bilimovich, A D *see* Ministerstvo finansov, 1802-1902

Bilingual research journal – Washington DC 1992+ – 1,5,9 – mf#18224.01 – us UMI ProQuest [400]

Bilingual review = La revista bilingue – Tempe AZ 1974+ – 1,5,9 – ISSN: 0094-5366 – mf#9878 – us UMI ProQuest [410]

The bilingual school : a study of bilingualism in south africa / Malherbe, Ernst Gideon – Johannesburg: Central News Agency, [1943] – 1 – us CRL [400]

Die bilin-sprache in nordost-afrika / Reinisch, L – Wien, [1883-1887] – 2mf – 9 – mf#NE-20258 – ne IDC [956]

A bill : as amended by the committee, to make temporary provision for the government of lower canada, prepared and brought in by lord john russell, lord viscount howick and sir george grey / Grande-Bretagne. Parliament. House of Commons – [London]: [s.n.] [1838] – 1mf – 9 – mf#SEM105P1142 – cn Bibl Nat [348]

Bill : an act to regulate the inspection and measurement of timber, masts, spars...in the ports of quebec and montreal / Canada (Province) – Kingston: R Stanton, 1841 [mf ed 1983] – 1mf – 9 – mf#SEM105P174 – cn Bibl Nat [324]

Bill : an act to repeal certain acts therein mentioned, and to provide for the further encouragement of elementary education in this province = Acte pour rappeler certains actes y mentionnes et pour pourvoir ulterieurement a l'encouragement de l'education elementaire en cette province / Bas-Canada – [s.l.]: Conseil legislatif, 1836 [mf ed 1982] – 2mf – 9 – mf#SEM105P133 – cn Bibl Nat [370]

Bill : an act to consolidate the acts respecting municipalities and roads in lower canada = Acte pour refondre les statuts relatifs aux municipalites et aux chemins dans le bas-Canada / Canada (Province) – [Quebec]: S Derbishire & G Desbarats [1859] (mf ed 1983) – 2mf – 9 – mf#SEM105P196 – cn Bibl Nat [380]

Bill : acte pour refondre les statuts relatifs aux municipalites et aux chemins dans le bas-canada / Canada (Province) – Toronto: [s.n.], [1859] (mf ed 1983) – 2mf – 9 – mf#SEM105P197 – cn Bibl Nat [350]

Bill : the lower canada municipalities act: (the lower canada municipal and road act) / Canada (Province) – [Quebec (Province)]: S Derbishire & G Desbarats, [1853] (mf ed 1996) – (= ser The lower Canada municipal and road act) – 1mf – 9 – mf#SEM105P2025 – cn Bibl Nat [348]
Bill, A *see* Zur erklaerung und textkritik des 1. buches tertullians, "adversus marcionem"
Bill (as amended by the committee) for uniting the legislatures of lower and upper canada / Grande-Bretagne. Parliament. House of Commons – Quebec: repr at the New Printing-Office, Free-Masons' Hall, 1822 [mf ed 1990] – 1mf – 9 – mf#SEM105P1146 – cn Bibl Nat [323]
Bill, August *see* Zur erklaerung und textkritik des 1. buches tertullians "adversus marcionem
Bill barnes : america's air ace comics – New York NY 1940-47 – 1 – mf#6129 – us UMI ProQuest [740]
Bill books of the u s house of representatives, 1814-1817 / U.S. House of Representatives – (= ser Records of the united states house of representatives) – 1r – 1 – (with printed guide) – mf#M1265 – us Nat Archives [324]
Bill books of the u s senate, 1795-1845 / U.S. Senate – (= ser Records of the united states senate) – 2r – 1 – (with printed guide) – mf#M1255 – us Nat Archives [324]
Bill for the restoration of the irish bishoprics – Dublin, Ireland. 1847 – 1r – us UF Libraries [240]
Bill, Ingraham Ebenezer *see* Fifty years with the baptist ministers and churches of the maritime provinces of canada
Bill intituled an act to explain and amend the laws relating to lands holden in free and common soccage in the province of lower canada – [S.l.] : [s.n.], [1831] (mf ed 1989) – 1mf – 9 – (other titles: an act to explain and amend the laws relating to lands holden in free and common soccage in the province of lower canada and: canada lands) – mf#SEM105P1141 – cn Bibl Nat [323]
Bill intituled an act to provide for the extinction of feudal and seigniorial rights...in the province of lower canada – [S.l: s.n, 1825] (mf ed 1997) – 1mf – 9 – (other title: an act to provide for the extinction of feudal and seigniorial rights and burthens on lands held a titre de fief and a titre de cens, in the province of lower canada) – mf#SEM105P2843 – cn Bibl Nat [323]
Bill introduit dans la chambre d'assemblee...pour mieux regler la milice de cette province = A bill introduced in the house of assembly...for better regulation of the militia of this province / Bas-Canada. Parlement. Chambre d'assemblee – Quebec: Imprime a la nouvelle impr, 1816 [mf ed 1977] – 1r – 5 – mf#SEM16P300 – cn Bibl Nat [348]
Bill, Ledyard *see* Winter in florida
Bill of rights in action – Los Angeles CA 1976+ – 1,5,9 – (cont: bill of rights newsletter) – ISSN: 0160-7731 – mf#10421.01 – us UMI ProQuest [322]
Bill of rights journal – New York NY 1979-96 – 1,5,9 – ISSN: 0006-2499 – mf#12036 – us UMI ProQuest [342]
Bill of rights newsletter – Los Angeles CA 1975-76 – 1,5,9 – (cont by: bill of rights in action) – ISSN: 0006-2502 – mf#10421.01 – us UMI ProQuest [322]
The bill of rights review : a quarterly – Bill of Rights Committee of the ABA. v1-2. 1940-42 (all publ) – 9mf – 9 – $13.50 – mf#LLMC 84-425 – us LLMC [322]
Bill of the play / Daly's Fifth Avenue Theatre. New York – 1879 1880-1896 1897 – 1 – us NY Public [790]
Le bill seigneurial expose sous son vrai jour par le journal "la patrie" refutation victorieuse du rapport soumis a la convention anti-seigneuriale / Rambau, Alfred-Xavier – Montreal: impr par Senecal & Daniel, 1855 [mf ed 1983] – 1mf – 9 – mf#SEM105P276 – cn Bibl Nat [323]
Bill to enforce the ancient laws of this province : compelling seigniors to concede their lands, subject only to rents and services... = Bill pour mettre en force les anciennes lois de cette province... – [S.l.]: [s.n.], [1825] (mf ed 1989) – 1mf – 9 – (in french and english) – mf#SEM105P1143 – cn Bibl Nat [340]
A bill to make provision with respect to the termination of his majesty's jurisdiction in palestine : and for purposes connected therewith – London, 1948 – 1mf – 9 – mf#J-28-101 – ne IDC [956]
Bill to provide for making and maintaining a rail road from the river st lawrence to the navigable waters of lake champlain = Bill pour pourvoir a la construction et a l'entretien d'un chemin a lisses, a partir du fleuve st laurent a aller jusqu'aux eaux navigables du lac champlain – [S.l.]: [s.n.], 1831 [mf ed 1990] – 1mf – 9 – (in french and english) – mf#SEM105P1222 – cn Bibl Nat [380]
A bill to provide for public elementary education in england and wales, 1870 : bill 33 – 1r 2mf – 1,9 – mf#1r 96784 2mf 86736 – uk Microform Academic [324]
Billard de Veaux, Robert *see* Memoires de billard de veaux (alexandre), ancien chef vendeen
Billard, Louis Phillip *see* Diary
Billard, Pierre *see*
– Cameroun physique
– La circulation dans le sud cameroun
Billaud-Varennes, Jacques *see* Memoires de billaud-varennes, ex-conventionnel, ecrits au port-au-prince en 1818
Billboard – New York NY 1894+ – 1,5,9 – ISSN: 0006-2510 – mf#6017 – us UMI ProQuest [780]
The billboard – Cincinnati, Ohio. v. 1-72. Nov 1 1894-Dec 31 1960 – 1 – us NY Public [071]
Billeb, Hermann *see* Die wichtigsten saetze der neueren alttestamentlichen kritik
Biller, Sarah *see*
– Memoir of the late hannah kilham
Billerbeg, F de *see* Epistola constantinopoli recens scripta. de praesenti turcici imperij statu, and gubernatoribus praecipuis, and de bello persico
Billerica 1627-1849 – Oxford, MA (mf ed 1995) – (= ser Massachusetts vital record transcripts to 1850) – 15mf – 9 – 0-87623-211-X – (mf 1t-3t: vital records 1627-1814. mf 3t: vital record indexes 1627-1847. mf 4t-8t: vital records 1627-1847. mf 8t: vital records index 1785-1848. mf 8t-10t: vital records 1785-1847. mf 11t: births & deaths 1800-48. mf 11t-12t: births 1843-49. mf 12t: marriages 1843-49; deaths 1843-49. mf 13t: out-of-town marriages 1674-1799. mf 14t-15t: baptisms 1747-1838) – us Archive [978]
Billet de loterie / Roger, Jean Francois – Paris, France. 1811 – 1r – us UF Libraries [440]
Billeter, Gustav *see* Wilhelm meisters theatralische sendung
Billeter, M et al *see* Berner beitraege zur geschichte der schweizerischen reformationskirchen
Billets en vers de m de saint-ussans – Paris: Chez la veuve de Claude Thiboust et Pierre Esclassan, 1688 – 4mf – 9 – mf#O-1892 – ne IDC [090]
Billette, J Emile *see* La cause des obligations et prestations
Billiard, Auguste *see* Voyage aux colonies orientales
Billiards for everybody / Roberts, Charles – London, England. 19–? – 1r – us UF Libraries [025]
Billiards simplified – London, England. 1889? – 1r – us UF Libraries [790]
Billiche antwurt joan : ecolampadij auff d martin luthers bericht des sacraments halb / Oecolampadius, J – Basel, Thomas Wolff, 1526 – 1mf – 9 – mf#PBU-368 – ne IDC [242]
Billing, Einar *see* Luthers laera om staten i dess samband med hans reformatoriska grundtankar och med tidigare kyrkliga laeror
Billing, Gottfrid *see* P waldenstroems uppsats, om foersoningens betydelse
Billing, John E *see* An analysis of selected attendance factors in the world league of american football
Billinger, Richard *see*
– Rauhnacht
– Rosse
– Sichel am himmel
Billings, Carolyn *see* Validation of the delta t mode of the omnisound 3000(tm)
Billings, Charles Towne *see* Movements and men of christian history
Billings county herald : [official county paper] – Medora, Billings Co, ND: Geo L Nelson, 1906; -v14 n25 aug 8 1919 (wkly) – 1 – (merged with: fryburg pioneer to form: billings county pioneer) – mf#00828-00831 – us North Dakota [071]
Billings county pioneer : [the official paper of billings county] – Fryburg, Billings Co, ND: Gerald P Nye. v1 n1 aug 15 1919- (wkly) – (= ser Fryburg pioneer; Billings county herald; Belfield herald) – 1 – (place of publ varies: fryburg, nd aug 15 1919-jan 11 1934; medora, nd jan 18 1934- : also bears numbering of the pioneer v6 n45-, and the herald v14 n26-, later dropped; formed by the union of: fryburg pioneer and: billings county herald; absorbed by: belfield herald; missing: 1992 jan 30; currently publ) – mf#02452 etc – us North Dakota [071]
Billings, Elkanah *see* The devonian fossils of canada west
Billings, John Shaw et al *see* The liquor problem
Billings, Mary DeWitt et al *see* American potpourri
Billings, Robert William *see*
– Architectural illustrations and account of the temple church
– Architectural illustrations and description of the cathedral church at durham
– Architectural illustrations of kettering church, northamptonshire
– An attempt to define geometric proportions of gothic architecture
– The baronial and ecclesiastical antiquities of scotland
Billings star – Billings MT. 1919 apr 20-1920 oct 16 – 1r – 1 – mf#852010 – us WHS [071]
Billings, William *see*
– The new england psalm-singer: or, american chorister
– The singing master's assistant, or a key to practical music
Billington – (= ser Bedfordshire parish register series) – 1mf – 9 – £3.00 – uk BedsFHS [929]
Billington, Mary Frances *see* Woman in india...
Billington, st michael and all angels monumental inscriptions – Bedfordshire Family HS 1985 – (= ser Bedfordshire parish register series) – 1mf – 9 – £1.25 – uk BedsFHS [929]
Billington, Thomas *see* Collection of works
Billon, Frederic Louis *see* Annals of st louis in its territorial days, from 1804 to 1821
Billon, Frederic Louis [comp] *see* Annals of st louis in its early days under the french and spanish dominations
Billot, Louis *see*
– De sacra traditione contra novam haeresim evolutionismi
– De verbo incarnato
Billroth, Gustav *see* Commentary on the epistles of paul to the corinthians
Bills and resolutions / U.S. Laws, Statutes, etc. (Bills) – Washington, DC. On film: 1st-72nd Congress; 1789-1933. LL-01 – 1 – us L of C Photodup [348]
The bills of exchange act, 1890 : an act to codify the laws relating to bills of exchange, cheques and promissory notes, passed by the parliament of canada, 53 vic, ch 33, with notes and comments / Girouard, Desire – Montreal: J Valois, 1891 – 8mf – 9 – (incl ind) – mf#03376 – cn Canadiana [348]
Billy bowlegs and the seminole war / Gifford, John Clayton – Coconut Grove, FL. 1925 – 1r – us UF Libraries [978]
Billy Graham Evangelistic Association *see* Decision
"Billy" sunday, the man and his message : with his own words which have won thousands for christ / Ellis, William Thomas – authorized ed Philadelphia: J C Winston, c1914 [mf ed 1990] – 432p/31pl on 2mf – 9 – 0-7905-8252-X – mf#1988-8115 – us ATLA [240]
Billy, Valmore-Armand de *see* Les etudiants tels qu'ils sont
Bilodeau, Francoise *see* Bibliographie du theatre canadien-francais de 1900-1955
Bilontra : orgao desafinado, monarquista republicano – Fortaleza, CE, 07 maio 1891 – (= ser Ps 19) – mf#P17,01,37 – bl Biblioteca [079]
O bilontra : litterario, critico, apimentado e galhofeiro – Fortaleza, CE. 12 out 1889 – (= ser Ps 19) – mf#P17,01,38 – bl Biblioteca [870]
O bilontra : periodico humoristico, litterario e noticioso – Cataguazes, MG. 23 abr 1885 – (= ser Ps 19) – bl Biblioteca [079]
Bilpin, Thomas Victor *see* To the banks of the zambezi
Bilse, Fritz Oswald *see*
– Aus einer kleinen garnison
– Dear fatherland
– Life in a garrison town
Bilson, T *see* The perpetuall government of christs church
Bilston herald – [West Midlands] Wolverhampton 7 oct 1871-29 sep 1906 [mf ed 2002] – 29r – 1 – (missing: 1872, jan-apr 1875, 1896, 1897; cont as: bilston weekly herald [jan 1883-dec 1887]; midland weekly herald [jan 1888-dec 1902]; midland herald [jan 1903-dec 1904]; midland weekly herald [jan 1905-sep 1906]) – uk Newsplan [072]
Bimbi sa ra : [short stories] / Ma Gha – Ran Kun: Ma Gha ca pe thut ve re tha na 1974 [mf ed 1990] – 1r with other items – 1 – (in burmese) – mf#mf-10289 seam reel 205/1 [§] – us CRL [830]
Bimeler, Joseph Michael *see*
– Sammlung auserlesner geistlicher lieder
– Die wahre separation
Bi-metsulot yam / Verne, Jules – Warsaw, Poland. 1876 – 1r – us UF Libraries [939]
Bimhah, G H *see* Mifananidzo yoruponseo
Bimilenario de la fundacion de la colonia norba caeserina / Barvo y Bravo Fernando – Caceres, 1967 – 1 – sp Bibl Santa Ana [946]
Bimilenario de la fundadcion romana de caceres – Madrid: Servicio de Publicaciones del Ministerio de Informacion y Turismo, 1967 – sp Bibl Santa Ana [946]
Bimler, Kurt *see* Die erste und zweite fassung von goethes "wanderjahren"
Bimstein, Emanuel *see* Gottfried der student
Bi'n fueer : geschichten un gedichten ut de lueneboerger heide / Freudenthal, Friedrich – 2. aufl. Norden: H Fischer 1883 [mf ed 1990] – 1r – 1 – (filmed with: ein glaubensbekenntniss / ferdinand freiligrath) – mf#7269 – us Wisconsin U Libr [880]
Binah la-'itim / Figo, Azariah – Warsaw, Poland. 1908 – 1r – us UF Libraries [939]
Binani, G D *see* India at a glance
Binayan, Narciso *see* Pedro henriquez urena
Bindarwish, Jamal *see* Social physique anxiety and exercise setting preferences among college students in a reqired pefwl course
Binder, Frauke *see* 'Der Zensierte Daemon'
Binder, Mathaeus *see* Predigten ueber die lauretanische litanei
Binder, Wilhelm *see* Allgemeine realencyclopaedie (ael1/16)
Bindery news labor / San Francisco Labor Council – v36 n13-v38 n5 [1984 may 11-1985 sep 13] – 1r – 1 – (cont by: gciu news local 583 labor) – mf#1099242 – us WHS [331]
Binding, Rudolf Georg *see*
– Ausgewaehlte und neue gedichte
– Coelestina
– Dichtungen
– Die geige
– Gesammeltes werk
– Groesse der natur; ruf des freien landes; vom inhalt des lebens
– Legenden der zeit
– Liebeskalender
– Moselfahrt aus liebeskummer
– Stolz und trauer
– Unsterblichkeit
– Unvergaengliche erinnerung
– Von der kraft deutschen worts als ausdruck der nation
– Von freiheit und vaterland
– Die waffenbrueder
Binding ties / Black Women's Community Development Foundation – 1973 jan-sep – 1r – 1 – mf#1053380 – us WHS [305]
Bindloss, Harold *see* Sunshine and snow
Bindungsstudien am mikrosomalen cytochrom p450 des rattenhodens mittels gleichgewichtsdialyse / Schuerer, Nanna Y – Duesseldorf 1986 (mf ed 1996) – 2mf – 9 – €40.00 – 3-8267-2350-3 – mf#DHS-AR 2350 – gw Frankfurter [612]
Bi-nedudim uva-mahteret / Kulkielko, Renya – 'En Harod, Israel. 1945 – 1r – us UF Libraries [939]
Bineman, I M *see* Kadry gosudartstvennogo i kooperativnogo apparata sssr
Binet, Etienne *see* Recueil des oeuvres spirituelles du p estienne binet...
Binet, Jacques *see* Budgets familiaux des planteurs de cacao au cameroun
Binet, Vincent le Cornu *see* Correspondence with the government, 1926-1931 and with dr clifford james on clothes, 1931
Bi-netivot ha-zeman veha-netsah / Girst, Judah Loeb – Jerusalem, Israel. 1955 – 1r – us UF Libraries [939]
Bing – Dakar, Senegal. n31-93 aug 1955-oct 1960; n120-227 1963-71 – 1 – us CRL [073]
Bingara advocate – Bingara. 1934-37, 1947-54 – at Pascoe [079]
Bingara advocate – Bingara, jan 1970-dec 1996 – at Pascoe [079]
Bingara telegraph – Bingara – 1r – at Pascoe [079]
Bingara telegraph – Bingara, aug 1897-dec 1907 – 3r – A$185.28 vesicular A$201.78 silver – at Pascoe [079]
Bingel, Horst *see*
– Deutsche lyrik
– Verband deutscher schriftsteller. phantasie und verantwortung
Bingen, Hildegard von *see* Liber scivias (cima50)
Binger, Louis Gustave *see*
– Du niger au golfe de guinee par le pays de kong et le mossi, 1887-1889
– Du niger au golfe du guinee par le pays de kong et le mossi
Bingham, Charles H *see* The story of naaman the syrian
Bingham, D *see* The bastille
Bingham, H A *see* A residence of twenty-one years in the sandwich islands
Bingham, Hiram *see* Residence of twenty-one years in the sandwich islands
Bingham, Joseph *see* Origenes ecclesiasticae
Bingham, Minerva Clarissa Brewster *see* Karaki aika baibara
Bingham, Peregrine *see*
– Bingham's new cases
– Bingham's reports
– Broderip and bingham's reports
Bingham's new cases : new cases in the court of common pleas... / Bingham, Peregrine – v1-6. 1834-40. London: Saunders & Benning, 1835-41 (all publ) – 54mf – 9 – $81.00 – mf#LLMC 84-750 – us LLMC [324]
Bingham's reports : reports of cases argued and determined in the court of common pleas... / Bingham, Peregrine – v1-10. 1822-34. London: J Butterworth/Saunders & Benning, 1824-34 (all publ) – 77mf – 9 – $115.00 – mf#LLMC 84-749 – us LLMC [324]

Bingley, Alfred Horsford *see* Notes on the warlike races of india and its frontiers...
Bingley guardian – [Yorkshire & Humberside] Bradford 28 sep 1928-dec 1950 [mf ed 2004] – 21r – 1 – uk Newsplan [072]
Bingley, William *see*
– Biographical conversations
– Travels in africa, from modern writers
– Travels in north america, from modern writers
Bingman, Mary Beth *see* "I've come a long way"
Binh, Nguyen Loc et al *see* Nhu-ng truyen ngan hay nhat cua que huong chung ta
Binh phu tan van – n2-29. Hue. 1er aout 1930-15 sept 1931 – 1 – (lacking: n11, 22, 28) – fr ACRPP [073]
Binnenlanden van het district nickerie / Cappelle, Herman Van – Baarn, Surinam. 1903 – 1r – us UF Libraries [972]
Binney, Amos *see*
– Binney's theological compend improved
– A theological compend
Binney, David M *see* Identification of selected attributes which predict competition climbing performance
Binney, Frederick Altona *see* Californian homes for educated englishmen
Binney, Hibbert, Lord Bishop of Nova Scotia *see*
– A charge delivered to the clergy
– A charge delivered to the clergy at the visitation held in the cathedral church of st luke
Binney, Horace *see* Opinion of horace binney, esq., upon the right of the city councils to subscribe for stock in the pennsylvania rail-road company
Binney, Joseph Getchell *see* The inaugural address of the rev j g binney
Binney, Juliette Pattison *see*
– Twenty-six years in burmah
Binney, Thomas *see*
– Great exhibition
– The great gorham case
– Is it possible to make the best of both worlds?
– Life and immortality brought to light through the gospel
– Lights and shadows of church-life in australia
– Sir thomas fowell buxton, bart
Binney's reports / Pennsylvania. Superior Court – v1-6. 1799-1814 (all publ) – (= ser Pre-nrs nominative reports) – 13mf – 9 – $58.00 – mf#LLMC 84-194 – us LLMC [340]
Binney's theological compend improved : containing a synopsis of the evidences, doctrines, morals, and institutions of christianity / Binney, Amos & Steele, Daniel – New York: Phillips & Hunt; Cincinnati: Walden & Stowe, c1875 [mf ed 1985] – 1mf – 9 – 0-8370-2705-5 – mf#1985-0705 – us ATLA [240]
Binnie, William *see*
– The church
– Proposed reconstruction of the old testament history
– The psalms
Binnie, William [comp] *see* 1841 census surname index
Binns, C T *see*
– Dinuzulu
– Last zulu king
Binns, Henry Bryan *see* A history of the adult school movement
Binns, Richard William *see* Worcester china
Binns, William *see* Lecture on theodore parker
O binoculo : jornal satyrico, chistoso e litterario – Bahia: Typ do Binoculo, 27 jun, 07 nov 1877 – (= ser Ps 19) – mf#P18B,02,58 – bl Biblioteca [410]
O binoculo – Rio de Janeiro, RJ: Typ Economica de J J Fontes, 05 out 1862 – (= ser Ps 19) – mf#P17,01,118 – bl Biblioteca [079]
O binoculo – Sao Paulo, SP. jun 1879 – (= ser Ps 19) – bl Biblioteca [079]
Binshtok, V I *see* Statisticheskii spravochnik po petrogradu
Bintang timur = Eastern star – 1865-68 [mf ed 2007] – (= ser Vernacular press in the netherlands indies, c1855-1925, unit 1) – 43mf – 9 – €525.00 – mf#mmp134/2 – ne Moran [079]
Bintang utara : surat kabar baru dari tanah seberang barat = Northern star – 1856-57 [mf ed 2007] – (= ser Vernacular press in the netherlands indies, c1855-1925, unit 1) – 5mf – 9 – €60.00 – mf#mmp134/1 – ne Moran [079]
Bintang-barat – Batawie: Ogilvie & Co (daily ex sun, wkly, semiwkly, 3 times/wk) [mf ed Jakarta: National Library of Indonesia 1989] – 1870-95 on 13r – 1 – (Began in 1869? vol designation begins jan 2 1872 with v3; in indonesian) – mf#mf-11208 seam – us CRL [079]
Bintang-djohar – Betawi: Bruining & Wijt [1873- (wkly) [mf ed Jakarta: National Library of Indonesia] – jan 11-dec 1873 (reel 1); oct 13-dec 22 1883; 1884; jan 4-mar 15 1886 (reel 2) – 1 – (iss for oct 13 1883-mar 15 1886 called: tahon jang ka 11, n42-tahon jang ka 14, n6; chiefly in indonesian, some in dutch; organ of: genootschap van in- en uitwendige zending, jul 12 1873-mar 15 1886; iss for 1873 filmed with: soerat chabar batawie, apr 3-jun 26 1858, and: bataviasche koloniale courant, jan 4-aug 2 1811; cont biang lala (jakarta, indonesia : 1867)) – mf#mf-7738 seam – us CRL [079]
Binterim, Anton Joseph *see* Pragmatische geschichte der deutschen concilien vom 4. jahrhundert bis zum concilium von trient
Binyon, Laurence *see* Songs of love and death
Binyon, Laurence, 1869-1943 *see* Akbar
Binyon, Robert Laurence *see*
– Dutch etchers of the seventeenth century
– Thomas girtin
Bio systems – Oxford UK 1967+ – 1,5,9 – ISSN: 0303-2647 – mf#42112 – us UMI ProQuest [574]
Bioastronautics : fundamental and practical problems – (= ser Advances in the astronautical sciences 17) – 9 – $20.00 – us Univelt [629]
Bioastronautics reports – Washington DC 1962-71 – 1 – ISSN: 0006-2901 – mf#1625 – us UMI ProQuest [629]
Biobehavioral reviews – Oxford UK 1977 – 1,5,9 – (cont by: neuroscience and biobehavioral reviews) – ISSN: 0147-7552 – mf#49529.01 – us UMI ProQuest [300]
Bio-bibliographie : la tres reverende mere, marie de saint-jean martin...prieure generale de l'u r / Marie-du-Perpetuel-Secours, mere – 1955 [mf ed 1978] – (= ser Bibliographies du cours...1947-66) – 2mf – 9 – (with ind; pref by ferdinand vandry) – mf#SEM105P4 – cn Bibl Nat [241]
Bio-bibliographie analytique, 1918 a 1961 inclusivement : de l'abbe ernest arsenault, missionnaire-colonisateur / Marie-de-Saint-Camille-de-Jesus, soeur – 1963 [mf ed 1978] – (= ser Bibliographies du cours...1947-66) – 3mf – 9 – (with ind; pref by guy hamel) – mf#SEM105P4 – cn Bibl Nat [241]
Bio-bibliographie analytique 1917-1941 de monsieur l'abbe pierre gravel cure de boischatel / Beatrice-du-Saint-Sacrement, soeur – 1961 [mf ed 1978] – (= ser Bibliographies du cours...1947-66) – 2mf – 9 – (with ind; pref by robert rumilly) – mf#SEM105P4 – cn Bibl Nat [241]
Bio-bibliographie analytique, 1941-1957 : de monsieur l'abbe pierre gravel, cure de boischatel / Georges-Andre, soeur – 1961 [mf ed 1978] – (= ser Bibliographies du cours...1947-66) – 2mf – 9 – (with ind;pref by robert rumilly) – mf#SEM105P4 – cn Bibl Nat [241]
Bio-bibliographie analytique de eddy boudreau : membre de la societe des ecrivains canadiens et de la societe des poetes / Vaillancourt, Emilienne – 1954 [mf ed 1978] – (= ser Bibliographies du cours...1947-66) – 1mf – 9 – (with ind; pref by alphonse desilets) – mf#SEM105P4 – cn Bibl Nat [410]
Bio-bibliographie analytique de f fitz osborne : geologue, titulaire de la chaire de petrologie a l'universite laval / Champagne, Andre – 1950 [mf ed 1979] – (= ser Bibliographies du cours...1947-66) – 1mf – 9 – (wit ind; pref by i w jones) – mf#SEM105P4 – cn Bibl Nat [550]
Bio-bibliographie analytique de gerard langlois : journaliste-directeur des editions du cactus / Prince, Madeleine – 1954 [mf ed 1979] – (= ser Bibliographies du cours...1947-66) – 2mf – 9 – (with ind; pref by rene labrecque) – mf#SEM105P4 – cn Bibl Nat [070]
Bio-bibliographie analytique de jean simard / Mainguy, Louise – 1959 [mf ed 1978] – (= ser Bibliographies du cours...1947-66) – 1mf – 9 – (with ind; pref by georges mainguy) – mf#SEM105P4 – cn Bibl Nat [920]
Bio-bibliographie analytique de l'oeuvre de l'abbe arthur maheux de la societe royale du canada : archiviste et professeur d'histoire des ameriques a l'universite laval / Boivin, Jean – [1954] (mf ed 1978) – (= ser Bibliographies du cours...1947-66) – 2mf – 9 – (with ind; pref by louis-albert vachon) – mf#SEM105P4 – cn Bibl Nat [378]
Bio-bibliographie analytique de l'oeuvre de marcel trudel : docteur es lettres / Voisine, Nive – 1959 [mf ed 1978] – (= ser Bibliographies du cours...1947-66) – 2mf – 9 – (with ind; pref by alphonse fortin) – mf#SEM105P4 – cn Bibl Nat [400]
Bio-bibliographie analytique de l'oeuvre du reverend pere ovila melancon / Sainte-Cecile, soeur – 1962 [i.e. 1963] (mf ed 1979) – (= ser Bibliographies du cours...1947-66) – 2mf – 9 – (with ind; pref by alexis paquet) – mf#SEM105P4 – cn Bibl Nat [241]
Bio-bibliographie analytique de m aime plamondon / Deslauriers, Francoise – 1948 [mf ed 1979] – (= ser Bibliographies du cours...1947-66) – 2mf – 9 – (with ind; pref by rene arthur) – mf#SEM105P4 – cn Bibl Nat [010]
Bio-bibliographie analytique de m elphege-j daignault / Bourget, Magdeleine – 1952 [mf ed 1978] – (= ser Bibliographies du cours...1947-66) – 1mf – 9 – (with ind) – mf#SEM105P4 – cn Bibl Nat [920]
Bio-bibliographie analytique de marcel trudel / Des Rochers, Guy – 1948 [mf ed 1978] – (= ser Bibliographies du cours...1947-66) – 1mf – 9 – (with ind) – mf#SEM105P4 – cn Bibl Nat [920]
Bio-bibliographie analytique de marthe bergeron-hogue / Bonin, Marie – 1964 [mf ed 1979] – (= ser Bibliographies du cours...1947-66) – 2mf – 9 – (with ind; pref by eugene l'heureux) – mf#SEM105P4 – cn Bibl Nat [920]
Bio-bibliographie analytique de me jean-charles bonenfant : membre de la societe royale du canada... / Saint-Emile, soeur – 1964 [mf ed 1979] – (= ser Bibliographies du cours...1947-66) – 2mf – 9 – (with ind) – mf#SEM105P4 – cn Bibl Nat [920]
Bio-bibliographie analytique de me marie-louis beaulieu : avocat a quebec / Hache, Patricia – 1957 [mf ed 1978] – (= ser Bibliographies du cours...1947-66) – 1mf – 9 – (with ind) – mf#SEM105P4 – cn Bibl Nat [340]
Bio-bibliographie analytique de mgr leonce boivin / Labbe, Wilfrid – 1961 [mf ed 1978] – (= ser Bibliographies du cours...1947-66) – 1mf – 9 – (with ind; pref by victor tremblay) – mf#SEM105P4 – cn Bibl Nat [241]
Bio-bibliographie analytique de monseigneur joseph ferland : cure de saint-roch de quebec / Saint-Joseph-de-la-Charite, soeur – 1960 [mf ed 1978] – (= ser Bibliographies du cours...1947-66) – 2mf – 9 – (with ind; pref by soeur saint-francois-de-l'alverne) – mf#SEM105P4 – cn Bibl Nat [241]
Bio-bibliographie analytique de monsieur gerald godin / Charest, Pauline – 1964 [mf ed 1979] – (= ser Bibliographies du cours...1947-66) – 1mf – 9 – (with ind;pref by herve biron) – mf#SEM105P4 – cn Bibl Nat [920]
Bio-bibliographie analytique de monsieur henri turgeon : notaire, professeur agrege a la faculte de droit de quebec / Cote, Berthe – 1956 [mf ed 1978] – (= ser Bibliographies du cours...1947-66) – 1mf – 9 – (with ind; pref by pierre-paul turgeon) – mf#SEM105P4 – cn Bibl Nat [340]
Bio-bibliographie analytique de monsieur herve biron : redacteur en chef au journal le nouvelliste des trois-rivieres... / Lemaire, Ghyslaine – [1963] (mf ed 1979) – (= ser Bibliographies du cours...1947-66) – 2mf – 9 – (with ind; pref by herve biron) – mf#SEM105P4 – cn Bibl Nat [070]
Bio-bibliographie analytique de monsieur l'abbe pascal potvin : principal de l'ecole normale de levis / Saint-Gerard, soeur – 1962 [mf ed 1978] – (= ser Bibliographies du cours...1947-66) – 2mf – 9 – (with ind; pref by arthur maheux) – mf#SEM105P4 – cn Bibl Nat [378]
Bio-bibliographie analytique de reine malouin : membre de la societe des ecrivains canadiens / Dubeau, Jean – 1963 [mf ed 1979] – (= ser Bibliographies du cours...1947-66) – 2mf – 9 – (with ind; pref by thomas-marie landry) – mf#SEM105P4 – cn Bibl Nat [410]
Bio-bibliographie analytique de reine malouin / Plamondon, Jocelyne – 1959 [mf ed 1978] – (= ser Bibliographies du cours...1947-66) – 1mf – 9 – (with ind; pref by soeur ste-claire-de-la-croix) – mf#SEM105P4 – cn Bibl Nat [410]
Bio-bibliographie analytique de rene ouvrard / Bergeron, Helene – 1957 [mf ed 1978] – (= ser Bibliographies du cours...1947-66) – 1mf – 9 – (with ind; pref by soeur ste-claire-de-la-croix) – mf#SEM105P4 – cn Bibl Nat [920]
Bio-bibliographie analytique de rodolphe laplante : regisseur et secretaire-general de l'office du credit agricole du quebec / Severien, frere – 1955 [i.e. 1956] (mf ed 1978) – (= ser Bibliographies du cours...1947-66) – 4mf – 9 – (wit ind; pref by albert rioux) – mf#SEM105P4 – cn Bibl Nat [920]
Bio-bibliographie analytique de roger chartier : directeur du personnel a l'hydro-quebec... / Miko, Eugenie – 1962 [mf ed 1979] – (= ser Bibliographies du cours...1947-66) – 1mf – 9 – (with ind; pref by marcel hudon) – mf#SEM105P4 – cn Bibl Nat [300]
Bio-bibliographie analytique de soeur saint-damase-de-rome : directrice du centre marguerite bourgeoys / Sainte-Marie-Gedeon, soeur – 1963 [i.e. 1964] (mf ed 1979) – (= ser Bibliographies du cours...1947-66) – 2mf – 9 – (with ind; pref by soeur sainte-marie-ernestine) – mf#SEM105P4 – cn Bibl Nat [241]
Bio-bibliographie analytique de soeur saint-francois-de-l'alverne : prefete provinciale des etudes / Sainte-Jeanne-d'Annecy, soeur – 1963 [mf ed 1978] – (= ser Bibliographies du cours...1947-66) – 2mf – 9 – (with ind; pref by soeur saint-jean) – mf#SEM105P4 – cn Bibl Nat [241]
Bio-bibliographie analytique de soeur saint-ignace-de-loyola / Sainte-Marie-Antonin, soeur – 1962 [mf ed 1978] – (= ser Bibliographies du cours...1947-66) – 1mf – 9 – (with ind; pref by mere sainte-madeleine-du-sacre-coeur) – mf#SEM105P4 – cn Bibl Nat [241]
Bio-bibliographie analytique des discours et conferences de l'honorable telesphore-damien bouchard / Morin, Maurice – 1964 [mf ed 1979] – (= ser Bibliographies du cours...1947-66) – 2mf – 9 – (with ind; pref by gustave morin) – mf#SEM105P4 – cn Bibl Nat [920]
Bio-bibliographie analytique des ecrits du docteur wilfrid leblond : professeur titulaire a l'universite laval / Ste-Suzanne, soeur – 1962 [mf ed 1978] – (= ser Bibliographies du cours...1947-66) – 2mf – 9 – (with ind; pref by broquerie fortier) – mf#SEM105P4 – cn Bibl Nat [378]
Bio-bibliographie analytique des imprimes des soeurs de la congregation de notre-dame de montreal / Sainte-Marie-de-Pontmain, soeur – 1952 [mf ed 1979] – (= ser Bibliographies du cours...1947-66) – 2mf – 9 – (with ind; pref by soeur sainte-madeleine-du-sacre-coeur) – mf#SEM105P4 – cn Bibl Nat [241]
Bio-bibliographie analytique du poete adolphe poisson / Richard, Marie-France – 1952 [mf ed 1979] – (= ser Bibliographies du cours...1947-66) – 1mf – 9 – (with ind) – mf#SEM105P4 – cn Bibl Nat [410]
Bio-bibliographie analytique du reverend frere marie-maximin : directeur-adjoint de la legion de marie, maison provinciale, quebec / Sainte-Marie-Rita, soeur – 1964 [mf ed 1979] – (= ser Bibliographies du cours...1947-66) – 2mf – 9 – (with ind) – mf#SEM105P4 – cn Bibl Nat [241]
Bio-bibliographie analytique du reverend pere laurent tremblay : missionnaire ecrivain / Marie de Ste-Marthe-de-la-Trinite, soeur – 1961 [mf ed 1978] – (= ser Bibliographies du cours...1947-66) – 1mf – 9 – (with ind; pref by victor tremblay) – mf#SEM105P4 – cn Bibl Nat [241]
Bio-bibliographie analytique du reverend pere paul-henri barabe : superieur de la province oblate notre-dame-du-tres-saint-rosaire, quebec / Sainte-Rollande-de-l'Immaculee, soeur – 1963 [mf ed 1978] – (= ser Bibliographies du cours...1947-66) – 2mf – 9 – (with ind; pref by gabriel bernier) – mf#SEM105P4 – cn Bibl Nat [241]
Bio-bibliographie analytique d'une religieuse educatrice : soeur sainte-madeleine-des-anges de la congregation de notre-dame / Sainte-Jeanne-de-Jesus, soeur – 1952 [mf ed 1979] – (= ser Bibliographies du cours...1947-66) – 1mf – 9 – (with ind; pref by marie-therese-du-sauveur) – mf#SEM105P4 – cn Bibl Nat [241]
Bio-bibliographie canadienne des oeuvres de mgr emile chartier, 1938-1962 / Levasseur, Georgette – 1963 [mf ed 1979] – (= ser Bibliographies du cours...1947-66) – 2mf – 9 – (with ind; pref by lionel groulx) – mf#SEM105P4 – cn Bibl Nat [440]
Bio-bibliographie critique d'anne hebert / Boutet, Odina – 1950 [mf ed 1978] – (= ser Bibliographies du cours...1947-66) – 1mf – 9 – (with ind) – mf#SEM105P4 – cn Bibl Nat [920]
Bio-bibliographie critique de l'avocat georges bellerive, 1859-1935 / Marie de St-Jean, soeur – 1950 [mf ed 1979] – (= ser Bibliographies du cours...1947-66) – 1mf – 9 – (with ind) – mf#SEM105P4 – cn Bibl Nat [340]
Bio-bibliographie de alcide fleury : historien et journaliste / Saint-Augustin, soeur – 1963 [mf ed 1979] – (= ser Bibliographies du cours...1947-66) – 2mf – 9 – (with ind; pref by joseph-walter houle) – mf#SEM105P4 – cn Bibl Nat [070]
Bio-bibliographie de beatrice clement / Mercier, Jeanne Mance – 1961 [mf ed 1978] – (= ser Bibliographies du cours...1947-66) – 2mf – 9 – (with ind; pref by jean-charles bonenfant) – mf#SEM105P4 – cn Bibl Nat [010]
Bio-bibliographie de damase potvin : ecrivain et journaliste / Raymond, frere – 1956 [mf ed 1979] – (= ser Bibliographies du cours...1947-66) – 2mf – 9 – (with ind) – mf#SEM105P4 – cn Bibl Nat [070]
Bio-bibliographie de damase potvin / Treffry, Philippe – 1947 [mf ed 1978] – (= ser Bibliographies du cours...1947-66) – 1mf – 9 – (with ind) – mf#SEM105P4 – cn Bibl Nat [241]
Bio-bibliographie de eugene achard / Tetreault, Madeleine – Montreal: Ecole de Bibliothecaires, Universite de Montreal, 1947 [mf ed 1994] – 1mf – 9 – (pref by juliette chabot) – mf#SEM105P2166 – cn Bibl Nat [010]

Bio-bibliographie de feu son eminence le cardinal jean-marie-rodrigue villeneuve : oblat de marie immaculee, archeveque de quebec / Houle, Rosaire – 1948 [mf ed 1979] – (= ser Bibliographies du cours...1947-66) – 1mf – 9 – (with ind; pref by g cote) – mf#SEM105P4 – cn Bibl Nat [241]

Bio-bibliographie de georgina lefaivre / Cote, Antonia – 1948 [mf ed 1979] – (= ser Bibliographies du cours...1947-66) – 2mf – 9 – (with ind) – mf#SEM105P4 – cn Bibl Nat [010]

Bio-bibliographie de gerard martin de la societe des ecrivains canadiens / Normand, Rita – 1953 [mf ed 1979] – (= ser Bibliographies du cours...1947-66) – 1mf – 9 – (with ind; pref by charles-marie boissonnault) – mf#SEM105P4 – cn Bibl Nat [410]

Bio-bibliographie de henri [i.e. hector] de saint-denys garneau (1953-1963) / Page, Jean-Pierre – 1964 [mf ed 1979] – (= ser Bibliographies du cours...1947-66) – 1mf – 9 – (with ind) – mf#SEM105P4 – cn Bibl Nat [920]

Bio-bibliographie de jules-s lesage / Bedard, Denyse – 1948 [mf ed 1979] – (= ser Bibliographies du cours...1947-66) – 1mf – 9 – (with ind) – mf#SEM105P4 – cn Bibl Nat [920]

Bio-bibliographie de la r s gabriel-lalemant : conseillere pedagogique au secretariat national de la j e c a montreal / Telmon, soeur – 1962 [mf ed 1978] – (= ser Bibliographies du cours...1947-66) – 1mf – 9 – (with ind. pref by soeur paul-emile) – mf#SEM105P4 – cn Bibl Nat [370]

Bio-bibliographie de l'abbe felix-antoine savard : professeur a la faculte des lettres de l'universite laval, quebec / Savard, Marcelle – 1947 [mf ed 1978] – (= ser Bibliographies du cours...1947-66) – 1mf – 9 – (with ind) – mf#SEM105P4 – cn Bibl Nat [400]

Bio-bibliographie de l'abbe joseph-william-ivanhoe caron, 1875-1941 / Pelletier, J-Antoine – 1947 [mf ed 1978] – (= ser Bibliographies du cours...1947-66) – 1mf – 9 – (with ind; pref by r f gareau) – mf#SEM105P4 – cn Bibl Nat [241]

Bio-bibliographie de m albert rioux / Carbonneau, Leopold – 1952 [mf ed 1978] – (= ser Bibliographies du cours...1947-66) – 1mf – 9 – (with ind; pref by pellerin lagloire) – mf#SEM105P4 – cn Bibl Nat [630]

Bio-bibliographie de m jean bruchesi : sous-secretaire et sous-registraire de la province de quebec / Boissonneault, Henri – 1948 [mf ed 1979] – (= ser Bibliographies du cours...1947-66) – 1mf – 9 – (with ind; pref by ferdinand vandry) – mf#SEM105P4 – cn Bibl Nat [350]

Bio-bibliographie de m l'abbe andre laliberte... : journaliste et educateur, 1926-32 / Sainte-Pauline, soeur – 1961 [mf ed 1978] – (= ser Bibliographies du cours...1947-66) – 2mf – 9 – (with ind; pref by eugene l'heureux) – mf#SEM105P4 – cn Bibl Nat [070]

Bio-bibliographie de mademoiselle louise marchand : bibliothecaire et ecrivain pour la jeunesse / Blanc, Claudette le – Ecole de bibliothécaires de l'Universite de Montreal, 1961 [mf ed [1973]] – 1r – 1 – (with ind; pref by guy boulizon) – mf#SEM35P51 – cn Bibl Nat [020]

Bio-bibliographie de me jean-charles bonenfant : membre de la societe royale du canada... / Jacques, Colette – 1954 [mf ed 1978] – (= ser Bibliographies du cours...1947-66) – 2mf – 9 – (with ind) – mf#SEM105P4 – cn Bibl Nat [020]

Bio-bibliographie de mere marguerite-marie lasalle : ursuline des trois-rivieres / Sainte-Catherine-de-Sienne, mere – 1964 [mf ed 1979] – (= ser Bibliographies du cours... 1947-66) – 1mf – 9 – (with ind; pref by mere j du saint-coeur de marie ferron) – mf#SEM105P4 – cn Bibl Nat [241]

Bio-bibliographie de mere marie-berthe thibault : bibliotheconomie, these presentee a l'universite laval / Bresoles, Judith de – 1962 [mf ed 1978] – (= ser Bibliographies du cours...1947-66) – 1mf – 9 – (with ind; pref by roch dancause) – mf#SEM105P4 – cn Bibl Nat [241]

Bio-bibliographie de mgr joseph-clovis k-laflamme / Trottier, Irenee – 1961 [mf ed 1978] – (= ser Bibliographies du cours...1947-66) – 1mf – 9 – (with ind; pref by rene bureau) – mf#SEM105P4 – cn Bibl Nat [241]

Bio-bibliographie de mme jeanne l'archeveque-duguay / Cote, Marielle – Ecole de bibliothecaires de l'Universite de Montreal, 1947 [mf ed [1973]] – 1r – 5 – (with ind) – mf#SEM16P247 – cn Bibl Nat [010]

Biobibliographie de mme marthe lemaire-duguay / Dery, Marie-Claire – 1962 [mf ed 1978] – (= ser Bibliographies du cours...1947-66) – 2mf – 9 – (with ind; pref by alphonse roux) – mf#SEM105P4 – cn Bibl Nat [010]

Bio-bibliographie de monseigneur alexandre vachon / Parent, Jean-Baptiste – 1947 [mf ed 1978] – (= ser Bibliographies du cours...1947-66) – 1mf – 9 – (with ind; pref by cyrias ouellet) – mf#SEM105P4 – cn Bibl Nat [971]

Bio-bibliographie de monseigneur elias roy : directeur national de l'union missionnaire du clerge et ancien superieur du college de levis / LaBrie, Hilda – 1954 [mf ed 1978] – (= ser Bibliographies du cours...1947-66) – 1mf – 9 – (with ind; pref by joseph ferland) – mf#SEM105P4 – cn Bibl Nat [241]

Bio-bibliographie de monseigneur louis-joseph-arthur melanson : premier archeveque de moncton, nb, 1879-1941 / Marie-Irene, soeur – 1963 [mf ed 1979] – (= ser Bibliographies du cours...1947-66) – 3mf – 9 – (with ind; pref by norbert robichaud) – mf#SEM105P4 – cn Bibl Nat [241]

Bio-bibliographie de monsieur albert gervais / Brousseau, Vincent – 1963 [mf ed 1979] – (= ser Bibliographies du cours...1947-66) – 3mf – 9 – (with ind) – mf#SEM105P4 – cn Bibl Nat [010]

Bio-bibliographie de monsieur carl faessler : professeur titulaire de mineralogie, faculte des sciences / Arteau, Jean-Marie – 1948 [mf ed 1979] – (= ser Bibliographies du cours...1947-66) – 1mf – 9 – (pref by j w laverdiere) – mf#SEM105P4 – cn Bibl Nat [540]

Bio-bibliographie de monsieur elphege bois : directeur du departement de biochimie de la faculte des sciences / Garant, J-Honorat – 1949 [mf ed 1979] – (= ser Bibliographies du cours...1947-66) – 1mf – 9 – (with ind; pref by louis cloutier) – mf#SEM105P4 – cn Bibl Nat [574]

Bio-bibliographie de monsieur gerard filion : journaliste, conferencier, ecrivain / Leveque, Isabella – 1964 [mf ed 1979] – (= ser Bibliographies du cours...1947-66) – 2mf – 9 – (with ind; pref by dominique beaudin) – mf#SEM105P4 – cn Bibl Nat [070]

Bio-bibliographie de monsieur gerard filteau : conseiller technique en pedagogie au departement de l'instruction publique / Maria-du-Sauveur, soeur – 1962 [mf ed 1978] – (= ser Bibliographies du cours...1947-66) – 1mf – 9 – (with ind; pref by lionel allard) – mf#SEM105P4 – cn Bibl Nat [370]

Bio-bibliographie de monsieur jean vallerand : licencie es lettres, diplome de l'universite de montreal en journalisme, critique musical / Juchereau-Duchesnay, Marguerite – 1953 [mf ed 1978] – (= ser Bibliographies du cours...1947-66) – 1mf – 9 – (pref by alice juchereau-duchesnay) – mf#SEM105P4 – cn Bibl Nat [780]

Bio-bibliographie de monsieur jean-charles bonenfant : attache a la bibliotheque de la legislature provinciale, charge de cours d'histoire politique a l'universite laval / Pouliot, Marcelle – 1947 [mf ed 1978] – (= ser Bibliographies du cours...1947-66) – 1mf – 9 – (with ind) – mf#SEM105P4 – cn Bibl Nat [320]

Bio-bibliographie de monsieur le chanoine victor tremblay : president de la societe historique du saguenay et membre de la societe des ecrivains canadiens-francais / Dandurand, Therese – 1956 [mf ed 1978] – (= ser Bibliographies du cours...1947-66) – 2mf – 9 – (with ind) – mf#SEM105P4 – cn Bibl Nat [920]

Bio-bibliographie de roger lemelin / Turgeon, Marguerite – 1949 [mf ed 1979] – (= ser Bibliographies du cours...1947-66) – 1mf – 9 – (with ind; pref by cyrille felteau) – mf#SEM105P4 – cn Bibl Nat [010]

Bio-bibliographie de s e le cardinal nicolas wiseman / Morissette, Lucy – 1947 [mf ed 1978] – (= ser Bibliographies du cours...1947-66) – 1mf – 9 – (with ind; pref by francis goyer) – mf#SEM105P4 – cn Bibl Nat [241]

Bio-bibliographie de yves leclerc, 1953-1961 / Denoncourt, Louise – 1964 [mf ed 1979] – (= ser Bibliographies du cours...1947-66) – 1mf – 9 – (with ind; pref by claire l. leclerc) – mf#SEM105P4 – cn Bibl Nat [010]

Bio-bibliographie des anciens eleves des freres maristes / Bolduc, Marcel – 1963 [mf ed 1979] – (= ser Bibliographies du cours...1947-66) – 2mf – 9 – (with ind; pref by frere georges-maurice) – mf#SEM105P4 – cn Bibl Nat [241]

Bio-bibliographie d'eugene rouillard : notaire, membre de la societe royale du canada... / Rouillard, Joseph – 1963 [mf ed 1979] – (= ser Bibliographies du cours...1947-66) – 2mf – 9 – (with ind) – mf#SEM105P4 – cn Bibl Nat [340]

Bio-bibliographie du chanoine lionel groulx : professeur a l'universite de montreal / Lefebvre, Marguerite – 1947 [mf ed 1978] – (= ser Bibliographies du cours...1947-66) – 1mf – 9 – (with ind) – mf#SEM105P4 – cn Bibl Nat [378]

Bio-bibliographie du colonel g e marquis : conservateur de la bibliotheque de la legislature du quebec / Magloire, frere – 1947 [mf ed 1978] – (= ser Bibliographies du cours...1947-66) – 1mf – 9 – (with ind) – mf#SEM105P4 – cn Bibl Nat [020]

Bio-bibliographie du docteur louis paul dugal / Potvin, Micheline – 1950 [mf ed 1979] – (= ser Bibliographies du cours...1947-66) – 1mf – 9 – (with ind; pref by rosario potvin) – mf#SEM105P4 – cn Bibl Nat [610]

Bio-bibliographie du docteur philippe hamel / Belzile, Marie-Paule – 1949 [i.e. 1950] (mf ed 1979) – (= ser Bibliographies du cours...1947-66) – 2mf – 9 – (with ind; pref by francis goyer) – mf#SEM105P4 – cn Bibl Nat [610]

Bio-bibliographie du frere robert : professeur au mont-saint-louis / Lippens-Giguere, Magdeleine – 1947 [mf ed 1978] – (= ser Bibliographies du cours...1947-66) – 1mf – 9 – (with ind; pref by paul-a giguere) – mf#SEM105P4 – cn Bibl Nat [378]

Biobibliographie du lieutenant ernest paichari [sic], soldat de france / Montreuil, Madeleine – 1949 [mf ed 1979] – (= ser Bibliographies du cours...1947-66) – 1mf – 9 – mf#SEM105P4 – cn Bibl Nat [355]

Bio-bibliographie du r p fernand porter / Caron, Alfred – 1952 [mf ed 1979] – (= ser Bibliographies du cours...1947-66) – 1mf – 9 – (with ind) – mf#SEM105P4 – cn Bibl Nat [010]

Bio-bibliographie du reverend pere francis goyer : consulteur provincial / Coulombe, Marguerite – 1947 [mf ed 1978] – (= ser Bibliographies du cours...1947-66) – 1mf – 9 – (with ind; pref by henri cloutier) – mf#SEM105P4 – cn Bibl Nat [241]

Bio-bibliographie du reverend pere gonzalve poulin : directeur de l'ecole sociale populaire, universite laval / Beaudoin, Gilles – 1955 [mf ed 1978] – (= ser Bibliographies du cours...1947-66) – 2mf – 9 – (with ind; pref by claude corrivault) – mf#SEM105P4 – cn Bibl Nat [241]

Bio-bibliographie du reverend pere joseph-francois richard / Toutant, Thomas – 1963 [mf ed 1979] – (= ser Bibliographies du cours...1947-66) – 2mf – 9 – (with ind) – mf#SEM105P4 – cn Bibl Nat [241]

Bio-bibliographie du reverend pere paul-emile breton : journaliste et ecrivain / Durocher, Georges – 1961 [mf ed 1978] – (= ser Bibliographies du cours...1947-66) – 2mf – 9 – (with ind) – mf#SEM105P4 – cn Bibl Nat [070]

Bio-bibliographie du t r p georges-henri levesque : doyen de la faculte des sciences sociales de l'universite laval / Ubald, frere – 1947 [mf ed 1979] – (= ser Bibliographies du cours...1947-66) – 1mf – 9 – (with ind; pref by stanislas) – mf#SEM105P4 – cn Bibl Nat [300]

Biobliographie de stanislas vachon : ses oeuvres, ses ecrits / Legare, Denise – 1955 [mf ed 1978] – (= ser Bibliographies du cours...1947-66) – 1mf – (with ind; pref by jacques legare) – mf#SEM105P4 – cn Bibl Nat [010]

A biochemical analysis of the exercise-induced dysfunction of the rat gastrocnemius sarcoplasmic reticulum ca2+-atpase protein / Luckin, Kristen A & Klug, Gary A – 1992 – 2mf – 9 – $8.00 – us Kinesology [612]

Biochemical education – Oxford UK 1972-1999 – 1,5,9 – (cont by: biochemistry and molecular biology education) – ISSN: 0307-4412 – mf#49286.01 – us UMI ProQuest [574]

Biochemical engineering journal – Oxford UK 1998+ – 1,5,9 – ISSN: 1369-703X – mf#42798 – us UMI ProQuest [660]

Biochemical genetics – Dordrecht, Netherlands 1967+ – 1,5,9 – ISSN: 0006-2928 – mf#10850 – us UMI ProQuest [574]

Biochemical journal [1906] – London UK 1906-72 – 1,5,9 – ISSN: 0006-2936 – mf#12983.02 – us UMI ProQuest [574]

Biochemical journal [1984] – London UK 1984+ – 1,5,9 – ISSN: 0264-6021 – mf#12983.02 – us UMI ProQuest [574]

Biochemical journal. cellular aspects – London UK 1973-83 – 1,5,9 – ISSN: 0306-3283 – mf#12984 – us UMI ProQuest [574]

Biochemical journal. molecular aspects – London UK 1973-83 – 1,5,9 – ISSN: 0306-3275 – mf#12983.02 – us UMI ProQuest [574]

Biochemical pharmacology – Oxford UK 1958+ – 1,5,9 – ISSN: 0006-2952 – mf#49020 – us UMI ProQuest [615]

Biochemical systematics and ecology – Oxford UK 1973+ – 1,5,9 – ISSN: 0305-1978 – mf#49021 – us UMI ProQuest [574]

Biochemische und molekularbiologische charakterisierung neuer mikrobakterieller nad(p)-abhaengiger alkoholhydrogenasen / Riebel, Bettina – (mf ed 1997) – 3mf – 9 – €49.00 – 3-8267-2478-X – mf#DHS 2478 – gw Frankfurter [574]

Biochemische zeitschrift – Berlin, J Springer. v[310-312]. 1942 – us CRL [660]

Biochemische zeitschrift beitraege zur chemischen physiologie und pathologie – Heidelberg, Germany 1906-48 – 1,5,9 – mf#1140 – us UMI ProQuest [540]

Biochemisches zentralblatt – v1-9. 1903-10 – 9 – $312.00 – (in german) – mf#0106 – us Brook [574]

Biochemistry – Dordrecht, Netherlands 1956+ – ISSN: 0006-2979 – mf#10816 – us UMI ProQuest [574]

Biochemistry – Washington DC 1962 – 1,5,9 – ISSN: 0006-2960 – mf#50002 – us UMI ProQuest [660]

Biochemistry and cell biology = Biochimie et biologie cellulaire – Ottawa ON 1986+ – 1,5,9 – (cont: canadian journal of biochemistry and cell biology = revue canadienne de biochimie et biologie cellulaire) – ISSN: 0829-8211 – mf#10947.02 – us UMI ProQuest [574]

Biochemistry and molecular biology education – Oxford UK 2001+ – 1,5,9 – (cont: biochemical education) – ISSN: 1470-8175 – mf#49286.01 – us UMI ProQuest [574]

Biochimica et biophysica acta = International journal of biochemistry and biophysics – Oxford UK 1947+ – 1,5,9 – ISSN: 0006-3002 – mf#42164 – us UMI ProQuest [574]

Biochimie – Oxford UK 1984+ – 1,5,9 – ISSN: 0300-9084 – mf#42404 – us UMI ProQuest [574]

Biocycle – Emmaus PA 1981+ – 1,5,9 – (cont: compost science/land utilization) – ISSN: 0276-5055 – mf#2712.02 – us UMI ProQuest [333]

Biodegradation – Dordrecht, Netherlands 1993+ – ISSN: 0923-9820 – mf#18596 – us UMI ProQuest [576]

Biodynamics – San Francisco CA 1949-2007 – 1,5,9 – ISSN: 0006-2863 – mf#453 – us UMI ProQuest [630]

Bioelectrochemistry – Oxford UK 2003+ – 1,5,9 – (cont: bioelectrochemistry and bioenergetics) – ISSN: 1567-5394 – mf#42115.01 – us UMI ProQuest [574]

Bioelectrochemistry and bioenergetics – Oxford UK 1974-99 – 1,5,9 – (cont by: bioelectrochemistry) – ISSN: 0302-4598 – mf#42115.01 – us UMI ProQuest [574]

Bioenergetics – Oxford UK 1967+ – 1,5,9 – ISSN: 0005-2728 – mf#42165 – us UMI ProQuest [612]

Bioessays – Cambridge UK 1990-96 – 1,5,9 – ISSN: 0265-9247 – mf#16519 – us UMI ProQuest [574]

Bioethics – Oxford UK 1987+ – 1,5,9 – ISSN: 0269-9702 – mf#17387 – us UMI ProQuest [575]

Bioethics northwest – Dordrecht, Netherlands 1976-77 – 1,5,9 – (cont by: bioethics quarterly) – ISSN: 0362-0824 – mf#12181.04 – us UMI ProQuest [170]

Bioethics quarterly – Dordrecht, Netherlands 1980-81 – 1,5,9 – (cont: bioethics northwest; cont by: journal of bioethics) – ISSN: 0163-9803 – mf#12181.04 – us UMI ProQuest [170]

Bio-factors – Ann Arbor MI 1988 – 1,5,9 – ISSN: 0887-1159 – mf#16445 – us UMI ProQuest [574]

Biofeedback and self-regulation – Dordrecht, Netherlands 1989-96 – 1,5,9 – (cont by: applied psychophysiology and biofeedback) – ISSN: 0363-3586 – mf#17657,01 – us UMI ProQuest [610]

Biogeochemistry – Dordrecht, Netherlands 1991+ – 1,5,9 – ISSN: 0168-2563 – mf#16769 – us UMI ProQuest [574]

Biogeographia dynamica / Sampaio, Alberto Jose De – Sao Paulo, Brazil. 1935 – 1r – us UF Libraries [972]

Biografia / Saldarriaga Betancur, Juan Manuel – Medellin, Colombia. 1954 – 1r – us UF Libraries [972]

Biografia de caceres / Agundez Fernandez, Antonio – Villanueva de la Serena, 1957 – 1 – sp Bibl Santa Ana [920]

Biografia de florentino castro soto / Segura, Rosalia De – San Jose, Costa Rica. 1954 – 1r – us UF Libraries [972]

Biografia de fr. luis de granada...demuestra... autor del libro de oracion / Cuervo, Fray usto – 1896 – 9 – sp Bibl Santa Ana [920]

Biografia de gregorio vasquez / Pizano Restrepo, Roberto – Bogota, Colombia. 1936 – 1r – us UF Libraries [972]

Biografia de joaquin de aguero / Aguero Y Estrada, Francisco – Habana, Cuba. 1935 – 1r – us UF Libraries [972]

Biografia de la esencia tragica en el impulso plen / Tur Canudas, Angel – Habana, Cuba. 1957 – 1r – us UF Libraries [972]

Biografia de la humildad / Vela, David – Guatemala, 1961 – 1r – us UF Libraries [972]

Biografia de lexcmo. d. vicente barrantes / Cortijo Valdes, A – 1873 – 9 – sp Bibl Santa Ana [920]

Biografia de lucas amadeo antomarchi en relacion... / Amadeo Gely, Teresa – San Juan, Puerto Rico. 1964 – 1r – us UF Libraries [972]

Biografia de miguel jeronimo gutierrez, 1822-1871 / Perez, Luis Marino – Habana, Cuba. 1912 – 1r – us UF Libraries [972]

Biografia de roberto g / Fernandez Molina, Antonio – Madrid, 1953 – 1 – sp Bibl Santa Ana [920]

Biografia del caribe / Arciniegas, German – Buenos Aires, Argentina. 1945 – 1r – us UF Libraries [972]

Biografia del dictador garcia moreno / Agramonte Y Pichardo, Roberto Daniel – Habana, Cuba. 1935 – 1r – us UF Libraries [972]

Biografia del doctor jose vargas / Villanueva, Laureano – Caracas, Venezuela. 1954 – 1r – us UF Libraries [972]

Biografia del dr y gral / Paredes, Lucas – Tegueigalpa, Honduras. 1938 – 1r – us UF Libraries [972]

Biografia del general pedro nel ospina / Vernaza, Jose Ignacio – Cali, Colombia. 1935 – 1r – us UF Libraries [972]

Biografia del general rafael urdaneta : ultimo pres... / Arbelaez Urdaneta, Carlos – Maracaibo, Venezuela. 1945 – 1r – us UF Libraries [972]

Biografia del ilmo s d fr ezequiel moreno y di... / Minguella Y Arnedo, Toribio – Barcelona, Spain. 1909 – 1r – us UF Libraries [972]

Biografia del padre reyes / Rosa, Ramon – Tegucigalpa, Mexico. 1955 – 1r – us UF Libraries [972]

Biografia del sir georges etienne cartier / Baillairge, Frederic-Alexandre – Ottawa: s.n, 1882 – 1mf – 9 – (text in italian) – mf#61705 – cn Canadiana [320]

Biografia do jornalismo carioca (1808-1908) / Fonseca, Gondin Da – Rio de Janeiro, Brazil. 1941 – 1r – us UF Libraries [070]

Biografia documentada / Lopez, Jose Maria et al; ed by Bayle, Constantino – Madrid: Razon y Fe, 1928 – 9 – sp Bibl Santa Ana [920]

Biografia y critica de d. jose espronceda isagoge / Munoz, Juan Antonio – Badajoz: Imprenta de la Diputacion Provincial, 1970 – sp Bibl Santa Ana [920]

Biografias. antonio maura, de taxonera, la emperatriz eugenia, de cabal y pedro de alvarado, de baron castro / Bayle, Constantino – Madrid: Razon y Fe, 1946 – 1 – sp Bibl Santa Ana [946]

Biografias bayle / Meseguer, Pedro – Madrid: Razon y Fe, 1944 – 1 – sp Bibl Santa Ana [920]

Biografias de los mandatarios y ministros de la re / Restrepo Saenz, Jose Maria – Bogota, Colombia. 1952 – 1r – us UF Libraries [972]

Biograficheskii slovar' studentov pervykh 27-mi kursov s.-peterburgskoi dukhovnoi akademii : k 100-letiiu s-peterburgskoi dukhovnoi akademii – St Petersburg: Tip I V Leonteva, 1907 – 1 – us CRL [947]

Biografisch archief van de benelux (bab1) = Biographical archive of the benelux countries (bab1) / Gorzny, Willi [comp] – [mf ed 1992-94] – 762mf (1:24) – 9 – diazo €10,060.00 (silver €11,080 isbn: 978-3-598-32630-1) – ISBN-10: 3-598-32610-6 – ISBN-13: 978-3-598-32610-3 – (with printed ind) – gw Saur [949]

Biografisch archief van de benelux. deel 2 (bab2) = Biographical archive of the benelux countries. series 2 (bab2) / Wispelwey, Berend [comp] – [mf ed 1999-2001] – 304mf – 9 – diazo €10,060.00 (silver €11,080 isbn: 978-3-598-34631-6) – ISBN-10: 3-598-34630-1 – ISBN-13: 978-3-598-34630-9 – (with printed ind) – gw Saur [949]

Biographiana : by the compiler of anecdotes of distinguished persons / Seward, William – London: printed for J Johnson 1799 [mf ed 1987] – 2v on 1r – 1 – (filmed with: my life / duncan, i) – mf#1980 – us Wisconsin U Libr [920]

Biographic register / U.S. Dept of State – 1869 70-1910 – 1 – us L of C Photodup [324]

A biographical and critical dictionary of painters and engravers / Bryan, Michael – London 1816 – (= ser 19th c art & architecture) – 17mf – 9 – mf#4.2.1457 – uk Chadwyck [700]

Biographical and historical memorial for flora k heebner, 1874-1947 : missionary of the home and foreign board of missions of the schwenckfelder church in the usa affiliated... – Norristown PA: Board of Pub of the Schwenckfelder Church, 1949 [mf ed 2003] – (= ser Schwenckfeldiana 2/1) – 1r – 1 – mf#2003-s008g – us ATLA [242]

Biographical and literary notices... / Carey, W – London, Northampton, 1886 – 2mf – 9 – mf#HTM-30 – ne IDC [910]

Biographical and literary studies / Little, Charles Joseph; ed by Stuart, Charles Macaulay – New York: Abingdon Press, 1916 – 1mf – 9 – 0-7905-9786-1 – mf#1989-1511 – us ATLA [410]

Biographical annals of jamaica / Cundall, Frank – Kingston, Jamaica. 1904 – 1r – us UF Libraries [972]

Biographical archive of the middle ages (bama) = Biographisches archiv des mittelalters (bama) / Wispelwey, Berend [comp] – [mf ed 2004-06] – 429mf (1:24) in 12 installments – 9 – diazo €10,060.00 (silver €11,080 isbn: 978-3-598-35411-3) – ISBN-10: 3-598-35410-X – ISBN-13: 978-3-598-35410-6 – (with printed ind) – gw Saur [931]

Biographical conversations : on celebrated travellers comprehending distinct narratives of their personal adventures; designed for the use of young persons / Bingley, William – London 1819 [mf ed Hildesheim 1995-98] – 3mf – 9 – €90.00 – 3-487-29889-9 – gw Olms [910]

Biographical cyclopaedia of the catholic hierarchy of the united states, 1784-1898 : a book for reference in the matter of dates, places and persons, in the records of our bishops, abbots and monsignori / Reuss, Francis X – Milwaukee: MH Wiltzius, 1898 – 1mf – 9 – 0-524-07038-5 – mf#1991-2891 – us ATLA [241]

Biographical data for elected officers and members of boards, commissions, and standing committees / Southern Baptist Convention – Comp. by Historical Comm. in co-op. with the Public Relations Dir., SBC, Exec. Comm. Feb 1961. 1348p – 1 – us Southern Baptist [242]

A biographical dictionary of eminent scotsmen / ed by Chambers, Robert – new rev ed. Edinburgh, London: Blackie & son, 1859 [mf ed 1987] – 5v on 1r – 1 – (with suppl vol continuing the biographies to the present time by the rev thos thomson) – mf#1977 – us Wisconsin U Libr [920]

A biographical dictionary of freethinkers of all ages and nations / Wheeler, Joseph Mazzini – London: Progressive Pub Co, 1889 [mf ed 1990] – 1mf – 9 – 0-7905-8254-6 – mf#1988-8117 – us ATLA [140]

Biographical directory of the u.s. congress, 1774-1989 : bicentennial edition – Washington: GPO, 1989 [all publ – 22mf – 9 – $33.00 – mf#llmc 95-029 – us LLMC [323]

Biographical essays / Meuller, Friedrich Max – New York: Charles Scribner, 1884 [mf ed 1995] – (= ser Yale coll) – 282p – 1 – 0-524-09853-0 – mf#1995-0853 – us ATLA [954]

Biographical file of african leaders / Adloff, Virginia McLean – [Stanford, CA: Hoover Library, Stanford Uni, 1965?] – us CRL [960]

Biographical history of primitive or old school baptist ministers of the united states : including a brief treatise on the subject of deacons, their duties, etc., with some personal mention of these officers / ed by Pittman, Reden Herbert – Anderson, Ind: Herald Pub Co, c1909 – 1mf – 9 – 0-524-08496-3 – mf#1993-3141 – us ATLA [242]

Biographical history of primitive or old school baptist ministers of the united states / Pittman, R H – 1909 – 1 – us Southern Baptist [242]

A biographical history of waterloo township and other townships of the county, vol 1 : being a history of the early settlers and their descendants, mostly all of pennsylvania dutch origin / Eby, Ezra E – Berlin Kitchener, ON: s.n, 1895 – v1 on 10mf – 9 – (incl ind) – mf#10019 – cn Canadiana [929]

A biographical history of waterloo township and other townships of the county, vol 2 : being a history of the early settlers and their descendants, mostly all of pennsylvania dutch origin / Eby, Ezra E – Berlin Kitchener, ON: s.n, 1896 – v2 on 10mf – 9 – (incl ind) – mf#10020 – cn Canadiana [929]

A biographical history of waterloo township and other townships of the county, vols 1 and 2 : being a history of the early settlers and their descendants, mostly all of pennsylvania dutch origin / Eby, Ezra E – Berlin Kitchener, on: s.n, 1895-1896 – 2v on 1mf – 9 – mf#10018 – cn Canadiana [929]

Biographical materials / Buck, William Calmes – 1790-1872. 184p – 1 – 6.44 – us Southern Baptist [920]

Biographical materials, correspondence, sermons / Eaton, Thomas Treadwell – 26,338p – 1 – $921.83 – us Southern Baptist [242]

A biographical memoir of samuel hartlib : milton's familiar friend: with bibliographical notices of works published by him... / Dircks, Henry – London: J R Smith [1865?] [mf ed 1990] – 1mf – 9 – 0-7905-5935-8 – (incl bibl ref) – mf#1988-1935 – us ATLA [920]

Biographical memoirs : being a record of the christian lives, experiences, and deaths of members of the religious society of friends from its rise to 1653 / Backhouse, Edward et al – London: W and F G Cash, 1854 – 1mf – 9 – 0-7905-6980-9 – mf#1988-2980 – us ATLA [240]

Biographical notices of members of the society of friends who were resident in ireland / Leadbeater, Mary – London: Harvey and Darton, 1823 – 1mf – 9 – 0-524-07017-2 – mf#1991-2870 – us ATLA [240]

Biographical notices of persian poets / Ousley, Gore – 1846 – (= ser Royal asiatic society oriental translation fund. old series) – 1r – 1 – mf#96417 – uk Microform Academic [490]

Biographical notices of some of the most distnguished jewish rabbies [sic] : and translations of portions of their commentaries, and other works / Turner, Samuel Hulbeart – New York: Stanford and Swords, 1847 – 1mf – 9 – 0-7905-0173-2 – mf#1987-0173 – us ATLA [270]

Biographical record of the san joaquin valley – San Joaquin Co, CA. 1905 – 1r – 1 – $50.00 – mf#B06104 – us Library Micro [929]

Biographical register of the officers and graduates of the us military academy at west point, ny since its establishment in 1802 / Cullum, George Washington – 3rd ed. 1891-1955 – 1 – us AMS Press [355]

Biographical sketch and writings of elder benjamin franklin / Franklin, Benjamin; ed by Rowe, John F & Rice, G W – 4th ed. Cincinnati: GW Rice, 1881 – 6mf – 9 – 0-524-07419-4 – mf#1991-3079 – us ATLA [240]

Biographical sketch of george e sebring, sr / Darcey, Barbara Berry – s.l, s.l? . 193-? – 1r – us UF Libraries [978]

A biographical sketch of george mercer dawson... / Ami, Henry Marc – Ottawa: [s.n.], 1901 – 1mf – 9 – 0-665-71025-9 – mf#71025 – cn Canadiana [920]

Biographical sketch of joanne bethune / Bethune, Joanne – 1768 – 1 – 5.00 – us Southern Baptist [242]

Biographical sketch of jose marti / Quesada Y Miranda, Gonzalo De – Habana, Cuba. 1948 – 1r – us UF Libraries [920]

Biographical sketch of major-general richard montgomery : of the continental army, who fell in the assault of quebec, december 31, 1775 / Cullum, George Washington – S.l: s.n, 1876 – 1mf – 9 – mf#24115 – cn Canadiana [920]

A biographical sketch of sir anthony panizzi... : late principal librarian, british museum / Cowtan, Robert – London: Asher, 1873 – 87p – 1 – mf#8599 – us Wisconsin U Libr [920]

Biographical sketch of tennessee baptist ministers / Borum, Joseph H – 1880 – 1 – us Southern Baptist [242]

Biographical sketch of the famous and brilliant canadian evangelist : robert kidd / Davidson, Judson France – [Toronto?: James & Williams], 1912 – 1mf – 9 – 0-665-87726-9 – mf#87726 – cn Canadiana [240]

A biographical sketch of the hon louis joseph papineau : speaker of the house of assembly of lower canada – Saratoga Springs, NY?: s.n, 1838 – 1mf – 9 – mf#53323 – cn Canadiana [920]

A biographical sketch of the late a f holmes... : including a summary history of medical department of mcgill college / Hall, Archibald – [Montreal?: University Medical Students' Association of McGee College, 1860 [mf ed 1983] – 1mf – 9 – 0-665-44936-4 – (incl bibl ref) – mf#44936 – cn Canadiana [610]

Biographical sketch of the rev. edward irving : with extracts from and remarks on his principal publications / Jones, William – London: John Bennett, 1835 – 1mf – 9 – 0-524-08839-X – mf#1993-1098 – us ATLA [240]

Biographical sketch of thomas chalmers, dd – Glasgow, Scotland. 1847 – 1r – us UF Libraries [240]

Biographical sketch of thomas sterry hunt / Douglas, James – S.l: s.n, 1892? – 1mf – 9 – mf#64581 – cn Canadiana [550]

Biographical sketch of w stanley hanson / Liddle, Carl – s.l, s.l? . 193-? – 1r – us UF Libraries [978]

Biographical sketches : being memorials of arthur penrhyn stanley...henry alford...mrs duncan stewart, etc / Hare, Augustus John Cuthbert – London: G Allen; New York: Dodd, Mead & Co 1895 [mf ed 1987] – 1r [ill] – 1 – (filmed with: historical memorials of westminster abbey / stanley, a p) – mf#1979 – us Wisconsin U Libr [920]

Biographical sketches / Paul, Charles Kegan – London, K Paul, Trench, 1883 – 1mf – 9 – 0-524-00077-8 – mf#1989-2777 – us ATLA [920]

Biographical sketches and anecdotes of members of the religious society of friends – Philadelphia: Tract Assoc of Friends, 1871 [mf ed 1993] – 1mf – 9 – 0-524-07554-9 – mf#1991-3174 – us ATLA [243]

Biographical sketches, funeral services, memorial sermon, addresses of condolence, resolutions of respect, etc etc : relating to charles albert massey: eldest son of mr and mrs h a massey, who died at his home in toronto, ont, february 12th 1884, aged 35 years, 4 months and 23 days – Toronto?: s.n, 1884? – 1mf – 9 – mf#07296 – cn Canadiana [080]

Biographical sketches of chinese communist military leaders / U.S. Office of the Chief of Military History – 1 – us L of C Photodup [951]

Biographical sketches of greeks in jacksonville – s.l, s.l? . 1939-1940 – 1r – us UF Libraries [978]

Biographical sketches of joshua marshman / Fenwick, John – Newcastle upon Tyne, England. 1843 – 1r – us UF Libraries [240]

Biographical sketches of loyalists of the american revolution : with an historical essay / Sabine, Lorenzo – Boston: Little, Brown, c1864 [mf ed 1990] – 2v on 3mf – 9 – 0-7905-7140-4 – mf#1988-3140 – us ATLA [975]

Biographical sketches of our pulpit / Carter, E R – 1856-88 – 1 – us Southern Baptist [242]

Biographical sketches of the fathers of new england / Clark, Mary – 1836 – 1 – $50.00 – us Presbyterian [920]

Biographical sketches of the founder and principal alumni of the log college : together with an account of the revivals of religion under their ministry / ed by Alexander, Archibald – Philadelphia: Presbyterian Board of Publ c1851 [mf ed 1990] – 1mf – 9 – 0-7905-6220-0 – (text probably written by archibald alexander yale) – mf#1988-2220 – us ATLA [242]

Biographical story of the constitution : a study of the growth of the american union / Elliot, Edward – New York, London: G P Putnam's Sons, 1910 – 5mf – 9 – $7.50 – mf#LLMC 95-091 – us LLMC [323]

Biographie avec portrait de m c o lenoir-rolland : ptre, s s et ancien directeur du college de montreal – Montreal: [s.n.], 1879 [mf ed 1983] – 1mf – 9 – mf#25915 – cn Canadiana [241]

Biographie avec portrait de m l'abbe mercier : cure de st-jacques de montreal / David, Laurent Olivier – [Montreal]: Typographie du journal "Le Bien public", 1875 [mf ed 1980] – 1mf – 9 – 0-665-04205-1 – mf#04205 – cn Canadiana [241]

Biographie d'abdel-kader : considerations qui l'ont amene a nous declarer la guerre – [Paris?]: Service historique de l'Armee, [19–] – us CRL [920]

Biographie de charles thibault, ecr – Quebec: L Brousseau, 1884 – 2mf – 9 – mf#08673 – cn Canadiana [920]

Biographie de gerin-lajoie : fragment / Casgrain, Henri Raymond – Montreal?: s.n, 1885? – 1mf – 9 – mf#01095 – cn Canadiana [440]

Biographie de jonathas granville – Paris, France. 1873 – 1r – us UF Libraries [920]

Biographie de joseph-francois perrault : ancien protonotaire de la cour du banc du roi, ancien depute – S.l: s.n, 1844? – 1mf – 9 – mf#53324 – cn Canadiana [340]

Biographie de la famille denechaud – Quebec: la Cie d'impr commerciale, 1895 [mf ed 1980] – 1mf – 9 – 0-665-04038-5 – mf#04038 – cn Canadiana [920]

Biographie de l'hon d b viger / Royal, Joseph – Montreal?: s.n, 1874? – (= ser Coll hommes d'etat) – 1mf – 9 – mf#12731 – cn Canadiana [971]

Biographie de m francois vezina : caissier de la banque nationale / Bechard, Auguste – St-Roche de Quebec: Ateliers du Nouvelliste, 1878 [mf ed 1980] – 1mf – 9 – 0-665-03028-2 – mf#03028 – cn Canadiana [920]

Biographie de samdech preah sanghareach chuon nath, superieur de l'ordre mahanikay = Biography of samdech preah sangharaja chuon nath, the chief mahanikaya order – Phnom-Penh: Le Comite Central [mf ed 1989] – 1r with other items – 1 – (in french & english) – mf#mf-10289 seam reel 005/11 [§] – us CRL [280]

Biographie de samdech preah sanghareach chuon-nath : superieur de l'ordre mahanikaya – [Phnom-Penh] Institut bouddhique [mf ed 1989] – (= ser Serie de culture et civilisation khmeres 6) – 1r with other items – 1 – mf#mf-10289 seam reel 005/10 [§] – us CRL [280]

Biographie de s.a.r. le prince norodom sihanouk / Cambodia. Ministere de l'information – [Phnom-Penh] 1959 [mf ed 1989] – 1r with other items – 1 – mf#mf-10289 seam reel 014/07 [§] – us CRL [959]

Biographie de s.a.r. le prince norodom sihanouk, chef d'etat du cambodge / Cambodia. Ministere de l'information – Phnom-Penh 1965] [mf ed 1989] – 1r with other items – 1 – (in french & english) – mf#mf-10289 seam reel 016/14 [§] – us CRL [959]

Biographie de sir n f belleau : chevalier commandeur de l'ordre de saint-michel et de saint-georges, et premier lieutenant-gouverneur de la province de quebec, sous la confederation des provinces de l'amerique du nord / Drapeau, Stanislas – [Quebec?: s.n.], 1883 – 1mf – 9 – 0-665-02759-1 – mf#02759 – cn Canadiana [971]

Biographie des hommes vivants : ou histoire par ordre alphabetique... – Paris. 5v. 1816-1819 – 31mf – 9 – mf#H-3028 – ne IDC [700]

Biographie du general montholon – [Paris 1849?] – us CRL [920]

Biographie et galerie historique des contemporains... / ed by Barthelmy, M P – Paris, 1822. 2v – 12mf – 9 – mf#H-3029 – ne IDC [700]

Biographie universelle / ed by Michaud, L G – Paris, 1843-1865. 45v – 340mf – 9 – mf#8839 – ne IDC [700]

Biographie universelle by micheud – 1843-65 [mf ed ProQuest] – 84v on 21r – 1 – (lists prominent personalities of the 18th & early 19th c, with emphasis on those living in continental europe) – us UMI ProQuest [940]

Biographie universelle des musiciens et bibliographie generale de la musique / Fetis, Francois J – 2nd ed. Paris. 8v+suppl. 1860-65 – 1 – $120.00 – (2v 1878-80) – mf#0203 – us Brook [780]

Biographies de a s falardeau et a e aubry / Casgrain, Henri Raymond – Montreal: Beauchemin & Valois, 1886 [mf ed 1980] – 2mf – 9 – 0-665-03021-5 – mf#03021 – cn Canadiana [920]

Biographies de a s falardeau et a e aubry / Casgrain, Henri-Raymond – Montreal: Beauchemin & Valois, 1886 [mf ed 1970] – 1r – 5 – mf#SEM16P10 – cn Bibl Nat [750]

Biographies de l'honorable barthelemi joliette et de m le grand vicaire a manseau / Bonin, Joseph – Montreal: E Senecal, 1874 [mf ed 1980] – 3mf – 9 – 0-665-02109-7 – mf#02109 – cn Canadiana [920]

Les biographies du manhal safi / Wiet, G – Cairo, 1932 – (= ser Memoires presentes a l'Institut d'Egypte) – 6mf – 9 – (memoires presentes a l'institut d'egypte v19) – mf#NE-20141 – ne IDC [956]

Biographies et portraits / David, Laurent-Olivier – Montreal: Beauchemin & Valois, 1876 – 4mf – 9 – mf#02511 – cn Canadiana [971]

Biographies et portraits d'ecrivains canadiens (1e serie) : etudes publiees dans le "propagateur", bulletin bibliographique de la librairie beauchemin – Montreal: Librairie Beauchemin, 1913 – 2mf – 9 – 0-665-76138-4 – mf#76138 – cn Canadiana [440]

Biographies of english catholics in the 18th century / Kirk, John – London: Burns & Oates, 1909 – 1mf – 9 – mf#1988-8045 – us ATLA [241]

Biographies of fellows, american academy of physical education / Clarke, Henry Harrison – 1953 – 3mf – 9 – $9.00 – us Kinesology [790]

Biographies of former greenville county physicians – South Carolina. 151p – 1 – us Southern Baptist [610]

Biographies of words and the home of the aryas / Mueller, Friedrich Max – London; New York: Longmans, Green, 1888 – 1mf – 9 – 0-7905-8860-9 – mf#1989-2085 – us ATLA [400]

Biographies of working men / Allen, Grant – London: SPCK, 1884 (London, Beccles: Clowes) – (= ser The people's library) – 3mf – 9 – (incl publ list) – mf#58890 – cn Canadiana [331]

Biographische charakterbilder aus der judischen geschichte / Katz, Albert – Berlin, Germany. 1922 – 1r – us UF Libraries [939]

Biographische denkmale / Varnhagen von Ense, Karl August – 2. verm verb aufl. Berlin: G Reimer 1845-46 – 5v on 1r – 1 – mf#7773 – us Wisconsin U Libr [943]

Biographische notizen ueber heinrich von kleist : in faksimilenachbildung / Schuetz, Wilhelm von; ed by Minde-Pouet, Georg – Berlin: Weidmann, 1936 [mf ed 1996] – (= ser Schriften der kleist-gesellschaft 16) – 19p – 1 – mf#8707 – us Wisconsin U Libr [920]

Biographisches archiv der antike (baa) = Biographical archive of the classical world / Schmuck, Hilmar [comp] – [mf ed 1996-99] – 656mf (1:24) – 9 – diazo €10,060.00 (silver €11,080 isbn: 978-3-598-33971-4) – ISBN-10: 3-598-33970-4 – ISBN-13: 978-3-598-33970-7 – (with guide & ind) – gw Saur [930]

Biographisches archiv der sowjetunion (1917-1991) (basu) = Biographical archive of the soviet union (1917-1991) / Frey, Axel [comp] – [mf ed 2000-03] – 525mf (1:24) in 12 installments – 9 – diazo €10,060.00 (silver €11,080 isbn: 978-3-598-34691-0) – ISBN-10: 3-598-34690-5 – ISBN-13: 978-3-598-34690-3 – (with printed ind) – gw Saur [947]

Biographisches jahrbuch und deutscher nekrolog – Berlin: G Reimer 1897-1917 [mf ed 1979] – 18v on 5r [ill] – 1 – (cont: biographische blaetter; cont by: deutsches biographisches jahrbuch) – mf#film mas c685 – us Harvard [920]

Biographisches lexikon / ed by Wurzbach, C von – Wien, 1856-1891. v1-60 – 299mf – 9 – mf#8762 – ne IDC [700]

Biographisches lexikon des kaisertums oesterreich / ed by Wurzbach, Constantin – Vienna. 60v. 1856-91 – 9 – $858.00 – (in german) – mf#0685 – us Brook [943]

Biographisches lexikon fuer das gebiet zwischen inn und salzach / Fuerst, Max – Muenchen 1901 [mf ed Hildesheim 1983] – (= ser Die schriftsteller- und gelehrtenlexika des 17., 18., und 19. jahrhunderts) – 1v on 3mf – 9 – diazo €19.80 silver €24.80 – gw Olms [920]

Biographisch-literarisches lexikon der katholischen deutschen dichter, volks-und jungendschriftsteller im 19. jahrhundert / Kehrein, Josef – Zuerich, Stuttgart, Wuerzburg 1868-71 [mf ed Hildesheim 1983] – (= ser Die schriftsteller- und gelehrtenlexika des 17., 18., und 19. jahrhunderts) – 2v on 8mf – 9 – diazo €37.80 silver €42.80 – gw Olms [430]

Biography – Honolulu HI 1987+ – 1,5,9 – ISSN: 0162-4962 – mf#16285 – us UMI ProQuest [920]

Biography of a new faith / Sen, Prasanta Kumar – Calcutta: Thacker, Spink & Co, 1950 – (= ser Samp: indian books) – us CRL [280]

Biography of a slave, 1875 / Thompson, Charles – 1r – 1 – mf#B29802 – us Ohio Hist [920]

Biography of baron dekalb gray / Seay, Warren Mosby – 1 – us Southern Baptist [242]

Biography of edwin james turpin : an earlier settler in fiji, 1971 / Diamond, A I – 1881-1971 – 1r – 1 – mf#pmb1183 – at Pacific Mss [980]

Biography of edwin james turpin, an earlier settler in fiji / Daimond, A I – 1971 – 1r – 1 – (available for ref) – mf#pmb1183 – at Pacific Mss [920]

The biography of eld. barton warren stone / Stone, Barton Warren – Cincinnati: J.A. & U.P. James, 1847, c1846 – 1mf – 9 – 0-7905-5976-5 – mf#1988-1976 – us ATLA [920]

The biography of elder james m. neff and his writings / Neff, James Monroe; ed by Neff, Florence – Elgin, IL: Brethren Pub House, 1913 – 1mf – 9 – 0-524-03559-8 – mf#1990-4754 – us ATLA [240]

Biography of elisha kent kane / Elder, William – Philadelphia: Childs & Peterson; London: Trubner, 1858 – 5mf – 9 – mf#53232 – cn Canadiana [617]

Biography of gospel song and hymn writers / Hall, Jacob Henry – New York: F.H. Revell, c1914 – 1mf – 9 – 0-7905-5048-2 – mf#1988-1048 – us ATLA [780]

Biography of his royal highness prince norodom sihanouk "upayuvareach," exking of cambodia – [n.p. 195-?] [mf ed 1989] – 1r with other items – 1 – mf#mf-10289 seam reel 014/03 [§] – us CRL [959]

Biography of iowa indians of kansas and nebraska from 1880-96 / Nuzum, George – 1 – us Kansas [978]

A biography of rev henry ward beecher / Beecher, William Constantine et al – New York: C L Webster, 1888 [mf ed 1991] – 2mf – 9 – 0-524-00505-2 – mf#1990-0005 – us ATLA [242]

Biography of rev. hosea ballou / Ballou, Maturin Murray – Boston: A. Tompkins, 1852 – 1mf – 9 – 0-7905-5750-9 – mf#1988-1750 – us ATLA [240]

Biography of samdech preah sanghareach chuon-nath, the chief of mahanikava order – [Phnom Penh]: Institut bouddhique [mf ed 1989] – (= ser Serie de culture et civilisation khmeres 7) – 1r with other items – 1 – mf#mf-10289 seam reel 017/31 [§] – us CRL [280]

Biography of william cullen bryant / Godwin, Parke – New York, NY. v1-2. 1883 – 1r – us UF Libraries [070]

Biography of william sherman wiley / Derigo, G A – 1867-1935 – 1 – $5.00 – us Southern Baptist [242]

Bioinformatics – Oxford UK 1998+ – 1,5,9 – (cont: computer applications in the biosciences: cabios) – ISSN: 1367-4803 – mf#16457,01 – us UMI ProQuest [500]

Biokhimiia – [Moskva]: Obedinennoe nauchno-tekhnicheskoe izd-vo. v15-16 (1950-1951). v17 n2 (1952) – us CRL [660]

[Biola-] biola broadcaster – CA. 1961-1973 – 3r – 1 – $180.00 – mf#R04007 – us Library Micro [071]

[Biola-] the biola chimes – Biola, CA. 1938-1990 – 7r – 1 – $420.00 – mf#B06015 – us Library Micro [071]

Biolaw : a legal and ethical reporter on medicine, health care, and engineering / ed by Campbell, Courtney S et al – 1986-2001 – backfiles per yr – 9 – $1330.00 – us UPA [344]

Biolley, Paul *see*
- Costa rica et son avenir
- Costa-rica und seine zukunft

Biologia contra la democracia : ensayo de solucion / Agramonte Y Pichardo, Roberto Daniel – Habana, Cuba. 1927 – 1r – us UF Libraries [304]

Biologia de la democracia : ensayo de solucion / Lamar Schweyer, Alberto – Habana, Cuba. 1927 – 1r – us UF Libraries [304]

Biologia hematologica elemental comparada / Picado Twight, Clodomiro – San Jose, Costa Rica. 1942 – 1r – us UF Libraries [616]

Biologia no brasil / Mello-Leitao, Candido De – Sao Paulo, Brazil. 1937 – 1r – us UF Libraries [574]

Biological aspects that influence the domestication potential of englerophytum natalense / Ech, H van – Stellenbosch: U of Stellenbosch 1998 [mf ed 1998] – 3mf – 9 – mf#mf.1283 – sa Stellenbosch [634]

Biological bulletin – Woods Hole MA 1899+ – 1,5,9 – ISSN: 0006-3185 – mf#8737 – us UMI ProQuest [574]

Biological chemistry – Berlin, Germany 1996+ – 1,5,9 – (cont: biological chemistry hoppe-seyler) – ISSN: 1431-6730 – mf#1143.02 – us UMI ProQuest [574]

Biological chemistry hoppe-seyler – Berlin; Germany 1985-96 – 1,5,9 – (cont: hoppe-seyler's zeitschrift fuer physiologische chemie; cont by: biological chemistry) – ISSN: 0177-3593 – mf#1143.02 – us UMI ProQuest [574]

Biological conservation – Oxford UK 1968+ – 1,5,9 – ISSN: 0006-3207 – mf#42113 – us UMI ProQuest [574]

Biological cybernetics – Dordrecht, Netherlands 1961+ – 1,5,9 – (cont: kybernetik) – ISSN: 0340-1200 – mf#13144,01 – us UMI ProQuest [000]

Biological mass spectrometry – Hoboken NJ 1991-94 – 1,5,9 – (cont: biomedical and environmental mass spectrometry) – ISSN: 1052-9306 – mf#13301,02 – us UMI ProQuest [540]

Biological oceanography – Abingdon, Oxfordshire 1983-85 – 1,5,9 – ISSN: 0196-5581 – mf#12421 – us UMI ProQuest [550]

Biological photographic association. journal – Atlanta GA 1932-79 – 1,5,9 – (cont by: journal of biological photography) – ISSN: 0006-3215 – mf#780.01 – us UMI ProQuest [574]

Biological psychiatry – Oxford UK 1969+ – 1,5,9 – ISSN: 0006-3223 – mf#42531 – us UMI ProQuest [616]

Biological psychology – Oxford UK 1973+ – 1,5,9 – ISSN: 0301-0511 – mf#42114 – us UMI ProQuest [150]

Biological research for nursing – Newbury Park CA 2000+ – 1,5,9 – ISSN: 1099-8004 – mf#31325 – us UMI ProQuest [610]

The biological review of ontario – Toronto: Biological Society of Ontario, [1894] – 9 – mf#P05155 – cn Canadiana [574]

Biological reviews – Cambridge UK 1998+ – 1,5,9 – (cont: biological reviews of the cambridge philosophical society) – ISSN: 0006-3231 – mf#14915.03 – us UMI ProQuest [574]

Biological reviews of the cambridge philosophical society – Cambridge UK 1985-97 – 1,5,9 – (cont by: biological reviews) – ISSN: 0006-3231 – mf#14915.03 – us UMI ProQuest [574]

Biological sciences curriculum study. bscs journal – Colorado Spings CO 1979-81 – 1,5,9 – (cont: biological sciences curriculum study journal) – mf#11727.01 – us UMI ProQuest [574]

Biological sciences curriculum study journal – Colorado Springs CO 1978-79 – 1,5,9 – (cont by: bscs journal) – ISSN: 0162-3613 – mf#11727.01 – us UMI ProQuest [574]

Biological sciences curriculum study newsletter *see* American institute of biological sciences. biological sciences curriculum study newsletter

Biological signals and receptors – Basel, Switzerland 2001+ – 1,5,9 – ISSN: 1422-4933 – mf#20575.02 – us UMI ProQuest [574]

Biological structures and morphogenesis – Paris, France 1988 – 1,5,9 – (cont: archives d'anatomie microscopique et de morphologie experimentale) – ISSN: 0989-8972 – mf#17270 – us UMI ProQuest [578]

Biological study of the tap water in the school of practical science, toronto / Acheson, George – [S.l: s.n, 1883?] [mf ed 1980] – 1mf – 9 – 0-665-02386-3 – mf#02386 – cn Canadiana [574]

Biological wastes – Oxford UK 1987-90 – 1,5,9 – (cont: agricultural wastes) – ISSN: 0269-7483 – mf#42009.01 – us UMI ProQuest [630]

Biological weapons convention protocol : status and implications: hearing...house of representatives, 107th congress, 1st session, jun 5 2001 / United States. Congress. House. Committee on Government Reform. Subcommittee on National Security, Veterans Affairs, and International Relations – Washington: US GPO 2002 [mf ed 2002] – 2mf – 9 – 0-16-068856-6 – us GPO [327]

Biologico : revista dos technicos do instituto biologico – Sao Paulo, Brazil 1973-77 – 1,5,9 – ISSN: 0366-0567 – mf#7974 – us UMI ProQuest [574]

Biologie und pathologie des weibes : ein handbuch der frauenheilkunde und geburtshilfe / ed by Halban, Josef et al – Berlin, Wien 1924-29 [mf ed 2006] – (= ser Hq 64) – 9v on 155mf – 9 – €820.00 – 3-89131-484-1 – gw Fischer [618]

Biologische verfahren des pflanzenschutzes im zierpflanzenbau : gezeigt am beispiel des einsatzes von raubmilben (phytoseiulus persimilis athias-henriot zur bekaempfung der spinnmilbe (tetranychus urticae koch) in hausrosen... / Hantke, Friedrich – (mf ed 1993) – 2mf – 9 – €40.00 – 3-89349-764-1 – mf#DHS 764 – gw Frankfurter [576]

Biologisches zentralblatt – Oxford UK 1972-91 – 1,5,9 – ISSN: 0006-3304 – mf#7244.01 – us UMI ProQuest [574]

Biologist – Charleston IL 1982-87 – 1,5,9 – ISSN: 0006-3339 – mf#12894 – us UMI ProQuest [574]

Biology : with preludes on current events / Cook, Joseph – Boston: James R Osgood 1877 [mf ed 1990] – 1mf – 9 – 0-7905-3931-4 – mf#1989-0424 – us ATLA [230]

Biology and fertility of soils – Dordrecht, Netherlands 1987+ – 1,5,9 – ISSN: 0178-2762 – mf#16978 – us UMI ProQuest [630]

Biology and life history of the palm-leaf skeletonizer / Creighton, John Thomas – s.l, s.l? . 1929 – 1r – us UF Libraries [630]

Biology digest – Medford NJ 1985+ – 1,5,9 – ISSN: 0095-2958 – mf#15203 – us UMI ProQuest [574]

Biology, life history, and control of the cotton leaf worm alabama argillacea (hubner) / Rowell, John Orian – s.l, s.l, s.l? 1933 – 1r – us UF Libraries [630]

Biology, life history, and control of the cross-striped cabbage worm / Cain, Thomas Leonard – s.l, s.l? . 1931 – 1r – us UF Libraries [630]

Biology, life history, and control of the diamond-back moth / Bess, Henry Alver – s.l, s.l? . 1931 – 1r – us UF Libraries [630]

Biology of aquatic and littoral insects / Usinger, Robert Leslie – Berkeley, CA. 1948 – 1r – us UF Libraries [590]

Biology of civilization / Walker, Cyril Charles – Toronto, ON. 1930 – 1r – us UF Libraries [572]

Biology of the cell – Lausanne, Switzerland 1986+ – 1,5,9 – ISSN: 0248-4900 – mf#42405 – us UMI ProQuest [578]

Biology of the neonate = Zeitschrift fuer die biologie des neugeborenen – Basel, Switzerland 1966+ – 1,5,9 – ISSN: 0006-3126 – mf#2048.01 – us UMI ProQuest [574]

Biology & philosophy – Dordrecht, Netherlands 1986+ – 1,5,9 – ISSN: 0169-3867 – mf#15257 – us UMI ProQuest [574]

Biomass – Oxford UK 1981-90 – 1,5,9 – ISSN: 0144-4565 – mf#42242 – us UMI ProQuest [630]

Biomass and bioenergy – Oxford UK 1991+ – 1,5,9 – ISSN: 0961-9534 – mf#49617 – us UMI ProQuest [333]

Biomaterials – Oxford UK 1980+ – 1,5,9 – ISSN: 0142-9612 – mf#13322 – us UMI ProQuest [610]

A biomechanical analysis of a sit-to-stand transfer among the elderly / Hughes, Lorraine C – 1999 – 2mf – 9 – $8.00 – mf#PE 3943 – us Kinesology [612]

A biomechanical analysis of canine gait before and after unilateral cemented total hip replacement / Dogan, Selami – 1989 – 264p 3mf – 9 – $12.00 – us Kinesology [619]

A biomechanical analysis of children's balance behavior : an investigation in balance theory / Kennedy, S O – 1991 – 3mf – 9 – $12.00 – us Kinesology [790]

Biomechanical analysis of forces and torques at the support lower extremity during two running-in-place exercises at three paces / Muniz, A E – 1990 – 4mf – 9 – $16.00 – us Kinesology [790]

A biomechanical analysis of patellofemoral stress syndrome / Moss, R I – 1989 – 2mf – 9 – $8.00 – us Kinesology [790]

A biomechanical analysis of the demi plie and grand plie / Buday, Marie T – 1989 – 83p 1mf – 9 – $4.00 – us Kinesology [612]

A biomechanical analysis of the effects of hand weights on the arm-swing while walking and running / Denny, Karen L – University of Wisconsin-La Crosse, 1995 – 1mf – 9 – $4.00 – mf#PE3588 – us Kinesology [612]

A biomechanical analysis of the prolonged effects on functional parameters of a test seating system for moderately involved cerebral palsied children / Boucher, George P – 1986 – 92p 1mf – 9 – $4.00 – us Kinesology [620]

A biomechanical analysis of the single arm versus the parallel double arm takeoffs in the triple jump / Larkins, Clifford – 1987 – 158p 2mf – 9 – $8.00 – us Kinesology [612]

A biomechanical and physiological analysis of efficiency during different running paces / Price, Kathleen M & Wilkerson, Jerry D – 1992 – 2mf – $8.00 – us Kinesology [612]
Biomechanical characteristics of the healthy and acl-reconstructed female knee / Marshall, Christina – 1999 – 2mf – 9 – $8.00 – mf#PE 3930 – us Kinesology [617]
Biomechanical comparison of support provided by the airstirrup ankle training brace'm pre- and post-exercise / Money, Sharon M & Kimura, Iris F – 1993 – 1mf – $4.00 – us Kinesology [612]
The biomechanical effects of crank arm length : on cycling mechanics / Sprules, Erica B – 2000 – 141 on 2mf – 9 – $10.00 – mf#PE 4120 – us Kinesology [612]
The biomechanical effects of prolotherapy on traumatized achilles tendons of male rats / Harrison, Maria E G – Brigham Young University, 1995 – 1mf – 9 – mf#PE 3651 – us Kinesology [617]
Biomechanical parameters influencing fourth grade children's free throw shooting / McKay, Laura L – 1997 – 2mf – 9 – $8.00 – mf#PE 3764 – us Kinesology [612]
Biomedical and environmental mass spectrometry – Hoboken NJ 1986-90 – 1,5,9 – (cont: biomedical mass spectrometry; cont by: biological mass spectrometry) – ISSN: 0887-6134 – mf#13301.02 – us UMI ProQuest [540]
Biomedical applications – Oxford UK 1992-93 – 1,5,9 – ISSN: 0378-4347 – mf#42722 – us UMI ProQuest [540]
Biomedical communications – Midland Park NJ 1977-84 – 1,5,9 – ISSN: 0092-8607 – mf#11693 – us UMI ProQuest [610]
Biomedical engineering – Dordrecht, Netherlands 1967+ – 1,5,9 – ISSN: 0006-3398 – mf#10876 – us UMI ProQuest [610]
Biomedical mass spectrometry – Hoboken NJ 1974-85 – 1,5,9 – (cont by: biomedical and environmental mass spectrometry) – ISSN: 0306-042X – mf#13301.02 – us UMI ProQuest [540]
Biomedical news – New York NY 1970-74 – 1 – ISSN: 0006-3401 – mf#9059 – us UMI ProQuest [574]
Biomedicine & pharmacotherapy = Biomedecine and pharmacotherapie – Oxford UK 1989+ – 1,5,9 – ISSN: 0753-3322 – mf#42644 – us UMI ProQuest [615]
Biomembranes – Oxford UK 1967+ – 1,5,9 – ISSN: 0005-2736 – mf#42167 – us UMI ProQuest [574]
Biometrics – Oxford UK 1945+ – 1,5,9 – ISSN: 0006-341X – mf#5828 – us UMI ProQuest [510]
Biometrika – Oxford UK 1901+ – 1,5,9 – ISSN: 0006-3444 – mf#11624 – us UMI ProQuest [310]
Biomolecular engineering – Oxford UK 2002+ – 1,5,9 – ISSN: 1389-0344 – mf#42450.03 – us UMI ProQuest [575]
Biomonitoring as a means to determine the pollution level in stellenbosch / Davis, Shaun A – U of the Western Cape 1991 [mf ed S.l: s.n. 1991] – 2mf – 9 – (abstract in afrikaans & english; incl bibl) – sa Misc Inst [628]
Biondi, A *see* Essequie della sacra cattolica real maesta del re di spagna don filippo 2. d'austria
[Biondo, M A] Ilg, A *see* Von der hochedlen malerei
Bioorganic & medicinal chemistry letters – Oxford UK 1991+ – 1,5,9 – ISSN: 0960-894X – mf#49607 – us UMI ProQuest [540]
Biopharmaceutics & drug disposition – Hoboken NJ 1979+ – 1,5,9 – ISSN: 0142-2782 – mf#11996 – us UMI ProQuest [615]
Biophysical chemistry – Lausanne, Switzerland 1973+ – 1,5,9 – ISSN: 0301-4622 – mf#42116 – us UMI ProQuest [612]
Biophysical journal – Bethesda MD 1960+ – 1,5,9 – ISSN: 0006-3495 – mf#12245 – us UMI ProQuest [612]
Biophysics of structure and mechanism – Dordrecht, Netherlands 1981-83 – 1,5,9 – ISSN: 0340-1057 – mf#13145.01 – us UMI ProQuest [612]
Biopolymers – Hoboken NJ 1963+ – 1,5,9 – ISSN: 0006-3525 – mf#11047 – us UMI ProQuest [540]
Biopuso ba basotho – Mafeteng, BCP Youth League. v[1]. jul 9-dec 19 1965 – us CRL [960]
Bioresource technology – Oxford UK 1991+ – 1,5,9 – ISSN: 0960-8524 – mf#42629 – us UMI ProQuest [630]
Bios – Ocean Grove NJ 1972+ – 1,5,9 – ISSN: 0005-3155 – mf#7754 – us UMI ProQuest [574]
Bioscience – Washington DC 1951+ – 1,5,9 – ISSN: 0006-3568 – mf#7098 – us UMI ProQuest [574]
The bioscope – 1908-1932 – 67r – 1 – £3200.00 – mf#BIS – uk World [790]
Biosensors – Oxford UK 1985-89 – 1,5,9 – (cont by: biosensors and bioelectronics) – ISSN: 0265-928X – mf#42561.01 – us UMI ProQuest [612]
Biosensors and bioelectronics – Oxford UK 1990+ – 1,5,9 – (cont: biosensors) – ISSN: 0956-5663 – mf#42561.01 – us UMI ProQuest [612]
Biot, E C *see* Dictionnaire des noms anciens et modernes des villes et arrondissements...dans l'empire chinois...
Biotechnic and histochemistry – v28-71. 1953-96 – (= ser Stain Technology) – 1,5,6,9 – $80.00r – (formerly: stain technology) – us Lippincott [540]
Biotechnologische verfahren in der sonnenblumenzuechtung / Wingender, Ruth – (mf ed 1999) – 1mf – 9 – €40.00 – 3-8267-2655-3 – mf#DHS 2655 – gw Frankfurter [574]
Bio/technology – London UK 1983-96 – 1,5,9 – (cont by: nature biotechnology) – ISSN: 0733-222X – mf#13372.01 – us UMI ProQuest [574]
Biotechnology advances – Oxford UK 1983+ – 1,5,9 – ISSN: 0734-9750 – mf#49442 – us UMI ProQuest [660]
Biotechnology and bioengineering – Hoboken NJ 1959+ – 1,5,9 – ISSN: 0006-3592 – mf#11048 – us UMI ProQuest [660]
Biotechnology investors' forum. worldwide ed – London UK 2001+ – 1,5,9 – ISSN: 1471-583X – mf#32373 – us UMI ProQuest [332]
Biotechnology progress [1985] – New York NY 1985-89 – 1,5,9 – ISSN: 8756-7938 – mf#14308 – us UMI ProQuest [660]
Biotechnology progress [1990] – Washington DC 1990+ – 1,5,9 – ISSN: 8756-7938 – mf#50030 – us UMI ProQuest [660]
Biotherapy – Dordrecht, Netherlands 1991-96 – 1,5,9 – ISSN: 0921-299X – mf#16770 – us UMI ProQuest [574]
Biotropica – Sarasota FL 1974+ – 1,5,9 – ISSN: 0006-3606 – mf#10068 – us UMI ProQuest [574]
Bip – snop – sop – 1971-82; 1985-88 – Inquire – 1 – (lacks some iss) – mf#ATLA S0456 – us ATLA [073]
Bi-perozdor / Wolfsberg, Oskar – Jerusalem, Israel. 1943 – 1r – us UF Libraries [939]
Bir avuc sacma / Halit, Refik [Karay] – Halep: Arakis Matbaasi, 1932 – (= ser Ottoman literature, writers and the arts) – 2mf – 9 – $40.00 – us MEDOC [470]
Bir cicek demeti / Resit, Mustafa – Istanbul: Matbaa-i Ebuezziya, 1304 [1877] – (= ser Ottoman literature, writers and the arts) – 1mf – 9 – $25.00 – us MEDOC [470]
Bir serencam / Kadri, Yakup [Karaosmanoglu] – Dersaadet [Istanbul]: Kitaphane-yi Islam ve Askeri, 1330 [1914] – (= ser Ottoman literature, writers and the arts) – 4mf – 9 – $60.00 – us MEDOC [470]
Birala, Ghanasyamadasa *see*
– Bapu ki prema prasadi
– The path to prosperity
Birch, George Henry *see* London churches of the 17th and 18th centuries
Birch, George W F *see* The presbyterian church in the united states of america against the rev. charles a. briggs, d.d
Birch, John *see*
– Country architecture
– Examples of labourers' cottages
– Picturesque lodges
Birch, John Grant *see* Travels in north and central china
Birch, S *see* Inscriptions in the hieratic and demotic character
Birch scroll : newsletter of the american-birkebeiner race at telemark / American Birkebeiner Ski Foundation – 1975 apr-1986 fall/winter – 1r – 1 – mf#1496155 – us WHS [790]
Birch, Walter de Gray *see*
– Early drawings and illuminations
– Fasti monastici aevi saxonici
– The history, art and palaeography of the manuscript styled the utrecht psalter
Bircher, Martin et al *see* Deutsche drucke des barock 1600-1720
Birch-Pfeiffer, Charlotte *see*
– Gesammelte novellen und erzaehlungen
– Der leiermann und sein pflegekind
– Nacht und morgen
– Steffen langer aus glogau
– Die waise aus lowood
Birchwood bulletin – Birchwood WI. 1918 feb 8-sep 27 – 1r – 1 – mf#959836 – us WHS [071]
Birchwood guardian – [NW England] Warrington Lib 5 nov 1982-17 jul 1987 – 1 – uk MLA; uk Newsplan [072]
Birckenstock, J *see* Sonate a violino solo e violoncello o basso continuo, opera prima
Bird, George W *see* Wanderings in burma
Bird, Isaac *see* Bible work in bible lands
Bird, Isabella *see* Leben einer dame in den felsengebirgen
Bird, Isabella Lucy *see* The yangtze valley and beyond
Bird, Isabella Lucy' *see* Among the tibetans
Bird island lighthouse letterbook, 1852-1875 – 6mf – 9 – mf#MSB 58 – sa National [380]
Bird, John *see*
– Annals of natal, 1495-1845
– Doctrine of justification briefly stated
Bird keeping in australia – Adelaide, Australia 1973 – 1 – ISSN: 0045-2076 – mf#7942 – us UMI ProQuest [590]
Bird, Laurice *see* Maxie mongoose
Bird, Mark Baker *see* Haiti
The bird of time : songs of life, death and the spring / Naidu, Sarojini – London: William Heinemann; New York: John Lane Co, 1912 – (= ser Samp: indian books) – (int by edmund gosse and portrait of the aut) – us CRL [780]
Bird, Phyllis T *see* Forts established in florida prior to 1700
Bird, R W *see* The spoliation of oudh
Bird, Robert *see* Jesus, the carpenter of nazareth
Bird study – Thetford UK 1980+ – 1,5,9 – ISSN: 0006-3657 – mf#15507 – us UMI ProQuest [590]
Bird, William *see* State of the cape of good hope, in 1822
Bird, William Hamilton [comp] *see* Oriental miscellany
Bird-banding – Statesboro GA 1944-79 – 1,5,9 – (cont by: journal of field ornithology) – ISSN: 0006-3630 – mf#3255.01 – us UMI ProQuest [590]
Birdland reasons / Cottam, John – [London, Ont?: s.n.], c1918 – 2mf – 9 – 0-665-77630-6 – mf#77630 – cn Canadiana [630]
Birds collected in cuba and haiti / Wetmore, Alexander – Washington, DC. 1933 – 1r – us UF Libraries [590]
Bird's creek baptist church : church records – 1854-1994 – 1 – $61.74 – (includes church minutes, cemetery lot records, membership records) – mf#6939 – us Southern Baptist [242]
Bird's eye view of british history in relation to papal claims / Paton, James – Edinburgh, Scotland. 1893 – 1r – us UF Libraries [941]
The birds of eastern north america known to occur east of the ninetieth meridian / Cory, Charles Barney – Chicago: Field Columbian Museum. 2v. 1899 – 1mf – 9 – mf#03603 – cn Canadiana [590]
The birds of eastern north america known to occur east of the ninetieth meridian, pt 1 : water birds; key to the family and species / Cory, Charles Barney – Chicago: Field Columbian Museum, 1899 – 2mf – 9 – mf#03604 – cn Canadiana [590]
The birds of eastern north america known to occur east of the ninetieth meridian, pt 2 : land birds; key to the family and species / Cory, Charles Barney – Chicago: Field Columbian Museum, 1899 – 3mf – 9 – mf#03605 – cn Canadiana [590]
Birds of prince edward island : their habits and characteristics / Bain, Francis – [S.l: s.n.], 1891 [mf ed 1980] – 1mf – 9 – 0-665-04171-3 – (incl ind) – mf#04171 – cn Canadiana [590]
Birds of the bahama islands / Riley, Joseph Harvey – Baltimore, MD. 1905 – 1r – us UF Libraries [590]
The birds of tunisia... / Whitaker, J I S – London, 1905. 2v – 22mf – 8 – mf#Z-1959 – ne IDC [590]
Birdseye view of indian policy / Commissioner of Indian Affairs – dec 1935 – (= ser American indian periodicals... 2) – 1mf – 9 – $95.00 – us UPA [305]
Bird's-eye view of life insurance : and mathematical and logical exposition of the level premium plan / Bruce, King – Toronto: K Bruce, 1888 [mf ed 1979] – 1mf – 9 – 0-665-00291-2 – mf#00291 – cn Canadiana [368]
Birdwood, Christopher Bromhead, Baron *see* A continent decides
Birdwood, George Christopher Molesworth *see*
– The industrial arts of india
– Report on the old records of the india office
Bire, Edmond *see* Memoires et souvenirs
Birebidzshan in 1935 un in 1936 yor... / Trotskii, B I – Moscow, Russia. 1936 – 1r – us UF Libraries [939]
Birgitta och reformationen : foeredrag i vadstene kyrka den 24 oktober 1916 / Soederblom, Nathan – Uppsala: Sveriges kristliga studentroerelses forlag [1916?] [mf ed 1991] – (= ser Sveriges kristliga studentroerelses skriftserie 64) – 1mf [ill] – 9 – 0-524-01896-0 – mf#1990-0523 – us ATLA [242]
Birgitta, Sancta *see* Revelaciones extravagantes
Birgitta-studier / Westman, Knut Bernhard – Uppsala: Akademiska boktryckeriet, 1911 – 1mf – 9 – 0-7905-6333-9 – mf#1988-2333 – us ATLA [240]
Birgys-barys = Aigle de paris – Paris. 1859-66 – 1 – (french title: aigle de paris. ed. arabe n1-179, ed. bilingue n1-25) – fr ACRPP [073]
O birimbau : orgao de cousa alguma – Baturite, CE: Typ do Seculo, 22 dez 1893 – (= ser Ps 19) – mf#P17,01,61 – bl Biblioteca [870]
Birin : novela / Benet Y Castellon, Eduardo – Santa Clara, Cuba. 1962 – 1r – us UF Libraries [830]
Biringer Gun Shop. Leavenworth, Kansas *see* Records and history
Biringer Gun Shop. Leavenworth. Kansas *see* Records and history
Biringuccio, V *see* De la pirotechnia libri 10...
Biriuch petrogradskikh gosudarstvennykh teatrov – Petrograd, 1918-19, 1919-20, 1920 – 10mf – 9 – us UMI ProQuest [780]
Biriukov, P *see* Ezhemesiachnoe obozrenie
Birk, A *see* Der suezkanal
Birk, Karl *see*
– Heinrich von kleist
Birkbeck, Morris *see*
– Letters from illinois
– Notes on a journey through france
Birkbeck, William John *see* Russia and the english church during the last fifty years
Birkeland, Harris *see* The lord guideth
Birkeland, Knut Bergesen *see* Light in the darkness, or, christianity and paganism
Die birken in den steinen : roman / Gerlach, Kurt – Prag: E Matthes, c1942 (mf ed 1990) – 1r – 1 – (filmed with: paul gerhardt's geistliche lieder) – us Wisconsin U Libr [830]
Birkenhead, 1901 surname/location index – (= ser 1901 census indexes [cheshire]) – 3mf – 9 – £4.50 – mf#356 – uk CheshireFHS [929]
Birkenhead and cheshire advertiser – [NW England] Birkenhead Lib 1866, 1864-1933 – 1 – uk MLA; uk Newsplan [072]
Birkenhead and tranmere areas – (= ser 1841 census surname/location indexes [cheshire] [cheshire]) – 1mf – 9 – £2.50 – mf#190 – uk CheshireFHS [929]
Birkenhead & cheshire advertiser – Birkenhead, England 5 jan 1861-6 jul 1940 – 1 – (wanting: jan-mar 1878, jul-dec 1910, jan-dec 1912; cont: birkenhead advertiser & tranmere telegraph [14 apr-29 dec 1860]; cont by: birkenhead advertiser & wallasey guardian [10 jul 1940-27 jun 1956]) – uk British Libr Newspaper [072]
Birkenhead, Earl of *see* Famous trials of history
Birkenhead news – Birkenhead, England 8 mar 1989- [mf 1986-] – 1 – (cont: wirral news [6 sep 1984-1 mar 1989]) – uk British Libr Newspaper [072]
Birkenhead news and wirral general advertiser – [NW England] Birkenhead 1 jun 1878-27 jun 1956 – 1 – (wanting: 1898, 1911; cont by: birkenhead news & advertiser [4 jul 1956-31 may 1968]) – uk Newsplan; uk British Libr Newspaper [072]
Birkenhead rd – (= ser 1891 census surname/location indexes [cheshire]) – 2mf – 9 – £4.00 – mf#247 – uk CheshireFHS [929]
Birkenhead, st mary – (= ser Cheshire monumental inscriptions) – 1mf – 9 – £2.50 – mf#382 – uk CheshireFHS [929]
Birkenhead, st mary 1712-1812 – (= ser Cheshire church registers) – 2mf – 9 – £4.00 – mf#376 – uk CheshireFHS [929]
Birkenhead surname & location index – (= ser 1861 census indexes [cheshire]) – 2mf – 9 – £4.00 – mf#227 – uk CheshireFHS [929]
Birkenhead surname & location index – (= ser 1871 census indexes [cheshire]) – 2mf – 9 – £3.00 – mf#246 – uk CheshireFHS [929]
Birkenmajer, A *see* Vermischte untersuchungen zur geschichte der mittelalterlichen philosophie
Birket-Smith, K *see* Preliminary report of the fifth thule expedition
Birkhaeuser, Jodocus Adolph *see* History of the church
Birkholz, Corie L *see* Nutritional knowledge and eating behaviors of phase 3 cardiac rehabilitation program participants
Birklein, Franz *see* Entwickelungsgeschichte des substantivierten Infinitivs
Birkmyre, William *see* The wealth of india and the hindrances to its increase
Birkner, Siegfried *see* Die mechanisierung des lebens im werk johann gottfried herders
Birkner, Thomas *see* Entwicklung, durchfuehrung und evaluation eines kurses 'gegenseitige ganzkoerperuntersuchung von medizinstudierenden' zur schulung der praktischen fertigkeiten im koerperlichen untersuchen
Birks, H A *see* God's champion, man's example
Birks, Herbert Alfred *see*
– Horae evangelicae
– The life and correspondence of thomas valpy french
Birks, John Betteley *see* Theory and practice of scintillation counting
Birks, Thomas Rawson *see*
– The bible and modern thought
– The difficulties of belief
– First elements of sacred prophecy
– Horae evangelicae
– Modern physical fatalism and the doctrine of evolution
– Modern rationalism and the inspiration of the scriptures
– Outlines of unfulfilled prophecy
– The pentateuch and its anatomists

- The scripture doctrine of creation
- Supernatural revelation
- Thoughts on the times and seasons of sacred prophecy

Birla, Ghanasyamadasa *see* In the shadow of the mahatma

Birlik – Cankiri [Turkey]: Birlik Matbaasi, nov 20 1950-aug 24 1953 (semiwkly) (gaps) [mf ed 1992] – 4r – 1 – us CRL [079]

Birlik – Galata [Turkey]: s.n. jan 2-aug 31 1950// (daily) (gaps) [mf ed 1992] – 1r – 1 – us CRL [079]

Birlik – Nicosia, Cyprus. Oct 1991-june 1992; sept-nov 1992 – 4r – 1 – us L of C Photodup [079]

Birlinger, Anton *see* Alemannia zeitschrift fuer sprache, litteratur und volkskunde des Elsasses und Oberrheins

Birmah, siam, and anam / Conder, Josiah – London 1826 [mf ed Hildesheim 1995-98] – 1v on 3mf [ill] – 9 – €90.00 – 3-487-27456-6 – gw Olms [915]

Birmann, Martin *see* Gesammelte schriften

Birmingham and lichfield chronicle – Birmingham, England 4 sep 1820-26 dec 1822 (wkly) – 1 – (cont: birmingham chronicle [9 sep 1819-7 sep 1820]; cont by: birmingham chronicle, & general advertiser of the midland counties (ns) 1 jan 1824-19 apr 1827) – uk British Libr Newspaper [072]

Birmingham argus : a weekly newspaper – Birmingham, England [31 oct 1818] – 1 – (ith single sheet of iss dated [16] jan 1819) – uk British Libr Newspaper [072]

Birmingham & aston chronicle – Birmingham, England 10 jan 1880-28 dec 1895 – 1 – (cont: aston chronicle [16 oct 1875-3 jan 1880]) – uk British Libr Newspaper [072]

Birmingham catholic news – Birmingham, England 16 nov 1895-14 jul 1934 – 1 – (wanting 1896,1912; incorp with: catholic herald [london]) – uk British Libr Newspaper [241]

Birmingham Chamber Of Commerce And Industry *see* South africa 1967

Birmingham chronicle [edgbaston & harborne] – Birmingham, England 8 sep 1989-23 mar 1990 – 1 – (cont by: chronicle (edgbaston & harborne) 30 mar 1990-29 dec 1995) – uk British Libr Newspaper [072]

Birmingham chronicle [kings heath & moseley ed] – Birmingham, England 12 jan-23 mar 1990 – 1 – (cont by: chronicle (kings heath & moseley) 30 mar 1990-18 sep 1992) – uk British Libr Newspaper [072]

Birmingham chronicle [northfield & kings norton ed] – Birmingham, England 12 jan-23 mar 1990 – 1 – (cont by: chronicle (northfield & kings norton) 30 mar 1990-18 sep 1992) – uk British Libr Newspaper [072]

Birmingham daily gazette – Birmingham, England 12 may 1862-29 jan 1904 [mf jan-jun 1871; jan-apr, jul-dec 1874; apr-jun 1875; may-dec 1897; may-dec 1898; sep-dec 1899; sep-dec 1900 – 1 – (cont by: birmingham gazette & express 1 feb 1904-16 nov 1912; wanting: may, jun 1874, jan-jun 1876, may-aug 1896, jan-apr 1897, jan-apr 1898, jan 1910-dec 1911) – uk British Libr Newspaper [072]

Birmingham daily mail – Birmingham, England 6 mar 1871-16 may 1918 (daily) – 1 – (cont by: birmingham mail [17 may 1918-8 apr 1963]) – uk British Libr Newspaper [072]

Birmingham daily news – Birmingham, England 5 nov 1985-7 aug 1987 [mf 1986-] – 1 – (cont: daily news [2 oct 1984-1 nov 1985]; cont by: daily news [11 aug 1987-17 may 1991]) – uk British Libr Newspaper [072]

Birmingham daily post – Birmingham, England 4 dec 1857-20 may 1918 (daily) – 1 – (cont by: birmingham post [21 may 1918-2 nov 1956]) – uk British Libr Newspaper [072]

Birmingham daily times – [West Midlands] Birmingham 4 nov 1885-31 mar 1890 [mf ed 2002] – 10r – 1 – (discontinued) – uk Newsplan; uk British Libr Newspaper [072]

Birmingham, David *see*
- Portuguese conquest of angola
- Trade and conflict in angola

Birmingham evening mail – Birmingham, England 7 aug 1967-1 oct 2005 – 1 – (title varies fr 1986-2005: birmingham evening mail / evening mail (birmingham); cont: birmingham evening mail & despatch [9 apr 1963-5 aug 1967]; cont by: birmingham mail [3 oct 2005-]) – uk British Libr Newspaper [072]

Birmingham exchange & sales record – Birmingham, England 3,10 jan 1891 – 1 – (cont by: birmingham & midland exchange & sales record [17 jan 1891]) – uk British Libr Newspaper [072]

Birmingham focus – Birmingham, England 24 may 1991-31 jan 1992 – 1 – (cont by: birmingham news [7 feb 1992-31 mar 1995]) – uk British Libr Newspaper [072]

Birmingham, G A *see* God's iron

Birmingham Genealogical Society *see* Pioneer trails

Birmingham graphic – Birmingham, England 7 mar-16 aug 1883, 28 mar- 13 jun 1884 [mf 1884] – 1 – (incorp with: owl) – uk British Libr Newspaper [072]

Birmingham illustrated weekly mercury – Birmingham, England [ns] 17 feb 1906-21 dec 1918 – 1 – (variant title: sunday mercury 1918-; wanting: 1898, dec 1910, jun-jul 1912, jan-mar, nov-dec 1975; cont: weekly mercury [7 mar 1903-10 feb 1906]; cont by: sunday mercury [ns] 29 dec 1918-18 apr 1920) – uk British Libr Newspaper [072]

Birmingham journal – Birmingham, England 26 nov 1859-13 feb 1869 (wkly) – 1 – (incorp with: birmingham daily post; cont: birmingham journal & general advertiser [4 jun 1825-19 nov 1859 (mf 1850)]) – uk British Libr Newspaper [072]

Birmingham labour party records, 1906-51 – (= ser Labour party in britain, origins and development at local level. series 1) – 5r – 1 – (int by peter d drake) – mf#97299 – uk Microform Academic [325]

Birmingham leader – Birmingham, England 26 apr-8 nov 1890 – 1 – uk British Libr Newspaper [072]

Birmingham ledger – Birmingham AL. 1912 jan ? ["anniversary & progress no, 1896-1912"] – 1r – 1 – (cont by: birmingham news) – mf#912104 – us WHS [071]

Birmingham liberal review : a journal of politics, literature & art – Birmingham, England 20 mar-9 oct 1880 – 1 – (discontinued) – uk British Libr Newspaper [072]

Birmingham mercury – Birmingham, England 30 dec 1848-27 mar 1858 [mf 1850] – 1 – (incorp with: birmingham weekly press) – uk British Libr Newspaper [072]

Birmingham metronews – Birmingham, England 16 sep 1993-27 dec 2001 [mf 1986-] – 1 – (publ once/wk fr 21 jun 1991-; cont: metro news [[ns] 21 may 1991-9 sep 1993; cont by: birmingham news [10 jan 2002-]) – uk British Libr Newspaper [072]

Birmingham & midland exchange & sales record – Birmingham, England 17 jan 1891) – 1 – (cont: birmingham exchange & sales record [3,10 jan 1891]; cont by: midland exchange & sales record [24 jan-26 dec 1891]) – uk British Libr Newspaper [072]

Birmingham morning news – Birmingham, England 2 jan-29 dec 1871, 8 feb [31 mar, 1 may 1872-15 jan 1876 [feb,mar,sep,dec 1872] – 1 – (cont by: birmingham morning & evening news [17 jan-27 may 1876]) – uk British Libr Newspaper [072]

Birmingham. Museum and Art Gallery *see*
- Catalogue of a special collection of works by david cox
- City of birmingham museum and art gallery catalogue...of...modern english animal painters
- Illustrated catalogue...of the permanent collection of paintings...at aston hall
- Illustrated handbook to the permanent collections of industrial art objects

Birmingham news – Birmingham, England 7 feb 1992-31 mar 1995 – 1 – (cont: birmingham focus [24 may 1991-31 jan 1992]; discontinued; replaced by: weekly observer (birmingham)) – uk British Libr Newspaper [072]

Birmingham news – Birmingham AL. 1999 jan 7-mar 25, apr-jun 24 – 2r – 1 – (cont; Daily news (Birmingham AL); Birmingham ledger) – mf#1391470 – us WHS [071]

Birmingham news [south ed] – Birmingham, England 6 feb 1892-20 jun 1959 [mf jan-dec 1893, jan-dec 1897, mar-dec 1910, jan-dec 1912] – 1 – (incorp with: solihull news; wanting: jan,feb 1910, jan-dec 1911, jan-dec 1919; cont: south birmingham news [3 mar 1888-30 jan 1892].) – uk British Libr Newspaper [072]

Birmingham, Peter *see* American art in the barbizon mood

Birmingham pictorial & dart – Birmingham, England 27 mar 1896-1 sep 1911 – 1 – (cont: dart & midland figaro [30 sep,18 nov 1881, 10 nov 1882, 12 jan 1883, 4 nov 1884-20 mar 1896]; discontinued) – uk British Libr Newspaper [072]

Birmingham post – Birmingham, England 24 sep 1964- [mf 1992-] – 1 – (cont: birmingham post & birmingham gazette [3 nov 1956-23 sep 1964]) – uk British Libr Newspaper [072]

Birmingham reference library catalogue 1879-1963 – [mf ed Chadwyck-Healey] – 341mf – 9 – (with special coll covering milton, cervantes, samuel johnson, war poetry & early and fine printing together with a vast range of material of interest to 19th c social & economic historians) – uk Chadwyck [020]

Birmingham society of artists : exhibition of modern works of art – Birmingham 1854 – (= ser 19th c art & architecture) – 1mf – 9 – mf#4.2.1691 – uk Chadwyck [700]

Birmingham stock & share list – Birmingham, England 2 jan 1891-27 dec 1912 [mf 1893] – 1 – uk British Libr Newspaper [332]

Birmingham suburban times – Birmingham, England 6 dec 1884-9 feb 1901 [mf 1888, 1897] – 1 – (discontinued) – uk British Libr Newspaper [072]

Birmingham sunday echo – [West Midlands] Birmingham 14 aug 1898-6 feb 1915 [mf ed 2002] – 15r – 1 – (cont by: birmingham echo [25 mar 1906-6 feb 1915 (mf 1912)]) – uk Newsplan [072]

Birmingham sunday mail – Birmingham, England 15 may 1898-28 may 1916 [mf may-dec 1898, jan-jun 1900, jan-apr 1911, feb [dec 1912] – 1 – (wanting jan-dec 1901, may 1911-jan 1912; discontinued) – uk British Libr Newspaper [072]

Birmingham sunday school union : church records – May 1848-Dec 1913 – 1 – 577.67 – us Southern Baptist [242]

Birmingham times – Birmingham AL. 1992 oct 29/nov 4-1999 oct 7/dec 23 – 24r – 1 – (with gaps) – mf#1288329 – us WHS [071]

Birmingham Trades Council et al *see* Labor advocate

Birmingham university chemical engineer – Birmingham UK 1976 – 1,5,9 – ISSN: 0006-3746 – mf#8191 – us UMI ProQuest [660]

Birmingham weekly mercury – Birmingham, England 8 nov 1884-28 feb 1903 – 1 – (wanting: 1898, dec 1910, jun-jul 1912, jan-mar, nov-dec 1975; cont by: weekly mercury [7 mar 1903-10 feb 1906]; birmingham illustrated weekly mercury [ns] 17 feb 1906-21 dec 1918; sunday mercury [29 dec 1918-18 apr 1920]) – uk British Libr Newspaper [072]

Birmingham weekly post – Birmingham, England 20 feb 1869-28 dec 1895, 1 jan 1898- 31 dec 1954 [mf 1888, 1910] – 1 – (wanting: 1911; cont: saturday evening post [5 dec 1857-13 feb 1869]; cont by: birmingham weekly post & midland pictorial [7 jan 1955-22 apr 1960]) – uk British Libr Newspaper [072]

Birmingham world – Birmingham AL. 1969 jun 28, aug 2, 1973 dec 29, 1974 jan 19 – 1r – 1 – mf#780622 – us WHS [071]

Birnamwood news – Birnamwood WI. 1900 may 9/1901 dec 25-1951 sep 26/1955 – 29r – 1 – (with gaps; cont by: wittenberg enterprise, wittenberg enterprise and birnamwood news (wittenberg wi: 1971)) – mf#952103 – us WHS [071]

Birnbach, Franz Bernhard *see* Heinrich federer. seine persoenlichkeit und seine kunstform

Birnbaum, Menachem *see* Schlemiel

Birnbaum, N *see* Die nationale wiedergeburt des juedischen volkes in seinem lande, als mittel zur loesung der judenfrage

Birnbaum, Nathan *see* Gots folk

Birney, Catherine H *see* The grimke sisters

Birney, Hoffman *see*
- Brothers of doom. the story of the pizarros of peru
- Los hermanos del destino (los pizarros y la conquista del peru)

Birney, William *see* James g birney and his times

Biro Dokumentasi Pers "Media" *see* Index nama penulis dalam kompas

Biro Pembangunan Masjarakat Desa *see* Swadaja-desa

Biro Penerangan *see* Yudhagama

Biro penerangan ekonomi : bulletin ekonomi-keuangan – Djakarta, 1962-1973 – 906mf – 9 – (missing: 1965; 1972, v17(4982-4983, 4990-4995, 4997); 1972, v18(5086, 5098, 5117, 5164-5165, 5209)) – mf#SE-276 – ne IDC [959]

Biro Penerangan Ekonomi (Djakarta) *see* Bulletin ekonomi keuangan

Biro penerangan sie vttiv : benteng negara – Semarang 1953-1956 – 2mf – 9 – (missing: 1953(5-12); 1956(2)) – mf#SE-584 – ne IDC [950]

Biro Public Relations Garuda Indonesian Airways *see* Radjawali

Biro Statistik dan Dokumentasi, Departemen Perindustrian Rakjat *see* Ekonomi dan industri

Biro Urusan Industrialisasi Ichtisar laporan unit2 Overheidsdienst Urusan Industrialisasi *see* Indonesia

Biro Urusan Industrialisasi Laporan tahunan *see* Indonesia

Biroat, Jacques *see* The eucharistic life of jesus christ

Birobidzhanskaia zvezda – Tikhonkaya (now called Birobidzhan), Russia 1973-88 [mf ed Norman Ross] – 4r – 1 – mf#nrp-1699 – us UMI ProQuest [077]

Birobidzshaner toyshvim / Gordon, Samuel – Moscow, Russia. 1947 – 1r – us UF Libraries [939]

Biron and bruckers sonntags-blatt / Free Thought League of North America – 1874 apr 5-dec 27 – 1r – 1 – (cont: milwaukee freidenker; cont by: freidenker (milwaukee wi: 1875)) – mf#1295888 – us WHS [210]

Birot, Pierre *see* Cycle d'erosion sous les differents climats

Birrell, Augustine *see* Emerson

Birrell, Charles Morton *see* The life of the rev. richard knill, of st. petersburg

Birsen, K *see* Devletler hususi hukuku

Birt, Henry Norbert *see*
- Benedictine pioneers in australia
- Downside
- The elizabethan religious settlement

Birt, John *see* Apostolical method of preaching the gospel

Birt, Th *see* Claudii claudiani carmina (mgh1:10.bd)

Birt, Theodor *see* Romische charakterkopfe

Birth – Oxford UK 1973+ – 1,5,9 – (cont: birth and the family journal) – ISSN: 0730-7659 – mf#10735.01 – us UMI ProQuest [362]

The birth and boyhood of jesus christ / Trench, George Henry – London: Skeffington, 1911 – 1mf – 9 – 0-7905-0357-3 – (incl ind) – mf#1987-0357 – us ATLA [920]

Birth and death of meaning / Becker, Ernest – New York, NY. 1962 – 1r – us UF Libraries [100]

The birth and development of ornament / Hulme, Frederick Edward – London 1893 – (= ser 19th c art & architecture) – 4mf – 9 – mf#4.2.1148 – uk Chadwyck [740]

The birth and infancy of jesus christ : according to the gospel narratives / Sweet, Louis Matthews – Philadelphia: Westminster Press, 1906 – 1mf – 9 – 0-8370-5471-0 – (incl ind) – mf#1985-3471 – us ATLA [240]

Birth and mortality statistics of the virgin islands / United States Navy Dept Bureau Of Medicine And S – Washington, DC. 1920 – 1r – us UF Libraries [304]

Birth and the family journal – Oxford UK 1974-81 – 1,5,9 – (cont by: birth) – ISSN: 0098-860X – mf#10735.01 – us UMI ProQuest [618]

Birth certificates from the presbyterian, independent and baptist registry : and from the wesleyan methodist metropolitan registry / Great Britain. General Register Office – 1742-1840 – 207 files – (original parchment & paper certificates fr wh entries in the registers of births fr the presbyterian, independent & baptist registry at dr williams' library were compiled, & the parchment certificates recorded in the registers of births & baptisms fr the wesleyan methodist metropolitan registry) – mf#gr5 – uk National [242]

Birth defects and hereditary disorders = Geboortegebreke en oorerflike afwykings / South Africa. Department of Health [Departement van Gesondheid] – Pretoria: Dept of Health [1975?] [mf ed Pretoria, RSA: State Library [199-]] – 13p [ill] on 1r with other items – 5 – (in english & afrikaans) – mf#op 06448 r23 – us CRL [616]

Birth of a dilemma / Mason, Philip – London, England. 1958 – 1r – us UF Libraries [960]

Birth of a plural society / Gann, Lewis H – Manchester, England. 1961, c1958 – 1r – us UF Libraries [960]

The birth of indian psychology and its development in buddhism / Davids, Caroline Augusta Foley Rhys – London: Luzac & Co, 1936 – (= ser Samp: indian books) – us CRL [280]

The birth of mormonism / Adams, John Quincy – Boston: Gorham Press, c1916 – (= ser Library of Religious Thought) – 1mf – 9 – 0-524-03209-2 – mf#1990-0837 – us ATLA [243]

Birth of new india / Besant, Annie Wood – Madras: Theosophical Pub House, 1917 – (= ser Samp: indian books) – us CRL [954]

The birth of the first southern baptist church of syracuse, 1957-58 – 1957-58.Formerly a mission of the Lasalle Baptist Church of Syracuse, NY. 98p – 1 – 5.00 – us Southern Baptist [242]

Birth-day – London, England. 18– – 1r – us UF Libraries [240]

The birthright church : a discourse / Judd, Sylvester – Augusta: William H Simpson, 1854 – 1mf – 9 – 0-524-04732-4 – mf#1991-2137 – us ATLA [240]

Births and deaths registers / Tonga. Ministry of Justice. Tongatapu Registry – 1867-1973 – 2r – 1 – (gaps, mainly 19th c registers. restricted access) – mf#PMB1095 – at Pacific Mss [920]

Births, deaths, and marriage record cards / Geary County, KS – undated – 1 – us Kansas [920]

Birtwell, Charles Wesley *see* The care of dependent, neglected, and wayward children microform

Biruni, Muhammad ibn Ahmad *see* Alberuni's india

Birven, Henri Clemens *see* Goethes faust und der geist der magie

Birzhevaia gazeta – city unknown, Russia 1877 [mf ed Norman Ross] – 1 – mf#nrp-87 – us UMI ProQuest [077]

Birzhevye vedomosti – St Petersburg, Russia 1880-1917 [mf ed Norman Ross] – 1 – mf#nrp-1579 – us UMI ProQuest [077]

Birzhevyia viedomosti – Petrograd: [s.n], jan-dec 1916 – 1 – us CRL [077]

"Bis hieher" : kurzgefasste geschichte der missouri-synode / Graebner, Augustus Lawrence – [s.l: s.n.] 1897 [mf ed 1993] – 1mf – 9 – 0-524-06587-X – mf#1990-5253 – us ATLA [242]

Bisbee daily review – Bisbee AZ. 1903 jul 19 – 1r – 1 – (cont: cochise review and arizona daily orb) – mf#914285 – us WHS [071]

Bisbee, Frederick Adelbert *see* A california pilgrimage

Bisbee, Frederick Adelbert et al *see* Good tidings

Die bischari-sprache tu-bedawie in nordost-afrika beschreibend und vergleichend dargestellt / Almkvist, H – Upsala, 1881 – 7mf – 9 – mf#NE-20177 – ne IDC [470]

Die bischoeflichen dioezesanbehoerden, insbesondere das bischoefliche ordinariat / Mueller, Joseph – Stuttgart, 1905 (mf ed 1995) – 2mf – 9 – €31.00 – 3-8267-3151-4 – mf#DHS-AR 3151 – gw Frankfurter [241]

Bischoff, Diedrich *see*
- Neuidealismus und freimaurerei
- Volkserziehungsgedanken eines deutschen freimaurers

Bischoff, Erich *see*
- Babylonisch-astrales im weltbilde des thalmud und midrasch
- Erlaeuterungen zu goethe's 'faust'
- Erlaeuterungen zu lessing's hamburgischer dramaturgie
- Genathliacon serenissimo neo-nato archiduci austriae leopoldo, augustissimi, romanorum imperatoris leopoldi primi...
- Kritische geschichte der thalmud-uebersetzungen aller zeiten und zungen
- Regium majestatis
- Der sieg der alchymie

Bischoff, H *see* Archiv fuer das studium der neueren sprachen

Bischoff, Heinrich *see* Nikolaus lenaus lyrik

Bischoff, James *see* Sketch of the history of van diemen's land

Bischoff, Karl *see* Gedenkschrift fuer ferdinand josef schneider, 1879-1954

Bischoff, Wilhelm *see* Reise durch die koenigreiche sachsen und boehmen in den jahren 1822 und 1823

Bischoff, Wilhelm Ferdinand *see* Deutsch-zigeunerisches woerterbuch

Ein bischofsbrief vom concil und eine deutsche antwort : ein beitrag zur unterscheidung von katholicismus und jesuitismus / [Nippold, Friedrich] – Berlin: C G Luederitz, 1870 [mf ed 1986] – 1mf – 9 – 0-8370-8464-4 – mf#1986-2464 – us ATLA [241]

Die bischofslisten und die apostolische nachfolge in der kirchengeschichte des eusebius / Overbeck, Franz – Basel: Fr Reinhardt, 1898 – 1mf – 9 – 0-7905-6714-8 – (incl bibl ref) – mf#1988-2714 – us ATLA [240]

Bischofswerdaer tageblatt *see* Der saechsische erzaehler

Biscuit maker – Croydon UK 1963-68 – 1 – ISSN: 0005-4151 – mf#1340 – us UMI ProQuest [660]

Bi-shenat ha-sheloshim / Histadrut Ha-Kelalit Shel Ha-'Ovdim Ha-'Ivrim Be-Erets-Yisra'el – Tel-Aviv, Israel. 1951/52 – 1r – us UF Libraries [939]

Eine bisher nicht erkannte schrift des papstes sixtus 2 / Harnack, Adolf von – Leipzig, 1895 – (= ser Tugal 1-13/1a) – 2mf – 9 – €5.00 – ne Slangenburg [240]

Eine bisher nicht erkannte schrift des papstes sixtus 2. vom jahre 257/8 / zur petrusapokalypse / patristisches zu luc. 16. 19 : drei abhandlungen / Harnack, Adolf von – Leipzig: J C Hinrichs, 1895 [mf ed 1989] – (= ser Tugal 13/1) – 1mf – 9 – 0-7905-1762-0 – (incl bibl ref) – mf#1987-1762 – us ATLA [225]

Eine bisher nicht erkannte schrift novatians / Harnack, Adolf von – Leipzig, 1895 – (= ser Tugal 1-13/4b) – 1mf – 9 – €3.00 – ne Slangenburg [240]

Eine bisher unbekannte version des ersten teiles der "apostellehre" / Iselin, L E – Leipzig, 1895 – (= ser Tugal 1-13/1b) – 1mf – 9 – €3.00 – ne Slangenburg [240]

Ein bisher unbekanntes werk des patriarchen eutychios von alexandrien (876-949) : mit zeugnissen ueber die heiligtuemer palaestinas / Graf, G – Koeln, 1911 – 1mf – 9 – mf#H-2876 – ne IDC [930]

Bishimi nebyakine – Kasempa, Zambia. 19-? – 1r – us UF Libraries [960]

Bishinik / Choctaw Nation – v1 n2-v4 n2 [1979 jul-1981 oct] – 1r – 1 – mf#675843 – us WHS [071]

Bishof, A *see* Kratkii obzor istorii i teorii bankov s prilozheniem ucheniia o birzhevykh operatsiiakh

Bishop amongst bananas / Bury, Herbert – London, England. 1911 – 1r – us UF Libraries [972]

A bishop and his flock / Hedley, John Cuthbert – London: Burns & Oates; New York: Benzinger, 1903 [mf ed 1990] – 1mf – 9 – 0-7905-3929-2 – mf#1989-0422 – us ATLA [241]

Bishop, Arthur Stanley *see* Ceylon buddhism

Bishop asbury / Lowrey, Asbury – Cincinnati: Published by the Cincinnati Annual Conference Historical Society; Curts & Jennings, [ca. 1898] – 1mf – 9 – 0-7905-5481-X – mf#1988-1481 – us ATLA [240]

Bishop Baraga Association *see* Bulletin apostle of the chippewas

Bishop butler's ethical discourses : to which are added some remains, hitherto unpublished = Ethical discourses / Butler, Joseph; ed by Passmore, Joseph Clarkson – Philadelphia: C Desilver, 1855 – 1mf – 9 – 0-7905-9162-6 – (incl bibl ref) – mf#1989-2387 – us ATLA [170]

Bishop, Carl Whiting *see* Origin of the far eastern civilizations

Bishop chase's reminiscences : an autobiography / Chase, Philander – 2nd ed. Boston: JB Dow, 1848, c1847 – 3mf – 9 – 0-7905-4617-5 – mf#1988-0617 – us ATLA [240]

Bishop colenso on the pentateuch reviewed / Porter, J L – Belfast, Northern Ireland. 1863 – 1r – us UF Libraries [240]

Bishop colenso utterly refuted : and categorically answered, by lord... / Burnand, F C – London, England. 1862 – 1r – us UF Libraries [240]

Bishop colenso's objections to the historical character of the pentateuch and the book of joshua (contained in pt 1) / Benisch, Abraham – London: Jewish Chronicle 1863 [mf ed 1985] – 1mf – 9 – 0-8370-2262-2 – mf#1985-0262 – us ATLA [221]

Bishop crowther's report of the overland journey : from lokaja to bida, on the river niger, and thence to lagos, on the sea coast, from november 10th 1871 to february 8th 1872 – London: Church Missionary House, 1872 – 1 – us CRL [916]

Bishop, E *see* Edward 6th and the book of common prayer

Bishop, Edmund *see*
- Edward 6 and the book of common prayer
- The genius of the roman rite
- Liturgica historica

Bishop, Farnham *see* Panama, past and present

Bishop for the hottentots / Simon, Jean Marie – New York, NY. 1959 – 1r – us UF Libraries [960]

Bishop, George Sayles *see* The doctrines of grace

Bishop gibson's three pastorel letters, to the people of his dioces – London, England. 1820 – 1r – us UF Libraries [240]

Bishop gore and the catholic claims / Chapman, John – London; New York: Longmans, Green, 1905 – 1mf – 9 – 0-8370-6656-5 – (incl bibl ref) – mf#1986-0656 – us ATLA [240]

Bishop gore's challenge to criticism : a reply to the bishop of oxford's open letter on the basis of anglican fellowship / Sanday, William – London; New York: Longmans, Green, 1914 – 1mf – 9 – 0-7905-0277-1 – mf#1987-0277 – us ATLA [241]

Bishop greene's four last things – London, England. 1820 – 1r – us UF Libraries [240]

Bishop hamilton's memorial : restoration of the choir of salisbury cathedral / Scott, George Gilbert – Salisbury [1870] – (= ser 19th c art & architecture) – 1mf – 9 – mf#4.1.403 – uk Chadwyck [720]

Bishop hannington : and the story of the uganda mission / Berry, William Grinton – New York: Revell [1908?] [mf ed 1992] – (= ser Anglican/episcopal coll) – 1mf – 9 – 0-524-04001-X – (incl bibl ref) – mf#1992-2001 – us ATLA [240]

Bishop harper and the canterbury settlement / Purchas, Henry Thomas – 2nd rev and enl ed. Christchurch, NZ: Whitcombe and Tombs, 1909 – 1mf – 9 – 0-524-01009-9 – mf#1990-0286 – us ATLA [241]

Bishop heber : poet and chief missionary to the east, second lord bishop of calcutta 1783-1826 / Smith, George – London: J Murray, 1895 – 1mf – 9 – 0-7905-5962-5 – (incl bibl ref) – mf#1988-1962 – us ATLA [240]

Bishop, Henry Halsall *see*
- Architecture
- Pictorial architecture of greece and italy

Bishop hoadly's celebrated sermon before george the first – London, England. 1840 – 1r – us UF Libraries [240]

Bishop hoadly's refutation of bishop sherlock's arguments against a... – London, England. 1790 – 1r – us UF Libraries [240]

Bishop Horne *see* Prevailing intercessor

Bishop, [J F] *see*
- Der goldene chersones
- Journeys in persia and kurdistan
- The yangtze valley and beyond

Bishop, James K *see* Samoa comes of age

Bishop, James Lord *see* A treatise on the common and statute law of the state of new york relating to insolvent debtors.

Bishop, Joel Prentiss *see*
- Commentaries on the the law of marriage and divorce.
- Commentaries on the written laws and their interpretation

Bishop john selwyn : a memoir / Selwyn, John Richardson – London: Isbister, 1899 – 1mf – 9 – 0-7905-4899-2 – mf#1988-0899 – us ATLA [240]

Bishop, Joseph Bucklin *see* Geothals, genius of the panama canal

Bishop joseph long : the peerless preacher of the evangelical association / Yeakel, Reuben – Cleveland, O[hio]: Thomas & Mattill, c1897 – 1mf – 9 – 0-7905-6918-3 – mf#1988-2918 – us ATLA [240]

Bishop, L C *see* "Massacre by indians..."

Bishop letter – v1 n1-v7 n1 [1981 feb-1987 feb] – 1r – 1 – mf#1611746 – us WHS [071]

Bishop, Nathaniel Holmes *see*
- En canot de papier de quebec au golfe du mexique
- Voyage of the paper canoe

Bishop of carlisle on the church of ireland – London, England. 1868? – 1r – us UF Libraries [240]

The bishop of oxford's open letter : an open letter in reply / Gwatkin, Henry Melvill – London: Longmans, Green, 1914 – 1mf – 9 – 0-524-08370-3 – mf#1993-3070 – us ATLA [241]

Bishop – owens valley herald – CA. 1908-1927 – 10r – 1 – $600.00 – mf#B06015 – us Library Micro [071]

Bishop patteson : the martyr of melanesia / Page, Jesse – New York: Fleming H Revell, [189-?] – 1mf – 9 – 0-8370-6591-7 – mf#1986-0591 – us ATLA [920]

Bishop potter : the people's friend / Keyser, Harriette A – New York: T Whittaker c1910 [mf ed 1991] – 1mf [ill] – 9 – 0-524-00996-1 – mf#1990-0273 – us ATLA [242]

The bishop potter memorial house : history of its origin, design, and operations, illustrating woman's spiritual mission in the christian church – Philadelphia: King & Baird, 1868. El Segundo, Ca: Micro Publication Systems, 1981 (1mf); Evanston: American Theol Lib Assoc, 1984 (1mf) – (= ser Women & the church in america) – 9 – 0-8370-1412-3 – mf#1984-2145 – us ATLA [240]

Bishop, Robert Hamilton *see* An outline of the history of the church in the state of kentucky

Bishop sarapion's prayer-book : an egyptian pontifical dated probably about a.d. 350-356 = Euchologion / Serapion of Thmuis, Saint – London: SPCK 1899 [mf ed 1992] – (= ser Early church classics) – 1mf – 9 – 0-524-04623-9 – (trans, int, notes and ind by john wordsworth) – mf#1990-1283 – us ATLA [240]

Bishop seabury and bishop provoost : an historical fragment / Perry, William Stevens – [S.l.: s.n.], 1862 – 1mf – 9 – 0-7905-6611-7 – mf#1988-2611 – us ATLA [240]

Bishop selwyn of new zealand and of lichfield : a sketch of his life and work, with some further gleanings from his letters, sermons, and speeches / Curteis, George Herbert – London: Kegan Paul, Trench 1889 [mf ed 1990] – 2mf [ill] – 9 – 0-7905-5644-8 – mf#1988-1644 – us ATLA [240]

Bishop ullathorne : the story of his life – London: Burns & Oates, [1886?] – 1mf – 9 – 0-8370-7015-5 – mf#1986-1015 – us ATLA [920]

Bishop Welles Brotherhood *see*
- Church scholiast
- Nashotah scholiast

Bishop White Prayer Book Society *see* The seventeenth anniversary of the bishop white prayer book society

Bishop, William Samuel *see* The development of trinitarian doctrine in the nicene and athanasian creeds

[Bishop/independence-] inyo register – CA. sep 1909-dec 1958; 1987- – 52+ r – 1 – $3120.00 – mf#BC02070 – us Library Micro [071]

Bishopric of the united church of england and ireland at jersualem / Hope-Scott, James Robert – London, England. 1841 – 1r – us UF Libraries [241]

The bishopric of truro : the first twenty-five years, 1877-1902 / Donaldson, Augustus Blair – London: Rivingtons, 1902 – 1mf – 9 – 0-524-03317-X – mf#1990-4677 – us ATLA [240]

The bishop's address at the opening of the general conference in adjourned session at napanee, january 9th, 1883 / Methodist Episcopal Church in Canada – [Napanee, Ont?: s.n, 1883?] – 1mf – 9 – 0-665-93352-5 – mf#93352 – cn Canadiana [242]

Bishops and clergy of other days : or, the lives of two reformers and three puritans / Ryle, John Charles – London: W Hunt, 1868 – 1mf – 9 – 0-524-01346-2 – mf#1990-0392 – us ATLA [240]

Bishops castle advertiser and clun news – [West Midlands] Shropshire jan 1901-dec 1950 [mf ed 2002] – 39r – 1 – (missing: 1907, 1921) – uk Newsplan [072]

Bishops castle chronicle & clun valley times – [West Midlands] Shropshire 14 jan-dec 1911 [mf ed 2003] – 1r – 1 – uk Newsplan [072]

Bishop's College Press, Calcutta *see* Specimen of printing types for book and other works, used at bishop's college press

The bishop's english : a series of criticisms on the right rev. bishop thornton's laudation of the revised version of the scriptures... / Moon, George Washington – 2nd ed. London: Swan Sonnenschein; New York: E P Dutton, 1904 – 1mf – 9 – 0-8370-9970-6 – mf#1986-3970 – us ATLA [220]

The bishops in the tower : a record of stirring events affecting the church and nonconformists from the restoration to the revolution / Luckock, Herbert Mortimer – London: Rivingtons, 1887 – 1mf – 9 – 0-7905-4769-4 – (incl bibl ref) – mf#1988-0769 – us ATLA [941]

The bishops of lindisfarne, hexham, chester-le-street, and durham, a.d. 635-1020 : being an introduction to the ecclesiastical history of northumbria / Miles, George – London: Wells Gardner, Darton, 1898 – 1mf – 9 – 0-524-02008-6 – mf#1990-0553 – us ATLA [240]

The bishops of scotland : being notes on the lives of all the bishops, under each of the sees, prior to the reformation / Dowden, John; ed by Thomson, J. Maitland – Glasgow: J. Maclehose, 1912 – 2mf – 9 – 0-7905-8004-7 – (incl bibl ref) – mf#1988-8004 – us ATLA [240]

The bishops of the american church : past and present / Perry, William Stevens – New York: Christian Literature Co, 1897 – 2mf – 9 – 0-524-01662-3 – (incl bibl ref) – mf#1990-0483 – us ATLA [240]

The bishops of the american church mission in china – Hartford: Church Missions Publ, [1906] [mf ed 1995] – (= ser Yale coll; Soldier and servant series 38) – 43p (ill) – 1 – 0-524-09961-8 – mf#1995-0961 – us ATLA [240]

Bishops' registers in the norwich diocese, 1299-1912 : from the ipswich & east suffolk record office – 10r – 1 – mf#97070 – uk Microform Academic [240]

Bishop's stortford gazette – Bishop's Stortford, England 25 Sep 1981-17 nov 1989 (wkly) – 1 – (cont by: bishop's stortford citizen gazette [24 nov 1989-23 mar 1990]) – uk British Libr Newspaper [072]

Bishop's stortford observer – Bishop's Stortford, England 7 jun 1979-25 feb 1982 (wkly) [mf 1980-] – 1 – (cont: herts & essex observer [5 jul 1862-31 may 1979]; cont by: herts & essex observer [4 mar 1982-]) – uk British Libr Newspaper [072]

Bishopston, glamorgan, parish church of st teilo : baptisms 1677-1891, burials 1677-1890, marriages 1678-1838 – 1mf – 9 – £1.25 – uk Glamorgan FHS [929]

Bishopston, st teilo & murton methodist, monumental inscriptions – 2mf – 9 – £2.50 – uk Glamorgan FHS [929]

Bishopville first baptist church. lee county : church records – 1848-1871 – 1 – 5.00 – us Southern Baptist [242]

Bismarck, Herbert von *see* Collection of correspondence of herbert von bismarck, 1881-1883

Bismarck, Otto, Fuerst von *see*
- Die gesammelten werke
- Die politischen berichte des fuersten bismarck aus petersburg und paris

Bismarck-erinnerungen des staatsministers / Lucius von Ballhausen, Robert Sigmund Maria Joseph, Freiherr – 1.-3. aufl. Stuttgart: Cotta 1920 [mf ed 1988] – 1r – 1 – (filmed with: memoirs of margaret / cooper, c h) – mf#2121 – us Wisconsin U Libr [943]

Bismarck-jahrbuch / ed by Kohl, Horst Ernst Arminius – Berlin: O Haering 1894-96; Leipzig: G J Goeschen 1897-99 [mf ed 1979] – 6v on 6r – 1 – (no more publ) – mf#film mas c682 – us Harvard [943]

Bismut, V *see* La nationalite des societes en tunisie

Bisping, August *see*
- Erklaerung der briefe an die ephesier, philipper, kolosser
- Erklaerung des briefes an die hebraeer
- Erklaerung des briefes an die roemer
- Erklaerung des ersten briefes an die korinther
- Erklaerung des zweiten briefes an die korinther, und des briefes an die galater
- Erklaerung des zweiten briefes an die thessalonicher, der drei pastoralbriefe und des briefs an philemon

Bissachere, Pierre J de la *see* Gegenwaertiger zustand von tunkin, cochinchina und der koenigreiche camboja, laos und lac-tho

Bissanui bhu ran ma / Mran U Ci pva re – Ran kun: Mon Krann Van 1976 [mf ed 1994] – on pt of 1r – 1 – mf#11052 r1709 n5 – us Cornell [959]

Bisschof, B *see* Mittelalterliche studien

Bisschop, W R *see* Rise of the london money market 1640-1826

Bissel, J *see* Ioannis bisselii e societate iesu, delicae aestatis

Bissell, Allen Page *see* The law of asylum in israel historically and critically examined

Bissell, Edwin Cone *see*
- The apocrypha of the old testament
- Biblical antiquities
- Historic origin of the bible
- The pentateuch
- Pentateuch analysis
- Pentateuch laws and the higher criticism
- A practical introductory hebrew grammar

Bissell, J H *see* Bissell's reports of cases in the seventh circuit, 1851-1883

Bissell's reports of cases in the seventh circuit, 1851-1883 / Bissell, J H – Philadelphia: Callaghan. v1-11. 1873-83 (all publ) – (= ser Early federal nominative reports) – 73mf – 9 – $109.00 – mf#LLMC 81-438 – us LLMC [340]

Bissing, F W von *see*
- Aegyptische kunstgeschichte von den aeltesten zeiten bis auf die eroberung durch die araber
- Catalogue general des antiquites egyptiennes du musee de caire
- Die mastaba des gem-ni-kai
- De oostersche grondslag der kunstgeschiedenis
- Untersuchungen zu den reliefs aus dem re-heiligtum des rathures

Bisso, Jose *see* Cronica de la provincia de murcia

Bisson, Andre *see* Rosaire

Bisson, Fernand *see* Bibliographie de certains maristes (peres, eveques et missionnaires) americains et neo-zelandais

Bissonnette, Antoine [comp] *see* Soixante ans de liberte, 1837-97

Bissoulet, J *see* [Correspondance entre pietro et alexandre verri pendant les annees 1774 et 1775]

Bissula / Dahn, Felix – Leipzig: Breitkopf & Haertel 1898 [mf ed 1993] – 4r – 1 – (filmed with: felix dahn's saemtliche werke poetischen inhalts) – mf#3468p – us Wisconsin U Libr [830]

Bissula : historischer roman aus der voelkerwanderung / Dahn, Felix – 7. aufl. Leipzig: Breitkopf & Haertel 1884 [mf ed 1979] – (= ser Kleine romane aus der voelkerwanderung 2) – 2r – 1 – (filmed with: attila, chlodovech, vom chiemgau & stilicho) – mf#film mas c494 – us Harvard [830]

Das bist du : ein spiel in 5 verwandlungen / Wolf, Friedrich – Dresden: R Kaemmerer 1920, c1919 [mf ed 1991] – (= ser Dramen der neuen schaubuehne 4) – 1r – 1 – (sketches by conrad felixmueller. filmed with: miniaturen / georg witkowski) – mf#3059p – us Wisconsin U Libr [820]

Bi-state reporter – Antioch IL, Kenosha WI. 1984 jan 19/jun-1990 apr 6/jun 29 – 1r – 1 – (cont by: kenosha county times) – mf#850302 – us WHS [071]

Bistoury, Andre F *see*
- Code et guide de l'etat civil a l'usage des minist...
- Face a la delinquance juvenile latente

Bistritzer deutsche zeitung – Bistritz (Bistrita RO), 1920 1 oct-1942 16 oct – 10r – 1 – (cont: siebenbuergisch-deutsche zeitung) – gw Misc Inst [077]

Bistritzer zeitung – Bistritz (Bistrita RO), 1891-1913 – 1 – gw Misc Inst [077]

Die bistumsgruendunen heinrichs des loewen (mgh schriften:3.bd) : untersuchungen zur geschichte des ostdeutschen kolonisation / Jordan, K – 1939 – (= ser Monumenta germaniae historica. schriften (mgh schriften)) – €11.00 – ne Slangenburg [931]

Bisu-yi azadi – London. shumarah-'i 1-4. farvardin-murdad 1357 [mar/apr-jul/aug 1978] – 1r – 1 – $53.00 – (r also incl: azaraksh; 19 bahman danishju i, and sitiz) – us MEDOC [079]

Bisu-yi azadi *see*
- 19 bahman danishju'i
- Azarakhsh
- Sitiz

Bisu-yi susyalism – Ittihad-i Mubarizan-i Kumunist, 1980-81. shumarah-'i 1-4. 1 murdad-bahman 1359 [23 jul 1980-jan 1981] – 1r – 1 – $53.00 – us MEDOC [079]

Bitburg skyblazer – v32 n16-v36 n17 [1981 may 8-1985 feb 8], 1992 jan 10-1993 dec 10 – 1r – 1 – (cont by: skyblazer (bitburg, germany)) – mf#1042972 – us WHS [071]

Bitburger kreis- und intelligenzblatt – Bitburg DE, 1854-67 – 1 – gw Misc Inst [074]

Bite of hunger / Kuper, Hilda – New York, NY. 1965 – 1r – us UF Libraries [960]

Bi-tehum ha-yamim / Snir, Mordecai – Tel-Aviv, Israel. 1953 – 1r – us UF Libraries [939]

Biter bit / Nobody, S – Oxford, England. 1804 – 1r – us UF Libraries [240]

Bitetti, Roque *see* Justicia social

Bithell, Jethro *see*
- An anthology of german poetry, 1830-1880
- Modern german literature, 1880-1950

Bitlis – (= ser Vilayet salnames) – 9 – (1310 [1892] 5mf $75; 1316h/1314 mali [1898] 4mf $60) – us MEDOC [956]

Bitovt, I *see* Redkie russkie knigi i letuchie izdaniia 18 veka s bibliograficheskimi primechaniiami, ukazaniem stepeni redkosti i tsenami antikvarov na nikh,

Bits and peaces : news and notes on the nuclear disarmament movement / Foundation for Global Peace & Operation Dismantle – n1-1925, 28-1943 [1987 jan 18, oct 9, nov 26, 1988 jan 28-nov 24, 1989 jan 13-mar 3] – 1r – 1 – mf#1840639 – us WHS [327]

Bits and pieces / Lincoln County Historical Association (NC) – 1980 apr/jun-1983 jan/mar – 1r – 1 – mf#1352997 – us WHS [071]

Bits of family history : paper read before the john bradford club, lexington, kentucky, december 8 1932 / Clay, Henrietta – [Lexington, KY: Clay Print Co, 1933] – us CRL [978]

Bittelman, Alex *see* Palestine

Bitter choice / Legum, Colin – Cleveland, OH. 1968 – 1r – us UF Libraries [960]

Die bitter gute *see* Ephrata codex

Bittere wahrheiten : eine unerwartete beleuchtung der "ernsten gedanken" des herrn oberstlieutenant v. egidy / Bornemann, Wilhelm – 3. unver. Aufl. Goettingen:Vandenhoeck & Ruprecht, 1891 – 1mf – 9 – 0-8370-2416-1 – mf#1985-0416 – us ATLA [240]

Bitteres leiden : oberammergauer passionsspiel: text von 1750 / Rosner, Ferdinand; ed by Mausser, Otto – Leipzig: K W Hiersemann 1934 [mf ed 1993] – (= ser Blvs 282) – 58r – 1 – (incl bibl ref. filmed with: coligny, gustav adolf, wallenstein / rhodius, narssius, vernuiaeus) – mf#3420p – us Wisconsin U Libr [820]

Bitting, A W *see*
- Big-head
- Leeches or leeching
- Liver fluke

Bitting, C C *see* Bible societies and the baptists

Bitting, Charles Carroll *see*
- Bible societies and the baptists
- Notes on the history of the strawberry baptist association of virginia for one hundred years, from 1776 to 1876

Bitting, Clarence R *see* Report on the everglades and contiguious areas

Bitting, William Coleman *see* A manual of the northern baptist convention

Bittinger, Benjamin Franklin *see* Manual of law and usage

Bittinger, Lucy Forney *see* German religious life in colonial times

Bittmann, Wilhelm *see* Eine studie ueber goethe's "iphigenie auf tauris"

Bittner, Andreas *see* Untersuchungen zum einsatz ir-spektrometrischer blutsubstratanalytik fuer die medizinische diagnostik

Bittner, Katharina *see* Untersuchungen zur parakrinen interaktion zwischen endothelzellen und chondrocyten bei der chondrocytenspaetdifferenzierung in vitro

Bitton, Nelson *see* The regeneration of new china

Bitton, William Nelson *see* Our heritage in china

Bittremieux, Leo *see*
- La societe secrete des bakhimba au mayombe

Bitullareaga, Mario *see* Cotorrona

Bituminous coal division, decisions and orders / U.S. Dept of the Interior – 1 oct 1940-june 30 1941 [all publ] – 14mf – 9 – $21.00 – mf#llmc 84-112 – us LLMC [622]

Bitzaron – (New York). 1939-54 – 1 – us AJPC [073]

Bitzius, Albert *see*
- Albert bitzius
- Die armennot / ein sylvestertraum / eines schweizers wort
- Bilder aus der schweiz
- Der emmentaler bauer bei jeremias gotthelf
- Erlebnisse eines schuldenbauers
- Familienbriefe jeremias gotthelfs
- Geld und geist
- Der geldstag
- Gotthelf
- Jakobs, des handwerksgesellen wanderungen durch die schweiz
- Jeremias gotthelf
- Jeremias gotthelf im kreise seiner amtsbrueder und als pfarrer
- Jeremias gotthelf in seinen beziehungen zu deutschland
- Jeremias gotthelf's ausgewaehlte werke
- Jeremias gotthelfs geld und geist
- Jeremias gotthelfs persoenlichkeit
- Jeremias metrik
- Jeremie gotthelf
- Die kaeserei in der vehfreude
- Kaethi, die grossmutter
- Kaethi die grossmutter
- Kalendergeschichten
- Kleinere erzaehlungen
- Leiden und freuden eines schulmeisters
- La maladie dans l'oeuvre de gotthelf
- Saemtliche werke in 24 baenden
- Die schwarze spinne
- Uli der paechter
- Ulric the farm servant
- Die vererbung bei den dichtern a bitzius, c f meyer und g keller
- Volksausgabe seiner werke im urtext
- Die wassernot im emmental / die armennot / eines schweizers wort
- Die wassernot im emmenthal / fuenf maedchen / dursli der branntweinlaeuter
- Wie anne baebi jowaeger
- Wie uli der knecht gluecklich wird
- Zeitgeist und bernergeist

The biu book : a collation and reference book on biu division (northern nigeria) / Davies, J G – Norla, Zaria: [s.n.] 1954-1956 – us CRL [960]

Biudzhet krestianskogo khoziaistva : rukovodstvo po schetovodnomu analizu krestianskogo khoziaistva dlia kooperatorov i agronomov / Studenskii, G A – 1923 – 124p 2mf – 9 – mf#COR-518 – ne IDC [335]

Biudzhetnye issledovaniia : istoriia i metody / Chaianov, A V – 1929 – 331p 4mf – 9 – mf#COR-228 – ne IDC [335]

Biudzhetnye obsledovaniia krest'ianskikh khoziaistv v dorevoliutsionnoi rossii / Korenevskaia, N N – M, 1954 – 4mf – 8 – mf#RZ-171 – ne IDC [314]

Biudzhety krest'ian starobel'skogo uezda – Khar'kov, 1915 – 9mf – 8 – mf#RZ-184 – ne IDC [314]

Biuleteni religiozno pedagogicheskago kabineta – Paris, France, 1928, 1929 – 1r – 1 – (russian periodical) – us IHRC [073]

Biuletyn / Polskie Towarzystwo Jezykoznawcze – Cracow. v. 1-13. 1927-19 – 1 – us NY Public [460]

Biuletyn / Zydowski Instytut Historyczny. Warsaw – nos 1-40 1951-1961 and index to nos 1-32 – 1 – us NY Public [947]

Biuletyn informacyjny – Warsaw, Poland 1941-44 [mf ed Norman Ross] – 1r – 1 – mf#nrp-2089 – us UMI ProQuest [077]

Biuletyn zbrodni hitlerowskich – Warsaw, Poland [mf ed Norman Ross] – 2r – 1 – (different yrs – inquire) – mf#nrp-2091 – us UMI ProQuest [077]

Biulleten' [Abastumnis astropizikuri observatoria] / Abastumnis astropizikuri observatoria – Tpilisi: SSRK mecnierebata akademiis sakartvelos pilialis gamomcemloga 1937- [irreg] [mf ed 2003] – 2r – 1 – (chiefly in georgian; some english & russian) – ISSN: 0375-6644 – mf#film mas c5529 – us Harvard [520]

Biulleten' astronomicheskikh institutov chekhoslovakii = Bulletin of the astronomical institutes of czechoslovakia / Ceskoslovenska akademie ved. Astrofysikalni observator – Praha: Astrofysikalni observator OndrejovvNakladetelstvi Ceskoslovenske akademie ved, 1953-66 [bimthly] [mf ed 2003] – 19v on 1r – 1 – (cont: bulletin of the central astronomical institutes of czechoslovakia; cont by: bulletin of the astronomical institutes of czechoslovakia [issn 0004-6248]; in english, russian, french & german) – mf#film mas c5570 – us Harvard [520]

Biulleten' astronomicheskoi observatorii im vp engel'gardta = Engelhardt observatory bulletin / Astronomicheskaia Engel'gardtovskaia observatoriia – Kazan: Izd-vo Kazanskogo gos universiteta 1934- [irreg] [mf ed 2001] – 1r – 1 – (n1-26 iss as suppl to: uchenye zapiski of the kazanskii gosudarstvennyi universitet; in russian, english or german) – mf#film mas c5084 – us Harvard [520]

Biulleten' [chita: 1919] : inform ch shtaba pomoshchnika komanduiushchego voiskami priamur voen okr – Chita [Zabaik obl]: [s n] 1919 [1919 22 sent-] – (= ser Asn 1-3) – n1-2 [1919] item 25, on reel n8 – 1 – mf#asn-1 025 – ne IDC [077]

Biulleten dlia gruppy sodeistviia rsdrp – n1. 1916 – 1mf – 9 – mf#R-18013 – ne IDC [077]

Biulleten' "doma pesni" – Moscow, 1912-17 – 1mf – 9 – (parallel texts in french, german & in pt russian) – us UMI ProQuest [780]

Biulleten' gazety "novoe slovo" / ed by Kossenko, M P – Semipalatinsk: [s n] 1919 [1919-] – (= ser Asn 1-3) – 4 nenum vyp [1919] item 26, on reel n8 – 1 – (suppl of: novoe slovo [asn-1 275]) – mf#asn-1 026 – ne IDC [077]

Biulleten' gosudarstvennogo astronomicheskogo instituta imeni p k shternberga pri moskovskom ordena lenina gosudarstvennom universitete imeni m v lomonosova / Gosudarstvennyi astronomicheskii institut im P K Shternberga – Moskva: Gos universitet im Lomonosova 1940-41 [irreg] [mf ed 2002] – 8v on 1r – 1 – (in russian & english; imprint varies) – mf#film mas c5416 – us Harvard [520]

Biulleten gubsoiuza – Saratov, 1921-1922(27) – 114mf – 9 – (cont as: golos kooperatora saratov 1922-1923(7);golos nizhne-volzhskogo kooperatora.saratov 1924-1928(1);golos kooperatora saratov,1928(2)-1929(21); missing:1921(1-12),1921(14)-1922(15),1922(19-20,22)) – mf#COR-578 – ne IDC [335]

Biulleten' inostrannoi kommercheskoi informatsii – Moscow, Russia 4 jan-31 dec 1955 – 1 – (in cyrillic) – uk British Libr Newspaper [380]

Biulleten' instituta teoreticheskoi astronomii / Institut teoreticheskoi astronomii (Akademiia nauk SSSR) – Leningrad: Izd-vo Akademii nauk SSSR 1947- [irreg] [mf ed 2005] – 1r – 1 – (ind: v1 (1924)-6 (1958); cont: biulleten' astronomicheskogo instituta akademii nauk sssr; suppl accompany some iss; chiefly in russian; occasional articles in french; publ: izd-vo "nauka," leningradskoe otd-nie 1967-; iss for 1947 have caption title also in french: bulletin de l'institut d'astronomie theorique) – ISSN: 0002-3302 – mf#film mas c5683 – us Harvard [520]

Biulleten... izdanie bunda – London, 1901. n1-26 – 5mf – 9 – (missing: 1901 no 5) – mf#R-18011 – ne IDC [072]

Biulleten kooperativnoi sektsii kominterna – 1924-1925 – 36mf – 9 – (cont as: mezhdunarodnaia kooperatsiia 1926-1930(6).missing:1924(1-10)) – mf#COR-650 – ne IDC [335]

Biulleten nizhegorodskogo gubernskogo kooperativnogo soveta – Nizhnyi novgorod, 1925-1926(12) – 109mf – 9 – (cont as: nizhgorodskii kooperator nizhnyi novgorod,1926(1-12)-1931(12);missing:1925(3)) – mf#COR-656 – ne IDC [335]

Biulleten' [omsk: 1919] / Osvedom upr Vost fronta Otd pechati – Omsk [Akmol obl]: Izd Osvedfronta 1919 [1919 20 iiunia-[noiab] – (= ser Asn 1-3) – n1-123 [1919] [gaps] item 24, on reel n8 – 1 – mf#asn-1 024 – ne IDC [077]

Biulleten' otdela tsk po rabote sredi zhenshchin – Moscow, 1921-25 [freq varies] [mf ed Norman Ross Publ] – 11mf – 9 – us UMI ProQuest [331]

Biulleten pravdy – Vienna, 1912. nos 1-4 – 1mf – 9 – mf#R-18016 – ne IDC [074]

Biulleten promkooperatsii – 1922-1923(5) – 7mf – 9 – (missing:1922(1-2),1922(5)-1923(1)) – mf#COR-551 – ne IDC [335]

Biulleten' soedinennogo ispolnitel'nogo biuro komiteta obschestvennoi bezopasnosti i soveta rabochikh, soldatskikh i krest'ianskikh deputatov / Krasnoiarskii komitet obshchestvennoj bezopasnosti. Soedinennoe ispolnitel'noe biuro – Krasnoyarsk, Russia 1917 [mf ed Norman Ross] – 1r – 1 – mf#nrp-796 – us UMI ProQuest [077]

Biulleten' sotsialisticheskogo bloka : organ sotsialisticheskogo bloka 12 armii – Russia 1917 [mf ed Norman Ross] – 1r – 1 – mf#nrp-88 – us UMI ProQuest [077]

Biulleten' sovieshchanii chlenov uchreditel'nago sobraniia – Parizh. n1-6 jan 12-feb 1 1921 [mf ed 1984] – 1v on 1r – 1 – mf#1103 – us Wisconsin U Libr [073]

Biulleten' tashkentskoi astronomicheskoi observatorii = Bulletin of the tashkent astronomical observatory / Toshkent astronomik observatoriiasi – [Tashkent: s.n.] 1933-49 [irreg] [mf ed 2003] – 2v on 1r – 1 – (chiefly in russian, with some english, french & german; imprint varies; iss for mart 1941-avg 1948 have also title in uzbek: toshkent astronomik observatoriiasining biulleteni) – mf#film mas c5597 – us Harvard [520]

Biulleten' teatra i muzyka – Elisavetgrad, 1912 [irreg 8 iss publ] – 1mf – 9 – us UMI ProQuest [780]

Biulleten 'tsentral' nogo statisticheskogo upravleniia – M, 1919-1926. n1-122 – 80mf – 9 – (missing: 1921(44, 46, 57)) – mf#RHS-2 – ne IDC [314]

Biulleten tsentralnogo komiteta partii sotsialistov-revoliutsionerov – [Geneva], 1906. n1 – 1mf – 9 – mf#R-18017 – ne IDC [074]

Biulleten' TsK partii levykh sotsialistov – revolyutsionerov (internatsionalistov) – Moscow, Russia 1919 [mf ed Norman Ross] – 1r – 1 – us UMI ProQuest [077]

Biulleten' uzakonenii i rasporiazhenii po sel'skomu i lesnomu khoziaistvu / Russia. (1917-R.S.F.S.R). Narodnyi Komissariat Zemledeliia – Moscow. 1928-30. On film: v1-3. LL-014 – 1 – us L of C Photodup [340]

Biulleten vserossiiskogo kooperativnogo banka – 1924-1926(4) – 9mf – 9 – (missing:1924(18)-1925(13,16-17(1))) – mf#COR-550 – ne IDC [335]

Biulleten' zasedaniia soveta federatsii – Moskva 1994-96 [irreg] – 129mf – 9 – $323.00/y – mf#m0787 – us East View [077]

Biulleteni diktatury tsentrokaspiia i prezidiuma vremennogo ispolnit k-ta / ed by Aiollo, G et al – Baku: [s n] 1918 [1918 fevr-] – (= ser Asn 1-3) – n5-18 [1918] [gaps] item 21, on reel n8 – 1 – mf#asn-1 021 – ne IDC [077]

Biulleteni gazety "golos stepi" / ed by Kustov, A V – Pavlodar [Semipalat obl]: T-vo kooperativov 1919 [1919 [ianv]-] – (= ser Asn 1-3) – n2-157 [1919] [gaps] item 20, on reel n7 – 1 – (suppl of: golos stepi [asn-097]) – mf#asn-1 020 – ne IDC [077]

Biulleteni knizhnykh i literaturnykh novostei – M., 1910-1911 – 298mf – 9 – (cont as: biulleteni literatury i zhizni. m., 1911-1917) – mf#R-4304, 1693 – ne IDC [077]

Biulleteni krasnoiarskogo soiuza sibirskikh oblastnikov-avtonomistov – Krasnoiarsk [Enis gub]: [s n] 1918 [1918 fevr-] – (= ser Asn 1-3) – n1 [1918] item 22, on reel n8 – 1 – mf#asn-1 022 – ne IDC [077]
Biulleteni shadrinskoi gorodskoi dumy – Shadrinsk [Perm gub]: [s n] 1918 [1918-] – (= ser Asn 1-3) – n2 [1918] item 23, on reel n8 – 1 – mf#asn-1 023 – ne IDC [077]
Bivero, P de *see*
– Sacrum oratorium piarum imaginum immaculatae mariae et animae creatae ac baptismo, poenitentia, et eucharistia innovatae
– Sacrum sanctuarium crucis et patientiae crucifixorum et cruciferorum, emblematicis imaginibus laborantium et aegrotantium ornatum
Bivort, Jean Baptiste *see* Code constitutionel de la belgique, ou commentaire sur la constitution, la loi electorale, la loi communale et la loi provinciale
Bivouacs en guyane / Quris, Bernard – Paris, France. 1956 – 1r – us UF Libraries [972]
Biweekly intelligence summaries, 1928-38 – (= ser U s military intelligence reports) – 6r – 1 – $1050.00 – 0-89093-659-5 – (with p/g) – us UPA [355]
The bixby gospels / Goodspeed, Edgar Johnson – Chicago: University of Chicago Press, 1915 – (= ser Historical and linguistic studies in literature related to the new testament) – 1mf – 9 – 0-8370-1994-X – mf#1987-6381 – us ATLA [226]
Bixby, James Thompson *see*
– The crisis in morals
– The new world and the new thought
– Religion and science as allies
Bixler, Marguerite Arthelda *see* Helpful hints on music
Bi-yeme bayit sheni / Klausner, Joseph – Berlin, Germany. 1923 – 1r – us UF Libraries [939]
Biz – Fairfield, jan 1970-jun 1972 – 1r – at Pascoe [079]
Biz – Fairfield, nov 1928-dec 1969 – 21r – 9 – A$1422.48 vesicular A$1306.98 silver – at Pascoe [079]
Biz – Toronto: S C Trethewey, [1893?-189- or 19–] [mf ed v1 n2 jan 1 1894] – 9 – mf#P04355 – cn Canadiana [650]
Biz beat – 1987 sep 25, 1988 jul – 1r – 1 – mf#4853345 – us WHS [071]
Bizari, P *see*
– Cyprivm bellvm, inter venetos, et selymvm tvrcarvm imperatorem...
– Histoire de la gverre qui c'est passee, entre les venitiens et la saincte ligue, contre les turcs, pour l'isle de cypre, es annees 1570, 1571 and 1572
Bizeul, Severe Jacques *see* Chinois et missionnaires
Bizim mecmua – Istanbul: Sanayi-i Nefise Matbassi, Evkaf Matbaasi, 1922-28. Mueduer-i Mes'ul: Huluesi, Mehmed Ali. n1-28. 5 nisan-12 tesrinievvel 1922 – (= ser O & t journals) – 8mf – 9 – $200.00 – us MEDOC [956]
Bizim pasa – Istanbul. v1 n1-3. 1949 [all publ] – (= ser O & t journals) – 1mf – 9 – $25.00 – (foll: malumpasa in the markopasa set; publ by: aziz nesin, sabahattin ali, and rifat ilgaz) – us MEDOC [956]
Bizimpasa *see* Markopasa
Bizzaron – New York. v1-65. 1939-1974 – 374mf – 9 – mf#J-91-3 – ne IDC [956]
Bjerre, Jens *see* Kalahari
Bjerregaard, Carl Henrik Andreas *see*
– The inner life and the tao-teh-king
– Lectures on mysticism and nature worship
Bjoerkman, Edwin *see* Scandinavia and the war
Bjog : an international journal of obstetrics and gynaecology – Oxford UK 2000+ – 1,5,9 – ISSN: 1470-0328 – mf#681.02 – us UMI ProQuest [618]
Bjornson, Bjornstjerne *see*
– Au dela des forces
– Ausgewaehlte werke
– Laboremus
– Ovind
Bju international – Oxford UK 1999+ – 1,5,9 – (cont: british journal of urology) – ISSN: 1464-4096 – mf#733.01 – us UMI ProQuest [616]
Bjurstrom, Per *see* Drawings of johan tobias sergel
Blaas, Erna *see* Ruehmung und klage
Blache, Robert *see* Spain's october
Blachere, R *see* Le coran
Black academy of arts and letters records : from the holdings of the schomburg center for research in black culture, manuscripts, archives and rare books division: the new york public library, astor, lenox and tilden foundations – 1995 – ca 10r – 1 – ca $850.00 – (guide sold separately for $20.00 covers all coll under "literature and the arts" d3305.g6) – mf#D3305P25 – Dist. us Scholarly Res – us L of C Photodup [700]
Black, Adam *see* Church its own enemy
Black advocate – 1992 may/jun, 1995 jan/feb-sep/oct – 1r – 1 – mf#2692245 – us WHS [071]
Black africa : wither? / Van Der Merwe, H J J M – Johannesburg, South Africa. 1963 – 1r – us UF Libraries [960]
Black agenda : news quarterly / National Black Political Assembly – v2 n1 [1980 spring] – 1r – 1 – mf#4775857 – us WHS [305]
Black, Alex *see*
– National blessings considered and improved
– Necessity of national reformation stated in a sermon
Black, Alexander *see* Exegetical study of the original scriptures considered
Black Allied Student Association [New York University] *see* Imani
Black america – 1970 jun, sep, 1971 may/jun, dec, 1972 jun-aug, 1973 aug – 1r – 1 – (cont: miss black america) – mf#4880779 – us WHS [640]
Black american literature forum – Terre Haute IN 1977-91 – 1,5,9 – (cont: negro american literature forum; cont by: african american review) – ISSN: 0148-6179 – mf#5044.02 – us UMI ProQuest [420]
Black american music review – v1 n1 [[1983]] – 1r – 1 – mf#4862704 – us WHS [780]
Black americans in congress 1870-1989 / Ragsdale, Bruce A & Treese, Joel D – Washington: GPO, 1990 [all publ] – 2mf – 9 – $3.00 – mf#llmc 95-031 – us LLMC [323]
Black and beautiful / Fortie, Marius – Indianapolis, IN. 1938 – 1r – us UF Libraries [025]
Black and dark night – London, England. 18– – 1r – us UF Libraries [240]
Black and Hispanic Images, Inc *see* Bhi newsletter
Black and third world periodicals: sample issues, 1844-1963 : from the holdings of the schomburg center for research in black culture, manuscripts, archives and rare books division: the new york public library, astor, lenox and tilden foundations – 1995 – 8r – 1 – $680.00 – (guide which covers all coll under "black periodicals" sold separately for $20.00 d3305.g4) – mf#D3305P14 – Dist. us Scholarly Res – us L of C Photodup [960]
Black and white / League of American Writers – v1-2 n2+5. 1939-40 [all publ] – (= ser Radical periodicals in the united states, 1881-1960. series 1) – 5mf – 9 – $200.00 – us UPA [335]
Black and white : unite and fight for a workers world – 1961 apr 14-1966 jun 16, 1966 jul 10-1970 dec 25 – 2r – 1 – (cont: colored and white, unite and fight for a workers world; cont by: workers world (new york ny: 1971)) – mf#1044254 – us WHS [331]
Black and white : a weekly illustrated record and review – London, England 6 feb 1891-13 jan 1912 (wkly) – 42r – 1 – (incorp with: sphere) – uk British Libr Newspaper [072]
Black and white africans / Strydom, Christiaan Johannes Scheepers – Cape Town, South Africa. 1967 – 1r – us UF Libraries [305]
Black and white in southeast africa : a study in sociology / Evans, Maurice Smethurst – London; New York: Longmans, Green, 1911 – 1mf – 9 – 0-7905-4904-2 – mf#1988-0904 – us ATLA [305]
Black and white in the southern states : a study of the race problem in the united states from a south african point of view / Evans, Maurice Smethurst – London, New York: Longmans, Green, 1915 [mf ed 1990] – 1mf – 9 – 0-7905-5389-9 – (incl bibl ref) – mf#1988-1389 – us ATLA [305]
Black and white magazine, port moresby, png / McCall, Grant – nov 1966-aug 1969 – 1r – 1 – mf#pmb doc462 – at Pacific Mss [870]
Black and white men together / san francisco – 1984 dec-1985 jan, may, jul-aug, oct, dec – 1r – 1 – mf#4863622 – us WHS [303]
Black arts new york : newsletter / Harlem Cultural Council – 1987 jun-oct, 1988 jan-mar, may-jun, sep-1989 jun, sep-1990 apr, jun, oct-dec, 1991 mar-jun, oct-1992 jun, oct, dec, 1993 jan, mar, may, nov, 1994 nov-dec/1995 jan – 1r – 1 – (cont: harlem cultural review) – mf#2847482 – us WHS [700]
Black arts quarterly – 1989 fall qtr – 1r – 1 – mf#5297608 – us WHS [700]
Black ascensions / Cuyahoga Community College – 1971 feb, dec, 1972 feb, 1973 winter I-spring I, 1974 winter – 1r – 1 – mf#2847504 – us WHS [373]
Black atlantic city magazine – 1986 mar/apr – 1r – 1 – (cont by: black new jersey magazine) – mf#5132314 – us WHS [071]
Black autonomy / Federation of Black Community Partisans – 1995 jan, mar, apr/may, jun/jul, aug/sep, 1996 jan/feb, mar/apr, may/jun, jul-sep, nov/dec, 1997 jan/feb, apr/may, jun/jul, aug/sep – 1r – 1 – mf#3202749 – us WHS [321]
Black bagdad / Craige, John Houston – New York, NY. 1933 – 1r – us UF Libraries [972]
Black Baptist Convention *see* Annuals
Black biographical dictionaries, 1790-1950 – [mf ed Chadwyck-Healey] – 297 titles on 1068mf – 9 – (suppl: 51 titles on 140mf. ea with printed handlists) – uk Chadwyck [305]
The black bishop : samuel adjai crowther / Page, Jesse – London: Hodder and Stoughton, 1908 – 2mf – 9 – 0-7905-8204-X – mf#1988-8087 – us ATLA [240]
The black book of paisley, and other manuscripts of the scotichronicon; with a note upon john de burdeus...and the pestilence / Murray, David – Paisley: A. Gardner, 1885 – 1 – us Wisconsin U Libr [941]
Black book of the admiralty (rs55) : monumenta juridica / ed by Twiss, Travers – (= ser The rolls series (rs)) – (with trans and app. v1 1871 €19. v2 1873 €19. v3 1874 €27. v4 1876 €25) – ne Slangenburg [355]
Black books bulletin – Chicago IL 1971-81 – 1,5,9 – ISSN: 0045-2114 – mf#8601 – us UMI ProQuest [070]
Black books bulletin / Institute of Positive Education (Chicago IL) – v1 n1-6/v2 n1 [1991 jul/aug-1992 [summer?] – 1r – 1 – mf#2406847 – us WHS [370]
Black Brigade *see* Do it loud
Black bulletin : black nationalist newsletter – 1952? – 1r – 1 – mf#3057878 – us WHS [322]
Black caesar : pirate enigma, had lair in keys belo... / Henry, Bruce L – s.l, s.l? . 193? – 1r – us UF Libraries [978]
Black caesar – s.l, s.l? . 193? – 1r – us UF Libraries [978]
Black Career Women, Inc *see* Charisma
Black careerst – v13 n1-v18 n2 [1977 jan/feb-1982 mar/apr – 1r – 1 – (cont: project, guidelines to equal opportunity employment) – mf#611490 – us WHS [331]
Black cat : a monthly magazine of original short stories – Killen TX 1895-1922 – 1 – mf#4645 – us UMI ProQuest [830]
Black cat – v1 n5 [1968 aug 25] – 1r – 1 – mf#765303 – us WHS [071]
Black, Catherine H *see* Influencing a broader understanding of jazz dance
Black caucus newsletter – v9 n1-v17 n3 [1983 sep-1988 dec] – 1r – 1 – (cont by: black caucus of ala newsletter) – mf#1053400 – us WHS [020]
Black caucus of ala newsletter – 1989 feb-1996 oct, 1997 feb-1998 oct – 2r – 1 – (cont: black caucus newsletter; cont by: newsletter of the black caucus of the american library association) – mf#1111728 – us WHS [020]
Black, Charles Clarke *see* Proof and pleadings in accident cases.
Black child – 1995 fall, 1996 jan-oct, fall, dec, 1997 feb-jul, fall, winter, 1998 spring, fall – 1r – 1 – mf#3466938 – us WHS [305]
Black church magazine – 1986 jun, 1990 jul – 1r – 1 – mf#4114387 – us WHS [240]
Black circles – n1-5 [1975 jan-1976 mar] – 1r – 1 – mf#354476 – us WHS [071]
The black code of georgia, u.s.a – (manuscripts). 1900 – 1 – us L of C Photodup [978]
The black code of the district of columbia, in force september 1st, 1848 / Snethen, Worthington Garrettson – New York, Harned, 1848. 61 p. LL-204 – 1 – us L of C Photodup [348]
Black, Colin *see* Lands and peoples of rhodesia and nyasaland
Black college radio news / Atlanta University Center (GA) – 1994 may/jun – 1r – 1 – mf#5327466 – us WHS [378]
Black college sports review – 1983 nov-97 jul – 1r – 1 – mf#2590200 – us WHS [790]
Black collegian – New Orleans LA 1970+ – 1,5,9 – ISSN: 0192-3757 – mf#6575 – us UMI ProQuest [305]
Black commentator – 1981 spring – 1r – 1 – mf#4798838 – us WHS [071]
Black Community Crusade for Children [US] *see* Necessary
Black congressional monitor – 1987 sep-1993 dec 30, 1994 jan 15-1996 dec 30, 1997 jan 15-1999 dec 30 – 3r – 1 – mf#1054683 – us WHS [071]
The black consciousness movement of south africa : material from the collection of gail m gerhart – [Nairobi: Kenyan Photographic Supply Co] for the Cooperative Africana Microform Project, 1979 – us CRL [322]
Black creek baptist church – New York. 1876-1919 (1) – 1 – $31.46 – mf#5910 – us Southern Baptist [242]
Black creek baptist church. dovesville (darlington), south carolina : church records – 1798-1896 – 1 – us Southern Baptist [242]
Black creek times – Black Creek WI. 1904-06, 1907-09, 1910-13, 1914-17, 1918-21, 1922-1928 apr 12 – 6r – 1 – mf#957324 – us WHS [071]
Black dawn / Steiner, Mia – Port-Au-Prince, Haiti. 1950 – 1r – us UF Libraries [972]
Black death: sources concerning the european plague, series 1 : rare printed sources from the herzog august bibliothek, wolfenbuttel, c1470-1822 – [mf ed Marlborough 1994] – 2pt – 1 – (pt1: sources, c1470-1621 [18r] £1650; pt2: sources, 1621-1822 [16r] £1500; with d/g) – uk Matthew [614]
Black democracy / Davis, Harold Palmer – New York, NY. 1936 – 1r – us UF Libraries [322]
Black dialogue – v2 n5 [1966 autumn] – 1r – 1 – mf#246321 – us WHS [071]
Black dwarf – La Salle IL 1817-24 – 1 – mf#4208 – us UMI ProQuest [073]
Black dwarf – v1-12. 1817-24 [all publ] – (= ser Radical periodicals of great britain, 1794-1914. period 1) – 101mf – 9 – $550.00 – us UPA [305]
Black eagle news – 1971 jul-1972 jan – 1r – 1 – mf#1106642 – us WHS [071]
Black earth advertiser – Black Earth WI. 1870 aug 11-1874, [1875-78], 1879-82, 1883-86, 1887 jan 7-1888 sep 15 – 5r – 1 – (cont: monthly advertiser; cont by: advertiser (black earth wi)) – mf#928153 – us WHS [071]
Black earth news – Black Earth WI. 1915 jun 4-aug 27 – 1r – 1 – (cont: black earth times; cont by: dane county news) – mf#938314 – us WHS [071]
Black economic times – 1993 mar 10-1996 oct 30 – 1r – 1 – mf#2760528 – us WHS [330]
Black economy, usa – v1-3 n3 [1973 may-1975 mar] – 1r – 1 – mf#156104 – us WHS [339]
Black employment and education journal – [1990 jan], [1996 winter], [1997 winter/spring, summer, fall] – 3r – 1 – mf#2405701 – us WHS [305]
Black enterprise – New York NY 1970+ – 1,5,9 – ISSN: 0006-4165 – mf#6177 – us UMI ProQuest [305]
Black ethnic collectibles – 1987 may/jun-1993 spring – 1r – 1 – mf#1579172 – us WHS [305]
Black excel news : the qtrly newsletter of black excel, college admissions and scholarship service – v1 n1, v2 n1-2/3, v3 n1/2 [1993 winter, 1994 spring-1995 spring, 1996 spring] – 1r – 1 – mf#2948414 – us WHS [378]
Black excellence – 1991 nov/dec, 1992 mar/apr-1994 nov/dec, 1995 jan/feb-1997 conference iss – 2r – 1 – mf#2683468 – us WHS [305]
Black excellence – v1 n2, v2 n3, v3 n5, 7 [1977 jun/aug, 1978 sep-nov, 1979 mar/may, sep/nov] – 1r – 1 – (cont by: excellence (milwaukee wi)) – mf#1699053 – us WHS [305]
Black exposition magazine – 1992 may – 1r – 1 – mf#3912588 – us WHS [305]
Black expression – v1 n1 [1968 fall], v2 n1 [[1969] fall] – 1r – 1 – mf#4992534 – us WHS [305]
Black Faculty and Staff Caucus (Baton Rouge LA) *see* Bf and sc news
Black family – 1982 dec, 1983 dec – 1r – 1 – mf#4025190 – us WHS [640]
Black fence – London, England. 1850 – 1r – us UF Libraries [240]
Black fire / Newcomb, Covelle – New York, NY. 1940 – 1r – us UF Libraries [972]
Black focus / Black Resources and Information Centre – 1979 jun – 1r – 1 – mf#4852943 – us WHS [305]
Black ghetto – 1969 aug-sep, 1972 jul/aug, 1973 feb – 1r – 1 – mf#4851589 – us WHS [307]
Black hair care – 1992 fall, 1994 aug, 1998 feb – 1r – 1 – mf#2695726 – us WHS [640]
Black hair digest / Word Up! Video Productions, Inc – v1 n1, v2 n2, v4 n1 [1993 nov, 1994 spring, 1995 spring] – 1r – 1 – mf#2901366 – us WHS [640]
Black hair styles – v2 n1 [1995 feb] – 1r – 1 – mf#3179532 – us WHS [640]
Black haiti / Niles, Blair – New York, NY. 1926 – 1r – us UF Libraries [972]
Black hawk *see* Gilpin county miscellaneous newspapers
Black hawk advetiser *see* Gilpin county miscellaneous newspapers
Black hawk times *see* Gilpin county miscellaneous newspapers
Black, Henry Campbell *see*
– Handbook on the construction and interpretation of the laws.
– A treatise on federal taxes
– A treatise on the law of judgments, including the doctrine of res judicata
– A treatise on the law of tax titles.
– A treatise on the laws regulating the manufacture and sale of intoxicating liquors
Black heritage : black heritage committee newsletter – 1988 may-oct – 1r – 1 – mf#5132125 – us WHS [321]
Black heritage – Orlando FL 1978-82 – 1,5,9 – (cont: negro heritage) – ISSN: 0197-8810 – mf#10790,01 – us UMI ProQuest [305]
Black Heritage Society of Washington State *see* Newsletter
Black Hills Alliance et al *see* Dead serious

Black hills alliance news, milwaukee – may 1979, [feb 1980] – 1r – 1 – mf#637227 – us WHS [071]
Black hills nuggets / Rapid City Society for Genealogical Research – 1983-89 – 1r – 1 – mf#1110265 – us WHS [929]
Black hills paha sapa report – v1 n1, 4-v3 n1 [1979 jul, 1980 feb-1982 mar/apr, 1982 aug] – 1r – 1 – mf#1220291 – us WHS [071]
Black history bulletin – Washington DC 2002+ – 1,5,9 – mf#877.01 – us UMI ProQuest [934]
Black hood – iss n9-18. win 1943-spr 1946 – 15 – mf#003MLJ-004MLJ – us MicroColour [740]
Black, Hugh *see* Friendship
Black interaction : newsletter of the center for black / University of California, Santa Barbara – 1971 nov 5-dec 9 – 1r – 1 – (cont: black vibrations) – mf#4851430 – us WHS [302]
The black interpreters : notes on african writing / Gordimer, Nadine – [Johannesburg] SPRO-CAS/Ravan [1973] – us CRL [490]
Black issues book review – Fairfax VA 2000+ – 1,5,9 – ISSN: 1522-0524 – mf#28164 – us UMI ProQuest [321]
Black issues in higher education – Fairfax VA 1987+ – 1,5,9 – ISSN: 0742-0277 – mf#16626.01 – us UMI ProQuest [321]
Black, J F *see* The bible way
Black jamaca : a study in evolution / Livingstone, W P – London, England. 1899 – 1r – us UF Libraries [972]
Black, Jeremiah S *see* Papers
Black, Jeremiah Sullivan *see* Mistakes of ingersoll and his answers complete
Black jewish dialog forum mini-newsletter – 1997 jun, sep, oct – 1r – 1 – mf#4023737 – us WHS [305]
Black, John *see*
– Cantate domino
– Presbyterianism in england in the eighteenth and nineteenth centuries
Black, John G *see* A comparative analysis of equivalent submaximal treadmil and bicycle ergometer exercise
Black, John Sutherland *see*
– The book of joshua
– The book of judges
– The christian consciousness
– Encyclopaedia biblica
– Lectures and essays of william robertson smith
– The life of william robertson smith
Black, Joseph *see* Industrial revolution: a documentary history, series 2
Black journal – 1970 5th season – 1r – 1 – mf#5266285 – us WHS [305]
Black journalism review [bjr] – 1976 fall – 1r – 1 – mf#4164338 – us WHS [071]
Black journals – 2ser – 1,9 – $10,340.00 coll ser1 $7735 ser2 $4710 – (50 periodicals, dating back to early 1800s, provide unparalleled historical record for afro-american and african studies. sold as complete coll & also by series & individual title; also listed separately) – us UPA [073]
Black justice exposed! : the most pressing domestic question of our time answered by an american expert on civil rights / Lynn, Conrad Joseph – [Philadelphia: Civil Liberties Department, Improved Benevolent Protective Order, Elks of the World], 1947 (mf ed 1976) – 1r – 1 – (incl bibl ref) – mf#ZZ-14306 – us NY Public [323]
Black, Kenneth Macleod *see* The scots churches in england
Black knight / yellow claw – iss n1-5 may 1955-apr 1956 (black knight); iss n1-4 oct 1956-apr 1957 (yellow claw) – (= ser Yellow Claw) – 15 – mf#043MV – us MicroColour [740]
Black lady – 1983 may/jun-1984 mar/apr – 1r – 1 – mf#4848556 – us WHS [305]
Black land news – 1971 may 1 – 1r – 1 – (cont by: black land news/magazine) – mf#3055245 – us WHS [333]
Black land news/magazine – 1972 sep 14/30 – 1r – 1 – (cont: black land news) – mf#3055253 – us WHS [333]
Black law journal – New York NY 1971-86 – 1,5,9 – (cont by: national black law journal) – ISSN: 0045-2181 – mf#6628.01 – us UMI ProQuest [340]
Black law journal *see* National black law journal
Black Liberation Party [US] *see* Liberation news
Black liberator / Alexis, Stephen – New York, NY. 1949 – 1r – us UF Libraries [972]
The black liberator : theoretical and discussion journal for black liberation – London: The Black Liberator Press 1972-78 – 1 – mf#914 – us Wisconsin U Libr [305]
Black lines – 1997 feb-1999 dec – 1r – 1 – (cont by: en la vida; identity (chicago il)) – mf#3797478 – us WHS [321]
Black lines : a journal of Black studies / University of Pittsburgh – 1970 fall – 1r – 1 – mf#151562 – us WHS [321]
Black lines – Pittsburgh PA 1970-72 – 1,5,9 – ISSN: 0045-2203 – mf#6375 – us UMI ProQuest [305]
Black literary players – 1993 jun-nov/dec, 1994 jan/feb-oct/nov, 1995 jan-nov, 1997 apr – 1r – 1 – mf#2775219 – us WHS [410]
Black literature, 1827-1940 / ed by Gates, Henry Louis Jr – [mf ed Chadwyck-Healey] – 200mf per y – 9 – (with cumulative ind) – uk Chadwyck [800]
Black living in westchester – 1984 jan/feb-1986 jan/feb – 1r – 1 – mf#4712827 – us WHS [307]
Black lung bulletin – [v1] n2-v3 n4 [1970 jul-1972 sep/oct] – 1r – 1 – mf#1053410 – us WHS [612]
Black, M *see* Rituale melchitarum
Black majesty / Vandercook, John W – New York, NY. 1928 – 1r – us UF Libraries [972]
Black man : a journal propagating the interests of workers throughout the african continent / Industrial and Commercial Workers Union of South Africa – Cape Town: Black Man Co Ltd 1920-[19–] (mthly) [mf ed Cape Town: SA Lib 1993] – 1r – 1 – (began in 1920; official organ of the industrial & commercial workers union of south africa) – mf#mp.1246 – sa National [331]
The black man : his antecedents, his genius, and his achievements / Brown, William Wells – New York: T Hamilton; Boston: R F Wallcut, 1863 (mf ed 1968) – 1r – 1 – mf#ZZ-6010 – us NY Public [920]
Black manifesto news – 1971 feb – 1r – 1 – mf#4851960 – us WHS [321]
Black man's portion / Reader, D H – Cape Town, South Africa. 1961 – 1r – us UF Libraries [305]
Black martinique, red gujana / Smith, Nicol – Indianapolis, IN. 1942 – 1r – us UF Libraries [972]
Black meetings and tourism – 1999 feb-jun – 1r – 1 – mf#3242717 – us WHS [338]
Black mother : the years of the african slave trade / Davidson, Basil – Boston, MA. 1961 – 1r – us UF Libraries [306]
The black mountain express *see* [Santa cruz-] miscellaneous titles
Black mountain review – nos. 1-7. 1954-57 – 1 – us AMS Press [800]
Black nation – 1982 fall/winter, 1983 summer/fall, 1984 summer/fall, 1986 summer/fall – 1r – 1 – mf#1193356 – us WHS [321]
Black nation information bulletin : the abc of islam – 1972 dec – 1r – 1 – mf#4851599 – us WHS [260]
Black nationalism in south africa : a short history / Walshe, Peter – Johannesburg, SPRO-CAS/Ravan 1973 – us CRL [322]
Black networking news – Washington DC. special preview iss 1989 jan, mar-apr, jun-aug, oct-nov, 1990 jan, jul-aug/sep – 1r – 1 – mf#3744685 – us WHS [070]
Black new ark – 1972 apr-1974 jan/feb – 1r – 1 – (cont: black news (newark nj)) – mf#859946 – us WHS [071]
Black newark – 1968 apr-nov – 1r – 1 – (cont by: black news (newark nj)) – mf#3362302 – us WHS [071]
Black news – 1970 jan 10,25, 1970 dec 10-1974 feb, 1974 mar-1978 jul, 1978 aug-1984 feb/mar – 1r – 1 – mf#355986 – us WHS [305]
Black news – v1 n4 1969 jan/mar – 1r – 1 – (cont: black newark; cont by: black new ark) – mf#3367921 – us WHS [305]
Black news – Columbia SC. 1987 aug 27-29, 1988 nov 3/1989 apr 27-1998 may 7/13-aug 27/sep 2 – 25r – 1 – (with gaps; cont: black on news) – mf#2304683 – us WHS [305]
Black news digest – Washington DC 1972-93 – 1,5,9 – ISSN: 0045-2238 – mf#7907 – us UMI ProQuest [305]
Black newspapers: sample issues, 1845-1966 : from the holdings of the schomburg center for research in black culture, manuscripts, archives and rare books division: the new york public library, astor, lenox and tilden foundations – 1995 – 4r – 1 – $340.00 – (guide which covers all coll under "black periodicals" sold separately for $20.00 d3305.g4) – mf#D3305P12 – Dist. us Scholarly Res – us L of C Photodup [071]
Black officer – 1986 jan/feb – 1r – 1 – mf#4882426 – us WHS [360]
Black orange – 1994, 1995 feb-1996 nov – 2r – 1 – mf#3107703 – us WHS [071]
Black orpheus – Ibadan, Nigeria: General Publ Section, Ministry of Education. n1-13 (sep 1957-nov 1963) – 1r – us CRL [073]
Black pages – 1971/72-1972/73, 1974/75-1975/76 – 1r – 1 – mf#4882543 – us WHS [071]
Black pages – 1987-89 – 1r – 1 – mf#4882560 – us WHS [071]
Black panther – 1967 may 15, jun 20-1970 mar 15, 1970 dec 19, 1970 mar 15-1971 jul 19, 1971 jul 24-1972 dec 30, 1973 jan-jun, 1973 oct 13-1974 dec 28, 1975 jan 4-1975 jul 28 – 7r – 1 – mf#29507 – us WHS [320]
Black panther – 1991 fall-1994 spring/summer – 1r – 1 – mf#2752687 – us WHS [320]
Black panther [1968] – Oakland CA 1968-80 – 1,5,9 – ISSN: 0523-7238 – mf#6048 – us UMI ProQuest [322]
Black Panther Party *see*
– People's community news
– Right on!: black community news service
Black parent / National Black Parents Organization (US) – 1976 jan – 1r – 1 – mf#5026604 – us WHS [640]
Black perspective in music – Cambria Heights NY 1973-90 – 1,5,9 – ISSN: 0090-7790 – mf#7891 – us UMI ProQuest [780]
Black politician – Los Angeles CA 1969-71 – 1 – ISSN: 0006-422X – mf#6615 – us UMI ProQuest [305]
Black post – Sumter SC. [1992 jan 9/15-apr 30/may 6]-[1996 jun 27/jul 3-sep 26/oct 2] – 14r – 1 – mf#2304816 – us WHS [071]
Black press / Montgomery Co. Dayton – jan-nov 1981 [wkly] – 1r – 1 – mf#B13555 – us Ohio Hist [976]
Black press / Montgomery Co. Dayton – v1 n1. (nov 1972-may 1980) very scattered [wkly] – 1r – 1 – mf#B34499 – us Ohio Hist [071]
Black pride – 1974 aug/sep-oct/nov, 1975 feb/mar-apr/may, aug/sep-dec, 1976 mar-apr – 1r – 1 – mf#2847467 – us WHS [071]
Black problem / Jabavu, Davidson Don Tengo – Lovedale, South Africa. 1921? – 1r – us UF Libraries [305]
Black progress review – 1992 jan-1997 winter – 1r – 1 – mf#2655483 – us WHS [071]
Black protest / Grant, Joanne – New York, NY. 1968 – 1r – us UF Libraries [303]
Black Psychiatrists of America *see* Newsletter
Black rap / Afrikan Students for Afrikan Liberation – v3 n1-2 [1970 early oct-mid nov], v4 n1 [1971 fall], v6 n5 [1974 nov] – 2r – 1 – mf#700591 – us WHS [321]
Black republican and office-holder's journal – New York, NY: Pluto Jumbo. n1 aug 10 1865; n2 aug 1865 – (= ser Negro Newspapers on Microfilm) – 1r – 1 – us L of C Photodup [071]
Black Resources and Information Centre *see* Black focus
Black review : 1972 – Durban, Black Community Programmes 1973 – us CRL [305]
Black river baptist church. black, missouri : church records – 1850-92 – 1 – 5.76 – us Southern Baptist [242]
Black river journal : notes from new orleans / Future Club of New Orleans – 1977 summer – 1r – 1 – mf#4877671 – us WHS [071]
Black sacred music – Durham NC 1990-95 – 1,5,9 – (cont: journal of black sacred music) – ISSN: 1043-9455 – mf#17596.01 – us UMI ProQuest [780]
Black sam – London, England. 18– – 1r – us UF Libraries [240]
Black, Samuel Charles *see*
– Building a working church
– Plain answers to religious questions modern men are asking
– Progress in christian culture
The black sash = Die swart serp – Johannesburg, Women's Defence of the Constitution League. v1-10 n3 1956; aug/oct 1966; v9-12 1965/66-1968/69 – us CRL [322]
Black Sash Society *see*
– [Paper, 3rd Series]
– Papers, 1955-1970
– Papers, 1955-1973
– Papers, 1955-1977
Black scholar – Oakland CA 1969+ – 1,5,9 – ISSN: 0006-4246 – mf#5983 – us UMI ProQuest [305]
Black secrets – 1998 dec-1999 nov, 2000 jan-nov – 2r – 1 – (cont: secrets) – mf#2847366 – us WHS [071]
Black sociologist – Piscataway NJ 1982 – 1,5,9 – ISSN: 0160-3566 – mf#12220.01 – us UMI ProQuest [305]
Black star – 1985 summer v3 n3-v5 n2 [1985 fall-1987: summer] – 1r – 1 – (cont: revista x) – mf#1546310 – us WHS [071]
Black star – v1-2 n1 [1975 dec-1978 jun?] – 1r – 1 – mf#499100 – us WHS [071]
Black star – Columbia, Greenville, Spartanburg SC. [1992 jan 9/15-apr 30/may 6]-[1996 jun 27/jul 3-sept 26/oct 2] – 14r – 1 – (with gaps; cont by: black news (columbia sc)) – mf#2304813 – us WHS [071]
Black star news – 1998 feb, jun, oct/nov-dec/1999 jan – 1r – 1 – mf#4145818 – us WHS [071]
Black stars – Chicago IL 1971-81 – 1,5,9 – ISSN: 0163-3007 – mf#6543 – us UMI ProQuest [790]
Black stars – 1973 jul, 1980 apr – 1r – 1 – (cont: tan (chicago il)) – mf#4852676 – us WHS [071]
Black Student League [Temple University] *see*
– Insight
– Maji-maji
Black Student Union (University of California, Santa Barbara) *see* Blackwatch
Black studies abstract / University of Michigan – 1973 – 1r – 1 – mf#5072181 – us WHS [321]
Black studies correspondence – 1989 mar 4 – 1r – 1 – mf#5319996 – us WHS [305]
Black swamp heritage – v7 n1-v8 n4 [1987 winter-1988 fall] – 1r – 1 – (cont: sandusky county heritage) – mf#1564632 – us WHS [071]
Black Teacher, Parent, Student Coalition *see* Communiviews
Black Teachers Workshop *see* Foresight
Black teen – 1986 jun – 1r – 1 – mf#4852602 – us WHS [305]
Black Theater Alliance *see* Blackstage
Black theater bulletin / Better Boys Foundation – v1 n2 [[1974] apr/may – 1r – 1 – mf#4992545 – us WHS [790]
Black theatre – Bronx NY 1968-72 – 1 – ISSN: 0006-4270 – mf#9985 – us UMI ProQuest [790]
Black Theatre Alliance *see* On stage
Black Theology Project [New York NY] *see* Bulletin of the black...
Black thought – 1997 feb-1998 dec, 1999 jan-2000 aug/sep – 2r – 1 – (cont by: thought (champaign il)) – mf#3844789 – us WHS [321]
Black times – 1975-1976 aug – 1r – 1 – mf#772450 – us WHS [071]
Black times – Charleston, Columbia SC. [1992 jan 9/15-apr 30/may 6]-1996 jun 27/jul 3-sep 26/oct 2 – 14r – 1 – (with gaps) – mf#2304713 – us WHS [071]
Black torch – 1972 jan 26 – 1r – 1 – mf#5076948 – us WHS [071]
Black track – v1 iss 2 [[1998]] – 1r – 1 – mf#4179097 – us WHS [071]
Black traveler – 1995 apr, 1996 jan, sep – 1r – 1 – (cont: black convention) – mf#3242894 – us WHS [071]
Black trek / Noble, Walter James – London, England. 1931 – 1r – us UF Libraries [025]
Black truth – Chicago IL. 1968 dec 20-1971 dec 20 – 1r – 1 – (cont: west side torch (chicago il: lawndale ed)) – mf#872740 – us WHS [071]
Black truth bulletin / National Joint Action Committee (Trinidad and Tobago) – n3 [1978] – 1r – 1 – mf#4867699 – us WHS [321]
Black vet / Black Veterans for Social Justice – 1988 jan/apr-1991 sep – 1r – 1 – mf#1054730 – us WHS [071]
Black Veterans for Social Justice *see* Black vet
Black vibrations / University of California, Santa Barbara – 1971 feb 5, mar 4, may 24-jun 10, 1972 apr 4-jun 6, aug 2, 1973 feb/mar-apr, oct-dec, 1974 may, 1975 feb – 1r – 1 – (cont by: black interaction) – mf#4798723 – us WHS [305]
Black viewpoint – 1972 dec 31, 1973 jan 31, mar 15 – 1r – 1 – mf#4867915 – us WHS [305]
Black viewpoint / ed by Biko, B S – Durban, Spro-Cas Black Community Programmes, 1972 – us CRL [305]
Black views – Columbia, Rock Hill SC. [1992 jan 16/22-apr 30/may 6]-[1996 jun 27/jul 3-sep 26/oct 2] – 14r – 1 – mf#2304829 – us WHS [321]
Black views / South Carolina Black Media Group – Rock Hill SC. 1988 dec 1/3-1989 sep 7 – 1r – 1 – mf#1663872 – us WHS [305]
Black visual arts notebook – 1992 jun 27-nov 7 – 1r – 1 – mf#4861661 – us WHS [700]
Black voice – Columbia, Orangeburg SC. [1992 jan 9/15-apr 30/may 6]-[1996 jun 27/jul 3-sep 26/oct 2] – 14r – 1 – mf#2304759 – us WHS [321]
Black voice : official organ of the côte des neiges black community development project – Montreal. v1-2 n10. may 25 1972-oct 1974// ? – 1r – 1 – Can$135.00 – cn McLaren [305]
Black voice news – Riverside CA. 1992 jan 2/jun 25-1998 jan 1/jun 25 – 13r – 1 – mf#1854149 – us WHS [305]
Black, William *see* Madcap violet
Black, William George *see* A handbook of the parochial ecclesiastical law of scotland
Black, William L *see* The rebirth of the historic old hole-in-the-rock trail as a recreational trail
Black woman in search of god / Brandel-Syrier, Mia – London, England. 1962 – 1r – us UF Libraries [305]
Black Women in Publishing, Inc *see* Interface
Black Women's Community Development Foundation *see* Binding ties
Black women's educational policy and research network newsletter – 1982 mar/apr-aug/sep – 2r – 1 – mf#4775523 – us WHS [376]
Black Workers Congress *see* Siege
Black Workers for Justice *see* Justice speaks
Black workers in the era of the great migration, 1916-1929 / ed by Grossman, James – (= ser Black studies research sources) – 25r – 1 – $4465.00 – 0-89093-740-0 – (with p/g) – us UPA [331]

Black works newsletter / Long Island University – 1973 mar 6-apr 10 – 1r – 1 – (cont: black works) – mf#4882502 – us WHS [071]

Black world – Chicago IL 1942-76 – 1,5,9 – ISSN: 0006-4319 – mf#5406 – us UMI ProQuest [305]

Black writer magazine – 1983 jun, 1986 apr – 1r – 1 – mf#4722054 – us WHS [305]

Black writers' news / International Black Writers' Conference – 1971 oct/nov, 1972 feb/mar, 1975 spring-summer – 1r – 1 – (cont: news (black writers' conference)) – mf#2882903 – us WHS [071]

Black x-press – Chicago IL. 1975 jun 30 – 1r – 1 – mf#4164341 – us WHS [071]

Blackall, Christopher Rubey *see* A story of six decades

Blackbook – 1979-96 – 1r – 1 – (cont: u s sports) – mf#2847588 – us WHS [071]

Blackburn and darwen mail – [NW England] Blackburn Lib 1982-aug 1989 – 1 – (title change: blackburn herald & post [24 aug 1989-22 aug 1991]) – uk MLA; uk Newsplan [072]

Blackburn evening express – [NW England] Blackburn Lib aug 1887-5 oct 1888 [mf ed 2003] – 1 – (title change: lancashire evening express & standard [12 oct 1888-1889]; evening express & standard [1890]; lancashire evening express [1891]; lancashire evening express & standard [1893-95]; lancashire daily express [1896]) – uk MLA; uk Newsplan [072]

Blackburn express – [NW England] Blackburn Lib 13 mar 1992-dec 1992 – 1 – uk MLA; uk Newsplan [072]

Blackburn, George *see* Poems, notes, and reports on the 1813 north carolina-south carolina boundary expedition, c1814

Blackburn, George Andrew *see*
- Discussions of philosophical questions
- Discussions of theological questions
- Life work of john l. girardeau...
- Sermons

Blackburn, Helen *see* Voices of the women's movement, 1850-1900

Blackburn, Henry *see*
- Academy notes
- The art of illustration
- English art in 1884
- Grosvenor notes
- Randolph caldecott
- [Royal academy. 1883]
- [Royal academy. 1884]
- [Royal academy. 1885]
- [Royal academy. 1886]
- [Royal academy. 1887]
- [Royal academy. 1888]
- [Royal academy. 1889]
- [Royal academy. 1890]
- [Royal academy. 1891]
- [Royal academy. 1892]

Blackburn herald – [NW England] Blackburn Lib 1930-35, 26 may 1949-5 jan 1956 – 1 – uk MLA; uk Newsplan [072]

Blackburn labour journal – [NW England] Blackburn Lib 1898-1909 – 1 – uk MLA; uk Newsplan [331]

Blackburn labour journal, 1898-1909 – 1r – 1 – mf#1375 – uk Microform Academic [072]

Blackburn mail – [NW England] Blackburn, Burnley Lib 29 may 1793-nov 1799, jul 1800-1825 – 1 – (title change: blackburn mail; and lancashire & yorkshire general advertiser [1826-29 (1829 incomplete)]) – uk MLA; uk Newsplan [072]

Blackburn researcher – v1 n1-v4 n4 [1982 sep-1985 dec] – 1r – 1 – mf#1009272 – us WHS [071]

Blackburn standard – [NW England] Blackburn 21 jan 1835-5 jun 1909 [mf ed 2002] – 55r – 1 – (missing: 1847, 1860, 1871-72, 1874, 1896-97, 1905; cont as: blackburn standard & north east lancashire advertiser [1876]; blackburn standard, darwen observer & nnorth east lancashire advertiser [jan 1877-dec 1888]; blackburn standard & weekly express [jan 1889-dec 1893]; weekly standard and express [jan 1895-jun 1909]) – uk Newsplan [072]

Blackburn standard – [NW England] Blackburn Lib 1837-38, 1840-53, 1855-59, 1861-7 – 1 – (title change: blackburn standard, darwen observer & north east lancashire advertiser [1877-88]; blackburn standard & weekly express [1890]; weekly standard & express [1894, 1897, 1899-1901]) – uk MLA; uk Newsplan [072]

Blackburn weekly telegraph – [NW England] Blackburn Lib 1900-01, 1907, 1910, 1912-13 – 1 – uk MLA; uk Newsplan [072]

Blackburn weekly times – [NW England] Blackburn Lib jun 1855-59, jan-14 sep 1861 – 1 – (title change: blackburn times [1861-64, mar 1865-dec 1865, mar 1866-1868, may 1869-dec 1869, 1871-1884, mar 1885-1903, 1906, 1909-15, 1921, 1923, oct-dec 1924, 1925, 1927-40, 1959-80*]) – uk MLA; uk Newsplan [072]

Blackburn, William Maxwell *see*
- Aonio paleario and his friends
- The college days of calvin
- History of the christian church

Blackburne, Edward Lushington *see*
- Sketches...for a history of the decorative painting applied to english architecture during the middle ages
- Suburban and rural architecture

Blacker, Irwir R *see* Cortes and the aztec conque consultent. gordon eckholm

Blacker, William *see* An essay on the improvement to be made in the cultivation of small farms by the introduction of green crops...

Blacket, John *see* Missionary triumphs among the settlers in australia and the savages of the south seas

Blacketer, Raymond Andrew *see* L'ecole de dieu

Blackett, Herbert Field *see* Two years in an indian mission

Blackfeet tribal news – v1 n1-v4 n3 [1982 jan 27-1985 mar 25] – 1r – 1 – mf#615497 – us WHS [307]

Blackfolk : journal of afro-american folklore – v1 n1-2 [1972 spring-73/1974 winter] – 1r – 1 – mf#4862520 – us WHS [390]

Blackford baptist church. hawesville, kentucky : church records – Feb 1853-Apr 1984 – 1 – 52.74 – us Southern Baptist [242]

Blackford, Dominique de *see* Precis de l'etat actuel des colonies angloises dans l'amerique septentrionale

Blackford's indiana reports – v1-8. 1817-47 – 53mf – 9 – $79.00 – mf#LLMC 84-133 – us LLMC [340]

Blackford's reports / Indiana. Supreme Court – v1-8. 1817-1847 (all publ) – (= ser Indiana Supreme Court Reports) – 53mf – 9 – $79.00 – (a pre-nrs title) – mf#LLMC 84-133 – us LLMC [347]

Blackham, Robert J *see* Wig and gown

Blackham, Robert James *see* Incomparable india

Blackhawk bulletin – Orfordville WI. 1976 feb 16-1979, 1980, 1981-1982 jan 25 – 3r – 1 – mf#959524 – us WHS [071]

Blackhawk Genealogical Society *see* Quarterly

Blackhawk talk / Parents Without Partners – 1976 jun-1978 dec – 1r – 1 – mf#618709 – us WHS [071]

Blackheath gazette, eltham, lee, and lewisham advertiser – Lewisham, London jan 1892-31 dec 1897 (daily) – 6r – 1 – uk British Libr Newspaper [072]

Blackhorse country – 1992 jan 27-1993 dec 29 – 1r – 1 – (cont: blackhorse) – mf#2365200 – us WHS [071]

Blackie, John Stuart *see*
- Four phases of morals
- Lay sermons
- The natural history of atheism
- On self-culture

Blackie, Walter Graham *see*
- Action of the free church commission ultra vires
- Observations on the report to her majesty by the commissioners...

Black-Jewish Information Center *see* Media project

Blackledge, William James *see* The legion of marching madmen

Blackletter journal *see* Harvard blackletter law journal

Blacklight – v3 n3-4 [c1982] – 1r – 1 – mf#4864112 – us WHS [071]

Blackman – Cape Town SA, 1920-21 – 1r – 1 – sa National [079]

Blackman, Ethan V *see* Miami and dade county, florida

Blackman, William Fremont *see*
- History of orange county
- Sugar and cane syrup in florida

Blackmar, Frank W *see*
- Spanish colonization in the southwest
- Spanish institutions in the southwest

Blackmill, paran baptist, monumental inscriptions – 1mf – 9 – £1.25 – uk Glamorgan FHS [929]

Blackmon, G H *see*
- Cover crop program for florida pecan orchards
- Fertilizer experiments with pecans
- Pecan growing in florida
- Pecan growing in florida and conditions suitable to maximum
- Top-working pecan trees

Black-polish conference newsletter – v2 n1-v3 n6 [1972 jan-1973 jun] – 1r – 1 – mf#1053415 – us WHS [327]

Blackpool and District Labour Representation Committee *see* Labour advocate

Blackpool and fleetwood gazette and fylde news – [NW England] Blackpool Lib 26 oct 1877-1890 – 1 – (title change: blackpool & fleetwood gazette, st anne's & fylde news [1891-92]) – uk MLA; uk Newsplan [072]

Blackpool and fylde citizen – [NW England] Blackpool Lib 18 dec 1987 – 1 – uk MLA; uk Newsplan [072]

Blackpool gazette – [NW England] Blackpool Lib 8 apr 1873-mar 1876 – 1 – (title change: blackpool gazette & fylde news 7 apr 1876-19 oct 1877; blackpool & fleetwood gazette & fylde news [26 oct 1877-1892]; gazette & news [1893-1919]) – uk MLA; uk Newsplan [072]

Blackpool herald – [NW England] Blackpool Lib 1876-12 may 1893 – 1 – (title change: blackpool herald & fylde advertiser [1915]; blackpool gazette & herald, fylde news & advertiser [1920-5 may 1972]; blackpool herald [5 may 1972-1979]; blackpool herald & times news [1980-mar 1983]; blackpool & fylde leader [mar 1983-mar 1985]) – uk MLA; uk Newsplan [072]

Blackpool labour advocate – [NW England] Blackpool Lib 1910-14 – 1 – uk MLA; uk Newsplan [072]

Blackpool times – 1 – (title change: blackpool times, st anne's times and fylde observer [1877-79, 1881, 1883-2nd oct 1920]; blackpool times [1879-96]; blackpool times, fleetwood express & st anne's visitor [9 oct 1920-1933]) – uk MLA; uk Newsplan [072]

Black-robes : or, sketches of missions and ministers in the wilderness / Nevin, Robert Peebles – Philadelphia: J B Lippincott, 1872 – 1mf – 9 – 0-8370-6232-2 – mf#1986-0232 – us ATLA [240]

Blacks and whites in south africa : an account of the past treatment and present condition of south african natives under british and boer control / Bourne, Henry Richard Fox – [2nd ed] London, 1900 – (= ser 19th c books on british colonization) – 2mf – 9 – mf#1.1.7985 – uk Chadwyck [322]

Blacks, boers, and british / Statham, Francis Reginald – New York, NY. 1969 – 1r – us UF Libraries [960]

Blacks in the railroad industry, 1946-1954 : from the holdings of the schomburg center for research in black culture, manuscripts, archives and rare books division: the new york public library, astor, lenox and tilden foundations – 1995 – 1r – 1 – $85.00 – (guide which covers all coll under "african-american organizations" sold separately for $20.00 d3305.g1) – mf#D3305P04 – Dist. us Scholarly Res – us L of C Photodup [380]

Blacks in the u.s. armed froces : basic documents, 1639-1973 / ed by Nalty, Bernard C & MacGregor, Morris J – 1994 – 5r – 1 – $425.00 – (comes with printed guide) – mf#S3304 – Dist. us Scholarly Res – us L of C Photodup [355]

Black's tourist guide to derbyshire / Adam And Charles Black (Firm) – Edinburgh, Scotland. 1864 – 1r – us UF Libraries [914]

Blacksburg first baptist church. cherokee county. south carolina : church records – 1900-1929, 1934-1970. History. 1889 – 1 – us Southern Baptist [242]

Blackshirt *see* Fascist week

Blackshirt, feb 1933-may 1939 – 3r – 1 – mf#97592 – uk Microform Academic [320]

Blackside, Inc *see* Eyes

Blacksmith's account book, 1842-1844 / Mitchell, Caleb – 1v on 1mf – 9 – mf#50/218 – us South Carolina Historical [680]

Blacksmiths, drop forgers and helpers journal / International Brotherhood of Blacksmiths, Drop Forgers, and Helpers – 1901-29 – (= ser Labor union periodicals, pt 1: the metal trades) – 6r – 1 – $1260.00 – 1-55655-229-7 – us UPA [680]

The blacksmiths guide : valuable instructions on forging, welding, hardening, tempering, casehardening, annealing, coloring, brazing, and general blacksmithing / Sallows, J F – 1st ed. Brattleboro, VT: The Technical Press 1907 – us CRL [670]

Blacksmiths journal : official organ / International Brotherhood of Blacksmiths and Helpers – 1901 mar-dec, 1903 feb, may-1908 oct, v9 n11-v15 n9 [1908 nov-1913 sep], v15 n10-v20 n6,8,10-1912 [1913 oct-1918 jun, aug, oct-dec], v21 n1-3,5-v22 n7 [1919 jan-mar, may-20 jul] – 1r – 1 – mf#1423319 – us WHS [680]

Blackstage / Black Theater Alliance – 1981 oct/nov, 1982 mar/apr-may/jun, 1983jun, 1984 mar/apr, 1989 spring – 1r – 1 – mf#4992548 – us WHS [790]

Blackstone 1844-1903 – Oxford, MA (mf ed 1992) – (= ser Massachusetts vital records) – 32mf – 9 – 0-87623-162-8 – (mf 1-4: military 1861-65. mf 5-7: births 1844-60. mf 8: marriages 1845-55. mf 8-9: deaths 1845-57. mf 10-14: births 1861-82. mf 15-19: marriages 1856-94. mf 20-24: deaths 1858-93. mf 25-29: births 1883-1903. mf 30: marriages 1895-1904. mf 31-32: deaths 1898-1903) – us Archive [978]

Blackstone, Frederick Elliot *see* Explanation of the system of the catalogue

Blackstone, Henry *see* H blackstone's reports

Blackstone Institute *see* Modern american law

Blackstone quizzer b. / Ellis, Griffith Ogden – 2d ed. Detroit, The Collector, 1895. 46 p. LL-869 – 1 – us L of C Photodup [340]

Blackstone, William *see*
- An analysis of the laws of england
- Blackstone's commentaries.
- Commentaries on the laws of england
- Commentaries on the laws of england applicable to real property
- Commentaries on the laws of england; in four books
- Reports of cases determined in the several courts of westminster-hall, from 1746-1779
- The student's blackstone

Blackstone, William E *see*
- The heart of the jewish problem
- How shall we know him?
- Jesus is coming
- Our god and his universe
- Satan, his kingdom and its overthrow
- Signs of the lord's coming
- The times of the gentiles

Blackstone's commentaries. / Blackstone, William – Philadelphia, Birch and Small, 1803. 5 v. LL-904 – 1 – us L of C Photodup [340]

Blackstone's commentaries on the laws of england : (american abridgements and extracts) – (= ser The yale law library blackstone coll) – 197mf – 9 – mf#LLMC 82-800 titles 137-178 – us LLMC [343]

Blackstone's commentaries on the laws of england : (english abridgements and extracts) – (= ser The yale law library blackstone coll) – 142mf – 9 – mf#LLMC 82-800 titles 46-79 – us LLMC [343]

Black-town – 1st iss [1997 dec?] – 1r – 1 – mf#4150775 – us WHS [071]

Blacktown advocate – Blacktown, jan 1991-jun 1993 – 5r – at Pascoe [079]

Blacktown advocate – Blacktown, may 1949-dec 1976 – 22r – A$1518.53 vesicular A$1639.53 silver – at Pascoe [079]

Blacktown district post – Blacktown, oct 1959-feb 1965 – 3r – A$192.37 vesicular A$208.87 silver – at Pascoe [079]

Blackview – 1984 jun – 1r – 1 – mf#4862630 – us WHS [071]

Blackwatch : a publication of the black student union of ucsb / Black Student Union (University of California, Santa Barbara) – [1991 feb?] – 1r – 1 – mf#4866809 – us WHS [378]

Blackwell, Antoinette Louisa Brown *see* Exegesis of 1 corinthians 14., 34,35

Blackwell, E B *see* An inventory for assessment of attitudes of high school students toward health-realted physical fitness

Blackwell, Elizabeth *see* The laws of life with special reference to the physical education of girls

Blackwell family *see* Papers

Blackwell, G F *see* Early adolescents' knowledge of and attitudes toward hiv and aids

Blackwell, Michael C *see* The ethical thought of carlyle marney

Blackwell newsletter – v1 n1/2-v7 [1979 jun-1985] – 1r – 1 – mf#1218245 – us WHS [071]

Blackwell, Ryan *see* Effect of vitamin e supplementation on delayed-onset muscle soreness

Blackwell, Thomas Evans *see* Repor...of the grand trunk railway company of canada for the year 1859

Blackwood, Andrew Watterson *see* The prophets

Blackwood & risca news – Blackwood, Wales 4-25 feb 1993 – 1 – (cont by: blackwood & risca news & weekly argus [4 mar 1993-27 aug 1998]) – uk British Libr Newspaper [072]

Blackwood's magazine – Edinburgh UK 1817-1980 – 1,5,9 – ISSN: 0006-436X – mf#515 – us UMI ProQuest [073]

En bladartikel / Kierkegaard, Soeren – Kobenhavn: C A Reitzel, 1859 [mf ed 1990] – 14p on 1mf – 9 – 0-7905-3787-7 – (in danish) – mf#1989-0280 – us ATLA [780]

Blade – Elkhorn WI. 1891 apr 17-1893, 1894-97, 1898-1901, 1902-1905 nov 28 – 4r – 1 – mf#962639 – us WHS [071]

Blade / Hamilton Co. Elmwood Place – (1928-sep 1929), mar 1931 [wkly] – 1r – 1 – mf#B12025 – us Ohio Hist [071]

Blade / Lucas Co. Toledo – v1 n1. dec 19, 1835, may 1837-1945 [daily, wkly, twice wkly, daily] – 394r – 1 – mf#B3507-3904 – us Ohio Hist [071]

Blade / Scioto Co. Portsmouth – (1890-1900, 1903, 1909-12) [wkly, semiwkly] – 11r – 1 – mf#B11172-11181 – us Ohio Hist [071]

Blade among the boys / Nzekwu, Onuora – London, England. 1964 – 1r – us UF Libraries [960]

Blade-atlas – Blanchardville WI. [1968 may 23-1969 apr 24]-[2000 jan 6-jun 1] – 21r – 1 – (cont: argyle atlas, blanchardville blade; cont by: argyle agenda (argyle wi: 1979), pecatonica valley leader) – mf#1008202 – us WHS [071]

Bladen enterprise – Bladen, NE: L E Spence. -v43 n12. mar 6 1936 (wkly) [mf ed v2 [n45 oct 24 1895]-mar 6 1936 (gaps)] – 10r – 1 – (absorbed by: blue hill leader) – us NE Hist [071]

Bladen union baptist church. fayetteville, north carolina : church records – 13 May 1956-10 Oct 1965 – 1 – us Southern Baptist [242]

Bladen union baptist church. fayetteville, north carolina : church records – 1859-1965 – 1 – us Southern Baptist [242]
Blades, Rowland H *see* Who was caxton?
Blades, William *see*
- How to tell a caxton
- A list of medals, jettons, tokens etc in connection with printers and the art of printing
- Numismata typographica
- The pentateuch of printing
- Proposals made by rev james kirkwood, (minister of minto) in 1699
- Shakspere and typography
- Some early type specimen books of england, holland, france, italy, and germany

Blaenavon express – [Wales] LLGC 7 oct 1871-22 feb 1873 [mf ed 2004] – 1r – 1 – (cont by: blaenavon & brynmawr express) – uk Newsplan [072]
Blaengwrach, st mary, monumental inscriptions – 1mf – 9 – £1.25 – uk Glamorgan FHS [929]
Blaeser, Edwin *see* Asymmetrische 1,3-dipolare cycloadditionen und hetero-diels-alder-reaktionen unter verwendung von (s)-prolinestern als chirale auxiliare
Blaeserstueck / Dopper, Cornelis – [s.l: s.n. 19–?] [mf ed 1988] – 1r – 1 – mf#pres. film 30 – us Sibley [780]
Blaetter aus dem werther-kreis / ed by Wolff, Eugen – Breslau: S Schottlaender 1894 [mf ed 1990] – 1r – 1 – (filmed with: goethes romische elegien / albert leitzmann) – mf#7371 – us Wisconsin U Libr [430]
Blaetter der erinnerung : meistens um und aus der paulskirche in frankfurt / Arndt, Ernst Moritz – Leipzig: Weidmann, 1849 [mf ed 1993] – 75p – 1 – mf#8459 – us Wisconsin U Libr [810]
Blaetter der juedischen buchvereinigung – Frankfurt/M DE, 1934-36 – 1 – gw Misc Inst [939]
Blaetter der platen-gesellschaft / ed by Praesidium und vom wissenschaftlichen Ausschuss der Platen-Gesellschaft e.V. – Berlin: Platen-Gesellschaft 1925-26 [mf ed 1992] – 1r [ill] – 1 – (incl bibl ref. filmed with: ausgewahlte fabeln und gedichte / gottlieb konrad pfeffel) – mf#2863p – us Wisconsin U Libr [430]
Blaetter der zeit – Braunschweig DE, 1848 2 oct-1855 25 mar – 2r – 1 – gw Misc Inst [074]
Blaetter des deutschen roten kreuzes – [1] 1922-[16] 1937 [mf ed 2005] – (= ser Freie wohlfahrtspflege 8) – 114mf – 9 – €750.00 – 3-89131-471-X – gw Fischer [360]
Blaetter des juedischen frauenbundes – Berlin DE, 1924 4 jul-1938 – 1 – gw Misc Inst [939]
Blaetter des nationalen vereins fuer deutschland – Kassel DE, 1849 5 feb-30 jun – 1r – 1 – gw Misc Inst [366]
Blaetter des operntheaters – Vienna 1919-20. 1 reel – 1 – us L of C Photodup [780]
Blaetter fuer alle – Berlin DE, 1927-1933 mar – 3r – 1 – gw Misc Inst [074]
Blaetter fuer das bayerische gymnasialschulwesen – Muenchen: Lindauer'sche Buchhandlung 1881-92 [mf ed 1978] – 11v on 17r – 1 – (cont: blaetter fuer das bayerische gymnasial- und real-schulwesen; ceased in 1891) – mf#film mas c396 – us Harvard [373]
Blaetter fuer das gymnasial-schulwesen – Muenchen: J Lindauer'sche Buchhandlung 1892-1916 [mf ed 1978] – 30v on 17r – 1 – (cont: blaetter fuer das bayerische gymnasialschulwesen (munich, germany); cont by: bayerische blaetter fuer das gymnasial-schulwesen; iss by: bayerischer gymnasiallehrerverein) – mf#film mas c396 – us Harvard [373]
Blaetter fuer demographie, statistik und wirtschaftskunde der juden – Berlin DE, 1923 feb-1925 jun – 1r – 1 – gw Misc Inst [939]
Blaetter fuer demographie, statistik und wirtschaftskunde der juden / ed by Brutzkus, Boris et al – Berlin. n1-5. 1923-25 [complete] – (= ser German-jewish periodicals...1768-1945, pt 1) – 1r – 1 – $125.00 – (in yiddish) – mf#B33 – us UPA [939]
Blaetter fuer den haeuslichen kreis – Stuttgart DE, 1872 – 1r – 1 – gw Misc Inst [640]
Blaetter fuer die juedische frau – Prag (CZ), 1932-36 [gaps] – 1 – gw Misc Inst [939]
Blaetter fuer die kunst : eine auslese aus den jahren 1892-1909 – Berlin, 1899-1909. 3v – 12mf – 8 – mf#H-442 – ne IDC [700]
Blaetter fuer die leipziger wohlfahrtspflege – Leipzig DE, 1924-28 – 1r – 1 – gw Misc Inst [366]
Blaetter fuer genossenschaftswesen. (innung der zukunft.) – v. 20-88. 1873-1941 – 1 – 581.00 – us L of C Photodup [331]
Blaetter fuer heimatliche geschichte *see* Zittauer stimmen
Blaetter fuer heimatpflege im kreise buetow *see* Buetower anzeiger
Blaetter fuer juedische geschichte und literatur – Mainz: Leopold Loewenstein. v1-5. 1899-1904 – (= ser German-jewish periodicals...1768-1945, pt 2) – 1r – 1 – $115.00 – (lacking: p177-184, dec 1902) – mf#B34 – us UPA [939]
Blaetter fuer juedische geschichte und literatur – Mainz DE, 1899-1900 n9, 1901 n1-8, 1902-1904 n9 – 1r – 1 – gw Misc Inst [939]
Blaetter fuer leipziger wohlfahrtspflege – Leipzig DE, 1924-28 – 1r – 1 – gw Misc Inst [366]
Blaetter fuer literarische unterhaltung – Leipzig: F A Brockhaus [1863- [n1-52 1863-98] (wkly) [mf ed 1979] – 24r – 1 – (cont: literarisches conversationsblatt; no more publ after 1898) – mf#film mas c628 – us Harvard [410]
Blaetter fuer literarische unterhaltung (klp17) / ed by Brockhaus, Heinrich et al – Leipzig: Friedrich Arnold Brockhaus 1826 jul-1898 [mf ed 2004] – (= ser Marbacher mikrofiche-editionen (mme) 17; Kultur – literatur – politik: deutsche zeitschriften des 19./20. jahrhunderts (klp)) – 955mf – 9 – €5900.00 – 3-89131-454-X – (with: literarischer anzeiger) – gw Fischer [430]
Blaetter fuer menschenrecht – Berlin DE, 1924 – 1 – (filmed by other misc inst: 1923 n1-1929 n9 (many gaps) [1r]; title varies: 1928 n12: menschenrecht) – gw Misc Inst [322]
Blaetter fuer polizei und kultur / ed by Niemann, A C H – Tuebingen 1801-03 – (= ser Dz. historisch-geographische abt) – 3jge[zu je 2bdn] on 25mf – 9 – €250.00 – mf#k/n1330 – gw Olms [360]
Blaetter fuer pommersche volkskunde – Stettin: J Burmeister. jahrg1-10 [okt 1892-sep 1902] [mf ed 198-] – 10v on 1r – 1 – mf#film mas 8368 – us Harvard [390]
Blaetter fuer scherz und ernst *see* Düsseldorfer zeitung 1814
Blaetter fuer sudetendeutsche sozialdemokraten – Malmo (s), 1944 feb-dec, 1945 jun, jul, oct, dec, 1946 feb & apr-jun, 1946 dec-1949 – 1r – 1 – (aka: may-nov 1944: mitteilungsblaetter fuer sudetendeutsche sozialdemokraten) – gw Misc Inst [325]
Blaetter fuer volksliteratur – v1-8. 1962-69 – 1 – us AMS Press [430]
Blaetter vermischten inhalts / ed by Gramberg, Gerhard Anton – Oldenburg 1787-97 [mf ed Hildesheim 1992-98] – (= ser Dz) – 6v on 24mf – 9 – €240.00 – mf#k/n5752 – gw Olms [074]
Blaetter vermischten inhalts – Oldenburg/Oldbg DE, 1787-88 (mpf), 1790-92 (mpf), 1797 (mpf) – 9 – gw Misc Inst [074]
Blaeu, J *see* Le theatre du monde
Blagdon, Francis *see* Paris as it was and as it is
Blagoi, D D *see* Istoriia russkoi literatury 18 veka
Blagoi, Dmitrii Dmitrievich *see* Istoriia russkoi literatury 18
Blagoveshchenskaia, M P *see* Evrei
Blagoveshchenskii *see* Izdanie olonetskogo gubernskogo statisticheskogo komiteta
Blagoveshchenskii, M *see* Kniga plach
Blagovestnik / ed by Fetler, Robert – Vladivostok. 1920-22. Incomplete – 1 – $124.80 – (publ. no. 3481-7d. one part of a four-part item) – us Southern Baptist [242]
Blagrove, George H *see* Marble decoration
Blaheta, Roman *see* Aussenohrcharakteristik bei verschiedenen nagern
Blaikie, Alexander *see*
- A history of presbyterianism in new england
- The philosophy of sectarianism

Blaikie, Francis [comps] *see* Holkham office cash accounts, 1808-1844
Blaikie, William Garden *see*
- After fifty years
- Better days for working people
- The book of joshua
- The colleges and theological institutions of america
- For the work of the ministry
- How to get strong and how to stay so
- A manual of bible history
- The preachers of scotland
- The public ministry and pastoral methods of our lord
- Questions on dr blaikie's bible history
- Right aim and spirit of the free church
- The second book of samuel
- Thomas chalmers
- William garden blaikie

Blaikie, William Garden et al *see* Is christianity true?
Blaine county beacon – Brewster, NE: Mr & Mrs F J Wengrzyn. 4v. 87th yr n12. dec 1 1971-90th yr n18. jan 8 1975 (wkly) [mf ed 1975] – 2r – 1 – (cont: nebraska beef producer) – us NE Hist [071]
Blaine county booster – Dunning, NE: Dopf Bros, 1909 (wkly) [mf ed n32. jun 25 1914-1955 (gaps) filmed 1972?] – 13r – 1 – (vol numbering begins with v6 n1 nov 19 1914. some irregularities in numbering) – us NE Hist [071]
Blaine, Ephraim *see* Papers of ephraim blaine
Blaine, James G *see* Papers
Blaine, Robert Gordon *see* Hydraulic machinery
Blainville, Charles Henri *see* Histoire generale, critique et philologique de la musique
Blair, A W *see*
- Composition of some of the concentrated feeding stuffs on sale in florida
- Pineapple culture vi
- Pineapple culture vii
- Soil studies 1
- Soil studies 2

Blair, Charles *see* Indian famines
Blair, Clay *see* Atomic submarine and admiral rickover
Blair courier – Blair, NE: Maynard and Hamilton. 19v. v1 n1. july 6 1889-v19 n11. aug 21 1907 (wkly) [mf ed with gaps] – 6r – 1 – (absorbed: herman cyclone. merged with: blair democrat to form: blair democrat and the blair courier) – us NE Hist [071]
Blair democrat – Blair, NE: Thos Osterman. v43 n26. dec 5 1912-v48 n23. nov 8 1917 (wkly) [mf ed with gaps)] – 6v on 2r – 1 – (cont: blair democrat and the blair courier. merged with: tribune to form: the tribune and blair democrat) – us NE Hist [071]
Blair democrat – Blair, NE: Thos T Osterman. 3v. v35 n37. mar 2 1905-v38 n10. aug 22 1907 (wkly) [mf ed with gaps] – 2r – 1 – (cont: blair republican. merged with: blair courier to form: blair democrat and the blair courier) – us NE Hist [071]
Blair democrat and the blair courier – Blair, NE: Thos Osterman. 6v. v38 n11. aug 29 1907-v43 n25. nov 28 1912 (wkly) – 3r – 1 – (formed by the union of: blair democrat and: blair courier. cont by: blair democrat (1912)) – us NE Hist [071]
Blair, emily n, family papers, ms 4342 – 1785-1972 – 17r – 1 – (correspondence, publ and unpubl writings, diaries and speeches, and family memorabilia of blair, a democratic party activist in the franklin roosevelt administration, and histories of the blair, newell, and mcdowell families) – us Western Res [320]
Blair, Emily Newell *see* Emily newell blair family papers, 1785-1972
Blair, Emma Helen *see* The philippine islands, 1493-1898
The blair family papers – 47r – 1 – $1,645.00 – Dist. us Scholarly Res – us L of C Photodup [975]
Blair, G W *see* Station and camp life in the bheel country
Blair & ketchum's country journal – Harrisburg PA 1974-86 – 1,5,9 – (cont by: country journal) – ISSN: 0094-0526 – mf#10275.01 – us UMI ProQuest [071]
The blair pilot – Blair, NE: Don C VanDeusen, jul 6 1927-v 57 n51. jan 30 1929 (wkly) [mf ed v56 n22. jul 13 1927-jan 30 1929] – 1r – 1 – (cont: pilot. merged with: tribune (1917) to form: pilot-tribune) – us NE Hist [071]
Blair press – Blair, Ettrick WI. 1908 oct 29-2002 – 54r – 1 – (with gaps; cont: ettrick advance) – mf#1004802 – us WHS [071]
[Blair-] press – NV. 1908-10 [wkly] – 1r – 1 – $60.00 – mf#U04421 – us Library Micro [071]
Blair republican – Blair, NE: C B Sprague, 1880-v35 n36. feb 23 1905 (wkly) [mf edv13 n[blank] jul 20 1882-feb 23 1905 (gaps)] – 3r – 1 – (cont: blair times; cont by: blair democrat) – us NE Hist [071]
Blair times – Blair, NE: V G Lantry, 1870-80// (wkly) [mf ed v3 n9. sep 26 1872-jun 12 1879 (gaps)] – 1r – 1 – (cont by: blair republican) – us NE Hist [071]
Blair, William *see* The united presbyterian church
Blair, William A *see* Journal of religion in disability and rehabilitation
Blair, William Newton *see* The korea pentecost
Blairgowrie advertiser and coupar-angus and alyth journal – [Scotland] Perthshire, Blairgowrie: D Christie jan 1858-dec 1859, jan 1936-dec 1950 (wkly) [mf ed 2003] – 16r – 1 – (cont by: advertiser for blairgowrie, rattray, coupar-angus, alyth, kirriemuir, strathmore and stormont) – uk Newsplan [072]
Blairgowrie advertiser and perth and forfar agriculturist – Blairgowrie: A Allen 1879-87 (wkly) – 398v on 9r – 1 – (cont: blairgowrie advertiser and strathmore and stormont news; cont by: blairgowrie advertiser (blairgowrie, scotland)) – uk Scotland NatLib [072]
Blairgowrie advertiser and strathmore and stormont news – Blairgowrie: D Christie 1861-79 (wkly) – 949v on 10r – 1 – (cont: advertiser for blairgowrie, rattray, coupar-angus, alyth, kirriemuir, strathmore and stormont; cont by: blairgowrie advertiser and perth and forfar agriculturist) – uk Scotland NatLib [072]
Blairgowrie advertiser (perth, scotland) – Perth: Scottish & Universal Newspapers Ltd 1981- (wkly) [mf ed 4 jan 1996-] – 1 – (cont: blairgowrie advertiser, east perthshire gazette & guardian) – ISSN: 0961-3803 – uk Scotland NatLib [072]
Blairgowrie news and alyth and couper-angus advertiser – [Scotland] Perthshire, Blairgowrie: Larg and Keir 25 may 1877, 22,29 mar 1878, & 29 mar 1878 & 6 apr 1878 (wkly) [mf ed 2004] – 2r – 1 – uk Newsplan [072]
Blairsville enterprise – Blairsville, PA. -w 1889-1912 – 13 – $25.00r – us IMR [071]
Blairsville. Presbytery (Pres. Church in the USA) *see* Minutes, 1830-1920
Blairsville record – Blairsville, PA., 1830 – 13 – $25.00r – us IMR [071]
Blais, Isidore *see* La gaspesie, la suisse canadienne
Blaisdell, James Joshua *see* The edgerton bible case
Blaise cendrars no brasil e os modernistas / Amaral, Aracy A – Sao Paulo, Brazil. 1970 – 1r – us UF Libraries [972]
Blaise pascal / Noel, Horace – London, England. 18– – 1r – us UF Libraries [025]
Blake, Buchanan *see*
- Joseph and moses, the founders of israel
- The problem of human suffering

The blake demonstration at the pavilion, horticultural gardens, monday evening september 19th, 1892 : full report of the proceedings – S.l: s.n, 1892? – 1mf – 9 – mf#38297 – cn Canadiana [320]
Blake, Edward *see*
- Address at the convocation of the university of toronto
- Address delivered in boston music hall, wednesday evening, january 31, 1894
- Canadian pacific resolutions
- Discours prononce par l'honorable m edward blake, m p
- The irish question
- "A national sentiment!"
- The situation
- Speech of hon mr blake, mp on the canadian pacific railway resolutions
- Speech of the hon mr blake on the address delivered in the house of commons, january 18, 1884

Blake, Euphemia Vale *see* Arctic experiences
Blake, Freeman N *see* Banquet to the hon f n blake, american consul
Blake, George *see* Barrie and the kailyard school
Blake, James Vila *see* Natural religion in sermons
Blake, Joaquin *see* La batalla de la albuera
Blake, John William *see* European beginnings in west africa, 1454-1578
Blake, Lillie Devereux *see* Woman's place to-day
Blake memorial baptist church. lake helen, florida : church records – 1891-Sep 1951 – 1 – us Southern Baptist [242]
Blake, Nancy E *see* The relationship between lactate and ventilatory thresholds in men with spinal cord injury
Blake, Sallie E *see* Tallahassee of yesterday
Blake, Samuel Hume *see* The knife of the higher critic / the judgment of the lord / the burial of an ass
Blake, Silas Leroy *see*
- The book
- The separates, or, strict congregationalists of new england

Blake studies – Memphis TN 1968-80 – 1,5,9 – ISSN: 0006-4548 – mf#3253 – us UMI ProQuest [420]
Blake, Thaddeus C *see* The old log house
Blake, Virgil *see* Public and access services quarterly
Blake, Warrenne [comp] *see* An irish beauty of the regency
Blake, Wilfrid Theodore *see*
- Central african survey
- Rhodesia and nyasaland journey

Blake, William *see*
- Engravings from the small and large books of designs
- Illustrations to blair's the grave
- Illustrations to bunyan's pilgrim's progress
- Illustrations to edward young's night thoughts, from british museum, dept. of prints and drawings
- Illustrations to edward young's "night thoughts" from sir john soane's museum, london
- Poems and prophecies
- Songs of experience
- Songs of innocence

Blake, william, milton (copy a) : from the british museum – 1r – 14 – mf#C545 – uk Microform Academic [810]
Blake-Hedges, Florence Edythe *see* The story of the catacombs
Blakeley sun : and alabama advertiser – Blakeley AL. v1 n43 [1819 may 7] – 1r – 1 – mf#851686 – us WHS [071]
Blakely baptist church. early county. georgia : church records – 1837-94 – 1 – us Southern Baptist [242]
Blakely, John *see* Golden vials full of odours
Blakeney, Richard Paul *see*
- Convocation
- Doctrine of reception
- Romanism, tridentine and vatican, refuted

Blakeslee, George Hubbard *see* Latin america
Blakesley, Joseph William *see* Seminaries of sound learning and religious education

Blakey, Robert *see*
– History of the philosophy of mind
– Lives of the primitive fathers
Blakiston, J F *see* The jami masjid at badaun
Blakiston, John *see* Twelve years' military adventure in three quarters of the globe
Blakney, Charles Philip *see*
– On 'banana' and 'iron'
Blakney, Raymond B *see* Yeh-su te sheng ping chiao hsun (ccm318)
Blampignon, Emile *see* De sancto cypriano et de primaeva carthaginiensi ecclesia
Blanc, Charles *see*
– Art in ornament and dress
– The history of the painters of all nations
Blanc, Claudette le *see* Bio-bibliographie de mademoiselle louise marchand
Blanc, H *see* A narrative of captivity in abyssinia
Blanc, Jean-Bernard Le *see* Le patriote anglois
Blanc, Joseph *see*
– L'agneau de dieu
– Taumua lelei
Blanc, Jules *see* Les martyrs d'aubenas
Blanc, Louis *see*
– Histoire de la revolution francaise
– Revolution francaise
Blanc, Ludwig G *see* Dr l g blanc's handbuch des wissenswuerdigsten aus der natur und geschichte der erde und ihrer bewohner
Blanca de borbon / Espronceda, Jose de; ed by Churman, Philip H – Extrait de la Revue Hispanique. Tome 8. New York, Paris, 1907 – 1 – sp Bibl Santa Ana [946]
Blancas y negras (cronicas) / Montoto de Sedas Castor – Madrid: Razon y Fe, 1927 – 1 – sp Bibl Santa Ana [305]
Blanch e Illa, Narciso *see* Cronica de la provincia de albacete
Blanchard, Albert Claude *see* Some aspects of mortality in florida, 1921-1930
Blanchard, Charles Albert *see* Modern secret societies
Blanchard, Emile *see* The transformations (or metamorphoses) of insects
Blanchard, Etienne *see* Vocabulaire bilingue par l'image
Blanchard, F M *see* Practical public speaking
Blanchard, Henri *see* Camille uesmoulins
Blanchard, Jon David *see* Florida conservation lands, 1998
Blanchard, Samuel Laman *see* George cruikshank's omnibus
Blanchard, William Gregg *see* What do you relly know of florida's petroleum poss...
Blanchardville blade – Blanchardville WI. [1896 jan 10-1897 sep 17]-[1967 jan 5-1968 may 16] – 40r – 1 – (cont by: argyle atlas, blade-atlas) – mf#1008189 – us WHS [071]
Blanche, Lenis *see* Histoire de la guadeloupe
Blanche parmi les noirs / Wannijn, Jeane – Leau, Belgium. 1939 – 1r – us UF Libraries [960]
Blanchet, Emilio *see* Historia y fantasia
Blanchet, Francois Xavier *see* Appel au parlement imperial et aux habitants des colonies angloises, dans l'amerique du nord
Blanchet, Jean Gervais Protais *see* Speech by the honourable j blanchet secretary of the province of quebec
Blanchet, Joseph *see* L'art
Blanchet, Jules *see*
– Destin de la jeune litterature
– Essais sur la culture
– Mission de l'institution communale
– Peint par lui-meme
– Politique etrangere et representation extertieure
Blanchet, Louis-Joseph-Napoleon *see* Une vie illustree de calixa lavallee
Blanchet Y Bitton, Emilio *see* Libro de las expiaciones
Blanchini, J *see* Opera omnia
Blanchon, Pierre *see* Jean guiton et le siege de la rochelle
Blanck, Karl *see* Der franzoesische einfluss im zweiten teil von gottscheds critischer dichtkunst
Blanck, Karl [comp] *see* Heine und die frau
Blanck Y Menocal, Guillermo De *see*
– Gotas de sangre
– Relaciones chino-cubanas y el tratado en estudio
– Reportaje
Blanckenburg, Curt *see* Studien ueber die sprache abrahams a santa clara
Blanckmeister, Franz *see* Goethe und die kirche seiner zeit
Blanco / Dario, Ruben – Paris, France. 1911 – 1r – us UF Libraries [972]
Blanco, Andres Eloy *see*
– Arbol de la noche alegre
– Poda
– Teatro
– Vargas, albacea de la angustia
Blanco, Antonio *see* Mosaicos romanos de merida. investigacion y ciencia
Blanco, Eduardo *see* Venezuela heroica
Blanco Fombona, Horacio *see* Crimenes del imperialismo norteamericano
Blanco Furniel, Armando *see* Mundo inoportuno
Blanco Garcia, Francisco *see*
– Adelardo lopez de ayala
– Antonio hurtado
– Bartolome jose gallardo...
– Carolina coronado
– Donoso cortes
– Espronceda
– Fernando de gabriel
– Gabino tejado
– Jose moreno nieto
– Jose sanchez arjona
– Juan justiniano y arribas
– Leandro herrero
– Nicolas diaz perez
– Vicente barrantes
Blanco, Indalecio *see* Coleccion...mejores poetas espanoles
Blanco, Jose Felix *see* Documentos para la historia de la vida publica del libertador de colombia, peru, y bolivia
Blanco, Jose Maria *see* Historia documentada de la vida y gloriosa muerte de los padres roque gonzalez de santa cruz, alonso rodriguez y juan del castillo...buenos aires, 1929
Blanco Sanchez, R *see* Para la historia del monasterio de guadalupe
Blanco Soto, P *see* Petri compostellani de consolatione rationis libri duo
Blanco, Tomas *see*
– Cinco sentidos
– Dragontea
– Prontuario historico de puerto rico
Blanco y negro : revista ilustrada. – Madrid. Ano 1-46. 1891-1936. (incomplete) – 1 – us NY Public [305]
Blanco-Fombona, Rufino *see*
– Hombre de hierro
– Letras y letrados de hispano-america
Blancs et noirs / Reboux, Paul – Paris, France. 1915 – 1r – us UF Libraries [305]
Bland, Desmond S *see* Early records of furnival's inn
Bland, John Otway Percy *see* Annals and memoirs of the court of peking
Blandford 1737-1860 – Oxford, MA (mf ed 1983) – (= ser Massachusetts vital records) – 51mf – 9 – 0-931248-36-1 – (mf 1-4: index to births 1741-1982. mf 5-10: index to marriages & deaths 1741-1842. mf 11-19: index to marriages & deaths 1843-1982. mf 20-22: miscellaneous records 1737-73. mf 23-29: miscellaneous records 1774-1802. mf 30-37: records 1803-34. mf 38-47: records 1835-60. mf 48-51: cemetery records) – us Archive [978]
Blandford, H [comp] *see* Catalogue of the archives of the moravian church, bristol
Blandon Berrio, Fidel *see* Que el cielo no perdona
Bland's chancery reports – v1-3. 1811-32 (all publ) – (= ser Maryland supreme court reports) – 9mf – 9 – $40.50 – mf#LLMC 84-147 – us LLMC [347]
Blane, William N *see* An excursion through the united states and canada during the years 1822-23
Blanes, Nilo *see* Desde colon a fidel
Blaney baptist church. elgin, south carolina : church records – 1905-75 – 1 – 8.24 – us Southern Baptist [242]
Blaney, Henry Robertson *see* Golden caribbean
Blank canvas : sargent johnson gallery quarterly / Western Addition Cultural Center – v1 n1 [1989 spring] – 1r – 1 – mf#394113 – us WHS [700]
Blank, Sally E *see* Versaclimber exercise elicits higher maximal oxygen uptake in women rowers than does treadmill exercise or rowing ergometry
Blankenburg, Quirinus van *see* Elementa musica
Blankenburg, Roland *see* Einfluesse impliziter eignungstheorien auf die beobachtungsgenauigkeit in assessment-centern
Blankenstein, Marcos Van *see* Suriname
Blanpied, Peter R *see* The comparison of active plantarflexor muscle stiffness between young and elderly human females
Blanqui, Adolphe *see*
– Voyage a madrid
– Voyage d'un jeune francais en angleterre et en ecosse, pendant l'automne de 1823
Blanton, Franklin S *see* Sand flies of florida
Blanton, Kelsey *see*
– Architecture
– Highlands county
– Highlands hammock
– Land reclamation at lakeland
– Sebring
Blaquiere, Edward *see*
– Briefe aus dem mittellaendischen meere
– Letters from greece
– Letters from the mediterranean
Blarney : ulster's only comic journal – Belfast, Ireland 6 mar-9 oct 1886 – 1/2r – 1 – uk British Libr Newspaper [072]
Blas de ledesma : un pintor recien descubierto / Torres Martin, Ramon – Badajoz: Imprenta Diputacion Prov., 1967 – sp Bibl Santa Ana [946]
Blaser, R E *see*
– Effect of fertilizer on growth and composition of carpet and other grasses
– Preliminary pasture clover studies
– Winter clover pastures for peninsular florida
Blashfield, De Witt Clinton *see* A treatise on instructions to juries in civil and criminal cases.
Blashfield, John Marriott *see*
– An account of the history and manufacture of...terra cotta
– A selection of vases, statues, busts, etc from terra-cotta
– Terra cotta chimney shafts, chimney pots, etc
Blasius, Johann *see* Reise im europaeischen russland in den jahren 1840 und 1841
Blasius, Richard *see* Appell
Blason de almas / Reyes Huertas, Antonio – Madrid: Editorial Paez, 1926 – 1 – sp Bibl Santa Ana [946]
Blason populaire de villedieu-les-poeles, arrondissement d'avranches (manche) / Brunet, Victor Armand – Sourdinopolis: L' Enclume 1888 [mf ed Bloomington IN: Indiana Uni Lib, Preservation Dept 1984] – 80p on 1r – 1 – us Indiana Preservation [390]
Blasphemous titles of the pope – Edinburgh, Scotland. 18– – 1r – us UF Libraries [240]
Blasphemy against the holy ghost / Miller, Samuel – Glasgow, Scotland. 1845 – 1r – us UF Libraries [240]
Blasphemy of the holy spirit / Reid, William – Edinburgh, Scotland. 18– – 1r – us UF Libraries [240]
Blasquez y Delgado-Aquilera, Antonio *see* Prehistoria de la region norte de marruecos
Blass, Fridericus *see* Acta apostolorum
Blass, Friedrich *see*
– [Barnabas] brief an die hebraeer
– Euangelium secundum iohannem
– Evangelium secundum matthaeum
– Religionsgeschichtliche parallelen zum alten testament – textkritische bemerkungen zu markus
Blass, Friedrich Wilhelm *see*
– Acta apostolorum
– Die entstehung und der charakter unserer evangelien
– Grammar of new testament greek
– Philology of the gospels
– Professor harnack und die schriften des lukas
– Textkritisches zu den korintherbriefen
Eine blassblaue frauenschrift / Werfel, Franz – Buenos Aires: Editorial Estrellas, c1941 – 1r – 1 – us Wisconsin U Libr [430]
Blast – London. v1-2. june 20 1914-july 1915 – 1 – us NY Public [073]
Blast – v1-2 n2,5. 1916-17 [all publ] – (= ser Radical periodicals in the united states, 1881-1960. series 1) – 3mf – 9 – $85.00 – us UPA [303]
Blast; a magazine of proletarian short stories – New York. v. 1 no. 1-5. Sept 1933-Nov 1934 – 1 – us NY Public [335]
Blatchford and howland's reports of cases in the southern district court of new york, 1827-1837 / Blatchford, S & Howland, F – New York: Halsted. 1v. 1937 (all publ) – (= ser Early federal nominative reports) – 7mf – 9 – $10.50 – (a selection of cases decided by judge samuel r betts, sometimes called "bett's decisions") – mf#LLMC 81-434 – us LLMC [324]
Blatchford, S *see*
– Blatchford and howland's reports of cases in the southern district court of new york, 1827-1837
– Blatchford's prize cases in the second circuit, 1861-1865
– Blatchford's reports of cases in the second circuit, 1845-1887
Blatchford's prize cases in the second circuit, 1861-1865 / Blatchford, S – New York: Baker-Voorhis, 1v. 1866 (all publ) – (= ser Early federal nominative reports) – 8mf – 9 – $12.00 – mf#LLMC 81-435 – us LLMC [340]
Blatchford's reports of cases in the second circuit, 1845-1887 / Blatchford, S – New York: Baker-Voorhis. v1-24. 1859-88 (all publ) – (= ser Early federal nominative reports) – 169mf – 9 – $253.00 – mf#LLMC 81-433 – us LLMC [340]
Blatchley, J S *see* Littleton and blatchley's digest of fire insurance decisions
Blatchley, W(illis) S(Tanley) *see* Nature wooing at ormond by the sea
Blathwayt, William *see* Accounts of british trade in america
Das blatt – Bogota (CO), 1943/44-1944/45 [gaps] – 1 – gw Misc Inst [079]
Das blatt der hausfrau *see* Dies blatt gehoert der hausfrau
Das blatt der hausfrau / wiener ausgabe *see* Ullsteins blatt der hausfrau / wiener ausgabe
Das blatt der wohltaetigkeit – Hamburg DE, 1806 1 nov-28 jun, 1808 2 jul-17 sep – 1r – 1 – gw Misc Inst [360]
Blatter der erinnerung zum 50jahrigen bestehen des israelitischen f... / Winter, D H – Lubeck, Germany. 1927 – 1r – us UF Libraries [939]
Blatter zur erinnerung an die einweihung der neuen synagoge in main... / Salfeld, Siegmund – Mainz, Germany. 1913 – 1r – us UF Libraries [939]
Blau, Bruno et al *see* Zeitschrift fuer demographie und statistik der juden
Blau, Georg *see* Ausdrucksbewegungen in heinrich von kleists werk
Blau ist das meer : eine erzaehlung aus der deutschen kriegsmarine / Zerkaulen, Heinrich – Leipzig: Quelle & Meyer [1936?] [mf ed 1991] – 1r – 1 – (filmed with: volk, ich breche deine kohle! / otto wohlgemuth) – mf#2964p – us Wisconsin U Libr [830]
Blau, Lajos *see* Zur einleitung in die heilige schrift
Blaue fernen : neue reisebilder / Hevesi, Lajos – Stuttgart: Adolf Bonz 1897 [1995] – 1r – 1 – (filmed with: herderbuch : reisejournal / ed by j loeber) – mf#3636p – us Wisconsin U Libr [914]
Die blaue mauritius : roman / Eckart-Helm, Martina – Neuaufl. Berlin: P Franke, 1944, c1939 (mf ed 1990) – 1r – 1 – (filmed with: peter moors fahrt nach suedwest) – us Wisconsin U Libr [830]
Der blaue tiger : roman / Doeblin, Alfred – Baden-Baden: P Keppler, 1947 [mf ed 1989] – (= ser Suedamerika-trilogie 2) – 423p – 1 – mf#7179 – us Wisconsin U Libr [830]
Blauenfeldt, Johanne *see* Hindoekinderen
Blaustein, Leopold *see* Das gotteserlebnis in hebbels dramen
Blaustein, Moses *see* Lerer redt zikh arop fun hartsn
Blau-weiss blaetter *see* Verein zur befoerderung der handwerke unter den inlaendischen israelitischer
Blau-weiss-blaetter – Berlin DE, 1916-18, 1924 – 1 – (with gaps) – gw Misc Inst [074]
Blavatsky, Helena Petrovna *see*
– From the caves and jungles of hindostan
– The key to theosophy
– A modern panarion
– The secret doctrine
– The theosophical glossary
Blaxall, Arthur William *see* Ten cameos from darkest africa
Blaxland, Bruce *see* The struggle with puritanism
[Blaxland, G] *see* A journal of a tour of discovery across the blue mountains in new south wales
Bla(y)lock genealogy news – 1987 nov/dec-1990 nov/dec – 1r – 1 – (cont: genealogy news of the blalock-blaylock clans) – mf#1056915 – us WHS [929]
Blayney advocate – Bayney, jan 1 1898-dec 28 1907 – 2r – 9 – A$111.10 vesicular A$122.10 silver – at Pascoe [079]
Blayney, Andrew *see* Narrative of a forced journey through spain and france, as a prisoner of war, in the years 1810 to 1814
Blaze, L E *see* The story of kandy
Blaze, M Henri *see* Le faust de goethe
Blazer news – Albion, Battle Creek etc MI. 1992 dec 23-1994 dec 28, 1995 jan 4-1996 jan 31/dec 4, 2000, 2001 jan 10/16-jun 27/jul 3 – 1r – 1 – mf#2633568 – us WHS [071]
Blazquez, Antonio *see*
– Extremadura en la guerra de la independencia (informe de gomez villafranca)
– Informe relativo a parte de la via romana num. 25 del itinerario de antonino
– Informe sobre declaracion de monumento nacional de puente de alcantara
– Via romana de merida a salamanca
– Vias romanas de la beturia de los turdulos, por don angel delgado
Blazquez del Barco, Juan *see*
– Explicacion...confesores..
– [Franciscano de la provincia de san miguel] en notas de bibliografia franciscana
– Relox del alma y oracion mental
– Trompeta evangelica...
– Trompeta evangelica, alfange..
Blazquez Izquierdo, Jose *see* 1st congreso sindical agrario de extremadura. ponencia 7. situacion de la, produccion ganadera
Blazquez Marcos, Jose *see*
– Por la vieja extremadura-provincia de caceres
– Transcendencia emocional de la lectura
Blazquez Prieto, Gabriel R *see*
– Resolucion de la incorporacion de olivenza..
– Resolucion...olivenza y las aldeas de..
– ...Se dirige a los curas...
Blazquez, Vidal *see* Epigrafia romana
Bleb – Tiburon CA 1970-71 – 1 – ISSN: 0006-467X – mf#7508 – us UMI ProQuest [810]
Bleby, Henry *see*
– Death struggles of slavery
– Jehovah's decree of predestination
Bleckmann, Friedrich *see* Griechische inschriften zur griechischen staatenkunde
Le bled en lumiere, folklore tunisien / Benattar, S C – Paris: J Tallandier, [c1923] – 1 – us CRL [390]
Bledsoe, Albert Taylor *see*
– Cotton is king, and pro-slavery arguments
– An essay on liberty and slavery
– A theodicy

Bledsoe, Anthony Jennings *see* Business law for business men, state of california
Bledsoe family quarterly – n1-3 [1985 jan-jul], n4-5 [1986 feb-oct], n6 [1987 may], n7 [1988 aug], n8-9/10 [1989 may], n11/12 [1990 dec], n13 [1991 jan] – 1r – 1 – mf#1728133 – us WHS [929]
Bleeding rose – 1968 may-1969 summer – 1r – 1 – mf#1053425 – us WHS [071]
Bleek, Dorothea Frances *see* Naron
Bleek, Friedrich *see*
– Beitraege zur evangelien-kritik
– Einleitung in das alte testament
– Einleitung in das neue testament
– Einleitung in die heilige schrift
– Der hebraeerbrief
– An introduction to the new testament
– Synoptische erklaerung der drei ersten evangelien
– Vorlesungen ueber die apokalypse
Bleek, Hermann *see* Die grundlagen der christologie schleiermachers
Bleek, Johannes Friedrich *see* Einleitung in das alte testament
Bleek, Wilhelm Heinrich Immanuel *see*
– African folk-lore
– Natal diaries of w h i bleek, 1855-1856
– On the origin of language
– Ueber den ursprung der sprache
– Zulu legends
Bleeker, Johannes Jacob *see* De polemiek der eerste christenen tegen de heidensche mythologie
Blegen, John Hansen *see* Zionsforeningens historie
Blei, Franz *see*
– Das grosse bestiarium der modernen literatur
– Karl henckell
Bleib stet! : vierzehn volksgeschichten / Rothacker, Gottfried – Muenchen: A Langen, G Mueller 1943, c1938 [mf ed 1991] – 1r – 1 – (filmed with: nietzsche und tolstoi / nikolaus grot) – mf#2851p – us Wisconsin U Libr [390]
Die bleibende bedeutung des alten testaments : ein konferenzvortrag / Kautzsch, Emil – 2. verm aufl. Tuebingen: J C B Mohr, 1903 – 1mf – 9 – 0-8370-9705-3 – mf#1986-3705 – us ATLA [221]
Die bleibende bedeutung des neutestamentlichen kanons fuer die kirche : vortrag auf der lutherischen pastoralkonferenz zu leipzig am 2. juni 1898 / Zahn, Theodor – Leipzig: A Deichert, 1898 – 1mf – 9 – 0-8370-9353-8 – (incl bibl ref) – mf#1986-3353 – us ATLA [225]
Bleiberg, German *see* La cancion petrarquista en la lirica espanola del siglo de oro, de enrique segura covarsi
Bleibtreu, Karl *see*
– Dies irae
– Dramatische werke
– Groessenwahn
– Lyrisches tagebuch
– Schlechte gesellschaft
– Zwei wackere helden
Bleibtreu, Karl et al *see* Die entscheidungsschlacht
Bleibtreu, Walther *see* Das geheimnis der froemmigkeit und die gottmenschheit christi
Blekinge folkblad – Karlskrona, 1903-21 – 9 – sw Kungliga [078]
Blekinge folkblad *see* Sydostra sveriges dagblad
Blekinge lans tidning – Karlskrona, Sweden. 1869- – 1 – (karlshamns allehanda; solvesborgstidningen) – sw Kungliga [078]
Blekinge lans tidning – Karlskrona, Sweden. 1869-1978 – 534r – 1 – (solvesborgstidningen, 1967-78, vaxjobladet 1967-75, karlshamns allehanda, 1976-78) – sw Kungliga [078]
Blekingen – Karlskrona, Sweden. 1885-92 – 3 – 1 – sw Kungliga [078]
Blekingeposten – Karlskrona, Sweden. 1909-10 – 1r – 1 – sw Kungliga [078]
Blekingeposten – Karlskrona, Sweden. 1945-78 – 32r – 1 – sw Kungliga [078]
Blekingeposten – Ronneby, Sweden. 1979- – 1 – sw Kungliga [078]
Blekingsposten – Karlskrona, Sweden. 1852-84 – 2r6 – 1 – sw Kungliga [078]
Bleklov, S M *see* Za faktami i tsiframi zapiski zemskogo statistika
Blemont, Emile *see* Mariage pour rire
Blemur, R M J de *see*
– L'annee benedictine
– Eloges de plusieurs personnes o s b
– Vie des saints
– Vie du r p pierre fournier
Blenck, Erna *see* Sudwest-afrika
Blencowe, Charles *see* Nature, operation, and reception of the gospel of christ
Blending lights : or, the relations of natural science, archaeology, and history to the bible / Fraser, William – New York:Robert Carter, 1874 – 1mf – 9 – 0-8370-3185-0 – mf#1985-1185 – us ATLA [210]
Blenerhasset, Thomas *see* A revelation of the true minerva
The blenheim papers : from the british library, london – 3-pt coll – 62r – 1 – (pt 1: 1st duke of marlborough (add mss 61101-61158 and 61160-61166 – 23r c39-16601. pt 2:...selected from add mss 61167-61303 – 18r c39-16602. pt 3:...add mss 61306-61315, 61319-21, 61363-73, 61378, 61380-61408, 61411-61413 – 21r c39-16603) – mf#C39-16600 – us Primary [941]
Blenk, James H *see* Fradryssa
Blennerhassett, Charlotte, Lady *see* John henry kardinal newman
Blennerhassett, Harman *see* Papers
Blennerhassett, R *see* Adventures in mashonaland, by two hospital nurses
Bles, Numa *see* Paris-montreal
Bleses, Peter *see* Das spd-konzept der "sozialen grundsicherung"
Blessed are the dead which die in the lord / Roberts, George – Monmouth, England. 1842 – 1r – us UF Libraries [240]
Blessed are they : or, thoughts on the beatitudes / Gilbert, Jesse Samuel – Paterson, NJ: Carleton M Herrick, 1890 – 1mf – 9 – 0-8370-3282-2 – mf#1985-1282 – us ATLA [240]
Blessed dead / Marsh, William Nathaniel Tilson – Leamington, England. 1843 – 1r – us UF Libraries [240]
Blessed is the nation whose god is the lord / Price, Thomas C – Bristol, England. 1879 – 1r – us UF Libraries [240]
Blessed joan the maid / Barnes, Arthur Stapylton – London: Burns & Oates; New York: Benziger, [1909?] – 1mf – 9 – 0-7905-6341-X – mf#1988-2341 – us ATLA [944]
Blessed martyrs of uganda / Streicher, Henri – Chishawasha, Zimbabwe. 1957 – 1r – us UF Libraries [960]
Blessed martyrs of uganda / Streicher, Henri – Mariannhill, South Africa. 1924 – 1r – us UF Libraries [960]
The blessed sacrament : or, the works and the ways of god / Faber, Frederick William – [3rd ed] London: Burns & Oates, [1861?] – 2mf – 9 – 0-7905-9195-2 – mf#1989-2420 – us ATLA [240]
Blessed sacrament the centre of immutable truth / Manning, Henry Edward – London, England. 1886 – 1r – us UF Libraries [240]
The blessed virgin – London: James Miller, [18–?] – 1mf – 9 – 0-8370-7984-5 – mf#1986-1984 – us ATLA [241]
The blessed virgin in the fathers of the first six centuries / Livius, Thomas – London: Burns and Oates, 1893 – 2mf – 9 – 0-8370-6143-1 – (incl bibl ref and ind) – mf#1986-0143 – us ATLA [240]
The blessed virgin in the nineteenth century : apparitions, revelations, graces / St-John, Bernard – London: Burns & Oates; New York: Benziger, [1903?] – 2mf – 9 – 0-8370-8621-3 – mf#1986-2621 – us ATLA [240]
Blessedness of departed believers / Macindoe, Peter – Kilmarnock, Scotland. 1841 – 1r – us UF Libraries [240]
Blessedness of dying in the lord / Hawtrey, C S – London, England. 1828 – 1r – us UF Libraries [240]
Blessedness of giving : greater than that of receiving / Stuart, John – Edinburgh, Scotland. 1809 – 1r – us UF Libraries [240]
"The blessedness of giving" and "perilous times" : two farewell sermons preached at tottenham court road chapel on sunday, nov 26 1876 / Bevan, Llewelyn D – London, England: F Davis 1876 – (= ser Penny pulpit [ns] 905, 906) – 1r – 1 – us UF Libraries [240]
Blessedness of those who die in the lord / Wood, Thomas – Bristol, England. 1818? – 1r – us UF Libraries [240]
Blessi, M *see* ...Nella rotta dell' armata de svltan selin, vltimo re de tvrchi
Blessing and ban from the cross of christ : meditations on the seven words on the cross / Dix, Morgan – New York: J Pott, 1898 [mf ed 1991] – 1v on 1mf – 9 – 0-7905-9186-3 – mf#1989-2411 – us ATLA [242]
Blessington, Marguerite *see*
– The idler in italy
– Journal of a tour through the netherlands to paris, in 1821
Bleter far geshikhte – Warsaw, Poland. 1934-38 – 1r – us UF Libraries [939]
Bletlekh tsu der geshikhte fun narsis leven / Schoijet, Jesekiel – Buenos Ayres, Argentina. 1953 – 1r – us UF Libraries [939]
Bletsoe – (= ser Bedfordshire parish register series) – 1mf – 9 – £3.00 – uk BedsFHS [929]
Bletz, Zacharias *see* Die dramatischen werke des luzerners zacharias bletz
Bleu et rouge – Port-au-Prince, Haiti: [s.n.] v2 n2-v4 n116 1916-1918. v7 n1-n38 1922 – 6 sheets – us CRL [972]
Bleuler, Sharon A *see* Nonverbal behavior of national figure skating coaches
Bleutler, Sharon A *see* Is there a relationship between prenatal exercise and postpartum depression
Blewett, George John *see*
– The christian view of the world
– The study of nature and the vision of god
Blewett, Jean *see*
– The cornflower
– Heart songs
– Heart stories
Bley *see* Dr buchner's report
Bley, Fritz *see* Horridoh!
Bley, Helmut *see* Kolonialherrschaft und sozialstruktur in deutsch-sudwestafrika
Bley, Wulf *see* Frau im wirbel
Blick *see* Schweriner blick
Ein blick auf russland / Melissander, Friedrich – Kiel: Lipsius und Tischer, 1892 – 1r – 1 – us Wisconsin U Libr [947]
Blick durch die wirtschaft – Frankfurt/M DE, 1965-97 – 62r – 1 – (1998 subsc [3r]. with suppl: monatsregister 1959-1980 jun [2r]) – mf#8726 – gw Mikropress [330]
Blick durch die wirtschaft – Frankfurt/Main DE, 1974-79 – 1 – (filmed by mikropress: 1965-1998 31 jul [until 1994 56r]; filmed by misc inst: 1958 nov-1964. with suppl: monatsregister 1959-jun 1980 [2r]) – gw Mikrofilm; gw Mikropress; gw Misc Inst [330]
Blick durch die wirtschaft – Monatsregister 1959-Juni 1980 – 2r – 1 – gw Mikropress [330]
Ein blick in die finsternisz / Reber, Geschrieben von J L – Lebanon PA: printed by John L Becker, for Von J A Sand & S Von Nieda 1850 – 1r – 1 – $35.00 – (in german) – mf^{s}-76 – us Commission [242]
Blick in die woche – Duesseldorf DE, jul 1952-may 1953 – 1r – 1 – (filmed by other misc inst: 1951 okt-1953 mai) – gw Misc Inst [074]
Blick in die zeit – Berlin DE, 1933-35 – 2r – 1 – gw Misc Inst [074]
Blicke in die geisteswelt der heidnischen kols : sammlung von sagen, maerchen und liedern der oraon in chota nagpur / Hahn, Ferdinand – Geutersloh: C Bertelsmann, 1906 [mf ed 1995] – (= ser Yale coll) – x/116p – 1 – 0-524-10093-4 – (in german) – mf#1995-1093 – us ATLA [390]
Blicke in die religionsgeschichte zu anfang des / Joel, Manuel – Breslau, Germany. v1-2. 1880-1883 – 1r – us UF Libraries [939]
Blicke in die zeit nach der schrift – Bremen DE, dec 25 1848-jul 14 1855 – 1r – 1 – gw Misc Inst [074]
Blicke in indisches heidentum – Basel: Verlag der Missionsbuchhandlung, 1906 [mf ed 1995] – (= ser Yale coll) – 24p (ill) – 1 – 0-524-09957-X – (in german) – mf#1995-0957 – us ATLA [954]
Blicke in schleiermachers theologie : vortrag... breslau am 21. feb 1873 / Treblin, Adolf – Berlin: F Henschel 1873 [mf ed 1991] – 1mf – 9 – 0-524-00179-0 – mf#1989-2879 – us ATLA [242]
Blickpunkt – Magdeburg DE, 1970 dec-1990 mar [gaps] – 4r – 1 – (autobahnbaukombinat) – gw Misc Inst [620]
Bliedner, Arno *see* Goethe und die urpflanze
Bliemetzrieder, Fr *see* Anselms von laon systematische sentenzen
Bliemetzrieder, Franz *see* Das generalkonzil im grossen abendlaendischen schisma
Bliesener, Johann *see* Trois quatuors pour deux violons, alto & violoncelle
Bliff, G Ripley *see* An address
Bligh, Harris Harding *see* Statutory annotations to the revised statutes of canada, 1906, and other canadian statutes (2d ed.) providing references to every change made by the annual statutes for 1907, 1908, 1909, 1910, 1911, 1912, 1913 and 1914
Bligh, Richard *see* Bligh's parliamentary reports
Bligh, W *see* A voyage to the south sea
Bligh watchman – Coonabarabran, jan 1898-dec 1910 – 2r – A$145.55 vesicular A$156.55 silver – at Pascoe [079]
Bligh, William *see*
– The mutiny on board h m s bounty
– Voyage a la mer du sud
Bligh's parliamentary reports : reports of cases heard in the house of lords on appeals and writs of error... / Bligh, Richard – v1-4 pt 1. 1819-21. London: J & W T Clarke and Baldwin, Cradock & Joy, 1823-? (all publ) – 26mf – 9 – $39.00 – mf#LLMC 84-752 – us LLMC [324]
Blik op het huidig bestuursbeleid in suriname – Paramaribo, Surinam. 1913 – 1r – us UF Libraries [972]
Blin, Jean Baptiste Nicolas *see* Vie de m jean hue
Blind apostle / Dowding, William Charles – Hannover, Germany. 1855 – 1r – us UF Libraries [240]
Blind betsey : or, comfort for the afflicted – London, England. 18– – 1r – us UF Libraries [240]
Blind justice / National Lawyers Guild – v3 n2 [1973 dec], v5 n3 [1975 jul], v7 n2 [1977 mar], v8 n2 [1978 may/jun]-v11 n2 [1981 may/jun], v11 n4 [1981 dec]-v18 n1 [1987 feb] – 1r – 1 – mf#1519079 – us WHS [340]
Blind man and the pedlar – London, England. 18– – 1r – us UF Libraries [240]
Blind man of jerusalem : a sermon / Fraser, Donald – London, England. 1879 – 1r – us UF Libraries [240]
Blind schoolmistress of devonshire : a true and interesting story – London, England. 18– – 1r – us UF Libraries [240]
Der blinde bruder *see* Aus nacht zum licht
Die blinde goettin : schauspiel in fuenf akten / Toller, Ernst – Berlin: G Kiepenheuer, 1933 – 1r – 1 – us Wisconsin U Libr [820]
Blinded eagle: a short life of edward irving / Whitely, H C – 9 – $10.00 – us IRC [920]
Blinken, Meir *see* Kortenshpiel
Blinkenberg, Christian *see* The thunderweapon in religion and folklore
Blinov, I A *see* Vsemirnyi ekonomicheskii, finansovyi i politicheskii spravochnik 1923 g
Blinovskii, P I *see* Zemstvo i kooperatsiia
Bliokh, I S *see* Finansy rossii 19 stoletiia
Bliss, Edwin Munsell *see*
– A concise history of missions
– The encyclopaedia of missions
– The missionary enterprise
– Organization and methods of mission work
– Turkey and the armenian atrocities
Bliss, F J *see* A mound of many cities (tell el hesy excavated)
Bliss, Frederick Jones *see*
– The development of palestine exploration
– Excavations at jerusalem 1894-97
– Excavations in palestine during the years 1898-1900
– A mound of many cities
Bliss, George Ripley *see* Commentary on the gospel of luke
Bliss not riches : love one another, or you will torment one another. colonisation on principles of pure christism... / King, Edward, of Blackthorn, Bicester – [London], 1845 – (= ser 19th c british colonization) – 1mf – 9 – mf#1.1.1780 – uk Chadwyck [240]
Bliss, P P *see* The charm of sunday schools
Bliss, Paul Franklin *see* A sociological survey of the unitarian churches in the united states
Bliss, PO *see* Gospel hymns and sacred songs
Bliss, Sylvester *see* Memoirs of william miller
Blitz : asia's foremost newsmagazine – Bombay, India: Blitz Publ, jan 3 1952 (wkly) [mf ed 1984] – 1r – 1 – (began in 1941; aka: blitz on sunday) – mf#8678 – us Wisconsin U Libr [079]
Blitz – Bombay, India: Blitz Publ, [1952-61] – 10r – 1 – us CRL [073]
Blitz – Hannover DE, 1953 29 oct-1954 16 mar – 1r – 1 – gw Mikrofilm [074]
Blitz : india's greatest weekly – Bombay. 1944-1951, 1961- – 1 – us L of C Photodup [073]
Blitz, Tzalel *see* Trit af san-martinisher erd
Blitzstrahl wider rom : die verfassung der christlichen kirche und der geist des christenthums / Baader, Franz von – 2. verb u erw Aufl. Wuerzburg: A Stuber, 1871 – 1mf – 9 – 0-8370-8963-8 – mf#1986-2963 – us ATLA [240]
Blium, A A *see* Istoriia kreditnykh uchrezhdenii i sovremennoe sostoianie kreditnoi sistemy v sssr
Blizhaishie ekonomicheskie zadachi / Miliutin, V – 1926 – 108p 2mf – 9 – mf#COR-173 – ne IDC [335]
Blizhaishie trebovaniia i konechnaia tselizdanie "iskry" – Geneva, 1905. nos 1-11 – 4mf – 9 – mf#R-18117 – ne IDC [077]
Blizzard : published by and for men of the 10th mountain division – Denver, CO: National Association of the 10th Mountain Division, 1971 [mf ed 1989] – 1r – 1 – mf#MF Bl619t – us Colorado Hist [355]
Blk – 1988 dec-1990 dec, 1991-1994 mar – 2r – 1 – mf#1831215 – us WHS [071]
Le bloc : organe de l'association generale des fonctionnaires de la guadeloupe et des dependances – Pointe-a-Pitre. n1-3. avr-juin 1926 – 1 – fr ACRPP [972]
Bloch, Armand *see* Phoenicisches glossar
Bloch, Ernest *see* Letters to edmond flegg
Bloch, I *see* Zeitschrift fuer sexualwissenschaft
Bloch, M E *see* Ichthyologie
Bloch, Marcus Elieser *see* Histoire naturelle des poissons
Bloch, Moses *see*
– Budapesti orszagos rabbikepzo-intezet ertesitoje
– Die civilprocess-ordnung nach mosaisch-rabbinischem rechte
– Das mosaisch-talmudische polizeirecht
Bloch, Philipp et al *see* Moses ben maimon
Bloch, Simson *see* Shevile 'olam
Blochet, Edgar *see* Le messianisme dans l'heterodoxie musulmane
Block book : western addition – San Francisco, 1890 – 1r – 1 – $50.00 – mf#B03723 – us Library Micro [978]
Block grants and indian tribes / Institute for the Development of Indian Law – v1 n1-v1 n8 [1983 mar-sep] – 1r – 1 – mf#717908 – us WHS [322]
Block, Louis *see* Mexican-american reports

Blockade of fort george, 1813 / Cruikshank, Ernest Alexander – Welland, Ont?: Niagara Historical Society?, 1898? – 1mf – 9 – mf#07382 – cn Canadiana [355]
The blockade of the port and harbour of hongkong by the hoppo, or farmer in canton of customs duties levied upon chinese vessels : proceedings...hongkong, on the 14th september 1874 – London: Kent & Co; Norwich: Fletcher & Son, [1875] – (= ser 19th c books on china) – 1mf – 9 – mf#7.1.38 – uk Chadwyck [951]
Der blockadebrecher – Bremen DE, 1924 sep-1925, 1926 18 jun-18 dec, 1927-30 [gaps], 1931 [many iss missing] – 3r – 1 – (suppl: neue wirtschaft 1926 [single iss]. special iss 1927-28) – gw Misc Inst [380]
Blodget, Lorin *see*
– The industries of philadelphia
– The revised statutes of the united states
Blodgett, Andrew D *see* Effects of carbohydrate supplementation on immune function with long endurance running and cycling
The blodgett collection of spanish civil war pamphlets / Harvard College. Library – 680 pamphlets published in 1936-39 in Spain, Europe, Latin America, and U.S. by agencies of the Spanish gov't., Spanish political parties, and organizations world-wide. Titles listed separately – 9 – us Harvard [080]
Blodgett, May Nellie *see* Character studies in genesis
Bloed over ambon – Amsterdam, 1951 – 1mf – 9 – mf#SE-1417 – ne IDC [950]
Bloem, Walter *see*
– Das eiserne jahr
– Das eiserne Jahr
– Faust in monbijou
– Held seines landes
– Komoediantinnen
– Die schmiede der zukunft
– Sohn seines landes
– Sonnenland
– Das verlorene vaterland
– Volk wider volk
– Wir werden ein volk
Bloem, Walter Julius *see* Heimkehr in die mannschaft
Bloemfontein post – Pretoria: The State Library, 15 mar 1900-30 jun 1915 – 74r – 1 – mf#MS00175 – sa National [079]
Bloemhof der doorluchtige voorbeelden : daer in door ware, vreemde en deftige geschiedenissen, leeringen en eygenschappen... / Heyns, Maria – t'Aemsteldam: J Lesaille, 1647 – 4mf – 9 – mf#O-3083 – ne IDC [090]
Bloemker, Friedrich *see* Das verhaeltnis von buergers lyrischer und episch-lyrischer dichtung zur englischen literatur
Bloemkrans van christelyke liefde- en zeededichten : nevens eenige christelyke gezangen met kunstplaaten / Willink, D – Amsteldam: J Oosterwyk, en H van de Gaete, 1714 – 3mf – 9 – mf#O-805 – ne IDC [090]
Bloem-tuyntje... / Schaep, J C – Amsterdam: Tymon Houthaak, 1660 – 3mf – 9 – mf#O-3163 – ne IDC [090]
Bloem-tuyntje... / Schaep, J C – t'Amsterdam: Jacob ter Beek, 1724 – 5mf – 9 – mf#O-749 – ne IDC [090]
Bloem-tuyntje... / Schaep, J C – t'Amsterdam: Jan Rieuwertsz., 1697 – 4mf – 9 – mf#O-3258 – ne IDC [090]
Bloem-tuyntje... / Schaep, J C – t'Amsterdam: Jan Rieuwertsz. de Jonge en Jacobus Deister, 1686 – 4mf – 9 – mf#O-3270 – ne IDC [090]
Bloem-tuyntje... / Schaep, J C – t'Amsterdam: Jan Rieuwertz, 1671 – 4mf – 9 – mf#O-3164 – ne IDC [090]
Bloesch, Emil *see* Geschichte der schweizerisch-reformierten kirchen
Bloesch, Hans *see*
– Am kachelofen
– Geld und geist
– Jeremias gotthelf
– Die kaeserei in der vehfreude
– Kalendergeschichten
– Saemtliche werke in 24 baenden
– Uli der paechter
– Zeitgeist und bernergeist
Blois, Charles N de *see* Les courants statiques induits de morton et quelques-unes de leurs applications en medecine
Blois, Louis de *see* Spiritual works of louis of blois, abbot of liesse
Blok, A *see* Poslednie dni imperatorskoi vlasti
Blok, Aleksandr Aleksandrovich *see* Stikhotvoreniia aleksandra bloka
Blom, Abraham Herman *see* De leer van het messiasrijk bij de eerste christenen
Blome, Hermann *see* Der rassengedanke in der deutschen romantik
Blome, Richard *see* A geographical description of the four parts of the world
Blome, Rud *see* Die kongregationalistische kirche
Blomfield, Alfred *see*
– A memoir of charles james blomfield...bishop of london
– The old testament and the new criticism
Blomfield, Charles James *see*
– Duty of family prayer
– First questions on religion
– Five lectures on the gospel of st john as bearing testimony to the...
– God's ancient people not cast away
– Manual of family prayers
– Manual of family prayers second series
– Manual of private devotion
– Sermon on the duty of family prayer
Blomfield, Reginald Theodore *see*
– The formal garden in england
– A history of renaissance architecture in england 1500-1800
Blomgren, Carl August *see* The elements of the christian religion
Blommaert, A *see*
– La forest des hermites et hermitesses d'egypte, et de la palestine...
– Sylva anachoretica aegypti et palaestinae
Blomquist, Melinda E *see* The effect of the reciprocal approach in teaching on the process of self-discovery for beginning modern dance students at the secondary level
Blonda, Maximo Aviles *see* Centro del mundo
Blondeau, P *see* Vollstaendiger bericht von allen sehens-wuerdigen freunden-festen
Blondel *see* Memorias de arquitectura...
Blondel, D *see*
– Actes authentiques des eglises reformees
– De la primaute en l'eglise
Blondel, F *see*
– L'art de jetter les bombes
– Uvelle maniere de fortifier les places
Blondel, Jacques Francois *see*
– Cours d'architecture enseigne dans l'academie royale d'architecture
– Cours d'architecture enseigné dans l'academie royale d'architecture
– De la distribution des maisons de plaisance et de la decoration desedifices en general
Blondin, Alphonse *see* Nouveau recueil de chansons comiques
Die blondjaeger : ein roman von negern, weissen maedchen, gentlemen und halunken / Leip, Hans – Berlin: Propylaeen-Verlag, c1929 – 1r – 1 – us Wisconsin U Libr [830]
Blood – Washington DC 1992+ – 1,5,9 – ISSN: 0006-4971 – mf#6541 – us UMI ProQuest [610]
Blood, Benjamin *see* Optimism
The blood covenant : a primitive rite and its bearings on scripture / Trumbull, Henry Clay – 2nd ed. Philadelphia: John D Wattles, 1893 – 1mf – 9 – 0-7905-0443-X – (incl bibl ref and indexes) – mf#1987-0443 – us ATLA [100]
Blood glutathione oxidation during human exercise / Viguie, Christine A – 1988 – 95p 1mf – 9 – $4.00 – us Kinesology [612]
Blood groups & blood group incompatibilities = Bloedgroepe & bloedgroeponverenigbaarhede / South Africa. Department of National Health and Population Development [Departement van Nasionale Gesondheid en Bevolkingsontwikkeling – 2nd ed. Pretoria: Dept of National Health & Population Development 1990 [mf ed Pretoria, RSA: State Library [199-]] – 22p [ill] on 1r with other items – 5 – (in english & afrikaans) – mf#op 09959 r23 – us CRL [616]
Blood Indian Reserve [Alta] *see* Kainai news
Blood lactate responses for three competitive swimming strokes / Chase, Lisa A – 1988 – 46p 1mf – 9 – $4.00 – us Kinesology [612]
Blood lipids and peak oxygen consumtion in young distance runners / Eisenmann, Joey C – 2000 – 3mf – 9 – $12.00 – mf#PH 1695 – us Kinesology [612]
The blood of jesus / Reid, William – Philadelphia: American Baptist Pub Soc [186-] [mf ed 1985] – 1mf – 9 – 0-8370-5444-3 – mf#1985-3444 – us ATLA [240]
The blood of jesus : what is its significance? / Waldenstroem, Paul – Chicago: J Martenson, 1888 – 1mf – 9 – 0-7905-7486-1 – mf#1989-0711 – us ATLA [220]
Blood of stones / Chattopadhyaya, Harindranath – Bombay: Padma Publ, 1944 – (= ser Samp: indian books) – us CRL [890]
Blood of the lamb and the union of the saints / Tregelles, Samuel Prideaux – London, England. 1851 – 1r – us UF Libraries [240]
Blood on the mercy-seat – London, England. 18– – 1r – us UF Libraries [240]
Blood royal : a novel / Allen, Grant – London: Chatto & Windus, 1893 – 4mf – 9 – (incl publ list) – mf#26647 – cn Canadiana [830]
Blood strain / Miller, Haiden – s.l, s.l? . 1924 – 1r – us UF Libraries [978]
Blood vessels – Basel, Switzerland 1974 – 1,5,9 – (cont: angiologica; cont by: journal of vascular research) – ISSN: 0303-6847 – mf#2046.02 – us UMI ProQuest [611]
Blood-horse – Lexington KY 1929+ – 1,5,9 – ISSN: 0006-4998 – mf#6043 – us UMI ProQuest [636]
The bloodhound : our best ally in ireland – [London], [1882] – (= ser 19th c ireland) – 1mf – 9 – mf#1.1.1877 – uk Chadwyck [941]
Bloodlines / Rural Organizing and Cultural Center (Lexington MS) – 1988 summer – 1r – 1 – mf#5294080 – us WHS [071]
Bloom, Debra *see* Locus of control, physical self-efficiacy and exercise frequency
Bloom, Michele L *see* Personals
Bloom, Solomon *see*
– A treatise on the law of mechanics' liens and building contracts with annotated forms
– Us constitution sesquicentennial commission
Bloomer advance – Bloomer WI. 1895 mar 21/1896 oct 29-2002 sep/dec – 101r – 1 – (with gaps) – mf#1133245 – us WHS [071]
Bloomer free press – Bloomer WI. 1923 sep 1-1925 jun 18 – 1r – 1 – mf#957448 – us WHS [071]
Bloomer workman – Bloomer WI. 1881 may 5-1883 mar, 1881 may 5-1883 mar 1, may 24, oct 11, dec 27, 1884 feb 21, jun 19, jul 24, 1885 mar 18, 1886 apr 8 – 2r – 1 – mf#957443 – us WHS [071]
Die bloomfield germania – Bloomfield, NE: Lohmann & Liewer. -jahrg 19 n32. mai 28 1914 (wkly) [mf ed jahrg 12 n30. 9 apr 1908-may 28 1914 (gaps)] – 2r – 1 – (in german. absorbed by: woechentliche omaha tribuene) – us NE Hist [071]
Bloomfield, J H *see* Cuban expedition
Bloomfield, John *see* Gospel glass
Bloomfield journal – Bloomfield, NE: Bloomfield Pub Co, 1892-v9 n3. oct 20 1921 (wkly) [mf ed v1 n27. nov 18 1892-oct 20 1921 (gaps)] – 7r – 1 – (absorbed by: bloomfield monitor. numbering dropped with sep 16 1898 issue and resumed with v6 n5 on feb 2 1899. issues for jun 11 1914-oct 20 1921 called v2 n13-v9 n3) – us NE Hist [071]
Bloomfield. Kansas. Public Schools *see* Teachers records
Bloomfield, Maurice *see*
– The atharva-veda and the gopatha-brahmana
– The life and stories of the jaina savior parcvanatha
– Religion of the veda
– Rig-veda repetitions
Bloomfield monitor – Bloomfield, NE: Needham Bros. -50th yr n53. nov 22 [ie 28] 1940 (wkly) [mf ed v2 n29. jun 24 1892-nov 28 1940 (gaps)] – 18r – 1 – (cont by: monitor) – us NE Hist [071]
Bloomfield monitor – Bloomfield NE: Wm A Skrivan. 60th yr n5. dec 29 1949- (wkly) – 1 – (cont: monitor) – us NE Hist [071]
Bloomfield, S T *see* Analytical view of the principal plans of church reform
Bloomfield, Samuel Thomas *see* He kaine diatheke
Bloomfield, Susan A et al *see* Site-specific changes in bone mass and alterations in calciotropic hormones with electrical stimulation exercise in individuals with chronic spinal cord injury
Blooming grove baptist church. mcleansboro, illinois : church records – 1850-68 – 1 – us Southern Baptist [242]
Blooming Grove [telephone directory : listing] – 1940-56 – 17r – 1 – (with gaps) – mf#2916537 – us WHS [917]
The bloomington advocate – Bloomington, NE: H M Crane. 5v. v49 n22. feb 5 1931-v53 n52. nov 30 1933 (wkly) – 2r – 1 – (cont: advocate-tribune; cont by: advocate-tribune (1933)) – us NE Hist [071]
Bloomington and Normal Trades and Labor Assembly et al *see* Livingston and mclean counties union news
The bloomington argus – Bloomington, NE: [I J Crane], ns: v1 n1. sep 21 1889- [mf ed sep 21 1889-sep 12 1890 (gaps)] – 1r – 1 – (ceased in 1890. absorbed by: rep valley echo. note iss for sep 21 1889-sep 12 1890 called also old ser v11-old ser v12) – us NE Hist [071]
Bloomington daily pantagraph – Bloomington IL. 1886 nov 4 – 1r – 1 – (cont: bloomington pantagraph (bloomington il: 1873 : daily); cont by: daily pantagraph (bloomington il)) – mf#991646 – us WHS [071]
Bloomington echo – Bloomington, NE: H M Crane. 6v. v15 n18. jan 1 1897-v20 n14. may 31 1901 (wkly) [mf ed with gaps] – 3r – 1 – (cont: rep valley echo. merged with: prickly pear to form bloomington advocate) – us NE Hist [071]
Bloomington free ryder – vn12 [1979 nov 1/19], v1 n22 [1980 jun 20/30], v1 n24, [1980 aug 1/18] – 1r – 1 – (cont: free ryder) – mf#664934 – us WHS [071]
Bloomington record – Bloomington WI. [1880 jul 29-1882 dec 14, 1888 jan 12-1892 feb 25] – 40r – 1 – mf#1004016 – us WHS [071]
Bloomington record – Bloomington WI. 1963 jul 11-1965, 1966-67, 1968-1969 oct 23, 1969 oct 30-1971 may 20, 1971 may 27-1973 feb 8 – 5r – 1 – (cont: record (bloomington wi: 1960); cont by: grant county herald independent) – mf#1003991 – us WHS [071]
Bloomington record – Bloomington WI. 1896 jun 4/1898 jun 9-1959-1960 jan 14 – 40r – 1 – (with gaps; cont: cassville index; cont by: record (bloomington wi: 1960)) – mf#1004004 – us WHS [071]
The bloomsbury review – Denver, CO. ill. -bim, -m. Has also occasional unnumbered suppls – 1 – us Wisconsin U Libr [420]
El bloque – Caceres 1907-19 – 5 – sp Bibl Santa Ana [074]
Bloqueo, rendicion y ocupacion de maracaibo por la... / Ortega Ricaurte, Enrique – Bogota, Colombia. 1947 – 1r – us UF Libraries [972]
Blore, Edward *see* The monumental remains of noble and eminent persons
Blore, Thomas *see* A guide to burghley house
Blossburg advertiser – Blossburg, PA. 1889-1910 – 13 – $25.00r – us IMR [071]
Blosser, raymond f, notes, ms 3273 – 1940-46 – 1r – 1 – (letters, clippings, and notes made during interviews with persons concerning oris p and mantis j van sweringen) – us Western Res [080]
Bloudy tenant of persecution for cause of conscience / Williams, Roger – London: J. Haddon, 1846-1854 – 1r – 1 – 0-8370-1699-1 – mf#1984-6079 – us ATLA [941]
Bloudy tenent of persecution *see* Publications
Blough, Jerome E *see* History of the church of the brethren of the western district of pennsylvania
Blouin, Egla Morales *see* Carne y sombra
Blouin, Gervaise *see* Bibliographie analytique de louis-alexandre belisle
Blouin, Gisele *see* Bibliographie analytique de joseph-thomas leblanc
Blount, George A *see* Materials
Blount, Godfrey *see* The science of symbols
Blount, Marie-Louise *see* Occupational therapy in mental health
Blount, Melesina Mary *see* God's jester
Blow, John *see* Amphon angelicus
Blow the trumpet / Galbraith, Richard – London, England. 1876 – 1r – us UF Libraries [240]
Blowers' report of the proceedings of the final wage conference : between the national glass bottle and vial manufacturers' association... / Glass Bottle Blowers Association of the United States and Canada & National Vial and Bottle Manufacturers' Association – 1907 – 1r – 1 – mf#3277443 – us WHS [331]
Blowers' report of the sessions of the final wage conference / Glass Bottle Blowers Association of the United States and Canada & National Vial and Bottle Manufacturers' Association – 1914 – 1r – 1 – mf#3277458 – us WHS [331]
Blowin / Forum for the Evolution of Progressive Arts – 1983 summer – 1r – 1 – mf#4992563 – us WHS [700]
Blowpipe practice : an outline of blowpipe manipulation and analysis, with original tables for determination of minerals / Chapman, Edward John – Toronto: Copp, Clark, 1893 – 4mf – 9 – (incl ind) – mf#26931 – cn Canadiana [550]
Bloxam, Matthew Holbeche *see* The principles of gothic architecture elucidated by question and answer
Bludau, August *see* Die beiden ersten erasmus-ausgaben des neuen testaments und ihre gegner
Bludeau, A *see* Die pilgerreise der aetheria
Blue and old gold – Cape Town, South Africa. 1953 – 1r – us UF Libraries [960]
The blue and white – Athens, Ohio. 1944-Nov 1945 – 1 – us AJPC [071]
The blue annals / 'Gos Lo-tsa-ba Gzon-nu-dpal – Calcutta: Royal Asiatic Society of Bengal, 1949- – (= ser Samp: indian books) – ([trans] by george n roerich) – us CRL [280]
Blue, Archibald *see* Colonel mahlon burwell
Blue banner faith and life : a quarterly publication devoted to expounding, defending, and applying the system of doctrine set forth in the word of god and summarized in the standards of the reformed presbyterian (covenanter) church – Clay Center KS: J G Vos. v1-34 [1946-79] [mf ed 1971-80] – 5r – 1 – mf#S0215 – us ATLA [242]
Blue banner faith and life / Reformed Presbyterian Church of North America – v32-1934 [1977 jan/mar-1979 oct/dec] – 1r – 1 – mf#354572 – us WHS [242]
Blue bells on the lea / Ewing, Juliana H – London, England. no date – 1r – us UF Libraries [025]
Blue cloud quarterly – Marvin SD 1972-88 – 1,5,9 – ISSN: 0006-5064 – mf#7302 – us UMI ProQuest [305]
Blue Cross Blue Shield United of Wisconsin *see* Images
Blue Cross of Wisconsin *see* Spotlight
The blue flag : or, the covenanters who contended for "christ's crown and covenant" / Kerr, Robert Pollok – Richmond, Va: Presbyterian Committee of Publication, 1905 – 1mf – 9 – 0-524-01654-2 – mf#1990-0475 – us ATLA [240]
Blue flame : journal of chicago blues and r and b – 14,16 – 1r – 1 – mf#4851572 – us WHS [780]
Blue grass roots / Kentucky Genealogical Society – 1975 fall-1980 winter, 1981 fall-1987 winter – 2r – 1 – mf#573444 – us WHS [929]

Blue hill leader – Blue Hill, NE: F P Shields. v5 n30. jun 25 1892 (wkly) [mf ed with gaps] – 1 – (absorbed: bladen enterprise) – us NE Hist [071]

Blue jay / Brown Co. Ripley – 1944-79 – 1r – 1 – (a school paper) – mf#B36625 – us Ohio Hist [071]

[Blue lake-] blue lake advocate – CA. may 1888-apr 3 1969 – 33r – 1 – $1980.00 – mf#BC02071 – us Library Micro [071]

The blue laws of connecticut : with an account of the persecution of witches and quakers in new england – New York: The Truth Seeker, 1899, c1898 – 1mf – 9 – 0-8370-7500-9 – mf#1986-1500 – us ATLA [340]

Blue mounds weekly news – Blue Mounds, Mount Horeb WI. 1883 jul 17-1886 jun 30 – 1r – 1 – (cont by: mount horeb weekly news) – mf#939409 – us WHS [071]

Blue mountain american – Sumpter OR: E E Young, 1899- [wkly] – 1 – (cont: sumpter miner (1899-1905)) – us Oregon Lib [071]

Blue Mountain College. Blue Mountain, Mississippi *see* Catalogs and college records

Blue mountain eagle – John Day OR: Grant County Blue Mountain Eagle Co, 1972- [wkly] – 1 – (cont: grant county blue mountain eagle) – us Oregon Lib; us Oregon Hist [071]

Blue mountain eagle – Canyon City OR: Patterson & Ward, -1948 [wkly] – 1 – (cont: grant county news; cont by: grant county blue mountain eagle) – us Oregon Hist [071]

Blue mountain eagle (canyon city, or) – Canyon City OR: Patterson & Ward, -1948 [wkly] – 1 – (absorbed: grant county news (canyon city, or), journal (prairie city, or); cont by: grant county blue mountain eagle (1948-72)) – us Oregon Lib [071]

Blue mountain times – LaGrande OR: Baker, Coggan & Co, [wkly] – 1 – (began in 1868) – us Oregon Lib [071]

Blue mountains advertiser – Katoomba. jan 1940-dec 1954, aug 1961-may 1978 – 12r – A$854.44 vesicular A$920.44 silver – at Pascoe [079]

Blue mountains courier – Katoomba, sep 1948-jul 1960 – 1r – A$82.98 vesicular A$88.48 silver – at Pascoe [079]

Blue mountains democrat – Katoomba, Apr 19 1961 – 9 – at Pascoe [079]

Blue mountains echo – Katoomba, mar 6 1909-dec 28 1928; may 12-oct 17 1939 – 8r – A$513.66 vesicular A$557.66 silver – at Pascoe [079]

Blue mountains gazette – Katoomba, jan 9 1903-dec 30 1904 – 1r – A$45.10 vesicular A$50.60 silver – at Pascoe [079]

Blue mountains gazette – Springwood, jan 1970-aug 1997 – at Pascoe [079]

Blue mountains misc newspapers – Blue Mountains – 2r – A$68.16 vesicular A$79.16 silver – at Pascoe [079]

Blue mountains star – Katoomba, jan 1929-feb 1931 – 1r – 9 – A$60.72 vesicular A$66.22 silver – at Pascoe [079]

Blue mountains times – Katoomba, oct 16 1931-nov 12 1937 – 1r – 9 – A$41.58 vesicular A$47.08 silver – at Pascoe [079]

Blue review : literature, drama, art, music – Killen TX 1913 – 1 – mf#4639 – us UMI ProQuest [073]

Blue ribbon – iss n1-10. nov 1939-mar 1941 – 15 – mf#005MLJ-006MLJ – us MicroColour [740]

The blue ribbon official gazette and gospel temperance herald the signal *see* Gospel temperance herald and blue ribbon official gazette

Blue ridger – 3rd anniv ed. [1945 jul 15], 1984 jun-1993 jul – 2r – 1 – mf#928432 – us WHS [071]

Blue river baptist church. washington county. indiana : church records – 1847-1869 – 1 – 6.93 – us Southern Baptist [242]

Blue sky : the life of harriet caswell-broad / Clark, Joseph Bourne – Boston: Pilgrim Press c1911 [mf ed 1989] – 1mf – 9 – 0-7905-4547-0 – mf#1988-0547 – us ATLA [920]

Blue sky news – National Assoc of Securities Commissioners, 1934-46 (all publ) – 12mf – 9 – $18.00 – (lacking: 1936-37. in 1947 the organization became "the national association of securities administrators") – mf#LLMC 84-426 – us LLMC [332]

Blue springs bee – Blue Springs, NE: C L Peckham (wkly) [mf ed v2 n15. apr 28 1927-sep 26 1946 (gaps)] – 4r – 1 – (absorbed: weekly arbor state) – us NE Hist [071]

Blue valley blade – Seward, NE: McCualley & Betzer. 70v. v2 n25. jul 24 1879-v71 n14. apr 3 1947 (wkly) [mf ed with gaps filmed [1974]] – 26r – 1 – (cont: seward advocate. absorbed : seward journal. cont by: seward blue valley blade) – us NE Hist [071]

Blue valley blade – Wilber, NE: luse & Meeker, jan 24 1884 (wkly) [mf ed v1 n3. feb 7 1884-feb 11 1886 (gaps) filmed [1974?]] – 1r – 1 – us NE Hist [071]

Blue valley journal – McCool Junction, NE: E C Gilliland. 46v. v1 n1. jun 18 1897-v46 n30. dec 31 1942 (wkly) [mf ed with gaps] – 30r – 1 – (cont: mccool junction record. absorbed by: york republican) – us NE Hist [071]

Blue valley record – Beatrice, NE: Howard & Nelson. v1 n1. jul 8 1868-oct 7 1868 (wkly) [mf ed with gaps] – 1r – 1 – (ceased 1869. cont by: beatrice clarion) – us NE Hist [071]

Blue winds talking leaves / Lac du Flambeau Band of Lake Superior Chippewa Indians – v1 n1-v2 n2 [1983 apr 13-1986 jun] – 1r – 1 – (cont: lac du flambeau tribal update) – mf#1354160 – us WHS [305]

Blueberry culture in florida / Mowry, Harold – Gainesville, FL. 1928 – 1r – us UF Libraries [634]

Bluefield College. Virginia *see* Catalogs

Bluefield first baptist church. bluefield, west virginia : church records – 1889-1903, 1912-46 – 1 – 84.69 – us Southern Baptist [242]

Bluefields sentinel – Mar. 10, 1892-Mar. 28, 1894 – 1 – us NY Public [970]

Der bluehende baum : erzaehlung / Sturm, Stefan – Karlsbad: A Kraft 1944 [mf ed 1991] – 1r – 1 – (filmed with: totenhorn-sudwand / karl hans strobl) – mf#2907p – us Wisconsin U Libr [880]

Der bluehende hammer : gedichte / Broeger, Karl – Berlin: Arbeiterjugend-Verlag, 1926 [mf ed 1989] – (= ser Die deutschen arbeiterdichter) – 52p – 1 – mf#7089 – us Wisconsin U Libr [810]

Das bluehende leben : roman / Eckmann, Heinrich – Braunschweig: G Westermann, c1939 [mf ed 1989] – 349p – 1 – mf#7201 – us Wisconsin U Libr [830]

Der bluehende stab : neun geschichten, neun holzschnitte / Schaumann, Ruth – Muenchen: J Koesel & F Pustet c1929 [mf ed 1991] – 1r [ill] – 1 – (filmed with: der krippenweg & other titles) – mf#2868p – us Wisconsin U Libr [830]

Bluejacket / Naval Air Station Memphis (TN) – v40 n48 [1982 dec 2], v42 n27-1928 [1984 jul 5-12], v43 n5-6 [1985 jan 31-feb 7, sec 2], v44 n33-35,37,43-1944,46-47,49 [1986 aug 14-28, sep 11, oct 23-30, nov 13-20, dec 4], v45 n1-8 [1987 jan 8-feb 26], 1987 mar/jul-1993 jan/jul 22 – 11r – 1 – mf#1544416 – us WHS [355]

Bluemel, Rudolf *see* Die deutsche schallform der letzten bluetezeit und ihrer auslaufer in dichtung und prosa

Bluemlein, Carl *see* Die floia und andere deutsche maccaronische gedichte

Bluemmer, H et al *see* Lehrbuch der griechischen antiquitaeten

Bluemner, Heinz Hubertus *see* Mexikanisches erlebnis

Bluemner, Hugo *see* Winckelmanns briefe an seine zuericher freunde

The bluenose – Halifax, N.S.: Imperial Pub. Co., 1900 – 9 – mf#P04162 – cn Canadiana [071]

Bluenose magazine of downeast canada – v1 n1-v4 n6 [1976 summer-1980 jun/jul] – 1r – 1 – mf#524782 – us WHS [071]

Blueprint / Columbus Air Force Base (MS) – 1981 may 1-1982 sep 3 – 1r – 1 – mf#1048993 – us WHS [071]

Blueprint for a british caribbean dominion / British Guiana Bureau Of Public Information – Georgetown, Guyana. 1950 – 1r – us UF Libraries [972]

Blueprint for social justice / Loyola University (New Orleans LA) – 1980 may-1989 nov – 1r – 1 – (cont: blueprint for the christian reshaping of society) – mf#1538780 – us WHS [230]

Blueprint for the christian reshaping of society / Loyola University (New Orleans LA) – 1964 sep-1979 may – 1r – 1 – (cont: christ's blueprint for the south; cont by: blueprint for social justice) – mf#1826635 – us WHS [230]

Blueprints / Wausau Homes, Inc – v13-1916 n2 [1976 spring-1979 may] – 1r – 1 – mf#499095 – us WHS [071]

Blues – Columbus, feb 1929-fall 1930 – 1 – (missing: n1-9.) – us NY Public [073]

Blues and soul, 1967-1986 – (= ser The Journals Of Popular Music) – 16r – 1 – (printed guide with complete bibliographic details available) – mf#C14R-11511 – us Primary [780]

Blues unlimited – Bexhill on Sea UK 1972-87 – 1,5,9 – ISSN: 0006-5153 – mf#8178 – us UMI ProQuest [780]

Der bluetenzweig : ein auswahl aus den gedichten / Hesse, Hermann – Zuerich: Fretz & Wasmuth 1945 [mf ed 1990] – 1r – 1 – (filmed with: das problem "volkstum und dichtung" bei herder / reta schmitz) – mf#2724p – us Wisconsin U Libr [810]

Die bluetezeit der deutschen politischen lyrik von 1840 bis 1850 : ein beitrag zur deutschen literatur und nationalgeschichte / Petzet, Christian – Muenchen, 1903 (mf ed 1994) – 6mf – 9 – €59.00 – 3-8267-3030-5 – mf#DHS-AR 3030 – gw Frankfurter [430]

Die bluetezeit der deutschen predigt im mittelalter, 1100-1400 / Albert, Felix Richard – Guetersloh: C Bertelsmann, 1896 [mf ed 1989] – (= ser Die geschichte der predigt in deutschland bis luther 3) – 1mf – 9 – 0-7905-4364-8 – (incl bibl ref) – mf#1988-0364 – us ATLA [240]

Bluethen des gefuehls gesprossen in meinem erholungsstunden / Reindahl, E Rullmann – Bremen: [J G Heyse] 1819 [mf ed 1995] – 2v in 1 on 1r – 1 – (filmed with: ferdinand raimunds saemtliche werke in drei teilen) – mf#3711p – us Wisconsin U Libr [800]

Bluethen und perlen : sammlung neuerer und aelterer gedichte auslaendischer und einheimischer dichter / ed by Warns, P F L – Milwaukee: im Selbstverlag des herausgebers 1886 [mf ed 1993] – 1r – 1 – (filmed with: neue deutsche gedichte / comp & int by richard bochinger) – mf#3344p – us Wisconsin U Libr [810]

Bluethezeit der romantik / Huch, Ricarda Octavia – Leipzig: H Haessel 1899 [mf ed 1992] – 1r – 1 – (filmed with: deutsche literaturgeschichte des neunzehnten jahrhunderts / friedrich kummer) – mf#3009p – us Wisconsin U Libr [430]

Bluethgen, Viktor *see*
- Badekuren
- Mama kommt!

Bluewater baptist church. dublin, georgia : church records – Oct 1942-Dec 1977. 416p – 1 – us Southern Baptist [242]

Bluff Point, New York. Bluff Point Baptist Church *see* Records

Bluffton general store account book, 1851-1860 – 1v – 9 – (with ind) – mf#34/341 – us South Carolina Historical [650]

Bluffview : news bulletin – 1988 mar-1990 jan/feb – 1r – 1 – (cont: Bluffviews; cont by: newsletter (bluffview acres, inc)) – mf#3183779 – us WHS [071]

Bluffview Acres, Inc *see* Newsletter

Bluffviews – 1963 apr 1-1969 dec 30, 1970 jan 7-1981 dec, 1982 jan-1988 feb – 3r – 1 – (cont by: bluffview) – mf#3183762 – us WHS [071]

Blum baptist church. texas : church records – 1886-1902 – 1 9.27 – us Southern Baptist [242]

[Blum, H] *see* Ein kunstreych buch von allerley antiquiteten

Blum, Hans *see*
- Des beruehmten meister hans blumen von lor am main nuezlichs seulenbuch
- Die deutsch revolution
- Qvinqve colvmnarvm exacta descriptio atque deliniatio

Blum, Karen M et al *see* Section 1983 litigation

Blum, Max *see* Krut un roeben

Blum, Otto *see* Wie erschliessen wir unsere kolonien?

Bluma, Daciano *see* De vita recessuoli...

Blumauer, Alois *see* Alois Blumauer's saemmtliche werke

Blumbergshof : geschichte einer kindheit / Vegesack, Siegfried von – Berlin: Universitas c1933 [mf ed 1991] – 1r – 1 – (filmed with: die gestohlene seele & other titles) – mf#2944p – us Wisconsin U Libr [943]

Blume, Clemens *see* Das apostolische glaubensbekenntniss

-Blume, J *see* Ein jahrtausend lateinischer hymnendichtung

Blume, Wilhelm von *see*
- Der deutsche militarismus
- Das familienrecht des buergerlichen gesetzbuchs
- Das familienrecht des buergerlichen gesetzbuchs; dritter abschnitt; vormundschaftsrecht
- Die wuerzeln der deutschen volkskraft

Blumenauer zeitung – Blumenau (BR), 1921-1938 2 dec [gaps] – 7r – 1 – gw Misc Inst [079]

Blumenfield, Samual N *see* Rashi ha-moreh

Blumen-Gesellschaft an der Pegnitz *see* Die betruebte pegnesis

Blumenlese aus den saemmtlichen werken von johann rudolf wyss dem juengern / Wyss, Johann Rudolf; ed by Greyerz, Otto von – Bern: K J Wyss 1872 [mf ed 1995] – 1r – 1 – (filmed with: der fahrende schueler / julius wolff) – mf#3765p – us Wisconsin U Libr [802]

Die blumensprache *see* Abendlied

Blumenstein, J *see* Die verschiedenen eidesarten nach mosaisch-talmudischem rechte und die faelle ihrer anwendung

Blumenstiel, Alexander *see*
- An act to establish a uniform system of bankruptcy throughout the united states
- The law and practice in bankruptcy.

Der blumenstrauss *see* Gedichtbuechelchen

Blumenthal, Harvey *see* Factors affecting the performance of intramural officials in competitive situations

Blumenthal, Lieselotte *see*
- Ein notizheft goethes von 1788
- Studien zur goethezeit

Blumenthal, M *see* Formen und motive in den apokryphen apostelgeschichten

Blumenthal, Oscar *see*
- Allerhand ungezogenheiten
- Buch der sprueche
- Die grosstadtluft
- Lebensschwaenke
- Nachdenkliche geschichten
- Paula's geheimnis
- Der schwur der treue
- Ein tropfen gift

Blumenthal, Oskar *see*
- Christian dietrich grabbe's saemmtliche werke und handschriftlicher nachlass
- Im weissen roessl
- Scherzgedichte

Blumer, J J *see* Die reformation im lande glarus

Blume's unreported opinions / Michigan. Supreme Court – 1v. 1836-1843 (all publ) – (= ser Michigan supreme court reports) – 3mf – 9 – $4.50 – (a pre-nrs title) – mf#LLMC 88-050 – us LLMC [347]

Blumhardt, James Fuller [comp] *see* Catalogue of the hindi, panjabi and hindustani manuscripts in the library of the british museum

Blumhardt, Johann Christoph *see*
- Handbuch der missionsgeschichte und missionsgeographie
- Von der nachfolge jesu christi

Blumner, Hugo *see* Romischen privataltertumer

Blumstein, Isaac *see* Bay di andn

Blumtritt, Walter *see* Der buecherwurm [klp9]

Blunck, Barthold *see* Der kapitaen

Blunck, Hans Friedrich *see*
- Eine auswahl aus dem dichterischen werk
- Bootsmann elbing
- Das brautboot
- Bruder und schwester
- Dammbruch
- Doerfliches leben
- Der feuerberg
- Der flammenbaum
- Frauen im garten
- Gestuehl der alten
- Gewalt ueber das feuer
- Glueckliche insel
- Die grosse fahrt
- Hein hoyer
- Italienisches abenteuer
- Die jaegerin
- Jungfern im nebel
- Kampf der gestirne
- Die kleine ferne stadt
- Koenig geiserich
- Der landsknecht
- Die luegenwette
- Mein leben
- Sage vom reich
- Schiffermaer
- Sicht des werkes
- Sommer im holmenland
- Spuk und luegen
- Streit mit den goettern
- Der trost der wittenfru
- Volkswende
- Vom ueberlisteten teufel
- Werdendes volk
- Wolter von plettenberg
- Der wundervogel

Blundell, Henry *see* Engravings and etchings of the principal statues, busts, bass-reliefs

"Blundering theology" : plain words addressed to mr thomas cooper, with notes and comments on his recent lectures & sermons, delivered in newcastle, july 12 to 18th, 1858 / Lucas, George – Gateshead, England: R Jackson 1858 – 1r – 1 – us UF Libraries [240]

Blunders and forgeries : historical essays / Bridgett, T E – London: Kegan Paul, Trench, Truebner, & Co, 1890 – 4mf – 9 – $6.00 – (allegations of anti-catholic bias in the press) – mf#LLMC 92-117 – us LLMC [320]

Blunders and forgeries : historical essays / Bridgett, T E – London: K. Paul, Trench, Truebner, 1890 – 1mf – 9 – 0-7905-5981-1 – mf#1988-1981 – us ATLA [240]

Blunham – (= ser Bedfordshire parish register series) – 2mf – 9 – £5.00 – uk BedsFHS [929]

Blunt, Anne *see* A pilgrimage to nejd, the cradle of the arab race

Blunt, Edward *see* Social service in india

Blunt, Henry *see* Lectures on the history of elisha

Blunt, J J *see* Introduction to a course of lectures on the early fathers

Blunt, John *see* Vestiges of ancient manners and customs, discoverable in modern italy and sicily

Blunt, John Henry *see*
- The annotated bible
- The atonement and the at-one-maker
- A companion to the new testament
- A companion to the old testament
- Dictionary of doctrinal and historical theology
- Dictionary of sects, heresies, ecclesiastical parties, and schools of religious thought
- A key to the knowledge and use of the holy bible
- The reformation of the church of england

Blunt, John James *see*
- The acquirements and principal obligations and duties of the parish priest
- The christian church during the first three centuries
- Principles for the proper understanding of the mosaic writings stated and applied
- Undesigned coincidences in the writings of both the old and new testament

Blunt, Walter *see* Confirmation
Blunt, Wilfrid Scawen *see*
- The future of islam
- Ideas about india

Bluntschli, H H *see* Memorabilia tigurina oder merkwuerdigkeiten der stadt und landschaft zuerich
Bluntschli, J R *see* Geschichte des schweizerischen bundesrechtes von den ersten ewigen buenden bis auf die gegenwart
Bluntschli, Johann Caspar *see*
- Alt-asiatische gottes- und weltideen in ihren wirkungen auf das gemeinleben der menschen
- Deutsches privatrecht
- Staats- und rechtsgeschichte der stadt und landschaft zuerich

Blut – Dordrecht, Netherlands 1981-84 – 1,5,9 – ISSN: 0006-5242 – mf#13107.01 – us UMI ProQuest [616]
Blut und eisen / Eyth, Max – Wiesbaden: Verlag des Volksbildungsvereins zu Wiesbaden [1909?] [mf ed 1989] – 1r – 1 – (pref by e eschertch; repr fr: hinter pflug und schraubstock) – mf#7228 – us Wisconsin U Libr [880]
Blut und rasse im deutschen dichter- und denkertum : eine auslese / Hoffmann, Paul Theodor [comp] – Hamburg: Hoffmann & Campe Verlag [c1934] [mf ed 1993] – 1r – 1 – (incl bibl ref. filmed with: sputnik contra bombe / ed by gerhard wolf) – mf#3336p – us Wisconsin U Libr [430]
Der blutaberglaube bei christen und juden / Strack, Hermann Leberecht – 3. Abdr. Muenchen: C.H. Beck (Oskar Beck), 1891 – (= ser Schriften Des Institutum Judaicum In Berlin) – 1mf – 9 – 0-8370-2027-1 – mf#1985-0027 – us ATLA [270]
Blutiker onhoyb / Grudzien, M N – Montevideo, Uruguay. 1945 – 1r – us UF Libraries [939]
Bluwstein, J *see* Spinozas briefwechsel und andere dokumente
Bly bulletin – Bly OR: Bulletin Press, [wkly] [mf ed 1964] – 1r – 1 – us Oregon Lib [071]
Blyde inkomst der allerdoorluchtighste koninginne : maria de medicis t'amsterdam / Baerle, K van – Amsterdam: Cornelis Blaev, 1639 – 4mf – 9 – mf#0-1113 – ne IDC [090]
Blyden, Edward Wilmot *see*
- Addresses and correspondence, 1857-1908
- The problems before liberia
- Writings, 1862?-1908

Blyden of liberia : an account of the life and labors of edward wilmot blyden, ll.d., as recorded in letters and in print / Holden, Edith – 1st ed. New York: Vantage Press, [1967, c1966] – us CRL [920]
Blyth citizen – Ontario, CN. jan-dec 1986 – 1r – 1 – cn Commonwealth Imaging [071]
Blyth, Frederic Cavan *see* Thoughts on the lord's prayer
[Blythe-] blythe herald – CA. jan 21 1915-1925 – 6r – 1 – $360.00 – mf#R03164 – us Library Micro [071]
Blythe, James *see* The death of the good man...rev john brown
[Blythe-] palo verde valley times – CA. jan 18 1925-dec 1989 – 71r – 1 – $4260.00 – mf#R02072 – us Library Micro [071]
Blythe spirit – Blytheville AR. 1981 may 1-1982, 1982-1985 jun, 1985 jul-1987 feb, 1987 mar-1988 mar, 1988 apr 7-1989 may 25 – 5r – 1 – mf#648442 – us WHS [071]
Blythe, Wayne T *see* The use of noye in the authentically pauline epistles
B'man family newsletter – v5 n2-v8 n2 [1981 oct-1984 nov] – 1r – 1 – (cont: b'man newsletter) – mf#871439 – us WHS [640]
B'man newsletter – 1980 jul-81 apr – 1r – 1 – (cont: beeman family newsletter (1977); cont by: b'man family newsletter (1981)) – mf#947569 – us WHS [071]
Bm/e [broadcast management/engineering] – Manhasset NY 1970-87 – 1,5,9 – (cont by: bme for technical and engineering management) – ISSN: 0005-3201 – mf#5914.02 – us UMI ProQuest [621]
Bme for technical and engineering management – Manhasset NY 1989 – 1,5,9 – (cont: bm/e: broadcast management/engineering; cont by: bme's television engineering) – ISSN: 1043-7487 – mf#5914.02 – us UMI ProQuest [380]
Bme marine engineer / Seafarers' International Union of North America – 1952 may-1958 dec, 1959 jan-jun – 2r – 1 – mf#1110046 – us WHS [623]
Bmi, the many worlds of music – New York NY 1973-87 – 1,5,9 – (cont by: musicworld) – ISSN: 0045-317X – mf#9498.01 – us UMI ProQuest [780]
Bmk-kurier *see* Kurier
Bmwe journal / Brotherhood of Maintenance of Way Employees – 1988 feb-1990 nov/dec, 1991 jan/dec-1994 nov/dec – 2r – 1 – mf#1766665 – us WHS [625]
B'nai b'rith anti-defamation league. adl bulletin – New York NY 1973-91 – 1,5,9 – (cont by: adl on the frontline) – ISSN: 0001-0936 – mf#8390 – us UMI ProQuest [320]
B'nai b'rith berlin : monatsschrift der berliner logen – Berlin DE, 1926-33 [gaps] – 1r – 1 – gw Misc Inst [270]
B'nai b'rith berlin : monatsschrift der berliner logen. uobb / ed by Baer, Karl M – Berlin. v1-13? 1921-1933? – (= ser German-jewish periodicals...1768-1945, pt 1) – 1r – 1 – $125.00 – (lacking: v1-5 1921-25 and misc iss in v6,8,9,12) – mf#B37 – us UPA [270]
B'nai b'rith covenant – Downsview, Ontario. Oct 1975-15 Dec 1987. Many issues missing – 1 – us AJPC [071]
B'nai b'rith deutschland – Berlin DE, 1891 apr-1937 – 3r – 1 – gw Misc Inst [270]
B'nai b'rith deutschland : der orden bne briss. mitteilungen der grossloge fuer deutschland 8 uobb – Berlin. 1891-1937 – (= ser German-jewish periodicals...1768-1945, pt 1) – 3r – 1 – $325.00 – (lacking: misc iss) – mf#B40 – us UPA [270]
B'nai B'rith District No 10 "Moravia" *see* Gedenkschrift zur feier des 25 jahr bestandes des israel
B'nai B'rith District No8 Silesia-Loge 36, Nr 477 *see* Gesetze, parlamentarische regeln und geschafts-ordnung
B'nai b'rith district two news – Cincinnati, OH. 1980-84 – 1 – us AJPC [071]
B'nai B'rith Grossloge Fur Deutschland 8 *see* Gedenkblatter fur die bruder ehrenvizegrossprasident hugo kuznitzky
B'nai B'rith Hillel Foundation (University of Wisconsin) *see*
- Hillel-o-gram
- Hillel-o-grams

B'nai b'rith international jewish monthly – Washington DC 1982+ – 1,5,9 – (cont: national jewish monthly) – ISSN: 0279-3415 – mf#10259.03 – us UMI ProQuest [939]
B'nai b'rith messenger – CA.1958 67 – 1 – us AJPC [939]
B'nai b'rith mitteilungen fuer oesterreich – Wien (A), 1924/25-1938 n1/2 – 1 – gw Misc Inst [074]
B'nai b'rith voice – San Antonio, TX.Jun 1960 – 1 – us AJPC [071]
B'nai brith voice – San Antonio, TX.May 1976-Apr 1987. Incomplete – 1 – us AJPC [071]
Bnai zion traveler – New york, NY. spring 1975-spring/summer/fall 1985 – 1 – (many iss lacking) – us AJPC [071]
Bna's union labor report / Bureau of National Affairs, Washington DC – 1955 jan-1956 jun 22 – 1r – 1 – (cont by: union labor report weekly newsletter) – mf#403984 – us WHS [331]
Bnchler, Johann *see* Land- und seereisen eines st gallischen kantonsbnrgers nach nordamerika und westindien
Bndd bulletin / U.S. Bureau of Narcotics and Dangerous Drugs – v1-4. 1969-73 (all publ) – 6mf – 9 – $9.00 – (lacking: v1 nos 2,4) – mf#LLMC 81-200 – us LLMC [360]
Bo luat dan-su' va thu'o'ng-su' to-tung / Nguyen Hung Tru'o'ng – Saigon: Khai-Tri 1973. 167p. LL-10031 – 1 – us L of C Photodup [340]
Bo, Tjan Ing *see* Terate mas
Boa, Myrtle J *see* Idylls of our island
Board of Arts and Manufactures for Lower Canada *see*
- Quarterly report of the sub-committee of the board of arts and manufactures for lower canada
- Report of the sub-committee of the board of arts and manufactures for lower canada

Board of arts and manufactures for upper canada. journal – Toronto. v1-8. jan 1861-feb 1868// – 3r – 1 – Can$330.00 – cn McLaren [600]
Board of contract appeals decisions / U.S. Dept of the Interior – v1-21. 1970-83 – 84mf – 9 – $126.00 – mf#llmc 82-202 – us LLMC [346]
Board Of County Commissioners Of Marion County, FL *see* Marion county, florida
Board of directors correspondence and committee materials, 1919-1955 – 2ser+suppl – (= ser Papers of the naacp 16) – 1 – (ser a: 1919-39 8r isbn 1-55655-477-X $1560. ser b: 1940-55 24r isbn 1-55655-478-8 $4640. suppl: board of directors files 1956-65 12r isbn 1-55655-877-5 $1750; 1966-70 isbn 1-55655-222-2 $1750. with p/g) – us UPA [322]
Board of directors minute books / Free Public Library, Ottawa, KS – 1872-1945 – 1 – us Kansas [020]
The board of directors of the young men's christian association of stratford, ont cordially invite your presence at luncheon in the new building in the market square : on the afternoon of thursday, march 3rd, from 4 to 6 o'clock, an early reply to the secretary is requested – S.l: s.n, 18–? – 1mf – 9 – mf#53228 – cn Canadiana [366]
Board of foreign missions records, 1908-1923 / Evangelical Lutheran Augustana Synod of North America. Board of Foreign Missions – 1r – 1 – (forms pt of: subgroup aug 24/5 china mission board; contains records fr the china mission board (cmb) from 1908-1923; comprise handwritten & typewritten minutes, correspondence, financial records, corporation documents, and other reports chronicling the work and evolution of the cmb; with finding aid) – mf#xa0114r – us ATLA [242]
Board of indian appeals decisions and orders / U.S. Dept of the Interior – v1-26. 1970-95 – (= ser Native american coll) – 131mf – 9 – $196.00 – (incl ind for v1-26. v1-8 pt of native american collection. updates available) – mf#llmc 82-203 – us LLMC [343]
Board of indian commissioners' annual reports / U.S. Dept of the Interior – 1869-1933 [all publ] – (= ser Native american coll) – 97mf – 9 – $145.00 – (lacking: reports n1, 18, 64) – mf#llmc 88-005 – us LLMC [343]
Board of land appeals decisions and orders / U.S. Dept of the Interior – v1-98. 1970-87 – 599mf – 9 – $898.00 – (updates available) – mf#llmc 82-204 – us LLMC [343]
Board of mine operations appeals decisions and orders / U.S. Dept of the Interior – v1-8. 1970-78 [all publ] – 39mf – 9 – $58.00 – mf#llmc 82-205 – us LLMC [622]
Board of National Missions (United Pres. Ch. in the USA) *see* Minutes, 1958-1972
Board of review, jag branch, china-burma-india/india-burma theater, holdings, opinions and reviews / U.S. Army. Judge Advocate General. Board of Review, CBI-IBT – v1-3. 1943-45. Washington: Office of the JAG, 1946? – 12mf – 9 – $18.00 – (covers: cm-cbi-15 to cm-cbi-249. cm-ibt-282 to cm-ibt-762) – mf#LLMC 84-224 – us LLMC [355]
Board of supervisors proceedings – Lake Co, CA. feb 6 1905-jul 9 1918 – 1r – 1 – $50.00 – mf#B40225 – us Library Micro [978]
Board of tax appeals reports – v1-47. 1924-1942 – 738mf – 9 – $1107.00 – mf#LLMC 79-420 – us LLMC [332]
Board of tax appeals, reports / U S Tax Court – v1-47. 1924-1942 – 9 – $975.00 set – (cont as: tax court reports) – mf#200081 – us Hein [336]
Board of trade journal, 1886-1960 – v1-179 – 146r – 1 – mf#95852 – uk Microform Academic [324]
Board of trade news – Toronto, Canada apr 1911-jul 1917, jan-nov 1918, feb 1919-dec 1921 – 2r – 1 – uk British Libr Newspaper [380]
Board of Trade of the City of Toronto *see* Monthly returns of imports and exports at the port of toronto for...
Board of trade: registrar general of shipping and seamen – 1 – (registers of births at sea of british nationals 1875-1891 [6v]; registers of deaths at sea of british nationals 1875-1888 [10v]) – uk National [380]
Board of trade: registrar general of shipping and seamen and predecessor: registers of births, deaths and marriages of passengers at sea – 1854-90 – 9v – 1 – uk National [380]
Board of trade: registrar general of shipping and seamen: register of seamen, central index – 1 – (alphabetical ser (cr1): 1921-41; numerical ser (cr2): 1921-41) – uk National [380]
Board of trade/overseas department economic surveys, 1921-1961 : economic surveys from 125 countries / Great Britain – [mf ed Chadwyck-Healey] – 806 reports on 1113mf – 9 – (individual rpts available separately: africa 90 surveys on 107mf. asia 85 surveys on 144mf. australasia 29 surveys on 51mf. europe & the near east 300 surveys on 422mf. latin america & the caribbean 206 surveys on 256mf. middle east 51 surveys on 61mf. north america 35 surveys on 70mf) – uk Chadwyck [330]
Board of trustees correspondence 1918 / General Synod of the Evangelical Lutheran Church in the United States. Woman's Home and Foreign Missionary Society – 1r – 1 – (with ind; forms pt of: subgroup gs 16/3 board of trustees; reel also incl xa0104r & xa0105r) – mf#xa0106r – us ATLA [242]
Board of trustees minutes 1884-1916 / General Synod of the Evangelical Lutheran Church in the United States. Woman's Home and Foreign Missionary Society – 1r – 1 – (with ind; forms pt of: subgroup gs 16/3 board of trustees) – mf#xa0094r – us ATLA [242]
Board of trustees, records / Baptist State Conventions (American Baptist). Vermont – 1824-1975 – 1 – (executive committee, records, 1964-75. commission on ordination, records, 1955-62. deeds and wills, 1886-1961) – us ABHS [240]
Board to morgon tidningen *see* Morgon tidningen social demokraten
Boarding and day school for young ladies and children : conducted by mrs s sinclair, (widow of the late samuel sinclair of montreal), and miss sinclair, (formerly of the church of england ladies' school, ottawa), no 119 o'connor street, ottawa – [Ottawa?: s.n, 18–] [mf ed 1983] – 1mf – 9 – 0-665-38371-1 – mf#38371 – cn Canadiana [370]
Boardman, George Dana *see*
- The church
- Ethics of the body
- Life and light
- Martin luther
- The problem of jesus
- Studies in the creative week
- Studies in the model prayer
- Unity of the church

Boardman, George Nye *see*
- Congregationalism
- Female education
- A history of new england theology
- Regeneration

Boardman, Henry Augustus *see*
- A discourse on the life and character of daniel webster
- The doctrine of election
- Earthly suffering and heavenly glory
- The general assembly of 1866
- The great question, will you consider the subject of personal religion?
- The prelatical doctrine of the apostolical succession examined
- A treatise on the scripture doctrine of original sin
- Two sermons
- The vanity of a life of fashionable pleasure

Boardman, James *see* America, and the americans
Boardman, John *see* The cretan collection in oxford and the dictean cave and iron age crete
Boardman mirror – Boardman OR: M A Cleveland, 1921-25 [wkly] [mf ed 1977] – 2r – 1 – (absorbed by: arlington bulletin (-1942)) – us Oregon Lib [071]
Boardman mirror *see* Arlington bulletin
Boardman, William Edwin *see*
- Faith work under dr. cullis in boston
- Gladness in jesus
- The higher christian life
- In the power of the spirit

Boardroom reports – Stamford CT 1975-95 – 1,5,9 – (cont by: bottom line business) – ISSN: 0045-2300 – mf#10136.01 – us UMI ProQuest [650]
The boards of trade general arbitrations act (1894) / Beatty, William Henry – Toronto, Hunter, Rose, 1894. 88 p. LL-2290 – 1 – us L of C Photodup [343]
Boas, Franz *see*
- Chinook texts
- Facial paintings of the indians of northern british columbia
- Indianische sagen von de nord-pacifischen kueste amerikas
- The mind of primitive man
- The mythology of the bella coola indians
- The professional correspondence of franz boas

Boas, Franz et al *see* Anthropology in north america
Boas, J *see* Buried with christ in baptism
Boase, Henry Samuel *see* A few words on evolution and creation
Boat trips / Goebel, Rubye K – s.l, s.l? . 1936 – 1r – us UF Libraries [910]
Boat who wouldn't float / Mowat, Farley – Toronto, ON. 1974 – 1r – us UF Libraries [025]
Boateng, Ernest Amano *see* A geographical study of human settlement in the eastern province of the gold coast colony west of the volta delta
Boating – New York NY 1956+ – 1,5,9 – ISSN: 0006-5374 – mf#1171 – us UMI ProQuest [790]
Boating safety newsletter – n1-32 [1976-1989] – 1r – 1 – mf#2461496 – us WHS [629]
Boatman of the padma / Bandyopadhyay, Manik – Bombay: Kutub, 1948 – (= ser Samp: indian books) – (transl from the bengali by hirendranath mukerjee) – us CRL [890]
Boatos – Rio de Janeiro, RJ: Typ J Paulo Hidelbradt, 1876 – (= ser Ps 19) – mf#P17,01,111 – bl Biblioteca [079]
Boaventura *see* Reencarnacionismo no brasil
Bob, der sonderling : seine geschichte und seine gedanken / Bertsch, Hugo – 4. aufl. Stuttgart; Berlin: J G Cotta, 1905 [mf ed 1989] – 227p – 1 – mf#7014 – us Wisconsin U Libr [830]

Bob Kastenmeier reports from the u s house of representatives – 1976 jan-1980 dec – 1r – 1 – (cont: your congressman bob kastenmeier reports from washington; cont by: congressman bob kastenmeier reports) – mf#1048491 – us WHS [323]
Bob kastenmeier reports to farmers from the u s house of representatives – 1979 mar-1982 mar – 1r – 1 – mf#655229 – us WHS [630]
Bobadilla, Emilio *see*
– A fuego lento
– Articulos periodisticos
– Muecas
Bobadilla, Perfecto H *see* Cartilla historica de honduras
Bobadilla Y Briones, Tomas *see* Discursos de bobadilla
Bobbin – New York NY 1993-2002 – 1,5,9 – ISSN: 0896-3991 – mf#16370.03 – us UMI ProQuest [670]
The bobbio missal (hbs61) : notes and studies / Wilmart, A et al – 1924 – (= ser Henry bradshaw society (hbs)) – 3mf – 8 – €7.00 – ne Slangenburg [241]
The bobbio missal, vol 1 (hbs53) : facsimile / Lowe, E A – 1917 – (= ser Henry bradshaw society (hbs)) – 9mf – 8 – €18.00 – ne Slangenburg [241]
The bobbio missal, vol 2 (hbs58) / Lowe, E A – 1920 – (= ser Henry bradshaw society (hbs)) – 4mf – 8 – €11.00 – ne Slangenburg [241]
Bobbsey twins : merry days indoors and out / Hope, Laura Lee – Racine, WI. 1950 – 1r – us UF Libraries [960]
Bobe, L *see*
– Christian lunds relation til kong frederik 3 om david danells tre rejser til gronland 1652-1654
– Dagboeger fra hans rejser i gronland 1739-1753
Bobea, Joaquin Maria *see* Hortaliza
Boberach, Heinz *see*
– Regimekritik, widerstand und verfolgung in deutschland und den besetzten gebieten
– Reichssicherheitshauptamt (bestand r 58) bd 22
Bobertag, Felix *see* Wielands romane
Bobo and Ray Families. South Carolina *see* Genealogy
Bobo, Rosalvo *see* A propos du centenaire
Bobrik, Eduard *see* Histoire de la franc-maconnerie
Bobrik, Johannes Eduard Guenther *see* Wielands don sylvio und oberon auf der deutschen singspielbuehne
Bobrishchev-Pushkin, Aleksandr M *see* Sud i raskoli' nikisektanty
Bobzin, Hartmut *see* Sammlung wagenseil
Boca da grota / Gusmao, Carlos De – Maceio, Brazil. 1970 – 1r – us UF Libraries [972]
Boca grande : lee county / Lamme, Corinne W – s.l, s.l? . 1936 – 1r – us UF Libraries [978]
Boca Raton Historical Society *see* Spanish river papers
Bocage : sua vida e epoca litteraria / Braga, Teofilo – Porto, Portugal. 1902 – 1r – us UF Libraries [025]
Bocanegra, Matias de *see* Auto general de la fee
Bocangelino, N *see* Libro de las enfermedades malignas y pestilentes, causas, pronosticos, curacion...
Boccaccio, Giovanni *see*
– De certaldo isignis opus de claris mulieribus
– De claris mulieribus
– Decameron
– Decamerone
– Il filocolo
– Opere minori
– Ein schoene cronica
Boccanera, Silio *see* Bahia epigraphica e iconographica
Boccherini, Luigi *see*
– Partition du stabat mater a trois voix avec deux violons, alto, violoncelle et contre basse [op 61]
– Quartet no 44 for 2 violins, viola and cello
– Sei quartetti concertanti per due violini, alto e violoncello, opera 27
– Sei quartetti, per due violini, alto, e violoncello obbligati...opera ia
– Sei quartetti per due violini, alto et basso, opera 26
– Sei quartetti per due violini, viola e violoncello, opera 33
– Sei quintetti concertanti, per due violini, alto et due violoncelle, opera 17, libro quarto di quintetti
– Sei quintetti, per due violini, alto et due violoncelli concertanti, opera 23
– Sei quintetti, per due violini, alto et due violoncelli concertanti...opera 20, libro terzo di quintetti
– Sei sextuor pour due violini, viola, due violoncelle e flauto, opera 15
– Sei sinfonie o sia quartetti per due violini, alto, e violoncello obbligati...opera 1a
– Sei trio per due violini, e violoncello obbligato...
– Six quartetto ou divertissement pour deux violons, taille et basse, oeuvre 7
– Six quatuors a deux violons, taille et basse obliges, oeuvre 7
– Six quintetti pour flute, deux violons, alto et violoncelle, oeuvre 21e
– Six quintettos for two violins, a tenor, and two violoncellos, 3d set, op 20
– Six sonates a violon seul et basse
– Six sonates pour le clavecin avec l'accompagnement d'un violon & violoncelle, oeuvre 12
– Trij 6 a due violini, e basso, muta 2da
Bocchi, A *see*
– Symbolicarum quaestionum de universo genere quas serio ludebat libri quinque
– Symbolicarum quaestionum, de universo genere, quas serio ludebat, libri quinque
Bocchi, F *see* Le bellezze della citta di firenze
Bocetos y recuerdos / Galarreta, Luis Adam – Habana, Cuba. 1915 – 1r – us UF Libraries [972]
Boch, Gustav *see* Neue berliner musikzeitung
Bochartus, Sam *see* Opera omnia
Bocher, Maxime *see* Introduction to the study of integral equations
Bochimanes! : khu de angola / Guerreiro, Manuel Viegas – Lisboa, Portugal. 1968 – 1r – us UF Libraries [960]
Bochinger, Richard [comp] *see* Neue deutsche gedichte
Bochius, J *see*
– Descriptio publicae gratulationis, spectaculorum et ludorum, in adventu sereniss principis ernesti archiducis austriae...
– Historica narratio profectionis et inaugurationis serenissimorum belgii principum alberti et isabellae, austriae archiducum
Bocholter-borkener volksblatt – Bocholt DE, 1972-76 – 26r – 1 – (filmed by misc inst: 1977- (ca 8r/yr)) – gw Mikrofilm; gw Misc Inst [074]
Bochsa, Robert Nicolas Charles *see*
– Deux nocturnes pour harpe & hautbois, oeuvre 50 no 1
– Judith
Der bochumer arzt dr carl arnold kortum, der dichter der jobsiade : sein leben und sein wirker / Tegeler, Ernst – Jena: Gustav Fischer 1931 [mf ed 1995] – (= ser Arbeiten zur kenntnis der geschichte der medizin im rheinland und in westfalen 7) – 1r [ill] – 1 – (incl bibl ref. filmed with: lebenserinnerungen des alten mannes in briefen an seinen bruder gerhard / wilhelm von kugelgen) – mf#3678p – us Wisconsin U Libr [610]
Bochumer kreisblatt 1842 – Bochum DE, 1830-39, 1848-49, 1887-98 – 15r – 1 – (title varies: 1 jul 1848: maerkischer sprecher, 18 sep 1909: bochumer zeitung. maerkischer sprecher; filmed by other misc inst: 1848 1 jan-24 jun) – gw Misc Inst [074]
Bochumer tageblatt *see* Heimat am mittag / bochumer tageblatt
Bock, Adolf *see* Die thueringische eisenbahn
Bock, Elfried *see* Holtschnitte des meisters ds
Bock, Ulrich Manfred *see* Hepatitis c bei haemodialysepatienten
Bock, W de *see* Materiaux pour servir a l'archeologie de l'egypte chretienne
Bocker Family *see* Diaries
Bocket, Thomas J *see* Differences in physical activity attitudes and fitness knowledge between health fitness standard, sex, and grade group
Bocklenberg, Ute *see* Moderne blockfloetentechniken und ihre vermittlung
Bocksgesang : in fuenf akten / Werfel, Franz – Muenchen: K Wolff 1921 [mf ed 1991] – 1r – 1 – (filmed with: der weg ins licht / gisela wenz-hartmann) – mf#3011p – us Wisconsin U Libr [820]
Bockum-hoeveler zeitung – Hamm (Westf) DE, 1957 1 oct-1959 – 1 – (bezirksausgabe von westfaelischer anzeiger und kurier) – gw Misc Inst [074]
Boclo, Ludwig *see* Der begleiter auf dem weser-dampfschiffe von muenden nach bremen
Bocquet pere et fils / Laurencin, M – Paris, France. 1840 – 1r – us UF Libraries [440]
Bod, Peter *see* Historia hungarorum ecclesiastica
Boda profunda / Orta Ruiz, Jesus – Habana, Cuba. 1957 – 1r – us UF Libraries [972]
Bodajbo. Priiskovyj Sovet rabochikh deputatov *see* Izvestiia soveta rabochikh deputatov priiskovogo rajona
Bodak, Shirley L *see*
– Brewer bulletin
– Dooley bulletin
– Grimes bulletin
– Howell-howel bulletin
– Page heritage
– Phillips bulletin
– Vanderslice bulletin
– Vanderslice newsletter
Bodarc record – Bodarc, NE: Slingerland & Hunter, 1886 (wkly) [mf ed v1 n4. nov 11 1886-apr 22 1887 (gaps) filmed 1975] – 1r – 1 – (absorbed by: sioux county herald) – us NE Hist [071]
Bodard, Auguste *see*
– Le cure labelle
– Emigration en canada
– En route pour le canada
– Exposition des faits et de la situation actuelle de la societe de colonisation du temiscamingue vis-a-vis des actionnaires francais
Bodas de plata misionales de la compania de maria... – Villavicencio, Colombia. 1929 – 1r – us UF Libraries [972]
Bodas y obras juveniles de zurbaran / Caturla, Maria Luisa – Granada: Imp. Francisco Roman Camacho, 1948 – sp Bibl Santa Ana [946]
Boddam Whetham, John Whetham *see* Across central america
Boddam-Whetham, John Whetham *see*
– Across central america
– Roraima and british guiana
Bodding, P O *see* Santal folk tales
Boddington, Mary *see* Slight reminiscences of the rhine, switzerland, and a corner of italy
Bode, G H *see* Scriptores rerum mythicarum tres romae nuper reperti
Bode, John Ernest *see* The absence of precision in the formularies of the church of england scriptural
Bode, Mabel Haynes *see* The pali literature of burma
Bode, Wilhelm *see*
– Der froehliche goethe
– Goethe in vertraulichen briefen seiner zeitgenossen
– Goethe ueber freunde und feinde
– Goethes gesundheitspflege
– Goethes leben
– Goethes lebenskunst
– Goethes persoenlichkeit
– Goethes weg zur hoehe
– Die tonkunst in goethes leben
– Weib und sittlichkeit in goethes leben und denken
– Der weisheit letzter schluss im faust
Bodecker, Albert *see* Reminiscences of life in butler county
[Bodega bay-] the bodega bay navigator – CA. oct 1987- – 2+ r – 1 – $120.00 (subs $50/y) – mf#B05029 – us Library Micro [071]
[Bodega bay-] the bodega bay signal – CA. apr 1985- – 12+ r – 1 – $720.00 (subs $50/y) – mf#B05028 – us Library Micro [071]
Bodemann, Eduard *see* Julie von bondeli und ihr freundeskreis
Bodemann, Friedrich Wilhelm *see* Der wichtigsten bekenntnisschriften der evangelisch-reformirten kirche
Boden – Lulea, Sweden. 1899-1905 – 2r – 1 – sw Kungliga [078]
Boden, August *see* Zur beurtheilung der christlichen glaubenslehre des dr. strauss
Bodenheimer, M J *see* Wohin mit den russischen juden?
Bodennutzung in venezuela : oekologische und oekonomische aspekte von markt- und subsistenzproduktion am beispiel des gemeindelandes von yaritagua, unter besonderer beruecksichtigung ihrer auswirkungen auf den boden / Jordan, Ronald – (mf ed 1991) – 1mf – 9 – €49.00 – 3-89349-450-2 – mf#DHS 450 – gw Frankfurter [333]
Bodenreuth, Friedrich *see*
– Alle wasser boehmens fliessen nach deutschland
– Das ende der eisernen schar
Bodens tidning – Boden, Sweden. 1986-88 – 1 – sw Kungliga [078]
Die bodenschaetze der kampfgebiete in ihrer bedeutung fuer uns und unsere feinde / Pompecki, Josef Felix – Tuebingen: Kloeres, 1915. 24p – 1 – us Wisconsin U Libr [943]
Der bodensee nebst dem rheinthale von s[ankt] luziensteig bis rheinegg : handbuch fuer reisende und freunde der natur, geschichte und poesie / Schwab, Gustav – Stuttgart [u.a.] 1827 [mf ed Hildesheim: 1995-98] – (= ser Fbc) – viii/550p on 4mf – 9 – €120.00 – 3-487-29490-7 – gw Olms [914]
Der bodenseher / Finckh, Ludwig – Muenchen: Deutscher Volksverlag [1914?] [mf ed 1989] – 1r – 1 – (filmed with: double, double, toil and trouble / lion feuchtwanger) – mf#7237 – us Wisconsin U Libr [830]
Die bodenstaendigkeit der synoptischen ueberlieferung vom werke jesu / Heinrici, Carl Friedrich Georg – Berlin: Edwin Runge 1913 [mf ed 1989] – (= ser Biblische zeit- und streitfragen 8/11) – 1mf – 9 – 0-7905-2593-3 – mf#1987-2593 – us ATLA [226]
Bodenstedt, F von *see*
– Tausend und ein tag im orient
– Die voelker der kaukasus und ihre freiheitskaempfe gegen die russen
Bodenstedt, Friedrich *see*
– Verschollenes und neues
– Vom atlantischen zum stillen ozean
Bodenstedt, Friedrich M von *see*
– Aus meinem leben
– Die voelker des kaukasus und ihre freiheitskaempfe gegen die russen
Bodenstedt, Friedrich Martin von *see*
– Aus morgenland und abendland
– Friedrich bodenstedt's gesammelte schriften
– Die lieder des mirza-schaffy
Bodette, Derek R *see* The perceptions of college physical education teaching majors with respect to teacher enthusiasm
Bodhicaryavatara : introduction a la pratique des futurs bouddhas / Santideva – Paris: Bloud, 1907 – 1mf – 9 – 0-524-06896-8 – mf#1991-0039 – us ATLA [280]
Bodhisattva, Asvaghosha *see* The fo-sho-hing-tsan-king (stbe19)
The bodhisattva ti-tsang (jizo) in china and japan / Visser, Marinus Willem de – Berlin: Oesterheld, 1914 [mf ed 1995] – (= ser Yale coll) – 181p (ill) – 1 – 0-524-10112-4 – (in german) – mf#1995-1112 – us ATLA [280]
Bodichon, Barbara Leigh *see* A brief summary, in plain language, of the most important laws of england concerning women
Bodie evening union – Bodie CA. v1 n21,35 [1879? oct 1,17] – 1r – 1 – mf#918726 – us WHS [071]
Bodin, Felix *see* Defense des resumes historiques
Bodington, Alice *see* Studies in evolution and biology
Bodington, Charles *see* Books of devotion
Bodington, George *see* Bodington on the deep-seated causes of irish adversity
Bodington on the deep-seated causes of irish adversity : and the appropriate remedial measures / Bodington, George – London, 1881 – (= ser 19th c ireland) – 1mf – 9 – mf#1.1.2193 – uk Chadwyck [330]
The Bodleian Library *see*
– Calendars of charters and rolls in the manuscript collections of the bodleian library
– Diaries of sir frederic madden
Bodleian library record – Oxford UK 1973+ – 1,5,9 – ISSN: 0067-9488 – mf#8119 – us UMI ProQuest [020]
Bodmer als parodist / Meissner, Erich – Naumburg: Sieling, 1904 [mf ed 1989] – 127p – 1 – mf#7042 – us Wisconsin U Libr [430]
Bodmer, Joh Jakob *see* Crito
Bodmer, Johann Jakob *see*
– Johann jakob bodmer
– Karl von burgund
– Schriften
– Vier kritische gedichte
Bodmer, Johann Jakob et al *see* Der mahler der sitten
Bod-mihi-ran dban = Tibetan freedom – Darjeeling, India. Jul 1965-1991 – 22r – 1 – us L of C Photodup [079]
Bodmin guardian and cornwall county chronicle – Bodmin, England 13 jan 1911-5 jan 1912 (wkly) – 1r – 1 – uk British Libr Newspaper [072]
Bodocnost – Milwaukee WI, 1913* – 1r – 1 – (slovenian newspaper) – us IHRC [071]
Bodrov-Poviraev, N I *see* Ishimskaia zhizn'
Body and mind : an inquiry into their connection and mutual influence, specially in reference to mental disorders / Maudsley, Henry – rev enld ed. New York: D Appleton & Co 1895 [mf ed 1987] – 1r – 1 – mf#1895 – us Wisconsin U Libr [150]
Body and soul : an enquiry into the effect of religion upon health / Dearmer, Percy – New York: E P Dutton, c1909 – 1mf – 9 – 0-7905-5692-8 – (incl bibl ref) – mf#1988-1692 – us ATLA [230]
The body builder: robert j. roberts / Brink, Benjamin D – 1916 – 4mf – 9 – $12.00 – us Kinesology [790]
Body, Charles William Edmund *see* The permanent value of the book of genesis as an integral part of the christian revelation
Body composition determination of older men / Latin, Richard W – 1982 – 1mf – 9 – $4.00 – us Kinesology [611]
Body composition of athletes assessed using a four-component model / Prior, Barry M – 1996 – 2mf – 9 – $8.00 – mf#PE 3807 – us Kinesology [612]
The body composition of masters women endurance athletes from 35 to 74 years of age / Riggs, Donna M & Wells, Christine L – 1990 – 2mf – 9 – $8.00 – us Kinesology [612]
Body, George *see* The atonement and the living christ
Body image and restricted eating patterns among female athletes / Walluk, Laura A – 1997 – 1mf – 9 – $4.00 – mf#PSY 1965 – us Kinesology [150]
"Body image by association" : women's interpretations of aerobics and the role of the fitness instructor / Vogel, Amanda E – 1998 – 2mf – 9 – $12.00 – mf#PSY 2027 – us Kinesology [150]
Body image, disordered eating, and obligatory exercise among women fitness instructors / Nardini, Maria – 1998 – 2mf – 9 – $8.00 – mf#PSY 2074 – us Kinesology [150]
Body mass scaling of endurance cycling performance / Heil, Daniel P – 1997 – 372p on 4mf – 9 – $20.00 – mf#PE 4212 – us Kinesology [612]

Body messages in popular men's and women's health and fashion magazines from 1991-95 : a content analysis / Oomen, Jody S – 1997 – 1mf – 9 – $4.00 – mf#PSY 2028 – us Kinesology [302]

The body of christ : an enquiry into the institution and doctrine of holy communion / Gore, Charles – New York: Scribner, 1901 – 1mf – 9 – 0-7905-9277-0 – mf#1989-2502 – us ATLA [240]

A body of divinity : contained in sermons upon the assembly's catechism / Watson, Thomas – new ed. London: printed & publ for the Pastors' College by Passmore & Alabaster, 1890 [mf ed 1991] – (= ser Congregational coll) – 2mf – 9 – 0-524-00738-1 – (rev and adapted by george rogers. pref and app by c h spurgeon) – mf#1990-4007 – us ATLA [242]

Body part identification and comprehension of spatial prepositions in handicapped and nonhandicapped preschool children / Toon, C J – 1991 – 2mf – 9 – $8.00 – us Kinesology [150]

The body politic : a newspaper for gay liberation – Toronto. n1-135. nov/dec 1971-feb 1987// (mthly) – 12r – 1 – Can$786.00 – cn McLaren [305]

The body politic – no.1-, Nov-Dec 1971-. -bim. Gay liberation newspaper. Continues in part: Our Image – 1 – us Wisconsin U Libr [072]

Body & soul – New York NY 2002+ – 1,5,9 – ISSN: 1539-0004 – mf#15049.03 – us UMI ProQuest [130]

Body temperature and capacity for work / Asmusson, Erling & Boje, Ore – 1945 – 1mf – 9 – $3.00 – us Kinesology [612]

Body therapy repatterning and the neuromotor system / Honka, Rita J M – University of Oregon, 1992 – 2mf – 9 – $8.00 – mf#PE3599 – us Kinesology [790]

Bod-youl ou tibet : (le paradis des moines) / Milloue, Leon de – Paris: Ernest Leroux, 1906 [mf ed 1995] – (= ser Yale coll; Annales du musee guimet. bibliotheque detudes t12) – 1 – 0-524-10027-6 – (in french) – mf#1995-1027 – us ATLA [951]

Boeah pikiran / Djamaloedin Bin Moh Rasad, B – 's-Gravenhage, 1910 – 2mf – 8 – mf#SE-1426 – ne IDC [959]

Boeck, F J de *see* Vier-honderd-jaerig jubil

Boeck, L B de *see* Manuel de lingala

Boeck, Thorvald Olaf *see* Efterretninger om geistlige embeder i norge

Boeck van rechten der stad kampen: dat gulden boeck / Kampen. Netherlands. Ordinances, Local Laws, etc – Zwolle, Willink, 1875. 279 p. LL-4009 – 1 – us L of C Photodup [348]

Boeckel, Otto *see* Das deutsche volkslied

Boeckh, Christian Gottfried *see*
- Allgemeine bibliothek fuer das schul- und erziehungswesen in deutschland
- Wochenschrift zum besten der erziehung der jugend

Boeckh, Richard *see* Der deutschen volkszahl und sprachgebiet in den europaeischen staaten

Boecklern, G A *see* Architectura curiosa va...

Boedder, Bernard *see* Theologia naturalis

Boeddicker, Otto *see* Veroeffentlichungen von der koeniglichen sternwarte zu goettingen

Boeer, Bjoern *see* Die betriebssteuer

Boege, Guenther *see* Nestroy als bearbeiter

Boegner, Alfred *see*
- Etude sur la jeunesse et la conversion de calvin
- Saintete de dieu dans l'ancien testament...

Boegner, Marc *see* The unity of the church

Boehl, Eduard *see*
- Dogmatik
- Forschungen nach einer volksbibel zur zeit jesu
- Die zweite helvetische confession
- Zwoelf messianische psalmen

Boehl, Franz Marius Theodor *see* Kanaanaeer und hebraeer

Boehlau, Helene *see*
- Altweimarische liebes- und ehegeschichten
- Im garten der frau maria strom
- Isebies
- Die kristallkugel
- Der rangierbahnhof
- Ratsmaedelgeschichten
- Das recht der mutter
- Sommerbuch
- Sommerseele / muttersehnsucht

Boehlaus, H *see* Luthers werke

Boehle, Uta-Regina *see*
- Molekulare phylogenie und evolution der gattung echium l
- Pflanzensoziologische analyse der halbtrockenrasen im suedlichen ringgau

Boehlen, Hippolytus *see* Die franziskaner in japan einst und jetzt

Boehlich, Ernst *see* Goethes propylaeen

Boehlig, Hans *see* Die geisteskultur von tarsos im augusteischen zeitalter

Boehm, Friedrich *see* Das alte testament im evangelischen religionsunterricht

Boehm, Gotthold *see* Statistische untersuchungen ueber die arbeitskaempfe

Boehm, Hans [comp] *see* Gedankendichtung der fruehromantik

Boehm, Karl *see* Der weg des georg freimarck

Boehm, Wilhelm *see*
- Englands einfluss auf georg rudolf weckherlin
- Faust, der nichtfaustische
- Gesammelte werke
- Hoelderlin und die schweiz
- Im kreuzfeuer zweier revolutionen

Boehme, E A *see* Handbuch der evangelisch-lutherischen synode von ohio und anderen staaten = manual of the evangelical lutheran joint synod of ohio and other states

Boehme, Herbert *see*
- Das deutsche gebet
- Gesaenge unter der fahne
- Der glaube lebt
- Das grossdeutsche reich
- Kampf und bekenntnis
- Der kirchgang des grosswendbauern

Boehme, Jakob *see*
- Jacob boehme's the way to christ
- Jakob boehme's saemmtliche werke
- Seraphinisch blumen-gaertlein

Boehme, Walther *see* Lessings minna von barnhelm

Boehmen / Gerle, Wolfgang – Pesth 1823 [mf ed Hildesheim: 1995-98] – (= ser Fbc) – 3v on 6mf – 9 – €120.00 – 3-487-29454-0 – gw Olms [943]

O boehmen! : roman / Watzlik, Hans – Leipzig: L Staackmann, 1923 – 1r – 1 – us Wisconsin U Libr [830]

Boehmen und maehren – Prag (CZ), 1940 n1, 1941 [gaps], 1943 n3/4 – 1 – gw Misc Inst [077]

Boehmen und seine nachbarlaender unter georg von podiebrad, 1458-61 : und des koenigs bewerbung um die deutsche krone / Bachmann, Adolf – Prag: J G Calve, 1878 [mf ed 1990] – xii/309/1p – 1 – mf#7438 – us Wisconsin U Libr [943]

Boehmer, E *see* Romanische studien

Boehmer, Eduard *see*
- Franzisca hernandez und frai franzisco ortiz
- Zwei reden an kaiser und reich

Boehmer, Edward *see*
- Juan de valdes' commentary upon the gospel of st matthew
- Spiritual milk, or, christian instruction for children

Boehmer, Elizabeth Wilhelmina *see* Dr jan bouws 1902-1978

Boehmer, G R *see* Bibliotheca scriptorum historiae naturalis

Boehmer, Guenter *see* Pan am fenster

Boehmer, Heinrich *see*
- Chronica fratris jordani
- Die faelschungen erzbischof lanfranks von canterbury
- Luther im lichte der neueren forschung
- Luthers romfahrt
- Studien zur geschichte der gesellschaft jesu
- Urkunden zur geschichte des bauernkrieges und der wiedertaeufer

Boehmer, Hinrich *see* Luther im lichte der neueren forschung

Boehmer, Julius *see*
- Der alttestamentliche unterbau des reiches gottes
- Gottesgedanken in israels koenigtum
- Kreuz und halbmond im nillande
- Reich gottes und menschensohn im buche daniel
- Reichgottesspuren in der voelkerwelt
- Der religionsgeschichtliche rahmen des reiches gottes

Boehme-zeitung – Soltau DE, 1978- – ca 5r/yr – 1 – gw Misc Inst [074]

Boehme-zeitung – Soltau DE, 1978- – ca 5r/yr – 1 – gw Misc Inst [074]

Boehmische korallen aus der goetterwelt / Krauss, Friedrich S – Wien: Rubinstein 1893 [mf ed Bloomington IN: Indiana Uni Lib, Preservation Dept 1984] – 1r – 1 – us Indiana Preservation [390]

Boehmische literatur / ed by Dobrowsky, Jos – Schoenfeld 1779 – (= ser Dz) – 1v on 3mf – 9 – €90.00 – mf#k/n5529 – gw Olms [430]

Boehmische und maehrische literatur / ed by Dobrowsky, Joseph – Prag 1780-81 – (= ser Dz) – 2,3v on 3mf – 9 – €90.00 – mf#k/n5548 – gw Olms [430]

Boehmisches wanderbuch : lieder und gedichte / Hoeller, Franz – Prag: Volk und Reich Verlag 1943 [mf ed 1990] – (= ser Prager feldpost-buecherei) – 1r – 1 – (filmed with: der dichter vor der geschichte : holderlin, novalis / reinhold schneider) – mf#2732p – us Wisconsin U Libr [810]

Boehn, Max Von *see* Menschen und moden im achtzehnten jahrhundert, nach bildern

Boehnke, Frieda *see* Die deutsche dichtung in der schule

Boehrer, George C A *see* Monarquia a republica

Boehtlingk, Arthur *see*
- Napoleon bonaparte, seine jugend und sein emporkommen
- Napoleon bonaparte und der rastatter gesandtenmord

Het boek der geestelijke gezangen / Mechelen, Lucas van – Amsterdam: Johannes Stichter, 1688 – 9mf – 8 – €18.00 – ne Slangenburg [240]

Boekan impian, boekan lamoenan / Jo, Boen Ek & Goerz – Batavia: Goedang Tjerita, 1948 [mf ed 1998] – (= ser Goedang tjerita 4) – 1r – 1 – (coll as pt of the colloquial malay coll; indonesian trans of chinese novel possibly entitled emei wei jianke, or the fierce sword-fighters from emei shan mountain [salmon, claudine. literature in malay by the chinese of indonesia. paris: editions de la maison des sciences de l'homme, c1981]; filmed with: lajangan biroe / im yang tjoe) – mf#10005 – us Wisconsin U Libr [830]

Boekhary, Mir Abdoel Kerim *see* Histoire de l'asie centrale (afghanistan. boekhara, khiva, khoquand)...

[Boekholt, J] *see* 'Tgeopende

Boeklen, Ernst *see*
- Adam und quain
- Die verwandtschaft der juedisch-christlichen mit der parsischen eschatologie

Boekoe beladjaran permoelaan bahasa nippon – Djakarta: "Asia-Raya" 2602 – vi/255p 3mf – 9 – mf#SE-2002 mf25-27 – ne IDC [480]

Boekoe peladjaran hoeroef kanji (permoelaan) : memoeat tjara menoelis, tjara membatja dan latihan / Ling, C S – Soerakarta, 2603 – 96p on 2mf – 9 – mf#SE-2002 mf101-102 – ne IDC [370]

Boekoe peladjaran ilmoe penjeberangan laoet / Pardi, M – (Djakarta: Kaidji Sookyoku, 2603) – 54p on 1mf – 9 – mf#SE-2002 mf126 – ne IDC [370]

Boekoe pengoempoelan oendang-oendang : disoesoen dengan peroebahan dan tambahan sampai penghabisan boelan 6, toehan syoowa 19 (2604) / Java. (Japanese Military Administration). Laws, statutes, etc – Djakarta, Kokumin Tosyokyoku, 2604 – 422p 5mf – 9 – (disoesoen oleh gunseikanbu) – mf#SE-2002 mf70-74 – ne IDC [355]

Boekoe petoendjoek praktek teknik bagi pemimpin seinendan / Java. (Japanese Military Administration) – ed 3. (Djakarta): Djawa Gunseikanbu(2604) – 101p 2mf – 9 – mf#SE-2002 mf67-68 – ne IDC [355]

De boekzaal van europa : te rotterdam / Slaart, Pieter van der – 1692-1670 – 113mf – 9 – €215.00 – ne Slangenburg [073]

Boell, Heinrich *see* Der zug war puenktlich

Boell, Paul Victor *see* Le protectorat des missions catholiques en chine

Boellenruecher, J *see* Gebete und hymnen an nergal

Boelsche, Wilhelm *see*
- Die mittagsgoettin
- Weltblick
- Wielands ausgewaehlte werke

Boelsing, Gottfried *see*
- Friedrich matthissons gedichte

Boemer, Aloys *see* Die pilgerfahrt des traeumenden moenchs

Boemus, J *see*
- Recueil de diverses histoires tovchant les sitvations de toutes regions and pays contenuz es trois parties du monde, auec les particulieres moeurs, loix et caeremonies de toutes nations and peuples y habitas
- Repertorivm librorvm trivm...de omnivm gentivm ritibvs

Boenigk, Otto, Freiherr von *see* Das urbild von goethes gretchen

Boenisch, Hermann Friedrich *see*
- Es reiten die chungusen
- Das tor in die freiheit

Boenneken, Margarete *see* Wilhelm raabes roman "die akten des vogelsangs"

Boennischer sitten-, staats- und geschichtslehrer – Bonn DE, 1772 [gaps] – 1 – gw Misc Inst [943]

Boennisches intelligenzblatt – Bonn DE, 1772 4 jul-1776 10 dec, 1780 1 apr-7 oct, 1781 [single iss], 1783 11 jan-1792, 1793 2 mar, 1795 28 jul-1796 17 jul – 5r – 1 – (iss missing. incl suppl: annalen jan 9-dec 25 1787 [gaps]; title varies: 28 jul 1795: boennische intelligenz-nachrichten, 2 aug 1795: bonner intelligenz-blatt) – gw Misc Inst [074]

Boennisches wochenblatt – Bonn DE, 1785-88 [gaps] – 1r – 1 – gw Misc Inst [074]

Boennisches wochenblatt *see* Wochenblatt des boennischen bezirks

Boer, Geert Egberts *see*
- Debat tusschen ds. e.l. meinders, herder en leeraar bij de ware hollandsche gereformeerde kerk te south holland, ill., en ds. g.e. boer, docent bij de theol. school van de holl. christ. geref. kerk te grand rapids, mich
- De gordel der waarheid
- Een man des volks

Boer, Richard Constant *see* Untersuchungen uber den ursprung und die entwicklung der...

Boer, Tjitze J de *see* The history of philosophy in islam

Boer war : miscellaneous pamphlets published in great britain and the united states., 1899-1902 – Chicago, IL: U of Chicago, Photoduplication Dept, 1972 (mf ed) – 1 – us CRL [960]

The boer war : its causes, and its interest to canadians: with a glossary of cape dutch and kafir terms / Biggar, Emerson Bristol – Toronto, Montreal: Biggar, Samuel, 1899 – 1mf – 9 – mf#26497 – cn Canadiana [960]

Boerde-echo – Wanzleben DE, 1962 2 aug-1965 9 oct – 1r – 1 – (later: mz am wochenende; aka: magdeburger zeitung am wochenende, publ in magdeburg) – gw Misc Inst [074]

Der boeren bode – Aliwal Noord SA, 1882-dec 29 1883 [wkly] [mf ed Cape Town: SA library 1986] – 1r – 1 – (filmed with: de boeren courant voor de noordelike districten) – mf#MS00376 – sa National [079]

Der boeren bode *see* De boeren courant

De boeren courant – Colesburg SA, aug 1 1871-jun 28 1873 [wkly] [mf ed Cape Town: SA library 1986] – 1r – 1 – (filmed with: der boeren bode) – mf#MS00376 – sa National [079]

Die boerenuus – Aliwal Noord SA, jan 4 1922-25 (wkly) [mf ed Cape Town: SA library 1986] – 2r – 1 – sa National [079]

Boerger, Willi *see* Vom deutschen wesen

Boerne der zeitgenosse : eine auswahl / ed by Kuh, Anton – Leipzig: Verlag der Wiener Graphischen Werkstaette, 1922 [mf ed 1993] – xxv/272p – 1 – mf#8524 – us Wisconsin U Libr [840]

Boerne, Ludwig *see*
- Boerne der zeitgenosse
- Boernes werke
- Briefe aus paris 1830-1831
- Fragments politiques et litteraires
- Gesammelte schriften
- Ludwig boerne
- Schriften zur deutschen literatur

Boerne und sein verhaeltnis zu goethe und jean paul / Stadtlaender, Wilhelm – Berlin: Junker und Duennhaupt, 1933 [mf ed 1989] – (= ser Neue forschung. arbeiten zur geistesgeschichte der germanischen und romanischen voelker 20) – 159p – 1 – (incl bibl) – mf#7053 – us Wisconsin U Libr [430]

Boerner, Imanuel Karl H et al *see* Oekonomische nachrichten der patriotischen gesellschaft in schlesien

Boerner, Klaus Erich *see*
- Gefaehrtin meines sommers
- Das unwandelbare herz
- Ursula

Boerner, Peter *see* Johann wolfgang von goethe in selbstzeugnissen und bilddokumenten

Boerne's leben / Gutzkow, Karl – Hamburg: Hoffmann und Campe, 1840 [mf ed 1993] – (= ser Ludwig boerne's gesammelte schriften. supplementband) – xxxvi/310p/[1]pl – 1 – mf#8524 – us Wisconsin U Libr [920]

Boernes werke : historisch-kritische ausgabe in zwoelf baenden / ed by Geiger, Ludwig et al – Berlin: Bong, [1912-1913] [mf ed 1989] – 12v – 1 – (incl bibl ref) – mf#7054 – us Wisconsin U Libr [802]

Boero, Joseph *see* The life of the blessed mary ann of jesus, de paredes y flores

Boersenblatt fuer den deutschen buchhandel – Leipzig. v. 53-82. 1886-1915 – 1 – us L of C Photodup [010]

Boersenblatt fuer den deutschen buchhandel, 1834-1945 = Journal for the german book trade – [mf ed 1979-81] – 3057mf (1:42) – 9 – diazo €10,430.00 – ISBN-10: 3-598-10177-5 – ISBN-13: 978-3-598-10177-9 – gw Saur [070]

Boersen-halle *see* Priviligirte liste der boersenhalle

Boersen-nachrichten *see* Ostsee-zeitung

Der boersenterminhandel und der dem reichstage am 19. febr. 1904 vorgelegte "entwurf eines gesetzes betr. die aenderung des abschnittes 4 des boersengesetzes" / Jung, Rudolf – Wuerzburg: J Seelmeyr, 1907 (mf ed 19–) – 86p – mf#ZT-TN pv12 n6 – us Harvard [332]

Boertige en ernstige minnezangen : nevens eenige puntdichten, en andere / [Sweerts, C & Alewijn, A] – Amsterdam: Strander, [1709] – 5mf – 9 – mf#O-3275 – ne IDC [090]

Boesak, Willem Andreas *see* Mag en geweld as teologies-etiese probleem

Boeschenstein, Bernhard *see* Leuchttuermen

Boese, Heinrich *see* Die glaubwuerdigkeit unserer evangelien

Boese, Karl *see* Geschichte der stadt schneidemuehl

Boeseken, A J *see* Nederlandsche commissarissen aan de kaap, 1657-1700

Boesen, Paul John *see* Regulae sancti benedicti index verborum

Boeser, P A A *see*
- Beschreibung der aegyptischen sammlung des niederlaendischen reichsmuseums der altertuemer in leiden
- Beschrijving van de egyptische verzameling in het rijksmuseum van oudheden te leiden
- Manuscrits coptes du musee d'antiquites des pays-bas a leyde

Boethius, Anicius Manlius Severinus *see*
- Consolacion de la filosofia
- De consolatione philosophiae
- King alfred's version of the consolations of boethius

Boethius de consolatione philosophiae / ed by Sehrt, Edward Henry & Starck, Taylor – Halle/S: M Niemeyer, 1933-34 [mf ed 1993] – (= ser Altdeutsche textbibliothek 32-34) – 3v – 1 – (latin text with old high german trans. incl bibl ref) – mf#8193 reel 3 – us Wisconsin U Libr [180]

Die dem boethius...zugeschriebene abhandlung des dominici gundisalvi de unitate / Correns, P – Muenster, 1891 – (= ser Bgphma 1/1) – 2mf – 9 – €5.00 – ne Slangenburg [100]

Boetjer basch : eine geschichte / Storm, Theodor – Berlin: Gebrueder Paetel 1887 [mf ed 1995] – 1r – 1 – (filmed with: ut'n knick / julius stinde) – mf#3752p – us Wisconsin U Libr [830]

Boets i pakhar' – Chita, Russia 1920-22 [mf ed Norman Ross] – 3r – 1 – mf#nrp-443 – us UMI ProQuest [077]

Boettcher, Alfred *see* Sprung ins kattegatt

Boettcher, Christoph *see* Elektronenmikroskopische untersuchungen zur stereoselektiven bildung mizellarer lipidfasern aus n-alkylaldonamiden

Boettcher, Friedrich *see*
- Ausfuehrliches lehrbuch der hebraeischen sprache
- Exegetisch-kritische aehrenlese zum alten testament

Boettcher, Kurt *see* Romanfuehrer a-z

Boettcher, Maximilian *see*
- Krach im vorderhaus
- Die wolfrechts

Boettger, Gustav *see* Topographisch-historisches lexicon zu den schriften des flavius josephus

Boetticher, Georg *see* Allotria

Boetticher, Gotthold *see*
- Denkmaeler der aelteren deutschen literatur fuer den litteraturgeschichtlichen unterricht an hoeheren lehranstalten
- Geschichte der deutschen literatur
- Das nibelungenlied im auszuge nach dem urtext

Boetticher, Gotthold [comp] *see*
- Die litteratur des achtzehnten jahrhunderts vor klopstock
- Die litteratur des siebzehnten jahrhunderts

Boetticher, Hans *see*
- Die flasche und mit ihr auf reisen
- Geheimes kinder-spiel-buch mit vielen bildern
- Liner roma
- Mein leben bis zum kriege
- Der nachlass
- Turngedichte

Boetticher, Otto *see* Das verhaeltnis des deuteronomiums zu 2. koen. 22. 23. und zur prophetie jeremia

Boetticher, Paul *see* Die anfaenge der reformation in den preussischen landen ehemals polnischen anteils bis zum krakauer frieden, 8. april 1525

Boetticher, Wilhelm *see* Los vom ultramontanismus

Boettiger, K W *see* Literarische zustaende und zeitgenossen

Boettiger, Karl August *see* Literarische zustaende und zeitgenossen

De boetveerdicheyt des levens... / Taffin, J – Ed 5. Haerlem, 1613 – 7mf – 9 – mf#PBA-314 – ne IDC [240]

Boeuf river baptist church. winnsboro, louisiana : church records – May 1911-Jan 1991 – 1 reel – 1 – $75.06 – (1,668p) – us Southern Baptist [242]

Boevoi udar – (city unknown), Russia 1941-45 [mf ed Norman Ross] – 1 – mf#nrp-89 – us UMI ProQuest [934]

Boevye predpriatiia sotsialistov-revoliutsionerov v osveshchenii okhranki – 1918 – 112p 2mf – 9 – mf#RPP-218 – ne IDC [325]

Bog i den'gi : epizody neskol'kikh zhizne'i [a novel] / Krymov, Vladimir – Berlin: gedr von Gebr Hirschbaum 1926 [mf ed 2004] – 2v on 1r – 1 – (no evidence other vols beyond v2 publ. filmed with: bog i den'gi) – mf#5544p – us Wisconsin U Libr [830]

Bogaert, A S *see* Schynvoets muntkabinet der roomsche keizers en keizerinnen

Bogaevskii, N N *see* Voennyi golos [novorossiisk: 1920]

Bogazici sirket-i hayriye tarihce, salname – 1330 [1914] – (= ser Ministry and special interest salnames) – 8mf – 9 – $130.00 – us MEDOC [956]

Bogbinder, Hilarius *see* Stadier paa livets vei

Bogdanov, B *see* Maslodelnye arteli v vologodskoi gubernii

Bogdanov, M *see* Rabochie deputaty v i-oi gosudarstvennoi dume

Bogdanovich, Savva *see* Missionerskaya beseda so shtundistami o kreste i o krestnom anamenii

Boge, Justin *see* Physical self-esteem across four phases of a cardiac rehabilitation program

Bogeat y Asuar, Antonio *see* Guia de villafranca de los barros

Bogens verden – Copenhagen, Denmark 1918-92 – 1,5,9 – ISSN: 0006-5692 – mf#3438 – us UMI ProQuest [070]

Boger, Margot *see* Christina mortens ehe

Bogerman, J *see*
- Praxis verae poenitentiae...
- Een schoon tractaet van de straffe
- Specimen conscientiae, candoris, veracitatis, simplicitatis et pietatis d. vorstii...
- Tractatus theologicus de salutari usu judiciorum dei...orationibus aliquot absolutus...

Bogert, George Gleason *see*
- The elements of business law
- Handbook of the law of trusts

Boggabri herald – Boggabri, jan-oct 1973 – 1r – at Pascoe [079]

Boggabri telegraph – Boggabri, 1981-82 – at Pascoe [079]

Boggie, Jeannie M *see* First steps in civilizing rhodesia

Boggs, Edna Garrido *see* Versiones dominicanas de romances espanoles

Boggs newsletter – 1976 may-1985 jun, 1985 sep-1987 jun – 2r – 1 – (cont by: boggs newsletter quarterly) – mf#813438 – us WHS [071]

Boggs, Stanley H *see* Tazumal en la arqueologia salvadorena

Boggs, William Bambrick *see*
- The baptists, who are they? and what do they believe?
- The needs of our foreign mission work

Bogler, Wilhelm *see* Harthmuth von kronberg

Bognar report – [Accra 1962] – us CRL [960]

Bognor observer and visitor's list – Bognor Regis, England 1 may 1872-26 mar 1930 – 1 – (wanting jan-jul 1900; cont by: bognor regis observer [2 apr 1930- (mf 1986-)]; variant ed of: chichester observer) – uk British Libr Newspaper [072]

Bognor regis post – Bognor Regis, England 10 aug 1929-12 sep 1985 [mf 1979-89, 1991-2] – 1 – (cont: bognor post [15 mar 1924-3 aug 1929]; cont by: chichester, littlehampton & bognor regis post [19 sep 1985-29 sep 1988]; bognor, chichester & littlehampton post [19 jan-18 may 1989]; bognor regis post [30 oct 1991-15 jan 1992]) – uk British Libr Newspaper [072]

Bogoiavlenskii, V A *see*
- Miasskaia novaia zhizn'
- Orenburgskii kazachii vestnik

Bogolepov, D P et al *see* Finansovaia entsiklopediia

Bogolepov, M I *see* Teoriia i praktika kommercheskogo banka

Bogoliubov, P A *see* Vestnik priural'ia

Bogomolov, N N *see* Narodnoe slovo [khar'kov: 1919: vech izd]

Bogor. Indonesia. Kebun Raja *see* Annales

Bogoslovskii, M M *see*
- Oblastnaia reforma petra velikogo
- Petr i materialy dlia biografii

Bogoslovskii, N *see* Izdanie novgorodskogo statisticheskogo komiteta

Bogoslovskii, S M *see* Zemskii meditsinskii biudzhet moskovskoi gubernii za 1883-1905 gg

Bogoslovskii vestnik, izdavaemyi moskovskoi dukhovnoi akademiei – New York. 1910-1996 (1) 1966-1996 (5) 1975-1996 (9) – 1539mf – 9 – mf#1426 – ne IDC [077]

O bogosluzhenii pravoslavnoi tserkvi / Germogen, Bishop of Pskov and Porkhov – Izd 9. S-Peterburg: Sinodal'naia tip, 1911 [mf ed 2002] – 1r – 1 – (filmed with: skazaniia ob antikhristie v slavianskikh perevodakh...(1874)) – mf#5225 – us Wisconsin U Libr [243]

Bogota : 8 (ie ocho) de junio / Vallejo, Alejandro – Bogota, Colombia. 1929 – 1r – us UF Libraries [972]

Bogota : la literatura colombiana a mediados del s. 19 / Gomez Restrepo, Antonio – Madrid: Razon y Fe, 1927 – 1 – sp Bibl Santa Ana [490]

Bogota 1538-1938 / Samper Ortega, Daniel – Bogota, Colombia. 1938 – 1r – us UF Libraries [972]

Bogota Concejo *see* Cabildos de santafe de bogota

[Bogota-] el tiempo – CK. 1971-1993 – 276r – 1 – $13,800.00 – mf#R060501 – us Library Micro [079]

[Bogota-] tribuna roja – CK. 1971-78 – 1r – 1 – $50.00 – mf#R04194 – us Library Micro [320]

Bogrov, Grigorii Isaakovich *see*
- Ketav-yad 'ivri
- Kinder khaper fun rusland
- Ma'asim she-hayu

Bogsamer zeitung – Bogsan (Bocsa-Montana RO), 1930 25 oct-1931 27 dec – 1r – 1 – gw Misc Inst [079]

Bogue, David *see*
- Diffusion of divine truth
- E au akoanga no nga tumu tuatua i kitea i roto i te tutua na te atua
- The history of dissenters
- On universal peace
- The theological lectures of rev david bogue

Boguslavskii, Mark Moiseevich *see* The legal status of foreigners in the u.s.s.r

Boguslavskoe obshchestvo trezvosti i bor'ba so shtundoyu / Skvortsov, Vasilii M – The Boguslav Society of Temperance and the Struggle with the Stunda. Kiev, 1895. One of 13 titles on reel – 1 – 86.44 – us Southern Baptist [242]

Bohannan, Paul *see* Justice and judgment among the tiv

Bohatec, Josef *see* Die cartesianische scholastik

Bohatec, Josef et al *see* Calvinstudien

Bohatta, Hans *see*
- Deutsches anonymen-lexikon, 1501-1910
- Deutsches pseudonymen-lexikon

La boheme : commedia lirica in quattro atti / Leoncavallo, Ruggiero – Milano: E Sonzogno 1897 [mf ed 19–] – 8mf – 9 – mf#fiche 603, 131 – us Sibley [780]

Bohemia – Prag, Czech Republic 1 jan-31 dec 1856; 19,20 aug, 24,25 oct 1916; 31 jan 1917-31 dec 1918 (imperfect) [mf 1916-18] – 1 – (filmed by misc inst: 1845 3 jan-1848, 1855, 1861-1938 [361r]) – uk British Libr Newspaper; gw Misc Inst [077]

Bohemia – Havana. v. 8-47. 1917-55 – 1 – us L of C Photodup [944]

Bohemia – Koeln DE, 1956 n57-1960 n93/94, 1962 n5-7 – 1 – gw Misc Inst [074]

Bohemia : list ceske narodni skupiny v nemecku – Munich, Germany 28 aug 1950-jul 1960 – 1 – uk British Libr Newspaper [700]

Bohemia from the earliest times to the fall of national independence in 1620 : with a short summary of later events / Maurice, Charles Edmund – London: T F Unwin [c1896] [mf ed 1986] – 1r [ill] – 1 – (filmed with: history of the commerce and town of liverpool.../ baines, thomas) – mf#6681 – us Wisconsin U Libr [943]

Bohemia nugget – Cottage Grove OR: C J Howard, 1899-1907 [wkly] – 1 – (absorbed by: cottage grove leader (1905-15)) – us Oregon Lib [071]

Bohemia. Zemsky Snem *see* Stenographische bericht. stenograficke zpravy

Bohemian / Franklin Co. Columbus – mar 1882-nov 1885 [wkly] – 2r – 1 – mf#B12014-12015 – us Ohio Hist [071]

Bohemian loan *see* Papers relating to the bohemian loan, 1620-1622

Bohemian voice – v1 n1-v3 n3 [1892 sep-1984 nov] – 1r – 1 – mf#1053457 – us WHS [071]

The bohemian voice / Bohemian-American National Committee & Lincoln. University of Nebraska. Libraries University Archives Special Collections Dept – Omaha, NE: Bohemian-American National Committee. 3v. v1 n1. sep 1 1892-v3 n3. nov 1894 (mthly) [mf ed 1985] – 1r – 1 – us NE Hist [071]

Bohemian-American National Committee *see* The bohemian voice

O bohemio – Filipeia, PA. 03 maio 1900 – (= ser Ps 19) – mf#P17,02,130 – bl Biblioteca [320]

O bohemio : folheto quinzenal de critica mansa – Sao Paulo, SP: Typ Internacional, 01 abr-01 maio 1889 – (= ser Ps 19) – mf#P17,02,221 – bl Biblioteca [079]

O bohemio : orgam crítico, humorístico e noticioso – Petropolis, RJ. 27 set-25 out 1903 – (= ser Ps 19) – mf#DIPER – bl Biblioteca [079]

O bohemio – Sao Paulo, SP: Typ e Lith Andrade & Comp, 04 abr 1896 – (= ser Ps 19) – mf#P18,01,78 – bl Biblioteca [410]

Bohic, O Carth *see* Chronica ordinis carthusiensis ab anno 1084 ad annum 1510

Bohl de Faber, Juan N *see* Teatro espanol anterior a lope de vega

Bohlen, Peter von *see*
- Bhartriharis sententiae et carmen quod chauri nomine circumfertur eroticum
- Introduction to the book of genesis

Bohlender, Jakob *see* Ortsgeschichte von der gemeinde ingenheim in der pfalz

Bohlendorff, Julius, Freiherr von *see* Hausbuch des herrn joachim von wedel auf krempzow schloss und blumberg erbgesessen

Bohlendorff, Julius, Freiherr von Bohlen *see* Hausbuch des herrn joachim von wedel auf krempzow schloss und blumberg erbgesessen

Bohlin, Karl *see* Recherches sur les perturbations de la comete de winnecke depuis 1809 a 1819

Bohlmann, Gerhard *see*
- Der vergessene kaiser
- Wallenstein ringt um das reich

Bohm, Erwin Herbert *see* The development of naturalism in german poetry

Bohmer, Walther *see* Pioniere

Bohn, Helmut *see* Sozialisten und die verteidigung

Bohn, Henry George *see* A guide to the knowledge of pottery, porcelain, and other objects of vertu

Bohn, John A *see*
- Administrative rules and regulations of the government of guam, 1975
- The civil and penal codes of the territory of guam, 1953
- The civil code of the territory of guam, 1970
- The code of civil procedure and probate code of guam, 1953
- The code of civil procedure and probate code of guam, 1970
- The government code of guam, 1952
- The government code of guam, 1961
- The government code of guam, 1970
- The penal code of the territory of guam, 1970
- Statutes and amendments to the codes of the territory of guam, 1951-1952

Bohnenblust, Gottfried *see*
- Goethe und die schweiz
- Goethe und pestalozzi
- Kaethi die grossmutter

Bohner, Theodor *see* Die negation bei goethe

Bohnert, Werner *see* Planung als durchsetzungsstrategie

Bo-hoa / Nguyen, Dinh Toan – [Gia Dinh, Vietnam]: Vo-Cau 1974 [mf ed 1993] – on pt of 1r – 1 – mf#11052 r469 n3 – us Cornell [959]

Der bohrkumpel – Gommern DE, 1956-57 [gaps], 1959-1990 26 jun [gaps] – 3r – 1 – (title varies: 1959: der erdoelpionier; 1963: im tempo der zeit) – gw Misc Inst [622]

Bohus-dals tidning – Uddevalla, 1936-40 – 1r – 1 – sw Kungliga [078]

Bohuskorrespondenten – Marstrand, 1873-76 – 9 – sw Kungliga [078]

Bohuskusten – Uddevalla, Sweden. 1941-43 – 1 reel – 1 – sw Kungliga [078]

Bohuslaningen – Uddevalla, Sweden. 1979- – 1 – (dals dagblad, 1983) – sw Kungliga [078]

Bohuslaningen – Uddevalla, Sweden. 1878-1978 – 419r – 1 – sw Kungliga [078]

Bohuslans allehanda – Uddevalla, Sweden. 1887-1891 – 4r – 1 – sw Kungliga [078]

Bohuslans annonsblad – Uddevalla, Sweden. 1891-95 – 2r – 1 – sw Kungliga [078]

Bohuslans nyhets och annonsblad – Uddevalla, Sweden. 1891 – 1r – 1 – sw Kungliga [074]

Bohuslans tidning – Uddevalla, Sweden. 1838-83 – 20r – 1 – sw Kungliga [078]

Bohuslans tidning uddevalla – Sweden, 1904-07 – 3r – 1 – sw Kungliga [078]

Bohusposten – Uddevalla, Sweden. 1910-52 – 100r – 1 – sw Kungliga [078]

Boi aru a / Jardim, Luis – Rio de Janeiro, Brazil. 1940 – 1r – us UF Libraries [972]

Boice, James Montgomery *see* Can you run away from god?

Boid, Edward *see* Travels through sicily and the lipari islands

Boie, Friedrich *see* Tagebuch gehalten auf einer reise durch norwegen im jahre 1817

Boie, Heinrich Christian *see*
- Deutsches museum
- Goettinger musenalmanach auf 1770
- Neues deutsches museum

Boieldieu, Francois Adrien *see*
- [Beniowksy]
- Ouverture du calife de bagdad
- Sonate pour le piano avec accompagnement de violon

Boigraphie [i.e. biographie] de monsieur l'abbe hermann plante : bachelier es arts de l'universite laval, licencie es lettres de l'universite de montreal, licencie en diction de l'ecole classique de montreal / Virginie Marie, soeur – 1964 [mf ed 1979] – (= ser Bibliographies du cours...1947-66) – 1mf – 9 – (with ind) – mf#SEM105P4 – cn Bibl Nat [378]

Boileau, Alexander Henry Edmondstone *see* Memorandum for reorganizing the indian army

Boiled-down essays / Crouter, John Wesley – [London, Ont?: s.n.], 1886 [mf ed 1980] – 1mf – 9 – 0-665-02224-7 – mf#02224 – cn Canadiana [360]

Boilermaker reporter / International Brotherhood of Boilermakers, Iron Shipbuilders etc – 1990-1994 nov/dec – 2r – 1 – (cont: boilermakers blacksmiths reporter) – mf#4208621 – us WHS [680]

Boilermakers blacksmiths reporter / International Brotherhood of Boilermakers, Iron Shipbuilders etc – 1962 jul, 1971 aug-1977 jul, 1977 sep-1981 dec, 1982 jan-1986 may, 1986 jun-1989 oct/dec – 5r – 1 – (cont: Boilermakers-blacksmiths record; Cont By: Boilermaker reporter) – mf#3238161 – us WHS [680]

Boilermakers-blacksmiths journal / International Brotherhood of Boilermakers, Iron Ship Builders, Blacksmiths, Forgers and Helpers – 1893-1955 – (= ser Labor union periodicals, pt 1: the metal trades) – 14r – 1 – $2925.00 – 1-55655-230-0 – us UPA [680]

Boiling springs baptist church. north carolina : church records – 1847-1963. Bulletins. 1949-63. Miscellaneous items. 1816-1959 – 1 – us Southern Baptist [242]

Boiling springs baptist church. spartanburg, south carolina : church records – Mar 1971-Oct 1976 – 1 – us Southern Baptist [242]

Boillet, Leon *see* Aux mines d'or du klondike

Boipuso ba basotho – [Mafeteng, Lesotho: BCP Youth League. v1 n1,5,7-13,16-23. jul 9-dec 19 1965 – 1r – us CRL [960]

Boirie, Jean-Bernard-Eugene Cantiran De *see* Jeunesse du grand frederic

Bois, Georges *see* Maconnerie nouvelle du grand-orient de france

Bois, Henri *see*
- Le dogme grec
- La poesie gnomique chez les hebreux et chez les grecs

Bois, Jacqueline *see* Bibliographie analytique de l'oeuvre de monseigneur arthur maheux de la societe royale du canada

Bois, Louis-Edouard *see*
- Le chevalier noel brulart de sillery
- Le coffret ou le tresor enfoui
- Le colonel dambourges
- La decouverte du mississipi
- Escaped from the gallows
- Esquisse de la vie et des travaux apostoliques de sa grandeur mgr fr xavier de laval-montmorency
- Etude biographique, m jean raimbault
- Etudes et recherches biographiques sur le chevalier noel brulart de sillery
- L'ile d'orleans
- Le juge a mabane
- Michel sarrasin
- Souvenir d'un prisonnier d'etat canadien en 1838

Bois, Patterson du *see* The natural way in moral training

Bois sinistre : roman canadien inedit / Lacerte, A B – Montreal: ed Edouard Garand, 1929 [mf ed 1982] – 2mf – 9 – (ill by albert fournier) – mf#SEM105P64 – cn Bibl Nat [830]

Bois, William Edward Burghardt du *see* The souls of black folk

Boise, James Robinson *see*
- The epistles of st paul written after he became a prisoner
- Notes critical and explanatory on paul's epistle to the galatians greek text of tischendorf
- Notes critical and explanatory on the greek text of paul's epistles to the ephesians, the colossians, philemon, and the philippians
- Notes, critical and explanatory, on the greek text of paul's epistles to the romans, the corinthians...

Bois-Melly, Charles du *see* Relations de la cour de sardaigne et de la republique de geneve depuis le traite de turin jusqn'a la fin de l'ancien regime, 1754-1792

Boisragon, Alan *see* The benin massacre

Boissard, J J *see*
- Emblematum liber
- Schawspiel menschliches lebens
- Theatrum vitae humanae
- Vitae et icones svltanorvm tvrcicorvm, principvm persarvm...

Boissard, L *see* L'eglise de russie

Boisseau, A *see* Supplement no 1 au catalogue de la bibliotheque de l'institut-canadien

Boisseau, Lionel *see* La mer qui meurt

Boisselier, Jean *see* Le cambodge

Boissevain, W T L *see* Memorie van overgave van madiun 1903-1907 door resident w t l boissevain

Boissier, Alfred *see*
- Documents assyriens relatifs aux presages. tome premier
- Mantique babylonienne et mantique hittite

Boissier de Sauvages, Pierre A *see* Dictionnaire languedocien-francois

Boissier, E P *see* Voyage botanique dans le midi de l'espagne pendant l'annee 1837

Boissier, Gaston *see*
- La fin du paganisme
- La religion romaine d'auguste aux antonins
- Rome and pompeii

Boissonnade, P *see* Saint-domingue a la veille de la revolution

Boissonnault, Charles-Marie *see* Histoire politique de la province de quebec

Boissonneault, Henri *see* Bio-bibliographie de m jean bruchesi

Boissy, Gabriel *see* Jules cesar

Boissy, Robert *see* Jupiter

Boiteux, Lucas Alexandre *see*
- Marinha de guerra brazileira nos reinados
- Marinha imperial versus cabanagem
- Pequena historia catharinense

Boivin, J *see* Byzantina historia (cbh23)

Boivin, Jean *see* Bio-bibliographie analytique de l'oeuvre de l'abbe arthur maheux de la societe royale du canada

Bojanowoer anzeiger : wochenblatt der posen-schlesischen grenze – Bojanowo (PL), 1926 7 aug-1932 30 jun – 1 – gw Misc Inst [077]

Bojarski, Gershon Meir *see* Regesh omarenu

Bojczyk, Kathryn E G *see* Object retrieval and interlimb coordination in the first year of life

Boje, Ore *see* Body temperature and capacity for work

Boje, Walter *see* Brand an der wolga

Bojovnik – Bratislava, Czechoslovakia. Jun-Oct 1945 – 1r – 1 – us L of C Photodup [077]

Bojovnik – Bratislava, Slovakia 3,9 aug, 12,13,16 dec 1945; 12 feb-15 may, 1 jun-2 nov, 14 nov-29 dec 1946 (imperfect) – 1 – uk British Libr Newspaper [077]

Bok, Marcia *see* Journal of hiv/aids prevention and education for adolescents and children

The boke of cokery / Salter, Elizabeth – (= ser Archives of the marquess of bath, longleat house, warminster, wiltshire) – 1r – 1 – mf#96821 – uk Microform Academic [640]

A boke or counseill against the disease called the sweate / Caius, John – 1522 – 9 – us Scholars Facs [616]

Boker, George Henry *see* Koenigsmark

Bokhanovskii, B *see* Ekonomicheskaia politika sssr

Bokvannen – v1-24. 1946-69 – 1 – us AMS Press [010]

Bokwe, J K *see* Letterbooks, 1882-9, 1894-7

Bokwe, John Knox *see* Amaculo ase lovedale

Bolaffey, Hayim Victa *see* An easy grammar of the primaeval language

Bolamba, Antoine Roger *see* Esanzo, chants pour mon pays

Bolanden, Conrad von *see* Luther's brautfahrt

Bolanos, Joaquin *see* La portentosa vida de la muerte

Bolayir *see* Ordunun defteri

El bolchevismo... / Gurian, Waldemar – Madrid: Razon y Fe, 1934 – 1 – (trad de emilio m martinez amador, barcelona 1932) – sp Bibl Santa Ana [946]

Bold, J D *see*
- Dictionary, grammar and phrase-book of fanagalo (kitchen kafir)
- Dictionary phrase-book and grammar of fanagalo

Boldness by the blood of christ / Tait, William – Edinburgh, Scotland. 18– – 1r – us UF Libraries [240]

Boldoni, Octavio *see* Theatrum temporaneum aeternitati caesaris montii s.r.e. cardinalis et archiep

Bolduc, Marcel *see* Bio-bibliographie des anciens eleves des freres maristes

Boldyrev, D *see* Russkoe delo

Bolecina razlike / Zizek, Slavoj – Maribor: Zalozba Obzorja 1972 – us CRL [999]

Bolenge : a story of gospel triumphs on the congo / Dye, Eva Nicols – [5th ed] Cincinnati, Ohio: Foreign Christian Missionary Society, 1911 – 1mf – 9 – 0-524-05948-9 – mf#1991-2348 – us ATLA [240]

Boleo, Jose De Oliveira *see*
- Mocambique

Bolero y plena / Arrivi, Francisco – San Juan, Puerto Rico. 1960 – 1r – us UF Libraries [972]

Boles, John B *see*
- The john pendleton kennedy papers
- The william wirt papers

Bolet, Julio C *see* San sebastian de los reyes. caracas, 1929

Boletim annunciador de benguella – Benguella: O Benguella, sep 28 1910 – us CRL [073]

Boletim anti-alccolico – Florianopolis, SC. out 1932 – (= ser Ps 19) – mf#UFSC/BPESC – bl Biblioteca [360]

Boletim da sociedade de geographia de lisboa – [Lisboa]: A Sociedade. v1-3, v5-53. dec 1876-1883,1885-1935 – 20r – us CRL [073]

Boletim de informacao / Conferencia das Organizacoes Nacionalistas das Colonias Portuguesas (CONCP) – Rabat: Conferencia das Organizacoes Nacionalistas das Colonias Portuguesas, Secretariado Permanente. n5. nov 1962 – us CRL [073]

Boletim de informacoes – Lisboa. 1945-53 – 1 – 92.00 – us L of C Photodup [330]

Boletim do expediente do governo : ministerio da justica – Rio de Janeiro, RJ: Typ Imperial e Constitucional de J Villeneuve & C, 1859-1862 – (= ser Ps 19) – mf#P07,01,16 n01 – bl Biblioteca [323]

Boletim do grande oriente do brasil : jornal official da maconaria brasileira – Rio de Janeiro, RJ: Typ da Grande Oriente e da Luz, dez 1871-dez1879; abr, set-dez 1880; jan-nov 1881; mar, jul, set-dez 1890; jan 1891-fev 1899 – (= ser Ps 19) – mf#P08,01,04-08 – bl Biblioteca [360]

Boletim do grande oriente unido e supremo conselho do brazil : jornal offical da maconaria brazileira – Rio de Janeiro, RJ: Typ do Grande Oriente Unido e Suprenmo Conselho do Brasil, jan 1873-dez 1877 – (= ser Ps 19) – mf#P08,01,02-03 – bl Biblioteca [360]

Boletim do militante / MPLA – [S.l.]: MPLA. n4. feb 1965 – us CRL [073]

Boletim do museu paraense de historia natural e ethnographia – Belem, PA: O Museu, set 1894-dez 1898; fev 1900; dez 1902; fev,dez 1904; mar 1906; fev 1908-1912; 1938; 1955; 1956 – (= ser Ps 19) – bl Biblioteca [500]

Boletim geografico – v. 1-11, no. 1-117. Apr 1943-Dec 1953 – 1 – us L of C Photodup [910]

Boletim liberal – Florianopolis, SC. 28 nov 1929-12 jan 1930 – (= ser Ps 19) – bl Biblioteca [079]

Boletim official : prefeitura do alto acre – Rio Branco, AC: Impresso nas officinas d'O Autonomista, 11 jul-set, dez 1915; jan-dez 1916; jan-maio, nov-dez 1917; 10,24 fev 1918 – (= ser Ps 19) – mf#P25,01,27 – bl Biblioteca [350]

Boletim oficial / Portugal. Direccao Geral das Alfandegas – 1892-1949 – 1 – 529.00 – us L of C Photodup [336]

Boletim oficial and supplements / Macao – 1957-1969- – 1 – us NY Public [324]

Boletim oficial dill / Timor – 1958-July 30, 1966 – 1 – us NY Public [946]

Boletim oficial do estado da india / Goa – 1942, 1951-Dec. 14, 1961 – 1 – us NY Public [954]

Boletim oficial do governo geral da provincia de angola – Luanda. series 1,2,3: 1935-1938 4r; 1957-1962 13r; series 1: 1954-1956 2r 1963-nov 10 1975 15r – us CRL [073]

Boletin / Academia de la Historia. Madrid – v1-137. 1877-1955 – 1 – $1458.00 – (v138-179 1956-82 $336 0005. in spanish) – mf#0004 – us Brook [946]

Boletin / Academia de la Historia. Madrid – v.1-76. Nov. 1877-June 1920 – 1 – us L of C Photodup [946]

Boletin / Argentine Republic. Secretaria de Comunicaciones – v. 1-24. 1929-52. Boletin Suplemento. N. 1-3386. 1932-48. (Scattered issues wanting) – 1 – 564.00 – us L of C Photodup [380]

Boletin / Asociacion Geofisica de Mexico – Mexico. v. 1-2 no. 3. July 1929-Sept 1930 – 1 – us NY Public [550]

Boletin : centro de espiritualidad de la compania de jesus – [Buenos Aires]: El Centro, [1968-1970]. [n2-4 (aug/sep1968-may/jun 1969); n6-7 (nov 1969-nov 1970)] (irreg) – 1r – 1 – us CRL [241]

Boletin / Chile. Ministerio de Hacienda – Santiago de Chile. t. 1-(27). 1888-1914. (Wanting t. 2, Jan-May 1889; t. 3-4, 1890-91; t. 6-13, 1893-1900; t. 23-24, 1910-11) – 1 – us L of C Photodup [350]

Boletin : interamerican children's institute / Interamerican Children's Institute. Montevideo – v15-30 1941-56 – 1 – $81.00 – us L of C Photodup [305]

Boletin / Mexico (City). Radiodifusora XELA – v. 1-12. Feb 4, 1952-Mar 4, 1963 – 1 – $46.00 – us L of C Photodup [380]

Boletin / Radiodifusora XELA. Mexico City – v. 1-12, N. 1-574. 4 Feb 1952-4 Mar 1963. (Scattered issues wanting) – 1 – us L of C Photodup [380]

Boletin astronomico del observatorio de madrid / Observatorio Astronomico de Madrid – Madrid Instituto Geografico y Catastral 1932- (irreg) [mf ed 2001] – 1-4(1932-55) on 1r [ill] – 1 – (v4 n1 misnumbered as v3 n9 [1949]; v4 n8 misnumbered as v5 n8; ind to v4; v3 n5 lacks numbering) – ISSN: 0373-7101 – mf#film mas c5063 – us Harvard [520]

Boletin bibliografico de antropologia americana – Mexico City 1977-79 – 1,5,9 – ISSN: 0067-9658 – mf#9187 – us UMI ProQuest [301]

Boletin bibliografico mexicano – Mexico City 1973+ – 1,5,9 – ISSN: 0185-2027 – mf#9327 – us UMI ProQuest [972]

El boletin catolico – Cebu: [s.n.], sep 26 1918-oct 17 1929 – us CRL [241]

Boletin celam – Santa fe de Bogota: Centro de Publicaciones CELAM. [n253-262 (apr/may 1993-jul /aug 1994); n264-290 (nov 1994-2000)] (mthly) – 4r – 1 – us CRL [972]

Boletin de artes visuales – Washington DC 1956-77 – 1,5,9 – ISSN: 0553-0571 – mf#9755 – us UMI ProQuest [700]

Boletin de el internacional – Tampa, FL. 1936 jun 27-dec 11; 1937 feb 19; mar 19; apr 9; MA – 1r – (1936 jul 24; aug 28; oct 30; nov 6, 20) – us UF Libraries [071]

Boletin de espiritualidad – [Buenos Aires]: Centro de Espiritualidad, Compania de Jesus [n8-174 (marzo 1971-nov/dic 1998)] (bimthly) – 4r – 1 – us CRL [241]

Boletin de estadistica peruana 1958-1962, 1964 / Peru. Direccion Nacional de Estadistica – (= ser Latin american & caribbean...1821-1982) – 44mf – 9 – (1964 not available) – uk Chadwyck [318]

Boletin de historia americana / Bayle, Constantino – Madrid: Razon y Fe, 1934 – 1 – sp Bibl Santa Ana [970]

Boletin de historias americanas / Bayle, Constantino – Madrid: Razon y Fe, 1933 – 1 – sp Bibl Santa Ana [970]

Boletin de informacion [federacion espanola de trabajadores de la ensenanza] = News bulletin – english ed. Valencia: La Federacion [1937] [semimthly] – 9 – (began in 1937?; title also in english & french with english text) – mf#w879 – us Harvard [320]

Boletin de la direccion general de archivos y bibliotecas, indices de los n. 1-62-99-104, 1952-1968 – 175mf – 9 – sp Cultura [020]

Boletin de la economica agricola de colombia / Varela Martinez, Raul – Bogota, Colombia. 1949 – 1r – us UF Libraries [630]

Boletin de la federacion / Federacion Cuban del Medio Oeste – adno 3 n2 [1977 apr], adno 5 n1 [1978 feb] – 1r – 1 – mf#669609 – us WHS [972]

Boletin de la real academia de la historia. informes / Coello, Francisco – Madrid, 1889 – 1 – sp Bibl Santa Ana [946]

Boletin de la real academia gallega – Coruna: Academia [mf ed 1985] – 6r [ill] – 1 – (began in 1906?. some issues are combined nos. filmed with: colecion de documentos historicos n23 [20 feb 1909]) – mf#1528 – us Wisconsin U Libr [360]

Boletin de la sociedad aragonesa de ciencias naturales – Zaragoza: Libreria editorial de Cecilio Gasca [mf ed 1986] – 1r – 1 – (began in 1902, ceased with t17 (1917); cont by: boletin de la sociedad iberica de ciencias naturales; some nos iss in combined form) – mf#6780 – us Wisconsin U Libr [500]

Boletin de la sociedad astronomica de barcelona / Sociedad Astronomica de Barcelona – Barcelona: Imp Moderna de Guinart y Pujolar 1910- (irreg) [mf ed 2001] – v1-4 (jul 1910-oct 1921) on 2r [ill] – 1 – (called also t1 [ano 1910-1912]-t4 [anos 1917-21]) – mf#film mas c5065 – us Harvard [520]

Boletin de la sociedad astronomica de mexico : revista mensual de astronomia, meteorologia, y fisica del globo / Sociedad Astronomica de Mexico – Mexico: Secretaria de la Sociedad, t1 n1 (abr 1902)- (mthly) mf ed 2006] – 2r – 1 – (ceased with v17 n159 (dec 1918)?; iss for jan 1903- called also ano 2-; some iss publ in combined form) – mf#film mas c6085 – us Harvard [520]

Boletin de leyes y decretos sobre ferrocarriles dictados / Chile Laws, Statutes, etc – Santiago, Chile. 1891 – 1r – us UF Libraries [972]

Boletin del archivo historico – Venezuela. v1-10. 1943-55 – 1 – us L of C Photodup [972]

Boletin del avuntamiento de madrid / Madrid. Ayuntamiento – 1956-1968 – 1 – us NY Public [946]

Boletin del ayuntamiento / Madrid. Ayuntamiento – Resumen Estadistico. 1959-1961 – 1 – us NY Public [946]

Boletin del gremio de obreros – [Habana]: Imprenta la Razon. v1 n5,8. aug 5, sep 20 1886 – 1mf – 9 – us CRL [073]

Boletin del segundo seminario sobre demografia / Seminario Sobre Demografia (2d : 1965) – Bogota, Colombia. 1965 – 1r – us UF Libraries [350]

Boletin eclesiastico de la diocesis de coria – 1856-1962 – 9 – sp Bibl Santa Ana [070]

Boletin eclesiastico del obispado de plasencia – 1859-1963 – 9 – sp Bibl Santa Ana [070]

Boletin eclesiastico del obispado del priorato de san marcos de leon – 1857-73 – 9 – sp Bibl Santa Ana [071]

Boletin especial / Dominican Republic Secretaria De Finanzas – Ciudad Trujillo, Dominican Republic. 1948 – 1r – us UF Libraries [336]

Boletin estadistico / Banco de Guatemala – July 1948-68 – 1 – us L of C Photodup [336]

Boletin estadistico 1964-1965, 1968, 1971 / Cuba. Direccion General de Estadistica – (= ser Latin american & caribbean...1821-1982) – 15mf – 9 – uk Chadwyck [318]

Boletin informativo del centro de espiritualidad – [Buenos Aires]: El Centro. [n1 (1 jul.1968)] (bimthly) – 1r – 1 – us CRL [241]

Boletin informativo extraordinario / Hermandad de Donates de Sangre – Caceres: Navidad, 1975. Caceres: Linea XXI, 1975 – 1 – sp Bibl Santa Ana [946]

Boletin judicial de la republica argentina / Argentine Republic. Courts – N. 4562-15313. Buenos Aires. 1910-39. (Wanting scattered issues and issues for Jan 1926-Jun 1937) – 1 – us L of C Photodup [972]

Boletín mensual del observatorio del ebro : serie a, heliofísica, meteorología, sismología / Observatorio del Ebro – Tortosa: Impr Moderna del Ebro de Alguero y Baiges [1937]- (mthly) [mf ed 2001] – v27-35(1936-47) on 2r [ill] – 1 – (with annual resumen; some iss accompanied by suppls; split into: boletín del observatorio del ebro. heliofísica; boletín del observatorio del ebro. meteorología; and boletín del observatorio del ebro. sismología; cont: boletin mensual del observatorio del ebro; ceased in dec 1947; v29 never publ; publ varies: imprenta de alguero y baiges 1913-35) – mf#film mas c4692 – us Harvard [520]

Boletin mensual del observatorio del ebro / observatorio de fisica cosmica del ebro = Bulletin de l'observatoire de l'ebre / Observatorio del Ebro – Barcelona: Impr Moderna de Guinart y Pujolar [1910]-1936 (mthly) [mf ed 2] – v1-26(1910-35) on 13r [ill] – 1 – (incl annual resumen, some iss accompanied by suppls; cont in part by: boletin mensual del observatorio del ebro. serie a, heliofisica, meteorologia, sismologia, in spanish & french 1910-19, imprint varies: tortosa: imprenta moderna del ebro de alguero y baiges 1913-35) – mf#film mas c4691 – us Harvard [520]

BOLETIN

Boletin municipal / Lima. (City). Consejo Provincal – 1959-66 – 1 – us NY Public [972]
Boletin municipal de zaragoza / Saragossa. Spain (City). Ayuntamiento – Zaragossa. 1960-1962 – 1 – us NY Public [946]
Boletin oficial / Argentine Republic – 1893-1956; 1971-79; 1980- – 1 – us L of C Photodup [972]
Boletin oficial / Argentine Republic – Buenos Aires. Feb. 24, 1899-1968 – 1 – us NY Public [324]
Boletin oficial / Baja California Sur. (Territory) – La Paz. Jan 1882-Dec 1912; Jan 1929-43. LL-02014 – 1 – 69.00 – us L of C Photodup [340]
Boletin oficial / California. (Lower). Southern territory – La Paz. 1945-1956 – 1 – us NY Public [324]
Boletin oficial / Chile. Direccion General de Correos y Telegrafos – v. 5-32, Jan 1926-June 1, 1960 – 1 – $207.00 – us L of C Photodup [380]
Boletin oficial, and supplements / Cape Verde Islands – (Praia). 1948-1967 – 1 – us NY Public [324]
Boletin oficial de la provincia de badajaz – 1836-1954 – 9 – sp Bibl Santa Ana [946]
Boletin oficial de la provincia de caceres – 1835-82 – 9 – sp Bibl Santa Ana [074]
Boletin oficial de la provincia de la habana – Habana: Imp del Gobierno y Capitania General. v1. 1879 – 55 sheets – us CRL [073]
Boletin oficial de la republica argentina – Bolsheviks Aires, jul 1 1893- – (seccion 3: feb 24 1899-1970) – us CRL [972]
Boletin oficial de ventas de bienes nacionales de la provincia de badajoz – Badajoz, 1870-72, 1875-1882, 1893-1899 y 1924 – 5 – sp Bibl Santa Ana [073]
Boletin oficial del estado / Spain – 1711-1986 5; 1984-1996 9 – 5,9 – sp Boletín [324]
Boletin oficial del gobierno constitucional del estado de sonara / Sonara. Mexico (State) – Hermosillo. On film: 1885-1955. LL-02035 – 1 – us L of C Photodup [340]
Boletin oficial del obispado de badajaz – 1855-1973 – 9 – sp Bibl Santa Ana [240]
Boletin oficial y judicial / Catamarca. Argentine Republic. (Province) – 1957-1968- – 1 – us NY Public [324]
Boletin parroquial de la diocesis de badajoz – Badajoz, 1914-1918 – 5 – sp Bibl Santa Ana [240]
Boletin revista del instituto de badajaz – 1881-82 – 9 – sp Bibl Santa Ana [070]
Boletin tecnico – Santiago, Chile 1972-73 – 1,5,9 – mf#7810 – us UMI ProQuest [660]
Boletin unido – Montivideo: [FIEU [dic 1981-dic 1998] (irreg) – 1r – 1 – (publicacion conjunta (de circulacion interna) en sustitucion de los boletines y circulares habituales de: congregaciones mennonites en el uruguay, iglesia evangelica metodista en el uruguay, iglesia evangelica del rio de la plata (ierp), iglesia evangelica valdense (area rioplatense) dic de 1981) – us CRL [242]
Bolin, Luis *see* Espana
Bolingbroke, Henry *see*
– Reise nach dem demerary
– Voyage to demerary
– A voyage to the demerary
Bolita / Richardson, Martin D – s.l, s.l? 1936 – 1r – 1 – us UF Libraries [978]
Bolivar – Bogota, Colombia. n1-48. 1951 jul-1957 oct – 6r – (gaps) – us UF Libraries [972]
Bolivar – Bogota, Colombia. n52-62. 1959 jul-1963 mar – 1r – (gaps) – us UF Libraries [972]
Bolivar a concha / Vega, Fernando De La – Bogota, Colombia. 1951 – 1r – us UF Libraries [972]
Bolivar Coronado, Rafael *see*
– Llanero (estudio de sociologia venezulana)
– Parnaso costarricense
Bolivar countries / Russell, William Richard – New York, NY. 1949 – 1r – us UF Libraries [972]
Bolivar e caxias : paralelo entre duas vidas / Monjardim, Adelpho Poli – Rio de Janeiro, Brazil. 1967 – 1r – us UF Libraries [972]
Bolivar first baptist church. bolivar, tennessee : church records – 1882-1937 – 1 – us Southern Baptist [242]
Bolivar, o brasil e os nossos vizinhos do prata / Mello, Arnaldo Vieira De – Rio de Janeiro, Brazil. 1963 – 1r – us UF Libraries [972]
Bolivar second baptist church. bolivar, missouri : church records – 1952-63 – 1 – 9.50 – us Southern Baptist [242]
Bolivar, Simon *see*
– America e o libertador
– Documentos
– Obras completas
Bolivar y dario / Rojas, Armando – Managua, Nicaragua. 1964 – 1r – us UF Libraries [972]
Bolivar y la emancipacion de sur-america / O'leary, Daniel Florencio – Madrid, Spain. v1-2. 1915 – 1r – us UF Libraries [972]
Bolivar y leon 12 / Leturia, Pedro S – Caracas, 1931; Madrid: Razon y Fe, 1931 – 1 – sp Bibl Santa Ana [946]
Bolivar y su obra / Gutierrez, Jose Fulgencio – Bogota, Colombia. 1953 – 1r – us UF Libraries [972]
Bolivariada / Rincon Y Serna, Jesus – Bogota, Colombia. 1953 – 1r – us UF Libraries [972]
Bolivia *see*
– Gaceta del gobierno
– Gaceta oficial
– Memoria de guerra y colonizacion
Bolivia. Departamento de Gobiern *see* Memoria...
Bolivia. Direccion General de Estadistica y Estudios Geograficos *see*
– Anuario geografico y estadistico de la republica de bolivia 1919
– Anuario nacional estadistico y geografico de bolivia 1917
Bolivia en cifras 1973 / Bolivia. Instituto Nacional de Estadistica – (= ser Latin american & caribbean...1821-1982) – 7mf – 9 – uk Chadwyck [318]
Bolivia. Instituto Nacional de Estadistica *see* Bolivia en cifras 1973
Bolivia. Laws, Statutes, etc *see*
– Codigo penal boliviano
– Coleccion oficial de leyes, decretos, ordenes, resoluciones.
Bolivia. Ministerio de Colonias y Agricultura *see* Memoria...
Bolivia. Ministerio de Colonizacion y Agricultura *see* Memoria...
Bolivia. Ministerio de Comunicaciones *see* Memoria...
Bolivia. Ministerio de Gobierno, Justicia y Relaciones Exteriores *see* Memoria...
Bolivia. Ministerio de Guerra *see*
– Informe [...]
– Memoria...
Bolivia. Ministerio de Guerra y Colonizacion *see*
– Memoria...
Bolivia. Ministerio de Hacienda *see*
– Informe del ministro de hacienda de bolivia al congreso ordinario de [...]
– Informe que el oficial mayor encargado del ministerio de hacienda presenta a la asamblea nacional ordinaria de [...]
– Memoria...
Bolivia. Ministerio de Hacienda e Industria *see*
– Informe [...]
– Informe del ministro de hacienda e industria a la asamblea ordinaria de [...]
– Memoria...
– Memoria que presenta al congreso ordinario de...
Bolivia. Ministerio de Hacienda i Culto *see* Memoria...
Bolivia. Ministerio de Hacienda y Estadistica *see*
– Informe que presenta a la legislatura ordinaria de [...]
– Memoria presentada al h. convencion nacional de...
Bolivia. Ministerio de Hacienda y Policia Material *see* Memoria que presenta al congreso constitucional de...
Bolivia. Ministerio de Justicia, Culto e Instruccion Publica *see* Memoria...
Bolivia. Ministerio de la Guerra *see* Informe presentado al congreso ordinario de [...]
Bolivia. Ministerio de Minas y Petroleo *see* Memoria presentada al h congreso nacional...
Bolivia. Ministerio de Obras Publicas y Comunicaciones *see* Memoria...
Bolivia. Ministerio de Relaciones Exteriores *see*
– Memoria...
Bolivia. Ministerio de Relaciones Exteriores y Colonizacion *see* Memoria...
Bolivia. Ministerio de Relaciones Exteriores y Culto *see*
– Informe [...]
– Memoria...
Bolivia. Ministerio del Culto e Instruccion Publica *see* Memoria...
Bolivia. Ministerio del Interior *see*
– Memoria...
Bolivia. Ministerio del Interior y Culto *see* Memoria...
Bolivia. Ministerio del Interior y Relaciones Exteriores *see*
– Memoria...
– Memoria que presenta al soberano congreso de bolivia...
Bolivia. Ministerio de Relaciones Exteriores y Colonizacion *see* Informe [...]
Bolivia President *see* Mensaje...
Bolkenstein, H *see* Wohltaetigkeit und armenpflege im vorchristlichen altertum
Boll, F *see* Vorlesungen und abhandlungen
Boll, Stefan *see*
– Segelsurfen im schulsport
– Vergleichende analyse der fernsehsportberichterstattung der ard aus der sicht von kommunikatoren und rezipienten
Boll, Uwe *see*
– Arbeiten zu film und fernsehen
– Bewertung der tv-serie auf empirischer grundlage
– Materialien zur medientheorie
Bolland en petrus / Schaepman, Herman Johan Aloysius Maria – 4th ed. Utrecht: Wed J R van Rossum 1899 [mf ed 1986] – 1mf – 9 – 0-8370-7103-8 – mf#1986-1103 – us ATLA [240]
Bolland, Gerardus Johannes Petrus Josephus *see*
– De pentateuch
– Gnosis en evangelie
– Rome en de geschiedenis
Bollandus, J *see*
– Acta sanctorum
– Januarii-novembris
Bollers, Henry J *see* Henry j bollers fortepiano book
Bolles, Albert Sidney *see*
– Everyman's lawyer...
– The law of the suspension of the power of alienation in the state of new york
– The law relating to banks and their depositors and to bank collections
Bolles, George S *see* Business man's commercial law library
Bolles, John Augustus *see* A treatise on usury and usury laws
Bolles school : san jose, south jacksonville / Shepherd, Rose – s.l, s.l? . 1937 – 1r – us UF Libraries [978]
Bolletino ufficiale / Trentino-Alto Adige. Italy – Trento. 1960-1968 – 1 – us NY Public [945]
Bollettino : official publication / Italian Catholic Federation – 1976 dec, 1982-84, 1985-87 – 3r – 1 – mf#592160 – us WHS [241]
Bollettino bibliografico musicale – Milan. v. 1-8 no. 4 5. Sept 1926-Apr May 1933 – 1 – us NY Public [780]
Bollettino bibliografico musicale – Milan. v1-8 n4-5. sep 1926-apr/may 1933 – 1 – $108.00 – mf#0110 – us Brook [780]
Bollettino della associazione degli africanisti italiani / Associazione Degli Africanisti Italiani – v1-2 1968-69 – 1 – us AMS Press [960]
Bollettino della societa di storia valdese – Torre Pellice IT: Tip Alpina. n61-63 [1934-35] [semiannual] [mf ed 2003] – 3v on 1r – 1 – mf#2003-s016 – us ATLA [242]
Bollettino della societa di studi valdesi – Torre Pellice IT: Tip Alpina. n64-185 [1935-99] [semiannual] [mf ed 2003] – 8r – 1 – (iss 88-89 lack collective title. in italian, english & french. summaries in italian) – mf#2003-s017 – us ATLA [242]
Bollettino della societa italiana di farmacia ospedaliera / Societa Italiana di Farmacia Ospedaliera – Torino Caselle, Italy 1975-80 – 1,5,9 – mf#10038 – us UMI ProQuest [615]
Bollettino dello sciopero dei sarti – Chicago IL, 1910 – 1r – 1 – (italian newspaper) – us IHRC [071]
Bollettino di psicologia applicata – Florence, Italy 1977-89 – 1,5,9 – ISSN: 0006-6761 – mf#1646 – us UMI ProQuest [150]
Bollettino di statistica e legislazione comparata / Italy. Ministero delle Finanze – Roma. Anno. 1-25. 1900-1926 1927 – 1 – us NY Public [336]
Bollettino di studi storico-religiosi – n1-2. 1921-22 [complete] – 1r – 1 – mf#ATLA 1994-S515 – us ATLA [200]
Bollettino officiale. ordine figli d'italia in america – New York NY, 1918-29 – 5r – 1 – (italian periodical) – us IHRC [073]
Bollettino storico della svizzera italiana – Bellinzona, Switzerland 1950-55 – 1 – ISSN: 0006-6869 – mf#466 – us UMI ProQuest [945]
Bollettino ufficiale / Italy. Dogane e Imposte Indirette. Direzione Generale delle – Milan. v. 1-79. 1862-1939. Incomplete – 1 – us NY Public [324]
Bolling beam – 1981 may 1-1982 jun, 1982 jul-1983 nov 18, 1984 aug 31-1986 feb 28, 1984 feb 17-1985 feb 22, 1985 mar-oct – 5r – 1 – mf#648341 – us WHS [071]
Bolling, Helmuth *see* A look into the past
Bollinger, Richard Amsey *see* Teaching and learning pastoral diagnosis
Bollington (prestbury), st gregory rc church : (and 3 small burial grounds) – [Macclesfield Ferrets] – (= ser Cheshire monumental inscriptions) – 1mf – 9 – £2.50 – mf#397 – uk CheshireFHS [929]
Bollington (prestbury), st john – [North Cheshire FHS] – (= ser Cheshire monumental inscriptions) – 3mf – 9 – £4.00 – mf#118 – uk CheshireFHS [929]
Bollington (prestbury), st john the baptist : burials 1835-1937 – [North Cheshire FHS] – (= ser Cheshire church registers) – 4mf – 9 – £5.50 – mf#422 – uk CheshireFHS [929]
Bollington (rostherne), holy trinity – [North Cheshire FHS] – (= ser Cheshire monumental inscriptions) – 1mf – 9 – £2.50 – mf#140 – uk CheshireFHS [929]
Bollnaes tidning – Gaevle, Sweden. 1876-80 – 2r – 1 – sw Kungliga [078]
Bollo, Sarah *see* Tres ensayos alemanes
Bollstandiges barburger – 1762 – 1 – us Southern Baptist [242]
Bolnhurst – (= ser Bedfordshire parish register series) – 1mf – 9 – £3.00 – uk BedsFHS [929]
Bol'nichnaia gazeta botkina – St Petersburg, Russia 1890-95 [mf ed Norman Ross] – 1 – mf#nrp-1580 – us UMI ProQuest [077]
Bolognetti, F *see* La christiana vittoria maritima
Bolona De Sierra, Concepcion *see* Pensamientos de coralia
Bolotov, P A *see*
– Sbornik polozhenii i instruktsii po bukhgalterskomu uchetu i otchetnosti soiuzov kustarno-promyslovoi i lesnoi kooperatsii
– Schetovodstvo proizvoditelno-trudovykh artelei v sviazi s osnovami obshchego schetovodstva
Bolshakov, A M *see*
– Ocherki derevni sssr, 1917-1927
– Sovremennaia derevnia v tsifrakh
Bol'shevik – (Kommunist). Moscow. Apr 1924-Oct 1952. Incomplete – 1 – us NY Public [335]
Bolshevik League of the US *see* Workers' tribune
Bolshevik Union of Canada *see* Proletarian revolution
Bol'shevik-chekist – (city unknown), Russia 1941-42 [mf ed Norman Ross] – 1 – mf#nrp-90 – us UMI ProQuest [934]
Bolsheviki, mensheviki i revoliutsionnaia sotsial-demokratiia / Lindov, G – 1917 – 46p 1mf – 9 – mf#RPP-24 – ne IDC [325]
Bolshevist – Cape Town SA, 1919-21 – 1r – 1 – sa National [079]
Bol'shevistskaia pechat' : publication of the central committee of the communist party = Bolshevik press – Moscow, Russia 1933 [mf ed Norman Ross] – mf#nrp-1059 – us UMI ProQuest [335]
Bol'shevistskii put' – Pavlodar, Kazakhstan 1973 [mf ed Norman Ross] – 4r – 1 – mf#nrp-1365 – us UMI ProQuest [077]
Bolt, Brian R *see* The influence of case discussions on physical education preservice teachers' reflection in an educational games class
Bolt, H M *see* Beurteilung des effektes organischer loesungsmittel auf das hoervermoegen
Bolt, Hermann M et al *see* Frueherkennung und vermeidung von arbeitsbedingten erkrankungen
Bolte, Johannes *see*
– Coligny – gustav adolf – wallenstein
– Drei schauspiele vom sterbenden menschen
– Gartengesellschaft
– Georg rollenhagens spiel vom reichen manne und armen lazaro
– Georg rollenhagens spiel von tobias
– Georg wickrams werke
– Jakob freys gartengesellschaft
– Martin montanus schwankbuecher
– Die reise der soehne giaffers
– Valentin schumanns nachtbuechlein
– Wallenstein
Bolten, Carl *see* Landgewinnungsarbeiten im bereich der halligen an der schleswigschen west kueste und ihre wirtschaftliche bedeutung
Bolton 1726-1905 – Oxford, MA (mf ed 1999) – (= ser Massachusetts vital records) – 102mf – 9 – 0-87623-402-3 – (mf 1-2,48: accounts 1781-91, 1810-46. mf 3-37: town records 1778-1872. mf 37: new town residents 1788-95. mf 37-58: payments 1782-1837, 1865-86. mf 59-63: poor farm 1827-45. mf64-73: mortgages 1806-62. mf 74: tax valuations 1860. mf 75,83: civil & revolutionary wars. mf 76: church members 1817-33. mf 77-79: births 1727-1849 a-w. mf 79,83: deaths 1823-34 a-w. mf 80-83: marriages 1742-1853 a-w. mf 84-86: births 1726-82. mf 85-86: deaths 1727, 1756-81. mf 86: marriages 1745-81. mf 87-89: 1st parish vitals 1759-1852. mf 90: births 1743-63, 1796-1848; deaths 1780, 1823-59. mf 91-92: births 1844-68 & index. mf 91-94: marriages 1844-56 & index. mf 93-94: deaths 1844-69 & index. mf 94-97: marriages 1856-1906 & index. mf 98-100: births 1864-1905 & index. mf 98,102: voters 1820, 1836, 1868-76. mf 100-102: deaths 1869-1906 & index. mf 102: non-resident burials 1855-1917) – us Archive [978]
Bolton 1728-1849 – Oxford, MA (mf ed 1995) – (= ser Massachusetts vital record transcripts to 1850) – 8mf – 9 – 0-87623-212-8 – (mf 1t-2t: births & deaths 1728-82. mf 3t-7t: births & deaths 1743-1849. mf 2t-6t: marriages 1746-1844. mf 3t-6t: intentions 1787-1850. mf 7t: out-of-town marriages 1739-1799. mf 8t: b,m,d 1844-49) – us Archive [978]
Bolton advertiser *see* Mackie's advertiser
Bolton and bury catholic herald – Bolton, England 19 sep 1908-14 jul 1934 (wkly) [mf 1912] – 1 – (wanting 1911; cont: bolton catholic herald [29 feb,1 aug-12 sep 1908]) – uk British Libr Newspaper [241]
Bolton, C J *see* Study of potato cooperative marketing associations in florida
Bolton, C S *see* Diary
Bolton catholic herald – [NW England] Bolton Lib 19 sep 1908-25 sep 1909 – 1 – (title change: bolton catholic herald [29 feb, 1 aug-12 sep 1908]) – uk MLA; uk Newsplan [241]

Bolton, Charles Knowles *see* Scotch irish pioneers in ulster and america
Bolton chronicle – [NW England] Bolton 8 jan 1831-21 dec 1917 [mf ed 2002-03] – 81r – 1 – (missing: 1842, 1861, 1886, 1894, 1911, 1891, 1896-97; cont: bolton chronicle and south lancashire advertiser [15 nov 1845-31 dec 1853]; bolton chronicle [7 jan 1854-21 dec 1917]) – uk Newsplan [072]
Bolton chronicle – [NW England] Bolton Lib 22 mar 1979-8 feb 1990 – 1 – (title change: bolton metro news [8 feb 1990-jun 1996]) – uk MLA; uk Newsplan [072]
Bolton chronicle, and register for wigan, blackburn, bury, rochdale – [NW England] Bolton Lib 22 jan-2 jul 1825, 30 sep 1826, 1827-8 nov 1845* – 1 – (cont by: bolton chronicle & south lancashire advertiser [15 nov 1845-31 dec 1853]; bolton chronicle [7 jan 1854-21 dec 1917]) – uk MLA; uk Newsplan [072]
Bolton daily chronicle – [NW England] Bolton 5 may 1873-31 dec 1895, 2 jan 1901-25 jul 1907 [mf jan-dec 1899; jan-apr, sep-dec 1910] – 1 – (cont by: bolton evening chronicle [1 aug 1907-21 dec 1917 (discontinued)]) – uk Newsplan [072]
Bolton evening echo – [NW England] Bolton 4 jun-16 aug 1894 (imperfect) [mf ed 2002] – 1r – 1 – uk MLA; uk Newsplan [072]
Bolton evening guardian – Bolton, England 29 sep 1873-20 may 1893 (daily) [mf 1859-93] – 1 – (incorp with: bolton evening news & weekly journal; fr 4 oct 1873-20 may 1893 sat morning ed were entitled: bolton weekly guardian; cont: bolton guardian [28 jan 1860-27 sep 1873]) – uk British Libr Newspaper [072]
Bolton evening news – [NW England] Bolton Lib 19 mar 1867-87 – 1 – (cont by: bolton evening news & district reporter [23 mar, 1 jul 1889-16 aug 1962 (gaps)]; evening news [17 aug 1962-31 jul 1976]; bolton evening news [3 aug 1976-9 sep 2006]; bolton news [11 sep 2006- (mf 1986-)]) – uk MLA; uk Newsplan [072]
Bolton express and county effective advertiser – [NW England] Bolton Lib 14 dec 1888-18 dec 1896* – 1 – (wanting: 1896,1897) – uk MLA; uk Newsplan [072]
Bolton express and lancashire advertiser – [NW England] Bolton Lib 5 jul 1823-26 jun 1824, 6 nov 1824, 4 jun 1825, 17 jun 1826 – 1 – (cont by: bolton express & lancashire general advertiser [6 nov 1824, 4 jun 1825, 17 jun 1826]) – uk MLA; uk Newsplan [072]
Bolton free press – [NW England] Bolton 21 nov 1935-24 dec 1847 (wkly) [mf 1842] – 1 – (discontinued) – uk MLA; uk Newsplan [072]
Bolton, Glorney *see* The tragedy of gandhi
Bolton independent – [NW England] Bolton 8 oct 1859-21 jan 1860 [mf 1859-93] [mf ed 2003] – 1 – (cont by: bolton guardian [28 jan 1860-27 sep 1873]; bolton evening guardian [29 sep 1873-20 may 1893]) – uk Newsplan; uk MLA [072]
Bolton journal – [NW England] Bolton 14 mar 1985-26 apr 1990 [mf 1986-] – 1 – (cont by: journal [3 may 1990-17 jun 1993]; bolton journal [24 jun 1993-]) – uk MLA; uk Newsplan [072]
Bolton mercury – [NW England] Bolton Lib 31 dec 1853-22 jul 1854 – 1 – uk MLA; uk Newsplan [072]
Bolton monthly advertiser – [NW England] Bolton may-aug, oct,dec 1854-jun 1855 [mf ed 2002] – 1 – uk Newsplan; uk MLA [072]
Bolton, Philip *see* Revival movement, and the way of salation explained...
Bolton, R *see* A discourse about the state of true happiness delivered in certaine sermons in oxford, and at st pauls crosse
Bolton, Robert *see* History of the protestant episcopal church in the county of westchester
Bolton, Robyn M *see* A study of the implications of the 1994 major league baseball players' strike and an analysis of the marketing strategies used by major league baseball and four teams in response to the strike
Bolton spectator – [NW England] Bolton 2 may 1857-27 feb 1858 – 1 – uk MLA; uk Newsplan [072]
Bolton star – [NW England] Bolton Lib 5 jun 1891-25 jun 1892 – 1 – uk MLA; uk Newsplan [072]
Bolton trades council records, 1875-1968 – (= ser Labour party in britain, origins and development at local level. series 2) – 5r – 1 – (with p/g. int by richard stevens) – mf#97573 – uk Microform Academic [331]
Bolton weekly guardian – [NW England] Bolton Lib 4 oct 1873-20 may 1893 – 1 – uk MLA; uk Newsplan [072]
Bolton weekly journal and district news – [NW England] Bolton 4 nov 1871-24 dec 1888, 6 jul 1889-20 may 1893 [mf 1871-7, 1879-85, 1887-97, 1899, 1901-50] [mf ed 2004] – 1 – (missing: 1878, 1886, 1898; cont by: bolton journal & guardian, & district news [may 1893-20 sep 1946]; bolton journal & guardian [jan 1947-dec 1950]) – uk Newsplan [072]
Bolton, William *see*
– North india
– The south india mission
– Travancore
Bolton-Smith, Robin *see* Portrait miniatures in the national museum of american art
Bolu – (= ser Vilayet salnames) – 9 – (1334 [1918] 6mf $90; 1337-38m [1921-22] 12mf $195) – us MEDOC [956]
Boman-Behram, B K *see* Educational controversies in india
Bomb hip-hop magazine – 1994-95 – 1r – 1 – mf#2947367 – us WHS [780]
Bombala herald – Bombala, jan 1899-aug 1911 – 2r – A$175.52 vesicular A$186.52 silver – at Pascoe [079]
Bombala times – Bombala, aug 1956-dec 1968 – 3r – A$226.95 vesicular A$243.45 silver – at Pascoe [079]
Bombala times – Bombala, jan 1899-dec 1905 – 2r – A$116.16 vesicular A$127.16 silver – at Pascoe [079]
Bombala times – Bombala, jan 1969-dec 1996 – at Pascoe [079]
Le bombardement des villes ouvertes – Paris, 1938? Fiche W 757. (Blodgett Collection of Spanish Civil War Pamphlets) – 9 – us Harvard [946]
Bombardements et agressions en espagne, juillet 1936-juillet 1938 / World Committee against War and Fascism – Paris 1938 – 9 – mf#w758 – us Harvard [946]
El bombardeo de almeria por la escuadra alemana – Madrid, 1937? – (= ser Blodgett coll) – 9 – mf#fiche w760 – us Harvard [946]
The bombardment of egyptian legation in madrid – n.p. 1937. Fiche W 759. (Blodgett Collection of Spanish Civil War Pamphlets) – 9 – us Harvard [946]
Bombardment of the british embassy in madrid = Bombardeo de la embajada inglesa en madrid – [Madrid]: Servicio Espanol de Informacion [1937?] – 9 – mf#w1512 – us Harvard [946]
Bombay / Sheppard, Samuel Townsend – Bombay: Times of India Press, 1932 – (= ser Samp: indian books) – us CRL [954]
Bombay, 1885 to 1890 : a study in indian administration / Hunter, William Wilson – London: H Frowde; Bombay: B M Malabari [1892] [mf ed 1987] – 1r – 1 – (filmed with: travels and researches in asia minor / fellows, c) – mf#1847 – us Wisconsin U Libr [954]
Bombay 1885 to 1890 a study in indian administration / Hunter, William Wilson – London, [1892] – (= ser 19th c british colonization) – 6mf – 9 – mf#1.1.4017 – uk Chadwyck [350]
Bombay and the sidis / Banaji, D R – London: Macmillan, 1932 – (= ser Samp: indian books) – us CRL [954]
Bombay courier – Bombay, India 5 jan 1793-29 dec 1846 (wkly) – 1 – (wanting: 1797) – uk British Libr Newspaper [079]
Bombay ducks / Dewar, Douglas – London, England. 1906 – 1r – us UF Libraries [590]
[Bombay-] economic and political weekly – II. 1973-92 – 36r – 1 – $1800.00 – mf#R63571 – us Library Micro [079]
Bombay gazette – Bombay, India 15 aug 1792, 7 apr 1813-31 dec 1814, 3 jan 1816-30 dec 1841, 1 jan 1850-31 dec 1868, 22 may 1911-7 mar 1914 [wkly] (imperfect) [mf 1792-1868] – 1 – uk British Libr Newspaper [079]
Bombay government gazette – 1931-1940; Apr. 1956-1960. Incomplete – 1 – us NY Public [324]
[Bombay-] illustrated weekly of india – II. 1972-85 – 21r – 1 – $1050.00 – mf#R63614 – us Library Micro [079]
Bombay in april, 1840 / Duff, Alexander – Edinburgh, Scotland. 1840 – 1r – us UF Libraries [240]
Bombay in the making : being mainly a history of the origin and growth of judicial institutions in the western presidency, 1661-1726 / Malabari, Phiroze Behramji Merwanji – London: T Fisher Unwin, 1910 – (= ser Samp: indian books) – us CRL [340]
The bombay quarterly review – Bombay: Smith, Taylor; London: Smith, Elder. v1-[7] (n1-14). 1855-sep 1858 – us CRL [073]
Bombay samachar – Bombay, India. Apr 1944-Jul 1995 – 278+ r – 1 – (cont as: mumbai samacara) – us L of C Photodup [079]
Bombay times *see* Times of india, 1861-1889
Die bombe – Wien (A) 1889 [mf ed 2004] – 1r – 1 – gw Mikrofilm [074]
Die bombenflieger *see* Roter marsch / der leichnam auf dem thron / die bombenflieger
Bomber offensive / Harris, Arthur Travers – London: Collins 1947 [mf ed 1984] – 1r – 1 – (filmed together: greek life and thought / larue van hook & other titles) – mf#11115 – us Wisconsin U Libr [934]
Bomberger, John Henry Augustus *see*
– Infant salvation in its relation to infant depravity, infant regeneration and infant baptism
– Reformed, not ritualistic, apostolic, not patristic
– The revised liturgy
– Selected works
– Text-book of church history
Bombes sur le guatemala / Desinor, Yvan M – Port-Au-Prince, Haiti. 1960 – 1r – us UF Libraries [972]
Bombs over barcelona / Medical Bureau and North American Committee to Aid Spanish Democracy – New York 1938 – 9 – mf#w1034 – us Harvard [946]
Bomfim, Manoel Jose Do *see*
– America latine
– Brasil
Der bommeraner – Witten DE, 1985 aug-1997 – 2r – 1 – gw Misc Inst [074]
Bompas, William Carpenter *see*
– Diocese of mackenzie river
– Lessons and prayers in the tenni or slavi language of the indians of mackenzie river in the north-west territory of canada
Bompiani, S *see* Italian explorers in africa
Bompiani, Valentino *see* Albertina
O bom-successo – Bom Sucesso, MG: Typ d'O Bom-Successo, 30 abr-23 jul 1893 – (= ser Ps 19) – mf#P11B,03,80 – bl Biblioteca [079]
Bon, Antoine *see* Brazil
Le bon apotre / Soupault, Philippe – Paris: Editions du Sagittaire c1923 – 1 – us Wisconsin U Libr [830]
Bon appetit – New York NY 1988+ – 1,5,9 – ISSN: 0006-6990 – mf#17081 – us UMI ProQuest [640]
Le bon francais – Paris. 1817-23 fevr 1818 – 1 – fr ACRPP [073]
Le bon francais – Paris. n1-7. 15 fevr-15 mai 1890 – 1 – fr ACRPP [073]
Bon numero / Barde, Andre – Paris, France. 1905 – 1r – us UF Libraries [440]
Le bon pasteur : (s jean, ch 10), meditation sacerdotale, noel / Emard, Joseph-Medard – Valleyfield [Quebec]: Bureaux de la chancellerie, 1920 [mf ed 1995] – 1mf – 9 – 0-665-74165-0 – mf#74165 – cn Canadiana [241]
Un bon patriote d'autrefois : le docteur labrie / Gosselin, Auguste – [Quebec?: Dussault & Proulx], 1903 – 3mf – 9 – 0-665-74001-8 – mf#74001 – cn Canadiana [610]
Bon petit menage / Landay, Maurice – Paris, France. 1928 – 1r – us UF Libraries [440]
Bon roi dagobert / Samuel-Rousseau, Marcel – Paris, France. 1927 – 1r – us UF Libraries [440]
Le bon sens – Port-au-Prince, Haiti: Imp H Amblard, jul 1909-apr 19 1910 – 25 sheets – us CRL [972]
Le bon sens du cure j meslier *see* Superstition in all ages
Le bon sens republicain – Paris. n1-69. 2 aout-10 oct 1881 – 1 – (mq no. 28-29, 66, 68) – fr ACRPP [073]
Bon valet / Pompigny, Maurin De – Paris, France. 1809 – 1r – us UF Libraries [440]
Bona, Giovanni *see* Ascetical treatise on the sacrifice of the mass / a letter on the great importance of the divine
Bona, Johannes *see* Rerum liturgicarum libri duo
Bon-accord [aberdeen, scotland : 1880] : the illustrated news of the north – [Scotland] Aberdeen: W J Clark 3 jan 1880-13 aug 1914 (mthly) [mf ed 2003] – 83r – 1 – (subtitle varies; cont by: bon accord and northern pictorial [3 apr 1926-4 jun 1927]) – uk Newsplan [072]
Bon-accord and northern pictorial – [Scotland] Aberdeen: H Munro 3 apr 1926-ns: dec 1950 (mthly) [mf ed 2003] – 42r – 1 – (cont: bon-accord (aberdeen, scotland : 1880); cont by: aberdeen bon-accord and northern pictorial: the weekend family journal of the north east with which is incorporated "the bailie" [11 jun 1927-18 jun 1959]) – uk Newsplan [072]
Bon-accord reporter – [Scotland] Aberdeen: R Edward & Co jul 1842-ap 1844 (mthly) [mf ed 2003] – 1r – 1 – uk Newsplan [072]
Bonaccorsi, Giuseppe *see*
– Harnack e loisy
– I tre primi vangeli e la critica letteraria, ossia, la questione sinottica
– Psalterium latinum cum graeco et hebraeo comparatum
Bonaeret, Benedictius de *see* Colophons de manuscrits...
Bonaire / Brenneker, Father – Willenstad, Curacao. 1947 – 1r – us UF Libraries [972]
Bonald, Louis Gabriel Ambroise, vicomte de *see* Recherches philosophiques sur les premiers objects des connaissances morales – demonstration philosophique du principe constitutif de la societe – meditations politiques tirees de l'evangile
Bonan, Jules *see* Comptabilite des affaires a enzel et leurs consequences juridiques pour les commercants
Bonaparte : ou l'homme du destintablettes historiques et chronologiques, presentant le precis de la vie entiere de cet homme extraordinaire... / Cuisin, J P – Paris 1821 [mf ed Hildesheim 1995-98] – 1v on 2mf [ill] – 9 – €60.00 – 3-487-26228-2 – gw Olms [944]
Bonar, Andrew Alexander *see*
– Christ and his church in the book of psalms
– A commentary on the book of leviticus
– Narrative of a mission of inquiry to the jews from the church of scotland in 1839
– Office of deacon
Bonar, Horatius *see*
– Catechisms of the scottish reformation
– The desert of sinai
– God's way of holiness
– Old gospel
– Words to the winners of souls
Bonar, Marjory *see* Reminiscences of andrew a. bonar, d.d
Bonardi, Carlo *see* Enrico heine nella letteratura italiana
Bonaventura *see* Nachtwachen
Bonaventura Argonensis *see* De optima legendorum ecclesiae patrum methodo
Bonaventure / Dupeuty, Charles – Paris, France. 1840 – 1r – us UF Libraries [440]
Bonavides, Paulo *see* Crise politica brasileira
Bonch-Bruevich, V D *see* Literaturnyi, nauchnyi i politicheskii zhurnal
De bond : alguns da civilisacao brazileira / Chagas, Joao Pinheiro – Lisboa, Portugal: Livraria moderna 1897 – 1r – 1 – us UF Libraries [972]
Bond / American Servicemen's Union. Committee for GI Rights – v1 n1-v6 n2 [1967 jun 23-1972 jan 27], v1 n5-v8 n4 [1967 aug-1974 sep/oct], v1 n1-v8 n4 [1967 jun 23-1974 sep/oct] – 3r – 1 – mf#720819 – us WHS [355]
O bond : orgao social – Rio de Janeiro, RJ. 02 nov 1881 – (= ser Ps 19) – mf#P17,01,123 – bl Biblioteca [079]
O bond : periodico faceto critico e noticioso – Bahia: Typ Bahiana, 12 jul 1876 – (= ser Ps 19) – mf#P17,01,29 – bl Biblioteca [870]
Bond baptist church. preston, mississippi : church records – 1839-1950. Formerly: New Hope Baptist Church (until 1951) – 1 – us Southern Baptist [242]
Bond, Carol *see* Practice and belief barriers to diabetes management in cambodian non-insulin-dependent diabetes mellitus patients
Bond, Dale *see* An evaluation of the importance of moderate exercise, t'ai chi, and problem solving in relation to psychological distress
Bond, E A *see* Chronica monasterii de melsa, a fundatione usque ad annum 1396 (rs43)
Bond, Francis *see*
– Dedications and patron saints of english churches
– Dedications and patron saints of english churches: ecclesiastical symbolism
– Dedications and patron saints of english churches, ecclesiastical symbolism, saints and their emblems
– English cathedrals illustrated
– An introduction to english church architecture
– Wood carvings in english churches
Bond, George J *see* Our share in china and what we are doing with it
Bond, George John *see*
– Our share in china and what we are doing with it
– Skipper george netman
Bond, Gregory *see* Whipped curs and real men
Bond, Horace Mann *see* The horace main bond papers
Bond, John *see* They were south africans
Bond, L H *see* Bond's reports of cases in the sixth circuit, 1856-1871
Bond lake and the highlands of york, thornhill, richmond hill, aurora, newmarket, and intermediate points, via the metropolitan railway co : and bond lake to schomberg via the schomberg and aurora ry co: guide and time table / Metropolitan Railway Co, Toronto – Toronto: Bryan Publ, 1904 – 1mf – 9 – 0-665-87737-4 – mf#87737 – cn Canadiana [380]
Bond, S *see* Church membership
Bond, T M *see* Republication of the minutes of the mississippi baptist association, a, 1806-47
Bond, Thomas E *see* An appeal to the methodists in opposition to the changes proposed in their church government
A bond to save from bondage : a few suggestions for a state tenants' defence association / Cleary, Thomas – [Ennis], 1879 – (= ser 19th c ireland) – 1mf – 9 – mf#1.1.1936 – uk Chadwyck [333]
Bond & Weigley *see* The legal, bank, and reporting directory.
Bondar, Gregorio Gregorievitch *see* Cacao
The bondelswarts rebellion of 1922 / Lewis, Gavin Llewellyn Mackenzie – Grahamstown: [s.n.] 1977 – us CRL [960]
Bondfield, G H et al *see* History of union church
Bondi, Georg *see* Erinnerungen an stefan george

Bondi, J H *see* Dem hebraeisch-phoenizischen sprachzweige angehoerige lehnwoerter in hieroglyphischen und hieratischen texten
Bondi, Simon *see* Or ester
Bonds of disunion : or english misrule in the colonies / Rowe, Charles James – London 1883 – (= ser 19th c british colonization) – 4mf – 9 – mf#1.1.3694 – uk Chadwyck [320]
Bonds of the state of tennessee : first mortgage liens on railroads in that state: opinion of charles o'conor upon statement of e l andrews / O'Conor, Charles – New York: American Bank Note Co, Type Dept, 1879 (mf ed 19–) – 133p – mf#ZV-TPG pv61 n16; ZV-TPR pv18 n7 – us Harvard [380]
Bond's reports of cases in the sixth circuit, 1856-1871 / Bond, L H – Cincinnati: Clarke. v1-2. 1879 (all publ) – (= ser Early federal nominative reports) – 13mf – 9 – $19.50 – mf#LLMC 81-436 – us LLMC [340]
Bonduel times – Bonduel WI. 1909 mar 1/1910-1970 mar 19/1971 jul 29 – 19r – 1 – (cont by: times-press (seymour wi: 1977), press (seymour wi)) – mf#967240 – us WHS [071]
Bone – Oxford UK 1978+ – 1,5,9 – ISSN: 8756-3282 – mf#49429 – us UMI ProQuest [617]
Bone and mineral – Oxford UK 1988-92 – 1,5,9 – ISSN: 0169-6009 – mf#42497 – us UMI ProQuest [610]
Bone density patterns in adult females with a history of anorexia nervosa / Siemers, Beverly J & Gench, Barbara E – 1992 – 3mf – $12.00 – us Kinesology [612]
Bone mineral and menstrual cycle status in competitive female athletes : a longitudinal study / Robinson, Tracey L – Oregon State University, 1994 – 3mf – 9 – $12.00 – mf#PH 1505 – us Kinesology [612]
Bone, Renee Yolande *see* Exchange control as an instrument to regulate south africa's external equilibrium
Bonekemper, Johannes *see* Diaries
Bonel, M *see* Tour du sud
Bonelli, Aurelio *see* Il primo libro de ricercari et canzoni a quattro voci
Boner, Jose M *see* El p juan de maldonado
Boner, Kathleen *see* Dr f c kolbe
Boner, Ulrich *see*
– Der edelstein
– Der edelstein / des teufels netz / sibyllenweissagung [cima7]
Bo'ness journal : an edition of the linlithgowshire journal and gazette – Grangemouth: Johnston (Falkirk) Ltd 1990-(wkly) [mf ed 2 jan 1998-] – 1 – (cont in pt: linlithgowshire journal & gazette) – ISSN: 0964-3540 – uk Scotland NatLib [072]
Bo'ness journal and linlithgow advertiser – [Scotland] Bo'ness: F W Broome 31 may 1879-dec 1950 (wkly) [mf ed 2003] – 65v on 59r – 1 – (cont by: bo'ness journal and linlithgowshire advertiser) – uk Newsplan [072]
Bonet-Maury, Gaston *see*
– Le congres des religions a chicago en 1893
– Croyances et legendes du moyen age
– Early sources of english unitarian christianity
– France
– L'islamisme et le christianisme en afrique
– La liberte de conscience en france depuis l'edit de nantes jusqu' a la separation
– Les precurseurs de la reforme et de la liberte de conscience dans les pays latins du 12e au 15e siecle
– L'unite morale des religions
[Bonfanti collection of materials relating to the southern sudan 1956-71] – Chicago: Uni of Chicago, Photodup Dept, 1973 – us CRL [960]
Bonfatti, Emilio et al *see* Momenti di cultura tedesca
Bonfiglio *see* Aerobic fitness testing and feeling states among 9 to 11 year old students
Bonfire / Carneiro, Cecilio J – New York, NY. 1944 – 1r – us UF Libraries [972]
Bong bhaanga omaamaana kwa jesus christ omaan ka oyagbin – High Wycombe, England. 19–? – 1r – us UF Libraries [960]
Bong, Hok Sioe *see* Lelie berdoeri
Bong, Kok No *see* Doea lobang pelor / tjoe bo kim so
Bong thuyen say : truyen dai / Nguyen, Mong Giac – Saigon: Nam-Giao 1974 [mf ed 1992] – on pt of 1r – 1 – mf#11052 r88 n9 – us Cornell [830]
Bonghi, Ruggiero *see* La santa sede
Bonham, Milledge L *see* Papers
Le bonheur au foyer domestique / Pietremont, Maria – Paris: Garnier, 1891 – (= ser Les femmes [coll]) – 4mf – 9 – mf#12827 – fr Bibl Nationale [640]
Bonhomme, Colbert *see* Revolution et contre-revolution en haiti de 1946 a...
Bonhomme jadis / Murger, Henri – Paris, France. 18–? – 1r – us UF Libraries [440]
Bonhomme jadis / Murger, Henri – Paris, France. 1852 – 1r – us UF Libraries [440]
Bonhomme jadis / Murger, Henri – Paris, France. 1859 – 1r – us UF Libraries [440]
Bonhomme job / Souvestre, Emile – Paris, France. 1835 – 1r – us UF Libraries [440]
Bonhomme, Joseph *see* Noir or
Bonhomme, Pierre *see* Melodiae sacrae...
Le bonhomme richard – 6avr 1832-31 aout 1833. devenu: L'Impartial. Paris. 1er sept 1833-6 oct 1836 – 1 – fr ACRPP [970]
Le bonhomme richard – Journal de Franklin. no. 1-3. Paris. 1848 – 1 – fr ACRPP [970]
Boniface / Smith, Isaac Gregory – London: SPCK; New York: E & J B Young 1896 [mf ed 1990] – (= ser The fathers for english readers) – 1mf – 9 – 0-7905-5964-1 – mf#1988-1964 – us ATLA [241]
Boniface, Alexandre *see* Geographie elementaire descriptive
Boniface of crediton and his companions / Browne, George Forrest – London: SPCK, 1910 – 1mf – 9 – 0-7905-4162-9 – mf#1988-0162 – us ATLA [240]
Bonifacius amerbach und die reformation / Burckhardt-Biedermann, Theophil – Basel: R. Reich, 1894 – 1mf – 9 – 0-7905-4251-X – (incl bibl ref) – mf#1988-0251 – us ATLA [242]
Bonifas, Francois *see*
– La doctrine de la redemption dans schleiermacher
– Histoire des dogmes de l'eglise chretienne
Bonifatius : der zerstoerer des columbanischen kirchentums auf dem festlande: ein nachtrag zu dem werke die iroschottische missionskirche / Ebrard, Johannes Heinrich August – Guetersloh: C Bertelsmann 1882 [mf ed 1992] – 1mf – 9 – 0-524-03281-5 – (incl bibl ref) – mf#1990-0892 – us ATLA [241]
Bonifatius, der apostel der deutschen, und die slavenapostel, konstantinos (cyrillus) und methodios : eine historische parallele / Hoefler, Karl Adolf Constantin, Ritter von – Prag: H Dominicus, 1887 [mf ed 1990] – 1mf – 9 – 0-7905-5605-7 – (incl bibl ref) – mf#1988-1605 – us ATLA [240]
Bonilla, Abelardo *see*
– Historia y antologia de la literatura costarricens
– Letras costarricenses
Bonilla, Albelardo *see* Valle nublado
Bonilla Atiles, Jose A *see* Mocion y un discurso
Bonilla Atiles, Jose Antonio *see* Discursos y conferencias enjuiciando la...
Bonilla, Manuel Antonio *see* Caro y su obra
Bonilla, Marcelina *see* Diccionario historico-geografico de las poblacione
Bonilla Naar, Alfonso *see* 21 [i.e. veintiun] anos de poesia colombiana, 1942-1963
Bonilla Samaniego, A *see* Exercitation medica phylosophica sobre la essencia del morbo gallico
Bonilla y San Martin, Adolfo *see*
– Fuero de usagre (siglo 18)
– Las leyendas de wagner en la literatura espanola
Bonin, Daniel *see* Johann georg zimmermann u johann gottfried herder
Bonin, Joseph *see* Biographies de l'honorable barthelemi joliette et de m le grand vicaire a manseau
Bonin, Marie *see* Bio-bibliographie analytique de marthe bergeron-hogue
Bonivard, Francois *see*
– Advis et devis de la sovrce de lidolatrie et tyrannie papale
– Advis et devis des lengues
Bonjoannes, Berardus *see* Compendium of the summa theologica of st thomas aquinas, pars prima
Bonjour : hebdomadaire bilingue de grande information. – Strasbourg. 1964-sept 1972 – 1 – fr ACRPP [073]
Bonjour – New York NY 1967-81 – 1,5,9 – ISSN: 0006-7121 – mf#5841 – us UMI ProQuest [370]
Bonn, Alfred *see* Ein jahrhundert rheinische mission
[Bonn-] vorwarts – DE. 1976-86 – 22r – 1 – $1100.00 – mf#R04234 – us Library Micro [074]
Bonnard, Auguste *see* Thomas eraste (1524-1583) et la discipline ecclesiastique
Bonnard, F *see* Histoire de l'abbaye royale et de l'ordre des chanoines reguliers de st victor de paris
Bonnard, Jean *see* Les traductions de la bible en vers francais au moyen age
Bonnaud, Dominique *see* D'ocean a ocean
Bonnaud, L *see* Apostolat en haiti
La bonne fermiere : revue trimestrielle d'economie domestique et d'agriculture feminine – Quebec: les Cercles. v1 n1 janv 1920- (qrtly) [mf ed 1979] – 1r – 5 – (with bibl; ceased 1930?) – mf#SEM16P317 – cn Bibl Nat [640]
La bonne litterature francaise – Montreal: Leprohon & Leprohon, [1894?-1900?] – 9 – mf#P04238 – cn Canadiana [440]
La bonne sainte : ou, l'histoire de la devotion a sainte anne / Charland, Paul-Victor – Quebec: [s.n.], 1904 – 3mf – 9 – 0-665-71128-X – mf#71128 – cn Canadiana [241]
La bonne ste-anne : sa vie, ses miracles, ses sanctuaires / Frederic, de Ghyvelde, pere – Quebec: [s.n.] 1900 [mf ed 1985] – 5mf – 9 – mf#SEM105P533 – cn Bibl Nat [920]
Bonneau *see* Exercices francais
Bonneau, B *see* Abrege de la grammaire selon l'academie
Bonnechose, Charles de *see* Montcalm et le canada francais
Bonnefon, Jean de *see* Lourdes et ses tenanciers
Bonnell, E *see* Geschichte der christlichen kirche
Bonnemant, Emile *see* Projet pour l'etablissement d'une sucrerie de betteraves au canada
Bonner and middleton's bristol journal – Bristol, England 24 dec 1774; 7,14 jan, 11 mar, 1 apr 1775; 6,13 jan 1776; 19 jul, 6 sep-25 oct, 15 nov, 6,20, 27 dec 1783; 10 jan-7,28 feb, 13,20 mar, 3-17 apr, 24 jul 1784-26 jul 1788; 21 aug 1790; 19 mar 1791; 15 mar 1794; 7 jan 1797-27 dec 1800 – 1 – (cont by: fenley & sheppard's bristol journal [7 jan 1804]) – uk British Libr Newspaper [072]
Bonner anzeiger – Bonn DE, 1850 28 jun-8 nov – 1 – gw Misc Inst [074]
Bonner depesche : funktionaersorgan der spd – Bonn DE, 1961-65 – 1r – 1 – mf#3772 – gw Mikropress [325]
Bonner, Hypatia Bradlaugh *see* Penalties upon opinion, or, some records of the laws of heresy and blasphemy
Bonner intelligenz-blatt *see* Boennisches intelligenzblatt
Bonner jahrbuch – Bonn DE, 1895 s18-155, 1896 s54-163 – 1r – 1 – mf#3596 – gw Mikropress [943]
Bonner, John *see* An essay on the registry laws of lower canada
Bonner nachrichts- und anzeige-blatt : feuille d'affiches – Bonn DE, 1812 5 jan-1814 14 jan – 1r – 1 – mf#4888 – gw Mikropress [074]
Bonner nachrichts- und anzeigeblatt *see* Wochenblatt des boennischen bezirks
Bonner rundschau : ausgabe der koelnischen rundschau fuer die bundeshauptstadt – Bonn, Germany 1 nov 1950-1 dec 1952 – 1 – uk British Libr Newspaper [074]
Bonner rundschau – Bonn DE, 1946 19 mar-1969 – 83r – 1 – (until sep 1949 regional ed of koelnische rundschau / bonn-land, fr 1 oct 1949 regional ed of koelnische rundschau, koeln) – gw Misc Inst [074]
Bonner volksblatt – Bonn DE, 1962 23 sep & 1 oct-30 nov – 1 – gw Misc Inst [074]
Bonner zeitschrift fuer theologie und seelsorge – 1(1924)-8(1931) – 56mf – 9 – €115.00 – ne Slangenburg [240]
Bonner zeitschrift fuer theologie und seelsorge – Duesseldorf, 1924-31 [mf ed 2001] – (= ser Christianity's encounter with world religions, 1850-1950) – 2r – 1 – (in german) – mf#2001-s116 – us ATLA [240]
Bonner zeitung 1824 – Bonn DE, 1824 2 oct-30 oct, 1830 – 1 – gw Misc Inst [074]
Bonner zeitung 1848 – Bonn DE, 1848 7 mai-1850 30 jun – 2r – 1 – (incl suppl: spartacus, jan 15-jul 9 1849; title varies: 1849 neue bonner zeitung) – gw Misc Inst [074]
Bonnes d'enfans : ou, une soiree aux boulevards-neu / Brazier, Nicholas – Paris, France. 1825 – 1r – us UF Libraries [440]
Bonnet, Jacques *see* Histoire de la musique, et de ses effets
Bonnet, Jules *see*
– Aonio paleario
– Calvin au val d'aoste
– Notice sur la vie et les ecrits de m. merle d'aubigne
Bonnet, Max *see*
– Acta apostolorum apocrypha
– Le latin de gregoire de tours
Bonnet, Rene *see* Veillee limousine
Le bonnet rouge – Paris: Impr Le Douarin, apr 10-16, 19, 22 1871 – (filmed as pt of: commune de paris newspapers. newspapers on these reels are filmed chronologically, not alphabetically) – us CRL [074]
Le bonnet rouge – [Paris]: Bonaventure et Ducessois, jun 11-15/18 1848 – us CRL [074]
Le bonnet rouge – Paris. 6 dec 1913-10 juil 1917, 18 oct, 22 nov 1922 [wkly] – 1 – (quotidien republicain du soir) – fr ACRPP [073]
Bonnet y Reveron, Buenoventura *see* Las canarias y la conquista franco-normanda
Bonnetain, Paul (Madame) *see*
– Dans le brousse
– Une francaise au soudan sur la route de tombouctou, du senegal au niger
Bonnet-t-e's and kin – v6-1913 [1978 mar/jun-1985] – 1r – 1 – mf#627920 – us WHS [071]
Bonneville dam chronicle – Bonneville OR: Dam Pub Co, 1934-39 [wkly] – 1 – (cont: dam chronicle (1934); cont by: cascade locks chronicle and bonneville dam chronicle (1939)) – us Oregon Lib [071]
Bonney, Charles Carroll *see* The present conflict of labor and capital
Bonney, Edwin *see* Life and letters of john lingard, 1771-1851
Bonney, Thomas George *see*
– Christian doctrines and modern thought
– Influence of science on theology
– Old truths in modern lights
– The present relations of science and religion
Bonnichon, Andre *see* Psychologie de l'art dramatique
Bonnier, Gaetan *see* L'occupation de tombouctou
Bonniot, Joseph de *see* Le miracle et ses contrefacons
Bonnycastle, Richard *see* Spanish america
Bonomi, G F *see* Io. francisci bonomij bononiensis chiron achillis
Bonomi, Joseph *see* Nineveh and its palaces
Bonomo, Joe *see* Bossa nova
Bononcini, Giovanni Maria *see* Mvsico prattico, che breuemente dimostra il modo di giungere alla perfetta cognizione di tutte quelle cose... opera ottava
Bonorden, Hermann Friedrich *see* Die erkenntniss des christenthumes vom naturwissenschaftlichen standpuncte
Bonpland, A J A *see* Voyage aux regions equinoxiales du nouveau continent
Os bons exemplos : jornal da congregacao das filhas de maria e das familias catholicas – Rio de Janeiro, RJ: Typ Americana, nov 1870 – (= ser Ps 19) – mf#P17,01,112 – bl Biblioteca [241]
Les bons romans illustres – Montreal: J H A Lamarre, [1887-18– ou 19–] – 9 – mf#P04065 – cn Canadiana [700]
Bonsal, Stephen *see* Fight for santiago
Bonsels, Waldemar *see*
– Brasilianische tage und nachte
– Eros und die evangelien
– Der hueter der schwelle
– Mario und gisela
Bonsels, Waldemar et al *see* Die erde
Bonsignore, S *see* Nelle solenni esequie per la sacra cesarea reale apostolica maest...di leopoldo secondo...orazione funebre...xxi marzo 1792
Bonsmann, Th *see* Gregor 1. der grosse
Bonstetten, Karl V von *see*
– Briefe von karl viktor von bonstetten an friederike brun
– La scandinavie et les alpes
Bonsu Kyeretwie, K *see* Ashanti heroes
La bonte : journal philosophique – Paris: Impr Ed Proux etc [aug 24-sep 10 1849] – (= ser Periodicals relating to the French Revolution of 1848) – 1r – 1 – us CRL [100]
Bontier, Pierre *see* Canarian [the]
Bonus, Albert *see*
– Collatio codicis lewisiani rescripti evangeliorum sacrorum syriacorum
– Collatio codicis lewisiani rescripti evangeliorum sacrorum syriacorum cum codice curetoniano (mus. brit. add. 14,451)
Bonus, Arthur et al *see* Der moderne mensch und das christentum
Bonus Expeditionary Force *see* Heffernan b e f news
Bonus, John *see* Shadows of the rood
Bonvilston, glamorgan, parish church of st mary : baptisms 1696-1900, burials 1696-1901, marriages 1696-1837 – 1mf – 9 – £1.25 – uk Glamorgan FHS [929]
Bonvilston, st mary; carmel united reformed; zoar presbyterian, monumental inscriptions – 1mf – 9 – £1.25 – uk Glamorgan FHS [929]
Bonwell, James *see* Perishing in the gainsaying of core
Bonwetsch, G N *see*
– Die buecher der geheimnisse henochs
– Drei georgisch erhaltene schriften von hippolytus
– Die geschichte des montanismus
– Hippolyts kommentar zum hohenlied
– Methodius von olympus, vol 1
– Studien zu den kommentaren hippolyts zum buche daniel und hohen liede
– Die unter hippolyts namen ueberlieferte schrift ueber den glauben
Bonwetsch, Gottlieb Nathanael *see*
– Cyrill und methodius
– Die dogmengeschichte der alten kirche
– Exegetische und homiletische schriften
– Die geschichte des montanismus
– Jesus christus im bewusstsein und in der froemmigkeit der kirche
– Das religioese erlebnis fuehrender persoenlichkeiten in der erweckungszeit des 19. jahrhunderts
– Die schriften tertullians nach der zeit ihrer abfassung
– Studien zu den kommentaren hippolyts zum buche daniel und hohen liede
– Texte zur geschichte des montanismus
– Theologische studien
– Die unter hippolyts namen ueberlieferte schrift ueber den glauben
Bonwetsch, N *see* Methodius (gcsej12)
Bonwick, James *see*
– Australia's first preacher
– The british colonies and their resources
– Egyptian belief and modern thought
– First twenty years of australia
– Irish druids and old irish religions
– The mormons and the silver mines

Bonzen und rebellen : geschichte eines unbekannten freiwilligen der nation / Weller, Anton Friedrich Tuedel – Muenchen: Zentralverlag der NSDAP: F Eher Nachf 1939 [mf ed 1992] – 1r – 1 – (filmed with: diederich von dem werder / georg witkowski) – mf#7936 – us Wisconsin U Libr [943]

Boochs, Wolfgang *see* Steuervorteile durch vereinbarungen zwischen familienangehoerigen

Boodin, John Elof *see*
- Realistic universe
- Truth and reality

Boodt, A de *see*
- Symbola varia diversorum principum, sacrosanc ecclesiae et sacri imperij romani

The book : its history and development / Davenport, Cyril – New York: D Van Nostrand, 1914 – (= ser Westminster Series (D. Van Nostrand)) – 1mf – 9 – 0-524-03893-7 – (incl bibl ref) – mf#1990-1152 – us ATLA [000]

The book : or, when and by whom the bible was written / Blake, Silas Leroy – Boston: Congregational Sunday-School and Pub Society, c1886 – 1mf – 9 – 0-524-04088-5 – (incl bibl ref) – mf#1992-0046 – us ATLA [220]

A book about fans : the history of fans and fan-painting / Fletcher, Banister Flight – New York 1895 – (= ser 19th c art & architecture) – 2mf – 9 – mf#4.2.1669 – uk Chadwyck [740]

A book about lawyers / Jeaffreson, John C – New York: C W Carleton & Co. 2v in 1. 1867 – 5mf – 9 – $7.50 – mf#LLMC 91-065 – us LLMC [340]

The book and its story : a narrative for the young – New York: Robert Carter, 1861 – 2mf – 9 – 0-524-05743-5 – mf#1992-0586 – us ATLA [220]

Book and magazine production – Northbrook IL 1981-82 – 1,5,9 – (cont: book production industry and magazine production) – ISSN: 0273-8724 – mf#1702.02 – us UMI ProQuest [680]

The book and the message / Alleman, Herbert Christian – Philadelphia: Lutheran Publ Soc, c1909 – (= ser Lutheran Teacher-Training Series for the Sunday School) – 1mf – 9 – 0-524-05592-0 – mf#1992-0447 – us ATLA [220]

Book arts : architecture, applied arts, studio arts – (= ser Art exhibition catalogues on microfiche) – 67 catalogues on 81mf – 9 – £595.00 – (individual titles not listed separately) – uk Chadwyck [740]

The book buyer – New york. v. 9-10; n.s. 1-2; 3rd series 37-39. nov 1875-dec 1877; sep 1878-sep 1880; 1912-1915 – 1 – us NY Public [410]

The book buyer : a review and record of current literature – New York: [Charles Scribner's Sons, 1867-1903]. v7-10 n3 oct 1873-77 – 1 – us CRL [410]

Book by book : popular studies on the canon of scripture / Carpenter, William Boyd – London: Isbister, 1892 – 2mf – 9 – 0-8370-9451-8 – mf#1986-3451 – us ATLA [220]

Book catalogs of american law libraries – 1805-1903 – 9 – $690.00 – mf#0111 – us Brook [340]

Book for all nations, and all times / Miller, J C – Birmingham, England. 1851 – 1r – us UF Libraries [240]

Book for beginners / Patrick, S – London, England. 1841 – 1r – us UF Libraries [240]

Book hoa matthew / Matthaeus Apostolus – London: Tilling & Hughes 1816 – (= ser Whsb) – 3mf – 9 – €40.00 – mf#Hu 357 – gw Fischer [225]

Book indexes to boston passenger lists, 1899-1940 – (= ser Records of the immigration and naturalization service, 1891-1957) – 107r – 1 – (no book indexes for 1901) – mf#T790 – us Nat Archives [975]

Book indexes to new york passenger lists, 1906-1942 – (= ser Records of the immigration and naturalization service, 1891-1957) – 807r – 1 – mf#T612 – us Nat Archives [975]

Book indexes to philadelphia passenger lists, 1906-1926 – (= ser Records of the immigration and naturalization service, 1891-1957) – 23r – 1 – mf#T791 – us Nat Archives [975]

Book indexes to portland, maine, passenger lists, 1907-1930 – (= ser Records of the immigration and naturalization service, 1891-1957) – 12r – 1 – mf#T793 – us Nat Archives [975]

Book indexes to providence passenger lists, 1911-1934 – (= ser Records of the immigration and naturalization service, 1891-1957) – 15r – 1 – mf#T792 – us Nat Archives [975]

Book links – Chicago IL 1996+ – 1,5,9 – ISSN: 1055-4742 – mf#21854 – us UMI ProQuest [070]

Book lore : a magazine devoted to old time literature – Killen TX 1884-87 – 1 – mf#2860 – us UMI ProQuest [420]

Book news monthly – Killen TX 1882-1918 – 1 – mf#2862 – us UMI ProQuest [070]

Book notes : newsletter of black classic press – 1993 winter, 1994 spring, 1996 winter – 1r – 1 – mf#3006691 – us WHS [070]

The book of adam and eve : also called the conflict of adam and eve with satan – London: Williams & Norgate, 1882 [mf ed 1989] – 1mf – 9 – 0-7905-2172-5 – (incl ind) – mf#1987-2172 – us ATLA [221]

A book of american explorers / Higginson, Thomas Wentworth – Boston: Lee and Shepard; New York: C T Dillingham, c1887 – (= ser Young folks' series) – 5mf – 9 – mf#01327 – cn Canadiana [917]

Book of american explorers / Higginson, Thomas Wentworth – Boston: Lee & Shepard, c1877 (mf ed 19–) – (alt title: young folks' book of american explorers) – mf#ZH-544 – us NY Public [917]

The book of amos / ed by Cooke, George Albert – London: Methuen, 1914 [mf ed 1989] – (= ser Westminster commentaries) – 1mf – 9 – 0-7905-1939-9 – (notes by ernest arthur edghill. incl ind) – mf#1987-1939 – us ATLA [221]

The book of amos / Schmoller, Otto – New York: Charles Scribner, c1874 [mf ed 1986] – (= ser A commentary on the holy scriptures. old testament 14/4) – 1mf – 9 – 0-8370-6236-5 – (trans and enl by talbot w chambers) – mf#1986-0236 – us ATLA [221]

[Book of anglican chant...in manuscript] – [c1842?] [mf ed 19–] – 3mf – 9 – mf#fiche 30 – us Sibley [780]

The book of art : cartoons, frescoes, sculpture, and decorative art / Hunt, Frederick Knight – London 1846 – (= ser 19th c art & architecture) – 3mf – 9 – mf#4.1.14 – uk Chadwyck [700]

The book of art : cartoons, frescoes, sculpture, and decorative art, as applied to the new houses of parliament and to buildings in general / ed by Hunt, Frederick Knight – London: Jeremiah How, 1846 – (= ser 19th c art & architecture) – 2mf – 9 – mf#4.1.14 – uk Chadwyck [700]

Book of benjamin – London, England. 1879? – 1r – us UF Libraries [240]

The book of books : a brief introduction to the bible for christian teachers and readers = Kurze bibelkunde / Schaller, John – St Louis, MO: Concordia Pub House, 1918 – 1mf – 9 – 0-524-05937-3 – (in english) – mf#1992-0694 – us ATLA [220]

The book of books : a study of the bible / Ragg, Lonsdale – London: Edward Arnold, 1910 – 1mf – 9 – 0-7905-0195-3 – (incl ind) – mf#1987-0195 – us ATLA [220]

The book of books : what it is, how to study it / Evans, William – Chicago: Moody Press, c1902 – 1mf – 9 – 0-7905-1382-X – mf#1987-1382 – us ATLA [220]

Book of brome, the... : from the ipswich & east suffolk record office – 15th century – 1r – 1 – mf#65866 – uk Microform Academic [941]

The book of chinese poetry : being the collection of ballads, sagas, hymns, and other pieces known as the shih ching, or, classic of poetry – London: K Paul, Trench, Truebner, 1891 – 2mf – 9 – 0-524-07796-7 – mf#1991-0173 – us ATLA [480]

The book of common order of the church of scotland : commonly known as john knox's liturgy = Book of common order – Edinburgh: W Blackwood, 1901 – 1mf – 9 – 0-524-06367-2 – mf#1990-5237 – us ATLA [242]

The book of common order of the church of scotland – London, 1962 – 4mf – 8 – €11.00 – ne Slangenburg [242]

Book of common prayer : a national bond of peace – London, England. 1855 – 1r – us UF Libraries [240]

The book of common prayer : according to the use of the church of ireland: its history and sanction / Reeves, William, Bishop of Down, Connor, and Dromore – Dublin, 1871 – (= ser 19th c ireland) – 1mf – 9 – mf#1.1.2617 – uk Chadwyck [241]

The book of common prayer / Hart, Samuel – 2nd rev ed. Sewanee, TN: University Press at the University of the South, c1913 – (= ser Sewanee Theological Library) – 1mf – 9 – 0-524-07290-6 – mf#1990-5377 – us ATLA [240]

The book of common prayer : its origin and growth / Benton, Josiah Henry – Boston: Priv print, 1910 – 1mf – 9 – 0-524-03130-4 – mf#1990-4579 – us ATLA [240]

The book of common prayer among the nations of the world : a history of translations of the prayer book of the church of england and of the protestant episcopal church of america / Muss-Arnolt, William – London: SPCK; New York: E S Gorham, 1914 [mf ed 1990] – 2mf – 9 – 0-7905-8043-8 – (incl bibl ref) – mf#1988-6024 – us ATLA [242]

Book of common prayer examined in the light of the present age / Jevons, William – Ramsgate, England. v1. 1872 – 1r – us UF Libraries [240]

The book of common prayer of the reformed episcopal church : adopted and set forth for use by the second general council of the said church held in the city of new york in the month of may, 1874 – Philadelphia: James A Moore, 1874 – 2mf – 9 – 0-524-07587-5 – mf#1991-3207 – us ATLA [240]

Book of common worship *see* P'u t'ien ch'ung pai (ccm120)

The book of common worship / Presbyterian Church in the USA. General Assembly – Philadelphia: Presbyterian Board of Publ and Sabbath-School Work, 1906 [mf ed 1992] – (= ser Presbyterian coll) – 1mf – 9 – 0-524-04883-5 – mf#1990-5081 – us ATLA [242]

The book of concord, or, the symbolical books of the evangelical lutheran church : with historical introduction, notes, appendices and indices / Jacobs, Henry Eyster – Philadelphia: GW Frederick, 1893, c1882-c1883 – 3mf – 9 – 0-8370-8682-5 – (incl bibl ref and ind) – mf#1986-2682 – us ATLA [242]

The book of constitution of the grand lodge of ancient and free and accepted masons of canada / Freemasons. Grand Lodge (Canada) – [Hamilton, Ont?]: s.n, 1887 [mf ed 1983] – 2mf – 9 – 0-665-38146-8 – (with ind) – mf#38146 – cn Canadiana [366]

The book of constitution of the grand lodge of ancient, free and accepted masons of canada : in the province of ontario / Freemasons. Grand Lodge of Ontario – Toronto?: Hunter, Rose, 1891 – 2mf – 9 – (incl ind) – mf#26548 – cn Canadiana [366]

The book of constitution of the grand lodge of quebec, ancient free and accepted masons : as revised, amended and adopted by grand lodge...january 29th 1896, with all amendments to february, 1922 / Francs-macons. Grande loge de Quebec – Montreal: C R Corneil Ltd, 1922 [mf ed 1922] – 2mf – 9 – mf#SEM105P1637 – cn Bibl Nat [366]

The book of constitution of the grand lodge of quebec, ancient, free and accepted masons / Francs-macons. Grande loge de Quebec – Montreal: printed by John Wilson, 1882 [mf ed 2000] – 9 – (with ind) – cn Bibl Nat [366]

Book of crests / Fairbairn, J – Edinburgh. 2v. 1892 – 1r – 1 – mf#624 – uk Microform Academic [920]

The book of daniel : introduction, revised version with notes, index and map / ed by Charles, Robert Henry – New York: H Frowde; Edinburgh: T C & E C Jack, [1913?] – (= ser The New-Century Bible) – 1mf – 9 – 0-7905-3365-0 – (incl bibl ref) – mf#1987-3365 – us ATLA [221]

The book of daniel : or, the second volume of prophecy... / Murphy, James Gracey – Andover: Warren F Draper, 1885 – 1mf – 9 – 0-8370-4542-8 – mf#1985-2542 – us ATLA [221]

The book of daniel : with introduction and notes / Driver, Samuel Rolles – Cambridge: University Press, 1900 [mf ed 1985] – 1mf – 9 – 0-8370-2964-3 – (incl app and ind) – mf#1985-0964 – us ATLA [221]

The book of daniel : with notes and introduction / Wordsworth, Christopher – 2nd ed. London: Rivingtons, 1871 – 1mf – 9 – 0-8370-5744-2 – (incl bibl ref) – mf#1985-3744 – us ATLA [221]

The book of daniel and the minor prophets – London: J M Dent; Philadelphia: J B Lippincott 1902 [mf ed 1989] – (= ser The temple bible) – 1mf – 9 – 0-7905-1850-3 – mf#1987-1850 – us ATLA [221]

The book of daniel unlocked / Auchincloss, William Stuart – NY: D van Nostrand, 1905 – 1mf – 9 – 0-8370-2127-8 – (includes explanatory notes and chronological tables) – mf#1985-0127 – us ATLA [221]

Book of deuteronomy / Betteridge, Walter Robert – Philadelphia: American Baptist Publ Soc 1915 [mf ed 1990] – (= ser American commentary on the old testament) – 2mf – 9 – 0-7905-3363-4 – mf#1987-3363 – us ATLA [221]

The book of deuteronomy / Harper, Andrew – New York: A C Armstrong, 1895 – 2mf – 9 – 0-8370-2291-6 – (incl bibl ref) – mf#1985-0291 – us ATLA [221]

The book of deuteronomy : in the revised version / Smith, George Adam – Cambridge: University Press, 1918 [mf ed 1985] – (= ser The cambridge bible for schools and colleges) – 2mf – 9 – 0-8370-2887-6 – (incl bibl ref & ind) – mf#1985-0887 – us ATLA [221]

Book of discipline / Associate Presbyterian Church of North America – 1840 – 1 – $50.00 – us Presbyterian [241]

The book of discipline, in a revised form / Presbyterian Church in the USA. Committee on the Revision of the Book of Discipline – New York: SW Green, 1880 [mf ed 1992] – (= ser Presbyterian coll) – 1mf – 9 – 0-524-05550-5 – mf#1990-5154 – us ATLA [242]

Book of divine consolation of the blessed angela of foligno = Liber de vera fidelium experientia / Angela of Foligno – London: Chatto and Windus, 1909 – (= ser New medieval library) – 1mf – 9 – 0-524-05128-3 – mf#1990-1384 – us ATLA [240]

Book of ecclesiastes / Marshall, John Turner – Philadelphia: American Baptist Publ Soc 1904 [mf ed 1989] – (= ser American commentary on the old testament) – 1mf – 9 – 0-7905-1292-0 – mf#1987-1292 – us ATLA [221]

The book of ecclesiastes : its meaning and its lessons / Buchanan, Robert – London: Blackie, 1859 – 1mf – 9 – 0-7905-3366-9 – mf#1987-3366 – us ATLA [221]

The book of ecclesiastes : with a new translation / Cox, Samuel – New York: A C Armstrong, [1890] – 1mf – 9 – 0-8370-2760-8 – mf#1985-0760 – us ATLA [221]

The book of enlightenment for the instruction of the inquirer / Jacob, Son of Aaron; ed by Barton, William E – Sublette, IL: Puritan Press, 1913 – 1mf – 9 – 0-8370-9795-9 – mf#1986-3795 – us ATLA [240]

The book of enoch / Schodde, George Henry – Andover: W F Draper, 1882 – 1r – 1 – 0-8370-1531-6 – mf#1984-B394 – us ATLA [221]

The book of enoch : translated from professor dillmann's ethiopic text / ed by Charles, Robert Henry – Oxford: Clarendon Press, 1893 – 1mf – 9 – 0-8370-9535-2 – (incl ind) – mf#1986-3535 – us ATLA [221]

The book of enoch : with introduction, notes, appendices, and indices – Oxford: Clarendon Press 1893 – (translated from professor dillmann's ethiopic text, amended and revised in accordance with hitherto uncollated ethiopic mss. and with the gizeh and other greek and latin fragments which are here publ in full by r h charles) – us CRL [470]

A book of essays / Hirsch, Samuel Abraham – London: publ for the Jewish Historical Society of England by Macmillan, 1905 [mf ed 1990] – 1mf – 9 – 0-7905-5766-5 – mf#1988-1766 – us ATLA [470]

The book of esther : theologically and homiletically expounded = Das buch esther / Schultz, Friedrich Wilhelm; ed by Strong, James – New York: Scribner, c1877 [mf ed 1985] – (= ser A commentary on the holy scriptures. old testament 7[5/4]) – 1mf – 9 – 0-8370-4204-6 – (english trans by ed) – mf#1985-2204 – us ATLA [221]

The book of esther : illustrative of character and providence / McEwan, Thomas – Edinburgh:Andrew Elliot, 1877 – 1mf – 9 – 0-8370-4349-2 – mf#1985-2349 – us ATLA [221]

The book of esther : its practical lessons and dramatic scenes / Raleigh, Alexander – Edinburgh: Adam and Charles Black, 1880 – 1mf – 9 – 0-8370-4832-X – mf#1985-2832 – us ATLA [221]

The book of esther : with introduction and notes / Streane, A W – Cambridge: University Press, 1907 – (= ser The Cambridge Bible For Schools And Colleges) – 1mf – 9 – 0-8370-6841-X – (incl bibl ref and index) – mf#1986-0841 – us ATLA [221]

The book of exodus / Driver, Samuel Rolles – 1911 – 9 – $18.00 – us IRC [221]

The book of exodus : in the revised version: with introduction and notes / Driver, Samuel Rolles – Cambridge: University Press, 1911 – (= ser The Cambridge Bible For Schools And Colleges) – 2mf – 9 – 0-8370-6812-6 – (incl bibl ref and index) – mf#1986-0812 – us ATLA [221]

The book of exodus : with introduction and notes / McNeile, Alan Hugh – London: Methuen, 1908 – 1mf – 9 – 0-7905-1013-8 – (incl indes) – mf#1987-1013 – us ATLA [221]

The book of ezra : theologically and homiletically expounded = Das buch ezra / Schultz, Friedrich Wilhelm; ed by Briggs, Charles Augustus – New York: Scribner, Armstrong, c1877 [mf ed 1985] – (= ser A commentary on the holy scriptures. old testament 7[5/2]) – 1mf – 9 – 0-8370-4201-1 – (trans by ed) – mf#1985-2201 – us ATLA [221]

A book of facsimiles of monumental brasses on the continent of europe / Creeny, William Frederick – [London] 1884 – (= ser 19th c art & architecture) – 10mf – 9 – mf#4.2.1699 – uk Chadwyck [730]

A book of family worship – Philadelphia: Presbyterian Board of Publ and Sabbath-School Work, 1916 [mf ed 1992] – 1mf – 9 – 0-524-05364-2 – mf#1990-5115 – us ATLA [242]

A book of fifty drawings by aubrey beardsley / Beardsley, Aubrey Vincent – London 1897 – (= ser 19th c art & architecture) – 3mf – 9 – mf#4.2.1720 – uk Chadwyck [740]

The book of filial duty / Hsiao ching – London: John Murray 1908 [mf ed 1993] – (= ser Wisdom of the east series (london, england)) – 1mf – 9 – 0-524-07998-6 – mf#1991-0220 – us ATLA [180]

Book of forms, adapted to the code of procedure / New York. (State). Commissioners of the Code – Albany: Weed, Parsons, 1861. 273p. LL-1697 – 1 – us L of C Photodup [347]

The book of fortune : two hundred unpublished drawings = [liber fortunae] / Cousin, Jean – Paris: J Rouam, London: Remington and Co, 1883 – 7mf – 9 – mf#0-213 – ne IDC [090]

The book of gallant vagabonds / Beston, Henry – New York: G H Doran, [1925] – 1 – us CRL [960]

Book of genesis / Goodspeed, Calvin – Philadelphia: American Baptist Publ Soc 1909, c1908 [mf ed 1989] – (= ser American commentary on the old testament) – 1mf – 9 – 0-7905-4116-5 – (incl bibl ref) – mf#1988-0116 – us ATLA [221]

The book of genesis / Driver, Samuel Rolles – 1904 – 9 – $18.00 – us IRC [221]

The book of genesis – London: Methuen, 1907 – (= ser Westminster Commentaries) – 2mf – 9 – 0-524-06511-X – mf#1992-0895 – us ATLA [221]

The book of genesis / Wade, George Woosung – London, New York: Longmans, Green, 1896 – 1mf – 9 – 0-7905-0519-3 – (incl bibl ref and index) – mf#1987-0519 – us ATLA [221]

The book of genesis, and part of the book of exodus : a revised version, with marginal references, and an explanatory commentary / Alford, Henry – London: Daldy, Isbister, 1877 – 1mf – 9 – 0-8370-2071-9 – mf#1985-0071 – us ATLA [221]

The book of genesis in hebrew : with a critically revised text, various readings, and grammatical and critical notes / Wright, Charles Henry Hamilton – London: Williams and Norgate, 1859 – 3mf – 9 – 0-7905-2583-6 – mf#1987-2583 – us ATLA [221]

The book of genesis in the light of modern knowledge / Worcester, Elwood – New York: McClure, Phillips, 1901 – 2mf – 9 – 0-7905-0468-5 – (incl bibl ref and indexes) – mf#1987-0468 – us ATLA [221]

The book of habakkuk : introduction, translation, and notes on the hebrew text / Stonehouse, George Gordon Vigor – London: Rivingtons, 1911 – 1mf – 9 – 0-8370-9747-9 – (incl ind) – mf#1986-3747 – us ATLA [221]

The book of habakkuk : theologically and homiletically expounded / Kleinert, Paul – New York: Charles Scribner, c1874 [mf ed 1986] – (= ser A commentary on the holy scriptures. old testament 14/9) – 1mf – 9 – 0-8370-6194-6 – (english trans by charles elliott) – mf#1986-0194 – us ATLA [221]

The book of haggai / McCurdy, James Frederick – New York: Charles Scribner, c1874 [mf ed 1986] – (= ser A commentary on the holy scriptures. old testament 14/11) – 1mf – 9 – 0-8370-6075-3 – mf#1986-0075 – us ATLA [221]

The book of hosea / Schmoller, Otto – New York: Charles Scribner, c1874 [mf ed 1986] – (= ser A commentary on the holy scriptures. old testament 14/2) – 1mf – 9 – 0-8370-6237-3 – (english trans fr german with additions by james frederick mccurdy) – mf#1986-0237 – us ATLA [221]

The book of ighan = Kitab-i iqan / Bahar Allah – 2nd ed. Chicago, IL, USA: Bahai Pub Society, 1907 – 1mf – 9 – 0-524-02290-9 – (in english) – mf#1990-2913 – us ATLA [290]

Book of isaiah / Bannister, Henry – New York: Phillips & Hunt, Cincinnati: Cranston & Stowe 1886 [mf ed 1989] – (= ser Commentary on the old testament 7) – 2mf [ill] – 9 – 0-7905-2823-1 – (filmed with: books of jeremiah and of the lamentations by francis dana hemenway) – mf#1987-2823 – us ATLA [221]

The book of isaiah : and other historical studies / Wright, Charles Henry Hamilton – London: Francis Griffiths, 1906 – 2mf – 9 – 0-8370-9998-6 – mf#1986-3998 – us ATLA [240]

The book of isaiah / Smith, George Adam – New York: A C Armstrong, [1889]-98 [mf ed 1985] – (= ser The expositor's bible) – 4mf – 9 – 0-8370-2370-X – (incl bibl ref & ind) – mf#1985-0370 – us ATLA [221]

The book of jeremiah : chapters 21-52 / Bennett, William Henry – NY: A C Armstrong, 1895 [mf ed 1985] – (= ser The expositor's bible) – 1mf – 9 – 0-8370-2267-3 – (suppl: the prophecies of jeremiah: with a sketch of his life and times by c j ball. incl ind) – mf#1985-0267 – us ATLA [221]

The book of jeremiah : with introduction and notes / Douglas, George Cunningham Monteath – London: Hodder and Stoughton, 1903 – 1mf – 9 – 0-7905-1041-3 – mf#1987-1041 – us ATLA [221]

The book of jeremiah and lamentations / ed by Green, Edmund Tyrrell – London: J M Dent; Philadelphia: J B Lippincott 1902 [mf ed 1989] – (= ser The temple bible) – 1mf – 9 – 0-7905-1821-X – mf#1987-1821 – us ATLA [221]

Book of job / Froude, James Anthony – London, England. 1854 – 1r – us UF Libraries [221]

Book of job / Marshall, John Turner – Philadelphia: American Baptist Publ Soc 1904 [mf ed 1989] – (= ser American commentary on the old testament) – 1mf – 9 – 0-7905-1434-6 – mf#1987-1434 – us ATLA [221]

The book of job / Aitken, James – Edinburgh: T & T Clark, [1905?] – (= ser Handbooks for bible classes and private students) – 1mf – 9 – 0-524-04894-0 – mf#1992-0237 – us ATLA [221]

The book of job : a rhythmical version with introduction and annotations = Das buch job / Lewis, Tayler – New York: Charles Scribner, c1902 [mf ed – (= ser A commentary on the holy scriptures. old testament 8) – 2mf – 9 – 0-8370-3362-4 – (english trans fr german with additions by llewelyn joan evans. incl bibl) – mf#1985-1362 – us ATLA [221]

The book of job / Gibson, Edgar C S – London: Methuen, 1899 – (= ser Oxford Commentaries) – 1mf – 9 – 0-8370-3262-8 – mf#1985-1262 – us ATLA [221]

The book of job = Livre de job / Renan, Ernest – London: W M Thomson, [1889?] – 1mf – 9 – 0-8370-9413-5 – (in english. incl bibl ref) – mf#1986-3413 – us ATLA [221]

The book of job : a new critically revised translation / Wright, George Henry Bateson – London: Williams and Norgate, 1883 – 1mf – 9 – 0-7905-3426-6 – mf#1987-3426 – us ATLA [221]

The book of job / Strahan, James – Edinburgh: T & T Clark, 1913 – 1mf – 9 – 0-7905-0393-X – (incl bibl ref and index) – mf#1987-0393 – us ATLA [221]

The book of job : the text of the revised version adapted to modern printing, prepared in connection with the lectures of the people's institute, chicago – Chicago: Fleming H Revell, 1892 – 1mf – 9 – 0-8370-9806-8 – mf#1986-3806 – us ATLA [221]

The book of job and the book of ruth / ed by Addis, William Edward – London: J M Dent; Philadelphia: J B Lippincott 1902 [mf ed 1989] – (= ser The temple bible) – 1mf – 9 – 0-7905-1801-5 – mf#1987-1801 – us ATLA [221]

The book of job, and the prophets : translated from the vulgate, and diligently compared with the original text, being a revised edition of the douay version... – Baltimore: Kelly, Hedian & Piet, 1859 – 8mf – 9 – 0-8370-1916-8 – mf#1987-6303 – us ATLA [221]

The book of job in the revised version : edited with introductions and brief annotations / ed by Driver, Samuel Rolles – Oxford: Clarendon Press, 1906 – 1mf – 9 – 0-8370-2965-1 – (incl ind) – mf#1985-0965 – us ATLA [221]

The book of joel / Schmoller, Otto – New York: Charles Scribner, c1874 [mf ed 1986] – (= ser A commentary on the holy scriptures. old testament 14/3) – 1mf – 9 – 0-8370-6238-1 – (english trans fr german, with add notes and new version of hebrew text by john forsyth) – mf#1986-0238 – us ATLA [221]

The book of jonah / Kleinert, Paul – New York: Charles Scribner, c1874 [mf ed 1986] – (= ser A commentary on the holy scriptures. old testament 14/6) – 1mf – 9 – 0-8370-6195-4 – (trans and enl by charles elliott) – mf#1986-0195 – us ATLA [221]

The book of jonah : preceded by a treatise on the hebrew and the stranger / Kalisch, Marcus Moritz – London: Longmans, Green, 1878 – 1mf – 9 – 0-8370-3837-5 – (includes bibliography and index of authors cited) – mf#1985-1837 – us ATLA [221]

Book of joshua / Steele, Daniel – New York: Eaton & Mains, Cincinnati: Jennings & Graham c1873 [mf ed 1990] – (= ser Commentary on the old testament 3) – 2mf – 9 – 0-8370-1603-7 – (incl bibl ref; filmed with: books of judges to 2. samuel by milton spenser terry) – mf#1987-6082 – us ATLA [221]

The book of joshua / Blaikie, William Garden – New York: A C Armstrong, 1893 – (= ser The expositor's bible) – 1mf – 9 – 0-8370-2360-2 – mf#1985-0360 – us ATLA [221]

The book of joshua = Das buch josua / Fay, F R – New York: Scribner, Armstrong, 1876 [mf ed 1985] – (= ser A commentary on the holy scriptures. old testament 4) – 1mf – 9 – 0-8370-3105-2 – (trans fr german by george r bliss) – mf#1985-1105 – us ATLA [221]

The book of joshua : a critical and expository commentary of the hebrew text / Lloyd, John – London: Hodder and Stoughton, 1886 – 1mf – 9 – 0-7905-1223-8 – (incl bibl ref and index) – mf#1987-1223 – us ATLA [221]

The book of joshua / Douglas, George Cunningham Monteath – Edinburgh: T&T Clark, 1882 – 1mf – 9 – 0-8370-2954-6 – mf#1985-0954 – us ATLA [221]

The book of joshua : in the revised version: with introduction and notes / Cooke, George Albert – Cambridge: University Press; New York: G P Putnam [distributor], 1918 – 1mf – 9 – 0-8370-6729-4 – (incl bibl ref and index) – mf#1986-0729 – us ATLA [221]

The book of joshua : with map, introduction, and notes / Black, John Sutherland – Cambridge: University Press, 1891 – (= ser The Smaller Cambridge Bible for Schools) – 1mf – 9 – 0-8370-2355-6 – (incl ind) – mf#1985-0355 – us ATLA [221]

The book of joshua and the book of judges / ed by Kennedy, Archibald Robert Stirling – London: J M Dent; Philadelphia: J B Lippincott 1902 [mf ed 1989] – (= ser The temple bible) – 1mf – 9 – 0-7905-1828-7 – mf#1987-1828 – us ATLA [221]

The book of jubilees : or, the little genesis / ed by Charles, Robert Henry – London: Adam and Charles Black, 1902 – 1mf – 9 – 0-7905-0920-2 – (incl bibl ref and indexes) – mf#1987-0920 – us ATLA [240]

The book of jubilees or the little genesis / Charles, Robert Henry – London, 1902 – 7mf – 8 – €15.00 – (trans fr ed's ethiopic text) – ne Slangenburg [221]

The book of judges = Das buch der richter / Cassel, Paulus – New York: Scribner, Armstrong, 1876 [mf ed 1985] – (= ser A commentary on the holy scriptures. old testament 4/1) – 1mf – 9 – 0-8370-5987-9 – (trans by peter henry steenstra) – mf#1985-3987 – us ATLA [221]

The book of judges / Curtis, Edward Lewis – New York: Macmillan, 1913 – (= ser The bible for home and school) – 1mf – 9 – 0-7905-3325-1 – (incl bibl ref) – mf#1987-3325 – us ATLA [221]

The book of judges / Douglas, George Cunningham Monteath – Edinburgh: T&T Clark, 1881 – 1mf – 9 – 0-8370-2955-4 – mf#1985-0955 – us ATLA [221]

The book of judges : in the revised version: with introduction and notes / Cooke, George Albert – Cambridge: University Press; New York: G P Putnam [distributor], 1913 – 1mf – 9 – 0-8370-6730-8 – (incl bibl ref and index) – mf#1986-0730 – us ATLA [221]

The book of judges : with map, introduction, and notes / Black, John Sutherland – Cambridge: University Press, 1892 – (= ser The Smaller Cambridge Bible for Schools) – 1mf – 9 – 0-8370-2356-4 – (incl ind) – mf#1985-0356 – us ATLA [221]

The book of judges in greek : according to the text of codex alexandrinus / ed by Brooke, Alan England & McLean, Norman – Cambridge: University Press, 1897 – 1mf – 9 – 0-8370-1790-4 – mf#1987-6178 – us ATLA [221]

The book of kells / Westwood, John Obadiah – Dublin 1887 – (= ser 19th c art & architecture) – 1mf – 9 – mf#4.1.302 – uk Chadwyck [740]

The book of koheleth, commonly called ecclesiastes : considered in relation to modern criticism, and to the doctrines of modern pessimism / Wright, Charles Henry Hamilton – London: Hodder and Stoughton, 1883 – 2mf – 9 – 0-7905-0472-3 – (incl indes) – mf#1987-0472 – us ATLA [221]

Book of laws... : together with the proceedings of the annual / Kentucky state federation of labor – 15th [1919], 19th [1924] – 1r – 1 – (cont by: official proceedings...annual convention of the kentucky state federation of labor) – mf#3130085 – us WHS [348]

Book of laws... : together with the proceedings of the annual / Tennessee Federation of Labor – 23rd-24th [1919-20], 28th [1924], 31st [1927], 36th [1932] – 1r – 1 – mf#3130220 – us WHS [348]

Book of leviticus / Genung, George Frederick – Philadelphia: American Baptist Publ Soc 1906, c1905 [mf ed 1990] – (= ser American commentary on the old testament) – 1mf – 9 – 0-7905-3374-X – mf#1987-3374 – us ATLA [221]

The book of leviticus : english translation / Driver, Samuel Rolles – 1898 – 9 – $10.00 – us IRC [221]

The book of leviticus : in the revised version / Chapman, A T – Cambridge: University Press; New York: G.P. Putnam [distributor], 1914 – (= ser The Cambridge Bible For Schools And Colleges) – 1mf – 9 – 0-8370-6726-X – (includes appendixes on literary structure, priestly code, date of h compared with ezekiel, wave-offering, azazel and index) – mf#1986-0726 – us ATLA [221]

The book of leviticus / Kellogg, Samuel Henry – New York: A C Armstrong, 1891 – 2mf – 9 – 0-8370-2306-8 – mf#1985-0306 – us ATLA [221]

The book of malachi / Packard, Joseph – New York: Charles Scribner, c1874 [mf ed 1986] – (= ser A commentary on the holy scriptures. old testament 14/13) – 1mf – 9 – 0-8370-6076-1 – mf#1986-0076 – us ATLA [221]

The book of martyrs / Foxe, John – London: Cassell, Petter & Galpin [1866?] [mf ed 1992] – 2mf – 9 – 0-524-03397-8 – (rev, with notes & app by wiliam bramley-moore. incl bibl ref) – mf#1990-0951 – us ATLA [240]

The book of micah / Kleinert, Paul – New York: Charles Scribner, c1874 [mf ed 1986] – (= ser A commentary on the holy scriptures. old testament 14/7) – 1mf – 9 – 0-8370-6196-2 – (english trans fr german with additions by george ripley bliss) – mf#1986-0196 – us ATLA [221]

A book of modern german lyric verse, 1890-1955 / ed by Rose, William – Oxford: Clarendon Press, 1960 [mf ed 1993] – 284p – 1 – (incl bibl ref and ind) – mf#8352 – us Wisconsin U Libr [810]

The book of monographs : void execution, judicial and probate sales, etc – 2d ed. St. Louis, Central Law Journal, 1877. 144, vii, 42, xvi, 105, 52 p. LL-1621 – 1 – us L of C Photodup [340]

The book of mormon proved to be a fraud : and latter day saints shown to be building upon a false foundation / Cooper, William Henry – [Milverton, Ont?: s.n, 1920?] – 1mf – 9 – 0-665-88066-9 – mf#88066 – cn Canadiana [243]

The book of nahum / Kleinert, Paul – New York: Charles Scribner, c1874 [mf ed 1986] – (= ser A commentary on the holy scriptures. old testament 14/8) – 1mf – 9 – 0-8370-6197-0 – (trans and enl by charles elliott. incl bibl) – mf#1986-0197 – us ATLA [221]

The book of nehemiah : critically and theologically expounded, including the homiletical sections of dr schultz = Das buch nehemiah / Crosby, Howard – New York:Scribner, Armstrong, c1877 [mf ed 1985] – (= ser A commentary on the holy scriptures. old testament 5/3) – 1mf – 9 – 0-8370-3242-3 – (in english) – mf#1985-1242 – us ATLA [221]

A book of new england legends and folk lore : in prose and poetry / Drake, Samuel Adams – Boston: Roberts, 1884 – 6mf – 9 – (ill by f t merrill. incl ind) – mf#06241 – cn Canadiana [390]

Book of numbers / Genung, George Frederick – Philadelphia: American Baptist Publ Soc 1906 [mf ed 1990] – (= ser American commentary on the old testament) – 2mf – 9 – 0-7905-3375-8 – mf#1987-3375 – us ATLA [221]

The book of numbers : in the revised version: with introduction and notes / McNeile, Alan Hugh – Cambridge: University Press; New York: G P Putnam [distributor], 1911 – 1mf – 9 – 0-8370-6759-6 – (incl bibl ref and index) – mf#1986-0759 – us ATLA [221]

The book of obadiah / Kleinert, Paul – New York: Charles Scribner, c1874 [mf ed 1986] – (= ser A commentary on the holy scriptures. old testament 14/5) – 1mf – 9 – 0-8370-6198-9 – (english trans fr german with additions by george ripley bliss) – mf#1986-0198 – us ATLA [221]

A book of offices and prayers for priest and people / Addison, Charles Morris & Suter, John Wallace – 3rd ed. New York: Edwin S Gorham, 1899 [mf ed 1993] – (= ser Anglican/episcopal coll) – 1mf – 9 – 0-524-06691-4 – mf#1990-5262 – us ATLA [242]

The book of opening the mouth / Budge, Ernest Alfred Wallis – London, 1909 v1-2 – (= ser Books on egypt and chaldaea) – 6mf – 9 – (books on egypt and chaldaea. v26, 27) – mf#NE-20012 – ne IDC [956]

The book of opening the mouth : the egyptian texts with english translations / Budge, Ernest Alfred Wallis – London: Kegan Paul, Trench, Truebner 1909 [mf ed 1989] – (= ser Books on egypt and chaldaea 26-27) – 2v on 2mf – 9 – 0-8370-1177-9 – mf#1987-6013 – us ATLA [470]

The book of popery : a manual for protestants descriptive of the origin, progress, doctrines, rites, and ceremonies of the papal church / Cobbin, Ingram – Philadelphia: Presbyterian Board of Publ, [1840?] – 1mf – 9 – 0-524-02061-2 – mf#1990-0558 – us ATLA [242]

A book of prayer / Levy, Joseph Leonard – Pittsburgh: Publicity Press, 1902 [mf ed 1985] – 1mf – 9 – 0-8370-4095-7 – (incl ind) – mf#1985-2095 – us ATLA [270]

Book of prayers for israelitish congregations / Siddur – New York, NY. 1872 – 1r – us UF Libraries [270]

Book of proverbs / Berry, George Ricker – Philadelphia: American Baptist Publ Soc 1904 [mf ed 1989] – (= ser American commentary on the old testament) – 1mf – 9 – 0-7905-1262-9 – mf#1987-1262 – us ATLA [221]

The book of proverbs : critical edition of the hebrew text with notes / Mueller, August & Kautzsch, Emil – Leipzig: J C Hinrichs; Baltimore: Johns Hopkins Press, 1901 – 1mf – 9 – 0-8370-9258-2 – mf#1986-3258 – us ATLA [221]

The book of proverbs / Horton, Robert Forman – New York: A C Armstrong, 1891 – 1mf – 9 – 0-8370-3665-8 – mf#1985-1665 – us ATLA [221]

The book of proverbs : in an amended version / Muenscher, Joseph – Gambier, OH: Western Episcopalian Off, 1866 – 1mf – 9 – 0-8370-4534-7 – (with int & explanatory notes) – mf#1985-2534 – us ATLA [221]

The book of psalms : containing a free metrical rendering, a rhythmical translation, an extended introduction, and a tabular analysis of the entire book... – New York: Eaton & Mains; Cincinnati: Jennings & Graham, c1896 – (= ser The student's commentary) – 1mf – 9 – 0-7905-2432-5 – mf#1987-2432 – us ATLA [221]

The book of psalms : exegetically and practically considered / Thomas, David – (= ser Homilistic library) – 3v on 1r – 1 – 0-524-10527-8 – (incl ind) – mf#b00739 – us ATLA [221]

The book of psalms / Hibbard, Freeborn Garretson – New York: Phillips & Hunt; Cincinnati: Walden & Stowe 1882 [mf ed 1989] – (= ser Commentary on the old testament 5) – 2mf – 9 – 0-7905-2528-3 – mf#1987-2528 – us ATLA [221]

The book of psalms – 1450s-early 1500s – 7mf – 9 – (middle bulgarian version; most likely moldavian by origin) – us UMI ProQuest [090]

The book of psalms / Perowne, J J Stewart – 8th ed. 1892 – 9 – $39.00 – us IRC [220]

The book of psalms / Perowne, John James Stewart – 6th ed. London: George Bell; Cambridge: Deighton, Bell, 1888 – 2mf – 9 – 0-8370-1150-7 – mf#1987-6001 – us ATLA [221]

The book of psalms – late 1400s-early 1500s – 11mf – 9 – (russian version) – us UMI ProQuest [090]

The book of psalms / ed by Streane, Annesley William – London: J M Dent; Philadelphia: J B Lippincott 1902 [mf ed 1989] – (= ser The temple bible) – 1mf – 9 – 0-7905-1852-X – mf#1987-1852 – us ATLA [221]

The book of psalms : translated from a revised text with notes and introduction / Cheyne, Thomas Kelly – London: Kegan Paul, Trench, Truebner, 1904 – 2mf – 9 – 0-8370-6035-4 – (incl bibl ref and ind) – mf#1986-0035 – us ATLA [221]

The book of psalms : with introduction and notes / Kirkpatrick, Alexander Francis – stereotyped ed. Cambridge: University Press, 1891-1895 – (= ser The cambridge bible for schools and colleges) – 3mf – 9 – 0-8370-6132-6 – mf#1986-0132 – us ATLA [221]

The book of psalms [fragment] : bessarabskaia kollektsia [bessarabian collection] – 1350s-90s – 1mf – 9 – (russian version) – us UMI ProQuest [090]

The book of psalms in hebrew and english : arranged in parallelism – Andover: Warren F Draper, 1865, c1861 – 1mf – 9 – 0-8370-1852-8 – mf#1987-6239 – us ATLA [221]

A book of public prayer : compiled from the authorized formularies of worship of the presbyterian church... – New York: Scribner, 1859 [mf ed 1992] – 1mf – 9 – 0-524-03267-X – mf#1990-4670 – us ATLA [242]

Book of reference of the city of quebec and village of saint sauveur : accompanying the cadastral plan / Cousin, Paul – Quebec: P Cousin, 1875 [mf ed 1980] – 2mf – 9 – 0-665-00257-2 – mf#00257 – cn Canadiana [971]

Book of reference of the city of quebec and village of saint sauveur : accompanying the cadastral plan – Quebec: Paul Cousin, 1875 [mf ed 1980] – 2mf – 9 – 0-665-02829-6 – mf#02829 – cn Canadiana [307]

The book of remarkable trials and notorious characters / ed by Benson, L – From "Half-Hanged Smith," 1700, to Oxford Who Shot at the Queen, 1840.... Illus. by Phiz (pseud). London: J.C. Hotten (1871). iv, (9)/545p. With: The Germans by I.A.R. Wylie. 1 reel. 1260 – 1 – us Wisconsin U Libr [360]

The book of revelation / Dean, John Taylor – Edinburgh: T & T Clark, 1915 – (= ser Handbooks for bible classes and private students) – 1mf – 9 – 0-524-05977-2 – mf#1992-0714 – us ATLA [221]

The book of revelation : an exposition / Warren, Israel Perkins – New York: Funk & Wagnalls, 1886 – 1mf – 9 – 0-8370-5715-9 – mf#1985-3715 – us ATLA [221]

The book of revelation – London: Samuel Bagster, 1849 – 1mf – 9 – 0-8370-9751-7 – mf#1986-3751 – us ATLA [221]

The book of revelation / Milligan, William – New York:A C Armstrong, [1889] – 1mf – 9 – 0-8370-4439-1 – mf#1985-2439 – us ATLA [221]

The book of rules of tyconius / ed by Burkitt, Francis Crawford – Cambridge: University Press 1894 [mf ed 1989] – (= ser Texts and studies (cambridge, england) 3/1) – 1mf – 9 – 0-7905-3297-2 – (in latin; int in english & latin) – mf#1987-3297 – us ATLA [220]

The book of ruth = Das buch ruth / Cassel, Paulus – New York: Scribner, Armstrong, 1876 [mf ed 1985] – (= ser A commentary on the holy scriptures. old testament 4/2) – 1mf – 9 – 0-8370-5988-7 – (trans fr german by peter henry steenstra) – mf#1985-3988 – us ATLA [221]

The book of ruth : in the revised version: with introduction and notes / Cooke, George Albert – Cambridge: University Press; New York: G P Putnam [dist], 1913 – 1mf – 9 – 0-8370-6731-6 – (incl bibl ref and index) – mf#1986-0731 – us ATLA [221]

The book of ruth : a literal translation from the hebrew / Steuart, Robert Henry Joseph – London: David Nutt, 1912 – 1mf – 9 – 0-524-06828-3 – mf#1992-0970 – us ATLA [221]

The book of ruth in hebrew / Wright, Charles Henry Hamilton – London: Williams & Norgate; Leipzig: Rudolph Hartmann, 1864 – 1mf – 9 – 0-8370-6549-6 – (includes aramaic glossary) – mf#1986-0549 – us ATLA [221]

The book of saint basil the great, bishop of caesarea in cappadocia, on the holy spirit : written to amphilochius, bishop of iconium, against the pneumatomachi = on the holy spirit / Basil, Saint, Bishop of Caesarea – Oxford: Clarendon, 1892 – 1mf – 9 – 0-7905-3756-7 – mf#1989-0249 – us ATLA [240]

A book of south india / Molony, John Chartres – London: Methuen & Co, 1926 – (= ser Samp: indian books) – us CRL [915]

The book of the bee / ed by Budge, Ernest A Wallis – Oxford. v1-pt2. 1886 – (= ser Anecdota oxoniensia. semitic series) – 12mf – 8 – €23.00 – ne Slangenburg [221]

The book of the beginnings : a study of genesis / Newton, Richard Heber – New York: G P Putnam, 1884 – 1mf – 9 – 0-8370-4580-0 – (incl bibl ref) – mf#1985-2580 – us ATLA [221]

The book of the beginnings : a study of genesis with an introduction to the pentateuch / Newton, Richard Heber – New York, London: G.P. Putnam's sons, 1884. xv,311p. illus – 1 – us Wisconsin U Libr [240]

The book of the dead : facsimiles of the papyri of hunefer, anhai, kerasher and netchemet, with supplementary text from the papyrus of nu / Budge, Ernest Alfred Wallis – London, 1899 – 9mf – 9 – mf#NE-20015 – ne IDC [930]

The book of the discipline : vinaya-pitaka – London: Published for the Pali Text Society by Luzac & Co, 1940-1966 – (= ser Samp: indian books) – (trans by i b horner) – us CRL [280]

The book of the ganda clans / Kagwa, Apolo – [196-] – 1 – (transl of: ekitabo kye bika bya baganda) – us CRL [960]

The book of the holy rosary : a popular doctrinal exposition of the fifteen mysteries, mainly conveyed in select extracts from the fathers and doctors of the church, with an explanation of their corresponding types in the old testament / Formby, Henry – London: Burns, Oates, 1872 – 1mf – 9 – 0-524-06247-1 – mf#1990-5202 – us ATLA [240]

The book of the kindred sayings (sanyutta-nikaya) : or grouped suttas – London: Published for the Pali Text Society by The Oxford University Press, 1917- – (= ser Samp: indian books) – (trans by rhys davids; assisted by suriyagoda sumangala thera) – us CRL [280]

A book of the laws of washington relating to notaries public / Skinner, Joseph Osmun – San Francisco: Bancroft Whitney, 1911. 365p. LL-1344 – 1 – us L of C Photodup [340]

Book of the lodge [the] : and officer's manual; to which is added, a century of aphorisms, calculated for general instruction and the improvement of a masonic life / Oliver, George – [2nd ed]. London: R Spencer 1856 – 3mf – 9 – mf#vrl-86 – ne IDC [366]

The book of the mainyo-i-khard : also an old fragment of the bundehesh, both in the original pahlavi being a fascimile of a manuscript... / Andreas, F C – Kiel, 1882 – 1mf – 9 – mf#NE-20160 – ne IDC [470]

Book of the pageant of vancouver, june, 1914 : "from smoke to sunshine" / Vancouver Summer Festival Association – [Vancouver?]: The Association, [1914?] – 2mf – 9 – 0-665-77529-6 – mf#77529 – cn Canadiana [917]

The book of the patriarch job – London: James Duncan, 1837 – 2mf – 9 – 0-7905-2011-7 – mf#1987-2011 – us ATLA [221]

The book of the prophecies of isaiah / McFadyen, John Edgar – New York: Macmillan, 1910 – (= ser The bible for home and school) – 2mf – 9 – 0-7905-1460-5 – (incl ind) – mf#1987-1460 – us ATLA [221]

The book of the prophet daniel = Das buch daniel / Keil, Carl Friedrich – Edinburgh: T & T Clark 1877 [mf ed 1989] – (= ser Clark's foreign theological library. 4th series 34) – 2mf – 9 – 0-7905-1175-4 – (trans fr german by m g easton) – mf#1987-1175 – us ATLA [221]

The book of the prophet daniel = Der prophet daniel / Zoeckler, Otto; ed by Strong, James – New York: Charles Scribner 1890, c1876 [mf ed 1986] – (= ser A commentary on the holy scriptures. old testament 13/2) – 1mf – 9 – 0-8370-6879-7 – (trans fr german by ed) – mf#1986-0879 – us ATLA [221]

The book of the prophet ezekiel / Henderson, Ebenezer – Andover: Warren F Draper, 1870 – 1mf – 9 – 0-8370-3559-7 – mf#1985-1559 – us ATLA [221]

The book of the prophet ezekiel : in the revised version / Davidson, A B – Cambridge: University Press, 1916 – (= ser The Cambridge Bible For Schools And Colleges) – 2mf – 9 – 0-8370-6175-X – (incl ind) – mf#1986-0175 – us ATLA [221]

The book of the prophet ezekiel : theologically and homiletically expounded = Prophet hesekiel / Schroeder, Friedrich Wilhelm Julius; ed by Fairbairn, Patrick & Findlay, William – New York: Charles Scribner, 1890, c1876 [mf ed 1986] – (= ser A commentary on the holy scriptures. old testament 13/1) – 2mf – 9 – 0-8370-6774-X – (trans and enl by ed) – mf#1986-0774 – us ATLA [221]

The book of the prophet ezekiel / ed by Whitehouse, Owen Charles – London: J M Dent; Philadelphia: J B Lippincott 1902 [mf ed 1989] – (= ser The temple bible) – 1mf – 9 – 0-7905-1858-9 – mf#1987-1858 – us ATLA [221]

The book of the prophet ezekiel : with introduction and notes / Redpath, Henry Adeney – London: Methuen, 1907 – 1mf – 9 – 0-8370-7418-5 – (incl ind) – mf#1986-1418 – us ATLA [221]

The book of the prophet ezekiel : with notes and introduction / Davidson, Andrew Bruce – Cambridge: University Press; New York: Macmillan [distributor], 1892 – 1mf – 9 – 0-8370-6732-4 – (incl ind) – mf#1986-0732 – us ATLA [221]

The book of the prophet ezekiel: in the revised version / Davidson, Andrew Bruce – Cambridge: University Press, 1916. lxii,403p. 2 fiches – 9 – us ATLA [240]

The book of the prophet isaiah / ed by Davidson, Andrew Bruce – London: J M Dent; Philadelphia: J B Lippincott 1902 [mf ed 1989] – (= ser The temple bible) – 1mf – 9 – 0-7905-1812-0 – mf#1987-1812 – us ATLA [221]

The book of the prophet isaiah / Henderson, Ebenezer – London: Hamilton, Adams, 1840 – 2mf – 9 – 0-7905-1412-5 – mf#1987-1412 – us ATLA [221]

The book of the prophet isaiah : a new english translation printed in colors exhibiting the composite structure of the book – London: James Clarke, 1898 – (= ser The Sacred Books of the Old and New Testaments) – 1mf – 9 – 0-524-06917-4 – mf#1992-1010 – us ATLA [221]

The book of the prophet isaiah : with introduction and notes / Wade, George Woosung – New York: Edwin S Gorham; London: Methuen, [1911?] – (= ser Westminster Commentaries) – 2mf – 9 – 0-7905-2442-2 – (incl ind) – mf#1987-2442 – us ATLA [221]

The book of the prophet isaiah, chaps 1-39 : with introduction and notes / Skinner, John – stereotyped ed. Cambridge:University Press, 1896 – 1mf – 9 – 0-8370-2518-4 – (includes bibliography and index) – mf#1985-0518 – us ATLA [221]

The book of the prophet isaiah, chaps 40-66 : in the revised version / Skinner, John – Cambridge:University Press, 1917 – 1mf – 9 – 0-8370-2593-1 – (incl ind) – mf#1985-0593 – us ATLA [221]

Book of the prophet jeremiah / Brown, Charles Rufus – [Philadelphia]: American Baptist Publ Soc 1907 [mf ed 1985] – (= ser American commentary on the old testament) – 1mf – 9 – 0-8370-2473-0 – mf#1985-0473 – us ATLA [221]

The book of the prophet jeremiah : critical edition of the hebrew text – Leipzig: J C Hinrichs; Baltimore: Johns Hopkins Press, 1895 – 1mf – 9 – 0-8370-9215-9 – (incl indes) – mf#1986-3215 – us ATLA [221]

The book of the prophet jeremiah : theologically and homiletically expounded = Der prophet jeremia / Naegelsbach, Carl Wilhelm Eduard; ed by Asbury, Samuel Ralph – New York: Scribner, Armstrong, 1871 [mf ed 1985] – (= ser A commentary on the holy scriptures. old testament 12/1) – 2mf – 9 – 0-8370-5670-5 – mf#1985-3670 – us ATLA [221]

The book of the prophet jeremiah : a revised translation with introductions and short explanations / Driver, Samuel Rolles – London: Hodder & Stoughton, 1906 – 1mf – 9 – 0-8370-2966-X – (incl bibl ref and index) – mf#1985-0966 – us ATLA [221]

The book of the prophet jeremiah : together with the lamentations / Streane, Annesley William – Cambridge: University Press; London: CJ Clay, 1889 – (= ser The Cambridge Bible for Schools and Colleges) – 2mf – 9 – 0-8370-6842-8 – (incl bibl ref and ind) – mf#1986-0842 – us ATLA [221]

The book of the prophet jeremiah and that of the lamentations : translated from the original hebrew: with a commentary, critical, philological, and exegetical / Henderson, Ebenezer – Andover: Warren F Draper, 1868 – 1mf – 9 – 0-8370-4709-9 – mf#1985-2709 – us ATLA [221]

The book of the revelation / Scott, Charles Archibald Anderson – London, New York: Hodder and Stoughton, [1905?] – 1mf – 9 – 0-7905-0328-X – (incl bibl ref) – mf#1987-0328 – us ATLA [221]

The book of the saints of the ethiopian church : a translation of the ethiopic synaxarium made from the manuscripts oriental 660 and 661 in the british museum / ed by Budge, E A W – Cambridge, 1928. 4v – 18mf – 9 – mf#NE-20185 – ne IDC [243]

The book of the secrets of enoch / Charles, Robert Henry – Oxford: Clarendon Press, 1896 – 1mf – 9 – 0-7905-0921-0 – (incl indes) – mf#1987-0921 – us ATLA [240]

Book of the states / American Legislators' Association. Council of State Governments – v18 [1970-71] – 1r – 1 – mf#25132 – us WHS [323]

Book of the states – v1-33. 1935-2001 – 9 – $823.00 set – (with suppl) – ISSN: 87292-0763 – mf#401111 – us Hein [340]

The book of the ten masters / Singh, Puran – London: Selwyn & Blount, Ltd, 1926 – (= ser Samp: indian books) – (foreword by ernest rhys) – us CRL [280]

The book of the twelve minor prophets / Henderson, Ebenezer – 2nd ed. London: Hamilton, Adams, 1858 – 5mf – 9 – 0-8370-1670-3 – mf#1987-6100 – us ATLA [221]

The book of the twelve prophets : commonly called the minor / Smith, George Adam – New York: A C Armstrong, 1896 [mf ed 1985] – (= ser The expositor's bible) – 2v on 2mf – 9 – 0-8370-2371-8 – (incl bibl ref & ind) – mf#1985-0371 – us ATLA [221]

The book of the twelve prophets / Smith, George Adam – Armstrong. 1903 – 9 – $33.00 – us IRC [221]

Book of the victorian era ball : given at toronto on the twenty eighth of december 1897 – Toronto: Rowsell & Hutchison, 1898 [mf ed 1980] – 2mf – 9 – 0-665-02845-8 – mf#02845 – cn Canadiana [390]

The book of thekla / ed by Goodspeed, Edgar Johnson – Chicago: University of Chicago Press, 1901 – (= ser Historical and linguistic studies in literature related to the new testament; Ethiopic martyrdoms) – 1mf – 9 – 0-7905-3122-4 – mf#1987-3122 – us ATLA [221]

The book of trades : a circle of the arts and manufactures adapted for schools, colleges, and families / Wylde, James – Edinburgh: Gall & Inglis, 1870 – (= ser 19th c children's literature) – 6mf – 9 – mf#6.1.42 – uk Chadwyck [331]

Book of trinidad / Jackson, T – Port-of-Spain, Trinidad and Tobago. 1904 – 1r – us UF Libraries [972]

The book of virgins and lays and legends of the church and the world / Ross, William Stewart – London: W. Stewart & Co., 189-?. 224p – 1 – us Wisconsin U Libr [240]

The book of were-wolves : being an account of a terrible superstition / Baring-Gould, Sabine – London: Smith, Elder, 1865 – 1mf – 9 – 0-524-01160-5 – mf#1990-2236 – us ATLA [130]

A book of wines / Turner, William – 1568 – 9 – us Scholars Facs [640]

The book of wisdom : the greek text, the latin vulgate and the authorised english version: with an introduction, critical apparatus and a commentary / Deane, William John – Oxford: Clarendon Press, 1881 – 1mf – 9 – 0-7905-0938-5 – (in english, greek, and latin. incl indes) – mf#1987-0938 – us ATLA [220]

The book of wisdom : with introduction and notes / ed by Goodrick, Alfred Thomas Scrope – London: Rivingtons 1913 [mf ed 1989] – (= ser The oxford church bible commentary) – 2mf – 9 – 0-7905-1406-0 – (in english & greek; incl ind) – mf#1987-1406 – us ATLA [221]

The book of witches / Hueffer, Oliver Madox – London: Eveleigh Nash, 1908 – 1mf – 9 – 0-524-02021-3 – (incl bibliographic references) – mf#1990-2796 – us ATLA [130]

The book of zechariah / Chambers, Talbot Wilson – New York: Charles Scribner, c1874 [mf ed 1986] – (= ser A commentary on the holy scriptures. old testament 14/12) – 1mf – 9 – 0-8370-6033-8 – mf#1986-0033 – us ATLA [221]

The book of zephaniah / Kleinert, Paul – New York: Charles Scribner, c1874 [mf ed 1986] – (= ser A commentary on the holy scriptures. old testament 14/10) – 1mf – 9 – 0-8370-6199-7 – (trans and enl by charles elliott) – mf#1986-0199 – us ATLA [221]

A book on baptism / Pitts, F E – 1835 – 1 – $50.00 – us Presbyterian [242]

A book on building : civil and ecclesiastical / Grimthorpe, Edmund Beckett. 1st Baron – London 1876 – (= ser 19th c art & architecture) – 5mf – 9 – mf#4.2.1317 – uk Chadwyck [690]

Book on the physician himself : and things that concern his reputation and success / Cathell, D W – 10th rev enl ed. Philadelphia: F A Davis Co, 1895, c1892 – us CRL [610]

The book opened : or, an analysis of the bible / Nevin, Alfred – Indianapolis, IN: Religious Pub House, 1882 [mf ed 1993] – 1mf – 9 – 0-524-05816-4 – mf#1992-0643 – us ATLA [220]

Book plates in the british museum / British Museum. London – 10r – 1 – $1350.00 – 0-907006-88-4 – (over 25,000 family crests & heraldic documentation, british & european) – uk Mindata [929]

Book production industry – Northbrook IL 1926-76 – 1,5,9 – (cont by: book production industry and magazine production) – ISSN: 0006-7318 – mf#1702.02 – us UMI ProQuest [680]

Book production industry & magazine production – Northbrook IL 1978-79 – 1,5,9 – (cont: book production industry; cont by: book and magazine production) – ISSN: 0192-2874 – mf#1702.02 – us UMI ProQuest [680]

Book report – Worthington OH 1986-2002 – 1,5,9 – ISSN: 0731-4388 – mf#16156 – us UMI ProQuest [070]

Book research quarterly – Piscataway NJ 1985-92 – 1,5,9 – (cont by: publishing research quarterly) – ISSN: 0741-6148 – mf#14303.01 – us UMI ProQuest [070]

Book review [1893] : a monthly journal devoted to new and current publications – New York NY 1893-1901 – 1 – mf#2863 – us UMI ProQuest [070]

Book reviews of the month : an index to reviews appearing in selected theological journals – Fort Worth TX 1962-85 – 1,5,9 – ISSN: 0006-7342 – mf#6348 – us UMI ProQuest [200]

Book sales catalogues / Parke-Bernet Galleries, Inc – New York. 1937-June 18 1968 – 1 – us AMS Press [020]

Bookbinders' bulletin : official journal of local no 4, international brotherhood of bookbinders / Graphic Arts International Union – 1963 jan/feb-1972 jan/feb – 1r – 1 – mf#1048334 – us WHS [680]

Bookbinding / Cockerell, Douglas – New York, NY. 1908 – 1r – us UF Libraries [680]

Bookbird – Mansfield OH 1976+ – 1,5,9 – ISSN: 0006-7377 – mf#11323 – us UMI ProQuest [070]

Booker t washington papers – 388r – 1 – $13,580.00 – (with guide) – Dist. us Scholarly Res – us L of C Photodup [370]

Der bookesbeutel : lustspiel / Borkenstein, Hinrich; ed by Heitmueller, Franz Ferdinand – Leipzig: G J Goeschen, 1896 [mf ed 1993] – (= ser Deutsche litteraturdenkmale des 18. und 19. jahrhunderts 56-57, n f n6-7) – xxx/73p/1pl – 1 – (incl bibl ref) – mf#8676 reel 5 – us Wisconsin U Libr [820]

Bookfellow – Sydney, 1899-1925 – 2r – 1 – A$77.00 vesicular A$88.00 silver – at Pascoe [079]

Book-keeping by double and single entry : with an appendix on precis writing and indexing: designed for self-instruction and for use in schools and colleges – Kingston, Ont: Dominion Business College, 1887 [mf ed 1980] – 3mf – 9 – 0-665-02732-X – mf#02732 – cn Canadiana [650]

Book-keeping by single and double entry : designed for use in the public and high schools / Beatty, Samuel G – 6th ed. Toronto: W J Gage, 1881 [mf ed 1987] – (= ser W J Gage and Co's educational series) – 3mf – 9 – 0-665-09901-0 – (incl publ list) – mf#9901 – cn Canadiana [650]

Book-keeping, by single and double entry : designed for use in the public and high schools / Beatty, Samuel G – Toronto: A Miller, 1877 [mf ed 1982] – 3mf – 9 – (incl ind) – mf#11914 – cn Canadiana [650]

Book-keeping, by single and double entry : designed for use in the public and high schools / Beatty, Samuel G – Toronto; Winnipeg: Gage, 1882 [mf ed 1984] – (= ser W J Gage and Co's educational series) – 3mf – 9 – 0-665-38459-9 – (incl ind and publ list) – mf#38459 – cn Canadiana [650]

Bookletter – New York NY 1974-77 – 1,5,9 – mf#10345 – us UMI ProQuest [070]

Booklist – Chicago IL 1905+ – 1,5,9 – ISSN: 0006-7385 – mf#1911 – us UMI ProQuest [070]

Bookman – Sevenoaks UK 1891-1934 – 1 – mf#5866 – us UMI ProQuest [070]

Bookman: a review of books and life – La Salle IL 1895-1933 – 1 – mf#3876 – us UMI ProQuest [070]

Bookmark – Moscow ID 1948-98 – 1,5,9 – ISSN: 0735-0295 – mf#1927 – us UMI ProQuest [020]

The book-method of bible study / Evans, William – Chicago: Bible Institute Colportage Association, c1915 – 1mf – 9 – 0-524-05909-8 – mf#1992-0666 – us ATLA [220]

Bookplates in the news – Alhambra CA 1973-74 – 1 – ISSN: 0045-2521 – mf#7490 – us UMI ProQuest [790]

Books abroad – Norman OK 1927-76 – 1,5,9 – (cont by: world literature today) – ISSN: 0006-7431 – mf#834.01 – us UMI ProQuest [400]

Books and notions : official organ of the booksellers' and stationers' association of ontario – Toronto: [Bookseller's & Stationers' Assoc of Ontario, 1884-1895] – 9 – mf#P06027 – cn Canadiana [020]

Books and pamphlets : select, important, scarce, historical, descriptive, etc, collected by mr gooch, relating to british america, the united states, etc, great britain and ireland, france and other states and countries of europe... / Alfred Booker (Firm) – [Montreal?: s.n, 1869?] – 1mf – 9 – 0-665-89988-2 – mf#89988 – cn Canadiana [070]

Books and pamphlets by or about john leland / Leland, John – 1754-1841 – 1 – 63.73 – us Southern Baptist [242]

Books & arts – Washington DC 1979-80 – 1,5,9 – ISSN: 0193-4082 – mf#12004 – us UMI ProQuest [410]

Books concerning music printed before 1800 / U.S. Library of Congress. Music Division – 1 – 5060.00 – us L of C Photodup [780]

Books for old testament study : an annotated list for popular and professional use / Smith, John Merlin Powis – Chicago: University of Chicago Press, 1908 – 1mf – 9 – 0-7905-0342-5 – mf#1987-0342 – us ATLA [012]

Books for your children – Birmingham UK 1970-92 – 1,5,9 – ISSN: 0006-7482 – mf#7271 – us UMI ProQuest [070]

Books in canada : the independent book review magazine – Toronto: Canadian Review of Books. v1- . jul 1971- – 9 – Can$72.00y (1999+) – (back run v1-21 1971-98 can$1330) – cn McLaren [070]

Books in english / British Library. National Bibliographic Service – 1992-. Bi-monthly progressive cumulations of English-language titles – 17 – £425.00y + VAT, £549.00y overseas – (1971-80 cumulation £550. 1981-85 cumulation £350) – uk British Libr [010]

Books in manuscript : a short introduction to their study and use / Madan, Falconer – London: Kegan Paul, Trench, Truebner, 1893 [mf ed 1986] – (= ser Books about books) – 1mf – 9 – 0-8370-8126-2 – (incl ind) – mf#1986-2126 – us ATLA [090]

Books in review – New York NY 1976-78 – 1,5,9 – (cont: jewish bookland) – mf#7665.01 – us UMI ProQuest [070]

Books listed in aals law books recommended for libraries : administrative law – $735.00 – mf#409260 – us Hein [342]

Books listed in aals law books recommended for libraries : admiralty – 39v – 9 – $550.00 set – mf#408820 – us Hein [340]

Books listed in aals law books recommended for libraries : biography – 321v – 9 – $3,825.00 – mf#408910 – us Hein [340]

Books listed in aals law books recommended for libraries : business enterprises – 1999. 110v – 9 – mf#409190 – us Hein [346]

Books listed in aals law books recommended for libraries : comparative law – 1999. 34v – 9 – $500.00 set – mf#409170 – us Hein [340]

Books listed in aals law books recommended for libraries : conflicts – 51v – 9 – $765.00 – mf#408920 – us Hein [340]

Books listed in aals law books recommended for libraries : constitutional law – 111v – 9 – $875.00 – mf#408930 – us Hein [342]

Books listed in aals law books recommended for libraries : contracts – 132v – 9 – $895.00 set – mf#408940 – us Hein [346]

Books listed in aals law books recommended for libraries : criminal law and procedure – 88v – 9 – $695.00 – mf#408950 – us Hein [345]

Books listed in aals law books recommended for libraries : evidence – 37v – 9 – $495.00 – mf#418960 – us Hein [340]

Books listed in aals law books recommended for libraries : family law – $735.00 – mf#409270 – us Hein [346]

Books listed in aals law books recommended for libraries : intellectual and industrial property – 80v – 9 – $925.00 – mf#409000 – us Hein [340]

Books listed in aals law books recommended for libraries : judicial administration (part 1 monographs) – 263v – 9 – $775.00 – mf#409020 – us Hein [340]

Books listed in aals law books recommended for libraries : judicial administration, part 2 (serials) – 1327v – 9 – $4,235.00 set – (with 1990 supplement) – mf#409030 – us Hein [340]

Books listed in aals law books recommended for libraries : jurisprudence – 71v – 9 – $750.00 – mf#409060 – us Hein [340]

Books listed in aals law books recommended for libraries : labor law – 107v – 9 – $995.00 – mf#409080 – us Hein [344]

Books listed in aals law books recommended for libraries : legal history – 318v – 9 – $4,200.00 – mf#409090 – us Hein [340]

Books listed in aals law books recommended for libraries : legal profession – $725.00 – mf#409240 – us Hein [340]

Books listed in aals law books recommended for libraries : medical jurisprudence – 108v – 9 – $1,450.00 – mf#409180 – us Hein [340]

Books listed in aals law books recommended for libraries : property – 294v – 9 – $3,320.00 – mf#409100 – us Hein [340]

Books listed in aals law books recommended for libraries : roman law – 34v – 9 – $495.00 set – mf#409130 – us Hein [340]

Books listed in aals law books recommended for libraries : torts – 93v – 9 – $985.00 – mf#409110 – us Hein [340]

Books listed in aals law books recommended for libraries : trust and estates – 42v – 9 – $950.00 – mf#409120 – us Hein [340]

The books of chronicles / Bennett, William Henry – New York: A C Armstrong, 1894 – (= ser The expositor's bible) – 2mf – 9 – 0-8370-2340-8 – (incl bibl ref and ind) – mf#1985-0340 – us ATLA [221]

The books of chronicles : with maps, notes, and introduction / Barnes, William Emery – Cambridge [England]: University Press, 1899 – (= ser The Cambridge Bible for Schools and Colleges) – 1mf – 9 – 0-8370-2182-0 – (incl ind) – mf#1985-0182 – us ATLA [221]

The books of chronicles : with maps, notes and introduction / Elmslie, W A L – new ed. Cambridge: University Press, 1916 – (= ser The Cambridge Bible For Schools And Colleges) – 1mf – 9 – 0-8370-6813-4 – (incl ind) – mf#1986-0813 – us ATLA [221]

The books of chronicles in relation to the pentateuch and the "higher criticism" : five lectures / Hervey, Arthur Charles – London: SPCK; New York: E & J B Young, 1892 – 1mf – 9 – 0-7905-1096-0 – mf#1987-1096 – us ATLA [221]

Books of devotion / Bodington, Charles – London: Longmans, Green 1903 [mf ed 1992] – (= ser The oxford library of practical theology) – 1mf – 9 – 0-524-05135-6 – (incl bibl ref) – mf#1990-1391 – us ATLA [200]

Books of ezra, nehemiah, and esther / Keil, Carl Friedrich – Edinburgh, Scotland. 1879 – 1r – us UF Libraries [270]

The books of ezra, nehemiah, and esther / ed by Harper, James Wilson – London: J M Dent; Philadelphia: J B Lippincott 1902 [mf ed 1989] – (= ser The temple bible) – 1mf – 9 – 0-7905-1824-4 – mf#1987-1824 – us ATLA [221]

The books of ezra, nehemiah, and esther / Keil, Carl Friedrich – Edinburgh: T & T Clark 1873 [mf ed 1993] – (= ser Clark's foreign theological library. 4th series 38) – 1mf – 9 – 0-524-06209-9 – (trans fr german by sophia taylor) – mf#1992-0847 – us ATLA [221]

Books of jeremiah and of the lamentations *see* Book of isaiah

The books of job, psalms, proverbs, ecclesiastes, and the song of solomon according to the wycliffite version – Oxford: Clarendon Press, 1881 – 1mf – 9 – 0-7905-8282-1 – mf#1987-6387 – us ATLA [221]

The books of joel and amos / Driver, Samuel Rolles – Cambridge: University Press, 1897 [mf ed 1986] – (= ser The cambridge bible for schools and colleges) – 1mf – 9 – 0-8370-6110-5 – (incl bibl ref, ind, int & notes) – mf#1986-0110 – us ATLA [221]

Books of judges to 2. samuel *see* Book of joshua

The books of nahum, habakkuk and zephaniah : with introduction and notes / Davidson, Andrew Bruce – Cambridge: University Press; New York: Macmillan [distributor], 1896 – 1mf – 9 – 0-8370-6733-2 – (incl bibl ref and index) – mf#1986-0733 – us ATLA [221]

The books of nahum, habakkuk and zephaniah : with introduction and notes / Davidson, Andrew Bruce – rev ed. Cambridge: University Press, 1920 – 1mf – 9 – 0-8370-3707-7 – (incl ind) – mf#1985-1707 – us ATLA [221]

Books of prayer and healing / ed by Doane, A N – [mf ed Binghamton NY, 1994] – (= ser ASMMF) – 944 folios – 8 – $120.00v / £76.00v [institution] ($96v / £60v if part of subsc) – 0-86698-141-1 – mf#mr136 – us MRTS [090]

The books of samuel = Die buecher samuelis / Erdmann, David; ed by Toy, Crawford Howell & Broadus, John Albert – New York: Charles Scribner, c1877 [mf ed 1986] – (= ser A commentary on the holy scriptures. old testament 5) – 2mf – 9 – 0-8370-6735-9 – (incl app) – mf#1986-0735 – us ATLA [221]

Books of sermons / Winkler, E T – 1851-79. 528p – 1 – us Southern Baptist [242]

The books of the apocrypha : their origin, teaching and contents / Oesterley, William Oscar Emil – London: Robert Scott 1914 [mf ed 1990] – (= ser Library of historic theology) – 2mf – 9 – 0-7905-3393-6 – (incl bibl ref) – mf#1987-3393 – us ATLA [221]

Books of the bible : with relation to their place in history / Hazard, Marshall Custiss – Boston: Pilgrim Press, c1903 – 1mf – 9 – 0-7905-1091-X – (includes bibliographies) – mf#1987-1091 – us ATLA [220]

The books of the chronicles = Die buecher der chronik / Keil, Carl Friedrich – Edinburgh: T & T Clark; New York: C Scribner [dist] 1872 [mf ed 1989] – (= ser Clark's foreign theological library. 4th series 35) – 2mf – 9 – 0-7905-1416-8 – (trans fr german by andrew harper) – mf#1987-1416 – us ATLA [221]

The books of the chronicles = Die buecher der chronik / Zoeckler, Otto; ed by Murphy, James G – New York: Scribner, Armstrong [1876?] [mf ed 1985] – (= ser A commentary on the holy scriptures. old testament 5/1) – 1mf – 9 – 0-8370-5969-0 – mf#1985-3969 – us ATLA [221]

The books of the fairs : a collection of world's fair publications, 1834-1915 – 174r in 4 units – 1 – $18,270.00 coll $5,250.00 per unit – (drawn from the holdings of the smithsonian institution libraries. coll includes ca 2000 bks and pamphlets and covers a wide range of topics, including architecture, fine and decorative arts, technology etc. printed guide available. units 1,2,3 50r ea unit 4 24r) – us Primary [900]

The books of the kings = Die buecher der koenige / Baehr, Karl Christian Wilhelm Felix; ed by Harwood, Edwin & Sumner, William Graham – New York: Charles Scribner c1872 [mf ed 1986] – (= ser A commentary on the holy scriptures. old testament 6) – 2mf – 9 – 0-8370-6722-7 – (bk1 ed, trans & enl by edwin harwood, bk2 by william graham sumner) – mf#1986-0722 – us ATLA [221]

The books of the new testament / Pullan, Leighton – London: Rivingtons, 1901 – 1mf – 9 – 0-7905-1792-2 – (incl ind) – mf#1987-1792 – us ATLA [225]

The books of the old and new testaments canonical and inspired : with remarks on the apocrypha / Haldane, Robert – 1st american ed. Boston: American Doctrinal Tract Society, 1846 – 1mf – 9 – 0-524-06205-6 – mf#1992-0843 – us ATLA [220]

The books of the vaudois : the waldensian manuscripts preserved in the library of trinity college, dublin / Todd, James Henthorn – London: Macmillan, 1865 – 1mf – 9 – 0-524-00658-X – mf#1990-0158 – us ATLA [240]

Books & religion – New York NY 1985-92 – 1,5,9 – (cont: review of books and religion) – ISSN: 0890-0841 – mf#12987.01 – us UMI ProQuest [070]

Bookseller – New York NY 1976+ – 1,5,9 – ISSN: 0006-7539 – mf#11213 – us UMI ProQuest [070]

Bookseller and stationer – Toronto: MacLean, 1897-1907 – 9 – mf#P06029 – cn Canadiana [070]

Bookseller and stationer and canadian newsdealer – Montreal: MacLean, 1908-1910 – 9 – (cont: bookseller and stationer; cont by: bookseller and stationer and office equipment journal) – mf#P06030 – cn Canadiana [070]

The bookseller, newsdealer and stationer – New York: Excelsior Publ House, [1894- (v6-7 (mar 1897-feb 1898) – us CRL [070]

A bookseller of the last century : being some account of the life of john newbery and of the books he published / Welsh, Charles – [London], New York: printed for successors to Newbery & Harris and E P Dutton & Co, 1885 – (= ser 19th c publishing...) – 5mf – 9 – mf#3.1.40 – uk Chadwyck [920]

Bookseller & stationer and office equipment journal – Toronto, Canada mar 1913-sep 1916 – 3 1/2r – 1 – uk British Libr Newspaper [680]

Bookworm : an illustrated treasury of old-time literature – Killen TX 1888-94 – 1 – mf#2864 – us UMI ProQuest [420]

Boole, George *see* An investigation of the laws of thought

Boole, W H *see* "Shall our common school system be maintained as it is?

Boole, William H *see* Antidote to rev h j van dyke's pro-slavery discourse

Boom in orlando 1923-1936 / Allen, L – s.l, s.l? . 1936 – 1r – us UF Libraries [978]

Boom in paradise / Weigall, Theyre Hamilton – New York, NY. 1932 – 1r – us UF Libraries [978]

Boomer, Harriet Ann *see* Little miss ellerby and her big elephants

Booms, Hans *see*
- Liberale parteien
- Verein deutscher eisen- und stahlindustrieller / wirtschaftsgruppe eisenschaffende industrie (bestand r 13 I) bd 10

Boon, A *see* Pachomiana latina

Boon, Albertus Goswinus see Dissertatio historico-critica de dogmatices christianae fontibus eorumque usu publico omnium examini offert albertus goswinus boon
Boon, Martin James see The immortal history of south africa
Boondocks – Bethesda MD 1971-77 – 1,5,9 – mf#8513 – us UMI ProQuest [610]
Boone, Charles Theodore see Law of real property.
Boone companion – Albion, NE: News Print Co. 5v. v1 n1. oct 27 1958-v5 n30. may 14 1963 (wkly) [mf ed 1975] – 3r – 1 – (companion of: albion news; absosrbed in may 1963 by: albion news) – us NE Hist [071]
Boone county advance – St Edward, NE: S J Kennedy, 1900-nov 1929// (wkly) [mf ed 1908-29 (gaps) filmed 1974?] – 8r – 1 – (cont by: st edward advance) – us NE Hist [071]
Boone county argus – Albion, NE: Argus Publ Co, jun 30 1876 (wkly) – 1r – 1 – (cont by: albion argus; vol numbering ends with sep 16 1892) – us Bell [071]
Boone county blade – Albion, NE: J F Bixby. 2v. v1 n1. dec 2 1896-v2 n18. mar 30 1898 (wkly) – 1r – 1 – (absorbed by: albion argus) – us Bell [071]
Boone County Historical Society see Trail tales
Boone county outlook – Cedar Rapids, NE: Baird & Son. 3v. v11 n4. dec 13 1895-v13 n21. apr 8 1898 (wkly) – 1r – 1 – (cont: cedar rapids republican. absorbed: calliope (albion ne). cont by: cedar rapids outlook. issues for dec 11 1896-jan 8 1897 called v11 but constitute v12) – us Bell [071]
Boone enterprise – Boone, NE: A A Dodendorf; Albion, NE: Argus Print House (wkly) [mf ed v3 n42. mar 26 1908-jun 20 1912 (gaps)] – 2r – 1 – (suspended foll dec 29 1910 issue; resumed in 1911. issues for dec 14 1911-called v1 n1-) – us NE Hist [071]
Boone Family Association of Cal-Mont in Missouri see Cal-mont news
Boone family echoes – v4 n3-v6 n2 [1962 oct 21-64 apr] – 1r – 1 – (cont: boone pioneer echoes; cont by: boone pioneer echoes (1964)) – mf#1494720 – us WHS [071]
Boone, Ilsley see Elements in baptist development
Boone, Jerry Neal see Study of the effect of hearing loss of freshmen at the u of fl...
Boone pioneer echoes – v6 n3-v28 n1 [1964 jul-1986 jan] – 1r – 1 – (cont: boone family echoes (1964)) – mf#1494720 – us WHS [929]
Boone pioneer echoes – v1 n2-3 [1959 jul-oct], v2 n1-3 [1960 apr-oct 16], v3 n1-3 [1961 jan-jul], 1962 jan, v4 n1-2 [1962 apr 21-jul 21] – 1r – 1 – (cont: pioneer echoes; cont by: boone family echoes) – mf#1497392 – us WHS [929]
Boone, Susanna see On the efficacy of the grace of our lord and saviour jesus christ
Boone, William Jones see
– Address in behalf of the china mission
– The notions of the chinese concerning god and spirits
– A vindication of comments on the translation of ephesians 1
Boone's creek baptist church (formerly: boggs fork baptist church). lexington, kentucky : church records – 1795-1900. 730p – 1 – us Southern Baptist [242]
Boone's creek baptist church. pickens county. salem, south carolina : church records – 1912-1977 – 1 – 9.72 – us Southern Baptist [242]
Boone's sierra echos – 1968 jun-1980 oct – 1r – 1 – mf#555645 – us WHS [929]
Boor, C de see
– Chronographia
– Neue fragmente des papias, hegesippus und pierius
Boorowa news – Boorowa. jan 1969-dec 1974, sep 1983-dec 1996 – at Pascoe [079]
Boorowa news see Burrowa
Boos, Heinrich see
– Geschichte der freimaurerei
– Manuel de la franc-maconnerie
Booster / Franklin Co. Columbus – (jan 1971-dec 1983) [wkly] – 14r – 1 – mf#B35077-35090 – us Ohio Hist [071]
Booster / Licking Co. Granville – mar 1977-dec 1987 [biwkly, wkly] – 7r – 1 – mf#B29329-29335 – us Ohio Hist [071]
Booster / Naval Weapons Station (Yorktown VA) – 1981 feb-1989 dec, 1990-93 – 2r – 1 – mf#1110291 – us WHS [355]
"Booster" for trempealeau county schools – 1910 jan-17 mar – 1 – mf#1053502 – us WHS [071]
Boot / Parris Island (SC: Recruit depot) – Parris Island SC. 1970 nov 24/dec 23-1990 – 15r – 1 – (cont: paris island boot) – mf#703914 – us WHS [071]
Boot and shoe recorder – New York NY 1916-74 – 1 – ISSN: 0006-7628 – mf#950 – us UMI ProQuest [680]
Boot and Shoe Workers Union see
– Proceedings of the...convention of boot and shoe workers' union
– Union boot and shoe worker
Boot and Shoe Workers' Union see Shoe workers' journal
Boot camps as korrektiewe inrigting / Toit, Pauline du – Uni of South Africa 2001 [mf ed Pretoria: UNISA 2000] – 4mf – 9 – (incl bibl ref; summary in english) – mf#mfm14726 – sa Unisa [365]
Booth, Abraham see
– Pastoral cautions
– The reign of grace
Booth, Catherine Mumford see
– Papers on godliness
– The salvation army in relation to the church and state
Booth, Charles see England and ireland
Booth, Donald Carr see Study of the effect of florida tung oil on lacquer films
Booth, G Robert see The tradition and essential character of the evangelical congregational church
Booth, Lorenza see A series of original designs for decorative furniture
Booth, Meyrick see Collected essays of rudolf eucken
Booth, Robert Russell see Sermon preached at the funeral services of marshall s bidwell
Booth, Walter Sherman see
– The conveyancer's and notary's manual...in the state of minnesota
– The conveyancer's and notary's manual...in the state of south dakota
Booth, William see
– The general's letters, 1885
– In darkest england and the way out
Boothby, R see A breife discovery or description of the most famous island of madagascar or st lawrence in asia neare unto east-india
Boothe, Charles Octavius see The cyclopedia of the colored baptists of alabama
Booth-Tucker, Frederick de Latour see
– The consul
– Darkest india
– The life of catherine booth
Bootle gazette – [NW England] Bootle Lib 1887-88 – 1 – uk MLA; uk Newsplan [072]
Bootle herald, netherton herald – [NW England] Bootle nov 1949-1 may 1965 – 1 – uk MLA; uk Newsplan [072]
Bootle times – [NW England] Bootle, Crosby Lib 12 feb 1878-1896, 1898-21 may 1965 (wkly) – 1 – (title change: bootle times herald [28 may 1965-1969, jan-23 sep 1971]; bootle times [1971-77, 1979]) – uk MLA; uk Newsplan [072]
Bootsmann elbing / Blunck, Hans Friedrich – Wien: W Frick, 1943 [mf ed 1989] – (= ser Wiener buecherei 23) – 75p – 1 – (ill by olaf gulbransson) – mf#7036 – us Wisconsin U Libr [830]
Boot$trap / Interracial Council for Business Opportunity of New Jersey – 1983 aug/sep – 1r – 1 – mf#5306634 – us WHS [338]
Booysen, C Murray see Tales of south africa
Bopp, Advocat et al see Rechtslexikon fuer juristen aller teutschen staaten
Bopp, Franz see Vergleichende grammatik des sanskrit, send, armenischen, griechischen, lateinischen, litauischen, altslavischen, gothischen und deutschen
Bopp, Raul see
– Memorias de um embaixador
– Sol and banana
Boppe, Auguste et al see Les vignettes emblematiques sous la revolution
Boraisha, Menahem see
– Alekanndr kuprin
– Zavl rimer
Boramy, Leon see Cambodia
Boras nu – Boras, Sweden. 2006- – 1 – sw Kungliga [078]
Boras nya tidning see Boras tidning
Boras nyheter – Boras, Sweden. 1922-51; 1990-92 – 1 – sw Kungliga [078]
Boras tidning – Boras, Sweden. 1838 – 578r – 1 – (aka: boras nya tidning; 1834-35, 1837-38, boras tidning, 1838-1978) – sw Kungliga [078]
Boras tidning – Boras, Sweden. 1979- – 1 – sw Kungliga [078]
Boras weckoblad – Boras, Sweden. 1826-29, 1831-33 – 2r – 1 – sw Kungliga [078]
Borasposten – Boras, Sweden. 1893-1902 – 11r – 1 – sw Kungliga [078]
Borasposten veckoupplagan – Boras, Sweden. 1894-1901 – 3r – 1 – sw Kungliga [078]
Bor'ba : organ ekaterinburgskogo soveta rabochikh i soldatskikh deputatov – Ekaterinburg, Russia 1917 [mf ed Norman Ross] – 1r – 1 – mf#nrp-480 – us UMI ProQuest [077]
Borba – Belgrade, Yugoslavia. Jan-Jun 1955 – 2r – 1 – us L of C Photodup [079]
Borba – Beograd, 1976-1993ff – 77r – 1 – gw Mikropress [949]
Borba – Toronto, nov 1 1931-sep 19 1936// – 3r – 1 – Can$375.00 – (in serbo-croatian & english; cont by: pravda and slobodna misao) – cn McLaren [071]
Borba : organ komunisticke partije jugoslavije – Belgrade: [s.n, nov 15 1944-75] – 1 – us CRL [949]
Borba : organ socijalisticki saveza radnog naroda jugoslavije – Zagreb: "Borba", jan 3 1972-dec 31 1973/jan 1/2 1974 – 12r – us CRL [949]
Borba – Turnovo, Bulgaria. 1951-Dec 1986 – 32r – 1 – us L of C Photodup [077]
Borba – Zagreb, Yugoslavia. Jul 1949-1953; Aug-Dec 1956 – 9r – 1 – us L of C Photodup [072]
Borba see Nasha borba
Bor'ba bednoty : organ severo-dvinskogo gubkoma rkp(b) – Veliky Ustyug, Russia 1919 [mf ed Norman Ross] – 4r – 1 – mf#nrp-1916 – us UMI ProQuest [077]
Bor'ba i trud : organ sol'vychegodskogo uezdnogo ispolnitel'nogo komiteta i komiteta rkp(b) – Sol'vychegodsk, Russia 1919-20 [mf ed Norman Ross] – 2r – 1 – mf#nrp-1576 – us UMI ProQuest [077]
Borba, Jose Osorio De Morais see Comedia literaria
Bor'ba [khar'kov: 1920] : organ tsentr kom i khar'k gub kom ukr partii levykh sotsialistovrevoliutsionerov [bor'bistov] – Khar'kov: [s n] 1920 [1918 [dek]-1920 [?]] – (= ser Asn 1-3) – n124,142,143 [1920] item 3, on reel n1 – 1 – mf#asn-2 003 – ne IDC [335]
Borba klassov i partii v 1-i gosudarstvennoi dume / Tomsinskii, S G – Rostov n/D,Krasnodar, 1924 – 103p 2mf – 9 – mf#RPP-44 – ne IDC [325]
Borba klassov i partii vo vtoroi gosudarstvennoi dume / Tomsinskii, S G; ed by Pokrovskii, M M – 1924 – 173p 2mf – 9 – mf#RPP-45 – ne IDC [325]
Bor'ba [odessa: 1920] : organ iuzhn biuro tsk i odes gub kom ukr partii levykh sots[ialistov]-revol[iutsionerov] [bor'bistov] – Odessa: [s n] 1920 [1919 apr-1920, 15 avg – (= ser Asn 1-3) – n213 [1920] item 2, on reel n1 – 1 – mf#asn-2 002 – ne IDC [335]
Bor'ba [orenburg: 1918] : organ orenburg org partii sotsialistov-revoliutsionerov – Orenburg: [s n] 1918 [1918 8 avg-] – (= ser Asn 1-3) – n1-25 [1918] item 17, on reel n7 – 1 – (lacks n22; cont: sotsialist-revolutsioner) – mf#asn-1 017 – ne IDC [335]
Bor'ba so shtundoi = The struggle with the stunda / Ol'shevskii, I – Poltava, 1902 – 1 – 5.00 – us Southern Baptist [242]
Borba sotsialisticheskikh i burzhuaznykh tendentsii v russkom revoliutsionnom dvizhenii / Akselrod, P V – Spb, 1907 – 128p 2mf – 9 – mf#RPP-135 – ne IDC [325]
Bor'ba [tiflis: 1920-1921] : organ tsentr kom sots[ial]-dem[okrat] rabochei partii gruzii / ed by Sharashidze, D G – Tiflis: [s n] 1920-21 [1917 4 maia-1921] – (= ser Asn 1-3) – n2-300 [1920] n1-42 [1921] [gaps] item 18, on reel n7 – 1 – mf#asn-1 018 – ne IDC [325]
Borba za kachestvo v promkooperatsii / Kremianskii, I – 1931 – 79p 1mf – 9 – mf#COR-424 – ne IDC [335]
Borba za ustanovlenie i uprochenie sovetskoi vlasti v iakutii : sbornik dokumentov i materialov / Tebekin, D A & Nikolaeva, V V – Ikutsk: Ikutskoe knizhnoe izd-vo, 1957-1962 – us CRL [949]
Borbecker nachrichten – Essen-Borbeck DE, 1955 31 dec-1956 24 dec, 1958-60 [gaps] – 1 – gw Misc Inst [074]
Borberg, Allan see Clinical and genetic investigations into tuberous sclerosis and recklinghausen's neurofibromatosis
Borbis, Johannes R see
– Die evangelisch-lutherische kirche ungarns in ihrer geschichtlichen entwicklung
– Ueber den religions-unterricht auf dem k.k. evangelischen gymnasium zu teschen
Borboleta poetica : periodico politico e satyrico – Rio de Janeiro, RJ. 12 mar-02 abr 1849 – (= ser Ps 19) – mf#DIPER – bl Biblioteca [073]
Borbon, F see Medicina domestica...del medico caritativo...
Borchard, Edwin Montefiore see Convicting the innocent; errors of criminal justice
Borchardt, Bernard F see
– Animal tales
– Desoto
– Odd attractions near fort myers
– Personalities
– Piracy
– Springs of hillsborough county
– Thomas a edison
Borchardt, L see
– Das grabdenkmal des koenigs ne-user-re
– Die mittel zur zeitlichen festlegung von punkten der aegyptischen geschichte und ihre anwendung
– Das re-heiligtum des koenigs ne-woser-re
Borchardt, Rudolf see
– Das buch joram
– Der deutsche in der landschaft
– Das gespraech ueber formen
– Handlungen und abhandlungen
– Das hoffnungslose geschlecht
– Jugendgedichte
– Die paepstin iutta
– Vereinigung durch den feind hindurch
Borcherdt, Hans Heinrich see
– Die ersten ausgaben von grimmelshausens simplicissimus
– Geschichte des romans und der novelle in deutschland
– Goethes briefe an charlotte von stein
– Grimmelshausens werke in vier teilen
Borchgrevink, C see Das festland am suedpol
Borchmann, Johann Friedrich see Briefe zur erinnerung an merkwuerdige zeiten, und ruehmliche personen
Borcke-Stargordt, Henning, graf von see Der ostdeutsche landbau zwischen fortschritt, krise und politik
Borda, Andres de see Practica de confessores de monjas
Borda, Eugene see Ammonia toxicity in the fertilization of shade tobacco
Borda, Jose Joaquin see Historia de la compania de jesus en la nueva grana
La bordah du cheikh el bousiri : poeme en l'honneur de mohammed = Burdah / Busiri, Sharaf al-Din Muhammad ibn Said – Paris: E Leroux, 1894 – (= ser Bibliotheque orientale elzeverienne) – 1mf – 9 – 0-524-01768-9 – (incl bibl ref; in french) – mf#1990-2616 – us ATLA [470]
Bordas-Demoulin, Jean Baptiste see Le cartesianisme, ou, la veritable renovation des sciences
Borde, Pierre-Gustave-Louis see Histoire de l'ile de la trinidad
Bordeaux, Albert see
– Guyane inconnue
Bordeaux. France. Chambre de Commerce see Extraits des proces-verbaux
Bordeaux, Henry see Le mariage
Bordeleau, Daniele see Agir pour l'insertion
Borden citizen / Canadian Forces Base Borden – 1973 dec 19-1975 feb 26, v25 n49-v27 n6 [1975 mar 5-1975 dec 10] – 2r – 1 – mf#1053504 – us WHS [071]
Borden's review of nutrition research – Columbus OH 1940-71 – 5,9 – ISSN: 0006-7679 – mf#2388 – us UMI ProQuest [613]
Border beacon and advertiser for galashiels, hawick, melrose, selkirkshire, and the border counties – [Scotland] Galashiels: A W Lyall 3 jan 1863-13 may 1864 (wkly) [mf ed 2004] – 1r – 1 – uk Newsplan [072]
Border eagle – 1981 may 15-1982, 1983 jan-sep, 1983 oct-1984, 1985-1986 apr, 1986 may-1987 aug, 1987 sep-1988 sep, 1988 oct-1989, 1992, 1993 – 9r – 1 – mf#646376 – us WHS [071]
Border land – London, England. 18– – 1r – us UF Libraries [240]
Border mail and gazette for roxburgh, selkirk, berwick and northumberland – [Scotland] Kelso: printed by J C Thomson 19 apr 1934-26 oct 1949 (wkly) [mf 19 apr 1934-oct 1949] [mf ed 2004] – 11r – 1 – (cont: kelso mail & gazette for the counties of roxburgh, selkirk, berwick and northumberland [2 Jan 1854-12 Apr 1934]; cont by: kelso border mail & gazette for roxburgh, selkirk, berwick & northumberland [23 may 1945-26 oct 1949]: latter merged with: kelso chronicle and border pioneer to form: border counties & kelso chronicle & mail) – uk Newsplan [072]
Border morning mail – Albury, jan 1965-sep 1997 – (aka: border mail) – at Pascoe [079]
Border morning mail – Albury, nov 1903-dec 1964 – 157r – A$5181.00 vesicular A$6044.50 silver – at Pascoe [079]
Border news : charleston – Pretoria: State Library Corporate Communication, 18 jan 1910-12 apr 1910 – (= ser State library south africa newspaper microfilm project) – 1r – 1 – mf#MS00280 – sa National [079]
Border news see Charlestown mail / border news
Border post – Albury, nov 1856-oct 1902 – 26r – A$1001.00 vesicular A$1144.00 silver – at Pascoe [079]
Border standard – [Scotland] Galashiels: Border Standard Ltd 4 jan 1906-29 dec 1950 (wkly) [mf ed 2004] – 47r – 1 – (cont: scottish border record, for the counties of roxburgh, selkirk, peebles & berwick; merged with: southern reporter (1855) to form: southern reporter and the border standard; imprint varies) – uk Newsplan [072]
Border telegraph – Galashiels: A Walker & Son 1902- (wkly) [mf ed 1995-] – 1 – (cont: galashiels telegraph) – uk Newsplan; uk Scotland NatLib [072]
The border tour throughout the most important and interesting places in the counties of northumberland, berwick, roxburgh, and selkirk / Mason, John – Edinburgh 1826 [mf ed Hildesheim 1995-98] – 1v on 2mf – 9 – €60.00 – 3-487-27936-3 – gw Olms [914]
The border wars of new england : commonly called king william's and queen anne's wars / Drake, Samuel Adams – New York: C Scribner's Sons, 1897 – 4mf – 9 – mf#53419 – cn Canadiana [975]
Borderer – Sackville, NB: Edward Bowes, 1865-70 – 1r – 1 – cn Library Assoc [971]

Borders gazette – Alnwick, England 9 feb, 2 mar-22 jun, 24-31 aug, 19 oct 1989-14 jan 1999 – 1 – uk British Libr Newspaper [072]
Bordes / Giraudier, Antonio – Habana, Cuba. 1956 – 1r – us UF Libraries [972]
Bordes, Gabriel *see* Tableau synoptique de l'histoire de france et des principaux evenemens arrives en europe depuis la naissance de louis 14 jusqu'a l'epoque de la restauration de la monarchie francai...
Bordewijk, Hugo Willem Constantijn *see*
– Handelingen over de reglementen op het...
– Ontstaan en ontwikkeling van het staatsrecht van c...
Bordier, H L *see* Libri miraculorum aliaque opera minora
Bordier, Louis Charles *see* Traite de composition
Bordin, Ruth M *see* Temperance and prohibition papers, 1830-1933
Bordon de peregrino. poemas / Corredor Garcia, Antonio – Caceres: Ediciones Gruzada Mariana, 1965 – 1 – sp Bibl Santa Ana [810]
Bore – Stockholm, Sweden. 1848-51 – 3r – 1 – sw Kungliga [078]
Bore, E *see* Correspondance et memoires d'un voyageur en orient
Boreas – Abingdon, Oxfordshire 1986+ – 1,5,9 – ISSN: 0300-9483 – mf#13022 – us UMI ProQuest [550]
Borehamwood, elstree edgware post – Borehamwood, England 11 Sep 1975-17 Jul 1980 [mf 1977-] – 1 – (cont: borehamwood, elstree, barnet borough post [13 jun 1968-4 sep 1975]; cont by: borehamwood, elstree, radlett, edgware post [24 jul 1980-christmas iss 1986]) – uk British Libr Newspaper [072]
Borehamwood, elstree edgware post – Borehamwood, England 11 sep 1975-17 jul 1980 [mf 1977-] – 1 – (cont: borehamwood, elstree, barnet borough post [13 jun 1968-4 sep 1975]; cont by: borehamwood, elstree, radlett, edgware post [24 jul 1980-xmas iss 1986]) – uk British Libr Newspaper [072]
Borel, Eugene *see* Repartition des annuites de la dette publique ottomane, article 47 du traite de lausanne
Borel, Henri *see* Wu wei
Borelli, Giovanni Alfonso *see* De vi percussionis liber
Borelli, J *see* Ethiopie meridionale
Borely, N *see* La vie de messire christophle d'authier de sisgau, eveque de bethleem
Boret, Long *see*
– Khmer republic
– La lutte pour la survie
Borge, O F *see* Schwedisch-chinesische wissenschaftliche expedition nach den nordwestlichen provinzen chinas unter leitung von dr sven hedin und prof su ping-chang. algen
Borgeaud, Charles *see* Calvin
Borges Da Fonseca, Antonio Jose Victoriano *see* Nobiliarchia pernambucana
Borges Jacinto Del Castillo, Analola *see* Casa de austria en venezuela durante la guerra de
Borges, Milo Adrian *see*
– Compilacion ordenada y completa de la legislacion
– Manual de la legislacion colombiana
– Manual de la legislation colombiana
Borges, Pedro *see* Eugenio sarralbo aquareles, antonio correa y arturo alvarez. (ofm). inventario...
Borgese, Giuseppe Antonio *see* Mefistofele
Borget, A *see* Sketches of china and the chinese
Borghese, Antonio D R *see* New and general system of music
Borghini, R *see* Il riposo...
Borgia / Klabund – Wien, Austria. 1932 – 1r – us UF Libraries [025]
Borgius, Viktor Walther Paul *see* Der voelkerbund
Borgman, Jean Pawley *see* Faure requiem
Borgnet, A *see* Opera omnia
Borgonon, Helena Pastor *see* Eisenzeitliche keramik aus galilaa
Borhan-i taraqqi – Baku, Azerbaijan 1906-11 [mf ed Norman Ross] – 3r – 1 – mf#nrp-227 – us UMI ProQuest [077]
Borheck, August Chr et al *see* Versuch eine briefwechsels ueber das oeffentliche schul- und erziehungswesen
Boria, J de *see*
– Emblemata moralia
– Empresas morales
Borinquen Field, P R *see* Puerto rico
Borinqueneer = El borinqueno – 1986 sep, 1987 may-jun, 1988 jan, mar-may, 1993 spring – 1r – 1 – mf#1269808 – us WHS [071]
Boris, Otto *see*
– Der grenzbauer
– Masurens waelder rauschen
– Murzel
– Reiter fuer deutschlands ehre
Borisenko, F *see* Nasha rodina
Borisoglebsk Sovet rk i kd *see* Izvestiia borisoglebskogo soveta rabochikh, soldatskikh i krest'ianskikh deputatov
Borisovskaia kommuna – Borisovka, Russia 1930-41 [mf ed Norman Ross] – 20r – 1 – mf#nrp-1502 – us UMI ProQuest [077]
Borius wichart : roman aus der gegenreformation / Wegner, Max – 3. aufl. Stuttgart: G Truckenmueller c1939 [mf ed 1991] – 1r – 1 – (filmed with: wassermann : sein kampf um wahrheit / walter goldstein) – mf#3037p – us Wisconsin U Libr [830]
Borkener zeitung – Borken/Hessen DE, 1926-35 – 8r – 1 – gw Misc Inst [074]
Borkener zeitung – Borken/Westfalen DE, 1987- – 7r/yr – 1 – gw Misc Inst [074]
Borkenstein, Hinrich *see* Der bookesbeutel
Borkowski, Jerzy Roman *see* Artefizieller sphinkter 'as 800' am blasenhals
Borlaenge – Falun, Sweden. 1998-2001 – 37r – 1 – sw Kungliga [078]
Borlaenge tidning – Hedemora, Borlaenge, Falun, Sweden. 1885-1920, 1921-78, 1979- – 254r – 1 – sw Kungliga [078]
Borlaengeposten – Falun, Sweden. 1897-1902 – 5r – 1 – sw Kungliga [078]
Borland, C R *see* A descriptive catalogue of the western mediaeval manuscripts in edinburgh university library
Borland, John *see*
– An appeal to the montreal conference and the methodist church generally
– The assumptions of the seminary of st sulpice to be the owners of the seigniory of the lake of two mountains and the one adjoining examined and refuted
– Dialogues between two methodists, algernon newways and samuel oldpaths
– An examination of, and reply to, "a brief statement of facts
– Letters to a member of the wesleyan methodist church
– Observations on the moral agency of man and the nature and demerit of sin
– The reviewer reviewed
Borlase, William *see* Antiquities, historical and monumental of the county of cornwall
A born coquette / Hungerford, Margaret Wolfe – London: Spencer Blackett, 1890 – (= ser 19th c women writers) – 9mf – 9 – mf#5.1.112 – uk Chadwyck [420]
Born of water and spirit : a series of essays concerning regeneration and the new life / Hough, Samuel – New York: Sheldon, 1879 – 1mf – 9 – 0-8370-4860-5 – mf#1985-2860 – us ATLA [240]
Borne-blad – 1877-1878 – 1r – 1 – mf#1053503 – us WHS [071]
Bornemann, F W B *see* In investiganda monachatus origine...
Bornemann, Johann W *see* Einblicke in england und london im jahre 1818
Bornemann, Wilhelm *see*
– Die allegorie in kunst, wissenschaft und kirche
– Bittere wahrheiten
– Einfuehrung in die evangelische missionskunde
– Systematische darstellung des preussischen civilrechts mit benutzung der materialien des allgemeinen landrechts
– Unterricht im christentum
Bornemann, Wilhelm et al *see*
– Jesus
– Jesus as problem, teacher, personality and force
Borneo times – Sandakan, Borneo 6 mar 1962-14 sep 1963 (imperfect) – 4r – 1 – (english ed) – uk British Libr Newspaper [079]
Borneo-expedition : geological explorations in central borneo (1893-1894) / Molengraaff, G A F – New York. 1961-1971 (1) 1970-1971 (5) – 25mf – 9 – mf#2536 – ne IDC [915]
Bornhaeuser, Karl *see* Die vergottungslehre des athanasius und johannes damascenus – die grundwahrheiten der christlichen religion nach d. r. seeberg
Bornhausen, Karl *see*
– Die ethik pascals
– Faustisches christentum
– Religion in amerika
– Wandlungen in goethes religion
– Wir heissen's fromm sein
Bornier, Henri *see* Fille de roland
Bornitz, J *see*
– Emblematum sacrorum et civilium miscellaneorum sylloge prior
– Moralia bornitiana
Bornstedt, Louise von *see* Die legende der hl jungfrau und maertyrin sankt katharina
Bornstein, Paul *see* Hebbels herodes und mariamne
A bornu almanac for the year a d 1916 : (a h 1334 and part of 1335) – London, NY: Oxford UP, 1916 – 1 – us CRL [030]
Bornu province gazetteer – [Lagos: Govt Printer, 1929] – 1 – us CRL [960]
Borochov, B *see* Nationalism and the class struggle
Borochov, Ber *see*
– Keta'im mi-mishnato shel b borokhov
– Sozialismus und zionismus
– Yalkut borokhov
Borodaevskii, S V *see*
– Kak ustroit' melkii kredit v gorodakh
– Kooperatsii
– Kooperatsiia sredi slavian
– Kredit
– Sbornik po melkomu kreditu
– S-peterburgskoe otdelenie komiteta o sel'skikh ssudo-sberegatel'nykh i promyshlennykh tovarishchestvakh
Borodin, D N *see* Nauchno-populiarnyi illiustrirovannyi zhurnal
Borooloola inquest book, 28 december 1889 to 10 november 1930 – 4mf – 9 – A$16.50 – 0-949124-67-2 – (filmed with: anythony's lagoon mortuary book [1890-1948]; newcastle waters police station mortuary book [1893-1951]; adelaide river police day books [1946-58]; brock's creek police day books [1926-48]; rankine river police day books (incl a census) [1930-34]; pine creek police day books [1882-1948]) – at Northern [980]
Borot'ba : organ of the ukrainian working people – Vienna. v1 n1-16. jan 1 1920-oct 16 1920// – 1r – 1 – Can$85.00 – cn McLaren [331]
Borough of birmingham museum and art gallery : handbook with notes, to the collections of paintings / Watts, George Frederick & Burne-Jones, Edward Coley, 1st Baronet – Birmingham [1885] – (= ser 19th c art & architecture) – 1mf – 9 – mf#4.2.1772 – uk Chadwyck [750]
Borough of chelsea herald *see* Chelsea herald
Borough of hackney express and shoreditch observer *see* Shoreditch observer
Borough of hackney standard, bethnal green and shoreditch chronicle – [London & SE] Tower Hamlets 24 mar 1877-18 jul 1885 – 1 – (cont by: hackney standard, bethnal green & shoreditch chronicle [25 jul 1885-10 may 1907]) – uk Newsplan; uk British Libr Newspaper [072]
The borough of stoke on trent, staffordshire / Ward, J – 1r – 1 – mf#713 – uk Microform Academic [941]
Borough polytechnic news : organ of the borough polytechnic institute – Southwark, London 15 jan 1895-jul 1925 – 1/4r – 1 – (cont: borough polytechnic weekly [5 jan 1893-27 dec 1894]; cont by: borough polytechnic magazine (special interim ed) apr, oct 1926-oct 1935) – uk British Libr Newspaper [378]
The borough register *see* Ward lists and other records of the city of gloucester, 1843-86
Boroughs, R Z *see* Two booklets relative to the beginnings of southern baptist missions in niagara falls, ny, 1953-59
Borovoi, A *see* Anarkhizm
Borowski, Christian *see* Funktionelle charakterisierung von domaenen des elongationsfaktors g
Borowski, Felix *see* Ier quatuor a cordes
Borracha na politica economica do brasil / Mello Moraes, Trajano De – Rio de Janeiro, Brazil. 1943 – 1r – us UF Libraries [330]
Las borracheras y el problema de las conversaciones en indias / Bayle, Constantino – Madrid: Razon y Fe, 1943 – 1 – sp Bibl Santa Ana [954]
Borrallo Salgado, Teofilo *see* Ruero del baylio
Borrel, Eugene *see* L'interpretation de la musique francaise (de lully a la revolution)
Borrelli, Dina M *see* Examining the relationship among measures of anxiety, self-confidence, arousal, and performance of elite field hockey players
Borrero, Fernando *see* Descripcion de las provincias del rio de la plata
Borrero Y Pierra, Ana Maria *see*
– Crisis del lujo
– Quinto poder
Borris, Siegfried *see* Herbstaufbruch
Borrmann, A *see* Bethanien
Borromeo, Charles N *see* Reduction of sports injury morbidity with hyperbaric oxygen treatment
Borrow, George Henry *see*
– The bible in spain
– Romano lavo-lil, word-book of the romany
Borrowed times : alternative news for montanans / Montana Reconnaissance Project – 1974 nov 15-1978 jun, 1978-1980 summer – 2r – 1 – mf#492042 – us WHS [071]
Borrows, William *see* Salvation by christ, the grand object of christian missions
Borsa rehberi – Istanbul: Matbaa-yi Ebueziyya, 1928 – 15mf – 9 – $200.00 – us MEDOC [380]
Borst, A *see* Die katharen (mgh schriften:12.bd)
Borstidningen – Stockholm, Sweden. 1888-1900 – 7r – 1 – sw Kungliga [078]
Borst-Smith, Ernest Frank *see*
– Caught in the chinese revolution
– Mandarin and missionary in cathay
Borthwick castle : or, sketches of scottish history: with biographical notices of the chiefs of the house of argyll / Borthwick, John Douglas – Montreal: J M O'Loughlin, 1880 – 4mf – 9 – mf#10198 – cn Canadiana [941]
Borthwick, J D *see* Three years in california
Borthwick, Jane Laurie *see* The story of four centuries
Borthwick, John Douglas *see*
– The battles of the world
– Borthwick castle
– Cyclopdia of history and geography
– The elementary geography of canada
– Examples in historical and geographical antonomasia
– From darkness to light
– The harp of canaan
– Historical and biographical sketches from borthwick's gazetteer of montreal
– History of montreal and commercial register for 1885
– The history of scottish song
– History of the diocese of montreal, 1850-1910
– Rebellion de 37-38
– The tourist's pleasure book
Borthwick, John Douglas [comp] *see* Poems and songs on the south african war
Bortkevich, I *see* O denezhnoi reforme, proektirovannoi ministerstvom finansov
Borukh rekhovitski bukh / Rekhovitski, Borukh – Buenos Aires, Argentina. 1932 – 1r – us UF Libraries [939]
Borum, Jospeh H *see* Biographical sketch of tennessee baptist ministers
Boruttau, Carl *see* Julianus der abtruennige
Borwicz, Michal *see*
– Dokumenty zbrodni i meczenstwa
– Uniwersytet zbirow
Borwicz, Michal Maksymilian *see* Dokumenty zbrodni i meczenstwa, kolegium redakcyjne
Bory de Saint-Vincent, J B G M *see*
– Dictionnaire classique d'histoire naturelle
– Expedition scientifique de Moree
– Voyage dans les quatre principales oles des mers d'afrique...
Bory de Saint-Vincent, Jean *see*
– Lamuel ou le livre du seigneur
– Voyage dans les quatre principales iles des mers d'afrique
– Voyage souterrain
– Voyage to, and travels through the four principal islands of the african seas
Bory de Saint-Vincent, Jean B *see*
– Beitraege zur naturgeschichte der maskarenischen insel in die beiden organischen naturreiche und mehrere neue entdeckungen in denselben betreffend
– Geschichte und beschreibung der kanarien-inseln aus dem franzoesischen
– Reise nach den maskarenischen oder franzoesisch afrikanischen inseln ile de france und bourbon in den jahren 1801 und 1802
Bory de Saint-Vincent, Jean B de *see*
– Expedition scientifique de moree
– Guide du voyageur en espagne
Bory, Paul *see* Explorateurs de l'afrique
Bosanquet and puller's new reports : new reports of cases argued and determined in the court of common pleas... / Bosanquet, John B & Puller, Christopher – v1-2. 1804-07. London: J Butterworth, 1806-08 (all publ) – 5mf – 9 – $22.50 – mf#LLMC 84-754 – us LLMC [324]
Bosanquet and puller's reports : reports of cases argued and determined in the court of common pleas and exchequer chamber, and in the house of lords on appeal therefrom... / Bosanquet, John B & Puller, Christopher – v1-2. 1796-1801 – 15mf – 9 – $22.50 – (v1 printed for byrne & hudson, 1804. v2 printed for p byrne, 1803. both in philadelphia) – mf#LLMC 95-247 – us LLMC [324]
Bosanquet, Bernard *see*
– The civilization of christendom
– The essential of logic
– A history of aesthetic
– Knowledge and reality
– Logic
– Logic, or, the morphology of knowledge
– Metaphysic
– The philosophical theory of the state
– Psychology of the moral self
– The value and destiny of the individual
Bosanquet, John B *see*
– Bosanquet and puller's new reports
– Bosanquet and puller's reports
Bosanquet, Samuel Richard *see* "Vestiges of the natural history of creation"
Bosbach, Heinz *see* Fuerst bismarck und die kaiserin augusta
Bosbogaz : koyu cumhuiyetci siyasi gazete – Sahibi ve Muharriri: Mehmed Asaf (Borsaci). n6. 19 haziran 1930, 10-12,17-30. 16 tesrinievvel 1930 – (= ser O & t journals) – 2mf – 9 – $40.00 – us MEDOC [956]
Bosboom, S *see*
– Cort onderwys van de vijf colommen
– Cort onderwys van de vyf colomen door vinsent scamozzy...
Bosboom, [S] *see* De vyf colom-orden, met derzelver deuren en poorten...
Bosc, Louis Augustin Guillaume *see*
– Histoire naturelle des coquilles
– Histoire naturelle des crustaces
– Histoire naturelle des vers
Boscawen, William Saint Chad *see*
– The bible and the monuments
– The first of empires
Bosch, Bernardus de *see* Dichtlievende verlustigingen

Bosch, G B *see* Reizen in west indie
Bosch, J van den *see* Basilica, monasterium et le culte de st martin de tours
Bosch, J vn den Basilica *see* Monasterium et le culte de st martin de tours
Bosch, Juan *see*
– Dos pesos de agua, cuentos
– Espaldas a si mismo
Boschet, A *see* Le parfait missionnaire ou la vie du r p julien maunoir
Boschini, M *see* La carta del navegar pitoresco dialogo
Boschius, J *see*
– Symbolographia sive de arte symbolica sermones septem
Boschma, Anne L C *see* Breast support for the active women
Boscobel appeal – Boscobel WI. 1868 jan 29-1869 feb 13 – 1r – 1 – (cont: appeal (boscobel wi); cont by: boscobel journal) – mf#957418 – us WHS [071]
Boscobel broad-axe – Boscobel WI. 1964 aug-1866 may 31 – 1r – 1 – (cont: boscobel hatchet) – mf#957428 – us WHS [071]
Boscobel dial – Boscobel WI. 1919 aug 27/1920 jan 14-2000 jul-dec – 90r – 1 – (cont: boscobel dial-enterprise) – mf#1008773 – us WHS [071]
Boscobel dial – Boscobel WI. 1873 apr 11, dec 26-1876, 1877-1878 may 17 – 2r – 1 – (cont by: dial (boscobel wi)) – mf#986199 – us WHS [071]
Boscobel dial – Boscobel WI. 1888 apr 5-1891, 1892-1895 sep 30, 1895 oct 3 – 3r – 1 – (cont: dial [boscobel wi]; cont by: dial-enterprise) – mf#1008757 – us WHS [071]
Boscobel dial-enterprise – Boscobel WI. 1908 aug 12/1909 sep 29-1918 aug 15/19 aug 20 – 8r – 1 – (cont: dial-enterprise, boscobel sentinel; cont by: boscobel dial [boscobel wi: 1919]) – mf#1008761 – us WHS [071]
Boscobel hatchet – Boscobel WI. 1864 jul 20 – 1r – 1 – (cont: national broad-axe; cont by: boscobel broad-axe) – mf#957426 – us WHS [071]
Boscobel journal – Boscobel WI. 1869 feb 20-1870 sep 16 – 1r – 1 – (cont: boscobel appeal) – mf#957420 – us WHS [071]
Boscobel sentinel – Boscobel WI. 1901-04, 1905-07, 1908-10, 1911-13, 1914-16, 1917-1919 aug 6 – 6r – 1 – (cont by: boscobel dial-enterprise) – mf#957707 – us WHS [071]
Boscobel weekly democrat – Boscobel WI. 1860 jan 28 – 1r – 1 – mf#957432 – us WHS [071]
Boscovic, Ruder *see* Journal d'un voyage de constantinople en pologne
Bose, Atindranath *see* Social and rural economy of northern india, cir 600 bc-200 ad
Bose, Buddhadeva *see* An acre of green grass
Bose, Chunilal *see* Sir gooroodass banerjee
Bose, Dakshina Ranjan *see* The cabinet mission in india
Bose, Eshan Chunder [comp] *see* The english works of raja ram mohun roy
Bose, George Mathias *see* Recherches sur la cause et sur la veritable teorie de l'electricite
Bose, Horace Mellard du *see* A history of methodism
Bose, Kheroth Mohini *see* The village of hope
Bose, Manindra Mohan *see* Sahajiya sahitya
Bose, Nirmal Kumar *see* Canons of orissan architecture
Bose, Phanindra Nath *see*
– The hindu colony of cambodia
– The indian colony of champa
Bose, Pramatha Nath *see* Swaraj, cultural, and political
Bose, Ram Chandra *see*
– Brahmoism
– Hindu philosophy popularly explained
Bose, S C *see* Buddha
Bose, Subhas Chandra *see*
– The mission of life
– Through congress eyes
Bose, Sudhansu Mohan *see* The working constitution in india
Bose, Suresh Chunder *see* The life of protap chunder mozoomdar
Bose, Vilmar Konrad *see* The struggle between conscience and law in times of war
Bosio, Ant *see* Roma sotteranea
Bosl, K *see* Die reichsministerialitaet der salier und staufer (mgh schriften:10.bd)
Bosley, methodist chapel – (= ser Cheshire monumental inscriptions) – 1mf – 9 – £2.50 – mf#12a – uk CheshireFHS [929]
Bosley, st mary – [North Cheshire FHS] – (= ser Cheshire monumental inscriptions) – 1mf – 9 – £2.50 – mf#117 – uk CheshireFHS [929]
Bosma, Menno John *see* Onderwijzing in de gereformeerde geloofsleer
Bosman, Willem *see*
– Description of the coast of guinea
– Nauwkeurige beschryving van de guinese goud-, tand-, en slave-kust...
Bosna – 1291 [1874] – (= ser Vilayet salnames) – 3mf – 9 – $55.00 – us MEDOC [956]
La bosnie consideree dans ses rapports avec l'empire ottoman / Pertusier, Charles – Paris 1822 [mf ed Hildesheim 1995-98] – 3mf – 9 – €90.00 – 3-487-29094-4 – gw Olms [910]
Bo's'n's whistle : published for the employees of the portland-vancouver area kaiser shipyards – Vancouver WA: Kaiser Co Inc, Portland [OR]: Oregon Shipbuilding Corp [1941-] [wkly] [mf ed v4-5 1944-45 (gaps)] – 1r – 1 – (absorbed: flat top flash 1944) – us Oregon Lib [623]
The bo's'n's whistle *see* Flat top flash
Le bosphore egyptien – Cairo, Egypt 3 oct 1884-14 jul 1886; 2 nov-2 dec 1894 (imperfect) – 3 3/4r – 1 – uk British Libr Newspaper [079]
Le bosphore egyptien – Cairo. Egypt. -d. 3 Oct 1884-4 Jul 1886, 2 Nov-2 Dec 1894. (Imperfect). (5 reels) – 1 – uk British Libr Newspaper [072]
Le bosphore et constantinople avec perspective des pays limitrophes / Tschichatscheff, P de – Paris, 1864 – 7mf – 9 – mf#AR-1634 – ne IDC [956]
Bosporus und attika : schilderungen / Reisewitz, Gustav – Berlin 1861 [mf ed Hildesheim 1995-98] – (= ser Fbc) – 2mf – 9 – €60.00 – 3-487-29105-3 – gw Olms [914]
Bosque de apolo / Rosales Y Rosales, Vicente – San Salvador, El Salvador. 1929? – 1r – us UF Libraries [910]
Bosquejo biografico del senor oidor juan antonio m... / Robledo, Emilio – Bogota, Colombia. v1-2. 1954 – 1r – us UF Libraries [972]
Bosquejo de la matematica espanola en los siglos de la decadencia / Penalver y Bachiller, P – Sevilla, 1930 – 2mf – 9 – sp Cultura [510]
Bosquejo economico politico de la isla de cuba / Torrente, Mariano – Madrid, Spain. v1-2. 1849-1853 – 1r – us UF Libraries [330]
Bosquejo fisico, politico e historico / Fernandez, Manuel – San Salvador, El Salvador. 1926 – 1r – us UF Libraries [972]
Bosquejo historico acerca de la virgen y monasterio de santa maria de guadalupe – Avila: Cayetano Gonzalez Hernandez – 1 – sp Bibl Santa Ana [240]
Bosquejo historico de honduras / Duron Y Gamero, Romulo Ernesto – Tegucigalpa, Mexico. 1956 – 1r – us UF Libraries [972]
Bosquejo historico de honduras, 1502 a 1921 / Duron Y Gamero, Romulo Ernesto – San Pedro Sula, Honduras. 1927 – 1r – us UF Libraries [972]
Bosquejo historico de la farmacia y la medicina en... / Reina Valenzuela, Jose – Tegucigalpa, Mexico. 1947 – 1r – us UF Libraries [972]
Bosquejo historico de la villa de ceclavin / Rosado, Joaquin – Caceres: Extremadura, 1927 – 1 – sp Bibl Santa Ana [946]
Bosquejo historico de las letras cubanas / Portuondo, Jose Antonio – Habana, Cuba. 1960 – 1r – us UF Libraries [972]
Bosquejo historico de las letras cubanas / Portuondo, Jose Antonio – Havana, Cuba. 1961 – 1r – us UF Libraries [972]
Bosquejo historico de las revoluciones de centro-america : desde 1811 hasta 1834 / Marure, Alejandro – 2nd ed. Guatemala: Tip de "El Progreso" 1877-78 [mf ed 1986] – 2v in 1 on 1r – 1 – (filmed with: resumen historico-critico de literatura colombiana / ruano, j m) – mf#7219 – us Wisconsin U Libr [972]
Bosquejo historico del brasil / Baez, Cecilio – Asuncion, Paraguay. 1940 – 1r – us UF Libraries [972]
Bosquejo para un curso de gimnasia para el uso de organizaciones juveniles de f.e.t. y de las jons de la provincia de caceres por... / O'Ferrall, Arturo – Caceres: Imprenta Moderna, 1938 – 1 – sp Bibl Santa Ana [946]
Bosquejo sobre el trabajo y la seguridad social en el dominio del canada / Sani Poblete, Margot – Santiago?, 1951. 66, 6p. LL-2397 – 1 – us L of C Photodup [340]
Bosquejo...literatura de asturias... / Fuertes Acevedo, Maximo – 1885 – 9 – sp Bibl Santa Ana [440]
Bosquejos cientificos / Fuertes Acevedo, Maximo – 1880 – 9 – sp Bibl Santa Ana [500]
Bosquejos, retratos, recuerdos / Pineyro, Enrique – Habana, Cuba. 1964 – 1r – us UF Libraries [972]
Boss tweed in court : a documentary history / ed by Hershkowitz, Leo – 6r – 1 – $1075.00 – 1-55655-167-3 – us UPA [364]
Bossa nova : the exciting new dance from brazil. new bonomo photo-step method / Bonomo, Joe – [New York, Bonomo Culture Institute, 1963?] – 1 – mf#*ZBD-*MGO pv28 – Located: NYPL – us Misc Inst [790]
Bossange, Gustave *see*
– Beautes de l'histoire du canada
– Il canada e l'emigrazione
– La nouvelle france
Bossano, Luis *see* Porblemas de la sociologia
Bossart, J J *see* Geschichte der mission der evangelischen brueder auf den caraibischen inseln s thomas, s croix und s jan
Bosschart, F *see* Troepenmacht in suriname
Bossdorf, Hermann *see*
– Bahnmeester dod
– De Faehrkrog
Bosse, Abraham *see*
– Sentimens sur la distinction des diverses manieres de peinture, dessein et graveure...
– Traicte des manieres de graver en taille-douce
– Traité des manieres de dessiner les ordres de l'architecture antique en toutes leurs parties...
Bosse, C L *see* Circular
Bossenbrook, William John *see* Justus moeser's approach to history
Bossert, Adolphe *see* Goethe et schiller
Bossert, Gustav *see*
– Das interim in wuerttemberg
– Wuerttemberg und janssen
Bossert, Helmuth Theodor *see* Altkreta
Bossert, Theodor Adolf *see* Friedrich heinrich jacobi und die fruehromantik
Bosshart, Jakob *see* Erzaehlungen
Bossi, Cesare *see*
– Bacchus & ariadne
– Les deux jumelles, ou, la meprise
– Favorite divertissement of le marchand de smyrne
– Hylas et temire
Bossi-Fedrigotti, Anton, Graf *see*
– Standschuetze bruggler
– Die tiroler kaiserjaeger am col di lana
Bossler, H Phil Karl *see* Musikalische realzeitung
Bossler, Heinrich Philipp *see* Bibliothek der grazien
Bosso, John A *see* Journal of infectious disease pharmacotherapy
Bossu, Jean *see* Nouveaux voyages aux indes occidentales
Bossu, N *see* Travels through that part of north america formerly called louisiana...
Bossuet and his contemporaries / Lear, H L Sidney – London: Rivingtons, 1877 – 2mf – 9 – 0-524-01884-7 – mf#1990-0511 – us ATLA [240]
Bossuet et la protestantisme : etude historique / Crousle, Leon – Paris: H Champion, 1901 – 1mf – 9 – 0-7905-7215-X – mf#1988-3215 – us ATLA [242]
Bossuet et les protestants / Julien, Eugene-Louis – Paris: G. Beauchesne, 1910 – 1mf – 9 – 0-7905-6235-9 – mf#1988-2235 – us ATLA [242]
Bossuet, Jacques Benigne *see*
– Devotion to the blessed virgin
– History of the variations of the protestant churches
– Oraisons fun ebres de bossuet
Bossuet, J-B *see*
– Conference avec m claude...sur la matiere de l'eglise
– Exposition de la doctrine de l'eglise catholique sur les matieres de controverse
Bossuit, F van *see* Beeld-snyders kunstkabinet...
Les bossus de quebec : bonne farce en trois petits actes / Sockeel, A – Paris: R Haton, [18–?] [mf ed 1984] – 1mf – 9 – 0-665-45089-3 – mf#45089 – cn Canadiana [820]
Bostock, John Knight *see* A handbook on old high german literature
Boston – Philadelphia PA 1973+ – 1,5,9 – ISSN: 0006-7989 – mf#8373 – us UMI ProQuest [071]
Boston 1630-1849 – Oxford MA (mf ed 1985) – (= ser Massachusetts vital records) – 104v on 540mf – 9 – 0-931248-76-0 – (mf1-3: births/deaths 1630-90. mf4-5: death index 1630-90. mf6-8: printed vitals 1630-99. mf9-23: county vitals 1630-66. mf24-32: vital records 1693-1820. mf33-43: typed births1630-1799. mf44-53: birth index 1630-1799. mf54-61: births 1635-1744. mf62-63: births 1810-49. mf64-69: birth index 1800-49. mf70-77: marriages 1646-1800. mf78-88: typed marriages 1646-1799. mf89-101: marr index 1646-1799. mf102-104: marriages 1651-62. mf102-107: marriages 1689-1720. mf108-109: marriages 1699-1751. mf110-111: marriages 1716-31. mf112-118: marriages 1720-1808. mf119-126: type marr 1720-1808. mf127-130: print marr 1700-51. mf131-134: marriages 1738-86. mf135-139: out-town marriages to 1800. mf140-149: marriages 1761-1809. mf150-158: marriages 1807-28. mf159-166: marriages 1825-40. mf167-174: marriages 1841-49. mf175-243: marriage index 1800-49. mf244-327: intentions 1707-1849. mf328-360: intentions index 1707-1849. mf361: deaths before 1700. mf362-363: deaths 1689-1720. mf364-377: death index 1700-1800. mf378-396: deaths 1800-24. mf397-402: death index 1810-24. mf403-422: deaths 1821-32. mf423-429: death index 1825-35. mf430-459: deaths 1833-48. mf460-466: death index 1835-46. mf467-486: death index 1810-48. mf487-540: death index 1801-48) – us Archive [978]
The boston advance – Boston, MA: Advance Pub Co, 1896-1907// – (= ser Negro Newspapers on Microfilm) – 1r – 1 – us L of C Photodup [071]
Boston advocate – Boston, Mass. 1905-52 – 1 – us AJPC [071]
Boston almanac and business directory – 1836-81 – 9 – $489.00 – mf#0112 – us Brook [978]
Boston bar journal – v1-45. 1957-2001 – 9 – $722.00 set – ISSN: 0524-1111 – mf#100971 – us Hein [340]
Boston beginnings 1630-1699 – Oxford MA (mf ed 1980) – 320p on 1mf – 9 – $6.00 – 0-931248-05-1 – (over 16,000 listings, arranged alphabetically, associate names with records of church, estate, indenture, land...) – us Archive [978]
Boston births 1849-1881 – Oxford MA (mf ed 1987) – (= ser Massachusetts vital records) – 41v on 234mf – 9 – 0-931248-77-9 – (mf1-177 1849-81. mf178-79 1852-69. mf180-196: corr & additions to birth records 1870-81. index 1849-69: mf197-203: vol a-g. mf 204-209: vol h-m. mf210-215: vol n-z. index 1870-81: mf216-222: vol a-g. mf223-228: vol h-m. mf229-234: vol n-z) – us Archive [978]
Boston births 1882-1895 – Oxford MA (mf ed 1988) – (= ser Massachusetts vital records) – 51v on 325mf – 9 – 0-931248-90-6 – (mf1-267: 1882-95. index 1882-91: mf268-313 a-z. mf314-325: index 1892-95) – us Archive [978]
Boston business journal – Charlotte NC 1989-98 – 1 – ISSN: 0746-4975 – mf#15204 – us UMI ProQuest [650]
Boston Chamber of Commerce. Bureau of Commercial and Industrial Affairs *see* Budgetary control for business
The boston chronicle – Boston. Mass. aug. 31, Dec. 21, 1940 – 1 – us NY Public [071]
The boston collection of instrumental music – Containing marches, quicksteps, waltzes, airs, cotillions, contra-dances, hornpipes, quadrilles, arranged with figures, Scotch and Irish jigs, reels, and strathspeys, arranged for brass, wooden & stringed instruments. Boston: O. Ditson 1850?. The music is in 2, 3 and 4 parts. MUSIC 1977 – 1 – us L of C Photodup [780]
Boston college environmental affairs law review – v1-27. 1971-2000 – 9 – $438.00 set – (title varies: v1-6 1971-78 as: environmental affairs) – ISSN: 0190-7034 – mf#100981 – us Hein [344]
Boston college industrial and commercial law review *see* Boston college law review
Boston college international and comparative law journal *see* Boston college international and comparative law review
Boston college international and comparative law review – v1-24. 1977-2 – 5,6,9 – $304.00 set – (v1-7 1977-84 in reel $80.00. v8-24 1985-2 in mf $224.00; title varies: v1-2 1977-79 as boston college international and comparative law journal) – ISSN: 0277-5778 – mf#101 – us Hein [341]
Boston college law review – v1-41. 1959-2000 – 1,5,6,9 – $1255.00 – (v1-36 1959-95 in reel or mf $1138. v37-41 1995-2000 in mf $117; title varies: v1-18 (1959-77) as boston college industrial and commercial law review) – ISSN: 0161-6587 – mf#100991 – us Hein [340]
Boston college. law school. boston college environmental affairs law review – Newton Centre MA 1979+ – 1,5,9 – (cont: environmental affairs) – ISSN: 0190-7034 – mf#10215.01 – us UMI ProQuest [333]
Boston college third world law journal – v1-21. 1980-2001 – 5,6,9 – $269.00 set – (v1-5 1980-85 in reel $60. v6-21 1986-2001 in mf $209) – ISSN: 0276-3583 – mf#102321 – us Hein [340]
Boston commonwealth – Killen TX 1862-96 – 1 – mf#3115 – us UMI ProQuest [071]
[Boston-] computerworld – MA. 1970-1980 – 24r – 1 – $1440.00 – mf#R04376 – us Library Micro [000]
The boston courant – Boston, MA: Courant Pub Co, 1890 (mf ed 1947) – (= ser Negro Newspapers on Microfilm) – 1r – 1 – us L of C Photodup [071]
Boston daily advertiser – Boston MA 3 mar 1813-30 dec 1820 – 1 – uk British Libr Newspaper [071]
Boston daily journal – Boston MA. 1845 jan 1-1847 dec 31 – 1r – 1 – (cont: evening mercantile journal [daily]; cont by: boston evening journal) – mf#780624 – us WHS [071]
Boston daily law bulletin – v1-2 n38. 1876-78 – 9 – (v2 called: boston daily law reporter. lacking: v1) – mf#LLMC 84-427 – us LLMC [340]
Boston deaths 1849-1890 – Oxford MA (mf ed 1987) – (= ser Massachusetts vital records) – 70v on 369mf – 9 – 0-931248-80-9 – (mf1-274: 1849-90. index 1849-69: mf275-280: vol a-g. mf281-285: vol n-z. index 1870-81: mf291-297: vol a-g-. mf298-302: vol h-m. mf303-307: vol n-z. index: mf308-325: 1882-90. mf326: index of removals 1823-59. mf327: deaths out of city 1850-54. mf328-329: deaths out of city 1853-98. mf3320-331: out-of-town index 1853-1901. mf332-338: record of deaths

1875. mf339-342: record of deaths 1876. mf343-348: record of deaths 1877. mf349-355: death index 1875-77. mf356-357: record of deaths 1878. mf358-359: death index 1878. mf360-363: stillborns 1854-81. mf364-365: stillborns 1882-96. mf366-369: index of still births 1875-81) – us Archive [978]

Le boston, double et triple boston : pour apprendre ou se perfectionner / Peter's, A – Paris: Editions Nilsson, [192-?] – 1 – mf#*ZB-56 – Located: NYPL – us Misc Inst [790]

Boston Draft Resistance Group *see* Newsletter

Boston evening transcript – Boston, MA: H W Dutton 1872-1941 – us CRL [071]

Boston evening transcript – Boston MA 2 jan1854-31 dec 1879, 8 feb 1900 – 1 – (cont: daily evening transcript [1 jan 1848-31 dec 1853]; filmed by wisconsin hs [1941 jan, feb, mar, apr]) – uk British Libr Newspaper; us WHS [071]

Boston firefighters digest / International Association of Fire Fighters – 1978 jan 2-1987 dec, 1988 jan-1994 nov 12 – 2r – 1 – mf#1322374 – us WHS [071]

Boston first baptist church. boston, massachusetts : church records – Orig. manuscript Minutes, 1665-1797) – 1 – 8.10 – us Southern Baptist [242]

Boston free press – 1st ed. [1968 may], n9 [1969] – 1r – 1 – mf#1582921 – us WHS [071]

Boston gazette, and country journal – Boston MA 12 apr 1756-30 dec 1793 – 1 – (cont: boston gazette, or country journal [7 apr 1755-5 apr 1756]; cont by: boston gazette, & republican weekly journal [6 jan 1794-17 sep 1798]) – uk British Libr Newspaper [071]

Boston gazette, commercial and political – Boston MA 9 oct 1800-28 dec 1815 – 1 – (cont: j russell's gazette, commercial & political [7 jun 1798-6 oct 1800]; cont by: boston commercial gazette [1 jan 1816-28 dec 1820]) – uk British Libr Newspaper [071]

Boston gazette [main ed] – [East Midlands] Boston 7 jan-31 mar 1860 [mf ed 2003] – 1 – (missing: 1861, 1870; cont by: boston gazette & lincolnshire commercial advertiser [7 apr-4 aug 1860]; boston gazette [11 aug 1860-21 oct 1891]; boston gazette & north holland advertiser [30 oct 1891-29 sep 1893]) – uk Newsplan [072]

Boston hebrew observer – Boston. Mass. 1883-86 – 1 – us AJPC [071]

Boston herald – Boston MA 1 may 1848-29 sep 1879 – 1 – uk British Libr Newspaper [071]

Boston historical edition of the jewish daily news – New York, NY. 1915 – 1 – us AJPC [071]

Boston Indian Council *see*
- Circle
- Knowledge of the circle
- News and views

Boston irish news – Boston 1981-90 – 4r – 1 – ie National [071]

Boston jewish times – Boston, MA. 1983-86 – 1 – us AJPC [071]

Boston journal of philosophy and the arts – La Salle IL 1823-26 – 1 – mf#3681 – us UMI ProQuest [190]

Boston law school magazine – 1v. 1896-97 (all publ) – 1mf – 9 – $1.50 – mf#LLMC 82-909 – us LLMC [340]

Boston literary magazine – La Salle IL 1832-33 – 1 – mf#3945 – us UMI ProQuest [420]

Boston lyceum – La Salle IL 1827 – 1 – mf#3946 – us UMI ProQuest [920]

Boston, Lyon *see* The u.s. as a creditor of insolvent debtors

Boston, MA *see*
- Christian endeavour world

Boston magazine – La Salle IL 1802-06 – 1 – mf#3561 – us UMI ProQuest [500]

Boston magazine, containing a collection of instructive and entertaining essays – La Salle IL 1783-86 – 1 – mf#3515 – us UMI ProQuest [071]

Boston marriages 1849-1890 – Oxford MA (mf ed 1986) – (= ser Massachusetts vital records) – 57v on 310mf – 9 – 0-931248-78-7 – (mf1-2: 1849. mf3-7: 1850. mf8-12: 1851. mf13-17: 1852. mf18-22: 1853. mf23-28: 1854. mf29-33: 1855. mf34-37: 1856. mf38-41: 1857. mf42-45: 1858. mf46-49: 1859. mf50-53: 1860. mf54-57: 1861 8-61: 1862. mf62-65: 1863. mf66-70: 1864. mf71-75: 1865. mf76-80: 1866. mf81-85: 1867. mf86-91: 1868. mf92-97: 1869. mf98-103: 1870. mf104-110: 1871. mf111-116: 1872. mf117-123: 1873. mf124-130: 1874. mf131-136: 1875. mf137-142: 1876. mf143-147: 1877. mf148-153: 1878. mf154-159: 1879. mf160-166: 1880. mf167-173: 1881. mf174-181: 1882. mf182-189: 1883. mf190-197: 1884. mf198-205: 1885. mf206-213: 1886. mf214-222: 1887. mf223-231: 1888. mf232-240: 1889. mf241-249: 1890. mf250-271: 1849-69. mf272-289: 1870-81. mf290-291: 1882. mf292-293: 1883. mf294-295: 1884. mf296-297: 1885. mf298-299: 1886. mf300-301: 1887. mf302-304: 1888. mf305-307: 1889. mf308-310: 1890) – us Archive [978]

Boston masonic mirror – La Salle IL 1824-34 – 1 – mf#4422 – us UMI ProQuest [366]

Boston (Mass). Registry Dept *see* Annual reports of the record commissioners of boston, 1876-1909

Boston massachusetts births 1700-1800 – Boston: Boston Record Commissioners 24th Report, 1894 – 379p on 1mf – 9 – $6.00 – us Archive [978]

Boston massachusetts births, baptisms, marriages, deaths 1630-1699 – Boston: Boston Record Commissioners 9th Report, 1900 – 281p on 1mf – 9 – $6.00 – us Archive [978]

Boston, Massachusetts. First Baptist Church *see* Records

Boston massachusetts marriages 1700-1751 – Boston: Boston Record Commissioners 28th Report, 1898 – 468p on 2mf – 9 – $12.00 – us Archive [978]

Boston mechanic, and journal of the useful arts and sciences – La Salle IL 1832-36 – 1 – mf#3947 – us UMI ProQuest [621]

Boston medical intelligencer – La Salle IL 1823-28 – 1 – mf#3709 – us UMI ProQuest [610]

Boston Men's Center *see* Men sharing

Boston miscellany of literature and fashion – La Salle IL 1842-43 – 1 – mf#3948 – us UMI ProQuest [740]

Boston monthly magazine – La Salle IL 1825-26 – 1 – mf#4058 – us UMI ProQuest [978]

Boston musical gazette – La Salle IL 1838-39 – 1,5,9 – mf#3949 – us UMI ProQuest [780]

Boston musical gazette : a semi-monthly journal devoted to the science of music – Boston: Otis, Broaders & Co 1838-39 [mf ed 19–] – 2v on 1r / 5mf – 1,9 – mf#film 1067 / fiche 918 – us Sibley [780]

Boston musical review – La Salle IL 1845 – 1 – mf#3725 – us UMI ProQuest [780]

[Boston-] new boston review – MA. 1975-1985 – 3r – 1 – $180.00 – mf#R04377 – us Library Micro [071]

Boston news-letter and city record – La Salle IL 1825-26 – 1 – mf#4423 – us UMI ProQuest [071]

Boston overseers of the poor records, 1733-1925 – ca 15r – 1 – ca $1,725.00 – (guide sold separately $25 d3478.g) – mf#D3478 – us MA Hist [978]

Boston phoenix – 1972 jun 7/sep 26-1981 nov/1982 jan – 45r – 1 – mf#1110298 – us WHS [071]

Boston phoenix – Boston MA 1973+ – 1 – ISSN: 0163-3015 – mf#8350 – us UMI ProQuest [917]

Boston post – Boston MA. 1857 jan 8,14, apr 8, may 14,30 [suppl], jun 29, jul 23, dec 4, 1867 aug 26 – 1r – 1 – (cont: boston morning post) – mf#882553 – us WHS [071]

Boston. Presbytery *see* Minutes

Boston press writer – v1 n12 [1903 may] – 1r – 1 – mf#1053511 – us WHS [071]

Boston quarterly review – La Salle IL 1838-42 – 1 – mf#3950 – us UMI ProQuest [190]

Boston review – Cambridge MA 1975+ – 1,5,9 – (cont: new boston review) – ISSN: 0734-2306 – mf#11619.01 – us UMI ProQuest [400]

Boston satirist : or, weekly museum – La Salle IL 1812 – 1 – mf#4083 – us UMI ProQuest [740]

The boston series, 1941-1945 (intelligence files, office of the director, oss) / U.S. Office of Strategic Services – (= ser Records Of The Office Of Strategic Services) – 3r – 1 – mf#M1740 – us Nat Archives [327]

Boston spectator : devoted to politicks and belles-lettres – La Salle IL 1814-15 – 1 – mf#3682 – us UMI ProQuest [071]

Boston standard – Boston, England 31 dec 1987-4 may 1989 – 1 – (cont: lincolnshire standard [10 feb 1984-24 dec 1987]; cont by: lincolnshire standard (boston) 11 may 1989-14 may 1992) – uk British Libr Newspaper [072]

Boston symphony orchestra. program book/ notes – Boston MA 1975+ – 1,5,9 – ISSN: 0006-8020 – mf#8591 – us UMI ProQuest [780]

Boston theological institute (b.t.i.) newsletter – 1968-88 – 2r – 1 – (lacks some iss) – mf#ATLA S0411 – us ATLA [200]

Boston Theological Institute et al *see* Placement news

Boston, Thomas *see*
- Pedwar cyflwr dyn, sef, ei gyflwr o ddiniweidrwydd, yn ein rhieni cyntaf yn mharadwys
- A soliloquy on the art of man-fishing

The boston times – Boston. Mass. feb. 24, Oct. 26, 1944 – 1 – us NY Public [071]

Boston union teacher : bulletin of boston teachers union, affiliated with the american federation of teachers – 1976 mar-1982, 1983-87 – 2r – 1 – mf#675042 – us WHS [071]

Boston unitarianism, 1820-1850 : a study of the life and work of nathaniel langdon frothingham / Frothingham, Octavius Brooks – New York: G P Putnam 1890 [mf ed 1990] – 1mf – 9 – 0-7905-5879-3 – mf#1988-1879 – us ATLA [243]

Boston University *see* Newsletter

Boston University. African Studies Library *see* Assorted rhodesian and south african pamphlets

Boston university africana libraries newsletter – Boston: Boston University, African Studies Library. n17-33. jun 1978-nov 1982 – 1r – us CRL [020]

Boston university international law journal – v1-19. 1982-2001 – 9 – $380.00 set – ISSN: 0737-8947 – mf#109371 – us Hein [341]

Boston university journal – Boston MA 1952-80 – 1,5,9 – ISSN: 0006-8039 – mf#8527 – us UMI ProQuest [378]

Boston university journal of science and technology law – v1-7. 1995-2 – 9 – $115.00 set – mf#118821 – us Hein [346]

Boston university law review – v1-6. 1921-25 – 21mf – 9 – $31.50 – mf#llmc97-240 – us LLMC [340]

Boston university law review – v1-81. 1921-2001 – 1,5,6,9 – $1431.00 set – (v1-74 1921-94 in reel or mf $1208. v75-81 1995-2001 in mf $223.00) – ISSN: 0006-8047 – mf#101021 – us Hein [340]

Boston weekly magazine [1743] – La Salle IL 1743 – 1 – mf#3512 – us UMI ProQuest [420]

Boston weekly magazine [1838] – La Salle IL 1838-41 – 1 – mf#3761 – us UMI ProQuest [390]

Boston weekly magazine, devoted to polite literature, useful science, biography, and dramatic criticism – La Salle IL 1816-24 – 1 – mf#3683 – us UMI ProQuest [071]

Boston weekly transcript – Boston, dec 1889-oct 1890 – us CRL [071]

Boston Women's Health Book Collective *see* Our bodies, ourselves

Bostonas latweeschu baptistu draudses darbibas pahrskats par 1940, gadu = Report of the bostonas latvian baptist church for 1940 – 1940. Publ. No. 6346,no. 6. One of sic items on reel. Total Pages 152 – 1 – us Southern Baptist [242]

Bostoner idishe shtimme = Boston jewish voice – Boston, MA. 1913-16 – 1 – us AJPC [071]

Boston-out-of-town marriages 1858-1895 – Oxford MA (mf ed 1988) – (= ser Massachusetts vital records) – 8v on 41mf – 9 – 0-931248-88-4 – (mf1-41: 1858-95. mf19-22: indexes 1857-83. mf32-36: indexes 1884-92. mf37: 1893. mf38: 1893-94. mf39: 1894-95. mf40-41: 1895) – us Archive [978]

Boston's awakening : a complete account of the great boston revival under the leadership of j wilbur chapman and charles m alexander, jan 26th to feb 21st 1909 / ed by Conrad, Arcturus Zodiac – Boston MA: King's Business Publ Co 1909 [mf ed 1986] – 1mf [ill] – 9 – 0-8370-6038-9 – mf#1986-0038 – us ATLA [242]

Bostrand, Torgerd *see* Diese deutschen

Bostroem, Christopher Jacob *see* Chr. jac. bostroems foerelaesningar i religionsfilosofi

Boswell, James [the Elder] *see* Etat de la corse

Boswell, Kasmin J *see* A review of the literature on self-efficacy and selected constructs

Bosworth baptist church. missouri : church records – 4Dec 1915-3 Jun 1964 – 1 – us Southern Baptist [242]

Bosworth, Edward Increase *see*
- Christ in everyday life
- New studies in acts
- Studies in the acts and epistles
- Studies in the life of jesus christ
- Studies in the teaching of jesus and his apostles

Bosworth, Francke Huntington *see* Study of architectural schools

Bosworth, Joseph *see* Compendious anglo-saxon and english dictionary

Bosworth, Mary C *see*
- Capitol building of tallahassee, florida
- Mayo, florida

Bosworth, Newton *see* Destruction of the last enemy considered and a tribute

The bosworth psalter : an account of a manuscript formerly belonging to o. turville-petre esq. of bosworth hall now addit. ms. 37517 at the british museum / Gasquet, Francis Aidan – London: George Bell, 1908 – 1mf – 9 – 0-7905-0082-5 – (incl bibl ref and index) – mf#1987-0082 – us ATLA [220]

Bot, S Pierre Njovk *see* Studies on basa customs

Bota-fogo – Desterro, SC: Typ Desterrense de Jose Joaquim Lopes, 24 out-12 dez 1858 – (= ser Ps 19) – bl Biblioteca [079]

Botana, Helvio Ildefonso *see* Vina y el grano

Botanica e agricultura no brasil no seculo 16 / Hoehne, Frederico Carlos – Sao Paulo, Brazil. 1937 – 1r – us UF Libraries [580]

Botanical and physiological memoirs / Henfrey, A – 1853 – (= ser Ray society publications 24) – 1r 14mf – 1,7 – mf#9/85976 – uk Microform Academic [580]

Botanical bulletin of academia sinica – Taipei, Taiwan 1973+ – 1,5,9 – ISSN: 0006-8063 – mf#8534.01 – us UMI ProQuest [580]

Botanical expedition to oregon / Murray, A – New York. 1974-1979 (1) 1974-1979 (5) 1974-1979 (9) – 1mf – 9 – mf#11185 – ne IDC [917]

Botanical gazette – Chicago IL 1875-1991 – 1,5,9 – (cont by: international journal of plant sciences) – ISSN: 0006-8071 – mf#135.01 – us UMI ProQuest [580]

Botanical review – Bronx NY 1949+ – 1,5,9 – ISSN: 0006-8101 – mf#314 – us UMI ProQuest [580]

Botanicheskii zhurnal / Akademiia Nauk. SSSR – Moskva, Leningrad: Izd-vo Akademii nauk SSSR. v35-37 n2 1950-mar/apr 1952. v60 n8-9 aug/sep 1975 – us CRL [580]

Botanische ergebnisse der schwedischen expedition nach patagonien und dem feuerlande 1907-1909 : pt 10: les mousses / Cardot, J & Brotherus, V F – Stockholm, 1923. v63 – 4mf – 9 – mf#H-542 – ne IDC [919]

Botanische ergebnisse der schwedischen expedition nach patagonien und dem feuerlande 1907-1909 : pt 6: die flechten / Zahlbruckner, A – Stockholm, 1917. v57 – 3mf – 9 – mf#H-542 – ne IDC [919]

Botanische ergebnisse der schwedischen expedition nach patagonien und dem feuerlande 1907-1909 / Skottsberg, C – Stockholm. v46, 50, 51, 56, 61, 63 – 30mf – 9 – mf#H-542 – ne IDC [919]

Botanische ergebnisse einer reise durch das oestliche transkaukasien und der aderbeidshan : ausgefuehrt in den jahren 1855 und 1856 / Seidlitz, N von – Dorpat: Gedruckt bei Schoenmann's Wwe & C Mattiesen, 1857 – 2mf – 9 – mf#BT-323 – ne IDC [915]

Botanische reisen in den hochgebirgen chinas und ost-tibets / Limpricht, H W – Dahlem bei Berlin, 1922. v12 – 10mf – 9 – mf#746 – ne IDC [915]

Botanische reisen in deutsch-suedwest-afrika / Dinter, K – Posen, 1918 [Dahlem bei Berlin, 1921]. v3 – 3mf – 9 – mf#746 – ne IDC [916]

Die botanischen ergebnisse der reise seiner koenigl hoheit des prinzen waldemar von preussen in den jahren 1845 und 1846 / Klotzsch, J F & Garcke, A – New York. 1973-1991 (1) 1974-1991 (5) 1974-1991 (9) – 7mf – 9 – mf#8390 – ne IDC [910]

Botanisk tidsskrift – Copenhagen, Denmark 1978 – 1,5,9 – ISSN: 0006-8187 – mf#9145 – us UMI ProQuest [580]

Botano-theology : an arranged compendium, chiefly from smith, keith, and thomson / Smith, James Edward – Oxford, 1825 – (= ser 19th c evolution & creation) – 2mf – 9 – mf#1.1.3059 – uk Chadwyck [580]

Botany. cryptogamia. filices : united states exploring expedition. during the years 1838-1842 under the command of charles wilkes / Brackenridge, W D – Philadelphia, 1854. v16 – 14mf – 9 – mf#6514 – ne IDC [910]

The botany of captain beechey's voyage : ...during the voyage to the pacific and beering's strait...in the years 1825-1828 / Hooker, W J & Arnott, G A W – London: Heny G Bohn, [1830-]1841 – 14mf – 9 – mf#5264 – ne IDC [910]

The botany of the antartic voyage of hm discovery ships erebus and terror in the years 1839-1843 : under the command of captain sir james clark ross / Hooker, J D – London: Reeve, 1844-1860 – 80mf – 9 – mf#457 – ne IDC [580]

The botany of the eastern coast of lake huron / Gibson, John & Macoun, John – S.l: s.n, 1876? – 1mf – 9 – mf#35147 – cn Canadiana [574]

The botany of the speke and grant expedition : an enumeration of the plants collected during the journey...from zanzibar to egypt / Grant, J A & Oliver, D – London. v29. 1875 – 12mf – 9 – mf#225 – ne IDC [580]

The botany of the voyage of hms herald : under the command of captain henry kellet... during the years 1845-1851 / Seemann, B C – Little Rock. 1957+ (1) 1970+ (5) 1977+ (9) – 22mf – 9 – mf#5739 – ne IDC [580]

Botany of the voyage of hms sulphur [the] : under the command of captain sir edward belcher...during the years 1836-1842 / Bentham, G – London: Smith, Elder & Co 1844[-1846] – 6pt on 10mf – 9 – mf#5375/2 – ne IDC [580]

Botany. phanerogamia : united states exploring expedition. during the years 1838-1842 under the command of charles wilkes / Gray, A – Indianapolis. 1947+ (1) 1970-1995 (5) 1975-1995 (9) – 37mf – 9 – mf#5938 – ne IDC [910]

Botataung – Rangoon, Burma. 1962; 1964; 1973-88 – 45r – 1 – us L of C Photodup [079]

Botchan (master darling) / Natsume, Soseki – Tokyo, Japan. 1918 – 1r – us UF Libraries [960]
Botchway, Samuel Asare *see* Towards people's participation and rural development
Der bote : ein mennonitisches familien- und gemeindeblatt – Saskatoon / Saskatchewan (CDN), 1972-76, 1977 [gaps] – 1 – gw Misc Inst [242]
Der bote – Winnipeg, Manitoba (CDN), 1971/72-1993 26 aug – 1 – gw Misc Inst [071]
Der bote *see* Mittheilungen von und fuer dippoldiswalde und umgegend
Bote an der inde – Eschweiler DE, 1957 2 nov-1959 30 jun [gaps] – 1 – (regional ed of aachener volkszeitung) – gw Misc Inst [074]
Bote an der weser – Minden/Westf DE, 1951-1953 31 oct, 1953 15 apr-1956 31 mar – 13r – 1 – (filmed by misc inst: 1956 1 oct-1959 [only local sect]) – gw Mikrofilm; gw Misc Inst [074]
Der bote aus dem riesengebirge / kriegsausgabe – Hirschberg (Jelenia Gora PL), 1914 8 aug-16 nov – 1r – 1 – gw Misc Inst [077]
Der bote aus dem riesengebuerge – Hirschberg [Jelenia Gora PL], 1818 & 1820, 1825-29, 1831, 1836-37, 1839-40, 1845 & 1847, 1865 jul-dec, 1929 jan-mar & jul-sep – 1 – (title varies: 22 apr 1813: der bote aus dem riesengebirge, 1 jul 1933: beobachter im iser- und riesengebirge; filmed by other misc inst: 1812 20 aug-1818, 1824, 1835-38, 1840-43, 1847 [12r]; war ed: 1914 8 aug-16 nov [11r]) – gw Misc Inst [077]
Der bote aus den sechs aemtern *see* Wochenblatt fuer den markt redwitz und umgegend
Der bote aus den vogesen – Annweiler DE, 1849 4 apr-29 dec [gaps] – 1r – 1 – gw Misc Inst [074]
Bote aus der heimat – Strasbourg, France. 1893-99 – 1 – fr ACRPP [074]
Der bote aus der heimat – Strassburg [Strasbourg F], 1893 7 jan-1899 24 sep – 1r – 1 – (title varies: 18 oct 1894: die heimat) – fr ACRPP; gw Misc Inst [074]
Der bote aus kurpfalz – Hockenheim DE, 1916-1933 12 mar – 17r – 1 – (title varies: 1919: "christliches volk" [publ in heidelberg], 1920: christliches volksblatt, 1925: "sonntagsblatt des arbeitenden volkes" [publ in karlsruhe], 1931: der religioese sozialist [publ in mannheim]) – gw Misc Inst [074]
Der bote aus thueringen / ed by Salzmann, Christian Gotth – Schnepfenthal 1788-1816 – (= ser Dz) – 102mf – 9 – €612.00 – mf#k/n5783 – gw Olms [074]
Der bote aus thueringen – Schnepfenthal (Waltershausen), 1791-93 [gaps] – 2r – 1 – gw Misc Inst [074]
Bote fuer stadt und land – Kaiserslautern DE, 1848-1849 28 jun – 2r – 1 – gw Misc Inst [074]
Bote, Hermann *see* Der koeker
Der bote vom allgaeu *see* Anzeiger von wurzach
Der bote vom geising und mueglitzthal-zeitung – Altenberg DE, 1845-1848 28 sep, 1866-1933 – 72r – 1 – (incl suppl: rund um den geisingberg, 1923-jan 1945 [1r]) – gw Misc Inst [074]
Bote vom muenstertal – Munster, France.1877-1914 – 1 – fr ACRPP [074]
Der bote vom muensterthal – Muenster, Elsass (Munster F), 1877 20 jul-1914 15 aug [gaps] – 1 – (fuer die cantone muenster & winzenheim) – fr ACRPP [074]
Der bote vom neckar und rhein – Heidelberg DE, 1822 1 jan-29 jun – 1r – 1 – gw Misc Inst [074]
Der bote vom niederrhein – Duisburg DE, oct 1 1865-jun 29 1866 – 1r – 1 – gw Misc Inst [074]
Der bote vom remtshale *see* Gemeinnuetziges wochenblatt
Bote vom unter-main – Miltenberg DE, 1983 1 jun- – ca 6r/yr – 1 – gw Misc Inst [074]
Der bote von aalen – Aalen DE, 1848 5 jan-1849 28 dec – 1r – 1 – gw Misc Inst [074]
Bote von der lahn – Marburg DE, 1853 2 jul-1854 29 mar – 1r – 1 – gw Misc Inst [074]
Boteler, Thomas *see* Narrative of a voyage of discovery to africa and arabia
Botelho De Magalhaes, Amilcar Armando *see*
– Impressoes da commissao rondon
– Pelos sertoes do brasil
Botello del Castillo, Carlos *see*
– Aritmetica para los alumnos
– Compendio de aritmetica..
– Compendio de geometria y trigonometria..
– Oracion...en el instituto provincial de badajoz..
Botenhagen, Kim A *see* Comparison of skinfold measurements under normally hydrated and dehydrated conditions in females ages to 54
Botero Isaza, Valerio *see* Regimen legal de aguas en colombia
Botero M, Jose Manuel *see*
– Geografia fisica y de la republica de colombia
Botero Y Botero, Ruben *see* Libro de oro de salamina
Botes, Petrus Johannes *see* Die suid-afrikaanse individuele skaal vir blindes
Both by land and by sea / Simons, Robert Bentham – [mf ed Spartanburg SC: Reprint Co [1981?]] – 2mf – 9 – (incl ind) – mf#51-154 – us South Carolina Historical [355]
Both sides now / Ohio Civil Service Employees Association – 1969 nov 29-1977 apr – 1r – 1 – mf#203638 – us WHS [366]
Both sides now / Ohio Civil Service Employees Association – v1 n8-v5 n6 [1975 nov-1979 oct] – 1r – 1 – mf#568777 – us WHS [331]
Both sides of the controversy between the roman and reformed churches : being 1. "a doctrinal catechism," etc... / Hughes, John – New York: Delisser & Proctor 1859 [mf ed 1986] – 2mf – 9 – 0-8370-9071-7 – mf#1986-3071 – us ATLA [230]
Botha, Andries Johannes *see* Die evolusie van 'n volksteologie
Botha, Christoffel Rudolph *see* Aspekte van die xhosakortverhaal
Botha, Colin Graham *see*
– Public archives of south africa, 1652-1910
– Social life in the cape colony with social customs in south africa
Botha, Daniel Jacobus Joubert *see* Urban taxation and land use
Botha, Jan Francois *see* Verwoerd is dead
Botha, Magda *see* Some general measures of repairable stochastic systems
Botha, Marika G *see* The influence of the home environment on the motor performance of preschool children
Botha, Marthinus Christoffel *see* Maskew miller's grammar of afrikaans
Botha, Matthys Izak *see* South africa's answer to un 'group of experts'
Botha, Philip Rudolph *see* Staatkundige ontwikkeling van die suid-afrikaanse republiek
Botha, smuts and south africa / Williams, Basil – London, England. 1946 – 1r – us UF Libraries [960]
Botha, Susanna Petronella Wilhelmina *see* Die gebruik van die opvoedkundig-sielkundige relasieteorie in die identifisering van'n middeljarekrisis
Bothma, C V *see* Ntshabeleng social structure
Bothne, Thrond *see* Kort udsigt over det lutherske kirkearbeide blandt nordmaendene i amerika
Bothner, Krisanne E *see* The development of a video-based motion analysis system
Bothner, Kristin E *see* Postural compensations to a disturbance of balance in humans
Boti y Barreiro, Regino Eladio et al *see* La lira cubana
Botlakhon rong 14 ong ruang namchai detdieo (ton thi 2) / Thotsiriwong, Prince – Phranakhon: s.n. 2423? [1980?] [mf ed 1994] – on pt of 1r – 1 – mf#11052 r1475 n1 – us Cornell [959]
Botly, William *see* Land tenure
O botocudo : jornal critico, litterario e recreativo – Rio de Janeiro, RJ: Typ Camoes, 01 jun, 01 ago 1887 – (= ser Ps 19) – mf#DIPER – bl Biblioteca [079]
Botschaft / Tuscarawas Co. Sugarcreek – v1 n1. jun 1975-jun 1976// [wkly] – 1r – 1 – (an amish-mennonite newspaper) – mf#B32393 – us Ohio Hist [071]
Botschafter – Schleisingerville WI. 1897 mar 27/1898 dec 31, 1899 jan-1902 dec, 1903 jan-1905 dec, 1906 jan-1909 dec, 1910 jan-1913 dec, 1914 jan-1917 aug 9 – 6r – 1 – mf#1093689 – us WHS [071]
Botschafter der wahrheit – v32-56. 1931-52 – (= ser Mennonite serials coll) – 2r – 1 – (lacks v35 n21, 23) – mf#ATLA 1993-S017 – us ATLA [242]
Botschafter des heils in christo *see* Die gemeinde unterm kreuz
Botschafter des heils in christo und zeichen dieser zeit – v1-3. 1889-91 – (= ser Mennonite serials coll) – 1r – 1 – (lacks v2 n1-2, 4, 6. v3 n6) – mf#ATLA 1994-S018 – us ATLA [242]
Botschko, R E *see* Jiskaur!
Botsford, Edmund *see* Memoirs
Botswana / Africa Institute Of South Africa – Pretoria, South Africa. 1968? – 1r – us UF Libraries [960]
Botswana / Great Britain Central Office Of Information Reference Division – New York, NY. 1966 – 1r – us UF Libraries [960]
Botswana : 'n studie in internasionele betrekkinge / Wolvaardt, Pieter Jacobus – 1968 – us CRL [960]
Botswana / Smit, Philippus – Pretoria, South Africa. 1970 – 1r – us UF Libraries [960]
Botswana *see*
– National development plan, 1968-73
– National development plan, 1970-75
– Statement on the luke report on localisation and training
Botswana. Central Statistics Office *see* Statistical abstract 1966-1976
Botswana daily news – [Gaborone, Botswana]: Botswana information service. oct 3 1966-may 11 1973; may 14 1973-sep 16 1975 – us CRL [079]
Botswana Information Services *see* This is botswana
Bott, Elisabeth *see* Ernst fries (1801-1833)
Botta, Carlo *see* Histoire de la guerre de l'independance des etats-unis d'amerique
Bottalla, Paul *see*
– The papacy and schism
– Pope honorius before the tribunal of reason and history
Bottari, G *see* Vite de' piu eccellenti pittori scultori e architetti
Bottari, G G *see* Raccolta di lettere sulla pittura, scultura ed architettura...
[Bottari, G G] *see* Raccolta di lettere sulla pittura scultura ed architettura...
Bottari, M *see* Il museo capitoli...
La botte de paille : suivie de le chapelet et la sentinelle, la cravate teinte de sang / Collin de Plancy, Jacques-Albin-Simon – Montreal: Librairie Saint-Joseph, Cadieux & Derome, [entre 1880 et 1910] – 1mf – 9 – 0-665-93513-7 – mf#93513 – cn Canadiana [390]
Bottego, Vittorio *see*
– L'esplorazione del guiba
– Il guiba esplorato
– Il guiba esporato
Bottens, Fulg *see*
– Het goddelyck herte
– Judicium pacifici salomonis christi domini nostri
Bottermann, Walther *see*
– Die beziehungen des dramatikers achim von arnim zur altdeutschen literatur
– Die beziehungen des dramatikers achim von arnim zur altdeutschen litteratur
Bottesch, Jessica M *see* The effects of magnetic therapy on physiological strength
Botticher, Gotthold *see* Hildebrandlied und waltharilied
Bottineau county herald – Bottineau, ND: Richard Costello. v28 n45 mar 10 1927-v31 n13 jul 25 1929 (wkly) – 1 – (cont: farmers advocate (bottineau, nd). merged with: bottineau courant (1895) to form: bottineau courant and bottineau county herald. missing: 1927 apr 7, may 19; 1929 jun 20-jul 4) – mf#06994-06995; 01631; 06995 – us North Dakota [071]
Bottineau county news : [official paper of bottineau county 1903-1909] – Bottineau Co, ND: F C Falkenstein. v5 n24 nov 12 1903-v19 n38 feb 1 1918 (wkly) – 1 – (cont: bottineau news. merged with: omemee herald to form: bottineau county news and omemee herald) – mf#06988-06992; 01624-01627 – us North Dakota [071]
Bottineau county news and omemee herald – [Bottineau, ND]: News Pub Co. v19 n39 feb 8 1918-v19 n46 mar 29 1918 (wkly) – 1 – (formed by the union of: bottineau county news and omemee herald. cont by: farmers advocate (bottineau nd).) – mf#01627; 06992 – us North Dakota [071]
Bottineau courant (1895) : [official paper for the city and county 1906-1929] – Bottineau, ND: J E Britton. v1 n1 may 4 1895-v18 n47 jul 25 1929 (wkly) – 1 – (carbury news, with a numbered masthead, publ as back page: v1 n1 dec 2 1915-v2 n19 apr 12 1917. numbering ceased and pg cont until jan 15 1920. cont: bottineau pioneer. absorbed: bottineau free lance. merged with: bottineau county herald to form: bottineau courant and bottineau county herald) – mf#06983++ – us North Dakota [071]
Bottineau courant (1969) : [official newspaper of bottineau county and city of bottineau 1969-1981] – Bottineau, ND: Bottineau Co-Operative Pub Co. v84 n5 dec 24 1969-v96 n51 dec 9 1981 (wkly) – 1 – (cont: bottineau courant and bottineau county herald. cont by: courant (bottineau, nd)) – mf#08593-08605 – us North Dakota [071]
The bottineau courant and bottineau county herald – us North Dakota [071]
Bottineau news : [county official paper 1903] – Bottineau, ND: F C Falkenstein, jun 9 1899; -v5 n23 nov 5 1903 (wkly) – 1 – (cont by: bottineau county news) – mf#06988 – us North Dakota [071]
Bottineau pioneer : [official paper of bottineau county 1886-april 1893] – Bottineau, ND: V B Nobel, 1885; -v10 n34 apr 27 1895 (wkly) – 1 – (cont by: bottineau courant (bottineau, nd: 1895)) – mf#06982-06983 – us North Dakota [071]
Bottisham 1561-1950 – (= ser Cambridgeshire parish register transcript) – 12mf – 9 – £15.00 – uk CambsFHS [929]
Botto, Antonio *see*
– Motivos de belleza
– Sonetos
– A verdade e nada mais
Bottom line – Bradford UK 2001+ – 1,5,9 – ISSN: 0888-045X – mf#28997 – us UMI ProQuest [020]
Bottom line [1993] – Austin TX 1993+ – 1,5,9 – ISSN: 0279-1889 – mf#19084 – us UMI ProQuest [650]
Bottom line business – Stamford CT 1995+ – 1,5,9 – (cont: boardroom reports) – ISSN: 1082-457X – mf#10136.01 – us UMI ProQuest [650]
Bottom rot and related diseases of cabbage caused by corticium / Weber, George F – Gainesville, FL. 1931 – 1r – us UF Libraries [630]
Bottome, Margaret et al *see* Women in the church
Bottome, Willard B *see* The stenographic expert
Bottomline – Washington DC 1983-92 – 1,5,9 – ISSN: 0740-5464 – mf#14283 – us UMI ProQuest [332]
Bottrall, Margaret Smith *see* William blake
Bottrigari, Ercole *see* Il desiderio, overo, de' concerti di varii strumenti musicali
Bottroper volkszeitung – Bottrop DE, 1883-91 [1889 many gaps], 1893-94, 1957 2 oct-13 dec, 1958-60 14 nov [in pts only local pgs] – 1 – (title varies: 1 mar 1949: ruhr-nachrichten, 22 oct 1949: bottroper volkszeitung, 1 oct 1953: ruhr-nachrichten; fr 1 mar 1949 ba v. ruhr-nachrichten, dortmund) – gw Misc Inst [074]
Botwood, Edward *see* Address delivered in st mary's church, st john's, nf
Boubacar, Barry *see* Le royaume du walo du traite de ngio en 1819 a la conquete en 1855
Boubee, Jean-Pierre Simon *see*
– Etudes historiques et philosophiques sur la franc-maconnerie ancienne et moderne
– Misraim, ou, les francs-macons
Bouchage, Fr *see* Saint gregoire le grand
Bouchard, Romeo *see* Deux pretres en colere
Bouchard, T-D [comp] *see* Catalogue de la bibliotheque de la legislature de la province de quebec
Bouchardy, Joseph *see*
– Fils du bravo
– Sonneur de saint-paul
– Vendredi
Bouchaud, Joseph *see* Cote du cameroun dans l'histoire et la cartographie des origines...
La bouche d'acier au marais – [Paris]: Impr Dondey-Dupre, sep 2 1848 – us CRL [944]
La bouche de fer – [Paris]: J Frey, aug 24 1848 – us CRL [944]
La bouche de fer – Paris. n26-27. aout-sept 1893 – 1 – fr ACRPP [320]
La bouche de fer – 1re serie: cercle social. lettre ire 67. 1790 (t. 1-2) – 1 – (publ. parallelement a la bouche de fer. 2e-3e serie: la bouche de fer. oct 1790-juil 1791 (t. i-iv). suivi de: annales de la confederation universelle des amis de la verite. supplement a la bouche de fer. 1791 (t. v, n1-2). suivi de: bulletin de la bouche de fer. juil 1790. paris) – fr ACRPP [320]
Bouche, Pierre Bertrand *see*
– Le cote des esclaves et le dahomey
– Sept ans en afrique occidentale.
Bouche-Leclercq, Auguste *see*
– Histoire de la divination dans l'antiquite
– Histoire des selucides (323-64 avant j.-c.)
– L'intolerance religieuse et la politique
Boucher and pratte's musical journal – Montreal: [s.n.], 1881-1882 – 9 – (cont: le canada musical. incl some text in french) – mf#P04169 – cn Canadiana [780]
Boucher, Andre *see* A travers les missions du togo et du dahomey
Boucher de la Richarderie, G *see* Bibliotheque universelle des voyages
Boucher, Edouard *see* Eloquence de la chaire
Boucher, George P *see* A biomechanical analysis of the prolonged effects on functional parameters of a test seating system for moderately involved cerebral palsied children
Boucher, Louis *see* Bibliographie des quotidiens de langue francaise parus dans la province de quebec depuis 1867
Boucher, Pierre *see* Histoire veritable et naturelle des moeurs et productions du pays de la nouvelle-france, vulgairement dite le canada
Boucher, Pierre, sieur de Boucherville *see*
– Canada in the seventeenth century
– Histoire veritable et naturelle des moeurs et productions du pays de la nouvelle france
Boucher-Belleville, Jean Baptiste *see* Les principes de la langue francaise
Bouchereau, Madeleine G Sylvain *see*
– Education des femmes en haiti
– Haiti, portrat eines freien landes
Bouchereau, Paul *see* Audience memorable au tribunal de cassation
Boucheron, Maxime *see* Miss helyett
Boucherville, Georges Boucher de *see* Une de perdue, deux de trouvees
Boucherville, Georges de *see*
– Le credit foncier
– Une de perdue, deux de trouvees
Bouchette, Errol *see*
– Emparons-nous de l'industrie
– Etudes sociales et economiques sur le canada
– L'independance economique du canada francais
Bouchette, Errol [comp] *see* Memoires...1805-1840
Bouchette, Joseph *see*
– Analyse chronologique relative a la concession du 25 fevrier 1661
– The british dominions in north america.
– The british dominions in north america, vol 1
– The british dominions in north america, vol 2

- The british dominions in north america, vols 1-2
- Description topographique de la province du bas-canada
- To his royal highness george augustus frederick...this topographical map of the province of lower canada
- A topographical description of the province of lower canada
- A topographical dictionary of the province of lower canada

Bouchette, Robert Shore Milnes *see* Memoires de robert-s-m bouchette, 1805-1840
Bouchez, E *see* Francs-macons septembriseurs
Bouchholtz, Fritz [comp] *see* Elsaessische stammeskunde
Bouchilloux, Helene *see* Apologetique et raison dans les pensees de pascal
Bouchor, Maurice *see* Conte de noel
Bouchot, Henri François Xavier Marie *see* The printed book, its history, illustration, and adornment
Bouclier de la foi : ou defense de la confession de foi des eglises reformees du royaume de france / Moulin, P du – Ed 2. Charenton, 1619 – 10mf – 9 – mf#PRS-142 – ne IDC [240]
Bouda, K *see* Die beziehungen des sumerischen zum baskischen, westkaukasischen und tibetischen
Boudaa, Azzedine *see* Die auswirkungen des bilingualismus in algerien (franzoesisch/ arabisch) auf das erlernen des deutschen als fremdsprache
Boudard, J B *see* Iconologie tir
Le bouddhisme au cambodge a l'epoque du nokor phnom / Pang-Khath – [Phnom-Penh: Universite bouddhique Preah Sihanouk Raj 196-?] [mf ed 1989] – (= ser Serie de culture et civilisation khmeres 6) – 1r with other items – 1 – mf#mf-10289 seam reel 002/11 [§] – us CRL [280]
Bouddhisme chinois : extraits du tripitaka, des commentaires, tracts, etc / Wieger, Leon – [Sienhsien (Hokienfu)]: Impr. de la Mission Catholique, 1910-1913. Chicago: Dep of Photodup, U of Chicago Lib, 1971 (1r); Evanston: American Theol Lib Assoc, 1984 (1r) – 1 – 0-8370-0551-5 – (incl ind) – mf#1984-B296 – us ATLA [280]
Le bouddhisme contemporain / Roussel, Alfred – Paris: P Tequi, 1916 – (= ser Religions Orientales) – 2mf – 9 – 0-524-04868-1 – (incl bibl ref) – mf#1990-3430 – us ATLA [280]
Bouddhisme, etudes et materiaux : theorie des douze causes / Vallee Poussin, Louis de la – Gand: E van Goethem; Londres: Luzac, 1913 [mf ed 1995] – (= ser Yale coll) – 1 – 0-524-09699-6 – (in french) – mf#1995-0699 – us ATLA [280]
Le bouddhisme japonais : doctrines et histoire des douze grandes sectes bouddhiques du japon / Fujishima, Ryauon – Paris: Maisonneuve et Ch Leclerc, 1889 – 1mf – 9 – 0-524-01356-X – mf#1990-2368 – us ATLA [280]
Le bouddhisme primitif / Roussel, Alfred – Paris: Pierre Tequi, 1911 – (= ser Religions Orientales) – 1mf – 9 – 0-524-02322-0 – (incl bibl ref) – mf#1990-2945 – us ATLA [280]
Le bouddhisme, son fondateur et ses ecritures / Neve, Felix – Paris: C Douniol: B Duprat, 1853 [mf ed 1991] – 1mf – 9 – 0-524-01070-6 – mf#1990-2218 – us ATLA [280]
Boudet, Marie *see* The art of dressmaking at home and in the workroom, vol 1
Boudinhon, Auguste *see*
- La nouvelle legislation de l'index
- Les proces de beatification et de canonisation

Boudinot, Elias *see*
- The man in a trance
- A star in the west

Boudot de Challaye *see* Petition presentee a l'assemble nationale constituante par m boudot-challaye
Boufflers, Stanislas-Jean, chevalier de *see* Journal inedit du second sejour au senegal (3 decembre 1786 – 25 decembre 1787)
Bouffonidor *see* Les fastes de Louis 15, de ses ministres, maitresses, gene'raux, et autres notables personnages de son regne
Bougainville *see* Voyage autour du monde, par la fregate du roi la boudeuse, et la fl-te l'etoile
Bougainville, Louis A de *see* Voyage autour du monde
Bougainville transitional and papua new guinea government newsletters : and related papers re the bougainville crisis, 1992-1995, 1997 – 1r – 1 – mf#pmb doc432 – at Pacific Mss [980]
Bougaud, Emile *see*
- Geschichte der heiligen monika
- History of st vincent de paul

Bougouin, E *see*
- La finance internationale et la guerre d'espagne
- La finanza internacional y la guerra de espana

Bouguer, M Pierre *see* Traite complet de la navigation
[Bouhours, D] *see*
- Les entretiens d'ariste et d'eugene

Bouhours, Dominique *see*
- The life of st. ignatius, founder of the society of jesus
- Vie de s francois xavier

Bouilhet, Louis Hyacinthe *see* Madame de montarcy
Bouillard, J *see* Histoire de l'abbaye de saint-germain-des-prez
Bouille, Francois C de *see* Memoires du marquis de bouille
Bouille, Louis J de *see* Memoires sur l'affaire de varennes
Bouillet, Jean Baptiste *see* Tablettes historiques de l'auvergne
Bouilly, Jean Nicolas *see*
- Fanchon la vielleuse
- Geschichtchen fuer meine tochter
- Haine aux femmes
- Jean-jacques rousseau, a ses derniers momens
- Vieillesse de piron

Bouinais, Albert *see* Le culte des morts dans le celeste empire et l'annam
Bouis, de *see* Le parterre geographique et historique, ou geographie-pratique
Boukari, Isbatou *see* Mineral analyses of selected western african foods with reference to nutritional status
Boulais, G *see* Manuel du code chinois
Boulangier, Edgar *see* Un hiver au cambodge
Boulbet, J et al *see* Les sites archeologiques de la region du bhnam gulen [phnom kulen]
[Boulder city-] backtrails – MI. feb 15 1972-mar 17 1987 – 1r – 1 – $110.00 – mf#U04910 – us Library Micro [071]
[Boulder city-] boulder city age – NV. may 1932; aug 1933 – 1r – 1 – $60.00 – mf#U04422 – us Library Micro [071]
[Boulder city-] boulder city citizen – NV. 1951 – 1 – $60.00 – mf#U04830 – us Library Micro [071]
[Boulder city-] boulder city daily reminder – NV. 1938-1940 – 4r – 1 – $240.00 – mf#U4831 – us Library Micro [071]
[Boulder city-] boulder city news – NV. 1941-[daily; wkly] – 65r – 1 – $3900.00 (subs $95y) – mf#UN04424 – us Library Micro [071]
[Boulder city-] desert shopper – NV. 1948 – 1r – 1 – $60.00 – mf#U04832 – us Library Micro [071]
[Boulder city-] the boulder city bulletin – NV. 1994- – 1r – 1 – $60.00 (subs $50y) – mf#U04907 – us Library Micro [071]
[Boulder city-] the desert scorpion – NV. 1941-1943 – 2r – 1 – $120.00 – (aka: sibert scorpion) – mf#U04425 – us Library Micro [071]
Boulder, CO *see* Charter of the city of boulder, state of colorado
Boulder county miscellaneous newspapers – Boulder, CO (mf ed 1991) – 2r – 1 – mf#MF Z99 B663 – us Colorado Hist [071]
[Boulder-] rolling stones – CO. 1970-79 – 14r – 1 – $840.00 – mf#R04000 – us Library Micro [071]
Boulding, J W *see* Expediency of christ's departure
Boule, Auguste Louis Desire *see*
- Prevot de paris,
- Trompettes de chamboran

Boulenger, Auguste *see* Manuel d'apologetique
Boulenger, G A *see*
- Catalogue of freshwater fishes of africa
- Catalogue of the lizards in the british museum (natural history)

Boulet, Jean Baptiste [comp] *see* Prayer book and catechism in the snohomish language
Boulet, Marie-Michele *see* Rapport 1 de l'etude exploratoire sur les possibilites d'utilisation d'un systeme-expert pour l'analyse de textes de conventions collectives
Les boulets rouges : le du club pacifique des droits de l'homme – [Paris]: F Malteste, jun 22-25 1848 – us CRL [944]
Le boulevard – Paris. dec 1861-juin 1863 – 1 – fr ACRPP [073]
O boulevard : jornal para senhoras – Bahia: Conde D'Eu de Candido Reinaldo da Rocha, 19 out 1870 – (= ser Ps 19) – mf#P17,1,30 – bl Biblioteca [079]
Boulevard baptist church. anderson, south carolina : church records – 1953-65 – 1 – $62.56 – us Southern Baptist [242]
The boulevardier – Paris. v1-6 n1. 1927-Jan 1932 – 1 – us NY Public [073]
Boulgakof, S *see*
- Du verbe incarne (agnus dei)
- Le paraclet

Boulger, Demetrius Charles de Kavanagh *see* England and russia in central asia
Boullon, Fernanda De *see* Lucha de razas
Boulnois, Helen Mary *see* Mystic india
Boulou : chacal du myombe / Trautmann, Rene – Bordeaux: Editions Delmas, 1944 – 1 – us CRL [960]
Boultbee, Thomas Pownall *see* An introduction to the theology of the church of england
Boulton, Mathew *see* Eight silver pattern books
Bouman, J C *see*
- De sociaal-psychologische aspecten van het zuid-molukse vraagstuk
- De sociaal-psychologische aspecten van het zuid-molukse vraagstuk

Bound records of the general land office relating to private land claims in louisiana, 1767-1892 / U.S. Bureau of Land Management – (= ser Records of the bureau of land management) – 8r – 1 – (with printed guide) – mf#M1382 – us Nat Archives [324]
Bound volumes of the general records of the united states consulate at yokohama, japan, 1936-1939 – (= ser Records of the foreign service posts of the department of state – consular posts) – 22r – 1 – (with printed guide) – mf#M1520 – us Nat Archives [327]
Bound volumes of the general records of the u.s. consulate at yokohama, japan, 1936-1939 – 22 rolls – 9 – $506.00 – Dist. us Scholarly Res – us L of C Photodup [324]
Boundary 2 – Durham NC 1988+ – 1,5,9 – ISSN: 0190-3659 – mf#17597 – us UMI ProQuest [400]
Boundary-layer meteorology – Dordrecht, Netherlands 1984+ – 1,5,9 – ISSN: 0006-8314 – mf#14741 – us UMI ProQuest [550]
Bounere, Benedictius du *see* Colephous de manuscrito
Bouniol, Bathild *see* Les marins francais
Bouniol, Joseph *see* White fathers and their missions
Bouquet, Alan Coates *see*
- An introduction to the study of efforts at christian reunion
- A man's pocket-book of religion
- Sacred books of the world

Bouquet du roi : ou le marche aux fleurs – Paris, France. 1815 – 1r – us UF Libraries [440]
Bouquet, Martin *see* Rerum gallicarum et francicarum scriptores (rgfs). recueil des historiens des gaules et de la france
Bouquetiere des champs-elysees / Kock, Paul De – Paris, France. 1843 – 1r – us UF Libraries [440]
Bouquillon, Thomas *see* Education, to whom does it belong?
Bourassa, Gustave *see*
- Conferences et discours
- Discours prononce au petit seminaire de montreal, le 2 fevrier 1890
- Les fables de la fontaine
- Mgr bourget
- La prophetie de malachie
- Les soldats du pape

Bourassa, Henri et al *see* Canadian nationalism and the war
Bourassa, Lucette *see* Le roman au canada-francais, 1925-1949
Bourassa, Napoleon *see*
- Causerie par m bourassa
- Melanges litteraires, vol 1
- Melanges litteraires, vol 2
- Reunion des anciens eleves du college de montreal, le 9 septembre 1885

Bourbon County, KS. Public School *see* Records
Bourbon del Monte Santa Maria, Giuseppe *see* L'islamismo e la confraternita dei senussi
Bourbon, J de *see* La grande et merueiileuse, and trescruelle oppugnatio de la noble cite de rhodes...
Bourbonniere, Joseph-Avila *see* Manuel pratique des ingenieurs, mecaniciens, chauffeurs, machinistes
Les bourbons : souvenirs et melanges / Conny de LaFay, Felix J de – Paris 1833 [mf ed Hildesheim 1995-98] – 1v on 2mf – 9 – €60.00 – ISBN-10: 3-487-26043-3 – ISBN-13: 978-3-487-26043-3 – gw Olms [940]
Bourdages, Jeanne *see* Bibliographie des ecrits du tres rev pere pascal d'ottawa, ex-prov de l'ordre des freres mineurs capucins, ex-prof de patrologie et d'histoire sainte
Bourdalque, Luis *see* Retiro espiritual para comunidades religiosas...
Bourdeaux, Jean *see* Les carrieres feminines intellectuelles
Bourdelot, Pierre *see* Histoire de la musique et de ses effets
Bourdes-de-peage et pis en sont! / Herment-Grenie – Paris, France. 1917 – 1r – us UF Libraries [440]
Bourdon, Hilaire *see* The church and the future
Bourg, Edme *see* Paris et ses environs
Bourg, Edme Theodore *see*
- Description historique des prisons de paris
- Napoleon

Bourgain, Louis *see* La chaire francaise au 12e siecle
Bourgeois, Charles-Edouard *see* Le service social diocesain
Bourgeois grand seigneur / Royer, Alphonse – Paris, France. 1842 – 1r – us UF Libraries [440]
Bourgeois, Nicolas *see* Voyages interessans dans differentes colonies francaises, espagnoles, anglaises, etc
Bourgeois, Phileas Frederic *see*
- Les anciens missionnaires de l'acadie
- Les anciens missionnaires de l'acadie devant l'histoire
- Henry wadsworth longfellow, sa vie, ses oeuvres litteraires, son poeme evangeline
- L'histoire du canada depuis sa decouverte jusqu'a nos jours
- L'histoire du canada en 200 lecons
- Petit resume de l'histoire du nouveau brunswick depuis quatre-vingts ans
- Vie de l'abbe francois-xavier lafrance

Bourgeois, R *see* Banyarwanda et barundi
[Bourges, Charles Doris des] *see* Geheimer briefwechsel zwischen dem kaiser napoleon und dem papst pius 7
Bourges pendant la guerre / Gignoux, Claude-Joseph – Paris: Les Presses Universitaires de France; New Haven: Yale UP, [1926] (mf ed 19–) – (= ser Histoire economique et sociale de la guerre mondiale. serie francaise) – xvi/64p – mf#Z-BTZO pv26 n1 – us NY Public [933]
Bourget, Ignace *see*
- Circulaire a messieurs les cures, missionnaires et autres pretres du diocese de montreal
- Circulaire a mm les cures et missionnaires du diocese de montreal
- Circulaire annoncant au clerge la retraite pastorale et le second synode diocesain
- Circulaire au clerge
- Circulaire au clerge, accompagnant le mandement de visite pour 1861 et 1862
- Circulaire au clerge concernant les 40 heures, l'ordo, l'indulgence des chapelets, l'annee religieuse, etc
- Circulaire au clerge de montreal
- Circulaire au clerge de montreal accompagnant le mandement du 8 dec 1862
- Circulaire au clerge du diocese de montreal
- Circulaire au clerge du diocese de Montreal
- Circulaire au clerge du diocese de montreal
- Circulaire au clerge du diocese de montreal accompagnant le mandement du 1 janvier 1865
- Circulaire au clerge du diocese de montreal, sur le cholera
- Circulaire au clerge du diocese de montreal sur le grand incendie du huit juillet
- [Lettre]
- Lettre pastorale de monseigneur l'eveque de montreal
- Lettre pastorale de monseigneur l'eveque de montreal sur le grand incendiedu 8 juillet
- Vie de saint viateur

Bourget, Magdeleine *see* Bio-bibliographie analytique de m elphege-j daignault
Bourgoin, J *see* Precis de l'art arabe
Bourgoing, Andre *see* Bibliographie de l'oeuvre de gerard tremblay
Bourgoing, Jean F de *see*
- Tableau de l'espagne moderne
- Travels in spain

Bourgois, Jean Jacques *see* Martinique et guadeloupe, terres francaises de ant...
Bourguignat, J R *see*
- Histoire malacologique de la regence de tunis
- Malacologie de l'algerie
- Testacea novissima quae cl de saulcy in itinere per orientem annis 1850 et 1851, collegit

Bourguignon, C *see* Vie du pere romillion, pretre del'oratoire de jesus
Bourignon, Antoinette *see*
- The light of the world
- Oeuvres

Bourinot, John George *see*
- Builders of nova scotia
- Canada
- Canada and the united states
- Canada as a home
- Canada during the victorian era
- Canada under british rule, 1760-1900
- Canada's marine and fisheries
- A canadian manual on the procedure at meetings of shareholders and directors of companies, conventions, societies and public assemblies generally
- Canadian materials for history, poetry, and romance
- Canadian studies in comparative politics
- Elected or appointed officials?
- Federal government in canada
- The fishery question
- Historical and descriptive account of the island of cape breton
- How canada is governed
- The island of cape breton
- Literature and art in canada
- A manual of the constitutional history of canada from the earliest period to 1901
- The national development of canada
- The national sentiment in canada
- Our intellectual strength and weakness
- Parliamentary procedure and practice in the dominion of canada
- Rules of order
- Social and economic conditions of the british provinces after the canadian rebellions, 1838-1840
- Statesmanship and letters

Bourke banner – 1899-1907 – 9r – 9 – A$586.92 vesicular A$636.42 silver – at Pascoe [079]

Bourke, John Gregory *see* Scatalogic rites of all nations
Bourke, Ulick Joseph *see*
– Ineffabilis deus
– The life and times of the most rev. john machale
Bourk-Rousseau, Adeline *see* Un heritage notice biographique
Bourlamaque, Francois-Charles de *see* Lettres de m de bourlamaque au chevalier de levis
Bourland bulletin – v2 n4-5 [1978 sep-dec], v3 n1-v4 n4, [1979 jan/feb-1980 oct/dec] – 1r – 1 – (cont: bourland family bulletin; cont by: loving letter; bourland bulletin and loving letter) – mf#1777609 – us WHS [071]
Bourland bulletin and loving letter / Harp and Thistle – 1981 spring/summer-1989 jun 12 – 1r – 1 – (cont: bourland bulletin, loving letter; cont by: bourland-loving bulletin) – mf#1777632 – us WHS [071]
Bourlon, I *see* Les assemblees du clergue et le jansenisme
Bourn 1563-1861 – (= ser Cambridgeshire parish register transcript) – 5mf – 9 – £6.25 – uk CambsFHS [929]
Bourn, Drew F *see* Gambling behavior among college student-athletes, non-athletes, and former athletes
Bourne, B A *see*
– Effects of freezing temperatures on sugarcane in the florida everglades
– Studies on the ring spot disease of sugarcane
Bourne, Edward Gaylord *see* Spain in america, 1450-1580
Bourne, Frederick William *see* The bible christians
Bourne, Henry Richard Fox *see*
– Blacks and whites in south africa
– The story of our colonies
Bourne, Hugh [comp] *see* An ecclesiastical history from the creation to the 18th century, a d
Bourne, John *see* Public works in india
Bourne, Kenneth *see* The papers of queen victoria on foreign affairs
Bourne, W Fitz G [comp] *see* Hindustani musalmans and musalmans of the eastern punjab
Bournemouth daily echo – Bournemouth, England 20 aug 1900-30 jun 1958 – 1 – (cont by: evening echo [1 jul 1958-4 apr 1997]) – uk British Libr Newspaper [072]
Bourneville *see* Le sabbat des sorciers
Bournonville, Antoine *see* Dagboger fra 1792
Bournonville, August *see* Cort adeler i venedig
Bourquin, Theodor *see* Grammatik der eskimo-sprache
Bourreaux et martyrs : conference donnee a l'institut canadien d'ottawa, le 12 fevrier 1891 – [Joliette, PQ?: s.n, 1891?] [mf ed 1980] – 1mf – 9 – 0-665-03815-1 – mf#03815 – cn Canadiana [944]
Bourret, Joseph-Christian Ernest *see* L'ecole chretienne de seville sous la monarchie des visigoths
Bourrienne, Louis A de *see* Memoires de m de bourrienne, ministre d'etat
Bourrit, Marc *see* Nouvelle description des glacieres, vallees de glace et glaciers qui forment la grande chaine des alpes de savoye, de suisse et d'italie
Bourse / Ponsard, Francois – Paris, France. 1856 – 1r – us UF Libraries [440]
Bourse de paris : copie du cours authentique. – Paris. avr-dec 1881 – 1 – fr ACRPP [332]
Bourse du travail de Paris *see* Bulletin officiel de la bourse du travail de paris
Bourse egyptienne – Cairo, Egypt. jul 1945-aug 1946; jul 1952-1956 – 18r – 1 – uk British Libr Newspaper [079]
La bourse egyptienne – Cairo, Egypt. -d. 15 June 1945-9 Aug 1946; 2 June 1952-31 Jan 1955; 2 May 1955-25 Oct 1956. 18 reels – 1 – uk British Libr Newspaper [072]
La bourse et la vie : recueil de renseignements utiles et d'informations exactes sur les cantons du nord et en particulier sur le territoire de la mantawa / Provost, Thomas Stanislas – Joliette, Quebec?: s.n, 1883 – 4mf – 9 – mf#12182 – cn Canadiana [917]
Bou-saada, cite de bonheur / Lehuraux, Leon – Alger, Soubiron, [1934?] – 1 – us CRL [960]
Bousfield, H B *see* Six years in the transvaal
Bousfield, William *see* The government of the empire
Boush, Christian Maximilian *see* Rulings by the civil courts governing religious societies
Boussenard, Louis *see*
– Le capitaine casse-cou
– Chasseurs canadiens
– Crusoes of guiana
Bousset, Jean Baptiste Drouart de *see* Recueil d'airs nouveaux serieux et a boire...
Bousset, Wilhelm *see*
– Der antichrist in der ueberlieferung des judentums, des neuen testaments und der alten kirche
– The antichrist legend
– Apophtegmata
– Der apostel paulus
– Die evangeliencitate justin des maertyrers in ihrem wert fuer die evangelienkritik von neuem untersucht
– The faith of a modern protestant
– Hauptprobleme der gnosis
– Jesu predigt in ihrem gegensatz zum judentum
– Jesus
– Die juedische apokalyptik
– Die offenbarung johannis
– Die religion des judentums im neutestamentlichen zeitalter
– Textkritische studien zum neuen testament
– Volksfroemmigkeit und schriftgelehrtentum
– Was wissen wir von jesus?
– Wesen der religion
– Das wesen der religion
Bousset's religion des judentums im neutestamentlichen zeitalter / Perles, Felix – Berlin: Wolf Peiser 1903 [mf ed 1989] – 1mf – 9 – 0-7905-3158-5 – (incl bibl ref) – mf#1987-3158 – us ATLA [270]
Boussod, Valadon and Co *see* Messrs. boussod, valadon and co....new and important publications
Boussuet, Jacques Benigne *see* Exposition of the doctrines of the catholic church
Boustany, W F *see* The palestine mandate
A bout de souffle : bulletin d'information mensuel prepare pour les pur-sang de l'outaouais – Hull: Les Pur-Sang de l'Outaouais. v1 n1 avril 1979- (mthly) [mf ed 1995-] – 1 – mf#SEM35P423 – cn Bibl Nat [073]
Boutard, Charles *see* Lamennais
Boutell, Charles *see*
– The arts and the artistic manufactures of denmark
– Christian monuments in england and wales
– Monumental brasses and slabs...of the middle ages
Bouterwek, F *see* Neues museum der philosophie und literatur
Bouterwek, Karl Wilhelm *see* Zur literatur und geschichte der wiedertaeufer
Boutet, Odina *see* Bio-bibliographie critique d'anne hebert
Bouthillier-Chavigny, Charles de *see* A travers le nord-ouest canadien
Bouthillier-Chavigny, Charles, vicomte de *see*
– A travers les grandes terres a ble du nord ouest canadien
– Address before the imperial institute of great britain on the 10th of march, 1898
– Le canada agricole et industriel
– Our land of promise
Boutillier, Thomas *see* Rapport des travaux de colonisation de l'annee 1855
Boutin, Rose-Anne *see*
– Bibliographie analytique de l'oeuvre de l'honorable senateur cyrille vaillancourt
– Bibliographie analytique de l'oeuvre de l'honorable senateur cyrille vaillancourt...
Boutroux, Emile *see*
– The beyond that is within
– The contingency of the laws of nature
– Education and ethics
– Historical studies in philosophy
– Pascal
– Science and religion in contemporary philosophy
– William james
Boutroux, Emile et al *see* Lectures
Boutwell, George S *see* The constitution of the united states at the end of the first century
Bouvard, Alexis *see* Tables astronomiques publiees par le bureau des longitudes de france
Bouverie-Pusey, Sidney Edward *see* Permanence and evolution
Bouvet de Cresse, Auguste *see* Voyage a reims
Bouvet, J *see* Voyage du pere joachim bouvet, jesuite, de peking...canton lorsqu'il fut envoye en europe par l'empereur kang-hi, en 1693
Bouvier, A *see* Henri bullinger...
Bouvier, Claude *see* La question michel servet
Bouvier, Roger *see* Commando ses sor
Bouwkonstige wercken, begrepen in 8 boeken... / Scamozzi, [V] – Amsterdam, 1661 – 7mf – 9 – mf#OA-80 – ne IDC [720]
Bouwman, M *see* Voetius over het gezag der synoden
Bouyer, Frederic *see* Guyane francaise
Bouzan, Ary *see* Aspectos legais e economicos da pequena empresa br...
Bouzon, Justin *see* Etudes historiques sur la presidence de faustin so...
[Bovard-] booster – NV. jan 1908 [wkly] – 1r – 1 – $60.00 – mf#U04426 – us Library Micro [071]
Bovea, A *see* Obra...de la grande y terrible mortandad
Bovee, Kristin K *see* Current trends in the reconstruction and rehabilitation of the anterior cruciate ligament
Bovell, James *see* Constitution and canons of the synod of the diocese of toronto
Bovenwindse eilanden : economische en sociale ophef / Netherlands Antilles Departement Sociale-En Economische Ophef – Willenstad, Curacao. 1955 – 1r – us UF Libraries [972]
Bovet, Felix *see*
– Egypt, palestine, and phoenicia
– Histoire du psautier
Bovine practice – Mission Viejo CA 1980-81 – 1,5,9 – ISSN: 0199-5456 – mf#12427.01 – us UMI ProQuest [636]
Bovini, G *see* I sarcofagi paleocristiani
Bovon, Jules *see*
– Dogmatique chretienne
– L'enseignement des apotres
– La vie et l'enseignement de jesus
Bovykin, V I *see*
– Formirovanie finansovogo kapitala v rossii, konets 19 v – 1908 g
– Organizatsionnye formy finansovogo kapitala v rossii
– Rossiia nakanune velikikh svershenii
– Zarozhdenie finansovogo kapitala v rossii
Bow and arrow news / North American Indian Association – 1973-74 – (= ser American indian periodicals... 2) – 1mf – 9 – $95.00 – us UPA [305]
Bow bells : a magazine of general literature and art, for family reading – La Salle IL 1864-65 – 1 – mf#4711 – us UMI ProQuest [073]
Bow, chelsea, and derby porcelain : being further information relating to these factories / Bemrose, William – London 1898 – (= ser 19th c art & architecture) – 4mf – 9 – mf#4.2.1535 – uk Chadwyck [730]
The bow in the cloud : fifteen discourses / Briggs, George Ware – 2d ed. Boston: Joseph Dowe, 1846. Beltsville, Md: NCR Corp, 1978 (3mf); Evanston: American Theol Lib Assoc, 1984 (3mf) – 9 – 0-8370-1076-4 – mf#1984-4429 – us ATLA [240]
Bow milkman – London, England. 18– – 1r – us UF Libraries [240]
Bowbells bulletin : [official paper of ward county] – Bowbells, Burke Co, ND: V F Snyder, 1903; -v13 n41 apr 22 1915 (wkly) – 1 – (publ as: bowbells bulletin=tribune nov 22 1907 to jun 25 1908 (due to fire at tribune plant). missing: 1907 may 17, jul 5, nov 8; 1908 jul 30, aug 6; 1914 sep 17, oct 8, dec 10) – mf#11087-11090 – us North Dakota [071]
Bowden, Henry Sebastian *see*
– Natural religion
– The religion of shakespeare
– Revealed religion
Bowden, James *see*
– Hand of god acknowledged in the loss of endeared relatives
– The history of the society of friends in america
Bowden, Jeanne J *see* A practical guide to revision of local court rules
Bowden, John Edward *see*
– The life and letters of frederick william faber, d.d
– Spiritual works of louis of blois, abbot of liesse
Bowden, Victoria L *see* The effect of training status on resting metabolic rate and substrate utilization in women
Bowdich, S *see* Excursions in madeira and porto santo...
Bowdich, Thomas *see*
– Mission der englisch-afrikanischen compagnie von cape coast castle nach ashantee
– Mission from cape coast castle to ashantee
Bowdich, Thomas E *see*
– An account of the discoveries of the portuguese in the interior of angola and mozambique
– An essay on the geography of north-western africa
Bowdich, Thomas Edward *see*
– The british and french expeditions to teembo
– Excursions in africa
– Mission from cape coast castle to ashantee...
Bowditch, Thomas Edward *see*
– An account of the discoveries of the portuguese in the interior of angola and mozambique
– Mission from cape coast castle to ashantee
Bowdler, Thomas *see* Letter addressed to the evangelical members of the church of england...
Bowdon, st mary: baptisms 1738-1812 – [North Cheshire FHS] – (= ser Cheshire church registers) – 5mf – 9 – £6.00 – mf#279 – uk CheshireFHS [929]
Bowdon, st mary: burials 1738-1841 – [North Cheshire FHS] – (= ser Cheshire church registers) – 9mf – 9 – £10.00 – mf#280 – uk CheshireFHS [929]
Bowdon, st mary: burials 1841-1908 – [North Cheshire FHS] – (= ser Cheshire church registers) – 11mf – 9 – £12.00 – mf#365 – uk CheshireFHS [929]
Bowe, William G Jr *see* A comparison of six personality factors between professional, college, and high school basketball players
Bowen, Anna Maude *see* Sources and text of richard wagner's opera "die meistersinger von nuernberg"
Bowen, Calvin *see* Guide to jamaica
Bowen, Clayton Raymond *see* The resurrection in the new testament
Bowen, Elias *see*
– History of the origin of the free methodist church
– Religious education of children
– Sermon on ministerial education
– Slavery in the methodist episcopal church
Bowen, Francis *see*
– Critical essays on a few subjects connected with the history and present condition of speculative philosophy
– A layman's study of the english bible
– Modern philosophy
– The principles of metaphysical and ethical science
– A treatise on logic
Bowen, George *see* The amens of christ
Bowen, George Ferguson *see* Thirty years of colonial government
Bowen, John *see*
– Journal of restaurant and foodservice marketing
– Memorials
– Memorials of john bowen
– Memorials of john bowen, ll.d., late bishop of sierra leone
Bowen or an evidence of grace / Roberson, Cecil F – 1969. Biography of Rev. Thomas Jefferson Bowen, SBC's first missionary to Nigeria, West Africa. Nigerian Bapt. Hist. Mat. 103p – 1 – us Southern Baptist [242]
Bowen, T J *see*
– Central africa
– Papers
– Papers of t.j. bowen
Bowen's virginia centinel and gazette : or, the winchester political repository – Winchester VA. 1790 may 26-1791 sep 10, 1795 sep, oct 12 – 2r – 1 – (cont: virginia centinel, or, the winchester mercury; cont by: bowen's virginia gazette: and the winchester centinel) – mf#882582 – us WHS [071]
Bower, Archibald *see* The history of the popes
Bower, F O *see* Wilhelm hofmeister – the works and life of a nineteenth century botanist
Bower, Ursula Graham *see* Naga path
Bowerman family newsletter – 1983 jun-1984 jan – 1r – 1 – (cont by: bowerman/bowman family newsletter) – mf#717874 – us WHS [071]
Bowerman, George Franklin *see* A selected bibliography of the religious denominations of the united states
Bowerman, Helen Cox *see* Roman sacrificial altars
Bowers, Alexander *see* Bhamo expedition
Bowers and Ruddy Galleries *see* Special coinletter
Bowers, Chester *see* Advanced tennis
Bowers, Faubion *see* The dance in india
Bowers, Richard *see* Plyometric training and its effects on speed, strength, and power of intercollegiate athletes
Bowerston, OH *see* Obituaries, 1879-1914
Bowes, James Lord *see*
– Japanese enamels
– Japanese pottery
Bowes, Michelle L *see* The development of weight-adjusted estimates of caloric expenditures for the nordic track
Bowhunting world – Plymouth MN 1989+ – 1,5,9 – (cont: archery world) – ISSN: 1043-5492 – mf#10676.01 – us UMI ProQuest [790]
Bowie, W Copeland *see* Liberal religious thought at the beginning of the twentieth century
Bowing in the name of jesus / Price, Thomas C – London, England. 1875 – 1r – us UF Libraries [240]
Bowker, Richard Rogers *see*
– Copyright, its history and its law
– The reader's guide in economic, social and political science
Bowler, Arthur *see*
– Conference sur haiti
– Haiti
Bowler, Ernest Constant *see* An album of the attorneys of rhode island
Bowles, Ada Chastina *see* Woman in the ministry
Bowles, Caroline *see* Nineteenth century literary manuscripts
Bowles, Samuel *see* Von ocean zu ocean
Bowles, William Lisle *see* Discourse, preached in salisbury cathedral, on king charles's marty
Bowling green exponent – Bowling Green, FL. 1938 nov-1939 dec – 1r – us UF Libraries [071]
Bowling green first baptist church. bowling green, kentucky : church records – 1833-Jan 1847 (incomplete), 1852-1946 – 2r – 1 – $76.95 – (1,710p) – us Southern Baptist [242]
Bowling green second baptist church. bowling green, missouri : church records – Dec 1890-Jul 1980. 1730p – 1 – 77.85 – us Southern Baptist [242]
Bowling news – 1948 sep 18-1954 may 22, 1954 jun 5-1960 jun 25 – 2r – 1 – mf#1222298 – us WHS [790]

Bowman, Amos *see*
- Preliminary report on field notes in cariboo district, b c
- Report on the geology of the mining district of cariboo, british columbia

Bowman, Anne *see* The bear-hunters of the rocky mountains

Bowman, Ariel *see* Hours of childhood and other poems

Bowman, Arthur Herbert *see* Christian thought and hindu philosophy

Bowman, Charles Victor *see* Missionsvaennerna i amerika

Bowman citizen – Bowman, ND: Citizens Pub Co. v4 n49 jan 5 1911-v11 n18 apr 30 1917 (wkly) – 1 – (an independent journal of real news and progressive views. official paper of bowman county 1911-1913. official paper of bowman village 1913. cont: bowman county news. merged with: bowman county pioneer (twin butte, nd) to form: bowman county pioneer and bowman citizen. missing: 1912 jul 25; 1914 feb 5; 1917 jan 11) – mf#09432-09433; 10370 – us North Dakota [071]

Bowman county leader – Bowman, ND: [s.n.] v9 n12 jan 20 1927-1928?// (wkly) – 1 – (cont: bowman county leader and farmer-labor monitor. missing: 1927 feb 10-17, apr 14, jul 19) – mf#10377 – us North Dakota [071]

Bowman county leader and farmer-labor monitor – Bowman, ND: H B French. v8 n35 jun 26 1925-v9 n11 jan 13 1927 (wkly) – 1 – (also bears whole numbering: n399-n451; n152-183; cont: farmer-labor monitor and farmers leader; cont by: bowman county leader; missing: 1925 dec 24; 1926 sep 16) – mf#10376-10377 – us North Dakota [071]

Bowman county news : [official paper of city and county 1908-1910] – Bowman, ND: McCann & Billyard, 1907; -v4 n48 dec 29 1910 (wkly) – 1 – (special harding co [sd] ed publ jun 10 1909. cont by: bowman citizen. missing: 1908 aug 6, oct 8, dec 22) – mf#10372 – us North Dakota [071]

Bowman county pioneer – Twin Butte, Bowman Co, ND: A L Lowden. v1 n1 may 16 1907-v9 n52 may 3 1917 (wkly) – 1 – (publ may 16-aug 7 1907 at twin butte, nd; aug 14 1907-may 3 1917 at bowman, nd. merged with: bowman citizen to form: bowman county pioneer and bowman citizen) – mf#00900-00903 – us North Dakota [071]

Bowman county pioneer (1929) : [official newspaper of bowman county] – Bowman, ND: H C Hagg. v22 n23 oct 10 1929-v38 n24 jun 19 1947 (wkly) – 1 – (cont: bowman county pioneer and bowman citizen. merged with: scranton star to form: bowman county pioneer and scranton star) – mf#00906-00913 – us North Dakota [071]

Bowman county pioneer (1949) – Bowman, ND: [s.n.] v41 n21 may 26 1949- (wkly) – 1 – (cont: bowman county pioneer and scranton star. currently publ) – mf#00913-00916++ – us North Dakota [071]

Bowman county pioneer and bowman citizen : [official newspaper of bowman county and bowman village] – Bowman, ND: Bowman Pub Co. v11 n1 may 10 1917-v22 n22 oct 2 [i.e. 3] 1929 (wkly) – 1 – (may 10-aug 30 1917 also bear vol numbering of the bowman citizen, v11 n19-v11 n35; bowman county pioneer (twin butte, nd) dropped v10 between title changes. formed by the union of: bowman county pioneer (twin butte, nd) and: bowman citizen. cont by: bowman county pioneer (bowman, nd: 1929)) – mf#00903-00906 – us North Dakota [071]

Bowman county pioneer and scranton star : [official paper of bowman county] – Bowman, ND: [s.n.] v38 n25 jun 26 1947-v41 n20 may 19 1949 (wkly) – 1 – (special county 40th anniv iss publ aug 28 1947; formed by the union of: scranton star and bowman county pioneer (bowman, nd: 1929); cont by: bowman county pioneer (bowman, nd: 1949)) – mf#00913 – us North Dakota [071]

Bowman, Elizabeth *see* Journal of trauma & dissociation

Bowman, Fred A *see* Some application of electric motors

Bowman, Frederick Charles *see* The reaction between bromic, hydriodic and arsenious acids

Bowman, George Ernest *see* The mayflower descendant 1620-1937

Bowman, George Ernest Bowman *see* Vital records of truro massachusetts to 1850

Bowman, Heath *see*
- Crusoe's island in the caribbean
- Westward from rio

Bowman, Hervey Meyer *see*
- Analysis of the gospels
- Die englische-franzosische friedensverhandlung
- Preliminary stages of the peace of amiens

Bowman, Isaac Daniel *see* A treatise on the lord's supper

Bowman, J N *see* Adobes in san mateo county

Bowman, Russell Keith *see* The connections of the geste des loherains with other french epics and mediaeval genres

Bowman, Shadrach Laycock *see* Historical evidence of the new testament

Bowman, Thomas *see* Historical review of the disturbance in the evangelical association

Bowman, Victor Virgil *see* Relative importance of the grade-lowering factors of citrus

Bowne, Borden Parker *see*
- The atonement
- The christian life
- The christian revelation
- The essence of religion
- Introduction to psychological theory
- Kant and spencer
- Metaphysics
- Personalism
- The philosophy of herbert spencer
- Philosophy of theism
- The principles of ethics
- Theism
- Theory of thought and knowledge

Bowral free press – Bowral, jul 1883-mar 1906 – 7r – A$457.20 vesicular A$495.70 silver – at Pascoe [079]

Bowraville guardian – Macksville, 1957-58 – (= ser Nambucca gazette) – (aka: nambucca gazette) – at Pascoe [079]

Bowring, Edgar Alfred *see* The poems of heine, complete

Bowring, J *see*
- The kingdom and people of siam
- A visit to the philippine islands

Bowring, John *see*
- Hymns
- On the religious progress beyond the christian pale

Bowser, Eileen *see* The merritt crawford papers

Bowsher, Charles A *see* Congressional oversight

Bowyer, George *see*
- Cardinal archbishop of westminster and the new hierarchy
- Observations on the arguments of dr twiss respecting the new roman

Box 551 / Transport Workers Union of America – 1975 mar-1976 may – 1r – 1 – mf#644309 – us WHS [331]

Box butte county rustler – Hemingford, NE: C A Burlew, 1886 (wkly) [mf ed v2 n15. oct 28 1887] – 1r – 1 – us NE Hist [071]

Box, George Herbert *see* A short introduction to the literature of the old testament

Boxare-upproret och foerfoeljelserna mot de kristna i kina 1900-1901 – Stockholm: Baptistmissionens Foerlagsexpedition, [1902] [mf ed 1995] – (= ser Yale coll) – 307p (ill) – 1 – 0-524-09226-5 – (in swedish) – mf#1995-0226 – us ATLA [951]

Boxberger, R *see* Klopstocks leben und werke / wielands leben und werke

Boxboard containers international – Shawnee Mission KS 1996+ – 1,5,9 – ISSN: 1084-5291 – mf#12383.01 – us UMI ProQuest [680]

Boxborough 1767-1849 – Oxford, MA (mf ed 1995) – (= ser Massachusetts vital record transcripts to 1850) – 3mf – 9 – 0-87623-213-6 – (mf 1t: births & deaths 1767-1844; intentions 1838-49; marriages 1841-44. mf 2t: intentions 1794-1844; marriages 1790-95, 1822-40+; births & deaths 1786-1843. mf 3t: births 1843-49; marriages 1844-49; out-of-town marriages 1784-1798; deaths 1844-49) – us Archive [978]

Boxborough 1767-1905 – Oxford, MA (mf ed 1996) – (= ser Massachusetts vital records) – 49mf – 9 – 0-87623-385-X – (mf 1-7: births & deaths 1767-1845. mf 1,5,7: town records 1755-1870. mf 5-7: marriage intentions 1794-1856; marriages 1790-97, 1822-44. mf 6-7: church members 1798, 1811-22. mf 8-13: town meetings 1783-1834. mf 14-27: town records 1835-66. mf 28-37: town records 1866-96. mf 38-44: town records 1896-1918. mf 45: births 1843-73. mf 46: marriages 1844-73; out-of-town marriages 1778-98. mf 46-47 deaths 1844-1905. mf 48-49: marriages & births 1874-1905) – us Archive [978]

Boxer, Charles Ralph *see*
- African eldorado
- Four centuries of portuguese expansion, 1415-1825
- Great luso-brazilian figure

Boxer, Frederick N *see*
- Hunter's hand book of the victoria bridge
- Reminiscences of the survey and cutting out of the boundary line between canada and the united states

Boxford 1666-1849 – Oxford, MA (mf ed 1995) – (= ser Massachusetts vital record transcripts to 1850) – 8mf – 9 – 0-87623-214-4 – (mf 1t-2t: births & deaths 1666-1757. mf 1t: marriages 1715-30. mf 2t: marriages & intentions 1690-1741. mf 3t-4t: births 1723-1822. mf 4t: deaths 1838-1843; marriages 1739-1843. mf 4t-6t: marriage publishments 1741-1849. mf 6t-7t: out-of-town marriages 1690-1799. mf 7t: marriages 1843-49; births 1843-49. mf 8t: deaths 1842-49) – us Archive [978]

Boxhorn, M Z *see*
- Emblemata politica
- Emblemata politica, et orationes
- Epistolae et poemata

Boxley, Robert F *see* Trends in double cropping

Boxworth 1588-1950 – (= ser Cambridgeshire parish register transcript) – 4mf – 9 – £5.00 – uk CambsFHS [929]

Boy comics – iss n6-25. oct 1942-dec 1945 – 15 – mf#002GL-005GL – us MicroColour [740]

Boy comics *see* Captain battle / boy comics

Boy in the bush / Lawrence, D H – New York, NY. 1924 – 1r – us UF Libraries [420]

The boy jesus and other sermons / Taylor, William Mackergo – New York: AC Armstrong, 1893 – 1mf – 9 – 0-7905-2606-9 – mf#1987-2606 – us ATLA [240]

Boy life on the prairie / Garland, Hamlin – rev ed. New York: Harper, c1899 [mf ed 1998] – 1r – 1 – (filmed with: ambitious man / ella wheeler wilcox & other titles) – mf#4390 – us Wisconsin U Libr [830]

The boy who would be king and other bible stories / Rosenberger, Elizabeth Delp – Elgin, IL: Brethren Pub House, 1906 – 1mf – 9 – 0-524-02751-X – mf#1990-4426 – us ATLA [220]

Boyaca / Penuela, Cayo Leonidas – Bogota, Colombia. 1936 – 1r – us UF Libraries [972]

Boyazoglu, Alexander J et al *see* Nungyeh hsin yung

Boyce, Edward Jacob *see* Catechetical hints and helps

Boyce, James Petigru *see*
- Abstract of systematic theology
- An inaugural address
- Special papers, 1850-1888

Boyce Thompson Institute for Plant Research *see* Contributions

[Boyce, W B] *see* Memoir of the rev william shaw, late general superintendent of the wesleyan missions in south-eastern africa

Boyce, William Binnington *see*
- Grammar of the kafir language
- The higher criticism and the bible
- Memoir of the rev. william shaw

Boyce's weekly – Chicago IL. 1903 jan 7-sep 2 – 1r – 1 – (cont by: saturday blade) – mf#870807 – us WHS [071]

Boyceville press – Boyceville WI. 1918 nov 15, 1923 apr 27, 1953 jun 26-1957 may 17 – 2r – 1 – (cont by: press-reporter) – mf#963578 – us WHS [071]

Boyceville press-reporter – Boyceville WI. 1975 jun 5-1975 jul 10, 1976 apr 22-1978 dec 28, 1979 jan 4-1979 dec 28, 1980-81, 1982, 1983-1984 jul 5 – 6r – 1 – (cont: press-reporter; cont by: glenwood city tribune, tribune press reporter) – mf#999062 – us WHS [071]

Boycott census – 1974 jan-1986 nov, 1986 jan/feb-1988 jul/aug – 2r – 1 – (cont: real paper; cont by: building economic alternatives) – mf#1546869 – us WHS [330]

Boycott update / Farm Labor Organizing Committee [Ohio] – v1 1 [1979 mar 26], v2 1-3 [1980 apr-nov], v3 2-5 [1981 mar, 27-oct], v4 1-3 [1982 feb-fall], 1983 jun-aug, 1984 dec, 1985 jul, 1986 feb-apr – 1r – 1 – (cont: update [farm labor organizing committee [ohio]: 1982]; cont by: update [farm labor organizing committee [ohio]: 1986]) – mf#1053529 – us WHS [071]

Boycott's news budget – La Crosse WI. 1893 may 30, 1898 sep 17, 1899 oct 14, nov11, 1900 mar 10, 31, apr 7, aug 4 – 1r – 1 – mf#931579 – us WHS [330]

Boyd, A K H *see* Early christian scotland, 400 to 1093 ad

Boyd, Andrew *see* Atlas of african affairs

Boyd, Andrew Kennedy Hutchion *see* Church life in scotland

Boyd, Archibald *see* National deliverance and national gratitude

Boyd bee – Boyd WI. 1959 nov6-1962 dec 28 – 1r – 1 – mf#1793828 – us WHS [071]

Boyd county advocate – Spencer, NE: Cal Moffet (wkly) [mf ed v4 n19. jan 3 1896-sep 3 1897 (gaps)] – 2r – 1 – (cont by: spencer advocate. issued with: boyd county register jan 29-sep 3 1897) – us NE Hist [071]

Boyd county democrat – Naper, NE: Vern Gibbens. v1 n14-20. jun 22-aug 3 1916 (wkly) – 1r – 1 – (cont: naper press. cont by: naper independent) – us NE Hist [071]

Boyd county register – Spencer, NE: C C Leonard. 14v. v1 n1. sep 17 1896-v14 n16. jan 7 1910 (wkly) [mf ed with gaps] – 2r – 1 – (absorbed by: butte gazette. publ at spencer ne, sep 17 1896-apr 22 1898; at butte ne, may 6 1898-jan 7 1910. issued with: boyd county advocate jan 29-sep 3 1897) – us NE Hist [071]

Boyd, James *see* Goethe's knowledge of english literature

Boyd, James Oscar *see* The octateuch in ethiopic

Boyd, James Robert *see*
- The being of god
- The communion table
- The westminster shorter catechism

Boyd, Julia *see* Bewick gleanings

Boyd, Robert *see* The lives and labors of moody and sankey

Boyd, Thomas Munford *see* Virginia bar examinations

Boyd, Thomas Parker *see* The how and why of the emmanuel movement

Boyd transcript – 1937-1943, 1943 dec 3-1946, 1947-1951, 1952-56, 1957-61, 1962-63, 1964-1966 jul 8 – 7r – 1 – mf#959834 – us WHS [071]

Boyd, William Andrew *see* Boyd's combined business directory for 1875-6

Boydell, Cary *see* Gender differences in sport centrality

Boydell, John *see* An alphabetical catalogue of plates

Boydell, Josiah *see* An alphabetical catalogue of plates

Boyds and their branches – v1 n1-v3 n3 [1985 summer-1987 winter] – 1r – 1 – mf#1671203 – us WHS [071]

Boyd's combined business directory for 1875-6 : containing an alphabetical list of all the merchants, manufacturers, tradesmen, etc of montreal, toronto...arranged under their proper headings... – Montreal: W Boyd, 1875 [mf ed 1980] – 4mf – 9 – 0-665-03714-7 – mf#03714 – cn Canadiana [380]

Boy-Ed, Ida *see* Das martyrium der charlotte von stein

Boyen, Herman von *see* Papers of herman von boyen, ca. 1787-1849

Boyer d'Agen *see* Introduction aux melodies gregoriennes

Boyer, Gaston *see* Un peuple de l'ouest africain

Boyer, Horace Clarence *see* Gospel song

Boyer, John Neely *see* The functional development of christian education in the church of the united brethren in christ

Boyer, Lucien *see* Paris-montreal

Boyer-Peyreleau, Eugene *see* Les antilles francaises, particulierement la guadeloupe

Der boyertown bauer – Boyertown, PA. 1851-1896. 1 roll – 13 – $25.00r – us IMR [071]

Boyertown democrat – Boyertown, PA. -w 1868-1915 – 13 – $25.00r – us IMR [071]

Boyesen, Hjalmar Hjorth *see*
- Essays on german literature
- A history of norway from the earliest times
- Ein kommentar zu goethes faust

Boyle, David *see*
- Some mental and social inheritances
- Uncle jim's canadian nursery rhymes
- The ups and downs of no 7, rexville

Boyle gazette, and roscommon reporter – Boyle, Ireland 14 feb-25 jul 1891 – 1 – (discontinued) – uk British Libr Newspaper [072]

Boyle, James Ernest *see* Marketing canada's wheat

Boyle, John *see* Plea for the episcopal church in scotland

Boyle, Leonard *see* Bibliotheca palatina

Boyle, Patrick *see* St vincent de paul and the vincentians in ireland, scotland, and england, a.d. 1638-1909

Boyle, Robert H et al *see* The waterhustlers

Boyle, Roger *see* Correspondence 1621-79

Boyles, Anne Mccollum *see* Story of orlando

Boylesve, Marin de *see* Le mois du precieux sang de n s jesus-christ

Boylston 1742-1905 – Oxford, MA (mf ed 1996) – (= ser Massachusetts vital records) – 87mf – 9 – (mf 1: vitals index 1768-1843. mf 2-3: intentions & marrs 1786-1835. mf 4: births & deaths 1768-1843. mf 4: out-of-town marriages 1788-99. mf 5: deaths 1776-1856. mf 6-7: intentions 1835-1917; marriages 1794-1843. mf 5,8-10: precinct 1742-86. mf 11-16: town records 1815-35. mf 17-24: town records 1835-62. mf 25-32: mortgages 1838-85. mf 33-41: store accounts 1802-25. mf 42-45: taxes 1797-1812. mf 45-51: taxes 1812-40. mf 52-56: taxes 1841-59. mf 57-61: taxes 1860-64. mf 62: rebellion rec 1861-65. mf 63-64: voters 1884-1939. mf 65-78: paupers 1846-1939. mf 79-81: paupers register 1907-43. mf 82-83: vital records 1844-66. mf 84-85: deaths 1867-1910. mf marriages 1867-1906. mf 87: births) – us Archive [978]

Boym, M P *see* Flora sinensis ou trait, des fleurs, des plantes et des animaux particuliers a la chine

Boynton, B *see* The physical growth of girls

Boynton, Charles Brandon *see* History of the navy during the rebellion

Boynton, Charles Luther *see* Notes on the chronological list of missionaries to china and the chinese, 1807-1942

Boynton, George Mills *see* The congregational way

Boynton, Richard Wilson *see* The vital issues of the war

Boyon, Jacques *see* Naissance d'un etat africain

Boy's adventures in the west indies / Ober, Frederick Albion – Boston, MA. 1888 – 1r – us UF Libraries [972]

The boys and girls clubs of nova scotia : one club's experience with take it e.a.s.y! / Davison, Carolyn J – 1998 – 2mf – 9 – $8.00 – mf#HE 634 – us Kinesology [370]

Boys' and girls' throwing development : a comparison of two cohorts twenty years apart / Pulito, Brenda – 2000 – 1mf – 9 – $4.00 – mf#PE 4060 – us Kinesology [611]

A boy's books, then and now, 1818, 1881 : a series of annotations from the "canada educational monthly" / Scadding, Henry – Toronto?: C B Robinson, 1882 – 2mf – 9 – mf#27984 – cn Canadiana [410]
Boys Choir of Harlem, Inc *see* Chorister
Boys, Edward *see* Narrative of a captivity and adventures in france and flanders
Boys, Ernest *see* The sure foundation
Boys' life – Irving TX 1911+ – 1,5,9 – ISSN: 0006-8608 – mf#2506 – us UMI ProQuest [370]
Boys' life of mark twain / Paine, Albert Bigelow – New York, NY. 1916 – 1r – us UF Libraries [420]
The boys of grand pre school / De Mille, James – Boston: Lee and Shepard, 1873 – (= ser The "bowc" series) – 4mf – 9 – 0-665-90791-5 – mf#90791 – cn Canadiana [830]
The boys of priors dean / Allen, Phoebe – London: John Hogg, [1891] – (= ser 19th c women writers) – 3mf – 9 – mf#5.1.130 – uk Chadwyck [830]
The boy's own book : a complete encyclopaedia of sports and pastimes, athletic, scientific, and recreative / Clarke, William – new ed. London: Crosby Lockwood & Son, 1889 – (= ser 19th c children's literature) – 8mf – 9 – mf#6.1.2 – uk Chadwyck [030]
The boy's own book : a complete encyclopedia of all the diversions, athletic, scientific, and recreative, of boyhood and youth / Clarke, William – London: Vizetelly, Branston & Co, 1828 – (= ser 19th c children's literature) – 5mf – 9 – mf#6.1.1 – uk Chadwyck [030]
Boys' own philatelist – Berlin [Kitchener], Ont.: Ontario Philatelic Co., [1897-1898] – 9 – (cont by: the canadian philatelic weekly) – mf#P05153 – cn Canadiana [760]
A boy's religion : from memory / Jones, Rufus Matthew – Philadelphia: Ferris & Leach, 1902 [mf ed 1990] – 1mf – 9 – 0-7905-7652-X – mf#1989-0877 – us ATLA [920]
A boy's religion from memory / Jones, Rufus M – Philadelphia: Ferris & Leach, 1902 – 141p/pl – 1 – mf#8314 – us Wisconsin U Libr [920]
Boys, Thomas *see*
– Christian dispensation miraculous
– A key to the psalms
Boys' weekly – 1922-29 – 1 – 57.96 – us Southern Baptist [242]
Boys, William Fuller Alves *see* A practical treatise on the office and duties of coroners in ontario and the other provinces and the territories of canada and in the colony of newfoundland
Boysen, Friedrich Eberhard *see* Allgemeines historisches magazin
Boysen van Nienkarken, Johannes Wilhelm *see* Leeder und stueckschen in ditmarscher platt
Boyton, Paul (Mrs) *see* A heroic priest
Boyvin, J G *see* Philosophia scoti a prolixitate et subtilitas ejus ab obscuritate libera et vindicata...
Boza Masvidal, Aurelio A *see* Evocaciones y reflexiones universitarias
Bozena : erzaehlung / Ebner-Eschenbach, Marie von – Stuttgart: J G Cotta 1920 [mf ed 1990] – 1r – 1 – (filmed with: hanchen und die kuechlein / a g eberhard & other titles) – mf#7268 – us Wisconsin U Libr [830]
Bozena : neue erzaehlungen / Ebner-Eschenbach, Marie von – Leipzig: H Fikentscher, H Schmidt & H Guenther [1928] [mf ed 1993] – 2r – 1 – mf#8570 reel 1 – us Wisconsin U Libr [830]
Bozkurt, Mahmut *see* Die waehrungspolitische kooperation in der europaeischen gemeinschaft
Bozner buergerspiele, alpendeutsche prang- und kranzfeste / ed by Doerrer, Anton – Leipzig: K W Hiersemann, 1941- [mf ed 1993] – (= ser Blvs 291) – 1 – (no more publ? incl bibl ref) – mf#8470 reel 57 – us Wisconsin U Libr [820]
Boznice starozytne w polsce – Warsaw, Poland 1912-13 [mf ed Norman Ross] – 1r – 1 – (with: przeglad judaistyczny: organ poswiecony nauce, literaturze i sztuce zydowskiej [poznan, poland] 1922) – mf#nrp-2092 – us UMI ProQuest [939]
Boztepe, Halil Nihad *see* Mahitap
Bpft, aft action / Bethel Park Federation of Teachers – v6 n1-5 [1978 jan-oct] – 1r – 1 – (cont: bft, aft news) – mf#647056 – us WHS [370]
Bpft news / Bethel Park Federation of Teachers – n7-11 [1974 dec-1975 may] – 1r – 1 – (cont by: bpft united teachers) – mf#647050 – us WHS [370]
Bpft united teachers / Bethel Park Federation of Teachers – v3 n1-8 [1975 oct-1976 jun] – 1r – 1 – (cont: bpft news; cont by: bft, aft news) – mf#647051 – us WHS [370]
Bpu pertani / Pertani – Djakarta, 1962(1-12) – 7mf – 9 – mf#SE-1895 – ne IDC [950]
Bqr 1821-1967 t 1-22 : [index] / Bibliotheque nationale du Quebec. Bureau de la bibliographie retrospective – Bibliographie du Quebec, 1821-1967 [mf ed 1990] – 35mf – 9 – cn Bibl Nat [019]
Braam, A E van *see* An authentic account of the embassy of the dutch east-india company
Braam Houckgeest, Andre Everard van *see* Voyage de l'ambassade de la compagnie des indes orientales hollandaises, vers l'empereur de la chine, en 1794 et 1795
Braam Houckgeest, Andreas E van *see* Voyage de l'ambassade de la compagnie des indes orientales hollandaises, vers l'empereur de la chine, en 1794 et 1795
Braasch, E F *see* Comparative darstellung des religionsbegriffes in den verschiedenen auflagen der schleiermacher'schen "reden"
Braasch, Theodor *see* Bilder aus der schweiz
Braatz, Donald Otto *see* Style of giovanni gabrieli
Braatz, Janelle S *see* The effect of a physical activity intervention based on the transtheoretical model in changing physical-activity-related behavior on low-income elderly volunteers
Brace, Charles Loring *see*
– Gesta christi
– The unknown god
Brace, David K *see* Measuring motor ability
Bracebridge gazette – 1955-1986 – cn Commonwealth Imaging [070]
Bracebridge herald-gazette – 1956-1986 – cn Commonwealth Imaging [070]
Le bracelet de fer : grand roman canadien inedit / Lacerte, A Bourgeois – Montreal: Ed E Garand, 1926 [mf ed 1982] – (= ser Le Roman canadien) – 2mf – 9 – (ill by: albert fournier) – mf#SEM105P65 – cn Bibl Nat [830]
Bracelete de safiras / Barroso, Gustavo – Rio de Janeiro, Brazil. 1931? – 1r – us UF Libraries [972]
Bracey, John H Jr *see* The bayard rustin papers
Bracey, John H, Jr *see* The papers of a philip randolph
Bracgrounder / Brotherhood of Railway, Airline and Steamship Clerks, Frieght Handlers, Express and Station Employees – v9 n1-3 [1987 mar/apr, aug, oct/nov] – 1r – 1 – mf#1601031 – us WHS [380]
Brachert, Thomas C *see* Klimasteuerung von karbonatsystemen
Brachmann, Friedrich *see* Christ-comoedia
Brachnaia gazeta / ed by Taratorin, I V – Irkutsk: [s n] 1919 [1919-] – (= ser Asn 1-3) – n2-8 [1919] item 19, on reel n7 – 1 – mf#asn-1 019 – ne IDC [077]
Brachvogel, Albert Emil *see*
– Ausgewaehlte werke
– Der deutsche michael
– Der fels von erz
– Lessings laokoon
– Narciss
– Object und methode der neutestamentlichen schriftlectuere in dem evangelischen religionsunterrichte der beiden oberen gymnasialklassen
Brachvogel, Carry *see* Der abtruennige
Brachvogel, Udo *see* Gedichte
Brackenbury, Henry *see* The river column
Brackenridge, Henry *see* Voyage to south america
Brackenridge, Henry M *see* Ansichten von louisiana
Brackenridge, W D *see*
– Botany. cryptogamia. filices
– Filices in charles wilkes' u.s. exploring expedition
Bracker, pastor *see*
– Die breklumer mission in indien
– Jeypur, land und leute
Bracket – v6 n3 [1978 oct], 1982 apr-spring, 1983 spring-1986 spring – 1r – 1 – (cont by: iowa historian) – mf#1132907 – us WHS [071]
Brackley observer & northamptonshire, bucks & oxon advertiser – Brackley, England 16 jul 1869-22 oct 1880 – 1 – (cont by: brackley observer & bicester advertiser [29 oct 1880-27 dec 1901]) – uk British Libr Newspaper [072]
Bracknell & ascot times – Bracknell, England 29 jul 1990- [mf 1972-81, 1983-] – 1 – (cont: bracknell times [25 may 1956-14 jun 1957, 6 jan 1972-31 dec 1981, 6 jan 1983-19 jul 1990) – uk British Libr Newspaper [072]
Bracknell news – Bracknell, England 5 feb 1959- [mf 1986-] – 1 – uk British Libr Newspaper [072]
Bracko, Michael R *see* Time motion analysis of the skating characteristics of professional ice hockey players
Braconnier, Edouard *see* Application de la geographie a l'histoire
Bradbury, James *see* India
Bradbury, John *see* Travels in the interior of america, in the years 1809, 1810, and 1811
Bradbury, William Batchelder *see*
– Fresh laurels for the sunday school
– Psalmist or choir melodies
– The victory: a new collection of sacred and secular music, comprising a great variety of tunes, anthems, glees, elementary exercises and social songs...
Bradbury's pleading and practice reports / New York. (State) – v1-5. 1910-19 – 36mf – 9 – $54.00 – mf#LLMC 80-018 – us LLMC [340]
Bradby, Eliza Dorothy *see* Short history of the french revolution, 1789-1795
Braddock tribune – Braddock, PA. -w 1889-1890; 1891-1893 – 13 – $25.00r – us IMR [071]
Braddon, Mary Elizabeth *see*
– Diavola
– Rough justice
– Rupert godwin
– Sensation fiction
Brade, William *see* Oedipus on the sphinx of the nineteenth century
Braden, Clark *see* Debate on the action of baptism
Braden, Roberta L *see* Sex role perceptions and defense mechanisms of female athletes
Braden, William *see* The beautiful gleaner
Bradenton herald – Bradenton, FL. 1941 jan-1942 apr – 2r – us UF Libraries [071]
Bradfield, William *see* Personality and fellowship
Bradford 1669-1849 – Oxford, MA (mf ed 1995) – (= ser Massachusetts vital record transcripts to 1850) – 15mf – 9 – 0-87623-215-2 – (mf 1t: births 1669-78. mf 1t-3t: vital records 1678-1739. mf 4t-5t: deaths 1736-74; marriages 1735-68. mf 4t-8t: births 1684-1796. mf 6t-8t: marriages 1679-1795. mf 7t-8t: deaths 1682-1796. mf 8t-11t: births 1772-1844. mf 11t-12t: marriages 1795-1848. mf 12t: out-of-town marriages 1684-1799. mf 12t-13t: deaths 1788-1847. mf 14t: births 1843-49. mf 15t: marriages & deaths 1843-49) – us Archive [978]
Bradford, Alden *see* Memoir of the life and writings of rev. jonathan mayhew, d.d
Bradford, Amory Howe *see*
– The age of faith
– The ascent of the soul
– Christ and the church
– Pilgrim in old england
Bradford argus – Towanda, PA. -w 1861-1912 – 13 – $25.00r – us IMR [071]
Bradford County Historical Society *see* Settler
Bradford county telegraph – Starke, FL. 1887-1998 jun – 80r – (gaps) – us UF Libraries [071]
Bradford county telegraph – Starke, FL. v118 n26-v119 n26. 1998 jan 08, jul 02-dec 31 – 1r – us UF Libraries [071]
Bradford daily argus – Bradford, England 16 jun 1892-14 jul 1923 – 1 – (wanting: jan-jun 1896, jul-dec 1897, dec 1899, jan-dec 1916) – uk British Libr Newspaper [072]
Bradford daily telegraph – Bradford, England 16 jul 1868-15 dec 1926 (daily) – 1 – (wanting: may-aug 1911, jan-apr, sep-dec 1912; cont by: bradford telegraph & argus [16 dec 1926-10 may 1930]) – uk British Libr Newspaper [072]
Bradford evening star and bradford daily record – Bradford, PA. -d 1894-1903; 1928-1931; 1931-1943; 1943-1946 – 13 – $25.00r – us IMR [071]
Bradford, Gamaliel *see* Types of american character
Bradford historical and antiquarian society : local record series – v1-4. 1929-53 – (= ser Publications of the english record societies, 1835-1972) – 9mf – 9 – uk Chadwyck [941]
Bradford, James C *see* Papers of john paul jones
Bradford, John *see* An address to the inhabitants of new brunswick, nova scotia, in north america
Bradford labour echo *see* Independent labour party newspapers
Bradford labour echo, 1895-99 : from bradford central library – 1r – 1 – mf#97017 – uk Microform Academic [072]
Bradford, Mary F *see* Side trips in jamaica
Bradford. New Hampshire. Christian Church *see* Church records, ms 1797
Bradford observer – Bradford, England 6 feb 1834-16 nov 1901 (wkly) [mf 1851, 15 may-dec 1871, 1872, 1896, apr-dec 1897, sep-dec 1900, sep-16 nov 1901, nov 1907-7 jan 1909, apr-jun 1910, 1911, jun-dec 1912, jan-mar 1917] – 1 – (cont by: yorkshire daily observer [18 nov 1901-15 jan 1909]) – uk British Libr Newspaper [072]
Bradford pioneer, the... 1913-35 : from bradford central library – 8r – 1 – mf#96918 – uk Microform Academic [072]
Bradford republican – Bradford IL. 1900 may 31, jun 21, 1901 apr 4, 1903 jul 9 – 1r – 1 – mf#1010682 – us WHS [071]
Bradford, Robert *see* Addresses delivered at agincourt, april 2nd and may 7th, 1878
Bradford, Simeon Briggs *see* Prohibition in kansas and the kansas prohibitory law
Bradford star – Bradford, PA., 1894-1903 – 13 – $25.00r – us IMR [071]
Bradford telegraph and argus – Bradford, England 16 dec 1926-10 may 1930 – 1 – (cont: bradford daily telegraph [16 jul 1868-15 dec 1926]; cont by: telegraph & argus [12 may 1930-]) – uk British Libr Newspaper [072]
Bradford, Thomas Lindsley *see* Bibliographer's manual of american history
Bradford trades council, 1867-1951 – (= ser Labour party in britain, origins and development at local level. series 1) – 5r – 1 – (int by john w boyle) – mf#97242 – uk Microform Academic [331]
Bradford weekly telegraph – Bradford, England 31 jul 1869-18 may 1878, 13 may 1882-12 apr 1884 (wkly) – 1 – (not publ between 18 may 1878-13 may 1882; cont by: illustrated weekly telegraph [19 apr 1884-29 jul 1899]) – uk British Libr Newspaper [072]
Bradford, William *see* History of plymouth plantation 1620-1647
Bradford's cases – Iowa. 1v. 1838-41 (all publ) – (= ser Iowa supreme court reports) – 2mf – 9 – $3.00 – (a pre-nrs title) – mf#LLMC 94-004 – us LLMC [347]
Bradish, Joseph Arno von *see*
– Der briefwechsel hofmannsthal-wildgans
– Goethe als erbe seiner ahnen
Bradke, P von *see* Dyaus asura, ahura mazda und die asuras
Bradlaugh Bonner, Hypatia *see* The reformer, 1897-1904
Bradlaugh, Charles *see*
– Atonement
– Cardinal's broken oath
– Few words about the devil
– Humanity's gain from unbelief
– Indian money matters
– Is there a god?
Bradlaugh versus besant / Bradlaugh, William Robert – London, England. 18– – 1r – us UF Libraries [240]
Bradlaugh, William Robert *see*
– Autobiography and conversion of w r bradlaugh
– Bradlaugh versus besant
– Christianity established by jewish and pagan testimony
– Is there a hell?
– Sceptic defeated with his own weapons
Bradle, T A *see* Levels of park satisfaction among florida state park visitors and their relationships to selected visitor characteristics
Bradley, A L *see* The wingate test
Bradley, Andrew Cecil *see*
– Philosophical lectures and remains of richard lewis nettleship
– Prolegomena to ethics
Bradley, Arthur Granville *see* The emigration of gentlemen's sons to the united states and canada
Bradley, B T *see* New etchings of old india
Bradley, Carolyn G *see* The effects of diet and exercise of varying intensities on the body composition of adult women
Bradley, Cornelius Beach *see* A half century among the siamese and the lao
Bradley County Historical Society *see* Views
Bradley, Dan Beach *see* Old testament history
Bradley, Ernest J *see* Distinctive plea of the disciples of christ
Bradley, Francis Herbert *see*
– Appearance and reality
– Essays on truth and reality
– Ethical studies
– The principles of logic
Bradley, George Granville *see* Lectures on ecclesiastes delivered in westminster abbey
Bradley, Henry *see* Story of the goths
Bradley, Hugh *see* Havana, cinderella's city
Bradley, James *see* The nettle creek church case or who are the regular baptists
Bradley, James et al *see* Astronomical observations made at the royal observatory at greenwich...
Bradley, John William *see*
– The life and works of giorgio giulio clovio
– A manual of illumination on paper and vellum
Bradley, John William [comp] *see* A dictionary of miniaturists, illuminators, calligraphers, and copyists
Bradley, Joshua [comp] *see* Accounts of religious revivals in many parts of the united states from 1815 to 1818
Bradley, Katharine *see* Michael field and fin-de-siecle culture and society
Bradley, Kenneth *see* The story of northern rhodesia
Bradley, Patrick *see* Irish convert
Bradley, Susan *see* Archives biographiques francaises (abf1)
Bradley tech – Peoria IL. v39 n18 1936 feb 6, v39 n19 1936 feb 13 – 1r – 1 – mf#1159559 – us WHS [071]
Bradley-Birt, Francis Bradley *see* Chota nagpore
O brado africano – Lourenco Marques: Empreza do Journal O Brado Africano, [dec 24 1918-nov 24 1956; 1966-jun 15 1974] – 25r – 1 – us CRL [079]
O brado americano – Rio de Janeiro, RJ: Typ de Nicolao Lobo Vianna & Filhos, 25 mar-12 abr 1859 – (= ser Ps 19) – mf#P25,03,08 n05 – bl Biblioteca [320]
O brado da liberdade : orgao de interesses politicos e sociais – Bahia: Typ do Diario, 15 set 1876 – (= ser Ps 19) – mf#P17,01,26 – bl Biblioteca [300]

O brado da patria – Sao Paulo, SP: Typ Alema, 03 fev-20 mar 1865 – (= ser Ps 19) – mf#P18,01,75 – bl Biblioteca [321]
O brado do amazonas – Rio de Janeiro, RJ: Typ Carioca de J I da Silva & Comp, 20 abr-20 jun 1849 – (= ser Ps 19) – mf#P15,01,26 – bl Biblioteca [321]
O brado do amazonas – Rio de Janeiro, RJ: Typ Francesa, 29 set-dez 1852; jan, abr, jun-jul, dez 1853; jun-dez 1854; jan-maio, jul-out 1855: jul-set 1856; jan-12 fev 1858 – (= ser Ps 19) – mf#P15,01,27 – bl Biblioteca [321]
O brado do amazonas – Rio de Janeiro, RJ: Typ Imparcial de P Brito, 05 abr-23 maio 1845 – (= ser Ps 19) – mf#P15,1,25 – bl Biblioteca [321]
O brado liberal da bahia : periodico politico, litterario e noticioso – Bahia: Typ de FA de Almeida, 22 dez 1869 – (= ser Ps 19) – mf#P17,01,32 – bl Biblioteca [321]
O brado natalence – Fortaleza, CE: Typ Americana, 21 jul, 21 ago 1849 – (= ser Ps 19) – mf#P17,02,198 – bl Biblioteca [320]
Bradshaw, James Daniel *see* A planned preaching program utilizing congregational involvement at first baptist church, mcdonough, georgia
Bradshaw, Maurice *see* Rucamiro
The bradshaw monitor – Bradshaw, NE: L D Beltzer. v13 n39. may 6 1909- (wkly) [mf ed -aug 26 1943 (gaps)] – 11r – 1 – (cont: bradshaw republican) – us NE Hist [071]
Bradshaw, W *see*
– English puritarisme
– A protestation of the kings supreamacie
– The unreasonablness of the separation
Bradshaws in america – v1-v9 n1 [1969 jun-1977 mar] – 1r – 1 – mf#354471 – us WHS [360]
Bradshaw's journal – La Salle IL 1841-43 – 1 – mf#4712 – us UMI ProQuest [073]
Bradshaw's railway manual – Shareholders' guide & official directory. London. v. 26-75. 1874-1923 – 1 – us NY Public [380]
Bradstreet, Anne *see* The tenth muse and from the manuscripts, meditations divine and morall, together with letters and occasional pieces
Bradt, Charles Edwin *see* Men and the modern missionary enterprise
Bradt, Charles Edwin et al *see* Around the world studies and stories of presbyterian foreign missions
Bradway-broadway bulletin – v1 n1-4=rev [1973 nov-1974 aug], v2 n1-4 [1974 nov-1975 jul] – 1r – 1 – (cont: br(o)adway bulletin; cont by: bradway-broadway bulletin and family research) – mf#626538 – us WHS [929]
Bradway-broadway bulletin and family research – 1975 nov-1983 summer – 1r – 1 – (cont: bradway-broadway bulletin; cont by: newsletter [bradway-broadway bulletin and family research]) – mf#626533 – us WHS [929]
Brady, Alexander *see*
– Democracy in the dominions
Brady, Charles B *see* An index to the arkansas reports, vols 1 to 31 inclusive
Brady, Christine P *see* Effects of acute resistive exercise on the resting metabolic rate of women
Brady, Cyrus Townsend *see*
– American fights and fighters
– A baby of the frontier
– A doctor of philosophy
– The fetters of freedom
– Gethsemane and after
– My lady's slipper
– Recollections of a missionary in the great west
– The records
– Under topsils and tents
Brady, F X *see* The great supper of god
Brady, Robert Alexander *see* Business as a system of power
Brady studio carte-de-visite portraits / U.S. Library of Congress. Prints and Photographs Division – 193 photographs. 1 reel. P&P2822 – 1 – us L of C Photodup [976]
Brady studio civil war views / U.S. Library of Congress. Prints and Photographs Division – 10,000 Civil War images. P&P1 – 1 – us L of C Photodup [976]
Brady vindicator – Brady, NE: Totter & Swancutt (wkly) [mf ed v2 n11. aug 13 1909-oct 30 1941 (gaps)] – 10r – 1 – (absorbed by: lincoln county tribune (north platte 1930)) – us NE Hist [071]
Brady, William Maziere *see*
– The episcopal succession in england, scotland and ireland, a.d. 1400 to 1875
– Essays on the english state church in ireland
– The irish reformation
– State papers concerning the irish church in the time of queen elizabeth
Braeker, Jakob *see* Der erzieherische gehalt in j j breitingers 'critischer dichtkunst'
Braeker, Ulrich *see*
– Der arme mann im tockenburg
– Das leben und die abenteuer des armen mannes im tockenburg
Det braendende sporgsmal – Copenhagen, Denmark. jan 1946-sep 1947 – 1/4r – 1 – uk British Libr Newspaper [074]
Braeuning-Oktavio, Hermann *see*
– Beitraege zur geschichte und frage nach den mitarbeitern der "frankfurter gelehrten anzeigen" vom jahre 1772
– Silhouetten aus der wertherzeit
Braeutigam, Harald *see* Die handelspolitik polens seit erlangung der selbstaendigkeit bis zum ablauf der genfer konvention am 15. juni 1925
Braga : vtoraia kniga stikhov 1921-1922 / Tikhonov, Nikolai – Moskva, Peterburg: Krug 1922 – us CRL [947]
Braga : vtoraia kniga stikhov, 1921-1922 / Tikhonov, Nikolai Semenovich – Moskva: "Krug", 1922 [mf ed 2002] – 1r – 1 – (filmed with: 255 stranitis maiakovskogo. knj 1 (1923)) – mf#5214 – us Wisconsin U Libr [800]
Braga, Teofilo *see*
– Bocage
– Garrett e o romantismo
O bragantino : jornal do povo – Bragança, SP: Typ do Bragantino, 23 dez 1876; abr 1877; maio 1879; maio-13 jun 1880 – (= ser Ps 19) – mf#P18,01,74 – bl Biblioteca [073]
Bragg briefs – Spring Lake NC 1975 – 1 – ISSN: 0006-8713 – mf#9055 – us UMI ProQuest [355]
Bragg briefs – [v1-v6 n2] 1969 jul 4-1973 feb – 1r – 1 – mf#765320 – us WHS [071]
Bragg, John *see* The diary, 1771-94
Bragg, Raymond Bennett *see* Principles and purposes of the free religious association
Bragg, Thomas *see* Diary
Braght, Thieleman Janszoon van *see* Martyrology of the churches of christ...
Bragin, M *see* Kooperativy
Braginskii, D *see* Bibliograficheskii ukazatel perevodnoi belletristiki v russkikh zhurnalakh za piat let 1897-1901 g
Braginskii, M et al *see* Obshchestva, tovarishchestva, tresty, arteli, kooperativy i drugie obedineniia
Brah nan indradevi : pralom lok knun qaksar sastr dak dan nyn pravattisastr [a novel] / Dik Gam – Bhnam Ben: Mau Can Thamn 2509 [1966] [mf ed 1990] – 1r with other items – 1 – (in khmer; incl bibl ref) – mf#mf-10289 seam reel 127/7 [§] – us CRL [830]
Brah nan jay devi : pralom lok phnaek pravattisastr / Ravivans Govid – Bhnam Ben: Pannagar Bejr Nil 2510 [1967] [mf ed 1990] – 1r with other items – 1 – (in khmer) – mf#mf-10289 seam reel 107/11 [§] – us CRL [959]
Brah pad padum raja : pralom lok phnaek pravattisastr / Hak Chai Huk – Bhnam Ben: Pannagar Sen Nuan Huat 2510 [1967] [mf ed 1990] – [ill] 1r with other items – 1 – mf#mf-10289 seam reel 097/09 [§] – us CRL [830]
Brah pad sri suriyobarn : pralom lok phnaek pravattisastr / Hak Chai Huk – Bhnam Ben: Pannalay Vannakamm Khmaer 2509 [1966] [mf ed 1990] – 1r with other items – 1 – (in khmer) – mf#mf-10289 seam reel 097/08 [§] – us CRL [480]
Brah qadity thmi rah loe phaen ti cas : [a novel] / Suan Surind – Bhnam Ben: Ron Bumb Viriyah 2504 [1961] [mf ed 1990] – 1r with other items – 1 – (in khmer) – mf#mf-10289 seam reel 103/08 [§] – us CRL [830]
Brah raj bidhi dvadasamas / Krasem – Bhnam Ben: Buddhasasan Pandity [1951-61] [mf ed 1990] – 3v on 1r with other items – 1 – (in khmer; [1] brah raj bidhi knun khae cetr; [2] brah raj bidhi knun khae bisakh nyn khae jesth; [3] brah raj bidhi knun khae qasadh-sraban-bhaddapad qassuj-kattik-migasir-puss-maghphalgun) – mf#mf-10289 seam reel 121/3 [§] – us CRL [390]
Brah sangh jajhloen maenr? – Bhnam Ben: Thac Saret 2515 [1971] [mf ed 1990] – 1r with other items – 1 – (in khmer) – mf#mf-10289 seam reel 116/2 [§] – us CRL [230]
Brah Silasamvara [Ras] *see* Gihi pratipatti bises bistar
Brahe, Tycho *see*
– Historias brasileiras
– Triangulorum planorum et sphaericorum praxis arithmetica
Brahler, C Jayne *see* Versaclimber exercise elicits higher maximal oxygen uptake in women rowers than does treadmill exercise or rowing ergometry
Brahm Kusal *see* Ras phcal maccuraj
Brahm, Otto *see*
– Gottfried keller
– Heinrich von kleist
Brahm, William Gerard de *see* Partial transcript report of the general survey in the southern district of north america
Brahma : ballo in sette atti e prologo. musica del maestro costantino dall' argine, da reppresentarsi al teatro comunale di bologna l'autunno 1868 / Monplaisir, Ippolito Giorgio – Bologna, Tip de G Vitali, 1868 – 1 – mf#*ZBD-*MGTZ pv7-Res – Located: NYPL – us Misc Inst [790]
Brahma, Nalinikanta *see* Philosophy of hindhu sadhana
Brahma und die brahmanen : vortrag in der oeffentlichen sitzung der k. akademie der wissenschaften am 28. maerz 1871 zur feier ihres einhundert und zwoelften stiftungstages / Haug, Martin – Muenchen: Koenigl Akademie, 1871 – 1mf – 9 – 0-524-01603-8 – mf#1990-2542 – us ATLA [280]
The brahmacharin : a monthly magazine devoted to hindu social, religious and moral reforms... – Jessore, India. 1901-07 [mf ed 2001] – (= ser Christianity's encounter with world religions, 1850-1950) – 1r – 1 – mf#2001-s098 – us ATLA [290]
Brahmadarsanam : or, intuition of the absolute: being an introduction to the study of hindu philosophy / Ananda Acharya – New York: Macmillan, 1917 [mf ed 1991] – 1mf – 9 – 0-524-00816-7 – (incl bibl ref) – mf#1990-2062 – us ATLA [280]
Brahmadarsanam : or, intuition of the absolute, being an introduction to the study of hindu philosophy / Ananda Acharya – New York: Macmillan, 1917 [mf ed 1995] – (= ser Yale coll) – xii/210p (ill) – 1 – 0-524-09598-1 – mf#1995-0598 – us ATLA [280]
Brahma-knowledge : an outline of the philosophy of the vedanta, as set forth by the upanishads and by sankara / Barnett, Lionel David – London: J Murray 1907 [mf ed 1991] – (= ser Wisdom of the east series (london, england)) – 1mf – 9 – 0-524-01039-0 – (incl english trans fr selections) – mf#1990-2187 – us ATLA [280]
Brahman : a study in the history of indian philosophy / Griswold, Hervey De Witt – New York: Macmillan 1900 [mf ed 1991] – (= ser Cornell studies in philosophy 2) – 1mf – 9 – 0-524-01489-2 – mf#1990-2465 – us ATLA [280]
Brahmananda keshub chunder sen : "testimonies in memoriam" / Banerji, G C [comp] – Allahabad: G C Banerji, 1937- – (= ser Samp: indian books) – us CRL [920]
The brahmanas of the vedas / Macdonald, Kenneth Somerled – 2nd ed. London: Christian Literature Society for India, 1901 – (= ser Sacred Books of the East Examined and Described) – 1mf – 9 – 0-524-01790-5 – mf#1990-2638 – us ATLA [280]
Brahmanical gods in burma : a chapter of indian art and iconography / Ray, Niharranjan – Calcutta: University of Calcutta, 1932 – (= ser Samp: indian books) – us CRL [280]
Brahmanism and hinduism : or, religious thought and life in india / Monier-Williams, Monier – 4th enl and improved ed. New York: Macmillan, 1891 – 2mf – 9 – 0-524-04345-0 – mf#1990-3329 – us ATLA [280]
Le brahmanisme / Godard, Charles – Paris: Librairie Bloud, 1904 – (= ser Science et Religion) – 1mf – 9 – 0-524-01176-1 – mf#1990-2252 – us ATLA [280]
The brahmans, theists and muslims of india : studies of goddess-worship in bengal, caste, brahmaism and social reform / Oman, John Campbell – 2nd ed. London: T Fisher Unwin, [1909?] – 1mf – 9 – 0-524-02100-7 – mf#1990-2864 – us ATLA [280]
The brahmans, theists, and muslims of india : studies of goddess-worship in bengal, caste, brahmaism and social reform, with descriptive sketches of curious festivals, ceremonies, and faquirs / Oman, John Campbell – London: T Fisher Unwin, 1907 – (= ser Samp: indian books) – us CRL [280]
Brahmins and pariahs : an appeal by the indigo manufacturers of bengal to the british government, parliament, and people, for protection against the lieut-governor of bengal... – London, 1861 – (= ser 19th c books on british colonization) – 3mf – 9 – mf#1.1.869 – uk Chadwyck [305]
The brahmo samaj : keshub chunder sen's lectures in india / Sen, Keshub Chunder – Calcutta: Brahmo Tract Society, 1883 – 1mf – 9 – 0-524-02233-X – mf#1990-2907 – us ATLA [280]
The brahmo samaj and arya samaj in their bearing upon christianity : a study in indian theism / Lillingston, Frank – London: Macmillan, 1901 – 1mf – 9 – 0-524-02091-4 – (incl bibl ref) – mf#1990-2855 – us ATLA [280]
The brahmo somaj : lectures and tracts / Sen, Keshub Chunder; ed by Collet, Sophia Dobson – London: Strahan, 1870 – (= ser Yale coll) – vii/288p – 1 – 0-524-09792-5 – mf#1995-0792 – us ATLA [280]
Brahmoism : or, history of reformed hinduism. from its origin in 1830, under rajah mohun roy, to the present time / Bose, Ram Chandra – New York: Funk and Wagnalls, 1884 – 1mf – 9 – 0-524-01254-7 – mf#1990-2290 – us ATLA [280]
Brahmopanisat-sara sangraha – Allahabad: Panini Office, Bhuvaneswari Asrama, 1916 – (= ser Samp: indian books) – (trans by vidyatilaka) – us CRL [280]
Brahms, Johannes *see*
– Oktaven und quinten, u.a
– Samtliche werke
Brahms, Johannes et al *see*
– Franz peter schubert
– Wolfgang a. mozart
Brahn Vira *see* Maha sankrant chnam khal naksatr
Braid, William David *see* Statement of the east india company's conduct towards the carnatic stipendiaries
Braiden, Russell W *see* The effect of cocaine on muscle carbohydrate metabolism and endurance during high intensity exercise in rats
Braidwood and araluen express – Braidwood, jan 1899-dec 1907 – 4r – 9 – A$266.24 vesicular A$288.24 silver – at Pascoe [079]
Braidwood dispatch – Braidwood. aug 1888-aug 1889, jan 1897-dec 1968 – 31r – A$2096.34 vesicular A$2266.84 silver – at Pascoe [079]
Braidwood dispatch – Braidwood, jan 1969-apr 1970 – 1r – at Pascoe [079]
Braidwood, J *see* True yoke-fellows in the mission field
Braidwood news – Braidwood, mar-sep 1862, jan-dec 1864 – 1r – A$41.80 vesicular A$47.30 silver – at Pascoe [079]
Braidwood observer – Braidwood, jan 1860-dec 1862 – 1r – A$66.75 vesicular A$72.25 silver – at Pascoe [079]
Brailsford, Henry Noel *see*
– Rebel india
– Subject india
Brailsford, Mabel Richmond *see* Quaker women, 1650-1690
Brain : a journal of neurology – Oxford UK 1878+ – 1,5,9 – ISSN: 0006-8950 – mf#3185 – us UMI ProQuest [616]
Brain as organ of the mind / Bastian, Henry – London: Kegan Paul, Trench, Trubner & Co, 1890 [mf ed 1987] – 708p – 1 – mf#1956 – us Wisconsin U Libr [150]
Brain, behavior and evolution – Basel, Switzerland 1968+ – 1,5,9 – ISSN: 0006-8977 – mf#2737 – us UMI ProQuest [500]
Brain, Belle Marvel *see*
– Holding the ropes
– The redemption of the red man
– The transformation of hawaii
Brain injury [bi] – Abingdon, Oxfordshire 1987+ – 1,5,9 – ISSN: 0269-9052 – mf#17295 – us UMI ProQuest [610]
Brain lesions and functional results / Clark, Daniel – Utica, NY?: s.n, 1881? – 1mf – 9 – mf#01668 – cn Canadiana [611]
Brain, mind & common sense – Los Angeles CA 1993-94 – 1,5,9 – (cont: new sense bulletin; cont by: brain, mind) – ISSN: 1064-671X – mf#11410.03 – us UMI ProQuest [616]
The brain of india / Ghose, Aurobindo – Chandernagore: Prabartak Pub House, 1923 – (= ser Samp: indian books) – us CRL [180]
Brain research bulletin – Oxford UK 1985+ – 1,5,9 – ISSN: 0361-9230 – mf#49531 – us UMI ProQuest [612]
Brain research. gene expression patterns – Oxford UK 2001+ – 1,5,9 – ISSN: 1567-133X – mf#42845 – us UMI ProQuest [616]
Brain research protocols – Oxford UK 1997+ – 1,5,9 – ISSN: 1385-299X – mf#42792 – us UMI ProQuest [616]
Brain research reviews – Oxford UK 1992+ – 1,5,9 – ISSN: 0165-0173 – mf#42682 – us UMI ProQuest [612]
Brain research series – Oxford UK 1966+ – 1,5,9 – ISSN: 0006-8993 – mf#42243 – us UMI ProQuest [612]
Brain stuffing and forcing / Clark, Daniel – Toronto?: Warwick, 1887 – 1mf – 9 – mf#01601 – cn Canadiana [150]
Brainard clipper – Brainard, NE: W H McGaffin, Jr. 54v v1 n1 jul 16 1897-v54 n6 sep 14 1950 (wkly) [mf ed with gaps filmed 1959] – 12r – 1 – (absorbed by: people's banner. suspended foll mar 25 1943 issue; resumed with jun 12 1947) – us NE Hist [071]
Brainard's musical world – Cleveland. v17, n198-v. 29, n347. Jun 1880-Nov 1892. Incomplete – 1 – us NY Public [780]
Brainerd, David *see* Memoirs of rev. david brainerd
Brainerd, Mary *see* Life of rev. thomas brainerd
Brainerd, Thomas *see* The life of john brainerd
Brainin, Reuven *see* Perez
Brainin, Ruben *see* Mimisrach umimaarabh
Brain.mind bulletin – Los Angeles CA 1977-92 – 1,5,9 – (cont by: new sense bulletin) – ISSN: 0273-8546 – mf#11410.03 – us UMI ProQuest [616]
Braithwaite, Joseph Bevan *see*
– Memoirs of anna braithwaite
– Memoirs of joseph john gurney
Braithwaite, Rock *see* The effects of attentional focus and trait anxiety between starting and nonstarting division 1 basketball players
Braithwaite, William Charles *see*
– The beginnings of quakerism
– The message and mission of quakerism
– Spiritual guidance in the experience of the society of friends

Braitmaier, Friedrich *see*
- Goethekult und goethephilologie
- Die poetische theorie gottsched's und der schweizer

Brake, Laurel *see* Nineteenth-century british periodicals

Brake, Peter H Vande *see* Divine passibility

Brake, Wilhelm *see* Nieder mit den sozialdemokraten

Brakel, Wessel J A van *see* Solving three-layer planar microwave structures with the method-of-lines

Brakespeare : or, the fortunes of a free lance / Lawrence, George Alfred – Toronto: McLeod & Allen, [1904?] – 6mf – 9 – 0-665-73555-3 – mf#73555 – cn Canadiana [830]

Brakte, E *see* Das sogenannte religionsgespraech am hof der sasaniden

Bralyn can biar : ryan [a novel] / Guy Lut – Bhnam Ben: Ron Bumb Syn Hen 2508 [1965] [mf ed 1990] – 1r with other items – 1 – (in khmer) – mf#mf-10289 seam reel 104/5 [§] – us CRL [830]

Bralyn puspa : ryan / Duc Sidym – Bhnam Ben: Krum h'un Vann Cand Bhayant 2508 [1965] [mf ed 1990] – 1r with other items – 1 – (in khmer) – mf#mf-10289 seam reel 107/1 [§] – us CRL [959]

Braman, Sidney T *see*
- Aviation bases, flying fields
- Historical barometer in miami

Braman, Wallis D *see* Use of silence in the instrumental works of representative composers

Brambach, Wilhelm *see* Ueber die betonungsweise in der deutschen lyrik

Brambauer zeitung – Luenen 1918 30 apr, 1926 28 jan-5 aug [mf ed 2005] – 1r – 1 – (vbg: (luenen-) brambauer, dortmund-brechten, dortmund-holthausen) – gw Mikrofilm [074]

Bramborski serbski casnik – Cottbus DE, 1848 5 jul-1933 29 jul – 13r – 1 – (sorbischer ortsname: chosebuz) – gw Misc Inst [074]

Brameld, Theodore B H *see* Remaking of a culture

Bramesfeld, Anke *see* Beduerfnisse alter menschen in psychiatrischer behandlung

Bramfeld-poppenbuettler zeitung – Hamburg DE, 1905-34 – 55r – 1 – gw Misc Inst [074]

Bramhall, baptist chapel : burials 1894-2000 – [North Cheshire FHS] – (= ser Cheshire church registers) – 1mf – 9 – £3.00 – mf#281 – uk CheshireFHS [929]

Bramhall, baptist chapel – [North Cheshire FHS] – (= ser Cheshire monumental inscriptions) – 1mf – 9 – £2.50 – mf#146 – uk CheshireFHS [929]

Brammeier, Michele R *see* The influence of the menstrual cycle and diet on metabolism during rest and exercise

Bramsen, John *see*
- Letters of a prussian traveller
- Remarks on the north of spain

Bramstedter anzeigenblatt *see* Bramstedter nachrichten

Bramstedter nachrichten – Bad Bramstedt, Bad Segeberg DE, 1881-99, 1902-1945 2 may, 1949 24 sep-1985 – 141r – 1 – (title varies: 13 may-23 sep 1949: bramstedter anzeigenblatt; later publ in bad segeberg) – gw Misc Inst [074]

Bramstedter tageblatt – Bad Bramstedt DE, 1905-06 – 2r – 1 – gw Misc Inst [074]

Bramston, Mary *see* Judaea and her rulers

Bran, Friedrich Alexander *see* Herder und die deutsche kulturanschauung

Bran, Telesphore *see* De l'etablissement en canada de la fabrication du sucre de betterave

Branas, Cesar *see* Viento negro

Branch 5 newsletter / National Association of Letter Carriers – 1989 jan-1979 oct – 1r – 1 – (not iss jul-aug; cont by: gate city news) – mf#665123 – us WHS [380]

Branch 193 nalc / National Association of Letter Carriers – v37 n10-v41 n9 [1985 feb 8-1989 jan 10] – 1r – 1 – (cont: letter carriers branch 193 labor; cont by: branch 193 bulletin) – mf#1497050 – us WHS [380]

Branch bulletin / National Association for the Advancement of Colored People – v2 n7/8, 11-v3 n1, 9-v5 n1 [1918 jun/jul nov-19 jan, sep-1921 jan – 1r – 1 – mf#778333 – us WHS [322]

Branch department files – 4ser – (= ser Papers of the naacp 25) – 1 – (ser a: regional files & special reports, 1941-55 25r isbn 1-55655-719-1 $4840. ser b: regional files & special reports, 1956-65 18r isbn 1-55655-735-3 $3475. ser c: branch newsletters & printed matter 11r isbn 1-55655-832-5 $2135. ser d: branch dept general subject files, 1956-65 40r isbn 1-55655-841-4 $7765. with p/g) – us UPA [322]

Branch department files, 1965-1972 – 4ser – (= ser Papers of the naacp 29) – 1 – (ser a: field staff files 18r isbn 1-55655-893-7 $3485. ser b: branch newsletters, annual branch activities reports, & selected branch dept subject files 15r isbn 1-55655-916-X $2905. ser c: branch newsletters & regional field office files, 1966-71 10r isbn 1-55655-924-0 $1935. ser d: branch dept general subject files, 1966-70 13r isbn 1-55655-928-3 $2520. with p/g) – us UPA [322]

Branch manager's correspondence and related papers / W R Carpenter & Ltd. Tulagi Branch – 1925-32 – 1r – 1 – mf#PMB1112 – at Pacific Mss [980]

Branch news / Ontario Genealogical Society – v1 n1-v11 n6 [1970 apr-1979 jun] – 1r – 1 – (cont by: ottawa branch news) – mf#1017000 – us WHS [929]

Branch reporter / Star Co. Canton – jun 1943-sep 1945, feb 1946-79 [mthly] – 3r – 1 – mf#B10120-10122 – us Ohio Hist [331]

Une branche de la famille amyot-larpiniere : m georges-elie amyot, manufacturier et brasseur de quebec, ses ancetres directs et ses enfants / Demers, Benjamin – [Quebec?: s.n.], 1906 – 1mf – 9 – 0-665-73921-4 – mf#73921 – cn Canadiana [929]

Branche des royaux lignages / Guillaume – Paris 1828 [mf ed Hildesheim 1995-98] – 2v on 6mf – 9 – €120.00 – 3-487-26318-1 – gw Olms [929]

Branco, Castello *see* Summary of the president's message to the 1965 nat...

Brancos e pretos na bahia : estudo de contacto rac / Pierson, Donald – Sao Paulo, Brazil. 1945 – 1r – us UF Libraries [972]

Brand – Stockholm, Sweden. 1898-1967 – 14r – 1 – sw Kungliga [078]

Brand : tillfallighetstidningar – Stockholm, Sweden. 1909-46 – 1r – 1 – sw Kungliga [078]

Brand, A *see*
- Brevis descriptio itineris sinensis...legatione moscovitica anno 1693, 94 et 95...
- Relation du voyage de mr evert isbrand / envoye de sa majeste czarienne...l'empereur de la chine, en 1692, 93 et 94

Brand an der wolga : historisch-politischer roman aus russlands juengster vergangenheit / Boje, Walter – Berlin: P J Oestergaard, 1936 [mf ed 1989] – 302p/1pl – 1 – mf#7050 – us Wisconsin U Libr [830]

Brand book – n11 [1964] – 1r – 1 – mf#794354 – us WHS [071]

Brand, C J J *see* Shumo

Brand, Charles *see* Journal of a voyage to peru

Brand, James *see*
- History of the first church, oberlin, ohio
- James brand

Brand, Joel *see* Advocate for the dead

Brand, Lucia *see* Dans- en bewegingsterapie as groepterapietegniek in die opvoedkundige sielkunde

Brand plucked out of the fire – London, England. 18– – 1r – us UF Libraries [240]

Brand, Robert Henry *see* Union of south africa

Brandao, Ambrosio Fernandes *see*
- Dialogos das grandezas do brasil
- Dialogos das grandezas do brasil pela primeira

Brandao, Claudio *see* Antologia contemporanea

Brandeis and brandeis : the reversible mind of louis d brandeis – [S.l: s.n, 1912?] (mf ed 19–) – 57p – mf#ZT-TN pv26 n10 – us Harvard [338]

Brandeis brief *see* Opinion and brief in the case of muller vs oregon

Brandeis, Louis Dembitz *see*
- Brandeis on zionism
- The louis d brandeis papers
- Opinion and brief in the case of muller vs oregon

Brandeis on zionism / Brandeis, Louis Dembitz – Washington, [1942] – 2mf – 9 – mf#J-28-4 – ne IDC [956]

Brandell, Jerrold R *see*
- Journal of analytic social work
- Psychoanalytic social work

Brandel-Syrier, Mia *see* Black woman in search of god

Brandenburg, Edward *see* Brandenburg's bankruptcy digest

Brandenburg, Edwin Charles *see* The law of bankruptcy, including the national bankruptcy law of 1898.

Brandenburg, Erich *see* Vortraege

Brandenburg, Hans *see*
- Joseph von eichendorff
- Die schatulle des grafen thruemmel und andere nachgelassene gedichte

Brandenburg, Werner *see* Das poetische genus personifizierter substantiva bei james thomson und edward young

Brandenburger anzeiger *see* Brandenburgischer anzeiger

Brandenburger, C L *see* Historia de polonia

Brandenburger sagen : sagen und geschichten / Eynatten, Carola, Freiin von – Leipzig: B Franke 1893 [mf ed 1989] – 1r – 1 – (filmed with: reise zu den demokraten / richard euringer) – mf#7227 – us Wisconsin U Libr [390]

Brandenburgische neueste nachrichten – Potsdam DE, 1954 jul-1990 – 65r – 1 – (filmed by other misc inst: 1992- [4r/yr]; title varies: 29 feb 1992: potsdamer neueste nachrichten) – gw Misc Inst [074]

Brandenburgische neueste nachrichten – Brandenburg, 1992 2 jan-2 jul 2 – 1r – 1 – (title varies: 29 feb 1992: potsdamer neueste nachrichten, stadtausgabe brandenburg) – gw Misc Inst [074]

Brandenburgischer anzeiger – Brandenburg DE, 1817-18, 1822, 1824, 1826-33, 1835 [gaps]-1851, 1853-56, 1858-59, 1861-62, 1917, 1922, 1924-1927 aug, 1927 nov-1931, 1932 mar-1937 feb, 1937 mai-1938 mar, 1938 jul-1939 jun, 1939 okt-1944 1 jan – 114r – 1 – (title varies: 1823?: brandenburgscher anzeiger; 30 jul 1845: brandenburger anzeiger) – gw Misc Inst [074]

Brandenburg's bankruptcy digest / Brandenburg, Edward – Chicago: Callaghan, 1899 – 6mf – 9 – $9.00 – (index-digest of bankruptcy decisions containing the decisions of the supreme court of the us from 1800-99, and of the federal and state courts of last resort under the act of 1867) – mf#LLMC 84-390 – us LLMC [346]

Brandenburgscher anzeiger *see* Brandenburgischer anzeiger

Brandende alleenspraak met god / Peters, Gerlach – Hasselt, 1947 – €7.00 – ne Slangenburg [240]

Brandes, Friedrich *see* John knox, der reformator schottlands

Brandes, Heinrich *see* Die koenigsreihen von juda und israel nach den biblischen berichten und den keilinschriften

Brandes, Heinrich K *see* Ausflug nach portugal im sommer 1863

Brandes, J L A *see* Rapporten van de commissie in nederlandsch-indie voor oudheidkundig onderzoek op java en mandoera

Brandes, Karl *see* Heilige petrus in rom und rom ohne petrum

Brandes, Mechthild *see* Bundespraesidialamt. amtszeit prof dr theodor heuss (bestand b 122) bd 38

Brandi, Salvatore Maria *see* The school question in the united states

Brandis, Cordt von *see* Der luchhof

Brandis, J Dietr et al *see* Technologisches taschenbuch fuer den kuenstler, fabrikanten und metallurgen auf das jahr 1786

Brandis, Johannes *see* Ueber den historischen gewinn aus der entzifferung der assyrischen inschriften

Brandl, Alois *see* Zwischen inn und themse

Brandl, Benedict *see* Lessings fragmentenstreit

Brandl, Johann Evangelist *see*
- Deuxieme quintuor pour la flute, violon, 2 altos et violoncelle, op 60
- Grand quatuor pour deux violons, alto & violoncelle, oeuvre 18me
- Notturno pour deux altos & violoncelle, oeuvre 19me
- Trois quatuors pour flute, violon, alto & violoncelle, oeuvre 40

Brandon daily mail – Brandon, MB. 1882-97 – 9r – 1 – cn Library Assoc [070]

Brandon daily news – Brandon, Canada 31 dec 1912-11 dec 1914 (daily) (imperfect) – 8r – 1 – uk British Libr Newspaper [071]

Brandon, Joshua Arthur *see*
- An analysis of gothick architecture
- Parish churches

Brandon, Raphael *see*
- An analysis of gothick architecture
- Parish churches

Brandon sun – Brandon, Manitoba, CN. 1959- – 12r per yr – 1 – cn Commonwealth Imaging [071]

Brandon times – Brandon, Fairwater WI. 1867 mar 9/1870 dec 28-1973 jan 18-1975 dec 25 – 38r – 1 – mf#986408 – us WHS [071]

Brandon. Vermont. Brandon Seminary *see* Catalogs and college records

Brandon weekly sun – Brandon, Canada 12 sep 1912-16 dec 1920 (wkly) (imperfect) – 11r – 1 – uk British Libr Newspaper [072]

Brands manadshafte – Stockholm, Sweden. 1908-09, 1913, 1915-16 – 1r – 1 – sw Kungliga [078]

Die brandschatzung zur franzosenzeit 1809-13 in illyrien, oder, die gestoerte see-idylle : melodram in 3 akten / Germonik, Ludwig – Neurode/Br-Schl: Leuschner & Tesch, [19–?] – 1r – 1 – us Wisconsin U Libr [820]

Brandstetter, J L *see* Repertorium ueber die in zeit- und sammelschriften der jahre 1812-1890 enthaltenen aufsaetze und mitteilungen schweizergeschichtlichen inhaltes

Brandstetter, R *see* An introduction to indonesian linguistics

Brandstetter, Renward *see* Tagalen und madagassen

Die brandstifter von karabanowka : roman aus der zeit der ersten russischen revolution, 1905/06 / Dohrmann, Hanns Arved – Heidelberg: Huethi, 1943 [mf ed 1989] – 286p – 1 – mf#7180 – us Wisconsin U Libr [830]

Brandt, August *see* Johann eccks predigttaetigkeit an u.l. frau zu ingolstadt, 1525-1542

Brandt, Dagmar *see* Gardariki

Brandt, G *see*
- Historie der reformatie
- Verantwoording ter saeke van sijne historie der reformatie tegens de beschuldigingen van d. henricus rulaeus

Brandt, Harry Alonzo *see* The widowed earth

Brandt, John Lincoln *see*
- America or rome
- The lord's supper
- Rome's attack on our public schools

Brand(t) names / Pence Publications – v1-6 [1985-1989] – 1r – 1 – mf#1861021 – us WHS [929]

Brandt, Rolf *see*
- Abschied von mariampol
- Liebe auf oesel

Brandt, Wilhelm *see*
- Elchasai
- Die evangelische geschichte und der ursprung des christenthums
- Juedische reinheitslehre und ihre beschreibung in den evangelien
- Die juedischen baptismen
- Die mandaeer
- Die mandaeische religion

Brandung : geschichten von de waterkant / Lau, Fritz – Hamburg: M Glogau 1921 [mf ed 1990] – 1r – 1 – (filmed with: wir tragen das leben / thor goote) – mf#2817p – us Wisconsin U Libr [830]

Brandung : novellen / Pauls, Eilhard Erich – Leipzig: C F Amelang 1918 [mf ed 1996] – 1r – 1 – (filmed with: der gott / rudolf pannwitz) – mf#3980p – us Wisconsin U Libr [830]

Die brandwag – Johannesburg SA, 1 sept-29 dec 1939 – 2r – 1, 16 diazo available at reduced price – sa National [079]

Brandweek – New York NY 1993+ – 1,5,9 – (cont: adweek's marketing week: national marketing ed) – ISSN: 1064-4318 – mf#16149.02 – us UMI ProQuest [650]

Brandywine baptist church. chadd's ford, pennsylvania : church records – 1699-1881(MS materials), 1791-1838, 1843-1904 – 1 – 53.55 – us Southern Baptist [242]

Branigan's chronicles and curiosities – Hamilton, C W [Ont]: T Branigan, [1858?-1859?] – 9 – mf#P04204 – cn Canadiana [073]

Braniss, Christlieb Julius *see* Ueber schleiermachers glaubenlehre

Branly, Roberto *see*
- Cisne
- Firme de sangre

Brann, Henry Athanasius *see*
- The age of unreason
- Curious questions
- Waifs and strays, vol 1

Brannan, Tori L *see* A comparison of anterior tibial-femoral laxity in female intercollegiate gymnasts to a normal population

Brannon, J D *see* History of cisco baptist association in texas

Bransdorfer, Alfred H *see* A kinematic analysis of the developmental sequence of kicking using a direct and angled approach

Branson leader *see* Miscellaneous newspapers of las animas county, reel 1

Branson news *see* Miscellaneous newspapers of las animas county, reel 1

Branson, Sheri W *see* A step by step guide to concert planning for dance in secondary education

Brant agriculturist and indian magazine – Brantford, Ont: [s.n, 1898?-19–?] [mf ed v5 n4 jan 1898] – 9 – mf#P04021 – cn Canadiana [630]

Brant, Sebastian *see*
- Das narrenschiff
- Sebastian brants narrenschiff
- Shyppe of fooles

Brant, Stefan *see* Aufstand

Branta, Crystal *see* A fieldwork study of how young children learn fundamental motor skills and how they progress in the development of striking

Brantford expositor – Ontario, CN. 1852- – 12r/y – 1 – Can$1065.00 – cn Commonwealth Imaging [071]

Brantome, Pierre de Bourdeille, seigneur de *see* Illustrious dames of the court of the valois kings

Brapa tjonto boeat mendjadi broentoeng / Pembantoe, bebrapa – Soerabaja: Tan's Drukkery, [1932] [mf ed 1998] – (= ser Penghidoepan 89) – 1r – 1 – (coll as pt of the colloquial malay collection. filmed with: nona olanda sebagi istri tionghoa / [njoo cheong seng]) – mf#10000 – us Wisconsin U Libr [920]

Brapa tjonto jang berfaeda / Pembantoe, bebrapa – Soerabaia: Tan's Drukkery, 1934 [mf ed 1998 – (= ser Penghidoepan 114) – 1r – 1 – (coll as pt of the colloquial malay collection. filmed with: poetri satrija dewi, atawa, resia madjapait / h s t) – mf#10001 – us Wisconsin U Libr [920]

Brasavola, Antonio Musa *see* Caelii calcagnini, ferrariensis, protonotarii apostolici, opera aliqvot

Brasch, Moritz *see*
- Moses mendelssohn
- Rudolf von gottschall

Braschi, Wilfredo *see* Neuvas tendencias en la literatura puertorriquena

Braselmann, Werner *see* Franz werfel

Brash, R R *see* The ecclesiastical architecture of ireland to the close of the 12th century
Brash, Richard Rolt *see*
– Ogam inscribed monuments of the gaedhil in the british island
– Ogam inscribed monuments of the gaedhil in the british islands
Brasil / Bomfim, Manoel Jose Do – Sao Paulo, Brazil. 1935 – 1r – us UF Libraries [972]
Brasil / Brazil. Departamento Nacional do Cafe – Rio de Janeiro, Brazil. 1940 – 1r – us UF Libraries [972]
Brasil / Cambolm, Natalicio – Madrid, Spain. 1929 – 1r – us UF Libraries [972]
Brasil : colonia de banqueiros / Barroso, Gustavo – Rio de Janeiro, Brazil. 1935 – 1r – us UF Libraries [972]
Brasil : hoy – Mexico City? Mexico. 1968 – 1r – us UF Libraries [972]
Brasil : integracao e desenvolvimento economico / Aguiar, Pinto De – Salvador, Brazil. 1958 – 1r – us UF Libraries [330]
Brasil / Martinez Amengual, Gumersindo – Habana, Cuba. 1964 – 1r – us UF Libraries [972]
Brasil / Melo, Luis Felipe De – Buenos Aires, Argentina. 1944 – 1r – us UF Libraries [972]
Brasil : pais del futuro / Zweig, Stefan – Buenos Aires, Mexico. 1944 – 1r – us UF Libraries [972]
Brasil / Reparaz, Gonzalo De – Madrid, Spain. 1892 – 1r – us UF Libraries [972]
Brasil : su vida, su trabajo, su futuro / Bernardez, Manuel – Buenos Aires, Argentina. 1908 – 1r – us UF Libraries [972]
Brasil : tempos modernos – Rio de Janeiro, Brazil. 1968 – 1r – us UF Libraries [972]
Brasil : terra lusiada / Cayolla, Julio – Lisboa, Portugal. 1942 – 1r – us UF Libraries [972]
Brasil : la tierra y el hombre / Deffontaines, Pierre – Barcelona, Spain. 1960 – 1r – us UF Libraries [972]
Brasil *see* Relatorios ministeriais
El brasil : la tierra y el hombre, seguido de un estudio historico de joaquina comas ros. barcelona / Deffontaines, Pierre – Madrid: Razon y Fe, 1946 – 1 – sp Bibl Santa Ana [946]
O brasil : diario da manha, independente – Sao Paulo, SP: [s.n.] 06, 18, 21, 26 mar 1899 – (= ser Ps 19) – mf#P11A,06,151 – bl Biblioteca [321]
O brasil : orgam critico, litterario e noticioso – Florianopolis, SCSC Gabinete: Typ Sul-Americano, 28 out 1901-19 jan 1902 – (= ser Ps 19) – bl Biblioteca [079]
O brasil : orgao constitucional – Rio de Janeiro, RJ: Typ da Luz, 15 mar-19 abr 1873 – (= ser Ps 19) – mf#P11,08,13 – bl Biblioteca [323]
O brasil – Rio de Janeiro, RJ: Typ Americana, 16 jun 1840-dez 1841; jan-abr,jul-dez 1842; jan 1843-jun 1845; ago,out 1846; ago-nov 1847; jan 1848-dez 1850; jan-set,dez 1851; jan-02 jun 1852 – (= ser Ps 19) – mf#P04A,04,12-21 – bl Biblioteca [323]
O brasil – Rio de Janeiro, RJ: Typ dos Annaes, 1822 – (= ser Ps 19) – mf#P01,04,03 – bl Biblioteca [321]
Brasil colonia e brasil imperio / Carvalho, Austricliano De – Rio de Janeiro, Brazil. v1-2. 1927 – 1r – us UF Libraries [972]
Brasil contemporaneo / Coelho, Jose Simoes – Lisboa, Portugal. 1915 – 1r – us UF Libraries [972]
Brasil de hoje / Morais, Alexandre De – Lisboa, Portugal. v1-2. 1943 – 1r – us UF Libraries [972]
Brasil de hontem / Moniz, Heitor – Rio de Janeiro, Brazil. 1928 – 1r – us UF Libraries [972]
Brasil de oeste / Achilles, Paula – Rio de Janeiro, Brazil. 1949 – 1r – us UF Libraries [972]
Brasil de ontem e o de hoje / Mattos Ibiapina, J De – Rio de Janeiro, Brazil. 1942 – 1r – us UF Libraries [972]
Brasil e a emigracao portuguesa / Simoes, Nuno – Coimbra, Portugal. 1934 – 1r – us UF Libraries [972]
Brasil e a raca / Baptista Pereira, Antonio – Sao Paulo, Brazil. 1928 – 1r – us UF Libraries [972]
Brasil e africa / Rodrigues, Jose Honorio – Rio de Janeiro, Brazil. v1-2. 1964 – 1r – us UF Libraries [972]
Brasil e o anti-semitismo / Pereira, Baptista – Rio de Janeiro, Brazil. 1933 – 1r – us UF Libraries [939]
Brasil e o mundo arabe / Lacerda, Carlos – Rio de Janeiro, Brazil. 1948 – 1r – us UF Libraries [972]
Brasil e o parana / Parana, Sebastiao – Curityba, Brazil. 1907 – 1r – us UF Libraries [972]
Brasil e od drama do petroleo / Maya, Emilio De – Rio de Janeiro, Brazil. 1938 – 1r – us UF Libraries [972]
Brasil e os brasileiros / Kidder, Daniel P – Sao Paulo, Brazil. v1-2. 1941 – 1r – us UF Libraries [972]
Brasil em face do prata / Barroso, Gustavo – Rio de Janeiro, Brazil. 1952 – 1r – us UF Libraries [972]
Brasil em perspectiva – Sao Paulo, Brazil. 1968 – 1r – us UF Libraries [972]
Brasil en la encrucijada historica / Furtado, Celso – Barcelona, Spain. 1966 – 1r – us UF Libraries [972]
O brasil historico : jornal historico, politico, litterario, scientifico e de propaganda... – Rio de Janeiro, RJ: Typ Brasileira, 29 maio 1864-jul 1865; jan 1866-dez 1868; ago 1873-jul 1874; fev-set 1882 – (= ser Ps 19) – mf#P03A,03,01-06 – bl Biblioteca [073]
O brasil illustrado : publicacao litteraria – Rio de Janeiro, RJ: Typ de N Lobo Vianna & Filhos, 14 mar 1835-31 dez 1856 – (= ser Ps 19) – mf#P03,02,01 – bl Biblioteca [440]
Brasil literario / Wolf, Ferdinand Joseph – Sao Paulo, Brazil. 1955 – 1r – us UF Libraries [972]
Brasil moderno / Saenz Hayes, Ricardo – Buenos Aires, Argentina. 1942 – 1r – us UF Libraries [972]
Brasil na crise actual / Amaral, Azevedo – Sao Paulo, Brazil. 1934 – 1r – us UF Libraries [972]
Brasil na ii grande guerra / Castello Branco, Mnoel Thomaz – Rio de Janeiro, Brazil. 1960 – 1r – us UF Libraries [972]
Brasil na lenda e na cartografia antiga / Barroso, Gustavo – Sao Paulo, Brazil. 1941 – 1r – us UF Libraries [972]
Brasil pitoresco / Ribeyrolles, Charles – Sao Paulo, Brazil. v1-2. 1941 – 1r – us UF Libraries [972]
Brasil post – Sao Paulo (BR), 1971- – 1 – (semanario brasileiro) – gw Misc Inst [079]
Brasil prosa e poesia – New York, NY. 1969 – 1r – us UF Libraries [440]
Brasil siglo 20 / Faco, Rui – Buenos Aires, Argentina. 1961 – 1r – us UF Libraries [972]
Brasil visto pelos ingleses / Mello-Leitao, Candido De – Sao Paulo, Brazil. 1937 – 1r – us UF Libraries [972]
Brasile / Malesani, Emilio – Roma, Italy. 1929 – 1r – us UF Libraries [972]
Il brasile com'e / Felici, Osea – Milano, 1923. 253p – 1 – us Wisconsin U Libr [972]
Brasilia e amazonia / Vaitsman, Mauricio – Rio de Janeiro, Brazil. 1959 – 1r – us UF Libraries [972]
Brasilianische tage und nachte / Bonsels, Waldemar – Berlin, Germany. 1931 – 1r – us UF Libraries [972]
Brasilianos and yankees / Lobo, Helio – Rio de Janeiro, Brazil. 1926 – 1r – us UF Libraries [972]
The brasilians – 1978-. New York: Jota Alves Enterprises. -m. To promote the interests of Brasil in the U.S.A – 1 – us Wisconsin U Libr [972]
Brasilien : ein land der zunkunft / Schuler, Heinrich – Stuttgart, Germany. 1924 – 1r – us UF Libraries [972]
Brasilien / Schuler, Heinrich – Stuttgart, Germany. 1919 – 1r – us UF Libraries [972]
Brasilien : volk und land / Schuck, Walter – Berlin, Germany. 1928 – 1r – us UF Libraries [972]
Brasilien als unabhaengiges reich in historischer, mercantilischer und politischer beziehung / Schaeffer, Georg A von – Altona 1824 [mf ed Hildesheim 1995-98] – 1v on 3mf – 9 – €90.00 – 3-487-26865-5 – gw Olms [972]
Brasilien nachtraege, berichtigungen und zusaetze zu der beschreibung meiner reise im oestlichen brasilien / Wied, Maximilian zu – Frankfurt am Main 1850 [mf ed Hildesheim 1995-98] – 1v on 1mf – 9 – €40.00 – 3-487-26857-4 – gw Olms [918]
Brasilien tag und nacht / Nohara, Komakichi – Berlin, Germany. 1938 – 1r – us UF Libraries [972]
Brasilien von heute / Schouler, Heinrich – Berlin, Germany. 1903? – 1r – us UF Libraries [972]
Brasiliens aufschwung in deutscher beleuchtung / Dettman, Eduard Johann Karl – Berlin, Germany. 1908 – 1r – us UF Libraries [972]
Brasilio machado (1848-1919) / Alcantara Machado, Jose De – Rio de Janeiro, Brazil. 1937 – 1r – us UF Libraries [972]
Brasis, brasil e brasilia / Freyre, Gilberto – Lisboa, Portugal. 1960 – 1r – us UF Libraries [972]
Braslavi, Joseph *see* Metsadah
Braslavsky, Mosheh *see*
– Aza iz undzer land
– Tenu'at ha-po'alim ha-erets-yisraelit
The brass band journal – A collection of new and beautiful marches, quicksteps, polkas, etc., arranged in an easy manner for brass bands of 6 to 12 instruments. First series. New York: Firth, Pond & Co., c1853-54. Twenty-four pieces in parts, No. 3 wanting. MUSIC 3031 – 1 – us L of C Photodup [780]
Brass ensemble : its history and music / Husted, Benjamin F – U of Rochester 1955 [mf ed 19–] – 2v on 1r – 1 – (with app & bibl) – mf#film 438 – us Sibley [780]
Brass worker *see* Machinists and blacksmiths' monthly journal, 1870-1875 / the brass worker, 1895-1896 / official journal, 1902-1904
Brassac, Augustus *see* The student's handbook to the study of the new testament
Brassard, Marc F *see* The effect of the airstirrup and a conventional method of strapping the ankle on agility and vertical jump performance
Brasseur de Bourbourg, abbe *see*
– Histoire du canada, de son eglise, et de ses missions
– Histoire du canada, de son eglise et de ses missions
Brassey, Annie A *see* A voyage in the "sunbeam"
Brassey, Annie, Baroness *see* A voyage in the 'sunbeam'
Brassey, Thomas, Earl *see*
– Grand trunk railway
– How best to improve and keep up the seamen of the country
– Papers and addresses
– Preferential duties, a council of the empire, the state of the navy
Brastow, Lewis Orsmond *see* Representative modern preachers
Brastvo – Beograd [etc] Drustvo sv Save [t]1-; 1887-19 – (= ser Kn[igi] drustva sv save) – 1 – mf#1117 – us Wisconsin U Libr [460]
Bratcher, Lewis M *see* The autobiography, longtime missionary to brazil
Bratcher, Robert *see* Translator's handbook on the gospel of mark
Bratia likhudy : opyt issledovaniia iz istorii tserkovnogo prosveshcheniia i tserkovnoi zhizni kontsa 17 i nachala 18 vekov / Smentsovskii, M N – 1899 – 9 – 8 – mf#R-7875 – ne IDC [243]
Brat'ia lugininy, pionery kreditnoi kooperatsii i pervyi kreditnyi kooperativ v rossii / Merkulov, A V – M, 1918 – 1mf – 9 – mf#COR-71 – ne IDC [332]
Bratke, Eduard *see*
– Luther's 95 thesen und ihre dogmenhistorischen voraussetzungen
– Das sogenannte religionsgespraech am hof der sasaniden
– Wegweiser zur quellen- und litteraturkunde der kirchengeschichte
Bratley, Homer Eells *see* Studies of fall webworm and walnut caterpillar
Bratranek, F T H *see* Goethes naturwissenschaftliche correspondenz
Bratranek, Franz Thomas *see*
– Goethe's briefwechsel mit den gebruedern von humboldt
– Goethes egmont und schillers wallenstein
Bratschi, Peter *see*
– Nacht ueber den bergen
– Was da klingt in der tiefe
Bratska sloga – Auckland, NZ. 1899 – 1r – 1 – mf#11.34 – nz Nat Libr [079]
Bratstvo – Brotherhood. v.17-19, 1940-1942 – 1 – us CRL [073]
Bratstvo – Calumet MI, 1926* – 1r – 1 – (slovenian periodical) – us IHRC [073]
Bratvold, Tyren J *see* A torn anterior cruciate ligament of the knee
Brau, Heinrich *see* Die neue gesellschaft
Brau, Salvador *see* Disquisiciones sociologicas, y otros ensyos
Brauchen wir christum, um gemeinschaft mit gott zu erlangen? *see* Do we need christ for communion with god?
Brauchen wir ein neues dogma? : vortrag:gehalten auf der leipziger pastoralkonferenz am 21. mai 1891 / Seeberg, Reinhold – Erlangen:Andr. Deichert (George Boehme), 1892 – 1mf – 9 – 0-8370-5207-6 – mf#1985-3207 – us ATLA [140]
Brauckmann, Sabine *see* Eine theorie fuer lebendes?
Braudel, Fernand *see* Navires et marchandises 'a l'entree du port de livourne
Brauer, Carl M *see* Civil rights during the kennedy administration
Brauer der Vereinigten Staaten *see* Protokoll der...jahres-konventiondes national-verbandes der brauer der vereinigtenstaaten
Brauer, Erich *see* Zuge aus der religion der herero
Brauer, Johann Hartwig *see* Die heidenboten friedrichs 4. von daenemark
Brauer, Karl *see* Die unionstaetigkeit john duries unter dem protektorat cromwells
Brauer, Sandra G *see* Mediolateral postural stability
Braun, Amanda *see* Twenty-five years of women's varsity intercollegiate basketball at duke university
Braun, Antoine *see*
– Une fleur du carmel
– Instructions dogmatiques sur le mariage chretien
– Memoire sur les biens des jesuites en canada
Braun, Bartholome *see* Carta del p bartholome braun
Braun, Carl *see*
– Amerikanismus, fortschritt, reform
– Distinguo
Braun, Daniel *see* Die milchkompositionen der saeugetiere im vergleich und die milch des menschen
Braun, F M *see*
– Jean le theologien
– Jean te theologien et son evangile dans l'eglise ancienne
Braun, Felix *see* Deutsche geister
Braun, Frank Xavier *see* Kulturelle ziele im werk gustav frenssens
Braun, G *see* Civitates orbis terrarvm
Braun, Hanns *see* Grillparzers verhaeltnis zu shakespeare
Braun, Holger *see* Evolution von samenglobulin-genen
Braun, J *see*
– Der christliche altar
– Die liturgische gewandung in occident und orient
– Die liturgische paramente in gegenwart und vergangenheit
Braun, J B *see* Statistique constitutionnelle de la chambre des deputes de 1814 a 1829
Braun, Joseph *see* Philosophia moralis
Braun, Karl *see*
– Landschafts- und staedtebilder
– Reisestudien
– Eine tuerkische reise
Braun, Lilli *see* Die neue gesellschaft
Braun, Lily *see*
– Gesammelte werke
– Mutter maria
– Le probleme de la femme
Braun, Max *see* Der junge schiller am rhein
Braun, Oscar *see* Wir deutsch-amerikaner
Braun, Otto *see*
– Abhandlungen
– Aus nachgelassenen schriften eines fruehvollendeten
– The diary of otto braun
– Von weimar zu hitler
– Werke
Braund, John *see* Illustrations of furniture candelabra musical instruments
Braune christen im hause des herrn : gottesdienstliche feiern in der tamulenmission / Gehring, Alwin – Leipzig: Evangelisch-Lutherische Mission, 1919 [mf ed 1995] – (= ser Yale coll) – [32]p – 1 – 0-524-09085-8 – (in german) – mf#1995-0085 – us ATLA [240]
Braune, Franz A von *see* Salzburg und berchtesgaden
Braune, Friedrich Wilhelm Otto *see* Caecilia
Braune, Karl *see*
– The epistle of paul to the ephesians
– The epistle of paul to the philippians
– The epistles general of john
Braune, Wilhelm *see*
– Aller praktik grossmutter
– An den christlichen adel deutscher nation von des christlichen standes besserung
– Auserlesene gedichte deutscher poeten
– Buch von der deutschen poeterei
– Die fabeln des erasmus alberus
– Der freund in der not
– Horribilicribrifax
– Peter squenz
Braune wirtschaftspost – Duesseldorf DE, 1932 1 jul-1939 18 mar 18 – 10r – 1 – gw Misc Inst [330]
Braunfels, Ludwig *see* Agnes
Braunkohle : zeitschrift fuer gewinnung und verwertung der braunkohle – Halle/Saale, Germany 5 apr 1901-22 mar 1910 – 6r – 1 – uk British Libr Newspaper [622]
Die braunkohle : bkw ammendorf – Halle S DE, 1950 nov-1953 [gaps], 1961-1968 3 jan [gaps] – 3r – 1 – gw Misc Inst [622]
Das braunkohlenkombinat – Lauchhammer DE, 1958 3 oct-1968 nov [gaps] – 4r – 1 – gw Misc Inst [622]
Braunkohlenkombinat geiseltal – Muecheln, Halle S DE, 1977 sep-1990 jun [gaps] – 4r – 1 – gw Misc Inst [622]
Braunkohlenkumpel – Unseburg DE, 1955 19 oct-1959 20 aug [gaps] – 1r – 1 – gw Misc Inst [622]
Brauns, Ernst *see* Skizzen von amerika
Braunschweig-Bevern, August Wilhelm, Herzog von *see* Papers of august wilhelm herzog von braunschweig-bevern, 1717-1781
Braunschweiger arbeiter zeitung : organ der dritten internationale – Braunschweig DE, 1920 13 nov-31 dec – 20r – 1 – (cont as: niedersaechsische arbeiter zeitung: organ der vereinigten kommunistischen partei deutschlands 1921-26 fr 1926 n256; cont as: neue arbeiter-zeitung 1927-26 feb 1933) – mf#4776 – gw Mikropress [335]
Braunschweiger arbeiter-zeitung : organ der dritten internationale braunschweig – Braunschweig DE, 1920 13 nov-1933 26 feb – 20r – 1 – (cont: jan 1921: niedersaechsische arbeiterzeitung [organ der vereinigten kommunistischen partei deutschlands]; 2 nov 1926: neue arbeiter-zeitung) – mf#4776 – gw Mikropress [331]

Braunschweiger, M *see* Die lehrer der mischnah
Braunschweiger neueste nachrichten – Braunschweig. v14-37. 1910-33 – 87r – 1 – gw Mikropress [074]
Braunschweiger neueste nachrichten *see* Neueste nachrichten
Braunschweiger stadtanzeiger – Braunschweig 1919 8 mar-31 dec [big gaps], 1920 31 jan-27 mai [big gaps] [mf ed 2004] – 1r – 1 – (23 aug 1888: braunschweigischer landesanzeiger; 23 jun 1905: braunschweigischer anzeiger; 25 mar 1906: braunschweiger allgemeiner anzeiger; filmed by misc inst: 1886 7 nov-1941 31 mai [160r]) – gw Mikrofilm; gw Misc Inst [074]
Braunschweiger volksfreund *see* Braunschweiger volksfreund 1871
Braunschweiger volksfreund 1871 – Braunschweig DE, 1871 15 may-1933 2 mar – 93r – 1 – (title varies: 2 nov 1878: braunschweigisches unterhaltungsblatt, 30 nov 1890: braunschweiger volksfreund, 1907: volksfreund; with suppls) – gw Misc Inst; gw Mikropress [074]
Braunschweiger zeitung [main edition] – Braunschweig DE, 1968- – ca 9r/yr – 1 – (filmed by mikropress: 1946 8 jan-1950 20 feb [4r] order#7367; regional ed available: salzgitter, salzgitter-zeitung [1977-], wolfsburg, wolfsburger nachrichten [1977-]) – gw Misc Inst; gw Mikropress [074]
Braunschweigische anzeigen 1745 – Braunschweig DE, 1906 3 jan-7 jun – 1r – 1 – (publ started jan 2 1745; cont by: oeffentliche anzeigen, feb 4 1809, braunschweigische anzeigen, nov 13 1813, braunschweigische staatszeitung (incl suppl: braunschweigisches magazin), oct 1 1923-jan 1 1934; filmed by other misc inst: 1789, 1794, 1796, 1848-49 [4r]) – gw Misc Inst [943]
Braunschweigische landeszeitung – Braunschweig 1914 1 aug-1916 31 jan, 1916 1 aug-1917 31 jul, 1917 1 nov-1918 31 jul, 1918 1 nov-1919 31 jan, 1919 8 mar-1920 27 mai [gaps] [mf ed 2004] – 12r – 1 – (3 jun 1941: braunschweiger landeszeitung; filmed by misc inst: 1881 16 sep-1936 15 feb, 1941 3 jun-1944 31 aug [169r]) – gw Mikrofilm; gw Misc Inst [074]
Braunschweigische morgenzeitung – Braunschweig 1919 21 nov-1922 25 mar [mf ed 2004] – 5r – 1 – (3 apr 1921: braunschweiger kurier) – gw Misc Inst [074]
Braunschweigische nachrichten – Braunschweig 1768-81, 1782-89 [single iss] [mf ed 2004] – 5r – 1 – gw Misc Inst [074]
Braunschweigische post – Braunschweig 1878 9 mar-29 sep [mf ed 2004] – 1r – 1 – gw Misc Inst [074]
Der braunschweigische post-bote – Braunschweig 1849 6 jan-30 jun [mf ed 2004] – 1r – 1 – gw Misc Inst [074]
Braunschweigische postzeitung – Braunschweig 1721 22 okt-1741 15 dec, 1742 3 nov-1767 11 dec [mf ed 2004] – 1 – ([n1-10 only in diazo]; 3 nov 1742: braunschweigische zeitung) – gw Misc Inst [074]
Braunschweigische sozialistische landeskorrespondenz – Braunschweig 1918 17 dec-1919 15 jan [mf ed 2004] – 1r – 1 – (arbeiter- und soldatenrat) – gw Misc Inst [335]
Braunschweigisches friedhofs- und bestattungsrecht / Ostmann, Hans – Leipzig, 1933 (mf ed 1994) – 1mf – €24.00 – 3-8267-3026-7 – mf#DHS 3026 – gw Frankfurter [340]
Braunschweigisches journal philosophischen, philologischen und paedagogischen inhalts / ed by Trapp, Ernst Christian et al – Braunschweig 1788-93 – (= ser Dz) – 3v on 39mf – 9 – €390.00 – mf#k/n419 – gw Olms [100]
Braunschweigisches magazin / ed by Eschenburg, Joh Joachim – Braunschweig 1788-1868 – (= ser Dz) – 81jge on 280mf – 9 – €1680.00 – mf#k/n4570 – gw Olms [074]
Braunschweigisches unterhaltungsblatt *see* Braunschweiger volksfreund 1871
Braunschweigwer tageszeitung *see* Niedersaechsische tageszeitung
Die braut des spaniers : ein drama in 5 acten / Becker, Carl – [S.l: s.n.], 1875 (New Orleans: Druck der 'Deutschen Zeitung') [mf ed 1989] – 123p – 1 – mf#7002 – us Wisconsin U Libr [820]
Das brautboot : und andere ernste und frohe geschichten aus aller welt / Blunck, Hans Friedrich – Berlin: G Grote, 1943 [mf ed 1989] – (= ser Grotes soldaten-ausgaben 27) – 112p – 1 – mf#7036 – us Wisconsin U Libr [830]
Brautlacht, Erich *see* Der spiegel der gerechtigkeit
Brava gente / Carvalho, Elisio De – Rio de Janeiro, Brazil. 1921 – 1r – us UF Libraries [972]
Brave francois / Leila, Hanoum – Paris, France. 1893 – 1r – us UF Libraries [440]
Brave leut' vom grund : volksstueck mit gesang in drei abteilungen / Anzengruber, Ludwig – Stuttgart: J G Cotta, 1892 [mf ed 1988] – 119p – 1 – mf#6947 – us Wisconsin U Libr [820]
Brave resena de las aguas sulfurado-sodi cas termales de montemayor o banos / Crespo y Escoriaza, Benito – Trujillo: Imp. Lib. y Enc. de Benito Pena y Pena, 1902 – 1 – sp Bibl Santa Ana [946]
Bravo – Muenchen DE, 1956 26 aug-1998 22 dec – 132r – 1 – gw Mikrofilm [305]
Bravo Aguilera, Francisco *see* Guia local comercial de managua
Bravo, Carlos M *see* Briznas
Bravo de Piedrahita, J *see* De hydrophobiae natura, causis atque medela...
Bravo, Luis *see* Cuando pinto zurbaran los cuadros de la cartuja de jerez de la frontera?
Bravo Murillo, Juan *see*
– Apuntes y documentos...administrativas
– De las deudas amortizables...cupones
– Discursos...congreso de los diputados...
– Opusculos
Bravo Riesco, Agustin *see*
– De la lamentacion de la virgen maria
– Flores de san bernardo. de la lamentacion de la virgen maria
Bravo sport – Muenchen DE, 1994 26 oct-1998 25 feb – 9r – 1 – gw Mikrofilm [790]
Brawer, A J *see* Arets
Brawer, Michael *see* Tsevi la-tsadik
[Brawley-] brawley wildcat – CA. nov 1929-jun 1986 – 4r – 1 – $240.00 – mf#R04008 – us Library Micro [071]
Brawley, Edward Macknight *see* The negro baptist pulpit
Brawley, Jodi *see* Assessment of factors which influence college students to participate in regular physical activity
[Brawley-] the brawley news – CA. sep 1904- – 252+ r – 1 – $15,120.00 (subs $480/y) – mf#RC02073 – us Library Micro [071]
Braxton bragg papers, 1833-1879 – [mf ed 1988] – 8r – 1 – (official and personal letters, and letter books, military reports and orders, telegrams, and memoranda relating to general bragg's service in the confederate army, and as an advisor to president jefferson davis) – mf#ms2000 – us Western Res [976]
Bray, Alfred James *see*
– Canada under the national policy
– Churches of christendom
– England and ireland
– Two discourses in review and criticism
Bray and south dublin herald – Bray, Ireland 6 jul 1901-26 aug 1922 [mf 1876-96, 1915] – 4r – 1 – (cont: bray herald & arklow reporter [17 jun 1893-29 jun 1901]; cont by: south dublin herald [2 sep 1922-27 jan 1923]) – uk British Libr Newspaper [072]
Bray, Anna Eliza (Kempe) Stothard *see* Life of thomas stothard
Bray, Caroline *see* The british empire
Bray, Charles *see*
– Illusion and delusion
– Modern protestantism
– Reign of law in mind as in matter...
– Reign of law in mind as well as in matter...
Bray, Corey T *see* The relationship between team cohesion and objective individual performance of high school basketball players
Bray, Francois G de *see* Voyage dans le tyrol, aux salines de salzbourg et de reichenhall
Bray herald and arklow reporter – Bray, Ireland 17 jun 1893-29 jun 1901 [mf 1876-96, 1915] – 1 – (cont: bray herald, & kingstown & dalkey advertiser [21 oct 1876-10 jun 1893]; cont by: bray & south dublin herald [6 jul 1901-26 aug 1922]) – uk British Libr Newspaper [072]
Bray herald and kingstown and dalkey advertiser – Bray, Wicklow. 1905-07, 1910-2 apr 1927 – 20r – 1 – (cont as: bray herald and arklow reporter [17 jun 1893-29 jun 1901]; cont as: bray and south dublin herald [6 jul 1901-26 aug 1922]; cont as: south dublin herald [2 sep 1922-27 jan 1923]; cont as: bray and south dublin herald [3 feb 1923-2 apr 1927]; incorp with: wicklow people) – ie National [072]
Bray herald, and kingstown and dalkey advertiser – Bray, Ireland 21 oct 1876-10 jun 1893 [mf 1876-96, 1915] – 1 – (cont by: bray herald & arklow reporter [17 jun 1893-29 jun 1901]) – uk British Libr Newspaper [072]
Bray, John *see* The indian princess, or la belle savage
Bray people – Bray, Ireland 13 may 1988- – 1 – uk British Libr Newspaper [072]
Bray, Roger *see*
– European music manuscripts, series 1
– Printed music before 1800
Bray, Warwick *see* Everyday life of the aztecs
Bray, Wayne D *see* The controversy over a new canal treaty between the u.s. and panama
Braye, Alfred Thomas Townshend Verney-Cave *see* The present state of the church in england
Brayer, Edith [comp] *see* Catalogue of french-language medieval manuscripts
Brayley, Edward W *see* Delineations, historical and topographical, of the isle of thanet and the cinque ports
Brayley, Edward Wedlake *see* Illustrations of her majesty's palace at brighton
Brayner, Floriano De Lima *see* Verdade sobre a feb
Brayon / Societe historique du Madawaska – v5 n1-v8 n3/4 [1976 sep-1980 oct/dec] – 1r – 1 – (cont by: revue de la societe historique du madawaska) – mf#573446 – us WHS [978]
Brayton's reports / Vermont. Supreme Court – 1v. 1815-1819 (all publ) – (= ser Vermont supreme court reports) – 3mf – 9 – $4.50 – (a pre-nrs title) – mf#LLMC 90-310 – us LLMC [347]
Brazao, Eduardo *see* Relacoes externas de portugal
Brazen serpent / Moore, Daniel – London, England. 1860 – 1r – us UF Libraries [240]
The brazen serpent : or, life coming through death / Erskine, Thomas – 3rd ed. Edinburgh: David Douglas, 1879 – 1mf – 9 – 0-7905-7729-1 – mf#1989-0954 – us ATLA [240]
Brazier, Nicholas *see*
– Begueule
– Bonnes d'enfans
– Coin de rue
– Cuisinieres
– Infidelites de lisette
– Madame frontin, ou, les deux duegnes
– Memoire de la blanchisseuse
– Pauvre de saint-roch
– Petites pensionnaires
– Sage et coquette
– Tony
'Brazil' / Oakenfull, J C – Freiburg, Germany. 1922 – 1r – 1 – us UF Libraries [972]
Brazil / Bon, Antoine – Sao Paulo, Brazil. 1950 – 1r – us UF Libraries [972]
Brazil : bulwark of inter-american relations / Phillips, Henry Albert – New York, NY. 1945 – 1r – us UF Libraries [972]
Brazil / Denis, Ferdinand – Paris, France. v1-2. no date – 1r – us UF Libraries [972]
Brazil / Freyre, Gilberto – Washington, DC. 1963 – 1r – us UF Libraries [972]
Brazil / Good, Reynolds E – Williamsport, PA. 1962 – 1r – us UF Libraries [972]
Brazil : an interpretation / Freyre, Gilberto – New York, NY. 1945 – 1r – us UF Libraries [972]
Brazil : its conditions and prospects / Andrews, C C – New York, NY. 1887 – 1r – us UF Libraries [972]
Brazil / Marshall, Andrew – New York, NY. 1966 – 1r – us UF Libraries [972]
Brazil / Momsen, Richard P – Princeton, NJ. 1968 – 1r – us UF Libraries [972]
Brazil : orchid of the tropics / Foster, Mulford Bateman – Lancaster, PA. 1946 – 1r – us UF Libraries [972]
Brazil / Orico, Osvaldo – New York, NY. 1957? – 1r – us UF Libraries [972]
Brazil : world frontier / Hunnicutt, Benjamin Harris – New York, NY. 1949 – 1r – us UF Libraries [972]
Brazil / Zweig, Stefan – New York, NY. 1941 – 1r – us UF Libraries [972]
Brazil *see*
– Codigo civil brasileiro
– Codigo de processo penal
– Codigo penal
– Codigo penal brasileiro (decreto-lei n2848...
– Constituicao de dez de novembro
– Diario oficial
– Direito do brasil
– Diretrizes e bases da educacao nacional
– Estatuto dos funcionarios publicos civis da uniao
– Legislacao brasileira de previdencia social
O brazil : jornal catholico, litterario e noticioso – Bahia: Typ de Camillo de Lellis Masson & C, 25 jan 1863 – (= ser Ps 19) – mf#P18B,02,13 – bl Biblioteca [241]
O brazil : jornal scientifico, litterario e artistico – Rio de Janeiro, RJ: Typ Imp de Brito & Irmaos, 25 jul-nov 1865; fev-1 set 1866 – (= ser Ps 19) – mf#P05,04,60 – bl Biblioteca [073]
O brazil *see* O catharinense
Brazil, 1938 / Instituto Brasileiro De Geografia E Estatistica – Rio de Janeiro, Brazil. 1939 – 1r – us UF Libraries [972]
Brazil after a century of independence / James, Herman Gerlach – New York, NY. 1925 – 1r – us UF Libraries [972]
Brazil and buenos ayres / Conder, Josiah – London 1825 [mf ed Hildesheim 1995-98] – 2v on 6mf – 9 – €120.00 – 3-487-26861-2 – gw Olms [918]
Brazil and her people of to-day / Winter, Nevin Otto – Boston, MA. 1910 – 1r – us UF Libraries [972]
Brazil and river plate mail : and south american mercantile journal – London, England 7 nov 1863-23 dec 1878 – 1 – (cont by: south american journal & brazil & river plate mail [8 jan 1879-aug 1955]) – uk British Libr Newspaper [072]
Brazil and the brazilians / Bruce, G J – London, England. 1915 – 1r – us UF Libraries [972]
Brazil and the brazilians / Kideer, Daniel P – Boston, MA. 1866 – 1r – us UF Libraries [972]
Brazil and the brazilians portrayed / Kidder, Daniel P – Philadelphia, PA. 1857 – 1r – us UF Libraries [972]
Brazil and the league of nations / Macedo Soares, Jose Carlos De – Paris, France. 1928 – 1r – us UF Libraries [972]
Brazil and the river plate in 1868 / Hadfield, Willilam – London, England. 1869 – 1r – us UF Libraries [972]
Brazil. Chefo do Governo Provisorio *see* Mensagem dirigida ao congresso nacional
Brazil. Comissao Brasileira dos Centenarios de Por... *see* Portugueses na marinha de guerra do brasil
Brazil. Comissao Exploradora do Planalto Central D... *see* Relatorio...
Brazil commercial – Rio de Janeiro, RJ. 14 mar-30 jul 1858 – (= ser Ps 19) – bl Biblioteca [079]
Brazil. Commissao Brazileira na Exposicao Universa... *see* Empire du bresil
Brazil. Commissao, Exposicao Universal, Philadelph... *see* Imperio do brazil na exposicao universal de 1876 e...
Brazil. Congresso Nacional *see* Emendas a constituicao de 1946
Brazil. Congresso Nacional Senado Federal Direto... *see* Congresso nacional e o programa de integracao soci...
Brazil. Conselho nacional de Estatistica *see* Revista brasileira de estatistica
Brazil. Courts *see* Diario da justica and apenso jurisprudencia
Brazil. Departamento Administrativo do Servico Pub... *see* Indicador da organizacao administrativa do executi...
Brazil. Departamento de Imprensa e Propaganda *see* Facts and information about brazil
Brazil. Departamento de Mprensa e Propaganda *see* Brazil in america
Brazil. Departamento Nacional do Cafe *see* Brasil
Brazil. Departamento Nacional do Cafe... *see* What brazil offers you
Brazil. Direcao Geral da Fazenda Nacional Assesso... *see* Diagnostico do sistema estatistico fazendario
Brazil. Divisao de Geologia e Mineralogia *see* Geologia historica do brazil
O brazil e a educacao popular / Leao, Antonio Carneiro – Rio de Janeiro: Jornal do Commercio, de Rodrigues, 1917 – 1 – us Wisconsin U Libr [972]
Brazil e a emigracao / Moreira Telles – Lisboa, Portugal. 1914 – 1r – us UF Libraries [972]
Brazil. Exercito *see* Revolucao de 31 de marco...
Brazil in america / Brazil. Departamento de Mprensa e Propaganda – Rio de Janeiro, Brazil. 1942 – 1r – us UF Libraries [972]
Brazil in capitals / Kelsey, Vera – New York, NY. 1942 – 1r – us UF Libraries [972]
Brazil in the making / Jobim, Jose – New York, NY. 1943 – 1r – us UF Libraries [972]
Brazil. Instituto Brasileiro de Geografia e Estatistica *see* Anuario estatistico de brasil 1908/1912-1969
Brazil. Instituto de Aposentadoria e Pensoes Dos I... *see* Seguro social
Brazil. Instituto de Expansao Commercial Antigo M... *see* Laranja no brasil
Brazil Laws, Statutes, Etc *see* Codigo penal brazileiro
Brazil looks forward / Hunnicutt, Benjamin Harris – Rio de Janeiro, Brazil. 1945 – 1r – us UF Libraries [972]
O brazil mental / Sampaio, Jose Pereira De – Porto, Portugal. 1898 – 1r – us UF Libraries [972]
Brazil. Ministerio da Agricultura *see*
– Relatorio [e annexos]
– Relatorios ministeriais, 1a republica, 1889-1929
Brazil. Ministerio da Educacao e Cultura Servico *see* Catalogo das publicacoes do servico de documentacao
Brazil. Ministerio da Educacao e Saude Publica *see* Exposicao machado de assis
Brazil. Ministerio da Fazenda *see*
– Panorama financeiro e economico da republica
– Relatorio apresentado ao presidente da republica dos estados unidos do brazil
– Relatorio apresentado ao vice-presidente da republica dos estados unidos do brazil
– Relatorio do ministro da fazenda...
Brazil. Ministerio da Guerra *see*
– Relatorios ministeriais, 1a republica
– Uniformes do exercito brasileiro
Brazil. Ministerio da Industria. Viacao e Obras Publicas *see* Relatorio...
Brazil. Ministerio da Instruccao Publica. Correios e Telegrafos *see* Relatorio...

BRAZIL

Brazil. Ministerio da Justica *see*
– Conta
– Exposicao apresentada ao chefe do governo provisorio da republica dos estados unidos do Brazil
– Relatorio apresentado a assemblea geral legislativa
– Relatorio apresentado ao presidente da republica dos estados unidos do brasil
– Relatorio da reparticao dos negocios da justica apresentado a assemblea geral legislativa
Brazil. Ministerio da Justica e Negocios Interiore *see* Constituicoes federal e estaduais
Brazil. Ministerio da Justica e Negocios Interiores *see*
– Relatorio...
– Relatorio appresentado ao vice-presidente da republica dos estados unidos do brasil
– Relatorio apresentado ao presidente da republica dos estados unidos do brasil
– Relatorio das atividades administrativas do exercicio de...
Brazil. Ministerio da marinha *see* Relatorio...
Brazil. Ministerio da Viacao e Obras Publicas *see* Relatorio...
Brazil. Ministerio das Relacoes Exteriores *see*
– Archivo diplomatico da independencia
– Relatorios ministeriais, 1a republica, 1890-1929
– Tratado de 8 de setembro de 1909 entre os estados
Brazil. Ministerio de Estado da Justica e Negocios Interiores *see* Relatorio appresentado ao vice-presidente da republica dos estados unidos do brasil
Brazil. Ministerio dos Transportes Servico de Doc... *see* Ministerio dos transportes na integracao e desenvo...
Brazil. Ministro da Justica e Negocios Interiores *see*
– Relatorio apresentado ao presidente da republica dos estados unidos do brasil
Brazil on the march / Cooke, Morris Llewellyn – New York, NY. 1944 – 1r – us UF Libraries [972]
Brazil on the move / Dos Passos, John – Garden City, NY. 1963 – 1r – us UF Libraries [972]
Brazil. Sao Paulo. Departamento do Archivo do Estado *see* Inventarios e testamentos
Brazil today and tomorrow / Joyce, Lilian Elwyn (Elliott) – New York, NY. 1921 – 1r – us UF Libraries [972]
Brazil under vargas / Loewenstein, Karl – New York, NY. 1942 – 1r – us UF Libraries [972]
Brazilian adventure / Fleming, Peter – New York, NY. 1942 – 1r – us UF Libraries [972]
Brazilian adventure / Fleming, Peter – New York, NY. no date – 1r – us UF Libraries [972]
Brazilian bulletin – New York NY 1944-77 – 1,5,9 – ISSN: 0006-9485 – mf#1507 – us UMI ProQuest [337]
Brazilian business – Rio de Janeiro, Brazil 1973-80 – 1,5,9 – ISSN: 0006-9493 – mf#8359 – us UMI ProQuest [338]
Brazilian coffee / Moreira, Nicolau Joaquim – New York: "O Novo Mundo" Printing Office, [1876] [mf ed 1989] – (= ser Books of the fairs) – 1 – mf#reel 47, item 4 – us Primary [640]
Brazilian colonization : from an european point of view / Assu, Jacare [pseud] – London, 1873 – (= ser 19th c books on british colonization) – 2mf – 9 – mf#1.1.5962 – uk Chadwyck [972]
Brazilian culture / Azevedo, Fernando De – New York, NY. 1950 – 1r – us UF Libraries [306]
Brazilian culture hearth / Schmeider, Oskar – Berkeley, CA. 1929 – 1r – us UF Libraries [390]
Brazilian economy / Spiegel, Henry William – Philadelphia, PA. 1949 – 1r – us UF Libraries [330]
Brazilian el dorado / Carvalho, J R De Sa – London, England. 1938 – 1r – us UF Libraries [972]
Brazilian portuguese self-taught / Ibarra, Francisco – New York, NY. 1943 – 1r – us UF Libraries [440]
Brazilian sketches / Ray, T Bronson – Louisville, KY: Baptist World Pub Co, 1912 – 2mf – 9 – 0-524-07458-5 – mf#1991-3118 – us ATLA [918]
Brazilian workers' party – 40r – 1 – (with printed guide; contents: 1. boletim mulheres do pt [pt's women's bulletin] mar/1993 a mai/1993 [1r]; 2. boletim nacional [national bulletin] out/1983 a out/1989; fev/1990 a out/1994 [2r]; 3. brasil agora [brasil now] (newspaper) set/1991 a mai/1993; mai/1993 a mai/1996 [2r]; 4. colecao de recortes de jornais e revistas [news clippings from newspapers and periodicals]: national by state, by (major) municipality, on lula, constituent meetings, elections, social movements, by member of parliament, impeachment of [pres.] collor de mello, the banking crisis, by personalities, miscellaneous, chronologies of events [14r]; 5. informe sorg [sorg informs] mar/2001 a mai/2003 [1r]; 6. jornal dos trabalhadores [workers' news] mar/1982 a mai/1983 [1r]; 7. linha aberta [open line] (diary eletronic bulletin) set/1995 a dez/1997, jan/1998 a out/1998 [2r]; 8. livros de atas [books of acts and resolutions] [1r]; 9. mulheres [women] ago/1993 a set/2000 [1r]; 10. pagina agraria (informativo san) [agrarian news] abr/1995 a set/1995 (informativo san), abr/1997 a jul/1999 (pagina agraria), jul/1999 a dez/2001 [2r]; 11. pagina internacional [international news] fev/2001 a set/2003 [1r]; 12. pt informa mulheres [pt informs women] ag/1992 a set/1999 [1r]; 13. pt noticias [pt news] (monthly newspaper) jun/1996 a nov/2003 [1r]; 14. publicacoes avulsas [special/single publ] (from the): national directorate national secretariat for political education, national secretariat for unions national secretariat for agriculture, national secretariat on finances, special pt, secretariat to monitor the 'fome zero' program etc etc [9r]; 15. smad net [smad net] mar/1998 a mai/2003 [1r]; 16. teoria e debate [theory and debate] dez/1987 a set/1991, out/1991 a nov/1994, dez/1994 a dez/1997, fev/1999 a nov/2003 [4r]) – ne IDC [335]
Brazilians / Tavares De Sa, Hernane – New York, NY. 1947 – 1r – us UF Libraries [306]
Brazilians and their country / Cooper, Clayton Sedgwick – New York, NY. 1917 – 1r – us UF Libraries [306]
Brazil's industrial evolution / Simonsen, Roberto Cochrane – Sao Paulo, Brazil. 1939 – 1r – us UF Libraries [338]
Brazil's popular groups : a microfilm collection of materials issued by socio-political, religious, labor and minority grass-roots organizations, 1987-1989 – Washington, DC: Library of Congress Preservation Microfilming Office, 1991 (mf ed) – 43r – 1 – (suppl to brazil's popular groups, 1966-1986; contains some materials issued pre-1987. reel 1: indexes. reels 2-6: agrarian reform and land issues. reels 7-8: blacks. reel 9: children. reel 10: ecology. reels 11-15: education. reels 16-18: human and minority rights. reels 19-21: indians. reels 22-25: labor and laboring classes. reels 26-30: political parties and issues. reels 31-35: religion and theology. reels 36-37: urban activism. reels 38-42: women. reel 43: posters) – us L of C Photodup [972]
Brazil's popular groups, 1966-1986 : [supplement 1, 1987-1989] – Washington DC: Library of Congress Preservation Microfilming Program [mf ed 1991] – 43r – 1 – (with guide) – us L of C Photodup [972]
Brazil's popular groups, 1966-1986 : [supplement 3, 1993] – Washington DC: Library of Congress Preservation Microfilming Program [mf ed 1995] – 32r – 1 – (with guide) – us L of C Photodup [972]
Brazil's popular groups, 1966-1986 : [supplement 4, 1994] – Washington DC: Library of Congress Preservation Microfilming Program [mf ed 1995] – 18r – 1 – (with guide) – us L of C Photodup [972]
Brazil's popular groups, 1966-1986 : [supplement 5, 1995] – Washington DC: Library of Congress Preservation Microfilming Program [mf ed 1996] – 28r – 1 – (with guide) – us L of C Photodup [972]
Brazil's popular groups, 1966-1986 : [supplement 6, 1996] – Washington DC: Library of Congress Preservation Microfilming Program [mf ed 1998] – 24r – 1 – (with guide) – us L of C Photodup [972]
Brazil's popular groups, 1966-1986 – Washington DC: Library of Congress Photodup Service, 1988 – 28 [i.e. 32]r – 1 – (with guide) – us L of C Photodup [972]
Brazil's popular groups, 1966-1986 / supplement 10, 2000 – [mf ed 2001] – 18r – 1 – us L of C Photodup [318]
Brazos. Synod (Cum. Pres. Ch.) *see* Minutes, 1849-1887
Bre : Black entertainment's premier magazine – 1992 dec/1993 mar-2000 jul/aug – 25r – 1 – (with gaps; cont: black radio exclusive) – mf#2599468 – us WHS [790]
BRE black music directory – 1993, 1997-2000 – 1r – 1 – (cont by: bre directory) – mf#3736659 – us WHS [780]
[Brea-] brea progress – CA. oct 31 1917- – 81r – 1 – $4860.00 (subs $50/y) – (aka: brea star) – mf#R02074 – us Library Micro [071]
Brea star *see* [Brea-] brea progress
The breach of promise trial : bardell v pickwick : adapted from "the pickwick papers" of charles dickens / Bengough, John Wilson – [Toronto?: s.n, 1907?] – 1mf – 9 – 0-665-73231-7 – mf#73231 – cn Canadiana [790]
Breach repaired in god's worship / Keach, Benjamin – or Singing of Psalms, Hymns and Spiritual Songs. London. 1691 – 1 – 6.72 – us Southern Baptist [242]
Bread and freedom – (Philadelphia). 1906 – 1 – us AJPC [830]
Bread and roses – v1 n1-v3 n1 [1977 sep-1982 winter], v3 n2 [1984] – 1r – 1 – mf#670402 – us WHS [071]
Bread and salt from the word of god : in sixteen sermons = Brot und salz aus gottes wort / Zahn, Theodor – Edinburgh: T & T Clark 1905 [mf ed 1989] – 1mf – 9 – 0-7905-0479-0 – (trans fr german by c s burn & andrew ewbank burn) – mf#1987-0479 – us ATLA [240]
The bread of life : or, st thomas aquinas on the adorable sacrament of the altar = de sacramento altaris / Rawes, Henry Augustus – New ed. London: Burns & Oates; New York: Benziger, [1879?] – 1mf – 9 – 0-8370-7028-7 – mf#1986-1028 – us ATLA [241]
Bread, peace, and land – 1970 dec, 1971 jan-feb, v1 n4, 9-10 [1971 mar, sep-oct], v2 n1-2 [1972 jan-feb], 1971 oct 19 – 1r – 1 – mf#1532468 – us WHS [334]
Bread & Roses Women's Health Center *see* Irregular periodical
Bread selling to the poor at half price / Hawker, Robert – London, England. 1820 – 1r – us UF Libraries [240]
O break : periodico chistoso e humoristico – Rio Claro, SP: Typ Rio Clarense, 30 dez 1877 – (= ser Ps 19) – bl Biblioteca [079]
Break free with 23 news / Yes on 23 [Organization] – v1 n1-v4 n4 [1979 jun-1982 jul] – 1r – 1 – (cont by: yes on 23/liberty amendment news) – mf#592647 – us WHS [071]
A break in the ocean cable / Baldwin, Maurice Scollard – Montreal: Dawson Bros, 1877 – 1mf – 9 – mf#56349 – cn Canadiana [240]
A break in the ocean cable / Baldwin, Maurice Scollard – Montreal: Dawson Bros, 1877 – 1mf – 9 – mf#00840 – cn Canadiana [240]
A break in the ocean cable / Baldwin, Maurice Scollard – Montreal: Dawson Bros, 1880 – 1mf – 9 – mf#13497 – cn Canadiana [240]
Breakdown – Klamath Falls OR: s.n, 1971- [wkly] – 1 – us Oregon Lib [071]
Breakers! : methodism adrift / Munhall, Leander Whitcomb – Philadelphia, PA: E & R Munhall, c1913 – 1mf – 9 – 0-524-02749-8 – mf#1990-4424 – us ATLA [240]
Breaking down chinese walls : from a doctor's viewpoint / Osgood, Elliott Irving – New York: Fleming H Revell, c1908 – 1mf – 9 – 0-8370-6590-9 – mf#1986-0590 – us ATLA [610]
Breaking the silence / Carleton University – 1984 fall-1989 mar/jul – 1r – 1 – mf#1053552 – us WHS [071]
The breaking waves dashed high : the pilgrim fathers / Hemans, Felicia – Boston: Lee & Shepard, 1880 – 1mf – 9 – $1.50 – mf#LLMC 91-012 – us LLMC [320]
Breakthrough / National Association of Commissions for Women – 1978 mar-1990 spring – 1r – 1 – (cont by: nacw's breakthrough) – mf#1803691 – us WHS [071]
Breakthrough / Prairie Fire Organizing Committee – 1977 mar-1980 winter – 1r – 1 – mf#499094 – us WHS [360]
Breakthrough / Recruitment and Training Program, Inc – 1974 oct-1981 apr – 1r – 1 – mf#4882323 – us WHS [071]
Breakthroughs in health and science – Palm Coast FL 1990-91 – 1,5,9 – (cont: science digest) – ISSN: 1050-6691 – mf#19444 – us UMI ProQuest [500]
Breaktime / American Postal Workers Union – 1984 oct-1990 apr/may – 1r – 1 – mf#1278575 – us WHS [380]
Break-up : effects and consequences on the two rhodesias – s.l, s.l? 1963 – 1r – us UF Libraries [960]
Break-up / Pearson, D S – Salisbury, Zimbabwe. 1963 – 1r – us UF Libraries [960]
Breakwall news / Lorain Co. Lorain – (1951-55, apr 1969-feb 1960) [irreg] – 1r – 1 – mf#B12098 – us Ohio Hist [331]
Breast cancer research and treatment – Dordrecht, Netherlands 1991+ – 1,5,9 – ISSN: 0167-6806 – mf#16771 – us UMI ProQuest [616]
Breast disease – Lausanne, Switzerland 1989-94 – 1,5,9 – ISSN: 0888-6008 – mf#42514 – us UMI ProQuest [616]
Breast self-examination, the health belief model and sexual orientation in women / Ellingson, Lyndall A – 1996 – 2mf – 9 – $8.00 – mf#HE 590 – us Kinesology [613]
Breast support for the active women : relationship to 3d kinematics of running / Boschma, Anne L C – Oregon State University, 1995 – 2mf – 9 – $8.00 – mf#PE 3632 – us Kinesology [612]
Breasted, James Henry *see*
– Ancient records of egypt
– Ancient times
– Development of religion and thought in ancient egypt
– First and second preliminary report of the egyptian expedition
– A history of the ancient egyptians
– Oriental forerunners of byzantine painting
Breasted, James Henry et al *see* Egypt
The breast-plate of faith and love : a treatise... / Preston, J – London: W I, 1630 – 7mf – 9 – mf#PW-20 – ne IDC [240]
The breath of god : a sketch: historical, critical and logical of the doctrine of inspiration / Hallam, Frank – New York: Thomas Whittaker, 1895 – 1mf – 9 – 0-8370-3453-1 – (incl bibl ref) – mf#1985-1453 – us ATLA [220]
A breathing after god : or a christians desire of gods presence / Sibbes, R – London: Iohn Dawson, 1639 – 5mf – 9 – mf#PW-80 – ne IDC [240]
Brechenmacher, Josef Karlmann *see* Deutsche sippennamen
Brechin advertiser – Brechin: D H Edwards [1879?]- (wkly) [mf ed 2 jan 1992-] – 1 – (not publ: 25 jun 1970; cont: brechin advertiser, and angus and mearns intelligencer) – uk Scotland NatLib [072]
Brecht, Bertolt *see*
– Baal
– Der dreigroschenroman
Brecht, Theodor *see* Papst leo 13. und der protestantismus
Breck, James Lloyd *see* The life of the reverend james lloyd breck
Breck, John *see*
– Family papers, ms 4675
– John breck family papers, 1782-1993
Breckenridge *see* Protest against the use of instrumental music in the stated...
Breckenridge, Roeliff Morton *see* The canadian banking system, 1817-1890
Breckerfelder zeitung – Breckerfeld 1917 2 jan-30 jun, 1918 2 jul-31 dec [gaps], 1920, 1933 8 apr & 15 apr [mf ed 2004] – 2r – 1 – (regional ed of: halversche zeitung, halver) – gw Mikrofilm [074]
Breckinridge, John *see* A discussion of the question, is the roman catholic religion, in any or in all its principles or doctrines, inimical to civil or religious liberty?
Breckinridge, Robert Jefferson *see*
– The knowledge of god
– Papism in the 19 century in the united states
Breckinridge, Sophonisba Preston *see*
– The family and the state
– Family welfare work in the metropolitan community
Brecknock beacon & general advertiser – Brecon, Wales 3 jul 1885-27 mar 1896 – 1 – (incorp with: brecon & radnor county times; cont: brecon free press, & general advertiser [20 oct 1883-26 jun 1885]) – uk British Libr Newspaper [072]
Brecksville. Ohio. Brecksville Congregational Church *see* Church records, ms 3168
Brecon and radnor express and carmarthen gazette – [Wales] Brecon 4 oct 1898-27 oct 1933 [mf 1897, 1898, 1909-50, 1986-] [mf ed 2004] – 1 – (cont by: brecon & radnor express & county times [2 nov 1933-2 nov 1972]; brecon & radnor express & powys county times (main ed) 16 nov 1972-]) – uk Newsplan [072]
Brecon county times – [Wales] Powys 5 may 1866-dec 1908; jan 1909-26 oct 1933 [mf ed 2003] – 61r – 1 – (missing: 1871, 1880, 1886-87, 1892, 1897-98; cont as: brecon & radnor county times [jan 1894-dec 1896]; brecon county times [jan 1899-oct 1933]) – uk Newsplan [072]
Brecon free press, and general advertiser – [Wales] Brecon 20 oct 1883-26 jun 1885 [mf ed 2003] – 1 – (cont by: brecknock beacon & general advertiser [3 jul 1885-27 mar 1896]) – uk Newsplan [072]
Brecon journal and county advertiser – [Wales] Brecon 27 jun 1857-9 may 1868 [mf ed 2003] – 1 – (cont: brecon journal, & town & country newspaper [6 oct 1855-20 jun 1857]; cont by: brecon journal & county advertiser [ns] 27 jun 1857-9 may 1868) – uk Newsplan [072]
Brecon reporter and south wales general advertiser – Brecon, Wales 12 sep 1863-16 nov 1867 (wkly) – 4r – 1 – (discontinued) – uk British Libr Newspaper [072]
Breda-street : ou, un ange dechu / Clairville, M – Paris, France. 1849? – 1r – us UF Libraries [440]
Bredbury, st mark – [North Cheshire FHS] – (= ser Cheshire monumental inscriptions) – 2mf – 9 – £3.25 – mf#147 – uk CheshireFHS [929]
Brede, Philipp *see* Reise durch teutschland, frankreich und holland im jahr 1806
Bredehoeft, Hermann *see*
– Preussischer herbst
– Die verirrten
Bredekamp, Horst *see* Kunst als medium sozialer konflikte
Bredekamp, Rosa *see* Individuele
Bredel, Willi *see* Die pruefung
Bredemeier, Brenda Jo Light *see* Goal orientation and moral atmosphere in youth sport
Bredenbeck, Anton *see* 1889
Bredenberg, Fritz von *see* Der kreis sensburg
Bredenkamp, Conrad Justus *see*
– Der prophet jesaia
– Der prophet sacharja
Brederek, Emil *see* Konkordanz zum targum onkelos

Bredetzky, Samuel see Reisebemerkungen ueber ungern und galizien
Bredikhin, Fedor Aleksandrovich see Etudes sur l'origine des meteores cosmiques et la formation de leurs courants
Bredius, A see Kuenstler-inventare
Bredthauer, Karl D et al see Ein baukran stuerzt um
Bree, Charles Robert see
- An exposition of fallacies in the hypothesis of mr darwin
- Species not transmutable, nor the result of secondary causes

Bree, Malwine see Leschetizky method
Breed, David Riddle see
- Abraham
- The history and use of hymns and hymn-tunes
- A history of the preparation of the world for christ
- Preparing to preach

Breed of barren metal, or, currency and interest : or, currency and interest: a study of social and industrial problems / Bennett, James William – Chicago: C H Kerr, c1895 (mf ed 19–) – (= ser Library of progress) – mf#ZT-54 – us NY Public [332]
Breed, William Pratt see
- Presbyterianism and its services in the revolution of 1776
- Presbyterianism three hundred years ago
- Presbyterians and the revolution
- Witherspoon

Breeder's gazette – Chicago IL 1949-64 – 1 – mf#54 – us UMI ProQuest [636]
Breen, Andrew Edward see
- A general introduction to the study of holy scripture
- A harmonized exposition of the four gospels

Breen, Hugh, Jr see On the corrections of bouvard's elements of the orbits of jupiter and saturn
Breen, J D see Continuity or collapse?
Breen, John Dunstan see Church of old england
Breevoort can ick vergeten niet / Hauer, H A – Aalten, 1956 – €7.00 – ne Slangenburg [890]
Breeze – Defuniak Springs, FL. 1936-1956 – 14r – (gaps) – us UF Libraries [071]
Breeze – Pardeeville, Wyocena WI. 1884 apr 9 – 1r – 1 – mf#960865 – us WHS [071]
The breeze – Peru, NE: E J Smith. 1v. 1884-v1 n16. mar 6 1885 (wkly) [mf ed v1 n9. jan 17-mar 6 1885 (lacks jan 24) filmed 1976] – 1r – 1 – us NE Hist [071]
The breeze see Miscellaneous newspapers of pueblo county
Breeze, James T see
- Canadian poems
- The fenian raid!!; the queen's own!

Bref expose historique des recherches en industrie laitiere faites dans la province de quebec / Allard, Gaston – 1950 [mf ed 2001] – 9 – cn Bibl Nat [338]
Breffort, Alexandre see Harengs terribles
Brega / Cruz, Carlos Manuel De La – Habana, Cuba. 1934 – 1r – us UF Libraries [972]
Bregel', E la see Denezhnoe obrashchenie i kredit sssr
Brehaut, Ernest see An encyclopaedist of the dark ages
Brehier, E see Les idees philosophiques et religieuses de philon d'alexandrie (ephm8)
Brehier, Emile see Les idees philosophiques et religieuses de philon d'alexandrie
Brehier, Louis see
- L'eglise et l'orient au moyen age
- Histoire anonyme de la premiere croisade.
- Le schisme oriental du 11e siecle

Brehm, Alfred see Reiseskizzen aus nord-ost-afrika oder den unter egyptischer herrschaft stehenden laendern egypten, nubien, sennahr, rosseeres und kordofahn
Brehm, Alfred Edmund see Tierreich nach brehm
Brehm, Bruno see
- Der abend ohne gefolge
- Apis und este
- Auf wiedersehn, susanne!
- Der duemmste sibiriak
- Das gelbe ahornblatt
- Die grenze mitten durch das herz
- Heimat ist arbeit
- Im grossdeutschen reiche
- Der koenig von ruecken
- Kuenstler
- Die sanfte gewalt
- Ein schloss in boehmen
- Tag der erfuellung
- They call it patriotism
- Weder kaiser noch koenig
- Die weisse adlerfeder
- Zu frueh und zu spaet

Breidenbach, Bernhard von see Aelterer deutscher 'macer' / ortolf von baierland: 'arzneibuch' / 'herbar' des bernhard von breidenbach / faerber- und maler-rezepte (cima13)
Breidenbaugh, Edward Swoyer see The pennsylvania college book, 1832-1882
A breife discovery or description of the most famous island of madagascar or st lawrence in asia neare unto east-india / Boothby, R – London, 1646 – 1mf – 9 – mf#HT-11 – ne IDC [916]
Breijo Hernandez, Ody see Nuevos poemas
Ein breiniad – [Wales] University of Wales-Bangor 21 medi 1878-rhag 1883 (various) [mf ed 2003] – 1r – 1 – uk Newsplan [072]
Breisgauer bote – Freiburg Br DE, 1849 1 nov-1934 20 jan [gaps] – 1 – (missing: 1916 jan-feb & jul-dec; title varies: 1 jan 1853: breisgauer zeitung; with suppl: extrablatt 1871 28 & 29 jan, 1877 24 jan, kreisverkuendigungs-blatt fuer den kreis freiburg 1840 1 nov-1862, 1869-79) – gw Misc Inst [074]
Breisgauer nachrichten see Badische zeitung [main ed]
Breitenbach, Marthinus Christofel see The effects of trade liberalisation on south african agriculture
Breitenkamp, Paul see Kuender deutscher einheit
Breitkopf, Gregor see Breuiuscula facilimaque commentatio in paruulum philosophiae moralis
Breitman, G N see Poslednie novosti
Breitscheid, Rudolf see Das freie volk
Breiz atao / Parti autonomiste breton – Rennes. 1919-aout 1939 – 1 – fr ACRPP [325]
Die breklumer mission in indien : ein reisebericht / Bracker, pastor – Breklum: Missionshauses [1919] [mf ed 1995] – (= ser Yale coll) – 653p – 1 – 0-524-10139-6 – (in german) – mf#1995-1139 – us ATLA [240]
Brelsford, William Vernon see Tribes of northern rhodesia
Brelvi, Mahmud see Islam in africa
Bremen, Carl von see
- Der deutsche berg im osten
- Die schifferwiege

The bremen lectures on great religious questions of to-day / Christlieb, Theodor et al – new impr ed. Philadelphia: American Baptist Pub Soc, 1898 [mf ed 1985] – 1mf – 9 – 0-8370-2443-9 – (english trans fr german by david heagle. originally trans fr: neun apologetische vortraege ueber einige fragen und wahrheiten des christentums) – mf#1985-0443 – us ATLA [240]
Bremen. Statistisches Amt see Jahrbuch fuer die amtliche statistik des bremischen staats
Bremen. Statistisches Landesamt see Jahrbuch fuer bremische statistik
Bremer arbeiter-zeitung – Bremen DE, dec 14 1918-sep 30 1922 – 5r – 1 – gw Misc Inst [331]
Bremer beobachter see Bremischer beobachter
Bremer beobachter 1911 – Bremen DE, jan 6 1911-mar 25 1917 – 1r – 1 – gw Misc Inst [074]
Bremer blaetter fuer jedermann – Bremen DE, aug 1917-oct 1918 – 1r – 1 – gw Misc Inst [074]
Bremer bote – Bremen DE, 1903-17 – 6r – 1 – gw Misc Inst [074]
Bremer buergerfreund see Der buergerfreund
Bremer buerger-zeitung – Bremen, Germany 20, 21 sep, 4, 21 dec 1916-7 aug 1919 [daily] (imperfect) – 5r – 1 – uk British Libr Newspaper [074]
Bremer buerger-zeitung – Bremen DE, 1890 1 mai-1919 3 feb, 1920-1933 12 mar – 96r – 1 – (title varies: 1919: bremer volksblatt, 2 oct 1922: bremer volkszeitung) – mf#1361 – gw Mikropress [074]
Bremer County argus – Waverly IA. 1860 aug 23 – 1r – 1 – mf#851242 – us WHS [071]
Bremer courier see Der courier an der weser
Bremer familienblatt see Bremer general-anzeiger
Bremer, Franz Peter see Franz von sickingens fehde gegen trier
Bremer freie zeitung – Bremen DE, jul 1 1876-oct 17 1878 – 2r – 1 – gw Misc Inst [074]
Bremer fremdenblatt 1855 – Bremen DE, 1855-60 – 5r – 1 – gw Misc Inst [074]
Bremer, Gabriele see Schleswig-holsteinisches kuenstlerlexikon des 20. jahrhunderts
Bremer general-anzeiger – Bremen DE, sep 24 1894-97 – 9r – 1 – (suppl: bremer familienblatt nov 10 1895-jun 20 1897) – gw Misc Inst [074]
Bremer handelsblatt – Bremen DE, 1866 6 jan-29 dec, 1867 5 jan-26 oct – 1 – gw Misc Inst [380]
Bremer handelsblatt – Bremen, Germany 11 oct 1851-29 sep 1883 – 17r – 1 – (discontinued) – uk British Libr Newspaper [380]
Bremer handels-zeitung – Bremen DE, 1883-84 – 1r – 1 – gw Misc Inst [380]
Bremer hausfrauen-zeitung – Bremen DE, 1922-23, 1925-28, 1930, 1932, 1934 – 4r – 1 – gw Misc Inst [640]
Bremer illustrierte woche – Bremen DE, 1931-33 – 2r – 1 – gw Misc Inst [074]
Bremer journal – Bremen DE, jul 1 1877-jun 17 1880 – 1r – 1 – gw Misc Inst [074]
Bremer montagsblatt 1874 – Bremen DE, jul 13 1874-jun 28 1875 – 1 – gw Misc Inst [074]
Bremer montagsblatt 1876 – Bremen DE, feb 21-may 29 1876 – 1 – gw Misc Inst [074]
Bremer morgenpost – Bremen DE, sep 1 1863-66, 1867 (many iss missing), 1868-sep 13 1870 – 13r – 1 – (suppl: morgenpost fuer die jugend, mar 29-dec 20 1868) – gw Misc Inst [074]
Bremer nachrichten – Bremen, Germany 1 mar 1942-2 aug 1944, 20 sep 1949-30 nov 1952 (imperfect) – 24r – 1 – uk British Libr Newspaper [074]
Bremer nachrichten see Bremer woechentliche nachrichten
Bremer nachrichten mit weser-zeitung see Bremer woechentliche nachrichten
Bremer nationalsozialistische zeitung – Bremen DE, 1941 2 jun-30 sep, 1942-1945 28 apr – 4r – 1 – (title varies: bremer zeitung, nov 1 1933; filmed by other misc inst: 1931 10 jan-1945 21 apr [43r]) – gw Misc Inst [320]
Bremer post – Bremen DE, 1856 n1-7, 1857 n8-12, 1858 n1-12, 1860 n1-12 – 1 – gw Misc Inst [074]
Bremer sonntagsblatt 1843 – Bremen DE, oct 8 1843-dec 29 1844 – 1 – gw Misc Inst [074]
Bremer sonntagsblatt 1853 – Bremen DE, 1853 2 jan-1866 25 mar – 7r – 1 – (suppl: literarischer wegweiser, 1864 jan-mar & mai-dez) – gw Misc Inst [074]
Bremer tageblatt 1855 – Bremen DE, dec 1855-mar 1859 – 1 – gw Misc Inst [074]
Bremer tageblatt 1895 – Bremen DE, nov 24 1895-jan 25 1896 – 1 – gw Misc Inst [074]
Bremer tageblatt 1897 – Bremen DE, nov 21 1897-1920 – 1 – (suppl: bremer tuermer, jul 21 1907-aug 2 1914) – gw Misc Inst [074]
Bremer tages-chronik see Tages-chronik
Bremer telegraph – Bremen DE, oct 2-dec 29 1846, dec 23 1847, jan 18-apr 25 1848 [many iss missing] – 1r – 1 – gw Misc Inst [074]
Bremer telegraph see Telegraph
Bremer volksblatt see Bremer buerger-zeitung
Bremer volksblatt 1875 – Bremen DE, 1875-sep 23 1878 – 4r – 1 – (title varies: 6 apr 1875: volks-blatt) – gw Misc Inst [074]
Bremer volks-zeitung – Bremen DE, apr 1-jun 10 1888 – 1r – 1 – gw Misc Inst [074]
Bremer volkszeitung see Bremer buerger-zeitung
Bremer weser-zeitung see
- Weser-zeitung

Bremer wochen-schrift – Bremen DE, dec 29 1749-dec 21 1750 – 1r – 1 – gw Misc Inst [074]
Bremer woechentliche nachrichten – Bremen DE, 1848-49 [4r], 1960 12 mar-1961 5 sep (tw. nur ausz.), 1961 20 oct-1962 8 mai, 1962 2 jun-1963 13 apr, 1963 29 jun-31 dec, 1964 11 jun-1966 – 1 – (title varies: 1 jan 1854: bremer nachrichten, 30 sep 1934: bremer nachrichten mit weser-zeitung, 20 sep 1949: bremer nachrichten; filmed by other misc inst: 1950-83, 1978 1 sep- [ca 10r/yr]) – gw Misc Inst [074]
Bremer zeitung see Bremer nationalsozialistische zeitung
Bremer zeitung 1741 – Bremen, Hannover DE, 1848 – 2r – 1 – (title varies: zeitung fuer norddeutschland, dec 26 1848; since then publ in hannover; filmed by other misc inst: 1741, 1742, 1752, 1765 [single iss], oct 1815-jun 30 1864, 1865-1871 feb-dec; filmed by mikropress: 1848 26 dec-1853 jun, 1854-1864 jun, 1865-70, 1871 27 feb-dec) – gw Misc Inst; gw Mikropress [074]
Bremer zeitung 1921 – Bremen DE, 1921-29 – 22r – 1 – (title varies: norddeutsche rundschau, sep 1 1923, nationale rundschau, sep 2 1924, neue bremer zeitung, apr 1 1926, bremer zeitung, aug 1 1926. incl suppl: bremen im weltverkehr, 1924, bremer schiffahrtszeitung, 1921-23, buehne und film, 1921-jun 25 1922, bz fuer kleinhandel und gewerbe, 1921-jun 12 1922, der deutsche arbeiter, 1921-24, der domshof, 1921-sep 1 1923, frauen-zeitung, 1921-aug 13 1922, die helling, 1921-23, kiekinnewelt, 1921-23, landwirtschaftliche rundschau, 1925-feb 8 1926, plattdeutsche waelte, 1922, der praktische ratgeber fuer haus und garten, 1921-23, wahl-ulk, 1921, weihnachtsanzeiger, 1923-24) – gw Misc Inst [074]
Bremerleher woechentliche anzeigen – Bremerhaven DE, 1859 5 jan-29 jun – 1r – 1 – (title varies: mai 1846: woechentliche anzeigen fuer lehe, umgegend und land wursten, 1848: der mittheiler an der unter-weser, 11 dec 1861: volksblatt an der nordsee, dez 1862: volksblatt an der weser; publ (bremerhaven-) lehe, later bremerhaven) – gw Misc Inst [074]
[Bremerton-] bremerton sun – WA. 1901- – 514r – 1 – $30,840.00 (subs $500y) – mf#R05400 – us Library Micro [071]
[Bremerton-] daily news searchlight – WA. 1922-45 – 53r – 1 – $3180.00 – mf#R05401 – us Library Micro [071]
[Bremerton-] kitsap county review – WA. 1903-25 – 10r – 1 – $600.00 – mf#R05402 – us Library Micro [071]
Bremervoerder volks-bote see Wochenblatt fuer die amtsgerichtsbezirke bremervoerde, beverstedt und zeven
Bremervoerder wochenblatt see Wochenblatt fuer die amtsgerichtsbezirke bremervoerde, beverstedt und zeven
Bremervoerder zeitung see Wochenblatt fuer die amtsgerichtsbezirke bremervoerde, beverstedt und zeven
Bremische blaetter – Bremen DE, 1835 n1-2, 1836 n3-5 – 1 – gw Misc Inst [074]
Bremische blaetter fuer unterhaltung, belehrung und witz – Bremen DE, apr 13-jun 1 1879 – 1 – gw Misc Inst [074]
Bremische correspondenz – Bremen DE, 1916 – 1r – 1 – gw Misc Inst [074]
Bremische Deutsche Gesellschaft see Versuch eines bremisch-niedersaechsischen woerterbuchs
Bremische polizeibeamten-zeitung – Bremen DE, 1924-34 – 1 – (title varies: mai? 1933: nachrichten der nsdap) – gw Misc Inst [074]
Bremische volkszeitung – Bremen DE, 18 oct 1878-25 feb 1879 – 1r – 1 – mf#2350 – gw Mikropress [074]
Bremischer beobachter – Bremen DE, 1849-55 – 2r – 1 – (title varies: 1 jan 1853: bremer beobachter) – gw Misc Inst [074]
Bremischer volksfreund – Bremen DE, 23 nov 1849-30 nov 1852 – 1r – 1 – gw Misc Inst [074]
Bremisches conversationsblatt – Bremen DE, 3 mai 1838-30 jun 1839 – 1r – 1 – gw Misc Inst [074]
Bremisches jahrbuch / ed by Kuenstlerverein [Bremen, Germany]. Historische Gesellschaft – Bremen: Verlag von G Gd Muller. v1-42 1864- (annual) [mf ed 1978] – 4r – 1 – (with gaps; ind in v24) – ISSN: 0341-9622 – mf#film mas c408 – us Harvard [943]
Bremisches magazin – Bremen DE, 1831-1834 n1-12 – 1r – 1 – gw Misc Inst [074]
Bremisches magazin zur ausbreitung der wissenschaften, kuenste und tugend / ed by Cassel, Johann Philipp – Hannover 1757-65 – (= ser Dz) – 7v[zu je 3st] on 33mf – 9 – €330.00 – (fr v2 publ in bremen & leipzig) – mf#k/n5253 – gw Olms [074]
Bremisches und verdisches theologisches magazin – Bremen DE, 1795-98 – 1r – 1 – gw Misc Inst [240]
Bremisches unterhaltungsblatt see Bremisches volksblatt
Bremisches volksblatt – Bremen DE, 1823-47, 1848 [single iss], 1855-59 – 12r – 1 – (title varies: 1 jan 1841: bremisches unterhaltungsblatt) – gw Misc Inst [074]
Bremmer, Rolf H, Jr see Manuscripts in the low countries
Bremner, Archie see City of london, ontario, canada
Bremner, Fred see Types of the indian army
Bremond, A see
- Bullarium ordinis ff praedicatorum
- Histoire du sentiment religieux en france

Bremond, Henri see
- L'inquietude religieuse
- The mystery of newman
- Sir thomas more (the blessed thomas more)

Bremsstrahlung niederenergetischer elektronen : experimentelle untersuchungen und simulationsrechnungen / Lindenstruth, Stefan – (mf ed 1994) – 1mf – 9 – €30.00 – 3-8267-2027-X – mf#DHS 2027 – gw Frankfurter [530]
Brenan, Michael John see An ecclesiastical history of ireland
Brenchley, Julius L see The brenchley papers
The brenchley papers / Brenchley, Julius L – 1840-65 – 1r – 1 – mf#PMB1050 – at Pacific Mss [920]
Brene, Jose R see
- Santa camila de la habana vieja
- Teatro

Brenendike brikn – Berlin, Germany. 1923 – 1r – us UF Libraries [939]
Brenes Cordoba, Alberto see Tratado de los bienes
Brenes La Roche, Santos see Fragua y fuelle
Brenes, Maria see Diez cuentos para un libro
Brenes Mesen, Roberto see
- Apologia del presidente roosevelt y un poema
- Critica americana
- Hacia nuevos embrales

Brengle branches – v1-6 n2 [1983 fall-1989 winter] – 1r – 1 – mf#1507376 – us WHS [071]
Brenier, Flavien see Evrei i talmud
Brenil, H see Glanes paleolithiques anciennes dans le bassin du guadiana
Brenk, B see Tradition und neuerung in der christlichen kunst des ersten jahrtausends (wbs3)
Brennan, Richard see A popular life of our holy father pope pius the ninth
Brennan, Robert Edward see History of psychology from the standpoint of a thomist
Brennan, sullivan and connelly scrapbooks – Cleveland, Cuyahoga, OH. 1844-1994 – 1r – 1 – (photographs, clippings, programs, letters, obituaries, memorials, certificates and memorabilia of the allied brennan, sullivan, and connelly families, irish immigrants to cleveland in the 19th century) – us Western Res [920]

Brenndorfer nachrichten – Brenndorf (Horineves CZ), 1929 5 jan-1937 – 2r – 1 – gw Misc Inst [077]
Brennecke, W *see*
– Forschungsreise s m s planet 1906/07
– Die ozeanographischen arbeiten der deutschen antarktischen expedition 1911-1912
Brenneker, Father *see* Bonaire
Der brennende baum : eine erzaehlung / Frenssen, Gustav – Berlin: G Grote 1931 [mf ed 1989] – (= ser Grote'sche sammlung von werken zeitgenoessischer schriftsteller 189) – 1r – 1 – (filmed with: ferdinand freiligrath / schmidt-weissenfels) – mf#7263 – us Wisconsin U Libr [880]
Brennende lieder und strophen / Biedenkapp, Georg – New York, NY: Im Selbstverlag, 1900 [mf ed 1989] – 128p – 1 – mf#7019 – us Wisconsin U Libr [810]
Brenner, D A I *see* Historia de las revoluciones de hungaria
Brenner, Joseph Hayyim *see*
– 'Arakhim
– Shekhol ve-khishalon
Brenner, Oskar *see* Altnordisches handbuch
Brenner, Stefan *see* Die entwicklung der frage der buendniszugehoerigkeit eines wiedervereinigten deutschlands von der maueroeffnung bis zum treffen von michail gorbatschow und helmut kohl in schelesnowodsk...
Brennerei-zeitung – Bonn 1896-1900 15 dec, 1902-05 [mf ed 2004] – 2mf=3df – 9 – gw Mikrofilm [074]
Die brennessel – Muenchen, Berlin DE, 1931-38 – 5r – 1 – gw Mikrofilm [074]
Brenning, Emil *see*
– Graf adolf friedrich von schack
– Wilhelm herbsts hilfsbuch fuer die deutsche litteraturgeschihte
Brennus – Berlin DE, 1802 – 1 – gw Misc Inst [074]
Brensa, Carel J *see* West-indie
Brent, Charles Henry *see*
– Adventure for god
– The conquest of trouble; and, the peace of god
– The consolations of the cross
– The inspiration of responsibility, and other papers
– Leadership
– Liberty and other sermons
– A master builder
– The mind of christ jesus in the church of the living god
– Presence
– Prisoners of hope and other sermons
– The revelation of discovery
– Sixth sense
– The splendor of the human body
– With god in prayer
– With god in the world
Brent leader – Brent, England 7 oct 1983-19 feb 1988 [mf 1986-8] – 3r – 1 – (replaced by local ed; cont: leader [brent ed] 13 may-30 sep 1983) – uk British Libr Newspaper [072]
Brent leader *see* Leader (brent ed)
Brent post – Brent, England 22 sep 1988-10 aug 1989 – 1 1/2r – 1 – (discontinued) – uk British Libr Newspaper [072]
Brentano als maerchenerzaehler / Gloeckner, Karl – Jena: E Diederich, [1937] [mf ed 1993] – (= ser Deutsche arbeiten der universitaet koeln 13) – 81p – (incl bibl ref) – mf#8215 reel 2 – us Wisconsin U Libr [430]
Brentano, Bernard von *see* Die ewigen gefuehle
Brentano, Bernhard von *see* Theodor chindler
Brentano, Christian *see* Der unglueckliche franzose
Brentano, Clemens *see*
– Brentanos werke
– Briefe
– Die chronika des fahrenden schuelers
– Chronika eines fahrenden schuelers
– Des knaben wunderhorn
– Gockel, hinkel und gackeleia
– Gustav wasa
– Romanzen vom rosenkranz
– Die schachtel mit der friedenspuppe
– Valeria, oder, vaterlist
Brentano, Franz Clemens *see*
– Ausgewaehlte werke
– The origin of the knowledge of right and wrong
– Der philister vor, in und nach der geschichte
– Die psychologie des aristoteles
Brentano, Lujo *see* Clemens brentanos liebesleben
Brentano, Maria Rafaela *see* Amalie fuerstin von gallitzin
Brentanos im rheingau : am urquell der rheinromantik / Doderer, Otto – 2. aufl. Ratingen [Germany]: A Henn 1955 [mf ed 1993] – (= ser Rheinische buecherei) – 1r [ill] – 1 – (filmed with: der erzieherische gehalt in j j breitingers "critischer dichtkunst" / comp by jakob braeker) – mf#8525 – us Wisconsin U Libr [430]
Brentanos jugenddichtungen : abschnitt 1: der ideengehalt des godwi / Kempner, Alfred – Halle-Wittenberg, 1914 (mf ed 1995) – 1mf – 9 – 3-8267-3132-8 – mf#DHS-AR 3132 – gw Frankfurter [430]
Brentanos werke / ed by Preitz, Max – crit ed. Leipzig: Bibliographisches Institut 1914 [mf ed 1993] – (= ser Meyers klassiker-ausgaben) – 3v on 1r – 1 – (incl bibl ref) – mf#8529 – us Wisconsin U Libr [800]
Brentanos werke / ed by Preitz, Max – Leipzig: Bibliographisches Institut 1914 [mf ed 1993] – (= ser Meyers klassiker-ausgaben) – 3v on 1r – 1 – (incl bibl ref) – mf#8529 – us Wisconsin U Libr [800]
Brentford and chiswick leader – Hounslow, England 4 jan 1996-20 apr 2001 – 1 – uk British Libr Newspaper [072]
Brenton, Edward Pelham *see* Life and correspondence of john, earl of st vincent...admiral of the fleet, etc
Brenton, Lancelot Charles Lee *see* The septuagint version of the old testament
Brents, Thomas Wesley *see* The gospel plan of salvation
Brentwood baptist church – Charleston Co, SC. 2308p. 1957-1988 – 2r – 1 – $103.86 – (deacons' minutes 1962-70, 1979; church council minutes 1986-88; history & by-laws 1957-88. financial records/reports) – mf#6650a – us Southern Baptist [242]
[Brentwood-] brentwood news – CA. mar 5 1937-1956; 1969-97 – 32+ r – 1 – $1920.00 – mf#BC02076 – us Library Micro [071]
Brentz, J *see* Tuerchen buchlein
[Brenz, Johannes] *see* Ordnung der kirchen inn eins erbarn raths zu schwaebischen hall oberkeit und gepiet gelegen
Brenztal-bote : ba der suedwestpresse – Giengen, Brenz DE, 1975- – 101r (1975-1990) – 1 – (bezirksausgabe von suedwest-presse, ulm) – gw Misc Inst [074]
Brepohl, Friedrich Wilhelm *see* Johannes calvin und seine bedeutung fuer unsere heutige kultur
Brereton, Charles David *see* The subordinate magistracy and parish system considered
Brereton, Frederick Sadleir *see* The hero of panama
Brereton, John *see* Doctrine of election considered with reference to the ministerial o...
Bres, G de *see*
– Le baston de la foy chrestienne...
– Le baston de la foy chrestienne
– Le baston de la foy chrestienne...
– Le baston de la foy chrestienne
– Le baston de la foy chrestienne...
– Confession de foy...
– Confession de foy
– Confession de foy...
– Confession de foy
– Correspondance
– Declaration sommaire dv faict de cevx de la ville de vallencienne
– Interrogatoires
– Oraison av seignevr
– Procedvres tenves...l'endroit de cevx de la religion dv pais bas
– La racine
– Remonstrance et svpplication de cevs de l'eglise reformee de la ville de valencenes
– Reqveste de cevs de l'eglise reformee de valencenes...
– De wortel
Brescia, Anthony M *see* The letters and papers of richard rush
Bresil / Dumon, Frederic – Bruxelles, Belgium. 1964 – 1r – us UF Libraries [972]
Bresil / Faust, Jean Jacques – Paris, France. 1966 – 1r – us UF Libraries [972]
Bresil : terre d'avenir / Zweig, Stefan – New York, NY. 1942 – 1r – us UF Libraries [972]
Le bresil : ou histoire, moeurs, usages et coutumes des habitans de ce royaume / Taunay, Hippolyte – Paris 1822 [mf ed Hildesheim 1995-98] – 14mf – 9 – €140.00 – ISBN-10: 3-487-26860-4 – ISBN-13: 978-3-487-26860-6 – gw Olms [918]
Bresil d'aujourd'hui / Burnichon, Joseph – Paris, France. 1910 – 1r – us UF Libraries [972]
Bresil d'aujourd'hui / CONSELHO NACIONAL DE ESTAT ISTICA – Rio de Janeiro, Brazil. 1956 – 1r – us UF Libraries [972]
Bresil litteraire / Wolf, Ferdinand Joseph – Berlin, Germany. 1863 – 1r – us UF Libraries [972]
Bresil meridional / Carvalho, Carlos Miguel Delgado De – Rio de Janeiro, Brazil. 1910 – 1r – us UF Libraries [972]
Breslau und umgebung / Markgraf, Hermann – Zuerich [1892] [mf ed Hildesheim: 1995-98] – (= ser Fbc) – 130p on 1mf [ill] – 9 – €40.00 – 3-487-29559-8 – gw Olms [914]
Breslauer anzeiger – Breslau (Wroclaw PL), 1854, 1856-60 [gaps], 1862 1 jan-30 mar & 1 jul-31 dec, 1873 1 jan-29 jun & 1 oct-31 dec, 1879 1 jul-31 dec, 1886 1 oct-31 dec, 1887 1 apr-30 jun, 1888 1 jul-30 sep, 1898 1 jan-31 mar & 1 jul-30 sep, 1907 1 sep-31 dec, 1908 1 mar-30 jun, 1908 1 sep-1909 30 apr, 1911 1 oct-31 dec, 1914 1 apr-30 jun – 34r – 1 – (with suppl; title varies: 29 mar 1853: kleine morgenzeitung, 1 jan 1858: morgenzeitung, 1 oct 1862: breslauer morgenzeitung; filmed by other misc inst: 1854 jan-5 aug, 1856, 1862 1 jan-23 apr & 17 jul-31 dec, 1873 [17r]) – gw Misc Inst [077]
Breslauer beitraege zur literaturgeschichte – Leipzig: M Hesse 1904-19 [mf ed 1992] – 1r – 1 – (iss 46 & 49 never publ) – mf#3100p – us Wisconsin U Libr [430]
Breslauer beitraege zur literaturgeschichte – Leipzig: M Hesse. heft1-50. 1904-19 [mf ed 1992] – 48v – 1 – (publ in leipzig, 1904-09 (heft1-18); in breslau (f hirt), 1910-12 (heft19-30); in stuttgart (j b metzler), 1912-1919 (heft31-50). heft11-30 also called neue folge 1-20; heft31-50 called neuere folge, but keep the numbering of the 1st ser (i.e, 31-50). heft46 and 49 never publ) – mf#8014 – us Wisconsin U Libr [430]
Der breslauer erzaehler – Breslau (WrocLaw PL), 1838, 1845-46 – 2r – 1 – gw Misc Inst [077]
Breslauer general-anzeiger – Breslau (WrocLaw PL), 1916 1 apr-dec, 1939-1944 29 feb – 16r – 1 – (title varies: breslauer neueste nachrichten, 15 apr 1918; filmed by other misc inst: 1931 1 mar-30 apr, 1935 1 sep-31 oct, 1938 1 jan- 28 feb, 1941 1 oct-1942 31 aug, 1943 1 jan-30 jun, 1943 1 sep-1944 29 feb [6r]) – gw Misc Inst [077]
Breslauer gerichts-zeitung – Breslau (WrocLaw PL), 1916 2 apr-1917 25 mar – 3r – 1 – gw Misc Inst [077]
Breslauer handelsblatt – Breslau (WrocLaw PL), 1854-56 [gaps], 1861-1863 30 sep [gaps], 1864 2 jan-31 oct, 1865, 1873, 1874 2 jul-31 dec – 12r – 1 – gw Misc Inst [380]
Breslauer hausblaetter *see* Breslauer hausblaetter fuer das volk 1863
Breslauer hausblaetter fuer das volk 1789 – Breslau (WrocLaw PL), 1789-1806 – 1 – gw Misc Inst [077]
Breslauer hausblaetter fuer das volk 1863 – Breslau (WrocLaw PL), 1943 [gaps], 1944 – 1 – (title varies: 1 apr 1869: breslauer hausblaetter; 1 jul 1871: schlesische volkszeitung) – gw Misc Inst [077]
Breslauer intelligenz-blatt *see* Breslausche auf das interesse der commerzien der schl. lande eingerichtete frag- und anzeigungs-nachrichten
Breslauer juedisches gemeindeblatt : amtliches blatt der synagogengemeinde breslau – Breslau. v1-15. 1924-38 – (= ser German-jewish periodicals...1768-1945, pt 1) – 2r – 1 – $220.00 – (lacking: misc iss) – mf#B435 – us UPA [939]
Breslauer juedisches gemeindeblatt – Breslau (WrocLaw PL), 1924 8 aug-1938 25 aug [gaps] – 2r – 1 – (title varies: 10 aug 1937: juedisches gemeindeblatt fuer die synagogengemeinde breslau) – gw Misc Inst [939]
Breslauer kreisblatt – Breslau (WrocLaw PL), 1928 – 1r – 1 – gw Misc Inst [077]
Breslauer montag – Breslau (WrocLaw PL), 1927 8 aug-5 dec – 1 – gw Misc Inst [077]
Breslauer neueste nachrichten *see* Breslauer general-anzeiger
Die breslauer ritualien / Jungniss, J – Breslau, 1892 – 1mf – 8 – €3.00 – ne Slangenburg [241]
Breslauer volksblaetter – Breslau (Wroclaw PL), 1904-06 – 2r – 1 – (with suppls) – gw Misc Inst [077]
Breslauer volksspiegel – Breslau (Wroclaw PL), 1846-47 – 2r – 1 – (title varies: 1847: volksspiegel) – gw Misc Inst [077]
Breslauer zeitung – Wroclaw, Poland 3 oct 1914, 12 jan 1916-10 aug 1919 (daily) (imperfect) – 1 – uk British Libr Newspaper [077]
Breslauer zeitung *see* Neue breslauer zeitung
Breslausche auf das interesse der commerzien der schl. lande eingerichtete frag- und anzeigungs-nachrichten – Breslau (WrocLaw PL), 1835 5 oct-28 dec, 1838 2 jul-24 dec – 2r – 1 – (title varies: 1816?: breslausches intelligenz-blatt, 1 sep? 1829: breslauer intelligenz-blatt; incl tw. subser: gemeinnuetziger anzeiger zum breslauer intelligenz-blatt; filmed by misc inst: 1811-12, 1816 nov, 1817 jul-aug, 1818 jan-feb, 1823-28, 1829 mai-jun & sep-okt, 1833 mai-jun, 1835 okt-dez, 1836 okt-dez, 1838 jul-dez) – gw Misc Inst [077]
Breslausches intelligenz-blatt *see* Breslausche auf das interesse der commerzien der schl. lande eingerichtete frag- und anzeigungs-nachrichten
Bresler, A *see* Perets smolenskin
Bresoles, Judith de *see* Bio-bibliographie de mere marie-berthe thibault
Bressani, Francisco Giuseppe *see* Les jesuites-martyrs du canada
Bresslau, H *see* Chronica heinrici surdi de selbach (mgh6:1.bd)
Bresslau, Marcus Heinrich *see* Shabthoth
Bresson / Semolue, Jean – Paris, France. 1959 – 1r – us UF Libraries [920]
Bresson, Jacques *see* Histoire financiere de la france
Brest on the quebec labrador / Combes, Le, sieur – Ottawa: J Hope & Sons, 1905 – 1mf – 9 – 0-665-72642-2 – (text in english and french) – mf#72642 – cn Canadiana [917]
La bretagne pittoresque et legendaire / Sebillot, Paul – Paris: H Daragon 1911 [mf ed Bloomington IN: Indiana Uni Lib, Preservation Dept 1984] – 1r – 1 – us Indiana Preservation [390]
La bretagne reelle – Merdrignac (Cotes-du-Nord). 1954-juin 1972 – 1 – fr ACRPP [073]
Bretau, Francisco *see* 3 anos de lucha resumen historico de la federacion sindical de plantas electricas, gas y agua
Bretholz, B *see* Cosmae pragensis chronica boemorum (mgh6:2.bd)
The brethren almanac for the year of our lord... – Ashland OH: publ by Brethren Book & Tract Committee, 1896 [annual] [mf ed 2003] – 1r – 1 – (reel incl earlier title: brethren's annual for the year of grace...) – mf1041 – us ATLA [242]
The brethren annual : or, church year book – Ashland OH: Brethren Publ Board, 1897-1913 [annual] [mf ed 2003] – 17v on 1r – 1 – (reel incl later title: brethren annual for...) – mf1042 – us ATLA [242]
The brethren annual for... – Ashland OH: Brethren Publ, 1914-16 [annual] [mf ed 2003] – 3v on 1r – 1 – (reel incl earlier title: brethren annual, or church year book) – mf1043 – us ATLA [242]
Brethren Church (Progressive Dunkers). Convention *see*
– Proceedings of the dayton convention
– Proceedings of the general convention of the brethren church
– Report of progressive convention
Brethren Church (Progressive Dunkers). General Conference *see*
– Minutes of the annual conference of the brethren church
– Minutes of the general conference of the brethren church for...and brethren annual for...and brethren annual for...
– Minutes of the...general conference of the brethren church
– [Report of the general conference of the brethren church]
The brethren hymn book : a collection of psalms, hymns and spiritual songs suited for song service in christian worship, for church service, social meetings and sunday schools – Elgin, IL: Brethren Pub House, 1901 – 2mf – 9 – 0-524-03610-1 – mf#1990-4770 – us ATLA [242]
The brethren hymnody, with tunes : for the sanctuary, sunday-school, prayer-meeting, and home circle / ed by Ewing, John Cook – Wilmington, Ohio: JC Ewing, 1884 – 3mf – 9 – 0-524-08755-5 – mf#1993-3260 – us ATLA [242]
Brethren in Christ Church *see* Handbook of missions
Brethren life and thought – Elizabethtown PA 1955+ – 1,5,9 – ISSN: 0006-9663 – mf#3294 – us UMI ProQuest [242]
The brethren's almanac – 1871-79 [complete] – 1r – 1 – mf#ATLA S0905 – us ATLA [030]
Brethren's almanac and directory – 1897-98 (complete) – (= ser Mennonite serials coll) – 1r – 1 – mf#ATLA 1993-S027 – us ATLA [030]
The brethren's annual for the year of grace... – Ashland OH: H R Holsinger, 1884-95 [annual] [mf ed 2003] – 12v on 1r – 1 – (lacks 1888 p7-8, 1889 p47-48. reel incl later title: brethren almanac for the year of our lord...) – mf1040 – us ATLA [242]
The brethren's annual for the year of grace 1885 : contains calendar for each month, biographical sketches... – Ashland OH: HR Holsinger [1885?] [mf ed 1992] – 1mf – 9 – 0-524-03064-2 – mf#1990-4553 – us ATLA [242]
The brethren's church manual : containing the declaration of faith, rules of order, how to conduct religious meetings etc / Brumbaugh, Henry Boyer – rev ed. Huntingdon, PA: JL Rupert, 1891 – 1mf – 9 – 0-524-02813-3 – mf#1990-4434 – us ATLA [242]
The brethren's family almanac – 1880-1917 [complete] – 2r – 1 – mf#ATLA S0906 – us ATLA [030]
The brethren's reasons for producing and adopting the resolutions of august 24th : consisting of a collection of petitions made to the annual meeting from year to year... / Kinsey, Samuel et al – Kinsey OH: Office of the Vindicator 1883 [mf ed 1992] – 1mf – 9 – 0-524-04227-6 – mf#1990-5018 – us ATLA [242]

The brethren's sunday-school song book : for use in sunday-schools, prayer and social meetings – Mt Morris, Ill: General Missionary and Tract Committee, c1894 – 3mf – 9 – 0-524-05541-6 – mf#1990-5145 – us ATLA [242]

The brethren's tracts and pamphlets : setting forth the claims of primitive christianity / Miller, Daniel Long et al – Dayton, Ohio: Brethren's Book and Tract Work, 1892 – 1mf – 9 – 0-524-02834-6 – mf#1990-4455 – us ATLA [242]

Breton de LaMartiniere, Jean *see*
- Aegypten, oder sitten, gebraeuche, trachten und denkmaeler der aegypter
- China
- La chine en miniature
- L'egypte et la syrie
- L'espagne et le portugal
- Le japon, ou moeurs, usages et costumes des habitans de cet empire
- La russie

Breton, Jose *see* Camino

Breton, Jules Adolphe Aime Louis *see* The life of an artist

Breton, Pierre Napoleon *see*
- Le collectionneur illustre des monnaies canadiennes
- Histoire illustree des monnaies et jetons du canada
- Popular illustrated guide to canadian coins, medals, etc
- Le secretaire canadien

Bretschneider, C G *see* Corpus reformatorum

Bretschneider, E *see* Mediaeval researches from eastern asiatic sources

Bretschneider, Emilii Vasil'evich *see* On the study and value of chinese botanical works

Bretschneider, Heinrich G von *see* Reise des herrn von bretschneider nach london und paris

Bretschneider, Horst *see* Aus meinem leben

Bretschneider, Karl Gottlieb *see*
- Aus meinem leben
- Handbuch der dogmatik der evangelisch-lutherischen kirche
- Henry and antonio
- A manual of religion and of the history of the christian church
- Der simonismus und das christenthum

Brett, Edwin John *see* A pictorial and descriptive record of the origin and development of arms and armour

Brett, George *see* Progress after death

Brett, George Sidney *see* A history of psychology, ancient and patristic

Brett, H J *see* Report on the industrial and economic situation of china in june, 1923

Brett, Henry *see* White wings

Brett, James Warden *see* Journal of a harpooner on board the whaling ship massachusetts

Brett, John Watkins *see* The illustrated catalogue of the valuable collection of pictures...coins and medals

Brett, Thomas *see*
- Honour of the christian priesthood
- Leading cases in modern equity

Brett, William Henry *see*
- Indian missions in guiana
- Indian tribes of guiana
- Mission work among the indian tribes in the forest
- Mission work among the indian tribes in the forests of guiana

Brettener sonntagszeitung – Bretten DE, 1913-19 [gaps] – 3r – 1 – gw Misc Inst [074]

Bretton woods agreement releases – n1-2. 9 jan 1950-19 jan 1955 (all publ) – (= ser The sec release series preceding the sec docket) – 1mf – 9 – $1.50 – mf#LLMC 89-004 – us LLMC [346]

Bretz, Adolf *see*
- Studien und texte zu asterios von amasea

Breuer, Isaac *see*
- Judenproblem
- Shpuren fun meshieh

Breuer, Joerg *see* Konzepte einer zentralen europaeischen verkehrswegeplanung

Breuer, Moses *see* Sophie bernhardi geb tieck als romantische dichterin

Breuer, Raphael *see*
- Gedankenwelt der halacha
- Zur abwehr

Breuer, Salomon *see* Juedische monatshefte

[Breughel, G H van] *see* Cupido's lusthof ende der amoureuse boogaert...

Breuiuscula facilimaque commentatio in paruulum philosophiae moralis : teneriori etati necessaria ad recte uirtuoseque uiuendum / Breitkopf, Gregor – [impressum Liptzik Per Jacobum Thanner Herbipolensem, anno nostre salutis 1504 quinta Junij] – (= ser Ethics in the early modern period) – 1mf – 9 – mf#pl-308 – ne IDC [170]

Breul, Karl [comp] *see* Romantic movement in german literature

Brev till henrik reuterdahl / Aulen, Gustaf – Stockholm: PA Norstedt, 1915 – 1mf – 9 – 0-524-03453-2 – mf#1990-0996 – us ATLA [240]

Brevard, Caroline Mays *see*
- History of florida
- History of florida from the treaty of 1763 to our...

Brevard's law reports / South Carolina. Supreme Court – v1-3. 1793-1816 (all publ) – (= ser Pre-nrs nominative law reports) – 20mf – 9 – $30.00 – mf#LLMC 94-010 – us LLMC [340]

Brevarium (siecle 15) – Valencia – 1r – 5,6 – sp Cultura [240]

Breve *see* Breve noticia de el origen que tuvo la devocion de la purisima concepcion de nuestra senora y de una santa imagen...

Breve antipologia al discurso nuevo... miguel fernandez de la pena...del uso de agua de la nieve en dia de purga / Perez Merino, I – Jaen, 1641 – 2mf – 9 – sp Cultura [615]

Breve antologia / Augier, Angel I – Santa Clara, Cuba. 1963 – 1r – us UF Libraries [972]

Breve antologia del cuento salvadoreno – San Salvador, El Salvador. 1962-1975 – 1r – us UF Libraries [972]

Breve biografia de antonio maceo / Porteil Vila, Herminio – Habana, Cuba. 1945 – 1r – us UF Libraries [920]

Breve compendio de todo lo que debe saber y entender el christiano : para poder logar, ver, conocer, y gozar de dios nuestro senor en el cielo eternamente / Ramírez, Antonio de Guadalupe – Mexico: impr D J de Jauregui 1784 – (= ser Books on religion...1543/44-c1800: doctrina cristiana, obras de devocion) – 2mf – 9 – mf#crl-57 – ne IDC [241]

Breve compendio della vita del famoso titiavecellio di cadore... / [Titian] TitiaVecellio – Venezia, 1622 – 1mf – 9 – mf#O-1072 – ne IDC [700]

Breve de n m s p clemente 14 en respuesta a la carta de el ill[ustrissi]mo s[eno]r arzobispo metropolitano de mexico : escrita a su santidad de resulta de su exaltacion al trono pontificio / Clement 14, Pope – [Mexico: s.n. 1770] – (= ser Books on religion... 1543/44-c1800: papas [cartas apostolicas, etc]) – 1mf – 9 – mf#crl-370 – ne IDC [241]

Breve descrittione dell'apparato funebre fatto per le sontuose esequie della serenissima reina isabella nel duomo di milano – Milano: Gio. Battista & Giuleo Cesare, [1644] – 1mf – 9 – mf#O-1988 – ne IDC [090]

Breve discurso em que se conta a conquista do reino do pegu abreu mousinho / Mousinho, Manuel de Abreu – Barcelos: Portugalense Editora 1936 [mf ed 1984] – 1r – 1 – (int by lopes d'almeida) – mf#1189 – us Wisconsin U Libr [959]

Breve ensayo sobre los nombres gentilicios usado en la alta extremadura / Gutierrez Macias, Valeriano – Badajoz: Dip. Provincial, 1970 – sp Bibl Santa Ana [946]

Breve estudio de la obra y personalidad del escultor y arquitecto don manuel tolsa. mexico / Escontria, Alfredo – Madrid: Razon y Fe, 1935 – 1 – sp Bibl Santa Ana [720]

Breve guia de trujillo, cuna de conquistadores / Moreno Lazaro, J – 1973 – 1 – sp Bibl Santa Ana [946]

Breve histoire de la petite armenie : l'armenie cilicienne / Iorga, N – Paris, 1930 – 2mf – 9 – mf#AR-1597 – ne IDC [915]

Breve historia constitucional y politica de colombia / Samper Bernal, Bustavo – Bogota, Colombia. 1957 – 1r – us UF Libraries [972]

Breve historia de guatemala / Contreras R, J Daniel – Guatemala, 1961 – 1r – us UF Libraries [972]

Breve historia de honduras / Barahona, Ruben – Tegucigalpa, Mexico. 1949 – 1r – us UF Libraries [972]

Breve historia de la poesia mexicana / Dauster, Frank N – Mexico City? Mexico. 1956 – 1r – us UF Libraries [440]

Breve historia de mexico / Bayle, Constantino & Vasconcelos, Jose – Madrid: Razon y Fe, 1944 – 1 – sp Bibl Santa Ana [972]

Breve historia de mexico / Vasconcelos, Jose – Madrid, Spain. 1952 – 1r – us UF Libraries [972]

Breve historia de mexico / Vasconcelos, Jose – Mexico City? Mexico. 1937 – 1r – us UF Libraries [972]

Breve historia del brasil / Mendonca, Renato – Madrid, Spain. 1950 – 1r – us UF Libraries [972]

Breve historia del modernismo / Henriquez Urena, Max – Mexico City? Mexico. 1962 – 1r – us UF Libraries [972]

Breve historial de las sagradas reliquias que se veneran en las parroquias de santa eulalia y sta. marta la mayor de merida / Gonzales y Gomez de Soto, Juan Jose – Merida: Tip.Lib.y Enc.Juan F.Rivera Silva, 1916 – 1 – sp Bibl Santa Ana [240]

Breve informe de la actividad de la oficina de prensa y propaganda de la representacion de espana en la argentina desde el mes de septiembre de 1937 hasta el de agosto de 1938 / Spain. Embajada (Argentina) – [Buenos Aires 1938] – 9 – mf#w1174 – us Harvard [946]

Breve metodo per fondatamente : e con facilita apprendere il canto fermo / Tettamanzi, Fabricio – Milano: Nelle stampe degl' Angelli 1706 [mf ed 19–] – 5mf – 9 – mf#fiche 903 – us Sibley [780]

Breve noticia de el origen que tuvo la devocion de la purisima concepcion de nuestra senora y de una santa imagen... / Breve – Madrid, 1706 – sp Bibl Santa Ana [946]

Breve noticia de la portentosa conversion : y admirable vida del venerable padre don martin de s cayetano, y jorganes, presbitero del oratorio del...senor san phelipe neri... / Vilaplana, Hermenegildo de – en Mexico:... Colegio de San Ildefonso, ano de 1760 – (= ser Books on religion...1543/44-c1800: biografias de religiosos) – 2mf – 9 – mf#crl-149 – ne IDC [241]

Breve noticia sobre o museu do dundo – Lisboa, Portugal. 1963 – 1r – us UF Libraries [960]

Breve practica, y regimen del confessonario de indios, en mexicano y castellano : para instruccion del confessor principiante, habilitacion, y examen del penitente / Velazquez de Cardenas, Carlos Celedonio – en Mexico: impr Bibliotheca Mexicana, ano de 1761 – (= ser Books on religion...1543/44-c1800: confesionarios) – 1mf – 9 – mf#crl-26 – ne IDC [241]

Breve puntual noticia de el modo, solemnidad y circunstancia con que se celebro la gloriosa aclamacion de nuestro inclito rey y senor luis primero...caceres...1724 – Madrid, 1724 – 1 – sp Bibl Santa Ana [946]

Breve racconto della trasportatione del corpo di papa paolo v dalla basilica di s pietro a'quella di s maria maggiore... / [Guidiccioni, L] – Roma, 1623 – 3mf – 9 – mf#O-1125 – ne IDC [700]

Breve relacion de la milagrosa y celestial imagen de santa domingo : patriarca de la orden de predicadores, trayda del cielo por mano de la virgen nuestra senora... – en Mexico: Con licencia, en la imprenta de Bernardo Calderon, ano de 1633 – (= ser Books on religion...1543/44-c1800: vidas y cultos de santos) – 1mf – 9 – mf#crl-110 – ne IDC [241]

Breve relacion de la procession del santissimo rosario, hecho en roma el dia dos de agosto del corriente ano de 1716 / Catholic Church Diocese of Rome [Italy] – en Madrid, Mexico, Manilla: [s.n. 1716] – (= ser Books on religion...1543/44-c1800: miscelanea) – 1mf – 9 – mf#crl-413 – ne IDC [241]

Breve relacion del origen, y fundacion de la insigne religion de los siervos maria santissimoa : y algunos prodigiosos favores de los innumerable, que la senora ha communicado a los devotos de sus dolores – en Mexico: Por dona Maria de Benavides en el Empedradillo, ano de 1699 – (= ser Books on religion...1543/44-c1800: servitas) – 3mf – 9 – mf#crl-251 – ne IDC [241]

Breve relacion del pleyto de orellana – S.l., s.i., s.a. – 1 – sp Bibl Santa Ana [946]

Breve Resena *see* Breve resena para gobierno de las que aspiran a ingresar en el instituto de hermanas de la caridad del sagrado corazon de jesus

Breve resena de badajoz / Direccion General de Turismo – Badajoz: Industrias Graficas, 1962 – sp Bibl Santa Ana [914]

Breve resena de las lineas...malpartida de plasencia a caceres.. – 1881 – 9 – sp Bibl Santa Ana [946]

Breve resena historica...san vicente de paul – 1862 – 9 – sp Bibl Santa Ana [240]

Breve resena para gobierno de las que aspiran a ingresar en el instituto de hermanas de la caridad del sagrado corazon de jesus / Breve Resena – S.L.Si.s.a. – 1 – sp Bibl Santa Ana [240]

Breve resumpta y tratado de la esencia, causas, pronostico..., y curacion de la peste / Barba, P – Madrid, 1648 – 69mf – 9 – sp Cultura [616]

Breve storia della nobile e celebre famiglia senese dei sozzini dalle sue origini fino alla sua estinzione (sec. 14-19) / Mazzei, Antonio – Siena: Giuntini & Bentivoglio, 1912 – 1mf – 9 – 0-524-08685-0 – mf#1993-3210 – us ATLA [240]

Breve tratado de peste con sus causas senales y curacion... / Perez, A – Madrid, 1589 – 2mf – 9 – sp Cultura [615]

Breve tratado de todo genero de bobedas asi regulares como irregulares... / Torija, J – Madrid, 1661 – 3mf – 9 – sp Cultura [720]

Breve y sumaria relacion de los senores de la nuev... / Zurita, Alonso De – Mexico City? Mexico. 1942 – 1r – us UF Libraries [972]

Breves advertencias para beber frio con nieve... / Porres, M – Lima, 1621 – 2mf – 9 – sp Cultura [610]

Breves apuntes de la vida de un patricio / Reynoso Garcia, Ulises – San Pedro de Macoris, Dominican Republic. 1945 – 1r – us UF Libraries [972]

Breves definiciones de historia general y de espana...sucesos...badajoz / Romero Morera, Joaquin – Badajoz: Imp. y libr. de Emilio Orduna, 1878 – 1 – sp Bibl Santa Ana [946]

Breves nouvelles d'indonesie – Berne, [1953]-1956. v1-4(16) – 8mf – 9 – (missing: [1953], v1-1954, v2(1-23); 1955, v3(15-19, 21, 24-28?); 1956, v4(3, 10, 11, 14)) – mf#SE-1363 – ne IDC [959]

Breves...aguas...banos de montamayor / Pesado Blanco, Sergio – 1897 – 9 – (1898 ed) – sp Bibl Santa Ana [890]

Breves-geografia astronomica / Guillen y Flores, Agustin – 1861 – 9 – sp Bibl Santa Ana [520]

Breves...isla de cuba / Fernandez Golfin, Luis – 1866 – 9 – sp Bibl Santa Ana [972]

Breve...universi / Urbano Octavo. Papa – 1632 – 9 – sp Bibl Santa Ana [240]

Les breviaires : manuscrits des bibliotheques publiques de france / Leroquais, Victor – Paris. tom 1-5+planches. 1934 – 81mf – 8 – €155.00 – ne Slangenburg [241]

Breviario – Calahorra – 1r – 5,6 – sp Cultura [240]

Il breviario ambrosiano / Cattaneo, E – Milano, 1943 – 7mf – 8 – €15.00 – ne Slangenburg [241]

Breviario critico / Morales, G Alfredo – Ciudad Trujillo, Dominican Republic. 1955 – 1r – us UF Libraries [972]

Breviario da bahia / Peixoto, Afranio – Rio de Janeiro, Brazil. 1945 – 1r – us UF Libraries [972]

Breviario de ciudadania... / Migueta, Juan – Madrid: Razon y Fe, 1927 – 1 – sp Bibl Santa Ana [680]

Breviario de la iglesia de calahorra (anno 1400) – Calahorra – 1r – 5,6 – sp Cultura [240]

Breviario de mi vida inutil / Cabrisas, Hilarion – Habana, Cuba. 1932 – 1r – us UF Libraries [972]

Breviario sentimental / Delgado Fernandez, Rufino – Caceres: Tip. El Noticiero, 1964 – 1 – sp Bibl Santa Ana [946]

Breviario (siecle 15) – Calahorra – 1r – 5,6 – sp Cultura [240]

Breviarium ad usum insignis ecclesiae eboracensis, 1-2 / Lawley, M – Edinburgh, 1883 – 18mf – 8 – €35.00 – ne Slangenburg [241]

Breviarium ad usum sarum / Procter, Fr & Wordsworth, Chr – Cambridge. v1-3. 1882-1886 – 3v – (v1: kalendarium-ordo temporalis-pica, 1882 €27. v2: psalterium-commune sanctorum, 1879 €14. v3: sanctorale-accentuarium, 1886 €27) – ne Slangenburg [241]

Breviarium chronographicum *see* Chronographia [cbh5]

Breviarium gothicum, el de silos : archivo monastico ms 6 / Cuesta, I F de la – MHS L. Madrid. 8. 1965 – €11.00 – ne Slangenburg [241]

Breviarium historiae metricum (cshb29) / Constantini Manassis; ed by Bekkerus, Imm – Bonnae, 1837 – (= ser Corpus scriptorum historiae byzantinae (cshb)) – €23.00 – (ioelis chronographia compendiaria. georgii acropolitae annales) – ne Slangenburg [241]

Breviarium historicum (cbh2,2) / S Nicephori, Patriarchae Constantinopolitani; ed by Petau, D – Parisiis, 1648 – (= ser Corpus byzantinae historiae (cbh)) – €12.00 – ne Slangenburg [241]

Breviarium historicum (cbh12,1) / Constantini Manassis; ed by Allatius, L & Fabrotus, C – Parisiis, 1655 – (= ser Corpus byzantinae historiae (cbh)) – €18.00 – ne Slangenburg [241]

Breviarium iuxta ritum ordinis cisterciensium... – Venetiis: luntas 1542 – (= ser Hqab. literatur des 16. jahrh.) – 11mf – 9 – €95.00 – mf#1542a – gw Fischer [241]

Breviarium iuxta ritum sacri ordinis cisterciensium : novissime castigatum, atque exactissime commodatum... – Venetiis: luntas 1579 – (= ser Hqab. literatur des 16. jahrh.) – 11mf – 9 – €95.00 – mf#1579a – gw Fischer [780]

Breviarium monasticum secundum ritum et morem monachorum ordinis s. benedicti de observantia casinensis congregationis alias s. justine : cum novo ac perutili repertorio ad quelibet facile in ipso breviario invenienda... – Venetiis: lunta 1573 – (= ser Hqab. literatur des 16. jahrh.) – 9mf – 9 – €85.00 – mf#1573a – gw Fischer [780]

Breviarium romanum a francisco cardinale quignonio. ed 1535 / Legg, J Wickham – 1e ed. Cambridge, 1888 – 5mf – 8 – €12.00 – ne Slangenburg [241]

Breviarium secundum usum cisterciensis ordinis per eundem monachum : qui duo precedentia maioris et minoris voluminis correxerat, sedula cura castigatum et tam...utiliter auctum et ad opus perfectum redactum – Parisiis: Marnef 1508 – (= ser Hqab. literatur des 16. jahrh.) – 8mf – 9 – €80.00 – mf#1508 – gw Fischer [780]

Breviarum lugdunense scriptum a stephano mantillarii sacerdotem parochiae sancti-symphoriani-castri anno 1487, cuius nomen legitur bis in codice, scilicet foliis 142v et 341v = Breviary. lyons – 1r – 1 – mf#1984-B364 – us ATLA [240]
Brevie, J *see* Islamisme contre "naturisme" au soudan francais
Brevier der lebenskunst : aus den briefen / Schirmer, Jo [comp] – Frankfurt a.M: Siegel-Verlag [1947] [mf ed 1995] – 1r – 1 – (filmed with: a morte de camoes / luis tieck) – mf#3754p – us Wisconsin U Libr [860]
Das brevier und der saekularklerus / Merk, K J – Stuttgart, 1950 – 3mf – 8 – €7.00 – ne Slangenburg [241]
Breviloquium...quator libros...sententiarum / Ovando Mogollon de Baredes, Francisco – 1584 – 9 – (1587 ed) – sp Bibl Santa Ana [880]
Brevis ac perspicua / Dathenus, P – n.p, 1558 – 1mf – 9 – mf#PBA-159 – ne IDC [240]
Brevis ac pia institutio christianae religionis, ad dispersos in hungaria...ministros... / Bullinger, Heinrich – Ovarini, 1559 – 1mf – 9 – mf#PBU-208 – ne IDC [240]
Brevis admonitio joannis calvini ad fratres polonos, ne triplicem in deo essentiam pro tribus personis imaginando, tres sibi deos fabricent / Calvin, J – Genevae: Ex officina Francisci Perrini, 1563 – 1mf – 9 – mf#CL-12 – ne IDC [242]
Brevis analysis tractatus de deo creatore / Jungmann, Bernardus – Ratisbonae [Regensburg], Neo Eboraci [New York]: Friderici Pustet, 1875 [mf ed 1985] – 1mf – 9 – 0-8370-4562-2 – mf#1985-2562 – us ATLA [241]
Brevis antibolè sive responsio secunda... ad...ioannis cochlei...replicam... / Bullinger, Heinrich – Tigvri, [Christoph] Froschoverus, 1544 – 1mf – 9 – mf#PBU-147 – ne IDC [240]
Brevis commentarius in facultates s. congregationis de propaganda fide / Paventi, Saverio M – Romae: Officium Libri Catholici 1944 [mf ed 1993] – (= ser Bibliotheca missionalis) – 1mf – 9 – 0-524-08129-8 – (incl bibl ref) – mf#1993-9035 – us ATLA [241]
Brevis defensio hieroglyphices : inventae a fr guil aug spohn et g seyffarth / Seyffarth, Gustav – Lipsiae: Barth 1827 – (= ser Whsb) – 1mf – 9 – €20.00 – mf#Hu 375 – gw Fischer [490]
Brevis descriptio itineris sinensis...legatione moscovitica anno 1693, 94 et 95... / Brand, A – 1mf – 9 – mf#HT-570 – ne IDC [915]
Brevis et expedita ratio... / Hitzenauer, Christoph – 1585 – (= ser Mssa) – lost – 9 – mfchl 59 – gw Fischer [780]
Brevis linguae chaldaicae : grammatica, litteratura, chrestomathia: cum glossario: in usum praelectionum et studiorum / Petermann, Julius Heinrich – Editio 2. Berolini: G Eichler, 1872 – 1mf – 9 – 0-8370-7183-6 – mf#1986-1183 – us ATLA [470]
Brevis linguae hebraicae : grammatica, litteratura, chrestomathia, cum glossario / Petermann, Julius Heinrich – Berolini [Berlin]: G Eichler, 1864 – 1mf – 9 – 0-524-06852-6 – mf#1992-0994 – us ATLA [470]
Brevis linguae syriacae grammatica *see* Syrische grammatik
Brevis linguae syriacae grammatica, litteratura, chrestomathia cum glossario : in usum praelectionum et studiorum privatorum / Nestle, Eberhard – Carolsruhae: H Reuther, 1881 – 1mf – 9 – 0-8370-8600-0 – (discussion in latin; texts in syriac) – mf#1986-2600 – us ATLA [470]
Brevis repetitio doctrinae orthodoxae de persona et officio christi / Pierius, U – Vvitebergae, 1591 – 1mf – 9 – mf#TH-1 mf 1276 – ne IDC [242]
Brevis responsio joannis calvini, ad diluendas nebulonis cuiusdam calumnias, quibus doctrinam de aeterna dei praedestinatione foedare conatus est / Calvin, J – Geneva: Excudit Crispinus, 1557 – 1mf – 9 – mf#CL-36 – ne IDC [242]
Brevissima rhetorices institutio / Evans, David – 1733. Also: Physica Compendiosa, 1737 – 1 – $50.00 – us Presbyterian [400]
Brevissima rudimenta musicae : pro incipientibus – Wratislaviae: Typis Georgi Bawman 1608 [mf ed 19–] – 1r – 1 – mf#pres. film 232 – us Sibley [780]
Brevis...tractatus de essentia...curatione et praecautione faucium et gutturis... / Perez de Herrera, C – Madrid, 1615 – 3mf – 9 – sp Cultura [610]
Brewarrina news – Brewarrina. jan-jun 1969, jan-dec 1972, jan 1974-oct 1975 – 1r – at Pascoe [079]
Brewer bulletin / Bodak, Shirley L – 1977 may 1, aug-dec – 1r – 1 – mf#637977 – us WHS [071]
Brewer, David J *see*
– The united states a christian nation
– The world's best orations
Brewer, Ebenezer Cobham *see* A dictionary of miracles
Brewer, J S *see*
– Giraldi cambrensis opera, vols 1-4
– Monumenta franciscana
– Registrum malmesburiense
– Rogeri bacon opera quaedam hactenus inedita
Brewer, James *see*
– The beauties of ireland
– The picture of england
Brewer, John Sherren *see*
– The athanasian origin of the athanasian creed
– The endowments and establishment of the church of england
– The reign of henry 8 from his accession to the death of wolsey
Brewer, Julia R *see* Coronary heart disease risk factors in children ages 9 to 11 years
Brewer, K D *see* The effects of intermittent compression and cold on edema in postacute ankle sprains
Brewer, Mandane Williamson *see* Diaries
Brewers' congress / United States Brewers' Association – 1868-78 – 640 – 1 – (cont by: proceedings of the...convention, united states brewers' association) – mf#3144855 – us WHS [071]
Brewer's guardian – Smithfield UK 1977-90 – 1,5,9 – ISSN: 0006-9728 – mf#10231 – us UMI ProQuest [660]
Brewery worker / International Union of Brewery, Flour, Cereal, Soft Drink and Distillery Workers of America – 1886-1973 – (= ser Labor union periodicals, pt 3: food and agricultural industries) – 17r – 1 – $3520.00 – 1-55655-615-2 – us UPA [660]
Brewery worker / International Union of United Brewery, Flour, Cereal, Soft Drink and Distillery Workers of America – 1886 oct 2/1892-1972-73 – 18r – 1 – mf#1110341 – us WHS [640]
Brewery Workers Local Union No 9 *see* Milwaukee brewery worker
Brewery Workers Local Union No 9, D A L U, AFL-CIO *see* Local union no 9 newsletter
Brewery workers news : local 9 – 1985 jan-1994 dec – 1r – 1 – (cont: newsletter: house of prayer church of god n1 1992 feb, 1995 jan-feb, apr-may, sep-oct, 1996 mar [1r]; national coalition for social change n2-7 [1974 jan-1976 fall] [1r]; library union caucus v1-3 n3 [1972 aug-1974 jul/sep] [1r]; daugherty family association 1984 mar-1985 dec [1r] cont by: daugherty family newsletter; grand rapids hist soc 1972 nov-1975 feb [1r]; withee wi 1943 mar-1970 dec 1943 mar-1977 oct [2r]; idaho bicentennial commission n1-35 [1973 jan-1976 spring] [1r]; citizen's constitutional committee 1974 sep 1-1975 apr 1 [1r]; wisconsin early childhood association 1978 jan 1 [1r] cont by: weca newsletter; national association for interdisciplinary ethnic studies [us] v8 n1-v9 n2 [1983 mar-1984 oct] [1r] cont: newsletter [national association of interdisciplinary ethnic studies [us]]; long island postal history soc 1980 aug-1983 dec [1r] cont by: long island postal history soc [series]; reservists committee to stop the war n1-2 [1r] cont by: redline; newfoundland status of women council 1974 jan/feb-1984 jun [1r] cont by: newsletter [st john's status of women council]; little big horn association v7 n1-v18 n8 [1973 jan-1984 nov] [1r]; national women's political caucus [us] 1971 dec-1975 nov [1r] cont by: quarterly report [national women's political caucus [us]]; kenosha area chamber of commerce v1 n5-v5 n31 [1974 dec-1979 dec] [1r] cont by: news [kenosha area chamber of commerce]; wisconsin council of agricultural cooperatives 1965 nov-1969 may [1r] cont: news letter [wisconsin council of agriculture cooperative : 1948] 1965 nov-1969 may; cont by: wac bulletin; williamson street grocery cooperative 1978 may-1983 jul [1r]; wisconsin citizens for right to work 1974 aug-1981 nov [1r] cont by: wisconsin right to work news; friends of micronesia 1971 sep/oct-nov/dec, 1973 fall-1974 winter, summer [v3 n4-v4 n1, v4 n3] [1r]; boston draft resistance group newsletter 1969 jan-1969 apr [1r]) – mf#1053560 – us WHS [640]
Brewin grant refuted / Cooper, Robert – London, England. 1853 – 1r – us UF Libraries [240]
Brewing world – v1 n1-v2 n2 [1883 jan-aug] – 1r – 1 – mf#3072974 – us WHS [640]
Brewington presbyterian church records, 1821-1975 – Philadelphia PA: Presbyterian Historical Soc – 1r – 1 – mf#45-341 – us South Carolina Historical [242]
Brewster 1745-1900 – Oxford, MA (mf ed 1994) – (= ser Massachusetts vital records) – 76mf – 9 – 0-87623-190-3 – (mf 1-3: births & deaths 1745-1838. mf births & deaths 1753-1892. mf 8,17: militia 1841-1864. mf 9-10: town records 1803-08. mf 10-11: marriages, intents 1802-26. mf 12: vital records 1769-1829. mf 11-17: town & vital 1803-31. mf 18-20: deeds 1829-45. mf 18-21: marriages, intents 1829-46. mf 21: dogs registered 1859-89. mf 22-28: town records 1829-71. mf 29-38: town records 1871-1918. mf 39-40: school 1834-69. mf 41-47: payments 1809-65. mf 48-49: poor records 1851-81. mf 50-57: mortgages 1846-1932. mf 58-60: valuation list 1890. mf 61: intentions index 1847-1904. mf 62-65: intentions 1847-1916. mf 66: birth index 1843-1916. mf 67: marriage index 1843-1916. mf 68: death index 1843-1916. mf 69-70: vital records 1843-51. mf 71-72: births 1852-1900. mf 73-74: marriages 1852-1900. mf 75-76: deaths 1852-1900) – us Archive [978]
Brewster, Chauncey Bunce *see*
– Aspects of revelation
– The kingdom of god and american life
Brewster, David *see* Martyrs of science
Brewster, Frederick Carroll *see*
– A treatise on practice in the courts of common pleas of pennsylvania
– A treatise on practice in the courts of pennsylvania
Brewster, James *see* Letter to the editor of the quarterly review
Brewster, Jonathan McDuffee et al *see* The centennial record of freewill baptists, 1780-1880
Brewster, William Nesbitt *see* The evolution of new china
Breyer, Ed A *see* R akiba
Breyer, Ralf *see* Beitrag zur geologie des "massif de ceze", oestlicher teil, gard (frankreich)
Breyfogel, Sylvanus Charles *see* Landmarks of the evangelical association
Breymann, Hermann *see*
– La dime de penitance
Breysig, Kurt *see* Die entstehung des gottesgedankens und der heilbringer
Breytenbach, Mariette *see* Clustering techniques
The brhad-devata, attributed to saunaka : a summary of the deities and myths of the rig-veda / ed by Macdonell, Arthur Anthony – Cambridge, Mass: Harvard University, 1904 – (= ser Harvard Oriental Series) – 7mf – 9 – 0-524-07383-X – mf#1991-0103 – us ATLA [280]
Brhajjataka = Brihat jataka / Varahamihira – Mysore: Govt Branch Press, 1929 – (= ser Samp: indian books) – (with an english transl and notes by v subrahmanya sastri) – us CRL [280]
Brian Walton *see* Biblia sacra polyglotta
Briand, Jean-Olivier *see* Lettre circulaire de monseigneur l'eveque, au clerge du diocese de quebec
Briand, Pierre *see* Les jeunes voyageurs en asie
Briano-Iragorry, Mario *see* Lecturas venezolanas
Brianskie chudotvortsy : materiali dlia russkoi agiologii – 1893 – 1mf – 9 – mf#R-18259 – ne IDC [243]
Brianskii, A M [comp] *see* Statisticheskii ezhegodnik 1923-1925 gg
Brianskii, N G *see* Moskovskie proizvoditelno-trudovye arteli
Brianskii rabochii – Bezhitsa, Russia 1917 [mf ed Norman Ross] – 1r – 1 – mf#nrp-333 – us UMI ProQuest [077]
Brianskii rabochii [bryansk] – Bryansk, Russia 1973-88 [mf ed Norman Ross] – 5r – 1 – mf#nrp-373 – us UMI ProQuest [077]
Briar patch – 1977-78, 1979-1981 feb, 1982-83, 1984-1985 dec/1986 jan – 5r – 1 – mf#498709 – us WHS [071]
Briarcliff quarterly – Briarcliff Manor, New York etc v. 1-3 no. 12. Jan 1944-Jan 1947 – 1 – us NY Public [378]
Bribery and boodling, fraud, hypocrisy and humbug : professional charges and pecuniary ethics: a paper read by c baillairge, chateau frontenac, quebec, oct 2, 1895 / Baillairge, Charles P Florent – Quebec: s.n, 1895 – 1mf – 9 – mf#02215 – cn Canadiana [170]
Bric-a-brac : devoted to amateur journalism and more especially, its extension in canada – Montreal: H W Robinson, [1885?-18–?] – 9 – (ceased 18–?) – mf#P04338 – cn Canadiana [070]
Brice and salnave / Vigoureux, Gustave – Jeremie, Haiti. 1932 – 1r – us UF Libraries [972]
Brice, Arthur John Hallam Montefiore *see* Henry m stanley
Briceno, Alfonso *see* Disputaciones metafisicas (1638)
Briceno, Manuel *see*
– Ilustres
Briceno Perozo, Mario *see* Causas de infidencia
Briceno Perozo, Ramon *see* De los hechos de la conquista durante la fundacion
Briceno Valero, Americo *see* Ciudad portatil
Briceno-Iragorry, Mario *see*
– Casa leon y su tiempo
– Ideario politico
– Pasion venezolana
– Tapices de historia patria
Briceno-Iregorry, Mario *see* Discurso de recepcion...en la academia de la historia
Brices weekly journal – Exeter, England 20 jun 1729-4 jun 1731 – 1 – (numeration irreg; cont by: brice's weekly collection of intelligence [8 dec 1738]) – uk British Libr Newspaper [072]
Brichford, Maynard *see* Guide to the university of illinois archives
Brick and clay record – Ann Arbor MI 1957-68 – 1 – ISSN: 0006-9760 – mf#1098 – us UMI ProQuest [730]
Brick and marble in the middle ages : notes of a tour in the north of italy / Street, George Edmund – London 1855 – (= ser 19th c art & architecture) – 4mf – 9 – mf#4.2.1024 – uk Chadwyck [720]
Brick yard at fort clinch 1865 – s.l, s.l? 193-? – 1r – us UF Libraries [978]
Bricklayer and mason – 1898 jul-03 jan – 1r – 1 – (cont by: bricklayer, mason and plasterer) – mf#3178665 – us WHS [690]
Bricklayer and mason – Indianapolis. v. 1-13. 1898-1910 – 1 – us NY Public [331]
Bricklayers, Masons, and Plasterers International Union of America *see* Biennial and report of the president, secretary and official auditor
Brickwork in italy / American Face Brick Association – Chicago, IL. 1925 – 1r – us UF Libraries [720]
Bricout, Joseph *see* Ou en est l'histoire des religions?
Bridal greetings : a marriage gift / Wise, Daniel – New York, NY: Carlton & Phillips, 1854 – 1r – 1 – (marriage manual) – us Western Res [390]
Bridel, Jean *see* Kleine fussreisen durch die schweiz
Bride's [1966] – New York NY 1966-92 – 1,5,9 – (cont by: bride's and your new home) – ISSN: 0161-1992 – mf#2214.02 – us UMI ProQuest [640]
Bride's [1996] – New York NY 1996+ – 1,5,9 – (cont: bride's and your new home) – ISSN: 1084-1628 – mf#2214.02 – us UMI ProQuest [640]
Bride's and your new home – New York NY 1992-95 – 1,5,9 – (cont: bride's; cont by: bride's) – ISSN: 1059-7476 – mf#2214.02 – us UMI ProQuest [640]
The bride's book of beauty / Anand, Mulk Raj & Hutheesing, Krishna – Bombay: Kutub Publishers, 1947 – (= ser Samp: indian books) – us CRL [640]
Brideshead revisited / Waugh, Evelyn – Boston, MA. 1945 – 1r – us UF Libraries [420]
Bridge – 1970 apr 24-1973 apr 5 – 1r – 1 – mf#1110345 – us WHS [071]
Bridge – 1978 sep/oct-1981 aug/sep – 1r – 1 – mf#671757 – us WHS [071]
Bridge – 1989 jan 23-1991 dec 6, 1992 mar 27 – 1r – 1 – mf#1055704 – us WHS [071]
Bridge / Citizens Alert [Organization : Chicago IL] – v4 n3 [1979 jul], v6 n1-v7 n1 [1981 jan-1982 jan] – 1r – 1 – mf#958051 – us WHS [360]
Bridge / Credit Union National Association – 1931 feb – 1r – 1 – (cont by: bridge [madison wi: 1936]) – mf#1420629 – us WHS [366]
Bridge / Indochina Resource Action Center [Washington DC] – 1984 jan-1992 spring – 1r – 1 – mf#2629378 – us WHS [071]
Bridge – Kondrook and Barham Bridge, jan 1978-dec 1992 – 6r – at Pascoe [079]
Bridge / Sandusky Co. Fremont – v1 n1. dec 1984-jul 1989 [wkly] – 4r – 1 – mf#B32919-32922 – us Ohio Hist [071]
Bridge – Sydney, 1964-73 (misc. yrs) – 1r – A$50.51 vesicular A$56.01 silver – at Pascoe [079]
Bridge [1924] – Ann Arbor MI 1924-34 – 1 – mf#2702 – us UMI ProQuest [332]
Bridge [1971] – New York NY 1971-85 – 1,5,9 – ISSN: 0045-2823 – mf#9183 – us UMI ProQuest [305]
Bridge, Bewick *see* An elementary treatise on algebra
Bridge bulletin – Eugene OR: [s.n.] -1963 [wkly] – 1 – (cont by: ferry street bridge bulletin) – us Oregon Lib [071]
Bridge launching / Forrest, Benjamin J – [S.l: s.n, 1904?] [mf ed 1991] – 1mf – 9 – 0-665-99513-X – mf#99513 – cn Canadiana [624]
Bridge of allan gazette – [Scotland] Stirling: Duncan & Jamieson jan 1892-dec 1950 (wkly) [mf ed 2004] – 72r – 1 – (ceased dec 31 1971; absorbed by: stirling observer; an edition of the stirling observer) – uk Newsplan [072]
The bridge of history over the gulf of time : a popular view of the historical evidence for the truth of christianity / Cooper, Thomas – London: Hodder & Stoughton, 1874 [mf ed 1985] – 1mf – 9 – 0-8370-2738-1 – mf#1985-0738 – us ATLA [240]
Bridge of weir advertiser : circulating in kilmacolm, houston, brookfield and kilbarchan – [Scotland] Renfrewshire, [Bridge of Weir] 18 aug 1934-30 mar 1935 (wkly) [mf ed 2003] – 1r – 1 – uk Newsplan [072]
Bridge, Stephen *see* Ascendancy of popery fatal to the truth of the gospel

Bridge to success – v12 n6, 8-9 [1986 jan/feb, jun/jul-oct/nov], v13 n 1-4, [1987 apr/may-nov/dec], v13 n6-1910 [1988 apr/may-nov/dec], v13 n11 [1989 jan/feb], v14 n12 [1989 may/jun], v15 n1-1912 [1989 jul-1990 jun], v16 n1-5 [1990 jul-dec], v16 n7-12 [1991 apr-nov], v17 n1-2 [1991 dec-1992 jan/feb] – 1r – 1 – (cont: volunteer [alameda ca]) – mf#1055765 – us WHS [360]
Bridgeburg review – Ontario, CN. jan 1930-dec 1931 – 2r – 1 – cn Commonwealth Imaging [071]
Bridgend – (= ser 1861 census returns [glamorgan]) – 8mf – 9 – £10.00 – (also covers llantrisant & st nicholas srd's) – uk Glamorgan FHS [314]
Bridgend county asylum burials 1915-1926 : and female case notes 1885-1902 – 5mf – 9 – £6.25 – uk Glamorgan FHS [929]
Bridgend, cowbridge & maesteg times – [Wales] LLGC apr 1859-23 mar 1860 [mf ed 2004] – 1r – 1 – uk Newsplan [072]
Bridgend, glamorgan: english weslyan methodist circuit : baptisms 1844-1925, marriages 1905-1920 – [Glamorgan]: GFHS [mf ed c2004] – 1mf – 9 – £1.25 – uk Glamorgan FHS [929]
Bridgend, glamorgan: ruhmah particular baptist, births 1800-1837 : tabernacle independent, baptisms 1785-1891, 1800-1837; burials 1861-1874 – [Glamorgan]: GFHS [mf ed 2003] – 1mf – 9 – £1.25 – uk Glamorgan FHS [929]
Bridgend, newcastle hill unitarian, monumental inscriptions – 1mf – 9 – £1.25 – uk Glamorgan FHS [929]
Bridgend nolton, st mary, monumental inscriptions – 1mf – 9 – £1.25 – uk Glamorgan FHS [929]
Bridgend, penyfai, all saints, monumental inscriptions – 1mf – 9 – £1.25 – uk Glamorgan FHS [929]
Bridgend, penyfai, smyrna baptist, monumental inscriptions – 1mf – 9 – £1.25 – uk Glamorgan FHS [929]
Bridgend r d (h.o.107/2461) – (= ser 1851 census returns [glamorgan]) – 4mf – 9 – £5.00 – uk Glamorgan FHS [314]
Bridgend reg dist – (= ser 1841 census returns [glamorgan]) – 4mf – 9 – £5.00 – uk Glamorgan FHS [314]
Bridgend, tabernacle congregational; ruhamah welsh baptist, monumental inscriptions – 1mf – 9 – £1.25 – uk Glamorgan FHS [929]
Bridgens, Richard *see*
– Costumes of italy switzerland and france
– Furniture with candelabra and interior decoration
Bridgen's surrogate's reports / New York. (State). Surrogate Court – 1v. 1825 (all publ) – 3mf – 9 – $4.50 – mf#LLMC 80-019 – us LLMC [340]
Bridgeport blade – Bridgeport, NE: Gary & Lowley, 1900-v9 n2. jul 17 1908 (wkly) [mf ed v1 n25. jan 11 1901-jul 17 1908 (gaps)] – 3r – 1 – (merged with: platte valley news to form: bridgeport news-blade) – us NE Hist [071]
[Bridgeport-] bridgeport chronicle union – CA. jul 1890-mar 1943; apr 1947-sep 1986 – 53r – 1 – $3180.00 – (cont: see mammoth lakes) – mf#C02077 – us Library Micro [071]
Bridgeport chronicle union *see* [Mammoth lakes-] review-herald
Bridgeport herald – Bridgeport, NE: Ray Ryason (wkly) [mf ed v15 n8. apr 9 1925-may 30 1929 (gaps)] – 1r – 1 – (absorbed by: bridgeport news-blade) – us NE Hist [071]
Bridgeport law review *see* Quinnipac law review
Bridgeport news-blade – Bridgeport, NE: J M Lynch. v9 n3. jul 24 1908- (wkly) [mf ed with gaps] – 1 – (formed by the union of: platte valley news and: bridgeport blade. absorbed: bridgeport herald (1929), dalton delegate (1951), broadwater news (1958) and: morrill county sun. issues for jul 24 1908-jun 18 1917 called also v6 n10-v11 n52) – us NE Hist [071]
[Bridgeport-] review herald – $60.00 – (cont: bridgeport chronicle union (see mammoth lakes)) – mf#B03585 – us Library Micro [071]
Bridges / Lithuanian-American Community, USA – v2 n1 [1978 jan], v5 n3-v10 n9 [1986 sep] – 1r – 1 – mf#934879 – us WHS [305]
Bridges, Calvin Blackman *see* Third-chromosome group of mutant characters of drosophila
Bridges, Charles *see* An exposition of the book of ecclesiastes
Bridges, Horace James *see* Some outlines of the religion of experience
Bridges in history and legend / Watson, Wilbur Jay – Cleveland, OH. 1937 – 1r – us UF Libraries [720]
Bridges, James *see*
– A letter to the right hon. robert pool, on the courts of law in scotland
– Patronage in the church of scotland considered
Bridges, John Henry *see* The home rule question eighteen years ago
Bridgetstow : some chronicles of a cornish parish / Pearse, Mark Guy – Cincinnati: Jennings & Graham [c1907] [mf ed 1984] – 3mf – 9 – 0-8370-0841-7 – mf#1984-4227 – us ATLA [941]
Bridgett, T E *see*
– Blunders and forgeries
– A history of the holy eucharist in great britain
– Our lady's dowry
– The ritual of the new testament
– The true story of the catholic hierarchy deposed by queen elizabeth
Bridgewater 1608-1849 – Oxford, MA (mf ed 1992) – (= ser Massachusetts vital record transcripts to 1850) – 21mf – 9 – 0-87623-216-0 – (mf 1t-4t: vital records 1645-1810. mf 3t-5t: marriages 1704-86. mf 5t-11t: vital records 1721-1846. mf 8t-12t: marriage bans 1762-1820. mf 13t-14t: marriages 1775-1831. mf 14t-19t: vital records 1608-1896. mf 17t-20t: marriage banns 1820-50. mf 18t: out-of-town marriages 1702-99; marriages 1830-33. mf 20t: marriages 1759-1843; births 1843-49. mf 21t: marriages & deaths 1843-49) – us Archive [978]
Bridgewater 1641-1900 – Oxford, MA (mf ed 1992) – (= ser Massachusetts vital records) – 173mf – 9 – 0-87623-160-1 – (mf 1-39: town & vitals 1641-1853. mf 40-64: town & vitals 1645-1887. mf 65-74: town & vitals: 1645-1881. mf 75: marriages 1788-1815. mf 76-77: births, deaths 1639-1882. mf 78-98: purchasers 1645-1847. mf 99-100: proprietors 1725-1835. mf 101-102: school book 1826-55. mf 103-109: town record 1789-1863. mf 110: town record 1703-28. mf 111-115: accounts 1821-59. mf 116-118: taxpayers 1837-49. mf 119-122: valuations 1840-50. mf 123-124: militia 1840-67. mf 125-126: rebellion 1861-65. mf 127: gar register 1890-1914. mf 128-131: paupers. mf 132-141: intentions 1835-1929. mf 142: marriages 1759-1843. mf 143-147: birth index 1843-1956. mf 148-151: marriage index 1843-1956. mf 152-156: deaths index 1843-1956. mf 157-159: vital records 1843-56. mf 160-164: deaths 1857-1904. mf 165-168: marriages 1857-1901. mf 169-173: births 1857-1902) – us Archive [978]
Bridgewater baptist church. virginia : church records – 1873-1919 – 1 – us Southern Baptist [242]
Bridgewater college : its past and present / ed by Wayland, John Walter et al – Elgin, Ill, USA: Printed by the Brethren Pub House, 1905 – 1mf – 9 – 0-524-02754-4 – mf#1990-4429 – us ATLA [378]
Bridgewater enterprise – Bridgewater VA. 1879 may 28 – 1r – 1 – (cont by: bridgewater journal) – mf#882372 – us WHS [071]
Bridgewater, New Hampshire. Free Will Baptist Church *see* Records
Bridgewater, Thomas *see* Seven years military life in southern india
Bridgham, Percy Albert *see* One thousand legal questions answered by the people's lawyer of the boston daily globe
Bridgman, E C *see*
– Brief memoir of the chinese evangelist leang afa...
– Description of the city of canton
Bridgman, E J Gillet *see* The life and labors of elijah coleman bridgman...
Bridgman, Eliza Jane Gillett *see* Daughters of china
Bridgnorth journal – Bridgnorth, England 9 feb 1973- [mf 1915-50, 1986-] – 1 – (cont: bridgnorth journal & south shropshire advertiser [5 may 1855-10 jul 1869, 22 oct 1887-2 feb 1973]; cont by: bridgnorth journal [9 feb 1973-]) – uk British Libr Newspaper [072]
Bridgnorth journal & south shropshire advertiser – [West Midlands] Shropshire jan 1901-dec 1950 [mf ed 2004] – 45r – 1 – (missing: 1910) – uk Newsplan [072]
Bridgnorth news – [West Midlands] Shropshire 11 sep 1936-24 nov 1939 [mf ed 2004] – 4r – 1 – (cont by: bridgnorth news & highley & alveley news [jan-nov 1939]) – uk Newsplan [072]
Bridgnorth weekly news – Bridgnorth, England 31 may-5 jul 1856 (wkly) – 1 – (discontinued) – uk British Libr Newspaper [072]
The bridgton record – Bridgeton, ME: Libby & Smith, nov 3 1915-sep 6 1916 – 1r – 1 – us CRL [071]
Bridgwater independent – [SW England] Bridgwater 11 jul 1885-10 jun 1933 [mf 1911] [mf ed 2003] – 46r – 1 – uk Newsplan [072]
Bridgwater mercury and western counties herald – [SW England] Somerset 25 jun 1857-dec 1930, jan 1948-dec 1950 [mf ed 2003] – 76r – 1 – (missing: 1865, 1873, 1886, 1897; cont as: western counties herald and bridgwater mercury [jan 1860-dec 1864]; bridgwater mercury [jan 1866-dec 1870]; bridgwater mercury and western counties herald [jan 1871-dec 1901]; bridgwater mercury and burnham, highbridge and weston-super-mare chronicle [jan 1902-dec 1930]; bridgwater mercury and county press [jan 1948-dec 1950]) – uk Newsplan [072]
Bridlington & quay advertiser & improved list of visitors – [Yorkshire & Humberside] East Riding 20 jul 1849 [mf ed 2004] – 1r – 1 – uk Newsplan [072]
Bridlington & quay advertiser & list of visitors – [Yorkshire & Humberside] East Riding 26 jun 1847-16 oct 1847 [mf ed 2004] – 1r – 1 – uk Newsplan [072]
Bridlington & quay advertiser & miscellany of useful and entertaining knowledge – [Yorkshire & Humberside] East Riding 7 jan 1854 [mf ed 2004] – 1r – 1 – uk Newsplan [072]
Bridlington-quay mercury & weekly list of visitors – [Yorkshire & Humberside] East Riding 20, 27 jul 1854, 6 sep 1859 [mf ed 2004] – 1r – 1 – uk Newsplan [072]
Bridlington-quay observer – [Yorkshire & Humberside] East Riding jul 1859-jul 1869; jun 1874-sep 1899 – 9r – 1 – (missing: 1860, 1876, 1883, 1897; publ summer season only) – uk Newsplan [072]
Bridwell, Arthur *see* Diaries
Brief : middle east highlights – Tel Aviv: Middle East Information Media 1971-1976. n1,3-4,8,13,15-20,22-117. jan 1-15, feb, apr 16-30, jul 1-15, aug-oct 1971, nov 16-20 1971-nov1/15 1975 – 1r – us CRL [956]
Brief / ed by Phi Delta Phi – v1-72. 1887-1978 (all publ) – 5,6 – $412.00 set – mf#108451 – us Hein [340]
The brief : a legal miscellany – Phi Delta Phi. v1-72. 1887-1976/78 – 322mf – 9 – $483.00 – (regular updates) – mf#LLMC 84-429 – us LLMC [340]
The brief : the solicitors' monthly review – London, UK. v1-3. 1894-95 (all publ) – 9mf – 9 – $13.50 – mf#LLMC 84-428 – us LLMC [340]
De brief aan de romeinen / Manen, Willem Christiaan van – Leiden: E J Brill, 1891 [mf ed 1989] – 308p on 1mf – 9 – 0-7905-1454-0 – (in dutch and greek. incl ind) – mf#1987-1454 – us ATLA [227]
Brief aan den hooggeleerden heer.. / Hofstede, Petrus – Rotterdam: J Bosch & R Arrenberg 1775 [mf ed 19–] – 1mf – 9 – mf#fiche 17 – us Sibley [780]
The brief (aba) – v8-30. 1978-2001 – 9 – $330.00 set – ISSN: 0273-0995 – mf#112111 – us Hein [340]
Brief account of a german minister – London, England. 18– – 1r – us UF Libraries [240]
Brief account of diocesan synods – Leeds, England. 1852 – 1r – us UF Libraries [240]
A brief account of the fenian raids on the missisquoi frontier in 1866 and 1870 – Montreal?: s.n, 1871 – 1mf – 9 – mf#02836 – cn Canadiana [971]
A brief account of the great revival in lawrence, kansas : feb, mar and apr 1872, in connection with the evangelistic labors of rev e payson hammond... – Lawrence, KN: Office of Republican Daily Journal, 1872 [mf ed 1991] – 1mf – 9 – 0-524-01243-1 – mf#1990-0382 – us ATLA [240]
Brief account of the jesuits – London, England. 1815 – 1r – us UF Libraries [241]
Brief account of the method of synodical action in the american chu... / Caswall, Henry – London, England. 1851 – 1r – us UF Libraries [240]
Brief account of the reasons which have induced the rev tc cowan... – Bristol, England. 1817 – 1r – us UF Libraries [240]
A brief account of the rise of the society of friends – Philadelphia: Friends' Book Store, 1878 [mf ed 1986] – 57p – 1 – mf#7433 – us Wisconsin U Libr [366]
Der brief an der hebraeer / Kurtz, Johann Heinrich – Mitau: Aug Neumann, 1869 – 2mf – 9 – 0-8370-4021-3 – mf#1985-2021 – us ATLA [227]
Der brief an die colosser / Kloepper, Albert – Berlin: G Reimer, 1882 – 2mf – 9 – 0-8370-9633-2 – (incl bibl ref) – mf#1986-3633 – us ATLA [227]
Der brief an die epheser als lehre von der gemeinde fuer die gemeinde / Stier, R – Berlin: Wilhelm Hertz, 1859 – 1mf – 9 – 0-7905-2147-4 – mf#1987-2147 – us ATLA [227]
Brief an die flora = Letter to flora / Ptolemy; ed by Harnack, Adolf von – Bonn: A Marcus & E Weber 1904 [mf ed 1992] – (= ser Kleine texte fuer theologische vorlesungen und uebungen 9) – 1mf – 9 – 0-524-04759-6 – (text in greek; pref & notes in german) – mf#1992-0201 – us ATLA [220]
Der brief an die galater / Sieffert, Friedrich – 9. aufl. Goettingen: Vandenhoeck und Ruprecht, 1899 – 1mf – 9 – 0-8370-5254-8 – mf#1985-3254 – us ATLA [227]
Der brief an die hebraeer : ein ermunterungsschreiben an zagende christen / Riggenbach, Eduard – Berlin-Lichterfelde: Edwin Runge 1916 [mf ed 1993] – (= ser Biblische zeit- und streitfragen 10/11-12) – 1mf – 9 – 0-524-07342-2 – mf#1992-1073 – us ATLA [225]
Der brief an die hebraeer : in sechs und dreissig betrachtungen / Stier, R – 2. neu bearb aufl. Braunschweig: C A Schwetschke, 1862 – 2mf – 9 – 0-7905-2148-2 – mf#1987-2148 – us ATLA [227]
Der brief an die hebraeer / Weiss, Bernhard – 6. verb aufl. Goettingen: Vandenhoeck und Ruprecht, 1897 – (= ser Kritisch Exegetischer Kommentar Ueber Das Neue Testament) – 1mf – 9 – 0-7905-2207-1 – mf#1987-2207 – us ATLA [227]
Der brief an die hebraeer / Zill, Leonhard – Mainz: Franz Kirchheim, 1879 – 7mf – 9 – 0-524-05765-6 – mf#1992-0608 – us ATLA [227]
Der brief an die hebraeer *see* The epistle to the hebrews
Der brief an diognetos : nebst beitraegen zur geschichte des lebens und der schriften des gregorios von neocaesarea / Draeseke, Johannes – Leipzig: J.A. Barth, 1881 – 1mf – 9 – 0-7905-6050-X – (incl bibl ref) – mf#1988-2050 – us ATLA [240]
Der brief an philemon *see* The epistle of paul to philemon
A brief analysis of sale / Landreth, Lucius Scott – Philadelphia, Welsh, 1880. 65 p. LL-386 – 1 – us L of C Photodup [346]
Brief analysis of the doctrine and argument in the case of gorham v... / Lindsay, Lord – London, England. 1850 – 1r – us UF Libraries [240]
Brief and authentic statement of the origin of an established church... / Procter, Payler Matthew – London, England. 1819 – 1r – us UF Libraries [240]
Brief authority / Hooper, Charles – London, England. 1960 – 1r – us UF Libraries [960]
Brief authority / Hooper, Charles – New York, NY. 1961 – 1r – us UF Libraries [960]
Brief baptist history : from the time of the apostles to the present, embracing every great movement, name and occurrence essential to the true story of the churches of christ, together with their doctrines and present statistics / Ford, Samuel Howard – 2nd enl and ill ed. St Louis, Mo: [s.n.], 1891 – 1mf – 9 – 0-524-04040-0 – mf#1990-4948 – us ATLA [242]
A brief biblical history : old testament / Foakes-Jackson, Frederick John – New York: Doran [1912?] [mf ed 1993] – 1mf – 9 – 0-524-08076-3 – mf#1992-1136 – us ATLA [221]
Brief biographical sketch of dr. thomas curtis founder for limestone college / Curtis, Thomas – Ed. by Dr. Elmer Douglas Johnson. From a MS. of Dr. R. W. Sanders. 16p – 1 – $5.00 – us Southern Baptist [920]
A brief biographical sketch of sir john william dawson / Ami, Henry Marc – Minneapolis: American Geologist, 1900 – 1mf – 9 – mf#00796 – cn Canadiana [920]
Brief biographical sketches of noted missionaries who labored on american soil – Columbus OH: Lutheran Book Concern [1873?] [mf ed 1992] – 1mf – 9 – 0-524-04067-2 – mf#1991-2012 – us ATLA [242]
Brief biographies of some members of the society of friends : showing their early religious exercises and experience in the work of regeneration / Walton, Joseph – Philadelphia: Friends' Book Store, [187-?] – 1mf – 9 – 0-524-03203-3 – mf#1990-4652 – us ATLA [240]
Brief candle – 1970 spring, v3 n1-2 [1970 fall-winter] – 1r – 1 – mf#1532474 – us WHS [071]
Brief case *see* Nlada briefcase
Brief comments on unusual happenings in early jack... / Clark, John – s.l, s.l? 193-? – 1r – us UF Libraries [978]
A brief compend of bible truth / Alexander, Archibald – Philadelphia: Presbyterian Board of Publ, c1846 [mf ed 1985] – 1mf – 9 – 0-8370-2252-5 – mf#1985-0252 – us ATLA [240]
Brief confutation of the errors of the church of rome / Secker, Thomas – London, England. 1796 – 1r – us UF Libraries [241]
Brief consideration of the present system of methodist episcopal church government : with a few suggestions towards its improvement. respectfully inscribed to the travelling ministers and the members of the methodist episcopal church. by a layman – 1824 – 1r – 1 – $35.00 – mf#um-15 – us Commission [242]
Brief considerations on the test laws – London, England. 1807 – 1r – us UF Libraries [240]
Brief defence of the "essays and reviews" / Wild, George John – London, England. 1861 – 1r – us UF Libraries [240]
Der brief des jakobus : in fuenfundzwanzig predigten / Koegel, Rudolf – Bremen: C E Mueller, 1889 – 1mf – 9 – 0-524-04802-9 – mf#1992-0222 – us ATLA [227]
Der brief des jakobus / Spitta, Friedrich – Goettingen: Vandenhoeck und Ruprecht, 1896 – 1mf – 9 – 0-8370-9581-6 – (incl bibl ref) – mf#1986-3581 – us ATLA [227]
Der brief des julius africanus an aristides : kritisch untersucht und hergestellt / Spitta, Friedrich – Halle: Waisenhaus, 1877 [mf ed 2003] – 1r – 1 – (incl bibl ref. discussion in german & greek. text in greek) – mf#b00654 – us ATLA [240]

Der brief des paulus an die philipper / Ewald, Paul – 1. und 2. aufl. Leipzig: A Deichert, 1908 – (= ser Kommentar zum Neuen Testament) – 1mf – 9 – 0-7905-3336-7 – mf#1987-3336 – us ATLA [227]

Der brief des paulus an die roemer / Kuehl, Ernst – Leipzig: Quelle & Meyer, 1913 – 2mf – 9 – 0-7905-1218-1 – (incl indes) – mf#1987-1218 – us ATLA [227]

A brief description of nova scotia : including a particular account of the island of grand manan / Lockwood, Anthony – London 1818 [mf ed Hildesheim 1995-98] – 1v on 2mf – 9 – €60.00 – 3-487-26625-3 – gw Olms [917]

Brief discourse concerning singing in public worship / Marlow, Isaac – London. 1690 – 1 – 5.00 – us Southern Baptist [242]

A brief discourse concerning the three chief principles of magnificent building : viz solidity, conveniency and ornament / Gerbier, B – London, 1664 – 1mf – 9 – mf#0A-286 – ne IDC [720]

A brief discussion of grace and good works : or, of the divine and the human agency in the work of human redemption / Milligan, Robert – St Louis: Christian Pub Co, 1889 [mf ed 1993] – 1mf – 9 – 0-524-06820-8 – mf#1991-2807 – us ATLA [240]

Brief documentary history of the translation of the scriptures into the arabic language / Smith, Eli & Van Dyck, Cornelius Von Alan – Beirut: American Presbyterian Mission Press, 1900 – 1r – 1 – 0-8370-0479-9 – mf#1984-B122 – us ATLA [220]

Brief drawing / Ringwalt, Ralph Curtis – New York: Longmans, Green 1929. 214p. LL-1262 – 1 – us L of C Photodup [340]

A brief examination of prevalent opinions on the inspiration of the scriptures of the old and new testaments / [Muir, John] – London: Longman, Green, Longman & Roberts, 1861 [mf ed 1986] – 1mf – 9 – 0-8370-9962-5 – (int by henry bristow wilson) – mf#1986-3962 – us ATLA [220]

Brief examination of professor keble's visitation sermon / Wilson, William – Oxford, England. 1837 – 1r – us UF Libraries [240]

A brief exposition of gospel differences given according to the divine law of progressive instruction / Horton, Mary B – rev ed. New York: Mary B Horton, 1892, c1891 [mf ed 1985] – 1mf – 9 – 0-8370-3664-X – mf#1985-1664 – us ATLA [226]

A brief for the trial of criminal cases / Abbott, Austin – New York, Diossy, 1889. 566p. LL-1271 – 1 – (2nd ed., rochester, lawyers co-op. pub. co., 1902. 814p. ll-1270) – us L of C Photodup [345]

A brief greek syntax and hints on greek accidence : with some reference to comparative philology, and with illustrations from various modern languages / Farrar, Frederic William – 8th ed. London: Longmans, Green, 1876 [mf ed 1988] – 1mf – 9 – 0-7905-0012-4 – (incl bibl ref and ind) – mf#1987-0012 – us ATLA [450]

A brief guide to the museum / Archeological Museum of Merida – Caceres: Edit. Extremadura, 1975 – 1 – sp Bibl Santa Ana [060]

A brief historical sketch of the catholic church on long island / Mulrenan, Patrick – New York: P O'Shea 1871 [mf ed 1993] – 1mf – 9 – 0-524-06755-4 – mf#1990-5281 – us ATLA [241]

A brief historical sketch of the grande-ligne mission from its beginning in 1835 to 1900, 65 years / Lafleur, Theodore – [Montreal?: D Bentley], 1900 – 1mf – 9 – 0-665-89718-9 – mf#89718 – cn Canadiana [242]

Brief historical sketch of the western baptist theological institute / Stevens, John – [preliminary ed] [S.l.: s.n., 1849?] – 1mf – 9 – 0-524-08589-7 – mf#1993-3174 – us ATLA [242]

Brief historical sketch of the western baptist theological institute, in covington, ky.: a reply – 1850 – 1 – 5.00 – us Southern Baptist [242]

Brief histories of the mission and its missionaries / Roman Catholic Mission, Fiji – 1815-1936 – 1r – 1 – mf#pmb453 – at Pacific Mss [241]

Brief histories of us army commands (army posts) and descriptions of their records / U.S. Army – (= ser Records of united states army continental commands, 1821-1920) – 1r – 1 – mf#T912 – us Nat Archives [355]

Brief history and tenth anniversary services / Truett Memorial Baptist Church – 1962 – 1 – 5.00 – us Southern Baptist [242]

A brief history from official sources of the legislation respecting separate schools since the year 1863 : in the united province of canada, and in the dominion since confederation – S.l: s.n, 18– – 1mf – 9 – mf#02835 – cn Canadiana [370]

A brief history of canada / Calkin, John Burgess – London: Nelson; Halifax, NS: A & W Mackinlay, [between 1905 and 1911] – 2mf – 9 – 0-665-73168-X – (incl ind) – mf#73168 – cn Canadiana [971]

A brief history of claar congregation / Adams, David M – Cleona, Lebanon Co, PA: Holzapfel Pub Co [1908?] [mf ed 1992] – 1mf – 9 – 0-524-03605-5 – mf#1990-4765 – us ATLA [242]

A brief history of congregation oheb shalom, baltimore, md / Rosenau, William – Baltimore: Guggenheimer, Wil, 1903 [mf ed 1985] – 1mf – 9 – 0-8370-4967-9 – mf#1985-2967 – us ATLA [270]

A brief history of early chinese philosophy / Suzuki, Daisetz Teitaro – London: Probsthain, 1914 [mf ed 1991] – (= ser Probsthain's oriental series 7) – 1mf – 9 – 0-524-01932-0 – mf#1990-2745 – us ATLA [180]

A brief history of german literature / Priest, George Madison – New York: C Scribner's Sons, 1928, c1909 [mf ed 1993] – xii/366p/2pl – 1 – (incl bibl ref and ind) – mf#8075 – us Wisconsin U Libr [430]

A brief history of great britain / Calkin, John Burgess – London: T Nelson; Halifax, NS: A & W Mackinlay, 1907 – 3mf – 9 – 0-665-71565-X – (incl ind) – mf#71565 – cn Canadiana [941]

A brief history of idaho and western montana : as settled and district organized by the church of the brethren / ed by Mow, A I – [a.l]: Mission Board of Idaho & Western Montana...1914 [mf ed 1992] – 1mf – 9 – 0-524-02748-X – mf#1990-4423 – us ATLA [242]

Brief history of manatee county / Liddle, Carl – s.l, s.l? 1936 – 1r – us UF Libraries [978]

A brief history of missionary enterprise in antient and modern times : lecture memoranda world missionary conference, edinburgh, 1910 – London, New York: Burroughs, Wellcome, [1910?] [mf ed 1986] – 1mf – 9 – 0-8370-7357-X – (incl ind) – mf#1986-1357 – us ATLA [240]

Brief history of my home town : bradenton / Tucker, Philip C – s.l, s.l? 193-? – 1r – us UF Libraries [978]

A brief history of nyasaland / Morris, Martin – London, New York; Longmans, Green 1952 – us CRL [960]

Brief history of south dakota / Robinson, Doane – New York: American Book Co, c1905 (mf ed 19–) – mf#ZH-495 – us NY Public [978]

A brief history of the albemarle baptist association : a discourse...chestnut grove church, aug 19 1891 / Turpin, John Broaddus – Richmond, VA: Virginia Baptist Historical Society [1892?] [mf ed 1993] – 1mf – 9 – 0-524-07987-0 – (incl bibl ref. int by wm w landrum) – mf#1990-5432 – us ATLA [242]

A brief history of the baptist church, hebden bridge, yorkshire, england / Williams, Charles – 1877 – 1 – $5.00 – us Southern Baptist [242]

Brief history of the baptist missionary society from its commenceme... – Calcutta? India. 1842 – 1r – us UF Libraries [242]

A brief history of the baptists and their distinctive principles and practices : from the beginning of the gospel to the present time / Duncan, William Cecil – New-York: E H Fletcher, 1855 [mf ed 1993] – 1mf – 9 – 0-524-06468-7 – (no more publ?) – mf#1990-5242 – us ATLA [242]

A brief history of the beginning of the mission work in nicomedia : by the american board of foreign missions / Nergararian, Garabed – Waynesboro, PA: Gazette Steam Printing House, 1885 [mf ed 1990] – 1mf – 9 – 0-7905-7123-4 – mf#1988-3123 – us ATLA [240]

A brief history of the christian church / Leonard, William Andrew – New York: E P Dutton, 1910 [mf ed 1992] – 1mf – 9 – 0-524-03236-X – (int by john williams) – mf#1990-0864 – us ATLA [240]

A brief history of the church of the brethren in china – Elgin, IL: Brethren Pub House [1915?] [mf ed 1992] – 1mf – 9 – 0-524-03929-1 – mf#1990-4923 – us ATLA [242]

A brief history of the court of customs and patent appeals / Rich, Giles S – Washington: GPO, 1981 (all publ) – (= ser US court of customs and patent appeals reports) – 3mf – 9 – $4.50 – (publ by authorization of the committee on the bicentennial of independence and the constitution of the judicial conference of the united states, 1980) – mf#LLMC 95-015 – us LLMC [340]

A brief history of the first 25 years' history of the ymca's in china *see* Chung-hua chi-tu chiao ch'ing nien hui er shih wu nien hsiao shih [ccm237]

Brief history of the florida east coast railway... / Florida East Coast Railway – St Augustine, FL. 1936 – 1r – us UF Libraries [380]

A brief history of the g-2 section, ghq, swpa and affiliated units / U.S. Army. Far East Command – 1948 – 1 – us L of C Photodup [355]

A brief history of the german baptist brethren church : showing the commencement of the work and line of progress in the city of lancaster...sep 1900 – Lancaster, PA: New Era Print, 1900 [mf ed 1992] – 1mf – 9 – 0-524-03692-6 – mf#1990-4797 – us ATLA [242]

A brief history of the indian peoples / Hunter, William Wilson – Oxford: Clarendon Press, 1903 – (= ser Samp: indian books) – us CRL [954]

Brief history of the indian peoples / Hunter, William Wilson; ed by Hutton, W H – 23rd ed. New York, Young People's Missionary Movement [1903] [mf ed 1995] – (= ser Yale coll; Mission study reference library 5(b)) – 1 – 0-524-09870-0 – mf#1995-0870 – us ATLA [954]

A brief history of the lutheran church in america = Kurzgefasste geschichte der lutherischen kirche amerikas / Neve, Juergen Ludwig – Burlington IA: German Literary Board 1904 – 1mf – 9 – 0-7905-5613-8 – (english trans fr german by joseph stump; 2nd ed publ in 1916, 3rd publ in 1934 as: history of the lutheran church in america) – mf#1988-1613 – us ATLA [242]

A brief history of the madison square presbyterian church and its activities / Parkhurst, Charles Henry – New York: [s.n.] 1906 [mf ed 1990] – 1mf – 9 – 0-7905-5786-X – mf#1988-1786 – us ATLA [242]

A brief history of the middle temple / Bedwell, Cyril E A – London: Butterworth, 1909 – 2mf – 9 – $3.00 – mf#LLMC 84-272 – us LLMC [323]

A brief history of the one hundredth regiment (roundheads) : to which is added short sketches of colonel leasure, and chaplain browne, with a few poems by h b durant / Bates, Samuel Penniman – New Castle, PA: J C Stevenson, 1884 (New Castle: W B Thomas) [mf ed 19–] – 1 – (repr with adds from s p bates' history of pennsylvania volunteers, 1861-5, v3 (1870) p[553]-563) – mf#*ZH-IAG pv260 n1 – us NY Public [976]

A brief history of the republic of china armed forces / U.S. Office of the Chief of Military History – 1971 – 1 – us L of C Photodup [951]

A brief history of the united states boundary question / James, George Payne Rainsford – London: Saunder and Otley, 1839 – 1mf – 9 – mf#36441 – cn Canadiana [327]

Brief history of the virgin islands / Jarvis, Jose Antonio – Charlotte Amalie, St Thomas. 1938 – 1r – us UF Libraries [972]

A brief inquiry into causes of the poetic element in the scottish mind : being a lecture delivered...city of kingston / George, James – Kingston, Ont?: J M Creighton, 1857 – 1mf – 9 – mf#35762 – cn Canadiana [390]

Brief inquiry into the law of the church of england with respect to... / Shaw, Benjamin – London, England. 1858 – 1r – us UF Libraries [241]

A brief intervention on environmental tobacco smoke and the attitudes and behaviors of childcare providers / Sydzyik, Robyn – 2000 – 57p on 1mf – 9 – $5.00 – mf#HE 671 – us Kinesology [150]

A brief introduction to modern philosophy / Rogers, Arthur Kenyon – New York: Macmillan, 1899 [mf ed 1991] – 1mf – 9 – 0-7905-9090-5 – mf#1989-2315 – us ATLA [190]

A brief introduction to new testament greek : with vocabularies and exercises / Green, Samuel Gosnell – New York: Fleming H Revell; London: Religious Tract Society [1894?] [mf ed 1986] – 1mf – 9 – 0-8370-9237-X – mf#1986-3237 – us ATLA [450]

A brief introduction to the bibliography of modern jewish history / Marcus, Jacob Rader – Cincinnati: Hebrew Union College, 1935 (mf ed 1995) – 1r – 1 – mf#*ZP-1485 – us NY Public [939]

Brief introduction to the study of the chinese language / Pettus, William Bacon – Shanghai: American Presbyterian Mission Press, 1915 [mf ed 1995] – (= ser Yale coll) – 26p (ill) – 1 – 0-524-09314-8 – mf#1995-0314 – us ATLA [480]

A brief introduction to the study of theology / Foster, Robert Verrell – Chicago: Fleming H Revell, c1899 [mf ed 1985] – 1mf – 9 – 0-8370-3168-0 – (incl app, bibl and ind) – mf#1985-1168 – us ATLA [240]

Der brief jakobi : in berichtigter lutherscher uebersetzung / Neander, August – Berlin: Karl Wiegandt, 1850 – 1mf – 9 – 0-8370-9569-7 – mf#1986-3569 – us ATLA [227]

Der brief jakobi : in zwei und dreissig betrachtungen / Stier, R – Barmen: W Langewiesche, 1845 – 1mf – 9 – 0-7905-2149-0 – mf#1987-2149 – us ATLA [227]

Der brief judae *see* The epistle general of jude

Der brief judae des apostels und bruders des herrn / Rampf, M F – Sulzbach: J E v Seidel, 1854 – 1mf – 9 – 0-7905-1375-7 – (in german, greek and latin. incl bibl ref) – mf#1987-1375 – us ATLA [227]

Der brief judae, des bruders des herrn : als prophetische mahnung allen glaeubigen unsrer zeit, die sich bewahren wollen / Stier, R – Berlin: Wilhelm Hertz, 1850 – 1mf – 9 – 0-7905-2150-4 – mf#1987-2150 – us ATLA [227]

Brief memoir : relative to the operations of the serampore missionaeries, bengal – London: Parbury, Allen, 1827 [mf ed 1995] – (= ser Yale coll) – 89p – 1 – 0-524-09300-8 – mf#1995-0300 – us ATLA [954]

Brief memoir of mr justice rokeby / Rokeby, Thomas – Durham, England. 1861? – 1r – us UF Libraries [240]

Brief memoir of mrs s parkinson : of sutton, in craven, yorkshire / Scott, P – Mytholmroyd, England. 1859 – 1r – us UF Libraries [240]

Brief memoir of the chinese evangelist leang afa... / Bridgman, E C – London, England. 1835? – 1r – us UF Libraries [242]

A brief memoir of the late honorable james william johnston : first judge in equity of nova scotia / Calnek, William Arthur – St John, NB: G Knodell, 1884 – 1mf – 9 – mf#00384 – cn Canadiana [340]

Brief memorial of the late rev james smart, chirnside... / Ritchie, William – Dunse, Scotland. 1854 – 1r – us UF Libraries [240]

Brief memorial of the lord's dealings with george picknell of chalv... / Forster, John – London, England. 1863 – 1r – us UF Libraries [240]

Brief memorials of alphonse fran+ois lacroix... : with brief memorials of mrs mullens, by her sister / Mullens, J – London, 1862 – 6mf – 9 – mf#HTM-144 – ne IDC [910]

A brief narrative of an unsuccessful attempt to reach repulse bay : through sir thomas rowe's "welcome", in his majesty's ship griper, in the year 1824 / Lyon, George F – London 1825 [mf ed Hildesheim 1995-98] – 1v on 2mf – 9 – €60.00 – 3-487-27009-9 – gw Olms [910]

A brief narrative of an unsuccessful attempt to reach repulse bay through sir thomas rowe's "welcome" in h m's ship griper in the year 1824 / Lyon, G F – London, 1825 – 5mf – 9 – mf#N-300 – ne IDC [917]

Brief narrative of the baptist mission in india – London, England. 1810 – 1r – us UF Libraries [242]

Brief narrative of the loss of the abeona etc... – Glasgow, Scotland. 1821 – 1r – us UF Libraries [240]

Brief narrative of the operations of the jaffna auxiliary bible society : in the preparation of a version of the tamil scriptures – Jaffna: [s.n.], 1868 [mf ed 1995] – (= ser Yale coll) – 172p – 1 – 0-524-09021-1 – mf#1995-0021 – us ATLA [220]

Brief notes on national construction in cambodia / Norodom Sihanouk, Prince – Phnom Penh: Impr Sangkum Reastr Niyum 1969 [mf ed 1989] – 1r with other items – 1 – mf#mf-10289 seam reel 027/14 [$] – us CRL [330]

Brief notes on pastoral theology / Hay, Charles Augustus – Gettysburg PA: W L Rutherford 1891 [mf ed 1994] – 1mf – 9 – 0-524-08866-7 – mf#1993-3330 – us ATLA [242]

Brief notes on the greek of the new testament / Trench, Francis Chenevix – London: Macmillan, 1864 – 1mf – 9 – 0-8370-9586-7 – (incl bibl ref) – mf#1986-3586 – us ATLA [225]

Brief notes on the island of anticosti : in the gulf of st lawrence, dominion of canada – [London?: s.n.], 1886 [mf ed 1980] – 1mf – 9 – 0-665-02523-8 – mf#02523 – cn Canadiana [917]

Brief notice of christian doctrine as held by the religious society – London, England. 1851 – 1r – us UF Libraries [240]

Brief notices of "the history and legislation of separate schools in upper canada" : by j george hodgins...of osgood hall, barrister-at-law – Toronto?: W Briggs?, 1889? – 1mf – 9 – mf#54254 – cn Canadiana [377]

A brief of the authorities upon the law of impeachable crimes and misdemeanors / Lawrence, William Beach – Washington, Govt. Print. Off., 1868. 27 p. LL-183 – 1 – us L of C Photodup [345]

A brief on the modes of proving the facts most frequently in issue. / Abbott, Austin – 3rd ed. Rochester, N.Y., The Lawyers' Cooperative Publishing Co., 1912. 1007 p. LL-934 – 1 – us L of C Photodup [340]

Brief outline and review of a work entitled "the principles of natu... / Chapman, John – London, England. 1847 – 1r – us UF Libraries [240]

Brief outline of christian unitarianism / Porter, J Scott – Belfast, Northern Ireland. 1871 – 1r – us UF Libraries [243]

Brief outline of the study of theology : drawn up to serve as the basis of introductory lectures = Kurze darstellung des theologischen studiums / Schleiermacher, Friedrich [Ernst Daniel] – Edinburgh: T & T Clark 1850 [mf ed 1990] – 1mf – 9 – 0-7905-9623-7 – (trans fr german by william farrer; reminiscences of schleiermacher by friedrich luecke, originally publ in german in theologische studien und kritiken) – mf#1989-1348 – us ATLA [240]

Brief outlines of christian doctrine : designed for senior epworth leagues and all bible students / Dewart, Edward Hartley – Toronto: W Briggs; Montreal: C W Coates, 1898 – (= ser Bibliotheque de theologie. serie 1. theologie dogmatique) – 1mf – 9 – mf#02673 – cn Canadiana [240]

Der brief pauli an der roemer *see* The epistle of paul to the romans

Der brief pauli an die galater / Schmoller, Otto – 2. durch aufl. Bielefeld: Velhagen und Klasing, 1865 – 1mf – 9 – 0-8370-5996-8 – mf#1985-3996 – us ATLA [227]

Brief pauli an die philipper = The epistle of paul to the pilippians / Neander, August – New York: Lewis Colby, 1851 – 1mf – 9 – 0-8370-9642-1 – (in english) – mf#1986-3642 – us ATLA [227]

Der brief pauli an die philipper : in berichtigter lutherscher uebersetzung / Neander, August – Berlin: Karl Wiegandt, 1849 – 1mf – 9 – 0-8370-9570-0 – mf#1986-3570 – us ATLA [227]

Der brief pauli an die roemer : forschenden bibellesern durch umschreibung und erlaeuterung erklaert und mit specieller einleitung, sowie mit den noetigen historischen, geographischen und antiquarischen anmerkungen / Couard, Hermann – 2. verb aufl. Potsdam: August Stein, 1895 – 1mf – 9 – 0-524-06833-X – mf#1992-0975 – us ATLA [227]

Brief plea for believers' baptism / Pengilly, Richard – London, England. 18– – 1r – us UF Libraries [242]

A brief record of the advance of the british egyptian expeditionary force in palestine, july 1917 to october 1918 – Ed 2. London, 1919 – 5mf – 9 – mf#J-28-166 – ne IDC [915]

Brief remarks on a pamphlet entitled "arguments to prove the policy and necessity of granting to newfoundland a constitutional government" : by p morris, an inhabitant of the colony of newfoundland – [Poole, England?: s.n.] 1828 [mf ed 1987] – 1mf – 9 – 0-665-63522-2 – mf#63522 – cn Canadiana [320]

Brief remarks on the anti-british effect of... criticism on modern art / Carey, William Paulet – [London? 1831] – (= ser 19th c art & architecture) – 2mf – 9 – mf#4.1.277 – uk Chadwyck [700]

Brief remarks on "the declaration of the catholic bishops... / Allwood, Philip – London, England. 1826 – 1r – us UF Libraries [241]

Brief remarks on the dispositions towards christianity / Rose, Hugh James – London, England. 1830 – 1r – us UF Libraries [240]

Brief remarks on the waste lands of the crown in the canadas : with reference to emigration and colonization / Forsyth, James Bell – [Quebec?: s.n.] 1848 [mf ed 1983] – 1mf – 9 – 0-665-29401-8 – mf#29401 – cn Canadiana [320]

Brief remarks upon the carnal and spiritual state of man / Allen, William – London, England. 1817 – 1r – us UF Libraries [240]

Brief review of all the texts in the new testament usually alleged – London, England. 1838 – 1r – us UF Libraries [240]

A brief review of criminal cases in the supreme court for the past year / Green, Frederick – Urbana: University of Illinois 1913. 24p. LL-543 – 1 – us L of C Photodup [345]

A brief review of doctor bond's " appeal to the methodists" / Shinn, Asa – Baltimore: Matchett 1827 – 1r – 1 – $35.00 – mf#um-15 – us Commission [242]

A brief review of ten years' missionary labour in india between 1852 and 1861 / Mullens, J – London, 1864 – 3mf – 9 – mf#HTM-140 – ne IDC [915]

Brief review of the ecclesiastical polity of great britain – Cambridge, England. 1818 – 1r – us UF Libraries [240]

Brief sketch of british honduras / Anderson, A H – Belize, Belize. 1948 – 1r – us UF Libraries [972]

Brief sketch of british honduras / Anderson, A H – Belize, Belize. 1958 – 1r – us UF Libraries [972]

A brief sketch of the brethren generally known as "dunkards" of northern indiana / Opperman, Owen – Goshen IN: News Printing Co 1897 [mf ed 1985] – 1mf – 9 – 0-524-03179-7 – mf#1990-4628 – us ATLA [242]

A brief sketch of the early history of the catholic church on the island of new york / Bayley, James Roosevelt – 2nd rev enl ed. New York: Catholic Pub Society, 1870 [mf ed 1993] – 1mf – 9 – 0-524-06240-4 – mf#1990-5195 – us ATLA [241]

A brief sketch of the establishment of the anglican church in india / Parlby, Brook Bridges – London 1851 – (= ser 19th c british colonization) – 2mf – 9 – mf#1.1.8780 – uk Chadwyck [242]

Brief sketch of the fate of 3000 indian pows in new guinea / Singh, Chint – 1943-45 – 1r – 1 – mf#pmb1249 – at Pacific Mss [934]

Brief sketch of the life and times of the late hon louis joseph papineau / Brown, Thomas Storrow – [Montreal?]: [s.n.], [1872?] [mf ed 1980] – 1mf – 9 – 0-665-00971-2 – mf#00971 – cn Canadiana [920]

Brief sketch of the life of charles, baron metcalfe, of fernhill, in berkshire... : to the period of his resigning the office of governor general of the british north american colonies, in 1845 / Erinensis [Walter Cavendish Crofton] – [Kingston, Ont?: s.n.] 1846 [mf ed 1983] – 1mf – 9 – 0-665-44467-2 – mf#44467 – cn Canadiana [971]

Brief sketch of the life of charles watson / Watson, Charles – Liverpool, England. 18– – 1r – us UF Libraries [240]

A brief sketch of the life of the rev father joseph henry tabaret : oblate of mary immaculate...founder and superior of the college of ottawa: died in ottawa, 28th february, 1886, aged 58 years – Ottawa?: s.n, 1886 – 1mf – 9 – mf#00868 – cn Canadiana [241]

A brief sketch of the morris movement : and of the firm founded by william morris to carry out his designs... – London: Priv print for Morris & Co, 1911 (mf ed 19–) – 63p – mf#ZM-3-MAR pv865 n7 – us Harvard [740]

A brief sketch of the present position of christian missions in northern india, and their progress during the year 1847 : compiled from recent missionary reports and letters... / Mullens, J – Calcutta, 1848 – 2mf – 9 – mf#HTM-141 – ne IDC [915]

A brief sketch of the waldenses / Strong, C H – Lawrence, KN: J S Boughton, 1893 [mf ed 1992] – 1mf – 9 – 0-524-03594-6 – mf#1990-1054 – us ATLA [242]

A brief sketch of the zoroastrian religion and customs : an essay written for the rahnumai mazdayasnan sabha of bombay / Bharucha, Sheriarji Dadabhai – 3rd rev enl ed. Bombay: D B Taraporevala Sons, 1928 [mf ed 1986] – 210p – 1 – (int by jivanji jamshedji modi) – mf#6903 – us Wisconsin U Libr [290]

A brief sketch of various attempts which have been made to diffuse a knowledge of the holy scriptures through the medium of the irish language – Dublin: Graisberry & Campbell, 1818 [mf ed 1989] – 1mf – 9 – 0-7905-0849-4 – (incl bibl ref) – mf#1987-0849 – us ATLA [220]

Brief sketches of floridas' outstanding and history – s.l, s.l? 193-? – 1r – us UF Libraries [978]

Der brief st pauli an die kolosser *see* The epistle of paul to the colossians

Der brief st pauli an die philipper *see* The epistle of paul to the philippians

Brief state of the contests that have lately arisen in the salisbury concert – Salisbury: Collins & Johnson 1781 [mf ed 19–] – 1mf – 9 – mf#fiche 59 – us Sibley [780]

A brief statement of objections to the policy of imposing export duties upon saw-logs, shingle bolts and stave bolts : and a few facts pertaining to the round timber trade in canada / Charlton, John – Lynedoch Ont: Leader Steam Press, 1869 – 1mf – 9 – mf#05955 – cn Canadiana [634]

Brief statement of reasons for bible societies in scotland / Thomas, William A – Edinburgh, Scotland. 1826 – 1r – us UF Libraries [240]

A brief statement of the reformed faith : according to the system of doctrine set forth in the standards of the presbyterian church of new zealand / Fraser, Philadelphus Bain [comp] – Dunedin, NZ: J Wilkie, 1909 [mf ed 1993] – 1mf – 9 – 0-524-07241-8 – mf#1991-2982 – us ATLA [242]

Brief statement of the rise, progress, and decline of the ancient c... / Jervis, John Jervis White – London, England. 1813 – 1r – us UF Libraries [240]

Brief statement of the tenets generaly held by the men reviled... / Reed, Thomas – London, England. 1822 – 1r – us UF Libraries [240]

A brief study of christian science / Sandt, George Washington – Philadelphia, PA: General Council Publ House, 1911 [mf ed 1990] – 1mf – 9 – 0-7905-6729-6 – mf#1988-2729 – us ATLA [240]

A brief summary, in plain language, of the most important laws of england concerning women / Bodichon, Barbara Leigh – 3rd ed. London, Trubner, 1869. 39p. LL-1700 – 1 – us L of C Photodup [340]

Brief summary of the recent controversy on infallibility / Ward, William George – London, England. 1868 – 1r – us UF Libraries [240]

A brief survey of equity jurisdiction / Langdell, Christopher Columbus – Cambridge, Harvard, 1904. 259 p. LL-1087 – 1 – us L of C Photodup [342]

Brief survey of the history of glass in the corning museum *see* History of glass

Brief survey of the ways of god to man / Wollaston, Francis – London, England. 1808? – 1r – us UF Libraries [240]

A brief text-book of moral philosophy / Coppens, Charles – New York: Schwartz, Kirwin & Fauss, c1895 [mf ed 1986] – 1mf – 9 – 0-8370-6173-3 – (incl ind. companion vol: brief text-book of logic and mental philosophy) – mf#1986-0173 – us ATLA [170]

Brief thoughts – Edinburgh, Scotland. 18– – 1r – us UF Libraries [240]

Brief thoughts and meditations on some passages in holy scripture / Trench, Richard Chenevix – London: Macmillan, 1884 – 1mf – 9 – 0-7905-2390-6 – mf#1987-2390 – us ATLA [220]

Brief traicte de la victoire que le compte charles de masfelt, ...a l'encontre du turc... l'an 1595 – Anvers, 1595 – 1mf – 9 – mf#H-8212 – ne IDC [956]

A brief treatise on the atonement / Kephart, Ezekiel Boring – Dayton, OH: United Brethren Pub House 1902 [mf ed 1989] – (= ser Doctrinal series (dayton, ohio)) – 1mf – 9 – 0-7905-2539-9 – mf#1987-2539 – us ATLA [240]

A brief treatise on the canon and interpretation of the holy scriptures : for the special benefit of junior theological students, but intended also for private christians in general / McClelland, Alexander – [rev ed] New York: Robert Carter, 1860 [mf ed 1985] – 1mf – 9 – 0-8370-4338-7 – (rev ed of: manual of sacred interpretation) – mf#1985-2338 – us ATLA [220]

A brief unpublished history of the baptists of south carolina, 1683-1937 / Allen, W C – 565p – 1 – $19.77 – us Southern Baptist [242]

Brief view of ecclesiastical history – Dublin, Ireland. 1844 – 1r – us UF Libraries [240]

Brief view of the baptist missions and translations : at the mission press, serampore... / Baptist Missionary Society – London: Printed by J Haddon, 1815 [mf ed 1995] – (= ser Yale coll) – 40p/4p – 1 – 0-524-09778-X – mf#1995-0778 – us ATLA [242]

A brief view of the laws of upper canada up to the present time : including a treatise on the law of executors and wills, and the law relative to landlord and tenant... / Keele, William Conway – Toronto?: s.n, 1884 – 3mf – 9 – (incl ind) – mf#10787 – cn Canadiana [340]

Brief view of the scriptural encouragements of the london society / Thelwall, Algernon Sydney – London, England. 1842 – 1r – us UF Libraries [240]

A brief view of the township laws up to the present time : with a treatise on the law and office of constable, the law relative to landlord and tenant, distress for rent, inn-keepers, etc – Toronto?: s.n, 1835 – 2mf – 9 – (incl ind) – mf#10778 – cn Canadiana [340]

Brief vindications of an essay to prove singing of psalms, etc / Allen, Richard – London. 1696 – 1 – us Southern Baptist [242]

Ein brief von gerolamo cardano an konrad gessner 1555 / Salzmann, C – (Gesnerus, 1956. v13(p53-60) – 1mf – 9 – mf#Z-2273 – ne IDC [590]

Briefe : eine auswahl / Holz, Arno; ed by Holz, Anita & Wagner, Max – Muenchen: R Piper, c1948 [mf ed 1995] – 308p/3pl (ill) – 1 – (incl bibl ref and ind. int by hans heinrich borcherdt) – mf#9163 – us Wisconsin U Libr [860]

Briefe / Brentano, Clemens – Leipzig: Insel-Verlag, 1941 [mf ed 1989] – 95p – 1 – (aft by hubert schiel) – mf#7084 – us Wisconsin U Libr [860]

Briefe / Brentano, Clemens; ed by Seebass, Friedrich – Nuernberg: H Carl, 1951 [mf ed 1989] – 2v – 1 – mf#7084 – us Wisconsin U Libr [860]

Briefe / Flex, Walter; ed by Eggert-Windegg, Walther – Muenchen: C H Beck [1927] [mf ed 1989] – 1r [ill] – 1 – (filmed with: lothar) – mf#7245 – us Wisconsin U Libr [860]

Briefe / Hoelderlin, Friedrich – Weimar: E Lichtenstein, 1922 [mf ed 1996] – 351p – 1 – mf#9685 – us Wisconsin U Libr [860]

Briefe / ed by Muschler, Reinhold Conrad – 3. stark verm aufl. Leipzig: F W Grunow, 1928 [mf ed 1989] – 327p – 1 – mf#7188 – us Wisconsin U Libr [860]

Briefe : die neueste litteratur betreffend / ed by Lessing, Gotthold Ephraim et al – Berlin 1759-66 – (= ser Dz. abt literatur) – 24pt on 71mf – 9 – €710.00 – (fr pt22ff publ in berlin & stettin) – mf#k/n4433 – gw Olms [860]

Briefe / Rosenzweig, Franz – Berlin, 1935 – 13mf – 8 – €25.00 – ne Slangenburg [140]

Briefe / Schleiermacher, Friedrich [Ernst Daniel] – Jena: E Diederichs 1906 [mf ed 1991] – 1mf – 9 – 0-7905-9111-1 – mf#1989-2336 – us ATLA [242]

Briefe : von der jugendzeit bis zum tode / Herzl, T – Zug, 1977 – 181mf – 9 – mf#J-72-101 – ne IDC [920]

Briefe, 2. bd (bdk60 1.reihe) / Cyprian, Saint (Cyprianus) – (= ser Bibliothek der kirchenvaeter. 1. reihe (bdk 1.reihe)) – €17.00 – ne Slangenburg [240]

Briefe an brinkmann, henriette v finckenstein, wilhelm v humboldt, rahel, friedrich tieck, ludwig tieck und wiesel / Burgsdorff, Wilhelm von; ed by Cohn, Alfons Fedor – Berlin: B Behr, 1907 [mf ed 1994] – (= ser Deutsche litteraturdenkmale des 18. und 19. jahrhunderts 139, 3. n f n19) – 1 – (incl bibl ref and ind) – mf#8676 reel 8 – us Wisconsin U Libr [860]

Briefe an bunsen von roemischen cardinaelen und praelaten, deutschen bischoefen und anderen katholiken aus den jahren 1818 bis 1837 / ed by Reusch, Heinrich – Leipzig 1897 (mf ed 1992) – 2mf – 9 – €24.00 – 3-89349-029-9 – mf#DHS-AR 66 – gw Frankfurter [241]

Briefe an deutsche freunde : von einer reise durch italien ueber sachsen, boehmen und oestreich 1820 und 1821 geschrieben und als skizzen zum gemaelde unserer zeit herausgegeben / Mueller, Wilhelm C – Altona 1824 [mf ed Hildesheim 1995-98] – (= ser Fbc) – 2v on 7mf – 9 – €140.00 – 3-487-29283-1 – gw Olms [860]

Briefe an eine christliche freundin ueber die grundwahrheiten des judenthums : mit einem biographischen vorwort / Izates, Esther – Leipzig: C L Morgenstern, 1883 – 1mf – 9 – 0-8370-2130-8 – mf#1985-0130 – us ATLA [920]

Briefe an freunde / Claudius, Matthias; ed by Jessen, Hans – Berlin-Steglitz: Eckartverlag, c1938 [mf ed 1989] – (= ser Matthias claudius correspondence 1) – 455p (ill) – 1 – mf#7152 – us Wisconsin U Libr [860]

Briefe an friedrich baron de la motte fouque von chamisso, chezy, collin...[et al] / Motte Fouque, Friedrich Heinrich Karl, Freiherr de la; ed by Motte-Fouque, Albertine Baronin de la – Berlin: W Adolf 1848 [mf ed 1992] – (= ser Bibliothek der deutschen literatur) – 1r – 1 – (iss in 2pt) – mf#7561 – us Wisconsin U Libr [920]

Briefe an ludwig tieck / Holtei, Karl von [comp] – Breslau: E Trewendt 1864 [mf ed 1991] – 4v on 1r – 1 – (incl ind) – mf#2950p – us Wisconsin U Libr [860]

Briefe an seine braut / Storm, Theodor; ed by Storm, Gertrud – Braunschweig: G Westermann 1922 [mf ed 1991] – 1r – 1 – (filmed with: auf leben und tod) – mf#2903p – us Wisconsin U Libr [860]

Briefe an seine freunde / Fontane, Theodor – 2.aufl. Berlin: S Fischer, 1925 [mf ed 1989] – 2v – 1 – mf#7075 – us Wisconsin U Libr [860]

Briefe an seine freunde hartmuth brinkmann und wilhelm petersen / Storm, Theodor; ed by Storm, Gertrud – Berlin: G Westermann, 1917 [mf ed 1993] – xii/226p – 1 – mf#7733 – us Wisconsin U Libr [860]

Briefe an seine gattin / Freytag, Gustav – 3. & 4. aufl. Berlin: W Borngraeber [1912] [mf ed 1989] – 1r – 1 – (filmed with: gustav freytags briefe an albrecht von stosch) – mf#7277 – us Wisconsin U Libr [860]

Briefe an seine kinder / Storm, Theodor; ed by Storm, Gertrud – Berlin: G Westermann c1916 [mf ed 1991] – 1r – 1 – (filmed with: briefe an seine freunde hartmuth brinkmann und wilhelm petersen) – mf#2904p – us Wisconsin U Libr [860]

Briefe an und von johann heinrich merck : eine selbstaendige folge der im jahr 1835 erschienenen briefe an j h merck / Merck, Johann Heinrich; ed by Wagner, Karl – Darmstadt: J P Diehl 1838 [mf ed 1993] – (= ser Bibliothek der deutschen literatur) – 1r [ill] – 1 – (sequel to: briefe an johann heinrich merck von goethe, herder, wieland...ed by karl wagner, publ by diehl in 1835; incl ind. filmed with: gedichte / alfred meissner) – mf#7606 – us Wisconsin U Libr [860]

A briefe and plaine declaration : concerning the desires of all those faithfull ministers, that have and do seeke for the discipline and reformation of the church of englande / Fulke, W – London: Robert Walde-graue, 1584 – 2mf – 9 – mf#PW-45 – ne IDC [241]

Briefe aus aegypten / Doyle, Charles W – Weimar 1805 [mf ed Hildesheim 1995-98] – 1v on 1mf – 9 – €40.00 – 3-487-26564-8 – gw Olms [916]

Briefe aus aegypten, aethiopien und der halbinsel des sinai : geschrieben in den jahren 1842-45: waehrend der auf befehl sr majestaet des koenigs friedrich wilhelm 4 von preussen ausgefuehrten wissenschaftlichen expedition / Lepsius, Carl R – Berlin 1852 [mf ed Hildesheim 1995-98] – 1v on 3mf [ill] – 9 – €90.00 – 3-487-27377-2 – gw Olms [916]

Briefe aus amerika fuer deutsche auswanderer / Koehler, Karl – Darmstadt 1852 [mf ed Hildesheim 1995-98] – 1v on 2mf [ill] – 9 – €60.00 – 3-487-27007-2 – gw Olms [860]

Briefe aus beiden hemisphaeren : ein sittengemaelde aus der tropenwelt / Schlichthorst, Carl – Celle 1833 [mf ed Hildesheim 1995-98] – 1v on 2mf – 9 – €60.00 – 3-487-26864-7 – gw Olms [910]

Briefe aus berlin : geschrieben im jahr 1832 / Steinmann, Friedrich A – Hanau 1832 [mf ed Hildesheim: 1995-98] – (= ser Fbc) – 4mf – 9 – €120.00 – 3-487-29564-4 – gw Olms [914]

Briefe aus china : als manuscript gedruckt zum besten der deutschen mission in canton / Jentzsch, Franz – Berlin: R Gaertner, 1883 [mf ed 1995] – (= ser Yale coll) – iv/245p – 1 – 0-524-09231-1 – (in german) – mf#1995-0231 – us ATLA [860]

Briefe aus columbien an seine freunde : geschrieben in dem jahre 1820 / Richard, Carl – Leipzig 1822 [mf ed Hildesheim 1995-98] – 1v on 2mf – 9 – €60.00 – 3-487-26900-7 – gw Olms [918]

Briefe aus dem mittellaendischen meere : enthaltend eine schilderung des buergerlichen und politischen zustandes von sicilien, tripoli, tunis und malta / Blaquiere, Edward – Weimar 1821 [mf ed Hildesheim 1995-98] – 2v on 5mf – 9 – €100.00 – 3-487-26501-X – gw Olms [910]

Briefe aus der hauptstadt und dem innern frankreichs / Meyer, Friedrich J – Tuebingen 1802 [mf ed Hildesheim 1995-98] – 2v on 4mf – 9 – €120.00 – 3-487-29632-2 – gw Olms [914]

Briefe aus hamburg : ein wort zur vertheidigung der kirche gegen die angriffe von sieben laeugnern [sic] der gottheit christi / [Pesch, Tilmann] – 3. rev aufl. Berlin: Verlag der Germania, 1889 [mf ed 1991] – (= ser Christ oder antichrist? 1) – 2mf – 9 – 0-7905-9437-4 – mf#1989-2662 – us ATLA [240]

Briefe aus indien : bilder aus der missionstaetigkeit der franziskanerinnen missionaerinnen mariens / Schlager, Patricius – Trier: Verlag der Paulinus-Druckerei, 1914 [mf ed 1995] – (= ser Yale coll; Aus allen zonen) – 1 – 0-524-10193-0 – (in german) – mf#1995-1193 – us ATLA [860]

Briefe aus indien / Hoffmeister, Werner – Braunschweig 1847 [mf ed Hildesheim 1995-98] – 1v on 3mf – 9 – €90.00 – 3-487-27428-0 – gw Olms [915]

Briefe aus italien : waehrend der jahre 1801, 1802, 1803, 1804, 1805 mit mancherlei beilagen / Rehfues, Philipp J von – Zuerich 1809-10 [mf ed Hildesheim 1995-98] – (= ser Fbc) – 10mf – 9 – €100.00 – 3-487-29308-0 – gw Olms [914]

Briefe aus meinem kloster / Ehrler, Hans Heinrich – 2. aufl. Stuttgart: Greiner & Pfeiffer c1922 [mf ed 1989] – 1r – 1 – (filmed with: menschen und affen / albert ehrenstein) – mf#7207 – us Wisconsin U Libr [860]

Briefe aus palaestina / Gordon, A D – Berlin, 1919 – 1mf – 9 – mf#J-28-5 – ne IDC [956]

Briefe aus paris : geschrieben in den monaten juli, aug, sept und oct 1815 / Demian, Johann A – Frankfurt am Main 1816 [mf ed Hildesheim 1995-98] – 2mf – 9 – €60.00 – 3-487-29675-6 – gw Olms [914]

Briefe aus paris 1830-1831 / Boerne, Ludwig – Hamburg 1832-33 [mf ed Hildesheim: 1995-98] – (= ser Fbc) – 4v on 9mf – 9 – €180.00 – 3-487-29671-3 – gw Olms [914]

Briefe aus paris geschrieben in den monaten sept, oct, nov 1830 / Held, Johann C – Sulzbach 1831 [mf ed Hildesheim 1995-98] – 2mf – 9 – €60.00 – 3-487-29691-8 – gw Olms [860]

Briefe aus paris und frankreich im jahre 1830 / Raumer, Friedrich L von – Leipzig 1831 [mf ed Hildesheim 1995-98] – 2v on 4mf – 9 – €120.00 – 3-487-29780-9 – gw Olms [860]

Briefe aus sizilien / Westphal, Johann H – Berlin [u. a.] 1825 [mf ed Hildesheim 1995-98] – (= ser Fbc) – 3mf [ill] – 9 – €90.00 – 3-487-29188-6 – gw Olms [914]

Briefe aus wien / Tuvora, Joseph – Hamburg 1844 [mf ed Hildesheim: 1995-98] – (= ser Fbc) – v1 on 2mf – 9 – €60.00 – 3-487-29441-9 – gw Olms [914]

A briefe confutation : of a popish discourse / Fulke, W – London: George Byshop, 1581 – 2mf – 9 – mf#PW-44 – ne IDC [240]

Briefe, depeschen, und berichte ueber luther vom wormser reichstage 1521 – Halle: Verein fuer Reformationsgeschichte 1898 [mf ed 1990] – (= ser Schriften des vereins fuer reformationsgeschichte 15/59) – 1mf – 9 – 0-7905-4828-3 – (incl bibl ref; trans by paul kalkoff) – mf#1988-0828 – us ATLA [240]

Die briefe der annette von droste-huelshoff / Droste-Huelshoff, Annette von; ed by Schulte Kemminghausen, Karl – Jena: E Diederichs. 2v. c1944 – 1r – 1 – us Wisconsin U Libr [860]

Die briefe der dichterin annette v droste-huelshoff / ed by Cardauns, Herman – Muenster/W: Aschendorff, 1909 [mf ed 1989] – xiii/443p – 1 – (incl bibl ref and ind) – mf#7188 – us Wisconsin U Libr [860]

Briefe der elisabeth stuart, koenigin von boehmen : an ihren sohn, den kurfuersten carl ludwig von der pfalz 1650-62 / Elizabeth, Queen, consort of Frederick 1, King of Bohemia; ed by Wendland, Anna – Stuttgart: Litterarischer Verein 1902 (Tuebingen: H Laupp, Jr) [mf ed 1993] – (= ser Blvs 228) – 58r – 1 – (english text with int in german) – mf#3420p – us Wisconsin U Libr [860]

Die briefe der frau rath goethe / ed by Koester, Albert – Leipzig: C E Poeschel, 1905 [mf ed 2000] – 2v – 1 – (incl bibl ref and ind) – mf#10479 – us Wisconsin U Libr [860]

Briefe der freiin annette von droste-huelshoff / Droste-Huelshoff, Annette von – Muenster: A Russell 1877 [mf ed 1993] – 1r – 1 – (incl bibl ref. filmed with: das geistliche jahr) – mf#8552 – us Wisconsin U Libr [860]

Briefe der herzogin elisabeth charlotte von orleans / ed by Holland, Wilhelm Ludwig – Stuttgart: Litterarischer Verein, 1867-1881 [mf ed 1993] – (= ser Blvs 87) – 6v – 1 – (incl bibl ref & ind) – mf#8470 reel 18 – us Wisconsin U Libr [860]

Briefe der prinzessin elisabeth charlotte von orleans an die raugraefin louise, 1676-1722 / ed by Menzel, Wolfgang – Stuttgart: Literarischer Verein, 1843 [mf ed 1993] – (= ser Blvs 6) – xviii/527p – 1 – mf#8470 reel 2 – us Wisconsin U Libr [860]

Die briefe des apostels paulus an timotheus und titus / Belser, Johannes Evangelist – Freiburg i B, St Louis MO: Herder, 1907 – 1mf – 9 – 0-8370-9528-X – mf#1986-3528 – us ATLA [227]

Die briefe des apostels paulus und die reden des herrn jesu : ein blick in den organischen zusammenhang der neutestamentlichen schriften / Roos, Friedrich – Ludwigsburg: Ad Neubert, 1887 – 1mf – 9 – 0-8370-4963-6 – mf#1985-2963 – us ATLA [227]

Die briefe des bischofs rather von verona (mgh epistolae 2:1.bd) – 1949 – (= ser Monumenta germaniae historica epistolae. 2. die briefe der deutschen kaiserzeit (mgh epistolae 2)) – €12.00 – ne Slangenburg [241]

Briefe des dichters ludwig zacharias werner – Muenchen: G Mueller 1914 [mf ed 1993] – 2v on 1r [ill] – 1 – (incl bibl ref & ind; filmed with: conversations of goethe with eckermann and soret) – mf#8551 – us Wisconsin U Libr [860]

Briefe des dreistoeckigen hausbesitzers mister schorsch dobbeljuh hutzelberger an die "mississippi blaetter" : "o tempora, o mores!": gedichte / Thiersch, Curt – St Louis: Im Selbstverlag des Verfassers 1899 [mf ed 1991] – 1r – 1 – (filmed with: gotti und gotteli / rudolf von tavel) – mf#2910p – us Wisconsin U Libr [810]

Die briefe des grossen apostels von indien und japan : des heiligen franz von xavier aus der gesellschaft jesu, als grundlage der missions-geschichte spaeterer zeiten – 2. aufl. Coblenz: Philipp Werle, 1845 [mf ed 1995] – (= ser Yale coll) – 3v in 2 (ill) – 1 – 0-524-09625-2 – (in german. trans and ann by joseph burg) – mf#1995-0625 – us ATLA [241]

Die briefe des heiligen johannes / Belser, Johannes Evangelist – Freiburg i B: Herder, 1906 – 1mf – 9 – 0-524-04393-0 – mf#1992-0086 – us ATLA [227]

Die briefe des hl bonifatius und lullus (mgh epistolae 4:1.bd) – 1916 – (= ser Monumenta germaniae historica epistolae 4. epistolae selectae (mgh epistolae 4)) – €15.00 – ne Slangenburg [241]

Die briefe des libanius / Seeck, Otto – Leipzig, 1906 – (= ser Tugal 2-30/1.2) – 8mf – 9 – €17.00 – ne Slangenburg [240]

Die briefe des libanius : zeitlich geordnet / Seeck, Otto – Leipzig: J C Hinrichs, 1906 [mf ed 1989] – (= ser Tugal 30/1-2) – 1mf – 9 – 0-7905-1678-0 – (incl in memoriam [for oskar von gebhardt]. incl ind) – mf#1987-1678 – us ATLA [180]

Die briefe des sextus julius africanus an aristides und origenes / Reichardt, W – Leipzig, 1909 – (= ser Tugal 3-34/3) – 2mf – 9 – €5.00 – ne Slangenburg [240]

Die briefe des sextus julius africanus an aristides und origenes / Reichardt, Walther – Leipzig: J C Hinrichs, 1909 – (= ser Tugal) – 1mf – 9 – 0-7905-1730-2 – (incl bibl ref and ind) – mf#1987-1730 – us ATLA [240]

Briefe deutscher philosophen (1750-1850) / ed by Henrichs, Norbert – [mf ed 1990] – 3141mf (1:24) – 9 – diazo €6600.00 (silver €7870 isbn 978-3-598-33020-9) – ISBN-10: 3-598-33010-3 – ISBN-13: 978-3-598-33010-0 – gw Saur [190]

Briefe, die ihn nicht erreichten / Heyking, Elisabeth von – Berlin: Paetel, 1903 [mf ed 1989] – 269p – 1 – mf#7086 – us Wisconsin U Libr [860]

Briefe, die neueste litteratur betreffend / ed by Lessing, Gotthold Ephraim et al – Berlin 1759-65 [mf ed Hildesheim 1977] – (= ser Allgemeinwissenschaftliche und literarische zeitschriften des 17. und 18. jahrhunderts) – 71mf – 9 – diazo €710.00 – (price for silver version on request) – gw Olms [430]

A briefe discourse of the troubles begun at frankeford in germany, an. dom. 1554 : about the booke of common prayer and ceremonies... / [Whittingham, W] – London: G Bishop and R White, 1642 – 2mf – 9 – mf#PW-59 – ne IDC [240]

Briefe eines aufmerksamen reisenden die musik betreffend / Reichardt, J F – Frankfurt / Breslau, Frankfurt / Leipzig, 1774, 1776. 2v – 3mf – 9 – mf#P-749 – ne IDC [910]

Briefe eines aufmerksamen reisenden die musik betreffend : an seine freunde geschrieben / Reichardt, Johann Friedrich – Frankfurt, Leipzig 1774-76 [mf ed 19–] – 2v on 5mf – 9 – (v2 publ in frankfurt & breslau) – mf#fiche 931, 933 – us Sibley [780]

Briefe eines deutschen kuenstlers aus italien : aus der nachgelassenen papieren / Speckter, Erwin – Leipzig 1846 [mf ed Hildesheim 1995-98] – 2v on 6mf – 9 – €120.00 – 3-487-29255-6 – gw Olms [860]

Briefe eines ehrlichen mannes bey einem wiederholten aufenthalt in weimar : deutschland 1800 – Leipzig: Xenien-Verlag, [1913?] [mf ed 1989] – 65p – 1 – mf#7086 – us Wisconsin U Libr [860]

Briefe eines lebenden / Foerster, Friedrich C – Berlin 1831 [mf ed Hildesheim 1995-98] – 2v on 6mf – 9 – €120.00 – 3-487-27740-9 – gw Olms [860]

Briefe eines reisenden franzosen ueber deutschland an seinen bruder zu paris / [Riesbeck, J K] – [Zuerich], 1783. 2v – 14mf – 9 – mf#HT-267 – ne IDC [914]

Briefe eines reisenden russen / Karamzin, Nikolaj M – Leipzig 1802 [mf ed Hildesheim 1995-98] – 6v on 12mf – 9 – €120.00 – 3-487-27821-9 – gw Olms [860]

Briefe eines suedlaenders / Fischer, Christian A – Leipzig 1805 [mf ed Hildesheim 1995-98] – 3mf – 9 – €90.00 – 3-487-29736-1 – gw Olms [860]

Briefe euber hinter-indien waehrend eines zehnjaehrigen aufenthalts daselbst an seine lieben freunde in europa / Roettger, E H – Berlin: In Commission der Enslinschen Buchhandlung, G W F Meuller, 1844 [mf ed 1995] – (= ser Yale coll) – xvi/320p (ill) – 1 – 0-524-09437-3 – (in german) – mf#1995-0437 – us ATLA [915]

Briefe friedrich leopolds grafen zu stolberg und der seinigen an johann heinrich voss : nach den originalen der muenchener hof- und staatsbibliothek / Stolberg, Friedrich Leopold, Graf – Muenster: Aschendorff 1891 [mf ed 1991] – 1r – 1 – (incl bibl ref. int, suppl & ann ed by otto hellinghaus. filmed with: storm : auswahl aus seinen werken & other titles) – mf#2955p – us Wisconsin U Libr [860]

Briefe friedrich schleiermachers an ehrenfried und henriette von willich geboren von muehlenfels, 1801-1806 – Berlin: Litteraturarchiv-Gesellschaft 1914 [mf ed 1991] – (= ser Mitteilungen aus dem litteraturarchive in berlin: neue folge) – 1mf – 9 – 0-7905-9624-5 – mf#1989-1349 – us ATLA [860]

Briefe (gcsej6) / Gregor von Nazianz (Gregory of Nazianzus, Saint); ed by Gallay, P – 1969 – (= ser Griechische christlichen schriftsteller der ersten jahr- hunderte (gcsej)) – €14.00 – ne Slangenburg [240]

Die briefe heinrichs 4 (mgh deutsches..: 1.bd) – 1937 – (= ser Monumenta germaniae historica. deutsches mittelalter. kritische studientexte (mgh deutsches...)) – €5.00 – ne Slangenburg [931]

Briefe in die heimat aus deutschland, der schweiz und italien / Hagen, Friedrich H von der – Breslau 1818-1821 [mf ed Hildesheim 1995-98] – 4v on 11mf – 9 – €110.00 – 3-487-29305-6 – gw Olms [860]

Briefe in die heimath : geschrieben auf einer reise nach england, italien, der schweiz und deutschland / Wolff, Ludwig – Hamburg 1833 [mf ed Hildesheim 1995-98] – 2v on 4mf – 9 – €120.00 – 3-487-27738-7 – gw Olms [914]

Briefe nach dem westwall : roman / Woerner, Hans – Berlin: Keil c1939 [mf ed 1993] – 1r – 1 – (filmed with: demetrius) – mf#7968 – us Wisconsin U Libr [830]

Die briefe pauli : ihre chronologie, entstehung, bedeutung und echtheit / Maier, Friedrich – 1. & 2. aufl. Muenster i W: Aschendorff 1909 [mf ed 1989] – (= ser Biblische zeitfragen 2/5-6) – 1mf – 9 – 0-7905-0505-3 – mf#1987-0505 – us ATLA [227]

Die briefe pauli an timotheus und titus / Weiss, Bernhard – 7. verb aufl. Goettingen: Vandenhoeck und Ruprecht, 1902 – (= ser Kritisch exegetischer Kommentar ueber das Neue Testament) – 1mf – 9 – 0-524-05065-1 – mf#1992-0318 – us ATLA [227]

Die briefe petri *see* The epistles general of peter

Die briefe petri und der brief judae : theologisch-homiletisch bearbeitet / Fronmueller, G F C – 2. verb. aufl. Bielefeld: Velhagen und Klasing, 1862 – (= ser Theologisch-homiletisches bibelwerk. neuen testamentes 14) – 1mf – 9 – 0-8370-3210-5 – mf#1985-1210 – us ATLA [227]

Briefe ueber das christliche dogma / Schlatter, Adolf von – Guetersloh: Bertelsmann, 1912 – (= ser Beitraege zur foerderung christlicher theologie) – 1mf – 9 – 0-7905-3219-0 – mf#1987-3219 – us ATLA [240]

Briefe ueber deutschland, frankreich, spanien, die balearischen inseln, das suedliche schottland und holland : geschrieben in den jahren 1809 bis 1814 / Holzenthal, Georg – Berlin 1817 [mf ed Hildesheim 1995-98] – 2mf – 9 – €60.00 – 3-487-29908-9 – gw Olms [860]

Briefe ueber die christliche religion / Mueller, F A – Stuttgart: J C Koetzle 1870 [mf ed 1985] – 1mf – 9 – 0-8370-4525-8 – (incl bibl ref) – mf#1985-2525 – us ATLA [225]

Briefe ueber die galanterien von berlin : auf einer reise gesammlet von einem oesterreichischen officier / [Friedel, J] – n.p, 1782 – 5mf – 9 – mf#HT-268 – ne IDC [914]

Briefe ueber die moralitaet der leiden des jungen werthers : eine verloren geglaubte schrift der sturm- und drangperiode / Lenz, Jakob Michael Reinhold; ed by Schmitz-Kallenberg, L – Muenster i.W: F Coppenrath, 1918 [mf ed 1990] – 50p – 1 – (incl bibl ref) – mf#7371 – us Wisconsin U Libr [430]

Briefe ueber einen theil von croatien und italien an caroline pichler / Artner, Maria T von – Halberstadt 1830 [mf ed Hildesheim 1995-98] – (= ser Fbc) – 1v on 2mf – 9 – €60.00 – 3-487-27784-0 – gw Olms [914]

Briefe ueber frankreich auf einer fussreise im jahre 1811 : durch das suedwestliche baiern, durch die schweiz... / Schultes, Joseph A – Leipzig 1815 [mf ed Hildesheim 1995-98] – 2v on 6mf – 9 – €120.00 – 3-487-29694-2 – gw Olms [914]

Briefe ueber freimaurerei : zur aufklaerung fuer alle kreise / Fischer, Robert – 2. veraend aufl. Gera: P Strebel 1875 – 2mf – 9 – mf#vrl-57 – ne IDC [366]

Briefe ueber hamburg und luebek / Merkel, Garlieb H – Leipzig 1801 [mf ed Hildesheim 1995-98] – 3mf – 9 – €90.00 – 3-487-29533-4 – gw Olms [914]

Briefe ueber merkwuerdigkeiten der litteratur / ed by Gerstenberg, Heinrich Wilhelm von et al – Schleswig/Leipzig 1766/67; Hamburg/Bremen 1770 [mf ed Hildesheim 1977] – (= ser Allgemeinwissenschaftliche und literarische zeitschriften des 17. und 18. jahrhunderts) – 8mf – 9 – diazo €42.80 silver €52.80 – gw Olms [860]

Briefe ueber merkwuerdigkeiten der litteratur / ed by Weilen, Alexander von – Heilbronn: Henninger, 1888-90 [mf ed 1993] – (= ser Deutsche litteraturdenkmale des 18. und 19. jahrhunderts 29-30) – 2v – 1 – (series of essays on modern literature (especially on shakespeare and english literature) on the german language, and esthetics) – mf#8676 reel 3-4 – us Wisconsin U Libr [410]

Briefe ueber polen, oesterreich, sachsen, bayern, italien...an die comtesse constance de s... : geschrieben auf einer reise vom monat mai 1807 bis zum monat feb 1808 / Uklanski, Carl T von – Nuernberg 1808 [mf ed Hildesheim 1995-98] – (= ser Fbc) – 7mf – 9 – €140.00 – 3-487-27794-8 – gw Olms [914]

Briefe ueber schweden im jahre 1812 : aus dem daenischen uebersetzt mit anmerkungen und zusaetzen des verfassers / Molbech, Christian – Altona 1818-20 [mf ed Hildesheim 1995-98] – (= ser Fbc) – 3v on 9mf – 9 – €180.00 – 3-487-28940-7 – gw Olms [914]

Briefe ueber seine werke / Flaubert, Gustave – Minden [Germany]: J C C Bruns' Verlag [1904] [mf ed 1999] – 5mf – 9 – 0-665-97407-8 – (in german; trans, int & ann by frederick philip greve) – mf#97407 – cn Canadiana [860]

Briefe ueber zustaende und begebenheiten in der tuerkei aus den jahren 1835-1839 / Moltke, H von – Berlin, 1841 – 5mf – 9 – mf#AR-2015 – ne IDC [956]

Briefe und erklaerung von j. von doellinger ueber die vaticanischen decrete, 1869-87 = Declarations and letters on the vatican decrees, 1869-87 / Doellinger, Johann Joseph Ignaz von – New York: Charles Scribner, 1891 – 1mf – 9 – 0-8370-8502-0 – mf#1986-2502 – us ATLA [240]

Briefe und erklaerungen von j. von doellinger ueber die vaticanischen decrete, 1869-87 / Doellinger, Johann Joseph Ignaz von; ed by Reusch, Franz Heinrich – Muenchen: C H Beck, 1890 – 1mf – 9 – 0-8370-8419-9 – (in german, french, english) – mf#1986-2419 – us ATLA [240]

Briefe und erzaehlungen aus amerika / Thilenius, Klara – Berlin 1849 [mf ed Hildesheim 1995-98] – 1v on 1mf – 9 – €40.00 – 3-487-27011-0 – gw Olms [860]

Briefe und gespraeche veranlasst durch die entfuehrung und gefangenschaftsreise des heiligen vaters pius des siebenten : von rom nach sawonna im juli und august 1809... / ed by Dewora, Viktor Joseph – Hadamar, Koblenz, 1816 (mf ed 1993) – 2mf – 9 – €31.00 – 3-89349-360-3 – mf#DHS-AR 360 – gw Frankfurter [240]

Briefe und schriften / Niebuhr, Barthold Georg; ed by Lorenz, Ludwig – Berlin [1918] (mf ed 1993) – 2mf – 9 – €24.00 – 3-89349-262-3 – mf#DHS-AR 119 – gw Frankfurter [800]

Briefe vom land : ein roman / Ehrler, Hans Heinrich – Stuttgart: Strecker & Schroeder c1918 [mf ed 1989] – 1r – 1 – (filmed with: menschen und affen / albert ehrenstein) – mf#7207 – us Wisconsin U Libr [830]

Briefe von adolf frey und carl spitteler / Frey, Adolf; ed by Frey, Lina – Frauenfeld: Huber 1933 [mf ed 1990] – 1r – 1 – (filmed with: unneren strohdack / friedrich freudenthal) – mf#7272 – us Wisconsin U Libr [860]

Briefe von alexander von humboldt an varnhagen von ense aus den jahren 1827 bis 1858 : nebst auszuegen aus varnhagen's tageuechern, und briefen von varnhagen und andern an humboldt / Humboldt, Alexander von – 3. aufl. Leipzig: F A Brockhaus 1860 [mf ed 1992] – 1r – 1 – (incl bibl ref. filmed with: die verkommenen / max kretzer & other titles) – mf#3183p – us Wisconsin U Libr [500]

Briefe von andreas masius und seinen freunde 1538-1573 (pgrg2) / ed by Lossen, Max – Leipzig, 1886 – (= ser Publikationen der gesellschaft fuer rheinische geschichtskunde (pgrg)) – €19.00 – ne Slangenburg [920]

Briefe von dorothea und friedrich schlegel an die familie paulus / Schlegel, Dorothea von & Schlegel, Friedrich; ed by Unger, Rudolf – Berlin: B Behr (F Feddersen), 1913 [mf ed 1993] – xxviii/192p – 1 – (incl bibl ref and ind) – mf#8676 reel 9 – us Wisconsin U Libr [920]

Briefe von goethes mutter an die herzogin anna amalia / ed by Burkhardt, C A H – Weimar: Goethe-Gesellschaft, 1885 [mf ed 1993] – (= ser Schriften der goethe-gesellschaft 1) – viii/151p – 1 – mf#8657 reel 1 – us Wisconsin U Libr [920]

Briefe von goethes mutter an ihren sohn, christiane und august v goethe – Weimar: Goethe-Gesellschaft, 1889 [mf ed 1994] – (= ser Schriften der goethe-gesellschaft 4) – 1 – (incl bibl ref & ind. int by bernhard suphan) – mf#8657 reel 1 – us Wisconsin U Libr [860]

Briefe von heinrich heine an heinrich laube / ed by Wolff, Eugen – Breslau: S Schottlaender; New York: G E Stechert 1893 [mf ed 1990] – 1r – 1 – (filmed with: akten uber die krankheit von heinrich heines vater & other titles) – mf#2707p – us Wisconsin U Libr [860]

Briefe von johann heinrich voss : nebst erlaeuternden beilagen / Voss, Johann Heinrich; ed by Voss, Abraham – Halberstadt: C Brueggemann 1829-33 [mf ed 1991] – (= ser Bibliothek der deutschen literatur) – 3v on 1r – 1 – mf#2976p – us Wisconsin U Libr [860]

Briefe von johann peter uz an einen freund : aus den jahren 1753-82 / Uz, Johann Peter; ed by Henneberger, August – Leipzig: F A Brockhaus 1866 [mf ed 1991] – 1r – 1 – (filmed with: das lied vom alten eisenhuth / wilhelm utermann) – mf#2933p – us Wisconsin U Libr [860]

Briefe von joseph von goerres an friedrich christoph perthes (1811-1827) / ed by Schellberg, Wilhelm – Koeln 1913 (mf ed 1992) – 2mf – 9 – €24.00 – 3-89349-063-9 – mf#DHS-AR 36 – gw Frankfurter [943]

Briefe von karl viktor von bonstetten an friederike brun – Frankfurt a M 1829 [mf ed Hildesheim 1995-98] – 2v on 6mf – 9 – €120.00 – 3-487-29335-8 – gw Olms [860]

Briefe von ludwig anzengruber : mit neuen beitraegen zu seiner biographie / ed by Bettelheim, Anton – Stuttgart: Cotta, 1902 [mf ed 1988] – 2v – 1 – mf#6951 – us Wisconsin U Libr [860]

Briefe von staegemann, metternich, heine und bettina von arnim : nebst briefen, anmerkungen und notizen von varnhagen von ense – Leipzig: F A Brockhaus 1865 [mf ed 1991] – 1r – 1 – (filmed with: das lied vom alten eisenhuth / wilhelm utermann) – mf#2933p – us Wisconsin U Libr [860]

Briefe von und an gottfried august buerger : ein beitrag zur literaturgeschichte seiner zeit... / Buerger, Gottfried August; ed by Strodtmann, Adolf – Berlin: Gebrueder Paetel 1874 [mf ed 1993] – (= ser Bibliothek der deutschen literatur) – 4v on 1r – 1 – (incl bibl ref & ind) – mf#8530 – us Wisconsin U Libr [860]

Briefe von und an michael bernays – Berlin: B Behr, 1907 [mf ed 1989] – xiv/220p – 1 – (incl bibl & ind) – mf#7010 – us Wisconsin U Libr [860]

Briefe von wilhelm von humboldt an friedrich heinrich jacobi / ed by Leitzmann, Albert – Halle a. S: M Niemeyer, 1892 [mf ed 1991] – viii/141p – 1 – mf#7490 – us Wisconsin U Libr [860]

Briefe waehrend meines aufenthalts in england und portugal an einen freund / Bernard, Esther – Hamburg 1802 [mf ed Hildesheim 1995-98] – (= ser Fbc) – 1v on 3mf [ill] – 9 – €90.00 – 3-487-27777-8 – gw Olms [914]

Briefe zu einer naehern verstaendigung ueber verschiedene meine thesen betreffende puncte : nebst einem namhaften briefe, an den herrn dr. schleiermacher / Harms, Claus – Kiel: Im Verlage der academischen Buchh, 1818 – 1mf – 9 – 0-524-00438-2 – (incl bibl ref) – mf#1989-3138 – us ATLA [240]

Briefe zur erinnerung an merkwuerdige zeiten, und ruehmliche personen : aus dem wichtigen zeitlaufe von 1740, bis 1778 / Borchmann, Johann Friedrich – Berlin: gedruckt mit Spenerschen Schriften 1778 [mf ed 19–] – 1r – 1 – mf#film 706, 1309, 1310 – us Sibley [943]

A brieff discours off the troubles begonne at frankford in germany anno domini 1554 : abowte the booke off off common prayer and ceremonies... / Travers, W – n.p., 1575 – 4mf – 9 – mf#PW-81 – ne IDC [240]

Briefings in bioinformatics – Oxford UK 2000+ – 1,5,9 – ISSN: 1467-5463 – mf#31696 – us UMI ProQuest [574]

Briefings in functional genomics and proteomics – London UK 2002+ – 1,5,9 – ISSN: 1473-9550 – mf#32087 – us UMI ProQuest [575]

Briefings in real estate finance – Hoboken NJ 2002+ – 1,5,9 – ISSN: 1473-1894 – mf#31755 – us UMI ProQuest [332]

Brief-making / Thompson, William Goodrich – Boston, Ellis Co., 1914. 25 p. LL-1103 – 1 – us L of C Photodup [340]

Briefs – Thorofare NJ 1970-83 – 1,5,9 – ISSN: 0007-0068 – mf#8732 – us UMI ProQuest [610]

Briefs on the law of insurance / Cooley, Roger William – St. Paul, West, 1905-19. 7 v. LL-1440 – 1 – us L of C Photodup [346]

Die briefsammlung gerberts von reims (mgh epistolae 2:2.bd) – 1966 – (= ser Monumenta germaniae historica epistolae. 2. die briefe der deutschen kaiserzeit (mgh epistolae 2)) – €17.00 – ne Slangenburg [241]

Briefsammlung trew = Trew letter collection – [mf ed 2006] – 3204mf – 9 – diazo €8900.00 silver €10,680.00 – 3-89131-477-9 – gw Fischer [610]

Briefsammlungen der zeit heinrichs 4 (mgh epistolae 2:5.bd) – 1950 – (= ser Monumenta germaniae historica epistolae. 2. die briefe der deutschen kaiserzeit (mgh epistolae 2)) – €18.00 – ne Slangenburg [241]

Brief.southern methodist university – Dallas TX 1965+ – 1,5,9 – ISSN: 0524-4684 – mf#7057 – us UMI ProQuest [378]

Briefue descriptio de la covrt dv grant tvrc et vng sommaire du regne des othmans auec vn abrege de leurs folles superstitions... / Geuffroy, A – Paris, 1546 – 2mf – 9 – mf#H-8276 – ne IDC [950]

Briefve histoire de la gverre de perse, faite l'an mil cinq cens septante huit... / Porsius, H – [Geneva], 1583 – 1mf – 9 – mf#H-8203 – ne IDC [956]

Briefwechsel : meist historischen und politischen inhalts / ed by Schloezer, August Ludwig – Goettingen 1776-82 – (= ser Dz. historisch-geographische abt) – 10pt on 30mf – 9 – €300.00 – mf#k/n1086 – gw Olms [860]

Briefwechsel balthasar paumgartners : des juengeren mit seiner gattin magdalena, geb behaim, 1582-98 / Paumgartner, Balthasar, der Juengere; ed by Steinhausen, Georg – Stuttgart: Litterarischer Verein 1895 (Tuebingen: H Laupp, Jr) [mf ed 1993] – (= ser Blvs 204) – 58r – 1 – mf#3420p – us Wisconsin U Libr [860]

Briefwechsel balthasar paumgartners, des juengeren : mit seiner gattin magdalena, geb behaim, 1582-1598 / ed by Steinhausen, Georg – Stuttgart: Litterarischer Verein, 1895 (Tuebingen: H Laupp, Jr) [mf ed 1993] – (= ser Blvs 204) – ix/304p – 1 – mf#8470 reel 42 – us Wisconsin U Libr [860]

Briefwechsel der brueder ambrosius und thomas blaurer 1509-1567 / ed by Schiess, T – Freiburg i Br, Ernst Fehsenfeld, 1908-1912. 3 v – 31mf – 9 – mf#PBU-444 – ne IDC [240]

Briefwechsel der familie des kinderfreundes / ed by Weisse, Chr Felix – Leipzig 1784-92 – (= ser Dz) – pt1-12 on 30mf – 9 – €300.00 – mf#k/n675 – gw Olms [370]

Der briefwechsel der schweizer mit den polen / Wotschke, T – Leipzig, M Heinsius Nf, 1908 – 5mf – 9 – mf#PBU-445 – ne IDC [240]

Briefwechsel des grossherzogs carl august von sachsen-weimar-eisenach mit goethe in den jahren von 1775 bis 1828 / Karl August, Grand Duke of Saxe-Weimar-Eisenach – Weimar: Landes-Industrie-Comptoir 1863 [mf ed 1991] – 2v in 1 on 1r – 1 – (filmed with: goethes briefwechsel mit seiner frau / ed by hans gerhard graf) – mf#2782p – us Wisconsin U Libr [920]

Der briefwechsel des mutianus rufus / Mutianus Rufus, Conradus; ed by Krause, Carl – Kassel: A Freyschmidt 1885 [mf ed 2005] – (= ser Zeitschrift des vereins fuer hessische geschichte und landeskunde; n f 9/9 suppl) – 1r – 1 – 0-524-10523-5 – (incl ind) – mf#b00735 – us ATLA [860]

Der briefwechsel hofmannsthal-wildgans / ed by Bradish, Joseph Arno von – ergaenz u. verb neudruck. Zuerich: Franklin Press 1935 [mf ed 1990] – 1r – 1 – (incl bibl ref. filmed with: gestern / hugo von hofmannsthal) – mf#2729p – us Wisconsin U Libr [860]

Der briefwechsel von emanuel geibel und paul heyse / ed by Petzet, Erich – Muenchen: J F Lehmann 1922 [mf ed 1989] – 1r – 1 – (filmed with: gedichte / august geib) – mf#7286 – us Wisconsin U Libr [860]

Briefwechsel zweier deutschen / ed by Pfizer, Paul Achatius – Berlin: B Behr, 1911- [mf ed 1993] – (= ser Deutsche litteraturdenkmale des 18. und 19. jahrhunderts 144, 3. folge n24) – 1 – (with: ziel und aufgaben des deutschen liberalismus by p a pfizer newly ed by georg kuentzel; no more publ?) – mf#8676 reel 9 – us Wisconsin U Libr [430]

Briefwechsel zwischen albrecht von haller und eberhard friedrich von gemmingen : nebst dem briefwechsel zwischen gemmingen und bodmer / ed by Fischer, Hermann – Stuttgart: Litterarischer Verein 189-? (Tuebingen; H Laupp, Jr) [mf ed 1993] – (= ser Blvs 219) – 58r – 1 – (incl bibl ref & ind) – mf#3420p – us Wisconsin U Libr [860]

Briefwechsel zwischen albrecht von haller und eberhard friedrich von gemmingen : nebst dem briefwechsel zwischen gemmingen und bodmer / ed by Fischer, Hermann – Stuttgart: Litterarischer Verein, 189-? (Tuebingen: H Laupp, Jr) [mf ed 1993] – (= ser Blvs 219) – ix/184p – 1 – mf#8470 reel 45 – us Wisconsin U Libr [860]

Briefwechsel zwischen christoph, herzog von wuerttemberg, und petrus paulus vergerius / Kausler, Eduard von & Schott, Theodor [comp] – Stuttgart: Litterarischer Verein 1875 (Tuebingen: H Laupp) [mf ed 1993] – (= ser Blvs 124) – 58r – 1 – (incl bibl ref & ind) – mf#3420p – us Wisconsin U Libr [860]

Briefwechsel zwischen christoph, herzog von wuerttemberg, und petrus paulus vergerius / Vergerio, Pietro Paolo; ed by Kausler, Eduard von & Schott, Theodor – Stuttgart: Litterarischer Verein, 1875 (Tuebingen: H Laupp) [mf ed 1993] – (= ser Blvs 124) – 517p – 1 – mf#8470 reel 26 – us Wisconsin U Libr [860]

Briefwechsel zwischen clemens brentano und sophie mereau / ed by Amelung, Heinz – Leipzig: Insel-Verlag, 1908 [mf ed 1989] – 2v – 1 – mf#7085 – us Wisconsin U Libr [860]

Briefwechsel zwischen george und hofmannsthal / George, Stefan Anton – Berlin: G Bondi [1938] [mf ed 1989] – 1r – 1 – (incl bibl ref. filmed with: der krieg) – mf#7294 – us Wisconsin U Libr [860]

Briefwechsel zwischen gleim und heinse / Gleim, Johann Wilhelm Ludewig; ed by Schueddekopf, Karl – Weimar: Emil Felber 1894-95 [mf ed 1993] – 2v on 1r – 1 – (filmed with: deutsche humoristen aus alter und neuer zeit / julius riffert) – mf#8582 – us Wisconsin U Libr [860]

Briefwechsel zwischen gleim und ramler / Gleim, Johann Wilhelm Ludewig; ed by Schueddekopf, Carl – Stuttgart: Litterarischer Verein 1906-07 (Tuebingen: H Laupp, Jr) [mf ed 1993] – (= ser Blvs 242,244) – 2v on 58r – 1 – (incl bibl ref) – mf#3420p – us Wisconsin U Libr [860]

Briefwechsel zwischen gleim und ramler / ed by Schueddekopf, Carl – Stuttgart: Litterarischer Verein, 1906-07 (Tuebingen: H Laupp, Jr) [mf ed 1993] – (= ser Blvs 242, 244) – 2v – 1 – (incl bibl ref. ann by ed) – mf#8470 reel 50 – us Wisconsin U Libr [860]

Briefwechsel zwischen gleim und uz / Gleim, Johann Wilhelm Ludwig; ed by Schueddekopf, Carl – Stuttgart: Litterarischer Verein, 1899 (Tuebingen: H Laupp, Jr) [mf ed 1993] – (= ser Blvs 218) – xv/553p – 1 – (incl bibl ref and ind) – mf#8470 reel 45 – us Wisconsin U Libr [860]

Briefwechsel zwischen gleim und uz / Gleim, Johann Wilhelm Ludwig; ed by Schueddekopf, Carl – Stuttgart: Litterarischer Verein 1899 (Tuebingen: H Laupp, Jr) [mf ed 1993] – (= ser Blvs 218) – 58r – 1 – (incl bibl ref & ind) – mf#3420p – us Wisconsin U Libr [860]

Briefwechsel zwischen goethe und k goettling in den jahren 1824-1831 / ed by Fischer, Kuno – Muenchen: F Bassermann, 1880 [mf ed 1990] – (= ser Bibliothek der deutschen literatur) – 1r – 1 – (filmed with: goethe als mensch und deutscher / gunther heyd) – mf#2664p – us Wisconsin U Libr [860]

Briefwechsel zwischen goethe und staatsrath schultz / ed by Duentzer, Heinrich – Leipzig: Dyk, 1853 [mf ed 1990] – x/410p/1pl – 1 – (incl bibl ref) – mf#7537 – us Wisconsin U Libr [920]

Briefwechsel zwischen goethe und zelter : 1799-1832 / ed by Fricke, Gerhard – 1. aufl. Nuernberg: H Carl, c1949 [mf ed 1993] – (= ser Geleit des geistes) – 219p/1pl – 1 – mf#8607 – us Wisconsin U Libr [860]

Briefwechsel zwischen h. l. martensen und j. a. dorner, 1839-1881 / Martensen, Hans & Dorner, Isaak August – Berlin: H. Reuther, 1888. Chicago: Dep of Photodup, U of Chicago Lib, 1970 (1r); Evanston: American Theol Lib Assoc, 1984 (1r) – 1 – 0-8370-0269-9 – mf#1984-B130 – us ATLA [920]

Briefwechsel zwischen joseph freiherrn von lassberg und ludwig uhland / Lassberg, Joseph Maria Christoph, Freiherr von; ed by Pfeiffer, Franz – Wien: W Braumueller 1870 [mf ed 1991] – 1r [ill] – 1 – (filmed by: das lied vom alten eisenhuth / wilhelm utermann) – mf#2933p – us Wisconsin U Libr [860]

Briefwechsel zwischen joseph victor von scheffel und paul heyse / Scheffel, Joseph Viktor von; ed by Hoeser, Conrad – Karlsruhe: [s.n.] 1932 [mf ed 1991] – 1r [ill] – 1 – (filmed with: frau aventiure) – mf#2869p – us Wisconsin U Libr [860]

Briefwechsel zwischen karl rosenkranz und varnhagen von ense / ed by Warda, Arthur – Koenigsberg 1926 (mf ed 1992) – 2mf – 9 – €24.00 – 3-89349-069-8 – mf#DHS-AR 35 – gw Frankfurter [943]

Briefwechsel zwischen schiller und goethe – 4. aufl. Stuttgart: J G Cotta 1881 [mf ed 1995] – 2v in 1 on 1r [ill] – 1 – (pref by wilhelm vollmer; incl ind) – mf#3732p – us Wisconsin U Libr [860]

Der briefwechsel zwischen theodor storm und gottfried keller / ed by Koester, Albert – 4th rev enl ed. Berlin: Gebrueder Paetel 1924 [mf ed 1991] – 1r – 1 – (incl ind. filmed with: briefe an seine kinder) – mf#2904p – us Wisconsin U Libr [860]

Briefwechsel zwischen varnhagen und rahel / Varnhagen von Ense, Karl August – Leipzig: F A Brockhaus 1874-75 [mf ed 1993] – (= ser Bibliothek der deutschen literatur) – 6v on 1r – 1 – mf#7774 – us Wisconsin U Libr [860]

Brief-writing and advocacy / Walter, Carroll Gibson – New York, Baker, Voorhis, 1931. 248 p. LL-1258 – 1 – us L of C Photodup [340]

Brieger, Auguste *see* Kain und abel in der deutschen dichtung

Brieger, Theodor *see*
- Constantin der grosse als religionspolitiker
- De formulae concordiae rabisbonensis origine atque indole
- Gasparo contarini und das regensburger concordienwerk des jahres 1541
- Der glaube luthers in seiner freiheit von menschlichen autoritaeten
- Martin luther und wir
- Die reformation
- Der speierer reichstag von 1526 und die religioese frage der zeit
- Die theologischen promotionen auf der universitaet leipzig 1428-1539
- Das wesen des ablasses am ausgange des mittelalters
- Zur geschichte des augsburger reichstages von 1530
- Zwei bisher unbekannte entwuerfe des wormser ediktes gegen luther

Briegisches wochenblatt – Brieg (Brzeg PL), 1794, 1825, 1829 2 oct-25 dec, 1833 1 jul-23 dec, 1845-46 – 3r – 1 – (title varies: 28 sep 1827: briegisches wochenblatt fuer leser aus allen staenden) – gw Misc Inst [077]

Briegisches wochenblatt fuer leser aus allen staenden *see* Briegisches wochenblatt

Briele, Wolfgang van der *see* Ausfahrt und landung

Brien, L *see* Miami valley, ohio families' genealogy, 1933-1939

Brier creek baptist church. wilkes county. north carolina : church records – 1783-1955 – 1 – us Southern Baptist [242]

Brier hill unionist / United Steelworkers of America – v2 n8-v7 n2 [1975 apr-1979 oct/nov] – 1r – 1 – mf#634125 – us WHS [660]

Brierley, Jonathan *see*
- Aspects of the spiritual
- Ourselves and the universe
- Rome from the inside, or, the priests' revolt

Briers, Steven *see* The development of an integrated model of risk

Brieue et claire exposition de la foy chrestienne annoncee par huldrich zwingle...escripte au roy chresten / Zwingli, H – 2mf – 9 – mf#PBU-532 – ne IDC [242]

Brieux, Eugene *see* La femme seule
Brieva, Matias *see*
- Coleccion de leyes...circulares...de la mesta desde el ano 1729 al de 1827
- Coleccion de leyes...ramo de la mesta

Brieve instruction : pour armer tous bons fideles contre les erreurs de la secte commune des anabaptistes / Calvin, J – Geneve: Par Jehan Girard, 1544 – 2mf – 9 – mf#CL-24 – ne IDC [242]
Brieve resolution sur les disputes qui ont este de nostre temps quant aux sacremens... / Calvin, J – Geneve: De l'imprimerie de Conrad Badius, 1555 – 1mf – 9 – mf#CL-52 – ne IDC [240]
De brieven aan de korinthiers / Manen, Willem Christiaan van – Leiden: E J Brill, 1896 [mf ed 1989] – 324p on 1mf – 9 – 0-7905-1455-9 – (in dutch and greek. incl ind) – mf#1987-1455 – us ATLA [227]
Brieven van de classis amsterdam en andere kerkelijke vergaderingen aan de kaapsche kerken (1651-1804) : en verdere archivalia op de geschiedenis van dit tijdvak betrekking hebbende / Spoelstra, Cornelis – Amsterdam: Hollandsch-Africaansche Uitgevers-Maatschappij 1907 [mf ed 1994] – (= ser Bouwstoffen voor de geschiedenis der nederduitsch-gereformeerde kerken in zuid-afrika 2) – 8mf – 9 – 0-524-08804-7 – mf#1993-3296 – us ATLA [242]
Brieven van de kaapsche kerken, hoofdzakelijk aan de classis amsterdam (1655-1804) / Spoelstra, Cornelis – Amsterdam: Hollandsch-Africaansche Uitgevers-Maatschappij 1906 [mf ed 1994] – (= ser Bouwstoffen voor de geschiedenis der nederduitsch-gereformeerde kerken in zuid-afrika 1) – 8mf – 9 – 0-524-08815-2 – mf#1993-3307 – us ATLA [242]
Brieven van...vermaerde en geleerde mannen deser eeuwe : o.a. van j arminius, j uytenbogaert, h de groot, s episcopius, n grevinchoven etc... / Wtenbogaert, J – Amsterdam, 1662. 2pts – 8mf – 9 – mf#PBA-366 – ne IDC [240]
Briffault, Robert *see* Mothers
Brig adventure : journal of voyage from salem towards st kitts – 1807 – 1r – 1 – cn Library Assoc [910]
Brigadeiros e generais de d joao 6 e d pedro 1 / Lago, Laurenio – Rio de Janeiro, Brazil. 1941 – 1r – us UF Libraries [972]
Brigadere, Anna *see* Kvelosa loka
Brigadier don juan sanchez ramirez / Troncoso De La Concha, M De J – Ciudad Trujillo, Dominican Republic. 1944 – 1r – us UF Libraries [972]
Le brigandage de la musique italienne / Goudar, Ange – Amsterdam 1780 [mf ed 19–] – 3mf – 9 – mf#fiche 485 – us Sibley [780]
Brigant, Jacques le *see* Elemens de la langue des celtes gomerites ou bretons
Brigenti, A *see* Villa burghesia vulgo pinciana, poetice descripta ab andrea brigentio patavino
Brigentius, A *see* Villa burghesia-vulgo pinciana poetice descripta ab a b patavino
Briggs, Charles Augustus *see*
- American presbyterianism
- The authority of holy scripture
- The bible, the church, and the reason
- Biblical history
- Biblical study
- The book of ezra
- The case against professor briggs
- The defence of professor briggs before the presbytery of new york, december 13, 14, 15, 19, and 22, 1892
- The ethical teaching of jesus
- The evidence submitted to the presbytery of new york
- The fundamental christian faith
- General introduction to the study of holy scripture
- The higher criticism of the hexateuch
- History of the study of theology
- The incarnation of the lord
- The messiah of the apostles
- The messiah of the gospels
- Messianic prophecy
- New light on the life of jesus
- Origin and history of premillenarianism
- The papal commission and the pentateuch
- Theological symbolics
- The virgin birth of our lord
- Whither?
- Who wrote the pentateuch?

Briggs, Charles Augustus et al *see* Inspiration and inerrancy
Briggs, George Ware *see* The bow in the cloud
Briggs, George Weston *see*
- The chamars
- Gorakhnath and the kanphata yogis

Briggs, Horace *see* Letters from alaska and the pacific coast
Briggs, John *see*
- India and europe compared
- A letter on the indian army
- Letters addressed to a young person in india
- The present land-tax in india considered as a measure of finance

Briggs, John C *see* Applegate's mineral springs
Briggs, Lawrence Palmer *see* Ancient khmer empire
Briggs, Martin Shaw *see*
- Homes of the pilgrim fathers in england and america
- Wren

Briggs, Robert Alexander *see* Bungalows and country residences
Briggs, S R *see* New notes for bible readings
Brigham, William Tufts *see* Guatemala
Brigham Young University *see*
- Buffalo hide
- Eagle's eye
- Mormon americana

Brigham young university education and law journal – v1-10. 1992-2 – 9 – $100.00 set – (title varies: v1-1992 as brigham young university journal of law and education) – mf#114631 – us Hein [344]
Brigham young university journal of law and education *see* Brigham young university education and law journal
Brigham young university journal of public law – v1-15. 1986-2 – 9 – $160.00 set – ISSN: 0896-2383 – mf#111971 – us Hein [342]
Brigham young university law review – 1975-2 – 9 – $496.00 set – ISSN: 0360-151X – mf#101061 – us Hein [344]
Brigham young university law review [1978] – Provo UT 1978+ – 1,5,9 – ISSN: 0360-151X – mf#11886 – us UMI ProQuest [340]
Brigham young university studies – Provo UT 1959-79 – 1,5,9 – (cont by: byu studies) – ISSN: 0277-7363 – mf#8942.02 – us UMI ProQuest [370]
Brigham young university studies [1984] – Provo UT 1984+ – 1,5,9 – (cont: byu studies) – mf#8942.02 – us UMI ProQuest [378]
Brighouse news – Brighouse, England 2 jul 1870-27 dec 1911 (wkly) [mf 1897] – 1 – (wanting: 1909; incorp with: brighouse observer) – uk British Libr Newspaper [072]
Brighouse & rastrick gazette – Brighouse, England 24 mar 1874-22 sep 1894 (wkly) [mf 1879, 1881, 1896] – 1 – (cont by: brighouse gazette & local railway guide [29 sep 1894-28 jan 1899]) – uk British Libr Newspaper [072]
Bright, J S *see* Before and after independence
Bright, Jagat S *see*
- Important speeches and writings of subhas bose
- Important speeches of jawaharlal nehru
- India on the march
- President kripalani and his ideas
- Subhas bose and his ideas
- The woman behind gandhi

Bright, James Wilson *see*
- The gospel of saint john in west-saxon
- The gospel of saint matthew in west-saxon

Bright, John *see*
- Production and markets
- Selected speeches of the rt. honble. john bright, m.p., on public questions
- The speech delivered by the right hon john bright, mp

Bright side / Forever His Ministries International [Jacksonville FL] – 1993 apr – 1r – 1 – mf#4114380 – us WHS [243]
Bright skies and dark shadows / Field, Henry M – New York, NY. 1890 – 1r – us UF Libraries [978]
Bright, William *see*
- The age of the fathers
- Ancient collects and other prayers
- The canons of the first four general councils of nicaea, constantinople, ephesus and chalcedon
- Chapters of early english church history
- Evening communions contrary to the church's mind, and why
- A history of the church
- Lessons from the lives of three great fathers
- Liber precum publicarum ecclesiae anglicanae
- The roman see in the early church
- Selected letters of william bright
- Some aspects of primitive church life
- Waymarks in church history

Brightly, Frederick C *see* Brightly's leading cases on the law of elections
Brightly's leading cases on the law of elections / Brightly, Frederick C – Philadelphia: Kay & Brother, 1871 (all publ) – 9mf – 9 – $13.50 – mf#LLMC 95-049 – us LLMC [340]
Brightly's reports / Pennsylvania. Nisi Prius. Supreme Court – 1v. 1809-1851 (all publ) – (= ser Pre-nrs nominative reports) – 6mf – 9 – $9.00 – mf#LLMC 95-053 – us LLMC [340]
Brightman, Frank Edward *see*
- The christian platonists of alexandria
- The english rite
- Liturgies, eastern and western
- Liturgies eastern and western, vol 1

Brightman, T *see* A revelation of the apocalyps
Brighton 1771-1873 – Oxford, MA (mf ed 1985) – (= ser Massachusetts vital records) – 33mf – 9 – 0-931248-86-8 – (mf 1: marriages, intents,...1771-1817. mf 2-7: b,m,d 1771-1873. mf 8-10: index to b,m,d 1771-1873. mf 11-20: b,m,d 1771-1874. mf 21: births & deaths 1817-45. mf 22-24: marriages & intentions 1817-58. mf 25-26: b,m,d 1843-55. mf 27-28: births 1855-73. mf 29-30: marriage intentions 1858-73. mf 31: marriages 1855-73. mf 32-33: deaths 1854-73) – us Archive [978]
Brighton convention and its doctrinal teaching – London, England. 1875 – 1r – us UF Libraries [240]
Brighton daily news – Brighton, England 1 jan 1870-31 may 1880 [mf jan-dec 1874, jun-dec 1875, may-aug 1876] – 1 – (incorp with: argus; cont: brighton daily news & south sussex gazette [2 nov 1868-31 dec 1869]) – uk British Libr Newspaper [072]
Brighton gazette – Brighton, England 6 jan 1825-1 may 1926, nov 1933-29 mar 1938 (wkly) – 1 – (wanting: 1897, 1898; cont by: brighton & hove gazette [2 apr 1938-12 nov 1965]) – uk British Libr Newspaper [072]
Brighton guardian – Brighton, England 31 jan 1827-25 dec 1895 (wkly) [mf 1871,1897] – 1 – (cont by: brighton & hove guardian & visitors' register [24 feb-15 sep 1897, 5 jan 1898-25 sep 1901]) – uk British Libr Newspaper [072]
Brighton institution, for promoting the fine arts : founded 1st of june...1820 – Brighton, 1820 – (= ser 19th c art & architecture) – 1mf – 9 – mf#4.2.1692 – uk Chadwyck [700]
Brighton, John George *see* Admiral of the fleet
Brighton patriot and lewes free press – [London & SE] East Sussex, Brighton Ref Lib 1835-39 – 2r – 1 – uk Newsplan [072]
Brigitta : erzaehlung / Auerbach, Berthold – Boston, MA: Ginn, c1908 [mf ed 1993] – (= ser International modern language series) – viii/165p – 1 – (int and ann by j howard gore) – mf#8464 – us Wisconsin U Libr [830]
Brigitta und andere erzaehlungen / Stifter, Adalbert – complete ed. Wiesbaden: Agrippina-Buecherei [19–?] [mf ed 1995] – 1r – 1 – (filmed with: abdias) – mf#3748p – us Wisconsin U Libr [830]
Brigitte – Hamburg, Germany 1968-73 – 1 – mf#5101 – us UMI ProQuest [770]
Brigitte *see* Dies blatt gehoert der hausfrau
Brigstocke, Frederick Hervey John *see*
- A paper on the revised version of the new testament
- Six branches of the missionary work of the church set forth as subjects for meditation during the week of intercession for missions, 1878
- Subjects for meditation during the week of intercession for missions 1877

Brigstocke, Frederick Hervey John [comp] *see* History of trinity church, saint john, new brunswick, 1791-1891
Brijon, C R *see* L'apollon moderne
Brill, Bernhard *see* Ein beitrag zur kritik von lessings laokoon
Brill, Hirsh *see* Shaul
Brill news – [Philadelphia]: J G Brill Co [v1 n1-13 (dec1913-1914)] (mthly) – 1r – 1 – us CRL [360]
Brill, Patricia A *see* Personality traits, cardiovascular fitness, and mortality in men
Brill, Robert H *see*
- Scientific investigations of ancient glasses and lead-isotope studies

Die brillantenkoenigin : original lebensbild mit gesang und tanz in 5 abtheilungen / Kaiser, Friedrich – Wien: Gustav Schoenwetter, 1874 [mf ed 1995] – 60p – 1 – mf#8797 – us Wisconsin U Libr [790]
Brillhart, John A *see* A pictorial history of the brillharts of america
Brillion news – Brillion, Forest Junction WI. 1894 sep 7/1895-2 jul/dec – 86r – 1 – (with gaps; cont: reedsville banner) – mf#1133666 – us WHS [071]
Bril's speshel = Bril's special – London, England 9 aug 1901-30 may 1902 – 1 – (in yiddish) – uk British Libr Newspaper [072]
Bril's telefon = Bril's telephone – London, England 6 aug 1901-8 jun 1902 – 1 – (in yiddish) – uk British Libr Newspaper [072]
Brimfield 1696-1893 – Oxford, MA (mf ed 1991) – (= ser Massachusetts vital records) – 83mf – 9 – 0-87623-134-2 – (mf 1-8: vital records 1696-1844. mf 9-12: vital records 1700-1850. mf 13-14: marr & birth 1724-1886. mf 15: marriage index 1724-1886. mf 15-26: proprietors 1730-1824. mf 27-41: town records 1730-1829: mf 42-48: selectmen 1766-1839. mf 49-52: property lists 1798. mf 53-59: tax valuations 1831-54. mf 60-61: school records 1825-61. mf 62-63: voter register 1847-82. mf 64: voter register 1877, 1889. mf 65-66: soldiers 1861-65. mf 67-70: paupers 1841-1917. mf 71-73: vital records 1843-55. mf 74-77: vitals index 1856-92. mf 78-79: births 1856-93. mf 80: intentions 1856-93. mf 81: marrs 1856-93, 1771-99. mf 82-83: deaths 1856-93) – us Archive [978]
Brimfield 1716-1849 – Oxford, MA (mf ed 1995) – (= ser Massachusetts vital record transcripts to 1850) – 9mf – 9 – 0-87623-217-9 – (mf 1t-2t: marriages 1726-1824. mf 2t-5t: births 1698-1838. mf 5t-6t: deaths 1730-1825. mf 6t: marriages 1814-46; out-of-town marriages 1727-98. mf 6t-7t: births 1795-1849. mf 7t: deaths 1813-44. mf 8t: publishments 1826-49; births 1844-49. mf 9t: marriages 1844-49; deaths 1844-49) – us Archive [978]
Brimstone : a journal / Ancient Brotherhood of Satan – 1989 apr-1991 jan – 1r – 1 – mf#1538797 – us WHS [130]
Brin Dumm *see*
- Nayopay paccuppann knun lok nin knun sruk khmaer
- Sangram nin santibhab
- Sangram pah por nyn pativattn

Brinckerhoff, Isaac W *see* The spirit of christ
Brinckman, Arthur *see*
- The controversial methods of romanism
- The controversial statistics of romanism
- Controversial statistics of romanism
- Notes on the papal claims

Brinckman, John *see*
- John brinckmans hoch- und niederdeutsche dichtungen
- John brinckmans plattdeutsche werke
- Kasper-ohm un ick
- Kleinere erzaehlungen

Brinckman-buch : john brinckmans leben und schaffen / Weltzien, Otto – Hamburg: R Hermes, 1914 – (= ser Niederdeutsche buecherei) – 112p (ill) – 1 – (incl bibl) – mf#7088 – us Wisconsin U Libr [920]
Brinckmann, John *see* John brinckmans saemtliche werke in fuenf teilen
Brinckmeier, Eduard *see* Louis napoleon bonaparte, praesident der franzoesischen republik
Brindley, William *see* Ancient sepulchral monuments...
Briner, Megan A *see* A comparison of perceived health and quality of life between cardiac rehabilitation participants and nonparticipants
Briney, J B *see* Otey-briney debate
Briney, John Benton *see*
- The form of baptism
- Instrumental music in christian worship
- The relation of baptism to the remission of alien sins
- The temptation of christ

Briney, John Benton et al *see* Churches of christ
Briney-taylor debate: "the church of the new testament." – 30 Mar-8 Apr 1881. Oakland Station, KY. Bowling Green District. 78p – 1 – us Southern Baptist [242]
Bring 'em back petrified / Brown, Lilian Maclaughlin – New York, NY. 1956 – 1r – us UF Libraries [972]
Bringer, Joy D *see* Psychosocial precursors to athletic injury in adolescent competitive athletes
Bringing criminal debt into balance : improving fine and restitution collection – Washington: FJC, n.d. (1992?) – 1mf – 9 – $1.50 – mf#LLMC 95-385 – us LLMC [345]
Bringing in sheaves / Earle, Absalom Backas – 10th thousand. Boston: James H Earle 1869 [mf ed 1984] – 5mf – 9 – 0-8370-0228-1 – mf#1984-0049 – us ATLA [242]
Bringing in the sheaves : gleanings from the mission fields of the christian and missionary alliance – [s.l: s.n.] 1898 [mf ed 1992] – (= ser Christian & missionary alliance coll) – 1mf – 9 – 0-524-02247-X – mf#1990-4254 – us ATLA [240]
Bringing the state back in – New York, NY. 1985 – 1r – us UF Libraries [025]
Brink, Benjamin D *see* The body builder: robert j. roberts
Brink, Bernhard Aegidius Konrad Ten *see* Geschichte der englischen litteratur
Brink, Carel Frederik *see* Journals of brink and rhenius
Brink, J W *see* Welke is de school voor onze kinderen?
Brink, Jan ten *see* Ostindische damen und herren
Brink, Jeanetha *see* Legal issues arising from the year 2000 problem in computer software
Brink, T L *see* Clinical gerontologist
Brink, Yvonne *see* Places of discourse and dialogue
Brinkerhoff *see* 30 ovi history
Brinkley 1599-1950 – (= ser Cambridgeshire parish register transcript) – 4mf – 9 – £5.00 – uk CambsFHS [929]
Brinkley, Harry John *see* Study of the effect of the length of the daily light period
Brinkley, John Richard *see*
- John richard brinkley vs. kansas state board of medical registration and examination, et al
- John richard brinkley vs. the kansas city star

Brinkmann, Friedrich *see* Studien und bilder aus sueddeutschem land und volk
Brinsley, John *see* A consolation for our grammar schooles

Brinsmade news – Brinsmade, ND: The Leeds News. v1 n1 sep 2 1926-jun 21 1928?// (wkly) – (= ser Leeds news) – 1 – (some iss misnumbered; iss with: leeds news (leeds, nd); missing: 1926 sep 16-30; 1927 jan 13, mar 24; 1928 may 17, jun 14) – mf#07376 – us North Dakota [071]
Brinsmade news *see* Leeds news
The brinsmade star – Brinsmade, ND: John Lindelien. v1 n1 may 31 1906-v2 n35 nov 6 1924 (wkly) – 1 – (suspended publ after oct 2 1919 (v14 n19 and resumed on mar 8 1923 (new v1 n1). missing: 1907 sep 12; 1908 nov 19; 1912 sep 5; 1913 dec 25; 1916 dec 28;1924 aug 7) – mf#11168-11171 – us North Dakota [071]
Brinton, Daniel Garrison *see*
– Aboriginal american authors and their productions
– American hero-myths
– The cradle of the semites
– Essays of an americanist
– Giordano bruno
– The language of palaeolithic man
– Lenape and their legends
– The myths of the new world
– Religions of primitive peoples
– The religious sentiment
Brinton, Maria *see* Effects of posture specific therapeutic exercise
Briny budget – Honolulu HI. 1901 aug 1 – 1r – 1 – mf#850511 – us WHS [071]
Brio – Edinburgh UK 1974+ – 1,5,9 – ISSN: 0007-0173 – mf#10198 – us UMI ProQuest [020]
Brion, Friederike-Elisabeth *see* Das haideroeslein von sesenheim
Briones, Mariano *see*
– La juventud anarquista: factor determinativo de la guerra y de la revolucion
– Totius terrae sanctae vrbivmqve et qvicqvid in eis memoria dignum actum gestumue fuit:...
Briquet, C M *see* Les filigraines
Briquet, Pierre de *see* Code militaire
Briquette – Scranton, ND: Scranton Pub Co. v10 n1 sep 13 1917-oct 20 1921?// (wkly) – 1 – (official paper of bowman county 1917-1919; official paper of scranton village (later: official city paper) 1917-1921; cont: scranton register; missing: 1920 apr 1) – mf#09479 – us North Dakota [071]
[Brisbane-] bee-democrat – CA. sep 28 1961-mar 13 1980 – 16r – 1 – $960.00 – (cont by: san bruno herald) – mf#B02078 – us Library Micro [071]
Brisbane courier – Brisbane, Australia 11 apr 1864-26 aug 1933 – 1 – (amalg with: daily mail & subsequently publ as: courier-mail) – uk British Libr Newspaper [079]
Brisbane courier – Brisbane, 1846-1996 – 864r – 1 – at Pascoe [079]
Brisbane daily mail *see*
– Daily mail
[Brisbane-] sun – CA. 1937-38 (wkly) – 1r – 1 – $60.00 – mf#B02079 – us Library Micro [071]
Brisbane, William *see*
– Account of my travels
– Receipt book
– Travel account
Brisbane's travels 1801-1807 *see* Account of my travels
Briscoe, John Potter *see* A contribution towards a bibliography of hosiery and lace, etc
Brisebarre, Edouard *see*
– Marie au second, garcon au cinquieme
– Menage de rigolette
Brisebois, Raymond *see*
– Decouvreurs et pionniers
– L'epopee canadienne
Briseno Sierra, Humberto *see*
– Arbitraje en el derecho privado
– Derecho procesal fiscal
Briseux, C E *see*
– L'aart de batir des maisons de campagne...
– Traite du beau essentiel dans les arts...avec un traite des proportions harmoniques...et les cinq ordres d'architecture
[Briseux, C E] *see* Architecture moderne o- l'art de bien batir...
Brisgovius, Huserus *see* Opera
Brissaud, Jean Baptiste *see* Manuel d'histoire du droit francais (sources, droit public, droit prive) a l'usage des etudiants en licence et en doctorat
Brisson papers : tahitian and other manuscripts formerly in the possession of captain victor brisson – 1862-1928 – 1r – 1 – mf#PMB1034 – at Pacific Mss [980]
Brisson, Pierre R de *see*
– Histoire du naufrage et de la captivite de m de brisson, officier de l'administration des colonies
– Perils and captivity
Brisson, Roger *see* Journal of internet cataloging
Brissot de Warville, Jacques *see* Nouveau voyage dans les etats-unis de l'amerique septentrionale, fait en 1788
Brissot de Warville, Jacques Pierre *see* Memoires de brissot
Bristed, John *see* America and her resources
Bristol adventurer and weekly news – England.5 May 1922-19 Oct 1923. -w. 1 reel – 1 – uk British Libr Newspaper [072]
Bristol advertiser evening telegram – England.Jul 1875-Jul 1876. -d. 2 reels – 1 – uk British Libr Newspaper [072]
Bristol advocate – England. -w. 17 Sep 1836-11 Feb 1837. (31 ft) – 1 – uk British Libr Newspaper [072]
Bristol and bath magazine : or instructive and entertaining miscellany – La Salle IL 1782-83 – 1 – mf#5262 – us UMI ProQuest [073]
Bristol and clifton amusements – England.Dec 1900-1903.-w. 2 reels – 1 – uk British Libr Newspaper [072]
Bristol and kingswood herald – London, UK. jul 1874-jan 1876 [wkly] – 1 – (aka: bristol district herald, feb 1875-jan 1876) – uk British Libr Newspaper [072]
Bristol and west of england advertiser *see* West of england advertiser
Bristol district herald *see* Bristol and kingswood herald
Bristol evening news – [SW England] Bristol 29 may 1877-jan 1932 [mf ed 2004] – 286r – 1 – (missing: jul-dec 1897, jul-dec 1898, 1911, jul-dec 1912; discontinued) – uk Newsplan [072]
Bristol evening post – Bristol, England 18 apr 1932-27 jan 1962 – 1 – (cont by: evening post [29 jan 1962-]) – uk British Libr Newspaper [072]
Bristol first – England.27 Nov 1923; 5 Jan-4 Apr 1925.-w. 1/2 reel – 1 – uk British Libr Newspaper [072]
Bristol, Frank Milton *see*
– The life of chaplain mccabe, bishop of the methodist episcopal church
– Providential epochs
Bristol free press – Bristol, FL. 1932-1989 – 47r – (gaps) – us UF Libraries [071]
Bristol gazette – [SW England] Bristol 24 dec 1767, 8 sep 1768 (imperfect) [mf ed 2003] – 1 – (missing: some early yrs; cont by: bristol gazette & public advertiser [28 mar-23 may, 6 jun 1771-11 jan 1781, 5 dec 1782, 20 feb 1783, 29 jan 1784, 23 feb, 4 may, 7 dec 1786, 18 jan 1787-25 dec 1800, 7 jan 1802-23 may 1872]) – uk Newsplan [072]
Bristol institution for the promotion of literature, science, and the fine arts... : third exhibition – Bristol 1826 – (= ser 19th c art & architecture) – 1mf – 9 – mf#4.2.1693 – uk Chadwyck [700]
Bristol liberal and west of england commercial and general advertiser – England. -w. 23 Jul 1831-3 Mar 1832. (21 ft) – 1 – uk British Libr Newspaper [072]
Bristol medico-chirurgical journal – Bristol UK 1883-1989 – 1,5,9 – (cont by: west of england medical journal) – ISSN: 0308-6356 – mf#2348.01 – us UMI ProQuest [617]
Bristol mercury – England. 1831-32.-w. 1 reel – 1 – uk British Libr Newspaper [072]
Bristol mercury and universal advertiser – [SW England] Bristol 1 mar 1790-30 nov 1909 [mf ed 2003] – 243r – 1 – (missing: sep-dec 1887, 1889, sep-dec 1891, 1897; cont as: bristol mercury [jan 1812-dec 1818]; bristol mercury and monmouthshire, south wales and west of england advertiser [jan 1819-dec 1877]; bristol mercury and daily post, western counties, monmouthshire and south wales advertiser [26 jan 1878-oct 1890]; bristol mercury, daily post, western counties and south wales advertiser [nov 1890-dec 1901]; bristol daily mercury, daily post, western counties and south wales advertiser [jan 1902-nov 1909]) – uk Newsplan [072]
Bristol observer – Bristol, England 12 jul 1862-1 jun 1962 [mf 1898] – 1 – (wanting: 1896,1897,1912; discontinued) – uk British Libr Newspaper [072]
Bristol paper – England.Feb 1932-Jan 1933. -w. 1 reel – 1 – uk British Libr Newspaper [072]
Bristol presentments, 1770-1917 : from the central reference library, college green, bristol – (= ser British records relating to america in microform) – 32r – 1 – (with guide. int by w e minchinton) – mf#97290 – uk Microform Academic [970]
Bristol record society – v1-19. 1930-55 – (= ser Publications of the english record societies, 1835-1972) – 65mf – 9 – uk Chadwyck [941]
Bristol standard – Bristol, England 23 jan 1839-27 jan 1842 (wkly) – 1r – 1 – (discontinued) – uk British Libr Newspaper [072]
Bristol times and bath advocate – [SW England] Bristol 2 mar 1839-26 mar 1853 [mf ed 2004] – 1 – (cont by: bristol times, & felix farley's bristol journal [2 apr 1853-31 dec 1864]) – uk Newsplan; uk British Libr Newspaper [072]
Bristol times, and felix farley's bristol journal – Bristol, England 2 apr 1853-31 dec 1864 – 1 – (cont: bristol times & bath advocate [2 mar 1839-26 mar 1853]; cont by: daily bristol times & mirror (ns) 5 jan 1865-31 dec 1883) – uk British Libr Newspaper [072]
Bristol, William W *see* Bristol's compilation; mechanics' lien law, torrens land title, builders' directory
Bristol's compilation; mechanics' lien law, torrens land title, builders' directory / Bristol, William W – Chicago, 1896. 189 p. LL-7 – 1 – us L of C Photodup [346]
Bristow enterprise – Bristow, NE: R O Willis. -v11 n54. feb 18 1932 (wkly) [mf ed v7 n47. apr 24 1908-feb 18 1932 (gaps)] – 8r – 1 – (suspended oct 1918; resumed with v1 n1 jan 6 1919. issued with: lynch herald feb 4-18 1932. merged with: lynch herald to form: herald-enterprise) – us NE Hist [071]
Bristow, Joseph L *see* Joseph l. bristow papers
Bristow, Lewis J *see* Scrapbook, southern baptist hospitals
Britain and america, the lost israelites : or, the ten tribes identified in the anglo-celtic race / McKillop, Peter S – St Albans, Vt: [s.n.], 1902 – 2mf – 9 – 0-524-08181-6 – mf#1992-1167 – us ATLA [572]
Britain and europe since 1945 – 2090mf coll – 9 – (comprising publ and documents of various groups and organizations. enlarged and updated annually. with guide. basic set 1945-72 296mf c39-27591) – mf#C39-27590 – us Primary [941]
Britain and her colonies / Hurlbert, Jesse Beaufort – London: E Stanford, 1865 [mf ed 1984] – 4mf – 9 – 0-665-45254-3 – (incl ind and bibl) – mf#45254 – cn Canadiana [320]
Britain and her colonies / Hurlbert, Jesse Beaufort – London, 1865 – (= ser 19th c british colonization) – 3mf – 9 – mf#1.1.3786 – uk Chadwyck [330]
Britain and her treaties on belize / Mendoza, Jose Luis – Guatemala, 1947 – 1r – us UF Libraries [972]
Britain and south africa / Austin, Dennis – London, England. 1966 – 1r – us UF Libraries [327]
Britain and the west indies / Whitson, Agnes Mary – London, England. 1948 – 1r – us UF Libraries [972]
Britain redeemed and canada preserved / Wilson, F A & Richards, Alfred Bate – London 1850 – (= ser 19th c british colonization) – 2pt on 7mf – 9 – mf#1.1.10031 – uk Chadwyck [337]
Britain's art paradise : or, notes on some pictures in the royal academy / Southesk, James Carnegie, earl of – Edinburgh 1871 – (= ser 19th c art & architecture) – 1mf – 9 – mf#4.2.1467 – uk Chadwyck [700]
Britain's legislation on education / Kerr, James – Greenock, Scotland. 1872 – 1r – us UF Libraries [240]
Britannia – [NW England] Manchester aug 1834-dec 1838 [mf ed 2002] – 2r – 1 – uk Newsplan [072]
Britannia : oder neue englische miszellen – Stuttgart 1825-27 [mf ed Hildesheim 1995-98] – 10v on 25mf – 9 – €250.00 – 3-487-27912-6 – gw Olms [914]
Britannia. 1-7. 18 oct 1912-20 dec 1918 / Women's Social and Political Union – 1 – 54.00 – us L of C Photodup [360]
Britannia and trades advocate – Hobart, TZ. 1846-55 – 3r – 1 – A$115.20 vesicular A$132.00 silver – at Pascoe [073]
Britannia and trades' advocate – Hobart, Australia 7 jan 1847-26 dec 1850 – 2r – 1 – uk British Libr Newspaper [380]
Britannia waives the rules / Culwick, Arthur Theodore – Cape Town, South Africa. 1963 – 1r – us UF Libraries [960]
Britannien und der krieg / Franz, Wilhelm – 2. verm aufl. Tuebingen: Kloeres 1915 [mf ed 1987] – (= ser Durch kampf zum frieden 12/13) – 1r – 1 – mf#6840 – us Wisconsin U Libr [941]
Britanny and the bible / Hope, I – London, England. 1852 – 1r – us UF Libraries [240]
Britian and the united states in the caribbean / Proudfoot, Mary Macdonald – London, England. 1954 – 1r – us UF Libraries [972]
Der britische feldzug nach abessinien / Hozier, H M – Berlin, 1870 – 3mf – 9 – mf#NE-20279 – ne IDC [956]
Britischen Militaerbehoerde *see* Neue rheinische zeitung
Britischen Weltnachrichtendienst *see* Weltpresse
British 20th century war art / Imperial War Museum. London – 122mf – 9 – $1010.00 – 0-907006-53-1 – (1500 works fr 1st ww coll; over 5000 works fr 2nd ww; each work has caption-details of artist, title, medium, dimensions & museum number; sequence is alphabetical by artist; with printed ind to artists) – uk Mindata [700]
The british achievement in india : a survey / Rawlinson, Hugh George – London: William Hodge & Co, 1948 – (= ser Samp: indian books) – us CRL [954]
British administration german new guinea government gazette – 15 oct 1914-27 nov 1919 – 1r – 1 – mf#pmb doc325 – at Pacific Mss [350]
British africa : internal affairs and foreign affairs, 1945-1959 / U.S. State Dept – (= ser Confidential u s state department central files) – 1 – $15,065.00 coll – (1945-49 14r isbn 1-55655-405-2 $2705. 1950-54 29r isbn 1-55655-406-0 $5610. 1955-59 39r isbn 1-55655-407-9 $7535. with p/g) – us UPA [327]
British aid statistics 1966-1973/77 – (= ser British government publications...1801-1977) – 16mf – 9 – uk Chadwyck [338]
British almanac companion – Killen TX 1828-88 – 1 – mf#4794 – us UMI ProQuest [941]
British america : arguments against a union of the provinces reviewed: with further reasons for confederation / McCully, Jonathan – London: F Algar, 1867 [mf ed 1981] – 1mf – 9 – mf#23468 – cn Canadiana [971]
British america : lectures delivered at the south place institute, finsbury, from 1895 to 1898 – London: K Paul, 1900 [mf ed 1979] – (= ser The british empire series) – 7mf – 9 – 0-665-00875-9 – mf#00875 – cn Canadiana [971]
British america : outline history of the grand lodge of canada, in the province of ontario / Robertson, John Ross – [S.l: s.n, 189-?] [mf ed 1983] – 1mf – 9 – mf#28894 – cn Canadiana [360]
British america assurance company, toronto, canada : incorporated 1833 – [Toronto?: s.n, 18–?] [mf ed 1982] – 1mf – 9 – mf#38064 – cn Canadiana [368]
British america, vol 1 / MacGregor, John – 2nd ed. Edinburgh; W Blackwood: London: T Cadell, 1833 [mf ed 1984] – 7mf – 9 – 0-665-42103-6 – mf#42103 – cn Canadiana [970]
British america, vol 1 / MacGregor, John – Edinburgh: W Blackwood; London: T Cadell, 1832 [mf ed 1983] – 6mf – 9 – 0-665-36845-3 – (incl bibl ref) – mf#36845 – cn Canadiana [917]
British america, vol 2 / MacGregor, John – 2nd ed. Edinburgh; W Blackwood: London: T Cadell, 1833 [mf ed 1984] – 7mf – 9 – 0-665-42104-4 – mf#42104 – cn Canadiana [971]
British america, vol 2 / MacGregor, John – Edinburgh: W Blackwood; London: T Cadell, 1832 [mf ed 1983] – 7mf – 9 – 0-665-36846-1 – mf#36846 – cn Canadiana [917]
British america, vols 1-2 / MacGregor, John – 2nd ed. Edinburgh; W Blackwood: London: T Cadell. 2v. 1833 – 1mf – 9 – 0-665-42102-8 – mf#42102 – cn Canadiana [970]
British america, vols 1-2 / MacGregor, John – Edinburgh: W Blackwood; London: T Cadell. 2v. 1832 – 1mf – 9 – mf#36844 – cn Canadiana [970]
British americain / Macgregor, John – Edinburgh 1832 [mf ed Hildesheim 1995-98] – 2v on 7mf – 9 – €140.00 – 3-487-27092-7 – gw Olms [910]
British american book and tract society reporter – Halifax, NS: [The Society, 1873-187- or 18–] [mf ed v1 n4 jul 1873] – 9 – mf#P05101 – cn Canadiana [240]
British american cultivator *see* The canada farmer
The british american cultivator – Toronto: J Eastwood & W G Edmundson, 1842-[1847] – 9 – mf#P04019 – cn Canadiana [630]
British American Friendly Society of Canada *see* Act of incorporation, bye-laws, rules and regulations
British american friendly society of canada : capital stock £100,000; head office, montreal, with branch offices and agencies in nearly every city and town in british north america – [Montreal?: s.n.], 1855 [mf ed 1984] – 1mf – 9 – 0-665-48278-7 – mf#48278 – cn Canadiana [366]
British american journal *see* Niagara peninsula newspapers, pt 2
The british american journal / ed by Hall, Archibald – Montreal: J Lovell, [1860-1862] [mf ed v1 [n1 jan 1860]-[v3 n12 dec 1862] – 9 – mf#P05183 – cn Canadiana [610]
British American Land Company *see* Lands for sale, in the eastern townships of lower canada
British American League Hamilton Branch *see* Address of the hamilton branch of the british american league
The british american medical and physical journal – Montreal: W Salter, [1850-1852?] [mf ed new ser: v6 n1 may 1850-v7 n8 [i.e. 9] jan 1852] – 9 – (incl some french text) – mf#P05181 – cn Canadiana [610]
British american presbyterian – Toronto: C B Robinson, [1872-1877] [mf ed v1 n1 feb 2 1872-v6 n299 oct 26 1877] – 9 – mf#P06066 – cn Canadiana [242]
British american union : a review of hon joseph howe's essay, entitled "confederation considered in relation to the interests of the empire" / Hamilton, Pierce Stevens – Halifax, NS?: A Grant, 1866 – 1mf – 9 – mf#23330 – cn Canadiana [323]

British and american diplomacy affecting canada, 1782-1899 : a chapter of canadian history / Hodgins, Thomas – Toronto: Publishers' Syndicate, 1900 – 2mf – 9 – mf#06805 – cn Canadiana [327]
The british and american mail – Rio de Janeiro, RJ: Typ Vivaldi, 17 ago 1877-24 mar 1879 – (= ser Ps 19) – mf#DIPER – bl Biblioteca [079]
British and Foreign Bible Society *see* Davids psalmer
British and foreign bible society, bible house, the upper canada bible society, 102 yonge street, toronto : the society keeps for sale all publications of the british and foreign bible society, consisting of bibles and testaments of all sizes and prices... – [Toronto?: s.n. 187-?] [mf ed 1983] – 1mf – 9 – 0-665-39731-3 – mf#39731 – cn Canadiana [220]
British and foreign bible society : review of the earl street commi – Edinburgh, Scotland. 1827 – 1r – us UF Libraries [240]
British and foreign evangelical review – Killen TX 1852-88 – 1 – mf#2866 – us UMI ProQuest [240]
British and foreign review : or european quarterly journal – La Salle IL 1835-44 – 1 – mf#3905 – us UMI ProQuest [240]
British and foreign school society, annual reports of the... 1814-1900 – 196mf – 9 – mf#87143 – uk Microform Academic [370]
British and foreign state papers, 1812-1968. v1-170 – 74r – 1 – $3,000.00 – us Trans-Media [343]
The british and french expeditions to teembo : with remarks on civilization in africa / Bowdich, Thomas Edward – Paris, 1821 – 8mf – 9 – mf#A-283 – ne IDC [916]
British and indian observer – London, England 14 dec 1823-11 jul 1824 (wkly) – 1 – uk British Libr Newspaper [072]
British and irish biographies, 1840-1945 / ed by Jones, David Lewis – [mf ed Chadwyck-Healey] – 14,287mf – 9 – (6pt collection available individually by title. only comprehensive reference work to the personalities of the victorian age and the 20th c to ww2) – uk Chadwyck [941]
British antarctic expedition 1907-1909 under the command of sir e h shackleton : reports on the scientific investigations, vol 1 – London. 1813-1828 (1) – 18mf – 9 – mf#2823 – ne IDC [919]
British antarctic expedition 1907-1909 under the command of sir e h shackleton : reports on the scientific investigations, vol 2 – Philadelphia. 1870-1882 (1) – 11mf – 9 – mf#2824 – ne IDC [919]
British apollo : or, curious amusements for the ingenious – La Salle IL 1708-11 – 1 – mf#4209 – us UMI ProQuest [870]
British archaeological discoveries in greece and crete, 1886-1936 / British School at Athens – London, England. 1936 – 1r – us UF Libraries [025]
British Architectural Library *see*
– The author/title and subject catalogue of books
– Comprehensive index to architectural periodicals
– Microfilmed collection of rare books
– Unpublished manuscripts collection
British archives of the international brigade to spain – 390mf – 9 – $2865.00 – 0-907006-79-5 – (articles, journals, pamphlets, correspondence, photos & ephemera relating to the spanish civil war, 1936-39, held at the marx memorial library, london; with p/g) – uk Mindata [946]
British art index : pictorial documentation on art in england, scotland and wales = Britischer kunst-index – bilddokumentation zur kunst in england, schottland und wales / ed by Bildarchiv Foto Marburg – Deutsches Dokumentationszentrum fuer Kunstgeschichte Philipps- Universitaet Marburg – [mf ed 2004-05] – 157mf (1:24) in 3 installments – 9 – silver €2055.00 – ISBN-10: 3-598-35613-7 – ISBN-13: 978-3-598-35613-1 – (sold only as set) – gw Saur [700]
British art, pictorial, decorative, and industrial / Wallis, George Harry – London [1882] – (= ser 19th c art & architecture) – 1mf – 9 – mf#4.2.1094 – uk Chadwyck [740]
British Assocation for the Advancement of Science *see* Journal of sectional proceedings
British Association for Labour Legislation *see* Report of a meeting held at the house of commons on thursday, mar 18, 1909
British association for the advancement of science : 67th meeting, toronto, 1897... – S.l: s.n, 1897? – 1mf – 9 – mf#53565 – cn Canadiana [500]
The british association for the advancement of science : a great world-educator / Bryce, George – [Winnipeg?: Manitoba Free Press], 1906 – 1mf – 9 – 0-665-74057-3 – mf#74057 – cn Canadiana [500]
British Association for the Advancement of Science. Canada *see*
– Canadian economics
– First report on conveyance as adopted by the executive committee
– Seventy-ninth annual meeting...winnipeg 1909
British attitude to german colonial development, 1880-1885 / Adams, M – 1935 – us CRL [943]
British australasian – London, England 2 oct 1884-14 feb 1924 – 1 – (cont by: british australian & new zealander [21 feb 1924-2 dec 1944]) – uk British Libr Newspaper [072]
British baker – Croydon UK 1963-73 – 1 – ISSN: 0007-0300 – mf#1339 – us UMI ProQuest [660]
The british bandsman and contest field – London. 1907-75. A weekly newspaper devoted entirely to bands. 7 reels – 1 – us L of C Photodup [780]
British banking statistics : with remarks on the bullion reserve and non-legal-tender note circulation of the united kingdom / Dun, John – London: E Stanford, 1876 (mf ed 19–) – ii/189p – mf#ZT-545 – us Harvard [332]
British baptist : 133 early english baptist titles – Filmed from the Regent's Park College Collection, Oxford, England. (Author Index furnished by request) – 1 – us Southern Baptist [242]
British baptist historical resource materials – Books, tracts, periodicals and sermons from the Libraries of English Baptist Colleges of Manchester and Regent's Park. 495 titles – 1 – 47.25r – us Southern Baptist [242]
British baptist materials – (43 books and pamphlets from British Library of the British Museum, London, England) – 1 – us Southern Baptist [242]
British baptist materials – (80 books and pamphlets from Angus Library, Regents Park College, Oxford, England) – 1 – us Southern Baptist [242]
British baptist materials : index – 5 – 5.04 – us Southern Baptist [242]
British baptist materials – Selected by faculty of Southwestern Baptist Theological Seminary from Whitley's Baptist bibliography, 1526-1837. (Title and author index furnished) – 1 – 47.25r – us Southern Baptist [242]
British baptist materials – Selected by professors of Southern Baptist Seminaries from Whitley's Baptist Bibliography, 1653-1862 – 1 – 552.86 – us Southern Baptist [242]
British baptist materials from angus library of regents park college – Oxford, England. Index. 3 reels – 1 – us Southern Baptist [242]
British baptist materials from angus library of regents park college, oxford, england – 17th-18th century items listed in Whitley's Baptist Bibliography. Five reels.Index available upon request – 1 – us Southern Baptist [242]
British baptist materials from the bodleian library, oxford, england – 17th-18th century selections from Whitley's Baptist Bibliography. Index available upon request – 1 – us Southern Baptist [242]
British baptist materials of 17th century – 1 – us Southern Baptist [242]
The british barbarians : a hill-top novel / Allen, Grant – London: J Lane; New York: G P Putnam's Sons, 1895 – 3mf – 9 – mf#17937 – cn Canadiana [830]
British battle fleet / Jane, Fred T – Boston, MA. v1-2. 1915 – 1r – us UF Libraries [025]
British beginnings in western india, 1579-1657 : an account of the early days of the british factory of surat / Rawlinson, Hugh George – Oxford: Clarendon Press, 1920 – (= ser Samp: indian books) – us CRL [954]
British biographical archive (bba1) = Britisches biographisches archiv (bba1) / Sieveking, Paul [comp] – [mf ed 1984-89] – 1236mf (1:24) – 9 – diazo €10,060.00 (silver €11,080 isbn: 978-3-598-30479-8) – ISBN-10: 3-598-30467-6 – ISBN-13: 978-3-598-30467-5 – (with printed ind) – gw Saur [941]
British biographical archive. series 2 (bba2) = Britisches biographisches archiv. neue folge (bba2) / ed by University of Glasgow – [mf ed 1992-94] – 632mf (1:24) – 9 – diazo €10,060.00 (silver €11,080 isbn: 978-3-598-33629-4) – ISBN-10: 3-598-33628-4 – ISBN-13: 978-3-598-33628-7 – (with printed ind) – gw Saur [941]
British biographical archive to 2002 (bba3) = Britisches biographisches archiv bis 2002 (bba3) / Nappo, Tommaso [comp] – [mf ed 2003-05] – 507mf (1:24) in 12 installments – 9 – diazo €10,060.00 (silver €11,080 isbn: 978-3-598-34781-8) – ISBN-10: 3-598-34780-4 – ISBN-13: 978-3-598-34780-1 – (with p/g) – gw Saur [941]
British birds – Icklesham UK 1960+ – 1,5,9 – ISSN: 0007-0335 – mf#1304 – us UMI ProQuest [590]
British birth control material 1800-1947 : at the british library of political and economic sciences / British Library. Political and Economic Sciences – 10r – 1 – £480.00 – mf#BCE – uk World [360]
British book news – London UK 1940-93 – 1,5,9 – ISSN: 0007-0343 – mf#8670 – us UMI ProQuest [070]
British Broadcasting Corporation *see*
– Bbc handbooks, annual reports and accounts, 1927-2002
British burma gazette – Rangoon: [s.n. 1875-86] [mf ed [London]: India Office Library & Records] – 12v on 40r – 1 – (iss in pt: pt1: extracts from the gazette of india [varies]; pt2: local gazette [varies]; pt3: notifications by the judicial commissioner & other departmental officers [varies]; pt4: advertisements [varies]; with ind; filmed with yr of coverage; iss for jan-sep 1886 filmed with: burma gazette oct-dec 1886 [pt1-3]; merged with: upper burma gazette to form: burma gazette) – us CRL [959]
British burmah and its people : being sketches of native manners, customs, and religion / Forbes, C J F S – London, 1878 – 5mf – 9 – mf#HT-47 – ne IDC [915]
British cabinet records – 2pt-coll – 18r – 1 – (pt 1: cabinet reports by prime ministers to the crown, 1837-67 5r c39-16101. pt 2:...1868-1916 13 r c39-16102) – mf#C39-16100 – us Primary [324]
British californian : official organ of the british societies – San Francisco CA jul, sep 1903; jun 1911; jan 1913-dec 1921 (mthly) (imperfect) – 1 – uk British Libr Newspaper [071]
The british captives in abyssinia / Beke, C T – Ed 2. London, 1867 – 5mf – 9 – mf#NE-20179 – ne IDC [916]
British carbonization research association. bcra review – Chesterfield UK 1977-78 – 1,5,9 – ISSN: 0305-8131 – mf#2756 – us UMI ProQuest [660]
British caribbean / TIMES, LONDON – London, England. 1950 – 1r – us UF Libraries [972]
British Catholic Association *see* Report of the committee of the british catholic association, and re...
British central africa : an attempt to give some account of a portion of the territories under british influence north of the zambezi / Johnston, Harry Hamilton – [London], 1897 – (= ser 19th c british colonization) – 7mf – 9 – mf#1.1.3392 – uk Chadwyck [916]
British Chamber of Commerce for the Netherlands East Indies *see* The java gazette
British Chamber of Commerce in Indonesia *see*
– General circular
– Review
– Weekly circular to members
The british chartered companies : 1877-1900, british north borneo, nigeria, british east africa, rhodesia 101=lund, franz edward – Madison, 1944 – us CRL [380]
British chautauquan – [Wales] LLGC 29 mar, 19 jul, 10,12,14,17,19,21,24,26,28 aug 1897 [mf ed 2004] – 1r – 1 – uk Newsplan [072]
British chronicle – [Scotland] 29 aug 1783 & 5 mar 1784 & 1 apr 1785-13 jun 1788 & 2 feb 1787 and 1 mar 1793 (wkly) [mf ed 2004] – 1r – 1 – (cont by: british chronicle or, union gazette) – uk Newsplan [072]
British chronicle *see* Hereford journal, 1770-1889
The british cicero : or a selection of the most admired speeches in the english language, arranged under three distinct heads of popular, parliamentary, and judicial oratory / Browne, Thomas – Philadelphia: Birch & Small. 3v. 1810 – 18mf – 9 – $27.00 – mf#LLMC 92-115 – us LLMC [340]
British colonial argus *see* Niagara peninsula newspapers, pt 2
British colonial office : palestine correspondence, 1927-1934 / British Colonial Office. Palestine – C.O. 733. All of the original correspondence for Palestine during the period of the British Mandate – 1 – (1927-1930 49r $6370 s0627-30. 1931-1934 103r $13390 s0631-34. registers 1927-1930 6r $780 s0527-30. registers 1931-1934 6r $780 s0531-34. with printed guides) – us Scholarly Res [941]
British Colonial Office. Palestine *see* British colonial office
British colonial policy / Wedderburn, David, 3rd Baronet – London 1881 – (= ser 19th c british colonization) – 1mf – 9 – mf#1.1.8271 – uk Chadwyck [320]
The british colonies : shall we have a colonial baronage? or, shall the colonial empire of great britain be resolved into republics? – London 1852 – (= ser 19th c british colonization) – 1mf – 9 – mf#1.1.534 – uk Chadwyck [327]
The british colonies : their history, extent, condition, and resources / Martin, Robert Montgomery – [London] [1851-1857] – (= ser 19th c british colonization) – 54mf – 9 – mf#1.1.6796 – uk Chadwyck [327]
The british colonies and their resources / Bonwick, James – London, 1886 – (= ser 19th c books on british colonization) – 8mf – 9 – mf#1.1.8334 – uk Chadwyck [333]
British colonies in north america – London: printed for SPCK, 1848 [mf ed 1983] – 3mf – 9 – mf#40284 – cn Canadiana [917]
British colonies in north america : the maritime provinces – London: printed for SPCK, 1848 [mf ed 1984] – 4mf – 9 – 0-665-41621-0 – mf#41621 – cn Canadiana [917]
British colonisation : a colonial want and an imperial necessity. a lecture delivered under the auspices of the balloon society of great britain in st james's hall, aug 7th, 1891 / Clayden, Arthur – London, [1891] – (= ser 19th c books on british colonization) – 1mf – 9 – mf#1.1.3806 – uk Chadwyck [941]
British colonist – Halifax, NS: Gran & Munroe, 1848-74 – 28r – 1 – cn Library Assoc [971]
British colonist – Stanstead, QC. 1823-31 – 2r – 1 – cn Library Assoc [971]
British colonist – Toronto, ON: H Scobie, 1838-54 – 10r – 1 – cn Library Assoc [971]
British colonist [halifax, canada] – Halifax, Canada 7 feb 1871-31 dec 1874 – 4r – 1 – uk British Libr Newspaper [071]
The british colonist in north america : a guide for intending emigrants – London, 1890 – (= ser 19th c books on british colonization) – 4mf – 9 – mf#1.1.8031 – uk Chadwyck [970]
The british colonist in north america : a guide for intending emigrants – London: S Sonnenschein, 1890 [mf ed 1980] – 4mf – 9 – 0-665-00294-7 – mf#00294 – cn Canadiana [304]
British colonization and coloured tribes / Bannister, Saxe [pseud] – London, 1838 – (= ser 19th c books on british colonization) – 4mf – 9 – mf#1.1.9534 – uk Chadwyck [941]
The british colony in russia / Johnstone, Catherine Laura – [Westminster], [1898?] – (= ser 19th c british colonization) – 1mf – 9 – mf#1.1.8235 – uk Chadwyck [940]
British columbia : an essay / Brown, Robert Christopher Lundin – New Westminster, BC?: Royal Engineer Press, 1867 – 2mf – 9 – mf#28209 – cn Canadiana [917]
British columbia : its resources and capabilities – Montreal: [s.n.], 1889 [mf ed 1980] – 1mf – 9 – 0-665-00861-9 – mf#00861 – cn Canadiana [971]
British columbia and vancouver island : comprising a historical sketch of the british settlements in the north-west coast of america and a survey of the physical character... / Hazlitt, William Carew [comp] – London; New York: G Routledge, 1858 [mf ed 1982] – 3mf – 9 – mf#35430 – cn Canadiana [971]
British columbia and vancouver's island / Albemarle, William Coutts Keppel, Earl of – London?: J Fraser?, 1858? – 1mf – 9 – mf#17961 – cn Canadiana [917]
B[ritish/] C[/olumbia/] Association of Non-Status Indians et al *see* Nesika
British Columbia Board of Trade *see* Annual report of the british columbia board of trade
British Columbia. Canada *see*
– British columbia law reports
– British columbia statutes, session laws and revisions
– Fell and langley's british columbia speaker's decisions
– Martin's mining cases
British columbia directories, 1860-1900 – Vancouver BC: microfilmed...for UBC... [mf ed 1997] – 21r – 1 – (filmed with: various british columbia directories; among them are: henderson's vancouver directory; henderson's victoria directory; henderson's british columbia gazetteer and directory etc) – cn UBC Preservation [971]
British columbia directories, 1900-1910 – Vancouver BC: microfilmed...for UBC... [mf ed 1993] – 16r – 1 – (filmed with: various british columbia directories; among them are: henderson's vancouver directory; henderson's victoria directory; henderson's british columbia gazetteer and directory etc) – cn UBC Preservation [971]
British columbia directories, 1911-1919 – Vancouver BC: microfilmed...for UBC... [mf ed 1995] – 21r – 1 – (filmed with: various british columbia directories; among them are: henderson's vancouver directory; henderson's victoria directory; henderson's british columbia gazetteer and directory etc) – cn UBC Preservation [971]
British columbia directories, 1920-1929 – Vancouver BC: microfilmed...for UBC... [mf ed 1996] – 21r – 1 – (filmed with: various british columbia directories; among them are: henderson's vancouver directory; henderson's victoria directory; henderson's british columbia gazetteer and directory etc) – cn UBC Preservation [971]
British columbia directories, 1948-1954 – Vancouver BC: microfilmed...for UBC... [mf ed 1998] – 18r – 1 – (incl items ind in: bond, mary e: canadian directories, 1790-1987) – cn UBC Preservation [971]
British columbia directories, 1955-1960 – Vancouver BC: microfilmed...for UBC... [mf ed 1999] – 22r – 1 – cn UBC Preservation [971]
British columbia directories, 1961-1965 – Vancouver BC: microfilmed...for UBC... [mf ed 2001] – 25r – 1 – cn UBC Preservation [971]

British columbia directories, 1966-1970 – Vancouver BC: microfilmed...for UBC... [mf ed 2003] – 30r – 1 – cn UBC Preservation [971]

British columbia directories, 1971-1975 – Vancouver BC: microfilmed...for UBC... [mf ed 2004] – 33r – 1 – cn UBC Preservation [971]

British columbia directories, 1976-1980 – Vancouver BC: microfilmed...for UBC... [mf ed 2005] – 36r – 1 – (incl items ind in: bond, mary e: canadian directories, 1790-1987) – cn UBC Preservation [971]

The british columbia directory for the years 1882-83 : embracing a business and general directory of the province, dominion and provincial official lists, reliable information about the country – Victoria, BC: R T Williams, 1882 – 6mf – 9 – (app by alexander caulfield anderson) – mf#24520 – cn Canadiana [030]

British columbia directory of mines – [Victoria, BC?: s.n.], 1897 [mf ed 1982] – 1mf – 9 – (with a synopsis of mining laws by archer martin) – mf#17438 – cn Canadiana [622]

British Columbia. Division of Vital Statistics *see* Special reports, 1954-1981

British columbia education history on microfilm. 1996-1997 series – Vancouver: UBC Libr, Facilities & Preservation Office – 5r/yr – 1 – (incl docs, reports, papers, etc fr b c dept & b c ministry of education) – cn UBC Preservation [370]

British columbia, emigration, and our colonies : considered practically, socially, and politically / Snow, William Parker – London 1858 – (= ser 19th c british colonization) – 2mf – 9 – mf#1.1.5629 – uk Chadwyck [304]

British columbia entomological society. bulletin – Saanichton BC 1906-08 – 1 – mf#8571 – us UMI ProQuest [590]

British columbia financial times – Vancouver, Canada 7 nov 1914-17 dec 1921 (wkly) – 3 1/2r – 1 – uk British Libr Newspaper [332]

British columbia for settlers : its mines, trade, and agriculture / Fraser, Agnes [pseud] – London, 1898 – (= ser 19th c books on british colonization) – 4mf – 9 – mf#1.1.8335 – uk Chadwyck [971]

British columbia for settlers : its mines trade and agriculture / Macnab, Frances – London: Chapman & Hall, 1898 [mf ed 1980] – 5mf – 9 – 0-665-03173-4 – (incl ind) – mf#03173 – cn Canadiana [917]

British Columbia Government Employees' Union *see* Provincial

British columbia government news – v14 n4-v20 n4/5 [1966 jul-1972 aug] – 1r – 1 – mf#1053582 – us WHS [350]

British Columbia Heritage Trust et al *see* Heritage west

The british columbia law notes – Victoria, BC?: s.n, 1894 – 9 – mf#P05039 – cn Canadiana [348]

British columbia law reports / British Columbia. Canada – v1-63. 1867-1947 (all publ) – 434mf – 9 – $651.00 – (cont by 2nd series. not offered by llmc) – mf#LLMC 81-019 – us LLMC [340]

British Columbia. Legislative Assembly *see*
- [British columbia] sessional papers
- Journals
- Sessional clipping books

British columbia library association bulletin – 1938-54 – 1r – 1 – cn Library Assoc [020]

British columbia library quarterly – Vancouver BC 1938-76 – 1,5,9 – ISSN: 0007-053X – mf#1944 – us UMI ProQuest [020]

British columbia mining critic – Vancouver: British Columbia Mining Critic Print & Pub Co, [1897-189- or 19–] – 9 – mf#P04194 – cn Canadiana [622]

B c mining exchange and investors' guide – Vancouver, Canada jun, jul 1899 – 1 – (cont by: mining tit-bits [aug, sep 1899]; british columbia mining exchange & investors' guide & mining tit-bits [oct 1899-jun/jul 1901]; b c mining exchange & investor's guide: (the b c mining exchange) [jul 1902-mar 1908]; cont by: b c mining exchange & engineering news [apr 1908-dec 1913, jun 1915-sep 1917]) – uk British Libr Newspaper [622]

British columbia mining prospectors' exchange and investors' guide – Vancouver: [s.n, 1899] [mf ed v1 n1 jan 1899-v1 n4 apr 1899] – 9 – mf#P04025 – cn Canadiana [622]

British columbia mining record – Victoria, BC: British Columbia Record, [1904-1908] [mf ed v11 n6 jul 1904-v15 n8 aug 1908] – 9 – (incl ind) – mf#P04963 – cn Canadiana [622]

British Columbia. Ministry of Education *see*
- Grade 12 provincial examination papers
- Provincial examination papers

The british columbia monthly and mining review – Victoria, BC: J M Leet, [1889?-18– or 19–] – 9 – mf#P04766 – cn Canadiana [622]

British Columbia Mountaineering Club *see* Constitution and by-laws

British Columbia Museums Association *see* Museum round-up

British columbia oddfellow : a monthly magazine devoted to the independent order, the elevation of human character and the good we can do – Vancouver: A Mackenzie; G L Center, [1895?-189- or 19–] [mf ed v1 n9 aug 1896] – 9 – ISSN: 1190-6405 – mf#P04658 – cn Canadiana [360]

British Columbia Rifle Association (Victoria, BC) *see* Constitution and by-laws

[British columbia] sessional papers / British Columbia. Legislative Assembly – Victoria BC: [Queen's Printer] [1872]-1982 [mf ed 1993] – 1 – (pt1: 1872-1920 [33r]; pt2: 1921-47 [34r]; 1872-1875, v1-4 appeared as app to the journals of the legislative assembly) – cn UBC Preservation [323]

British columbia statutes, session laws and revisions / British Columbia. Canada – Revised Statutes 1871-32nd Parliament 1st sess. 1871-1979 – 889 – 9 – $1,333.00 – (updates planned) – mf#LLMC 90-120 – us LLMC [348]

British Columbia Teachers' Federation *see* Teacher

British columbia, the most westerly province of canada : its position, advantages, resources and climates: new fields for mining, farming, fruit growing and ranching along the lines of the canadian pacific railway... – [Montreal?: Canadian Pacific Railway Co?], 1900 [mf ed 1980] – 1mf – 9 – mf#14576 – cn Canadiana [917]

British columbian – New Westminster, Canada 6 jan 1864-5 feb 1868 (imperfect) – 1 – (cont by: daily british columbian [16 mar, 2 jul 1869; 1 jun-4 jul 1888]) – uk British Libr Newspaper [071]

British columbian – New Westminster, BC: J Robson, 1861-69 – 6r – 1 – ISSN: 0841-7806 – cn Library Assoc [071]

British columbian *see* Weekly columbian

The british columbian and victoria guide and directory for 1863 : under the patronage of his excellency governor douglas, cb, and the executive of both colonies / Howard, Frederick P & Barnett, George [comp] – Victoria, VI [BC]: British Columbian and Victoria Directory, 1863 [mf ed 1983] – 3mf – 9 – 0-665-28206-0 – (incl ind) – mf#28206 – cn Canadiana [030]

The british columbian fancier – Nanaimo, BC: Nanaimo Poultry Society, [1894-189- or 19–] – 9 – ISSN: 1190-7363 – mf#P04511 – cn Canadiana [636]

The british columbian magazine – Victoria, BC: Dominion Magazine Co, [1889-18– or 19–] – 9 – mf#P04506 – cn Canadiana [971]

British columbia's blackout – n4-5,17-103 [1979 jun 23/jul 7-jul 2/21, 1979 jun 27/1980 jul 11-1984 mar 29/apr 6] – 1r – 1 – (cont by: bad british columbia blackout) – mf#592648 – us WHS [071]

The british command of the sea and what it means to canada / Wood, William – [Toronto: Toronto Branch of the Navy League, 1900] [mf ed 1981] – 1mf – 9 – mf#26052 – cn Canadiana [355]

British commerce and colonies : from elizabeth to victoria / Gibbins, Henry de Beltgens – [London], 1893 – (= ser 19th c books on british colonization) – 2mf – 9 – mf#1.1.8284 – uk Chadwyck [380]

British constitutional society [of] upper canada : at a meeting of a number of the members of the original constitutional society of york...held at morrison's tavern, on tuesday the lst day of july 1834 – [Toronto?: s.n, 1834?] [mf ed 1984] – 1mf – 9 – 0-665-45722-7 – mf#45722 – cn Canadiana [366]

The british consulate in jerusalem in relation to the jews of palestine / Hyamson, A M – London, 1939-1941. 2v – 13mf – 9 – mf#J-27-32 – ne IDC [956]

British contemporary press – ca 51r – 1 – us Primary [072]

British copyright : lord herschell's new bill: its effect upon canadian interests... / Lancefield, Richard T – [Hamilton, Ont?: s.n, 1898] [mf ed 1980] – 1mf – 9 – mf#08136 – cn Canadiana [346]

British corrosion journal – London UK 1989+ – 1,5,9 – ISSN: 0007-0599 – mf#15687.01 – us UMI ProQuest [660]

British council of churches. conference for world mission. handbook *see* Conference of missionary societies in great britain and ireland. reports and minutes of the annual conference / handbook

British critic : and quarterly theological review – La Salle IL 1793-1843 – 1 – mf#4210 – us UMI ProQuest [200]

British culture, series one and two : eighteenth and nineteenth centuries – 2 series – 16,957mf coll – 9 – (series 1: 4131mf c35-23310. series 2: 12,826mf c35-23320. series 1 covers 18th and 19th century english literature, and includes contemporary and retrospective histroy, biography and criticism. series 2 contains selected vols from the new cambridge bibliography of english literature 1800-1900. also includes significant number of religious works wh are available as separate colls as well. both series include printed guides) – mf#C35-23300 – us Primary [941]

British culture, series two : theology – 12,826mf – 9 – (titles from the new cambridge bibliography of english literature, v3 1800-1900) – us Primary [306]

British dental journal – London UK 1950+ – 1,5,9 – ISSN: 0007-0610 – mf#570 – us UMI ProQuest [617]

British dependencies in the caribbean and north atlantic, 1939-1952 – London, England. 1939-1952 – 1r – us UF Libraries [972]

British Diabetic Association *see* Diabetic medicine

British documents on the origin of the war, 1898-1914 / Gooch, G P & Temperley, Harold – v1-11. 1898-1914 – 1 – $240.00 – mf#0115 – us Brook [941]

The british dominions in north america. / Bouchette, Joseph – London, 1832. 2v – 1 – us CRL [971]

The british dominions in north america, vol 1 : or, a topographical and statistical description of the provinces of lower and upper canada, new brunswick, nova scotia, the islands of newfoundland, prince edward, and cape breton / Bouchette, Joseph – London: H Colburn & R Bentley, 1831 [mf ed 1983] – 7mf – 9 – 0-665-42807-3 – (incl bibl ref) – mf#42807 – cn Canadiana [917]

The british dominions in north america, vol 1 : or, a topographical and statistical description of the provinces of lower and upper canada, new brunswick, nova scotia, the islands of newfoundland, prince edward, and cape breton / Bouchette, Joseph – London: Longman, Rees, Orme, Brown, Green & Longman, 1832 [mf ed 1984] – 7mf – 9 – 0-665-48011-3 – (incl bibl ref) – mf#48011 – cn Canadiana [317]

The british dominions in north america, vol 2 : or, a topographical and statistical description of the provinces of lower and upper canada, new brunswick, nova scotia, the islands of newfoundland, prince edward, and cape breton / Bouchette, Joseph – London: H Colburn & R Bentley, 1831 [mf ed 1983] – 4mf – 9 – 0-665-42808-1 – (incl bibl ref) – mf#42808 – cn Canadiana [917]

The british dominions in north america, vol 2 : or, a topographical and statistical description of the provinces of lower and upper canada, new brunswick, nova scotia, the islands of newfoundland, prince edward, and cape breton / Bouchette, Joseph – London: Longman, Rees, Orme, Brown, Green & Longman, 1832 [mf ed 1984] – 4mf – 9 – 0-665-48012-1 – (incl bibl ref) – mf#48012 – cn Canadiana [317]

The british dominions in north america, vols 1-2 : or, a topographical and statistical description of the provinces of lower and upper canada, new brunswick, nova scotia, the islands of newfoundland, prince edward, and cape breton / Bouchette, Joseph – London: H Colburn & R Bentley. 2v. 1831 – 1mf – 9 – 0-665-42806-5 – mf#42806 – cn Canadiana [971]

The british dominions in north america, vols 1-2 : or, a topographical and statistical description of the provinces of lower and upper canada, new brunswick, nova scotia, the islands of newfoundland, prince edward, and cape breton / Bouchette, Joseph – London: Longman, Rees, Orme, Brown, Green & Longman. 2v. 1832 – 1mf – 9 – 0-665-48010-5 – mf#48010 – cn Canadiana [971]

British east africa : past, present and future / Hindlip, Charles Allsopp – London: T F Unwin, 1905 – 1 – us CRL [960]

British east africa or ibea : a history of the formation and work of the imperial british east africa company compiled with the authority of the directors... / Macdermott, P L – London 1893 – (= ser 19th c british colonization) – 5mf – 9 – mf#1.1.3717 – uk Chadwyck [960]

British Ecological Society *see* Journal of ecology

The british emigrant's advocate : being a manual for the use of emigrants and travellers in british america and the united states... / Duncumb, Thomas – London: Simpkin & Marshall, 1837 – 5mf – 9 – mf#61345 – cn Canadiana [304]

The british empire / Balough, Elemer – Halle (Saale) etc. Sack & Montanus, 1931-35. 2 v. LL-2326 – 1 – us L of C Photodup [340]

The british empire / Campbell, George – [London], [1887] – (= ser 19th c books on british colonization) – 3mf – 9 – mf#1.1.2798 – uk Chadwyck [941]

The british empire / Dilke, Charles Wentworth – London, 1899 [i.e. 1898] – (= ser 19th c books on british colonization) – 3mf – 9 – mf#1.1.3697 – uk Chadwyck [941]

The british empire : a sketch of the geography, growth, natural and political features of the united kingdom, its colonies and discrepancies / Bray, Caroline – London, 1863 – (= ser 19th c books on british colonization) – 7mf – 9 – mf#1.1.7389 – uk Chadwyck [941]

The british empire : a speech delivered at the banquet in boston, celebrating her majesty's diamond jubilee / Davin, Nicholas Flood – Winnipeg: Nor'-Wester, 1897 – 1mf – 9 – mf#03956 – cn Canadiana [941]

The british empire in 1820 : being a grammar of british geography in the four quarters of the world / Phillips, Richard – London 1820 [mf ed Hildesheim 1995-98] – 1v on 3mf – 9 – €90.00 – 3-487-28883-4 – gw Olms [900]

British Empire League *see*
- Canadian insolvency legislation
- Report of meeting at guildhall, 3rd december, 1896
- Report of the inaugural meeting of the league

British enterprise beyond the seas : or, the planting of our colonies / Fyfe, James Hamilton – London, 1863 – (= ser 19th c books on british colonization) – 3mf – 9 – mf#1.1.6388 – uk Chadwyck [338]

British federalism its rise and progress : a paper read before the royal colonial institute january 10, 1893 / Labilliere, Francis Peter de – [London, 1893] – (= ser 19th c british colonization) – 1mf – 9 – mf#1.1.3808 – uk Chadwyck [320]

British Film Institute *see* Monthly bulletin 1934-91

British film institute cinema pressbooks, 1920-1940 – [mf ed Chadwyck-Healey] – 1864mf – 9 – (incl printed list of titles in alphabetical order with the date, studio, director and main stars with 4 ind) – British Film Institute – uk Chadwyck [790]

British flag and christian sentinel : the official organ of the (united british) army scripture readers' & soldiers' friend society – London, England [ns] jan 1870-jul 1938 [mf 1894] – 1 – (cont: british flag: a journal for soldiers & sailors [ns] sep 1857-dec 1869; cont by: british flag [oct 1938-apr/jun 1950]) – uk British Libr Newspaper [240]

British food journal – Bradford UK 1994+ – 1,5,9 – ISSN: 0007-070X – mf#19300.03 – us UMI ProQuest [660]

British foreign missions, 1837-1897 / Thompson, Ralph Wardlaw & Johnson, Arthur N – London: Blackie 1899 [mf ed 1986] – (= ser The victorian era series) – 1mf – 9 – 0-8370-6781-2 – (incl bibl ref & ind) – mf#1986-0781 – us ATLA [941]

British Foreign Office *see*
- Japan correspondence, 1856-1905
- Japan correspondence, 1856-1948
- Japan correspondence, 1856-1951
- Russia correspondence, 1883-1948

British foreign office : us correspondence, 1930-48 – BFO file 371, 1978-81 – (1930-32: the early depression 36r $4680 s0830-32. 1933-34: fdr and the "hundred days" 38r $4940 s0833-34. 1935-36: period of social reform 41r $5330 s0835-36. 1937-38: the second new deal 51r $6630 s0837-38. 1939-40: the advent of war 46r $5980 s0839-40. 1941-44 world war 2 93r $12,090 s0841-44. 1945-46: dissolving the grand alliance 49r $6370 s0845-46. 1947-48: onset of the cold war 28r $3640 s0847-48. with printed guides) – us Scholarly Res [320]

The british friend / Friends House Library. The Religious Society of Friends – 1843-1913 – 15r – 1 – £720.00 – (monthly journal dealing with topical issues of the day. merged in 1913 with: the friend) – mf#BFR – uk World [073]

British galleries of painting and sculpture...catalogue / Westmacott, Charles Molloy – London 1824 – (= ser 19th c art & architecture) – 3mf – 9 – mf#4.2.353 – uk Chadwyck [700]

The british gazette – London: H M Stationery Off, may 5-13 1926 – 1r – 1 – (publ during general strike) – us CRL [074]

British gazette, and berwick advertiser – Berwick-upon-Tweed, England 2 jan 1808-16 dec 1820; 4 jan-15 nov 1823 – 1 – (cont by: berwick advertiser [22 nov-27 dec 1823; 1 jan 1825-23 feb 1983]) – uk British Libr Newspaper [072]

British government and the pope – London, England. 1889 – 1r – us UF Libraries [240]

British government in india : the story of viceroys and government houses / Curzon, George Nathaniel, Marquis of – London, New York: Cassell and Co, 1925 – (= ser Samp: indian books) – us CRL [954]

British government publications containing statistics, 1801-1977 – [mf ed Chadwyck-Healey] – 362r 1766mf – 1,9 – (individual titles also listed and may be purchased separately) – uk Chadwyck [941]

British governmental blue books of statistics for the caribbean and the americas : prior to independence – 319 – 1 – (int by d c dorward. 21 countries in the region are covered. individual countries are avail separately. apply for details) – mf#97485-97512 – uk Microform Academic [941]

British Grassland Society *see* Grass and forage science

British Guiana *see*
- British guiana
- British official gazette
- Building confidence
- Bulletin des actes administratifs de la prefecture de la guyane
- Gouvernementsblad van suriname

British guiana / British Guiana – London, England. 1924 – 1r – us UF Libraries [972]

British guiana / Crookall, L – London, England. 1898 – 1r – us UF Libraries [972]

British guiana / Smith, Raymond Thomas – London, England. 1962 – 1r – us UF Libraries [972]

British guiana archeology to 1945 / Osgood, Cornelius – New Haven, CT. 1946 – 1r – us UF Libraries [930]

British guiana bulletin – Georgetown, Guyana 15 jan, 15 mar, 15 apr-31 may 1958; 17 oct, 28 nov 1960 – 1 – uk British Libr Newspaper [079]

British Guiana Bureau Of Public Information *see*
- Blueprint for a british caribbean dominion
- Por and con

British guiana handbook, 1922 – Georgetown, Guyana. 1923 – 1r – us UF Libraries [972]

British Guiana Interior Development Committee *see* Handbook of natural resources of british guiana

British guiana, the land of six peoples / Swan, Michael – London, England. 1957 – 1r – us UF Libraries [972]

British heart journal – London UK 1939-95 – 1,5,9 – (cont by: heart) – ISSN: 0007-0769 – mf#1331.01 – us UMI ProQuest [616]

British heritage – 1979+ – 1,5,9 – ISSN: 0195-2633 – mf#12144 – us UMI ProQuest [941]

British history notes / Henderson, George E & Fraser, George A – Toronto: Educational Pub Co, [1897?] – 2mf – 9 – 0-665-92071-7 – mf#92071 – cn Canadiana [941]

British homoeopathic journal – London UK 1974-80 – 1,5,9 – ISSN: 0007-0785 – mf#9144 – us UMI ProQuest [615]

British honduras, past and present / Caiger, Stephen Langrish – London, England. 1951 – 1r – us UF Libraries [972]

The british impact on india / Griffiths, Percival Joseph – London: Macdonald, 1952 – (= ser Samp: indian books) – us CRL [954]

British in africa / Taylor, Don – London, England. 1962 – 1r – us UF Libraries [960]

The british in ireland, series 1 : dublin castle records, 1880-1921 – 7pt-coll – 136r – 1 – (pt 1: anti-government organisations, 1882-1921 (co 904/7-23, 27-29 and 157 14r c39-27581. pt 2: police reports, 1892-97 co 904/48-67 16r c39-27582. pt 3: police reports, 1898-1913 co 904/68-91 23r c39-27583. pt 4: police reports, 1914-21 co 904/92-122 and 148-156a 27r c39-27584. pt 5: public control and administration, 1884-1921 co 904/159-178 16r c39-27585. pt 6: judicial proceedings, enquiries and misc records, 1872-1926 co 904/30-35, 37-39, 45-47b and 180-189 16r c39-27586. pt 7: sinn fein and republican suspects, 1899-1921 co904/193-216 24r. comes with cumulative printed guide) – mf#C39-27580 – Public Record Office (PRO) – us Primary [941]

British india / Frazer, Robert Watson – London, 1896 – (= ser 19th c books on british colonization) – 5mf – 9 – mf#1.1.4479 – uk Chadwyck [954]

British india / Frazer, Robert Watson – London: T Fisher Unwin; New York: GP Putnam's Sons, 1898 – (= ser Samp: indian books) – us CRL [954]

British india : in its relation to the decline of hindooism and the progress of christianity... / Campbell, William – London: John Snow, 1839 [mf ed 1995] – (= ser Yale coll) – xii/596p (ill) – 1 – 0-524-09053-X – (ill with engravings on wood by g baxter) – mf#1995-0053 – us ATLA [954]

British india : in its relation to the decline of hindooism and the progress of christianity. containing remarks on the manners, customs, and literature of the people... / Campbell, W – London, 1858 – 7mf – 9 – mf#HT-23 – ne IDC [915]

British india : in its relation to the decline of hindooism and the progress of christianity: containing remarks on the manners, customs, and literature of the people / Campbell, William – London: John Snow, 1858 – 2mf – 9 – 0-8370-6568-2 – mf#1986-0568 – us ATLA [954]

British india and its rulers / Cunningham, Henry Stewart – London, 1881 – (= ser 19th c books on british colonization) – 4mf – 9 – mf#1.1.4016 – uk Chadwyck [954]

British industry, labour and trade unionism, 1887-1934 – 22r – 1 – us Primary [331]

British Institute of International Affairs *see* Journal of the british institute of international affairs

British Institution for Promoting the Fine Arts in the United Kingdom, London *see*
- Catalogue of a selection of the works of sir joshua reynolds
- [Catalogue of carlton house palace. 1826]
- [Catalogue of carlton house palace. 1827]
- [Catalogue of pictures by ancient and modern masters. 1806]
- [Catalogue of pictures by ancient and modern masters. 1807]
- [Catalogue of pictures by ancient and modern masters. 1808]
- [Catalogue of pictures by ancient and modern masters. 1809]
- [Catalogue of pictures by ancient and modern masters. 1810]
- [Catalogue of pictures by ancient and modern masters. 1811]
- [Catalogue of pictures by ancient and modern masters. 1812]
- [Catalogue of pictures by ancient and modern masters. 1813]
- [Catalogue of pictures by ancient and modern masters. 1814]
- [Catalogue of pictures by ancient and modern masters. 1815]
- [Catalogue of pictures by ancient and modern masters. 1816]
- [Catalogue of pictures by ancient and modern masters. 1817]
- [Catalogue of pictures by ancient and modern masters. 1818]
- [Catalogue of pictures by ancient and modern masters. 1819]
- [Catalogue of pictures by ancient and modern masters. 1820]
- [Catalogue of pictures by ancient and modern masters. 1821]
- [Catalogue of pictures by ancient and modern masters. 1822]
- [Catalogue of pictures by ancient and modern masters. 1823]
- [Catalogue of pictures by ancient and modern masters. 1824]
- [Catalogue of pictures by ancient masters. 1825]
- [Catalogue of pictures by ancient masters. 1828]
- [Catalogue of pictures by ancient masters. 1829 jun]
- [Catalogue of pictures by ancient masters. 1831 jun]
- [Catalogue of pictures by ancient masters. 1832 jul]
- [Catalogue of pictures by ancient masters. 1834]
- [Catalogue of pictures by ancient masters. 1835 may]
- [Catalogue of pictures by ancient masters. 1836 may]
- [Catalogue of pictures by ancient masters. 1837 may]
- [Catalogue of pictures by ancient masters. 1838 jun]
- [Catalogue of pictures by ancient masters. 1839 jun]
- [Catalogue of pictures by ancient masters. 1840 jun]
- [Catalogue of pictures by ancient masters. 1841 jun]
- [Catalogue of pictures by ancient masters. 1842 jun]
- [Catalogue of pictures by ancient masters. 1843 jun]
- [Catalogue of pictures by ancient masters. 1844 jun]
- [Catalogue of pictures by ancient masters. 1845 jun]
- [Catalogue of pictures by ancient masters. 1846 jun]
- [Catalogue of pictures by ancient masters. 1847 jun]
- [Catalogue of pictures by ancient masters. 1848 jun]
- [Catalogue of pictures by ancient masters. 1849 jun]
- [Catalogue of pictures by ancient masters. 1850 jun]
- [Catalogue of pictures by ancient masters. 1851 jun]
- [Catalogue of pictures by ancient masters. 1852 jun]
- [Catalogue of pictures by modern masters. 1830]
- [Catalogue of pictures by modern masters. 1832]
- [Catalogue of pictures by modern masters. 1835]
- [Catalogue of pictures by modern masters. 1836]
- [Catalogue of pictures by modern masters. 1837]
- [Catalogue of pictures by modern masters. 1838]
- [Catalogue of pictures by modern masters. 1839]
- [Catalogue of pictures by modern masters. 1841]
- [Catalogue of pictures by modern masters. 1845]
- [Catalogue of pictures by modern masters. 1846]
- [Catalogue of pictures by modern masters. 1847]
- [Catalogue of pictures by modern masters. 1849]
- [Catalogue of pictures by modern masters. 1850]
- [Catalogue of pictures by modern masters. 1851]
- [Catalogue of pictures by modern masters. 1852]
- Catalogue of the works of the late sir thomas lawrence
- Declaration issued in the preface to the catalogue

British Interplanetary Society *see*
- Journal of the british interplanetary society
- Realities of space travel

The british invasion from the north : the campaigns of generals carleton and burgoyne from canada, 1776-1777: with the journal of lieut. william digby, of the 53rd, or shropshire regiment of foot – Albany, NY: J Munsell's Sons, 1887 – 5mf – 9 – (ill by james phinney baxter) – mf#03506 – cn Canadiana [975]

British isles gazetteer – Bartholomew: 1904 – 8mf – 9 – NZ$32.00 – 0-908989-18-0 – (based on 1901 census index. incl. some maps) – nz BAB [941]

British journal for the history of science – Hants UK 1962+ – 1,5,9 – ISSN: 0007-0874 – mf#11398 – us UMI ProQuest [500]

British journal for the philosophy of science – Oxford UK 1989+ – 1,5,9 – ISSN: 0007-0882 – mf#17494 – us UMI ProQuest [500]

British journal of addiction – Abingdon, Oxfordshire 1982-85 – 1,5,9 – (cont by: addiction) – ISSN: 0952-0481 – mf#13420.03 – us UMI ProQuest [362]

British journal of aesthetics – Oxford UK 1960+ – 1,5,9 – ISSN: 0007-0904 – mf#10048 – us UMI ProQuest [700]

British journal of anaesthesia – Oxford UK 1923+ – 1,5,9 – ISSN: 0007-0912 – mf#1311 – us UMI ProQuest [617]

British journal of audiology – London UK 1967-80 – 1,5,9 – ISSN: 0300-5364 – mf#6943 – us UMI ProQuest [617]

British journal of biomedical science – Tunbridge Wells UK 1993+ – 1,5,9 – (cont: medical laboratory sciences) – ISSN: 0967-4845 – mf#15571.04 – us UMI ProQuest [619]

British journal of cancer – London UK 1947+ – 1,5,9 – ISSN: 0007-0920 – mf#1316 – us UMI ProQuest [616]

British journal of chiropody – Millom UK 1973-79 – 1,5,9 – ISSN: 0007-0939 – mf#8602 – us UMI ProQuest [617]

British journal of clinical pharmacology – Oxford UK 1974+ – 1,5,9 – ISSN: 0306-5251 – mf#11259 – us UMI ProQuest [615]

British journal of clinical psychology – Leicester UK 1994+ – 1,5,9 – ISSN: 0144-6657 – mf#14714 – us UMI ProQuest [616]

British journal of criminology – Oxford UK 1989+ – 1,5,9 – ISSN: 0007-0955 – mf#17493 – us UMI ProQuest [364]

British journal of dental science – London, Oxford House. v2 1858-1859; v3 n37-54 1859-1860; v5 n67-68 1862; v25 n335-358 1882; v26 n359-382 1883 – us CRL [617]

British journal of dermatology – Oxford UK 1980+ – 1,5,9 – ISSN: 0007-0963 – mf#15508.02 – us UMI ProQuest [616]

British journal of developmental psychology – Leicester UK 1994+ – 1,5,9 – ISSN: 0261-510X – mf#14716 – us UMI ProQuest [150]

British journal of diseases of the chest – Oxford UK 1907-88 – 1,5,9 – (cont by: respiratory medicine) – ISSN: 0007-0971 – mf#1313.01 – us UMI ProQuest [616]

British journal of disorders of communication – Abingdon, Oxfordshire 1991 – 1 – (cont by: european journal of disorders of communication) – ISSN: 0007-098X – mf#14161.02 – us UMI ProQuest [616]

British journal of educational studies – Oxford UK 1977+ – 1,5,9 – ISSN: 0007-1005 – mf#12172 – us UMI ProQuest [370]

British journal of educational studies, 1953-72 – v1-20 – 3r 9mf – 1,9 – mf#8/96856 – uk Microform Academic [370]

British journal of educational technology – Oxford UK 1970+ – 1,5,9 – ISSN: 0007-1013 – mf#6740 – us UMI ProQuest [370]

British journal of experimental pathology – Oxford UK 1920-89 – 1,5,9 – (cont by: journal of experimental pathology) – ISSN: 0007-1021 – mf#2517.02 – us UMI ProQuest [619]

British journal of haematology – Oxford UK 1980+ – 1,5,9 – ISSN: 0007-1048 – mf#15509 – us UMI ProQuest [616]

British journal of hospital medicine – London UK 1968+ – 1,5,9 – (cont by: hospital medicine) – ISSN: 0007-1064 – mf#6627.02 – us UMI ProQuest [610]

British journal of industrial medicine – London UK 1944-93 – 1,5,9 – (cont by: occupational and environmental medicine) – ISSN: 0007-1072 – mf#1332.01 – us UMI ProQuest [362]

British journal of industrial relations – Oxford UK 1988+ – 1,5,9 – ISSN: 0007-1080 – mf#15722 – us UMI ProQuest [331]

British journal of law and society *see* Journal of law and society

British journal of management – Oxford UK 1990+ – 1,5,9 – ISSN: 1045-3172 – mf#18143 – us UMI ProQuest [650]

British journal of medical hypnotism – Maidenhead UK 1949-59 – 1,5,9 – mf#1873 – us UMI ProQuest [615]

British journal of medical psychology – Leicester UK 1993+ – 1,5,9 – ISSN: 0007-1129 – mf#14710.02 – us UMI ProQuest [150]

British journal of music education [bjme] – Cambridge UK 1989+ – 1,5,9 – ISSN: 0265-0517 – mf#16520 – us UMI ProQuest [780]

British journal of non-destructive testing – Northampton UK 1980-93 – 1,5,9 – ISSN: 0007-1137 – mf#12782.01 – us UMI ProQuest [660]

British journal of nursing – London UK 1992+ – 1,5,9 – (cont: nursing) – ISSN: 0966-0461 – mf#19574 – us UMI ProQuest [610]

British journal of nutrition – Wallingford UK 1985+ – 1,5,9 – ISSN: 0007-1145 – mf#14913 – us UMI ProQuest [613]

British journal of obstetrics and gynaecology – Oxford UK 1975-99 – 1,5,9 – (cont: journal of obstetrics and gynaecology of the british commonwealth) – ISSN: 0306-5456 – mf#681.02 – us UMI ProQuest [618]

British journal of ophthalmology – London UK 1917+ – 1,5,9 – ISSN: 0007-1161 – mf#1327 – us UMI ProQuest [617]

British journal of oral and maxillofacial surgery – Oxford UK 1984+ – 1,5,9 – ISSN: 0266-4356 – mf#14918,01 – us UMI ProQuest [617]

British journal of orthodontics – Oxford UK 1985-99 – 1,5,9 – ISSN: 0301-228X – mf#13422.01 – us UMI ProQuest [617]

British journal of perioperative nursing – Harrogate UK 2001+ – 1 – ISSN: 1467-1026 – mf#26755.02 – us UMI ProQuest [610]

British journal of pharmacology – London UK 1946+ – 1,5,9 – ISSN: 0007-1188 – mf#1333 – us UMI ProQuest [615]

The british journal of photography – 1854-1999+ – 124r – 1 – £4950.00 – mf#BJP – uk World [770]

British journal of photography annual – 1860-1993 – 76r – 1 – £2650.00 – mf#BJA – uk World [770]

British journal of plastic surgery – Oxford UK 1982+ – 1,5,9 – ISSN: 0007-1226 – mf#13423.01 – us UMI ProQuest [617]

British journal of political science – Cambridge UK 1971+ – 1,5,9 – ISSN: 0007-1234 – mf#11033 – us UMI ProQuest [320]

British journal of preventive and social medicine – London UK 1947-79 – 1,5,9 – ISSN: 0007-1242 – mf#1334.01 – us UMI ProQuest [610]

British journal of psychology – London UK 1993+ – 1,5,9 – ISSN: 0007-1269 – mf#14711,02 – us UMI ProQuest [150]

British journal of psychology, 1904/5-83 : the journal of the british psychological society / British Psychology Society – 809mf – 7,9 – mf#2071 – uk Microform Academic [150]

British journal of radiology – London UK 1980+ – 1,5,9 – ISSN: 0007-1285 – mf#17212 – us UMI ProQuest [616]

British journal of religious education – Abingdon, Oxfordshire 1978+ – 1,5,9 – ISSN: 0141-6200 – mf#11895 – us UMI ProQuest [377]

British journal of rheumatology – Oxford UK 1983-98 – 1,5,9 – (cont: rheumatology and rehabilitation; cont by: rheumatology) – ISSN: 0263-7103 – mf#3457.03 – us UMI ProQuest [616]

British journal of social psychology – Leicester UK 1994+ – 1,5,9 – ISSN: 0144-6665 – mf#14713 – us UMI ProQuest [150]

British journal of social work – Oxford UK 1971+ – 1,5,9 – ISSN: 0045-3102 – mf#10804 – us UMI ProQuest [360]

British journal of sociology – 1950-99+ – 19r – 1 – £870.00 – mf#BJS – uk World [301]

British journal of special education – Oxford UK 1985+ – 1,5,9 – (cont: special education: forward trends) – ISSN: 0952-3383 – mf#11032.01 – us UMI ProQuest [370]

British journal of sports medicine – London UK 1980+ – 1,5,9 – ISSN: 0306-3674 – mf#17213.01 – us UMI ProQuest [617]

British journal of surgery – Hoboken NJ 1913+ – 1,5,9 – ISSN: 0007-1323 – mf#1292 – us UMI ProQuest [617]

British journal of urology – Oxford UK 1950-98 – 1,5,9 – (cont by: bju international) – ISSN: 0007-1331 – mf#733.01 – us UMI ProQuest [616]

British journal of venereal diseases – London UK 1964-84 – 1,5,9 – (cont by: genitourinary medicine) – ISSN: 0007-134X – mf#1354.02 – us UMI ProQuest [616]

British journal on alcohol and alcoholism – London UK 1977-82 – 1,5,9 – (cont: journal of alcoholism; cont by: alcohol and alcoholism: international journal of the medical council on alcoholism) – ISSN: 0309-1635 – mf#6588.01 – us UMI ProQuest [616]

British kinematograph, sound and television society. bksts journal – London UK 1974-81 – 1,5,9 – (cont: british kinematography, sound and television; cont by: image technology: journal of the bksts) – ISSN: 0305-6996 – mf#5327.02 – us UMI ProQuest [380]

British kinematography, sound and television – London UK 1969-73 – 1,5,9 – (cont by: bksts journal) – ISSN: 0373-109X – mf#5327.02 – us UMI ProQuest [380]

British labour history ephemera – 2 sects. 1880-1926 – 1 – £3,200.00 coll – (1880-1900 46r £2150 blh. 1900-26 22r £1050 ble) – uk World [331]

British labour statistics : historical abstract 1886-1968 – [mf ed Chadwyck-Healey] – (= ser British government publications...1801-1977) – 5mf – 9 – uk Chadwyck [331]

British labour statistics : yearbook 1969-1976 – [mf ed Chadwyck-Healey] – (= ser British government publications...1801-1977) – 36mf – 9 – uk Chadwyck [314]

British labourers' protector and factory child's friend – n1-31. 1832-33 [all publ] – (= ser Radical periodicals of great britain, 1794-1914. period 1) – 3mf – 9 – $55.00 – us UPA [331]

The british laws of the new hebrides : in force on 22 sep 1971 / Ballard, B C [comp] – v1-3. 1971 – 1r – 1 – (available for ref) – mf#pmb doc446 – at Pacific Mss [348]

British legislature / Gordon, James Edward – London, England. 1837 – 1r – us UF Libraries [240]

British librarian – La Salle IL 1737 – 1 – mf#4211 – us UMI ProQuest [020]

The british librarian : or handbook for students in divinity, a guide to the knowledge of theological works, in english, and in the learned and other foreign languages, classified under heads / Lowndes, William Thomas – [London], 1839 – (= ser 19th c publishing...) – 8mf – 9 – (text in two numbered clms) – mf#3.1.2 – uk Chadwyck [020]

British Library *see*
- Calendars of charters and rolls in the manucript collections of the british library
- Eighteenth century short title catalogue
- The samas religious texts

British Library. National Bibliographic Service *see*
- Books in english
- British national bibliography
- Fiction on fiche
- Name authority list
- Serials in the british library

British Library. Political and Economic Sciences *see* British birth control material 1800-1947

British lion : the only/official organ of the british fascisti/fascists – London, England [ns] jun 1926-[nov 1929] – 1 – (cont: fascist bulletin [13 jun 1925-12 jun 1926]; cont by: british fascism [ns]; special summer propaganda no; extra autumn iss; [ns] jun 1930-feb 1933, [jun,oct 1933], mar-jun 1934) – uk British Libr Newspaper [320]

British literary manuscripts from cambridge university library, series one : the medieval age, c1150-c1500 – 49r – 1 – (pt 1: medieval mss from mss dd-ff 17r. pt 2: medieval mss from mss gg-ii 17r. pt 3: medieval mss from mss kk-oo and additional 15r. with printed guide) – mf#C35-28220 – us Primary [420]

British literary manuscripts from cambridge university library, series two : the english renaissance, c1500-c1700 – 35r – 1 – (includes printed guide) – mf#C35-28221 – us Primary [420]

British literary manuscripts from the bodleian library, oxford : the english renaissance, c1500-c1700 – 43r – 1 – (includes printed guide) – mf#C35-28310 – us Primary [420]

British literary manuscripts from the british library, london, series 3 : the medieval age, c1150-c1500, pts 1 and 2 – 42r – 1 – (includes printed guide) – mf#C35-28230 – us Primary [420]

British literary manuscripts from the british library, london, series one : the english renaissance – literature from the tudor period to the restoration, c1500-c1700 – 109r – 1 – (includes printed guide) – mf#C35-28231 – us Primary [420]

British literary manuscripts from the british library, london, series two : the eighteenth century, c1700-c1800 – 50r – 1 – (includes printed guide) – mf#C35-28232 – us Primary [420]

British literary manuscripts from the folger shakespeare library, washington, d.c. : the english renaissance – literature from the tudor period to the restoration, c1500-c1700 – 30r – 1 – (includes printed guide) – mf#C35-28240 – us Primary [420]

British literary manuscripts from the national library of scotland, edinburgh : pt 1: medieval and renaissance literature, c1300-c1700 – 20r – 1 – (includes printed guide) – mf#C35-28250 – us Primary [420]

British literary manuscripts from the national library of scotland, edinburgh : pt 2: eighteenth century literary manuscripts c. 1700-c.1800 – 19r – 1 – (coll offers 165 literary mss from the age of the scottish enlightenment. three major writers featured are: allan ramsey, henry mackenzie and robert burns. with printed guide) – mf#C35-28251 – us Primary [420]

British literary manuscripts from the national library of scotland, edinburgh : pt 3 and 4: the nineteenth century, c.1800-1880 – 31r – 1 – (coll ranges from ballads, folk material and correspondence to the literary manuscripts of the great aristocratic colls) – mf#C35-28520 – us Primary [420]

British literary manuscripts from the princeton university library : the william cowper papers and other eighteenth century literary manuscripts – 10r – 1 – (includes printed guide) – mf#C35-28260 – us Primary [420]

British luminary and weekly intelligence – London, England 3 oct 1818-21 may 1820 – 1 – (cont by: british luminary, or weekly intelligencer [27 may-23 jul 1820]) – uk British Libr Newspaper [072]

British magazine – La Salle IL 1746-51 – 1 – mf#5266 – us UMI ProQuest [072]

British magazine and register of religious and ecclesiastical information – La Salle IL 1832-49 – 1 – mf#4213 – us UMI ProQuest [240]

British magazine; or monthly repository for gentlemen and ladies – La Salle IL 1760-67 – 1 – mf#4212 – us UMI ProQuest [790]

British masters of the albumen print : a selection of mid-nineteenth century victorian photography / International Museum of Photography at George Eastman House; ed by Sobieszek, Robert A – 1976 – 3 color mf – 15 – $55.00f – 0-226-69171-3 – us Chicago U Pr [770]

British Medical Association. Joint Committee on Psychiatry and the Law *see* The criminal law and sexual offenders; a report

British medical bulletin – Oxford UK 1943+ – 1,5,9 – ISSN: 0007-1420 – mf#2341 – us UMI ProQuest [610]

British medical journal [bmj] : international edition – London UK 1857+ – 1,5,9 – ISSN: 0959-8146 – mf#1205 – us UMI ProQuest [610]

British medicine – Oxford UK 1972-90 – 1,5,9 – ISSN: 0140-2722 – mf#49373 – us UMI ProQuest [610]

The british mercury : and wednesday's evening post (london) – 7 jan 1824-29 dec 1824; 5 jan 1825-13 jul 1825 – (= ser 19th c british periodicals) – r8 – 1 – us Primary [073]

The british mercury – Hamburg DE, 1787 apr-dec, 1789 jul-1790 sep – 3r – 1 – gw Misc Inst [074]

British mercury (london) : or, wednesday evening post – 30 apr 1806-31 dec 1806; 6'7 jan 1807-30 dec 1807 – (= ser 19th c british periodicals) – 2r – 1 – (6 jan 1808-28 dec 1808, 4 jan 1809-27 dec 1809 [3r]; 3 jan 1810-26 dec 1810, 2 jan 1811-25 dec 1811, 1 jan 1812-5 feb 1812 [4r]; 30 mar 1814, 12 oct 1814-19 oct 1814, 7 jan 1818-30 dec 1818, 6 jan 1819-29 dec 1819 [5r]; 5 jan 1820-27 dec 1820, 3 jan 1821-26 dec 1821 [6r]) – us Primary [072]

British mercury (london) : or, wednesday evening post – 2 jan 1822-25 dec 1822, 1 jan 1823-31 dec 1823 – (= ser 19th c british periodicals) – 7r – 1 – (cont as: the british mercury, and wednesday's evening post fr 14 may 1823) – us Primary [072]

British messenger – [Scotland] Stirling: P Drummond jan 1854 (mthly) [mf ed 2004] – 1r – 1 – (a monthly journal devoted to the diffusion of scriptural knowledge, the promotion of vital religion and the advancement of social reformation) – uk Newsplan [240]

British methodism / Hurst, John Fletcher – New York: Eaton & Mains, 1902 – 4mf – 9 – 0-524-04213-6 – mf#1990-5004 – us ATLA [242]

British mezzotint portraits : being a descriptive catalogue / Smith, John Chaloner – London 1878-84 – (= ser 19th c art & architecture) – 27mf – 9 – mf#4.2.264 – uk Chadwyck [760]

British minstrel : and musical and literary miscellany – La Salle IL 1843-45 – 1 – mf#5267 – us UMI ProQuest [780]

British moralists : being selections from writers principally of the 18th century / ed by Selby-Bigge, Lewis Amherst – Oxford: Clarendon Press 1897 [mf ed 1990] – 3mf – 9 – 0-7905-9634-2 – mf#1989-1359 – us ATLA [170]

British morning news – Vienna, Austria 3,7 feb 1946 – 1 – (cont by: morning news [22,28 jan 1947-19 jun 1949]) – uk British Libr Newspaper [072]

British morning news – Vienna, Austria nov 1945-jun 1949 [mf ed Norman Ross] – 4r – 1 – mf#nrp-1934 – us UMI ProQuest [074]

British Museum *see* Materials on the early educational life of the english baptists

British museum collections of natural history specimens : and drawings from the "endeavour" voyage of captain cook, 1768-1771 – [mf ed Chadwyck-Healey] – 3pts on 38 colour + 18 b/w mf – 15,9 – (with catalogues) – uk Chadwyck [574]

British Museum. Dept of Printed Books *see*
- Catalogue of books in the library of the british museum printed in england, scotland, and ireland to the year 1640
- Catalogue of books printed in the 15th century now in the british museum

British Museum. Dept of Prints & Drawings *see*
- Crace collection of london views in the british museum
- Historical prints in the british museum

British museum entomological literature, 1800-1864 / British Museum (Natural History); ed by Gilbert, Pamela – 6,003mf – 9 – £21,940.00 – (guide included £42) – uk Chadwyck [590]

British museum karaite mss : descriptions and collation of six karaite manuscripts of portions of the hebrew bible in arabic characters / Hoerning, Reinhart – London: Williams and Norgate, 1889 – 2mf – 9 – 0-8370-9393-7 – mf#1986-3393 – us ATLA [090]

British Museum. London *see*
- Book plates in the british museum
- Costume prints in the british museum
- National photographic record and survey
- Trade cards in the british museum

British Museum, London. Dept of British and Mediaeval Antiquities *see* Antiquities from the city of benin

British Museum, London. Dept of Coins and Medals *see* A catalogue of english coins in the british museum

British Museum (Natural History) *see*
- British museum entomological literature, 1800-1864
- Early american herbaria

British national bibliography / British Library. National Bibliographic Service – 1993-. Annual volume – 17 – £155.00y + VAT, £199.00y overseas – (1950-84 cumulation. £520. 1981-85 full cumulation £350) – uk British Libr [010]

British national catalogue – 1963-83 – 7r – 1 – £350.00 – mf#BNF – uk World [790]

British, natives and boers in the transvaal... : the appeal of the swazi people / Bartlett, Ellis Ashmead – London, 1894 – (= ser 19th c books on british colonization) – 1mf – 9 – mf#1.1.4709 – uk Chadwyck [960]

The british nepos : consisting of the lives of illustrious britons, who have distinguished themselves by their virtues, talents, or remarkable advancement in life / Mavor, William Fordyce – 13th rev enl ed. London: printed for Longman, Hurst...1819 – (= ser 19th c children's literature) – 6mf – 9 – mf#6.1.3 – uk Chadwyck [941]

British new guinea annual reports – 1886-30 jun 1906 – 2r – 1 – mf#pmb doc312 – at Pacific Mss [324]

British new guinea deaths 1888-1906 : date of publication, name, place, occupation, cause of death, date of death – [mf ed 1986] – 1mf – 9 – A$4.40 – 0-949124-18-4 – (filmed with: land owners papua new guinea 1891-1906 [date, number, name, place]) – at Northern [929]

British new guinea government gazette – 3 jan 1903-6 aug 1906 – 1r – 1 – mf#pmb doc314 – at Pacific Mss [323]

British news of canada – Montreal, Canada 13 jan-21 sep 1912 [wkly] – 1 – (cont by: canadian & british news of canada [28 sep 1912-19 apr 1913]) – uk British Libr Newspaper [071]

British north america : reports of progress together with a preliminary and general report, on the assiniboine and saskatchewan exploring expedition... / Hind, Henry Youle – London: printed by George Edward Eyre & William Spottiswoode...1860 [mf ed 1983] – 3mf – 9 – (with ind) – mf#SEM105P285 – cn Bibl Nat [917]

The british north american arithmetic : containing elementary lessons for the younger classes in common schools, prepared expressly for the british provinces – Stanstead, LC [Quebec]: Walton & Gaylord, 1833 [mf ed 1984] – 1mf – 9 – 0-665-43140-6 – mf#43140 – cn Canadiana [510]

The british north american magazine and colonial journal – [Halifax, NS?: E Ward, 1831-183-?] – 9 – mf#P04913 – cn Canadiana [420]

British Officer Of Sir John Clottworthy's Regiment *see* History of the war of ireland from 1641 to 1653

British official gazette / British Guiana – Georgetown. 1952-May 21, 1966. For later file See: Guyana. Official gazette – 1 – us NY Public [324]

British official publications not published by hmso document delivery service – [mf ed Chadwyck-Healey, 1980] – 2 major coll + 13 subject coll – 9 – (also available by sects. major coll: science & technology 1111mf. social sciences 1473mf. subject coll: annual reports 151mf. local govt 61mf. background briefs & foreign policy documents. agriculture, forestry, fisheries 444mf. environment. education 256mf. employment & working conditions 259mf. food 20mf. health & medicine 308mf. health & safety. library information science & bibliographies 114mf. local govt. scotland 19mf. transport 96mf) – uk Chadwyck [324]

British opium policy and its results to india and china / Turner, Frederick Storrs – London 1876 – (= ser 19th c british colonization) – 4mf – 9 – mf#1.1.5283 – uk Chadwyck [380]

British oribatidae / Michael, A D – v1-2. 1884, 1888 – (= ser Ray society publications 61, 65) – 1,7 – (v1 1r or 10mf 9/86313. v2 9mf 86368) – uk Microform Academic [580]

British paintings 1500 to 1850 – (= ser Sotheby's pictorial archive) – 235mf – 9 – $1635.00 – 1-900853-85-X – (over 20,000 reproductions) – uk Mindata [750]

British planning history, 1900-1952 / ed by Simpson, Michael et al – Printed and archival records of the National Housing and Town Planning Council, The Town and Country Planning Association and the Royal Town Planning Institute – 38r – 1 – uk Microform Academic [710]

British playbills, 1736-1900 : from the harvard theatre collection – 6pts – (= ser Playbills from the Harvard Theatre coll) – 100r coll – 1 – (previous title: playbills from the harvard theatre collection. british playbills from the mid-18th-20th century. pt 1: adelphi, astley's surrey and globe theatres 13r c35-12210. pt 2: grecian, royal colburg, olympic, st james and strand theatres 11r c35-12211. pt 3: princess', sadler's wells, english opera house theatres 20r c35-12212. pt 4: theatres royal, covent garden 19r c35-12213. pt 5: theatre royal, drury lane 25r c35-12214. pt 6: theatre royal, haymarket 14r c35-12215) – mf#C35-12200 – us Primary [790]

British poetry since 1970 – Manchester, England. 1980 – 1r – us UF Libraries [810]

British poets – Riverside ed. Boston. 122 v. (lacks v. 14-17, 26, 61) – 1 – us L of C Photodup [810]

British policy in changing africa / Cohen, Andrew – Evanston, IL. 1959 – 1r – us UF Libraries [960]

British policy in china : is our war with the tartars or the chinese? / Scarth, John – London: Smith, Elder & Co; Edinburgh: Edmonston & Douglas, 1860 – (= ser 19th c books on china) – 1mf – 9 – mf#7.1.34 – uk Chadwyck [951]

British policy in relation to the gold coast 1815-1850 / James, Philip Gilbert – [London 1935] – us CRL [960]

British policy in the sudan 1882-1902 / Shibikah, Makki – London, New York, Oxford University Press, 1952 – us CRL [960]

British policy towards sindh : upto the annexation, 1843 / Khera, P N – Lahore: Minerva Book Shop, 1941 – (= ser Samp: indian books) – us CRL [954]

British political and social cartoons / U.S. Library of Congress. Prints and Photographs Division – 2200 prints from 1650-1832. 4 reels. P&P12022 – 1 – us L of C Photodup [941]

British political party general election addresses – 28r – 1 – (pt 1: general election addresses, 1892-1922 12r c39-20401. pt 2:...1923-31 16r c39-20402) – mf#C39-20400 – us Primary [941]

British press : or morning literary advertiser – London, England [1,26 jan-7 jul 1803]-[1 jan 1818-31 oct 1826] (imperfect) – 1 – uk British Libr Newspaper [072]

The british protectorates and the union of south africa 1930-1950 / Thayer, Ralph Noyes – Madison, 1951 – us CRL [960]

British Psychology Society *see*
- British journal of psychology, 1904/5-83

British public record office archival material at stanford university : with an addendum of other significant british microform holdings / Rozkuszka, W David – Stanford, CA: The Libraries, 1983 (mf ed 1988) – 1mf – 9 – mf#*XM-17,226 – us NY Public [941]

British Record Society Ltd *see* Index library

British relations with the nagpur state in the 18th century : an account, mainly based on contemporary english records / Wills, Cecil Upton – Nagpur: Central Provinces Govt Press, 1926 – (= ser Samp: indian books) – us CRL [954]

British review and london critical journal – La Salle IL 1811-25 – 1 – mf#4214 – us UMI ProQuest [410]

British review and national observer *see* Scots observer

British rule in india : a historical sketch / Martineau, Harriet – London 1857 – (= ser 19th c british colonization) – 4mf – 9 – mf#1.1.8381 – uk Chadwyck [954]

British rule in south africa / Holden, William Clifford – Pretoria, South Africa. 1969 – 1r – us UF Libraries [960]

British ruling cases / Great Britain. Courts – 1900-31.6 reels – 1 – $300.00 – us Trans-Media [340]

British school : paintings and watercolours – (= ser Christie's pictorial archive: painting and graphic art) – 170mf – 9 – $1280.00 – 0-907006-02-7 – (over 1500 artists, over 10,00 reproductions) – uk Mindata [750]

British School at Athens *see* British archaeological discoveries in greece and crete, 1886-1936

British school at athens. annual – London: MacMillan. v1-16. 1894-1910 – 1 – $144.00 – (v17-75 1910/11-1979 $652 [0117]) – mf#0116 – us Brook [450]

British School at Athens. Bulletin *see* [Of archaeology at athens]

British School of Archaeology, Egypt *see* Ancient egypt and the east

The british school of sculpture illustrated by twenty engravings / Scott, William Bell – London [1871?] – (= ser 19th c art & architecture) – 4mf – 9 – mf#4.2.936 – uk Chadwyck [730]

British Science Museum. London *see* Pictorial history of science, industry and medicine

British seafarer : the official organ of the british seafarers' union – Southampton, England jan 1913-jan 1922 (mthly) (imperfect) – 2r – 1 – uk British Libr Newspaper [331]

British seaman – [Wales] Cardiff 31 may-21 jun 1884 [mf ed 2002] – 1r – 1 – uk Newsplan [072]

British settlement of natal / Hattersley, Alan Frederick – Cambridge, England. 1950 – 1r – us UF Libraries [960]

British socialist : a monthly socialist review – v1-2. 1912-13 [all publ] – (= ser Radical periodicals of great britain, 1794-1914. period 2) – 1r – 1 – $115.00 – us UPA [335]

British Society of Dowsers *see* Journal of the british society of dowsers

British society of franciscan studies – Aberdeniae. v1-9. 1908-1920 – 34mf – 8 – mf#H-765 – ne IDC [241]

The british soldier in india / Mouat, Frederic John – [London] 1859 – (= ser 19th c british colonization) – 1mf – 9 – mf#1.1.4129 – uk Chadwyck [355]

British Solomon Islands Protectorate *see*
- Agricultural gazette
- News sheet

British somaliland and its tribes – [n.p.] Military Govt of British Somaliland, 1945 – 1 – us CRL [960]

British Somaliland. Customs and Excise Dept *see* Annual trade report

British Somaliland. Survey Dept [London] *see* Report on general survey of british somaliland

British south africa and the zulu war : a paper read before the royal colonial institute, with the discussion, feb 18 1879 / Noble, John – London 1879 – (= ser 19th c british colonization) – 1mf – 9 – mf#1.1.4944 – uk Chadwyck [960]

British South Africa Co *see* Reports on the native disturbances in rhodesia, 1896-1897

British South Africa Company *see* Rhodesia

British sovereignty in india / Wilson, John – Edinburgh, Scotland. 1837 – 1r – us UF Libraries [240]

British Soviet Friendship Society *see* The british soviet friendship society presents soviet dancers in britain in action photographs

The british soviet friendship society presents soviet dancers in britain in action photographs / British Soviet Friendship Society – [London, 1954?] – 1 – mf#*ZBD-*MGO pv21 – Located: NYPL – us Misc Inst [790]

The british species of angiocarpus lichens elucidated by their sporidia / Leighton, W A – 1851 – (= ser Ray society publications 20) – 1r 4mf – 1,7 – mf#9/86072 – uk Microform Academic [580]

British stage and literary cabinet – La Salle IL 1817-22 – 1 – mf#4215 – us UMI ProQuest [790]

British statistical blue books – 1821-1947 – 5789mf – 9 – (sect: africa [1 title on 188mf]; australia [3 titles on 703mf]; east asia [1 title on 616mf]; europe [3 titles on 1446mf]; latin america [2 titles on 403mf]; middle east/ north africa [2 titles on 548mf]; south asia [1 title on 883mf]; south east asia: general [1 title on 1002mf]; with printed catalogue) – ne IDC [310]

British steelmaker, 1935-58 – 35r – 1 – mf#526 – uk Microform Academic [670]

British tax review – Andover UK 1993+ – 1,5,9 – ISSN: 0007-1870 – mf#18028 – us UMI ProQuest [336]

British territories in east and central africa, 1945-1950 / GREAT BRITAIN COLONIAL OFFICE – London, England. 1950 – 1r – us UF Libraries [960]

British theorists of the nineteenth century / Finney, Charles Herbert – U of Rochester 1957 [mf ed 19–] – 1r – 1 – (with app & bibl) – mf#film 581 – us Sibley [780]

British theses relating to british history, 1688-1715 / McLeod, W Reynolds [comp] – 31r – 1 – mf#97183-97212 – uk Microform Academic [941]

The british tourists : or traveller's pocket companion, through england, wales, scotland, and ireland; comprehending the most celebrated tours in the british islands / Mavor, William – London 1800 [mf ed Hildesheim 1995-98] – 6v on 12mf – 9 – €120.00 – 3-487-28898-2 – gw Olms [914]

British trade union history collection – 50r – 5 – £1,800.00 – (also available in mf) – mf#BTU – uk World [331]

British transport commission: annual report *see* Government control of railways; estimates of the pooled revenue, receipts and expenses and resultant net revenue 1939/40-1947

British treasury, economic depression, and international finance, 1916-1943 – 38r (coll) – 1 – (pt 1: international finance situation and policy, 1916-1943, 20r. pt 2: domestic, monetary and unemployment policy, 1919-1943, 18r) – us Primary [330]

British trials, 1660-1900 : detailed first-hand accounts of thousands of trials – [mf ed Chadwyck-Healey] – 1:24 [complete] – 9 – (incl ind & printed title listing) – uk Chadwyck [345]

British underwriter and insurance advocate – London, England oct 1906, jan 1908-feb 1913 (irreg) – 1 – (cont by: insurance advocate & british underwriter [mar 1913-dec 1928]) – uk British Libr Newspaper [368]

The british union jack : a short history of our national flag for the children of our public shcools / Howell, Henry Spencer – Cambridge, Ont?: s.n, 1897 – 1mf – 9 – mf#07022 – cn Canadiana [941]

British versus american civilization : a lecture delivered in shaftesbury hall, toronto, 19th april, 1873 / Davin, Nicholas Flood – Toronto: Adam, Stevenson, 1873 – mf#23813 – cn Canadiana [900]

British veterinary journal – Oxford UK 1875-1996 – 1,5,9 – (cont by: veterinary journal) – ISSN: 0007-1935 – mf#1312.01 – us UMI ProQuest [636]

British Virgin Islands. Statistics Office *see* Statistical abstract number 1, 1974

British vogue 1916-1939 – 1382mf – 9 – $7800.00 set – 0-907006-24-8 – (available in 4 sects: 1916-23 [175 iss on 366mf] $2280 [isbn: 0-907006-04-3]; 1924-29 [152 iss on 316mf] $1950 [isbn: 0-907006-09-4]; 1930-34 [162 iss on 332mf] $2060 [isbn: 0-907006-14-0]; 1935-39 [170 iss on 368mf] $2290 [isbn: 0-907006-19-1]) – uk Mindata [740]

British volunteer and manchester weekly express – [NW England] Manchester 25 aug 1804, 1 feb, 22 mar, 1 nov 1806, jan 1807; jan 1810-dec 1819, 5 feb 1820; 1821; 2,30 mar 1822, 13,27 jul 1822, 3 aug 1822, 11 jan, 1 feb, 21 jun 1823, 24 apr, 15 may, 14 aug 1824, 5 feb 1825 [mf ed 2003] – 7r – 1 – uk Newsplan [072]

British war office : american revolution, 1773-1783 – 1r – 1 – $130.00 – mf#S1291 – us Scholarly Res [941]

British watercolours and drawings to 1850 – (= ser Sotheby's pictorial archive) – 217mf – 9 – $1645.00 – 1-900853-90-6 – (over 20,000 reproductions) – uk Mindata [700]

The british weekly : a journal of social and christian progess – v1-157. 5 nov 1886-1990 – 90r – 1 – (lacks some pp) – ISSN: 0007-1951 – mf#atla s0052 – us ATLA [073]

British west indies and the sugar industry / Root, John William – Liverpool, England. 1899 – 1r – us UF Libraries [972]

British work in india / Carstairs, Robert – [Edinburgh], 1891 – (= ser 19th c books on british colonization) – 4mf – 9 – mf#1.1.7820 – uk Chadwyck [954]

British worker (manchester ed) – [NW England] Manchester 5-17 may 1926 [mf ed 2003] – 1r – 1 – uk Newsplan [331]

British workman – London, England apr 1856-feb 1921 – 1 – (cont: british workman & friend of the sons of toil [feb 1855-mar 1856]; cont by: british workman & home monthly [mar-sep 1921]) – uk British Libr Newspaper [331]

British world – London, England nov 1906 (irreg) – 1 – (founded for the promotion of trade & intercourse with & between the states of the british empire; discontinued) – uk British Libr Newspaper [380]

The british world in the east : a guide historical, moral, and commercial, to india, china, australia, south africa, and the other possessions or connexions of great britain in the eastern and southern seas / Ritchie, Leitch – London 1847 – (= ser 19th c british colonization) – 2v on 12mf – 9 – mf#1.1.5000 – uk Chadwyck [900]

British yearbook of international law – v1-44. 1920-70 – 1,5,6 – $545.00 – ISSN: 0068-2691 – mf#101111 – us Hein [341]

The british-american – Philadelphia. 10 Dec 1887-Jul 1918 (imperfect).-w,m. 30 reels – 1 – uk British Libr Newspaper [071]

The british-american reader – Montreal: J Lovell; Toronto: H & A Miller, 1860 – 4mf – 9 – mf#42958 – cn Canadiana [910]

The british-canadian gold fields exploration, development and investment company – Toronto: The Company, 1896 [mf ed 1979] – 1mf – 9 – 0-665-00284-X – mf#00284 – cn Canadiana [622]

Britisyha mran ma khet ca krann tuik mya a khre a ne, 1826-1947 / Ae Kru, U – Ran kun: A thak tan Panna U ci Thana, Bhasa pran nhan Ca aup Thut ve re Thana 1979 [mf ed 1990] – (= ser Mran ma mhu ca can 3) – [pl/ill] 1r with other items – 1 – (in burmese; with bibl & ind) – mf#mf-10289 seam reel 177/4 [§] – us CRL [020]

Brit'ns are comming – v1 n1-v3 n4 [1987 apr-1989 oct] – 1r – 1 – mf#1578548 – us WHS [071]

Brito Conde, Herminio De *see* Tragedia ocular de machado de assis

Brito, Manuel Carlos de *see* European music manuscripts, series 2

Brito, Rodrigues De *see* Economia brasileira no alvorecer do seculo 19

Briton – La Salle IL 1762-63 – 1 – mf#4216 – us UMI ProQuest [073]

Briton – London, England 25 sep-20 nov 1819 (irreg) – 1 – uk British Libr Newspaper [072]

Briton ferry, giants grave cemetery, monumental inscriptions – 1mf – 9 – £1.25 – uk Glamorgan FHS [929]

Briton ferry / llansawel, glamorgan, parish churches of st mary & st clement : baptisms 1668-1925, burials 1686-1924, marriages 1668-1837 – [Glamorgan]: GFHS [mf ed c2000] – 2mf – 9 – £2.50 – uk Glamorgan FHS [929]

Briton ferry, st mary, monumental inscriptions – 1mf – 9 – £1.25 – uk Glamorgan FHS [929]

The briton in india / George, T J – Madras: TJ George, 1935 – (= ser Samp: indian books) – us CRL [954]

Brits, Rudolf Marthinus *see* The singular spectrum in a banach algebra h

Brittain, John *see*
- Elementary agriculture and nature study
- A first course in chemistry
- Nature study and agriculture

Brittan, Harriette G *see* Kardoo

Britten, Emma Hardinge *see*
- Ghost land
- Modern american spiritualism

Britten, James *see* Remaines of gentilisme and judaisme

Britten, William *see* Ghost land

Britting, Georg *see*
- Der bekraentze weiher
- Lebenslauf eines dicken mannes, der hamlet hiess

Brittische bibliothek / ed by Mueller, Karl Wilh – Leipzig 1757-67 – (= ser Dz) – 6v on 24mf – 9 – €240.00 – mf#k/n208 – gw Olms [020]

Brittisches museum fuer die deutschen / ed by Eschenburg, Joh Joachim – Leipzig [Aachen] 1777-80 – (= ser Dz. abt literatur) – 6v on 19mf – 9 – €190.00 – mf#k/n4504 – gw Olms [060]

Brittisches theologisches magazin / ed by Bamberger, Johann Peter – Halle 1769-74 – (= ser Dz. abt theologie) – 4v[zu je 4st] on 24mf – 9 – €240.00 – mf#k/n2088 – gw Olms [240]

Britto, Jose Gabriel De Lemos *see* Gloriosa sotaina do primeira imperio

Britton, John *see*
- The architectural antiquities of great britain
- Beauties of wiltshire
- Catalogue raisonne of the pictures belonging to the...marquis of stafford
- The fine arts of the english school
- Graphic illustrations
- Graphical and literary illustrations of fonthill abbey, wiltshire
- An historical and architectural essay relating to redcliffe church, bristol
- Historical and descriptive accounts...of...english cathedrals
- The history and antiquities of bath abbey church
- The history and description, with graphic illustrations, of cassiobury park, hertfordshire
- Illustrations of the public buildings of london
- Picturesque antiquities of the english cities
- Restoration of the church of saint mary, redcliffe, bristol
- The union of architecture, sculpture, and painting

Britton, Mignon *see* A group dynamic interpretation of a teambuilding event

Britton, Nathaniel Lord *see* Flora of bermuda

Brittonia – Bronx NY 1973+ – 1,5,9 – ISSN: 0007-196X – mf#8093 – us UMI ProQuest [580]

Briuk, D I et al *see* Slavianskie knigi kirillovskoi pechati 15-18 vv

Briva zeme – Riga, Latvia 1919-40 [mf ed Norman Ross] – 46r – 1 – mf#nrp-1454 – us UMI ProQuest [077]

Briva zeme – Riga, Latvia 3 jan 1938-27 apr 1940 (daily) – 28r – 1 – uk British Libr Newspaper [077]

Brivais strelnieks – Riga, Latvia 1917-18 [mf ed Norman Ross] – 1r – 1 – mf#nrp-1455 – us UMI ProQuest [077]

Brivibas talcinieks : a[merikas] l[atviesu] j[aunatnes] a[pvienibas] publiskas informacijas nozares biletens / Amerikas Latviesu Jaunatnes Apvieniba – n1-30 [1957 nov-1983 jun] – 1r – 1 – mf#2892227 – us WHS [350]

Brix, Fritz *see* Tilsit-ragnit

Brixham advertiser and directory – [SW England] Torbay aug 1932-may 1940 [mf ed 2004] – 3r – 1 – uk Newsplan [072]

Brixton citizen – London, UK. Oct-dec 1950; feb-mar 1952; 1953; oct-dec 1960 – 1/4r – 1 – uk British Libr Newspaper [072]

Brixton, clapham and streatham post – Lambeth, London 25 mar 1871-11 oct 1873 [mf 1873] – 1 – (cont by: brixton & clapham post [oct 1873-8 may 1875]) – uk British Libr Newspaper [072]

Brixton free press : emergency news bulletin – Lambeth, London 5 may-14 may 1926 – 1 – (publ during the general strike) – uk British Libr Newspaper [072]

Brixton streatham and norwood times and south london gazette *see* South london gazette and brixton streatham and norwood times

Brizna en el oleaje / Pou, Angel Neovildo – Guanabacoa, Cuba. 1956 – 1r – us UF Libraries [972]

Briznas / Bravo, Carlos M – Habana, Cuba. 1951 – 1r – us UF Libraries [972]

Brno 65 [etc]...[international trade fair] – Brno, Czech Republic 1965-90 (imperfect) – 1 – uk British Libr Newspaper [380]

Broad, Charlie Dunbar *see*
- Perception, physics, and reality

The broad church : or, what is coming / Haweis, Hugh Reginald – London: S Low, Marston, Searle, & Rivington, 1891. Beltsville, MD: NCR Corp, 1978 (4mf); Evanston: American Theol Lib Assoc, 1984 (4mf) – 9 – 0-8370-0927-8 – mf#1984-4245 – us ATLA [240]

The broad church : or, what is coming? / Haweis, Hugh Reginald – London: Sampson Low, Marston, Searle & Rivington, 1891 – 1mf – 9 – 0-8370-5005-7 – mf#1985-3005 – us ATLA [240]

The broad highway / Farnol, Jeffery – Toronto: McClelland & Goodchild, 1913 [mf ed 1999] – 6mf – 9 – 0-659-90266-4 – mf#9-90266 – cn Canadiana [830]

Broad river baptist church. cherokee county. gaffney, south carolina : church records – 1931, 1934-39, 1953-72 – 1 – 5.00 – us Southern Baptist [242]

Broad River Genealogical Society *see* Eswau huppeday

The broad-axe – Charlottetown, PEI: [s.n, 1871] – 9 – mf#P04160 – cn Canadiana [350]

Broad-axe (eugene, or) – Eugene OR: Amis & Son, [wkly] – 1 – (cont: broad-axe tribune; absorbed: weekly record (eugene, or)) – us Oregon Lib [071]

Broad-axe (eugene, or) *see* Weekly record (eugene, or)

Broad-axe tribune – Eugene OR: J F Amis, [wkly] – 1 – (cont by: broad-axe (eugene, or)) – us Oregon Lib [071]

Broadbent, E H *see* The pilgrim church

Broadbent, William *see* Progress of gentile error

Broadcast engineering – Shawnee Mission KS 1965+ – 1,5,9 – ISSN: 0007-1994 – mf#1728 – us UMI ProQuest [621]

Broadcast engineering. world edition – Shawnee Mission KS 2002+ – 1,5,9 – mf#20377.03 – us UMI ProQuest [621]

Broadcaster / Brotherhood of Railway, Airline and Steamship Clerks, Freight Handlers, Express and Station Employees – 1977 jan/feb-oct/nov – 1r – 1 – mf#379645 – us WHS [380]

Broadcaster – Sydney, dec 1923-dec 1927 – 2r – 9 – A$98.82 vesicular A$109.82 silver – at Pascoe [079]
Broadcasting – New York NY 1931-92 – 1,5,9 – (cont by: broadcasting & cable) – ISSN: 0007-2028 – mf#2970.01 – us UMI ProQuest [380]
Broadcasting and television – Sydney, Australia 9 jan-24 dec 1959 – 1r – 1 – uk British Libr Newspaper [380]
Broadcasting & cable – New York NY 1994+ – 1,5,9 – (cont: broadcasting) – ISSN: 1068-6827 – mf#2970.01 – us UMI ProQuest [380]
Broadcasting for the integration of women in development : the atrcw perspective – Addis Ababa: s.n, 1977? – (= ser African training and research centre for women publications on microfilm) – 1mf – 9 – us CRL [376]
Broadcasting Service, Ministry of Information *see* Voice of indonesia
Broadchalke sermon-essays on nature, mediation, atonement, absolution, etc / Williams, Rowland – London: Williams & Norgate 1867 [mf ed 1991] – 1mf – 9 – 0-7905-9762-4 – mf#1989-1487 – us ATLA [242]
Broaddus, Andrew *see* The extra examined
Broadhurst, Cyrus Napoleon *see* Personal work
Broadmead Church *see* The records, bristol, england
Broadmead church: the records *see* Publications
Broadmoor baptist church – Shreveport, LA. 3216p. 1930-89 – 1 – $144.72 – mf#6680 – us Southern Baptist [242]
Broadmouth baptist church. abbeville, south carolina : church records – 1837-1966 – 1 – 53.82 – us Southern Baptist [242]
Broadsheet – jul 1972-oct 1976; mar 1995-jul 1997 – 8r – mf#ZB 35 – nz Nat Libr [079]
Broadside – 1980 sep-1989 feb – 1r – 1 – mf#962312 – us WHS [071]
Broadside and the free press – v8 n23-1924 [1969/1970 dec 31/jan 13-1970 jan 14/27], v9 n15 [1970 sep 9/22] – 1r – 1 – (cont: broadside [cambridge ma: 1965]; free press) – mf#1532484 – us WHS [071]
Broadstairs and st peter's mail, thanet news and east kent times – [London & SE] Kent Arts & Libraries, Margate Lib 1912 – 1 – uk Newsplan [072]
Broadus, Andrew *see* The dover selection of spiritual songs
Broadus, John Albert *see*
– Addresses, essays, lectures
– The books of samuel
– A catechism of bible teaching
– Commentary on the gospel of matthew
– A harmony of the gospels in the revised version
– Immersion essential to christian baptism
– Jesus of nazareth
– Lectures on the history of preaching
– Materials
– Memoir of james petigru boyce, d. d., ll. d
– Memoir of james petigru boyce, d.d., ll.d.
– Sermons and addresses
Broadwater news – Broadwater, NE: W J Eby (wkly) [mf ed v4 n21. aug 6 1914-v46 n48. jan 30 1958 (gaps)] – 12r – 1 – (absorbed by: bridgeport news-blade) – us NE Hist [071]
Broadway – Killen TX 1867-73 – 1 – mf#5338 – us UMI ProQuest [790]
Broadway baptist church. louisville, kentucky : church records – 1878-1975 – 1 – us Southern Baptist [242]
Broadway baptist church. memphis, tennessee : church records – 1944-63 – 1 – us Southern Baptist [242]
Broadway baptist church. woman's missionary society. louisville, kentucky : church records – 1887-1966. Lacks 1896-1900. 5046p – 1 – us Southern Baptist [242]
Broadway journal – La Salle IL 1845-46 – 1 – mf#4361 – us UMI ProQuest [790]
The broadway journal – v1-2. 1845-46 – 1 – us AMS Press [410]
Broadwell, J S et al *see* Methodism and literature
The broadwood archive : the business records of john broadwood and sons, piano-makers – 1794-1901 – 95r – 1 – £4650.00 – (with the lucy broadwood folksong archive wh may be purchased separately) – mf#BRO – uk World [780]
El brocense : conferencia...en la casa de salamanca de madrid el dia 16 de mayo de 1958 / Alamillo Salgado, Ildefonso – Madrid: summ.grag, 1958 – 1 – sp Bibl Santa Ana [946]
Brochet, Henri *see* Saint felix et ses pommes de terre
Brochet, Maurice *see* Reseau des chemins de fer colombiens de bogota 'a
Brochmand, Erasmus Johannes *see* Ethices historicae specimen
Brochu, Andre *see* Le reel, le realisme et la litterature quebecoise
La brochure : napoleon 1 et napoleon 3 et l'opinion – [s.l.]: [s.n.], [s.d.] (mf ed 1984) – 1mf – 9 – mf#SEM105P443 – cn Bibl Nat [944]
La brochure populaire mensuelle – n1-24. Paris. 1934-35 – 1 – fr ACRPP [073]
Brochure-souvenir et historique du 75eme anniversaire de la fondation de sacre-coeur de jesus, east-broughton, bce, 1871-1946 – [Quebec (Province): s.n, 1946] (mf ed 1996) – 2mf – 9 – (with ind; int by j-odina roy) – mf#SEM105P2539 – cn Bibl Nat [241]
Brochure-souvenir et historique du 100e anniversaire de la fondation de la paroisse de st-eugene, 1867-1967 – [Quebec (Province): s.n.], impr 1967 [mf ed 2000] – 2mf – 9 – mf#SEM105P3280 – cn Bibl Nat [971]
The brock bugle – [Brock, NE: Claudia Dougherty], aug 1965 (wkly ex aug) [mf ed n18. dec 9 1965- (gaps) filmed 1977] – 1 – (vol numbering begins with v2 n1 aug 25 1966) – us NE Hist [071]
Brock bulletin – Brock, NE: A K Ovenden (wkly) [mf ed v4 n39. may 20 1898-v62 n45. mar 23 1944 (gaps)] – 9r – 1 – (publ in talmage ne, sep 23 1926-jul 14 1927. issues for may 27 1898- called v3 n4- . issued with: talmage tribune sep 23 1926-mar 3 1927) – us NE Hist [071]
The brock champion – Brock, NE: A L Ogden & W C Ogdon, 1895 (wkly) [mf ed v1 n11. nov 1 1895-dec 17 1897 (gaps)] – 1r – 1 – us NE Hist [071]
Brock, Erich *see* Das weltbild ernst juengers
Brock, Isaac *see* A lecture delivered in the city hall on tuesday evening march 8, 1870
Brock, M J *see* History of placer and nevada counties
Brock, Mourant *see*
– Justification by faith only
– Lord's coming
– Sacrament of the lord's supper
The brock news – Brock News, NE: T J Shelton, 1891 [mf ed v1 n14. oct 3 1891-jul 29 1892 (gaps)] – 1r – 1 – us NE Hist [071]
Brock, Paul *see*
– Die auf den morgen warten!
– Das opfer der unbekannten
Brock, Robert Kincaid *see* Needham law school, 1821-1842, cumberland county, virginia
Brock, Stephan *see*
– Caroline von wolzogens "agnes von lilien"
Brock, William *see*
– Infidelity in high places
– Prodigal's return
– Sacramental religion subversive of vital christianity
Brockelmann, Carl *see*
– Altturkestanische volksweisheit
– Geschichte der arabischen litteratur
– Grundriss der vergleichenden grammatik der semitischen sprachen
– Lexicon syriacum
– Mahmud al-kasgharis darstellung des tuerkischen verbalbaus
– Semitische sprachwissenschaft
– Syrische grammatik
– Volkskundliches aus altturkestan
Brockelmann, Carl et al *see* Geschichte der christlichen litteraturen des orients
Brockenbrough and holmes' cases – 1v. 1789-1814 – (= ser A pre-nrs title) – 4mf – 9 – $6.00 – mf#llmc90-312 – us LLMC [347]
Brockenbrough, J W *see* Brockenbrough's reports of cases in the fourth circuit, 1802-1833
Brockenbrough's reports of cases in the fourth circuit, 1802-1833 / Brockenbrough, J W – Philadelphia: J Kay. v1-2. 1837 (all publ) – (= ser Early federal nominative reports) – 13mf – 9 – $19.50 – (sometimes known as: chief justice marshall's decisions) – mf#LLMC 81-437 – us LLMC [347]
Die brockensammlung : [zeitschrift fuer angewandten buddhismus] – Berlin: Neu-Buddhistischer Verlag 1924-38 [mf ed 2005] – 14v on 1r – 1 – (none publ in 1928?; iss as doppelheft 1924-27) – mf#2005c-s033 – us ATLA [280]
Brockes, Barthold Heinrich *see* Ein gelegenheitsgedicht von brockes
Brockett, Linus Pierpont *see*
– Our western empire
– The philanthropic results of the war in america
– The story of the karen mission in bassein, 1838-1890
Brockhaus, Clemens Friedrich *see* Nicolai cusani de concilii universalis potestate sententia explicatur
Das brockhaus conversations-lexikon 1796-1898 (ael1/35) : gesamtausgabe der ersten 14 auflagen und der dazugehoerigen supplemente – [mf ed 1997] – (= ser Archiv der europaeischen lexikographie, abt 1: enzyklopaedien) – 1400mf – 9 – €9670.00 set – 3-89131-250-4 – (vols listed separately) – gw Fischer [030]
Brockhaus' conversations-lexikon (ael1/35.22) : allgemeine real-encyclopaedie – 13th ed. Leipzig 1882-87 [mf ed 1997] – (= ser Das brockhaus conversations-lexikon 1796-1898 (ael1/35)) – 16v on 161mf – 9 – €770.00 – 3-89131-271-7 – gw Fischer [030]
Brockhaus' conversations-lexikon (ael1/35.23) : allgemeine deutsche real-encyclopaedie – suppl vol to 13th ed. Leipzig 1887 [mf ed 1997] – (= ser Das brockhaus conversations-lexikon 1796-1898 (ael1/35)) – 1v on 11mf – 9 – €80.00 – 3-89131-272-5 – gw Fischer [030]
Brockhaus' conversations-lexikon (ael1/35.24) – 14th ed. Leipzig/Berlin/Wien. 16v+suppl= v17 1897). 1892-95 [mf ed 1997] – (= ser Das brockhaus conversations-lexikon 1796-1898 (ael1/35)) – 191mf – 9 – €920.00 – 3-89131-273-3 – gw Fischer [030]
Brockhaus, Heinrich *see* Die kunst in den athos-kloestern
Brockhaus, Heinrich et al *see* Blaetter fuer literarische unterhaltung (klp17)
Brockington, Alfred Allen *see*
– Old testament miracles in the light of the gospel
– The parables of the way
Brockkhaus' conversations-lexikon (ael1/35.25) – rev jubilee ed. 14th ed. Leipzig/Berlin/Wien 1898 [mf ed 1997] – (= ser Das brockhaus conversations-lexikon 1796-1898 (ael1/35)) – 17v on 191mf – 9 – €920.00 – 3-89131-274-1 – gw Fischer [030]
Brockley news, new cross and hatcham review – Lewisham, England 19 sep 1891-29 apr 1931 (wkly) [mf 1894] – 1 – (cont: brockley news & hatcham & new cross review [3 may 1890-12 sep 1891]) – uk British Libr Newspaper [072]
Brockman, H J *see* Letter to the woman of england
Brockmeier, Wolfram *see*
– Ewiges deutschland
– Die ravensburger fahnentraeger
Brockmeyer, Gretchen A *see*
– The construct validity of a scale to measure teacher enthusiasm in secondary physical education
– Gender differences in overt coaching behaviors of high school soccer coaches
– Perceptions of the importance and achievement of student teaching objectives
– Perceptions of "utility" and "likeability" dimensions of a physical education fitness curriculum
Brockport, New York. Brockport Baptist Church *see* Records
Brockton 1795-1849 – Oxford, MA (mf ed 1995) – (= ser Massachusetts vital record transcripts to 1850) – 7mf – 9 – 0-87623-218-7 – (mf 1t: births 1795-1850. mf 1t-2t: publishments 1821-49. mf 2t-3t: marriages 1821-47. mf 3t-6t: births 1843-49. mf 6t: marriages 1843-49. mf 6t-7t: deaths 1843-49) – us Archive [978]
Brockville gazette – Brockville, ON. 1828-32 – 1r – 1 – ISSN: 1181-5590 – cn Library Assoc [071]
Brockville recorder – Brockville, ON. 1830-49 – 6r – 1 – cn Library Assoc [071]
Brockway, A Fenner *see* The indian crisis
Brockway, Alice Pickford *see* Letters from the far east
Brockway, Fenner *see* The truth about barcelona
Brockway, Josephus *see* Mr. brockway's apology to the rev. nathan s.a. beman..
Brockway, K Nora *see* A larger way for women
Brockwayville weekly record – Brockwayville, PA. -w 1887-1973. 31 rolls – 13 – $25.00r – us IMR [071]
Brockwood-norwalk-ontario county line connection – Norwalk, Ontario WI, 1983 nov 17-dec 1 – 1r – 1 – mf#1028732 – us WHS [071]
Brod, A *see* Russkaia armiia [omsk: 1919]
Brod, Max *see*
– Abenteuer in japan
– Das buch der liebe
– Erloeserin
– Die erste stunde nach dem tode
– Die hoehe des gefuehls
– The master
Brodbeck, Adolf *see* Einleitung in die philosophie
[Broderick-] the independent – may 1946-jun 1948 – 1r – 1 – $60.00 – mf#C02080 – us Library Micro [071]
[Broderick-] yolo independent – CA. 1911-jan 27 1922 – 6r – 1 – $360.00 – mf#C03586 – us Library Micro [071]
Broderip and bingham's reports : reports of cases argued and determined in the court of common pleas... / Broderip, William J & Bingham, Peregrine – v1-3. 1819-22. London: A Strahan, 1820-22 (all publ) – 7mf – 9 – $31.50 – mf#LLMC 84-755 – us LLMC [324]
Broderip, Robert *see* Short introduction to the art of playing the harpsichord
Broderip, William J *see* Broderip and bingham's reports
Brodersen, Arvid *see* Stefan george
Broderskap – Stockholm, Sweden. 1980- – 1 – sw Kungliga [078]
Brodhead independent – Brodhead WI. 1861 feb 27-1864 sep 30, 1864 jan 15, 1864 oct 7-1867 feb 22 – 3r – 1 – (cont by: independent [brodhead wi: 1867]) – mf#1047281 – us WHS [071]
Brodhead independent – Brodhead WI. 1871 may 12-1871 sep 8, 1872 jan 5-1872 aug 16, 1872 aug 23-1872 dec 13, 1873 feb 7, 1874 aug 14-oct 14, 1875 jun, 25-aug 13 – 2r – 1 – (cont by: independent [brodhead wi: 1875]) – mf#1032362 – us WHS [071]
Brodhead independent – Brodhead, Orfordville W: 1881 oct 7/dec 23-1908 jun 11/1909 jul 15 – 16r – 1 – (cont: independent [brodhead wi: 1875]; cont by: brodhead register, independent-register) – mf#1047290 – us WHS [071]
Brodhead, Jane Napier *see* The religious persecution in france, 1900-1906
Brodhead news – Brodhead WI. 1909 sep 16/dec 9-1929 jan 3-1930 nov 27 – 13r – 1 – mf#964300 – us WHS [071]
Brodhead register – Brodhead WI. 1883 nov 1-1887 may 21, 1887 may 28-1890 dec 27, 1891 jan 3-1894 apr 11 – 3r – 1 – (cont by: register [brodhead wi]) – mf#963583 – us WHS [071]
Brodhead register – Brodhead WI. 1898 jan 5-1899 aug 23, 1899 aug 30-1901 apr 17, 1901 apr 23-1902 dec 17, 1902 dec 24-1904 sep 14, 1904 sep 21-1906 may 23, 1906 may 30-1908 mar 11, 1908 mar 18-1909 jul 14 – 1r – 1 – (cont: register [brodhead wi]; cont by: independent-register [brodhead wi]) – mf#963586 – us WHS [071]
Brodhead weekly independent – Brodhead WI. 1868 nov 24-1870 dec 23 – 1r – 1 – (cont: independent [brodhead wi: 1867]) – mf#1047284 – us WHS [071]
Brodhead weekly reporter – Brodhead WI. 1859 jun 24-1862 apr 15 – 1r – 1 – mf#957894 – us WHS [071]
Brodhull *see* The white and black books of the cinque ports from 1433
Brodie-Brockwell, Charles Alexander *see* The bible and the critic
Brodie-Innes, John William *see* Scottish witchcraft trials
Brodnitz, Kaethe *see* Der junge tieck und seine maerchenkomoedien
Brodovikov, A M *see* Amurskoe ekho
Brodovskii, I *see* Evreiskaia nishcheta v odessie
Brodribb, William Jackson *see* Constantinople
Brodrick, Alan Houghton *see* Little vehicle
Brodrick, George Charles *see*
– A collection of the judgments of the judicial committee of the privy council
– Home rule and justice to ireland
Brodrick-Cloete, W *see* The history of the great boer trek and the origins of south african republics
Brodruch, Karl *see* Der kamps um badajoz un fruhjahr
Brody, Heinrich *see* Shaar ha-shir
Broeckaert, Joseph *see* The fact divine
Broecker, A v *see* Moderner christusglaube
Broeckhoff, J P *see* Dicht- en zedekundige zinnebeelden en bespiegelingen
Broeger, Karl *see*
– Der bluehende hammer
– Nuernberg
– Der ritter eppelein
– Sturz und erhebung
– Volk, ich leb aus dir
Broehl-Delhaes, Christel *see* Ein soldat schrieb an ein kleines maedchen
Broehmer, Heinrich *see* Die einwirkungen der reformation auf die organisation und besetzung des reichskammergerichts
Broek, J A van den *see* De cheribonsche opstand van 1806
Broek, J O M *see* Place names in 16th and 17th century borneo
Broeker, Heinz *see*
– Alarm ueber tage
– Die tapferen tage
Broemsen, Karl M von *see* Russland und das russische reich
Brogan, O *see* The roman frontier settlement at ghirza
Brogden, James *see* Catholic safeguards against the errors, corruptions, and novelties of the church of rome
Broglie, Albert, Duc de *see*
– King's secret
– Saint ambrose
Broglie, Auguste Theodore Paul, Abbe de *see*
– Monotheisme, henotheisme, polytheisme
– Le present et l'avenir du catholicisme en france
– Problemes et conclusions de l'histoire des religions
Broglie, Charles F de *see* Politique de tous les cabinets de l'europe
Broglie, E de *see* Bernard de montfaucon et les bernardins 1715-1750
Broglie, M l'abbe de [Auguste Theodore Paul] *see* Preuves psychologiques de l'existence de dieu
Broh tae prak r tammnak dyk bhnaek gra kroy pamphut : pralom lok khan manosancetana / Kuy Yak Hu – Pat Tam Pan: Pannagar Q'yn H'un 2502 [1959] [mf ed 1990] – 1r with other items – 1 – (in khmer) – mf#mf-10289 seam reel 109/2 [§] – us CRL [959]

Broiler industry – Mount Morris IL 1974-99 – 1,5,9 – ISSN: 0007-2176 – mf#9647 – us UMI ProQuest [636]
Brokaw, George Lewis *see*
- Doctrine and life
- The lord's supper
Broken arcs : a west country chronicle by christopher hare / Andrews, Marian [pseud Christopher Hare] – London, New York: Harper & Bros, 1898[1897] – (= ser 19th c women writers) – 4mf – 9 – mf#5.1.84 – uk Chadwyck [420]
Broken arrow / Selfridge Air Force Base – 1969 aug 4-1971 apr 5, 1969 aug 4-1971 apr 5 – 2r – 1 – mf#720772 – us WHS [071]
Broken barriers – v1 n1-v5 n3 [1975 jun 14/jul 20-1980 early spring] – 1r – 1 – mf#637335 – us WHS [071]
Broken bits of byzantium by c g curtis... : lithographed...with some additions, by mary a walker – Constantinople: Lorentz & Keil Libraires de S M I le Sultan [1887-91] – (= ser 19th c art & architecture) – 2mf – 9 – mf#4.1.58 – uk Chadwyck [930]
Broken bow daily republican – Broken Bow, NE: D M Amsberry, mar 1888 (daily ex sun) [mf ed mar 21-dec181888 (gaps) filmed [1989-93]] – 2r – 1 – (cont by: daily republican) – us NE Hist [071]
Broken bow daily republican – Broken Bow, NE: J Pigma. v1 n1. feb 6 1911- (daily) [mf ed feb 6, 28 1911 filmed 1999] – 1r – 1 – (cont: daily evening republican) – us NE Hist [071]
Broken bow daily republican – Broken Bow, NE: Republican Pub Co (daily ex sun) [mf ed v5 n70. jun 10 1892] – 1r – 1 – (cont: daily republican. cont by: daily evening republican) – us NE Hist [071]
Broken bow free press – Broken Bow, NE: J G Painter (wkly) [mf ed v1 n12. jan 18-mar 21 1912 (lacks feb 1 1912) filmed 1999] – 1r – 1 – us NE Hist [071]
The broken bow times – Broken Bow, NE: Times Pub Co (wkly) [mf ed v2 n25. may 25 1888 filmed [1989]] – 1r – 1 – us NE Hist [071]
Broken hill age – Broken Hill, Australia 3 oct 1893-13 may 1895 (wkly) (very imperfect) – 1 – (discontinued) – uk British Libr Newspaper [079]
Broken hopes and perfect life – London, England. no date – 1r – us UF Libraries [240]
The broken platform : or, a brief defence of our symbolical books against recent charges of alleged errors / Hoffman, John N – Philadelphia: Lindsay & Blakiston 1856 [mf ed 1992] – 1mf – 9 – 0-524-04769-3 – (incl bibl ref) – mf#1991-2155 – us ATLA [242]
Broken ties and other stories / Tagore, Rabindranath – London: Macmillan and Co, 1925 – (= ser Samp: indian books) – us CRL [830]
The broken title of episcopal inheritance : or, a discovery of the weake reply... / Burgess, C – London: John Bellamie, and Ralph Smith, 1642 – 1mf – 9 – mf#PW-65 – ne IDC [240]
The broken wing : songs of love, death and destiny, 1915-1916 / Naidu, Sarojini – London: William Heinemann, 1917 – (= ser Samp: indian books) – us CRL [780]
Brokensha, Miles *see* Fourth of july raids
Broker magazine – New York NY 2001+ – 1,5,9 – ISSN: 1540-0824 – mf#29631 – us UMI ProQuest [332]
Brokhage, J D *see* Francis patrick kenrick's opinion on slavery
Brokiga blad – Stockholm, 1908-30 – 23r – 1 – sw Kungliga [078]
Brokmiller, I F *see* Novosti zhizni
Broling, Gustav *see* Bemerkungen auf einer reise durch england
Bromage, Richard Raikes *see* The holy catechism of nicolas bulgaris
Bromberger tageblatt – Bromberg (Bydgoszcz PL), 1942 apr-dec, 1944 may-sep [gaps] – 2r – 1 – (with gaps; title varies: 1 oct 1919: ostdeutsche rundschau, 1 jan 1921?: deutsche rundschau in polen, 2 sep 1939: deutsche rundschau; filmed by misc inst: 1920, 1921, 1922-23, 1924-39, 1940, 1941-42, 1943, 1918 3 apr-1919 30 mar [2r]) – uk British Libr Newspaper; gw Misc Inst [077]
Bromborough, st barnabas – (= ser Cheshire monumental inscriptions) – 2mf – 9 – £4.00 – mf#9 – uk CheshireFHS [929]
The bromfield gazette – Bromfield, NE: C H Israel, 1892 (wkly) [mf ed v1 n4. jun 17 1892] – 1r – 1 – us NE Hist [071]
Bromfield, Louis *see* The rains came
Bromham – (= ser Bedfordshire parish register series) – 1mf – 9 – £3.00 – uk BedsFHS [929]
Bromham, st owen monumental inscriptions – Arthur Weight Matthews 1907-16 – (= ser Bedfordshire parish register series) – 1mf – 9 – £1.25 – uk BedsFHS [929]
Bromley and hayes news shopper – Bromley, England 10 feb 1983-27 feb 1991 [mf 1986-] – 1 – (variant ed of: beckenham & penge record; cont: news shopper [12 dec 1968-3 feb 1983]; cont by: news shopper (bromley & hayes) 6 mar 1991-) – uk British Libr Newspaper [072]
Bromley and west kent telegraph – Bromley, England 25 dec 1886-31 dec 1896 – 1 – (cont: bromley telegraph, st mary cray, sevenoaks & local journal [1 feb 1868-25 may 1872]; cont by: bromley telegraph: (bromley telegraph & west kent herald) 8 jan-10 sep 1898) – uk British Libr Newspaper [072]
Bromley & beckenham advertiser – Bromley, England 17 nov 1989-2 aug 1990 – 1 – (cont, in pt, of: beckenham & penge advertiser; variant ed of: croydon advertiser; discontinued) – uk British Libr Newspaper [072]
Bromley, beckenham, chislehurst times – Bromley, England 2 nov 1989-31 may 1990 – 1 – (amalg of: beckenham times bromley times & chislehurst times [2 nov 1989-31 may 1990] and of: beckenham times & bromley times [7 jun 1990-]) – uk British Libr Newspaper [072]
Bromley, beckenham, chislehurst times – Bromley, England 2 nov 1989-31 may 1990 – 1 – (amalg of: beckenham times, bromley times & chislehurst times [2 nov 1989-31 may 1990] & of: beckenham times & bromley times [7 jun 1990-]; cont by: bromley & beckenham times-7 jun 1990-28 sep 2000) – uk British Libr Newspaper [072]
Bromley & beckenham express – Bromley, England 10 jan 2001-15 sep 2004 – 1 – (cont by: express (bromley) 23 sep 2004-19 jan 2006) – uk British Libr Newspaper [072]
Bromley & beckenham leader – Bromley, England 14 jun 1996-3 mar 2000 [mf 1986-2000] – 1 – (cont: leader [3 mar 1995-7 jun 1996]; cont by: bromley & district leader [10 mar-22 dec 2000]) – uk British Libr Newspaper [072]
Bromley & beckenham times – Bromley, England 7 jun 1990-28 sep 2000 – 1 – (cont: bromley, beckenham, chislehurst times [2 nov 1989-31 may 1990]; cont by: kentish times (bromley & beckenham) 5 oct 2000-20 sep 2002) – uk British Libr Newspaper [072]
Bromley [& beckenham] times leader – Bromley, England 30 oct 1992-24 feb 1995 [mf 1986-2000] – 1 – (cont: bromley leader [6 dec 1990-23 oct 1992]; cont by: leader (bromley & beckenham) 3 mar 1995-7 jun 1996) – uk British Libr Newspaper [072]
Bromley borough news – Bromley, England 6 jan 1939-26 dec 1944 – 1 – (cont: bromley & hayes guide & borough news [6 nov 1936-30 dec 1938]) – uk British Libr Newspaper [072]
Bromley borough news [biggin hill news] – Bromley, England 7 mar 1981-29 jun 1995 [mf 1986-] – 1 – (numeration cont that of biggin hill news; cont by: bromley news [6 jul 1995-]) – uk British Libr Newspaper [072]
Bromley chronicle – Bromley, England 10 sep 1891-2 jun 1921 (wkly) – 1 – (incorp with: bromley mercury; wanting: 1897) – uk British Libr Newspaper [072]
Bromley & county independent – Bromley, England 4 sep 1889 – 1/4r – 1 – uk British Libr Newspaper [072]
Bromley & district times – Bromley, England 4 jan 1889-25 may 1928 – 1 – (cont by: bromley & kentish times [1 jun 1928-13 nov 1970]) – uk British Libr Newspaper [072]
Bromley, James *see* The romish inquisition as adopted by the wesleyan conference
Bromley journal and west kent herald – [London & SE] Bromley LSL 21 may 1869-3 may 1912 (wkly) – 31r – 1 – (lacking: 1886) – uk Newsplan [072]
Bromley leader – Bromley, England 6 dec 1990-23 oct 1992 [mf 1986-2000] – 1 – (cont: bromley comet leader [8 feb-15 nov 1990]; cont by: bromley [and beckenham] times leader [30 oct 1992-24 feb 1995]) – uk British Libr Newspaper [072]
Bromley local guide and advertiser – Bromley, England 28 feb 1903-31 oct 1936 [mf 1904] – 1r – 1 – (cont by: bromley & hayes guide & borough news [6 nov 1936-30 dec 1938]) – uk British Libr Newspaper [072]
Bromley record and monthly advertiser – Bromley, England jun 1858-dec 1913 [mf 1858-62,1885,1887,1888,1890,1904] – 1 – uk British Libr Newspaper [072]
Bromley, Scott *see* The relationship of the congruence of perceived and preferred cohesion to sport performance and satisfaction
Bromley times – Bromley, England 20 nov 1970-26 oct 1989 [mf 1986-oct 1989] – 1 – (cont: bromley & kentish times [1 jun 1928-13 nov 1970]) – uk British Libr Newspaper [072]
Bromma nyheter – Stockholm, Sweden. 1946-54 – 4r – 1 – sw Kungliga [078]
Bromme, Traugott *see* Reisen durch die vereinigten staaten und ober-canada
Bromwell, Henrietta Elizabeth *see* Colorado portrait and biography index
Bromwell, William J *see* A digest of the military and naval laws of the confederate states
Bronces de mexico / Jinesta, Carlos – Mexico City? Mexico. 1949 – 1r – us UF Libraries [972]
Bronces y llamas / Montagu Y Vivero, Guillermo De – Habana, Cuba. 1941 – 1r – us UF Libraries [972]
Brondsted, J *see* Handel og samfaerdsel i oldtiden
Bronevskii, B A *see* Osvobozhdenie rossii
Bronkhurst, H V P *see*
- Among the hindus and creoles of british guyana
- Colony of british guyana and its labouring population
Bronn, H G *see* System der urweltlichen pflanzenthiere...
Bronn, Heinrich *see* Ergebnisse meiner naturhistorisch-oeconomischen reisen
Bronnen, Arnolt *see*
- Katalaunische schlacht
- Napoleons fall
- Ostpolzug
- Rheinische rebellen
- O s
Bronnen en grondslagen van het godsdienstig geloof : formeel gedeelte van de geloofsleer, op het standpunt van de moderne wetenschap / Hoekstra, Sytze – Amsterdam:P.N. van Kampen, 1864 – 1mf – 9 – 0-8370-3613-5 – mf#1985-1613 – us ATLA [210]
Bronnen tot de kennis van hret leven en de werken van d van coornhert / Becker, Bernhard – Den Haag, 1928 – 7mf – 8 – €15.00 – ne Slangenburg [100]
De bronnen van carel van mander voor "het leven der doorluchtighe nederlandtsche en hoogduytsche schilders" / Greve, H E – Haag, 1903 – 2v on 5mf – 9 – mf#O-518 – ne IDC [700]
Bronnitskii uezd : ispolnitel'nyj komitet sovetov. izvestiia bronnitskogo uispolkoma i uezdkoma rkp – Bronnitsy, Russia 1918 [mf ed Norman Ross] – 1r – 1 – mf#nrp-369 – us UMI ProQuest [077]
Brons, Anna *see* Ursprung, entwicklung und schicksale der taufgesinnten oder mennoniten
Bronson alcott's fruitlands / Sears, Clara Endicott [comp] – Boston: Houghton Mifflin, 1915 [mf ed 1990] – 1mf – 9 – 0-7905-6951-5 – (with: transcendental wild oats by louisa may alcott) – mf#1988-2951 – us ATLA [190]
Bronson, Edgar Beecher *see* The vanguard
Bronson, Walter C *see* History of brown university, 1764-1914
Bronte, Charlotte *see* Jane eyre
Bronte manuscripts : literary manuscripts and correspondence of the bronte family from the bronte parsonage, haworth and the british library, london – 12r – 1 – (includes printed guide) – mf#C35-14500 – us Primary [420]
Bronterre's national reformer : in government, law, property, religion and morals – London, England 7 jan-18 mar 1837 – 1 – uk British Libr Newspaper [320]
Bronterre's national reformer *see* Herald of the rights of industry, 1834
Bronwley, K A *see* The effects of music on psychophysiological stress responses to graded exercise
Bronwood baptist church – Terrell Co, GA. 2332p. 1898-1992 – 1 – $104.94 – (lacking: sep 1976-aug 1977) – mf#6711 – us Southern Baptist [242]
Bronx county historical society. journal – Bronx NY 1987-89 – 1,5,9 – ISSN: 0007-2249 – mf#16444 – us UMI ProQuest [978]
Bronze texan news – Fort Worth TX. 1969 may 2, aug 21, oct 2,16 – 1r – 1 – (cont: bronze news) – mf#4164348 – us WHS [071]
The bronzes of nalanda and hindu-javanese art / Bernet Kempers, August Johan – Leiden: EJ Brill, [1933] – (= ser Samp: indian books) – us CRL [730]
Der brook : [A novel] / Tuegel, Ludwig – 2. aufl. Hamburg: Hanseatische Verlagsanstalt 1943, c1938 [mf ed 1991] – 1r – 1 – (filmed with: leuchtendes land / luis trenker) – mf#2917p – us Wisconsin U Libr [830]
Brook, Benjamin *see* Memoir of the life and writings of thomas cartwright, b.d., the distinguished puritan reformer
Brook farm : its members, scholars, and visitors / Swift, Lindsay – New York: Macmillan, 1900 [mf ed 1990] – (= ser National studies in american letters) – 1mf – 9 – 0-7905-6690-7 – (incl bibl ref) – mf#1988-2690 – us ATLA [975]
The brook farm papers, 1842-1901 – [mf ed 1978] – 1r – 1 – (with p/g. coll contains unique & complete record of one of america's most noted 19th-c intellectual communes) – us MA Hist [420]
Brook, W Carr *see* Reason versus authority
Brook, William M *see* A history of the eastern defense command
Brookbank, Joseph *see* Well-tuned organ
Brookdale center on aging newsletter – 1988 fall – 1r – 1 – mf#5308383 – us WHS [618]
Brooke, Alan E *see*
- The fragments of heracleon
- The old testament in greek
Brooke, Alan England *see* The book of judges in greek
Brooke, Arthur de *see*
- Sketches in spain and morocco
- Ein winter in lappland und schweden
Brooke, Donald Lloyd *see* Citrus-grove cooperative caretaking
Brooke finchley's daughter / Albert, Mary – London: Chatto & Windus, 1891 – (= ser 19th c women writers) – 4mf – 9 – mf#5.1.103 – uk Chadwyck [830]
Brooke, Frances *see* The history of emily montague
Brooke, George J *see* Allegro qumran collection on microfiche
Brooke, Henry *see* The fool of quality
Brooke, John R *see*
- Civil report of major john r brooke
- Civil report of major-general john r brooke
Brooke, John T *see* The legal profession: its moral nature, and practical connection with civil society
Brooke, R *see* A discourse
Brooke, Stopford Augustus *see*
- Christ in modern life
- The development of theology as illustrated in english poetry from 1780 to 1830
- The early life of jesus
- The fight of faith
- Freedom in the church of england
- God and christ
- The gospel of joy
- The history of early english literature
- The kingship of love
- The late rev. f.d. maurice
- Life, letters, lectures, and addresses of frederick w robertson
- The life superlative
- Milton
- The old testament and modern life
- The ology in the english poets
- The onward cry and other sermons
- Religion in literature
- Sermons preached in st james's chapel, york street, london
- Short sermons
- The spirit of christian life
- Tennyson
- The unity of god and man, and other sermons
Brooke, Thomas *see* History of the island of st helena
Brooke, William Graham *see* The public worship regulation act, 1874
Brooker, Marvin A *see*
- Farm tenancy in jackson county, florida
- Farmers' cooperative associations in florida
- Study of the cost of transportation of florida citrus fruits with comparative costs
Brookes, Edgar Harry *see*
- City of god and the city of man in africa
- South africa in a changing world
Brookes, Henry *see* The bank act of 1844
Brookes, Iveson *see* Abolition and emancipation
Brookes, Iveson L *see* A discourse
Brookes, James Hall *see*
- The christ
- God spake all these words
- Is the bible inspired?
- Is the bible true?
- Salvation
Brookfield 1696-1849 – Oxford, MA (mf ed 1995) – (= ser Massachusetts vital record transcripts to 1850) – 20mf – 9 – 0-87623-219-5 – (mf 1t,3t: births 1701-96. mf 1t-3t: marriages 1718-93. mf 3t: deaths 1758-1806. mf 4t-9t: births & deaths 1696-1814. mf 9t: intentions & marriages 1766-77. mf 9t-12t: births & deaths 1696-1847. mf 12t-14t: intents & marriages 1792-1825. mf 15t-16t: marriages 1816-43. mf 16t-17t: intents & marriages 1840-45. mf 17t-18t: births 1839-49. mf 18t: marriages 1842-49. mf 19t: deaths 1843-49. mf 20t: out-of-town marriages 1696-1799) – us Archive [978]
Brookfield 1700-1895 – Oxford, MA (mf ed 1987) – (= ser Massachusetts vital records) – 44mf – 9 – 0-87623-019-2 – (mf 1-9: births & deaths 1700-1818. mf 10-14: births & deaths 1768-1847. mf 15-22: marriages, publishments, misc. 1793-1844. mf 23-26: b,m,d 1844-60. mf 23-24: births 1843-58. mf 25: births 1859-60; marriages 1843-57. mf 26: deaths 1843-57. mf 27: marriage intentions 1845-49; marriages 1850-72. mf 28: marrigaes 1873-95. mf 29-31: town records 1845-1918. mf 32-34: births 1861-95. mf 34-37: marriages 1722-1895. mf 37-39: deaths 1857-95. mf 40-44: index to vital records 1857-1911) – us Archive [978]
Brookfield, Frances Mary *see* The cambridge "apostles"
Brookfield news – Brookfield WI. 1955 aug 18/1956-2003 nov-dec – 132r – 1 – (with gaps) – mf#1002773 – us WHS [071]

Brookfield [telephone directory] : listing – 1948 jun, 1949 pt, 1949 pt, 1953 nov pt, 1953 nov pt, 1954 pt, 1954 pt, 1956 pt, 1956 pt 1958 – 10r – 1 – mf#2862324 – us WHS [917]

Brookfield-elm grove post – Brookfield, Elm Grove, West Allis WI. 1977 mar/jun-1980 sep-1981 jan – 13r – 1 – (with gaps) – mf#1049819 – us WHS [071]

Brookgreen bulletin – v1 n1-v15 n1 [1971 summer-1985] – 1r – 1 – (cont by: brookgreen journal) – mf#1095461 – us WHS [071]

Brookhill baptist church. etowah, tennessee : church records – Apr 1946-Dec 1982. Formerly New Hope Baptist Church, name changed to West Etowah in 1949, name changed to Brookhill in 1965 – 1 – us Southern Baptist [242]

Brookings bulletin – Washington DC 1962-82 – 1,5,9 – (cont by: brookings review) – ISSN: 0007-229X – mf#6984 – us UMI ProQuest [338]

Brookings papers on economic activity – Washington DC 1970+ – 1,5,9 – ISSN: 0007-2303 – mf#6160 – us UMI ProQuest [330]

Brookings review – Washington DC 1987+ – 1,5,9 – (cont: brookings bulletin) – ISSN: 0745-1253 – mf#15723 – us UMI ProQuest [338]

Brookings, Robert Somers *see* Big business and the public

Brookings-harbor pilot – Brookings OR: D Akers & D Holman, 1946-78 [wkly] [mf ed 1964-79] – 20r – 1 – (cont by: curry coastal pilot (1978-)) – us Oregon Lib [071]

Brookline 1655-1849 – (mf ed 1995) – (= ser Massachusetts vital record transcripts to 1850) – 11mf – 9 – 0-87623-220-9 – (mf 1t-3t: b,d,m 1683-1758. mf 2t-4t: births & deaths 1707-1847. mf 3t-4t: publishments & marriages 1753-1842. mf 4t-6t: b,d,m 1680-1845. mf 6t-7t: out-of-town marriages 1655-1799. mf 7t-8t: marriages 1839-45. mf 7t: publishments 1842-46. mf 8t: vital records 1772-1849. mf 9t: intentions 1846-49. mf 9t-10t: births 1844-49. mf 10t-11t: marriages 1844-49. mf 11t: deaths 1844-49) – us Archive [978]

Brookline 1748-1928 – Oxford, MA (mf ed 1999) – (= ser New hampshire town and vital records) – 87mf – 9 – 0-87623-175-X – (mf 1-3: births 1748-1896; deaths 1764-1921. mf 2: marriages 1778-1849. mf 3-6: marriages & intents 1850-1923. mf 7-16: town and tax record 1769-1831. mf 9,12: marriages 1810-13. mf 14: births 1764-79. mf 14-15: marriages 1791-1841. mf 17-25: town & tax recs 1821-48. mf 26-29: town payments 1782-1840. mf 29-31: militia 1827-37, 1877. mf 32-48: tax invoices 1836-57. mf 49-54: town rayments 1864-94. mf 55-62: town records 1848-67. mf 63-70: town records 1867-82. mf 70-79: town records 1882-95. mf 80,82: deaths 1838-96. mf 80-81: marriages 1879, 1882-97. mf 80-82: births 1850-97. mf 83-84: marriages & index 1897-1928. mf 85: births & index 1897-1901. mf 85-87: deaths & index 1897-1928) – us Archive [978]

Brooklyn barrister – Brooklyn NY 1950-94 – 1,5,9 – ISSN: 0007-232X – mf#2382 – us UMI ProQuest [340]

Brooklyn barrister – v1-52. 1950-2000 – 9 – $620.00 set – ISSN: 0007-232X – mf#101121 – us Hein [340]

Brooklyn jewish examiner – Brooklyn, NY. 1932-37 – 1 – us AJPC [071]

Brooklyn journal of international law – v1-26. 1975-2001 – 5,6,9 – $390.00 set – (v1-10 1975-84 in reel $88. v11-26 1985-2001 in mf $302) – ISSN: 0740-4824 – mf#101131 – us Hein [341]

Brooklyn law review – v1-66. 1932-2001 – 1,5,6 – $1694.00 – (v1-61 1932-95 in reel $1534. v62-66 1996-2001 in mf $160) – ISSN: 0007-2362 – mf#101141 – us Hein [340]

Brooklyn life – Brooklyn. v1-83. 1890-1931 – 1 – us L of C Photodup [073]

Brooklyn longshoreman / International Longshoremen's Association – special merger iss, 1973 dec-1978 aug – 1r – 1 – mf#498712 – us WHS [639]

Brooklyn. New York. Siloam Presbyterian Church *see* Semi-centennial

Brooklyn news – Brooklyn WI. 1898 jan 20-1900 jun 20, 1900 jun 27-1902 feb 26, 1902 mar 5-1903 nov 4, 1903 nov 11-1905 nov 1 – 4r – 1 – mf#943151 – us WHS [071]

Brooklyn, Presbytery (Pres. Church in the USA) *see* Records, 1838-1918

Brooklyn record – Brooklyn WI. 1882 sep 7-1883 apr 26 – 1r – 1 – mf#939411 – us WHS [071]

Brooklyn teller – Brooklyn WI. 1915 jan 13/1916 sep 20-1953/56 – 23r – 1 – (with gaps) – mf#943155 – us WHS [071]

Brooks, A D *see* History of ellis county baptist association

Brooks, A M *see* Unwritten history of old st augustine

Brooks, A N *see*
– Crimp
– Strawberries in florida

Brooks, Abbie M *see*
– Petals plucked from sunny climes

Brooks, Alfred Mansfield *see* Architecture

Brooks, Arthur *see* Phillips brooks

Brooks, Charles Wolcott *see* Early migrations

Brooks, E W *see* The syriac chronicle

Brooks, Edgar Harry *see* The bantu in south african life

Brooks, Elbridge Gerry *see*
– Our new departure
– Universalism, a practical power

Brooks, Elbridge Streeter *see* The life-work of elbridge gerry brooks, minister in the universalist church

Brooks, Elizabeth Harper *see* Java and its challenge

Brooks, Ernest Walter *see* Joseph and asenath

Brooks, Geo B *see* A chapter from the north-west rebellion

Brooks, Gladys C *see* History of lexington baptist church, oglethorpe county, georgia, 1847-1974

Brooks, Henry M *see* Olden-time music

Brooks, Henry Mason *see* Olden-time music

Brooks, J W *see* A new arrangement of the proverbs of solomon

Brooks, James *see* Two duetts for one performer on the violin...op 4

Brooks, John A *see* A debate on the beginning of messiah's reign, the abrogation of the mosaic law, and first proclamation of the gospel

Brooks, John Cotton *see*
– Essays and addresses
– Sermons for the principal festivals and fasts of the church year

Brooks, John Graham *see* As others see us

Brooks, John Rives *see* Scriptural sanctification

Brooks, Joseph S *see* Laws the detective should know

Brooks, Michael John *see* A program of new church member orientation in the fairfax baptist church, valley, alabama

Brooks, Philip Coolidge *see* Diplomacy and the borderlands

Brooks, Phillips *see*
– Addresses
– Baptism and confirmation
– The candle of the lord and other sermons
– Essays and addresses
– The law of growth and other sermons
– Lectures on preaching
– Letters of travel
– The light of the world, and other sermons
– New starts in life and other sermons
– Phillips brooks' addresses
– Seeking life and other sermons
– Sermons
– Sermons for the principal festivals and fasts of the church year
– The spiritual man, and other sermons
– Tolerance
– Twenty sermons

Brooks, Richard L *see* Comparison of citrus fruit grown on various rootstocks

Brooks, Robert Clarkson *see* Government and politics of switzerland

Brooks, Samuel H *see*
– Designs for cottage and villa architecture
– Modern architecture
– Rudimentary treatise on the erection of dwelling-houses

Brooks, Thomas *see* Letter from a good and happy father to his daughter

Brooks, Walter Rollin *see* God in nature and life

Brooksiana : or, the controversy between senator brooks and archbishop hughes: growing out of the recently enacted church property bill – New York: T W Strong 1870 [mf ed 1986] – 1mf – 9 – 0-8370-7128-3 – (int by archbishop of new york) – mf#1986-1128 – us ATLA [241]

Brooksville herald – Brooksville, FL. v25 n1-v40 n94. 1926 jan 07-1928 nov 23 – 1r – (missing: 1926 jun 4-8, 18, aug 31; 1927 jan 04, aug 26-sep 27, oct 07; 1928 jan 24, jun 29) – us UF Libraries [071]

Brooksville journal – Brooksville, FL. 1928 jan 05-1959 – 25r – (gaps) – us UF Libraries [071]

Brooksville sun – Brooksville, FL. v1 n1-v24 n44. 1932 mar 04-1959 apr 16 – 11r – (gaps) – us UF Libraries [071]

Brooksville sun-journal – Brooksville, FL. 1960-1980 – 32r – (gaps) – us UF Libraries [071]

Brookville american – Brookville, PA. -w 1981-1982 – 13 – $25.00r – us IMR [071]

Brookville american – Jefferson, PA., 1918-1981 – 13 – $25.00r – us IMR [071]

Brookville democrat – Brookville, PA., 1879-1884 – 13 – $25.00r – us IMR [071]

Brookville democrat – Jefferson, PA., 1879-1883 – 13 – $25.00r – us IMR [071]

Brookville republican – Brookville, PA., 1826-1837, 1873 – 13 – $25.00r – us IMR [071]

Brookwood – Brookwood, Katonah, New York. v. 1-14. mar 20 1923-july 1936. (incomplete) – 1 – us NY Public [073]

Broom : an international magazine of the arts – Rome. v. 1-6 n1. Nov 1921-Jan 1924 – 1 – us NY Public [700]

The broom : san diego's progressive weekly / ed by Aryan, C Leon de – San Diego, CA: C Leon de Aryan [mar 14 1932-dec 21 1953] (wkly) – 5r – 1 – us CRL [071]

Broom maker : official journal / International Broom and Brush Makers' Union – 1903 sep-1904 dec – 1r – 1 – mf#3240136 – us WHS [680]

Broom, Robert *see* Finding the missing link

Broomcorn growing – Tallahassee, FL. 1941 – 1r – us UF Libraries [630]

Broome, Gordon *see* Report on the canadian phosphates

Broome, Mary A *see* Ein jahr aus dem leben einer hausfrau in sued-afrika

Broome, Mary Anne (Stewart) Barker, lady *see* The bedroom and boudoir

Broomhall, B *see* The evangelisation of the world

Broomhall, Benjamin *see* The evangelisation of the world

Broomhall, Marshall *see*
– The chinese empire
– Chuan chiao wei jen ma-li-hsun
– Doctor lee
– Heirs together of the grace of life
– Islam in china
– The jubilee story of the china inland mission
– Last letters and further records of martyred missionaries of the china inland mission
– Martyred missionaries of the china inland mission
– Pioneer work in hunan by adam dorward
– Present day conditions in china

Broomstick / Options for Women Over Forty [Organization]. San Francisco Women's Centers – 1978 dec-1983, 1984 jan/feb-1988 may/jun – 2r – 1 – mf#642638 – us WHS [305]

Bror leonard grondal papers, 1908-1974 [bulk 1945-1974] / Grondal, Bror Leonard – 3 cubic ft [3 boxes] – (some biogr material on mf; genealogical materials are in swedish; with finding aid) – us UW Libraries [634]

Brosch, Hermann Josef *see*
– Der seinsbegriff bei boethius

Brosch, Moritz *see* Papst julius 2. und die gruendung des kirchenstaates

Brossard, Sebastien de *see*
– Dictionaire de musique
– Elevations et motets a 2. et 3. voix, et a voix seule, deux dessus de violon, ou deux flutes

Brosses, Charles de *see* Du culte des dieux fetiches

Brosset, M *see*
– ...De patriarche armenien de constantinople
– Rapport sur la 2me partie du voyage du p sargis dchalaliants dans la grande-armenie
– Rapports sur un voyage archeologique dans la georgie et dans l'armenie, execute en 1847-1848
– Relation du pays de ta ouan...

Brosur / Wanita Perhimpunan Sardjana Hukum Indonesia – Djakarta, 1960/1961. v1 – 2mf – 9 – mf#SE-1985 – ne IDC [959]

Brot : roman / Waggerl, Karl Heinrich – Leipzig: Inselverlag 1938 [mf ed 1991] – 1r – 1 – (filmed with: werke und briefe / wilhelm heinrich wackenroder) – mf#3021p – us Wisconsin U Libr [830]

Brot und salz aus gottes wort *see* Bread and salt from the word of god

Brotes liricos / Romano, R Clodomiro – Ciudad Trujillo, Dominican Republic. 1958 – 1r – us UF Libraries [972]

Brother azarias : the life story of an american monk / Smith, John Talbot – New York: William H Young, 1897 – 1mf – 9 – 0-8370-6776-6 – mf#1986-0776 – us ATLA [920]

Brother john's canaan in carolina, 1872-1956 / Washburn, W Wyan – 1 – us Southern Baptist [242]

Brother jonathan – La Salle IL 1842-43 – 1 – mf#3951 – us UMI ProQuest [420]

The brother of girls : the life story of charles n. crittenton / Crittenton, Charles Nelson – Chicago: World's Events, 1910 – 1mf – 9 – 0-7905-8240-6 – mf#1988-8103 – us ATLA [240]

Brotherhood = Bratstvo – Cleveland, OH: American Russian National Brotherhood, v17-18. 1940-1941; v14 1942 – us CRL [060]

Brotherhood : a monthly magazine designed to help the peaceful evolution of a juster and happier social order. london – May 1895-Apr. 1900. (incomplete) – 1 – us NY Public [073]

Brotherhood *see* County derry liberal

Brotherhood commission publication guide – oct 1966-67 – 1 – us Southern Baptist [242]

Brotherhood eyes – 1936 oct 31 – 1r – 1 – mf#5259177 – us WHS [360]

Brotherhood journal – Memphis. Tenn. 1940-67 – 1 – us Southern Baptist [242]

Brotherhood, nature's law / Harding, Burcham – New York: B Harding, 1897 – 1mf – 9 – 0-524-00884-1 – mf#1990-2107 – us ATLA [100]

Brotherhood [northern counties, ireland ed] – Limavardy, Ireland 11 may 1889-31 may 1890 – 1r – 1 – (incorp with: belfast weekly star; fr 14 sep 1889 publ in belfast) – uk British Libr Newspaper [072]

Brotherhood of locomotive engineer's monthly journal : devoted to the interests of the locomotive department of railroads – 1872-75, 1876-79, 1880-83, 1884-86, 1887-89, 1890-92, 1893-95, 1896-98, 1899-1900, 1901-02 – 10r – 1 – (cont: locomotive engineers' monthly journal; cont by: brotherhood of locomotive engineers journal) – mf#1411044 – us WHS [380]

Brotherhood of Locomotive Engineers [US] *see*
– Locomotive engineer
– Locomotive engineer's monthly journal
– Report of the legislative representatives of the order of railroad conductors, brotherhood of locomotive firemenand enginemen, brotherhood of railway

Brotherhood of Locomotive Engineers [US] et al *see*
– Joint legislative report of wisconsin state legislative representatives of the brotherhood of locomotive engineers, order of railway conductors, broth
– Joint report of wisconsinstate legislative representatives of the brotherhood of locomotive engineers, order of railway conductors, brotherhood of railway trainmen

Brotherhood of Locomotive Firemen *see* Locomotive firemen's magazine

Brotherhood of Locomotive Firemen and Enginemen *see*
– Directory
– Enginemen's press

Brotherhood of Locomotive Firemen [US] *see*
– Firemen's magazine
– Locomotive firemen's monthly magazine

Brotherhood of locomotive firemen's magazine – 1876 dec-1878 nov – 1r – 1 – (cont by: locomotive firemen's monthly magazine) – mf#2596356 – us WHS [360]

Brotherhood of locomotive firemen's magazine – Peoria IL. 1901 jan-oct, 1901 nov-1902 dec – 2r – 1 – (cont: locomotive firemen's magazine; cont by: brotherhood of locomotive firemen and enginemen's magazine) – mf#1416970 – us WHS [360]

Brotherhood of Machinery Molders *see* Machinery molders journal, 1888-1892 / vulcan record, 1868-1875

Brotherhood of Maintenance of Way Employees *see* Bmwe journal

Brotherhood of Marine Officers *see* Bell

Brotherhood of Metal Workers *see* Metal workers bulletin, 1910-1914 / weldors' journal, 1938-1941

Brotherhood of Painters and Decorators of America *see* Painters journal

Brotherhood of Painters, Decorators and Paperhangers of America *see*
– D c 9 newsletter
– Official journal of the brotherhood of painters, decorators and paperhangers of america
– Painter

Brotherhood of Painters, Decorators, and Paperhangers of America *see* Glassworker

Brotherhood of Railroad Trainmen *see*
– Railroad brakemen's journal
– Railroad trainman
– Railroad trainmen's journal
– Torpedo
– Trainman news

Brotherhood of Railway, Airline and Steamship Clerks, Freight Handlers, Express and Station Employees *see* Broadcaster

Brotherhood of Railway, Airline, and Steamship Clerks, Freight Handlers, Express, and Station Employes et al *see* Interchange

Brotherhood of Railway, Airline and Steamship Clerks, Frieght Handlers, Express and Station Employees *see* Bracgrounder

Brotherhood of Railway and Steamship Clerks, Freight Handlers, Express, and Station Employees *see* Railway clerk

Brotherhood of Railway Postal Clerks *see* Harpoon

The brotherhood of religions / Wadia, Sophia – Bombay, India: International Book House Ltd, 1944 – (= ser Samp: indian books) – us CRL [230]

Brotherhood of rr signalman / Hamilton Co. Cincinnati – (1967-jun 1980) [irreg, qrtly] – 1r – 1 – mf#B10318 – us Ohio Hist [331]

Brotherhood Railway Carmen of America *see*
– Railway carmen's journal

Brotherhood Rally of American Veterans Organizations *see* Veterans outlook

Brotherly love / Cox, John Edmund – London, England. 1850 – 1r – us UF Libraries [240]

Brotherly-kindness and unity essential to the christian character / Aitken, Roger – Aberdeen, Scotland. 1813 – 1r – us UF Libraries [240]

The brothers : from the bengali of svarnalata, a novel / Gangopadhyaya, Tarakanatha – London: India Society, 1928 – (= ser Samp: indian books) – (trans by edward thompson) – us CRL [830]

Brothers and Sisters for African Unity *see* Habari barua
Brothers, Barbara Jo *see* Journal of couples therapy
The brothers d'amours / Hannay, James – [S.l: s.n, 1898?] – 1mf – 9 – 0-665-93344-4 – mf#93344 – cn Canadiana [929]
Brothers Hospitallers of St John of God *see*
– Descripcion que hace la ilustre provincia del sr s raphael del peru, y reyno de chile
– Regla de n p s augustin, obispo, y doctor de la iglesia
Brothers Of Christian Instruction Of Ploermel *see* Manuel d'histoire d'haiti
Brothers of doom. the story of the pizarros of peru / Birney, Hoffman – New York: G.P. Potnams Sons, 1942 – sp Bibl Santa Ana [350]
The brothers of holy cross / Trahey, James Joseph – Notre Dame, Ind: University Press, [1904?] – 1mf – 9 – 0-524-04242-X – mf#1990-5033 – us ATLA [240]
Brothers of the Christian Schools [comp] *see*
– The third book of reading lessons
Brothers, Richard *see* Revealed knowledge of the prophecies and times, particularly of the present time
The brothers wiffen : memoirs and miscellanies / ed by Pattison, Samuel Rowles – London: Hodder and Stoughton, 1880 – 1mf – 9 – 0-524-06730-9 – mf#1991-2760 – us ATLA [920]
Brotherton messenger – v1 n1-v6 n6 [1981 nov-1986 dec] – 1r – 1 – mf#1508841 – us WHS [071]
Brotherus, V F *see* Botanische ergebnisse der schwedischen expedition nach patagonien und dem feuerlande 1907-1909
Brotteroder anzeiger – Brotterode DE, 1911 1 apr-8 apr [gaps], 1912 2 mar-1923 24 oct [gaps], 1924-1937 29 jun – 15r – 1 – gw Misc Inst [074]
Bro'ty advertiser – [Scotland] Broughty Ferry: A Bowman 22 jan 1915-7 feb 1919 (wkly) [mf ed 2003] – 3r – 1 – (cont by: carnoustie herald and bro'ty advertiser [jan 1918-feb 1919; split into: bro'ty advertiser & carnoustie herald) – uk Newsplan [072]
Bro'ty advertiser & carnoustie herald – [Scotland] Angus, Broughty Ferry: A Bowman 14 jul 1916-7 feb 1919 (wkly) [mf ed 2004] – 1r – 1 – (cont in pt: bro'ty advertiser) – uk Newsplan [072]
Brou, Louis *see*
– Antifonario visigotico...
– The monastic ordinale of st vedast's abbey arras, vol 1-2
– The psalter collects
Brou y J Vives, L *see* Antifonario visigotico mozarabe de la catadral de leon
Brouerius van Niedek, Matheus *see*
– Zederyke zinnebeelden der tonge
Brough, William *see* The natural law of money
Brougham : a weekly paper of useful and entertaining knowledge – [Scotland] Glasgow: printed...by A Colville 28 apr-11 aug 1832 (wkly) [mf ed 2004] – 1r – 1 – uk Newsplan [072]
Brougham and Vaux, Henry Brougham, Baron *see* Lives of men of letters of the time of george 3
Brougham, Henry Peter *see*
– A letter from the right hon lord brougham to the right hon sir james graham, bt, mp
– The life and times of henry, lord brougham, written by himself
Broughton baptist church. fulton county. arkansas : church records – 1893-1915 – 1 – us Southern Baptist [242]
Broughton, Len Gaston *see*
– The plain man and his bible
– The revival of a dead church
Broughton, Morris *see* Press and politics of south africa
Broughton, Rhoda *see* Nancy
Broughton, Samuel *see* Letters from portugal, spain, and france
Broughton, Thomas *see*
– Christian soldier
– Les marattes
Broughton, William *see* Voyage de decouvertes dans la partie septentrionale de l'ocean pacifique
Brouillette, Benoit *see*
– Le canada par l'image
– Geographie economique
Broullion, Nicolas *see* Missions de chine
Brousseau freres (Firme) *see* Les soirees canadiennes
Brousseau, Georges *see* Souvenirs de la mission savorgnan de brazza
Brousseau, Serge *see* Le beau roman d'amour de rolande desormeaux
Brousseau, Vincent *see* Bio-bibliographie de monsieur albert gervais
[Brousson, C] *see* Estat des reformez en france
Brousson, Jean-Jacques *see* Conversion de figaro
Brouwer, Anneus Marinus *see* Onze verhouding tot indie
Brouwer, Dirk *see* Diskussie van de waarnemingen van satellieten 1, 2 en 3 van jupiter, gedaan te johannesburg door r t a innes in de jaren 1908-1925
Brouwer, Johannes *see*
– Famous dutch writer denounces rebel atrocities
– The last days of unomuno
– Words of indignation and of truth
Brovender, Samuel J *see* Effectiveness of an abdominal training protcol on an unstable surface
Brovkin, T M *see* Bibliograficheskii ukazatel kooperativnoi literatury za 1927 g
Broward County Library *see* Building bridges
Broward jewish world – Boca Raton, Florida. v10 n44 (oct 25-31 1991)-v13 n7 (feb 25-mar 3 1994) – 100ft – (merged with: miami jewish tribune and palm beach jewish world to form: south florida jewish tribune) – us AJPC [939]
Broward jewish world *see* Palm beach jewish world
Broward latino – Hollywood, FL. 1983 jul-1989 feb – 1r – us UF Libraries [071]
Broward times – Pompano Beach FL. 1993 jul 9-dec 31, 1994 jan 7-apr 29, 1994 may 6-aug 26, 1994 sep 2-dec 30, 1995 jan 6-jun 30, 1995 jul-dec, 1996 jan 5-may 3 – 7r – 1 – mf#2733571 – us WHS [071]
Browarzik, Ulrich *see* Glaube, historie und sittlichkeit
Browder, Earl *see* Next steps to win the war in spain
Brower, David *see* Wilderness
Brower family circle – v1 n3 [1976 oct], v2 n1-2 [1977 apr-jul], v2 n4-v8 n2 [1978 jan-1983 jul] – 1r – 1 – mf#653595 – us WHS [640]
Brower, Jacob V *see* Field notebooks
Brower, Jacob Vradenberg *see* The mississippi river and its source
Brown, Abel J *see* The lutheran church built on the only true foundation
Brown, Alec *see* The juryman's handbook
Brown Alexander *see* Genesis of the united states
Brown, Alexander *see* Prosperity of the soul
Brown, Alfred W *see* Evesham friends in the olden time
Brown american / National Association of Negroes in American Industries – Philadelphia, 1936-45 [all publ] – (= ser Black journals, series 1) – 12mf – 9 – $125.00 – (v1-5 n8 cont as yrs 1941-45) – us UPA [305]
Brown, Archibald G *see* Lion-killing on a snowy day
Brown, Arthur Judson *see*
– Foreign missionary
– The foreign missionary
– The mastery of the far east
– The nearer and farther east
– New forces in old china
– Report of a visitation of the china missions
– Report of a visitation of the china missions of the presbyterian board of foreign missions
– Report of a visitation of the korea mission of the presbyterian board of foreign missions
– Report of a visitation of the philippine islands
– Report of a visitation of the siam and laos missions of the presbyterian board of foreign missions
– Report of a visitation of the syria mission of the presbyterian board of foreign missions
– Rising churches in non-christian lands
– Unity and missions
– The why and how of foreign missions
Brown, Benjamin F *see*
– A concise statement of the law of partnership
– Early religious history of maryland
Brown, Brian *see* The wisdom of the hindus
Brown, Brian et al *see* Divine or civil obedience?
Brown, Charles Barrington *see* Canoe and camp life in british guiana
Brown, Charles Brockden *see* The rhapsodist
Brown, Charles et al *see* Youth and life
Brown, Charles J *see*
– Church establishments defended
– Church of rome brought to the test of the epistle to the romans
– The divine glory of christ
– Last enemy
– Marriage affinity question
– Preaching, its properties, place, and power
Brown, Charles John *see* Disruption question stated
Brown, Charles Philip *see*
– English and telugu dictionary
– Telugu-english dictionary
– Vakyavali
Brown, Charles Reynolds *see*
– The cap and gown
– Faith and health
– The latent energies in life
– The main points
– The modern man's religion
– The quest of life
– The social message of the modern pulpit
– The young man's affairs
Brown, Charles Rufus *see*
– An aramaic method
– Book of the prophet jeremiah
Brown, Charles Stagmaier *see* Art of chorale-preluding and chorale accompaniment as presented in kittel's der angehende praktische organist
Brown, Clark *see* Sermon preached at wareham (massachusetts), 31 mar 1695
Brown Co. Georgetown *see*
– Brown county news
– Castigator
– Democratic standard
– Gazette
– News democrat
– News-democrat
– Southern ohio argus
– Western aegis series
Brown Co. Manchester *see* Signal series
Brown Co. Mount Orab *see* Brown county press
Brown Co. Ripley *see*
– Bee
– Blue jay
– Castigator
– Ripley bee
– Times
Brown, Colin Campbell *see*
– China in legend and story
– A chinese st francis
Brown county democrat – Ainsworth, NE: Clarence C Jones, 1906-v39 n22. jun 1 1945 (wkly) – 24r – 1 – (merged with: ainsworth star-journal (1893) to form: ainsworth star-journal and brown county democrat; suppls accompany some iss) – us Bell [071]
Brown county democrat – De Pere, Little Chute, Wrightstown WI,1891 mar 5/dec 24-1918 apr 12/1919 mar 13 – 24r – 1 – (with gaps; cont by: brown county journal-news, de pere journal-democrat) – mf#1006350 – us WHS [071]
Brown county herald – Green Bay WI. 1878 mar 7-sep 30 – 1r – 1 – (cont: fort howard herald, and general advertiser for brown county; cont by: de pere news and brown county herald) – mf#920574 – us WHS [071]
Brown County Historical Society et al *see* Green bay historical bulletin
Brown county journal – 1916 nov10-17 dec 14, 1917 dec 21-1918 apr 12 – 2r – 1 – (cont by: brown county journal-news) – mf#965400 – us WHS [071]
Brown county journal-news – De Pere WI. 1918 apr 19-19 mar 13 – 1r – 1 – (cont: de pere news; cont by: brown county journal) – mf#965396 – us WHS [071]
Brown county news / Brown Co. Georgetown – dec 1877-81, 1883-86 (fire damaged) [wkly] – 4r – 1 – mf#B9768-9771 – us Ohio Hist [071]
Brown county news / Brown Co. Georgetown – jul-oct 1864, jan-dec 1865 [wkly] – 1r – 1 – mf#B156 – us Ohio Hist [071]
Brown county press / Brown Co. Mount Orab – sep 1977-dec 1980 [wkly] – 3r – 1 – mf#B29707-29709 – us Ohio Hist [071]
Brown Daily Herald Voluntary Publishing Association *see* Fresh fruit
Brown, David *see*
– The apocalypse
– Christ's second coming
– Commentary on the epistle to the romans
– The epistle to the romans
Brown, David Stevens *see* Colonization
Brown, DE *see* Predicting blood pressure from activity, fitness, and body composition of borderline hypertensive individuals
Brown deer herald – Brown Deer, Shorewood, Whitefish Bay WI. 1960 nov 17/1961-1977 may-aug 25 – 23r – 1 – (cont by: north shore herald [brown deer wi ed]) – mf#1159249 – us WHS [071]
Brown deer herald – New Berlin WI. feb-mar, apr-jun, jul-sep, oct-dec – 4r – 1 – (cont: herald [brown deer wi ed: 1989]; cont by: fox point-bayside-river hills herald [shorewood wi: 1965], glendale herald [new berlin wi], shorewood herald [west allis wi], whitefish bay herald [west allis wi], north shore herald [west allis wi]) – mf#5466350 – us WHS [071]
Brown deer [telephone directory] : listing – 1948 jun, 1949 pt, 1953 nov pt, 1954 pt, 1956 pt, 1958 – 10r – 1 – mf#2862243 – us WHS [917]
Brown, Donald Mackenzie *see* The white umbrella
Brown, Douglas *see* Against the world
Brown, Edgar G *see* The civilian conservation corps and colored youth
Brown, Edmund Woodward *see* The divine indwelling
Brown, Elijah P *see* From nowhere to beulahland
Brown, Elizabeth Baldwin *see* Stoics and saints
Brown, Elizabeth W *see* The whole world kin
Brown, Elmer E *see* Die stellung des staates zur kirche in bezug auf den religionsunterricht in der schule in preussen, england und den vereinigten staaten von nordamerika
Brown, ephraim, papers, ms 1872 – 1790-1887 – 22r – 1 – (correspondence, land documents and deeds, business and financial records, and court docket books of ephraim brown, an early western reserve settler, and his descendants, a prominent cleveland industrialist family) – us Western Res [978]
Brown, Ernest Faulkner *see* The pastoral epistles
Brown, Fortune Charles *see* Christ on the throne of power and antichrist
Brown, Francis *see*
– Assyriology
– Church unity
– Teaching of the twelve apostles
Brown, Frederick *see* Religion in tientsin
Brown, G A *see* Financial and economic survey
Brown, Gamaliel *see* Appeal to the preachers of all the creeds
Brown, George *see*
– [Circular]
– George brown
– Melanesians and polynesians
– Pacific island culture and society
– The volume of creation
Brown, George Alexander *see* The diaries and memoirs, 1811-70
Brown, George Boylston *see* Survey of iberian folk-song and a study of the jota aragonesa
Brown, George M *see* Ponce de leon land and florida war record
Brown, George Stayley *see* Yarmouth, nova scotia
Brown, George William *see*
– The economic history of liberia
– The relation of the legal profession to society
Brown, Gerald F X *see* War dairy, patrol reports and personal papers
Brown, Gerard Baldwin *see*
– Fine art as a branch of university study
– From schola to cathedral
Brown, Gordon D et al *see* A comparison of skeletal muscle responsiveness to exercise in male and female sprague-dawley rats treated with an anabolic steroid
Brown, Gwethalyn Graham Erichsen *see* Earth and high heaven
Brown, H B *see* Brown's reports of admiralty and revenue cases in the sixth circuit
Brown, Hamlin L *see* Marketing florida tomatoes
Brown, Harvey *see* The doctrine of eternal punishment refuted upon natural principles. the reign of a thousand years
Brown, Henriette *see*
– Affaire guibord
Brown, Henry *see* Arminian inconsistencies and errors
Brown, Hiram Chellis *see* The historical bases of religions
Brown, Howard Nicholson *see* A life of jesus for young people
Brown, Hubert William *see* Latin america
Brown, Hugh Stowell *see* Baptism according to st paul
Brown, Isaac Van Arsdale *see* A historical vindication of the abrogation of the plan of union by the presbyterian church in the united states of america
Brown, J Coggin *see* Catalogue raisonne of the prehistoric antiquities in the indian museum at calcutta
Brown, J Colvin *see* Development of agricultural education in the state of florida from 1918-1928
Brown, J Newton *see* Encyclopedia of religious knowledge
Brown, James *see*
– The life of john eadie, d.d., ll.d
– Papers, 1843-51
Brown, James A et al *see* Lectures on the augsburg confession
Brown, James Baldwin *see*
– The christian policy of life
– The divine life in man
– The divine mysteries
– The doctrine of annihilation in the light of the gospel of love
– The doctrine of the divine fatherhood in relation to the atonement
– The higher life
– Idolatries, old and new
– Misread passages of scripture
– The risen christ, the king of men
– The soul's exodus and pilgrimage
– Stoics and saints
Brown, James Bryce *see* Views of canada and the colonists
Brown, James W *see* Fanti confederation
Brown, John *see*
– Apology for the more frequent administration of the lord's supper
– Apostolical succession in the light of history and fact
– Commonwealth england
– Danger of opposing christianity and the certainty of its final triu...
– The english puritans
– Five discourses preached before the university of cambridge
– From the restoration of 1660 to the revolution of 1688
– Historical account of the rise and progress of the secession
– The history of the english bible
– John brown letters
– Letter to the rev dr chalmers
– Letters, manuscripts, articles, etc
– Memorandum book
– of the light of nature

- On religion, and the means of its attainment
- On the character, duty, and danger, of those who forget god
- On the state of scotland in reference to the means of religious ins...
- Papers
- The pilgrim fathers of new england and their puritan successors
- Puritan preaching in england
- Selected papers from the kshs collection

Brown, John Crombie *see* Pastoral discourses
Brown, John F *see* The self-proving accounting system
Brown, John Newton *see* The obligation of the sabbath
Brown, John Porter *see* The dervishes, or, oriental spiritualism
Brown, John S *see* John s. brown papers
Brown, John T *see* A list of legal fees
Brown, John Thomas *see* Churches of christ
Brown, John Tod *see* Posthumous testimony of true faith
Brown, John Tom *see*
- Among the bantu nomads
- Setswana dictionary

Brown, Joseph *see* Sabbath-school missions in wisconsin
Brown, Joseph D *see* The effects of cooperative and individualistic goal structures on the learning domains of beginning tennis students
Brown, Josiah *see*
- Brown's parliamentary cases
- Reports of cases upon appeals and writs of error determined in the high court of parliament

Brown, L E G *see*
- The hereford breviary, vol 1
- The hereford breviary, vol 2
- The hereford breviary, vol 3

Brown, Lilian Mabel Alice (Roussel) *see* Unknown tribes, uncharted seas
Brown, Lilian Maclaughlin *see* Bring 'em back petrified
Brown, Lorna Joan *see* A therapeutic relationship
Brown, Louise Fargo *see* The political activities of the baptists and fifth monarchy men in england during the interregnum
Brown, Lucinda White *see* Lucinda white brown diary
Brown, Marc A *see* Relation of body fatness and blood lipids as risk factors for coronary heart disease in african american women
Brown, Marianna C *see* Sunday-school movements in america
Brown, Marianne Hawke *see* Orchestral style of gustav mahler as illustrated in the second symphony in c minor
Brown, Marshall *see* Wit and humor of bench and bar
Brown, Maurice Henry *see* Mortgages
Brown, Nelson Courtlandt *see* Preliminary examination of the forest conditions of florida
Brown, O E *see* Women preachers
Brown, Oliver May *see* Bible chronology vindicated by its own internal evidence
Brown, Otis, Mrs *see* Catholics in florida
Brown, P Hume *see* John knox
Brown, P M *see* Foreigners in turkey
Brown papers / National Institute for Women of Color [US] – n1-2 [1984-85] – 1r – 1 – mf#1219174 – us WHS [305]
Brown, Percy *see*
- Indian architecture
- Indian painting
- Indian painting under the mughals, ad 1550 to ad 1750

Brown, Peter Hume *see*
- George buchanan, humanist and reformer
- John knox
- Life of goethe

Brown, Ralph W *see* [Letters]
Brown, Richard *see*
- The coal fields and coal trade of the island of cape breton
- Domestic architecture
- On the geology of cape breton
- The principles of practical perspective
- The rudiments of drawing cabinet and upholstery furniture

Brown, Richard L *see* [31]p metabolic responses to activity of nonspecifically trained muscle tissue
Brown, Robert *see*
- The great dionysiak myth
- Miscellaneous botanical works, 1886-88
- Researches into the origin of the primitive constellations of the greeks, phoenicians and babylonians
- Semitic influence in hellenic mythology

Brown, Robert Allen *see* List of indiana lawyers, judges and prosecutors
Brown, Robert Christopher Lundin *see* British columbia
Brown, Robert K *see* Mission work in british columbia
Brown, Ronald Keith *see* The use of systems theory of organization in analyzing, planning and administering the work of first baptist church, columbia, tennessee
Brown, Rose (Johnston) *see*
- American emperor
- Two children and their jungle zoo

Brown rot of irish potatoes and its control / Eddins, A H – Gainesville, FL. 1936 – 1r – us UF Libraries [630]
Brown, Samuel *see* Lectures on the atomic theory
Brown, Samuel Windsor *see* The secularization of american education
Brown, Sanger *see* The sex worship and symbolism of primitive races
Brown, Seth E *see* The effects of internet-based instructional lesson planning on teacher trainee performance
Brown side out – v1 n8,12 [1990 nov 9, dec 7] – 1r – 1 – mf#1834381 – us WHS [071]
Brown, Stefani *see* Fat intake of university students
Brown, T J *see* Letter to the very rev archdeacon daubeny
Brown, Theron *see* The story of the hymns and tunes
Brown, Thomas *see*
- Annals of the disruption
- Church and state in scotland
- Priveleges of those to whom are committed the oracles of god

Brown, Thomas I *see* Economic cooperation among the negroes of georgia
Brown, Thomas Richard *see* The essentials of sanscrit grammar
Brown, Thomas Storrow *see*
- 1837
- Brief sketch of the life and times of the late hon louis joseph papineau
- Canada correspondence
- From the new york evening express, tuesday, april 22, 1873 canada correspondence
- My escape in 1837

Brown, Thomas Storrow [comp] *see* A history of the grand trunk railway of canada
Brown, Timothy *see* Commentaries on the jurisdiction of courts
Brown, Tom *see* Oil on ice
Brown, W G E *see* Report on the forest inventory of the island of do...
Brown, W Jethro *see* The austinian theory of law
Brown, W Kennedy *see* Gunethics, or, the ethical status of woman
Brown, W L *see* Letter to george hill
Brown, W Norman *see* India, pakistan, ceylon
Brown, Walter H *see* Notes on the smelting processes at freiberg
Brown, Warner *see* Judgement of very weak sensory stimuli
Brown, Wenzell *see* Angry men, laughing men
Brown, William *see*
- The history of the christian missions of the sixteenth, seventeenth, eighteenth, and nineteenth centuries
- Silver in its relation to industry and trade
- Three rondos for the piano forte or harpsichord

Brown, William Adams *see*
- The christian hope
- Christian theology in outline
- The essence of christianity
- Is christianity practicable?
- Modern missions in the far east
- Modern theology and the preaching of the gospel
- Morris ketchum jesup

Brown, William Alden *see* Portland cement industry
Brown, William Barrick *see* The history of a famous court house located at carlinville, illinois
Brown, William Bryant *see*
- The early history of congregationalism in new jersey and the middle provinces
- The problem of final destiny

Brown, William Carlos *see* An address before the commercial exchange of des moines, iowa, thursday, december 20, 1894
Brown, William Harvey *see*
- On the south african frontier

Brown, William John *see* Jamaican journey
Brown, William Laurence *see* Nature, the causes, and the effects of indifference with regard to...
Brown, William Linton *see* A treatise on free trade addressed to the farmers of canada
Brown, William Montgomery *see*
- The church for americans
- The crucial race question, or, where and how shall the color line be drawn
- The level plan for church union

Brown, William Norman *see*
- Art of enamelling on metal
- Manuscript illustrations of the uttaradhyana sutra
- A practical manual of wood engraving
- The united states and india and pakistan

Brown, William Oscar *see* Race relations in the american south and in south africa
Brown, William Wells *see* The black man
Brown, Winifred R *see* Federal rulemaking
Brown, Wm Holmes *see* House practice
Brown, Wm M, 3rd *see* Sediment transport and turbidity in the eel river basin
Brown, Wrisley *see* Impeachment: a monograph on the impeachment of the federal judiciary
Browne, Albert G *see* The utah expedition
Browne, Arthur S *see* Origin of the protestant episcopal church in the district of columbia
Browne, Chad *see* Memorial
Browne, Edward Granville *see*
- Kitab-i nuqtatu'l-kaf
- A literary history of persia
- Traveller's narrative written to illustrate the episode of the bab
- A year amongst the persians

Browne, Edward Harold *see*
- An exposition of the thirty-nine articles
- Life in knowledge of god
- The pentateuch and the elohistic psalms
- The pentateuch and the elohistic psalms in reply to bishop colenso
- Sermons on the atonement

Browne, George Forrest *see*
- An address on the anglo-saxon coronation forms
- Alcuin of york
- Anglican orders
- Augustine and his companions
- Boniface of crediton and his companions
- The christian church in these islands before the coming of augustine
- The continuity of possession at the reformation
- Continuity of possession at the reformation
- The continuity of the holy catholic church in england
- Continuity of the holy catholic church in england
- The conversion of the heptarchy
- The election, confirmation and homage of bishops of the church of england
- Glastonbury
- The history of the british and foreign bible society
- The odore and wilfrith
- On what are modern papal claims founded?
- The recollections of a bishop
- St aldhelm
- The st augustine commemoration
- Theodore and wilfrith
- The venerable bede

Browne, George Forrest et al *see*
- The church and life of to-day
- Lectures
- Lectures. third series

Browne, Henry *see*
- A handbook of hebrew antiquities
- Handbook of homeric study

Browne, Irving *see*
- Browne's index/digest to the new york court of appeals reports
- Humorous phases of the law
- Short studies in evidence
- Short studies of great lawyers
- A treatise on the admissibility of parole evidence in respect to written instruments

Browne, J H *see* On justification
Browne, J Ross *see* Crusoe's island
Browne, John *see* Sermons preached before the university of oxford
Browne, John Hutton Balfour *see* South africa
Browne, Joseph Vincent *see* The improvement of the harbor of quebec
Browne, knight and a day. leipzig 1928 / Bayle, Constantino – Madrid: Razon y Fe, 1928 – 9 – sp Bibl Santa Ana [999]
Browne, Lewis *see* This believing world
Browne [Miss] *see* Evening song to the virgin [at sea]
Browne, N E *see* A(merican) l(ibrary) a(ssociation) portrait index
Browne, P *see*
- The civil and natural history of jamaica
- The civil and natural history of jamaica in three parts

Browne, P D *see* The seventh and james street baptist church of waco, texas
Browne, R H C *see* The canadian polar expedition
Browne, Robert *see*
- A "new years guift"
- The "retractation" of robert browne
- The "retractation" of robert browne

Browne, St John *see* Our day-schools
Browne, Thomas *see* The british cicero
Browne, W H *see* The catholicos of the east and his people...
Browne, Walter R *see* Present aspect of the conflict with atheism
Browne, Walter Raleigh *see* The inspiration of the new testament
Browne, William *see*
- Nouveau voyage dans la haute et basse egypte, la syrie, le dar-four
- W g brown's reisen in afrika, egypten und syrien

Browne, William Hand *see* George calvert and cecilius calvert
Browne, William Hardcastle *see* Famous women of history
Browne, William Henry *see*
- The catholicos of the east and his people
- Ten coloured views taken during the arctic expedition of her majesty's ships "enterprise" and "investigator"

Brownell, Charles De Wolf *see* The indian races of america
Brownell, Henry Howard *see* The english in america
Browne's index/digest to the new york court of appeals reports : vols 1-95 / Browne, Irving – Albany: Parsons. 1v. 1884 (all publ) – 7mf – 9 – $10.50 – mf#LLMC 79-522 – us LLMC [347]
Browne's national bank cases *see* National bank cases
Brownfield baptist church. golconda, illinois : church records – 1860-1950 – 1 – us Southern Baptist [242]
Brownfield, J M *see* Diary
Brownfield, Maine. Brownfield Free Will Baptist Church *see* Records
Brownfield, Richard Charles *see* The church covenant committing
Brownhills & chasetown post – Brownhills, England 11 feb 1888-12 sep 1890 (wkly) – 2 1/2r – 1 – uk British Libr Newspaper [072]
The brownies abroad / Cox, Palmer – New York: Century, c1899 – 2mf – 9 – mf#17971 – cn Canadiana [830]
The brownies around the world / Cox, Palmer – New York: Century Co, c1894 – 2mf – 9 – mf#17972 – cn Canadiana [830]
The brownies at home / Cox, Palmer – New York: Century, c1891 – 2mf – 9 – mf#17004 – cn Canadiana [830]
Brownies' book – 1920 jan-1921 dec – 1r – 1 – mf#1053633 – us WHS [071]
The brownies, their book / Cox, Palmer – New York: Century, c1887 – 2mf – 9 – mf#17005 – cn Canadiana [830]
The brownies through the union / Cox, Palmer – New York: Century Co, c1894 – 2mf – 9 – mf#34487 – cn Canadiana [830]
The browning, eliot, thackeray and trollope manuscripts *see* Nineteenth century literary manuscripts
Browning, Elizabeth Barrett *see* Poetical works of elizabeth barrett browning
Browning engineering company bulletin – Cleveland, Cuyahoga, OH. 1902-12 – 1r – 1 – (serial publication advertising this cleveland-based company's material handling products. also included is the first issue of hoisting machinery march, 1912) – us Western Res [620]
Browning, Henry B *see* The new theology
Browning, king and co's illustrated monthly – v10-15 [1895-97] – 1r – 1 – (cont by: browning, king & co's monthly magazine) – mf#3104581 – us WHS [071]
Browning, king and co's monthly magazine – 1901 apr-1902 jun – 1r – 1 – (cont: browning, king and co's illustrated monthly; cont by: browning's magazine) – mf#3104613 – us WHS [071]
Browning, Oscar *see*
- Aspects of education
- Impressions of indian travel
- An introduction to the history of educational theories

Browning, Robert *see*
- Inn album
- Nineteenth century literary manuscripts

Browning, Robert Franklin *see* A program of preparation for marriage for the youth of the rich pond baptist church
Browning, Thomas Blair *see* Chart of elocutionary drill
Browning, William *see* History of the huguenots during the sixteenth century
Brownlee, C *see* Reminiscences of kaffir life and history, and other papers...
Brownlee, Frank *see* Transkeian native territories
Brownlee, William Craig *see*
- Romanism in the light of prophecy and history
- Secret instructions of the jesuits

Brownlow, John *see* Liberty of conscience
Brownlow, John T *see* Mr brownlow's speech in the house of commons
Brownlow north – Edinburgh, Scotland. 1879 – 1r – us UF Libraries [240]
Brownlow north, esq / Sinclair, George – Edinburgh, Scotland. 1858 – 1r – us UF Libraries [240]
Brownrigg, Abraham [Ossory, Bishop of] et al *see* Should clergymen criticize the bible?
Brown's baptist church. tar river association. warren county. north carolina : church records – 1830-84 – 1 – 6.75 – us Southern Baptist [242]
Brown's literary omnibus : news, books entire, sketches, reviews, tales, miscellaneous intelligence – La Salle IL 1837-38 – 1 – mf#5270 – us UMI ProQuest [420]
Browns literary omnibus – Philadelphia, PA., 1838 – 13 – $25.00r – us IMR [071]
Brown's nisi prius reports – Michigan. v1-2. 1869-71 (all publ) – (= ser Michigan appellate reports) – 11mf – 9 – $16.50 – mf#LLMC 81-305 – us LLMC [340]

Brown's parliamentary cases : reports of cases upon appeals and writs of error determined in the high court of parliament / Brown, Josiah – 2nd ed. v1-8. 1702-1800. London: J Butterworth, 1803 – 54mf – 9 – $81.00 – (with notes and additional cases brought down to 1800. first 3v completed during j brown's lifetime. t e tomlins was responsible for v4-8. 8th vol is called an appendix and contains additional cases reported by tomlins) – mf#LLMC 84-756 – us LLMC [324]
Brown's reports of admiralty and revenue cases in the sixth circuit / Brown, H B – New York: Baker-Voorhis. 1v. 1876 (all publ) – (= ser Early federal nominative reports) – 7mf – 9 – $10.50 – mf#LLMC 81-446 – us LLMC [340]
Brownson, Henry Francis *see*
– The convert
– An essay in refutation of atheism
– Faith and science, or, how revelation agrees with reason, and assists it
– Orestes a. brownson's ... life ...
Brownson, Lydia B et al *see* Genealogical notes on cape cod families, 1620-1901
Brownson, Orestes Augustus *see*
– The american republic, its constitution, tendencies and destiny
– Conversations on liberalism and the church
– The convert
– An essay in refutation of atheism
– Essays and reviews
– New views of christianity, society, and the church
– Papers orestes a brownson
– The spirit-rapper
Brownson's quarterly review – (formerly The Boston Quarterly Review, 1838-1842; in 1844 resumed publication as Brownson's). v1-29. 1838-75 – (= ser The boston quarterly review) – 1 – us AMS Press [800]
Brownson's quarterly review – La Salle IL 1844-75 – 1 – mf#3952 – us UMI ProQuest [240]
Brownstone Revival Committee of New York City *see* Brownstoner
Brownstoner / Brownstone Revival Committee of New York City – 1973 oct-1980 dec, 1981 feb-dec – 2r – 1 – mf#555542 – us WHS [071]
Brownsville advertiser – North Brownsville OR: G A Dyson, 1878-79 [biwkly] [mf ed 1971] – 1r – 1 – us Oregon Lib [071]
Brownsville banner – Brownsville OR: G C Blakely, 1882- – 1r – 1 – us Oregon Lib [071]
Brownsville clipper – Brownsville, PA. -w 1889-1912 – 13 – $25.00r – us IMR [071]
Brownsville times – Brownsville OR: McDonald & Cavender, -1960 [wkly] – 1 – (began in 1889; cont by: times (1960-)) – us Oregon Lib [071]
Brownsville United Trades Council *see* Union worker
Brownville biograph – Brownville, NE: Biograph Pub Co, 1902 (wkly) [mf ed v1 n5. nov 14 1902-mar 27 1903 (gaps) filmed [1973]] – 1r – 1 – us NE Hist [071]
Brownville courier – Brownville, NE: G W Fairbrother & Co, 1888 (wkly) [mf ed v1 n7. jun 8 1888 filmed [1973]] – 1r – 1 – us NE Hist [071]
Brownville democrat – Brownville, NE: Whitehead & Porter, jul 11 1868-jan 1874 (wkly) [mf ed v5 n13. sep 27 1872 (gaps)] – 1r – 1 – (cont by: nemaha county granger) – us NE Hist [071]
Brownville Fine Arts Association *see* Bulletin of the brownville...
Brownville Historical Society *see* Bulletin of the brownville...
The brownville letter – Brownville, NE: C E & MR Witherow, 1904 (wkly) [mf ed v1 n19. oct 14 1904-jan 19 1906 (gaps) filmed [1973]] – 1r – 1 – us NE Hist [071]
The brownville news – Brownville, NE: P H Drennen (wkly) [mf ed v1 n32. sep 20 1889-dec 5 1890 (gaps)] – 1r – 1 – us NE Hist [071]
The brownville record – Brownville, NE: W E Moore & Co (wkly) [mf ed v1 n23. jun 9 1894 filmed [1973]] – 1r – 1 – us NE Hist [071]
Brownville republican – Brownville, NE: John C Thompson. v1 n1. apr 27 1882- (wkly) – 1r – 1 – us Bell [071]
The brownville sun – Brownville, NE: J G Sanders (wkly) v1 n1. dec 10 1897- (wkly) [mf ed -feb 17 1900] – 1r – 1 – us NE Hist [071]
Brownwood, David O *see* Selected east african cases on commercial law
Brozas. la encomienda mayor : conferencia pronunciada en brozas con motivo de "el carro de la alegria". organizado por informacion y turismo el 12 de diciembre de 1969 / Conde de Canilieros – Caceres: Tip. Extremadura, 1970 – sp Bibl Santa Ana [338]
Bruan mas : [a novel] / O Phu – Bhnam Ben: Ron Bumb Khmaer [195-?] [mf ed 1990] – [ill] 1r with other items – 1 – (in khmer) – mf#mf-10289 seam reel 096/08 [§] – us CRL [830]
Bruant, Aristide *see* Chansons et monologues
Brubaker, Ella Miller *see* A personal testimony of the love of jesus
Brucciani, Dominico *see* Catalogue of casts for sale
Bruce, Alexander B *see* The gospel history
Bruce, Alexander Balmain *see*
– Apologetics
– The chief end of revelation
– The epistle to the hebrews
– The galilean gospel
– The humiliation of christ in its physical, ethical, and official aspects
– The miraculous element in the gospels
– The moral order of the world
– The parabolic teaching of christ
– The providential order of the world
– St paul's conception of christianity
– The training of the twelve
Bruce, Alexander Balmain et al *see* The expositor's greek testament
Bruce, Charles *see* Graphic scenes in african story
Bruce cooperator / Bruce Publishing Co – v20 n1-v29 n12 [1941 mar 16-1951 may/jun] – 1r – 1 – mf#630453 – us WHS [071]
Bruce, G J *see* Brazil and the brazilians
Bruce, George *see* Printed traps
Bruce, Henry James *see* Letters from india
Bruce herald – Milton, New Zealand 21 mar 1893-19 jun 1896 (semiwkly) (imperfect) – 1 – uk British Libr Newspaper [079]
Bruce, Herbert *see* The age of schism
Bruce, James *see*
– Lives of the eminent men of aberdeen
– Lives of the eminent men of fife
– National element in the scottish episcopal church
– Travels in abyssinia and nubia, 1768-1773
– Travels to discover the source of the nile
– Travels to discover the source of the nile, in the years 1768, 1769, 1770, 1771, 1772 and 1773
– Voyage aux sources du nil, en nubie et en abyssinie
– Voyage en nubie et en abyssinie, entrepris pour decouvrir les sources du nil
Bruce, James Douglas *see* The anglo-saxon version of the book of psalms
Bruce, John *see*
– Discourse preached in the new north church, edinburgh
– Great disruption principle
– Lecture on the lawfulness of the church accepting an endowmen from...
– Testimony and remonstrance regarding the moderatorship of next gene...
Bruce, John Edward *see* Concentration of energy
Bruce, King *see* Bird's-eye view of life insurance
Bruce news=letter – Bruce WI. 1905 oct 27/1907 jul 12-1969 dec 4/1971 oct 28 – 23r – 1 – (cont by: ladysmith news) – mf#1004147 – us WHS [071]
Bruce, Peter Henry *see* Bahamian interlude
Bruce Publishing Co *see*
– Bruce cooperator
– Bruce's weekly buzzer
– City council journal
Bruce, Robert *see* Apostolic order and unity
Bruce, William Straton *see*
– The ethics of the old testament
– The formation of christian character
– Social aspects of christian morality
Bruce's weekly buzzer / Bruce Publishing Co – 1920 nov 27-1929 dec 28, 1930 nov 15-1940 dec 21 – 2r – 1 – mf#467098 – us WHS [071]
Bruch, Max *see*
– [Erste] sinfonie, (es-dur), op 28
– Phantasie, d moll, fuer 2 klaviere zu 4 haenden, op 11
Bruchesi, Louis Joseph Paul Napoleon *see*
– Deuxieme centenaire de la fondation de l'institut des freres des ecoles chretiennes
– Imposition du pallium a mgr l'archeveque duhamel
Bruchhaus, K *see* Im banne der goetzen
Bruchsaler rundschau – Bruchsal DE, 1983 1 jun- – ca 10r/yr – 1 – (ba v. badische neueste nachrichten, karlsruhe) – gw Misc Inst [074]
Bruchstuecke aus einer reise durch einen theil italiens : im herbst und winter 1798 und 1799 / Arndt, Ernst M – Leipzig 1801 [mf ed Hildesheim 1995-98] – 2v on 6mf – 9 – €120.00 – 3-487-29317-X – gw Olms [914]
Bruchstuecke aus einer reise von baireuth bis wien im sommer 1798 / Arndt, Ernst M – Leipzig 1801 [mf ed Hildesheim: 1995-98] – (= ser Fbc) – 397p on 3mf – 9 – €90.00 – 3-487-29409-5 – gw Olms [914]
Bruchstuecke aus einigen reisen nach dem suedlichen russland : in den jahren 1822 bis 1828 mit besonderer ruecksicht auf die nogayen-tataren am asowschen meere / Schlatter, Daniel – St Gallen 1830 [mf ed Hildesheim 1995-98] – (= ser Fbc) – 4mf [ill] – 9 – €120.00 – 3-487-28988-1 – gw Olms [914]
Bruchstuecke der sahidischen bibeluebersetzung : nach handschriften der kaiserlichen oeffentlichen bibliothek zu st. petersburg / ed by Lemm, Oskar Eduardovich – Leipzig: JC Hinrichs, 1885 – 1mf – 9 – 0-8370-1791-2 – mf#1987-6179 – us ATLA [220]
Die bruchstuecke der skeireins / ed by Dietrich, Ernst – Strassburg: KJ Truebner, 1903 – (= ser Texte und Untersuchungen zur altgermanischen Religionsgeschichte) – 2mf – 9 – 0-524-02788-9 – mf#1987-6482 – us ATLA [240]
Bruchstuecke des ersten clemensbriefes : nach dem achmimischen papyrus der strassburger universitaets- und landesbibliothek, mit biblischen texten derselben handschrift = First epistle of clement to the corinthians. selections / Clement 1, Pope; ed by Roesch, Friedrich – Strassburg i. E: Schlesier & Schweikhardt, 1910 – 1mf – 9 – 0-7905-4256-0 – (in german and coptic) – mf#1988-0256 – us ATLA [227]
Bruchstuecke des evangeliums und der apokalypse des petrus / Harnack, Adolf von – 2. verb erw aufl. Leipzig: J C Hinrichs, 1893 [mf ed 1989] – (= ser Tugal) – 1mf – 9 – 0-7905-1709-4 – (incl ind) – mf#1987-1709 – us ATLA [226]
Bruchstuecke des evangeliums und der apokalypse des petrus / Harnack, Adolf von – Leipzig, 1893 – (= ser Tugal 1-9/2) – 2mf – 9 – €5.00 – ne Slangenburg [240]
Bruchstuecke einer reise durch das suedliche frankreich, spanien und portugal [im jahr 1802] / Jariges, Karl F von – Leipzig 1810 [mf ed Hildesheim 1995-98] – (= ser Fbc) – 1v on 2mf – 9 – €60.00 – 3-487-27775-1 – gw Olms [914]
Bruchstuecke einer reise durch frankreich im fruehling und sommer 1799 / Arndt, Ernst M – Leipzig 1802-03 [mf ed Hildesheim 1995-98] – 3v on 9mf – 9 – €180.00 – 3-487-29781-7 – gw Olms [914]
Bruchstuecke eines tagebuches gehalten in groenland / Saabye, H E – Hamburg, 1817 – 5mf – 9 – mf#N-379 – ne IDC [917]
Bruck, J *see*
– Emblemata moralia et bellica
– Emblemata politica
– Emblemata pro toga et sago
Bruck, Julius *see* Bunte bluethen
Brucker, Joseph *see*
– L'eglise et la critique biblique (ancien testament)
– Jacques marquette et la decouverte de la vallee du mississipi sic
Bruckner : der roman der sinfonie / Bachmann, Luise George – 2. aufl. Paderborn: F Schoeningh, c1938 [mf ed 1989] – 480p – 1 – mf#6971 – us Wisconsin U Libr [830]
Bruckner, Albert *see*
– Der alte weg zum alten gott
– Julian von eclanum
Bruckner, Ferdinand *see* Elisabeth von england
Bruder, Carl Hermann *see* Tamieion ton tes kaines diathekes lexeon
Bruder hansens marienlieder / ed by Batts, Michael S – Tuebingen: M Niemeyer, 1963 [mf ed 1993] – (= ser Altdeutsche textbibliothek n58) – xvii/273p/[7pl] – 1 – (middle high german text; int in german; incl bibl ref & ind) – mf#8193 reel 5 – us Wisconsin U Libr [780]
Bruder hermanns klause / Ehrler, Hans Heinrich – Stuttgart: Fleischhauer & Spohn c1927 [mf ed 1989] – 1r – 1 – (filmed with: menschen und affen / albert ehrenstein) – mf#7207 – us Wisconsin U Libr [830]
Bruder, K *see* Die philosophische elemente in den opuscula sacra des boethius
Bruder lustig : fuer die maerchenspiele der kuenstlerischen volksbuehne schlicht und getreu nach grimms maerchen in handlung und rede gesetzt / Guembel-Seiling, Max – Leipzig: Breitkopf & Haertel [19-?] [mf ed 1990] – (= ser Deutsche maerchenspiele 3) – 1r – 1 – (filmed with: guerillaskrieg : versprengte lieder) – mf#2694p – us Wisconsin U Libr [820]
Bruder philipps des carthaeusers marienleben / ed by Rueckert, Heinrich – Quedlinburg, Leipzig: G Basse, 1853 [mf ed 1993] – (= ser Bibliothek der gesammten deutschen national-literatur von der aeltesten bis auf die neuere zeit sect1/34) – viii/391p – 1 – (incl bibl ref) – mf#8438 reel 7 – us Wisconsin U Libr [810]
Bruder und schwester : novelle / Blunck, Hans Friedrich – Leipzig: P Reclam, c1928 [mf ed 1989] – 74p – 1 – (aft by paul wittko) – mf#7036 – us Wisconsin U Libr [830]
Bruder und schwester : novelle / Blunck, Hans Friedrich – Leipzig: P Reclam c1928 [mf ed 1989] – (= ser Reclams universal-bibliothek 6831) – 1r – 1 – (aft by paul wittko. filmed with: allerhand ungezogenheiten / oscar blumenthal) – mf#7036 – us Wisconsin U Libr [830]
Bruechner, Kathrin *see* Konfrontative untersuchungen zur lexikalischen dimension der fachlichkeit von texten
Brueck, Heinrich *see*
– Geschichte der katholischen kirche in deutschland im neunzehnten jahrhundert
– History of the catholic church
– Lehrbuch der kirchengeschichte
Die bruecke – Bad Liebenwerda DE, 1951-1958 11 aug [gaps] – 2r – 1 – gw Misc Inst [074]
Die bruecke – Danzig (Gdansk PL), 1919 4 oct-1921 – 1r – 1 – gw Misc Inst [077]
Die bruecke : deutsche wochenzeitschrift fuer ostasien – Schanghai (VR), 1925-1933 jun – 3r – 1 – gw Misc Inst [079]
Die bruecke : deutsche wochenzeitschrift fuer ostasien – Schanghai (VR), 1925-1933 jun – 3r – 1 – gw Misc Inst [079]
Die bruecke : halbmonatszeitschrift fuer politik, kultur, wirtschaft – Frankfurt/M DE, 1946-dec 1 1948 – 1 – gw Misc Inst [074]
Die bruecke – [Welkom, South Africa?: s.n.], [Middelburg, Transvaal, South Africa: O H Schultz] [qrterly] [mf ed 2004] – 5r – 1 – (mf: v7-88 1929/30-1988, 1989-90 [gaps]. began in 1924, ceased in 1990? publ suspended 1940-46. some iss not publ, some iss combined. some iss have suppl called: wissenschaftliche beilage) – mf0643 – us ATLA [240]
Die bruecke : schauspiel in vier aufzuegen / Kolbenheyer, Erwin Guido – Muenchen: G Mueller, 1933 – 1r – 1 – us Wisconsin U Libr [820]
Die bruecke – Flensburg, Kiel, Heide, Holst DE, 1924-33 – 1r – 1 – (title varies: jan 1925: deutsche zukunft. with suppl) – gw Misc Inst [074]
Die bruecke – Warschau (PL), 1948 n1-1949 n25, 1975 [special ed] – 1r – 1 – gw Misc Inst [077]
Die bruecken : eine auswahl aus seinem schaffen / Zerkaulen, Heinrich – Berlin: Die Heimbuecherei, 1942 – 1r – 1 – us Wisconsin U Libr [943]
Bruecken des lebens : das leben des menschen in zeit und gesellschaft, widergespiegelt in deutschen gedichten von walther von der vogelweide bis zur gegenwart / Czechowski, Heinz [comp] – Halle (Saale): Mitteldeutscher Verlag 1969 [mf ed 1993] – 1r – 1 – (filmed with: deutsche lyrik seit 1850 / ed by heinrich spiero) – mf#3340p – us Wisconsin U Libr [810]
Bruecken-bau / Walter, C – Augsburg, 1766 – 3mf – 9 – mf#OA-124 – ne IDC [720]
Brueckenkopf : novelle / Truestedt, Haro – [Berlin: Wiking Verlag c1944] [mf ed 1991] – 1r – 1 – (ill by heinz raebiger; filmed with: leuchtendes land / luis trenker) – mf#2917p – us Wisconsin U Libr [830]
Brueckmann, Arthur *see* Nachgelassene schriften
Brueckner, Aleksander *see* Dzieje kultury polskiej
Brueckner, Alexander *see* Katharina die zweite
Brueckner, Gustav *see* Hebraeisches lesebuch fuer anfaenger und geuebtere
Brueckner, Martin *see*
– Die entstehung der paulinischen christologie
– Das fuenfte evangelium (das heilige land)
– Die komposition des buches jes. c. 28-33
– Der sterbende und auferstehende gottheiland
Brueckner, Wilhelm *see* Die chronologische reihenfolge
Die brueder : aus vergangenheit und gegenwart der bruedergemeine / ed by Uttendoerfer, Otto & Schmidt, Walther E – Herrnhut: Verlag des Vereins fuer Bruedergeschichte, in Kommission der Unitaetsbuchh in Gnadau, 1914 – 5mf – 9 – 0-524-07366-X – mf#1990-5403 – us ATLA [240]
Die brueder : eine erzaehlung / Frenssen, Gustav – Berlin: G Grote, 1920 (mf ed 1990) – 1r – 1 – (filmed with: ferdinand freiligrath) – us Wisconsin U Libr [830]
Die brueder alfonso und juan de valdes : zwei lebensbilder aus der geschichte der reformation in spanien und italien / Schlatter, Wilhelm – Basel: R Reich, 1901 – 1mf – 9 – 0-524-07716-9 – mf#1991-3301 – us ATLA [240]
Brueder im sturm : roman / Bartsch, Rudolf Hans – Graz: L Stocker, 1940 [mf ed 1989] – 400p – 1 – mf#6981 – us Wisconsin U Libr [830]
Die brueder von st bernhard : schauspiel in fuenf aufzuegen / Ohorn, Anton – 8. Aufl. Berlin: Vita Deutsches Verlagshaus [between 1906 and 1926?] – 1r – 1 – us Wisconsin U Libr [820]
Der bruederbote : monatszeitschrift des bessarabischen gemeinschaftsverbandes – v4-9. 1951-55; v18-44. 1964-91* – 6r – 1 – mf#ATLA S0423 – us ATLA [242]
Das bruederliche jahr : gedichte / Moeller, Eberhard Wolfgang – new enl ed. Wien: Wiener Verlag 1943, c1941 [mf ed 1990] – 1r – 1 – (filmed with: ende gut, alles gut / melchior meyr) – mf#2835p – us Wisconsin U Libr [810]
Bruees, Otto *see*
– Die affen des grossen friedrich
– Fliegt der blaufuss?
– Das gauklerzelt
– Heilandsflur

– Heiterkeit des herzens
– Das maedchen von utrecht
– Der schlaue herr vaz
– An den vier waellen
– Weites feld der liebe
– Die wiederkehr
Brueggemann, Diethelm *see* Vom herzen direkt in die feder
Brueggemann, Fritz *see* Gellerts schwedische graefin
Brueggemann, Joseph *see* Ludwig tieck als uebersetzer mittelhochdeutscher dichtung
Bruehl, I A *see* Geschichte der katholischen literatur deutschlands vom 17. jahrhundert bis zur gegenwart
Brueil, J du *see* L'art universel des fortifications...
Bruel, A *see* Recueil des chartes de l'abbaye de cluny
Bruel, Georges *see* Bibliographie de l'afrique equatoriale francaise
Bruell, Adolf *see* Adolf bruell's populaerwissenschaftliche monatsblaetter [...]
Bruell, Andreas *see* Der hirt des hermas
Bruell, Nehemiah *see* Zentral-anzeiger fuer juedische literatur
Bruell, Nehemias *see* Jahrbuecher fuer juedische geschichte und literatur
Bruen, Edward Tunis *see* Outlines for the management of diet
Bruen, Matthias *see* Essays, descriptive and moral
Bruenner montagsblatt – Brünn (Brno CZ), 1940 2 dec-1941 31 may, 1941 1 jul-1944 30 jun [gaps] – 4r – 1 – gw Misc Inst [077]
Bruenner tagblatt *see* Tagesbote aus maehren und schlesien
Bruenner tagespost *see* Tagespost
Bruennow, R E *see* Die provincia arabia
Bruesseler zeitung – Brussels, Belgium 18 feb-21 mar, 8 jun, 5 jul 1941-2 sep 1944 – 1 – uk British Libr Newspaper [074]
Bruesseler zeitung 1936 – Bruessel (B), 1936 4 jul-2 aug – 1 – gw Misc Inst [074]
Bruesseler zeitung 1940 – Brussels (B), 1940 1 jul-1944 2 sep – 9r – 1 – mf#3424 – gw Mikropress [074]
Bruestle, Wilhelm *see* Klopstock und schubart
Brueys *see*
– Avocat patelin
Brueys, Claude *see* Iardin deys mvsos provensalos
Brueys, D-A *see*
– Defense du culte exterieur de l'eglise
– Reponse au livre de monsieur de condom
Die brug : tussen protestant en katoliek – v16-18. 1967-69 [complete] – 1r – 1 – mf#ATLA S0531 – us ATLA [240]
Brugensis, Galbertus notarius *see* De multro, traditione et occisione gloriosi karoli comitis flandriarum
Bruger, Ferdinand *see* Das herz befiehlt!
Brugere, Lud-Fred *see*
– De ecclesia christi
– De vera religione
Bruges book of hours – 15th c – (= ser Holkham library manuscript books 48) – 1 col r – 14 – (illuminated at bruges by vrelant) – mf#C503 – uk Microform Academic [090]
Bruges, Roger *see* Reiseskizzen aus west-indien, mexico und nord-amerika
Bruggeboes, W *see* Die fraterherren im luechtenhofe zu hildesheim
Brugghen, Guillaume Anne van den *see* Calvijn, 10 juli 1509-27 mei 1564
Brugghen, Guillaume Anne van der *see* Een merkwaardig chinees
Brugsch, Heinrich *see* Kleine hieroglyphen-grammatik (auszug) nach dem werk h. prof. heinrich brugsch, berling, handschrift
Brugsch, Heinrich Karl *see*
– Die aegyptologie
– Der bau des tempels salomo's nach der koptischen bibelversion
– Dictionnaire geographique de l'ancienne egypte
– Egypt under the pharaohs
– Hieroglyphisch-demotisches woerterbuch
– Die neue weltordnung nach vernichtung des suendigen menschengeschlechtes
– Religion und mythologie der alten aegypter
– Thesaurus inscriptionum aegyptiacarum
– The true story of the exodus of israel
Bruguera, O *see* Novae ac infestae destillationis... quae civitati barcinonesis anni 1562 accidit brevis enaratio
Bruhat, L *see* Le monachisme en saintonge et en aunis. (11 et 12 siecles)
Bruhn, Wilhelm *see* Theosophie und theologie
Bruhns, Carl *see* Life of alexander von humboldt
Bruied treasure : new smyrna, fla / Sweett, Zelia Wilson – s.l, s.l? 1936? – 1r – us UF Libraries [978]
Bruin, Cornelis de *see*
– Aanmerkingen op otto van veens zinnebeelden der goddelijke liefde
– Uitbreiding
Bruington baptist church. gaston association. gaston county. north carolina : church records – 1853-72 – 1 – 5.00 – us Southern Baptist [242]
Bruinier, Johannes Weijgardus *see*
– Das engelsche volksschauspiel doctor johann faust als faelschung
Bruinwold Riedel, J *see* Goethes faust als levensbeeld
Bruist, B *see* Beginzelen der vesting-bouw
Brukliner idishe shtimme = Brooklyn jewish voice – Brooklyn, NY. 1932-34 – 1 – us AJPC [071]
The brule citizen – Brule, NE: Roy R Barnard. -v13 n41. apr 10 1941 (wkly) [mf ed v9 n32. jan 28 1937-apr 10 1941 (lacks mar 9 1939)] – 1r – 1 – (absorbed by: keith county news) – us NE Hist [071]
Brulotte, Marie-Berthe *see* Bibliographie analytique de mme paule develuy
Brum, Baltasar *see* Paz de america
Brum, Cathrina de [comp] *see* Selected decisions and digests of decisions for the period apr 1982 to mar 1985
Brumbaugh, Henry Boyer *see*
– The brethren's church manual
– The church manual
Brumbaugh, Martin Grove *see*
– A history of the german baptist brethren in europe and america
– Juniata bible lectures
– The life and works of christopher dock
– Limitations of leadership
– Rose day address
– Stories of pennsylvania
Brumbaugh, Martin Grove et al *see* Two centuries of the church of the brethren
Brumell, Henry Peareth H *see*
– The mineral waters of canada
– Notes on manganese in canada
– On the geology of natural gas and petroleum in southwestern ontario
The brumer catalog of rabbinic manuscripts – Clearwater Publ Co – (= ser Hebrew manuscript catalogs from the jewish theological seminary; Research colls in judaica) – 34mf (24:1) – 9 – $250.00 – us UPA [090]
Brumley, Frank Warner *see* Labor requirements of florida crops
Brummerloh, Carsten *see* Rettungshubschrauber im vergleich
Brummet, Stephan D H *see* Die spanische politik im westsahara-konflikt
Brumund, Jan Frederik Gerrit *see* Bijdragen tot de kennis van het hindoeisme op java
Brun, Amedee *see*
– Deux amours
– Pages retrouvees
– Sans pardon
Brun, B *see* Cronica johannis vitodurani (mgh6:3.bd)
Brun, C le *see* Voyages de corneille le brun par la moscovie, en perse, et aux indes orientales
Brun, Guillaume, abbe *see* L'abbe j-p lapauze
Brun, J *see* Dictionarium syriaco-latinum
Brun, Jean *see* Empedocle
Brun, Lyder *see* Jesu evangelium
Brun, Regis *see* Les acadiens a moncton
Brun, Sophie *see* Roemisches leben
Brun von schonebeck / ed by Fischer, Arwed – Stuttgart: Litterarischer Verein, 1893 (Tuebingen: H Laupp, Jr) [mf ed 1993] – (= ser Blvs 198) – lxii/443p – 1 – (incl bibl ref and ind) – mf#8470 reel 41 – us Wisconsin U Libr [430]
Brun von schonebeck / ed by Fischer, Arwed – Stuttgart: Litterarischer Verein 1893 (Tuebingen: H Laupp, Jr) [mf ed 1993] – (= ser Blvs 158) – 58r – 1 – (incl bibl ref & ind. filmed with: bibliothek des litterarischen vereins in stuttgart) – mf#3420p – us Wisconsin U Libr [430]
Brun, Xavier *see* Adelbert de chamisso de boncourt
Brundage, Burr Cartwright *see* Rain of darts
Brune, J de *see*
– Emblemata of zinnewerck
Bruneau, Alfred *see* Jardin du paradis
Bruneau, Arthur-Aime *see* Resume du proces bolduc et son execution a sorel, le 5 avril 1918
Brunel, Ismael-Matthieu *see* Le general faidherbe
Brunel, J *see* La femme mariee et les charges du menage
Brunel, Nore *see* Petrus
Brunet, Gustave *see* Curiosites theologiques
Brunet, Jacques Charles *see*
– Manuel de libraire et de l'amateur de livres
– Manuel du libraire et de l'amateur de livres
Brunet, Ovide *see*
– Catalogue des plantes canadiennes
– Elements de botanique et de physiologie vegetale
– Enumeration des genres de plantes de la flore du canada
– Histoire des picea qui se rencontrent dans les limites du canada
– Michaux and his journey in canada
– Notes sur les plantes
– Notice sur le musee botanique de l'universite laval
– Notice sur les plantes de michaux, et sur son voyage au canada et a la baie d'hudson
– Voyage d'andre michaux en canada
Brunet, Pierre *see* Voyage a l'ile de france, dans l'inde et en angleterre
Brunet, Roger *see* L'annexion du congo a la belgique et le droit international
Brunet, Victor Armand *see* Blason populaire de villedieu-les-poeles, arrondissement d'avranches (manche)
Brunetere, Ferdinand *see* Evolution de la poesie lyrique en france au dix-neuvieme siecle
Brunetiere, Ferdinand *see* Les difficultes de croire
Brunet-P Tremblay *see* La renaissance du 12e siecle. les ecoles et l'enseignement
Brunetti, Gaetano *see*
– [Sinfonia con violini, oboe, corni, viola, fagotto e basso, no 25. partition]
– [Sinfonia con violini, oboe, corni, viola, fagotto e basso, no 29. partition]
Brunfels, Otto *see*
– Onomastikon medicinae
– Reformation der apotecken
– Theses seu communes, loci totius rei medicae
Bruni, A *see* Libro della gverra de ghotti composto da misser leonardo aretino...
Bruni, Antonio Bartolomeo *see* Six trios concertants pour deux violons et alto
Bruni Celli, Blas *see*
– Estudios historicos
– Secuestros en la guerra de independencia
Bruni, Leonardo [Leonardo Aretino] *see*
– Bellum punicum 1...
– Livius, books 31-40/dictys...
– Opuscula 30...
Brunier, Ludwig *see* Kurland
Bruning banner – Bruning, NE: R J Epp. v1 n1. may 3 1918-v47 n2. jun 2 1966 (wkly) – 7r – 1 – (cont by: thayer county banner-journal; iss for oct 24 1930-dec 10 1942 accompanied a separately numbered sect: belvidere news; iss for oct 8 1964-jun 2 1966 accompanied by a separately numbered sect: davenport people's journal) – us Bell [071]
Bruning banner – Bruning, NE: R J Epp. v1 n1. may 3 1918-v47 n2. jun 2 1966 (wkly) – 3r – 1 – (cont by: thayer county banner-journal. issues for oct 24 1930-dec 10 1942 accompanied by separately numbered sect: belvidere news; for oct 8 1964-jun 2 1966: davenport people's journal) – us NE Hist [071]
Bruning booster – Bruning, NE: A S Pettit. v1 n1. mar 21 1913-v4 n41. dec 22 1916 (wkly) – 1r – 1 – (chiefly in english with some articles in german) – us NE Hist [071]
Bruning courier – Bruning, NE: McGrew & Boyd. v1 n1. jun 30 1899-apr 19 1907 (gaps) – 1r – 1 – us NE Hist [071]
Brunk, Max E *see*
– Celery harvesting methods in florida
– Economic study of celery marketing
– Factors affecting farming returns in jackson county, florida
– Labor and material requirements for crops and livestock
Brunken, David L *see* Carolina balance index
Brunnell, John *see* Demerara after 15 years of freedom
Brunnenrauschen : kalendergeschichten / Leppa, Karl Franz – Karlsbad: A Kraft 1942 [mf ed 1990] – 1r – 1 – (filmed with: die freunde machen den philosophen, der englander, der waldbruder von jakob michael reinhold lenz / comp by ilse kaiser) – mf#2823p – us Wisconsin U Libr [390]
Brunner, A *see* Brunner's reports of cases in the circuit courts of the u.s., 1789-1879
Brunner abendblatt – Brno, Czechoslovakia. Dec 1940-Aug 1944 – 3r – 1 – us L of C Photodup [077]
Brunner, F *see* La doctrine de la matiere chez avicebron
Brunner, Heinrich *see* Das anglonormannische erbfolgesystem
Brunner, Samuel *see*
– Ausflug ueber constantinopel nach taurien im sommer 1831
– Streifzug durch das oestliche ligurien, elba, die ostkueste siciliens, und malta
Brunner, Sebastian *see*
– Die hofschranzen des dichterfuersten
– Rom und babylon
– Unter lebendigen und todten
Brunner tagblatt – Brno, Czechoslovakia. Dec 1940-Feb 1945 – 6r – 1 – us L of C Photodup [077]
Brunner, Thomas *see* Jacob und seine zwoelf soehne
Brunner's reports of cases in the circuit courts of the u.s., 1789-1879 / Brunner, A – San Francisco: Sumner-Whitney. 1v. 1884 (all publ) – (= ser Early federal nominative reports) – 8mf – 9 – $12.00 – mf#LLMC 81-439 – us LLMC [347]
Brunngraber, Rudolf *see*
– Opiumkrieg
– Zucker aus cuba
Brunnquell, Paul [comp] *see* Dialoge in poetischer und prosaischer form
Bruno, Anibal *see* Nova gramatica da lingua portuguesa
Bruno brehm zum fuenfzigsten geburtstag / Buch des dankes – Karlsbad, Leipzig: A Kraft, c1942 [mf ed 1989] – 371p (ill) – 1 – mf#7082 – us Wisconsin U Libr [430]
Bruno, Camille *see* Panegyrique d'edmond paul
Bruno Cartusiense *see* Opera et vita...
Bruno chap books – New York. v1-3 1915-1916 – 1 – us NY Public [800]
Bruno di Segni, Saint *see* Sancti brunonis carthusianorum institutoris expositio in psalmos
Bruno, Giordano *see*
– De gl'eroici furori
– Le opere italiane di giordano bruno
Brunonis de bello saxonico liber (mgh7:15.bd) – 1880 – (= ser Monumenta germaniae historica 7: scriptores rerum germanicarum in usum scholarum (mgh7)) – €7.00 – ne Slangenburg [240]
Bruno's bohemia – Killen TX 1918 – 5 – mf#4649 – us UMI ProQuest [400]
Brunos buch vom sachsenkrieg (mgh deutsches..: 2.bd) – 1937 – (= ser Monumenta germaniae historica. deutsches mittelalter. kritische studientexte (mgh deutsches...)) – €7.00 – ne Slangenburg [931]
Bruno's weekly – New York. v. 1-3. July 26 1915-Dec 30 1916. Incomplete – 1 – us NY Public [800]
Bruns, Friedrich *see*
– Friedrich hebel und otto ludwig
– Goethe's poems and aphorisms
– Die lese der deutschen lyrik
Bruns, Marianne et al *see* Deutsche stimmen 1956
Bruns, Paul *see* Neuster bericht ueber eine reise von hamburg nach sued-australien sowie ueber das land selbst
Bruns, Paul Jakob *see* Neue systematische erdbeschreibung von afrika
Brunschvig, R *see* La tunisie dans le haut moyen age
Brunschwig, Hieronymus *see*
– Apoteck fuer den gemainen man
– Das buch zu distilieren die zusamen gethonen ding
Brunson, Alfred *see* A western pioneer
Brunswick & byron advocate – Mullumbimby, jan 1969-dec 1972 – 4r – at Pascoe [079]
Brunswick byron advocate – Mullumbimby, 1965-68 – at Pascoe [079]
Brunswick, Maine. Baptist Church *see* Records
Brunton, E *see* A grammar and vocabulary of the susco language
Brunton, G *see* Sedment 1 and 2
Brunton, Guy *see*
– Lahun 2
– Lahun i: the treasure
Brunton, Paul *see*
– A hermit in the himalayas
– The hidden teaching beyond yoga
– Indian philosophy and modern culture
– A message from arunachala
– A search in secret india
Brunton, Thomas Lauder *see* The bible and science
Brunton, William *see* Messiah's exhortation to his people
Bruny, Chevalier de *see* Lettre sur j j rousseau
Bruschi, Antonio Filippo *see* Regole per il contrapunto
Brush and pencil – 1897-1907 [mf ed Chadwyck-Healey] – (= ser Rare 19th Century American Art Journals) – 110mf – 9 – uk Chadwyck [700]
Brush and pencil : an illustrated magazine of arts of today – Killen TX 1897-1907 – 1 – mf#2868 – us UMI ProQuest [700]
Brush creek baptist church. tennessee : church records – 1828-1984 – 1 – 75.69 – us Southern Baptist [242]
Brush, Florence C *see* A comparison of selected neuromuscular and kinematic variables before and after learning an aiming task
Brushy creek baptist church. copiah county. mississippi : church records – 1875-1882 – 1 – 5.00 – us Southern Baptist [242]
Brushy creek baptist church. greenville county. south carolina : church records – June 1795-1969; Deacons' Minutes, 1961-72 – 1 – 77.00 – us Southern Baptist [242]
Brussel, Pierre *see* La promenade utile et recreative de deux parisiens en cent soixante cinq jours
Brussius, G *see* ...De tartaris diarivm
Brussolo, Armando *see* Tudo pelo brasil!
Brust, Alfred *see* Spiele
Bruston, Charles *see*
– La descente du christ aux enfers
– Etudes sur daniel et l'apocalypse
– Histoire critique de la litterature prophetique des hebreux
– La vie future d'apres saint paul
Bruston, Edouard *see* Ignace d'antioche, ses epitres, sa vie, sa theologie
Brut, A A *see* Ust'e amura
Brut y tywysogion (rs17) : or the chronicle of the princes of wales (681-1281) / ed by Williams, J, ab Ithel – 1860 – (= ser The rolls series (rs)) – €19.00 – ne Slangenburg [931]

BRUTA

La bruta (heroes de ahora) / Trigo, Felipe – Madrid: Renacimiento, 6th ed 1907 – sp Bibl Santa Ana [946]
Brutal mandate / Lowenstein, Allard K – New York, NY. 1962 – 1r – us UF Libraries [960]
Brutalidad de los negros / Labra Y Cadrana, Rafael Maria De – Madrid, Spain. 1876 – 1r – us UF Libraries [972]
Brutalitaeten : skizzen und studien / Conradi, Hermann – Zuerich: verlags-magazin, 1886 [mf ed 1989] – 88p – 1 – mf#7160 – us Wisconsin U Libr [880]
The brute / Kummer, Frederick Arnold – Toronto: McLeod & Allen, c1912 [mf ed 1994] – 4mf – 9 – 0-665-73552-9 – mf#73552 – cn Canadiana [830]
Bru-Thiellay, Paul *see* Intrigue au bal
Brutus : oder der tyrannenfeind / ed by Biergans, F Th – Koeln 1795 – (= ser Dz. historisch-politische abt) – 5mf – 9 – €100.00 – mf#k/n1738 – gw Olms [930]
Brutus : trauerspiel / Kruse, Heinrich – 2. aufl. Leipzig: S Hirzel 1882 [mf ed 1990] – 1r – 1 – (filmed with: hein wieck / timm kroger) – mf#2777p – us Wisconsin U Libr [820]
Brutus, Edner *see* Instruction publique en haiti
Brutus, lache cesar! / Rosier, Joseph-Bernard – Paris, France. 1849 – 1r – us UF Libraries [440]
Brutus! schlaefst du? : zeitgedichte / Strodtmann, Adolf – London: Truebner; Hamburg: J P F E Richter [18–?] [mf ed 1990] – (= ser Bibliothek der deutschen literatur) – 1r [ill] – 1 – (filmed with: goethes faust in urspruenglicher gestalt / ed by erich schmidt) – mf#7324 – us Wisconsin U Libr [810]
Brutzkus, Boris et al *see* Blaetter fuer demographie, statistik und wirtschaftskunde der juden
Bruun, Geoffrey *see* Nineteenth-century european civilization, 1815-1914
Bruun, Laurids *see* Van zanten's happy days
Bruwer, J P Van S *see* South west africa
Bruxelles. Academie Royale des Sciences Coloniales. Classe des Sciences Morales et Politiques *see* Memoires in-80
Bruxelles et ses environsguide de l'etranger dans cette capitale : contenant l'histoire abregee de la ville de bruxelles, la description de ses monuments... / Wauters, Alphonse – Bruxelles 1845 [mf ed Hildesheim 1995-98] – 2mf [ill] – 9 – €60.00 – 3-487-29614-4 – gw Olms [914]
Bruxelles. Institut Royal Colonial Belge. Section des Sciences Morales et Politiques *see* Memoires
Le bruxellois – Brussels, Belgium 25 aug 1916-3 apr, 9 dec 1917; 2 feb-10 mar 1918 (very imperfect) – 1r – 1 – uk British Libr Newspaper [074]
Bruyere, Jean Marie *see* Controversy between dr ryerson, chief superintendent of education in upper canada, and rev j m bruyere, rector of st michael's cathedral, toronto
Bruylants, P *see* Les oraisons du missel romain
Bruylofts-kost : bestaande in verscheyden zedighe en boertighe echts-gezangen... – Aemsteldam: [Smeerbol], n.d. – 4mf – 9 – mf#O-3277 – ne IDC [090]
Bruyn, Karlheinz de *see* Italien im deutschen gedicht
Bruzzi, Nilo *see* Casimiro de abreu
Bry, J I de *see* Emblemata saecularia
Bry, J Th de *see*
– Emblemata nobilitati et vulgo scitu digna
– Emblemata saecularia
– Emblemata secularia
Bry, J Theodor de *see* Emblemata nobilitatis
Bryan and darrow at dayton : the record and documents of the "bible-evolution trial" / Allen, Leslie Henri [comp] – New York: A Lee & Co c1925 [mf ed 1986] – 1r [ill] – 1 – ("account of the case of the state of tennessee against john thomas scopes". filmed with: japan in world politics / kawakami, k k) – mf#1675p – us Wisconsin U Libr [347]
The bryan campaign for the american people's money / Donnelly, Ignatius – Chicago: Laird & Lee, 1896 (mf ed 19–) – xxv/186p – mf#ZT-545 – us Harvard [332]
Bryan, Carlton H Jr *see* An analysis of selected attendance factors in the world league of american football
The bryan democrat *see* Miscellaneous newspapers of pitkin county
Bryan, F Macdonald *see* Home building and beautification
Bryan, George *see* The imperialism of john marshall: a study in expediency
Bryan, Joseph Harris *see*
– The art of questioning
– The organized adult bible class
– The what, why, and how of sunday-school work
Bryan, Lindsay M *see*
– Cigarmakers' union dispute in tampa
– Lakes in hillsborough county
– Mystery of the golden tarpon
– Sulphur springs
Bryan, Michael *see* A biographical and critical dictionary of painters and engravers
Bryan newsletter – 1982 jan 1-1987 jan 1 – 1r – 1 – (cont by: bryan/bryant newsletter) – mf#1321041 – us WHS [071]
Bryan, R G *see* God's witness to his own word
Bryan, William Jennings *see*
– The commoner
– Weekly world-herald
Bryans creek baptist church. lincoln county. missouri : church records – 1831-1948. 468p – 1 – us Southern Baptist [242]
Bryant, Alfred T *see*
– Incwadi yesisngisi nesizulu
– Olden times in zululand and natal
– Zulu people as they were before the white man came
– Zulu-english dictionary with notes on pronunciation
Bryant backtrails – v1 n1-v4 n4 [1977 jan/mar-1980 dec] – 1r – 1 – (cont by: kenneth g lindsay report) – mf#569609 – us WHS [071]
Bryant, D *see*
– Igirama lesingisi
Bryant, Edwin Eustace *see*
– The constitution of the united states with notes of the decisions of the supreme court thereon.
– Forms in civil actions and proceedings in the courts of record of wisconsin
– A selection of forms to accompany the volume on wisconsin code practice.
Bryant Family *see* Letters
Bryant, John Ebenezer *see* Agriculture in public schools
Bryant, Joshua *see* Account of an insurrection of the negro slaves in the colony of demarara
The bryant memorial meeting of the goethe club of the city of new york : wednesday, october 30th, 1878 – New York: GP Putnam, 1879 – 1mf – 9 – 0-524-04760-X – mf#1992-2040 – us ATLA [975]
Bryant, Stratton and Odell's Business College *see* The index
Bryant, William C *see*
– L'amerique du nord pittoresque
– The embargo
Bryant, Wm M *see* Ethics and the "new education"
Bryce, George *see*
– The british association for the advancement of science
– Educational thoughts for the diamond jubilee year
– Great britain as seen by canadian eyes
– The old settlers of red river
– Original letters and other documents relating to the selkirk settlement
– Recorder adam thom
– The remarkable history of hudson's bay company
– A short history of the canadian people
– University education
Bryce, J *see*
– Impressions of south africa
– Transcaucasia and ararat
Bryce, James Bruce, viscount *see* Handbook of home rule
Bryce, James Bryce, Viscount *see*
– America del sud
– The american commonwealth
– American correspondence of james bryce, 1871-1922
– Circa sacra
– The holy roman empire
– Modern democracies
– Second letter on the present position of the church of scotland, addressed to george cook, dd...
– Studies in contemporary biography
– Ten years of the church of scotland, from 1833 to 1843
– University and historical addresses
Bryce on american democracy : selections from the american commonwealth and the hindrances to good citizenship / ed by Fulton, Maurice G – New York: Macmillan, 1919 – 5mf – 9 – $7.50 – mf#LLMC 95-086 – us LLMC [323]
Bryce, Peter Henderson *see*
– The illumination of joseph keeler, esq
– Saving canadians from the degeneracy due to the industrialism in cities of older civilization
Brychner, Hans *see* Om det religiyse i dets enhed med det humane
Bryden, Henry Anderson *see* Gun and camera in southern africa
Brydges, Harford J *see* An account of the transactions of his majesty's mission to the court of persia
Brydges, Samuel *see* Letters from the continent
Brydie, Andrew *see* Death the last enemy
Brydone, Patrick *see*
– A tour through sicily and malta
– Voyage en sicile et a malthe
Bryers, Fred *see* The cyclists' road guide of canada
Brymner, Douglas *see*
– Church of scotland's endowment
– The jamaica maroons
– Property and civil rights
– The twa mongrels
Bryn Dumm *see* Dassanah khnum
Brynmenin, betharan congregational, monumental inscriptions – 1mf – 9 – £1.25 – uk Glamorgan FHS [929]
Brynna & peterston s montem, st peter, monumental inscriptions – 1mf – 9 – £1.25 – uk Glamorgan FHS [929]
Bryologist – Lewiston ME 1957+ – 1,5,9 – ISSN: 0007-2745 – mf#959 – us UMI ProQuest [580]
Bryson, Mary Isabella *see*
– Cross and crown
– James gilmour and john horden
– John kenneth mackenzie
Y brython = Briton – [Wales] Flintshire 8 chw 1906-23 chw 193 [mf ed 2002] – 34r – 1 – (incorp with: y cymro; 1907,1908 imperfect; 1911 very imperfect) – uk Newsplan [072]
Brython cymreig – [Wales] LLGC ion 1892-medi 1901 [mf ed 2004] – 5r – 1 – uk Newsplan [072]
Brzezinski, Zbigniew K *see*
– Ideology and power in soviet politics
Brzoska, Maria *see* Anthropomorphe auffassung des gebaeudes und seiner teile
Bs. berliner sozialdemokrat *see* Der sozialdemokrat 1946
Bsl aktuell – Schkopau DE, 1998, 14 jan-2000 may – 1r – 1 – (buna-werke) – gw Misc Inst [074]
BSSR: Ekonomiko-statisticheskii spravochnik *see* Sektor narodno-khoziaistvennogo ucheta gosplana
Bssr k 11 sezdu sovetov – Minsk, 1935. 128p – 2mf – 9 – mf#RHS-14 – ne IDC [314]
B-troop news – 1970 may- jun, v1 n3-4 [1970 may-jun] – 2r – 1 – mf#721051 – us WHS [355]
Bubastis / Naville, E – London, 1891 – (= ser Mees 8) – 8mf – 8 – €17.00 – ne Slangenburg [930]
Bubb, LoriAnn K *see* The predictive power of different methods of measuring body composition
Bubbles from the brunnens of nassau / Head, Francis B – London 1834 [mf ed Hildesheim: 1995-98] – (= ser Fbc) – iv/406p on 3mf – 9 – €90.00 – 3-487-29608-X – gw Olms [918]
Bubbles of the foam – London: Medici Society, 1914 – (= ser Samp: indian books) – (trans fr original mss by f w bain) – us CRL [830]
Buber, Martin *see*
– Arab-jewish unity
– Daniel
– Hasidism
– Die juedische bewegung
– Kampf um israel
– Die legende des baal-schem
– Reden uber das judentum
– Vom geist des judentums
Buber, Solomon *see*
– Anshe shem
– Midrasch tehillim
– Pesiktah de rav kahana
Bubnov, Nikolai Mikhailovich *see* Friedrich nietzsches kulturphilosophie und umwertungslehre
Bubonic plague in cuba / Guiteras, Juan – Havana, Cuba. 1915 – 1r – us UF Libraries [972]
Buc, G *see* Institutiones theologicae seu locorum communium christianae religionis analysis
Buccaeus, I *see* Het necrologium dioecesis harlemensis
Buccaneer islands / Cochran, Hamilton – New York, NY. 1941 – 1r – us UF Libraries [972]
The buccaneers and marooners of america / ed by Pyle, Howard – New illus. ed. London: T. Fisher Unwin; New York: Macmillan, 1891. 403p. illus – 1 – us Wisconsin U Libr [970]
Buccaneers in the west indies / Haring, Clarence Henry – New York, NY. 1910 – 1r – us UF Libraries [972]
Buccaneers of america / Exquemelin, A O – London, England. 1924 – 1r – us UF Libraries [972]
Bucci, A *see* ...Oratione della pace, and della guerra contra turchi, a'i prencipi christiani
Bucelini, Gabr *see*
– Menologium benedictinum
– Sacrarium benedictinum
Bucer, G(?) *see* Dissertatio de gubernatione ecclesiae...
Bucer, Martin *see*
– Gesprechbiechlin neuew karsthans
– Metaphrases et enarrationes perpetuae epistolarum d pauli apostoli. tomus primus
– Psalmorum libri quinque ad hebraicam veritatem
Buceta Facorro, Luis *see* El sociograma de la estructura informal
Das buch : zeitschrift fuer die unabhaengige deutsche literatur – Paris (F), 1938-40 – 1r – 1 – gw Misc Inst [430]
Buch al-chazari / Judah, Ha-Levi – Breslau, Germany. 1885 – 1r – us UF Libraries [939]
Das buch baruch : geschichte und kritik, uebersetzung und erklaerung: auf grund des wiederhergestellten hebraeischen urtextes / Kneucker, Johann Jacob – Leipzig: FA Brockhaus, 1879 [mf ed 1985] – 1mf – 9 – 0-8370-3941-X – mf#1985-1941 – us ATLA [221]
Das buch bei den griechen und roemern / Schubert, W – Berlin, 1921 – €7.00 – ne Slangenburg [450]
Das buch daniel / Behrmann, Georg – Goettingen: Vandenhoeck & Ruprecht, 1894 – (= ser Handkommentar zum alten testament) – 2mf – 9 – 0-7905-2883-5 – mf#1987-2883 – us ATLA [221]
Das buch daniel / Kranichfeld, Rudolph – Berlin: Gustav Schlawitz, 1868 – 1mf – 9 – 0-8370-3992-4 – mf#1985-1992 – us ATLA [221]
Das buch daniel / Marti, Karl – Tuebingen: J C B Mohr (Paul Siebeck), 1901 – 2mf – 9 – 0-7905-0436-7 – (incl ind) – mf#1987-0436 – us ATLA [221]
Das buch daniel : text-kritische untersuchung / Riessler, Paul – Stuttgart: Jos Roth, 1899 – 1mf – 9 – 0-8370-4901-6 – (incl bibl ref) – mf#1985-2901 – us ATLA [221]
Ein buch, das gern ein volksbuch werden moechte / Ebner-Eschenbach, Marie von – Berlin: Gebrueder Paetel, 1911 – 1r – 1 – us Wisconsin U Libr [430]
Das buch der beispiele der alten weisen / ed by Holland, Wilhelm Ludwig – Stuttgart: Litterarischer Verein, 1860]mf ed 1993] – (= ser Blvs 56) – vi/261p – 1 – (trans fr latin of giovanni da capua's directorium humanae vitae by anthonius von pforr. incl bibl ref) – mf#8470 reel 11 – us Wisconsin U Libr [450]
Buch der einheit / Ibn Ezra, Abraham Ben Meir – Berlin, Germany. 1921 – 1r – us UF Libraries [939]
Das buch der jubiaeen und sein verhaeltniss zu den midraschim : ein beitrag zur orientalischen sagen- und alterthumskunde / Beer, B – Leipzig: W. Gerhard, [1856?] – 1mf – 9 – 0-7905-1922-4 – (incl bibl ref) – mf#1987-1922 – us ATLA [270]
Das buch der jubilaeen : oder, die kleine genesis / Dillmann, August; ed by Roensch, Hermann – Leipzig: Fues, 1874. Chicago: Department of Photodup, U of Chicago Lib, 1969 (1r); Evanston: American Theol Lib Assoc, 1984 (1r) – 1 – 0-8370-0588-4 – (incl bibl ref and ind) – mf#1984-B104 – us ATLA [221]
Das buch der liebe : gedichte / Brod, Max – Muenchen: Kurt Wolff c1921 [mf ed 1995] – 1r – 1 – (filmed with: der tod des vergil / hermann broch) – mf#3808p – us Wisconsin U Libr [810]
Das buch der liebe : liebenswuerdiges und verliebtes von zeitgenoessischen autoren: mit vieler alten und neuen bildern / Hochstetter, Gustav – 2nd ed. Berlin: Eysler [1916?] [mf ed 1993] – 1r – 1 – (incl ind. filmed with: sputnik contra bombe / gerhard wolf [ed]) – mf#3336p – us Wisconsin U Libr [800]
Buch der lieder / Heine, Heinrich – Muenchen, Germany. 1920 – 1r – us UF Libraries [025]
Buch der lyrik : auswahl deutscher dichtung / ed by Maurer, Friedrich – Berlin: Cornelsen Verlag 1947 [mf ed 1993] – 1r – 1 – (filmed with: das neue lied / ed by wolf kornatzki & other titles) – mf#3341p – us Wisconsin U Libr [810]
Das buch der maccabaeer in mitteldeutscher bearbeitung / ed by Helm, Karl – Stuttgart: Litterarischer Verein, 1904 (Tuebingen: H Laupp, Jr) [mf ed 1993] – (= ser Blvs 233) – xcvi/432p – 1 – (middle high german poem. incl bibl ref) – mf#8470 reel 48 – us Wisconsin U Libr [810]
Das buch der makkabaeer in mitteldeutscher bearbeitung / ed by Helm, Karl – Stuttgart: Litterarischer Verein 1904 (Tuebingen: H Laupp, Jr) [mf ed 1993] – (= ser Blvs 233) – 58r – 1 – (incl bibl ref & ind) – mf#3420p – us Wisconsin U Libr [810]
Das buch der malerzeche in prag / Pangeri, M – Wien, 1878. v13 – 2mf – 9 – mf#O-517 – ne IDC [700]
Das buch der natur [cima33] : farbmikrofiche-edition der handschrift heidelberg, universitaetsbibliothek, cod pal germ 311 und der bilder aus cod pal germ 300 / Megenberg, Konrad von – [mf ed 1997] – (= ser Codices illuminati medii aevi [cima] 33) – 47p on 15 color mf – 15 – €470.00 – 3-89219-033-X – (filmed with: johannes hartlieb: kraeuterbuch; int & description by gerold hayer) – gw Lengenfelder [090]
Buch der pastoralregel (bdk4 2.reihe) / Gregor der Grosse – (= ser Bibliothek der kirchenvaeter. 2. reihe (bdk 2.reihe)) – €14.00 – ne Slangenburg [240]
Das buch der psalmen in neuer und treuer uebersetzung : nach der vulgata, mit fortwaehrender beruecksichtigung des urtextes / ed by Langer, J – 3. aufl. Freiburg i.B, St Louis, MO: Herder, 1889 – 2mf – 9 – 0-7905-0045-0 – mf#1987-0045 – us ATLA [220]
Das buch der reisendie interessantesten und neuesten reiseabenteuer / Wachenhusen, Hans – Berlin [1860] [mf ed Hildesheim 1995-98] – 2v on 11mf – 9 – €110.00 – 3-487-26743-8 – gw Olms [910]
Das buch der richter *see* The book of judges

Das buch der richter. erster band : mit besonderer ruecksicht auf die geschichte seiner auslegung und kirchlichen verwendung / Bachmann, Johannes – Berlin: Wiegandt & Grieben. 2v. 1868-69 – 2mf – 9 – 0-7905-0245-3 – (issued in two pts. no more publ. incl bibl ref) – mf#1987-0245 – us ATLA [221]

Das buch der ringsteine farabis / Horten, M – Muenster, 1906 – (= ser Bgphma 5/3) – 10mf – 8 – €19.00 – ne Slangenburg [190]

Buch der sprueche : [poems] / Blumenthal, Oscar – 2. aufl. Berlin: Concordia Deutsche Verlags-Anstalt, c1909 [mf ed 1993] – 247p – 1 – mf#8520 – us Wisconsin U Libr [810]

Das buch der weisheit / Gutberlet, Constantin – Muenster: Coppenrath, 1874 – 2mf – 9 – 0-7905-0997-0 – (in german and greek) – mf#1987-0997 – us ATLA [221]

Das buch der weisheit / Heinisch, Paul – Muenster in Westf: Aschendorff, 1912 – (= ser Exegetisches Handbuch zum Alten Testament) – 1mf – 9 – 0-7905-3258-1 – (incl bibl ref) – mf#1987-3258 – us ATLA [221]

Das buch der weisheit des jesus sirach (josua ben sira) in seinem verhaeltniss zu den salomonischen spruechen und seiner historischen bedeutung / Seligmann, Caesar – Halle (Saale): [s.n.], 1883 (Breslau [Wroclaw]: Th Schatzky) – 1mf – 9 – 0-8370-9820-3 – (incl bibl ref) – mf#1986-3820 – us ATLA [221]

Das buch der welt – Stuttgart DE, 1842-44, 1848-50 – 1 – gw Misc Inst [074]

Buch des dankes *see*
- Bruno brehm zum fuenfzigsten geburtstag
- Kuenstler

Buch des dankes fuer hans carossa : dem 15. dezember 1928 – Leipzig: Insel-Verlag [1928?] [mf ed 1989] – 1r [ill] – 1 – (filmed with: ungleiche welten) – mf#7145 – us Wisconsin U Libr [880]

Das buch des marco polo als quelle fuer die religionsgeschichte / Witte, Johannes – Berlin: Hutten-Verlag [1916] [mf ed 1995] – (= ser Yale coll) – 126p – 1 – 0-524-09584-1 – (in german) – mf#1995-0584 – us ATLA [200]

Das buch des propheten daniel / Rohling, August – Mainz: Franz Kirchheim, 1876 – 1mf – 9 – 0-8370-9897-1 – (incl bibl ref) – mf#1986-3897 – us ATLA [221]

Das buch des propheten ezechiel / Cornill, Carl Heinrich – Leipzig: J C Hinrichs, 1886 – 2mf – 9 – 0-7905-0931-8 – (text in german and hebrew; critical apparatus in german, latin, ethiopic, greek, hebrew, and syriac) – mf#1987-0931 – us ATLA [221]

Das buch des propheten habackuk / Happel, Otto – Wuerzburg: Andreas Goebel, 1900 – 1mf – 9 – 0-8370-3468-X – mf#1985-1468 – us ATLA [221]

Das buch des propheten sophonias / Lippl, Joseph – Freiburg im Breisgau, St Louis MO: Herder 1910 [mf ed 1989] – (= ser Biblische studien 15/3) – 1mf – 9 – 0-7905-2018-4 – (in german, hebrew & greek) – mf#1987-2018 – us ATLA [221]

Das buch deutscher briefe / ed by Heynen, Walter – Wiesbaden: Insel-Verlag 1957 [mf ed 1993] – 1r – 1 – (incl bibl ref & ind. filmed with: concordanz der deutschen national-literatur / h a berlepsch [ed]) – mf#8503 – us Wisconsin U Libr [860]

Das buch ester : nach der septuaginta hergestellt, uebersetzt und kritisch erklaert – Leiden: E J Brill, 1901 – 1mf – 9 – 0-8370-3760-3 – mf#1985-1760 – us ATLA [221]

Das buch esther : auf seine geschichtlichkeit / Jampel, Sigmund – Frankfurt (Main): J Kauffmann, 1907 – 1mf – 9 – 0-8370-3767-0 – (incl bibl ref) – mf#1985-1767 – us ATLA [221]

Das buch esther *see*
- The book of esther
- An explanatory commentary on esther

Das buch exodus / Eerdmans, Bernadus Dirk – Giessen: A Toepelmann, 1910 [mf ed 1989] – (= ser Alttestamentliche studien 3) – 1mf – 9 – 0-7905-0762-5 – (incl bibl ref) – mf#1987-0762 – us ATLA [221]

Das buch ezechiel : auf grund der septuaginta hergestellt – Leipzig: Eduard Pfeiffer, 1905 – 1mf – 9 – 0-8370-3761-1 – (incl ind of hebrew, german and greek words) – mf#1985-1761 – us ATLA [221]

Das buch ezechiel / Kraetzschmar, Richard – Goettingen: Vandenhoeck & Ruprecht, 1900 – (= ser Handkommentar zum alten testament) – 3mf – 9 – 0-7905-2915-7 – (incl bibl ref) – mf#1987-2915 – us ATLA [221]

Das buch ezechiel / Schmalzl, Peter – Wien: Mayer, 1901 – (= ser Kurzgefasster Wissenschaftlicher Kommentar Zu Den Heiligen Schriften Des Alten Testamentes) – 2mf – 9 – 0-7905-2063-X – (incl ind) – mf#1987-2063 – us ATLA [221]

Das buch ezra *see* The book of ezra

Das buch fuer alle – Stuttgart/Berlin/Leipzig DE, 1893, 1894, 1896, 1911-13 – 1 – (filmed by other misc inst: 1895, 1899, 1900, 1904, 1909; 1872 & 1877, 1897-98 [2r]) – gw Misc Inst [074]

Das buch henoch : aethiopischer text / Ethiopic book of Enoch – Leipzig: J C Hinrichs, 1902 – (= ser Tugal) – 1mf – 9 – 0-7905-1699-3 – mf#1987-1699 – us ATLA [240]

Das buch henoch : aethiopischer text / Flemming, J – Leipzig, 1902 – (= ser Tugal 2-22/1) – 3mf – 9 – €7.00 – ne Slangenburg [221]

Das buch henoch : aus dem aethiopischen in die urspruenglich hebraeische abfassungssprache / zurueckuebersetzt – Berlin: Richard Heinrich, 1892 – 1r – 1 – 0-8370-3332-2 – mf#1985-1332 – us ATLA [470]

Das buch henoch / Dillmann, August – Leipzig: Fr Chr Wilh Vogel, 1853 – 1mf – 9 – 0-7905-0879-6 – (incl bibl ref) – mf#1987-0879 – us ATLA [221]

Das buch henoch : sein zeitalter und sein verhaeltniss zum judasbriefe / Philippi, Ferdinand – Stuttgart: SG Liesching, 1868 – 1mf – 9 – 0-7905-0319-0 – (incl bibl ref) – mf#1987-0319 – us ATLA [221]

Das buch henoch (gcsej8) / ed by Flemming, J & Radermacher, L – 1901 – (= ser Griechische christlichen schriftsteller der ersten jahr- hunderte (gcsej)) – €12.00 – ne Slangenburg [221]

Das buch hiob / Budde, Karl – 2. neubearb aufl. Goettingen: Vandenhoeck & Ruprecht, 1913 – (= ser Goettinger handkommentar zum alten testament) – 1mf – 9 – 0-8370-9447-X – mf#1986-3447 – us ATLA [221]

Das buch hiob / Duhm, Bernhard – Freiburg i.B: J C B Mohr, 1897 – 1mf – 9 – 0-8370-2991-0 – (includes subject index) – mf#1985-0991 – us ATLA [221]

Das buch hiob / Hengstenberg, Ernst Wilhelm – Leipzig: J C Hinrichs, 1875 – 2mf – 9 – 0-7905-2911-4 – mf#1987-2911 – us ATLA [221]

Das buch hiob : nach seinem inhalt, seiner kunstgestaltung und religioesen bedeutung / Ley, Julius – Halle a S: Verlag der Buchh des Waisenhauses, 1903 – 1mf – 9 – 0-7905-1967-4 – mf#1987-1967 – us ATLA [221]

Das buch hiob uebersetzt und ausgelegt / Hitzig, Ferdinand – Leipzig: C F Winter, 1874 – 1mf – 9 – 0-8370-3596-1 – (incl bibl ref and index) – mf#1985-1596 – us ATLA [221]

Das buch jeremia / Cornill, Carl Heinrich – Leipzig: Chr Herm Tauchnitz, 1905 – 2mf – 9 – 0-8370-9691-X – (incl ind) – mf#1986-3691 – us ATLA [221]

Das buch jeremia / Giesebrecht, Friedrich – Goettingen: Vandenhoeck & Ruprecht, 1907 – (= ser Handkommentar zum alten testament) – 1mf – 9 – 0-8370-3271-7 – mf#1985-1271 – us ATLA [221]

Das buch jeremia – Tuebingen: J C B Mohr, 1903. Chicago: Dep of Photodup, U of Chicago Lib, 1971 (1r); Evanston: American Theol Lib Assoc, 1984 (1r) – (= ser HIS die poetischen und prophetischen buecher des alten testaments) – 1 – 0-8370-0442-X – mf#1984-B201 – us ATLA [221]

Das buch jesaia / Duhm, Bernhard – 3. verb verm aufl. Goettingen: Vandenhoeck & Ruprecht, 1914 – (= ser Goettinger handkommentar zum alten testament) – 2mf – 9 – 0-7905-3191-7 – mf#1987-3191 – us ATLA [221]

Das buch jesaja / Marti, Karl – Tuebingen: J C B Mohr (Paul Siebeck), 1900 – 2mf – 9 – 0-7905-1233-5 – (incl ind) – mf#1987-1233 – us ATLA [221]

Das buch jesaja / Marti, Karl – Tuebingen: J C B Mohr (Paul Siebeck), 1900 [mf ed 2003] – (= ser Kurzer hand-commentar zum alten testament 5/10) – 1r – 1 – (incl ind) – mf#b00652 – us ATLA [221]

Das buch job : nach anleitung der strophik und der septuaginta – Wien: Carl Gerold, 1894 – 1mf – 9 – 0-7905-0555-X – mf#1987-0555 – us ATLA [410]

Das buch job : uebersetzt und erklaert / Zschokke, Hermann – Wien: Wilhelm Braumueller, 1875 – 1mf – 9 – 0-8370-5973-9 – mf#1985-3973 – us ATLA [221]

Das buch job *see* The book of job

Das buch job als strophisches kunstwerk nachgewiesen / Hontheim, Joseph – Freiburg i Breisgau, St Louis MO.: Herder 1904 [mf ed 1989] – (= ser Biblische studien 9/1-3) – 1mf – 9 – 0-7905-2474-0 – mf#1987-2474 – us ATLA [221]

Das buch job aussgelegt unnd erklaert in 141 predigen... / Lavater, L – Zuerych: Christoffel Froschouer, 1582 – 6mf – 9 – mf#PBU-321 – ne IDC [240]

Das buch joram / Borchardt, Rudolf – Leipzig: Insel-Verlag, 1907 [mf ed 1989] – 51p – 1 – mf#7052 – us Wisconsin U Libr [880]

Das buch josua / Holzinger, Heinrich – Tuebingen: J C B Mohr, 1901 – 1mf – 9 – 0-8370-3645-3 – (includes subject index) – mf#1985-1645 – us ATLA [221]

Das buch josua *see* The book of joshua

Das buch judith als geschichtliche urkunde : vertheidigt und erklaert nebst eingehenden untersuchungen ueber dauer und ausdehnung der assyrischen obmacht in asien und aegypten, ueber die hyksos, ueber die ursitze der chaldaeer und deren zusammenhang mit den skythen, ueber phud, lud, elam, chna / Wolff, O – Leipzig: Doerffling und Franke, 1861 – 1mf – 9 – 0-8370-9754-1 – (incl bibl ref) – mf#1986-3754 – us ATLA [221]

Das buch kohelet : kritisch und metrisch untersucht / Zapletal, Vincenz – Freiburg (Schweiz): O Gschwend, 1905 – 1mf – 9 – 0-8370-7439-8 – (commentary in german; text in hebrew and german. incl bibl ref) – mf#1986-1439 – us ATLA [221]

Das buch kohelet : nach der auffassung der weisen des talmud und midrasch und der juedischen erklaerer des mittelalters. theil 1, von der mischna bis zum abschluss des babyl. talmud von 200-500 n. d. g. z / Schiffer, Sinai – Frankfurt a M: J Kauffmann, [1884] – 1mf – 9 – 0-8370-5090-1 – mf#1985-3090 – us ATLA [270]

Das buch kohelet im talmud und midrasch / Schiffer, Sinai – Hannover: Arnold Weichelt, 1884 – 1mf – 9 – 0-7905-0293-3 – mf#1987-0293 – us ATLA [270]

Buch kusari des jehuda ha-levi / Judah – Leipzig, Germany. 1869 – 1r – us UF Libraries [939]

Buch, L von *see* Reise durch norwegen und lappland

Buch, Leopold von *see* Reise durch norwegen und lappland

Das buch leviticus / Eerdmans, Bernadus Dirk – Giessen: A Toepelmann, 1912 [mf ed 1989] – (= ser Alttestamentliche studien 4) – 1mf – 9 – 0-7905-0763-3 – mf#1987-0763 – us ATLA [221]

Buch, Maganlal Amritlal *see* Rise and growth of indian liberalism

Das buch nehemiah *see* The book of nehemiah

Das buch ruth *see* The book of ruth

Das buch ruth in der midrasch-litteratur : ein beitrag zur geschichte der bibelexegese / Hartmann, David – Leipzig: Baer & Hermann, 1901 [mf ed 1985] – 1mf – 9 – 0-8370-3506-6 – (in german. incl app) – mf#1985-1506 – us ATLA [221]

Das buch sidrach / ed by Jellinghaus, H – Stuttgart: Litterarischer Verein, 1904 (Tuebingen: H Laupp, Jr) [mf ed 1993] – (= ser Blvs 235) – xii/240p – 1 – (middle low german text. int in german. incl bibl ref) – mf#8470 reel 48 – us Wisconsin U Libr [890]

Das buch sidrach : nach der kopenhagener mittelniederdeutschen handschrift 5. j. 1479 / ed by Jellinghaus, Hermann Friedrich – Stuttgart: Litterarischer Verein 1904 (Tuebingen: H Laupp, Jr) [mf ed 1993] – (= ser Blvs 235) – 58r – 1 – (incl bibl ref & ind. middle low german text; int in german) – mf#3420p – us Wisconsin U Libr [270]

Das buch tobias / Reusch, Franz Heinrich – Freiburg i.B: Herder, 1857 – 1mf – 9 – 0-8370-9730-4 – (in german, latin, and greek) – mf#1986-3730 – us ATLA [226]

Das buch tobit / Sengelmann, H – Hamburg: Perthes-Besser & Mauke, 1857 – 1mf – 9 – 0-7905-0334-4 – mf#1987-0334 – us ATLA [221]

Das buch um anton wildgans / Soyka, Josef – Leipzig: L Staackmann 1932 [mf ed 1991] – 1r [ill] – 1 – (filmed with: armut) – mf#3052p – us Wisconsin U Libr [430]

Das buch von den vier quellen / Wibbelt, Augustin – Warendorf: J Schnellschen Buchhandlung, 1912 [mf ed 1992] – 213p (ill) – 1 – (ill by balduin) – mf#7937 – us Wisconsin U Libr [430]

Buch von der deutschen poeterei / Opitz, Martin; ed by Braune, Wilhelm – Halle a.S: Max Niemeyer 1882 [mf ed 1993] – (= ser Neudrucke deutscher litteraturwerke des 16. und 17. jahrhunderts 1) – 11r [ill] – 1 – mf#3387p – us Wisconsin U Libr [810]

Buch von der deutschen poeterey *see* Martin opitzen's buch von der deutschen poeterei

Das buch von der erkenntniss der wahrheit, oder, der ursache aller ursachen : nach den syrischen handschriften zu berlin, rom, paris und oxford / ed by Kayser, C – Leipzig: JC Hinrichs, 1889 – 2mf – 9 – 0-8370-7297-2 – (incl ind) – mf#1986-1297 – us ATLA [470]

Ein buch von guter speise = The book of good food – Stuttgart: Literarischer Verein, 1844 [mf ed 1993] – (= ser Blvs 9/2) – vi/29p – 1 – mf#8470 reel 2 – us Wisconsin U Libr [640]

Buch, Walter *see* Niedergang und aufstieg der deutschen familie

Das buch wanderschaft / Winnig, August – Hamburg: Hanseatische Verlagsanstalt c1941 [mf ed 1991] – 1r [ill] – 1 – (filmed with: dichterische arbeiten / eugen gottlob winkler) – mf#3058p – us Wisconsin U Libr [910]

Das buch weinsberg, bd 2 1552-1577 (pgrg4) / ed by Hoehlbaum, K – Leipzig, 1887 – (= ser Publikationen der gesellschaft fuer rheinische geschichtskunde (pgrg)) – €17.00 – ne Slangenburg [931]

Das buch zu distilieren die zusamen gethonen ding : composita genant durch die einzingen ding und das buch thesaurus pauperum genant... / Brunschwig, Hieronymus – Strassburg: Bartholeme & Grueniger 1532 [mf ed 19–] – (= ser German books before 1601) – 1r [ill] – 1 – us Misc Inst [615]

Buchan, George *see* A narrative of the loss of the winterton east indiaman, wrecked on the coast of madagascar in 1792

Buchan observer – Peterhead: P Scrogie Ltd 27 sep 1988- (wkly) [mf ed 3 may 1994-] – 1 – (cont: buchan observer & east aberdeenshire advertiser [14 mar 1893-20 sep 1988]) – uk Scotland NatLib [072]

Buchanan, Agnes *see* The treasures of hassan

Buchanan, Alexander Carlisle *see*
- Canada 1863
- Rapport de a c buchanan
- Rapport de l'agent en chef de l'immigration, (a c buchanan, ecr,) pour l'annee 1860

Buchanan, Arthur William Patrick *see* The buchanan book

The buchanan book : the life of alexander buchanan, qc of montreal, followed by an account of the family of buchanan / Buchanan, Arthur William Patrick – Montreal: [s.n.], 1911 – 7mf – 9 – 0-665-72268-0 – (incl bibl ref) – mf#72268 – cn Canadiana [920]

Buchanan, Briggs *see* Catalogue of near eastern seals in the ashmolean museum

Buchanan, C *see* Christian researches in india...

Buchanan county journal – Independence, Jesup IA. 1887 jan 13-20 – 1r – 1 – (cont by: buchanan county bulletin, bulletin-journal [independence ia: 1891]) – mf#851246 – us WHS [071]

Buchanan, Daniel Houston *see* The development of capitalistic enterprise in india

Buchanan, Dr *see* Considerations, explanatory and recommendatory...

Buchanan, Dugald *see* Laoidhean spioradail

Buchanan, Edgar Simmons *see*
- The epistles of s paul from the codex laudianus
- The four gospels
- The four gospels from the codex corbeiensis (ff [or ff2])
- Syllabus of a course of four lectures on the history and authority of the holy scriptures in the church

Buchanan, Francis *see*
- An account of the kingdom of nepal
- A journey from madras through the countries of mysore, canara, and malabar

Buchanan, James *see*
- Analogy considered as a guide to truth
- The doctrine of justification
- Faith in god and modern atheism compared
- Improvement of affliction
- Letter to his excellency sir francis bond head
- Message of the president of the united states
- The office and work of the holy spirit
- Papers
- Sketches of the history, manners, and customs of the north american indians
- The uses of creeds and confessions of faith

Buchanan, John Lane *see* Johann lane buchanans, missionars der schottischen kirche, reisen durch die westlichen hebriden, waehrend der jahre 1782 bis 1790

Buchanan, Joseph Rodes *see* Moral education

Buchanan, Robert *see*
- The book of ecclesiastes
- Christ and casar
- Church of scripture and the church of the disruption
- Close of sermon preached in free north church, stirling,30th september...
- God to be obeyed rather than men
- Principles and position of the free church of scotland
- Reply to an attack on the general assembly's church accommodation c...
- The ten years' conflict

Buchanan, Robert J *see* Canada

Buchanan, Sara Louise *see* Legal status of women in the united states of amer...

Buchanan, William *see* Memoirs of painting

Buchanan, William I *see* Central american peace conference held at washington

Buchanan-Gould, Vera *see* Vast heritage

Buchauer wochenblatt *see* Wochenblatt fuer die fuerstlich thurn und taxischen besitzungen im donaukreis buchau

Buchauer wochenblatt vom federsee *see* Wochenblatt fuer die fuerstlich thurn und taxischen besitzungen im donaukreis buchau

Buchauer zeitung *see* Wochenblatt fuer die fuerstlich thurn und taxischen besitzungen im donaukreis buchau

Buchbauer, Oskar *see* Jungen der fernen grenze

Buchbinder-zeitung – Stuttgart DE, 1885-1904 oct, 1905-32 – 1 – gw Misc Inst [380]

Die buchdrucker-familie froschauer in zuerich 1521-1595 / Rudolphi, E C – Zuerich, 1869 – 2mf – 9 – mf#ZWI-97 – ne IDC [240]

Buchdrucker-wacht – Leipzig DE, 1896-1902 – 1 – gw Misc Inst [074]

Buchdrucker-wacht – Leipzig DE, 1896-1902 – 1 – gw Misc Inst [680]
Buchdrucker-zeitung – 1873-1940, 1887-1940 – (= ser Labor union periodicals, pt 2: the printing trades) – 5r – 1 – $1050.00 – 1-55655-304-8 – us UPA [680]
Buchdrucker-zeitung / Deutsch-Amerikanische Typographia – v55-67 [1927 jul-1940 jul] – 1r – 1 – (cont: deutsch-amerikanische buchdrucker-zeitung) – mf#3189544 – us WHS [680]
Buchdrucker-zeitung – New York NY (USA), 1927-40 – 1r – 1 – gw Misc Inst [680]
Bucher, George *see* The garb law
Buchet, Edmond Edouard *see* Children of wrath
Bucheum : the history and archaeology of the site / Mond, R & Myers, O L – London, 1934 – (= ser Mees 41/1 – 10mf – 8 – €19.00 – ne Slangenburg [930]
Bucheum : the inscriptions / Mond, R & Myers, OL – London, 1934 – (= ser Mees 41/2 – 5mf – 8 – €12.00 – ne Slangenburg [930]
Bucheum : the plates / Mond, R & Myers, OL – London, 1934 – (= ser Mees 41/3 – 16mf – 8 – €31.00 – ne Slangenburg [930]
Buchgemeinschaften in deutschland 1918-1933 / Scholl, Bernadette – (mf ed 1994) – 4mf – 9 – €56.00 – 3-89349-873-7 – mf#DHS 873 – gw Frankfurter [430]
Buchhaendler im neuen reich – Berlin DE, 1936-40 [gaps] – 1 – gw Misc Inst [070]
Buchhaendler palm : ein deutsches heldenschicksal aus dem jahre 1806 / May, Werner – Breslau: H Handel 1935 [mf ed 1990] – 1r – 1 – (incl bibl ref. filmed with: wittenberg und rom / gustav kuhne) – mf#2832p – us Wisconsin U Libr [830]
Die buchhalterin roman / Kretzer, Max – 3. aufl. Leipzig: P List [191-?] [mf ed 1995] – 1r – 1 – (filmed with: berliner skizzen / max kretzer) – mf#3910p – us Wisconsin U Libr [830]
Buchheim, C A *see* Harzreise
Buchheim, Carl Adolf *see* First principles of the reformation
Buchheit, V *see* Studien zu methodios von olympos
Buchheld, Kurt *see* Der deichgraf
Buchholtz, Arend [comp] *see* Die geschichte der familie lessing
Buchholtz, Ludwig *see* Die christliche lehre auf heilsgeschichtlichem grunde
Buchholz gallery-curt valentin catalogue – New York, 1937-1955 – (= ser Art exhibition catalogues on microfiche) – 144 catalogues on 145mf – 9 – £915.00 – (individual titles not listed separately) – uk Chadwyck [700]
Buchholz, Paul Ferdinand Friedrich *see* Geschichte napoleon bonaparte's
Buchholzer anzeiger zu noerdlicher anzeiger *see* Wochenblatt fuer niederschoenhausen, schoenholz, pankow, rosenthal, nordend und wilhelmsruh
Buchi = bushman / Moscoso, Antonio – Panama, 1961 – 1r – us UF Libraries [972]
Buchka, Gerhard von *see*
- Landesprivatrecht der grossherzogtuemer mecklenburg-schwerin und mecklenburg-strelitz
- Vergleichende darstellung des buergerlichen gesetzbuches fuer das deutsche reich und des gemeinen rechts

Buchmann, Jacob *see*
- Kirchliche autoritaet und macht der wissenschaft
- Krumme wege zur unfehlbarkeit
- Ein missionsbischof aus laengst vergangener zeit
- Populaersymbolik, oder, vergleichende darstellung der glaubensgegensaetze zwischen katholiken und protestanten nach ihren bekenntnisschriften
- Die unfreie und die freie kirche
- Von palaestrina nach anagni
- Zaghafte und entschlossene politik

Buchmann, Klaere *see*
- The odor fontane
- Theodor fontane
- Der widerschein

Buchner, Otto *see* Die feuermeteore
Buchner, R *see* Textkritische untersuchungen zur lex ribvaria (mgh schriften..: 5.bd)
Buchon, Jean A *see* Collection des chroniques nationales francaises
Buchon, Jean Alexandre C *see* Chronique de la conquete de constantinople et de l'etablissement des francais en moree
Buchsbaum, Ralph Morris *see* Animals without backbones
Buchta, Aegidius *see* Das religioese in clemens brentanos werken
Buchtenkirch, Gustav *see* Kleists lustspiel "der zerbrochene krug" auf der buehne
Buchwald, Georg *see* Wittenberger ordiniertenbuch
Buchwald, Reinhard *see*
- Das leben goethes
- Schiller und beethoven
- Das vermaechtnis der deutschen klassiker

Das buchwesen im altertum und im byzantinischen mittelalter / Gardthausen, Viktor – 2. Aufl. Leipzig: Veit, 1911 – 1mf – 9 – 0-7905-1393-5 – (in german, greek and latin. includes bibliographies and index) – mf#1987-1393 – us ATLA [760]
Buck, Adriaan de *see*
- The egyptian coffin texts, vol 2
- The egyptian coffin texts, vol 6

Buck, Carl Darling *see* Introduction to the study of the greek dialects
Buck, Cecil Henry *see* Faiths, fairs, and festivals of india
Buck creek baptist church. calhoun, kentucky : church records – 1824-Aug 1973 – 1 – 81.00 – us Southern Baptist [242]
Buck creek baptist church. spartanburg, south carolina : church records – 1851-1966 – 1 – us Southern Baptist [242]
Buck, Daniel Dana *see*
- An original harmony and exposition of the 24th chapter of matthew and the parallel passages in mark and luke
- Our lord's great prophecy and its parallels throughout the bible, harmonized and expounded

Buck, Dudley *see* The centennial meditation of columbia
Buck, Edward *see* Massachusetts ecclesiastical law
Buck, Gladys *see* Indian legend
Buck, Herbert *see* Kassel und ahnaberg
Buck, Jirah Dewey *see* The new avatar and the destiny of the soul
Buck, John Lossing *see* Land utilization in china
Buck, Leffert Lefferts *see* A few remarks about the niagara gorge
Buck, Michael Richard *see* Ulrichs von richental chronik des constanzer concils
Buck, Pearl Sydenstricker *see* House of earth
Buck, Richard *see* Ulrichs von richental chronik des constanzer consils 1414-18
Buck, Victor de *see* Les saints martyrs japonais de la compagnie de jesus
Buck, William Calmes *see*
- The baptist hymn book
- Biographical materials
- Theology: the philosophy of religion

Buckbee, Charles A *see* Discussion on the necessity of revising king james' version of the holy scriptures
Bucke, Richard Maurice *see*
- Alcohol in health and disease
- Calamus
- The correlation of the vital and physical forces
- Cosmic consciousness
- Notes and fragments
- Surgery among the insane in canada
- The wound dresser

Buckel, A *see* Die gottesbezeichnungen in den liturgien der ostkirchen
Buckeye / Fulton Co. Archbold – dec 1970-dec 1982 [wkly] – 13r – 1 – mf#B13001-13013 – us Ohio Hist [071]
Buckeye / Highland Co. Leesburg – apr 1899-jun 1900, sep 1900-12 [wkly] – 5r – 1 – mf#B12378-12382 – us Ohio Hist [071]
Buckeye / Logan Co. DeGraff – (mar 1879-jun 1899) scattered [wkly] – 1r – 1 – mf#B1478 – us Ohio Hist [071]
Buckeye american : weekly ku klux klan newspaper – Warren, OH. 11 Sept 1923 – 1r – 1 – us Western Res [071]
Buckeye banner : weekly temperance newspaper – Warren, OH. 27 July and 3 Aug 1849 – 1r – 1 – us Western Res [071]
Buckeye engineer : official publication of local unions 18, 18-a, 18-b, 18c, 18-ra / International Union of Operating Engineers – 1988 jan-1993 dec, v11 n6-v21 n12 [1978 jun-1987 dec] – 2r – 1 – mf#1573047 – us WHS [620]
Buckeye first southern baptist church. buckeye, arizona : church records – 1925-90 – 1 – $88.38 – us Southern Baptist [242]
Buckeye flyer – 1981 may-1993 dec – 1r – 1 – mf#5486791 – us WHS [071]
Buckeye guard – 1984 sep/oct, 1986 jul/aug, 1988 win, sum, 1991 fall, 1992 sum, 1993 spr-sum – 1r – 1 – mf#1053652 – us WHS [071]
Buckeye news / Franklin Co. Canal Winches – 1915-17, 1940-41, 1959-65 – 6r – 1 – mf#B36222-36227 – us Ohio Hist [071]
Buckeye review / Mahoning Co. Youngstown – aug 1967-dec 1976 [wkly] – 6r – 1 – mf#B4992-4997 – us Ohio Hist [071]
Buckeye review / Trumbull Co. Youngstown – aug 1967-dec 1976 [wkly] – 6r – 1 – (a black newspaper) – mf#B4992-4997 – us Ohio Hist [071]
Buckeye review – Youngstown OH. v31 n46 [[1968 sep 6]-1971 may 14]-[1998 jan 7/13-dec 23/jan 5] – 22r – 1 – mf#801225 – us WHS [071]
Buckeye smoke signals / North American Indian Association – jan-feb 1969 – (= ser American indian periodicals... 2) – 1mf – 9 – $95.00 – us UPA [305]
Buckeye state / Columbiana Co. Lisbon – 1875-78, sep 1879-1920 [wkly, semimthly] – 20r – 1 – mf#B27618-27637 – us Ohio Hist [071]
Buckfastleigh western guardian – [SW England] Devon jan 1903-dec 1907 [mf ed 2003] – 5r – 1 – uk Newsplan [072]
Buckham, John Wright *see*
- Christ and the eternal order
- Personality and the christian ideal
- Religious progress on the pacific slope

Buckhead baptist church : minutes, membership rolls, wms – Buckhead, GA. 1894-1995 – 1 – $122.40 – (some yrs missing) – mf#6895 – us Southern Baptist [242]
Bucking chute gazette – 1988 jun – 1r – 1 – mf#4798735 – us WHS [071]
Buckingham, James *see*
- The buried city of the east
- Reisen durch syrien und palestina
- Tea-garden coolies in assam
- Travels among the arab tribes inhabiting the countries east of syria and palestine
- Travels in assyria, media, and persia
- Travels in mesopotamia
- Travels in palestine, through the countries of bashan and gilead, east of the river jordan

Buckingham, James Silk *see*
- Autobiography...including his voyages, travels, adventure, speculations, successes and failures, faithfully and frankly narrated
- Outline sketch of the voyages, travels, writings, and public labours of james silk buckingham
- Travels in palestine

Buckingham : lee county / Hanson, W Stanley – s.l, s.l? 1936 – 1r – us UF Libraries [978]
Buckingham, Samuel Giles *see* A memorial of the pilgrim fathers
Buckinghamshire and adjacent counties advertiser – Amersham, England 15 nov 1853-26 dec 1854 – 1 – (cont by: buckinghamshire advertiser, & middlesex, herts, berks, beds, & oxon gazette [2 jan-19 jun 1855]) – uk British Libr Newspaper [072]
Buckinghamshire Association Of Baptist Churches *see* Common errors respecting christian experience
Buckinghamshire record society – v1-11. 1937-56 – (= ser Publications of the english record societies, 1835-1972) – 25mf – 9 – uk Chadwyck [941]
Buckland 1841-1849 – Oxford, MA (mf ed 1995) – (= ser Massachusetts vital record transcripts to 1850) – 1mf – 9 – 0-87623-221-7 – (mf 1t: births & marriages 1843-49; deaths 1841-49) – us Archive [978]
Buckland 1873-1895 – Oxford, MA (mf ed 1988) – (= ser Massachusetts vital records) – 5mf – 9 – 0-87623-046-X – (mf 1: births 1873-89. mf 2: births 1889-95; marriages 1869-91. mf 3: marriages 1882-95. mf 4: deaths 1873-90. mf 5: deaths 1891-95) – us Archive [978]
Buckland, Augustus Robert *see*
- James gilmour and john horden
- Women in the mission field

Buckland, C E *see*
- Bengal under the lieutenant-governors
- Dictionary of indian biography

Buckland, William *see*
- Geology and mineralogy considered with reference to natural theology
- Reliquiae diluvianae

Buckland, William Warwick *see* Inquiry whether the sentence of death pronounced at the fall of man
Buckle, Henry Thomas *see*
- The miscellaneous and posthumous works of henry thomas buckle
- The miscellaneous and posthumous works of henry thomas buckle, vol 1
- The miscellaneous and posthumous works of henry thomas buckle, vol 2

Buckle, Mary *see* Art in needlework
Buckler, John Chessell *see* A history of the architecture of the abbey church of st alban
Buckley, Christopher *see* Five ventures
Buckley, Edmund *see* Phallicism in japan
Buckley, Homer John *see* Science of marketing by mail
Buckley, James Monroe *see*
- An address on supposed miracles
- Constitutional and parliamentary history of the methodist episcopal church
- Faith-healing, christian science and kindred phenomena
- The fundamentals and their contrasts
- History of methodists in the united states
- A history of methodists in the united states
- Perfect love
- What methodism owes to women

Buckley, John *see* An impartial account of the late debate at lyme in the colony of connecticut
Buckley, Katharine C *see* Florida and mexico competition for the winter fresh vegetable market
Buckley, Peter F *see* Journal of dual diagnosis
Buckley, Robert Burton *see* Irrigation works in india and egypt
Buckley, Theodore Alois *see* A history of the council of trent
Buckley-mathew collection, 1850-1856 – Liverpool Public Library – (= ser BRRAM series) – 1r – 1 – £67 / $134 – (int by h e s fisher) – mf#95805 – uk Microform Academic [975]
Buckminster, Joseph Stevens *see*
- Notice of griesbach's edition of the new testament
- Sermons
- The works of joseph stevens buckminster

Bucknell review – Cranbury NJ 1954+ – 1,5,9 – (cont: bucknell university studies) – ISSN: 0007-2869 – mf#12135.01 – us UMI ProQuest [378]
Bucknell university studies – Cranbury NJ 1941-70 – 1,5,9 – (cont by: bucknell review) – mf#12135.01 – us UMI ProQuest [378]
Bucknell world – Lewisburg, PA., 1984-1987 – 13 – $25.00r – us IMR [071]
Bucknellan paper – Lewisburg, PA., 1984-1987 – 13 – $25.00r – us IMR [071]
Buckner, Heike *see* Lohn- und tarifpolitik in der metallindustrie 1918 bis 1933
Bucknew, H F *see* A grammar of the maskoke
Bucknill, George *see* He being dead yet speaketh
Buckower lokal-anzeiger – Buckow DE, 1933-1934 30 jun, 1935-1937 30 jun, 1938-1941 28 jun, 1943-1944 29 jun – 1 – gw Misc Inst [074]
Bucks advertiser & aylesbury news – Aylesbury, England 2 jan 1847-27 jun 1980 – 1 – (wanting 1881, jan 1896-feb 1897, 1903, 1911; cont: aylesbury news & advertiser for bucks & the surrounding counties [3 dec 1836-30 dec 1846]; cont by: bucks advertiser [4 jul 1980-]) – uk British Libr Newspaper [072]
Bucks county gazette – New Hope, PA. -w 1972 – 13 – $25.00r – us IMR [071]
Bucks County Historical Society *see* Collection of papers read before the...
Bucks county intelligencer – Doylestown, PA. -w 1898-1912; 1918-1927 – 13 – $25.00r – us IMR [071]
Bucks County. Pennsylvania *see* Tax lists
Bucks county press – Levittown, PA. -w 1952-1953 – 13 – $25.00r – us IMR [071]
Bucks examiner – Chesham, England 5 jan 1906-16 aug 2002 [mf 1986-] – 1 – (wanting: 1896, 1911; cont: chesham examiner, amersham & rickmansworth times [24 jul 1889-29 dec 1905]; cont by: buckinghamshire examiner [22 aug 2002-]) – uk British Libr Newspaper [072]
Bucks gazette – Aylesbury, England 17 nov 1821-6 oct 1849 – 1 – (incorp with: bucks chronicle) – uk British Libr Newspaper [072]
Buckskin – 1970-71 – (= ser American indian periodicals... 1) – 7mf – 9 – $95.00 – us UPA [305]
Buckskin bulletin – v6-8 [1971 fall-1974 fall] – 1r – 1 – mf#223751 – us WHS [071]
Bucolica... / Siculus, Calpurnius – 15th c – (= ser Holkham library manuscript books 334) – 1r – 1 – (filmed with: bucolica by vergil; achilleis by statius) – mf#96936 – uk Microform Academic [450]
Bucolica et georgica / Vergil – 15th c – (= ser Holkham library manuscript books 307) – 1 col r – 14 – mf#96936 – uk Microform Academic [810]
Bucolica, georgica, aeneis(cima23) : farbmikrofiche-edition der handschrift valencia, biblioteca general i historica de la universitat, ms 837 / Vergilius Maro, Publius – (mf ed 1992) – (= ser Codices illuminati medii aevi (cima) 23) – 42p on 19 color mf – 15 – €360.00 – 3-89219-023-2 – (int by antonie wlosok) – gw Lengenfelder [090]
Bucquoi, Erdmann Friedrich et al *see* Bunzlauische monathschrift zum nutzen und vergnuegen
Bud Nan, Bhikkhu *see* Ranasirs pram yan
Budagesti orszagos rabbikepzo-intezet ertesitoje az 1883- 84-iki ta – Budapest, Hungary. 1884 – 1r – us UF Libraries [939]
Budaja djaja. Madjalah kebudajaan umum *see* Dewan kesenian djakarta
Budapest orszagos rabbikepzo-intezet ertesitoje az 1890- 91-iki tan – Budapest, Hungary. 1891 – 1r – us UF Libraries [939]
Budapester rundschau – Budapest (H), 1976-1992 6 jan [gaps] – 16r – 1 – gw Mikropress [380]
Budapester rundschau : wochenzeitung fuer politik, wirtschaft und kultur – Budapest, Hungary 3 mar 1967-6 jan 1992 [mf 1967-75] – 1 – (specimen iss dated 11 feb 1967) – uk British Libr Newspaper [077]
Budapester zeitung – Budapest (H), 1999 17 jul-2001 23 dec – 1 – gw Misc Inst [077]
Budapesti hirlap – Budapest. 1926-1930, 1935-1936, Jan-Apr 1939 – 1 – us NY Public [947]
Budapesti hirlap – Budapest, Hungary 1 jan 1905-30 apr 1939 – 1 – (lacking: sept, oct 1937) – uk British Libr Newspaper [077]
Budapesti orszagos rabbikepzo-intezet ertesitoje / Bloch, Moses – Budapest, Hungary. 1882 – 1r – us UF Libraries [939]
Budapesti szemle – ser. 1, v. 1-ser. 3, v. 177, no. 505. 1857-1919. (incomplete) – 1 – us NY Public [073]
Buday, Laszlo *see* Dismembered hungary

Buday, Marie T *see* A biomechanical analysis of the demi plie and grand plie
Budbaereren / Norwegian Lutheran Church of America – Everett, Kent, Seattle, Tacoma WA. 1918 mar 1-1922 jun 21, 1922 jul 5-1928 dec 29, 1929 jan 2-jun 19 – 3r – 1 – (cont: pacific herald, budstikken; cont by: pacific lutheran herald) – mf#1425699 – us WHS [242]
Budbaereren *see* Lutheraneren
Budberg, Leonhard G von *see* Gallerie der neuesten reisen, von russen durch russland und fremde laender unternommen
Budd, Henry *see*
– The 39 articles of our established church – 1571
– Leading cases in the american law of real property
– Petition proposed to be presented respectively to the three estates
– Scriptural education ilustrated according to the formularies of the church of england
Budde, Karl *see*
– Auf dem wege zum monotheismus
– Das buch hiob
– Die buecher richter und samuel
– Eduard reuss' briefwechsel mit seinem schueler und freunde karl heinrich graf
– Geschichte der althebraeischen litteratur
– Der kanon des alten testaments
– Das prophetische schrifttum
– Die religion des volkes israel bis zur verbannung
– Die schaetzung des koenigtums im alten testament
Buddeberg *see* Ueber das bei dem hebraeischen unterricht zu grunde zu legende uebungsbuch – die schulnachrichten
Buddelmeyer-zeitung – Berlin DE, 1849 2 apr-1851 29 dec – 1r – 1 – gw Misc Inst [074]
Buddensieg, Rudolf *see*
– Johann wiclif und seine zeit
Buddeus, Johann Franz *see*
– Allgemeines historisches lexikon
Buddh pravatti sankhep / Suvannajoto, Bhikkhu – Bhnam Ben: Buddhasasan Pandity 2512 [1969] [mf ed 1990] – 1r with other items – 1 – (in khmer) – mf#mf-10289 seam reel 104/6 [§] – us CRL [280]
Buddha : being a dramatised version of sir edwin arnold's "the light of asia" / Bose, S C – London: Kegan Paul, Trench, Treubner, [1916] [mf ed 1995] – (= ser Yale coll) – 31p – 1 – 0-524-09234-6 – mf#1995-0234 – us ATLA [280]
Buddha : his life, his doctrine, his order = Buddha / Oldenberg, Hermann – London: Williams and Norgate, 1882 – 2mf – 9 – 0-524-02355-7 – (incl bibl ref. in english) – mf#1990-2966 – us ATLA [280]
Buddha : ein culturbild des ostens / Dahlmann, Joseph – Berlin: FL Dames, 1898 – 1mf – 9 – 0-524-01424-8 – (incl bibl ref) – mf#1990-2419 – us ATLA [280]
Buddha : epische dichtung in zwanzig gesaengen / Widmann, Joseph Viktor – Bern: Dalp (C Schmid) 1869 [mf ed 1995] – 1r – 1 – (filmed with: die gnaedige frau von paretz / ernst wichert) – mf#3759p – us Wisconsin U Libr [810]
Buddha / Hardy, Edmund – Leipzig: G J Goeschen 1903 [mf ed 1991] – (= ser Sammlung goeschen) – 1mf – 9 – 0-524-01602-X – mf#1990-2541 – us ATLA [280]
Buddha : his part in human evolution / Laffitte, M Pierre – Yokohama; Shanghai: Kelly & Walsh, [1901] [mf ed 1995] – (= ser Yale coll; Humanity's great exemplars) – 1 – 0-524-10039-X – (fr french of m pierre laffitte) – mf#1995-1039 – us ATLA [280]
Buddha : sein evangelium und seine auslegung / Held, Hans Ludwig – Muenchen: Hans Sachs, 1912 [mf ed 1991] – 2v on 2mf – 9 – 0-524-01831-6 – (iss in pts; v2 lief 11-14 only; no more publ? with bibl ref) – mf#1990-2666 – us ATLA [280]
Buddha abhidhamma khyup / So Jan, U – [Ran kun: Mit sac] 1974 [mf ed 1994] – on pt of 1r – 1 – mf#11052 r1826 n3 – us Cornell [959]
Buddha and buddhism / Lillie, Arthur – New York: Scribner 1900 [mf ed 1991] – (= ser The world's epoch-makers) – 1mf – 9 – 0-524-01841-3 – mf#1990-2676 – us ATLA [280]
Buddha and early buddhism / Lillie, Arthur – London: Truebner, 1881 – 1mf – 9 – 0-524-01786-7 – mf#1990-2634 – us ATLA [280]
The buddha and his religion = Bouddha et sa religion / Barthelemy Saint-Hilaire, Jules – London: G Routledge, 1895 – (= ser Sir John Lubbock's Hundred Books) – 1mf – 9 – 0-524-02291-7 – (incl bibl ref. in english) – mf#1990-2914 – us ATLA [280]
Buddha and his sayings : with comments on re-incarnation, karma, nirvana, etc / Syama Sankara – London: F Griffiths, 1914 – 1mf – 9 – 0-524-03375-7 – mf#1990-3209 – us ATLA [280]
Buddha and the gospel of buddhism / Coomaraswamy, Ananda Kentish – London: George G Harrap & Co, 1916 – (= ser Samp: indian books) – (with ill in colour by abanindro nath tagore & nanda lal bose and 32 reproductions in black and white from photographs) – us CRL [280]
Buddha bhasa kamma vada nahn mranma yna kye mhu / Thavan Mran, U, Dagun – Yan Kun: Ca Yan Pan Cape 1977 [mf ed 1990] – 1r with other items – 1 – (in burmese) – mf#mf-10289 seam reel 133/9 [§] – us CRL [280]
Buddha e upama ta thon / Le Mruin – Ran kun: Hansa Ca pe 1975 [mf ed 1995] – on pt of 1r – 1 – mf#11052 r1947 n4 – us Cornell [280]
Buddha e vi nann vatthu to kri / Janeyyabhivamsa, a rhan – Ran kun: Kaya Sukha Pitakat Pum nhip tuik: Phran khyi re thana, Sacca Manduin Pitakat Ca up chuin 1976 [mf ed 1995] – on pt of 1r – 1 – mf#11052 r1957 n4 – us Cornell [280]
Buddha nikay na rap / Can nve, Da gun U – Ton Ukkalapa [Ran kun]: Da gun Ca pe 1975 [mf ed 1994] – on pt of 1r – 1 – mf#11052 r1825 n2 – us Cornell [280]
The buddha of christendom : a book for the present crisis / Anderson, Robert – London: Hodder & Stoughton, 1899 – 1mf – 9 – 0-8370-2493-5 – (incl indes) – mf#1985-0493 – us ATLA [240]
Buddha ron khrann nhan pu gam mrui / Cui Mran, U – Ran Kun: Mra Ca pay Ca pe: Prann lum kyvat phran khyi ra thana, Ca pay U 1983 [mf ed 1990] – 1r with other items – 1 – (in burmese) – mf#mf-10289 seam reel 133/2 [§] – us CRL [280]
Buddha und die frauen / Schreiber, Max – Tuebingen: JCB Mohr, 1903 – 1mf – 9 – 0-524-02365-4 – (incl bibl ref) – mf#1990-2976 – us ATLA [280]
Buddha vada a me a phre ta thon "2" / Man Canda – Ran kun: Ta kon Ca up tuik 1975 [mf ed 1994] – on pt of 1r – 1 – mf#11052 r1733 n10 – us Cornell [280]
Buddhabhasa sui ku pron khran atthuppatti kyam / Nnui, U – Ran kun: Thin Van Ca pe phran khyi re 1962 [mf ed 1990] – 1r with other items – 1 – (in burmese; autobiography of u nnui, a burmese government official who converted to christianity and later reverted to buddhism; cover title: van thok kri u nnui e atthuppatti bhava a lan pra sankhepa kyam) – mf#mf-10289 seam reel 133/5 [§] – us CRL [280]
Buddhaghosa *see*
– Buddhaghosha's parables
– Buddhist legends
– Visuddhimagg bhag 3
Buddhaghosha's parables / Buddhaghosa – London: Truebner, 1870 – 1mf – 9 – 0-524-08402-5 – mf#1993-4012 – us ATLA [280]
Buddhaghosuppatti : or, the historical romance of the rise and career of buddhaghosa / Mahamangala; ed by Gray, James – London: Luzac, 1892 – 2mf – 9 – 0-524-07216-7 – mf#1991-0078 – us ATLA [490]
The buddha-karita of asvaghosha / ed by Cowell, E B – Oxford, 1893 – €11.00 – (ed fr 3 mss) – ne Slangenburg [280]
Die buddha-legende in den skulpturen des tempels von boro-budur / Pleyte, Cornelis Marinus – Amsterdam: J H de Bussy, 1901 – (= ser Yale coll) – xvi/183p (ill) – 1 – (in german) – mf#1996-1241 – us ATLA [730]
Die buddha-legende und das leben jesu nach den evangelien : erneute pruefung ihres gegenseitigen verhaeltnisses / Seydel, Rudolf – 2. Aufl. mit ergaenzenden Anmerkungen von Martin Seydel. Weimar: Emil Felber, 1897 – 1mf – 9 – 0-7905-2135-0 – (incl bibl ref and index) – mf#1987-2135 – us ATLA [240]
Buddharakkhita, Mahathera *see* Jinaalankaara
Buddha's teachings : being the sutta-nipata or discourse-collection – Cambridge MA: Harvard University Press, 1932 – (= ser Samp: indian books) – (ed in the original pali text with an english version facing it by lord chalmers) – us CRL [280]
The buddha's way of virtue : a translation of the dhammapada from the pali text / Wagiswara, W D C & Saunders, K J – New York: E P Dutton 1912 [mf ed 1993] – (= ser The wisdom of the east) – 1mf – 9 – 0-524-06898-4 – mf#1991-0041 – us ATLA [280]
Buddhasasana prajadhipateyy sadharanaratth / Khiav Jum, Bhikkhu dhammapal – Bhnam Ben: Pannagar Banly Jati 2515 [1971] [mf ed 1990] – 1r with other items – 1 – (in khmer) – mf#mf-10289 seam reel 128/1 [§] – us CRL [230]
Buddhasasanapandity [Institute: Phnom Penh, Cambodia] *see* Inscriptions modernes d'angkor
Buddhism : being a sketch of the life and teachings of gautama, the buddha / Davids, Thomas William Rhys – new rev ed. London: SPCK 1910 [mf ed 1992] – (= ser Non-christian religious systems) – 1mf [ill] – 9 – 0-524-04157-1 – (incl bibl ref) – mf#1990-3287 – us ATLA [280]
Buddhism : in its connexion with brahmanism and hinduism, and in its contrast with christianity / Monier-Williams, Monier – New York: Macmillan 1889 [mf ed 1992] – (= ser Duff missionary lectures 1888) – 2mf – 9 – 0-524-02315-8 – mf#1990-2938 – us ATLA [280]
Buddhism : its doctrines and its methods / David-Neel, Alexandra – London: John Lane the Bodley Head, 1939 – (= ser Samp: indian books) – us CRL [280]
Buddhism : its essence and development / Conze, Edward – Oxford: Bruno Cassirer, 1951 – (= ser Samp: indian books) – us CRL [280]
Buddhism : its historical, theoretical and popular aspects / Eitel, Ernest John – 3rd rev ed. Hongkong: Lane, Crawford 1884 [mf ed 1991] – 1mf – 9 – 0-524-01480-9 – mf#1990-2456 – us ATLA [280]
Buddhism : its history and literature / Davids, Thomas William Rhys – London, New York: GP Putnam's Sons, 1926 – (= ser Samp: indian books) – us CRL [280]
Buddhism : its history and literature / Davids, Thomas William Rhys – New York: G P Putnam 1896 [mf ed 1991] – (= ser American lectures on the history of religions. 1st series 1894-95) – 1mf – 9 – 0-524-00714-4 – mf#1990-2042 – us ATLA [280]
Buddhism / Small, Annie H – London: J M Dent; New York: E P Putton 1905 [mf ed 1995] – (= ser Yale coll; Studies in the faiths 1) – 1r [ill] – 1 – 0-524-09177-3 – (filmed with other works) – mf#1995-0177 – us ATLA [280]
Buddhism : a study of the buddhist norm / Davids, Caroline Augusta Foley Rhys – New York: H Holt; London: Williams & Norgate [1912?] [mf ed 1991] – (= ser Home university library of modern knowledge 44) – 1mf – 9 – 0-524-00712-8 – mf#1990-2040 – us ATLA; us CRL [280]
Buddhism and asoka / Gokhale, Balkrishna Govind – Baroda: Padmaja Publ; Bombay: Sole distributors, Padma Publ, [1948] – (= ser Samp: indian books) – us CRL [280]
Buddhism and buddhist pilgrims : a review of m stanislas julien's "voyages des pelerins bouddhistes" / Mueller, Friedrich Max – London: Williams & Norgate 1857 [mf ed 1991] – 1mf – 9 – 0-524-01854-5 – mf#1990-2689 – us ATLA [280]
Buddhism and christianity : a parallel and a contrast / Scott, Archibald – Edinburgh: D Douglas 1890 [mf ed 1991] – (= ser Croall lectures 1889-90) – 1mf – 9 – 0-7905-9879-5 – (incl bibl ref) – mf#1989-1604 – us ATLA [230]
Buddhism and immortality / Bigelow, William Sturgis – Boston: Houghton Mifflin 1908 [mf ed 1991] – (= ser The ingersoll lecture 1908) – 1mf – 9 – 0-524-00692-X – mf#1990-2020 – us ATLA [280]
Buddhism and its christian critics / Carus, Paul – Chicago: The Open Court Publ Co 1897 [mf ed 1996] – 1r – 1 – (filmed with: history of the wesleyan methodist church of south africa / whiteside, joseph) – mf#1737 – us Wisconsin U Libr [280]
Buddhism and its place in the mental life of mankind / Dahlke, Paul – London: Macmillan and Co, 1927 – (= ser Samp: indian books) – us CRL [280]
Buddhism and science / Dahlke, Paul – London: Macmillan 1913 [mf ed 1991] – 1mf – 9 – 0-524-01423-X – (trans fr german by bhikkhu silacara) – mf#1990-2418 – us ATLA [230]
Buddhism as a religion : its historical development and its present conditions / Hackmann, Heinrich Friedrich – 2nd ed. London: Probsthain 1910 [mf ed 1991] – (= ser Probsthain's oriental series 2) – 1mf – 9 – 0-524-00880-9 – (in english; incl bibl ref) – mf#1990-2103 – us ATLA [280]
Buddhism in cambodia = Le bouddhisme au cambodge / Pang-Khath – [Phnom-Penh]: Institut Bouddhique 1970 [mf ed 1989] – (= ser Serie de culture et civilisation khmeres 8) – [Pl/ill] 1r with other items – 1 – (also available in french) – mf#mf-10289 seam reel 020/12 [§] – us CRL [280]
Buddhism in china / Beal, Samuel – London: SPCK; New York: E & JB Young 1884 [mf ed 1991] – (= ser Non-christian religious systems) – 1mf [ill] – 9 – 0-524-00689-X – mf#1990-2017 – us ATLA; uk Chadwyck [280]
Buddhism in its relationship with hinduism / Dharmapala, Anagarika – Calcutta: Anagarika Brahmachari Dharmapala, Maha Bodhi Society, 1918 [mf ed 1995] – (= ser Yale coll) – 29p – 1 – 0-524-09253-2 – mf#1995-0253 – us ATLA [280]
Buddhism in kerala / Alexander, Padinjarethalakal Cherian – Annamalainagar: Annamalai University, 1949 – (= ser Samp: indian books) – us CRL [280]
Buddhism in translations : passages selected from the buddhist sacred books and translated from the original pali into english – Cambridge MA: Harvard University Press, 1953, c1896 – (= ser Samp: indian books) – us CRL [280]
The buddhism of tibet : or lamaism: with its mystic cults, symbolism and mythology... / Waddell, Laurence Austine – London: W H Allen 1895 [mf ed 1992] – 2mf [ill] – 9 – 0-524-05352-9 – (incl bibl ref) – mf#1990-3473 – us ATLA [280]
Buddhism, primitive and present : in magadha and in ceylon / Copleston, Reginald Stephen – 2nd ed. London, New York: Longmans, Green 1908 [mf ed 1991] – 1mf – 9 – 0-524-01422-1 – (incl bibl ref; first printed 1892) – mf#1990-2417 – us ATLA [280]
Le buddhisme : precede d'un essai sur le vedisme et le brahmanisme / Lafont, Gaston de – Paris: Chamuel, 1895 – 1mf – 9 – 0-524-01776-X – (incl bibl ref) – mf#1990-2624 – us ATLA [280]
Le buddhisme au cambodge / Leclere, Adhemard – Paris: E. Leroux, 1899. xxxi,535p. front.,illus.,plates – 1 – us Wisconsin U Libr [280]
Der buddhismus / Hackmann, Heinrich Friedrich – Halle a. S: Gebauer-Schwetschke 1905 [mf ed 1995] – (= ser Religionsgeschichtliche volksbuecher fuer die deutsche christliche gegenwart 3/4-5,7) – 3v on 1r – 1 – 0-524-10171-X – mf#1995-1171 – us ATLA [280]
Der buddhismus : oder, der vorchristliche versuch einer erloesenden universalreligion / Wurm, Paul – Guetersloh: C Bertelsmann, 1880 – 1mf – 9 – 0-524-03604-7 – mf#1990-3248 – us ATLA [280]
Buddhismus (buddha und seine lehre) / Beckh, Hermann – 2. aufl. Berlin, Leipzig: G J Goeschen, 1919 [mf ed 1995] – (= ser Yale coll; Sammlung goeschen) – 2v – 1 – 0-524-10264-3 – (in german) – mf#1996-1264 – us ATLA [280]
Der buddhismus in china : eine religionsgeschichtliche studie / Piton, Charles – Basel: Missionsbuchh, 1902 – (= ser Basler missionsstudien) – 1mf – 9 – 0-524-02226-7 – mf#1990-2900 – us ATLA [280]
Der buddhismus in seiner psychologie / Bastian, Adolf – Berlin: Ferd Duemmler, 1882 – 1mf – 9 – 0-524-01161-3 – mf#1990-2237 – us ATLA [280]
Der buddhismus nach aeltern paali-werken / Hardy, Edmund – Muenster i W: Aschendorff 1890 [mf ed 1991] – (= ser Darstellungen aus dem gebiete der nichtchristlichen religionsgeschichte 1) – 1mf – 9 – 0-524-01445-0 – (incl bibl ref) – mf#1990-2440 – us ATLA [280]
Buddhismus und christenthum : was sie gemein haben und was sie unterscheidet / Schroeder, Leopold von – 2. verm Aufl. Reval [Tallinn]: F Kluge, 1898 – 1mf – 9 – 0-524-02541-X – mf#1990-3036 – us ATLA [230]
Buddhismus und christentum / Bertholet, Alfred – 2. durchg aufl. Tuebingen: J C B Mohr 1909 [mf ed 1991] – (= ser Sammlung gemeinverstaendlicher vortraege und schriften aus dem gebiet der theologie und religionsgeschichte 28) – 1mf – 9 – 0-524-01253-9 – (incl bibl ref; first printed 1902) – mf#1990-2289 – us ATLA [230]
Der buddhismus und das christentum vor dem forum des philosophischen and ethischen denkens : in verschiedenen gymnasien und anderen oeffentlichen saelen gehaltener vortrag / Bernstein, P – Esslingen: S Mayer, 1911 – 1mf – 9 – 0-524-02068-X – mf#1990-2832 – us ATLA [170]
Buddhist and christian gospels : now first compared from the originals / Edmunds, Albert Joseph; ed by Anesaki, Masaharu – 4th ed., being the Tokyo ed., rev. and enl. Philadelphia: Innes, 1908-1909 – 2mf – 9 – 0-8370-1771-8 – mf#1987-6160 – us ATLA [230]
The buddhist antiquities of nagarjunakonda, madras presidency / Longhurst, Albert Henry – Delhi: Manager of Publications, 1938 – (= ser Samp: indian books) – us CRL [280]
Buddhist art in india, ceylon, and java / Vogel, Jean Philippe – Oxford: Clarendon Press, 1936 – (= ser Samp: indian books) – (transl from the dutch by a j barnouw) – us CRL [700]
Buddhist art in its relation to buddhist ideals : with special reference to buddhism in japan / Anesaki, Masaharu – Boston: Houghton Mifflin, 1915 – 1mf – 9 – 0-524-04634-4 – mf#1990-3377 – us ATLA [700]
A buddhist bible / ed by Goddard, Dwight – Vermont, USA: Dwight Goddard, 1938 – (= ser Samp: indian books) – us CRL [280]
Buddhist cave temples of india / Wauchope, Robert Stuart – [Calcutta: Calcutta General Print Co, 1933] – (= ser Samp: indian books) – us CRL [280]
Buddhist china / Johnston, Reginald Fleming – London: John Murray, 1913 – 2mf – 9 – 0-524-00907-4 – (incl bibl ref) – mf#1990-2130 – us ATLA [280]
Buddhist Churches of America *see* Wheel of dharma
Buddhist churches of america newsletter – 1967-1974 mar – 1r – 1 – (cont: buddhist newsletter) – mf#1064709 – us WHS [280]

Buddhist economics and the modern world / Jha, Hari Bansh – Kathmandu: Dharmakirti Baudha Adhyayan Gosthi, 1979 – us CRL [330]

Buddhist essays = Aufsaetze zum verstaendnis des buddhismus / Dahlke, Paul – London: Macmillan 1908 [mf ed 1992] – 1mf – 9 – 0-524-02076-0 – (trans fr german by bhikkhu silacara) – mf#1990-2840 – us ATLA [280]

Buddhist hymns : versified translations from the dhammapada and various other sources / Carus, Paul – Chicago: Open Court, 1911 – 1mf – 9 – 0-524-02242-9 – mf#1990-2909 – us ATLA [780]

Buddhist hymns : versified translations from the dhammapada and various other sources; adapted to modern music / Carus, Paul – Chicago: Open Court Publ; London: Kegan Paul, Trench, Treubner, 1911 [mf ed 1995] – (= ser Yale coll) – 40p – 1 – 0-524-09731-3 – mf#1995-0731 – us ATLA [280]

Buddhist ideals : [a study in comparative religion] / Saunders, Kenneth James – Madras: Christian Literature Soc for India 1912 [mf ed 1988] – 1r – 1 – (filmed with: (julian the apostate) the death of the gods / merezhkovsky, kmitry s) – mf#2218 – us Wisconsin U Libr [280]

Buddhist ideals : a study in comparative religion / Saunders, Kenneth James – Madras: Christian Literature Society for India, 1912 – 1mf – 9 – 0-524-01455-8 – (incl bibl ref) – mf#1990-2450 – us ATLA [280]

Buddhist india / Davids, Thomas William Rhys – Calcutta: Susil Gupta (India) Ltd, 1950 – (= ser Samp: indian books) – us CRL [954]

Buddhist india / Davids, Thomas William Rhys – New York: GP Putnam, 1903 – (= ser The story of the nations) – 1mf – 9 – 0-524-01690-9 – mf#1990-2592 – us ATLA [954]

Buddhist india – Calcutta: S C Talukder, "Buddhist India" Office, printed at the Banik Press [1927]- [v1-5 n16-18 (1927-apr/may 1935)] [biwkly] [mf ed 2005] – 1r with other item – 1 – (v1 n2-4 publ by rangoon printing works, rangoon, burma; none publ in 1930 & 1932?; with suppl; official organ of: all-india buddhist conference and buddhist india society, 1931-1935; filmed with: the buddhist world) – mf#2005c-s019 – us ATLA [280]

Buddhist legends : [translated from the original pali text by eugene watson burlingame] / Buddhaghosa – Cambridge MA: Harvard University Press, 1921 – (= ser Samp: indian books) – us CRL [280]

Buddhist manual of psychological ethics of the 4th century b c : being a translation, now made for the first time, from the original pali, of the first book in the abhidhamma pitaka, entitled dhamma-sangani (compendium of states or phenomena) – London: Royal Asiatic Society, 1900 [mf ed 1995] – (= ser Yale coll; Oriental translation fund. new series 12) – xcv/393p – 1 – 0-524-09148-X – (with int essays and notes by caroline augusta foley rhys) – mf#1995-0148 – us ATLA [280]

Buddhist meditations from the japanese : with an introductory chapter on modern japanese buddhism / Fukyo Taikwan – Tokyo: Rikkyo Gakuin Press, 1905 – 1mf – 9 – 0-524-01286-5 – mf#1990-2322 – us ATLA [290]

Buddhist newsletter – 1963 dec-1966 dec – 1r – 1 – (cont by: buddhist churches of america newsletter) – mf#1110378 – us WHS [280]

Buddhist nirvana : a review of max mueller's dhammapada / De Alwis, James – Colombo: William Skeen, 1871 – 1mf – 9 – 0-524-07604-9 – mf#1991-0130 – us ATLA [280]

Buddhist parables – New Haven: Yale University Press, 1922 – (= ser Samp: indian books) – (trans fr original pali by eugene watson burlingame) – us CRL [280]

Buddhist philosophy in india and ceylon / Keith, Arthur Berriedale – Oxford: Clarendon Press, 1923 – (= ser Samp: indian books) – us CRL [280]

Buddhist popular lectures : delivered in ceylon in 1907 / Besant, Annie Wood – Adyar, Madras, S India: Theosophist Office, 1908 [mf ed 1991] – 1mf – 9 – 0-524-01167-2 – mf#1990-2243 – us ATLA [280]

The buddhist praying-wheel : a collection of material bearing upon the symbolism of the wheel and circular movements in custom and religious ritual / Simpson, William – London: Macmillan, 1896 – 1mf – 9 – 0-524-01922-3 – (incl bibl ref) – mf#1990-2735 – us ATLA [280]

Buddhist psychology : an inquiry into the analysis and theory of mind in pali literature / Davids, Caroline Augusta Foley Rhys – London: G Bell 1914 [mf ed 1991] – (= ser The quest series) – 1mf – 9 – 0-524-00713-6 – (incl bibl ref) – mf#1990-2041 – us ATLA [280]

Buddhist remains in andhra and the history of andhra between 225 and 610 ad / Subramanian, K R – Madras: Diocesan Press, 1932 – (= ser Samp: indian books) – us CRL [930]

Buddhist scriptures : a selection – New York: E P Dutton 1913 [mf ed 1993] – (= ser Wisdom of the east series (new york, ny)) – 1mf – 9 – 0-524-05769-9 – mf#1991-0012 – us ATLA [280]

Buddhist shrines in india – [New Delhi?]: Issued by the Publ Division, Ministry of Information and Broadcasting, Govt of India, 1951 – (= ser Samp: indian books) – us CRL [280]

Buddhist shrines in india / Valisinha, Devapriya – Colombo: Maha Bodhi Society of Ceylon 1948 [mf ed 1981] – 1r [ill] – 1 – (incl ind) – mf#123 – us Wisconsin U Libr [280]

Buddhist stories = Buddhistische erzaehlungen / Dahlke, Paul – London: K Paul, Trench, Truebner, 1913 – 1mf – 9 – 0-524-02348-4 – (in english) – mf#1990-2959 – us ATLA [280]

The buddhist stupas of amaravati and jaggayyapeta in the krishna district, madras / Burgess, James – London 1887 – (= ser 19th c art & architecture) – 3mf – 9 – mf#4.1.324 – uk Chadwyck [720]

Buddhist suttas (stbe11) – 1881 – (= ser Sacred book of the east (sbte)) – €15.00 – (trans fr pali by t w rhys davids. 1: the mahaparinibbana suttanta. 2: the dhamma-kakka-ppvattana sutta. 3: the tevigga suttanta. 4: the akkankheyya sutta. 5: the ketokhila sutta. 6: the maha-sudassana suttanta. 7: the sabbasava sutta) – ne Slangenburg [280]

Buddhist texts as recommended by asoka : with an english translation by vidhushekhara bhattacharya – [Calcutta]: University of Calcutta, 1948 – (= ser Samp: indian books) – us CRL [280]

Buddhist texts quoted as scripture by the gospel of john : a discovery in the lower criticism (john 7, 38; 12, 34) / Edmunds, Albert Joseph – 2nd ed. Philadelphia: Innes, 1911 – 1mf – 9 – 0-524-01479-5 – mf#1990-2455 – us ATLA [230]

The buddhist way of life : its philosophy and history / Smith, Frederick Harold – London: Hutchinson's University Library, 1951 – (= ser Samp: indian books) – us CRL [280]

The buddhist world – Bangalore City: B B D Power Press [v1 n8 (dec 1933)] [mthly] [mf ed 2005] – 1r with other item – 1 – (official organ of: maha sangha raja sabha of the united buddhist world and the united buddha society; filmed with: buddhist india) – mf#2005c-s020 – us ATLA [280]

Buddhistische anthologie : texte – Leiden: EJ Brill, 1892 – 1mf – 9 – 0-524-07666-9 – mf#1991-0143 – us ATLA [280]

Buddhistische und neutestamentliche erzaehlungen : das problem ihrer gegenseitigen beeinflussung / Faber, Georg – Leipzig: J C Hinrichs 1913 [mf ed 1991] – (= ser Untersuchungen zum neuen testament 4) – 1mf – 9 – 0-524-00833-7 – (incl bibl ref) – mf#1990-2079 – us ATLA [230]

Buddingh, D *see* S[/ain/]t jan, de dooper en de evangelist

Buddist mahayana texts, pt 1 (stbe49) : the buddha-karita of asvaghosha – Oxford, 1894 – (= ser Sacred book of the east (sbte)) – 4mf – 8 – €11.00 – (trans fr sanskrit by e b cowell) – ne Slangenburg [280]

Buddist mahayana texts, pt 2 (stbe50) : the larger sukhavati-vyuha... – Oxford, 1894 – (= ser Sacred book of the east (sbte)) – 5mf – 8 – €12.00 – ne Slangenburg [280]

Bude, Eugene de *see*
- Lettres inedites adressees de 1686 a 1737 a j.-a. turrettini
- Vie de benedict pictet, theologien genevois, 1655-1724
- Vie de j.-a. turrettini, theologien genevois 1671-1737

Budeau, Georges *see* Poder executivo na franca

Buder, Christian Gottlieb *see* Burcardi gotthelffii struvii...corpus historiae germanicae

Budge, E A W *see* The book of the saints of the ethiopian church

Budge, Ernest A Wallis *see* The book of the bee

Budge, Ernest Alfred Wallis *see*
- Assyrian texts
- Babylonian life and history
- The book of opening the mouth
- The book of the dead
- The chapters of coming forth by day
- Cleopatra's needles and other egyptian obelisks
- The contendings of the apostles
- The contendings of the apostles, vol 2, the english translation
- The dwellers on the nile
- The egyptian heaven and hell
- Egyptian ideas of the future life
- Egyptian magic
- The egyptian saudaan
- Facsimiles of egyptian hieratic papyri in the british museum
- The gods of the egyptians
- The greenfield papyrus in the british museum
- A hieroglyphic vocabulary to the theban recension of the book of the dead
- A history of egypt
- The history of esarhaddon (son of sennacherib) king of assyria, b.c. 681-668
- The life and exploits of alexander the great
- The life of takla haymanot...
- The literature of the ancient egyptians
- The lives of maba' syon and gabra krestos
- The mummy
- The nile
- Osiris and the egyptian resurrection
- The papyrus of ani
- The rosetta stone
- A short history of the egyptian people

Budge, Ernest Alfred Wallis, Sir *see*
- Coptic biblical texts in the dialect of upper egypt
- The earliest known coptic psalter
- The history of the blessed virgin mary and the history of the likeness of christ which the jews of tiberias made to mock at

Budge, Jane *see* Glimpses of george fox and his friends

Budge, Sir Ernest Alfred Thompson Wallis *see* One hundred and ten miracles of our lady mary, translated from ethiopic manuscripts for the most part in the british museum..

Budgell, E A *see* Letter to the merchants and tradesmen of great britain

Budget / Tuscarawas Co. Sugarcreek – jan 1950-dec 1992 [wkly] – 60r – 1 – (an amish-mennonite newspaper) – mf#B32543-32602 – us Ohio Hist [071]

Budget / Tuscarawas Co. Sugarcreek – mar-may 1897, jun 1898-may 1900 [wkly] – 1r – 1 – (an amish-mennonite newspaper) – mf#B32542 – us Ohio Hist [071]

Budget... / Cho lon [Vietnam] – Saigon: Impr commerciale Rey, Curiol & cie 1893, 1896, 1900, 1901, 1904, 1909, 1911, 1914, 1917 – 1r – 1 – mf#mf-12636 seam – us CRL [079]

The budget – Toronto: W Campbell, [1881?-1893] – 9 – ISSN: 1190-6545 – mf#P04579 – cn Canadiana [360]

The budget *see* The budget and taranaki weekly herald

The budget and taranaki weekly herald – jan 1875-jan 1877; 19 may 1877-1887; jan 1892-1898; jan 1900-jun 1905; jan 1906-1909; jan 1911-32 – 1 – (aka: the budget (taranaki)) – mf#21.8 – nz Nat Libr [079]

Budget control : what it does and how to do it, prepared and published in the interest of better business / Ernst and Ernst – [S.l.]: Ernst & Ernst, 1929 (mf ed 19–) – 38p – mf#Z-1688 – us Harvard [650]

Le budget de henri 3 : ou les premiers etats de blois, comedie historique / Roederer, Antoine – Paris 1830 [mf ed Hildesheim 1995-98] – 1v on 3mf – 9 – €90.00 – ISBN-10: 3-487-26113-8 – ISBN-13: 978-3-487-26113-3 – gw Olms [820]

Budget de l'arsenal de saigon... / Republique francaise, gouvernement general de l'Indo-chine francaise – Hanoi-Haiphong: Impr d'Extreme-Orient 1923, 1925-26 – 1r – 1 – mf#mf-12635 seam – us CRL [079]

Budget de l'exploitation des chemins de fer / Republique francaise, gouvernement general de l'Indo-chine francaise, inspection generale des travaux publics – Hanoi-Haiphong: Impr d'Extreme-Orient 1914, 1916-18, 1920, 1921, 1924-40, 1943 – 2r – 1 – (mf-12636 seam [1r] 1945) – mf#mf-12637 seam – us CRL [350]

Budget des recettes et des depenses departementales / Oise. France. (Dept) – Beauvais. Exercise 1945-1949 & Supplements – 1 – us NY Public [350]

Budget du bresil / Straten-Ponthoz, Gabriel Auguste Van Der – Paris, France. v1-3. 1854 – 1r – us UF Libraries [972]

Budget extraordinaire pour l'exercice... / Protectorat de l'Annam et du Tonkin – Hanoi: F H Schneider 1897, 1900 – 1r – 1 – mf#mf-12362 seam – us CRL [350]

Budget general pour l'exercice... / Republique francaise, gouvernement general de l'indo-chine francaise – Saigon: Impr Coloniale 1900, 1907-08, 1910, 1914, 1916, 1923-24, 1926-34, 1936 – 8r – 1 – mf#mf-12274 seam – us CRL [350]

Budget local de l'annam pour l'exercice... / Republique francaise, gouvernement general de l'indo-chine francaise – Hanoi: Impr Typo-Lithographique F H Schneider 1899, 1901-45 – 8r – 1 – mf#mf-12273 seam – us CRL [350]

Budget local du laos de l'exercice... / Republique francaise, gouvernement general de l'Indochine – Hanoi-Haiphong: Impr d'Extreme-Orient 1942 [911, 1913, 1916-31, 1933-43] – 5r – 1 – mf#mf-12620 seam – us CRL [350]

Budget local du tonkin pour l'exercice... / Republique francaise, gouvernement general de l'Indo-chine – Hanoi-Haiphong: Impr d'Extreme-Orient 1909, 1910, 1912, 1914, 1923, 1927, 1930-36, 1939, 1944 – 3r – 1 – mf#mf-12639 seam – us CRL [350]

Budget local pour l'exercice... / Cochinchine, Direction de l'interiur – Saigon: Impr nationale, Gouvernement general de l'Indochine 1875, 1879-82, 1898, 1901-09, 1911, 1915-18, 1924-25, 1927-28, 1933, 1936-38, 1940-42, 1944 – 8r – 1 – mf#mf-12624 seam – us CRL [350]

Budget minicipal de la ville de hanoi – Hanoi 1894-1896, 1905-05, 1909-10, 1913-36, 1939, 1941 – 3r – 1 – us CRL [350]

Budget municipal pour l'exercice... / Gouvernement general de l'Indochine, Concession francaise de Tourane – [Tourane]: Impr de Quinhon 1922-45 – 1r – 1 – mf#mf-12643 seam – us CRL [350]

Budget of letters from japan : reminiscences of work and travel in japan / Maclay, Arthur Collins – New York: A C Armstrong, 1886 [mf ed 1995] – (= ser Yale coll) – viii/391p (ill) – 1 – 0-524-09979-0 – mf#1995-0979 – us ATLA [915]

Budget pour l'exercice... / Protectorat francais du Cambodge – Phnom-Penh: Impr Typo-Lithographique 1890-1892, 1894, 1896-1910, 1912-26, 1928-29, 1931-34, 1940, 1942, 1944, 1945 – 8r – 1 – mf#mf-12699 seam – us CRL [350]

Budget speech : delivered...on friday february 22nd, 1878 / Cartwright, Richard – Ottawa: s.n, 1878 – 1mf – 9 – mf#04185 – cn Canadiana [336]

Budget speech, delivered by honorable john s hall, treasurer of the province : in the legislative assembly of quebec, on friday, may 20th, 1892 – [Quebec?: Morning Chronicle], 1892 – 1mf – 9 – 0-665-94068-8 – mf#94068 – cn Canadiana [336]

Budget speech delivered by the honourable a w atwater, treasurer of the province : in the legislative assembly of quebec on wednesday 9th december, 1896 – Quebec?: s.n, 1896 – 1mf – 9 – mf#04092 – cn Canadiana [336]

Budget speech delivered in the house of commons of canada : friday, 14th march, 1879 / Tilley, Samuel Leonard – [S.l.: s.n, 1879?] [mf ed 1987] – 1mf – 9 – 0-665-62203-1 – mf#62203 – cn Canadiana [336]

Budget speech delivered in the house of commons of canada : on friday, february 25, 1876 / Cartwright, Richard – [Ottawa?: s.n.], 1876 [mf ed 1985] – 1mf – 9 – 0-665-53950-9 – mf#53950 – cn Canadiana [336]

Budget speech delivered in the house of commons of canada : on tuesday, february 20th, 1877 / Cartwright, Richard – Ottawa: [s.n.], 1877 [mf ed 1980] – 1mf – 9 – 0-665-05710-5 – mf#05710 – cn Canadiana [336]

Budget speech delivered in the house of commons of canada : on tuesday, the 1st april, 1873 / Tilley, Samuel Leonard – [S.l: s.n, 1873] [mf ed 1984] – 1mf – 9 – mf#27709 – cn Canadiana [336]

Budget speech delivered in the house of commons of canada : on tuesday, the 1st april, 1873 / Tilley, Samuel Leonard – [S.l: s.n, 1873?] [mf ed 1986] – 1mf – 9 – 0-665-53899-5 – mf#53899 – cn Canadiana [336]

Budget speech delivered in the house of commons of canada : tuesday, 9th march, 1880 / Tilley, Samuel Leonard – Ottawa: [s.n.], 1880 [mf ed 1985] – 1mf – 9 – 0-665-53969-X – mf#53969 – cn Canadiana [336]

Budget speech delivered in the legislative assembly of quebec on the 21st february, 1890 / speech delivered in the legislative assembly of quebec, on the 21st february, 1890 / Shehyn, Joseph & Mercier, Honore – Quebec: [s.n.], 1890 [mf ed 1981] – 1mf – 9 – mf#13414 – cn Canadiana [336]

Budget speech delivered in the provincial legislature : on monday, march 29th, 1886 / Duck, Simeon – [Victoria, BC?: s.n.], 1886 [mf ed 1981] – 1mf – 9 – mf#14934 – cn Canadiana [336]

Budget speech of hon j h turner – [Victoria, BC?: s.n, 1894?] [mf ed 1981] – 1mf – 9 – mf#16282 – cn Canadiana [336]

The budget speech of hon mr wuertele, treasurer of the province of quebec : delivered on the 15th may, 1882 – [Quebec?: s.n.], 1882 [mf ed 1986] – 1mf – 9 – 0-665-48753-3 – (also available in french) – mf#48753 – cn Canadiana [336]

The budget speech of hon mr wuertele, treasurer of the province of quebec : delivered on the 16th february, 1883 – Quebec: [s.n.], 1883 [mf ed 1982] – 2mf – 9 – mf#28667 – cn Canadiana [336]

The budget speech of the hon j g robertson, treasurer of the province of quebec : legislative assembly, quebec, 24th march, 1885 – [Montreal?: s.n.], 1885 [mf ed 1981] – 1mf – 9 – mf#12644 – cn Canadiana [336]

Budgetary control for business / McKinsey, James Oscar – Boston: Boston Chamber of Commerce, c1924 (mf ed 19–) – 47p – mf#ZT-TN pv78 n6 – us Harvard [338]

The budgetary impact of possible changes in diversity jurisdiction / Partridge, Anthony – Washington: FJC, 1988 – 1mf – 9 – $1.50 – mf#LLMC 95-340 – us LLMC [340]

Budgets familiaux des planteurs de cacao au cameroun / Binet, Jacques – Paris, France. 1956 – 1r – us UF Libraries [960]

BUERGERBLATT

Budha bhasa pui karan pe ca samuin [1851-1970] / Bhun Mran, U – Yan Kun: Sapre U Cape 1975 [mf ed 1990] – 1r with other items – 1 – (in burmese, karen or english) – mf#mf-10289 seam reel 177/6 [§] – us CRL [480]
Budha tara nhan nhlum svan rhu thon / Van Tan, U – Yan Kun: Vijodaya 1976 [mf ed 1990] – 1r with other items – 1 – (in burmese) – mf#mf-10289 seam reel 143/2 [§] – us CRL [280]
Budhal, Rishichand Sookai *see* The impact of the principal's instructional leadership on the culture of teaching and learning in the school
Budil'nik – St Petersburg, Russia 1865-1917 [mf ed Norman Ross] – 649mf – 9 – $3,800.00 – mf#nrp-1581 – us UMI ProQuest [790]
Budissiner nachrichten *see* Budissinische woechentliche nachrichten
Budissinische nachrichten *see* Budissinische woechentliche nachrichten
Budissinische woechentliche nachrichten – Bautzen DE, 1921-40 – ca 50r – 1 – (title varies: 7 jan 1809: budissinische nachrichten, 5 jan 1828: budissiner nachrichten, 9 jun 1868: bautzener nachrichten, 1 aug 1934: der freiheitskampf ba v. der freiheitskampf, dresden, 2 jan 1935: ns-tageszeitung fuer bautzen) – gw Misc Inst [074]
Budkaflen – Stockholm, Sweden. 1883-98 – 5r – 1 – sw Kungliga [078]
Buds and blossoms / Thayers, M J – [Toronto?: s.n.], 1894 [mf ed 1981] – 3mf – 9 – mf#24684 – cn Canadiana [810]
Budstikken / Valdres Samband – 1970 dec-1980 dec – 1r – 1 – (cont: valdres samband budstikken) – mf#555790 – us WHS [280]
Budushchee rossii – Moscow, Russia. n 1 (1993), n3(1994), n2[8] (1997), n1[9](1998) – 1 – mf#mf-12248 (reel 2) – us CRL [077]
Budushchnost *see* Lavenir
Budweiser zeitung – Budweis (Ceske Budejovice CZ), 1938 24 aug-1942 27 oct – 1 – gw Misc Inst [077]
Budzinski, Karen M *see* A comparison of ratings of perceived exertion during stairmaster and treadmill exercise
Die buecher der bibel / ed by Rahlwes, F – Berlin, Wien [1923] (mf ed 1996) – (= ser Monographien zur wissenschaft des judentums) – 9 – (v1: ueberlieferung und gesetz. das fuenfbuch mose und das buch josua 6mf €59 isbn: 3-8267-3184-0; v6: die liederdichtung. die psalmen, die klagelieder, das hohelied 4mf €45 isbn: 3-8267-3186-7; v7: die lehrdichtung. die sprueche, hiob, der prediger, ruth, jona, esther, daniel 4mf €45 isbn: 3-8267-3190-5. v2-5 not publ) – gw Frankfurter [220]
Die buecher der chronik *see*
– The books of the chronicles
Die buecher der chronik der vulgata und des hebraeischen textes / Neteler, Bernhard – Muenster i. W: Theissing, 1899 – 1mf – 9 – 0-8370-3874-X – mf#1985-1874 – us ATLA [221]
Die buecher der geheimnisse henochs / Bonwetsch, G N – Leipzig, 1922 – (= ser Tugal 3-44/2) – 3mf – 9 – €7.00 – ne Slangenburg [221]
Die buecher der hirten- und preisgedichte, der sagen und saenge und der haengenden gaerten / George, Stefan Anton – 3. Aufl. Berlin: G Bondi, 1907 – 1r – 1 – us Wisconsin U Libr [880]
Die buecher der hirten- und preisgedichte, der sagen und saenge, und der haengenden gaerten / George, Stefan Anton – Einzelausg. Godesberg: H Kupper vormals G Bondi, 1950 – 1 – us Wisconsin U Libr [880]
Die buecher der koenige : mit neun abbildungen im text, einem plan des alten jerusalem und einer geschichtstabelle / Benzinger, Immanuel – Freiburg i. B: J C B Mohr (Paul Siebeck), 1899 – (= ser Kurzer Hand-Commentar Zum Alten Testament, Abt. 9) – 1mf – 9 – 0-8370-2280-0 – (includes chronological table and index) – mf#1985-0280 – us ATLA [220]
Die buecher der koenige *see* The books of the kings
Die buecher esra (a und b) und nehemja : text-kritisch und historisch-kritisch untersucht mit erklaerung der einschlaegigen prophetenstellen und einem anhang ueber hebraeische eigennamen / Jahn, Gustav – Leiden: E J Brill, 1909 – 1mf – 9 – 0-8370-3762-X – (incl ind and appendix) – mf#1985-1762 – us ATLA [221]
Die buecher esra, nechemia und ester / Bertheau, Ernst – In 2. aufl. Leipzig: S. Hirzel, 1887 – (= ser Kurzgefasstes Exegetisches Handbuch Zum Alten Testament) – 2mf – 9 – 0-7905-3305-7 – mf#1987-3305 – us ATLA [221]
Die buecher exodus, leviticus, numeri *see* Introduction to the three middle books of the pentateuch
Die buecher exodus und leviticus / Dillmann, August; ed by Ryssel, Victor – 3. aufl. Leipzig: S Hirzel, 1897 – 2mf – 9 – 0-8370-9460-7 – mf#1986-3460 – us ATLA [221]
Die buecher josua, der richter, samuelis und der koenige : in uebersichtlicher nebeneinanderstellung des urtextes, der septuaginta, vulgata und luther-uebersetzung, so wie der wichtigsten varianten der vornehmsten deutschen uebersetzungen fuer den praktischen handgebrauch – Bielefeld: Velhagen & Klasing, 1851 – 9mf – 9 – 0-524-08208-1 – mf#1993-0003 – us ATLA [221]
Die buecher moses und josua : eine einfuehrung fuer laien / Merx, Adalbert – Tuebingen: J C B Mohr 1907, c1905 [mf ed 1989] – (= ser Religionsgeschichtliche volksbuecher fuer die deutsche christliche gegenwart 2/3,1-2) – 1mf – 9 – 0-7905-1466-4 – mf#1987-1466 – us ATLA [221]
Die buecher richter und samuel : ihre quellen und ihr aufbau / Budde, Karl – Giessen: J Ricker, 1890 – 1mf – 9 – 0-8370-2509-5 – mf#1985-0509 – us ATLA [221]
Die buecher samuelis *see* The books of samuel
Buecher, Wilhelm *see* Grillparzers verhaeltnis zur politik seiner zeit
Die buecher-kommentare : vierteljahreshefte der deutschen kommentare – Stuttgart DE, 1952 jul-1967 – 1 – (comparison: deutsche kommentare, heidelberg) – gw Misc Inst [430]
Der buecherwurm [klp9] : eine monatsschrift fuer buecherfreunde / ed by Weichardt, Walter [d.i. Walter Blumtritt] – Muenchen 1910/11-1942/43 [mf ed 2003] – (= ser Kultur – literatur – politik: deutsche zeitschriften des 19./20. jahrhunderts [klp] 9) – 102mf – 9 – €590.00 – 3-89131-367-5 – gw Fischer [070]
Das buechlein jesu sirach : in gesaenge verfasst / Heymair, Magdalena – s.l. 1578 – (= ser Hqab. literatur des 16. jahrh.) – 3mf – 9 – €40.00 – (newly corr by greogorium sunderreuetter) – mf#1578b – gw Fischer [780]
Das buechlein vom leben nach dem tode *see* The little book of life after death
Ein buechlein, von dem banne, vnd andern kirchenstraffen, aus gottes wort / Sarcerius, E – Eisleben, [1555] – 2mf – 9 – mf#TH-1 mf 1313-1314 – ne IDC [242]
Buechler, Adolf *see*
– The political and the social leaders of the jewish community of sepphoris in the second and third centuries
– Die priester und der cultus im letzten jahrzehnt des jerusalemischen tempels
– Das synedrion in jerusalem und das grosse beth-din in der quaderkammer des jerusalemischen tempels
– Die tobiaden und die oniaden im 2. makkabaeerbuche und in der verwandten juedisch-hellenistischen litteratur
– Types of jewish-palestinian piety from 70 b c e to 70 c e
Buechler, Franz *see* August der starke
Buechner, Georg *see*
– Das eherne gesetz
– Georg buechners saemtliche poetische werke
– Plays of georg buechner
– Saemmtliche werke und handschriftlicher nachlass
– Saemtliche poetische werke
Buechner, Ludwig *see*
– Aus dem geistesleben der thiere
– Gott und die wissenschaft
– Kraft und stoff
– Man in the past, present and future
– Natur und geist
Buechner, Wilhelm *see*
– Fauststudien
– Goethes faust
Buechner-preis-reden, 1951-1971 – Stuttgart: P Reclam 1972 [mf ed 1992] – 1r – 1 – (incl bibl ref; pref by ernst johann. filmed with: stilwandel / emil staiger) – mf#3207p – us Wisconsin U Libr [430]
Buechsel, Friedrich *see* Der begriff der wahrheit in dem evangelium und den briefen des johannes
Bueck, Friedrich *see* Wegweiser durch hamburg und die umliegende gegend
Buecker, Bernd *see* Der herzog und sein kumpan
Buehel, Hans von *see* Dyocletianus leben
Buehl, Eduard *see* Christologie des alten testamentes
Buehlbaecker, Alexander *see* Zur verwendbarkeit von stutenmilch, kumyss und eselmilch als diaetetika und heilmittel unter besonderer beruecksichtigung der beduerfnisse des saeuglings und des fruehgeborenen
Buehler, Adolph *see* Theokrisis
Buehler bote *see* Acher-bote
Buehler, Christian *see* Der altkatholicismus
Buehler, Georg *see*
– The laws of manu
– On the indian sect of the jainas
Buehlmann, Heinrich *see* Goethes faust
Buehne und bildende kunst : ein epilog zur faust-auffuehrung am muenchener kuenstler-theater 1908 / Oberlaender, Hans – Koeln (Rhein): A Ahn [1908] [mf ed 1990] – 1r – 1 – (filmed with: goethes faust: eine evangelische auslegung / firso melzer) – mf#7356 – us Wisconsin U Libr [790]
Das buehnenfestspiel in bayreuth : eine studie ueber richard wagners "ring des nibelungen" / Porges, Heinrich – 2. durchges aufl. Muenchen: C Merhoff 1877 – 1mf – 9 – mf#wa-129 – ne IDC [780]
Das buehnenfestspiel zu bayreuth : eine kritische studie / Kalbeck, Max – Breslau: Schletter'sche Buchhandlung [E Franck] 1877 – 2mf – 9 – mf#wa-52 – ne IDC [780]
Die buehnenfestspiele in bayreuth : authentischer beitrag zur geschichte ihrer entstehung und entwickelung / Heckel, Karl – Leipzig: E W Fritzsch [1891?] – 1mf – 9 – mf#wa-41 – ne IDC [780]
Die buehnenfestspiele in bayreuth : ihre gegner und ihre zukunft / Plueddemann, Martin – 2. ausg. Leipzig: Senf [1881] – 1mf – 9 – mf#wa-130 – ne IDC [780]
Die buehnengerechten einrichtungen der schillerschen dramen fuer das koenigliche national-theater zu berlin / Schmieden, Alfred – Berlin: E Fleischel, 1906 – 1 – (incl bibl ref) – us Wisconsin U Libr [790]
Die buehnengeschichte des goethe'schen faust / Creizenach, Wilhelm Michael Anton – Frankfurt a/M: Ruetten & Loening, 1881 (mf ed 1990) – 1r – 1 – (filmed with: goethes faust in seiner haltesten gestalt) – us Wisconsin U Libr [790]
Die buehnenwirksamkeit der symbole in hugo von hofmannsthals und richard strauss' zauberoper die frau ohne schatten : ein einblick in die inszenierungsgeschichte des 20. jahrhunderts / Obermaier, Gerlinde – (mf ed 2000) – 4mf – 9 – €56.00 – 3-8267-2743-6 – mf#DHS 2743 – gw Frankfurter [790]
Buehner, Karl Hans *see* Hermann hesse und gottfried keller
Buehrlen, Friedrich L *see* Bilder aus dem schwarzwald
Buel, Frederick *see* Discussion on the necessity of revising king james' version of the holy scriptures
Buel, James William *see*
– America's wonderlands
Buel, Samuel *see*
– The apostolical system of the church defended
– Eucharistic presence, eucharistic sacrifice, and eucharistic adoration
– A treatise of dogmatic theology
Buell collection of historical documents relating to the corps of engineers, 1801-1819 / U.S. War Dept. Office of the Chief of Engineers – (= ser Records of the office of the chief of engineers) – 3r – 1 – (with printed guide) – mf#M417 – us Nat Archives [355]
Buellingen, Ludwig von *see* Annales typographici colonienses
Buelow, Eduard *see* Der arme mann im tockenburg
Buelow, Eduard von *see* Christian legends
Buelow, Frieda, Freiin von *see*
– Im lande der verheissung
– Irdische liebe
Buelow, G *see* Des dominicus gundissalinus schrift "von der unsterblichkeit der seele"
Buelow, Hans von *see* Ueber richard wagners faust-ouverture
Buelow, Wilhelm *see* Das knappschaftswesen im ruhrkohlenbezirk bis zum allgemeinen preussischen berggesetz vom 24.6.1865
Buen ladron / Veloz Maggiolo, Marcio – Ciudad Trujillo, Dominican Republic. 1960 – 1r – us UF Libraries [972]
Buen Lozano, Nestor De *see* Decadencia del contrato
Buen vecino / Vegas, Jose De La – Bogota, Colombia. 1941 – 1r – us UF Libraries [972]
Buena greetings – Chicago IL. 1909 jul-1920 nov – 1r – 1 – mf#1053662 – us WHS [071]
[Buena park-] buena park news – CA. feb 1993-jun 1994 – 6r – 1 – $360.00 – mf#R04009 – us Library Micro [071]
Buena vista advocate – Buena Vista VA. 1890 may 23, oct 17, 1891 jan 2, feb 13, 27-mar 6, 20, apr 3-17, may 1,15-29, jun 19-dec 11, 1892 jan 8-jul 29, aug 12, 1895 dec 18, 1900 jun 29-aug 3, 1901 jan 25, jun 14 – 1r – 1 – mf#883988 – us WHS [071]
Buena vista baptist church. owensboro, kentucky : church records – 1920-Sept 1984 – 1 – 77.85 – us Southern Baptist [242]
Buenaventura, San *see*
– Liber de profectu religiosorum
– Meditaciones de la vida de cristo...
Buender general-anzeiger – Buende/Westf DE, 1938-40 – 1 – (later: buender zeitung, publ in herford) – gw Misc Inst [074]
Buender tageblatt – Buende Westf DE, 1901 21 jan-1905, 1910-11, 1920 19 jan-30 dec, 1922 2 jan-19 dec, 1923 & 1924, 1926 5 jan-6 sep, 1927 2 mar-30 nov, 1928-1930 30 apr, 1933-35, 1936 1 apr-1937 31 mar, 1938 1 feb-31 aug, 1942 1 aug-30 nov, 1955 25 jun-3 dec – 1 – (suppls: illustrirtes sonntags-blatt 1901-05, 1910-13, 1916 [with gaps]; wort und bild 1924, 1926-28 [gaps]) – gw Misc Inst [074]
Buender zeitung *see* Buender general-anzeiger
Das buendnis – Rossleben DE, 1957 31 jan-dec, 1960 jan-12 aug, 1966-90 – 3r – 1 – (with gaps; title varies: 1 jul 1963) – gw Misc Inst [074]
Das buendnis – Zeitz DE, 1957-65 [gaps] – 1r – 1 – gw Misc Inst [074]
Das buendnis der parteien : herausbildung und rolle des mehrparteiensystems in der europaeischen soziolistischen laendern / Grosser, Guenther – Berlin: Buchverlag Der Morgen 1967 – (= ser Schriften der ldpd 3) – 1 – mf#1561 – us Wisconsin U Libr [325]
Buendnis und bekenntnis 1529/1530 : der toleranzgedanke im reformationszeitalter / Schubert, Hans von & Hermelink, Heinrich – Leipzig: Verein fuer Reformationsgeschichte, 1908. (Schriften des Vereins fuer Reformationsgeschichte; 26. Jahrg., Schrift 98) – 1mf – us ATLA [240]
Buendnis und bekenntnis 1529/1530 / der toleranzgedanke im reformationszeitalter : vortraege gehalten auf der 25. generalversammlung des vereins fuer reformationsgeschichte zu bretten... / Schubert, Hans von – Leipzig: Verein fuer Reformationsgeschichte 1908 [mf ed 1990] – (= ser Schriften des vereins fuer reformationsgeschichte 26/98) – 1mf – 9 – 0-7905-4710-4 – (incl bibl ref) – mf#1988-0710 – us ATLA [242]
Buendnispolitik und revolutionaere krise : zu einigen aspekten der britisch-russischen beziehungen 1917 / Maier, Lothar August – Heidelberg, 1975 – 3mf – 9 – 3-89349-687-4 – gw Frankfurter [327]
Buenner, D *see* L'ancienne liturgie romaine
Bueno, Francisco Da Silveira *see* Estilistica brasileira
Bueno, Ramon *see* Apuntes sobre la provincia misional del orinoco...caracas, 1933
Bueno, Salvador *see*
– Antologia del cuento en cuba
– Enrique pineyro y la critica literaria
– Historia de la literatura cubana
– Letra como testigo
– Policromia y sabor de costumbristas cubanos
– Temas y personajes de la literatura cubana
Buenos Aires. Centro de Investigacion y Accion Social *see* Cias
Buenos aires herald – Buenos Aires, Argentina 15 jun 1878-31 dec 1976, 2 jan-27 apr 1979, 1 jun-31 oct 1982, 2 jan- 30 apr 1983, 2 jan 1984- [mf 1876-1985, jan-apr 2005] – 1 – (cont: herald [15 sep 1876-14 jun 1878]) – uk British Libr Newspaper [079]
[Buenos aires-] mundo peronista – AG. 1951-1955 – (= ser Latin american perspectives coll) – 4r – $200.00 – mf#R04160 – us Library Micro [079]
Buenos aires musical – [Buenos Aires, s.n.] v25-34. 1970-79 (irr) [19–] – 2r – 1 – (began mar 1946; title varies: bam) – mf#828 – us Wisconsin U Libr [780]
Buenos aires musical – Buenos Aires. 1954-75. 3 reels – 1 – 47.00 – us L of C Photodup [780]
Buenos aires musical – Buenos Aires, Argentina 16 aug 1956-15 dec 1975, mar 1976-may 1978 – 1 – uk British Libr Newspaper [780]
Buenos ayres et le paraguay : ou histoire, moeurs, usages et costumes des habitans de cette partie de l'amerique / Denis, Jean F – Paris 1823 [mf ed Hildesheim 1995-98] – 2v on 4mf – 9 – €120.00 – 3-487-26846-9 – gw Olms [972]
Buerckstuemmer, Christian *see* Geschichte der reformation und gegenreformation in der ehemaligen freien reichstadt dinkelsbuehl (1524-1648)
Buerener zeitung – Bueren DE, 3 nov 1896-97, 1900, 4 jan 1902-10, 2 oct-24 dec 1912, 1914-15, 1920-44, 1950-54, 1 apr, 30 jun 1955, 2 jan, 29 sep 1956, 1957-31 mar 1959, 1 jul, 30 dec 1959 – 1 – (with gaps; title varies: fr 1925 kopfblatt v: der patriot, lippstadt; 1 nov 1935: der patriot, 12 dec 1952: westfalenpost) – gw Misc Inst [074]
Buergeler nachrichten – (Offenbach-) Buergel, Fechenheim, 1900 4 apr-1902 30 jun [gaps] – 2r – 1 – gw Misc Inst [074]
Der buerger – Goettingen DE, 1732 may-sep – 1r – 1 – gw Misc Inst [074]
Der buerger *see* Rheinisch-westfaelische frauen-zeitung
Buerger, Gottfried August *see*
– Akademie der schoenen redekuenste
– Briefe von und an gottfried august buerger
– Buergers gedichte in zwei teilen
– G a buergers ausgewaehlte werke
– Saemmtliche gedichte
Buerger, Mark Claus *see* Computertomographische kieferschnittbilder zur evaluation von knochenneubildung und regenerierten implantatlokalisationen
Buerger- und bauern-zeitung – Potsdam, Berlin DE, 1849 5 jul-1850 22 may – 1r – 1 – gw Misc Inst [630]
Buergerblatt *see* Niederrheinische heimatblaetter

Der buergerfreund / ed by Metternich, M – Mainz 1792-93 – (= ser Dz. historisch-politische abt) – 1v on 3mf – 9 – €90.00 – mf#k/n1717 – gw Olms [074]

Der buergerfreund – Bremen DE, 1816 1 apr-1866 – 37r – 1 – (title varies: 25 jan 1857: bremer buergerfreund) – gw Misc Inst [074]

Der buergerfreund : eine wochenschrift fuer fabrikanten, manufakturisten, handwerker und buerger / ed by Nenke, Karl Christoph – Berlin 1784 – (= ser Dz) – 24st on 3mf – 9 – €90.00 – mf#k/n2943 – gw Olms [330]

Buergerliche baukunst : darinnen gezeiget wird wie die innerliche einrichtung der buergerlichen wohngebaeude, damit sie den absichten des bauherrn gemaess seye / Voch, L – Augsburg. 4v. 1780-1782 – 12mf – 9 – mf#OA-123 – ne IDC [720]

Der buergerliche baumeister... / Schmidt, F C – Gotha, 1790-1799. 4v – 40mf – 9 – mf#OA-270 – ne IDC [720]

Das buergerliche gesetzbuch fuer das deutsche reich : nebst dem einfuehrungsgesetze; text; ausgabe mit sachregister – Berlin: R v Decker, (1896?) – (= ser Civil law 3 coll) – 7mf – 9 – (incl ind) – mf#LLMC 96-506 – us LLMC [348]

Die buergerliche und politische gleichberechtigung aller confessionen : die unbeschraenkte freiheit der sektenbildung und die trennung der kirche vom staat / Ullmann, Karl – Stuttgart: JG Cotta, 1848 – 1mf – 9 – 0-524-08627-3 – mf#1993-1077 – us ATLA [323]

Buergerliches gesetzbuch, allgemeiner teil / Oertmann, Paul – 3., umgearb Aufl. Berlin: C Heymann, 1927 – (= ser Civil law 3 coll; Kommentar zum buergerlichen gesetzbuch und seinen nebengesetzen, buergerliches gesetzbuch) – 10mf – 9 – (incl bibl ref) – mf#LLMC 96-555 – us LLMC [348]

Buergerliches lebensgefuehl in grillparzers dramen / Weissbart, Gertrud – Bonn: L Roehrscheid 1929 [mf ed 1992] – (= ser Mnemosyne 3) – 1r – 1 – (incl bibl ref. filmed with: mnemosyne / oskar walzel) – mf#3113p – us Wisconsin U Libr [430]

Buergerliches naturgefuehl und offizielle landschaftsmalerei in frankreich 1753-1824 / Rosenthal, Gisela – Heidelberg – 3mf – 9 – 3-89349-397-2 – gw Frankfurter [750]

Buergermeisterblatt – Duesseldorf DE, 1850 31 jul-1869, 1871-92 – 33r – 1 – (title varies: 2 jul 1871: duesseldorfer volkszeitung) – gw Misc Inst [350]

Buergernaehe der verwaltung als thema der ausbildung von beamten des gehobenen nichttechnischen dienstes / Moellers, Martin – (mf ed 1992) – 3mf – 9 – €49.00 – 3-89349-556-8 – mf#DHS 556 – gw Frankfurter [350]

Buergers beziehungen zu herder / Peveling, Adolfine – [S.l.: s.n.], 1917 (Weimar: Druck von R Wagner) [mf ed 1989] – 61p – 1 – (incl bibl ref) – mf#7095 – us Wisconsin U Libr [430]

Buergers gedicht die nachtfeier der venus / ed by Stammler, Wolfgang – Bonn: A Marcus und E Weber, 1914 [mf ed 1989] – (= ser Kleine texte fuer vorlesungen und uebungen 128) – 56p – 1 – mf#7095 – us Wisconsin U Libr [430]

Buergers gedichte in zwei teilen / Buerger, Gottfried August; ed by Consentius, Ernst – Berlin: Bong [1914] [mf ed 1993] – (= ser Bongs goldene klassiker bibliothek) – 2v – 1 – (incl bibl ref and ind. with engravings etc) – mf#8527 – us Wisconsin U Libr [430]

Buergers gedichte in zwei teilen : kritisch durchgesehene und erlaeuterte ausgabe / Buerger, Gottfried August; ed by Consentius, Ernst – Berlin: Bong [1914] [mf ed 1993] – 1r – 1 – (filmed with: saemmtliche gedichte / g a buerger) – mf#8527 – us Wisconsin U Libr [810]

Buergers verskunst / Zaunert, Paul – Marburg a.L: N G Elwert 1911 [mf ed 1992] – (= ser Beitraege zur deutschen literaturwissenschaft 13) – 1r – 1 – (incl bibl ref. filmed with: gustav freytag und das junge deutschland / otto mayrhofer) – mf#3091p – us Wisconsin U Libr [430]

Buerger-zeitung – Memel (Klaipeda LT), 1859-61 [gaps] – 2r – 1 – gw Misc Inst [077]

Buerger-zeitung *see*
- Duesseldorfer buerger-zeitung
- Xanthippus

Buergerzeitung – Perjamosch (Perjamos/Lovrin RO), 1933 15 jan-1942 27 dec – 1 – gw Misc Inst [077]

Buergerzeitung *see* Hamburg-altonaer volksblatt

Buergschaft und garantievertrag unter besonderer beruecksichtigung der bankpraxis / Dudenhausen, Wolfgang – Leipzig, 1936 (mf ed 1994) – 1mf – 9 – €24.00 – 3-8267-3003-8 – mf#DHS-AR 3003 – gw Frankfurter [346]

Buerki, Jakob *see* A der heiteri

Buerki, Robert A *see*
- Journal of pharmacy teaching
- Pharmacy practice – the challenge of ethics

Buerkle, Veit *see*
- Heimat ohne ende
- Lasst das fruehjahr kommen!
- Der schelmensack

Buerokratie als lernende organisation / Maetzing, Ortwin – (mf ed 1994) – 2mf – 9 – €40.00 – 3-8267-2035-0 – mf#DHS 2035 – gw Frankfurter [370]

Buersche volkszeitung – Gelsenkirchen DE, 1922 18 feb, 1932 1 oct-1934 30 jun, 1934 1 oct-31 dec – 3mf=6df – 9 – (1934: vestische neueste nachrichten [gelsenkirchen-] buer) – gw Mikrofilm [074]

Buersche zeitung 1881 – Gelsenkirchen DE, 1909 19 may-1940 30 jun [gaps] – 31r – 1 – (buer, fr 1 apr 1928 to gelsenkirchen) – gw Misc Inst [074]

Buesa, Jose Angel *see*
- Alegria de proteo
- Canto final
- Nuevo oasis
- Poeta enamorado
- Versos de amor

Buesch, Joh Georg et al *see* Handlungsbibliothek

Buescher, Bradley R *see* The segmental energetics in the take-off and landing phases

Buesching, Ant Friedr *see* Gelehrte abhandlungen und nachrichten aus und von russland

Buesching, Anton *see*
- Dr anton friedrich bueschings neue erdbeschreibung
- Geographie universelle
- Nouveau traite de geographie

Buesching, Anton Friedr *see* Magazin fuer die neue historie und geographie

Buesching, Anton Friedrich *see* Dr anton friedrich bueschings neue erdbeschreibung

Buesching, Johann Gustav Gottlieb et al *see* Pantheon

Buescu, Mircea *see* Historia economica do brasil pesquisas e analises

Buesing, Georg *see* Vom tapferen leben

Buess, Eduard *see* Jeremias gotthelf

Der buesser : eine erzaehlung aus dem bergmannsleben / Deschmann, Ida Maria – Leipzig: P Reclam 1941 [mf ed 1989] – (= ser Reclams universal-bibliothek 7490) – 1 – mf#7174 – us Wisconsin U Libr [830]

Buetower anzeiger – Buetow (Bytow PL), apr 1926-apr 1927 – 1r – 1 – (incl suppl: blaetter fuer heimatpflege im kreise buetow; filmed by other misc inst: 1889 8 jan-1894, 1905 3 jan-29 jun, 1908-09, 1912 1 jul-31 dec, 1916 1 jul-30 dec, 1918 [7r]) – gw Misc Inst [077]

Buettner, H E et al *see* Ansbachische monatsschrift

Buettner, Heinrich Christoph et al *see* Fraenkisches archiv

Buetzler, Carl *see* Untersuchungen zu den melodien walthers von der vogelweide

Bueyuek donanma mueneccimi – 1330M [(1329) 1913] 4. sene – (= ser Ministry and special interest salnames) – 3mf – 9 – $55.00 – us MEDOC [956]

Bueyuek duygu – Istanbul: Cemiyet Kuetuephanesi, Mesai Matbaasi, Resimli Kitap Matbaasi, 1913-14. Mueduer ve Mueessisleri: Duendar Alp, S. Ulug: Mueduer-i Mesul: Tevfik. n1-26. 2 mart-18 kanunisani 1329 [1913-14] – (= ser O & t journals) – 9mf – 9 – $150.00 – us MEDOC [956]

Bueyuek mecmu'a – n1-17. 1919 [all publ] – (= ser O & t journals) – 9mf – 9 – $150.00 – us MEDOC [956]

Bueyuek salnamesi – 1341-42 [1925-26] – (= ser Ministry and special interest salnames) – 4mf – 9 – $70.00 – us MEDOC [956]

Buez ar saint, gant reflexionon spirituel : var o c'haera actionou, pere a ell servichout da veditation evit pep deiz eus ar bloas, hac un instruction evit disqui aez ha facil ober meditation pe oraeson,... / Fontaine, Nicolas – E Sant-Briec: Prud'homme 1824 – (= ser Whsb) – 9mf – 9 – €85.00 – mf#Hu 119 – gw Fischer [410]

Buezo, Rodolfo *see* Sangre de hermanos

Bufe, Franz Gustav *see* Licht und schatten

Buffalo – 1981 jul 31, 1982 jan 8-1984 jun 29, 1984 jul 6-1986 may 16 – 2r – 1 – mf#713030 – us WHS [071]

Buffalo – v1 n1-v3 n1 [1980 nov-1982 jan] – 1r – 1 – mf#615608 – us WHS [071]

Buffalo baptist church. cherokee county. south carolina : church records – 1805-1910, 1858-1917, 1919-23, 1940-41, 1945-60 – 1 – us Southern Baptist [242]

Buffalo baptist church. rutledge, tennessee : church records – 1859-Jul 1953 – 1 – us Southern Baptist [242]

Buffalo bill : king of scouts; a narrative of thrilling adventures and graphic description of frontier life / Hawkeye, Harry – Baltimore, MD: I & M Ottenheimer, 1908 [mf 197-] – 1r – 1 – us NY Public [920]

Buffalo bill picture stories – New York NY 1949 – 1 – mf#6146 – us UMI ProQuest [740]

Buffalo bill's wild west courier & international horticultural exhibition gazette – London, England 7 may 1892 – 1 – (iss at an exhibition held at earl's court) – uk British Libr Newspaper [072]

Buffalo charger – 1986 aug-1993 – 1r – 1 – (cont: charger [buffalo ny]; cont by: charger [buffalo ny: 1997]) – mf#1056108 – us WHS [071]

Buffalo county beacon – Gibbon, NE: Andrew J Price, jul 1872-mar 1873// (wkly) [mf ed v1 n3. jul 27 1872 filmed 1985] – 1r – 1 – (absorbed by: central nebraska press) – us NE Hist [071]

Buffalo county beacon – Gibbon, NE: G W Read. v1 n1. sep 15 1882- (wkly) [mf ed -1896 (gaps)] – 7r – 1 – (publ as buffalo co. beacon sep 15 1882-jun 22 1883. suspended with v5 n16 nov 20 1896) – us NE Hist [071]

Buffalo county herald – Mondovi WI. 1876 sep 21, oct 5-1878 aug 2, 1878 aug 9-1879 aug 30, 1880 may 15, 1882 jan 13-1883 aug 3, 1881 oct 8, 1883 aug 10-1887 mar 11, 1887 mar 18-1890 jun 6 – 5r – 1 – (cont by: mondovi herald) – mf#1131891 – us WHS [071]

Buffalo county journal – Alma, Pepin WI. 1879 jun 5/1882 oct 5-1994 [gaps] – 61r – 1 – (cont: alma weekly express [pepin lake]; cont by: buffalo-pepin county journal) – mf#1001346 – us WHS [071]

Buffalo county journal – Alma WI. 1987 jan 29-dec, 1988-94 – 8r – 1 – (cont by: buffalo-pepin county journal) – us WHS [071]

Buffalo county journal – Alma WI. v1 n1-v2 n52 [1861 apr 27-1863 jun 18] – 1r – 1 – (cont by: alma journal) – mf#1001366 – us WHS [071]

Buffalo county journal – Kearney, NE: L B Cunningham, 1880-93// (wkly) [mf ed apr 6 1881] – 1r – 1 – us NE Hist [071]

Buffalo county pilot – Kearney, NE: A P Salgren, 1898 (wkly) [mf ed v3 n1. apr 8 1898-nov 1 1900 (gaps)] – 1r – 1 – (formed by the union of: elm creek pilot and: elm creek times. cont numbering of: elm creek times) – us NE Hist [071]

Buffalo county republican – Fountain City W: 1924 feb 7-1925, 1926-27, 1928-29, 1929 aug 1-dec 26, 1930/32-1957/1959 jan 3 – 11r – 1 – (cont: buffalo county republikaner [fountain city w: 1894]; cont by: cochrane-fountain city recorder) – mf#1043733 – us WHS [071]

Buffalo county republikaner – Alma, Fountain City WI. 1862 sep 3-1863 sep 5, [1866 sep 16-1871]-1923/1924 jan 31 – 18r – 1 – (cont: buffalo county republikaner und alma blaetter [fountain city wi]; cont by: buffalo county republican) – mf#1043728 – us WHS [071]

Buffalo county republikaner – Fountain City W. 1862 sep 3-1863 sep 5 – 1r – 1 – (cont by: alma blaetter [fountain city wi: 1888]; buffalo county republikaner und alma blaetter) – mf#915759 – us WHS [071]

Buffalo county republikaner – Fountain City WI (USA), aug 26 1920-dec 27 1923 – 2r – 1 – gw Misc Inst [071]

Buffalo county sun – Kearney, NE: Geo J Shepard, 1894-v4 n46. oct 23 1897 (wkly) [mf ed jul 27 1895-oct 23 1897 (gaps)] – 1r – 1 – (cont by: sun) – us NE Hist [071]

Buffalo criminal law review – Buffalo Criminal Law Center. v1-4. 1997-2001 – 9 – $101.00 set – mf#117311 – us Hein [345]

Buffalo criterion – Buffalo NY. 1991 jul 11/17-dec 26/jan 1 1992, 1992 jan 4/10-dec 26/jan 1 1993, 1993 jan 2/8-dec 25/31, 1994 jan 1/7-dec 31/jan 6 1995, 1995 jan 7/13-dec 30/jan 5 1996, 1996 jan 6/12-dec 28/jan 3 1997, 1997 jan 4/10-dec 27/jan 2 1998, 1998 jan 3/9-dec 26/jan 1 1999 – 8r – 1 – (cont: metropolitan buffalo criterion) – mf#4030934 – us WHS [071]

The buffalo criterion – Buffalo. N.Y. Aug. 16, Oct. 11, 1941 – 1 – us NY Public [071]

Buffalo grove baptist church. jefferson city, tennessee : church records – Jul 1905-Mar 1979 – 1 – us Southern Baptist [242]

Buffalo hide / Brigham Young University – v5 n3-v10 n2 [1982 jul-1988 aug] – 1r – 1 – mf#1508365 – us WHS [378]

Buffalo intellectual property law journal – v1-2000- – 9 – (inquire for info) – mf#118941 – us Hein [346]

Buffalo jewish review – Buffalo. N.Y. 1963-67 – 1 – us AJPC [071]

Buffalo law review – v1-49. 1951-2001 – 5,6,9 – $840.00 set – (v1-33 1951-84 in reel $418. v34-49 1985-2001 in mf $422) – ISSN: 0023-9356 – mf#101151 – us Hein [340]

Buffalo leader – 1939 aug 10-sep – 1r – 1 – (cont by: buffalo union leader) – mf#2821157 – us WHS [071]

Buffalo new times – Syracuse NY 1973-74 – 1 – mf#8771 – us UMI ProQuest [071]

Buffalo. New York. Superior Court *see* Sheldon's superior court reports

Buffalo public interest law journal – State University at Buffalo. v1-18. 1980-2000 – 9 – $112.00 set – (title varies: v1-16 1980-98 as: in the public interest) – ISSN: 1140-4707 – mf#114611 – us Hein [342]

Buffalo republic and progress – reel 1,2,3 – 3r – 1 – (cont: progress) – mf#483106 – us WHS [071]

Buffalo ridge baptist church. washington county. tennessee : church records – 1827-74 – 1 – us Southern Baptist [242]

Buffalo springs sentinel – Buffalo Springs, Bowman Co, ND: George D Skinner. v1 n1 nov 29 1907)- (wkly) – 1 – (cont by: scranton register) – mf#09489 – us North Dakota [071]

Buffalo. Synod (Pres. Ch. in the U.S.A. Old School) *see* Minutes, 1843-1870

Buffalo trails / Cowley County Genealogical Society – v1 n2 [1989 sep], v2 n1-2 [1990 nov] – 1r – 1 – mf#1743430 – us WHS [929]

Buffalo union leader – 1939 sep 21-1940 apr 19 – 1r – 1 – (cont: buffalo leader; cont by: union leader [buffalo ny]) – mf#2821145 – us WHS [071]

Buffalo volksfreund – Buffalo NY (USA), sep 1 1920-may 16 1939 (incomplete) – 30r – 1 – gw Misc Inst [071]

Buffaloer arbeiter-zeitung – Buffalo NY (USA), 1898-1918 2 apr [gaps] – 11r – 1 – (title varies: 6 nov 1898: arbeiter-zeitung) – gw Misc Inst [071]

Buffalo-pepin county journal – Alma WI. 1985 dec 5/1986 dec-1987 jan 1-22 – 2r – 1 – (cont: buffalo county journal [alma wi: 1879]; cont by: buffalo county journal [alma wi: 1987]) – mf#1001341 – us WHS [071]

Buffet, Edward Payson *see* The layman revato

Buffon, G L L de *see*
- Histoire naturelle des oiseaux
- Histoire naturelle generale et particuliere

Buffon, Georges Louis Leclerc, comte de *see* Histoire naturelle, generale et particuliere

O bufo : folha humoristica, noticiosa, critica e commercial – Sao Paulo, SP. 07 abr 1898 – (= ser Ps 19) – mf#P17,02,235 – bl Biblioteca [079]

Bufon escarlata / Velasquez, Rolando – San Miguel, El Salvador. 1943 – 1r – us UF Libraries [972]

Bufonadas del instituto de literatura / Castro, Tomas De Jesus – Bayamon, Puerto Rico. 1957 – 1r – us UF Libraries [972]

Buganda and king : a royal history of buganda / Zimbe, Bartolomayo Musoke – Chicago: Uni of Chicago, Photodup [19–] – us CRL [960]

Bugarin, Jose *see* Diccionario ibanag-espanol

Bugarski, Georg M *see* Die natur und der determinisumus des willens bei leibniz

Bugenhagen, J *see*
- Der 29 psalm ausgelegt
- Ain christlicher sendprieff an frauw anna
- Annotationes io bvgenhagij pomerani in epistolas pauli
- Bekentnis von seinem glauben vnd lere/ geschrieben an eynen widderteuffer
- Eyn christlich vnterricht eynes gottseligen lebens
- Eyn sendebrieff
- Eyn sermon von der eygenschafft vnd weyse des sacraments der tauff
- In 4 priora capita euangelij secundum matthaeum
- In ieremiam prophetam commentarium
- In regvm dvos vltimos libros, annotationes post samuelem iam primu emissae
- Indices qvidam ionnis bvgenhagij pomerani in euangelia
- Ioan bvgenhagii pomerani in hiob annotationes
- Ioannis bvgenhagii pomerani commentarius
- Ioannis bvgenhagii publica
- Ioannis bvgenhague pomerani annotationes ab ipso iam emissae in deuteronomium in samuelem propheta, id est duos libros regu
- Ionas propheta
- Von dem konigreych vnd priesterthum christi der hundert vnd zehende psalm dauids
- Von der closter keuescheyt vnnd christenlicher beicht

Bugenhagen, Johann *see*
- Dr. johannes bugenhagens briefwechsel
- Johannes bugenhagens braunschweiger kirchenordnung, 1528

Bugenhagen, Johannes *see* [Ehrbaren stadt braunschweig christenliche ordnung] der erbarn stadt braunschwyg christenliche ordenung, zu dienst dem heiligen evangelio christlicher lieb, zucht, fride und eynigkeit

Bugge, Christian August *see* Das gesetz und christus im evangelium

Bugge, Sophus *see* Studier over de nordiske gude- og heltesagns oprindelse

Buglawton, st john the evangelist – (= ser Cheshire monumental inscriptions) – 2mf – 9 – £4.00 – mf#357 – uk CheshireFHS [929]

Buglawton, st john the evangelist : burials – (= ser Cheshire church registers) – 2mf – 9 – £4.00 – mf#358 – uk CheshireFHS [929]

Bugle – Chambers, NE: Wry & Sackett (wkly) [mf ed 1892, 1895-1902 (gaps)] – 12r – 1 – (cont by: chambers bugle) – us NE Hist [071]

Bugle – Turtle Lake WI. 1901 oct 17-1904 may 19, 1904 may 26-1906 nov 8, 1906 nov 15-1909 dec 31, 1910 jan 6-1913 oct 9 – 4r – 1 – (cont by: turtle lake times) – mf#1009763 – us WHS [071]

Bugle / Erie Co. Vermillion – v1 n1. sep 1876-jan 1877 [wkly] – 1r – 1 – mf#B33275 – us Ohio Hist [071]

Bugle – Redbridge, England 13 dec 1907-20 jan 1917, 7 mar 1919-5 may 1922 – 1 – (set consists of eds for wanstead & woodford) – uk British Libr Newspaper [072]

The bugle – Birkenhead, NZ. jun 1981-aug 1982 – 1r – 1 – mf#11.44 – nz Nat Libr [079]

Bugle american – 1970 sep 16/1972-1977 feb/1978 dec 7 – 10r – 1 – (with gaps) – mf#780630 – us WHS [071]

Bugle blast – Lake Mills WI. 1863 oct, dec, 1864 apr, oct-1865 apr – 1r – 1 – mf#933213 – us WHS [071]

Bugler, Jeremy *see* Polluting britain

Bugul'minskaia gazeta : organ komiteta kul'turnoprosvetitel'skogo otdela – Bugul'ma, Russia 1918 [mf ed Norman Ross] – 1r – 1 – mf#nrp-385 – us UMI ProQuest [077]

Bug-zeitung : nachrichtenblatt des grossen soldatenrates der etappe bug – ZZ=o.O/s.l. 1917 17 sep-1918 12 nov [mf ed 2004] – 1 – gw Mikrofilm [355]

Buhay na maogma nin princesa sampaguita asin princesa ardelisa – [Nueva Caceres?: Libreria Mariana 190-?] [mf ed Bloomington IN: Indiana Uni Lib, Preservation Dept 1984] – (= ser Coll...in the bikol language) – 1r – 1 – us Indiana Preservation [490]

Buhay na pig-aguihan ni manrique na hinoclob – [Nueva Caceres: Libreria Mariana 190-?] [mf ed Bloomington IN: Indiana Uni Lib, Preservation Dept 1984] – (= ser Coll...in the bikol language) – 2v on 1r – 1 – us Indiana Preservation [490]

Buhay na pinagdaanan ni d juan, hari sa mundo ng austria at ni dona maria sa kaharian nang murcia – Maynila: Sayo ni Soriano 1919 [mf ed Bloomington IN: Indiana Uni Lib, Preservation Dept 1984] – (= ser Coll...in the tagalog language 1) – 1r – 1 – us Indiana Preservation [490]

Buhayri, Marwan *see* Al-hilf al-atlasi wa-al-sharq al-awsat

Buheiry, Marwan *see* Us threats of intervention against arab oil, 1973-79

Buhl, Frants *see* Geographie des alten palaestina

Buhle, Joh Gottlieb Gerh et al *see* Goettingisches philosophisches museum

Buhler, Georg *see* On the indian sect of the jainas

El buho del ribero / Delgado Solis, Sebastian – Caceres: imp. y enc. vda. de floriano, 1957 – 1 – sp Bibl Santa Ana [946]

Buhre, U T *see* The effect of a 60-minute duration exercise, at the intensities of the lactate and the individual anaerobic thresholds, on the cardiovascular drift

Buhren, Frank *see* Sozial- und moralphilosophie in der 'alten und neuen kritischen theorie'

Bui, Duc Tinh *see* Van-hoc va ngu hoc

Bui Ke, U *see* Pu gam su te sa na lam nnvhan

Bui phan dai guong / Nghiem, Le Quan – [Saigon] Thuy Chung [1974] [mf ed 1992] – on pt of 1r – 1 – mf#11052 r359 n8 – us Cornell [959]

Buies, Arthur *see*
- L'ancient et le futur quebec
- Anglicismes et canadianismes
- Animals of canada
- Au portique des laurentides
- Le chemin de fer du lac saint-jean ses origines
- Chroniques canadiennes
- Chroniques, vol 1
- Chroniques, vol 2
- Conferences
- Une evocation
- La lanterne
- Lecture sur l'entreprise du chemin de fer du nord
- Lettres sur le canada
- L'outaouais superieur
- Les poissons et les animaux a fourrures au canada
- La province de quebec
- Question franco-canadienne
- Recits de voyages
- La region du lac saint-jean, le grenier de la province de quebec
- Reminiscences; les jeunes barbares
- Reponse a un ordre de l'assemblee legislative, en date du 11 decembre 1890
- Le saguenay et la vallee du lac saint-jean
- Le saguenay et le bassin du lac saint-jean
- La vallee de la matapedia

Buies, Arthur et al *see* Reports on the counties of rimouski, matane and temiscouata

The buik of the chronicles of scotland (rs6) : or: a metrical version of the history of hector boece, by william stewart / ed by Turnbull, W B – 1858 – (= ser The rolls series (rs)) – 3v – €23.00v – ne Slangenburg [941]

Buil Ba Kui *see* Bha kri rvhan

Buil Bakui *see* Tarup prann sac khari

Buil khup aon chan e lvat lap re krui pam mhu / Ne Van, Takkasuil – Ran Kun: Lan Yun ca pe 1982 [mf ed 1990] – 1r with other items – 1 – (in burmese) – mf#mf-10289 seam reel 201/3 [§] – us CRL [959]

Buil khyup aon chan nhan a janann kri mya atthuppatthi – [s.l: s.n. 1972] [[Rangoon]: Pran kra re nhan A sam lvhan U ci Thana] – [ill] 1r with other items – 1 – (in burmese; biographies of general on chan, 1915-47, burmese statesman, and six other prominent statesmen assassinated while attending a cabinet meeting in july 1947) – mf#mf-10289 seam reel 137/5 [§] – us CRL [959]

Buil mhu bha rhan e a no ra tha a ran ka mran ma nuin nam / Bha Rhan Bhuil mhu – Ran kun: Chve Chve Mo Ca pe 1975 [mf ed 1994] – on pt of 1r – 1 – mf#11052 r1709 n1 – us Cornell [959]

Buil ta thon = Vanguard daily – Rangoon: [s.n. (daily) [mf ed Ithaca NY: [John M Echols Collection] Cornell University 2004-] – reel 1-8 – 1 – (in burmese; lacking many iss) – mf#mf-10871 seam – us CRL [079]

Build to serve – v5 n1-2,4 [1980 jan 13-apr 20, jul 27], v6 n1,3-4, [1980 oct 9, 1981 apr 26-jul 15], v7 n1 [1981 oct 25], v8 n1 [1982: nov 21], v9 n1-4 1983 jan 23-nov 27], v10 n5-7 [1984 mar 24-oct], v12 n3 [1986 jul], v13 n1-3 [1986 nov-1987 may], v14 n1-16 [1987 dec-1990 fall], 1990 win-1993 win – 1r – 1 – mf#1053679 – us WHS [071]

Builder – Coos Bay OR: J F Kutch, 1975-76 [wkly] – 1 – (cont: empire builder (1966-75); cont by: coos bay empire builder (1976-77)) – us Oregon Lib [071]

Builder – Washington DC 1983+ – 1,5,9 – ISSN: 0744-1193 – mf#13573.02 – us UMI ProQuest [690]

"The builder album" of royal academy architecture – London 1891-93 – (= ser 19th c art & architecture) – 13mf – 9 – mf#4.1.193 – uk Chadwyck [720]

"The builder album" of royal academy architecture – London, 1891-93 – (= ser 19th c visual arts & architecture) – 3v on 13mf – 9 – mf#4.1.193 – uk Chadwyck [720]

Builders / AFL-CIO [American Federation of Labor-Congress of Industrial Organizations] – 1979 mar 12-1980 aug 25, 1980 jul 21-1983 mar 14, 1983 jan 10-1994 dec – 3r – 1 – mf#625994 – us WHS [690]

The builders of babel / M'Causland, Dominick – London: Richard Bentley 1874 [mf ed 1985] – 1mf – 9 – 0-8370-4216-X – mf#1985-2216 – us ATLA [221]

Builders of latin america / Stewart, Watt – New York, NY. 1942 – 1r – us UF Libraries [972]

Builders of nova scotia : a historical review, with an appendix containing copies of rare documents relating to the early days of the province / Bourinot, John George – [S.l: s.n, 1899?] [mf ed 1979] – 3mf – 9 – (incl bibl ref) – mf#00220 – cn Canadiana [920]

Builders of nova scotia : a historical review with an appendix containing copies of rare documents relating to the early days of the province / Bourinot, John George – Toronto: Copp-Clark, 1900 [mf ed 1981] – 3mf – 9 – (incl ind and bibl ref) – mf#26585 – cn Canadiana [920]

Builders of united italy / Holland, Rupert Sargent – New York: H Holt & Co 1908 [mf ed 1988] – 1r [ill] – 1 – mf#2192 – us Wisconsin U Lib [945]

The builders' portfolio : of street architecture / Collis, James – London 1831 – (= ser 19th c art & architecture) – 1mf – 9 – mf#4.2.1058 – uk Chadwyck [720]

The builder's practical director : or buildings for all classes containing plans, sections and elevations for the erection of cottages, villas, farm buildings, dispensaries, public schools etc with detailed estimates, quantities prices etc – Leipzig, Dresden: A H Payne; London: J Hagger. 2v. [1855-1857?] – (= ser 19th c art & architecture) – 9mf – 9 – (ill by numerous plates and diagrams) – mf#4.1.207 – uk Chadwyck [690]

Building – London UK 1842+ – 1,5,9 – ISSN: 0007-3318 – mf#1271 – us UMI ProQuest [720]

Building a bridge between athletics and academics / Kilbourne, John R – 1994 – 2mf – $8.00 – us Kinesology [790]

Building a working church / Black, Samuel Charles – New York: Revell c1911 [mf ed 1991] – 1mf – 9 – 0-7905-7686-4 – mf#1989-0911 – us ATLA [240]

Building and Construction Trades Council of Alameda County *see* Journeyman

Building and engineering journal – Sydney, jun 1888-dec 1905 – 7r – A$433.05 vesicular A$471.55 silver – at Pascoe [079]

Building and environment – Oxford UK 1965+ – 1,5,9 – ISSN: 0360-1323 – mf#49024 – us UMI ProQuest [690]

Building bridges : a newsletter about the african-american research library and cultural center / Broward County Library – 1997 jul – 1r – 1 – mf#5296690 – us WHS [020]

Building confidence / British Guiana – East Demarara, Guyana. 1955 – 1r – us UF Libraries [972]

Building construction – New York NY 1964-67 – 1 – (cont by: building design and construction) – mf#1640.01 – us UMI ProQuest [690]

Building design – London UK 1985-91 – 1,5,9 – ISSN: 0007-3423 – mf#11355 – us UMI ProQuest [720]

Building design & construction – New York NY 1986+ – 1,5,9 – (cont: building construction) – ISSN: 0007-3407 – mf#1640.01 – us UMI ProQuest [690]

Building eras in religion / Bushnell, Horace – New York: Scribner 1881 [mf ed 1991] – 2mf – 9 – 0-7905-9251-7 – mf#1989-2476 – us ATLA [240]

Building for peace : or, gandhi's ideas on social (adult) education / Nayyar, Dev Parkash – Delhi: Atma Ram & Sons, 1952 – (= ser Samp: indian books) – us CRL [327]

Building news *see* Freehold land times building news

Building news and engineering journal – London, 1854-1926. v1-130 – 1764mf – 9 – mf#OA-303 – ne IDC [720]

The building news and engineering journal – London, 1856-1926 – (= ser Architectural periodicals at avery library, columbia university) – 122r – 1 – $16,510.00 – us UPA [690]

The building of character / Miller, James Russell – New York: Thomas Y Crowell, c1894 – 1mf – 9 – 0-8370-7179-8 – mf#1986-1179 – us ATLA [240]

Building of gold and of stubble / Woodford, James Russell – London, England. 1855 – 1r – us UF Libraries [240]

The building of the church / Jefferson, Charles Edward – New York: Macmillan, 1910 – (= ser Lyman Beecher Lectures) – 1mf – 9 – 0-7905-7787-9 – mf#1989-1012 – us ATLA [240]

Building of the house of god / Wilberforce, Henry William – Southampton, England. 1839? – 1r – us UF Libraries [240]

The building of the kosmos and other lectures / Besant, Annie Wood – Madras: Theosophist, 1894 – (= ser Samp: indian books) – us CRL [520]

Building of the tabernacle, or, the duty and privilege of contribut / Roxburgh, John – Glasgow, Scotland. 1847 – 1r – us UF Libraries [240]

Building of the walls of jerusalem – Ashby-de-la-Zouch, England. 1843 – 1r – us UF Libraries [240]

Building operating management – Milwaukee WI 1972+ – 1,5,9 – ISSN: 0007-3490 – mf#7497 – us UMI ProQuest [690]

Building record – British Columbia, CN. 1912-51 – 32r – 1 – (some missing iss) – cn Commonwealth Imaging [690]

Building research establishment digest – Norwich UK 1973-81 – 1,5,9 – ISSN: 0144-8536 – mf#8087 – us UMI ProQuest [690]

Building service employee – 1942-43, 1952 feb-1956 dec, 1955-51 – 3r – 1 – (cont by: service employee) – mf#1336414 – us WHS [331]

Building Service Employees' International Union *see*
- Pennsylvania labor record
- Report to locals
- Report to locals newsletter
- Service employee
- Service union reporter

Building services engineer – London UK 1974-78 – 1,5,9 – ISSN: 0301-6536 – mf#8338 – us UMI ProQuest [690]

Building specialties – Bolinas CA 1950-54 – 1 – mf#716 – us UMI ProQuest [690]

Building supply business – New York NY 1996 – 1 – (cont: building supply home centers: bshc national ed) – ISSN: 1086-2943 – mf#14866.04 – us UMI ProQuest [690]

Building supply home centers – New York NY 1989-93 – 1 – (cont by: building supply home centers [bshc] national ed) – ISSN: 0890-9008 – mf#14866.04 – us UMI ProQuest [690]

Building supply home centers [bshc] national ed – New York NY 1994-95 – 1,5,9 – (cont: building supply home centers; cont by: building supply business) – mf#14866.04 – us UMI ProQuest [690]

Building systems design – Lilburn GA 1904-80 – 1,5,9 – (cont by: energy engineering: journal of the association of energy engineers) – ISSN: 0002-2284 – mf#756.01 – us UMI ProQuest [690]

Building technology and management – Ascot UK 1976-89 – 1,5,9 – ISSN: 0007-3709 – mf#11269 – us UMI ProQuest [690]

Building Trades Council [FL] et al *see* Miami citizen

Building Trades Council of Alameda County et al *see* East bay labor journal

Building Trades Council of Cincinnati and Vicinity *see* Labor advocate

Building tradesman / Detroit Building Trades Council – 1960/62-1991 jan/1992 dec – 16r – 1 – (with gaps; cont: detroit michigan building tradesman) – mf#3320906 – us WHS [690]

The building up of the old testament / Girdlestone, Robert Baker – London: Robert Scott 1912 [mf ed 1989] – (= ser Library of historic theology) – 1mf – 9 – 0-7905-2901-7 – mf#1987-2901 – us ATLA [221]

Buildings – Cedar Rapids IA 1972+ – 1,5,9 – ISSN: 0007-3725 – mf#6370 – us UMI ProQuest [690]

Buildings and monuments : modern and mediaeval / Godwin, George – London 1850 – (= ser 19th c art & architecture) – 3mf – 9 – mf#4.2.737 – uk Chadwyck [720]

The buildings at samaria / Crowfoot, J W – PEF. 1942 – 9 – $10.00 – us IRC [930]

Builes G, Miguel Angel *see*
- Cronicas misionales del excmo y revmo sr dr dn...
- Cuarenta dias en el vaupes

Buimov, I M *see* Novyi luch

Buiskool, Johannes Ate Eildert *see* Suriname nu en straks

Buisson, Ferdinand Edouard *see*
- Libre-pensee et protestantisme liberal
- La religion, la morale et la science
- Sebastien castellion

Buisson, J C du *see* The gospel according to st mark

Buitenbezittingen / Politiek verslag, 1852 – 25mf – 8 – mf#SD-100 mf 26-50 – ne IDC [959]

Buitenlandse sending kwartaalblad van die ned. geref. kerk in s.a. – Mkhoma, Nyasaland: Sending Drukkery. v1 1923; v2 n2-4 apr/jun-oct/dec 1924; v3 n2-4 apr/jun-oct/dec 1925; v4-8 1926-30 – 1r – us CRL [960]

Die buiteposte in die ekonomie van die kaapse verversingstasie : 1652-1795 / Sleigh, Dan – Stellenbosch: U van Stellenbosch 1987 [mf ed 1987] – 8mf – 9 – mf#mf.735 – sa Stellenbosch [330]

Bujak, Franciszek *see* Stan gospodarczy polski

Buk in al kab tun ko : non ro dri ailin in marshall / Pease, E M – New York: Biglow & Main, 1889 [mf ed 1995] – (= ser Yale coll) – 126p – 1 – 0-524-09452-7 – (in marshall) – mf#1995-0452 – us ATLA [490]

Buka re munamato wevese / Church Of The Province Of Central Africa – London, England. 1963 – 1r – us UF Libraries [960]

Bukacz, Franz *see* Die deutschen schutzgebiete in afrika

Bukareshter zamibikher – Bukaresht, Romania. v1. 1947 – 1r – us UF Libraries [939]

Bukarest und stambul : skizzen aus ungarn, rumunien und der tuerkei / Kunisch, Richard – Berlin 1861 [mf ed Hildesheim 1995-98] – 3mf – 9 – €90.00 – 3-487-29146-0 – gw Olms [910]

Bukarester deutsche tagespost – Bukarest (RO), 1924-1925 15 nov – 3r – 1 – gw Misc Inst [077]

Bukarester deutsches tageblatt – Bukarest (RO), 1941-1943 mar, 1943 jul-1944 30 apr – 1 – (title varies: 1927: bukarester tageblatt; filmed by other misc inst: 1926-43 [30r]) – gw Misc Inst [077]

Bukarester lloyd – Bukarest (RO), 1023 2 jan-1924 8 mar – 1 – gw Misc Inst [077]

Bukarester post – Bukarest (RO), 1933 8 jan-1940 1 jan – 5r – 1 – gw Misc Inst [077]

Bukarester tageblatt – Bukarest, Romania 1-31 jul, 1 sep 1942-6 jan 1944; 31 jan-22 aug 1944 (imperfect) – 1 – uk British Libr Newspaper [077]

Bukarester tageblatt *see* Bukarester deutsches tageblatt

Bukarester tageblatt 1880 – Bukarest (RO), 1889 3 sep-1896 4 aug, 1918 1 jul-7 nov – 13r – 1 – (later: rumaenischer lloyd) – gw Misc Inst [077]

Buket : zhurnal shit'ia, vyshivaniia, mod, domashnego khoziaistva, literat i mod novostei – St Petersburg, jan-apr 1860 [mf ed Norman Ross Publ] – 3mf – 9 – us UMI ProQuest [640]

Bukh, Niels E *see* Fundamental gymnastics

Bukh un der lezer – Warsaw, Poland. 1910/11 – 1r – us UF Libraries [939]

Bukhari von Johor *see* De kroon aller koningen

Bukhari, Muhammad ibn Ismail *see*
- Selections from the sahih of al-buhari
- Les traditions islamiques

Bukharin, Nikolai Ivanovich *see*
- Leninizm i problema kul'turnoi revoliutsii
- Partiia i oppozitsionnyi blok

Bukhbinder, N A *see* Materialy dlia istorii evreiskogo rabochego dvizheniia v rossii

Bukhov, Arkadii Sergeevich *see* Chortovo koleso

Bukkyo shoshi / Fujii, Sensho – Kyoto: Otani Shintaido, Meiji 29 [1896] – 2mf – 9 – 0-524-08298-7 – mf#1993-4003 – us ATLA [280]

Bukovetskii, A I *see* Materialy po denezhnoi reforme 1895-1897 gg

Bukovina. Landtag *see* Stenographische protokolle

Bukowinaer provinzbote – Storoynez (RO), 1931-32 – 1 – gw Misc Inst [077]

Buku alamat dagang – Surabaja, 1961 – 5mf – 9 – mf#SE-1367 – ne IDC [950]

Buku duku re masoko anoyera / Knecht, Friedrich Justus – Mariannhill, South Africa. 1915 – 1r – us UF Libraries [960]

Buku kita : madjalah untuk buku dan pembatja – Djakarta. 2v. 1955-1956 – 23mf – 9 – mf#SE-631 – ne IDC [370]

Buku la mapemphero – Issy-les Moulineaux, France. 1951 – 1r – us UF Libraries [960]

Buku pedoman / Universitas baperki – Djakarta, 1961-1963 – 4mf – 9 – mf#SE-1974 – ne IDC [959]

Buku ra vana – Cape Town, South Africa. 1918 – 1r – us UF Libraries [960]

Buku re masoko anoyera e chirangano che kare ne chipswa / Gilmour, R – Mariannhill, South Africa. 1917 – 1r – us UF Libraries [960]

Bukvar' tatarskago i arabskago pis'ma: s priloz. slov so znakami, pokazuvajuscimi ich vygovor / Atnometev, Nijat Baku – Sanktpeterburge: Imperatorskaja Akademia Nauki 1802 – (= ser Whsb) – 1mf – 9 – €20.00 – (fibel der tatar. u. arab. schrift mit beifuegung der woerter mit zeichen, die ihre aussprache angeben) – mf#Hu 233 – gw Fischer [410]

Bula in apostolatus culmine led papa paulo iii / Paul 3, Pope – Ciudad Trujillo, Dominican Republic. 1944 – 1r – us UF Libraries [972]

La bula "inter graviores curas" de pio 7 en la orden franciscana y ulterior regimen general de la orden en espana (1804-1904) / Barrado Manzano, Arcangel – Madrid: Archivo Ibero-Americano, 1964 – 1 – sp Bibl Santa Ana [240]

Bula matari / Wasserman, Jakob – New York, NY. 1933 – 1r – us UF Libraries [960]

Bulaeus (du Boulay), Caesar Egassius *see* Historia universitatis parisiensis

Bulak, 'Arif *see* The divan project

Bulanzhe *see* Dvukhnedelnoe obozrenie, posviashchennoe voprosam bratskoi zhizni, kak ikh obiasnial liudiam khristos i kak napominaet teper l n tolstoi

Las bulas alejandrinas de 1493 referentes a las indias / Bayle, Constantino – Madrid: Razon y Fe, 1945 – 1 – sp Bibl Santa Ana [970]

Bulawayo / Ransford, Oliver – Cape Town, South Africa. 1968 – 1r – us UF Libraries [960]

Bulawayo express : weekly edition – Bulawayo, Zimbabwe 24 dec 1904-25 mar 1905 – 1 – uk British Libr Newspaper [079]

Bulck, Gaston Van *see* Manual de linguistique bantoue

Buletinul demografic al romaniei – Bucuresti: Ministerului Muncii, Sanatatatii si ocrotirilor 1932- [mf ed 2002] – 1r – 1 – (filming in process. library has: anul 1-anul 3 (1932-34); anul 5 (1936); anul 6 n4 (1937: apr); anul 6 n6 (1937: iunie); anul 6 n10-12 (1937: oct-dec); anul 8 n6-10 (1939: iunie-oct); anul 9 n2 (1940: feb); anul 9 n4 (1940: apr); anul 9 n7 (1940: iulie); anul 11 n1 (1940: ian); anul 11 n3-12 (1940: mar dec); anul 12 (1943)) – mf#?? – us Wisconsin U Libr [314]

Buletinul oficial / Romania – Bucharest. Pt. I Annul I-Pt. III Annul IV. Aug. 21, 1965-Dec. 31, 1968 – 1 – us NY Public [324]

Buley, Ernest Charles *see* South brazil

Bulfinch, Stephen Greenleaf *see* Romanism

Bulgakov, Afanasii Ivanovich *see* The question of anglican orders

Bulgakov, F I *see* Illiustrirovannaia istoriia knigopechataniia i tipografskogo iskusstva

Bulgakov, P G *see* Arabskie rukopisi sobraniia leningradskogo gosudarstvennogo universiteta

Bulgaria *see* Mouvement de la population

Bulgaria. Darzhavno Upravlenie za Informatsiya *see*
- Statisticheski godishnik na narodna republika bulgaria 1956-1970
- Statistickeski godishnik na narodna republika bulgaria 1956-1970

Bulgaria. Glavna Direktsiia na Statistikata *see*
- Mesechni statisticheski izvestiia
- Statisticheski godishnik na bulgarskoto tsarstvo

Bulgaria. Laws, Statutes, etc *see*
- Law of administrative procedure
- Law of administrative violations and punishments

Bulgaria. Obiknoveno Narodno Subraniye *see* Dnevnitzi

Bulgaria past and present : historical, political, and descriptive / Samuelson, James – London: Truebner & Co 1888 [mf ed 1986] – 1r [ill] – 1 – mf#1575 – us Wisconsin U Libr [949]

The bulgarian exarchate : its history and the extent of its authority in turkey = Machtbereich des bulgarischen exarchats in der tuerkei / Mach, Richard von – London: T F Unwin; Neuchatel: Attinger, 1907 – 1mf – 9 – 0-7905-6487-4 – (incl bibl ref. in english) – mf#1988-2487 – us ATLA [949]

Bulgarian review – Rio de Janeiro, Brazil 1961-78 – 1,5,9 – ISSN: 0007-3946 – mf#7671 – us UMI ProQuest [077]

Bulgarian-british review – 1928-40 – 1 – us L of C Photodup [949]

La bulgarie – Sofia, Bulgaria. 1892; 1926-35 – 11r – 1 – us L of C Photodup [949]

La bulgarie – Sofia, Bulgaria. 2 jan 1924-27 dec 1935 – 1 – mf#m.f.683 – uk British Libr Newspaper [077]

La bulgarie – Sofia, Bulgaria. -d. 2 Jan 1924-27 Dec 1935. 12 reels – 1 – uk British Libr Newspaper [949]

Bulgariia – Sofia, Bulgaria 2 dec 1918-14 jul 1919 (imperfect) – 1 – (in cyrillic) – uk British Libr Newspaper [077]

Bulgaris, Nicolas *see* The holy catechism of nicolas bulgaris

Bulgarische wochenschau – Sofia, Bulgaria. Jul 1940-1942 – 1r – 1 – us L of C Photodup [077]

Bulgaro-sovetsko edinstvo – Sofia, Bulgaria. 1951-54 – 1r – 1 – us L of C Photodup [949]

Bulgarski knigopis – Sofia. v. 1-48. 1897-1945 – 1 – 93.00 – us L of C Photodup [949]

Bulgarski turgovski viestnik = Bulgarische handelszeitung – Sofia, Bulgaria 10 dec 1917-5 jun 1918 (imperfect) – 1 – (in cyrillic (bulgarian) & german) – uk British Libr Newspaper [077]

Bulger, Andrew H *see*
- An autobiographical sketch of the services of the late captain andrew bulger
- Papers referring to red river settlement

Bulgin, Robert *see* Exiled jesuits on trespass in england

Bulitin-i bahs-i dakhili – London, mar 1976-oct 1977 – 1r – 1 – $53.00 – (minutes of meetings of an iranian trotskyite organization based in london) – us MEDOC [320]

Bulkeley, Owen T *see* Lesser antilles

Bulkley, Charles *see* The signs of the times

The bull apostolicae curae and the edwardine ordinal / Puller, Frederick William – London: SPCK 1896 [mf ed 1992] – (= ser Church historical society (series) 16) – 1mf – 9 – 0-524-03326-9 – (incl bibl ref) – mf#1990-4686 – us ATLA [242]

Bull, Bartle E *see* Christ and his apostles

Bull board – n1-129 [1944 jul 7-1945 aug 3] – 1r – 1 – mf#2892701 – us WHS [071]

Bull, Canon *see* Centennial poem

Bull, Earl *see* Bernard jean bettelheim

Bull, Edvard *see* Folk og kirke i middelalderen

Bull, Edvard et al *see* Byer og bybebyggelse

Bull fight / Youngblood, Alice P – s.l, s.l? 1937? – 1r – us UF Libraries [978]

Bull, George *see* Corruptions of the church of rome

Bull, John *see* Plain appeal to the common sense of all the men and women of great britain and ireland

Bull, Lucien *see* La cinematographie

Bull moose / Association of Political Items Collectors – v1-4 n6 [1975 aug-1978] – 1r – 1 – mf#499090 – us WHS [320]

Bull of pope pius the ninth and the ancient british church / Harington, Edward Charles – London, England. 1850 – 1r – us UF Libraries [240]

Bull, Paul Bertie *see* The sacramental principle

Bull sheet / 135 Medical Regiment [Organization] – 1974-82 – 1r – 1 – (cont: chaplain's bulletin [west de pere wi]; cont by: christmas bulletin [madison wi]) – mf#1497281 – us WHS [355]

Bull sheet – v2 n11, 17-18 [1941 nov 28, 1942 jan 16-23], v3 n3 [1942 feb 13], 1945 aug 11-sep 11 – 1r – 1 – mf#645697 – us WHS [071]

Bull swamp baptist church : centennial address by g w gardner on history of church – Dublin. 1972-78 [1,5] 1975-78 [9] – 1 – $10.00 – mf#6503 – us Southern Baptist [242]

Bulla confirmationis et nouae concessionis priuilegiorum omnium ordinum mendicantium : cum certis declarationibus decretis [et] inhibitio[n]ibus s d n d pij papae 5. motu p[ro]prio / Pius 5, Pope – Mexici: Apud Antonium de Spinosa, anno 1568 – (= ser Books on religion...1543/44-c1800: ordenes mendicantes) – 1mf – 9 – mf#crl-248 – ne IDC [241]

Bulla de su santidad : de gregorio 15 en razon de la sugrecion, y subordinacion, que los regulares... / Gregory 15, Pope – [Mexico s.n. 1638?] – (= ser Books on religion...1543/44-c1800: historia ecclesiastical) – 1mf – 9 – mf#crl-427 – ne IDC [241]

Bulla erectionis sanctae metropolitanae ecclesiae mexiceae / Clement 7, Pope – [Mexico City: s.n. 1735?] – (= ser Books on religion...1543/44-c1800: iglesias, catedrales) – 1mf – 9 – mf#crl-351; crl-463 – ne IDC [241]

Bullae, motus propis ac breviaque.. – 1704 – 9 – sp Bibl Santa Ana [240]

Bullard *see* Barbary coast

Bullard, Arthur *see* Panama

Bullard, J M *see* The growth hormone response to exercise at different times of the day

Bullarii romani continuatio / ed by Barberi, A & Spetia, A – Romae. v1-19. 1835-1857 – 455mf – 8 – €868.00 – ne Slangenburg [240]

Bullarium canonicorum regularium congregationis sanctissimi salvatoris – Romae, 1733 – €57.00 – ne Slangenburg [241]

Bullarium carmelitanum / Monsignano, Eliseo – 9 – €278.00 – (pars prima duplici indice exornata, romae 1715. pars secunda duplici indice instructa, romae 1718. a fratre josepho alberto ximenez: pars tertia, romae 1768. pars quarta a clemente 11 usque ad clementem 13, romae 1768) – ne Slangenburg [241]

Bullarium casinense seu constitutiones...pro congregatione casinense / ed by Margarinus, G – Venetiis-Tuderti. v1-2. 1650-70 – €107.00 – ne Slangenburg [241]

Bullarium equestris...iacobi de spatha / Lopez Agurleta, Jose – 1719 – 9 – sp Bibl Santa Ana [790]

Bullarium lateranense : collectio privilegiorum apostolicorum a sancta sede canonicis regularibus ordinis sancti augustini congregationis salvatoris lateranensis concessorum – Romae, 1727 – €40.00 – ne Slangenburg [241]

Bullarium ordinis ff praedicatorum / ed by Bremond, A – Romae. v1-8. 1729-40 – 8v on 246mf – 8 – €469.00 – ne Slangenburg [241]

Bullarium ordinis recollectorum sancti augustini / Fernandez de S Corde, J – Madrid: Archivo Ibero Americano, 1963 – 1 – sp Bibl Santa Ana [240]

Bullarium...alcantara...pereiro / Ortega y Cotes, Ignacio – 1759 – 9 – sp Bibl Santa Ana [946]

Bulldozer / Prism Solidarity Collective – v1 n1-8 [1980 aug-1985 sum] – 1r – 1 – mf#1266027 – us WHS [334]

Bulldozer Group [Toronto, Ontario] *see* Prison news service

Bulle, Konstantin *see* Geschichte des zweiten kaiserreiches und des koenigreiches italien

Die bullen der paepste : bis zum ende des zwoelften jahrhunderts / Pflugk-Harttung, Julius von – Gotha: FA Perthes, 1901 – 1mf – 9 – 0-7905-7070-X – mf#1988-3070 – us ATLA [240]

Bullen, Frank Thomas *see*
- Back to sunny seas
- Creatures of the sea
- The cruise of the "cachalot"
- Denizens of the deep
- Idylls of the sea
- Sea puritans
- With christ at sea

Buller, Frank *see* The influence of certain ocular defects in causing headache

Buller, James *see* Forty years in new zealand

Buller's campaign / Symons, Julian – London, England. 1963 – 1r – us UF Libraries [960]

Bullet, P *see* L'architecture pratique

Bulletin / Aeronautical Society of America – New York. Jul. 1908-Dec. 1909; July 1911 – 1 – us NY Public [073]

Bulletin / The American Astronomical Society – v1-. 1969- – 1,5,6 – us AIP [520]

Bulletin / American Iron and Steel Association – v1-46. 1886-1912 – 1 – $241.00 – us L of C Photodup [670]

Bulletin / The American Physical Society – Series 1. v1-30. 1925-55 – 1,5,6,9 – us AIP [530]

Bulletin / American Physical Society – Series 2. v1-. 1956- – 1,5,6,9 – us AIP [530]

Bulletin / Arctic Club of America – New York. n1-27. 1907-1911 – 1 – us NY Public [366]

Bulletin / L'Association canadienne des parents des prisonniers de guerre – Montreal: Association canadienne des parents des prisonniers de guerre, [ca 1942]- (irreg) [mf ed 1987] – 1mf – 9 – mf#SEM105P798 – cn Bibl Nat [366]

Bulletin / Association des Amis de Romain Rolland – no. 1-102. Paris. aout 1946-72 – 1 – fr ACRPP [440]

Bulletin / L'Association Emile-Zola – no. 1-7. Paris. 1910-12 – 1 – fr ACRPP [440]

Bulletin – Sydney, jan 1880-jun 1997 – 9 – (available on subsc. apply for details) – at Pascoe [073]

Bulletin / Bank Indonesia – Djakarta, 1953-1960(20) – 30mf – 9 – (missing: 1953(1); 1956(12)) – mf#SE-260 – ne IDC [959]

Bulletin / Comite de l'Afrique francaise – devenu: L'Afrique francaise. Bulletin mensuel du Comite de l'Afrique francaise du Maroc. Paris. 1891-1940, 1952-janv mars 1960 – 1 – fr ACRPP [960]

Bulletin / Comite d'Etudes Historiques et Scientifiques de l'Afrique Occidentale Francaise – v1-21 1918-38 – 1 – us CRL [960]

Bulletin – n7-12 – 1r – 1 – (cont by: bulletin [united states. bureau of forestry]) – mf#2463272 – us WHS [634]

Bulletin – Chicago IL. 1968 sep 11-1969 apr 2 – 1r – 1 – (cont by: south side bulletin) – mf#868982 – us WHS [071]

Bulletin – San Francisco CA. 1915 feb 20, 1918 jul 1-6 – 1r – 1 – (cont: daily evening bulletin) – mf#931332 – us WHS [071]

Bulletin / Ecole Francaise d'Extreme-Orient – v. 1-51. 1901-63 – 1 – 618.00 – us L of C Photodup [950]

Bulletin / Ethnological Society – 1 – us AMS Press [306]

Bulletin / Federation radicale et radicale-socialiste de Guyane. Comite Executif – Cayenne. dec 1908-sept 1910 – 1 – fr ACRPP [073]

Bulletin / Florida Geological Survey – Tallahassee, FL. n1-11. 1908-1933 – 2r – us UF Libraries [500]

Bulletin / Florida Geological Survey – Tallahassee, FL. n54-58. 1972 – 1r – us UF Libraries [500]

Bulletin / Florida Geological Survey – Tallahassee, FL. n59-63. 1988-1991 – 1r – us UF Libraries [500]

Bulletin / France. Agence Generale des Colonies – v. 1-27. 1908-34. (v. 16, no. 192; v. 17, no. 200, 201 wanting) – 1 – us L of C Photodup [944]

Bulletin / France. Conseil Economique – 1948-52 – 1 – fr ACRPP [330]

Bulletin / France. Convention Nationale – Paris. 21 sept 1794-28 aout 1795 – 1 – fr ACRPP [944]

Bulletin / France. L'Assemblee nationale – Paris. juil 1789-janv 1790 – 1 – fr ACRPP [324]

Bulletin / Gallia Co. Gallipolis – (1868-74), 75-nov 77, 1880-94 [wkly] – 8r – 1 – mf#B6117-6124 – us Ohio Hist [071]

Bulletin / Gallia Co. Gallipolis – 1895-99, 11/01-11/18, 19-jan 1920 [wkly] – 9r – 1 – mf#B10948-10956 – us Ohio Hist [071]

Bulletin / Groupe du Bas-Languedoc de l'Association Sully – Montpellier, Nimes. dec 1933-aout 1939, 1943-juin 1944 – 1 – (devenu: sully) – fr ACRPP [073]

Bulletin / Hancock Co. VanBuren – feb 1945-nov 1948 [mthly, biwkly] – 1r – 1 – mf#B5441 – us Ohio Hist [071]

Bulletin / Institut de recherches scientifiques au Congo – Brazzaville, 1962-1963 – us CRL [073]

Bulletin / L'Institut Francais d'Afrique Noire – Paris. 1939-65 – 1 – fr ACRPP [073]

Bulletin / L'Institut Francais d'Afrique Noire – T. 1-15. Jan 1939-Oct 1953, and Bulletin. Series A, Sciences naturelles. T. 16-27. 1954-65 – 1 – us L of C Photodup [960]

Bulletin / Institut General Psychologique. Paris – v. 1-33. 1900-33 – 1 – 93.00 – us L of C Photodup [150]

Bulletin / Inter-African Labour Institute – London: Commission for Technical Co-operation in Africa South of the Sahara, 1960-v12 n2. may 1965 (qrtly) [mf ed aug 1953-mar 1955 filmed [19–]] – 1r – 1 – (in english and french) – mf#Sc 331.096-I – us NY Public [331]

Bulletin / International Association of Jewish Lawyers and Jurists – Tel Aviv, Israel 1978-81 – 1,5,9 – mf#11705 – us UMI ProQuest [340]

Bulletin / International Typographical Union – 1956 jan-1959 jun, 1959 jul-1962 jan, 1962 feb-1964 jul, 1964 aug-1966 nov, 1966 dec-1968 dec, 1969 jan-1970 aug, 1970 sep-1973 mar, 1973 apr 11-1975 oct, 1976-79 – 9r – 1 – us WHS [680]

Bulletin / Jewish Telegraphic Agency – New York. N.Y. 1956-67 – 1 – us AJPC [073]

Bulletin / Komite Olympiade Indonesia – Djakarta, 1953, v1(1-3); 1954, v2(1-12); 1955, v3(1-6/7) – 17mf – 9 – mf#SE-1753 – ne IDC [959]

Bulletin / L'Ecole francaise d'Extreme-Orient – Hanoi. 1901-08, 1910-11, 1914-18 – 1 – fr ACRPP [930]

Bulletin / Library Association. China – Peking. v. 1-14, no. 5. June 1925-Mar Apr 1940 – 1 – us NY Public [020]

Bulletin / Ligue de defense des libertes publiques (Haiti) – [Port-au-Prince]: la Ligue. n1. 29 dec 1953 – us CRL [073]

Bulletin / L'Institut francais de sociologie – Paris. 1931-33 (1-3) – 1 – fr ACRPP [073]

Bulletin / Literary and Historical Society of Quebec – [S.l: The Society?, 1900-1904] [mf ed n1 apr 14 1900-n2 victoria day, 1904] – 9 – mf#P05116 – cn Canadiana [410]

Bulletin / Madjelis Ilmu Pengetahuan Indonesia – Djakarta, 1960-1969. v1-12 – 31mf – 9 – (missing: 1960 v1) – mf#SE-493 – ne IDC [959]

Bulletin / Maine Archaeological Society – 1973-80 – 1r – 1 – (cont: maine archaeological society: [bulletin]) – us WHS [930]
Bulletin – Milwaukee WI. 1971 apr 11-may 23 – 1r – 1 – mf#1166108 – us WHS [071]
Bulletin / Montgomery Co. Miamisburg – jan 1868-jan 1869,jul 1870-aug 1872 [wkly] – 1r – 1 – mf#B5453 – us Ohio Hist [071]
Bulletin – New Orleans LA. 1884 sep 6-oct 11,18, nov 1, 5,12-19,26 – 1r – 1 – mf#861288 – us WHS [071]
Bulletin / Societe de l'Histoire de France – puis Annuaire-bulletin. Paris. 1834-69 – 1 – fr ACRPP [944]
Bulletin / Societe de l'histoire du Protestantisme Francais. Paris – v1-52, 94-98. 1852-1902, 1947-51. Lacking Oct-Dec 1948 and Jan-Jun 1951 – 1 – $368.00 – us L of C Photodup [240]
Bulletin / Societe des Amis de Georges Bernanos – Paris. n1-58. dec 1949-janv 1966 – 1 – (mq n31, 49) – fr ACRPP [073]
Bulletin / Societe des compositeurs de musique – Paris. 1863-70 – 1 – fr ACRPP [780]
Bulletin / Societe des Etudes Indochinoises de Saigon – 1883-1923 – 1 – 69.00 – us L of C Photodup [959]
Bulletin / Societe des Etudes Indochinoises de Saigon – Saigon. 1926-59 – 1 – fr ACRPP [490]
Bulletin / Societe des Gens de Lettres – Paris. 1847 (t. 3) – 1 – fr ACRPP [800]
Bulletin / Societe des sciences historiques et naturelles de la Corse. Bastia – v(1)-46/49. 1881-1929 – 1 – $119.00 – us L of C Photodup [944]
Bulletin / Societe Ethnologique de Paris – Paris. 1846-47 – 1 – fr ACRPP [306]
Bulletin / Societe Historique du VIe Arondissement de Paris – Paris. v1-30 1898-1929 – 1 – us NY Public [944]
Bulletin / Societe litteraire des Amis d'Emile Zola – no. 1-24. Paris. 1922-38 – 1 – fr ACRPP [830]
Bulletin / Societe Paul Claudel – no. 1-32, no. special. Paris. 1958-68 – 1 – fr ACRPP [810]
Bulletin / Society for Applied Spectroscopy – Bound Brook, N.J. v1-6 1946-Nov 1952 – 1 – us NY Public [621]
Bulletin / Summit Co. Twinsburg – (jul 1959-dec 1991) [wkly] – 35r – 1 – mf#B32034-32068 – us Ohio Hist [071]
Bulletin / Transport Workers Union of America – 1943 apr-dec, 1944-48 – 2r – 1 – (cont: transport bulletin; cont by: twu express) – us WHS [380]
Bulletin / U.S. Treasury Dept. Bureau of Customs – v1-20. 1966-86.Continues: Treasury Decisions.78-234B – 9 – $132.00 – us LLMC [336]
Bulletin / Wisconsin Conference of Social Work – 1940 mar-1941 oct – 1r – 1 – (cont: news from the wisconsin conference of social work; cont by: wisconsin welfare [madison wi: 1941]) – us WHS [360]
Bulletin / Yerkes Observatory of the University of Chicago – Chicago: The University of Chicago Press 1896-1903 (irreg) [n1-19 (1896-1903)] [mf ed 1999] – 1r – 1 – (n3 publ as v6 n3 of the: astrophysical journal, but was not iss separately in the series: bulletin; n1-19 by george e hale; n5 contains: proceedings of the conferences held at the yerkes observatory oct 18-21 1897) – mf#film mas c4166 – us Harvard [520]
Bulletin *see*
– Correspondance
– Gold hill nugget
Le bulletin – St-Jerome [Quebec]: J J Grignon, 1892 – 9 – ISSN: 1181-215X – mf#P04167 – cn Canadiana [380]
The bulletin – 1880- – 4r per y – 1 – us UMI ProQuest [072]
The bulletin : a journal devoted to the mineral industry of british columbia – Rossland, BC: Collins, [1900-19–] – 9 – mf#P04061 – cn Canadiana [622]
The bulletin – Sydney, Australia. 2 April-29 Oct 1941; 25 Feb 1942-18 Nov 1953 – 23r – 1 – uk British Libr Newspaper [980]
The bulletin / Taylor Chapel A M E Church (Bowling Green KY) – Bowling Green KY: Taylor Chapel [wkly] [mf ed 2004] – v1 n2-31 (jan 13-nov 2 1924); ns: v1 n2-32 (jan 9-aug 7 1949) on 1r – 1 – (lacks v1 n7,16,21,26; [ns]: v1 n1,3,18,20,22; damaged: v1 (1924) n12 p2; "men's day program" was printed in place of regular bulletin for jul 3 1949, "ladies' day" program was printed for jul 31 1949; iss for 1949 begin new series; cont by: taylor chapel a m e church (bowling green, ky). monthly bulletin [2004-s070] filmed on same reel) – mf#2004-s069 – us ATLA [242]
The bulletin – Toronto: Bulletin Pub Co, [1893-1950] – 9 – mf#P05132 – cn Canadiana [360]
The bulletin *see* The western pilot
Bulletin: a monthly newsletter / Confederated Tribes of the Colville Reservation – v3 n4 [1981 jul] – 1r – 1 – mf#639797 – us WHS [305]
Bulletin administratif / Cambodia – 1921-23, 1927 – 1 – fr ACRPP [959]
Bulletin administratif / Congo. Belgian – Leopoldville. Pt1: 1958-59, Pt2: 1958-59 – 1 – us NY Public [960]
Bulletin administratif / Indochina. French – 1921 – 1 – fr ACRPP [073]
Bulletin administratif / Laos – 1919, 1921-22, 1927 – 1 – fr ACRPP [959]
Bulletin administratif du tonkin – Hanoi: [s.n.] v5 [1906], v10 [1911], v14-15 [1915-16], v33 n1-6 [jan-mar 1934], v39 n7-12 [apr-jun 1940] [mf ed Hanoi, Vietnam: National Library of Vietnam 1998] – 6r – 1 – (master neg held by crl) – mf#mf-11750 seam – us CRL [350]
Bulletin administratif et commercial = Bestuurs-en handelsblad van belgisch-congo feb 10 1915 / Belgian Congo. Secretariat general – Boma, Belgian Congo: Secretariat General 1912-27 [semimthly] – 1 – (cont by: belgian congo. bulletin administratif / congo belge = bestuurlijk blad / belgisch-conge; some nos accompanied also by suppls, called "supplements," and/or also by "annexes") – mf#1152 – us Wisconsin U Libr [960]
Bulletin administratif et judiciaire des annales forestieres – Paris: Bureau des Annales Forestieeres 1844- [t1-4 (1842/43-1848/49)] [mf ed 2005] – 1r – 1 – mf#film mas 99999 – us Harvard [634]
Bulletin analytique de linguistique francaise – Paris, France 1988 – 1,5,9 – ISSN: 0007-408X – mf#16566 – us UMI ProQuest [440]
Bulletin and aurora news / Summit Co. Twinsburg – (may 12 1956-may 21 1959) [semiwkly, wkly] – 1r – 1 – mf#B31710 – us Ohio Hist [071]
Bulletin annexe au journal officiel / France – 1905-12 – 1 – fr ACRPP [323]
Bulletin apostle of the chippewas : quarterly bulletin for the cause of bishop frederick baraga / Bishop Baraga Association – v1 n1-v10 n4 [1946 jul-1956 mar] – 1r – 1 – (cont by: baraga bulletin) – mf#614804 – us WHS [241]
Bulletin archeologique du comite des travaux historiques et scientifiques – Paris. 1955 56-1963 64 – 1 – fr ACRPP [930]
Bulletin argus / Corporation des bibliothecaires professionnels du Quebec – Montreal: la Corporation. n73 nov/dec 1985-n83 nov/dec 1987 [mf ed 1987-1989] – 5 – (cont: argus journal; cont by: corpo clip) – mf#SEM16P365 – cn Bibl Nat [020]
Bulletin astronomique / Universite de Besancon. Observatoire astronomique, chronometrique et meteorologique – Besancon: P Jacquin [1900] (annual) [mf ed 2000] – v1-11(1886-96) on 1r – 1 – mf#film mas c4653 – us Harvard [520]
Bulletin badan kerdja sama tani-militer – Djakarta, 1960 – 2mf – 9 – mf#SE-585 – ne IDC [959]
Bulletin – bank of finland – Helsinki, Finland 1975+ – 1,5,9 – ISSN: 0784-6509 – mf#10648 – us UMI ProQuest [332]
Bulletin [bend, or] – Bend OR: Bend Bulletin Inc, 1963- [daily ex sat] – 1 – (cont: bend bulletin [1917]) – us Oregon Lib [071]
Bulletin bibliographique / Association francaise des observateurs d'etoiles variables – [Saint-Geris-Laval [Rhone]]: Assoc francaise d'observateurs d'etoiles variables [1932]- [mf ed 2003] – 1r – 1 – (link: bulletin de l'association francaise des observateurs d'etoiles variables [issn 0153-9949]) – mf#film mas c5591 – us Harvard [520]
Bulletin board – Molalla OR: Molalla Printing & Graphics, [wkly] – 1 – (began in 1972; ceased in 1972? cont by: bulletin [molalla or]) – us Oregon Lib [071]
Bulletin board / Dane County Childcare Union, District 65, UAW – v1 n1-5 [1985 jun-1986 mar] – 1r – 1 – (cont by: union bond) – mf#1898061 – us WHS [360]
Le bulletin (buckingham, quebec) : la revue de la lievre – Buckingham: [s.n.] v5 n27 12 mai 1986- (wkly) [mf ed 1988-] – 1 – mf#SEM35P283 – cn Bibl Nat [071]
Bulletin bulanan industri minjak dan gas bumi indonesia / Direktorat Djenderal Minjak dan Gas Bumi – Djakarta, 1965-1969 – 3mf – 9 – (missing: 1965-1969(jan-mar)) – mf#SE-1369 – ne IDC [959]
Bulletin canadian sociology and anthropology association / Canadian Sociology and Anthropology Association – Montreal. 1976-1976 (1) 1976-1976 (5) 1976-1976 (9) – ISSN: 0008-5049 – mf#7260 – us UMI ProQuest [301]
Bulletin (carothers observatory) / Carothers Observatory; ed by Carothers, Warren Fay – Houston TX: Carothers Observatory [1913]- [mf ed 2007] – 1r – 1 – mf#film mas 37496 – us Harvard [520]
Bulletin (chicago edition) / Mahoning Co. Youngstown – v1 n1. nov 1933-dec 1967 [mthly] – 7r – 1 – mf#B11807-11813 – us Ohio Hist [331]
Bulletin (chicago edition) / Trumbull Co. Youngstown – v1 n1. nov 1933-dec 1967 [mthly] – 7r – 1 – mf#B11807-11813 – us Ohio Hist [331]
Bulletin chronometrique / Universite de Besancon. Observatoire astronomique, chronometrique et meteorologique; ed by Gruey, Louis Jules & Lebeuf, A – Besancon: Impr Millot Freres 1889-1925 (irreg) [mf ed 2000] – v1-26/29 (1889-1913/1924) on 2r [ill] – 1 – (cont by: annales francaises de chronometrie; some nos iss in 2pt: pt1: chronometrie; pt2: memoires; imprint varies) – mf#film mas c4665 – us Harvard [520]
Bulletin colonial / Communist Party. France – Paris. oct 1933-juin 1936 – 1 – fr ACRPP [335]
Bulletin commercial agricole – Paris. 27 janv-11 juil 1848 – 1 – fr ACRPP [630]
Bulletin communiste / Comite de la Troisieme Internationale – Paris. mars 1920-30, 1933 – 1 – fr ACRPP [335]
Bulletin communiste : organe du comite de la troisieme internationale – Paris. v1 n1-v14 n32-33. mar 1920-jul 1933 – (= ser Communist international periodicals from the feltrinelli archives) – 60mf – 9 – $250.00 – us UPA [335]
Bulletin communiste – Paris. v. 1-5. June 10 1920-Nov 14 1924. Incomplete – 1 – us NY Public [335]
Bulletin: connecticut labor department / Connecticut. Employment Security Division – v.1, n.1, Jan. 1936-78. 94 fiches. (Harvard Law School Library Collection.) – 9 – us Harvard Law [336]
Bulletin – council for research in music education / Council for Research in Music Education – Urbana. 1963+ (1) 1971+ (5) 1976+ (9) – ISSN: 0010-9894 – mf#3236 – us UMI ProQuest [780]
Le bulletin de buckingham – Buckingham: [s.n.] v1 n1 24 juil 1958-v22 n9 2 oct 1979 (wkly) [mf ed 1987-91] – 10r – 1 – mf#SEM35P282 – cn Bibl Nat [073]
Bulletin de correspondance africaine – Alger. 1882, 1884-86 (I, III-V) – 1 – fr ACRPP [960]
Bulletin de correspondance hellenique – v1-89. 1877-1965 – 1 – $1110.00 – (v90-106 1966-82 $336 [0121]) – mf#0119; 0120 – us Brook [450]
Bulletin de la bibliotheque nationale – [Montreal] v1 n3 sep 1967-v6 n1 nov 1972 (irreg) [mf ed 1977] – 1r – 5 – (cont: bulletin de la bibliotheque saint-sulpice; cont by: bulletin de la bibliotheque nationale du quebec 0045-1967) – mf#SEM16P296 – cn Bibl Nat [020]
Bulletin de la bibliotheque nationale du quebec – Montreal. v7 n1 mars 1973-v17 special no dec 1983 [qrtly] [mf ed 1977-1984] – 1r – 5 – (cont by: incunable) – mf#SEM16P296 – cn Bibl Nat [020]
Bulletin de la bouche de fer *see* La bouche de fer
Bulletin de la caisse nationale d'economie – Montreal: la Caisse: Association Saint-Jean-Baptiste de Montreal. v1 n1 juin 1904-v10 n9 sep 1913 (mthly) [mf ed 2001] – 1 – cn Bibl Nat [360]
Bulletin de la chambre de commerce francaise de grande-bretagne *see* Communique de la chambre de commerce francaise de londres
Bulletin de la classe historico-philologique de lacademie imperiale des sciences de saint-petersbourg – Spb., Leipzig, 1844-1859. S 2. v1-16. Tabs – 127mf – 9 – mf#R-1702 – ne IDC [077]
Bulletin de la classe physico-mathematique de lacademie imperiale des sciences de st.-petersbourg – Spb., Leipzig, 1843-1856. v1-14 – 141mf – 9 – mf#R-5820 – ne IDC [077]
Bulletin de la commission archeologique de l'indochine / France. Commission archeologique de l'Indo-chine – Paris: Impr nationale 1908- [mf ed New Haven CT: SEAsia Coll, Yale Uni Library 1991] – 1r – 1 – (at head of title: ministere de l'instruction publique et des beaux-arts; "bibliographie raisonnee des travaux relatifs a l'archeologie du cambodge et du champa, par m g coedes": annee 1909) – us CRL [930]
Bulletin de la commission de toponymie et dialectologie – Liege. v1-3. 1927-1929 – 10mf – 9 – mf#H-10026 – ne IDC [400]
Bulletin de la commission internationale penale et penitentiaire / International Penal and Prison Commission – v. 1-21. 1880-1930. Publication suspended from 1906-09 and 1911-24 – 1 – us L of C Photodup [360]
Bulletin de la commune de port-au-prince / Port-Au-Prince (Haiti) – Port-Au-Prince, Haiti. 1932 – 1r – us UF Libraries [972]
Bulletin de la federation des societes de gynecologie... / Federation des societes de gynecologie et d'obstetrique de langue francaise – Paris, France 1969-71 – 1 – ISSN: 0046-3515 – mf#3429 – us UMI ProQuest [618]
Bulletin de la federation jurassienne de l'association internationale des travailleurs / L'Association internationale des Travailleurs. La Federation jurassienne – Sonvillier, Locle, La Chaux-de-Fonds. v1 n1-v7 n12. feb 1872-mar 1878 – (= ser Important periodicals of italian and international socialism, 1868-1917) – 1r – 1 – $150.00 – us UPA [335]
Le bulletin de la ferme – Quebec. v1 n1 sep 1913-v24 n39 24 sep 1936 (mthly) [mf ed 1982] – 12r – 1 – mf#SEM35P182 – cn Bibl Nat [630]
Bulletin de la milice canadienne – Quebec: A l'Impr canadienne, [1807-18–] – 9 – (ceased 180-?) – mf#P05007 – cn Canadiana [355]
Bulletin de la semaine politique, sociale et religieuse – Paris. 1905-aout 1914 – 1 – fr ACRPP [944]
Bulletin de la societ imperiale des naturalistes de moscou – Moscow 1829-1886(-87) v1-62; n.s. 1887-1917 v1-30 – 1972mf – 9 – (cont as: moskovskoe obshchestvo ispytatelei prirody. biulletene. otdel biologicheskii. n.s., 1922/1923-1968, v31-73) – mf#7118/2 – ne IDC [580]
Bulletin de la societe bienveillante st-roch – Quebec: La Societe, [1893-1903?] – 9 – ISSN: 1190-7819 – mf#P04135 – cn Canadiana [366]
Bulletin de la societe centrale forestiere de belgique / Societe centrale forestiere de Belgique – Bruxelles: Impr Vanbuggenhoudt 1893- (mthly) [9-11 annee sociale (1902-04)] [mf ed 2005] – 2r [ill] – 1 – (cont by: bulletin de la societe royale forestiere de belgique; ceased with dec 1950; publ suspended aug 1914-dec 1918, jun 1940-nov 1944) – mf#film mas c6035 – us Harvard [634]
Bulletin de la societe chimique de france / Societe chimique de France – Lausanne MA 1991-97 – 1,5,9 – ISSN: 0037-8968 – mf#42664 – us UMI ProQuest [540]
Bulletin de la societe de l'histoire du theatre – Geneve, 1902-22 – 1 – $60.00 – mf#0123 – us Brook [790]
Bulletin de la societe des anciens textes francais / Societe des anciens textes francais (Paris, France) – Paris: Librairie Firmin-Didot. v1-40 [1875-1914], v47-62 [1921-36] [mf ed 1985] – 2r – 1 – (v41-46 1915-20 never publ; v51-52, 53-55, 56-59, 60-62 all iss combined) – mf#6724 – us Wisconsin U Libr [440]
Bulletin de la societe des artisans canadiens-francais de la cite de montreal : organe des societes canadiennes de secours mutuels – Montreal: La Societe, [1891-19–] – 9 – (ceased 1899?; cont by: artisan) – mf#P04054 – cn Canadiana [740]
Bulletin de la societe des professeurs de langues vivantes : langues modernes – Paris, 1903-1946 – 370mf – 8 – (missing: 1941-1944) – mf#H-444c – ne IDC [410]
Bulletin de la societe des recherches congolaises – Brazzaville: Impr du gouvernement general. n1-27. 1922-39 – 1 – us CRL [960]
Bulletin de la societe d'etudes camerounaises – Doula: Institut francais d'afrique noire, Centre local Cameroun. n1-19/20. dec 1935-sep/dec 1947 – 1 – us CRL [960]
Bulletin de la societe d'histoire et de geographie d'haiti – [Port-au-Prince]: Imp Cheraquit. v1 n1. may 1925 – r1 – us CRL [073]
Bulletin de la societe d'histoire vaudoise – Pignerol: Chiantore & Mascarelli, 1884-1933 [semiannual] [mf ed 2003] – 60v on 4r – 1 – (iss 6,15,31 & 57-58 lack collective title. in french & italian) – mf#2003-s015 – us ATLA [242]
Bulletin de la societe d'industrie laitiere de la province de quebec = Bulletin of the dairymen's association of the province of quebec – [St-Hyacinthe, Quebec]: La Societe, [1891?]- – 9 – (ceased 189-?) – mf#P04004 – cn Canadiana [630]
Bulletin de la societe entomologique de france – Paris, 1898-1940 – 9 – $1467.00 – mf#0122 – us Brook [590]
Bulletin de la societe forestiere – [Paris]: Au Secretariat de la Societe forestiere [1859?]- (irreg) [annee 5 n16/17-anee 7 n24 (1859-61)] [mf ed 2005] – 1r [ill] – 1 – (cont: bulletin trimestriel de la societe forestiere) – mf#film mas 99999 – us Harvard [634]
Bulletin de la societe francaise de dermatologie et de syphiligraphie – Paris, France 1890-1972 – 1,5 – ISSN: 0049-1071 – mf#3407 – us UMI ProQuest [616]
Bulletin de la societe francaise de philosophie / Societe Francaise de Philosophie – Paris. 1901-71 – 5 – fr ACRPP [100]

Bulletin de la societe francaise de statistique universelle : compte rendu des seances, rapports et arretes de la societe et de son conseil et des articles de melanges / Societe Francaise de Statistique Universelle – Paris 1831-32 [mf ed Hildesheim 1995-98] – 2v on 5mf – 9 – €100.00 – 3-487-29955-0 – gw Olms [910]

Bulletin de la societe historique et archeologique du perigord / Societe historique et archeologique du Perigord – Perigueux. 1950-1954 (1) – ISSN: 0037-9425 – mf#603 – us UMI ProQuest [930]

Bulletin de la societe lepidopterologique – Geneve 1905/1909-1918 – v1-4 on 22mf – 9 – mf#z-928/2 – ne IDC [590]

Bulletin de la societe malacologique de france – Paris 1884-90 – v1-7 on 51mf – 9 – mf#z-893/2 – ne IDC [590]

Bulletin de la societe ornithologique suisse – Geneve, Bale 1865-70 – v1-2 on 10mf – 9 – mf#z-2015/2 – ne IDC [590]

Bulletin de la societe zoologique de france – Paris 1876-1920 – v1-45 on 286mf – 9 – mf#6234/2 – ne IDC [590]

Bulletin de l'abq = Qla newsletter / Association des bibliothecaires du Quebec & Quebec Library Association – Montreal. v13 n2 nov 1971-(bimthly) [mf ed 1985-89] – 1r – 5 – (cont: bulletin de nouvelles) – mf#SEM16P350 – cn Bibl Nat [020]

Bulletin de l'abq = QLA newsletter / Association des bibliothecaires du Quebec & Quebec Library Association – Montreal, 1971- (bimthly) [mf ed 1992-] – (= ser QLA newsletter) – 1r – 5 – mf#SEM105P1687 – cn Bibl Nat [020]

Bulletin de lacadamie imperiale des sciences de st-petersbourg – Spb. 1860-88 – v1-32 on 332mf – 9 – mf#r-5821/2 – ne IDC [580]

Bulletin de lacademie imperiale des sciences de st-petersbourg – n.s. Spb. 1890-94 – v1-4 on 81mf – 9 – mf#r-1701/2 – ne IDC [580]

Bulletin de l'agence de presse libre du quebec / Agence de presse libre du Quebec – Montreal. n1 18/25 mars 1971-n118 28 juin/4 juil 1973 (wkly) [mf ed 1976] – 2r – 1 – (cont by: bulletin populaire) – mf#SEM35P114 – cn Bibl Nat [070]

Le bulletin de l'amicale *see* La vie ecoliere

Bulletin de l'amicale des polices de l'indochine [api] – Hanoi: Impr Mac-Dinh-Tu [1928-] n1-35 [1928-35] – 1r – 1 – mf#mf-12647 seam – us CRL [360]

Le bulletin de l'art ancien et moderne – Suppl. de: La Revue de l'art ancien et moderne. no. 1-819. Paris. 1899-1935 – 1 – fr ACRPP [700]

Bulletin de l'association des bibliothecaires du quebec / Association des bibliothecaires du Quebec – Montreal: l'association. n1 mar 1939-n34 1950/1952? (biannual) [mf ed 1985] – (= ser The QLA bulletin) – 1r – 5 – cn Bibl Nat [020]

Bulletin de l'association francaise des observateurs d'etoiles variables / Association francaise des observateurs d'etoiles variables – [Saint-Geris-Laval [Rhone]]: Assoc francaise d'observateurs d'etoiles variables [1932]- [mf ed 2003] – 2r – 1 – (cont: bulletin de l'observatoire de lyon; cont in pt by: bibliographie mensuelle de l'astronomie 1933; link: bulletin bibliographique [association francaise des observateurs d'etoiles variables]) – ISSN: 0153-9949 – mf#film mas c5590 – us Harvard [520]

Bulletin de l'ecole polytechnique de montreal / Ecole polytechnique [Montreal, Quebec] – Montreal. v1 n1 janv 1913-v2 n8 aout 1914 [mthly] [mf ed 1985] – 1r – 5 – (cont by: la revue trimestrielle canadienne) – mf#SEM16P356 – cn Bibl Nat [378]

Bulletin de legislation et du jurisprudence egyptienne – Alexandria. On film: v1-8; 1941-49. (incomplete). LL-020 – 1 – us L of C Photodup [340]

Bulletin de l'herbier boissier. *see* Memoires de l'herbier boissier

Bulletin de liaison saharienne – no. 1-32. Alger. oct 1950-58 – 1 – fr ACRPP [960]

Bulletin de l'institut botanique de buitenzorg – Buitenzorg, 1898-1905. v1-22 – 188mf – 9 – (cont as: bulletin du departement de l'agriculture aux ind neerlandaises buitenzorg, 1906-1911 v1-47; bulletin du jardin botanique de buitenzorg buitenzorg, s2, 1911-1918, v1-28; s3, 1918-1949/1950, v1-18+suppl v1-3) – mf#7715 – ne IDC [580]

Bulletin de l'institut francais d'archeologie orientale – Le Caire. v1-41. 1901-1942 – 174mf – 9 – mf#NE-490 – ne IDC [930]

Bulletin de l'institut pasteur / Institut Pasteur, Paris, France – Paris. 1968+ (1) 1971+ (5) 1971+ (9) – ISSN: 0020-2452 – mf#5104 – us UMI ProQuest [610]

Bulletin de l'observatoire astronomique de beograd / Astronomska opservatorija u Beogradu – Beograd: Impr nationale du Royaume de Yougoslavie 1936-91 [irreg] [mf ed 2006] – 3r – 1 – (merged with: publications of the department of astronomy, to become: bulletin astronomique de belgrade; chiefly in french, some articles in english; imprint varies) – ISSN: 0373-3734 – us Harvard [520]

Bulletin de l'observatoire astronomique de beograd / Astronomska opservatorija u Beogradu – Beograd: Impr nationale du Royaume de Yougoslavie 1936-91 [mf ed 2006] – (= ser Publications de l'observatoire astronomique de l'universite de belgrade 1936-1942; Publications de l'observatoire astronomique de belgrade 1949-1950) – 1r – 1 – (merged with: publications of the department of astronomy, to become: bulletin astronomique de belgrade; iss for 1936-41 called also 1 n1-[6], n3-12; iss for 1949-1950 called also 14-15; 1936-1982 chiefly in french, some articles in english; 1983-1988 english only; summaries in english & serbo-croatian (cyrillic), -n144; imprint varies) – ISSN: 0373-3734 – mf#film mas 37493 – us Harvard [520]

Bulletin de l'observatoire de lyon / Observatoire de Lyon [France] – Lyon: H. Georg, [1913-1931] (mthly) [mf ed 2000] – n1-t13(1913-31) [gaps?] on 5r – 1 – (1921 incomplete? cont by: bulletin de l'association francaise des observateurs d'etoiles variables; some iss have suppls, a & b. sect b called: bibliographie rapide de l'astronomie et de la physique du globe; imprint varies) – mf#film mas c4662 – us Harvard [520]

Bulletin de l'opposition bolcheviks-leninistes – no. 1 2-72. Paris. juil 1929-38. mq no. 20, 33, 40-41 – 1 – fr ACRPP [335]

Bulletin de l'union coloniale francaise – Paris: L'Union. v1. nov 1894-1895 – us CRL [320]

Bulletin de madagascar – [Tananarive: Haut Commissariat de la Republique francaise a Madagascar et dependances, Service general de l'information. n89 oct 1953; n120-121 may-jun 1956; n131-295 apr 1957-70 – 13r – 1 – (lacks: n136 sep 1957. n145-151 jun-dec 1958. n162 nov 1959. n254-255 jul-aug 1967. n258 nov 1967. n260-261 jan-feb 1968. n266-267 jul-aug 1968) – us CRL [960]

Bulletin de nouvelles = News bulletin / Corporation des bibliothecaires professionnels du Quebec – Montreal: la Corporation. n14 19 juin 1969-n24 sep/oct 1971 [mf ed 1977] – 1r – 5 – (in french & english; cont: bulletin special aux bibliothecaires; cont by: argus) – mf#SEM16P290 – cn Bibl Nat [020]

Bulletin de nouvelles = Newsletter / Association des bibliothecaires du Quebec & Quebec Library Association – Montreal. v1 n1 aut 1954-v13 n1 janv./mars 1971 (irreg) [mf ed 1985] – 1r – 5 – (cont: qla bulletin; cont by: bulletin de l'abq) – mf#SEM16P349 – cn Bibl Nat [020]

Bulletin de nouvelles anti-repression : organe du comite pour les droits democratiques du peuple ou cddp – Montreal: [s.n.] v1 n1-(irreg) [mf ed 1972] – 1r – 1 – (ceased 12 janv 1970?) – mf#SEM35P14 – cn Bibl Nat [360]

Bulletin de nouvelles d'afrique – [oct-dec 1960] – us CRL [073]

Bulletin de nouvelles du congo – [oct-dec 1960] – us CRL [073]

Bulletin de paris – Paris. 1955-57 – 1 – fr ACRPP [944]

Bulletin de police criminelle... / Police de l'Indochine, Service de la surete du Tonkin, Section judiciaire – Hanoi: Impr Tan-Dan, n159-260 [1940-41] – 2r – 1 – mf#mf-12644 seam – us CRL [364]

Bulletin de presse / Inforcongo. 1ere Direction. Presse et relations publiques – [Brussels]: Inforcongo, sep 10 1956-jan 21 1957; feb 1957-dec 22 1958; jan 25-sep 19 1960 – 1r – us CRL [070]

Bulletin de ste anne de la pointe-au-pere – Rimouski: P Sylvain, [1882-1883] – 9 – (cont by: messager de sainte anne) – mf#P04850 – cn Canadiana [241]

Bulletin de ste anne de la pointe-au-pere *see* Le messager de sainte anne

Bulletin de theologie ancienne et medievale – 1(1929)-4(1945) – 45mf – 9 – €86.00 – ne Slangenburg [240]

Bulletin de theologie ancienne et medievale – Louvain, 1929-1943. v1-4 – 42mf – 8 – (missing: 1937-1940 v3(p189-360)) – mf#H-309c – ne IDC [240]

Bulletin departemen penerangan, humas pimpinan pusat partai muslimin indonesia / Partai Muslimin Indonesia – Djakarta, 1968-1970 – 6mf – 9 – (missing: 1968, v1(1-4); 1969(feb-sep); 1970(feb-may)) – mf#SE-1867 – ne IDC [959]

Bulletin des 3. kongresses der kommunistischen internationale – Moscow. n1-24, 24 jun-20 jul 1921 – (= ser Communist international periodicals from the feltrinelli archives) – 6mf – 9 – $95.00 – us UPA [335]

Bulletin des 4. kongresses der kommunistischen internationale – Moscow. n1-31, 11 nov-12 dec 1922 – (= ser Communist international periodicals from the feltrinelli archives) – 8mf – 9 – $95.00 – us UPA [335]

Bulletin des actes administratifs de la prefecture de la guyane / British Guiana – Cayene. 1957-July 4, 1966. Incomplete – 1 – us NY Public [324]

Bulletin des amis de la verite – no. 1-121. Paris. janv-avr 1793 – 1 – fr ACRPP [073]

Bulletin des amis du laos – Hanoi. juil 1937-aout 1940 – 1 – fr ACRPP [959]

Bulletin des amis du vieux hue – Hanoi, Haiphong. 1914-avr juin 1944 (1-31) – 1 – fr ACRPP [959]

Bulletin des annonces legales obligatoires (b.a.l.o.) / France – 1912-1993 – 1 – fr ACRPP [323]

Bulletin des annonces legales obligatoires (balo) – 1974- – 9 – €99.09y – fr Journal Officiel [340]

Bulletin des arrets / France. Cour de Cassation. Chambre criminelle – Paris. 1798-1816, 1818-19, 1821-40, 1943-86, tb: 1798-1986 – 1 – fr ACRPP [960]

Bulletin des arrets / France. Cour de Cassation. Chambres civiles – Paris. 1798-1942, 1947-86, tb: 1958-86 – 1 fr ACRPP [960]

Bulletin des commissions royales d'art et d'archeologie / Belgium. Ministere de l'Interieur – Bruxelles, 1921-42 – 9 – $300.00 – (in french) – mf#0098 – us Brook [930]

Bulletin des commissions royales d'art et d'archeologie – Brussels, 1862-93 – (= ser Architectural periodicals at avery library, columbia university) – 10r – 1 – $1330.00 – us UPA [720]

Bulletin des constructeurs – Paris, 1895-99, 1904-05 – (= ser Architectural periodicals at avery library, columbia university) – 3r – 1 – $395.00 – us UPA [690]

Bulletin des ecrivains proletaires – no. 2, 4. Vincennes. avr-juin 1932 – 1 – fr ACRPP [335]

Bulletin des fabricants de papier : organe pratique & revue technique universelle de la papeterie & des industries qui s'y rattachent – Paris. 15 jul 1890-jul 1914 – 11 1/2r – 1 – uk British Libr Newspaper [670]

Bulletin des freres et amis : ou l'echo de l'opinion. – n1-19. Paris. juil 1797 – 1 – fr ACRPP [073]

Bulletin des groupes plans *see* Plans

La bulletin des halles – Paris, France 14 jun-9 sep 1882 (imperfect) – 1 – (cont by: bulletin des halles, bourses et marches [1 jun 1917-27 mar 1918]) – uk British Libr Newspaper [074]

Bulletin des lois, dea1crets et documents officiels / Exposition Universelle 1900. Paris – Paris. v2-7 Dec 5 1895-Nov 16 1900. Incomplete – 1 – us NY Public [900]

Bulletin des metiers d'art *see* Art and decoration

Bulletin des musees – Paris, 1890-93 – (= ser Architectural periodicals at avery library, columbia university) – 1r – 1 – $155.00 – us UPA [060]

Bulletin des nouvelles – Cap-Haitien: Imp du Progres – (sheets 1-3 dec 1900-jan 16,19-22, 24-feb 1,4 16-26,28 1901; sheet 4 mar 9 1901 filmed out of sequence; sheets 5-6 mar 11,16-19,29, apr 13-15,18,23 1901; sheet 5 may 2-7 1901 filmed out of sequence) – us CRL [073]

Bulletin des presse- und informationsamtes der bundesregierung – Bonn DE, 1951 27 oct-1956 – 5r – 1 – mf#6458 – gw Mikropress [350]

Bulletin des relations industrielles – Quebec: Departement des relations industrielles...v1 n1 15 sep 1945-v5 n10 juil 1950 (mthly) [mf ed 1982] – 1r – 5 – (cont by: relations industrielles; in french & english) – mf#SEM16P311 – cn Bibl Nat [331]

Bulletin des seances de la societe malacologique de belgique – Bruxelles 1863/1865-1875. v1-10 – 130mf – 9 – (cont as: proces-verbaux des seances de la societe malacologique de belgique [bruxelles 1871-98] v1-27) – mf#z-892/2 – ne IDC [590]

Bulletin des statistiques de la republique d'haiti / Laforest, Antoine – Port-Au-Prince, Haiti. 1913 – 1r – us UF Libraries [972]

Bulletin des usines electriques : organe du syndicat professionel des usines de l'electricite – Paris, France dec 1896-15 jan 1907 – 3r – 1 – uk British Libr Newspaper [621]

Bulletin d'etudes orientales / Institut francais de Damas – Le Caire, Paris. 1931-66 (I-XIX) – 1 – fr ACRPP [950]

Bulletin d'information de l'a.t.e.n / ATEN – Paris. 1966-73 – 5 – fr ACRPP [073]

Bulletin d'information du bureau permanent du front de liberation de mozambique a alger – Alger: Le Bureau. n1-3 jan-apr/may 1964; n4? jul 1964; n3-5 mar/may 1965; sep 1966 – us CRL [960]

Bulletin d'information ouvriere – no. 2-8. Paris. mars-juin 1940 – 1 – fr ACRPP [331]

Bulletin d'information religieuse – En cambodgien. Cambodge. 1932-34, 1936-42 – 1 – fr ACRPP [240]

Bulletin d'informations / Institut Francais d'Opinion Publique – Paris. oct 1944-53, 1955 – 1 – (devenu: sondages. revue francaise de l'opinion publique) – fr ACRPP [303]

Bulletin d'informations du secretariat permanent de la conference des organizations nationalistes des colonies portugaises (concp) – Rabat: Le Secretariat. n1-3 dec 30 1961-may 10 1962; n6 jan 1963 – us CRL [946]

Bulletin du bureau medical du college des medecins et chirurgiens de la province de quebec – Montreal: Le College, 1894-[19–?] – 9 – (ceased 189-?) – mf#P05127 – cn Canadiana [610]

Bulletin du cancer – Lausanne MA 1908+ – 1,5,9 – ISSN: 0007-4551 – mf#5704 – us UMI ProQuest [616]

Le bulletin du club cartier – Montreal: Berthiaume & Sabourin, [1880] – 9 – ISSN: 1190-7746 – mf#P04136 – cn Canadiana [320]

Bulletin du comite de vigilance des intellectuels antifascistes *see* Vigilance

Bulletin du comite d'etudes historiques et scientifiques de l'afrique occidentale francaise – Paris [etc]: Librairie Larose [etc]. v1-21. 1918-1938 – 11r – us CRL [960]

Le bulletin du commerce – Noumea, New Caledonia 3 jan-30 mar 1940 – 1/4r – 1 – uk British Libr Newspaper [380]

Bulletin du livre – Paris, France 1976-77 – 1,5,9 – ISSN: 0007-456X – mf#10233 – us UMI ProQuest [410]

Le bulletin du travail : organe des interets temporels de l'artisan et du laboureur – Quebec: [s.n.] 1re annee n4 15 dec 1900-(wkly) [mf ed 1988] – 9 – (ceased 1903?) – mf#SEM105P928 – cn Bibl Nat [331]

Bulletin du vicariat / Roman Catholic Mission Fiji – jun 1891-aug 1894 – 1r – 1 – mf#pmb doc210 – at Pacific Mss [241]

Bulletin economique / Madagascar – 1901-15, 1919-20 – 1 – fr ACRPP [330]

Bulletin ekonomi keuangan / Biro Penerangan Ekonomi (Djakarta) – Djakarta, 1967(jul)-1972 – 649mf – 9 – (missing: 1968 v2(102, 119, 202, 203); 1970 v3(241-246); 1972 v6(29, 40, 53, 60)) – mf#SE-1370 – ne IDC [330]

Bulletin elementaire franco – Phnom Penh [1925-29; 1931-39] – 6r – 1 – us CRL [079]

Bulletin elementaire franco-khmere – Phnom Penh [1925-39] – 7r – 1 – us CRL [079]

Le bulletin et journal des journaux, reviseur impartial du pour et du contre – devenu: Le Reviseur impartial et universel de tous les journaux pour et contre et Bulletin de Madame de Beaumont. Paris. 1791-juil 1792 (I-III) – 1 – fr ACRPP [944]

Bulletin eucharistique – Montreal: [s.n, 1896?-1955?] – 9 – (incl suppl; ceased 1955?; cont by: bulletin (congregation du tres saint-sacrement)) – mf#P04589 – cn Canadiana [241]

Bulletin fakta-fakta ekonomi daerah istimewa atjah / Darussalam, Banda Atjah, Lembaga Penjelidikan Ekonomi dan Sosial, Fakultas Ekonomi Universitas Sjiah Kuala – Darussalam. v1-4. 1968-1971 – 5mf – 9 – (missing: 1986/1969 v1(3-4); 1969/1970 v2(7-8)) – mf#SE-1371 – ne IDC [330]

Bulletin financier de la bangque de bruxelles – Brussels, Belgium 9 oct 1964-15 jan 1965 – 1/4r – 1 – uk British Libr Newspaper [332]

Bulletin for... / John Birch Society – 1963 sep, 1964 nov, 1965 oct, 1966 nov-1970 may, 1970 jun-1975 mar – 2r – 1 – (cont by: john birch society bulletin) – mf#515684 – us WHS [366]

A bulletin for the promotion of the italo-indonesian trade and cultural relations / Il Marco Polo – Djakarta, 1963 – 7mf – 9 – mf#SE-10825 – ne IDC [959]

Bulletin from the monastic interreligious dialogue – 1978-94 [complete]] – 1r – 1 – (title varies) – mf#ATLA S0931A-E – us ATLA [230]

Bulletin ge'ez : dirige par sylvian grebaut / Aethiops – Paris, 1922-1938. 6v – 6mf – 9 – (missing: 1936 v5(3-4)) – mf#NE-20176 – ne IDC [956]

Bulletin (grants pass, or: 1964) *see* Illinois valley news

Bulletin [grants pass, or: 1964] – Grants Pass OR: Grants Pass Bulletin Pub Co, 1949-1960 [wkly] – 1 – (formed by the union of: grants pass bulletin [grants pass or] and: rogue river record, and: gold hill nugget; cont by: grants pass bulletin [grants pass or: 1960]) – us Oregon Lib [071]

Bulletin historique et archeologique du vaucluse et des departements limitrophes – Avignon, 1879-83 – (= ser Architectural periodicals at avery library, columbia university) – 2r – 1 – $270.00 – us UPA [720]

Bulletin horaire du bureau international de l'heure [bih] / Bureau international de l'heure – Paris: Le Bureau [1922- (irreg) [mf ed 2001] – 1922 jan-1950 jul/dec [gaps?] on 2r [ill] – 1 – (cont by 3. ser 1951-54; 4. ser 1955-58; 5. ser 1959-62; 6. ser 1963-; iss for janv-avril 1945 through juil-dec 1950 called 2. ser; imprint varies) – ISSN: 1010-2264 – mf#film mas c4681 – us Harvard [520]
Bulletin in defense of marxism / Fourth Internationalist Tendency [Group] – 1983 dec-1986 dec, 1987 jan-1989 oct – 2r – 1 – (cont by: labor standard [tucson az]) – mf#1611887 – us WHS [335]
Bulletin – indian library association / Indian Library Association – New Delhi. 1965-80 – 1,5,9 – ISSN: 0019-5782 – mf#7142 – us UMI ProQuest [020]
Bulletin indonesian institute of sciences / Lembaga Geologi dan Pertambangan Nasional – Bandung, 1968, v1(1); 1969, v2(1-3); 1970, v3(1-2) – 6mf – 9 – mf#SE-1762 – ne IDC [959]
Bulletin international de l'electricite – Paris, France. 4 jan 1886-26 aug 1895; 6 jan 1900-25 dec 1910 – 4r – 1 – uk British Libr Newspaper [621]
Bulletin international du mouvement syndicaliste – Paris, France [ns] 6 oct 1907, 5 apr 1908-19 jul 1914 (imperfect) – 1 1/4r – 1 – (fr 1/5 apr 1914 onward publ in amsterdam) – uk British Libr Newspaper [335]
Bulletin kantor berita pab – Djakarta: AKRAB Institution (daily ex sun) [mf ed Ithaca NY: [John M Echols Collection] Cornell University [1998]] – 13r – 1 – (began in jun 1972? cont: news for the general public, armed forces functionaries, servicemen and defense and security officials) – mf#mf-11587 seam – us CRL [079]
Bulletin karya ilmiah permias / Persatuan Mahasiswa Indonesia di Amerika Serikat – Washington, 1964 – 1mf – 9 – (missing: 1964 v1(1-2)) – mf#SE-747 – ne IDC [959]
Bulletin [kodaikanal observatory [kodaikanal, india]] / Kodaikanal Observatory [Kodaikanal, India] – Madras: Printed by the Supt, Govt Press. 177v. 1905- (irreg) [mf ed 2001] – n1-131(1905-51) on 2r [ill] – 1 – (cont by series a, b, c; cont by: kodaikanal observatory bulletins. series a; iss called also v1-14; imprint varies) – mf#film mas c4679 – us Harvard [520]
Bulletin koperasi / Direktorat Djenderal Koperasi – Djakarta, 1966-1972 – 29mf – 9 – (missing: 1966, v1; 1967, v2(1-12, 17-20); 1968, v3(1-2, 5/6, 9-20); 1969, v4(1-7/8, 15); 1971, v6(7-12)-1972(1-3)) – mf#SE-1372 – ne IDC [959]
Bulletin [lowell observatory] / Lowell Observatory – [Flagstaff AZ]: The Observatory [1903- (irreg) [mf ed 2000] – n1-88(1903-35) on 1r [ill] – 1 – (n167- called also v9 n1-; n1-50 called also v1 [iss for 1903-1911]; n51-75 called also v2 [iss for 1909-16]; n76-88 called also v3 [iss for 1916-35]; n87 never publ) – mf#film mas c4511 – us Harvard [520]
Bulletin mensuel de l'union patriotique – Port-au-Prince: Comite de Port-au-Prince. n1-2/3. dec 1920-jan/feb 1921 – 2 sheets – us CRL [325]
Bulletin mensuel d'information des groupes d'etudes *see* Masses
Bulletin mensuel du comite de l'afrique francaise et du comite du maroc – Paris, 1891-1940 – 1230mf – 8 – (missing: 1896, 1908, 1910-12, 1914) – mf#A-387c – ne IDC [956]
Le bulletin mensuel du travail : organe des interets temporels de l'artisan et du laboureur – Quebec: [s.n.] 1re annee n1 1er sep 1900-1re annee n3 1er nov 1900 (mthly) [mf ed 1988] – 1mf – 9 – mf#SEM105P925 – cn Bibl Nat [331]
Bulletin midweek shopper – Kenosha WI. 1981 nov 21-1982, 1983-1984 apr 3 – 2r – 1 – (cont by: midweek bulletin) – mf#1793928 – us WHS [071]
Bulletin [molalla, or] – Molalla OR: [s.n.] [wkly] – 1 – (cont: bulletin board [molalla or]) – us Oregon Lib [071]
Bulletin monumental – 1834-1920. v1-79 – 676mf – 9 – (with ind) – mf#O-1219 – ne IDC [700]
Bulletin municipal / Dijon – 1959-1968 – 1 – us NY Public [944]
Bulletin municipal officiel / Lyons. France – 1940-1968 – 1 – us NY Public [944]
Bulletin municipal officiel / Marseille – 1958-1968 – 1 – us NY Public [944]
Bulletin municipal officiel / Paris. Conseil Municipal – 1959-1968 – 1 – us NY Public [944]
Bulletin municipal officiel / Versailles. France – Dec. 1955-1967 – 1 – us NY Public [944]
Bulletin municipal officiel de la ville de paris – Paris, France 1 aug 1943-9 may 1944 (imperfect) – 1/2r – 1 – uk British Libr Newspaper [350]
Bulletin oder taegliche nachrichten des national convents – Strassburg (Strasbourg F), 1792 28 sep-1793 29 jul [gaps] – 1 – fr ACRPP [074]
Bulletin oepp eppo bulletin / European and Mediterranean Plant Protection Organisation – Paris. 1985-1996 (1,5,9) – ISSN: 0250-8052 – mf#15523 – us UMI ProQuest [630]
Bulletin of alloy phase diagrams – Novelty OH 1980-90 – 1,5,9 – ISSN: 0197-0216 – mf#12934.02 – us UMI ProQuest [660]
Bulletin of amnesty international usa / Amnesty International USA – n1-7 [1984 jun-1985 dec] – 1r – 1 – mf#1131360 – us WHS [341]
Bulletin of applied botany and plant breeding – L, 1908-31, v1-27(5); 1957 v30(3); 1958 v33 – 628mf – 9 – mf#13260 – ne IDC [630]
Bulletin of beloit college / Beloit College – 1946 jan-1951 feb – 2r – 1 – (cont: beloit college bulletin. the alumnus; cont by: bulletin of beloit college [1951]) – mf#1053151 – us WHS [378]
Bulletin of beloit college / Beloit College – 1951 spr-1953 sum, 1953 sep-1969 jun – 2r – 1 – (cont: bulletin of beloit college. the alumnus; cont by: beloit, bulletin issue; beloit, magazine issue) – mf#1053150 – us WHS [378]
Bulletin of bibliography – Ann Arbor MI 1979+ – 1,5,9 – (cont: bulletin of bibliography and magazine notes) – ISSN: 0190-745X – mf#2355.01 – us UMI ProQuest [020]
Bulletin of bibliography & magazine notes – Ann Arbor MI 1897-1978 – 1,5,9 – (cont by: bulletin of bibliography) – ISSN: 0007-4780 – mf#2355.01 – us UMI ProQuest [020]
Bulletin of business research – Columbus OH 1926-80 – 1,5,9 – ISSN: 0007-4799 – mf#6178 – us UMI ProQuest [338]
Bulletin of club activities / Wisconsin Federation of Stamp Clubs – 1944 oct-1958 may – 1r – 1 – (cont: philatelic news bulletin; cont by: bulletin [wisconsin federation of stamp clubs]) – mf#3579553 – us WHS [760]
Bulletin of colgate rochester, bexley hall, crozer theological seminary – v43-51 n4. 1970-79 (gaps) – Inquire – 1 – mf#ATLA 1994-S527 – us ATLA [200]
Bulletin of commercial law league of america *see* Commercial law journal
Bulletin of concerned asian scholars – Abingdon, Oxfordshire 1971-2000 – 1,5,9 – ISSN: 0007-4810 – mf#6049.01 – us UMI ProQuest [073]
Bulletin of economic research – Oxford UK 1975+ – 1,5,9 – ISSN: 0307-3378 – mf#9140 – us UMI ProQuest [330]
Bulletin of entomological research – London 1910-20 – v1-10 on 137mf – 9 – mf#z-924/2 – ne IDC [590]
Bulletin of entomological research – v1-60. 1910-70 – 9 – $1098.00 – mf#0124 – us Brook [590]
Bulletin of environmental contamination and toxicology – Dordrecht, Netherlands 1966+ – 1,5,9 – ISSN: 0007-4861 – mf#13109 – us UMI ProQuest [360]
Bulletin of experimental biology and medicine – Dordrecht, Netherlands 1957+ – 1,5,9 – ISSN: 0007-4888 – mf#10815 – us UMI ProQuest [619]
Bulletin of far eastern bibliography – v. 1-5. 1936-40 – 1 – us L of C Photodup [950]
The bulletin of far eastern bibliography / American Council of Learned Societies. (Committee on Far Eastern Studies) – v1-5. 1936-40 – 1 – us AMS Press [010]
Bulletin of harvard international law club *see* Harvard international law journal
Bulletin of international news – Ann Arbor MI 1925-45 – 1 – mf#1197 – us UMI ProQuest [934]
Bulletin of international socialism / American Committee for the Fourth International Workers League [US] – v2 n4-v4 n11 [1965 feb 22-1968 feb 19] – 1r – 1 – (cont by: bulletin [workers league [us]]) – mf#661872 – us WHS [335]
Bulletin of international socialism – New York. v. 1, no. 2-4, no. 11. 28 Sep 1964-19 Feb 1968. Incomplete – 1 – us NY Public [335]
Bulletin of latin american research – Oxford UK 1981+ – 1,5,9 – ISSN: 0261-3050 – mf#49497 – us UMI ProQuest [972]
Bulletin of mathematical biology – Oxford UK 1973+ – 1,5,9 – ISSN: 0092-8240 – mf#49025 – us UMI ProQuest [510]
Bulletin of notes and queries / Institute of American Genealogy – 1935 dec-1936 jun, sep-oct, 1937 apr, dec, 1938 feb, apr-sep, dec, 1939 mar, may, nov, 1940 feb, aug, 1942 jan-jul – 1r – 1 – mf#1112553 – us WHS [929]
Bulletin of occupational education – Washington DC 1966-74 – 1,5,9 – mf#8089 – us UMI ProQuest [370]
Bulletin of peace proposals – London UK 1981-92 – 1,5,9 – (cont by: security dialogue) – ISSN: 0007-5035 – mf#13023.01 – us UMI ProQuest [327]
Bulletin of prosthetics research – Washington DC 1974-83 – 1,5,9 – ISSN: 0007-506X – mf#9149.01 – us UMI ProQuest [617]
Bulletin of solar phenomena / Tokyo Tenmondai [Tokyo Astronomical Observatory] – Tokyo: The Observatory, Showa 25 [1950-70] [qrtly] [mf ed 2001] – 22v on 1r – 1 – (Chiefly tables & charts) – mf#film mas c5307 – us Harvard [520]
Bulletin of the 4 asian games – Djakarta, 1962(1-5) – 7mf – 9 – mf#SE-1319 – ne IDC [950]
Bulletin of the academy of sciences of the ussr... / Academy of Sciences of the USSR. Division of Chemical Sciences – Dordrecht, Netherlands 1952-77 – 1,5 – ISSN: 0568-5230 – mf#10814.02 – us UMI ProQuest [540]
Bulletin of the adyar library – Madras, 1937-1965. v1-29 – 279mf – 8 – (missing: 1965 v29) – mf#I-365 – ne IDC [240]
Bulletin of the all-peoples congress / All-Peoples Congress – 1981 sep 25-1982 dec – 1r – 1 – (cont by: newsletter [all-peoples congress]) – mf#697245 – us WHS [325]
Bulletin of the amalgamated meat... : hebrew butcher workers union, local 234 / Amalgamated Meat Cutters and Butcher Workmen of North America – v2 n2-v3 n3 [1973 sep/oct-1977 dec] – 1r – 1 – mf#502077 – us WHS [636]
Bulletin of the american academy of psychiatry and the law *see* Journal of the american academy of psychiatry and the law
Bulletin of the american civil... / American Civil Liberties Union – n781-1264 [1937 sep 11-1947 jan 20] – 1r – 1 – (cont by: weekly news bulletin [american civil liberties union]) – mf#645698 – us WHS [071]
Bulletin of the american college of chest physicians / American College of Chest Physicians – Northbrook IL 1962-73 – 1,5 – (cont by: cardiopulmonary medicine) – ISSN: 0002-7960 – mf#1736.01 – us UMI ProQuest [616]
Bulletin of the american institute of accountants / American Institute of Accountants – Jersey City NJ 1924-36 – 1 – mf#1607 – us UMI ProQuest [650]
Bulletin of the american museum of natural history / American Museum of Natural History – New York NY 1983+ – 1,5,9 – ISSN: 0003-0090 – mf#12598 – us UMI ProQuest [060]
Bulletin of the american postal... / American Postal Workers Union et al – n1-9 [1981 apr 13-jul 15] – 1r – 1 – (cont by: bulletin [joint bargaining committee [us]: 1984]) – mf#634169 – us WHS [071]
Bulletin of the american society for information science and technology / American Society for Information Science and Technology – Silver Spring MD 2001+ – 1,5,9 – ISSN: 1931-6550 – mf#11872.01 – us UMI ProQuest [020]
Bulletin of the anglo-saxon... / Anglo-Saxon Federation of America – 1930 jan-1932 dec – 1r – 1 – (cont by: messenger of the covenant) – mf#505638 – us WHS [071]
Bulletin of the asian-african conference – Djakarta, 1955 – 5mf – 9 – mf#SE-1318 – ne IDC [950]
Bulletin of the association... / Association of Wisconsin School Administrators – 1979 sep – 1r – 1 – (cont: bulletin [association of wisconsin school administrators. elementary section]; cont by: awsa bulletin) – mf#813433 – us WHS [370]
Bulletin of the association for business communication / Association for Business Communication (US) – 1985-94 – 1,5,9 – (cont: abca bulletin; cont by: business communication quarterly) – ISSN: 8756-1972 – mf#8090.02 – us UMI ProQuest [650]
Bulletin of the astronomical institutes of czechoslovakia [prague, czech republic: 1949] / Statni hvezdarna [czechoslovakia] – Praha: [National Observatory] 1949-50 [irreg] [mf ed 2003] – 2v on 1r – 1 – (cont by: bulletin of the central astronomical institute of czechoslovakia; chiefly in english; some also in french) – mf#film mas c5568 – us Harvard [520]
Bulletin of the astronomical observatory in torun = Biuletyn obserwatorium astronomicznego w toruniu – Torun: Wydawn. Uniwersytetu Mikolaja Kopernika c1946- [irreg] [mf ed 2002] – (= ser Studia societatis scientiarum torunensis) – 1r – 1 – (ceased with v70 in 1986; in polish, english or french; title varies slightly) – mf#film mas c5404 – us Harvard [520]
Bulletin of the atomic scientists – Chicago IL 1945+ – 1,5,9 – ISSN: 0096-3402 – mf#900 – us UMI ProQuest [300]
Bulletin of the avery institute... / Avery Institute of Afro-American History and Culture – v1 n1-v4 n1 [1981 spr-1985 spr], v6 n2 [1986 fall], v8 n2-v9 n1 [1988 fall-1989 spr] – 1r – 1 – mf#1507709 – us WHS [305]
Bulletin of the badger state... / Badger State Matchcover Club – n1-64 [1978 jan-1988 jul/aug] – 1r – 1 – mf#1780095 – us WHS [790]
Bulletin of the bar association of the district of columbia *see* Journal of the bar association of the district of columbia
Bulletin of the black... / Black Theology Project [New York NY] – 1985 apr – 1r – 1 – mf#4864171 – us WHS [240]
Bulletin of the british association for american studies – ns: n1-2. 1960-66 – 1r – 1 – mf#96744 – uk Microform Academic [073]
Bulletin of the british ornithologists' club – London 1892/1893-1920 – v1-40 on 128mf – 9 – (ind v16-39) – mf#z-2012/2 – ne IDC [590]
Bulletin of the british psychological society – London: The Society, [1953- v30 (1977) – 1r – us CRL [150]
Bulletin of the brownville... / Brownville Fine Arts Association – 1977 spr-1980 – 1r – 1 – (cont: bulletin of the brownville historical society, inc) – mf#676369 – us WHS [978]
Bulletin of the brownville... / Brownville Historical Society – v14 n1-v20 n4 [1970 jan-1976:fall] – 1r – 1 – (cont: bulletin of the brownville, nebraska historical society; cont by: bulletin of the brownville historical society and the brownville fine arts association) – mf#676350 – us WHS [978]
Bulletin of the california... / California Central Coast Genealogical Society – 1985 spr-1987 win, index 1980-82, 1980 win-1984 – 2r – 1 – (cont: bulletin of california central coast genealogical society, inc of san luis obispo county california; cont by: san luis obispo county genealogical society, inc) – mf#467097 – us WHS [929]
Bulletin of the california... / California Central Coast Genealogical Society – 1968 jun-1972 sep, 1972 oct-1978 win, 1979 – 3r – 1 – (cont: california central coast genealogical society bulletin; cont by: bulletin of california central coast genealogical society, inc of san luis obispo county california) – mf#1806957 – us WHS [929]
Bulletin of the canadian association... / Canadian Association in Support of the Native Peoples – 1972 oct-1978 fall – 1r – 1 – (cont: casnp bulletin) – mf#470891 – us WHS [322]
Bulletin of the canadian manufacturers' association – Toronto: Canadian Manufacturer Pub Co, [1898-189- or 19–] – 9 – mf#P04062 – cn Canadiana [670]
Bulletin of the canadian oral... / Canadian Oral History Association – v2 n3-v3 n3 [1976 spr-1977 fall] – 1r – 1 – (cont: bulletin [canadian aural/oral history association]; cont by: journal [canadian oral history association]) – mf#289059 – us WHS [390]
Bulletin of the canadian society for immunology = Bulletin de la societe canadienne d'immunologie / Canadian Society for Immunology – Downsview. 1972-1977 (1) 1976-1979 (5) 1976-1977 (9) – ISSN: 0068-9653 – mf#6996 – us UMI ProQuest [616]
Bulletin of the center... / Center for Public Representation – v2 n2,6 [1976 jan/feb, sep/oct], v3 n1-6 [1977 jan/feb-dec] – 1r – 1 – (cont by: public 1) – mf#384127 – us WHS [350]
Bulletin of the center for children's books – Champaign IL 1947+ – 1,5,9 – ISSN: 0008-9036 – mf#2743 – us UMI ProQuest [020]
Bulletin of the central astronomical institute of czechoslovakia / Ustredni ustav astronomicky [Czechoslovakia] – Prague: The Institute, 1951[irreg]-1952[bimthly] [mf ed 2003] – 2v on 1r – 1 – (cont: bulletin of the astronomical institutes of czechoslovakia [prague, czech republic: 1949]; cont by: biulletin' astronomicheskikh institutovchekhoslovakii; chiefly in english; some also in french & german) – mf#film mas c5569 – us Harvard [520]
Bulletin of the chemical society of japan / Nippon Kagakukai & Chemical Society of Japan – Tokyo, Japan 1926-91 – 1,5,9 – ISSN: 0009-2673 – mf#3487 – us UMI ProQuest [540]
Bulletin of the chester county... / Chester County Genealogical Society – 1978 spr-1984 dec – 1r – 1 – mf#966277 – us WHS [929]
Bulletin of the circus... / Circus World Museum [Baraboo WI] – 1963 apr 22-1963 may 1 – 1r – 1 – mf#3564513 – us WHS [790]
Bulletin of the citizens'... / Citizens' Governmental Research Bureau [WI] – v51 n1-2 [1963 jan 12-feb 18], v55 n3,12-13,13-v72 n6 [1967 feb 25, jul 12-22, sep 23-1982 apr 24 – 1r – 1 – (cont by: in fact [milwaukee wi]) – mf#1405316 – us WHS [350]
Bulletin of the citizens... / Citizens Energy Council – n9,13,17-18,20-23 [1986 jul 8, aug 25, oct 5-16, nov 13-dec 26], n24-28 [1987 jan 5-feb 26], 1987 apr 15, may 11-23, jun 22, jul 7, aug 6-dec 28, n52-59 [1988 jan 20-may 10] – 1r – 1 – (cont: energy news digest) – mf#1519935 – us WHS [360]

Bulletin of the cleveland museum of art / Cleveland Museum of Art – Cleveland OH 1914-94 – 1,5,9 – ISSN: 0009-8841 – mf#5811 – us UMI ProQuest [060]

Bulletin of the college art association of america – New York. v1. 1913-1918 – 289mf – 9 – (cont as: the art bulletin. new york, 1919-1945. v2-27) – mf#O-492c – ne IDC [700]

Bulletin of the commercial law league – v1-28. 1895-1923 (all publ) – 109mf – 9 – $163.00 – (v10 no 8 was never issued. lacking: v1-5,14,20,21,23,27) – mf#LLMC 84-4301 – us LLMC [346]

Bulletin of the committee... / Committee on Canadian Labour History – n1-7 [1976 spr-1979 spr] – 1r – 1 – (cont: canadian labour history; cont by: labour) – mf#646916 – us WHS [331]

Bulletin of the congregation... / Congregation Anshai Lebowitz [Milwaukee WI] – 1954 feb-1964 dec – 1r – 1 – mf#405103 – us WHS [071]

Bulletin of the consumers'... / Consumers' League of Cincinnati – n1-5 [1915 jun-1917 may] – 1r – 1 – (cont by: consumers' league bulletin) – mf#3144562 – us WHS [380]

Bulletin of the consumers'... / Consumers' League of Massachusetts – n14 [1917 nov], n17, 18-19 [1919 jan, jan-mar], n20, [1920 mar], n22 [1921 jun], n23 [1922 jan], n24, 25-27, [1924 jan, jan-oct], n28-30 [1925 jan-oct], n31 [1926 mar], n33-34 [1927:may-dec] – 1r – 1 – mf#3194651 – us WHS [380]

Bulletin of the consumers'... / Consumers' League of New York City – v1 n2 [1922 feb], v2 n7-8 [1923 oct-nov], v3 n1-7, 9 [1924 jan-oct, dec], v4 n1-9 [1925 jan-dec], v5 n1-5 [1926 jan-jun] – 1r – 1 – mf#3193025 – us WHS [380]

Bulletin of the cooper ornithological club of california / Cooper Ornithological Club – Santa Clara. 1899-1899 (1) 1899-1899 (5) 1899-1899 (9) – (cont by: condor) – mf#12347 – us UMI ProQuest [590]

Bulletin of the coos... / Coos Genealogical Forum – v15 n1-v22 n2 [1979 mar-1987 spr] – 1r – 1 – mf#1238635 – us WHS [929]

Bulletin of the council for democratic germany – New York NY (USA), 1944 sep-1945 may – 1r – 1 – gw Misc Inst [943]

Bulletin of the croatian... / Croatian Genealogical Society et al – n3-4-1 [1980 dec 1-1982 mar 1] – 1r – 1 – (cont: quarterly bulletin [croatian-american academic association of the pacific]; cont by: ragusan research bulletin) – mf#652327 – us WHS [929]

Bulletin of the dairymen's association of the province of quebec – [St-Hyacinthe, Quebec]: The Association, [1891?]- [mf ed 1988] – 1mf – 9 – mf#P04005 – cn Canadiana [630]

Bulletin of the deccan college research institute. – Poona, 1939-1957. V.1-18. – 150= – 8 – mf#I-236 – ne IDC [240]

Bulletin of the dental guidance council for cerebral palsy / Dental Guidance Council for Cerebral Palsy – New York NY 1961-72 – 1 – (cont by: dental guidance council on the handicapped journal) – ISSN: 0011-8591 – mf#5061.01 – us UMI ProQuest [617]

Bulletin of the dunn... / Dunn County School of Agriculture and Domestic Economy – 1902-12 – 1r – 1 – mf#2625939 – us WHS [630]

Bulletin of the early... / Early Sites Research Society – 1974 sep-1986 dec – 1r – 1 – mf#1712351 – us WHS [930]

Bulletin of the ecological society of america / Ecological Society of America – Durham. 1970-1996 (1) 1970-1996 (5) 1976-1996 (9) – ISSN: 0012-9623 – mf#6113 – us UMI ProQuest [574]

Bulletin of the entomological society of america / Entomological Society of America – Lanham MD 1955-89 – 1,5,9 – (cont by: american entomologist) – ISSN: 0013-8754 – mf#9056.01 – us UMI ProQuest [590]

Bulletin of the evangelical theological society / Evangelical Theological Society – Wheaton. 1958-1968 (1,5,9) – (cont by: journal of the evangelical theological society) – ISSN: 0361-5138 – mf#11887 – us UMI ProQuest [242]

Bulletin of the evansville... : devoted to interests of evansville seminary / Evansville Seminary [Evansville WI] – 1900 mar-1906 jan/feb – 1r – 1 – mf#1854013 – us WHS [240]

Bulletin of the executive committee of the state grange of wisconsin, p of h / Patrons of Husbandry – v1 [1875] – 1r – 1 – (cont by: bulletin) – mf#1067479 – us WHS [636]

Bulletin of the frisbie-frisbee... / Frisbie-Frisbee Family Association of America – 1975 jan-1984 jan – 1r – 1 – (cont: bulletin of the frisbie-frisbee-frisby family of america; cont by: frisbie-frisbee family bulletin) – mf#711724 – us WHS [929]

Bulletin of the genealogical... / Genealogical Society of South Brevard – 1987 sep-1989 aug – 1r – 1 – mf#1053713 – us WHS [929]

Bulletin of the georgia academy of science / Georgia Academy of Science – Lawrenceville GA 1943-67 – 1 – ISSN: 0016-8114 – mf#8797.01 – us UMI ProQuest [500]

Bulletin of the giles county... / Giles County Historical Society – v1-3 [1974-76], v6-9 [1979 jul 22-1982 oct 24] – 1r – 1 – (cont by: giles county historical society bulletin) – mf#637831 – us WHS [978]

Bulletin of the golfiana... / Golfiana Collectors Club – n1 [1970 sep] – 1r – 1 – (cont by: bulletin [golf collectors' society]) – mf#956163 – us WHS [790]

Bulletin of the harvard college observatory / Harvard College Observatory – Cambridge MA: The Observatory (irreg) [n1-921(1898-1952)] [mf ed 1999] – 3r – 1 – (iss n1-500 are copies of handwritten accounts of telegrams received by the harvard college observatory; title varies) – ISSN: 0891-3943 – mf#film mas c4167 – us Harvard [520]

Bulletin of the historical... / Historical Society of Decatur County [IN] – v4 n93-108 [1982 dec-1986 oct], v4 n110-111 [1987 apr-jul] – 1r – 1 – mf#1058289 – us WHS [978]

Bulletin of the history of dentistry – Baltimore MD 1953-95 – 1,5,9 – (cont by: journal of the history of dentistry) – ISSN: 0007-5132 – mf#7280.01 – us UMI ProQuest [617]

Bulletin of the history of medicine – Baltimore MD 1933+ – 1,5,9 – ISSN: 0007-5140 – mf#7632 – us UMI ProQuest [610]

Bulletin of the indonesian organization for afro-asian people's solidarity – Peking, 1967-1972. v1-6(1) – 8mf – 9 – (missing: 1967, v1; 1968, v2; 1969, v3(1-4, 9); 1970, v4(3-[10]); 1971/1972, v5(4-[10])) – mf#SE-1854 – ne IDC [959]

Bulletin of the institute of paper chemistry / Institute of Paper Chemistry – Appleton. 1948-1954 (1) – ISSN: 0096-9680 – mf#172 – us UMI ProQuest [660]

Bulletin of the international bureau of education / International Bureau of Education – Geneva, Switzerland 1985-93 – 1,5,9 – (cont: educational documentation and information: bulletin of the international bureau of education) – ISSN: 1019-3189 – mf#461.01 – us UMI ProQuest [370]

Bulletin of the john rylands university library of manchester / John Rylands University Library of Manchester – Manchester UK 1985+ – 1,5,9 – ISSN: 0301-102X – mf#15298.01 – us UMI ProQuest [020]

Bulletin of the kenosha... / Kenosha County Historical Society and Museum – 1962 jan [v17 n1]-1973 mar/apr – 1r – 1 – (cont by: southport newsletter) – mf#586062 – us WHS [978]

Bulletin of the lamesa area... / Lamesa Area Genealogical Society – n1,3 [1972 dec, 1973 aug 13] – 1r – 1 – (cont by: threads of life) – mf#575701 – us WHS [929]

Bulletin of the league... / League of Women Voters of Dane County [WI] – 1971 may-1975 apr, 1975 may-1982 oct, 1982 nov-1988 nov – 3r – 1 – (cont: bulletin [league of women voters of madison [wi]: 1969]) – mf#706950 – us WHS [325]

Bulletin of the league... / League of Women Voters of Greater Milwaukee – 1971 may-1973 jun – 1r – 1 – (cont: bulletin [league of women voters of milwaukee]; cont by: league bulletin [milwaukee wi]) – mf#625799 – us WHS [325]

Bulletin of the league... / League of Women Voters of Madison [WI] – 1969 oct-1971 apr – 1r – 1 – (cont: league bulletin [madison wi]; cont by: bulletin [league of women voters of dane county]) – mf#706953 – us WHS [325]

Bulletin of the league... / League of Women Voters of Madison [WI] – 1954 jun-1956 jun – 1r – 1 – (cont: league of women voters of madison [series]; cont by: league bulletin [madison wi]) – mf#706955 – us WHS [325]

Bulletin of the league... / League of Women Voters of Milwaukee – 1970 nov-1971 mar – 1r – 1 – (cont by: bulletin) – mf#625710 – us WHS [071]

Bulletin of the library / Foundation for Reformation Research Library – St. Louis. 1972-1973 (1) 1966-1971 (5) (9) – ISSN: 0015-8941 – mf#6399 – us UMI ProQuest [020]

Bulletin of the livesay... / Livesay Family Association – v9 n1-4 [1965 oct 1-1966 jun 1] – 1r – 1 – (cont by: bulletin [livesay patriotic and historical society]) – mf#683103 – us WHS [929]

Bulletin of the livesay patriotic... / Livesay Patriotic and Historical Society – v10 n1-15 [i.e. 14] n3 [1966 oct 1-1971 jul] – 1r – 1 – (cont: bulletin [livesay family association]; cont by: livesay bulletin) – mf#683108 – us WHS [978]

Bulletin of the livonia... / Livonia Education Association – v13 n9-v15 n9 [1976 oct 5-1977 nov 8] – 1r – 1 – (cont by: lea bulletin) – mf#675832 – us WHS [370]

Bulletin of the london mathematical society / London Mathematical Society – London. 1989-1996 (1) – ISSN: 0024-6093 – mf#14632 – us UMI ProQuest [510]

Bulletin of the los angeles... / Los Angeles Union Label Council – 1983 mar 15-1984 nov 11 – 1r – 1 – mf#1224280 – us WHS [331]

Bulletin of the lower cape... / Lower Cape Fear Historical Society – v1 n1-v31 n3 [1957 oct-1988 may] – 1r – 1 – (cont by: lower cape fear historical society journal; newsletter) – mf#1840886 – us WHS [978]

Bulletin of the maritime... / Maritime Federation of the Pacific Coast – 1940 feb-may – 1r – 1 – mf#3184570 – us WHS [360]

Bulletin of the maritime library association / Maritime Library Association – Maritme Library Assoc, 1936-57 – 1r – 1 – ISSN: 0317-6665 – cn Library Assoc [020]

Bulletin of the maritime museum... / Maritime Museum of British Columbia – n2 of 1965 [1965 mar]-n3 of 1968 [1968 nov], 1971 oct, n16-55 [1972 may-1982 sum/fall] – 1r – 1 – mf#622869 – us WHS [060]

Bulletin of the massachusetts... / Massachusetts Bay Tercentenary – n6-30 [1928 feb-1930 jun] – 1r – 1 – (cont: bulletin [massachusetts bay celebration committee]) – mf#923905 – us WHS [978]

Bulletin of the massachusetts audubon society / Massachusetts Audubon Society – Boston. 1950-1955 (1) – ISSN: 0275-472X – mf#449 – us UMI ProQuest [574]

Bulletin of the massachusetts bay... / Massachusetts Bay Celebration Committee – n1-4 [1927 feb-jun] – 1r – 1 – (cont by: bulletin [massachusetts bay tercentenary]) – mf#923905 – us WHS [978]

Bulletin of the medical library association / Medical Library Association – Chicago IL 1911+ – 1,5,9 – ISSN: 0025-7338 – mf#1833.01 – us UMI ProQuest [610]

Bulletin of the menninger clinic – New York NY 1936+ – 1,5,9 – ISSN: 0025-9284 – mf#1484 – us UMI ProQuest [616]

Bulletin of the meter... / Meter Stamp Society – n4-6 [1947 nov-1949 jan] – 1r – 1 – (cont by: monthly bulletin) – mf#956233 – us WHS [760]

Bulletin of the meter... / Meter Stamp Society – 1960 jan/feb-1984 – 1r – 1 – (cont: Monthly bulletin; cont by: Quarterly bulletin) – mf#956080 – us WHS [760]

Bulletin of the national association... / National Association of Builders of the USA – v2 n2, v3 n2-v4 n5 [1895 oct, 1897-98] – 1r – 1 – mf#1113259 – us WHS [690]

Bulletin of the national consumers league / National Consumers League – 1961 may-1983 may/jun – 1r – 1 – (cont by: ncl bulletin) – mf#1385481 – us WHS [380]

Bulletin of the national council... / National Council of Jewish Women – 1943 feb-1977 aug – 1r – 1 – mf#390656 – us WHS [939]

Bulletin of the national federation... / National Federation of Federal Employees – 1976 jan-1982 oct – 1r – 1 – mf#1094178 – us WHS [331]

Bulletin of the national league... / National League of Women Voters [US] – v1 n2-v4 n3 [1927 nov-1930 jul/aug] – 1r – 1 – (cont by: league news) – mf#1004134 – us WHS [325]

Bulletin of the national popular... / National Popular Government League – n4,60,62-69,81-100,102-110,112,115-125,127,153,157 – 1r – 1 – mf#1431558 – us WHS [350]

The bulletin of the national tax association – v1-33. 1915-47 (all publ) – 33mf – 9 – $49.50 – mf#LLMC 84-369 – us LLMC [336]

Bulletin of the national treasury... / National Treasury Employees Union – 1975 mar 1/20-1980 mar 17 – 1r – 1 – mf#773253 – us WHS [331]

Bulletin of the nevada county... / Nevada County Historical Society [CA] – v32 2-v35 4 [1978 apr-1981 oct] – 1r – 1 – (cont: nevada county historical society [ca: series]) – mf#622275 – us WHS [978]

Bulletin of the new hebrides chamber of commerce, industry and agriculture / bulletin of the vanuatu chamber of commerce / New Hebrides Chamber of Commerce & Vanuatu Chamber of Commerce – Port Vila, 1976-86 – 1r – 1 – mf#PMB Doc425 – at Pacific Mss [338]

Bulletin of the new york... / New York Genealogical and Biographical Society – v1 n1 [1869 dec] – 1r – 1 – (cont by: new york genealogical and biographical record) – mf#1833262 – us WHS [929]

Bulletin of the new york academy of medicine / New York Academy of Medicine – Oxford. 1925-1997 (1) 1965-1997 (5) 1970-1997 (9) – ISSN: 0028-7091 – mf#258 – us UMI ProQuest [610]

Bulletin of the newark... / Newark Teachers Union [NJ] – 1984 dec-1992 oct – 1r – 1 – (cont: ntu bulletin) – mf#1066042 – us WHS [370]

Bulletin of the newberry... / Newberry County Historical Society – v1 n1 [1970 jun], v3 n2 [1982 dec]-v15 n1 [1984 dec] – 1r – 1 – mf#1222512 – us WHS [978]

Bulletin of the newfoundland... / Newfoundland Teachers' Association – 1982 sep 15-1988 jun, 1988 sep-1992 mar – 2r – 1 – (cont: n t a bulletin) – mf#1066053 – us WHS [370]

Bulletin of the newspaper... / Newspaper and Mail Deliverers' Union [NY] – v74 n8-v89 n4 [1977 oct-1991 sep/oct] – 1r – 1 – mf#2493243 – us WHS [331]

Bulletin of the northwest indian... / Northwest Indian Fisheries Commission – v1 n2-3 [1975 jun 19-aug] – 1r – 1 – (cont by: newsletter [northwest indian fisheries commission]) – mf#819077 – us WHS [071]

Bulletin of the nuttall ornithological club / ed by Allen, J A – Cambridge 1876-83. v1-8 – 393mf – 9 – (cont as: auk. a quarterly journal of ornithology [boston etc 1884-1920] v1-37) – mf#z-2014/2 – ne IDC [590]

Bulletin of the ontario libertarian party : newsletter of the... / Ontario Libertarian Party – v10 n3-v13 n3 [1984 apr/may-1987 dec] – 1r – 1 – (cont by: libertarian bulletin [toronto ont]) – mf#1533286 – us WHS [325]

Bulletin of the oregon... / Oregon Genealogical Society – 1976 sep-1981 sum – 1r – 1 – (cont: oregon genealogical bulletin; cont by: quarterly) – mf#603324 – us WHS [929]

Bulletin of the orton society / Orton Society – Baltimore MD 1980-81 – 1,5,9 – (cont by: annals of dyslexia) – ISSN: 0474-7534 – mf#12837.01 – us UMI ProQuest [370]

Bulletin of the overseas... / Overseas Press Club of America – 1972 aug 15-1974 dec 15 – 1r – 1 – (cont: overseas press bulletin; cont by: opc bulletin [1975]) – mf#1814293 – us WHS [070]

Bulletin of the overseas press... / Overseas Press Club of America – 1950 may 27-1952 dec 27, 1953 jan 3-1956 sep 29 – 2r – 1 – (cont: opc bulletin; cont by: overseas press bulletin) – mf#1807274 – us WHS [070]

Bulletin of the pace... / Pace Society of America – 1968 mar-1985 dec – 1r – 1 – mf#1139362 – us WHS [366]

Bulletin of the pan american union / Pan American Union – Washington DC 1893-1948 – 1 – mf#6243 – us UMI ProQuest [323]

Bulletin of the patrons of husbandry / Patrons of Husbandry – v2-7 [1876 jan/feb-1882 jan 15], v8 n1,14 [1882 feb 1, sep 4], v10 n2,4 [1884 jan 15, feb 18] – 1r – 1 – (cont: bulletin of the executive committee of the state grange of wisconsin, p of h) – mf#3523064 – us WHS [636]

Bulletin of the people... / People, Food and Land Foundation – 1983 may-aug, oct/Nov, 1984 jan /Feb-apr, jun-dec, 1985 feb-mar, jun – 1r – 1 – mf#1222703 – us WHS [333]

Bulletin of the psychonomic society – Austin TX 1973-93 – 1,5,9 – (cont by: psychonomic bulletin and review) – ISSN: 0090-5054 – mf#7026 – us UMI ProQuest [150]

Bulletin of the racine county... / Racine County Historical Society – v1 n1-v4 n1 [1980 feb-1983 feb] – 1r – 1 – (cont by: quarterly [racine county historical society and museum]) – mf#670232 – us WHS [978]

Bulletin of the railroad... / Railroad Station Historical Society [Crete NE] – 1968 jan/feb, 1977 jul-1989 dec – 1r – 1 – mf#271537 – us WHS [380]

Bulletin of the retail clerks... / Retail Clerks Union, Local 197 [Stockton CA] – 1976 jul-1979 may/jun – 1r – 1 – (cont by: bulletin [united food and commercial workers international union. local 197 [stockton ca]]) – mf#657645 – us WHS [331]

Bulletin of the saskatchewan... / Saskatchewan Genealogical Society – 1977-81, 1982-86, 1987 mar-1990 dec – 3r – 1 – mf#606179 – us WHS [929]

Bulletin of the school of oriental studies – London, 1917-1939. v1-9 – 196mf – 8 – (cont as: bulletin of the school of oriental and african studies, london 1940/42-1946 v10-11) – mf#A-398c – ne IDC [956]

Bulletin of the science fiction writers of america / Science Fiction Writers of America – Chesterton MD 1974-91 – 1,5,9 – (cont by: bulletin – science fiction and fantasy writers of america) – ISSN: 0192-2424 – mf#9835.01 – us UMI ProQuest [420]

Bulletin of the scottish institute of missionary studies / Scottish Institute of Missionary Studies – Edinburgh UK 1976-91 – 1,5,9 – ISSN: 0048-9778 – mf#10636 – us UMI ProQuest [240]

Bulletin of the scottish rite [masonic order] / Scottish Rite [Masonic order] – 1975 aug-1996 fall – 1r – 1 – mf#5699645 – us WHS [360]

Bulletin of the second... / Second Division Association. United States – 1920 jan 1 – 1r – 1 – (cont: indian [neuwied, germany]; cont by: indian head [washington dc]) – mf#1896053 – us WHS [366]

Bulletin of the service employees... / Service Employees International Union – n60 [holidays ed, 1970/71], n70-72 [1973 autumn-1974 feb/mar] – 1r – 1 – (cont by: seiu local 11 bulletin) – mf#676214 – us WHS [331]

Bulletin of the society for american music / Society for American Music – Wayne PA 2001+ – 1,5,9 – mf#16415.02 – us UMI ProQuest [780]

Bulletin of the southern states... / Southern States Industrial Council – 1970 dec 1-1973 jul 1 – 1r – 1 – (cont by: bulletin [united states industrial council]) – mf#688897 – us WHS [331]

Bulletin of the spencer... / Spencer Family Association – v1 n1 [1978 jul] – 1r – 1 – (cont by: despencer) – mf#1573058 – us WHS [929]

Bulletin of the state bar of wisconsin *see* Wisconsin lawyer

Bulletin of the temple... / Temple Beth El [Madison WI] – 1977 may-1986 jul/aug – 1r – 1 – (cont by: temple beth el) – mf#1330352 – us WHS [270]

Bulletin of the toronto hospital for the insane – Toronto: Warwick Bros's & Rutter, [1907-1916] – 9 – mf#P05213 – cn Canadiana [616]

Bulletin of the torrey botanical club / Torrey Botanical Club – New York. 1870-1996 (1) 1970-1996 (5) 1977-1996 (9) – (cont by: journal of the torrey botanical society) – ISSN: 0040-9618 – mf#1472 – us UMI ProQuest [580]

Bulletin of the united food... / United Food and Commercial Workers International Union – 1979 jul/aug-1982 jul/sep – 1r – 1 – (cont: bulletin [retail clerks union, local 197 [stockton ca]]) – mf#657650 – us WHS [331]

Bulletin of the united states... / United States Industrial Council – 1973 aug 1-1980 dec, 1981 feb 15-1983 jun 15 – 2r – 1 – (cont: bulletin [southern states industrial council]; cont by: bulletin [united states business and industrial council]) – mf#688898 – us WHS [331]

Bulletin of the united steelworkers of america : official newsletter of local 7896 uswa, simonds cutting tools / United Steelworkers of America – v4 n6-12 [1979 jun 18-dec 17], v5 n1-v10 n12[1980 jan 14-1985 dec], v11 n1-3,5 [1986 jan-mar, may] – 1r – 1 – mf#1782778 – us WHS [660]

Bulletin of the university of kansas *see* Studies in bergson's philosophy

Bulletin of the university of nebraska state museum – Lincoln NE 1975-91 – 1,5,9 – mf#10351 – us UMI ProQuest [060]

Bulletin of the van brocklin... / Van Brocklin Family Association – 1961 oct-1979 oct – 1r – 1 – mf#1074036 – us WHS [929]

Bulletin of the va-nc... / VA-NC Piedmont Genealogical Society – 1979 feb-1984 – 1r – 1 – (cont by: piedmont lineages) – mf#962307 – us WHS [929]

Bulletin of the vermont... / Vermont Old Cemetery Association – 1966 jan-1978 fall – 1r – 1 – (cont: president's new letter; cont by: voca) – mf#1313928 – us WHS [366]

Bulletin of the vorpagel... / Vorpagel Family Association – v4-5 n2 [1975 spr-1976 fall] – 1r – 1 – mf#384133 – us WHS [929]

Bulletin of the whitley county... / Whitley County Historical Society [IN] – 1986 feb-1990 dec – 1r – 1 – mf#1074752 – us WHS [978]

Bulletin of the windsor jewish community council – (Windsor, Ont.) September 1968 – June 1984 – us AJPC [270]

Bulletin of the wisconsin building... / Wisconsin Building and Loan League – v1 n1-v6 n5 [1939 feb-1944 sep] – 1r – 1 – (cont by: bulletin [wisconsin saving and loan league]) – mf#665811 – us WHS [071]

Bulletin of the wisconsin canners association / Wisconsin Canners Association – 1939 jan 4-1947 dec 23, 1948 jan 9-1953 dec 22, 1954 jan 6-1958 dec 20, 1959-61, 1962-64 – 5r – 1 – (cont by: wisconsin canners and freezers bulletin) – mf#1656563 – us WHS [660]

Bulletin of the wisconsin conference... / Wisconsin Conference of Social Work – v1 n1-5,8-9, n17,19 – 1r – 1 – (cont: bulletin of wisconsin conference of social work; cont by: publication) – mf#920609 – us WHS [360]

Bulletin of the wisconsin federation... / Wisconsin Federation of Music Clubs – v15 [1938]-v16 [1939], v18 [1941]-v35 [1958/59] – 1r – 1 – mf#923895 – us WHS [780]

Bulletin of the wisconsin federation... / Wisconsin Federation of Stamp Clubs – 1936 apr-1958 apr – 1r – 1 – (cont: bulletin of club activities) – mf#3579525 – us WHS [760]

Bulletin of the wisconsin interscholastic... / Wisconsin Interscholastic Athletic Association – 1932 sep-1944 mar, 1944 apr-1950 may, 1950 sep-1956 dec, 1957 jan-1964 may 15 – 4r – 1 – (cont by: wiaa bulletin) – mf#682866 – us WHS [790]

Bulletin of the wisconsin recreation association / Wisconsin Recreation Association – 1959 dec-1961 feb – 1r – 1 – (cont: wisconsin recreation association: [newsletter]; cont by: wisconsin recreation bulletin) – mf#3562772 – us WHS [790]

Bulletin of the wisconsin savings... / Wisconsin Savings and Loan League – 1944 oct-1960 dec, 1961 jan-1971 jan, 1971 feb-1974 feb – 3r – 1 – (cont: bulletin [wisconsin building and loan league]; cont by: executive [wisconsin savings and loan league]) – mf#665812 – us WHS [071]

Bulletin of the wisconsin state... / Wisconsin State Brewers' Association – 1955 feb 15-dec 9, 1961 jan 14-1965 dec 24, 1968 jan 24-1969 mar 24 – 1r – 1 – mf#439333 – us WHS [660]

Bulletin of the wisconsin state... / Wisconsin State College, River Falls – 1951-60 – 1r – 1 – (cont: bulletin – river falls state teachers colleges...series 2) – mf#479155 – us WHS [378]

Bulletin of the wisconsin state... / Wisconsin State Telephone Association – 1969 oct 28-1985 oct 18 – 1r – 1 – mf#1296377 – us WHS [380]

Bulletin of the workers league [us] : bi-weekly organ of the workers league / Workers League [US] – 1968 mar 4-1970 oct 19, 1970 oct 26-1972 dec 25 – 2r – 1 – (cont: bulletin of international socialism; cont by: bulletin [workers league [us]. central committee]) – mf#661878 – us WHS [331]

Bulletin of the workers league [us] : twice-weekly organ of the central committee / Workers League [US] – 1974 feb 12/1975 nov 28-1992 jan 10/1993 mar 26 – 20r – 1 – (with gaps; cont: bulletin [workers league [us]]; cont by: international workers bulletin) – mf#661882 – us WHS [331]

Bulletin of the world health organization = Bulletin de l'organisation mondiale de la sante / World Health Organization – Geneva. 1982+ (1,5,9) – ISSN: 0042-9686 – mf#14939 – us UMI ProQuest [360]

Bulletin of the yell county... / Yell County Historical and Genealogical Association [AR] – 1981 jan-1984 feb, 1984 apr-1987 dec – 2r – 1 – mf#550789 – us WHS [929]

Bulletin official / Morocco – 1970-79. (Part I-Textes legislatifs). (Part II-Annonces Legales) – 32r – 1 – $675.00; outside North America add $1.25r – (1980-. ca $85.00y) – us L of C Photodup [324]

Bulletin officiel = Ambtelijk blad / Ruanda-Urundi – Usumbura [mf ed 1942-juin 30 1962 filmed [19–]] – 1 – (superseded by: rwanda. journal officiel, and by: burundi. bulletin officiel) – mf#*ZAN-565 – us NY Public [960]

Bulletin officiel / Cambodia – Protectorat francais. 1883-86 – 1 – fr ACRPP [959]

Bulletin officiel / Congo. Belgian – Brussels. 1949. 1952-1959 – 1 – us NY Public [960]

Bulletin officiel / Expedition de Cochinchine. Puis, de la Cochinchine francaise – 1862, 1864-69, 1872-79, 1881-87; TB. 1861-77 – 1 – fr ACRPP [959]

Bulletin officiel / Federation Nationale des Ouvriers Metallurgistes de France – Paris. 1891-1901. Continued as: L'Ouvrier Metallurgiste – 1 – fr ACRPP [660]

Bulletin officiel / Gabon – 1885 – 1 – fr ACRPP [960]

Bulletin officiel / Indochina. French – 1887-1902, 1904-05, 1911 – 1 – fr ACRPP [324]

Bulletin officiel / Indochina. French. Commissariat – 1947-51 – 1 – fr ACRPP [324]

Bulletin officiel / Indochina. French. Commissariat – mars 1948-50 – 1 – fr ACRPP [073]

Bulletin officiel / Protectorate of Annam and Tonkin – 1883-86 – 1 – fr ACRPP [324]

Bulletin officiel / Syndicat regional unitaire des mineurs – Federation unitaire du sous-sol. 10 no. Lille. 1927-30 ? – 1 – fr ACRPP [331]

Bulletin officiel / Union Saint-Joseph d'Ottawa – Ottawa: Le Bureau, [1895-189- ou 19–] [mf ed 1re annee n1 15 mai 1895] – 9 – ISSN: 1190-7630 – mf#P04105 – cn Canadiana [360]

Bulletin officiel de la bourse du travail de paris / Bourse du travail de Paris – Union des syndicats du department de la Seine. no. 1-16. Paris. juil 1904-oct 1905 – 1 – fr ACRPP [331]

Bulletin officiel de la concurrence, de la consommation et de la repression des fraudes – €35.90y – (backfile: 1941-1980 €121.96) – fr Journal Officiel [360]

Bulletin officiel de la nouvelle caledonie – 1853-64 – 1r – 1 – mf#pmb doc47 – at Pacific Mss [079]

Bulletin officiel de la nouvelle caledonie – 1853-64 – 1r – 1 – mf#pmb doc46 – at Pacific Mss [079]

Bulletin officiel de la nouvelle caledonie – 1872-73 – 1r – 1 – mf#pmb doc48 – at Pacific Mss [079]

Bulletin officiel de la nouvelle caledonie – 1874-75 – 1r – 1 – mf#pmb doc49 – at Pacific Mss [079]

Bulletin officiel de la nouvelle caledonie – 1876-79 – 1r – 1 – mf#pmb doc50 – at Pacific Mss [079]

Bulletin officiel de la nouvelle caledonie – 1880-83 – 1r – 1 – mf#pmb doc51 – at Pacific Mss [079]

Bulletin officiel de la nouvelle caledonie – 1884-86 – 1r – 1 – mf#pmb doc52 – at Pacific Mss [079]

Bulletin officiel de la nouvelle caledonie – 1887-90 – 1r – 1 – mf#pmb doc53 – at Pacific Mss [079]

Bulletin officiel de la nouvelle caledonie – 1891-94 – 1r – 1 – mf#pmb doc54 – at Pacific Mss [079]

Bulletin officiel de la nouvelle caledonie – 1895-97 – 1r – 1 – mf#pmb doc53 – at Pacific Mss [079]

Bulletin officiel de la nouvelle caledonie, 1895-1897 – Noumea: Impr du Gouvernement 1871 (?) – 1r – 1 – mf#pmb doc55 – at Pacific Mss [079]

Bulletin officiel de la nouvelle caledonie, 1898-1900 – Noumea: Impr du Gouvernement 1871 (?) – 1r – 1 – mf#pmb doc56 – at Pacific Mss [079]

Bulletin officiel de la nouvelle caledonie, 1901-1902 – Noumea: Impr du Gouvernement 1871 (?) – 1r – 1 – mf#pmb doc57 – at Pacific Mss [079]

Bulletin officiel de la nouvelle caledonie, 1903-1904 – Noumea: Impr du Gouvernement 1871 (?) – 1r – 1 – mf#pmb doc58 – at Pacific Mss [079]

Bulletin officiel de la nouvelle caledonie, 1905-1907 – Noumea: Impr du Gouvernement 1871 (?) – 1r – 1 – mf#pmb doc59 – at Pacific Mss [079]

Bulletin officiel de la solidarite-sante – 1988- (wkly) – €134.80y – fr Journal Officiel [614]

Bulletin officiel de l'education nationale – Suite de: Bulletin officiel du Ministere de l'Education nationale. Paris.oct 1944-64 – 1 – fr ACRPP [370]

Bulletin officiel des annonces commerciales / France – B.O.D.A.C.. 1972 – 1 – fr ACRPP [323]

Bulletin officiel des forces francaises libres / France – no. 1. Londres. 15 aout 1940. suivi de Journal officiel du Haut Commissariat de France puis du Commandement en chef EN Afrique. no. 1-24. Alger. janv-30 mai 1943. suivi de: Journal officiel de la France libre puis de la France combattante. Londres. 20 janv 1941-16 sept 1943. suivi de: Journal officiel de la Republique francaise. Alger. 10 juin 1943-31 aout 1944 – 1 – fr ACRPP [323]

Bulletin officiel du departement de l'instruction publique – Port-au-Prince: Imp de la jeunesse McDonald Dugue & H. Archer. sheets 1-5 [dec 1895-mar 2 1898]; sheets 5-21 [oct/nov/dec 1908/jan 1909-jan/sep 1919]; sheets 22-43 [dec/jan 1923-july/aug/sep 1931] – 43 sheets – (scattered issues wanting) – us CRL [370]

Bulletin officiel du gouvernement militaire de bade – Freiburg Br DE, 1945 28 may-1946 30 nov – 1 – gw Misc Inst [355]

Bulletin officiel du travail, de l'emploi et de la formation professionelle – 1988- – €76.30y – fr Journal Officiel [331]

Bulletin on graphite / Ells, Robert Wheelock – Ottawa: S E Dawson, 1904 – 1mf – 9 – 0-665-99741-8 – mf#99741 – cn Canadiana [550]

Bulletin on narcotics – New York NY 1949-91 – 1,5,9 – ISSN: 0007-523X – mf#11332 – us UMI ProQuest [615]

Bulletin periodique des actes administratifs / Indochina. French – juil 1921 – 1 – fr ACRPP [959]

Bulletin periodique du bureau socialiste international = Periodical bulletin of the international socialist bureau – Brussels: [s.n.] 1909-13 [mf ed 1981-] – 1 – (in french, german & english; some nos accompanied by suppls) – mf#129 – us Wisconsin U Libr [335]

Bulletin pertamina / Dinas Humas Pusat – Djakarta, 1965-1971 – 31mf – 9 – (missing: 1965-1968, v1-4(1-19, 24-25, 27-52); 1969, v5(1-7, 19, 21, 24, 27, 33-34); 1970, v6(7-9, 34, 50); 1971, v7(4, 6-7, 11, 15, 22-23, 25, 27, 31, 41)) – mf#SE-1373 – ne IDC [959]

Bulletin populaire / Agence de presse libre du Quebec – Montreal. n1 1/7 dec 1973-n59 22 avril/5 mai 1976 (wkly) [mf ed 1976] – 2r – 1 – mf#SEM35P114 – cn Bibl Nat [073]

Bulletin pour les prisonniers francais en allemagne *see* General-anzeiger fuer wesel

Bulletin publie par la societe linguistique turque – Ankara, 1933-1968 – 403mf – 8 – (missing: s1 v34; s2 v21; s4 v3 5 (partly)) – mf#NE-113 – ne IDC [956]

Bulletin quotidien d'informations – [Ouagadougou], [mar 5 1956-1966] – us CRL [073]

Bulletin – science fiction and fantasy writers of america / Science Fiction and Fantasy Writers of America – Chestertown MD 1992+ – 1,5,9 – (cont: bulletin of the science fiction writers of america) – mf#9835.01 – us UMI ProQuest [420]

Bulletin scientifique de lacademie imperiale des sciences de saint-petersbourg – Spb. 1836-42 – v1-10 on 76mf – 9 – mf#r-5819/2 – ne IDC [580]

Bulletin sekretariat agit : prop dp partai murba – Djakarta, 1964. v1(1-5) – 1mf – 9 – (missing: 1964 v1(1-4)) – mf#SE-1776 – ne IDC [950]

Bulletin socialiste – no. 125, 141-397. Paris. sept 1933, 1934-aout 1939 – 1 – fr ACRPP [335]

Bulletin socialiste – no. 1-373. Paris. mars 1970-71. fait suite a: Le Populaire. no. 12719. 28 fevr 1970. Remplace par: L'Unite voir a ce titre – 1 – fr ACRPP [335]

Bulletin socialiste *see* L'unite

Bulletin special aux bibliothecaires = Special bulletin to librarians – Montreal: Comite... bibliothecaires professionnels du Quebec. v1 n1 janv 1966-v1 n13 oct 1968 (irreg) [mf ed 1977] – 1r – 1 – (cont by: bulletin de nouvelles (corporation des bibliothecaires professionnels du quebec)) – mf#SEM16P290 – cn Bibl Nat [020]

Bulletin statistique – Ouagadougou. v[1-6]. 1960-1966 – us CRL [310]

Bulletin – strike paper / Allen Co. Lima – may-jun 1957 [daily] – 1r – 1 – (incl mss matl) – mf#B10145 – us Ohio Hist [071]

Bulletin (superseded by the shakespeare quarterly) / The Shakespeare Association of America – v. 1-24. 1924-49 – 1 – us AMS Press [420]

Bulletin sur les chemins / Camirand, J A – [Quebec?]: Dep de l'agriculture, 1897 [mf ed 1980] – 1mf – 9 – 0-665-02828-8 – mf#02828 – cn Canadiana [380]

Bulletin [sydney] – Sydney, Australia 31 jan 1880-30 jun 1998 – 1 – (with ind dated fr 1880-1962) – uk British Libr Newspaper [079]

Bulletin technique du bureau veritas – Paris. 1919-72; 1985-1992 – 5 – fr ACRPP [600]

Bulletin theosophique – Saint-Amand (Cher): Destenay, 1900-48 [mthly, qrterly] [mf ed 2003] – 49v on 4r – 1 – (organ of: section francaise de la societe theosophique, 1900-oct 1908; societe theosophique de france, nov 1908-48. no iss publ aug 1940-mar 1945. mf: v3-49 [1902-48] lacks v3 n17-23, v7 n52-54,56, v32 p247-256, v41 n2) – mf#2003-s099 – us ATLA [290]

Bulletin thomiste – Le Saulchoir, 1(1924)-33(1956) – 121mf – 9 – €236.00 – (lacking: 9(1932). 14(1937)-16(1939)) – ne Slangenburg [241]

Bulletin trimestriel de la societe forestiere – [Paris]: Au Secretariat de la Societe forestiere (4nos/yr) [annee 1 n2-annee 5 (1856-59)] [mf ed 2005] – 1r – 1 – (lacks annee 3 n9); cont by: bulletin de la societe forestiere) – mf#film mas 99999 – us Harvard [634]

Bulletin with which is incorporated the st thomas commercial and shipping gazette – Saint Thomas VI. v26 n135-136 [1901 jun 13-14] – 1r – 1 – (cont: bulletin [saint thomas vi: 1875]; cont by: bulletin [saint thomas vi: 1916]) – mf#871343 – us WHS [380]

Bulletin (youngstown edition) / Mahoning Co. Youngstown – (sep 1919-77) [mthly, irreg] – 9r – 1 – mf#B11729-11737 – us Ohio Hist [331]

Bulletin (youngstown edition) / Trumbull Co. Youngstown – (sep 1919-77) [mthly, irreg] – 9r – 1 – mf#B11729-11737 – us Ohio Hist [331]

Bulletino di archaeologica cristiana *see* Christian art 2

Bulletinof Beloit College – 1946 jan-1951 feb, – 1r – 1 – mf#1053151 – us WHS [071]

Bulletin...of san luis obispo county california / California Central Coast Genealogical Society – v13 n1-2 [1980 spr-sum], index 1980 – 1r – 1 – (cont: bulletin of the california central coast genealogical society; cont by: bulletin) – mf#1806964 – us WHS [929]

Bulletin'of the american geographical... / American Geographical Society of New York – v47 [1915] – 1r – 1 – (cont: journal of the american geographical society of new york; cont by: geographical review) – mf#190822 – us WHS [917]

Bulletin...of the astronomical observatory of the university of illinois, urbana, illinois / University of Illinois [Urbana-Champaign campus]. Astronomical Observatory – Cambridge, USA: Printed for the University of Illinois 1898 [mf ed 2000] – n1(1898) on 1r [ill] – 1 – (no more publ) – mf#film mas c4597 – us Harvard [520]

Bulletins / American Baptist Theological Seminary – 1924-78 – 1 – $80.15 – us Southern Baptist [242]

Bulletins / Baptist Southern Convention – 1950-70 – 1 – $14.07 – us Southern Baptist [242]

Bulletins / Wisconsin. Dept. of Revenue. Bureau of Local Financial Assistance – 76 fiches. (Harvard Law School Library Collection.) – 9 – $ – (indebtedness. 1948-79. municipal resources provided and expended. 1921-78. property tax. 1937-78. taxes and aids. 1920-68. town, village and city taxes. 1920-78) – us Harvard Law [336]
Bulletins and other ephemera relating to the fourth international, 1930-1940 – 5r – 1 – (with guide. in english, french, spanish, dutch and german) – mf#97524 – uk Microform Academic [335]
Bulletins, catalogs and other material / Golden Gate Baptist Theological Seminary – 1 – us Southern Baptist [242]
[Bulletins de vote au nom de louis-napoleon bonaparte] – Paris, 1848 – us CRL [944]
Bulletins des seances de la societe vaudoise des sciences naturelles – Lausanne 1842-1942/45 – v1-62 – 9 – (ind [1842-1939] v1-60) – mf#z-247c/2 – ne IDC [590]
Bulletins et travaux compte rendu des seances / Institut Indochinois pour l'Etude de l'Homme. Hanoi – v. 1-6. 1938-43 – 1 – $25.00 – us L of C Photodup [930]
Bulletins of the laws observatory / University of Missouri. Laws Observatory – Columbia: E W Stephens Pub Co 1902-21 (irreg) [v1-2(1902/05-1911/21)] [mf ed 1999] – 2v on 1r ill] – 1 – (cont by: publ of the university of missouri observatory) – mf#film mas c4262 – us Harvard [520]
Bulletins of the society for the study of labour history, 1960-1982 – 42mf – 9 – mf#87285 – uk Microform Academic [331]
Bulletins of the u s bureau of labor and the u s bureau of labor statistics, 1895-1919 – 838mf – 9 – $8090.00 – 1-55655-469-9 – (with p/g) – us UPA [331]
Bulletins officiels de la grande armee dictes par l'empereur napoleon / ed by Goujon, Alexandre M – Paris 1822 [mf ed Hildesheim 1995-98] – 2v on 6mf – 9 – €120.00 – 3-487-26357-2 – gw Olms [355]
Bullettino della societa entomologica italiana – Firenze 1869-1920 – v1-52 on 309mf – 9 – mf#z-930/2 – ne IDC [590]
Bullettino malacologico italiano – Pisa 1868-74. v1-7 – 133mf – 9 – (cont as: bullettino della societa malacologica italiana [pisa 1875-1895/96] v1.1-20) – mf#z-894/2 – ne IDC [590]
Bullettino settimanale delle leggi e dei decreti del regno d'italia / Italy. Laws, Statutes, etc – Napoli. On film: v1-19; 1894-1912. LL-0251 – 1 – us L of C Photodup [348]
Bulley, Agnes Amy *see* Women's work
Bulley, M W *see* Manual of nyanja (as spoken on the shores of lake nyasa)
Bullialdus, Ism *see* Historia byzantina (cbh15)
Bullinger, Ethelbert William *see* A key to the psalms
Bullinger, H *see* Handlung oder acta gehaltner disputatio vnd gespraech zu zoffingen...
Bullinger, Heinrich *see*
– Abbrege de la doctrine evangelique et papistique
– Absolvta de christi...sacramentis tractatio
– L'accord passe et conclvd tovchant la matiere des sacremens...
– Ad ioannis cochlei de canonicae scriptvrae... authoritate libellum...responsio
– Ad libros commetariorum d. joannis oecolampadii...praefatio
– Ad magnificos...ministros...in polonia...praefatio
– Ad septem accvsationis capita...responsio
– Ad testamentvm d ioannis brentii...responsio
– Adam melchior
– Adhortatio ad omnes...verbi dei ministros
– Adversus omnia catabaptistarum prava dogmata
– Adversvs anabaptistas libri 6
– Der alt gloub
– Der alt gloub...
– An den durchlaeuchtigesten...herrn allbrechten...
– Anklag vnd ernstliches ermanen gottes...
– Antiqvissima fides et vera religio
– Antithesis et compendivm evangelicae et papisticae doctrinae
– Antwort der dieneren der kyrchen zuo zuerych vff d. jacoben anderesen...widerlegen...antwort... vff d. jacoben andresen...erinnerung...
– Apologetica expositio
– Apologie...en laquelle est demonstrè...
– Bedencken ob der verraehter judas auch an dem tisch desz herren gesessen...
– Bekanntnusz desz waaren gloubens...
– Bericht der krancken
– Bericht wie die, so...mit...fragen versuocht werdend, antworten...moegind
– Bewiligung vnd confirmation eines burgermeisters...
– Bewilligung vnd confirmation...ueber die restitution vnd verbesserung ettlicher maenglen vnd miszbruechen...
– Brevis ac pia institutio christianae religionis, ad dispersos in hungaria...ministros...
– Brevis antibolè sive responsio secunda...ad... ioannis cochlei...replicam...
– Bvllae papisticae...contra...reginam elizabetham... promulgatae, refutatio
– Catechesis pro advltioribvs scripta
– Catechismus
– Cent sermons svr l'apocalypse
– The christen state of matrimony...
– The christen state of matrymonye
– Der christenheit rechte vollkommenheit
– Christennlich ordnung vnd bruch der kirchen zuerich
– Der christlich eestand
– Christliches baettbuechlin
– Les cinq decades
– Commentarii in omnes pauli epist et epist catholicas. de testamento..dei...- de utraque in christo natura
– Commentarii in omnes pauli epistolas
– Commentariorum libri 10 in...evangelium secundum ioannem
– Commentariorum libri 12 in...evangelium secundum matthaeum
– Compendium christianae religionis
– Concilium tridentinum non institutum esse ad inquirendam...veritatem...demonstratio
– Confessio et expositio simplex orthodoxae fidei
– Confession et simple exposition de la vraye foy
– A confession of fayth
– La confessione elvetica...
– Confessiun da la vera cardienscha...
– Consensio mvtva in re sacramentaria
– Daniel...expositvs homilijs 66. epitome temporvm
– Das die evangelischen kilchen weder kaetzerische noch abtruennige...syend gruntliche erwysung
– De coena domini sermo
– De conciliis
– De conciliis...in primitiva ecclesia
– De fine seculi and iudicio...
– De gratia dei iustificante nos propter christum
– De gratia dei ivstificante
– De hebdomadis, qvae apvd danielem sunt, opusculum
– De la sevle foy en christ ivstifiante
– De omnibvs sanctae scriptvrae libris...
– De origine erroris et de conciliis
– De origine erroris, in divorum ac simvlachrorvm cvltv
– De origine erroris, in negocio evcharistiae...
– De origine erroris libri duo
– De origine erroris libri dvo...
– De persecvtionibvs ecclesiae christianae
– De prophetae officio...
– De sacrosancta coena domini nostri iesv christi
– De scriptvrae sanctae avthoritate...deque episcoporum...institutione
– De scriptvrae sanctae praestantia, dignitate...
– De testamento sev foedere...expositio
– De vera hominis christiani iustificatione
– The decades
– Dispositio et perioche historiae evangelicae
– Ecclesiae scholaeque tigurinae, de iisdem thesibus [zanchii] iudicium
– Ecclesias evangelicas neqve haereticas neqve schismaticas...esse...apodixis
– Einhaelligkeit der dienern der kirhen zuo zuerich vnd herren joannis caluinj
– Epistolae dvae ad ecclesias polonicas
– Epistolarum fasciculus
– Erzaehlung des sempacher krieges...
– An exhortation to the ministers of gods woord
– Festorvm diervm...sermones
– Fiftie godlie and learned sermons
– Freuntliche ermanung zur grechtigheit...
– Fvndamentvm firmvm
– Gaegenbericht...vff den bericht herren johansen brentzen
– Gaegensatz vnnd kurtzer begriff der euangelischen vnd baepstischen leer
– Gottsaeliger vnd grundtlicher bericht von der hochheit...heiliger goettlicher geschrifft...
– Hauszbuch...
– Heinrich bullingers diarium
– Histoire des persecvtions de l'eglise...
– Hoffnung der gloeubigen
– Huysboeck
– Huysboeck, vijf decades
– A hvndred sermons vpon the apocalips
– Hvysboec
– Ieremias...propheta, expositus...concionibus 170
– In acta apostolorvm...commentariorvm libri 6
– In apocalypsim...conciones centum
– In d apostoli pauli ad thessalonicenses... epistolas commentarii...
– In d apostoli pavli ad galatas, ephesios, philippen...
– In d apostoli pavli ad thessalonicenses, timotheum, titum & philemonem epistolas...commentarij
– In d petri apostoli epistolametranqe... commentarius
– In divinvm...euangelium secundum ioannem, commentariorum libri 10
– In epistolam [primam] ioannis...expositio
– In luculentum et sacrosanctum evengelium... secundum lucam commentariorum lib 9
– In lvcvlentvm...euangeliu...secundum lucam, commentariorum lib 9
– In omnes apostolicas epistolas, divi videlicet pavli 14. et 8
– In posteriorem d pavli ad corinthios epistolam... commentarius
– In priorem d pavli ad corinthios epistolam... commentarius
– In sacrosanctum evangelium domini nostri iesu christi sec marcum commentariorum lib 6
– In sacrosanctvm euangelium...secundum marcu, commentariorum lib 6
– In sacrosanctvm...euangelium secundum matthaeum, commentariorum libri 12
– In...pavli ad hebraeos epistolam...commentarius
– In...pavli ad romanos epistolam...commentarius
– Institvtio eorvm qui...de fide examinantur...
– Isaias...expositus homilijs 190
– Das juengste gericht
– Kerckelycke sermoenen over de feestdaghen...
– Ministrorvm tigvrinae ecclesiae, ad confutationem d iacobi andreae, apologia
– A most excellent sermon of the lordes supper...
– Die offenbarung jesu christi
– De openbaringhe jesu christi
– Ordnung synodi...yetz widerumb erneuwert vnd verbessert
– Orthodoxa tigvrinae ecclesiae ministrorum confessio
– Het oude gelooue
– Perfectio christianorvm
– La perfection des chrestiens
– Qvo pacto cvm aegrotantibus...agendu sit:...
– Ratio stvdiorvm
– Die rechten opffer der christenheit
– Reformationsgeschichte
– Repetitio et...explicatio...de inconfusis proprietatibus naturarum christi...
– Resolvtion de tovs les poincts de la religion chrestienne
– Responsio [ad ioannem brentium]
– Resvrrectio
– Salz zum salat
– Ein schoen spil von der geschicht...lucretiae...
– Een seer schoon troostelick boeck...
– Series et digestio temporvm et rervm...in actis apostolorum
– Sermonum decades quinque
– Sermonvm decades duae
– Sermonvm decades quinque
– Sermonvm decas quarta
– Sermonvm decas quinta
– Sermonvm decas tertia
– La sovrce d'errevr: redige en devx livres
– Studiorum ratio...
– Summa christenlicher religion
– Summa christlicher religion...
– Teghens de vvederdoopers ses boecken...
– Threnorvm sev lamentationvm...ieremiae... explicatio
– Tractatio verborvm domini, in domo patris...
– The tragedies of tyrantes
– Der tuergg
– Utrivsque in christo natvrae...assertio
– Uvaarachtighe bekentnis van de dienaars der kerken tot zurigh
– Vande ghenade gods...
– Vanden oorspronc der dvvalinghe...van de concilien
– Verglichung der vralten vnd vnser zyten kaetzeryen
– Vermanung an alle diener des worts gottes
– Veruolgung
– Vester grund
– Vff etliche scharpffe vnnd bittere buechle verantwortung
– Vff herren johannsen brentzen testament... antwort
– Vff johannsen [fabri] wyenischen bischoffs trostbuechlin... verantwurtung...
– Vff siben klagartickel...verantwortung
– Vo rechter buosz oder besserung desz suendigen menschens
– Vom antichrist vnd seinem reich...
– Von allen buecheren heiliger vnd goettlicher gschrifft...
– Von dem einigen vnnd ewigen testament oder pundt gottes...
– Von dem heil der gloeubigen
– Von dem heiligen nachtmal
– Von dem himmel vnd der graechten gottes
– Von dem vnuerschampte fraefel der...widertoeuffern...
– Von den concilijs
– Von der bekerung desz menschen zu gott
– Von der verklaerung jesu christi
– Von hoechster froeud vnd groestem leyd desz...juengsten tags
– Von raechter hilff vnd erretung in noeten
– Von warem bestaendigem glouben in aller not vnd anfaechtung
– Von warer rechtfertigung...
– Warhaffte bekanntnuss der dieneren der kilchen zu zuerych...
– Wider die schwartzen kuenst
– Widerlegung der bullen dess papst pij 5. ...wider...elizabetham, koenigin in engelland... aussgegangen
– Der widertoufferen ursprung/furgang/section/ wasen/furnem me und gemine jrer leer artickle
– Wjr burgermeyster vnnd rath... [reformationsmandat]
– Zwo predigen ueber den 130. ouch 133. psalmen
Bullinger, Heinricheinrich *see* Lucretia-dramen
Bullinger, Heinricheinrich et al *see* Miscellanea tigurina
Bullingers briefwechsel mit vadian / Schiess, T – Zuerich, Faesi & Beer [vorm D Haehr], 1906 – (= ser Jahrbuch fuer Schweizerische Geschichte) – 1mf – 9 – (jahrbuch fuer schweizerische geschichte, 1906. v31(p 23-68)) – mf#PBU-443 – ne IDC [240]
Bullingers korrespondenz mit den graubuendnern / ed by Schiess, T – Basel, Basler Buch- und Antiquariatshandlung, vorm. Adolf Geering, 1904-1906. 3 v – 23mf – 9 – mf#PBU-441 – ne IDC [240]
[Bullion-] district miner – NV. jul 1907 [wkly] – 1r – 1 – $60.00 – mf#U04427 – us Library Micro [071]
Bullion ledgers of the united states mint at philadelphia, 1794-1802 / U.S. Bureau of the Mint – (= ser Records of the united states mint) – 1r – 1 – mf#T587 – us Nat Archives [332]
Bulletin quotidien d'informatio ouagadougou – [sep 20-nov 8 1965] – (filmed with: upper volta. service de l'information. bulletin quotidien d'information) – us CRL [073]
Bulloch County. Georgia. Macedonia Baptist Church *see* Souvenir program and 100-year history
Bulloch, John *see* George jameson
Bullock, Charles *see*
– Mashona and the matabele
– Mashona (the indigenous native of s rhodesia)
Bullock, George E *see* Ice hockey injuries
Bullock, Hannah *see* History of the isle of man
Bullock, Miles Gaylord *see* What christians believe
Bullock report *see* Language for life
Bullock, William *see* Six months' residence and travels in mexico
Bullon de Mendoza, Alfonso *see* Las ordenes militares en la reconquista de la provincia de badajoz
Bullon Ramirez, Francisco *see* Canalizacion del ahorro provincial. conferencia. colegio o. de secretarios, interventores y depositarios de administracion local de la provincia de caceres
Bull's-eye – 1981 may 15/1982 jan-1993 jul 2/dec 17 – 11r – 1 – (with gaps) – mf#646600 – us WHS [071]
Bullseye – 1981 aug, 1982 mar, sep-1993 nov, 1994 feb, jun, oct-1986: jan, apr-aug, oct-dec, 1987 feb, 1988 aug, 1989 jan-may, aug-1993 jul/aug=92 feet – 1r – 1 – mf#661587 – us WHS [071]
Bulova, Josef Ad *see* Die einheitslehre (monismus) als religion
Bulovec, Stefka *see* Bibliografija edvarda kardelja
Bulow, Franz Josef von *see* Deutsch-sudwestafrika
Bulow, Franz von *see* Im felde gegen die hereros
Bulow, Heinrich von *see* Deutsch-sudwestafrika seit der besitzergreifung, die zuge und kriege gegen die eingeborenen
Bulpett, C W L *see* A picnic party in wildest africa
Bulpin, Thomas Victor *see*
– Golden republic
– Hunter is death
– Lost trails of the transvaal
– Lost trails on the low veld
– Natal and the zulu country
– Rhodesia and nysaland
– Shaka's country
– Southern africa
– Storm over the transvaal
– To the banks of the zambezi
– To the shores of natal
Bulstrode, Richard *see* Newsletters of richard bulstrode, 1667-1689
Bulteel, H B *see*
– Letter to the nine clergymen of oxford who voted against their fell
– Sermon on i corinthians ii
Bulteel, Henry Bellenden *see* Reply to dr burton's remarks upon a sermon
Bultema, Harry *see*
– Maranatha
– De twee gewraakte punten
Bulter, Rhea S *see* Inclusive physical education
Bulthaupt, Heinrich Alfred *see*
– Die arbeiter
– Ueber den einfluss des zeitungswesens auf litteratur und leben
Bultmann. Fritz A *see* Das kleingedruckte
Buluwayo (bulawayo) chronicle – Bulawayo, Zimbabwe 12 oct 1894-29 mar 1940, 31 jan 1941-8 jun 1951 (very imperfect) – 1 – (1941, 42 very imperfect; cont by: chronicle [15 jun 1951-30 jan, 1 may 1953, 2 oct 1953-28 nov 1980, 1 jan 2000-]) – uk British Libr Newspaper [079]
Bulwark : or reformation journal: a monthly journal on behalf of reformation principles – [Scotland] Glasgow: printed...by Aird & Coghill jul 1851-dec 1924 (mthly) [mf ed 2004] – 41r – 1 – (cont: bulwark, or, reformation journal; iss by the scottish reformation society) – uk Newsplan [242]
The bulwarks of the faith : a brief and popular treatise on the evidences of christianity, or the authenticity, truth and inspiration of the holy scriptures / Gray, James Martin – Nyack: Christian Alliance, c1899 [mf ed 1992] – (= ser Christian & missionary alliance coll) – 2mf – 9 – 0-524-02254-2 – mf#1990-4261 – us ATLA [220]
Bulwer Lytton, Edward G *see* England and the english

Bulwer, William H *see* An autumn in greece
Bumazhnye denezhnye znaki rossii i sssr / Malyshev, A I et al – M, 1991 – 9mf – 9 – mf#REF-186 – ne IDC [332]
Bumblebee – Empire City OR: [s.n.] [wkly] – 1 – us Oregon Lib [071]
Bumgarner, Simeon Columbus *see* Bumgarner's annotated pocket code of tennessee.
Bumgarner's annotated pocket code of tennessee. / Bumgarner, Simeon Columbus – Harrisburg, Pa., United Evangelical Church, 1915. 451 p. LL-488 – 1 – us L of C Photodup [348]
Bump, Orlando Franklin *see*
- Composition in bankruptcy
- Decisions constitucionales de los tribunales federales de estados unidos desde 1789.

Bunau-Varilla, Philippe *see* Great adventure of panama
Bunavestire – Bucharest, 1940-41 – 1 – us CRL [077]
Bunbury, Charles James Fox *see* Journal of residence at the cape of good hope
Bunbury, Robert Shirley *see* Effects of prudence on the temporal and spiritual welfare of man
Bunbury, st boniface: baptisms 1848-66 : including tilstone fearnall – (= ser Cheshire church registers) – 3mf – 9 – £4.50 – mf#319 – uk CheshireFHS [929]
Bunbury, st boniface (interior) – (= ser Cheshire monumental inscriptions) – 3mf – 9 – £4.50 – mf#377 – uk CheshireFHS [929]
Bunce, C R *see* Catalogue of the archives of the dean and chapter of canterbury, 1805-1806
Bunce, Dearl Linwood I *see* Coggins memorial baptist church's involvement in a meaningful world hunger ministry
Bunce, John Thackray *see*
- Fairy tales
- History of the corporation of birmingham

Bunch velvet beans to control root-knot / Watson, J R – Gainesville, FL. 1922 – 1r – us UF Libraries [630]
Der bund – Bern (CH), 1963-66, 1974 1-30 apr – 1r – 1 – (filmed by misc inst: 1914 1 jul-1919 30 apr [gaps]; 1940-1977 jul, 1979-91; 1981-83; 1915 jan-okt, 1916-1920 mar, 1921 dez) – gw Mikrofilm; gw Misc Inst [074]
Der bund : das gewerkschaftsblatt der britischen zone – Koeln DE, 1947 22 apr-1949 21 dec – 1r – 1 – (cont: welt der arbeit; filmed by misc inst: suppl: wirtschaft und wissen 1949 [publ in duesseldorf 1r]) – mf#3865 – gw Mikropress; gw Misc Inst [331]
Der bund – Nuernberg DE, 1921 – 1 – gw Misc Inst [074]
Der bund – Vienna, Austria jan 1905-dec 1919 [mf ed Norman Ross] – 1r – 9 – mf#nrp-1940 – us UMI ProQuest [074]
Bund der Bau-, Maurer- und Zimmermeister zu Berlin *see* Jahrbuch
Der bund der freimaurer / Horneffer, August – Jena: Verlegt bei Eugen Diederichs 1913 – 3mf – 9 – mf#vrl-202 – ne IDC [366]
Bund der technischen Angestellten und Bund der technischen Angestellten und Beamten *see* Schriften
Bund der technisch-industriellen Beamten. Berlin *see* Protokoll des ordentlichen bundestages
Bund Deutscher Offizier *see* Ehren-rangliste des ehemaligen deutschen heeres..
Bund Evangelischer Pfarrer in der Deutschen Demokratischen Republik *see* Evangelisches pfarrerblatt
Der bund (hq41) : zentralblatt des bundes oesterreichischer frauenvereine – 1905-19 [mf ed 1999] – (= ser Hq 41) – 14v on 31mf – 9 – €160.00 – 3-89131-352-7 – gw Fischer [305]
Bund i sionizm / Zhabotinskii, V – Odessa, 1906 – 1mf – 9 – mf#RPP-97 – ne IDC [325]
Bundarra & tingha advocate – Bundurra, dec 1900-dec 1906 – 2r – A$118.45 vesicular A$129.45 silver – at Pascoe [079]
Der bundehesh / Justi, F – Leipzig, 1868 – 6mf – 9 – mf#NE-20151 – ne IDC [956]
Der bundehesh / ed by Justi, Ferdinand – Leipzig: FCW Vogel, 1868 – 2mf – 9 – 0-524-02418-9 – mf#1990-3002 – us ATLA [280]
Bundehesh, liber pehlvicus e vetustissimo codice havniensi descripsit / Westergaard, N L – Havniae, 1851 – 1mf – 9 – mf#NE-20150 – ne IDC [956]
Bundes anzeiger – Bonn, Germany 8 feb 1950-15 aug 1951, 24 jan 1952-31 dec 1955 – 59r – 1 – uk British Libr Newspaper [074]
Bundes-anzeiger – Bonn, Frankfurt/M, Koeln DE, 1958 4 jan-1 oct, 1983-87 – 1 – (beginning also: oeffentlicher anzeiger fuer das vereinigte wirtschaftsgebiet) – gw Mikrofilm [343]
Bundesbote – Teplitz (Teplice CZ), 1930 – 1r – 1 – gw Misc Inst [077]
Bundesbote-kalender – 1886-1947 [complete] – (= ser Mennonite serials coll) – 2r – 1 – mf#ATLA 1993-S024 – us ATLA [242]
Bundesdenkmalamt *see* Mittheilungen der k k centralkommission zur erforschung und erhaltung der baudenkmale
Bundesforschungsamt fuer naturschutz und landschaftsoekologie (bestand b 245) bd 44 : teilfindbuch reichsstelle fuer naturschutz / zentralstelle fuer naturschutz und landschaftspflege / bundesanstalt fuer naturschutz und landschaftspflege / ed by Wettengel, Michael – 1993 – xiii/87p – €5.50 – ISBN-13: 978-3-89192-040-4 – gw Bundesarchiv [943]
Bundesgesetz uber das verfahren bei dem bundesgerichte in burgerlichen rechtsstretigkeiten / Switzerland. Laws, Statutes, etc – (Vom 22. November 1850) 54 p. LL-4024 – 1 – us L of C Photodup [348]
Bundesministerium der justiz (bestand b 141) bd 69 – (t1: strafrecht (1949-70) ed by vera huebel [1999 614p €15.50] isbn: 978-3-89192-080-0; t2: zivilrecht (1946-) 1949-74 (-1997) ed by reinhold bauer [2004 2v on 932p €34.50] isbn: 978-86509-146-8) – gw Bundesarchiv [342]
Bundespraesidialamt. amtszeit heinrich luebke 1959-1969 (bestand b 122) bd 74 / ed by Werner, Wolfram – 2000 – 258p – €11.00 – ISBN-13: 978-3-89192-091-6 – gw Bundesarchiv [342]
Bundespraesidialamt. amtszeit prof dr theodor heuss (bestand b 122) bd 38 / ed by Brandes, Mechthild – 1990 – xv/220p – €7.50 – ISBN-13: 978-3-89192-028-2 – gw Bundesarchiv [943]
Bundessekretariat des Deutschen Kulturbundes, Sektor Publikationen *see* Hoelderlin, friedrich
Bundesstaat und bundeskrieg in nordamerika : mit einem abriss der colonialgeschichte als einleitung / Hopp, Ernst Otto – Berlin: G Grote 1886 – (= ser Allgemeine geschichte in einzeldarstellungen 4/4) – [pl/ill] – 1 – mf#film mas c604 – us Harvard [975]
Die bundesvorstellung im alten testament in ihrer geschichtlichen entwickelung / Kraetzschmar, Richard – Marburg: N G Elwert, 1896 – 1mf – 9 – 0-8370-3987-8 – (incl bibl ref, appendixes on the etymology of the hebrew word berit and words used with it and index of biblical citations) – mf#1985-1987 – us ATLA [221]
Die bundeswehr – Dortmund DE, 1956 dec-1966 – 1 – gw Misc Inst [355]
Bundeto, C *see* El espejo de la muerte
A bundle of memories / Holland, Henry Scott – London: Wells Gardner, Darton [1915?] [mf ed 1990] – 1mf – 9 – 0-7905-4975-1 – mf#1988-0975 – us ATLA [920]
Bundling / Stiles, Henry Reed – New York, NY. 1934 – 1r – us UF Libraries [025]
Bundschuh, Joh Kaspar *see*
- Der fraenkische merkur
- Neuer fraenkischer merkur

Bundschuh, Joh Kaspar et al *see* Journal von und fuer franken
Bundy, Charles Smith *see* The justices' manual of statute, judicial and elementary law, with appropriate forms.
Bungalows and country residences : a series of designs and examples / Briggs, Robert Alexander – London 1891 – (= ser 19th c art & architecture) – 1mf – 9 – mf#4.2.430 – uk Chadwyck [720]
Bunge, N *see* Russkie bumazhnye den'gi
Bungei sosho meiji-jidai – The Diplomatic Record Office, Ministry of Foreign Affairs of Japan – 9 – Y1,000,000 (Y1300/sheet) – (in japanese) – ja Yushodo [480]
Bungendore mirror – Bungendore, oct 1887-aug 1888 – 1r – A$35.02 vesicular A$40.52 silver – at Pascoe [079]
Bungener, Felix *see*
- Calvin
- Christ et le siecle
- History of the council of trent
- Pape et concile au 19e siecle
- Saint paul

Bungener, Laurence Louis Felix *see* History of the council of trent
The bunhill memorials / Jones, J A – 1849 – 1 – us Southern Baptist [242]
Bunimovitsh, Yisra'el *see* Memuarn fun yisra'el bunimovitsh
Bunimovitz, Avraham Eliyahu *see* Sefer katuv
Bunin, Ivan Alekseevich *see* Temnye allei
Bunker, A *see* Sketches from the karen hills...
Bunker, Alonzo *see*
- Sketches from the karen hills
- Soo thah

Bunker Hill Monument Association *see*
- Proceedings of the bunker hill monument association

Bunkergeschichten / ed by Mungenast, Ernst Moritz – Stuttgart: Verlag Deutsche Volksbuecher [1943] [mf ed 1993] – 1r – 1 – (filmed with: traum und sendung / comp & aft by heinz kindermann) – mf#3362p – us Wisconsin U Libr [830]
Bunkyo kyoku nichi ma jiten : kamoes bahasa nippon-indonesia / Java. (Japanese Military Administration). Naimubu – (Djakarta?): Djawa Gunseikanbu (2603) – vi/950p 11mf – 9 – mf#SE-2002 mf75-85 – ne IDC [355]
Bunkyokyoku : didikan boedi pekerti / Java. (Japanese Military Administration) – Djakarta: Gunseikanbu Kanrikojo "Kolff" 2602 23 (no daftar 751) – 1mf – 9 – mf#SE-2002 mf58 – ne IDC [959]
Bunnell / Scoville, Dorothy R – s.l, s.l? 1936 – 1r – us UF Libraries [978]
Bunols, J Esteban *see* Por cuenta del estado
Bunpo, T S *see* Un peu de poesie et surtout de la morale
Bunpodo [comp] *see* Bunpodo zassan
Bunpodo zassan : divers documents compiled by bunpodo, a publisher in the late edo era. in the holdings of the national archive – 625 items on 32r – 1 – Y480,000 – (with 50p guide; in japanese) – ja Yushodo [950]
Bunsen, Christian Karl Josias, Freiherr von *see*
- Analecta ante-nicaena
- Bibelgeschichte
- Bibelurkunden
- God in history
- Hippolytus and his age
- Outlines of the philosophy of universal history applied to language and religion
- Signs of the times

Bunsen, Ernest de *see*
- The angel-messiah of buddhists, essenes, and christians
- The chronology of the bible
- The keys of saint peter
- Das symbol des kreuzes bei allen nationen und die entstehung des kreuz-symbols der christlichen kirche
- Die ueberlieferung

Bunsen, Frances *see* A memoir of baron bunsen
Bunster, Arthur *see* Speech delivered by mr bunster, mp
Buntar *see* Organ russkikh anarkhistov-kommunistov
Bunte blaetter zur unterhaltung und belehrung *see* Neusser intelligenzblatt
Bunte bluethen : scherz und ernst in versen / Bruck, Julius – 2nd ed. New York: S Zickel 1880 [mf ed 1998] – 1r – 1 – (filmed with: clemens brentano und die brueder grimm / reinhold steig) – mf#9966 – us Wisconsin U Libr [810]
Bunte bluethen / Steinlein, A – La Crosse WI: J Ulrich 1884 [mf ed 1992] – 1r – 1 – (poems, chiefly in german (some in english). filmed with: briefe von alexander von humboldt an varnhagen von ense aus den jahren 1827 bis 1858) – mf#3183p – us Wisconsin U Libr [810]
Buntetojogi dontvenytar. – Budapest. On film: v1-27; 1908-35. LC set imperfect: v25 wanting. LL-0263 – 1 – us L of C Photodup [340]
Bunthorne abroad : or, the lass that loved a pirate / Bengough, John Wilson – [S.l: s.n, 1883?] [mf ed 1980] – 1mf – 9 – 0-665-00124-X – mf#00124 – cn Canadiana [830]
Bunting, Brian Percy *see* Rise of the south african reich
Bunting, Ian David *see* The consensus tigurinus and john calvin
Bunting, Jabez *see* Memoir of the late thomas holy, esq...
Bunting, Thomas Percival *see* The life of jabez bunting, dd
A bunvadi perrendtartas zsebkonyve, kiegeszitve az ujabb bunvadi eljarasi szabalyokkal; irtak edvi illes karoly es vargha ferenc / Hungary. Laws, Statutes, etc – 7. kiad. Budapest: Grill K., 1919. 687p. LL-4106 – 1 – us L of C Photodup [348]
Bunyan / Froude, James Anthony – London: Macmillan 1880 [mf ed 1990] – (= ser English men of letters (london, england)) – 1mf – 9 – 0-7905-5991-9 – mf#1988-1991 – us ATLA [420]
Bunyan / Froude, James Anthony – New York: Harper & Bros 1880 [mf ed 1980] – 1r – 1 – mf#67 – us Wisconsin U Libr [420]
Bunyan, John *see*
- Bunyan's awakening works
- Bunyan's consoling works
- Bunyan's devotional works
- Bunyan's directing works
- Bunyan's doctrinal works
- Bunyan's experimental works
- Bunyan's inviting works
- Bunyan's searching works
- Kuhamba kwomuhambi
- Leeto la mokreste
- Life of john bunyan
- Lwendo lwa muendi
- Pilgrim's progress
- The pilgrim's progress
- The pilgrim's progress from this world to that which is to come
- Pilgrim's robe
- Ugwalo lu ka bunyana ogutiwa uguhamba gwomhambi
- Ugwalo lu ka bunyane ogutiwa uguhamba gwomhambi
- The works of john bunyan

Bunyan's awakening works = Awakening works / Bunyan, John – Philadelphia: American Baptist Publication Society, 1851, c1850 – 1mf – 9 – 0-524-08334-7 – mf#1993-2024 – us ATLA [240]
Bunyan's consoling works = Consoling works / Bunyan, John – Philadelphia: American Baptist Publication Society, c1851 – 1mf – 9 – 0-524-08455-6 – mf#1993-3100 – us ATLA [240]
Bunyan's devotional works = Devotional works / Bunyan, John – Philadelphia: American Baptist Publiation Society, c1850 – 1mf – 9 – 0-524-08456-4 – mf#1993-3101 – us ATLA [240]
Bunyan's directing works = Directing works / Bunyan, John – Philadelphia: American Baptist Publication Society, c1851 – 1mf – 9 – 0-524-08457-2 – mf#1993-3102 – us ATLA [240]
Bunyan's doctrinal works = Doctrinal works / Bunyan, John – Philadelphia: American Baptist Publication Society, c1852 – 1mf – 9 – 0-524-08458-0 – mf#1993-3103 – us ATLA [240]
Bunyan's experimental works = Experimental works / Bunyan, John – Philadelphia: American Baptist Publication Society, c1852 – 1mf – 9 – 0-524-08459-9 – mf#1993-3104 – us ATLA [240]
Bunyan's inviting works = Inviting works / Bunyan, John – Philadelphia: American Baptist Publication Society, c1850 – 1mf – 9 – 0-524-08460-2 – mf#1993-3105 – us ATLA [240]
Bunyan's searching works = Searching works / Bunyan, John – Philadelphia: American Baptist Publication Society, c1851 – 1mf – 9 – 0-524-08566-8 – mf#1993-3151 – us ATLA [240]
Bunzel, Ulrich *see* Schlesien lebt
Bunzlauer sonntagsblatt – Bunzlau (Boleslawiec PL), 1843 31 dec-1844 – 1r – 1 – gw Misc Inst [077]
Bunzlauer stadtblatt – Bunzlau (Boleslawiec PL), 1915 1 jan-30 jun, 1916 1 jan-30 jun – 3r – 1 – gw Misc Inst [077]
Bunzlauische monathschrift zum nutzen und vergnuegen / ed by Bucquoi, Erdmann Friedrich et al – Bunzlau 1774-83 – (= ser Dz. abt literatur) – 10jge on 30mf – 9 – €300.00 – mf#k/n4478 – gw Olms [400]
[Buommattei, B] *see* Descrizion delle feste fatte in firenze per la canonizzazione di s. to andrea corsini
Buon lau / Hoang Ha – [Saigon] Ve Tinh 1974 [mf ed 1992] – on pt of 1r – 1 – mf#11052 r130 n11 – us Cornell [959]
Buon nhu doi nguoi : truyen dai / Nguyen, Thi Hoang – [Saigon] Doi Moi [1974] [mf ed 1993] – on pt of 1r – 1 – mf#11052 r453 n8 – us Cornell [830]
Buonagurio, Joseph J *see* The relationship between perceived stressful life events and occurrence of athletic injury among college level gymnasts
Buonaiuti, Ernesto *see* The programme of modernism
Buonanni, Filippo *see*
- Descrizione degl'istromenti armonici d'ogni genere, del padre bonanni
- Gabinetto armonico pieno d'istromenti sonori

Buonaparte et sa famille : ou confidences d'un de leurs anciens amis – Paris 1816 [mf ed Hildesheim 1995-98] – 4mf – 9 – €120.00 – 3-487-26354-8 – gw Olms [944]
Buonaparte und die bourbons : oder ueber die nothwendigkeit, dass sich frankreich, zu seinem eignen und ganz europa's glueck, mit seinen rechtmaessigen fuersten wieder vereinige / Chateaubriand, F A de – Berlin, 1814 [mf ed 1993] – 1mf – 9 – €24.00 – 3-89349-121-X – mf#DHS-AR 90 – gw Frankfurter [944]
Buonarroti, Filippo *see* Giornale patriottico di corsica
Buonarroti, Filippo M *see* Conspiration pour l'egalite dite de babeuf
Buondelmonti, Christophorus *see* Liber insularum archipelagi
Buonpensiere, Enrico *see* Commentaria in 1. p. summae theologicae s. thomae aquinatis, o.p., a q. 1. ad q. 23 (de deo uno)
Burachek, S *see* Trudy uchenykh i literatorov russkikh i inostrannykh
Burbank – 1972-75 – (= ser California telephone directory coll) – 22r – 1 – $1100.00 – mf#P00010 – us Library Micro [917]
[Burbank-] burbank leader – CA. mar 1909; oct 1910-jun 1911, dec 1911, 1912 (scats), 1913-14, 1916, 1923, 1927, 1927-28, 1931-81, 1984- – 504+ r – 1 – $30,240.00 (subs $2300/y) – (aka: burbank daily review) – mf#H03168 – us Library Micro [071]
Burbank daily review *see* [Burbank-] burbank leader
Burbidge, George Wheelock *see*
- A digest of criminal law of canada (crimes and punishments
- A general index to the statutes of new brunswick now in force, other than those contained in the consolidated statutes

Burbuja en el limbo / Dobles, Fabian – San Jose, Costa Rica. 1946 – 1r – us UF Libraries [972]
Burcardi gotthelffii struvii...corpus historiae germanicae / Struve, Burkhard Gotthelf & Buder, Christian Gottlieb – Ienae: Sumptibus I F Bielckii, 1730 [mf ed 1988] – (= ser Corpus historiae germanicae) – 1r – 1 – mf#SEM35P322 – cn Bibl Nat [943]
Burch, Derek George *see* Checklist of the woody cultivated plants of florida
Burchard 1. von worms und die deutsche kirche seiner zeit (1000-1025) : ein kirchen- und sittengeschichtliches zeitbild / Koeniger, Albert Michael – Muenchen: J J Lentner, 1905 [mf ed 1990] – (= ser Veroeffentlichungen aus dem kirchenhistorischen seminar muenchen 2 ser/6) – 1mf – 9 – 0-7905-6302-9 – (incl bibl ref) – mf#1988-2302 – us ATLA [240]
Burchard news – Burchard, NE: F A Harrison (wkly) [mf ed v2 n21. jul 3 1885-dec 4 1886 (gaps) filmed 1957] – 1r – 1 – (cont by: pawnee county news) – us NE Hist [071]
Burchard news – Burchard, NE: Swallow & Helmes, 1887 (wkly) [mf ed v4 n190. oct 0 [ie 8]-22 1887 filmed 1957] – 1r – 1 – (cont: pawnee county news) – us NE Hist [071]
Burchard times – Burchard, NE: A H Smith & W D Smith. v3 n24. jul 31 1896-apr 1936// (wkly) [mf ed -feb 19 1931 (gaps) filmed -1981] – 3r – 1 – (cont: pawnee county times. merged with: nebraska citizen to form: nebraska citizen and burchard times) – us NE Hist [071]
The burchard times – Burchard, NE: J C Hester & J P Swallow (wkly) [mf ed v4 n4. jun 10 1892=whole n160] – 1r – 1 – (cont by: pawnee county times) – us NE Hist [071]
Burchardi praepositi urspergensis chronicon (mgh7:16.bd) – ed 2a – (= ser Monumenta germaniae historica 7: scriptores rerum germanicarum in usum scholarum (mgh7)) – €7.00 – ne Slangenburg [221]
Burchardt, M *see* Die altkanaanaeischen fremdworte und eigennamen im aegyptischen
Burchardus, Johannes *see* Johannis burchardi, argentinensis, capelle pontificie sacrorum rituum magistri diarium, sive, rerum urbanarum commentarii (1483-1506)
Burchardus, U *see* Chronicvm...
Burchelati, B *see* Commentariorum memorabilium multiplicis hystoriae tarvisinae locuples promptuarium libris quatuor distributum...
Burchell, W J *see* Travels in the interior of southern africa
Burchell, William *see*
– Reisen in das innere von sued-afrika
– Travels in the interior of southern africa
Burchell, William John *see*
– Travels in the interior of southern africa
– Woodcut vignettes
Burchill, James Frederick *see* Analytical study of nielsen's commotio
Burchinal, Mary Cacy *see* Hans sachs and goethe
Burcio, Humberto F *see* La ceca de la villa imperial de potosi y la moneda colonial buenos aires, 1945
Burck, Joachim a *see* Zwantzig deutsche liedlein mit vier stimmen
Burckhard, C *see* Le mandat francais en syrie et au liban
Burckhard, Max *see* Das nibelungenlied
Burckhardt Heinrich *see* Aus dem walde
Burckhardt, Jacob *see* Die zeit constantin's des grossen
Burckhardt, Jakob *see* Griechische-kulturgeschichte
Burckhardt, Jakob Christoph *see* The cicerone
Burckhardt, John *see*
– Notes on the bedouins and wahabys
– Travels in arabia
– Travels in nubia
– Travels in syria and the holy land
Burckhardt, John L *see* Bemerkungen ueber die beduinen und wahaby
Burckhardt, Paul *see* Huldreich zwingli
Burckhardt-Biedermann, Theophil *see* Bonifacius amerbach und die reformation
Burdach, Konrad *see*
– Goethes eigenhaendige reinschrift des west-oestlichen divan
– Die schluss-szene in goethes faust
Burdach, Konrad et al *see* Kleine schriften
Burden, Harold Nelson *see* Life in algoma
The burden of the lord : aspects of jeremiah's personality, mission, and age / Thomson, W R – London, 1919 – 5mf – 8 – €12.00 – ne Slangenburg [221]
Burder, Henry Forster *see* Sermon delivered at hoxton chapel, on thursday, august 15, 1811, on occasion of the death of the...
Burder, Samuel *see* Memoirs of eminently pious women
Burdett, Staunton S *see*
– The baptist harmony
Burdett's official intelligence, 1882-1898 : from the guildhall library, london – 26r – 1 – mf#97541 – uk Microform Academic [941]
Burdick, Charles Kellog *see* Cases on the law of public service
Burdick, Francis Marion *see* The law of torts
Burdick, Lewis Dayton *see*
– Foundation rites, with some kindred ceremonies
– The hand
Burditt, B A *see*
– The germania; a collection of the most favorite operatic airs, marches, polkas, waltzes, dances, and melodies of the day
– The new germania.
Burdon, J A *see* Historical notes on certain emirates and tribes
Bureau d'amenagement de l'Est du Quebec *see* Rapport [de travail]
Bureau de statistique / Denmark. Statistiske Bureau – (= ser European official statistical serials, 1841-1984) – 2mf – 9 – uk Chadwyck [314]
Bureau der Katholischen Indianer-Missionen *see* Annalen der katholischen indianer-missionen von amerika
Bureau des commissaires d'ecoles catholiques romains de la cite de Montreal *see*
– [Aux honorables] membres du conseil executif, du conseil legislatif [et de l'as]semblee legislative de la province de Quebec
– Memoire presente au gouvernement de la province de quebec
Bureau des commissaires d'ecoles catholiques romains de Montreal *see* Notice sur les ecoles relevant du bureau...de la cite de montreal
Bureau drawer / Greater Madison Convention and Visitors Bureau – 1975 sum-1980 nov – 1r – 1 – mf#653737 – us WHS [978]
Bureau, Gabriel *see* Guyane meconnue
Bureau, Helene *see* Guide pratique pour le choix des professions feminines
Bureau international de l'heure *see* Bulletin horaire du bureau international de l'heure [bih]
Bureau memorandum – v2-v21 n1 [1960 nov-1980] – 1r – 1 – (cont: m r memorandum) – mf#349673 – us WHS [350]
Bureau memorandum – Washington DC 1976-79 – 1,5,9 – mf#11126 – us UMI ProQuest [370]
Bureau of american ethnology bulletins and annual reports, smithsonian institution : the culture and history of north and south american indian tribes – 1897-1971 [mf ed Mircofilming Corp of America] – 248v on 42r – 1 – (coll divided into 2 sect: bulletins (v1-200, 1887-1971). annual reports (v1-48, 1879/1880-1930/1931)) – us UMI ProQuest [305]
Bureau of Catholic Indian Missions (US) *see*
– Report of the bureau of catholic indian missions
– Report of the director of the bureau of catholic indian missions
Bureau Of Census And Statistics *see* Population census, 1951
Bureau of Foreign Trade, Ministry of Industry *see* Tsui chin san shih ssu nien lai chung-kuo t'ung shang k'ou an tui wai mao i t'ung chi
Bureau of indian affairs education research bulletin – 1973-76 – (= ser American indian periodicals... 1) – $95.00 – us UPA [350]
Bureau of indian affairs records created by the santa fe indian school, 1890-1918 – (= ser Records of the bureau of indian affairs) – 38r – 1 – (with printed guide) – mf#M1473 – us Nat Archives [350]
Bureau of Justice Statistics *see* Sourcebook of criminal justice statistics
Bureau of National Affairs *see*
– What's new in collective bargaining negotiations and contracts
– White collar report
Bureau of National Affairs [Washington DC] *see* Daily labor report
Bureau of National Affairs, Washington DC *see*
– Bna's union labor report
– Union labor report weekly newsletter
Bureau of Railway Economics, Washington, DC *see* Catalogue of railroad mortgages
Bureau of Social Affairs, the City Government of Greater Shanghai *see*
– Shang-hai shih chih kung tzu lu
– Shang-hai shih kung jen sheng huo fei chih shu
Bureau of Social Affairs, the City of Government of Greater Shanghai *see* Shang-hai shih lao tzu chiu fen t'ung chi
Bureau of social hygiene project and research files, 1913-1940 / Rockefeller University. Archives – 1980 – 31r – 1 – $4030.00 – (with printed guide) – mf#S1846 – Rockefeller Archive Center – us Scholarly Res [360]
Bureau of the budget bill reports : public laws, 87th and 88th congresses – (= ser Presidential documents series) – 23r – 9 – $3590.00 – 0-89093-363-4 – (with p/g) – us UPA [336]
Bureau Zuid-Molukken *see*
– Ambon beroept zich op recht en trouw
– De stem der ambonnezen
– Erkenning van ambon
– Rondom de affaire kapitein andi abdul azis
– De stem der ambonnezen rede
Bureaucracy a la mode / Latimer, Joseph W – 1924-25 – (= ser American indian periodicals... 2) – 1mf – 9 – $95.00 – us UPA [305]
Bureaucracy a la mode – n2-11 [1924 apr 15-1925 dec] – 1r – 1 – mf#1002661 – us WHS [350]
Bureaucracy in ghana : the civil service / Tiger, Lionel Samuel – London, 1962 – us CRL [350]
Bureaucrat – McLean VA 1972-93 – 1,5,9 – (cont by: public manager) – ISSN: 0045-3544 – mf#11101.01 – us UMI ProQuest [350]
Buren, J van *see* The new reformation
Buren, Martin van *see* Papers of martin van buren
Burenin-Petrov, N A *see* Ural'skii maiak
Buresch, K *see* Aus lydien
Burevestnik *see* Organ russkikh anarkhistov-kommunistov.
Burford, J B *see* Computer input microfilm (cim) feasibility study
Burford, R *see*
– Description of a view of canton, the river tigress, and the surrounding country...
– Description of a view of macao in china...
– Description of a view of the island and bay of hong kong...
Burford, Robert *see*
– Description of a view of the city of cairo
– Description of a view of the city of mexico
– Description of a view of the city of quebec
– Description of a view of the continent of boothia
– Description of a view of the falls of niagara
– Description of summer and winter views of the polar regions
Die burg – Krakau (Krakow PL), 1941 n3-1944 n2 – 1 – gw Misc Inst [077]
Die burg : vierteljahresschrift des instituts fuer deutsche ostarbeit krakau – Krakow, Poland 1940-44 [mf ed Norman Ross] – 2r – 1 – mf#nrp-781 – us UMI ProQuest [939]
Burg, H van den *see*
– Verzameling van uitgekorene zin-spreuken
– Zedige byschriften
Burg neideck / Riehl, Wilhelm Heinrich; ed by Jonas, Johannes Benoni Eduard – Boston: D C Heath c1907 [mf ed 1995] – (= ser Heath's modern language series) – 1r – 1 – (german text, int etc in english. filmed with: jean paul / richard benz) – mf#3715p – us Wisconsin U Libr [430]
Burgazki far – Burgas, Bulgaria. 4 Jul-15 Aug 1944 – 1r – 1 – us L of C Photodup [077]
Burgdorfer kreisblatt *see* Burgdorfer wochenblatt
Burgdorfer wochenblatt – Burgdorf b Lehrte [Kr Hannover DE], 1978 1 sep-1986 30 apr – 40r – 1 – (title varies: 1897: burgdorfer kreisblatt; filmed by other misc inst: 1867 2 jan-1885, 1887-1945 4 apr [77r]) – gw Misc Inst [074]
Burge, Lorenzo *see*
– Aryas, semites and jews
– Origin and formation of the hebrew scriptures
– Pre-glacial man and the aryan race
Burge, William *see* Observations on the supreme appellate jurisdiction of great britain.
Burgenlaendische freiheit : landesorgan der sozialistischen partei burgenlands – Eisenstadt, Austria 27 jul 1946-15 feb 1948 (imperfect) – 1r – 1 – uk British Libr Newspaper [325]
De burger – Bloemfontein. South Africa. 1894-97 – 3r – 1 – sa National [079]
Die burger : all ct eds – Cape Town, South Africa: Nasionale Pers, 1 Jul 1915- current – 553+ r – 1 (diazo also available at reduced price) – sa National [079]
Die burger – Kaapstad, South Africa: [Die nasionale pers beperk] jul 26 1915- (daily ex sun) [mf ed 19–] – 2r – 1 – (based on: 16 mei 1969; iss for 1969- also called: 54. jaarg-; in afrikaans) – us CRL [079]
Die burger = Oos-kaap uitgawe – Port Elizabeth SA, 1993- – 85+ r – 1,16 – sa National [960]
Burger, Heinz Otto *see*
– Annalen der deutschen literatur
– "Dasein heisst eine rolle spielen"
– Gedicht und gedanke
Burger, J P *see* English-lozi vocabulary
Burger, Magdalen Keith *see* Our india story
Burger, O F *see*
– Cucumber rot
– Lettuce drop
– Preliminary report on controlling melanose
Burger, Troy *see* Complex training compared to a combined weight training and plyometric training program
Burger- und realschule der israelitischen gemeinde zu frankfurt / Hess, Michael – Frankfurt am Main, Germany. 1857 – 1r – us UF Libraries [939]
Burger zeitung – Burg, Dithmarschen DE, 1950 18 feb-1853 19 nov – 2mf=3df – 9 – (publ in wilster. filmed with suppl) – gw Mikrofilm [074]
Burgersdijck, Franco *see* Idea philosophiae moralis
Die burger-senaat, 1795-1828 / Oberholster, Jacobus Johannes – U of Stellenbosch 1936 [mf ed S.l: s.n.] 1936] – 4mf – 9 – (incl bibl) – mf#mf.784 – sa Stellenbosch [323]
Burges, H *see* Address delivered by the rev h burges
Burges, Tom King *see* Orchestral development of the french horn to the time of johann sebastian bach
Burges, William *see*
– Art applied to industry
– The designs of william burges
Burgess a weekly review of cardiff – [Wales] Cardiff 13 feb 1879-20 feb 1880 [mf ed 2003] – 1r – 1 – uk Newsplan [072]
Burgess, C *see* The broken title of episcopal inheritance
Burgess, Charles F *see* Dancing
Burgess, Ernest Watson *see* Personality and the social group
Burgess, Frederick William *see* Chats on old coins
Burgess, G A *see* Free baptist cyclopaedia
Burgess, George *see*
– The lowliness of the episcopate
– Pages from the ecclesiastical history of new england during the century between 1740 and 1840
Burgess, Henry *see* Bible and lord shaftesbury
Burgess, Isaac Bronson *see* The life of christ
Burgess, James *see*
– The ancient monuments, temples and sculptures of india
– The buddhist stupas of amaravati and jaggayyapeta in the krishna district, madras
– The cave temples of India
– The chronology of modern india
– On the indian sect of the jainas
– On the muhammadan architecture...in gujarat
– Report on the antiquities in the bidar and aurangabad districts
– Report on the buddhist cave temples and their inscriptions
– The rock-temples of elephanta or gharapuri
Burgess, Jas *see* On the indian sect of the jainas
Burgess, John W *see* Recent changes in american constitutional theory
Burgess, O O *see* Consciousness, being, immortality; divine healing and christian science
Burgess, Otis Asa *see* A debate on total depravity, election, the polity or church government of the regular baptist church, free moral agency...
Burgess, Richard *see* Lectures on the insufficiency of unrevealed religion
Burgess, Thomas *see*
– Charge delivered to the clergy of the diocese of salisbury
– Greek original of the new testament asserted
– Greeks in america
– The people of the eastern orthodox churches, the separated churches of the east, and other slavs
– Sermon, preached at the anniversary of the royal humane society
Burgess, Thomas Joseph Workman *see*
– Abstract of a historical sketch of canadian institutions for the insane
– A historical sketch of our canadian institutions for the insane
– The lake erie shore as a botanizing ground
– Memorial notice richard maurice bucke
– Notes on the genus rhus
– Notes on the history of botany
– Obituary
– Thyroid feeding
– Two cases of ephemeral mania
– Valedictory address delivered to the graduates in medicine
Burgess, Walter H *see* John smith, the se-baptist and the pilgrim fathers helwys and baptist origins
Burgess, Walter Herbert *see* John smith the se-baptist, thomas helwys, and the first baptist church in england
Burgess, William *see*
– The bible in shakspeare [sic]
– The liquor traffic and compensation
– Nature and necessity of sober-mindedness
– Reciprocal duties of a minister and his people
– The religion of ruskin
Der burgfried – Duesseldorf DE, 1950-65 – 5r – 1 – (title varies: jul 1952: die heimat) – gw Misc Inst [074]
Der burggraf von nuernberg : oder, der hohenzollern weltgeschichtlicher beruf: historisches schauspiel in fuenf acten und einem vorspiel / Bauer, Hugo – 6. aufl. Berlin: H Bauer, 1888 [mf ed 1993] – 108p – 1 – mf#7780 – us Wisconsin U Libr [820]
Burgh, William *see*
– An exposition of the book of revelation
– Scriptural observance of the "first day of the week"
Burghaber, A *see* Theologia polemica...
Burghardt, W J *see* The image of god in man according to cyril of alexandria (sca14)
The burghersdorp gazette – Burgersdorp SA, 1872-73 (wkly) [mf ed Cape Town: SA library 1986] – 1r – 1 – (absorbed by: albert times) – mf#MS00436 – sa National [079]

Burghley, William Cecil *see* The execution of justice in england

Burghs' reformer – [Scotland] Leith: W L Rollo 19 mar 1859-27 jun 1863 (wkly) [mf ed 2004] – 2r – 1 – (merged with: leith herald and commercial advertiser to form: herald & reformer, and general advertiser) – uk Newsplan [072]

Burgis, D S *see* Investigation of some uncultivated native shrubs to determine metho...

Burgman, Charles F *see* Port orange, florida

Burgoa, Francisco de *see*
- Geografica descripcion
- Geografica descripcion de la parte septentrional del polo artico de la america
- Geografica descripcion de la parte septentrional del polo artico de la america y nueva iglesia de las indias occidentales...2 vol. mexico, 1934

Burgon, John William *see*
- The causes of the corruption of the traditional text of the holy gospels
- England and rome
- Inspiration and interpretation
- Last 12 verses of the bible according to st mark vindicated
- The last twelve verses of the gospel according to s. mark
- Lives of twelve good men
- A plain commentary on the four holy gospels
- Prophecy – not "forecast"
- The revision revised
- The servants of scripture
- The traditional text of the holy gospels

Burgos. archivo historico provincial. los protocolos del archivo historico provincial / Castro, P – Burgos, 1986 – 13mf – 9 – sp Cultura [946]

Burgos, Julia *see* Mar y tu

Burgos, Julia De *see* Obra poetica

Burgoyne, Charles G *see* Burgoyne's directory of lawyers practicing in new york city, 1883

Burgoyne, John Charles *see* Chronological account of india

Burgoyne's campaign, june-october, 1777 : justice to schuyler / Peyster, John Watts de – S.l: s.n, 1868? – 1mf – 9 – mf#32174 – cn Canadiana [975]

Burgoyne's directory of lawyers practicing in new york city, 1883 / Burgoyne, Charles G – New York, 1883 36 p. LL-1102 – 1 – us L of C Photodup [340]

Burgoyne's invasion of 1777 : with an outline sketch of the american invasion of canada, 1775-76 / Drake, Samuel Adams – Boston: Lee & Shepard, 1889 [mf ed 1980] – (= ser Decisive events in American history) – 2mf – 9 – 0-665-02755-9 – (incl ind) – mf#02755 – cn Canadiana [971]

Burgsdorff, Wilhelm von *see* Briefe an brinkmann, henriette v finckenstein, wilhelm v humboldt, rahel, friedrich tieck, ludwig tieck und wiesel

Burgt, Joannes Michael M van der *see* Het kruis geplant in een onbekend negerland van midden-afrika..

Burgt, Johannes Michael M van der *see* Un grand peuple de l'afrique equatoriale

Burguesito recien pescado / Gomez, Francisco Gregorio – Habana, Cuba. 1951 – 1r – us UF Libraries [972]

Burguete, Ricardo *see* Guerra! cuba

Burguieres, Jules M *see* Some notes on florida

Burguillos, aldea y basilica del siglo 17 / Martinez Martinez, Matias Ramon – Caceres: Tip. Enc. y Lib. Jimenez, 1904 – 1 – sp Bibl Santa Ana [946]

Burguillos, Pedro *see* Novena a jesus nazareno en su..

Burgundia, Antonium *see*
- Des wereldts proef-steen ofte de ydelheydt door de waerheyd beschuldight ende overtuyght van valscheydt
- Linguae vitia et remedia emblematice expressa...
- Mundi lapis lydius
- Mundi lapis lydius sive vanitas per veritatem falsi accusata et convicta

Burgus, Heinrich von *see* Heinrich von burgus der seele rat

Burhaneddin, Kazi *see* The divan project

Buri, Fritz *see* Gottfried kellers glaube

The burial chamber of the treasurer sobk-mose / Hayes, William C – MMA. 1939 – 9 – $10.00 – us IRC [930]

Burial registers for military posts, camps, and stations, 1768-1921 / U.S. Office of the Quartermaster General – (= ser Records of the office of the quartermaster general) – 1r – 1 – (with printed guide) – mf#M2014 – us Nat Archives [355]

Burial registers of the fulbourn asylum 1903-1950 – 1mf – 9 – £1.25 – uk CambsFHS [929]

The burial service : musical setting / Archer, Harry Glasier & Reed, Luther Dotterer – Philadelphia: General Council Publication Board, 1910 – 1mf – 9 – 0-524-07974-9 – mf#1990-5419 – us ATLA [240]

The burial service : a reply to an article by professor st george mivart...in the jan number of the "nineteenth century" review / Legg, John Wickham – London: SPCK 1897 [mf ed 1993] – (= ser Church historical society (series) 21) – 1mf – 9 – 0-524-07204-3 – mf#1990-5362 – us ATLA [242]

Buriatiia / Izdanie Verkhovnogo Soveta i Soveta Ministrov Buriatskoi SSR – Ulan-Ude: "Buriatiia" 1990- [dec 19 1990-95] [mf ed Minneapolis MN: East View Publ [199-] – 10r – 1 – mf#mf-12328 seemp – us CRL [077]

Buried cities and bible countries / St Clair, George – London: Kegan Paul, Trench, Truebner, 1891 – 2mf – 9 – 0-524-08836-5 – (incl bibl ref) – mf#1993-0059 – us ATLA [220]

The buried city of the east : nineveh / Buckingham, James – London [1851] [mf ed Hildesheim 1995-98] – 1v on 2mf – 9 – €60.00 – 3-487-27631-3 – gw Olms [930]

'Buried in cheshire?', vol 1-3 – (= ser Cheshire church registers) – 9 – (v1 [5mf] £5.50; v2 [5mf] £5.50; v3 [3mf] £4.50) – mf#74, 76, 250 – uk CheshireFHS [929]

Buried treasure / Saunders, H J – s.l, s.l? 1936 – 1r – us UF Libraries [978]

Buried treasure from aceto genealogical files – v1 n1-v3 n6 [1984 jan-1986 nov/dec] – 1r – 1 – mf#1238883 – us WHS [929]

Buried with christ in baptism : or, an essay on romans 6:3-4 / Boas, J – Carlisle PA: Steam Book & Job Print Est, [1876?] [mf ed 1993] – 1mf – 9 – 0-524-07280-9 – mf#1992-1057 – us ATLA [225]

Burk, Frederic *see* From fundamental to accessory in the development of the nervous system and of movements

Burk, William Herbert *see* The church handbook for teacher training classes

Burkart, Joseph *see* Aufenthalt und reisen in mexico in den jahren 1825 bis 1834

Burke, A J C *see* Resurrection in the acts of the apostles

Burke, Arthur Meredyth *see* Indexes to the ancient testamentary records of westminster

Burke, Barlow *see* Discovery problems in civil cases

Burke, Darren *see* A comparison of manual and machine assisted proprioceptive neuromuscular facilitation flexibility techniques

[Burke, E] *see* A philosophical inquiry into the origin of our ideas of the sublime and beautiful

Burke, Edmund *see*
- Politics in the age of revolution, 1715-1848
- Speech of edmund burke, esq member of parliament for the city of bristol
- Speeches of the right hon edmund burke
- Thoughts on the cause of the present discontents
- Works of edmund burke

Burke, J *see* Changes in spinal excitability preceding a voluntary movement in young and old adults

Burke, J B *see*
- Dormant and extinct peerages
- Encyclopaedia of heraldry
- Extinct and dormant baronetcies of england

Burke, Junius Jessel *see* Letters to a law student

Burke, Thomas *see* Thomas burke papers

Burke, Thomas Nicholas *see*
- Ireland's case stated in reply to mr. froude
- Lectures and sermons
- Lectures and sermons. second series

Burkes connaught journal *see* Connaught journal

Burkesville baptist church. burkesville, kentucky : church records – 1893-1966 – 1 – us Southern Baptist [242]

Burkhard, H *see* Schlagwortkatalog der monographien und periodika bis erwerbungsjahr 1964

Burkhard, Werner *see* Grimmelshausen, erloesung und barocker geist

Burkhardt, C A H *see* Briefe von goethes mutter an die herzogin anna amalia

Burkhardt, Carl August Hugo *see* Der historische hans kohlhase und heinrich von kleist's michael kohlhaas

Burkhardt, Gustav Emil *see* Die entwickelung der evangelischen mission

Burkhardt, Max *see* Heustecher

Burkitt, F C *see*
- Ephraim's quotations from the gospel
- The old latin and the itala
- The rules of tyconius

Burkitt, F Crawford *see*
- Evangelion da-mepharreshe

Burkitt, Francis Crawford *see*
- The book of rules of tyconius
- The earliest sources for the life of jesus
- Early christianity outside the roman empire
- Early eastern christianity
- Evangelion da-mepharreshe
- The failure of liberal christianity
- The gospel history and its transmission
- The old latin and the itala
- Two lectures on the gospels

Burkitt, Lemuel *see* A concise history of the kehukee baptist association

Burkitt, Miles Crawford *see* South africa's past in stone and paint

Burkitt, William *see* Help and guide to christian families

Burkley, Renee L *see* The effectiveness of behavioral contracts in promoting the maintenance of cancer risk reduction behavior while utilized in a college cancer avoidance course

Burla burlando / Marquez Sterling, Manuel – Habana, Cuba. 1907 – 1r – us UF Libraries [972]

Burla, Yehuda *see* Sipurim

Burlador / Lilar, Suzanne – Bruxelles, Belgium. 1945 – 1r – us UF Libraries [440]

Burlaton, Louis *see* Le venerable geronimo

Burleigh, B *see* Two campaigns

Burleigh, shrewsbury u leicester in schillers maria stuart / Kennel, Albert – Speyer: Dr Jaeger 1906 [mf ed 1991] – 1r – 1 – (filmed with: don karlos in der geschichte und in der poesie / richard pappritz) – mf#2873p – us Wisconsin U Libr [430]

Burleson family bulletin – v1 n1-v7 n4 [1981 aug-1988 feb] – 1r – 1 – mf#1544381 – us WHS [929]

Burlingame advance – CA. jun 1910-dec 1917; 1920-mar 1926 (wkly) – 10r – 1 – $600.00 – mf#B02081 – us Library Micro [071]

[Burlingame-] boutique and villager – CA. may 1965- – 37+ r – 1 – $2220.00 (subs $90/y) – (cont: burlingame villager) – mf#B02085 – us Library Micro [071]

[Burlingame-] burlingame advance star – CA. apr 1926-dec 1932; apr-sep 1946; sep 1949-feb 1950; jan 19 1951; 1967-69 (wkly) – 44r – 1 – $2640.00 – mf#B02082 – us Library Micro [071]

Burlingame Cemetery *see* Lot record book

[Burlingame-] the editor of burlingame – CA. jan 1948-nov 1949 – 1r – 1 – $60.00 – mf#C02084 – us Library Micro [071]

Burlingame-broadway editor – CA. oct 1944-dec 1947 – 1r – 1 – $60.00 – mf#C02083 – us Library Micro [071]

Burlington 1752-1849 – Oxford, MA (mf ed 1995) – (= ser Massachusetts vital record transcripts to 1850) – 9 – 0-87623-222-5 – (mf 1t-2t: marriages & intentions 1799-1849; family births & deaths 1752-1851. mf 2t: births & deaths 1809-46; b,m,d 1843-49) – us Archive [978]

The burlington advertiser – Burlington, N.J. Apr 13 1790-Dec 13 1791 – 1 – us NY Public [071]

Burlington & east-riding advertiser – [Yorkshire & Humberside] East Riding 8 jul 1843-26 sep 1844 [mf ed 2004] – 1r – 1 – uk Newsplan [072]

Burlington Fine Arts Club *see* Catalogue of specimens of japanese lacquer and metal work exhibited in 1894

Burlington Fine Arts Club, London *see*
- Catalogue of prints and books illustrating the history of engraving in japan
- Exhibition illustrative of the french revival of etching
- Exhibition of drawings and studies by sir edward burne-jones
- Illustrated catalogue
- Illustrated catalogue of specimens of persian and arab art

Burlington Fine Arts Club. London *see*
- Early german art
- English mezzotint portraits

Burlington free press – 1881 feb 22/1883-1954/1955 may 24 – 23r – 1 – (with gaps; cont by: burlington standard-press) – mf#964301 – us WHS [071]

Burlington gazette – Burlington Wl. 1859 may 14-jun 21 – 1r – 1 – (cont: rockton gazette; cont by: burlington weekly gazette) – mf#922774 – us WHS [071]

Burlington magazine – London UK 1984+ – 1,5,9 – ISSN: 0007-6287 – mf#14728.01 – us UMI ProQuest [700]

Burlington magazine for conisseurs – London. v1-37. 1903-1920 – 293mf – 9 – mf#O-497 – ne IDC [720]

Burlington reporter & burlington-quay fashionable advertiser – [Yorkshire & Humberside] East Riding jun 1843-6 sep 1844 [mf ed 2004] – 1r – 1 – uk Newsplan [072]

Burlington standard – Burlington Wl. 1863 oct 14-1965 oct 4, 1865 oct 11-1867 nov 27, 1867 dec 3-1872 jan 4, 1873 jan 2-1875 aug 26, 1875 sep 2-1877 jul 19, 1877 jul 26-1879 may 10, 1879 may 17-1882 sep 9, 1882 sep 16-1886 apr 3 – 8r – 1 – (cont by: standard democrat) – mf#1133828 – us WHS [071]

Burlington standard press – Burlington Wl. 1967 aug 3/sep 21-2001 nov/dec – 178r – 1 – (with gaps; cont: burlington standard-press) – mf#1133665 – us WHS [071]

Burlington standard-press – Lancaster Wl. 1955 jun 2/dec-1967 jun 15/jul 27 – 18r – 1 – (with gaps; cont: standard democrat, burlington free press; cont by: burlington standard democrat) – mf#1139536 – us WHS [071]

Burlington, Vermont. Gove Hill Christian Association *see* Trustees records

Burlington, Vermont. Vermont Baptist Home, Inc *see* Records

Burlington weekly gazette – Burlington Wl. 1859 jun 28-oct 18 – 1r – 1 – (cont: burlington gazette [burlington wi]; cont by: weekly burlington gazette) – mf#922780 – us WHS [071]

Burma – Rangoon: Office of the Supt Govt Printing. pt3. 1911 – (= ser Census of india) – 1 – us CRL [315]

Burma : a short study of its people and religion / Trotman, F E – Westminster: Society for the Propagation of the Gospel in Foreign Parts, 1917 [mf ed 1995] – (= ser Yale coll) – viii/151p (ill) – 1 – 0-524-09243-5 – mf#1995-0243 – us ATLA [280]

Burma. Bhun To Kri Kron Mya Tvan Tuin Ran Bhasa Panna San Kra Mhu Cum Cam Re Aphvai *see* Aci ran kham ca nhan ovada cariya chara to mya i ovada katha

Burma. Customs Dept *see* Annual statement of the sea-borne trade and navigation of burma with foreign countries and indian ports

Burma. Dept of Agriculture *see* Report on the operations of the dept of agriculture, burma, for the year ended...

Burma. Dept of Land Records and Agriculture *see* Report on the department of land records and agriculture, burma, for the year...

Burma. Excise Dept *see* Report on the administration of the excise department in burma

The burma gazette – Rangoon: [s.n. 1886-[wkly] [mf ed [London]: India Office Library & Records 1976-] – 1 – (iss in pt: pt1: local gazette, notifications by government [varies]; pt2: extracts from the gazette of india other than acts and bills of the governor-general in council [varies]; pt3: acts and bills of the governor-general in council [varies]; pt4: notifications by the courts and heads of departments [varies]; later iss incl additional pt: acts, bills, proceedings of the burma legislative council; mandalay suppl also publ; some iss accompanied by various other suppl; in english & burmese jan 6 1951- ; with ind; iss for oct-dec 1886 [pt1-3] filmed with: british burma gazette jan-sep 1886; cont: british burma gazette; absorbd by: upper burma gazette) – us CRL [959]

The burma gazette – Rangoon: [s.n. 1886-[wkly [mf ed sep 13 1913-dec 29 1923 New York: NYPL 1969] – 1 – (with ind; some iss incomplete or missing; iss in pt: pt1: local gazette, notifications by government [varies]; pt2: extracts from the gazette of india other than acts and bills of the governor-general in council [varies]; pt3: acts and bills of the governor-general in council [varies]; pt4: notifications by the courts and heads of departments [varies]; later iss incl additional pt: acts, bills, proceedings of the burma legislative council; mandalay suppl also publ; some iss accompanied by various other suppl; in english & burmese jan 6 1951- ; cont: british burma gazette and: upper burma gazette) – us CRL [959]

Burma. Judicial Dept *see* Report on the administration of civil justice in burma

Burma. Laws, Statutes, etc *see* The land nationalization bill, 1948

The burma mission herald – Rangoon, Burma: Burma Mission of the Methodist Episcopal Church, 1904 [semiannual] [mf v1-35 1904-41] – 1r – 1 – (lacks: several iss) – mf#2003-s029 – us ATLA [242]

Burma past and present with personal reminiscences of the country / Fytche, A – London: C Kegan Paul & Co, 1878. 2v – 9mf – 9 – mf#SE-20141 – ne IDC [915]

Burma. Tuin Ran Bhasa Panna nhan Asak Mve Vam Kron Panna Re Pru Pran Ci Mam Mhu Komiti *see* Aci ran kham ca

Burmah Baptist Missionary Convention *see*
- Annual report of the...1873-
- Annual report of the...1866-1871
- Minutes of a missionary convention at which was formed the burmah baptist missionary convention

The burman – Rangoon, Burma. Dec 1945-sept 1963 – 53r – 1 – us L of C Photodup [079]

Burman, Debajyoti *see* The english works of raja rammohun roy

Burman, F *see*
- Orationes
- Synopsis theologiae

Burman, Jose *see* Who really discovered south africa?

Burmann, Karl *see* Im dunklen erdteil

Burmeister, H *see* Reise durch die la plata-staaten, mit besonderer ruecksicht auf die physische beschaffenheit und den culturzustand der argentinischen republik

Burmeister, Hermann *see*
- Reise durch die la plata-staaten
- Viagem ao brasil

Burmese Baptist Missionary Convention *see* Minutes of the...annual meeting of the...

The burmese empire a hundred years ago / Sangermano, V – Westminster, 1893 – 4mf – 9 – mf#SE-20209 – ne IDC [915]

The burmese review and monday new times – Rangoon: "Burmese Review" Press, [1948-1960]. [jan 9 1956-jul 25 1960] – 10r – us CRL [079]

Burn, A E *see*
- The athanasian creed and its early commentaries
- Niceta of remesiana

Burn, A U *see* Facsimiles of the creeds from early manuscripts (hbs36)

Burn, Andrew Eubank *see* An introduction to the creeds and to the te deum

Burn, Andrew Ewbank *see*
- The apostles' creed
- The athanasian creed and its early commentaries
- Niceta of remesiana

Burn, David *see* Vindication of van diemen's land

Burn, George *see* Modern science

Burn, John Ilderton *see* Familiar letters on population, emigration, home colonization etc

Burn, Richard *see*
- Le juge a paix, et officier de paroisse
- A new law dictionary

Burn, Robert *see* Kita-b al-qudus...

Burn, Robert Scott *see*
- The colonist's and emigrant's handbook of the mechanical arts
- Ornamental drawing, and architectural design

Burn, William Scott *see* The connection between literature and commerce

Burnaby advertisor – Burnaby, British Columbia, CN. 1929-64 – 8r – 1 – cn Commonwealth Imaging [071]

Burnaby broadcast – Burnaby, British Columbia, CN. 1926-35 – 2r – 1 – cn Commonwealth Imaging [073]

Burnaby news – British Columbia, CN. 1987- – 2r per yr – 1 – cn Commonwealth Imaging [071]

Burnaby news courier – Burnaby, British Columbia, CN. jan 1944-aug 1953 – 5r – 1 – cn Commonwealth Imaging [071]

Burnaby now – Burnaby, British Columbia, CN. nov 1983-dec 1987 – 5r – 1 – cn Commonwealth Imaging [071]

Burnaby post – British Columbia, CN. may 1934-nov 1937 – 1r – 1 – cn Commonwealth Imaging [071]

Burnaby sunday news – Burnaby, British Columbia, CN. jan 1991-dec 1992 – 4r – 1 – cn Commonwealth Imaging [071]

Burnaby today – British Columbia, CN. oct 1979-dec 1981 – 2r – 1 – cn Commonwealth Imaging [071]

Burnam *see* Vitruvius pollio, 1556

Burnand, F C *see* Bishop colenso utterly refuted

Burnap, George Washington *see* Lectures on the sphere and duties of woman, and other subjects

Burnap, George Washington et al *see* [Unitarian doctrines]

Burnat, Eugene *see* Lelio socin

Burne-Jones, Edward Coley, 1st Baronet *see* Borough of birmingham museum and art gallery

Burnes, A *see* Travels into bokhara

Burnes, Alexander *see* Travels into bokhara

Burnes, James *see* A narrative of a visit to the court of sindea

Burnes, William *see* A manual of religious belief

Burnet, Gilbert *see* History of the reformation

Burnet, J *see* Authentic report of the discussion which took place between the rev...

Burnet, John *see*
- An essay on the education of the eye
- The ethics of aristotle
- Practical essays on various branches of the fine arts
- The progress of a painter in the nineteenth century

The burnett blade – Burnett, NE: A E Shelton. v1 n1. aug 14 1884- (wkly) – 1r – 1 – (publ also in battle creek apr 9 1885-) – us Bell [071]

Burnett county enterprise – Webster WI. 1917-21, 1922-25, 1926-1929 nov 7, 1946-48, 1949 jan 7-1951 sep 7 – 5r – 1 – (cont by: journal of burnett county and the times; journal of burnett county and the times and burnett county enterprise) – mf#948838 – us WHS [071]

Burnett county leader – Siren WI. 1944-46, 1947-50, 1951-52, 1953 jan-1955 mar – 4r – 1 – (cont by: inter-county leader [frederic wi]) – mf#931289 – us WHS [071]

Burnett county sentinel – Alpha, Branstad, Danbury etc WI. 1962 nov 22/1964 jun 10-1995 jan-apr [gaps] – 68r – 1 – mf#1144804 – us WHS [071]

Burnett county sentinel – Grantsburg WI. 1875 aug 20, 1876 jan 14/1877 mar 16-1908 jan 16-1910 jan 27 – 15r – 1 – (cont by: journal of burnett county; journal of burnett county and burnett county sentinel) – mf#923552 – us WHS [071]

Burnett, Daniel Frederick *see* Cases on the law of private corporations.

Burnett, Edmund C *see* Letters of members of the continental congress

Burnett, James J *see* Tennessee pioneer baptist preachers

Burnett, Madeline L *see* The development of american hymnody, 1620-1900

Burnett, Peter Hardemann *see*
- The path which led a protestant lawyer to the catholic church
- Reasons why we should believe in god, love god, and obey god

Burnett, Thomas R *see* Center shots

Burnett's reports / Wisconsin. Supreme Court – 1v. 1839-1843 (all publ) – (= ser Wisconsin supreme court reports; Pinney's reports) – 3mf – 9 – $4.50 – (for purposes of pre-nrs coverage, all of burnett's cases are incl in pinney) – mf#LLMC 91-302 – us LLMC [347]

Burney, C *see*
- Carl burney's der musik doctors tagebuch seiner musikalischen reisen
- The present state of music in france and italy
- The present state of music in germany, the netherlands, and united provinces

Burney, C F *see* Israel's hope of immortality

Burney, Charles *see*
- 6 cornet pieces
- Account of the musical performances in westminster abbey and the pantheon, may 26th, 27th, 29th
- Dr karl burney's nachricht von georg haendel's lebensumstaenden
- Four sonatas or duets for two performers on one pianoforte or harpsichord
- Memoirs of the life and writings of the abate metastasio
- Present state of music in france and italy
- Present state of music in germany, the netherlands, and united provinces
- Second set of four sonatas or duets

Burney, Fanny *see*
- Diary and letters of madame d'arblay
- Diary and letters of madame d'arblay ed by ger niece [charlotte barrett]

[Burney-] inter-mountain news – CA. 1963; 1966; 1970; 1977; 1979- – 15+ r – 1 – $900.00 (subs $50/y) – mf#B02086 – us Library Micro [071]

Burney, James *see* A chronological history of the north-eastern voyages of discovery

Burnham, Sherburne Wesley *see* General catalogue of double stars within 121° of the north pole

Burnham, Sylvester et al *see* Gospel from two testaments

Burnham-on-sea gazette and highbridge express – Burnham-on-Sea, England 5 sep 1947-31 may 1983 – 3r – 1 – (cont: burnham-on-sea gazette, residents' & visitors' list [24 feb 1917-30 aug 1947]; cont by: burnham-on-sea gazette & highbridge express [5 sep 1947-31 may 1983]; burnham & highbridge gazette [7 jun 1983-12 mar 1986, 24 sep 1986-10 oct 1989]) – uk British Libr Newspaper [072]

Burnichon, Joseph *see* Bresil d'aujourd'hui

Burnier, L *see* Histoire litteraire de l'education morale et religieuse en france et dans la suisse romande

Burnier, Raymond *see* Hindu medieval sculpture

Burnier, Theophile *see* Ames primitives

Burning at santiago but an accident in the mariolatry of the church / Macklin, Thomas – Glasgow, Scotland. 1881 – 1r – us UF Libraries [240]

Burning bush / Metropolitan Church Association – v75 n1 [1976 jan/feb], v77 n1-v88 n6, [1978 jan/feb-1989 nov/dec – 1r – 1 – mf#1700770 – us WHS [242]

Burning bush not consumed / Dickerson, Philip – London, England. 1862? – 1r – us UF Libraries [240]

Burning of st pierre and the eruption of mont pel / Royce, Frederick – Chicago, IL. 1902 – 1r – us UF Libraries [972]

Burning of the reichstag / Lubbe, Marinus Van Der – London, England. 1934 – 1r – us UF Libraries [025]

Burning questions of the life that now is and of that which is to come / Gladden, Washington – New York: Century, 1890 – 1mf – 9 – 0-7905-3444-4 – mf#1987-3444 – us ATLA [240]

Burning river – Cleveland OH. 1970 apr 15/28-jun 18/jul 1 – 1r – 1 – (cont: burning river oracle) – mf#1053739 – us WHS [071]

Burning river news – Cleveland OH. 1969 oct 28-1970 feb 6/mar 6 – 1r – 1 – (cont: big us; cont by: burning river oracle) – mf#765348 – us WHS [071]

Burning river oracle – Cleveland OH. 1970 mar 3/17-mar 31/apr 13 – 1r – 1 – (cont: burning river news; cont by: burning river [cleveland oh]) – mf#1056240 – us WHS [071]

Burning spear / African People's Socialist Party et al – v1 n10-v2 n5 [1970 oct 13/27-1971 jul], 1971 nov, v3 n1-9 [1973 nov-1974 sep 15/oct 15], v4 n12 [1977 jun], v5 n8-v6 n11 [1978 jul-1979 dec], 1980-83, 1984 jan-1986 sep, 1986 oct-1990 dec, [1990 feb-1998 sep/oct] – 6r – 1 – mf#400606 – us WHS [325]

Burnley advertiser – [NW England] Burnley Lib 1852-80 – 1 – uk MLA; uk Newsplan [072]

Burnley and east lancashire mid-weekly gazette – [NW England] Burnley Lib 5 mar 1884-1888 – 1 – uk MLA; uk Newsplan [072]

Burnley citizen – [NW England] Burnley Lib feb 1983-mar 1984 – 1 – uk MLA; uk Newsplan [072]

Burnley evening star : rossendale front – [NW England] Burnley Lib oct 1965-aug 1967 – 1 – (title change: evening star: rossendale front [9 feb 1970-30 jun 1971; 1 jul 1971-aug 1983]) – uk MLA; uk Newsplan [072]

Burnley express and clitheroe division advertiser – [NW England] Burnley Lib 1877-1995 – 1 – uk MLA; uk Newsplan [072]

Burnley free press and general advertiser for east lancashire – [NW England] Burnley Lib 1863-jan 1864 – 1 – uk MLA; uk Newsplan [072]

Burnley gazette and east lancashire advertiser – [NW England] Burnley Lib 1864-1915 – 1 – uk MLA; uk Newsplan [072]

Burnley mentor – [NW England] Burnley Lib nov 1852-oct 1853 – 1 – uk MLA; uk Newsplan [072]

Burnley news – [NW England] Burnley Lib 16 nov 1912-1933 – 1 – uk MLA; uk Newsplan [072]

Burnouf, Emile *see* The science of religions

Burnouf, Eugene *see*
- Commentaire sur le yacna, l'un des livres liturgiques des parses
- Essai sur le pali
- Extrait d'un commentaire et d'une traduction nouvelle du vendidad sade
- Legends of indian buddhism
- Observations grammaticales sur quelques passages de l'essai sur le pali de mm e burnouf et lassen
- Observations sur la partie de la grammaire comparative de m f bopp

Burns – Oxford UK 1974+ – 1,5,9 – ISSN: 0305-4179 – mf#13961 – us UMI ProQuest [617]

Burns and Elliott *see* Calgary, alberta, canada, her industries and resources

Burns, Arthur Lee *see* Peace-keeping by un forces

Burns, Emile *see* Spain

Burns, George *see* National church a national treasure

Burns, I *see* Memoir of the rev wm c burns, missionary to china

Burns, Islay *see*
- Catholicism and secretarianism
- The first three christian centuries
- The history of the church of christ
- Memoir of the rev wm c burns

Burns, James Aloysius *see* The catholic school system in the united states

Burns, James C *see* How the spirit of god may be quenched

Burns, John *see*
- A sermon preached in the presbyterian church in stamford, upper canada, on the 3rd day of june, 1814
- True patriotism

Burns, Julia A *see* The impact of the la crosse wellness project on the health promotion involvement of college students residing on the campus of the university of wisconsin-la crosse

Burns, Martha J *see* Tourism marketing on the world wide web

Burns, Nelson *see* Autobiography of the late rev nelson burns

Burns news *see* Burns times-herald

Burns news (burns, or) *see* Times-herald (burns, or)

Burns news [burns, or] – Burns OR: D Mullarky, 1926-29 [wkly] [mf ed 1967-72] – 1r – 1 – (merged with: times-herald [burns, or] to form: burns times-herald [1930-]; cont: harney county news [burns, or: 1913]; 1929 incl newspaper publ...by burns high school students) – us Oregon Lib [071]

Burns Paiute Indian Reservation *see*
- Tu kwa hone newsletter

Burns Philp (South Sea) Co Ltd, Labasa Branch, Fiji *see*
- Business correspondence
- Managers' reports on annual balances
- Miscellaneous correspondence

Burns press – Burns OR: S D Pierce, 1931-32 [wkly] [mf ed 1965] – 1r – 1 – (cont: free press [1930-31]; cont by: free press [1932-40]) – us Oregon Lib [071]

Burns, Robert *see*
- Lecture on the use of the episcopal liturgy in presbyterian churche...
- Poetical works of robert burns
- Scottish voluntaryism the atheist's ally
- The songs of robert burns
- The works of robert burns
- The works of robert burns

Burns, Robert Ferrier *see*
- The maine liquor law
- Maple leaves from canada, for the grave of abraham lincoln

Burns times *see*
- East oregon herald
- Times-herald (burns, or)

Burns times-herald – Burns OR: Burns Times-Herald Inc, 1930- [wkly] – 1 – (merger of: times-herald [1896-1929]; burns news [1926-29]; 1930 incl newspaper publ...by burns high school students) – us Oregon Lib [071]

Burns times-herald *see* Times-herald (burns, or)

Burns, W *see* Sons of the soil

Burns watchmaker and jeweller, north side of the market-square, saint john, new brunswick : has always on hand a choice selection of the following articles... – S.l: s.n, 18–? – 1mf – 9 – mf#53051 – cn Canadiana [680]

Burns, William John *see* The masked war

Burnside, Helen Marion *see*
- Robinson crusoe

Burnside, Robert *see* Fruits of the spirit, the ornaments of christians

Burnside, William *see* Theory of groups of finite order

Burnsview baptist church. spartanburg county. south carolina : church records – 1921-72 – 1 – us Southern Baptist [242]

The burnt offering / Cotes, Everard, mrs [Sara Jeanette Duncan] – London: Methuen, 1909 – 4mf – 9 – 0-665-77191-6 – (first publ in 1909; incl aut's and publ's list) – mf#77191 – cn Canadiana [830]

Burokratischer verwaltungsstaat und soziale demokratie / Sultan, Herbert – Hannover, Germany. 1955 – 1r – us UF Libraries [025]

Burpee, Lawrence J *see* Flowers from a canadian garden

Burque, Francois-Xavier *see* Elevations poetiques

Burr, Aaron *see* Reports of the trials of colonel aaron burr, (late vice president of the united states), for treason.

Burr, Agnes Rush *see* Russell h. conwell

Burr, Anna Robeson Brown *see*
- The autobiography
- Religious confessions and confessants

Burr, Enoch Fitch *see* Universal beliefs

Burr, Frank A *see* A new, original and authentic record of the life and deeds of general u s grant

Burr, Jonathan Kelsey et al *see* Job, proverbs, ecclesiastes, and solomon's song

The burr mcintosh monthly guide *see* American theatre periodicals of the nineteenth and early twentieth centuries

Burr oak – 1853 oct 7-1854 dec 29 – 1r – 1 – (cont: dodge county gazette) – mf#927159 – us WHS [071]

The burr oak – Burr, NE: John Cornell Lee, 1896 (wkly) [mf ed v1 n5. jun 26-dec 11 1896] – 1r – 1 – us NE Hist [071]

Burr star – Burr, NE: S W McCoy. v1 n1. mar 20 1897-1897// (wkly) [mf ed -jul 30 1897 (lacks jul 16 1897)] – 1r – 1 – (absorbed by: sterling eagle) – us NE Hist [071]

Burr, William Henry *see* The doctrine of hell

Burrage, Champlin *see*
- The church cevenant idea
- The church covenant idea
- The early english dissenters in the light of recent research
- John penry
- Nazareth and the beginnings of christianity
- New facts concerning john robinson
- A "new years guift"
- The "retractation" of robert browne
- The true story of robert browne (1550?-1633), father of congregationalism

Burrage, Henry S *see*
- The act of baptism in the history of the christian church
- Baptist hymn writers and their hymns

Burrage, Henry Sweetser *see*
- A history of the anabaptists in switzerland
- History of the baptists in maine
- A history of the baptists in new england

Burrangong argus – Burrangong, apr 1865-des 1907 – 12r – 9 – A$761.46 vesicular A$827.46 silver – at Pascoe [079]

Burrangong chronicle – Young, dec 1873-jul 1876 – 1r – at Pascoe [079]

Burrard, S G *see* A sketch of the geography and geology of the himalaya mountains and tibet

Burre, Paul *see* Es reiten die wilden jaeger

Burrell collection – 104mf – 9 – $630.00 – 0-907006-48-5 – (visual catalogue of complete coll of fine & applied art in the important new museum opened in 1983; 7000 captioned photos with printed ind) – uk Mindata [700]

Burrell, David James *see*
- The religions of the world
- The teaching of jesus concerning the scriptures
- The wonderful teacher and what he taught

Burrell, Mary Banks [comp] *see* Richard wagner

Burress/burrows/burroughs researchers – v1 n1-4 [1986 win-1987 fall] – 1r – 1 – mf#1609408 – us WHS [071]

Burrill, Alexander *see* New law dictionary and glossary
Burrill, Alexander Mansfield *see* A treatise on the law and practice of voluntary assignments for the benefit of creditors.
Burris, F Holiday *see* The trinity
Burritt, Elihu *see* Chips from many blocks
Burritt, Elijah Hinsdale *see* The geography of the heavens and class book of astronomy
O burro magro : jogando de garupa nas placas russianas – Rio de Janeiro, RJ: Typ Brasiliense, 23 nov 1833-10 jan 1834 – (= ser Ps 19) – mf#P15,01,57 n02 – bl Biblioteca [321]
Burrough green 1571-1950 – (= ser Cambridgeshire parish register transcript) – 5mf – 9 – £6.25 – uk CambsFHS [929]
Burroughs clearing house – Detroit MI 1916-80 – 1,5,9 – ISSN: 0007-6341 – mf#1766 – us UMI ProQuest [332]
Burroughs, Henry *see* A historical account of christ church, boston
Burroughs, John *see*
– The light of day
– Sharp eyes
– Time and change
– Whitman
Burroughs, Joseph Washington *see* Farm journals
Burroughs, P E *see*
– The convention system of teacher training
– Meet b. h. carroll
Burroughs, Stephen *see* His excellency lord gosford, the governor-general of the canadas etc etc
Burroughs Wellcome and Co *see* Crown and realm
Burroughs, William Henry *see* A treatise on the law of taxation as imposed by the states and their municipalities.
Burrow, James *see* Burrow's reports
Burrowa – Boorowa, jun 1874-dec 1968 – 29r – A$1748.67 vesicular A$1908.17 silver – (aka: boorowa news) – at Pascoe [079]
Burrowa – Boorowa, jun 13 1874-dec 19 1968 – (= ser Burrowa news) – 29r – 9 – A$1748.67 vesicular A$1908.17 silver – (aka: burrowa news) – at Pascoe [079]
Burrows, Charles Acton *see*
– The canadian pacific railway telegraph
– North western canada, its climate, soil and productions
Burrows, G *see*
– The curse of central africa
– The land of the pigmies
Burrows, H. Lansing *see* Miscellaneous manuscript works
Burrows, J L *see* A christian merchant
Burrows, J Lansing *see* American baptist register for 1852
Burrows, John Lansing *see* What baptists believe
Burrows, Montagu *see* Wiclif's place in history
Burrow's reports : reports of cases argued and adjudged in the court of kin's bench during the time of lord mansfield... / Burrow, James – 4th ed. v1-5. 1756-72. London: A Strahan, 1790 (all publ) – (= ser Burrow's reports tempore mansfield) – 33mf – 9 – $49.50 – (all 5v were publ by strahan in 1790, but v1-3 are called "4th ed corr". v4-5 are called "2nd ed corr". title also known as: burrow's reports tempore mansfield) – mf#LLMC 84-757 – us LLMC [324]
Burr-Reynaud, Frederic *see*
– Anacaona
– Visages d'arbres et de fruits haitiens
Burry, Frederic W *see* Twelve essays
Bursa mecmuasi – Bursa: Vilayet Matbaas, 1916-20. Yayimliyan: Bursa Muhibleri Cemiyeti; Mueduerue: Muhiddin Baha [pars] n1. 1 kanuniewel 1917-6,8. 15 mart 1918 – (= ser O & t journals) – 2mf – 9 – $40.00 – us MEDOC [956]
Burschen heraus! : roman aus der zeit unserer tiefsten erniedrigung / Sperl, August – 11. aufl. Muenchen: C H Beck 1922 [mf ed 1991] – 1r – 1 – (filmed with: die alten und die jungen / conrad alberti) – mf#2954p – us Wisconsin U Libr [830]
Bursics, Zoltan *see* A magyar bunvadi eljarasi jog vazlata. 3. kiad
Bursma, A *see* Het profetische woord
Burssens, Amaat *see*
– Manuel de tshiluba (kasayi, congo belge)
– Tonologische schets van het tshiluba (kayasi, belgisch kongo)
Burssens, Amaat F S *see* Inleiding tot de studie van de kongolese bantoetalen
Burssens, Amaat Fs *see* Manuel de tshiluba (kasayi, congo belge)
Burt, A W *see* Lessons in literature for high school entrance examinations 1892-1893
Burt, Armistead *see* Letters, 1847-1848
Burt county herald – Tekamah, NE: W H Korns, 1884-aug 1942// (wkly) [mf ed 1892,1895-1942 (gaps) filmed in 1970] – 28r – 1 – (absorbed: tekamah journal (1903), 1922; burt county tribune 1929. issues for jul 4 1929-may 15 1930 called also tribune v3 n50-v4 n43) – us NE Hist [071]
Burt county news – Craig, NE: L H Warner (wkly) [mf ed 1892-1900 (gaps)] – 3r – 1 – (merged with: craig advertiser to form: craig advertiser and the burt county news) – us NE Hist [071]
Burt county plaindealer – Tekamah, NE: Plaindealer Pub Co. v1 n1. apr 19 1934- (wkly) – 1 – (lacks: nov 25 1948-feb 10 1949 and mar 31-dec 29 1966. cont: tekamah news. absorbed: decatur advertiser) – us NE Hist [071]
Burt, Edward *see* Letters from a gentleman in the north of scotland to his friend in london
Burt, Henry Martyn *see* Burt's illustrated guide of the connecticut valley
Burt letters, 1847-1848 *see* Letters, 1847-1848
Burte, Hermann *see* Sieben reden
Burtner, Patricia A *see* Mechanical and muscle activation characteristics during crouch stance balance in children with spastic cerebral palsy
Burton, Catharine *see* An english carmelite
Burton, Catherine J *see* The heart rates of elementary children during physical education classes
Burton chronicle and barton etc and tutbury journal – Burton on Trent 18 oct 1860-96, 1898-1908, 1910-n5032 20 jun 1957 [mf ed only 1872, 1898] – 1 – (incorp with: burton observer) – uk Newsplan [072]
Burton, Clarence Monroe *see* A sketch of the life of antoine de la mothe cadillac
Burton, Clarence Monroe [comp] *see* "Cadillac's village", or, "detroit under cadillac"
Burton, E *see* Three primers put forth in the reign of henry 8th
Burton, Edward *see*
– A description of the antiquities and other curiosities of rome
– Remarks upon a sermon, preached at st mary's on sunday
Burton, Edwin Hubert *see* The life and times of bishop challoner (1691-1781)
Burton, Ernest De Witt *see*
– Four letters of the apostle paul
– The life of christ
– Notes on new testament grammar
– The origin and teaching of the new testament books
– An outline handbook of the life of christ
– The purpose and plan of the gospel of john
– The purpose and plan of the gospel of luke
– The purpose and plan of the gospel of mark
– The purpose and plan of the gospel of matthew
– Report on christian education in china
– A short introduction to the gospels
– Some principles of literary criticism and their application to the synoptic problem
– Studies in the gospel according to mark
– Syntax of the moods and tenses in new testament greek
– Syntax of the moods and tenses of new testament greek
Burton, Ernest DeWitt *see*
– A handbook of the life of the apostle paul
– A harmony of the gospels for historical study
Burton, Ernest DeWitt et al *see* Biblical ideas of atonement
Burton Family *see* Reunion records
Burton, Francis Nathaniel *see* Extraits
Burton independent – Burton, NE: Frank W McManis. -v10 n36. aug 16 1917 (wkly) [mf ed v5 n1. dec 7 1911-aug 16 1917 (gaps) filmed 1979] – 2r – 1 – (cont: springview independent) – us NE Hist [071]
Burton, Isabel, Lady *see* Arabia, egypt, india
Burton, J E *see* Essay on comparative agriculture
Burton, J W *see* The call of the pacific
Burton, John *see*
– The duty and reward of propagating principles of religion and virtue exemplified in the history of abraham
– The french canadian, imperium in imperio
Burton, John Hill *see* Life and correspondence of david hume
Burton, John Wear *see* The fiji of to-day
Burton, Joseph *see* Ministrial usefulness
Burton, Joseph F et al *see* Letters and instructions for church officers
Burton mail – Burton-upon-Trent, England 2 may-26 sep 1898, 2 jan 1899-30 apr 1907 – 1 – (cont by: burton daily mail [2 may 1907-21 mar 1981]; burton mail [23 mar 1981- (mf 1986-)]) – uk British Libr Newspaper [072]
Burton, Margaret E *see*
– The education of women in china
– The education of women in japan
Burton, Margaret Ernestine *see* Comrades in service
Burton, Marion Le Roy *see* The problem of evil
Burton, Nathanael *see* History of the royal hospital, kilmainham, near dublin
Burton, R *see* First footsteps in east africa
Burton, R F *see*
– The lake regions of central africa, a picture of exploration
– A mission to gelele, king of dahome
– Two trips to gorilla land and the cataracts of the congo
[Burton, R F] *see* Wanderings in west africa from liverpool to fernando po
Burton, Richard *see*
– Colonial discourses, series 2
– Life of sir richard burton
Burton, Richard Francis *see*
– Letters from the battle-fields of paraguay
– Sind revisited
– Unexplored syria
– Vikram and the vampire, or, tales of hindu devilry
Burton, Robert or Richard [pseud of: Nathaniel Crouch] *see* The history of the house of orange
Burton, Theodore Elijah *see* Theodore e burton papers, 1869-1958
Burton, William *see* Handbook of marks on pottery and porcelain
Burton, William Frederick Padwick *see* Luba religion and magic in custom and belief
Burtonian – Tekamah, NE: Burt County Pub Co. v23 n41. jul 9 1896-dec 1901// (wkly) [mf ed with gaps filmed [1967]] – 4r – 1 – (cont: weekly burtonian (1883). merged with: tekamah journal to form: tekamah journal, the weekly burtonian) – us NE Hist [071]
Burt's illustrated guide of the connecticut valley : containing descriptions of mount holyoke, mount mansfield, white mountain... / Burt, Henry Martyn – Northampton, MA: New England Pub Co, 1867 [mf ed 1980] – 4mf – 9 – 0-665-05693-1 – mf#05693 – cn Canadiana [917]
Burtsev, A E *see*
– Bibliograficheskoe obozrenie drevneslavianskoi i russkoi pismennosti i drugikh pamiatnikov ot 16 do nachala 20 v
– Khudozhestvenno-bibliograficheskii zhurnal
– Opisanie starykh i redkikh russkikh gazet, zhurnalov, raznykh letuchikh listkov i lubochnykh kartinok
– Russkiie knizhnye redkosti
– Slovar redkikh knig i gravirovannykh portretov
Burtsev, V *see*
– Ezhenedelnaia gazeta
– Obshchee delo
– Sotsialno-politicheskoe obozrenie
– Svobodnaia rossiia
Burtsev, V L *see*
– Istoriko-revoliutsionnyi sbornik
– Zhurnal, izdavavshiisia pod redaktsiei v l burtseva
Burtt, Scott L *see* Physical activity, calcium intake, body composition and stature as predictors of bone indices in college-aged men
Burty, Philippe *see*
– Charles meryon
– Chefs-d'oeuvre of the industrial arts
Burundi. Departement des Etudes et Statistiques *see* Annuaire statistique 1969-1975
Burwash, Edward Moore *see*
– The geology of michipicoten island
– The geology of vancouver and vicinity
– "The new theology"
– The pleistocene volcanoes of the coast range of british columbia
Burwash, Nathanael *see*
– Inductive studies in theology
– Jane clement jones
– Letter to b e walker
– Manual of christian theology on the inductive method
– Some further facts concerning federation
– Wesley's doctrinal standards, pt 1
Burwell 1561-1950 – (= ser Cambridgeshire parish register transcript) – 15mf – 9 – £18.75 – uk CambsFHS [929]
The burwell bell – Burwell, NE: L M Hart. v1 n1. mar 6 1885- [mf ed filmed 1999] – 1r – 1 – us NE Hist [071]
Burwell, Cynthia B *see* The relationship between adolescent smoking behavior and peer influence
Burwell mascot – Burwell, NE: Mascot Pub Co, 1896-v15 n41. oct 8 1903 (wkly) [mf ed 9th yr n1. jan 7 1897-oct 8 1903 (gaps)] – 2r – 1 – (cont: burwell progress. absorbed: eye may 1898 and: burwell tribune (1898), aug 1902. cont by: burwell tribune) – us NE Hist [071]
Burwell progress – Burwell, NE: L J Harris, 1894-dec 1896// (wkly) [mf ed v7 n33. aug 15 1895-dec 17 1896 (gaps)] – 1r – 1 – (cont: garfield enterprise. cont by: burwell mascot) – us NE Hist [071]
Burwell tribune – Burwell, NE: Burwell Print Co, 1898-aug 1902// (wkly) [mf ed feb 2 1899-aug 14 1902 (gaps)] – 1r – 1 – (absorbed by: burwell mascot) – us NE Hist [071]
Burwell tribune – Burwell, NE: W Z Todd. v15 n42. oct 15 1903- (wkly) [mf ed with gaps] – 1 – (cont: burwell mascot) – us NE Hist [071]
Bury free press – Bury St Edmunds 14 jul 1855-to date [mf ed only 1871-10 feb 1872, 1897] – 1 – (lacking: 14 jul 1855) – uk Newsplan [072]
Bury guardian – [NW England] Bury 4 jul 1857-dec 1935 [mf ed 2003] – 101r – 1 – (missing: 1862; title change: bury guardian & general advertiser for south & east lancashire [1860-76*]; bury guardian & radcliffe, ramsbottom, heywood, haslingden & rossendale standard [1877-85]; bury guardian [1886-1935]) – uk Newsplan; uk MLA [072]
Bury, Herbert *see* Bishop amongst bananas
Bury, John Bagnell *see*
– The constitution of the later roman empire
– A history of freedom of thought
– A history of the eastern roman empire
– A history of the later roman empire
– An inaugural lecture
– The life of st patrick and his place in history
Bury, Karin *see* Der bildhauer kurt lehmann
Bury, Richard de *see*
– Histoire de saint louis, roi de france
– Philobiblon
Bury st edmunds mercury – Ipswich, England 7 aug 1986-17 jan 1997 – 1 – (cont by: bury mercury [24 jan 1997-]) – uk British Libr Newspaper [072]
Bury, Thomas Talbot *see* Remains of ecclesiastical woodwork
Bury times – [NW England] Bury Lib jul 1855-1884, 1886-95, 1897-1902, 1904-07, 1909-12, 1914-85 – 1 – uk MLA; uk Newsplan [072]
Buryshkin, P A *see* Moskva kupecheskaia
Burzenlaender bote – Kronstadt (Brasov RO), 1921 20 oct-12 nov – 1r – 1 – gw Misc Inst [077]
Burzhuaiia i pomeshchiki v 1917 godu : chastnye soveshchaniia chlenov gosudarstvennoi dumy / ed by Drezen, A K – 1932 – 328p 4mf – 9 – mf#RPP-7 – ne IDC [325]
Burzhuaziia i tsarizm v pervoi russkoi revoliutsii / Chermenskii, E D – 1970 – 448p 5mf – 9 – mf#RPP-52 – ne IDC [325]
Burzhuaznye i melkoburzhuaznye partii rossii v oktiabrskoi revoliutsii i grazdanskoi voine : materialy simpoziuma v g kalinine i g tskhaltubo / ed by Komin, V V – 1980 – 156p 2mf – 9 – mf#RPP-8 – ne IDC [325]
Bus transportation – New York NY 1950-56 – 1 – mf#365 – us UMI ProQuest [380]
Bus & truck transport – Toronto ON 1976-86 – 1,5,9 – (cont by: truck fleet) – ISSN: 0007-635X – mf#10763.01 – us UMI ProQuest [380]
Busby, C A *see* A series of designs for villas and country houses
Busby, James *see* Journal of a recent visit to the principal vineyards of spain and france
Busca, G *see* Della espugatione et difesa delle fortezze libri due
Buscadores de diamentes en la guayana venezolana / Canellas Casals, Jose – Madrid, Spain. 1958 – 1r – us UF Libraries [972]
Busch, Ioannes *see* Des augustinerpropstes ioannes busch
Busch, Karl August *see* William james als religionsphilosoph
Busch, M *see* Ridala arkamise ajalugu-the history of the revival in ridala
Busch, Wilhelm *see*
– Edward's dream
– Gesammelte werke
– Hans huckebein, der unglueckselrabe
– Pater filucius
– Schein und sein
– Zu guter letzt
Buschatz, Dirk *see* Interkorrelation von epidemiologischen und polysomnographischen risikofaktoren des ploetzlichen saeuglingstodes
Buschbecker, Karl Matthias *see*
– Und doch schlaegt das herz an den grenzen
– Wie unser gesetz es befahl
Buscher, Heide *see* Die funktion der nebenfiguren in fontanes romanen
Buschleben in australien / Haygarth, Henry W – Dresden [u a] 1849 [mf ed Hildesheim 1995-98] – 1v on 2mf – 9 – €60.00 – 3-487-26813-2 – gw Olms [307]
Die buschmaenner der kalahari / Passarge, Siegfried – Berlin: Dietrich Reimer [Ernst Vohsen] 1907 – (= ser [Travel descriptions from south africa, 1711-1938]) – 2mf – 9 – mf#zah-46 – ne IDC [307]
Busenbaum, H *see* Theologia moralis...
Buser, Deborah E *see* Occupational exposure characterization of vacuum pump maintenance technicians in a semiconductor manufacturing environment
Busetti, Franco Raoul *see* Metaheuristic approaches to realistic portfolio optimisation
Bush advocate – may 1888-93, 16 may-30 dec 1899, jan 1901-jun 1901, jul 1901-apr 1903, jan 1904-jun 1909, jan 1904-jun 1909, jan 1910-apr 1912 – 59r – 1 – (cont as: dannevirke advocate fr jul 1901) – mf#35.5 – nz Nat Libr [079]
Bush and boma / Cairns, John – London, England. 1959 – 1r – us UF Libraries [960]
Bush, Annie Forbes *see* Memoirs of the queens of france
Bush, George *see*
– Anastasis
– Notes, critical and practical, on the book of exodus
– Notes, critical and practical, on the book of genesis
– Notes, critical and practical, on the book of joshua
– The resurrection of christ
– The soul
– The valley of vision

Bush, George Gary *see* History of education in florida
Bush mama – 1980 harvest – 1r – 1 – mf#4848544 – us WHS [071]
Bush master / Smith, Nicol – Indianapolis, IN. 1941 – 1r – us UF Libraries [972]
Bush negro art / Dark, Phillip John Crosskey – [New Haven] 1950 – us CRL [700]
Bush, P S et al *see* Reply to: brief historical sketch of the western baptist theological institute in covington, kentucky by the board of trustees, p.s.bush, j.m. frost and lewis roach 1850
Bush river baptist church : reedy river association – Newberry Co, SC. 865p. 1791-1945; History 1771-1933 – 1 – $38.93 – mf#0900-1 – us Southern Baptist [242]
Bush river, south carolina : church records – 1791-1945 – 1 – us Southern Baptist [242]
Bush, Sam Stone *see* The rush to the klondike
Bush telegraph – Pahiatua, NZ. 1978-87 – 16r – 1 – (previous title: north wairarapa news) – mf#48.13 – nz Nat Libr [079]
Bush telegraph *see* North wairarapa news
Bush, Wendell T *see* Avenarius and the standpoint of pure experience
Bushchik, L P *see* Istoriia sssr
Bush-conant file relating to the development of the atomic bomb, 1940-1945 / U.S. Office of Scientific Research and Development – (= ser Records Of The Office Of Scientific Research And Development) – 14r – 1 – (with printed file) – mf#M1392 – us Nat Archives [324]
Bushe commission, report, minutes of evidence and memoranda of the... may 1933 – Cmd 4623. 1937 – 1r – 1 – (with int by h f morris) – mf#96621 – uk Microform Academic [960]
Bushell, John J *see* Picturesque bermuda in picture
Bushell, Stephen Wootton *see* On yuan chwang's travels in india, 629-645 a.d
Bushido / Nitobe, Inazo – Tokyo, Japan. 1938 – 1r – us UF Libraries [025]
Bushido in the past and in the present / Imai, John Tashimichi – Tokyo: Kanazashi, Kanda, [1906] [mf ed 1995] – (= ser Yale coll) – 73p – 1 – 0-524-09373-3 – mf#1995-0373 – us ATLA [170]
Bushido, the soul of japan : an exposition of japanese thought / Nitobe, Inazo – 10th rev enl ed. New York: GP Putnam, c1905 – 1mf – 9 – 0-524-07732-0 – mf#1991-0154 – us ATLA [180]
Bushman speaks / Phillips, Mary – Cape Town, South Africa. 1961 – 1r – us UF Libraries [960]
Bushmen and other non-bantu peoples of angola / Almeida, Antonio De – Johannesburg, South Africa. 1965 – 1r – us UF Libraries [960]
Bushmen of the southern kalahari / Jones, John David Rheinallt – Johannesburg, South Africa. 1937 – 1r – us UF Libraries [960]
Bushnell, Horace *see*
- Building eras in religion
- The character of jesus
- Christ in theology
- Christian nurture
- Discourses on christian nurture
- Forgiveness and law
- God in christ
- God's thoughts fit bread for children
- Moral uses of dark things
- Nature and the supernatural
- Sermons on christ and his salvation
- Sermons on living subjects
- The spirit in man
- The vicarious sacrifice, grounded in principles of universal obligation
- Views of christian nurture, and of subjects adjacent thereto
- Women's suffrage
- Work and play

Bushnell. Katharine C *see* The queen's daughters in india
The bushnell record – Bushnell, NE: J G Todd. -v28 n43. nov 22 1944 (wkly) [mf ed v3 n25. jul 24 1919-nov 22 1924] – 8r – 1 – us NE Hist [071]
Busia, Kofi Abrefa *see* Report on a social survey of sekondi-takoradi
Business – Atlanta GA 1979-90 – 1,5,9 – (cont: atlanta economic review) – ISSN: 0163-531X – mf#7592.01 – us UMI ProQuest [338]
Business administration – Singapore 1971-76 – 1,5,9 – (cont by: chief executive monthly) – ISSN: 0007-6414 – mf#6020.02 – us UMI ProQuest [650]
Business after the war / Roberts, George Evans – [New York: s.n, 1916] (mf ed 19–) – 21p – (an address before the michigan bankers' assoc, flint mi, jun 13 1916) – mf#Z-BTZE pv225 n11 – us Harvard [338]
Business america – Washington DC 1978-98 – 1,5,9 – ISSN: 0190-6275 – mf#11745 – us UMI ProQuest [338]
Business and commercial aviation – New York NY 1971-93 – 1,5,9 – ISSN: 0191-4642 – mf#5984.02 – us UMI ProQuest [629]
Business and economic dimensions – Gainesville FL 1965-83 – 1,5,9 – ISSN: 0007-6457 – mf#8794 – us UMI ProQuest [338]
Business and economic review – Columbia SC 1983+ – 1,5,9 – ISSN: 0007-6465 – mf#14444 – us UMI ProQuest [650]
Business and family papers re activities in the new hebrides / Zeitler, Adolf – 1899-1935 – 1r – 1 – mf#PMB1091 – at Pacific Mss [920]
Business and finance division newsletter. special libraries association – Alexandria VA 1978-80 – 1,5,9 – ISSN: 0038-674X – mf#11236.02 – us UMI ProQuest [020]
Business and financial papers, 1780-1939 : selected titles from the bodleian library, oxford and the british library newspaper library, london – 3 ser – 1 – (ser1: international trade – pt1: the anglo-japanese gazette 1902-09 & 10 other titles [30r] £2800, pt2: african review 1892-1904 [23r] £2150; ser2: economic impact of scientific & technical change – pt1: mechanical engineer 1897-1907 [22r] £2050, pt2: mechanical engineer 1908-17 [16r] £1500; ser3: industrial enterprise – pt1: oil news 1912-39 [24r] £2250; with d/g) – uk Matthew [338]
Business and health – Montvale NJ 1985-2001 – 1,5,9 – ISSN: 0739-9413 – mf#15282 – us UMI ProQuest [360]
Business and home tv screen – New York NY 1978 – 1,5,9 – (cont: business screen) – ISSN: 0160-7294 – mf#195.01 – us UMI ProQuest [790]
Business and law, or the careful man's guide : a complete legal and business compendium / Roe, E T & Loomis, Elihu G – Boston/Chicago: Hertel, Jenkins & Co, 1907 – 8mf – 9 – $12.00 – mf#LLMC 92-192 – us LLMC [346]
Business and politics / Vanderlip, Frank Arthur – [New York: s.n], 1914 (mf ed 19–) – 16p – (address before the new york state bankers assoc convention, new london ct, june 11 1914) – mf#ZT-TLH pv55 n3 – us Harvard [332]
Business and society – Newbury Park CA 1960+ – 1,5,9 – ISSN: 0007-6503 – mf#1796 – us UMI ProQuest [306]
Business and society review – Oxford UK 1972+ – 1,5,9 – ISSN: 0045-3609 – mf#6964 – us UMI ProQuest [306]
Business as a system of power / Brady, Robert Alexander – New York: Columbia UP 1943 [mf ed 1988] – 1r – 1 – mf#2121 – us Wisconsin U Libr [338]
Business asia – Sydney – 1r – A$27.50 vesicular A$33.00 silver – at Pascoe [079]
Business atlanta – Atlanta GA 1980-94 – 1,5,9 – ISSN: 0192-0855 – mf#12240.02 – us UMI ProQuest [338]
Business bankruptcy / Warren, Elizabeth – Washington: FJC, 1993 – 2mf – 9 – $3.00 – mf#LLMC 95-383 – us LLMC [346]
Business bookkeeping and practice / Sadler, W H & Rowe, H M – Baltimore: Sadler-Rowe Co, 1894 – 3mf – 9 – $4.50 – (incl ind) – mf#LLMC 92-128 – us LLMC [346]
Business by phone in the federal courts / Meierhofer, Barbara S – 1983 – 1mf – 9 – $1.50 – mf#llmc99-013 – us LLMC [347]
Business & commercial aviation – New York NY 1999+ – 1,5,9 – (cont: business & commercial aviation international) – mf#5984.02 – us UMI ProQuest [380]
Business & commercial aviation international – New York NY 1995-97 – 1,5,9 – (cont: business and commercial aviation; cont by: business and commercial aviation) – mf#5984.02 – us UMI ProQuest [380]
Business communication quarterly / Association for Business Communication (US) – 1995+ – 1,5,9 – (cont: bulletin of the association for business communication) – ISSN: 1080-5699 – mf#8090.02 – us UMI ProQuest [650]
Business communications review – 1971+ – 1,5,9 – ISSN: 0162-3885 – mf#6796 – us UMI ProQuest [650]
Business computer systems – New York NY 1982-86 – 1,5,9 – ISSN: 0745-0745 – mf#13319 – us UMI ProQuest [000]
Business computing – Tulsa OK 1984-85 – 1,5,9 – ISSN: 0741-4641 – mf#13621.01 – us UMI ProQuest [000]
Business conditions – Chicago IL 1919-76 – 1,5,9 – (cont by: economic perspectives) – ISSN: 0007-6589 – mf#5156 – us UMI ProQuest [338]
Business conditions digest [bcd] – Washington DC 1972-90 – 1,5,9 – ISSN: 0146-7735 – mf#6292 – us UMI ProQuest [332]
The business corporations law. / Jones, Dwight Arven – 5th ed. New York, Baker, Voorhis, 1897. 129 p. LL-1327 – 1 – us L of C Photodup [346]
The business corporations law...and other laws concerning business corporations in the state of new york / Jones, Dwight Arven – 3rd ed. New York: Baker, Voorhis, 1893.120p. LL-689 – 1 – us L of C Photodup [346]
Business correspondence / Burns Philp (South Sea) Co Ltd, Labasa Branch, Fiji – 1924-1950 – 1r – 1 – mf#pmb152 – at Pacific Mss [650]
Business correspondence 1915-19 : accounts and wage books 1935-43 / Ussher, J – Nguna, New Hebrides – 1r – 1 – mf#PMB1131 – at Pacific Mss [650]
Business courier – Charlotte NC 1998+ – 1 – (cont: cincinnati business courier) – ISSN: 1096-8636 – mf#16673.01 – us UMI ProQuest [650]
Business credit – Columbia MD 1987+ – 1,5,9 – (cont: credit and financial management: c&fm) – ISSN: 0897-0181 – mf#1980.02 – us UMI ProQuest [332]
Business economics – Washington DC 1965+ – 1,5,9 – ISSN: 0007-666X – mf#2510 – us UMI ProQuest [338]
Business education forum – Reston VA 1947+ – 1,5,9 – ISSN: 0007-6678 – mf#5059 – us UMI ProQuest [338]
Business education index – Little Rock AR 1920+ – 1,5,9 – ISSN: 0068-4414 – mf#2531 – us UMI ProQuest [370]
Business education world – Hightstown NJ 1920-90 – 1,5,9 – ISSN: 0007-6694 – mf#785 – us UMI ProQuest [370]
Business entities – New York NY 1999+ – 1,5,9 – ISSN: 1524-3583 – mf#29263 – us UMI ProQuest [336]
Business ethics – Oxford UK 1992+ – 1,5,9 – ISSN: 0962-8770 – mf#18769 – us UMI ProQuest [170]
Business first [1984] – Charlotte NC 1984+ – 1,5,9 – ISSN: 0748-6146 – mf#16674 – us UMI ProQuest [650]
Business first [1987] – Charlotte NC 1987-99 – 1,5,9 – ISSN: 0748-6138 – mf#16682 – us UMI ProQuest [650]
Business forms, labels and systems – Philadelphia PA 1990+ – 1,5,9 – (cont: business forms & systems) – ISSN: 1044-758X – mf#2208.02 – us UMI ProQuest [680]
Business forms reporter – Philadelphia PA 1966-81 – 1,5,9 – (cont by: business forms & systems) – ISSN: 0007-6767 – mf#2208.02 – us UMI ProQuest [680]
Business forms & systems – Philadelphia PA 1983-89 – 1,5,9 – (cont: business forms reporter; cont by: business forms, labels and systems) – ISSN: 0745-3914 – mf#2208.02 – us UMI ProQuest [680]
Business forum – Los Angeles CA 1981+ – 1,5,9 – (cont: los angeles business and economics) – ISSN: 0733-2408 – mf#12340.01 – us UMI ProQuest [338]
Business graduate – Hoboken NJ 1984-87 – 1,5,9 – ISSN: 0306-3895 – mf#14803 – us UMI ProQuest [650]
Business graphics – New York NY 1967-75 – 1,5,9 – ISSN: 0007-6775 – mf#3026 – us UMI ProQuest [740]
Business guide book to djakarta, 1969-1971 – 5mf – 9 – (missing: 1969) – mf#SE-1374 – ne IDC [338]
Business history – Abingdon, Oxfordshire 1989+ – 1,5,9 – ISSN: 0007-6791 – mf#17466 – us UMI ProQuest [338]
Business history collection, 1916-1975 – 2233mf (coll) – 1 – us Primary [330]
Business history review – Boston MA 1926+ – 1,5,9 – ISSN: 0007-6805 – mf#687 – us UMI ProQuest [338]
Business horizons – Oxford UK 1958+ – 1,5,9 – ISSN: 0007-6813 – mf#1871 – us UMI ProQuest [650]
Business ideas and facts – Ypsilanti MI 1967-73 – 1,5,9 – ISSN: 0045-3633 – mf#7990 – us UMI ProQuest [650]
Business in brief – New York NY 1953-80 – 1,5,9 – ISSN: 0007-6821 – mf#5799 – us UMI ProQuest [650]
Business insurance – Detroit MI 1967+ – 1,5,9 – ISSN: 0007-6864 – mf#2750 – us UMI ProQuest [368]
Business International Corporation *see* New brazil
Business japan – Tokyo, Japan 1982-91 – 1,5,9 – (cont by: japan 21st) – ISSN: 0300-4341 – mf#13095.02 – us UMI ProQuest [338]
Business journal – 1983 oct 24/1984 apr 30-2003 oct/dec – 67r – 1 – (with gaps) – mf#671661 – us WHS [650]
Business journal – Portland OR: The Business Journal of Portland Inc, c1984?- [wkly] – 1 – (with suppl) – us Oregon Lib [071]
Business journal [1990] – Charlotte NC 1990-96 – 1 – ISSN: 0895-1632 – mf#16684.02 – us UMI ProQuest [650]
Business journal [1991] – Charlotte NC 1991+ – 1,5,9 – ISSN: 0887-5588 – mf#16672 – us UMI ProQuest [650]
Business korea – Seoul, South Korea 1991+ – 1,5,9 – mf#18349 – us UMI ProQuest [338]
Business law / Hirschl, Samuel D – Chicago: LaSalle Extension University, 1927 – 6mf – 9 – $9.00 – mf#LLMC 96-050 – us LLMC [346]
Business law / Pomeroy, Dwight Abel – 2d ed. Cincinnati: South-Western 1939. 906p. L.C. copy imperfect: p. 93-98 wanting. LL-1006 – 1 – (manual to accompany business law, 2nd ed. cincinnati, 1939. 236p. ll-1006) – us L of C Photodup [346]
Business law case method. / Commerce Clearing House – Chicago, 1915 7 v. LL-885 – 1 – us L of C Photodup [346]
Business law for business men, state of california / Bledsoe, Anthony Jennings – 9th ed. San Francisco, Business Law Publishing Co., 1912. 1021 p. LL-889 – 1 – us L of C Photodup [346]
Business law journal *see* University of miami business law review
Business law review – Ann Arbor MI 1979-81 – 1,5,9 – ISSN: 0145-9074 – mf#12285.04 – us UMI ProQuest [346]
Business law review – v1-15. 1960-94 – 9 – $577.00 set – ISSN: 0143-6295 – mf#114561 – us Hein [346]
Business lawyer (aba) – v1-56. 1946-2001 – 1,5,6 – $1760.00 – ISSN: 0007-6899 – mf#101241 – us Hein [340]
Business ledger 1827-1874 / Archibald, Smith – 1r – 1 – mf#B41469 – us Ohio Hist [650]
Business management [1950] – London UK 1950-69 – 1 – mf#477 – us UMI ProQuest [650]
Business management [1951] – Stamford CT 1951-71 – 1,5,9 – ISSN: 0007-6910 – mf#1511 – us UMI ProQuest [650]
Business management training pedoman – Bandung, 1962 – 4mf – 9 – (missing: 1963-70) – mf#SE-450 – ne IDC [959]
The business man's adviser. / Butts, Isaac Ridler – Boston, Butts, 1854. 10, 132, vi, 120, iv, 108 p. LL-246 – 1 – us L of C Photodup [346]
Business man's commercial law library / Bolles, George S – NY: Double-Day/Collier. 6v. 1924 – 18mf – 9 – $27.00 – mf#LLMC 92-173 – us LLMC [346]
The business man's encyclopedia : the business man's brain, partners, contracts, commercial usage, business law, foreign trade, saving systems – 16th rev ed. Chicago etc: A W Shaw Co, 1927 – 6mf – 9 – $9.00 – mf#LLMC 96-058 – us LLMC [650]
Business manual : comprising twenty-five chapters on business law... / Anger, William Henry – [St Catharines [Ont]: s.n, 1892?] [mf ed 1980] – 2mf – 9 – 0-665-02406-1 – (incl ind) – mf#02406 – cn Canadiana [346]
Business marketing – Detroit MI 1984-93 – 1,5,9 – (cont: industrial marketing; cont by: advertising age's business marketing) – ISSN: 0745-5933 – mf#348.03 – us UMI ProQuest [650]
Business mexico – Mexico City 1991+ – 1,5,9 – ISSN: 0187-1455 – mf#18381 – us UMI ProQuest [338]
Business month – Boston MA 1988-90 – 1,5,9 – (cont: dun's business month) – ISSN: 0892-4090 – mf#202.02 – us UMI ProQuest [338]
The business of salvation / Otten, Bernard John – St Louis, Mo: B Herder, 1911 – 1mf – 9 – 0-7905-9431-5 – mf#1989-2656 – us ATLA [240]
The business of the supreme court: a study in the federal judicial system / Frankfurter, Felix – New York: Macmillan, 1927. 349p. LL-1174 – 1 – us L of C Photodup [347]
The business of travel: a fifty years' record of progress / Rae, William Fraser – 1841-Leicester to Loughborough (12 miles) 1891-all over the globe. London, New York: T. Cook and Son, 1891. History of the origin and progress of Thomas Cook and Son company, 22 July 1897 – 1 – us Wisconsin U Libr [910]
Business perspectives – Memphis TN 1987+ – 1,5,9 – ISSN: 0896-3703 – mf#16713 – us UMI ProQuest [650]
Business process management journal – Bradford UK 2001+ – 1,5,9 – ISSN: 1463-7154 – mf#31582.01 – us UMI ProQuest [650]
Business publishing – New York NY 1992-93 – 1,5,9 – (cont: personal publishing) – ISSN: 1060-2208 – mf#17364.01 – us UMI ProQuest [070]
Business quarterly – London ON 1952-98 – 1,5,9 – (cont: quarterly review of commerce; cont by: ivey business quarterly) – ISSN: 0007-6996 – mf#12002.03 – us UMI ProQuest [338]
Business records / Clark, Gabriel Penn – 1853-1920, Relate to Platte county, Missouri, and Shawnee county, KS – 1 – us Kansas [380]
Business review [1950] / Federal Reserve Bank of Philadelphia – Philadelphia PA 1950+ – 1,5,9 – ISSN: 0007-7011 – mf#8193 – us UMI ProQuest [332]
Business review [1973] / Federal Reserve Bank of San Francisco – San Francisco CA 1973-75 – 1,5 – (cont by: economic review federal reserve bank of san francisco; cont: monthly review federal reserve bank of san francisco) – ISSN: 0093-8262 – mf#335.02 – us UMI ProQuest [332]

Business review [1987] / Wells Fargo Bank – San Francisco CA 1987-91 – 1,5,9 – (cont by: wells fargo economic monitor: california) – ISSN: 0883-9670 – mf#15912 – us UMI ProQuest [332]
Business rules and secretariat instructions / Andhra Pradesh (India) – [s.l: s.n] 1959 (Hyderabad: Director, Govt Stamps Press, Mint Compound) – 1r – 1 – us CRL [650]
Business s a – Johannesburg, South Africa 1977-78 – 1,5,9 – (cont: business south africa) – mf#5328.01 – us UMI ProQuest [338]
Business screen – New York NY 1938-77 – 1,5,9 – (cont by: business and home tv screen) – ISSN: 0007-7046 – mf#195.01 – us UMI ProQuest [790]
Business south africa – Johannesburg, South Africa 1969-73 – 1,5,9 – (cont by: business s a) – ISSN: 0007-7070 – mf#5328.01 – us UMI ProQuest [338]
Business statistics – Washington DC 1979-92 – 1,5,9 – ISSN: 0083-2545 – mf#12410 – us UMI ProQuest [338]
The business tax – Ottawa: s.n, 1891? – 1mf – 9 – mf#03022 – cn Canadiana [336]
Business telegraph = Tigawane – [Blantyre: s.n, [nov 27/dec3 1997-jun 30/jul 2 1998] (semiwkly) – 1r – 1 – us CRL [650]
Business times – Kuala Lumpur, Malaysia. 1977-92 – 83r – 1 – (continuation offered) – us L of C Photodup [079]
Business times – Dar es Salaam: Business Times Ltd, nov 4 1988-jul 12/18 1996 – us CRL [650]
The business times – Singapore, 1984-2003 [daily] – 381r – 1 – (covers both european and local business news) – mf#97308 – uk Microform Academic [338]
Business today – Princeton NJ 1968+ – 1,5,9 – ISSN: 0007-7100 – mf#5113 – us UMI ProQuest [650]
Business tokyo – Tokyo, Japan 1989-92 – 5,9 – mf#16313 – us UMI ProQuest [650]
Business traveller asia-pacific – London UK 2001+ – 1,5,9 – ISSN: 0255-7312 – mf#32372.01 – us UMI ProQuest [650]
Business traveller germany – London UK 2002+ – 1,5,9 – mf#33090 – us UMI ProQuest [650]
Business trends – Petaluma CA 1973-84 – 1,5,9 – (cont: news front) – ISSN: 0194-9225 – mf#9099.01 – us UMI ProQuest [338]
Business venezuela – Caracas, Venezuela 1971-81 – 1,5,9 – ISSN: 0045-3641 – mf#7989 – us UMI ProQuest [338]
Business week : [industrial/technology ed] – New York NY 1932+ – 1,5,9 – ISSN: 0739-8395 – mf#36 – us UMI ProQuest [650]
Business woman – Toronto. v. 1-4, n10. nov 1926-oct 1929. (incomplete) – 1 – us NY Public [305]
Business world – Clifton NJ 1964-69 – 1 – ISSN: 0007-7143 – mf#5329 – us UMI ProQuest [650]
Business world / WMC Foundation, Inc – 1984 jan-jul, oct-nov, 1985 feb-jun/jul, oct-nov, 1986 jan-may, sep-dec, 1987 jan-may, nov-dec, 1988 jan-mar, sep – 1r – 1 – mf#1548589 – us WHS [650]
Business-Industry Political Action Committee *see* Politikit
BusinessWeek careers – New York NY 1986-88 – 1,5,9 – (cont: businessweek's guide to careers) – ISSN: 0891-6578 – mf#15653.01 – us UMI ProQuest [331]
BusinessWeek's guide to careers – New York NY 1985-86 – 1,5,9 – (cont by: businessweek careers) – ISSN: 8756-9116 – mf#15653.01 – us UMI ProQuest [331]
Businger, Lucas Caspar *see* History of the catholic church
Busiri, Sharaf al-Din Muhammad ibn Said *see* La bordah du cheikh el bousiri
Buskirk, Philip Clayton van *see* Diaries, 1851-1902
Buslaev, Fedor Ivanovich *see* Narodnaia poeziia
Buslett's – 1922 jan-oct, 1923 nov – 1r – 1 – mf#830576 – us WHS [071]
Busnach, William *see* Assommoir
Busnelli, Juan *see* Manual de teosofia o breve estudio critico de las doctrinas teosoficas. mejico, 1929
Busqueda / Sarusky, Jaime – Habana, Cuba. 1961 – 1r – us UF Libraries [972]
Busqueda / Vazquez Rodriguez, Benigno – Habana, Cuba. 1957? – 1r – us UF Libraries [972]
Busqueda pastoral – La Paz, Bolivia: Busqueda Pastoral, [n9-101 (1970-1989)] (bimthly) – 2r – 1 – us CRL [242]
Busqueda pastoral (la paz, bolivia) : documentacion – La Paz: [Conferencia Episcopal Boliviana, Subsecretariado de Pastoral, [n5-6 (marzo-jun 1979)] – 1r – 1 – us CRL [242]
Busqueda y plasmacion de nuestra personalidad / Robles De Cardona, Mariana – San Juan, Puerto Rico. 1958 – 1r – us UF Libraries [972]
Buss, Claude Albert *see* Southeast asia and the world today
Das die buss rewe oder erkentnis des zorns vnd der suenden eigentlich allein aus dem gesetz / Flacius Illyricus d A, M – [Jhena, 1559] – 1mf – 9 – mf#TH-1 mf 464 – ne IDC [242]
Buss, Robert William *see*
– The almanack of the fine arts for the year 1850
– The almanack of the fine arts for the year 1852
– English graphic satire
Buss, Septimus *see* The trial of jesus illustrated from talmud and roman law
Die bussbuecher und das kanonische bussverfahren / Schmitz, H J – Duesseldorf, 1898 – €43.00 – ne Slangenburg [241]
Die bussbuecher und das kanonische bussverfahren / Schmitz, Hermann Joseph – Duesseldorf: L Schwann, 1898 – (= ser Bussbuecher und die bussdisciplin der kirche) – 2mf – 9 – 0-524-01667-4 – (incl bibl ref) – mf#1990-0488 – us ATLA [240]
Die bussbuecher und die bussdisciplin der kirche / Schmitz, H J – Mainz, 1883 – €49.00 – ne Slangenburg [241]
Die bussbuecher und die bussdisciplin der kirche / Schmitz, Hermann Joseph – Mainz: Franz Kirchheim, 1883 – 2mf – 9 – 0-8370-7424-X – (incl ind) – mf#1986-1424 – us ATLA [240]
Busschere, C Gonsalve Marie de *see* Le rosaire de marie
Die bussdisciplin der kirche von den apostelzeiten bis zum siebenten jahrhundert / Frank, Friedrich – Mainz: F Kirchheim, 1867 – 3mf – 9 – 0-7905-9203-7 – mf#1989-2428 – us ATLA [240]
Busse, Carl *see*
– Gedichte
– Neue gedichte
– Novalis' lyrik
– Ueber zeit und dichtung
– Vagabunden
Die busse, die letzte oelung, die priesterweihe und die ehe / Oswald, Johann Heinrich – 2. verb Aufl. Muenster: Aschendorff, 1864 – 1mf – 9 – 0-524-04557-7 – mf#1991-2121 – us ATLA [240]
Busse, Hermann Eris *see*
– Fides
– Grimmelshausen
– Mein leben
– Spiel des lebens
Bussell, Frederick William *see*
– Christian theology and social progress
– Marcus aurelius and the later stoics
– The school of plato
Bussell, FW *see* The school of plato
Bussierre, Marie Theodore Renouard, vicomte de *see* Histoire de schisme portugais dans les indes
Die busslehre luthers und ihre darstellung in neuster zeit / Galley, Alfred – Guetersloh: C Bertelsmann, 1900 – (= ser Beitraege zur foerderung christlicher theologie) – 1mf – 9 – 0-7905-9275-4 – mf#1989-2500 – us ATLA [242]
Busson, H *see* Les sources et le developpement du rationalisme dans la litterature francaise de la renaissance (1533-1601)
Bussstufen und katechumenatsklassen / Schwartz, E – Strassburg, 1911 – €5.00 – ne Slangenburg [241]
Bussstufen und katechumenatsklassen / Schwartz, Eduard – Strassburg: K J Truebner 1911 [mf ed 1990] – (= ser Schriften der wissenschaftlichen gesellschaft in strassburg 7) – 1mf – 9 – 0-7905-6884-5 – (in german & greek; incl bibl ref) – mf#1988-2884 – us ATLA [240]
Bussy-Rabutin, Roger de *see* Histoire amoureuse des gaules
Bustamante Arellana, Carlos *see* Amor y caridad
Bustamante, Gregorio *see* Historia militar de el salvador
Bustamante Y Montoro, Antonio Sanchez De *see* Ironia y generacion
Bustamente Bustillo, Antonio *see* El doctor d antonio bustamante bustillo pable fernandez
Bustamente, Coton *see* Cocotologia
Bustamente Yepez, Marco A *see* America y la 'hilea amazonica'
Bustami, Perang., sutan *see* Hikajat sitti rabihatoen
The bustan al-ukul / Nathanael ibn al Fayyumi; ed by Levine, David – New York: s.n, 1908, c1907 [mf ed 1985] – 1mf – 9 – 0-8370-4553-3 – (in english & arabic) – mf#1985-2553 – us ATLA [270]
Busteed, Henry Elmsley *see* Echoes from old calcutta
El busto de elisa / Hurtado, Antonio – 1871 – 9 – sp Bibl Santa Ana [830]
Bu-ston Rin-chen-grub *see* History of buddhism
Buston, Thomas Fowell *see* The african slave trade
Bustos de Olmedilla, G *see* El monstruo horrible de grecia mortal enemigo del hombre
Busy body – La Salle IL 1759 – 1 – mf#4217 – us UMI ProQuest [420]
Busy citizen – Brodhead WI. 1893 oct 10-1895 aug 20, 1895 aug 27-1897 may 13, 1897 may 20-1898 dec 29 – 3r – 1 – mf#957414 – us WHS [071]
Busy pastor's guide *see* Baptist standard church directory and busy pastor's guide
But god : the resources and sufficiency of god / Simpson, Albert B – Brooklyn NY: Christian Alliance c1899 [mf ed 1992] – (= ser Christian & missionary alliance coll) – 1mf – 9 – 0-524-02141-4 – mf#1990-4207 – us ATLA [210]
But in our lives' : a romance of the indian frontier / Younghusband, Francis Edward – London: J Murray, 1926 – (= ser Samp: indian books) – us CRL [830]
But mau : tap truyen ngan cua nhieu tac gia vung do thi – [s.l]: Van Nghe Giai Phong 1973 [mf ed 1992] – on pt of 1r – 1 – mf#11052 r171 n3 – us Cornell [830]
Butane-propane news [1939] – New York NY 1939-69 – 1 – (cont by: butane-propane news) – ISSN: 0007-7259 – mf#1038 – us UMI ProQuest [550]
Butane-propane news [1969] – Arcadia CA 1969+ – 1,5,9 – (cont: butane-propane news) – ISSN: 0007-7259 – mf#5979 – us UMI ProQuest [550]
Butcher, Edith Louisa *see* The story of the church of egypt
Butcher, Samuel *see*
– Claims of the additional curates' fund society
– Conservative character of the english reformation...
– Few thoughts on the supreme authority of the word of god
– Relative value and importance of divine and human knowledge
– Reunion with rome, as advocated in the eirenicon of dr pusey
Butcher, W W *see* W w butcher's canadian newspaper directory
Butcher worker / Hebrew Butcher Workers' Union – 1937-1942 – (= ser Labor union periodicals, pt 3: food and agricultural industries) – 3r – 1 – $605.00 – 1-55655-625-X – us UPA [660]
Butcher workman / Amalgamated Meat Cutters and Butcher Workmen of North America – 1915-60 – (= ser Labor union periodicals, pt 3: food and agricultural industries) – 8r – 1 – $1645.00 – 1-55655-620-9 – us UPA [660]
Butchers' 532 review / Amalgamated Meat Cutters and Butcher Workmen of North America – 1976 mar-1980 aug – 1r – 1 – (cont by: local 532 butchers' review) – mf#1278033 – us WHS [660]
Butchers local #532 review / United Food and Commercial Workers International Union – 1981 jul-1987 jun – 1r – 1 – (cont: local 532 butchers' review; cont by: local 532 review) – mf#1268956 – us WHS [660]
Butchers' union local n120 / Amalgamated Meat Cutters and Butcher Workmen of North America – v1 n1 [1979 aug] – 1r – 1 – mf#679061 – us WHS [660]
Bute broadsides : from the houghton library, harvard university – 1r – 1 – (contains 466 broadsides and other ephemera collected by john patrick crichton stuart. documents many political and social aspects of 17th century life) – mf#C35-14600 – us Primary [090]
Buteman : incorporating the rothesay express – Rothesay: Bute Newspapers Ltd 1955- (wkly) [mf ed 1995-] – 1 – (cont: buteman & the rothesay express) – uk Scotland NatLib [072]
Buteman and general advertiser for the western isles – [Scotland] Argyll & Bute, Rothesay: R M'Fie jan 1855-dec 1878, jan 1890-dec 1950 (wkly) [mf ed 2003] – 56r – 1 – (cont as: buteman advertiser for the western isles and visitors list [jan 1896-dec 1913]; buteman and west coast chronicle [jan 1914-dec 1950]) – uk Newsplan [072]
Un buten singt de nachtigall : un annere beller un geschichten up moensterlaennsk blatt / Wagenfeld, Karl – Essen-Ruhr: Fredebeul und Koenen, 1912 – 1r – 1 – us Wisconsin U Libr [430]
Butenschon, Peter *see* Material concerning the nigeria-biafra conflict, 1967-1970
Butki, Brian D *see* The relationship between physical activity and multidimensional self-concept among adolescents
Butler, A J *see* The churches and monasteries of egypt and some neighbouring countries
Butler, Alford Augustus *see* How to study the life of christ
Butler, Alfred Joshua *see*
– The ancient coptic churches of egypt
– The arab conquest of egypt and the last thirty years of the roman dominion
Butler area news – Butler WI. 1980 apr-1981 dec, 1982 jan-1987 sep – 2r – 1 – (cont: butler chamber news) – mf#3374157 – us WHS [338]
Butler bulletin – 1969 mar-1970 jan – 1r – 1 – (cont by: butler chamber news) – mf#4106163 – us WHS [338]
Butler, C *see* The lausiac history of palladius (ts6/1-2)
Butler, C E *see* Old testament word studies
Butler, C M *see* Reformation in sweden
Butler, Carlos A *see* The temple in the time of christ as restored by herod
Butler chamber news – Butler WI. 1970 jun, 1970 jun-1975 mar, 1975 apr-1980 mar – 3r – 1 – (cont: butler bulletin [butler wi]; cont by: butler area news) – mf#3374146 – us WHS [338]
Butler, Charles *see*
– The feminin 'monarchi', or the histori of the bee's
– Letter to the right reverend c j blomfield
– Principles of musik, in singing and setting
Butler, Charles Henry *see* Century at the bar of the supreme court of the us
Butler, Clement Moore *see*
– An ecclesiastical history from the 1st to the 13th century
– An ecclesiastical history from the 13th to the 19th century
– Inner rome
– Ritualism of law in the protestant episcopal church of the united states
– St paul in rome
Butler Co. Col.Corner *see* College corner news
Butler Co. Fairfield *see*
– Echo
– Sun-press series
Butler Co. Hamilton *see*
– Advertiser series
– American series
– Butler county press
– Deutsch amerikaner
– Gazette and miami register
– Hamilton true telegraph
– Intelligencer
– Miami herald
– Miami intelligencer
– Philanthropist
Butler Co. Oxford *see*
– Citizen
– Citizen / town / news
– Press
– Press (mid-century edition)
– Press series
Butler Co. Rossville *see* Miami democrat
Butler Co. Trenton *see*
– Edgewood this week
Butler Co. West Chester *see* Union times
Butler, Colin Gasking *see* World of the honeybee
Butler county citizen – Pennsylvania. -d and -w 1876-1919 – 13 – $25.00r – us IMR [071]
Butler county democrat – Hamilton, OH. 18 aug 1887 – 1r – 1 – (daily democratic newspaper) – us Western Res [071]
Butler County Historical Society [AL] *see* Quarterly publication of the butler county historical society
Butler county press / Butler Co. Hamilton – apr 1913-aug 1946 – 9r – 1 – mf#B35149-35157 – us Ohio Hist [331]
Butler county press – David City, NE: C D Casper & Co, 1873-81st yr n17. dec 3 1953 [wkly] [mf ed sep 4 1891-dec 3 1953 [gaps]] – 21r – 1 – (merged with: people's banner to form: banner-press) – us NE Hist [071]
Butler county republican – David City, NE: O A Keith. v1 n1. feb 5 1897-1898// (wkly) [mf ed with gaps filmed [1974?] – 1r – 1 – us NE Hist [071]
Butler county telegraph – Hamilton, OH, apr 29-nov 11 1847 – 1r – 1 – (weekly democratic newspaper) – us Western Res [071]
Butler democratic – Butler, PA., 1898-1901 – 13 – $25.00r – us IMR [071]
Butler, Edmund *see* The apartments of the house
Butler, Edward Cuthbert *see*
– Authorship of the dialogus de vita crysostomi
– The lausiac history of palladius
Butler, Eleanor, Lady *see* Ladies of llangollen
Butler, Elizabeth *see* Letters from the holy land
Butler, H C *see* Architecture and other arts...
Butler herald – Butler, PA., 1901-1912 – 13 – $25.00r – us IMR [071]
Butler, J *see* Travels and adventures in the province of assam
Butler, J M *see* Scrapbook
Butler, James, 1st Duke of Ormonde *see* Letters and papers, 1610-88 and 1665-1745
Butler, James, 2nd Duke of Ormonde *see* Letters and papers, 1610-88 and 1665-1745
Butler, James Glentworth *see* Vital truths respecting god and man
Butler, Jeffrey *see* Liberal party and the jameson raid
Butler, John Jay *see* Lectures on systematic theology
Butler, John P *see* Index to the papers of the continental congress, 1774-1789
Butler, Joseph *see*
– The analogy of religion natural and revealed
– Bishop butler's ethical discourses
Butler, Josephine Elizabeth Grey *see* Catharine of siena
Butler, Josephine Elizabeth Grey et al *see* Woman's place in church work

Butler journal – Butler WI. 1946 jan 24-apr 4, 1949 jan 2-1952 apr 30, 1949 oct 27 – 3r – 1 – mf#957435 – us WHS [071]

Butler news / Butler Paper Co – v12 n3-v18 n2 [1976 may-1982 spr] – 1r – 1 – mf#618041 – us WHS [071]

Butler, Nicholas Murray *see*
- Aspects of education
- New outlook

Butler opinion / Montgomery Co. Vandalia – aug 1950-sep 1951 [wkly] – 1r – 1 – mf#B5303 – us Ohio Hist [071]

Butler Paper Co *see* Butler news

Butler plantation papers : the papers of pierce butler (1744-1822) and successors from the historical society of pennsylvania – [mf ed Marlborough 1996] – 22r – 1 – £2050.00 – (with d/g) – uk Matthew [976]

Butler, Robert N *see* Love and sex after sixty

Butler, Samuel *see*
- Evolution, old and new
- Luck, or cunning, as the main means of organic modification?

Butler special – 1982 jul 30-1988 jul 15 – 1r – 1 – (cont by: community) – mf#3475146 – us WHS [071]

Butler, Stanley B *see* Physical education and nonphysical education majors

Butler sun – Butler, Hartland WI. 1987 oct 6-1988 dec 27, 1989 jan 1-dec 26, 1990 jan 2-dec 25, 1991 jan 1-dec 31, 1992 jan-jun, 1992 jul-dec, 1993 jan-jun, 1993 jul-dec – 8r – 1 – mf#2256788 – us WHS [071]

Butler [telephone directory] : listing – 1948 jun, 1949 pt, 1953 nov pt, 1954 pt, 1956 pt, 1958 – 10r – 1 – mf#2861789 – us WHS [917]

[Butler-] the tonopah miner – NV. jul-dec 1906 – 1r – 1 – $60.00 – mf#U04428 – us Library Micro [071]

Butler township school district 5, ms 908 – 1846-61 – (= ser Records of the butler township school district 5) – 1r – 1 – (attendance and grading records of students in the butler township, columbiana co, ohio, school district 5) – us Western Res [978]

Butler, W Archer *see* Primitive church principles not inconsistent with universal christi...

Butler, William *see*
- From boston to bareilly and back
- The land of the veda
- Mexico in transition

Butler, William A *see* Two millions

Butler, William Archer *see*
- Letters on romanism
- Sermons, doctrinal and practical. first series
- Sermons, doctrinal and practical. second series

Butler, William Edward *see* Igor stravinsky

Butler, William Francis *see*
- Far out
- From naboth's vineyard
- The great lone land
- The hero of pine ridge
- Red cloud, the solitary sioux
- Report by lieutenant butler (69th regt) of his journey from fort garrey to rocky mountain house and back
- The wild north land

Butler's guide to better shopping – 1962 may-jul – 1r – 1 – mf#4106162 – us WHS [380]

Butler's journal – Fredericton, NB: Martin Bo, 1890-95, 1898-1903 – 2r – 1 – cn Library Assoc [071]

Butron, I de *see* Discursos apologeticos, en que se defiende la ingenuidad del arte de la pintura...

Butt, Abdullah *see* Aspects of abul kalam azad

Butt, Isaac *see* The irish deep sea fisheries

Butt, K L *see* Four middle school physical education teachers' experiences during a collaborative action research staff development project

Buttari Guanaurd, J *see* Discursos y conferencias

Butte city freie presse – Butte MT. 1888 apr 28, aug 4 – 1r – 1 – mf#1010685 – us WHS [071]

Butte county – 1992- – (= ser California telephone directory coll) – 2r – 1 – $100.00 – mf#P00011 – us Library Micro [917]

[Butte county-] butte, colusa, glenn, nevada, placer, shasta, sutter, tehama and yuba counties – CA. 1892-1894 – 1r – 1 – $50.00 – mf#D008 – us Library Micro [978]

[Butte county-] butte, colusa, sutter, tehama and yuba counties – CA. 1881; 1884-1885 – 4r – 1 – $200.00 – mf#D007 – us Library Micro [978]

[Butte county-] chico and oroville city directories – CA. 1904-1905; 1913-1914; 1921-1928; 1937-1938; 1948-1950 – 9r – 1 – $450.00 – mf#D006 – us Library Micro [917]

Butte county historical society "diggin's" / McIntosh, Patricia & Parker, Virginia – v1-3. 1957-69 – 3r – 1 – $150.00 – mf#B40205 – us Library Micro [978]

[Butte county-] history of butte county / Wells, Harry L – CA. 1882 – 1r – 1 – $50.00 – mf#B40209 – us Library Micro [978]

[Butte county-] history of butte county, oroville, california / McGee, Joseph F – CA. 1956 – 1r – 1 – $50.00 – mf#B40208 – us Library Micro [978]

[Butte county-] history of wyandotte, butte county, california / Dunstane, William – CA. 1884 – 1r – 1 – $50.00 – mf#B40210 – us Library Micro [978]

Butte evening news – Butte MT 1910 jan-apr – 1r – 1 – mf#852013 – us WHS [071]

Butte gazette – Butte, NE: T S Armstrong, 1892 (wkly) [mf ed v1 n24. dec 3 1892 with gaps] – 1 – (absorbed: boyd county register) – us NE Hist [071]

Butte weekly miner – Butte MT. 1896 jan 2-1897 feb 27, 1897 mar 4-1898 feb 24, 1898 mar 3-oct 27, 1898 nov 3-1899 jun 29, 1899 jul 6-1900 feb 22, 1900 mar 1-sep 27, 1900 oct 4-1901 may 23, 1901 may 30-dec 26, 1902 jan-aug, 1902 sep-dec – 10r – 1 – mf#852011 – us WHS [622]

Buttenwieser, Moses *see*
- Die hebraeische elias-apokalypse
- Outline of the neo-hebraic apocalyptic literature
- The prophets of israel from the eighth to the fifth century
- Psalms

Butterfield, Consul Willshire *see* History of the discovery of the northwest by john nicolet in 1634

Butterfield, David L *see* The effects of high-volt pulsed current electrical stimulation on delayed onset muscle scoreness

Butterfield, Jane M *see* Problem of diction in choral technique

Butterfield, Kenyon Leech *see* The country church and the rural problem

Butterfield overland mailbag / Overland Mail Centennial – 1958 feb 16-1960 apr – 1r – 1 – mf#1053746 – us WHS [071]

Butterflies of california / Comstock, John adams – Gainesville: Scientific Publ 1989 [mf ed Ithaca NY: Cornell University, A R Mann Libr...; Bethlehem Pa: Micrographic Preservation Service (MAPS) Inc 1993] – 1r – 1 – (int, biogr & rev checklist by thomas c emmel & john f emmel) – mf#5087 r18 n5 – us Cornell [590]

Butterfly – Killen TX 1899-1900 – 1 – mf#3198 – us UMI ProQuest [073]

Butterfly; a humorous and artistic magazine – Killen TX 1893-94 – 1 – mf#3179 – us UMI ProQuest [073]

Butternut bulletin – Butternut WI. 1922 feb 1-1925, 1928-43, 1943 oct 7-1947, 1948-62, 1963-1966 feb 23, 1967 – 13r – 1 – mf#999734 – us WHS [071]

Butternut eagle – Butternut WI 1904 apr 9 – 1r – 1 – mf#1221710 – us WHS [071]

Butterworth, Alan *see*
- A collection of the inscriptions on copper-plates and stones in the nellore district
- The southlands of siva

Butterworth, F Edward *see* Sons of the sea

Butterworth, Hezekiah *see*
- The log school-house on the columbia
- Lost in nicaragua
- The story of the hymns
- Zigzag journeys in the great northwest

Butterworth, John *see* A new concordance to the holy scriptures

Buttery, John A *see* Why kruger made war

Buttmann, Alexander *see* A grammar of the new testament greek

Buttmann, Philipp *see*
- Die christliche heilslehre
- Lexilogus
- Recensus omnium lectionum quibus codex sinaiticus descrepat a textu editionis novi testamenti cui est titulus

Button, Charles P *see* The general law of partnership as applied to commercial and business liabilities

Buttonworld magazine – [vl n1]-v4 n5 [1971 sep/oct-1976 oct] – 1r – 1 – mf#669368 – us WHS [071]

Butts, Isaac Ridler *see*
- The business man's adviser.
- Directions and forms for the execution and acknowledgement of deeds to be used or recorded in other states
- The trader's guide, and business man's legal companion, containing the laws of trade.

Butts, John L *see* Program for agricultural education in dade county

Butts, Nancy Kay *see*
- Aerobic responses to 12 weeks of exerstriding or walking training in sedentary adult women
- A comparison of rating of perceived exertion in treadmill vs track walking and running
- The energy cost of women walking with and without hand weights while performing rhythmic arm movements
- A five-mile mountain bicycle test to predict vo2max
- Handrail assisted versus nonhandrail assisted stairmaster gauntlet ergometry
- Physiological responses obtained during exercise on the stairmaster gauntlet with and without the use of hands
- Psychological profiles before and after 12 weeks of walking or exerstrider training in adult women
- Validation of equations to predict lactate threshold, fixed blood lactate concentrations, and peak values from a 3200 meter performance time
- A validation study of the q-plex 1 cardiopulmonary exercise system

Buttstett, J H *see* Ut, mi, sol, re, fa, la, tota musica et harmonia aeterna

Butzbacher zeitung, wetterauer bote – Butzbach DE, 1988- – 6r/yr – 1 – gw Misc Inst [074]

Butzlaff, Martin E *see* Vergleich der gastralen saeuresekretion bei diabetikern mit und ohne autonome neuropathie

Buurman, Ulrich *see* Erlaeuterungen und aufsaetze zur einfuehrung in goethes faust fuer lehrer und den gebildeten

Buxbaum, Heinrich *see* Geschichte der israel

Buxbaum, Philipp *see* Wildhecken

Buxoyo, Simon Benito *see* Historia de caceres y su patrona

Buxtehuder tagblatt *see* Buxtehuder wochenblatt

Buxtehuder wochenblatt – Buxtehude DE, 1988- – 6r/yr – 1 – (title varies: 2 jan 1926: buxtehuder tagblatt) – gw Misc Inst [074]

Buxton chronicle, north derbyshire and north cheshire reporter – Buxton then Glossop, Derbyshire, jul 1888-26 jan 1906 [mf ed 2004] – 16r – 1 – (previously: buxton advertiser & high peak chronicle; after 26 jan 1906 incorp in: high peak chronicle) – uk Newsplan [072]

Buxton eagle – Buxton IA. 1903 oct 10 – 1r – 1 – mf#851119 – us WHS [071]

Buxtorf, Johann *see* Otsar sharshi leshon ha-kodesh

Buyers, W *see* Recollections of northern india

Buyers, William *see*
- Letters on india
- Recollections of northern india

Buyouts & acquisitions – Seattle WA 1986-88 – 1,5,9 – (cont: journal of buyouts and acquisitions; cont by: corporate growth) – ISSN: 1045-1161 – mf#14431.05 – us UMI ProQuest [332]

Buys, E *see* A new and complete dictionary of terms of art

Buyser y Aquino, Fernando *see*
- Ang laa sa bugay: dulang inawitan: duha ka acto ug tulo ka cuadro
- Haring gangis ug haring leon
- Si kristo gikawat

Buytendijk, S H *see* Het vijfentwintigjarig bestaan van het christelijk nationaal zendingsfeest

Buzo Gomes, Sinoforiano *see* Indice de la poesia paraguaya

Buzonniere, Leon de *see*
- Le touriste ecossais, ou itineraire general de l'ecosse
- Voyage en ecosse

Buzshe raynarski, 1897-1914 / Forem, Leon – New York, NY. 1938 – 1r – us UF Libraries [939]

Buzulukskii uezd : ispolnitel'nyj komitet sovetov. izvestiia buzulukskogo uezdnogo ispolnitel'nogo komiteta sovete rabochikh i krest'ianskikh deputatov – Buzuluk, Russia 1918 [mf ed Norman Ross] – 1r – 1 – mf#nrp-391 – us UMI ProQuest [077]

Buzy, Odette *see* La notion de congregation, sa portee en droit civil francais.

Buzzell, John M G *see*
- The life of elder benjamin randal
- Religious magazine containing an account of the united churches of christ commonly called freewill baptist

Buzzworm – Boulder CO 1988-92 – 1,5,9 – (cont by: earth journal) – ISSN: 0898-2996 – mf#18477.01 – us UMI ProQuest [639]

Bvllae papisticae...contra...reginam elizabetham... promulgatae, refutatio / Bullinger, Heinrich – Londini, Iohannes Day, 1571 – 2mf – 9 – mf#PBU-236 – ne IDC [240]

Bwebwenato in baibel / Whitney, J F [comp] – New York: American Tract Society, [1883] [mf ed 1995] – (= ser Yale coll) – 336p (ill) – 1 – 0-524-09475-6 – (in marshall) – mf#1995-0475 – us ATLA [220]

By any means necessary / New Afrikan People's Organization [US] – Birmingham AL, Jackson MS. 1987? sep/oct, 1992 jun/jul, 1993 jan, aug/sep – 1r – 1 – mf#2691799 – us WHS [320]

By canoe and dog train : among the cree and salteaux indians / Young, Egerton Ryerson – London: C H Kelly, 1890 – 4mf – 9 – (int by mark guy pearse) – mf#54506 – cn Canadiana [242]

By canoe and dog train : among the cree and salteaux indians / Young, Egerton Ryerson – Toronto: W Briggs; Montreal: C W Coates, 1890? – 3mf – 9 – (int by mark guy pearse) – mf#30579 – cn Canadiana [920]

By canoe and dog-train among the cree and salteaux indians / Young, E R – London, 1894 – 5mf – 9 – mf#N-458 – ne IDC [917]

By canoe and dog-train among the cree and salteaux indians / Young, Egerton Ryerson – New York: Hunt & Eaton; Cincinnati: Cranston & Stowe, 1891 [mf ed 1986] – 1mf – 9 – 0-8370-6638-7 – mf#1986-0638 – us ATLA [917]

By horse, canoe and float through the wilderness... / Cook, William Azel – Akron, OH. 1909 – 1r – us UF Libraries [972]

By intervention of providence / Mckenna, Stephen – Boston, MA. 1923 – 1r – us UF Libraries [972]

By nile and euphrates : a record of discovery and adventure / Geere, Henry Valentine – Edinburgh: T & T Clark, 1904 – 1mf – 9 – 0-524-00836-1 – mf#1990-2082 – us ATLA [910]

By order of the czar : a novel / Hatton, Joseph – New York: J W Lovell, c1890 – 5mf – 9 – 0-665-05208-1 – mf#05208 – cn Canadiana [830]

By pu. booster – Louisville. Kentucky. Oct 1920-Oct 1922. (The Reflector. Jan 1928. Junior Baptist. Aug 1922, Oct 1922). Baptist Young Adults. 1956-61 – 1 – us Southern Baptist [242]

By temple shrine and lotus pool / Robinson, William; ed by Smith, George – London: Morgan and Scott, 1910 [mf ed 1995] – (= ser Yale coll; Morgan and scott's missionary series) – xvi/295p – 1 – 0-524-10049-7 – mf#1995-1049 – us ATLA [240]

By the aurelian wall : and other elegies / Carman, Bliss – Boston: Lamson, Wolffe, 1898 [mf ed 1980] – 2mf – 9 – 0-665-00484-2 – mf#00484 – cn Canadiana [810]

By the equator's snowy peak : a record of medical missionary work and travel in british east africa / Crawford, E May Grimes – London: Church Missionary Society, 1913 – 1 – us CRL [240]

By the great wall : letters from china, the selected correspondence of isabella riggs williams, missionary of the american board to china, 1866-97 / Williams, Isabella Riggs – New York: Fleming H Revell, [1909] [mf ed 1995] – (= ser Yale coll) – 400p (ill) – 1 – 0-524-10076-4 – (int by arthur h smith) – mf#1995-1076 – us ATLA [951]

By the mill born / Milbourne & Tull Teserach Center – iss n1-34 & 35 [1976 apr-1984 may/aug] – 1r – 1 – mf#354473 – us WHS [071]

By the president of the united states of america : a proclamation – Washington?: s.n, 1812? – 1mf – 9 – mf#60362 – cn Canadiana [975]

By the river chebar : some applications of ezekiel's visions / Lewis, Howell Elvet – London: Hodder and Stoughton, 1903 – 1mf – 9 – 0-8370-4102-3 – (incl ind of biblical passages cited) – mf#1985-2102 – us ATLA [920]

By the still waters : a meditation on the twenty-third psalm / Miller, James Russell – New York:Thomas Y Crowell, c1897 – 1mf – 9 – 0-8370-4430-8 – mf#1985-2430 – us ATLA [220]

By waysides in india / Frost, Adelaide Gail – 2nd ed. [S.l.: s.n., 1902?] – 1mf – 9 – 0-524-04373-6 – mf#1991-2077 – us ATLA [240]

B'yachad – Northridge, CA. Sept 1977-Sept 1984 – 1 – us AJPC [071]

Byalistoker leksikon – Bialystok, Poland. 1935 – 1r – us UF Libraries [939]

Bybel- en zededichten : bestaende in zinnebeelden... / Schim, H – Delft: Reinier Boitet, 1726 – 8mf – 9 – mf#O-750 – ne IDC [090]

Bychkov, A F *see*
- Opisanie slavianskikh i russkikh rukopisnykh sbornikov imperatorskoi publichnoi biblioteki
- Opisanie tserkovno-slavianskikh i russkikh rukopisei imperatorskoi publichnoi biblioteki
- Pisma petra velikogo, khraniashchiesia v imperatorskoi publichnoi biblioteke i opisanie nakhodiashchikhsia v nei rukopisei, soderzhashchikh materialy dlia istorii ego tsarstvovannia

Die bydrae van die kaapse maleier tot die afrikaanse volkslied / Du Plessis, Izak David – Kaapstad: Nasionale Pers 1935 [mf ed Cape Town: SA Lib 1977] – 4mf – 9 – mf#mf.957 – sa National [780]

Bye-laws of the municipal council of the district of london : passed at the 1st session of 1847 – [s.l: s.n.] 1847 [mf ed 1984] – 1mf – 9 – 0-665-45064-8 – mf#45064 – cn Canadiana [348]

Bye-laws of the municipal council of the district of london : passed at the 1st session of 1849 – [London, Ont? s.n.] 1849 [mf ed 1984] – 1mf – 9 – 0-665-45121-0 – mf#45121 – cn Canadiana [348]

Bye-laws, rules and orders of the trinity house, quebec : 10th november, 1820, and 19th april 1821 / Trinity House of Quebec – [s.l: s.n, 1821?] [mf ed 1984] – 1mf – 9 – 0-665-44246-7 – mf#44246 – cn Canadiana [380]

Byelorussian voice *see* Belaruski holas

Byen : junker friklover : nutidsroman / Claussen, Sophus – Kobenhavn: Det Schubotheske Forlag 1900 [mf ed 1984] – 1r – 1 – mf#962 – us Wisconsin U Libr [830]

De byencorf der h roomsche kercke / Marnix van S Aldegonde, P van – [Emden, 1569] – 6mf – 9 – mf#PBA-255 – ne IDC [241]

Bye-paths in baptist history : a collection of interesting, instructive, and curious information, not generally known, concerning the baptist denomination / Goadby, Joseph Jackson – London: Elliot Stock, 1871 – 1mf – 9 – 0-524-07981-1 – mf#1990-5426 – us ATLA [242]

Byer og bybebyggelse / ed by Bull, Edvard et al – Stockholm: A Bonnier 1933 [mf ed Bloomington IN: Indiana Uni Lib, Preservation Dept 1984] – 99p on 1r [ill] – 1 – us Indiana Preservation [390]

Byern, E von *see* Bilder aus griechenland und der levante

Byers, Charles Francis *see* Contribution to the knowledge of florida odonata

Byers, George D *see* Diaries

Byers newsletter – n13-15 [1968-75] – 1r – 1 – mf#354475 – us WHS [071]

Byers, W C *see* Soil survey of bradford county, florida

Byford, Charles T *see* Peasants and prophets

Byford, Charles Thomas *see* The soul of russia

Byford, Charles Thomas et al *see* Modern baptist heroes and martyrs

Byggnadskultur / ed by Erixon, Sigurd – Stockholm: A Bonnier 1953 [mf ed Bloomington IN: Indiana Uni Lib, Preservation Dept 1984] – 416p on 1r [ill] – 1 – us Indiana Preservation [390]

Bygone church life in scotland / Tyack, George Smith et al; ed by Andrews, William – London: W Andrews, 1899 – 1mf – 9 – 0-524-03451-6 – mf#1990-0994 – us ATLA [240]

Byington, Ezra Hoyt *see* The puritan as a colonist and reformer

Bykhovskii, N I *see* Dlia chego nuzhny sovety krestianskikh deputatov

De bykorf des gemoeds : honing zaamelende uit allerley bloemen / Luyken, Jan – Amsteldam: Wed P Arentz, en K vander Sys, 1711 – 8mf – 9 – mf#O-354 – ne IDC [090]

Bylae tot die oosterlig, die volksblad, die volkstem : mikrofilm deur staatsbiblioteek pretoria – 1953-65 [mf ed Pretoria: Staatsbiblioteek Korporatiewe Kommunikasie, [1982]] – 1 – mf#MS00215 – sa National [079]

By-laws and constitution : minutes / Ladies' Enterprise Association. Mound City, Kansas – 1864-69 – 1 – us Kansas [060]

By-laws and regulations of the trinity house : concerning pilots and others, and the navigation of the river st lawrence / Trinity House of Quebec – [Quebec?: s.n.] 1843 [mf ed 1985] – 1mf – 9 – 0-665-28076-9 – mf#28076 – cn Canadiana [343]

By-laws and rules as revised 1899 / Association of Ontario Land Surveyors – Toronto?: s.n, 1899? – 1mf – 9 – mf#17070 – cn Canadiana [520]

By-laws for the management of the affairs of the bank of montreal / Bank of Montreal – Montreal?: J Lovell, 1856 – 1mf – 9 – mf#48732 – cn Canadiana [332]

By-laws of maple leaf lodge of ancient free and accepted masons, no, st catharines, c w : constituted by dispensation of the grand lodge of canada, bearing date may 17... / Freemasons. Maple Leaf Lodge (St Catharines, Ont) – St Catharines, Ont?, 1859 – 1mf – 9 – mf#47217 – cn Canadiana [360]

By-laws of st francis lodge of ancient, free, and accepted masons, n24 grc : instituted a l 5839 at smith's falls, ontario – [Smith's Falls, Ont?: s.n.], 1895 [mf ed 1980] – 1mf – 9 – 0-665-03253-6 – mf#03253 – cn Canadiana [360]

By-laws of the... : to which is prefixed, the act of incorporation, 8th victoria, cap 93, passed by the provincial parliament, 1845 / Mechanics' Institute of Montreal – [Montreal?: s.n.] 1847 [mf ed 1984] – 9 – 0-665-43096-5 – mf#43096 – cn Canadiana [360]

By-laws of the artisans' permanent building society : adopted at the general meeting held the 15th january 1875 – [Quebec?: s.n.], 1875 [mf ed 1980] – 1mf – 9 – mf#07063 – cn Canadiana [366]

By-laws of the association of ontario land surveyors / Association of Ontario Land Surveyors – S.l: s.n, 1892? – 1mf – 9 – mf#54745 – cn Canadiana [366]

By-laws of the montreal protestant house of industry and refuge : as amended and finally passed by the governors of the corporation, on 9th march, 1864... / Montreal Protestant House of Industry and Refuge – Montreal?: Herald Steam Press, 1864 – 1mf – 9 – mf#67787 – cn Canadiana [360]

By-laws of the mutual marriage aid association of canada : (incorporated under chapter 167, revised statutes of canada), head office, hamilton, ont / Mutual Marriage Aid Association of Canada – Hamilton, Ont: J A Griffin, 1881? – 1mf – 9 – mf#11188 – cn Canadiana [366]

By-laws, rules and regulations of the beechwood cemetery company, ottawa : incorporated 1873 – Ottawa: J Durie, 1886 [mf ed 1980] – 1mf – 9 – 0-665-03303-6 – mf#03303 – cn Canadiana [360]

By-laws, rules, special rules, regulations and orders : for the use and guidance of the servants, employes [sic] and officers of the grand trunk railway of canada / Grand Trunk Railway Company of Canada – [s.l: s.n, 1865?] [mf ed 1983] – 2mf – 9 – 0-665-44877-5 – mf#44877 – cn Canadiana [380]

The byles family papers, 1757-1837 – [mf ed 1984] – 2r – 1 – (with p/g. the letters of mather byles, jr. wh comprise the main body of the coll, provide personal accounts of a family torn apart over the revolutionary conflict) – us MA Hist [975]

Byles, John Barnard *see* A treatise of the law of bills of exchange, promissory notes, bank-notes and checks

Byley, st john – (= ser Cheshire monumental inscriptions) – 2mf – 9 – £4.00 – mf#108a – uk CheshireFHS [929]

Byloe : sbornik po istorii russkago osvoditel nago dvisheniia – Paris. 57nos. 1900-33 – 1 – $432.00 – (in russian) – mf#0126 – us Brook [460]

Byloff, Fritz *see* Das verbrechen der zauberei (crimen magiae)

Byposten – Oslo, Norway. 322p. 1907-1909 – 1r – 1 – (orders outside usa via universitetsbiblioteket i oslo) – mf#6620 – us Southern Baptist [242]

By-products of the study of law / Northwestern University. School of Law – Chicago, Ill.: School of Law 1939 14p. LL-1362 – 1 – us L of C Photodup [340]

The b.y.p.u. booster – Louisville, KY. Oct 1920-Oct 1922. The Reflector, Jan 1928. Junior Baptist, Aug, Oct 1922 – 1 – us Southern Baptist [242]

Bypu quarterly *see* Baptist young people

Byr, Robert [Bayer, Robert von] *see*
- Auf abschuessiger bahn
- Auf der station
- Der kampf um's dasein
- Nomaden
- Sesam
- Sphinx
- Der weg zum glueck

Byrd, B J *see* The relationship of history of stressors, personality, and coping resources, with the incidence of athletic injuries

Byrd, Daniel Ellis *see* Papers, 1947-1981

Byrd, Marcia J *see* Drug and alcohol use by freshman at siuslaw high school and their opinions regarding potentially effective drug and alcohol education programs

Byrd, Simpson Lesley *see* Studies in the administration of the indians in new spain. berkeley, 1934

Byrd, William *see* Manuscripts

Byrne, Donn *see*
- Crusade
- Power of the dog

Byrne, James *see* Naturalism and spiritualism

Byrne, James et al *see* Essays on the Irish church

Byrne, John Elliott *see* Federal criminal procedure, with forms for the defense.

Byrne, William *see* The catholic doctrine of faith and morals, gathered from sacred scripture; decrees of councils, and approved catechisms

Byrne's emigrants / journal – Pretoria: State Library Corporate Communication, feb 1850-mar 1850 – (= ser State library south africa newspaper microfilm project) – 1r – 1 – mf#MS00281 – sa National [079]

Byrnes, John *see* The relationship between elementary classroom teachers' perceptions of school health education and their level of health teaching

Byrom hall / Houghton, J – Liverpool, England. 1881 – 1r – us UF Libraries [240]

Byron and espronceda / Churchman, Philip H – Extrait de la Revue Hispanique. Tome 20. New York, Paris, 1909 – 1 – sp Bibl Santa Ana [946]

The byron blade – Byron, NE: P J George (wkly) [mf ed v4 n8. jan 3 1908-dec 20 1912 (gaps)] – 2r – 1 – (publ as: weekly byron blade oct 20 1911-jan 19 1912) – us NE Hist [071]

[Byron-] byron times – CA. jul 19 1907-jun 1935 – 13r – 1 – $780.00 – mf#BC03169 – us Library Micro [071]

Byron et le romantisme francais / Esteve, Edmond – Paris, France. 1929 – 1r – us UF Libraries [025]

Byron, George Gordon, 6th Baron *see*
- Kayin
- Poems and dramas of lord byron

The byron gleaner – Byron, NE: Cyrus Black (wkly) [mf ed v4 n52. jul 28-nov 24 1892 (gaps) filmed [1973]] – 1r – 1 – us NE Hist [071]

The byron herald – Chester, NE: H A Brainerd, nov 20 1896-nov 1905// (wkly) [mf ed v1 n6. dec 25 1896=whole n6-apr 29 1904 (gaps)] – 1r – 1 – (absorbed by: chester herald) – us NE Hist [071]

Byrum, Enoch Edwin *see* Divine healing of soul and body

Bysh's edition of the life of robinson crusoe – London, England. 183- – 1r – us UF Libraries [830]

Bysshe, Edward *see* Visitation of the county of essex

Bystander *see* Annual report of the port royal relief committee

Bystritskii, K *see* Sel'skaia zhizn'

Bystrov, F M *see* Ural'skaia zhizn'

Bystrov, I *see* Opyt alfavitnogo ukazatelia k russkim periodicheskim izdaniiam

Byte – Manhasset NY 1975-98 – 1,5,9 – ISSN: 0360-5280 – mf#11392 – us UMI ProQuest [000]

Bytie i soznanie / Russkoe natsional'noe dvizhenie – Troitsk, Russia. n1(7/92)-2(1993), n1[3]-2[4](1993) – 1 – mf#mf-12248 (reel 2) – us CRL [077]

Bytown gazette – Ottawa, ON. 1836-45 – 3r – 1 – cn Library Assoc [071]

Byu studies – Provo UT 1981-83 – 1,5,9 – (cont: brigham young university studies; cont by: brigham young university studies) – ISSN: 0278-1980 – mf#8942.02 – us UMI ProQuest [378]

Byzantina historia (cbh23) / Nicephori Gregorae; ed by Boivin, J – Parisiis, 1702 – (= ser Corpus byzantinae historiae (cbh)) – €75.00 – ne Slangenburg [240]

Byzantina historia (cshb6,7,8) : graece e latine / Nicephori Gregorae; ed by Schopeni, Lud – Bonnae. v1. 1829 – (= ser Corpus scriptorum historiae byzantinae (cshb)) – €23.00 – (cum annotationibus h wolfii, c ducangii, io boivini et c capperonnerii. v2 bonnae 1830 €27. v3 bonnae 1855 €19) – ne Slangenburg [240]

The byzantine and romanesque court in the crystal palace / Wyatt, Matthew Digby & Waring, John Burley – London 1854 – (= ser 19th c art & architecture) – 2mf – 9 – mf#4.2.448 – uk Chadwyck [720]

Byzantine architecture / Texier, Charles Felix Marie & Pullan, Richard Popplewell – London 1864 – (= ser 19th c art & architecture) – 8mf – 9 – mf#4.2.864 – uk Chadwyck [720]

Byzantine bibliography : based on byzantislavica, prague – Zug, 1985 – 24mf – 9 – mf#O-1768 – ne IDC [720]

Byzantine catholic world – Pittsburgh PA 1973-79 – 1 – mf#8235 – us UMI ProQuest [241]

The byzantine catholic world – Pittsburgh, PA: Pittsburgh Byzantine Catholic Press Associates, 1956-74 – 1 – us CRL [241]

Byzantine portraits / Diehl, Charles – New York: A A Knopf 1927 [mf ed 1986] – 1r – 1 – (originally iss as figures byzantines [paris: armand colin 1906]) – mf#10365 – us Wisconsin U Libr [931]

Byzantine studies – Tempe AZ 1974-79 – 1,5,9 – mf#6968 – us UMI ProQuest [949]

Byzantinische kulturgeschichte / Gelzer, Heinrich – Tuebingen: J C B Mohr, 1909 – 1mf – 9 – 0-7905-4588-8 – mf#1988-0588 – us ATLA [930]

Byzantinische legenden – Jena: E Diederichs, 1911 – 1mf – 9 – 0-7905-7249-4 – mf#1988-3249 – us ATLA [240]

Byzantinische zeitschrift – 1(1892)-42(1943/49) – 687mf – 9 – €1310.00 – ne Slangenburg [931]

Byzantislavica *see* Recueil pour l'etude des relations byzanti-slaves

Byzanz und persien in ihren diplomatisch-voelkerrechtlichen beziehungen im zeitalter justinians / Gueterbock, K – Berlin, 1906 – 2mf – 9 – mf#AR-1894 – ne IDC [956]

Bz am abend – Berlin DE, 1949 15 jul-1954 31 jul – 9r – 1 – (filmed by misc inst: 1954 2 aug-1990 [68r], 1992-, 1949 15 jul-1954. title varies: 3 dec 1990: berliner kurier am abend, 3 aug 1992: berliner kurier) – gw Mikrofilm; gw Misc Inst [074]

Bz am abend – Pressburg (Bratislava SK), 1924 – 1r – 1 – Dist. gw Mikrofilm – gw Misc Inst [077]

C *see* Coquille herald

C and o canaller / Chesapeake and Ohio Canal Association – 1969 aug – 1r – 1 – (cont: level walker; cont by: along the towpath) – mf#681372 – us WHS [380]

C B Fisk (Firm) *see* Memoranda concerning government bonds, for the information of investors...

C b i roundup – 1944 jun 29, jul 13,27, aug 17 – 1r – 1 – mf#2892897 – us WHS [338]

C c n y black alumni news / City University of New York – 1986 may – 1r – 1 – mf#4851569 – us WHS [378]

The c d acts in india : official report of mr maclaren's speech in the house of commons, on june the 5th, 1888. reprinted from the "crewe & nantwich chronicle," of saturday, june 16th, 1888 / MacLaren, Duncan – London 1889 – (= ser 19th c british colonization) – 1mf – 9 – mf#1.1.5898 – uk Chadwyck [324]

C F gellerts briefstilreform : tradition, entwicklung und wirkungsgeschichte der gellertschen epistolartheorie / Arto-Haumacher, Rafael – (mf ed 1995) – 2mf – 9 – €40.00 – 3-8267-2126-8 – mf#DHS 2126 – gw Frankfurter [430]

C f gellert's saemtliche schriften / Gellert, Christian Fuerchtegott – Berlin: Weidmann; Leipzig: Hahn 1867 [mf ed 1993] – 10v in 5 on 2r – 1 – mf#8583 – us Wisconsin U Libr [802]

C f meyers "angela borgia" / Weishaar, Friedrich – Marburg a.L: N G Elwert 1928 [mf ed 1992] – (= ser Beitraege zur deutschen literaturwissenschaft 30) – 1r – 1 – (incl bibl ref. filmed with: wortsinn und wortschoepfung bei meister eckehart / rudolf fahrner & other titles) – mf#3099p – us Wisconsin U Libr [430]

C f stegemann's wanderung durch deutschland, polen, russland, caucasien, aegypten und persien nach jerusalem : in den jahren 1814 bis 1821 – Danzig 1824 [mf ed Hildesheim 1995-98] – 1v on 1mf – 9 – €40.00 – 3-487-26671-7 – gw Olms [910]

C f tombe's bataillonschefs reise in ostindien in den jahren 1802 bis 1806 / Tombe, Charles F – Leipzig 1811 [mf ed Hildesheim 1995-98] – 1v on 3mf [ill] – 9 – €90.00 – 3-487-27484-1 – (with ann by c s sonnini) – gw Olms [915]

C fu jen hsiao hsiang : san mu chu / Tung, Mei-k'an – Hsiang-kang: Mei chou chu she, Min kuo 28 [1939] – (= ser P-k&k period) – us CRL [820]

C G T *see* La bataille syndicaliste

C G T -F O *see* Force ouvriere

C h s bandwagon / Circus Historical Society – 1951 autumn-1957 feb – 1r – 1 – (cont: hobby bandwagon; cont by: bandwagon [richmond in]) – mf#932433 – us WHS [978]

C h spurgeon : a biography / Fullerton, William Young – London: W Williams & Norgate 1920 [mf ed 1988] – 1r [ill] – 1 – mf#2174 – us Wisconsin U Libr [242]

C i manigault's travels 1822-1848 *see* Travel notes

C i o news / Federation of Flat Glass Workers of America – v1 n6-v3 n31 [1938 jan 14-1940 jul 29] – 1r – 1 – (cont: Flat glass worker; cont by: cio news [glass, ceramic and silica sand ed]) – mf#659728 – us WHS [660]

C j stewart's catalogue of ecclesiastical law and polity – London: CJ Stewart, [1851?] – 1mf – 9 – 0-524-08616-8 – mf#1993-1066 – us ATLA [012]

C: jet communication: journalism education today – Manhattan KS 1972+ – 1,5,9 – ISSN: 0010-3535 – mf#8115.01 – us UMI ProQuest [070]

C m schrebian's aufenthalt in morea, attika und mehreren inseln des archipelagus – Leipzig 1825 [mf ed Hildesheim 1995-98] – 2mf – 9 – €60.00 – 3-487-29062-6 – gw Olms [914]

C olevianus und z ursinus : leben und ausgewaehlte schriften / Sudhoff, Karl – Elberfeld: R L Friderichs 1857 [mf ed 1991] – (= ser Leben und ausgewaehlte schriften der vaeter und begruender der reformirten kirche 8) – 2mf – 9 – 0-524-00609-1 – mf#1990-0109 – us ATLA [242]

C p u journal / Canadian Paperworkers Union – v2 n3-v4 n3 [1977 apr-1980 jan] – 1r – 1 – (cont: canadian paperworker journal) – mf#505503 – us WHS [670]

C S *see* Joseph

C s a o news / Civil Service Association of Ontario – 1970 may-1975 oct – 1r – 1 – (cont by: opseu news) – mf#697508 – us WHS [350]

C s lewis : exponent of tradition and prophet of postmodernism? / Moodie, Charles Anthony Edward – Uni of South Africa 2000 [mf ed Johannesburg 2000] – 6mf – 9 – (incl bibl ref) – mf#mfm14862 – sa Unisa [230]

C s p a a bulletin – New York NY 1941-81 – 1,5,9 – ISSN: 0010-1990 – mf#7698 – us UMI ProQuest [070]

C s p newsletter/ – 1990 spr/summer – 1r – 1 – mf#4851391 – us WHS [378]

C suetoni tranquilli quae supersunt omnia / Suetonius – Lipsiae, Germany. 1893 – 1r – us UF Libraries [960]

C w a c newsletter / Canadian Women's Army Corps – Ottawa: Dept National Defence. v1-2 n7. jan 1944-oct 1945// (mthly) – 1r – 1 – Can$110.00 – cn McLaren [355]

C w hufelands journal der practischen heilkunde *see* Journal der practischen arzneykunde und wundarzneykunst

Ca: a cancer journal for clinicians – Baltimore MD 1950- – 1,5,9 – ISSN: 0007-9235 – mf#7285 – us UMI ProQuest [616]
Ca aup ca pe – Ran Kun: Ca pe Biman 1973 [mf ed 1990] – 1r with other items – 1 – (in burmese) – mf#mf-10289 seam reel 160/1 [§] – us CRL [480]
Ca bu ton / Can nve, Da gun U – Ran kun Mrui': Da gun u Can nve 1975 [mf ed 1993] – on pt of 1r – 1 – mf#11052 r613 n9 – us Cornell [959]
CA charter – Sydney, Australia 2002+ – 1,5,9 – ISSN: 1446-4543 – mf#18680.05 – us UMI ProQuest [650]
Ca chemissues – Current Chemical Abstracts on Microfiche distributed weekly plus backfile of Chemical Abstracts on 16mm microfilm casettes or microfiches – 6,9 – us Chemical [540]
Ca chui panna rhi kalidasa e bha va a tve a krum mya / Lha, Lu thu U – Mantale: Kri pva re Pum nhip tuik 1961- [mf ed 1990] – 1r with other items – 1 – (in burmese) – mf#mf-10289 seam reel 201/5 [§] – us CRL [490]
Ca chui to nan tvan vatthu kri : [a novel] / Khan Khan Le, Da gun – Ran kun: Ba ma khet Sa tan ca Tuik 1951-52 [mf ed 1990] – 1r with other items – 1 – (in burmese) – mf#mf-10289 seam reel 187/2 [§] – us CRL [830]
Ca cuam ait chon sat pum / Cui Van, U – Ran Kun: Chan Nnvhan U pum nhip tuik 1969 [mf ed 1990] – 1r with other items – 1 – (in burmese) – mf#mf-10289 seam reel 148/1 [§] – us CRL [480]
Ca dao lao dong – [Ha-noi]: Pho Thong 1974 [mf ed 1992] – on pt of 1r – 1 – mf#11052 r98 n5 – us Cornell [959]
Ca et la / Baillairge, Frederic Alexandre – Montreal: Cadieux & Derome, 1881 [mf ed 1979] – 1mf – 9 – 0-665-00073-1 – mf#00073 – cn Canadiana [241]
Ca et la : cochinchine et cambodge; l'ame khmere; angkor / Reveilliere, Paul Emile Marie [Paul Branda: pseud] – Paris: Fischbacher 1886 [mf ed 1989] – 1r with other items – 1 – mf#mf-10289 seam reel 009/09 [§] – us CRL [915]
Ca ira – Anvers. n1-20. avr 1920-janv 1923 – 1 – fr ACRPP [073]
Ca ira! : or, danton in the french revolution; a study / Gronlund, Laurence – Boston: Lee & Shepard; New York: C T Dillingham 1888 [mf ed 1987] – 1r – 1 – mf#8599 – us Wisconsin U Libr [944]
Ca journal of poetry – nos. 1-13. 1963-66 – 1 – us AMS Press [810]
Ca ka lum krvay ra nann padesa = How to increase your word power in burmese / Lha Samin – Ran kun: Tan Chve Ae Ca aup chuin [19]69 [mf ed 1990] – 1r with other items – 1 – (in burmese) – mf#mf-10289 seam reel 147/6 [§] – us CRL [480]
Ca ka pum tara – Ran kun: On lam to Ca pe 1985 [mf ed 1990] – [ill] 1r with other items – 1 – mf#mf-10289 seam reel 180/7 [§] – us CRL [390]
Ca kui ca ka kui – Ton Ukkalapa [Ran kun]: Ri Le Ca pe 1974 [mf ed 1993] – on pt of 1r – 1 – mf#11052 r603 n3 – us Cornell [959]
Ca lum, che cak, con krui, nhan kattipa ka lip / Da gun Tara – Kan to ka le, Ran kun: Prannlumkyvat Phran khyi re acay dami Ca pe tuik 1974 [mf ed 1993] – on pt of 1r – 1 – mf#11052 r606 n1 – us Cornell [959]
Ca lum pon sat pum nnhi nhuin khyak, rhan lan khyak, chum phrat khyak / Mon Mon Kri, Takkasuil – Ran Kun: Nha lum lha ca pe phran khyi re 1975 [mf ed 1990] – 1r with other items – 1 – (in burmese) – mf#mf-10289 seam reel 197/2 [§] – us CRL [480]
CA magazine – Edinburgh. 1993+ (1,5,9) – (cont: accountant's magazine) – mf#11283,01 – us UMI ProQuest [650]
CA magazine – Toronto ON 1937+ – 1,5,9 – ISSN: 0317-6878 – mf#9858 – us UMI ProQuest [650]
Ca mrin vuin / Ae Kruin, Do – Ran Kun: Kam ko mruin ca pe 1984 [mf ed 1990] – [ill/pl] 1r with other items – 1 – (in burmese) – mf#mf-10289 seam reel 175/6 [§] – us CRL [640]
Ca myak rhu / Mon Mon Kri, Takkasuil – Ran Kun: 'Alimma Ca pe 1966 [mf ed 1990] – (= ser Nha lum Lha Ca 'up 62) – 1r with other items – 1 – (in burmese) – mf#mf-10289 seam reel 196/3 [§] – us CRL [480]
Ca nay jan sa muin : nidan / Kri Mon, Kre mum – Ran Kun: Khyac Mruin Ca pe 1974 [mf ed 1990] – (= ser Khyac mruin ca can 17) – 1r with other items – 1 – mf#mf-10289 seam reel 163/4 [§] – us CRL [070]
Ca nay jan samuin ca tan mya / Ba San et al – Yan Kun: Cape Biman 1978 [mf ed 1990] – (= ser Prynn su lak cavai cacan) – 2v on 1r with other items – 1 – (in burmese) – mf#mf-10289 seam reel 164/1 [§] – us CRL [070]
Ca on microfilm – v1-101. 1907 – 6 – us Chemical [540]
Ca pay ni ka bya mya : [poems] / Mon Chann Cya et al – Ran Kun: Takkasuil rip sa [195-?] [mf ed 1990] – 1r with other items – 1 – (in burmese) – mf#mf-10289 seam reel 183/3 [§] – us CRL [810]
Ca pe a svan nhan kyvan to a mran / Mon Mon Kri, Takkasuil – Ran kun: Ca khyac su 1975 [mf ed 1993] – on pt of 1r – 1 – mf#11052 r595 n4 – us Cornell [959]
Ca pe biman thut-ca aup mya – Ran Kun: Ca pe biman ron vay phok ka re tha na 1963 [mf ed 1990] – 1r with other items – 1 – (in burmese) – mf#mf-10289 seam reel 177/8 [§] – us CRL [070]
Ca pe chu ra vatthu tui mya = Short stories. selections / Sota Chve – Ran kun: Sota Chve Ca pe 1974 [mf ed 1993] – on pt of 1r – 1 – mf#11052 r1251 n11 – us Cornell [830]
Ca pe chve nve pvai / Lha on, a thok to – [Ran kun: a Mran Sai Ca pe 1975 [mf ed 1993] – on pt of 1r – 1 – mf#11052 r594 n3 – us Cornell [959]
Ca pe chve nve pvai / Sin Phe Mran – Ran kun: U Sin Phe Mran 1975 [mf ed 1993] – on pt of 1r – 1 – mf#11052 r1243 n1 – us Cornell [959]
Ca pe chve nve pvai / Sota Chve – Ran kun: Sota Chve Ca pe 1975 [mf ed 1993] – on pt of 1r – 1 – mf#11052 r1239 n3 – us Cornell [959]
Ca pe gita nhan van kri padesaraja (1047-1114) – Ran kun: Cin Pan Mruin Ca pe tuik 1974 [mf ed 1993] – on pt of 1r – 1 – mf#11052 r1218 n7 – us Cornell [959]
Ca pe nhan lu mhu lak tve / Mra San, Sa Khan – Ran Kun: Na va rat ca pe 1967 [mf ed 1990] – 1r with other items – 1 – (in burmese) – mf#mf-10289 seam reel 132/4 [§] – us CRL [180]
Ca pe nhan yan kye mhu / Lha Bhe, Dagun U – Ran Kun: Ca pe pon ku pum nhip tuik 1976 [mf ed 1990] – 1r with other items – 1 – (in burmese) – mf#mf-10289 seam reel 147/8 [§] – us CRL [480]
Ca pe suta, ca pe rasa / Ne La, Mon – Ran Kun: Amara Ca pe 1987 [mf ed 1990] – 1r with other items – 1 – (in burmese) – mf#mf-10289 seam reel 151/3 [§] – us CRL [480]
Ca pe van kui kye kye pvan pvan tham kra cui – Ran Kun: U Bhui 'Han, Ya mu na Ca pe 1328 [1966] – 1r with other items – 1 – (incl bibl ref) – mf#mf-10289 seam reel 203/8 [§] – us CRL [959]
Ca to khyak kyam : rhe mran ma bhu ran a chak chak tui a sum to kham ca to chak ca phui sann tui pru ci nam khai sann a ra sa pho nann nissa yann – Ran Kun: Hamsavati pum nhip tuik 1965 [mf ed 1990] – 1r with other items – 1 – (in burmese) – mf#mf-10289 seam reel 165/3 [§] – us CRL [640]
Ca va – New York NY 1969-72 – 1,5,9 – ISSN: 0007-9243 – mf#5842 – us UMI ProQuest [370]
Caa journal – Washington DC 1940-52 – 1 – mf#5760 – us UMI ProQuest [629]
Caalogo de imprevistos / Matas, Julio – Habana, Cuba. 1963 – 1r – us UF Libraries [972]
Caatingas e chapadoes / Iglesias, Francisco Da Assis – Sao Paulo, Brazil. v1-2. 1958 – 1r – us UF Libraries [972]
The cab *see* Political pamphlets... 19th c
Caba, Pedro *see*
- Algunos rasgos del hombre extremeno
- La ciencia, la naturaleza y el milagro
- Eugenio noel
- Filosofia de la presencia humana
- La filosofia del conocimiento de san agustin
- La filosofia del no-ser en el pensamiento griego
- El hombre ante el hombre
- El hombre contra la naturaleza
- Metafisica de los sexos humanos
- La metafisica del espacio
- Misterio y poesia
- Nostalgia de dios en agustin y en pascal
- Los sexos, el amor y la historia
- Sobre la vida y la muerte
- Tierra y mujer o lazara la profetisa

Caba, Ruben *see* Impetu, pasion y fuga
Cabada, Carlos *see* Cuentos de ciencia-ficcion
Cabal Cabal, Camilo J *see* Gestion oficial en agricultura
Cabal, Juan *see*
- Balboa, descubridor del pacifico
- Carabelas de espana...pinzon, juan de la cosa, diaz de solis juan sebastian elcano

The cabala : its influence on judaism and christianity / Pick, Bernhard – Chicago: Open Court, 1913 – 1mf – 9 – 0-7905-1780-9 – (incl ind) – mf#1987-1780 – us ATLA [270]
Caballero Calderon, Eduardo *see*
- Cristo de espaldas
- Diario de tipacoque
- Historia privada de los colombianos
- Siervo sin tierra

Caballero Calderon, Lucas *see* Figuras politicas de colombia
Caballero de el dorado / Arciniegas, German – Bogota, Colombia. 1958? – 1r – us UF Libraries [972]
Caballero de jamaica / Siri, Eros Nicola – Buenos Aires, Argentina. 1944 – 1r – us UF Libraries [972]
El caballero de la gloria / Vial Solar, Javier – Santiago de Chile: imprenta y litografia la ilustracion, 1916 – 1 – sp Bibl Santa Ana [946]
Caballero, Fernan *see*
- Gaviota
- Obras completas

Caballero, Fernando *see* La casa en que murio hernan cortes en castilleja de la cuesta
Caballero, Jose Agustin *see* Philosophia electiva
Caballero, Jose Maria *see* Particularidades de santafe
Caballero y Tejerina, Antonio *see* Memorial ajustado...del pleyto...de la aliseda...
Caballeros clerigos extremenos del orden y caballeria de alcantara / Velo Nieto, Gervasio – Madrid: Imp. Accasor, 1953 – sp Bibl Santa Ana [240]
Caballeros de espuela dorada (descubrimiento y conquista del peru) / Funes, Jorge Ernesto – Buenos Aires: Emece, 1980 – 1 – sp Bibl Santa Ana [972]
Caballeros de las ordenes militares en mexico / Martinez Cosio, Leopoldo – Mexico City? Mexico. 1946 – 1r – us UF Libraries [972]
Los caballeros de las ordenes militares en mexico / Martinez Cosio, Leopoldo – Madrid: Razon y Fe, 1947 – 1 – sp Bibl Santa Ana [355]
Los caballeros de nuestra senora de salor / Munoz de San Pedro, Miguel – Madrid: Diana Artes Graficas, 1954. Hidalguia II pp. 449-460 – 1 – sp Bibl Santa Ana [946]
Caballeros del monasterio de yuste – Caceres: Tip. La Minerva, 1971 – 1 – sp Bibl Santa Ana [340]
Caballito verde / Arroyo, Anita – Habana, Cuba. 1956 – 1r – us UF Libraries [972]
El caballo de batalla de los nuevos cruzados en la america espanola / Bayle, Constantino – Madrid: Razon y Fe, 1926 – 1 – sp Bibl Santa Ana [972]
Caballo rojo / Barrett, Maca – Guatemala, 1959 – 1r – us UF Libraries [972]
Caban Soler, Jose *see* De norte a sur
Cabanas Ventura, Felipe *see* Badajoz taurino
Cabanis, J *see* Ornithologisches centralblatt
Cabanis, J L *see* Museum heineanum. verzeichniss der ornithologischen sammlung des oberamtmann ferdinand heine...
An cabayong tabla ni don juan de valencia na agom ni d a maria de asturia – [Nueva Caceres?: Libreria Mariana? 190-?] [mf ed Bloomington IN: Indiana Uni Lib, Preservation Dept 1984] – 1r – 1 – us Indiana Preservation [490]
Cabeen, Violet Abbott *see* Changes in the documents of british india caused by the government of india act 1935
Cabel record – Milton, WV. 1954+ (1) – mf#67368 – us UMI ProQuest [071]
Cabell county press – Guyandotte, WV. 1869-1873 (1) – mf#67305 – us UMI ProQuest [071]
Cabell, James Branch *see* Music from behind the moon
Cabell, James Lawrence *see* The testimony of modern science to the unity of mankind
Cabellian – Garden City. 1968-1972 (1) 1970-1972 (5) – ISSN: 0007-926X – mf#3451 – us UMI ProQuest [073]
Cabello de Balboa, Miguel *see* Obras
Cabello de Carbonera, Mercedes *see* Sacrificio y recompensa
Cabet, Etienne *see* Revolution de 1830, et situation presente (juillet 1832)
Cabeus, N *see* Philosophia magnetica...
Cabeza del Buey. Ayuntamiento *see*
- Feria y fiestas, 1947. cabeza del buey
- Ferias y fiestas en honor a san miguel. 1979

Cabezas, Juan Antonio *see* Ruben dario
Cabiati, Attilio *see* Problemi commerciali e finanziari dell'italia
Cabildos de indios en la america espanola / Bayle, Constantino – Madrid: Missionalia Hispanica, 1951 – 1 – sp Bibl Santa Ana [972]
Cabildos de las ciudades de nuestra senora de la c / Piedrahita, Diogenes – Cali, Colombia. 1962 – 1r – us UF Libraries [972]
Cabildos de santafe de bogota / Bogota Concejo – Bogota, Colombia. 1957 – 1r – us UF Libraries [972]
Cabinet : or, monthly report of polite literature – London. 1807-1809 (1) – mf#4218 – us UMI ProQuest [920]
Cabinet : a repository of polite literature – Boston. 1811-1811 (1) – mf#3684 – us UMI ProQuest [420]
Le cabinet de lecture – Paris. 4 oct 1829-34, 1836-45 – 1 – fr ACRPP [073]
Le cabinet de toilette d'une honnete femme / Gence, comtesse de – Paris: Pancier, 1909 – (= ser Les femmes [coll]) – 5mf – 9 – mf#10717 – fr Bibl Nationale [390]
Le cabinet des beaux arts... / Perrault – Paris, 1690 – 3mf – 9 – mf#O-1086 – ne IDC [700]
Le cabinet des beaux arts... / Perrault – Paris, 1699 – 3mf – 9 – mf#O-1085 – ne IDC [700]
Cabinet des fees : ou, collection choisie des contes des fees – v1-41. 1785-89 – 9 – $462.00 – (in french) – mf#0128 – us Brook [440]
Cabinet des singularitez d'architecture, peinture, sculpture, et graveure... / Comte, F le – Paris. 3v. 1699-1700 – 29mf – 9 – mf#O-206 – ne IDC [700]
Le cabinet des tuileries sous le consulat et sous l'empire : ou memoires pour servir a la vie de napoleon / Vauban, Jacques A de – Paris 1827 [mf ed Hildesheim 1995-98] – 346 S. 2mf – 9 – €60.00 – ISBN-10: 3-487-26256-8 – ISBN-13: 978-3-487-26256-7 – gw Olms [944]
The cabinet maker : a journal of designs. for the use of upholsterers, decorators, carvers, gilders / Charles, Richard – London 1868 – (= ser 19th c art & architecture) – 4mf – 9 – mf#4.2.1426 – uk Chadwyck [740]
The cabinet maker's sketch book / King, Thomas – London [1835-36] – (= ser 19th c art & architecture) – 2mf – 9 – mf#4.2.1278 – uk Chadwyck [740]
The cabinet mission in india / Banerjee, Anil Chandra & Bose, Dakshina Ranjan – Calcutta: A Mukherjee & Co, 1946 – (= ser Samp: indian books) – us CRL [954]
Cabinet newspaper – London, England. -w. 27 Nov 1858-Feb 1860. 1 reel – 1 – uk British Libr Newspaper [072]
Cabinet of catholic information : a collection of lectures and writings of eminent prelates and priests of the catholic church in america and europe – New York: Murphy & McCarthy, c1903 – 2mf – 9 – 0-8370-8987-5 – mf#1986-2987 – us ATLA [241]
Cabinet of jade / O'neil, David – Boston, MA. 1918 – 1r – us UF Libraries [960]
Cabinet of literature – [Toronto?: s.n.], 1838-[18–?] [mf ed v1 n1 nov 1838-[v1 n12 sep 1839]] – 9 – mf#P05006 – cn Canadiana [420]
The cabinet of scientific industry : being essays by working men and others / Marples, John [comp] – [Hamilton, Ont?: s.n.], 1872 (Hamilton [Ont]: R Raw) – 1mf – 9 – 0-665-91744-9 – mf#91744 – cn Canadiana [331]
Cabinet papers, series 1, prem 3 : papers concerning defence & operational subjects, 1940-45, winston churchill, minister of defence, secretariat papers – [mf ed Marlborough 1994] – (= ser Complete classes from the cab and prem series in the public record office) – 11pt – 1 – (pt1: prem 3/1-51: subject files covering acrobat, aden, aegean, aerodromes... [15r] £1450; pt2: prem3/52-111: covering army, artillery, atlantic – battle of australasian forces, balkans, blockade... [16r] £1550; pt3: prem3/112-169: covering cyprus, czechoslovakia... [20r] £1900; pt4: prem3/170-198: covering finland, fleet air arm... [13r] £1250; pt5: prem 3/199-238: covering gibraltar, greece, hess... [15r] £1450; pt6: prem 3/239-276: covering italy, japan, joint intelligence committee... [17r] £1650; pt7: prem 3/277-325: covering mexico, middle-east... [18r] £1750; pt8: prem3/326-358: covering netherlands east indies, norway... [16r] £1550; pt9: prem 3/359-403: covering police, ports, portugal... [20r] £1900; pt10: prem 3/404-450: covering soe, spain, submarines & anti-submarine warfare... [19r] £1800; pt11: prem 3/451-515: covering usa, venezuela, vulcan... [22r] £2100; with d/g) – uk Matthew [324]
Cabinet papers, series 2 : cab 50 & cab 51 – the papers of the committee of imperial defence – papers of the oil board, 1925-1939, & middle east questions, 1930-1939 – (= ser Complete classes from the cab and prem series in the public record office) – 13r – 1 – £1250.00 – (with d/g) – uk Matthew [324]
Cabinet papers, series 3 : cab 128 and cab 129 – cabinet conclusions and cabinet memoranda, 1945 and following – (= ser Complete classes from the cab and prem series in the public record office) – 7pt – 1 – (pt1: attlee government, aug 1945-oct 1951 (cab 128/1-13 & cab 129/1-20) [15r] £1550; pt2: attlee government, aug 1945-oct 1951 (cab 128/14-22 & cab 129/21-47) [16r] £1550; pt3: churchill/eden governments, oct 1951-jan 1957 [17r] £1650; pt4: macmillan/home govts, jan 1957-oct 1964 (cab 128/31-38 & cab 129/85-118) [21r] £2000; pt5: wilson govt, oct 1964-dec 1968 (cab 128/39-43, 46 & cab 129/119-139) [11r] £1050; pt6: the wilson govt, jan 1969-may 1970 (cab 128/44, 46 & cab 129/140-146) [6r] £575; pt7: heath govt, jun 1969-mar 1974 [15r] £1450; with d/g) – uk Matthew [324]

The cabinet-maker and upholsterer's drawing-book / Sheraton, Thomas – [3rd ed]. London 1802 – (= ser 19th c art & architecture) – 8mf – 9 – mf#4.2.1429 – uk Chadwyck [740]
The cabinet-maker, upholsterer, and general artist's encyclopedia / Sheraton, Thomas – [London? 1804-07?] – (= ser 19th c art & architecture) – 6mf – 9 – mf#4.2.1246 – uk Chadwyck [740]
The cabinet-maker's assistant : a series of original designs for modern furniture – Glasgow 1853 – (= ser 19th c art & architecture) – 8mf – 9 – mf#4.1.459 – uk Chadwyck [740]
Cable / Communications Workers of America – 1943 jan-1951 mar – 1r – 1 – mf#1110438 – us WHS [380]
Cable, A Mildred *see* The fulfilment of a dream of pastor hsi's
Cable car days : clippings – San Francisco Public Lib, 1880-89 – 1r – 1 – $50.00 – mf#B40312 – us Library Micro [978]
Cable, Dale *see* Jackie robinson and the integration of organized baseball
Cable, John Levi *see* Loss of citizenship, denaturalization, the alien in wartime
Cable television business – Englewood. 1982-1991 (1) 1982-1991 (5) 1982-1991 (9) – (cont: tvc) – ISSN: 0745-2802 – mf#8072,02 – us UMI ProQuest [380]
Cable tv station distribution file / U.S. National Technical Information Service – Lists cable TV stations (sorted by call sign) carried by different cable communities – 9 – us NTIS [000]
Cable vision – Denver. 1983+ (1,5,9) – ISSN: 0361-8374 – mf#14108 – us UMI ProQuest [380]
Cablegrams exchanged between general headquarters, american expeditionary forces, and the war department, 1917-1919 / U.S. Army. American Expeditionary Forces – (= ser Records of the american expeditionary forces (world war 1), 1917-1923) – 19r – 1 – (with printed guide) – mf#M930 – us Nat Archives [355]
O cabloco : orgam critico – Cachoeiro de Itapemirim, ES. 27 out-nov 1901; 27 abr 1902 – (= ser Ps 19) – mf#P11B,05,16 – bl Biblioteca [079]
Caboclo brasileiro / Cearense, Catullo De Paixao – Rio de Janeiro, Brazil. 1939 – 1r – us UF Libraries [972]
Cabon, Adolphe *see* Mgr alexis-jean-marie guilloux
Cabos sueltos para la historia de chiguinguira / Mesanza, Andres – Madrid: Razon y Fe, 1935 – 1 – sp Bibl Santa Ana [972]
Cabot, Ella Lyman *see* Ethics for children
Cabot, James Elliot *see* A memoir of ralph waldo emerson
Cabot, John Moors *see* Diplomatic papers of john moors cabot, 1929-1978
Cabot, Samuel *see* The papers of samuel cabot, 1713-1858
Cabot's discovery of north america : the dates connected with the voyage of the matthew of bristol: mr g e weare's further reply to mr henry harrisse – London: Privately printed for the author, 1897 – 1mf – 9 – mf#25431 – cn Canadiana [910]
Cabral, Antonio Augusto Pereira *see* Indigenas da colonia de mocambique
Cabral, Carlos Castilho *see* Tempos de janio e outros tempos
Cabral, Cid Pinheiro *see* Senador de ferro
Cabral De Moncada, Francisco Xavier *see* Campanha do bailundo em 1902
Cabral e as origens do brasil / Cortesao, Jaime – Rio de Janeiro, Brazil. 1944 – 1r – us UF Libraries [972]
Cabral, Luis Gonzaga *see* Jesuitas no brasil
Cabral, Manuel Del *see*
– Compadre mon
– De este lado del mar
– Sangre mayor
Cabral, Manuel del *see* 20 cuentos de manuel del cabral
Cabral, Oswaldo R *see*
– Assuntos insulanos
– Historia de santa catarina
– Santa catharina
Cabrales, Gonzalo *see* Epistolario de heroes
Cabrales, Luis Alberto *see*
– Politica de estados unidos y poesia de hispano ame...
– Ruben dario
Cabranes, Diego de *see* Abito y armadura espiritual...con privilegio imperial. 1544
Cabrera de la Rocha, Juan *see* Alegato de buena prueba presentado
Cabrera, F *see* Remedios espirituales y corporales para curar y preservar del mal de peste...
Cabrera, Francisco De Asis *see* Razon y fuerza
Cabrera, Francisco Manrique *see*
– Apuntes para la historia literaria de puerto rico
– Historia de la literatura puertorriquena
Cabrera, J F *see* Poetas de puerto rico
Cabrera, Lydia *see* Pourquoi
Cabrera, Miguel *see* Maravilla americana
Cabrera Munoz, Rosalinda *see* Derecho de familia y la legislacion guatemalteca
Cabrera Navarro, Victor Manuel *see* Seguro de desocupacion y de retiro
Cabrera, Pablo *see*
– Los aborigenes del pais de cuyo
– La antigua biblioteca jesuitica de cordoba
– Ensayos sobre etnologia argentina
– Introduccion a la historia eclesiastica del tucuman, 1535 a 1590. buenos aires, 1934
– La segunda imprenta de cordoba
– Tesoros del pasado argentino...
Cabrera, pablo... / ed by Bayle, Constantino – Madrid: Razon y Fe, 1926 – 1 – sp Bibl Santa Ana [920]
Cabrera, Raimundo *see*
– Cuba and the cubans
– Mis malos tiempos
– Sombras eternas
Cabrera y Quintero, Cayetano *see* Escudo de armas de mexico
Cabrisas, Hilarion *see* Breviario de mi vida inutil
Cabrol, F *see* Les eglises de jeruzalem
Cabrol, Fernand *see*
– L'angleterre chretienne avant les normands
– Les eglises de jerusalem
– Introduction aux etudes liturgiques
– Le livre de la priere antique
– Les origines liturgiques
Cabuhi nga nadangat sang principe igmidio cag sang princesa clauriana sa guinharian sa gran cairo – [Iloilo]: La Editorial 1908 [mf ed Bloomington IN: Indiana Uni Lib, Preservation Dept 1984] – (= ser Coll...in the bisaya language 1) – 1r – 1 – us Indiana Preservation [490]
Cac a khyac nhan thon : [a novel] / Lha, Lu thu U – Ran Kun: Kri pva re pum nhip tuik 1960 [mf ed 1990] – 1r with other items – 1 – (in burmese) – mf#mf-10289 seam reel 185/1 [§] – us CRL [830]
Cac a tvan nhan cac pri ca pra jat mya / San Nvay, Prann – Ran Kun: Ca pe Biman 1985 [mf ed 1990] – (= ser Ca pe Biman thut prann su lak cvai ca can) – 1r with other items – 1 – (in burmese) – mf#mf-10289 seam reel 171/1 [§] – us CRL [790]
Cac Kuin Lah Rlvhe *see* Rup rhn gita nahn kyvan to
Cac kuin u bhui san / aphum pan khyi u cam tui / Bhui San, Cac kuin U – Ran Kun: Ca pe Biman 1979 [mf ed 1990] – 1r with other items – 1 – (in burmese) – mf#mf-10289 seam reel 135/3 [§] – us CRL [959]
Cac rhum su : [a novel] / Rvhe U Don – Ran Kun: Mala Mruin Pum nhip tuik 1961 [mf ed 1990] – (= ser Nve ta ri ca can 9) – 1r with other items – 1 – (in burmese) – mf#mf-10289 seam reel 203/6 [§] – us CRL [830]
Cac su kri mahosadha / Chan Mon, U – Yan Kun: Kyan Tvan Cape 1976 [mf ed 1990] – 1r with other items – 1 – (in burmese) – mf#mf-10289 seam reel 152/6 [§] – us CRL [959]
Caca modesa / Warschaver, Fina – Buenos Aires, Argentina. 1947 – 1r – us UF Libraries [972]
[Cacace, G B] *see* Theatrum omnium scientiarum sive apparatus quo exerptus fuit excimus princeps d innicus de guevara, et tassis...a professoribus gymnasij neapolitani, decretus ab illustrissimo domi d ioanne de salamanca...
Cacao / Bondar, Gregorio Gregorievitch – Bahia, Brazil. pt1-2. 1924 – 1r – us UF Libraries [025]
Cacciatore, Niccolo *see* Della cometa apparsa in luglio del 1819
Caceres / Caceres. Junta Provincial de Turismo – Madrid: Sucesos de Rivadeneyra – (texto tambien en frances, ingles y aleman) – sp Bibl Santa Ana [914]
Caceres / Junta Provincial de Turismo – Vitoria: Tip. Fourner, s.a. – 1 – (texto en ingles) – sp Bibl Santa Ana [338]
Caceres *see*
– Comision especial de aguas. dictamen...
– Comision especial de aguas. dictamen...6.11.1934
– Constitucion de la corporacion municipal el 6 de febrero de 1949
– Constitucion de la corporacion municipal el 31 de julio de 1948
– Constituido...2 de febrero de 1958
– Contaduria. memoria del ano 1910
– Folleto informativo sobre el servicio municipal de limpieza
– Liquidacion del presupuesto de 1914, deudores y acreedores para resultas
– Ordenanzas 1 al 6. presupuesto extraordinario de obras, servicios y urbanismo, vigencia indefinida desde 1957
– Ordenanzas fiscales que agregando o rectificando algunas de 1939 forman
– Ordenanzas formadas para el cobro de los impuestos y arbitrios del presupuesto 1915-26
– Ordenanzas formadas para el cobro de los impuestos y arbitrios del presupuesto de 1959
– Ordenanzas municipales...de caceres...aprobadas en 1912
– Ordenanzas para el cobro de los impuestos y arbitrios del presupuesto
– Patronato local de cantinas escolares
– Presupuesto de gastos e ingresos para el ano
– Presupuesto ordinario de gastos e ingresos para 1948. memoria justificada ordenanzas y tarifas nuevas o modificadas. bases de ejecucion
– Presupuesto ordinario de ingresos y gastos ano
– Presupuesto ordinario de ingresos y gastos para el ano
– Presupuesto ordinario para ingresos y gastos para el ano 1910
– Rectificaciones y adiciones a las ordenanzas mineras 2, 4, 5, 8, 9, 12, 13, 14 bis, 16 bis, 18, 20, 24, 25-26, 28, 31, 35, 47, 48, 51, 52 y 65 formuladas para el cobro de impuestos y arbitrios del presupuesto, vigentes para el de – 1958 y sucesivas
– Reglamentacion municipal contra el paro forzoso...1931
– Reglamento de la academia de musica y banda municipal de caceres
– Reglamento de la mutualidad de prevision para funcionarios del ayuntamiento de caceres, 1955
– Reglamento de procedimineto administrativo y de obras y servicios...1955
– Reglamento de regimen interior dela guardia municipal de la ciudad de caceres
– Reglamento de sanidad e higiene...1959
– Reglamento del servicio municipal de mercados 1955...
– Reglamento municipal contra el paro forzoso
– Reglamento para el regimen interior de las oficinas del excmo
– Reglamento para el regimen interior de las oficinas del excmo. ayuntamiento de caceres
– Reglamento para elregimen interior del cementerio publico municipal de esta capital
– ...Reglamento...del matadero municipal de caceres...1919
Caceres. 34th Asamblea de la Federacion Espanola de Centros de Iniciativas y Turismo *see* A celebrar en caceres del 5 al 11 de octubre de 1969
Caceres ante la historia. la cuestion critica de la fundacion y el nombre de caceres / Floriano Cumbreno, Antonio C – Caceres: Imprenta, Libreria y Encuadernacion de Garcia Floriano, 1931 – sp Bibl Santa Ana [946]
Caceres, Antonio de *see* Sermones varios de diversos asuntos panegiricos
Caceres. Asociacion Cultural *see* Estatutos
Caceres. Asociacion de Padres de Alumnos del Colegio San Antonio de Padua *see* Estatutos de la...caceres
Caceres. Asociacion Nacional de Invalidos Civiles *see* Memoria de las actividades desarrolladas por la junta directiva...
Caceres. Ayuntamiento *see*
– Exposicion historia de la feria de mayo en caceres (documentacion del archivo municipal) catalogo 1973
– Feria de san miguel, 1954
– Feria y fiestas mayo de 1952. guia comercial
– Feria y fiestas mayo-junio, 1953
– Ferias y fiestas 1952
– Ferias y fiestas. 1973. guia oficial
– Ferias y fiestas de mayo 1954
– Ferias y fiestas de mayo de 1950 guia comercial
– Ferias y fiestas mayo 1947
– Ferias y fiestas mayo 1963
– Ferias y fiestas patronales. san jorge, 1960
– Fiesta de san jorge, 1953
– Fiestas patronales san jorge, 1955
– Fiestas patronales san jorge 1959
– Guia comercial de ferias y fiestas mayo, 1945
– Presupuesto ordinario de gastos e ingresos para 1951
– Revista oficial ferias y fiestas 1972
Caceres bajo la reina catolica y su camarero sancho paredes golfin / Orti Belmonte, Miguel Angel – Badajoz: Dip. Provincial, 1955 – 1 – sp Bibl Santa Ana [241]
Caceres. centro extremeno – Caceres: Rime Publicidad, 1973 – 1 – sp Bibl Santa Ana [946]
Caceres. Club de tenis "Cabeza Rubia" *see* Estatutos sociales del...
Caceres. Comision Semana Santa *see* Semana santa en caceres, 1957
Caceres. Delegacion Provincial de Organizaciones del Movimiento *see* 2 juegos deportivos
Caceres. Delegacion Provincial del Ministerio, de Educacion *see* Desde la formacion profesional a las facultades o escuelas universitarias e industrias y servicios
Caceres, Diego de *see*
– Cargos que resultan contra...el p. diego de caceres, general de...s. geronimo..
– Conciliabulo de basilea
– Dilucidatio in generali capitulo ordino d. hyeronymi a..
– Dioscoro patriarcha alexandrino
– Elucidationes de potestate papae
– Exposicion al nuncio
– Memorial al rey pidiendole se remita la causa al nuncio y a otros prelados y da cuenta de su vida
– Parecer del reverendo...debe ser cabeza de la dicha religion
– Sentencia que nrpmf, general de la orden de nps geronimo, dio y pronuncio en el monasterio de san bartolome...
Caceres. Diputacion Provincial *see* Presupuesto ordinario de ingresos y gastos para el ejercicio de 1972
Caceres en 1828. datos historicos, estadisticos y otras curiosidades tomados de...instituto de segunda ensenanza de la misma – 1874 – 9 – sp Bibl Santa Ana [946]
Caceres en tiempo de los romanos / Huebner, Emilio – Caceres: Tip. y Lib. de N.M. Jimenez, 1899 – 1 – sp Bibl Santa Ana [946]
Caceres. Espana en Paz *see* Chronica de veinticinco anos
Caceres. Hermandad de Donantes de Sangre. San Pedro de Alcantara *see* Memoria 1980
Caceres. Hermandad Nacional de Alfereces Provisionales *see* Relacion alfabetica de los alfereces provisionales de la provincia de caceres
Caceres. iglesia del convento de san francisco... / Meseguer Fernandez, Juan – Madrid: Graf. Calleja, 1970 – 1 – sp Bibl Santa Ana [240]
Caceres. Junta Provincial de Turismo *see*
– Caceres
– Caceres. plano guia
– Trujillo
Caceres. Junta Provincial del Turismo. Caceres *see* Plano-guia
Caceres Lara, Victor *see*
– Fechas de la historia de honduras
– Humus
Caceres. mapa de comunicaciones de la provincia – Arqueros, 1950 – 1 – sp Bibl Santa Ana [380]
Caceres. mapa guia turistico provincial / Junta Provincial de Informacion, Turismo y Educacion Popular – San Sebastian: Valverde S.A, 1959 – 1 – sp Bibl Santa Ana [338]
Caceres monumental / Callejo Serrano, Carlos – Madrid: Editorial Plus Ultra, 1960 – 1 – sp Bibl Santa Ana [946]
Caceres. Mutua Aseguradora de Transportes *see* Memoria. ejercicio 1964
Caceres. Mutua Aseguradora de transportistas *see* Memoria. ejercicio 1960
Caceres. Mutua Aseguradora de Transportistas de la Provincia de Caceres *see*
– Estatutos aprobados por orden ministerial hacienda de 1960
– Reglamento del ministerio de hacienda de 17 de junio de 1960
– Reglamento del ramo de accidentes aprobado por orden del ministerio de hacienda de 17 de junio de 1960
Caceres. Mutua Cerealistica *see*
– Estatutos
– Memoria. ejercicio 1959
Caceres. Mutua extremena de Vehiculos *see* Memoria. ejercicio 1973
Caceres. Mutualidad General Deportiva *see* Normas de procedimiento en caso de accidente deportivo
Caceres. Obra Sindical de Educacion y Descanso *see* Navidad 74
Caceres. Parroquia de San Jose *see* Iglesia parroquial de san jose
Caceres. Pena. Amigos del Flamenco de Extremadura *see* Estatutos...
Caceres. plano guia / Caceres. Junta Provincial de Turismo – Alicante: Graf. Gutenberg, 1948 – sp Bibl Santa Ana [914]
Caceres primera cuna de la orden militar de santiago / Munoz Gallardo, Juan Antonio – Badajoz: Imprenta Diputacion Provincial, 1974. Separata Rev. Estudios Extremenos – 1 – sp Bibl Santa Ana [350]
Caceres, Rafael de *see* Perjuicios de la separacion de la medicina y la cirujia..
Caceres. resena de los festejos celebrados en esta capital para solemnizar la promulgacion de la ley fundamental del estado...en 1869 – Caceres: Imp. Nicolas M. Jimenez, 1869 – 1 – sp Bibl Santa Ana [946]
Caceres. (Spain) *see* Ordenanzas municipales
Caceres. Spain *see*
– Consejo provincial de agricultura, industria y energia de caceres
– Resena de los festejos..
Caceres. Terpresa *see* Memoria 1974
Caceres Tinoco De Giron, Ela *see* Desarrollo del programa oficial
Caceres: Universidad Laboral Hispanoamericana *see* Orquesta filarmonica morava de olomone-sinfonica del estado
Caceres y Sotomayor, Antonio *see*
– Paraphasis de los psalmos de david...y modo de hablar de la lengua espanola
– Sermones y discursos de tiempo...predicados por..
– Tercera parte de los sermones y discursos que contiene desde el...sermon en la feria segunda..
Cach mang gouc gia – Ho Chi Minh City, Vietnam. 1960-1961 (1) – mf#68404 – us UMI ProQuest [079]
Cach mang quoc gia – Saigon: [s.n. feb 5-dec 31, 1958 – 1r – 1 – (lacks: feb 12-22,27, mar 7-10,12,17-19, apr 15,29, may 2,6-7,12-20,24, 29-31, jun 10, 13,16,28,30, nov 2) – mf#mf-4132 seam – us CRL [079]
Cacharron, Francisco de Paula *see* Ocios y versos

Cachon / Monclus, Miguel Angel – Ciudad Trujillo, Dominican Republic. 1958 – 1r – us UF Libraries [972]

Les cachots d'haldimand : grand roman canadien historique inedit / Feron, Jean – Montreal: editions Edouard Garand, 1925 [mf ed 1987] – 1mf – 9 – (ill by albert fournier) – mf#SEM105P841 – cn Bibl Nat [830]

O cacique : jornal noticioso e recreativo – Desterro, SC: Typ de J A do Livramento, 02 ago 1870-29 abr 1871 – (= ser Ps 19) – mf#UFSC/BPESC – bl Biblioteca [073]

Cacique de marien / Peraza De Zell, Rosa L – Habana, Cuba. 1926 – 1r – us UF Libraries [972]

Cacique de turmeque y su epoca / Rojas, Ulises – Tunja, Colombia. 1965 – 1r – us UF Libraries [972]

Caciques aborigenes venezolanos / Reyes, Antonio – Caracas, Venezuela. 1952 – 1r – us UF Libraries [972]

Cacosh health and safety news / Chicago Area Committee on Occupational Safety and Health – v2 n1-v7 n3 [1975 jun-1979 nov/dec], 1980 apr/may-1984 may/jun, 1986 jul/aug – 1r – 1 – mf#1313385 – us WHS [360]

Cactaceas de la flora de santo domingo / Moscoso, Rafael M – Ciudad Trujillo, Dominican Republic. 1941 – 1r – us UF Libraries [580]

Cactos / Zea Ruano, Rafael – Guatemala, 1952 – 1r – us UF Libraries [972]

Cactus and succulent journal – Santa Barbara. 1921-1989 (1) 1972-1989 (5) 1976-1989 (9) – ISSN: 0007-9367 – mf#6707 – us UMI ProQuest [580]

Cactus comet – Yuma Marine Corps Air Station AZ. 1981 apr 30-1983, 1984, 1985-1986 mar, 1986 apr-1987jun, 1987, 1987 jul-1988 sep, 1988, 1989, 1990 – 9r – 1 – mf#703347 – us WHS [355]

Cactus patch – v22 n4 [1983 win], v23 n5-6 [1983 fall-1984 win], v24, n2 [1984 sum], v25 n4 [1986 win], v26 n1,4 [1986 spr, oct], v27 n3-4 [1987 sum-oct], v27 n1-4, [1988 apr-dec], v28 n1-2 [1989 jun-aug],v28 n1,3-4 [1990 mar, nov-dec], v28 n1-2,4 [1991 may-jul, dec], v28 n1-2,4 [1992 feb-jun, dec], v29 n1 [1993 may] – 1r – 1 – mf#1053832 – us WHS [071]

Cacua Prada, Antonio *see* Libertad de prensa en colombia

Cada dia tiene su afan / Lindo, Hugo – San Salvador, El Salvador. 1965 – 1r – us UF Libraries [972]

Cadaloz – Istanbul: Tercueman-i Hakikat Matbaasi, 1911. Sahib-i Imtiyaz ve Sermuharriri: Nureddin Ruesdi. n20,24. 10, 25 haziran 1911 – (= ser O & t journals) – 1mf – 9 – $25.00 – us MEDOC [956]

Cadalso en colombia – Bucaramanga, Colombia. 1925 – 1r – us UF Libraries [972]

Cadams progress – Cadams, NE: J Ord Cresap, may 25 1916 [wkly] [mf ed v1 n2. jun 1-dec 14 1916 [lacks oct 19] filmed 1979] – 1r – 1 – (absorbed by: nelson gazette) – us NE Hist [071]

An cadanayan na lacao nin quinaban napipinta sa buhay ni nicolas asin crispina na mag-amang macalolodoc / Ariate, Nicolas – [Nueva Caceres?: Libreria Mariana? 190-?] [mf ed Bloomington IN: Indiana Uni Lib, Preservation Dept 1984] – (= ser Coll...in the bikol language) – 1r – 1 – us Indiana Preservation [490]

Cadastral plans : city of montreal / Sicotte, Louis-Wilfrid – [Montreal]: [Dept of Public Works], [ca 1874] (mf ed 1974) – 1r – 1 – mf#SEM35P118 – cn Bibl Nat [917]

Cadastre abrege de la partie de la seigneurie de bourchemin est... / Judah, Henry – Quebec: impr par Stewart Derbishire & George Desbarats, 1861 [mf ed 2000] – 1mf – 9 – mf#SEM105P1183 – cn Bibl Nat [350]

Cadastre abrege de la partie de la seigneurie de l'islet st jean... / Lelievre, Simeon – [Quebec: impr par Stewart Derbishire & George Desbarats, 1863] (mf ed 2000) – 1mf – 9 – mf#SEM105P1162 – cn Bibl Nat [350]

Cadastre abrege de la partie sud-ouest de la seigneurie de bourg louis... / Lelievre, Simeon – Quebec: Stewart Derbishire & George Desbarats, 1861 [mf ed 2000] – 1mf – 9 – mf#SEM105P1168 – cn Bibl Nat [350]

Cadastre abrege de la seigneurie d'auteuil... / Lelievre, Simeon – Quebec: impr par Stewart Derbishire & George Desbarats, 1862 [mf ed 2000] – 1mf – 9 – mf#SEM105P1185 – cn Bibl Nat [350]

Cadastre abrege de la seigneurie de beauharnois... / Judah, Henry – Quebec: impr par Stewart Derbishire & George Desbarats, 1861 [mf ed 2000] – 9 – cn Bibl Nat [350]

Cadastre abrege de la seigneurie de deschambault... / Lelievre, Simeon – [Quebec: impr par Stewart Derbishire & George Desbarats, 1862] (mf ed 2000) – 1mf – 9 – mf#SEM105P1176 – cn Bibl Nat [350]

Cadastre abrege de la seigneurie de la grande vallee des monts... / Le Bel, J G – [Quebec: impr par George Desbarats, 1863] (mf ed 2000) – 1mf – 9 – mf#SEM105P1179 – cn Bibl Nat [350]

Cadastre abrege de la seigneurie de la nouvelle longueuil... / Judah, Henry – [Quebec: impr par George Desbarats, 1863] (mf ed 2000) – 1mf – 9 – mf#SEM105P1122 – cn Bibl Nat [350]

Cadastre abrege de la seigneurie de levrard (ou st pierre les becquets)... / Turcotte, Joseph-Edouard – Quebec: Stewart Derbishire & George Desbarats, 1861 [mf ed 1992] – 1mf – 9 – (with ind) – mf#SEM105P1526 – cn Bibl Nat [350]

Cadastre abrege de la seigneurie de rocquetaillade... / Turcotte, Joseph-Edouard – Quebec: impr par Stewart Derbishire & George Desbarats, 1861 [mf ed 1992] – 1mf – 9 – mf#SEM105P1525 – cn Bibl Nat [350]

Cadastre abrege de la seigneurie de st valier... / Lelievre, Simeon – [Québec?: s.n, 1862?] (mf ed 1992) – 1mf – 9 – mf#SEM105P1524 – cn Bibl Nat [350]

Cadastre abrege de la seigneurie delery... / Judah, Henry – Quebec: impr par Stewart Derbishire & George Desbarats, 1861 [mf ed 2000] – 1mf – 9 – mf#SEM105P1110 – cn Bibl Nat [350]

Cadastre abrege du fief coteau st louis : appartenant ci-devant a l'ordre des jesuites, fait le 16 mars, 1864 / Judah, Henry – [Quebec: impr par George Desbarats, 1864] (mf ed 1992) – 1mf – 9 – mf#SEM105P1167 – cn Bibl Nat [241]

Cadastre abrege du fief st joseph ou l'epinay... / Lelievre, Simeon – Quebec: impr par Stewart Derbishire & George Desbarats, 1861 [i.e. 1862?] (mf ed 1992) – 1mf – 9 – mf#SEM105P1371 – cn Bibl Nat [350]

Cadastre abrege du fief vieuxpont... / Dumas, Norbert – Quebec: impr par Stewart Derbishire & George Desbarats, 1861 [mf ed 2000] – 1mf – 9 – mf#SEM105P1066 – cn Bibl Nat [350]

Cadastres abreges des seigneuries appartenant a la couronne... – Quebec: impr par George Desbarats, 1863 [mf ed 1985] – 1r – 1 – (with ind) – mf#SEM35P226 – cn Bibl Nat [333]

Cadastres abreges des seigneuries du district de montreal... – Quebec: impr par Stewart Derbishire & George Desbarats, 1863 [mf ed 1985] – 2r – 1 – (with ind) – mf#SEM35P207 – cn Bibl Nat [333]

Cadastres abreges des seigneuries du district de quebec... – Quebec: impr par George Desbarats. 2v. 1863 [mf ed 1985] – 2r – 1 – (with ind) – mf#SEM35P224 – cn Bibl Nat [333]

Cadastres abreges des seigneuries du district des trois-rivieres... – Quebec: Stewart Derbishire & Georges Desbarats...1863 [mf ed 1985] – 1r – 1 – (with ind) – mf#SEM35P225 – cn Bibl Nat [333]

Cadavid G, J Ivan *see* Fueros de la iglesia ante el liberalismo y el cons...

Cadbury, M Christabel *see* Robert barclay

Cad/cam technology – Dearborn. 1982-1984 (1,5,9) – (cont by: cim technology: casa, sme's magazine of computers in design and manufacturing) – ISSN: 0737-660X – mf#13437 – us UMI ProQuest [000]

Caddell, Cecilia Mary *see*
- Hidden saints
- A history of the missions in japan and paraguay

Cadden, Joseph *see* Spain

"Caddie woodlawn", adapted by greg gunning from the novel by carol ryrie brink : cue sheet for students / Aguirre-Sacasa, Roberto – [Washington DC: Kennedy Center]: US Dept of Education, Office of Educ Research & Improvement...[c2000] [mf ed 2001] – 1mf – 9 – us GPO [790]

Caddington – (= ser Bedfordshire parish register series) – 2mf – 9 – £5.00 – uk BedsFHS [929]

Caddington, all saints monumental inscriptions monumental inscriptions – Arthur Weight Matthews 1907-16 – (= ser Bedfordshire parish register series) – 1mf – 9 – £1.25 – uk BedsFHS [929]

Caddo free press – Shreveport LA. 1840 apr 30 – 1r – 1 – mf#860870 – us WHS [071]

Cadell, Robert *see* Nineteenth century literary manuscripts

Cadell, William A *see* A journey in carniola, italy, and france

Cadena De Vilhasanti, Pedro *see* Relacao diaria do cerco da baia de 1638

Cadence of the clinical laboratory – Bellaire. 1970-1976 [1]; 1975-1976 [5,9] – mf#8618 – us UMI ProQuest [619]

The cadenza – Kansas City. 1894-1924. A music magazine. 7 reels – 1 – us L of C Photodup [780]

Cadenza in the piano concerto / Howe, Richard Esmond, Jr – U of Rochester 1956 [mf ed 19–] – 2v on 6mf – 9 – mf#fiche 177 – us Sibley [780]

Cadernos cedi – Rio de Janeiro: Centro Ecumenico de Documentacao e Informacao, [1980?- n1 1980] – 1r – us CRL [972]

Cadernos do ceas / Centro de Estudos e Acao Social – Salvador, Bahia, Brasil: Centro de Estudos e Acao Social, [n5-184 (fev1970-1999)] (bimthly) – 10r – 1 – us CRL [300]

Cadernos do cedi – Rio de Janeiro: Centro Ecumenico de Documentacao e Informacao, [1980?- n2-20 (1980-1990) – 1r – us CRL [972]

Cades baptist church – New York. 1972+ (1) 1952+ (5) 1974+ (9) – 1r – 1 – $13.91 – mf#6501 – us Southern Baptist [242]

Cadet – Fort Bragg NC, v2 n1-14 [1979 nov 15/1980 oct 30]-v13 n5-8 [1976 jul 16/oct 15] – 2r – 1 – mf#1520883 – us WHS [071]

The cadet – Montreal: J C Becket, [1852-1854] – 9 – ISSN: 1190-6855 – mf#P04208 – cn Canadiana [360]

The cadet – [St John, NB]: Grand Section [1867?] – 9 – ISSN: 1190-6626 – mf#P04536 – cn Canadiana [360]

Cadet buteux legislateur, ou, la constitution en v – Paris, France. 1815 – 1r – us UF Libraries [440]

Cadet de Gassicourt, Charles *see* Voyage en autriche, en moravie et en baviere

Cadet, Ernest *see* Le mariage en france

Cadetes / Cunha, Rui Vieira Da – Rio de Janeiro, Brazil. 1966 – 1r – us UF Libraries [972]

The cadets' trumpet – Windsor, NS: Cadets of Temperance, [1880] – 9 – mf#P04295 – cn Canadiana [360]

Cadier, Albert *see* La lampe sous le boisseau

Cadilla, Arturo *see* Oro de la dicha

Cadilla De Martinez, Maria *see*
- Costumbres y tradicionalismos de mi tierra...
- Hitos de la raza
- Poesia popular en puerto rico
- Rememorando el pasado heroico

Cadilla De Ruibal, Carmen Alicia *see* Antologia poetica

Cadilla Ruibal, Carmen *see* Cien sinrazones

"Cadillac's village", or, "detroit under cadillac" : with list of property owners and a history of the settlement, 1701 to 1710 / Burton, Clarence Monroe [comp] – Detroit: [s.n.], 1896 [mf ed 1980] – 1mf – 9 – 0-665-02297-2 – mf#02297 – cn Canadiana [978]

Cadiz to cathay / Duval, Miles Percy – Palo Alto, CA. 1940 – 1r – us UF Libraries [972]

Cadman, Harry W *see* The christian unity of capital and labor

Cadman, James Piper *see* Christ in the gospels

Cadman, Samuel Parkes *see*
- Answers to everyday questions
- The three religious leaders of oxford and their movements

Cadmus et hermione : tragedie mise en musique...partition generale... / Lully, Jean Baptiste – Paris: C Ballard 1719 [mf ed 19–] – 1r – 1 – (libretto by philippe quinault) – mf#film 513 – us Sibley [780]

Cadott blade – Cadott WI. 1893 dec 22, 1894 jul 13, 1900 may 25, 1902 may 17 [special ill ed], 1904 feb 19, 1910 oct 7 – 1r – 1 – mf#957898 – us WHS [071]

Cadott sentinel – Cadott WI. 1914 apr 10/1915 mar 26-2000 jul/dec – 73r – 1 – (cont: cadott blade) – mf#995049 – us WHS [071]

Cadoxton-j-barry, glamorgan, parish church of st cadoc : baptisms 1724-1889, burials 1724-1901, marriages 1724-1837 – 1mf – 9 – £1.25 – uk Glamorgan FHS [929]

Cadoxton-j-neath, glamorgan, parish church of st catwg : baptisms 1721-1899, burials 1721-1899, marriages 1721-1837 – 7mf – 9 – £8.75 – uk Glamorgan FHS [929]

Cadoxton-juxta-barry, st cadoc; philadelphia baptist, monumental inscriptions – 1mf – 9 – £1.25 – uk Glamorgan FHS [929]

Caducean / Tripler Army Medical Center – Honolulu HI. v38 n12-v43 n1 [1983 jul 1-1988 sep 10] – 1r – 1 – mf#1344634 – us WHS [355]

Cadwalader, J *see* Cadwalader's cases in the district court of pennsylvania, 1858-1879

Cadwalader's cases in the district court of pennsylvania, 1858-1879 / Cadwalader, J – Philadelphia: Welsh & Co. v1-2. 1907 – (= ser Early federal nominative reports) – 15mf – 9 – $22.50 – mf#LLMC 81-440 – us LLMC [347]

Cadwell baptist church – Cadwell, GA. 903p. aug 1941-sep 1962, oct 1965-1990 – 1 – $40.64 – mf#5188 – us Southern Baptist [242]

Cady, H Emilie *see* Lessons in truth

Cady, Lyman van Law *see* Tsung chiao hsin li hsueh (ccm5)

Cady, Philander Kinney *see* The rewards of the profession of the law

Caecilia / American Society of Saint Caecilia – 1886 sep 1, 1887 jan 1-1887 feb 1, 1893 apr – 1r – 1 – (cont by: catholic choirmaster; sacred music) – mf#969479 – us WHS [780]

Caecilia : [collection of compositions of old italian masters. original manuscript of the editor, o braune] / ed by Braune, Friedrich Wilhelm Otto – [185-?] [mf ed 1986] – 2v on 15mf / 1r – 9,1 – mf#fiche1229 / pres. film 63 – us Sibley [780]

Caecilia : eine zeitschrift fuer die musikalische welt – Mainz. 1824-48. 4 reels – 1 – 75.00 – us L of C Photodup [780]

Caecilia : eine zeitschrift fuer die musikalische welt / ed by Weber, Gottfried & Dehn, Siegfried Wilhelm – Mainz 1824-48 [mf ed Hildesheim 1979] – 143mf – 9 – diazo €748.00 silver €888.00 – gw Olms [780]

Caecilia. eine zeitschrift fuer die musikalische welt – Paris-Mainz. v. 1-27. 1824-1848 – 1 – us NY Public [780]

Caecilia von albano : dramatisches gedicht in fuenf aufzuegen / Mosenthal, Salomon Hermann – Wien: [s. n.], 1849 (U Klopf Senior & A Eurich) [mf ed 1995] – 1r – 1 – mf#3698p – us Wisconsin U Libr [820]

Caedmon manuscript, the... : bodleian library, ms. junius ii – 1,14 – mf#1r 525; 1 col r C525 – uk Microform Academic [810]

Caeiro, Jose *see* Primeira publicacao apos 160 anos do manuscrito

Caelii avgvstini cvrionis sarracenicae historiae libri tres... / Curio, C A – Francofvrdi, 1596 – 4mf – 9 – mf#H-8387 – ne IDC [956]

Caelii calcagnini, ferrariensis, protonotarii apostolici, opera aliqvot / Calcagnini, Celio; ed by Brasavola, Antonio Musa – 1st ed. Basileae 1544 [mf ed 1984] – 1 – (incl ind) – mf#7182 – us Wisconsin U Libr [520]

Caelii sedulii opera / Arevalo, Faustino – Roma: Antonio Fulgonium, 1794 – 1 – sp Bibl Santa Ana [780]

Caementaria hibernica : being the public constitutions that have served to hold together the freemasons of ireland / Crawley, William John Chetwode – Dublin: W M'Gee 1895-1900 – 3v on 9mf – 9 – (int w j chetwode crawley) – mf#vrl-21 – ne IDC [366]

Caeneghem, E P R van *see*
- Hekserij bij de baluba van kasai
- Studie over de gewoontelijke strafbepalingen tegen het overspel bij de baluba en ba lulua van kasai

Caerau, glamorgan, parish church of st mary : baptisms 1725-1903, burials 1720-1903, marriages 1727-1836; & michaelston-s-ely, st michael, baptisms 1721-1905, burials 1724-1909, marriages 1724-1835 – 1mf – 9 – £1.25 – uk Glamorgan FHS [929]

Caerau, st mary; and michaelston-s-ely, st michael, monumental inscriptions – 1mf – 9 – £1.25 – uk Glamorgan FHS [929]

Caeremoniale iuxta ritum romanum / Aloysio Maria a Carpo – ed tertia Romana, auctior et emendatior. Romae: SC de Propaganda Fide, 1874 – 2mf – 9 – 0-524-08856-X – mf#1993-3320 – us ATLA [240]

Caernarvon advertiser of political, literary, scientific and commecial intelligence – [Wales] University of Wales-Bangor jan-23 mar 1822 [mf ed 2003] – 1r – 1 – uk Newsplan [072]

Caerphilly, glamorgan, parish church of st martin : baptisms 1813-1924, burials 1713-1923, marriages 1813-1820 – [Glamorgan]: GFHS [mf ed c2000] – 2mf – 9 – £2.50 – uk Glamorgan FHS [929]

Caerphilly groeswen, welsh congregational chapel, monumental inscriptions – 2mf – 9 – £2.50 – uk Glamorgan FHS [929]

Caerphilly, st martin, monumental inscriptions – 3mf – 9 – £3.75 – uk Glamorgan FHS [929]

Caerphilly, tonyfelin welsh baptist, monumental inscriptions – 1mf – 9 – £1.25 – uk Glamorgan FHS [929]

Caerphilly, watford congregational, monumental inscriptions – 1mf – 9 – £1.25 – uk Glamorgan FHS [929]

Caesar, Carl Adolph *see* Denkwuerdigkeiten aus der philosophischen welt

Caesar flaischlen : ein essay / Thiess, Frank – Berlin: E Fleischel 1914 [mf ed 1989] – 1r – 1 – (filmed with: der tod vor dem spiegel / edmund finke) – mf#7242 – us Wisconsin U Libr [840]

Caesar, G J *see* Les commentaires de cesar

Caesar, Julius *see*
- Caesar's gallic war
- Fifteenth century italian manuscripts
- Opera...
- Selections

Caesareis virtutib et symbolis adornatum... / Cenotaphium piis manibus Ferdinandi 3... – Augustae Vindel: Melchiore et Matthaeo Kuesell, 1657 – 2mf – 9 – mf#O-1541 – ne IDC [090]

Caesarii Heisterbacensis (Caesarius of Heisterbach) *see* Dialogus miraculorum

Caesarii Heisterbacensis [Caesarius of Heisterbach] *see* Dialogus miraculorum

Caesarii Heisterbasensis *see* Homiliae

Caesarius of Heisterbach *see*
- Dialogus miraculorum
- Johann hartliebs uebersetzung des dialogus miraculorum von caesarius von heisterbach

Caesarius, S *see* Arelatensis episcopus, regula sanctarum virginum aliaque opuscula ad sanctimoniales directa

Caesarius von arelate und die gallische kirche seiner zeit / Arnold, Carl Franklin – Leipzig: J.C. Hinrichs, 1894 – 2mf – 9 – 0-7905-5680-4 – (incl bibl ref) – mf#1988-1680 – us ATLA [240]
Caesar's gallic war / Caesar, Julius – Chicago, IL. 1907 – 1r – us UF Libraries [025]
Caetani, L *see* Annali dell'islam, 1-10
O caeteense : orgao patriota, litterario, industrial – Caete, PA. 25 mar 1888 – (= ser Ps 19) – bl Biblioteca [079]
Caetes / Ramos, Graciliano – Rio de Janeiro, Brazil. 1955 – 1r – us UF Libraries [972]
Cafasso, Arthur *see* Das bild in der dramatischen sprache grillparzers
Cafeteria call / Hotel and Restaurant Employees and Bartenders International Union – 1942 jan-1966, 1967 jan-1974 sum – 2r – 1 – mf#354481 – us WHS [640]
Caffaro – Genoa, Italy. -d. 27 Jan 1918-5 Aug 1919. Imperfect. 4 reels – 1 – uk British Libr Newspaper [072]
Caffeine : effects on blood pressure, heart rate and short term muscular endurance in static exercise of muscle groups of varying mass / Bailey, Mark L – 1988 – 147p 2mf – 9 – $8.00 – us Kinesology [612]
Caffeine, carbohydrate loading, and physical performance / O'Connor, MaryLou & McMurray, Robert G – 1992 – 1mf – 9 – $4.00 – us Kinesology [613]
Cafferata, Juan F *see* Labor parlamentaria 1912-1916, 1920-1924, 1924-1928. buenos aires, 1928
Caffese, Maria E *see* Mayo en la bibliografia. buenos aires, 1961
Cafres / Lacerda, Francisco Gavicho De – Lisboa, Portugal. 1944 – 1r – us UF Libraries [960]
Cagigal, Jose *see* La muerte de luis 16
Cagigal, Juan Manuel *see* Escritos literarios y cientificos
Caglayan – Balikesir: Karesi Matbaasi, 1925-26. Mueduerue: Orhan Saik [Goekyay] n1-2,4-15. 20 tesrinievvel 1341 [1925]-15 mayis 1926 – (= ser O & t journals) – 3mf – 9 – $65.00 – us MEDOC [956]
Cagliostro, Alessandro, conte di *see* Memoire pour le comte de cagliostro, accuse
Cagnolo, C *see* The akikuyu
Cahangahangang buhay ni santa margarita de cortona na taga toscana sa nayon ng diocesis at limang virgenes, at apat na puong soldados na pauang mga martires at ibang nadamay / Ignacio, Cleto R – Maynila: P Sayo 1916 [mf ed Bloomington IN: Indiana Uni Lib, Preservation Dept 1984] – (= ser Coll...in the tagalog language 1) – 1r – 1 – us Indiana Preservation [490]
Cahaya india = Light of the indies – 1885- [ed 2007] – (= ser Vernacular press in the netherlands indies, c1855-1925, unit 1) – 13mf – 9 – €160.00 – mf#mmp134/3 – ne Moran [079]
Cahete : orgao republicano nativista – Maceio, AL. 12 out-26 nov 1896 – (= ser Ps 19) – bl Biblioteca [079]
Cahid, Burhan [Morkaya] *see*
– Goenuel yuvasi
– Harb doenuesue
Cahier / Societe d'histoire des Pays d'en Haut – n1-11 [1979 printemps-1981 sep] – 1r – 1 – (cont by: cahiers d'histoire des pays-d'en-haut) – mf#962308 – us WHS [971]
Le cahier – Revue mensuelle des lettres et des arts. Paris. 1929-avr 1940 – 1 – fr ACRPP [800]
Cahier de spicileges / Lasnier, Roseline [comp] – 1r – 1 – mf#SEM35P310 – cn Bibl Nat [355]
Les cahiers algeriens – Paris. n2. juil aout 1950 – 1 – fr ACRPP [073]
Cahiers bleus – Paris. n1-119. aout 1928-mai 1932 – 1 – fr ACRPP [073]
Cahiers bleus *see* Chantiers cooperatifs
Les cahiers communistes – Paris. n1-6. 9 nov-21 dec 1922 – 1 – (deux n6 different) – fr ACRPP [335]
Cahiers congolais d'anthropologie et d'histoire – [Brazzaville, Congo: s.n.] 1976- – 1 – (in french, with some summaries in english) – mf#976 – us Wisconsin U Libr [301]
Les cahiers d'amani-y – Saint-Marc, [Haiti: s.n. [v1 n2-9]. (1953-1954) 1955 – 3 sheets – us CRL [972]
Les cahiers d'aujourd'hui – no.1-10. oct 1912-avr 1914; n.s. nos.1-15. nov 1920-24. [irregular] – 1 – us Wisconsin U Libr [073]
Les cahiers d'aujourd'hui – Paris. no. 1-9. oct 1912-fevr 1914 – 1 – fr ACRPP [073]
Les cahiers de contre-enseignement proletarien – Paris. dec 1931-mai 1937 – 1 – fr ACRPP [073]
Cahiers de droit – Quebec. 1974+ (1) 1974+ (5) 1974+ (9) – ISSN: 0007-974X – mf#9538 – us UMI ProQuest [340]
Les cahiers de droit – Quebec: Faculte de droit. v1 n1- dec 1954- [mf ed 1978-] – 5 – mf#SEM16P310 – cn Bibl Nat [323]
Cahiers de geographie de quebec / Universite Laval (Quebec). Institut de geographie – Quebec: Les Presses universitaires Laval. v1 n1 oct 1956 – [mf ed 1979-1991] – 7r – 5 – (merger of: cahiers de geographie with: notes de geographie) – mf#SEM16P313 – cn Bibl Nat [917]
Cahiers de geographie de quebec / Universite Laval (Quebec). Institut de geographie – Quebec: Presses universitaires Laval. v1 n1 oct 1956- [mf ed 1992-] – 9 – (merger of: cahiers de geographie with: notes de geographie) – mf#SEM105P1706 – cn Bibl Nat [917]
Cahiers de geographie du quebec – Quebec. 1976+ [1,5,9] – ISSN: 0007-9766 – mf#11203 – us UMI ProQuest [971]
Cahiers de la democratie – Paris. n1-54. juin 1933-avr 1939 – 1 – (lacking: n13) – fr ACRPP [325]
Les cahiers de la jeunesse – Revue des jeunes de notre temps. Paris. n1-21.juil 1937-avr 1939 – 1 – fr ACRPP [305]
Cahiers de la nouvelle journee – Paris. 1924-avr 1932 (1-21) – 1 – fr ACRPP [073]
Cahiers de la quinzaine – Paris. 1900-juil 1914, 1925-34 – 1 – fr ACRPP [073]
Cahiers de la quinzaine – Paris. ser. 1-15. Jan 5 1900-July 7 1914; ser. 16-27, n2. Jan 1925-1936.(incomplete) – 1 – us NY Public [073]
Les cahiers de la republique des lettres, des sciences et des arts – Paris. n1-12, avec un suppl. de mai-juil 1926.avr 1926-28 – 1 – fr ACRPP [073]
Cahiers de l'academie canadienne-francaise / Barbeau, Victor – Montreal: l'Academie. 1 (1956)-v15 (1977) – 1r – 5 – mf#SEM16P117 – cn Bibl Nat [440]
Cahiers de l'afrique et de l'asie – Paris. v 1-3 n.d – 1 – us NY Public [950]
Les cahiers de l'enfance inadaptee – Paris. 1952-May 1973 – 5 – fr ACRPP [305]
Les cahiers de l'hexagone – Paris. n1-56 57. avr 1961-juin 1972 – 5 – fr ACRPP [073]
Cahiers de l'iep – Beyrouth-Liban: Institut d'etudes palestiniennes. n10. 1980 – 1r – us CRL [956]
Cahiers de l'institut maurice thorez – no. 1-24. Paris. avr 1966-71 – 5 – fr ACRPP [944]
Cahiers des comites de prevention du batiment et des travaux publics – Issy-les-Moulineaux. sept oct 1968-73 – 1 – fr ACRPP [073]
Les cahiers des droits de l'homme – Paris. 1920-fevr 1940 – 1 – fr ACRPP [322]
Cahiers des quatre saisons – Paris. n1-49. aout 1955-67 – 1 – (devenu: cahiers des saisons) – fr ACRPP [073]
Cahiers des saisons *see* Cahiers des quatre saisons
Cahiers d'etudes africaines – n1-28. 1960-67 – 1 – us AMS Press [970]
Cahiers d'etudes de radio-television – n1-27 28. Paris. 1954-sept dec 1960 – 1 – fr ACRPP [380]
Cahiers d'etudes revolutionnaires – no. 1-6. Marseille. nov dec 1963-juil 1965 – 1 – fr ACRPP [320]
Cahiers d'histoire des Pays-d'en-Haut / Societe d'histoire des Pays d'en Haut – 1981 nov-1984 – 1r – 1 – (cont: cahier [societe d'histoire des pays d'en haut]; cont by: cahier [societe d'histoire des pays-d'en-haut: 1989]) – mf#962310 – us WHS [071]
Les cahiers d'occident – Paris.1, n1-10.1926-27; 2e s., n1-10. 1928-30 – 1 – (fusionne avec: latinite. revue des pays d'occident et la reaction pour l'ordre pour former: la revue du siecles) – fr ACRPP [073]
Cahiers du bolchevisme – Paris. nov 1924-aout 1939 – 1 – (a partir de juil 1939, parait sous le titre de: cahiers du communisme. pour les suppl. voir les dossiers de l'agitateur et bulletin colonial) – fr ACRPP [335]
Cahiers du cercle proudhon – Paris. 1912, 1914 – 1 – fr ACRPP [320]
Cahiers du cinema – v1-27. 1951-64 – 1 – us AMS Press [790]
Cahiers du college de pataphysique – Paris. 1952-63 – 1 – (devenu: college de pataphysique. dossiers) – fr ACRPP [140]
Cahiers du communisme – Paris. v.3-25. Dec 31, 1927-Dec 1948. Incomplete – 1 – 69.00 – us L of C Photodup [335]
Cahiers du communisme *see* Cahiers du bolchevisme
Les cahiers du contadour – Paris. puis Saint-Paul. ete 1936-fevr 1939 – 1 – fr ACRPP [073]
Cahiers du groupe francoise minkowska – Paris. juil 1958-65 – 1 – fr ACRPP [073]
Les cahiers du jazz – Paris. n1-16/17. nov 1959(?)-1968 [all publ] – (= ser Jazz periodicals, 1914-1977) – 1r – 1 – $220.00 – us UPA [780]
Les cahiers du militant / Parti communiste francais (S F IC) – no. 1, 4, 9-10. Paris. mai 1924-juil 1925 – 1 – fr ACRPP [335]
Les cahiers du mois – Paris. n1-25 26. mai 1924-juin 1927 – 1 – fr ACRPP [073]
Les cahiers du plateau – Passy (Haute-Savoie). n17-18, 20. 1938-juin 1939 – 1 – fr ACRPP [073]
Les cahiers du sud – Marseille. 1924-66 – 1 – fr ACRPP [073]
Les cahiers du theatre – no. 1-10. Paris. mars 1926-avr 1932, sept 1936 – 1 – fr ACRPP [790]
Cahiers francais – London, UK. Jul 1944 – 1 – uk British Libr Newspaper [072]
Cahiers g.l.m – Paris. mai 1936-mars 1939, 1954-56 – 1 – fr ACRPP [073]
Les cahiers jaunes – Paris. n1-2,4. 1932-33 – 1 – fr ACRPP [073]
Cahiers leon bloy – La Rochelle. sept oct 1924-mai aout 1939, mars avr 1952 – 5 – fr ACRPP [073]
Les cahiers naturalistes – Paris. n1-36. 1955-68 – 1 – fr ACRPP [073]
Cahiers pierre loti – Paris. juin 1950-72 – 5 – fr ACRPP [440]
Les cahiers rationalistes – No.1, 1931-. Paris: Union Rationaliste. -m. -irr. Indexes 1931-65. Includes index to Courrier Rationaliste. 1v. 1289 – 1 – us Wisconsin U Libr [140]
Les cahiers rouges – Paris. n3-12. aout sept 1937-juin juil 1938 – 1 – (mq n 6) – fr ACRPP [073]
Cahiers victoriens and edouardiens – Montpellier. 1989-1989 (1) – ISSN: 0220-5610 – mf#16503,01 – us UMI ProQuest [400]
Cahill, Daniel William *see* First american edition of the works of the rev d w cahill
Cahill, M *see* A scottish knight-errant
Cahoba valley news – Birmingham, AL. 1963-1963 (1) – mf#61982 – us UMI ProQuest [071]
Cahper journal / California Association for Health, Physical Education and Recreation – Los Angeles. 1972-1978 (1) 1972-1978 (5) 1975-1978 (9) – ISSN: 0007-7763 – mf#7586 – us UMI ProQuest [790]
Cahper journal times / California Association for Health, Physical Education, and Recreation – Danville. 1978-1980 (1) 1978-1980 (5) 1978-1980 (9) – (cont by: cahperd journal times) – ISSN: 0194-8261 – mf#7586,01 – us UMI ProQuest [790]
Cahper journal times / California Association for Health, Physical Education, Recreation, and Dance – Danville. 1980+ (1) 1980+ (5) 1980+ (9) – ISSN: 0273-6896 – mf#7586,02 – us UMI ProQuest [790]
Cahuide: revista nacional – Lima. 2-22, 1939-59. Incomplete – 1 – 69.00 – us L of C Photodup [972]
Cai tao – Ha Noi: Pham-van-thu -1954 [1950 v3 n84-97 [jan 7-may 13] n101-126 [jul 1-dec 30]; 1952: v5 n177/178-181 [jan 12-23] n183-9 [mar 8-jun 28]; 1953: v6 n223-260 [jan 3-dec 19]; 1954: v6 n261-281 [jan 2-jul 24] – 1r – 1 – mf#mf-4172 seam – us CRL [079]
Caicedo Montua, Francisco A *see* Banzay
Caida de adan / Ulloa, Juan – San Salvador, El Salvador. 1961 – 1r – us UF Libraries [972]
Caida de jorge ubico – Guatemala, 1944 – 1r – us UF Libraries [972]
Caida de una tirania / Montufar, Rafael – Guatemala, 1923- – 1r – us UF Libraries [972]
Caida del gobierno constitucional en costa rica / Oreamuno, Jose Rafael – New York, NY. 1919 – 1r – us UF Libraries [972]
Caidin, Martin *see* Night hamburg died
Caiger, Stephen Langrish *see* British honduras, past and present
Caigniez, Louis-Charles *see*
– Amans en poste
– Petite bohemienne
– Pie voleuse
Caillaud, Frederic *see* Voyage a meroe, au fleuve blanc, au-dela de f'azoql dans le midi du royaume de sennar, a syouah et dans cinq autres oasis
Caille, Nicolas Louis de la *see* Journal historique du voyage fait au cap de bonne-esperance
Caillemer, Exupere *see* Le credit foncier a athenes
Cailler-Bois, Ricardo *see*
– Ensayo sobre el rio de la plata y la revolucion francesa
– Nuestros corsarios 1. brown y bouchard en el pacifico, 1815-1816
Cailleux, Andre *see* Application a la geographie des methodes d'etude d...
Cailleville, Jacques De *see* Cercle inutile
Cailliatte, Charles *see* 1re these
Cailliaud, F *see* Voyage...meroe
Caillie, R *see* Travels through central africa to timbuctoo
Caillie, Rene *see* Journal d'un voyage a tembouctou et a jenne, dans l'afrique centrale
Caillot, Antoine *see*
– Beautes de la marine
– Memoires pour servir a l'histoire des moeurs et usages des francais
– Tuileur portatif des trente-trois degres de l'ecossisme du rit ancien et accepte
Caim news / Council for American Indian Ministry – 1973 oct/dec-1984 mar, 1974 apr/may/jun, v2 2 qtr [1974 apr] – 3r – 1 – mf#819420 – us WHS [360]
Caiman sonoro / Riveron Hernandez, Francisco – Habana, Cuba. 1959 – 1r – us UF Libraries [972]
Caimaw review / Canadian Association of Industrial, Mechanical and Allied Workers – 1983 jul-1991 nov – 1r – 1 – mf#1053754 – us WHS [600]
Caimo, Norberto *see* Voyage d'espagne, fait en l'annee 1755 [mil sept cent cinquantecinq]
Cain, J Byron *see* With the nea in florida
Cain, Thomas Leonard *see* Biology, life history, and control of the cross-striped cabbage worm
Caine, Caesar *see*
– Capella de gerardegile
– The story of mashonaland and the missionary pioneers
Caine, John Thomas *see* The legislative commission scheme
Caine, William Ralph Hall *see* Cruise of the 'port kingston'
Caines, George *see* Practical forms of the supreme court, taken from tidd's appendix of the forms of the court of king's bench.
Caird, Edward *see*
– A critical account of the philosophy of kant
– The evolution of theology in the greek philosophers
– Hegel
– Lay sermons and addresses
– Lectures and essays on natural theology and ethics
– The social philosophy and religion of comte
Caird, James *see*
– India
– The plantation scheme
Caird, John *see*
– Essays for sunday reading
– The fundamental ideas of christianity
– An introduction to the philosophy of religion
– Religion in common life
– Spinoza
– University addresses
– University sermons
Caird, John et al *see*
– The faiths of the world
– Oriental religions
Caird, William Renny *see*
– Christi worte ueber vollendung der wege gottes
– Letter to the rev r h story, rosneath
Caire et ses environs – Paris, France. 1909 – 1/4r – 1 – uk British Libr Newspaper [072]
Cairnes, John Elliott *see*
– The slave power
– University education in ireland
Cairns, Adam *see* The inauguration of the political independence of victoria
Cairns argus – Cairns, Australia 3 nov 1891-10 dec 1897 (very imperfect) (wkly) – 6r – 1 – uk British Libr Newspaper [079]
Cairns Collection of American Women Writers *see*
– Godey's lady's book
– Woman's home companion
Cairns, David *see* Christ, the morning star
Cairns, David Smith *see* Christianity in the modern world
Cairns, John *see*
– Bush and boma
– Christ the central evidence of christianity
– Christ, the morning star
– An examination of professor ferrier's theory of knowing and being
– False christs and the true
– The hongkong register
– The jews in relation to the church and the world
– Memoir of john brown, d.d
– Moral greatness of the temperance enterprise
– Outlines of apologetical theology
– Oxford rationalism and english christianity
– The scottish philosophy
– Unbelief in the eighteenth century
Cairns, John et al *see* The presbyterian church of england
Cairns martyn st monumental inscriptions – 12mf – 9 – A$55.00 set – at Cairns [929]
Cairns, William *see* Christ, the morning star
Cairo baptist church. missouri : church records – 1888-1956 – 1 – us Southern Baptist [242]
Cairo city weekly gazette – Cairo, IL. 1861-1862 (1) – mf#68993 – us UMI ProQuest [071]
Cairo daily democrat – Cairo IL. 1865 apr 16 – 1r – 1 – (cont by: cairo democrat) – mf#984202 – us WHS [071]
Cairo farm record – Cairo, NE: Richard J and Jeanne Mohanna. 8v. v63 n44, mar 24 1967-v70 n13. aug 9 1973 [mf ed mar 24 1967-aug 9 1973 (lacks dec 20 1968) filmed 1969-77] – 4r – 1 – (cont: cairo record. cont by: cairo record (1973)) – us NE Hist [071]
Cairo press review – sept-dec 1962; may 1963-dec 1982 – 1 – us L of C Photodup [073]

Cairo record – Cairo, NE: Elliott Harrison, 1902- v63 n43. mar 17 1967 (wkly) [mf ed v5 n4. apr 26 1907-09, 1920-67 (gaps) filmed 1969] – 17r – 1 – (cont by: cairo farm record. publ as: hall county record jan 3-jun 27 1923) – us NE Hist [071]
Cairo record – Cairo, NE: Richard J and Jeanne Mohanna. v70 n14. aug 16 1973- (wkly) [mf ed filmed 1977-] – 1 – (cont: cairo farm record) – us NE Hist [071]
Cairo standard – Harrisville, WV. 1947-1949 (1) – mf#67312 – us UMI ProQuest [071]
Cairo to cape town / Reynolds, Reginald – Garden City, NY. 1955 – 1r – us UF Libraries [960]
Cairu / Cairu, Jose Da Silva Lisboa – Rio de Janeiro, Brazil. 1958 – 1r – us UF Libraries [972]
Cairu, Jose Da Silva Lisboa *see* Cairu
Caissa : revista argentina de ajedrez. buenos aires. – v1-14. 1937-1950. (incomplete) – 1 – us NY Public [073]
Caisse d'epargne et l'ecole en haiti / Janvier, Louis Joseph – Port-Au-Prince, Haiti. 1906 – 1r – us UF Libraries [972]
Caisse nationale d'economie, fondee le 1er janvier 1899 : arthur gagnon, secretaire-tresorier / Association Saint-Jean-Baptiste de Montreal. Caisse nationale d'economie – [Montreal?: s.n, 1899?] [mf ed 1980] – 1mf – 9 – 0-665-00378-1 – mf#00378 – cn Canadiana [366]
Caisse nationale d'economie, fondee le 1er janvier 1899 – [Montreal?]: [s.n.], [1899?] [mf ed 1981] – 1mf – 9 – 0-665-13778-8 – mf#13778 – cn Canadiana [360]
Caisse Nationale des Monuments Historiques et des Sites. Paris *see*
– Antiquities
– Decorative art
– Drawings in provincial and other museums
– Drawings in the louvre and national museums
– Indexes
– Manuscripts
– Paintings in provincial and other museums
– Paintings in the louvre
– Sculpture
Caisse Nationale des Monuments Historiques et des Sites. Paris. Archives Photographiques *see* Architecture and early photography in france
Caithness chronicle and advertiser for the northern counties – [Scotland] Thurso: Wm Manson 2 apr-16 dec 1847 (wkly) [mf ed 2003] – 17v on 1r – 1 – uk Newsplan [072]
Caithness courier : and incorporating the john o' groat journal – Thurso: Highland News Group of Newspapers (Caithness Division) 1977- (wkly) [mf ed 2 jan 1991-] – 1 – (cont: caithness courier, and weekly advertiser for the northern counties; nov 9 1977- : sister publ to: john o' groat journal & caithness monthly miscellany; imprint varies) – ISSN: 1354-9642 – uk Scotland NatLib [072]
Caithness, Marie Sinclair, Countess of *see* Old truths in a new light
Caius, John *see* A boke or counseill against the disease called the sweate
Caix de Saint-Aymour, A de *see* La france en ethiopie
Caixa auxiliar : orgam de propaganda da caixa auxiliar da ponte hercilio luz limitada – Florianopolis, SC: Imprensa Official, 22 mar, jun 1927; 23 jun 1928 – (= ser Ps 19) – mf#UFSC/BPESC – bl Biblioteca [079]
O caixeiro : hebdomadario republicano – Natal, RN: Typ d'A Republica, 01 ago 1892-14 mar 1894 – (= ser Ps 19) – mf#P22B,04,198 – bl Biblioteca [079]
O caixeiro : publicacao litteraria, scientifica e noticiosa – Bahia: [s.n.] 01 mar 1878 – (= ser Ps 19) – mf#P18B,02,21 – bl Biblioteca [079]
O caixeiro nacional – Bahia: Typ Liberal do Argos Bahiano, 24 ago 1855 – (= ser Ps 19) – bl Biblioteca [079]
Caja Ahorros y Monte de Piedad de Caceres *see*
– Memoria y datos estadisticos correspondientes al ano 1954
– Reglamento provisional de procedimiento electoral
Caja Ahorros y Monte de Piedad de Caceres. *see* Estatutos generales
Caja de Ahorros de Plasencia *see* Trujillo en fiestas
Caja de ahorros, prestamos y socorros fontes. memoria, balance y estadistica del ejercicio 1st de su funcionamiento, cerrado el 31 de agosto de 1909 – Almendralejo: Imp. Juan Bote, 1909 – 1 – sp Bibl Santa Ana [946]
Caja de Ahorros y Monte de Piedad *see*
– Guion de conferencias pronunciadas en el aula de cultura de la caja de ahorros y monte de piedad de plasencia. abril-mayo 1971
– Memoria y datos estadisticos correspondientes al ejercicio 1958
– Memoria y datos estadisticos correspondientes al ejercicio de 1957
Caja de Ahorros y Monte de Piedad. Caceres *see*
– Estatutos generales
– Memoria correspondiente al ejercicio de 1961
– Memoria correspondiente al ejercicio de 1962
– Memoria correspondiente al ejercicio de 1963
– Memoria correspondiente al ejercicio de 1964
– Memoria de 1976
– Memoria y datos estadisticos correspondientes al ano 1952
– Memoria y datos estadisticos correspondientes al ano de 1955
Caja de Ahorros y Monte de Piedad. Plasencia *see*
– Memoria. 1958
– Memoria 1972
Caja extremena de prevision social – Madrid: Imp. de Minuesa, 1930 – 1 – sp Bibl Santa Ana [946]
Caja General Frexnense *see* Ejercicio social del ano 1912. memoria y balance leidos y aprobados en junta general de socios...1913
Caja Postal de Ahorros *see* 15th certamen nacional del ahorro
Caja real de filipinas (anno 1565-1701) – Sevilla – 1r – 5,6 – sp Cultura [959]
Caja real de mejico (anno 1553-1563) – Sevilla – 1r – 5,6 – sp Cultura [972]
Caja real de nueva espana (anno 1533-1553) – Sevilla – 1r – 5,6 – sp Cultura [946]
Caja real de nueva espana (anno 1540-1549) – Sevilla – 1r – 5,6 – sp Cultura [946]
Caja real de panama. cuentas de descargas (anno 1550-1576) – Sevilla – 1r – 5,6 – sp Cultura [972]
Caja real de puerto rico (anno 1627-1633) – Sevilla – 1r – 5,6 – sp Cultura [972]
Caja real de san francisco de quito (anno 1549-1590) – Sevilla – 1r – 5,6 – sp Cultura [977]
Caja reales de indias – Sevilla – 17r – 5,6 – sp Cultura [970]
Caja Rural de Ahorros y Prestamos de Almendralejo *see*
– Estatutos y reglamentos
– Memoria correspondiente al ejercicio de 1967 aprobada en junta general de accionistas celebrada el 18-2-1968
– Memoria y balance del tercer ejercicio social, leida y aprobada en la junta general de socios...1909
Caja Rural de Ahorros y Prestamos de Fuente de Cantos *see* Memoria leida en la junta general de socios
Caja Rural de Ahorros y Prestamos de los Santos de Maimona *see* Estatutos
Caja Rural de Ahorros y Prestamos Nuestra Senora de Botoa *see*
– Memoria. ejercicio de 1966
– Memoria. ejercicios de 1956-57
– Memoria y ejercicio de 1961
Caja Rural de Almendralejo *see* Memoria 1975
Cajas reales de panama y portobelo (anno 1514-1760) – Sevilla – 24r – 5,6 – sp Cultura [972]
Cajetan, T *see* Opera omnia quotquot in sacrae scripturae expositionum reperiuntur
Cajetan, Tommaso de Vio Gaetani *see*
– Commentarii illustres...in quinque mosaicos libros
– Commentarii in summam theologicam thomae aquinatis
– De nominum analogia
– Opuscula omnia
– Opuscula omnia 1558
Cajigal, Juan Manuel De *see* Memorias sobre la revolucion de venezuela
Il cajo fabrizio : opera en 3 atti del sig jomelli / Jommelli, Nicolo – [1772] [mf ed 1988] – 1r – 1 – (libretto by mattia verazi) – mf#pres. film 17 – us Sibley [780]
[El cajon-] california express line – CA. 1989 – 61+ r – 1 – $3660.00 (subs $765/y) – mf#R04023 – us Library Micro [071]
[El cajon-] daily californian – CA. 1974- – 192r – 1 – $11,520.00 (subs $480/y) – (aka: inland empire daily californian; formerly: valley news) – mf#RC02193 – us Library Micro [071]
Cajon de sastre / Sanchez Arjona, Vicente – Sevilla: Imprenta Zambrano, 1956 – 1 – sp Bibl Santa Ana [946]
[El cajon-] heartland news – CA. 1950-52 – 1r – 1 – $60.00 – mf#C02194 – us Library Micro [071]
[El cajon-] valley news – CA. 1927-61; 1964-65 – 40r – 1 – $2400.00 – mf#C02195 – us Library Micro [071]
Cajun cable – 1991 jan, mar-jun, aug – 1r – 1 – mf#2253581 – us WHS [071]
Caka pum ka pro so dassana / Canda Chve – Ran Kun: Cin Krann ca pum nhip tuik 1977 [mf ed 1990] – 1r with other items – 1 – (proverbs in burmese; bibl also incl items in english; glos in english & burmese; with bibl) – mf#mf-10289 seam reel 162/2 [§] – us CRL [390]
Cakchiquel texts vocabularies and miscellaneous notes / Rosales, Juan de Dios – Chicago: University of Chicago Library, 1976 (mf ed) – (= ser Microfilm coll of manuscripts on cultural anthropology) – 1r – 1 – us Chicago U Pr [490]
Cake walks / Clendenen, Frank Leslie – [St Louis? 190-?] – 1 – mf#*ZBD-*MGO pv10 – Located: NYPL – us Misc Inst [790]
Cakrasakha : the companion of god / Acharya, Ananda – Scandinavia [ie Alvdal, Norway]: Brahmakul Gaurisankar, [1922?] – (= ser Samp: indian books) – us CRL [280]
Cala newsletter / Community Action on Latin America – 1971 nov-1981 apr – 1r – 1 – mf#31999 – us WHS [360]
Calabria during a military residence of three years in a series of letters / Duret de Tavel – London 1832 [mf ed Hildesheim 1995-98] – 3mf – 9 – €90.00 – 3-487-29213-0 – gw Olms [860]
Calado, Manoel *see* Valeroso lucideno e triunfo da liberdade
Calado, Raphael Ruiz *see* El d raphael ruiz calado
Calahan first baptist church. calahan, florida : church records – 1841-1943 – 1 – us Southern Baptist [242]
Calahan, Harold Augustin *see* So you're going to buy a boat
Calamus : a series of letters written during the years 1868-1880 by walt whitman to a young friend (peter doyle) / ed by Bucke, Richard Maurice – Boston: L Maynard, 1897 – 2mf – 9 – (int by ed) – mf#37422 – cn Canadiana [860]
Calamy, Edmund *see*
– Four speeches delivered in guild-hall, 1643
– Inspiration of the holy writings of the old and new testaments
Calancha, Antonio de la *see* Coronica moralizada del orden de san augustin en el peru
Caland, Willem *see* L'agnistoma
Calasibeta, M *see* La rosa de palermo, antidoto de la peste y de todo mal contagioso
[Calaveras county-] calaveras county – 1893 – 1r – 1 – $50.00 – mf#D009 – us Library Micro [978]
[Calaveras county-] calaveras, san joaquin, stanislaus and tuolumne counties – CA. 1856 – 1r – 1 – $50.00 – mf#D010 – us Library Micro [978]
Calaveras/los gatos/saratoga – 1992 – (= ser California telephone directory coll) – 1r – 1 – $50.00 – mf#P00012 – us Library Micro [917]
Calaveras/tuolumne – 1910-38; 1992- – (= ser California telephone directory coll) – 31r – 1 – $1550.00 – mf#P00013 – us Library Micro [917]
Calc report / Clergy and Laity Concerned [US] – 1975 sep 14, 1977 jan-1985 jun – 1r – 1 – mf#1081669 – us WHS [240]
Calcagni, F *see* Antiquarum statuarum urbis romae
Calcagnini, Celio *see* Caelii calcagnini, ferrariensis, protonotarii apostolici, opera aliqvot
Calcagno, Francisco *see* Diccionario biografico cubano
Calcified tissue international – Heidelberg. 1981-1992 (1,5,9) – ISSN: 0171-967X – mf#13147,01 – us UMI ProQuest [574]
Calcium activated neutral protease (calpain) and the neutrophil : their relationship and association with the acute inflammatory response to exercise / Raj, Daniel Adelbert – 1997 – 2mf – 9 – $8.00 – mf#PH 1601 – us Kinesology [612]
Calcographia : or, the art of multiplying... drawings, after the manner of chalk, [etc] / Hassell, John – London 1811 – (= ser 19th c art & architecture) – 1mf – 9 – mf#4.2.1566 – uk Chadwyck [740]
Calcoin news / California State Numismatic Association – 1976 win-1986 fall – 1r – 1 – (cont by: n a s c quarterly; california numismatist) – mf#1084332 – us WHS [730]
Calculations to determine at what point in the side of a hill its attraction will be the greatest etc / Hutton, Charles – London: J Nichols 1780 [mf ed 1998] – 1r – 1 – mf#film mas 28292 – us Harvard [520]
The calculus of observations : a treatise on numerical mathematics / Whittaker, Edmund Taylor et al – 2nd ed. London: Blackie & Son 1926 – 1r [pl/ill] – 1 – (incl ind) – mf#film mas 28407 – us Harvard [510]
Calcutta. British Indian Association *see*
– A half-yearly general meeting of the british indian association...tuesday the 31st july, 1866...
– Letter addressed to the board of revenue
– Petition of the british indian association
– Petition of the british indian association to the house of commons, 1859
– Petition to parliament from the members of the british indian association
– Petitions and letters of the british indian association
– Petitions of the british indian association
– Proceedings of the british indian association, the 3rd june, 1853
– The twelfth annual general meeting of the british indian association...wednesday, the 24th february 1864...
Calcutta christian herald – Calcutta, India. jul 1844-dec 1845 [wkly] – 1r – 1 – uk British Libr Newspaper [079]
Calcutta gazette – India. -w. Jan 1792-March 1800. 5 reels – 1 – uk British Libr Newspaper [072]
The calcutta gazette – Alipore. 1903-1940; 1957-1966 – 1 – us NY Public [324]
Calcutta. General Association of Missionaries *see* A statement respecting a central institution or college
Calcutta government gazette, 1815-32 – 27r – 1 – mf#3635 – uk Microform Academic [954]
Calcutta. India *see* Municipal gazette
Calcutta journal : or political, commercial and literary gazette – Calcutta. 1819-1823 (1) – mf#4845 – us UMI ProQuest [320]
Calcutta journal of natural history and miscellany of the arts and sciences in india – Calcutta 1841-47 – v1-8 on 90mf – 9 – mf#1-276/2 – ne IDC [079]
Calcutta mathematical society. bulletin – v1-43. 1909-51 – 9 – $252.00 – mf#0129 – us Brook [510]
Calcutta Missionary Conference *see* Statistical tables of protestant missions in india, burma and ceylon
Calcutta morning post, 1812-13 – 1r – 1 – mf#4898 – uk Microform Academic [079]
Calcutta review, 5.3 – Calcutta 1921-56 – v1-141 on 1011mf – 9 – (missing: 1943/1944 [v87(4-12)]) – mf#1-109/2 – ne IDC [590]
Calcutta review, new series – Calcutta 1913-20 – v1-7 on 73mf – 9 – mf#1-108/2 – ne IDC [590]
Calcutta review, series 1 – Calcutta 1844-1912 – v1-135 on 1157mf – 9 – (missing: 1853 v20) – mf#1-107/2 – ne IDC [580]
Calcutta weekly notes – Calcutta, 1896-1965. v1-69 – 1861mf – 8 – (lacks) – mf#I-537 – ne IDC [079]
Caldas Barbos, Domingos *see* Viola de lereno
Caldcleugh, Alexander *see*
– Reisen in sued-amerika waehrend der jahre 1819, 1820, 1821
– Travels in south america, during the years 1819-20-21
Caldecote 1649-1949 – (= ser Cambridgeshire parish register transcript) – 2mf – 9 – £2.50 – uk CambsFHS [929]
Caldecott, Alfred *see*
– The philosophy of religion in england and america
– Selections from the literature of theism
Caldecott, William Shaw *see*
– The second temple in jerusalem
– Solomon's temple
– The tabernacle
Caldeira, Clovis *see* Mutirao
Calder, Frederick *see* Memoirs of simon episcopius
Calder, William R *see*
– The seven sayings on the cross
– The thief and the cross
Calderio, Francisco *see*
– Catolicos y comunistas
– Experiencias de cuba
– Por la igualdad de todos los cubanos
Calder-Marshall, Arthur *see* Glory dead
Calderon : poemita dramatico / Baumgartner, Alexander – Madrid: Libreria de San Jose, 1882 [mf ed 1995] – 106p – 1 – (trans fr german by la ciencia cristiana) – mf#8972 – us Wisconsin U Libr [820]
Calderon Altamirano de Chaves Hinojosa y Paredes, Luis Francisco *see* Opusculos de oro, virtudes morales christianas
Calderon, Cesareo *see* Apuntes biograficas del m.i. sr. d. francisco de paula soto y mancera, arcipreste de la santa y apostolica iglesia de santiago de compostela, natural de zafra, provincia de badajoz
Calderon de la barca / Frutos Cortes, Eugenio – Barcelona: Labor S.A., 1949 – sp Bibl Santa Ana [946]
Calderon de la barca. autos sacramentales. antologia / Frutos Cortes, Eugenio – Madrid: Editora Nacional, 1947 – sp Bibl Santa Ana [440]
Calderon de la Barca, Pedro *see*
– El alcalde de zalamea
– la aurora en copacavana
Calderon de Robles, Juan *see* Privilegia selectiora militiae sancti iuliani..
Calderon, Diego *see* Tiernos afectos de amor, temor, humildad, y confianza
Calderon guardia, lider y caudillo / Fernandez Mora, Carlos – San Jose, Costa Rica. 1939? – 1r – us UF Libraries [972]
Calderon, Jose Tomas *see*
– Anhelos de un ciudadano
– Prontuario geografico – comercial – estadistico y...
Calderon Quijano, Jose Antonio *see* Belice, 1663 (?)-1821
Calderon Ramirez, Salvador *see*
– Cuentos para mi carmencita
– De adentro

Calderon S *see* Meteorito de guarena
Calderon, S *see* Estudio petrografico del meteorito de guarena
Calderwood, Henry *see*
- David hume
- Evolution and man's place in nature
- Handbook of moral philosophy
- The parables of our lord
- Philosophy of the infinite
- The relations of science and religion
Caldesi, Blanford & Co *see* The royal collection of pictures in the gallery at buckingham palace
Caldesi, Leonida *see*
- The national gallery
- Photographs by cave. leonida caldesi
Caldicott, Thomas Ford *see* Hannah corcoran
Caldlish, Robert Smith *see* Narrative relating to certain recent negotiations for the settlemen...
Caldwell, John Henderson *see* A history of presbyterian education in east tennessee
Caldwell. Kansas *see*
- Police docket
- Records
Caldwell, Louis Goldsborough *see* A suggested method of compliance with the davis amendment to the radio law
Caldwell, R *see* Reminiscences of bishop caldwell
Caldwell, R E *see* Spectrographic study of certain everglades soils...
Caldwell, Robert *see* Chi-nyanja simplified
Caldwell, Robert, 1814-1891 *see* A comparative grammar of the dravidian
Caldwell, Robert Granville *see* Lopez expeditions to cuba 1848-1851
Caldwell, Samuel Lunt *see*
- Cities of our faith
- A discourse preached in warren at the completion of the first century of the warren association, september 11, 1867
Caldwell-sweaney researcher – 1983 spr-fall – 1r – 1 – (cont by: caldwell-looney trader) – mf#717444 – us WHS [071]
Cale, Walter *see* Nachgelassene schriften
Caleb, C C *see* The song divine
The caleb emerson family papers, 1795-1905 / Emerson, Caleb – [mf ed 1974] – 6r – 1 – (primarily correspondence, & legal & financial papers of Caleb Emerson (1779-1853), & other family members. Emerson lived in Marietta, Ohio, was active in publishing, abolitionism, & politics) – mf#ms830 – us Western Res [360]
The caleb strong papers, 1657-1818 – [mf ed 1978] – 1r – 1 – (with p/g) – us MA Hist [978]
Caleb, the collier – London, England. 18– – 1r – us UF Libraries [240]
Caledonia / Berlepsch, Emilie von – Hamburg 1802-04 [mf ed Hildesheim 1995-98] – 4v on 8mf – 9 – €160.00 – 3-487-27908-8 – gw Olms [914]
Caledonia : or, an account, historical and topographic of north britain: from the most ancient to the present times / Chalmers, George – London 1807-24 [mf ed Hildesheim 1995-98] – 4v on 31mf – 9 – €310.00 – 3-487-27548-1 – gw Olms [914]
The caledonia interchange – Nass River, BC: Printed...by...J B McCullagh's Indian Boys, [1899?-19–] – 9 – ISSN: 1190-7533 – mf#P04502 – cn Canadiana [242]
Caledonia pictorial – Caledonia, Oak Creek WI. 1977 sep 23-1978 feb 24, 1978 mar-jun, jul-dec, 1979 jan-feb 15 – 4r – 1 – (cont: oak creek, caledonia pictorial; cont by: caledonia-raymond pictorial) – mf#965438 – us WHS [071]
Caledonia pictorial – Caledonia, Oak Creek WI. 1970 jan 1/aug 27-1977 apr 1/may 20 – 16r – 1 – (with gaps; cont by: oak creek, caledonia pictorial) – mf#965468 – us WHS [071]
Caledonia romana : a descriptive account of the roman antiquities of scotland / Stuart, R – Edinburgh, London, 1845 – 14mf – 8 – mf#H-1135 – ne IDC [700]
Caledonian – [Scotland] Glasgow: Niven, Napier & Khull 11 apr-7 nov 1807 (wkly) [mf ed 2003] – 31v on 1r – 1 – (cont by: western star) – uk Newsplan [072]
Caledonian : an illustrated family magazine – v1-23. 1901-23 – 1 – us L of C Photodup [073]
Caledonian gazetteer – [Scotland] Edinburgh: printed for & by John Robertson 31 may-28 jun 1776 (3 times/wk) [mf ed 2004] – 13v on 1r – 1 – (companion paper: caledonian mercury, appeared on mon, wed & sat; incl: london news & foreign news notes, edinburgh news, essays on current topics, local theater advertising, real estate auctions & public notices; ceased with: n13 (1776)) – uk Newsplan [072]
Caledonian mercury : "being a short account of the most considerable news foreign and domestick..." – [Scotland] Edinburgh: printed for W Rolland by William Adams, Junior 28 apr 1720-1867 [mf ed 2005] – 142r – 1 – (imprint varies; merged with: daily express (edinburgh, scotland) to form: caledonian mercury and daily express; cont: caledonian mercury and daily express; absorbed by: weekly scotsman) – uk Newsplan [072]
Caledonian mercury – Edinburgh. 1722-1799 – 1 – mf#5215 – us UMI ProQuest [073]
Caledonian mercury – Edinburgh, Scotland, UK.1832. -w. 2 reels – 1 – uk British Libr Newspaper [072]
Caledonia-raymond pictorial – Caledonia, Oak Creek, Raymond WI 1979 feb 22-jun, jul-dec, 1980 jan-feb 28 – 3r – 1 – (cont: caledonia pictorial [oak creek wi: 1977]; cont by: pictorial [oak creek wi: 1980]) – mf#965434 – us WHS [071]
Calef and chuter letter book, 1783-1796 – Oxford: Rhodes House Library – (= ser BRRAM series) – 1r – 1 – £67 / $134 – (int by oscar tapper) – mf#r95911 – uk Microform Academic [337]
Calef, Robert *see* Salem witchcraft
Calegari, Francesco Antonio *see* Ampla dimostrazione degli armoniali musicale tuoni trattato teorico-prattico de fra frances calegari min
[calendar] *see* Democratic party of dane county
Calendar: an anthology of poetry – Prairie City, Ill. 1940-1942. 3 v – 1 – us NY Public [810]
Calendar of ancient records of dublin : in the possession of the municipal corporation of that city / Gilbert, John Thomas – Dublin: J Dollard 1889-1919 [mf ed 1980] – 1r [ill] – 1 – (v8- ed by rosa m gilbert) – mf#6171 – us Wisconsin U Libr [941]
Calendar of business – 1975 aug 1/dec 19-1992 oct – 70r – 1 – mf#1938492 – us WHS [650]
Calendar of entries in the papal registers relating to great britain and ireland. papal letters / Great Britain. Public Record Office – v. 1-11. 1893-1921 – 1 – 85.00 – us L of C Photodup [240]
Calendar of events / Greater Madison Convention and Visitors Bureau – 1982 apr/may, aug/sep-oct/dec, 1983 jan/mar, oct/nov-1985 nov/dec, 1986 jul/sep, 1987 dec, 1988 jan-dec, 1989 jan-apr/jul – 1r – 1 – (lacks: 1984 jun/aug, 1985 mar/may) – mf#1053904 – us WHS [060]
Calendar of events / Twin Cities Peace and Justice Coalition [MN] – 1985 aug-1987 sep – 1r – 1 – (cont by: calendar of events [minnesota peace and justice coalition]) – mf#1278081 – us WHS [340]
Calendar of items microfilmed at the india office, london / [Crane, Robert I] – [Ann Arbor: Center for South and Southeast Asian Studies? 196-] – us CRL [954]
Calendar of middlesex and westminster sessions book, 1638-1751 – 5r – 1 – (incl orders of court fr 1716) – mf#96494 – uk Microform Academic [340]
Calendar of official papers, (1803-1878) / Governors of Ohio – 3r – 1 – mf#B26520-26522 – us Ohio Hist [324]
The calendar of st willibrord / Wilson, H Austin – 1919 – (= ser Henry bradshaw society (hbs)) – 4mf – 8 – €11.00 – ne Slangenburg [241]
Calendar of state papers, domestic series of the reign of charles i, 1625-1649 / Great Britain. Public Record Office – 1858-97 – 1 – us L of C Photodup [941]
Calendar of the... / Madison Art Center – 1986 oct/nov-1979 may/jun – 1r – 1 – (cont: newsletter of the madison art center; cont by: madison art center newsletter) – mf#1609104 – us WHS [060]
Calendar of the... / Smithsonian Institution – 1972 jan-1978 apr – 1r – 1 – mf#228380 – us WHS [060]
Calendar of the ecclesiastical dignitaries of st paul's cathedral, from the year 1800 to / Simpson, W Sparrow – London, England. 1877 – 1r – us UF Libraries [240]
Calendar of the gods in china / Richard, Timothy – [2nd ed] Shanghai: Commercial Press, 1916 – 1mf – 9 – 0-524-02042-6 – mf#1990-2817 – us ATLA [290]
A calendar of the...records, 1505-1750 – London: Masters of the Bench, 1896 – 20mf – 9 – $30.00 – (v1-3 ed by f a inderwick. v4 ed by r a roberts) – mf#LLMC 84-297 – us LLMC [520]
Calendar of virginia state papers and other manuscripts / Virginia – Richmond. 1875-93 – 1 – us L of C Photodup [978]
Calendar of wisconsin's bicentennial events / American Revolution Bicentennial Commission of Wisconsin – 1975 sep-1976 win – 1r – 1 – mf#352113 – us WHS [978]
Calendar to montreal gazette / Lower, A R M – Montreal, QC. 1778-1841 – 1r – 1 – cn Library Assoc [971]
Calendaring practices of the eastern district of north carolina / Olson, Susan M – Washington: FJC, 1987 – 1mf – 9 – $1.50 – mf#LLMC 95-354 – us LLMC [340]
Calendario : nome hebdomadario do 1.dia do mez de janeiro de cada ano, desde 1582, epoca da reforma gregoriana, ate o anno 4000 – Pernambuco: Typ de Pinheiro & Faria, 1835 – (= ser Ps 19) – bl Biblioteca [079]
Calendario civico-cultural, 1969 / RIO DE JANEIRO INSTITUTO NACIONAL DO LIVRO – Rio de Janeiro, Brazil. 1969 – 1r – us UF Libraries [972]
Calendario de Extremadura *see* 1851
Calendario de extremadura para el ano 1863...y aumentado con el calendario portugues de barda d'agua / Extremadura. Spain (Province.) – 1863 – 9 – sp Bibl Santa Ana [520]
Calendario del campeonato 1972-73 / Federacion Extremena de Futbol – Caceres: Tip. Extremadura, 1972 – 1 – sp Bibl Santa Ana [790]
Calendario escolar 1968/69 / Seminario Diocesano de San Anton – Badajoz: Graficas Tejado, 1968 – sp Bibl Santa Ana [240]
Calendario escolar 1969/70 / Seminario Diocesano de San Anton – Badajoz: Graficas Tejado, 1970 – sp Bibl Santa Ana [240]
Calendario folclorico do distrito federal / Lira, Mariza – Rio de Janeiro, Brazil. 1957? – 1r – us UF Libraries [390]
Calendario, manual, y guia de foresteros de las islas filipinas, para el ano de... – Manila: Impreso en Sto Tomas a cargo de D Candido Lopez, 1839-40 – 1r – 1 – us CRL [520]
Calendario para la provincia de extremadura.. / Extremadura. Spain (Province.) – 1827 – 9 – sp Bibl Santa Ana [520]
Calendario y plan de estudios del curso academico 1972-73. seminario diocesano / Coria-Caceres – Caceres: Edit. Extremadura, 1972 – 1 – sp Bibl Santa Ana [370]
Calendario y plan de estudios del curso academico 1975-76. seminario mayor diocesano – Caceres: Edit. Extremadura, 1975 – 1 – sp Bibl Santa Ana [370]
Calendars of charters and rolls in the manucript collections of the british library / British Library – -1911 [mf ed Chadwyck-Healey] – 19r – 1 – uk Chadwyck [090]
Calendars of charters and rolls in the manuscript collections of the bodleian library / The Bodleian Library – 12th-20th c [mf ed Chadwyck-Healey] – 327mf – 9 – uk Chadwyck [090]
Calendarul ortodox credinta – Detroit MI, 1966-71 – 2r – 1 – (romanian periodical) – us IHRC [073]
Calendarul zialurui desteptarea – Detroit MI, 1922-64 – 5r – 1 – (romanian periodical) – us IHRC [073]
Calender fuer das volk – Hannover DE, 1788 & 1805 – 1r – 1 – gw Misc Inst [943]
Calendrier de l'eglise catholique d'haiti / Maisonneuve, Gerard F – Port-Au-Prince, Haiti. 1962 – 1r – us UF Libraries [972]
Calendrier pour les chantres pour l'annee... – Saint-Philippe: Joseph Hebert, impr [ca 182-] (mf ed 1971) – 1r – 5 – mf#SEM16P8 – cn Bibl Nat [241]
La calentura mesenterica... / Lloret y Marti, F – Madrid, 1730 – 6mf – 9 – sp Cultura [616]
Calepinus, A *see* Dictionarivm latino lvsitanicvm...
Calera de Leon *see*
- Fiestas de tentudia, 1969
- Fiestas de tentudia, 1970
- Tentudia. fiestas de septiembre 1973
- Tentudia. revista de ferias y fiestas de 1975
- Tentudia. revista oficial de fiestas, 1972
Calera de Leon, Ayuntamiento *see* Tentudia
Calero Orozco, Adolfo *see*
- Cuentos nicaraguenses
- Cuentos pinoleros
Caleta, joya arqueologica antillana / Herrera Fritot, Rene – Habana, Cuba. 1946 – 1r – us UF Libraries [972]
[Calexico-] calexico chronicle – CA. jan 1906-71+ r – 1 – $4260.00 (subs $50/y) – mf#R02087 – us Library Micro [071]
Caley, Llewellyn Neville *see* The church handbook for teacher training classes
Calgary, alberta, canada, her industries and resources / ed by Burns and Elliott – Calgary: Burns & Elliott, 1885 – 2mf – 9 – mf#30063 – cn Canadiana [917]
Calgary albertan *see* Morning albertan
Calgary daily herald – Canada. -d. April 1911-Dec 1920. 59 1 2 reels – 1 – uk British Libr Newspaper [072]
The calgary diocesan magazine – [Innisfail, Alta: Free Lance Print, 1899-19–] – 9 – mf#P05977 – cn Canadiana [242]
Calgary eye opener – Calgary, Alberta, CN. jan 1902-dec 1922 – 1r – 1 – cn Commonwealth Imaging [071]
Calgary eye-opener – High River, Calgary, AB. 1902-22 – 1r – 1 – cn Library Assoc [071]
Calgary herald – Calgary, Alberta, CN. 1888-36r/y – 1 – Can$2660.00 silver Can$2500.00 vesicular – cn Commonwealth Imaging [071]
Calgary Indian Friendship Centre *see* Elbow drums
Calgary jewish times – Calgary, Albert, Canada.1980 – 1 – us AJPC [071]
Calgary mirror – Alberta, CN. jan 1983-dec 1987 – 20r – 1 – cn Commonwealth Imaging [071]
Calgary news telegram – Alberta, CN. mar 1907-sept 1918 – 45r – 1 – cn Commonwealth Imaging [071]
Calgary optimist – Calgary. Alberta. CN. nov 27 1909-feb 12 1910 – 1 – cn Commonwealth Imaging [071]
Calgary periodicals – Alberta, CN. jan 1884-dec 1894 – 1r – 1 – cn Commonwealth Imaging [073]
Calgary real estate and live stock bulletin – Calgary [Alta]: Fitz Gerald & Lucas, [1892?-189- or 19–] [mf ed v1 n1 jan 1892] – 9 – mf#P04053 – cn Canadiana [333]
Calgary rebel – Calgary. Alberta. CN. 1937-38 – 1 – cn Commonwealth Imaging [071]
Calgary standard *see* Provincial standard
Calgary sunday standard *see* Western standard illustrated weekly
Calgary tribune – Calgary, Alberta, CN. jan 1885-dec 1899 – 1r – 1 – cn Commonwealth Imaging [071]
Calgary weekly herald – Calgary, Alberta, CN. aug 1883-dec 1901 – 4r – 1 – cn Commonwealth Imaging [071]
Calgary weekly herald *see* Weekly herald
Calgary western standard illustrated weekly *see* Western standard illustrated weekly
Calgary women's newspaper – v3 n9-v7 n3 [1977 oct-1981 apr/may] – 1r – 1 – mf#675632 – us WHS [305]
Calhan news and ramah record *see* El paso county miscellaneous newspapers, reel 2
Calhoun, Alfred R *see* A soldier's story
Calhoun baptist church – Calhoun, KY. 1889-1992 – 1 – $205.92 – mf#2713 – us Southern Baptist [242]
Calhoun chronicle – Grantsville, WV. 1893+ (1) – mf#67303 – us UMI ProQuest [071]
Calhoun city first baptist church. calhoun city, mississippi : church records – 1904-83. 840p – 1 – us Southern Baptist [242]
Calhoun County AFL-CIO Council *see* Square deal
Calhoun County Genealogical Society *see*
- Karankawa kountry
- Karankawa kountry quarterly
Calhoun county times – Blountstown, FL. 1947-1948 – 1r – (1947 jan 16-dec; 1948 jan 22, aug 27, oct 23, nov 20) – us UF Libraries [071]
Calhoun, Frederick S *see*
- Letters received by the attorney general
Calhoun, John C *see*
- A disquisition on government
- Papers
Calhoun, John Caldwell *see* Letters, 1824, 1831, 1844-1850
Calhoun journal – Bruce, MS. 1977-1983 (1) – mf#68266 – us UMI ProQuest [071]
Calhoun letters, 1824-1850 *see* Letters, 1824, 1831, 1844-1850
Calhoun, Robert Lowry *see* The dilemma of humanitarian modernism
Calhoun-liberty journal – Bristol, FL. v12 n1-v17 n1-53. 1992-1997 – 6r – (gaps) – us UF Libraries [071]
Caliban – Paris. n1-55. fevr 1947-51 – 1 – fr ACRPP [073]
Caliber / Libertarian Party of California – v5 n1-v11 n1 [1977 mar, 1978 jan/feb-1983 sum] – 1r – 1 – mf#1043251 – us WHS [325]
Calice, F *see* Grundlagen der aegyptisch-semitischen wortvergleichung
Calico print – 1950 nov-1952 apr – 1r – 1 – (cont: calico print) – mf#1053913 – us WHS [071]
Calico print – Yerma, CA. 1951-1953 (1) – mf#62309 – us UMI ProQuest [071]
Calidoscopio de haiti / Monclus, Miguel Angel – Buenos Aires, Argentina. 1953 – 1r – us UF Libraries [972]
[Caliente-] caliente express – NV. 1905 – 1r – 1 – $60.00 – mf#U04429 – us Library Micro [071]
[Caliente-] herald – NV. 1928-68 [wkly] – 20r – 1 – $1200.00 – mf#U04430 – us Library Micro [071]
[Caliente-] lincoln county record – NV. 1900-1905; 1926-1931; 1968- – 27r – 1 – $1620.00 (subs $50y) – mf#UN04663 – us Library Micro [071]
[Caliente-] lode-express – NV. 1905; 1906-08 – 1r – 1 – $60.00 – mf#U04431 – us Library Micro [071]
[Caliente-] the caliente progress – NV. 1904 – 1r – 1 – $60.00 – mf#U04833 – us Library Micro [071]
[Caliente-] the prospector – NV. 1909-13 [wkly] – 7r – 1 – $420.00 – mf#U04432 – us Library Micro [071]
[Caliente-] weekly news – NV. 1920-25 – 1r – 1 – $60.00 – mf#U04433 – us Library Micro [071]
Calife de bagdad / Saint-Just – Paris, France. 1812 – 1r – us UF Libraries [440]
California – Beverly Hills. 1981-1991 (1) 1981-1991 (5) 1981-1991 (9) – ISSN: 0747-4563 – mf#12436,01 – us UMI ProQuest [073]

California / Evangelical Mission Covenant Association of California – 1944 mar 30/1944 sep 12-1956 aug 2/1958 jun 12 – 8r – 1 – (with gaps; cont: missionstidningen california) – mf#921521 – us WHS [242]

California – Rio de Janeiro, RJ: Typ Brasiliense, 30 mar-05 abr 1849 – (= ser Ps 19) – mf#P14,02,34 n01 – bl Biblioteca [321]

California : session laws of american states and territories – 1849-1994 – 9 – $5397.00 set – mf#402530 – us Hein [348]

California : west's annotated california code – St Paul: West Publ Co, 1954-apr 2002 update – 9 – $11,464.00 set – mf#400961 – us Hein [348]

California *see*
- Coffey's probate reports
- Labatt's district reports
- Myrick's probate decisions
- Ragland's superior court decisions
- Reports and opinions
- Reports, post-nrs
- Reports, pre-nrs

California academy of sciences proceedings – San Francisco. 1974-1988 (1) 1974-1988 (5) 1974-1988 (9) – mf#9008 – us UMI ProQuest [500]

California academy of sciences vascular plant type collection – [mf ed Chadwyck-Healey] – 263mf – 9 – (families can be purchased separately. with ind) – uk Chadwyck [580]

California activist '83 : bulletin of the libertarian party of california / Libertarian Party of California – 1983 apr-1984 nov – 1r – 1 – mf#1771915 – us WHS [325]

California administrative code – 1945-79 – 9 – $8500.00 set – mf#401121 – us Hein [348]

California advocate – Fresno CA. 1975 jan 5/1978 apr 28 [scat iss]-1998 jan 16/dec 11 – 11r – 1 – (with gaps) – mf#873364 – us WHS [071]

California afl-cio news / California Labor Federation, AFL-CIO – 1975 mar 7-1978 jun 2, 1978 jun 9-1984, 1985 jan 11-1989 jun – 3r – 1 – (cont: weekly news letter, california labor federation, afl-cio; cont by: california labor news) – mf#593450 – us WHS [331]

California aft teacher / California Federation of Teachers – 1973 sep-dec, 1974 mar-may, dec, 1975 jan, apr, jun, sep-dec – 1r – 1 – (cont by: california teacher) – mf#989718 – us WHS [370]

California appellate court briefs – 3rd series: v1-235. 4th series: v1-10 – 9 – $14,500.00 set 3rd ser. $800.00 4th ser ($80.00v update service) – mf#B50570 3rd ser. B50571 4th ser – us Library Micro [347]

California appellate reports – 1st series: v1-72. 1905-25 – 66mf (1:42) 500mf (1:24) – 9 – $1047.00 – (vols after v40 will be filmed as they fall out of copyright) – mf#LLMC 84-124 – us LLMC [340]

The california architect and building news – 1879-99 – 5r – 1 – £250.00 – mf#CBN – uk World [720]

The california architect and building news – San Francisco, CA. 1879-99 – 5r – 1 – $250.00 – (aka: california architect and building review; quarterly architect) – mf#B40303 – us Library Micro [720]

California architect and building review *see*
- The california architect and building news
- [San francisco-] california architect and building news

California Association for Health, Physical Education and Recreation *see* Cahper journal

California Association for Health, Physical Education, and Recreation *see* Cahper journal times

California Association for Health, Physical Education, Recreation, and Dance *see* Cahper journal times

California attorney general reports and opinions – 1852-2001 – 9 – $3006.00 – (reports 1852-1958 ind + tables 1953-72 on reel $1785. 1943-2001 on mf $1221) – mf#408140 – us Hein [348]

California bar journal – v1-2001. 1994-2001 – 9 – $183.00 set – mf#115931 – us Hein [340]

California birds – Del Mar. 1970-1972 (1) – (cont by: western birds) – ISSN: 0045-3897 – mf#7318 – us UMI ProQuest [590]

California Black Faculty and Staff Association *see* Cbfsa news

California business – Los Angeles. 1979-1990 (1,5,9) – ISSN: 0008-0926 – mf#12382 – us UMI ProQuest [338]

California census of 1852 : counties of sierra, solano, trinity and tulare – 1r – 1 – $50.00 – mf#B50004 – us Library Micro [317]

California Center for Research and Education in Government *see* Tax revolt digest

California Central Coast Genealogical Society *see*
- Bulletin of the california...
- Bulletin...of san luis obispo county california

California Central Coast Genealogical Society bulletin – v1 n1,3-5 [1968 jan, mar-may] – 1r – 1 – (cont by: bulletin of the california central coast geological society) – mf#1806952 – us WHS [929]

California city and county directories : 1852 – present – 1 – $50.00r – (individual counties listed separately) – us Library Micro [978]

California city press – California City, CA. 1966-1967 (1) – mf#62108 – us UMI ProQuest [071]

California county agricultural commission reports, 1920-1981 – CA. 1920-81 – 9 – $300.00 – (years vary for each county) – mf#B50524 – us Library Micro [630]

California cpa / California Society of Certified Public Accountants – Redwood City. 2000+ (1) 2000+ (5) 2000+ (9) – (cont: outlook) – ISSN: 1530-4035 – mf#7218,02 – us UMI ProQuest [650]

California cpa quarterly – Palo Alto. 1972-1980 (1) 1972-1980 (5) 1975-1980 (9) – (cont by: outlook) – ISSN: 0008-0934 – mf#7218 – us UMI ProQuest [650]

California democrat *see* [San francisco-] california staats zeitung

California demokrat – San Francisco CA (USA), 1923 7 jan-1939 15 dec [gaps] – 6r – 1 – gw Misc Inst [071]

California. Dept. of Employment *see* Research series bulletins

California. Dept. of Industrial Relations *see* Report

California. Dept. of Insurance *see* Report of the insurance commissioner of california

California. Dept. of Savings and Loan *see* Report: savings and loan commissioner of california

California. Division of Industrial Welfare *see* Reports: industrial welfare commissioners

California education – Sacramento. 1963-1966 – 1 – ISSN: 0575-5603 – mf#1635 – us UMI ProQuest [370]

California election returns compilation by district of all primary and general election returns federal state and county – 1849-may 1916 – 24r – 1 – $1200.00 – mf#B50520 – us Library Micro [340]

California elementary administrator – Burlingame. 1969-1971 – 1 – ISSN: 0008-1019 – mf#2151 – us UMI ProQuest [370]

California. Employment Development Dept *see* Operations reports

California. Fair Employment Practice Commission *see* Reports

California Federation of Teachers *see* California aft teacher

California fish and game – Long Beach. 1972+ (1) 1972+ (5) 1975+ (9) – ISSN: 0008-1078 – mf#6385 – us UMI ProQuest [639]

California. Franchise Tax Board *see* Reports

California freie presse – San Francisco CA [USA], 1972-82 – 1 – (cont: amerika-woche, chicago) – gw Misc Inst [071]

California geology – Sacramento. 1972+ (1) 1972+ (5) 1975+ (9) – ISSN: 0026-4555 – mf#6937 – us UMI ProQuest [550]

California governor's commission on the los angeles riots [mccone report watts riots] – 1966 – 1 – $270.00 – mf#0130 – us Brook [303]

California gp – San Francisco. 1972-1973 (1) 1972-1973 (5) (9) – ISSN: 0410-2894 – mf#6579 – us UMI ProQuest [610]

The california great register of voters indexes, 1900-1944 – California State Library – 1 – $60.00r (1-20r) $55.00r (21r or more) (Entire coll – inquire for price) – (coll of over 1700v contains records of all 58 california counties. individual counties also listed separately) – mf#R03720 – us Library Micro [917]

California historian – Carmel, CA. 1954-1970 (1) – mf#61968 – us UMI ProQuest [071]

California historical courier – 1973 jul/aug-1987 sep/oct – 1r – 1 – (cont: notes [california historical society]; cont by: california chronicle) – mf#1518990 – us WHS [978]

California historical materials concerning the negro baptist – 1899-1964. 194p – 1 – 6.79 – us Southern Baptist [242]

California history center foundation newsletter / De Anza College – v1 iss 1-v3 iss 2 [1977 fall-1980 win] – 1r – 1 – (cont: trianon reflections; cont by: californian) – mf#669259 – us WHS [978]

California indian herald – 1923-24 – (= ser American indian periodicals... 1) – 2mf – 9 – $95.00 – us UPA [305]

California indian legal services newsletter – v9 n1-v10 n1 [1979 dec-1980 jul] – 1r – 1 – mf#678889 – us WHS [340]

California industry – San Francisco. 1942-1972 (1) 1972-1972 (5) (9) – ISSN: 0043-390X – mf#7163 – us UMI ProQuest [600]

California Institute of International Studies *see* World affairs report

California institute of international studies report – Stanford. 1970-1972 (1) – (cont by: world affairs report) – ISSN: 0068-564X – mf#8213 – us UMI ProQuest [327]

California jewish life – Los Angeles. Calif. 1950-51 – 1 – us AJPC [071]

California jewish press – Los Angeles. Calif. 1957-62 – 1 – us AJPC [071]

California jewish review – Los Angeles, CA.1922-29 – 1 – us AJPC [071]

California jewish review *see* [Los angeles-] california jewish bulletin

California jewish voice – Los Angeles. Calif. 1960 67 – 1 – us AJPC [071]

California journal – San Francisco CA (USA), 1920 20 feb-1940 2 may [many gaps] – 6r – 1 – gw Misc Inst [071]

California journal of educational research – San Francisco. 1950-1975 (1) 1971-1975 (5) – ISSN: 0008-1213 – mf#2188 – us UMI ProQuest [370]

California journal of elementary education – Sacramento. 1932-1963 – 1 – mf#967 – us UMI ProQuest [370]

California journal of teacher education (cjte) – Claremont. 1976-1983 – 1,5,9 – (cont by: teacher education quarterly) – ISSN: 0278-6052 – mf#11026 – us UMI ProQuest [370]

California Labor Federation, AFL-CIO *see*
- California afl-cio news
- Force for progress
- Labor legislation
- Sacramento story

California law journal and literary review – v1-2. 1862-63 (all publ) – 9 – mf#LLMC 82-910 – us LLMC [340]

California law review – Berkeley. 1985+ (1,5,9) – ISSN: 0008-1221 – mf#15673 – us UMI ProQuest [340]

California law review – v1-14. 1912-1925/26 (all publ) – 103mf – 9 – $154.00 – mf#LLMC 90-317 – us LLMC [340]

California law review – v1-88. 1912-2000 – 1,5,6,9 – $2592.00 set – (v1-84 1912-96 in reel $2425. v 85-88 1997-2000 in mf $167) – ISSN: 0008-1221 – mf#101261 – us Hein [340]

California lawyer – v1-21. 1981-2001 – 9 – $563.00 – (cont: california state bar journal) – ISSN: 0279-4063 – mf#101911 – us Hein [340]

California lawyer / State Bar of California – San Francisco. 1987-1991 (1) 1987-1991 (5) 1987-1991 (9) – (cont: california state bar journal) – ISSN: 0279-4063 – mf#14109 – us UMI ProQuest [340]

California legal studies journal – v1-11 (1985-95) – 9 – $55.00 set – (v8 never publ) – mf#112281 – us Hein [340]

California legionaire : official publication of the american legion, department of california – 1929 nov-1935 feb, 1935 mar-1938 dec 15 – 2r – 1 – (cont: legion news) – mf#702188 – us WHS [350]

California legislative bills – 1963- – over 1600mf ea 2yr session – 9,5 ($950.00/update B50513) – us Library Micro [323]

California legislature selected reports – Sacramento. 1913-1964 (1) – mf#3119 – us UMI ProQuest [350]

California librarian – Sacramento. 1939-1978 (1) 1973-1978 (5) 1975-1978 (9) – ISSN: 0008-123X – mf#7009 – us UMI ProQuest [020]

California. (Lower). Southern territory *see* Boletin oficial

California management review – Berkeley. 1958+ (1) 1970+ (5) 1975+ (9) – ISSN: 0008-1256 – mf#1578 – us UMI ProQuest [650]

California Medical Association *see* California medicine

California Medical Facility *see* Speaking leaves

California medicine / California Medical Association – San Francisco. 1902-1973 (1) 1971-1973 (5) – (cont by: western journal of medicine) – ISSN: 0008-1264 – mf#2239 – us UMI ProQuest [610]

California missionary baptist – 1 May 1940-15 Apr 1973, 15 Jun 1973 – 1 – 54.04 – us Southern Baptist [242]

California monitor of education – v8 n5-v9 n10[1985 jan-1986 jun] – 1r – 1 – (cont by: national monitor of education) – mf#1123052 – us WHS [370]

California nurse – San Francisco. 1983+ (1,5,9) – ISSN: 0008-1310 – mf#14136,07 – us UMI ProQuest [610]

California Office Of State Engineer *see* Irrigation development

California oil fields – San Francisco. 1949-1950 (1) – mf#325 – us UMI ProQuest [622]

California oil worker – Bakersfield, CA. 1922-1922 (1) – mf#62088 – us UMI ProQuest [071]

California oil worker : official newspaper / Union Oil Workers of California, District Council No 1 – v5 n25-29, 33-v6 n21 [1923 sep 6-oct 4, nov 8-1924 may 22] – 1r – 1 – (cont by: oil worker [long beach ca: 1924]) – mf#1008422 – us WHS [622]

California, oregon, washington – 1899 – (= ser California telephone directory coll) – 1r – 1 – $50.00 – mf#P00145 – us Library Micro [917]

A california pilgrimage / Bisbee, Frederick Adelbert – Boston: Murray Press, 1915 [mf ed 1991] – (= ser Unitarian/universalist coll) – 1mf – 9 – 0-524-01714-X – mf#1990-4106 – us ATLA [917]

California. Pooled Money Investment Board *see* – Report

California post – Fresno CA (USA), 1920 13 may-1924 1 may, 1925 19 nov-1927 3 nov – 2r – 1 – gw Misc Inst [071]

California preservation – v7 n1-v8 n3 [1982 spr-1984 jul] – 1r – 1 – (cont: newsletter [californians for preservation action]) – mf#668294 – us WHS [978]

California presse – Los Angeles CA (USA), 1926 8 jan-1930 24 apr – 2r – 1 – gw Misc Inst [071]

California. Public Utilities Commission *see*
- Decisions
- Reports

California school employee – San Jose. 1973-1974 – 1 – ISSN: 0008-1515 – mf#8590 – us UMI ProQuest [370]

California school libraries – Burlingame. 1929-1977 [1]; 1970-1977 [5]; 1976-1977 [9] – ISSN: 0008-1523 – mf#2007 – us UMI ProQuest [020]

California School Library Association *see* Journal – california school library association

California schools – Sacramento. 1930-1963 – 1 – mf#968 – us UMI ProQuest [370]

California senate and assembly daily journals : legislature of the state of california – 1850- – 9 – $1150.00 per update – (updates every two yr) – mf#B50582- – us Library Micro [325]

California serials collection : 1800's-present – 1995 – 1 – apply for price – (a collaboration with california state library and local california libraries to preserve a piece of california's history and provide a source for researchers. project is entitled "preserving the dream". available as a coll or on a per reel basis) – us Library Micro [978]

California social democrat – v1-6 n1-263. 1911-16 [all publ] – (= ser Radical periodicals in the united states, 1881-1960. series 2) – 1r – 1 – $200.00 – us UPA [325]

California socialist – 1973 jun-1985 apr – 1r – 1 – (cont by: socialist tribune [milwaukee [wi]]; california socialist and socialist tribune) – mf#583615 – us WHS [335]

California socialist and socialist tribune – v9 n4-5 [1985 may-jun=whole n122-123] – 1r – 1 – (cont: california socialist; cont by: socialist [los angeles ca]) – mf#1122404 – us WHS [071]

California Society of Certified Public Accountants *see* California cpa

California Society of Friends of Russian Freedom *see* Russian review

California southern baptist – Nov 1941-1991 – 1 – $1,086.16 – us Southern Baptist [242]

California staats-zeitung – Los Angeles CA (USA), 1917 3 may-1918 4 apr, 1920 27 feb-1938 8 jul [gaps], 1972- – 1 – gw Misc Inst [071]

California staats-zeitung – San Francisco CA (USA), 1917 3 may-1918 4 apr – 1 – gw Misc Inst [071]

The california state assembly file analysis – 1975-2000 – 9 – $600.00 set ($50.00/update B50543) – mf#B50542 – us Library Micro [324]

California. State Banking Dept. Superintendant of Banks *see* Report

California. State Bar Association *see* Proceedings

California state bar journal / State Bar of California – Los Angeles. 1975-1981 (1,5,9) – (cont by: california lawyer) – ISSN: 0161-9241 – mf#10530 – us UMI ProQuest [340]

California state bar journal – v1-56. 1926-1981 [all publ] – 9 – $688.00 set – (title varies: v1-17 1926-1942 as state bar journal of the state bar of california; v18-46 1942-1971 as journal of the state bar of california; cont as: california lawyer) – mf#101271 – us Hein [340]

California. State Board of Equalization *see* Reports

California State College [PA] *see* Southwestern pennsylvania

California. State Conciliation Service *see* Adjustment of labor-management disputes in california

California State Council of Cannery Unions, Teamsters *see* Afl cannery reporter, 1949-1954 / united dairy farmer, 1941, 1944-1945

California state employee – 1967 sep 15-1970 jun 12, 1970 jun 26-1972 dec 22, 1973 jan 12-1975 dec 24, 1976 jan 21-1979 dec 19, v51 1-v57 6 [1980 jan 16-1986 dec] – 5r – 1 – (cont: bulletin [california state employees association]; cont by: california pride) – mf#1053954 – us WHS [350]

California State Govt *see* Teachers guide to the education of spanish speaking children

California State Grange *see* Journal of proceedings of the...annual session of the california state grange, patrons of husbandry

California State Library *see* News notes of california libraries

California State Numismatic Association *see* Calcoin news

California State Prison at San Quentin *see* San quentin news

California State University, Northridge *see* Popo

California statesman – 1973 feb-1974 oct, 1975 aug – 1r – 1 – mf#355241 – us WHS [071]

California. Superior Court. San Francisco Probate Dept *see* Reports of decisions in probate

California. Supreme Court *see*
- California supreme court reports
- California unreported cases
- Late political decisions.

California supreme court briefs – 3rd series: v1-54. oct 1969-94. 4th series: v1- – 9 – $4000.00 set 3rd ser. $95.00v 4th ser – (update and addendum service. incl special index) – mf#B50535 3rd ser. B50536 4th ser – us Library Micro [347]

California supreme court reports / California. Supreme Court – v1-198. 1850-1926 – 1798mf – 9 – $2697.00 – (pre-nrs: v1-63 1850-83 497mf $745.00. updates planned) – mf#LLMC 80-800 – us LLMC [347]

California teacher – 1976 feb-1984 dec – 1r – 1 – (cont: california aft teacher) – mf#999596 – us WHS [370]

California Teachers' Association *see* Cta journal

California Teamsters Public Affairs Council *see* Teamstergrams

California tomorrow – San Francisco. 1982-1983 (1) 1982-1983 (5) 1982-1983 (9) – (cont: cry california) – ISSN: 0744-8686 – mf#6916,01 – us UMI ProQuest [639]

California university chronicle – Berkeley. 1898-1906 – 1 – mf#5708 – us UMI ProQuest [378]

California university. publications in botany – v1-14. 1902-29 – 1 – $216.00 – mf#0132 – us Brook [580]

California university. publications in zoology – v1-20. 1902-23 – 9 – $990.00 – mf#0131 – us Brook [590]

California unreported cases / California. Supreme Court – v1-7 (all publ) – 74mf – 9 – $111.00 – (a pre-nrs title) – mf#LLMC 90-003 – us LLMC [340]

California Urban Indian Health Council *see*
- Clinician's letter
- Indian alcohol times

California v savio, et al / Meiklejohn Civil Liberties Library – 1964-67 – 1 – us AMS Press [321]

California veckoblad – Los Angeles CA. 1912 mar 8/1914 may 22-1957 apr 4/1958 dec 25 – 22r – 1 – mf#851233 – us WHS [071]

California veteran – v1 n2 [veterans day iss], v1 n2 [veterans day iss] – 2r – 1 – mf#720855 – us WHS [305]

California veterinarian – Sacramento. 1947-1981 (1) 1972-1981 (5) 1975-1981 (9) – ISSN: 0008-1612 – mf#7221 – us UMI ProQuest [636]

California voice – Berkeley etc CA. 1973 dec 13, 1974 jan 3-10,24 – 1r – 1 – mf#897059 – us WHS [071]

California voice – San Francisco, CA. 1978-1986 (1) – mf#67984 – us UMI ProQuest [071]

California voice *see* [Berkeley-] voice

California vorwaerts – Fresno CA (USA), 1922 2 aug-1938 25 aug [gaps] – 4r – 1 – gw Misc Inst [071]

California weckruf – Los Angeles CA (USA), 1936 13 jul-1937 9 dec [gaps] – 1r – 1 – gw Misc Inst [071]

California western international law journal – v1-31. 1970-2001 – 1,5,6 – $425.00 set – (v1-24 1970-94 in reel $297. v25-31 1994-2001 in mf $128) – ISSN: 0886-3210 – mf#101281 – us Hein [341]

California western law review – v1-37. 1965-2001 – 1,5,6 – $574.00 set – (v1-30 1965-94 in reel $445. v31-37 1994-2001 in mf $129.) – ISSN: 0008-1639 – mf#101291 – us Hein [340]

California wild – San Francisco. 1997+ (1) 1997+ (5) 1997+ (9) – (cont: pacific discovery) – ISSN: 1094-365X – mf#6859,01 – us UMI ProQuest [500]

California youth authority quarterly – Sacramento. 1977-1981 (1,5,9) – ISSN: 0008-1671 – mf#11537 – us UMI ProQuest [350]

Californian – Carmel, CA. 1936-1937 (1) – mf#62112 – us UMI ProQuest [071]

Californian – San Francisco: Burton H Wolfe. v1 n1 jan 1960-v3 n6 jun 1962 (mf ed 1963) – 1r – 1 – (cont by: american liberal. former title: california liberal jan-feb 1960) – mf#MN *ZAN-3181 – us NY Public [071]

Californian – 1846 aug 29-1848 sep 9, 1848 mar 15 – 2r – 1 – (cont by: california star; californa star and californian) – mf#840802 – us WHS [071]

Californian : a western monthly magazine – San Francisco. 1880-1882 – 1 – mf#5271 – us UMI ProQuest [073]

Californian homes for educated englishmen : a practical suggestion for a model colony: congenial english society, a glorious climate, lovely scenery, and the most fertile of soils / Binney, Frederick Altona – London, 1875 – (= ser 19th c books on british colonization) – 1mf – 9 – mf#1.1.8207 – uk Chadwyck [950]

Californian illustrated magazine – San Francisco. 1891-1894 (1) – mf#5272 – us UMI ProQuest [978]

California-Nevada Token Society *see* Cal-neva token ledger

Californians for Preservation Action *see* Newsletter

California's health – Sacramento. 1943-1973 (1) – ISSN: 0008-168X – mf#6790 – us UMI ProQuest [360]

La californie et les routes interoceaniques / Holynski, Aleksander – Bruxelles 1853 [mf ed Hildesheim 1995-98] – 1v on 3mf – 9 – €90.00 – 3-487-27158-3 – gw Olms [917]

Les californies, l'oregon, et l'amerique russe / Denis, Ferdinand – [s.l: s.n, 1849?] [mf ed 1984] – 2mf – 9 – 0-665-44854-6 – (incl app and bibl ref) – mf#44854 – cn Canadiana [917]

Calinich, Hermann Julius Robert *see* Luther und die augsburgische confession

[Calipatria-] calipatria herald – CA. mar 1920-apr 1922 – 1r – 1 – $60.00 – mf#R02088 – us Library Micro [071]

The caliph haroun alraschid and saracen civilization / Palmer, Edward Henry – New York: G.P. Putnam's Sons, 1881. 228p. fold. geneal. tab – 1 – us Wisconsin U Libr [900]

The caliphate : its rise, decline, and fall / Muir, William – new and rev ed. Edinburgh: J Grant, 1915 – 2mf – 9 – 0-524-02026-4 – (incl bibliographic references) – mf#1990-2801 – us ATLA [956]

Calippe, Charles *see* L'education chretienne de la democratie

Calispell valley times – Cusick, WA. 1909-1910 (1) – mf#66981 – us UMI ProQuest [071]

[Calistoga-] independent calistogian – CA. dec 26 1877-dec 1892; dec 1894-aug 1896 (wkly) – 5r – 1 – $300.00 – mf#B02089 – us Library Micro [071]

[Calistoga-] the weekly calistogan – CA. sep 11 1896-dec 1961; 1974-77; 1989- – 42+ r – 1 – $2520.00 (subs $50/y) – mf#B02090 – us Library Micro [071]

Calixte, Demosthenes Petrus *see* Haiti

Calixte, Nyll F *see* Fort-liberte d'hier et d'aujourd'hui

Calixto, Benedicto De Jesus *see* Capitanias paulistas

Calkin, John Burgess *see*
- A brief history of canada
- A brief history of great britain
- Calkin's new introductory geography
- The geography and history of nova scotia
- Historical geography of bible lands
- History of british america
- A history of the dominion of canada
- Notes on education
- Old time customs, memories and traditions
- School geography
- School geography of the world
- The world

Calkins, Harvey Reeves *see* Ganga dass

Calkins, Mary Whiton *see*
- Association
- The metaphysical system of hobbes as contained in twelve chapters from his elements of philosophy concerning body...and human nature...and leviathan
- The persistent problems of philosophy

Calkin's new introductory geography : with outlines of history / Calkin, John Burgess – London: T Nelson; Halifax, NS: A & W MacKinlay, 1891 – (= ser Nova scotia school series) – 2mf – 9 – 0-665-92072-5 – mf#92072 – cn Canadiana [910]

Calkins, R *see* Jeremiah the prophet

Calkins, Wolcott *see*
- Keystones of faith
- Parables for our times

Call – 1919 oct 10-1920 may 10 – 1r – 1 – mf#3574958 – us WHS [071]

Call – Baltimore. 1970-1974 (1) – ISSN: 0045-4028 – mf#7978 – us UMI ProQuest [574]

Call / Communist Party [Marxist-Leninist] – 1972 oct-1976 jun 28, 1976 jul 5-1977 oct 31, 1977 nov 7-1979 jan 29, 1979 feb-dec, 1980-1982 mar/apr – 5r – 1 – mf#355242 – us WHS [335]

Call – Kansas City MS. 1972 mar 17/23, 1975 oct 31/nov 6 – 1r – 1 – (cont: kansas city call; call [kansas city mo: 1933: tulsa ed]) – mf#1880067 – us WHS [071]

Call – Drummond, MT. 1905-1906 (1) – mf#64360 – us UMI ProQuest [071]

Call – Great Falls, MT. 1921-1922 (1) – mf#64416 – us UMI ProQuest [071]

Call – Kalispell, MT. 1896-1896 (1) – mf#64501 – us UMI ProQuest [071]

Call – Kansas City, MO. 1954-1998 (1) – mf#64179 – us UMI ProQuest [071]

Call – Lafayette, IN. 1897-1901 (1) – mf#62869 – us UMI ProQuest [071]

Call – Philipsburg, MT. 1902-1903 (1) – mf#64594 – us UMI ProQuest [071]

Call – Seattle, WA. 1885-1898 (1) – mf#67100 – us UMI ProQuest [071]

Call / Servosse Association – 1928 mar 19 – 1r – 1 – mf#5012858 – us WHS [071]

Call / Students for a Democratic Society [US] – 1967 feb-1968 apr – 1r – 1 – mf#1110486 – us WHS [320]

Call – Whiting, IN. 1906-1921 (1) – mf#63005 – us UMI ProQuest [071]

Call *see*
- Central oregonian
- Crook county journal

The call – London, England. An organ of international socialism. -w. 24 Feb 1916-29 July 1920. 2 reels – 1 – uk British Libr Newspaper [072]

The call : qualifications and preparation of candidates for foreign missionary service / Speer, Robert Robert Elliott et al – New York: Student Volunteer Movement for Foreign Missions, 1901 – 1mf – 9 – 0-8370-6235-7 – mf#1986-0235 – us ATLA [240]

The call – Schuylkill, PA., 1981-1983, 1952-1985 – 13 – $25.00r – us IMR [071]

Call advertiser *see* [Norwalk-] call

Call africa 999 / Nugent, John Peer – New York, NY. 1965 – 1r – us UF Libraries [960]

The call and challenge of mongolia / Sturt, Reginald W – Glasgow: Pickering & Inglis, [1917?] [mf ed 1995] – (= ser Yale coll) – 32p (ill) – 1 – 0-524-10185-X – (foreword by g h bondfield) – mf#1995-1185 – us ATLA [951]

Call and post – Cleveland, Columbus OH. 1974 mar 9-30 – 1r – 1 – mf#904330 – us WHS [071]

Call and post : (cleveland edition) – Cleveland, OH. 1934+ (1) – mf#61023 – us UMI ProQuest [071]

Call and post – Cleveland OH. 1975 oct 4 – 1r – 1 – (cont: cleveland call and post) – mf#780636 – us WHS [071]

Call and post *see* [San francisco-] the daily morning call

Call and post (cincinnati edition) – Cleveland, OH. 1972-1999 (1) – mf#68007 – us UMI ProQuest [071]

Call and post (columbus edition) – Cleveland, OH. 1972-1999 (1) – mf#68006 – us UMI ProQuest [071]

Call and post (state edition) – Cleveland, OH. 1990+ (1) – mf#68836 – us UMI ProQuest [071]

Call and response : newsletter of the african american studies program / University of Alabama at Birmingham – 1994 win-fall, 1995 spr, 1995 fall-1996 spr/summer, 1997 fall/winter – 1r – 1 – (cont by: uab african american studies newsletter) – mf#2901371 – us WHS [305]

Call bulletin – San Francisco, CA. 1956-1959 (1) – mf#62267 – us UMI ProQuest [071]

Call bulletin *see* [San francisco-] the bulletin

Call center crm solutions – Norwalk. 1999-1999 (1,5,9) – (cont: call center solutions) – ISSN: 1529-1782 – mf#18353,03 – us UMI ProQuest [380]

Call center product news – Cleveland. 1998+ (1) – ISSN: 1098-1667 – mf#28081 – us UMI ProQuest [380]

Call center solutions – Norwalk. 1998-1999 (1,5,9) – (cont: telemarketing and call center solutions. cont by: call center crm solutions) – ISSN: 1521-0774 – mf#18353,02 – us UMI ProQuest [380]

A call for a convention to effect a national organization for the improvement of religious and moral education through the sunday school and other agencies : to be held in chicago in feb or mar 1903 – Chicago: under the auspices of the Council of Seventy...[1902?] [mf ed 1986] – (= ser Official document (council of seventy) 1) – 1mf – 9 – 0-8370-7537-8 – (repr fr: biblical world, nov 1902) – mf#1986-1537 – us ATLA [377]

Call for nmu democracy – 1966 aug-1970 jul – 1r – 1 – mf#1053962 – us WHS [320]

Call from god = Shen Chao – v7. n6 sep 1932; v9 n4 jul 1934* – (= ser Chinese christian coll 60) – 1r – 1 – (in chinese) – mf#ATLA S0296P – us ATLA [240]

A call from macedonia : a sermon preached at hingham in new-england, oct 12 1768. at the ordination of the reverend mr caleb gannet... / Gay, Ebenezer – Boston, New England: printed by Richard Draper, and Thomas & John Fleet, 1768 [mf ed 1984] – 1mf – 9 – 0-665-44276-9 – mf#44276 – cn Canadiana [240]

Call it sleep / Roth, Henry – New York, NY. 1965, c1934 – 1r – us UF Libraries [025]

Call leader – Elwood, IN. 1995-2000 (1) – mf#61381 – us UMI ProQuest [071]

Call, Nancy June *see* Symbolic use of color in the novels of miguel ange...

The call of a child *see* Tung hsin te hu huan (ccm83)

A call of attention to the behaists or babists of america / Stenstrand, August J – [s.l: s.n, 1907?] [mf ed 1992] – 1mf – 9 – 0-524-02615-7 – mf#1990-3065 – us ATLA [290]

The call of cathay : a study in missionary work and opportunity in china old and new / Cornaby, William Arthur – London: Wesleyan Methodist Missionary Society, 1910 – 1mf – 9 – 0-8370-6483-X – (incl ind) – mf#1986-0483 – us ATLA [240]

The call of india : a study in conditions, methods and opportunities of missionary work among hindus / Thompson, Edgar Wesley – London: Wesleyan Methodist Missionary Society, 1912 [mf ed 1995] – (= ser Yale coll) – xv/319p (ill) – 1 – 0-524-09998-7 – mf#1995-0998 – us ATLA [240]

The call of korea : political, social, religious / Underwood, Horace Grant – New York: Fleming H Revell, c1908 – 1mf – 9 – 0-8370-6433-3 – (includes bibliographies) – mf#1986-0433 – us ATLA [951]

Call of the carpenter / White, Bouck – Garden City, NY. 1915 – 1r – us UF Libraries [960]

The call of the christ : a study of the challenge of jesus to the present century / Willett, Herbert Lockwood – New York: Fleming H Revell, c1912 – 1mf – 9 – 0-524-06565-9 – mf#1991-2649 – us ATLA [240]

Call of the church to women of leisure / Muir, Pearson Madam – Edinburgh, Scotland. 1894 – 1r – us UF Libraries [240]

The call of the east : sketches from the history of the irish mission to manchuria, 1869-1919 / O'Neill, Frederick William Scott – London: James Clarke, [1919] [mf ed 1995] – (= ser Yale coll) – 129p (ill) – 1 – 0-524-09826-3 – mf#1995-0826 – us ATLA [240]

The call of the harvest / McKay, Charles L – Convention Press. 1956 – 1 – 5.00 – us Southern Baptist [242]

The call of the home land : a study in home missions / Phillips, Alexander Lacy – 3rd ed., rev. and enl. Richmond, Va.: Presbyterian Committee of Publication, 1910, c1906 – 1mf – 9 – 0-8370-6596-8 – (includes bibliographies and index) – mf#1986-0596 – us ATLA [240]

The call of the minaret / Cragg, Kenneth – New York, 1956 – 7mf – 8 – €15.00 – ne Slangenburg [260]

The call of the new day to the old church / Stelzle, Charles – New York: Fleming H Revell, c1915 – 1mf – 9 – 0-524-03050-2 – mf#1990-0807 – us ATLA [240]

The call of the new era : its opportunities and responsibilities / Muir, William – London: Morgan & Scott, 1910 – 1mf – 9 – 0-8370-6223-3 – (incl ind) – mf#1986-0223 – us ATLA [240]

The call of the pacific / Burton, J W – London: Charles H Kelly, [1912] [mf ed 1995] – (= ser Yale coll) – xiv/286p (ill) – 1 – 0-524-09547-7 – mf#1995-0547 – us ATLA [980]

Call to christians to consider their ways / Sheppard, Henry W – London, England. 1840 – 1r – us UF Libraries [240]

A call to commitment : fathers' involvement in children's learning / US. Dept of Education and US Dept of Health and Human Services – [Washington DC]: Partnership for Family Involvement in Education, US Dept of Education, Office of Educ Research & Improvement...[c2000] [mf ed 2000] – 1mf – 9 – (incl bibl ref) – us GPO [370]

A call to prayer / Simpson, Albert B – South Nyack, NY: Christian Alliance Pub Co, c1897 [mf ed 1992] – (= ser Christian & missionary alliance coll) – 1mf – 9 – 0-524-03738-8 – mf#1990-4843 – us ATLA [240]

Call to the christians and the hebrews / Theaetetus – London, England. 1819 – 1r – us UF Libraries [240]

Call to union on the principles of the english reformation / Hook, Walter Farquhar – London, England. 1839 – 1r – us UF Libraries [242]

The call to young india / Lajpat Rai, Lala – Madras: S Ganesan & Co, [1920?] – (= ser Samp: indian books) – us CRL [954]

Call up! / Soldiers for Democratc Action – v1 n1 [1970 sep 26] – 2r – 1 – mf#720854 – us WHS [355]

A call upon the unemployed talent of the church : a sermon in behalf of the american sunday-school union / Potts, George – Philadelphia: American Sunday-School Union [1853?] [mf ed 1993] – 1mf – 9 – 0-524-08497-1 – mf#1993-3142 – us ATLA [240]

Callage, Fernando *see* Sociologia catholica e o materialismo

Callaghan, Thomas *see* Acts and ordinances of the governor and council of new south wales

Callaloo – Baltimore. 1989+ (1,5,9) – ISSN: 0161-2492 – mf#18005 – us UMI ProQuest [400]

Callan, Edward *see*
- Albert john luthuli and the south african race conflict

Callander advertiser – [Scotland] Stirling: Duncan & Jamieson jan 1912-dec 1950 (wkly) [mf ed 2004] – 52r – 1 – (ceased dec 31 1971; cont: callander advertiser and trossachs, strathyre, lochearnhead, killin, dalmally and oban observer; absorbed by: stirling observer) – uk Newsplan [072]

Callander advertiser – Stirling: Duncan & Jamieson 1893-1971 (wkly) [mf ed 1893-1911] – 1 – (ceased dec 31 1971; cont: callander advertiser & trossachs, strathyre, lochearnhead, killin, dalmally & oban observer; absorbed by: stirling observer) – uk Scotland NatLib; uk Newsplan [072]
Callander advertiser and trossachs, strathyre, killin, dalmally and oban observer – Stirling: Duncan & Jamieson 1884-1993 (wkly) – 1 – (cont by: callander advertiser) – uk Scotland NatLib [072]
Callaway courier – Callaway, NE: Geo B Mair. -v17 n36. feb 24 1905 (wkly) [mf ed v6 n4. jun 24 1892, 1894-97, 1900-05 (gaps)] – 4r – 1 – (merged with: weekly tribune (1903) to form: courier-tribune) – us NE Hist [071]
The callaway courier – Callaway, NE: M L Chaloupka. v1 n1. mar 27 1968- (wkly) [mf ed filmed 1977-] – 1 – us NE Hist [071]
Callaway, Godfrey *see*
– Fellowship of the veld
– Pioneers in pondoland
Callaway, Henry *see*
– Nursery tales, traditions, and histories of the zulus
– Religious system of the amazulu in the zulu language
Callaway, John *see*
– Hints on the cingalese and english languages
– A school dictionary
Callaway Jr., M *see* Studies in the syntax of the lindisfarne gospels
The callaway standard – Callaway, NE: C A Sherwood (wkly) [mf ed v2 n1. aug 18 1887-oct 11 1888 (gaps) filmed 1999] – 1r – 1 – us NE Hist [071]
Callaway weekly tribune – Callaway, NE: [Frank W Conly] 2v. v18 whole n19. feb 8 1902-v19 whole n18. feb 7 1903 (wkly) [mf ed with gaps] – 1r – 1 – (cont: weekly tribune. cont by: weekly tribune (1903)) – us NE Hist [071]
Call-bulletin – San Francisco CA. v161 n108 [1937 may 26] – 1r – 1 – (cont: san francisco call; san francisco bulletin) – mf#846293 – us WHS [071]
Callcott, Maria (Dundas) Graham *see* Memoirs of the life of nicholas poussin
Callcott, Maria (Dundas) Graham, lady *see*
– Continuation of essays towards the history of painting
– Description of the chapel of the annunziata dell'arena...padua
– Essays towards a history of painting
Callcott, Wilfrid Hardy *see* Caribbean policy of the united states, 1890-1920
Calle oscura / Ozores, Renato – Panama, 1955 – 1r – us UF Libraries [972]
Calle Restrepo, Arturo *see* Conflictos familiares y problemas humanos
Called of god / Davidson, Andrew Bruce – New York: Scribner, (1902). 1 fiche – 9 – us ATLA [240]
The called of god / Davidson, A B; ed by Paterson, James Alexander – New York: Imported by Charles Scribner, [ca. 1902] – 1mf – 9 – 0-8370-2832-9 – mf#1985-0832 – us ATLA [220]
Callejero y guia historica de badajoz / Lopez e Sosoaga y Borinaga, Benigno – Badajoz: La Minerva Extremena, 1963 – sp Bibl Santa Ana [946]
Callejo, Carlos *see* El monasterio de guadalupe
Callejo Serrano, Carlos *see*
– La arqueologia de norba cesarina
– Caceres monumental
– Catalogo de las pinturas de la cueva de maltravieso
– Cedulas epigraficas del campo norbense
– Nuevo repertorio epigrafico de la provincia de caceres
– El origen y el nombre de caceres
[Callender 1] fayssoux collection of william walker papers, 1856-1860 : the only extant documents covering william walkers controversial activities in latin america – [mf ed Norman Ross Publ] – 4r – 1 – (with p/g) – Latin American Library, Tulane University – us UMI ProQuest [972]
Callender, Clarence N *see* Selected cases on contracts
Callender, John *see* An historical discourse on the civil and religious affairs of the colony of rhode-island
Calles de la habana – Habana, Cuba. 1936 – 1r – us UF Libraries [972]
Calles Mariscal, Alfredo *see* Ganado porcino extremeno
Calles Mariscal, Juan *see* Ganado porcino extremeno
Callewaert, C *see* Sacris erudiri
Callie self baptist church. greenwood county. south carolina : church records – 1942-71 – 1 – us Southern Baptist [242]
Callimachus, P *see* ...De bello turcis inferendo, oratio grauissima...
Calling our nation / Aryan Nations-Teutonic Unity Publ et al – n38-67, n4-10,12-14,16-44 [1978-84] – 2r – 1 – mf#718414 – us WHS [243]
Callinicus / Haldane, J B S – New York, NY. 1925 – 1r – us UF Libraries [500]
Calliope – Albion, NE: Calliope Print Co, dec 6 1895 (wkly) – 1r – 1 – (absorbed by: boone county outlook) – us Bell [071]
The calliope – [Trois Rivieres, Quebec: s.n, 1859] – 9 – mf#P04887 – cn Canadiana [073]
The calliopean – Hamilton, C W [Ont]: P Ruthven, [1847-18-?] – 9 – mf#P05963 – cn Canadiana [640]
Callisen, Adolph Carl Peter *see* Medicnisches schriftsteller-lexikon der jetzt lebenden aerzte, wundaerzte und geburtshelfer, apotheker und naturforscher aller gebildeten voelker (ael3/18)
Callista : a tale of the third century / Newman, John Henry – new ed. London, New York: Longmans, Green, 1895 [mf ed 1991] – (= ser The works of cardinal newman) – 1mf – 9 – 0-7905-8534-0 – (1st printed 1856) – mf#1989-1759 – us ATLA [830]
Calliste *see* Etude sur les origines de la penitence chretienne
Callsign index to non-government master frequency data base / U.S. National Technical Information Service – Annual. Includes two suppl. issued in May and Oct of each year – 9 – us NTIS [000]
Callwell, Jm *see* Little curiosity
Calm, Felix *see* Richard wagners "ring des nibelungen"
A calm review of the inaugural address of prof charles a briggs / Morris, Edward D – 1891 – 9 – $50.00 – us Presbyterian [240]
Calma la pena amara : sung by sigr rubinelli, in the opera of armida / Mortellari, Michele – London: Longman & Broderip [1786] [mf ed 19–] – 1r – 1 – (in italian) – mf#pres. film 73 – us Sibley [780]
Calman, A L *see* Life and labors of john ashworth 1871-1875
Calman, David *see* Calman's code time-table
Calman's code time-table / Calman, David – 3rd ed. New York: Strouse, 1891. 143p. LL-622 – 1 – us L of C Photodup [348]
Calmes, Th *see*
– Comment se sont formes les evangiles
– L'evangile selon saint jean
– Qu'est-ce que l'ecriture sainte?
Calmet, Aug *see*
– Commentaire litteral, historique et moral
– Commentarius literalis, historico-moralis in regulam s p benedicti
Calmeyn, Maurice *see* Au congo belge
Calmon, Pedro *see*
– Crime de antonio vieira
– Espirito da sociedade colonial
– Estado e o direito n'os lusiadas
– Figuras de azulejo, perfis e cenas da historia do...
– Gomes carneiro
– Historia da bahia
– Historia da casa da torre
– Historia da civilizacao brasileira
– Historia da fundacao da bahia
– Historia de castro alves
– Historia de la civilizacion brasilena
– Historia diplomatica do brasil
– Historia do brasil
– Historia do brasil na poesia do povo
– Pequena historia da civilizacao brasileira para e...
– Rei filosofo
– Segredo das minas de prata
Cal-mont news / Boone Family Association of Cal-Mont in Missouri – v1 n1,3-4,11-20,22-23,25-36, v2 n1-3 [1979 aug, 1980 feb-may, 1982 feb-1984 may, 1984 nov-1985 feb, 1985 aug-1988 aug, 1988 nov-1989 may] – 1r – 1 – (cont by: boone family association of cal-mont in mo: [newsletter]) – mf#1551062 – us WHS [929]
Calnek, William Arthur *see* A brief memoir of the late honorable james william johnston
Cal-neva token ledger / California-Nevada Token Society – v4-v6 n2 [1974 feb-1976 apr] – 1r – 1 – mf#355987 – us WHS [366]
Calnin, R J *see* The effect of moderate exercise on lipid profiles in a healthy college age population
Calogeras / Gontijo De Carvalho, Antonio – Sao Paulo, Brazil. 1935 – 1r – us UF Libraries [972]
Calogeras, Joao Pandia *see*
– Estudos historicos e politicos
– Formacao historica do brasil
– History of brazil
– Marquez de barbacena
– Problemas de administracao
– Problemas de governo
Les calomniateurs confondus / Frechette, Louis – Quebec?: s.n, 1872? – 1mf – 9 – mf#23800 – cn Canadiana [440]
Calomnies anti-protestantes / Doumergue, Emile – Paris: Bureaux de foi & vie; Lausanne: G Bridel, 1912 [mf ed 1990] – 1mf – 9 – 0-7905-6806-3 – (incl bibl ref. no more publ?) – mf#1988-2806 – us ATLA [242]
Calonne, Charles Alexandre de *see*
– De l'etat de la france
– Reponse de m de calonne a l'ecrit de m necker, publie en avril 1787
– Requete au roi
Calonne-Beaufaict, Adolphe de *see* Etudes bakongo
Caloosa belle – Labelle, FL. 1973-1997 – 25r – us UF Libraries [071]
Caloosahatchee / Gonzalez, Thomas A – Estero, FL. 1932 – 1r – us UF Libraries [978]
Caloosahatchee current – Labelle, FL. 1922 mar 3; 1923 may 10, 18, 25; jun 8; jul 13; SE – 1r – us UF Libraries [071]
Caloosahatchee river and lake okeechobee drainage / OKEECHOBEE FLOOD CONTROL DISTRICT, FLA – Washington, DC. 1930 – 1r – us UF Libraries [978]
Calphad : computer coupling of phase diagrams and thermochemistry – Elmsford. 1981-1994 (1) 1977-1994 (5,9) – ISSN: 0364-5916 – mf#49252 – us UMI ProQuest [500]
Calpin, G H *see* There are no south africans
Calspan news – Buffalo. (1) 1973-1973 (5) (9) – mf#8730 – us UMI ProQuest [500]
Calthrop, Gordon *see*
– God's works made to be remembered
– How to interpret "accidents"
– Only partial knowledge possible now
The calumet – v1-2. apr 1831-35 – (= ser The library of world peace studies) – 10mf – 9 – $105.00 – us UPA [305]
Calumet Baking Powder Co *see* The truth about baking powder
Calumet city burnham star – Chicago Heights, IL. 1987-1989 (1) – mf#68364 – us UMI ProQuest [071]
Calumet county reporter – New Holstein WI. 1908 sep 2-1910 mar 30, 1910 apr 6-1911 oct 25, 1911 nov 1-1914 apr 24, 1914 may 1-1915 nov 26, 1915 dec 3-1917 aug 31, 1917 sep 14-1919 jun 27, 1919 jul-sep 26 – 7r – 1 – (cont by: new holstein reporter) – mf#968240 – us WHS [071]
Calumet day – Whiting, IN. 1980-1989 (1) – mf#69032 – us UMI ProQuest [071]
Calumet republican – Gravesville WI. 1859 aug 4-1861 dec 16 – 1r – 1 – mf#918698 – us WHS [071]
Calumet world – Chicago, IL. 1924-1928 (1) – mf#62536 – us UMI ProQuest [071]
Calumniae nebulonis cuiusdam, quibus odio et invidia gravare conatus est doctrinam joh calvini de occulta dei providentia : johannis calvini ad easdem responsio – [Genevae]: Ex officina Conradi Badii, 1558 – 2mf – 9 – mf#CL-9 – ne IDC [242]
Calumnies refuted / Sinnott, J – York, England. 1843 – 1r – us UF Libraries [240]
Calunga / Lima, Jorge De – Buenos Aires, Argentina. 1941 – 1r – us UF Libraries [972]
Calungasinho : orgao do novo club terpsychore – Rio de Janeiro, RJ. 24 jan-dez 1886; mar-ago, dez 1887; jan-28 abr 1888 – (= ser Ps 19) – mf#P17,01,84 – bl Biblioteca [079]
Calvario de guatemala / Comite De Estudiantes Universitarios Anticomunista – Guatemala, 1955 – 1r – us UF Libraries [972]
Calvary Assembly *see* Charisma
Calvary baptist church : church minutes – jun 1961-sep 1996 (positive only) – 1 – $33.03 – mf#6944 – us Southern Baptist [242]
Calvary baptist church. aiken county. south carolina : church records – 1962-68, Sept 1975-1983.342p – 1 – us Southern Baptist [242]
Calvary baptist church. dallas, texas : church records – Nov 1914-39 – 1 – us Southern Baptist [242]
Calvary baptist church. florence county. florence, south carolina : church records – 1949-72 – 1 – us Southern Baptist [242]
Calvary baptist church. hannibal, missouri : church records – 1887-1976 – 1 – us Southern Baptist [242]
Calvary baptist church. klamath falls, oregon : church records – 1947-May 1973 – 1 – us Southern Baptist [242]
Calvary baptist church. lancaster county. south carolina : church records – 1924-43, 1951-54. 509p – 1 – us Southern Baptist [242]
Calvary baptist church. mcpherson, kansas : church records – 8Nov 1953-Dec 1962 – 1 – 5.00 – us Southern Baptist [242]
Calvary baptist church. renton, washington : church records – 2Jun 1950-71 – 1 – us Southern Baptist [242]
Calvary baptist church. shelby, north carolina : church records – 1935-63 – 1 – 52.74 – us Southern Baptist [242]
Calvary baptist church. tuscaloosa, alabama : church records – 1Jan 1911-Oct 1958 – 1 – 48.78 – us Southern Baptist [242]
Calvary baptist church. vancouver, washington : church records – 30 Dec 1948-71 – 1 – us Southern Baptist [242]
Calvary baptist church. ville platte, louisiana : church records – 1846-1947 – 1 – us Southern Baptist [242]
Calvary crier – Calvary, Fond Du Lac, Mount Calvary WI. 1985 jun 4-1987 dec 15, 1988-1990 jun, 1990 jul 17-1992 dec, 1993-94 – 4r – 1 – mf#1108161 – us WHS [071]
Calvary Gospel Church *see* Prophecy update
Calve, Adolphe (Azal pseud) *see* Sylves noires
Calvert, Albert Frederick *see* South-west africa
Calvert, Amelia Catherine Smith *see* Year of costa rican natural history
The calvert papers / ed by Cox, Richard J – 27r – 1 – $3510.00 – (with printed guide) – mf#S1611 – us Scholarly Res [920]
Calvert, Samuel *see* A memoir of edward calvert
Calvert, Stephen R *see* Agenda for the commonwealth
Calvert, Thomas *see* Established church
Calvert, William *see* Preparation for death
Calvete de Estrella, Juan Cristobal *see* Rebelion de pizarro en el peru
Calvijn : een strijder voor de anti-revolutionaire beginselen, toegelicht vooral uit zijne worsteling voor de vrijmaking der kerk / Proosdij, Cornelius van – Leiden: Donner, 1899 – 1mf – 9 – 0-524-07911-0 – (incl bibliographic references) – mf#1991-3456 – us ATLA [242]
Calvijn, 10 juli 1509-27 mei 1564 : ter eere van den grootsten christen der 16de eeuw / Brugghen, Guillaume Anne van den – [s.l]: Neerbosch' Boekhandel 1909 [mf ed 1993] – 1mf [ill] – 9 – 0-524-07853-X – mf#1991-3398 – us ATLA [242]
Calvijn als bedienaar des woords / Biesterveld, Petrus – Kampen: JH Bos, 1897 – 3mf – 9 – 0-524-08732-6 – mf#1993-3237 – us ATLA [242]
Calvijn en de economie / Diepenhorst, Pieter Arie – Wageningen: Naamlooze Venootschap drukkerij "Vada" 1904 [mf ed 1993] – 4mf – 9 – 0-524-07410-0 – (incl bibl ref) – mf#1991-3070 – us ATLA [230]
Calvijn in het strijdperk / Doumergue, Emile – Amsterdam: W Kirchner, 1904 – 6mf – 9 – 0-524-08749-0 – mf#1993-3254 – us ATLA [242]
Calvijn's beschouwing over kerk en staat / Schoch, Samuel – Groningen: JB Wolters, 1902 – 2mf – 9 – 0-524-07464-X – mf#1991-3124 – us ATLA [242]
Calvijn's jeugd, jongelingsjaren, omzwervingen, bekeering, en eerste optreden als reformator / Doumergue, Emile – Amsterdam: W Kirchner, 1903 – 6mf – 9 – 0-524-08750-4 – mf#1993-3255 – us ATLA [242]
Calvin : fondateur de l'academie de geneve / Borgeaud, Charles – Paris, Armand Colin, 1897 [mf ed 1993] – 1mf – 9 – 0-524-08736-9 – mf#1993-3241 – us ATLA [242]
Calvin : rede bei der akademischen calvin-gedaechtnisfeier in der gr. aula der universitaet heidelberg am 11. juli 1909 / Schubert, Hans von – Tuebingen: JCB Mohr, 1909 – 1mf – 9 – 0-7905-7666-X – (incl bibl ref) – mf#1989-0891 – us ATLA [242]
Calvin : sa vie, son oeuvre et ses ecrits / Bungener, Felix – Paris: J Cherbuliez, 1862 – 6mf – 9 – 0-524-08738-5 – mf#1993-3243 – us ATLA [242]
Calvin als unionsmann / Reichel, Gerhard – Tuebingen: JCB Mohr, 1909 – 1mf – 9 – 0-7905-7132-3 – mf#1988-3132 – us ATLA [242]
Calvin and his enemies : a memoir of the life, character, and principles of john calvin / Smyth, Thomas – new ed. Philadelphia: Presbyterian Board of Publ, c1856 – 1mf – 9 – 0-524-04893-2 – mf#1991-2175 – us ATLA [242]
Calvin and servetus : the reformer's share in the trial of michael servetus historically ascertained = Relation du proces criminel intente a geneve, en 1553, contre michel servet / Rilliet, Albert – Edinburgh: John Johnstone 1846 [mf ed 1992] – 1mf – 9 – 0-524-03592-X – (english trans fr french, notes & additions by w k tweedie) – mf#1990-1052 – us ATLA [920]
Calvin and the reformation : four studies / Doumergue, Emile et al – New York: Fleming H Revell, c1909 – 1mf – 9 – 0-7905-4353-2 – (incl bibl ref) – mf#1988-0353 – us ATLA [242]
Calvin au val d'aoste / Bonnet, Jules – Paris: Grassart 1861 [mf ed 1993] – 1mf – 9 – 0-524-08713-X – mf#1993-1083 – us ATLA [914]
Calvin aujourd'hui : allocution adressee aux proposants devant messieurs de la venerable compagnie le 17 decembre 1909 / Berguer, Henry – Geneve: Wyss et Duchaene, 1910 – 1mf – 9 – 0-524-08711-3 – mf#1993-1081 – us ATLA [242]
Calvin defended : a memoir of the life, character and principles of john calvin / Smyth, Thomas – new ed. Philadelphia: Presbyterian Board of Publication, 1909 – 1mf – 9 – 0-524-06499-7 – mf#1991-2599 – us ATLA [242]

Calvin et l'eloquence francaise / Lefranc, Abel – Paris: Fischbacher, 1934 – (= ser Publications de la societe calviniste de france 4) – 1 – (filmed with: love and marriage / montaigne, m e) – mf#1986 – us Wisconsin U Libr [242]

Calvin et les genevois : ou, la verite sur calvin – Paris: A Berthoud, 1907 – 1mf – 9 – 0-524-08714-8 – mf#1993-1084 – us ATLA [242]

Calvin et servet : lecon publique. faite a l'ouverture des cours de l'ecole de theologie evangelique de geneve... / Ruffet, Louis – Geneve: Chez les principaux libraires, 1910 (Montbeliard: Societe Anonyme d'Imprimerie Montbeliardaise) – 1mf – 9 – 0-524-02600-9 – mf#1990-0652 – us ATLA [242]

Calvin et servet (1509-1511-1553-1564) : esquisse biographique / Magnin, Jean Pierre – Wiesbaden: Carl Ritter, 1886 – 1mf – 9 – 0-524-07828-9 – mf#1991-3375 – us ATLA [242]

Calvin et son ideal theocratique / Monod, Leopold – Lyon: Royer, 1909 – 1mf – 9 – 0-524-08716-4 – mf#1993-1086 – us ATLA [242]

Calvin hebraisant et interprete de l'ancien testament / Baumgartner, Anton Jean – Paris: Librairie Fischbacher, 1889 – 1mf – 9 – 0-524-08730-X – mf#1993-3235 – us ATLA [242]

Calvin in his letters / Henderson, Henry F – London: J.M. Dent, 1909 – 1mf – 9 – 0-7905-5152-7 – (incl bibl ref) – mf#1988-1152 – us ATLA [242]

Calvin, J *see*
- Acta synodi tridentinae
- Les actes du concile de trente
- Advertissement contre l'astrologie, qu'on appelle judiciaire
- Advertissement tresutile du grand proffit qui reviendroit a la chrestiente s'il se faisoit inventoire de tous les corps sainct...
- Brevis admonitio joannis calvini ad fratres polonos, ne triplicem in deo essentiam pro tribus personis imaginando, tres sibi deos fabricent
- Brevis responsio joannis calvini, ad diluendas nebulonis cuiusdam calumnias, quibus doctrinam de aeterna dei praedestinatione foedare conatus est
- Brieve instruction
- Brieve resolution sur les disputes qui ont este de nostre temps quant aux sacremens...
- Christianae religionis institutio, totam fere pietatis summam, et quicquid est in doctrina salutis cognitu necessarium, complectens
- Commentariorum joannis calvini in acta apostolorum
- Commentariorum joannis calvini in acta apostolorum, liber posterior
- Congratulation a venerable prestre messire gabriel de saconnay precenteur de l'eglise de lyon, touchant la belle preface & mignonne, dont il a rempare le livre du roy d'angleterre
- Contre la secte phantastique et furieuse des libertins
- Creed rebellion alias bible rebellion
- De aeterna dei praedestinatione, qua in salutem alios ex hominibus elegit, alios suo exitio reliquit
- De scandalis, quibus hodie plerique absterrentur, nonnulli etiam alienantur a pura evangelii doctrina
- Declaration pour maintenir la vraye foy que tiennent tous chrestiens de la trinite des personnes en un seul dieu...
- Defensio orthodoxae fidei de sacra trinitate, contra prodigiosos errores michaelis serveti hispani
- Defensio sanae et orthodoxae doctrinae de sacramentis, eorumque natura, vi, fine, usu, et fructu
- Defensio sanae et orthodoxae doctrinae de servitute et liberatione humani arbitrii, adversus calumnias alberti pighii campensis
- Des scandales qui empeschent aujourd'huy beaucoup de gens de venir a la pure doctrine de l'evangile, et en desbauchent d'autres
- Dilucida explicatio sanae doctrinae de vera participatione carnis et sanguinis christi in sacra coena, ad discutiendas heshusii nebulas...
- Epinicion, christo cantatum ab ioanne calvino, calendis januarii, anno 1541
- Epistola ad senatum populumque genevensem, qua in obedientiam romani pontificis eos reducere conatur. joannis calvini responsio
- Epistola joannis calvini, qua fidem admonitionis ab eo nuper editae, apud polonos confirmat
- Epistre de jaques sadolet cardinal, envoyee au senat et peuple de geneve
- Excuse de jehan calvin a messieurs les nicodemites
- Expositions of the epistles of paul to the philippians and colossians
- La forme des prieres et chants ecclesiastiques
- Harmonia ex tribus evangelistis composita, matthaeo, marco et luca
- In librum psalmorum, johannis calvini commentarius
- In omnes pauli apostoli epistolas, atque etiam in epistolam ad hebraeos, item in canonicas petri, johannis, jacobi, et judae, quae etiam catholicae vocantur, joh. calvini commentarii
- Institutio christianae religionis, in libros quatuor nunc primum digesta, certisque distincta capitibus, ad aptissimam methodum
- Institutio christianae religionis nunc vere demum suo titulo respondens
- Institutio christianae religionis nunc vere demum suo titulo respondes
- Institutio totius christianae religionis, nunc ex postrema authoris recognitione, quibusdam locis auctior, infinitis vero castigatior
- Institution de la religion chrestienne
- Institution de la religion chrestienne: composee en latin par jehan calvin, et translatee en francoys par luymesme
- Interim adultero-germanum
- L'interim, c'est a dire, provision faicte sur les differens de la religion, en quelques villes et pais d'allemagne
- Ioannis calvin commentarii in epistolam pauli ad romanos
- Joannis calvini commentarii in isaiam prophetam...
- Joannis calvini in librum josue brevis commentarius, quem paulo ante mortem absolvit
- Joannis calvini praelectiones
- Joannis calvini praelectiones in duodecim prophetas (quos vocant) minores
- Joannis calvini praelectiones in librum prophetiarum danielis, joannis budaei et caroli jonvillaei labore et industria exceptae
- Joannis calvini responsio ad balduini convicia
- Joannis calvini, sacrarum literarum in ecclesia genevensi professoris, epistolae duae, de rebus hoc saeculo cognitu apprime necessariis
- La manyere de faire prieres aux eglises francoyses, tant devant la predication comme apres, ensemble pseaulmes et canticques francoys qu'on chante aus dictes eglises, apres s'ensuyt l'ordre...
- Mosis libri 5, cum johannis calvini commentariis
- Petit traicte de la saincte cene de nostre seigneur jesus christ
- Petit traicte, monstrant que c'est que doit faire un homme fidele congnoissant la verite de l'evangile
- Petit traicte monstrant que doit faire un homme fidele congnoissant la verite de l'evangelie quand il est entre les papistes...
- Pro g farello et collegis eius, adversus petri caroli theologastri calumnias, defensio nicolai gallasii
- Quatre sermons de m jehan calvin, traictans des matieres fort utiles pour nostre temps, comme on pourra veoir par la preface
- Responsio ad versipellem quendam mediatorem, qui pacificandi specie rectum evangelii cursum in gallia abrumpere molitus est
- Secunda defensio piae et orthodoxae de sacramentis fidei, contra joachimi westphali calumnias...
- Supplex exhortatio, ad invictiss
- Ultima admonitio joannis calvini ad joachimum westphalum, cui nisi obtemperet, eo loco posthac habendus erit, quo pertinaces haereticos haberi jubet paulus
- Vivere apud christum non dormire animis sanctos, qui in fide christi decedunt

Calvin, J. *see* Sermons de m jean calvin sur le livre de job. recueillis fidelement de sa bouche selon qu'il les preschoit

[Calvin, J] *see*
- Les actes de la journee imperiale, tenue en la cite de regespourg, aultrement dicte ratispone...
- Admonitio paterna pauli 3 romani pontificis ad invictiss
- Advertissement sur la censure qu'ont faicte les bestes de sorbonne, touchant les livres qu'ilz appellent heretiques
- Les articles de la sacree faculte de theologie de paris, concernans nostre foy et religion chrestienne, et forme de prescher
- Articuli a facultate sacrae theologiae parisiensi determinati super materiis fidei nostrae hodie controversis
- De la predestination eternelle de dieu
- Excuse de noble seigneur, jaques de bourgoigne, s de fallez et bredam
- Gratulatio ad venerabilem presbyterum, dominum gabrielem de saconay, praecentorem ecclesiae lugdunensis, de pulchra et eleganti praefatione quam libro regis angliae inscripsit
- Histoire d'un meurtre execrable
- Impietas valentini gentilis detecta, et palam traducta, qui christum non sine sacrilega blasphemia deum essentiatum esse fingit
- Response a un cauteleux et ruse moyenneur, qui souz couleur d'appaiser les troubles touchant le faict de la religion, a tente tous les moyens d'empescher et rompre le cours de l'evangile par la france...
- Supplication et remonstrance

Calvin, Jean *see*
- Commentaires de m jean calvin, sur les cinq livres de moyse...
- Contre la secte phantastique et furieuse des libertins

Calvin, John *see*
- Appendix libelli adversus interim adultero-germanum
- Institutes of the christian religion
- Opera omnia
- Treatise on relics

Calvin, le predicateur de geneve : conference faite dans la cathedrale de saint-pierre, a geneve / Doumergue, Emile – Geneve: Atar, [1909?] [mf ed 1990] – 1mf – 9 – 0-7905-6046-1 – (in french) – mf#1988-2046 – us ATLA [242]

Calvin memorial addresses : delivered...at savannah, ga, may, 1909 / Reed, Richard Clark et al – Richmond, VA: Presbyterian Cttee of Pub, c1909 [mf ed 1990] – 1mf – 9 – 0-7905-5799-1 – mf#1988-1799 – us ATLA [242]

Calvin, servet, guillaume de trie et le tribunal de vienne : reponse a l'action radicale / Weiss, Nathanael – Geneve: Imprimerie Nationale, 1908 – 1mf – 9 – 0-524-02608-4 – mf#1990-0660 – us ATLA [242]

Calvin theological journal – Grand Rapids. 1966+ (1) 1972+ (5) 1975+ (9) – ISSN: 0008-1795 – mf#6373 – us UMI ProQuest [242]

Calvin, twisse and edwards on the universal salvation of those dying in infancy / Stagg, John Weldon – Richmond, VA: Presbyterian Comm of Publ, c1902 – 1mf – 9 – 0-8370-5518-0 – (incl bibl ref) – mf#1985-3518 – us ATLA [242]

Calvin und basel bis zum tode des myconius, 1535-1552 / Wernle, Paul – Basel: F Reinhardt, 1909 – 2mf – 9 – 0-7905-6911-6 – (incl bibl ref) – mf#1988-2911 – us ATLA [242]

Calvin und montaigne : rede zum vierhundertjaehrigen jubilaeum calvins / Lobstein, Paul – Strassburg: E van Hauten, 1909 [mf ed 1990] – 1mf – 9 – 0-7905-7659-7 – (in german; incl bibl ref) – mf#1989-0884 – us ATLA [242]

Calvin und servet : vortrag / Barth, Fritz – Bern: A Francke, 1909 – 1mf – 9 – 0-524-02638-6 – mf#1990-0662 – us ATLA [242]

Calviniana religio : oder calvinisterey, so faelschlich die reformirte religion genennet wird / Gedik, S – Leipzig, 1615 – 8mf – 9 – mf#TH-1 mf 526-533 – ne IDC [242]

Die calvinische und die altstrassburgische gottesdienstordnung : ein beitrag zur geschichte der liturgie in der evangelischen kirche / Erichson, Alfred – Strassburg: JH Ed Heitz, 1894 – 1mf – 9 – 0-524-02388-3 – (incl bibl ref) – mf#1990-4290 – us ATLA [242]

Calvinischer betlersmantel darin angezeiget wird mit was kleider sie sich bekapen den schalck verbergen vnd zudecken koennen / Engel, A – np, 1598 – 1mf – 9 – mf#TH-1 mf 400 – ne IDC [242]

Calvinism : an address delivered at st andrew's, march 17, 1871 / Froude, James Anthony – New York: Charles Scribner 1871 [mf ed 1986] – 1mf – 9 – 0-8370-8671-X – mf#1986-2671 – us ATLA [242]

Calvinism / Froude, James Anthony – London, England. 1871 – 1r – us UF Libraries [242]

Calvinism : six lectures / Kuyper, Abraham – New York: F H Revell, [1899?] [mf ed 1990] – (= ser Stone lectures) – 1mf – 9 – 0-7905-6071-2 – mf#1988-2071 – us ATLA [242]

Calvinism and evangelical arminianism : compared as to election, reprobation, justification, and related doctrines / Girardeau, John Lafayette – Columbia SC: W J Duffie; New York: Baker & Taylor 1890 [mf ed 1986] – 2mf – 9 – 0-8370-8672-8 – (incl bibl ref & ind) – mf#1986-2672 – us ATLA [242]

Calvinism in history / McFetridge, Nathaniel S – Philadelphia: Presbyterian Board of Publ & Sabbath-School Work 1912 [mf ed 1992] – 1mf – 9 – 0-524-04843-6 – (incl bibl ref) – mf#1990-1335 – us ATLA [242]

Calvinism in its relations to scripture and reason : or, an examination into the nature and consequences of calvinistic principles / Munro, Alexander – Glasgow: Hugh Margey, 1856 [mf ed 1993] – 3mf – 9 – 0-524-07444-5 – mf#1991-3104 – us ATLA [242]

Calvinism, pure and mixed : a defence of the westminster standards / Shedd, William Greenough Thayer – New York: Charles Scribner 1893 [mf ed 1986] – 1mf – 9 – 0-8370-8788-0 – mf#1986-2788 – us ATLA [242]

Calvinism taught in the thirty-nine articles / Crawford, John Howard – Edinburgh, Scotland. 1878 – 1r – us UF Libraries [242]

Het calvinisme : zes stone-lezingen in october 1898 te princeton (n-j) gehouden / Kuyper, Abraham – Amsterdam: Boekhandel voorheen Hoeveker & Wormser [c1898] [mf ed 1986] – 1mf – 9 – 0-8370-8757-0 – (incl bibl ref) – mf#1986-2757 – us ATLA [242]

Le calvinisme de l'avenir / Berthoud, Aloys – Geneve: Wyss et Duchene, 1890 – 1mf – 9 – 0-524-08712-1 – mf#1993-1082 – us ATLA [242]

Het calvinisme en de kunst : rede bij de overdracht van het rectoraat der vrije universiteit op 20 oct 1888 / Kuyper, Abraham – Amsterdam: J A Wormser, 1888 [mf ed 1990] – 1mf – 9 – 0-7905-6200-6 – (incl bibl ref) – mf#1988-2200 – us ATLA [242]

Calvinisme en schriftstudie : een woord van verweer tegen g.a. v. d. bergh van eysinga / Leeuwen, Jacobus Adrien Cornelius van – Utrecht: Ruys 1909 [mf ed 1993] – 1mf – 9 – 0-524-07253-1 – mf#1991-2994 – us ATLA [242]

Calvinisme en socialisme : een woord voor onzen tijd / Rudolph, Roelof Jan Willem – Kampen: Kok, 1901 – 1mf – 9 – 0-524-07265-5 – (incl bibliographic references) – mf#1991-3006 – us ATLA [242]

Calvinisme et liberte / Goumaz, Louis – Geneve: [s.n.], 1951 – 1r – 1 – 0-8370-1602-9 – mf#1984-T024 – us ATLA [242]

Calvinistarum vera, viva et genuina descriptio / Hoe von Hoenegg, M – Lipsiae, 1620 – 2mf – 9 – mf#TH-1 mf 673-674 – ne IDC [242]

The calvinistic conception in lutheran theology : an examination as to the confessional character of the doctrine of the synod of missouri on eternal election / Cronenwett, Emanuel – Columbus, Ohio: [s.n.], 1883 – 1mf – 9 – 0-524-06401-6 – mf#1991-2523 – us ATLA [242]

Calvinistic Conference on Psychology and Psychiatry *see* Proceedings of the calvinistic conference on psychology and psychiatry

The calvinistic doctrine of election and reprobation no part of st paul's teachings : bible-study / Harris, John Andrews – Philadelphia: Porter & Coates, 1890 – 1mf – 9 – 0-8370-5015-4 – mf#1985-3015 – us ATLA [225]

Calvinistic Methodist Church in the U.S.A. General Assembly *see* Minutes

Calvins bedeutung fuer die geschichte und das leben der protestantischen kirche / Hadorn, Wilhelm – Neukirchen: Verlag der Buchhandlung des Erziehungsvereins, [1910?] – 1mf – 9 – 0-7905-7642-2 – mf#1989-0867 – us ATLA [242]

Calvin's hermeneutics of the imprecations of the psalter / Mpindi, Paul Mbunga – Calvin Theological Seminary, 2003 [mf ed 2004] – 1r – 1 – 0-524-10481-6 – mf#d00007 – us ATLA [221]

Calvins jenseits-christentum : in seinem verhaeltnisse zu den religioesen schriften des erasmus / Schulze, Martin – Goerlitz: Rudolf Duelfer, 1902 – 1mf – 9 – 0-7905-8882-X – mf#1989-2107 – us ATLA [242]

Calvins persoenlichkeit und ihre wirkungen auf das geistige leben der neuzeit : festrede / Barth, Fritz – Bern: A Francke, 1909 – 1mf – 9 – 0-7905-7682-1 – mf#1989-0907 – us ATLA [242]

Calvins praedestinationslehre : ein beitrag zur wuerdigung der eigenart seiner theologie und religiositaet / Scheibe, Max – Halle a. S: Ehrhardt Karras, 1897 – 1mf – 9 – 0-8370-5139-8 – (incl bibl ref) – mf#1985-3139 – us ATLA [242]

Calvinstudien : festschrift zum 400. geburtstage johann calvins / Bohatec, Josef et al; ed by Reformierte Gemeinde Elberfeld – Leipzig: Rudolf Haupt, 1909 – 2mf – 9 – 0-524-04188-1 – (incl bibl ref) – mf#1990-1227 – us ATLA [242]

Calvinvs ivdaizans, hoc est : ivdaicae glossae et corrvptele, qvibvs iohannes calvinvs illustrissima / Hunnius, A – Vvitebergae, 1593 – 2mf – 9 – mf#TH-1 mf 758-759 – ne IDC [242]

Calvisius, Seth *see* Harmonia cantionum ecclesiasticarum

Calvisius, Sethus *see* Harmonia cantionum ecclesiasticarum

Calvo Abeytar, Fernando *see* Libro de albayteriade

Calvo, Carlos *see*
- Annales historiques de la revolution de l'amerique latine
- Recueil complet des traites, conventions, capitulations, armistices et autres actes diplomatiques de tous les etats de l'amerique latine

Calvo Flores, Josquin *see*
- Poesias
- Vientos alegres, vientos desolados

Calvo, Joaquin Bernardo *see* Campana nacional contra los filibusteros en 1856 y

Calvo Serer, Rafael *see* Europa en 1949

Calvocoressi, Peter *see* South africa and world opinion

Calwell, James *see* Campsie case

Calwer Verlagsverein *see* Der christliche glaube in acht buechern

Calya parva – Calcutta: Bharata Press, 1889 – 1mf – 9 – 0-524-08010-0 – mf#1991-0232 – us ATLA [280]

Calypso and carnival of long ago and today / Jones, Charles – Port-Au-Prince, Haiti. 1947 – 1r – us UF Libraries [972]

La calzada de oropesa, su santo cristo y sus monjas / Ayape, Eugenio – Madrid: Editorial Augustinus, 1976 – sp Bibl Santa Ana [240]

Calzadilla, Rafael S De *see* Cuba para los cubanos

Calzado Pedrilla, Felipe *see*
- Defensa de d. luis calderon...presos de confesion
- Opinion formada en el asunto de los ferrocarriles de la provincia de caceres

Cam Sa On, U *see* Ananda candra [rhc ra cu rakhuin vesali man]

Cam Sa on, U *see* E di khrok ra cu nhan yan ma tuin mi ra khuin prannsum akkhara

Cam Su, Mon, Da gun *see* Pro ra khak tay

Cam Tha, U *see* Mran ma alanka kyam

Cam Thvan Aon *see* Prann thon cu kye lak pyo rvhan mhu mya

Camacho, Carlos S *see*
- Governor's report on commonwealth affairs
- Gubernatorial inauguration address

Camacho Carrizosa, Guillermo *see* Critica historica

Camacho Leyva, Ernesto *see* Policia en los territorios nacionales

Camacho Macias, Aquilino *see* Don francisco de navarra, obispo de badajoz (1545-1556). sus interveneiones en trento sobre la obligacion episcopal de residir

Camacho Montoya, Guillermo *see* Santander

Camacho, Panfilo D *see* Marti un genio creador

Camacho, Panfilo Daniel *see* Varona

Camacho Perea, Miguel *see* Geografia e historia del departamento del valle de...

Camacho Roldan, Salvador *see* Memorias

Camacho, Roldan, Salvador *see* Notas de un viaje (colombia y estados unidos)

Camagueyano – 1979 sep-1982 oct, 1982 nov-1985 sep, 1985 oct-1988 nov/dec – 3r – 1 – mf#620986 – us WHS [071]

Camara Cascudo, Luis Da *see* Marquez de olinda e seu tempo (1793-1870)

Camara de comercio latino de hialeah – Hialeah, FL. 1974 mar – 1r – us UF Libraries [071]

Camara, Ezequiel Enrique *see* Tablas de reducciones y equivalencias del antiguo sistema de pesas

Camara lenta / Lainfiesta, Margot – Tegucigalpa, Mexico. 1935 – 1r – us UF Libraries [972]

Camara Manoel, Jeronymo Pinheiro d'Almeida *see* Missoes dos jesuitas no oriente nos seculos 16 e 17

Camara Oficial de Comercio e Industria Badajoz *see* Memoria comercial y estadistica relativa al estado de los negocios en la provincia de badajoz

Camara Oficial de Comercio e Industria de Badajoz *see*
- Decreto y resolucion de la direccion general de expansion comercial sobre organizacion del registro general de exportadores y de los registros especiales
- Memoria acerca del movimiento de los negocios en los anos 1962-1963
- Memoria acerca del movimiento de los negocios en los anos 1964-1965
- Memoria comercial y estadistica relativa al estado de los negocios en la provincia de badajoz

Camara Oficial de Comercio e Industria de Caceres *see*
- Informe de los resultados de la encuesta sobre la situacion del comercio, industria y servicios
- Reglamento de regimen interior de la...

Camara Oficial de la Propiedad Urbana *see* Memoria de los trabajos realizados durante el ejercicio de 1931, que se eleva al excmo. sr. ministro de trabajo y prevision

Camara Oficial de la Propiedad Urbana de la Provincia de Caceres *see* Memoria de los trabajos realizados durante el ano 1940...

Camara Oficial Sindical Agraria *see*
- Memoria de actividades
- Memoria de actividades, 1967
- Memoria de actividades, 1968
- Reglamento de trabajo agricola para la provincia de badajoz

Camara Oficial Sindical Agraria. Caceres *see*
- Orden de 31 de enero de 1958...la incoacion para las hermandades sindicales de labradores y ganaderos...exaccion por la via de apremio de los cupones impagados
- Reglamento para el regimen y funcionamiento de las secciones sociales de las entidades sindicales menores, 12 abril, 1950

Camara Privada de Compensacion Bancaria de Caceres *see*
- Memoria 1977
- Reglamento de la...

O camaradinha – Rio de Janeiro, RJ: Typ de FA de Almeida, ago 1851 – (= ser Ps 19) – mf#P15,01,76 – bl Biblioteca [321]

Camargo, Joao Ayres De *see* Patriotas paulistas na columna sul

Camargo, Paulo Florio Da Silveira *see* Historia eclesiastica do brasil

Camargo Perez, Gabriel *see* Barro al acero en la roma de los chibchas

[Camarillo/port huneme-] the daily news – CA. 1967-1976; apr 1977-sep 1981; 1982- – 179+ r – 1 – $10,740.00 (subs $360/y) – mf#H03171 – us Library Micro [071]

Camas hot springs exchange – Hot Springs, MT. 1938-1957 (1) – mf#64482 – us UMI ProQuest [071]

Camas post and washougal record – Washougal, WA. 1932-1942 (1) – mf#67174 – us UMI ProQuest [071]

[Camas-washougal-] the camas-washougal record and shoppers guide – WA. 1982-1984; 1987- – 12r – 1 – $720.00 (subs $80y) – mf#B05407 – us Library Micro [071]

Cambefort, Gaston *see* Introduction au cambodgien

Cambell, Arthur *see* The mystery of martha warne

Cambell, Jacques *see* I fiori dei tre compagni... milano, 1967

Cambell, W *see* Materials for a history of the reign of henry 7 (rs60)

Camberos de Yegros, Fernando *see*
- El heroe serafico de san pedro de alcantara
- Verdad ilustrada contra las imposturas que ha escrito el r.p. fr. marcos de alcala

Camberwell and peckham express *see* Camberwell peckham and dulwich express

Camberwell and peckham times – London, UK. apr 1870-27 feb 1969 [wkly] – 121r – 1 – (aka: camberwell & peckham time lambeth & south london observer; south london observer camberwell and peckham times and, south london observer) – uk British Libr Newspaper [072]

Camberwell news south london advertiser – London, UK. 9 dec 1876-1881 – 5r – 1 – (aka: camberwell news peckham and south london advertiser; south london gazette and camberwell news) – uk British Libr Newspaper [072]

Camberwell peckham and dulwich express – London, UK. mar-dec 1871, 1873 – 1 1/2r – 1 – (aka: camberwell and peckham express) – uk British Libr Newspaper [072]

Cambiagi, G *see* Descrizione dell'imperiale giardidi boboli...

Cambiasi, Pompeo *see* La scala 1778-1906; note storiche e statistiche di pompeo cambiasi...

Cambie, Henry John *see* An unrecorded property of clay

Cambier, Enrique *see* Sonetos mios

Cambini, A *see*
- Commentario...della origine de tvrchi, et imperio della casa ottomanna
- Libro...della origine de tvrchi et imperio delli ottomanni

Cambini, Giuseppe Maria *see*
- Six nouveaux quatuors pour deux violons, alto et basse
- Six quatuor d'airs connus dialogues et varies
- Trois quintetti pour deux violons, alto, violoncelle et basse...3me. livre de qiomtetto

Cambio – Oakland CA. 1986 jan-1994 mar – 1r – 1 – mf#2883781 – us WHS [071]

Cambio 16 – Madrid. nov 22, 1971- – 1 – us L of C Photodup [073]

Cambio diez y seis *see* Cambio 16

Cambios en la concepcion y estructura de la narrativa mexicana de7 / Passafari De Gutierrez, Clara – Rosario, Argentina. 1968 – 1r – us UF Libraries [972]

Cambodge : an 12 de l'independance – [Hongkong: Continental Print Co 196-] [mf ed 1989] – 1r with other items – 1 – mf#mf-10289 seam reel 014/14 [§] – us CRL [959]

Cambodge / Cambodia. Ministere de l'information – Phnom-Penh [mf ed 1989] – 1r with other items – 1 – (with bibl) – mf#mf-10289 seam reel 002/07 [§] – us CRL [959]

Cambodge / Dannaud, Jean Pierre – Lausanne [Switzerland]: Ed Clairefontaine [1956] [mf ed 1989] – [ill/pl] 1r with other items – 1 – (photos by cahery et al) – mf#mf-10289 seam reel 003/04 [§] – us CRL [306]

Cambodge : fetes civiles et religieuses / Leclere, Adhemard – Paris: Impr nationale 1916 [mf ed 1992] – (= ser Annales du musee guimet 42) – 2mf [ill] – 9 – 0-524-04646-8 – (incl bibl ref) – mf#1990-3389 – us ATLA [390]

Cambodge / Lacouture, Simonne – Lausanne: Editions Rencontre [1963] [mf ed 1989] – (= ser L'atlas des voyages) – 1r with other items – 1 – mf#mf-10289 seam reel 006/04 [§] – us CRL [959]

Cambodge – Hongkong: printed by Contenental Print Co 1961?] [mf ed 1989] – 1r with other items – 1 – (photos by raymond cauchetier [et al]) – mf#mf-10289 seam reel 003/03 [§] – us CRL [915]

Le cambodge / Aymonier, Etienne – Paris: E Leroux 1900-04 [mf ed 1989] – 3v [ill/facs] on 1r with other items – 1 – (1: le royaume actuel; 2: les provinces siamoises; 3: le group d'angkor et l'histoire) – mf#mf-10289 seam reel 011/01 [§] – us CRL [959]

Le cambodge / Boisselier, Jean – Paris: A & J Picard & Cie [mf ed 1989] – (= ser Manuel d'archeologie d'extreme-orient) – 1r with other items – 1 – (with bibl) – mf#mf-10289 seam reel 021/08 [§] – us CRL [930]

Le cambodge : guide franco khmer: avec la transcription phonetique romanisee des statistiques sur le cambodge moderne, des cartes et renseignements touristiques, des notions elementaires de grammaire / Sonolet, Jean-Francois – Phnom-Penh: Bouth-Neang [mf ed 1989] – 1r with other items – 1 – (with bibl) – mf#mf-10289 seam reel 025/07 [§] – us CRL [480]

Le cambodge – Paris: F Maspero [mf ed 1989] – (= ser Front solidarite indochine 8) – 1r with other items – 1 – (incl bibl ref) – mf#mf-10289 seam reel 017/29 [§] – us CRL [959]

Le cambodge de sihanouk : ou, de la difficulte d'etre neutre. lettre preface de norodom sihanouk / Dauphin-Meunier, Achille – Paris: nouvelle ed latines [1965] [mf ed 1989] – (= ser Coll survol du monde) – 1r with other items – 1 – (with bibl) – mf#mf-10289 seam reel 009/01 [§] – us CRL [959]

Cambodge d'hier et d'aujord'hui : une monarchie millenaire, une democratie socialiste / Cambodia. Ministere de l'information – [Phnom-Penh?]: Secretariat d'Etat a l'information du Cambodge [ca 1956?] [mf ed 1989] – 1r with other items – 1 – mf#mf-10289 seam reel 017/38 [§] – us CRL [915]

Le cambodge economique / Dreyfus, Pierre – Paris: V Giard & E Briere [mf ed 1989] – 1r with other items – 1 – mf#mf-10289 seam reel 026/28 [§] – us CRL [330]

Le cambodge en lutte / Chinese Journalists Delegation to Cambodia – Pekin: Ed en langues etrangeres 1975 [mf ed 1989] – [ill] 1r with other items – 1 – mf#mf-10289 seam reel 017/42 [§] – us CRL [335]

Cambodge et cambodgiens : metamorphose du royaume khmer par une methode francaise de protectorat / Collard, Paul Maria Alexandre – Paris: Societe d'editions geographiques, maritimes et coloniales 1925 [mf ed 1989] – [ill] 1r with other items – 1 – mf#mf-10289 seam reel 008/12 [§] – us CRL [959]

Cambodge et java : ruines khmeres et javanaises 1893-1894 / Tissandier, Albert – Paris: G Masson [mf ed 1989] – [ill/pl] 1r with other items – 1 – mf#mf-10289 seam reel 007/01 [§] – us CRL [930]

Cambodge et la france – [Hongkong: Continental Print Co 1967?] [mf ed 1989] – [ill] 1r with other items – 1 – (with: le cambodge et la france by norodom sihanouk, chef de l'etat [p1-3]) – mf#mf-10289 seam reel 014/05 [§] – us CRL [327]

Le cambodge et la politique de neutralite / Stein, Renee – Dijon [mf ed 1989] – 1r with other items – 1 – mf#mf-10289 seam reel 005/01 [§] – us CRL [327]

Le cambodge et ses relations avec ses voisins / Norodom Sihanouk, Prince – [Phnom-Penh]: Le Ministere de l'information [1962?] [mf ed 1989] – 1r with other items – 1 – mf#mf-10289 seam reel 003/09 [§] – us CRL [327]

Cambodge et siam : voyage et sejour aux ruines des monuments kmers / Filoz, Auguste Achille Hippolyte – Paris: Gedalge [1896?] [mf ed 1989] – [ill] 1r with other items – 1 – (notes du capitaine a filoz) – mf#mf-10289 seam reel 008/01 [§] – us CRL [915]

Cambodge et siam : voyage et sejour aux ruines des monuments kmers / Filoz, Auguste Achille Hippolyte – Thonon: A Dubouloz [1876] [mf ed 1989] – 1r with other items – 1 – mf#mf-10289 seam reel 008/03 [§] – us CRL [915]

Le cambodge [kampuchea] : apercu geographique et historique – [Phnom-Penh?: s.n. 1972?] [mf ed 1989] – 1r with other items – 1 – mf#mf-10289 seam reel 008/08 [§] – us CRL [959]

Le cambodge, passe, present, avenir / Testoin, Edouard – Tours: Impr E Mazereau 1886 [mf ed 1989] – 1r with other items – 1 – mf#mf-10289 seam reel 005/02 [§] – us CRL [959]

Cambodge (Phnom Penh, Cambodia) : [quotidien d'information khmer] – Phnom-Penh: Association Sutharot -1971 (daily ex sun) [mf] – [mf ed Ithaca NY: Cornell University] – 1r – 1 – (in french; publ varies) – mf#mf-11157 seam – us CRL [079]

Cambodge terre de travail et oasis de paix / Cambodia. Ministere de l'information – Phnom-Penh 1963 [mf ed 1989] – 1r with other items – 1 – mf#mf-10289 seam reel 002/06 [§] – us CRL [959]

Le cambodge, un ilot de paix et de stabilite en asie du sud-est : une serie d'articles publies dans le monde diplomatique de septembre 1966 – Phnom-Penh?]: Le Ministere de l'information 1966 [mf ed 1989] – [ill] 1r with other items – 1 – mf#mf-10289 seam reel 003/02 [§] – us CRL [959]

Le cambodgien / Monod, Guillaume-Henri – Paris: Ed Larose 1931 [mf ed 1989] – 1r with other items – 1 – mf#mf-10289 seam reel 130/2 [§] – us CRL [930]

Le cambodgien sans maitre : methode progressive, rapide & pratique a la portee de tous / Tiw-Oll – 1st ed. [Phnom-Penh?] 1957 [mf ed 1989] – 1r with other items – 1 – mf#mf-10289 seam reel 025/08 [§] – us CRL [480]

Cambodia : the administration's version and the historical record / Kahin, George McTurnan – [Ithaca NY: Glad Day Press 1970] [mf ed 1989] – 1r with other items – 1 – (paper contains replies to prepared question & answer materials distributed by the administration on may 1 in regard to the cambodian invasion...) – mf#mf-10289 seam reel 017/32 [§] – us CRL [327]

Cambodia / Boramy, Leon – London: Conflict Education Library Trust 1970 [mf ed 1989] – (= ser Peace press: an international information service 6/3-4) – 1r with other items – 1 – mf#mf-10289 seam reel 017/24 [§] – us CRL [959]

Cambodia : land of contrasts. illustrated with photos / Tooze, Ruth – New York: Viking Press [1962] [mf ed 1989] – [ill] 1r with other items – 1 – mf#mf-10289 seam reel 015/34 [§] – us CRL [959]

Cambodia : march 1970 – [n.p. 1970] [mf ed 1989] – [ill] 1r with other items – 1 – mf#mf-10289 seam reel 016/17 [§] – us CRL [959]

Cambodia : the search for security / Leifer, Michael – New York: Praeger [1967] [mf ed 1989] – 1r with other items – 1 – (with bibl ref incl in notes) – mf#mf-10289 seam reel 020/10 [§] – us CRL [327]

Cambodia : the widening war in indochina / ed by Grant, Jonathan S et al – New York: Washington Square Press [1971] [mf ed 1989] – 1r with other items – 1 – (with bibl) – mf#mf-10289 seam reel 019/03 [§] – us CRL [959]

Cambodia *see*
- Bulletin administratif
- Bulletin officiel

Cambodia 1972 / leng Sary – [Phnom Penh?] Royal Govt of National Union of Cambodia [1972?] [mf ed 1989] – [ill] 1r with other items – 1 – mf#mf-10289 seam reel 017/10 [§] – us CRL [959]

Cambodia. Commission des murs et coutumes *see*
- Chamrieng cheat niyum ou chant patriotique
- Jeux populaires au cambodge

Cambodia. Departement de l'information *see* Les agressions sud-vietnamiennes des 7 et 8 mai 1964 contre le territoire cambodgien

Cambodia. Departement du tourisme *see* Guide du cambodge

Cambodia. Ministere de l'information *see*
- Biographie de s.a.r. le prince norodom sihanouk
- Biographie de s.a.r. le prince norodom sihanouk, chef d'etat du cambodge
- Cambodge
- Cambodge d'hier et d'aujord'hui
- Cambodge terre de travail et oasis de paix
- Cambodia today and yesterday
- Documents relatifs a la suspension des relations diplomatiques entre le cambodge et la thailande
- Documents sur l'agression vietcong et nord-vietnamienne contre le cambodge [1970]
- La femme cambodgienne a l'ere du sangkum
- Images de dix congres nationaux, septembre 1955-decembre 1960
- Livre blanc sur la rupture des relations diplomatiques entre le cambodge et la thailande le 23 octobre 1961
- Livre blanc sur l'agression vietcong et nord-vietnamienne contre la republique khmere [1970-71]
- Massacres commis par les agresseurs nord-vietnamiens et vietcong en territoire khmer
- Les progres du cambodge, 1954-1964
- Quand sihanouk denoncait le communisme asiatique...[1966-1969]
- Les refugies khmers, victimes de l'invasion nord-vietnamienne
- Le retour de l'independance nationale, 9 novembre 1953
- Self-aid for cambodia
- Les troupes nord-vietnamiennes et vietcong envahissent la republique khmere
- Violations de l'article 20 des accords de paris par le nord-vietnam

Cambodia. Ministere du plan *see* Le plan quinquennal preah norodom sihanouk, 1960-1964

Cambodia. Ministry of Foreign Affairs *see* Dossier des agressions des forces americano-sud-vietnamiennes a chantrea, province de svayrieng [cambodge] le 19 mars 1964, soumis au conseil de securite de l'o n u

Cambodia. Office national du tourisme *see* A preface to angkor

Cambodia. Office of Exchange Control *see* Concerning the import and export of funds and holdings by travellers

Cambodia. Permanent Mission to the United Nations *see*
- Discours de son excellence monsieur l'ambassadeur huot sambath, president de la delegation du cambodge
- The kingdom of cambodia
- Statement made by h e mr huot sambath

Cambodia, problems of neutrality and independence may 1970 / International Documentation and Information Centre, Hague – The Hague [1970] [mf ed 1989] – 1r with other items – 1 – mf#mf-10289 seam reel 017/03 [§] – us CRL [327]

Cambodia. Residence superieure *see* Rapport sur la situation du cambodge [octobre 1902-juillet 1903]

Cambodia. Sangkum Reastr Niyum. Comite executif *see* Politique economique du sangkum reastr niyum

Cambodia today and yesterday = Cambodge d'hier et d'aujourd'hui / Cambodia. Ministere de l'information – [Phnom-Penh 196-?] [mf ed 1989] – 1r with other items – 1 – (in english & french) – mf#mf-10289 seam reel 009/08 [§] – us CRL [959]

Cambodian : basic course units 1-12 / Foreign Service Institute [US] – Washington DC 1959 [mf ed 1989] – 1r with other items – 1 – mf#mf-10289 seam reel 029/06 [§] – us CRL [480]

Cambodian : language familiarization manual. prepared for the dept of defense / Educational Services [Washington DC] – [Washington c1960] [mf ed 1989] – 1r with other items – 1 – mf#mf-10289 seam reel 024/10 [§] – us CRL [480]

Cambodian chronicles and the siamese invasions of angkor / Vickery, Michael – [s.l: s.n. 196-] [mf ed 1989] – 1r with other items – 1 – (incl bibl ref) – mf#mf-10289 seam reel 021/09 [§] – us CRL [959]

Cambodian documents – [1976-80] [mf ed 1981] – 3r – 1 – us L of C Photodup [959]

Cambodian glory : the mystery of the deserted khmer cities and their vanished splendour; and a description of life in cambodia today / Ponder, H W – London: T Butterworth ltd [1936] – 1r with other items – 1 – (1st publ 1936; with bibl) – mf#mf-10289 seam reel 010/07 [§] – us CRL [915]

Cambodian names and titles / Huffman, Franklin Eugene – [New Haven: Institute of Far Eastern Languages, Yale Uni 1968] [mf ed 1989] – 1r with other items – 1 – (incl transcription system for standard cambodian) – mf#mf-10289 seam reel 015/15 [§] – us CRL [480]

Cambodian names and titles / Huffman, Franklin Eugene – [New Haven: Institute of Far Eastern Languages, Yale Uni 1968] [mf ed 1989] – 1r with other items – 1 – (incl transcription system for standard cambodian) – mf#mf-10289 seam reel 015/15 [§] – us CRL [480]

The cambodian resistance / Norodom Sihanouk, Prince et al – Auckland: Auckland Vietnam Cttee [mf ed 1990] – 1r with other items – 1 – (incl bibl) – mf#mf-10289 seam reel 124/4 [§] – us CRL [959]

Cambodian system of writing and beginning reader with drills and glossary / Huffman, Franklin E et al – New Haven: Yale UP [mf ed 1989] – 1r with other items – 1 – (added title in cambodian; with bibl) – mf#mf-10289 seam reel 026/02 [§] – us CRL [480]

Cambodian tipitaka / Phnom-Penh. Institut Bouddhique – 1931-69 (mf ed) – 1142mf – 9 – $800.00 set – (chrieng script with khmer translation on facing pages and footnote ref to publ burmese and thai eds) – us IASWR [090]

Cambodian-american relations : a story in pictures / United States Information Service, Phom Penh – [Phnom-Penh: USIS & USOM 1955] [mf ed 1989] – [ill] 1r with other items – 1 – mf#mf-10289 seam reel 026/25 [§] – us CRL [382]

Cambodia's quest for survival / Poole, Peter A – [New York: American-Asian Educational Exchange 1969] [mf ed 1989] – 1r with other items – 1 – (incl bibl ref) – mf#mf-10289 seam reel 015/18 [§] – us CRL [959]

Cambodja / Claassen, Antoon – Amsterdam [Netherlands]: Sjaloom-Odijk 1974 [mf ed 1989] – 1r with other items – 1 – mf#mf-10289 seam reel 002/19 [§] – us CRL [959]

Cambolm, Natalicio *see* Brasil

Los Camborios *see* Estatutos por los que se rige la pena "los camborios"

Cambourne-redruth packet – England.20 Sept 1955-18 Dec 1956; 1957-59. -w. 2 reels – 1 – uk British Libr Newspaper [072]

Cambrai, Gui de *see* Barlaam und josaphat

Cambria freeman – Ebensburg, PA. -w 1899-1912 – 13 – $25.00r – us IMR [071]

Cambria news – Cambria WI. 1893 oct 27/1898-1943 dec 24/1949 may 26 – 15r – 1 – (with gaps; cont by: pardeeville-wyocena times) – mf#964304 – us WHS [071]

[Cambria-] the cambrian – CA. sep 1931-1973; 1979; 1981-82 – 21r – 1 – $1260.00 – mf#B02091 – us Library Micro [071]

Cambrian and caledonian quarterly magazine and celtic repertory – London. 1829-1833 – 1 – mf#4220 – us UMI ProQuest [073]

Cambrian and general weekly advertiser for the principality of wales – [Wales] Swansea jan 1804-14 mar 1930 [mf ed 2003] – 100r – 1 – (cont by: cambrian and weekly general advertiser for swansea and the principality of wales [jan 1870-mar 1930]) – uk Newsplan [072]

Cambrian bibliography : containing an account of the books printed in the welsh language, or relating to wales, from the year 1546 to the end of the eighteenth century / Rowlands, William; ed by Evans, Daniel Silvan – Llanidloes: printed & publ by John Pryse, 1869 – (= ser 19th c publishing...) – 9mf – 9 – mf#3.1.4 – uk Chadwyck [680]

Cambrian law review – v1-31. 1970-2000 – 5,6,9 – $341.00 set – (v1-15 1970-84 in reel $100. v16-31 1985-2000 in mf $241) – ISSN: 0084-8328 – mf#101301 – us Hein [340]

Cambrian times – [Wales] LLGC 19 sep 1936-dec 1938 [mf ed 2004] – 2r – 1 – uk Newsplan [072]

Cambridge 1635-1849 – Oxford, MA (mf ed 1995) – (= ser Massachusetts vital record transcripts to 1850) – 29mf – 9 – 0-87623-223-3 – (mf 1t-2t: vital records 1635-92. mf 2t-4t,29t: births 1688-1822. mf 4t-5t: marriage intentions 1800-21. mf 5t-7t: marriages 1700-1821. mf 7t-8t: deaths 1699-1819. mf 8t-15t: marriages & intentions 1814-45. mf 13t-17t: deaths & burials 1825-64. mf 14t-15t: births 1794-1852. mf 17t-18t: marriage intentions 1845-49. mf 18t-19t: out-of-town marriages 1686-1799. mf 19t-27t: vital records 1843-49. mf 27t-28t: births & deaths 1849. mf 28t-29t: marriage intentions 1772-80) – us Archive [978]

Cambridge, all saints 1538-1950 – 12mf – 9 – £16.00 – uk CambsFHS [929]

[Cambridge-] american poetry review – MA. 1972-1979 – 2r – 1 – $120.00 – mf#R04378 – us Library Micro [420]

Cambridge and dublin mathematical journal – Cambridge. 1837-1854 (1) – mf#5147 – us UMI ProQuest [510]

The cambridge and saybrook platforms of church discipline : with the confession of faith of the new england churches adopted in 1680, and the heads of agreement assented to by the presbyterians and congregationalists in england in 1690 – Boston: TR Marvin, 1829 – 1mf – 9 – 0-524-07558-1 – mf#1991-3178 – us ATLA [242]

The cambridge "apostles" / Brookfield, Frances Mary – New York: Scribner, 1906 – 1mf – 9 – 0-7905-9154-5 – mf#1989-2379 – us ATLA [920]

Cambridge Camden Society *see*
- Church enlargement and church arrangement
- Churches of cambridgeshire and the isle of ely
- A few hints on...study of ecclesiastical architecture
- A few words to church builders
- Instrumenta ecclesiastica

Cambridge characteristics in the seventeenth century : or, the studies of the university and their influence on the character and writings of the most distinguished graduates during that period / Mullinger, James Bass – London: Macmillan, 1867 – 1mf – 9 – 0-7905-5076-8 – (incl bibl ref) – mf#1988-1076 – us ATLA [941]

Cambridge chronicle – Cambridge, NE: Chronicle Publ Co, 1887 (wkly) [mf ed 1888-89 (gaps) filmed 1995] – 1r – 1 – us NE Hist [071]

Cambridge chronicle, the... 1770-1934 – 107r – 1 – mf#13136 – uk Microform Academic [072]

Cambridge city labour party records, 1906-49 / cambridgeshire labour party records, 1918-51 – (= ser Labour party in britain, origins and development at local level. series 1) – 5r – 1 – (int by christopher j howard) – mf#97130 – uk Microform Academic [325]

Cambridge clarion – Cambridge, NE: Clarion Pub Co. v1 n1. feb 3 1899- (wkly) [mf ed with gaps] – 1 – (cont: cambridge kaleidoscope (1898). absorbed: wilsonville review 1966, combine edition 1952) – us NE Hist [071]

Cambridge daily news – Cambridge, England. -d. Jan-Dec 1893. 3 reels – 1 – uk British Libr Newspaper [072]

Cambridge edition – 1984-88 – 7r – 1 – mf#15.38 – nz Nat Libr [079]

Cambridge gazette – Cambridge WI. 1867 sep 6 – 1r – 1 – mf#936868 – us WHS [071]

Cambridge general advertiser – Cambridge, England, 9 jan 1839-25 dec 1850 – 6r – 1 – (fr 31 jul 1839 cambridge advertiser etc; fr 22 jun 1850 new cambridge advertiser; suppls 6 jan-13 oct 1847, 19 jan-20 dec 1848) – uk Newsplan [072]

Cambridge, George *see* The military papers of george cambridge, 2nd duke, 1838-1900

Cambridge guardian, university intelligencer : literary & general advertiser – [East Midlands] Cambridgeshire, 6-3 mar 1838 [mf ed 2004] – 1r – 1 – uk Newsplan [072]

The cambridge history of india – Cambridge, [England]: University Press, 1928- – (= ser Samp: indian books) – us CRL [954]

Cambridge, holy sepulchre 1567-1950 – (= ser Cambridgeshire parish register transcript) – 9mf – 9 – £11.25 – uk CambsFHS [929]

Cambridge, holy trinity 1566-1950 – (= ser Cambridgeshire parish register transcript) – 19mf – 9 – £23.75 – uk CambsFHS [929]

Cambridge independent – 9 jun 1967-16 may 1968; 21 may 1968-24 dec 1968; 21 jan 1969-19 may 1969; 28 oct 1969-9 apr 1970; 14 apr 1970-24 dec 1970; 14 jan 1971-29 apr 1971; 4 may 1971-14 sep 1971; 16 sep 1971-25 jul 1972; aug 1972-dec 1987 – 1 – mf#15.21 – nz Nat Libr [079]

Cambridge intelligencer – Cambridge, England, 27 jul 1793-18 apr 1795, 13, 20 jun, 1 aug 1795-9 sep 1797, 6 jan 1798-17 aug, 2 nov-28 dec 1799, 15 feb-27 dec 1800 (wkly) – 2r – 1 – (cont by: politics radical) – uk Newsplan [072]

Cambridge jeffersonian – Cambridge, OH. may 19, 1870-may 15, 1873 – 2r – 1 – (weekly democratic newspaper) – us Western Res [071]

Cambridge journal – London. 1947-1954 – 1 – mf#2152 – us UMI ProQuest [073]

Cambridge journal of education – Cambridge. 1975-1987 (1) 1975-1987 (5) 1975-1987 (9) – ISSN: 0305-764X – mf#10487 – us UMI ProQuest [370]

Cambridge kaleidoscope – Cambridge, NE: John C Harlan. -v12 n18. nov 6 1896 (wkly) [mf ed 1887-96 (gaps)] – 2r – 1 – (cont by: kaleidoscope. issues for jun 3 1892-nov 6 1896 called also whole n361-591) – us NE Hist [071]

Cambridge kaleidoscope – Cambridge, NE: Kaleidoscope Power Print Co. v13 n27. jan 7 1898-v14 n30. jan 27 1899=whole n652-707 (wkly) [mf ed with gaps] – 1r – 1 – (cont: kaleidoscope. cont by: cambridge clarion) – us NE Hist [071]

Cambridge, Lark A *see* Expressing universal themes through storydance choreography

Cambridge law journal – London. 1990-1991 (1) – ISSN: 0008-1973 – mf#16521 – us UMI ProQuest [340]

Cambridge mission to north india (delhi). report / Cambridge mission to delhi. report and annual reports – 1879-1966 [mf ed 2001] – (= ser Christianity's encounter with world religions, 1850-1950) – 9r – 1 – (a mission of the church of england) – mf#2001-s021-028 – us ATLA [242]

Cambridge news – Cambridge WI. 1886 apr 30/1895 aug 30-1998 jul-dec – 75r – 1 – mf#891563 – us WHS [071]

Cambridge opera journal – Cambridge. 1991+ (1) – ISSN: 0954-5867 – mf#17115 – us UMI ProQuest [780]

Cambridge, our lady and the english martyrs (roman catholic) 1841-1890 – (= ser Cambridgeshire parish register transcript) – 2mf – 9 – £2.50 – uk CambsFHS [929]

The cambridge paragraph bible of the authorized english version – Cambridge [Eng]: University Press 1873 [mf ed 1990] – 14mf – 9 – 0-8370-1853-6 – mf#1987-6240 – us ATLA [220]

Cambridge Philological Society *see* Proceedings of the cambridge philological society

Cambridge Philosophical Society *see*
- Mathematical proceedings of the cambridge philosophical society
- Transactions of the cambridge philosophical society, 1822-1928

The cambridge platform of church discipline : adopted in 1648. and, the confession of faith: adopted in 1680 / Emmons, Nathanael – Boston: Congregational Board of Pub, 1855 [mf ed 1993] – (= ser Congregational coll) – 1mf – 9 – 0-524-06766-X – mf#1991-2773 – us ATLA [242]

The cambridge platonists : being selections from the writings of benjamin whichcote, john smith and nathanael culverwel / Whichcote, Benjamin et al – Oxford: Clarendon Press, 1901 [mf ed 1990] – 1mf – 9 – 0-7905-3806-7 – mf#1989-0299 – us ATLA [180]

Cambridge quarterly – Cambridge. 1990+ (1,5,9) – ISSN: 0008-199X – mf#18487 – us UMI ProQuest [400]

Cambridge quarterly of healthcare ethics: cq – New York. 1995+ (1,5,9) – ISSN: 0963-1801 – mf#21010 – us UMI ProQuest [170]

Cambridge quarterly, the... 1965/6-1971 – v1-8 – 1r – 1 – (with ind) – mf#96912 – uk Microform Academic [420]

Cambridge sermons / Abbott, Edwin Abbott – 2nd ed. London: Macmillan, 1875 – 1mf – 9 – 0-7905-3121-6 – mf#1987-3121 – us ATLA [240]

Cambridge sermons / Lightfoot, Joseph Barber – London; New York: Macmillan, 1890 – 1mf – 9 – 0-7905-7522-1 – mf#1989-0747 – us ATLA [240]

Cambridge sermons / Westcott, Brooke Foss et al; ed by Prior, C H – London: Methuen, 1893 – 1mf – 9 – 0-7905-0378-6 – (incl bibl ref) – mf#1987-0378 – us ATLA [240]

The cambridge shorter history of india / Allan, John et al; ed by Dodwell, H H – Cambridge: University Press, 1934 – (= ser Samp: indian books) – us CRL [954]

Cambridge sporting news – [East Midlands] Cambridgeshire 11 apr-7 dec 1929 [mf ed 2004] – 2r – 1 – uk Newsplan [790]

Cambridge, st andrew the great 1600-1950 – (= ser Cambridgeshire parish register transcript) – 18mf – 9 – £22.50 – uk CambsFHS [929]

Cambridge, st andrew the less 1599-1950 – (= ser Cambridgeshire parish register transcript) – 24mf – 9 – £36.25 – (baptisms only 1599-1950 [9mf] £11.25; marriages & banns only 1599-1950 [11mf] £13.75; burials only 1599-1950 [6mf] £7.50; ind only 1599-1950 [3mf] £3.75) – uk CambsFHS [929]

Cambridge, st barnabas 1878-1950 – 8mf – 9 – £10.00 – uk CambsFHS [929]

Cambridge, st benedict (aka st bene'ts) 1539-1950 – 12mf – 9 – £15.00 – uk CambsFHS [929]

Cambridge, st botolph 1564-1950 – 10mf – 9 – £12.50 – uk CambsFHS [929]

Cambridge, st clement 1560-1992 – (= ser Cambridgeshire parish register transcript) – 11mf – 9 – £13.75 – uk CambsFHS [929]

Cambridge, st edwards 1588-1950 – (= ser Cambridgeshire parish register transcript) – 13mf – 9 – £16.25 – uk CambsFHS [929]

Cambridge, st giles 1585-1950 – (= ser Cambridgeshire parish register transcript) – 21mf – 9 – £26.25 – uk CambsFHS [929]

Cambridge, st mary the great 1559-1950 – (= ser Cambridgeshire parish register transcript) – 13mf – 9 – £16.25 – uk CambsFHS [929]

Cambridge: st mary the less 1557-1950 – (= ser Cambridgeshire parish register transcript) – 11mf – 9 – £14.75 – uk CambsFHS [929]

Cambridge, st michael's 1538-1950 – 6mf – 9 – £7.50 – uk CambsFHS [929]

Cambridge, st paul's 1845-1950 – 14mf – 9 – £17.50 – uk CambsFHS [929]

Cambridge, st peter's 1586-1908 – (= ser Cambridgeshire parish register transcript) – 5mf – 9 – £6.25 – uk CambsFHS [929]

Cambridge, st philip's 1903-1950 – (= ser Cambridgeshire parish register transcript) – 7mf – 9 – £8.75 – uk CambsFHS [929]

Cambridge Tenants Organizing Committee *see* Tenants' newsletter

Cambridge University. Library *see*
- A catalog of the manuscripts preserved in the library of the university of cambridge
- Early english printed books 1475-1640

Cambridge university magazine – London. 1840-1843 – 1 – mf#4221 – us UMI ProQuest [378]

Cambridgeshire and eastern counties weekly gazette – [East Midlands] Cambridgeshire 3 mar-dec 1899 [mf ed 2004] – 2r – 1 – (ceased publ) – uk Newsplan [072]

Cambridgeshire and huntingdonshire births & baptisms register at the wesleyan methodists metropolitan registry 1812-1837 – 1mf – 9 – £1.25 – uk CambsFHS [242]

Cambridgeshire and huntingdonshire women in the british lying-in-hospital 1750-1867 – 1mf – 9 – £1.25 – uk CambsFHS [941]

Cambridgeshire baptism index 1801-1837 – 42mf, 2mf per set of surnames – 9 – £36.00 set £2.50 per set of 2mf – (surnames: abi – barker, james; barker, john r – bond, rachel; bond, reason – butler, mary; butler, mary – claypole, mary; claypole, mary – croxon, mary; croxon, mary a – edwards, thomas; edwards, thomas – frost, jane; frost, jemima – green, mary; green, mary – heigho, james; heigho, john c – iseson, elizabeth; isgate, john – larkins, samuel; larkins, samuel – marshall, william; marshall, william – naylor, mary a; naylor, mary a – patman, sarah; patman, sarah – purr, ezra; purr, george – rutherford, robert g; ruttleton, thomas – smith, hannah; smith, hannah – tarlton, john; tarndley, samuel – waller, john; waller, john – williams, harriet; williams, harriet – yules, thomas) – uk CambsFHS [929]

Cambridgeshire baptisms registered in dr william's library – 4mf – 9 – £5.00 – (incl cambridgeshire and huntingdonshire) – uk CambsFHS [243]

Cambridgeshire census 1841 – 1mf for ea parish unless otherwise stated – 9 – £1.25mf – (abington, great & little; abington, little [o/n1841/20]; abington pigotts & litlington [o/n1841/50]; arrington, longstowe & whaddon [o/n1841/21]; ashley & cheveley [o/n1841/1]; babraham & stapleford [o/n1841/22]; balsham [o/n1841/51]; barrington & meldreth [o/n1841/53]; bartlow, hildersham & shudy camps [o/n1841/52]; barton, coton & grantchester [o/n1841/23]; bassingbourn & royston 2mf [o/n1841/54]; benwick [o/n1841/55]; bottisham [o/n1841/56]; bourn & knapwell [o/n1841/57]; boxworth, conington & lolworth [o/n1841/24]; brinkley & stetchworth [o/n1841/9]; burrough green & dullingham [o/n1841/25]; burwell

2mf [o/n1841/26]; caldecote, childerley & dry drayton [o/n1841/27]; cambridge all saints & st giles 2mf [o/n1841/94]; cambridge holy sepulchre & st botolph [o/n1841/58]; cambridge, holy trinity & st edward 2mf [[order n1841/81]; cambridge st andrew the great 2mf [o/n1841/82]; cambridge st andrew the less 5mf [o/n1841/28]; cambridge st benet with addenbrookes [o/n1841/59]; cambridge st mary the great & st mary the less [o/n1841/60]; cambridge st michael & st peter [o/n1841/61]; cambridge st clement [o/n1841/10]; cambridge university [o/n1841/29]; carlton cum willingham, westley waterless & weston colville [o/n1841/95]; castle camps & horseheath [o/n1841/2]; caxton & papworths [o/n1841/3]; chatteris 3mf [o/n1841/62]; cherry hinton & fen ditton [o/n1841/63]; chesterton [o/n1841/64]; chippenham, kennett, landwade & snailwell [o/n1841/65]; comberton [o/n1841/66]; cottenham [o/n1841/67]; coveney & manea [o/n1841/96]; croxton & eltisley [o/n1841/30]; croydon cum clopton, shingay, tadlow & wendy [o/n1841/11]; enquire for further details regarding parishes d-w) – uk CambsFHS [350]

Cambridgeshire census 1851 – 188mf set – 9 – £190.00 set £1.25mf – (enquire for further details regarding parishes a-w; overall surname ind [4mf] £5) – uk CambsFHS [350]

Cambridgeshire census 1861 – 1mf for ea parish unless otherwise stated – 9 – £1.25mf – (abington pigotts [o/n1861/68]; arrington [o/n1861/1]; ashley, snailwell, landwade & kennett [o/n1861/2]; babraham [o/n1861/48]; balsham [o/n1861/54]; barrington [o/n1861/73]; bartlow [o/n1861/96]; barton [o/n1861/97]; bassingbourn-cum-kneesworth 2mf [o/n1861/74]; benwick [o/n1861/129]; bottisham [o/n1861/109]; bourn [o/n1861/9]; boxworth [o/n1861/146]; brinkley [o/n1861/10]; burrough green & westley waterless [o/n1861/92]; burwell & reach [o/n1861/3]; enquire for further details regarding parishes c-y) – uk CambsFHS [350]

Cambridgeshire census 1871 – 1mf for ea parish unless otherwise stated – 9 – £1.25mf – (abington pigotts [o/n1871/37]; arrington [o/n1871/46]; ashley-cum-silverly[o/n1871/9]; babraham [o/n1871/123]; balsham [o/n1871/117]; barrington [o/n1871/36]; bartlow [o/n1871/61]; barton [o/n1871/47]; bassingbourn 2mf [o/n1871/62]; benwick [o/n1871/65]; bottisham 2mf [o/n1871/130]; bourn [o/n1871/48]; boxworth [o/n1871/10]; brinkley [o/n1871/142]; burrough green [o/n1871/104]; burwell & swaffham prior 3mf [o/n1871/138]; enquire for further details regarding parishes c-w) – uk CambsFHS [350]

Cambridgeshire census 1891 – 1mf for ea parish unless otherwise stated – 9 – £1.25mf – (cambridge st andrew the less 10mf [o/n1891/1]; cambridge st clements [o/n1891/2]; cambridge (chesterton) st luke's 4mf [o/n1891/6]; castle camps [o/n1891/9]; hardwick [o/n1891/4]; histon [o/n1891/5]; impington [o/n1891/3]; linton 2mf [o/n1891/8]; horseheath [o/n1891/11]; shudy camps [o/n1891/10]; swavesey [o/n1891/7]) – uk CambsFHS [350]

Cambridgeshire census 1901 – 1mf for ea parish unless otherwise stated – 9 – £1.25mf – (boxworth [o/n1901/2]; castle camps [o/n1901/15]; conington [o/n1901/3]; elsworth [o/n1901/6]; fen drayton [o/n1901/4]; hardwick [o/n1901/11]; histon [o/n1901/1]; horseheath [o/n1901/16]; impington [o/n1901/12]; kirtling [o/n1901/13]; knapwell [o/n1901/6]; lolworth [o/n1901/9]; longstanton [o/n1901/5]; over [o/n1901/8]; papworth everard & st agnes [o/n1901/7]; shudy camps [o/n1901/14]; swavesey [o/n1901/10]) – uk CambsFHS [350]

Cambridgeshire census strays 1998 – 6mf – 9 – £7.50 – uk CambsFHS [929]

Cambridgeshire deserters – 1mf – 9 – £1.25 – (transcr fr the police gazette 1828-1845) – uk CambsFHS [941]

Cambridgeshire index 1801-1837 – 28mf set 2mf per set of surnames – 9 – £25.00 set £2.50 per set of 2mf – (surnames: abbess, godfrey -beckett, william; beckett, william – butcher, elizabeth; butcher, elizabeth – cornwell, elizabeth; cornwell, elizath – ellingham, charles; ellingham, david – golding, william; golding, william – hilger, rose; hilger, william – kerridge, william; kerry, elizabeth – masters, susan; masters, thomas – oyston, sarah; pace, mary rawlings, james; rawlings, james – shippey, henry; shippey, james – theake, susan; theake, william – westwood, peter; westwood, rachel – yules, thomas) – uk CambsFHS [929]

Cambridgeshire labour party records, 1918-51 *see* Cambridge city labour party records, 1906-49 / cambridgeshire labour party records, 1918-51

Cambridgeshire marriage index 1626-1675 – 9 – (men [8mf] £10; women [8mf] £10) – uk CambsFHS [929]

Cambridgeshire masters and their apprentices – 6mf – 9 – £7.50 – (transcr fr apprenticeship records for the years 1763-1811) – uk CambsFHS [941]

Cambridgeshire monumental inscriptions – 1mf for ea parish unless otherwise stated – 9 – £1.25mf – (enquire for further details regarding parishes a-w) – uk CambsFHS [941]

Cambridgeshire poor law papers – 6mf – 9 – £7.50 – (ind of settlement papers, certificates, examinations, removal orders, vagrancy passes, apprenticeships and bastardy orders fr 1697) – uk CambsFHS [941]

Cambridgeshire tax on male servants 1780 – 1mf – 9 – £1.25 – uk CambsFHS [336]

The cambro-briton : and general celtic repository (london) – sep 1819-jun 1822 – (= ser 19th c british periodicals) – r24 – 1 – us Primary [073]

Cambry, Jacques *see* Reise durch einen theil des westlichen frankreichs

O cambucyense – Rio de Janeiro, RJ. 02-10 jun 1904 – (= ser Ps 19) – mf#DIPER – bl Biblioteca [079]

Cambuslang advertiser : circulating in parliamentary constituency of rutherglen – [Scotland] South Lanarkshire, Cambuslang: J Lithgow & Sons 28 may 1898-dec 1950 (wkly) [mf ed 2004] – 32r – 1 – (absorbed by: east kilbride news; subtitle varies) – uk Newsplan [072]

Cambuslang and rutherglen express – [Scotland] Rutherglen: G Hutcheson & Co 7 sep 1898-19 nov 1902 (wkly) [mf ed 2003] – 220v on 2r – 1 – uk Newsplan [072]

Cambuslang pilot – [Scotland] South Lanarkshire, Hamilton: Hamilton Herald Printing & Publ Co Ltd 2 jan 1920-25 dec 1925 (wkly) [mf ed 2004] – 2r – 1 – (cont by: pilot and cambuslang district news [12 jul 1940-23 feb 1945]; circulated in cambuslang, newton, halfway, westburn, silverbanks, kirkhill...etc) – uk Newsplan [072]

Camden advertiser – Camden, 1936-57 (misc iss) – 3r – at Pascoe [079]

Camden and hampstead and highgate record and chronicle *see* Hampstead record

Camden and holborn and finsbury guardian *see* Holborn guardian and bloomsbury chronicle

Camden and s[/ain/]t pancras chronicle – London UK, 1964, 1986-92 – 18 1/2r – 1 – (aka: peoples advertiser; peoples advertiser sale and exchange gazette; s[ain]t pancras chronicle peoples advertiser sale and exchange gazette) – uk British Libr Newspaper [072]

The camden colony : or, the seed of the righteous: a story of the united empire loyalists, with genealogical tables / Tucker, William Bowman – Montreal: J Lovell, 1908 – 3mf – 9 – 0-665-74572-9 – mf#74572 – cn Canadiana [929]

Camden Defense Committee *see* Viewpoints

Camden haven courier – Laurieton, jan-dec 1969 – 1r – at Pascoe [079]

Camden journal and hampstead news – London. -w. 1973-24 nov 1978; 26 Jan 1979-1980 16 r – 1 – uk British Libr Newspaper [072]

Camden labor – Camden NJ. 1905 jan 8 – 1r – 1 – mf#868986 – us WHS [071]

Camden new journal – London UK – 1 – uk British Libr Newspaper [072]

Camden news – Camden, jan 1969-jun 1982 – 14r – 9 – at Pascoe [079]

Camden news – Camden, jun 1895-dec 1968 – 24r – A$1805.80 vesicular A$1937.80 silver – at Pascoe [079]

Camden republican – United States. 26 Oct 1839-27 May 1843.-w. 1 reel – 1 – uk British Libr Newspaper [071]

Camden Society *see* Publications

Camden society. publications – v1-62 1871-1901 – 1 – $390.00 – mf#0133 – us Brook [941]

Camden, W *see* A chronographical description of england, scotland, ireland and islands adjacent

Camellias in california – S.l., S.L., S.l? . 1930 – 1r – us UF Libraries [630]

Camelot, P Th *see* Foi et gnose

Cameo glass *see* History of glass

Cameos from the life of george fox / Taylor, Ernest Edwin – London: Headley Bros, [1907?] – 1mf – 9 – 0-524-01216-4 – mf#1990-4074 – us ATLA [920]

Cameos of a chinese city / Darley, Mary – London: Church of England Zenana missionary Society; Marshall Bros, 1917 [mf ed 1995] – (= ser Yale coll) – 210p (ill) – 1 – 0-524-09517-5 – mf#1995-0517 – us ATLA [915]

Camera : madjalah film dan umum / Camera Press – Djakarta, 1963-1965 – 13mf – 9 – (missing: 1963/1964, v1(9-end); 1964, v2(11-end); 1965, v3(1-2)) – mf#SE-877 – ne IDC [959]

Camera – Munich. 1950-1981 (1) 1971-1981 (5) 1976-1981 (9) – ISSN: 0366-7073 – mf#524 – us UMI ProQuest [770]

Camera and darkroom – Beverly Hills. 1990-1995 (1,5,9) – (cont: darkroom photography) – ISSN: 1056-8484 – mf#11830,01 – us UMI ProQuest [770]

Camera craftsman – Denver. 1955-1980 (1) 1955-1980 (5) 1955-1980 (9) – ISSN: 0527-3919 – mf#10227 – us UMI ProQuest [770]

Camera notes – New York: Camera Club. v1-6. 1897-1903 – 2r – 1 – us CRL [760]

Camera obscura – Berkeley. 1989-1996 (1,5,9) – ISSN: 0270-5346 – mf#18006 – us UMI ProQuest [770]

Camera obscura : or, life in glasgow – [Scotland] Glasgow : Printed by W. Clark & Co 28 aug-25 dec 1830 & 30 apr-7 may 1831 (wkly) [mf ed 2004] – 1r – 1 – (containing literature, sporting intelligence, and police reports) – uk Newsplan [072]

Camera Press *see* Camera

Camera thirty-five – Palisades. 1971-1982 (1) 1957-1982 (5) 1974-1982 (9) – ISSN: 0008-2171 – mf#6090 – us UMI ProQuest [770]

The cameralists : the pioneers of german social polity / Small, Albion Woodbury – Chicago: University of Chicago Press [etc], 1909 [mf ed 1970] – (= ser Library of american civilization 10353) – xxv/606p on 1mf – 9 – us Chicago U Pr [943]

Camerarius, J *see*
- Capita pietatis et religionis christianae
- De philippi melanchthonis ortv, totivs vitae cvrricvlo et morte
- De rebvs tvrcicis commentarii dvo accvratissimi
- Homiliae qvi svnt sermones habiti de iis, qvae in christianis ecclesiis legvntvr
- Symbolorum ac emblematum ethico-politicorum centuriae quatuor...
- Symbolorum et emblematum...
- Symbolorum et emblematum
- Symbolorum et emblematum centuriae...
- Symbolorum et emblematum centuriae tres
- Symbolorum et emblematum ex animalibus quadrupedibus desumtorum centuria altera collecta...
- Symbolorum et emblematum ex aquatilibus et reptilibus desumpforum
- Symbolorum et emblematum ex re herbaria desumtorum centuria una collecta a...
- Vier hundert wahl-sprueche und sinnenbilder durch welche beygebracht und aussgelegt werden

Camerer, Theodor *see* Spinoza und schleiermacher

Camerer, W *see* Eduard moerike und klara neuffer

Camerlynck, Achille *see* Commentarius in actus apostolorum

Cameron, Alexander Mackenzie *see* New south wales

Cameron, Angus deMille *see* John white chadwick

Cameron, C R *see* Nature and design of the church

Cameron, Charles Hay *see* An address to parliament on the duties of great britain to india

Cameron, Charles Innes *see* Poems and hymns

Cameron, Charles Richard *see* Considerations on the divine authority of the lord's day

Cameron county echo – Emporium, PA. -w 1972-1980. 7 rolls – 13 – $25.00r – us IMR [071]

Cameron county press – Emporium, PA. -w 1889-1928 – 13 – $25.00r – us IMR [071]

Cameron county press and emporium independent – Emporium, PA. 1879-1971 (1) – mf#68914 – us UMI ProQuest [071]

Cameron, D *see* Discourse delivered before the synod of ross

Cameron, D R *see* Remarks upon the true location of the international boundary line at the mouth of the river st clair

Cameron, Donald *see* To the canadian public...

Cameron, Donald Roderick *see*
- An aid to national defence
- Correspondence relating to the eastern boundary of the province

Cameron echo – Cameron WI. 1931 mar 11-1935 aug 29, 1935 sep 5-1939 jun 29, 1939 jul 6-1941 dec 25, 1946 apr 18-1951 jun 13 – 4r – 1 – (cont by: barron county leader) – mf#966483 – us WHS [071]

Cameron, George Frederick *see*
- An entirely new and original military opera in three acts, entitled
- Leo, the royal cadet

Cameron, George G *see* Persepolis treasury tablets

Cameron, Henry Clay *see* Jonathan dickinson and the college of new jersey

Cameron, James *see* African revolution

Cameron, James Chalmers et al *see* An american text-book of obstetrics for practioners and students

Cameron, James Robertson *see* The renascence of jesus

Cameron, John *see* Lectures on "infant church membership, etc"

Cameron, John Hillyard *see* The digest of cases determined in the court of queen's bench from michaelmas term, tenth george 4, to hilary term, third victoria

Cameron, John Home *see* Le voyage de monsieur perrichon

Cameron, John Kennedy *see* The free church of scotland, 1843-1910

Cameron, Malcolm Graeme *see*
- The ditches and watercourses acts of ontario
- A treatise on the law of dower

Cameron, Mrs *see*
- Bee-hive cottage
- Coronation
- Honest penny is worth a silver shilling

Cameron, Norman Eustace *see* History of the queen's college of british guiana

Cameron review – Cameron WI 1915 apr 2 – 1r – 1 – mf#966159 – us WHS [071]

Cameron, Robert *see* The doctrine of the ages

Cameron, Simon *see* Papers

Cameron, VL *see* Across africa

Cameron's directory and railway and steamship guide to toronto – [Toronto?: Cameron?, 1892-189- or 19–] [mf ed n1 jun 1892] – 9 – mf#P04856 – cn Canadiana [380]

Cameron's legal opinions / Canada. General – 1v. 1859-76 (all publ) – 3mf – 9 – $4.50 – mf#LLMC 81-006 – us LLMC [340]

Cameron's supreme court cases / Canada. Supreme Court – 1v. 1880-1900 (all publ) – 7mf – 9 – $10.50 – mf#LLMC 81-007 – us LLMC [347]

Cameroon / Gardinier, David E – London, England. 1963 – 1r – us UF Libraries [960]

Cameroon. Direction de la Statistique et de la Comptabilite Nationale *see* Note annuelle statistique 1973-1975

Cameroon tribune : english edition – Yaounde, Cameroon. Sept 5 1984-Feb 7 1992 – 13r – 1 – (scattered issues lacking) – us L of C Photodup [079]

Cameroons champion – Victoria, Cameroon. 26 nov-24 dec 1960; 4 feb-1 sep 1961 (imperfect) – 1/4r – 1 – uk British Libr Newspaper [079]

Cameroons under united kingdom administration / Great Britain. Colonial Office – 1947-49 – 1 – us CRL [960]

Cameroun / Wilbois, Joseph – Paris, France. 1934 – 1r – us UF Libraries [025]

Cameroun *see* Agence economique des territoires francais sous mandat

Cameroun libre – Yaounde, Cameroon. 20 feb 1942-15 jul 1945 – 1r – 1 – uk British Libr Newspaper [079]

Cameroun physique / Billard, Pierre – Lyon, France. 1962 – 1r – us UF Libraries [960]

Cameroun political ephemera, 1952-1961 – St Louis, MO: Washington Uni Libraries, 1964 – 1 – us CRL [080]

Cameroun political ephemera, 1954-1963 / Welch, Claude Emerson – Palo Alato, CA: Stanford Uni Photographic Dept, 1965 – 1 – us CRL [960]

Cami *see* Trablusgarbdan sahra-i kebire dogru

Camille, Roussan *see* Gerbe pour deux amis

Camille uesmoulins / Blanchard, Henri – Paris, France. 1850? – 1r – us UF Libraries [440]

Camilli, C *see* Imprese illustri di diversi con discorsi di camillo camilli

Camilo torres / Forero, Manuel Jose – Bogota, Colombia. 1952 – 1r – us UF Libraries [972]

Camin, Alfonso *see*
- El adelantado de la florida pedro menendez de aviles
- Alabastros

Camina de justicia – v1 n2 [1970 apr] – 1r – 1 – mf#720852 – us WHS [071]

Caminando por la literatura hispanica, 1948-1964 / Fernandez Spencer, Antonio – Santo Domingo, Dominican Republic. 1964 – 1r – us UF Libraries [972]

Caminero J, Luis Augusto *see* Letras de mi jardin

Camingos cruzados / Verissimo, Erico – Lisboa, Portugal. 1947 – 1r – us UF Libraries [972]

O caminho da escola : luta popular pela escola publica / Ribero, Vera Masagao – Sao Paulo: Centro Ecumenico de Documentacao e Informacao, 1986 – (texto baseado nos depoimentos de adriano diogo...[et al]) – us CRL [972]

Caminho de ferro de quelimane eo futuro da colonia portugueza / Stuchy, Joseph E – Lisboa, Portugal. 1904 – 1r – us UF Libraries [960]

Caminhos antigos e povoamento do brasil / Abreu, Joao Capistrano De – Rio de Janeiro, Brazil. 1960 – 1r – us UF Libraries [972]

Caminhos do novo mundo / Fernandes, Jose Fonseca – Rio de Janeiro, Brazil. 1965 – 1r – us UF Libraries [972]

Caminhos historicos de invasao / Souza, Antonio De – Rio de Janeiro, Brazil. 1950 – 1r – us UF Libraries [972]

Camino / Breton, Jose – Santiago de los Caballeros, Dominican Republic. 1937 – 1r – us UF Libraries [972]

Camino Burgos, Luis G *see* Los besos bajo tierra (poemas del amor y del desamos)

El camino de la unidad : un discurso socialista / Lamoneda, Ramon – Madrid? 1937? – (= ser Blodgett coll) – 9 – mf#fiche w984 – us Harvard [946]

El camino de la victoria : ilamamiento del partido comunista a todos los pueblos de espana y a cuantos aman la paz, el progreso y la libertad – [Valencia: s.n. 193-?] [mf ed 1977] – (= ser Blodgett coll) – 1mf – 9 – mf#w771 – us Harvard [946]
Camino de marti / Lizaso, Felix – Habana, Cuba. 1953 – 1r – us UF Libraries [972]
Camino de sombras (poemas) / Jimenez Malaret, Rene – San Juan, Puerto Rico. 1956 – 1r – us UF Libraries [972]
Camino del cielo / Purificacion y Tornavacas, Pedro de – 1818 – 9 – sp Bibl Santa Ana [240]
Camino del cielo / Suarez de Figueroa, Diego – 1739. 3v – 9 – sp Bibl Santa Ana [946]
Camino del cielo en lengua mexicana : con todos los requisitos necessarios para conseguir este fin, co[n] todo lo que vn [christ]iano deue creer, saber, y obrar, desde el punto que tiene vso de razon, hasta que muere / Leon, Martin de – en Mexico: impr Diego Lopez Danalos...ano de 1611 – (= ser Books on religion...1543/44-c1800: doctrina cristiana, obras de devocion) – 4mf – 9 – mf#crl-43 – ne IDC [241]
Un camino en la selva / Barreda, Ernesto Maria – Buenos Aires. 1916 – 1 – us CRL [830]
Camino espiritual / San Pedro de Alcantara, Domingo de – Caceres: Tip. Extremadura, 1962 – 1 – sp Bibl Santa Ana [240]
Caminos / Diaz Martinez, Manuel – Habana, Cuba. 1962 – 1r – us UF Libraries [972]
Caminos de espana : plasencia 2 – Madrid: Compania Espanola de Penicilina, S.A., 1951 – sp Bibl Santa Ana [914]
Caminos de espana : zafra 1 – Madrid: Compania Espanola de Penicilina, S.A., 1952 – sp Bibl Santa Ana [914]
Caminos de espana : zafra 2 – Madrid: Compania Espanola de Penicilina, S.A., 1958 – sp Bibl Santa Ana [914]
Caminos de espana. merida. ruta 44 – Madrid: Imp. Lonja y Cia, 1960 – 1 – sp Bibl Santa Ana [914]
Caminos de espana: merida-zafra – Madrid: Compania Espanola de Penicilina, 1960 – 1 – sp Bibl Santa Ana [946]
Caminos de guerra y conspiracion / Rodriguez, Amadeo – Barcelona, Spain. 1955 – 1r – us UF Libraries [972]
Caminos de renunciacion / Zepeda Turcios, Roberto – Tegucigalpa, Mexico. 1947 – 1r – us UF Libraries [972]
Caminos de servidumbre / Romero Mendoza, Pedro – Madrid: Imp. G. Hernandez y Galo Saez, 1926 – 1 – sp Bibl Santa Ana [830]
Caminos del agro / Castillo, Moises – Panama, 1959 – 1r – us UF Libraries [972]
Caminos del aire / UNIVERSITY OF PUERTO RICO (RIO PIEDRAS CAMPUS) – Rio Piedras, Puerto Rico. 1951 – 1r – us UF Libraries [972]
Caminos del alma / Merchan Cantisan, Felisa – Madrid: Tip. Rubio y Castro, S.L. 1975 – 1 – sp Bibl Santa Ana [946]
Caminos del alma / Perez Alonso, Jaime – Managua, Nicaragua. 1960 – 1r – us UF Libraries [972]
Caminos y luchas por la independencia / Valdes Oliva, Arturo – Guatemala, 1956 – 1r – us UF Libraries [972]
Camirand, J A *see* Bulletin sur les chemins
The camisards : a sequel to the huguenots in the seventeenth century / Tylor, Charles – London: Simpkin, Marshall, Hamilton, Kent: E Hicks, 1893 – 2mf – 9 – 0-7905-6377-0 – (incl bibl ref) – mf#1988-2377 – us ATLA [240]
The camisards : a sequel to the huguenots in the seventeenth century / Tylor, Charles – London: Simpkin, Marshall, Hamilton, Kent: E. Hicks, 1893 – 2mf – us ATLA [305]
Camlibel, Faruk Nafiz *see*
– Canavar
– Suda halkalar
Camloey dyk bhnaek : pralom lok jivit bit knun paccuppann / Hak Chai Huk – Bhnam Ben: Pannalay Vannakamm Khmaer 2509 [1966] [mf ed 1990] – 1r with other items – 1 – (in khmer) – mf#mf-10289 seam reel 097/06 [§] – us CRL [480]
Camm, Bede *see*
– Courtier, monk and martyr
– Tyburn conferences
– William cardinal allen
Cammann, Henry J *see* The charities of new york, brooklyn, and staten island
Camocio, G F *see* L'ordine delle galere et insegne loro, con li fano, nomi, et cognomi delli magnifici, et generosi patroni de esse...
Camoenopaedia sacra concertus vocant 2. 3. 4. 5. 6. 8. vocum / Kraf(f), Michael – 1627 – (= ser Mssa) – 4mf – 9 – €60.00 – mfchl 274 – gw Fischer [780]
Camoens / Tapia Y Rivera, Alejandro – San Juan, Puerto Rico. 1944 – 1r – us UF Libraries [972]
Camoens a calderon en el centenario de este / Coronado, Carolina – 1881 – 9 – sp Bibl Santa Ana [440]
Camoes, Luis De *see* Lusiadas
Camp, A F *see*
– Citrus propagation
– Japanese persimmon in florida
– Soil temperature studies with cotton
– Some symptoms of citrus malnutrition in florida
Camp adair sentry – Camp Adair OR: Sentry Pub, 1942-44 [wkly] – 1 – us Oregon Lib [355]
Camp and cantonment : a journal of life in india in 1857-1859, with some account of the way thither...to which is added a short narrative of the pursuit of the rebels in central india / Paget, Georgiana Theodosia (Fitzmoor-Halsey) – London 1865 – (= ser 19th c british colonization) – 5mf – 9 – mf#1.1.7237 – uk Chadwyck [954]
Camp and lamp : rambles in realms of sport, story, song / Baylis, Samuel Mathewson – Montreal: W Drysdale, 1897 [mf ed 1980] – 4mf – 9 – (incl ind) – mf#03508 – cn Canadiana [790]
Camp and mill news / Woodworkers' Industrial Union of Canada – v1 n1 [1949 apr 26] – 1r – 1 – mf#681694 – us WHS [634]
Camp cleghorn assembly herald – 1899 sep-1907jun, 1899-1908 – 2r – 1 – mf#2892868 – us WHS [071]
Camp crier – 1971-76 – (= ser American indian periodicals... 1) – 33mf – 9 – $210.00 – us UPA [305]
'The camp doctor and other stories' / Young, Egerton Ryerson – Toronto: Musson, [1909?] – 4mf – 9 – 0-659-91941-9 – mf#9-91941 – cn Canadiana [830]
Camp, Edgar W et al *see* Encyclopaedia of evidence
The camp fire : a monthly record and advocate of the temperance reform – Toronto: F S Spence, [1894-189- or 19–] – 9 – mf#P04052 – cn Canadiana [360]
Camp, Hugh N *see* The charities of new york, brooklyn, and staten island
Camp, John Perlin *see* Effect of copper sulfate and potassium arsenate on the accumulatio...
Camp lejeune globe – Camp Lejeune NC. 1987, 1988, 1990 – 3r – 1 – (cont: globe [camp lejeune nc: 1976]; cont by: globe [camp lejeune nc: 1994]) – mf#3627483 – us WHS [071]
Camp lejeune globe – Camp Lejeune NC. 1956/57-1976 jan/jul 15 – 15r – 1 – (cont: globe [camp lejeune nc]; cont by: globe [camp lejeune nc: 1976]) – mf#3627466 – us WHS [071]
Camp lejeune globe – Camp Lejeune, New River NC. 1944 feb 23-1945 jan, 1945-46, 1947-1948 sep 2 – 3r – 1 – (cont: new river pioneer; cont by: globe [camp lejeune nc]) – mf#704409 – us WHS [071]
Camp libertad periodico – 1980 may 14-sep 24 – 1r – 1 – mf#512957 – us WHS [320]
Camp life and sport in south africa : experiences of kaffir warfare with the cape mounted rifles / Lucas, Thomas J – London, 1878 – (= ser 19th c british colonization) – 4mf – 9 – mf#1.1.5689 – uk Chadwyck [790]
Camp meeting manual : a practical book for the camp ground, in two parts / Gorham, Barlow Weed – Boston: H V Degen, 1854 [mf ed 1984] – 2mf – 9 – 0-8370-0946-4 – mf#1984-4299 – us ATLA [242]
Camp New Amsterdam [Huls ter Heide, Netherlands] *see* Interceptor
Camp news / Chicago Area Military Project – v1 n1-v4 n8 [1970 feb 24-1973 aug 15] – 1r – 1 – mf#1336843 – us WHS [355]
Camp niagara : with a historical sketch of niagara-on-the-lake and niagara camp / Cruikshank, Ernest Alexander – Niagara Falls [Ont]: F H Leslie, [1906?] – 1mf – 9 – 0-665-72124-2 – mf#72124 – cn Canadiana [355]
Camp, Paul D *see* Study of range cattle management in alachua county
Camp wallace trainer / Antiaircraft Replacement Training Center [TX] – 1943 jul 30-sep 3, sep 17, oct 1-8, oct 22 – 1r – 1 – mf#928469 – us WHS [355]
Camp, Walter Chauncey *see* How to play football
Camp, Walter Mason *see* Notes on track
Campa Y Caraveda, Miguel Angel *see*
– Cenizas gloriosas
– Politica regional del caribe
Campagne a deux / Dupeuty, Charles – Paris, France. 1843 – 1r – us UF Libraries [440]
Campagne contre le papier-monnaie / Bijou, Cajuste – Port-Au-Prince, Haiti. 1898 – 1r – us UF Libraries [972]
Campagne dans le haut senegal et dans le haut niger, 1885-1886 / Frey, Henri Nicolas – Paris: E Plon, Nourrit, 1888 – 1 – us CRL [960]
Campagne electorale : le toryisme, voila l'ennemi!: importance de la prochaine election et l'avenir du pays / Liberal Party of Canada – [Ottawa?: s.n, 1900?] – 1mf – 9 – 0-665-91746-5 – (incl english text) – mf#91746 – cn Canadiana [325]
Campagne in frankreich / Goethe, Johann Wolfgang von; ed by Direction de l'education publique, G.M.Z.F.O. – Offenburg/Baden: Lehrmittel-Verlag [1946] [mf ed 1990] – (= ser Klassiker der weltliteratur) – 1r – 1 – (filmed with: goethes campagne in frankreich, 1792) – mf#7321 – us Wisconsin U Libr [933]
La campagne politico-religieuse de 1896-1897 / Landry, Philippe – Quebec?: s.n, 1897 – 2mf – 9 – mf#30307 – cn Canadiana [370]
Campagnoli, Bartolomeo *see* L'arte d'inventer a l'improviste des fantaisies et cadences pour le violon
Campaign... / United Way of Greater Milwaukee – 1987 oct 30, nov 16 – 1r – 1 – mf#2174435 – us WHS [071]
Campaign act amendments, 1974, 1976 and 1979 – Washington, 1977, 1983 – 33mf – 9 – $49.50 – mf#llmc 90-370 – us LLMC [342]
The campaign against residential segregation, 1914-1955 – (= ser Papers of the naacp 5) – 23r – 1 – $4455.00 – 0-89093-968-3 – (suppl: residential segregation, general office files, 1956-65 16r isbn 1-55655-545-8 $3115. with p/g) – us UPA [322]
Campaign Against Utility Service Exploitation *see* News from cause
Campaign constitution – Washington DC. 1860 aug 23 – 1r – 1 – mf#933962 – us WHS [325]
Campaign finance law / Federal Election Commission – 1984, 1986, 1988, 1990, 1992 – 24mf – 9 – $36.00 – mf#LLMC 95-027 – us LLMC [346]
Campaign for Economic Democracy (CA) *see* Economic democrat
Campaign for Economic Democracy [CA] *see* Ced news
The campaign for educational equality, 1913-1965 – 4ser – (= ser Papers of the naacp 3) – 1 – (ser a: legal dept & central office records, 1913-40 24r isbn 0-89093-893-8 $4640. ser b: legal dept & central office records, 1940-50 19r isbn 0-89093-894-6 $3685. ser c: legal dept & central office records, 1951-55 23r isbn 1-55655-543-1 $4455. ser d: general office files, 1956-65 13r isbn 1-55655-553-9 $2520. with p/g) – us UPA [322]
Campaign for Political Rights [US] et al *see* Organizing notes
Campaign for U N Reform [Organization] *see* U n reform campaigner
The campaign in halton, mr macdougall's record : speech delivered by mr pattullo, secretary of the reform association of the province of ontario, at georgetown, friday evening, august 30th, 1878 – [Toronto?: s.n.], 1878 [mf ed 1987] – 1mf – 9 – 0-665-63103-0 – mf#63103 – cn Canadiana [325]
Campaign of 1813 on the ohio frontier : sortie at fort meigs, may 1813, address of thomas christian, a volunteer in col dudley's regiment – s.l: s.n, 1870? – 1mf – 9 – mf#58985 – cn Canadiana [975]
The campaign of adowa and the rise of menelik / Berkeley, G F H – London, 1902 – 5mf – 9 – mf#NE-20238 – ne IDC [956]
Campaign text books / Democratic Party. National Committee – v1-17. 1876-1940 [all publ] – 9 – $198.00 – mf#0178 – us Brook [325]
Campaign update / National Peace Academy Campaign [US] – v1 n1-v7 n1 [1978 spr-1984 spr] – 1r – 1 – (cont by: peace institute report) – mf#1096459 – us WHS [327]
Campaigner / Prohibition Party [WI] – 1915 may-1928 aug – 1r – 1 – (cont: prohibition journal; cont by: forward press) – mf#3500375 – us WHS [325]
Campaigner for economic democracy – v1 n1 [1977 mar] – 1r – 1 – (cont by: ced news) – mf#667561 – us WHS [325]
Campaigning for christ in japan / Wainright, Samuel Hayman – Nashville; Dallas: Publ House of the M E Church, South, 1915 [mf ed 1995] – (= ser Yale coll) – 170p – 1 – 0-524-09523-X – mf#1995-0523 – us ATLA [240]
Campaigning in cuba / Kennan, George – New York, NY. 1899 – 1r – us UF Libraries [972]
Campaigning in south africa and egypt / Molyneux, W C F – London, 1896 – 4mf – 9 – mf#HT-93 – ne IDC [916]
Campaigns and elections – Washington. 1980+ (1,5,9) – ISSN: 0197-0771 – mf#12264 – us UMI ProQuest [325]
The campaigns of 1812, 1813, and 1814 : also the causes and consequences of the french revolution, to which is added the french confiscations, contributions, requisitions etc from 1793, till 1814 / M'Queen, James – Glasgow: printed...for W Summerville, A Fullarton, J Blackie...1815 [mf ed 1984] – 11mf – 9 – 0-665-46133-X – mf#46133 – cn Canadiana [355]
Campaigns of 1812-1814 : contemporary narratives by captain w h merritt, colonel william claus, lieut-colonel matthew elliott and captain john norton / ed by Cruikshank, Ernest Alexander – [Niagara, Ont?: s.n.], 1902 – 1mf – 9 – 0-665-76087-6 – mf#76087 – cn Canadiana [355]
The campaigns of 'ala'u'd-din khilji : being the khaza'inul futuh (treasures of victory) of hazrat amir khusrau of delhi / Amir Khusraw Dihlavi – Bombay: DB Taraporewala, Sons, 1931 – (= ser Samp: indian books) – (trans into english with notes and parallel passages from other persian writers by muhammad habib; and with an historical int by s krishnaswami aiyangar) – us CRL [954]
Campaigns of the civil war – New York. v1-13. 1881-82 – 9 – $466.00 – mf#0134 – us Brook [976]
Campaigns of the war of 1812-15, against great britain, sketched and criticised : with brief biographies of the american engineers / Cullum, George Washington – New York: J Miller, 1879 [mf ed 1980] – 5mf – 9 – mf#03637 – cn Canadiana [355]
Campaigns protesting against nuclear testing in the pacific : press cuttings and scrapbooks / Greenpeace New Zealand – 1973-1975, 1985 – 1r – 1 – mf#pmb1238 – at Pacific Mss [333]
Campamentos 1971. caceres / Delegacion Provincial de la Juventudes – Caceres: Imp. M. Sergio Dorado, 1971 – 1 – sp Bibl Santa Ana [946]
Los campamentos de quinto cecilio metelo pio / Lumbreras Valiente, Pedro – Caceres: Edit. Extremadura, 1973 – 1 – sp Bibl Santa Ana [946]
Campamn sasana / Lon Nol – Bhnam Ben: [s.n.] 2514 [1970] [mf ed 1990] – 1r with other items – 1 – (in khmer) – mf#mf-10289 seam reel 115/2 [§] – us CRL [230]
Campan, Jeanne Louise Henriette Genest *see*
– Journal anecdotique de mme campan
– Memoires sur la vie privee de marie-antoinette, reine de france et de navarre
La campana – Paris. 5 janv-8 juil 1898, 1 avr-5 aout 1900 – 1 – fr ACRPP [073]
Campana admirable – Caracas, Venezuela. 1965 – 1r – us UF Libraries [972]
Campana, C *see*
– Compendio historico, delle gverre vltimamente successe tra christiani, and turchi, and tra turchi, and persiani...
– Della guerra...fatta per difesa di religione...
Campana de campamentos 1974 / Delegacion Provincial de la Juventudes – Caceres: Imp. M. Sergio Dorado, 1974 – 1 – sp Bibl Santa Ana [350]
Campana de portugal...extremadura / Garcia, Miguel – 1762 – 9 – sp Bibl Santa Ana [946]
Campana de portugal...extremadura / Mascarenas, Geronimo – 1662 – 9 – sp Bibl Santa Ana [914]
Campana del brasil. antecedentes coloniales. tomo 1 : (documentos referentes a la... independencia...de...argentina...segunda serie) / Bayle, Constantino – Buenos Aires, 1931; Madrid: Razon y Fe, 1932 – 1 – sp Bibl Santa Ana [972]
Campana libertadora de 1821 / Florez Alvarez, Leonidas – Bogota, Colombia. 1921 – 1r – us UF Libraries [972]
Campana nacional contra los filibusteros en 1856 y / Calvo, Joaquin Bernardo – San Jose, Costa Rica. 1909 – 1r – us UF Libraries [972]
La campana protestante en america / Bayle, Constantino – Madrid: Razon y Fe, 1928 – 9 – sp Bibl Santa Ana [242]
Campana reaundada / Bayle, Constantino – Madrid: Razon y Fe, 1928 – 9 – sp Bibl Santa Ana [999]
Campanario *see* Programa oficial de feria y fiestas en honor de la santisima virgen de piedra-escrita
Campanario. Ayuntamiento *see* Ordenanzas municipales
Campanas de maceo en la ultima guerra de independe / Piedra Martel, Manuel – Habana, Cuba. 1946 – 1r – us UF Libraries [972]
Campanas en el rif y gebala, del general berenguer / Bayle, Constantino – Madrid: Razon y Fe, 1923 – 1 – sp Bibl Santa Ana [355]
Campanas que doblaron solas / Mendez Merida, Virgilio – Panama, 1963 – 1r – us UF Libraries [972]
Campanella, Tommaso *see*
– The sonnets of michael angelo buonarroti and tommaso campanella
– Sonnets of michaelangel buonarroti and tommaso campanella
The campaner thal, and other writings / Richter, Johann Paul Friedrich – New York: United States Book Co., n.d. 383p – 1 – (from the german) – us Wisconsin U Libr [430]
A campanha – Campanha, MG, 19 maio, 29 ago 1908 – (= ser Ps 19) – 1,5,6 – bl Biblioteca [079]

Campanha de 1923 / Cunha, Jose Antonio Flores Da – Rio de Janeiro, Brazil. 1942? – 1r – us UF Libraries [972]
Campanha de libertacao [discursos] / Gomes, Eduardo – Sao Paulo, Brazil. 1946 – 1r – us UF Libraries [972]
Campanha do bailundo em 1902 / Cabral De Moncada, Francisco Xavier – Loanda, Angola. 1903 – 1r – us UF Libraries [025]
Campanha do sul de angola em 1915 / Pereira De Eca, Antonio Julio Da Costa – Lisboa, Portugal. 1921 – 1r – us UF Libraries [960]
Campanha presidencial / Vargas, Getulio – Rio de Janeiro, Brazil. 1951 – 1r – us UF Libraries [972]
Campanhas de imprensa / Nabuco, Joaquim – Sao Paulo, Brazil. 1949 – 1r – us UF Libraries [972]
Campanulas / Bermejo, Fernando – Ciudad Trujillo, Dominican Republic. 1946 – 1r – us UF Libraries [972]
Campanus, J *see* Tvrcicorvm tyrannorvm qvi inde vsqve ab otomanno rebus turcicis praefuerunt...
Campbel, Charles *see* Conversations with a ranter
Campbell, A et al *see* Psalms, hymns and spiritual songs
Campbell, A J *see* Sermon on the fulness of christ
Campbell, A, mrs *see* The inner life
Campbell, Alexander *see*
– The bible and wine
– The campbell year book
– Canadian pacific railway
– Christian baptism
– The christian preacher's companion
– Debate between rev a campbell and rev n l rice
– A debate on christian baptism
– A debate on the roman catholic religion
– A discussion of the doctrines of endless misery and universal salvation
– Familiar lectures on the pentateuch
– Heart of africa
– In the case of louis riel, convicted of treason, and executed therefor
– The memorable sermon on the law
– Report of the minister of justice
– Report on the subject of the red river of the north
– Songs, hymns, and spiritual songs
– The true greenback
Campbell, Alexander Colin *see* The making of a dollar bill
Campbell, Alexander Duncan *see*
– A dictionary of the teloogoo language
– Dictionary of the teloogoo language, commonly termed the gentoo
Campbell, Archer Stuart *see* Cigar industry of tampa, florida
Campbell, Archibald *see* A voyage round the world
Campbell, Archibald, Lord *see* Highland dress, arms and ornament
Campbell Area Genealogical and Historical Society *see* Pipeline
Campbell, Arthur *see* Songs of the pinewoods
Campbell, Belle McPherson *see* Madagascar
Campbell, C *see* Vitruvius britannicus...
Campbell, Charles *see* Canada
Campbell citizen – Campbell, NE: Will R Burr, jan 1901-v44 n30. feb 24 1944 (wkly) [mf ed v8 n8. mar 13 1908-14, 1916-17,1919-25,1927-39,1941-44 (gaps)] – 16r – 1 – (absorbed by: franklin county sentinel) – us NE Hist [071]
Campbell, Colin *see* Marriage vow
Campbell contacts – v7 iss 3-v9 iss 1 [1985 nov-1987 may] – 1r – 1 – (cont: campbell contacts in america) – mf#1495201 – us WHS [071]
Campbell contacts in america – 1979 may-1983, 1984 feb-1985 aug, – 1r – 1 – mf#689452 – us WHS [071]
Campbell contacts in america – 1987 aug-1989 feb – 1r – 1 – mf#1497393 – us WHS [071]
Campbell, Courtney S et al *see* Biolaw
Campbell, Dana M *see* Harmonic equipment evidenced in the works of claudio monteverdi (1567-1643)
Campbell, David *see* The judgment period preparatory to the establishment of the kingdom of heaven
Campbell, David Stephen *see* Geography of the coffee industry of puerto rico
Campbell, Donald *see*
– Aventures de donald campbell
– Memorials of john mcleod campbell, d.d
– Memorials of john mcleod campell, d.d.
– Reminiscences and reflections
Campbell, Donald Alexander *see* Pioneers of medicine in nova scotia
Campbell, Douglas *see* The puritan in holland, england, and america
Campbell, Douglas Graves *see* George w chadwick
Campbell, Dugald *see* In the heart of bantuland
Campbell, Elizabeth M *see* The land of morning calm
Campbell Family *see* Papers of campbell, preston and floyed families
Campbell, Francis *see* Index-catalogue of indian official publications in the library, british museum
Campbell, Francis Wayland *see*
– The fenian invasions of canada of 1866 and 1870
– History of the formation of the medical faculty, university of bishop's college, montreal
– Introductory lecture delivered at the opening of the second session of the medical faculty of the university of bishop's college, october 2nd, 1872
– "War"
Campbell, Frank *see* Catalogue of official reports relating to india issued as english parliamentary papers (and in connection with the india office) during the year 1892
Campbell, Frank Carter *see* Polychoral motets of adam gumpelzhaimer (1559-1625)
Campbell, George *see*
– Alarms in regard to popery
– The british empire
Campbell, H J *see* The canadian cricketer's guide
Campbell, Henry D *see* The congo
Campbell, J *see* Travels in south africa
Campbell, J Baxter *see* Charter of the city of quincy, florida
Campbell, James et al *see* The church of scotland, past and present
Campbell, James Mann *see*
– After pentecost, what?
– The heart of the gospel
– Paul the mystic
– Paul, the mystic
– The teachings of the books
Campbell, James Mannann *see* The indwelling christ
Campbell, James R *see* Missions in hindustan
Campbell, John *see*
– The american indian
– The coptic element in languages of the indo-european family
– The cumberland coal fields, nova-scotia
– Faithful minister's character and reward
– John angell james
– The land of robert burns
– Life of john, lord campbell, lord high chancellor of great britain
– Lives of the lord chancellors
– The martyr of erromanga
– Perpetual duration of christianity established by historic proof
– The primitive history of the ionians
– Reisen in sued-afrika
– A short history of the non-subscribing presbyterian church of ireland
– Travels in south africa
Campbell, John Campbell *see* Law of church rates
Campbell, John Francis *see*
– My circular notes, vol 1
– My circular notes, vol 2
– My circular notes, vols 1 and 2
– A short american tramp in the fall of 1864
Campbell, John Gregorson *see* Witchcraft and second sight in the highlands and islands of scotland
Campbell, John Lorne *see* The patmos letters
Campbell, John McLeod *see*
– Christ the bread of life
– Memorials of john mcleod campbell, d.d
– The nature of the atonement
– Reminiscences and reflections
– Thoughts on revelation
Campbell, John Mcleod *see* Everlasting gospel
Campbell, John P *see* Vindex or the doctrines of the strictures vindicated
Campbell, John Quincy Adams *see* John quincy adams campbell diaries, 1861-1864
Campbell, John Roy *see* An answer to some strictures in brown's sequel to campbell's history of yarmouth
Campbell, Joseph *see*
– Myths and symbols in indian art and civilization
– Philosophies of india
Campbell law review – v1-23. 1979-2001 – 9 – $348.00 set – ISSN: 0198-8174 – mf#101321 – us Hein [340]
Campbell, Lewis *see*
– The epistles of saint paul to the thessalonians, galatians and romans
– The epistles of st paul to the thessalonians, galatians and romans
– Religion in greek literature
– Theological essays of the late benjamin jowett
Campbell, Mary J *see*
– Daughters of india
– The power-house at pathankot
Campbell, ML *see* Rating of college courses
Campbell, N W *see* Why am i a presbyterian?
The campbell news – Campbell,NE: Robert Emil Hartman. 30v. v1 n1. nov 24 1944-v30 n11. dec 20 1973) (wkly) [mf ed with gaps filmed -1975] – 10r – 1 – us NE Hist [071]
Campbell, Peter Colin *see* The theory of ruling eldership
Campbell press – CA. 1895-1948; jul 1973-may 31 1985 – 30r – 1 – $1800.00 – mf#B02092 – us Library Micro [071]
The campbell press – Campbell, NE: B A Simpson, -1903// (wkly) [mf ed v6 n22 (mar 18 1892, 1895-1903 (gaps)] – 2r – 1 – us NE Hist [071]
Campbell, Ralph J *see* Local lyrics
Campbell, Reginald John *see*
– Christianity and the social order
– City temple sermons
– A faith for to-day
– The new theology
– New theology sermons
– The restored innocence
– A spiritual pilgrimage
– The war and the soul
Campbell river courier – British Columbia, CN. jan 1946-dec 1972; jan 1973-dec 1986 – 34r – 1 – cn Commonwealth Imaging [071]
Campbell, Robert *see*
– The discovery and exploration of the pelly (yukon) river
– The discovery and exploration of the youcon (pelly) river
– A flag of distress
– The flora of montreal island
– The flora of the rocky mountains
– James johnstone vs the minister and trustees of st andrew's church
– Reasons why there are so few candidates for the holy ministry
– The relations of the christian churches to one another
– Supplemental notes on the flora of cap-a-l'aigle
– Union or co-operation – which?
Campbell, Robert Fishbourn *see* Mission work among the mountain whites in asheville presbytery, north carolina
Campbell, Roderick *see*
– The importer's guide
– The importers' guide
Campbell, Roy *see* Adamastor
Campbell, Samuel Miner *see* Across the desert
Campbell, Selina Huntington *see* Home life and reminiscences of alexander campbell
Campbell, Th L *see* Dionysius the ps-areopagite
Campbell, Thomas *see* Memoirs of elder thomas campbell
Campbell, Thomas Joseph *see*
– Jesuits, 1534-1921
– Out of the grave
– Pioneer priests of north america, 1642-1710
– Pioneer priests of north america, 1642-1710, vol 1
– Various discourses
Campbell, Thomas Moody *see* Hebbel, Ibsen and the analytic exposition
Campbell Thompson, R *see* Late babylonian letters
Campbell, W *see* British india
Campbell, W Graham *see* The new world, or, recent visit to america
Campbell, Wilfred *see*
– A beautiful rebel
– The beauty, history, romance and mystery of the canadian lake region
– Beyond the hills of dream
– Canada
– [Canada's responsibility to the empire and the race]
– Departure
– The dread voyage
– Healing
– Ian of the orcades
– Langemarck
– A list of members of the house of assembly for upper canada
– The lyre degenerate
– Lyrics of the dread redoubt
– Morning
– Night / The house of dreams
– Our heritage / sublimity
– The poems of wilfred campbell
– Poetical tragedies
– Quebec tercentenary
– Sagas of vaster britain
– Snowflakes and sunbeams
– Songs of yoho
– The vanguard
– Victoria
Campbell, William *see*
– British india
– Campbell's third collection of the newest and most favorite country dances & cotillons, for the violin, harp, harpsichord, and german flute
– The crown lands of australia
– Formosa under the dutch
– Handbook of the english presbyterian mission in south formosa
– Sketches from formosa
Campbell, William Cecil *see* Preparation for church growth in the miami springs baptist church, miami springs, florida
Campbell, William J *see* Chronological list of the practicing members of the philadelphia bar
Campbell, William S *see* Forms of code pleading for nebraska, kansas and oklahoma, fully annotated
The campbell year book : choice selections for every day in the year from the writings of alexander campbell / Campbell, Alexander – [S.l.]: W Burleigh, c1909 – 1mf – 9 – 0-524-06985-9 – mf#1991-2838 – us ATLA [240]
Campbellism examined / Jeter, Jeremiah Bell – New York: Sheldon, Lamport, & Blakeman; Boston: Gould & Lincoln, 1855, c1854 – 1mf – 9 – 0-7905-7790-9 – mf#1989-1015 – us ATLA [240]
Campbellism exposed : in an examination of lard's review of jeter / Williams, Alvin Peter – Nashville TN: Southwestern Pub House 1860 [mf ed 1991] – 1mf – 9 – 0-524-01218-0 – (int by j b jeter) – mf#1990-4076 – us ATLA [240]
Campbellism re-examined / Jeter, Jeremiah Bell – New York: Sheldon, Blakeman, [1856?] – 1mf – 9 – 0-524-00045-X – mf#1989-2745 – us ATLA [240]
Campbellism, what is it? : a series of lectures on that which is commonly called campbellism by religious teachers who oppose the teachings of the word of god / Chism, J W – Nashville, Tenn: Gospel Advocate Pub, 1901 – 1mf – 9 – 0-524-06014-2 – mf#1991-2374 – us ATLA [220]
Campbell-Johnson, Alan *see* Mission with mountbatten
Campbell/los gatos/saratoga – 1993- – (= ser California telephone directory coll) – 3r – 1 – $150.00 – mf#P00014 – us Library Micro [917]
Campbell's foreign semi-monthly magazine : or, select miscellany of european literature and art – Philadelphia. 1842-1844 – 1 – mf#3954 – us UMI ProQuest [073]
Campbell's third collection of the newest and most favorite country dances & cotillons, for the violin, harp, harpsichord, and german flute : with their proper figures as performed at court, bath and all public assemblys etc / Campbell, William – London: William Campbell [1786?] [mf ed 19–] – 1r – 1 – mf#pres. film 116 – us Sibley [780]
Campbell/saratoga – 1942 – (= ser California telephone directory coll) – 1r – 1 – $50.00 – mf#P00015 – us Library Micro [917]
Campbellsport news – Campbellsport WI. 1908 jun 11/1909 dec 16-1998 – 56r – 1 – mf#1009197 – us WHS [071]
Campbellsville baptist church. kentucky : church records – (records beginning with Church of Christ on Pitman's Creek). 1802-28, 1885-1955, 1961-67 – 1 – us Southern Baptist [242]
Campbelltown district star – Campbelltown, jul 1975-aug 1986 – 11r – at Pascoe [079]
Campbelltown herald – Campelltown, feb 14 1880-sep 7 1881; jan 5 1889-oct 11 1919 – 4r – 9 – A$184.49 vesicular A$206.49 silver – at Pascoe [079]
Campbelltown news – Campelltown, jan 2 1920-dec 23 1968 – 26r – 9 – A$1594.03 vesicular A$1737.03 silver – at Pascoe [079]
Campbeltown courier – [Scotland] Argyll & Bute, Campbeltown: R Wilson jan 1875-dec 1879, jan 1893-dec 1950 (wkly) [mf ed 2004] – 34r – 1 – (began in 1873) – uk Newsplan [072]
Campbeltown courier – Campbeltown: R Wilson 1873- (wkly) [mf ed 1995-] – 1 – (began in 1873; imprint varies) – uk Newsplan; uk Scotland NatLib [072]
Campbeltown journal and argyle and buteshire advertiser – [Scotland] Argyll & Bute, Campbeltown : Thomas Clark and Son 1 mar 1851-24 feb 1855 (wkly) [mf ed 2003] – 90v on 2r – 1 – (cont by: campbeltown journal and argyle, bute and dumbartonshire advertiser [jan 1854-feb 1855]) – uk Newsplan [072]
Campe, Joachim *see* Voyage au spitzberg et a la nouvelle-zemble, entrepris, en 1596, par j heemskerk, dans le dessein de trouver, par le nord, un passage aux indes orientales
Campe, Joachim H *see* Bibliotheque geographique et instructive des jeunes gens
Campe, Joachim Heinrich *see*
– Beitraege zur weitern ausbildung der deutschen sprache
– Histoire de la decouverte de l'amerique
– Histoire de la decouverte et de la conquete de l'amerique
Campe, Julius *see* Zwei bisher unveroeffentlichte zeit-dokumente des verlags-archives ueber heines "reisebilder" und "buch der lieder"
O campeao : periodico politico, noticioso, social, critico e facetas – Recife, PE. 21 out 1863 – (= ser Ps 19) – bl Biblioteca [079]
Campeche. Mexico (State) *see* Periodico oficial del gobierno constitucional del estado de campeche
Campell, Jennifer A *see* Metabolic and cardiovascular responses to shallow water exercise in younger and older women
Campello, Enrico di *see* Count campello
Campelltown camden chronicle – Campelltown, nov 1984-jun 1997 – (= ser MacArthur chronicle) – (aka: macarthur chronicle) – at Pascoe [079]
Campen, J van *see* Afbeelding van 't stadt huys van amsterdam...
Campenhausen, H von *see* Ambrosius von mailand als kirchenpolitiker
Campenhausen, Leyon *see* Travels through several provinces of the russian empire

Campeonato de Pesca 1, 1975 *see* 1st campeonato de pesca en la modalidad de c grupo iberduero. embalse de alcantara en aguas del rio tajo. finca la carrascosa. dia 21 de junio de 1975. reglamento y programa de actos

Campeonato de pesca 2 1976. ...en la modalidad de carpidos...dia 19 de junio 1976. reglamento y programa de actos – Caceres: Imp. Rodriguez, 1976 – 1 – sp Bibl Santa Ana [350]

Camper – Kerr Lake, NC. 1975-1979 (1) – mf#69223 – us UMI ProQuest [071]

Campesinado colombiano / Perez Ramirez, Gustavo – Bogota, Colombia. 1959 – 1r – us UF Libraries [972]

Campesinado colombiano / Perez Ramirez, Gustavo – Friburgo, Switzerland. 1962 – 1r – us UF Libraries [972]

El campesino / Gonzalez, Valentin – Paris: Plon, 1950 – 1 – sp Bibl Santa Ana [946]

Los campesinos y la republica / Uribe, Vicente – Conferencia pronunciada el dia 22 de enero en el Teatro Apolo, de Valencia.Valencia, 1938? Fiches W1239, 1240. (Blodgett Collection of Spanish Civil War Pamphlets) – 9 – us Harvard [946]

Campfires of the afro-americans : or, the colored man as a patriot, soldier, sailor and hero in the cause of free america... / Guthrie, Jas M – Philadelphia: Afro-American Publ Co, 1899 – us CRL [305]

Camphuysen, D R *see*
– Stichtelycke rymen
– Theologische werken

Campian, Edmond *see* Appeal to the members of the two universities presenting ten reason...

Campiliiensis, Christanus *see* Opera poetica

Campina interiorana / Tejeira, Gil Blas – Mexico City? Mexico. 1956 – 1r – us UF Libraries [972]

Camping – Cambridge. 1926-1929 (1) – mf#5133 – us UMI ProQuest [790]

Camping and cruising in florida / Henshall, James A – Cincinnati, OH. 1884 – 1r – us UF Libraries [790]

Camping at the pole / Krenkel, Ernst Teodorovich – Moscow: Foreign Languages Pub House, 1939 [mf ed 2000] – 1r – 1 – mf#29077 – us Harvard [919]

Camping facilities : a selected bibliography / White, Anthony G – Monticello, IL: Vance Bibliographies, [1983] (mf ed 1986) – 1mf – 9 – mf#*XM-16128 – us NY Public [790]

Camping in the canadian rockies : an account of camp life in the wilder parts of the canadian rocky mountains, together with a description of the region about banff, lake louise, and glacier, and a sketch of the early explorations / Wilcox, Walter Dwight – New York; London: G P Putnam's Sons, 1896 [mf ed 1981] – 5mf – 9 – (incl ind; with full-page photogravures, and many text ill fr photos by aut) – mf#16459 – cn Canadiana [790]

Camping in the muskoka region / Dickson, James – Toronto: C B Robinson, 1886 [mf ed 1980] – 2mf – 9 – 0-665-02660-9 – mf#02660 – cn Canadiana [790]

Camping journal – New York. 1973-1981 (1) 1976-1981 (5) 1976-1981 (9) – ISSN: 0527-4478 – mf#8370 – us UMI ProQuest [790]

Camping magazine – Martinsville. 1930+ (1) 1969+ (5) 1970+ (9) – ISSN: 0740-4131 – mf#960 – us UMI ProQuest [790]

Campion, Edmund *see* A historie of ireland

Campioni, Carlo Antonio *see* 6 duo, a deux violons. oeuvre 8

Campitelli, Michael A *see* Why student at eastern washington university choose to participate or not in intermural sports activites while at ewu

Camp-meeting : the reply of spectator [i.e. charles walker] to the rev j draper's pretended review of his strictures on the late camp-meeting which appeared in the people's press / Walker, Charles – [Fonthill ON?: s.n.] 1863 [mf ed 1987] – 1mf – 9 – 0-665-63236-3 – mf#63236 – cn Canadiana [240]

Camp-meetings : their origin, history, and utility: also their perversion, and how to correct it / Swallow, S C – New York: Nelson and Phillips; Cincinnati: Hitchcock and Walden, 1879 – 1mf – 9 – 0-524-01134-6 – mf#1990-0348 – us ATLA [240]

Campo adentro / Rodriguez, Mario Augusto – Panama, 1947 – 1r – us UF Libraries [972]

Campo Cardona, Antonio del *see*
– Estudio demografico comparativo de espana y la provincia de caceres (decenio 1921-1930)
– La lucha antipaneidica en la provincia de caceres

Campo, Cupertino Del *see* Vibraciones y reflejos

El campo de gibraltar – Algeciras, Spain. 22 aug 1916-7 march 1919 [daily] – 3r – 1 – uk British Libr Newspaper [074]

Campo Lacasa, Cristina *see* Notas generales sobre la historia eclesiastica de...

Campo misionero – Buenos Aires: Fundacion Cristiana de Evangelizacion. v34 n7-v49 n588. jan 1978-jun 1993 – 8r – 1 – us CRL [240]

El campo propio del sacerdote secular en la evangelizacion americana / Bayle, Constantino – Madrid: Missionalia Hispanica, 1946 – 1 – sp Bibl Santa Ana [240]

Campomanes, Conde de *see* Memorial ajustado al expediente de concordia...mesta con la diputacion provincial...extremadura

O camponez : semanario para defesa das classes agrarias – Urucanga, SC. 20 nov, 11 dez 1932 – (= ser Ps 19) – mf#UFSC/BPESC – bl Biblioteca [079]

Campori, G *see* Memorie biografiche degli scultori, architetti, pittori ecc nativi di carrara e di altri luoghi della provincia di massa...

Campos, Augusto De *see* Balanco da bossa

Campos, Camilo *see*
– Normas supremas

Campos, Camilo D *see* Normas supremas

Campos, Ernesto De Souza *see* Educacao superior no brasil

Campos, Francisco *see*
– Estado nacional
– Problemas do brasil e as grandes solucoes do novo

Campos gerais, estruturas agrarias – Curitiba, Brazil. 1968 – 1r – us UF Libraries [972]

Campos, Humberto De *see*
– Conceito e a imagem na poesia brasileira
– Critica
– Memorias inacabadas
– Notas de um diarista
– Sepultando os meus mortos
– Sombra das tamareiras
– Sonho de pobre
– Tonel de diogenes
– Vale de josaphat

Campos, Humberto de *see* Sombras que sofrem

Campos, Joachim Joseph A *see* History of the portuguese in bengal

Campos, Joaquim Pinto De *see* Vida do grande cidadao brasileiro, luiz alves de l...

Campos, Jorge *see*
– Antologia hispano-americana
– Vida y trabajos de un libro viejo

Campos, Paulo Mendes *see* Testamento do brasil

Campos, Pedro Dias De *see* Incola e o bandeirante na historia de sao paulo

Campos Porto, Manuel Ernesto De *see* Apontamentos para a historia da republica dos esta

Campos, Roberto De Oliveira *see* Temas e sistemas

Campos, Sabino De *see* Catimbo

Campra, Andre *see* Idomenee

Camprubi De Jimenez, Zenobia *see* Monumento de amor

Camps' answer to dr forbes : the church of rome vindicated from every calumny: with epistle dedicatory to father dayman – New York: F A Brady, 1859 [mf ed 1986] – 1mf – 9 – 0-8370-7923-3 – mf#1986-1923 – us ATLA [241]

Camps, David *see* Balance

Camps in the rockies : being a narrative of life on the frontier, and sport in the rocky mountains, with an account of the cattle ranches of the west / Baillie-Grohman, William Adolph – 3rd ed. London: S Low, Marston, Searle & Rivington, 1882 [mf ed 1980] – 5mf – 9 – 0-665-03971-9 – (incl ind) – mf#03971 – cn Canadiana [790]

Campsie case / Calwell, James – Glasgow, Scotland. 1881 – 1r – us UF Libraries [240]

Campsie news – Campsie, jan 1969-79 – 9r – at Pascoe [079]

Campsie news – Lakemba, jan 1963-dec 1968 – 4r – at Pascoe [079]

Campton, New Hampshire. Campton Baptist Church *see* Records

Campton with shefford – (= ser Bedfordshire parish register series) – 3mf – 9 – £7.50 – uk BedsFHS [929]

Campus : poemas 1954-1958 / Hernandez Sanchez, Jesus – Barcelona, Spain. 1958 – 1r – us UF Libraries [810]

Campus – Rio de Janeiro: OPUS da Junta de Mocidade da Convencao Batista Brasileira. v1-9 n36. 1977-90 – 1r – us CRL [972]

Campus capitalist – 1988 nov-1990 mar 16 – 1r – 1 – mf#1579168 – us WHS [071]

Campus echo / North Carolina Central University – v1 n1 [1971 sep 16], n44 [1995 jan 20], n60 [1996 oct 19] – 1r – 1 – mf#3693564 – us WHS [378]

Campus law enforcement journal – Athens. 1979+ (1,5,9) – ISSN: 0739-0394 – mf#12242,01 – us UMI ProQuest [360]

Campus life – Carol Stream. 1972+ (1) 1972+ (5) 1976+ (9) – ISSN: 0008-2538 – mf#7532 – us UMI ProQuest [400]

Campus ministry women newsletter – 1981 nov, 1982 may/jun-nov/dec, 1983:summer-fall, 1984 feb, jul, nov, 1985 mar-may, sep-nov, 1986:winter, sum, fall, dec,1987 jan/feb-may/jun, oct-dec, 1988 feb-apr – 1r – 1 – (cont: interim [dayton oh]) – mf#1053984 – us WHS [305]

Campus underground – v1-2 n5. 1968-69 – (= ser The new prairie primer) – 1 – (superseded by: the new prairie primer) – us AMS Press [378]

Camrian pi pad: : [qatthapad samramn] / Som Samq'un – Bhnam Ben: Pannagar Banly vijja 1972 [mf ed 1990] – 1r with other items – 1 – (in khmer) – mf#mf-10289 seam reel 108/5 [§] – us CRL [480]

Camrose canadian – Alberta, CN. 1898- – 2r/y – 1 – Can$93.00r – cn Commonwealth Imaging [071]

Camus, Albert *see* Fall, and exile and the kingdom

Camus, J P *see* Reparties succinctes a l'abrege des controverses de m charles drelincourt... ensembles les antitheses protestantes...

Camus, Raoul *see* National tune index

Can altered body position alleviate post-exercise pulmonary diffusing capacity impairment? / Stewart, Ian Braidwood – 1997 – 1mf – 9 – $4.00 – mf#PH 1594 – us Kinesology [612]

Can american parents coalition for quality education / African American Parents Coalition for Quality Education – 1991 may 6, jun 28, jul/dec – 1r – 1 – mf#4841755 – us WHS [370]

Can critical thinking processes enhance classroom practice? / Phala, Melomong Augustinos – Pretoria: Vista University 2001 [mf ed 2001] – 2mf – 9 – (incl bibl ref) – mf#mfm15186 – sa Unisa [370]

Can do : a national publicationfor all seabees / Seabee Veterans of America – 1968 nov-1970 aug – 1r – 1 – (cont by: seabee's can do) – mf#1842418 – us WHS [305]

Can do review / Naval Construction Training Center [Port Hueneme CA] – 1990 dec/ jan – 1r – 1 – (cont: bee in the know) – mf#1698601 – us WHS [623]

Can extreme voluntaryism be made an open question? / Anderson, William – Aberdeen, Scotland. 1871 – 1r – us UF Libraries [240]

Can Lvan, U *see* Mran ma ca ka pum mya a phvan

Can man know god? / Strong, Thomas – London, England. 1886 – 1r – us UF Libraries [240]

Can, Ngo Hu *see* Culture and cytological development of psilocybe cubensis

Can nve, Da gun U *see*
– Buddha nikay na rap
– Ca bu ton
– Sadda a mran ca a mran

Can Nve, U *see* De va dat a ja ta sat

Can one be saved without baptism? – London, England. 18– – 1r – us UF Libraries [242]

Can Sa min *see*
– Mran ma sve mran ma dha mran ma panna
– Prann su khyac tai a no ra tha

Can salyang aqui na si cacaseno *see* Historia ni bertoldino

Can the old faith live with the new? : or, the problem of evolution and revelation / Matheson, George – 3rd ed. Edinburgh: William Blackwood, 1889 – 1mf – 9 – 0-8370-4308-5 – mf#1985-2308 – us ATLA [210]

Can the old faith live with the new? : or the problem of evolution and revelation / Matheson, George – [Edinburgh] 1885 – (= ser 19th c evolution & creation) – 5mf – 9 – mf#1.1.11439 – uk Chadwyck [210]

Can vermezler tekkesi / Kamil, Ahmet – Istanbul: Kitaphane-yi Sudi, 1922 – (= ser Ottoman literature, writers and the arts) – 2mf – 9 – $40.00 – us MEDOC [470]

Can we believe in miracles? / Warington, George – London: SPCK, [1871] – 1mf – 9 – 0-8370-5711-6 – mf#1985-3711 – us ATLA [210]

Can we still be christians? = Koennen wir noch christen sein? / Eucken, Rudolf – New York: Macmillan, 1914 – 1mf – 9 – 0-7905-3669-2 – (in english) – mf#1989-0162 – us ATLA [240]

Can we still follow jesus? : a study of the teachings of jesus in its modern applications / Garvie, Alfred Ernest – London, New York: Cassell, 1913 [mf ed 1989] – 1mf – 9 – 0-7905-0834-6 – mf#1987-0834 – us ATLA [240]

Can we trust the bible? : chapters on biblical criticism / Ballard, Frank et al – London: Religious Tract Society, 1908 – 1mf – 9 – 0-524-05655-2 – mf#1992-0505 – us ATLA [220]

Can you run away from god? / Boice, James Montgomery – Wheaton IL: Victor Books, c1977 [mf ed 2003] – 1r – 1 – mf#b00667 – us ATLA [221]

Can you speak english? / SAMBO – Bulawayo, Zimbabwe. 19– – 1r – us UF Libraries [960]

Cana, Frank Richardson *see* South africa

Canaan / Aranha, Graca – Santiago, Chile. 1935 – 1r – us UF Libraries [972]

Canaan / Aranha, Jose Pereira Da Graca – Mexico City? Mexico. 1954 – 1r – us UF Libraries [972]

Canaan : or, the land of promise – London, England. 1843 – 1r – us UF Libraries [240]

Canaan d'apres l'exploration recente / Vincent, Hugues – Paris: Victor Lecoffre 1907 [mf ed 1989] – (= ser Etudes bibliques) – 2mf – 9 – 0-7905-2392-2 – (incl bibl ref & ind) – mf#1987-2392 – us ATLA [930]

Canaan Fernandez, Euridice *see* Depravados

Canaan, Taufik *see* Aberglaube und volksmedizin im lande der bibel

Canaanite myths and legends / Driver, G R – T. and T. Clark, 1956 – 9 – $10.00 – us IRC [390]

Canada – Benton, NB: M R Knight, [1891?-1892] – 9 – mf#P04112 – cn Canadiana [420]

Canada / Bourinot, John George – London, 1897 – (= ser 19th c books on british colonization) – 6mf – 9 – mf#1.1.4551 – uk Chadwyck [971]

Canada : a brief outline of her geographical position, productions, climate, capabilities, educational and municipal institutions, etc – Toronto, Canada West: [s.n.], 1857 [mf ed 1999] – 1mf – 9 – mf#SEM105P3147 – cn Bibl Nat [917]

Canada : a brief outline of her geographical position, productions, climate, capabilities, educational and municipal institutions, fisheries, railroads, etc / Canada (Province). Departement de l'agriculture – 2nd ed. Quebec: printed by John Lovell, 1860 [mf ed 1983] – 1mf – 9 – mf#SEM105P226 – cn Bibl Nat [917]

Canada / Campbell, Wilfred – Toronto: Macmillan, [1906?] – 7mf – 9 – 0-665-71323-1 – (ill by t mower martin) – mf#71323 – cn Canadiana [917]

Canada : correspondence relative to emigration to canada – London: printed by W Clowes & sons, 1841 [mf ed 1983] – 2mf – 9 – mf#SEM105P232 – cn Bibl Nat [304]

Canada : the country, its people, religions, politics, rulers, and its apparent future: being a compendium of travel from the atlantic to the pacific, the great lakes, manitoba, the north-west, and british columbia... / Captain Mac – Montreal: [s.n.], 1892 [mf ed 1980] – 5mf – 9 – mf#15533 – cn Canadiana [917]

Canada : an essay: to which was awarded the first prize by the paris exhibition committee of canada / Hogan, John Sheridan – Montreal: B Dawson, 1855 [mf ed 1981] – 1mf – 9 – 0-665-45101-6 – mf#45101 – cn Canadiana [917]

Canada : further papers relative to the affairs of canada – London: printed by W Clowes & sons, 1849 [mf ed 1984] – 1mf – 9 – mf#SEM105P406 – cn Bibl Nat [971]

Canada : further papers relative to the affairs of canada – London: printed by W Clowes & sons...1849 [mf ed 1984] – 1mf – 9 – mf#SEM105P406 – cn Bibl Nat [971]

Canada / Gahan, James Joseph – Quebec?: P G Delisle, 1877 – 1mf – 9 – mf#24174 – cn Canadiana [810]

Canada : le guide du colon francais, belge et suisse / Drapeau, Stanislas – Ottawa: [s.n.], 1896 [mf ed 1986] – 1mf – 9 – 0-665-56033-8 – mf#56033 – cn Canadiana [304]

Canada : le guide du colon francais, belge, suisse, etc: brochue preparee et publiee / Drapeau, Stanislas – Ottawa: [s.n.], 1887 [mf ed 1980] – 2mf – 9 – mf#02760 – cn Canadiana [304]

Canada – Hampton, N.B: M R Knight, [1892-189- or 19–] – 9 – mf#P04060 – cn Canadiana [420]

Canada : instructions to the earl of gosford, and the commissioners appointed to inquire into the grievances complained of in lower canada – [s.l]: House of Commons, 1836 [mf ed 1984] – 1mf – 9 – mf#SEM105P410 – cn Bibl Nat [971]

Canada : its commerce, its colleges, and its churches / Lindsay, James – London: Athenaeum, 1900 [mf ed 1985] – 1mf – 9 – 0-665-35286-7 – mf#35286 – cn Canadiana [370]

Canada : its rise and progress / Smith, George Barnett – London 1898 – (= ser 19th c british colonization) – 4mf – 9 – mf#1.1.7477 – uk Chadwyck [971]

Canada : a metrical story / Campbell, Charles – Toronto: W Briggs, 1897 [mf ed 1980] – 1mf – 9 – 0-665-00404-4 – mf#00404 – cn Canadiana [810]

Canada : a modern nation / Lighthall, William Douw – Montreal: Witness Print House, 1904 [mf ed 1997] – 1mf – 9 – 0-665-83527-2 – mf#83527 – cn Canadiana [333]

Canada : papers relative to the affairs of canada – London: printed by William Clowes & sons...1849 [mf ed 1984] – 1mf – 9 – mf#SEM105P441 – cn Bibl Nat [971]

Canada : a patriotic address / Buchanan, Robert J – [Hamilton, Ont?: s.n.], 1907 – 1mf – 9 – 0-665-99474-5 – mf#99474 – cn Canadiana [971]

Canada : physical, economic and social / Lillie, Adam – Toronto: Maclear, 1855 [mf ed 1982] – 4mf – 9 – (incl bibl ref) – mf#37267 – cn Canadiana [550]

Canada : a portfolio of original photographic views of our country... – Toronto: Art Pub Co, 1894? – 2mf – 9 – mf#54455 – cn Canadiana [770]

Canada : present and future, a patriotic poem / Awde, Robert – [S.l: s.n, 1889?] – 1mf – 9 – 0-665-01006-0 – mf#01006 – cn Canadiana [810]

Canada : return of the names and quality or station of the several persons arrested and placed in confinement in the prisons of toronto, etc – [s.l.]: the House of Commons, 1839 [mf ed 1984] – 1mf – 9 – mf#SEM105P388 – cn Bibl Nat [360]

Canada : session laws of canada – 1792-2001 – 9 – $3643.00 set – mf#408050 – us Hein [348]

Canada : a short history of the dominion of canada / Archer, Andrew – St John, NB: J & A McMillan, 1889 – (= ser New Brunswick school series) – 3mf – 9 – mf#26064 – cn Canadiana [971]

Canada / Vekeman, Gustave – Brussels?: "Journal populaire", 1887 – 1mf – 9 – (also available in french) – mf#40501 – cn Canadiana [304]

Canada / Willson, Beckles – London: T C & E C Jack; Toronto: Copp Clark, 1907 – 4mf – 9 – 0-665-73891-9 – (with 12 reproductions fr original coloured drawings by henry sandham) – mf#73891 – cn Canadiana [971]

Canada : zijn de belgen goede landverhuizers? / Vekeman, Gustave – Brussels?: J van Gompel-Trion, 1884 – 1mf – 9 – mf#62218 – cn Canadiana [304]

Canada *see*
- Acte pour pourvoir plus amplement a l'incorporation de la ville de st hyacinthe
- Reponse a une adresse de l'assemblee legislative en date du 22 juin dernier

Le canada – 2e ed. Quebec: impr par John Lovell, 1860 [mf ed 1999] – 1mf – 9 – cn Bibl Nat [333]

Le canada / Clapin, Sylvia – Paris: Plon, 1897? – 1mf – 9 – mf#04167 – cn Canadiana [917]

Le canada – Ottawa, ON. 1865-69 – 4r – 1 – cn Library Assoc [071]

Le canada : ou notes d'un colon / Vekeman, Gustave – Sherbrooke, Quebec?: s.n, 1884 – 1mf – 9 – mf#16985 – cn Canadiana [971]

Le canada / Vekeman, Gustave – Bruxelles: "Journal populaire", 1887 – 1mf – 9 – mf#25322 – cn Canadiana [304]

O canada : mon pays, mes amours! chant patriotique / Labelle, Jean Baptiste – nouv ed. Montreal: A J Boucher, [entre 1905 et 1924] – 1mf – 9 – mf#SEM105P908 – cn Bibl Nat [780]

Canada 1862 : pour l'information des immigrants / Canada (Province). Bureau d'agriculture et d'immigration – Quebec: Impr de Leger Brousseau, 1862 [mf ed 1983] – 1mf – 9 – mf#SEM105P233 – cn Bibl Nat [971]

Canada 1862 : pour l'information des immigrants – [Quebec?: s.n.], 1862 [mf ed 1985] – 1mf – 9 – 0-665-51160-4 – mf#51160 – cn Canadiana [304]

Canada 1863 : for the information of immigrants / Buchanan, Alexander Carlisle – [Quebec?: s.n.], 1863 [mf ed 1985] – 1mf – 9 – 0-665-43252-6 – mf#43252 – cn Canadiana [304]

Canada, a memorial volume : general reference book on canada, describing the dominion at large, and its various provinces and territories... – Montreal: E Biggar, 1889 [mf ed 1980] – 12mf – 9 – 0-665-02302-2 – (pref by e b biggar; incl poem: fair canada by a h wingfield; incl ind) – mf#02302 – cn Canadiana [917]

Canada, a short history of the dominion of canada / Archer, Andrew – St John, NB: J & A McMillan, 1884 – (= ser New Brunswick school series) – 3mf – 9 – mf#06061 – cn Canadiana [971]

Le canada agricole et industriel : les "ranchs canadiens" / Bouthillier-Chavigny, Charles, vicomte de – Montreal: E Senecal, 1888 – 1mf – 9 – mf#51020 – cn Canadiana [338]

Canada and great britain : report of erastus wiman on the congress of the chambers of commerce of the british empire, held in london, june, 1892 / Wiman, Erastus – [S.l: s.n, 1892?] [mf ed 1981] – 1mf – 9 – mf#25988 – cn Canadiana [380]

Canada and her commerce : from the time of the first settlers to that of the representative men of to-day who have shaped the destiny of our country / ed by Hedley, James – Montreal: Sabiston Litho & Pub Co, 1894 [mf ed 1980] – 4mf – 9 – 0-665-05533-1 – (incl the official history of the dominion commercial travellers' association comp by h w wadsworth) – mf#05533 – cn Canadiana [380]

Canada and her relations to the empire / Dennison, George Taylor – Toronto: Week Pub, 1895 [mf ed 1980] – 1mf – 9 – 0-665-02635-8 – (repr fr: the westminster review) – mf#02635 – cn Canadiana [971]

Canada and her resources / Armstrong, Charles Newhouse – London: Metchim, [1885?] [mf ed 1980] – 1mf – 9 – 0-665-02469-X – mf#02469 – cn Canadiana [917]

Canada and her resources : an essay to which, upon a reference from the paris exhibition committee of canada was awarded by his excellency sir edmund walker head, bart, governor general of british north america, etc, etc, etc, the second prize / Morris, Alexander – Montreal: B Dawson; London: S Low, 1855 [mf ed 1984] – 2mf – 9 – 0-665-40772-6 – mf#40772 – cn Canadiana [917]

Canada and her resources : an essay to which, upon a reference from the paris exhibition committee of canada was awarded by his excellency sir edmund walker head, bart, governor general of british north america, etc, etc, etc, the second prize / Morris, Alexander – Montreal?: s.n, 1855 [mf ed 1983] – 3mf – 9 – mf#38243 – cn Canadiana [917]

Canada and its capital : with sketches of political and social life at ottawa / Edgar, James David – Toronto: G Morang, 1898 [mf ed 1980] – 3mf – 9 – 0-665-02878-4 – mf#02878 – cn Canadiana [971]

Canada and its relations to the empire : an address / Flavelle, Joseph – [Toronto?: s.n, 1917?] [mf ed 1995] – 1mf – 9 – 0-665-77044-8 – mf#77044 – cn Canadiana [327]

Canada and mr goldwin smith / Hincks, Francis – [S.l: s.n. 1881?] [mf ed 1985] – 1mf – 9 – 0-665-24953-5 – mf#24953 – cn Canadiana [380]

Canada and newfoundland / ed by Ami, Henry Marc – London: E Stanford, 1915 – 13mf – 9 – 0-665-75947-9 – (original iss in series: stanford's compendium of geography and travel (new issue) north america) – mf#75947 – cn Canadiana [917]

Canada and newfoundland, etc- chronology : dioceses, vicariates-apostolic, prefectures-apostolic, cardinal...1508 to 1891 – s.l: s.n, 1891? – 1mf – 9 – mf#02489 – cn Canadiana [240]

Canada, and other poems / Herbin, John Frederic – Windsor, NS: J Anslow, 1891 [mf ed 1980] – 1mf – 9 – 0-665-05548-X – mf#05548 – cn Canadiana [810]

Canada and the empire : address delivered by robert meighen at the complimentary banquet given to george e drummond at the canada club, july 21 1904 – [Canada?: s.n, 1904?] – 1mf – 9 – 0-665-86428-0 – mf#86428 – cn Canadiana [380]

Canada and the united states : an address on the american conflict...december 22, 1864 / Cordner, John – Manchester: A Ireland, 1865 – 1mf – 9 – mf#33340 – cn Canadiana [976]

Canada and the united states : an historical retrospect / Bourinot, John George – [S.l: s.n, 1891?] [mf ed 1980] – 1mf – 9 – 0-665-02530-0 – (incl bibl ref) – mf#02530 – cn Canadiana [327]

Canada and the united states : a study in comparative politics / Bourinot, John George – [S.l: s.n, 1890] [mf ed 1980] – 1mf – 9 – 0-665-02531-9 – (fr: annals of the american academy of political and social science, jul 1890) – mf#02531 – cn Canadiana [327]

Canada and the united states : their past and present relations / Bourinot, John George – [S.l: s.n, 1891?] [mf ed 1980] – 1mf – 9 – (fr: quarterly review, apr 1891; incl bibl ref) – mf#05952 – cn Canadiana [327]

Canada and the united states : their past and present relations – [S.l: s.n, 1891] [mf ed 1980] – 1mf – 9 – 0-665-00413-3 – mf#00413 – cn Canadiana [327]

Canada and the united states compared : with practical notes on commercial union, unrestricted reciprocity and annexation / Facktz, P N – Toronto: Toronto News Co, 1889 [mf ed 1981] – 1mf – 9 – mf#01269 – cn Canadiana [337]

Canada and the world – Toronto. 1975-1994 (1) 1975-1994 (5) 1976-1994 (9) – (cont by: canada and the world backgrounder) – ISSN: 0043-8170 – mf#10764 – us UMI ProQuest [300]

Canada and the world backgrounder – Waterloo. 1994+ (1,5,9) – (cont: canada and the world) – ISSN: 1189-2102 – mf#10764,01 – us UMI ProQuest [300]

Canada as a home / Bourinot, John George – London: Truebner, 1882 [mf ed 1980] – 1mf – 9 – 0-665-03812-7 – mf#03812 – cn Canadiana [360]

Canada as a home / Bourinot, John George – Toronto?: s.n, 1882 [mf ed 1982] – 1mf – 9 – 0-665-26583-2 – (repr fr: the westminster review, jul 1882) – mf#26583 – cn Canadiana [360]

Canada, as it is : comprising details relating to the domestic policy, commerce and agriculture, of the upper and lower provinces / Hume, George Henry – New York: W Stodart, 1832 [mf ed 1984] – 3mf – 9 – 0-665-45100-8 – mf#45100 – cn Canadiana [917]

Canada at the colonial and indian exhibition – [London?: s.n.], 1886 [mf ed 1980] – 1mf – 9 – 0-665-00414-1 – mf#00414 – cn Canadiana [971]

Canada at the universal exhibition of 1855 = Le canada et l'exposition universelle de 1855 / Comite executif canadien de l'Exposition universelle a Paris (1855) – Toronto: printed by John Lovell, 1856 [mf ed 1983] – 5mf – 9 – mf#SEM105P154 – cn Bibl Nat [330]

The canada baptist magazine and missionary register – Montreal: Pub by W Greig...[1837-1841] – 9 – mf#P05067 – cn Canadiana [242]

Canada. Bas-Canada *see* Code civil du bas canada

Canada. Bas-Canada. Parlement. Chambre d'Assemblee *see* Rapport

Canada bookseller – Toronto: Adam, Stevenson, [1870-1871?] – 9 – (cont by: the canada bookseller miscellany and advertiser) – mf#P05008 – cn Canadiana [070]

The canada bookseller – Toronto: Rollo & Adam, [1865?-1869?] – 9 – mf#P06003 – cn Canadiana [070]

Canada bookseller and stationer – Toronto: MacLean, 1896-1897 [mf ed v12 n1 jan 1896-v13 n4 apr 1897] – 9 – mf#P06028 – cn Canadiana [020]

Canada bookseller and stationer *see* Bookseller and stationer

Canada bookseller miscellany and advertiser – [Toronto]: Adam, Stevenson, [1872?-18–] – 9 – (cont: the canada bookseller) – mf#P05011 – cn Canadiana [070]

Canada. Canadian Army. Scarboro' Volunteer Rifle Company *see* Rules and regulations of the scarborough volunteer rifle company

Canada. Canadian Army. Volunteer Militia Rifle Company of Toronto, 1st *see* Rules and regulations of the 1st volunteer militia rifle company, of toronto

Canada canal communication : return to an address to his majesty, dated 4 february 1831 for copies of the correspondence between the treasury, the secretary of state for the colonies and ordnance... – [s.l.]: the House of Commons, 1831 [mf ed 1984] – 2mf – 9 – mf#SEM105P387 – cn Bibl Nat [380]

Canada. Central Board of Health *see* Regulations etc adopted by the...under the act 12 vict cap 8

Le canada chante, vol 1 : les horizons / Ferland, Albert – Montreal: Deom Frere, 1908-1910 – 4v on 4mf – 9 – 0-665-75051-X – mf#75051 – cn Canadiana [810]

Le canada chante, vol 2 : le terroir / Ferland, Albert – Montreal: Deom Frere, 1908-1910 – 4v on 4mf – 9 – 0-665-75052-8 – mf#75052 – cn Canadiana [810]

Le canada chante, vol 3 : l'ame des bois / Ferland, Albert – Montreal: Deom Frere, 1908-1910 – 4v on 4mf – 9 – 0-665-75053-6 – mf#75053 – cn Canadiana [810]

Le canada chante, vol 4 : la fete du christ a ville-marie / Ferland, Albert – Montreal: Deom Frere, 1908-1910 – 4v on 4mf – 9 – 0-665-75054-4 – mf#75054 – cn Canadiana [810]

The canada christian monthly : a review and record of christian thought, christian life and christian work – Chatsworth [Ont]: J Morrison, [1873-1878?] – 9 – ISSN: 1190-7606 – mf#P04174 – cn Canadiana [240]

The canada citizen and temperance herald : a journal devoted to the advocacy of prohibition and the promotion of social progress and moral reform – Toronto: Citizen Pub Co, [1879?-18– or 19–] – 9 – mf#P06073 – cn Canadiana [230]

Canada (clergy reserves) : copy of a letter from captain pringle to the secretary of state for the colonies, dated 9th may 1840, relative to the extent and value of the clergy reserves in upper canada – [S.l: s.m, 1840?] [mf ed 1987] – 1mf – 9 – 0-665-64019-6 – mf#64019 – cn Canadiana [340]

Canada. Commission of Conservation. Committee on Fisheries, Game and Fur-Bearing Animals *see* Fur-farming in canada

Canada. Commissioners to Revise the Public General Statutes *see* Special report.

Canada company : letter, dated 9th january 1840, soliciting appropriation of sum due to the crown to the encouragement of emigration to upper canada – [s.l.]: the House of Commons, 1840 [mf ed 1984] – 1mf – 9 – mf#SEM105P390 – cn Bibl Nat [971]

Canada, Conde de la *see*
- Instituciones practicas de los juicios civiles
- Observaciones practicas sobre los recursos de fuerza.... 1794

Canada constellation *see* Niagara peninsula newspapers, pt 1

Canada co-operative supply association (limited), albert building, victoria square, montreal : capital $150,000 in 30,000 shares of $5 each – [Montreal?: s.n, 1881?] [mf ed 1980] – 1mf – 9 – 0-665-05698-2 – mf#05698 – cn Canadiana [334]

The canada corn bill : lord stanley's speech, in the house of commons, on friday, may 19 – [London], [1843?] – (= ser 19th c books on british colonization) – 1mf – 9 – mf#1.1.4961 – uk Chadwyck [348]

Canada corporations act 1968; chap. 53, r.s.c. 1952, as amended / Canada. Laws, Statutes, etc – 2nd ed. Don Mills, Ont.: CCH Canadian 1968 144p. LL-2373 – 1 – us L of C Photodup [348]

Canada correspondence / Brown, Thomas Storrow – Montreal?: s.n, 1873? – 1mf – 9 – mf#03728 – cn Canadiana [971]

Canada. Court of Inquiry on Lime Ridge Engagement *see* Proceedings and report of the court of inquiry

Le canada de l'atlantique au pacifique et a la mer polaire, expeditions arctiques et voyages de decouverte au nord, etc, etc / Baillairge, George Frederick – Ottawa?: s.n, 1891? – 3mf – 9 – mf#00930 – cn Canadiana [919]

Canada. Departement de l'Instruction Publique pour le Haut-Canada *see* Law of separate schools in upper canada

Canada. Departement des finances *see* Report of the minister of finance on the reciprocity treaty with the united states

Canada. Dept of Indian Affairs *see* Annual report

Canada. Dept of Marine and Fisheries *see* Survey of tides and currents in canadian waters

Canada. Dept of National Health and Welfare *see* Mental health legislation in canada, 1959

Canada. Dept of National Revenue *see* Report, containing statements relative to customs-excise revenue and other services

Canada. Dominion Bureau of Statistics *see*
- Census of canada
- The maritime provinces since confederation

Canada during the victorian era : a historical review / Bourinot, John George – Ottawa: J Durie; Toronto: Copp, Clark, 1897 – 1mf – 9 – (incl ill) – mf#00984 – cn Canadiana [971]

Canada during the victorian era : a short historical review in two parts / Bourinot, John George – S.l: s.n, 1887? – 1mf – 9 – mf#26676 – cn Canadiana [971]

Il canada e l'emigrazione / Bossange, Gustave – Paris: Symonds, 1872? – 1mf – 9 – mf#03711 – cn Canadiana [304]

Canada educational monthly – Toronto: Canada Educational Monthly Pub Co, 1897-1902 – 9 – (cont: the canada educational monthly and school chronicle; cont by: the educational monthly of canada) – mf#P04044 – cn Canadiana [370]

Canada educational monthly and school chronicle – Toronto: Educational Monthly Pub Co, 1879?-1897 – 9 – (absorbed: school magazine; cont by: canada educational monthly) – mf#P04023 – cn Canadiana [370]

Canada. Emergency Measures Organization *see* Emo national digest

Le canada et la marine : discours prononce par le tres honorable sir wilfrid laurier...chef de l'opposition, en presentant l'amendement au bill de l'aide navale a la chambre de communes le 12 decembre 1912 – Ottawa: [s.n.], 1913 – 1mf – 9 – 0-665-74882-5 – (also available in english) – mf#74882 – cn Canadiana [355]

Le canada et l'exposition universelle de 1855 = Canada at the universal exhibition of 1855 / Comite executif canadien de l'Exposition universelle a Paris (1855) – Toronto: des presses a vapeur de John Lovell, 1856 [mf ed 1983] – 6mf – 9 – mf#SEM105P153 – cn Bibl Nat [330]

Canada et terreneuve etc : chronologie...statistiques eclesiastiques: 1508 a 1891? / Baillarge, George Frederick – [S.l: s.n, 1891?] [mf ed 1980] – 1mf – 9 – 0-665-04357-0 – mf#04357 – cn Canadiana [971]

Canada. Exchequer Court *see* Exchequer reports of canada

Canada. Exploration Geologique *see* Descriptive catalogue of a collection of the economic minerals of canada, and of its crystalline rocks

The canada farmer – Hamilton [Ont]: J E Force, [1855?-18–] – 9 – mf#P04884 – cn Canadiana [630]

The canada farmer – Toronto: R Brewer, 1847 – 9 – (merged with: british american cultivator. merged to become: the agriculturalist and canadian journal) – mf#P04205 – cn Canadiana [630]

The canada farmer (toronto, ont: 1864) – Toronto: G Brown, 1864-1876 – 9 – mf#P04206 – cn Canadiana [630]

Canada first movement – 1r – 1 – (william alexander foster scrapbook) – cn Library Assoc [971]

Canada for canadians : extracts from the election law, with directions for voting: liberal-conservative candidate for the city and county, charles a everett – s.l: s.n, 1885 [mf ed 1987] – 1mf – 9 – mf#56236 – cn Canadiana [342]

Canada for gentlemen : being letters from james seton cockburn – London?: Army & Navy Co-operative Society, 1884? – 1mf – 9 – mf#27082 – cn Canadiana [971]

"Canada for the canadians" : political pointers for the campaign of 1896 / Hague, John – [Ottawa?: s.n, 1896?] – 1mf – 9 – 0-665-29431-X – mf#29431 – cn Canadiana [337]

Canada foundry company limited : ornamental iron department bulletin no 2, jun 1st 1901 – [Toronto?: s.n, 1901?] [mf ed 1991] – 1mf – 9 – 0-665-99525-3 – mf#99525 – cn Canadiana [660]

Canada from the atlantic to the pacific and arctic oceans, arctic voyages of discovery in the north and public works, etc, etc / Baillarge, George Frederick – [Ottawa?: s.n, 1891?] [mf ed 1980] – 3mf – 9 – 0-665-03315-X – (incl ind) – mf#03315 – cn Canadiana [917]

Canada from the atlantic to the pacific and arctic oceans, arctic voyages of discovery in the north and public works, etc, etc / Baillarge, George Frederick – [S.l: s.n, 1890?] [mf ed 1984] – 3mf – 9 – (incl ind) – mf#28830 – cn Canadiana [917]

Canada gazette – 1841-69 – 35r – 1 – cn Library Assoc [971]

Canada gazette : second series – Ottawa, ON: Queen's Printer, 1867-92 – 27r – 1 – cn Library Assoc [971]

Canada. General *see*
- Cameron's legal opinions
- Canada statutes, session laws and revisions
- Canadian law times
- Canadian reports
- Cartwright's cases on the british north american act
- Eastern law reporter
- Hunter's torrens cases
- Laperriere's speaker's decisions
- Western law reporter

Canada. geological survey. memoirs – v1-34. 1910-19 – 1 – $108.00 – (v35-321 1919-63 $1050 [0137]) – mf#0136 – us Brook [550]

Canada go bragh : being an inaugural address to the young liberal club, of seaforth, ont, on the 27th october, 1886 / Cartwright, Richard – Toronto: [s.n.], 1886 [mf ed 1980] – 1mf – 9 – 0-665-00499-0 – mf#00499 – cn Canadiana [325]

Canada goose – aug 23/sep 5-1969 mar 21/apr 3 – 1r – 1 – mf#1110496 – us WHS [071]

Canada. Gouverneur general *see* Further correspondence relative to the projected railway from halifax to quebec

The canada grande ligne mission : with a photographic sketch of the french-canadian hut where in madame feller, the self-denying founder of the said mission began her benevolent enterprise, by instructing a few french-canadian children – [Canada: s.n, 19–?] – 1mf – 9 – 0-665-76238-0 – mf#76238 – cn Canadiana [810]

Canada (halifax, etc railway), railways (british north america) : copy of official communications...on the subject of a proposed communication by railway between the port of halifax and those provinces – [S.l: s.n, 1862?] [mf ed 1984] – 1mf – 9 – 0-665-38168-9 – mf#38168 – cn Canadiana [380]

Canada health journal – London, Ont: J Cameron, [1870] [mf ed v1 n1 jan 1870-v1 n5 may 1870] – 9 – mf#P05195 – cn Canadiana [614]

Canada health journal – Ottawa: Canada Health Journal, [1890?-189- or 19–] [mf ed v12 n1 jan 1890-v13 n12 dec 1891] – 9 – mf#P04588 – cn Canadiana [614]

The canada health journal – [Ottawa: Health Journal, 1886-1888] [mf ed v8 n10 [i.e. 9] sep 1886-v10 n6 jun 1888] – 9 – mf#P04586 – cn Canadiana [614]

The canada herd book : containing the pedigrees of improved short horned cattle, vol 1 – [Toronto?: s.n.] 1867 [mf ed 1984] – 7mf – 9 – 0-665-22495-8 – (incl ind) – mf#22495 – cn Canadiana [636]

Canada, historical and descriptive : from sea to sea / Adam, Graeme Mercer – Toronto: W Bryce, 1888 [mf ed 1979] – 2mf – 9 – 0-665-00771-X – mf#00771 – cn Canadiana [917]

Canada in 1871 : or, our empire in the west: a lecture, delivered at the russell institution, london, 22nd january, 1872 / Duncan, Francis – London : W Mitchel, 1872 – 1mf – 9 – mf#07106 – cn Canadiana [917]

Canada in flanders / Beaverbrook, Max Aitken, Baron – London: Hodder & Stoughton 1916- [mf ed 1985] – 1r – 1 – (pref by a bonar law; int by robert borden; with maps & app) – us UMI ProQuest [933]

Canada in memoriam, 1812-14 : her duty in the erection of monuments in memory of her distinguished sons and daughters / Curzon, Sarah Anne – Welland [Ont]: Telegraph, 1891 [mf ed 1980] – 1mf – 9 – mf#03641 – cn Canadiana [971]

Canada in the seventeenth century / Boucher, Pierre, sieur de Boucherville – Montreal: printed by George E Desbarats & Co, 1883 [mf ed 1979] – (= ser Histoire veritable et naturelle des moeurs et productions du pays de la nouvelle france vulgairement dite le Canada) – 1mf – 9 – 0-665-00874-0 – (trans by by edward louis montizambert) – mf#00874 – cn Canadiana [971]

Canada Iron Mining and Manufacturing Co *see* Prospectus of the...

Canada, is she prepared for war? : or, a few remarks on the state of her defences / Denison, George Taylor – Toronto?: s.n, 1861 – 1mf – 9 – mf#22896 – cn Canadiana [355]

Canada, its growth and prospects : two lectures delivered before the mechanics' institute, toronto, on the 13th and 27th february, 1852 / Lillie, Adam – Toronto: T Maclear, 1852 [mf ed 1984] – 1mf – 9 – (repr fr: journal of education for upper canada, mar, 1852; incl bibl ref) – mf#22312 – cn Canadiana [317]

Canada, its growth and prospects : two lectures delivered before the mechanics' institute, toronto, on the 13th and 27th february, 1852 / Lillie, Adam – [Brockville, Ont?: s.n.], 1852 [mf ed 1986] – 1mf – 9 – 0-665-48225-6 – (with app; incl bibl ref) – mf#48225 – cn Canadiana [317]

Canada japan trade council newsletter – Ottawa. 1972-1981 (1) 1972-1981 (5) 1980-1981 (9) – ISSN: 0045-4214 – mf#7840 – us UMI ProQuest [380]

The canada journal of dental science – Hamilton [Ont]: W G Beers, C S Chittenden, R Trotter, [1868-1879?] – 9 – mf#P04218 – cn Canadiana [617]

Canada kalender – Berlin, ON: Rittinger & Motz, 1882-1920//? – 52mf – 9 – Can$275.00 – (in german. lacking: 1883, 1887, 1895, 1901 and 1905) – cn McLaren [071]

Canada land amendment association : prospectus and constitution together with some remarks on the present system of land transfer in ontario... – Toronto: The Association, 1883 – 1mf – 9 – mf#02021 – cn Canadiana [343]

Canada law journal / Canada. Ontario – v4-58. 1868-1922 (all publ) – 358mf – 9 – $537.00 – (cont: v3 of upper canada law journal, new series. aka: canada law journal, new series) – mf#LLMC 81-015 – us LLMC [340]

Canada law journal – Montreal: J Lovell, 1867-[1868] – 9 – (cont: the lower canada law journal) – mf#P05019 – cn Canadiana [348]

Canada law journal – Toronto: W C Chewett, 1868-[1922] – 9 – (cont: the upper canada law journal and local courts gazette; merged with: the canadian law times, to become: the canadian bar review) – mf#P04911 – cn Canadiana [347]

Canada law journal – Toronto. 68v. 1855-1922 – 1 – $610.00 – (title varies: os: as upper canada law journal; os: v1-10 1855-64, ns: 1-58 1865-1922) – mf#408850 – us Hein [340]

Canada. Laws, Statutes, etc *see*
- An act for the abolition of feudal rights and duties in lower canada
- Acte d'amendement des municipalites et des chemins du bas-canada, de 1856
- Canada corporations act 1968; chap. 53, r.s.c. 1952, as amended
- Index to dominion statute amendments
- Index to eastern provinces and dominion statute amendments to 1926.
- An index to the statutes of canada
- Index to western provinces and dominion statute amendments to 1926.
- The provincial laws of the customs, and a collection of those parts of the imperial acts on the same subject.
- The seigniorial acts: viz: the seigniorial act of 1854, 16 vict. cap. 3
- A synoptical index of the consolidated statutes of canada and upper canada, with notices of the later acts which affect them; including the session of 1864

The canada life assurance company bill is stated to be listed to be heard by the banking and commerce committee : on wednesday the 10th day of march, 1909... / Laidlaw, William – [Toronto?: s.n, 1909?] [mf ed 1996] – 1mf – 9 – 0-665-78958-0 – mf#78958 – cn Canadiana [368]

Canada lumberman – Toronto: C H Mortimer, [1895-1905] – 9 – mf#P04954 – cn Canadiana [634]

The canada lumberman and millers', manufacturers' and miners gazette – Toronto: A Begg, [1880-1887] – 9 – mf#P04420 – cn Canadiana [634]

Canada lumberman and woodworker – Toronto, Canada. -w. 1 may, 1 nov, 15 dec 1919; 1920; 1921 – 7r – 1 – uk British Libr Newspaper [670]

The canada lumberman (monthly ed) – Peterborough [Ont]: A G Mortimer, [1887-1904] [mf ed v7 n10 oct 1887-v24 [i.e. 25], n12 dec 1904] – 9 – mf#P04953 – cn Canadiana [634]

Canada medical and surgical journal – Montreal: G E Desbarats, [1872-1888] [mf ed v1[n1 jul 1872]-[v16 n12 jun 1888] – 9 – mf#P05177 – cn Canadiana [610]

Canada medical association : montreal meeting, tuesday 26th aug 1884 – [s.l: s.n, 1884?] [mf ed 1985] – 1mf – 9 – 0-665-01766-9 – mf#01766 – cn Canadiana [610]

Canada medical journal and monthly record of medical and surgical science / ed by Fenwick, George E & Wayland, Francis – Montreal: Dawson Bros, [1864-1872] [mf ed v1[n1 jul 1864]-[v8 n12 jun 1872] – 9 – (in english and french; cont: the british american journal, dec 1862) – mf#P05176 – cn Canadiana [610]

Canada medical journal and monthly record of medical and surgical science – Montreal: printed by J Lovell, 1852-1853 [mf ed vl n1 mar 1852-v1 n12 feb 1853] – 9 – (incl ind) – mf#P04196 – cn Canadiana [610]

The canada medical record : a monthly journal of medicine and surgery – [Montreal?: s.n, 1872-1904] – 9 – mf#P05185 – cn Canadiana [610]

Canada military gazette – Ottawa: D Kerr, [1857-18–?] [mf ed v1 n1 feb 3 1857-v1 n14 may 15, 1857] – 9 – mf#P04359 – cn Canadiana [355]

Canada (militia bills) : return to an address of the house of lords, dated 10th february 1863... in reference to the militia bills proposed and passed in the canadian parliament – [London?: s.n, 1863?] [mf ed 1984] – 1mf – 9 – 0-665-45287-X – mf#45287 – cn Canadiana [323]

The canada (monthly) general railway and steam navigation guide – Toronto: Pub...by MacLear, [1856-18–?] – 9 – mf#P04367 – cn Canadiana [380]

Canada must have prohibition / Watkins, Thomas C – [Hamilton, Ont?: s.n, 188-?] [mf ed 1994] – 1mf – 9 – 0-665-94625-2 – (original iss in ser: prohibition series. incl bibl ref) – mf#94625 – cn Canadiana [170]

Canada. Natural Resources Intelligence Service *see* Natural resources of the prairie provinces

Canada. Ontario *see*
- Canada law journal
- Upper canada law journal

Canada, ontario, the british flag and other poems / Gray, Nelson Cockburn – [Montreal: s.n.], c1908 – 1mf – 9 – 0-665-97291-1 – mf#97291 – cn Canadiana [810]

Canada organ and piano co, limited : capital under charter, $250,000: organs and melodeons – [Toronto?: s.n, 187-?] [mf ed 1983] – 1mf – 9 – 0-665-39746-1 – mf#39746 – cn Canadiana [780]

Canada, our frozen frontier – [s.l: s.n, 1862?] [mf ed 1984] – 2mf – 9 – 0-665-32088-4 – (contains 3 essays bound together and publ in jan & feb 1862) – mf#32088 – cn Canadiana [327]

Canada pacific railway : elements for a prospectus / Waddington, Alfred – [Ottawa: s.n, 1870?] [mf ed 1980] – 1mf – 9 – 0-665-02043-0 – mf#02043 – cn Canadiana [380]

Canada, papers relating to the removal of the seat of government, and to the annexation movement : presented to both houses of parliament by command of her majesty, 15th april,1850 / Grande-Bretagne. Colonial Office – London: W Clowes, 1850 [mf ed 1984] – 1mf – 9 – 0-665-45282-9 – mf#45282 – cn Canadiana [971]

Le canada par l'image / Brouillette, Benoit – 4e ed. [Montreal]: Librairie Beauchemin, 1946 [mf ed 1986] – 3mf – 9 – mf#SEM105P681 – cn Bibl Nat [971]

Canada. Parlement. Chambre des communes *see*
- Rapport et temoignage sur la derniere election pour le district electoral de kamouraska
- Rapport sur les avantages et la necessite d'etablir un reseau de telegraphe sous-marin dans le fleuve et le golfe st laurent

Canada. Parlement. Conseil legislatif *see* Extract from a return dated the 2nd of december, 1854

Canada penitentiary : answers to the questions proposed by messr mondelet and neilson, lower canada commissioners – [s.l: [s.n.], 1842 [mf ed 1983] – 1mf – 9 – mf#SEM105P347 – cn Bibl Nat [360]

Canada Permanent Building and Savings Society *see* Annual report

Canada Permanent Loan and Savings Company *see* Annual report

Canada poultry journal – Brooklin, Ont: [H M Thomas & E R Grant], 1875-[18– or 19–] [mf ed v1 n1 sep 15 1875-v1 n12 aug 15, 1876] – 9 – mf#P04983 – cn Canadiana [636]

The canada presbyterian – Toronto: C B Robinson, [1877-189- or 19–] [mf ed new ser: v1 n1 nov 2 1877-v25 n52 dec 23 1896] – 9 – ISSN: 1191-2057 – mf#P04796 – cn Canadiana [242]

Canada (Province) *see*
- An act concerning bankrupts and the administration of their effects
- An act for limiting the time of service in the army
- An act for the better establishment and maintenance of public schools in upper-canada
- An act further to amend the judicature acts of lower canada
- An act respecting the militia, extracted from consolidated statutes of canada
- An act respecting the preservation of the public health
- An act to abolish imprisonment for debt and for the punishment of fraudulent debtors in lower canada and for other purposes
- An act to amend the acts relating to the grand trunk railway company of canada
- An act to authorize the grand trunk railway company of canada to construct a bridge over the river st clair at sarnia
- An act to define seigniorial rights in lower canada
- An act to grant additional aid to the grand trunk railway company of canada
- An act to regulate the inspection and measurement of timber, masts, spars, deals, staves and other articles of a like nature in the ports of quebec and montreal
- Acte des municipalites et des chemins de 1855...
- L'acte municipal du bas canada de 1860
- Acte pour abroger certaines lois y mentionnees pour mieux pourvoir a la defense de cette province et pour en regler la milice
- Acte pour abroger certains actes y mentionnes
- Acte pour abroger certains actes y mentionnes et etablir de meilleures dispositions relativement a l'admission des arpenteurs et a l'arpentage des terres en cette province
- Acte pour amender et consolider les dispositions de l'ordonnance pour incorporer la cite et ville de montreal
- Acte pour amender et refondre les differents actes concernant le notariat
- Acte pour amender l'acte municipal refondu du bas-canada
- Acte pour amender les actes de judicature du bas-canada
- Acte pour amender les lois en force
- Acte pour amender les lois relatives a la milice de cette province, et les rendre parmanentes
- Acte pour augmenter la representation du peuple de cette province en parlement
- Acte pour faire de plus amples dispositions pour l'incorporation de la ville des trois-rivieres
- Acte pour l'abolition des droits et devoirs feodeaux dans le bas-canada
- Acte pour pourvoir a la decision sommaire des petites causes, dans le bas-canada
- Acte pour regler la milice de cette province et pour abroger les actes maintenant en force a cette fin
- Actes concernant l'education et les ecoles dans le bas-canada
- Actes d'education elementaire
- Actes des municipalites du bas-canada
- Les actes et ordonnances revises du bas-canada
- Actes pour promouvoir l'education dans le bas-canada
- Actes relatifs aux chemins a barrieres et ponts dans et pres quebec
- Actes relatifs aux pouvoirs
- Acts relating to the grand trunk railway and for the prevention of accidents on railways
- Acts relating to the powers, duties and protection of justices of the peace in lower canada
- Anno vicesimo-tertio victoriae reginae
- Bill
- Code civil du bas canada
- Code de procedure civile
- Collection de plusieurs des actes et ordonnances les plus utiles en force dans le bas-canada
- A collection of some of the most useful acts and ordinances in force in lower canada
- Communication du greffier de la couronne en chancellerie
- The consolidated statutes for lower canada
- The consolidated statutes for upper canada
- The consolidated statutes of canada
- Consolidated statutes respecting the militia
- Copies de correspondances entre le surintendant-en-chef des ecoles pour le haut-canada et autres personnes
- Copies of correspondence between members of the government and the chief superintendent of schools
- Correspondence, documents, evidence and proceedings in the enquiry of messrs lafrenaye and doherty...
- Customs, excise and commercial laws of canada
- Etat et avenir du canada en 1854
- First report of the commissioners appointed to enquire into the losses occasioned by the troubles during the years 1837 and 1838
- Lois d'education, bc
- Lower canada municipal and road act of 1855...

- The lower canada municipal and road amendment act of 1856
- The lower canada municipal and road amendment act of 1857
- Municipal act of lower canada, 10 and 11 vict 1847
- The provincial laws of the customs
- Railway clauses consolidation acts and acts
- Rapport annuel du maitre-general des postes... 1852-1856
- Rapport des commissaires de l'amerique britannique du nord
- Rapport des commissaires enqueteurs dans l'affaire du meurtre de corrigan
- Rapport des commissaires nommes pour faire une enquete sur la conduite des autorites de police lors de l'emeute de l'eglise chalmers, le 6 juin 1853...
- Rapport des commissaires nommes pour faire une enquete sur les affaires du department des postes dans l'amerique septentrionale britannique
- Rapport des commissaires nommes pour preparer un projet afin de mieux organiser le departement de l'adjudant-general de la milice
- Rapport des commissaires nommes pour preparer un projet afin de mieux organiser le departement de l'adjutant-general [sic] de la milice
- Rapport des commissaires nommes pour s'enquerir de la cause de l'incendie qui a detruit l'hotel du parlement
- Rapport des commissaires nommes pour s'enquerir de l'etat des lois et autres circonstances qui se rattachent a la tenure seigneuriale dans le bas-canada et appendice
- Rapport des commissaires nommes pour s'enquerir de l'origine et des causes de l'incendie qui a consume l'hospice des soeurs de la charite
- Rapport des commissaires nommes pour s'enquerir et faire rapport des meilleurs moyens de reorganiser la milice en canada
- Rapport des commissaires nommes pour s'enquerir et faire rapport des meilleurs moyens de reorganiser la milice en canada et d'etablir un systeme efficace et economique de defense publique
- Rapport des commissaires speciaux, nommes le 8 septembre 1856
- Rapport du bureau central de la sante
- Rapport du maitre-general des postes pour l'annee expiree le...
- Rapport du ministre des finances sur le traite de reciprocite avec les etats-unis
- Rapport du surintendant d'education du bas-canada, pour l'annee 1846
- Rapport final des commissaires
- Rapport preliminaire du bureau des inspecteurs d'asiles, prisons, etc, 1859
- Rapport sur l'etat de la milice de la province
- Rapport sur l'etat de la milice de la province du canada
- Reponse a une adresse de l'assemblee legislative
- Reponse a une adresse de l'assemblee legislative, a son excellence le gouverneur general, datee le 16 du mois dernier
- Reponse a une adresse de l'assemblee legislative a son excellence le gouverneur general en date du 10 septembre
- Reponse a une adresse de l'assemblee legislative, a son excellence le gouverneur-general, datee le 4 juin 1850
- Reponse a une adresse de l'assemblee legislative, en date du 23 du mois dernier
- Reponse a une adresse de l'assemblee legislative, en date du 28 ultimo
- Reponse a une adresse de l'assemblee legislative priant son excellence de vouloir bien mettre devant cette chambre
- Reponse a une adresse...en date du 13 avril 1853
- Report of the central board of health
- Report of the commission appointed to inquire into matters connected with the public buildings at ottawa
- Report of the commission appointed to inquire into the affairs of the grand trunk railway
- Report of the commissioners appointed to enquire into the affairs and financial condition of toronto university and university college, upper canada
- Report of the commissioners appointed to inquire into the cause of the fire at the parliament buildings
- Report of the commissioners appointed to inquire into the conduct of the police authorities
- Report of the commissioners appointed to inquire into the state of the laws and other circumstances connected with the seigniorial tenure in lower canada
- Report of the commissioners appointed to investigate and report upon the best means of re-organizing the militia of canada
- Report of the special commissioners appointed on the 8th of september, 1856
- Return to an address from the legislative assembly
- The revised acts and ordinances of lower-canada
- Statutes regulating the judicature of lower canada
- Statutes relating to elementary education
- Statutes relating to the duties of justices of the peace in lower canada
- Statuts concernant les devoirs des juges de paix dans le bas canada
- Statuts de la province du canada...1852-1866
- Les statuts provinciaux du canada...
- Statuts provinciaux du canada
- Les statuts refondus du canada
- Les statuts refondus pour le bas-canada

Canada. (Province) *see* Rapport des commissaires nommes pour s'enquerir de la cause de l'incendie qui a detruit l'hotel du parlement

Canada (Province). Assemblee legislative *see* Rapport...auquel a ete renvoye le sujet de la formation d'un pont de glace sur le saint-laurent, devant quebec

Canada (Province). Bureau d'agriculture et des statistiques *see* Memoire sur le cholera

Canada (Province). Bureau d'agriculture et d'immigration *see* Canada 1862

Canada (Province). Bureau d'enregistrement et de statistiques *see*
- Recapitulation
- Recensement des canadas, 1851-2
- Recensement des canadas, 1860-61
- Recensement du canada, 1861

Canada (Province) Bureau des brevets *see* Patents of canada

Canada (Province). Bureau du secretaire *see* Correspondence relative to the accounts of the indian department in canada west

Canada (Province). Commissaire sous l'Acte seigneurial refondu *see*
- Cadastre abrege de la partie de la seigneurie de bourchemin est...
- Cadastre abrege de la partie de la seigneurie de l'islet st jean...
- Cadastre abrege de la partie sud-ouest de la seigneurie de bourg louis...
- Cadastre abrege de la seigneurie d'auteuil...
- Cadastre abrege de la seigneurie de beauharnois...
- Cadastre abrege de la seigneurie de deschambault...
- Cadastre abrege de la seigneurie de la grande vallee des monts...
- Cadastre abrege de la seigneurie de la nouvelle longueuil...
- Cadastre abrege de la seigneurie de levrard (ou st pierre les becquets)...
- Cadastre abrege de la seigneurie de rocquetaillade...
- Cadastre abrege de la seigneurie de st valier...
- Cadastre abrege de la seigneurie delery...
- Cadastre abrege du fief coteau st louis
- Cadastre abrege du fief st joseph ou l'epinay...
- Cadastre abrege du fief vieuxpont...

Canada (Province). Commissaires charges de codifier les lois du Bas Canada, en matieres civiles *see*
- Cedule
- Code de procedure civile du bas canada

Canada (Province). Commissaires charges de reviser les actes et ordonnances du Bas-Canada *see* Tables relative to the acts and ordinances of lower-canada

Canada (Province). Commissaires nommes pour s'enquerir d'une serie d'accidents et retardements sur le Grand chemin de fer occidental, Canada Ouest *see* Rapports... en vertu d'une commission en date du 3 novembre 1854

Canada (Province). Cour superieure (Bas-Canada) *see*
- Rules and orders of practice of the superior court, lower canada
- Rules, orders, and tariff of fees in insolvency, 1864

Canada (Province). Departement de la Milice *see* List of officers of the sedentary militia of lower canada, 1862

Canada (Province). Departement de la milice *see*
- The annual volunteer and service militia list of canada
- Return to an address of the legislative assembly, of 1st september, 1863

Canada (Province). Departement de l'agriculture *see* Canada

Canada (Province). Departement des finances *see*
- Discours prononce par l'honorable a t galt...en presentant le budget
- Speech of the hon a t galt
- Speech of the hon a t galt...
- Speech of the honorable a t galt...in introducing the budget

Canada (Province). Departement des terres de la couronne *see*
- Etat des sommes depensees a meme l'octroi de £30, 000 vote dans le but d'aider a l'etablissement des terres vacantes de la couronne dans le bas-canada
- Remarks on upper canada surveys, and extracts from the surveyors' reports

Canada (Province). Departement des travaux publics *see* Documents relating to the construction of the parliamentary and departemental buildings at ottawa

Canada (Province). Dept of Public Instruction for Upper Canada *see* The educational museum and school of art and design for upper canada

Canada (Province). Gouverneur general *see*
- Correspondence relating to the civil list and military expenditure in canada
- Halifax railway and public works
- Return to two addresses of the honorable the legislative assembly to his excellency the governor general dated 28th february, 1856...

Canada (Province). Gouverneur general (1847-1854: Elgin) *see* Chemin de fer de halifax et de quebec et travaux publics

Canada (Province). Parlement *see*
- Debats parlementaires sur la question de la confederation des provinces de l'amerique britannique du nord
- Parliamentary debates on the subject of the confederation of the british north american provinces

Canada (Province). Parlement. Assemblee legislative *see*
- Annual revenue and expenditure of lower canada
- Chemin de fer de quebec et halifax
- Constitutions, regles et reglements de l'assemblee legislative du canada
- Debats dans l'assemblee legislative sur la tenure seigneuriale
- Depeches du secretaire de sa majeste pour les colonies et autres documents relatifs a l'union federale des colonies britanniques de l'amerique du nord...
- Depeches du secretaire de sa majeste pour les colonies, et autres documents relatifs au siege du gouvernement...
- First and second reports of the select committee of the legislative assembly
- First report
- First report of the special committee appointed to inquire into the causes which retard the settlement of the eastern townships of lower canada
- Parochial and township subdivisions of lower canada
- Premier et second rapports du comite special
- Premier rapport du comite auquel a ete renvoye la consideration de l'etat des peches exploitees par les habitants de cette province dans le golfe st laurent et sur la cote du labrador...
- Premier rapport du comite special
- Procedes
- Proceedings of the standing committee on railroads, etc
- Rapport
- Rapport du comite charge de s'enquerir de la cause des desastres eprouves par les batiments et paquebots transportant les passagers du royaume-uni et d'ailleurs au canada
- Rapport du comite nomme pour s'enquerir du tarif d'honoraires de la cour d'amiraute
- Rapport du comite nomme pour s'enquerir et faire rapport des actes qui
- Rapport et deliberations...sur les accusations contre la derniere administration
- Rapport sur la petition de wm l mackenzie
- Rapport sur la petition du rev m j destroismaisons
- Rapport sur les terrains auriferes du canada
- Rapport...auquel a ete renvoye le rapport annuel du principal agent de l'emigration
- Rapport...auquel a ete renvoye le sujet de la formation d'un pont de glace sur le fleuve st laurent au-dessus des rapids du richelieu
- Rapport...charge de rechercher s'il est possible d'adopter des mesures legislatives
- Rapport...charge de s'enquerir de l'etat du bureau du surintendant des mesureurs de bois
- Rapport...charge de s'enquerir de quelle maniere ont ete depenses les fonds votes en 1855
- Rapport...nomme pour prendre en consideration la colonisation des terres incultes du bas-canada
- Rapport...nomme pour s'enquerir de l'etat de l'education et du fonctionnement de la loi des ecoles dans le bas canada
- Rapport...nomme pour s'enquerir des causes de l'emigration du canada aux etats-unis d'amerique ou ailleurs
- Rapport...nomme pour s'enquerir des causes et de l'importance de l'emigration qui a lieu tous les ans du bas-canada vers les etats-unis
- Rapport...nomme pour s'enquerir et faire rapport sur l'etat, l'administration et l'avenir de la compagnie du chemin de fer grand tronc
- Rapport...pour s'enquerir des transactions de la compagnie du chemin de fer de montreal et bytown
- Rapport...sur la colonisation
- Rapport...sur la convenance de defendre le travail du dimanche dans les departements publics de la province
- Rapport...sur la culture de la vigne au canada
- Rapport...sur la loi pour prohiber la vente des liqueurs fortes
- Rapport...sur les comptes publics
- Rapport...sur l'etat de l'agriculture du bas-canada
- Rapport...sur l'opportunite d'attirer l'emigration francaise, belge et suisse en canada
- Reponse a une adresse de l'assemblee legislative
- Reponse a une adresse de l'assemblee legislative du 23 ultimo
- Report of select committee on georgian bay and lake ontario ship canal
- Report of the select committee appointed to enquire into the causes of emigration from canada to the united states of america or elsewhere
- Report of the select committee appointed to inquire and reporn (sic) upon the present system of management of the public lands
- Report of the select committee of the legislative assembly
- Report of the select committee of the legislative assembly to whom was referred the subject of the formation of an ice bridge over the st lawrence at quebec
- Report of the select committee on the geological survey
- Report of the select committee to whom was referred the annual report of the chief emigration agent
- Report of the select committee...appointed to enquire into the state of education and the working of the school laws in lower canada
- Report of the special committee on colonisation
- Report on the canadian gold fields
- Report...to enquire into the admiralty tariff of fees
- Return to an address from...for copies of all transactions, sales or contracts...
- Return to an address of the legislative assembly for copies of certain seigniorial documents
- Subdivisions du bas canada en paroisses et townships depuis 1853
- Subdivisions du bas-canada en paroisses et townships
- Trente-septieme rapport du comite des bills prives
- Troisieme rapport et deliberations...auquel ont ete renvoyees les resolutions adoptees...le seize juin mil huit cent cinquante, au sujet de la tenure seigneuriale

Canada. (Province). Parlement. Assemblee legislative *see* Rapport du comite charge de s'enquerir des circonstances relatives a la reduction recente des droits sur le pin rouge

Canada (Province). Parlement. Assemblee legislative. Bibliotheque *see*
- Catalogue of books in the library of the legislative assembly of canada
- Catalogue of books relating to the history of america
- Supplementary catalogue of book added to the collection on the history of america

Canada (Province). Parlement. Assemblee legislative. Comite nomme pour s'enquerir des relations commerciales entre le Canada et la Grande-Bretagne... *see*
- Rapport sur le commerce
- Report on trade and commerce

Canada (Province). Parlement. Bibliotheque *see*
- Alphabetical catalogue of the library of parliament
- Catalogue of books in the library of parliament

Canada (Province). Parlement. Chambre d'assemblee *see* Rapport...sur les iles de la magdeleine et sur la partie ouest de cette province au-dessus du lac huron

Canada (Province). Parlement. Conseil legislatif *see*
- Constitutions, regles et reglements du conseil legislatif du canada
- Regles et reglements permanents du conseil legislatif du canada
- Report from the select committee of the legislative council on the accusations made against the members of the late administration
- Report of the select committee on immigration
- Rules, orders and forms of proceeding of the legislative council of canada
- Rules, orders, and forms of proceeding of the upper house of parliament of canada
- The select committee appointed on the 21st day of september last
- Standing orders of the legislative council

Canada (Province). Parlement. Conseil legislatif. Bibliotheque *see* Alphabetical catalogue of the library of the hon the legislative council of canada

Canada (Province). Surintendant de l'education pour le Bas-Canada *see*
- Rapport du surintendant de l'education pour le bas-canada, pour...
- Rapport du surintendant d'education pour le bas-canada pour l'annee...
- Rapport special...sur les finances de son departement
- Rapport sur l'education dans le bas canada suivi de tableaux statistiques pour l'annee scolaire...
- Report of the superintendent of education for lower canada, for the year 1846
- Report of the superintendent of education for lower canada for...1850/1851-1866
- Report on education in lower canada
- Tableau...indiquant les comtes qui ont recu des sommes d'argent pour la construction de maisons d'ecole, etc...

– Table...shewing the counties which have received sums of money for the construction of school houses, etc and the municipalities which have received their proportion of the common school fund...

Canada (Province). Surintendant de l'education pour le bas-Canada *see* Education, bas-canada

Canada (Province). Surintendant des ecoles du Haut-Canada *see*
– Rapport special sur les dispositions des ecoles separees de la loi des ecoles du haut-canada
– Rapport special sur les mesures qui ont ete adoptees pour l'etablissement d'une ecole normale
– Special report of the measures which have been adopted for the establishment of a normal school
– Special report on the separate school provisions of the school law of upper canada

Canada (Province). Surintendant des ecoles du haut-Canada *see* Extracts from the chief superintendent's report on education in upper canada for the year 1857

Canada. Quebec *see*
– Desjardin's speaker's decisions
– Dorion's decisions on appeal
– Judgment de conseil souverain
– Lower canada jurist
– Lower canada law journal
– Lower canada reports
– Montreal condensed reports
– Montreal law reports
– Perrault's conseil superieur
– Perrault's prevoste de quebec
– Pyke's lower canada reports
– Quebec law reports
– Quebec practice reports
– Quebec revised reports
– Ramsay's appeal cases
– Rapports judiciaires de quebec
– La revue de jurisprudence
– Revue de legislation et de la jurisprudence
– La revue legale
– Stuart's lower canada court king's bench reports
– Stuart's lower canada vice-admiralty reports

Canada. Quebec. (Province) *see*
– Beaubien
– Cook's lower canada admiralty court cases

Le canada reconquis par la france / Barthe, Joseph Guillaume – Paris?: Ledoyen, 1855 – 5mf – 9 – mf#29602 – cn Canadiana [971]

Canada school journal – Toronto: [A Miller, 1877-1885] – 9 – (cont by: canada school journal and weekly review) – mf#P04030 – cn Canadiana [370]

Canada school journal – Toronto: Canada School Journal Pub Co, [1886-1887] – 9 – (cont: the canada school journal and weekly review, merged with: educational weekly. merged to become: educational journal) – mf#P04032 – cn Canadiana [370]

Canada school journal *see* The educational weekly

Canada school journal and weekly review – Toronto: Canada School Journal Pub Co, [1885-1886?] – 9 – (cont: canada school journal) – mf#P04031 – cn Canadiana [370]

Canada. Service de l'environnement atmospherique. Bureau des previsions du Quebec *see* Cartes

Le canada, son present et son avenir : politique et finances / Fournier, Jules – Montreal?: s.n, 1865 – 1mf – 9 – mf#35174 – cn Canadiana [971]

Le canada sous la domination francaise : d'apres les archives de la marine et de la guerre / Dussieux, Louis – Paris: Librairie V Lecoffre, 1883 – 4mf – 9 – mf#12589 – cn Canadiana [971]

The canada spelling book : intended as an introduction to the english language, consisting of a variety of lessons progressively arranged in three acts... / Davidson, Alexander – Toronto: printed & publ for the author by H Rowsell, 1840 – 3mf – 9 – (with app) – mf#34668 – cn Canadiana [420]

The canada spelling book : intended as an introduction to the english language, consisting of a variety of lessons progressively arranged in three acts... / Davidson, Alexander – Toronto: R McPhail, 1864 – 2mf – 9 – (with app) – mf#48479 – cn Canadiana [420]

The canada spelling book : intended as an introduction to the english language, consisting of a variety of lessons, progressively arranged in three parts... / Davidson, Alexander – Niagara Ont: A Davidson, 1845 – 2mf – 9 – (with app) – mf#35058 – cn Canadiana [420]

Canada stamp and coin journal – Halifax, NS: J R Findlay, [1888-1889] [mf ed v1 n1 jul 1888-v1 n11 may 1889] – 9 – mf#P04564 – cn Canadiana [760]

Canada stamp sheet – Quebec: W G L Paxman, [1899?-1901?] – 9 – (merged with: energy to become: canada stamp sheet and energy) – mf#P04563 – cn Canadiana [760]

The canada stamp sheet *see* Energy

Canada stamp sheet and energy *see*
– Energy
– The philatelic advocate

Canada statutes, session laws and revisions / Canada. General – 1st Parliament-32nd Parliament 2nd sess. 1867-1984 – 1420mf – 9 – $2,130.00 – (updates planned) – mf#LLMC 90-100 – us LLMC [348]

Le canada stenographique – Montreal: M Gabard, [1900-19–] – 9 – ISSN: 1190-7843 – mf#P04092 – cn Canadiana [650]

Canada sunday school advocate – Toronto: S Rose, [1865?-18– or 19–] [mf ed v12 n3 nov 12 1876; v13 n1 oct 12 1867 [i.e. 1877] – 9 – mf#P04918 – cn Canadiana [242]

Canada. Supreme Court *see*
– Cameron's supreme court cases
– Supreme court reports

The canada temperance advocate – [Montreal: R Campbell, 1835-18–] – 9 – mf#P05000 – cn Canadiana [230]

The canada temperance manual and prohibitionist's handbook / Foster, George Eulas – Montreal?: Witness, 1884 – 2mf – 9 – mf#27395 – cn Canadiana [360]

Canada, the printed record : a bibliographic register with indexes to the microfiche series of the canadian institute for historical microreproductions = Catalogue d'imprimes canadiens: repertoire bibliographique avec index de la collection – 9th ed. Ottawa: the Institute, 1997 – 9 – 0-665-90295-6 – (incl: notices bibliographiques completes (57mf); index a: auteurs, titres, collections (63mf); index b: vedette-matieres anglaises (54mf); index c: vedettes-matieres francaises (54mf); index d: indice dewey (29mf); index e: lieux de publication (31mf); index f: date de publication (29mf); index g: numero de collection de l'icmh (29mf)) – mf#99966 – cn Canadiana [010]

Canada, the resources and future greatness of her great north-west prairie lands : with information for all, of interest to the intending settler and the capitalist seeking profitable and safe investments / Spence, Thomas – Ottawa: Dept of Agriculture, 1886 – 1mf – 9 – mf#33542 – cn Canadiana [917]

Canada times – Toronto. v1-apr 1982- (semiwkly) – 1 (subs Can$140/y) – (in japanese & english. supersedes: continental times, toronto, 1948-82. back run (1982-97) 15r can$1675) – cn McLaren [071]

Canada / transvaal : dedie aux diplomates francais qui ont du bon sens / Aron, Joseph – Paris: [s.n.], 1896 [mf ed 1981] – 2mf – 9 – (incl english text) – mf#30031 – cn Canadiana [327]

Canada under british rule, 1760-1900 / Bourinot, John George – Cambridge: University Press, 1900 [mf ed 1980] – (= ser Cambridge historical series) – 4mf – 9 – 0-665-03344-3 – (incl ind and bibl ref) – mf#03344 – cn Canadiana [971]

Canada under the administration of lord lorne / Collins, Joseph Edmund – Toronto: Rose, 1884 [mf ed 1980] – (= ser Roses' canadian national series) – 7mf – 9 – 0-665-00715-9 – (incl app and ind) – mf#00715 – cn Canadiana [971]

Canada under the national policy : arts and manufactures, 1883 / Bray, Alfred James – Montreal: Industrial Pub Co, 1883 – 2mf – 9 – mf#03718 – cn Canadiana [330]

Canada. Upper Canada. House of Assembly *see* Journals and appendices

The canada visitor : or, monthly magazine – Montreal: Pub...by W Greig, [1837-18–] – 9 – mf#P05135 – cn Canadiana [240]

Canada weekly – 1973 jan 10-1976 dec 29, 1977-79, 1980-82, 1983-1985 jul 24 – 4r – 1 – (cont: canadian weekly bulletin) – mf#646369 – us WHS [071]

Canada west and the hudson's-bay company : a political and humane question of vital importance to the honour of great britain, to the prosperity of canada, and to the existence of the native tribes / Aborigines Protection Society, London – [London], 1856 – (= ser 19th c books on british colonization) – 1mf – 9 – mf#1.1.3932 – uk Chadwyck [330]

Le canada-francais – Quebec: L-J Demers, 1888-1891 – 9 – (incl bibl) – mf#P04940 – cn Canadiana [440]

Le canada-francais et la providence / Masson, Philippe – Quebec?: L Brousseau, 1875 – 1mf – 9 – mf#24053 – cn Canadiana [210]

Canada-normandie : organe officiel de la federation de l'association canada-normandie / Association Canada-Normandie. Section de Montreal – Montreal. v1 n1 oct 1970-v4 n1 janv 1974 (irreg) [mf ed 1973-75] – 1r – 1 – mf#SEM35P15 – cn Bibl Nat [071]

Canada's actual condition / Bender, Prosper – [Boston: s.n, 1886] [mf ed 1980] – 1mf – 9 – mf#03569 – cn Canadiana [330]

Canada's approaching peril : the forest a vital necessity in regulating water powers and sustaining agriculture... / Biggar, Emerson Bristol – Toronto: Biggar-Wilson, [1908?] – 1mf – 9 – 0-665-76263-1 – (also available in french) – mf#76263 – cn Canadiana [634]

Canada's brave sons off to the war : for the canadian patriotic fund... – [Winnipeg?]: Pollard, Daniels, [1900] [mf ed 1982] – 1mf – 9 – mf#17542 – cn Canadiana [790]

Canada's canal problem and its solution : a reply to the toronto board of trade / Federation of Boards of Trade and Municipalities – [Ottawa?: s.n, 1912?] [mf ed 1994] – 1mf – 9 – 0-665-72833-6 – mf#72833 – cn Canadiana [380]

Canada's crisis : political, commercial, and industrial relations with the united states and other countries; our transportation problems and the railway rule of canada / Biggar, Emerson Bristol – Toronto: Biggar-Wilson, [1911?] – 1mf – 9 – 0-665-73635-5 – mf#73635 – cn Canadiana [337]

Canada's jews / Rosenberg, Louis – Montreal, Quebec. 1939 – 1r – us UF Libraries [939]

Canada's late premier / Aberdeen and Temair, Ishbel Gordon, marchioness of – [S.l: s.n, 1895?] [mf ed 1979] – 1mf – 9 – 0-665-00766-3 – (repr fr: the outlook, jan 26 1895) – mf#00766 – cn Canadiana [971]

Canada's marine and fisheries / Bourinot, John George – [S.l: s.n, 1872?] [mf ed 1979] – 1mf – 9 – 0-665-00201-7 – mf#00201 – cn Canadiana [639]

Canada's mental health – Ottawa, CN. 1953-88 – 9r – 1 – (with ind. french ed also available) – cn Commonwealth Imaging [360]

Canada's metals : a lecture delivered at the toronto meeting...august 20, 1897 / Roberts-Austen, William Chandler – London, New York: Macmillan, 1898 – 1mf – 9 – mf#11195 – cn Canadiana [660]

Canada's missionary congress : addresses delivered at the canadian national missionary congress, held in toronto... – Toronto: Canadian Council Laymen's Missionary Movement, [1909?] – 1mf – 9 – 0-524-03575-X – mf#1990-1035 – us ATLA [240]

Canada's national policy : mr c c colby's great speech on tariff revision, house of commons, march, 1878... / Colby, Charles Carroll – [Ottawa?: s.n, 1878?] [mf ed 1980] – 1mf – 9 – 0-665-00698-5 – mf#00698 – cn Canadiana [336]

Canada's patriot statesman : the life and career of the right honorable sir john a macdonald...etc: based on the work of edmund collins / Collins, Joseph Edmund – London: McDermid & Logan, 1891 – 8mf – 9 – mf#08397 – cn Canadiana [920]

[Canada's responsibility to the empire and the race] / Campbell, Wilfred – [Canada?: s.n, 1915?] – 1mf – 9 – 0-665-74880-9 – mf#74880 – cn Canadiana [933]

Canada's wool and woolens : the problem of clothing the canadian people with canadian wool manufactured by canadian woolen mills – Toronto: Biggar-Wilson, c1908 – 1mf – 9 – 0-665-77222-X – mf#77222 – cn Canadiana [636]

Canada-sir f b head : copy of a despatch from sir f b head, in answer to charges preferred against him by dr c duncombe, in petition presented to the house of commons on the august 1836 – [s.l.]: House of Commons, 1837 [mf ed 1984] – 2mf – 9 – mf#SEM105P411 – cn Bibl Nat [971]

Canada-united states law journal – Case Western Reserve University. v1-27. 1978-2001 – 5,6,9 – $387.00 set – (v1-8 1978-84 in reel $90. v9-27 1985-2001 in mf $297) – ISSN: 0163-6391 – mf#101341 – us Hein [340]

Canadia, ode : epinikios / Belsham, Jacobus – Londini: ...apud J Clarke...R & J...Dodsley...et J Buckland, 1760 [mf ed 1984] – 1mf – 9 – 0-665-20296-2 – mf#20296 – cn Canadiana [810]

The canadian – London, Ont: Grand Council of the CMBA of Canada, [1895-19–] – 9 – mf#P04723 – cn Canadiana [360]

The canadian accountant : a practical system of book-keeping, containing a complete elucidation of the science of accounts by the latest and most approved methods... / Beatty, Samuel G & Johnson, John Wesley – Belleville, Ont: Ontario Business College, 1894 – 4mf – 9 – (incl ind) – mf#39843 – cn Canadiana [650]

Canadian accounting perspectives – Toronto. 2002+ (1,5,9) – ISSN: 1499-8653 – mf#32800 – us UMI ProQuest [650]

Canadian administrator – Edmonton. 1979-1998 – 1,5,9 – ISSN: 0008-2813 – mf#12233 – us UMI ProQuest [370]

The canadian advertiser – Toronto: Canadian Advertiser Pub Co, 1893-[189- or 19–] – 9 – mf#P04057 – cn Canadiana [650]

Canadian advertising rates and data / Maclean-Hunter Research Bureau – v40 n1-v41 n2 [1967 jan-feb] – 1r – 1 – (cont: canadian advertising; canadian media rates and data; cont by: card [toronto ont]) – mf#1110497 – us WHS [650]

Canadian Aeronautics and Space Institute *see* Casi transactions

Canadian aeronautics and space journal – Ottawa. 1955+ (1) 1972+ (5) 1977+ (9) – ISSN: 0008-2821 – mf#2198 – us UMI ProQuest [629]

Canadian agricultural journal – Montreal: Lovell & Gibson [1844?-1847] – 9 – (cont by: agricultural journal and transactions of the lower canada agricultural society) – mf#P05005 – cn Canadiana [630]

Canadian agriculture, pt 1 : the prairie / Fream, William – [London?: s.n.], 1885 [mf ed 1981] – 2mf – 9 – 0-665-09761-1 – (incl bibl ref) – mf#09761 – cn Canadiana [630]

Canadian agriculture, pt 2 : the eastern provinces / Fream, William – [London?: s.n.], 1885 [mf ed 1981] – 1mf – 9 – 0-665-09762-X – (incl bibl ref) – mf#09762 – cn Canadiana [630]

Canadian agriculturist – Toronto: [W McDougall, G Buckland, 1848?-1863] [mf ed v1 n1 jan 1 1849-v15 n12 dec 1863] – 9 – mf#P04016 – cn Canadiana [630]

Canadian Air Line Flight Attendants Association *see* Unity

The canadian almanac and miscellaneous directory for the year... – Toronto: Copp, Clark, [18—19–] – (containing full and authentic commercial, statistical, astronomical, departmental, ecclesiastical, educational, financial and general information" tables. incl ind) – mf#A02669 – cn Canadiana [030]

The canadian almanac and repository of useful knowledge for the year 1857 : being the first after leap year... – Toronto: Maclear, [1857?] [mf ed 1984] – 2mf – 9 – 0-665-32378-6 – mf#32378 – cn Canadiana [030]

The canadian almanac and repository of useful knowledge for the year 1858 : being the second after leap year... – Toronto: Maclear, [1858?] [mf ed 1984] – 2mf – 9 – 0-665-32379-4 – mf#32379 – cn Canadiana [030]

The canadian almanac and repository of useful knowledge for the year 1872 : being the second after leap year: containing full and authentic commercial, statistical, astronomical... and general information – Toronto: Copp, Clark, [1872?] [mf ed 1983] – 2mf – 9 – 0-665-32697-1 – mf#32697 – cn Canadiana [030]

The canadian almanac and repository of useful knowledge for the year 1874 : being the second after leap year: containing full and authentic commercial, statistical, astronomical...and general information – Toronto: Copp, Clark, [1874?] [mf ed 1983] – 3mf – 9 – 0-665-32698-X – (incl ind) – mf#32698 – cn Canadiana [030]

Canadian anaesthetists' society journal = Journal de la societe canadienne des anesthesistes – Toronto. 1954-1986 (1) 1972-1986 (5) 1976-1986 (9) – (cont by: canadian journal of anesthesia) – ISSN: 0008-2856 – mf#7177 – us UMI ProQuest [617]

Canadian annual review of public affairs – v1-38. 1901-38 – 1 – $486.00 – mf#0138 – us Brook [350]

Canadian architect – Don Mills. 1991-1996 (1) 1991-1996 (5) 1991-1996 (9) – ISSN: 0008-2872 – mf#3025 – us UMI ProQuest [720]

Canadian architect and builder : a journal of constructive and decorative art – Toronto. v1-21 n19. jan 1888-apr 1908// (mthly) – 12r – 1 – Can$795.00 – cn McLaren [720]

Canadian architect and builder – Toronto: C H Mortimer. v1 n1 jan 1888-v22 n4 apr 1908 [mf ed 1974] – 1r – 1 – mf#SEM35P99 – cn Bibl Nat [720]

The canadian artillery team at shoeburyness, 1896 / Cole, Frederick Minden – Montreal?: s.n, 1897? – 1mf – 9 – (incl extracts from the british press) – mf#03212 – cn Canadiana [790]

Canadian Association for Health, Physical Education and Recreation *see* Canadian journal of history of sport and physical

Canadian Association in Support of the Native Peoples *see* Bulletin of the canadian association...

Canadian Association of Industrial, Mechanical and Allied Workers *see* Caimaw review

Canadian Association of Radiologists *see* Journal of the canadian association of radiologists

Canadian association of radiologists journal = Journal l'association canadienne des radiologistes – Montreal. 1986-1996 (1) 1986-1996 (5) 1986-1996 (9) – (cont: journal of the canadian association of radiologists = journal de l'association canadienne des radiologistes) – ISSN: 0846-5371 – mf#3032,01 – us UMI ProQuest [616]

The canadian atlantic telegraph – S.l: s.n, 18– – 1mf – 9 – mf#05701 – cn Canadiana [380]

Canadian author and bookmen – Toronto, Ontario, CN. 1922-75 – 7r – 1 – cn Commonwealth Imaging [420]

Canadian aviation – v. 1-36. 1928-63 – 1 – 640.00 – us L of C Photodup [629]

Canadian banker – Toronto. 1983-2000 (1) 1983-2000 (5) 1983-2000 (9) – (cont: canadian banker and icb review) – ISSN: 0822-6830 – mf#6871,02 – us UMI ProQuest [332]

Canadian banker – Toronto. 1972-1973 (1) 1972-1972 (5) (9) – ISSN: 0008-297X – mf#6871 – us UMI ProQuest [332]
Canadian banker and icb review – Toronto. 1974-1983 (1) 1975-1983 (5) 1975-1983 (9) – (cont by: canadian banker) – ISSN: 0315-6230 – mf#6871,01 – us UMI ProQuest [332]
The canadian banking system, 1817-1890 / Breckenridge, Roeliff Morton – New York: Publ for the American Economic Association by Macmillan; London: Swann, Sonnenschein, 1895 – (= ser Publications of the american economic association 10) – 6mf – 9 – mf#26695 – cn Canadiana [332]
The "canadian baptist" and dr ryerson – Toronto: s.n, 1872 – 1mf – 9 – mf#04996 – cn Canadiana [242]
Canadian baptist telegu missions. report – 1878-1911 [mf ed 2001] – (= ser Christianity's encounter with world religions, 1850-1950) – 6r – 1 – (filmed with: canadian baptist mission [india]. report [1913-26]; report of the canadian baptist mission among the telegus, oriyas and savaras [1927-40]; among the telegus and bolivians) – mf#2001-s021-028 – us ATLA [242]
Canadian bar association journal – v1-4. 1970-73 – 9 – $34.00 set – (cont: canadian bar journal) – mf#115371 – us Hein [340]
Canadian bar association yearbook – 1896/97; 1915-85 – 108mf – 9 – $486.00 – (regular updates planned) – mf#LLMC 84-431 – us LLMC [340]
Canadian bar journal – v1-12. 1958-69 – 9 – $245.00 set – (cont by: canadian bar association journal) – mf#115381 – us Hein [340]
Canadian bar journal *see* Canadian bar association journal
Canadian bar review – v1-79. 1923-2000 – 5,6,9 – $1358.00 set – (v1-61 1923-83 in reel or mf $891. v62-79 1984-2000 in mf $467) – ISSN: 0008-3003 – mf#101351 – us Hein [340]
Canadian bee journal – Beeton, Ont: D A Jones, [1885-1913] – 9 – (cont by: canadian horticulturist and beekeeper) – mf#P04191 – cn Canadiana [630]
The canadian bibliographer and library record – Hamilton [Ont: Griffin & Kidner, 1889-1890] – 9 – mf#P04050 – cn Canadiana [020]
Canadian bicentenary papers : no 1: the history of nonconformity in england in 1662, by w f clark, no 2: the reasons for nonconformity in canada, by f h marling – Toronto?: s.n, 1862 – 1mf – 9 – (incl bibl ref) – mf#18769 – cn Canadiana [242]
Canadian biographical archive (caba) = Archives biographiques canadiennes (caba) / Baillie, Laureen [comp] – [mf ed 2001-03] – 421mf (1:24) in 12 installments – 9 – diazo €10,060.00 (silver €11,080 isbn: 978-3-598-34721-4) – ISBN-10: 3-598-34720-0 – ISBN-13: 978-3-598-34720-7 – (with printed ind) – gw Saur [971]
The canadian biographical dictionary : and portrait gallery of eminent and self-made men – Toronto, Chicago: American Biographical Pub Co, 1881 – 2v on 1mf – 9 – (individual vols also available) – mf#08544 – cn Canadiana [920]
Canadian biographies : [artists, authors and musicians] – 1948-52 – 1r – 1 – cn Library Assoc [920]
The canadian birthday book : with poetical selections for everyday in the year from canadian writers, english and french / Seranus [comp] – Toronto: C B Robinson, 1887 – 5mf – 9 – mf#06400 – cn Canadiana [810]
Canadian bookman : devoted to literature and the creative arts – Toronto. v1-22. jan 1919-oct/nov 1939// – 5r – 1 – Can$675.00 – (hardcopy ind available: index to canadian bookman comp & ed by grace heggie & anne mcgaughey) – cn McLaren [700]
Canadian bookman : a monthly review of contemporary literature devoted to the interests of the canadian book buyer – Toronto. v1-2 n6. jan 1909-jun 1910// – 1r – 1 – Can$110.00 – cn McLaren [070]
The canadian boy : a magazine for young canada – Guelph, Ont: Canadian Boy Pub Co, [1900?-19–] – 9 – mf#P04997 – cn Canadiana [305]
Canadian boy scouts : report of officer commanding the canadian boy scouts' contingent to england, 1911; with introduction respecting the growth of the movement in canada to 1912 / Cole, Frederick Minden – [Montreal?: s.n, 1912?] – 1mf – 9 – 0-665-77596-2 – mf#77596 – cn Canadiana [360]
Canadian breeder and agricultural review – Toronto: [s.n, 1884?]- [mf ed v2 n1 jan 2 1885-v2 n51 dec 31 1885] – 9 – ISSN: 1190-7274 – mf#P04033 – cn Canadiana [630]
Canadian broadcaster – Toronto, Canada. 8 jan-24 dec 1959; 1960-1 oct 1964 – 4r – 1 – uk British Libr Newspaper [072]
Canadian broadcaster – Toronto, CN. 1942-79 – 36r – 1 – cn Commonwealth Imaging [073]
Canadian Brotherhood of Railway, Transport and General Workers *see* Canadian transport
Canadian building – Toronto. 1975-1991 (1) 1975-1991 (5) 1975-1991 (9) – (cont by: building) – ISSN: 0008-3070 – mf#10766 – us UMI ProQuest [690]
Canadian business – Montreal, Canada. 1957-nov 1961; 1962 – 6r – 1 – uk British Libr Newspaper [071]
Canadian business – Toronto. 1976+ (1,5,9) – ISSN: 0008-3100 – mf#11292 – us UMI ProQuest [338]
Canadian business review – Ottawa. 1980-1996 (1,5,9) – ISSN: 0317-4026 – mf#12873 – us UMI ProQuest [338]
Canadian camp life / Herring, Frances Elizabeth – London: T F Unwin, 1900 [mf ed 1980] – 3mf – 9 – 0-665-05560-9 – mf#05560 – cn Canadiana [390]
Canadian Campaign for Nuclear Disarmament *see* Sanity
The canadian casket – Hamilton, UC [Ont]: A Crosman, [1831-18–?] – 9 – mf#P04356 – cn Canadiana [420]
The canadian cattle agitation : mr gardner's policy denounced: plain speaking by farmers at public meeting in town hall, dundee, on 1st august, 1893 – Dundee Scotland: W & D C Thomson, 1893 – 1mf – 9 – mf#02034 – cn Canadiana [636]
Canadian cattleman – Winnipeg, CN. 1938-87 – 33r – 1 – cn Commonwealth Imaging [636]
The canadian census / Colmer, Joseph Grose – S.l: s.n, 1891? – 1mf – 9 – mf#17963 – cn Canadiana [317]
The canadian census of 1871 : remarks on mr harvey's paper published in the february number of "the canadian monthly" / Tache, Joseph-Charles – S.l: s.n, 1872? – 1mf – 9 – mf#23743 – cn Canadiana [317]
Canadian century – Montreal, Canada. 8 jan 1910-18 nov 1911 – 4r – 1 – (aka: canadian century and canadian life and resources; canadian life and resources) – uk British Libr Newspaper [072]
Canadian century *see* Canadian life and resources
Canadian champion – Milton, ON: James Campbell, 1862-73 – 3r – 1 – ISSN: 0834-6925 – cn Library Assoc [971]
Canadian champion – Milton, Ontario, CN. jan 1870-jun 1874; 1949-51; 1984 – 4r – 1 – cn Commonwealth Imaging [071]
Canadian checkerist – Toronto: W H Darlington, [1888-18–] [mf ed v1 n1 feb 14 1888] – 9 – ISSN: 1190-6227 – mf#P04329 – cn Canadiana [790]
The canadian cheese and butter maker – Williamstown, Ont: G F Brown, 1898? – 9 – (iss for dec 1898 also publ in french) – ISSN: 1190-7010 – mf#P04006 – cn Canadiana [630]
Canadian Chiropractic Association *see* Journal of the canadian chiropractic association
The canadian christian examiner and presbyterian review – Niagara, UC [Ont]: W D Miller, 1837-[1840] – 9 – mf#P04998 – cn Canadiana [242]
Canadian christmas song / Evans, Walter Norton – S.l: s.n, 1894? – 1mf – 9 – mf#63034 – cn Canadiana [780]
Canadian chronicle *see* The new era
The canadian church juvenile – Toronto: Board of Management of the Domestic and Foreign Missionary Society, [1893?-191-?] – 9 – mf#P05980 – cn Canadiana [240]
Canadian church magazine and mission news – Hamilton, Ont: Domestic and Foreign Missionary Society of the Church of England in Canada, [1887-1898] – 9 – (cont: our mission news) – mf#P04040 – cn Canadiana [242]
The canadian church magazine and mission news *see* Wellington deanery magazine
The canadian church missionary gleaner – Toronto: Canadian Church Missionary Society, [1896?-1903?] – 9 – (incl: the church missionary gleaner) – mf#P05982 – cn Canadiana [242]
The canadian church press : a journal of ecclesiastical, literary and general intelligence – Toronto: Lovell & Gibson, 1860 – 9 – mf#P04069 – cn Canadiana [242]
Canadian churchman / Anglican Church of Canada – 1959-64, 1965-1970 jun, 1975 jan-1979 mar, 1979 apr-1982 – 4r – 1 – (cont: dominion churchman) – mf#969889 – us WHS [242]
Canadian churchman – Toronto, ON: Anglican Church of Canada, 1876-1926 – 36r – 1 – ISSN: 0008-3216 – cn Library Assoc [240]
Canadian citizen – Toronto, Canada. 19 nov 1948-21 dec 1951, 1952-9 jul 1954, 10 sep-17 sep 1954, 19 may-14 jul 1955 – 4 3/4r – 1 – (aka: ukrainian christian worker; ukrainian toiler) – uk British Libr Newspaper [071]
Canadian coach magazine – Toronto. 1973-1973 (1) – ISSN: 0045-4559 – mf#8113 – us UMI ProQuest [380]
Canadian collector – Berlin [Kitchener], Ont: F I Weaver, [1898] [mf ed v1 n1 sep 1898] – 9 – mf#P04570 – cn Canadiana [760]
The canadian collector and philatelic punch – Berlin [Kitchener], Ont: E A Giller, [1899] [mf ed v1 n2 apr 1899-v1 n3 jun 1899] – 9 – mf#P04571 – cn Canadiana [760]
Canadian colliery guardian and critic *see* Critic
Canadian colonization – [s.l: s.n, 1853?] [mf ed 1984] – 1mf – 9 – 0-665-32296-8 – mf#32296 – cn Canadiana [320]
Canadian comment : interpretative articles and summaries of world news – Toronto: Current Publ. v1-7 n5. feb 1932-may 1938// – 3r – 1 – Can$215.00 – cn McLaren [320]
Canadian Communist League [Marxist-Leninist] et al *see* Forge
Canadian confederation : the case of nova scotia – London?: s.n, 1868? – 1mf – 9 – mf#54072 – cn Canadiana [323]
The canadian constitution; a study of the written and unwritten features of our system of government / Lawson, W J – Ottawa, Queen's Printer, 1960. 29 p. LL-2398 – 1 – us L of C Photodup [342]
Canadian constitutional development : shown by selected speeches and despatches, with introductions, and explanatory notes / Egerton, Hugh Edward & Grant, William Lawson – Toronto: Musson, [1907?] – 6mf – 9 – 0-665-74148-0 – mf#74148 – cn Canadiana [323]
Canadian constitutional history and law / Hassard, Albert Richard – Toronto: Carswell, 1900 [mf ed 1980] – 3mf – 9 – 0-665-05207-3 – (incl ind and bibl ref) – mf#05207 – cn Canadiana [323]
Canadian consumer – Ottawa. 1977-1993 (1,5,9) – ISSN: 0008-3275 – mf#11721 – us UMI ProQuest [380]
The canadian contingents and canadian imperialism : a story and a study / Evans, William Sanford – Toronto: Publishers' Syndicate, 1901 [mf ed 2000] – 5mf – 9 – 0-659-91667-3 – mf#9-91667 – cn Canadiana [971]
Canadian contract record – Toronto: C H Mortimer, [1889-1908] [mf ed v1 n1 nov 27 1889-v19 n19 may 6 1908] – 9 – (merged with: canadian architect and builder to become: contract record) – mf#P06062 – cn Canadiana [690]
Canadian contractor – Toronto. 2000+ (1,5,9) – ISSN: 1498-8941 – mf#32863 – us UMI ProQuest [690]
Canadian controls and instrumentation – Toronto. 1975-1977 (1) 1975-1977 (5) 1975-1977 (9) – (cont by: canadian controls + instruments) – ISSN: 0008-3283 – mf#10767 – us UMI ProQuest [621]
Canadian controls and instrumentation – Toronto. 1983-1985 (1) – (cont: canadian controls + instruments) – ISSN: 0705-3193 – mf#10767,02 – us UMI ProQuest [621]
Canadian controls + instruments – Toronto. 1978-1982 (1,5,9) – (cont by: canadian controls and instrumentation) – ISSN: 0705-3193 – mf#10767,01 – us UMI ProQuest [621]
Canadian controls + instruments – Toronto. 1978-1982 [1,5,9] – (cont: canadian controls and instrumentation) – ISSN: 0705-3193 – mf#10767,01 – us UMI ProQuest [621]
The canadian co-operator and patron – Owen Sound [Ont]: R J Doyle, [1882-1900] – 9 – mf#P04433 – cn Canadiana [630]
Canadian copyright : the following editorials have appeared in the toronto telegram... – [S.l: Canadian Copyright Association, 1884?] [mf ed 1980] – 1mf – 9 – 0-665-05702-4 – mf#05702 – cn Canadiana [346]
Canadian copyright – London [Ont]: J S Virtue, [1889?] [mf ed 1986] – 1mf – 9 – 0-665-56735-9 – mf#56735 – cn Canadiana [346]
Canadian copyright / Wilson, Daniel – [S.l: s.n, 1892?] [mf ed 1981] – 1mf – 9 – mf#14254 – cn Canadiana [346]
Canadian counsellor = Conseiller canadien – Kanata. 1976-1985 – 1,5,9 – (cont by: canadian journal of counselling=revue canadienne de counseling) – ISSN: 0008-333X – mf#11409 – us UMI ProQuest [370]
Canadian countryman – Toronto. Canada. -w. Apr 1913-Dec 1939; 22 Mar 1941-22 Sep 1951. (62mqn reels) – 1 – uk British Libr Newspaper [072]
Canadian countrymen – Toronto, Canada. 19 apr 1913-1939; 22 mar 1941-1950; 13 jan-22 sep 1951 – 54 1/2r – 1 – uk British Libr Newspaper [071]
Canadian county connections – 1978 feb 15-1987 – 1r – 1 – mf#626153 – us WHS [929]
Canadian County Genealogical Society *see* Newsletter
Canadian courant and montreal advertiser – Montreal, QC. 1807-34 – 9r – 1 – cn Library Assoc [071]
Canadian craftsman and masonic record – Port Hope, Ont: J B Trayes, 1877-[1898?] – 9 – (cont: the craftsman and canadian masonic record) – mf#P04075 – cn Canadiana [360]
The canadian cricket field : a journal devoted to the interests of cricket in canada – [Toronto]: A G Brown, G G S Lindsey, [1882-18–] – 9 – (in dble clms) – mf#P04070 – cn Canadiana [790]
The canadian cricketer's guide : containing photographs and biographical sketch of two prominent cricketers, character of the game, hints for playing, the clubs of canada... / Phillipps, T D & Campbell, H J – Ottawa?: C W Mitchell, 1877 – 2mf – 9 – mf#12029 – cn Canadiana [790]
Canadian cultivator and household magazine – Sherbrooke: G H Bradford, 1890-[1892?] [mf ed 1989] – 1mf – 9 – (ceased 1892?) – mf#P04056 – cn Canadiana [635]
Canadian dairyman and farming world – Peterboro [Peterborough], Ont: Dairyman Pub Co and Farming World, [1908] – 9 – (cont by: farm and dairy and rural home) – mf#P05025 – cn Canadiana [630]
Canadian datasystems – Toronto. 1975-1992 (1) 1976-1992 (5) 1976-1992 (9) – (cont by: it magazine) – ISSN: 0008-3364 – mf#10768 – us UMI ProQuest [000]
The canadian day-star : a monthly magazine devoted to the exhibition of the gospel in its glorious fulness and unfettered freeness – [Montreal: J Lovell, 1861-1864?] – 9 – (incl ind) – mf#P04360 – cn Canadiana [240]
Canadian directory of shopping centres – Toronto. 2003+ (1,5,9) – ISSN: 0822-7799 – mf#32864 – us UMI ProQuest [650]
Canadian Disarmament Information Service *see*
– Peace calendar
– Peace magazine
Canadian domestic lawyer : with plain and simple instructions for the merchant, farmer, and mechanic, to enable them to transact their business according to law / Whitley, John – Stratford [Ont]: Vivian & Maddocks, 1864 [mf ed 1982] – 5mf – 9 – (incl ind) – mf#34143 – cn Canadiana [346]
The canadian dry goods review : the organ of the canadian dry goods, hats, caps and furs, millinery and clothing trades – Toronto: Dry Goods Review Co, [1891?-1933] – 9 – mf#P04470 – cn Canadiana [680]
The canadian ecclesiastical gazette : or, monthly church register for the dioceses of quebec, toronto, and montreal – Toronto: H Rowsell, [1854?-1862] – 9 – mf#P04459 – cn Canadiana [242]
The canadian ecclesiastical gazette – Quebec: G Stanley, [1850-1853?] – 9 – mf#P04308 – cn Canadiana [242]
The canadian eclectic magazine of foreign literature, science and art – [Toronto: s.n, 1871-1872] – 9 – mf#P04839 – cn Canadiana [073]
Canadian economics : being papers prepared for reading before the economical section, with an introductory report: montreal meeting, 1884 / British Association for the Advancement of Science. Canada – Montreal: Dawson, 1885 [mf ed 1980] – 5mf – 9 – 0-665-03723-6 – mf#03723 – cn Canadiana [330]
The canadian economist : a book of tried and tested receipts / Ladies' Association of Bank Street Church, Ottawa [comp] – Ottawa: A Mortimer; Toronto: Hunter, Rose, 1881 – 7mf – 9 – (incl ind) – mf#07261 – cn Canadiana [640]
The canadian economist – Montreal: Printed for the Committee of the Montreal Free Trade Association, Donoghue and Mantz, [1846-1847] – 9 – mf#P05050 – cn Canadiana [380]
Canadian educator for home and school use / ed by McLaughlin, Sara B – Toronto: Iroquois Press, 1920 – 9mf – 9 – 0-665-87964-4 – mf#87964 – cn Canadiana [917]
The canadian electrical news – Montreal: Hart Bros, [1884-18– or 19–] – 9 – mf#P04058 – cn Canadiana [621]
Canadian electrical news and engineering journal – Toronto: C H Mortimer, [1899-1910] [mf ed new ser v9 n1 jan 1899-new ser v10 n12 dec 1900] – 9 – (incl ind) – mf#P04469 – cn Canadiana [621]
Canadian electrical news and engineering journal [electrical news] – Toronto, Canada. dec 1908-15 jun 1922 – 21r – 1 – uk British Libr Newspaper [621]
Canadian electrical news and steam engineering journal – Toronto: C H Mortimer, [1891?-1898] [mf ed v1 n1 jan 1891-v2 n12 dec 1892; new ser v3 n1 jan 1893-new ser v8 n12 dec 1898] – 9 – mf#P04468 – cn Canadiana [621]
Canadian electronics – Willowdale. 1992-1996 (1,5,9) – ISSN: 1187-6026 – mf#19536 – us UMI ProQuest [621]
Canadian electronics engineering (cee) – Toronto. 1957-1990 (1) 1957-1990 (5) 1957-1990 (9) – ISSN: 0008-3461 – mf#10769 – us UMI ProQuest [621]
Canadian emigrant and western district commercial and general advertiser – Sandwich, ON. 1831-36 – 1r – 1 – cn Library Assoc [071]

The canadian emigrant housekeeper's guide / Traill, Catherine Parr – Montreal: J Lovell, 1861 [mf ed 1983]] – 2mf – 9 – 0-665-41581-8 – mf#41581 – cn Canadiana [640]

Canadian engineer – Toronto, Canada. May 1893-1909, feb 1912-26 may 1921, 20 oct 1921-9 may 1922 – 41 1/2r – 1 – (aka: canadian engineer weekly) – uk British Libr Newspaper [620]

The canadian engineer – Toronto: Canadian Engineer Co, [1893-1939] – 9 – (incl ind) – mf#P04084 – cn Canadiana [620]

Canadian engineer weekly *see* Canadian engineer

The canadian engineering news – Montreal: W E Gower, [1893-189- or 19–] – 9 – mf#P04051 – cn Canadiana [620]

Canadian entomologist – London, Toronto 1869-1920 – v1-52 on 321mf – 9 – mf#z-931/2 – ne IDC [590]

The canadian entomologist – Toronto: Copp, Clark, [1868]- – 9 – (incl ind) – mf#P05087 – cn Canadiana [590]

The canadian epworth era – Toronto: W. Briggs, [1899-1915] – 9 – mf#P04326 – cn Canadiana [242]

Canadian equipment rentals – Toronto. 2000+ (1,5,9) – mf#33086 – us UMI ProQuest [690]

Canadian essays and addresses / Peterson, William – London: Longmans, Green, 1915 – 5mf – 9 – 0-665-75600-3 – mf#75600 – cn Canadiana [370]

Canadian estate tax and succession duties acts; including all amendments to december 1, 1966 / Commerce Clearing House Canadian Limited – 6th ed. Don Mills, Ont. 1966 vii, 215 p. LL-2370 – 1 – us L of C Photodup [343]

Canadian evangelist – Toronto: Evangelist Pub Co, [1890-1895] – 9 – (cont by: the disciple of christ and canadian evangelist) – mf#P04635 – cn Canadiana [242]

Canadian evangelist and disciple of christ – Hamilton, Ont: G Munro, [1896] – 9 – (cont: the disciple of christ and canadian evangelist) – mf#P04637 – cn Canadiana [242]

Canadian exhibitor – Toronto: Trades Pub Co, [1886-18– or 19–] [mf ed july 1 1886] – 9 – ISSN: 1190-6286 – mf#P04172 – cn Canadiana [060]

Canadian expeditionary force / Ottawa. Dept of Militia and Defence – 13v. 1914-17 (nominal rolls) – 6r – 1 – Can$675.00 – (with printed guide) – cn McLaren [355]

The canadian family herald – Toronto: Printed for D McDougall by J Stephens, [1851-18–?] – 9 – (incl ind) – ISSN: 1190-7185 – mf#P04337 – cn Canadiana [073]

Canadian farmer – Montreal: J Smith, [1851-18–] [mf ed 1989] – 1mf – 9 – (ceased 1851?) – ISSN: 1190-6936 – mf#P04127 – cn Canadiana [630]

Canadian farmer – Winnipeg/Manitoba, Canada. 11 apr-26 dec 1928; 1929-25 dec 1940 – 13r – 1 – uk British Libr Newspaper [072]

The canadian farmer : a weekly paper, established 1878... – [S.l: s.n, 1878?] [mf ed 1985] – 1mf – 9 – 0-665-53736-0 – mf#53736 – cn Canadiana [630]

The canadian farmer and grange record and organ of the ontario bee-keepers' association – Welland, Ont: N B Colcock, [1880?-1884] – 9 – mf#P06055 – cn Canadiana [630]

The canadian farmer and mechanic : to promote the country's wealth and the people's good – Kingston [Ont]: Garfield & Good, 1841 – 9 – ISSN: 1190-6952 – mf#P04109 – cn Canadiana [630]

The canadian farmer, the general consumer and the wool tariff / Biggar, Emerson Bristol – Toronto: Biggar-Wilson, [1910?] – 1mf – 9 – 0-665-73634-7 – mf#73634 – cn Canadiana [636]

The canadian farmer's almanac and general memorandum-book for the year 1824 : being the first after bissextile or leap year: the calculations for the meridian of york... – York, UC [Toronto]: C Fothergill, [1824?] [mf ed 1985] – 1mf – 9 – 0-665-43557-6 – mf#43557 – cn Canadiana [520]

The canadian farmer's almanac and general memorandum-book for the year 1825 : being the first after bissextile or leap year: the calculations for the meridian of york... – York, UC [Toronto]: C Fothergill, [1825?] [mf ed 1985] – 1mf – 9 – 0-665-43558-4 – mf#43558 – cn Canadiana [520]

The canadian farmer's almanac and memorandum book for the year of our lord 1850 : being the second after bissextile or leap year and till the 20th day of june,the thirteenth year of the reign of her most gracious majesty queen victoria: calculated for the meridian of sherbrooke... – Sherbrooke, CE [Quebec]: W Brooks, [1850?] [mf ed 1985] – 1mf – 9 – 0-665-43573-8 – mf#43573 – cn Canadiana [520]

The canadian farmers' almanac for the year of our lord... – Montreal: R Miller, 1873?-1880? – 9 – mf#A00179 – cn Canadiana [630]

Canadian farmers' gazette – Brantford, CW [Ont]: A Webber, [1861?-18–?] [mf ed v1 n2 feb 1861] – 9 – ISSN: 1190-6731 – mf#P04345 – cn Canadiana [630]

The canadian farmer's manual of agriculture : the principles and practice of mixed husbandry as adapted to canadian soils and climate... / Whitcombe, Charles Edward – Toronto: W R Burrage, 1876 [mf ed 1984] – 7mf – 9 – 0-665-37367-8 – (incl ind) – mf#37367 – cn Canadiana [636]

The canadian farmer's manual of agriculture : the principles and practice of mixed husbandry as adapted to canadian soils and climate... / Whitcombe, Charles Edward – Toronto: W[illin]g & Williamson, [1879?] [mf ed 1984] – 7mf – 9 – 0-665-41563-3 – (incl ind) – mf#41561 – cn Canadiana [636]

The canadian farmer's manual of agriculture : the principles and practice of mixed husbandry as adapted to canadian soils and climate... / Whitcombe, Charles Edward – Toronto: J Adam, 1874 [mf ed 1983] – 7mf – 9 – mf#25785 – cn Canadiana [636]

A canadian farmer's report : minnesota and dakota compared with manitoba and the canadian north-west: the facts as personally seen by a canadian farmer / Webster, W A – [Ottawa?: s.n.], 1888 [mf ed 1982] – 1mf – 9 – (with app) – mf#30541 – cn Canadiana [917]

The canadian farmer's travels in the united states of america : in which remarks are made on the arbitrary colonial policy practised in canada and the free and equal rights and happy effects of the liberal institutions and astonishing enterprise of the united states / Davis, Robert – Buffalo: printed [by] Steele's Press, 1837 [mf ed 1983] – 2mf – 9 – mf#21585 – cn Canadiana [917]

Canadian fiction – Kingston. 1998+ (1) 1998+ (5) 1998+ (9) – (cont: canadian fiction magazine) – mf#9676,01 – us UMI ProQuest [420]

Canadian fiction magazine – Kingston. 1971-1997 (1) 1975-1997 (5) 1975-1997 (9) – (cont by: canadian fiction) – ISSN: 0045-477X – mf#9676 – us UMI ProQuest [420]

The canadian field-naturalist / Ottawa Field-Naturalists' Club – [Ottawa?: The Club?, 1919] – 9 – mf#P05001 – cn Canadiana [500]

Canadian finance – Winnipeg, Canada. 21 jun 1911-21 dec 1921 – 12r – 1 – uk British Libr Newspaper [332]

The canadian fireside : an entertaining magazine for the leisure hour – [Montreal: W Bennet, 1888-188- or 189-] – 9 – ISSN: 1190-6987 – mf#P04071 – cn Canadiana [870]

Canadian fisherman – Sainte Anne de Bellevue, Canada. sep 1919-jan 1923 – 2r – 1 – uk British Libr Newspaper [630]

The canadian florist and cottage gardener : devoted to the cultivation of flowers, vegetables and fruits – Peterborough [Ont]: F Mason, [1885-18– or 19–] – 9 – ISSN: 1190-7312 – mf#P04115 – cn Canadiana [635]

Canadian folk song and handicraft festival... quebec, may 20-22 1927 : fifth concert, may 22 – [Quebec?]: [s.n.], [1927] (mf ed 1991) – 1mf – 9 – (in english and french) – mf#SEM105P1354 – cn Bibl Nat [780]

Canadian folk song and handicraft festival... quebec, may 20-22 1927 : first concert, may 20 – [Quebec?]: [s.n.], [1927] (mf ed 1991) – 1mf – 9 – (in english and french) – mf#SEM105P1351 – cn Bibl Nat [780]

Canadian folk song and handicraft festival... quebec, may 20-22 1927 : fourth concert, may 22 – [Quebec?]: [s.n.], [1927] (mf ed 1991) – 1mf – 9 – (in english and french) – mf#SEM105P1353 – cn Bibl Nat [780]

Canadian folk song and handicraft festival... quebec, may 20-22 1927 : second concert, may 21 – [Quebec?]: [s.n.], [1927] (mf ed 1991) – 1mf – 9 – (in french and english) – mf#SEM105P1363 – cn Bibl Nat [780]

Canadian folk song and handicraft festival... quebec, may 20-22 1927 : third concert, may 21 – [Quebec?]: [s.n.], [1927] (mf ed 1991) – 1mf – 9 – (in english and french) – mf#SEM105P1352 – cn Bibl Nat [780]

Canadian folk songs for american schools / Davis, Reginald Charles – U of Rochester 1950 [mf ed 19–] – 4mf – 9 – mf#fiche216, 715 – us Sibley [780]

Canadian Forces Base Borden *see* Borden citizen

Canadian forest industries – Ste-Anne-de-Bellevue. 1991-1991 (1,5,9) – ISSN: 0318-4277 – mf#18758 – us UMI ProQuest [634]

The canadian forester's illustrated guide / Chapais, Jean Charles – Montreal: E Senecal, 1885 – 3mf – 9 – (incl ind) – mf#03590 – cn Canadiana [634]

Canadian forests, forest trees, timber and forest products / Small, Henry Beaumont – Montreal: Dawson, 1884 [mf ed 1981] – 1mf – 9 – mf#13683 – cn Canadiana [634]

Canadian forests, forest trees, timber and forest products / Small, Henry Beaumont – Montreal: Dawson, 1884 [mf ed 1982] – 1mf – 9 – mf#33289 – cn Canadiana [634]

Canadian forum – Toronto: Canadian Forum 1920- [mthly] [mf ed UMI] – (incl ind) – ISSN: 0008-3631 – mf#543 – us UMI ProQuest [071]

Canadian forward – n341-347 [1915 jul 15-dec 16] – 1r – 1 – (cont: cotton's weekly) – mf#700498 – us WHS [071]

Canadian franchise and election laws : a manual for the use of revising officers, municipal officers, candidates, agents, and electors: with supplement containing the amending acts of 1886 / Ermatinger, Charles Oakes – Toronto: Carswell, 1886 [mf ed 1980] – 5mf – 9 – 0-665-02920-9 – (incl ind and bibl ref) – mf#02920 – cn Canadiana [325]

Canadian free press / Argyle Co-operative House – 1967 feb 15/mar 15-1968 jul 19, 1967 feb 25/mar 15-1968 jul 19 – 2r – 1 – (cont by: octopus [ottawa ont]) – mf#786891 – us WHS [334]

Canadian freeman – Toronto, ON: Mallon & Maylan, 1862-73 – 5r – 1 – cn Library Assoc [071]

Canadian freeman – Toronto, ON, (York, Upper Canada) 1825-34 – 1r – 1 – ISSN: 1181-1811 – cn Library Assoc [071]

Canadian freeman *see* Chatham newspapers, pt 1

The canadian freemason – Montreal: Hill's Book Store, [1860-1861] – 9 – mf#P04079 – cn Canadiana [366]

The canadian freemason – Toronto: Aldrich & Co, [1874-18–?] – 9 – mf#P04365 – cn Canadiana [366]

Canadian Friends Historical Association *see* Canadian quaker history newsletter

Canadian fruit, flower, and kitchen gardener : a guide in all matters relating to the cultivation of fruits, flowers and vegetables and their value for cultivation in this climate / Beadle, Delos White – Toronto: J Campbell, 1872 [mf ed 1980] – 5mf – 9 – 0-665-02995-0 – mf#02995 – cn Canadiana [635]

The canadian fruit-culturist : or letters to an intending fruit-grower, on the proper location, soil, preparation, planting, and after-cultivation of orchards, vineyards, and gardens... / Dougall, James – Montreal: John Dougall & Son, publ, 1867 [mf ed 1980] – 1mf – 9 – 0-665-06238-9 – mf#06238 – cn Canadiana [634]

Canadian Fund for Relief of Distress in Ireland *see* Report of the joint committee selected from the committees of the duchess of marlborough relief fund...

Canadian garland : a semi-monthly literary journal – Hamilton, UC: W Smyth. v1 n1-26. sep 15 1832-aug 31 1833// – 1r – 1 – Can$110.00 – cn McLaren [410]

Canadian genealogist – 1979 3-1987 – 1r – 1 – mf#615766 – us WHS [929]

The canadian gentleman's journal and sporting times – Toronto: P. Collins, [187-?-188-?] – 9 – mf#P04325 – cn Canadiana [790]

Canadian geographer = Geographe canadien – Toronto. 1951+ (1) 1951+ (5) 1951+ (9) – ISSN: 0008-3658 – mf#13850 – us UMI ProQuest [917]

Canadian geographic – Ottawa. 1978+ (1) 1978+ (5) 1978+ (9) – (cont: canadian geographical journal) – ISSN: 0706-2168 – mf#1582,01 – us UMI ProQuest [917]

Canadian geographical journal – Ottawa. 1930-1978 (1) 1969-1978 (5) 1975-1978 (9) – (cont by: canadian geographic) – ISSN: 0315-1824 – mf#1582 – us UMI ProQuest [917]

Canadian geotechnical journal = Revue canadienne de geotechnique – Ottawa. 1963+ (1) 1977+ (5) 1977+ (9) – ISSN: 0008-3674 – mf#10946 – us UMI ProQuest [550]

Canadian gleaner – Huntingdon, QC. 1863-1900 – 12r – 1 – ISSN: 0845-793X – cn Library Assoc [071]

The canadian gold fields and farm lands : how to get there – Liverpool England: Allan Line Offices 1899 – 1mf – 9 – mf#00036 – cn Canadiana [917]

Canadian granger : devoted to the interests of patrons of husbandry – London, Ont: [s.n, 1876-18– or 19–] – 9 – (cont: the granger) – mf#P04059 – cn Canadiana [636]

The canadian green bag : an entertaining magazine for lawyers / ed by Longueville, F – Montreal: J Lovell, [1895] – 9 – mf#P04171 – cn Canadiana [340]

Canadian grocer – Toronto. 1985-1996 (1,5,9) – ISSN: 0008-3704 – mf#15475 – us UMI ProQuest [380]

Canadian hardware and metal merchant – Toronto: J B McLean Pub Co, [1889 or 189–1904?] [mf ed v6 n41 oct 13 1894] – 9 – mf#P04959 – cn Canadiana [680]

The canadian herbal : or botanic family physician: comprising a variety of the indian remedies and medicinal plants of this country, and adapted to various forms of disease / Stewart, Schuyler – Hamilton, Ont?: Canada Christian Advocate, 1851 – 1mf – 9 – mf#53609 – cn Canadiana [615]

Canadian historical dates and events : 1492-1915 / Audet, Francis-Joseph – [Ottawa?: G Beauregard], 1917 – 3mf – 9 – 0-665-72289-3 – mf#72289 – cn Canadiana [971]

The canadian historical exhibition, 1897 / Howland, Oliver Aiken – Toronto?: s.n, 1896? – 1mf – 9 – mf#12991 – cn Canadiana [060]

Canadian historical quarterly – [Toronto?]: Hunter, Rose, [1899-19–] [mf ed v1 n1 dec 1899] – 9 – mf#P04076 – cn Canadiana [971]

Canadian historical review – North York. 1920+ (1) 1973+ (5) 1976+ (9) – ISSN: 0008-3755 – mf#8523 – us UMI ProQuest [971]

Canadian history / Hughes, James Laughlin – New York: Phillips & Hunt; Cincinnati: Hitchcock & Walden, 1880 [mf ed 1980] – (= ser The Chataugua Text-books) – 1mf – 9 – mf#07042 – cn Canadiana [971]

Canadian history / Hughes, James Laughlin – Toronto: W J Gage, 1881 [mf ed 1986] – (= ser W J Gage and Co's educational series) – 1mf – 9 – 0-665-32595-9 – mf#32595 – cn Canadiana [971]

Canadian history : the siege and blockade of quebec, by generals montgomery and arnold, in 1775-6: a paper read...march 6th, 1872 / Anderson, William James – Quebec: Middleton & Dawson, 1872 [mf ed 1979] – 1mf – 9 – 0-665-00034-0 – mf#00034 – cn Canadiana [971]

Canadian history and biography : and passages in the lives of a british prince and a canadian seigneur, the father of the queen and the hero of chateauguay... / Anderson, William James – Quebec: printed by Middleton & Dawson...1867 [mf ed 1983] – 1mf – 9 – 0-665-42778-6 – mf#42778 – cn Canadiana [971]

Canadian history readings, vol 1 : for schools, libraries, and general readers: embracing seventy-two topics, treated by twenty-six writers, including well-known specialists / ed by Hay, George Upham – Saint John, NB: Barnes, 1900 [mf ed 1980] – 4mf – 9 – 0-665-05520-X – (incl ind) – mf#05520 – cn Canadiana [971]

Canadian home economics journal = Revue canadienne d'economie familiale – Ottawa. 1980+ [1,5,9] – ISSN: 0008-3763 – mf#12451 – us UMI ProQuest [640]

The canadian home, farm and business cyclopaedia : a treasury of useful and entertaining knowledge... – Toronto, Whitby Ont: J S Robertson, 1884 – 10mf – 9 – mf#03837 – cn Canadiana [630]

Canadian home journal – Toronto: Home Journal Pub Co, [1895-1958] – 9 – mf#P04431 – cn Canadiana [640]

Canadian home rule herald – Parkdale, Ont: G D Griffin, [1890-189- or 19-] [mf ed n1 1890] – 9 – mf#P04185 – cn Canadiana [330]

The canadian honey producer – Brantford, Ont: E L Goold, [1887-1889] – 9 – mf#P04986 – cn Canadiana [630]

The canadian horticultural magazine – [Montreal]: Montreal Horticultural Society, [1897-1899?] – 9 – (ceased 1899?) – mf#P04158 – cn Canadiana [635]

The canadian horticulturist – St. Catharines [Ont.]: Fruit Growers' Association of Ontario, 1878-[1914] – 9 – (incl ind) – mf#P04048 – cn Canadiana [635]

Canadian hospital – Toronto. 1925-1973 (1) 1971-1973 (5) – (cont by: dimensions in health service) – ISSN: 0008-3798 – mf#562 – us UMI ProQuest [360]

Canadian hostility to annexation / Hopkins, John Castell – [S.l: s.n, 1892?] [mf ed 1982] – 1mf – 9 – mf#17804 – cn Canadiana [327]

Canadian illustrated news – v1-2 n1 [1976 may-1977 spr] – 1r – 1 – (cont: canadian treasure) – mf#242740 – us WHS [071]

Canadian illustrated news – Hamilton, CW. nov 8 1862-feb 13 1864// – 2r – 1 – Can$198.00 – cn McLaren [071]

Canadian illustrated news – Montreal, Canada. 4 jan 1873-30 oct 1875; 23 sep-23 dec 1876; 6 jan 1877-15 feb 1879; 2 apr-25 jun 1881; 1 jul-30 dec, 27 jan 1883 (imperfect) – 5 1/2r – 1 – uk British Libr Newspaper [071]

The canadian illustrated news portfolio and dominion guide for 1873 – Montreal: Canadian Illustrated News, 1873? – 13mf – 9 – mf#02038 – cn Canadiana [720]

Canadian illustrated shorthand writer – Toronto: Bengough Bros, [1880-1881] [mf ed v1 n7 nov 1880] – 9 – mf#P04442 – cn Canadiana [650]

Canadian independence, annexation and british imperial federation / Douglas, James – New York: G P Putnam, 1894 [mf ed 1980] – 2mf – 9 – 0-665-02744-3 – (incl ind) – mf#02744 – cn Canadiana [971]

The canadian independent – Toronto: MacLear, [1854-1894] – 9 – mf#P04946 – cn Canadiana [200]

The canadian indian – Owen Sound, ON: J Rutherford. v1-12. oct 1890-sep 1891// – 1r – 1 – Can$110.00 – (publ under the auspices of the canadian indian research & aid society) – cn McLaren [305]

Canadian indian art crafts / National Indian Arts and Crafts Corporation – v1 n4-v5 n1 [1975 aug-1979] – 1r – 1 – mf#462825 – us WHS [740]

Canadian Indian Centre of Toronto *see* Native times

Canadian Indian Voice Society et al *see* Indian voice

Canadian insolvency legislation : report of meeting of the league held on wednesday, december 4th, 1895 / British Empire League – London: Commerce Print & Pub Co, [1895?] [mf ed 1979] – 1mf – 9 – 0-665-00296-3 – mf#00296 – cn Canadiana [323]

Canadian Institute *see*
– Proceedings of the canadian institute
– Transactions of the canadian institute

Canadian institute : first lecture course, session 1882-83 – [s.l: s.n. 1882?] [mf ed 1983] – 1mf – 9 – 0-665-39739-9 – mf#39739 – cn Canadiana [090]

Canadian institute, established 1849, incorporated by royal charter : president, hon chief justice robinson, first vice-president, professor croft – [Toronto?: s.n.] 1854 [mf ed 1987] – 1mf – 9 – 0-665-57986-1 – mf#57986 – cn Canadiana [366]

Canadian Institute for Ukrainian Studies. University of Alberta *see* The canadian ukrainian collection

Canadian institute of food science and technology journal – Toronto. 1989-1991 (1) – (cont by: food research international) – ISSN: 0315-5463 – mf#16635,01 – us UMI ProQuest [660]

Canadian Institute of Mining and Metallurgy *see* Cim bulletin

Canadian institute's caxton celebration, toronto, wednesday, june 13th 1877, at 8 pm : exhibition of early specimens of typography, etc – [Toronto?: s.n. 1877?] [mf ed 1983] – 1mf – 9 – 0-665-39755-0 – mf#39755 – cn Canadiana [680]

Canadian insurance – Toronto. 1994-1996 (1) 1994-1996 (5) 1994-1996 (9) – ISSN: 0008-3879 – mf#15917 – us UMI ProQuest [368]

Canadian interiors – Toronto. 1975-1990 (1) 1975-1990 (5) 1975-1990 (9) – ISSN: 0008-3887 – mf#10770 – us UMI ProQuest [740]

The canadian iron and steel industry : a study in the economic history of a protected industry / Donald, William John – Boston: Houghton Mifflin, 1915 – (= ser Hart, schaffner and marx prize essays) – 5mf – 9 – 0-665-98518-5 – (incl app) – mf#98518 – cn Canadiana [338]

Canadian jewish historical society journal – 1978 spr-1984 fall – 1r – 1 – (cont: jewish historical society of canada journal) – mf#806693 – us WHS [939]

Canadian jewish news : an independent community newspaper serving as a forum for diverse viewpoints – Toronto. v1- . jan 1, 1960- (wkly) – 1 (subs Can$335/y) – (back run (1960-98) 65r can$6285; some iss incl suppls viewpoints (canadian jewish congress) & beginning in 1986, in touch (jewish women's federation); extensive coverage of local, national & world news affecting the jewish community) – cn McLaren [939]

Canadian jewish news [montreal edition] – Toronto. v1- . 1976- (wkly) – 1 (subs Can$340/y) – (back run (1996-97) 6r can$680) – cn McLaren [071]

Canadian jewish outlook : canada's progressive jewish magazine – Toronto, Vancouver. v1-23 n11. oct 1963-dec 1985// – 6r – 1 – Can$435.00 – (cont as: outlook; political and social coverage in the tradition of toronto yiddish-language newspapers der veg and vochenblatt) – cn McLaren [939]

Canadian jewish weekly – Canada. jan 1924-dec 1978 – 55r – 1 – (in yiddish) – cn Commonwealth Imaging [071]

Canadian journal : a repertory of industry, science, and art, and a record of the proceedings of the canadian institute / ed by Hind, Henry Youle – Toronto: Publ by H Scobie for the Council of the Canadian Institute, 1852-1855 – 9 – (cont by: the canadian journal of industry, science and art. incl ind) – mf#P04982 – cn Canadiana [073]

The canadian journal : canada for canadians – Toronto: W R Haight, [1882-18–] – 9 – mf#P05995 – cn Canadiana [971]

Canadian journal of agricultural economics = Revue canadienne d'economie rurale – Ottawa. 1977+ (1,5,9) – ISSN: 0008-3976 – mf#11432 – us UMI ProQuest [630]

Canadian journal of anesthesia = Journal canadien d'anesthesie – Toronto. 1987+ (1) 1987+ (5) 1987+ (9) – (cont: canadian anaesthetists' society journal) – ISSN: 0832-610X – mf#7177,01 – us UMI ProQuest [617]

Canadian journal of animal science – Ottawa. 1976+ (1,5,9) – ISSN: 0008-3984 – mf#10988 – us UMI ProQuest [590]

Canadian journal of behavioural science – Ottawa. 1969+ (1) 1975+ (5) 1976+ (9) – ISSN: 0008-400X – mf#10439 – us UMI ProQuest [150]

Canadian journal of biochemistry = Journal canadien de biochimie – Ottawa. 1964-1982 (1) 1976-1982 (5) 1976-1982 (9) – (cont by: canadian journal of biochemistry and cell biology=revue canadienne de biochimie et biologie cellulaire) – ISSN: 0008-4018 – mf#10947 – us UMI ProQuest [574]

Canadian journal of biochemistry and cell biology = Revue canadienne de biochimie et biologie cellulaire – Ottawa. 1983-1985 (1,5,9) – (cont: canadian journal of biochemistry. cont by: biochemistry and cell biology) – ISSN: 0714-7511 – mf#10947,01 – us UMI ProQuest [574]

Canadian journal of biochemistry and physiology – Ottawa. 1944-1963 (1) – ISSN: 0576-5544 – mf#10948 – us UMI ProQuest [574]

Canadian journal of botany – Ottawa. 1935+ (1) 1977+ (5) 1977+ (9) – ISSN: 0008-4026 – mf#10949 – us UMI ProQuest [580]

Canadian journal of botany – Ottawa. 1935+ [1]; 1977+ [5,9] – ISSN: 0008-4026 – mf#10949 – us UMI ProQuest [580]

Canadian journal of chemical engineering – Ottawa. 1944+ (1) 1976+ (5) 1976+ (9) – ISSN: 0008-4034 – mf#11366 – us UMI ProQuest [660]

Canadian journal of chemistry – Ottawa. 1935+ (1) 1975+ (5) 1976+ (9) – ISSN: 0008-4042 – mf#10950 – us UMI ProQuest [540]

Canadian journal of chemistry – Ottawa. 1935+ [1]; 1975+ [5]; 1976+ [9] – ISSN: 0008-4042 – mf#10950 – us UMI ProQuest [540]

Canadian journal of civil engineering = Revue canadienne de genie civil – Ottawa. 1974-1996 (1) 1976-1996 (5) 1976-1996 (9) – ISSN: 0315-1468 – mf#10951 – us UMI ProQuest [624]

Canadian journal of comparative medicine = Revue canadienne de medecine comparee – Ottawa. 1984-1985 (1,5,9) – (cont by: canadian journal of veterinary research=revue canadienne de recherche veterinaire) – ISSN: 0008-4050 – mf#14067,02 – us UMI ProQuest [610]

Canadian journal of counselling = Revue canadienne de counseling – Ottawa. 1986+ – 1,5,9 – (cont: canadian counsellor) – ISSN: 0828-3893 – mf#11409,01 – us UMI ProQuest [370]

Canadian journal of criminology = Revue canadienne de criminologie – Ottawa. 1978+ (1) 1978+ (5) 1978+ (9) – (cont: canadian journal of criminology and corrections) – ISSN: 0704-9722 – mf#3278,01 – us UMI ProQuest [360]

Canadian journal of criminology and corrections = Revue canadienne de criminologie – Ottawa. 1958-1977 (1) 1971-1977 (5) 1977-1977 (9) – (cont by: canadian journal of criminology) – ISSN: 0315-5390 – mf#3278 – us UMI ProQuest [360]

Canadian journal of dietetic practice and research – Toronto. 1998+ (1) – ISSN: 1486-3847 – mf#14841,01 – us UMI ProQuest [613]

Canadian journal of earth sciences – Ottawa. 1964+ (1) 1977+ (5) 1977+ (9) – ISSN: 0008-4077 – mf#10952 – us UMI ProQuest [550]

Canadian journal of earth sciences – Ottawa. 1964+ [1]; 1977+ [5,9] – ISSN: 0008-4077 – mf#10952 – us UMI ProQuest [550]

Canadian journal of economics = Revue canadienne d'economique – Malden. 1968+ (1) 1974+ (5) 1975+ (9) – ISSN: 0008-4085 – mf#9908 – us UMI ProQuest [330]

Canadian journal of education = Revue canadienne de l'education – Toronto. 1983+ – 1,5,9 – ISSN: 0380-2361 – mf#14309 – us UMI ProQuest [370]

Canadian journal of experimental psychology = Revue canadienne de psychologie experimentale – Ottawa. 1993+ (1) 1993+ (5) 1993+ (9) – (cont: canadian journal of psychology) – ISSN: 1196-1961 – mf#9909,01 – us UMI ProQuest [150]

Canadian journal of fabrics – Montreal: E B Biggar, [1883-1906?] – 9 – (cont by: canadian textile journal; ceased 1906?) – mf#P04945 – cn Canadiana [670]

Canadian journal of family law = Revue canadienne de droit familial – Agincourt. 1978+ (1,5,9) – ISSN: 0704-1225 – mf#11857 – us UMI ProQuest [346]

Canadian journal of family law – v1-17. 1978-2000 – 9 – $395.00 set – ISSN: 0704-1225 – mf#110021 – us Hein [346]

Canadian journal of fisheries and aquatic sciences = Journal canadien des sciences halieutiques et aquatiques – Ottawa. 1980+ (1) 1980+ (5) 1980+ (9) – ISSN: 0706-652X – mf#7099,01 – us UMI ProQuest [639]

Canadian journal of forest research – Ottawa. 1971+ (1) 1977+ (5) 1977+ (9) – ISSN: 0045-5067 – mf#10953 – us UMI ProQuest [634]

Canadian journal of forest research – Ottawa. 1971+ [1]; 1977+ [5,9] – ISSN: 0045-5067 – mf#10953 – us UMI ProQuest [634]

The canadian journal of health – Toronto: H Fox, [1894-189- or 19–] – 9 – mf#P04450 – cn Canadiana [613]

Canadian journal of higher education = La revue canadienne d'enseignement superieur – Toronto. 1979+ – 1,5,9 – ISSN: 0316-1218 – mf#12305,01 – us UMI ProQuest [378]

Canadian journal of history = Annales canadiennes d'histoire – Saskatoon. 1966+ (1) 1971+ (5) 1977+ (9) – ISSN: 0008-4107 – mf#2683 – us UMI ProQuest [971]

Canadian journal of history of sport = Revue canadienne de l'histoire des sports – 1981 dec-1984, 1985 may-1990 dec – 1r – 1 – mf#835035 – us WHS [790]

Canadian journal of history of sport and physical / Canadian Association for Health, Physical Education and Recreation – v4 n2-v11 n2 [1973 dec-1980 dec] – 2r – 1 – (cont: canadian journal of history of sport and physical education; cont by: sport history review) – mf#774505 – us WHS [790]

Canadian journal of hospital pharmacy – Hamilton. 1982+ (1,5,9) – ISSN: 0008-4123 – mf#12346,01 – us UMI ProQuest [615]

Canadian journal of industry, science, and art – Toronto: Printed for the Canadian Institute by Lovell & Gibson, [1856?-1867] – 9 – (cont by: the canadian journal of science, literature, and history) – mf#P05122 – cn Canadiana [073]

Canadian journal of law and jurisprudence – University of Western Ontario. v1-14. 1988-2001 – 9 – $415.00 set – (cont: university of western ontario law review) – ISSN: 0841-8209 – mf#111841 – us Hein [340]

Canadian journal of linguistics – Montreal. 1961-1995 (1) 1961-1995 (5) 1961-1995 (9) – (cont: journal of the canadian linguistic association) – ISSN: 0008-4131 – mf#12024,01 – us UMI ProQuest [400]

Canadian journal of mathematics = Journal canadien de mathematiques – Toronto. 1949+ (1) 1949+ (5) 1949+ (9) – ISSN: 0008-414X – mf#13851 – us UMI ProQuest [510]

Canadian journal of medical radiation technology = Journal canadien des techniques en radiation medicale – Ottawa. 1987+ (1) 1987+ (5) 1987+ (9) – (cont: canadian journal of radiography, radiotherapy, nuclear medicine) – ISSN: 0820-5930 – mf#7078,02 – us UMI ProQuest [616]

The canadian journal of medical science : a monthly journal of british and foreign medical science, criticism and news – [Toronto?: Guardian Book & Job Print, 1876?-1883] – 9 – mf#P05186 – cn Canadiana [610]

The canadian journal of medicine and surgery – Toronto: [s.n, 1897?-1936] – 9 – mf#P05193 – cn Canadiana [610]

Canadian journal of microbiology – Ottawa. 1954+ (1) 1977+ (5) 1977+ (9) – ISSN: 0008-4166 – mf#10954 – us UMI ProQuest [576]

Canadian journal of microbiology – Ottawa. 1954+ [1]; 1977+ [5,9] – ISSN: 0008-4166 – mf#10954 – us UMI ProQuest [576]

Canadian journal of neurological sciences – Calgary. 1979+ (1,5,9) – ISSN: 0317-1671 – mf#12025 – us UMI ProQuest [616]

Canadian journal of occupational therapy = Revue canadienne d'ergotherapie – Ottawa. 1983+ (1,5,9) – ISSN: 0008-4174 – mf#14186,02 – us UMI ProQuest [610]

The canadian journal of odd-fellowship – Stratford [Ont.]: Odd-Fellows' Print and Pub Assoc, [1875?-1876] – 9 – (incl ind) – mf#P04336 – cn Canadiana [420]

Canadian journal of ophthalmology = Journal canadien d'ophtalmologie – Ottawa. 1966+ (1) 1972+ (5) 1974+ (9) – ISSN: 0008-4182 – mf#6997 – us UMI ProQuest [617]

Canadian journal of otolaryngology = Journal canadien d'otolaryngologie – Toronto. 1972-1975 (1) 1972-1974 (5) (9) – (cont by: journal of otolaryngology) – ISSN: 0045-5083 – mf#7640 – us UMI ProQuest [617]

The canadian journal of philately : a monthly magazine devoted to the science of philately – Toronto: H A Fowler, [1893] – 9 – ISSN: 1190-6448 – mf#P04548 – cn Canadiana [760]

The canadian journal of philately *see* The international philatelist

Canadian journal of photography – Toronto: Ewing & Co, mar 1870-mar 1871; jan 1-sep 1 1875// – 1r – 1 – Can$75.00 – (first canadian photographic magazine) – cn McLaren [770]

Canadian journal of physics – Ottawa. 1935+ (1) 1976+ (5) 1976+ (9) – ISSN: 0008-4204 – mf#10955 – us UMI ProQuest [530]

Canadian journal of physics – Ottawa. 1935+ [1]; 1976+ [5,9] – ISSN: 0008-4204 – mf#10955 – us UMI ProQuest [530]

Canadian journal of physiology and pharmacology – Ottawa. 1964+ (1) 1977+ (5) 1977+ (9) – ISSN: 0008-4212 – mf#10956 – us UMI ProQuest [612]

Canadian journal of plant science = Revue canadienne de phytotechnie – Ottawa. 1957+ (1) 1976+ (5) 1976+ (9) – ISSN: 0008-4220 – mf#10989 – us UMI ProQuest [580]

Canadian journal of psychiatric nursing – Winnipeg. 1979-1990 (1,5,9) – ISSN: 0008-4247 – mf#12065,01 – us UMI ProQuest [610]

Canadian journal of psychiatry = Revue canadienne de psychiatrie – Ottawa. 1979+ (1) 1979+ (5) 1979+ (9) – (cont: canadian psychiatric association journal) – ISSN: 0706-7437 – mf#5875,01 – us UMI ProQuest [616]

Canadian journal of psychology = Revue canadienne de psychologie – Old Chelsea. 1947-1992 (1) 1974-1992 (5) 1976-1992 (9) – (cont by: canadian journal of experimental psychology) – ISSN: 0008-4255 – mf#9909 – us UMI ProQuest [150]

Canadian journal of public health = Revue canadienne de sante publique – Ottawa. 1913+ (1) 1972+ (5) 1973+ (9) – ISSN: 0008-4263 – mf#6846 – us UMI ProQuest [360]

Canadian journal of public health – v1-54. 1909-63 – 1 – us AMS Press [614]

Canadian journal of radiography, radiotherapy, nuclear medicine – Ottawa. 1974-1986 (1) 1974-1986 (5) 1974-1986 (9) – (cont: canadian journal of radiography, radiotherapy, nucleography. cont by: canadian journal of medical radiation technology) – ISSN: 0319-4434 – mf#7078,01 – us UMI ProQuest [616]

Canadian journal of radiography, radiotherapy, nucleography – Ottawa. 1970-1974 (1) 1970-1974 (5) 1973-1974 (9) – (cont by: canadian journal of radiography, radiotherapy, nuclear medicine) – ISSN: 0015-4938 – mf#7078 – us UMI ProQuest [616]

Canadian journal of religious thought – v1-9. 1924-32 [complete] – Inquire – 1 – ISSN: 0382-6589 – mf#ATLA 1993-S512 – us ATLA [200]

Canadian journal of remote sensing – Ottawa. 1977+ (1,5,9) – ISSN: 0703-8992 – mf#11728 – us UMI ProQuest [629]

Canadian journal of research – Ottawa. 1929-1935 (1) – mf#10957 – us UMI ProQuest [500]

Canadian journal of science, literature and history – Toronto. 1852-1878 – 1 – ISSN: 0381-8624 – mf#2780 – us UMI ProQuest [073]

Canadian journal of science, literature, and history – Toronto: Canadian Institute, [1868-1878] – 9 – (cont by: canadian institute. proceedings of the canadian institute) – mf#P05012 – cn Canadiana [073]

Canadian journal of soil science – Ottawa. 1957+ (1) 1976+ (5) 1976+ (9) – ISSN: 0008-4271 – mf#10990 – us UMI ProQuest [630]

Canadian journal of sport sciences = Journal canadien des sciences du sport – Downsview. 1989-1990 (1,5,9) – ISSN: 0833-1235 – mf#16439,01 – us UMI ProQuest [790]

Canadian journal of surgery = Journal canadien de chirurgie – Ottawa. 1957+ (1) 1966+ (5) 1970+ (9) – ISSN: 0008-428X – mf#2405 – us UMI ProQuest [617]

Canadian journal of veterinary research = Revue canadienne de recherche veterinaire – Ottawa. 1986+ (1,5,9) – (cont: canadian journal of comparative medicine revue canadienne de medecine comparee) – ISSN: 0830-9000 – mf#14067,03 – us UMI ProQuest [610]

Canadian journal of zoology – Ottawa. 1935+ (1) 1976+ (5) 1976+ (9) – ISSN: 0008-4301 – mf#10958 – us UMI ProQuest [590]

Canadian journal of zoology – Ottawa. 1935+ [1]; 1976+ [5,9] – ISSN: 0008-4301 – mf#10958 – us UMI ProQuest [590]

Canadian Journalism Foundation *see* Last post

Canadian Kodak Co *see* How to make good pictures

Canadian laboratory – Willowdale. 1989-1990 (1,5,9) – mf#17473 – us UMI ProQuest [500]

Canadian labour / Canadian Labour Congress – 1984-1989 spr, v24-v25 [1979 mar 30-1980 dec] – 2r – 1 – (cont: canadian unionist; trades and labour congress journal; cont by: clc today) – mf#1253529 – us WHS [331]

Canadian labour comment / Canadian Labour Congress – 1973 jun 15-1977 feb 24, 1977 mar 11-1979 mar 2 – 2r – 1 – (cont by: canadian labour) – mf#219379 – us WHS [331]

Canadian Labour Congress *see*
– Canadian labour
– Canadian labour comment
– Clc today

Canadian land advertiser : issued for distribution in canada and in great britain and ireland amongst british emigrants: containing descriptions, prices and terms of purchase for over five million dollars worth of improved farms... – Toronto: W J Fenton, 1883? – 2mf – 9 – mf#01123 – cn Canadiana [333]

The canadian law of banks and banking : the clearing house, currency and dominion notes, bills, notes, cheques and other negotiable instruments / Falconbridge, John Delatre – Toronto: Canada Law Book Co, 1913 [mf ed 1998] – 10mf – 9 – 0-665-97720-4 – mf#97720 – cn Canadiana [346]

The canadian law of fixtures / Manning, Harold Ernest – Toronto: Canada Law Book Co. 1927. 289-364p. LL-2378 – 1 – us L of C Photodup [348]

Canadian law review – Toronto, CN: Canadian Law Review Co. v1-6. 1901-07 (all publ) – 14mf – 9 – $63.00 – mf#LLMC 84-432 – us LLMC [340]

Canadian law review – Toronto. v1-6. 1901-07 – (= ser Historical legal periodical series) – 1 – $85.00 set – mf#408860 – us Hein [340]

Canadian law times / Canada. General – v1-42. 1881-1922 (all publ) – 467mf – 9 – $700.00 – (includes 2v of digest/index. cont by: the canadian bar review which is not offered by llmc) – mf#LLMC 81-016 – us LLMC [340]

Canadian law times – Toronto. v1-42. 1881-1922 – 1 – $605.00 set – mf#408870 – us Hein [340]

Canadian lawyer – Aurora. 1979-1996 (1,5,9) – ISSN: 0703-2129 – mf#11981 – us UMI ProQuest [340]

Canadian lawyer – v1-25. 1977-2001 – 9 – $378.00 set – ISSN: 0703-2129 – mf#112701 – us Hein [340]

Canadian leaves : history, art, science, literature, commerce: a series of new papers read before the canadian club of new york / ed by Fairchild, George Moore – New York: N Thompson, 1887 [mf ed 1980] – 4mf – 9 – 0-665-00428-1 – (ill by thomson willing) – mf#00428 – cn Canadiana [410]

Canadian Liberation Movement *see* New canada

Canadian Library Association *see* Cm

Canadian library journal – Ottawa, ON: Canadian Library Association, 1944-92 – 17r – 1 – ISSN: 008-4352 – cn Library Assoc [020]

Canadian life and resources – Montreal, Canada. feb 1906-nov 1910 – 2r – 1 – (incorp with: canadian century fr nov 1910) – uk British Libr Newspaper [071]

Canadian life and resources *see* Canadian century

Canadian life and scenery : with hints to intending emigrants and settlers / Argyll, John Douglas Sutherland Campbell, duke of – [London]: The Religious Tract Society, 1891 [mf ed 1981] – (= ser The RST library, illustrated) – 3mf – 9 – mf#26269 – cn Canadiana [917]

Canadian life and scenery : with hints to intending immigrants and settlers / Argyll, John Douglas Sutherland Campbell, Duke of – London: Religious Tract Society, 1886 [mf ed 1980] – (= ser The rst library, illustrated) – 3mf – 9 – mf#00048 – cn Canadiana [971]

Canadian Linguistic Association *see* Journal of the canadian linguistic association

Canadian literary journal – Toronto: Flint & Van Norman, [1870-1871] – 9 – (cont by: the canadian magazine) – mf#P04226 – cn Canadiana [420]

The canadian literary magazine – York [Toronto]: G Gurnett, [1883] – 9 – mf#P04225 – cn Canadiana [400]

The canadian literary news letter and booksellers' advertiser – Montreal: H Ramsey, [1855] – 9 – ISSN: 1190-6804 – mf#P04235 – cn Canadiana [410]

Canadian literature – Vancouver. 1959+ (1) 1972+ (5) 1975+ (9) – ISSN: 0008-4360 – mf#6978 – us UMI ProQuest [410]

Canadian live-stock and farm journal – Hamilton [Ont]: Stock Journal Co, [1886-1895?] – 9 – (cont: canadian live-stock journal) – mf#P04046 – cn Canadiana [636]

Canadian live-stock journal – Hamilton [Ont]: Stock Journal Co, [1885?-1886] – 9 – (cont: canadian stock-raisers' journal; cont by: the canadian live-stock and farm journal) – mf#P04045 – cn Canadiana [636]

Canadian local history : the first gazetteer of upper canada / Scadding, Henry – [S.l: s.n, 188-?] [mf ed 1981] – 2mf – 9 – mf#13247 – cn Canadiana [971]

Canadian local history : the first gazetteer of upper canada, with annotations / Scadding, Henry – [S.l: s.n, 1875?] [mf ed 1984] – 1mf – 9 – 0-665-33642-X – mf#33642 – cn Canadiana [971]

Canadian machinery and manufacturing news – Toronto, Canada. 1909-12; 9 jan 1919-17 jul 1912; 6 dec 1920-1927; 12 jan-22 mar 1928 – 31r – 1 – uk British Libr Newspaper [670]

Canadian machinery and metalworking – Toronto. 1975-1996 (1,5,9) – ISSN: 0008-4379 – mf#10771 – us UMI ProQuest [660]

Canadian magazine – Toronto, ON: Ontario Publ Co, 1904-39 – 25r – 1 – ISSN: 1181-3458 – cn Library Assoc [071]

Canadian magazine – York [Toronto]: R Stanton, [1883] [mf ed v1 n1 jan 1833-v1 n4 apr 1833] – 9 – ISSN: 1190-6812 – mf#P04229 – cn Canadiana [410]

Canadian magazine of politics – Toronto, Canada. Mar 1893-1904 – 18r – 1 – uk British Libr Newspaper [071]

The canadian magazine of politics, art, science and literature – 1905-39 – 25r – 1 – cn Library Assoc [073]

Canadian magazine of politics, science, art and literature – Toronto. 1893-1906 – 1 – mf#2969 – us UMI ProQuest [073]

Canadian magazine of science and the industrial arts, patent office record – Montreal?: s.n, 1883-1891? (mthly) [mf ed 1990] – 12mf – 9 – (cont: scientific canadian mechanics' magazine and patent office record; ceased 1891?) – mf#P04873 – cn Canadiana [600]

Canadian manager – Toronto. 1988-1996 (1,5,9) – ISSN: 0045-5156 – mf#14385,01 – us UMI ProQuest [650]

A canadian manual on the procedure at meetings of shareholders and directors of companies, conventions, societies and public assemblies generally / Bourinot, John George – Toronto: Carswell, 1894 – 2mf – 9 – mf#26677 – cn Canadiana [350]

Canadian manufacturer – Toronto: Canadian Manufacturer Pub Co, [1908-19–] – 9 – (cont: the canadian manufacturer and industrial world) – mf#P05040 – cn Canadiana [670]

Canadian manufacturer and industrial world – Toronto: Canadian Manufacturer Co, [1882-1908] – 9 – (cont by: canadian manufacturer; cont: industrial world and national economist) – mf#P04937 – cn Canadiana [670]

Canadian Manufacturers' Association *see* Industrial canada

The canadian maple leaf song book – Toronto: A S Irving, [between 1867 and 1872] – 2mf – 9 – 0-665-89182-2 – (incl ind) – mf#89182 – cn Canadiana [780]

Canadian masonic pioneer – Montreal: Owler & Stevenson, [1856-1857] – 9 – ISSN: 1190-6979 – mf#P04078 – cn Canadiana [360]

Canadian materials for history, poetry, and romance / Bourinot, John George – [S.l: s.n], 1871 [mf ed 1980] – 1mf – 9 – 0-665-06482-9 – mf#06482 – cn Canadiana [410]

Canadian mathematical bulletin – Montreal. 1989-1996 (1) – ISSN: 0008-4395 – mf#16458 – us UMI ProQuest [510]

Canadian matrimonial news – Toronto: M Watson, [1892-189- or 19–] [mf ed v1 n1 may 14 1892] – 9 – mf#P04366 – cn Canadiana [306]

The canadian mecca / Beers, William George – [New York?: Century Co, c1882?] [mf ed 1980] – 1mf – 9 – 0-665-01012-5 – (fr: the century magazine, may 1882) – mf#01012 – cn Canadiana [639]

Canadian mechanics magazine and patent office record – [Ottawa: Burland-Desbarats Litho Co, 1876-1878] [mf ed v4 n1 jan 1876-v6 n12 dec 1878] – 9 – mf#P04864 – cn Canadiana [600]

The canadian mechanics ready reckoner : or, tables for converting english lineal, square and solid measures into french, and the contrary / Pigott, I – Three-Rivers Quebec: G Stobbs, 1832 – 1mf – 9 – mf#39595 – cn Canadiana [510]

Canadian media rates and data / Standard Rate & Data Service – 1953 jan/sep-1965 jul 9/1966 jul 9 – 20r – 1 – mf#861228 – us WHS [000]

Canadian Medical Association *see* Origin and organization of the canadian medical association

Canadian medical association journal (cmaj) = Journal de l'association medicale canadienne – Ottawa. 1911+ (1) 1966+ (5) 1970+ (9) – ISSN: 0820-3946 – mf#2086 – us UMI ProQuest [610]

The canadian mercantile test – Toronto: [s.n., 18—18– or 19–] – 9 – ISSN: 1190-7568 – mf#P04330 – cn Canadiana [332]

Canadian messenger – [Montreal: s.n., 1892-1899] – 9 – (cont: messenger of the sacred heart; cont by: the canadian messenger of the sacred heart. incl ind) – mf#P04901 – cn Canadiana [240]

The canadian messenger *see* Le messager canadien du sacre-coeur de jesus

Canadian messenger and journal of missions *see* The montreal witness

Canadian messenger of the sacred heart – [Montreal: s.n, 1899-1961] – 9 – (cont: the canadian messenger) – mf#P04902 – cn Canadiana [240]

Canadian methodism, its epochs and characteristics : written at the request of the london, toronto and montreal conferences / Ryerson, Egerton – Toronto: W Briggs, 1882 [mf ed 1981] – 5mf – 9 – (repr with add matter fr: canadian methodist magazine; incl bibl ref) – mf#12794 – cn Canadiana [242]

Canadian methodist magazine – Toronto: S Rose, 1875-1888 – 9 – (absorbed: earnest christianity; cont by: the methodist magazine) – mf#P04086 – cn Canadiana [242]

The canadian methodist pulpit : a collection of original sermons from living ministers of the wesleyan methodist church in canada / ed by Phillips, Samuel G – Toronto: Hunter, Rose; Montreal: Drysdale, 1875 [mf ed 1981] – 5mf – 9 – (incl ind; int by edward hartley dewart) – mf#11846 – cn Canadiana [242]

Canadian methodist quarterly : a review devoted to theology, philosophy, sociology, science, and christian work – Toronto: ...auspices of the Theological Union, [1889-1893] – 9 – (cont by: the canadian methodist review) – mf#P04088 – cn Canadiana [242]

Canadian methodist review – Toronto: ...auspices of the Theological Union, 1894-1895 – 9 – (cont: the canadian methodist quarterly; absorbed by: methodist magazine and review) – mf#P04087 – cn Canadiana [242]

The canadian military gazette – Montreal: [s.n., 1878-18–?] – 9 – mf#P04977 – cn Canadiana [355]

Canadian military review – Quebec: [s.n, 1880-1881] – 9 – (ceased 1881? incl some preliminary text in french) – mf#P04148 – cn Canadiana [355]

The canadian military review – [Ottawa?: A S Woodburn, 1877-18– or 19–] [mthly] – 9 – mf#P04186 – cn Canadiana [355]

Canadian military review, (quebec, quebec) *see* Partie francaise de la revue militaire canadienne

The canadian militia : an historical sketch: a lecture delivered...montreal, on 8th march, 1886 / Oswald, William Robert – S.l: s.n, 1886? – 1mf – 9 – mf#11558 – cn Canadiana [971]

The canadian militia / Wicksteed, Richard John – Ottawa?: MacLean, Roger, 1875 – 2mf – 9 – mf#23984 – cn Canadiana [355]

The canadian miller and grain trade review – Toronto: A G Mortimer, [188–1894] – 9 – mf#P06064 – cn Canadiana [630]

The canadian miner – Toronto: Canadian Miner Pub Co, 1897-[189- or 19–] – 9 – mf#P04956 – cn Canadiana [622]

Canadian mines and reciprocity : being a paper read before the commercial union club / Ledyard, Thomas D – Toronto: Hunter, Rose, 1888 [mf ed 1980] – 1mf – 9 – mf#08670 – cn Canadiana [622]

Canadian mining and mechanical review – Ottawa: [Review Pub Co, 1891-1894] [mf ed v10 n1 jan 1891-v13 n7 jul 1894] – 9 – mf#P04199 – cn Canadiana [622]

The canadian mining gazette – Toronto: [s.n, 1899-19–] – 9 – mf#P04184 – cn Canadiana [622]

Canadian mining handbook – Toronto, Ontario, CN. 1931-80 – 40r – 1 – cn Commonwealth Imaging [622]

Canadian Mining Institute *see* The journal of the canadian mining institute

Canadian mining journal – Toronto, Canada. 1 jul-15 dec 1912; 1913-nov 1935; 1936-1938; jul-dec 1939; apr 1941-oct 1947; 1948-sep 1951 (imperfect) – 69r – 1 – uk British Libr Newspaper [622]

Canadian mining law : a paper read before the american institute of mining engineers, at the wilkes-barre meeting, june, 1911 / Clark, John Murray – [S.l: s.n.], 1911 – 1mf – 9 – 0-665-71337-1 – mf#71337 – cn Canadiana [343]

Canadian mining review – Ottawa, Canada. -m. Jan 1889-may 1890; may 1891; jul, aug, dec 1892; feb 1893-1901; 31 jan-30 jun 1902 – 8 1/2r – 1 – uk British Libr Newspaper [622]

Canadian mining review – Ottawa: [Review Pub Co, 1882?-1890] [mf ed v1 n7 may 1883-v3 n8 nov 1885; v4 n1 jan 1886-v9 n12 dec 1890] – 9 – mf#P04197 – cn Canadiana [622]

The canadian mining review – Ottawa: [Review Pub Co, 1894-1907] [mf ed v13 n8 aug 1894-v28 n2 feb 1907] – 9 – mf#P04198 – cn Canadiana [622]

Canadian missionary link : in the interests of the baptist foreign mission societies of canada – Toronto: [Dudley & Burns, 1878-1927] – 9 – (merged with: baptist visitor to become: the link and visitor) – mf#P04082 – cn Canadiana [242]

Canadian modern language review = La revue canadienne des langues vivantes – North York. 1944+ (1) 1972+ (5) 1972+ (9) – ISSN: 0008-4506 – mf#9104 – us UMI ProQuest [410]

Canadian monetary times and insurance chronicle *see* The trade review and intercolonial journal of commerce

Canadian monthly and national review – Toronto: Adam, Stevenson [1872-78] – 9 – (merged with: belford's monthly magazine to become: rose-belford's canadian monthly and national review. incl ind) – mf#P05010 – cn Canadiana [971]

Canadian monthly and national review – Toronto. 1872-1878 (1) – mf#2781 – us UMI ProQuest [971]

The canadian monthly free press – Montreal: Devins & Bolton, [1877-18–?] – 9 – mf#P04368 – cn Canadiana [073]

Canadian municipal journal – Canada. 1905-feb 1914; may 1915-17 – 6r – 1 – uk British Libr Newspaper [073]

The canadian municipal journal : devoted more particularly to the exposition of municipal, school and other legislative enactments relating to local municipalities – Toronto: A.L. Willson, [1891?-1892] – 9 – (incl ind) – mf#P04741 – cn Canadiana [350]

The canadian music and drama : a monthly journal devoted to the interests of local and universal news – Kingston [Ont]: Canadian Music and Drama, [1895-189- or 19–] – 9 – mf#P04849 – cn Canadiana [790]

Canadian music and trades journal – Toronto: D C Nixon, [1900-1930?] – 9 – mf#P04203 – cn Canadiana [780]

Canadian music folio – Toronto: Canadian Music Folio Co, [18–?-18– or 19–] [mf ed nov 1892] – 9 – mf#P06072 – cn Canadiana [780]

Canadian musician – Toronto: Whaley, Royce, [1889?-1899?] – 9 – (cont by: musician (toronto, ont)) – mf#P04427 – cn Canadiana [780]

The canadian mute – Belleville, Ont: Institution for the Deaf and Dumb, [1892-19–] – 9 – mf#P04409 – cn Canadiana [360]

The canadian nation – Toronto: [s.n, 1890?-189- or 19–] – 9 – mf#P06009 – cn Canadiana [071]

Canadian National League *see* Home market and farm

Canadian national magazine – Montreal. 1957-1957 (1) – ISSN: 0703-5306 – mf#1131 – us UMI ProQuest [380]

Canadian nationalism and the war / Bourassa, Henri et al – Montreal: [s.n.], 1916 [mf ed 1984] – 1mf – 9 – mf#SEM105P392 – cn Bibl Nat [971]

Canadian nationality : the cry of labor, and other essays / Hatheway, Warren Frank – Toronto: W Briggs, 1906 – 3mf – 9 – 0-665-74433-1 – mf#74433 – cn Canadiana [320]

Canadian nationality : its growth and development / Canniff, William – Toronto: Hart & Rawlinson, 1875 – 1mf – 9 – mf#23990 – cn Canadiana [971]

Canadian Native Friendship Center *see* Edmonton native news

Canadian Native Friendship Centre (Edmonton AB) *see* Edmonton native news

The canadian naturalist : a series of conversations on the natural history of lower canada / Gosse, Philip Henry – London: J van Voorst, 1840 [mf ed 1982] – 5mf – 9 – (incl ind) – mf#37210 – cn Canadiana [500]

Canadian naturalist and geologist – Montreal: B Dawson, [1856-1868] – 9 – (cont by: the canadian naturalist and quarterly journal of science) – mf#P04260 – cn Canadiana [500]

Canadian naturalist and quarterly journal of science – Montreal: Dawson Bros., 1869-1883 – 9 – (cont: the canadian naturalist and geologist; cont by: the canadian record of science) – mf#P04261 – cn Canadiana [500]

Canadian negro – Toronto. v1-4. jun 1953-dec 1956// (mthly) – 1r – 1 – Can$135.00 – cn McLaren [305]

Canadian news – London, UK. 1856-75 – 9r – 1 – cn Library Assoc [072]

Canadian news – London. -w. 11 Jun 1856-16 Mar 1876. (16 reels) – 1 – uk British Libr Newspaper [072]

Canadian News Synthesis Project et al *see* Latin america and caribbean inside report

The canadian north west : a speech delivered by his excellency the marquis of lorne, governor general of canada, winnipeg / Argyll, John Douglas Sutherland Campbell, Duke of – Ottawa: Dept of Agriculture, 1881 – 1mf – 9 – mf#55864 – cn Canadiana [917]

The canadian north-west : its history and its troubles, from the early days of the fur trade to the era of the railway and the settler; with incidents of travel in the region, and the narrative of three insurrections / Adam, Graeme Mercer – Toronto: Rose Pub Co; Whitby [Ont]: J S Robertson, 1885 [mf ed 1982] – 5mf – 9 – mf#30264 – cn Canadiana [971]

Canadian notabilities / Dent, John Charles – Toronto: J B Magurn, 1880 – 2v on 1mf – 9 – (individuals vols also available separately) – mf#03600 – cn Canadiana [971]

Canadian notabilities, vol 1 / Dent, John Charles – Toronto: J B Magurn, 1880 [mf ed 1980] – 2mf – 9 – 0-665-03601-9 – mf#03601 – cn Canadiana [920]

Canadian notabilities, vol 2 / Dent, John Charles – Toronto: J B Magurn, 1880 [mf ed 1980] – 2mf – 9 – 0-665-03602-7 – mf#03602 – cn Canadiana [920]

Canadian notabilities, vols 1-2 / Dent, John Charles – Toronto: J B Magurn. 2v. 1880 – 1mf – 9 – 0-665-03600-0 – mf#03600 – cn Canadiana [920]

Canadian numismatic bibliography : a review of mr r w mclachlan's "canadian numismatics", and other books and pamphlets describing canadian coins and medals – Montreal: repr fr 'The Gazette,' 1886 [mf ed 1980] – 1mf – 9 – 0-665-02826-1 – mf#02826 – cn Canadiana [730]

Canadian numismatics : a descriptive catalogue of coins, tokens and medals issued in or relating to the dominion of canada and newfoundland... / McLachlan, Robert Wallace – [Montreal?: R W McLachlan], 1886 [mf ed 1980] – 2mf – 9 – mf#09447 – cn Canadiana [730]

Canadian nurse – Ottawa. 1905+ (1) 1971+ (5) 1976+ (9) – ISSN: 0008-4581 – mf#1776 – us UMI ProQuest [610]

Canadian opinions on the bill introduced into the dominion parliament by desire girouard, esq, mp, (jacques-cartier) : legalizing marriage with the sister of a deceased wife, and with the widow of a brother – [S.l: s.n, 1880?] [mf ed 1980] – 1mf – 9 – 0-665-02825-3 – mf#02825 – cn Canadiana [340]

Canadian Oral History Association *see* Bulletin of the canadian oral...

The canadian orange minstrel, for 1860 : contains nine new and original songs, mostly all of them showing some wrong that effects the order or the true course of protestant loyalty to the british crown / McBride, Robert – [London, Ont?: s.n.] 1860 [mf ed 1983] – 1mf – 9 – 0-665-38220-0 – mf#38220 – cn Canadiana [780]

The canadian orange minstrel for 1870 : written for the purpose of keeping in remembrance the dark doings and designs of popery in this country: an antidote for pamphile lemay's songs, etc / McBride, Robert – Toronto?: P H Stewart, 1870 – 1mf – 9 – mf#01499 – cn Canadiana [870]

Canadian Order of Chosen Friends. Eureka Council *see* Constitution and by-laws of eureka council, no 13

Canadian Order of Foresters *see* Preamble, constitution, endowment law, and rules of order of the right worthy high court

The canadian ornithologist – Toronto: Willing & Williamson, 1973 – 9 – mf#P04740 – cn Canadiana [590]

The canadian pacific : the new highway to the orient across the mountains, prairies and rivers of canada – [Montreal?: s.n, 1886?] [mf ed 1981] – 1mf – 9 – mf#14587 – cn Canadiana [380]

The canadian pacific and north shore railways : correspondence relating to the efforts of the canadian pacific railway co to reach quebec / Canadian Pacific Railway Co – Montreal?: s.n, 1885 – 1mf – 9 – mf#00450 – cn Canadiana [380]

Canadian Pacific Railway *see*
– Annotated time table

Canadian pacific railway : contract with the syndicate / Campbell, Alexander – [S.l: s.n, 1881?] [mf ed 1980] – 1mf – 9 – 0-665-00396-X – mf#00396 – cn Canadiana [380]

Canadian pacific railway : contract with the syndicate: speech / Campbell, Alexander – Ottawa?: s.n, 1881? – 1mf – 9 – mf#05695 – cn Canadiana [380]

Canadian pacific railway : hon sir john a macdonald's speech, ottawa, 17th january 1881 – [Ottawa?: s.n, 1881?] [mf ed 1980] – 1mf – 9 – mf#09332 – cn Canadiana [380]

Canadian pacific railway : reports in reference to location of second section west of red river / Fleming, Sandford – [Ottawa?: s.n.], 1880 [mf ed 1980] – 1mf – 9 – 0-665-05454-8 – mf#05454 – cn Canadiana [380]

The canadian pacific railway : address at the annual convention at milwaukee, wisconsin, june 28, 1888 / Keefer, Thomas C – New York: s.n, 1888? – 1mf – 9 – mf#07800 – cn Canadiana [625]

Canadian pacific railway and the new northwest – [S.l: s.n, 1882?] [mf ed 1981] – 1mf – 9 – mf#15375 – cn Canadiana [380]

Canadian pacific railway annotated time table : with information as to cpr transcontinental routes / Compagnie du chemin de fer canadien du Pacifique – [S.l: s.n, 1897?] [mf ed 1987] – 1mf – 9 – 0-665-63826-4 – mf#63826 – cn Canadiana [380]

Canadian Pacific Railway Co *see*
– The canadian pacific and north shore railways
– A time-table with notes of the transcontinental trains, the great lakes route, and the montreal and toronto line
– A time-table, with notes of the westbound transcontinental train
– A time-table with notes of the westbound transcontinental train, the great lakes route and the boston and toronto lines

Canadian Pacific Railway Company *see* Joint passenger tariff to the canadian north-west, northern minnesota, dakota and transcontinental points via canadian routes

Canadian pacific railway company : to the shareholders / Mount Stephen, George Stephen, Baron – S.l: s.n, 1887? – 1mf – 9 – mf#26757 – cn Canadiana [380]

The canadian pacific railway company and its extraordinary telegraphic and telephonic privileges : a letter to the hon sir charles tupper, from the representatives of telegraph companies in canada / Crawford, John & Wiman, Erastus – [S.l: s.n, 1884?] [mf ed 1980] – 1mf – 9 – mf#03625 – cn Canadiana [380]

Canadian pacific railway, ottawa, 1st july, 1880 / Fleming, Sandford – [Ottawa?: s.n, 1880?] [mf ed 1980] – 1mf – 9 – 0-665-03127-0 – mf#03127 – cn Canadiana [380]

Canadian pacific railway summer tours, vol 4 : western tours – S.l: Canadian Pacific Railway Co, 1898 – 1mf – 9 – (incl ind) – mf#55905 – cn Canadiana [380]

The canadian pacific railway telegraph : remarks on its present condition and the necessity for an immediate change of location / Burrows, Charles Acton – Ottawa: Citizen Print & Pub Co, 1880 – 1mf – 9 – mf#03828 – cn Canadiana [380]

Canadian pacific resolutions : thorough sifting of a great scandal: astounding discrepancies in government estimates: convincing argument for rejection of terms / Blake, Edward – [S.l: s.n, 1884?] [mf ed 1979] – 1mf – 9 – 0-665-00158-4 – mf#00158 – cn Canadiana [380]

Canadian Pacific Telegraph *see* List of offices in canada and tariff

Canadian packaging – Toronto. 1975-1996 (1) 1975-1995 (5) 1975-1995 (9) – ISSN: 0008-4654 – mf#10772 – us UMI ProQuest [680]

Canadian paint and finishing – Toronto. 1975-1978 (1) 1975-1978 (5) 1975-1978 (9) – (cont by: coatings in canada) – ISSN: 0008-4662 – mf#10773 – us UMI ProQuest [660]

Canadian papermaker – Toronto. 1992-1996 (1,5,9) – (cont: pulp and paper journal) – ISSN: 1191-887X – mf#10776,02 – us UMI ProQuest [670]

Canadian paperworker journal – v1 n7-v2 n2 [1976 mar-1977 mar] – 1r – 1 – (cont by: c p u journal, canadian paperworkers union) – mf#505506 – us WHS [071]

Canadian Paperworkers Union *see*
– C p u journal
– Intercom

The canadian parliamentary companion, 1883 / ed by Gemmill, John Alexander – Ottawa: J Durie, 1883 – 5mf – 9 – (established 1862. incl ind) – mf#32957 – cn Canadiana [971]

Canadian Party of Labour *see* Worker

Canadian patent law and practice / Fisher, Harold & Smart, Russel Sutherland – Toronto: Canada Law Book Co, 1914 [mf ed 1995] – 6mf – 9 – 0-665-76834-6 – (app by william joseph lynch) – mf#76834 – cn Canadiana [346]

The canadian patent office record – [Montreal: G E Desbarats, 1873-1960] – 9 – (some text in french. incl ind) – mf#P04863 – cn Canadiana [346]

The canadian patent office record and mechanics' magazine – Montreal: G E Desbarats. v1 n1 mar 1873-v3 n12 dec 1875 (mthly) [mf ed 1990] – 33mf – 9 – mf#P04862 – cn Canadiana [600]

Canadian pen and ink sketches / Fraser, John – [Montreal?: s.n.], 1890 [mf ed 1980] – 5mf – 9 – 0-665-03177-7 – mf#03177 – cn Canadiana [971]

Canadian periodical index – 1938-47, 1948-59 – 2r – 1 – cn Library Assoc [030]

Canadian personnel and industrial relations journal – Toronto. 1954-1981 (1) 1971-1981 (5) 1975-1981 (9) – ISSN: 0008-4727 – mf#2221 – us UMI ProQuest [650]

Canadian pharmaceutical journal – Toronto: J M Trout, [1868-1984] – 9 – (cont in pt by: canadian journal of pharmaceutical sciences; cont by: cpj: canadian pharmaceutical journal. incl ind) – mf#P05106 – cn Canadiana [615]

The canadian philatelic and curio advertiser – Montreal: A L Hamilton, [1886?-188-?] – 9 – ISSN: 1190-6561 – mf#P04572 – cn Canadiana [760]

Canadian philatelic journal – [St. Catharines, Ont] H E French, [1888-188-?] – 9 – (cont: the canadian philatelist (niagara falls, ont); absorbed by: the niagara falls philatelist) – mf#P04152 – cn Canadiana [760]

Canadian philatelic journal *see* The niagara falls philatelist

The canadian philatelic journal : published on the 25th of each month in the interests of stamp collectors – Merritton, Ont: Canadian Pub Co, [1894] – 9 – ISSN: 1190-7037 – mf#P04561 – cn Canadiana [760]

The canadian philatelic magazine – Halifax, N.S: A M Muirhead, [1893-1901?] – 9 – mf#P04554 – cn Canadiana [760]

Canadian philatelic review – Berlin [Kitchener], Ont: F I Weaver, [1899] – 9 – (cont: the canadian philatelic weekly) – mf#P04939 – cn Canadiana [760]

Canadian Philatelic Society of Great Britain *see* Maple leaves

Canadian philatelic weekly – Berlin [Kitchener], Ont: F I Weaver, [1898-1899] – 9 – (cont: the boys own philatelist; cont by: the canadian philatelic review) – mf#P05152 – cn Canadiana [760]

The canadian philatelic weekly – London [Ont]: L M Staebler, [1894] – 9 – mf#P04556 – cn Canadiana [760]

The canadian philatelic weekly – Toronto: Canadian Philatelic Weekly, [1898-189- or 19–] – 9 – mf#P04565 – cn Canadiana [760]

Canadian philatelist – Niagara Falls, Ont: Canadian Philatelic Co, [1888] – 9 – (cont by: canadian philatelic journal (st catherines, ont)) – mf#P04566 – cn Canadiana [760]

Canadian philatelist – Whitby, Ont: L F Barker, [1884?-1885] – 9 – (cont by: the canadian philatelist and numismatist) – mf#P04559 – cn Canadiana [760]

The canadian philatelist : an illustrated monthly magazine devoted to stamp collecting – Quebec: International Stamp Co, [1872-1873?] – 9 – mf#P04557 – cn Canadiana [760]

The canadian philatelist : a monthly magazine for stamp collectors – Toronto: G A Lowe, [1886] – 9 – ISSN: 1190-7029 – mf#P04555 – cn Canadiana [760]

The canadian philatelist : official organ of the philatelic society of canada – London, Ont: L M Stoebler, [1891-1896] – 9 – ISSN: 1190-626X – mf#P04550 – cn Canadiana [760]

The canadian philatelist – Quebec: Birt, Williams & Co, [1872] – 9 – ISSN: 0701-3590 – mf#P04562 – cn Canadiana [760]

Canadian philatelist and numismatist – Whitby, Ont: L F Barker, [1885] – 9 – (cont: canadian philatelist (whitby, ont)) – mf#P04558 – cn Canadiana [760]

The canadian phonetic pioneer : a monthly journal, devoted to the spread of the writing, printing, and spelling reform – Oshawa, C W [Ont]: W H Orr, [1858-18–] – 9 – mf#P04974 – cn Canadiana [400]

Canadian photographic journal – Toronto: Geo W Wilson. v1-6. feb 1892-feb 1897// – (= ser Professional photographer) – 3r – 1 – Can$260.00 – (absorbed by: professional photographer, buffalo, ny) – cn McLaren [770]

Canadian photographic standard – Montreal: D H Hogg, [1893?-1899?] – 9 – (cont: canadian photo standard; ceased 1899?) – mf#P04128 – cn Canadiana [770]

Canadian photography – Toronto. 1975-1983 (1,5,9) – ISSN: 0031-8582 – mf#10774 – us UMI ProQuest [770]

Canadian phrenological and psychological magazine – [S.l: s.n, 1891-189- or 19–] [mf ed [jul 1891]-[dec 1891] – 9 – mf#P04364 – cn Canadiana [150]

Canadian pictorial – Montreal, Canada. -m. 1908-nov 1916 – 4r – 1 – uk British Libr Newspaper [072]

Canadian pictorial and illustrated war news – Toronto: Grip Printing & Pub Co. v1 n1-18. apr 4-aug 1 1885// (wkly) – 1r – 1 – Can$85.00 – cn McLaren [971]

Canadian pictures : drawn with pen and pencil / Argyll, John Douglas Sutherland Campbell, duke of – London: Religious Tract Society, [1882?] [mf ed 1986] – 3mf – 9 – 0-665-52541-9 – (incl ind) – mf#52541 – cn Canadiana [740]

Canadian pictures : drawn with pen and pencil / Argyll, John Douglas Sutherland Campbell, duke of – London: Religious Tract Society, [1884?] [mf ed 1980] – 3mf – 9 – 0-665-02222-0 – mf#02222 – cn Canadiana [740]

Canadian pictures, drawn with pen and pencil / Argyll, John Douglas Sutherland Campbell, duke of – London: Religious Tract Society, 1885 [mf ed 1981] – 3mf – 9 – (incl ind) – mf#26270 – cn Canadiana [740]

The canadian pioneers / Casgrain, Henri Raymond – Montreal?: C O Beauchemin, 1896 – 1mf – 9 – (trans fr french by a w l gompertz) – mf#10423 – cn Canadiana [971]

Canadian plastics – Don Mills. 1991-1992 (1,5,9) – ISSN: 0008-4778 – mf#18760 – us UMI ProQuest [660]

Canadian pleistocene / Dawson, John William – London: Trubner, 1883 [mf ed 1987] – 1mf – 9 – 0-665-64664-X – (incl bibl ref) – mf#64664 – cn Canadiana [550]

Canadian poems : respectfully dedicated to w s griffin, wesleyan methodist minister, port hope / Breeze, James T – [Port Hope, Ont?: s.n, 1866?] [mf ed 1985] – 1mf – 9 – 0-665-50557-4 – mf#50557 – cn Canadiana [810]

Canadian poets in miniature / Clio – [S.l: s.n, 1892?] [mf ed 1980] – 1mf – 9 – 0-665-00685-3 – (fr: dominion illustrated monthly, nov 1892) – mf#00685 – cn Canadiana [810]

The canadian polar expedition : or, will canada claim her own / Browne, R H C – Ottawa, Canada: [s.n.], 1901 [mf ed 1985] – 1mf – 9 – mf#SEM105P476 – cn Bibl Nat [919]

Canadian political history : outlines of a course of ten lectures delivered in connection with the educational work of the young men's christian association of montreal during the autumn of 1894 / Ames, Herbert Brown – [Montreal]: The Association, [1894] – 1mf – 9 – 0-665-04012-1 – mf#04012 – cn Canadiana [323]

Canadian politics in war and peace : letter from the ex-minister of finance... / Fielding, William Stevens – [Halifax NS?: s.n, 1918?] [mf ed 1995] – 1mf – 9 – 0-665-74251-7 – mf#74251 – cn Canadiana [325]

The canadian portland cement co, limited : deseronto, ontario, canada, works at marlbank and strathcona, ontario – [Montreal?: s.n, 1901?] [mf ed 1991] – 1mf – 9 – 0-665-99515-6 – mf#99515 – cn Canadiana [660]

The canadian portrait gallery / Dent, John Charles – Toronto: J B Margurn, 1880-1881 – 4v on 1mf – 9 – (individuals vols also available separately) – mf#07403 – cn Canadiana [971]

The canadian poultry chronicle : a monthly journal devoted to poultry and pigeon breeding – Toronto: Publ...by the Globe Print Co, [1870-1872?] – 9 – (incl ind) – mf#P04322 – cn Canadiana [636]

Canadian poultry review – Strathroy, Ont: J Fullerton, 1877-1975 – 9 – mf#P04175 – cn Canadiana [636]

Canadian poultry review *see* Pigeon fancier

The canadian presbyter – Montreal: J Lovell, 1857-1858 – 9 – mf#P04077 – cn Canadiana [242]

The canadian presbyterian magazine : especially devoted to the interests of the united presbyterian church – Toronto: [J Cleland, 1851-1854] – 9 – mf#P04398 – cn Canadiana [242]

Canadian Press Association *see* Journal of proceedings at 39th annual meeting held at toronto, february 4th and 5th, 1897

Canadian printer – Toronto. 1989-1994 (1,5,9) – (cont: canadian printer and publisher) – ISSN: 0849-0767 – mf#10775,01 – us UMI ProQuest [680]

The canadian printer – Montreal: C T Palsgrave, [1860?-18–?] – 9 – mf#P05117 – cn Canadiana [680]

Canadian printer and publisher – Toronto. 1975-1989(1,5,9) – (cont by: canadian printer) – ISSN: 0008-4816 – mf#10775 – us UMI ProQuest [680]

The canadian prison sunday – Toronto: Prisoners' Aid Association of Canada, [1892-189- or 19–] – 9 – mf#P04224 – cn Canadiana [360]

Canadian Protesting Committee *see* An account of the canadian protest

Canadian psychiatric association journal = Revue de l'association des psychiatres du canada / Association des psychiatres du Canada – Ottawa. 1956-1978 (1) 1970-1978 (5) 1976-1978 (9) – (cont by: canadian journal of psychiatry) – ISSN: 0008-4824 – mf#5875 – us UMI ProQuest [616]

Canadian psychological review = Psychologie canadienne – Montreal. 1975-1979 (1) 1975-1979 (5) 1977-1979 (9) – (cont: canadian psychologist = psychologie canadienne. cont by: canadian psychology = psychologie canadienne) – ISSN: 0318-2096 – mf#1449,01 – us UMI ProQuest [150]

Canadian psychologist psychologie canadienne – Calgary. 1960-1974 (1) 1971-1974 (5) – (cont by: canadian psychological review = psychologie canadienne) – ISSN: 0008-4832 – mf#1449 – us UMI ProQuest [150]

Canadian psychology = Psychologie canadienne – Ottawa. 1980+ (1,5,9) – (cont: canadian psychological review = psychologie canadienne) – ISSN: 0708-5591 – mf#1449,02 – us UMI ProQuest [150]

Canadian public lands : speech delivered in the house of commons by mr j b plumb, mp, on monday, 5th april, 1880 / Plumb, Josiah Burr – [S.l: s.n, 1880?] [mf ed 1980] – 1mf – 9 – 0-665-02294-8 – mf#02294 – cn Canadiana [333]

Canadian pulp and paper industry – Toronto. 1975-1981 (1,5,9) – (cont by: pulp and paper journal) – ISSN: 0008-4867 – mf#10776 – us UMI ProQuest [670]

Canadian quaker history newsletter / Canadian Friends Historical Association – n6-9,11-40,41-42 [1973 dec-1974 sep, 1975 mar-1986 dec, 1987 sum-1988 win] – 1r – 1 – (cont by: canadian quaker history journal) – mf#1611546 – us WHS [243]

The canadian quarterly review – Toronto?: s.n, 1856?-18– or 19– – 9 – mf#P04451 – cn Canadiana [400]

The canadian quarterly review and family magazine – Hamilton, Ont?: Pub for G D Griffin...by Donnelley & Lawson, [1863-1866] – 9 – mf#P04230 – cn Canadiana [073]

The canadian queen : a magazine of fashion, art, literature, etc – Toronto. v2-5. sep 1890-jun 1892 (mthly) – 1r – 1 – Can$82.00 – cn McLaren [360]

Canadian radio guide – Toronto. v1-2 n17. dec 19 1931-feb 18 1933//? – 1r – 1 – Can$130.00 – cn McLaren [790]

Canadian rail / Canadian Railroad Historical Association – 1949-60, 1977-79, 1980-82, 1983-88 – 3r – 1 – (cont: crha news report) – mf#1054020 – us WHS [380]

Canadian rail / Canadian Railroad Historical Association, Inc – Montreal: the Association. n135 jul/aug 1962- (mthly) [mf ed 1995] – 1 – (cont: crha news report; has suppl: association news, 197-?-1973; has suppl: crha communications, 1974-1979) – mf#SEM35P417 – cn Bibl Nat [380]

Canadian Railroad Historical Association *see* Canadian rail

Canadian Railroad Historical Association, Inc *see* Canadian rail

The canadian railroad historical association, inc – [Montreal: the Association] oct 1949-May 1951 (mthly) [mf ed 1995] – 1r – 1 – mf#SEM35P414 – cn Bibl Nat [380]

Canadian railway and marine world – Toronto, Canada. jul 1914-1921 – 10 1/2r – 1 – uk British Libr Newspaper [071]

Canadian railway and marine world *see* Railway and shipping world

Canadian railway and steam navigation guide *see* Robertson's canadian railway and steam navigation guide

Canadian railway and steamboat guide – Montreal: publ by H Rose, [1855?-18– or 19–] – 9 – (ceased 18–?) – ISSN: 1190-755X – mf#P04331 – cn Canadiana [380]

The canadian railway problem / Biggar, Emerson Bristol – Toronto: Macmillan Co of Canada, c1917 – 3mf – 9 – 0-665-73580-4 – (incl app) – mf#73580 – cn Canadiana [380]

Canadian railwayman / International Non-operating Railway Unions in Canada – 1976 mar-1986 sep – 1r – 1 – mf#1230866 – us WHS [380]

Canadian reader – Toronto. 1959-1979 [1]; 1971-1979 [5]; 1978-1979 [9] – ISSN: 0008-4891 – mf#1821 – us UMI ProQuest [070]

Canadian Real Estate Association *see* Crea reporter

Canadian realtor – Toronto. 1955-1971 (1) – ISSN: 0008-4905 – mf#7180 – us UMI ProQuest [333]

Canadian reciprocity : remarks of hon n s townshend of ohio in the house of representatives, february 24, 1853, on the bill establishing reciprocrl [sic] trade with the british north american provinces, on certain conditions / Townshend, Norton Strange – [Washington?: s.n.], 1853 [mf ed 1987] – 1mf – 9 – 0-665-39228-1 – mf#39228 – cn Canadiana [337]

Canadian reciprocity : why some canadians want reciprocity: why englishmen want it: why we don't want it – Philadelphia: American Iron & Steel Association, [18–] [mf ed 1979] – 1mf – 9 – 0-665-00789-2 – mf#00789 – cn Canadiana [337]

Canadian reciprocity treaty : remarks of hon george f edmunds, of vermont, in the senate of the united states, january 22, 1875 / Edmunds, George Franklin – Washington: GPO, 1875 [mf ed 1982] – 1mf – 9 – mf#32198 – cn Canadiana [337]

Canadian record of science : including the proceedings of the natural history society of montreal and replacing the canadian naturalist – Montreal: Natural History Society, [1884?-1916] – 9 – (cont: canadian naturalist and quarterly journal of science. suspended: 1898, and in 1905-1913. incl ind) – mf#P04195 – cn Canadiana [500]

Canadian records of the united society for the propagation of the gospel : e series reports c1901-1952 [brram] / United Society for the Propagation of the Gospel. Archives – 14r – 1 – (int by peter lyon) – mf#97369 – uk Microform Academic [220]

The canadian remembrancer : a loyal sermon preached on st george's day, april 23, 1826, at the episcopal church in york / Phillips, T – York Toronto: R Stanton, 1826 – 1mf – 9 – mf#37203 – cn Canadiana [240]

The canadian repealer's almanac : for the year 1856, being leap year: containing statistics, essays, and memoranda... / Mackenzie, William Lyon [comp] – Toronto: Printed and publ by compiler, 1856? – 1mf – 9 – (incl ind) – mf#43391 – cn Canadiana [971]

Canadian reports : appeal cases / Canada. General – v1-14. 1807-1905; 10v. 1906-13 (all publ) – 140mf – 9 – $210.00 – mf#LLMC 81-008 – us LLMC [324]

Canadian research – Toronto. 1975-1989(1,5,9) – (cont: canadian research and development) – ISSN: 0319-1974 – mf#10777,01 – us UMI ProQuest [574]

Canadian research and development – Toronto. 1975-1975 (1) 1975-1975 (5) 1975-1975 (9) – (cont by: canadian research) – ISSN: 0008-493X – mf#10777 – us UMI ProQuest [574]

Canadian revenues : a bill intituled an act to amend an act...for establishing a fund towards defraying the charges of the administration of justice and support of the civil government within the province of quebec in america – [s.l.]: [s.n.], 1831 [mf ed 1984] – 1mf – 9 – mf#SEM105P433 – cn Bibl Nat [336]

Canadian review and journal of literature – Montreal: [s.n.] (Montreal: John C Becket) v1 n1 jul 1855-v1 n6 dec 1855 [197-] – 1r – 1 – (cont: literary supplement to the montreal witness; cont by: canadian review and literary supplement to montreal witness; suppl to: montreal witness) – mf#SEM35P6 – cn Bibl Nat [410]

Canadian review and journal of literature *see* The montreal witness

Canadian review and literary and historical journal – Montreal: H H Cunningham, [1824-1825] – 9 – (cont by: canadian review and magazine) – mf#P04968 – cn Canadiana [971]

Canadian review and literary supplement to Montreal witness *see* The montreal witness

Canadian review and literary supplement to montreal witness – Montreal: [s.n.] (Montreal: J C Becket. v2 n1. jan 1856- 1856? [mf ed 197-] – 1r – 1 – (cont: canadian review and journal of literature; supplement to: montreal witness) – mf#SEM35P6 – cn Bibl Nat [410]

Canadian review and magazine – Montreal: Printed for the Proprietor at the Office of the Montreal Gazette, [1826] – 9 – (cont: canadian review and literary and historical journal; ceased 1826?) – mf#P04987 – cn Canadiana [971]

Canadian review of music and art – Toronto. v1-6. feb 1942-jan 1948// – (= ser Northern Review) – 2r – 1 – Can$155.00 – (absorbed by: northern review) – cn McLaren [780]

Canadian review of sociology and anthropology = Revue canadienne de sociologie et d'anthropologie – Toronto. 1964+ (1) 1964+ (5) 1964+ (9) – ISSN: 0008-4948 – mf#5887 – us UMI ProQuest [301]

The canadian revolt : a short review of its causes, progress, and probable consequences – [s.l: s.n, 1838?] [mf ed 1984] – 1mf – 9 – 0-665-44248-3 – mf#44248 – cn Canadiana [971]

Canadian rockies : new and old trails / Coleman, Arthur Philemon – Toronto: H Frowde, 1911 – 6mf – 9 – 0-665-71139-5 – (incl ind) – mf#71139 – cn Canadiana [917]

The canadian rockies : new and old trails / Coleman, Arthur Philemon – Toronto: H Frowde, 1912 – 6mf – 9 – 0-665-74131-6 – mf#74131 – cn Canadiana [917]

The canadian royal arcanum journal : devoted to the interests of the royal arcanum in canada – Toronto: [s.n, 1894-189- or 19–] – 9 – mf#P05973 – cn Canadiana [060]

Canadian rural education : a social study / Sutherland, John Campbell – Montreal: Montreal News Co, 1913 – 1mf – 9 – 0-665-73419-0 – mf#73419 – cn Canadiana [370]

Canadian ruthenian = Kanady-isky-i rusyn – Winnipeg, MB. 1911-30 – 14r – 1 – (in ukrainian) – ISSN: 0845-9517 – cn Library Assoc [071]

The canadian school geography / Ewing, Thomas – Montreal: Armour & Ramsay, 1843 [mf ed 1984] – 1mf – 9 – 0-665-44860-0 – mf#44860 – cn Canadiana [910]

Canadian science monthly : devoted to the interests of the canadian postal college, teachers and naturalists – Wolfville, NS: A J Pineo, [1884-1885] – 9 – (cont: the acadian scientist) – mf#P04165 – cn Canadiana [500]

The canadian senator : or, a romance of love and politics / Oakes, Christopher – Toronto: National Pub Co, 1890? – 3mf – 9 – mf#30422 – cn Canadiana [830]

Canadian service employee – 1977 jan-1985 jan, 1985 feb-1988 feb – 2r – 1 – (cont: canada works) – mf#963454 – us WHS [331]

Canadian shoe and leather journal – Toronto: J Acton, 1888-1910 – 9 – (cont by: canadian footwear journal) – mf#P04108 – cn Canadiana [680]

Canadian social studies – North York. 1991+ (1,5,9) – (cont: history and social science teacher) – ISSN: 1191-162X – mf#11569,02 – us UMI ProQuest [300]

Canadian Society for Immunology *see* Bulletin of the canadian society for immunology

Canadian Society of Authors *see* Report on copyright

Canadian Society of Civil Engineers *see*
– Report of committee on a standard specification for portland cement
– Standard portland cement tests

Canadian Sociology and Anthropology Association *see* Bulletin canadian sociology and anthropology association

Canadian son of temperance – Toronto: [s.n, 1852] [mf ed v2 n1 jan 5 1852-v2 n30 dec 20 1852] – 9 – (cont: canadian son of temperance and literary gem; cont by: the canadian son of temperance[and] literary gem) – mf#P04333 – cn Canadiana [360]

Canadian son of temperance [and] literary gem – Toronto: [s.n, 1853] – 9 – (cont by: the son of temperance and canadian literary gem; cont: canadian son of temperance) – mf#P04334 – cn Canadiana [230]

Canadian son of temperance [and] literary gem – Toronto: [s.n, 1853] [mf ed v3 n1 jan 3 1853-v3 n52 dec 27 1853] – 9 – (cont: canadian son of temperance) – mf#P04334 – cn Canadiana [360]

Canadian son of temperance and literary gem – Toronto: [s.n, 1851] [mf ed v1 n1 feb 26 1851-v1 n24 dec 29 1851] – 9 – (cont by: canadian son of temperance) – mf#P06070 – cn Canadiana [360]

Canadian sportsman and livestock journal – Toronto: E K Dodds, [187-?-19–] [mf ed v8 n116 oct 12 1883; christmas no dec 23 1898] – 9 – mf#P05090 – cn Canadiana [636]

Canadian statesman – Bowmanville, ON. 1868-1900 – 16r – 1 – ISSN: 0834-5651 – cn Library Assoc [971]

Canadian stock-raisers' journal – Hamilton, Ont: Stock Journal Co, [1883?-1884?] [mf ed v1 n4 feb 1884; v1 n8 jun 1884] – 9 – mf#P04578 – cn Canadiana [636]

The canadian student / Dawson, John William – Montreal?: s.n, 1892? – 1mf – 9 – mf#03959 – cn Canadiana [378]

Canadian studies in comparative politics / Bourinot, John George – Montreal: Dawson Bros, 1890 [mf ed 1979] – 1mf – 9 – 0-665-00226-2 – mf#00226 – cn Canadiana [320]

Canadian summer resort guide : illustrated souvenir and guide book of some of the principal fishing, hunting, health and pleasure resorts and tourist and excursion routes of canada... – 7th ed. Toronto: F Smily, 1900 [mf ed 1987] – 2mf – 9 – 0-665-61946-4 – mf#61946 – cn Canadiana [639]

Canadian summer resorts : illustrated souvenir and guide book of some of the principal resorts of ontario... / ed by Smily, Frederick – 2nd ed. Toronto: F Smily, 1895 [mf ed 1981] – 2mf – 9 – mf#13695 – cn Canadiana [917]

Canadian Synod. (Pres. Church in the USA) *see* Minutes, 1907-55

The canadian system of banking and the national banking system of the united states : a comparison with reference to the banking requirements of canada / Walker, Byron Edmund – Toronto?: Trout & Todd, 1890 – 1mf – 9 – mf#25419 – cn Canadiana [332]

The canadian tariff / Galt, Alexander Tilloch – London?: Women's Printing Society, 1879? – 1mf – 9 – mf#28315 – cn Canadiana [336]

Canadian ten cent ball-room companion and guide to dancing : comprising rules of etiquette, hints on private parties, toilettes for the ball-room, etc – Toronto: W Warwick, 1871 – 1mf – 9 – mf#01094 – cn Canadiana [790]

Canadian theosophist / Theosophical Society in Canada – 1920 may 15 – 1r – 1 – mf#3910420 – us WHS [290]

Canadian thresherman and farmer : canada's farm machinery magazine, winnipeg, canada – Winnipeg: E H Heath, 1[1902-1919] – 9 – (cont by: canadian power farmer) – mf#P04973 – cn Canadiana [630]

Canadian timber trees : their distribution and preservation / Drummond, Andrew Thomas – [Montreal?: s.n.], 1879 [mf ed 1980] – 1mf – 9 – 0-665-02766-4 – mf#02766 – cn Canadiana [634]

Canadian tit-bits – [Toronto?: s.n, 1891-189- or 19–] [mf ed v1 n1 may 23 1891] – 9 – mf#P04454 – cn Canadiana [071]

Canadian token – 1983-85, 1986-1987 sep – 2r – 1 – (cont by: transactions of the canadian numismatic research society, numismatica canada) – mf#1238439 – us WHS [730]

Canadian trade progress : a series of articles reprinted from the columns of the "journal of commerce" of montreal... – Montreal: s.n, 1895 – 1mf – 9 – mf#63079 – cn Canadiana [380]

The canadian trade review – Montreal: H Harvey, [1885-1906?] – 9 – mf#P04949 – cn Canadiana [670]

Canadian transport / Canadian Brotherhood of Railway, Transport and General Workers – 1963 sep 16-1971 oct 15, 1971 nov-1980 dec, 1981-88, 1989 jan-1994 jun – 4r – 1 – (cont: canadian railway employees' monthly; cont by: transport canadien) – mf#1054025 – us WHS [380]

Canadian transport = Transport canadien – Ottawa. 1973-1993 (1) – ISSN: 0045-5466 – mf#8445 – us UMI ProQuest [380]

Canadian travel courier – Toronto. 1976-1989 (1) 1976-1989 (5) 1976-1989 (9) – (cont by: travel courier) – ISSN: 0008-5219 – mf#10778 – us UMI ProQuest [917]

Canadian treasure – v1-3 [collector's n1-7], 1973 sum-1975 – 1r – 1 – (cont by: canadian illustrated news) – mf#177166 – us WHS [071]

Canadian tribune : canadian communist paper – Toronto, Ontario, CN. jan 1940-dec 1989 – 33r – 1 – cn Commonwealth Imaging [071]

Canadian tribune – v67 n2579-2652 [1988 jan 11-jun 26] – 1r – 1 – mf#1581247 – us WHS [071]

Canadian ufo report – Duncan. 1969-1970 (1) 1969-1970 (5) (9) – ISSN: 0008-5243 – mf#7580 – us UMI ProQuest [000]

Canadian ukrainian – Winnipeg, Canada. 1 may 1920-15 feb 1922 – 1 1/2r – 1 – uk British Libr Newspaper [072]

The canadian ukrainian collection / Canadian Institute for Ukrainian Studies. University of Alberta – 1906-70 – 170r – 1 – cn Commonwealth Imaging [080]

Canadian union news / Oil, Chemical and Atomic Workers International Union – 1976 aug-1979 may – 1r – 1 – mf#499088 – us WHS [331]

Canadian Union of Postal Workers *see* Cupw perspective

Canadian united presbyterian magazine – Toronto: C Fletcher, [1854?-1861] [mf ed v1 n1 jan 1854-v8 n12 dec 1861] – 9 – (incl ind) – mf#P04980 – cn Canadiana [242]

The canadian united service magazine – [Ottawa?]: United Service Club, [1895?-189- or 19–] – 9 – ISSN: 1190-7282 – mf#P04130 – cn Canadiana [355]

Canadian university music review = Revue de musique des universites canadiennes – Ottawa. 1980+ (1,5,9) – ISSN: 0710-0353 – mf#12815 – us UMI ProQuest [780]

Canadian uutiset – Port Arthur, ON. v1- . nov 11 1915- (wkly) [mf ed jan 3 1918-dec 29 1927] – 10r – 1 – Can$795.00 – (many issues from 1923-27 have damaged or missing pages. the longest history of continuous publ of any finnish canadian newspaper. publ suspended oct 17-nov 28 1918 under order-in-council. english translation appears in columns adjoining finnish text, dec 5 1918-apr 10 1919) – cn McLaren [071]

Canadian veterinary journal = Revue veterinaire canadienne – Ottawa. 1968+ (1) 1971+ (5) 1975+ (9) – ISSN: 0008-5286 – mf#2754 – us UMI ProQuest [636]

A canadian view of annexation / Bender, Prosper – Boston?; New York?: s.n, 1883? – 1mf – 9 – mf#07991 – cn Canadiana [327]

Canadian vocational journal = Journal de l'association canadienne de la formation professionnelle – Ottawa. 1984-1996 – 1,5,9 – ISSN: 0045-5520 – mf#14235,01 – us UMI ProQuest [370]

The canadian weekly stamp news – Toronto: W R Adams, [1896-1897?] – 9 – ISSN: 1190-7495 – mf#P04568 – cn Canadiana [760]

Canadian wesleyan – Hamilton, UC [Ont]: H Ryan, [183-18–] [mf ed v2 n5 nov 8 1832] – 9 – mf#P06063 – cn Canadiana [242]

The canadian wesleyan hymn book : or, mr wesley's hymn book republished... – York [Toronto]: printed for the Canadian Wesleyan Methodists, 1831 – 6mf – 9 – 0-665-89701-4 – (incl ind) – mf#89701 – cn Canadiana [242]

Canadian Wesleyan Methodist Church *see* Minutes of the...annual conference

Canadian Wesleyan Methodist New Connexion Church *see* Minutes of the...annual conference

The canadian west : its discovery by the sieur de la verendrye: its development by the fur-trading companies, down to the year 1822 / Dugas, Georges – Montreal: Librairie Beauchemin, 1905 – 4mf – 9 – 0-665-74199-5 – (also available in french) – mf#74199 – cn Canadiana [971]

Canadian west india trading association (limited) – [Halifax, NS: s.n.], 1893 [mf ed 1980] – 1mf – 9 – 0-665-00455-9 – mf#00455 – cn Canadiana [380]

The canadian wheelman – London [Ont: Canadian Wheelmen's Association, 1883-189-?] – 9 – mf#P04231 – cn Canadiana [790]

Canadian wild flowers – Montreal: J Lovell, 1868 [mf ed 1981] – 2mf – 9 – (painted & litho by agnes fitz gibbon, with botanical descriptions by c p traill) – mf#06559 – cn Canadiana [580]

Canadian wild flowers – Montreal: J Lovell, 1869 [mf ed 1982] – 2mf – 9 – (painted & litho by agnes fitz gibbon, with botanical descriptions by c p traill) – mf#26830 – cn Canadiana [580]

Canadian wild flowers / Ross, John Hugh – [Montreal?: s.n.], 1893 [mf ed 1980] – 1mf – 9 – 0-665-00459-1 – mf#00459 – cn Canadiana [580]

Canadian wild flowers : selections from the writings of miss helen m johnson of magog, pq, canada – Boston: J M Orrock, 1884 [mf ed 1980] – 3mf – 9 – 0-665-07623-1 – (with a sketch of her life by j m orrock) – mf#07623 – cn Canadiana [810]

Canadian woman and her work – Toronto: Canadian Suffrage Association, [19–] – 1mf – 9 – 0-665-65393-X – mf#65393 – cn Canadiana [305]

Canadian Women's Army Corps *see* C w a c newsletter

Canadian woodworker – Toronto, Canada. 1911-28. -w – 18 1/2r – 1 – uk British Libr Newspaper [640]

Canadian woodworker / Woodworkers' Industrial Union of Canada – v1 n1-8 [1948 nov 3-1949 mar 2] – 1r – 1 – (cont: b c lumber worker; cont by: union woodworker [1949]) – mf#618443 – us WHS [634]

Canadian workmen's compensation acts and cases : containing comparative tables and references to the acts of british columbia, alberta, manitoba and saskatchewan / Dale, Edgar Thorniley – Winnipeg, Butterworth, 1915. 162 p. LL-2333 – 1 – us L of C Photodup [344]

Canadian yearbook of international law – v1-34. 1963-96 – 1,5,6 – $653.00 set – (v1-27 1963-89 in reel $330. v28-34 1990-96 in mf $323) – ISSN: 0069-0058 – mf#101391 – us Hein [341]

Canadiana : a collection of canadian notes – [Montreal?: Gazette Print Co], 1889-1890 – 9 – (incl french text) – mf#P04190 – cn Canadiana [971]

Canadiana : containing sketches of upper canada and the crisis in its political affairs / Wells, William Benjamin – London: printed by C & W Reynell, 1837 [mf ed 1982] – 3mf – 9 – mf#34142 – cn Canadiana [971]

Canadiana authorities = Canadiana vedettes d'autorite / National Library of Canada – [Ottawa: National Library of Canada] 1979-[2000] [trimthly] (0225-1574) – (cont: bibliotheque nationale du canada c a n / m a r c: authorities [ottawa: bibliotheque nationale du canada] 0701-810x; canadiana authorities [ottawa: national library of canada] 1979-[2000] 0225-1574) – cn Library and Archives [025]

Canadiana germanica / German-Canadian Historical Association – n21-46 [1979 feb-1985 jun] – 1r – 1 – (cont: mitteilungsblatt der historical society of mecklenburg upper canada) – mf#963392 – us WHS [971]

Canadiana-americana : an unusually important offering – [S.l: s.n, 1898?] [mf ed 1986] – 1mf – 9 – 0-665-54458-8 – mf#54458 – cn Canadiana [971]

Canadian-american law journal – v1-4. 1982-1988 (all publ) – 9 – $40.00 set – mf#110381 – us Hein [340]

Canadian-american slavic studies = Revue canadienne-americaine d'etudes slaves – Pittsburgh. 1967-1979 (1) 1970-1979 (5) 1975-1979 (9) – ISSN: 0090-8290 – mf#3450 – us UMI ProQuest [327]

Canadianco-operator : a magazine of social and economic progress / Cooperative Union of Canada – 1932 nov – 1r – 1 – mf#3910406 – us WHS [334]

Canadians all – Toronto, aut 943-sum/aut 1946// – 1r – 1 – Can$55.00 – (cont: poles in canada) – cn McLaren [305]

The canadian's right the same as the englishman's : a dialogue between a barrister at law, and a juryman / Hawles, John – York, UC Toronto: C Fothergill, 1823 [mf ed 1984] – 1mf – 9 – 0-665-44953-4 – (first publ in london, 1680, under title: the englishman's right. incl bibl ref) – mf#44953 – cn Canadiana [347]

Canadicae missionis relatio ab anno 1611 usque ad annum 1613 : cum statu ejusdem missionis, annis 1703 & 1710 / Jouvancy, Joseph de – Romae: Ex typographia Georgii Plachi, 1710 [mf ed 1984] – 1mf – 9 – 0-665-20065-X – (original ed: romae : ex typographia georgii plachi, 1710) – mf#20065 – cn Canadiana [241]

Le canadien – Quebec, QC. 1806-25 – 3r – 1 – cn Library Assoc [071]

La canadienne : samedi, 8e janvier, 1825 – [s.l: s.n, 1825?] [mf ed 1984] – 1mf – 9 – 0-665-43057-4 – mf#43057 – cn Canadiana [320]

Les canadiens de france / Gourmont, Remy de – Paris: Firmin-Didot, 1893? – 3mf – 9 – mf#03490 – cn Canadiana [720]

Les canadiens de l'ouest / Tasse, Joseph – 4e ed. [Montreal?: s.n.] 2v. 1882 [mf ed 1985] – 2v on 1mf – 9 – mf#49000 – cn Canadiana [920]

Les canadiens des etats-unis : leon 13 aux eveques d'amerique relativement aux immigres italiens / Goesbriand, Louis de – Burlington, VT?: s.n, 1889? – 1mf – 9 – mf#03456 – cn Canadiana [975]

Canadiens, mefiez-vous : une experience de vignt ans / Gagnon, Ernest – Montreal: Revue canadienne, 1900 – 1mf – 9 – mf#05754 – cn Canadiana [230]

Les canadiens-francais de lowell, mass : recensement, valeur commerciale, valeur immobiliere, condition religieuse, civile et politique... – Lowell, MA?: A Bourbonniere, 1896 – 3mf – 9 – mf#00987 – cn Canadiana [978]

Canaima / Gallegos, Romulo – Buenos Aires, Mexico. 1944 – 1r – us UF Libraries [972]

Canakkale *see* Yeni mecmua

Canakkale muharebesi – Istanbul: Askeri Matbaasi, 1927 – (= ser Ottoman histories and historical sources) – 3mf – 9 – $58.00 – us MEDOC [956]

Canal Barrachina, Avelino *see* Historia y destino

Canal boatman's magazine, the... 1829-32 – 1r – 1 – mf#96665 – uk Microform Academic [380]

Canal de panama / Rebolledo, Alvaro – Cali, Colombia. 1957 – 1r – us UF Libraries [972]

Canal de panama / Wyse, Lucien Napoleon Bonaparte – Paris, France. 1886 – 1r – us UF Libraries [972]

Canal de panama completement acheve pour quatre ce... / Sautereau, Gustave – Paris, France. 1889 – 1r – us UF Libraries [972]

Canal de panama, el istmo americano / Wyse, Lucien Napoleon Bonaparte – Panama, 1959 – 1r – us UF Libraries [972]

El canal de panama en las guerras futuras / Olmedo, Alfaro – Guayaquil, 1930; Madrid: Razon y Fe, 1931 – 1 – sp Bibl Santa Ana [380]

Le canal de suez / Saint Victor, G de – Paris, 1934 – 4mf – 9 – mf#ILM-2337 – ne IDC [956]

Le canal de suez / Voisin-Bey, Francois Philippe – v1-6. Paris 1902-1906 – 1 – us NY Public [627]

Le canal de suez... / Yeghem, F – Dijon, 1927 – 3mf – 9 – mf#ILM-2212 – ne IDC [956]

Canal oceanique de panama / Renaut, Francis Paul – Paris, France. 1915 – 1r – us UF Libraries [972]

Canal Ramirez, GonzalO *see* Del 13 ie trece de junio al 10 ie diez de

Canal Ramirez, Gonzalo *see*
– El 13 [i.e. trece] de junio en 33 numeros de ya
– Estado cristiano y boliveriano del 13 de junio

Canal Rosado, Jose *see* Viento amarrado

Canal treaties : executive documents presented to the u.s. senate, together with the proceedings of the senate thereon relative to the panama canal – Senate doc no 456. 63rd Congress 2nd sess. Washington: GPO, 1914 – 1mf – 9 – $1.50 – mf#LLMC 82-100D Title 18 – us LLMC [324]

Canal zone code, 1934 : as approved by congress on june 19 1934, with an appendix containing treaties and laws of the u.s. relating to the canal zone or the panama canal / Panama Canal Zone; ed by Bentz, Paul A – Washington: GPO, 1934-37 – 15mf – 9 – $22.50 – (with suppl 1 covering 19 jun 1934-2 sep 1937) – mf#LLMC 82-100D Title 4 – us LLMC [348]

Canal zone code, 1962 : an act to revise and codify the general and permanent laws relating to and in force in the canal zone and to enact the canal zone code: p.l.87-845, 87th congress, approved oct 18 1962 / Panama Canal Zone – Washington: GPO, 1962 – 9mf – 9 – $13.50 – mf#LLMC 82-100D Title 5 – us LLMC [348]

Canal zone code annotated, 1962 / Panama Canal Zone – Oxford, NH: Equity Publ Co. 1963 – 29mf – 9 – $43.50 – (with cumulative suppl to 1976) – mf#LLMC 82-100D Title 6 – us LLMC [348]

Canal zone pilot / Haskins, William C – Panama, 1908 – 1r – us UF Libraries [972]

Canal zone reports, vol 3 : u.s. district court for the canal zone, may 1 1914-january 1 1926 / Panama Canal Zone – Mount Hope, CZ: Panama Canal Press, 1927 – 7mf – 9 – $10.50 – (no publ decisions between 1926-46. partial coverage for 1946-82 provided in the federal suppl series of west publ co's national reporter system) – mf#LLMC 82-100D Title 9 – us LLMC [347]

Canal zone rules and regulations, 1966 / Panama Canal Zone – Washington: GPO, 1966 – 2mf – 9 – $3.00 – (p12202-12351 of the federal register v31 no 180 pt 2 16 sept 1966) – mf#LLMC 82-100D Title 10 – us LLMC [348]

Canal zone supreme court reports, vols 1-2 : cases adjudged in the supreme court of the canal zone / Panama Canal Zone. Supreme Court – v1 jul term 1905-oct term 1908. Ancon, CZ: Isthmian Canal Commission, 1909. v2 oct 1908-jun 1914. Mount Hope, CZ: Panama Canal Press, 1915 – 6mf – 9 – $9.00 – (supreme court of the cz ceased to exist 1 july 1914. succeeded by us district court for the cz, with appellate jurisdiction being exercised by the us circuit court of appeals for the 5th circuit in new orleans) – mf#LLMC 82-100D Title 8 – us LLMC [347]

Il canale di suez – Milano, nd – 2mf – 9 – mf#ILM-2897 – ne IDC [956]

Canale, Floriano *see* Canzoni da sonare a quattro et otto voci

Canales de riego de cataluna y reino de valencia... / Jaubert de Passa, M – Valencia, 1844 – 25mf – 9 – sp Cultura [627]

Canales, Nemesio R *see* Paliques

Canaletti, Giovanni Battista *see* 7 trii per violino due e cetra

Canalizacion del ahorro provincial. conferencia. colegio o. de secretarios, interventores y depositarios de administracion local de la provincia de caceres / Bullon Ramirez, Francisco – Caceres: s.i., 1951 – sp Bibl Santa Ana [628]

Canals and railroads, ship canals and ship railways : a discussion of the paper of e sweet, the radical enlargement of the erie canal / Corthell, Elmer Lawrence – S.l: s.n, 1885? – 1mf – 9 – mf#03614 – cn Canadiana [627]

Canals, Angel Maria *see* Sendas de apostolado

Canape-vert / Thoby-Marcelin, Philippe – New York, NY. 1944 – 1r – us UF Libraries [972]

Canard en vacances *see* Canard enchaine

Canard enchaine – Paris, France. 1 aug 1956, 2 jan, 10 apr, 8 may, 26 jun, 21 aug, 2 oct 1957, dec 1959, 10 aug 1960, 2 aug 1961, 8 aug 1962, 2 jan 1963, 7 jan-30 dec 1970 – 1 1/4r – 1 – (aka: canard en vacances) – uk British Libr Newspaper [072]

Canard enchaine – Paris. 1989-1991 – 1 – ISSN: 0008-5405 – mf#10077 – us UMI ProQuest [073]

Le canard enchaine – Paris. sept 1915-juin 1940, sept 1944-1993 [wkly] – 1 – (journal satirique) – fr ACRPP [870]

Le canard enchaini – 1989- – 2r per y – 5 – enquire for prices – us UMI ProQuest [070]

Canard, M *see* La relation du voyage d'ibn fadln chez les bulgares de la volga

Canard suavage – Algiers. 14 nov 1943-20 oct 1944 – 1r – 1 – uk British Libr Newspaper [072]

Canarian [the] : or, book of the conquest and conversion of the canarians in the year 1402 by messire jean de bethencourt, kt / Bontier, Pierre – London: Printed for the Hakluyt Society 1872 – (= ser [Travel descriptions from south africa, 1711-1938]) – 3mf – 9 – (trans & ed, with notes & int by richard henry major) – mf#zah-4 – ne IDC [915]

Las canarias y la conquista franco-normanda / Bonnet y Reveron, Buenoventura – Laguna de Tenerife, 1944-54 – 2v – 1 – us CRL [946]

Canaries vs chickens : or, money in canaries / Cottam Bird Seed (Firm) – London, Ont: Cottam Bird Seed, c1906 – 1mf – 9 – 0-665-79420-7 – mf#79420 – cn Canadiana [338]

O canario : semanario critico – Bagagem, MG: Typ Allianca, 21 fev"jun, 22 set 1891 – (= ser Ps 19) – mf#P17,02,71 – bl Biblioteca [079]

Canas, Alberto F *see* Luto robado

Canas y bueyes / Moscoso Puello, Francisco Eujenio – Santo Domingo, Dominican Republic. 1935 – 1r – us UF Libraries [972]

Canastra / Alves, Raul – Rio de Janeiro, Brazil. 1936 – 1r – us UF Libraries [972]

O canastra : periodico critico, chistoso e litterario – Bahia: Typ S Quintanilha, 31 out 1867 – (= ser Ps 19) – bl Biblioteca [079]

Canavar / Camlibel, Faruk Nafiz – s.l: Necm-i Istikbal Matbaasi, 1926 – (= ser Ottoman literature, writers and the arts) – 1mf – 9 – $25.00 – us MEDOC [470]

Canaveral. Ayuntamiento *see*
– Feria y fiestas de primavera 1974
– Feria y fiestas de primavera 1975
– Feria y fiestas mayo 1970

Canaviais e engenhos na vida politica do brasil / Azevedo, Fernando De – Rio de Janeiro, Brazil. 1948 – 1r – us UF Libraries [972]

Canavieiros em greve : campanhas salariais e sindicalismo – Sao Paulo: Centro Ecumenico de Documentacao e Informacao, 1985 – us CRL [972]

Canberra times – Canberra, sep 1926-jul 1997 – 368r – 1 – at Pascoe [079]

The canberra times – 1926- – 12r per y – 1 – enquire for prices – us UMI ProQuest [079]

Canbronero Salazar, Miguel Angel *see* Escuela secundaria guatemalteca, problemas y soluc...

Canby herald *see* Canby herald and wilsonville spokesman

Canby herald and clackamas county news – Canby OR: M R Boehmer, [wkly] [mf ed 1968] – 2r – 1 – (ceased in 1924; merger of: canby herald [1914-22]; clackamas county news [1916-22]; cont by: canby herald [1924-]) – us Oregon Lib [071]

Canby herald and clackamas county news *see* Clackamas county news

Canby herald and wilsonville spokesman – Canby OR: T R Dillon, 1985 [wkly] – 1 – (related to: canby herald [canby or: 1924]; cont by: wilsonville spokesman) – us Oregon Lib [071]

Canby herald (canby or: 1914) *see* Clackamas county news

Canby herald [canby, or: 1914] – Canby OR: C P Leonard, 1914- [wkly] [mf ed 1968] – 1r – 1 – (merged with: clackamas county news, to form: canby herald and clackamas county news; cont: canby irrigator. ceased in 1922) – us Oregon Lib [071]

Canby herald [canby, or: 1924] – Canby OR: W C Culbertson, [wkly] – 1 – (began in 1924; cont by: canby herald & wilsonville spokesman; cont: canby herald and clackamas county news; 1924 incl newspaper publ by canby high school) – us Oregon Lib [071]

Canby irrigator – Canby OR: H P Bennett, 1911-14 [wkly] [mf ed 1968] – 2r – 1 – (cont: canby tribune [1911]; cont by: canby herald [1914-22]) – us Oregon Lib [071]

Canby tribune and willamette valley irrigator – Canby OR: Valley Pub Co, 1909-11 [wkly] [1968] – 1r – 1 – (merger of: canby and willamette valley irrigator; tribune [canby or]; cont by: canby tribune [canby or: 1911]) – us Oregon Lib [071]

Canby tribune [canby, or: 1908] – Canby OR: G W Dixon, -1909 [wkly] [mf ed 1968] – 1r – 1 – (cont by: tribune [canby, or]) – us Oregon Lib [071]

Canby tribune [canby, or: 1911] – Canby OR: F M Roth, 1911 [wkly] [mf ed 1968] – 1r – 1 – (cont: canby tribune and williamette valley irrigator [1909-11]; cont by: canby irrigator [1911-14]) – us Oregon Lib [071]

Cancelacion de una mision diplomatica / Dominican Republic Secretaria De Relaciones Exter... – Ciudad Trujillo, Dominican Republic. 1946 – 1r – us UF Libraries [972]

Cancellieri, F *see* Storia de' solenni possessi de' sommi pontefici detti anticamente processi o processioni dopo la loro coronazione dalla basilica vaticana alla lateranense...

Cancer – Philadelphia. 1948+ (1) 1973+ (5) 1975+ (9) – ISSN: 0008-543X – mf#9614 – us UMI ProQuest [616]

Cancer and metastasis reviews – Dordrecht. 1988-1996 (1,5,9) – ISSN: 0167-7659 – mf#16772,01 – us UMI ProQuest [616]

Cancer chemotherapy and pharmacology – Heidelberg. 1981-1991 (1) 1981-1991 (5) 1981-1991 (9) – ISSN: 0344-5704 – mf#13148 – us UMI ProQuest [615]

Cancer chemotherapy reports – Bethesda. 1972-1975 (1) 1959-1975 (5) 1972-1975 (9) – (cont by: cancer treatment reports) – ISSN: 0576-6559 – mf#6386 – us UMI ProQuest [615]

Cancer chemotherapy reports – v1-13. 1956-68 – 1 – us AMS Press [616]

Cancer communications – New York. 1989-1991 (1,5,9) – (cont by: oncology research) – ISSN: 0955-3541 – mf#49567 – us UMI ProQuest [616]

Cancer genetics and cytogenetics – New York. 1979-1996 (1) 1979-1996 (5) 1987-1996 (9) – ISSN: 0165-4608 – mf#42117 – us UMI ProQuest [575]

Cancer immunology and immunotherapy – Heidelberg. 1981-1992 (1,5,9) – ISSN: 0340-7004 – mf#13149 – us UMI ProQuest [616]

Cancer letters – Amsterdam. 1975-1993 (1) 1975-1993 (5) 1987-1993 (9) – ISSN: 0304-3835 – mf#42118 – us UMI ProQuest [616]

Cancer nursing – Philadelphia. 1992+ (1,5,9) – ISSN: 0162-220X – mf#18698 – us UMI ProQuest [610]

Cancer radiotherapie – Paris. 1997+ (1) – ISSN: 1278-3218 – mf#42784 – us UMI ProQuest [616]

Cancer research – Baltimore. 1941+ [1]; 1966+ [5]; 1970+ [9] – ISSN: 0008-5472 – mf#1406 – us UMI ProQuest [616]

Cancer treatment reports – Washington. 1976-1987 (1) 1976-1987 (5) 1976-1987 (9) – (cont: cancer chemotherapy reports) – ISSN: 0361-5960 – mf#6386,01 – us UMI ProQuest [615]

Cancio Villa-Amil, Mariano *see* Cuba

Una cancion de amor / Hurtado, Antonio – 1874 – 9 – sp Bibl Santa Ana [810]

Cancion de cuna / Chavarria Flores, Manuel – Guatemala, 1952 – 1r – us UF Libraries [972]

Cancion de cuna / Urunuela Ortiz, Juan – sp Bibl Santa Ana [780]

Cancion de la hora / Lomar, Martha – San Juan, Puerto Rico. 1959 – 1r – us UF Libraries [972]

Cancion de marti / Lazaro, Angel – Habana, Cuba. 1953 – 1r – us UF Libraries [972]

Cancion de una vida : poesias / Fiallo, Fabio – Madrid, Spain. 1926 – 1r – us UF Libraries [972]

Cancion del caminante / Villegas, Silvio – Bogota, Colombia. 1960? – 1r – us UF Libraries [972]

Cancion del camino / Castillo, Moises – Panama, 1954 – 1r – us UF Libraries [972]

La cancion petrarquista en la lirica espanola del siglo de oro, de enrique segura covarsi / Bleiberg, German – Madrid: Arbor, 1949 – 1 – sp Bibl Santa Ana [780]

Cancion redonda / Lars, Claudia – San Jose, Costa Rica. 1937 – 1r – us UF Libraries [972]

Cancion y poesia de scanlan / Scanlan, Eduardo – Ciudad Trujillo, Dominican Republic. 1946 – 1r – us UF Libraries [972]

Cancioneiro da biblioteca nacional, cancioneiro...antigo colocci-brancuti – Lisboa. v1-8. 1949-56 – 1 – $108.00 – mf#0139 – us Brook [440]
Cancioneiro geral : altportugiesische liedersammlung des edeln garcia de resende / ed by Kausler, E H von – Stuttgart: Litterarischer Verein, 1846-52 [mf ed 1993] – (= ser Blvs 15,17,26) – 3v – 1 – mf#8470 reels 4, 6 – us Wisconsin U Libr [780]
Cancionero *see* Cancionero de la virgen de la aurora
Cancionero de la restauracion / Mota, Fabio A – Santo Domingo, Dominican Republic. 1963 – 1r – us UF Libraries [972]
Cancionero de la virgen de la aurora / Cancionero – Parroquia de Zarza Capilla:Toledo Editorial Catolico-Toledana, 1951 – sp Bibl Santa Ana [946]
Cancionero folklorico / Macau, Miguel Angel – Havana, Cuba. 1956 – 1r – us UF Libraries [390]
Cancionero popular cuyano – Mendoza [Argentina]: Best Hermanos 1938 [mf ed Bloomington IN: Indiana Uni Lib, Preservation Dept 1984] – 1r – 1 – us Indiana Preservation [390]
Cancionero popular de extremadura. contribucion al folklore musical de la region / Gil Garcia, Bonifacio – (Cataluna): E. Castells, impresor, 1931.-v1 – 1 – sp Bibl Santa Ana [390]
Cancionero popular de la rioja – Buenos Aires: A Baiocco 1942 [mf ed Bloomington IN: Indiana Uni Lib, Preservation Dept 1984] – 3v on 1r [ill] – 1 – us Indiana Preservation [390]
Cancionero popular venezolano – 2nd ed. Caracas: L Puig Ros & P Almenar 1922 [mf ed Bloomington IN: Indiana Uni Lib, Preservation Dept 1984] – 1r – 1 – (with musical suppl) – us Indiana Preservation [780]
Canciones / Lars, Claudia – San Salvador, El Salvador. 1960 – 1r – us UF Libraries [972]
Canciones / Marre, Luis – Habana, Cuba. 1964 – 1r – us UF Libraries [972]
Canciones *see*
- Almendralejo asociacion de adoradores de jesus sacramentado y practicas religiosas
- Canciones devotos que acostumbran cantar en sus misiones los padres misioneros del colegio seminario de ntra.sra. de aguas santas

Canciones de ruta y sueno / Zapata Acosta, Ramon – San Juan, Puerto Rico. 1954 – 1r – us UF Libraries [972]
Canciones del senor. canciones y oraciones biblicas – Seleccion de L. Fanlo. Dibujos de Froma. Don Benito, Imp. Sanchez Trejo, 1968 – sp Bibl Santa Ana [780]
Canciones devotos que acostumbran cantar en sus misiones los padres misioneros del colegio seminario de ntra.sra. de aguas santas / Canciones – Almendralejo: Luciano Carballar, 1901 – 1 – sp Bibl Santa Ana [240]
Canciones en carne viva / Alvarez Lencero, H – (Bilbao: Zero S.A., 1973) – sp Bibl Santa Ana [780]
Canciones en la sombra : poemas / Ramirez Brau, Enrique – San Juan, Puerto Rico. 1954 – 1r – us UF Libraries [972]
Canciones guerreras / Alonso de la Avecilla, Pablo – 1834 – 9 – sp Bibl Santa Ana [780]
Canciones para tu historia (1936-1939) / Augier, Angel I – Habana, Cuba. 1941 – 1r – us UF Libraries [972]
Canciones quadragesimales quadruplices / Trujillo, Fray Thomas – Barcelona, 1591 – 1 – sp Bibl Santa Ana [780]
Canciones y otros poemas : manuscrito / Garcia Lorca, Federico – 2mf – 9 – sp Cultura [810]
Canciones...guadalupe...adventu / Trujillo, Thomas de – 1591 – 9 – sp Bibl Santa Ana [810]
Canclini, Santiago *see* Scrapbooks
Cand rah jivit ralat : ryan / Tryn Thac Saen – Bhnam Ben: Ron Bumb Rasmi 2510 [1967] [mf ed 1990] – 1r with other items – 1 – (in khmer) – mf#mf-10289 seam reel 107/10 [§] – us CRL [959]
Canda Chve *see* Caka pum ka pro so dassana
Canda devi / Rvhe U Don – Ran Kun: Mran Ma prann Ca up tuik 1960 [mf ed 1990] – 1r with other items – 1 – (in burmese) – mf#mf-10289 seam reel 182/5 [§] – us CRL [959]
Candado / Sanz Lajara, J M – Ciudad Trujillo, Dominican Republic. 1959 – 1r – us UF Libraries [972]
Candanedo, Cesar A *see* Clandestinos
Candelabros del tropico / Rivas, Nicolas – San Juan, Puerto Rico. 1950 – 1r – us UF Libraries [972]
[Candelaria-] chloride belt – NV. 1890-92 [wkly] – 1r – 1 – $60.00 – mf#U04434 – us Library Micro [071]
[Candelaria-] true fissure – NV. 1881-86 [wkly] – 2r – 1 – $120.00 – mf#U04435 – us Library Micro [071]
Candelo / eden union / southern auckland advocate – Candelo, jan 189-dec 1904 – 1r – A$72.16 vesicular A$77.66 silver – (aka: eden union; southern auckland advocate) – at Pascoe [079]
A candid examination of the question whether the pope of rome is the great antichrist of scripture / Hopkins, John Henry – New York: Hurd & Houghton, 1868 [mf ed 1986] – 1mf – 9 – 0-8370-8352-4 – (incl bibl ref) – mf#1986-2352 – us ATLA [241]
A candid examination of theism : by physicus / Romanes, George John – London, 1878 – (= ser 19th c evolution & creation) – 3mf – 9 – mf#1.1.10904 – uk Chadwyck [140]
A candid examination of theism / Romanes, George John – 3rd ed. London: Kegan Paul, Trench, Truebner, 1892 [mf ed 1985] – 1mf – 9 – 0-8370-4959-8 – mf#1985-2959 – us ATLA [210]
Candid examiner – Montrose. 1826-1827 (1) – mf#4362 – us UMI ProQuest [240]
A candid history of the jesuits / McCabe, Joseph – London: E Nash, 1913 [mf ed 1990] – 2mf – 9 – 0-7905-5257-4 – mf#1988-1257 – us ATLA [241]
Candid reasons for declining to become a member of temperance socie... / Fraser, William – Edinburgh, Scotland. 1832 – 1r – us UF Libraries [240]
Candid reflections on the report (as published by authority) of the general-officers : appointed by his majesty's warrant of the first of november last, to enquire into the causes of the failure of the late expedition to the coasts of france / Holland, Henry Fox, Baron – 3rd ed. London: printed for S Hooper & A Morley...1758 [mf ed 1984] – 1mf – 9 – 0-665-44100-2 – mf#44100 – cn Canadiana [941]
Candid warning to public men in a series of letters / Hancock, Edward – London, England. 1836? – 1r – us UF Libraries [240]
Candidates and referenda / League of Women Voters of Madison [WI] – 1974 apr 7 – 1r – 1 – (cont: important election information; cont by: candidates questions and answers) – mf#623411 – us WHS [325]
Candidates answers / League of Women Voters of Dane County [WI] et al – 1970 apr 7-1982 apr 6 – 1r – 1 – (cont: candidates questions and answers) – mf#623425 – us WHS [325]
Candidates questions and answers / League of Women Voters of Madison [WI] – 1965 mar 9, 1966 mar 8, 1966 apr 5 – 1r – 1 – (cont: candidates and referenda; cont by: candidates answers) – mf#623412 – us WHS [325]
Candide – Grand hebdomadaire parisien et litteraire. Paris. 20 mars 1924-9 aout 1944 – 1 – fr ACRPP [820]
Candidius, G *see* A short account of the island of formosa, in the indies, situated near the coast of china
Candido de Rivera, J *see* Memorial ajustado...
Candidus *see*
- Observations on a letter by lucius to the rev andrew thomson
- Plain truth addressed to the inhabitants of america

Candidus, V *see* Illustres disquisitiones morales
Candle by night: a history of woman's missionary union auxiliary to the baptist general convention of texas / Patterson, Roberta Turner – 1800-1955. 1955 – 1 – 7.07 – us Southern Baptist [242]
The candle of the lord and other sermons / Brooks, Phillips – New York: EP Dutton, 1881 – 1mf – 9 – 0-7905-3617-X – mf#1989-0110 – us ATLA [240]
Candler, Allen D *see* Confederate records of the state of georgia
Candler, Edmund *see*
- The mantle of the east
- On the edge of the world

Candler, Isaac *see* A summary view of america
Candler, Warren A *see*
- Practical studies in the fourth gospel
- Wesley and his work

Candler, Warren Akin *see*
- Great men and great movements
- Great revivals and the great republic

Candlin, George T *see* Chinese fiction
Candlish, James S *see*
- The christian sacraments
- The epistle of paul to the ephesians
- The work of the holy spirit

Candlish, Robert S *see*
- Four letters to the rev e b elliott on some passages in his hora...
- Principle of free inquiry and private judgment

Candlish, Robert Smith *see*
- The atonement
- The christian's sacrifice and service of praise, or, the two great commandments
- Church and state
- Church's unity in diversity
- Contributions towards the exposition of the book of genesis
- Discourses bearing upon the sonship and brotherhood of believers
- Examination of mr maurice's theological essays
- The fatherhood of god
- The first epistle of john
- The gospel of forgiveness
- John know and his 'devout imagination'
- John knox
- Lectures on foreign churches
- Lord's short work on the earth
- Reason and revelation
- Report of the speech delivered at a meeting...
- Scripture characters
- Sermons

Candlish, Rs *see* Reason insufficient without revelation
Candlish, William James *see* The illinois law of voluntary assignments for the benefit of creditors
Candray, Jose Eulalio *see* Acuarelas
Candy industry – New York. 1993+ (1,5,9) – ISSN: 0745-1032 – mf#18813,04 – us UMI ProQuest [640]
Cane branch baptist church. horry county. loris, south carolina : church records – 1907-82.658p – 1 – us Southern Baptist [242]
Cane cape vauin : [1975: mandalay, burma] – Man tale: Hamsavati 1976 [mf ed 1990] – 1r with other items – 1 – (in burmese; incl bibl ref) – mf#mf-10289 seam reel 136/1 [§] – us CRL [480]
Cane creek baptist church. union, south carolina : church records – 1835-1890 – 1 – 6.53 – us Southern Baptist [242]
Cane growers' quarterly bulletin – Brisbane. 1933-1981 (1) 1972-1981 (5) 1980-1981 (9) – ISSN: 0008-5553 – mf#7154 – us UMI ProQuest [630]
Cane, Miguel *see* En viaje, 1881-1882
The cane ridge meeting-house : to which is appended, the autobiography of b. w. stone. and, a sketch of david purviance / Rogers, James Richard et al – Cincinnati: Standard Pub Co, c1910 – 1mf – 9 – 0-524-01011-0 – mf#1990-0288 – us ATLA [920]
Cane syrup in infant feeding / Townsend, Ruth O – Gainesville, FL. 1944 – 1r – us UF Libraries [630]
Cane, syrup, sugar / Stockbridge, Horace E – Lake City, FL. 1898 – 1r – us UF Libraries [630]
Canedo, Lino Gomez *see* Archivos historicos de puerto rico
Canellas, Angel *see* Coleccion diplomatica de san andres de faulo (958-1270). zaragoza, 1964
Canellas Casals, Jose *see* Buscadores de diamentes en la guayana venezolana
Canestrelli, Philip *see*
- Lu skuskuests lu t st marie
- Yakasinkinmiki

Canetti y Alvarez de Gades, Liborio *see* El mago de logrosan
Caneville / Van Den Berghe, Pierre L – Middletown, CT. 1964 – 1r – us UF Libraries [960]
Caney, harrisonville, and spring hill church records / Caney. Kansas. Methodist Episcopal Church – 1870-89 – 1 – us Kansas [240]
Caney. Kansas. Methodist Episcopal Church *see* Caney, harrisonville, and spring hill church records
Canfield, George L *see* The law of the sea
Canfield, James Hulme *see* Religion and public education
Canfield, Leon Hardy *see* The early persecutions of the christians
Canfield. Ohio. Presbyterian Church of Christ *see* Presbyterian church of christ, canfield, ohio records, 1804-1860
Cangaceiros e fanaticos / Faco, Rui – Rio de Janeiro, Brazil. 1965 – 1r – us UF Libraries [972]
Cange, C du *see*
- Annales
- Chronicon paschale
- De imperatorum constantinopolitanorum seu de inferioris aevi
- Historia byzantina duplici commentario illustrata
- Imperatorii grammatici historiarum libri seu de rebus gestis a joanne et mannuele gommensis impp

Cange, Ch Dufresne Du *see* Glossarium ad scriptores mediae et infimae graecitatis duos in tomos digestum
Cange, Ch. Dufresne Du *see* Glossarium mediae et infimae latinitatis
Canh hoa truoc gio : truyen dai tinh cam xa hoi / Lan Phuong – ed: In lan 1. Saigon: Anh Loc 1974 [mf ed 1992] – on pt of 1r – 1 – mf#11052 r152 n2 – us Cornell [480]
Canh man trang / Canh Tra – Ha-noi: Van Hoc 1975 [mf ed 1992] – on pt of 1r – 1 – mf#11052 r14 n18 – us Cornell [959]
Canh nong luan – Saigon. 24 aout 1929-4 avr 1931 – 1 – fr ACRPP [073]
Canh Tra *see* Canh man trang
Canihuante, Gustavo *see* Revolucion chilena
Canina, Luigi et al *see* Illustrations, architectural and pictorial
Canine practice – Santa Barbara. 1974-1987 (1) 1975-1987 (5) 1975-1987 (9) – ISSN: 0094-4904 – mf#9792 – us UMI ProQuest [636]
Canine practice – Santa Barbara. 1990-1994 (1,5,9) – mf#17818 – us UMI ProQuest [636]
Canini, G A *see* Iconografia cioa disegni d'imagini de famosissimi monarchi, regi, filosofi, poeti ed oratori dell' antichit...
Canini, M A *see* Iconografia cioa disegni d'imagini de famosissimi monarchi, regi, filosofi, poeti ed oratori dell' antichit...
Canivez, J-M *see*
- Auctarium d c de visch ad bibliothecam scriptorum s o cisterciensis
- L'ordre de citeaux en belgique des origines (1132) au 20m siecle
- Statuta capitulorum generalium ordinis cisterciensis

Canizo Gomez, Jose del *see* Ideas actuales sobre las plagas de langosta
Cankaya – n1. 1 mayis 1928 [all publ] – (= ser O & t journals) – 2mf – 9 – $40.00 – us MEDOC [956]
Canna e o assucar nas antilhas / Dias Filho, Manoel A Santos – Rio de Janeiro, Brazil. 1908 – 1r – us UF Libraries [972]
Cannabich, Christian *see* Periodical overture in eight parts...no 1
Cannabich, Johann *see*
- Lehrbuch der geographie nach den neuesten friedensbestimmungen
- Statistisch-geographische beschreibung des koenigreichs preussen

Cannabrava Filho, Paulo *see* Militarismo e imperialismo en brasil
Canne, John *see* A necessity of separation from the church of england
Cannecattim, Bernardo Maria de *see* Colleccao de observacoes grammaticaes sobre a lingua bunda
Cannegieter, Tjeerd *see* Degodsdienst uit plichtbesef en de geloofsvoorstelling uit dichtende verbeelding geboren?
Cannella, Felix *see* Dr phillipe
Canner packer – Chicago. 1904-1977 (1) 1969-1977 (5) 1976-1977 (9) – (cont by: processed prepared food) – ISSN: 0190-8731 – mf#840 – us UMI ProQuest [660]
Cannery and field union news, 1937 / cio news cannery workers edition, 1938-1939 / ucapawa news, 1939-1944 / fta news, 1945-1950 / United Cannery, Agricultural, Packing and Allied Workers of AmericaFood, Tobacco, and Agricultural and Allied Workers Union of America – (= ser Labor union periodicals, pt 3: food and agricultural industries) – 1r – 1 – $210.00 – 1-55655-610-1 – us UPA [660]
Cannery Workers' Union of the Pacific, Los Angeles County Harbor District *see* Fishery worker
Canney, Maurice Arthur *see* Essays on the social gospel
Cannibal cousins / Craige, John Houston – New York, NY. 1934 – 1r – us UF Libraries [972]
Canniff, C M *see* Pocket manual of mining
Canniff, William *see*
- Canadian nationality
- History of the province of ontario (upper canada)
- History of the settlement of upper canada
- A manual of the principles of surgery
- The medical profession in upper canada, 1783-1850

Canning, Albert Stratford George *see* Words on existing religions
Canning, George *see* Corrected report of the speech
Cannington gleaner – Ontario, CN. jan 1888-dec 1977 – 31r – 1 – (some iss missing) – cn Commonwealth Imaging [071]
The cann-leighton official theatrical guide *see* American theatre periodicals of the nineteenth and early twentieth centuries
Il cannocchiale aristotelico... / Tesauro, E – Torino: Bartolomeo Zavatta, 1670 – 14mf – 9 – mf#0-1957 – ne IDC [090]
Cannock advertiser – England. -w. Aug 1893-Dec 1924. (Wanting 1896, 1911). (19 reels) – 1 – uk British Libr Newspaper [072]
Cannock chase courier *see* Cannock chase news
Cannock chase courier & west staffordshire councillor – [West Midlands] Staffordshire jan 1929-dec 1950 [mf ed 2003] – 22r – 1 – uk Newsplan [072]
Cannock chase examiner – Hednesford. England. -w. 2 May 1874-5 Oct 1877. (2 reels) – 1 – uk British Libr Newspaper [072]
Cannock chase news – England. may 1889-dec 1928 – 30r – 1 – (aka: cannock chase courier; wanting 1911) – uk British Libr Newspaper [072]
Cannon beach gazette – Cannon Beach OR: V Hawkins, 1977- [semimthly] [mf ed 1995] – 1 – (cont: cannon gazette [1976-77]) – us Oregon Lib [071]
Cannon, Clarence *see*
- Cannon's precedents of the house of representatives
- Cannon's procedure in the house
- Cannon's procedure in the house of representatives

Cannon, Edward W *see* The relationship between an increased aerobic power and the excess post exercise oxygen consumption

Cannon gazette – Cannon Beach OR: V Hawkins, 1976-77 [semimthly] [mf ed 1995] – 1r – 1 – (cont by: cannon beach gazette [1977-]) – us Oregon Lib [071]

Cannon, Harry Sharp *see* Sudermann's treatment of verse

Cannon, Richard *see* Historical record of the thirty-sixth or the herefordshire regiment of foot

Cannon, Walter Bradford *see* Address delivered april 18, 1937

Cannoneer – 1980 sep 18/1981 sep-1993 mar 25/sep 23 – 16r – 1 – mf#584817 – us WHS [071]

Cannon's precedents of the house of representatives / Cannon, Clarence – Washington: GPO. 11v. 1935-41 [all publ] – 56mf – 9 – $84.00 – mf#llmc 84-105 – us LLMC [323]

Cannon's procedure in the house / Cannon, Clarence – Washington: GPO, 1959 [all publ] – 6mf – 9 – $9.00 – (new ed with notes and add by william t roy) – mf#llmc 84-107 – us LLMC [323]

Cannon's procedure in the house of representatives / ed by Cannon, Clarence – 3rd ed. Washington: GPO, 1939 (all publ) – 6mf – 9 – $9.00 – mf#llmc 84-106 – us LLMC [323]

Cannot and can fall from grace / Peebles, Isaac Lockhart – Nashville, Tenn: Publishing House of the ME Church, South, 1914 – 1mf – 9 – 0-7905-9055-7 – mf#1989-2280 – us ATLA [240]

Cano, M *see* Opera

Canoe – Kirkland. 1978-1993 (1,5,9) – (cont by: canoe and kayak) – ISSN: 0360-7496 – mf#11770 – us UMI ProQuest [790]

Canoe and boat building : a complete manual for amateurs: containing plain and comprehensive directions for the construction of canoes, rowing and sailing boats and hunting craft / Stephens, W P – 5th rev enl ed. New York: Forest & Stream Publ Co, 1891 – us CRL [790]

Canoe and boat building / Stephens, William Picard – New York, NY. 1885 – 1r – us UF Libraries [790]

Canoe and camp life in british guiana / Brown, Charles Barrington – London, England. 1876 – 1r – us UF Libraries [920]

Canoe and kayak – Harrisburg. 1994+ (1,5,9) – (cont: canoe) – ISSN: 1077-3258 – mf#11770,01 – us UMI ProQuest [790]

A canoe trip through temagaming the peerless in the land of hiawatha / Armstrong, Louis Olivier – Montreal?: Canadian Pacific Railway, 1900 – 1mf – 9 – mf#03883 – cn Canadiana [917]

Canoeing on the columbia / Coleman, Arthur Philemon – S.l: s.n, 1889? – 1mf – 9 – mf#18004 – cn Canadiana [790]

Canoemates / Munroe, Kirk – New York, NY. 1892 – 1r – us UF Libraries [978]

Canoemates / Munroe, Kirk – New York, NY. 1905 – 1r – us UF Libraries [978]

Canoga park – 1946-47; 1982-87 – (= ser California telephone directory coll) – 9r – 1 – $450.00 – mf#P00016 – us Library Micro [917]

The canol project / U.S. Army. Office of the Chief of Military History. Eastern Defense Command – 1945. 7 v. illus., charts, maps, photos. 1 reel – 1 – us L of C Photodup [977]

Canon and text of the new testament / Gregory, Caspar Rene – New York: Charles Scribner, 1907 – 2mf – 9 – 0-7905-0314-X – (incl ind) – mf#1987-0314 – us ATLA [225]

Canon and text of the new testament / Gregory, Caspar Rene – New York: C Scribner's Sons 1907 [mf ed 1988] – (= ser International theological library) – 1r – 1 – mf#1749 – us Wisconsin U Libr [225]

The canon and text of the new testament *see* Hsin yueh cheng ching cheng li shih (ccm15)

Canon law : a basic collection – initial ed. Honolulu, 1987 – (= ser Llmc canon law catalogs on fiche) – 1mf – 9 – mf#LLMC 87-000A – us LLMC [348]

Canon law : a basic collection – permanent ed. Kaneohe, 1996 – (= ser Llmc canon law catalogs on fiche) – 1mf – 9 – mf#LLMC 87-000B – us LLMC [348]

Canon law : a basic collection / ed by Reynolds, Thomas H – From the Berkeley Law Library, Uni of California. Additional materials from the law libraries of Yale, Uni of Michigan and other law schools. Complete index – 222 titles 5730mf – 9 – $8,055.00 – (titles are too numerous to list separately. free catalog available on request. full cataloguing as a major microform collection available from oclc) – us LLMC [348]

Canon law / Mcneile, Hugh – London, England. 1850 – 1r – us UF Libraries [240]

Canon law abstracts – Hove. 1988-1988 (1) – ISSN: 0008-5650 – mf#15936 – us UMI ProQuest [200]

Canon muratorianus : the earliest catalogue of the books of the new testament / ed by Tregelles, Samuel Prideaux – Oxford: Clarendon Press, 1867 – 1mf – 9 – 0-8370-5566-0 – (incl bibl ref) – mf#1985-3566 – us ATLA [220]

The canon of the bible : its formation, history, and fluctuations / Davidson, Samuel – New York: Peter Eckler, [1899] – 1mf – 9 – 0-8370-2836-1 – (from the 3rd rev enl ed of this same work publ in 1878) – mf#1985-0836 – us ATLA [220]

The canon of the holy scriptures from the double point of view of science and of faith = Canon des saintes ecritures / Gaussen, Samuel Robert Louis – London: James Nisbet, 1862 – 2mf – 9 – 0-7905-1396-X – (incl bibl ref and index) – mf#1987-1396 – us ATLA [210]

The canon of the old and new testaments ascertained : or, the bible complete without the apocrypha and unwritten traditions / Alexander, Archibald Browning Drysdale – new ed. Philadelphia: Presbyterian Board of Publ, c1851 – 1mf – 9 – 0-7905-0842-7 – (incl bibl ref) – mf#1987-0842 – us ATLA [220]

The canon of the old testament : an essay on the gradual growth and formation of the hebrew canon of scripture / Ryle, Herbert Edward – London, New York: Macmillan, 1892 – 1mf – 9 – 0-8370-9415-1 – (in english and greek. incl ind) – mf#1986-3415 – us ATLA [221]

The canon of the old testament / Mullen, Tobias – New York: Fr Pustet, 1892, c1888 – 8mf – 9 – 0-7905-8310-0 – mf#1987-6415 – us ATLA [221]

Canones et decreta : des hochheiligen, oekumenischen und allgemeinen concils von trient / Smets, Wilhelm – 5. Aufl. Bielefeld: Velhagen & Klasing, 1858 – 2mf – 9 – 0-8370-8330-3 – (in german and latin) – mf#1986-2330 – us ATLA [240]

Canones et decreta = The doctrinal decrees and canons of the council of trent – New-York: American and Foreign Christian Union, 1854 – 1mf – 9 – (in english) – mf#1986-1453 – us ATLA [240]

Canones et decreta concilii tridentini : ex editione romana a. 1834 repetiti = Canones et decreta / ed by Richter, Aemelius Ludwig – Lipsiae [Leipzig]: Typis et sumptibus B Tauchnitii, 1853 – 2mf – 9 – 0-524-03502-4 – mf#1990-4724 – us ATLA [240]

Canones, et echo sex vocibus / Agostini, Lodovico – 1572 – (= ser Mssa) – 2mf – 9 – €35.00 – mfchl 170 – gw Fischer [780]

Die canones hippolyti / Achelis, Hans – Leipzig: J C Hinrichs, 1891 – (= ser Die aeltesten quellen des orientalischen kirchenrechtes; Texte und untersuchungen zur geschichte der altchristlichen literatur) – 1mf – 9 – 0-7905-1920-8 – (incl bibl ref) – mf#1987-1920 – us ATLA [240]

Canonical and uncanonical gospels : with a translation of the recently discovered fragment of the gospel of peter... / Barnes, William Emery – London: Longmans, Green, 1893 – 1mf – 9 – 0-8370-2183-9 – (appendix contains translation of the gospel of peter) – mf#1985-0183 – us ATLA [226]

Canonicarum quaestionum / Gutierrez, Juan – Liber Primus. 1730 – 9 – (liber secundus 1729. liber tertius 1730) – sp Bibl Santa Ana [240]

Canonicarum utrusque fori liber primus / Gutierrez, Juan – 1587 – 9 – (liber secundus 1608. liber tertius 1617) – sp Bibl Santa Ana [240]

Canonici, Luciano *see* La porziencula nei piu antichi documenti francescani...

Canonici regularis scti victoris parisiensis opera omnia / Hugh of Saint-Victor – Rothomagi. v1-3. 1648 – 87mf – 8 – €166.00 – ne Slangenburg [241]

Canonicity : a collection of early testimonies to the canonical books of the new testament / Charteris, Archibald Hamilton – Edinburgh: William Blackwood 1880 [mf ed 1989] – 2mf – 9 – 0-7905-1635-7 – (incl bibl ref & ind) – mf#1987-1635 – us ATLA [225]

Canonis misse expositio ediderunt : oberman-courtenay / Biel, G – Wiesbaden, 1963 – 7mf – 8 – ne Slangenburg [241]

The canonisation of saints / Macken, Thomas F – Dublin: MH Gill, 1910 – 1mf – 9 – 0-8370-6916-5 – mf#1986-0916 – us ATLA [240]

The canons and decrees of the sacred and oecumenical council of trent : celebrated under the sovereign pontiffs, paul 3, julius 3 and pius 4 = Canones et decreta – London: C Dolman, 1848 – 2mf – 9 – 0-524-04998-X – (in english) – mf#1990-5086 – us ATLA [240]

Canons, by-laws and resolutions adopted by the synod of the diocese of toronto : with an historical digest of the proceedings from 1851 to 1872 inclusive / Church of England. Diocese of Toronto – [Toronto: s.n.], 1873 [mf ed 1980] – 5mf – 9 – (incl ind) – mf#05723 – cn Canadiana [242]

The canons of 1571 in english and latin – London: SPCK 1899 [mf ed 1993] – (= ser Church historical society (series) 40) – 2mf – 9 – 0-524-05556-4 – (in english & latin; notes by william edward collins) – mf#1990-5160 – us ATLA [242]

The canons of athanasius of alexandria : the arabic and coptic versions / ed by Riedel, Wilhelm & Crum, Walter Ewing – London: Williams and Norgate, 1904 – 1mf – 9 – 0-7905-3752-4 – mf#1989-0245 – us ATLA [240]

Canons of indigenous traditions and western values : the voice of african women writers / Kwatsha, Linda Loretta – Pretoria: Vista University 2002 [mf ed 2002] – 5mf – 9 – (incl bibl ref) – mf#mfm15235 – sa Unisa [305]

Canons of orissan architecture / Bose, Nirmal Kumar – Calcutta: R Chatterjee, 1932 – (= ser Samp: indian books) – us CRL [720]

Canons of professional ethics / American Bar Association – Chicago, 1947. 54p. LL-2246 – 1 – us L of C Photodup [340]

The canons of the first four general councils of nicaea, constantinople, ephesus and chalcedon / Bright, William – 2nd ed. Oxford: Clarendon Press, 1892 – 1mf – 9 – 0-7905-5511-5 – (incl bibl ref) – mf#1988-1511 – us ATLA [240]

The canons of the first four general councils of the church : and those of the early local greek synods... / ed by Lambert, William – London: RD Dickinson, [1868?] – 1mf – 9 – 0-524-01583-X – mf#1990-0449 – us ATLA [240]

Canons of the synod of the diocese of ontario : with acts of parliament affecting ecclesiastical rights, and forms of grants, requests and trusts for church purposes / Church of England. Diocese of Ontario – [Kingston, Ont?: s.n.], 1873 [mf ed 1980] – 1mf – 9 – 0-665-05720-2 – (incl ind) – mf#05720 – cn Canadiana [242]

Canons of the synod of the diocese of ontario and of the provincial synod of canada : with a collection of statutes affecting ecclesiastical rights and form of grants, requests and trusts for church purposes / Eglise d'Angleterre en Canada. Diocese of Ontario – [Kingston Ont?: s.n, 1891] [mf ed 1980] – 3mf – 9 – 0-665-00639-X – mf#00639 – cn Canadiana [242]

Canonsburg herald – Canonsburg, PA. -w 1872-1888 – 13 – $25.00r – us IMR [071]

Canonsburg notes weekly – Canonsburg, PA. -w 1902-1904 – 13 – $25.00r – us IMR [071]

Canova, Antonio *see* The works of antonio canova

Canovas Del Castillo, Antonio *see* Paz de cuba

Canovas del Castillo, Antonio *see* Discurso pronunciado 8 noviembre de 1888

Canowindra news – Canowindra, jan 1983-dec 1996 – at Pascoe [079]

Canowindra star – Canowindra, apr 1900-dec 1907, jan 1910-dec 1959 – 17r – A$1126.09 vesicular A$1219.59 silver – at Pascoe [079]

Canowindra star – Canowindra, jan 1969-feb 1971 – 1r – at Pascoe [079]

Canright, Dudley Marvin *see* Seventh-day adventism renounced

Canseco, Manuel *see*
- Circular en la que el director..
- Circular recomendando la puntualidad del ayuntamiento..

Cansinos Assens, Rafael *see* Estetica y erotismo de la pena de muerte

Canstatt, Oskar *see* Republikanische brasilien in vergangenheit und geg...

Cant, Gourlay and Co *see* Price list of cant, gourlay and co, "galt machine works," galt, ontario, canada

Cantamos por la herida / Juarez Toledo, Enrique – Guatemala, 1962 – 1r – us UF Libraries [972]

Cantares de juventud / Sanchez Arjona, Vicente – Sevilla: Imprenta Alvarez, Tomo 1-9 y 10-29. 1957 – 1 – sp Bibl Santa Ana [780]

Cantares viejos / Sanchez Arjona, Vicente – Sevilla: Imprenta Carlos Acuna, s.a. – 1 – sp Bibl Santa Ana [780]

Cantata composed in honor of h r h the prince of wales' visit to canada : sung by the montreal musical union, at the grand musical festival, august 1860 = [Cantate en l'honneur du prince de galles. libretto] / Sabatier, Charles Wugk – [Montreal?: s.n, 1860?] [mf ed 1984] – 1mf – 9 – 0-665-22818-X – (original french title: cantate en l'honneur de son altesse royale le prince de galles a l'occasion de son voyage au canada) – mf#22818 – cn Canadiana [780]

Cantata con istromenti : [for solo voice accompanied by 2 violins, viola obbligato and basso] / Holzbauer, Ignaz – [17-?] [mf ed 19–] – 1r – 1 – mf#film 1414 – us Sibley [780]

Cantatas / Valdes Machuca, Ignacio – Habana, Cuba. 1829 – 1r – us UF Libraries [972]

Cantate : la confederation dediee a l'hon george etienne cartier, ministre de la milice / Achintre, Auguste – [Montreal]: [s.n.], [1868] [mf ed 1980] – 1mf – 9 – 0-665-04514-X – mf#04514 – cn Canadiana [780]

Cantate : les cygnes malades – Montreal: [s.n.], 1879 [mf ed 1980] – 1mf – 9 – 0-665-02571-8 – mf#02571 – cn Canadiana [780]

Cantate domino : a hymnal and chants for public worship / Black, John – Toronto: Copp, Clark, 1874 [mf ed 1981] – 2mf – 9 – (incl ind) – mf#11906 – cn Canadiana [780]

Cantate, la confederation / Achintre, Auguste & Labelle, Jean-Baptiste – [S.l: s.n, 1868?] [mf ed 1980] – 1mf – 9 – mf#02387 – cn Canadiana [780]

Cantate morali a voce sola del conte pirro albergati. opera terza / Albergati Capacelli, Pirro – Bologna: G Monti 1685 [mf ed 1988] – 1r – 1 – mf#pres. film 35 – us Sibley [780]

Cantates a voix seule : et avec simfonie... premier livre qui contient six cantates francoise et deux cantates italiennes / Monteclair, Michel Pignolet de – Paris: author [1709] [mf ed 19–] – 1r – 1 – mf#pres. film 34 – us Sibley [780]

Cantates francoises a 1 et 2 voix : avec simphonie, et sans simphonie...livre premier / Clerambault, Louis Nicolas – Paris: chez l'auteur, Sr Foucault 1710 [mf ed 19–] – 1r – 1 – mf#film 1611 – us Sibley [780]

Cantave, Philippe *see* Vrai visage d'haiti

Canteen – v4 n10,11, v5 n6 [1942 oct 1, dec 1, 1943 jun] – 1r – 1 – mf#964481 – us WHS [071]

Cantemus domino / Plasencia. Secretariado Catequistico – Plasencia: Imp. La Victoria, 1960 – sp Bibl Santa Ana [946]

Cantera, Eugenio *see* Historia del santisimo cristo de la victoria que se venera en la villa de serradilla (caceres)

Cantera y Burgos, Francisco *see* Alvar garcia de santa maria

Canterbury / Jenkins, Robert Charles – London: SPCK 1880 [mf ed 1992] – (= ser Diocesan histories) – 1mf – 9 – 0-524-03385-4 – mf#1990-4697 – us ATLA [242]

Canterbury and otago almanac – 1887 – 1r – 1 – mf#ZB 12 – nz Nat Libr [079]

The canterbury benedictional (hbs51) / Woolley, R M – 1917 – (= ser Henry bradshaw society (hbs)) – 4mf – 8 – €11.00 – ne Slangenburg [241]

Canterbury cathedral library catalogue of pre-1801 books – [mf ed Marlborough 1996] – 17mf – 9 – £115.00 – (with d/g; also incl: rochester cathedral library catalogue of pre-1901 books) – uk Matthew [020]

Canterbury. Christ Church Priory *see* Literae cantuarienses (rs85)

Canterbury farmer – Christchurch, NZ. may 1981-nov 1984 – 1r – 1 – mf#70.31 – nz Nat Libr [079]

Canterbury, New Hampshire. Canterbury Free Will Baptist Church *see* Records

Canterbury, New Hampshire. First Free Will Baptist Society *see* Records

Canterbury provincial roll 1868-69 – 2mf – 9 – NZ$9.00 – 0-908797-65-6 – nz BAB [325]

Canterbury provincial roll 1870-71 – 2mf – 9 – NZ$9.00 – 0-908797-66-4 – nz BAB [325]

Canterbury provincial roll 1872-73 – 3mf – 9 – NZ$14.00 – 0-908797-28-1 – nz BAB [325]

Canterbury provincial roll 1873-74 – 4mf – 9 – NZ$18.00 – 0-908797-29-X – nz BAB [325]

Canterbury psalter, the... : trinity college, cambridge, ms. r.17.1 – 12th c – 1r – 14 – mf#C561 – uk Microform Academic [780]

Canterbury standard – Christchurch, NZ. 1854-60 – 5r – 1 – mf#70.15 – nz Nat Libr [079]

Canterbury tales / Chaucer, Geoffrey – New York, NY. 1931 – 1r – us UF Libraries [420]

Canterbury tales / [geoffrey chaucer] and faerie queene / [edmund spenser] / Chaucer, Geoffrey & Spenser, Edmund – Boston: Lee & Shepard [pref 1869] [mf ed 1984] – 1r – 1 – (ed for popular perusal with current ill & explanatory notes by d laing purves) – mf#954 – us Wisconsin U Libr [810]

Canti parva – Calcutta: Bharata Press, 1890 – 4mf – 9 – 0-524-08011-9 – mf#1991-0233 – us ATLA [280]

Cantica : ex sacris literis...in usum pastorum, diaconorum, & iuventutis scholasticae, iam postremum recognita et aucta – s.l: Rihelius 1575 – (= ser Hqab. literatur des 16. jahrh.) – 2mf – 9 – €30.00 – mf#1575a – gw Fischer [780]

Cantica canticorum : eighty-six sermons on the song of solomon = Sermones super cantica canticorum / Bernard of Clairvaux, Saint; ed by Eales, Samuel John – London: Elliot Stock, 1895 – 2mf – 9 – 0-524-01220-2 – (in english) – mf#1990-0359 – us ATLA [220]

Cantica sacra : an aid to devotion / Silloway, Thomas William – Boston: NE Universalist Pub House, 1865 – 1mf – 9 – 0-524-02964-4 – mf#1990-4516 – us ATLA [240]
Cantica sacra : partim ex sacris literis desumta, partim ab orthodoxis patribus et piis ecclesiae doctoribus composita... / ed by Ulysseo, Francisco Elero – Hamburgi: Wolff 1588 – (= ser Hqab. literatur des 16. jahrh.) – 5mf – 9 – €60.00 – mf#1588a – gw Fischer [780]
Cantica selecta veteris novique testamenti : cum hymnis et collectis, seu orationib. purioribus, quae in orthodoxa atque catholica ecclesia cantari solent / addita dispositione & familiari expositione christophori corneri – Lipsiae: Voegelianus 1571 – (= ser Hqab. literatur des 16. jahrh.) – 3mf – 9 – €40.00 – mf#1571b – gw Fischer [780]
Cantica selecta veteris novique testamenti : cum hymnis et collectis, seu orationibus purioribus, quae in orthodoxa atque catholica ecclesia cantari solent – Lipsia: Voegelin 1568 – (= ser Hqab. literatur des 16. jahrh.) – 3mf – 9 – €40.00 – mf#1568a – gw Fischer [780]
Cantica selecta veteris novique testamenti : cum hymnis et collectis, seu orationibus purioribus, quae in orthodoxa atque catholica ecclesia cantari solent – Lipsiae: Schneider 1575 – (= ser Hqab. literatur des 16. jahrh.) – 3mf – 9 – €40.00 – mf#1575b – gw Fischer [780]
Cantica selecta veteris noviqve testamenti : cvm hymnis et collectis, sev orationibus purioribus, quae in orthodoxa atque catholica ecclesia cantari solent – Lipsiae: Steinmannus 1588 – (= ser Hqab. literatur des 16. jahrh.) – 3mf – 9 – €40.00 – mf#1588b – gw Fischer [780]
The canticles of the christian church : eastern and western, in early and medieval times / Mearns, James – Cambridge: University Press, 1914 – 1mf – 9 – 0-7905-5488-7 – mf#1988-1488 – us ATLA [240]
Cantico mortal a julia de burgos / Gonzalez, Josemilio – Yauco, Puerto Rico. 1956 – 1r – us UF Libraries [972]
Canticorum liber primus cum quinque vocibus / Corteccia, Francesco – 1571 – (= ser Mssa) – 3mf – 9 – €50.00 – mfchl 197 – gw Fischer [780]
Canticorum liber primus cum sex vocibus / Corteccia, Francesco – 1571 – (= ser Mssa) – 3mf – 9 – €50.00 – mfchl 198 – gw Fischer [780]
Canticos, liturgia, ngoma – Vila Pery, Mozambique. 1966 – 1r – us UF Libraries [960]
Canticum beatae mariae quod magnificat nuncupatur / Guerrero, Francisco – 1563 – (= ser Mssa) – 4mf – 9 – €60.00 – mfchl 243 – gw Fischer [780]
Canticum beatae mariae virginis – 1557 – (= ser Mssa) – 1mf – 9 – €20.00 – mfchl 122 – gw Fischer [780]
Canticum gratulatorium... / Faber, Benedikt – 1607 – (= ser Mssa) – 1mf – 9 – €20.00 – mfchl 212 – gw Fischer [780]
Canticum virginis seu magnificat... / Holzner, Anton – 1625 – (= ser Mssa) – 4mf – 9 – €60.00 – mfchl 258 – gw Fischer [780]
Cantigas de santa maria / Alfonso 10 el Sabio, Rey de Castilla – Madrid: Editorial Patrimonio Nacional, 1974 – 1 – sp Bibl Santa Ana [240]
Cantillon, Philippe de *see* Delices du brabant et de ses campagnes
Cantilupe society – v1-20. 1906-32 – (= ser Publications of the english record societies, 1835-1972) – 74mf – 9 – uk Chadwyck [941]
Cantimpre, Thomas de *see* Liber de natura rerum (cima55)
Cantin, Louise *see* Bibliographie analytique de l'oeuvre de madame marcelle lepage-thibaudeau
Cantional, oder gesangbuch augspurgischer confession : in welchem des herrn d martini lutheri vnd anderer frommen christan / Schein, Johann Hermann – [Leipzig]: in Verlegung des Austoris 1627 [mf ed 1994] – 1r – 1 – mf#pres. film 131 (pm 1103) – us Sibley [780]
Cantiones aliquot quinque vocum / Lasso, Orlando di – 1569 – (= ser Mssa) – 2mf – 9 – €35.00 – mfchl 284 – gw Fischer [780]
Cantiones bohemicae : leiche, lieder und rufe des 13., 14. und 15. jahrhunderts nach handschriften aus prag... / ed by Dreves, Guido Maria – Leipzig: Fues, 1886 [mf ed 1986] – (= ser Analecta hymnica medii aevi 1) – 1mf – 9 – 0-8370-7454-1 – (text of hymns primarily in latin. int in german. incl bibl ref) – mf#1986-1454 – us ATLA [220]
Cantiones ecclesiasticae latinae, dominicis et festis diebvs : in commemoratione cenae domini, per totius anni circulum cantandae... / Spangenberg, Johann – [Magdeburg]: [Michael Lotther] 1545 – (= ser Hqab. literatur des 16. jahrh.) – 9mf – 9 – €85.00 – mf#1545c – gw Fischer [780]
Cantiones quinque vocum selectissimae – 1539 – (= ser Mssa) – 5mf – 9 – €70.00 – mfchl 86 – gw Fischer [780]
Cantiones sacrae / Pevernage, Andreas – 1602 – (= ser Mssa) – 8mf – 9 – €100.00 – mfchl 370 – gw Fischer [780]
Cantiones sacrae... / Lasso, Ferdinando di – 1588 – (= ser Mssa) – 3mf – 9 – €50.00 – mfchl 310 – gw Fischer [780]
Cantiones sacrae... / Victoria, Tomas Luis de – 1589 – (= ser Mssa) – 5mf – 9 – €70.00 – mfchl 438 – gw Fischer [780]
Cantiones sacrae de festis... / Hassler, Hans Leo – 1597 – (= ser Mssa) – 5mf – 9 – €70.00 – mfchl 253 – gw Fischer [780]
Cantiones sacrae quinque vocum / Flori, Jakob – 1599 – (= ser Mssa) – 4mf – 9 – €60.00 – mfchl 222 – gw Fischer [780]
Cantiones sacrae...tomus primus / Praetorius, Hieronymus – 1607 – (= ser Mssa) – 7mf – 9 – €90.00 – mfchl 375 – gw Fischer [780]
Cantiones selectissimae quatuor vocum. liber secundus – 1549 – (= ser Mssa) – 2mf – 9 – €35.00 – mfchl 95 – gw Fischer [780]
Cantiones septem, sex & quinque vocum – 1546 – (= ser Mssa) – 5mf – 9 – €70.00 – mfchl 93 – gw Fischer [780]
Le cantique de debora : etude exegetique et critique / Segond, Albert – Geneve: W Kuendig, 1900 – 1mf – 9 – 0-8370-7335-9 – mf#1986-1335 – us ATLA [220]
Le cantique des cantiques : commentaire philologique et exegetique / Joueon, Paul – 2e ed. Paris: G Beauchesne, 1909 – 1mf – 9 – 0-524-05616-1 – (incl bibl ref) – mf#1992-0471 – us ATLA [220]
Le cantique des cantiques / Renan, Ernest – 2e rev corr ed. Paris: Michel Levy, 1861 – 1mf – 9 – 0-8370-9414-3 – (incl bibl ref) – mf#1986-3414 – us ATLA [220]
Le cantique des cantiques (etb) / Robert, A et al – Paris, 1963 – 8mf – 8 – €17.00 – ne Slangenburg [220]
Cantiques de marseilles accommodes a des airs vulgaires / Durand, Laurent – Quebec: Impr a la Nouvelle Imprimerie, 1819. [mf ed 1984] – 4mf – 9 – 0-665-44858-9 – mf#44858 – cn Canadiana [241]
Cantiques et prieres : extraits de la priere chantee: manuel complet pour le chant de tous, a l'usage des paroisses, des maisons d'education et des oeuvres, conforme aux instructions pontificales / Dubois, Emile [comp] – nouv ed. Paris [etc]: Societe de S Jean l'evangeliste, Desclee & cie, [1948?] (mf ed 1998) – 3mf – 9 – (with ind) – mf#SEM105P2919 – cn Bibl Nat [780]
Cantiques populaires du canada francais – Quebec?: L Brousseau, 1897 – 1mf – 9 – mf#03276 – cn Canadiana [780]
Cantiunculae pascales / Rasch, Johann – 1572 – (= ser Mssa) – 1mf – 9 – €20.00 – mfchl 384 – gw Fischer [780]
Canto a extremadura / Quijano Quijano, Adolfo – Cadiz: Libreria Universal de Morillas, s.a. – sp Bibl Santa Ana [780]
Canto a juan delgado – Habana, Cuba. 1954 – 1r – us UF Libraries [972]
Canto a la argentina / Dario, Ruben – Buenos Aires, Argentina. 1949 – 1r – us UF Libraries [972]
Canto a la ceiba de colon / Moreno Jimenes, Domingo – San Cristobal, Venezuela. 1958 – 1r – us UF Libraries [972]
Canto a la encontrada patria y su heroe / Suarez, Clementina – Tegucigalpa, Mexico. 1958 – 1r – us UF Libraries [972]
Canto a la provincia trujillo y otros poemas / Sanchez Lamouth, Juan – Ciudad Trujillo, Dominican Republic. 1960 – 1r – us UF Libraries [972]
Canto a la vera / Verde, Josefina – Plasencia: Sanguino offset y Tip., 1979 – 1 – sp Bibl Santa Ana [780]
Canto a los angeles / Joglar Cacho, Manuel – San Juan, Puerto Rico. 1958 – 1r – us UF Libraries [972]
Canto a los argonautas y otros poemas / Espada Rodriguez, Jose – Yauco, Puerto Rico. 1958 – 1r – us UF Libraries [972]
Canto al amor profundo / Rivel, Isa De – Mayaguez, Puerto Rico. 1956 – 1r – us UF Libraries [972]
Canto da saudade / Soares, Amandio – Rio de Janeiro, Brazil. 1932 – 1r – us UF Libraries [972]
Canto de amor para la patria novia / Rodriguez, Mario Augusto – Panama, 1957 – 1r – us UF Libraries [972]
Canto de bronce : poemas / Derpich Aguilar, Juan – Havana, Cuba. 1963 – 1r – us UF Libraries [972]
Canto de fe universal al benefactor de la patria / Pena Santana, Santiago De – Ciudad Trujillo, Dominican Republic. 1955 – 1r – us UF Libraries [972]
Canto de la locura / Matos Paoli, Francisco – San Juan, Puerto Rico. 1962 – 1r – us UF Libraries [972]
Canto de los olvidos / Palma, Marigloria – Barcelona, Spain. 1965 – 1r – us UF Libraries [972]
El canto de relacion en el folklore infantil de extremadura / Gil Garcia, Bonifacio – Badajoz, 1963 – 1 – sp Bibl Santa Ana [390]
Canto de soledad y doce poemas crepusculares / Jordan Diaz, Alfredo Alberto – Habana, Cuba. 1954 – 1r – us UF Libraries [972]
Canto de tierra adentro / Geigel Polanco, Vicente – New York, NY. 1965 – 1r – us UF Libraries [972]
Canto del amor infinito / Geigel Polanco, Vicente – San Juan, Puerto Rico. 1962 – 1r – us UF Libraries [972]
Canto di baldassare donato il primo libro di madrigali a cinque e a sei voci con tre dialoghi a sette novamente per antonio gardano / Donato, Baldassare – Venice, 1560. 6v – 1 – us L of C Photodup [780]
Canto eterno, poesias / Hernandez-Santana, Gilberta – Habana, Cuba. 1934 – 1r – us UF Libraries [972]
Il canto fermo in prattica per uso di qual si voglia sorte di religiosi, e religiose : che bramano con un metodo facilissimo, e breve imparare il canto gregoriano per il buon concerto, e regolamento de'cori / Caselli, Domenico Antonio – Roma: Nella stamperia del Chracas 1724 [mf ed 1983?] – 1mf – 9 – mf#fiche 865 – us Sibley [780]
Canto final / Buesa, Jose Angel – Habana, Cuba. 1938 – 1r – us UF Libraries [972]
Canto funebre – 1821 – 9 – sp Bibl Santa Ana [810]
Canto llano, 1954-1955 / Vitier, Cintio – Habana, Cuba. 1956 – 1r – us UF Libraries [972]
Canto vivo / Ovalle Lopez, Werner – Guatemala, 1952 – 1r – us UF Libraries [972]
Canto y saloma / Gonzalez Bazan, Carlos R – Panama, 1958 – 1r – us UF Libraries [972]
Cantoclarus, Car *see* Excerpta de legationibus (cbh1,2)
Canton, cardiff, glamorgan, the parish church of st john's : baptisms 1858-1924 – [Glamorgan]: GFHS [mf ed c2005] – 15mf – 9 – £18.75 – uk Glamorgan FHS [929]
Canton christian college : its growth and outlook = Ling naam hok hau / Ling nan ta hseueh (Canton, China) – New York: Trustees of the Canton Christian College, [1919] [mf ed 1995] – (= ser Yale coll) – 66p (ill) – 1 – 0-524-10084-5 – mf#1995-1084 – us ATLA [377]
Canton daily ledger – Canton IL. 1917 jan 26 – 1r – 1 – mf#1159494 – us WHS [071]
Canton estremenofel (sic), plasencia – 1884-89.No. sueltos – 9 – sp Bibl Santa Ana [070]
El canton extremeno – Plasencia, 1887-1889 – 5 – sp Bibl Santa Ana [073]
Canton general price current – Canton, China. -w. 7 jan 1834-26 dec 1837 – 1r – 1 – uk British Libr Newspaper [079]
Canton ind. sentinel – Canton, PA., 1939-1979 – 13 – $25.00r – us IMR [071]
Canton, OH *see* Selections (1820-1928)
Canton. OH. (Elkton Circuit). Methodist-Episcopal Church *see* Church records, ms 2822
Canton press – Macao, China. 12 sep 1835-30 mar 1844 – 4r – 1 – uk British Libr Newspaper [072]
The canton press – Canton: [s.n.] sep 12 1835-mar 30 1944 – 4r – 1 – us CRL [079]
Canton register – Canton, China. 1835-1837; 21 may 1839; jun 1840; 8 jun 1841; 8 nov 1827-30 dec 1834 (very imperfect) – 2r – 1 – uk British Libr Newspaper [072]
Canton register – Canton: [James Matheson], nov 8 1827-jun 10 1843 – 4r – 1 – (filmed with: hongkong, late canton, register jun 20-dec 26 1843) – us CRL [079]
Canton sentinel – Canton, PA. 1875-1939. 41 rolls – 13 – $25.00r – us IMR [071]
Canton telephone books : 1925, 33-52, 54, 56, 63-73, 75-77,79-1989 – 17r – 1 – mf#B31469-31485 – us Ohio Hist [978]
Canton, William *see* A history of the british and foreign bible society
Canton world – Canton, PA. 1909-15. 2 rolls – 13 – $25.00r – us IMR [071]
Cantone, Serafino *see* Sacrae cantiones...octonis vocibus...
Cantonese union church bulletin *see* Shanghai kuang-tung chung-hua chi-tu-chiao-hui yueh pao (ccs32)
Les cantons de la province de quebec : nomenclature / Fafard, Francois-Xavier [comp] – Quebec: [s.n.] 1913 [mf ed 1997] – 1mf – 9 – 0-665-81878-5 – mf#81878 – cn Canadiana [971]
Der cantor : [collected aphorisms] / Wolff, Gottfried August Benedict – [s.l: s.n, 1872?] [mf ed 1991] – 1r [ill] – 1 – (filmed with: volk, ich breche deine kohle! / otto wohlgemuth) – mf#2964p – us Wisconsin U Libr [390]
Cantor del niagara en santo domingo / Rodriguez Demorizi, Emilio – Ciudad Trujillo, Dominican Republic. 1939 – 1r – us UF Libraries [972]
Cantor lectures : the decorative treatment of natural foliage / Stannus, Hugh Hutton – London 1891 – (= ser 19th c art & architecture) – 1mf – 9 – mf#4.2.1052 – uk Chadwyck [740]
Cantor lectures on the art of lace-making / Cole, Alan Summerly – London 1881 – (= ser 19th c art & architecture) – 1mf – 9 – mf#4.2.577 – uk Chadwyck [740]
Cantos a la naturaleza cubana del siglo 19 / Feijoo, Samuel – Santa Clara, Cuba. 1964 – 1r – us UF Libraries [972]
Cantos de amanecer / X, Marilola – Habana, Cuba. 1934 – 1r – us UF Libraries [972]
Cantos de amor y de dolor / Gabulli, Plorio A – Montevideo, Uruguay. 1951 – 1r – us UF Libraries [972]
Cantos de apolo / Perdomo, Apolinar – Ciudad Trujillo, Dominican Republic. 1943 – 1r – us UF Libraries [972]
Cantos de pitirre / Diego, Jose De – Palma de Mallorca, Spain. 1950 – 1r – us UF Libraries [972]
Cantos de vida y esperanza / Dario, Ruben – Buenos Aires, Argentina. 1946 – 1r – us UF Libraries [972]
Cantos de vida y esperanza / Dario, Ruben – Buenos Aires, Argentina. 1952 – 1r – us UF Libraries [972]
Cantos del pueblo de dios – Caceres: Tip. El Noticiero, 1975 – 1 – sp Bibl Santa Ana [780]
Cantos en negro y esperanza / Montero Monago, Nemesio E – Madrid: Altamira Tip, 1980 – 1 – sp Bibl Santa Ana [780]
Cantos liturgicos / Colegio San Francisco Javier. Fuente de Cantos (Badajoz) – Zafra: Industrias Tipgraficas Extremenas, 1969 – sp Bibl Santa Ana [780]
Cantos liturgicos / Iglesia – Badajoz: Tip. Manuel Barrena, 1969 – 1 – sp Bibl Santa Ana [240]
Cantos para la cuaresma y semana santa *see* Lamentacion y sagrada pasion de nuestro senor jesucristo, stabat y perdon
Cantos para soldados y sones para turistas / Guillen, Nicolas – Buenos Aires, Argentina. 1952 – 1r – us UF Libraries [972]
Cantos partioticos / Heredia, Jose Maria – Habana, Cuba. 1916 – 1r – us UF Libraries [972]
Cantos y cuentos / Sanchez Arjona y Sanchez Arjona, Jose – 1877 – 9 – sp Bibl Santa Ana [810]
Cantos y rumbos / Inda Hernandez, Jose – Ciego de Avila, Cuba. 1939 – 1r – us UF Libraries [972]
Cantu, Cesare *see*
– Gli eretici d'italia
– Histoire universelle
Cantu Corro, Jose *see* Mujer a traves de los siglos
Cantum ecclesiasticum praecibus apud deum animas juvandi, corporaque humandi defunctorum officium, missam et stationes juxta ritum sacrosanctae romanae ecclesiae omnium ecclesiarum matris et magistrae : juxta breviarij, missalique romani novissimam recognitionem – Antverpine: H Aertssens 1691 [mf ed 1992] – 1r – 1 – mf#pres. film 117 – us Sibley [780]
Cantwell, John S et al *see* The winchester centennial, 1803-1903
Canudos (diario de uma expedicao) / Cunha, Euclydes Da – Rio de Janeiro, Brazil. 1939 – 1r – us UF Libraries [972]
Canyon cinemanews – Sausalito. 1972-1976 (1) 1972-1976 (5) 1976-1976 (9) – (cont by: cinema news) – ISSN: 0008-5758 – mf#7615 – us UMI ProQuest [790]
Canyon city news – Azusa, CA. 1951-1954 (1) – mf#62085 – us UMI ProQuest [071]
Canyon creek current – Canyonville OR: S & B Eller, 1974-82 [wkly] – 6r – 1 – (merged with: mail [myrtle creek or] to form: umpqua free pressa) – us Oregon Lib [071]
[Canyon lake-] canyon lake community news – CA. sep 1986- – 1r – 1 – $60.00 (sub $50/y) – mf#R04010 – us Library Micro [071]
[Canyon lake-] canyon lake menifee valley news – CA. aug 1989- – 1r – 1 – $60.00 (subs $50/y) – mf#R04012 – us Library Micro [071]
[Canyon lake-] canyon lake weekly – CA. sep 1990- – 1+ r – 1 – $60.00 (subs $50/y) – mf#R04011 – us Library Micro [071]
[Canyon lake-] friday flyer – CA. dec 1990- – 1+ r – 1 – $60.00 (subs $50/y) – mf#R04013 – us Library Micro [071]
Canzler, Friedrich Gottlob *see*
– Allgemeines litteraturarchiv fuer geschichte, geographie und statistik
– Neues magazin fuer die neuere geschichte, erd- und voelkerkunde
Canzler, Karl Christian et al *see* Fuer aeltere litteratur und neuere lectuere
Canzon a selin imperator di turchi : in desperation della sua armata, e gente persa – 1mf – 9 – mf#H-8181 – ne IDC [956]

Canzone a ballo composte dal magnifico lorenzo de medici et da m agnolo politiano, e altri autori insiem con la nencia da barberino, e la beca da dicomacompote dal medesimo lorenzo... / Medici, L de et al – N p, n d – 2mf – 9 – mf#O-1108 – ne IDC [700]
Canzone nella felicissima vittoria christiana contra infideli al sereniss d gio d'avstria / Guarnello, A – [Venice, 1571] – 1mf – 9 – mf#H-8313 – ne IDC [956]
Canzone nella nativita di nostro signor giesv christo : nella allegrezza della vittoria hauuta contra turchi / Forzanini, G P – Venetia, 1572 – 1mf – 9 – mf#H-8325 – ne IDC [956]
Canzone nella vittoria dell' armata della santissima lega contra la turchesca – [Venice, 1571] – 1mf – 9 – mf#H-8174 – ne IDC [956]
Canzone sopra la vittoria ottenvta dall' armata de' prencipi christiani contra la turchesca – Venetia, 1571 – 1mf – 9 – mf#H-8175 – ne IDC [956]
Canzone...per la felicissima vittoria nauale contra turchi / Tiepolo, G – Vinegia, 1572 – 1mf – 9 – mf#H-8334 – ne IDC [956]
Canzonette a quatro voci libro primo / Hassler, Hans Leo – 1590 – (= ser Mssa) – 2mf – 9 – €35.00 – mfchl 250 – gw Fischer [780]
Canzonette a quattro voci / Croce, Giovanni – 1604 – (= ser Mssa) – 2mf – 9 – €35.00 – mfchl 202 – gw Fischer [780]
Canzonette a quattro voci, composte de diuversi eccti musici, con l'intauolatura del cimbalo et del liuto – Roma 1591 [mf ed 19–] – 1r – 1 – mf#film 315 – us Sibley [780]
Canzonette a sei voci / Vecchi, Orazio – 1587 – (= ser Mssa) – 2mf – 9 – €35.00 – mfchl 425 – gw Fischer [780]
Canzonette a tre voci – 1604 – (= ser Mssa) – 2mf – 9 – €35.00 – mfchl 151 – gw Fischer [780]
Canzonette a tre voci / Pecci, Tomaso – 1604 – (= ser Mssa) – 2mf – 9 – €35.00 – mfchl 368 – gw Fischer [780]
Canzonette a tre voci di horatio vecchi... – 1597 – (= ser Mssa) – 1mf – 9 – €20.00 – mfchl 148 – gw Fischer [780]
Canzonette a tre voci. libro secondo – 1604 – (= ser Mssa) – 2mf – 9 – €35.00 – mfchl 152 – gw Fischer [780]
Canzonette amorose spirituali a tre voci / Ballis, Oliviero – 1607 – (= ser Mssa) – 1mf – 9 – €20.00 – mfchl 173 – gw Fischer [780]
Canzonette spirituali : divise in tre parti, la prima serve per ogni tempo / Casini, Giovanni Maria – In Firenze: Nella Stamperia di Sua Altezza Reale...1703 [mf ed 198-] – 1r – 1 – mf#pres. film 36 – us Sibley [780]
Canzoni a 4 & 8 voci / Taeggio, Giovanni Domenico Rognoni – 1605 – (= ser Mssa) – 4mf – 9 – €60.00 – mfchl 391 – gw Fischer [780]
Canzoni alla francese et ricercari ariosi – 1605 – (= ser Mssa) – 1mf – 9 – €20.00 – mfchl 465 – gw Fischer [780]
Canzoni da sonare a quattro et otto voci / Canale, Floriano – 1600 – (= ser Mssa) – 2mf – 9 – €35.00 – mfchl 186 – gw Fischer [780]
Canzoni et sonate... / Gabrieli, Giovanni – 1615 – (= ser Mssa) – 4mf – 9 – €60.00 – mfchl 234 – gw Fischer [780]
Canzoni per sonare con ogni sorte di stromenti – 1608 – (= ser Mssa) – 5mf – 9 – €70.00 – mfchl 153 – gw Fischer [780]
Cao Huy Dinh *see* Tim hieu tien trinh van hoc dan gian viet nam
Caonabo : seigneur de la maguana / Corvington, Hermann – Port-Au-Prince, Haiti. 1944 – 1r – us UF Libraries [972]
Caonex : novela / Sanz-Lajara, J M – Buenos Aires, Argentina. 1949 – 1r – us UF Libraries [830]
Caos / Herrera, Flavio – Guatemala, 1949 – 1r – us UF Libraries [972]
El caos de religiones nuevas / Bayle, Constantino – Madrid: Razon y Fe, 1929 – 9 – sp Bibl Santa Ana [240]
Caouette, Jean Baptiste *see* Le vieux muet
Caoutchouc et la gutta-percha – Paris, France. 16 mar 1904-15 dec 1906; 1907-09 – 4r – 1 – uk British Libr Newspaper [072]
Le caoutchouc et la gutta-percha – Paris, France. -m. March 1904-Dec 1909. 4 reels – 1 – uk British Libr Newspaper [410]
Cap alert / National Federation of Republican Women – v1 n1 [p1-4]-v2 n8 [p31-34] [1978 oct 8-1979 jul 31], v3 n1 [p1-2]-v3 n6 [p23-26] [1980 jan 14-jun 30] – 1r – 1 – (cont by: comprehensive advocacy program) – mf#999832 – us WHS [305]
The cap and gown / Brown, Charles Reynolds – New York: Pilgrim Press, c1910 – 1mf – 9 – 0-7905-9158-8 – mf#1989-2383 – us ATLA [240]
Le cap au diable / Deguise, Charles – Ste Anne de la Pocatiere, Quebec?: F H Proulx, 1863 – 1mf – 9 – mf#23046 – cn Canadiana [390]
Cap Bin, Ghun Uttamaprija *see* Prasna vinayapitak sankhep
Le cap de bonne-esperance au 17e siecle : l'escale maritime – johan van riebeeck – les colons europeens... – Paris: Hatchette, 1909 – 1 – us CRL [960]
Le cap eternite : poeme suivi des etoiles filantes / Gill, Charles – Montreal: Edition du Devoir, 1919 – 2mf – 9 – 0-665-71504-8 – (pref by albert lozeau) – mf#71504 – cn Canadiana [810]
Cap francais vu par une americaine / Hassel, Mary – Port-Au-Prince, Haiti. 1936 – 1r – us UF Libraries [972]
Cap of liberty – London, UK. 1819-20. -irr. 13 feet – 1 – uk British Libr Newspaper [072]
Cap tien – Ho Chi Minh City, Vietnam. 1969-1972 (1) – mf#67821 – us UMI ProQuest [079]
Capaccio, G C *see* Delle imprese, trattato di giulio cesare capaccio
La capacidad cambiaria en el derecho internacional privado : estudio comparativo de las legislaciones americanas / Garcia Calderon K, Manuel – Lima: Libreria e Impr. Gil, 1951. 267p. LL-8008 – 1 – us L of C Photodup [340]
Capacidad de la republica dominicana / Dominican Republic Comision Para El Estudio Del I – Ciudad Trujillo, Dominican Republic. 1946 – 1r – us UF Libraries [972]
Capacity building for educators as a way to improve classroom performance / Mvula, Shadrack Hanania – Pretoria: Vista University 2001 [mf ed 2001] – 2mf – 9 – (incl bibl ref) – mf#mfm15253 – sa Unisa [370]
Capacity management review – Phoenix. 1990-1998 (1) 1990-1998 (5) 1990-1998 (9) – (cont: edp performance review) – ISSN: 1049-2194 – mf#15753,01 – us UMI ProQuest [650]
Capacity of the dominican republic to absorb refug... / Dominican Republic Comision Para El Estudio Del I – Trujillo City, Dominican Republic. 1945 – 1r – us UF Libraries [972]
Caparraso, Carlos Arturo *see* Ciclos de lirismo colombiano
Caparroso, Carlos Arturo *see* Antologia lirica
Capart, J *see* Une rue de tombeaux a saqqarah
The cape, and canada : viewed as to their eligibility for british emigration – London: Kent & Richards, 1848 – (= ser 19th c economics) – 1mf – 9 – mf#1.1.107 – uk Chadwyck [304]
The cape, and canada : viewed as to their eligibility for british emigration. giving ample details to meet the inquiries of all classes – London, 1848 – (= ser 19th c books on british colonization) – 1mf – 9 – mf#1.1.107 – uk Chadwyck [971]
The cape and south africa / Noble, John – Cape Town 1878 – (= ser 19th c british colonization) – 3mf – 9 – mf#1.1.3935 – uk Chadwyck [916]
Cape and south africa [the] / Noble, John – London: Longmans, Green & Co 1878 – (= ser [Travel descriptions from south africa, 1711-1938]) – 3mf – 9 – mf#zah-75 – ne IDC [916]
Cape argus – Cape Town SA: printed...by S Solomon & Co 3 jan 1857- – 1 – (semiwkly [jan 3 1857-]; triwkly [mar 30 1858-]; has suppl: cape argus weekly ed; cont by: argus [dec 1969-]) – sa National [079]
Cape breton intererts sic sacrificed : its catholic clergy affronted and its french population ignored by adption sic of the central route / Chisholm, Murdoch – Halifax, NS?: Holloway, 1887 – 1mf – 9 – mf#03862 – cn Canadiana [380]
Cape breton railway : specification for the construction of the work – [Ottawa?: s.n, 1886?] [mf ed 1995] – 1mf – 9 – 0-665-94772-0 – mf#94772 – cn Canadiana [625]
Cape breton railway extension company of canada, 1890 – [Halifax, NS: s.n, 1890?] [mf ed 1980] – 1mf – 9 – 0-665-02068-6 – mf#02068 – cn Canadiana [380]
Cape breton's magazine – n25-38 [1979-84] – 1r – 1 – mf#177742 – us WHS [071]
Cape chave nave pavai / Sauin Chauin, Mon – Yan Kun: Khayn Tavn Tauik 1975 [mf ed 1990] – 1r with other items – 1 – (in burmese) – mf#mf-10289 seam reel 173/3 [§] – us CRL [400]
Cape chronicle – Cape Town SA, W.F. Mathew, 1860-62 – 1r – 1 – (cont: the cape weekly chronicle) – sa National [079]
The cape chronicle – Cape Town SA, W Foster 1870-71 (wkly) [mf ed Cape Town: SA library 1986] – 1r – 1 – mf#MS00437 – sa National [079]
Cape chu ra vatthu tui mya : [short stories] / Sao Ta Chave – Yan Kun: Sao Ta Cape 1974 [mf ed 1990] – 1r with other items – 1 – (in burmese) – mf#mf-10289 seam reel 170/2 [§] – us CRL [830]
Cape coast castle record book, 1777-1803 – [s.l: s.n, 19–?] – 1 – us CRL [960]
Cape cod sounding – 1989 oct 21-dec 22, v1 n1-2,27,28,30,32 [1990 jan 5-12, jul, 20 [lacks p1-2], jul 27, aug 3,24] – 1r – 1 – mf#1759159 – us WHS [071]
Cape colour question / Macmillan, William Miller – Cape Town, South Africa. 1968 – 1r – us UF Libraries [960]
Cape colour question / Macmillan, William Miller – London, England. 1927 – 1r – us UF Libraries [960]
Cape coloured franchise / Thompson, Leonard Monteath – Johannesburg, South Africa. 1949 – 1r – us UF Libraries [960]
The cape daily telegraph – Port Elizabeth SA, 1898-1908 – 1 – sa National [079]
Cape directory 1800 – Cape Town, South Africa. 1969 – 1r – us UF Libraries [960]
Cape fear record – Wilmington, NC. 1818-1832 (1) – mf#65348 – us UMI ProQuest [071]
Cape florida lighthouse / Clifford, William G – s.l, s.l? 193-? – 1r – us UF Libraries [978]
Cape frontier times – Grahamstown SA, 1840-64 – 6r – 1 – (title varies: colonial times) – mf#MS00249 – sa National [079]
Cape Girardeau. Missouri *see* Proceedings of the bethel church
Cape guardian – Cape Town: Stewart Printing Co, feb 19-jun 11 1937 – 1r – 1 – us CRL [079]
The cape hornet : an illustrated weekly journal – Port Elizabeth SA, 1879 (wkly) [mf ed Cape Town: SA library 1985] – 1r – 1 – mf#MS00378 – sa National [079]
Cape in mid-eighteenth century [the] : being the biography of rudolf siegfried allemann, captain of the military forces and commander of the castle in the service of the dutch east india company at the cape of good hope / Mentzel, O F – Cape Town: T Maskew Miller 1920 – (= ser [Travel descriptions from south africa, 1711-1938]) – 2mf – 9 – (trans fr german by margaret greenlees) – mf#zah-64 – ne IDC [960]
Cape law journal – v1-17. 1984-1900 – 9 – $235.00 – (cont by: south african law journal) – mf#101401 – us Hein [340]
Cape law journal *see* South african law journal
Cape librarian = Kaapse bibliotekaris – Cape Town. 1957-1996 (1) 1972-1996 (5) 1976-1996 (9) – ISSN: 0008-5790 – mf#7038 – us UMI ProQuest [020]
Cape maclear / Cole-King, Pa – Zomba, Malawi. 1968 – 1r – us UF Libraries [960]
Cape malays / Du Plessis, Izak David – Cape Town, South Africa. 1947 – 1r – us UF Libraries [960]
Cape may county times – Sea Isle City, NJ. 1974-1976 (1) – mf#64835 – us UMI ProQuest [071]
The cape mercury – King William's Town SA, 1875-1947 – 123r – 1 (diazo also available at reduced price) – sa National [079]
The cape mercury and weekly magazine – Cape Town SA, 1859 (wkly) [mf ed Cape Town: SA library 1986] – 1r – 1 – (fr jan 1859 as: cape town weekly magazine) – mf#MS00439 – sa National [079]
The cape monitor – Cape Town SA, 1850-62 – 6r – 1 – mf#MS00250 – sa National [079]
The cape monthly magazine – Cape Town: J C Juta. 1857-81 – 1 – (v11 jan-jun 1862) – us CRL [073]
Cape of Good Hope *see* Statistical blue books 1821-1885
Cape of good hope – Cape Town, South Africa. 1911 – 1r – us UF Libraries [960]
Cape of good hope and its dependencies : an accurate and truly interesting description of those delightful regions, situated five hundred miles north of the cape... / Stout, Benjamin – London 1820 [mf ed Hildesheim 1995-98] – 1v on 1mf – 9 – €40.00 – 3-487-27291-1 – gw Olms [916]
Cape of good hope and its dependencies : an accurate and truly interesting description of those delightful regions, situated five hundred miles north of the cape, formerly ingland... / Stout, Benjamin – London: Edwards & Knibb 1820 – (= ser [Travel descriptions from south africa, 1711-1938]) – 2mf – 9 – mf#zah-44 – ne IDC [916]
Cape of good hope and the eastern province f algoa bay / Chase, John Centlivres – Cape Town, South Africa. 1967 – 1r – us UF Libraries [960]
Cape of good hope and the eastern province of algoa bay [the] : with statistics of the colony / Chase, John Centlivres; ed by Christophers, Joseph S – London: Pelham, Richardson 1843 – (= ser [Travel descriptions from south africa, 1711-1938]) – 4mf – 9 – mf#zah-81 – ne IDC [916]
Cape of Good Hope (Colony). Parliament *see* Official publications of the cape of good hope 1854-1910
Cape of good hope government gazette – Cape Town SA, Government Printer 1800-1910 – 155r – 1 – sa National [324]
Cape of good hope impartial observer *see* De meditator
Cape of Good Hope. Laws, Statutes, etc *see* Statutes...1652-1905
The cape of good hope literary magazine – Cape Town: [s.n.], 1847- – 1 – (filmed with: the cape monthly magazine v1 1847. v2 n5-10 feb-dec 1848) – us CRL [800]
Cape Of Good Hope Native Affairs Commission *see* Reports (interim and final) 1910
Cape of good hope official publications, 1854-1910 : cape colonial parliamentary papers, 1854-1910 – [mf ed Cape Town: SA library 1992] – 4358mf – 9 – mf#MFM11595 – sa National [324]
Cape of Good Hope. Parliament. House *see* Index to the annexures and printed papers of the house of assembly... 1854-1897
Cape of good hope / port natal shipping / mercantile gazette – London: BLNL, 5 may 1844-27 dec 1861? – 2r – 1 – (cont by: cape of good hope shipping list) – uk British Libr [380]
Cape of good hope shipping list – Cape Town, South Africa. Cape of Good Hope & Port Natal Shipping & Mercantile Gazette. -w. 7 Jan 1840-28 Dec 1855. Imperfect. – 212r – 1 – £24.00r – uk British Libr Newspaper [380]
Cape Of Good Hope (South Africa) Census Office *see* Results of a census of the colony of the cape of good hope
Cape Of Good Hope (South Africa) Commission On Native Laws *see* Report and proceedings, with appendices, of the government
Cape of Good Hope. South Africa. Parliament *see*
– Papers 1854-1910
– Report of the select committee 1909
Cape of Good Hope. South Africa. Parliament. Legislative Council *see* Report of the select committee 1856-1909
Cape of Good Hope. Surveyor-General's Office *see* Report...with appendices
The cape register – Cape Town SA, 25 apr 1890-4 dec 1903 – 15r – 1 – sa National [079]
Cape standard – Cape Town SA, 1866-69 – 8r – 1 – mf#mp.1119 – sa National [079]
Cape standard – Cape Town: Stewart Printing Co, may 11 1936-nov 25 1947 – 10r – 1 – us CRL [079]
Cape standard *see* South african advertiser and mail
The cape standard – Cape Town SA, 1936-47 – 13r – 1 – sa National [079]
Cape times – Cape Town: R W Murray & F Y St Leger, jul 1938-jan 1986; mar-may 1986 – 1 – us CRL [079]
Cape times – Cape Town SA, 27 mar 1876- (daily) [mf ed Cape Town: SA library c1989] – 1 – sa National [079]
Cape times – Cape Town, South Africa. 1913-1935 (1) – mf#67819 – us UMI ProQuest [079]
Cape times – Cape Town: Times Media Ltd 1876- (daily) [mf ed Johannesburg: Microfile [195-?]- – 1 – (ind in: cape town press index 1876-) – mf#mp.1025 – sa National [079]
The cape times weekly edition – Cape Town SA, 4 jan 1887-26 jan 1917 – 39r – 1 – sa National [960]
Cape town daily news / the general advertiser – SA, 4 jan 1875-30 mar 1878 – 7r – 1 – sa National [079]
Cape town diocesan magazine – v1-10. aug 1939-may 1949 – 3r – 1 – (some orig pp damaged) – mf#atla s0689 – us ATLA [240]
Cape town english press index – Cape Town: SA Library – (= ser South african library index series) – 9 – 0-86968-046-3 – (with: suppl 1871 (13mf) isbn: 0-86968-046-3 and suppl 1872 (12mf) isbn: 0-86968-047-1) – sa National [079]
The cape town english press index 1871-75 / Coates, Peter Ralph – Cape Town: South African Library – 64mf – 11 – sa National [010]
Cape town guide [the] : an illustrated volume of reference for travellers containing information of every character for visitors and residents... / Edwards, Dennis [comp] – 4th ed. Cape Town: Dennis Edwards & Co Publ [1904] – (= ser [Travel descriptions from south africa, 1711-1938]) – 6mf – 9 – mf#zah-16 – ne IDC [916]
The cape town mail – Cape Town SA, 6 mar 1841-53 – 1 – mf#MS00251 – sa National [079]
The cape town mirror – Cape Town SA, 5 sept 1848-26 jun 1849 – 1r – 1 – sa National [079]
Cape Town. South Africa. Civil Rights League *see* Newsletter
[Cape Town]. University of Cape Town [1951] *see* A preliminary survey of the turkana
Cape town weekly magazine *see* The cape mercury and weekly magazine
Cape Verde Islands *see* Boletin oficial, and supplements
Cape Verde Islands. Seccao de Estatistica *see* Anuario estatistico 1933-1952
The cape weekly chronicle – Cape Town SA, W F Mathew, feb 4 1859-jan 27 1860 (wkly) [mf ed Cape Town: SA library 1986] – 1r – 1 – mf#MS00438 – sa National [079]
Capeau, Charles *see* La convention collective de travail (loi du 24 juin 1936) et l'arbitrage obligatoire

Capecelatro, Alfonso *see*
- Christ, the church, and man
- Der heilige philippus neri

Cape-fear recorder – Wilmington NC. 1818 nov 28, 1827 apr 11 – 1r – 1 – mf#858847 – us WHS [071]

Capefigue, Jean Baptiste Honore Raymond *see*
- Diplomatie de la france et de l'espagne depuis l'avenement de la maison de bourbon
- Essai sur les invasions maritimes des normands dans les gaules
- L'europe depuis l'avenement du roi louis-philippe
- Histoire constitutionnelle et administrative de la france depuis la mort de philippe-auguste
- Histoire de la reforme, de la ligue, et du regne de henri 4
- Histoire de la restauration et des causes qui ont amene la chute de la branche ainee des bourbons
- Histoire de philippe-auguste
- La societe et les gouvernements de l'europe
- Trois siecles de l'histoire de france

Capek, Karel *see* Epoque ou nous vivons

Capel llanilterne, glamorgan, chapelry of st ellteyrn : baptisms 1717-1925, burials 1724-1999, marriages 1727-1950 – [Glamorgan]: GFHS [mf ed c2000] – 1mf – 9 – £1.25 – uk Glamorgan FHS [929]

Capel llanilterne, st ellteyrn, monumental inscriptions – 1mf – 9 – £1.25 – uk Glamorgan FHS [929]

Capel-Cure, Edward *see*
- From grace to grace
- Good and faithful service

Capell, Frank J [comp] *see* Compiled ordinances of the city of council bluffs, iowa

Capella de gerardegile / Caine, Caesar – Haltwhistle, England. 1908 – 1r – us UF Libraries [939]

Capelle, Paul *see* Le texte du psautier latin en afrique

Capelletti, Giuseppe *see* Scotch ghost

Capello, H *see* From benguella to the territory of yacca

Capello, Hermenegildo *see* De benguella as terras do iacca...

Capello, Hermenegildo Carlos De Brito *see* From benguella to the territory of yacca

Capenhurst, holy trinity – (= ser Cheshire monumental inscriptions) – 1mf – 9 – £2.50 – mf#10 – uk CheshireFHS [929]

Capers, John G *see* Federal laws governing licensed dealers

Capes, John Moore *see*
- The church of the apostles
- To rome and back
- What can be certainly known of god and of jesus of nazareth?

Capes, Mary Reginald *see* Richard of wyche

Capes, William Wolfe *see*
- The english church in the 14th and 15th centuries
- Roman history, the early empire

Capesius, Josef Franz *see* Das religioese in goethes faust

Capesthorne, holy trinity : burials 1725-1989 – [North Cheshire FHS] – (= ser Cheshire church registers) – 1mf – 9 – £3.00 – mf#423 – uk CheshireFHS [929]

Capetown (south africa) diocesan records / United Society for the Propagation of the Gospel. Archives – 19th c – 15r – 1 – (with int by isobel pridmore) – mf#96722 – uk Microform Academic [025]

Capetown to stockholm / Makepeace, Gordon – Port Elizabeth, South Africa. 1929 – 1r – us UF Libraries [025]

Capgrave, John *see*
- Chronicle of england
- Liber de illustribus henricis

Cap-haitien – Port-Au-Prince, Haiti. 1953 – 1r – us UF Libraries [972]

Capharnauem et ses ruines / Orfali, G – Paris, 1922 – 3mf – 9 – mf#H-2856 – ne IDC [956]

Caphat san kra nann a myui myui = Teaching reading / Kvan, U – Ran Kun: Aon mit chak 1972 [mf ed 1990] – 1r with other items – 1 – (with bibl) – mf#mf-10289 seam reel 145/1 [§] – us CRL [370]

Capilano review – North Vancouver. 1972-1987 (1) 1972-1987 (5) 1979-1987 (9) – ISSN: 0315-3754 – mf#9949 – us UMI ProQuest [420]

Capistrano de abreu / Vianna, Helio – Rio de Janeiro, Brazil. 1955 – 1r – us UF Libraries [972]

Capita pietatis et religionis christianae / Camerarius, J – Lipsiae, 1551 – 1mf – 9 – mf#TH-1 mf 188 – ne IDC [242]

Capitaine belronde / Picard, Louis-Benoit – Paris, France. 1817 – 1r – us UF Libraries [440]

Le capitaine casse-cou / Boussenard, Louis – Montreal: Montreal Printing and Publ, 1903 [mf ed 1985] – 3mf – 9 – mf#SEM105P527 – cn Bibl Nat [830]

Capitaine charlotte / Bayard, Jean-Francois-Alfred – Paris, France. 1842 – 1r – us UF Libraries [440]

Capitaine de voleurs / Xavier – s.l, s.l? 1846 – 1r – us UF Libraries [025]

A capital – Lisbon, Portugal 2 sep 1916-7 aug 1919; 30 apr, 4 may, 25 jun 1974; 13,21 jun 1975 [mf 1916-19] – 1 – uk British Libr Newspaper [074]

Capital – Ellensburg, WA. 1888-1940 (1) – mf#66985 – us UMI ProQuest [071]

Capital – Fredericton, NB. 1880-89 – 8r – 1 – cn Library Assoc [071]

Capital – Hamburg. 1969-1972 (1) 1971-1972 (5) (9) – ISSN: 0008-5847 – mf#5100 – us UMI ProQuest [338]

La capital – Rosario, Argentina: [s.n.] 1945-aug 1948; oct 1948-mar 1949; may 1949-feb 1950; july-oct, 1950; 1951-jun 1953 – 1 – us CRL [079]

Le capital – Paris. 1922-juin 1940 – 1 – fr ACRPP [073]

Capital and class – London. 1990+ (1,5,9) – ISSN: 0309-8168 – mf#18284 – us UMI ProQuest [330]

Capital and labor : containing the views of eminent men of the united states and canada on the labor question, social reform and other economic subjects / ed by Keys, William – Montreal: Dominion Assembly Knights of Labor [1904?] [mf ed 1996] – 3mf – 9 – 0-665-81336-8 – (with pref) – mf#81336 – cn Canadiana [331]

Capital and labour : archives of the employers and workers organizations, 1850-1939 – 37r – 1 – (series 1: archives of the federation of biritish industries minutes, 1916-1939, 11r. series 2: trade union archives, 26r.) – us Primary [331]

Capital and labour – London, UK. 25 Feb 1874-20 Dec 1882. -w. 9 reels – 1 – uk British Libr Newspaper [072]

Capital area ruralist – Madison WI. 1943 jan 28-1946, 1947-1950 aug 10 – 2r – 1 – mf#921457 – us WHS [071]

Capital baptist – New York. 1945+ (1) 1969+ (5) 1975+ (9) – 1 – mf#1094 – us Southern Baptist [242]

Capital chronicle – Salem OR: J H Upton & A Noltner, 1867- [wkly] – 1 – (ceased with v1 n30 (mar 16 1868)?) – us Oregon Lib [071]

Capital city collegian – Helena, MT. 1925-1933 (1) – mf#64454 – us UMI ProQuest [071]

Capital city courier – Lincoln, NE: Wessel & Dobbins, 1885-v8 n27. jun 11 1893 (wkly) [mf ed 1887-93 (gaps)] – 2r – 1 – (cont by: sunday morning courier) – us NE Hist [071]

Capital city letter carrier / National Association of Letter Carriers – 1979 apr-1983, 1984 feb-1990 dec – 2r – 1 – mf#709276 – us WHS [380]

The capital city of canada and its surroundings : the most picturesque capital in the world – Ottawa: H E Dickson, 1892 [mf ed 1980] – 1mf – 9 – 0-665-02067-8 – mf#02067 – cn Canadiana [917]

Capital city quarterly / Madison Urban League [WI] – 1985 aug 29-1988 3rd qtr – 1r – 1 – (cont: quarterly [madison urban league [wi]]; cont by: capitol city quarterly) – mf#1330604 – us WHS [071]

Capital city sun – Lincoln, NE: Sun Newspapers of Lincoln. 12v. v51 n44. nov 23 1961-; -v12 n3. dec 27 1972) (wkly) [mf ed with gaps filmed 1970-78] – 12r – 1 – (split from: lincolnland sun. cont: misc (all issues on filmstrip reel 6). split into: northwest lincoln sun and: southwest lincoln sun. first issue adopts its numbering from lincolnland sun) – us NE Hist [071]

Capital comment – v1-5 n28 [1947 may 3-1951 sep] – 1r – 1 – mf#1054073 – us WHS [071]

Capital de la gran colombia / Castro, Luis Gabriel – Cucuta, Colombia. 1943 – 1r – us UF Libraries [972]

Capital district business review – Albany. 1989-1995 (1) – ISSN: 0747-3699 – mf#16675,01 – us UMI ProQuest [650]

Capital extranjero en la america latina / Cuba Comision Nacional De La Unesco – Habana, Cuba. 1962 – 1r – us UF Libraries [972]

Capital farm and home news – Havelock Sta. (Lincoln), NE: Albert W Ballenger. 3v. v45 n17. feb 29 1936-v47 n16. feb 23 1938 (wkly) [mf ed with gaps] – 1r – 1 – (cont: lancaster county weekly. cont by: lincoln farm and home news) – us NE Hist [071]

Capital flyer – 1981 may/dec-1993 jul/dec – 16r – 1 – (with gaps) – mf#627592 – us WHS [071]

Capital forum – Richmond, VA. 1992-1994 (1) – mf#68918 – us UMI ProQuest [071]

Capital investment sector circular mod / Commercial Advisory Foundation in Indonesia – Djakarta, 1970-1972. nos 1-84 – 11mf – 9 – (missing: 1970-1971(1-40, 42, 43, 58, 83)) – mf#SE-1385 – ne IDC [959]

Capital investments in canada : some facts and figures respecting one of the most attractive investment fields in the world / Field, Frederick William – Montreal: Monetary Times of Canada, c1911 [mf ed 2000] – 3mf – 9 – 0-659-91724-6 – mf#9-91724 – cn Canadiana [332]

Capital journal – Salem, OR. 1955-1980 (1) – mf#60565 – us UMI ProQuest [071]

Capital journal – Topeka, KS. 1980+ (1) – mf#60475 – us UMI ProQuest [071]

Capital journal am – Topeka, KS. 1948-1980 (1) – mf#60474 – us UMI ProQuest [071]

Capital journal (salem, or) *see* Statesman journal

Capital journal [salem, or: 1893] – Salem OR: Capital Journal Pub Co, 1893-95 [daily ex sun] – 1 – (began with feb 10 or feb 11 1893; cont: evening capital journal; cont by: daily capital journal [salem, or: 1896]) – us Oregon Lib [071]

Capital journal [salem, or: 1919] – Salem OR: G Putnam, 1919-80 [daily ex sun] – 1 – (cont: daily capital journal [salem or: 1903]; merged with: oregon statesman [salem or: 1916]; statesman journal) – us Oregon Lib [071]

Capital labor news – Tallahassee, FL. 1960; 1963-1964 [scattered] – 1r – us UF Libraries [071]

Capital market theories and pricing models : evaluation and consolidation of the available body of knowledge / Laubscher, Eugene Rudolph – Uni of South Africa 2001 [mf ed Johannesburg 2001] – 8mf – 9 – (incl bibl ref) – mf#mfm14991 – sa Unisa [332]

Capital punishment / Hartmann, Franz – London, England. 1890 – 1r – us UF Libraries [240]

Capital times – Lincoln, NE: Capital Times, jul 13 1988-jan 24 1991 (wkly) – 5r – 1 – (cont: sun (1987)) – us NE Hist [071]

Capital times – Madison WI. 1928 jan 1/jan 27-1982 jan 2/15 – 394r – 1 – (with gaps) – mf#832537 – us WHS [071]

Capital university law review – v1-28. 1972-2000 – 9 – $560.00 set – ISSN: 0198-9693 – mf#101411 – us Hein [340]

Capital xtra! : ottawa's lesbian and gay monthly – Ottawa. n1- . sep 24 1993- – 1 – (back run (1993-97) 2r can$265) – cn McLaren [305]

La capitale – Denver, CO: La Capitale Publ Co, [dec 29 1917-oct 1923] – us CRL [071]

La capitale – Sacramento, CA: V Panattoni, [dec 8 1917-apr 1944] – 22r – us CRL [071]

Los capitales yanquis en la argentina / Sommi, Luis Victor – Buenos Aires: Editorial Monteagudo, 1949. 212p.Illus.Incl. bibliog. 1 reel. 1273 – 1 – us Wisconsin U Libr [336]

Capitalism, nature, socialism – Santa Cruz. 1992-1996 (1) – ISSN: 1045-5752 – mf#18384 – us UMI ProQuest [333]

Capitalism, socialism, or villagism? / Kumarappa, Bharatan – Madras: Shakti Karyalayam, 1946 – (= ser Samp: indian books) – us CRL [320]

Capitalismo del centavo / Tax, Sol – Guatemala, v1-2. 1964 – 1r – us UF Libraries [972]

Capitals of jamaica / Roberts, Walter Adolphe – Kingston, Jamaica. 1955 – 1r – us UF Libraries [972]

El capitan diego de caceres ovando paladin extremeno de los reyes catolicos / Munoz de San Pedro, Miguel – Badajoz: imp. de la dipt.prov., 1952 – 1 – sp Bibl Santa Ana [946]

El capitan don gonzalo pizarro, padre de pizarro hernando, juan y gonzalo pizarro, conquistadores del peru / Cuneo-Vidal, Romulo – Madrid: rev. arch. bibliot. y museos, 1926 – 1 – sp Bibl Santa Ana [920]

El capitan general marques de monsalud / Monsalud, Marques de – Madrid: suc, de rivadeneyra, 1909 – 1 – sp Bibl Santa Ana [920]

Capitanes generales. cartas florida y luisiana (anno 1764-1823) – Sevilla – 43r – 5,6 – sp Cultura [977]

Capitanes generales de la habana – Sevilla – 34r – 5,6 – sp Cultura [972]

Capitania das minas gerais / Lima Junior, Augusto De – Rio de Janeiro, Brazil. 1943 – 1r – us UF Libraries [972]

Capitania de sao paulo / Luiz Pereira De Souza, Washington – Sao Paulo, Brazil. 1938 – 1r – us UF Libraries [972]

Capitanias paulistas / Calixto, Benedicto De Jesus – Sao Paulo, Brazil. 1927 – 1r – us UF Libraries [972]

Capito und butzer, strassburgs reformatoren : nach ihrem handschriftlichen briefschatze, ihren gedruckten schriften und anderen gleichzeitigen quellen / Baum, Johann Wilhelm – Elberfeld: R L Friderichs 1860 [mf ed 1989] – (= ser Leben und ausgewaehlte schriften der vaeter und begruender der reformirten kirche 3) – 2mf – 9 – 0-7905-4065-7 – mf#1988-0065 – us ATLA [242]

Capito und butzer, strassburgs reformatoren : nach ihrem handschriftlichen briefschatze, ihren gedruckten schriften und anderen gleichzeitigen quellen / Baum, Johann Wilhelm – Elberfeld: R.L. Friderichs, 1860. (Leben und ausgewaehlte Schriften der Vaeter und Begruender der reformirten Kirche; 3. T.) – 2mf – us ATLA [240]

Capitol : woman: a newsletter of the house committee onconstitutional revision and women's rights – v7 n1-v10 n6 [1983 jan-1986 dec] – 1r – 1 – (cont by: capitol women) – mf#1277535 – us WHS [305]

Capitol building of tallahassee, florida / Bosworth, Mary C – s.l, s.l? 193-? – 1r – us UF Libraries [978]

Capitol building of tallahassee, florida – s.l, s.l? 193-? – 1r – us UF Libraries [978]

Capitol bulletin / Minnesota Women's Consortium – n1-379 [1981 feb 2-1988 dec 21] – 1r – 1 – mf#1054076 – us WHS [305]

Capitol city fed / United Federation of Postal Clerks – 1971 mar – 1r – 1 – mf#635206 – us WHS [380]

Capitol comment / Wisconsin Hospital Association – v1 n1-v3 n1 [1979 jan 5-1981 jan 16,1981 feb-1984 mar 23] – 1r – 1 – (cont: legislative lookout) – mf#957564 – us WHS [360]

Capitol communal / Capitol Senior High School [Baton Rouge LA] – v11 n1-4 [1995 sep-1996 may] – 1r – 1 – mf#3912563 – us WHS [373]

Capitol drumbeat / Arizona Commission of Indian Affairs – 1979-91 – 1r – 1 – (cont by: capitol drumbeat newsletter) – mf#470221 – us WHS [306]

Capitol headline – 1977 feb 10/aug 26-1994 sep/dec – 55r – 1 – (with gaps; cont by: capitol headlines from the legislative reference bureau) – mf#467096 – us WHS [071]

Capitol hill beacon – Oklahoma City, OK. 1948-1956 (1) – mf#65788 – us UMI ProQuest [071]

Capitol notes / National Association of Letter Carriers [US] – v2 n1-v6 n4 [1980 feb-1984 nov] – 1r – 1 – (cont by: postmark washington) – mf#853644 – us WHS [380]

Capitol report / Texas AFL-CIO [American Federation of Labor and Congress of Industrial Organizations] – 1981 jan 19-1986 apr 3 [v1 n1-v6 n1, i.e. 2], 1987 feb – 1r – 1 – mf#1289069 – us WHS [331]

Capitol Senior High School [Baton Rouge LA] *see* Capitol communal

Capitol studies – Washington. 1972-1978 (1) 1972-1978 (5) 1976-1978 (9) – (cont by: congressional studies) – ISSN: 0045-5687 – mf#7498 – us UMI ProQuest [320]

[Capitola-] mid-county post – CA. 1990-1994 – 5r – 1 – $300.00 – mf#B02094 – us Library Micro [071]

The capitolan *see* [Santa cruz-] miscellaneous titles

Capiton, W *see*
- Hexemeron dei opus
- In habakuk prophetam enarrationes
- In hoseam prophetam commentarius
- Institutionum hebraicarum, libri duo
- Responsio, de missa...

Capitularia regum francorum *see* Monumenta germaniae historica leges 2. leges in quarto. a legum sectio 2 (mgh leges 2b)

Capitularia regum francorum (mgh leges. 1:1.bd) – 1835 – (= ser Monumenta germaniae historica leges 1. leges in folio (mgh leges 1)) – €31.00 – ne Slangenburg [240]

Capitulo de la autobiografia de marti / Carbonell, Nestor – Habana, Cuba. 1946 – 1r – us UF Libraries [972]

Capitulos da historia social de s paulo / Ellis Junior, Alfredo – Sao Paulo, Brazil. 1944 – 1r – us UF Libraries [972]

Capitulos da sociologia brasileira / Dornas, Joao – Rio de Janeiro, Brazil. 1955 – 1r – us UF Libraries [972]

Capitulos de historia colonial, 1500-1800 / Abreu, Joao Capistrano De – Rio de Janeiro, Brazil. 1954 – 1r – us UF Libraries [972]

Capitulos de un libro sobre historia financiera de / Gonzalez Viquez, Cleto – San Jose, Costa Rica. 1965 – 1r – us UF Libraries [972]

Capitulos escogidos de la geografia fisica / Pittier, Henri – San Jose, Costa Rica. 1942 – 1r – us UF Libraries [972]

Caplan, Gerald L *see* Elites of barotseland, 1878-1969

Caplow, Theodore *see* Urban ambience

Cap'n warren's wards / Lincoln. Joseph Crosby – Toronto: McLeod & Allen, c1911 [mf ed 1995] – 5mf – 9 – 0-665-74854-X – (ill by edmund frederick) – mf#74854 – cn Canadiana [830]

Capo, Jose Maria *see* Tres dictadores negros

Capo-Bianco, A *see* Corona e palma militaire di artiglieria et fortificatione

Capomazza, Ilario *see*
- La lingua degli afar
- La lingua degli afar; vocabolario italiano-dankalo e dankalo-italiano

Le caporal *see* La sentinelle du peuple

Capot! : ou, les adieux au pouvoir, chanson nouvelle – Paris [1848?] – us CRL [944]

Cappa, Ricardo S J *see*
- Estudios criticos acerca de la dominacion espanola en america
- Historia de peru

Cappadelta, Luigi *see* Luther

Cappel, L *see*
- Critica sacra...
- Critica sacra
- Syntagma thesium theologicarum in academia salmurensi disputatarum

Cappelle, Herman Van *see*
- Au travers des forets vierges de la guyane holland
- Binnenlanden van het district nickerie
- Essai sur la constitution geologique de...

Cappelletti, G *see* Storia dell'isola di san lazzaro e della congregazione de'monaci armeni, unita alla storia delle magistrature venete

Cappello, Felice M *see* De visitatione ss. liminum et dioeceseon ac de relatione s. sedi exhibenda

Capper, Arthur *see* Letters and speeches

Capper, Charles *see* Port and trade of london

Capra, Daniel J *see*
- Advisory committee notes to the federal rules of evidence that may require clarification
- Case law divergence from the federal rules of evidence

Caprara, A *see*
- Funebris pompa serenissimi ranutii farnesii parmae et placentiae ducis 4...
- Insegnamenti del vivere del conte alberto caprara

Capreolus, Johannes (Capreolus, Jean) *see* Defensiones theologiae divi thomae aquinatis

Caprice brilliant pour piano et violon / Saint-Saens, Camille – [1859] [mf ed 19–] – 1r – 1 – mf#film 1325 – us Sibley [780]

Capricioso / Dolin, Anton – New York, 1941 – 1r – 1 – (pictures from the ballets presented by the ballet theatre at the majestic theatre, new york, mar 6th 1941) – mf#*ZC-2 – Located: NYPL – us Misc Inst [790]

Capricorn africa – [London?: s.n. 1953?] – us CRL [960]

Capricorn Africa Society *see*
- Handbook for speakers
- Newsletter

The capricorn convention : the world press on the capricorn africa society's declarations – [London: Neane, 1953] – 1 – us CRL [960]

The capricorn convention : the world press on the capricorn africa society's declarations – [London: Neane, 1953] – us CRL [960]

Capricornian – Rockhampton, Australia. 3 jan 1885-1929 – 104r – 1 – uk British Libr Newspaper [072]

The capricornian – Rockhampton, Australia. 3 Jan 1885-26 Dec 1929.-d. 105 reels – 1 – uk British Libr Newspaper [072]

Capriotti, Paul V *see* The effects of acute dietary creatine supplementation on power output indices and blood lactate concentrations during high-intensity intermittent cycling exercise

Capron, Frederick Hugh *see* The conflict of truth

Capron trail / Comstock, Bertha – s.l, s.l? 1936 – 1r – us UF Libraries [978]

Capron trail / Huss, Veronica E – s.l, s.l? 193-? – 1r – us UF Libraries [978]

Caps and taps : house organ of adolph coors company – Golden, CO: Adolph Coors Company (mf ed 1990) – 1r – 1 – (vol for 1973 incl special centennial iss. spring iss fr 1955-60, 1968 called "award edition") – mf#MF Ca174 – us Colorado Hist [660]

Capstone / Howard University – 1986 mar 3, apr 28-1997 jan 22, feb 4,25, mar 28, apr 18, may 6, jun 16, aug 11,26, sep 2,29, nov 17, dec 16, [1998 jan 27-2000 may 29] – 2r – 1 – (cont by: capstone online) – mf#2540197 – us WHS [378]

Capsulas gelatinosas / Gonzalez, Antonio – 1 – sp Bibl Santa Ana [946]

Capt, Louis *see* Gellerts lustspiele

Captain america – iss n1-60. mar 1941-jan 1947 – 15 – mf#007MV-016MV; 052MV-053MV – us MicroColour [740]

Captain andrew jackson lea : madison county – s.l, s.l? 193-? – 1r – us UF Libraries [978]

Captain battle / boy comics – iss n1-2 (capt battle); iss n3-5 (boy comics) sum 1941-aug 1942 – 15 – mf#001GL – us MicroColour [740]

Captain billy's whiz bang – December 1920 – August 1936 – 5r – 1 – $150.00 $30.00r – us Minn Hist [870]

Captain brand of the schooner 'centipede' / Wise, Henry Augustus – New York, NY. 1894 – 1r – us UF Libraries [972]

Captain clapperton's last expedition to africa : from the royal commonwealth society library / Lander, Richard – 2v. 1830 – 9mf – 7 – mf#2986 – uk Microform Academic [916]

Captain, Gwendolyn *see* Social, religious, and leisure pursuits of northern california's african american population

Captain lightfoot : the last of the new england highwaymen / G FD – Topsfield, MA: The Wayside Press, 1926 – 2mf – 9 – $3.00 – mf#LLMC 92-125 – us LLMC [975]

Captain Mac *see* Canada

Captain robert percival's, verfassers einer beschreibung von ceylon beschreibung des vorgebirgs der guten hoffnung : nach seinem ehemaligen und jetzigen zustande, in historischer, geographischer, typographischer, statistischer und kommerzieller hinsicht; aus dem englischen / Percival, Robert – Weimar 1805 [mf ed Hildesheim 1995-98] – 1v on 4mf – 9 – €120.00 – 3-487-26578-8 – gw Olms [915]

Captain sir john ross zweite entdeckungsreise nach den gegenden des nordpols : 1829-1833 – Berlin 1835-[36] [mf ed Hildesheim 1995-98] – 4v on 11mf – 9 – €110.00 – 3-487-27079-X – gw Olms [919]

Captain tracy b. kittredge's "the evolution of global strategy" / U.S. Joint Chiefs of Staff – (= ser Records Of The U.S. Joint Chiefs Of Staff) – 1r – 1 – mf#T1174 – us Nat Archives [355]

Captains and comrades in the faith : sermons historical and biographical / Davidson, Randall Thomas – London: John Murray, 1911 – 1mf – 9 – 0-7905-4288-9 – mf#1988-0288 – us ATLA [240]

Captains of brazil / Sanceau, Elaine – Porto, Portugal. 1965 – 1r – us UF Libraries [972]

Captive / Mere, Charles – Paris, France. 1920 – 1r – us UF Libraries [440]

The captive city of god : or, the churches seen in the light of the democratic ideal / Heath, Richard – London: Headley 1905 [mf ed 1990] – 1mf – 9 – 0-7905-6178-6 – (original ed publ 1904) – mf#1988-2178 – us ATLA [230]

The captive missionary : being an account of the country and people of abyssinia / Stern, Henry Aaron – London, New York: Cassell, Petter and Galpin, [1869]. Chicago: Dep of Photodup, U of Chicago Lib, 1971 (1r); Evanston: American Theol Lib Assoc, 1984 (1r) – 1 – 0-8370-0310-5 – mf#1984-B270 – us ATLA [240]

The captive missionary : being an account of the country and people of abyssinia embracing a narrative of king theodore's life, and his treatment of political and religious missions / Stern, H A – London, [1868] – 5mf – 9 – mf#HTM-185 – ne IDC [916]

Captive of the simbas / Hayes, Margaret – New York, NY. 1966 – 1r – us UF Libraries [960]

Captives of capitalism / International Workers Aid. Committee – Chicago: International Workers' Aid, [192-?] (mf ed 19–) – 16p – mf#ZT-167 – us Harvard [335]

Captives of tipu : survivors' narratives / ed by Lawrence, A W – London: Jonathan Cape, 1929 – (= ser Samp: indian books) – us CRL [954]

The captivity and the pastoral epistles : with introduction and notes / Strahan, James – New York: Fleming H Revell; London: Andrew Melrose [191-?] – 1mf – 9 – 0-7905-1379-X – (incl ind) – mf#1987-1379 – us ATLA [227]

The captivity, sufferings, and escape, of james scurry : who was detained a prisoner during ten years, in the dominions of hyder ali and tippoo saib / Scurry, James – London 1824 [mf ed Hildesheim 1995-98] – 1v on 3mf – 9 – €90.00 – 3-487-27421-3 – gw Olms [920]

Capture : roman d'amour / Telpail, Prosper – [Berthierville?: s.n, entre 1940 et 1967] (mf ed 1993) – 1mf – 9 – mf#SEM105P1897 – cn Bibl Nat [830]

Capture of havana in 1762 by the forces of george – Cambridge, MA. 1898 – 1r – us UF Libraries [972]

The capture of mount washington : november 16th, 1776, the result of treason / De Lancey, Edward Floyd – New York: s.n, 1877 – 1mf – 9 – mf#03790 – cn Canadiana [975]

Captured by franco / Ornitz, Lou – New York: Friends of the Abraham Lincoln Brigade 1939 – 9 – mf#w1091 – us Harvard [946]

Captured german documents filmed at berlin, 1960 / American Historical Association – (= ser National archives coll of foreign records seized, 1941-) – 986r – 1 – mf#T580 – us Nat Archives [943]

Captured german records filmed at berlin / University of Nebraska – (= ser Records Of The Headquarters Of The German Navy High Command (Okm).) – 49r – 1 – mf#T611 – us Nat Archives [943]

Captured japanese ships' plans and design data – (= ser Records of the bureau of ships) – 10r – 1 – mf#M1176 – us Nat Archives [355]

Capus, Alfred *see* Le personnel feminin des p t t pendant la guerre

Caputo, Jennifer L *see* Psychosocial stress and abominal fat patterning in black premenopausal women

Car and driver – New York. 1955+ (1) 1971+ (5) 1969+ (9) – ISSN: 0008-6002 – mf#1645 – us UMI ProQuest [380]

Car craft – Los Angeles. 1953+ (1) 1971+ (5) 1974+ (9) – ISSN: 0008-6010 – mf#3058 – us UMI ProQuest [790]

Car exchange – 1979 apr-1979, 1980, 1981, 1982, 1983, 1984 jan-apr, 1986 jun-1987 sep – 6r – 1 – mf#661575 – us WHS [380]

Car model – Phoenix. 1962-1973 (1) 1971-1972 (5) – ISSN: 0008-6045 – mf#2409 – us UMI ProQuest [790]

Car review – London, UK. Aug, Nov, Dec 1906.-irr. 13 feet – 1 – uk British Libr Newspaper [072]

Car trust securities / Rawle, Francis – A paper read at the eighth annual meeting of the American Bar Association at Saratoga Springs, New York, August 20th, 1885. Philadelphia, Dando, 1885. 48 p. LL-1208 – 1 – us L of C Photodup [340]

Car wheel / Committee Against Racism – 1974 feb-1975 sum – 1r – 1 – mf#354482 – us WHS [305]

Car window glimpses : en route to quebec by daylight via quebec central railway – New York?: Leve & Alden's Publication Dept, 18–? – 1mf – 9 – mf#28184 – cn Canadiana [380]

Cara a cara / Spanish Speaking Catholic Commission – 1976 mayo/unio-1980 nov – 1r – 1 – mf#620786 – us WHS [241]

Caraballo, Vicente *see* Negro obeso

Carabelas de espana...pinzon, juan de la cosa, diaz de solis juan sebastian elcano / Bayle, Constantino & Cabal, Juan – Madrid: Razon y Fe, 1944 – 1 – sp Bibl Santa Ana [946]

An carabhan = Karavane / Hauff, Wilhelm & O Moghrain, Padraic – [Baile Atha Cliath]: O Fallamhain i Gcomhar le hOifig an tSolathair 1930 [mf ed 1990] – 1r – 1 – (filmed with: der frosch / otto erich hartleben) – mf#2699p – us Wisconsin U Libr [830]

Carabinades / Choquette, Ernest – Montreal: Deom Freres, 1900 – 3mf – 9 – mf#06045 – cn Canadiana [610]

Caracas, 1935 / Parra, Caracciolo – Madrid: Razon y Fe, 1935 – 1 – sp Bibl Santa Ana [946]

Caracas politica, intelectual y mundana. caracas, 1966 / Parra Marquez, Hector – Madrid: Graf. Calleja, 1968 – 1 – sp Bibl Santa Ana [321]

Caracci, A *see* Le arti di bologna disegnate da annibale caracci ed intagliate da simone guillini coll'assistenza di alessandro algardi

Caracciolo, Henrietta *see* Memoirs of henrietta caracciolo

El caracter : definicion, importancia, ideal, origenes... / Guibert, J; ed by Bayle, Constantino – Madrid: Razon y Fe, 1928 – 9 – sp Bibl Santa Ana [150]

Caracter de la literatura hebrea / Halevy, Fabian S – Buenos Aires, Argentina. 1928 – 1r – us UF Libraries [939]

Caracter de la revolucion guatemalteca / Diaz Rozzotto, Jaime – Mexico City? Mexico. 1958 – 1r – us UF Libraries [972]

Caractere, culture, vodou / Derose, Rodolphe – Port-Au-Prince, Haiti. 1955 – 1r – us UF Libraries [390]

Caracteres : pages choisies / La Bruyere, Jean de – London: J M Dent; New York: G P Putnam 1907 [mf ed 1984] – 1r – 1 – mf#946 – us Wisconsin U Libr [440]

Caracteres chinois / Wieger, Leon – 3rd ed [Hien-hien]: [s.n.], 1916 [mf ed 1995] – (= ser Yale coll) – 1200p (ill) – 0-524-09592-2 – (in french) – mf#1995-0592 – us ATLA [480]

Caracteres constantes en las letras cubanas / Estenger, Rafael – Havana, Cuba. 1954 – 1r – us UF Libraries [972]

Les caracteres des passions... / La Chambre, Marin Cureau de – 2e rev corr ed. Paris: Chez Jacques d'Allin. 5v. 1662 [mf ed 1978] – 1r – 1 – mf#SEM35P159 – cn Bibl Nat [150]

Caracterisation du milieu et utilisation des donnees numeriques de teledetection pour la cartographie de l'evapotranspiration : un exemple sur le senegal / Mbaye, Constance – (mf ed 2000) – 2mf – 9 – €40.00 – 3-8267-2726-6 – mf#DHS 2726 – gw Frankfurter [960]

Caracteristicas de la actividad agropecuaria en co... / Costa Rica Oficina De Planificacion – San Jose? Costa Rica. 1965 – 1r – us UF Libraries [972]

Caracteristicas de la carta preliminar / Martorell Otzet, Ramon – Ciudad Trujillo, Dominican Republic. 1947 – 1r – us UF Libraries [972]

Las caracteristicas de la revolucion espanola / Togliatti, Palmiro – Barcelona, 193? – (= ser Blodgett coll) – 9 – mf#fiche w1229 – us Harvard [946]

Caracterizacion y propiedades de una vermiculita de badajoz / Gonzalez, F et al – Madrid: CSIC, 1954. Sep Ana. Edaf. y Fisio. Veg. Tomo 13. no 2. 1954 – 1 – sp Bibl Santa Ana [946]

Le caraeme / Ermoni, Vincent – Paris: Bloud, 1907 – (= ser Science et religion) – 1mf – 9 – 0-524-03463-X – (incl bibl ref) – mf#1990-1006 – us ATLA [240]

Carafa, Vincent *see* Elevations a dieu

The carafas of maddaloni: naples under spanish dominion / Reumont, Alfred von – Trans. from the German of Alfred von Reumont. London: H.G. Bohn, 1854. xiv,465p. Incl. geneal. tables. 1 reel. 1262 – 1 – us Wisconsin U Libr [945]

Caragoli / Pirch, Otto von – Berlin 1832 [mf ed Hildesheim 1995-98] – 2v on 4mf – 9 – €1200.00 – 3-487-29150-9 – gw Olms [914]

Caraja...kou trois ans chez les indiens du brasil / Falaise, Rayliane de la – Madrid: Razon y Fe, 1940 – sp Bibl Santa Ana [306]

Caramel apple – [1983] dec/jan-1984 feb – 1r – 1 – mf#4848421 – us WHS [071]

Caramuel, J *see* [P]raecursor logicus...cuius partes tres

Carande, Bernardo Victor *see* Manuel conmigo, ilustraciones de enrique sopena scapardini

O carangolense : orgao litterario, noticioso e agricola – Carangola, MG: Typ do Carangolense, 10 ago 1884 – (= ser Ps 19) – bl Biblioteca [079]

A carapuca – Rio de Janeiro, RJ: Typ Carioca de J I da Silva & Comp, 27 fev 1850 – (= ser Ps 19) – 1,5,6 – mf#P15,01,65 n.02 – bl Biblioteca [079]

O carapuceiro : periodico sempre moral, e so por accidens politico – Pernambuco: Typ Fidedigna, 1832-1834,1837-1840,1842 – (= ser Ps 19) – mf#P19,02,37-42 – bl Biblioteca [320]

Caras y caretas – Buenos Aires. v. 6-42. 26 dec 1903-7 oct 1939 – 1 – us NY Public [073]

Caravalho, Nelson R *see* Operacao brasil

The caravan – Brooklyn NY, 1953-61 – 4r – 1 – (arabic newspaper) – us IHRC [071]

Caravan, Ronald L *see* Counterpoint in the music of claude debussy

Caravana pasa / Dario, Ruben – Paris, France. 1919? – 1r – us UF Libraries [972]

Caravanner – Bakersfield, CA. 1959-1969 (1) – mf#62089 – us UMI ProQuest [071]

Caravasios, Peter *see* Greek proverbs from mrs peter caravasios

Caravel – Marjorca. n1-5. summer 1934-mar 1936 – 1 – us NY Public [073]

Caravelle : cahiers du monde hispanique et luso-bresilien – Toulouse. n1-17. 1963-71 – 1 – fr ACRPP [972]

Carayon, A *see* Relations inedites des missions de la compagnie de jesus...constantinople et dans le levant au 17e siecle

Carballeyra, Leopoldo *see* Poemas revolucionarios

Carballido Rey, Jose M *see* Gallo pinto

Carballo, Julio *see* Teatro infantil

Carbohydrate polymers – London. 1981+ (1) 1981+ (5) 1987+ (9) – ISSN: 0144-8617 – mf#42244 – us UMI ProQuest [540]

Carbohydrate research – Amsterdam. 1965+ (1) 1965+ (5) 1987+ (9) – ISSN: 0008-6215 – mf#42178 – us UMI ProQuest [540]

Carbon – New York. 1963+ (1,5,9) – ISSN: 0008-6223 – mf#49026 – us UMI ProQuest [530]

The carbon advocate – Mauch Chunk, PA. 1890-94. 1 roll – 13 – $25.00r – us IMR [071]

Carbon county chronicle – Red Lodge, MT. 1903-1925 (1) – mf#64618 – us UMI ProQuest [071]

Carbon county democrat – Mauch Chunk, PA., 1847-1873 – 13 – $25.00r – us IMR [071]

Carbon county democrat – Red Lodge, MT. 1899-1902 (1) – mf#64619 – us UMI ProQuest [071]

Carbon county democrat and mauch – Mauch Chunk, PA., 1848-1849 – 13 – $25.00r – us IMR [071]

Carbon county gazette – Mauch Chunk, PA., 1844-1847 – 13 – $25.00r – us IMR [071]

Carbon county gazette – Red Lodge, MT. 1905-1907 (1) – mf#64620 – us UMI ProQuest [071]

Carbon county gazette and mauch – Mauch Chunk, PA., 1849-1852 – 13 – $25.00r – us IMR [071]

Carbon county journal – Red Lodge, MT. 1909-1918 (1) – mf#64621 – us UMI ProQuest [071]

Carbon county sentinel – Gebo, MT. 1898-1902 (1) – mf#64397 – us UMI ProQuest [071]

Carbon county transit – Mauch Chunk, PA., 1843-1844 – 13 – $25.00r – us IMR [071]

Carbon county transit & gazette – Mauch Chunk, PA. 1843-1846. Also incl. Carbon County Gazette (Mauch Chunk), 1844-47; Carbon Democrat (Mauch Chunk), 1847-48; Carbon County Gazette & Mauch Chunk Courier, 1847-48. 1 roll – 13 – $25.00r – us IMR [071]

The carbon democrat – Mauch Chunk, PA. 1853-1869. Also incl. Mauch Chunk Gazette, 1856-58. 1 roll – 13 – $25.00r – us IMR [071]

Carbon news – Alberta, CN. jan 1927-dec 1960 – 12r – 1 – cn Commonwealth Imaging [071]

Carbon star *see* Miscellaneous newspapers of weld county

Carbone, Caesar *see* De modernistarum doctrinis

Carbonell, Abel *see* Por la doctrina
Carbonell, Abel Francisco *see* Quincena politica
Carbonell Barberan, Ramiro *see* Legislacion notarial
Carbonell, Diego *see* Lo morboso en ruben dario
Carbonell, Jose Manuel *see*
– Carlos a boissier y diaz
– Evolucion de la cultura cubana
– Juan clemente zenea, poeta y martir
– Manuel sanguily, adalid, tribuno y pensador
– Pedro angel castellon
Carbonell, Miguel Angel *see*
– Elogio de los fundadores
– Sanguily que yo conoci
Carbonell, Nestor *see*
– Capitulo de la autobiografia de marti
– General ramon leocadio bonachea
– Marques [salvador cisneros betancourt]
– Marti
– Prosas oratorias
Carbonell Y Rivero, Miguel Angel *see* Varona que yo conoci
Carbonero Bravo, D *see* Ganado karakul
Carbones encendidos / Ulloa, Juan – San Salvador, El Salvador. 1946 – 1r – us UF Libraries [972]
Carbonneau, Fred *see* The girl i can't forget
Carbonneau, Leopold *see*
– Bibliographie analytique de l'oeuvre de m albert rioux
– Bio-bibliographie de m albert rioux
Carbonneau, Louis *see* Fievres d'afrique
Carbra, Romulo D *see* La cronica oficial de las indias occidentales. la plata 1934
Carcamaya vilakkam : selections from tamil literature relating to religion and morals = Satsamaya vilakkam / ed by Popley, Herbert A – Madras: Christian Literature Soc for India 1915 [mf ed 1995] – (= ser Samp: indian books 40824) – 1r – 1 – (master mf held by crl) – mf#mf-10452 r041 – us CRL [490]
Carcanet – nos. 1-5. 1969-70 – 1 – us AMS Press [810]
Carcassonne and company / Robinson, Holland – Binghampton, NY. 1926 – 1r – us UF Libraries [720]
La carcel de mujeres de madrid / Bayle, Constantino – Burgos: Razon y Fe, 1938 – 1 – sp Bibl Santa Ana [946]
Carcinogenesis – Oxford. 1988-1996 (1,5,9) – ISSN: 0143-3334 – mf#16446 – us UMI ProQuest [616]
Carco, Francis *see* Prisons de femmes
Carcoar chronicle – Carcoar, 1863-1943 (misc issues) – 5r – A$290.58 vesicular A$318.08 silver – at Pascoe [079]
Card catalog of foreign publications – 895mf – 9 – $6,000.00 coll – us UMI ProQuest [020]
Card catalog of gubernskie, oblastnye and voiskovye vedomosti from the national library of russia, st petersburg – [mf ed Norman Ross Publ] – 87mf – 9 – (87 vedomosti listed, comprising the biggest coll in a single library) – us UMI ProQuest [077]
Card catalog of hermitagiana – (mf ed 2000) – 117mf – 9 – $499.00 – us UMI ProQuest [060]
Card catalog of russian books and serials – 480mf – 9 – $3,000.00 coll – us UMI ProQuest [020]
Card catalog of russian personalities (b l modzalevskii collection) : from the manuscript department of the institute of russian literature of the russian academy of science (pushkinskii dom) / Modzalevskii, B L – 364mf – 9 – $2,200.00 coll – (a coll of biographical materials from the 18th to early 20th century, reflecting little-known facts and background material relating to nearly 100,000 influential people in russian art and society. a large number of cards pertain to less well known people but whose contribution to the development of russian culture was essential) – us UMI ProQuest [920]
Card catalog of the department of the literature of the nationalities of the former soviet union / Russian National Library – 2703mf – 9 – $13,500.00 coll – us UMI ProQuest [020]
Card catalog of the former library of the russkii zagranichnyi istoricheskii arkhiv (rzia) : records from the former library of the prague archive of the slovanska knihovna – 1945-90 [mf ed Norman Ross Publ] – 267mf [576cards/mf] – 9 – (in russian. contains books, periodicals, & newspapers from emigre communities in the ussr & around the world. int by richard kneeley) – us UMI ProQuest [020]
Card catalog of the g w blunt white research library at mystic seaport museum : a card catalog of maritime and nautical history – [mf ed Chadwyck-Healey] – 133mf – 9 – uk Chadwyck [380]
Card catalog of the library of the state hermitage museum – (mf ed 2000) – 2 card catalogs – 9 – $7300.00 set – (consists of: russian books and serials 480mf $2700. foreign publs 897mf $5000) – us UMI ProQuest [020]
The card catalog of the music library of the st petersburg state conservatory (rimsky-korsakov) – 312mf – 9 – $2,000.00 coll – us UMI ProQuest [780]
The card catalog of the peace palace library, the hague / Peace Palace Library. The Hague – Clearwater Publ Co – 1814mf (24:1) – 9 – $12,140.00 – (periodicals ref guide $4580. universal bibl catalogue $7995. suppl through 1984 $2615 317mf) – us UPA [020]
Card catalog of the slavic collection of the library of the academy of sciences, st petersburg, 16th century to 1930 / Russian National Library – (mf ed 1996) – 434mf – 9 – $2500.00 – us UMI ProQuest [460]
Card catalogs of the harvard law school library, 1817-1981 – [mf ed 1984-85] – 2422mf (42:1) – 9 – $8315.00 – (author-title catalog 1750,000 cards on 1085mf $4375. anglo-american subject catalog 470,000 cards on 288mf $1155. foreign & comparative law subject catalog 520,000 cards on 330mf $1330. catalog of international law & relations 525,000 cards on 342mf $1365. jurisdictional shelf-list 550,000 cards on 288mf $1155. international shelf-list 200,000 cards on 89mf $350) – us UPA [020]
Card catalogs of the harvard university fine arts library, 1895-1981 – [mf ed 1984] – 516mf [42:1] – 9 – $4895.00 – (sold as complete set wh incl foll: dictionary catalog 355mf, catalog of auction sales catalogs 21mf, shelf-list catalog 107mf, catalog of the ruebel asiatic research collection 33mf) – us UPA [020]
[Card catalogue (authors and titles) of the institute's library] / South African Institute of Race Relations Library – Johannesburg, Microfile, 1969 – us CRL [020]
Card catalogue file / Southern Baptist Theological Seminary. Library – 240,000p – 1 – us Southern Baptist [242]
Card catalogue of the working class movement : books and pamphlets. 1. author. 2. subject – 30r – 5 – £1,200.00 – (also on mf) – mf#CCW – uk World [331]
Card file of baptist materials / Southern Baptist Theological Seminary. Library – 6,000p – 1 – us Southern Baptist [242]
Card, Henry *see* Historical outlines of the rise and establishment of the papal power
Card index to 'old loan' ledgers of the bureau of the public debt, 1790-1836 / U.S. Treasury Dept. Bureau of the Public Debt – (= ser Records of the bureau of the public debt) – 15r – 5 – (with printed guide) – mf#M521 – us Nat Archives [336]
Card index to pictures collected by the george washington bicentennial commission – (= ser Records of exposition, anniversary, and memorial commissions) – 1r – 1 – mf#T271 – us Nat Archives [020]
Card manifests (alphabetical) of entries through the port of detroit, michigan, 1906-1954 – (= ser Records of the immigration and naturalization service, 1891-1957) – 117r – 1 – mf#M1478 – us Nat Archives [975]
Card records of headstones provided for deceased union civil war veterans, ca 1879-ca 1903 – 22r – 1 – mf#M1845 – us Nat Archives [976]
Cardaire, Michel *see* L'islam et le terroir africaine
Cardano, Girolamo *see* De rerum varietate libri 17...
Cardauns, Herman *see* Die briefe der dichterin annette v droste-huelshoff
Cardauns, Hermann *see*
– Aus luise hensels jugendzeit
– Die goerres-gesellschaft, 1876-1901
– Klemens brentano
– Die kommende romantik philipp veit und ernst lieber
Cardauns, Ludwig *see* Zur geschichte der kirchlichen unions- und reformbestrebungen von 1538 bis 1542
Cardecera, Valentin *see* Retrato de don pedro de valdivia. informe
Car-del digest – v1 n1-4 [1982 feb-aug] – 1r – 1 – (cont: chedwato dispatch) – mf#622377 – us WHS [071]
Car-del scribe / Chedwato Service – 1979 jan-1984 may – 1r – 1 – (cont: missing links, ancestral notes) – mf#802606 – us WHS [929]
Carden, Allen D *see*
– The missouri harmony
– Missouri harmony, 1840
– United states harmony
Carden, Andrew *see* An answer to mr dalton's pamphlet on the irish question
Cardenal antonio caggiano, obispo de rosario. la figura de san francisco solano y su actuacion en el locuman... / Bayle, Constantino – Madrid: Missionalia Hispanica, 1951 – 1 – sp Bibl Santa Ana [240]
Cardenal, Ernesto *see* Mayapan
El cardenal goma, primado de espana. madrid, 1969 / Granados, Anastasio – Madrid: graf. calleja, 1970 – 1 – sp Bibl Santa Ana [240]
Cardenal Iracheta, Manuel *see* Vida de gonzalo pizarro
Cardenas Acosta, Pablo E (Pablo Enrique) *see* Movimiento comunal de 1781 en el nuevo reino de gr...
Cardenas Acosta, Pablo Enrique *see*
– Comuneros
– Vasallaje a la insurreccion de los comuneros
Cardenas, Daniel *see* El espanol de jalisco
Cardenas Garcia, Jorge *see* Frente nacional y los partidos politicos
Cardenas, Joaquin E *see* Sucesos miguelenos
Cardenas Salazar, Manuel *see* Cardenas salazar y la republica dominicana
Cardenas salazar y la republica dominicana / Cardenas Salazar, Manuel – Santiago de los Caballeros, Dominican Republic. 1947 – 1r – us UF Libraries [972]
Cardenas Y Echarte, Raul De *see* Recurso de inconstitucionalidad
Cardenas Y Rodriguez, Jose Maria De *see* Coleccion de articulos satiricos y de costumbres
Cardiac rehabilitaion exercise adherence : the influence of exercise benefits, barriers, locus of control, and intrinsic motivation / Gregory, Arden R – 1998 – 219p on 3mf – 9 – $15.00 – mf#PSY 2167 – us Kinesology [617]
Cardiac rehabilitation – New York. 1970-1980 (1) 1970-1980 (5) 1976-1980 (9) – ISSN: 0147-3875 – mf#7747 – us UMI ProQuest [616]
Cardiff, adamsdown cemetery, monumental inscriptions – 1mf – 9 – £1.25 – uk Glamorgan FHS [929]
Cardiff advertiser – [Wales] Cardiff 9 jul 1858-23 jul 1859 [mf ed 2003] – 1r – 1 – uk Newsplan [072]
Cardiff advertiser and local exchange and mart – [Wales] Cardiff 15 oct 1891-13 mar 1894 [mf ed 2003] – 2r – 1 – (cont as: cardiff advertiser [jan 1893-mar 1894]) – uk Newsplan [072]
Cardiff & caerphilly srd's – (= ser 1861 census returns [glamorgan]) – 10mf – 9 – £12.50 – uk Glamorgan FHS [314]
Cardiff chronicle – [Wales] Cardiff 3 aug 1866-7 mar 1868 [mf ed 2003] – 2r – 1 – uk Newsplan [072]
Cardiff examiner – [Wales] Cardiff apr 1873-jun 1874 [mf ed 2003] – 2r – 1 – (missing: apr-2 aug 1873) – uk Newsplan [072]
Cardiff free press and district intelligencer – [Wales] Cardiff 4 nov 1876-dec 1877; 22 mar 1878-5 jul 1884 [mf ed 2004] – 7r – 1 – uk Newsplan [072]
Cardiff free press and south wales advertiser – [Wales] Cardiff may/jun 1924 [mf ed 2002] – 1r – 1 – uk Newsplan [072]
Cardiff free press and south wales journal – [Wales] Cardiff 18 oct 1884 [mf ed 2002] – 1r – 1 – uk Newsplan [072]
Cardiff freemen – 1777-1993 – 1mf – 9 – £1.25 – uk Glamorgan FHS [941]
Cardiff, glamorgan non conformist register : ebenezer welsh wesleyan 1799-1837, ebenezer independent 1817-1837, welsh wesleyan circuit 1818-1837, ebenezer welsh congress members 1826-1854 – [Glamorgan]: GFHS [mf ed c2006] – 4mf – 9 – £5.00 – uk Glamorgan FHS [242]
Cardiff, glamorgan, parish church of all saints : baptisms 1867-1893, burials 1867-1944 – [Glamorgan]: GFHS [mf ed c2003] – 1mf – 9 – £1.25 – uk Glamorgan FHS [929]
Cardiff, glamorgan, parish church of bethany baptist : baptisms 1807-1837, burials 1807-1837; cardiff, charles st cong + members baptisms 1853-1925, burials 1854-1947, marriages 1857-1925; & windsor place presbyterian, baptisms 1903-1925 – 1mf – 9 – £1.25 – uk Glamorgan FHS [929]
Cardiff, glamorgan, parish church of bethany english baptist st mary street : baptisms 1804-1837, burials 1807-1837; & membership registration 1806-1857 – 2mf – 9 – £2.50 – uk Glamorgan FHS [929]
Cardiff, glamorgan, parish church of guildford street & newport road : baptisms 1868-1933; guildford street, baptisms 1864-1868 & membership registration 1869-1893 – 3mf – 9 – £3.75 – uk Glamorgan FHS [929]
Cardiff, glamorgan, parish church of st andrew : baptisms 1863-1925, burials 1863-1911 – [Glamorgan]: GFHS [mf ed c2003] – 1mf – 9 – £1.25 – uk Glamorgan FHS [929]
Cardiff, glamorgan, parish church of st dyfrig : baptisms 1885-1894, burials 1895-1927 – [Glamorgan]: GFHS [mf ed c2003] – 1mf – 9 – £1.25 – uk Glamorgan FHS [929]
Cardiff, glamorgan, parish church of st illtyd : baptisms 1890-1925; & st samson baptisms 1904-1918; & st stephen baptisms 1878-1925 – 1mf – 9 – £1.25 – uk Glamorgan FHS [929]
Cardiff, glamorgan, parish church of st john baptist : baptisms 1669-1840, burials 1669-1841, marriages 1669-1837 – 6mf – 9 – £7.50 – uk Glamorgan FHS [929]
Cardiff, glamorgan, parish church of st mary : baptisms 1843-1925, burials 1848-1918 – [Glamorgan]: GFHS [mf ed c2001] – 7mf – 9 – £8.75 – uk Glamorgan FHS [929]
Cardiff, glamorgan, parish church of st teilo : baptisms 1884-1925 – [Glamorgan]: GFHS [mf ed c2003] – 2mf – 9 – £2.50 – uk Glamorgan FHS [929]
Cardiff, glamorgan, parish church of tabernacle (the hayes) : baptisms 1813-1847 + transcript membership registration – 1mf – 9 – £1.25 – uk Glamorgan FHS [929]
Cardiff, glamorgan, r c parish church of st david : baptisms 1836-1915, burials 1839-1873, marriages 1839-1915 – [Glamorgan]: GFHS [mf ed c2003] – 6mf – 9 – £7.50 – uk Glamorgan FHS [929]
Cardiff, glamorgan, r c parish church of st david : burials 1839-1873 – [Glamorgan]: GFHS [mf ed c2005] – 1mf – 9 – £1.25 – uk Glamorgan FHS [929]
Cardiff, glamorgan, r c parish church of st paul [newtown] : baptisms 1881-1915, marriages 1882-1915 – [Glamorgan]: GFHS [mf ed c2005] – 1mf – 9 – £1.25 – uk Glamorgan FHS [929]
Cardiff, glamorgan, r c parish church of st peter : baptisms 1861-1915, marriages 1862-1915 – [Glamorgan]: GFHS [mf ed 2004] – 3mf – 9 – £3.75 – uk Glamorgan FHS [929]
Cardiff independent – [Wales] Cardiff 23 oct-13 nov 1875 [mf ed 2002] – 1r – 1 – uk Newsplan [072]
Cardiff mercury – [Wales] Cardiff 8 jun 1861-25 jan 1862 [mf ed 2003] – 1r – 1 – uk Newsplan [072]
Cardiff monthly stock and share list – [Wales] Cardiff oct 1893-dec 1894 [mf ed 2003] – 1r – 1 – uk Newsplan [072]
Cardiff news – [Wales] Cardiff 12 feb-jun 1864 [mf ed 2003] – 1r – 1 – uk Newsplan [072]
Cardiff port & channel pilots – 1810-1979 – 1mf – 9 – £1.25 – uk Glamorgan FHS [380]
Cardiff r d with caerphilly, llantrisant & st nicholas – (= ser 1851 census returns [glamorgan]) – 8mf – 9 – £10.00 – uk Glamorgan FHS [314]
Cardiff reg dist (incl caerphilly & llantrisant) – (= ser 1841 census returns [glamorgan]) – 6mf – 9 – £7.50 – uk Glamorgan FHS [314]
Cardiff roath, glamorgan, parish church of st german : baptism 1884-1924, marriages 1887-1918 – [Glamorgan]: GFHS [mf ed c2005] – 8mf – 9 – £11.25 – uk Glamorgan FHS [929]
Cardiff settlement examinations – 1735-86 – 2mf – 9 – £2.50 – uk Glamorgan FHS [941]
Cardiff shipping and mercantile gazette – [Wales] Cardiff 12 jul 1869-dec 1911 [mf ed 2003] – 20r – 1 – (missing: 1871, 1888, 1890-91, 1893) – uk Newsplan [072]
Cardiff, slater's directory (facsimile) – 1882 – 2mf – 9 – £2.50 – uk Glamorgan FHS [380]
Cardiff, st john baptist, monumental inscriptions – 1mf – 9 – £1.25 – uk Glamorgan FHS [929]
Cardiff standard and county chronicle – [Wales] Cardiff 6 apr 1864-11 jan 1865 [mf ed 2003] – 1r – 1 – uk Newsplan [072]
Cardiff suburban news – [Wales] Cardiff 9 feb 1924-dec 1950 [mf ed 2004] – 23r – 1 – (cont by: cardiff & suburban news jan 1930-dec 1950]) – uk Newsplan [072]
Cardiff, tabernacle welsh baptist, monumental inscriptions – 1mf – 9 – £1.25 – uk Glamorgan FHS [929]
Cardiff times – [Wales] Cardiff 10 oct 1857-21 jun 1930 [mf ed 2003] – 96r – 1 – (cont as: cardiff times and newport and south wales advertiser [jan 1859-dec 1861]; cardiff times, merthyr, aberdare and pontypridd gazette [jan 1862-dec 1867]; cardiff times, south wales, monmouthshire and western counties advertiser [jan 1868-dec 1875]; cardiff times and south wales weekly news [dec 1876-dec 1886]; cardiff times [1887]; cardiff times and south wales weekly news [jan 1889-dec 1928]; south wales weekly news and cardiff times [jan 1929-jun 1930]) – uk Newsplan [072]
Cardiff times – Wales. -w. 1858, 1863, 1868. 2 1 2 reels – 1 – uk British Libr Newspaper [072]
Cardiff, western mail directory (facsimile) – 1902 – 4mf – 9 – £5.00 – uk Glamorgan FHS [380]
Cardiff, wright's directory (facsimile) – 1891 – 9 – £3.75 – uk Glamorgan FHS [380]
Cardigan county times and shropshire and mid-wales advertiser – [Wales] Powys 10 jul 1897-15 jun 1907 [mf ed 2003] – 11r – 1 – uk Newsplan [072]
Cardigan observer – [Wales] Ceredigion jan 1877-25 sep 1898 [mf ed 2003] – 14r – 1 – (missing: 1893) – uk Newsplan [072]
Cardigan & tivy-side advertiser – [Wales] Ceredigion jan 1909-dec 1950 [mf ed 2004] – 42r – 1 – uk Newsplan [072]
Cardiganshire & merioneth herald and cambrian visitor / merioneth news, etc – [Wales] Gwynedd jan 1885-jun 1920 [mf ed 2003] – 26r – 1 – (missing: 1897, 1907, 1911; cont as: merioneth news and herald and barmouth record [jan 1889-jun 1920]) – uk Newsplan [072]

CARDILLAC

Cardillac / Barr, Robert – Toronto: McLeod & Allen, c1909 – 5mf – 9 – 0-665-76356-5 – mf#76356 – cn Canadiana [830]
Cardillo, Cheryl M *see* Effects of a 30-minute walk on ground reaction forces
Cardillo, Giacomo Antonio *see* Sacrarum modulationum. liber secundus
Cardim, Fernao *see* Tratados da terra e gente do brasil
Cardinal archbishop of westminster and the new hierarchy / Bowyer, George – London, England. 1850 – 1r – us UF Libraries [241]
Cardinal, Bradley J *see* The effectiveness of the stages of change model and experimental exercise prescriptions in increasing female adults' physical activity and exercise behavior
The cardinal democrat, henry edward manning / Taylor, Ida Ashworth – London: Kegan Paul, Trench, Truebner, 1908 – 1mf – 9 – 0-524-04125-3 – (incl bibl ref) – mf#1992-2011 – us ATLA [240]
Cardinal elements of the christian faith / Adam, David Stow – London, New York: Hodder & Stoughton [1911?] [mf ed 1990] – 1mf – 9 – 0-7905-3508-4 – mf#1989-0001 – us ATLA [240]
The cardinal facts of canadian history : carefully gathered from the most trustworthy sources / Taylor, James P – [Toronto?: s.n.], 1899 [mf ed 1986] – 3mf – 9 – 0-665-24659-5 – (incl ind and bibl ref) – mf#24659 – cn Canadiana [971]
Cardinal, Jeffrey S *see* Effects of coach interactions on college soccer players' behavior and perception
Cardinal lavigerie; and, the african slave trade / ed by Clarke, Richard Frederick – London: Longmans, Green, 1889 – 1mf – 9 – 0-524-03276-9 – mf#1990-0887 – us ATLA [240]
Cardinal manning / Hutton, Arthur Wollaston – London: Methuen, 1892 – 1mf – 9 – 0-7905-6187-5 – (incl bibl ref) – mf#1988-2187 – us ATLA [240]
Cardinal manning as represented in his own letters and notes / Manning, Henry Edward – London: E Stock, 1896 – 1mf – 9 – 0-7905-8173-6 – mf#1988-8056 – us ATLA [240]
Le cardinal manning et son action sociale / Lemire, Jules – Paris: V Lecoffre, 1893 – 1mf – 9 – 0-7905-6814-4 – mf#1988-2814 – us ATLA [240]
Cardinal, Marita K *see* A survey analysis of dance wellnessrelated curricula in american higher education
Cardinal mercier's retreat to his priests = Retraite pastorale / Mercier, Desire – Bruges: Ch Beyaert, 1912 – 2mf – 9 – 0-7905-8851-X – (in english) – mf#1989-2076 – us ATLA [240]
Cardinal Mindszenty Foundation *see* Red line
Cardinal newman / Meynell, Wilfrid – 6th ed. rev. London: Burns and Oates, 1907 – 1mf – 9 – 0-7905-4837-2 – mf#1988-0837 – us ATLA [240]
Cardinal newman : reminiscences of fifty years since / Lockhart, William – London: Burns & Oates; New York: Catholic Publication Society, 1891 – 1mf – 9 – 0-7905-5059-8 – mf#1988-1059 – us ATLA [240]
Cardinal newman : reminiscences of fifty years since / Lockhart, William – London: Burns & Oates; New York: Catholic Publication Society, 1891 – 1mf – us ATLA [240]
Cardinal newman : the story of his life / Jennings, Henry James – Birmingham: Houghton; London: Simpkin, Marshall, 1882 – 1mf – 9 – 0-7905-4931-X – mf#1988-0931 – us ATLA [240]
Cardinal newman and the encyclical pascendi dominici gregis : an essay / O'Dwyer, Edward Thomas – London; New York: Longmans, Green, 1908 – 1mf – 9 – 0-8370-8367-2 – mf#1986-2367 – us ATLA [240]
Le cardinal nicolas de cues (1401-1464) : l'action – la pensee / Vansteenberghe, E – Paris, 1920 – 14mf – 8 – €27.00 – ne Slangenburg [110]
Cardinal truths of the gospel / Halfyard, Samuel Follet – New York: Methodist Book Concern, c1915 – 1mf – 9 – 0-7905-7643-0 – (incl bibl ref) – mf#1989-0868 – us ATLA [226]
Cardinal von geissel : aus seinem handschriftlichen nachlass geschildert / Pfuelf, Otto – Freiburg i.B., 1895 (mf ed 1993) – 2pts 8mf – 9 – €99.00 – 3-89349-208-9 – mf#DHS-AR 98 – gw Frankfurter [240]
Cardinal wiseman's appeal – London, England. 18– – 1r – us UF Libraries [240]
Cardinal wolsey / Creighton, Mandell – London, New York: Macmillan 1888 [mf ed 1990] – (= ser Twelve english statesmen) – 1mf – 9 – 0-7905-5458-5 – mf#1988-1458 – us ATLA [941]
Cardinal wolsey / Martin, Samuel – London, England. 1849? – 1r – us UF Libraries [240]
Cardinal ximenes : statesman, ecclesiastic, soldier and man of letters / Lyell, James Patrick Ronaldson – London: Grafton, 1917 – 1mf – 9 – 0-524-03555-5 – (incl bibl ref) – mf#1990-4750 – us ATLA [220]
Il cardinale raffaele...del val...roma / Cenci, Pio – Torino, 1933; Madrid: Razon y Fe, 1933 – 1 – sp Bibl Santa Ana [240]
Cardinall, Allan Wolsey *see* The natives of the northern territories of the gold coast
Die cardinalpunkte der franz baader'schen philosophie / Hamberger, Julius – Stuttgart: JF Steinkopf, 1855 – 1mf – 9 – 0-524-08631-1 – mf#1993-2091 – us ATLA [190]
Cardinal's broken oath / Bradlaugh, Charles – London, England. 1882 – 1r – us UF Libraries [240]
Cardington – (= ser Bedfordshire parish register series) – 2mf – 9 – £5.00 – uk BedsFHS [929]
Cardington, cemetery monumental inscriptions – Bedfordshire Family HS 1978 – (= ser Bedfordshire parish register series) – 1mf – 9 – £1.25 – uk BedsFHS [929]
Cardington, st mary monumental inscriptions monumental inscriptions – Arthur Weight Matthews 1914 – (= ser Bedfordshire parish register series) – 1mf – 9 – £1.25 – uk BedsFHS [929]
Cardiology – Basel. 1966-1996 (1) 1966-1996 (5) 1994-1996 (9) – ISSN: 0008-6312 – mf#2049 – us UMI ProQuest [616]
Cardiology clinics – Philadelphia. 1983+ (1,5,9) – ISSN: 0733-8651 – mf#13376 – us UMI ProQuest [616]
Cardiology in review – v1-4. 1993-1996 – 4r – 1,5,6,9 – $65.00r – us Lippincott [616]
Cardiomorphoseos sive ex corde desumpta emblemata sacra / Pona, F – Veronae, 1645 – 3mf – 9 – mf#O-857 – ne IDC [090]
Cardiopulmonary medicine – Park Ridge. 1975-1980 (1) 1976-1980 (5) 1976-1980 (9) – (cont: bulletin of the american college of chest physicians) – ISSN: 0149-6719 – mf#1736,01 – us UMI ProQuest [610]
Cardiopulmonary responses to unsupported and supported arm exercise in normal subjects and patients with obstructive pulmonary disease / Lebzelter, Joseph – Temple University, 1996 – 3mf – 9 – $12.00 – mf#PH 1500 – us Kinesology [612]
Cardio-pulmonary resuscitation knowledge of registered nurses working in private hospital wards / Hutchings, Pauline Linda Joan – Uni of South Africa 2001 [mf ed Johannesburg 2001] – 9 – (incl bibl ref) – mf#mfm14814 – sa Unisa [610]
Cardio-respiratory response to upright and aero-posture cycling / Origenes, M M – 1991 – 1mf – 9 – $4.00 – us Kinesology [612]
Cardiorespiratory responses : following an 8 week deep water running trail program in elderly women / Hu, Kelly S – 2000 – 150 p on 2mf – 9 – $10.00 – mf#PH 1710 – us Kinesology [612]
Cardiorespiratory responses of controlled frequency breathing during submaximal exercise / Tracy, Michael L – 1980 – 1mf – 9 – $4.00 – us Kinesology [790]
Cardiorespiratory responses of world class whitewater slalom paddlers / Law, R Craig – 1988 – 88p 1mf – 9 – $4.00 – us Kinesology [612]
Cardiorespiratory responses to circuit weight training as measured by a biokinetic swim-bench test and a treadmill run test / Chiang, J – 1989 – 2mf – 9 – $8.00 – us Kinesology [612]
Cardiovascular and body composition responses to aerobic dance training of varying frequencies and total program lengths / Ipsen, Lillas F & Roundy, Elmo S – 1990 – 2mf – $8.00 – us Kinesology [612]
Cardiovascular and interventional radiology – Heidelberg. 1980-1996 (1,5,9) – ISSN: 0174-1551 – mf#13152,01 – us UMI ProQuest [616]
Cardiovascular and metabolic responses and alternations in selected measures of mood with a single bout of dynamic tae kwon do exercise / Toskovic, Nebojsa N – 2000 – 243p on 3mf – 9 – $15.00 – mf#PSY 2170 – us Kinesology [612]
The cardiovascular and metabolic responses of men with cardiovascular disease to aqua dynamic exercise / Miller, K A – 1990 – 1mf – 9 – $4.00 – us Kinesology [612]
Cardiovascular disease risk in adults with mental retardation and down syndrome / Draheim, Christopher C – 2000 – 2mf – 9 – $8.00 – mf#HE 660 – us Kinesology [612]
Cardiovascular drugs and therapy – Norwell. 1987-1996 (1,5,9) – ISSN: 0920-3206 – mf#16773 – us UMI ProQuest [616]
Cardiovascular endurance effects of a required college health, physical education, and recreation class / Fitzgerald, Dani J – 1997 – 1mf – 9 – $4.00 – mf#PH 1588 – us Kinesology [612]
Cardiovascular nursing – Dallas. 1965-1996 (1) 1965-1996 (5) 1965-1996 (9) – ISSN: 0008-6355 – mf#8457 – us UMI ProQuest [610]
Cardiovascular research – London. 1972+ (1) 1972+ (5) 1972+ (9) – ISSN: 0008-6363 – mf#6593 – us UMI ProQuest [616]
Cardiovascular surgery – Kidlington. 1993+ (1,5,9) – ISSN: 0967-2109 – mf#19660 – us UMI ProQuest [617]
Cardona, Faust *see* Lenguaje de los tambores africanos
Cardona, Jenaro *see* Del calor hogareno
Cardona Rossell, Mariano *see* Aspectos economicos de nuestra revolucion
Cardoni, Giuseppe *see* Elucubratio de dogmatica romani pontificis infallibilitate eiusque definibilitate
Cardosa, Onelio Jorge *see* Cuentero
Cardoso, Clodoaldo *see* Municipios maranhenses
Cardoso, F *see* Utilidades del agua i de la nieve, del bever frio i caliente...
Cardoso, Joaquin *see* Sangre en los tapehuanes
Cardoso, Manuel Da Costa Lobo *see* Sao paulo da assumpcao de luanda
Cardoso, Onelio Jorge *see*
- Cuentos completos
- Otra muerte del gato
- Perro
- Pueblo cuenta
- Taita, diga usted como

Cardoso, Vicente Licinio *see* Pensamentos brasileiros
Cardot, J *see* Botanische ergebnisse der schwedischen expedition nach patagonien und dem feuerlande 1907-1909
Cardoza Y Aragon, Luis *see*
- Apolo y coatlicue
- Gutemala

Cardozo arts and entertainment law journal – Yeshiva University. v1-19. 1982-2001 – 9 – $358.00 set – ISSN: 0736-7694 – mf#109081 – us Hein [340]
Cardozo, Benjamin Nathan *see* An address delivered in chancellors hall, state education building, albany, ny
Cardozo journal of international and comparative law – v1-8. 1992-2000 – 9 – $140.00 set – (title varies: v1-2 n1 1992-93 as: new europe law review) – ISSN: 1069-3181 – mf#114011 – us Hein [341]
Cardozo law review – v1-22. 1979-2001 – 9 – $862.00 set – ISSN: 0270-5192 – mf#101421 – us Hein [340]
Cardozo, Michael H *see* Exchange of patent rights and technical information under mutual aid programs
Cardozo women's law journal – v1-6. 1993-1999 – 9 – $185.00 set – (cont: women's annotated legal bibliography) – ISSN: 1074-5785 – mf#114781 – us Hein [342]
Cardross case : lord jerviswoode's decision – Edinburgh, Scotland. 18– – 1r – us UF Libraries [240]
Cardross case / Robertson, Andrew – Edinburgh, Scotland. 1861 – 1r – us UF Libraries [240]
Cardross case and the spiritual independence of non-established chu... – Edinburgh, Scotland. 1875 – 1r – us UF Libraries [240]
Cardross case in relation to the civil rights of the community – Edinburgh, Scotland. 1860 – 1r – us UF Libraries [240]
Cardston news – Alberta, CN. sept 1925-jun 1958 – 14r – 1 – cn Commonwealth Imaging [071]
Carducho, V *see* Dialogos de la pintura...
Cardwell, Edward *see*
- Documentary annals of the reformed church of england
- History of conferences
- A history of conferences and other proceedings connected with the revision of the book of common prayer

The care and cataloguing of manuscripts: as practiced by the minnesota historical society / Nute, Grace Lee – 1936 – 1r – 1 – $5.00 – us Minn Hist [025]
The care of dependent, neglected, and wayward children microform : being a report of the second section of the international congress of charities, correction and philanthropy, chicago, june, 1893 / ed by Spencer, Anna Garlin & Birtwell, Charles Wesley – Baltimore: Johns Hopkins Press, 1894 [mf ed 1984] – (= ser Women & the church in america 164) – 1mf – 9 – 0-8370-1464-6 – mf#1984-2164 – us ATLA [362]
Care of the elderly = Bejaardesorg / South Africa. Department of Health [Departement van Gesondheid] – Pretoria: Dept of Health 1979 [mf ed Pretoria, RSA: State Library [199-]] – 1r with other items – 5 – (incl bibl ref) – mf#op 06843 r24 – us CRL [362]
Care of the elderly / South Africa. Department of National Health and Population Development [Departement van Nasionale Gesondheid en Bevolkingsontwikkeling – 2nd ed. Pretoria: Dept of National Health & Population Development 1986 [mf ed Pretoria, RSA: State Library [199-]] – 100p [ill] on 1r with other items – 5 – mf#op 08223 r24 – us CRL [362]
The care of the soul / Fuller, Andrew – 1805 – 1 – 5.00 – us Southern Baptist [242]
Career / United Office and Professional Workers of America – 1948 oct-1950 jun 15 – 1r – 1 – (cont: insurance career, office and professional news; cont by: champion [new york ny]) – mf#3564493 – us WHS [650]
Career advantage – 1987 spr-nov/dec – 1r – 1 – mf#4882465 – us WHS [331]
Career development for exceptional individuals – Reston. 1978+ [1,5,9] – ISSN: 0885-7288 – mf#12791 – us UMI ProQuest [331]
Career development international – Bradford. 2001+ (1,5,9) – mf#31276 – us UMI ProQuest [331]
Career development quarterly – Tulsa. 1986+ (1) 1986+ (5) 1986+ (9) – (cont: vocational guidance quarterly) – ISSN: 0889-4019 – mf#3178,01 – us UMI ProQuest [331]
Career education quarterly : an official publication of the national association for career education / National Association for Career Education – Glassboro. 1977-1979 (1,5,9) – ISSN: 0276-7848 – mf#11636 – us UMI ProQuest [331]
Career Guidance Foundation *see*
- College catalog collections: national
- College catalog collections: regional
- The international collection
- Special collection
- State education directories

Career mobility patterns of head coaches in the national basketball association / Gibbs, E Nathan – 1997 – 1mf – 9 – $4.00 – mf#PE 3838 – us Kinesology [790]
The career of the god-idea in history / Tuttle, Hudson – Boston: Adams and Co, c1869 – 1mf – 9 – 0-524-01517-1 – mf#1990-2493 – us ATLA [230]
Career satisfaction of dental hygienists performing expanded functions as compared to dental hygienists performing only traditional duties / Sylvis, Robin – 1981 – 1mf – 9 – $4.00 – us Kinesology [617]
Career world – Highland Park. 1985+ (1,5,9) – ISSN: 0744-1002 – mf#11650,01 – us UMI ProQuest [331]
Career world 1 – Highwood. 1980-1981 (1,5,9) – ISSN: 0198-7615 – mf#11649 – us UMI ProQuest [331]
Career world 2 – Highwood. 1978-1981 (1) 1978-1981 (5) 1974-1975 (9) – ISSN: 0198-7623 – mf#11650 – us UMI ProQuest [331]
Careers in space [aasms49] – 1984 – (= ser Aasms 1968) – 6papers on 2mf – 9 – $12.00 – 0-87703-206-8 – us Univelt [331]
Careful and strict inquiry into the pretensions and designs of dr h... – Glasgow, Scotland. 1833 – 1r – us UF Libraries [240]
Carel, Auguste *see* La france ancienne et moderne
Careless church-goers – London, England. 1837 – 1r – us UF Libraries [240]
Carell, Paul *see* Foxes of the desert
Carencro news – Opelousas, LA. 1989-2000 (1) – mf#68899 – us UMI ProQuest [071]
Carette *see* Souvenirs intimes de la cour des tuileries
Carew Hunt, R N *see* Theory and practice of communism
Carew poyntz book of hours : fitzwilliam museum, cambridge ms. 48 – 14th c – 1r – 14 – mf#C590 – uk Microform Academic [240]
Carew, Thomas *see*
- The poems of thomas carew

Carey, Annie *see* The history of a book
Carey, E *see* Memoir of william carey, late missionary to bengal...
Carey, Eustace *see* Memoir of william carey
Carey, Frances *see* Journal of a tour in france
Carey, Henry *see* The geography, history, and statistics, of america, and the west indies
Carey, Henry Charles *see*
- Financial crises
- The slave trade, domestic and foreign

Carey, John *see*
- farewell sermon, preached in the episcopal churches, st john, n b
- Rideau canal

Carey Jones, N S *see* Pattern of a dependent economy
Carey, Mathew *see*
- An address to william tudor, esq author of letters on the eastern states
- The olive branch
- Sketch of the irish code

Carey, Robert *see* Memoirs of robert cary, earl of monmouth
Carey, T H *see* Christian baptism (illustrated)
Carey, Thomas Joseph *see*
- Law at a glance
- The legal advisor

Carey, W *see* Biographical and literary notices...
Carey, W H [comp] *see* The good old days of honorable john company
Carey, Walter Julius *see* Have you understood christianity?

Carey, William *see*
- Adventures in tibet
- College library: catalogue of early indian imprints, 1714-1850
- Dialogues intended to facilitate the acquiring of the bengalee language
- A dictionary of the bengalee language, vol 1
- A dictionary of the bhotanta, or boutan language
- An enquiry into the obligations of christians
- Fifty-two letters to dr. john ryland
- Grammar of the punjabee language
- Letters from the rev dr carey
- Missionary tour in the hucli and howrah districts, lower bengal – india
- Ramayuna of valmeeki

Carey, William: *see* Grammar of the mahratta language

Carey, William et al *see* Garo jungle book

Carey, William Paulet *see*
- Brief remarks on the anti-british effect of... criticism on modern art
- Critical description and analytical review of "death on the pale horse" painted by benjamin west
- Cursory thoughts
- Desultory exposition of an anti-british system of incendiary publication
- The national obstacle to the national public style
- Observations on the primary object of the british institution
- Ridolfi's critical letters on the style of wm etty...
- Some memoirs of the patronage and progress of the fine arts

Carey, William paulet *see* Critical description of the procession of chaucer's pilgrims to canterbury, painted by thomas stothard

Carey's library of choice literature – Philadelphia. 1835-1836 (1) – mf#4619 – us UMI ProQuest [420]

Carey's manitoba reports / Manitoba. Canada – 1v. 1875 (all publ) – 2mf – 9 – $3.00 – mf#LLMC 81-024 – us LLMC [340]

Cargo airlift – New York. 1942-1976 (1) 1971-1976 (5) 1976-1976 (9) – (cont by: air cargo magazine) – ISSN: 0002-2217 – mf#242 – us UMI ProQuest [380]

Cargo courier – 1989 jan 7-1993 dec 11 – 1r – 1 – (cont: phantom's eye) – mf#1057911 – us WHS [071]

Cargo of the "wilhelmina" / american trade in munitions of war / sinking of the "frye" – Boston: World Peace Foundation 1915 [mf ed 1992] – (= ser World peace foundation pamphlet series 5/4/5) – 1mf – 9 – 0-524-03225-4 – mf#1990-0853 – us ATLA [933]

Cargos / Comision de Monumentos – Madrid: Ed. Reus, 1922. B.R.A.H. 80. p. 304 – 1 – sp Bibl Santa Ana [946]

Cargos que resultan contra...el p. diego de caceres, general de...s. geronimo.. / Caceres, Diego de – 1641 – 9 – sp Bibl Santa Ana [240]

Cari figli un' altro amplesso : sung by sigr marchesi in the opera of giulio sabino / Sarti, Giuseppe – London: Longman & Broderip [1778?] [mf ed 1989] – 1r – 1 – (words by pietro giovannini) – mf#pres. film 42 – us Sibley [780]

Carias Reyes, Marcos *see*
- Germinal, cuentos
- Heredad
- Hombres de pensamiento
- Juan ramon molina

Caribbean : contemporary colombia / Conference On The Caribbean (12th : 1961) – Gainesville, FL. 1962 – 1r – us UF Libraries [972]

Caribbean : its health problems / Wilgus, A Curtis – Gainesville, FL. 1965 – 1r – us UF Libraries [972]

Caribbean / Roberts, Walter Adolphe – Indianapolis, IN. 1940 – 1r – us UF Libraries [972]

Caribbean : sea of the new world / Arciniegas, German – New York, NY. 1946 – 1r – us UF Libraries [972]

Caribbean : venezuelan development / Conference On The Caribbean (13th : 1962) – Gainesville, FL. 1963 – 1r – us UF Libraries [972]

Caribbean area / George Washington University Seminar Conference – Washington, DC. 1934 – 1r – us UF Libraries [972]

Caribbean area, 1941-1943 / Walsh Construction Company – s.l, s.l? no date – 1r – us UF Libraries [972]

Caribbean backgrounds and prospects / Jones, Chester Lloyd – New York, NY. 1931 – 1r – us UF Libraries [972]

Caribbean business news – Toronto. 1972-1980 (1) 1978-1980 (5) 1978-1980 (9) – ISSN: 0045-5792 – mf#7999 – us UMI ProQuest [338]

Caribbean circuit / Luke, Harry Charles Joseph – London, England. 1950 – 1r – us UF Libraries [972]

Caribbean Commission *see*
- Caribbean islands and the war
- Caribbean tourist trade
- Guide to commercial shark fishing in the caribbean area
- Industrial development of puerto rico and the virg...
- Promotion of industrial development in the caribbe

Caribbean contact – Bridgetown, Barbados. 1988 mar-1990 dec and 1991 jan-1994 aug – 2r – (1988 apr, jul, aug; 1989 jan-mar, oct-nov; 1992 jan-feb, sep-nov; aug 1993) – us UF Libraries [079]

Caribbean cruise / Bertram, Kate – New York, NY. 1948 – 1r – us UF Libraries [918]

Caribbean cruise / Foster, Henry La Tourette – New York, NY. 1928 – 1r – us UF Libraries [918]

Caribbean danger zone / Rippy, James Fred – New York, NY. 1940 – 1r – us UF Libraries [972]

Caribbean daylight – 1994 may 22 [v3 n9], 1994 oct 9 [v3 n29]/dec 25-2000 jan 7/jun 30 – 11r – 1 – (with gaps) – mf#3006580 – us WHS [071]

Caribbean islands and the war / Caribbean Commission – Washington, DC. 1943 – 1r – us UF Libraries [972]

Caribbean journal of education – Kingston. 1980+ – 1,5,9 – ISSN: 0376-7701 – mf#12625 – us UMI ProQuest [370]

Caribbean journal of religious studies – Kingston, Jamaica: United Theological College of the West Indies. v1-12. sep 1975-sep 1991 – 1r – 1 – us CRL [240]

Caribbean labour congress, 1947-49 – [mthly] – (= ser The private coll of richard hart) – 3mf – 9 – mf#87551 – uk Microform Academic [325]

Caribbean lands / Carpenter, Frances – New York, NY. 1955 – 1r – us UF Libraries [972]

Caribbean lands / Macpherson, John – London, England. 1963 – 1r – us UF Libraries [972]

Caribbean lands : mexico, central america and the w... / Carpenter, Frances – New York, NY. 1950 – 1r – us UF Libraries [972]

Caribbean newsletter / Friends for Jamaica – 1991-99 – 1r – 1 – (cont: friends for jamaica newsletter) – mf#1352575 – us WHS [071]

Caribbean policy of the united states, 1890-1920 / Callcott, Wilfrid Hardy – Baltimore, MD. 1942 – 1r – us UF Libraries [972]

Caribbean quarterly – Mona. 1949+ (1) 1975+ (5) 1977+ (9) – ISSN: 0008-6495 – mf#8948 – us UMI ProQuest [073]

Caribbean readers / Newman, Arthur James – London, England. bk1 introd-bk5. 1937-1953 – 1r – us UF Libraries [972]

Caribbean Research Center focus / City University of New York – 1989 sep, 1990 jan – 1r – 1 – mf#5294558 – us WHS [972]

Caribbean Research Council Committee On Agricultu... *see* Livestock in the caribbean

Caribbean review – Miami. 1969-1989 (1) 1972-1989 (5) 1975-1989 (9) – ISSN: 0008-6525 – mf#6381 – us UMI ProQuest [073]

Caribbean since 1900 / Jones, Chester Lloyd – New York, NY. 1936 – 1r – us UF Libraries [972]

Caribbean tourist trade / Caribbean Commission – Washington, DC. 1945 – 1r – us UF Libraries [338]

Caribbeana – London, England. v1-6. 1910-19 – 2r – us UF Libraries [972]

Caribbee cruise / Vandercook, John W – New York, NY. 1938 – 1r – us UF Libraries [972]

Caribbee islands under the proprietary patents / Williamson, James Alexander – London, England. 1926 – 1r – us UF Libraries [972]

Caribe – Santo Domingo, Dominican Republic. 1978 sepT-1999 dec – 117r – (gaps) – us UF Libraries [079]

Caribe – Santo Domingo, Dominican Republic. 18 dec 1954-28 jan 1955 – 1r – 1 – uk British Libr Newspaper [079]

El caribe – Ciudad Trujillo, Dominican Republic: editora del caribe, 1956-57 – 1 – us CRL [079]

El caribe – Santo domingo, dominican republic. 1948-1987 (1) – mf#67691 – us UMI ProQuest [079]

Caribou : the voice of the newfoundland micmac / Newfoundland Federation of Indians – 1982 aug 31, nov 30-1986 mar 15, dec 23-1987 mar 30 – 1r – 1 – mf#1095633 – us WHS [307]

Caribou Indian Education and Training Centre *see* Coyoti prints

Caribou shooting in newfoundland : with a history of england's oldest colony from 1001 to 1895 / Davis, Samuel T – [S.l: s.n], 1895 [mf ed 1980] – 3mf – 9 – 0-665-02600-5 – mf#02600 – cn Canadiana [639]

Carica papaya = Papaya farm / Trainor, A W – s.l, s.l? 1936 – 1r – us UF Libraries [978]

Caricaturas / Rendon, Ricardo – Bogota, Colombia. v1-2. 1931 – 1r – us UF Libraries [972]

A caricature history of canadian politics : events from the union of 1841, as illustrated by cartoons from "grip", and various other sources / Bengough, John Wilson – Toronto: Grip Print & Pub Co, 1886 – 2v on 1mf – 9 – (int by principal grant) – mf#07441 – cn Canadiana [971]

Caricature politique au canada = Free lance political caricature in canada / Ryan, Alonzo [ill] – Montreal: Dominion Pub Co A T Chapman, 1904 [mf ed 1980] – 2mf – 9 – (int by by lucien lasalle and h m williams; in french and english) – mf#SEM105P57 – cn Bibl Nat [760]

Caricias de lumbre / Nolasco Cordero, Francisco – Ciudad Trujillo, Dominican Republic. 1961 – 1r – us UF Libraries [972]

La caridad cristiana / Fernandez Fernandez, Juan – Badajoz: Tip. Arqueros, 1953 – 1 – sp Bibl Santa Ana [946]

La caridad en los primeros siglos del cristianismo / Cicognani, H J – Madrid, 1931; Madrid: Razon y Fe, 1931 – 1 – sp Bibl Santa Ana [240]

La caridad misional y la epistola de san pablo a los filipenses. badajoz / Vera, Emilio de & Fernandez y Fernandez, Juan – Madrid: Razon y Fe, 1947 – 1 – sp Bibl Santa Ana [240]

Caries research – Basel. 1967-1996 (1) 1967-1996 (5) 1970-1996 (9) – ISSN: 0008-6568 – mf#3148 – us UMI ProQuest [617]

Carilla, Emilio *see*
- Literatura de la independencia hispanoamericana
- Olvidado poeta colonial
- Romanticismo en la america hispanica

Carillo Y Anacona, Crescencio *see* Obispado de yucatan historia de su fundacion y sus obispos

Cario, Louis *see* L'exotisme

O carioca – Rio de Janeiro, RJ: Typ de Silva Santos & Cia, 04-25 abr 1853 – (= ser Ps 19) – mf#P15,01,54 n01 – bl Biblioteca [321]

Cariocas e paulistas / Correa, Antonio Augusto Mendes – Porto, Portugal. 1935 – 1r – us UF Libraries [972]

Carissimi, Giacomo *see* Ars cantandi

Caristas diocesanas / Caritas Diocesana de Coria – Caceres: Tip. Extremadura, S.A. 1955 – sp Bibl Santa Ana [240]

Carit khmaer / Punnacand Mul – Bhnam Ben: Ron Bumb Nagar Dham 2517 [1973] [mf ed 1990] – 1r with other items – 1 – (in khmer) – mf#mf-10289 seam reel 122/10 [§] – us CRL [959]

Caritas : erzaehlungen fuer das deutsche haus / Gerhardt, Dagobert von – Leipzig: G Fock [18–?] [mf ed 1993] – 1r – 1 – (filmed with: gerke suteminne / gerhard von amyntor [dagobert von gerhardt] & other titles) – mf#8584 – us Wisconsin U Libr [830]

Caritas *see* Una caritas parroquial sencilla

Caritas anglicana : or, an historical inquiry into those religious and philanthropical societies that flourished in england between the years 1678 and 1740 / Portus, Garnet Vere – London: AR Mowbray 1912 [mf ed 1989] – 1mf – 9 – 0-7905-7185-4 – (int by w h hutton) – mf#1988-3185 – us ATLA [360]

Caritas Diocesana *see*
- Caritas diocesana de accion catolica. coria-caceres. 1955
- Memoria 1957
- Memoria-informe 1973. reconciliacion? ser justo y fraternal con todos

Caritas diocesana de accion catolica. coria-caceres. 1955 / Caritas Diocesana – s.l., s.i., s.a. – 1 – sp Bibl Santa Ana [241]

Caritas Diocesana de Coria *see* Caristas diocesanas

Una caritas parroquial sencilla / Caritas – Badajoz: Tip. A. Mangas, 1969 – sp Bibl Santa Ana [946]

Carl august im niederlaendischen feldzug 1814 / Egloffstein, Hermann, Freiherr von – Weimar: Goethe-Gesellschaft, 1927 [mf ed 1993] – (= ser Schriften der goethe-gesellschaft v40) – viii/248p/[2pl] (ill) – 1 – (incl bibl ref and ind) – mf#8657 reel 10 – us Wisconsin U Libr [943]

Carl august im niederlaendischen feldzug 1814 / Egloffstein, Hermann, Freiherr von und zu – Weimar: Goethe-Gesellschaft, 1927 [mf ed 1993] – (= ser Schriften der goethe-gesellschaft 40) – viii/248p/2pl (ill) – 1 – (incl bibl ref and ind) – mf#8657 reel 10 – us Wisconsin U Libr [430]

Carl burney's der musik doctors tagebuch seiner musikalischen reisen : v2: durch flandern, die niederlande und am rhein bis wien / Burney, C – Hamburg, 1773 – 4mf – 9 – mf#P-656 – ne IDC [780]

Carl burney's der musik doctors tagebuch seiner musikalischen reisen : v3: durch boehmen, sachsen, brandenburg, hamburg und holland... / Burney, C – Hamburg, 1773 – 4mf – 9 – mf#P-657 – ne IDC [780]

Carl friedrich von ledebour's...reise durch das altai-gebirge und die soongorische kirgisen-steppe : auf kosten der kaiserlichen universitaet dorpat unternommen im jahre 1826 in begleitung der herren d carl anton meyer und d alexander von bunge, mit kupfern und karten – Berlin 1829-30 [mf ed Hildesheim 1995-98] – 2v on 14mf – 9 – €140.00 – 3-487-27614-3 – gw Olms [912]

Carl friedrich von naegelsbach's homerische theologie = Homerische theologie / Naegelsbach, Carl Friedrich – 3. Aufl. Nuernberg: C Geiger, 1884 – 2mf – 9 – 0-524-02223-2 – (incl bibl ref) – mf#1990-2897 – us ATLA [250]

Carl friedrich zelters darstellungen seines lebens / ed by Schottlaender, Johann-Wolfgang – Weimar: Verlag der Goethe-Gesellschaft, 1931 [mf ed 1994] – (= ser Schriften der goethe-gesellschaft 44) – xxvii/403p/10pl [ill] – 1 – (incl bibl ref & ind) – mf#3562P – us Wisconsin U Libr [880]

Carl gustav carus als erbe und deuter goethes / Wilhelmsmeyer, Hans – Berlin: Junker & Duennhaupt 1936 [mf ed 1992] – (= ser Neue deutsche forschungen. abteilung neuere deutsche literaturgeschichte 82) – 2r – 1 – (incl bibl ref) – mf#3185p – us Wisconsin U Libr [430]

Carl hildebrand freiherr v. canstein : zum theil nach handschriftlichen quellen, mit portrait und facsimile / Plath, Karl Heinrich Christian – Halle: Verlag der Buchh des Waisenhauses, 1861 – 1mf – 9 – 0-524-03293-9 – (incl bibl ref) – mf#1990-0904 – us ATLA [240]

Carl loewes (1796-1867) werke : gesamtausgabe der balladen, legenden, lieder und gesaenge fuer eine singstimme, im auftrag der loeweschen familie = Carl loewe's works. complete edition of the ballads, legends, songs, and arias for solo voice by commission of the loewe family / ed by Runze, Max – Leipzig: Breitkopf & Haertel. 17v. 1899-1904 – 11 – $135.00 set – us Univ Music [780]

Carl m marcy, senate service 1950-1973 : chief of staff, senate foreign relations committee – (= ser Us senate historical office oral history coll) – 4mf – 9 – $20.00 – us Scholarly Res [327]

Carl philipp bach and the growth of the sonata form / Frank, Lawrence Stroup – U of Rochester 1933 [mf ed 19–] – 1r – 1 – mf#film 477 – us Sibley [780]

Carl philipp emanuel bach's concept of the free fantasia / Elder, Elinor Goertz – U of Rochester 1980 [mf ed 19–] – 3mf – 9 – mf#fiche 1171 – us Sibley [780]

Carl t curtis health news / Omaha Tribe of Nebraska – 1986 dec, 1987 jan/feb, apr/may, sep/oct – 1r – 1 – (cont: newsletter [omaha tribe of nebraska]) – mf#1054084 – us WHS [306]

Carl Vinson [Ship] *see* Eagle

Carlberg, Gustav *see*
- Glimpses from central honan
- Honan glimpses

Carlblom, August *see* Zur lehre von der christlichen gewissheit

Carle, Erwin *see* Allen gewalten zum trotz

Carlebach, David *see* Biblische koenigsdramen in der franzoesischen tragoedie des 16. und 17. jahrhunderts

Carlebach, Salomon *see*
- Geschichte der juden in lubeck
- Ratgeber fur das judische haus

The carleton enterprise – Carleton, NE: W H McCurdy. 22v. v1 n1. nov 21 1919-v22 n29. may 22 1941 (wkly) [mf ed with gaps] – 6r – 1 – us NE Hist [071]

Carleton, George Washington *see* Our artist in cuba fifty drawings on wood

Carleton island in the revolution : the old fort and its builders : with notes and brief biographical sketches / Durham, J H – Syracuse, NY: Bardeen, 1889 – 2mf – 9 – mf#05203 – cn Canadiana [971]

Carleton, James G *see* The bible of our lord and his apostles

Carleton journalism review – v1 n1-v3 n1 [1977 spr-1980 win] – 1r – 1 – mf#666105 – us WHS [070]

Carleton leader – Carleton, NE: Chas W Eisenbise (wkly) [mf ed v13 n4. dec 2 1905-07,1910-11,1913-14 (gaps) filmed 1989] – 2r – 1 – (vol numbering irregular nov 23-dec 7 1911) – us NE Hist [071]

Carleton miscellany – Northfield. 1960-1980 [1]; 1971-1980 [5]; 1976-1980 [9] – ISSN: 0008-6649 – mf#1608 – us UMI ProQuest [400]

Carleton University *see* Breaking the silence

The carleton weekly – Carleton, NE: S M Figge (wkly) [mf ed v1 n34. jul 22 1892 filmed 1973] – 1r – 1 – us NE Hist [071]

Carleton, Will *see*
- Drifted in
- Farm ballads

Carletti, Tomaso *see* Attraverso il benadir

Carli, Gian Rinaldo *see* Lettres americaines

Carli, Gileno De *see* Anatomia da renuncia

Carlieri, I *see* Notizie varie dell' imperio della china...
Carlile, John Charles *see* The story of the english baptists
Carlile, Richard *see* Manual of freemasonry: in three parts
Carlile, Warrand *see* Christ tempted in all points like as we are
[Carlin-] carlin express – NV. 1993- – 2r – 1 – $120.00 (subs $50y) – mf#U04835 – us Library Micro [071]
[Carlin-] courier – NV. 1976 – 1r – 1 – $60.00 – mf#N04436 – us Library Micro [071]
[Carlin-] nevada democrat – NV. feb, apr 1917; oct-nov 1914 [wkly] – 1r – 1 – $60.00 – mf#U04436 – us Library Micro [071]
[Carlin-] western home builder – NV. 1914-19 (scats) [wkly] – 2r – 1 – $120.00 – mf#U04437 – us Library Micro [071]
Carling, Jon *see* The effect of transverse pedal spacing on cycling efficiency
Carlisle 1732-1849 – Oxford, MA (mf ed 1995) – (= ser Massachusetts vital record transcripts to 1850) – 6mf – 9 – 0-87623-224-1 – (mf 1t: births & deaths 1741-56. mf 1t-4t,6t: marriage intentions 1780-1849. mf 1t-4t: marriages 1780-1843. mf 4t: out-of-town marriages 1755-98. mf 4t-6t: births & deaths 1732-1843. mf 6t: births & deaths 1843-49; marriages 1844-49) – us Archive [978]
Carlisle american – Carlisle, PA. -w 1856-64. 3 rolls – 13 – $25.00r – us IMR [071]
Carlisle baptist church. stewart county. tennessee : church records – 1913-Aug 1966 – 1 – 9.36 – us Southern Baptist [242]
Carlisle express and cumberland advertiser – [NE England] Cumbria 16 mar 1861-25 mar 1870 [mf ed 2003] – 10r – 1 – uk Newsplan [072]
Carlisle express and examiner – [NE England] Cumbria jan 1897-23 aug 1913 [mf ed 2003] – 16r – 1 – (missing: 1896) – uk Newsplan [072]
Carlisle gazette – Carlisle, PA. -w 1823-1897; 1897-1900. 4 rolls – 13 – $25.00r – us IMR [071]
Carlisle, George William Frederick Howard, 7th earl of *see* Secular education
Carlisle herald – Carlisle, PA. -d 1802-1920 – 58 rolls – 13 – $25.00r – us IMR [071]
Carlisle herald and expositer – Carlisle, PA., 1837-1847 – 13 – $25.00r – us IMR [071]
Carlisle mirror – Carlisle, PA. -w 1875-79. 2 rolls – 13 – $25.00r – us IMR [071]
Carlisle, Nicholas *see*
- Hints on rural residences
- A memoir of...william wyon

Carlisle patriot – [NE England] Cumbria 3 jun 1815-jun 1910 [mf ed 2003] – 80r – 1 – uk Newsplan [072]
Carlisle. Presbytery (Pres. Church in the USA) *see* Minutes
Carlisle, Ralph C *see* Making a long time program in vocational agriculture for sneads com...
Carlisle rep and framers mecn. – Carlisle, PA., 1830-1931 – 13 – $25.00r – us IMR [071]
Carlisle republican – Carlisle, PA. -w 1831-38; 1890-91. 3 rolls – 13 – $25.00r – us IMR [071]
Carlisle volunteer – Carlisle, PA. -w 1905-13. 6 rolls – 13 – $25.00r – us IMR [071]
Carlisle whig & various papers – Carlisle, PA. -w 1822-23. 2 rolls – 13 – $25.00r – us IMR [071]
Carlos 4 y maria luisa, de juan perez de guzman y gallo / Godoy, Manuel & Fernandez de Bethencourt, Francisco – Madrid: Fortanet, 1913. B.R.A.H. 62. pp. 460-464 – 1 – sp Bibl Santa Ana [946]
Carlos a boissier y diaz / Carbonell, Jose Manuel – Habana, Cuba. 1958 – 1r – us UF Libraries [972]
Carlos alban / Vernaza, Jose Ignacio – Cali, Colombia. 1948 – 1r – us UF Libraries [972]
Carlos manuel de cespedes / Torriente Y Peraza, Cosme De La – Habana, Cuba. 1946 – 1r – us UF Libraries [972]
Carlos mendieta / Marcos Suarez, Miguel De – Habana, Cuba. 1923 – 1r – us UF Libraries [972]
Carlota joaquina / Cheke, Marcus – Rio de Janeiro, Brazil. 1949 – 1r – us UF Libraries [972]
Carlow independent and carlow post *see* Carlow independent and leinster agricultural journal
Carlow independent and leinster agricultural journal – Carlow, Ireland. 28 jun 1879-jun 1882 – 1 1/4r – 1 – (aka: carlow independent and carlow post) – uk British Libr Newspaper [072]
Carlow journal : or leinster chronicle – Carlow, 27 mar 1784, 12 feb 1785 – 0.25r – 1 – ie National [072]
Carlow mercury : or leinster advertiser – Carlow, 25 oct 1788 – 0.25r – 1 – ie National [072]
Carlow morning post – Ireland. -d. 3 Jan 1828-27 May 1833, 30 Nov 1833-24 Jan 1835. (6 reels) – 1 – uk British Libr Newspaper [072]
Carlow nationalist, and leinster times – Ireland. The Nationalist, and Leinster Times. -w. 22 Sept 1883-1923; 1927; 1986-1992; 1994. 55 reels – 1 – uk British Libr Newspaper [072]
Carlow post – Carlow, Ireland. 15 oct 1853-11 may 1878 – 8 1/2r – 1 – uk British Libr Newspaper [072]
Carlow sentinel – Ireland. -w. 1832-Oct 1920. 30 1/2 reels – 1 – uk British Libr Newspaper [072]
Carlow standard – Ireland. -w. 2 Jan-19 Apr 1832. (1/4 reel) – 1 – uk British Libr Newspaper [072]
Carlow vindicator and leinster standard – Carlow, Ireland. 1892 – 1/2r – 1 – uk British Libr Newspaper [072]
Carlow weekly news and general advertiser – Carlow, Ireland. 27 mar 1858-24 oct 1863 – 3r – 1 – uk British Libr Newspaper [072]
[Carlsbad-] carlsbad journal – CA. 1926-28, 1930- – 68+ r – 1 – $4080.00 (subs $50/y) – (aka: carlsbad champion) – mf#H03173 – us Library Micro [073]
Carlsbad champion *see* [Carlsbad-] carlsbad journal
Carlscronas wekloblad *see* Karlskrona weckoblad
Carlshafener zeitung – Bad Karlshafen DE, 1913-1914 30 sep – 1r – 1 – gw Misc Inst [074]
Carlshamn – Karlshamn, Sweden. 1864-67 – 4r – 1 – sw Kungliga [078]
Carlson, Frank *see*
- Papers
- Selected papers

Carlson, Fred Albert *see* Geography of latin america
Carlson, Gerald A *see* Double-cropping wheat and soybeans in the southeast
Carlson, PD *see* The effect of tactile and whole/part drill on the acquisition of opposition in a successful basketball lay-up
Carlsruher beytraege zu den schoenen wissenschaften / ed by Molter, Friedr – Frankfurt, Leipzig 1760-65 – (= ser Dz. abt literatur) – 3v on 11mf – 9 – €110.00 – mf#k/n4437 – gw Olms [500]
Carlsruher wochenblatt – Karlsruhe DE, 1756 dec [single iss], 1757-59 – 1r – 1 – (filmed by other misc inst: 1756-58, 1774-75 [1r]) – gw Misc Inst [074]
Carlsruher zeitung – Karlsruhe DE, 1848-49 – 3r – 1 – (filmed by other misc inst: 1784-1933 [gaps] [109r]; title varies: 1 jan 1811: grossherzoglich badische staats-zeitung, 1 jan 1817: karlsruher zeitung, between 15 may-24 jun 1849: organ der provisorischen regierung; with several suppls) – gw Misc Inst [074]
Carlsruher zeitung – Karlsruhe DE, 1784-1933 [gaps] – 109r – 1 – (filmed by other misc inst: 1848-49'[3r]. title varies: 1 jan 1811: grossherzoglich badische staats-zeitung; 1 jan 1817: karlsruher zeitung; fr 15 may-24 jun 1849: organ der provisorischen regierung. with suppls) – gw Misc Inst [074]
Carlstads allehanda – Karlstad, Sweden. 1854-55 – 1r – 1 – sw Kungliga [078]
Carlton – (= ser Bedfordshire parish register series) – 1mf – 9 – £3.00 – uk BedsFHS [929]
Carlton, Frank T *see* History and problems of organized labor
Carlton news – British Columbia, CN. jan 1938-dec 1943 – 1r – 1 – cn Commonwealth Imaging [071]
Carlton news – Carlton OR: G R Knapton, 193?-1946 [wkly] – 1 – (cont by: yamhill county news [1946-19-?]) – us Oregon Lib [071]
Carlton sentinel – Carlton OR: J D Burt, -1931 [wkly] [mf ed 1967] – 2r – 1 – (cont by: newberg scribe and carlton sentinel [1931-32]) – us Oregon Lib [071]
Carlton-cum-willingham 1588-1950 – (= ser Cambridgeshire parish register transcript) – 5mf – 9 – £6.25 – uk CambsFHS [929]
Carlton-yamhill review – Carlton OR: N K Stewart, 1946- [wkly] – 1r – us Oregon Lib [071]
Carlucci, Joseph Barry *see* Analytical study of published clarinet sonatas by american composers
Carluke chronicle, and strathclyde advertiser – [Scotland] South Lanarkshire, Carluke: J Cossar 5 mar 1870 [mf ed 2003] – 1r – 1 – uk Newsplan [072]
Carlyle, Alexander James *see*
- Christianity in history
- The influence of christianity upon social and political ideas

Carlyle, Gavin *see* The light of all ages
Carlyle, Rev. G *see* The collected writings of edward irving
Carlyle, Thomas *see*
- Correspondence of thomas carlyle and janet welsh
- The correspondence of thomas carlyle and ralph waldo emerson, 1834-1872
- Essays on the greater german poets and writers
- French revolution
- Goethe
- The life of john sterling
- Pleadings with my mother
- Scottish and other miscellanies

Carlyles einfluss auf kingsley in sozialpolitischer und religioes-ethischer hinsicht / Meyer, Maria – Leipzig, 1914 (mf ed 1994) – 2mf – 9 – €31.00 – 3-8267-3074-7 – mf#DHS-AR 3074 – gw Frankfurter [170]
Carlyles stellung zu christentum und revolution / Schultze-Gaevernitz, G von – Leipzig, 1891 (mf ed 1993) – 1mf – 9 – €24.00 – 3-89349-261-5 – mf#DHS-AR 118 – gw Frankfurter [240]
Carlyle's translation of wilhelm meister / Marx, Olga – Baltimore: Waverly Press 1925 [mf ed 1990] – 1r – 1 – (incl bibl ref) – mf#7371 – us Wisconsin U Libr [430]
Carlyon, H C *see* Work among the jats of the rohtak district
Car-madison newsletter / International Committee Against Racism – 1974 apr 1-1978 jun – 1r – 1 – mf#665395 – us WHS [305]
La carmagnole – Au Marais [Paris]: Dondey-Dupre, jun 1-11/15 1848 – us CRL [074]
Carman, Albert *see* The supernatural
Carman, Albert Richardson *see*
- The pensionnaires
- The preparation of ryerson embury

Carman, Bliss *see*
- Address to the graduating class, 1911, of the unitrinian school of personal harmonizing
- An apostle of personal harmonizing
- April airs
- At michaelmas
- Ballads of lost haven
- Behind the arras
- By the aurelian wall
- Christmas eve
- Christmas eve at s kavin's
- Corydon
- Echoes from vagabondia
- Four sonnets
- The friendship of art
- From the book of myths
- The gate of peace
- The grave-tree / the wind and the tree / seven wind songs / overlord
- In the heart of the hills
- James whitcomb riley
- The kinship of nature
- Low tide on grand pre and ballads of lost haven
- Marian drurie
- The master of the isles / an afterword / a robin song / the tragedy of willow / the faithless lover / the faithful love
- Moonshine, songs and ballads
- More songs from vagabondia
- Ode on the coronation of king edward
- "An open letter" from bliss carman
- A pagan's prayer
- A painter's holiday
- The path to sankoty
- Pipes of pan
- The poetry of life
- The rough rider
- Sappho
- Songs from vagabondia
- Songs of sappho, vol 1
- The trail of the bugles
- The vengeance of noel brassard
- The white gull
- A windflower
- A winter holiday
- The word at st kavin's

Carmarthen express and general advertiser, etc – [Wales] LLGC 12 may 1876-12 dec 1878 [mf ed 2004] – 3r – 1 – uk Newsplan [072]
Carmarthen journal – Wales: The Journal, 1821-22; 1832-35; 1841-43; 1845-65; 1867-68; 1871; 1876-78; 1880; 1886; 1889; 1893-96; Jun 1910-Dec 1911; 1925; 1950-51; 1976- – 95+ r – 1 – uk British Libr Newspaper [072]
Carmarthen times, and joint counties weekly gazette – [Wales] LLGC 30 jan-21 aug 1875 [mf ed 2003] – 1r – 1 – uk Newsplan [072]
Carmarthen weekly reporter – [Wales] LLGC jan 1896-9 sep 1921 [mf ed 2004] – 22r – 1 – uk Newsplan [072]
Carmarthen weekly reporter etc – Wales, UK. 2 Sept 1860-1870; 8 Apr 1871-1878; 3 Jan 1879-1895; 189?-1899 – 13 1/2r – 1 – (missing: 1872) – uk British Libr Newspaper [072]
Carmel : allgemeine illustrierte judenzeitung – Pest: Josef Baermann, W A Meisel. v1-2? 1860-61? [complete?] – (= ser German-jewish periodicals...1768-1945, pt 1) – 1r – 1 – $125.00 – (cont as: allegemeine illustrierte judenzeitung) – mf#B49 – us UPA [270]
Carmel : une legende de la tribu des cris / Prud'homme, Louis Arthur – Ottawa: impr pour la Societe Royale du Canada, 1920 – 1mf – 9 – 0-665-75327-6 – mf#75327 – cn Canadiana [390]
Carmel baptist church. mansfield, georgia : church records – 30 Nov 1835-3 Jan 1943 – 1 – us Southern Baptist [242]
Carmel baptist church. ruther glen, virginia : church records – 1799-1819, 1864-72, 1872-89, 1889-1901, 1902-35. WMS records. 1885-1933 – 1 – us Southern Baptist [242]
[Carmel-] carmel pine cone – CA. 1941-59 – 14r – 1 – $840.00 – mf#C02095 – us Library Micro [071]
Carmel in america : a centennial history of the discalced carmelites in the united states / Currier, Charles Warren – Baltimore: John Murphy, 1890 – 2mf – 9 – 0-524-03141-X – mf#1990-4590 – us ATLA [240]
Carmel in ireland : a narrative of the irish province of teresian, or, discalced carmelites, a.d. 1625-1896: with a supplement chiefly from letters of irish missionaries of the seventeenth century / Rushe, James P – Dublin: Sealy, Bryers and Walker, M H Gill; New York: Benziger, 1903 – 1mf – 9 – 0-8370-7101-1 – (incl indes) – mf#1986-1101 – us ATLA [240]
Le carmel. paris, 1929 / Vaussard, M-M – Madrid: Razon y Fe, 1930 – 1 – sp Bibl Santa Ana [944]
Carmelite – Carmel, CA. 1928-1932 (1) – mf#62118 – us UMI ProQuest [071]
Carmelite review – Falls View [Niagara Falls, Ont.]: Carmelite Fathers of North America, 1893-1903 – 9 – (cont by: the new carmelite review) – mf#P04324 – cn Canadiana [241]
Carmen acadium : ode for the jubilee year of the reign of queen victoria / Dole, William Peters – St John, NB: s.n, 1887 – 1mf – 9 – mf#06002 – cn Canadiana [941]
Carmen de bello parthico *see* Exposite in terentium...
Carmen de bello saxinico (mgh7:17.bd) – 1889 – (= ser Monumenta germaniae historica 7: scriptores rerum germanicarum in usum scholarum (mgh7)) – €3.00 – (accedit conquestio heinrici 4 imperatoris) – ne Slangenburg [240]
Carmen des gestis frederici 1. imperatoris in lombardia (mgh7:62.bd) – 1965 – (= ser Monumenta germaniae historica 7: scriptores rerum germanicarum in usum scholarum (mgh7)) – €12.00 – ne Slangenburg [240]
The carmen la rosa home course in ballet and toe dancing for beginners / La Rosa, Carmen – New York: Carmen La Rosa School of Ballet [c1944] – 1 – (la rosa, carmen) – mf#*ZBD-*MGO pv19 – Located: NYPL – us Misc Inst [790]
Carmen Natalia *see* Llanto si termino por el hijo nunca llegado
Carmen paschale *see* Historia...
Carmenes de oro malva / Padro, Humberto – San Juan, Puerto Rico. 1947 – 1r – us UF Libraries [972]
Carmichael, A C *see* Domestic manners and social condition of the white, coloured, and negro population of the west indies
Carmichael, A Wilson- *see* From sunrise land, letters from japan
Carmichael, Amy *see*
- From sunrise land
- From the fight
- Lotus buds
- Overweights of joy
- Things as they are

[Carmichael-] carmichael courier – CA. 1952-may 1975 – 12r – 1 – $720.00 – mf#R02096 – us Library Micro [071]
[Carmichael-] carmichael times – CA. sep 1981-dec 1994 – 4r – 1 – $240.00 – mf#B02093 – us Library Micro [071]
Carmichael, Gertrude *see* History of the west indian islands of trinidad and...
Carmichael, Hartley *see* One holy catholic and apostolic church
Carmichael, J Kevin *see* The effect of cranklength on oxygen consumption when cycling at a constant work rate
Carmichael, James *see*
- Church of england teaching
- Design and darwinism
- The errors of the plymouth brethren
- Essay on the character of jesus christ
- Is there a god for man to know?
- The kingdom and the church
- The kingdom of god, or, kingdom
- Precis of the wars in canada
- A sermon preached by the very rev the dean of montreal
- The tares and the wheat
- Why some fairly intelligent persons do not endorse the hypothesis of evolution

Carmichael, James Wilson *see* The art of marine painting in water-colours
Carmichael, William Miller *see* The early christian fathers
Carmina : formae tplila 109 / Petrus Blesensis – [mf ed 2002] – (= ser ILL – ser a; Cccm 128) – 2mf+viii/35 – 9 – €30.00 – 2-503-64282-9 – be Brepols [400]
Carmina / Sanchez Arjona, Vicente – Sevilla: Imprenta Alvarez, 1957 – 1 – sp Bibl Santa Ana [810]
Carmina *see* Exposite in terentium...
Carmina... / Mussatus, Albertinus [Mussato, Albertino] – 14th c – (= ser Holkham library manuscript books 425) – 1r – 1 – mf#95900 – uk Microform Academic [450]

Carmina burana : lateinische und deutsche lieder und gedichte einer handschrift des 13. jahrhunderts aus benedictbeuern auf der k bibliothek zu muenchen – Stuttgart: Literarischer Verein, 1847 [mf ed 1993] – (= ser Blvs 16/1) – xiv/275p – 1 – mf#8470 reel 4 – us Wisconsin U Libr [780]
Carmina cantabrigiensia (mgh7:40.bd) – 1926 – (= ser Monumenta germaniae historica 7: scriptores rerum germanicarum in usum scholarum (mgh7)) – €7.00 – ne Slangenburg [240]
Carmina crucis / Greenwell, Dora – Boston: Roberts, 1869 – 1mf – 9 – 0-7905-7632-5 – mf#1989-0857 – us ATLA [240]
Carmina scripturarum / Marbach, C – Strasbourg, 1907 – 13mf – 8 – €25.00 – ne Slangenburg [221]
Carmina varia *see* Metamorphoses...
Carmona Alonso, Miguel *see* Primeras jornadas de comercio exterior. camara oficial de comercio e industria de caceres. ponente d. ...
Carmona, Dario *see* Prohibida la sombra
Carmona Guillen, Juan *see* Libro del maestro
Carmona, Juan *see*
– Tractatus de peste ac febribus cum puncticulis vulgo tavardillo
– Tractatus de peste...ac febribus cum puncticulis vulgo tabardillo
Carmona, Miguel *see* Excavaciones de america
Carmouche, M *see* Sac a charbon, ou, le pere jean
Carmouche, M (Pierre-Frederic-Adolphe) *see* N, i, ni
Carmouche, Pierre-Frederic-Adolphe *see*
– Cricri et ses mitrons
– Maris a vendre, ou les dispenses anglaises
– Vieillesse de frontin
Carnapas, Anna Macdonald *see* The gospel in its native land
Carnarvon and denbigh herald *see* Carnarvon herald
Carnarvon, Henry Howard Molyneux, Earl of *see* Recollections of the druses of the lebanon, and notes on their religion
Carnarvon herald – Caernarvon, Wales. 1831-82 – 40r – 1 – uk British Libr Newspaper [072]
Carnarvon herald and north wales advertiser – [Wales] Gwynedd jan 1831-dec 1950 [mf ed 2003] – 111r – 1 – (missing: 1862, 1864, 1873, 1886; cont by: carnarvon herald and north and south wales advertiser [jan 1835-dec 1836]; carnarvon and denbigh herald and north and south wales independent [jan 1837-dec 1878]; carnarvon and denbigh herald and north and south wales advertiser [jan 1879-dec 1887]; carnarvon and denbigh herald and north and south wales independent [jan 1888-dec 1920]; carnarvon and denbigh herald and merioneth news [jan 1921-dec 1922]; carnarvon and denbigh herald [jan 1923-jun 1937]; caernarvon and denbigh herald and north wales observer [jul 1937-dec 1950]) – uk Newsplan [072]
Der carnaval und die somnambuele / Immermann, Karl Leberecht; ed by Gerz, Alfred – Potsdam: Ruetten & Loening, [1944?] [mf ed 1991] – (= ser Troesteinsamkeit, eine sammlung deutscher meistererzaehlungen) – 145p – 1 – mf#7500 – us Wisconsin U Libr [830]
Carne / Ribeiro, Julio – Rio de Janeiro, Brazil. 1964 – 1r – us UF Libraries [972]
Carne de quimera / Labrador Ruiz, Enrique – Habana, Cuba. 1947 – 1r – us UF Libraries [972]
Carne, John *see*
– Letters from switzerland and italy, during a late tour
– Letters from the east
– Recollections of the travels in the east
Carne, Louis de *see* Vues sur l'histoire contemporaine
Carne y alma / Gonzalez, Graciela – Managua, Nicaragua. 1952 – 1r – us UF Libraries [972]
Carne y sombra / Blouin, Egla Morales – New York, NY. 1957 – 1r – us UF Libraries [972]
Carnegie – Pittsburgh, 1998+ [1,5,9] – (cont: carnegie magazine) – mf#9629,01 – us UMI ProQuest [700]
Carnegie classics of international law / Carnegie Endowment for International Law – 8r – 1 – $350.00 – us Trans-Media [341]
Carnegie corporation of new york. report – 1922-69 – 9 – $192.00 – mf#0141 – us Brook [360]
Carnegie, D *see* Among the matabele
Carnegie Endowment for International Law *see* Carnegie classics of international law
Carnegie Endowment for International Peace. Division of International Law *see* Pamphlets
Division ofCarnegie Endowment For International Peace Monograph *see* American foreign policy
Carnegie foundation for the advancement of teaching. annual reports – v1-60. 1906-65 – 9 – $431.00 – mf#0142 – us Brook [370]
Carnegie Institution of Washington *see* Contributions from the mount wilson observatory
Carnegie Institution of Washington, Mount Wilson Solar Observatory *see* Communications to the national academy of sciences
Carnegie institution of washington publication *see*
– General catalogue of double stars within 121° of the north pole
– Papers of the mount wilson observatory
– Papers of the mount wilson solar observatory
Carnegie magazine – Pittsburgh. 1927-1997 (1) 1973-1997 (5) 1976-1997 (9) – ISSN: 0008-6681 – mf#9629 – us UMI ProQuest [700]
Carnegie quarterly – New York. 1953-1996 (1) 1975-1996 (5) 1976-1996 (9) – ISSN: 0576-7954 – mf#10436 – us UMI ProQuest [370]
The carnegie survey of the architecture of the south : photographs / Johnston, Frances Benjamin – [mf ed Chadwyck-Healey, 1984] – 132mf – 9 – uk Chadwyck [720]
Carnegie, William Hartley *see*
– Churchmanship and character
– Democracy and christian doctrines
Carnegie-rochester conference series on public policy – Amsterdam. 1978+ (1) 1978+ (5) 1987+ (9) – ISSN: 0167-2231 – mf#42210 – us UMI ProQuest [338]
Carneiro, Cecilio J *see* Bonfire
Carneiro Da Silva, Jose Juliao *see* Memoria topographica e historica sobre os campos d...
Carneiro, David *see*
– Cerco da lapa e seus herois
– Fuzilamentos de 1894 no parana
– Historia da guerra cisplatina
– Historia do periodo provincial do parana
– Parana e a revolucao federalista
– Problema da federacao brasileira
– Trofeus na historia do brasil
Carneiro, Edison *see*
– Quilombo do palmares, 1630-1695
– Quilombo dos palmares
Carneiro, J Fernando *see*
– Imagracao e colonizacao no brasil
Carneiro Leao, Antonio *see*
– Sentido de la evolucion cultural del brasil
– Sociedade rural, seus problemas e sua educacao
Carneiro, Levi *see* Dois arautos da democracia
Carneiro, Milton *see* Filmando janio
Carne-Marcein, Louis Joseph Marie de Carne, Comte de *see* Travels in indo-china and the chinese empire
Carnero / Rodriguez Freyle, Juan – Bogota, Colombia. 1935 – 1r – us UF Libraries [972]
Carnero / Rodriguez Freyle, Juan – Bogota, Colombia. 1942 – 1r – us UF Libraries [972]
Carnet-agenda du forestier pour... / Societe forestiere de Franche-Comte & Belfort – Besancon: P Jacquin (annual) [1903-05] [mf ed 2005] – 1r [ill] – 1 – (lacks 1904; cont: agenda du forestier pour...) – us Harvard [634]
Les carnets d'un curieux : collaboration speciale a "la patrie" / Fauteux, Aegidius – [mf ed 1971] – 1r – 1 – (with ind) – mf#SEM35P47 – cn Bibl Nat [971]
Carnevali, Luigi *see*
– Il ghetto di mantova
Carney, Colleen M *see* The effects of acute and chronic exercise on serum potassium in hemodialysis patients
Carney, Deborah A *see* The effects of a six-month exercise maintenance program on the cardiovascular fitness levels of participants
Carney, Thomas *see* Letters
Carney, William Harrison Bruce *see* History of the alleghany evangelical lutheran synod of pennsylvania
Carney, William P *see* No democratic government in spain. russia's part in spain's civil war. murder and antireligion in spain
Carnforth guardian – [NW England] Carnforth, Lancaster Lib 1933, jan-jun 1935, 1936, jan-jun 1937, 1939-41 – 1 – uk MLA; uk Newsplan [072]
Carnival / Wisconsin Society of the Children of the American Revolution – 1972 sum-1982 – 1r – 1 – (cont by: wisconsin c.a.r.es) – mf#645271 – us WHS [975]
Carnival glass encore – 1975 oct-1979 dec, 1980 feb-1983 apr – 2r – 1 – (cont by: encore [kansas city mo]) – mf#1494888 – us WHS [071]
Carnochan, J *see*
– Igbo revision course for gce, wasc and similar examinations
Carnochan, Janet *see*
– Centennial poem
– Centennial st andrew's, niagara, 1794-1894
– Centennial, st mark's church, niagara
Carnotes – 1990 sep/oct – 1r – 1 – mf#3123861 – us WHS [071]
Carnoy, H *see* Folklore de constantinople
Caro Baroja, Julio *see* Los pueblos de espana. ensayo de etnologia
Caro, Carl *see* Gudrun
Caro, Elme *see*
– Essai sur la vie et la doctrine de saint-martin, le philosophe inconnu
– Etudes morales sur le temps present
– Le materialisme et la science
– Nouvelles etudes morales sur le temps present
Caro, Elme Marie *see* Le pessimisme au 19e siecle
Caro, Georg *see* Sozial- und wirtschaftsgeschichte der juden im...
Caro Grau, Francisco *see* Parnaso colombiano
Caro, Hugo de S *see* Opera omnia in universum vetus et novum testamentum
Caro, Isaac Ben Joseph *see* Toldot yitshak
Caro, Jose Eusebio *see* Antologia
Caro, Miguel Antonio *see*
– Estudios constitucionales
– Poesias latinas
– Versiones latinas
Caro mio ben : a celebrated song, sung by sigr pacchierotti... / Giordani, Giuseppe – London: I Preston [1785?] [mf ed 19–] – 1r – 1 – (english & italian words) – mf#pres. film 96 – us Sibley [780]
Caro ncinonono / Chiume, M W Kanyama – London, England. 1957 – 1r – us UF Libraries [960]
Caro, Nestor *see* Cielo negro
Caro y su obra / Bonilla, Manuel Antonio – Bogota, Colombia. 1947 – 1r – us UF Libraries [972]
Carocciolo, F *see* I commentari i delle gverre fatto co' turchi da d giovanni d'avstria...
Caroli a linne species plantarum : exhibentes plantas rite cognitas ad genera relatas... / Linne, Carl von – ed quarta. Berolini [Berlin]: Impensis G C Nauk. 9v. 1797 [mf ed 1986] – 9 – 0-665-55334-X – mf#55334 – cn Canadiana [580]
Caroli lachmanni in t lucretii cari de rerum natura libros / Lachmann, Karl – Berolini, Germany. 1850 – 1r – us UF Libraries [025]
Caroli linnai systema natura / Linne, Carl Von – Lipsiae, Germany. 1894 – 1r – us UF Libraries [500]
Caroli ruaei e societate jesu carminum libri quatuor... / La Rue, Ch de – Lutetiae Parisiorum: Apud Simonem Benard, 1680 – 4mf – 9 – mf#0-1350 – ne IDC [090]
Carolina balance index : a multiple regression analysis of four balance/postural stability index systems / Brunken, David L – 1999 – 86p on 1mf – 9 – $5.00 – mf#PE 4157 – us Kinesology [613]
Carolina comments – Raleigh. 1972+ (1) 1972+ (5) 1977+ (9) – ISSN: 0576-808X – mf#6349 – us UMI ProQuest [978]
Carolina coronado / Blanco Garcia, Francisco – Madrid: Saenz de Jubera, 1909 – sp Bibl Santa Ana [440]
Carolina coronado / Munoz de San Pedro, Miguel – Madrid, 1953. Sep.Ind. nº 64 Junio, 1953 – 1 – (notas y papales ineditos) – sp Bibl Santa Ana [946]
Carolina flyer – Fayetteville, NC. 1999-2000 (1) – mf#69382 – us UMI ProQuest [071]
Carolina gazette – Charleston SC. 1798 jan-dec 27, 1799 jan 31-1800 dec 25 – 1r – 1 – mf#858671 – us WHS [071]
Carolina genealogist – n33-52 [1978/79 win-1984 fall] – 1r – 1 – mf#780422 – us WHS [929]
Carolina indian voice – 1979 may 24, aug 30-1980 dec, 1981, 1982, 1983 jan-sep, 1983 oct-1984 jul, 1984 aug-1985 sep, 1985 oct-1986, 1987, 1988 jan-1989 jun – 9r – 1 – mf#572101 – us WHS [307]
Carolina israelite – Charlotte. N.C. 1944-68 – 1 – us AJPC [071]
The carolina israelite – Charlotte. v. 1-v. 3, no. 4, v. 12 no. 7-v. 16 no. 6. Feb 1944-May 1946; Mar 1954-Nov Dec 1958* – 1 – us NY Public [073]
Carolina journal of medicine, science, and agriculture – Charleston. 1825-1826 (1) – mf#3955 – us UMI ProQuest [610]
Carolina labor news / Durham Central Labor Union – 1964 mar 19-1965 may 7 – 1r – 1 – (cont: durham labor journal; cont by: labor news [durham nc]) – mf#1223730 – us WHS [331]
Carolina law journal – Columbia. 1830-1831 (1) – mf#3956 – us UMI ProQuest [323]
Carolina law journal – Columbia, S.C.: 1v. 1830-31 (all publ) – 2mf – 9 – $9.00 – mf#LLMC 82-911 – us LLMC [340]
Carolina law repository – Raleigh. 1813-1816 (1) – mf#3685 – us UMI ProQuest [323]
Carolina law repository – Raleigh. v1-2. 1813-1816 (all publ) – (= ser Historical legal periodical series) – 1,5,6 – $40.00 set – mf#408880 – us Hein [340]
Carolina law repository – (= ser Supreme court reports) – 9 – (title renumbered into official run of the north carolina supreme court reports as 4 n.c.) – mf#LLMC 95-266 – us LLMC [340]
Carolina Media Project *see* Protean radish
Carolina news – Chapin, SC. 1897-1899 (1) – mf#66467 – us UMI ProQuest [071]
Carolina peacemaker – Greensboro NC. 1970 oct 10, dec 5-12, dec 26-2003 jul/dec – 41r – 1 – (with gaps) – mf#655080 – us WHS [071]
Carolina quarterly – Chapel Hill. 1948+ (1) 1975+ (5) 1976+ (9) – ISSN: 0008-6797 – mf#10382 – us UMI ProQuest [073]
Carolina Rifle Club *see* Minutes of the carolina rifle club
Carolina spartan – Spartanburg, SC. 1849-1893 (1) – mf#66519 – us UMI ProQuest [071]
Carolina State *see* Fundamental constitution
Carolina tams quarterly / Carolina Token and Medal Society [Greensboro NC] – 1982 mar-1988 nov – 1r – 1 – mf#1054089 – us WHS [730]
Carolina times – Durham, NC. 1937-1994 (1) – mf#65302 – us UMI ProQuest [071]
Carolina times – Durham NC. 1960 feb 20, 1972 apr 29-may 6, 1975 aug 30, sep 20, oct 11 – 1r – 1 – mf#780641 – us WHS [071]
Carolina times – Durham, North Carolina, 1963-1969 – 7r – (gaps) – us UF Libraries [071]
Carolina Token and Medal Society [Greensboro NC] *see* Carolina tams quarterly
Carolinas genealogical society bulletin – v15 1 [1978 sum], v16 1-v17 1 [1979 sum-1980 sum] – 1r – 1 – mf#666370 – us WHS [929]
Caroline : ou, le tableau / Roger, Francois – Paris, France. 1810 – 1r – us UF Libraries [440]
The caroline h dall papers, 1811-1917 – [mf ed 1981] – 45r – 1 – (with p/g. coll provides insight into women's studies, 19th-century religion, literature, and social and political history) – us MA Hist [322]
Caroline progress – Bowling Green, VA. 1999-2000 (1) – mf#66677 – us UMI ProQuest [071]
Caroline und dorothea schlegel in briefen / ed by Wieneke, Ernst – Weimar: G Kiepenheuer, 1914 [mf ed 1988] – 596p/pl – 1 – mf#2146 – us Wisconsin U Libr [920]
Caroline von wolzogens "agnes von lilien" (1798) : ein beitrag zur geschichte des frauenromans / Brock, Stephan – Berlin, 1914 (mf ed 1995) – 2mf – 9 – €31.00 – 3-8267-3112-3 – mf#DHS-AR 3112 – gw Frankfurter [430]
Caroline von wolzogens "agnes von lilien" (1798) : ein beitrag zur geschichte des frauenromans / Brock, Stephan – Berlin: H Blanke 1914 [mf ed 1992] – 1r – 1 – (incl bibl ref. filmed with: joseph von lassberg / ed by karl s bader) – mf#3144p – us Wisconsin U Libr [430]
Carolinian – Raleigh NC. 1962 mar 31/1963-1999 jan 4/mar 29 – 80r – 1 – (with gaps; cont: carolina tribune) – mf#780642 – us WHS [071]
The carolinian florist : as adapted (in english) for the more ready use of the flora caroliniana of thomas walter / Drayton, John; ed by Meriwether, Margaret Babcock – South Caroliniana Library, Uni of South Carolina, 1943 [mf ed Charleston SC, 1981] – 2mf – 9 – (english with latin ind) – mf#51-500 – us South Carolina Historical [580]
Carollo, James J *see* A model of gait performance on functional sub-system quantification
Carolus magnus redivivvs, hoc est caroli magni... : cum henrico m gallorum & nauarrorum rege...comparatio... / Stucki, J W – [Tigvri, Ioannes Vvolph], 1592 – 2mf – 9 – mf#PBU-505 – ne IDC [240]
Caron, Adolphe *see*
– Aux electeurs du comte de quebec
– Catalogue of the private collection of books belonging to the estate of the late sir a p caron
– Discours de sir adolphe caron sur l'execution de louis riel
– Discours sur la question riel, prononce le 17 mars 1886, a la chambre des communes
– Protest against amendment to ex-ministers' pension bill
– Speech of sir adolphe caron, mp on the remedial bill
Caron, Alfred *see* Bio-bibliographie du r p fernand porter
Caron, Francois *see* A true description of the mighty kingdoms of japan and siam
Caron, Louis Bonaventure *see* Opinion of the honorable mr justice caron and judgment of the superior court
Caron, Max *see* Jesus, doctor
Caron, mere *see*
– Directions diverses donnees en 1878 par la reverende mere caron
Caron, Mother *see* Directions diverses donnees en 1878 par la rev mere caron
Caron, Napoleon *see*
– Deux voyages sur le saint maurice
– Histoire de la paroisse d'yamachiche
– Legendes et revenants
– Petit vocabulaire a l'usage des canadiens-francais

Caros en colombia : su fe, su patriotismo, su amor / Holguin Y Caro, Margarita – Bogota, Colombia. 1953 – 1r – us UF Libraries [972]
Caroso, Fabritio *see*
– Il ballarino di m fabritio caroso da sermoneta, diuiso in due trattati
– Raccolta di varij balli fatti in occorrenze di nozze
Carossa, Hans *see*
– Doctor gion
– Fuehrung und geleit
– Gedichte
– Geheimnisse des reifen lebens
– Gesammelte gedichte
– Das jahr der schoenen taeuschungen
– Eine kindheit
– Eine kindheit und verwandlungen einer jugend
– Die schicksale doktor buergers / die flucht
– Tag in terracina
– Ungleiche welten
– Wirkungen goethes in der gegenwart
Carothers Observatory *see*
– Auxiliary bulletin. solar / carothers observatory (private astronomical)
– Auxiliary bulletin. weather / carothers observatory (private astronomical)
– Bulletin (carothers observatory)
Carothers, Warren Fay *see*
– Auxiliary bulletin. solar / carothers observatory (private astronomical)
– Auxiliary bulletin. weather / carothers observatory (private astronomical)
– Bulletin (carothers observatory)
Carotte d'or / Melesville, M – Paris, France. 1846 – 1r – us UF Libraries [440]
Carp, Matatias *see* Cartea neagra
The carpathian – [Pittsburgh, PA: Carpathian Pub Co. v1 n1-v3 n7-9. oct 1941-jul-sep 1943] – 1 – us CRL [073]
The carpatho-russian american – Yonkers, NY: Lemko Association of the United States and Canada, 1968-jan 1969 – us CRL [305]
The carpatho-russian youth – Stanford, CT: [s.n.] v1 n1-v2 n6 jan 20 1938-nov 1940; n1-2 apr,aug 1941 – 1 – us CRL [305]
[Carpeau du Saussay] *see* Voyage de madagascar, connu aussi sous le nom de l'isle de st laurant
Carpenter / United Brotherhood of Carpenters and Joiners of America – 1981 sep-1992 apr – 1r – 1 – (cont: los angeles county carpenter; cont by: southern california carpenter) – mf#2541660 – us WHS [690]
Carpenter and related family historical journal – 1981 jan/mar-1985 dec/oct – 1r – 1 – (cont: carpenter and related family paper) – mf#1095455 – us WHS [929]
Carpenter and related family paper – n1-39 [1977 mar-1980 dec] – 1r – 1 – (cont by: carpenter and related family historical journal) – mf#524516 – us WHS [929]
Carpenter, Edmund James *see* Roger williams
Carpenter, Edward *see*
– Civilisation
– Fabian economic and social thought, series 1
– Modern science
– Pagan and christian creeds
Carpenter, Edward Childs *see* Romeo and–jane
Carpenter family courier – v1 n1-v2 n4 [1985 apr-1987 jan] – 1r – 1 – mf#1085385 – us WHS [929]
Carpenter, Frances *see*
– Caribbean lands
Carpenter, Frank George *see*
– Land of the caribbean
– Uganda to the cape
Carpenter, George Herbert *see* Insect transformation
Carpenter, Henry *see* Church visible, and the church invisible
Carpenter, Henry Barrett *see* Introduction to the history of architecture
Carpenter, James C *see* I crossed the plains in the '50's
Carpenter, Joseph Estlin *see*
– The bible in the nineteenth century
– Comparative religion
– The composition of the hexateuch
– Ethical and religious problems of the war
– The hexateuch according to the revised version
– James martineau, theologian and teacher
– The life and work of mary carpenter
– Life in palestine when jesus lived
– Personal and social christianity
– Phases of early christianity
– The place of christianity among the religions of the world
Carpenter, Kenneth E *see* The harvard university library
Carpenter, Lant *see*
– Discourse on divine influences and conversion
– On the beneficial tendency of unitarianism
– Primitive christian faith
Carpenter, Mason B *see* Mining code
Carpenter, Maurice *see* Indifferent horseman
Carpenter, Rhys *see* Land beyond mexico
Carpenter, Russell Lant *see*
– Personal and social christianity
– Six lectures on the scriptural doctrine of reconciliation or atonement, and connected subjects
Carpenter, Thomas *see* The scholar's spelling assistant
Carpenter, William *see* Political letters and pamphlets by william carpenter
Carpenter, William B *see* Mesmerism, spiritualism, and c.: historically and scientifically considered
Carpenter, William Benjamin *see*
– Mesmerism, spiritualism, &c
– Principles of general and comparative physiology
Carpenter, William Benjamin. *see* Is man an automaton?
Carpenter, William Boyd *see*
– Book by book
– The great charter of christ
– An introduction to the study of the scriptures
– The permanent elements of religion
– A popular history of the church of england
– Some thoughts on christian reunion
– The son of man among the sons of men
– The wisdom of james the just
– The witness of religious experience
– The witness of the heart to christ
– The witnesses to the influence of christ
Carpenter, William Hookham *see* Pictorial notices
[Carpenteria-] carpenteria herald – CA. dec 8 1911-nov 28 1913, dec 4 1914-oct 1954, 1976-80 – 19r – 1 – $1140.00 – (aka: valley news) – mf#H03176 – us Library Micro [071]
Carpenters' District Council of Western Pennsylvania *see* Western pennsylvania carpenter
Carpenter's monthly political magazine – v1-2 n2,2. 1831-32 [all publ] – (= ser Radical periodicals of great britain, 1794-1914. period 1) – 6mf – 9 – $115.00 – us UPA [331]
Carpentersville countryside – Barrington, IL. 1982-1983 (1) – mf#68644 – us UMI ProQuest [071]
Carpentier, Alejo *see*
– Ecue-yamba-o!
– Guerra del tiempo
– Kingdom of the world
– Lost steps
– Musica en cuba
– Pasos perdidos
– Reino de este mundo
– Royaume de ce monde
– Siglo de las luces
– Tres relatos
Carpentier Alting, A S *see*
– Loge ritualen voor den leerling-graad
– Woordenboek voor vrijmetselaren
Carpentier, Denyse *see* Bibliographie analytique de louis hemon
Carpentry and building – New York: David Williams Co, 1879-1909. v3 1881. v14-15 1892-93 – us CRL [690]
Carpet bag rule in florida / Wallace, John – Jacksonville, FL. 1888 – 1r – us UF Libraries [978]
O carpinteiro joze *see* O mestre joze
Carpmael, A *see* Patent laws of the world
Carpmael, Charles *see*
– On the reduction of the barometer to sea level
– Report of the canadian observations of the transit of venus
Carpmael, E *see* Patent laws of the world
Carr, Arthur *see*
– The general epistle of james
– The gospel according to saint matthew
– The gospel according to st luke
– Horae biblicae
Carr, B *see* The gentlemens amusement
Carr, Benjamin *see*
– Dead march and monody
– Manuscript collection of pianoforte music
– Manuscript collection of vocal music
– Masses, vespers, litanies, hymns, psalms, anthems & motets
– Six imitations of english, scotch, irish, welch, spanish and german airs
– Three divertimentos (for the piano)
Carr, Edward Hallett *see* New society
Carr, Herbert Wildon *see*
– Leibniz
– The philosophy of change
Carr, J H *see* The effect of arm movement on the biomechanics of standing up
Carr, James Anderson *see* The life and times of james ussher
Carr, John *see* L'ete du nord, ou voyage autour de la baltique
Carr, John David *see* From the cam to the cays
Carr, Joseph William Comyns *see*
– Art in provincial france...1882
– Essays on art
– Examples of contemporary art
– Frederick walker
Carr, Marilyn *see* Appropriate technology for african women
Carr, Marilyn. *see* Appropriate technology
Carr, Michael W *see* A history of catholicity in northern ohio and the diocese of cleveland
Carr milestones – v1 iss 1-v5 iss 19 [1983 oct-1988 apr] – 1r – 1 – mf#1322429 – us WHS [071]
Carr, Ralph *see* American papers of ralph carr, 1741-1778
Carr, Simon Joseph *see* Thomae edesseni tractatus de nativitate domini nostri christi
Carr, T W *see* Another gospel
Carr, W David *see* Observations and perceptions of the physical presence, cooperation, and communication between athletic training clinical and classroom instructors
Carr, William G *see* Scriptural outlines by books and themes
Carra de Vaux, Bernard, Baron *see*
– Avicenne
– Gazali
Carracedo, San Salvador de *see* Registro de documentos siecle 11-16 (anno 1792)
Carradine, Beverly *see*
– Are secret societies a blessing or a curse?
– Sanctification
– The sanctified life
– The second blessing in symbol
Carradori, Arcangelo *see*
– Arcangelo carradori's ditionario della lingua italiana e nubiana
– ...Dizionario della lingua italiana e nubiana
Carral Oviedo, Don Benigno *see* Memorias de un loco
An carranach : the general interest magazine of locharron, applecross and torridon districts – Locharron: An Carranach Society 1985-97 (mthly) – 1 – uk Scotland NatLib [072]
Carranca Y Trujillo, Raul *see* Evolucion politica de ibero-america
Carranza, Arturo Bartolome *see* Digesto constitucional americano
Carranza, Jesus E *see* General justo rufino barrios
[Carrara-] carrara miner – NV. jul 1929 – 1r – 1 – $60.00 – mf#U04438 – us Library Micro [071]
[Carrara-] obelisk – NV. 1914-16 [wkly] – 1r – 1 – $60.00 – mf#U04439 – us Library Micro [071]
Carrasco, Adolfo *see* Descubrimiento y conquista de chile
Carrasco Alvarez, Antonio *see*
– Comentario al articulo 1361 del codigo civil
– Incongruencias legales de las faltas contra la propiedad de corchero y compania
Carrasco, Antonio *see* Documentos de 1584 a 1595, relativos a don luis zapata de chaves, existentes en el archivo municipal de llerena
Carrasco Canales, Jose *see* Pasatiempos arroyanos
Carrasco, Castulo *see*
– Correspondencia con juan alcaide sanchez
– Peliculas de aventuras
– Por los blancos caminos del margen. notas para una psicologia del lector
Carrasco Lianes, Virgilio *see*
– Documentos y monumentos epigraficos del museo provincial de badajoz
– Los pueblos tras su historia. bienvenida
Carrasco Montero, Gregorio *see* Novena al santisimo cristo de la salud. brozas
Carrasquilla, Rafael Maria *see* Sermones y discurses escogidos
Carrasquilla, Tomas *see*
– Marquesa de yolomba
– Marquesa de yolombo
– Salve, regina
– Seis cuentos
– Sus mejores cuentos
Carrasquillo, Pedro *see* Requinto
Carratraca en extremadura – 1842 – 9 – sp Bibl Santa Ana [946]
Carrau, Ludovic *see* La philosophie religieuse en angleterre
Carrazzoni, Andre *see* Getulio vargas
Carre de Chambon, Barthelemy *see* Voyage des indes orientales
Carre, Jean Marie *see*
– Goethe
– Goethe en angleterre
– Velikii iazychnik
Carre, William H *see*
– Art work on british columbia, canada
– Art work on hamilton, canada
– Art work on montreal, canada
– Art work on ottawa, canada
– Art work – quebec, canada
Carred, Henri *see* La guerre
Carrefour – Paris. aout 1944-15 oct 1977; 1978-nov 1986 – 1 – fr ACRPP [073]
Carrefour africain – Ouagadougou: [s.n, mar 10 1960-may 1967] – 2r – 1 – us CRL [079]
Carrefour chretien – Montreal: Les Buissonnets inc, [ca 1962]-1985. -v37 n6 nov/dec 1985 (mthly) [mf ed 1991] – 1 – (cont: sourire (montreal, quebec); cont by: presse chretienne) – mf#SEM35P84 – cn Bibl Nat [241]
Carrel, Frank *see*
– Guide to the city of quebec
– Our french canadian friends
– The quebec tercentenary commemorative history
Carreno, Alberto Maria *see*
– Un desconocido cedulario del siglo 16. mexico, 1944
– Isabela
– Mexico y los estados unidos de america
Carrera, Carlos *see* De mi barrio y otros cuentos
Carrera Damas, German *see* Tres temas de historia
Carrera De Wever, Margarita *see* Tematica y romanticismo en la poesia de juan diego
Carrera Justiz, Francisco *see* Orientaciones necesarias
Carreras, Carlos Noriega *see* Sortija de agua
Carrere, Frederic *see* De la senegambie francaise
Carreta / Marques, Rene – Rio Piedras, Puerto Rico. 1961 – 1r – us UF Libraries [972]
Carretas, J *see* Indice de los papeles de la junta central suprema gubernativa del reino y...
The carriage and implement journal (of canada) : devoted to the interests of the manufacturers and dealers in carriages, implements, wagons and harness – Toronto: W H Miln, [1900-19–] – 9 – mf#P04193 – cn Canadiana [680]
Carriage and wagon workers journal, 1899-1908 / official journal, 1912-1915 / mesa educator, 1944-1951 – (= ser Labor union periodicals, pt 1: the metal trades) – 1r – 1 – $210.00 – 1-55655-231-9 – us UPA [621]
Carriage tax 1754-1766 – 2mf – 9 – £2.50 – uk CambsFHS [336]
Carriage, Wagon, and Automobile Workers International Union of North America *see* Spark plug
Carrick democrat etc – Carrick, Ireland. 13 Sept 1883 – 1/4r – 1 – uk British Libr Newspaper [072]
Carrick gazette – Girvan: Carrick gazette & Girvan News 1981- (wkly) [mf ed 5 jan 1995-] – 1 – (cont: carrick gazette and girvan news) – uk Scotland NatLib [072]
Carrick herald and ayr advertiser – Girvan: A Guthrie & Sons Ltd 1974- (wkly) [mf ed 5 jan 1995-] – 1 – (not publ: mar 13 1980; cont: carrick herald and south ayrshire advertiser; title & imprint vary) – ISSN: 1358-4294 – uk Scotland NatLib [072]
Carrick herald and south ayrshire advertiser – [Scotland] South Ayrshire, Girvan: H Wallace jul 1909-dec 1950 (wkly) [mf ed 2004] – 22r – 1 – (incorp with: carrick echo n1441 (22 sep 1939)-n1442 (29 sep 1939); cont by: carrick herald and ayr advertiser) – uk Newsplan [072]
Carrick, John James *see* What you ought to know about mariday park, port arthur
Carrick opinion – Clonmel, Tipperary. 20 oct 1978-27 nov 1981 – 2r – 1 – ie National [072]
Carrick times – Lurgan, Ireland. 14 may 1987-98 – 39 1/2r – 1 – (aka: carrick times and east antrim times) – uk British Libr Newspaper [072]
Carrick times and east antrim times *see* Carrick times
Carrickfergus advertiser and east antrim gazette – Carrickfergus, Ireland 11 sep 1891-9 jan 1931, 25 oct 1946-9 may 1990, 25 sep 1992- [mf 1986-] – 1 – (missing: 1923, 1925; cont: carrickfergus advertiser & county gazette [4 apr 1884-4 sep 1891; cont by: carrickfergus advertiser & guardian [23 may 1990-27 feb 1991]; cont: carrickfergus guardian & advertiser [6 mar 1991-12 aug 1992]) – uk British Libr Newspaper [072]
Carrickfergus advertiser and guardian – Carrickfergus, Ireland 23 may 1990-27 feb 1991 – 1 – (cont: carrickfergus advertiser & east antrim gazette [11 sep 1891-9 jan 1931: 25 oct 1946-9 may 1990]; cont by: carrickfergus guardian & advertiser [6 mar 1991-12 aug 1992]; carrickfergus advertiser & east antrim gazette [25 sep 1992-]) – uk British Libr Newspaper [072]
Carrickfergus freeman – Ireland, 22 Apr 1865-28 Jul 1866 – 1/2r – 1 – uk British Libr Newspaper [072]
Carrick's daily advertiser – Dublin, 19 oct 1812-1813, 12, 16 feb 1814; 22 feb 1814-31 – 39r – 1 – (cont as: carrick's morning post [25 apr 1814-23 mar 1821]; cont as: dublin morning post [24 mar 1821]-ca 5 may 1832; incorp with: dublin times [7 may 1832]) – ie National [072]
Carrie, Pierre *see* Crepuscule
Carrier / Naval Air Station [Alameda CA] – Alameda CA. v44 n49 [1982 dec 3], v45 n9,13,21 [1983 mar 4, apr 1, 27], 1984 jul 6-27, aug 24, sep 5-28, oct 12, nov 2, 16, dec 7-14, 1985 feb 1, 1986 aug 29, oct 10,24-31, nov 5, 1988 jan 8-29, feb 12-26, mar 11-18, apr 8-15, oct 7-14, 28-dec 16, 1989 jan 6-sep 22, oct 6-nov 3,24-dec 15, 1990 jan 5-1991 dec 20, 1992 jan 10-1993 apr 2 – 3r – 1 – mf#1054099 – us WHS [355]
Carrier : official publicationof national association of letter carriers, george t russell branch 576 / National Association of Letter Carriers [US] – 1983 jan-1988 apr – 1r – 1 – mf#1671194 – us WHS [380]
Carrier, Albert *see* Coutumier du 11 [sic] siecle de l'ordre de saint-ruf (chanoines reguliers de saint-augustin) en usage a la cathedrale de maguelone
Carrier, Augustus Stiles *see* The hebrew verb
Carrier, Gaston Marcel *see* Samuel mcchord crothers

[Carrier indian mission paper, 1891-94] / Morice, Adrien Gabriel – [Stuart Lake, BC: s.n, 1894] – 2mf – 9 – 0-665-15665-0 – mf#15665 – cn Canadiana [241]
Carrier, Joseph C *see* Histoire physiologique et chimique d'un flambeau ou bougie de cire
Carrier, Joseph Celestin *see* Histoire physiologique et chimique
Carrier, Nicole *see* Almanachs et annuaires de la ville de quebec de 1780 a 1900
Carriere, Gaston *see* Histoire documentaire de la congregation des missionnaires...otawa, 1963
Carriere, Moriz *see*
– Lebensbilder
– Die philosophische weltanschauung der reformationszeit
– Die philosophische weltanschauung der reformationszeit in ihren beziehungen zur gegenwart
– Die poesie, ihr wesen und ihre formen
– Religioese reden und betrachtungen fuer das deutsche volk
Les carrieres feminines intellectuelles / Bourdeaux, Jean – France-ed. Paris, 1923 – (= ser Les femmes [coll]) – 3mf – 9 – mf#8351 – fr Bibl Nationale [305]
Carriers' flash / National Association of Letter Carriers – n1-[25?] [1975 may 27-jul 30], 1978 aug 30 – 1r – 1 – mf#355985 – us WHS [380]
Carrier's news / National Association of Letter Carriers [US] – 1983 mar-1989 nov/dec – 1r – 1 – mf#1054100 – us WHS [380]
Carrier's voice / National Association of Letter Carriers [US] – v17 n1-v19 n12 [1980 feb-1982 dec], v21 n9-10, [1984 sep-nov], v22 n1-v30 n12 [1985 jan-1988 dec] – 1r – 1 – mf#1058127 – us WHS [350]
Carrighan, Terentius *see* The chancery student's guide in the form of a didactic poem.
Carriker, Robert C *see* The pacific northwest tribes missions collection of the oregon province archives of the society of jesus, 1853-1960
O carril : jornal para a distracao dos viajantes – Bahia: Typ de J G Tourinho, 01 dec 1870 – (= ser Ps 19) – bl Biblioteca [073]
Carrillo, Alfonso *see* Algunos aspectos juridicos de la controversia...
Carrillo Chumacero, Fernando *see* Epistola de laudibus paetriae nostrae
Carrillo de Albornoz, A *see* Die spanische inquisition und die alumbrados. berlin-bonn, 1934
Carrillo Lopez, Ignacio *see* Pensil americano florido...maria de guadalupe de mexico. 1797
Carrillo, Mario *see* In the saddle with gomez
Carrillo, Rafael *see* Ambiente axiologico de la teoria pura del derecho
Carrillo, Santiago *see*
– En marcha hacia la victoria
– La juventud: factor de la victoria
– Por la republica y la legalidad constitucional: todos unidos a la lucha
– Unidad y lucha
Carrillo y Perez, Ignacio *see* Pensil americano florido en el rigor del invierno
Carrington and kirwan's reports : reports of cases argued and ruled at nisi prius in the courts of queen's bench, common pleas and exchequer, together with cases tried on the circuits, and in the central criminal court, also the crown cases reserved / Carrington, F A & Kirwan, A V – v1-3. 1843-53. London: S Sweet, 1845-52 – 25mf – 9 – $37.50 – mf#LLMC 84-759 – us LLMC [324]
Carrington and marshman's reports : reports of cases argued and ruled at nisi prius in the courts of queen's bench, common pleas and exchequer, together with cases tried on the circuits, and in the central criminal court... / Carrington, F A & Marshman, J R – v1-2 in 1bk. 1841-42. London: S Sweet, 1843 (all publ) – 8mf – 9 – $12.00 – mf#LLMC 84-758 – us LLMC [324]
Carrington and marshman's reports *see* Carrington and payne's reports
Carrington and payne's reports : reports of cases argued and ruled at nisi prius in the courts of king's bench and common pleas...and on the oxford summer circuit / Carrington, Frederick A & Payne, J – v1-9. 1823-41. London: S Sweet, 1825-41 [all publ] – 74mf – 9 – $111.00 – (with v9 title is described as: carrington and marshman's reports) – mf#llmc84-760 – us LLMC [324]
Carrington, F A *see*
– Carrington and kirwan's reports
– Carrington and marshman's reports
Carrington, Frederick A *see* Carrington and payne's reports
Carrington, Henry *see* On marriage with the sister of a deceased wife
Carrington, Henry Beebee *see* Battles of the american revolution, 1775-1781
Carrington, John F *see*
– Comparative study of some central african gong-languages
– Talking drums of africa
Carrington, Philip *see* The primitive christian catechism
Carrington Smith, Herbert *see* On the frontier of british guiana and brazil
Carrington, st george – (= ser Cheshire monumental inscriptions) – 1mf – 9 – £2.50 – mf#416 – uk CheshireFHS [929]
Carrion, C *see* Biblioteca manual medico-practica...
Carrion Marquez, Jesus *see* Defensa de la naturaleza
Carrion, Miguel De *see*
– Esfinge
– Honradas
– Milagro
Carrion Y Cardenas, Miguel De *see* Impuras
Carrizo, Juan Alfonso *see* Antecedentes hispano-medioevales de la poesia tradicional argentina
Carrizosa Pardo, Hernando *see* Sucesiones
El carro de la alegria, 1967 / Los Santos de Maimona – Madrid: industrias graficas m.s.a., 1967 – sp Bibl Santa Ana [946]
Carro de la alegria en las localidades de hervas, jarandilla... / Delegacion Provincial de Ministerio de Informacion y Turismo – Febrero, 1970. Caceres: La Minerva, 1969 – 1 – sp Bibl Santa Ana [946]
Carro, P *see* Gramatica ilocana
Carroll, Ana Ella *see* The star of the west
Carroll, Anna E *see* The romish church opposed to the liberties..
Carroll, Austin *see* A catholic history of alabama and the floridas
Carroll, Benajah Harvey *see*
– Baptists and their doctrines
– Course in the english bible
– The genesis of american anti-missionism
– Opening of the course in the english bible
Carroll, BH *see*
– Ecclesia the church, bible class lectures
– The genesis of american anti-missionism
Carroll, Catherine Agnes *see* Percy goetschius
Carroll, Charles *see*
– The charles carroll papers
– Unpublished letters of charles carroll of carrollton and of his father, charles carroll of doughoregan
Carroll chronicle series / Carroll Co. Carrollton – aug 1875-dec 1934 [wkly] – 26r – 1 – mf#B9724-9749 – us Ohio Hist [071]
Carroll Co. Carrollton *see*
– Carroll chronicle series
– Carroll county chronicle
– Carroll county union
– Carroll democrat
– Carroll free press
– Carroll journal
– Carroll union press
– Citizen democrat
– Democratic companion
– Free press standard
– Free press standard series
– Ohio picayune
– Republican series
Carroll Co. Leesville *see* Connotton valley times
Carroll Co. Malvern *see* Community news
Carroll Co. Minerva *see*
– Leader
Carroll Co. Sherodsville *see* Standard
Carroll College *see* Quarterly report
Carroll College [Waukesha WI] *see* Pioneer
Carroll College [Waukesha, WI] *see* Quarterly
Carroll county chronicle / Carroll Co. Carrollton – apr 1871-jul 1875 [wkly] – 2r – 1 – mf#B3984-3985 – us Ohio Hist [071]
Carroll county genealogical quarterly – 1982 spr-1987 fall – 1r – 1 – mf#1231134 – us WHS [929]
Carroll County Genealogical Society [OH] *see*
– Carroll cousins
– Newsletter
Carroll county times – Westminster, MD. 1995+ (1) – mf#61189 – us UMI ProQuest [071]
Carroll county union / Carroll Co. Carrollton – sep 1861-aug 1862 [wkly] – 1r – 1 – mf#B3983 – us Ohio Hist [071]
Carroll cousins / Carroll County Genealogical Society [OH] – 1982 jan-1988 nov/dec – 1r – 1 – (cont: newsletter) – mf#1544945 – us WHS [929]
Carroll democrat / Carroll Co. Carrollton – jan 1860-sep 1861 [wkly] – 1r – 1 – mf#B3983 – us Ohio Hist [071]
Carroll echo – Waukesha WI. 1943 sep 30-1951, 1951-61, 1961 sep 28-1967 may 12, 1967 sep 22-1976 may 7 – 4r – 1 – mf#1110547 – us WHS [378]
Carroll, Edward, Jr *see* Law printing laws.
Carroll, Frank *see* Introduction to antonio soler
Carroll free press / Carroll Co. Carrollton – aug 1875-92, 1897-1905 [wkly] – 11r – 1 – mf#B8972-8982 – us Ohio Hist [071]
Carroll free press / Carroll Co. Carrollton – (sep 1835-36,42-48,52-61,69-73) [wkly] – 6r – 1 – mf#B13053-13058 – us Ohio Hist [071]
Carroll, Henry K *see* Religious forces of the united staes
Carroll, Henry King *see*
– Proceedings of the fourth ecumenical methodist conference
– Statistics of the churches of the united states of america for 1914
The carroll index – Carroll, NE: Arthur P Childs, 1901-v28 n19. may 30 1928 (wkly) [mf ed v1 n26. sep 20 1901-11,1925-28 (gaps) filmed [1972-86]] – 5r – 1 – (absorbed by: wayne herald) – us NE Hist [071]
Carroll, John *see*
– The besiegers' prayer
– Case and his cotemporaries
– The "exposition" expounded, defended and supplemented
– "Father corson"
– A needed exposition
– Past and present
– Reasons for wesleyan belief and practice, relative to water baptism
– The school of the prophets
– The stripling preacher
Carroll, Joseph *see* Our missionary life in india
Carroll journal / Carroll Co. Carrollton – 1957-66, dec 1970-jun 1971 [wkly] – 5r – 1 – mf#B11566-11569 – us Ohio Hist [071]
Carroll journal / Carroll Co. Carrollton – apr 1935-dec 1945 [wkly] – 5r – 1 – mf#B8983-8987 – us Ohio Hist [071]
Carroll journal / Carroll Co. Carrollton – jan 1946-dec 1956 [wkly] – 5r – 1 – mf#B11214-11218 – us Ohio Hist [071]
Carroll, K K *see* Development of a predictive equation for maximal oxygen consumption on the steptreadmill
Carroll, Lewis *see*
– Alice in wonderland
– Alice's adventures in wonderland
– The hunting of the snark
Carroll, New Hampshire. Carroll Baptist Church *see* Records
Carroll news – Hillsville, VA. 1980-1984 (1) – mf#68184 – us UMI ProQuest [071]
Carroll, Phidellia Patton *see* Soul-winning
Carroll union press / Carroll Co. Carrollton – sep-dec 1862, jan 1866-dec 1868 [wkly] – 2r – 1 – mf#B3983-3984 – us Ohio Hist [071]
Carrollton star – New Orleans, LA. 1851-1856 (1) – mf#68746 – us UMI ProQuest [071]
Carrona / Halftermeyer, Gratus – Leon, Nicaragua. 1944 – 1r – us UF Libraries [972]
Carrothers, Julia D *see* The sunrise kingdom
Carrottoman baptist church. ottoman, virginia : church records – 1886-1967 – 1 – us Southern Baptist [242]
Le carrousel : journal de la cour, de la ville et des departements. – Paris. n1-32. mars 1836-juil 1837 – 1 – fr ACRPP [073]
Carrousel art – 1978 apr-1984 feb – 1r – 1 – mf#807883 – us WHS [071]
Carrs hill baptist church. transylvania county. north carolina : church records – 1882-1927 – 1 – 8.19 – us Southern Baptist [242]
Carruaje bajo la lluvia / Figueroa, Carlos Alberto – Guatemala, 1959 – 1r – us UF Libraries [972]
Carruth, Hayden *see* Track's end
Carruthers, Robert *see* Life of alexander pope
Carruthers, SW *see* The westminster confession of faith
Carry, John *see* An exposure of the mischievious perversions of holy scripture in the national temperance society's publications
Carrying the gospel to all the non-christian world : with supplement, presentation and discussion of the report in the conference on 15th june 1910 – Edinburgh: Publ for the World Missionary Conference by Oliphant, Anderson & Ferrier; New York: Fleming H Revell, [1910?] – 2mf – 9 – 0-8370-6472-4 – (incl indes) – mf#1986-0472 – us ATLA [240]
Cars and trucks – McLean. 1965-1979 [1]; 1971-1979 [5]; 1975-1979 [9] – ISSN: 0027-5778 – mf#1975 – us UMI ProQuest [380]
Carson, Alexander *see*
– Examination of the principles of biblical interpretation of ernesti, ammon, stuart
– Knowledge of jesus, the most excellent of the sciences
– Refutation of the review in the christain guardian for january 1832
[Carson-] chronicle – NV. 1935-52, 1927-48 (incomplete), feb-dec 1967, 1969 – 9r – 1 – $540.00 – (aka: morning chronicle) – mf#UN04442 – us Library Micro [071]
[Carson city-] capital news – NV. dec 1950 – 1r – 1 – $60.00 – mf#U04440 – us Library Micro [071]
[Carson city-] carson boys – NV. 1885-1886 – 1r – 1 – $60.00 – mf#U04441 – us Library Micro [071]
[Carson city-] carson daily appeal – NV. 1865-1877; 1877-1946 – 142r – 1 – $8520.00 – (aka: morning appeal. cont by: nevada appeal) – mf#N04443 – us Library Micro [071]
[Carson city-] carson daily bee – NV. oct-nov 1882, 14 jul 1883 (scats) – 1r – 1 – $60.00 – mf#U04449 – us Library Micro [071]
[Carson city-] carson daily times – NV. 1880-81 – 1r – 1 – $60.00 – mf#10037 – us Library Micro [071]
[Carson city-] carson evening gazette – NV. 21-23 jul 1914 – 1r – 1 – $60.00 – mf#U04444 – us Library Micro [071]
[Carson city-] carson free lance – MI. mar 1885-nov 1886 [wkly] – 1r – 1 – $110.00 – mf#U04445 – us Library Micro [071]
[Carson city-] carson review – NV. 1972-1973 – 2r – 1 – $120.00 – mf#N4447 – us Library Micro [071]
[Carson city-] chronicle – NV. 1935-52; 1927-48 (incomplete) [wkly] – 7r – 1 – $420.00 – mf#U04448 – us Library Micro [071]
[Carson city-] daily evening herald – NV. aug-sep 1875 – 1r – 1 – $60.00 – mf#U04451 – us Library Micro [071]
[Carson city-] daily index – NV. 1863-64 (scats); 1880-87 – 8r – 1 – $480.00 – mf#U04450 – us Library Micro [071]
[Carson city-] daily morning post – NV. mar, apr 1865 – 1r – 1 – $60.00 – mf#U04452 – us Library Micro [071]
[Carson city-] daily state register – NV. dec 1870-72 – 3r – 1 – $180.00 – mf#U04454 – us Library Micro [071]
[Carson city-] enlightener – NV. aug-sep 1914 – 1r – 1 – $60.00 – mf#U04455 – us Library Micro [071]
[Carson city-] independent – NV. 1863-1864 – 1r – 1 – $60.00 – mf#U04457 – us Library Micro [071]
[Carson city-] morning news – NV. 1891-1930, jan-apr 1961 [daily] – 41r – 1 – $2460.00 – (aka: carson city news) – mf#U04459 – us Library Micro [071]
[Carson city-] nevada appeal – NV. mar-apr 1877, 1947-71 – 81r – 1 – $4860.00 – (cont by: carson daily appeal; morning appeal) – mf#UN04460 – us Library Micro [071]
[Carson city-] nevada capitol news – NV. 1950 – 1r – 1 – $60.00 – mf#U04461 – us Library Micro [071]
[Carson city-] nevada index-union – NV. 1887-88 [daily] – 2r – 1 – $120.00 – mf#U04462 – us Library Micro [071]
[Carson city-] nevada patriot – NV. jul-aug 1876 (scattered issues) – 1r – 1 – $60.00 – mf#U04463 – us Library Micro [071]
[Carson city-] nevada state journal – NV. may 1886 – 1r – 1 – $60.00 – mf#U04464 – us Library Micro [073]
[Carson city-] nevada state recorder – NV. 1984-1986 – 3r – 1 – $180.00 – mf#N03703 – us Library Micro [071]
[Carson city-] nevada tribune – NV. 1875-96 [daily] – 22r – 1 – $1320.00 – mf#U04467 – us Library Micro [071]
[Carson city-] nevada union – NV. 1886-87 [daily] – 1r – 1 – $60.00 – mf#U04468 – us Library Micro [071]
[Carson city-] nevadian times – NV. mar-jul 1935 [wkly] – 1r – 1 – $60.00 – mf#U04469 – us Library Micro [071]
[Carson city-] new indian – NV. sep 1899; 1903-04 – 1r – 1 – $60.00 – mf#U04470 – us Library Micro [071]
Carson city news *see* [Carson city-] morning news
[carson city, nv-] the indian advance – sep 1899; jan 1901-sep 1903 – 1r – 1 – mf#U04457 – us Library Micro [071]
[Carson city-] parish rubric – NV. feb 1898; mar-apr 1901; jun 1904 – 1r – 1 – $60.00 – mf#U04471 – us Library Micro [071]
[Carson city-] range magazine – NV. 1994- – 2r – 1 – $120.00 (subs $90y) – mf#U04836 – us Library Micro [071]
[Carson city-] republican principals – NV. 1888 – 1r – 1 – $60.00 – mf#U04472 – us Library Micro [071]
[Carson city-] silver age – NV. jul-oct 1861 (scats) – 1r – 1 – $60.00 – mf#U04473 – us Library Micro [071]
[Carson city-] the flash – NV. 1969-1970 – 1r – 1 – $60.00 – mf#N04456 – us Library Micro [071]
[Carson city-] the indian advance – NV. sep 1899; 1901-03 – 1r – 1 – $60.00 – mf#U04458 – us Library Micro [071]
[Carson city-] the nevada state veteran – NV. 17, 30 apr 1946; mar-dec 1949 – 1r – 1 – $60.00 – mf#U04465 – us Library Micro [071]
[Carson city-] the nevada statesman – NV. 1960-66 – 1r – 1 – $60.00 – mf#U04466 – us Library Micro [071]
[Carson city-] the weekly – NV. 1891-1918 – 19r – 1 – $1140.00 – mf#U04475 – us Library Micro [071]
[Carson city-] white ribbon – NV. jul 1894 (scats) – 1r – 1 – $60.00 – mf#U04476 – us Library Micro [071]
Carson free lance – Carson City, NV. 1885-86 – 2r – 1 – $100.00 – mf#N04447 – us Library Micro [071]
Carson, George Stephen *see* A primary catechism for religious instruction in the home and sabbath school
Carson, James Crawford Ledlie *see* The heresies of the plymouth brethren
Carson, Rachel *see* Edge of the sea

Carson valley news – Genoa, NV. 1875-76, 1883-89 – 2r – 1 – $90.00 – mf#U04536 – us Library Micro [071]

[Carson valley-] territorial enterprise – NV. 1859 – 1r – 1 – $60.00 – mf#U04474 – us Library Micro [071]

[Carson valley-] territorial enterprise – UT. jan 1859 – 1r – 1 – $60.00 – mf#U05300 – us Library Micro [071]

Carson, William Robert *see* Reunion essays

Carstairs journal – Alberta, CN. jan 1907-dec 1923 – 5r – 1 – cn Commonwealth Imaging [071]

Carstairs news – Alberta. CN. aug 1924-dec 1924, 1926 – 1 – cn Commonwealth Imaging [071]

Carstairs, Robert *see* British work in india

Carstens, Margret *see* Indigene land- und selbstbestimmungsrechte in australien und kanada unter besonderer beruecksichtigung des internationalen rechts

Carswell chronicle / Carswell Family Association – v1 n2, v4 n1-v8 n1/2 [1972 aug, 1975 feb-1979 aug] – 1r – 1 – mf#626461 – us WHS [929]

Carswell Family Association *see* Carswell chronicle

Carswell sentinel – 1981 may 15/1983 apr-1987 apr 3/1988 mar 11, v34 n1-4,15-26...[1992 jan 10/31, apr 17/jul 10...], v35 n4-6,9-11,17 [1993 feb 5-12, mar 5-19, apr 30] – 6r – 1 – mf#660870 – us WHS [071]

Cart, Jacques *see*
- Histoire des cinquante premieres annees de l'eglise evangelique libre du canton de vaud
- Histoire du movement religieux et ecclesiastique dans le canton de vaud
- Pierre viret

Carta / Cortes, Hernando – 1865 – 9 – sp Bibl Santa Ana [912]

Carta al autor de la oracion apologetica por la espena y su merito literario / Conchudo, J – 1787 – 9 – sp Bibl Santa Ana [840]

Carta al marques de san simon / Godoy, Manuel – Madrid: Fortanet, 1891. B.R.A.H. 18, pp. 470-472 – sp Bibl Santa Ana [946]

Carta apologetica de la sentencia del sumo pontifice benedicto 14 : sobre que el sacerdote que al consagrar el pan... / Perez Calama, Joseph – en Mexico: por D Felipe de Zuniga y Ontiveros, ano de 1780 – (= ser Books on religion...1543/44-c1800: obispos: obispos de quito) – 1mf – 9 – mf#crl-402 – ne IDC [241]

Carta circular – 1967-72, 1978-89* – 1 – (cont by: espana evangelica) – mf#atla s0365 – us ATLA [240]

Carta circular, o edicto, de el ilustrisimo : y reverendisimo senor d fr josef antonio de s alberto, del consejo de s m y obispo de cordova del tucuman / Jose Antonio de San Alberto, archbishop of Rio de la Plata – Buenos Ayres: impr de los Ninos Expositos 1781 – (= ser Books on religion...1543/44-c1800: arzobispos: arzobispos de la plata) – 2mf – 9 – mf#crl-433 – ne IDC [241]

Carta colectiva de los obispos espanoles a los de todo el mundo con motivo de la guerra en espana – Pamplona: Graficas Bescansa 1937 – 9 – mf#w780 – us Harvard [946]

Carta de a...al partir para roma / Ramirez Vazquez, Fernando – 1867 – 9 – (1866 ed) – sp Bibl Santa Ana [440]

Carta de bartolo / Ipnocausto, Paulo – 1790 – 9 – sp Bibl Santa Ana [830]

Carta de bartolo sobrino de don fernando perez / Forner Segarra, Juan Pablo – 1790 – 9 – sp Bibl Santa Ana [946]

Carta de don...en que demuestra quan inaccesibles han sido los esfuerzos de d. bernardo arayo para defender que no que phtisis pulmonar... / Herrero, Antonio Maria – Madrid, 1757 – 1mf – 9 – sp Cultura [616]

Carta de edificacion : en que el p juan antionio baltasar, provincial de esta provincia de nueva espana... / Baltasar, Juan Antonio – en Mexico: ...Joseph Bernardo de Hogal, ano de 1751 – (= ser Books on religion...1543/44-c1800: jesuitas) – 1mf – 9 – mf#crl-227 – ne IDC [241]

Carta de edificacion : en que el p juan antonio baltasat, de la compania de jesus, rector del colegio de san gregorio de este ciudad de mexico da noticia a todos los superiores de las casas... / Baltasar, Juan Antonio – en Mexico: Por Joseph Bernardo de Hogal, ano de 1737 – (= ser Books on religion...1543/44-c1800: jesuitas) – 1mf – 9 – mf#crl-222 – ne IDC [241]

Carta de fernando el catolico comunicando la toma de granada (anno 1492) – Cordoba – 1r – 5,6 – sp Cultura [946]

Carta de paracuellos / Perez, Fernando – 1789 – 9 – sp Bibl Santa Ana [946]

Carta de pedro ponce de leon, obispo de plasencia, a felipe 2, sobre las reliquias y librerias de su obispado y sus actividades literarias / Andres Martinez, Gregorio – Badajoz: Dip. Provincial, 1967 – sp Bibl Santa Ana [240]

Carta de privilegio de los reyes catolicos a la ciudad de badajoz, fechada en el campamento real "sobre toro" el dia 21 de julio de 1475 / Guerra Guerra, Arcadio – Badajoz: Dip. Provincial, 1974. Sep. REE – 1 – sp Bibl Santa Ana [240]

Carta de un juez / Hurtado, Oscar – Habana, Cuba. 1963 – 1r – us UF Libraries [972]

Carta del doctor mariano seguer...a un erudito y sabio... / Seguer, Mariano – SL, SA – 1mf – 9 – sp Cultura [616]

La carta del navegar pitoresco dialogo : opera de marco boschini / Boschini, M – Venetia, 1660 – 13mf – 9 – mf#O-167 – ne IDC [700]

Carta del p bartholome braun : visitador de la provincia tarahumara a los pp superiores de esta provincia de nueva espana, sobre la apostolica vida: virtudes, y santa muerte del p francisco hermano glandorff / Braun, Bartholome – [Mexico City]:...Colegio de San Ildefonso de Mexico, ano de 1764 – (= ser Books on religion...1543/44-c1800: biografias de religiosos) – 1mf – 9 – mf#crl-152 – ne IDC [241]

Carta del p joseph de arjo, de la compania de jesus, preposito de la casa professa de esta ciudad de mexico : en que da noticia a todos los superiores de esta provincia de nueva-espana... / Arjo, Jose – en Mexico: por Joseph Bernardo de Hogal...ano de 1727 – (= ser Books on religion...1543/44-c1800: biografias de religiosos) – 1mf – 9 – mf#crl-138 – ne IDC [241]

Carta del padre pedro de morales de la compania de iesus : para el muy reverendo padre everardo mercuriano, general de la misma compania / Morales, Pedro de – en Mexico: Por Antonio Ricardo, ano 1579 – (= ser Books on religion...1543/44-c1800: vidas y cultos de santos) – 5mf – 9 – mf#crl-108 – ne IDC [241]

Carta del padre provincial francisco zevallos sobre la apostolica vida : y virtudes del p fernando konsag, insigne missionero de la california / Zevallos, Francisco – [Mexico City]:...Colegio de San Ildefonso de Mexico, ano de 1764 – (= ser Books on religion...1543/44-c1800: biografias de religiosos) – 1mf – 9 – mf#crl-153 – ne IDC [241]

Carta Diocesanas *see* Memoria 1959

Carta edificante : en que el p antonio de paredes de la compana de jesus rector del colegio del espiritu santo da noticia a su provincia mexicana de las solidas virtudes... / Paredes, Antonio de – [Mexico City: s.n. 1777] – (= ser Books on religion...1543/44-c1800: biografias de religiosos) – 1mf – 9 – mf#crl-157 – ne IDC [241]

Carta edificante : en que el p antonio de paredes de la extinguida compania de jesus refiere la vida vida exemplar de la hermana salvadora de los santos... / Paredes, Antonio de – 2nd ed. Mexico:...D Joseph de Jauregui... ano de 1784 – (= ser Books on religion... 1543/44-c1800: biografias de religiosos) – 2mf – 9 – mf#crl-159 – ne IDC [241]

Carta edificante del h augustin de valenziaga : coadjutor temporal formado de la compania de jesus, defuncto en el colegio del espiritu santo de puebla a 13, de enero de 1738 / Ansaldo, Matheo – [Mexico City: s.n. 1742] – (= ser Books on religion...1543/44-c1800: jesuitas) – 1mf – 9 – mf#crl-223 – ne IDC [241]

Carta edificativa : en que el p andres xavier garcia, preposito de la casa professa de la sagrada compania de jesus... / Garcia, Andres Javier – en Mexico: impr Bibliotheca Mexicana, ano de 1763 – (= ser Books on religion...1543/44-c1800: jesuitas) – 1mf – 9 – mf#crl-234 – ne IDC [241]

Carta en otono / Feijoo, Samuel – Habana, Cuba. 1957 – 1r – us UF Libraries [972]

Carta familiar de un sacerdote, respuesta a un colegial amigo suyo : en que le da cuenta de la admirable conquista espiritual del vasto imperio del gran thibet... / Francisco, de Ajofrin, fray – en Mexico: impr bibliotheca Mexicana...1765 – (= ser Books on religion... 1543/44-c1800: ordenes, etc: capuchinos) – 1mf – 9 – mf#crl-172 – ne IDC [241]

Carta inedita de jose marti – Habana, Cuba. 1934 – 1r – us UF Libraries [972]

Carta medita de la duquesa de plasencia dona leonor pimentel. donando a los dominicos el convento de san vicente ferrer de la ciudad de plasencia (22 de agosto y 10 de octubre de 1484) / Palomo Iglesias Crescencio – Badajoz: Imprenta de la Diputacion Prov., 1975 – sp Bibl Santa Ana [920]

Carta pastora / Ortiz y Gutierrez, Luis Felipe – 1886 – 9 – sp Bibl Santa Ana [240]

Carta pastoral : 1st asamblea diocesana de accion de catolica / Alcazar Alenda, Jose Maria – Badajoz: Tipografia Espanola, 1933 – 1 – sp Bibl Santa Ana [240]

Carta pastoral / Becerra y Valcarzel, Diego – 1694 – 9 – sp Bibl Santa Ana [240]

Carta pastoral / Garcia Gil, Manuel – 1854 – 9 – sp Bibl Santa Ana [830]

Carta pastoral / Rodriguez de Rivas y Velasco, Diego – [Guadalajara: s.n. 1768] – (= ser Books on religion...1543/44-c1800: obispos: obispos de guadalajara) – 1mf – 9 – mf#crl-394 – ne IDC [241]

Carta pastoral / Rodriguez de Rivas y Velasco, Diego – [Guadalajara s.n. 1769] – (= ser Books on religion...1543/44-c1800: obispos: obispos de guadalajara) – 2mf – 9 – mf#crl-395 – ne IDC [241]

Carta pastoral / Torrijos y Gomez, Ramon – 1895 – 9 – sp Bibl Santa Ana [240]

Carta pastoral / Valero y Lossa, Francisco – 1759 – 9 – sp Bibl Santa Ana [240]

Carta pastoral al inaugurar su pontificado / Conde y Corral, Bernardo – Plasencia, 1858 – 1 – sp Bibl Santa Ana [240]

Carta pastoral. confirmacion / Garcia Gil, Manuel – 1856 – 9 – sp Bibl Santa Ana [830]

Carta pastoral del illustrissimo senor d don manuel abad yllana : del consejo de su magestad obispo de arequipa &c... / Abad Yllana, Manuel – en Lima: [s.n.] ano de 1777 – (= ser Books on religion...1543/44-c1800: obispos: obispos de arequipa) – 2mf – 9 – mf#crl-390; crl-391 – ne IDC [241]

Carta pastoral del illustrissimo y reuerendissimo, senor doctor d augustin rodrugues : delgado por la gracia de dios, y de la santo sede apostolica... / Rodriguez Delgado, Agustin – en Lima: Por Juan Joseph Gonzalez de Cossio, ano de 1735 – (= ser Books on religion...1543/44-c1800: obispos: obispos de la paz) – 3mf – 9 – mf#crl-396 – ne IDC [241]

Carta pastoral del illvstrissimo senor obispo de la pvebla de los angeles, d ivan de palafox, y mendoza : previniendo los animos de los fieles de sv obispado... / Palafox y Mendoza, Juan de – en Mexico:...de Bernardo Calderon 1649 – (= ser Books on religion...1543/44-c1800: iglesias, catedrales) – 1mf – 9 – mf#crl-343 – ne IDC [241]

Carta pastoral del ilmo. sr. obispo de plasencia al clero y fieles de su diocesis / Casas Souto, Pedro – 1876 – 9 – sp Bibl Santa Ana [240]

Carta pastoral del...con ocasion del 4th centenario de la muerte de hernando cortes y del homenaje de espana a nuestra senora de guadalupe / Alcazar Alenda, Jose Maria – Badajoz: Imp. Provincial, 1947 – 1 – sp Bibl Santa Ana [240]

Carta pastoral, que dirige a los parrocos, sacerdotes y demas fieles de su diocesi / Jose Antonio de San Alberto, archbishop of Rio de la Plata – Buenos-Ayres: impr de los Ninos Expositos 1781 – (= ser Books on religion...1543/44-c1800: arzobispos: arzobispos de la plata) – 1mf – 9 – mf#crl-434 – ne IDC [241]

Carta pastoral que el excmo...adolfo perez munoz dirige al clero y fieles de su diocesis / Perez Munoz, Adolfo – Badajoz: Tip.Uceda Hnos, 1915 – 1 – sp Bibl Santa Ana [240]

Carta pastoral que el illustrissimo senor d fr joseph antonio de san alberto, arzobispo de la plata : dirige a sus amados hijos los curas a la entrada de su gobierno en el arzobispado / Jose Antonio de San Alberto, archbishop of Rio de la Plata – Buenos-Ayres: impr de los Ninos Expositos 1784 – (= ser Books on religion...1543/44-c1800: arzobispos: arzobispos de la plata) – 3mf – 9 – mf#crl-435 – ne IDC [241]

Carta pastoral que el illustrissimo senor d manuel rubio salninas arzobispo de mexico dirige al clero y pueblo de su diocesi : con motivo de las noticias, que ultimamente se han recibo de espana... / Rubio Salinas, Manuel – en Mexico: impr Bibliotheca Mexicana, ano de 1756 – (= ser Books on religion...1543/44-c1800: arzobispos: arzobispos de mexico) – 1mf – 9 – mf#crl-376 – ne IDC [241]

Carta pastoral que el illustrissimo senor don fray joseph antonio de san alberto, arzobispo de la plata : dirige a todos los que en el pasado concurso han sido nombrados... / Jose Antonio de San Alberto, archbishop of Rio de la Plata – en Buenos Ayres: impr de los Ninos Expositos, ano de 1791 – (= ser Books on religion...1543/44-c1800: arzobispos: arzobispos de la plata) – 7mf – 9 – mf#crl-437 – ne IDC [241]

Carta pastoral que...al...dean y cabildo / Ramirez Vazquez, Fernando – 1866 – 9 – sp Bibl Santa Ana [240]

Carta pastoral que...dirige a sus diocesanos en su pontificado en marzo de 1864 / Lopez y Zaragoza, Gregorio Mª. – Madrid: Imp. y Lib. de D. Eusebio Aguado, 1864 – 1 – sp Bibl Santa Ana [240]

Carta pastoral que...obispo de badajoz dirige al clero y fieles de su diocesis / Perez Munoz, Adolfo – Badajoz: Tip. Uceda Hermanos, 1920 – 1 – sp Bibl Santa Ana [240]

Carta pastoral...a todos los prelados y religiosos de dicha provincia... / Molina, Gaspar de – 1 – sp Bibl Santa Ana [240]

Carta pastoral...al clero / Ramirez Vazquez, Fernando – 1867 – 9 – sp Bibl Santa Ana [240]

Carta pastoral...al clero / Ramirez Vazquez, Fernando – 1879 – 9 – sp Bibl Santa Ana [240]

Carta pastoral...quenta cura / Hernandez y Herrero, Joaquin – 1865 – 9 – sp Bibl Santa Ana [240]

Carta politica del ciudadano juan jose arevalo / Marroquin Rojas, Clemente – Guatemala, 1965 – 1r – us UF Libraries [972]

Carta que el p francisco xavier, rector del colegio maximo de s pablo : y al preferente preposito provincial de la provincia de peru remitio a los padres rectores de los colegios... / Xavier, Francisco de – en Lima: Por loseph de Contreras, ano 1689 – (= ser Books on religion...1543/44-c1800: biografias de religiosos) – 2mf – 9 – mf#crl-135 – ne IDC [241]

Carta, que sobre la vida : y muerte de el padre doctor francisco xavier lazcano, dirige a los padres de la compania de jesus de la provincia de mexico / Gandara, Salvador de la – [Mexico City]:...Colegio de S Ildefonso de Mexico, ano de 1763 – (= ser Books on religion...1543/44-c1800: biografias de religiosos) – 2mf – 9 – mf#crl-150 – ne IDC [241]

Carta real por la que se exime a serradilla de la jurisdiccion de plasencia (24 de noviembre de 1557) / Archivo Municipal, Serradilla – Ayuntamiento de Serradilla, Plasencia: Tip. La Victoria, 1956 – sp Bibl Santa Ana [946]

Carta segunda pastoral que el illustrissimo senor d fr joseph antonio de san alberto, arzobispo de la plata : dirige a los curas, tenientes y sacerdotes de su diocesi / Jose Antonio de San Alberto, archbishop of Rio de la Plata – en Buenos-Ayres: impr de los Ninos Expositos, ano de 1786 – (= ser Books on religion...1543/44-c1800: arzobispos: arzobispos de la plata) – 2mf – 9 – mf#crl-436 – ne IDC [241]

Carta sin sobre escrito al sr. ossorio y gallardo / Salazar Alonso, Rafael – Madrid: Roca, Impresor, 1929 – 1 – sp Bibl Santa Ana [946]

Carta y otros documentos de hernando cortes / Becker, Jeronimo – Madrid: Fortanet, 1916. B.R.A.H. lxix/pp. 313-316 – 1 – sp Bibl Santa Ana [350]

Cartagena, A *see* Liber de peste, de signis febrium et de diebus criticis

Cartagena de indias / Marco Dorta, Enrique – Cartagena, Colombia. 1960 – 1r – us UF Libraries [972]

Cartagena, Donaro *see* Semana de miedo

Cartagena hispanica, 1533 a 1810 / Porras Tronconis, Gabriel – Bogota, Colombia. 1954 – 1r – us UF Libraries [972]

Cartagena y su gente / Manrique, Ramon – Cartagena, Colombia. 1945 – 1r – us UF Libraries [972]

Cartagena y sus cercanias / Urueta, Jose P – Cartagena, Colombia. 1912 – 1r – us UF Libraries [972]

Cartaphilus : or, the wandering jew / Reed, Orville Sibbitt – Cincinnati: Standard Pub Co, 1902 [mf ed 1993] – (= ser Christian church (disciples of christ) coll) – 1mf – 9 – 0-524-07035-0 – mf#1991-2888 – us ATLA [220]

Cartari, V *see*
- Imagines deorum...
- Le imagini de i dei de gli antichi...
- Imagini delli dei de gli antichi...
- Seconda novissima editione delle imagini degli dei delli antichi...

Cartas : a cerca da provincia de santa catharina – Desterro, SC: Typ de Jose Joaquim Lopes, 20 jan 1857-06 out 1858 – (= ser Ps 19) – mf#UFSC/BPESC – bl Biblioteca [079]

Cartas a amigos / Nabuco, Joaquim – Sao Paulo, Brazil. v1-2. 1949 – 1r – us UF Libraries [972]

Cartas a elpidio / Varela, Felix – Habana, Cuba. 1960 – 1r – us UF Libraries [972]

Cartas a evelina / Moscoso Puello, Francisco E – Ciudad Trujillo, Dominican Republic. 1941 – 1r – us UF Libraries [972]

Cartas a fidel castro / Betancourt Agramonte, Oscar – Habana, Cuba. 1960 – 1r – us UF Libraries [972]

Cartas a florinda / Alegria, Jose S – San Juan, Puerto Rico. 1958 – 1r – us UF Libraries [972]

Cartas a floro : sobre primera ensenanza y educacion / Codina, Luis – 1864 – 9 – sp Bibl Santa Ana [370]

Cartas a la novia / Aradillas Agudo, Antonio – Madrid: Ediciones Stadium, 1962 – sp Bibl Santa Ana [946]

Cartas a lopez prudencio / Guerra Guerra, Arcadio – Badajoz: Imp. Dip. Provincial, 1966. Sep. REE – sp Bibl Santa Ana [946]

Cartas a luz caballero / Entralgo, Elias Jose – Habana, Cuba. 1949 – 1r – us UF Libraries [972]

Cartas a nestor ponce de leon / Marti, Jose – Habana, Cuba. 1952 – 1r – us UF Libraries [972]
Cartas a un ciudadano / Figueres Ferrer, Jose – San Jose, Costa Rica. 1956 – 1r – us UF Libraries [972]
Cartas a un esceptico en materia de religion *see* Letters to a sceptic on religious matters
Cartas al pueblo americano sobre cuba / Casas, Antonio De Las – Buenos Aires, Argentina. 1897 – 1r – us UF Libraries [972]
Cartas al rey acerca de la isla de cuba / Bas Y Cortes, Vincente – Habana, Cuba. 1871 – 1r – us UF Libraries [972]
Cartas ao amigo ausente / Rio Branco, Jose Maria Da Silva Paranhos – Rio de Janeiro, Brazil. 1953 – 1r – us UF Libraries [972]
Cartas apocrifas sobre la conferencia de guayaquil / Lecuna, Vicente – Caracas, Venezuela. 1945 – 1r – us UF Libraries [972]
Cartas boca arriba / Sanchez Felipe, Juan A – Caceres: Tip. Editorial Extremadura, s.a. 1954? Anaquel de Forja no 8 – sp Bibl Santa Ana [946]
Cartas confidenciales de la reina maria luisa y de don manuel godoy – Madrid: M. Aguilar – 1 – sp Bibl Santa Ana [946]
Cartas d'africa / Ornellas De Vasconcellos, Ayres D' – Lisboa, Portugal. 1930 – 1r – us UF Libraries [960]
Cartas de america / Delgado, Luis Humberto – Lima, Peru. 1940 – 1r – us UF Libraries [972]
Cartas de arturo gazul a fernando villalba / Segura Otano, Enrique – Badajoz: Imp. Dipt. Provincial, 1971 – sp Bibl Santa Ana [912]
Cartas de barolome jose gallardo. noticia / Fita, Fidel – Madrid: Fortanet, 1913. B.R.A.H. 62. p. 182 – 1 – sp Bibl Santa Ana [946]
Cartas de china / Maas, Otto – Sevilla: J Santigosa, 1917 [mf ed 1995] – (= ser Yale coll) – 2v – 1 – 0-524-09822-0 – (in spanish) – mf#1995-0822 – us ATLA [951]
Cartas de d...que tratan del descubrimiento y conquista de chile / Valdivia, Pedro de – Sevilla: Est. tip. de M. Carmona, 1929 – 1 – sp Bibl Santa Ana [350]
Cartas de godoy a ferrer del rio / Valgoma y Diaz-Varela, Dalmiro de la – Madrid: Imp. y Edit. Maestre, 1968. B.R.A.H. 163. Cuaderno I. pp. 57-58 – 1 – sp Bibl Santa Ana [946]
Cartas de la habana / Cartas, Francisco – Habana, Cuba. 1856 – 1r – us UF Libraries [972]
Cartas de los padres de la compania de la mision de filipinas / Society of Jesus. Philippine Islands – Letters from missions. Manila. 10v. 1877-95 – 1 – us L of C Photodup [240]
Cartas de maximo gomez / Gomez, Maximo – Ciudad Trujillo, Dominican Republic. 1936 – 1r – us UF Libraries [972]
Cartas de relacion de la conquista de mejico / Hernan Cortes – Madrid: Espasa Calpe, 5th ed. tomo 1. 1942 – 1 – (tambien tomo 2, 5th ed) – sp Bibl Santa Ana [350]
Cartas de ruben dario / Dario, Ruben – Madrid, Spain. 1963 – 1r – us UF Libraries [972]
Cartas del libertador. tomos 1 a 10 / Lecuna, Vicente – Caracas, 1929-30; Madrid: Razon y Fe, 1931 – 1 – sp Bibl Santa Ana [946]
Cartas diocesanas. memoria 1957 – S.L., s.i., 1957 – 1 – sp Bibl Santa Ana [240]
Cartas do imperador d pedro 2 ao barao de cotegi – Sao Paulo, Brazil. 1933 – 1r – us UF Libraries [972]
Cartas do padre antonio vieira – Coimbra, Portugal. v1-3. 1925 – 1r – us UF Libraries [972]
Cartas do solitario / Tavares Bastos, Aureliano Candido – Sao Paulo, Brazil. 1938 – 1r – us UF Libraries [972]
Cartas edificantes de los misioneros de la compania de jesus en filipinas, 1898-1902 – Barcelona: Henrich y Compania en Comandita, 1903 [mf ed 1995] – (= ser Yale coll) – xiii/379p (ill) – 1 – 0-524-09081-5 – (in spanish) – mf#1995-0081 – us ATLA [241]
Cartas escriptas da india e da china nos annos de 1815 a 1835... / Andrade, J I de – Lisboa: Na Imprensa Nacional, 1843. 2v – 6mf – 9 – mf#HT-582 – ne IDC [915]
Cartas escritas a los muy nobles doctores...se dize que el sal azidoy alcali... / Juanini, J – Madrid, 1691 – 2mf – 9 – sp Cultura [610]
Cartas familiares / Marti, Jose – Habana, Cuba. 1953 – 1r – us UF Libraries [972]
Cartas familiares de don bartolome jose gallardo / Perez de Guzman, Juan – Madrid: Fortanet, 1920. B.R.A.H. 77. pp. 312-318 – 1 – sp Bibl Santa Ana [946]
Cartas, Francisco *see* Cartas de la habana
Cartas ineditas y semi-ineditas de donoso cortes / Valle, Antonio – Madrid: Razon y Fe, 1936 – 1 – sp Bibl Santa Ana [946]
Cartas literarias...sobre gregorio silvestre / Fernandez Almuzara, E – Madrid: Razon y Fe, 1940 – 1 – sp Bibl Santa Ana [440]
Cartas para o brasil / Grave, Joao – Porto, Portugal. 1929 – 1r – us UF Libraries [972]
Cartas pastorales. bahia, 1928 / Schumacher, Pedro – Madrid: Razon y Fe, 1930 – 1 – sp Bibl Santa Ana [240]
Cartas pastorales, y edictos / Lorenzana, Francisco Antonio – en Mexico: impr Joseph Antonio de Hogal, ano de 1770 – (= ser Books on religion...1543/44-c1800: arzobispos: arzobispos de mexico) – 3mf – 9 – mf#crl-380 – ne IDC [241]
Cartas pastorales y otras exhortaciones... doctor don pedro casas souto...obispo de plasencia / Casas y Gonzalez, Juan Bautista – 1898 – 9 – sp Bibl Santa Ana [240]
Cartas pedagogicas / Saiz Otero, Concepcion y Urbano Gonzales Serrano – 1895 – 9 – sp Bibl Santa Ana [370]
Cartas sertanjas / Ribeiro, Julio – Lisboa, Portugal. 1908 – 1r – us UF Libraries [972]
Cartas (siecle 16) / Cruz, San Juan de la – Andujar – 1r – 5,6 – sp Cultura [946]
Cartas sobre quintos / Delicado, Juan M – 1827 – 9 – sp Bibl Santa Ana [946]
Cartas y documentos / Cortes, Hernando – Mexico City? Mexico. 1963 – 1r – us UF Libraries [972]
Cartas y documentos de las misiones de los p.p. capuchinos en venezuela 1781-1788 / Rionegro, Friolan – Vigo, 1931; Madrid: Razon y Fe, 1933 – 1 – sp Bibl Santa Ana [240]
Cartas y extasis de gema galgani. barcelona, 1933 / San Estanislao, German de – Madrid: Razon y Fe, 1934 – 1 – sp Bibl Santa Ana [946]
Cartas y memorial al rey. manila, 23 junio 1584 / Plasencia, Juan de – Archivo Ibero-Americano, 1916 – 1 – sp Bibl Santa Ana [946]
Cartas y mensajes / Santander, Francisco De Paula – Bogota, Colombia. v1-10. 1953-1956 – 3r – us UF Libraries [972]
Cartas y otros documentos novisimamente descubiertos en el archivo general de indias de sevilla / Cortes, Hernando – Sevilla: Tipografia de F. Diaz y Compania, 1915 – 1 – sp Bibl Santa Ana [946]
Cartas y relaciones de hernan cortes al emperador – Paris, France. 1866 – 1r – us UF Libraries [972]
Cartas y relaciones de hernando cortes al emperador carlos v / Cortes, Hernando – 1866 – 9 – sp Bibl Santa Ana [946]
Cartas y testamento / Maroquin, Francisco – Guatemala, 1963 – 1r – us UF Libraries [972]
Cartas...bartolo gallar.. / Zapatilla, Lupianejo (pseud. de Adolfo de Castro) – 1851 – 9 – sp Bibl Santa Ana [920]
O cartaz : folha humoristica, satirica – Bahia: [s.n.] 27 fev 1890 – (= ser Ps 19) – mf#P18B,02,29 – bl Biblioteca [870]
Carte des prefectures de chine et de leur population chretienne en 1911 / Moidrey, Joseph de – Chang-hai: Imprimerie de la Mission Catholique, 1913 [mf ed 1995] – (= ser Yale coll; Varietes sinologiques 35) – 1 – 0-524-09803-4 – (in french) – mf#1995-0803 – us ATLA [241]
Carte generale de la monarchie francoise : contenant l'histoire militaire, depuis clovis premier roy chretien, jusqu'a la quinzieme annee accomplie du regne de louis 15 / Lemau de la Jaisse, Pierre – [Paris]: [s.n.], 1733 [mf ed 1984] – 1r – 1 – mf#SEM35P205 – cn Bibl Nat [944]
Carte linguistique du congo belge / Hulstaert, G – Bruxelles, Belgium. 1950 – 1r – us UF Libraries [470]
La carte postale : saynete enfantine / Dandurand, Josephine – Montreal: C-O Beauchemin, 1896? – 1mf – 9 – mf#04889 – cn Canadiana [830]
Carte topographique de la lune : sur 615 millimetres de diametre, divisee en 9 planches dressee et dessinee / Chemla-Lamech, Felix – Toulouse: V Cazelles 1934 [mf ed 1998] – 1r [ill/pl] – 1 – (incl bibl ref) – mf#film mas 28405 – us Harvard [520]
Cartea neagra : suferintele evreilor din romania, 1940-1944 / Carp, Matatias – Bucuresti: Atelierele grafice Socec, 1946- [mf ed 1980, 1998] – 3v on 1r – 1 – (v1: legionarii si rebeliunea; v2a: pogromul dela iasi; v3: transnistria; incl bibl ref & ind; imprint varies) – mf#27207 – us Harvard [934]
Il carteggio del comitato di emigrazione di rimini (1859-60) / Nicoletti, Luigi – Fabriano: Premiata tip. economica 1925 [mf ed 1990] – 1r – 1 – mf#20306 – us Harvard [945]
Carteggio inedito d'artisti dei secoli 14, 15, 16... / Gaye, J W – Firenze, 1839-1840. 3v – 24mf – 9 – mf#O-263 – ne IDC [700]
Cartelas – Habana. v1-41, 1919-60 – 1 – us L of C Photodup [073]
Carter, Alfred George Washington *see* The old court house
Carter, Charles Sydney *see*
– The english church and the reformation
– The english church in the 17th century
– The english church in the 18th century
Carter, Clarence E *see* The territorial papers of the united states
Carter, E R *see* Biographical sketches of our pulpit
Carter, Elizabeth (Simerwell) *see* Diary
The carter family papers, 1659-1797 : in the sabine hall collection / University of Virginia Library – 4r – 1 – $340.00 – (with printed guide) – mf#D3182 – us Virginia U Pr [920]
Carter, Gavin *see* Patrol reports, field journals, photographs and related papers
Carter, George Robert *see* Journal of a canoe voyage along the kauai palis, made in 1845
Carter, Gwendolen Margaret *see*
– Five african states
– Government and politics in the twentieth century
– Independence for africa
– National unity and regionalism in eight african states
– Politics of inequality
– South africa's transkei
– Transition in africa
Carter, Hazel *see* Notes on the tonal system of northern rhodesian plateau tonga
Carter, Howard *see* The tomb of tut-ankh-amen, 3
Carter, J F M *see* Life and work of the rev. t.t. carter
Carter, James Coolidge *see*
– Law
– The proposed codification of our common law
Carter, James Treat *see* The nature of the corporation as a legal entity, with especial reference to the law of maryland
Carter, Jane Frances Mary *see*
– The life and times of john kettlewell
– Life and work of the rev. t. t. carter
Carter, Jesse Benedict *see*
– De deorum romanorum cognominibus
– Religion of numa
– The religion of numa, and other essays on the religion of ancient rome
– The religious life of ancient rome
Carter, John *see*
– Journal and account book
– Specimens of gothic architecture
Carter, Mary Eddie *see* Polymerization studies on beta-nitrostyrene derivatives
Carter, Merle *see* Solomon islands diaries and correspondence
Carter, Russell Kelso *see*
– Divine healing
– The supernatural gifts of the spirit
Carter, Ruth C *see*
– Cataloging and classification quarterly
– Journal of internet cataloging
Carter, T T *see* Grace of the apostolic priesthood
Carter, Thomas *see*
– French mission life
– Shakespeare, puritan and recusant
Carter, Thomas Fortescue *see* A narrative of the boer war
Carter, Thomas H *see* Papers
Carter, Thomas Thellusson *see*
– Life of penitence
– A memoir of john armstrong
– Rome catholic and rome papal
Carter, Tony *see* Journal of hospital marketing and public relations
Carter watch – v1 n1-v4 n12 [1977 oct-1981 jan] – 1r – 1 – mf#635569 – us WHS [071]
Carter, William *see* A memorial of the congregational ministers and churches of the illinois association
Cartera del coronel conde de adlercreutz – Paris, France. 1928 – 1r – us UF Libraries [972]
Carteret, Leopold *see* Le tresor du bibliophile romantique et moderne, 1801-1875
Carteret/granville correspondence, c1615-1727 – (= ser Archives of the marquess of bath, longleat house, warminster, wiltshire) – 1r – 1 – mf#96840 – uk Microform Academic [860]
Carteret-Hill, P *see* Attendance of protestant children at roman catholic schools
Carter-karis collection of south african political materials – Chicago: Uni of Chicago, Photodup Dept, 1974 (mf ed) – 1 – us CRL [960]
Carter's ford baptist church. colleton county. south carolina : church records – 1855-1933, 1945-1979 – 1 reel – 1 – $41.22 – (membership rolls 1855-1979. total, 916p) – us Southern Baptist [242]
Carter's ford baptist church. lodge, south carolina : church records – 1855-1953 – 1 – us Southern Baptist [242]
Cartersville baptist church. cartersville, georgia : church records/wmu book – 1873-94; Manuscript letters from Lottie Moon. 1874 – 1 – $12.51 – us Southern Baptist [242]
Cartes : surface / Canada. Service de l'environnement atmospherique. Bureau des previsions du Quebec – Cartes meteorologiques manuscrites, avr 1972-juil 1975 – 14r – 1 – mf#SEM35P139 – cn Bibl Nat [550]
Cartes des provinces et des missions de la compagnie avant la suppression (1763-1773) / Pfister, L – n.p, n.d. – 3mf – 9 – mf#HTM-228 – ne IDC [912]
Cartes marines a l'usage des armees du roy de la grande-bretagne / Hooghe, Romein de – Amsterdam: Chez Pierre Mortier...1693 [mf ed 1981] – 1r – 1 – mf#SEM35P172 – cn Bibl Nat [914]
Die cartesianische scholastik : in der philosophie und reformierten dogmatik des 17. jahrhunderts / Bohatec, Josef – Leipzig: A Deichert, 1912 – 1mf – 9 – 0-524-00863-9 – (incl bibl ref) – mf#1990-0248 – us ATLA [240]
Le cartesianisme chez les benedictins : dom robert desgabets / Lemaire, Paul – Paris: Felix Alcan, 1901 – 1mf – 9 – 0-7905-9302-5 – mf#1989-2527 – us ATLA [100]
Le cartesianisme, ou, la veritable renovation des sciences : ouvrage couronne par l'institut / Bordas-Demoulin, Jean Baptiste – Paris: J Hetzel, 1843 – 3mf – 9 – 0-524-00361-0 – (incl bibl ref) – mf#1989-3061 – us ATLA [190]
Cartheuser, Fr A *see* Vermischte schriften aus der naturwissenschaft, chymie und arzeneygelahrtheit
Cartier and hochelaga : maisonneuve and ville-marie: two historic poems of montreal / Evans, Walter Norton – Montreal: W Drysdale & Co, 1895 – (= ser Early canadiana) – 9 – mf#02928 – cn Canadiana [810]
Cartier et son temps / DeCelles, Alfred Duclos – Montreal: Librairie Beauchemin, 1907 – 3mf – 9 – 0-665-72895-6 – (incl ind, app and bibl ref) – mf#72895 – cn Canadiana [320]
Cartier et son temps / DeCelles, Alfred Duclos – [2e ed]. Montreal: Librairie Beauchemin ltee, 1913 [mf ed 1985] – (= ser Bibliotheque canadienne. coll champlain 703b) – 3mf – 9 – (with ind) – mf#SEM105P514 – cn Bibl Nat [920]
Cartier et son temps / DeCelles, Alfred Duclos – [3e ed]. Montreal: Librairie Beauchemin ltee, 1925 [mf ed 1985] – (= ser Bibliotheque canadienne. coll champlain 703b) – 3mf – 9 – (with ind) – mf#SEM105P532 – cn Bibl Nat [920]
Cartier et son temps / DeCelles, Alfred Duclos – Montreal: Librairie Beauchemin ltee, 1907 [mf ed 1985] – 3mf – 9 – (with ind) – mf#SEM105P513 – cn Bibl Nat [920]
Cartier, Etienne *see* Lumiere et tenebres
Cartier, George Etienne *see* Cantate
Cartilla agraria en verso para uso de las escuelas de primera ensenanza / Cuadrado Retamosa, Joaquin – 1887 – 9 – sp Bibl Santa Ana [630]
Cartilla de correspondencia y legislacion mercanti / Fernandez Bolandi, Tomas – San Jose, Costa Rica. 1931 – 1r – us UF Libraries [972]
Cartilla divulgadora. sobre explotacion ovina en su faceta de lana / Direccion General de Ganaderia. Junta Provincial de Fomento Pecuario de Badajoz – Badajoz: Graficas Iberia, 1945 – 1 – sp Bibl Santa Ana [350]
Cartilla forestal cubana para uso de autoridades y... – Habana, Cuba. 1924 – 1r – us UF Libraries [972]
Cartilla historica de costa rica / Fernandez Guardia, Ricardo – San Jose, Costa Rica. 1927 – 1r – us UF Libraries [972]
Cartilla historica de honduras / Bobadilla, Perfecto H – San Pedro Sula, Honduras. 1938 – 1r – us UF Libraries [972]
Cartilla historico-politica / Moreno, Francisco – 1871 – 9 – sp Bibl Santa Ana [946]
Cartilla politica donosiana / Becerro de Bengoa, Ricardo – Caceres: Imp. y Enc. Vda. de Garcia Floriano, 1948 – sp Bibl Santa Ana [320]
Cartilla redactada para dar a conocer los trabajos que se realizan en la granja escuela practica de agricultura de badajoz : publicada e expensas del consejo provincial de fomento / Consejo Provincial de Fomento. Badajoz – Badajoz: Tip. y Enc. de Uceda Hermanos, 1913 – sp Bibl Santa Ana [630]
Cartilla, y doctrina espiritual, para la crianza : y educacion de los novicios, que tomaren el habito en la orden de nuestro padre san francisco... – en Mexico:...D Felipe de Zuniga y Ontiveros, ano de 1775 – (= ser Books on religion...1543/44-c1800: franciscanos) – 2mf – 9 – mf#crl-211 – ne IDC [241]
Cartland, Fernando Gale *see* Southern heroes
Cartmel, cartmell, cartmill family quarterly – n1-n8 [1979 fall-1981 fall], n11 [1982 sum] – 1r – 1 – mf#569148 – us WHS [929]
Cartografia jesuistica del rio de la plata... / Furlong Cardiff, Guillermo – Madrid: Razon y Fe, 1940 – 1 – sp Bibl Santa Ana [241]
Cartographica – North York. 1985+ (1,5,9) – ISSN: 0317-7173 – mf#13852,02 – us UMI ProQuest [520]
Cartography and geographic information science – Bethesda. 1999+ (1,5,9) – (cont: cartography and geographic information systems) – ISSN: 1523-0406 – mf#12484,02 – us UMI ProQuest [917]

Cartography and geographic information systems – Bethesda. 1990-1998 (1,5,9) – (cont: american cartographer. cont by: cartography and geographic information science) – ISSN: 1050-9844 – mf#12484,01 – us UMI ProQuest [520]
Carton, R *see* La synthese doctrinale de roger bacon
Cartones de la frontera / Miro, Baltasar – Trujillo, Peru. 1945 – 1r – us UF Libraries [972]
The cartoon : a serio-comic and illustrated journal – St John, NB: R & E Armstrong, [1878] – 9 – mf#P04534 – cn Canadiana [870]
Cartoonist profiles – Fairfield. 1969-1996 (1) 1969-1996 (5) 1969-1996 (9) – ISSN: 0008-7068 – mf#12014 – us UMI ProQuest [740]
Cartoons and satire : subject collections – (= ser Art exhibition catalogues on microfiche) – 9 catalogues on 9mf – 9 – £75.00 – (individual titles not listed separately) – uk Chadwyck [700]
Cartoons for the cause, 1886-1896 : a souvenir of the international socialist workers and trade union congress, 1896 / Crane, Walter – London: Twentieth Century Press 1896 [mf ed 1981] – 1r [ill] – 1 – (repr fr various journals, with some accompanying verses & a fable) – mf#8267 – us Wisconsin U Libr [740]
Cartoons for the legal profession – Crane Paper Co, n.d. – 1mf – 9 – $1.50 – mf#LLMC 91-081 – us LLMC [740]
Cartoons from punch / Tenniel, John – London [1868?] – (= ser 19th c art & architecture) – 4mf – 9 – mf#4.2.1571 – uk Chadwyck [740]
The cartoons of st mark / Horton, Robert Forman – New York: Fleming H Revell, 1894 – 1mf – 9 – 0-8370-3666-6 – mf#1985-1666 – us ATLA [225]
Cartoons of the campaign : dominion of canada general elections, 1900 / Bengough, John Wilson – Toronto: Poole,1900 – 2mf – 9 – mf#03572 – cn Canadiana [325]
Cartouche / Ennery, Adolphe D' – Paris, France. 1859 – 1r – us UF Libraries [440]
Cartouche / Theodore – Paris, France. 1840 – 1r – us UF Libraries [440]
Le cartulaire de cormery. / Cormery. France. Benedictine Abbey – Tours. 1861 – 1 – us CRL [090]
Cartulaire de la chartreuse du val de ste-aldegonde pres saint-omer / Pas, J de – Saint-Omer, 1805 – €19.00 – ne Slangenburg [241]
Cartulaire de l'abbaye de bonneval en rouergue / Verlaguet, P A & Rigal, J L – Rodez, 1938 – 17mf – 8 – €32.00 – ne Slangenburg [241]
Cartulaire de l'abbaye de cambron / Smet, J J de – Bruxelles. v1-2. 1869 – 17mf per v – 8 – €65.00 – ne Slangenburg [241]
Cartulaire de l'abbaye de flines / Hautcoeur, E – Lille. v1-2. 1873 – v1 11mf v2 12mf – 8 – €44.00 – ne Slangenburg [241]
Cartulaire de l'abbaye de notre dame des vaux de gernay / Merlet, L & Moutie, A – Paris. v1-2. 1857 – v1 15mf v2 17mf – 8 – €61.00 – ne Slangenburg [241]
Cartulaire de l'abbaye de notre dame d'ourscamp / Peigne-Delacourt, M – Amiens, 1865 – 20mf – 8 – €38.00 – ne Slangenburg [241]
Cartulaire de l'abbaye de saint trond / Piot, Ch – Brussel v1 1870; Brussel v2 1874 – v1 22mf; v2 25mf – 8 – €90.00 – ne Slangenburg [241]
Cartulaire de l'abbaye de silvanes / Verlaguet, P A – Rodez, 1910 – 16mf – 8 – €31.00 – ne Slangenburg [241]
Cartulaire de l'abbaye d'orval / Goffinet, Hippolyte – Brussel, 1879 – 27mf – 8 – €52.00 – ne Slangenburg [241]
Cartulaire de l'abbaye du val-benoit / Cuvelier, J – Brussel, 1906 – 31mf – 8 – €60.00 – ne Slangenburg [241]
Cartulaire de l'ancien consulat d'espagne a bruges / Gilliodts-Van Severen, Louis – Bruges. 1901-02 – 1 – us CRL [949]
Cartulaire de l'ordre des hospitaliers de saint-jean de jerusalem (1100-1310) / Delaville le Roulx, Joseph Marie Antoine – Paris: E Leroux. 4v. 1894-1906 [mf ed 1975] – 4r – 1 – mf#SEM35P60 – cn Bibl Nat [360]
Cartulaire de marcigny-sur-loire, 1045-1144 / Richard, Jean – Dijon, 1962 – 8mf – 8 – €17.00 – ne Slangenburg [241]
Cartulaire de marmoutier pour le perche : [n.-d. du vieux-chateau, collegiale de saint leonard de belleme, et prieure de st.-martin-du-vieux-belleme / Barret, M l'Abbe – Mortagne: Impr Georges Meaux, 1894 – 1 – us CRL [090]
Cartulaire de notre dame de prouille / Guiraud, Jean – Paris. v1-2. 1907 – v1 21mf v2 18mf – 8 – €75.00 – ne Slangenburg [241]
Cartulaire des abbayes saint-pierre de la couture et de saint-pierre de solesmes – Le Mans, 1881 – €48.00 – ne Slangenburg [241]
Cartulaire du chapitre de saint-laud d'angers. / Angers. France. St. Laud (Church) – Angers. 1903 – 1 – us CRL [090]
Cartulaire du prieure de saint marcel-les-chalon : publis d'apres les manuscrits de marcel canat de chizy par paul canat de chizy / St Marcel-les-Chalon, France. (Benedictine priory) – Chalon-sur-Saone: L Marceau 1894 [mf ed 1978?] – 1r – 1 – mf#19 – us Wisconsin U Libr [241]
Cartulaire noir de la cathedrale d'angers. / Angers. France. Cathedrale – Angers. 1908 – 1 – us CRL [090]
Cartulario del archivo con documentos de los anos 1362-1618 – Albarracin – 1r – 5,6 – sp Cultura [946]
Cartulario (siecle 13-17) – Albarracin – 1r – 5,6 – sp Cultura [946]
Cartularium monasterii de rameseia (rs79) / Ramsey Abbey; ed by Hart, W H & Lyons, P A – (= ser The rolls series (rs)) – (v1 1884 €18. v2 1886 €15. v3 1893 €21) – ne Slangenburg [241]
Cartwright, Alan Patrick *see*
– Gold paved the way
– Golden age
– This is south africa
Cartwright, Conway Edward *see* Lena, a legend of niagara
Cartwright, Cyril G F *see* Letters from ocean island [banaba] and the gilbert islands [kiribati]
Cartwright, Otho Grandford *see* Middle west side
Cartwright, Peter *see*
– Autobiography of peter cartwright
– Fifty years as a presiding elder
Cartwright, Richard *see*
– Budget speech
– Budget speech delivered in the house of commons of canada
– Canada go bragh
– Discours sur le budget prononce a la chambre des communes du canada
– The economic condition of canada and her trade policy
– Memories of confederation
– Speech by the rt hon sir richard cartwright
– Speech of sir richard cartwright, mp, on the budget
Cartwright, Robert David *see* The first and last words of a pastor to his people
Cartwright, T *see*
– A full and plaine declaration of ecclesiasticall discipline owt off the word off god...
– A replye to an ansvvere made of m doctor whitgifte
– The second replie of thomas cartwright
– A seconde admonition to the parliament
Cartwright, Thomas *see* A commentary upon the epistle of st paul written to the colossians
Cartwright's cases on the british north american act / Canada. General – v1-5. 1868-96 (all publ) – 41mf – 9 – $61.00 – mf#LLMC 81-009 – us LLMC [340]
Carus, Carl Gustav *see* Natur und idee
Carus, Paul *see*
– Buddhism and its christian critics
– Buddhist hymns
– Chinese philosophy
– Chinese thought
– The dawn of a new religious era
– The dharma
– Edward's dream
– God
– Goethe and schiller's xenions
– The gospel of buddha according to old records
– Helgi und sigrun
– The history of the devil and the idea of evil
– Ein leben in liedern
– The mechanistic principle and the non-mechanical
– Nietzsche and other exponents of individualism
– The oracle of yahveh
– Philosophy as a science
– The pleroma
– The story of samson and its place in the religious development of mankind
– Tai-shang kan-ying pien
– Whence and whither
– Yin chih wen
Carus, Victor A *see* Das altarwerk zu lauenstein und die anfaenge des barock in sachsen
Caruso, Christina M *see* Psychological and physiological changes associated with a period of increased training
Carus-Wilson, Ashley [Mrs] *see* Irene petrie
Carus-Wilson, Ashley, Mrs *see* Clews to holy writ
Caruthers, Abraham *see* History of a lawsuit
Carvajal, Gaspar De *see*
– Descobrimentos do rio das amazonas
– Relacion del nuevo descubrimiento del famoso rio...
Carvajal, Gaspar de *see* Descubrimiento del rio amazonas
Carvajal, Jacinto De *see* Relacion del descubrimiento del rio apure hasta su...
Carvajal, Micael *see*
– Cortes de la muerte
– Tragedia josephine
Carvajal, Pedro de *see* Constituciones synodales del obispado de coria
Carvajal Rodriguez, Dora *see* Seis villancicos cubanos
Carvajal Y Bello, Juan Eduardo Fernandez *see* Obra lirica
Carvajal y Mendoza, Luisa de *see* Poesias espirituales
Carvalho, Affonso De *see*
– Caxias
– Poetica de olavo bilac
Carvalho, Alfredo De *see* Aventuras e aventureiros no brasil
Carvalho, Antonio Feliciano de Santa Rita *see* Pastoral do arcebispo eleito de goa, primaz do oriente, governador, e vigario capitular do mesmo arcebispado metropolitano
Carvalho, Antonio Feliciano de Santa Rita, Archbishop *see* Resposta ao folhetinho, que tem por titulo
Carvalho, Austricliano De *see* Brasil colonia e brasil imperio
Carvalho, Carlos Miguel Delgado De *see*
– Bresil meridional
– Historia da cidade do rio de janeiro
– Organizacao social e politica brasileira
Carvalho E Menezes, Vasco Guedes De *see* Apontamentos para a historia d'angola
Carvalho, Elisio De *see*
– Brava gente
– Principes del espiritu americano
Carvalho, Estevao Leitao De *see* Servico do brasil na segunda guerra mundial
Carvalho, Fernando Setembrino De *see* Memorias, dados para a historia do brasil
Carvalho Franco, Francisco De Assis *see* Dicionario de bandeirantes e sertanistas do brasil
Carvalho, Henrique Augusto Dias De *see* Lubuco
Carvalho, Hernani De *see* Sociologia da vida rural brasileira
Carvalho, J R De Sa *see* Brazilian el dorado
Carvalho, Luiz Antonio Da Costa *see* Realizacoes do governo getulio vargas no campo do...
Carvalho, Maria Da Conceicao Vicente De *see* Vicente de carvalho
Carvalho, Menelick De *see* Revolucao de 30 e o municipio
Carvalho, Orlando M *see*
– Problemas fundamentaes do municipio
– Rio da unidade nacional
Carvalho, Osvaldo Ferraro *see* Ensaio sobre a problematica dos transportes
Carvalho, Ronald De *see*
– Pequena historia da literatura brasileira
– Pequena historia de literatura brasileira
Carvalho Soares Brandao, Ulysses De *see* Pernambuco de outr'ora
Carvalho, Vicente Augusto De *see* Poemas e cancoes
Carvallo Arvelo, Salvador *see* Historia de un proceso
Carvell, Alice Maude *see* In jungle depths
Carver 1733-1900 – Oxford, MA (mf ed 1992) – (= ser Massachusetts vital records) – 46mf – 9 – 0-87623-145-8 – (mf 1-7: town & vital records 1733-1847. mf 8-12: vital records index 1733-1847. mf 13-18: town records 1814-52. mf 18-19: marriage intentions 1814-52. mf 19-22: misc town records 1834-56: mf 23-32: town record copy 1790-1854. mf 32-34: intentions transcript 1791-1852. mf 34-35: marriage transcript 1788-1844. mf 36-37: vital records 1843-57. mf 38: vital records index 1858-1905. mf 39-41: births 1859-1900. mf 42-43: marriages 1859-1900. mf 44-46: deaths 1895-1905) – us Archive [978]
Carver 1748-1849 – Oxford, MA (mf ed 1995) – (= ser Massachusetts vital record transcripts to 1850) – 6mf – 9 – 0-87623-225-X – (mf 1t: index to vitals 1748-1844. mf 1t-4t: births & deaths 1748-1844. mf 1t-2t: marriages 1788-1844. mf 2t: intentions 1791-1814. mf 4t-5t: intentions 1814-49. mf 5t-6t: births 1843-49. mf 6t: marriages & deaths 1843-49) – us Archive [978]
Carver, Jonathan *see* Voyage dans les parties interieures de l'amerique septentrionale
Carver journal / Carver Vocational Technical High School [Baltimore MD] – [1982 apr] – 1r – 1 – mf#4863520 – us WHS [331]
Carver research news and reviews / Tuskegee Institute – 1969 jun 1 – 1r – 1 – mf#4841788 – us WHS [370]
Carver School of Missions and Social Work (Woman's Missionary Union Training School for Christian Workers). Louisville, Ky *see* Catalogs and college records
Carver, T A *see* Comparison of bilateral normal tibial rotation in adult males
Carver Vocational Technical High School [Baltimore MD] *see* Carver journal
Carver, W A *see* Cotton varieties for florida
Carver, W O *see* Collection
Carver, William Owen *see*
– Missions and modern thought
– Missions in the plan of the ages
Carville, F E *see* A collection of ye old-fashioned dances of 1850
Carwithen, John Bayly Sommers *see*
– History of the christian church
– A view of the brahminical religion
Cary, George Lovell *see* The synoptic gospels
Cary grove countryside – Barrington, IL. 1982-1984 (1) – mf#68643 – us UMI ProQuest [071]
Cary, Henry *see* A collection of statutes affecting new south wales
Cary, Orland R *see* Correspondence
Cary, Otis *see*
– A history of christianity in japan
– Japan and its regeneration
Carzo, Jose M *see* La parte subjetiva en el conocimiento intelectual segun santo tomas de aquino
Cas – Banska Bystrica, Czechoslovakia. Sept 1944-Feb 1948 – 12r – 1 – us L of C Photodup [077]
Un cas de conscience / Diana, Pierre – Paris, 193? Fiche W 1505. (Blodgett Collection of Spanish Civil War Pamphlets) – 9 – us Harvard [946]
Le cas des catholiques basques / Hiriartia, J de – Paris, 193? Fiche W945. (Blodgett Collection of Spanish Civil War Pamphlets) – 9 – us Harvard [946]
Cas min dan dum : [a novel] / Gan Sukhum – Bhnam Ben: Samagam qnak nibandh Khmaer 2516 [1972] [mf ed 1990] – 1r with other items – 1 – (in khmer) – mf#mf-10289 seam reel 098/03 [§] – us CRL [830]
Cas registry handbook-common names – 6,9 – $1,190.00 – us Chemical [540]
Casa colonial venezolana / Gasparini, Graziano – Caracas, Venezuela. 1962 – 1r – us UF Libraries [972]
Casa de austria en venezuela durante la guerra de / Borges Jacinto Del Castillo, Analola – Salzburgo, Austria. 1963 – 1r – us UF Libraries [972]
La casa de bernarda alba / Garcia Lorca, Federico – MS DE 1936 – 2mf – 9 – sp Cultura [820]
Casa de la Cultura. Cine Club de la O.S. de E. y D. *see* Noviembre 1975. sesiones cinematograficas de arte y ensayo
Casa de los ladrillos rojos / Zachrisson, Boris A – Panama, 1958 – 1r – us UF Libraries [972]
La casa de montalvo – Ecuador. 1-24, 1931-56 (incomplete) – 1 – us L of C Photodup [073]
Casa de pensao / Azevedo, Aluisio – Rio de Janeiro, Brazil. 1944 – 1r – us UF Libraries [972]
Casa de sao clemente / Pereira, Edgard Baptista – Rio de Janeiro, Brazil. 1949 – 1r – us UF Libraries [972]
Casa de vidrio / Lars, Claudia – Santiago, Chile. 1942 – 1r – us UF Libraries [972]
La casa donde nacio san francisco de asis, patronato del estado espanol / Barrado, Angel – Madrid, 1944. Sep. Rev. Verda y vida no 7, 1944 – sp Bibl Santa Ana [946]
Casa El Salvador "Farabundo Marti" et al *see* El salvador's link
La casa en que murio hernan cortes en castilleja de la cuesta / Caballero, Fernando – 1844 – 9 – sp Bibl Santa Ana [946]
Casa, Jose Joaquin *see* Semblanzas
Casa leon y su tiempo : aventura de un anti-heroe / Briceno-Iragorry, Mario – Caracas, Venezuela. 1954 – 1r – us UF Libraries [972]
La casa natale di s. francesco...secolo 13. roma (1966) / Abate, Giuseppe – Madrid: Graf. Calleja, 1966 – 1 – sp Bibl Santa Ana [946]
Casa solariega / Bauza, Obdulio – San Juan, Puerto Rico. 1954 – 1r – us UF Libraries [972]
Casacion en lo civil / Martinez Escobar, Manuel – Habana, Cuba. 1936 – 1r – us UF Libraries [972]
Casa-grande and senzala / Freyre, Gilberto – Brasilia, Brazil. 1963 – 1r – us UF Libraries [972]
Casa-grande and senzala / Freyre, Gilberto – Rio de Janeiro, Brazil. v1-2. 1950 – 1r – us UF Libraries [972]
Casa-grande and senzala / Freyre, Gilberto – Rio de Janeiro, Brazil. v1-2. 1958 – 1r – us UF Libraries [972]
Casal feliz – [Rio de Janeiro: s.n., 1985-] n1 – us CRL [972]
Casal, G *see*
– Historia natural y medica del principado de asturias
– Mal de la rosa
Casal, Julian Del *see*
– Cronicas habaneras
– Julian del casal
– Poesias
– Selected prose of julian del casal
Casalduero, Joaquin *see* Sentido y forma del quijote
Casali, Lodovico *see* Mottectorum octonis vocibus. liber primus
Casalis, A *see* English-sesuto vocabulary
Casalis, E *see* The basutos
Casalis, Eugene Arnaud *see* Basutos
Casalog – Monterey CA. 1945 apr 18-dec 14 – 1r – 1 – mf#2892895 – us WHS [071]
Casals Llorente, Jorge *see* Epopeya de marti desda paula hasta dos rios

Casanova di Seingalt, Giacomo G *see* Amours et aventures de casanova

Casanova in wien : komoedie, drei akte in versen / Auernheimer, Raoul – Muenchen: Drei Masken Verlag, 1924 [mf ed 1995] – 151p – 1 – mf#8920 – us Wisconsin U Libr [820]

Casanova, Jose Manuel *see* Cuban economic standard

Casanova, Silvio di *see* Lieder der liebe und einsamkeit

Casanovas, Martin *see* Orbita de la revista de avance

Casar baptist church. casar, north carolina : church records – 1901-63 – 1 – us Southern Baptist [242]

Casas, Alvaro Maria De Las *see* Sonetos brasilenos

Casas, Antonio De Las *see* Cartas al pueblo americano sobre cuba

Casas, Bartolome de las *see*
- Del unico modo de atraer a todos los pueblos a la verdadera religion. advertencia...mexico, 1942
- Del unico modo de atraer todos los pueblos a la verdadera religion

Casas, Lucas de *see* Monte de las glorias de dios en las de la exaltacion canonica de el inclyto patriarcha de la carmelitana descalzez s juan de la cruz

Casas Souto, Pedro *see*
- Carta pastoral del ilmo. sr. obispo de plasencia al clero y fieles de su diocesis
- Constituciones sinodales del obispado de plasencia
- Pastoral del venerable obispo de plasencia

Casas y Gonzalez, Juan Bautista *see* Cartas pastorales y otras exhortaciones...doctor don pedro casas souto...obispo de plasencia

Casas y Souto, Pedro *see* Pastoral

Casasus, Juan J E *see* Por la abolicion del castigo capital

Casasus, Juan Jose Exposito *see* Mariano aramburo

Casatejada. Ayuntamiento *see*
- 136th feria de santiago para toda clase de ganados 1971
- 142nd feria de santiago para toda clase de ganados
- 144th feria de santiaago. julio de 1979
- Feria de santiago de toda clase de ganado. durante...julio, 1961
- Feria de santiago de toda clase de ganados. durante...julio, 1962
- Fiestas de la soledad, 1961
- Fiestas de la soledad...1960
- Grandes fiestas de santiago de toda clase de ganados y generos de comercio...24, 25 y 26 de julio, 1960
- Revista anual de cultura 1980

Casati, G *see* Ten years in equatoria and the return with emin pasha

Casati, Gaetano *see* Ten years in equatoria and the return with emin pasha

Casati, Tomaso *see* Arioaldo, re de' longobardi

Casaubon, E D *see* Le nouveau contrat social

Casaysayan ng catotohanang buhay ng haring clodeveo at reyna clotilde sa reyno nang francia na tinula sa lubos na catiagaan / Ignacio, Cleto R – Manila, I F: P Sayo 1917 [mf ed Bloomington IN: Indiana Uni Lib, Preservation Dept 1984] – (= ser Coll...in the tagalog language 2) – 1r – 1 – us Indiana Preservation [490]

El cascabel – Madrid, Spain. oct 1863-apr 1877 – 4r – 1 – uk British Libr Newspaper [072]

Cascade comix monthly – n1-n23 [1978 mar-1981 apr] – 1r – 1 – mf#669539 – us WHS [740]

Cascade roarer / Summit Co. Akron – jun 1845-jul 1846 [wkly] – 1r – 1 – mf#B6810 – us Ohio Hist [071]

Cascaden, Gordon *see* Shall unionism die?

Cascales Munoz, Jose *see*
- Apuntes para la historia de villafranca de los barros
- Apuntes y materiales para la biografia de don jose de espronceda
- El autentico espronceda pornografico y el apocrifo en general
- Las bellas artes plasticas en sevilla, tomo 1
- Las bellas artes plasticas en sevilla...desde el siglo 13 hasta nuestros dias...tomo 2
- La confederacion de las clases. el programa de un nuevo partido
- Los conflictos del proletariado. el movimiento social contemporaneo: por que, cuando y como ha nacido el problema obrero
- De sevilla a batalha. excursion...de sevilla a merida y badajoz..
- Democracia colectivista. lecciones de sociologia sobre una nueva politica a la antigua espanola...por...
- Espronceda su epoca su vida y sus obras
- Francisco de zurbaran. su epoca, su vida y sus obras
- Historia de a cuerda granadina contada por algunos de sus nudos, apuntes para la misma recopilados por...
- Los primeros frutos de mi huerta. (versos muy malos)
- Rasgos de nuestra epopeya (episodios y personajes)
- Sevilla intelectual, sus ecritores y artistas contemporaneos
- Solo dios es grande

O cascalho : jornal politico, joco-serio – Rio de Janeiro, RJ: Typ Liberal de F F & Ramalho, 11 mar-02 jul 1849 – (= ser Ps 19) – mf#P14,02,33 n05 – bl Biblioteca [320]

Casco bay breeze – South Harpswell, ME. 1901-1916 (1) – mf#63574 – us UMI ProQuest [071]

Cascudo, Luis Da Camara *see*
- Antologia do folclore brasileiro
- Coisas que o povo diaz
- Conde d'eu
- Contos tradicionais do brasil
- Literatura oral
- Vaqueiros e cantadores

Case : as to the legal force of the judgment of the privy council in... – Oxford, England. 1864 – 1r – us UF Libraries [240]

Case against disestablishment / Odom, William – London, England. 18– – 1r – us UF Libraries [240]

The case against professor briggs / Briggs, Charles Augustus – New York: Scribner, 1892-1893 – 2mf – 9 – 0-8370-2603-2 – mf#1985-0603 – us ATLA [240]

The case against tax-exempt bonds : open letters to...sir robert borden...prime minister of canada, and to...sir thomas white...minister of finance / Killam, Izaak Walton – [Montreal?: s.n, 1918?] [mf ed 1996] – 1mf – 9 – 0-665-81326-0 – mf#81326 – cn Canadiana [336]

The case against the nazi war criminals : opening statement for the u.s.a., and other documents / Jackson, Robert Houghwout – 1st ed. New York: A.A. Knopf, 1946. xiii,216p. plates – 1 – us Wisconsin U Libr [345]

Case, Alan J *see* An exploration of the opinions of recreation and parks/leisure studies faculty and public sector practitioners concerning the computer competency skills of recreation and parks/leisure studies bacca-laureate students

Case, Alden Buell *see* Thirty years with the mexicans

Case and comment – v1-28. 1894-1922 – 152mf – 9 – $228.00 – (lacking: v2 no 5. v13, v14. updates planned) – mf#LLMC 84-434 – us LLMC [150]

Case and comment – Rochester. 1894-1990 (1) 1975-1990 (5) 1975-1990 (9) – ISSN: 0008-7238 – mf#2807 – us UMI ProQuest [340]

Case and his cotemporaries : or, the canadian itinerants' memorial: constituting a biographical history of methodism in canada, from its introduction into the province, till the death of the rev wm case in 1855 / Carroll, John – Toronto: S Rose, 1867 – 5mf – 9 – (incl ind) – mf#05316 – cn Canadiana [242]

Case and his cotemporaries : or, the canadian itinerants' memorial: constituting a biographical history of methodism in canada, from its introduction into the province, till the death of the rev wm case in 1855 / Carroll, John – Toronto: Wesleyan Conference Office, 1869 – 6mf – 9 – (incl ind) – mf#05317 – cn Canadiana [242]

Case and his cotemporaries : or, the canadian itinerants' memorial: constituting a biographical history of methodism in canada, from its introduction into the province, till the death of the rev wm case in 1855 / Carroll, John – Toronto: Methodist Conference Office, 1877 – 5mf – 9 – mf#05320 – cn Canadiana [242]

Case and his cotemporaries : or, the canadian itinerants' memorial: constituting a biographical history of methodism in canada, from its introduction into the province, till the death of the rev wm case in 1855 / Carroll, John – Toronto: S Rose, 1867 – 1mf – 9 – mf#05315 – cn Canadiana [242]

Case and his cotemporaries : or, the canadian itinerants' memorial: constituting a biographical history of methodism in canada, from its introduction into the province, till the death of the rev wm case in 1855 / Carroll, John – Toronto: Wesleyan Conference Office, 1871 – 6mf – 9 – mf#05318 – cn Canadiana [242]

Case and his cotemporaries : or, the canadian itinerants' memorial: constituting a biographical history of methodism in canada, from its introduction into the province, till the death of the rev wm case in 1855 / Carroll, John – Toronto: Wesleyan Conference Office, 1874 – 6mf – 9 – mf#05319 – cn Canadiana [242]

Case and opinion on the will of the reverend george powell, deceased, ex-parte the president and governors of the radcliffe infirmary, oxford / Sewell, Richard Clarke – Oxford: Trash, 1840. 20p. LL-2311 – 1 – us L of C Photodup [340]

Case as it is : or, a documented detail of the occurrences in the pe... – Edinburgh, Scotland. 1821 – 1r – us UF Libraries [240]

Case as it is : or, a reply to the letter of dr pusey to his grace / Goode, William – London, England. 1842 – 1r – us UF Libraries [240]

Case, Carl Delos *see* The incarnation and modern thought

Case concerning the northern cameroons : cameroon v united kingdom – Hague, Netherlands. 1963 – 1r – us UF Libraries [960]

Case currents / Council for Advancement and Support of Education – Washington. 1975-1983 (1) 1976-1983 (5) 1976-1983 (9) – (cont by: currents) – ISSN: 0360-862X – mf#10671 – us UMI ProQuest [378]

Case examples of educators' creative strategies in solving teaching and learning problems in rural kzn schools / Mthabela, Muzi Sandy Sadler – Pretoria: Vista University 2000 [mf ed 2000] – 2mf – 9 – (incl bibl ref) – mf#mfm15251 – sa Unisa [370]

Case file 35-30 / O'Hare, Kate Richards – Washington, DC: National Archives and Records Service [19–] – us CRL [324]

Case file on the rebellion of lares, 1868-1869 *see* Expediente sobre la rebelion de lares, 1868-1869

Case files in suits involving consuls and vice consuls and the repeal of patents of the u.s. district court for the southern district of new york, 1806-1860 / U.S. District Court – (= ser Records of district courts of the united states) – 2r – 1 – (with printed guide) – mf#M965 – us Nat Archives [346]

Case files of approved pension applications of widows and other dependents of civil war and later navy veterans ("navy widows' certificates"), 1861-1910 / U.S. War Dept. – (= ser Records of the veterans administration) – ca 40,000mf – 9 – (with printed guide) – mf#M1279 – us Nat Archives [355]

Case files of chinese immigrants, 1895-1920, from district no.4 (philadelphia) of the immigration and naturalization service / U.S. Immigration and Naturalization Service – (= ser Records Of The Immigration And Naturalization Service) – 51r – 1 – (with printed guide) – mf#M1144 – us Nat Archives [975]

Case files of disapproved pension applications of widows and other dependents of civil war and later navy veterans ("navy widows' originals"), 1861-1910 / U.S. War Dept. – (= ser Records of the veterans administration) – ca 8500mf – 9 – (with printed guide) – mf#M1274 – us Nat Archives [355]

Case files of investigations by levi c. turner and lafayette c. baker, 1861-1866 / U.S. War Dept. Adjutant General's Office – (= ser Records of the adjutant general's office, 1780's-1917) – 137r – 1 – (with printed guide) – mf#M797 – us Nat Archives [355]

The case for canada *see* Advantages of imperial federation

The case for india / Durant, Will – New York: Simon and Schuster; Dodballapur, Mysore State, India: Distributed in India by Taluk Congress Committee, 1930 – (= ser Samp: indian books) – us CRL [954]

The case for the government / Langdon-Davies, John – NY, 1939 – (= ser Blodgett coll) – 9 – mf#fiche w986 – us Harvard [946]

Case for the society in scotland for propagating christian knowledg – Edinburgh, Scotland. 1843 – 1r – us UF Libraries [240]

Case for tithes simply stated in a few plain notes / Price, Thomas – Rhyl, England. 1887 – 1r – us UF Libraries [240]

Case, H W *see* On sea and land, on creek and river

Case law and index; a complete series of condensed reports, federal, state, and english, including canadian, australian, new zealand and hawaiian reports. vol. i. banks and banking – New York: Case Law Co. 1903. 1432p. LL-331 – 1 – (index to vol. i. new york 1903. 207p) – us L of C Photodup [342]

Case law divergence from the federal rules of evidence / Capra, Daniel J – 2000 – 1mf – 9 – $1.50 – mf#llmc99-038 – us LLMC [347]

Case management and court management in u.s. district courts / Flanders, Steven – Washington: FJC, Sept 1977 – 2mf – 9 – $3.00 – mf#LLMC 95-813 – us LLMC [347]

Case management procedures in the federal court of appeals / McKenna, Judith A et al – 2000 – 3mf – 9 – $4.50 – mf#llmc99-048 – us LLMC [347]

Case manager – Little Rock. 1998+ (1,5,9) – ISSN: 1061-9259 – mf#21574 – us UMI ProQuest [610]

Case, Nelson *see* Copies of speeches

The case of arthur ernest hatheway : a british subject, who, induced by the promises of quick profits in the west, settled at big horn city, wyoming territory, us, october 6, 1884 and...arrrested by united states soldiers... – S.l: s.n, 1885? – 1mf – 9 – mf#02548 – cn Canadiana [355]

Case of catholic subscription to the thirty-nine articles considered / Keble, John – London, England. 1841 – 1r – us UF Libraries [241]

Case of conscience solved / Milner, John – London, England. 1801 – 1r – us UF Libraries [240]

Case of cuba / Sherwood, John D – New York, NY. 1869 – 1r – us UF Libraries [972]

The case of dr marcus dods correctly stated : in answer to recent mis-statements of it / Scrymgeour, William – Glasgow: J. Maclehose, 1878 – 1mf – 9 – 0-7905-3415-0 – mf#1987-3415 – us ATLA [220]

The case of england and western australia in respect to transportation / Grellet, Henry Robert – London, 1864 – (= ser 19th c books on british colonization) – 1mf – 9 – mf#1.1.7089 – uk Chadwyck [348]

The case of henry ward beecher : opening address / Tracy, Benjamin Franklin – New York: George W Smith, 1875 – 1mf – 9 – 0-524-08599-4 – mf#1993-3184 – us ATLA [240]

The case of peter du calvet, esq of montreal in the province of quebeck : containing...an account of the long and severe imprisonment he suffered in the said province by the order of general haldimand... / Du Calvet, Pierre – London: [s.n.], 1784 [mf ed 1973] – 1r – 5 – mf#SEM16P24 – cn Bibl Nat [971]

Case of pharaoh – London, England. 18– – 1r – us UF Libraries [240]

Case of the black warrior / United States Dept Of State – Washington, DC. 1854 – 1r – us UF Libraries [972]

Case of the church in wales / Bevan, W L – London, England. 1886 – 1r – us UF Libraries [240]

Case of the colonists of the eastern frontier of the cape of good hope : in reference to the kaffir wars of 1835-36 and 1846 / Godlonton, Robert – Grahamstown, 1879 – (= ser 19th c books on british colonization) – 2mf – 9 – mf#1.1.3695 – uk Chadwyck [960]

Case of the dissenters : in a letter addressed to the lord chancello – London, England. 1834 – 1r – us UF Libraries [240]

The case of the rev e b fairfield... : being an examination of his "review of the case of henry ward beecher," together with his "reply" and a rejoinder / Raymond, Robert Raikes – [2nd ed] New York: [s.n.], 1874 [mf ed 1992] – 1mf – 9 – 0-524-02987-3 – (with app containing letters etc by rossiter w raymond) – mf#1990-0774 – us ATLA [345]

The case of the rev g c gorham against the bishop of exeter : as heard and determined by the judicial committee of the privy council on appeal from the arches court of canterbury / Gorham, George Cornelius – London: V & R Stevens and G S Norton, 1852 – 2mf – 9 – 0-524-05178-X – mf#1990-5097 – us ATLA [241]

Case of the rev mr shore / Phillpotts, Henry – London, England. 1849 – 1r – us UF Libraries [240]

Case of the rev walter c smith / Freer, James – Glasgow, Scotland. 1867 – 1r – us UF Libraries [240]

Case of thomas pooley : the cornish well-sinker / Holyoake, George Jacob – London, England. 1857? – 1r – us UF Libraries [240]

Case of william robertson smith in the free church / Smith, William Robertson – [S.l.: s.n., 18–] – 1r – 1 – 0-8370-0783-6 – mf#1984-T085 – us ATLA [240]

The Case Of William Robertson Smith In The Free Church Of Scotland *see* Uncritical criticism

Case papers of the court of admiralty of the state of new york, 1784-1788 – (= ser Records of district courts of the united states) – 1r – 1 – (with printed guide) – mf#M948 – us Nat Archives [347]

Case papers of the u.s. district court for the eastern district of virginia, 1863-1865, relating to the confiscation of property / U.S. District Court – (= ser Records of district courts of the united states) – 1r – 1 – mf#M435 – us Nat Archives [347]

Case respecting the maintenance of the london-clergy / Moore, John – London, England. 1802 – 1r – us UF Libraries [240]

Case respecting the maintenance of the london-clergy / Moore, John – London, England. 1812 – 1r – us UF Libraries [240]

Case, Shirley Jackson *see* The historicity of jesus

Case study analysis of teacher change with the sport education model / Dayton, Danielle M – 1999 – 2mf – 9 – $8.00 – mf#PE 3941 – us Kinesology [790]

Case study approach to some features of cross-cultural social work practice with indian families / Gower, Myrna Zoe – Johannesburg: U of the Witwatersrand 1978 [mf ed 1978] – 4mf – 9 – sa Misc Inst [362]

A case study of a multiple-joint resistance exercise for an individual with cerebral palsy / Cohen, Jenna S – 1999 – 1mf – 9 – $4.00 – mf#PE 4000 – us Kinesology [617]

A case study of selected effects of an organized summer residential camp upon staff memebers / Glick, Jeffrey – 1980 – 4mf – 9 – $16.00 – us Kinesology [790]
A case study of the impact of a sequential swim program an behaviors of one young child with autism and his mother / Ostlund, Linda D – 1999 – 2mf – 9 – $8.00 – mf#PSY 2121 – us Kinesology [150]
A case study of the process of tourism development in rural communities in the state of indiana / Lewis, James B – 1996 – 4mf – 9 – $16.00 – mf#RC 505 – us Kinesology [338]
A case study of the school development functions of a school governing body in a historically disadvantage secondary school / Kani, Bennett Zolile – Vista University 2000 [mf ed Johannesburg 2000] – 3mf – 9 – (incl bibl ref) – mf#mfm14755 – sa Unisa [373]
Case, Thomas et al *see* Lectures on the method of science
Case western reserve journal of international law – Cleveland. 1991-1996 (1,5,9) – ISSN: 0008-7254 – mf#16372 – us UMI ProQuest [341]
Case western reserve journal of international law – v1-32. 1968-2000 – 9 – $399.00 set – ISSN: 0008-7254 – mf#101431 – us Hein [341]
Case western reserve law review – v1-51. 1949-2001 – 1,5,6 – $1135.00 – (title varies: v1-18 [1949-67] as western reserve law review) – ISSN: 0008-7262 – mf#101441 – us Hein [340]
Case Western Reserve University *see* Ju-ju
Casel, Odo *see* Vom christlichen mysterium
Caselius, Johannes *see* In ethicorum aristotelis interpretationem prolegomena
Caselli, Domenico Antonio *see* Il canto fermo in prattica per uso di qual si voglia sorte di religiosi, e religiose
The caseload experiences of the district courts from 1972 to 1983 : a preliminary analysis / Meierhoefer, Barbara S & Armen, Eric V – Washington: FJC, 1985 – 1mf – 9 – $1.50 – mf#LLMC 95-836 – us LLMC [347]
Casemate – Fort Monroe VA. 1980 jan-1983 jun, 1983 jul-1987 oct, 1987 oct-1990 dec, 1991 feb-1993 sep – 4r – 1 – mf#663539 – us WHS [071]
Casentini, Marsilio *see* Tirsi e clori. terzo libro de' madrigali a cinque voci
Caserio del carmen : cuentos y cuadros / Espendez Navarro, Juan – Humacao, Puerto Rico. 1937 – 1r – us UF Libraries [972]
La caserne – Paris. nov 1924-fevr 1929 – 1 – (puis organe de defense des matelots) – fr ACRPP [073]
Caseron del cerro / Pogolotti, Marcelo – Santa Clara, Cuba. 1961 – 1r – us UF Libraries [972]
Cases adjudged in the u.s. circuit courts of appeal – New York: Banks. v1-63. 1893-99 (al publ) – (= ser U s circuit & district court reports) – 612mf – 9 – $918.00 – (with 2 index vols. each title page carries the legend "official edition") – mf#LLMC 79-422 – us LLMC [347]
Cases and materials on legislation / Parkinson, Thomas Ignatius – Rev. 1936. New York, 1936. 2v. in 3. LL-1132 – 1 – us L of C Photodup [340]
Cases and materials on security transactions / Maloney, John Philip – New York: St. John's Univ. Press, 1947. 748p. LL-300 – 1 – us L of C Photodup [340]
Cases and materials on the law of sales / Llewellyn, Karl N – Chicago, Callaghan, 1930. 1081 p. LL-290 – 1 – us L of C Photodup [346]
Cases and materials on the law of vendor and purchase / Handler, Milton – St. Paul: West, 1933. 238p. LL-173 – 1 – us L of C Photodup [346]
Cases, Cesare *see* Stichworte zur deutschen literatur
Cases decided in the united states court of claims / United States Court of Claims – Washington. 1975-1977 (1) 1976-1977 (5) 1976-1977 (9) – ISSN: 0149-2810 – mf#6240 – us UMI ProQuest [347]
Cases in common law actions. / Keigwin, Charles Albert – Rochester, N.Y., The Lawyers Co-Operating Publishing Co., 1928. 302 p. LL-522 – 1 – us L of C Photodup [346]
Cases in georgia reports that have been overruled, doubted, criticised, or modified / Downing, Hugh Urquhart – Columbia, Ga. 1922. LL-1306 – 1 – us L of C Photodup [340]
Cases of the law of bills and notes selected from decisions of english and american courts / Smith, Howard Leslie – St. Paul: West, 1910. 756p. LL-1117 – 1 – us L of C Photodup [347]
The cases of the u.s. court of appeals for the d.c. circuit / Beremant, Gordon et al – Washington: FJC, July 1982 – 1mf – 9 – $1.50 – mf#LLMC 95-349 – us LLMC [347]
Cases on bailments and carriers / Roberts, John Stuart – Chicago: Thompson 1911. 233p. LL-1455 – 1 – us L of C Photodup [340]
Cases on certain equitable doctrines and remedies / Lloyd, William Henry – Philadelphia, International Printing Co. 1917 418 p. LL-807 – 1 – us L of C Photodup [340]
Cases on common law pleading. 2nd ed / Sunderland, Edson Read – Chicago, Callaghan, 1932. 693 p. LL-1564 – 1 – us L of C Photodup [340]
Cases on constitutional law / Thayer, James Bradley – Cambridge Mass. Sever, 1895. 2 v. LL-1315 – 1 – us L of C Photodup [342]
Cases on criminal law / Mikell, William Ephraim – Philadelphia: International Printing Co., 1903. 983p. LL-509 – 1 – (3rd ed. st. paul: west, 1933. 775p. ll-738) – us L of C Photodup [345]
Cases on damages selected from decisions of english and american courts / Mechem, Floyd Russell – St. Paul: West, 1909. 626p. LL-875 – 1 – us L of C Photodup [347]
Cases on equitable relief against defamation and injuries to personality. supplementary to ames's cases in equity jurisdiction / Pound, Roscoe – v.1. Cambridge, Mass., 1920. 77p. LL-1205 – 1 – us L of C Photodup [342]
Cases on equity jurisdiction; restraint of infringement of incorporeal rights. part 1. a collection of cases with notes / Lewis, William Draper – Philadelphia, International Printing Co., 1904. 200 p. LL-1617 – 1 – us L of C Photodup [342]
Cases on federal jurisdiction and procedure. / Medina, Harold Raymond – St. Paul: West, 1926. 674p. LL-1463 – 1 – us L of C Photodup [340]
Cases on labor law / Landis, James McCauley – 2nd ed. Chicago, Foundation Press, 1942. v. 1-2. LL-1130 – 1 – (1947. supplement. brooklyn, 1948. 181 p. ll-1130) – us L of C Photodup [344]
Cases on personal property. / Griffin, Levi Thomas – St. Paul: West, 1895. 202p. LL-259 – 1 – us L of C Photodup [346]
Cases on persons and domestic relations : selected from decisions of english and american courts / Kales, Albert Martin – St. Paul, West, 1911. 654 p. LL-1544 – 1 – us L of C Photodup [347]
Cases on restraint of trade / Wyman, Bruce – Cambridge, Harvard, 1902-?4. 5 pt. LL-1625 – 1 – us L of C Photodup [343]
Cases on the federal anti-trust laws of the united states / MacLachlan, James Angell – New York, Ad Press 1930 684 p. LL-237 – 1 – us L of C Photodup [346]
Cases on the law of admiralty / Lord, George de Forest – 2nd ed. St. Paul, West, 1939. 1044p. LL-215 – 1 – us L of C Photodup [355]
Cases on the law of admiralty / Lord, George de Forest – St. Paul, West, 1926. 837p. LL-216 – 1 – us L of C Photodup [355]
Cases on the law of agency : including the law of principal and agent and the law of master and servant / Huffcut, Ernest Wilson – 2d ed. Boston: Little, Brown, 1907. 837p. LL-409 – 1 – us L of C Photodup [340]
Cases on the law of evidence : selected from decisions of english and american courts / Hinton, Edward Wilcox – St. Paul: West, 1919. 1098p. LL-1182 – 1 – us L of C Photodup [347]
Cases on the law of evidence. / Hughes, Thomas Welburn – St. Paul: West, 1896. 141p. LL-1285 – 1 – (chicago: callaghan, 1921. 922p. ll-1332) – us L of C Photodup [347]
Cases on the law of executors and administrators / Vosseler, Edward Adolph – Brooklyn, 1948. 208 p. LL-1555 – 1 – us L of C Photodup [340]
Cases on the law of insurance...2nd ed / Vance, William Reynolds – St. Paul, West, 1931. 1020 p. LL-1254 – 1 – us L of C Photodup [336]
Cases on the law of municipal corporations / Tooke, Charles Wesley – 1931 ed. New York, Commerce Clearing House, 1931. 896 p. LL-1249 – 1 – us L of C Photodup [346]
Cases on the law of partnership / Mechem, Floyd Russell – 2d ed. by Floyd R. Mechem and Frank L. Sage...3d ed. Chicago: Callaghan, 1905. 1104, 209-224p. LL-526 – 1 – us L of C Photodup [346]
Cases on the law of private corporations. / Burnett, Daniel Frederick – Boston, Little, Brown, 1917. 828 p. LL-1272 – 1 – us L of C Photodup [346]
Cases on the law of public service / Burdick, Charles Kellog – Boston, Little, Brown, 1916. 544 p. LL-234 – 1 – us L of C Photodup [340]
Cases on the law of succession to property after the death of the owner. / Mechem, Floyd Russell – St. Paul: West, 1895. 184p. LL-93 – 1 – us L of C Photodup [346]
Cases on the law of taxation: parts 1, 2 and 3...4 and 5 / Maguire, John MacArthur – New York: Commerce Clearing House. 1931. 950p LL-265 – 1 – us L of C Photodup [343]
Cases on the law of wills / Schmid, John Henry – Brooklyn, 1924. 365p. LL-1325 – 1 – us L of C Photodup [346]
Cases submitted to the house of lords on appeal from the courts of england, scotland, and northern ireland / Great Britain. Parliament. House of Lords – London, etc. On film: cases 1-1027. LL-052 – 1 – us L of C Photodup [347]
La caseta de la cordialidad. feria de san miguel, 1972. bailes, actos culturales. concursos. cena de gala / Zafara 72 – Caceres: Edit. Extremadura, 1972 – 1 – sp Bibl Santa Ana [390]
Casey, Calvert *see*
– Memorias de una isla
– Regreso
Casey, Elizabeth *see* Illustrious irishwomen
Casey, George Elliott *see* Speech of g e casey, mp on the remedial bill
Casey, Kevin M *see* Concentric and eccentric strength differences in the lead and back legs of division 1 college level fencers
Casey, Patrick H *see* The bible and its interpreter
Casey, Robert E *see* The declaration of independence
Casey, Thomas Lincoln *see* Revision of the cucujidae of america north of mexico
Casey, Timothy *see* Circular
Casgrain, Eugene *see* Le mouton
Casgrain, Henri Raymond *see*
– Biographie de gerin-lajoie
– Biographies de a s falardeau et a e aubry
– The canadian pioneers
– Champlain
– F x garneau et francis parkman
– The french-war papers of the marechal de levis
– Montcalm et levis
– Une paroisse canadienne au 17e siecle
– Un pelerinage au pays d'evangeline
– Les quarante dernieres annees, le canada depuis l'union de 1841, par john charles dent
– Voyage au canada dans le nord de l'amerique septentrionale fait depuis l'an 1751 a 1761
Casgrain, Henri-Raymond *see*
– A s falardeau et a e aubry
– Biographies de a s falardeau et a e aubry
– De gaspe et garneau
– Une excursion a l'ile aux coudres
– Legendes canadiennes
– Oeuvres completes de l'abbe casgrain
– A s falardeau et a e aubry
Casgrain, Thomas Chase *see*
– Address
– Aux electeurs du comte de montmorency
– The courts of quebec
Casgrain, Thomas Chase et al *see*
– Deuxieme rapport de la commission chargee de reviser et de modifier le code de procedure civile du bas-canada
– Premier rapport de la commission chargee de la revision et de la modification du code de procedure civile du bas-canada
– Quatrieme rapport de la commission chargee de reviser et de modifier le code de procedure civile du bas-canada
– Troisieme rapport de la commission chargee de reviser et de modifier le code de procedure civile du bas-canada
Cash and glory : the commercialization of major league baseball as a sports spectacular, 1865-1892 / Voigt, David Q – 1962 – 7mf – 9 – $28.00 – mf#PE 4005 – us Kinesology [790]
Cash, Tamra L *see* Effects of different exercise promotion strategies and stage of exercise on reported physical activity, self-motivation, and stages of exercise in worksite employees
Cash, William Thomas *see* Story of florida
Cashaway baptist church. darlington district. south carolina : church records – 1767-1805 – 1 – 7.20 – us Southern Baptist [242]
Cashbook of the department of state, 1785-1795 / U.S. Dept of State – (= ser General records of the department of state) – 1r – 1 – mf#T904 – us Nat Archives [324]
Cashel gazette and weekly advertiser – Cashel, Ireland. 14 may 1864-6 may 1865, 3 jun 1865-7 jul 1866, 30 oct-24 dec 1868, 2 jan-25 sep 1869, 1870-18 dec 1886, 1887-8 jul 1893 – 11 1/4r – 1 – (aka: cashel gazette tipperary reporter and weekly advertiser) – uk British Libr Newspaper [072]
Cashel gazette tipperary reporter and weekly advertiser *see* Cashel gazette and weekly advertiser
Cashel sentinel – Cashel, Ireland. 12 jan 1889-1896, 1899-1901, 1904 – 5 3/4r – 1 – (aka: cashel sentinel and weekly general advertiser) – uk British Libr Newspaper [072]
Cashel sentinel and weekly general advertiser *see* Cashel sentinel
Cashflow – Overland Park. 1986-1988 (1,5,9) – (cont by: corporate cashflow) – ISSN: 0196-6227 – mf#15757 – us UMI ProQuest [332]
Cashie baptist church. windson association. berie county. north carolina : church records – 1791-1924 – 1 – us Southern Baptist [242]
Cashton record – Cashton WI. 1900 jan 4/1901 oct 18-2000 – 58r – 1 – (with gaps) – mf#1005489 – us WHS [071]
Casi transactions / Canadian Aeronautics and Space Institute – Ottawa. 1972-1972 (1) 1972-1972 (5) (9) – ISSN: 0007-7852 – mf#6869 – us UMI ProQuest [629]
[Casie Chitty, Simon] *see* Sketch of the rise and progress of the catholic church in ceylon
Casimir, : ou, le premier tete-a-tete / Desnoyer, Charles – Paris, France. 1831 – 1r – us UF Libraries [440]
Casimiro de abreu / Bruzzi, Nilo – Rio de Janeiro, Brazil. 1949 – 1r – us UF Libraries [972]
Casimirus emblematico anagrammaticus reverendissimo et eminentissimo dno d. anselmo casimiro sacrae sedis moguntinae archiepiscopo... / Marx, J R – Moguntiae: Typographia Meresiana, apud Ioannem Cratonum Schmidt, 1636 – 1mf – 9 – mf#O-37 – ne IDC [090]
Casini, Giovanni Maria *see* Canzonette spirituali
Casino kyogle courier – Casino, 1905-32 – at Pascoe [079]
Casket : devoted to literature, science, the arts, news, etc – Cincinnati. 1846-1846 – 1 – mf#3957 – us UMI ProQuest [073]
Casket – Hudson. 1811-1812 (1) – mf#3686 – us UMI ProQuest [420]
The casket, or musical pocket companion; a collection of the most popular songs, duetts, marches, waltzes, dances &c – Carefully arranged for the flute, violin, Kent bugle, or flageolet. New York: James L. Hewitt 183-. Includes: "Old King Cole" and "Tis the Last Rose of Summer." MUSIC 1988 – 1 – us L of C Photodup [780]
Un caso curioso de derecho y de anatomia mineral / Bayle, Constantino – Madrid: Razon y Fe, 1925 – 1 – sp Bibl Santa Ana [611]
Caso de angola / Ventura, Reis – Braga, Portugal. 1964 – 1r – us UF Libraries [960]
Caso de belice a la luz de la historia / Santiso Galvez, Gustavo – Guatemala, 1941 – 1r – us UF Libraries [972]
Caso de belice ante la conciencia de america / INTERNATIONAL AMERICAN CONFERENCE – Guatemala, 1948 – 1r – us UF Libraries [972]
Un caso de extirpacion de la laringe...de esta operacion / Cisneros, Juan – 1890 – 9 – sp Bibl Santa Ana [610]
El caso del judaizante jeronimo fray diego de marchena / Sicioff, A A – Madrid: Castalia, 1966 – 1 – sp Bibl Santa Ana [946]
El caso del obispo marcial de merida / Garcia de la Fuente, P Arturo – Badajoz: tip. y enc. la alianza, 1933 – 1 – sp Bibl Santa Ana [946]
Caso palmer / Gomez, C R A – Santiago, Dominican Republic. 1932 – 1r – us UF Libraries [972]
Casopis macicy serbskeje *see* Casopis towarstwa macicy serbskeje
Casopis towarstwa macicy serbskeje – Bautzen DE, 1848-1937 – 11r – 1 – (title varies: 1873: casopis macicy serbskeje sorbisch) – gw Misc Inst [074]
Casos e coisas da bahia / Vianna, Antonio – Salvador, Brazil. 1950 – 1r – us UF Libraries [972]
Casos para el estudio de los derechos reales / Rodriguez Ramos, Manuel – San Juan, Puerto Rico. 1956 – 1r – us UF Libraries [972]
Casos y cosas de la politica / Olavarria Bravo, Arturo – Santiago, Chile. 1950 – 1r – us UF Libraries [972]
Caspar cruciger : nach gleichzeitigen quellen / Pressel, Theodor – Elberfeld: R L Friderichs 1862 [mf ed 1991] – (= ser Leben und ausgewaehlte schriften der vaeter und begruender der lutherischen kirche 8/2) – 1mf – 9 – 0-524-00584-2 – (incl bibl ref) – mf#1990-0084 – us ATLA [242]
Caspar hauser : oder, die traegheit des herzens: roman / Wassermann, Jakob – 1.-4. aufl. Stuttgart: Deutsche Verlags-Anstalt 1908, c1905 [mf ed 1991] – 1r – 1 – (filmed with: richard wagner / hans von wolzogen) – mf#2974p – us Wisconsin U Libr [830]
Caspari, Carl Paul *see*
– Alte und neue quellen zur geschichte des taufsymbols und der glaubensregel
– A grammar of the arabic language
– Kirchenhistorische anecdota
– Konkordiebogen
– Populaere foredrag over profeten daniel
– Ueber den syrisch-ephraimitischen krieg unter jotham und ahas
– Ueber micha den morasthiten und seine prophetische schrift
– Ungedruckte, unbeachtete und wenig beachtete quellen zur geschichte des taufsymbols und der glaubensregel
Caspari, Chretian Edouard *see* Chronological and geographical introduction to the life of christ
Caspari, Wilhelm *see*
– Die bedeutung der wortsippe kvd im hebraeischen
– Die bedeutungen der wortsippe "kbd" im hebraeischen

- Echtheit, hauptbegriff und gedankengang der messianischen weissagung, jes. 9, 1-6
- Erd- oder feuerbestattung
- Die pharisaeer bis an die schwelle des neuen testaments
- Die religion in den assyrisch-babylonischen busspsalmen
- Vorstellung und wort friede im alten testament

Caspary, Eugen *see* Nachrichtendienst
Caspary, Eugen et al *see* Zedakah
Le casque a meche – [Paris]: Impr de Beaule et Maignand, may 1849 – us CRL [944]
Casquete, Antonio *see* El cristo de la reja
Casquete Hernando, Antonio *see* Noticias de la villa de segura de leon
La casquette du pere duchene : pamphlet socialiste – [Paris]: Impr Bonaventure et Ducessois, [1848?] – us CRL [325]
Casrilho Barreto e Noronha, Augusto Vidal de *see* O districto de lourenco marques, no presento e no futuro
Cass county democrat – Plattsmouth, NE: Fellows & Kirkham. v4 n12. jun 7 1901-[wkly] [mf ed -1902 [gaps] filmed 1979] – 1r – 1 – (cont: weekly post) – us NE Hist [071]
Cass county echo – [Plattsmouth, NE]: Call Print Co, aug 16 1943-v3 n46. jun 28 1946 (wkly) [mf ed with gaps] – 2r – 1 – us NE Hist [071]
Cass county herald – Nehawka, NE: D L Hamilton (wkly) [mf ed jan 8-22 1970 filmed 1972] – 1r – 1 – (absorbed: louisville courier (1963). foll absorption cont numbering of: louisville courier) – us NE Hist [071]
The cass county sentinel – Plattsmouth, NE: Geo H Thompson (wkly) [mf ed v1 n36. nov 1 1879 filmed 1979] – 1r – 1 – us NE Hist [071]
The cass county tribune – Plattsmouth, NE: G F S Burton, 1895 (wkly) [mf ed v1 n16. oct 11 1895-97 (gaps) filmed 1979] – 1r – 1 – us NE Hist [071]
Cassandra – Rio de Janeiro, RJ. ago 1874 – (= ser Ps 19) – mf#P17,01,95 – bl Biblioteca [321]
Cassandre – Brussels Belgium. 16 feb-29 jun, 2 nov 1941; 3 jan 1943-9 jul 1944 – 2r – 1 – uk British Libr Newspaper [074]
Cassandre-agamemnon et colombine-cassandre, parodi / Barre, M – Paris, France. 1804 – 1r – us UF Libraries [440]
Cassandri, Georgii (Cassander, George) *see* Opera quae reperiri potuerunt omnia
Cassar, Francisco del *see* Respuesta que da el m.r.p...fr. j. torrubia...sobre la legitimidad del libro de oracion..
Cassava as a money crop / Stockbridge, Horace E – Lake City, FL. 1899 – 1r – us UF Libraries [634]
Cassava, the velvet bean, prickly comfrey, taro, chinese yam, canaigre, alfalfa, flat pea, sachaline / Clute, O – Lake City, FL. 1896 – 1r – us UF Libraries [634]
Casseday, Morton M *see* Land of manatee
Cassegrain, Arthur *see* La grande troniciade ou itineraire de quebec a la riviere-du-loup
Cassel, Daniel Kolb *see*
- Geschichte der mennoniten
- History of the mennonites

Cassel, Johann Philipp *see* Bremisches magazin zur ausbreitung der wissenschaften, kuenste und tugend
Cassel, Paulus *see*
- The book of judges
- The book of ruth
- An explanatory commentary on esther
- Vom nil zum ganges

Casseler fremden-verkehrs-zeitung *see* Fremden-verkehrs-zeitung
Casseler grundstuecks- und hypotheken-boerse – Kassel DE, 1907 27 sep-1 nov – 1r – 1 – gw Misc Inst [332]
Casseler stadt-anzeiger – Kassel DE, 1960 2 sep-20 oct & 31 oct-30 dec, 1961 28 jan-1963 11 jul [gaps], 1963 10 sep-29 sep – 1 – (filmed by other misc inst: 1889-1904 jun, 1906-08, 1912, 1914-28, 1930-34 [gaps], 1935 mar-1939 mar, 1939 jul-1943 mar [gaps], 1949 nov-1956 feb, 1956 apr-1969 25 feb [178r]; title varies: 11 may 1897: hessische post und casseler stadtanzeiger, 1916: hessische post. kasseler stadtanzeiger, 10 jan 1923: kasseler post / stadtausgabe) – gw Misc Inst [074]
Casseler tages-post – Kassel DE, 1861 21 sep-1866 25 mar – 11r – 1 – gw Misc Inst [074]
Die casseler woche – Kassel DE, 1925 17 oct-1926 – 1r – 1 – gw Misc Inst [074]
Casselische zeitung von policey-, commercien und andern dem publico dienlichen sachen – Kassel DE, 1733 2 may-1808 27 jun, 1814-21 – 65r – 1 – (title varies: 1751: casselsche policey-, gelehrte und commercien-zeitung, later: casselische policey- und commercien-zeitung, later: casselsche policey- und commercien-zeitung) – gw Misc Inst [380]
Cassell & Co Ltd *see*
- Cassell's illustrated family exhibitor
- The illustrated exhibitor

Cassell, John *see* The works of eminent masters
Cassell's gazetteer of great britain and ireland – London. 6v. 1894-98 – 1 – us L of C Photodup [914]
Cassell's illustrated family exhibitor / Cassell & Co Ltd – London 1862 – (= ser 19th c art & architecture) – 3mf – 9 – mf#4.2.894 – uk Chadwyck [700]
Cassels, Robert *see*
- A digest of cases
- A digest of cases decided by the supreme court of canada from the organization of the court, in 1875, to the 1st day of may 1886
- Manual of procedure in the supreme and exchequer courts of canada
- Report of robert cassels, esq

Cassels, Samuel Jones *see* Christ and antichrist
Cassels, Walter Richard *see*
- Letter on a gold currency for india
- A reply to dr lightfoot's essays
- Supernatural religion

Casselsche policey-, gelehrte und commercien-zeitung *see* Casselische zeitung von policey-, commercien und andern dem publico dienlichen sachen
Casser, Paul *see* Die westfaelischen musenalmanache und poetischen taschenbuecher
Casset, A *see* Citonga grammar and vocabulary for the use of the settlers
Casseus, Maurice A *see* Mambo
Cassi kwoc – 9 – $100.00 – (keyword out of context ind) – us Chemical [540]
Cassimir, Heinrich *see* Ludwig ganghofer als buehnendichter
Cassiodoris senatoris variae (mgh1:12.bd) / ed by Mommsen, Theodor – 1894 – (= ser Monumenta germaniae historica 1: scriptores – auctores antiquissimi) – €38.00 – (accedunt 1: epistolae theodoricianae variae ed th mommsen. 2: acta synodorum habitarum romae a 498 ed th mommsen. 3: cassiodori orationum reliquiae ed l traube) – ne Slangenburg [240]
Cassirer, Ernst *see* Heinrich von kleist und die kantische philosophie
Cassius, Johann Ludwig *see* Lehrgebaeude der polnischen sprachlehre
Cassville american – Cassville, Potosi, Tennyson WI. 1941 nov 14/1945 jan 25-1971 jul 15/1973 feb 8 – 11r – 1 – (with gaps) – mf#966178 – us WHS [071]
Cassville current – Cassville WI. 1885 dec 12 – 1r – 1 – mf#958035 – us WHS [071]
Cassville index – Cassville WI. 1888 mar 8/1891 aug 20-1913 oct 2/1917 aug 9 – 10r – 1 – mf#966181 – us WHS [071]
Cast metals research journal – Des Plaines. 1971-1975 (1) 1965-1975 (5,9) – ISSN: 0008-7467 – mf#6208 – us UMI ProQuest [660]
Cast thy bread upon the waters / Sadler, Thomas – London, England. 1846 – 1r – us UF Libraries [240]
Castalleda, Vicente *see* Fallecimiento
Castanea – Charlotte. 1949-1954 (1) – ISSN: 0008-7475 – mf#390 – us UMI ProQuest [580]
Castaneda, Carlos Eduardo *see* Lands of middle america
Castaneda, Gabriel Angel *see* Roman cero
Castaneda, Gloria *see* Piedra
Castaneda, P *see* Firmes
Castaneda S, Gustavo A *see*
- Combate de el obrajuelo
- Dominio insular de honduras

Castaneda, Vicente *see*
- Bibliofilia sentimental
- Muerte de d. francisco barrado y font
- Trujillo. declaracion de monumento historico-artistico de su castillo
- Viniegra vera, virgilio. correspondiente de la real acad. de la historia en santa marta

Castaneyra, Isidro Alphonso de *see* Manual summa de las ceremonias de la provincia de el santo evangelio de mexico
Castanis, C Plato *see* The greek boy and the sunday-school
Las castas del mexico colonial...1924 / Leon, Nicolas; ed by Bayle, Constantino – Madrid: Razon y Fe, 1928 – 9 – sp Bibl Santa Ana [972]
Caste and credit in the rural area : a survey by s s nehru / Nehru, Shri Shridhar – Calcutta: Longmans, Green & Co, 1932 – (= ser Samp: indian books) – us CRL [305]
Caste and outcast / Mukerji, Dhan Gopal – London: J M Dent & Sons, 1923 – (= ser Samp: indian books) – us CRL [305]
Caste and outcaste / Sanjana, J E – Bombay: Thacker & Co, 1946 – (= ser Samp: indian books) – us CRL [305]
Caste and race in india / Ghurye, Govind Sadashiv – London: Kegan Paul, Trench, Truebner & Co, 1932 – (= ser Samp: indian books) – us CRL [305]
Caste in a peasant society / Tumin, Melvin Marvin – Princeton, NJ. 1952 – 1r – us UF Libraries [306]
Caste in india : the facts and the system / Senart, Emile – London: Methuen & Co, 1930 – (= ser Samp: indian books) – (trans by e denison ross) – us CRL [305]
Caste in india : its nature, function, and origins / Hutton, John Henry – London, New York: Oxford University Press, 1951 – (= ser Samp: indian books) – us CRL [305]
Caste or christ? : sketches of indian life / Hodge, John Zimmerman & Hicks, George Elgar – London: Morgan & Scott; Regions beyond missionary union, [1906] [mf ed 1995] – (= ser Yale coll) – 127p (ill) – 1 – 0-524-10033-0 – (pref by harry guinness. ill fr photos by alexander l banks) – mf#1995-1033 – us ATLA [306]
Castel, Elie *see* Les huguenots et la constitution de l'eglise reformee de france en 1559
Castel, Joaquin *see*
- Algunas ideas sobre el engrandecimiento de caceres
- Cuestion de actualidad
- Influencia del manantial de marco en el desarrolla material de caceres
- Replica al folleto de don francisco galan castillo titulado "al publico"

The castel of helth / Elyot, Thomas – 1541 – 9 – us Scholars Facs [610]
Castel, Rene-Richard *see* Histoire naturelle de buffon
Castelao, Fernanda *see* Monografia historico del castillo de jarndilla
Castelar, Emilio *see* Dona carolina coronado
Castelhun, Friedrich Karl *see* Gedichte
Castellan, Antoine *see*
- Lettres sur la moree, l'hellespont et constantinople
- Lettres sur l'italie
- Sitten, gebraeuche und trachten der osmanen
- Turkey

Castellan, Antoine L *see* A l castellan's briefe ueber morea und die inseln cerigo, hydra und zante
La castellana de ribera del fresno. leyenda. / Antunez Toriblo, Manuel – 1865 – 9 – sp Bibl Santa Ana [830]
Castellani, Alessandro *see*
- Antique jewellery and its revival
- Italian jewellery as worn by the peasants of italy

Castellani, Ch *see* Vers le nil francais avec la mission marchand
Castellani, Charles Jules *see* Marchand l'africain
Castellano, Conde de *see* Un complot terrorista en el siglo 15th. madrid, 1927
Castellano, P *see* Vitae illustrium medicorum...
Castellanos, Francisco Jose *see* Ensayos y dialogos
Castellanos G, Gerardo *see* Motivos de cayo hueso (contribution) a la historia
Castellanos Garcia, Gerardo *see*
- Discursos leidos en la recepcion publica
- Raices del 10 de octubre de 1868
- Soldado y conspirador
- Viajando por los mares de trinidad

Castellanos, Jesus *see*
- Conjura
- Optimistas

Castellanos, Joaquin *see* El doctor alem y el radicalismo
Castellanos, Juan De *see* Elegias de varones ilustres de indias
Castellanos, Juan de *see* Obras. tomo 1
Castellanos Romero, Carlos *see*
- Curso de procediminetos penales
- Primer -segundo curso de procedimientos civiles

Castelle, Friedrich *see*
- Dichtungen der droste
- Gustav falke

Castelle, Friedrich [comp] *see* Dichtungen der droste
Castelli, Bartholommeo *see* Lexicon medicum graeco-latinum (ael3/10)
Castelliunculus, Lapus *see* Bellum punicum 1...
Castello branco : revolucao o democracia / Wamberto, Jose – Rio de Janeiro, Brazil. 1970 – 1r – us UF Libraries [972]
Castello Branco, Mnoel Thomaz *see* Brasil na ii grande guerra
Castello, J Aderaldo *see* Aspectos do romance brasileiro
Castellon, Hildebrando A *see* Resumen de la geografia de nicaragua
Castelnau *see*
- Expedicao as regioes centrais da america do sul. t. 1
- Expedicao as regioes centrais da america do sul. t. 2

Castelnau, Francis de *see* Renseignements sur l'afrique centrale et sur une nation d'hommes a queue qui s'y trouverait
Castelo Branco, Camilo *see* Polemicas em portugal e no brasil
Castelo Branco, Renato *see* Civilizacao do couro
Castelo, Garcia *see* Trozos de literatura de autores extremenos
Castelo, Placido Aderaldo *see* Historia do ensino no ceara
Castelo-Branco, Fernando A *see* Actividades dos missionarios...
Castelpoggi, Atilio Jorge *see* Miguel angel asturias
Castes and tribes of southern india / Thurston, Edgar – Madras: Govt Press 1909 – (= ser Samp: indian books) – us CRL [305]
Casti, Giovanni Battista *see* La papesse
Castigatissimi annali con la loro copiosa tavola della eccelse and illustrissima repubblica di genoa, da fideli and approuati scrittore... / Giustiniani, A – Genoa, 1537 – 11mf – 9 – mf#H-8250 – ne IDC [950]
Castigator / Brown Co. Georgetown – v1 n1. (jun 1824-sep 28, jul 32-apr 1837) [wkly] – 2r – 1 – mf#B12408-12409 – us Ohio Hist [071]
Castigator / Brown Co. Ripley – v1 n1. (jun 1824-sep 28-jul 32-apr 1837) [wkly] – 2r – 1 – mf#B12408-12409 – us Ohio Hist [071]
Castile. Laws, Statutes, etc *see* Extracto de las siete partidas
Castilho, Augusto Ferreira De *see* Democracia no brasil
Castilian days / Hay, John – Boston: J R Osgood & Co 1871 [mf ed 1987] – 1r – 1 – mf#1821 – us Wisconsin U Libr [914]
Castilla agricola para la ensenanza de la agricultura...caceres / Quintanilla, Guillermo & Arche, Jose Vicente – Badajoz: Ciudad Real y Albacete; Madrid: Imp. de los Hijos de M.G. Hern'andez, 1905 – sp Bibl Santa Ana [630]
Castilla, Juan de *see* La justicia revolucionaria en espana
Castille, Cheryl L *see* The perceived importance of a leisure education component in outpatient weight management programs serving adult women
Castillero R, Ernesto J *see*
- Causa inmediata de la emancipacion de panama
- Historia de la comunicacion interoceanica
- Universidad interamericana

Castillo Armas, Carlos *see* Discursos del presidente de guatemala
Castillo, Balthasar del *see* Luz, y guia de los ministros evangelicos
Castillo de Bovadilla, J *see* Politica para coregidores y senores de vassalos en tiempos de paz y de guerra...
El castillo de castellar : datos para la historia de zafra / Salazar Fernandez, Antinio – Zafra: imp segedana, 1955 – 1 – sp Bibl Santa Ana [946]
Castillo de guadamuz / Velo Nieto, Gervasio – Madrid, s.i., 1956 – 1 – sp Bibl Santa Ana [946]
Castillo de la alta extremadura : eljas (con noticias historicas de la encomienda de su nombre) / Velo Nieto, Gervasio – Badajoz: Imprenta Diputacion Provincial, 1968 – sp Bibl Santa Ana [946]
El castillo de loarre (informe) / Monsalud, Marques de – Madrid: Est. Tip. Fortanet, 1905. B.R.A.H. 47, 1905, pp. 448-451 – 1 – sp Bibl Santa Ana [946]
El castillo de medellin en la ruta del turismo / Garcia Sanchez, Francisco – Don Benito, Sanchez Trejo, 1969 – sp Bibl Santa Ana [946]
Castillo de oro / Ruben, Carlos – Habana, Cuba. 1951 – 1r – us UF Libraries [972]
El castillo de piedrabuena / Escobar Prieto, Eugenio – Caceres: imp luciano jim'enez, 1908 – 1 – sp Bibl Santa Ana [946]
El castillo de santibanez el alto / Velo Nieto, Gervasio – Madrid: Accasor, 1956 – 1 – sp Bibl Santa Ana [946]
El castillo e los marqueses de las navas / Perez Minguez, Fidel – Madrid: tip arch, bibl y mus, 1930 – sp Bibl Santa Ana [946]
Castillo, Jose Leon *see* Geografia general nacionalista de la america del c...
Castillo, Manuel *see*
- Extremadura
- Gramatica castellana
- Programa de la asignatura de castellano.(primo curso)

Castillo, Marciano *see* La federacion
Castillo, Moises *see*
- Caminos del agro
- Cancion del camino

Castillo Puche, Jose Luis *see* Sin camino, novelo
Castillo R *see* Ahuizote
Castillo Y Guevara, Francisca Josefa De *see*
- Afectos espirituales de la venerable madre y obser
- Mi vida
- Su vida

El castillo y plaza fuerte de alcantara / Velo Nieto, Gervasio – Madrid, 1963 – (sep boletin asociacion espanola de amigos de los castillos) – sp Bibl Santa Ana [946]
Un castillo y varios castellanos... / Perez Minguez, Fidel – Madrid: Razon y Fe, 1927 – 1 – sp Bibl Santa Ana [946]
Castillos de la alta extremadura : penafiel, con breves noticias de la encomienda de su nombre / Velo Nieto, Gervasio – Madrid, 1957 – 1 – (aparte hidalguia nov-dic 1957 n25 p1-22) – sp Bibl Santa Ana [946]
Castillos, torres y casas fuertes de la provincia de caceres / Hurtado de Mendoza, Publio – Caceres: Imprenta y libreria Catolica de Santos Floriano, 1912 – sp Bibl Santa Ana [946]

Castle – 1985 feb/mar-1993 aug – 1r – 1 – mf#1058327 – us WHS [071]
Castle – 1980 aug 22-1981 jul, 1981 aug-1982 apr, 1982 may 14-oct 29, 1982 nov 5-1983 jun 24, 1983 jul 1-1984 feb 24, 1984 aug-1985 feb, 1984 mar-aug 3 – 14r – 1 – (cont by: Centerpiece [Fort Belvoir (VA)]; belvoir eagle) – mf#570167 – us WHS [071]
Castle camps 1563-1950 – (= ser Cambridgeshire parish register transcript) – 7mf – 9 – £8.75 – uk CambsFHS [929]
Castle camps congregational church registers 1817-1933 – (= ser Cambridgeshire parish register transcript) – 1mf – 9 – £1.25 – uk CambsFHS [929]
Castle comments – 1984 jul/aug, 1987 sep-1993 sep – 1r – 1 – mf#1058336 – us WHS [071]
Castle corner – v6 n10 [1983 oct], v7 n3-4 [1984 jun-aug], v9 n5, [1987 oct/nov], v10 n1,4,6 [1988 feb/mar, aug/sep, dec/1989 jan, v11 n1-4,6 [1989 feb/mar-aug/sep, dec-1990/jan], v12 n4 [1990 may/jun] – 1r – 1 – mf#1054111 – us WHS [071]
Castle donington weekly express – [East Midlands] Leicestershire 26 jun 1858-30 nov 1867 [mf ed 2002] – 5r – 1 – (cont as: castle donnington telegraph & leicestershire & derbyshire advertiser [jan 1861-nov 1867]) – uk Newsplan [072]
Castle, Eduard *see*
– Ferdinand raimunds saemtliche werke in drei teilen
– Gespraeche mit goethe in den letzten jahren seines lebens
– Lenau und die familie loewenthal
– Lenaus leben
– Nikolaus lenau
– Saemtliche werke und briefe in sechs baenden
Castle, Egerton *see* Schools and masters of fence
A castle in spain : a novel / De Mille, James – London: Chatto and Windus, 1885 – 4mf – 9 – mf#06960 – cn Canadiana [830]
Castle lite : newlsetter / Passaic County Historical Society – v6 n1 [1975 spr], v7 n1 [1977 spr] – 1r – 1 – (cont: passaic county historical society newsletter) – mf#1875793 – us WHS [071]
Castle, Nicholas *see* The exalted life
Castle, Nicolas *see* The witness of the spirit
The castle of love 1549?, a translation by john bourchier / San Pedro, Diego de – Lord Berners, of Carcel de Amor. 1492 by Diego de San Pedro – 9 – us Scholars Facs [830]
The castle of otranto / Walpole, Horace; ed by Doughty, Oswald – London: The Scholartis Press, 1929. lxxx,111p. 2 pl – 1 – us Wisconsin U Libr [830]
The castle st louis, quebec, 1759-1834 / LeMoine, James McPherson – Toronto: Ontario Pub Co, [1896?] [mf ed 1980] – 1mf – 9 – 0-665-08610-5 – cn Canadiana [720]
Castleacre deeds, 1300-1400 – bundle 2, n7a-11 – (= ser Holkham library early estate records) – 1r – 1 – mf#97232 – uk Microform Academic [343]
Castleacre manor court rolls, 1300-1400 – (= ser Holkham library early estate records) – 1r – 1 – mf#6510 – uk Microform Academic [343]
Castleford chronicle and knottingley advertiser – [Yorkshire & Humberside] Wakefield 24 apr 1858-27 sep 1862 [mf ed 2003] – 4r – 1 – uk Newsplan [072]
Castleford gazette and district advertiser – [Yorkshire & Humberside] Wakefield jan 1874-14 feb 1902 [mf ed 2004] – 20r – 1 – (missing: 1880, 1897; cont by: castleford gazette and normanton, whitwood, methley, allerton, kippax, fairburn, ferrybridge, knottingley and pontefract chronicle [jan 1896-feb 1902]) – uk Newsplan [072]
Castleford star – [Yorkshire & Humberside] Wakefield 10 jul 1869-30 mar 1872 [mf ed 2004] – 2r – 1 – (cont by: castleford star and free press [jan 1871-mar 1872]) – uk Newsplan [072]
Castleford telegraph and whitwood, kippax, methley, and featherstone chronicle – [Yorkshire & Humberside] Wakefield 10 jun 1892-dec 1898 [mf ed 2004] – 6r – 1 – uk Newsplan [072]
The castles, palaces and prisons of mary of scotland / Mackie, Charles – London, 1849. 480p. illus – 1 – us Wisconsin U Libr [941]
Castletown, Bernard Edward Barnaby Fitzpatrick, 2nd Baron *see*
– The abc of the irish land question
– Ireland's brighter prospects
Caston, Alfred de *see* La turquie en 1873
Caston, M *see* Independency in warwickshire
Castonnet des Fosses, Henri *see* L'abyssinie et les italiens
Castonnet Des Fosses, Henri Louis *see* Perte d'une colonie
Castonnet des Fosses, Henri Louis *see* Madagascar
Castrametatio : dat is legermeting / Stevin, S – Leyden, 1633 – 1mf – 9 – mf#0A-176 – ne IDC [720]
Castrametatio : dat is legermeting / Stevin, S – Rotterdam, 1617 – 1mf – 9 – mf#0A-174 – ne IDC [720]
La castrametation... / Stevin, S – Rotterdam, 1618 – 1mf – 9 – mf#0A-175 – ne IDC [720]
La castreida / Salas, Francisco Gregorio de – 1838 – 9 – sp Bibl Santa Ana [830]
Castren, M A *see* Reiseberichte und briefe aus den jahren 1845-1849
Castries, H de *see* Sources inedites de l'histoire du maroc de 1530 a 1845
Castro Albarran, A de *see*
– Este es el cortejo...salamanca 1938
– Guerra santa: el sentido catolico de la guerra espanola. burgos, 1938
– Polvo de sus sandalias
Castro, Alf A *see* Adversus omnes haereses libri 14
Castro alves : conferencias / Neiva, Venancio De Figueiredo – Rio de Janeiro, Brazil. 1947 – 1r – us UF Libraries [972]
Castro alves / Peixoto, Afranio – Sao Paulo, Brazil. 1942 – 1r – us UF Libraries [972]
Castro, Antonio de *see*
– Peticion escrito de conclusiones al nuncio por el p. caceres con el p. juan de la serena y otros
– Peticion...al nuncio por el p. caceres..
– Por...fr. diego de caceres, general...de s. geronimo...y demas diputados que se confirma la sentencia
Castro Bajo, Julian *see* Flores y espinas
Castro de Torres *see* Panegirico al chocolate
Castro, Eduardo Gomes De Albuquerque *see* Angola
Castro, Eugenio De *see* Ensaios de geographia linguistica
Castro, Eugenio de *see* Obras poeticas
Castro Fernandez, Hector Alfredo *see*
– Pounette
– Vitral
Castro, Ferreira De *see* Selva
Castro, Francisco de *see* La octava maravilla
Castro, Gabriel *see* Salvacion de colombia
Castro, J *see* Historia de las virtudes y propiedades del tabaco y de los modos de tomarse...
Castro, Jean de *see* Novae cantiones sacrae...
Castro, Jesus *see* Antologia de poetas hondurenos
Castro, Joannes a *see*
– De on-ghemaskerde liefde des hemels
– Zedighe sinne-belden (sic) ghetrocken uyt...
Castro, Jose Agustin de *see* El triunfo del silencio
Castro, Jose de *see*
– Elogio...antonio mendes correia
– Primera regla de la fecunda madre santa clara de assis
Castro, Josue De *see*
– Alimentacion en los tropicos
– Documentario do nordeste
– Geografia da fome
– Problema da alimentacao no brasil
Castro, Juan Francisco *see* Geografia elemental de la republica del salvador
Castro, Justino M *see* Vida civil y militar de don hermenegildo galeana
Castro, Luis Gabriel *see* Capital de la gran colombia
Castro, Luiz Paiva De *see* Guia poetica da cidade do rio de janeiro
Castro, Manuel *see*
– Baltasar cuartero y huerta y antonio vargas zuniga y montero de espinosa. marques de siete iglesias. indice...
– Cristobal de san antonio, ofm...en notas bibliografia franciscana
– Francisco pizarroso, ofm, en notasde bibliografia franciscana...
– Jeronimo zapata, natural de azuaga, en notas de bibliografia franciscana
Castro, Manuel de *see* Meridion
Castro Noboa, H B De *see* Antologia poetica trujillista
Castro, P *see* Burgos. archivo historico provincial. los protocolos del archivo historico provincial
Castro, Pedro Andres de *see* Ortografia y reglas de lengua tagalag, ordenada por...
Castro Ramirez, Manuel *see*
– Derecho panal salvadoreno
– Lecciones de logica judicial
Castro, Ricardo *see* Paginas historicas colombianas
El castro romano de caceres el viejo. nuevas inscripciones / Fita, Fidel – Madrid: Fortanet, 1911 – sp Bibl Santa Ana [946]
Castro Saavedra, Carlos *see* Rios navegados
Castro, Salomon G *see* Enciclopedia colombiana
Castro Sampaio, Manuel de *see* Ensaios poeticos
Castro Seoane, Jose *see* El p bartolome de olmedo
Castro, Therezinha De *see* Historia documental do brasil
Castro, Tomas De Jesus *see* Bufonadas del instituto de literatura
[Castro valley-] reporter – CA. jun 3 1931-feb 1951; jan 1955-1979 – 11r – 1 – $660.00 – mf#B02097 – us Library Micro [071]
Castro Y Calvo, Jose Maria *see* Ruben dario y el modernismo en la literatura hispa
Castro y Castro, Manuel *see* Union misional franciscana. su naturaleza y organizacion
Castrofuerte, Marques de *see* Noticias (hallazgos por el marques de castrofuerte en caceres
Castrovido, Roberto *see* Las dos republicas: el 11 de febrero y el 14 de abril
[Castroville-] times – CA. 1959-74 – 6r – 1 – $360.00 – (cont: times journal) – mf#B02099 – us Library Micro [071]
[Castroville-] times journal – CA. jun 1950-52; 1954-58 – 2r – 1 – $120.00 – mf#B02098 – us Library Micro [071]
Castroville times/moss landing harbor news *see* Moss landing
Castroville times/north county news – CA. 1975-77 – 1r – 1 – $60.00 – (cont by: north county news, salinas) – mf#B02100 – us Library Micro [071]
Casus papales et episcopales (i. zaragoza, 1479-1484) – Burgos – 1r – 5,6 – sp Cultura [220]
Caswall, Henry *see*
– Brief account of the method of synodical action in the american chu...
– Mormonism and its author
Cat : truyen dai / Thao Truong – [Saigon]: Nhu Y 1974 [mf ed 1992] – on pt of 1r – 1 – mf#11052 r172 n3 – us Cornell [830]
Cataclismo / Desnoes, Edmundo – Habana, Cuba. 1965 – 1r – us UF Libraries [972]
Catacomb – London SA, 1 jan 1949-31 dec 1952 – 1r – 1, 16 diazo available at reduced price – sa National [079]
Le catacombe romane : secondo gli ultimi studi e le pi u recenti scoperte / Marucchi, Orazio – Roma: Desclee, Lefebvre, 1903 – 2mf – 9 – 0-7905-6760-1 – (incl bibl ref) – mf#1988-2760 – us ATLA [930]
Les catacombes de rome : histoire de l'art et des croyances religieuses pendant les premiers siecles du christianisme / Roller, Theophile – Paris: A Morel, [1881?] – 1r – 1 – 0-524-03669-1 – (incl bibl ref) – mf#1990-B000 – us ATLA [240]
Les catacombes de rome – souvenirs de rome / Abelous, Louis David et al – Paris: Agence de la Societe des ecoles du dimanche, 1860 – 1mf – 9 – 0-524-02969-5 – (incl bibl ref) – mf#1990-0756 – us ATLA [914]
The catacombs of rome as illustrating the church of the first three centuries / Kip, William Ingraham – New York: Redfield, 1854, c1853 – 1mf – 9 – 0-7905-5354-6 – mf#1988-1354 – us ATLA [240]
Los catalanes en grecia... / Rubio y Lluch, A; ed by Bayle, Constantino – Madrid: Razon y Fe, 1928 – 9 – sp Bibl Santa Ana [946]
Catalanus, Josepho *see* De codice sancti evangelii, libri 3
Catalina – 1992- – (= ser California telephone directory coll) – 3r – 1 – $150.00 – mf#P00017 – us Library Micro [917]
Catalog / Amherst College. Amherst, MA – 1822-26, 1829, 1830, 1832, 1835, 1840 41, 1845 46 – 1 – us CRL [378]
Catalog – n42 [1982 sum] – 1r – 1 – (cont: krupp dealers' catalog) – mf#660344 – us WHS [621]
Catalog / Georgetown College. Georgetown, Ky – 1845-75 – 1 – us Southern Baptist [242]
Catalog / Regent's Park College. Angus Library – (Annotated v. 1908 – 1 – us Southern Baptist [242]
Catalog / U.S. Bureau of the Census – 1 – us AMS Press [317]
[Catalog 1873] / Society of Lady Artists – London 1873 – (= ser 19th c art & architecture) – 1mf – 9 – mf#4.2.675 – uk Chadwyck [700]
[Catalog 1874] / Society of Lady Artists – London 1874 – (= ser 19th c art & architecture) – 1mf – 9 – mf#4.2.676 – uk Chadwyck [700]
[Catalog 1875] / Society of Lady Artists – London 1875 – (= ser 19th c art & architecture) – 1mf – 9 – mf#4.2.677 – uk Chadwyck [700]
[Catalog 1877] / Society of Lady Artists – London 1877 – (= ser 19th c art & architecture) – 1mf – 9 – mf#4.2.678 – uk Chadwyck [700]
[Catalog 1878] / Society of Lady Artists – London 1878 – (= ser 19th c art & architecture) – 1mf – 9 – mf#4.2.679 – uk Chadwyck [700]
[Catalog 1879] / Society of Lady Artists – London 1879 – (= ser 19th c art & architecture) – 1mf – 9 – mf#4.2.680 – uk Chadwyck [700]
[Catalog 1880] / Society of Lady Artists – London 1880 – (= ser 19th c art & architecture) – 1mf – 9 – mf#4.2.681 – uk Chadwyck [700]
[Catalog 1881] / Society of Lady Artists – [3rd ed]. London 1881 – (= ser 19th c art & architecture) – 1mf – 9 – mf#4.2.686 – uk Chadwyck [700]
[Catalog 1881] / Society of Lady Artists – London 1881 – (= ser 19th c art & architecture) – 1mf – 9 – mf#4.2.682 – uk Chadwyck [700]
[Catalog 1884] / Society of Lady Artists – [2nd ed]. London 1884 – (= ser 19th c art & architecture) – 1mf – 9 – mf#4.2.695 – uk Chadwyck [700]
[Catalog 1885] / Society of Lady Artists – London 1885 – (= ser 19th c art & architecture) – 1mf – 9 – mf#4.2.683 – uk Chadwyck [700]
[Catalog 1886] / Society of Lady Artists – London 1886 – (= ser 19th c art & architecture) – 1mf – 9 – mf#4.2.684 – uk Chadwyck [700]
[Catalog 1887] / Society of Lady Artists – [London] 1887 – (= ser 19th c art & architecture) – 1mf – 9 – mf#4.2.685 – uk Chadwyck [700]
[Catalog 1889] / Society of Lady Artists – [London] 1889 – (= ser 19th c art & architecture) – 1mf – 9 – mf#4.2.687 – uk Chadwyck [700]
[Catalog 1890] / Society of Lady Artists – [London] 1890 – (= ser 19th c art & architecture) – 1mf – 9 – mf#4.2.688 – uk Chadwyck [700]
[Catalog 1893] / Society of Lady Artists – [London] 1893 – (= ser 19th c art & architecture) – 1mf – 9 – mf#4.2.689 – uk Chadwyck [700]
[Catalog 1894] / Society of Lady Artists – [London] 1894 – (= ser 19th c art & architecture) – 1mf – 9 – mf#4.2.690 – uk Chadwyck [700]
[Catalog 1895] / Society of Lady Artists – [London] 1895 – (= ser 19th c art & architecture) – 1mf – 9 – mf#4.2.691 – uk Chadwyck [700]
[Catalog 1896] / Society of Lady Artists – [London] 1896 – (= ser 19th c art & architecture) – 1mf – 9 – mf#4.2.692 – uk Chadwyck [700]
[Catalog 1897] / Society of Lady Artists – [London] 1897 – (= ser 19th c art & architecture) – 1mf – 9 – mf#4.2.693 – uk Chadwyck [700]
[Catalog 1898] / Society of Lady Artists – [London] 1898 – (= ser 19th c art & architecture) – 1mf – 9 – mf#4.2.694 – uk Chadwyck [700]
Catalog age – Overland Park. 1988+ (1,5,9) – ISSN: 0740-3119 – mf#16475 – us UMI ProQuest [650]
Catalog der hebraeischen bibelhandschriften der kaiserlichen oeffentlichen bibliothek in st petersburg / Harkavy, Albert & Strack, Hermann Leberecht – St Petersburg: C Ricker; Leipzig: J C Hinrichs 1875 [mf ed 1989] – 1mf – 9 – 0-7905-2780-4 – mf#1987-2780 – us ATLA [090]
Catalog of books – Allahabad, Supt, Govt Press, United Provinces, mar 1923; sep 1924 – us CRL [020]
Catalog of copyright entries / U.S. Copyright Office – no. 1-782. 1 11 Jul 1891-28 Jun 1906 – 1 – us L of C Photodup [010]
Catalog of copyright entries / U.S. Copyright Office – Third series. v. 1-21. 1947-67 – (part 5: music. 1. complete. parts 1-13. 1) – us AMS Press [070]
Catalog of copyright entries / U.S. Copyright Office – New series. v. 1-43. 1906-46 – (pt3: musical compositions; complete pts 1-4) – us AMS Press [070]
Catalog of copyright entries / U.S. Copyright Office – v. 1-43 and v. 1-21. Total new series and third series, complete. 1906-67 – 1 – us AMS Press [070]
Catalog of copyright entries: musical compositions / U.S. Copyright Office – 1891-1946 – 1 – us L of C Photodup [780]
Catalog of copyrighted dramas, 1870-1916 / U.S. Copyright Office – Washington, D.C. G.P.O., 1918. 2 v. 2 reels – 1 – us L of C Photodup [070]
A catalog of files and microfilms of the german foreign ministry archives, 1867-1920 / Germany. Foreign Ministry – (= ser National archives coll of foreign records seized, 1941-) – 1r – 1 – mf#T322 – us Nat Archives [943]
A catalog of long island newspapers on microfilm – Mineola, NY: Nassau County Historical Museum, 1970 (mf ed 1984) – (= ser Booklet series (nassau county historical museum)) – 12p – mf#FSN 39,591 – us NY Public [071]
Catalog of publications issued by the government of the united provinces and obtainable from the book depot, government press, allahabad – Allahabad, Supt, Govt Press, United Provinces, jun 1928-sep 1931; jun 1932-sep 1933; jun-sep 1934; jun-sep 1936; jun 1938 – us CRL [350]
A catalog of the descendants of thomas watkins / Watkins, Francis N – 1852 – 1 – $50.00 – us Presbyterian [920]

A catalog of the manuscripts preserved in the library of the university of cambridge / Cambridge University. Library – v. 1-5. 1866-67 – 1 – us L of C Photodup [090]

Catalog of the official publications of the florida agricultural experiment station / University Of Florida Agricultural Experiment Station – Gainesville, FL. 1938 – 1r – us UF Libraries [630]

A catalog of the printed books.. / Middle Temple. London. Library – Glasgow: Maclehose & Co. 3v. 1914 – 26mf – 9 – $39.00 – (alphabetically arranged, with an index of subjects, by c e a bedwell) – mf#LLMC 84-307 – us LLMC [340]

Catalogacion de leyes y disposiciones de trabajo d... / Bauer Paiz, Alfonso – Guatemala, 1965 – 1r – us UF Libraries [972]

Cataloging and classification quarterly / ed by Carter, Ruth C – v1- 1980- – 1, 9 ($175.00 in US $245.00 outside hardcopy subsc) – us Haworth [020]

La catalogne / Catalonia. Comissariat de Propaganda – Barcelona. 1963? Fiche W 783. (Blodgett Collection of Spanish Civil War Pamphlets) – 9 – us Harvard [946]

Catalogne, Gedeon de *see* Manuscript relating to the early history of canada (from the archives of the literary and historical society)

Catalogne, Gerard De *see*
– Haiti a l'heure du tiers-monde
– Haiti devant son destin...
– Nostalgies de san francisco

Catalogne, Gerard de *see* Dialogue entre deux mondes

Catalogne, Gideon de *see* Recueil de ce qui s'est passe en canada au sujet de la guerre

Catalogo / Colegio San Jose – 1938-1939. Villafranca. Badajoz – 1 – sp Bibl Santa Ana [946]

Catalogo / Museo provincial de Bellas artes – Badajoz: Imp. Provincial, 1974 – sp Bibl Santa Ana [060]

Catalogo da exposicao de etnografia angolana / Lisbon Exposicao De Etnografia Angolana, 1946 – Lisboa, Portugal. 1946 – 1r – us UF Libraries [960]

Catalogo das publicacoes do servico de documentaca / Brazil. Ministerio da Educacao e Cultura Servico – Rio de Janeiro, Brazil. 1965 – 1r – us UF Libraries [350]

Catalogo de documentos relativos a las islas filipinas existentes en el archivo de indias de sevilla / Bayle, Constantino – Madrid: Razon y Fe, 1926 – 1 – sp Bibl Santa Ana [954]

I catalogo de la biblioteca circulante a utilizar por los asegurados y beneficiarios internados en este centro sanitario, consta de un total de seiscientos volumenes / Ministerio de Trabajo – Badajoz: Tip. Barrena, 1962 – 1 – sp Bibl Santa Ana [020]

Catalogo de la coleccion gomez-imaz : a large collection of documents relating to the spanish war of independence / Biblioteca Nacional, Madrid – 1808-14 [mf ed Chadwyck-Healey] – 13mf – 9 – (in english) – uk Chadwyck [946]

Catalogo de la exposicion fortografica vida de mat... / Archivo Nacional De Cuba – La Habana, Cuba. 1945 – 1r – us UF Libraries [770]

Catalogo de las labras heraldicas de la ciudad de villanueva de la serena (badajoz) / Cotta y Marquez de Prado, Fernando de et al – Madrid, s.i. 1958 – 1 – sp Bibl Santa Ana [946]

Catalogo de las obras / Davila y Figueroa, Marino – 1898 – 9 – sp Bibl Santa Ana [946]

Catalogo de las pinturas de la cueva de maltravieso / Callejo Serrano, Carlos – Tirada aparte de la Cronica del 11th Congreso Nacional de Arqueologia – 1 – sp Bibl Santa Ana [930]

Catalogo de libros y revistas donados por el gobie / Ciudad Trujillo Universidad De Santo Domingo – Ciudad Trujillo, Dominican Republic. 1948 – 1r – us UF Libraries [040]

Catalogo de los alumnos del colegio de san jose. 1904-1905 / Colegio San Jose – Madrid: Imp. Aurial, 1905 – 1 – (tambien 1905-1906; 1906-1907; 1908-1909; 1910-1911; 1914-1915; 1915-1916; 1909-1910; 1916-1917; 1917-1918; 1918-1919; 1929-1930; 1933-1934) – sp Bibl Santa Ana [240]

Catalogo de los colegiales del insigne : viejo y mayor de santa maria de todos santos, que el illmo dr don francisco rodriguez santos...fundo en mexico a 15 de agosto de 1573 anos... / Arechederreta y Escalada, Juan Bautista de – en Mexico: Por don Mariano Joseph de Zuniga y Ontiveros 1796 – (= ser Books on religion... 1543/44-c1800: colegios religiosos) – 1mf – 9 – mf#crl-363 – ne IDC [241]

Catalogo de los documentos relativo a las islas filipinas existentes en el archivo de indias de sevilla. tomo 7 / Bayle, Constantino – Barcelona, 1932; Madrid: Razon y Fe, 1932. 2v – 1 – sp Bibl Santa Ana [950]

Catalogo de los documentos relativos a las islas filipinas... / Torres y Lanzas, Pedro; ed by Bayle, Constantino – Madrid: Razon y Fe, 1928 – 9 – sp Bibl Santa Ana [959]

Catalogo de los documentos relativos a las islas filipinas... / Torres y Lanzas, Pedro – Madrid: Razon y Fe, 1927 – 1 – sp Bibl Santa Ana [959]

Catalogo de los documentos relativos a las islas filipinas existentes en el archivo de indias de sevilla (1592-1602). barcelona, 1928 / Torres Lanzas, Pedro & Pastells, Pablo – Madrid: Razon y Fe, 1930 – 1 – sp Bibl Santa Ana [959]

Catalogo de los fondos / Archivo Nacional De Cuba – Habana, Cuba. 1944 – 1r – us UF Libraries [972]

Catalogo de los fondos americanos del archivo de protocolos de sevilla... / Bayle, Constantino – Burgos: Razon y Fe, 1939 – 1 – sp Bibl Santa Ana [240]

Catalogo de los fondos cubanos / Archivo General De Indias – Madrid, Spain. 1929- – 1r – us UF Libraries [972]

Catalogo de los libros...caceres – 1871 – 9 – sp Bibl Santa Ana [010]

Catalogo de los materiales codigologicos....de auspach 1966... / Fernandez Caton, Jose Maria – Madrid: Graf. Calleja, 1966 – 1 – sp Bibl Santa Ana [946]

Catalogo de los obispos de cordoba. 1st parte / Gomez Bravo, Juan – Cordoba: Simon Ortega y Leon, 1739 – 1 – sp Bibl Santa Ana [240]

Catalogo de los objetos...exposicion / Diaz Perez, Nicolas – 1883 – 9 – sp Bibl Santa Ana [900]

Catalogo de modelacion impresa 1958 / Minerva Extremena – Badajoz, 1958 – 1 – sp Bibl Santa Ana [020]

Catalogo de obras que existen en esta biblioteca en 30 de marzo de 1952 / Biblioteca Publica Municipal – Trujillo: Imp. Sobrino de Benito Pena, s.a. – 1 – sp Bibl Santa Ana [020]

Catalogo de pasajeros a indias durante los siglos 16, 17 y 18 / Archivo General De Indias – Sevilla, Spain. v1-2. 1940- – 1r – us UF Libraries [972]

Catalogo de pasajeros a indias durante los siglos 16, 17 y 18. vol 2 (1535-1538) / Bayle, Constantino & Bermudez Plata, Cristobal – Madrid: Razon y Fe, 1944 – 1 – sp Bibl Santa Ana [920]

Catalogo de publicaciones periodicas : a comprehensive inventory of all periodicals, spanish and foreign from the national library / Biblioteca Nacional, Madrid – [mf ed Chadwyck-Healey] – 133mf – 9 – (in spanish) – uk Chadwyck [020]

Catalogo de tomos varios (b.n. departamento de manuscritos) / Paz, J – Madrid, 1938 – 9 – sp Cultura [020]

Catalogo de una serie miscelanea procedente del convento de san antonio del prado y colegios jesuiticos... / Hernandez Andres, J M – Madrid. v2. 1967-1968 – 1 – sp Bibl Santa Ana [240]

Catalogo de varios especiales : 56,000 records, covering the 16th-19th century / Biblioteca Nacional, Madrid – [mf ed Chadwyck-Healey] – 153mf – 9 – (in spanish) – uk Chadwyck [946]

Catalogo dei molluschi raccolti dalla missione italiana in persia / Issel, A – (= ser Mem Reale Accad Sc S 2 Torino) – 2mf – 8 – (mem reale accad sc, s2 torino 1866 v23) – mf#Z-583 – ne IDC [956]

Catalogo del archivo de la casa del sol / Rubio Merino, Pedro – Badajoz: Dip. Provincial, 1979 – 1 – sp Bibl Santa Ana [020]

Catalogo del archivo de la diputacion provincial de teruel / Floriano Cumbreno, Antonio C – Madrid: Tip. de Archivos, 1930 – sp Bibl Santa Ana [020]

Catalogo del archivo de musica de la real capilla de palacio / Garcia, Marcellan, Jose – Madrid: Editorial del Patrimonic Nacional [19–] [mf ed 1980] – 1 – mf#103 – us Wisconsin U Libr [780]

Catalogo del concurso-exposicion de fotografias sobre temas cacerenos / Junta Provincial de Turismo – Caceres, 1957 – sp Bibl Santa Ana [338]

Catalogo del museo de ciencias naturales / Janer, F – 1,279mf – 9 – sp Cultura [500]

Catalogo delle lingue conosciute e notizia della loro affinita, e diversita : opera / Hervas y Panduro, Lorenzo – Cesena: Biasini 1784 – (= ser Whsb) – 3mf – 9 – €40.00 – mf#Hu 001 – gw Fischer [410]

Catalogo exposicion de trofeos de caza mayor 1970 / Jefatura provincial Servicio Pesca Continental, Caza y Parques Nacionales – Caceres: Tip. Extremadura, 1970 – 1 – sp Bibl Santa Ana [946]

Catalogo exposicion international de artesania – Caceres, 4 a 11 Enero, 1953 – sp Bibl Santa Ana [700]

Catalogo formado por...de los principales articulos que componen la selecta libreria de d.j. boehl de faber / Gallardo, Bartolome Jose – Madrid: Ed. Reus, 1922. B.R.A.H. 81. pp. 478-494; y 82, pp. 69-94, 165-190 y 248-267 – 1 – sp Bibl Santa Ana [946]

Catalogo general de la exposicion betico extremena celebrada en el alcazar de sevilla – 1874 – 9 – sp Bibl Santa Ana [900]

Catalogo general de la libreria espanola e hispanoamericana – Anos 1901-1930. 5v. 1932-51 – 1,9 – us AMS Press [010]

Catalogo general de libros impresos, 1982-1987 *see* Author catalogues

Catalogo general de libros impresos, hasta 1981 *see* Author catalogues

Catalogo general de productos fitosanitarios – Badajoz: Agrotecnica Extremena, 1970 – sp Bibl Santa Ana [630]

Catalogo generale dei musei di antichita...regio museo di torino : antichita egizie / Fabretti, A, Rossi, F and Lanzone, R – Roma, Torino, 1882-1888. 2v – 15mf – 9 – mf#NE-397 – ne IDC [956]

Catalogo geral das publicacoes da comissao rondon / Conselho Nacional De Protecao Aos Indios (Brazil) – Rio de Janeiro, Brazil. 1950 – 1r – us UF Libraries [972]

Catalogo manuscrito de la biblioteca de la universidad central de madrid / Villaamil y Castro, J – Madrid, 1878 – 762mf – 9 – sp Cultura [020]

Catalogo monumental de espana : caceres (1914-1916) / Melida, Jose Ramon – Madrid: Razon y Fe, 1926 – 1 – sp Bibl Santa Ana [946]

Catalogo monumental de espana. provincia de badajoz. texto 2 / Melida, Jose Ramon – Madrid: Imp. de la Ciudad Lineal, 1926 – 1 – sp Bibl Santa Ana [946]

Catalogo monumental de espana. provincia de caceres. (1914-1916) / Melida, Jose Ramon – Madrid: Nº de Instruccion Publica y Bellas Artes, Texto 2. 1924 – 1 – sp Bibl Santa Ana [946]

Catalogo monumental de espana. provincia de caceres. texto 1 / Melida, Jose Ramon – Madrid: Imp. de la Ciudad Lineal, 1924 – 1 – sp Bibl Santa Ana [946]

Catalogo... obispos de cordoba / Gomez Bravo, Juan – 1739 – 9 – sp Bibl Santa Ana [240]

Catalogo oficial ilustrado de la exposicion de las obras de francisco de zurbaran / Viniegra, Salvador – Madrid: J. Lacoste, 1905 – 1 – sp Bibl Santa Ana [946]

Catalogo razonado de las leyes de guatemala / Guatemala Laws, Statutes, Etc (Indexes) – Guatemala, 1945 – 1r – us UF Libraries [972]

Catalogo razonado de obras anonimas y sendonimas de autores de la compania de jesus / Uriarte, Jose E – Madrid. 1904-16. 5v – 1 – us L of C Photodup [010]

Catalogo razonado y critico...extremadura / Barrantes Moreno, Vicente – 1865 – 9 – sp Bibl Santa Ana [946]

Catalogo y guia de la riqueza de extremadura – 1 – sp Bibl Santa Ana [946]

Catalogo y guia de la riqueza de extremadura – 1 – (dibujos) – sp Bibl Santa Ana [946]

Catalogo...biblioteca instituto...caceres / Lopez Sanchez, Eulogio – 1871 – 9 – sp Bibl Santa Ana [020]

Catalog...of gammon theological seminary / Gammon Theological Seminary – Atlanta GA: [s.n.] 1919-21 [mf ed 2006] – v27-28 (1919-1921) [complete] on 1r – 1 – (iss for jul 1921 also called: quadrennial – alumni ed; iss for jul 1919 & 1920 both called v27 n1; cont: gammon theological seminary. quarterly bulletin; cont by: gammon theological seminary. annual catalog) – mf#2006-s009 – us ATLA [242]

Catalogo-guia 1950 / Mahizflor. Museo Taurino. Aceuchal – Badajoz: Tip. Arqueros, 1950 – 1 – sp Bibl Santa Ana [020]

Catalogo...informaciones genealogicas de los pretendientes a cargos de santo oficio – Valladolid, 1928 – 9 – sp Cultura [920]

Catalogs / American Baptist Theological Seminary. Nashville, Tennessee – 1980-85 – 1 – $7.70 – us Southern Baptist [242]

Catalogs / Bluefield College. Virginia – 1922-55 – 1 – $109.06 – us Southern Baptist [242]

Catalogs / Florida Memorial College – 1910-80 – 1 – $47.80 – us ABHS [378]

Catalogs and college records / Baylor University College of Medicine. Houston, Texas – 1900-57 – 1 – $127.54 – us Southern Baptist [610]

Catalogs and college records / Blue Mountain College. Blue Mountain, Mississippi – 1873-Apr 1955 – 1 – $202.58 – us Southern Baptist [242]

Catalogs and college records / Brandon. Vermont. Brandon Seminary – 1832-1866 – 1 – $5.00 – us Southern Baptist [242]

Catalogs and college records / Carver School of Missions and Social Work (Woman's Missionary Union Training School for Christian Workers). Louisville, Ky – 1908-55 – 1 – $91.28 – us Southern Baptist [242]

Catalogs and college records / Clear Creek Baptist School. Pineville, Kentucky – 1936-82 – 1 – $28.80 – us Southern Baptist [378]

Catalogs and college records / Decatur Baptist College. Decatur, Texas – 1899-1906 – 1 – $32.76 – us Southern Baptist [242]

Catalogs and college records / Grand Canyon College. Phoenix, Arizona – 1949-55 – 1 – $21.35 – us Southern Baptist [242]

Catalogs and college records / Greenville Woman's College. South Carolina – 1857-1937 – 1 – $173.74 – us Southern Baptist [242]

Catalogs and college records / Murfreesboro. North Carolina. Chowan College – 1886-1916 – 1 – $89.04 – us Southern Baptist [242]

Catalogs and college records / New Orleans Baptist Theological Seminary. (Formerly: Baptist Bible Institute). New Orleans, Louisiana – 1918-54 – 1 – $110.95 – us Southern Baptist [242]

Catalogs and college records / Norman College. Norman Park, Georgia – 1903-46 – 1 – $79.94 – (lacking 1904-06) – us Southern Baptist [242]

Catalogs and college records / Roger Williams University. Nashville, Tenn – 1881-93 – 1 – $12.95 – us Southern Baptist [378]

Catalogs and college records / Southern Baptist College. Walnut Ridge, Ark – 1942-54 – 1 – $30.73 – us Southern Baptist [242]

Catalogs and college records / Southern Baptist Theological Seminary – 1859-89 – 1 – $23.80 – (also: history of the establishment and organization of the southern baptist theological seminary, greenville, s.c., 1860) – us Southern Baptist [242]

Catalogs and college records / Stetson University. DeLand, Florida – 1885-1953 – 1 – $466.06 – us Southern Baptist [242]

The catalogs of the art exhibition catalog collection *see* Art exhibition catalogs subject index, 1977-1990

Catalogs of vocal music / Ditson, Oliver – Boston. 1879, 1882, 1898, 1900, 1901, 1903-04, 1913, 1915, 1924 – 1 – $23.00 – us L of C Photodup [780]

Catalogue : grand encan de livres francais et anglais par oct lemieux et cie d'une partie de la magnifique bibliotheque de son excellence le comte de premio-real, consul general d'espagne... / Oct Lemieux et cie – [S.l: s.n, 1883?] [mf ed 1983] – 1mf – 9 – mf#35541 – cn Canadiana [020]

Catalogue / Montgomery Ward – [Baltimore, MD] : Montgomery Ward, 1935/36 (mf ed 1982) – 6r – 1 – mf#MF M766s – us Northeast [978]

Catalogue : poetry collection, lockwood memorial library – Buffalo: State University of New York, [1972?] – us CRL [810]

Catalogue : vente a l'encan de la bibliotheque de feu l'hon elzear gerin-lajoie...comprenant pres de 3,000 volumes – Quebec?: s.n, 1888? – 1mf – 9 – mf#03333 – cn Canadiana [030]

Catalogue : vente a l'encan par mm oct lemieux & cie de la bibliotheque de m p le may, comprenant plus de 1,000 volumes...vendredi, le 27 sep 1889... / Oct Lemieux & Cie – Quebec: C Darveau, 1889 [mf ed 1994] – 1mf – 9 – 0-665-94644-9 – mf#94644 – cn Canadiana [010]

Catalogue : vente a l'encan par oct lemieux et cie de la bibliotheque de feu son excellence le comte de premio-real comprenant pres de 2,000 volumes, droit, litterature... – [Quebec?: s.n.], 1888 [mf ed 1986] – 1mf – 9 – 0-665-54398-0 – mf#54398 – cn Canadiana [020]

Catalogue. / American Law Association – Chicago?. 1893. 3-24 numb. LL-923 – 1 – us L of C Photodup [340]

[Catalogue] : 10th season. re-organized 1865 / Society of Female Artists, afterwards Society of Lady Artists – London 1866 – (= ser 19th c art & architecture) – 1mf – 9 – mf#4.2.471 – uk Chadwyck [700]

[Catalogue] : 11th season. re-organized, jan 1865 / Society of Female Artists, afterwards Society of Lady Artists – London 1867 – (= ser 19th c art & architecture) – 1mf – 9 – mf#4.2.472 – uk Chadwyck [700]

[Catalogue] : 12th season. re-organized jan 1865 / Society of Female Artists, afterwards Society of Lady Artists – London 1868 – (= ser 19th c art & architecture) – 1mf – 9 – mf#4.2.473 – uk Chadwyck [700]

[Catalogue] : 13th season. re-organized jan 1865 / Society of Female Artists, afterwards Society of Lady Artists – London 1869 – (= ser 19th c art & architecture) – 1mf – 9 – mf#4.2.474 – uk Chadwyck [700]

[Catalogue] : 14th season. re-organized, jan 1865 / Society of Female Artists, afterwards Society of Lady Artists – London 1870 – (= ser 19th c art & architecture) – 1mf – 9 – mf#4.2.475 – uk Chadwyck [700]

[Catalogue] : 1st exhibition 1857 / Society of Female Artists, afterwards Society of Lady Artists – London [1857] – (= ser 19th c art & architecture) – 1mf – 9 – mf#4.2.462 – uk Chadwyck [700]

[Catalogue] : 2nd exhibition 1858 / Society of Female Artists, afterwards Society of Lady Artists – London 1858 – (= ser 19th c art & architecture) – 1mf – 9 – mf#4.2.463 – uk Chadwyck [700]

[Catalogue] : 3rd exhibition 1859 / Society of Female Artists, afterwards Society of Lady Artists – London 1859 – (= ser 19th c art & architecture) – 1mf – 9 – mf#4.2.464 – uk Chadwyck [700]

[Catalogue] : 4th exhibition 1860 / Society of Female Artists, afterwards Society of Lady Artists – London 1860 – (= ser 19th c art & architecture) – 1mf – 9 – mf#4.2.465 – uk Chadwyck [700]

[Catalogue] : 5th exhibition 1861 / Society of Female Artists, afterwards Society of Lady Artists – London 1861 – (= ser 19th c art & architecture) – 1mf – 9 – mf#4.2.466 – uk Chadwyck [700]

[Catalogue] : 6th exhibition 1862 / Society of Female Artists, afterwards Society of Lady Artists – London 1862 – (= ser 19th c art & architecture) – 1mf – 9 – mf#4.2.467 – uk Chadwyck [700]

[Catalogue] : 7th exhibition 1863 / Society of Female Artists, afterwards Society of Lady Artists – London 1863 – (= ser 19th c art & architecture) – 1mf – 9 – mf#4.2.468 – uk Chadwyck [700]

[Catalogue] : 8th exhibition 1864 / Society of Female Artists, afterwards Society of Lady Artists – London 1864 – (= ser 19th c art & architecture) – 1mf – 9 – mf#4.2.469 – uk Chadwyck [700]

[Catalogue] : re-organized, jan 1865 / Society of Female Artists, afterwards Society of Lady Artists – London 1865, 1871, 1872 – (= ser 19th c art & architecture) – 3mf – 9 – mf#4.2.470; 4.2.476; 4.2.477 – uk Chadwyck [700]

Catalogue 1887 : exhibition held at the galleries of the art association, phillips square, montreal, open to the public on wednesday, april 20th, at 9 am / Royal Canadian Academy of Arts – S.l: s.n, 1887? – 1mf – 9 – mf#46523 – cn Canadiana [700]

Catalogue 1900 : twenty-first annual exhibition, opened on the 15th february, 1900 in the national gallery, ottawa / Academie royale des arts du Canada – [Toronto?: s.n, 1900?] [mf ed 1984] – 1mf – 9 – 0-665-46707-9 – mf#46707 – cn Canadiana [700]

Catalogue and announcement of the texas deaf and blind institute for colored youths / Texas Deaf and Blind Institute for Colored Youths – Austin TX: Texas Deaf & Blind Institute for Colored Youths [mf ed 2006] – 1919/1920 [complete] on 1r – 1 – (cont by: texas deaf and dumb and blind institute for colored youths. catalogue and announcement of the texas deaf and dumb and blind institute for colored youths [2005-s131] filmed on same reel; reel also incl: the freedman's friend [2005-s132]) – mf#2005-s130 – us ATLA [362]

Catalogue and announcement of the texas deaf and blind institute for colored youths / Texas Deaf and Blind Institute for Colored Youths – Austin TX: Texas Deaf & Blind Institute for Colored Youths 1919/20 [mf ed 2006] – 1r – 1 – (cont by: texas deaf and dumb and blind institute for colored youths. catalogue and announcement of the texas deaf and dumb and blind institute for colored youths; reel incl other titles: 2005-s131 & 2005-s132) – mf#2005-s130 – us ATLA [362]

Catalogue and announcement of the texas deaf and dumb and blind institute for colored youths / Texas Deaf and Blind Institute for Colored Youths – Austin TX: Texas Deaf & Dumb and Blind Institute for Colored Youths 1922/1923-1933/1938 [mf ed 2006] – 1r – 1 – (cont: texas deaf and blind institute for colored youths. catalogue and announcement of the texas deaf and blind institute for colored youths; lacks: 1923/24-1926/27; reel incl other titles: 2005-s130 & 2005-s131) – mf#2005-s131 – us ATLA [362]

Catalogue and announcement of the texas deaf and dumb and blind institute for colored youths / Texas Deaf and Blind Institute for Colored Youths – Austin TX: Texas Deaf & Blind Institute for Colored Youths [mf ed 2006] – 1922/23-1933/38 – 1 – (lacks: 1923/24-1926/27]; cont: texas deaf and blind institute for colored youths. catalogue and announcement of the texas deaf and blind institute for colored youths [2005-s130] filmed on same reel; reel also incl: the freedman's friend [2005-s132]) – mf#2005-s131 – us ATLA [362]

Catalogue and index – Enfield. 1975-1996 (1) 1976-1996 (5) 1976-1996 (9) – ISSN: 0008-7629 – mf#10181 – us UMI ProQuest [020]

Catalogue and price list of armstrong patent tool holders for turning, planing and boring metals : over 50,000 in use: for sale by aikenhead hardware co, 6 adelaide st, east, toronto, ont, canada / Aikenhead Hardware Co – [Toronto?: s.n, 1895?] [mf ed 1984] – 1mf – 9 – mf#02451 – cn Canadiana [680]

Catalogue de la bibliotheque de la legislature de la province de quebec / Bouchard, T-D [comp] – Quebec: R Paradis, Impr du roi. 2v. 1932-1933 [mf ed 1993] – 6mf – 9 – mf#SEM105P1986 – cn Bibl Nat [020]

Catalogue de la bibliotheque de l'apostolat des bons livres / Apostolat des bons livres. Bibliotheque – Quebec: Typ Laflamme & Proulx, 1910 [mf ed 1992] – 2mf – 9 – (with ind) – mf#SEM105P1608 – cn Bibl Nat [020]

Catalogue de la bibliotheque de l'oeuvre des bons livres, erigee a montreal / Bibliotheque paroissiale de Notre-Dame – Montreal: Impr de Louis Perrault, 1845 [mf ed 1992] – 1mf – 9 – mf#SEM105P1577 – cn Bibl Nat [020]

Catalogue de la bibliotheque historique et scientifique de feu m le docteur j court / Court, Juergen – Paris: C Leclerc, 1884 [mf ed 1980] – 3mf – 9 – 0-665-03599-3 – mf#03599 – cn Canadiana [016]

Catalogue de la librairie de j b rolland et fils a montreal division du catalogue : histoire, litterature, theologie, etc etc / J B Rolland & fils – Montreal: J B Rolland et fils, libraires, [1878?] (mf ed 2001) – 9 – cn Bibl Nat [020]

Catalogue de la precieuse bibliotheque de feu m le docteur j court : comprenant une collection unique de voyageurs et d'historiens relatifs a l'amerique. / Court, Juergen – Paris: C Leclerc, 1884 [mf ed 1980] – 2mf – 9 – 0-665-03598-5 – mf#03598 – cn Canadiana [016]

Catalogue de l'herbier de syrie / Puel, T & Gaillardot, C – Paris, [1854] – 1mf – 8 – mf#1047 – ne IDC [956]

Catalogue de l'histoire de france / Bibliotheque Nationale. France – [mf ed Chadwyck-Healey] – 1500mf – 9 – (with p/g & ind. in french. catalogue is the largest subject catalogue of the library containing all works on the history of france to the end of 1987) – uk Chadwyck [944]

Catalogue de l'histoire de l'afrique / Bibliotheque National. France. Dept des Imprimes – Paris, 1895 – 1 – us CRL [960]

Catalogue de manuscrits arabes chretiens con-serves au caire / Graf, Georg – Citta del Vaticano, 1934 – 9mf – 8 – €18.00 – ne Slangenburg [240]

Catalogue des francs-macons suisses 1910-1911 : Deuxieme partie, comprenant toutes les loges sauf celles de Geneve / ed by Vogt, William – 3mf – 9 – (Geneve: [s.n. 1912]) – mf#vrl-208 – ne IDC [366]

Catalogue des gentilshommes de normandie... / La Roque, Louis de & Barthelemy, Edouard de – Paris: E Dentu...: Aug Aubry...1864 [mf ed 1982] – 2mf – 9 – mf#SEM105P85 – cn Bibl Nat [929]

Catalogue des livres appartenant a feu j a n provencher, ecr : ...vendredi, le 30 mai 1890... / Marcotte et Ecrement (Firme) – [Montreal?: s.n, 1890?] [mf ed 1994] – 1mf – 9 – 0-665-94657-0 – mf#94657 – cn Canadiana [010]

Catalogue des livres appartenant a la bibliotheque de la chambre d'assemblee = Catalogue of books in the library of the house of assembly / Bas-Canada. Parlement. Chambre d'Assemblee. Bibliotheque – Quebec: Impr par Frechette & Cie, 1835 [mf ed 1982] – 1mf – 9 – mf#SEM105P132 – cn Bibl Nat [020]

Catalogue des livres composant la bibliotheque de feu m. le baron james de rothschild / Rothschild, Nathan J E – 5v. 1884-1920 – 1,9 – us AMS Press [010]

Catalogue des livres orientaux et autres composant la bibliotheque de feu m. garcin de tassy : suivi du catalogue des manuscrits hindoustanis, persans, arabes, turcs 111=deloncle, m f – Paris: A Labitte, 1879 – us CRL [950]

Catalogue des manuscrits ethiopiens de la collection antoine d'abbadie / Chaine, M – Paris, 1921 – 2mf – 9 – mf#NE-20268 – ne IDC [960]

Catalogue des mineraux, roches et fossiles du canada : avec notes descriptives et explicatives / Harrington, Bernard James & Selwyn, Alfred Richard Cecil – Londres: G E Eyre & W Spottiswoode, 1878 – 2mf – 9 – (trans by paul de cazes) – mf#55197 – cn Canadiana [550]

Catalogue des plantes canadiennes : contenues dans l'herbier de l'universite laval et recueillies pendant les annees 1858-65 / Brunet, Ovide – Quebec?: C Darveau, 1865 – 1mf – 9 – mf#15588 – cn Canadiana [580]

Catalogue des plantes du maroc (spermatophytes et pteridophytes / Jahandiez, E – Alger, 1931-1941. 4v – 16mf – 9 – mf#9392 – ne IDC [956]

Catalogue des vegetaux ligneux du canada : pour servir a l'intelligence des collections de bois economiques envoyees a l'exposition universelle de paris, 1867 – Quebec: C Darveau, 1867 – 1mf – 9 – (incl ind) – mf#23464 – cn Canadiana [634]

Catalogue d'une bibliotheque canadienne : ouvrages sur l'amerique et en particulier sur le canada / Dunn, Oscar – [Quebec: s.n.], 1880 [mf ed 1980] – 1mf – 9 – mf#05951 – cn Canadiana [019]

Catalogue d'une bibliotheque canadienne : ouvrages sur l'amerique et en particulier sur le canada, droit, litterature, science, poesies, etc, etc – [Quebec?: s.n.], 1885 [mf ed 1980] – 1mf – 9 – mf#07107 – cn Canadiana [020]

Catalogue general de la librairie garneau / Librairie Garneau – Quebec: J-P Garneau, [1914?] [mf ed 1995] – 3mf – 9 – 0-665-75116-8 – mf#75116 – cn Canadiana [017]

Catalogue general des antiquites egyptiennes du musee de caire : fayencegefaesse / Bissing, F W von – Wien, 1902 – 3mf – 9 – mf#NE-20415 – ne IDC [930]

Catalogue general des antiquites egyptiennes du musee de caire : metallgefaesse / Bissing, F W von – Wien, 1901 – 2mf – 9 – mf#NE-20414 – ne IDC [930]

Catalogue general des antiquites egyptiennes du musee de caire : steingefaesse / Bissing, F W von – Wien, 1907 – 5mf – 9 – mf#NE-20416 – ne IDC [930]

Catalogue general des antiquites egyptiennes du musee de caire : tongefaesse / Bissing, F W von – Wien, 1913 – 2mf – 9 – mf#NE-20417 – ne IDC [930]

Catalogue general des monuments d'abydos decouverts pendant les fouilles de cette ville / Mariette, A – Paris, 1880 – 11mf – 9 – mf#NE-362 – ne IDC [956]

Catalogue general des periodiques des origines a 1959 / Bibliotheque Nationale. France – [mf ed Chadwyck-Healey] – 1344mf – 9 – (in french. incl addendum [44mf]. with p/g. periodicals coll in the library is one of the most diverse & wide-ranging in the world) – uk Chadwyck [073]

Catalogue no 1 des volumes a etre vendus a l'encan par marcotte et ecrement, jeudi, vendredi et samedi, 17, 18, et 19 janvier 1889... : 9,000 volumes, art, litterature, science medecine... / Marcotte et Ecrement (Firme) – [Montreal?]: Marcotte & Ecrement, [1889?] [mf ed 1994] – 1mf – 9 – 0-665-94647-3 – mf#94647 – cn Canadiana [010]

Catalogue no 2 comprenant partie des 10,000 volumes devant etre vendus a l'enchere : ...montreal...jeudi, vendredi et samedi, 17, 18, 19 janv 1889... / Marcotte et Ecrement (Firme) – [Montreal?: s.n, 1889?] [mf ed 1994] – 1mf – 9 – 0-665-94659-7 – mf#94659 – cn Canadiana [010]

Catalogue of 340 specimens from the collection of the historical and scientific society, winnipeg : comprising geology, mineralogy, ethnology, and history of the canadian northwest...dominion and centennial exhibition held at st john, new brunswick, october 1883 / Manitoba Historical and Scientific Society – Winnipeg: s.n, 1883 – 1mf – 9 – mf#17362 – cn Canadiana [550]

Catalogue of a collection of privately printed books / Dobell, Bertram – London, 1891-93 – (= ser 19th c publishing...) – 3mf – 9 – mf#3.1.9 – uk Chadwyck [070]

Catalogue of a large and valuable collection of english and french books : belonging to a private gentleman, among which are to be found a considerable number of scarce and rare books relating to the early history of america... – S.l: Mercury Office, 1860? – 1mf – 9 – mf#47198 – cn Canadiana [000]

A catalogue of a magnificent and superlatively elegant assemblage of parisian furniture – [London?] 1816 – (= ser 19th c art & architecture) – 1mf – 9 – mf#4.2.1009 – uk Chadwyck [740]

A catalogue of a miscellaneous collection of second hand, french, italian, german, latin and greek books, dictionaries, classical translations, and mathematical works : on sale at prices affixed at philip naughten's book store, 178 rideau street, ottawa, canada – [Ottawa?: s.n, between 1875 and 1879] (Ottawa: J C Wilson) – 1mf – 9 – 0-665-90790-7 – mf#90790 – cn Canadiana [070]

Catalogue of a part of the library of the late professor joseph henry thayer : of harvard university – [s.l: s.n] 1902 [mf ed 1989] – 1mf – 9 – 0-7905-2745-6 – (in english, german, french, latin & greek) – mf#1987-2745 – us ATLA [012]

Catalogue of a selection of the works of sir joshua reynolds / British Institution for Promoting the Fine Arts in the United Kingdom, London – London 1833 – (= ser 19th c art & architecture) – 1mf – 9 – mf#4.2.650 – uk Chadwyck [750]

Catalogue of a special collection of works by david cox : with descriptive notes and illustrations / Birmingham. Museum and Art Gallery – Birmingham 1890 – (= ser 19th c art & architecture) – 1mf – 9 – mf#4.2.1411 – uk Chadwyck [700]

Catalogue of a valuable assemblage of paintings : including some historical portraits... / Hodgson London – London, 1856 – (= ser 19th c visual arts & architecture) – 1mf – 9 – mf#4.1.119 – uk Chadwyck [750]

Catalogue of an exhibition of bibles : in commemoration of the tercentenary of the authorized version, 1611-1911 – provisional issue, under revision. Glasgow: J Maclehose, 1911 – 1mf – 9 – 0-7905-3151-8 – mf#1987-3151 – us ATLA [220]

Catalogue of an exhibition of books, portraits, and facsimiles illustrating the history of the english translation of the bible : in commemoration of the tercentenary anniversary of the king james version, 1611, at the yale university library, new haven, connecticut, april, 1911 – [s.l: s.n, 1911?] [mf ed 1990] – 1mf – 9 – 0-7905-3505-X – mf#1987-3505 – us ATLA [220]

Catalogue of an exhibition of manuscript and printed copies of the scriptures : illustrating the history of the transmission of the bible... mar to dec 1911 / John Rylands Library Bible Tercentenary Exhibition – Manchester: University Press 1911 [mf ed 1989] – 1mf – 9 – 0-7905-2119-9 – mf#1987-2119 – us ATLA [220]

Catalogue of an exhibition of portraits by charles willson and... / Pennsylvania Academy Of Fine Arts – Philadelphia, PA. 1923 – 1r – us UF Libraries [720]

Catalogue of ancient persian bronzes in the ashmolean museum, oxford / Moorey, P S – 5mf – 9 – mf#87383 – uk Microform Academic [060]

Catalogue of arabic manuscripts from ghana and adjacent territories from institute of african studies : university of ghana, legon / Wilks, Ivor & Ferguson, Phyllis – Chicago, IL: Uni of Chicago, Photodup Dept, 1974 – 1 – us CRL [010]

Catalogue of architectural design for churches and parsonages 1889-1890 – 1r – 1 – $35.00 – mf#um-311 – us Commission [720]

Catalogue of articles shewn at the provincial exhibition : held at montreal and inaugurated by his royal highness the prince of wales, august, 1860 – Montreal: printed by M Longmoore & co...1860 [mf ed 1983] – 1mf – 9 – mf#SEM105P365 – cn Bibl Nat [060]

Catalogue of bihar and orissa government publications – Patna, Govt Print, 1929-1936, 1939 – us CRL [324]

Catalogue of books and an abridgement of the constitution and rules... : with the names of the officers and members / Farmers' Institute and Subscription Library Society – [Toronto?: s.n.] 1840 [mf ed 1987] – 1mf – 9 – 0-665-59309-0 – mf#59309 – cn Canadiana [016]

A catalogue of books and manuscripts : presented to the wesleyan theological institution in the year 1859 by james heald, esq / [Jackson, Thomas] – [s.l: s.n, 1859?] (London: James Nichols) [mf ed 1993] – 1mf – 9 – 0-524-08701-6 – mf#1993-3226 – us ATLA [242]

Catalogue of books contained in the library of the american bible society : embracing editions of the holy scriptures in various languages and other biblical and miscellaneous works – New York: American Bible Society's Press, 1863 [mf ed 1992] – 1mf – 9 – 0-524-05204-2 – mf#1992-0337 – us ATLA [012]

Catalogue of books, imported from London : and for sale at j neilson's shop, no 3, mountain street, quebec / Neilson, John – [Quebec?: s.n.] 1811 [mf ed 1994] – 1mf – 9 – 0-665-94697-X – mf#94697 – cn Canadiana [010]

Catalogue of books, in every department of oriental literature : including the philology, religion, history, etc of eastern nations: the holy scriptures in hebrew, and in the various oriental versions...together with a collection of oriental manuscripts – London: Howell & Stewart 1826 – (= ser Whsb) – 3mf – 9 – €40.00 – mf#Hu 473 – gw Fischer [470]

Catalogue of books in the library of parliament / Canada (Province). Parlement. Bibliotheque – Quebec: printed at John Lovell's steam printing establishment, 1852 [mf ed 1983] – 2mf – 9 – mf#SEM105P223 – cn Bibl Nat [020]

Catalogue of books in the library of the british museum printed in england, scotland, and ireland to the year 1640 / British Museum. Dept of Printed Books – 3v. 1884 – 1,9 – us AMS Press [010]

Catalogue of books in the library of the house of assembly = Catalogue des livres appartenant a la bibliotheque de la chambre d'assemblee / Bas-Canada. Parlement. Chambre d'Assemblee. Bibliotheque – Quebec: printed by Neilson & Cowan, 1831 [mf ed 1982] – 1mf – 9 – mf#SEM105P147 – cn Bibl Nat [020]

Catalogue of books in the library of the legislative assembly of canada : printed by order of the legislative assembly / Canada (Province). Parlement. Assemblee legislative. Bibliotheque – Kingston: Desbarats & Cary, 1842 [mf ed 1982] – 1mf – 9 – mf#SEM105P144 – cn Bibl Nat [020]

Catalogue of books in the library of the mechanics' institute, of montreal : with the rules of the library and reading room / Mechanics' Institute of Montreal. Library – [Montreal?: s.n.] 1884 [mf ed 1984] – 3mf – 9 – 0-665-01435-X – (incl ind) – mf#01435 – cn Canadiana [020]

Catalogue of books in the piot collection / Victoria and Albert Museum – London. 2v – 10mf – 9 – mf#O-1089 – ne IDC [720]

Catalogue of books in the pwd secretariat library of the government of bombay – [3rd corr ed]. Bombay, Govt Central Press, 1st nov 1927 – us CRL [020]

Catalogue of books in the secretariat libary of the government of bombay 150=bombay, govt central press, 31st jul 1938 – us CRL [020]

A catalogue of books on art and architecture in mcgill university library and the gordon home blackader library of architecture / McGill University. Library – 2nd rev ed. Montreal: McGill University Library, 1926 [mf ed 1991] – 3mf – 9 – (incl ind; pref by gerhard richard lomer) – mf#SEM105P1490 – cn Bibl Nat [700]

A catalogue of books on history, biography, topography, heraldry and family history, old poetry, and the drama, philology, bibliography, fine arts, divinity, etc etc : published or sold by john russell smith, 36, soho square, london – London?: s.n, 1865? – 1mf – 9 – mf#47180 – cn Canadiana [000]

Catalogue of books printed in the 15th century now in the british museum / British Museum. Dept of Printed Books – 7v. 1912-49 – 1,9 – us AMS Press [010]

A catalogue of books relating to the discovery and early history of north and south america / Church, Elihu Dwight – 5v. 1907 – 1,9 – us AMS Press [019]

Catalogue of books relating to the history of america : forming part of the library of the legislative assembly of canada / Canada (Province). Parlement. Assemblee legislative. Bibliotheque – Quebec: William Cowan & Son, 1845 [mf ed 1983] – 1mf – 9 – mf#SEM105P218 – cn Bibl Nat [970]

[Catalogue of carlton house palace. 1826] / British Institution for Promoting the Fine Arts in the United Kingdom, London – London 1826 – (= ser 19th c art & architecture) – 1mf – 9 – mf#4.2.643 – uk Chadwyck [720]

[Catalogue of carlton house palace. 1827] / British Institution for Promoting the Fine Arts in the United Kingdom, London – London 1827 – (= ser 19th c art & architecture) – 1mf – 9 – mf#4.2.644 – uk Chadwyck [720]

Catalogue of casts for sale / Brucciani, Dominico – London [1870?] – (= ser 19th c art & architecture) – 1mf – 9 – mf#4.2.285 – uk Chadwyck [730]

Catalogue of chinese objects in the south kensington museum / South Kensington Museum, London – London 1872 – (= ser 19th c art & architecture) – 1mf – 9 – mf#4.1.344 – uk Chadwyck [700]

Catalogue of civil publications relating to agriculture, forestry, civic, commerce, finance, legislation, industry, public health, railways, science, trade, etc – Delhi: Govt Publ Branch, [1925-mar 1970] – us CRL [350]

Catalogue of compact discs to december 1993 – 9 – NZ$22.50 – nz Nat Libr [020]

[Catalogue of designs for ecclesiastical metalwork etc] / Hardman and Co, John – Birmingham 1879] – (= ser 19th c art & architecture) – 3mf – 9 – mf#4.2.145 – uk Chadwyck [730]

Catalogue of early indian imprints / William Carey College. Library – 1714-1850. 1,296p – 1 – us Southern Baptist [242]

A catalogue of english and foreign theology, sermons, discourses and lectures – Manchester: James & Joseph Thomson, 1852 [mf ed 1993] – 1mf – 9 – 0-524-08539-0 – mf#1993-2064 – us ATLA [012]

Catalogue of english and french books in the quebec library : at the bishop's palace, where the rules may be seen / Quebec Library – Quebec: Printed at the New Printing Office, 1808 [mf ed 1971] – 1r – 5 – mf#SEM16P13 – cn Bibl Nat [020]

Catalogue of english and french books in the quebec library / Quebec Library – Quebec: S Neilson, 1792 [mf ed 1971] – 1r – 5 – mf#SEM16P11 – cn Bibl Nat [020]

Catalogue of english and french books in the quebec library at the bishop's palace : where the rules may be seen... / Quebec Library – Quebec: Printed at the New Printing Office, 1801 [mf ed 1971] – 1r – 5 – mf#SEM16P12 – cn Bibl Nat [020]

A catalogue of english coins in the british museum / British Museum, London. Dept of Coins and Medals – London 1887 – (= ser 19th c art & architecture) – 13mf – 9 – mf#4.1.409 – uk Chadwyck [730]

Catalogue of english garden and flower seeds for sale by p robert inches : druggist and apothecary... – [S.l: s.n, 18–?] [mf ed 1986] – 1mf – 9 – 0-665-53053-6 – mf#53053 – cn Canadiana [635]

Catalogue of field, garden and flower seeds, fruit and ornamental trees, shrubs, roses, etc for sale by h mitchell : senior partner of the late firm of mitchell and johnston, grower, importer and dealer in seeds... – [Victoria, BC?: s.n.], 1878 [mf ed 1981] – 1mf – 9 – mf#15653 – cn Canadiana [635]

The catalogue of first annual loan and sale exhibition of the newspaper artists' association : held at art association gallery, phillips square, june 29th, 1903 / Art Association of Montreal – [Montreal?: s.n, 1903?] (Montreal: J Fortier) – 1mf – 9 – 0-665-74783-7 – mf#74783 – cn Canadiana [700]

Catalogue of first editions for american authors, poets / Leon & Brother – New York, NY. 1885 – 1r – us UF Libraries [025]

Catalogue of french-language medieval manuscripts : in the koninklijke bibliotheek [royal library] of the netherlands and meermanno-westreenianum museum, the hague / Brayer, Edith [comp] – 18mf – 9 – €325.00 – (with p/g; int by anne s korteweg) – mf#mmp102 – ne Moran [090]

Catalogue of freshwater fishes of africa / Boulenger, G A – v1-4. 1909-16 – 4r – 5 – mf#9/85979-80 – uk Microform Academic [590]

Catalogue of fruit and ornamental trees, flowering shrubs and plants, green-house shrubs and plants, bulbous flower roots, american and indigenous trees and plants, etc... : cultivated and for sale at guilbaut's botanic garden, coteau-baron, st lawrence street, montreal – [Montreal?: s.n.], 1834 [mf ed 1984] – 1mf – 9 – 0-665-47494-6 – mf#47494 – cn Canadiana [635]

Catalogue of fruit and ornamental trees, flowering shrubs, garden seeds and green-house plants, bulbous roots and flower seeds : cultivated and for sale at the toronto nursery, dundas street, near york / Custead, William W – York [Toronto]: printed by W L Mackenzie, 1827 [mf ed 1987] – 1mf – 9 – 0-665-58249-8 – mf#58249 – cn Canadiana [635]

Catalogue of gammon school of theology / Gammon Theological Seminary – Atlanta GA: Clark UP 1888-91 [annual] [mf ed 2006] – 1888-1891 [complete] on 1r – 1 – (no iss publ 1889; with suppls entitled: an address on the occasion of the laying of the corner-stone; dedication of the library building, gammon theological seminary, may 26th 1869 and: address at the dedication of the new library building of gammon theological seminary, atlanta, georgia; cont: gammon theological seminary. circular of the gammon school of theology; cont by: gammon theological seminary. quarterly bulletin; reel also contains 3 monographs: two addresses & a dedication) – mf#2006-s007 – us ATLA [242]

Catalogue of garden, agricultural and flower seeds for sale by james fleming, seedsman and florist, yonge street, toronto : general remarks in arranging material for this catalogue... – [Toronto?: s.n.], 1855 [mf ed 1987] – 1mf – 9 – 0-665-68163-1 – mf#68163 – cn Canadiana [635]

Catalogue of government publications – Madras: Printed by the Director of Stationery and Printing, 1950-1951, jul 1954, jun 1958, jan 1961, jan 1963 – us CRL [350]

Catalogue of illustrations of the artistic supply company / Artistic Supply Co Ltd – London [1895?] – (= ser 19th c art & architecture) – 1mf – 9 – mf#4.2.1255 – uk Chadwyck [740]

Catalogue of law forms : published and for sale by d h doust, r carswell, law bookseller, law stationer, lithographer, etc / R Carswell (Firm) – Toronto: Carswell, 1876 [mf ed 1983] – 1mf – 9 – mf#10545 – cn Canadiana [340]

Catalogue of maps, charts and books furnished by james spencer, publisher, 65 colborne street, toronto on – [s.l: s.n. 1840?] [mf ed 1987] – 1mf – 9 – 0-665-26984-6 – mf#26984 – cn Canadiana [912]

Catalogue of miscellaneous books by auction : at the sale rooms, 361 notre dame st on thursday, 10th march / Alfred Booker (Firm) – Montreal: A Booker?, 1870? – 1mf – 9 – mf#06451 – cn Canadiana [070]

Catalogue of moravian archives : original typescript from fairfield moravian church, manchester – 1mf – 7 – mf#376 – uk Microform Academic [240]

Catalogue of near eastern seals in the ashmolean museum / Buchanan, Briggs – 4mf – 9 – mf#87382 – uk Microform Academic [060]

A catalogue of notable middle temple templars / Hutchinson, John – London: Butterworth, 1902 – 4mf – 9 – $6.00 – mf#LLMC 84-296 – us LLMC [920]

Catalogue of official publications in english and kannada available for sale at the government central book depot, bangalore – Bangalore: Printed by the Supt of the Govt Press, 1936-1968 – us CRL [324]

Catalogue of official reports relating to india issued as english parliamentary papers (and in connection with the india office) during the year 1892 / Campbell, Frank – London: [Truslove & Bray], 1893 – us CRL [324]

A catalogue of official reports upon geological surveys of the united states and territories : and of british north america / Prime, Frederick – Philadelphia?: Sherman, 1879 – 1mf – 9 – mf#24779 – cn Canadiana [550]

A catalogue of old and new books : including many curious and rare works relating to america, canada, etc, all in good order... / Johnston, William – [Toronto?: s.n, 1886?] [mf ed 1995] – 1mf – 9 – 0-665-94789-5 – (in dble clms) – mf#94789 – cn Canadiana [010]

Catalogue of oriental coins in the british museum / Lane-Poole, S – London, 1875-1883. 8 v+suppl 1889-1890 – 68mf – 8 – (missing: 1881 v6) – mf#H-372 – ne IDC [956]

Catalogue of pastors of baptist churches of new hampshire / New Hampshire – 1892-1934. By D. Donovan. Unpublished mss, 1934 – 1 – us ABHS [242]

Catalogue of photographic views of quebec and vicinities : most respectfully presented to the tourist visiting quebec by l p vallee, portrait and landscape photographer... / L P Vallee (Photographer) – [Quebec?: L Brousseau, between 1890 and 1901] – 1mf – 9 – 0-665-94560-4 – mf#94560 – cn Canadiana [770]

Catalogue of pictures and sculpture by members of the canadian art club – [Toronto?: s.n.], c1910 – 1mf – 9 – 0-665-73378-X – mf#73378 – cn Canadiana [700]

Catalogue of pictures at longford castle and categorical list of family portraits / Radnor, Helen Matilda (Chaplin), countess of – 2nd ed. [London] 1898 – (= ser 19th c art & architecture) – 1mf – 9 – mf#4.1.93 – uk Chadwyck [750]

Catalogue of pictures belonging to the earl of ilchester – [London?] 1883 – (= ser 19th c art & architecture) – 3mf – 9 – mf#4.2.354 – uk Chadwyck [700]

[Catalogue of pictures by ancient and modern masters. 1806] / British Institution for Promoting the Fine Arts in the United Kingdom, London – London 1806 – (= ser 19th c art & architecture) – 1mf – 9 – mf#4.2.973 – uk Chadwyck [750]

[Catalogue of pictures by ancient and modern masters. 1807] / British Institution for Promoting the Fine Arts in the United Kingdom, London – London 1807 – (= ser 19th c art & architecture) – 1mf – 9 – mf#4.2.974 – uk Chadwyck [750]

[Catalogue of pictures by ancient and modern masters. 1808] / British Institution for Promoting the Fine Arts in the United Kingdom, London – London 1808 – (= ser 19th c art & architecture) – 1mf – 9 – mf#4.2.975 – uk Chadwyck [750]

[Catalogue of pictures by ancient and modern masters. 1809] / British Institution for Promoting the Fine Arts in the United Kingdom, London – London 1809 – (= ser 19th c art & architecture) – 1mf – 9 – mf#4.2.976 – uk Chadwyck [750]

[Catalogue of pictures by ancient and modern masters. 1810] / British Institution for Promoting the Fine Arts in the United Kingdom, London – London 1810 – (= ser 19th c art & architecture) – 1mf – 9 – mf#4.2.977 – uk Chadwyck [750]

[Catalogue of pictures by ancient and modern masters. 1811] / British Institution for Promoting the Fine Arts in the United Kingdom, London – London 1811 – (= ser 19th c art & architecture) – 1mf – 9 – mf#4.2.978 – uk Chadwyck [750]

[Catalogue of pictures by ancient and modern masters. 1812] / British Institution for Promoting the Fine Arts in the United Kingdom, London – London 1812 – (= ser 19th c art & architecture) – 1mf – 9 – mf#4.2.979 – uk Chadwyck [750]

[Catalogue of pictures by ancient and modern masters. 1813] / British Institution for Promoting the Fine Arts in the United Kingdom, London – London 1813 – (= ser 19th c art & architecture) – 1mf – 9 – mf#4.2.980 – uk Chadwyck [750]

[Catalogue of pictures by ancient and modern masters. 1814] / British Institution for Promoting the Fine Arts in the United Kingdom, London – London 1814 – (= ser 19th c art & architecture) – 1mf – 9 – mf#4.2.981 – uk Chadwyck [750]

[Catalogue of pictures by ancient and modern masters. 1815] / British Institution for Promoting the Fine Arts in the United Kingdom, London – London 1815 – (= ser 19th c art & architecture) – 1mf – 9 – mf#4.2.982 – uk Chadwyck [750]

[Catalogue of pictures by ancient and modern masters. 1816] / British Institution for Promoting the Fine Arts in the United Kingdom, London – London 1816 – (= ser 19th c art & architecture) – 1mf – 9 – mf#4.2.983 – uk Chadwyck [750]

[Catalogue of pictures by ancient and modern masters. 1817] / British Institution for Promoting the Fine Arts in the United Kingdom, London – London 1817 – (= ser 19th c art & architecture) – 1mf – 9 – mf#4.2.984 – uk Chadwyck [750]

[Catalogue of pictures by ancient and modern masters. 1818] / British Institution for Promoting the Fine Arts in the United Kingdom, London – London 1818 – (= ser 19th c art & architecture) – 1mf – 9 – mf#4.2.985 – uk Chadwyck [750]

[Catalogue of pictures by ancient and modern masters. 1819] / British Institution for Promoting the Fine Arts in the United Kingdom, London – London 1819 – (= ser 19th c art & architecture) – 1mf – 9 – mf#4.2.986 – uk Chadwyck [750]

[Catalogue of pictures by ancient and modern masters. 1820] / British Institution for Promoting the Fine Arts in the United Kingdom, London – London 1820 – (= ser 19th c art & architecture) – 1mf – 9 – mf#4.2.987 – uk Chadwyck [750]

[Catalogue of pictures by ancient and modern masters. 1821] / British Institution for Promoting the Fine Arts in the United Kingdom, London – London 1821 – (= ser 19th c art & architecture) – 1mf – 9 – mf#4.2.988 – uk Chadwyck [750]

[Catalogue of pictures by ancient and modern masters. 1822] / British Institution for Promoting the Fine Arts in the United Kingdom, London – London 1822 – (= ser 19th c art & architecture) – 1mf – 9 – mf#4.2.989 – uk Chadwyck [750]

[Catalogue of pictures by ancient and modern masters. 1823] / British Institution for Promoting the Fine Arts in the United Kingdom, London – London 1823 – (= ser 19th c art & architecture) – 1mf – 9 – mf#4.2.990 – uk Chadwyck [750]

[Catalogue of pictures by ancient and modern masters. 1824] / British Institution for Promoting the Fine Arts in the United Kingdom, London – London 1824 – (= ser 19th c art & architecture) – 1mf – 9 – mf#4.2.991 – uk Chadwyck [750]

[Catalogue of pictures by ancient masters. 1825] / British Institution for Promoting the Fine Arts in the United Kingdom, London – London 1825 – (= ser 19th c art & architecture) – 1mf – 9 – mf#4.2.642 – uk Chadwyck [750]

[Catalogue of pictures by ancient masters. 1828] / British Institution for Promoting the Fine Arts in the United Kingdom, London – London 1828 – (= ser 19th c art & architecture) – 1mf – 9 – mf#4.2.645 – uk Chadwyck [750]

[Catalogue of pictures by ancient masters. 1829 jun] / British Institution for Promoting the Fine Arts in the United Kingdom, London – London 1829 – (= ser 19th c art & architecture) – 1mf – 9 – mf#4.2.646 – uk Chadwyck [750]

[Catalogue of pictures by ancient masters. 1831 jun] / British Institution for Promoting the Fine Arts in the United Kingdom, London – London 1831 – (= ser 19th c art & architecture) – 1mf – 9 – mf#4.2.648 – uk Chadwyck [750]

[Catalogue of pictures by ancient masters. 1832 jul] / British Institution for Promoting the Fine Arts in the United Kingdom, London – London 1832 – (= ser 19th c art & architecture) – 1mf – 9 – mf#4.2.649 – uk Chadwyck [750]

[Catalogue of pictures by ancient masters. 1834] / British Institution for Promoting the Fine Arts in the United Kingdom, London – London 1834 – (= ser 19th c art & architecture) – 1mf – 9 – mf#4.2.651 – uk Chadwyck [750]

[Catalogue of pictures by ancient masters. 1835 may] / British Institution for Promoting the Fine Arts in the United Kingdom, London – London 1835 – (= ser 19th c art & architecture) – 1mf – 9 – mf#4.2.652 – uk Chadwyck [750]

[Catalogue of pictures by ancient masters. 1836 may] / British Institution for Promoting the Fine Arts in the United Kingdom, London – London 1836 – (= ser 19th c art & architecture) – 1mf – 9 – mf#4.2.653 – uk Chadwyck [750]

[Catalogue of pictures by ancient masters. 1837 may] / British Institution for Promoting the Fine Arts in the United Kingdom, London – London 1837 – (= ser 19th c art & architecture) – 1mf – 9 – mf#4.2.654 – uk Chadwyck [750]

[Catalogue of pictures by ancient masters. 1838 jun] / British Institution for Promoting the Fine Arts in the United Kingdom, London – London 1838 – (= ser 19th c art & architecture) – 1mf – 9 – mf#4.2.655 – uk Chadwyck [750]

[Catalogue of pictures by ancient masters. 1839 jun] / British Institution for Promoting the Fine Arts in the United Kingdom, London – London 1839 – (= ser 19th c art & architecture) – 1mf – 9 – mf#4.2.656 – uk Chadwyck [750]

[Catalogue of pictures by ancient masters. 1840 jun] / British Institution for Promoting the Fine Arts in the United Kingdom, London – London 1840 – (= ser 19th c art & architecture) – 1mf – 9 – mf#4.2.657 – uk Chadwyck [750]

[Catalogue of pictures by ancient masters. 1841 jun] / British Institution for Promoting the Fine Arts in the United Kingdom, London – London 1841 – (= ser 19th c art & architecture) – 1mf – 9 – mf#4.2.658 – uk Chadwyck [750]

[Catalogue of pictures by ancient masters. 1842 jun] / British Institution for Promoting the Fine Arts in the United Kingdom, London – London 1842 – (= ser 19th c art & architecture) – 1mf – 9 – mf#4.2.659 – uk Chadwyck [750]

[Catalogue of pictures by ancient masters. 1843 jun] / British Institution for Promoting the Fine Arts in the United Kingdom, London – London 1843 – (= ser 19th c art & architecture) – 1mf – 9 – mf#4.2.660 – uk Chadwyck [750]

[Catalogue of pictures by ancient masters. 1844 jun] / British Institution for Promoting the Fine Arts in the United Kingdom, London – London 1844 – (= ser 19th c art & architecture) – 1mf – 9 – mf#4.2.661 – uk Chadwyck [750]

[Catalogue of pictures by ancient masters. 1845 jun] / British Institution for Promoting the Fine Arts in the United Kingdom, London – London 1845 – (= ser 19th c art & architecture) – 1mf – 9 – mf#4.2.662 – uk Chadwyck [750]

[Catalogue of pictures by ancient masters. 1846 jun] / British Institution for Promoting the Fine Arts in the United Kingdom, London – London 1846 – (= ser 19th c art & architecture) – 1mf – 9 – mf#4.2.663 – uk Chadwyck [750]

[Catalogue of pictures by ancient masters. 1847 jun] / British Institution for Promoting the Fine Arts in the United Kingdom, London – London 1847 – (= ser 19th c art & architecture) – 1mf – 9 – mf#4.2.664 – uk Chadwyck [750]

[Catalogue of pictures by ancient masters. 1848 jun] / British Institution for Promoting the Fine Arts in the United Kingdom, London – London 1848 – (= ser 19th c art & architecture) – 1mf – 9 – mf#4.2.665 – uk Chadwyck [750]

[Catalogue of pictures by ancient masters. 1849 jun] / British Institution for Promoting the Fine Arts in the United Kingdom, London – London 1849 – (= ser 19th c art & architecture) – 1mf – 9 – mf#4.2.666 – uk Chadwyck [750]

[Catalogue of pictures by ancient masters. 1850 jun] / British Institution for Promoting the Fine Arts in the United Kingdom, London – London 1850 – (= ser 19th c art & architecture) – 1mf – 9 – mf#4.2.667 – uk Chadwyck [750]

[Catalogue of pictures by ancient masters. 1851 jun] / British Institution for Promoting the Fine Arts in the United Kingdom, London – London 1851 – (= ser 19th c art & architecture) – 1mf – 9 – mf#4.2.668 – uk Chadwyck [750]

[Catalogue of pictures by ancient masters. 1852 jun] / British Institution for Promoting the Fine Arts in the United Kingdom, London – London 1852 – (= ser 19th c art & architecture) – 1mf – 9 – mf#4.2.669 – uk Chadwyck [750]

A catalogue of pictures by british artists...at tabley house / Young, John [comp] – London 1825 – (= ser 19th c art & architecture) – 2mf – 9 – mf#4.2.318 – uk Chadwyck [700]

[Catalogue of pictures by modern masters. 1830] / British Institution for Promoting the Fine Arts in the United Kingdom, London – London 1830 – (= ser 19th c art & architecture) – 1mf – 9 – mf#4.2.627 – uk Chadwyck [750]

[Catalogue of pictures by modern masters. 1832] / British Institution for Promoting the Fine Arts in the United Kingdom, London – London 1832 – (= ser 19th c art & architecture) – 1mf – 9 – mf#4.2.628 – uk Chadwyck [750]

[Catalogue of pictures by modern masters. 1835] / British Institution for Promoting the Fine Arts in the United Kingdom, London – London 1835 – (= ser 19th c art & architecture) – 1mf – 9 – mf#4.2.629 – uk Chadwyck [750]

[Catalogue of pictures by modern masters. 1836] / British Institution for Promoting the Fine Arts in the United Kingdom, London – London 1836 – (= ser 19th c art & architecture) – 1mf – 9 – mf#4.2.630 – uk Chadwyck [750]

[Catalogue of pictures by modern masters. 1837] / British Institution for Promoting the Fine Arts in the United Kingdom, London – London 1837 – (= ser 19th c art & architecture) – 1mf – 9 – mf#4.2.631 – uk Chadwyck [750]

[Catalogue of pictures by modern masters. 1838] / British Institution for Promoting the Fine Arts in the United Kingdom, London – London 1838 – (= ser 19th c art & architecture) – 1mf – 9 – mf#4.2.632 – uk Chadwyck [750]

[Catalogue of pictures by modern masters. 1839] / British Institution for Promoting the Fine Arts in the United Kingdom, London – London 1839 – (= ser 19th c art & architecture) – 1mf – 9 – mf#4.2.633 – uk Chadwyck [750]

[Catalogue of pictures by modern masters. 1841] / British Institution for Promoting the Fine Arts in the United Kingdom, London – London 1841 – (= ser 19th c art & architecture) – 1mf – 9 – mf#4.2.634 – uk Chadwyck [750]

[Catalogue of pictures by modern masters. 1845] / British Institution for Promoting the Fine Arts in the United Kingdom, London – London 1845 – (= ser 19th c art & architecture) – 1mf – 9 – mf#4.2.635 – uk Chadwyck [750]

[Catalogue of pictures by modern masters. 1846] / British Institution for Promoting the Fine Arts in the United Kingdom, London – London 1846 – (= ser 19th c art & architecture) – 1mf – 9 – mf#4.2.636 – uk Chadwyck [750]

[Catalogue of pictures by modern masters. 1847] / British Institution for Promoting the Fine Arts in the United Kingdom, London – London 1847 – (= ser 19th c art & architecture) – 1mf – 9 – mf#4.2.637 – uk Chadwyck [750]

[Catalogue of pictures by modern masters. 1849] / British Institution for Promoting the Fine Arts in the United Kingdom, London – London 1849 – (= ser 19th c art & architecture) – 1mf – 9 – mf#4.2.638 – uk Chadwyck [750]

[Catalogue of pictures by modern masters. 1850] / British Institution for Promoting the Fine Arts in the United Kingdom, London – London 1850 – (= ser 19th c art & architecture) – 1mf – 9 – mf#4.2.639 – uk Chadwyck [750]

[Catalogue of pictures by modern masters. 1851] / British Institution for Promoting the Fine Arts in the United Kingdom, London – London 1851 – (= ser 19th c art & architecture) – 1mf – 9 – mf#4.2.640 – uk Chadwyck [750]

[Catalogue of pictures by modern masters. 1852] / British Institution for Promoting the Fine Arts in the United Kingdom, London – London 1852 – (= ser 19th c art & architecture) – 1mf – 9 – mf#4.2.641 – uk Chadwyck [750]

Catalogue of pictures by the ancient masters : and the works of modern british artists, in the gallery of the northern society for the encouragement of the fine arts / Northern Society for the Encouragement of the Fine Arts, Leeds – Leeds: printed by Hernaman & Perring, 1830 – (= ser 19th c visual arts & architecture) – 1mf – 9 – mf#4.1.8 – uk Chadwyck [700]

A catalogue of pictures, statues, busts...at hendersyde park / Waldie, John – [London], 1859 – (= ser 19th c art & architecture) – 3mf – 9 – mf#4.2.341 – uk Chadwyck [700]

Catalogue of pre-1650 manuscript maps held by county record offices in england and wales / Fowkes, Dudley [comp] – 8mf – 9 – mf#87517 – uk Microform Academic [914]

Catalogue of prints and books illustrating the history of engraving in japan / Burlington Fine Arts Club, London – [London?] 1888 – (= ser 19th c art & architecture) – 2mf – 9 – mf#4.2.1290 – uk Chadwyck [760]

Catalogue of publications – Cuttack, Supt, Orissa Govt Press, 1941, 1943, 1950, 1964, 1967, 1969 – us CRL [350]

Catalogue of publications – Delhi: Ministry of Education, 1956, 1962, 1965 – us CRL [350]

Catalogue of publications – Karachi: Govt Book Depot & Record Office, nov 1939; Suppl: aug 1940, jan 1941, 1942 – us CRL [350]

Catalogue of publications – Patna: Supt, Govt Print, 1939 – us CRL [350]

Catalogue of railroad mortgages / Princeton University. Pliny-Fisk Statistical Library & Bureau of Railway Economics, Washington, DC – Washington, DC: [s.n], 1919-22 (mf ed 19–) – iv/163/40p – (accompanied by: supplement to catalogue of railroad mortgages) – mf#ZV-TPG pv145 n8 – us Harvard [380]

A catalogue of religious, scientific, illustrated, juvenile, and miscellaneous books (including educational works) : constantly kept for sale by john bennett strong, bookseller, stationer and news agent, hollis street, halifax, 1860 / John B Strong (Firm) – [Halifax, NS?: s.n, 1860?] [mf ed 1987] – 1mf – 9 – 0-665-12950-5 – mf#12950 – cn Canadiana [020]

Catalogue of school books stationery, etc, etc / D and J Sadlier and Co – Montreal: The author, [1886] (mf ed 1976) – 1r – 5 – mf#SEM16P268 – cn Bibl Nat [010]

Catalogue of seals in the department of manuscripts in the british museum – London, 1887-1900. 6v – 94mf – 8 – mf#H-1371 – ne IDC [929]

Catalogue of silurian fossils from arisaig, nova scotia / Ami, Henry Marc – S.l: s.n, 1892? – 1mf – 9 – mf#38486 – cn Canadiana [560]

A catalogue of some marbles, bronzes, pictures, and gems, at the hyde, near ingatestone, essex / Disney, John – [London?] 1809 – (= ser 19th c art & architecture) – 2mf – 9 – mf#4.2.1526 – uk Chadwyck [730]

Catalogue of specimens of japanese lacquer and metal work exhibited in 1894 / Burlington Fine Arts Club – London: printed for the Burlington Fine Arts Club, 1894 – (= ser 19th c art & architecture) – 3mf – 9 – mf#4.1.170 – uk Chadwyck [740]

Catalogue of stars observed at the united states naval observatory : during the years 1845 to 1877 / Yarnall, Mordecai [comp] – 2nd rev ed. Washington: GPO 1878 [mf ed 1998] – 1r – 1 – mf#film mas 28421 – us Harvard [520]

A catalogue of the arabic books and manuscripts : in the library of the asiatic society of bengal / Mirza Ashraf 'Ali, Shams-ul-'Ulama – Calcutta, 1899-1904 [mf ed 1969] – 153p on 1r – 1 – mf#3210 – us Wisconsin U Libr [090]

Catalogue of the arabic manuscripts preserved in the university library, ibadan, nigeria / Kensdale, W E N – Ibadan, 1955-1958 – us CRL [090]

Catalogue of the arabic mss in the convent of s catharine on mount sinai / Gibson, Margaret Dunlop [comp] – London: C J Clay 1894 [mf ed 1990] – (= ser Studia sinaitica 3) – 1mf – 9 – 0-8370-1843-9 – (in arabic, greek & english) – mf#1987-6231 – us ATLA [090]

Catalogue of the archiepiscopal manuscripts in lambeth palace library / Todd, H J [comp] – 1r – 1 – mf#710 – uk Microform Academic [090]

Catalogue of the archives of the dean and chapter of canterbury, 1805-1806 / Bunce, C R – (= ser Canterbury cathedral. archives) – 3r – 1 – mf#96833 – uk Microform Academic [242]

Catalogue of the archives of the moravian church, bristol / Blandford, H [comp] – 1r – 1 – mf#4294 – uk Microform Academic [240]

A catalogue of the articles of ornamental art / Great Britain. Department of Practical Art – London [1851] – (= ser 19th c art & architecture) – 2mf – 9 – mf#4.2.908 – uk Chadwyck [740]

A catalogue of the books in the bangor cathedral library : containing upwards of 1400 volumes on various subjects, arranged and edited with annotations / Jones, Charles William Frederick – Bangor: printed & publ by Nixon & Jarvis, Booksellers, 1872 – (= ser 19th c publishing...) – 1mf – 9 – mf#3.1.5 – uk Chadwyck [020]

Catalogue of the books in the library of the law society of upper canada / Law Society of Upper Canada Library; ed by Adam, Graeme Mercer – Toronto: printed by C B Robinson, 1880 [mf ed 1982] – 7mf – 9 – mf#10657 – cn Canadiana [020]

Catalogue of the books in the library of the law society of upper canada : with an index of subjects / Law Society of Upper Canada Library; ed by Adam, Graeme Mercer – Toronto: printed by C B Robinson, 1880 [mf ed 1981] – 5mf – 9 – mf#10656 – cn Canadiana [020]

Catalogue of the books, in the montreal library / Montreal Library – [Montreal?: s.n.] 1824 [mf ed 1984] – 2mf – 9 – 0-665-44281-5 – mf#44281 – cn Canadiana [020]

Catalogue of the bronzes, greek, roman, and etruscan, in the...british museum / Walters, Henry Beauchamp – London 1899 – (= ser 19th c art & architecture) – 6mf – 9 – mf#4.2.1556 – uk Chadwyck [730]

Catalogue of the celebrated collection of...ralph bernal / Christie, Manson and Woods, Ltd, London – [London] 1855 – (= ser 19th c art & architecture) – 6mf – 9 – mf#4.2.390 – uk Chadwyck [740]

Catalogue of the celebrated collection...british india / Christie, Manson and Woods, Ltd, London – London [1857] – (= ser 19th c art & architecture) – 2mf – 9 – mf#4.2.501 – uk Chadwyck [700]

Catalogue of the celebrated fontaine collection / Christie, Manson and Woods, Ltd, London – [London] 1884 – (= ser 19th c art & architecture) – 2mf – 9 – mf#4.2.391 – uk Chadwyck [700]

Catalogue of the chateau ramezay museum – [Montreal?]: Women's Branch of the Numismatic & Antiquarian Society of Montreal, 1898 [mf ed 1980] – 1mf – 9 – 0-665-02123-2 – mf#02123 – cn Canadiana [700]

A catalogue of the chinese translation of the buddhist tripitaka : the sacred canon of the buddhists in china and japan / Nanjio, Bunyiu – Oxford: Clarendon Press, 1883 [mf ed 1992] – 1mf – 9 – 0-524-05346-4 – (incl bibl ref. added in 1930: japanese alphabetical index of nanjio's catalogue of the buddhist tripitaka...ed by daijo tokiwa and unrai ogiwara) – mf#1990-3467 – us ATLA [280]

Catalogue of the choice collection...from blenheim palace / Christie, Manson and Woods, Ltd, London – London [1883] – (= ser 19th c art & architecture) – 2mf – 9 – mf#4.2.392 – uk Chadwyck [700]

Catalogue of the christiansburg industrial institute / Christiansburg Industrial Institute – [Cambria VA?]: Institute Press Print 1897/1998-1907/1908 [mf ed 2005] – 1r – 1 – (lacks: 1899/1900,1901/02,1904/05,1906/07; cont by: christiansburg industrial institute. annual catalogue) – mf#2005-s104 – us ATLA [366]

Catalogue of the civil and mechanical engineering designs : from the library of the royal society, london / Smeaton, John – 1741-92 – (= ser Newcomen society extra publication 5) – 6mf – 7 – mf#86577 – uk Microform Academic [621]

Catalogue of the collection of japanese works of art / Huish, Marcus Bourne – London 1895 – (= ser 19th c art & architecture) – 4mf – 9 – mf#4.2.490 – uk Chadwyck [700]

Catalogue of the collection of...his grace the duke of hamilton / Christie, Manson and Woods, Ltd, London – London [1882] – (= ser 19th c art & architecture) – 3mf – 9 – mf#4.2.563 – uk Chadwyck [700]

Catalogue of the copinger collection of editions of the latin bible : with bibliographical particulars / Copinger, Walter Arthur – Manchester: [WA Copinger?], 1893 – 1mf – 9 – 0-7905-1747-7 – mf#1987-1747 – us ATLA [012]

Catalogue of the coptic manuscripts in the british museum / Crum, Walter Ewing – London, 1905 – 60mf – 8 – €115.00 – ne Slangenburg [090]

Catalogue of the dante collection presented by willard fiske / Cornell University Libraries – Ithaca, NY. v1-2. 1898-1899 – 1r – us UF Libraries [025]

Catalogue of the drawings of the riba – 4r – 1 – £200.00 – mf#RCD – uk World [740]

Catalogue of the eastlake library in the national gallery / Green, George & Molini, Morgan – London: printed...for HMSO, 1872 – (= ser 19th c publishing...) – 2mf – 9 – mf#3.1.17 – uk Chadwyck [020]

A catalogue of the english books printed before 1601 / Sinker, Robert – London: George Bell & Sons, 1885 – (= ser 19th c publishing...) – 6mf – 9 – (with list of abbr & errata list) – mf#3.1.15 – uk Chadwyck [070]

Catalogue of the ethiopic mss in the british museum / Wright, W – London, 1877 – 8mf – 9 – mf#NE-20269 – ne IDC [960]

Catalogue of the first annual exhibition of the association of canadian etchers / Association of Canadian Etchers – Toronto: [s.n.], 1885 [mf ed 1980] – 1mf – 9 – 0-665-00847-3 – mf#00847 – cn Canadiana [760]

A catalogue of the first circulating collection of water-colour paintings of the british school / Victoria and Albert Museum, South Kensington – London 1900 – (= ser 19th c art & architecture) – 1mf – 9 – mf#4.1.340 – uk Chadwyck [750]

Catalogue of the first portion of the extensive and varied collections of rare books and manuscripts relating chiefly to the history and literature of america : comprising the great collections of voyages and travels of de bry (in latin and german), hulsius, thenevot, purchas and hakluyt... / Stevens, Henry – London: s.n, 1881 – 3mf – 9 – mf#13960 – cn Canadiana [000]

Catalogue of the florida department of the confede... / Confederate Memorial Literary Society, Richmond – Richmond, VA. 1914 – 1r – us UF Libraries [978]

Catalogue of the foreign and commonwealth office library : a major historical reference – 1977-1980 [mf ed Chadwyck-Healey] – 67mf – 9 – uk Chadwyck [320]

Catalogue of the greek manuscripts on mount athos / Lambros, Spyr P – Cambridge. v1-2. 1895-1900 – 28mf – 8 – €54.00 – ne Slangenburg [450]

Catalogue of the guildhall library's major archive and manuscript holdings / Guildhall Library. London – 10r – 1 – £470.00 – mfΛC – uk World [941]

Catalogue of the hebrew manuscripts in the bodleian library and in the college libraries of oxford : including mss. in other languages which are written with hebrew characters, or relating to the hebrew language or literature, and a few samaritan mss / Neubauer, Adolf [comp] – Oxford: Clarendon Press 1886-1906 – (= ser Catalogi codd mss bibliothecae bodleianae 12/1-2) – 2v – 1 – (facs illustrating various forms of rabbinical characters with transcriptions portfolio) – mf#132p – us Wisconsin U Libr [470]

Catalogue of the highly important collection of...james price / Christie, Manson and Woods, Ltd, London – [London] 1895 – (= ser 19th c art & architecture) – 2mf – 9 – mf#4.2.393 – uk Chadwyck [700]

Catalogue of the hindi, panjabi and hindustani manuscripts in the library of the british museum / Blumhardt, James Fuller [comp] – London: [Gilbert & Rivington], 1899 [mf ed 1996] – (= ser Yale coll) – xii/84p – 1 – 0-524-10238-4 – mf#1996-1238 – us ATLA [090]

Catalogue of the household furniture, books and other effects and property, belonging to david chisholme, esq : to be sold without reserve, by public auction at three rivers, on thursday, 5th january, 1837 ... / Chisholme, David – [Trois-Rivières, Quebec?: s.n, 1836?] [mf ed 1994] – 1mf – 9 – 0-665-94684-8 – mf#94684 – cn Canadiana [640]

Catalogue of the important collection of 360 modern paintings and water colour drawings / Eadon, W H and J A Auctioneers – [Sheffield? 1895?] – (= ser 19th c art & architecture) – 1mf – 9 – mf#4.2.1462 – uk Chadwyck [700]

Catalogue of the important historical collection of coins and medals made by gerald e hart, esq : comprising ancient coins of greece, rome and judaea, mediaeval and modern coins, chiefly of france and england... / Frossard, Edouard – [Boston?: s.n.], 1888 [mf ed 1981] – 2mf – 9 – (incl ind) – mf#11766 – cn Canadiana [730]

A catalogue of the kenya national archive collection on microfilm at syracuse university / Fedha, Nathan W & Webster, John B [comp] – Syracuse, NY: Bibliographic Section, Program of Eastern African Studies, Maxwell Graduate School for Citizenship and Public Affairs, Syracuse University, 1967 (mf ed 19–) – [10] leaves – mf#Z-1951 – us NY Public [960]

Catalogue of the law and classical library of the late j a tailhaides, esq, advocate : to be sold...on wednesday, 22nd december, 1852 – S.l: J Lovell, 1852? – 1mf – 9 – mf#54392 – cn Canadiana [340]

Catalogue of the law library of the late r a ramsay, esq, advocate : to be sold...on thursday, 26th may, 1887 / Arnton, William H – [Montreal?: s.n, 1887?] [mf ed 1992] – 1mf – 9 – 0-665-94646-5 – mf#94646 – cn Canadiana [013]

Catalogue of the law library of the new-york life insurance co, montreal building / New York Life Insurance Company. Law Library – S.l: s.n, 1889? – 2mf – 9 – mf#54076 – cn Canadiana [340]

Catalogue of the library of charles darwin now in the botany school, cambridge / Rutherford, H W [comp] – Cambridge: University Press 1908 [mf ed 1985] – 1r – 1 – (int by francis darwin) – mf#6604 – us Wisconsin U Libr [575]

Catalogue of the library of h macnab stuart, esq, advocate, 5,000 volumes : extremely rich, rare and complete collection of english, american and canadian authors...to be sold...thursday, july 3rd, 1884 – [Quebec?: C Darveau, 1884?] – 1mf – 9 – 0-665-89139-3 – mf#89139 – cn Canadiana [020]

Catalogue of the library of the late hon sir james stuart, bart, chief justice of lower canada – Quebec: printed by Lovell & Lamoureux...1854 [mf ed 1983] – 2mf – 9 – mf#SEM105P224 – cn Bibl Nat [340]

Catalogue of the library of the right honourable the earl of yarmouth, lately deceased : containing a large and curious collection of books... – [London: s.n, 1734] [mf ed 1987] – 50/16p on 1r – 1 – mf#2124 – us Wisconsin U Libr [010]

Catalogue of the lizards in the british museum (natural history) / Boulenger, G A – London. 1974-1978 (1) 1974-1976 (5) 1974-1976 (9) – 26mf – 9 – mf#8338 – ne IDC [590]

Catalogue of the magnificent contents of alton towers... : seat of the earls of shrewsbury / Christie, Manson and Woods, Ltd, London – London [1857] – (= ser 19th c art & architecture) – 3mf – 9 – mf#4.2.1483 – uk Chadwyck [700]

Catalogue of the manuscripts and some early printed books in the library at holkham – printed by Thomas Philips – (= ser Holkham library, the house, park & art colls 749) – 1r – 1 – (filmed with: accounts of the kitchen gardens, woods and plantations, water boats etc, 1743-1759) – mf#97108 – uk Microform Academic [090]

Catalogue of the manuscripts in the library of john alexander thynne, (fourth) marquess of bath : mss 1864 – (= ser Archives of the marquess of bath, longleat house, warminster, wiltshire) – 1r – 1 – mf#96889 – uk Microform Academic [090]

Catalogue of the manuscripts in the library of the earl of leicester, holkham hall, 1816-1828 / Roscoe, William & Madden, Frederick – (= ser Holkham library, the house, park & art colls 770) – 3r – 1 – mf#2919 – uk Microform Academic [090]

Catalogue of the marlborough gems : being a collection of works in cameo and intaglio / Christie, Manson and Woods, Ltd, London – [London? 1899?] – (= ser 19th c art & architecture) – 2mf – 9 – mf#4.2.1494 – uk Chadwyck [730]

Catalogue of the miscellaneous collection of manuscript books : miscellaneous mss 1-34 – (= ser Archives of the marquess of bath, longleat house, warminster, wiltshire) – 1r – 1 – mf#96889 – uk Microform Academic [090]

Catalogue of the museum of archaeology at sarnath / Sahni, Daya Ram – Calcutta: Supt Govt Print, India, 1914 – (= ser Samp: indian books) – us CRL [060]

Catalogue of the museum of mediaeval art / Cottingham, Lewis Nockalls – London 1850 – (= ser 19th c art & architecture) – 1mf – 9 – mf#4.2.1348 – uk Chadwyck [700]

Catalogue of the nineteenth exhibition of the norwich society of artists – Norwich [1823] – (= ser 19th c art & architecture) – 1mf – 9 – mf#4.2.1696 – uk Chadwyck [700]

A catalogue of the original works of john wyclif / Shirley, Walter Waddington – Oxford: Clarendon Press, 1865 [mf ed 1992] – 2mf – 9 – 0-524-02241-0 – (rev in pt as: shirley's catalogue of the extant latin works of john wyclif rev by johann loserth [london: the wycliff society [1924]]) – mf#1990-0581 – us ATLA [012]

A catalogue of the paintings at hatfield house / Salisbury, Mary Catherine Cecil, marchioness of – London 1865 – (= ser 19th c art & architecture) – 1mf – 9 – mf#4.2.342 – uk Chadwyck [750]

Catalogue of the parliamentary papers of southern rhodesia / Willson, Francis Michael Glenn – Salisbury, Zimbabwe. 1965 – 1r – us UF Libraries [960]

A catalogue of the pictures at grosvenor house, london / Young, John [comp] – London 1821 – (= ser 19th c art & architecture) – 2mf – 9 – mf#4.2.316 – uk Chadwyck [700]

A catalogue of the pictures at leigh court, near bristol / Young, John [comp] – London 1822 – (= ser 19th c art & architecture) – 1mf – 9 – mf#4.2.317 – uk Chadwyck [700]

Catalogue of the portraits in the jamaica / Cundall, Frank – Kingston, Jamaica. 1914 – 1r – us UF Libraries [972]

A catalogue of the portraits painted by sir joshua reynolds / Cotton, William – London 1857 – (= ser 19th c art & architecture) – 1mf – 9 – mf#4.2.1551 – uk Chadwyck [750]

Catalogue of the printed books in the library of the hon. society of lincoln's inn / Nicholson, John – London: printed by C F Roworth, 1890 – (= ser 19th c publishing...) – 5mf – 9 – (suppl vol containing the additions fr 1859-1890 by john nicholson) – mf#3.1.14 – uk Chadwyck [020]

Catalogue of the printed books in the library of the hon. society of lincoln's inn / Spilsbury, William Holden – London: printed by C Roworth & Sons, 1859 – (= ser 19th c publishing...) – 11mf – 9 – mf#3.1.13 – uk Chadwyck [020]

Catalogue of the printed books in the royal asiatic society's society's library : and third report of the oriental translation committee / Royal Asiatic Society of Great Britain and Ireland – London: Cox 1830 – (= ser Whsb) – 2mf – 9 – €30.00 – mf#Hu 476 – gw Fischer [470]

Catalogue of the private collection of books belonging to the estate of the late sir a p caron : to be sold by public auction...the 13th and 14th of january, 1910 at...ottawa... william a cole, auctioneer / William A Cole [Firm] – [Ottawa?: s.n.] 1909 [Ottawa: A Bureau & Freres] – 2mf – 9 – 0-665-75844-8 – mf#75844 – cn Canadiana [070]

Catalogue of the publications of the government of bengal available for sale at the bengal secretariat book depot – Calcutta, Secretariat. pt 1-2 1926; pt 1 1927; pts 1-2 1936 – us CRL [350]

Catalogue of the queen square methodist sunday school library, saint john, nb – St John, NB?: s.n, 1881 – 1mf – 9 – mf#54617 – cn Canadiana [020]

Catalogue of the renowned collection of... hollingworth magniac / Christie, Manson and Woods, Ltd, London – London [1892] – (= ser 19th c art & architecture) – 4mf – 9 – mf#4.2.388 – uk Chadwyck [700]

A catalogue of the second circulating collection of water-colour paintings of the british school / Victoria and Albert Museum, South Kensington – London 1900 – (= ser 19th c art & architecture) – 1mf – 9 – mf#4.1.341 – uk Chadwyck [750]

Catalogue of the special loan exhibition of spanish and portuguese ornamental art / South Kensington Museum, London – London [1881] – (= ser 19th c art & architecture) – 3mf – 9 – mf#4.1.353 – uk Chadwyck [740]

Catalogue of the specimens of lizards in the collection of the british museum / Gray, J E – London, 1845 – 3mf – 9 – mf#Z-2256 – ne IDC [590]

A catalogue of the students of the wesleyan academy, mount allison, sackville, nb : for the three years ending december, 1851 / Mount Allison Wesleyan Academy – [Halifax, NS?: s.n.] 1851 [mf ed 1983] – 1mf – 9 – 0-665-43448-0 – mf#43448 – cn Canadiana [378]

Catalogue of the syriac mss in the convent of s catharine on mount sinai / Lewis, Agnes Smith [comp] – London: C J Clay 1894 [mf ed 1990] – (= ser Studia sinaitica 1) – 1mf – 9 – 0-8370-1844-7 – (in syriac, greek & english; app comp by j rendel harris) – mf#1987-6232 – us ATLA [090]

Catalogue of the various articles of antiquity : to be disposed of, at the egyptian tomb / Belzoni, Giovanni Battista – London 1822 – (= ser 19th c art & architecture) – 1mf – 9 – mf#4.2.1169 – uk Chadwyck [700]

Catalogue of the various works of art forming the collection of matthew uzielli – London 1860 – (= ser 19th c art & architecture) – 4mf – 9 – mf#4.1.411 – uk Chadwyck [700]

Catalogue of the works of art forming the collection of robert napier : of west shandon, dumbartonshire / Robinson, John Charles [comp] – London, 1865 – (= ser 19th c art & architecture) – 4mf – 9 – mf#4.1.76 – uk Chadwyck [700]

Catalogue of the works of the late sir edwin landseer / Graves, Algernon – London 1874 – (= ser 19th c art & architecture) – 1mf – 9 – mf#4.2.1782 – uk Chadwyck [750]

Catalogue of the works of the late sir thomas lawrence / British Institution for Promoting the Fine Arts in the United Kingdom, London – London 1830 – (= ser 19th c art & architecture) – 1mf – 9 – mf#4.2.647 – uk Chadwyck [750]

A catalogue of the...collection of...john julius angerstein / Young, John [comp] – London 1823 – (= ser 19th c art & architecture) – 3mf – 9 – mf#4.2.355 – uk Chadwyck [700]

Catalogue of the...collection...by the late adrian hope, esq / Christie, Manson and Woods, Ltd, London – London [1894] – (= ser 19th c art & architecture) – 2mf – 9 – mf#4.2.394 – uk Chadwyck [700]

Catalogue of thematic incipits of copenhagen, kg bibl ms gl kgl sml 1872 40 / Runyan, William Edward – U of Rochester 1973 [mf ed 1984] – 5mf – 9 – (with bibl) – mf#fiche 1136 – us Sibley [780]

Catalogue of works of industry and art : sent from japan / Alcock, Rutherford – London [1862] – (= ser 19th c art & architecture) – 1mf – 9 – mf#4.2.843 – uk Chadwyck [700]

Catalogue officiel – 2e ed. [Paris]: E Panis, [1855] – us CRL [324]

Catalogue officiel / Exposition Universelle. Paris, 1878 – Tome 1, Oeuvres d'art, classes 1 a 5. 1878 – 1 – us CRL [900]

Catalogue officiel : tome 1. groupe 1, oeuvres d'art, classes 1 a 5 / Exposition universelle internationale de 1878, a Paris. Commissariat General – 2e ed. Paris: Impr national 1878 – us CRL [324]

Catalogue of...modern pictures and drawings : and ancient and modern engravings / Grundy, John Clowes – [London? 1867] – (= ser 19th c art & architecture) – 3mf – 9 – mf#4.1.412 – uk Chadwyck [700]

A catalogue of...paintings : including some historical portraits / Hodgson and Co, Auctioneers – London [1856] – (= ser 19th c art & architecture) – 1mf – 9 – mf#4.1.119 – uk Chadwyck [750]

Catalogue, picture gallery, 1888 / Hamilton Art Exposition – [Hamilton, Ont?: s.n, 1888?] [mf ed 1994] – 1mf – 9 – 0-665-94604-X – mf#94604 – cn Canadiana [700]

Catalogue – poetry collection, lockwood memorial library / State University of New York at Buffalo. University Libraries – Buffalo: State University of New York, [1972?] – 1 – us CRL [020]

Catalogue provisoire des manuscripts mauritaniens en langue arabe preserves en mauritanie / Hamidoun, Mokhtar ould & Heymowski, Adam – Stockholm, 1965-66 [i.e. 1966?] – us CRL [960]

Catalogue raisonne : or, a list of the pictures in blenheim palace / Scharf, George – London 1862 – (= ser 19th c art & architecture) – 3mf – 9 – mf#4.2.434 – uk Chadwyck [700]

Catalogue raisonne de manuscrits ethiopiens / Abbadie, A d' – Paris, 1859 – 5mf – 9 – mf#NE-20278 – ne IDC [956]

Catalogue raisonne des tableaux du roy : avec une abrege de la vie des peintres / Lepicie, B – Paris. 2v. 1752-1754 – 8mf – 9 – mf#O-1080 – ne IDC [700]

A catalogue raisonne of the engraved works of william woollett / Fagan, Louis Alexander – London 1885 – (= ser 19th c art & architecture) – 2mf – 9 – mf#4.2.262 – uk Chadwyck [760]

Catalogue raisonne of the pictures belonging to the...marquis of stafford / Britton, John – London 1808 – (= ser 19th c art & architecture) – 2mf – 9 – mf#4.2.340 – uk Chadwyck [750]

Catalogue raisonne of the prehistoric antiquities in the indian museum at calcutta / Brown, J Coggin; ed by Marshall, John – Simla: Govt Central Press, 1917 – (= ser Samp: indian books) – us CRL [060]

A catalogue raisonne of the works of the most eminent dutch, flemish, and french painters / Smith, John – London 1829-42 – (= ser 19th c art & architecture) – 55mf – 9 – mf#4.2.1222 – uk Chadwyck [750]

A catalogue raisonne...works of sir joshua reynolds / Hamilton, Edward – new ed. London 1884 – (= ser 19th c art & architecture) – 3mf – 9 – mf#4.2.263 – uk Chadwyck [750]

Catalogue synonymique des coleopteres d'europe et d'algerie / Gaubil, J – Paris, 1849 – 6mf – 8 – mf#Z-1269 – ne IDC [956]

Catalogued manuscripts, 1847-19? / Catholic Archdiocese of Papeete – 3r – 1 – mf#pmb1082 – at Pacific Mss [241]

Catalogue...from 1793 to 1827 inclusive / Litchfield Law School – Litchfield, CT: Smith, 1828. 27p. LL-2340 – 1 – us L of C Photodup [340]

Catalogues / Roger Williams University – 1873-1929 – 1 – $81.69 – us Southern Baptist [020]

Catalogues des manuscrits syriaques et sabeens (mandaites) de la bibliotheque nationale / Zoterberg, H – Paris, 1874 – 15mf – 8 – €53.00 – ne Slangenburg [470]

Catalogues du departement des arts du spectacle / Bibliotheque Nationale. France – [mf ed Chadwyck-Healey] – 1020mf – 9 – (in french) – uk Chadwyck [790]

Catalogus codicum astrologorum graecorum – Bruxellis. v1-5. 1898-1940 – 30mf – 8 – mf#H-435 – ne IDC [450]

Catalogus codicum copticorum manuscriptorum / Zoega, G – Romae, 1808; Leipzig, 1903 – 39mf – 8 – €75.00 – ne Slangenburg [240]

Catalogus codicum graecorum sinaiticorum / Gardthausen, V – Oxonii, 1886 – 6mf – 8 – €14.00 – ne Slangenburg [450]

Catalogus codicum hagiographicorum latinorum antiquorum saecula 16... – Bruxelles, etc, 1889-1893. 3v – 38mf – 8 – mf#H-299 – ne IDC [700]

Catalogus codicum hagiographicorum latinorum bibliothecarum romanarum praeter quam vaticanae / Poncelet, Albertus – Bruxelles: Apud Editores, 1909 [mf ed 2004] – (= ser Subsidia hagiographica 9) – 1r – 1 – 0-524-10502-2 – (incl bibl ref & ind) – mf#b00639 – us ATLA [450]

Catalogus codicum manuscriptorum bibliothecae bodleianae oxoniensis / Dillmann, A – Oxonii, 1848 – 2mf – 9 – mf#NE-20275 – ne IDC [956]

Catalogus historico-cirticus librorum rariorum / Vogt, J – Hamburg, 1753 – 13mf – 8 – €25.00 – ne Slangenburg [020]

Catalogus librorum / Rask, Rasmus Kristian – Havniae: Poppiana 1833 – (= ser Whsb) – 1mf – 9 – €20.00 – mf#Hu 023 – gw Fischer [410]

Catalogus of naamlijst van schilderijen... / Hoet, G – 's-Gravenhage, 1752. 3v – 30mf – 9 – mf#O-294 – ne IDC [700]

Catalogus personarum, & domiciliorum : in quibus sub a r p societatis jesu praeposito generali 16. p petro zespedes hispaniarum assistente. p joanne antonio balthazar provinciae mexicanae praeposito provinciali 66... / Jesuits. Provincia de Mexico – Mexici: Ex Regalis, & antiquioris divi Ildephonsi Collegii typographia, anno 1751 – (= ser Books on religion... 1543/44-c1800: jesuitas) – 1mf – 9 – mf#crl-228 – ne IDC [241]

Catalogus plantarum in algeria sponte nascentium / Munby, G – Oran, 1859 – 1mf – 9 – mf#11183 – ne IDC [956]

Catalogus testium veritatis : qui ante nostram aetatem reclamarunt papae / Flacius, M – Basileae, 1556 – 12mf – 8 – €23.00 – ne Slangenburg [241]

Catalogus testivm veritatis, qvi ante nostram aetatem reclamarunt papae 101=[flacius illyricus d a, m] – Basileae, [1556] – 12mf – 9 – mf#TH-1 mf 441-452 – ne IDC [242]

Catalogus van boeken voor studie en ontwikkeling / Willemstad (Curacao) Gouvernements-Bibliotheek – Willenstad, Curacao. 1950 – 1r – us UF Libraries [972]

Catalonia. Comissariat de Propaganda *see*
- La catalogne
- El fascismo pretende encarcelar espana

Catalysis communications – Amsterdam, 2000+ [1,5,9] – ISSN: 1566-7367 – mf#42842 – us UMI ProQuest [660]

Catalysis today – Amsterdam. 1987-1993 (1,5,9) – ISSN: 0920-5861 – mf#42499 – us UMI ProQuest [540]

Catalyst – 1972 aug-1974 nov – 1r – 1 – mf#416356 – us WHS [071]

Catalyst : action newsletter / Union of Concerned Scientists – 1985 feb-mar, jun-dec, 1986 feb-aug, nov, 1987 mar, sep, dec, 1988 mar, jun, sep – 1r – 1 – mf#1110552 – us WHS [500]

Catalyst / American Freedom from Hunger Foundation – v1 n5-6 [1972 sep-oct/nov] – 1r – 1 – mf#1582998 – us WHS [360]

Catalyst – Amherst. 1965-1985 (1) 1970-1985 (5) 1977-1985 (9) – ISSN: 0008-7661 – mf#2247 – us UMI ProQuest [300]

Catalyst / Atlanta-Fulton Public Library – 1987 spr-1991 sum – 1r – 1 – mf#1287597 – us WHS [020]

Catalyst – Blacksburg. 1995+ – 1,5,9 – (cont: community services catalyst) – mf#14618,01 – us UMI ProQuest [374]

Catalyst – Dublin. 2002+ (1,5,9) – mf#33062 – us UMI ProQuest [650]

Catalyst – Gary IN. 1971 dec 20 – 1r – 1 – mf#926638 – us WHS [071]

Catalyst – v2 n7 [1970 feb 3-17] – 1r – 1 – mf#1583017 – us WHS [071]

Catalyst for change – Commerce. 1971+ (1) 1974+ (5) 1974+ (9) – ISSN: 0739-2532 – mf#10438 – us UMI ProQuest [073]

Catalyst for environmental quality – New York. 1970-1978 (1) 1974-1978 (5) 1975-1978 (9) – (cont by: catalyst for environment/energy) – ISSN: 0008-7688 – mf#9944 – us UMI ProQuest [333]

Catalyst for environment/energy – New York. 1978-1982 (1) 1978-1982 (5) 1978-1982 (9) – (cont: catalyst for environmental quality) – ISSN: 0194-1445 – mf#9944,01 – us UMI ProQuest [333]

Catamarca. Argentine Republic. (Province) *see* Boletin oficial y judicial

Cataneo, G *see*
- Avertimenti et essamini intor a quelle cose che richiede a un bombardiero
- Libro nuove di fortificare...

Cataneo, P *see*
- L'architettura...
- I quattro primi libri di architettura

Catani, Baldo *see* La pompa funerale fatta dall' ill. mo & r. mo cardinale montalto nella traportatione (sic)...

Catanzariti, Jason C *see* A comparison of two methods for teaching three-ball juggling

Catasauqua independent – Catasauqua, PA., 1892-1897 – 13 – $25.00r – us IMR [071]

Catasauqua valley record – Catasauqua, PA. -w 1889-1892 – 13 – $25.00r – us IMR [071]

La catastrofe de barcelona / Escudero Gonzalez, Jose – Merida: Imp. Juan Rejas Lopez, 1969 – sp Bibl Santa Ana [946]

Catastrophe model of anxiety and performance : application to field hockey / Mills, Brett D & Gray, Marvin – 1992 – 1mf – 9 – $4.00 – us Kinesology [150]

The catastrophe of the presbyterian church, in 1837 : including a full view of the recent theological controversies in new england / Crocker, Zebulon – New Haven: B and W Noyes, 1838 – 1mf – 9 – 0-524-01720-4 – (incl bibl ref) – mf#1990-4112 – us ATLA [242]

Catastrophic injuries in junior high and high school wrestling : a five-season study / Laudermilk, Julie I – 1988 – 49p 1mf – 9 – $4.00 – us Kinesology [617]

Catat, L *see* Voyage...madagascar (1889-1890)

Catawba baptist church. york county. south carolina : church records – 1949-72 – 1 – us Southern Baptist [242]

Catawba. Synod (Pres. Church in the USA) *see* Minutes, 1887-1913

Catc helpline / Cincinnati Area Teacher Center – v2 n1-v3 n16 [1979 sep 5-1981 may] – 1r – 1 – mf#630956 – us WHS [370]

Catch-all gazette : local 1 newsletter / Wisconsin State Employees Union – 1982 jul-1994 oct – 1r – 1 – mf#1054116 – us WHS [331]

Cate, Steven Blaupot ten *see*
- Geschiedenis der doopsgezinden in friesland
- Geschiedenis der doopsgezinden in holland, zeeland, utrecht en gelderland
- Geschiedkundig onderzoek naar den waldenzischen oorsprong van de nederlandsche doopsgezinden

Catechesis davidis chytraei postremo recognita / Chytraeus, D – Magdeburgae, 1578 – 3mf – 9 – mf#TH-1 mf 328-330 – ne IDC [242]

Catechesis pro advltioribvs scripta / Bullinger, Heinrich – Tigvri, [Christoph] Frosch[auer], 1559 – 2mf – 9 – mf#PBU-206 – ne IDC [240]

Catechetical hints and helps : a manual for parents and teachers on giving instruction in the catechism of the church of england / Boyce, Edward Jacob – 3rd rev enl ed. London: George Bell, 1875 – 1mf – 9 – 0-524-07088-1 – mf#1991-2911 – us ATLA [241]

The catechetical oration of gregory of nyssa = Great catechesis / Gregory of Nyssa, Saint; ed by Srawley, James Herbert – Cambridge: University Press, 1903 – (= ser Cambridge Patristic Texts) – 1mf – 9 – 0-7905-9945-7 – (incl bibl ref) – mf#1989-1670 – us ATLA [240]

Catechetics : historical, theoretical, and practical / Ziegler, Henry – Philadelphia: Lutheran Board of Publication, 1876 – 1mf – 9 – 0-524-06282-X – (incl bibl ref) – mf#1991-2473 – us ATLA [240]

Le catechime des commencants / Gosselin, David – Ste Anne de la Pocatiere: F Proulx, 1886 – 1mf – 9 – mf#03480 – cn Canadiana [241]

Catechism : explanatory of the leading truths of the gospel / Bagot, Daniel – London, England. 18– – 1r – us UF Libraries [240]

Catechism : in which the principal testimonies in proof of the divi... / Gray, Robert – London, England. 1820 – 1r – us UF Libraries [240]

Catechism : shewing the real difference bewtween the established church... – Cheltenham? England. 1843 – 1r – us UF Libraries [240]

Catechism *see* Chiao yu wen ta [ccm158]

Catechism about typhoid or enteric fever = Kategismus oor tifeuse koors of ingewandskoors / South Africa. Department of Public Health – Pretoria: Dept of Public Health 1936 [mf ed Pretoria, RSA: State Library [199-]] – 5p on 1r with other items – 5 – mf#op 12825 r25 – us CRL [616]

The catechism explained : an exhaustive exposition of the christian religion, with special reference to the present state of society and the spirit of the age / Spirago, Francis; ed by Clarke, Richard Frederick – New York: Benziger, 1899 – 2mf – 9 – 0-524-05198-4 – mf#1991-2234 – us ATLA [240]

A catechism for sunday schools and families : in fifty two lessons / Schaff, Philip – Philadelphia: Lindsay & Blakiston, 1862 [mf ed 1992] – 1mf – 9 – 0-524-03104-5 – mf#1990-0829 – us ATLA [242]

Catechism for the instruction and direction of young communicants / Colquhoun, John – Edinburgh, Scotland. 1821 – 1r – us UF Libraries [240]

Catechism for the instruction of communicants in the nature and use... / Thomson, Andrew – Edinburgh, Scotland. 18– – 1r – us UF Libraries [240]

A catechism of baptism / Currie, Duncan Dunbar – enl ed. Toronto: S Rose, 1877 [mf ed 1984] – 2mf – 9 – 0-665-08341-6 – mf#08341 – cn Canadiana [240]

A catechism of bible teaching / Broadus, John Albert – Philadelphia: American Baptist Publ Society; Nashville: Sunday-School Board of Southern Baptist Convention [c1890] [mf ed 1989] – 1mf – 9 – 0-7905-3313-8 – mf#1987-3313 – us ATLA [242]

A catechism of christian doctrine – 2nd ed. Malta: [s.n.] 1911 [mf ed 1993] – 1mf – 9 – 0-524-05706-0 – mf#1991-2320 – us ATLA [241]

Catechism of christian doctrine / Deharbe, Joseph – new rev ed. New York, NY: Fr Pustet, c1901 – 1mf – 9 – 0-524-04069-9 – mf#1991-2014 – us ATLA [240]

Catechism of christian doctrine : prepared and enjoined by order of the third plenary council of baltimore – Woodstock, MD: Woodstock College, 1891 – 2mf – 9 – (trans by philip canestrelli) – mf#29164 – cn Canadiana [241]

Catechism of christian doctrine – Rome, Italy. 1913 – 1r – us UF Libraries [240]

Catechism of christian doctrine in english and chiswina – Mariannhill, South Africa. 1915 – 1r – us UF Libraries [240]

A catechism of church government : with special reference to that of the methodist episcopal church, south / McTyeire, Holland Nimmons – Nashville, TN: Publishing House of the M E Church, South, 1894, c1878 [mf ed 1990] – 1mf – 9 – mf#1988-1487 – us ATLA [242]

The catechism of positive religion = Catechisme positiviste / Comte, Auguste – 3rd ed, rev and corr. London: Kegan Paul, Trench, Truebner, 1891 – 1mf – 9 – 0-7905-7437-3 – (in english) – mf#1989-0662 – us ATLA [200]

Catechism of private and public hygiene / Desroches, Joseph Israel; ed by Wright, Alexander – Montreal: A Wright, 1899 [mf ed 1980] – 1mf – 9 – 0-665-02666-8 – (trans fr french by ed) – mf#02666 – cn Canadiana [613]

The catechism of rodez explained in form of sermons : a work equally useful to the clergy, religious communities, and faithful = Catechisme de rodez explique en forme de prones / Luche – 6th ed. St Louis, MO: B Herder, 1917 – 2mf – 9 – 0-524-07572-7 – (in english) – mf#1991-3192 – us ATLA [240]

Catechism of the christian religion : being, with some small changes, a compendium of the catechism of montpellier... / Keenan, Stephen – Boston: Patrick Donahoe, 1852 – 2mf – 9 – 0-524-06548-9 – mf#1991-2632 – us ATLA [240]

Catechism of the church of england – London, England. 1818 – 1r – us UF Libraries [241]

Catechism of the history of newfoundland : with an introductory chapter on the discovery of america by the ancient scandinavians / St John, William Charles – rev ed. Boston: G C Rand, 1855 [mf ed 1983] – 1mf – 9 – 0-665-40653-3 – mf#40653 – cn Canadiana [971]

The catechism of the orthodox, catholic, eastern church : examined and approved by the most holy governing synod, and published for the use of schools and of all orthodox christians – San Francisco, CA: Murdock Press, 1901 – 1mf – 9 – 0-8370-7501-7 – mf#1986-1501 – us ATLA [240]

A catechism of the shaiva religion / Sabhapati Mudaliyar & Sadashiva Mudaliyar – London: Williams & Norgate, 1861 [mf ed 1992] – 1mf – 9 – 0-524-03253-X – (trans fr tamil by thomas foulkas) – mf#1990-3183 – us ATLA [280]

Catechism on modernism : according to the encyclical "pascendi dominici gregis" of his holiness, pius 10 = Catechisme sur le modernisme / Lemius, Jean Baptiste – London: R & T Washbourne; New York: Benziger, 1908 – 1mf – 9 – 0-8370-8527-6 – (incl bibl ref) – mf#1986-2527 – us ATLA [240]

Catechism on the doctrines of the plymouth brethren / Croskery, Thomas – London, England. 1868 – 1r – us UF Libraries [242]

Catechism on the doctrines of the plymouth brethren / Croskery, Thomas – London, England. 1878 – 1r – us UF Libraries [242]

Catechism on the government and discipline of the presbyterian church – Glasgow, Scotland. 1842 – 1r – us UF Libraries [242]

Catechism on the voluntary church association – Edinburgh, Scotland. 1833 – 1r – us UF Libraries [240]

Catechism, prayers and hymns in sindebele as spoken in the mangwe – Mariannhill, South Africa. 1900 – 1r – us UF Libraries [470]

Catechism wherein the christian principles and doctrines of the soc... / Barclay, Robert – Manchester, England. 1871 – 1r – us UF Libraries [240]

Catechisme a l'usage du diocese de quebec : imprime par l'ordre de monseigneur jean olivier briand, eveque de quebec / Eglise catholique. Diocese de Quebec – 7e ed. Quebec: de la Nouvelle impr, 1822 [mf ed 2000] – 9 – cn Bibl Nat [241]

Catechisme a l'usage du diocese de quebec : imprime par l'ordre de monseigneur jean olivier briand, eveque de quebec / Eglise catholique. Diocese de Quebec – Saint Philippe: a l'Impr ecclesiastique, 1827 [mf ed 1998] – 9 – cn Bibl Nat [241]

Catechisme abrege en la langue de madagascar : pour instruire sommairement ces peuples, les inviter et les disposer au bapteme – [Rome] [1785] – (= ser Whsb) – 1mf – 9 – €20.00 – (latin & malagassy in parallel columns) – mf#Hu 335 – gw Fischer [240]

Catechisme creole / Kersuzan, Francoise Marie – Vannes, France. 1922 – 1r – us UF Libraries [240]

Catechisme de controverse premiere partie / Begin, Louis-Nazaire – Quebec: [J P Garneau libraire-editeur], 1902 [mf ed 1994] – 9 – cn Bibl Nat [241]

Catechisme de la venerable mere marie de l'incarnation, fondratrice des ursulines de quebec : ou explication familiere de la doctrine chretienne – 3e ed. Paris: Lib Internationale-Catholique, Leipzig: L A Kittler, Tournai France: H Casterman, 1878 [mf ed 1985] – 4mf – 9 – 0-665-09900-2 – mf#09900 – cn Canadiana [241]

Le catechisme des electeurs d'apres l'ouvrage de a gerin-lajoie – 15e mille. Montreal: J B Thivierge & fils, editeurs, 1936 [mf ed 1990] – 2mf – 9 – mf#SEM105P1273 – cn Bibl Nat [325]

Le catechisme des electeurs d'apres l'ouvrage de a gerin-lajoie – 15e mille. Montreal: J B Thivierge & fils, editeurs, 1936 [mf ed 1993] – 2mf – 9 – mf#SEM105P1242 – cn Bibl Nat [325]

Le catechisme des electeurs d'apres l'ouvrage de a gerin-lajoie – nouv ed. 10e mille. Montreal: J-B Thivierge & fils, editeurs, [1935 ?] (mf ed 1992) – 2mf – 9 – mf#SEM105P1660 – cn Bibl Nat [325]

Le catechisme des provinces ecclesiastiques de quebec, montreal, ottawa / Eglise Catholique. Province de Quebec – nouv ed. Quebec: A O Pruneau, editeur, [1908?] (mf ed 1990) – 2mf – 9 – (in latin and french) – mf#SEM105P1217 – cn Bibl Nat [241]

Catechisme d'hygiene privee / Desroches, Joseph Israel – Montreal: W F Daniel, 1889 [mf ed 1980] – 1mf – 9 – 0-665-02667-6 – mf#02667 – cn Canadiana [613]

Catechisme d'hygiene privee et publique / Desroches, Joseph Israel – Montreal: Cadieux & Derome, 1897 [mf ed 1980] – 2mf – 9 – 0-665-02668-4 – mf#02668 – cn Canadiana [613]

Catechisme du diocese de sens / Languet de Gergy, Jean-Joseph – a Quebec: chez Brown & Gilmore, impr, 1765 [mf ed 1988] – 2mf – 9 – mf#SEM105P887 – cn Bibl Nat [241]

Catechisme du diocese de sens / Languet, Jean-Joseph – Quebec: Chez Brown & Gilmore...1765 [mf ed 1984] – 2mf – 9 – 0-665-45451-1 – mf#45451 – cn Canadiana [241]

Catechisme du libre-penseur / Monteil, Edgar – Paris: C Marpon et E Flammarion [1876?] – 3mf – 9 – mf#vrl-79 – ne IDC [366]

Catechisme historique : contenant en abrege l'histoire sainte et la doctrine chretienne / Fleury, Claude – nouv ed. Quebec: Nouvelle impr, 1807 [mf ed 1971] – 1r – 5 – mf#SEM16P39 – cn Bibl Nat [241]

Catechisme ou cours abrege de l'histoire sainte, de l'histoire du canada et des autres provinces de l'amerique britannique du nord – Montreal: s.n, 1873? – 3mf – 9 – mf#33015 – cn Canadiana [220]

Catechisme politique ou elemens du droit public et constitutionnel du canada : mis a la portee du peuple / Gerin-Lajoie, Antoine – Montreal?: s.n, 1851 – 2mf – 9 – mf#10788 – cn Canadiana [323]

Catechisme populaire de la lettre encyclique de notre t saint-pere leon 13 / Gosselin, David – Quebec: A Cote, 1891 – 1mf – 9 – mf#06433 – cn Canadiana [241]

Le catechisme romain : ou, l'enseignement de la doctrine chretienne. explication nouvelle / Bareille, Georges – Montrejeau: J-M Soubiron, 1906-1910 – 10mf – 9 – 0-8370-8244-7 – mf#1986-2244 – us ATLA [240]

Catechismi tres systematice coordinati pro plena juventutis christianae instructione / Weninger, Francis Xavier – Cincinnati, OH: Roberti Clarke, 1871 – 1mf – 9 – 0-8370-6714-6 – mf#1986-0714 – us ATLA [240]

Catechismo sul modernismo : secondo l'enciclica pascendi dominici gregis di sua santit a pio x = Catechisme sur le modernisme / Lemius, Jean Baptiste – Roma: Tipografia Vaticana, 1908 – 1mf – 9 – 0-8370-8528-4 – (incl bibl ref) – mf#1986-2528 – us ATLA [240]

Catechismo y examen para los que comulgan en lengua castellana y timuquana : en el qual co[n]tiene el respecto o que se deue rener a los templos, con algunos similes del santissimo sacramento, y sus effectos... / Pareja, Francisco de – en Mexico: impr Iuan Ruyz, ano de 1627 – (= ser Books on religion...1543/44-c1800: catecismos) – 7mf – 9 – mf#crl-4 – ne IDC [241]

Catechisms – Baptist Catechsim, 1866; Baptist Catechism, a Catechism for little children; Boyce, James P., A brief Catechism of Bible doctrine, 1864; Broadus, John A., A Catechism of Bible teachings, 1892; Dayton, A. C., A catechism for the little children; Graves, A. D., The child's scripture catechism in rhyme, v. 1, 1861; Sunday School primer, 1864; Winkler, Edwin T., Notes and questions for the oral instruction of colored people, 1857 – 1 – us Southern Baptist [242]

Catechisms of the scottish reformation / ed by Bonar, Horatius – London: J Nisbet, 1866 – 1mf – 9 – 0-524-06385-0 – mf#1991-2507 – us ATLA [242]
Catechisms of the second reformation : with historical introduction and biographical notices / Mitchell, Alexander F – London: James Nisbet, 1886 – 1mf – 9 – 0-7905-5491-7 – mf#1988-1491 – us ATLA [242]
Catechismus / Bullinger, Heinrich – Zuerych, Johann Wolff, 1597 – 1mf1mf – 9 – mf#PBU-207 – ne IDC [240]
Catechismus / Jud, L – [Zuerich, Christoph Froschauer, 1534] – 3mf – 9 – mf#PBU-469 – ne IDC [240]
Catechismus : oder kinderlehr von den fuernemmen haeuptpuncten christlicher religion als da sind / Hunnius, A – Franckfort am Mayn, 1596 – 1mf0 – 9 – mf#TH-1 mf 760-769 – ne IDC [242]
Catechismus : oder, kurtzer unterricht christlicher lehre, wie der in kirchen und schulen der chur-fuerstlichen pfaltz getrieben wird. mit nuetzlichen randfragen – Berlin: Gedruckt bey Christoft Runge 1657 [mf ed 19–] – 1r – 1 – mf#pres. film 120 – us Sibley [240]
Catechismus: ain kurtze christliche leer und underweysung fuer die jugent... / Meckhart, Johann – Augspurg: Ulhart 1557 – (= ser Hqab. literatur des 16. jahrh.) – 1mf – 9 – €20.00 – mf#1557c – gw Fischer [780]
Catechismus, brevissima christianae religionis formula instituendae iuuentuti tigurinae... / Jud, L – Tigvri, Christoph Froschouer, [1539] – 1mf – 9 – mf#PBU-533 – ne IDC [240]
Catechismus, das ist : trostreiche vnd nuetzliche auslegung vber die fuenff heubtstueck der christliche lehre / Mathesius, J – Leipzig, 1586 – 6mf – 9 – mf#TH-1 mf 1015-1020 – ne IDC [242]
Catechismus in kurtze gebetlein verfasset / Musculus, A – [Erffurdt, 1505] – 2mf – 9 – mf#TH-1 mf 1213-1214 – ne IDC [242]
De catechismus oft kinderleere... / [Utenhoven, J] – Embden, 1558 – 2mf – 9 – mf#PBA-378 – ne IDC [240]
Catechismus romanus = Catechism of the council of trent – Dublin: James Duffy, 1914 – 2mf – 9 – 0-8370-8249-8 – (incl bibl ref. in english) – mf#1986-2249 – us ATLA [240]
Catechismuspredigten / Eber, P – [Nuernberg], 1578 – 4mf – 9 – mf#TH-1 mf 385-388 – ne IDC [242]
Catechismvs, hoc est, christianae doctrinae methodvs item, obiectiones in evndem / Lossius, L – Vvitebergae, 1560 – 5mf – 9 – mf#TH-1 mf 849-853 – ne IDC [242]
Catechismvs per omnes quaestiones et circumstantias, quae in iustam tractationem incidere possunt, in usum praedicatorum diligenter ac pie absolutus / Sarcerius, E – Marpurgi, 1537 – 2mf – 9 – mf#TH-1 mf 1315-1316 – ne IDC [242]
Catechismvs predigsweise gestelt fuer die kirche zu regenspurg / Gallus, N – [Regenspurg], 1554 – 5mf – 9 – mf#TH-1 mf 480-484 – ne IDC [242]
Catechist – Dayton. 1976+ (1,5,9) – ISSN: 0008-7726 – mf#10691 – us UMI ProQuest [240]
O catechista : folha commercial, noticiosa e analytica – Manaus, AM. 12 jul 1862; ago 1863-jun 1865; mar, nov-dez 1869; jan-26 set 1871 – (= ser Ps 19) – mf#P11B,06,14 – bl Biblioteca [079]
A catechist's manual : seven lessons on the church catechism / Norris, John Pilkington – new ed. London: Longmans, Green, 1875 [mf ed 1993] – (= ser Anglican/episcopal coll) – 1mf – 9 – 0-524-06490-3 – mf#1991-2590 – us ATLA [242]
De catechizandis rudibus / Augustine, Saint, Bishop of Hippo; ed by Wolfhard, Adolf – 2. vollst neubearb ausg. Freiburg i B: J C B Mohr, 1893 [mf ed 1992] – (= ser Sammlung ausgewaehlter kirchen- und dogmengeschichtlicher quellenschriften 4) – xv/76p on 1mf – 9 – 0-524-02637-8 – mf#1990-0661 – us ATLA [241]
Catechumen / Davidson, P – Edinburgh, Scotland. 1847 – 1r – us UF Libraries [240]
Catecismo breve da doutrina crista em portugues e chichangane / Barbosa, Martinho Da Rocha – Beira, Mozambique. 1929 – 1r – us UF Libraries [960]
Catecismo breve en lengua otomi / Miranda, Francisco de – en Mexico: impr Bibliotheca Mexicana...ano de 1759 – (= ser Books on religion...1543/44-c1800: catecismos) – 1mf – 9 – mf#crl-14 – ne IDC [241]
Catecismo da doutrina crista – Rome, Italy. 1956 – 1r – us UF Libraries [240]
Catecismo de la doctrina cristiana y nociones de urbanidad / Llanes Garcia, Enrique – Don Benito: Tip. de Trejo, 1926 – sp Bibl Santa Ana [240]
Catecismo mexicano : que contiene toda la doctrina christiana con todas sus declaraciones: en que el ministro de almas hallara, lo que a ellas debe ensenar: y estas hallaran lo que, para salvarle, deben saber, creer y observar / Ripalda, Geronimo de – en Mexico: impr Bibliotheca Mexicana...ano de 1758 – (= ser Books on religion...1543/44-c1800: catecismos) – 3mf – 9 – mf#crl-13 – ne IDC [241]
Catecismo para uso de los parrocos : hecho por el 4. concilio provincial mexicano, celebrado ano de 1771 / Catholic Church. Province of Mexico City [Mexico] – en Mexico: impr D J de Jauregui...ano de 1772 – (= ser Books on religion...1543/44-c1800: catecismos) – 6mf – 9 – mf#crl-16 – ne IDC [241]
Catecismo patriotico espanol / Gonzalez Menendez-Reigada, Albino – 3rd ed. Salamanca: ...M P Criado 1939 – 9 – mf#w1043 – us Harvard [440]
Catecismo politico...monarquia espanola – 1820 – 9 – sp Bibl Santa Ana [946]
Catecismo social / Fernandez Santana, Ezequiel – Portada de Orduna: Huelva, Imp Munoz, 1947 – 1 – sp Bibl Santa Ana [240]
Catecismo social / Sanchez Ruiz, Valentin M – Madrid: Apostolado de la Prensa, 2nd ed 1935 – 1 – sp Bibl Santa Ana [240]
Catecismo social... / Sanchez Ruiz, Valentin M – Madrid: Apostolado de la Prensa, 1933 – 1 – sp Bibl Santa Ana [240]
Catecismo social a sea la enciclica "rerum novarum" del papa leon 13 puesta en preguntas y respuetas para su mejor inteligencia / Crespo, Manuel Maria – Don Benito: Colegio del Corazon de Maria, 1920 – sp Bibl Santa Ana [946]
La catedral de caracas y sus funciones de culto / Navarro, Nicolas E – Caracas, 1931; Madrid: Razon y Fe, 1931 – 1 – sp Bibl Santa Ana [240]
La catedral de oviedo : perfiles historico-arqueologicos / Alvarez Amandi, Justo – Oviedo: Impr Region, 1929 (mf ed 19–) – (= ser Harvard Western European local history preservation microfilm project) – 126p [25] pl – mf#ZM-3-MAR pv238 n3 – us NY Public [241]
Catedral primada de america / Lara Fernandez, Carmen – Ciudad Trujillo, Dominican Republic. 1950 – 1r – us UF Libraries [972]
Categories grammaticales et distribution : les limites entre preposition, conjonction, adverbe / Xatard, Veronique – 1mf – 9 – (10092) – fr Atelier National [440]
Catel, Albert *see* Chartes et documents de l'abbaye cistercienne de preuilly
Catel, Charles-Simon *see* Premiere suite d'harmonie a huit parties
Catena aurea : commentary on the four gospels / Thomas, Aquinas, Saint – Oxford: J H Parker, 1842-1845 – 7mf – 9 – 0-524-06055-X – mf#1992-0768 – us ATLA [220]
Catena in acta ss apostolorum e cod nov coll / ed by Cramer, John Anthony – Oxonii: E Typographeo Academico, 1838 [mf ed 1992] – 2mf – 9 – 0-524-05399-5 – mf#1992-0409 – us ATLA [226]
A catena of buddhist scriptures from the chinese / Beal, Samuel – London: Truebner, 1871 [mf ed 1991] – 2mf – 9 – 0-524-00690-3 – (incl selections fr buddhist scriptures in english trans. with bibl ref) – mf#1990-2018 – us ATLA [280]
Catenae graecorum patrum in novum testamentum / ed by Cramer, J A – London. v1-8. 1839-1844 – 8v on 73mf – 8 – €140.00 – ne Slangenburg [225]
Catenen : mitteilungen ueber ihre geschichte und handschriftliche ueberlieferung / Lietzmann, H – Freiburg, 1897 – €7.00 – ne Slangenburg [090]
Catenen : mitteilungen ueber ihre geschicte und handschriftliche ueberlieferung / Lietzmann, Hans & Usener, Hermann – Freiburg i B: Mohr, 1897 – 1mf – 9 – 0-7905-3352-9 – mf#1987-3352 – us ATLA [220]
Cateretes do sul de minas gerais / Alvarenga, Oneyda – Sao Paulo: Departamento de Cultura, 1937 – mf#*ZBD-*MGO pv9 – Located: NYPL – us Misc Inst [790]
Caterpillar : a gathering of the tribes – n1-7. 1967-69 – 1 – us AMS Press [800]
Caterpillar – Pasadena. 1970-1973 (1) 1970-1973 (5) (9) – ISSN: 0008-784X – mf#6076 – us UMI ProQuest [400]
Caterpillars and their moths / Eliot, Ida Mitchell et al – New York: The Century Co 1902 [mf ed Ithaca NY: Cornell University, A R Mann Libr...; Bethlehem Pa: Micrographic Preservation Service (MAPS) Inc 1993 – 1r – 1 – (with ill fr photos of living caterpillars & spread moths by edith eliot) – mf#5087 r5 n4 – us Cornell [590]
Cates, J M D *see*
– The companion
– The sacred harp
– The voice of truth
Catesby, M *see* The natural history of carolina, florida and the bahama islands
Catfish creek baptist church. dillon county. langley, south carolina : church records – 1802-1971 – 1 – us Southern Baptist [242]
Cathala, Pierre Adolphe Juste *see* Face aux realites, la direction des finances francaises sous l'occupation
Catharina von georgien / Gryphius, Andreas; ed by Flemming, Willi – Tuebingen: M Niemeyer, 1955 [mf ed 1993] – xiii/107p – 1 – (fr 1663 and 1657 eds. incl bibl ref) – mf#8452 – us Wisconsin U Libr [820]
The catharine maria sedgwick papers, 1798-1908 – [mf ed 1984] – 18r – 1 – (with p/g) – us MA Hist [420]
Catharine of aragon and the sources of the english reformation = Catherine d'aragon et les origines du schisme anglican / Du Boys, Albert; ed by Yonge, Charlotte Mary – London: Hurst and Blackett, 1881 – 2mf – 9 – 0-7905-4850-X – (incl bibl ref. in english) – mf#1988-0850 – us ATLA [941]
Catharine of siena : a biography / Butler, Josephine Elizabeth Grey – London: Dyer, 1878 – 1mf – 9 – 0-7905-6922-1 – mf#1988-2922 – us ATLA [240]
O catharinense – Desterro, SC. 28 jul-ago 1831; 25 jan 1832 – (= ser Ps 19) – bl Biblioteca [079]
O catharinense : jornal politico e noticioso – Desterro, SC: Typ Catharinense, 31 out 1860-27 mar 1861 – (= ser Ps 19) – bl Biblioteca [320]
Catharri suffocativi ejesque curationes historia / Rotundis, P – Madrid, 1728 – 9 – sp Cultura [615]
Catharsis / Quonset-Davisville GI's for Peace – n1 [1970 aug, nov 5], n1 [1970 aug, nov 5] – 2r – 1 – mf#720848 – us WHS [355]
Cathay and the way thither : being a collection of medieval notices of china / Yule, H; ed by Cordier, H – 4v – 28mf – 9 – mf#U-668 – ne IDC [915]
Cathcart, Charles Murray Cathcart, Earl [Canada (Province). Governor general) *see* Copy of the speech of the governor-general to the legislative assembly of canada
Cathcart, Wallace Daniel *see* Commercial law questions
Cathcart, William *see*
– Ancient british and irish churches
– The ancient british and irish churches
– The baptism of the ages and of the nations
– The baptist encyclopaedia
– The baptists and the american revolution
– The papal system
Cathechismus ex decreto concilii tridentini ad parochos pii 5. et clementis 13 pont. max : jussu editus ad editionem romae a.d. 1845. publici iuris factam accuratissime expressus = Cathechismus romanus – Ratisbonae: GJ Manz, 1905 – 2mf – 9 – 0-8370-8326-5 – mf#1986-2326 – us ATLA [240]
Cathechismvs pro ijs, qui volunt suscipere baptismvm in octo dies diuisus / Rhodes, Alexandre de – Romae, typis sacrae Congregationis de propaganda fide [1651?] [mf ed 1995] – (= ser Yale coll) – 1 – 0-524-09838-7 – (in latin) – mf#1995-0838 – us ATLA [241]
Cathecismo de la doctrina christiana traducido en lengua cahita / Yapuguay, Nicolas – en Mexico: Por Francisco Xavier Sanchez...ano de 1737 – (= ser Books on religion...1543/44-c1800: catecismos) – 1mf – 9 – mf#crl-11 – ne IDC [241]
Cathecismo romano – en Mexico: Por Francisco de Rivera Calderon, ano de 1723 – (= ser Books on religion...1543/44-c1800: catecismos) – 3mf – 9 – (in spanish & nahuatl; trans by p f manuel perez) – mf#crl-9 – ne IDC [241]
Cathedra petri : a political history / Greenwood, Thomas – London: C.J. Stewart, 1856-1872 – 2r – 1 – 0-8370-0680-5 – mf#1984-S025 – us ATLA [900]
The cathedral : its necessary place in the life and work of the church / Benson, Edward White – London: J. Murray, 1878 – 1mf – 9 – 0-7905-5566-2 – mf#1988-1566 – us ATLA [240]
Cathedral age – Washington. 1925-1973 (1) – ISSN: 0008-7874 – mf#8795 – us UMI ProQuest [700]
Cathedral and university sermons / Reichel, Charles Parsons – London; New York: Macmillan, 1891 – 1mf – 9 – 0-7905-8564-2 – mf#1989-1789 – us ATLA [240]
Cathedral and university sermons / Salmon, George – 2nd ed. London: John Murray, 1901 – 1mf – 9 – 0-7905-9864-7 – mf#1989-1589 – us ATLA [240]
The cathedral builders : the story of a great masonic guild / Baxter, Lucy E (Barnes) [pseud: Leader Scott] – London 1899 – (= ser 19th c art & architecture) – 7mf – 9 – mf#4.2.1407 – uk Chadwyck [720]
The cathedral churches of ireland / Fallow, Thomas McCall – London [1894] – (= ser 19th c art & architecture) – 2mf – 9 – mf#4.2.1382 – uk Chadwyck [720]
[Cathedral city-] cathedral citizen – CA. mar 1981-1983 – 6r – 1 – $360.00 – mf#R02101 – us Library Micro [071]
The cathedral monthly – [Fredericton, NB: Christ Church Cathedral, 1888?-189- or 19–] – 9 – (incl: the church monthly, pub in london, england) – mf#P05096 – cn Canadiana [242]
The cathedral of santiago de compostella / Thompson, Thurston – London 1868 – (= ser 19th c art & architecture) – 2mf – 9 – mf#4.2.592 – uk Chadwyck [720]
The cathedral of santiago de compostella in spain... : especially...the portico de la gloria / Thompson, Thurston – London 1868 – (= ser 19th c art & architecture) – 3mf – 9 – mf#4.2.1452 – uk Chadwyck [720]
The cathedral of the holy trinity...dublin / Street, George Edmund – Dublin 1882 – (= ser 19th c art & architecture) – 16mf – 9 – mf#4.2.1574 – uk Chadwyck [720]
Les cathedrals : prelude [pour le poeme dramatique de m eugene morand]. version avec ou sans choeurs [pour orchestre] / Pierne, Gabriel – Paris: Rouart, Leroille & Cie 1916 [mf ed 1991] – 1r – 1 – mf#pres. film 103 – us Sibley [780]
Cathell, D W *see* Book on the physician himself
Cather, Willa *see* Sapphira and the slave girl
Catherine-de-Saint-Augustin, soeur *see* Bibliographie des biographies des religieuses decedees a l'hotel-dieu du sacre-coeur de jesus de quebec
Catherwood, Frederick *see* Views of ancient monuments in central america
Catherwood, Mary Hartwell *see*
– The lady of fort st john
– Old caravan days
Catheterization and cardiovascular interventions – New York, 1999+ [1,5,9] – ISSN: 1522-1946 – mf#22245,01 – us UMI ProQuest [616]
The catholic : letters addressed by a jurist to a young kinsman proposing to join the church of rome / Derby, Elias Hasket – Boston: John P Jewett, 1856 – 1mf – 9 – 0-8370-8499-7 – (incl bibl ref) – mf#1986-2499 – us ATLA [241]
The catholic : a religious weekly periodical – Kingston [Ont]: Patriot & Farmer's Monitor, 1830-[184-?] – 9 – (incl some ind) – mf#P04110 – cn Canadiana [241]
Catholic action : a national monthly – Washington. 1949-1953 (1) – mf#292 – us UMI ProQuest [241]
Catholic advocate – Newark, NJ. 1986-1996 (1) – mf#68100 – us UMI ProQuest [071]
Catholic African Congress (1st : 1952 Aug : Chishawasha) *see* Mharidzo dzekongress
Catholic African Congress (2nd : 1952 Aug : Chishawasha) *see* Mharidzo dzekongress
Catholic African Congress (3rd : 1952 Aug : Chishawasha) *see* Mharidzo dzekongress
Catholic agitator / Ammon Hennacy House of Hospitality – 1971-84 – 1r – 1 – mf#962754 – us WHS [241]
Catholic and protestant / Kinsman, Frederick Joseph – New York: Longmans, Green, 1913 – 1mf – 9 – 0-524-03584-9 – mf#1990-1044 – us ATLA [240]
Catholic and protestant countries compared : in civilization, popular happiness, general intelligence, and morality / Young, Alfred – 9th ed. New York: Catholic Book Exchange, 1898, c1895 – 2mf – 9 – 0-8370-7039-2 – (incl ind) – mf#1986-1039 – us ATLA [230]
Catholic and protestant nations compared : in their threefold relations to wealth, knowledge, and morality / Roussel, Napoleon – Boston: John P. Jewett, 1855. Chicago: Dep of Photodup, U of Chicago Lib, 1971 (1r); Evanston: American Theol Lib Assoc, 1984 (1r) – 1 – 0-8370-0493-4 – mf#1984-B235 – us ATLA [240]
Catholic and protestant priests, freemasons and liberals shot by the rebels / Spain. Embajada. Great Britain – London: [Press Dept of Spanish Embassy] 1937 – 9 – mf#w1171 – us Harvard [946]
The catholic and tolerant character of the church of england, is it to be maintained? : being the substance of an address...on sunday, the 2nd of july, 1871 / Wood, Edmund – Montreal?: J Lovell, 1871 – 1mf – 9 – mf#26046 – cn Canadiana [242]
Catholic answer – Huntington, IN. 1987-2000 (1) – mf#68421 – us UMI ProQuest [071]
Catholic Archdiocese of Papeete *see*
– Administrative archives, 1833-1969
– Catalogued manuscripts, 1847-19?
– Miscellaneous manuscripts, 1968-1983
A catholic atlas : or, digest of catholic theology... / Grafton, Charles Chapman – New York: Longmans, Green, 1908 [mf ed 1986] – 1mf – 9 – 0-8370-8742-2 – mf#1986-2742 – us ATLA [241]
Catholic belief : or, a short and simple exposition of catholic doctrine / Di Bruno, Joseph Faa; ed by Lambert, Louis Aloisius – american ed. New York: Benziger c1912 [mf ed 1986] – (= ser Library of popular instruction) – 1mf – 9 – 0-8370-8337-0 – (incl ind) – mf#1986-2337 – us ATLA [241]

Catholic biblical quarterly – 1(1939)-8(1946) – 67mf – 9 – €128.00 – ne Slangenburg [241]

Catholic biblical quarterly – Washington. 1949+ (1) 1971+ (5) 1975+ (9) – ISSN: 0008-7912 – mf#199 – us UMI ProQuest [220]

A catholic catechism for the parochial and sunday schools of the united states / Groenings, Jakob – [large ed] New York: Benziger Bros 1900 [mf ed 1993] – 1mf – 9 – 0-524-07624-3 – mf#1991-3231 – us ATLA [241]

Catholic Central Verein of America *see*
– Fest-zeitung

Catholic charismatic – Mahwah. 1976-1980 (1,5,9) – ISSN: 0145-9368 – mf#11396 – us UMI ProQuest [241]

A catholic christian church the want of our time / Tayler, John James – London: Williams & Norgate, 1867 [mf ed 1991] – 1mf – 9 – 0-524-00393-9 – mf#1989-3093 – us ATLA [241]

Catholic christian instructed in the sacraments, sacrifice, ceremon... / Challoner, Richard – Dublin, Ireland. 1831 – 1r – us UF Libraries [241]

Catholic christianity : or, an essay toward lessening the number of... / Synge, Edward – London, England. 1790 – 1r – us UF Libraries [241]

Catholic chronicle / Lucas Co. Toledo – dec 1934-dec 1974 [wkly] – 29r – 1 – mf#B4078-4107 – us Ohio Hist [241]

Catholic chronicle / Lucas Co. Toledo – jan 1975-dec 1983 [wkly, biwkly] – 8r – 1 – mf#B14886-14893 – us Ohio Hist [071]

Catholic chronicle / Lucas Co. Toledo – jan 1984-dec 1986 [biwkly] – 3r – 1 – mf#B29156-29158 – us Ohio Hist [071]

Catholic chronicle / Lucas Co. Toledo – jan 1987-jun 1993 [biwkly, semimthly] – 4r – 1 – mf#B33016-33019 – us Ohio Hist [071]

Catholic Church *see*
– Cerimonial y rubricas generales
– Hore presentes ad usum sarum
– Louenge de dieu de de sa tressaincte
– Manual para administrar a los Indios del idioma cahita los santos sacramentos
– Manuale ad usum patrum societatis iesu qui in reductionibus paraqvariae versantur ex rituali romano ac toletano decerptum anno domini 1721
– Manuale sacrame[n]to[r]u[m]
– Missale ad vsum insignis ecclesie sarisburiensis nunc recens typis elegantioribus exaratum, historijs nouis, varijs ac proprijs insignitum
– New world
– Report in the archdiocese of new york
– Spirit
– Tablet
– Times-review
– Zvidzidzo zvesangano dzvene

Catholic church and christian state : a series of essays on the relation of the church to the civil power = Katholische kirche und christlicher staat / Hergenroether, Joseph – London: Burns and Oates, 1876 – 2mf – 9 – 0-7905-4973-5 – (incl bibl ref. in english) – mf#1988-0973 – us ATLA [240]

Catholic church and the holy bible, protestantism and its variation – York, England. 1852? – 1r – us UF Libraries [240]

The catholic church and the race question : a unesco educational studies publication / Congar, Y M J – 2mf – 7 – mf#3405 – uk Microform Academic [241]

Catholic Church. Archdiocese of Omaha (NE) *see* Catholic voice

Catholic Church. Archidiocese de Quebec *see* Circulaire

Catholic Church. Archidiocese de Quebec. Archeveque (1850-1867 : Turgeon) *see* Mandement de l'archeveque et des eveques [sic] de la province ecclesiastique de quebec

Catholic Church. Diocese de Montreal. Eveque *see*
– Circulaire au clerge

Catholic Church. Diocese de Montreal Eveque (1836-1840: Lartigue) *see* Jean jacques lartigue

Catholic Church. Diocese de Montreal. Eveque (1836-1840: Lartigue) *see*
– Circulaire a messieurs les pretres et autres ecclesiastiques du diocese de montreal
– Circulaire a mrs les cures du diocese de montreal
– Circulaire au clerge du diocese de montreal

Catholic Church. Diocese de Montreal. Eveque (1840-1876 : Bourget) *see*
– Circulaire annoncant la celebration du troisieme concile provincial de quebec...
– Lettre pastorale de monseigneur l'eveque de montreal a l'occasion de la nouvelle annee

Catholic Church. Diocese de Quebec Eveque (1825-1833 : Panet) *see* Mandement du 12 mai 1830

Catholic Church. Diocese de Quebec. Eveque (1833-1844: Signay) *see* Lettre circulaire a mm les cures et vicaires

Catholic Church. Diocese of Chatham. Conference of the Clergy (1882: Chatham, NB) *see* Report of the conference of the clergy of the diocese of chatham on october 19th 1882

Catholic Church Diocese of Rome [Italy] *see* Breve relacion de la procession del santissimo rosario, hecho en roma el dia dos de agosto del corriente ano de 1716

Catholic Church Diocese of Santiago [Chile]. Synod [1688] *see* Synodo diocesana, con la carta pastoral convocatoria para ella

Catholic Church Diocese of Santiago [Chile]. Synod [1763] *see*
– Al exc[ellentissi]mo senor don christoval protocarrero, guzman, luna, enriquez de almanfa, pacheco, acuna, funes de villalpando...
– Synodo diocesana

The catholic church from within – London; New York: Longmans, Green, 1901 – 1mf – 9 – 0-8370-6939-4 – (incl bibl ref and index) – mf#1986-0939 – us ATLA [241]

The catholic church in china from 1860 to 1907 / Wolferstan, Bertram – London: Sands; St Louis: B Herder, 1909 [mf ed 1995] – (= ser Yale coll) – xxxvii/470p – 1 – 0-524-09804-2 – mf#1995-0804 – us ATLA [241]

The catholic church in colonial days : the thirteen colonies, the ottawa and illinois country, louisiana, florida, texas, new mexico and arizona, 1521-1763 / Shea, John Dawson Gilmary – New York: J G Shea, 1886 [mf ed 1990] – 2mf – 9 – 0-7905-8075-6 – (incl bibl ref) – mf#1988-6056 – us ATLA [241]

Catholic church in indonesia : archives of the archbishopric of batavia/jakarta, 1807-1949 – 3302mf – 9 – €12,315.00 – (with p/g in english) – mf#m301 – ne MMF Publ [241]

The catholic church in new york : a history of the new york diocese from its establishment in 1808 to the present time / Smith, John Talbot – New York: Hall & Locke, c1905 – 2mf – 9 – 0-7905-8227-9 – mf#1988-6127 – us ATLA [241]

The catholic church in the niagara peninsula, 1626-1895 / Harris, William Richard – Toronto: W Briggs, 1895 – 5mf – 9 – mf#05379 – cn Canadiana [241]

The catholic church in the united states of america : undertaken to celebrate the golden jubilee of his holiness, pope pius 10 – New York: Catholic Editing Co, c1912-c1914 – 4mf – 9 – 0-524-06366-4 – mf#1990-5236 – us ATLA [241]

Catholic Church. Plenary Council of Baltimore *see* Decreta concilii plenarii baltimorensis tertii

Catholic Church. Pope [1775-1799: Pius 6] *see* Alocucion

Catholic Church. Province of Calcutta (India). Concilium Provinciale (1st: 1894) *see* Acta et decreta

Catholic Church. Province of Lima. Concilio Provincial *see* Tercero cathecismo y exposicion de la doctrina christiana, por sermones

Catholic Church. Province of Mexico City [Mexico] *see* Catecismo para uso de los parrocos

Catholic Church Province of Mexico City [Mexico]. Concilio Provincial [1st: 1555] *see* Concilios provinciales primero, y segundo

Catholic Church Province of Mexico City [Mexico]. Concilio Provincial [3rd: 1585] *see* Concilium mexicanum provinciale 3. celebratum mexici anno 1585

Catholic Church. Province of Peru. Concilio Provincial *see*
– Confessionario para los curas de indios
– Doctrina christiana, o cartilla

The catholic church, the renaissance and protestantism : lectures given at the catholic institute of paris, jan to mar 1904 = Eglise catholique, la renaissance, le protestantisme / Baudrillart, Alfred – London: Kegan Paul, Trench, Truebner, 1908 [mf ed 1990] – (= ser The international catholic library 4) – 1mf – 9 – 0-7905-5562-X – (trans fr french into english by mrs philip gibbs. with pref letter by cardinal perraud. incl bibl ref) – mf#1988-1562 – us ATLA [241]

Catholic churchmen in science. first series : sketches of the lives of catholic ecclesiatics who were among the great founders in science / Walsh, James Joseph – 2nd ed. Philadelphia: American Ecclesiastical Review, 1910 – 1mf – 9 – 0-8370-7034-1 – mf#1986-1034 – us ATLA [920]

Catholic citizen – 1878 dec 21/1880 apr 10-1933 apr 8/1934 aug 18 – 26r – 1 – (cont: catholic vindicator [milwaukee wi], catholic review [new york ny]; catholic american [new york ny]; cont by: catholic herald of wisconsin; catholic herald citizen) – mf#1094425 – us WHS [071]

Catholic colonial missions, 1803-27 : from westminster cathedral archives – 2r – 1 – (int by b fisher) – mf#96296 – uk Microform Academic [241]

Catholic columbian / Franklin Co. Columbus – jan 1875-dec 1876 [wkly] – 1r – 1 – mf#B1455 – us Ohio Hist [241]

Catholic columbian / Franklin Co. Columbus – jul 2 1898; apr 1918-aug 1939 [wkly] – 10r – 1 – mf#B35104-35113 – us Ohio Hist [241]

The catholic conception of the church : a study of the traditional idea of the nature and constitution of the church / Sparrow-Simpson, William John – London: Robert Scott; New York: S R Leland [1914?] [mf ed 1991] – (= ser Library of historic theology) – 1mf – 9 – 0-7905-9674-1 – mf#1989-1399 – us ATLA [241]

Catholic courier – Rochester, NY. 1935-1945 (1) – mf#65184 – us UMI ProQuest [071]

Catholic courier journal – Rochester, NY. 1945-2000 (1) – mf#61650 – us UMI ProQuest [071]

Catholic daily tribune – Iowa. 1933 oct 6-dec 31, 1934 jan 3-mar 10, 1934 jun 9-sep 5, 1934 mar 11-jun 8, 1934 sep 6-dec 30 – 5r – 1 – (cont: daily american tribune; cont by: daily tribune [dubuque ia]) – mf#854106 – us WHS [241]

Catholic democracy : individualism and socialism / Day, Henry Cyril – London: Heath, Cranton & Ouseley, 1914 – 1mf – 9 – 0-524-04609-3 – mf#1990-1269 – us ATLA [335]

Catholic digest – St. Paul. 1936+ (1) 1970+ (5) 1976+ (9) – ISSN: 0008-7998 – mf#287 – us UMI ProQuest [241]

Catholic doctrine of a trinity proved by above an hundred short and... / Jones, William – London, England. 1802 – 1r – us UF Libraries [241]

The catholic doctrine of faith and morals, gathered from sacred scripture; decrees of councils, and approved catechisms / Byrne, William – Boston: Cashman, Keating, 1892 – 2mf – 9 – 0-8370-8323-0 – (incl ind) – mf#1986-2323 – us ATLA [241]

The catholic doctrine of the atonement : an historical review / Oxenham, Henry Nutcombe – 3rd ed. London: W H Allen 1881 [mf ed 1991] – 2mf – 9 – 0-524-00074-3 – (incl bibl ref; first printed in 1865) – mf#1989-2774 – us ATLA [241]

Catholic dogma : the fundamental truths of revealed religion / Littlejohn, Abram Newkirk et al – New York: E & J B Young 1892 [mf ed 1985] – (= ser The church club lectures 1891) – 1mf – 9 – 0-8370-3199-0 – mf#1985-1199 – us ATLA [241]

Catholic education today – Twickenham. 1967-1980 (1) 1976-1980 (5) 1976-1980 (9) – ISSN: 0008-8013 – mf#7682 – us UMI ProQuest [377]

Catholic educational conditions in the united states / Macksey, Charles – Columbus, OH: Catholic Educational Association, 1913 – 1mf – 9 – 0-8370-8590-X – mf#1986-2590 – us ATLA [241]

Catholic educational review – Washington. 1911-1969 – 1 – ISSN: 0884-0598 – mf#415 – us UMI ProQuest [377]

Catholic educator – New York. 1931-1970 [1,5,9] – mf#1888 – us UMI ProQuest [241]

The catholic educator : a library of catholic instruction and devotion: / ed by Shea, John Dawson Gilmary – New York: Thomas Kelly, c1888? – 3mf – 9 – 0-8370-7263-8 – (incl bibl ref) – mf#1986-1263 – us ATLA [052]

Catholic emancipation : considered on protestant principles / Monteagle, Thomas Spring-Rice, 1st Baron – London, 1827 – (= ser 19th c ireland) – 1mf – 9 – mf#1.1.1888 – uk Chadwyck [241]

Catholic emancipation – London, England. 1805 – 1r – us UF Libraries [241]

Catholic eschatology and universalism : an essay on the doctrine of future retribution / Oxenham, Henry Nutcombe – 2nd ed, rev and enl. London: WH Allen, 1878 – 1mf – 9 – 0-7905-8544-8 – (incl bibl ref) – mf#1989-1769 – us ATLA [241]

Catholic exponent / Mahoning Co. Youngstown – jan 1944-dec 1974 [wkly] – 24r – 1 – mf#B4408-4431 – us Ohio Hist [241]

Catholic exponent / Mahoning Co. Youngstown – jan 1975-dec 1983 [wkly, biwkly] – 7r – 1 – mf#B278-284 – us Ohio Hist [241]

Catholic exponent / Mahoning Co. Youngstown – jan 1984-dec 1986 [biwkly] – 3r – 1 – mf#B29159-29161 – us Ohio Hist [071]

Catholic exponent / Trumbull Co. Youngstown – jan 1944-dec 1974 [wkly] – 24r – 1 – mf#B4408-4431 – us Ohio Hist [241]

Catholic exponent / Trumbull Co. Youngstown – jan 1975-dec 1983 [wkly, biwkly] – 7r – 1 – mf#B278-284 – us Ohio Hist [241]

The catholic faith : or, doctrines of the church of rome contrary to scripture and the teaching of the primitive church / Treat, John Harvey – Nashotah, WI: Bishop Welles Brotherhood, 1888, c1886 – 2mf – 9 – 0-8370-8555-1 – (in english, greek and latin. incl ind) – mf#1986-2555 – us ATLA [241]

Catholic film newsletter / National Center for Film Study – v33 n8-v40 n24 [1968 jan-1975 dec 30] – 1r – 1 – (cont by: film and broadcasting review) – mf#152985 – us WHS [241]

Catholic gems : or, treasures of the church. a repository of catholic instruction and devotion / DeLigney, Francis & Shea, John Gilmary – New York: Office of Catholic Publ, c1887 – 3mf – 9 – 0-8370-7134-8 – (incl bibl ref) – mf#1986-1134 – us ATLA [241]

Catholic herald – 1981 dec 17-1981 dec 31, 1982, 1983-1984 mar, 1984 apr-1985 feb, 1985 mar-dec, 1986-89 – 9r – 1 – mf#1094433 – us WHS [241]

Catholic herald – 1981 dec 17-31, 1982-89 – 9r – 1 – (cont: catholic herald citizen [milwaukee wi: madison ed]) – mf#1012349 – us WHS [241]

The catholic herald – 1888-1997+ – 83r – 1 – £4000.00 – mf#CHE – uk World [241]

The catholic herald – Philadelphia. Pennsylvania. v. 1-12. 1833-44. Scattered issues wanting – 1 – 65.00 – us L of C Photodup [241]

Catholic herald citizen – 1955-1981 jan 3/dec 10 – 29r – 1 – (cont by: catholic herald [madison wi]) – mf#1012391 – us WHS [241]

Catholic herald citizen – 1953 dec 5/1954-1981 jan 3/dec 10 – 25r – 1 – (cont by: catholic herald [milwaukee wi: superior ed]) – mf#1094432 – us WHS [071]

Catholic herald citizen – 1936 jan 4/mar 21-1954 – 19r – 1 – (with gaps; cont: catholic citizen [milwaukee wi]; catholic herald of wisconsin; cont by: catholic herald [milwaukee wi]) – mf#1012391 – us WHS [241]

Catholic herald of wisconsin – 1925 oct 28 jubilee ed, 1926 educational suppl, 1925 sep 24/1926 oct 7-1934 nov 15/1935 sep 19 – 9r – 1 – (with gaps; cont by: catholic citizen [milwaukee wi]; catholic herald citizen [milwaukee wi]) – mf#1012536 – us WHS [071]

Catholic historical review – Washington. 1915+ [1]; 1970+ [5]; 1976+ [9] – ISSN: 0008-8080 – mf#416 – us UMI ProQuest [241]

A catholic history of alabama and the floridas / Carroll, Austin – New York: P J Kenedy 1908 [mf ed 1993] – 1mf – 9 – 0-524-06365-6 – (no more publ) – mf#1990-5235 – us ATLA [241]

The catholic hospital – Patna City: Medical Mission Sisters, feb 1945-apr/jun 1954 [mf ed 2005] – 2r – 1 – (official journal of: catholic hospitals association (india, burma, and ceylon), 1945-1948; catholic hospitals assocation (india, pakistan, burma, ceylon), 1949-april-june 1954; cont by: medical service) – mf#2005C-S022 – us ATLA [241]

Catholic interests in the nineteenth century = Des interaests catholiques au 19e siecle / Montalembert, Charles Forbes, comte de – London: C Dolman, 1852 – 1mf – 9 – 0-7905-6768-7 – (incl bibl ref. in english) – mf#1988-2768 – us ATLA [241]

Catholic journalist – Ronkonkoma. 1973-1973 (1) – ISSN: 0008-8129 – mf#7516 – us UMI ProQuest [070]

Catholic Junior Leagues of Wisconsin *see* Mantle

Catholic junior leagues of wisconsin – v1 n1-3 [1945 sep-nov] – 1r – 1 – (cont by: mantle) – mf#679990 – us WHS [241]

Catholic Knights of St George [US] *see* Knight of st george

Catholic lawyer – v1-39. 1955-2000 – 1,5,6 – $462.00 set – (v1-32 1955-89 in reel $330. v33-39 1990-2000 in mf $132) – ISSN: 0008-8137 – mf#101451 – us Hein [340]

Catholic layman – Dublin, Ireland. jan-nov 1852; jan-jun, aug, oct-nov 1853; mar-may, jul-aug 1854; oct 1854-1858 – 1r – 1 – uk British Libr Newspaper [072]

Catholic laymen (supplement) – Dublin, Ireland. 1862 – 1/4r – 1 – (containing general ind analytical digest and chronological tables of the councils popes fathers and ecclesiastical writes) – uk British Libr Newspaper [072]

Catholic League for Religious and Civil Rights [US] *see* Catholic league newsletter

Catholic league newsletter / Catholic League for Religious and Civil Rights [US] – 1973 dec-1981 – 1r – 1 – (cont by: catalyst [milwaukee wi]) – mf#609158 – us WHS [322]

Catholic library world – Bryn Mawr. 1929+ (1) 1970+ (5) 1977+ (9) – ISSN: 0008-820X – mf#1559 – us UMI ProQuest [241]

Catholic life – Detroit. 1954-1980 (1) 1972-1980 (5) 1975-1980 (9) – ISSN: 0008-8218 – mf#7092 – us UMI ProQuest [241]

Catholic life and letters of cardinal newman : with notes on the oxford movement and its men / Oldcastle, John – 3rd ed. London: Burns & Oates; New York: Catholic Publ Soc [1885][mf ed 1986] – 1mf [ill] – 9 – 0-8370-6923-8 – mf#1986-0923 – us ATLA [241]

Catholic light – Scranton. 1972+ (1) – ISSN: 0164-9418 – mf#9710 – us UMI ProQuest [241]

Catholic london a century ago / Ward, Bernard – London: Catholic Truth Soc 1905 [mf ed 1990] – 1mf – 9 – 0-7905-6906-X – mf#1988-2906 – us ATLA [241]

Catholic looks at spain / Semprun Gurrea, Jose Maria – London: Labour Publ Dept [1937] – 9 – mf#w787 – us Harvard [241]

Catholic mind – New York. 1903-1982 [1]; 1971-1982 [5]; 1975-1982 [9] – ISSN: 0008-8242 – mf#1805 – us UMI ProQuest [241]
The catholic mission in australasia / Ullathorne, William Bernard – 3rd ed. London: Keating & Brown; Booker & Dolman, 1838 [mf ed 1995] – (= ser Yale coll) – 57p – 1 – 0-524-09246-X – mf#1995-0246 – us ATLA [241]
Catholic Mission, Solomon Islands *see* Na turupatu na lotu katolika
Catholic Mission, Wallis Island *see*
- Correspondence with french high commissioner, noumea
- 'Notes sur la mission' by father jean-marie bazin

Catholic missions in southern india to 1865 / Strickland, William & Marshall, Thomas William M – London: Longmans, Green, 1865 [mf ed 1995] – (= ser Yale coll) – viii/240p – 1 – 0-524-10001-2 – mf#1995-1001 – us ATLA [241]
The catholic monthly calendar – Toronto: G M Rose, [1898-189- or 19–] – 9 – mf#P04187 – cn Canadiana [241]
Catholic moral teaching and its antagonists : viewed in the light of principle and of contemporaneous history = Katholische moral und ihre gegner / Mausbach, Joseph – New York: Joseph F Wagner, c1914 – 2mf – 9 – 0-524-07576-X – (in english) – mf#1991-3196 – us ATLA [241]
Catholic Negro-American Mission Board *see* Educating in faith
Catholic news – New Rochelle. 1970-1981 (1) – ISSN: 0008-8250 – mf#5861 – us UMI ProQuest [241]
Catholic oath : the temporalities of the established church and the... / Creagh, Pierse – Dublin, Ireland. 1856 – 1r – us UF Libraries [241]
Catholic opinion – London, 10 Jul 1869-31 Dec 1870 – 1r – 1 – uk British Libr Newspaper [072]
Catholic Order of Foresters *see*
- Rituel
- Rituel de l'ordre des forestiers catholiques

Catholic orthodoxy and anglo-catholicism : a word about intercommunion between the english and the orthodox churches / Overbeck, Julian Joseph – London: N Truebner, 1866 – 1mf – 9 – 0-7905-6663-X – mf#1988-2663 – us ATLA [241]
Catholic parent – Huntington, IN. 1995-2000 (1) – mf#69071 – us UMI ProQuest [071]
Catholic poor-school committee, annual reports of the... 1848-1900 : roman catholic voluntary schools – 71mf – 7 – mf#87144 – uk Microform Academic [377]
Catholic press – Sydney, nov 1895-dec 1911 – 27r – A$1614.18 vesicular A$1762.68 silver – at Pascoe [079]
Catholic principles : as illustrated in the doctrine, history, and organization of the american catholic church in the united states commonly called the protestant episcopal church / Westcott, Frank Nash – Milwaukee: Young Churchman, c1902 – 1mf – 9 – 0-8370-8799-6 – mf#1986-2799 – us ATLA [241]
Catholic principles of allegiance illustrated / Gillow, Thomas – Newcastle upon Tyne, England. 1807 – 1r – us UF Libraries [241]
Catholic psychological record – v1-6, no. 2. 1963-68 – 1 – us AMS Press [241]
Catholic question – Bristol, England. 18– – 1r – us UF Libraries [241]
Catholic reasons for rejecting the modern pretensions and doctrines / Wray, Cecil – London, England. 1846 – 1r – us UF Libraries [241]
Catholic record – Ontario Prov., Canada. 1874-1947 – 1 – cn Commonwealth Imaging [241]
Catholic reform : letters, fragments, discourses / Hyacinthe, Father – London: Macmillan, 1874 – 1mf – 9 – 0-8370-9072-5 – mf#1986-3072 – us ATLA [241]
Catholic register – Toronto: Catholic Register Print & Pub Co, [1893-1908?] – 9 – (cont by: catholic register and canadian extension) – mf#P04934 – cn Canadiana [241]
Catholic register *see* The catholic weekly review
The catholic religion : a manual of instruction for members of the anglican church / Staley, Vernon – 4th ed. Oxford: Mowbray, 1894 [mf ed 1992] – (= ser Anglican/episcopal coll) – 1mf – 9 – 0-524-05021-X – (incl bibl ref) – mf#1991-2191 – us ATLA [242]
Catholic safeguards against the errors, corruptions, and novelties of the church of rome : being discourses and tracts, selected from the works of eminent divines of the church of england, who lived during the seventeenth century / Brogden, James – London: John Murray. 3v. 1851 – 6mf – 9 – 0-8370-9069-5 – (incl bibl ref and index) – mf#1986-3069 – us ATLA [230]
Catholic school – London. -irr. Oct 1853, 14 Oct 1854, 14 Mar 1855. (8 ft) – 1 – uk British Libr Newspaper [377]
The catholic school book : containing easy and familiar lessons for the instruction of youth of both sexes in the english language and the paths of true religion and virtue / Andrews, William Eusebius – Montreal: R Miller, 1864 – 2mf – 9 – mf#41538 – cn Canadiana [241]
Catholic school journal – Stamford. 1901-1970 – 1,5,9 – mf#1896 – us UMI ProQuest [377]
The catholic school system in the united states : its principles, origin, and establishment / Burns, James Aloysius – New York: Benziger Brothers, 1908 – 1mf – 9 – 0-8370-7533-5 – (incl ind) – mf#1986-1533 – us ATLA [377]
Catholic scripture manual atlas : specially prepared with reference to the catholic scripture manuals / ed by Cecilia, Madame – London: K Paul, Trench, Truebner 1905 [mf ed 1991] – 1mf – 9 – 0-8370-1922-2 – mf#1987-6309 – us ATLA [220]
Catholic sentinel – Portland OR: Herman & Atkinson, 1870- * [wkly slightly irreg] – 1 – (official organ of: archdiocese of oregon city (sometimes called oregon) 1878-jan 17 1929, with its suffragan dioceses, 1878-, with diocese of boise, jun 9 1921-jan 17 1929; archdiocese of portland in oregon, jan 24 1929- . suspended publ jul 3-oct 8 1884. incl suppl. related to: catholic sentinel (portland or)) – mf#451 – us Oregon Lib [241]
Catholic sentinel – Chippewa Falls WI. 1891 jan 29, mar 19-apr 2, 1892 jan 21/1893 jul 6-1914 jul 16/1916 sep 14 – 16r – 1 – (with gaps; cont: chippewa sentinel) – mf#921381 – us WHS [241]
The catholic shield : a monthly chronicle and general review – Ottawa: [A Bureau), 1881-1882] – 9 – (incl ind) – mf#P04975 – cn Canadiana [241]
Catholic socialism = Studi sul socialismo contemporaneo / Nitti, Francesco Saverio – London: S Sonnenschein, New York: Macmillan 1895 [mf ed 1990] – (= ser Half guinea international library) – 2mf – 9 – 0-7905-6004-6 – (incl bibl ref, trans fr 2nd italian ed by mary mackintosh; int by david g ritchie) – mf#1988-2004 – us ATLA [335]
Catholic standard – London. 20 oct 1849-dec 1870 [wkly] – 43r – 1 – (aka: weekly register and catholic standard) – uk British Libr Newspaper [241]
Catholic standard – Georgetown, Guyana. Jan 17 1954-Aug 4 1967; Sept 5 1969-Dec 18 1977; Jan 7 1979-1992 – 15r – 1 – (incomplete) – us L of C Photodup [241]
A catholic sunday-school hymn book : consisting of hymns contained in the manual of the sodality and a selection of other hymns adapted to children – 4th enl ed. Philadelphia: Henry McGrath 1850 [mf ed 1993] – 1mf – 9 – 0-524-05648-X – mf#1991-2317 – us ATLA [241]
Catholic telegraph / Montgomery Co. Dayton – jan 1968-dec 1972 [wkly] – 3r – 1 – mf#B5243-5245 – us Ohio Hist [241]
Catholic telegraph *see* Weekly telegraph
Catholic telegraph register / Montgomery Co. Dayton – mar 1940-51, 54-56, 58-may 1959 [wkly] – 8r – 1 – mf#B5246-5253 – us Ohio Hist [241]
Catholic the same in meaning as sovereign / Laing, Francis Henry – London, England. 18– – 1r – us UF Libraries [241]
Catholic theatre – Washington. 1937-1959 (1) – mf#267 – us UMI ProQuest [790]
Catholic Theological Society of America *see* Proceedings of the annual convention
Catholic thoughts on the bible and theology / Myers, Frederic – London: Daldy, Isbister, 1879 – 1mf – 9 – 0-8370-3856-1 – mf#1985-1856 – us ATLA [220]
Catholic thoughts on the church of christ and the church of england / Myers, Frederic – London: W Isbister 1874 [mf ed 1992] – (= ser Present-day papers on prominent questions in theology) – 2mf – 9 – 0-524-05089-9 – mf#1991-2213 – us ATLA [242]
Catholic times / Franklin Co. Columbus – jan 1982-dec 1987 [wkly] – 6r – 1 – mf#B29166-29171 – us Ohio Hist [241]
Catholic times / Franklin Co. Columbus – jan 1988-dec 1989 [wkly] – 2r – 1 – mf#B34874-34875 – us Ohio Hist [241]
Catholic times / Franklin Co. Columbus – v1 n1. oct 1951-dec 1981 [wkly] – 24r – 1 – mf#B11877-11900 – us Ohio Hist [241]
The catholic times – v1-3. 3 dec 1892-23 nov 1895* – 1r – 1 – (incorp by: universe) – ISSN: 0041-8226 – mf#ATLA S0267 – us ATLA [241]
Catholic times of south africa – Johannesburg, [South Africa]: Catholic Times Ltd, v1-15 (1936-58) [mf ed 2005] – (= ser Christianity's encounter with world religions, 1850-1950) – 23v on 4r – 1 – (publ by: brothers of the society of st vincent de paul, 1936-oct 1937; vicor apostolic of the transvaal, nov 1937-1950; lacks: v1 n6 (jun 1936) p8-9; some pgs damaged) – mf#2006c-s054 – us ATLA [241]
Catholic Traditionalist Movement *see* Quote – unquote
Catholic transcript : (bridgeport edition) – Hartford, CT. 1963-1971 (1) – mf#62346 – us UMI ProQuest [071]
Catholic transcript – Hartford, CT. 1898-2000 (1) – mf#61247 – us UMI ProQuest [071]
Catholic truth and historical truth / Coulton, George Gordon – Cairo: Nile Mission Press, [ca 1906] – 1mf – 9 – 0-8370-7929-2 – (incl bibl ref) – mf#1986-1929 – us ATLA [230]
The catholic truth society : its aims and objects – Ottawa?: The Society, 1891 or 1892 – 1mf – 9 – mf#00517 – cn Canadiana [241]
Catholic Truth Society of Ottawa *see* Annual report for...
Catholic twin circle – 1980-81, 1982 jan-aug, 1982 sep-1983 mar, 1983 aug-1984 may, 1984 jun-1985 mar, 1985 apr-1986 jan, 1986 feb-dec, 1987-93 – 15r – 1 – (cont: twin circle; cont by: catholic faith and family) – mf#570162 – us WHS [241]
Catholic unity – Ottawa: J Durie, [18–?] [mf ed 1994] – 1mf – 9 – 0-665-94717-8 – (original iss in ser: tracts by canadian laymen n3) – mf#94717 – cn Canadiana [241]
Catholic universe / Cuyahoga Co. Cleveland – jan 1901-may 1926 [wkly] – 16r – 1 – mf#B1322-1337 – us Ohio Hist [241]
Catholic universe / Cuyahoga Co. Cleveland – jul 1874-92, sep 1895-1900 [wkly] – 10r – 1 – mf#B3208-3217 – us Ohio Hist [241]
Catholic universe bulletin / Cuyahoga Co. Cleveland – jun 1926-dec 1974 [wkly] – 56r – 1 – mf#B1338-1393 – us Ohio Hist [241]
Catholic university bulletin – Washington. 1895-1908 (1) – mf#2870 – us UMI ProQuest [241]
Catholic university law review – v1-50. 1950-2001 – 5,6,9 – $908.00 set – (v1-34 1959-85 in reel $440; v35-50 1985-2001 in mf $468; title varies: v1-19 1950-70 catholic university of america law review) – ISSN: 0008-8390 – mf#101461 – us Hein [340]
Catholic University of America *see* Jurist
Catholic university of america law review *see* Catholic university law review
The Catholic University of America Studies in Sacred Theology *see* Dionysius the ps-areopagite
Catholic univrs bulletin / Cuyahoga Co. Cleveland – jan 1975-dec 1989 [wkly, biwkly] – 17r – 1 – mf#B34876-34892 – us Ohio Hist [241]
Catholic vet / Catholic War Veterans of the United States of America – 1944 sep-1952 apr – 1r – 1 – mf#1054128 – us WHS [071]
Catholic vet / Catholic War Veterans of the United States of America – v1 n1,3-4,7,10-11,14 [1947 feb, may-jun, nov, 1948 feb-mar, sep] – 1r – 1 – (cont by: wisconsin catholic vet) – mf#3564592 – us WHS [305]
The catholic view of the public school question : a lecture delivered in the hall of the cooper institute, sunday evening, january 16, 1870 / Preston, Thomas Scott – New York: Robert Coddington, 1870, c1869 – 1mf – 9 – 0-8370-7821-0 – mf#1986-1821 – us ATLA [377]
Catholic vindicator – v4 n12-33 [1874 jan 22-jun 20], v4 n20 [1876 mar 25] – 2r – 1 – (cont: catholic vindicator and star of bethlehem; cont by: catholic citizen [milwaukee wi]) – mf#1013106 – us WHS [241]
The catholic visitor – Quebec: F Belanger, [1874-1875] – 9 – ISSN: 1190-6820 – mf#P04214 – cn Canadiana [241]
Catholic voice : omaha archdiocesan newspaper / Catholic Church. Archdiocese of Omaha (NE) – Omaha, NE: The Archdiocese. v71 n9. sep 14 1973- (biwkly) [mf ed 1977-] – 1 – (cont: true voice (1955). not publ last week of dec 1982-) – us NE Hist [071]
Catholic voice – 1968 aug 28/1970 jan 21-1986 dec/1988 dec 19 – 14r – 1 – (with gaps) – mf#1054129 – us WHS [241]
The catholic voice – [Oakland CA]: The Diocese [biwkly, wkly] [mf ed 2004] – 1 – (mf: v41-2003-) – mf1051 – us ATLA [241]
Catholic war veteran / Catholic War Veterans of the United States – 1968 mar/apr-1981 jul/aug – 1r – 1 – mf#584052 – us WHS [305]
Catholic War Veterans of the United States *see* Catholic war veteran
Catholic War Veterans of the United States of America *see*
- Catholic vet
- Wisconsin catholic vet
- Wisconsin catholic vet newsletter

The catholic weekly review : a journal devoted to the interests of the catholic church in canada – Toronto: A C Macdonell and F W G Fitzgerald [1887-1892] – 9 – (merged with: irish canadian to become: catholic register) – mf#P04950 – cn Canadiana [241]
Catholic witness – Harrisburg, PA., 1966 – 13 – $25.00r – us IMR [071]
Catholic worker – 1933 may-1951, 1952-63, 1964-66, 1967-69, 1970-1973 jan, 1973 feb-1986, 1987 jan-1994 dec – 8r – 1 – mf#765731 – us WHS [241]
Catholic worker – New York. 1970+ (1) 1979+ (5) 1979+ (9) – ISSN: 0008-8463 – mf#7588 – us UMI ProQuest [241]
Catholic worker – v1-27. 1933-61 – (= ser Radical periodicals in the united states, 1881-1960. series 1) – 2r – 1 – $365.00 – us UPA [241]
Catholic world – Mahwah. 1989-1996 (1) 1989-1996 (5) 1989-1996 (9) – (cont: new catholic world) – ISSN: 1042-3494 – mf#813,01 – us UMI ProQuest [241]
Catholic Young Mens' Society of Fiji *see* Constitution, correspondence with bishop c j nicolas, sm
Catholic zulu testimony / Wanger, W – Mariannhill, South Africa. 1913 – 1r – us UF Libraries [241]
Catholica – 1(1932)-25(1971) – 142mf – 9 – €271.00 – ne Slangenburg [241]
Catholicism and independence : being studies in spiritual liberty / Petre, Maude Dominica – London; New York: Longmans, Green, 1907 – 1mf – 9 – 0-8370-8778-3 – mf#1986-2778 – us ATLA [241]
Catholicism and secretarianism / Burns, Islay – Edinburgh, Scotland. 1864 – 1r – us UF Libraries [241]
Catholicism and the vatican : with a narrative of the old catholic congress at munich / Whittle, James Lowry – London: Henry S King, 1872 – 1mf – 9 – 0-524-05832-6 – mf#1990-1527 – us ATLA [241]
Catholicism, roman and anglican / Fairbairn, Andrew Martin – New York: Scribner 1899 [mf ed 1990] – 2mf – 9 – 0-7905-3736-2 – mf#1989-0229 – us ATLA [241]
Un catholicisme americain / Delattre, Alphonse J – Namur: Auguste Godenne, 1898 – 1mf – 9 – 0-8370-8417-2 – (incl bibl ref) – mf#1986-2417 – us ATLA [241]
Le catholicisme dans les temps modernes / ed by Gibier, abbe – Paris: P Lethielleux, [1904?] – 4mf – 9 – 0-8370-7946-2 – (incl ind) – mf#1986-1946 – us ATLA [241]
Catholicisme en angleterre au 19e siecle = The english catholic revival in the nineteenth century / Thureau-Dangin, Paul; ed by Wilberforce, Wilfred – New York: E P Dutton. 2v. [19–] – 4mf – 9 – 0-8370-7030-9 – (in english. incl bibl ref and index) – mf#1986-1030 – us ATLA [241]
Le catholicisme en chine au 8 siecle de notre ere : avec une nouvelle traduction de l'inscription de sy-ngan-fou accompagnee d'une grande planche / Dabry de Thiersant, Philibert – Paris: Ernest Leroux, 1877 [mf ed 1995] – (= ser Yale coll) – 58p (ill) – 1 – 0-524-10228-7 – (in french) – mf#1996-1228 – us ATLA [241]
Catholicisme et critique : reflexions d'un profane sur l'affaire loisy / Desjardins, Paul – Paris: Libres entretiens, 1905 – 1mf – 9 – 0-8370-8734-1 – mf#1986-2734 – us ATLA [241]
Catholicisme et loyalisme / Moreno, Enrique – Paris: Ed des Archives espagnoles [1937] [mf ed 1983] – 1mf – 9 – (incl bibl ref) – us NY Public [241]
Catholicisme et papaute / Batiffol, Pierre – Paris, France. 1925 – 1r – us UF Libraries [241]
Catholicisme et rebellion / Martin-Chauffier, Louis – Paris: Comite franco-espagnol 1936 [mf ed 1977] – (= ser Blodgett coll) – 1mf – 9 – mf#w1030 – us Harvard [241]
Catholicity and pantheism : all truth or no truth / Concilio, Januarius de – New York: D & J Sadlier, 1874 [mf ed 1985] – 1mf – 9 – 0-8370-2858-2 – mf#1985-0858 – us ATLA [241]
Catholicity in its relationship to protestantism and romanism : being six conferences. delivered at newark, n. j... / Ewer, Ferdinand Cartwright – new rev ed. New York: E & JB Young, c1878 – 1mf – 9 – 0-8370-8737-6 – mf#1986-2737 – us ATLA [241]
Catholicity in philadelphia : from the earliest missionaries down to the present time / Kirlin, Joseph Louis J – Philadelphia: John Jos McVey, 1909 – 2mf – 9 – 0-524-03936-4 – mf#1990-4930 – us ATLA [241]
Catholicity, protestantism and infidelity : an appeal to candid americans / Weninger, Francis Xavier – 13th ed. New York: Sadlier; Cincinnati: John P Walsh, 1869, c1861 – 1mf – 9 – 0-8370-6856-8 – (also issued under title: protestantism and infidelity) – mf#1986-0856 – us ATLA [240]
O catholico : periodico academico – Sao Paulo, SP: Typ da Tribuna Liberal, 22 jun 1876 – (= ser Ps 19) – bl Biblioteca [377]
Catholicon (ael2/12) : ou dictionnaire universel de la langue francoise / catholicon oder franzoesisch-deutsches universalwoerterbuch der franzoesischen sprache / Schmidlin, Johann Josef – Hamburg 1771-79 [mf ed 1995] – (= ser Archiv der europaeischen lexikographie: woerterbuecher) – 44mf – 9 – €360.00 – 3-89131-195-8 – (int by manfred hoefler) – gw Fischer [054]

The catholicos of the east and his people : being the impressions of five years' work in the "archbishop of canterbury's assyrian mission"... / Maclean, Arthur John & Browne, William Henry – London: SPCK; New York: E & JB Young, 1892 – 1mf – 9 – 0-7905-4952-2 – mf#1988-0952 – us ATLA [390]
The catholicos of the east and his people : being the impressions of five years' work in the "archbishop of canterbury's assyrian mission" / Maclean, Arthur John & Browne, William Henry – London: S.P.C.K.; New York: E.& J.B. Young, 1892 – 1mf – us ATLA [241]
The catholicos of the east and his people... : an account of the religious and secular life...of the eastern syrian christians of kurdistan and northern persia... / Maclean, A J & Browne, W H – London, 1892 – 5mf – 9 – mf#HT-169 – ne IDC [243]
Catholics and the american revolution / Griffin, Martin Ignatius Joseph – Ridley Park PA: M I J Griffin 1907-11 [mf ed 1991] – 3v on 3mf – 9 – 0-524-03153-3 – (incl bibl ref. no more publ) – mf#1990-4602 – us ATLA [241]
Catholics and the civil war in spain : a collection of statements by world-famous catholic leaders on the events in spain – New York NY: Workers Libr Publ [1936?] – 9 – mf#w788 – us Harvard [241]
Catholics and the spanish state / Moreno, Enrique – London: [Friends of Spain] 1937 – 9 – mf#w1063 – us Harvard [230]
Catholics in florida / Brown, Otis, Mrs – s.l, s.l? 193-? – 1r – us UF Libraries [978]
The catholics of ireland under the penal laws in the 18th century / Moran, Patrick Francis – London: Catholic Truth Society, 1900 [mf ed 1986] – 1mf – 9 – 0-8370-7004-X – mf#1986-1004 – us ATLA [241]
The catholics of scotland : from 1593, and the extinction of the hierarchy in 1603, till the death of bishop carruthers in 1852 / Dawson, Aeneas McDonell – London, Ont: T Coffey, 1890 – 10mf – 9 – mf#02602 – cn Canadiana [241]
The catholics of scotland : from 1593, and the extinction of the hierarchy in 1603, till the death of bishop carruthers in 1852 / Dawson, Aeneas McDonell – London, Ont: T Coffey, 1890 – 10mf – 9 – mf#02602 – cn Canadiana [241]
Catholics reply to open letter of 150 protestant signatories on spain – New York NY: America Press [1937] – 9 – (repr fr "catholic mind" nov 22 1937) – mf#w1221 – us Harvard [241]
Catholics speak for spain / North American Committee to Aid Spanish Democracy – New York 1937 – 9 – mf#w789 – us Harvard [241]
Catholicus reformatus : hoc est, expositio et declaratio... / ed by Perkins, William – Hanoviae: Apud Guilielmum Antonium, 1601. Chicago: Dep of Photodup, U of Chicago Lib, 1973 (1r); Evanston: American Theol Lib Assoc, 1984 (1r) – 1 – 0-8370-0009-2 – mf#1984-B383 – us ATLA [241]
Le catholique – v1, no. 1-53. Paris. 5 nov 1881-20 mai 1883. – 1 – (Lacking:no. 14, 32, 50; v2. no. 1-20) – fr ACRPP [241]
Le catholique canadien : etes vous du nombre de ceux qui soutiennent le principe qu'il ne faut pas distribuer la sainte bible au peuple... / Reeves, James – S.l: s.n, 18–? – 1mf – 9 – mf#47652 – cn Canadiana [241]
Le catholique d'action = Catolico de accion / Palau, Gabriel – 3e ed. Paris: Casterman, [1905?] – 1mf – 9 – 0-8370-7091-0 – mf#1986-1091 – us ATLA [241]
Le catholique orthodoxe oppose au catholique papiste... / Rivet, A – Saumur, 1616 – 15mf – 9 – mf#CA-148 – ne IDC [240]
Catholisch gesangbuch voller geistlicher lieder vnd psalmen der alten apostolischer recht vnd war-glaubiger christlicher kirchen / Leisentrit, Johann – Budissin: Wolrab 1584 – (= ser Hqab. literatur des 16. jahrh.) – 8mf – 9 – €80.00 – (missing: p1,152,192) – mf#1584a – gw Fischer [780]
Catholisch gesangbuechlein : in fuenff underschiedliche theil abgetheilt; bey dem catechismo, auch fuernemmen festen, in processionen, creutzgaengen und kirchenfaerten sehr nutzlich zu gebrauchen... – Costantz: Kalt 1600 – (= ser Hqab. literatur des 16. jahrh.) – 6mf – 9 – €70.00 – mf#1600a – gw Fischer [780]
The catholographer : or universal writer – Rockwood, Ont: E Collom, 1868-[18– or 19–] (mthly) – 9 – mf#P05149 – cn Canadiana [400]
Cathrein, Victor *see*
– Die frauenfrage
– Socialism
– Socialism exposed and refuted
Catilina et jugurtha / Sallustius [Gaius Sallustius Crispus] – 15th c – (= ser Holkham library manuscript books 539) – 1r – 1 – (italian trans) – mf#96893 – uk Microform Academic [450]
Catilina et jugurtha *see* Commoediae...
Catilina et jugurtha... / Sallustius [Gaius Sallustius Crispus] – 15th c – (= ser Holkham library manuscript books 339) – 1r – 1 – (filmed with: orationes homeri et eulogium othonis by l aretinus) – mf#96587 – uk Microform Academic [450]
Catimbo / Campos, Sabino De – Rio de Janeiro, Brazil. 1946 – 1r – us UF Libraries [972]
Catinella, Salvatore *see* La corte suprema federale nel sistema costituzionale degli stati uniti d'america
Catley, Delwyn *see* Psychological antecedents of the frequency and intensity of flow in golfers
Catlin, George Edward Gordon *see* In the path of mahatma gandhi
Catlin, Louise E *see* Marjory and her neighbors
Catlow, Agnes *see* Popular conchology
Cato journal – Washington. 1981+ (1,5,9) – ISSN: 0273-3072 – mf#12662 – us UMI ProQuest [350]
Cato von eisen : lustspiel in drei acten / Laube, Heinrich – 2. aufl. Leipzig: J J Weber 1892 [mf ed 1995] – 1r – 1 – (filmed with: heinrich laubes meisterdramen) – mf#3680p – us Wisconsin U Libr [820]
El catolicismo liberal / Tejado, Gabino – 1875 – 9 – sp Bibl Santa Ana [241]
El catolico filipino – Manila: [s.n.], dec 13-17,20-21,24 1898 – us CRL [241]
Catolico o krausista? / Fernandez Valbuena, Ramiro – 1882 – 9 – sp Bibl Santa Ana [241]
Un catolico va al cine / Perez Lozano, Jose Maria – Barcelona: Editorial Juan Flors, 1956 – 1 – sp Bibl Santa Ana [241]
Catolicos y comunistas / Calderio, Francisco – Habana, Cuba. 1940 – 1r – us UF Libraries [972]
Catolocismo de la juventud colombiana / Fermoso Estebanez, Paciano – Bogota, Colombia. 1961 – 1r – us UF Libraries [972]
Caton, James R *see* Legislative chronicles of the city of alexandria
Caton, John Dean *see* A summer in norway
Caton-Thompson, Gertrude *see* Zimbabwe culture
Catoosa county news – Ringgold, GA. 1991-2000 (1) – mf#68700 – us UMI ProQuest [071]
Cator, Leonce *see* Precis elementaire de droit commercial.
Catorce pecados de humor y una vida descabellada / Llorens, Washington – San Juan, Puerto Rico. 1959 – 1r – us UF Libraries [972]
Catriona : a sequel to "kidnapped": being memoirs of the further adventures of david balfour at home and abroad / Stevenson, Robert Louis – Toronto: Musson Book Co; London: Cassell, 1892? – 5mf – 9 – mf#33588 – cn Canadiana [830]
Catrou, Fran ois *see* History of the mogul dynasty in india
Catrou, Francois *see* The general history of the mogol empire
Catrufo, Giuseppe *see*
– Duo de felicie
– Rondeau de l'intrigue au chateau
Cats, Jacob *see*
– Maechden-plicht ofte ampt der jonck-vrouwen
– Maegden-plicht ofte ampt der ionghvrouwen
– Monita amoris virginei
– Proteus ofte minne-beelden verandert in sinne-beelden
– Silenus alcibiadis
– Silenus alcibiadis sive proteus
– Spiegel van den ouden en nieuwen tyd...
– Spiegel van den ouden en nieuwen tydt...
– Spiegel van den ouden ende nieuwen tijdt...
– Zinne- en minne-beelden
[Cats, Jacob] *see* Silenus alcibiadis
Catt, Carrie Clinton Chapman *see* Papers
Cattaneo, Carlo *see* Terre italiane
Cattaneo, E *see* Il breviario ambrosiano
Cattaneo, Raffaele *see* Architecture in italy from the sixth to the eleventh century
Cattano, Carlo *see* Scritti economici
Cattaraugus republican – Ellicottville, NY. 1844-1854 (1) – mf#64954 – us UMI ProQuest [071]
Catteau-Calleville, Jean *see*
– Gemaelde der ostsee in physischer, geographischer, historischer und merkantilischer ruecksicht
– Voyage en allemagne et en suede
Cattell, Raymond B *see* Subjective character of cognition and the pres-sensational developm...
Catterall, Ralph Charles Henry *see* The second bank of the united states
Cattle and kinship among the gogo / Rigby, Peter – Ithaca, NY. 1969 – 1r – us UF Libraries [307]
Cattle feeding in southern florida / Kidder, Ralph W – Gainesville, FL. 1941 – 1r – us UF Libraries [636]
Cattle plague : a warning voice to britain from the king of nations / Waldegrave, Samuel – London, England. 1866 – 1r – us UF Libraries [240]
Cattleman – Fort Worth. 1949+ (1) 1970+ (5) 1976+ (9) – ISSN: 0008-8552 – mf#235 – us UMI ProQuest [636]
Catto, Octavius V *see* Our alma mater
Il cattolicismo rosso: studio sul presente movimento di riforma nel cattolicismo / Prezzolini, Giuseppe – Napoli: R. Ricciardi, 1908. xx,348p – 1 – us Wisconsin U Libr [241]
Cattopadhyaya, Basantakumara *see* The teachings of the upanishads
Cattopadhyaya, Saratcandra *see*
– The deliverance
– Srikanta
Catuaba – Fortaleza, CE. 06 nov 1890 – (= ser Ps 19) – mf#P17,01,39 – bl Biblioteca [079]
Catulli, tubulli, propertii, carmina / Mauricio Hauptio, A – Lipsiae, Germany. 1912 – 1r – us UF Libraries [960]
Catullus – London, England. 1926 – 1r – us UF Libraries [450]
Catullus, Gaius Valerius *see* Select poems of catullus edited
Caturla, Maria Luisa *see*
– Bodas y obras juveniles de zurbaran
– Zurbaran en san pablo de sevilla
Catv : newsweekly of catv and pay-cable – Atlanta. 1967-1976 (1) 1972-1976 (5) 1974-1976 (9) – (cont by: vue: news magazine of catv and pay-cable) – ISSN: 0574-9204 – mf#8073 – us UMI ProQuest [380]
"Cat-wagon trails." / Clugston, W G – 1 – us Kansas [978]
Cau Sam Qan *see* Pratidin paramn-khmaer
Cau Sau *see* Ganapaks prajadhipateyy 1972-73
Cau sradap cek / Yas Nin – Bhnam Ben: Pannagar Jyn Nuan Huat 2502 [1959] [mf ed 1990] – 1r with other items – 1 – mf#mf-10289 seam reel 107/9 [§] – us CRL [959]
Cauca (Colombia) *see* Codigo de leyes y decretos del estado s del cauca
Cauca (Colombia : Dept) *see* Recopilacion de leyes del estado soberano del cauc...
Le caucase, nouvelles impressions de voyage / Dumas, A [pere] – Leipzig: A Durr, 1859. 3v – 12mf – 9 – mf#AR-2087 – ne IDC [914]
Cauce – Badajoz, 1958-1964 – 5 – sp Bibl Santa Ana [073]
Cauce hondo / Ramirez De Arellano De Nolla, Olga – San Juan, Puerto Rico. 1947 – 1r – us UF Libraries [972]
Le cauchemar des intrigants politiques – [Paris]: Typographie Benard et Comp. n1. oct 9 1848 – us CRL [320]
Cauchois, H *see* Cours oral de franc-maconnerie symbolique
Cauchois-Lemaire, Louis Augustin *see* Lettres sur les cent-jours
Cauchon, Alphonse *see* Lac megantic, la compagnie nantaise, le chemin de fer 1879-1936
Cauchon, Joseph *see* Discours de l'hon jos cauchon sur la question de la confederation
Caucus calendar / Wisconsin Women's Political Caucus – v1 n2-v2 n1 [1973 mar-1974 mar] – 1r – 1 – (cont by: caucus news) – mf#697323 – us WHS [325]
Caudillism and militarism in venezuela, 1810-1910 / Gilmore, Robert L – Athens, OH. 1964 – 1r – us UF Libraries [972]
Caudillismo en la republica dominicana / Monclus, Miguel Angel – Ciudad Trujillo, Dominican Republic. 1948 – 1r – us UF Libraries [972]
Caudillo y gobernante / Navia Varon, Hernando – Cali, Colombia. 1964 – 1r – us UF Libraries [972]
El caudillo y los combatientes – Bilbao, 1938 – (= ser Blodgett coll) – 9 – mf#fiche w807 – us Harvard [946]
Caufeild, Miss *see* Martyrs omitted by foxe
Caught in the chinese revolution : a record of risks and rescue / Borst-Smith, Ernest Frank – London; Leipsic: T Fisher Unwin, 1912 [mf ed 1995] – (= ser Yale coll) – 125p (ill) – 1 – 0-524-09221-4 – mf#1995-0221 – us ATLA [951]
Cauiliuiling dalauang pu at dalauang cuento : nang magcapatid na si leandro at si orencio at ng naguucol na amang si rubin ng canicaniyang mga carapatan / Ignacio, Cleto R – [Maynila]: P Sayo 1916 [mf ed Bloomington IN: Indiana Uni Lib, Preservation Dept 1984] – (= ser Coll...in the tagalog language 2) – v2 on 1r – 1 – us Indiana Preservation [490]
Caula, Giacomo Alessandro *see* Baltazarini e il "balet comique de la royne"
Caulfeild, Sophia Frances Anne *see* The dictionary of needlework
Cauliflower / Hume, H Harold – Lake City, FL. 1901 – 1r – us UF Libraries [634]
Caumette, Ch *see* Eclaircissemens des antiquites de la ville de nismes
Caumont, A de *see* Cours d'antiquitees monumentales
Caumont, Armand *see* Goethe et la litterature francaise
Caumont de LaForce, Charlotte R de *see* Anecdotes du seizieme siecle
Caurenta anos de vida de la academia / Santovenia Y Echaide, Emeterio Santiago – Habana, Cuba. 1950 – 1r – us UF Libraries [972]
Caus, Salomon de *see* La perspective avec la raison des ombres et miroirs
Causa – 1981 jul-1988 – 1r – 1 – (cont by: american leadership) – mf#929336 – us WHS [243]
Causa inmediata de la emancipacion de panama / Castillero R, Ernesto J – Panama, Panama. 1933 – 1r – us UF Libraries [972]
Causa usa report – v1 n1-6 [1984 apr-nov/dec], v2 n3,5-8 [1985 jan/feb-apr, jun-aug], v3 n2,4-7 [1986 jul, sep-dec], v4 n1-11 [1987 jan-dec], v5 n1-10 [1988 jan-oct] – 1r – 1 – mf#1054137 – us WHS [243]
Causal attributions and task persistence of learned-helpless and mastery-oriented sixth graders : in math, physical education, and reading / Griffith, Joseph B – 1994 – 2mf – $8.00 – us Kinesology [370]
Die causalbetrachtung in den geisteswissenschaften / Ritschl, Otto – Bonn: A. Marcus u. E. Weber, 1901 – 1mf – 9 – 0-7905-6356-8 – mf#1988-2356 – us ATLA [100]
Causalite et creation : le continu et le discontinu dans l'oeuvre d'henri bergson / Anastassopoulou, Itheoni – 2mf – 9 – (10531) – fr Atelier National [110]
Causas de infidencia / Briceno Perozo, Mario – Madrid, Spain. 1961 – 1r – us UF Libraries [972]
Causas y efectos de una dictadura / Paredes Cruz, Joaquin – Cali, Colombia. 1957 – 1r – us UF Libraries [972]
Causation : and, freedom in willing; together with, man a creative first cause, and kindred papers / Hazard, Rowland Gibson; ed by Hazard, Caroline – Boston: Houghton, Mifflin 1889 [mf ed 1991] – 1mf – 9 – 0-7905-8656-8 – mf#1989-1881 – us ATLA [120]
Cause – Dublin, Ireland. 3 jan-17 jan 1880 – 1/4r – 1 – uk British Libr Newspaper [072]
The cause and circumstances of mr bidwell's banishment by sir f b head : correctly stated and proved / Ryerson, Egerton – Kingston Ont?: s.n, 1838 (Kingston Ont: T H Bentley) – 1mf – 9 – mf#47493 – cn Canadiana [971]
The cause and cure of infidelity : including a notice of the author's unbelief and the means of his rescue / Nelson, David – New York: American Tract Society, c1841 [mf ed 1994] – 1mf – 9 – 0-524-08849-7 – mf#1993-2134 – us ATLA [210]
Cause and cure of social evil considered – Edinburgh, Scotland. 1861 – 1r – us UF Libraries [240]
Cause and remedy for national distress / Stewart, James Haldane – London, England. 1826 – 1r – us UF Libraries [240]
La cause catholique : discours destine a la seance de cloture du congres catholique reuni a malines en 1863 / Dechamps, Victor Auguste – Tournai: H Castermann, 1863 – 1mf – 9 – 0-524-03790-6 – mf#1990-4862 – us ATLA [241]
La cause des obligations et prestations / Billette, J Emile – Montreal, 1933. 157, 2 p. LL-2325 – 1 – us L of C Photodup [340]
La cause immorale, etude de jurisprudence / Dorat des Monts, Roger – Paris, Librairie Rousseau, 1956. 173 p. LL-4097 – 1 – us L of C Photodup [340]
Cause masson-prevost : memoire presentee a monseigneur zotique racicot, p a, juge delegue en cette cause par l'avocat du demandeur / Auclair, Elie-Joseph – Montreal: Arbour & Laperle, 1900 – 1mf – 9 – (incl bibl ref) – mf#61708 – cn Canadiana [346]
Cause of christ and the cause of satan / Duff, Alexander – Edinburgh, Scotland. 1843 – 1r – us UF Libraries [240]
The cause of god and truth : in four parts / Gill, John – new ed. London: WH Collingridge, 1855 [mf ed 1994] – 4mf – 9 – 0-524-08788-1 – (incl bibl ref) – mf#1993-3280 – us ATLA [230]
The cause of the degradation of man / Adams, Henry – St John, NB: E J Armstrong, 1895 – 1mf – 9 – mf#06151 – cn Canadiana [230]
The cause of the distress at present prevailing in great britain and ireland / Barker, Joseph – [London? 1845?] – (= ser 19th c ireland) – 1mf – 9 – mf#1.1.9495 – uk Chadwyck [941]
Cause of the lord's sufferings : and the true nature of the atonement / Sibly, Manoah – London, England. 1796 – 1r – us UF Libraries [240]
The cause of the operatives of ireland : advocated in a letter to sir robert kane / Naper, James Lenox William – Dublin, 1853 – (= ser 19th c ireland) – 1mf – 9 – mf#1.1.2216 – uk Chadwyck [338]
Cause of the people *see* New moral world, 1845

The cause of the war / Jefferson, Charles Edward – New York: Church Peace Union, [1914?] – (= ser The Church and International Peace) – 1mf – 9 – 0-7905-9233-9 – mf#1989-2458 – us ATLA [940]
Cause/effect – Boulder. 1980-1999 (1) 1980-1999 (5) 1980-1999 (9) – ISSN: 0164-534X – mf#12570 – us UMI ProQuest [378]
La causerie – Paris. 1859-avr 1862 – 1 – fr ACRPP [073]
Causerie a propos de lohengrin / Jullien, Adolphe – Paris: Librarie E Sagot, – 1re annee n1 [mars 1887], p4-14 – 1mf – 9 – (in: l'independance musicale et dramatique) – mf#wa-50 – ne IDC [780]
Une causerie agricole – St Hyacinthe Quebec: Courrier, 1872 – 1mf – 9 – mf#28042 – cn Canadiana [630]
Causerie par m bourassa : a la chapelle notre-dame de lourdes de montreal, le 22 juin 1880 – Montreal: s.n, 1880? – 1mf – 9 – mf#04048 – cn Canadiana [720]
Causeries : la liberte vs la mode et al, et la politique – Saint-Hyacinthe, Quebec: [s.n.], 1886 [mf ed 1980] – 1mf – 9 – 0-665-04360-0 – mf#04360 – cn Canadiana [120]
Causeries congolaises / Torday, E – Bruxelles: A Dewit, 1925 – 1 – us CRL [490]
Causeries litteraires et historiques / Janin, Jules Gabriel – Paris, France. 1894 – 1r – us UF Libraries [960]
Causeries sur le protestantisme d'aujourd'hui. english *see* Plain talk about the protestantism of to-day
The causes and cure of unbelief = Pourquoi l'on ne croit pas / Laforet, Nicolas Joseph; ed by Gibbons, James – Philadelphia: HL Kilner, c1909 [mf ed 1991] – 1mf – 9 – 0-7905-8821-8 – (in english) – mf#1989-2046 – us ATLA [210]
The causes and effects of war : a sermon, delivered in salem, aug 20 1812, the day of national humiliation and prayer / Emerson, Brown, 1778-1872 – Salem [MA]: printed by Joshua Cushing, 1812 [mf ed 1984] – 1mf – 9 – 0-665-44469-9 – mf#44469 – cn Canadiana [230]
Causes and reduction of mechanical vibration in an electric drive system / Zhao, Jian – Stellenbosch: U of Stellenbosch 1998 [mf ed 1998] – 3mf – 9 – mf#mf.1309 – sa Stellenbosch [621]
The causes and remedies of pauperism in the united kingdom considered : pt 1: being a defence of the principles and conduct of the emigration committee, against the charges of mr sadler / Wilmot-Horton, Robert John – London 1829 – (= ser 19th c british colonization) – 2mf – 9 – mf#1.1.2250 – uk Chadwyck [304]
Causes, consequences, and remedy of intemperance / Sherman, James – London, England. 1841? – 1r – us UF Libraries [240]
Causes et precis des troubles, des crimes, des desordres dans le departement du gard et dans d'autres lieux du midi de la france, en 1815 et en 1816 : faisant suite aux eclaircissemens historiques en reponse aux calomnies dont les protestans du gard sont l'objet / Lauze de Peret, Pierre – Paris 1819 [mf ed Hildesheim 1995-98] – 1v on 3mf – 9 – €90.00 – 3-487-26322-X – gw Olms [944]
The causes leading to the organization of the cumberland presbyterian church / Stephens, John Vant – Nashville: Cumberland Presbyterian Pub. House, 1898 – 1mf – 9 – 0-7905-6367-3 – (incl bibl ref) – mf#1988-2367 – us ATLA [242]
Causes of declension in christian churches / Arundel, John – London, England. 1830 – 1r – us UF Libraries [240]
The causes of foreign invasion in spain – n.p. 193? Fiche W 790. (Blodgett Collection of Spanish Civil War Pamphlets) – 9 – us Harvard [946]
The causes of the corruption of the traditional text of the holy gospels : being the sequel to the traditional text of the holy gospels / Burgon, John William; ed by Miller, Edward – London: George Bell; Cambridge: Deighton, Bell, 1896 – 1mf – 9 – 0-8370-3055-2 – (incl bibl ref, general index, and index of biblical passages cited) – mf#1985-1055 – us ATLA [226]
Causes of the decline of interest in critical theology : an address delivered...cambridge theological school," jul 16 1847 / Noyes, George Rapall – [Boston: William Crosby; London: John Chapman, 1847] [mf ed 1984] – (= ser Biblical crit – us & gb 78/1) – 1mf – 9 – 0-8370-1578-2 – mf#1984-6251 – us ATLA [220]
Causes of the failure of the cement pipe used in sub-irrigation / Hubbard, Donald – s.l, s.l? 1924 – 1r – us UF Libraries [630]
Causes of the increases of the churches / Williams, William R – (A discourse). 1842 – 1 – 5.00 – us Southern Baptist [242]
Caussede, Jean Pierre De *see* Self-abandonment to divine providence
Caussin, N *see*
– De symbolica aegyptiorum sapientia in quo symbola, parabolae, historiae selectae...
– Electorum symbolorum et parabolarum historicarum syntagmata
Cautereels, Peter Jozef *see* Piet de franc-macon
Caution against enthusiasm – London, England. 1794 – 1r – us UF Libraries [240]
Caution against ill company / Ellesby, James – London, England. 1792 – 1r – us UF Libraries [240]
Caution against infidelity and atheism – Newcastle upon Tyne, England. 1851 – 1r – us UF Libraries [240]
Caution against irreligion and anarchy / Keith, George Skene – Edinburgh, Scotland. 1794 – 1r – us UF Libraries [240]
Cautions and counsels to new converts – London, England. 18– – 1r – us UF Libraries [240]
Cautions for the times : addressed to the parishioners of a parish in england by their former rector / Whately, Richard & Fitzgerald, William – 2nd ed. London: John W Parker, 1854 – 2mf – 9 – 0-524-00418-8 – mf#1989-3118 – us ATLA [240]
Cautions to continental travellers / Cunningham, J W – London, England. 1823 – 1r – us UF Libraries [240]
Cautus *see* Letter to the right honourable father in god, william skinner
Cauvain, Henri *see* Le grand vaincu
Cauvin, Leger *see*
– Affaire maunder
– Discours sur la constitution de 1889
Cauwenbergh, P van *see* Etudes sur les moines d'egypte
Caux de Cappeval *see* Apologie du gout francois, relativement a l'opera
Cavagnoli, Stefania *see* Il linguaggio della medicina con funzione divulgativa
Cavaignac et louis-napoleon devant le pays – Paris, [1848?] – us CRL [944]
Cavaille, Jean Pierre *see* Le monde de descartes
Cavaille-Coll, Aristide *see* De l'orgue et de son architecture
Cavalacanti De Carvalho, M *see* Evolucao do estado brasileiro
Cavalcanti, A *see* Esequie del serenissimo principe francesco
Cavalcanti, Araujo *see* Desenvovimento economico e social dos municipios
Cavalcanti De Albuquerque Mello, Felix *see* Memorias de um cavalcanti
Cavalcaselle, Giovanni Battista *see*
– The early flemish painters
– A history of painting in north italy...
– A new history of painting in italy from the 2nd-16th century
– Raphael
– Titian
Cavaleiro da esperanca / Amado, Jorge – Rio de Janeiro, Brazil. 1956 – 1r – us UF Libraries [972]
Cavalerie signale – [18–?] [mf ed 1989] – 1r – 1 – mf#pres. film 62, 33 – us Sibley [780]
Cavalier, Anthony Ramsen *see* In northern india
Cavalieri, J M *see* Opera omnia liturgica
Cavalieri, Pio Franchi de *see* Gli atti dei ss montano
Cavaliers and roundheads of barbados / Davis, Nicholas Darnell – Georgetown, Guyana. 1887 – 1r – us UF Libraries [972]
Cavallera, F *see* Le schisme d'antioche (4e-5e siecle)
Cavallo, J A *see* Vollstaendiger bericht von allen sehens-wuerdigen freunden-festen
Cavalo na formacao do brasil / Goulart, Jose Alipio – Rio de Janeiro, Brazil. 1964 – 1r – us UF Libraries [972]
Cavalry tactics, u.s. army, assimilated to the tactics of infantry and artillery (new york, 1874) / U.S. War Dept. Adjutant General's Office – (= ser Records of the adjutant general's office, 1780's-1917) – 1r – 1 – mf#T1109 – us Nat Archives [355]
Cavan herald and inland general advertiser – Cavan, 31 aug-23 nov 1824, mar-2 aug 1825 – 1r – 1 – ie National [072]
Cavan herald and inland general advertiser – Cavan, Ireland. 14 jul 1818 – 1/4r – 1 – uk British Libr Newspaper [072]
Cavan observer – Cavan, Ireland. 11 jul 1857-29 oct 1864 – 3r – 1 – uk British Libr Newspaper [072]
Cavan weekly news – Ireland. -w. 1893. 30 ft – 1 – uk British Libr Newspaper [072]
Cavan weekly news and general advertiser – Cavan, 1901-02, 1905-07, 10 jul-18 dec 1909 – 5.5r – 1 – ie National [072]
Cavan weekly news and general advertiser – Cavan, Ireland. 16 dec 1864-1896; 1900 (jun 1873 missing) – 13r – 1 – uk British Libr Newspaper [072]
Cavanagh, William Henry *see* The word protestant in literature, history and legislation
Cavanillas y Munoz, Juan Alonso *see* Articulos de costumbres
Cavazzi, G A *see* Istoria descrizione de tre regni congo, matamba et angola
Cavazzi, Giovanni Antonio *see* Relation historique de l'ethiopie occidentale
Cave, Alfred *see*
– The battle of standpoints
– Battle of the standpoints
– The inspiration of the old testament inductively considered
– An introduction to theology
– The scriptural doctrine of sacrifice and atonement
– The spiritual world
Cave, Charles John Philip *see* Roof bosses in medieval churches
Cave, Henry *see* The ruined cities of ceylon
Cave spring baptist church. roanoke, virginia : church records – 1899-1935, 1948-59. Bulletins. 1947-57 – 1 – 77.18 – us Southern Baptist [242]
The cave temples of India / Fergusson, James & Burgess, James – London 1880 – (= ser 19th c art & architecture) – 8mf – 9 – mf#4.2.488 – uk Chadwyck [720]
Caveat emptor – West Orange. 1975-1982 (1) 1975-1982 (5) 1975-1982 (9) – ISSN: 0045-6004 – mf#9655 – us UMI ProQuest [650]
Caveeshar, Sardul Singh *see* The sikh studies
Caveler, William *see* Select specimens of gothic architecture
Cavelier, German *see* Statement of the laws of colombia in matters affec...
Cavelti, Sigisbert *see* Angelomontana
Cavelti, Sigisbert et al *see* Angelomontana
Caven, William *see*
– Christ's teaching concerning the last things
– The divine foundation of the lord's day
– The scripture readings
– The testimony of christ to the old testament
Cavenagh, Orfeur *see* Reminiscences of an indian official
Cavendish, George *see*
– The life and death of cardinal wolsey
– The life of cardinal wolsey
Cavendish, Henry *see*
– Debates of the house of commons in the year 1774 on the bill for making more effectual provision for the government of province of quebec
– Experiments to determine the density of the earth
Cavens/cavins newsletter – v1 n1-v5 n3 [1982 jul-1987 mar] – 1r – 1 – mf#1220581 – us WHS [071]
Caverno, Charles *see* A narrow ax in biblical criticism
Caviedes, Juan del Valle y *see*
– [Poesias]
Cavling, Henrik *see* Danske vestindien
[Cavriuolo, A] *see* Il sontuoso apparato, fatta dalla magnifica citta di brescia, nel felice ritordell'illu e reverendiss vescovo suo, il cardinale morosini
Caw tca union : the national magazine of caw-canada – 1988 fall-1992 spr/summer – 1r – 1 – (cont: national union magazine) – mf#1870803 – us WHS [331]
CAW-Canada *see* National union magazine
Cawdrey, Robert *see* A table alphabeticall of hard usual english wordes
Cawood, John *see*
– Christian watchman
– Church of england and dissent
Caxias / Carvalho, Affonso De – Rio de Janeiro, Brazil. 1944 – 1r – us UF Libraries [972]
Caxias e o problema militar brasileiro / Raposo, Amerino – Rio de Janeiro, Brazil. 1969 – 1r – us UF Libraries [972]
Caxton 1599-1950 – (= ser Cambridgeshire parish register transcript) – 5mf – 9 – £6.25 – uk CambsFHS [929]
Caxtons / Lytton, Edward Bulwer Lytton, Baron – Boston, MA. v1-2. 1898? – 1r – us UF Libraries [025]
Cayce, Claudius Hopkins *see* Is salvation conditional or unconditional?
Caycedo, Bernardo J *see* Grandezas y miserias de dos victorias
Cayet, P *see* Paradigmata de qvator lingvis orientalibvs, praecipvis arabica, armena, syra, aethiopica
Cayet, Pierre-Victor *see*
– Chronologie septenaire de l'histoire de la paix entre les roys de france et d'espagne
Cayetano, Luis *see* Promptuario llenerense gramatico-latino
Cayetano Rosado, Moises *see* He tenido sujeta la palabra entre los dientes
Cayla, Jean-Mamert *see*
– Le 89 du clerge
– Jesuites hors la loi
Cayley, George *see* Aeronautical and miscellaneous notebooks, c1799-1826
Caylus, M de *see* Memoire sur la peinture... l'encaustique et sur la peinture...la cire
Caylus, Marthe Margu'erite LeValois de Villette de Marcay *see* Les souvenirs de m le comte de caylus
Caymanian compass : or the daily caymanian compass – George town, Cayman Islands. jan-sept 16 1986, mar 1987-aug 1990 – 18r – 1 – us L of C Photodup [079]
Cayolla, Julio *see* Brasil
Cayton's weekly – Seattle, WA. 1917-1921 (1) – mf#67101 – us UMI ProQuest [071]
Cayuaga chief – Weedsport, NY. 1877-1953 (1) – mf#65278 – us UMI ProQuest [071]
Cayuaga chief chronicle – Weedsport, NY. 1954-1975 (1) – mf#69013 – us UMI ProQuest [071]
Cayuga chief – Auburn NY, Fort Atkinson WI. 1853 jan 4-1855 sep 25, 1853 may 3-1855 jan 30, 1855 oct 2-1857 jun 17, 1857 jan 28-jun 17 – 4r – 1 – (cont by: wisconsin chief) – mf#856285 – us WHS [071]
Cayuga republican – Auburn NY, 1823 aug 20, 1824 oct 20, 1825 jan 5-mar 30, 1828 sep 3, 1829 may 27 – 1r – 1 – (cont: auburn gazette; cont by: free press) – mf#855767 – us WHS [071]
La caza de la perdiz con reclamo / Gonzales Borreguero, G – Caceres: Tip. de El Noticiero – 1 – sp Bibl Santa Ana [790]
La caza del perdigon / Vela Hidalgo y Burriel, Angel – Madrid, 1920 – 1 – sp Bibl Santa Ana [630]
Cazadores de cabezas del amazonas / Up De Graff, Fritz W – Madrid, Spain. 1928 – 1r – us UF Libraries [972]
Caze, Jean *see* Refutation de la vie de napoleon de sir walter scott
Cazenave De La Roche, Jean *see* Tension aux antilles francaises
Cazenove, John Gibson *see* Historic aspects of the priori argument concerning the being and atrributes of god
Cazenovia leader – Cazenovia WI. 1938 nov 3-1940 dec 5 – 1r – 1 – (cont by: la valle weekly) – mf#955077 – us WHS [071]
Cazenovia reporter – Cazenovia WI. 1911 mar 3-1914, 1915-32, 1933-1937 jan 27 – 8r – 1 – mf#958041 – us WHS [071]
Cazes, Paul de *see*
– Code scolaire de la province de quebec
– Deux points d'histoire
– L'episode de l'ile de sable
– La frontiere nord de la province de quebec
– L'instruction publique dans la province de quebec
– La langue que nous parlons
– Manuel de l'instituteur catholique de la province de quebec
– Le masque de fer n'etait pas matthioli
– Notes sur le canada
– Le petit manuel canadien
Cazes, Paul de [comp] *see*
– Code de l'instruction publique de la province de quebec
– Code of public instruction of the province of quebec
Cazet, Cl *see* Du mode de filiation des racines semitiques et de l'inversion
Cazneau, Jane Maria Mcmanus *see* Life in santo donimgo
Cazzaniga, Ignazio *see* Problemi intorno alla farsaglia
Cba builder / Conservative Baptist Association of America – Oct 1957-Aug 1970 – 1 – us Southern Baptist [242]
Cba record – Chicago. 1987-1996 (1,5,9) – ISSN: 0892-1822 – mf#16710 – us UMI ProQuest [340]
Cba record – v1-15. 1987-2001 – 9 – $274.00 set – (cont: chicago bar record) – ISSN: 0892-1822 – mf#110991 – us Hein [340]
Cbc quarterly/ / Citizens Budget Commission [New York NY] – v1 n1-v3 n1 [1981 jul-1983:winter] – 1r – 1 – (cont by: quarterly [citizens budget commission [new york ny]) – mf#681350 – us WHS [350]
Cbfsa news / California Black Faculty and Staff Association – 1982 sep/oct, 1983 jan/feb, nov/dec, 1985 jan/mar, sep/oct, 1986 apr/jun, 1988 jan/mar, 1989 dec/1990 mar – 1r – 1 – mf#3946027 – us WHS [305]
Cbia news / Connecticut Business and Industry Association – Hartford. 1980-1980 (1) – ISSN: 0199-686X – mf#9810,02 – us UMI ProQuest [338]
CBMR digest / Columbia College – Chicago IL. 1995 fall, 1996 fall – 1r – 1 – (cont: inside cbmr) – mf#1355931 – us WHS [378]
Cbnl : [manuscripts] – [mf ed Phnom Penh, Cambodia: Cambodia Microfilm Project, Ministry of Information and Culture; Ithaca NY: [John M Echols Collection] Cornell University 1990] – 39r – 1 – (mf of mss held in the cambodia national library; in khmer & pali) – mf#mf-11613. – us CRL [480]
Cbnm : [manuscripts] – [mf ed Phnom Penh, Cambodia: Cambodia Microfilm Project, Ministry of Information and Culture; Ithaca NY: [John M Echols Collection] Cornell University 1990] – 7r – 1 – (mf of mss held in the library of the cambodia national museum; chiefly in khmer) – mf#mf-11612 – us CRL [480]
Cbo role and performance : enhancing accuracy, reliability, and responsiveness in budget and economic estimates: hearing...washington dc, may 2 2002 / United States. Congress. House. Committee on the Budget – Washington: US GPO 2002 [mf ed 2002] – 1mf – 9 – 0-16-068659-8 – (incl bibl ref) – us GPO [339]

Cbu drumbeat : ...newsletter / Coalition for Black Unity – 1994 mar – 1r – 1 – mf#3912551 – us WHS [321]
Ccar journal / Central Conference of American Rabbis – New York. 1972-1978 (1) 1972-1978 (5) 1976-1978 (9) – (cont by: journal of reform judaism) – ISSN: 0007-7976 – mf#6963 – us UMI ProQuest [270]
Ccar journal – New York. 1991+ (1) 1991+ (5) 1991+ (9) – (cont: journal of reform judaism) – ISSN: 1058-8760 – mf#6963,02 – us UMI ProQuest [270]
The ccc and wildlife – Washington, DC: GPO, 1938 – us CRL [630]
Ccca newsletter / Concerned Citizens for Choice on Abortion [Vancouver BC] – n1-22 [1981 dec-1988 apr] – 1r – 1 – mf#1053760 – us WHS [170]
Ccco news notes / Central Committee for Conscientious Objectors – 1951 oct-1969 – 1r – 1 – (cont: news notes of the central committee for conscientious objectors; cont by: objector [san francisco ca]) – mf#632208 – us WHS [355]
Ccco news notes / Central Committee for Conscientious Objectors – Philadelphia. 1949-1993 (1) 1970-1993 (5) 1976-1993 (9) – ISSN: 0008-5952 – mf#3241 – us UMI ProQuest [320]
Ccea newsletter / East China Christian Education Association – v2-3. 1948-49* – 1r – 1 – ISSN: 0310-1878 – mf#ATLA S0707A – us ATLA [240]
Ccf news for british columbia and the yukon – Vancouver, British Columbia. v. 10, no. 22-v. 25, no. 9. Jan 3 1946-Sept 19 1961. Incomplete – 1 – us NY Public [971]
Cclp : contents of current legal periodicals – Wilmington. 1976-1980 (1) 1976-1980 (5) 1976-1980 (9) – (cont: contents of current legal periodicals. cont by: legal contents (lc)) – ISSN: 0147-0493 – mf#9481,01 – us UMI ProQuest [340]
Ccq : critical care quarterly – Rockville. 1978-1986 (1,5,9) – (cont by: critical care nursing quarterly) – ISSN: 0160-2551 – mf#12728 – us UMI ProQuest [610]
Ccssq community college social science quarterly – El Cajon. 1970-1976 (1) 1970-1976 (5) 1970-1976 (9) – ISSN: 0045-7728 – mf#10973 – us UMI ProQuest [300]
Ccwhp : [newsletter] / Coordinating Committee on Women in the Historical Profession – 1982 mar-1983 dec – 1r – 1 – (cont: ccwhp newsletter [1974]; cont by: ccwhp newsletter [1984]) – mf#1222454 – us WHS [305]
Ccwhp newsletter / Conference Group in Women's History – 1974 jan-1981 fall – 1r – 1 – (cont by: ccwhp: [newsletter]) – mf#1221661 – us WHS [305]
Ccwhp newsletter / Coordinating Committee on Women in the Historical Profession – 1984 feb-1987 jan – 1r – 1 – (cont: ccwhp; cont by: cgwh newsletter [manhattan ks]; ccwh newsletter) – mf#1222455 – us WHS [305]
Cd review – Hancock. 1989-1992 (1) 1989-1989 (5) 1989-1989 (9) – ISSN: 1044-1700 – mf#16388,03 – us UMI ProQuest [621]
Cd4- und cxcr4-vermittelte apoptose als moeglicher mechanismus der t-zell-depletion bei aids / Berndt, Christina – (mf ed 1998) – 2mf – 9 – €40.00 – 3-8267-2575-1 – mf#DHS 2575 – gw Frankfurter [574]
Cdl report / Christian Defense League [US] – 1980 jul/aug, 1981-85 – 1r – 1 – (cont: christian vanguard [metairie la]) – mf#1043306 – us WHS [240]
Cdl reporter / Citizens for Decency through Law [US] – 1986 jan/feb-1989 may/jun – 1r – 1 – (cont: national decency reporter; cont by: clf reporter) – mf#1609435 – us WHS [343]
Cdm weekly bulletin *see* Oranjemund newsletter and weekly bulletin
Cd-rom professional – Wilton. 1990-1996 (1) 1990-1996 (5) 1990-1996 (9) – (cont: laserdisk professional. cont by: e media professional) – ISSN: 1049-0833 – mf#16703,01 – us UMI ProQuest [020]
Cds review – Chicago. 1973-1991 (1) 1975-1980 (5) 1975-1980 (9) – (cont: fortnightly review of the chicago dental society) – ISSN: 0091-1666 – mf#3307,01 – us UMI ProQuest [617]
Cdu-informationsdienst : informationsdienst des zonenausschusses der cdu – Frankfurt/M DE, 1947 n1-24, 1948 12 aug-1952 [gaps] – 2r – 1 – (title change: 17 may 1950: informationsdienst der cdu deutschlands: 1950 17 may-23 dec [n23-87/88], 1951 5 jan-10 feb [n1/2-12], 3 mar-29 dec [n18-98/99], 1952 5 jan-31 dec [n1/2-104/105], union im wahlkampf: 1949 20 may-13 aug [n2-25], union in deutschland: 1949 1 sep-31 dec [n168-201], 1950 4 jan-30 dec [n1-101], 1951 3 jan-28 feb [n1-17]) – mf#1959 – gw Mikropress [325]
Cdu-informationsdienst : informationsdienst des zonenausschusses der cdu – Koeln, 1947-jan 1950 – 2r – 1 – (title change: informationsdienst der cdu deutschlands 1950-52; union im wahlkampf 1949; union in deutschland 1949-51) – gw Mikropress [325]
Cdu-informationsdienst fuer die britische zone – Koeln DE, 1947-51 [gaps] – 1 – mf#1959 – gw Mikropress [074]
Ce contrat et marche conclu entre sa majeste la reine agissant en ce qui concerne la puissance du canada : et a ces fins representee par l'honorable sir charles tupper, kcmg, ministre des chemins de fer et canaux... – [S.l: s.n, 1880?] [mf ed 1980] – 1mf – 9 – mf#03413 – cn Canadiana [380]
Ce que je ferais si j'etais pretre – Montreal: D Bently & cie, impr, 1880 [mf ed 1980] – 1mf – 9 – 0-665-04122-5 – mf#04122 – cn Canadiana [241]
Ce que le gouvernement a fait pour quebec / Federation liberale nationale du Canada – [Canada: s.n, 1908?] [mf ed 1995] – 1mf – 9 – 0-665-77300-5 – mf#77300 – cn Canadiana [325]
Ce que pensent les fleurs : saynete enfantine / Dandurand, Josephine – Montreal: Beauchemin, [1895] [mf ed 1980] – 1mf – 9 – 0-665-05416-5 – mf#05416 – cn Canadiana [820]
Ce que tout canadien devrait savoir concernant les filatures de laine et les tricoteries du canada – [Quebec (Province)]: [s.n.], 1925 [mf ed 1992] – 1mf – 9 – mf#SEM105P1541 – cn Bibl Nat [670]
Ce qui sauve / Cougnard, J – Nimes, France. 1867 – 1r – us UF Libraries [240]
Ce qui se passe au concile – 3e ed. Paris: Henri Plon, 1870 – 1mf – 9 – 0-8370-8997-2 – (incl bibl ref) – mf#1986-2997 – us ATLA [240]
Ce qu'il faut dire – Paris. avr 1916-17 – 1 – fr ACRPP [073]
Ce soir – Paris. 2 mars 1937-25 aout 1939, 22 aout 1944-1 2 mars 1953 – 1 – fr ACRPP [073]
Ce soir – Paris, France. 6 sep 1944-22 nov 1945; 19 nov 1946-1952; 1 jan-2 mar 1963 – 17 1/2r – 1 – uk British Libr Newspaper [072]
Cea critic / College English Association – College Station. 1939+ (1) 1971+ (5) 1976+ (9) – ISSN: 0007-8069 – mf#6909 – us UMI ProQuest [420]
Cea forum / College English Association – Shreveport. 1972+ (1) 1972+ (5) 1976+ (9) – ISSN: 0007-8034 – mf#7749 – us UMI ProQuest [378]
Cea, Jose Roberto *see* Poetas jovenes de el salvador
Cean Bermudez, J A *see* Diccion rio historico de los m s illustres profesores de las bellas artes en espana
Ceara (Brazil) Governor *see* Relatorios dos presidentes, 1a republica, 1891-1930
Ceara (Brazil) President *see* Relatorios dos presidentes, epoca do imperio, 1836-1889
Cearense / Barroso, Parsifal – Rio de Janeiro, Brazil. 1969 – 1r – us UF Libraries [972]
Cearense, Catullo Da Paixao *see*
– Fabulas e alegorias
– Poemas bravios
– Poemas escolhidos
Cearense, Catullo De Paixao *see* Caboclo brasileiro
Cearense de jacuna – Ceara, 04 set 1833 – (= ser Ps 19) – mf#P17,01,40 – bl Biblioteca [321]
Ceasarian – Canberra, 1972-92 – 3r – at Pascoe [079]
Cebel-i luebnan – (= ser Vilayet salnames) – 9 – (1304 [1887] 2mf $40; 1309 [1892] def'a 6 – 2mf $75) – us MEDOC [956]
Cebrian De Quesada, Arnaldo *see* Clarinadas
La ceca de la villa imperial de potosi y la moneda colonial buenos aires, 1945 / Burcio, Humberto F – Madrid: Razon y Fe, 1947 – 1 – sp Bibl Santa Ana [972]
Cecchi, A *see*
– L'abissinia settentrionale e le strade che vi conducono da massaua
– Da zeila alle frontiere del caffa
Cecco, John P De *see* Journal of homosexuality
Cece guide : a bimonthly guide to cultural events in the african american community / Community Economics and Cultural Exchange – 1992 oct/dec, 1993 feb-mar – 1r – 1 – mf#2681051 – us WHS [302]
Cech – Prague, Czechoslovakia. -d. 1 feb-23 jul 1919 (imperfect) – 1r – 1 – uk British Libr Newspaper [072]
Cechoslovak – Milwaukee, WI: John V Klabough (semimthly) [mf ed 1949] – 5r – 1 – (cheifly in czech; some also in english) – mf#1166130 – us WHS [071]
Cechoslovak – Rosenberg TX: Vydavatelska, Spolecnost Cechoslovak =Cechoslovak Pub Co, 1917-dec 1919// (wkly) [mf ed 1983] – 1r – 1 – (chiefly in czech; some also in english; merged with: westske noviny to form: cechoslovak and westske noviny) – us Barker [071]
Cechoslovak = The czechoslovak / Czechoslovakia. Ministerstvo zahranicnich veci – Londyn : Vychazi peci Ministerstva zahranicnich veci CSR. 6v. roc2 cis29. 19 cerven 1940-roc7 cis14. 6 dub 1945 (wkly) [mf ed with gaps filmed 1986] – 2r – 1 – (cont: cechoslovak v anglii. cont by: cechoslovak v zahranici) – us NE Hist [071]
Cechoslovak – Milwaukee WI. 1935 dec 21-1937 apr 24, 1937 may 1-1938 sep 24, 1938 oct 1-1940 feb 17, 1940 mar 2-1941 dec 20, 1943 dec 4-1948 apr 3 – 5r – 1 – mf#1166130 – us WHS [071]
Cechoslovak *see* Cechoslovak v anglii
Cechoslovak and westske noviny – West TX: Cechoslovak Pub Co, roc9 cis1. jan 6 [1920]-roc42 cis27. 7 cerven 1961 (wkly) [mf ed 1983] – 34v on 11r – 1 – (chiefly in czech; some also in english; formed by the union of: westske noviny and: cechoslovak (rosenbery tx); absorbed by: hospodar) – us Barker [071]
Cechoslovak v anglii – London, UK. 20 oct 1939-12 jul 1940 – 1 – (aka: cechoslovak, 26 jul 1940-6 apr 1945; cechoslovak v zahranici, 13 apr-31 dec 1945, 24 apr 1953-29 nov 1967) – uk British Libr Newspaper [072]
Cechoslovak v zahranici = The czechoslovak abroad / Czechoslovakia. Ministerstvo zahranicnich veci – Londyn : [Publ R J Stursa for the Czechoslovak Ministry of Foreign Affairs] 38v. roc7 cis15. 13 dub 1945-roc7 cis52. 31 pros 1945 (wkly) [mf ed 1986] – 1r – 1 – (cont: cechoslovak) – us NE Hist [071]
Cechoslovak v zahranici *see* Cechoslovak v anglii
Cecil, Algernon *see* Six oxford thinkers
Cecil, David *see* Stricken deer
Cecil, Florence, Lady *see* Changing china
Cecil, Herbert Myron *see*
– Fundamental principles of the organization, management, and teaching of the school band
– Treatise on the trombone
Cecil, Joe S *see*
– Administration of justice in a large appellate court
– Deciding cases without argument
– The role of staff attorneys and face-to-face conferencing in non-argument decisionmaking
– Summary judgement practice in three district courts
Cecil, Joe S et al *see* Jury service in lengthly civil trials
Cecil of Chelwood, Robert Gascoyne-Cecil, Viscount *see* Our national church
Cecil rhodes / Baker, Herbert – London, England. 1934 – 1r – us UF Libraries [960]
Cecil rhodes / Lockhart, John Gilbert – London, England. 1933 – 1r – us UF Libraries [960]
Cecil rhodes / Maurois, Andre – New York, NY. 1953 – 1r – us UF Libraries [960]
Cecil rhodes / Millin, Sarah Gertrude Liebson – New York, NY. 1933 – 1r – us UF Libraries [960]
Cecil rhodes / Williams, Basil – New York, NY. 1921 – 1r – us UF Libraries [960]
Cecil, Richard *see*
– Friendly visit to the house of mourning
– A friendly visit to the house of mourning
Cecil, William Rupert Ernest Gascoyne, Lord *see* Changing china
Cecile staub genhart : her biography and her concepts of piano playing / Gordon, Stewart – U of Rochester 1965 [mf ed 19–] – 8mf – 9 – mf#fiche 208 – us Sibley [780]
Cecilia baptist church (formerly: rudes creek). kentucky : church records – 1819-1968 – 1 – 60.21 – us Southern Baptist [242]
Cecilia, Madame *see* Catholic scripture manual atlas
Cecilia valdes / Rodriguez Herrera, Esteban – Habana, Cuba. 1953 – 1r – us UF Libraries [972]
Cecilia valdes / Villaverde, Cirilo – Habana, Cuba. 1941 – 1r – us UF Libraries [972]
Ced news / Campaign for Economic Democracy [CA] – 1977 jun/jul-1980 aug – 1r – 1 – (cont: campaigner for economic democracy; cont by: economic democrat) – mf#667571 – us WHS [330]
Cedar bluff baptist church – Albany. 1972-1999 (1) 1969-1999 (5) 1973-1999 (9) – 1r – 1 – $25.74 – mf#6506 – us Southern Baptist [242]
Cedar bluff's opinion – Cedar Bluffs, NE: F C Yenny (wkly) [mf ed v1 n38. nov 24 1892 filmed 1973] – 1r – 1 – us NE Hist [071]
Cedar bluffs standard – Cedar Bluffs, NE: C A Sherwood. -v73 n47. oct 8 1964 (wkly) [mf ed v5 n32. oct 25 1895-1964 (gaps) filmed -1993] – 21r – 1 – (cont by: new cedar bluffs standard) – us NE Hist [071]
Cedar county leader – Hartington, NE: Z M Baird, 18894-98// (wkly) [mf ed aug 9-oct 18 1895 (lacks oct 4 1895) filmed [1965?]] – 1r – 1 – (cont: hartington leader. absorbed by: hartington herald) – us NE Hist [071]
Cedar county news – Hartington, NE: Z M Baird, jan 13 1898 (wkly) [mf ed v1 n7. feb 24 1898 filmed [1969?]-] – 1 – (absorbed: wynot tribune and: obert times jan 4 1934) – us NE Hist [071]
Cedar county waechter – Hartington, NE: Chas Weiss. 20v. 28 aug 1898-jahrg 20 n52. 22 aug 1918 (wkly) [mf ed jahrg 10 n19. 2 jan 1908-18 lacks 29 jun 1916 filmed [1965?]] – 1r – 1 – (absorbed by: woechentliche omaha tribuene) – us NE Hist [071]
Cedar creek baptist church *see* Belleview baptist church
Cedar creek pilot – Athens, TX. 1996-1999 (1) – mf#68077 – us UMI ProQuest [071]
Cedar falls gazette – Cedar Falls IA. 1878 jul 19 – 1r – 1 – (cont: recorder [cedar falls ia]) – mf#851125 – us WHS [071]
Cedar grove baptist church. maryville, tennessee : church records – 1890-Apr 1962. Lacking: Feb 1930-Jul 1941 – 1 – us Southern Baptist [242]
Cedar grove baptist church. south carolina : church records – 1825-1943 – 1 – us Southern Baptist [242]
Cedar rapids commercial – Cedar Rapids, NE: N Fodrea (wkly) [mf ed v3 n2. aug 24 1894 filmed 1973] – 1r – 1 – us NE Hist [071]
Cedar rapids leader – Cedar Rapids, NE: Edwin Van Ackeren, jul 1933-v4 n39. apr 8 1937 (wkly) – 1r – 1 – (merged with: cedar rapids outlook to form: cedar rapids leader-outlook) – us Bell [071]
Cedar rapids leader-outlook – Cedar Rapids, NE: Edwin E Van Ackeren. 7v. v53 n30. apr 15 1937-v59 n28. mar 25 1943 (wkly) – 2r – 1 – (absorbed: primrose press; formed by the union of: cedar rapids leader and: cedar rapids outlook; absorbed by: albion weekly news) – us Bell [071]
Cedar rapids press – Cedar Rapids, NE: G M Cox. v1 n1. nov 21 1957-v1 n2. nov 28 1957; v10 n3. dec 5 1957- (wkly) [mf ed 1957- filmed 1976-] – 1 – (cont: cedar valley press. v2-9 not publ) – us NE Hist [071]
Cedar rapids republican – Cedar Rapids, NE: Baird & Son, 1885-dec 6 1895// (wkly) – 1r – 1 – (cont by: boone county outlook. issue for nov 25 1892 called v8 n1) – us Bell [071]
Cedar rapids review – Cedar Rapids, NE: G A Mayfield, mar 19902 (wkly) [mf ed v1 n4. apr 5-may 17 1902 (gaps) filmed 1973] – 1r – 1 – us NE Hist [071]
Cedar rapids standard – Cedar Rapids IA. 1885 sep 10 – 1r – 1 – (cont by: new standard [cedar rapids ia]) – mf#880426 – us WHS [071]
Cedar rapids star – [Greeley, NE: s.n.] v1 n1. apr 9 1948- (wkly) [mf ed -apr 16 1948 filmed 1973] – 1r – 1 – us NE Hist [071]
Cedar spring baptist church. spartanburg county. south carolina : church records – 1794-1972 – 1 – 85.14 – us Southern Baptist [242]
Cedar valley news – Bradford IA. 1860 jan 26, sep 20, oct 11, 1861 dec 14 – 1r – 1 – mf#851120 – us WHS [071]
Cedar valley press – Cedar Rapids, NE: G M Cox. v9 n26. nov 14 1957 (wkly) – 1r – 1 – (cont: cedar valley promoter. cont by: cedar rapids press) – us NE Hist [071]
Cedar valley promoter – Cedar Rapids, NE: A C Leonard and Maries Carlson. 9v. v1 n1. jun 2 1949-v9 n25. nov 7 1959 (wkly) – 3r – 1 – (cont by: cedar valley press) – us NE Hist [071]
Cedarburg enterprise – Cedarburg WI. 1880 mar 10-dec 29 – 1r – 1 – (cont by: ozaukee county enterprise) – mf#958037 – us WHS [071]
Cedarburg news – Cedarburg, Grafton...WI. 1894 jan 3/1896 mar 4-1959/1960 jun 8 – 41r – 1 – (with gaps; cont: cedarburg weekly news; cont by: news graphic [cedarburg wi]) – mf#966480 – us WHS [071]
Cedarburg weekly news – Cedarburg WI. 1883 jan 17-1886 apr 21, 1886 apr 28-1889 aug 28, 1889 sep 4-1892 dec 7, 1892 dec 14-1893 dec 27 – 4r – 1 – (cont by: cedarburg news) – mf#966479 – us WHS [071]
Cedarville. Ohio. Reformed Presbyterian Church *see* Centennial souvenir, 1809-1909
[Cedarville-] surprise valley journal – CA. aug 21 1952-feb 13 1958 – 2r – 1 – $120.00 – mf#B02102 – us Library Micro [071]
[Cedarville-] surprise valley record – CA. apr 25 1906-feb 4 1937 – 7r – 1 – $420.00 – mf#B02103 – us Library Micro [071]
Ceddia, Michael A *see* The effects of four consecutive days of acute exercise on macrophage antigen presentation
CEDI, Centro Ecumenico de Documentacao e Informacao, Programa de Assessoriaa Pastoral Protestante *see* Onze de abril
Cedr quarterly / Phi Delta Kappa Center on Evaluation, Development and Research – Bloomington. 1980-1982 – 1,5,9 – ISSN: 0147-9741 – mf#12601,01 – us UMI ProQuest [330]
Cedula y el sufragio / Montoya, Hernan – s.l, s.l? 1938 – 1r – us UF Libraries [972]
Cedulario cubano (los origenes de la colonizacion 1. (1493-1512)...tomo 6 / Chacon Y Calvo, Jose Maria – Madrid: Razon y Fe, 1929 – 1 – sp Bibl Santa Ana [946]
Cedulario de la monarquia espanola relativo a la i... / Charles 1 Of Spain, King – Caracas, Venezuela. v1-2. 1961 – 1r – us UF Libraries [972]
Cedulario del peru : siglos 16, 17 y 18 / Porras Barrenechea, Paul – Lima: Dept de Relaciones Culturales del Ministerio de Relciones Exteriores del Peru. v1-2. 1944-48 – 1 – us CRL [972]

Cedularios de la monarquia espanola relativos a la... / Charles 1 Of Spain, King – Caracas, Venezuela. v1-2. 1959 – 1r – us UF Libraries [972]
Cedulas epigraficas del campo norbense / Callejo Serrano, Carlos – Salamanca, 1968. Sep. Zephyrus 18th, 1967, pp. 85-120 – 1 – sp Bibl Santa Ana [946]
Cedule : resolutions contenant les amendements qui doivent etre faits au role imprime du code de procedure civile du bas canada = Schedule: resolutions containing the amendments to be made in the printed roll of the code of civil procedure / Canada (Province). Commissaires charges de codifier les lois du Bas Canada, en matieres civiles – Ottawa: impr par G E Desbarats, [1866?] (mf ed 1998) – 1mf – 9 – (in english and french) – mf#SEM105P2906; SEM105P2907 – cn Bibl Nat [348]
Cedule de certains etats relatifs a l'importation et exportation de la province du canada – Montreal: impr par Lovell & Gibson, [1846] (mf ed 1992) – 1mf – 9 – mf#SEM105P1734 – cn Bibl Nat [380]
Cedule des diverses compagnies incorporees pour la construction de chemins de fer en canada : depuis la date de la premiere charte jusqu'a la cloture de la session de 1852-53 classees... – [Quebec: Lovell et Lamoureux, 1853 ?] (mf ed 1992) – 1mf – 9 – mf#SEM105P1760 – cn Bibl Nat [690]
Cee – chemical engineering education – Gainesville. 1972-1996 (1) 1972-1996 (5) 1974-1996 (9) – ISSN: 0009-2479 – mf#7166 – us UMI ProQuest [660]
Cefn chronicle = Cronicl y cefn – [Wales] LLGC jul 1913-dec 1950 [mf ed 2004] – 30r – 1 – uk Newsplan [072]
Cefn cribbwr, nebo welsh baptist; and kenfig hill, pisgah welsh baptist, monumental inscriptions – 1mf – 9 – £1.25 – uk Glamorgan FHS [929]
Cefn cribbwr, siloam congregational, monumental inscriptions – 1mf – 9 – £1.25 – uk Glamorgan FHS [929]
Cefn-coed-y-cymmer, carmel baptist; ebenezer congregational, monumental inscriptions – 1mf – 9 – £1.25 – uk Glamorgan FHS [929]
Cefn-coed-y-cymmer, hen-dy-cwrdd unitarian, monumental inscriptions – 2mf – 9 – £2.50 – uk Glamorgan FHS [929]
Cefn-coed-y-cymmer tabor, monumental inscriptions – 2mf – 9 – £2.50 – uk Glamorgan FHS [929]
Cefp journal / Council of Educational Facility Planners – Columbus. 1962-1988 (1) 1975-1988 (5) 1975-1988 (9) – (cont by: educational facility planner) – ISSN: 0007-8220 – mf#10304 – us UMI ProQuest [370]
Cehila : boletin informativo – Quito, Ecuador: Comision de Estudios de Historia de la Iglesia en Latinoamerica. [n1-52 (1973-1996)] (irreg) – 1 – us CRL [240]
Cei – Rio de Janeiro: Centro Ecumenico de Informacao. [n17-149/150]. oct 1966-apr/may 1979 – 1r – us CRL [972]
Cei documento – [Rio de Janeiro: Tempo e Presenca. [n58-92]. nov 1974-may 1979 – 1r – us CRL [972]
Ceiba en el tiesto / Laguerre, Enrique A – San Juan, Puerto Rico. 1956 – 1r – us UF Libraries [972]
Ceida, Amelia *see* Puertas
Ceide, Amelia *see* Cuando el cielo sonrie
Ceillier, R *see* Histoire generale des auteurs sacres et ecclesiastiques
Ceirim, pismo mlodych ruchu agudat hanoar haiwri "akiba" – Krakow, L'viv, various, 1934-35 – 1r – 1 – us UMI ProQuest [939]
Cejador y Frauca, Julio *see* Historia de la lengua y literatura castellana desde los origenes hasta carlos 5
Cela biedrs – 1956-60, 1961-65, 1966-70 – 3r – 1 – mf#681853 – us WHS [071]
Cela, Camilo Jose *see* Historias de venezuela
Celal, Mehmed *see* The divan project
Celaleddin *see* The divan project
Celam : [boletin] – [Bogota: Consejo de Redaccion, Secretariado General del CELAM, -1993]. [n56-n252 (apr 1972-marzo 1993)] (mthly) – 2r – 1 – us CRL [073]
Celaya, J de *see* Expositio in octo libros phisicorum aristotelis...
Celebi, Ibrahim Cevri *see* Tarih-i cevri celebi
Celebi, Katib *see* Fezkeke-i tarih
Celebi, Solakzade Mehmed Hemdemi *see* Solakzade tarihi
Celebrated pictures exhibited at the glasgow international exhibition / Armstrong, Walter – London 1888 – (= ser 19th c art & architecture) – 3mf – 9 – mf#4.2.1283 – uk Chadwyck [700]
"Celebrated sentence" : or, a calm review of the controversy it has – London, England: Jackson & Walford 1839? – 1r – 1 – (repr fr congregational magazine for feb 1839) – us UF Libraries [240]
Celebration des noces d'or de m le chanoine archambeault – [Montreal?: s.n.], 1887 [mf ed 1980] – 1mf – 9 – mf#07852 – cn Canadiana [920]
Celebration des noces d'or de m le chanoine archambeault a st hugues, 13 janvier 1887 – Montreal: Cie d'impr & de lith Gebhardt-Berthiaume, 1887 [mf ed 1980] – 1mf – 9 – 0-665-00751-5 – mf#00751 – cn Canadiana [241]
Celebration of the 50th anniversary of the appointment of professor william henry green : as an instructor in princeton theological seminary, may 5 1896 – New York: Charles Scribner 1896 [mf ed 1989] – 1mf [ill] – 9 – 0-7905-1530-X – mf#1987-1530 – us ATLA [242]
Celebration of the lord's supper every lord's day – Edinburgh, Scotland. 1802 – 1r – us UF Libraries [240]
Celebrations and amusements among negroes of florida / Muse, Viola B – s.l, s.l? 1937 – 1r – us UF Libraries [305]
Celebrity doll club – 1966 nov-1972 may – 1r – 1 – (cont by: celebrity doll journal) – mf#515147 – us WHS [071]
Celebrity doll journal – 1972 aug-1980 aug – 1r – 1 – (cont: celebrity doll club) – mf#515358 – us WHS [071]
Celery diseases in florida / Foster, Arthur C – Gainesville, FL. 1924 – 1r – us UF Libraries [634]
Celery harvesting methods in florida / Brunk, Max E – Gainesville, FL. 1944 – 1r – us UF Libraries [634]
The celestial and his religions : or, the religious aspect in china. being a series of lectures on the religions of the chinese / Ball, James Dyer – Hongkong: Kelly and Walsh, 1906 – 1mf – 9 – 0-524-01159-1 – mf#1990-2235 – us ATLA [290]
Celestial empire – Shanghai. 4 jul 1874-1883; 25 apr 1884-1897; 17 oct 1908; 1918-mar 1927 – 54 1/2r – 1 – uk British Libr Newspaper [072]
Celestial empire – Shanghai. China. -w. 4 Jul 1874-26 Dec 1883, 25 Apr 1884-31 Dec 1897, 17 Oct 1908, 5 Jan 1918-28 Mar 1927. (56 reels) – 1 – uk British Libr Newspaper [079]
The celestial empire – Shanghai: Loureiro & Co, jul 4 1874-1897; 1902-sep 1905; 1906-13; 1922-mar 1927 – 66r – 1 – us CRL [079]
The celestial keys / Sibbrena, Ireleda – London: Kegan Paul, Trench & Truebner, 1909 – 1mf – 9 – 0-524-02370-0 – mf#1990-2981 – us ATLA [200]
Celestial lirio / Domingo de San Pedro de Alcantara – 1755 – 9 – sp Bibl Santa Ana [810]
Celestial mechanics – Dordrecht. 1984-1988 (1,5,9) – (cont by: celestial mechanics and dynamical astronomy) – ISSN: 0008-8714 – mf#14742 – us UMI ProQuest [520]
Celestial mechanics and dynamical astronomy – Dordrecht. 1989+ (1,5,9) – (cont: celestial mechanics) – ISSN: 0923-2958 – mf#14742,01 – us UMI ProQuest [520]
Celestin, Clement *see*
– Idees et opinions, la reforme de l'etat
Celestine / Achard, Paul – Paris, France. 1942 – 1r – us UF Libraries [440]
Celestine / Achard, Paul – Paris, France. 1946, c1942 – 1r – us UF Libraries [440]
Celestine and sallie : or, two dolls and two homes / Armstrong, Jessie F – London: Chas Kelly, 1890 – (= ser 19th c children's literature) – 1mf – 9 – mf#6.1.35 – uk Chadwyck [830]
Celestrin, Heliodoro G *see* Bajareque
Celibate woman – v1 n1-v4 n2 [1982 jul-1988 oct] – 1r – 1 – mf#1519151 – us WHS [305]
Celie / Lafontant, Delorme – Port-Au-Prince, Haiti. 1939 – 1r – us UF Libraries [972]
Celine : ou une autre magdeleine a l'asile du bon pasteur de quebec / Amicus – [Sherbrooke, Quebec?: s.n.], 1886 [mf ed 1980] – 1mf – 9 – 0-665-02563-7 – mf#02563 – cn Canadiana [241]
Celine, : ou, la famille de l'absent / Fournier, Narcisse – Paris, France. 1842? – 1r – us UF Libraries [440]
Celine, Louis-Ferdinand *see*
– Bagatelles pour un massacre
– Mort a credit
Celine-de-la-Presentation, Soeur *see* Bibliographie de l'oeuvre de monsieur l'abbe andre jobin
Cell and tissue kinetics – Oxford. 1980-1990 (1) 1980-1990 (5) 1980-1990 (9) – (cont by: cell proliferation) – ISSN: 0008-8730 – mf#15510 – us UMI ProQuest [574]
Cell and tissue research – Heidelberg. 1978-1990 (1,5,9) – ISSN: 0302-766X – mf#13110,03 – us UMI ProQuest [574]
Cell biochemistry and function – Chichester. 1983-1994 (1,5,9) – ISSN: 0263-6484 – mf#16097 – us UMI ProQuest [574]
Cell biology and molecular basis of liver transport : 2nd international ringberg conference on hepatic transport / ed by Wehner, Frank & Petzinger, Ernst – Dortmund: projekt vlg. 1995 (mf ed 1996) – 4mf – 9 – €45.00 – 3-8267-9709-4 – mf#DHS 9709 – gw Frankfurter [574]
Cell calcium – Edinburgh. 1987-1996 (1,5,9) – ISSN: 0143-4160 – mf#13424 – us UMI ProQuest [611]
Cell differentiation – Shannon. 1972-1988 (1) 1972-1988 (5) 1987-1988 (9) – (cont by: cell differentiation and development) – ISSN: 0045-6039 – mf#42179 – us UMI ProQuest [574]
Cell differentiation and development – Shannon. 1989-1990 (1,5,9) – (cont: cell differentiation. cont by: mechanisms of development) – ISSN: 0922-3371 – mf#42179,01 – us UMI ProQuest [574]
Cell migration in three-dimensional collagen lattices. integrins, cell-matrix-interactions and migration strategies : fundamental differences in t lymphocytes and tumor cells / Friedl, Peter – (mf ed 1998) – 2mf – 9 – €40.00 – 3-8267-2533-6 – mf#DHS 2533 – gw Frankfurter [616]
Cell proliferation – Oxford. 1991-1994 (1) 1991-1994 (5) 1991-1994 (9) – (cont: cell and tissue kinetics) – ISSN: 0960-7722 – mf#15510,01 – us UMI ProQuest [574]
Cellensis, Petrus *see* Commentaria in ruth. tractatus de tabernaculo
Celler beobachter – Celle DE, 1932 20 aug-1934, 1938-1942 30 jun [gaps] – 7r – 1 – (suppl of: niedersaechsische tageszeitung, hannover until oct 1937) – gw Misc Inst [074]
Celler volkszeitung – Celle DE, 1928 nov 1-1932 – 12r – 1 – gw Misc Inst [074]
Cellerier, Jacob Elisee *see* Manuel d'hermeneutique biblique
Cellesche anzeigen *see* Zellescher anzeiger 1817
Cellesche zeitung *see* Zellescher anzeiger 1817
Cellesche zeitung und anzeigen *see* Zellescher anzeiger 1817
Cellini, B *see*
– Due trattati uintoralle otto principali arti dell' oreficeria
– Vita...da lui medesimo scritta
Cellini, Benvenuto *see* Memoirs
Cells tissues organs : in vivo, in vitro – Basel. 1999+ (1) – (cont: acta anatomica) – ISSN: 1422-6405 – mf#2043,01 – us UMI ProQuest [574]
Cellular and molecular biology – Oxford. 1977-1992 (1,5,9) – (cont by: cellular and molecular biology research) – ISSN: 0145-5680 – mf#49291 – us UMI ProQuest [574]
Cellular and molecular biology research – Oxford. 1993-1995 (1,5,9) – (cont: cellular and molecular biology) – ISSN: 0968-8773 – mf#49291,01 – us UMI ProQuest [574]
Cellular and molecular life sciences – Basel. 1997+ (1) 1997+ (5) 1997+ (9) – (cont: experientia) – ISSN: 1420-682X – mf#1377,01 – us UMI ProQuest [500]
Cellular business – Overland Park. 1984-1997 (1,5,9) – ISSN: 0741-6520 – mf#14720 – us UMI ProQuest [380]
Cellular signalling – Oxford. 1989-1994 (1,5,9) – ISSN: 0898-6568 – mf#49559 – us UMI ProQuest [574]
La cellule : seminaire de joliette – [Joliette] :le Seminaire. v1 n1 oct 1973-1982?// (irreg) [mf ed 1990] – 9 – (cont: estudiant) – mf#SEM105P1211 – cn Bibl Nat [200]
Die celluloid-industrie : beilage der gummi-zeitung, berlin – Berlin, 22 mar 1912-24 jul 1914 – 1 – uk British Libr Newspaper [670]
Die celluloid-industrie *see* Gummi-zeitung
Celorio Y Cobo, Alfonso *see* Vibraciones distintas
Y celt – [Wales] LLGC 19 ebrill 1878-mehefin 1906 [mf ed 2002] – 44r – 1 – (cont by: y celt newydd [1903-06]) – uk Newsplan [072]
Y celt – Bala, Gwynedd, Wales 26 apr-25 dec 1902* [mf apr-may 1878; jan-may 1879; jan-dec 1882; jan-dec 1884; jan 1890-dec 1891; oct-dec 1893] – 1 – (cont by: y celt newydd [2 jan 1903-7 jul 1905]; fr 3 jun 1881-30 jun 1882 publ at rhyl; fr 7 jul 1882-28 sep 1894 at bangor; fr 5 oct 1894-27 dec 1895 at aberavon; fr 6 mar 1896-29 jun 1900 at llanelly; fr 6 jul 1900-22 jan 1904 at aberdare; fr 29 jan 1904-7 jul 1905 at ystalyfera; fr 14 jul 1905 onward at llanelly) – uk British Libr Newspaper [072]
Celt – Waterford, Ireland. 7 oct 1876-21 jul 1877 – 1r – 1 – (incorp into: munster express 1877) – uk British Libr Newspaper [072]
The celtic church in britain and ireland = Keltische kirche in britannien und irland / Zimmer, Heinrich – London: David Nutt, 1902 – 1mf – 9 – 0-7905-7202-8 – (incl bibl ref. in english) – mf#1988-3202 – us ATLA [240]
The celtic church in ireland : the story of ireland and irish christianity from before the time of st. patrick to the reformation / Heron, James – London: Service & Paton, 1898 – 1mf – 9 – 0-7905-4864-X – mf#1988-0864 – us ATLA [240]
The celtic church in scotland : being an introduction to the history of the christian church in scotland down to the death of saint margaret / Dowden, John – London: SPCK; New York: E & J B Young, 1894 – 1mf – 9 – 0-7905-4400-8 – (incl bibl ref) – mf#1988-0400 – us ATLA [240]
The celtic church of wales / Willis Bund, John William – London: D. Nutt, 1897. vii,533p – 1 – us Wisconsin U Libr [243]
Celtic magazine : a monthly periodical devoted to the literature, history, antiquities, folklore, traditions. of the celt – Inverness. 1875-1888 (1) – mf#4713 – us UMI ProQuest [490]
Celtic mythology and religion : with a chapter on the "druid" circles / Macbain, Alexander – New York: EP Dutton, 1917 – 1mf – 9 – 0-524-01969-X – mf#1990-2760 – us ATLA [290]
Celtic news & west wales general advertiser – [Wales] LLGC 2 mar 1923-28 oct 1937 [mf ed 2004] – 15r – 1 – (cont by: cymric times [jan 1928-dec 1936]; cymric times & telegraph – the cymric times & western telegraph [jan-oct 1937]) – uk Newsplan [072]
Celtic religion in pre-christian times / Anwyl, Edward – London: Archibald Constable, 1906 [mf ed 1991] – (= ser Religions ancient and modern) – 1mf – 9 – 0-524-00678-4 – (incl bibl ref) – mf#1990-2006 – us ATLA [290]
Celtic researches, on the origin, traditions & language of the ancient britons : with some introductory sketches on primitive society / Davies, Edward – London: Booth 1804 – (= ser Whsb) – 7mf – 9 – €75.00 – mf#Hu 098 – gw Fischer [490]
The celtic tragedy, vol 1 : british races, languages, and religions, the anglo-saxon myth and orange fanaticism / Murray, Norman – Montreal: N Murray, [1919-1921] – 5v on 5mf – 9 – (v2 79422 isbn: 0-665-79423 v3 79423 isbn: 0-665-79423-1 v4 79424 isbn: 0-665-79424-X v5 79425 isbn: 0-665-79425-8. incl some text in french) – mf#79421 – cn Canadiana [306]
The celts / Maclear, George Frederick – London: S.P.C.K.; New York: Pott, Young, [1878?] – (= ser Conversion Of The West) – 1mf – 9 – 0-7905-5428-3 – (incl bibl ref) – mf#1988-1428 – us ATLA [290]
Celulosas *see* Celulosas de extremadura
Celulosas de extremadura / Celulosas – Merida: Sarrio Compania papelera de Leiza, S.A. – 1 – sp Bibl Santa Ana [946]
Cem – Istanbul. n34-39. 12 kanunievvel 1928-2 mayis 1929 [12 dec 1928-2 may 1929] – (= ser 0 & t journals) – 4mf – 9 – $60.00 – (cont: djem) – us MEDOC [956]
Cem anos de ensino secundario no brasil (1826-1926 / Dodsworth, Henrique De Toledo – Rio de Janeiro, Brazil. 1968 – 1r – us UF Libraries [972]
Cemal, Mehmed *see* Anadolu
Cement and concrete composites – Essex. 1990-1996 (1,5,9) – (cont: international journal of cement composites and lightweight concrete) – ISSN: 0958-9465 – mf#42582,02 – us UMI ProQuest [690]
Cement and concrete research – Elmsford. 1971+ (1,5,9) – ISSN: 0008-8846 – mf#49027 – us UMI ProQuest [690]
Cement and lime manufacture – London. 1928-1969 (1) – mf#2462 – us UMI ProQuest [690]
Cement, concrete, and aggregates – Conshohocken. 1979-1994 (1,5,9) – ISSN: 0149-6123 – mf#11878 – us UMI ProQuest [690]
The cemeteries of abydos : pt 1: the mixed cemetery and umm el-ga'ab / Naville, E et al – London, 1914 – (= ser Mees 33) – 5mf – 8 – €12.00 – ne Slangenburg [930]
The cemeteries of abydos : pt 2 / Peet, T E – London, 1914 – (= ser Mees 34) – 10mf – 8 – €19.00 – ne Slangenburg [930]
The cemeteries of abydos : pt 3 / Peet, T E & Loat, W LS – London, 1913 – (= ser Mees 35) – 5mf – 8 – €12.00 – ne Slangenburg [930]
Cemetery cards / St. Mary's Cemetery, Geary County, KS – undated – 1 – us Kansas [920]
Cemetery inscriptions fairview park, ohio : tombstone records of fairview park (old rockport) cemetery / Daughters of the American Revolution. Lakewood. Ohio Chapter – 1r – 1 – us Western Res [920]
Cemetery records / Cherokee County, KS – 1916-1924, Galena, KS, Undertakers' Records – 1 – us Kansas [920]
Cemetery records, 1849-1929 : and early history of town / Germantown, OH – 1r – 1 – mf#B26151 – us Ohio Hist [978]
Cemetery tombstone and obituary cards / Jefferson County, KS – undated – 1 – us Kansas [920]
Cemiyet-i tedrisiye-yi islamiye salnamesi – 1332 [1913] – (= ser Ministry and special interest salnames) – 4mf – 9 – $60.00 – us MEDOC [956]
Cen Gim Jhuan *see* Vacananukram sadisasabd
Cenci, Pio *see* Il cardinale raffaele...del val...roma
Cendrillon / Maximilien, M – Paris, France. 1838 – 1r – us UF Libraries [440]
Cenitagoya, Vicente *see* Los machiguengas. lima, 1944
Cenizas del alma / Arroyo, Angel Manuel – New York, NY. 1949 – 1r – us UF Libraries [972]

Cenizas gloriosas / Campa Y Caraveda, Miguel Angel – Habana, Cuba. 1945 – 1r – us UF Libraries [972]
Cennini, C *see*
– Traite de la peinture
– Trattato della pittura
– A treatise on painting
[Cennini, C] Ilg, A *see* Kunst oder tractat der malerei des cennicennini da colle di valdelsa
Cennini, Cennino *see* A treatise on painting
Cenotaph / Hutchinson County Genealogical Society – 1968 spr-1972 – 1r – 1 – mf#1054154 – us WHS [929]
Cenotaphium piis manibus Ferdinandi 3... *see* Caesareis virtuatib et symbolis adornatum...
Cenove zpravy / Czechoslovakia. Statni Urad Statisticky – Prague. v1-19, 26-28. 1921-39, 1946-48* – 1 – us NY Public [314]
Le censeur des journaux – Paris. nov 1795-sept 1797 – 1 – fr ACRPP [073]
Le censeur europeen – Paris. 1-XII. Fevr 1817-17 avr 1819; 1-172, 15 juin 1819-20 juin 1820 – 1 – fr ACRPP [073]
Le censeur hebdomadaire – Utrecht, Paris. 1760-aout 1761 – 1 – fr ACRPP [073]
Censo agropecuario, 1963 / Costa Rica Direccion General De Estadistica Y Cen... – San Jose, Costa Rica. 1965 – 1r – us UF Libraries [972]
Censo agropecuario de 1950 / Costa Rica Direccion General De Estadistica Y Cen... – San Jose, Costa Rica. 1953 – 1r – us UF Libraries [972]
Censo da populacao em 1940 – Lourenco Marques. v5. 1942 – us CRL [310]
Censo de 1943 / Cuba Direccion General Del Censo – Habana, Cuba. 1945 – 1r – us UF Libraries [972]
Censo de la poblacion de espana : region de extremadura. cuaderno n_o 8 / Ministerio de Trabajo – Madrid: Sucesores de Rivadeneyra, S.A. 1941 – 1 – sp Bibl Santa Ana [314]
Censo de la poblacion de espana / Spain. Direction general de estadistica – 1797-1940 – 1 – (lacks 1860) – us CRL [314]
Censo de la republica de cuba / Cuba Direccion General Del Censo – Habana, Cuba. 1920? – 1r – us UF Libraries [972]
Censo de la republica de cuba / Cuba Oficina Del Censo – Washington, DC. 1908 – 1r – us UF Libraries [972]
Censo de las poblaciones de espana segun la inscripcion de 31 de diciembre de 1940 / Direccion General de Estadistica, Ministerio de Trabajo – Madrid: Barranco, 1945 – 1 – sp Bibl Santa Ana [304]
Censo de poblacion de espana segun el empadronamiento de...1887, tomo 1-3 / Instituto Geografico – Madrid, 1887 – 43mf – 9 – sp Cultura [946]
Censo de poblacion de espana segun el recuento de 21 de mayo de 1857 / Comision de Estadistica – Madrid, 1858 – 16mf – 9 – sp Cultura [946]
Censo de poblacion de las provincias y partidos de la corona de aragon en el siglo 16 / Gonzalez, T – Madrid, 1829 – 7mf – 9 – sp Cultura [946]
Censo escolar correspondiente a fines de 1883, principios de 1884-1885, tomo 1-3 – Buenos Aires, 1885 – 27mf – 9 – sp Cultura [972]
Censo escolar nacional. resumenes generales y preliminares levantado a fines de 1883, principios de 1884 – Buenos Aires, 1884 – 1,093mf – 9 – sp Cultura [972]
Censo espanol / Floridablanca, Conde de – Madrid, 1787 – 2mf – 9 – sp Cultura [946]
Censo espanol executado por orden del rey en el ano 1787 / Floridablanca, Conde de – Madrid, 1787 – 4mf – 9 – sp Cultura [946]
Censo general de buenos aires de 1887, tomo 1-2 – Buenos Aires, 1889 – 23mf – 9 – sp Cultura [972]
Censo general de la republica de guatemala en 1893 – Guatemala, 1894 – 5mf – 9 – sp Cultura [972]
Censo general de la republica de guatemala levantado el ano 1880 – Guatemala, 1881 – 9mf – 9 – sp Cultura [972]
Censo general de la republica mexicana, 1895, 1900, 1910, 1921, 1930 / Mexico. Direccion general de estadistica – 1897-1936 – 1 – us L of C Photodup [318]
Censo general de poblacion / Mexico. Direccion General de Estadistica – 1 – (7th: v1-32 1950 $120 [0359]. 8th: v1-32 1960 $432 [0358]) – us Brook [318]
Censo oficial de los senores agentes comerciales en 1st de junio de 1961 / Colegio Oficial de Agentes Comerciales – Caceres: Imp. Sanguino, 1961 – 1 – sp Bibl Santa Ana [304]
Censor – Boston. 1771-1772 (1) – mf#3513 – us UMI ProQuest [320]
Censor – Los Angeles CA. v6 n52, 2-4,12,18-19,28 [1888 jul 19, aug 2-16, oct 11, nov 22-29, 1889 feb 7] – 1r – 1 – (cont: western wave) – mf#919845 – us WHS [071]
Censor : an entirely original work devoted to literature, poetry, and the drama – London. 1828-1829 (1) – mf#4223 – us UMI ProQuest [420]
Censor – London. 1715-1717 (1) – mf#4222 – us UMI ProQuest [420]
Censor de la revolucion – Santiago, Chile. 1960 – 1r – us UF Libraries [972]
Die censoriade : fuenf buecher censorenlieder / Sehring, Wilhelm – Strassburg: G L Schuler, 1943 – 1r – 1 – us Wisconsin U Libr [780]
The censorship of hebrew books / Popper, William – New York: Knickerbocker Press, 1899 – 2mf – 9 – 0-524-07477-1 – (incl bibl ref) – mf#1992-1080 – us ATLA [470]
The censorship of the church of rome : and its influence upon the production and distribution of literature / Putnam, George Haven – New York: GP Putnam, 1906-1907 – 3mf – 9 – 0-524-00779-9 – mf#1990-0211 – us ATLA [240]
Censur und confiscation hebraischer bucher im kircenstaate / Berliner, Abraham – Frankfurt am Main, Germany. 1891 – 1r – us UF Libraries [939]
Censura del senor doctor d pedro joseph bermudez de la torre y solier, alguacil mayor de corte desta real audiencia de lima : en respuesta a un papel, con que el autor desta obra le remitio a su examen, paraque le dixesse su parecer sobre ella / Bermudez de la Torre y Solier, Pedro Jose – [Lima: s.n.] diciembre 11 de 1715 anos – (= ser Books on religion...1543/44-c1800: miscelanea) – 1mf – 9 – mf#crl-412 – ne IDC [241]
Censura sencilla del papel que publico en esta corte el reverendo fray buenaventura angeleres : con el titulo de desengano de la filosofia real y desempeno de la medicina sanativa / Gamez, A – Madrid, S.A. – 1mf – 9 – sp Cultura [610]
Census commissioner's notes on census arrangements in individual provinces and states / India. Census Commissioner – 1st series. 1912? – 1 – us CRL [315]
Census data with maps for small areas of new york city, 1910-1960 – 10r – 1 – (with guide) – us Primary [317]
Census enumeration district descriptions, 1900 / United States. Census Office – Washington, DC: National Archives, 1977 (mf ed) – (= ser National Archives microfilm publications) – 10r – 1 – (reel 1: alabama-connecticut. reel 2: delaware-illinois. reel 3: indian territory-kansas. reel 4: kentucky-massachusetts. reel 5: michigan-montana. reel 6: nebraska-new york (districts 1-3). reel 7: new york (districts 4-19)-ohio. reel 8: oklahoma-pennsylvania. reel 9: rhode island-vermont. reel 10: virginia-wyoming) – us Nat Archives [317]
Census of 1897 : pervaia vseobshchaia prepis' naseleniia rossiiskoi imperii 1897 / Russia – Spb, 1897-1904. v1-89 – 852mf – 8 – mf#3000 – ne IDC [314]
Census of 1926 : vsesoiuznaia perepis' naseleniia 1926 goda / Russia – M, 1926 – 664mf – 8 – (with ind) – mf#R-160 – ne IDC [314]
Census of 1939 : pod'iachikh, p g vsesoiuznaia perepis' naseleniia 1939 goda / metodologiia i organizatsiia provedeniia perepisi i razrabotki itogov / Russia – 2nd ed. M, 1957 – 2mf – 9 – mf#R-18,327 – ne IDC [314]
Census of 1959 : chislennost', sostav i razmeshchenie naseleniia sssr / kratkie itogi vsesoiuznoi perepisi naseleniia 1959 goda / Russia – M, 1961 – 1mf – 9 – mf#R-18,331 – ne IDC [314]
Census of 1959 : isupov, a a natsional'nyi sostav naseleniia ssr (po itogam perepisi 1959 goda) / Russia; ed by Pod'iachikh, P G – M, 1964 – 1mf – 9 – mf#R-18,330 – ne IDC [314]
Census of 1959 : itogi vsesoiuznoi perepisi naseleniia 1959 goda / Russia – M, 1962-1963. 15v – 36mf – 9 – mf#R-18329 – ne IDC [314]
Census of 1970 : itogi vsesoiuznoi perepisi naseleniia 1970 goda / Russia – M, 1972-1974 7v – 36mf – 9 – mf#R-18,332 – ne IDC [314]
Census of 1970 : o predvaritel'nykh itogakh vsesoiuznoi perepisi naseleniia 1970 goda / Russia – M, 1970 – 1mf – 9 – mf#R-18,333 – ne IDC [314]
Census of canada / Canada. Dominion Bureau of Statistics – 1851-91 – 1 – $216.00 – mf#0135 – us Brook [317]
The census of ceylon – Colombo: Govt Press. v1 pts1-2, v2-3 1946; v4 – 1 – us CRL [315]
The census of ceylon – [s.l., s.n.] v3 1891; v1,3 1901 – 1 – us CRL [315]
Census of creek indians taken by parsons and abbott in 1832 / U.S. Bureau of Indian Affairs – (= ser Records relating to census rolls and other enrollments) – 1r – 1 – mf#T275 – us Nat Archives [317]
Census of england and wales, 1871 / Great Britain. Census Office – 1872-73 – 1 – us L of C Photodup [941]
Census of governments *see* Us bureau of the census. census of governments
Census of great britain on education, report of royal commissioners for taking a... 1852-3 : command n1692 – 6mf – 9 – mf#87112 – uk Microform Academic [370]
Census of india : price list of publications / India. Ministry of Home Affairs. Office of the Registrar General – New Delhi, 1957 – us CRL [317]
Census of industrial production, 1961 / Tanzania Maktaba Ya Takwimu – Dar es Salaam, Tanzania. 1964 – 1r – us UF Libraries [960]
Census of palestine 1931 – Alexandria, 1933. 2v – 28mf – 9 – mf#J-28-178 – ne IDC [956]
Census of population, 1790 – Philadelphia: J Gales, [1791?]-1908 – 3mf – 1 – us Misc Inst [310]
Census of population, 1840 – Washington DC: Blair and Rives, 1841 – 1mf – 1 – us Misc Inst [310]
Census of population, 1850 – Washington DC: Robert Armstrong, Public Printer, 1853-1854 – 1mf – 1 – us Misc Inst [310]
Census of population, 1860 – Washington DC: US Govt Print Office, 1862-1866 – 1mf – 1 – us Misc Inst [317]
Census of population, 1870 – Washington: US GPO. 1872-1874 – 2mf – 1 – us Misc Inst [317]
Census of population, 1880 – Washington: US GPO, 1881-1888 – 4mf – 1 – us Misc Inst [317]
Census of population, 1890 – Washington: US GPO, 1892-1898 – 10mf – 1 – us Misc Inst [317]
Census of population, 1900 – Washington: US Govt Census Office, 1900-1907 – 8mf – 1 – us Misc Inst [317]
Census of population, 1910 – Washington: US GPO, 1910-1915 – 5mf – 1 – us Misc Inst [317]
Census of population, 1920 – Washington: US GPO, 1920-1931 – 10mf – 1 – us Misc Inst [317]
Census of population, 1800-1830 – Washington DC: [s.n.], 1801-1835 – 1mf – 1 – us Misc Inst [310]
Census of religions / Addington, John Gellibrand Hubbard – London, England. 1882 – 1r – us UF Libraries [240]
Census of san mateo and santa cruz counties – San Mateo Co, CA. 1860 – 1r – 1 – $50.00 – mf#B40258 – us Library Micro [978]
Census of the philippine islands. 1918 / Philippines. Census Office – 1-4 v. 1920-21 – 1 – 54.00 – us L of C Photodup [915]
Census of the philippines, 1939 / Philippines. (Commonwealth). Commission of the Census – 5v. 1940-43 – 1 – $161.00 – us L of C Photodup [315]
Census of the republic of cuba 1919 / Cuba Direccion General Del Censo – Havana, Cuba. 1920? – 1r – us UF Libraries [972]
Census of travancore : code of procedure – Trivandrum. pt1. 1911 – 1 – us CRL [315]
[Census reports] – 13,856mf 133r – 9,1 – (sects: africa [1 title on 4mf]; east asia [20 titles on 348mf]; eastern europe [9 titles on 1600mf]; europe [24 titles on 972mf]; latin america [33 titles on 598mf]; middle east/north africa [65 titles on 608mf]; south asia [21 titles on 4731mf]; south-east asia [32 titles on 553mf]; india population census 1872-1951 [1 title on 4442mf]; india population census 1961-71 [133r]; with p/g) – ne IDC [310]
Census returns : filed october 15 1870 / Ohio. Ashtabula Co – (mf ed 1974) – 1r – 1 – (filmed by genealogical society of utah, 1974) – us Western Res [978]
Census returns, 1841-1881, on microfilm : a directory to local holdings / Gibson, Jeremy Sumner Wycherley [comp] – Plymouth, England: Fed of Family History Soc, 1986 (mf ed 1987) – (= ser Guides for genealogists, family, and local historians) – 1mf – 9 – mf#FSN-46530 – us NY Public [314]
Census returns for 1850 / Portage Co. Ohio – 1r – 1 – (filmed by genealogical society of utah, 1974) – us Western Res [978]
Census roll, 1835, of the cherokee indians east of the mississippi / U.S. Bureau of Indian Affairs – (= ser Records relating to census rolls and other enrollments) – 1r – 1 – (with index) – mf#T496 – us Nat Archives [317]
Cent fables : choisies des anciens auteurs, mises en vers latin... / Faerno, G – Londres: Guill. Darres & Claude Du Bosc, 1743 (1744). – 4mf – 9 – (in latin and french) – mf#O-1856 – ne IDC [090]
Cent fleurs de mon herbier : etudes sur le monde vege tal a la portee de tous / Massicotte, Edouard Zotique – Montreal: Beauchemin, 1906 – 3mf – 9 – 0-665-65361-1 – mf#65361 – cn Canadiana [580]
Cent per cent swadeshi : or, the economics of village industries / Gandhi, Mahatma – Ahmedabad: Navajivan Press, 1938 – (= ser Samp: indian books) – us CRL [338]
Cent sermons svr l'apocalypse / Bullinger, Heinrich – [Geneve], Jean Crespin, Pour Nicolas Barbier, & Thomas Courteau, 1558 – 1mf0mf – 9 – mf#PBU-198 – ne IDC [240]
Cent trente-cinq ans apres ou la renaissance acadienne : suivi de notules historiques et anecdotiques, et d'un petit discours prononcé par l'auteur le 15 aout 1890 a annapolis, nouvelle-ecosse (ancienne acadie) / Fontaine, L Urgele – Montreal: Gebhardt-Berthiaume, 1890 [mf ed 1980] – 1mf – 9 – 0-665-03140-8 – mf#03140 – cn Canadiana [971]
Centamilc celvi – Tirunelveli: Tirunelveli Tennintiya Caivacittanta Nurpatippuk Kalakam. [v1-14 (1923-1936)] (mthly) – 4r – 1 – us CRL [490]
Centanerio de colon y ferias / Badajoz – 1892 – 9 – sp Bibl Santa Ana [946]
Le centaure – Paris. 1896-97 – 1 – fr ACRPP [073]
Centenaire de la ville et de la paroisse de st-hubert : comte de chambly, province de quebec, 1862-1962 = Centenary of the town and parish of st-hubert, chambly county, province of Quebec, 1862-1962 – [St-Hubert?: s.n, 1962?] (mf ed 1992) – 1mf – 9 – mf#SEM105P1568 – cn Bibl Nat [971]
Centenaire de l'ecole des langues oriental vivantes 1795-1895 – Paris, 1895 – 6mf – 9 – mf#AR-1798 – ne IDC [956]
Centenaire de l'independance nationale d'haiti / Devot, Justin – Paris, France. 1901 – 1r – us UF Libraries [972]
Centenario de don fray juan de zumarraga, el 4th / Bayle, Constantino – Madrid: Missionalia Hispanica, 1948 – 1 – sp Bibl Santa Ana [240]
Centenario de la congregacion de misioneros de la preciosa sangre. recuerdos de las fiestas de su celebracion en caceres, 15 agosto 1815-15 de agosto 1915... – Caceres: Tip. Jimenez Merino, 1915? – 1 – sp Bibl Santa Ana [240]
Centenario de la guerra nacional de nicaragua cont... / Aleman Bolanos, Gustavo – Guatemala, . 1956 – 1r – us UF Libraries [972]
Centenario de la independencia espanola. noticia genealogica y biografica y biografica del mariscal campo... / Croquer Cabezas, Emilio – Cadiz: Tipografia Comercial, 1912 – 1 – sp Bibl Santa Ana [920]
Centenario de la universidad de antioquia, 1822-19 / Universidad De Antioquia – Medellin, Colombia. 1922 – 1r – us UF Libraries [378]
El centenario de magallanes / Bayle, Constantino – Madrid: Razon y Fe, 1920 – 1 – sp Bibl Santa Ana [946]
Centenario de un episodio de la guerra de la independencia ocurrido el 21 de agosto de 1809 / Jerte, Ayuntamiento de – Madrid: Asilo de Huerfanos de S.C. Jesus, 1909 – 1 – sp Bibl Santa Ana [320]
Centenario de varona / Pan American Union Division Of Philosophy, Letter – Washington, DC. 1950 – 1r – us UF Libraries [972]
Centenario do conselheiro rodrigues alves – Sao Paulo, Brazil. v1-2. 1951 – 1r – us UF Libraries [972]
Centenario do nascimento do almirante julio cesar / Santos, Noronha – Rio de Janeiro, Brazil. 1945 – 1r – us UF Libraries [972]
Centenario y panegirico / Santovenia Y Echaide, Emeterio Santiago – Habana, Cuba. 1948 – 1r – us UF Libraries [972]
The centenary celebration of the baptist missionary society, 1892-3 : reports of the commemoration services held at nottingham, leicester, kettering, london, and northampton, and list of contributions to thanksgiving fund / ed by Myers, John Brown – Holborn: Baptist Missionary Society, 1893 – 8mf – 9 – 0-524-08729-6 – (incl ind) – mf#1993-3234 – us ATLA [242]
The centenary commemoration of the birth of dr. william ellery channing, april 7th, 1880 : reports of the meetings in london, belfast, aberdeen, tavistock, manchester, and liverpool – London: British & Foreign Unitarian Association, 1880 – 1mf – 9 – 0-524-07744-4 – mf#1991-3312 – us ATLA [240]
Centenary history and handbooks of british guiana / Webber, Albert Raymond Forbes – Georgetown, Guyana. 1931 – 1r – us UF Libraries [972]
Centenary memorial of the planting and growth of presbyterianism in western pennsylvania and parts adjacent : containing the historical discourses...dec 7-9 1875 / Junkin, David Xavier et al – Pittsburgh: printed...by B Singerly 1876 [mf ed 1992] – 2mf [ill] – 9 – 0-524-02475-8 – mf#1990-4334 – us ATLA [242]
The centenary of american methodism : a sketch of its history, theology, practical system, and success / Stevens, Abel – New York: Carlton & Porter, 1866, c1865 – 1mf – 9 – 0-7905-6205-7 – mf#1988-2205 – us ATLA [242]

The centenary of catholicity in kentucky / Webb, Benedict Joseph – Louisville: C A Rogers 1884 [mf ed 1993] – 2mf – 9 – 0-524-06297-8 – mf#1990-5226 – us ATLA [241]

Centenary of methodism in eastern british america, 1782-1882 – Halifax, NS: S F Huestis, [1882?] [mf ed 1980] – 2mf – 9 – 0-665-02171-2 – mf#02171 – cn Canadiana [242]

Centenary of robert burns : a lecture delivered by the rev w mckenzie, before the mechanics' institute at ramsay – [Montreal?: s.n.] 1859 [mf ed 1984] – 9 – 0-665-45537-2 – mf#45537 – cn Canadiana [420]

Centenary of saint peter and the general council / Manning, Henry Edward – London, England. 1867 – 1r – us UF Libraries [240]

The centenary of saint peter and the general council : a pastoral letter to the clergy etc / Manning, Henry Edward – London: Longmans, Green, 1867 – 1mf – 9 – 0-8370-8727-9 – (incl bibl ref) – mf#1986-2727 – us ATLA [240]

The centenary of the birth of ralph waldo emerson : as observed in concord, may 25, 1903 – [Boston?]: Printed at the Riverside Press for the Social Circle in Concord, 1903 – 1mf – 9 – 0-524-01093-5 – mf#1990-4058 – us ATLA [420]

The centenary of the methodist new connexion, 1797-1897 / Crothers, Thomas Davison et al; ed by Packer, George – London: Geo Burroughs, [1897?] – 1mf – 9 – 0-524-06368-0 – (incl bibl ref) – mf#1990-5238 – us ATLA [242]

The centenary of the society of mary / Garvin, John E – Dayton, Ohio: Brothers of Mary, c1917 – 1mf – 9 – 0-524-03842-2 – mf#1990-4889 – us ATLA [240]

The centenary of wesleyan methodism : a brief sketch of the rise, progress, and present state of the wesleyan methodist societies throughout the world / Jackson, Thomas – New York: T Mason and G Lane, 1839 – 1mf – 9 – 0-524-01815-4 – mf#1990-4153 – us ATLA [242]

Centenary pictorial album : being contributions of the early history of methodism in the state of maryland / Roberts, George C M – Baltimore: JW Woods, 1866 – 1mf – 9 – 0-524-08580-3 – mf#1993-3165 – us ATLA [240]

Centenary souvenir, 1851-1951 / Fernando, J S A & Jayewardene, Gustavus – [Colombo: Lanka Trading Co, 1951] – 1 – us CRL [954]

Centenary thoughts for the pew and pulpit of methodism in eighteen hundred and eighty-four / Foster, Randolph Sinks – New York: Phillips and Hunt; Cincinnati: Cranston and Stowe, 1884 – 1mf – 9 – 0-524-00968-6 – mf#1990-4026 – us ATLA [242]

Centenary voices : or, a part of the work of the women of the universalist church: from its centenary to the present time – Philadelphia: Woman's Centenary Assoc 1886 [mf ed 1992] – (= ser Unitarian/universalist coll) – 1mf – 9 – 0-524-04279-9 – mf#1991-2063 – us ATLA [240]

The centenary volume of the baptist missionary society, 1792-1892 / Henderson, William John et al; ed by Myers, John Brown – 2nd ed. [London]: Baptist Missionary Society, 1892 – 5mf – 9 – 0-524-08827-6 – (incl bibl ref and ind) – mf#1993-3319 – us ATLA [242]

The centenary volume of the church missionary society for africa and the east, 1799-1899 – London: Church Missionary Society, 1902 [mf ed 1990] – 3mf – 9 – 0-7905-5517-4 – mf#1988-1517 – us ATLA [242]

Centennial *see* Clear creek county miscellaneous newspapers

The centennial : an old canadian fort / Garrett, John C – Niagara, Ont?: s.n, 1900 – 1mf – 9 – mf#09099 – cn Canadiana [720]

The centennial : a poem written on the centenary of st mark's church, niagara, ont (1792-1892) / Garrett, John C – S.l: s.n, 1892 – 1mf – 9 – mf#01400 – cn Canadiana [810]

Centennial addresses / Hayden, Warren Luce – Indianapolis, IN: [s.n.] c1909 [mf ed 1992] – (= ser Christian church (disciples of christ) coll) – 1mf – 9 – 0-524-03157-6 – mf#1990-4606 – us ATLA [240]

Centennial addresses : synod of north carolina... / Crawford, A W et al – [s.l: s.n, 1913?] [mf ed 1992] – (= ser Presbyterian coll) – 1mf – 9 – 0-524-02570-3 – mf#1990-4382 – us ATLA [242]

Centennial anniversary of the death of john wesley and of the foundation of methodism in canada : held in centenary church, hamilton, on monday evening, march 2nd, 1891... – [S.l: s.n, 1891?] [mf ed 1987] – 1mf – 9 – 0-665-46614-5 – mf#46614 – cn Canadiana [242]

The centennial anniversary of the elevation of john marshall to the office of chief justice of the supreme court of the united states of america. / Marshall, John – Philadelphia: Buchanan, 1901. 68p. LL-485 – 1 – us L of C Photodup [347]

Centennial Association for Connecticut *see* Spirit of 'seventy-six

Centennial avenue school headliner cash – Roosevelt NY. 1993 spr – 1r – 1 – mf#4864006 – us WHS [370]

Centennial baptist church. nashville, tennessee : church records – Feb 1894-Mar 1982 – 1 – 60.66 – us Southern Baptist [242]

The centennial campfire / Craig, Laura Gerould – 2nd ed. Indianapolis, Ind., USA: Christian Woman's Board of Missions, 1909 – 2mf – 9 – 0-524-02250-X – mf#1990-4257 – us ATLA [240]

The centennial celebration of the evacuation of detroit by the british, july 11, 1796-july 11, 1896 : report of the proceedings... – Detroit: J F Eby, 1896 – 1mf – 9 – mf#27431 – cn Canadiana [978]

Centennial celebration of the settlement of this province by the u e loyalists : will be held at niagara under the patronage of the county council of lincoln – S.l: s.n, 1884? – 9 – mf#59700 – cn Canadiana [971]

The centennial celebration of the theological seminary of the presbyterian church in the united states of america, at princeton, new jersey : may fifth, may sixth, may seventh, nineteen hundred and twelve – Princeton: At the Theological seminary, 1912 – 2mf – 9 – 0-7905-6547-1 – mf#1988-2547 – us ATLA [242]

The centennial celebration of the theological seminary of the presbyterian church in the united states of america, at princeton, new jersey : may fifth, may sixth, may seventh, nineteen hundred and twelve – Princeton: At the Theological seminary, 1912 – 2mf – us ATLA [242]

The centennial celebration of tolarsville baptist church – St Pauls, NC. 1874-1974 – 1 – $5.00 – us Southern Baptist [242]

Centennial convention report : one hundredth anniversary of the disciples of christ, pittsburg [sic], october 11-19, 1909 – Cincinnati, O[hio]: Standard Pub Co, [1909?] – 2mf – 9 – 0-524-07675-8 – mf#1991-3260 – us ATLA [240]

Centennial edition of the baptist denomination : being the past and present of the baptist church throughout the world / Haynes, Dudley C – New York: Sheldon, 1875 – 1mf – 9 – 0-8370-8907-7 – (incl ind) – mf#1986-2907 – us ATLA [242]

Centennial historical discourses : delivered in the city of philadelphia, june, 1876 / McGill, Alexander Taggart et al – Philadelphia: Presbyterian Board of Publ c1876 [mf ed 1992] – 1mf – 9 – 0-524-02127-9 – mf#1990-4193 – us ATLA [242]

Centennial history of american methodism : inclusive of its ecclesiastical organization in 1784 and its subsequent development under the superintendency of francis asbury / Atkinson, John – New York: Phillips & Hunt; Cincinnati: Cranston & Stowe, 1884 [mf ed 1989] – 2mf – 9 – 0-7905-4371-0 – (incl bibl ref) – mf#1988-0371 – us ATLA [242]

The centennial history of the associate reformed presbyterian church, 1803-1903 – Charleston, SC: Walker, Evans & Cogswell, 1905 – 2mf – 9 – 0-524-06360-5 – mf#1990-5230 – us ATLA [242]

Centennial hymn, op 27 / Paine, John Knowles – [1876] [mf ed 1988] – 1r – 1 – (words by [john greenleaf] whittier) – mf#pres. film 7 – us Sibley [780]

Centennial living-link souvenir – Cincinnati: Foreign Christian Missionary Society, 1909 – 1mf – 9 – 0-524-06483-0 – mf#1991-2583 – us ATLA [240]

The centennial meditation of columbia / Buck, Dudley – A cantata for the inaugural ceremonies at Philadelphia, May 10, 1876. Poem by Sidney Lanier, of Georgia. Music by Dudley Buck, of Connecticut. New York: G. Schirmer, 1876. Piano-vocal score. music 902 – 1 – us L of C Photodup [780]

The centennial memorial of the presbytery of carlisle : series of papers, historical and biographical, relating to the origin and growth of presbyterianism in the central and eastern part of southern pennsylvania / Chambers, Talbot Wilson et al – Harrisburg: Meyers Print and Pub House, 1889 – 3mf – 9 – 0-524-02486-3 – mf#1990-4345 – us ATLA [242]

The centennial northwest : an illustrated history of this great section of the united states from its earliest settlement to the present time / Tuttle, Charles Richard & Pennock, Ames Castle – Madison, WI: Inter-state Book Co, 1876 – 8mf – 9 – mf#16288 – cn Canadiana [978]

Centennial of home missions : in connection with the 114th general assembly of the presbyterian church in the united states of america, new york city, may 16-20 1902 / McCook, Henry Christopher et al – Philadelphia: Presbyterian Board of Publ & Sabbath-School Work 1902 [mf ed 1986] – 1mf – 9 – 0-8370-6319-1 – mf#1986-0319 – us ATLA [242]

The centennial of religious journalism / ed by Barrett, John Pressley – 2nd ed. Dayton, Ohio: Christian Pub Association, 1908 – 2mf – 9 – 0-524-06238-2 – mf#1990-5193 – us ATLA [070]

Centennial of the province of upper canada, 1792-1892 : proceedings at the gathering held at niagara-on-the-lake, july 16, 1892... – [S.l.]: printed... by Arbuthnot & Adamson, 1893 [mf ed 1980] – 1mf – 9 – 0-665-02562-9 – mf#02562 – cn Canadiana [971]

Centennial of upper canada and the province of ontario – [Toronto?: s.n, 1892?] [mf ed 1981] – 1mf – 9 – mf#13229 – cn Canadiana [325]

Centennial papers / Fowler, William Chauncey et al; ed by General Conference of the Congregational Churches of Connecticut – Hartford: Case, Lockwood, & Brainard, 1877 – 1mf – 9 – 0-524-03151-7 – mf#1990-4600 – us ATLA [240]

Centennial poem / Curzon, Sarah Anne – [Niagara, Ont?: s.n.], 1897 [mf ed 1980] – (= ser Niagara Historical Society) – 1mf – 9 – (incl: fort niagara, ny, 1783-1796 by rev canon bull; slave rescue in niagara, sixty years ago miss carnochan) – mf#06181 – cn Canadiana [971]

The centennial record of freewill baptists, 1780-1880 / Brewster, Jonathan McDuffee et al – Dover, NH: Printing Establishment, 1881 – 1mf – 9 – 0-524-03786-8 – (incl bibl ref) – mf#1990-4858 – us ATLA [242]

A centennial review of the bowdoinham association of baptist churches in maine / Small, Edwin S – 1887 – 1 – $5.00 – us Southern Baptist [242]

Centennial sermons and papers : delivered at the one hundredth anniversary of the organization of the cumberland presbyterian church before the eightieth general assembly, dickson, tenn... – Nashville, Tenn: Cumberland Press, 1911 – 4mf – 9 – 0-524-07968-4 – mf#1990-5413 – us ATLA [242]

Centennial souvenir, 1809-1909 / Cedarville. Ohio. Reformed Presbyterian Church – 9 – $50.00 – us Presbyterian [240]

Centennial souvenir of the new hampshire yearly meeting of free baptists, 1792-1892 / ed by Wiley, Frederick Levi – Laconia, NH: Pub by the Board of Directors, [1892?] – 1mf – 9 – 0-524-06871-2 – mf#1990-5290 – us ATLA [241]

Centennial st andrew's, niagara, 1794-1894 / Carnochan, Janet – Toronto: W Briggs, 1895 [mf ed 1979] – 1mf – 9 – 0-665-00994-1 – mf#00994 – cn Canadiana [240]

Centennial, st mark's church, niagara / Carnochan, Janet – Toronto: J Bain, 1892 [mf ed 1979] – 1mf – 9 – 0-665-00461-3 – mf#00461 – cn Canadiana [240]

Centennial survey of foreign missions : a statistical supplement to "christian missions and social progress" / Dennis, James Shepard – New York: Fleming H Revell, 1902 – 2mf – 9 – 0-8370-7212-3 – (incl indes) – mf#1986-1212 – us ATLA [240]

Centeno, A *see* Historia de cosas del oriente primera y segunda parte

Centeno, Christopher J *see* Journal of whiplash and related disorders

Centeno Guell, Fernando *see*
- Angel y las imagenes
- Evocacion de zande
- Rapsodia de aglae

Center baptist church. marion, illinois : church records – Feb 1887-Jan 1950. 606p – 1 – us Southern Baptist [242]

Center Creek. Baptist Church, Jefferson County, MO *see* Records

Center first baptist church. center, texas : church records – Sep 1889-Sep 1941. 998p. Lacks May 1908-Dec 1917 – 1 – us Southern Baptist [242]

Center for Alternative Media [US] *see* Moore's weekly

Center for black culture and research bulletin / West Virginia University – 1993 apr-may – 1r – 1 – (cont by: bulletin [west virginia university. center for black culture and research]) – mf#2918436 – us WHS [305]

Center for Changes [Detroit MI] *see* Changes socialist monthly

Center for Community Technology *see* New and views

Center for Community Technology [Madison WI] *see*
- Human scale
- Newsletter

Center for Constitutional Rights *see* Rights

Center for Coordination of Research on Social Indicators [US] *see* Social indicators newsletter

Center for Defense Information [Washington DC] *see* Defense monitor

Center for Media Literacy. University of Minnesota *see* Enclitic

Center for Migration Studies [US] *see* Migration today

Center for National Security Studies [Washington, DC] et al *see* First principles

Center for Public Representation *see*
- Bulletin of the center...
- Consumer and responsive government news
- Public eye
- Public i
- Transportation news

Center for Public Representation. United States *see* Clearing the air

Center for Public Representation [US] *see* Making waves

Center for reformation research newsletter – St. Louis. 1975-1992 (1) 1975-1976 (5) 1975-1976 (9) – (cont: foundation for reformation research newsletter) – ISSN: 0362-563X – mf#6400,01 – us UMI ProQuest [242]

Center for southern folklore – v2 n2-v4 n2 [1979 spr-1982 win] – 1r – 1 – (cont: center for southern folklore newsletter) – mf#676053 – us WHS [390]

Center for southern folklore newsletter – v1 n1-v2 n1 [1978 spr-1979 win] – 1r – 1 – (cont by: center for southern folklore) – mf#676060 – us WHS [390]

Center for the study of democratic institutions center report – Santa Barbara. 1967-1976 (1) 1972-1976 (5) 1975-1976 (9) – mf#7242 – us UMI ProQuest [322]

Center for the Study of Practical Politics *see* Practical politics

Center for United Labor Action *see*
- Newsletter
- United labor action

Center for Urban Missions *see* Reaching the city

Center for Women's Studies and Services *see* Longest revolution

Center for Women's Studies and Services [San Diego, CA] *see* Feminist bulletin

Center gallery newsletter / Art Place/Center Gallery [Madison WI] – 1982 jan-1983 nov/dec – 1r – 1 – (cont by: art place) – mf#706261 – us WHS [700]

Center hill baptist church. neshoba county. mississippi : church records – 1866-78 – 1 – 5.00 – us Southern Baptist [242]

Center house bulletin – New York. 1971-1974 (1) 1971-1974 (5) 1971-1974 (9) – (cont by: presidential studies quarterly) – ISSN: 0098-809X – mf#10622 – us UMI ProQuest [320]

Center line / Naval Training Equipment Center [US] – 1966 sep 21-1968 sep 19, 1968 sep 26-1970 sep 17, 1970 sep 24-1973 may 31, 1973 jun-1976 jan, 1976 feb 2-1979 jan 18, 1979 feb-1984 sep 20, 1984 oct 4-1985 sep 27 – 7r – 1 – (cont by: centerline [orlando fl]) – mf#1497451 – us WHS [355]

Center magazine – Santa Barbara. 1967-1987 (1) 1971-1987 (5) 1967-1987 (9) – ISSN: 0008-9125 – mf#3362 – us UMI ProQuest [320]

Center news / African American Catholic Pastoral Center [Oakland CA] – 1992 dec, 1993 jan, mar, may, aug, 1994 jan, aug, dec, 1995 jan, jul, dec, 1996 mar, jun, oct – 1r – 1 – mf#2978755 – us WHS [241]

Center news – Milwaukee WI. 1938 sep 9-1964 jun 12, 1966 nov 4-1971 dec 10 – 2r – 1 – mf#1054160 – us WHS [071]

Center news – Rochester, N.Y. – 1 – (v36, no29, (10 apr. 1975)-v36, no32, (19 may 1975)) – us AJPC [071]

Center on Law and Pacifism [Colorado sprs CO] *see* Center peace

Center peace / Center on Law and Pacifism [Colorado sprs CO] – v3 n3-v8 n1 [1980 sep/oct-1985 fall] – 1r – 1 – mf#1042669 – us WHS [303]

Center register – Center, NE: O R Robinson, -oct 1906// (wkly) [mf ed v2 n16. mar 24 1905-06 (gaps) filmed 1971] – 1r – 1 – (cont: knox county broad ax. absorbed by: crofton journal) – us NE Hist [071]

Center relay – v70 n6,9-10 [1983 mar 25, may 6-20], v73 n8 [1984 oct 19], v76 n3-4,8,10 [1986 aug 8-22, oct 31, nov 28], v79 n8 [1988 apr 15], v80 n5-12 [1988 sep 2-dec 23], v81 n1-4,6-13 [1989 jan n6-feb 17, mar 17-jun 23], v82 n1-12 [1989 jul 7-dec 22], v83 n1-7,9 [1990 jan 5-mar 30, apr 27], 1990 may 11-1993 dec – 2r – 1 – mf#1054162 – us WHS [071]

Center report / Hastings Center Hastings – 1971+ (1) 1973+ (5) 1975+ (9) – ISSN: 0093-0334 – mf#8639 – us UMI ProQuest [610]

Center reporter *see* Miscellaneous newspapers of saguache county

Center scope / Good Samaritan Medical Center [Milwaukee WI] – v1 n1-v2 n1 [1981 feb-1982 jan], v3 n1-v6 [i.e. 5] n2 [1983 may-1985 jun] – 1r – 1 – (cont: deacon lite) – mf#647016 – us WHS [360]

Center shots : a brief discussion of many religious subjects in which both sides are heard / Burnett, Thomas R – Austin, Tex: Firm Foundation Pub House, [1912?] – 1mf – 9 – 0-524-06479-2 – mf#1991-2579 – us ATLA [240]

Center township, records, ms 566 – 1818-21 – 1r – 1 – (docket book for various justices of the peace and constables from this township) – us Western Res [978]

Center white creek baptist church (little white creek; white creek). cambridge, new york : church records – 1779-84, 1806-13, 1784-1816, 1827-94 – 1 – us Southern Baptist [242]

Centerfold : artists' news magazine – Calgary, Toronto. v1-4 n1. aug 1976-nov 1979// – 1r – 1 – Can$85.00 – cn McLaren [700]

Centerline / Naval Training Systems Center [US] – 1983 mar-1993 sep, 1985 oct-1989 jun, 1985 oct-1993 sep – 3r – 1 – (cont: center line [orlando fl]) – mf#1546483 – us WHS [355]

Centerpiese – Palo Alto, California – (= ser Center News) – 1 – (vol. 1, no. 2 (winter 1983)-v. 1, no. 3 (early spr. 1984); v. 2, no. 5 (fall 1985)-v. 3, no. 6 (mar. 1986); v. 5, no. 9 (may 1988)-v. 6, no. 2 (oct. 1988)) – us AJPC [978]

Centers of the southern struggle : fbi files on selma, memphis, montgomery, albany, and st augustine / ed by Garrow, David – (= ser Black studies research sources) – 21r – 1 – $3745.00 – 1-55655-047-2 – (with p/g; filmed fr dr garrow's personal holdings of released fbi files) – us UPA [322]

Centerview *see* Miscellaneous newspapers of saguache county

Centerville-bellbrook times / Montgomery Co. Centerville/Kettering – jan 1990-dec 1992 [semiwkly] – 4r – 1 – mf#B33444-33447 – us Ohio Hist [071]

Centerville-bellbrook times / Montgomery Co. Kettering – dec 1973-mar 1980 [wkly] – 22r – 1 – mf#B33448-33469 – us Ohio Hist [071]

[Centifolium stultorum] hundert weniger eine thorheit in eben so vielen kupfern vorgestellt... / Abraham...Sancta Clara – Wien, 1782 – 7mf – 9 – mf#O-1508 – ne IDC [090]

Centi-folium stultorum in quarto : oder hundert aussbaendige narren in folio / [Abraham...Sancta Clara] – Wien: Lercher/Seyinger, [1709-]1713 – 14mf – 9 – mf#O-1820 – ne IDC [090]

Centinel of the northwest territory – Cincinnati, OH. 1793-1796 (1) – mf#65410 – us UMI ProQuest [071]

Centinela – Almendralejo. 1884-86. No. sueltos – 9 – sp Bibl Santa Ana [074]

El centinela – [Habana]: El centinela. v12 n1356-1378,1380-1396,1399-1402. mar 9-may 1, may 6-jun 12, jun 19-26 1897; n1405-1411,1413-1417,1419-1422,1425-1450. jul 3-17, jul 22-31, aug 5-14, aug 19-oct 16 1897 – us CRL [079]

El centinela – Panama city, Panama. 12 jul 1857-26 dec 1858 – 1r – 1 – us L of C Photodup [079]

Centinela contra judios / Torrejoncillo, Francisco de – 1691 – 9 – sp Bibl Santa Ana [270]

El centinela de israel – 1873 – 9 – sp Bibl Santa Ana [830]

Centinela de la libertad – Miami, FL. 1963 apr 06-1965 may – 1r – us UF Libraries [071]

La centinela de la patria – Cadiz, Spain. 21 Jun-22 Aug 1810.-w. 6 ft – 1 – uk British Libr Newspaper [072]

Centinela dogmatico-moral con oportunos avisos al confessor, y penitente vigilias apostolicas : en que daniel, y maximino, sacerdotes missioneros, proponen, y resuelven algunas dudas, especialmente sobre el uso de las opiniones, tratos, y contratos / Vilaplana, Hermenegildo de – en Mexico: impr Bibliotheca mexicana, en el Puente del Espiritu Santo, ano de 1767 – (= ser Books on religion...1543/44-c1800: confesionarios) – 3mf – 9 – mf#crl-29 – ne IDC [241]

Cento concerti ecclesiastici... / Viadana, Lodovico da – 1605 – (= ser Mssa) – 5mf – 9 – €70.00 – mf#435 – gw Fischer [780]

Cento, Fernando *see* Peginas escogidas

Cento salmi – Roma: Tipografia poliglotta della SC di Propaganda Fide, 1875 – 1mf – 9 – 0-524-07184-5 – mf#1992-1054 – us ATLA [220]

Centon epistolario del bachiller fernan gomez de cibdadreal. generaciones... / Vera y Figueroa, Juan Antonio – Madrid, 1775 – 4mf – 9 – sp Cultura [946]

Central africa : adventures and missionary labors in several countries in the interior of africa, from 1849 to 1856 / Bowen, T J – Charleston, 1857 – 4mf – 9 – mf#HTM-19 – ne IDC [916]

Central africa : a record of the work of the universities' mission to central africa – v1-82. 1883-1964 – 11r – 1 – (lacks some iss) – mf#ATLA S0224 – us ATLA [240]

Central africa *see* Political party, trade union and pressure group materials

Central africa, 1883-1964 : the universities' mission to central africa monthly magazine. from the archives of the u s p g – 9r – 1 – (with int by r g stuart) – mf#96838 – uk Microform Academic [240]

Central africa, japan, and fiji : a story of missionary enterprise, trials and triumphs / Pitman, Emma Raymond – London: Hodder & Stoughton 1882 [mf ed 1987] – 1r – 1 – mf#8472 – us Wisconsin U Libr [240]

Central african emergency / Sanger, Clyde – London, England. 1960 – 1r – us UF Libraries [960]

Central african mail – Lusaka, Zambia 27 feb 1962-25 may, 1/3 jun, 18 oct 1963 – 1 – (wanting: jun, jul 1961; cont: african mail [8 mar 1960-20 feb 1962]) – uk British Libr Newspaper [079]

Central african post – Lusaka, Zambia. -tw. 30 March 1950-3 May 1951; 2 Aug 1951-22 May 1952; 2 Jan 1954-31 Dec 1958. 18 reels – 1 – uk British Libr Newspaper [072]

Central African Republic *see* Journal officiel de la republique centrafricaine

Central African Republic. Direction de la Statistique et de la Conjoncture *see* Annuaire statistique de la republique centrafricaine 1962

Central african survey / Blake, Wilfrid Theodore – London, England. 1961 – 1r – us UF Libraries [960]

Central african times – Blantyre, Malawi 22 jul 1899-27 jun 1908 (wkly) – 10r – 1 – (cont by: nyasaland times [5 jan 1911-18 sep 1939; 6 jan-30 jun, 18 aug 1947-27 apr 1954; 2 jan 1957-28 jun 1963) – uk British Libr Newspaper [079]

Central america / Joyce, Lilian Elwyn – London, England. 1924 – 1r – us UF Libraries [972]

Central america / Koebel, William Henry – London, England. 1917 – 1r – us UF Libraries [972]

Central america / Starr, Frederick – Chicago, IL. 1930 – 1r – us UF Libraries [972]

Central america and the caribbean, 1930-1945 – (= ser Confidential u s diplomatic post records) – 1 – $38,365.00 coll – (cuba: pt1: 1930-39 61r isbn 0-89093-635-8 $10,620; pt2: 1940-45 63r isbn 0-89093-641-2 $10,960. el salvador, 1930-45 28r isbn 0-89093-629-3 $4875. honduras, 1930-45 42r isbn 0-89093-631-5 $7300. nicaragua, 1930-45 38r isbn 0-89093-630-7 $6620. with p/g) – us UPA [327]

Central America Education Fund [Cambridge MA] *see* Central america report

Central america report / Central America Education Fund [Cambridge MA] – 1984 oct-dec, 1985 jan, mar-jul, sep-dec, 1986 jan-jul, sep-dec, 1987 jan-jul, sep-dec, 1988 jan-apr, jun-jul – 1r – 1 – (cont by: central america reporter) – mf#1534237 – us WHS [972]

Central America Resource Center [Minneapolis MN] *see* Minnesota central america connection

Central american currency and finance / Young, John – Princeton, NJ. 1925 – 1r – us UF Libraries [972]

Central american journey / Babson, Roger Ward – Yonkers-on-Hudson, NY. 1920 – 1r – us UF Libraries [972]

Central american peace conference held at washington / Buchanan, William I – Washington, DC. 1908 – 1r – us UF Libraries [972]

Central americans / Ruhl, Arthur Brown – New York, NY. 1928 – 1r – us UF Libraries [972]

Central and autonomic nervous system activity during self-paced motor performance: a study of the activation construct in marksmen / Hatfield, Bradley D – 1982 – 2mf – 9 – $8.00 – us Kinesology [790]

Central and south america / Keane, A H – London, England. v1-2. 1901 – 1r – us UF Libraries [972]

Central Arizona Labor Council *see* Labor journal

Central asia and the anglo-russian frontier question : a series of political papers / Vambery, Armin – London: Smith, Elder & Co 1874 [mf ed 1985] – 1r – 1 – (first publ in "unsere zeit", 1867-73; trans by fanny elizabeth bunnett) – mf#6966 – us Wisconsin U Libr [950]

Central asian fragments of the ashtadasasahasrika prajnaparamita and of an unidentified text / ed by Konow, Sten – Delhi: Manager of Publ, 1942 – (= ser Samp: indian books) – us CRL [490]

Central asian serials – late 19th- to early 20th-century : a core collection of newspapers and serials from the imperial period in russia – 1871-1991 [mf ed Norman Ross Publ 1994] – 98 titles on 622r – 1 – (in russian; individual titles listed separately; foll 4 serials incl: shark iulduzi, tashkent 1931-91 [122r]; kaspii, baku 1881-1917 [108r]; molla nasredin, 1906-31 [5r]; turkistan vilayatin gazeti, 1870-84, 1910-17 [9r]) – us UMI ProQuest [077]

Central asian survey – Oxford. 1993-1996 (1) 1993-1996 (5) 1993-1995 (9) – ISSN: 0263-4937 – mf#20929 – us UMI ProQuest [320]

Central avenue baptist church. memphis, tennessee : church records – 1873-1937 – 1 – 5.00 – us Southern Baptist [242]

Central baptist – Missouri. Aug 1868-1912. Lacking 1873-76 – 1 – 825.75 – us Southern Baptist [242]

Central baptist church – New York. 1972-1972 (1) 1954-1972 (5) (9) – 1r – 1 – $82.08 – (formerly: tabernacle baptist church) – mf#6507 – us Southern Baptist [242]

Central baptist church : minutes – Paris, KY. 1946-96 – 1 – mf#7048 – us Southern Baptist [242]

Central baptist church. alcoa, tennessee : church records – Sept 1943-Jun 1984 – 1 – us Southern Baptist [242]

Central baptist church. greenville county. south carolina : church records – 1893-Oct 1982. Includes mission committee minutes, 1974-75. 1713p – 1 – 77.09 – us Southern Baptist [242]

Central baptist church. morgan county. decatur, alabama : church records – 1892-1956 – 1 – us Southern Baptist [242]

Central baptist church. nashville, tennessee : church records – 20 Oct 1858-May 1946 – 1 – 79.29 – us Southern Baptist [242]

Central baptist church. pampa, texas : church records – 1931-60 – 1 – us Southern Baptist [242]

Central baptist church. washington, dc : church records – August 1826-February 1835 – 1 reel – 1 – $5.00 – us Southern Baptist [242]

Central blatt fuer die gesamte etc – Vienna, Austria. 5 jan 1899-19 dec 1901, 1908-20 dec 1909 – 2r – 1 – (aka: maschinen und metall industrie zeitung; internationales zentralblatt fuer bau keramik etc) – uk British Libr Newspaper [074]

Central blatt fuer glas industrie und keramik – Vienna, Austria. jan-20 dec 1897, 1899-20 dec 1900, 1901-15 dec 1904, 5 jan 1905-15 dec 1907 – 3r – 1 – (aka: internationales zentralblatt fuer baukeramik und glasindustrie) – uk British Libr Newspaper [074]

Central blatt fuer maschinen industrie und eisengiesserei – Vienna, Austria. 27 sep 1896-4 dec 1898 – 1r – 1 – uk British Libr Newspaper [072]

Central Branch Union Pacific Railroad Company *see* Report to secretary of interior

Central California Newspaper Guild, Local 92 *see* Guild journal

Central California Non-partisan Alliance et al *see* Kern county union labor journal

Central Canada Chamber of Mines *see* Constitution of the central canada chamber of mines, winnipeg, canada

Central canterbury news – may 1983-88 – 6r – 1 – (commenced publ may 1983) – mf#70.24 – nz Nat Libr [079]

Central canterbury news *see* Malvern record

Central china baptist mission minutes – Scattered yrs. 1907-48. Also misc. items, scattered years 1889-1940; Executive Committee minutes, scattered years, 1910-42 – 1 – us Southern Baptist [242]

Central china famine relief committee, shanghai, china : report and accounts from oct 1 1911 to jun 30 1912 – Shanghai: North-China Daily News & Herald, 1912 [mf ed 1995] – (= ser Yale coll) – 79p (ill) – 1 – 0-524-10082-9 – mf#1995-1082 – us ATLA [360]

Central christian advocate / Methodist Episcopal Church – 1872 oct 9/1880 mar 31-1909 sep 1/dec 22 – 38r – 1 – mf#518184 – us WHS [242]

Central chronik – Omaha, NE: John P Mueller, 1896 (wkly) [mf ed v2 n7. 15 apr 1897-98 (gaps) filmed 1981] – 1r – 1 – (in german) – us NE Hist [071]

Central city courier – Central City, NE: Bowerman & Steele. v1 n49. mar 11 1875-v21 n39. dec 27 1894 [wkly] [mf ed with gaps filmed -1974] – 5r – 1 – (cont: lone tree courier) – us NE Hist [071]

Central city democrat – Central City, NE: E Lena Spear. -v12 n33. jan 12 1905 (wkly) [mf ed 1895-1905 (gaps)] – 4r – 1 – (cont by: central city record) – us NE Hist [071]

Central city herald – Stevens Point WI. 1935 mar 22-apr 19 – 1r – 1 – (cont by: central wisconsin herald) – mf#935892 – us WHS [071]

Central city nonpareil – Central City, NE: H G Taylor. 50v. v22 n2. jan 8 1903-v71 n10. jan 29 1953 (wkly) [mf ed with gaps] – 1 – (cont: nonpareil. absorbed clarks enterprise) – us NE Hist [071]

Central city record – Central City, NE: Fitch Bros. v12 n34. jan 19 1905-v19 n26. nov 25 1909 [wkly] [mf ed lacks feb 1 1906] – 3r – 1 – (cont: central city democrat; absorbed by: central city republican) – us NE Hist [071]

Central city republican – Central City, NE: O D Henyan. 59v. v1 n1. jul 15 1893-v59 n41. feb 19 1953 (wkly) [mf ed with gaps] – 20r – 1 – (absorbed: central city record. merged with: central city nonpareil (1903) to form: central city republican-nonpareil. issues for feb 5-19 1953 called also v71 n11-v71 n13) – us NE Hist [071]

Central city republican-nonpareil – Central City, NE: Elgin O White. v59 n42. feb 26 1953-=v71 n14- (wkly) – 1 – (formed by the union of: central city republican+central city nonpareil (1903)) – us NE Hist [071]

Central coast express – Gosford, jan 1952-sep 1963 (misc periods) – 6r – A$426.10 vesicular A$459.10 silver – at Pascoe [079]

Central coast express – Gosford. sep 1974-sep 1988, jan-mar 1989, apr-sep 1990, apr-jun 1991, jan-mar 1992, oct-dec 1992 – 68r – at Pascoe [079]

Central Coast Railway Club *see* Ferroequinologist

Central Committee for Conscientious Objectors *see*
– Ccco news notes
– Counter pentagon
– Draft counselor's newsletter
– News notes of the central committee for conscientious objectors

Central Committee of Indonesian Independence *see* Free indonesia

The central conception of buddhism and the meaning of the word "dharma" / Shcherbatskoi, Fedor Ippolitovich – London: Royal Asiatic Society, 1923 – (= ser Samp: indian books) – us CRL [280]

The central conception of buddhism and the meaning of the word dharma / Shcherbatsky, T – 1923 – (= ser Royal asiatic society prize publication fund) – 1r – 1 – mf#664 – uk Microform Academic [280]

Central Conference of American Rabbis *see*
– Ccar journal
– Views on the synod
– Yearbook

Central Co-operative Wholesale [US] *see* Co-operative builder

Central counties advertiser and stirling, mentieth, lennox, and bridge of allan gazette – [Scotland] Stirling: C Rogers 24 jan 1862-11 mar 1863 (wkly) [mf ed 2004] – 1r – 1 – (cont by: stirling gazette and central counties advertiser) – uk Newsplan [072]

Central courier – jul 1979-mar 1981 – 7r – 1 – (incl south auckland ed) – mf#11.39 – nz Nat Libr [079]

Central criminal sessional papers / Great Britain. Courts – Old Bailey. 1816-1913. 49 reels – 1 – $3,000.00 – us Trans-Media [345]

Central daily news *see* Zhong yang ri bao

Central district times *see* Taihape times

Central district times [taihape] – jan 10 1953-feb 1959 – 1 – (title changed to: taihape times mar 1959-oct 1960, jan 5 1961-mar 1979, jan 1980-1987) – mf#42.1 – nz Nat Libr [079]

Central districts farmer – Palmerston North, NZ. 7 apr 1982-87 – 5r – 1 – mf#45.10 – nz Nat Libr [079]

Central faith of christianity *see*
– Chi-tu chiao ti chung hsin hsin yang [ccm69]

Central farmer – Omaha, NE: C Vincent. n1080. mar 28 1901-n1201. jul 30 1903 – 2r – 1 – (cont: central farmer and the nonconformist. absorbed by: farmers advocate (topeka, ks)) – us NE Hist [071]

Central fife gazette – [Scotland] Fife, Cowdenbeath: J Parker 17 oct 1902-26 dec 1905 (wkly) [mf ed 2004] – 167v on 2r – 1 – uk Newsplan [072]

Central fife times and advertiser – Lochgelly: A Romanes & Son Ltd (Dunfermline Press) 1973-(wkly) [mf ed 1994-] – 1 – (formed by union of: cowdenbeath advertiser & kelty news and: times for lochgelly, bowhill, dundonald, cardenden, glencraig & lochore) – ISSN: 1358-4731 – uk Scotland NatLib [072]

Central first baptist church. central, south carolina : church records – 1890-1957 – 1 – us Southern Baptist [242]

Central five / Greenbie, Sydney – Evanston, IL. 1943 – 1r – us UF Libraries [972]

Central florida advocate – Orlando FL. v5 n27,45 [1997 jul 4-10, nov 7-13] – 1r – 1 – mf#4024999 – us WHS [071]

Central Florida Christian School et al *see* Faith and freedom

Central florida exposition / Harold, William G – s.l, s.l? 1936 – 1r – us UF Libraries [978]

Central florida times – Ocala, FL. 1926 apr-aug 3 – 1r – us UF Libraries [071]

Central glamorgan gazette – Bridgend, Wales, UK. 29 Jun 1866-Mar 1894. -w.15 reels – 1 – uk British Libr Newspaper [072]

Central glamorgan gazette and general, commercial and agricultural advertiser – [Wales] Bridgend 29 jun 1866-mar 1894 [mf ed 2003] – 19r – 1 – uk Newsplan [072]

Central hawkes bay press – Waipukurau, NZ. nov 1971-mar 1972; may 1974-nov 1980 – 1 – mf#35.3 – nz Nat Libr [079]

The central idea of christianity / Peck, Jesse Truesdell – Rev ed. New York: Nelson & Phillips; Cincinnati: Hitchcock & Walden, 1876 – 1mf – 9 – 0-524-00303-3 – mf#1989-3003 – us ATLA [240]

Central Ikanagon Lands Ltd *see* Fruit farming at kelowna

Central illinoian – Beardstown IL. 1864 may 19 – 1r – 1 – mf#874019 – us WHS [071]
Central illinois wochenblatt – La Salle, IL: H E Hagenbach, 1921-sep 25 1925 – 5r – us CRL [074]
Central india – Lucknow. pt4. 1901 – (= ser Census of india) – 1 – us CRL [315]
Central india / Whitley, Edward Hamilton – Westminster: Society for the Propagation of the Gospel in Foreign Parts, 1933 [mf ed 1995] – (= ser Yale coll; Spg handbooks. new series) – 1 – 0-524-09241-9 – mf#1995-0241 – us ATLA [240]
Central india in 1857 : being an answer to sir john kaye's criticisms on the conduct of the late sir henry marion durand, whilst in charge of central india during the mutiny / Durand, Henry Mortimer – London, 1876 – (= ser 19th c books on british colonization) – 1mf – 9 – mf#1.1.2516 – uk Chadwyck [954]
Central Institute of Research in Indigenous Systems of Medicine *see* The jamnagar experiment
Central Islamic Association of the Khmer Republic et al *see* The martyrdom of khmers muslims
Central kentucky researcher / Taylor County Historical Society – 1970 sep-1978 – 1r – 1 – mf#524149 – us WHS [978]
Central Labor Council of Cincinnati and Vicinity *see* Chronicle
Central Labor Council of San Mateo County *see* San mateo county labor
Central Labor Council of Seattle and Vicinity et al *see* Seattle union record
Central Labor Council [Portsmouth OH] *see* Labor review
Central Labor Union *see*
– Labor journal
– Penna. labor news
– S[/ain/]t petersburg advocate
– Toledo union
– Toledo union leader
– Union leader
Central Labor Union [Charlotte, NC] *see* Charlotte labor journal and dixie farm news
Central Labor Union [Harrisburg PA] *see* Central penna. labor news
Central labor union news – Gary, IN. 1919-1920 (1) – mf#62786 – us UMI ProQuest [071]
Central Labor Union of Augusta *see* Labor review
Central Labor Union of Indianapolis *see* Workingman
Central Labor Union of Pittston [PA] *see* Industrial advocate
Central Labor Union [Toledo, OH] *see* Labor union
Central law journal – St. Louis. 1874-1927 (1) – mf#5275 – us UMI ProQuest [340]
Central law journal – St Louis. v1-100. 1874-1927 (all publ) – (= ser Historical legal periodical series) – 1,5,6 – $1095.00 set – mf#408890 – us Hein [340]
Central law journal – v1-100, 1824-1927 + index/digest for v1-54 + index for v1-30 (all publ) – 608mf – 9 – $912.00 – mf#LLMC 82-912 – us LLMC [340]
Central law monthly – v1-3. 1880-82 [all publ] – 4mf – 9 – $18.00 – (cont by: the chicago law journal) – mf#llmc84-435 – us LLMC [340]
Central leader – Auckland, NZ. jan 1976-dec 1988 – 41r – 1 – mf#11.22 – nz Nat Libr [079]
Central Lunatic Asylum for Colored Insane, Virginia *see* Report of the board of directors and medical superintendent of the central lunatic asylum for colored insane, virginia, for the year...
Central Maine Theological Circle *see* Records
Central methodist – Catlettsburg, KY. 1872-1901 (1) – mf#63454 – us UMI ProQuest [071]
Central michigan life – Mount Pleasant. 1976-1977 – 1 – ISSN: 0008-9451 – mf#9053 – us UMI ProQuest [378]
Central montana wagon trails / Lewistown Genealogy Society – v1 n1-v7 n4 [1979 aug-1986 aug] – 1r – 1 – mf#1080808 – us WHS [929]
Central Nacional Sindicalista. Caceres *see* Asamblea ejecutiva de la c.n.s. de caceres
Central nebraska press – Kearney, NE: Webster Eaton, 1873 (wkly) [mf ed -1881 (gaps)] – 1r – 1 – (absorbed: buffalo county beacon (1872). cont by: kearney weekly hub and central nebraska press) – us NE Hist [071]
Central nebraska republican – Grand Island, NE: Seth P Mobley & Sister, 1894-jan 6 1900// (wkly) [mf ed 1895-99 (gaps)] – 3r – 1 – (absorbed by: free press. issues for sep 14 1895-nov 21 1896 also called old ser v21. issues for new ser v3 n1-new ser v6 n4 also called old ser v22-old ser v25. daily ed: grand island daily republican) – us NE Hist [071]
Central nebraskan – Hastings, NE: A D Williams, 1878 (semiwkly) [mf ed -1879 filmed [1973?]-75] – 2r – 1 – (cont: kenesaw times. cont by: hastings weekly nebraskan) – us NE Hist [071]
Central new jersey times – Plainfield, NJ. 1868-1880 (1) – mf#64841 – us UMI ProQuest [071]
Central New York. Synod. (Pres. Church in the USA) *see* Minutes
Central Obrera Nacional Sindicalista *see* – Estatutos
Central Old Settlers Union, Jefferson County, KS *see* Records
Central oregon enterprise – Prineville OR: A M Byrd, [wkly] – 1 – (cont by: call [prineville or]; ceased in 1920?) – us Oregon Lib [071]
Central oregon midstatesman – Bend OR: Central Oregon Midstatesman Inc, 1957- [wkly] [mf ed 1959] – 1r – 1 – (cont: midstatesman [1954-57]) – us Oregon Lib [071]
Central oregon press – Bend OR: Bend Press Pub Co, -1926 [daily ex mon] [mf ed 1968] – 4r – 1 – (cont: bend press; absorbed by: bend bulletin [bend or: 1917]) – us Oregon Lib [071]
Central oregonian – Prineville OR: Crook County Pub Co, 1921- [semiwkly] – 1 – (merger of: call [prineville, or]; crook county journal; absorbed: crook county news [prineville, or]; tribune [prineville, or]; iss for dec 7 1939-apr 4 1940 called: central oregonian with which is consolidated crook county news. iss for oct 15 1953-mar 25 1954 called: central oregonian and tribune) – us Oregon Lib [071]
Central oregonian – Prineville, OR. jul 14 1921-jun 1999 – 1 – us Oregon Hist [071]
Central oregonian *see*
– Crook county journal
– Crook county news
– Tribune (prineville, or)
Central Organization of US Marxist-Leninists *see* Workers' daily
Central Organization of U.S. Marxist-Leninists et al *see* Proletarian internationalism
Central Organization of US Marxist-Leninists et al *see* Internacionalismo proletario
Central otago news – Alexandra, NZ. jan 1974-dec 1988 – 1 – mf#83.1 – nz Nat Libr [079]
Central penna. labor news / Central Labor Union [Harrisburg PA] – v8 n32-v9 n19 [1942 may 29-1943 feb 26] – 1r – 1 – (cont by: Penna. labor news) – mf#357722 – us WHS [071]
Central penna. labor news / Greater Harrisburg Region Central Labor Council – 1967 sep 8-1968 aug 23, 1968 aug 30-1971 jan 29, 1971 feb 5-1972 dec 22, 1973 jan 12-1975 jun 29, 1978 jan-1985 aug – 5r – 1 – (cont: penna. labor news) – mf#1054174 – us WHS [331]
The central plains record – Kenesaw, NE: James L Kistner. v1 n1. oct 5 1956-57// (wkly) [mf ed -jun 28 1957 (gaps)] – 1r – 1 – us NE Hist [071]
Central point american – Central Point OR: J B Sheley & N B Sheley, -1927 [wkly] – 1 – (cont by: ashland american [1927-27]) – us Oregon Lib [071]
Central point american [central point, or] – Central Point OR: A E Powell, 1936- [wkly] – 1 – (cont: american [central point, or]) – us Oregon Lib [071]
Central point herald – Central Point OR: Lancaster & Pattison, 1906-17 [wkly] – 1 – (merged with: southern oregon news, to form: central point herald and southern oregon news) – us Oregon Lib [071]
Central point herald *see* Central point herald and southern oregon news
Central point herald and southern oregon news – Central Point OR: Herald Pub Co, 1917- [wkly] – 1 – (merger of: southern oregon news, central point herald) – us Oregon Lib [071]
Central point herald and southern oregon news *see* Central point herald
Central point star – Gold Hill OR: Mac's Printing Co, [wkly] – 1 – us Oregon Lib [071]
Central point times – Central Point OR: J Anderson, 1964- [wkly] – 1 – us Oregon Lib [071]
Central press – London, UK. 18 Jan-28 Feb 1871; Nov 1873-29 Jun 1874. -d. 5 reels – 1 – uk British Libr Newspaper [072]
Central Provinces (India). Department of Public Instruction *see*
– Hindi first reader
– Hindi second reader
– Hindi third reader
Central queensland herald – Rockhampton, Australia. -w. 2 Jan-31 July 1930. 2 reels – 1 – uk British Libr Newspaper [072]
Central Railway and Engineering Club of Canada *see* Official proceedings
Central region limited / National Railway Historical Society – n2-45 [1973 feb-1981 jan] – 1r – 1 – (cont by: public relations express) – mf#664938 – us WHS [380]
Central reguladora de adquisicion de patatas : reglamento de regimen interior para la... – Caceres: Tip. El Noticiero, S.A., 1946 – 1 – sp Bibl Santa Ana [060]
The central reporter – New York: Lawyers's Co-op. v1-13. 1885-88 (all publ) – 135mf – 9 – $202.00 – (covers ny, nj, pa, md, de and dc) – mf#LLMC 81-413 – us LLMC [340]
Central republican – Faribault, MN. 1858-1870 (1) – mf#63919 – us UMI ProQuest [071]
Central review – Strum WI. 1978 dec 7, 21-1980 may 2 – 1r – 1 – mf#955189 – us WHS [071]
Central s[/ain/]t croix news – Hammond, Roberts WI. 1975 apr 1/jun 24-1994 – 21r – 1 – (with small gaps) – mf#951309 – us WHS [071]
Central service yearbook from hospital topics – Chicago. 1956-1973 (1) 1973-1973 (5) (9) – ISSN: 0577-0947 – mf#8791 – us UMI ProQuest [610]
Central shopper – Hammond, Roberts WI. 1973-1974 sep 24, 1974 oct-1975 mar 25 – 2r – 1 – (cont: cowles shopper; cont by: central st croix news) – mf#1278405 – us WHS [071]
Central somerset gazette – Glastonbury & Wells, England. 25 Oct 1862-1868; 2-10 Jul 1869; 22 Jan 1870-1981. -w. 128 reels – 1 – uk British Libr Newspaper [072]
Central south sider / 3d Ward Chamber of Commerce [Chicago IL] – Chicago IL. 1929 jul 6 – 1r – 1 – mf#5012852 – us WHS [380]
Central star of empire – Kearney, NE: Moses H Sydenham (mthly) [mf ed 1896,1899-1906 (gaps) filmed 1975-83] – 3r – 1 – (cont: star of empire. issued in newspaper format 1896-1902. beginning with mar 1906 numbering starts over with v1 n1) – us NE Hist [071]
Central State Hospital (Petersburg VA) *see* Annual report
Central State Hospital [WI] *see*
– Grapevine
– Ideas and insights
– Patient's pen
Central state pointer / Wisconsin State College [Stevens Point] – 1951 dec 13-1958 may 15 – 1r – 1 – (cont: pointer [stevens point wi: 1919]; cont by: pointer [stevens point wi: 1958]) – mf#601400 – us WHS [378]
Central State Teachers College [Stevens Point WI] *see*
– Normal pointer
– Pointer
Central State University *see* Tower
Central states news – Belington, WV. 1930-1943 (1) – mf#67203 – us UMI ProQuest [071]
Central states speech journal – West Lafayette. 1984-1988 – 1,5,9 – (cont by: communication studies) – ISSN: 0008-9575 – mf#14901 – us UMI ProQuest [302]
Central sun – Aexandra, NZ. sep-dec 1982 – 1r – 1 – mf#83.13 – nz Nat Libr [079]
The central teaching of jesus christ : a study and exposition of the five chapters of the gospel according to st. john, 13 to 17 inclusive / Bernard, Thomas Dehany – New York: Macmillan, 1892 – 1mf – 9 – 0-8370-2293-2 – mf#1985-0293 – us ATLA [220]
Central Trades and Labor Assembly of Tampa *see* Union label
Central treasury records of the continental and confederation governments, 1775-1789 / U.S. Treasury Dept – (= ser Blotters of the office of the register of the treasury, 1782-1810) – 23r – 1 – (with printed guide. vols 8-17 were filmed as: blotters of the office of the register of the treasury, 1782-1810) – mf#M1014 – us Nat Archives [336]
Central treasury records of the continental and confederation governments relating to foreign affairs, 1775-1787 / U.S. Treasury Dept – 3r – 1 – (with printed guide) – mf#M1004 – us Nat Archives [336]
Central treasury records of the continental and confederation governments relating to military affairs, 1775-1789 / U.S. Treasury Dept – 7r – 1 – (with printed guide) – mf#M1015 – us Nat Archives [336]
Central treasury records relating to the loan of 1790 / U.S. Treasury Dept. Bureau of the Public Debt – (= ser Records of the bureau of the public debt) – 1r – 1 – mf#T786 – us Nat Archives [336]
Central union – Westfield WI. 1878 jun 19-1959 jan/1960 aug – 25r – 1 – (with gaps; cont: oxford times [oxford wi]; cont by: marquette county tribune) – mf#948688 – us WHS [071]
Central United Presbyterian Church, Topeka, KS *see* Records
Central Utah Relocation Project *see* Topaz times
Central valley voice – Merced, Winton CA. 1992 aug 7-1995 jan 26, 1995 feb 6-1997 dec, 1998: jan-1999 dec, 2000 jan-dec – 4r – 1 – mf#2624636 – us WHS [071]
Central virginian – Louisa, VA. 1928-2000 (1) – mf#66752 – us UMI ProQuest [071]
Central washington agworld – Wenatchee, WA. 1996-1996 (1) – mf#68922 – us UMI ProQuest [071]
Central weekly times and tullamore commercial and agricultural advertiser – Tullamore, Ireland. 4 jan-26 jul 1859 – 1/4r – 1 – (incorp with: leinster reporter) – uk British Libr Newspaper [072]
Central western daily – Bathurst, 1965-66 – at Pascoe [079]
Central western daily – Orange, jan 1969-jul 1997 – at Pascoe [079]
Central wisconsin – Wausau WI. 1868 oct 14/1871 dec 31-1907 jan 1/1910 feb 5 – 19r – 1 – (cont by: sun [wausau wi]) – mf#937178 – us WHS [071]
Central wisconsin – Wausau WI. 1857 apr 22-1861 apr 18, 1861 may 2-1863 nov 12 – 2r – 1 – mf#956121 – us WHS [071]
Central Wisconsin Center for the Developmentally Disabled *see* Coupon clipper
Central Wisconsin Community Action Council *see* Head start newsletter
Central wisconsin farmer – 1960 sep 2-1961 mar 24 – 1r – 1 – (cont by: southern wisconsin farmer) – mf#3503653 – us WHS [630]
Central wisconsin herald – Stevens Point WI. 1935 apr 26-1936 dec 31, 1937 jan 1-1938 mar 11 – 2r – 1 – (cont: central city herald) – mf#935896 – us WHS [071]
Central wisconsin resorter – 1978 mar 10-1978 aug 31, 1979 may 24-1980 aug 28, 1981 may 21-1981 sep 3, 1982 may 27-1984 aug 30, 1985 may 14-aug 29, 1987-95 – 14r – 1 – mf#4184468 – us WHS [071]
Central-anzeiger fuer elsass : indicateur central d'alsace – Strassburg (Strasbourg F), 1874 28 nov -1877 2 jun – 1 – fr ACRPP [074]
Central-blatt / Deutscher Romisch-Katholischer Central-Verein von Nord Amerika – v1 [1908 apr-1909 mar] – 1r – 1 – (cont by: central-blatt and social justice) – mf#4881927 – us WHS [241]
Central-blatt fuer die deutsche papier-fabrikation – Dresden DE, 1876-1902 [mnthly] – 16r – 1 – uk British Libr Newspaper [670]
Centralblatt fuer die gesamte eisenstahl industrie *see* Maschinen und metallindustrie zeitung
Centralblatt fuer die oesterreichische ungarische papierindustrie – Wien (A), 1883 15 oct-1909 20 dec – 26r – 1 – uk British Libr Newspaper [670]
Central-blatt fuer glas-industrie und keramik : internationales zentralblatt fuer baukeramik und glasindustrie – Wien (A), 1897 jan-1909 dec – 4r – 1 – uk British Libr Newspaper [660]
Centralblatt fuer maschinen industrie und eisengiesserei – Wien [A], 1896 27 sep-1912 19 dec – 7r – 1 – (title varies: 1902: maschinen- und metallindustrie-zeitung; 1905: oesterreichische eisenhaendler-zeitung) – uk British Libr Newspaper [600]
Centrale de l'enseignement du Quebec *see* Magazine ceq
Centrale des bibliotheques. Bibliotheque nationale du Quebec *see* Point de repere
Centralia enterprise – Wisconsin Rapids WI. [1882 jan 5-1885 feb 26] scattered iss, 1879 may 21-dec 24, 1880 apr 29-1881 dec 22, 1880 jan 4-1881 jul 28, 1883 jan 4-1884 may 8, 1884 may 15-1886 jan 28, 1885 apr 9, may 28, 1886 feb 4-1887 may 26 – 8r – 1 – (cont by: grand rapids tribune [wisconsin rapids wi: 1873]; centralia enterprise and tribune) – mf#951857 – us WHS [071]
Centralia enterprise and tribune – Wisconsin Rapids WI. 1887 jun 4-1887 jul 9, 1887 jun 4-1888, [1887 jul 16-1898 oct 29], [1889 jan-1893 aug 5], 1893 aug 12-1896 sep 12, 1896 sep 19-1899 aug 26, 1899 sep 1-1900 apr 14 – 7r – 1 – (cont: centralia enterprise [wisconsin rapids wi]; grand rapids tribune [wisconsin rapids wi: 1873 : weekly]; cont by: grand rapids tribune [wisconsin rapids wi: 1900 : weekly]) – mf#951860 – us WHS [071]
Centralian advocate – Alice Springs. may 1947-dec 1979, jan-jun 1982 – 28r – at Pascoe [079]
Die central-karpathen mit den naechsten voralpen : handbuch fuer gebirgsreisende / Fuchs, Friedrich – Pest 1863 [mf ed Hildesheim 1995-98] – 2mf – 9 – €60.00 – 3-487-29156-8 – gw Olms [914]
Central-verein deutscher staatsbuerger juedischen glaubens : c-v zeitung – Berlin. v1-17. 1922-38 [complete] – (= ser German-jewish periodicals...1768-1945, pt 2) – 7r – 1 – $760.00 – mf#B51 – us UPA [270]
Central-vereins-dienst – Berlin DE, 1924-27 – 1 – gw Misc Inst [360]
Central-volksblatt fuer den regierungsbezirk arnsberg *see* Central-volksblatt fuer die kreise soest, arnsberg, iserlohn, hamm
Central-volksblatt fuer die kreise soest, arnsberg, iserlohn, hamm – Arnsberg DE, 1856-57 – 1r – 1 – (n50 1856: central-volksblatt fuer den regierungsbezirk arnsberg) – Dist. gw Mikrofilm – gw Misc Inst [074]
Centre afrique – Bukavu: [s.n., dec 27/28 1959. – (issues filmed with: bartlett, robert e: collection of african newspapers) – us CRL [079]
Centre broadsheet : a political monthly / ed by Barclay, Rachel Mary – [Scotland] Edinburgh: printed at the Darien Press jun 1928-dec 1929 (mthly) [mf ed 2004] – 1r – 1 – uk Newsplan [320]
Centre catholique du cinema de Montreal. Commission des cine-clubs et al *see*
– Sequences

Centre court news – v1 n1-v3 n1 [1985 may/jun-1987 jan/feb] – 1r – 1 – mf#1312948 – us WHS [071]
Centre daily times : (am edition) – State College, PA. 1985-1986 (1) – mf#66089 – us UMI ProQuest [071]
Centre daily times – State College, PA. 1961-2000 (1) – mf#60573 – us UMI ProQuest [071]
Le centre de l'amour : decouvert soubs divers emblesmes galans et facetieux – Paris: Cupidon, 1680 – 5mf – 9 – mf#0-71 – ne IDC [090]
Centre De Recherche Et D'information Socio-Politiques *see* Rwanda politique
Centre de recherches metallurgiques *see* Metallurgical reports
Centre democrat – Bellefonte, PA. -w 1971-1981 – 13 – $25.00r – us IMR [071]
Centre for the history of european expansion, university of leiden – Leiden, 1975-1987 – 60mf – 9 – mf#H-15 – ne IDC [914]
Centre island news – Hicksville, NY. 1963-1965 (1) – mf#64997 – us UMI ProQuest [071]
Centre national de la recherche scientifique (France) *see* Constantes fondamentales de l'astronomie
Centre Party Southern Rhodesia *see* Papers, 1959-1976
Centre point – Salisbury (Zimbabwe): Centre Party, nov 1970-feb 1973 – 1r – us CRL [960]
Centre polytechnicien d'etudes economiques – Paris. 1933-aout 1939. Les no. 1-6 ont paru sous le titre de: X Crise – 1 – fr ACRPP [330]
Centre Protestant d'Etudes *see* Centre protestant d'etudes
Centre protestant d'etudes : bulletin / Centre Protestant d'Etudes – Geneva, SZ. 1967-90 – 4r – 1 – mf#ATLA S0391 – us ATLA [242]
Centre Protestant d'Etudes et de Documentation *see* Centre protestant d'etudes et de documentation
Centre protestant d'etudes et de documentation : bulletin / Centre Protestant d'Etudes et de Documentation – Paris, FR. n38-356. 1959-90* – 12r – 1 – ISSN: 0008-9842 – mf#ATLA S0415 – us ATLA [242]
Centre Syndical d'Action Contre la Guerre *see* Feuille bimensuelle d'informations syndicales
Centreville times – Centreville, VA. 1992-2000 (1) – mf#68920 – us UMI ProQuest [071]
Centripetal/hakol – Palo Alto, CA. 1980-81 – 1 – us AJPC [071]
Centro Cultural Chicano *see* Visiones de la raza
Centro De Artilleria (Bogota, Colombia) *see* Artilleria colombiana
Centro de Educacion Especial *see* Jornadas de educacion especial
Centro de Estudos Angolanos *see* Angola
Centro de Estudos e Acao Social *see* Cadernos do ceas
Centro De Geografia Do Ultramar *see* Relacao dos nomes geograficos de s tome e principe
Centro de Iniciativas Turisticas *see*
– Plasencia en ferias 1977
– Plasencia. navidad-76
Centro de Iniciativas Turisticas de Plasencia *see* Valle del jerte
Centro del mundo / Blonda, Maximo Aviles – Santa Domingo, Dominican Republic. 1962 – 1r – us UF Libraries [972]
Centro Ecumenico de Documentacao e Informacao *see* Povos indigenas no brasil
Centro Ecumenico de Documentacao e Informacao. Rio de Janeiro *see* Aconteceu
[El centro-] imperial valley press – CA. 1901- – 268r – 1 – $16,080.00 (subs $390/y) – mf#RC02196 – us Library Micro [071]
Centro tecnico dos eletricistas brasileiros – Rio de Janeiro, RJ: Typ Penna de Ouro, 30 nov 1889 – (= ser Ps 19) – mf#P17,01,94 – bl Biblioteca [621]
Centroamerica : es tu nombre / Arreola, Eduardo – Guatemala, . 1957 – 1r – us UF Libraries [972]
Centro-amerika : nach den gegenwaertigen zustaenden des landes und volkes, in beziehung auf die verbindung der beiden oceane, und im interesse der deutschen auswanderung bearbeitet / Reichardt, C F – Braunschweig 1851 [mf ed Hildesheim 1995-98] – 1v on 2mf – 9 – €60.00 – 3-487-26996-1 – gw Olms [972]
Cents jours / Forzano, Giovacchino – Paris, France. 1932? – 1r – us UF Libraries [440]
Centum fabulae ex antiquis... / Faerno, G – Lugdunum Batavorum: Apud Christophorum Raphelencium, 1600 – 2mf – 9 – mf#0-29 – ne IDC [090]
Centuria outlook – Centuria WI. 1902 feb 14-19, 1906-1909 oct 8 – 2r – 1 – mf#963577 – us WHS [071]
Centuria similitudinum...hundert gleichnussen in welchen durch vorstellung leiblicher figuren... lehren fuergebildet werden : vornehmlich auss h schrift und der alten lehrer monumenten... / Sudermann, D – [Argentorati]: Gedruckt in Verlegung Jacobs von der Heyden Chalgographi, 1624 – 3mf – 9 – mf#0-866 – ne IDC [700]
Les centuries de magdebourg, ou, la renaissance de l'historiographie ecclesiastique au seizieme siecle : lecon d'ouverture / Jundt, Auguste – Paris: Fischbacher, 1883 – 1mf – 9 – 0-7905-7053-X – (incl bibl ref) – mf#1988-3053 – us ATLA [240]
Centurion – 1982 may/jun-1993 aug – 1r – 1 – mf#1054184 – us WHS [071]
Centurion : official publicationof the manitoba centennial corporation – Manitoba. 1965 mar-1967 dec – 1r – 1 – mf#1054185 – us WHS [338]
Century – 1968-76 – at Pascoe [079]
Century 2 / Pendleton District Historical and Recreational Commission – [v1 n1]-v2 n4 [1974 sum-1976 spr] – 1r – 1 – mf#357717 – us WHS [978]
Century 21 (Firm) *see* Homes
Century and a half of jewish history / Emanuel, Charles Herbert Lewis – London, England. 1910 – 1r – us UF Libraries [939]
Century at the bar of the supreme court of the us / Butler, Charles Henry – New York, NY. 1942 – 1r – us UF Libraries [347]
Century gazette : heritage of the nation – 1960 jan 9-mar 5 – 1r – 1 – mf#1054186 – us WHS [975]
Century illustrated monthly magazine – New York. 1870-1906 – 1 – mf#5276 – us UMI ProQuest [073]
A century in the pacific : scientific, sociological, historical, missionary, general / ed by Colwell, James – London: C H Kelly, 1914 [mf ed 1990] – 1v on 2mf – 9 – 0-7905-5519-0 – (incl bibl ref) – mf#1988-1519 – us ATLA [980]
Century minutes of the philadelphia baptist association – 1707-1807 – 1 – $16.94 – us Southern Baptist [242]
A century of american diplomacy : being a brief review of the foreign relations of the united states, 1776-1876 / Foster, John W – Boston, New York: Houghton, Mifflin & Co: 1901 – 6mf – 9 – $9.00 – mf#LLMC 92-187 – us LLMC [327]
Century of archaeological discoveries = Archaeologische entdeckungen des neunzehnten jahrhunderts / Michaelis, Adolf – New York: E P Dutton 1908 [mf ed 1992] – 1mf – 9 – 0-524-03601-2 – (english trans fr german by bettina kahnweiler; pref by percy gardner) – mf#1990-3245 – us ATLA [933]
A century of artists : a memorial of the glasgow international exhibition 1888 / Henley, William Ernest – Glasgow 1889 – (= ser 19th c art & architecture) – 5mf – 9 – mf#4.2.516 – uk Chadwyck [700]
A century of baptist achievement / ed by Newman, Albert Henry – Philadelphia: American Baptist Publ Society, 1901 [mf ed 1986] – 2mf – 9 – 0-8370-9087-3 – (incl bibl ref) – mf#1986-3087 – us ATLA [242]
A century of bibles : or, the authorised version from 1611 to 1711... / Loftie, William John – London: Basil Montague Pickering, 1872 [mf ed 1988] – 1mf – 9 – 0-7905-0138-4 – mf#1987-0138 – us ATLA [220]
Century of books / NEW YORK TIMES – New York, NY. 1951 – 1r – us UF Libraries [025]
Century of brazilian history since 1865 / Graham, Richard – New York, NY. 1969 – 1r – us UF Libraries [972]
A century of christian progress and its lessons / Johnston, James – London: James Nisbet, 1888 [mf ed 1986] – 1mf – 9 – 0-8370-6743-X – (incl ind) – mf#1986-0743 – us ATLA [240]
A century of dishonor : a sketch of the united states government's dealings with some of the indian tribes / Jackson, Helen Hunt – new enl ed. Boston: Roberts Bros, 1889 [mf ed 1993] – (= ser Society of friends (quakers) coll) – 2mf – 9 – 0-524-05891-1 – mf#1991-2341 – us ATLA [975]
A century of episcopacy in portland : a sketch of the history of the episcopal church in portland, maine, from the organization of st paul's church, falmouth, nov 4 1763, to the present time / Perry, William Stevens – Portland: [s.n.] 1863 [mf ed 1993] – 1mf – 9 – 0-524-08579-X – mf#1993-3164 – us ATLA [242]
A century of faith / White, Charles L – 1932 – 1 – $11.48 – us Southern Baptist [242]
A century of german lyrics / Kroeker, Kate Freiligrath [comp] – London: W Heinemann, 1894 [mf ed 1993] – xiv/225p – 1 – (trans fr german) – mf#8349 – us Wisconsin U Libr [810]
A century of jewish miss[i]ons / Thompson, Albert Edward – Chicago: Fleming H Revell, c1902 [mf ed 1986] – 1mf – 9 – 0-8370-6624-7 – mf#1986-0624 – us ATLA [230]
The century of life : the niti shataka of bhartrihari freely rendered into english verse / Ghose, Aurobindo – Madras: Shama'a Pub House, 1924 – (= ser Samp: indian books) – us CRL [180]
Century of mission work in basutoland (1833-1933) / Ellenberger, Victor – Morija, Zimbabwe. 1938 – 1r – us UF Libraries [240]
A century of municipal history : pt 2: 1841-1893 / Cruikshank, Ernest Alexander [comp] – Welland, Ont: publ by authority of the County Council, 1893? – 2mf – 9 – mf#03634 – cn Canadiana [350]
A century of municipal history, 1792-1841 / Cruikshank, Ernest Alexander [comp] – Welland, Ont: publ by authority of the County Council. 2v. 1892? – 1mf – 9 – mf#03632 – cn Canadiana [350]
A century of municipal history, 1792-1892 : pt 1: 1792-1841 / Cruikshank, Ernest Alexander [comp] – Welland, Ont: publ by authority of the County Council, 1892? – 1mf – 9 – mf#03633 – cn Canadiana [350]
The century of preparation and the means and time of fulfillment : a sermon. delivered before the foreign missionary society of new-york and brooklyn... / Cheever, George Barrell – New-York: Almon Merwin, 1854 – 1mf – 9 – 0-524-02519-3 – mf#1990-0619 – us ATLA [240]
Century of protestant missions in china *see* China christian yearbook / the china mission handbook / a century of protestant missions in china
A century of protestant missions in china (1807-1907) : being the centenary conference historical volume / ed by MacGillivray, Donald – Shanghai: printed at the American Presbyterian Mission Press, 1907 [mf ed 1990] – 2mf – 9 – 0-7905-4950-6 – (incl bibl ref) – mf#1988-0950 – us ATLA [242]
A century of science : and other essays / Fiske, John – Bosto, New York: Houghton Mifflin & Co, 1900 (c1899) [mf ed 1986] – vii/477/[1]p – 1 – mf#8757 – us Wisconsin U Libr [840]
A century's change in religion / Harris, George – Boston: Houghton Mifflin, 1914 [mf ed 1990] – 1mf – 9 – 0-7905-3942-X – mf#1989-0435 – us ATLA [240]
A century's progress in religious life and thought / Adeney, Walter Frederic – London: James Clark, 1901 [mf ed 1989] – 1mf – 9 – 0-7905-1500-8 – mf#1987-1500 – us ATLA [240]
Cent-vingt jours de service actif : recit historique tres complet de la campagne du 65ieme au nord-ouest / Daoust, Charles-Roger – Montreal: E Senecal et fils...1886 [mf ed 1982] – 3mf – 9 – mf#SEM105P76 – cn Bibl Nat [971]
Ceol – Dublin. 1963-1981 (1) – ISSN: 0009-0174 – mf#8916 – us UMI ProQuest [941]
Ceos e terras do brasil / Taunay, Alfredo D'escragnolle Taunay – Sao Paulo, Brazil. 1904 – 1r – us UF Libraries [972]
Cep – chemical engineering progress – New York. 1950-1986 (1) 1950-1986 (5) 1975-1986 (9) – (cont by: chemical engineering progress) – ISSN: 0360-7275 – mf#717 – us UMI ProQuest [660]
Cepari, Virgilio *see*
– Life of saint aloysius gonzaga
– Vita del beato giovanni berchmans della compagnia di gesu
Cepari, Virgilio et al *see* Vie de ste. francoise romaine
Cepeda, Joaquin *see* Famossisimos romances
Cepero, Alfredo *see* Poemas del exilio
Cephalalgia – London. 1989+ (1,5,9) – ISSN: 0333-1024 – mf#13024 – us UMI ProQuest [616]
Cepkova, Drahoslava *see*
– Mitteldeutsche reimfassung der interrogatio sancti anshelmi
Cepy – Chicago, IL: Cepy Publ Co. v19 n44-v23 n31. nov 1917-jul 1921 – us CRL [071]
Cepy i nowy swiat – Chicago, IL; New York: [s.n.]: [v23 n32-v26 n12]. aug 1921-apr 1 1923 – us CRL [071]
Cera kings of the sangam period / Sesha Aiyar, K C – London: Luzac & Co, 1937 – (= ser Samp: indian books) – us CRL [930]
The ceramic art of great britain : from pre-historic times down to the present day: being a history of the ancient and modern pottery and porcelain works of the kingdom and of their productions of every class / Jewitt, Llewellynn Frederick William – London: Virtue & Co Ltd, 1878 – (= ser 19th c art & architecture) – 13mf – 9 – mf#4.1.96 – uk Chadwyck [730]
Ceramic arts and crafts – Detroit. 1966-1990 (1) 1971-1981 (5) 1977-1981 (9) – ISSN: 0009-0190 – mf#2127 – us UMI ProQuest [730]
Ceramic industry – Troy. 1979+ (1,5,9) – ISSN: 0009-0220 – mf#12088 – us UMI ProQuest [660]
Ceramics : architecture, applied arts, studio arts – (= ser Art exhibition catalogues on microfiche) – 68 catalogues on 86mf – 9 – £632.00 – (individual titles not listed separately) – uk Chadwyck [730]
Ceramics – London. 1950-1955 (1) – ISSN: 0009-0301 – mf#600 – us UMI ProQuest [730]
Ceramics – (= ser Christie's pictorial archive: decorative and applied art) – 110mf – 9 – $840.00 – 0-907006-32-9 – (over 6500 reproductions of english & continental items, glass & art nouveau) – uk Mindata [730]
Ceramics and glass *see* The index of american design (tiam)
Ceramics collection / Victoria and Albert Museum. London – (= ser Decorative art in the victoria and albert museum) – 237mf – 9 – $1760.00 – 0-907006-15-9 – (over 14,000 reproductions) – uk Mindata [730]
Ceramics for the potter / Home, Ruth M – Peoria, IL. 1952 – 1r – us UF Libraries [730]
Ceramics monthly – Columbus. 1953+ [1]; 1968+ [5]; 1970+ [9] – ISSN: 0009-0328 – mf#1799 – us UMI ProQuest [660]
La ceramique d'asie-mineure et de constantinople du 14 au 18 siecle / Sakisian, A B & Migeon, G – Paris, 1923 – 1mf – 9 – mf#AR-1848 – ne IDC [956]
Cerca / Giraudier, Antonio – Habana, Cuba. 1958 – 1r – us UF Libraries [972]
Cercado ajeno / Otero Silva, Miguel – Caracas, Venezuela. 1961 – 1r – us UF Libraries [972]
Cerceau, J A du *see* Histoire de la derniere revolution de perse
Cercheor of la crosse – v1 n1-7 [1981 jan 15-feb 28] – 1r – 1 – mf#632126 – us WHS [071]
Cercle catholique de Quebec *see* Annuaire...
Cercle et carre – n1-3. Paris. mars-juin 1930 – 1 – fr ACRPP [700]
Cercle inutile / Cailleville, Jacques De – Paris, France. 1932 – 1r – us UF Libraries [440]
Cercle linguistique de prague – Travaux. v1-8. 1929-1939 – 36mf – 8 – mf#2271 – ne IDC [460]
Cercle Ville-Marie (Montreal, Quebec) *see* Historique du cercle et rapport general du secretaire pour l'annee 1886-1887
Cercles agricoles : instructions pour l'organisation et la direction des cercles agricoles / Barnard, Edouard-Andre – S.l: s.n, 1893? – 1mf – 9 – mf#55098 – cn Canadiana [630]
Cerco da lapa e seus herois / Carneiro, David – Rio de Janeiro, Brazil. 1934 – 1r – us UF Libraries [972]
El cerco de zamora / Martinez Abeytia, Mateo – 1833 – 9 – sp Bibl Santa Ana [830]
Cerdan, F *see*
– Discursos physico-medico, politico-moral que tratan ser toda calentura hectica contagiosa
– Disertacion physico-medica de las virtudes medicinales, uso y abuso de las aguas termales de la villa de archena
– Naturaleza triunfante y crisol de mesentericas...
– Tuta...medicatio pro lientericis...
El cerdo de tipo iberico en la provincia de badajoz / Juana Sardon, Amalio de la – Cordoba: Imprenta Moderna, 1954 – sp Bibl Santa Ana [946]
Cereada, F *see* Don juan de carvajal. un espanol al servicio de la santa sede. madrid, 1947
Cereal chemistry – St. Paul. 1924+ (1) 1970+ (5) 1976+ (9) – ISSN: 0009-0352 – mf#261 – us UMI ProQuest [630]
Cereal foods world – St. Paul. 1975+ (1) 1975+ (5) 1976+ (9) – ISSN: 0146-6283 – mf#2557,01 – us UMI ProQuest [660]
Cereal science today – St. Paul. 1956-1974 (1) 1971-1974 (5) – (cont by: cereal foods world) – ISSN: 0009-0360 – mf#2557 – us UMI ProQuest [660]
Cereal workers world : official publication of local union n110... / Retail, Wholesale and Department Store Union – 1975 mar-1986 mar – 1r – 1 – (cont: local 110 tele news; cont by: cereal worker) – mf#358240 – us WHS [331]
Cerebral blood flow responses to a cognitive challenge in an older population / Albo, Jamy M – 1999 – 2mf – 9 – $8.00 – mf#PSY 2118 – us Kinesology [612]
Cereceda, F *see*
– Gonzalez, julio. alfonso 9th. madrid, 1944
– Los origenes de europa. traduccion de francisco elias de tejada. madrid, 1945
Ceremonia del...santa maria de tudia / Grima y Villa-Senor, Gabriel de – 1705 – 9 – sp Bibl Santa Ana [240]
Ceremonial del altar...san gabriel / Santano de Membrio, Juan – 1710 – 9 – sp Bibl Santa Ana [240]
Ceremonial del recuerdo / Ros-Zanet, Jose Guillermo – Panama, Panama. 1955? – 1r – us UF Libraries [972]
Ceremonial du sacre des rois de france : precede d'un discours preliminaire sur l'ancienneté de cet acte de religion... / Alletz, Pons A – Paris 1775 [mf ed Hildesheim 1995-98] – 1v on 3mf – 9 – €90.00 – 3-487-25901-X – gw Olms [944]
Ceremonial guide to low mass : or, plain directions for the consecration and administration of the sacrament of the holy communion – London: Pickering, 1883 [mf ed 1992] – (= ser Anglican/episcopal coll) – 1mf – 9 – 0-524-03199-1 – mf#1990-4648 – us ATLA [242]

Ceremonial in the church of england – London: Rivingtons, 1881 – 1mf – 9 – 0-524-05181-X – mf#1990-5100 – us ATLA [241]

Ceremonial of the altar : a guide to low mass according to the ancient customs of the church of england – London: Swan Sonnenschein 1888 [mf ed 1993] – 1mf – 9 – 0-524-08253-7 – (incl bibl ref) – mf#1993-3008 – us ATLA [242]

The ceremonial of the english church / Staley, Vernon – Oxford: AR Mowbray, [1899?] – 1mf – 9 – 0-524-03192-4 – mf#1990-4641 – us ATLA [240]

Das ceremonial-gesetz des alten testamentes : darstellung desselben und nachweis seiner erfuellung im neuen testament / Lisco, Friedrich Gustav – Berlin: Enslin, 1842 – 1mf – 9 – 0-524-05406-1 – (incl ind) – mf#1992-0416 – us ATLA [270]

Ceremonie de distribution des prix aux laureats au concours litteraire en langue nationale, session de 1959, sous la tres haute presidente de sa majeste la reine, le 25 novembre 1959 – [Phnom-Penh? 1959] [mf ed 1989] – 1r with other items – 1 – mf#mf-10289 seam reel 026/03 [§] – us CRL [480]

Ceremonies et coutumes religieuses de tous les peuples du monde – nouv ed. Amsterdam, Paris: Chez Laporte...4v. 1783 [mf ed 1985] – 4v on 1mf – 9 – mf#48776 – cn Canadiana [390]

Ceremonies et coutumes religieuses de tous les peuples du monde, representees par des figures dessinees... / Picart, B – Amsterdam, 1723-1743. 9 v – 49mf – 9 – mf#AR-1685 – ne IDC [956]

Ceremonies usitees au japon pour les mariages et les funerailles : suivies de details sur la poudre dosia, de la preface d'un livre de confoutzee sur la piete filiale / Titsingh, Isaac – Paris 0 [mf ed Hildesheim 1995-98] – 2v in 8 on 3mf – 9 – €90.00 – 3-487-27541-4 – gw Olms [390]

Ceremorias en el colegio militar de nuestra sra. sta. maria de tudia del orden de santiago de la universidad de salamanca / Grima y Villa-Senor Gabriel de – Salamanca, 1705 – 1 – sp Bibl Santa Ana [350]

Ceres : english edition – Rome. 1968-1996 [1,5,9] – ISSN: 0009-0379 – mf#6126 – us UMI ProQuest [320]

[Ceres-] ceres courier – CA. 1921-56 – 19r – 1 – $1140.00 – mf#C02104 – us Library Micro [071]

The ceresco courier – Ceresco, NE: Inter-State Newspaper Co, 1892 (wkly) [mf ed 1892,1894-98,1908-25 (gaps)] – 10r – 1 – us NE Hist [071]

Ceresco news – Ceresco, NE: Otto F Olsen, dec 1924-v32 n31. jun 28 1956 (wkly) [mf ed 1925-56 (gaps)] – 10r – 1 – us NE Hist [071]

Ceresini, Giovanni *see* Il primo libro de'madrigali a quattro voci

Ceresoles, Mauricio *see*
- Analisis de la contestacion del diputado de provincia don antonio concha y del libelo informativo...d joaquin rodriguez leal...
- Demonstracion y vindicacion de las injusticias... por acusacion del delator don joaquin rodriguez leal
- Resultado de la vista del articulo...en la causa que sigue a don mauricio ceresoles

Cerezo Moreno, Antonio *see* Una manana gris

Cerf, Bennett A *see* Great german short novels and stories

Cerfberr, Anatole *see* Repertoire de la comedie humaine de h de balzac

Cerff, Karl *see* "In uns ist alles"

Ceria, E *see* Don boscocon dios

Ceriani, Antonio Maria *see*
- Codex syro-hexaplaris ambrosianus
- Fragmenta latina evangelii s lucae, parvae genesis et assumptionis mosis, baruch, threni et epistola jeremiae versionis syriacae pauli telensis
- Missale ambrosianum
- Opuscula et fragmenta miscella magnam partem apocrypha
- Pentateuchi et josue
- Pentateuchi syro-hexaplaris
- Translatio syra pescitto veteris testamenti

Ceride-i resmiyye *see* T c resmi gazete

Cerimonial y rubricas generales : con la orden de celebrar las missas, y auisos para los defectos q[ue] acerca dellas pueden acontecer. sacados del nueuo missal tridentino / Catholic Church – en Mexico: Por Pedro Balli 1579 – (= ser Books on religion...1543/44-c1800: manuales de rito) – 120lea on 3mf – 9 – (trans by iuan ozcariz) – mf#crl-66 – ne IDC [241]

Cerkovnaja nauka – Church adviser.1, 1903 – 1 – us CRL [240]

Cerkovnyj vistnik – Church messenger. 1946-1970, 1973-1974 – 1 – us CRL [240]

Le cerneen – Port Louis, Mauritius 20 jul 1839, 11 jul 1855, 1 jan 1864-31 dec 1888, 6 jan 1930-30 jul 1938, 1 dec 1938-31 mar 1942, 3 nov 1949-31 dec 1951, 1 oct 1953-29 feb 1968 – 1 – (fr 1942-49 amalg with: le mauricien & advance & publ as: le cerneen-le mauricien-advance) – uk British Libr Newspaper [079]

Le cerneen-le mauricien-advance – Port Louis, Mauritius 1 apr 1942-27 nov 1948 – 1 – (cont by: mauricien-le cerneen [1 dec 1948-30 sep 1949]) – uk British Libr Newspaper [079]

Cerniki, Zihni *see* The divan project

Cerone, Pietro *see* El melopeo y maestro

Cerqueira, Ivo Benjamin De *see* Vida social indigena na colonia de angola (usos e costumes)

Cerrado e retiro / Costa, Esdras Borges – Rio de Janeiro, Brazil. 1960 – 1r – us UF Libraries [972]

Cerralbo, Marques de *see* Una cronica de los moctezuma, madrid...1954

[El cerrito-] the contra costan – CA. 1917-1924 – 1r – 1 – $60.00 – mf#B03592 – us Library Micro [071]

Cerro Sanchezherrera, Eduardo *see* Aportacion al estudio del fuero de baylio

Cerro y Contreras, Felix *see* Nociones de aritmetica y su metrico decimal...elementales de ninos... sexos

Cerro y llanura / Diaz Montero, Anibal – San Juan, Puerto Rico. 1964 – 1r – us UF Libraries [972]

Cert report / Council of Energy Resource Tribes – v4 n4-12 [1982 mar 23-oct 8], v5 n1-2 [1983 apr 25-aug 8] – 1r – 1 – mf#412786 – us WHS [333]

Certad, Leonardo *see* Proteccion posesoria

Certain and sufficient maintenance the right of christ's ministers / Benson, Christopher – London, England. 1834 – 1r – us UF Libraries [240]

Certain difficulties felt by anglicans in catholic teaching considered : in a letter addressed to the rev e b pusey...on occasion of his eirenicon of 1864: and in a letter addressed to the duke of norfolk, on occasion of mr. gladstone's expostulation of 1874 / Newman, John Henry – London: Basil M Pickering, 1876 – 1mf – 9 – 0-7905-9539-7 – mf#1989-1244 – us ATLA [241]

Certain difficulties felt by anglicans in catholic teaching considered : in twelve lectures addressed to the party of the religious movement of 1833 / Newman, John Henry – London: Burns & Oates 1879 [mf ed 1990] – 1mf – 9 – 0-7905-9540-0 – mf#1989-1245 – us ATLA [241]

Certain legumes of major importance of north central florida with... / Struthers, Orville W – s.l, s.l? 1940 – 1r – us UF Libraries [630]

Certain reasons of a private christian against conformitie to kneeling in the very act of receiving the lord's supper... / Dighton, T – n.p., 1618 – 2mf – 9 – mf#PW-42 – ne IDC [240]

Certain river mounds of duval county, florida / Moore, Clarence Bloomfield – Philadelphia, PA. 1895 – 1r – us UF Libraries [550]

Certain sand mounds of the st john's river, florida / Moore, Clarence Bloomfield – Philadelphia, PA. v1-2. 1894 – 1r – us UF Libraries [550]

A certain way to save our country : and make us a more happy and flourishing people, than at any former period of our history / Edwards, George – Newcastle: printed by S Hodgson, 1807 – (= ser 19th c economics) – 1mf – 9 – mf#1.1.47 – uk Chadwyck [339]

Certainties in religion / Monk, Henry Wentworth – [Ottawa?: s.n, 1886?] [mf ed 1995] – 1mf – 9 – 0-665-94758-5 – mf#94758 – cn Canadiana [210]

Certainties in religion / Williams, John Aethuruld, 1817-1889 – Toronto [ON]: Briggs 1882 [mf ed 1987] – 1mf – 9 – 0-665-39904-9 – mf#39904 – cn Canadiana [200]

Certainty in religion / Wyman, Henry H – New York: Columbus Press, 1905 – 1mf – 9 – 0-8370-8239-0 – mf#1986-2239 – us ATLA [240]

Certainty of saving truth / Mcneile, Hugh – London, England. 1867 – 1r – us UF Libraries [240]

Certainty of the future punishment of the wicked – Glasgow, Scotland. 18– – 1r – us UF Libraries [240]

Certamen del patriotismo / COOPERATIVA BANANERA COSTARRICENSE – San Jose, Costa Rica. 1928 – 1r – us UF Libraries [972]

Certamen pharmaceutico-galenico in quo tres continentur dissertationes... / Arnau, J – Valencia, 1727 – 2mf – 9 – sp Cultura [615]

Certamen poetico...por la real academia...x de su instalacion.. – 1872 – 9 – sp Bibl Santa Ana [810]

Certeyne preceptes gathered by hulrichus zuinglius declaring howe the ingenious youth ought to be instructed and brought vnto christ / Zwingli, H – Jppswich, Anthony Scoloker, 1548 – 1mf – 9 – mf#PBU-670 – ne IDC [242]

Certidumbre de america / Arrom, Jose Juan – Habana, Cuba. 1959 – 1r – us UF Libraries [972]

Le certificat d'heritier dans les departements du bas-rhin, du haut-rhin et de la moselle d'apres la loi d'introduction du droit francais du 1er juin 1924 / Krug, Camille – Paris, Recueil Sirey, 1939. 175 p. LL-4109 – 1 – us L of C Photodup [340]

Certificates of enrollment issued for merchant vessels at [...] / U.S. Bureau of Marine Inspection and Navigation – (= ser Records Of The Bureau Of Marine Inspection And Navigation) – 1 – (galveston, texas 1846-60 and 1865-70, and master abstracts of enrollments issued for merchant vessels at all texas ports, 1846-60 and 1865-june 1911 2r m1857. buffalo, new york may 1816-nov 1896 13r m1861. cleveland, ohio apr 1829-may 1915 14r m1862. oswego, new york 1815-1911 6r m1864) – us Nat Archives [380]

Certificates of enrollment issued for merchant vessels at buffalo, new york, may 1816-november 1896 / U.S. Bureau of Marine Inspection and Navigation – (= ser Records Of The Bureau Of Marine Inspection And Navigation) – 13r – 1 – mf#M1861 – us Nat Archives [380]

Certificates of enrollment issued for merchant vessels at cleveland, ohio, april 1829-may 1915 / U.S. Bureau of Marine Inspection and Navigation – (= ser Records Of The Bureau Of Marine Inspection And Navigation) – 14r – 1 – mf#M1862 – us Nat Archives [380]

Certificates of head tax paid by aliens arriving at seattle, wa, from foreign contiguous territory, 1917-1924 – (= ser Records Of The Immigration And Naturalization Service) – 10r – 5 – mf#M1365 – us Nat Archives [975]

Certificates of ratification of the constitution and the bill of rights : including related correspondence and rejections of proposed amendments, 1787-1792 / U.S. Constitutional Convention – (= ser General records of the united states government) – 1r – 1 – mf#M338 – us Nat Archives [323]

Certificates of registry, enrollment, and license issued at edgartown, massachusetts, 1815-1913 / U.S. Bureau of Marine Inspection and Navigation – (= ser Records Of The Bureau Of Marine Inspection And Navigation) – 9r – 1 – (with printed guide) – mf#M130 – us Nat Archives [380]

Certified accountant, 1905-87 : the journal of the association of certified accountants – 21r 139mf – 1,9 – mf#9/86657 – uk Microform Academic [650]

Certified copy / Greater Cleveland Genealogical Society – v4 n1-5 [1975 jun-nov], v5 n1-v12 n4 [1976 jan-1983/84 win] – 1r – 1 – (cont: greater cleveland genealogical society) – mf#1532683 – us WHS [929]

Certified copy of dr fleming's sermon – Glasgow, Scotland. 1836 – 1r – us UF Libraries [240]

Certifying questions of state law : experience of federal judges / Seron, Carroll – Washington: FJC, Jan 1983 – 1mf – 9 – $1.50 – mf#LLMC 95-833 – us LLMC [340]

La certitude de la foi et la certitude historique : etude sur le probleme du fondement de la vie religieuse / Menegoz, Fernand – Bale: E Finckh; Paris: Fischbacher, 1906 [mf ed 1990] – 1mf – 9 – 0-7905-7533-7 – mf#1989-0758 – us ATLA [230]

Certitude, providence, and prayer / McCosh, James – New York: Scribner 1884, c1883 [mf ed 1991] – 1mf – 9 – 0-7905-9803-5 – mf#1989-1528 – us ATLA [200]

Cerulean baptist church. cerulean, kentucky : church records – 1887-1979. Scattered Minutes: 1904-07; 1928-38. Lacking Minutes: Sept 1948-Aug 1973 – 1 – us Southern Baptist [242]

Cerulli, E *see*
- Etiopia occidentale dallo scioa alla frontiera del sudan
- Folk-literature of the galla of southern abyssinia

Cerulli, Enrico *see*
- La costituzione etiopica
- Il libro etiopico dei miracoli di maria e le sue fonti nelle letterature del medio evo latino
- Il libro etiopico del miracolo di maria e le sue fonti nelle letterature del medio evo latino
- Storia della letteratura etiopica
- Studi etiopici...

Cervantes : ein roman / Frank, Bruno – Stockholm: Bermann-Fischer 1944 [mf ed 1989] – 1r – 1 – (filmed with: totaliter aliter / hans franck) – mf#7255 – us Wisconsin U Libr [830]

Cervantes, Augustin *see* Duelos en cuba

Cervantes Bermudez de Cana, Tomas *see* Compendio de las instituciones de derecho canonigo segun el metodo de domingo cavallario

Cervantes en el ateneo de badajoz / Ateneo – Badajoz: Tip. Correo de la Manana, 1918 – 1 – sp Bibl Santa Ana [946]

Cervantes en el pais de fausto / Bertrand, Jean Jacques Achille – Madrid, Spain. 1950 – 1r – us UF Libraries [960]

Cervantes Saavedra, Miguel de *see* La galatea

Cervantines : y otros ensayos / Astrana Marin, Luis – Madrid, Spain. 1944 – 1r – us UF Libraries [025]

Cervarius Tubero, L *see* ...De turcarum origine, moribus, and rebus gestis commentarius

Cervati, Raphael C *see*
- Annuaire oriental du commerce de l'industrie, de l'administration et de la magistrature 9me annee 1889-1890
- Annuaire-almanach du commerce de l'industrie, de l'administration et de la magistrature 4me annee 1883

Cervenka, Zdenek *see* Bibliography on the nigerian civil war

Cervi, Emilio *see* Comparaciones historicas

Cerwin, Herbert *see* These are the mexicans

Cesaire, Aime *see*
- Toussaint l'ouverture

Cesalpino, Andrea *see* De metallicis libri tres andrea caesalpino aretino

Cesar : ou, le chien du chateau / Scribe, Eugene – Paris, France. 1839 – 1r – us UF Libraries [440]

Cesar, A *see* Ahuizote

Cesar cortes / Alcala, Manuel – Mexico; Editorial Jus, 1950 – sp Bibl Santa Ana [920]

Cesare in egitto : ballo eroico storico in cinque atti [la musica del ballo e di varii migliori maestri] da rappresentarsi nel regio teatro di parma il carnevale 1861-62 / Massini, Federico – Parma: Dalla Stamperia di A Stocchi, 1861 – 1 – mf#*ZBD-*MGTZ pv3-Res – us Misc Inst [790]

Cesare, Raffaele de *see* Antonio scialoja

Cesarismo democratico / Vallenilla Lanz, Laureano – Caracas, Venezuela. 1961 – 1r – us UF Libraries [972]

Cesarski Uniwersytet Warszawski. Obserwatorium Astronomiczne *see* Observations faites au cercle meridien de l'observatoire de l'universite imperiale de varsovie

Ces-Caupenne, Octave *see* Anvers

Cesiex en cifras / Consejo Economico Sindical – Badajoz: CESIEX, 1970 – sp Bibl Santa Ana [946]

Ceske slovo – Prague. jul 1938-jun 5 1940 – 13r – 1 – (filmed with: a-zet pondelnik jul 1938-nov 1939) – us CRL [077]

Ceske slovo – Prague, Czechoslovakia. May 1923-1933; May 1940-Mar 1942 – 2r – 1 – (scattered issues) – us L of C Photodup [077]

Cesko-slovansky obzor – Oklahoma City, OK: [s.n.], -roc4 cis13. 17.list.1910 (wkly) [mf ed 1908-10 filmed 1978] – 1r – 1 – (publ in oklahoma city, ok apr 30 1908-may 21 1908; in omaha, ne and oklahoma city, ok, may 28,1908-nov 17 1910. issues for sep 24 1908-nov 17 1910 called also cis. 81-cis. 193. chiefly in czech with some engish) – us NE Hist [071]

Ceskoslovenska akademie ved. Astrofysikalni observator *see* Biulleten' astronomicheskikh institutov chekhoslovakii

Ceskoslovenska republika – Prague, Czechoslovakia. 7 Jul 1921; 1926-Sept 1932 – 15r – 1 – us L of C Photodup [077]

Ceskoslovensky boj – Paris, France. 12 jan-7 jun 1940 – 1r – 1 – uk British Libr Newspaper [072]

Ceskoslovensky hornik – Prague, Czechoslovakia. Aug 1955-1958 – 1r – 1 – us L of C Photodup [077]

Cesky boj – London, UK. 1 May 1948-8 Jan 1951; 28 Sept 1957-12 Dec 1958; Aug 1959 – 1 – uk British Libr Newspaper [072]

Cesky denik – 1999- – 1r per y – 1 – (1995 1r $75.00. backfile 6r per y $510y. as of jan 1995: cesky tydenik) – us UMI ProQuest [070]

Cesky denik – Plzen, Czechoslovakia. -d. 9 feb-22 jul 1919 (imperfect) – 1r – 1 – uk British Libr Newspaper [072]

Cesky literarni spolek v St Louis, MO *see* Hlas

Cesky obzor – Omaha, Schuyler, NE: Bohemian-American Newspaper Union. (wkly) [mf ed roc2 cis9. 5 srp 1903-11, 1913-14 (gaps) filmed 1978] – 2r – 1 – (in czech) – us NE Hist [071]

Cesky zapas – Prague, Czechoslovakia. -w. 19 oct 1933-28 may 1936; 11 jun 1936-24 aug 1939 – 2r – 1 – uk British Libr Newspaper [072]

Cespedes, A *see* Libro de instrumentos nuevos de geometria

Cespedes Y Quesada, Carlos Manuel De *see* Discursos leidos en la recepcion publica

Cessnock eagle – Cessnock, dec 1969-dec 1982 – 9r – at Pascoe [079]

Cessnock eagle – Cessnock, nov 1913-dec 1963 – 28r – A$1853.02 vesicular A$2007.02 silver – at Pascoe [079]

C'est la faute a cremieux / Cremieux, A – Paris, France. 1879 – 1r – us UF Libraries [440]

Cesta – V Praze: Tiskem Prazske akciove tiskarny 1918-1930 [mf ed Bloomington IN: Indiana Uni Lib, Preservation Dept 1989] – 5r – 1 – us Indiana Preservation [460]

Cesta miru – Liberec, Czechoslovakia. 1953-Mar 1960 – 4r – 1 – us L of C Photodup [077]

Cestero Burgos, Tulio A *see* Epistolas a mi amigo aristos telasca
Cestero, Tulio Manuel *see*
- Ciudad romantica
- Estados unidos y las antillas
- Hostos
- Sangre

Cetnicke novine = Chetniks newspaper – Milwaukee WI. 1976 jan-1982 nov – 1r – 1 – mf#683995 – us WHS [071]
La cetra sonora : Sonata a tre, doi violini e violone o arcileuto col basso per l'organo, opera prima / Franchi, Giovan Pietro – Amsterdam: E Roger [171-?] [mf ed 1988] – 4pt on 1r – 1 – mf#pres. film 39 – us Sibley [780]
Cetreria del titere / Garcia Vega, Lorenzo – Santa Clara, Cuba. 1960 – 1r – us UF Libraries [972]
Cetshwayo's dutchman : being the private journal of a white trader in zululand during the british invasion / Vijn, Cornelius – London: Longmans, Green & Co 1880 – (= ser [Travel descriptions from south africa, 1711-1938]) – 3mf – 9 – (trans fr dutch; ed with pref & notes, by j w colenso) – mf#zah-59 – ne IDC [916]
Cetshwayo's dutchman / Vijn, Cornelius – New York, NY. 1969 – 1r – us UF Libraries [960]
Cette afrique-la! / Ikelle-Matiba, Jean – Paris, France. 1963 – 1r – us UF Libraries [960]
Cetto, Gitta von *see* Die kleine welt
Ceuerio de Vera, J *see* Viaie de la tierra santa, y descripcion de ierusalen, y del santo monte libano, con son relacion de cosas marauillosas...
Ceux qui souffrent : piece en un acte / Coupal, Louis – [Montreal?: s.n, 1918?] – 1mf – 9 – 0-659-90905-7 – mf#9-90905 – cn Canadiana [820]
Cevallos, Jose Antonio *see* Recuerdos salvadorenos
Cevat Pasa, Ahmet *see*
- Tarih-i askeri-yi osmani
- Tarih-i osmani

Cevdet, Ahmet *see* Mecelle-i ahkamn adliye
Cevdet, Mehmed *see* Duenyaya ikinci gelis yahut istanbul'da neler olmus
Cew – chemical engineering world – Bombay. 1972-1989 (1) 1972-1980 (5) 1975-1980 (9) – ISSN: 0009-2517 – mf#7340 – us UMI ProQuest [660]
Ceylan : ou recherches sur l'histoire, la litterature, les moeurs et les usages des chingulais / Gauttier d'Arc, Edouard – Paris 1823 [mf ed Hildesheim 1995-98] – 1v on 2mf [ill] – 9 – €60.00 – 3-487-27452-3 – gw Olms [954]
Ceylon : an account of the island physical, historical, and topographical... / Tennent, J E – London: Longman, Green, Longman and Roberts, 1859. 2v – 24mf – 9 – mf#Z-291 – ne IDC [915]
Ceylon *see*
- The ceylon government gazette
- Statistical blue books 1862-1938

Ceylon a key to india, and an open letter to the constituency of the american board / Leitch, Mary & Leitch, Margaret W – Boston: American Board of Commissioners for Foreign Missions, 1898 [mf ed 1995] – (= ser Yale coll) – 80p – 1 – 0-524-09620-1 – mf#1995-0620 – us ATLA [954]
Ceylon almanac and compendium of useful information – Colombo, 1814-39 – 132mf – 8 – (cont as: ceylon calendar and compendium of useful information [colombo, 1840-43]; cont as: ceylon almanac and compendium of useful information [colombo, 1844-50]; cont as: ceylon almanac and annual register [colombo,1851-62]) – mf#I-900 – ne IDC [954]
Ceylon and its capabilities : an account of its natural resources, indigenous productions, and commercial facilities... / Bennett, John Whitchurch – London, 1843 – (= ser 19th c books on british colonization) – 6mf – 9 – mf#1.1.8347 – uk Chadwyck [954]
Ceylon and the hollanders, 1658-1796 / Pieris, Paulus Edward – Tellippalai, Ceylon: American Ceylon Mission Press, 1918 [mf ed 1995] – (= ser Yale coll) – xvi/181p – 1 – 0-524-10249-X – (with glos) – mf#1996-1249 – us ATLA [954]
Ceylon antiquary and literary register – v. 1-10, no. 2. Jul 1915-Oct 1924 – 1 – us NY Public [490]
Ceylon buddhism / Gogerly, Daniel John; ed by Bishop, Arthur Stanley – Colombo: Wesleyan Methodist Book Room 1908 [mf ed 1992] – 2v on 2mf [ill] – 9 – 0-524-04540-2 – mf#1990-3374 – us ATLA [280]
Ceylon daily news – Colombo, Ceylon: Assoc Newspapers of Ceylon, 1956-66 (mf ed 1981) – 1 – us CRL [079]
Ceylon daily news – Colombo, Sri Lanka. -d. Jan 1919-Dec 1921. 12 reels – 1 – uk British Libr Newspaper [072]
The ceylon government gazette / Ceylon – 1902-45 – 1 – us NY Public [954]
Ceylon Laymen's Missionary Movement *see* Progress report of the general committee of the ceylon laymen's missionary movement
Ceylon. Legislative Council *see*
- Ceylon sessional papers, 1860-july 1931
- Sessional papers

Ceylon sessional papers, 1860-july 1931 / Ceylon. Legislative Council – Papers relating to agriculture, etc – 1 – us CRL [954]
Ceylon. Supreme Court *see* Decisions...on appeal from courts of requests. pt. 1
Ceza kanunu / Nuri, Ibnuerrefik Ahmet – Istanbul: Orhaniye Matbaasi, 1924 – (= ser Ottoman histories and historical sources) – 2mf – 9 – $40.00 – us MEDOC [956]
Cezair-i bahri-i sefid – (= ser Vilayet salnames) – 9 – (1292 [1875] def'a 6 2mf $275; 1301 [1894] def'a 8, 1311 [1893] def'a 11 5mf $75; 1312 [1894] def'a 12 3mf $55; 1313 [1895] def'a 13 5mf $75; 1318 [1900] def'a 18 3mf $150; 1321 [1903] def'a 20 5mf $75) – us MEDOC [956]
CF TC *see* Le syndicalisme chretien
Cfo – Boston. 1985-1996 (1,5,9) – ISSN: 8756-7113 – mf#14825 – us UMI ProQuest [338]
Cfo europe – London. 1998+ (1,5,9) – ISSN: 1462-5601 – mf#31860 – us UMI ProQuest [650]
Cfo journal / Colorado Field Ornithologists – Boulder. 1977-1979 (1) 1977-1979 (5) 1977-1979 (9) – (cont: colorado field ornithologist) – ISSN: 0362-9902 – mf#8388,01 – us UMI ProQuest [590]
Cfo.com – Boston. 2001+ (1,5,9) – mf#31861 – us UMI ProQuest [330]
Cfr : carpet, flooring, retail – Tonbridge. 2000+ (1,5,9) – ISSN: 1471-8162 – mf#26390,01 – us UMI ProQuest [740]
Cgt organo oficial de la confederacion general del... – Buenos Aires, Argentina. 1968- – 1r – us UF Libraries [079]
Ch. d. lippe's bibliographisches lexicon der gesammten juedischen literatur der gegenwart und adress-anzeiger : ein lexicalisch geordnetes schema mit adressen von rabbinen... in chronologischer anordnung und reihenfolge dargestellt = Bibliographisches lexicon der gesammten juedischen literatur der gegenwart und adress-anzeiger / Lippe, Chayim David – Wien: D Loewy, 1881 – 2mf – 9 – 0-524-08302-9 – mf#1993-4007 – us ATLA [014]
Ch'a hua chi – Shang-hai: Hsin Chung-kuo shu chu, Min kuo 20 [1931] – (= ser P-k&k period) – us CRL [840]
Cha ra kri sa khan kuiy to mhuin tan ca pro kra khai sann pum pran mya / Mhuin, Sa khan Kuiy to – Ran Kun: Van Mo U Ca pe phran khyi re 1967 [mf ed 1990] – 1r with other items – 1 – (in burmese) – mf#mf-10289 seam reel 161/5 [§] – us CRL [870]
Ch'a yeh pang tzu / Ch'en, Pai-ch'en – Shang-hai: K'ai ming shu tien, 1937 – (= ser P-k&k period) – us CRL [480]
Ch'a yen hsieh / Fan, Yen-ch'iao – Shang-hai: Chung fu shu chu, 1934 – (= ser P-k&k period) – us CRL [480]
Ch'a yu sui pi – Shang-hai: Hui wen t'ang hsin chi shu chu, 1932 – (= ser P-k&k period) – us CRL [480]
Chaadaeva, O *see* Pomeshchiki i ikh organizatsii v 1917 g
Chaane, Monyane Elizabeth *see* Pedagogical relevancy of inclusive education in the foundation phase of education
Chabad Lubavich House *see*
- Chabad voice
- L'chaim! to life!

Chabad lubavitch press – Montreal, Quebec. 1979-1985 – 1 – us AJPC [071]
Chabad times – Portland, Oregon, Mar. 1986; Oct. 1986; Dec. 1986; Apr. 1987 – us AJPC [071]
The chabad times – Cincinnati, OH. 1978-83 – 1 – us AJPC [071]
Chabad voice / Chabad Lubavich House – Madison WI. v1 n1-3 [1982 aug/sep-1983 jan/feb] – 1r – 1 – mf#718103 – us WHS [270]
Chaban, D *see* Golos sibiriaka
Chabar hindia-olanda – Betawi: Karsseboom & Co -1889] (daily) [mf ed Jakarta: National Library of Indonesia] – reel 2 [jan 2-apr 1888], 2r [may 1888-89] – 1 – (began with jan 1 1888 iss – tahon 2 n299 (31 dec 1889); some iss damaged; iss for jan 2-apr 1888 filmed with: hindi-nederland, 1883-1886, and: soerat chabar batawie, apr 5-jun 26 1858; iss for apr-dec 1889 filmed with: hindia-olanda, jan-mar 15, 1890; cont'd by hindia-olanda) – mf#mf-7739 seam – us CRL [079]
Chablani, S P *see* Economic conditions in sind, 1592-1843
Chaboillez, Augustin *see* Questions sur le gouvernement ecclesiastique du distric de montreal
Chace : selected from the celebrated poem of william somervile, esqr, set to musick for a voice...to which is added, rosalinda / Flackton, William – London: printed for aut & sold by Mr Walsh [1738?] [mf ed 19–] – 1mf – 9 – mf#fiche 1132 – us Sibley [780]
Der chacham kohelet als philosoph und politiker : ein kommentar zum biblischen buche kohelet, zugleich ein studie zu religioesen und politischen entwicklung des volkes israel im zeitalter herodes des grossen / Gerson, Adolf – Frankfurt a. M: J Kauffmann, 1905 – 1mf – 9 – 0-8370-3257-1 – (incl ind) – mf#1985-1257 – us ATLA [220]
Chacko, Charuvil Padeettathil Chacko *see* The nature and measurement of environmental literacy for sustainability
Chacon, A *see* Vitae gesta summorum pontificum a christo domiusque ad clementum 8...
Chacon Ferral, Antonio *see* Latigo de jesus
Chacon Miriam E *see* The relationship of trunk and hip muscle strength to low back pain
Chacon Trejos, Gonzalo *see* Tradiciones costarricenses
Chacon, Vamireh *see* Cooperativismo e comunitarismo
Chacon Villareso, Carmen *see* Coleccion de poesias
Chacon Y Calvo, Jose Maria *see*
- Cedulario cubano (los origenes de la colonizacion 1. (1493-1512)...tomo 6
- Criticismo y libertad
- Discursos leidos en la recepcion publica
- Documento y la reconstruccion historica
- Hermanito menor

Chacon y Calvo, Jose Maria *see* Ensayos de literatura cubana
Chacon Zamora, Fernando *see* La pedagogia juridica norteamericana
Chacun chez soi / Leonce – Paris, France. 1845 – 1r – us UF Libraries [440]
Chacun son tour : ou, l'echo de paris / Desaugiers, Marc-Antoine – Paris, France. 1816 – 1r – us UF Libraries [440]
Chad *see* Journal officiel du tchad
Chad browne memorial – 1638-1888. 280p – 1 – us Southern Baptist [242]
Chad. Sous Direction de la Statistique et des Etudes Economiques *see* Annuaire statistique 1966-1975
Chadbourne, Paul A *see* Lectures on natural theology
Chadbourne, Paul Ansel *see* Prominence of the religious element in education
Chadderton and middleton chronicle – [NW England] Chadderton 1959-25 may 1963 – 1 – uk MLA; uk Newsplan [072]
Chadds Ford, Pennsylvania. Brandywine Baptist Church *see* Records
Chadman, Charles Erehart *see*
- Constitutional law.
- Illustrative cases on personal rights and the domestic relations including teacher and pupil
- Negotiable instruments and principal and surety.
- Principles of the law of private and public corporations.
- Principles of the law of real property and the law of pleading and practice at common law
- A treatise on criminal law and criminal procedure
- Wills

Chadron advocate – Chadron, NE: A E & J D Sheldon. v2 n39. aug 9 1889-92// (wkly) [mf ed with gaps] – 1r – 1 – (cont: northwestern temperance advocate. absorbed by: dawes county journal) – us NE Hist [071]
Chadron chronicle – Chadron, NE: C H Pollard. -v34 n33. nov 25 1943 [wkly] [mf ed 1919-43 [gaps]] – 9r – 1 – (merged with: chadron journal [1900] to form: chadron record) – us NE Hist [071]
The chadron citizen – Chadron, NE: Bailey & Hill. v6 n31. mar 19 1891- (wkly) [mf ed -1895 (gaps)] – 1r – 1 – (cont: chadron democrat) – us NE Hist [071]
Chadron democrat – Chadron, NE: Democrat Pub Co. 6v. v1 n1. aug 27 1885-v6 n30. mar 12 1891 (wkly) – 3r – 1 – (cont by: chadron citizen) – us NE Hist [071]
Chadron journal – Chadron, NE: D S Efner. v17 n3. nov 9 1900-v60 n12. nov 26 1943 (wkly) [mf ed with gaps] – 18r – 1 – (cont: dawes county journal. merged with: chadron chronicle to form: chadron record) – us NE Hist [071]
Chadron record – Chadron, NE: Lester J Mann. v105 n1. sep 13 1988- [semiwkly] [mf ed 1989-] – 1 – (cont: chadron record and crawford tribune) – us NE Hist [071]
Chadron record – Chadron, NE: Chadron Print Co.36v. v60 n13. dec 2 1943-v95 n87. sep 19 1979 (semiwkly) [mf ed with gaps filmed -1980] – 50r – 1 – (formed by the union of: chadron chronicle and: chadron journal. absorbed: panhandle digest. merged with: crawford tribune to form: chadron record and crawford tribune) – us NE Hist [071]
Chadron record and crawford tribune – Chadron, NE: R G Dietz. 10v. v95 n88. sep 20 [ie 22] 1979-v104 n103. sep 9 1988 (semiwkly) [mf ed filmed 1980-89] – 19r – 1 – (formed by the union of: chadron record and: crawford tribune. cont by: chadron record (1988). sep 22 1979 issue misdated sep 20 1979) – us NE Hist [071]
Chadron recorder – Chadron, NE: Claude T Taylor. 5th yr n31. mar 25 1897- (wkly) [mf ed apr 8-aug 27 1897 (gaps)] – 1r – 1 – (cont: signal-recorder) – us NE Hist [071]
Chadron recorder – Chadron, NE: John O Taylor. 3v. 1st yr n1. jul 22 1893-3rd yr n41. apr 23 1896 (wkly) [mf ed with gaps] – 1r – 1 – (merged with: chadron signal to form: signal-recorder) – us NE Hist [071]
Chadron signal – Chadron, NE: A E & J D Sheldon, 1892-apr 1896 (wkly) [mf ed with gaps] – 1r – 1 – (merged with: chadron recorder to form: signal-recorder) – us NE Hist [071]
Chadron times – Chadron, NE: Clark & Ricker. 3v. os: v9 n1 jan 8 1903-v11 n31. jul 28 1905=ns: v1 n1-v3 n31 (wkly) – 1 – (cont: chadronian) – us NE Hist [071]
Chadron weekly republican – Chadron, NE: A W Davison. v1 n1. apr 16 1890- (wkly) [mf ed -may 2 1890 filmed 1973] – 1r – 1 – us NE Hist [071]
Chadronian – Chadron, NE: Phipps Bros, jan 4 1901-ns: v2 n52. dec 31 1902 (wkly) [mf ed with gaps] – 1r – 1 – (cont: news. cont by: chadron times. issues for aug 13-dec 31 1902 called also old ser v6 n32-51) – us NE Hist [071]
Chadwick, Edward Marion *see* The ontarian genealogist and family historian
Chadwick, George Alexander *see* Christ bearing witness to himself
Chadwick, George Whitefield *see* Judith, lyric drama
Chadwick, Hector Munro *see* The cult of othin
Chadwick, J W et al *see* [National conference controversy, 1865-1894]
Chadwick, John White *see*
- The faith of reason
- The man jesus
- The odore parker
- Old and new unitarian belief
- William ellery channing

Chadwick, Nora Kershaw *see* Poetry and letters in early christian gaul
Chadwick, O *see* John cassien
Chadwick, William Edward *see*
- The church, the state, and the poor
- The pastoral teaching of st paul
- Social relationships in the light of christianity

Chadwick, William Sydney *see* Mother africa hits back
Chafee, Zechariah, Jr *see* The zechariah chafee, jr papers
Chafer, Lewis Sperry *see*
- The kingdom in history and prophecy
- Satan
- True evangelism

[Chafey-] news – NV. 1908-09 [wkly] – 1r – 1 – $60.00 – mf#U04477 – us Library Micro [071]
Chaffanjon, Jean *see* Orenoque et le caura
Chaffee county miscellaneous newspapers – Chaffee County, Poncha Springs, Salida, CO [mf ed 1991] – 1r – 1 – (mine, stack and rail [nov 1 1885]; granite mining journal [sep 12 1910]; poncha springs herald [feb 5 1881]; the chronicle [sep 1 1904-sep 30 1904]; salida record [apr 17 1928]) – ISSN: 0 – mf#MF Z99 C346 – us Colorado Hist [071]
Chaffeegram – v1 n1-30, 32-43 [1980 may 4-jun 25, 27-jul 23] – 1r – 1 – mf#512238 – us WHS [071]
Chaffer, H J *see* Orlando
Chaffers, William *see* The keramic gallery
Chafik, H *see* Statut juridique international de l'egypte
Chafulumira, E W *see*
- Gwaza
- Kantini
- Kazitape
- Kuphika
- Mbiri ya amang'anja
- Mfumu watsopano
- Mtendere
- Wopambana ndani?

Chagas, Joao Pinheiro *see* De bond
Chagas, Paulo Pinheiro *see* Teofilo otoni
Chagla, Mahomedali Currim *see* Law, liberty, and life
Chagnon, Joseph Antoine *see* Etude sur la loi criminelle du canada en rapport avec les lois penales de la province de quebec...
Chagres : river of westward passage / Minter, John Easter – New York, NY. 1948 – 1r – us UF Libraries [972]
Chagrin falls township minutes – Chagrin Falls, Cuyahoga, OH. 1845-1991 – 4r – 1 – us Western Res [978]
Chagrin falls village cemetery deeds – Chagrin Falls, Cuyahoga, OH. 1881-1989 – 3r – 1 – us Western Res [978]
Chagrin falls village minutes : minutes of the council of the village of chagrin falls – Cuyahoga, OH. 1844-1992 – 11r – 1 – us Western Res [978]
Ch'a-ha-erh sheng chuan shui chien li wei yuan hui hui k'an – [China]: Ch'a-ha-erh sheng chuan shui chien li wei yuan hui, 1936 – (= ser P-k&k period) – us CRL [336]
Chahan de Cirbied, Jacques *see* Grammaire de la langue armenienne
Chahta anumpa = The choctaw times / Southeastern Indian Antiquities Survey, Inc – v1-v2 n7 [1968 may-1971 may] – 1r – 1 – mf#1054255 – us WHS [307]

Chai, Cheng Sien *see* Mirip satoe impian
Chai, Chu *see* The humanist way in ancient china
Chai, Fang *see* Lun cheng tang wen ta
Chai fu tzu yu t'an / Pu-ch'u-t'ing-ts'ao-chai-fu – Shang-hai: Shen pao kuan, 1932 – (= ser P-k&k period) – us CRL [840]
Chai kung ch'ien shuo / Chou, Chung-jen – Ch'ung-ch'ing: Ch'ing chen ssu, Min kuo 32 [1943] – (= ser P-k&k period) – us CRL [390]
Chai lights – Shreveport, Louisiana – 1 – (feb 1984; nov 1984-mar 1985; may 1985; sep 1985-oct 1985; dec 1985; apr 1986-may 1986; sep 1986-feb 1987) – us AJPC [270]
Chai men chi / Pa, Jen, 1897- – Hsiang-kang: Hai yen shu tien, Min kuo 30 [1941] – (= ser P-k&k period) – us CRL [480]
Chai, Winberg *see* The humanist way in ancient china
Chaianov, A *see*
– Istoriia biudzhetnykh issledovanii
– Materialy po voprosam organizatsii prodovolstvennogo dela
– Metody kolichestvennogo ucheta effekta zemleustroistva
Chaianov, A V *see*
– Biudzhetnye issledovaniia
– Kak organizovat krestianskoe khoziaistvo v nechernozemnoi polose
– Kapitaly krestianskogo khoziaistva i ego kreditovanie pri agrarnoi reforme
– Kratkii kurs kooperatsii
– Ocherki po teorii trudovogo khoziaistva
– Optimalnye razmery selsko-khoziaistvennykh predpriiatii
– Opyt anketnogo issledovaniia denezhnykh elementov krestianskogo khoziaistva moskovskoi gubernii
– Organizatsiia krestianskogo khoziaistva
– Organizatsiia severnogo krestianskogo khoziaistva
– Osnovnye idei i metody raboty obshchestvennoi agronomii
– Osnovnye usloviia uspekha kooperativnogo sbyta produktov selskogo khoziaistva
– Pamiatka Inovoda-kooperatora
– Prodovolstvennyi vopros
– Russkoe Inovodstvo, Inianoi rynok i Inianaia kooperatsiia
– Soiuznoe stroitelstvo kooperativov
Chaikovskii, B K *see* Kievskaia mysl' [vech vypusk]
Chailley, marcel et al *see* Notes et etudes sur l'islam en afrique noire
Chailley-Bert, Joseph *see*
– Administrative problems of british india
– The colonisation of indo-china
Chaillou, B *see* Lasthenie
Chaillu, P B du *see*
– Explorations and adventures in equatorial africa
– A journey to ashango-land
Chaim / Vidal, Ioan – San Jose, Costa Rica. 1960 – 1r – us UF Libraries [972]
L'chaim! to life! / Chabad Lubavich House – Milwaukee WI. v1 n1-8 [1982 aug/sep=elul 5742-1985 mar=adar 5745] – 1r – 1 – mf#1131009 – us WHS [270]
Chaim weizmann decade, 1952-1962 / Weizmann, Chaim – Rehovoth, Israel. 1962? – 1r – us UF Libraries [939]
Chaima, Nelson J *see* Nthano va kasimu
Chain leader : winning the chain restraurant game – Des Plaines. 2002+ (1,5,9) – ISSN: 1528-4999 – mf#32049 – us UMI ProQuest [640]
Chain of fathers / Husenbeth, Frederick Charles – London, England. 18– – 1r – us UF Libraries [240]
Chain saw age – Portland. 1972-1981 [1]; 1977-1981 [5,9] – ISSN: 0009-093X – mf#8605 – us UMI ProQuest [634]
Chain store age – New York. 1995+ (1) 1995+ (5) 1995+ (9) – (cont: chain store age executive with shopping center age) – ISSN: 1087-0601 – mf#8667,02 – us UMI ProQuest [650]
Chain store age : drug edition – New York. 1974-1978 (1) 1975-1978 (5) 1976-1978 (9) – mf#8665,01 – us UMI ProQuest [650]
Chain store age : executives edition/including shopping center age – New York. 1974-1974 (1) 1974-1974 (5) 1974-1974 (9) – (cont by: chain store age executive with shopping center age) – ISSN: 0885-1425 – mf#8667 – us UMI ProQuest [650]
Chain store age : general merchandise group – New York. 1976-1979 (1) 1976-1979 (5) 1976-1979 (9) – (cont: chain store age: newsmagazine of the general merchandise group. cont by: chain store age: general merchandise ed) – ISSN: 0193-1350 – mf#8664,02 – us UMI ProQuest [650]
Chain store age : general merchandise trends – New York. 1985-1988 (1) 1985-1988 (5) 1985-1988 (9) – (cont: chain store age: general merchandise ed) – ISSN: 0885-050X – mf#8664,04 – us UMI ProQuest [650]
Chain store age : newsmagazine of the general merchandise group – New York. 1975-1976 (1) 1975-1976 (5) 1976-1976 (9) – (cont by: chain store age: general merchandise group) – mf#8664,01 – us UMI ProQuest [650]
Chain store age : supermarkets – New York. 1975-1983 (1) 1975-1983 (5) 1975-1983 (9) – ISSN: 0193-1369 – mf#8668,01 – us UMI ProQuest [650]
Chain store age executive with shopping center age – New York. 1975-1995 (1) 1975-1995 (5) 1975-1995 (9) – (cont: chain store age: executives edition/including shopping center age. cont by: chain store age) – ISSN: 0193-1199 – mf#8667,01 – us UMI ProQuest [650]
Chain store age general merchandise ed – New York. 1979-1984 (1) 1979-1984 (5) 1979-1984 (9) – (cont: chain store age: general merchandise group. cont by: chain store age: general merchandise trends) – mf#8664,03 – us UMI ProQuest [650]
Chaine electrique / Gabriel, M – Paris, France. 1842 – 1r – us UF Libraries [440]
Chaine, Leon *see* Menus propos d'un catholique liberal
Chaine, M *see*
– Catalogue des manuscrits ethiopiens de la collection antoine d'abbadie
– La chronologie des temps chretiens de l'egypte et de l'aethiopie
La chaine symbolique : origine, developpement et tendances de l'idee maconnique...mythe, traditions, documents, histoire et dogmes de la franc-maconnerie moderne... / Galiffe, John-Barthelemy-Gaifre – Geneve: Elie Carey 1852 – 6mf – 9 – mf#vrl-95 – ne IDC [366]
Chair of peter / Collette, Charles Hastings – London, England. 1887? – 1r – us UF Libraries [240]
La chaire francaise au 12e siecle : d'apres les manuscrits / Bourgain, Louis – Paris: Societe generale de librairie catholique, 1879 – 1mf – 9 – 0-7905-7206-0 – (incl bibl ref) – mf#1988-3206 – us ATLA [240]
Chairman's address at the annual meeting, 1870 / Baptist Tract Society – London, England. 1870? – 1r – us UF Libraries [240]
Chairman's correspondence / New Zealand. Methodist Church. Methodist Overseas Mission. Solomon Islands Mission – 1952-61 – 6r – 1 – (ind) – mf#PMB1111 – at Pacific Mss [240]
Chaitanya and his age / Sen, Dineshchandra – Calcutta: University of Calcutta, 1922 – (= ser Samp: indian books) – us CRL [920]
Chaitanya and his companions : with two tri-colour illustrations: being lectures delivered at the university of calcutta as ramtanu lahiri research fellow for 1913-14 / Sen, Dineshchandra – Calcutta: University of Calcutta, 1917 – (= ser Samp: indian books) – us CRL [280]
Chaitanya's pilgrimages and teachings : from his contemporary bengali biography, the chaitanya-charit-amrita: madhya-lila = Srisricaitanyacaritamrta / Krsnadasa Kaviraja Gosvami – Calcutta: MC Sarkar, 1913 – 1mf – 9 – 0-524-01509-0 – (in english) – mf#1990-2485 – us ATLA [470]
Chajes, Hirsch Perez *see* Markus-studien
Chaka / Mofolo, Thomas – London, England. 1967 – 1r – us UF Libraries [960]
Chaka / Mofolo, Thomas – Morija, Zimbabwe. 1948 – 1r – us UF Libraries [960]
Chaka / Mofolo, Thomas – Paris, France. 1940 – 1r – us UF Libraries [960]
Chakaipa, Patrick *see*
– Karikoga gumiremiseve
– Rudo ibofu
– Spear of blood
Chakladar, Haran Chandra *see* Social life in ancient india
Chakravarti, Amiya et al *see* Rabindranath
Chakravarti, Chandra *see* The ratnavali
Chakravarti, Jnan Saran *see* Rai bahadur biresvar chakravarti's translation of the bhagavad gita in english rhyme
Chakravarti, Prabhat Chandra *see* The linguistic speculations of the hindus
Chakravarty, Amiya Chandra *see*
– The dynasts and the post-war age in poetry
– Mahatma gandhi and the modern world
Chakravarty, Apurba Kumar *see* Origin and development of indian calendrical science
Chakravorty, Ramendranath *see* Abanindranath tagore
Chalcedon Presbyterian Church [Atlanta GA] *see* Counsel of chalcedon
The chalcedonian decree : or, historical christianity, misrepresented by modern theology, confirmed by modern science, and untouched by modern criticism / Fulton, John – New York: T Whittaker, 1892 [mf ed 1990] – (= ser Charlotte wood slocum lectures) – 1mf – 9 – 0-7905-5042-3 – mf#1988-1042 – us ATLA [240]
Das chalcidius kommentar zu platos timaeus / Switalski, B W – Muenster, 1902 – (= ser Bgphma 3/6) – 3mf – 9 – €7.00 – ne Slangenburg [180]
Chalcocondylas, L *see* L'histoire de la decadence de l'empire grec, et establissement de celvy des turcs...
Chaldaeisches lesebuch : aus den targumin des alten testaments / Winer, Georg Benedikt – 2. durchaus verb aufl. Leipzig: Im Tr Woeller 1864 [mf ed 1991] – 1mf – 9 – 0-7905-8320-8 – (rev by julius fuerst) – mf#1987-6425 – us ATLA [470]
Chaldaeisches woeterbuch : ueber die targum und einen grossen theil des rabbinischen schriftthums / Levy, J – Leipzig. v1-2. 1867-68 – 8 – €54.00 – (1. band aleph-lamed, leipzig 1867 12mf; 2. band mem-tav., leipzig 1868 16mf) – ne Slangenburg [270]
La chalde chretienne / Avril, Adolphe d' – Paris: Aux Bureaux de l'oeuvre des coles d'orient, 1892 [mf ed 1986] – (= ser Bibliotheque de l'association pour l'union des eglises) – 1mf – 9 – 0-8370-8001-0 – mf#1986-2001 – us ATLA [240]
Chaldeae sev aethiopicae lingvae institvtiones : nunquam antea a latinis visae, opus vtile, ac eruditum... / Vittorio, M – Romae, 1552 – 1mf – 9 – mf#NE-20294 – ne IDC [956]
Chaldean account of genesis / Denys, George Williams – London, England. 1876? – 1r – us UF Libraries [240]
The chaldean account of genesis : containing the description of the creation, the deluge, the tower of babel, the destruction of sodom, the times of the patriarchs, and nimrod, babylonian fables, and legends of the gods / Smith, George – new rev corr ed. New York: Scribner, [1880?] – 1mf – 9 – 0-7905-3287-5 – mf#1987-3287 – us ATLA [221]
Chaldean magic : its origin and development / Lenormant, F – London, 1877 – 5mf – 9 – mf#NE-446 – ne IDC [956]
Chaldean magic : its origin and development = Magie chez les chaldeens et les origines accadiennes / Lenormant, Francois – London: Samuel Bagster, [1877?] – 1mf – 9 – 0-7905-2014-1 – (incl bibl ref and index. in english) – mf#1987-2014 – us ATLA [930]
Chalet-des-brises : cinquieme reunion de bonzes qui se prennent au serieux! / Morin, Victor – Montreal: impr pour l'auteur par Adj Menard, 1919 [mf ed 1987] – 1mf – 9 – mf#SEM105P781 – cn Bibl Nat [820]
Chalgrave – (= ser Bedfordshire parish register series) – 2mf – 9 – £5.00 – uk BedsFHS [929]
Chalif Russian Normal School of Dancing *see* [Announcement]
Chalkley, Lyman *see* Scotch-irish settlement in virginia, 1745-1800
Chalkley, Thomas *see* On the great love of god to mankind, through jesus christ our lord
Challen, James *see*
– Baptism in spirit and in fire
– The gospel and its elements
Challen typescripts, the... : from the guildhall library, london – 87v. 16th, 17th and 18th c – 21r 3mf – 1,9 – (with ind) – mf#96647/85678 – uk Microform Academic [920]
Challener, Richard *see* The correspondence series and speeches series of the personal papers of john foster dulles (1888-1959)
A challenge : am i my brother's keeper: sequel to the schwenckfelder migration (1734) – Norristown PA: Board of Pub of the Schwenckfelder Church, 1951 [mf ed 2003] – (= ser Schwenckfeldiana 2/2) – 1r – 1 – mf#2003-s008h – us ATLA [242]
Challenge – 1971-1972 sep – 1r – 1 – mf#1110593 – us WHS [071]
Challenge / A Philip Randolph Senior Center [New York NY] – 1972 nov-1973 jan – 1r – 1 – mf#4881946 – us WHS [071]
Challenge – Armonk. 1982+ (1,5,9) – ISSN: 0577-5132 – mf#13345,01 – us UMI ProQuest [338]
Challenge – Boston. v1-2. mar 1 1934-spring fall 1937 – 1 – us NY Public [073]
Challenge – Brooklyn. 1979+ (1) 1985+ (5) 1985+ (9) – ISSN: 0009-1049 – mf#12129 – us UMI ProQuest [320]
Challenge – 1972 feb-1975 dec – 1r – 1 – (cont by: winning spirit) – mf#555901 – us WHS [071]
Challenge – Washington. 1978-1981 (1) 1978-1981 (5) 1978-1981 (9) – (cont: hud challenge) – ISSN: 0196-1969 – mf#6300,01 – us UMI ProQuest [360]
Challenge – Peru, NE: Maverick Media. 3v. v6 n6. mar 16 1978-v8 n9. apr 3 1980 (wkly) [mf ed filmed 1980] – 3r – 1 – (cont: peru challenge. absorbed by: syracuse journal-democrat) – us NE Hist [071]
Challenge = Desafio / Progressive Labor Party – New York NY. 1964 oct 13/1970 feb-1988 nov/1989 jun – 18r – 1 – mf#1110596 – us WHS [325]
Challenge – Gainesville, FL.Fall 1980-Spring 1981 – 1 – us AJPC [071]
Challenge – Kingstown. St. Vincent. -f. 3 Jan 1959-15 Dec 1960, 6 Jan 1962-15 Jan 1964. (1 reel) – 1 – uk British Libr Newspaper [072]
Challenge : a libertarian weekly – v1-2 n2,18. 1938-39 [all publ] – (= ser Radical periodicals in the united states, 1881-1960. series 1) – 1r – 1 – $115.00 – us UPA [320]
Challenge / National Federation of Republican Women – 1972 jul/aug-1980 – 1r – 1 – mf#671207 – us WHS [325]
Challenge / National Federation of Republican Women – 1978 jan-1984 dec – 1r – 1 – (cont: winning spirit; cont by: new challenge) – mf#555895 – us WHS [325]
Challenge : newsletter / Fisk University – 1969 apr – 1r – 1 – mf#5026498 – us WHS [071]
Challenge / Sierra Army Depot – 1983 dec-1993 sep – 1r – 1 – mf#1110594 – us WHS [071]
Challenge / Wisconsin Federation of Cooperatives – 1982 sep/oct-1984 jul/aug – 1r – 1 – (cont: newsletter [wisconsin federation of cooperatives]; cont by: wfchallenge) – mf#929635 – us WHS [334]
Challenge *see* Fascist and anti-fascist newspapers
Challenge! : ypsl / Young People's Socialist League – v1-4 n2,3. 1943-46 [all publ] – 5mf – 9 – $95.00 – us UPA [335]
The challenge : agent, wm payne, bicycle and importer, po box 304 london, ontario / William Payne (Firm) – Coventry, England?: Iliffe & Son, 1883 – 1mf – 9 – mf#54754 – cn Canadiana [790]
Challenge, 1935-39 – (= ser Fascism and reactions to fascism in britain 1918-1989) – 2r – 1 – mf#97593 – uk Microform Academic [072]
Challenge in educational administration – Edmonton. 1961+ (1) 1974+ (5) 1974+ (9) – ISSN: 0045-625X – mf#10094 – us UMI ProQuest [370]
Challenge of amazon's indians / Tylee, Ethel Canary – Chicago, IL. 1931 – 1r – us UF Libraries [972]
Challenge of change / Steward, Alexander – London, England. 1962 – 1r – us UF Libraries [960]
The challenge of christ / Masterman, John Howard Bertram – London: Robert Scott; New York: George H Doran, 1913 – 1mf – 9 – 0-7905-1435-4 – mf#1987-1435 – us ATLA [240]
The challenge of christianity to a world at war / Griffith-Jones, Ebenezer – London: Duckworth, 1915 – 1mf – 9 – 0-7905-3851-2 – mf#1989-0344 – us ATLA [240]
Challenge of conservative baptist home missions – Apr 1954-Nov 1972 – 1 – us Southern Baptist [242]
The challenge of german literature / ed by Daemmrich & Haenicke, Diether H – Detroit: Wayne State University Press, 1971 – 1 – (incl bibl ref) – us Wisconsin U Libr [430]
The challenge of the city / Strong, Josiah – New York: Eaton & Mains; Cincinnati: Jennings & Graham c1907 [mf ed 1990] – (= ser Forward mission study courses) – 1mf – 9 – 0-7905-7146-3 – (incl bibl ref) – mf#1988-3146 – us ATLA [360]
The challenge of the north-west frontier : a contribution to world peace / Andrews, Charles Freer – London: George Allen & Unwin Ltd, 1937 – (= ser Samp: indian books) – us CRL [320]
Challenge to cardinal wiseman : or, lectures by the rev john cumming...and the rev r j m'ghee...at the great meetings held in exeter hall to which cardinal wiseman was invited in march and april, 1853 – 2nd ed. London: Arthur Hall, Virtue 1854 [mf ed 1986] – 1mf – 9 – 0-8370-8011-8 – mf#1986-2011 – us ATLA [241]
The challenge to christian missions : missionary questions and the modern mind / Welsh, Robert Ethol – New York: Young People's Missionary Movement, 1908 [mf ed 1986] – 1mf – 9 – 0-8370-6459-7 – (incl bibl ref) – mf#1986-0459 – us ATLA [240]
Challenge to women / Amrit Kaur, Rajkumari – Allahabad: New Literature, 1946 – (= ser Samp: indian books) – us CRL [305]
Challenger – Buffalo NY. 1991 jan 2/jun 26-2001 jan 3/jun 27 – 17r – 1 – mf#1839633 – us WHS [071]
Challenger – Columbus, OH. 1963-1971 (1) – mf#65442 – us UMI ProQuest [071]
Challenger – Wilmington NC. [1994 jan 6/12-jun 30/jul 6]-[1998 jul 2/8-dec 24/30] – 10r – 1 – (cont by: greater diversity news) – mf#2848791 – us WHS [071]
Challenger / Trumbull Co. Youngstown – feb-aug 1969, jan-dec 1972 [wkly] – 1r – 1 – (black press) – mf#B29905 – us Ohio Hist [071]
Challenger / United Steelworkers of America – 1978 jan/feb-1985 win, 1986 spr/summer-autumn/winter, 1988 spr, 1989 spr, 1989 fall, 1990 win – 1r – 1 – mf#1888916 – us WHS [660]
The challenger – Lilongwe: United Printers Ltd [sep 24-oct 11 1993] – 1r – 1 – us CRL [079]
"Challenger" commission p c numbered documents, 1986 / U S Temporary Committees, Commissions and Boards – 73r – 5 – mf#M1496 – us Nat Archives [324]

Challenges – Middletown. 1991-1992 (1,5,9) – (cont: current consumer and lifestudies) – ISSN: 1058-4773 – mf#11651,02 – us UMI ProQuest [338]
Challenges associated with the implementation of curriculum 2005 : a comparison between ivory park and midrand primary schools / Manyisa, Parafin Louis – Pretoria: Vista University 2002 [mf ed 2002] – 2mf [ill] – 9 – (incl bibl; abstract in english & afrikaans) – mf#mfm15163 – sa Unisa [370]
Challenges to press freedom in india, 1947 to 1963 / Irani, Behram S – Madison, 1965 – us CRL [360]
Challes, Robert *see* Journal d'un voyage fait aux indes orientales
Challis, James *see* A translation of the epistle of the apostle paul to the romans
Challoner, Richard *see*
– An abstract of the history of the old and new testaments
– Catholic christian instructed in the sacraments, sacrifice, ceremon...
– Memoirs of missionary priests
Chalmel, Jean *see* Histoire de touraine
Chalmers, Alexander *see* General biographical dictionary
Chalmers, Andrew *see* Transylvanian recollections
Chalmers church watchman : a monthly record of christian work in chalmers church – [Montreal?]: Young Peoples' Society of Christian Endeavor, [1890?-189- or 19–] – 9 – (ceased 189-?) – mf#P05047 – cn Canadiana [240]
Chalmers, George *see* Caledonia
Chalmers, J *see*
– Pioneer life and work in new guinea, 1877-1894
– Work and adventure in new guinea 1877 to 1885
Chalmers, James *see* Adventures in new guinea
Chalmers, John *see* The origin of the chinese
Chalmers, John Aitken *see* Tiyo soga
Chalmers, Thomas *see*
– Alexander campbell's tour in scotland
– Attempt to point out the duty which the church owes to the people...
– Christian union
– Churches and chapels
– Conference with certain ministers and elders of the church of scotland
– Considerations on the economics and platform of the free church of...
– Evidence given before the select committee of the house of commons
– Fulness and freeness of the gospel message
– Importance of civil government to society and the duty of christian
– Importance of civil government to society and the duty of christian...
– Influence of bible societies on the temporal necessities of the poo...
– Lectures on the epistle of paul the apostle to the romans
– Lectures on the establishment and extension of national churches
– On natural theology
– On preaching to the common people
– On the evangelical alliance
– On the evils which the established church in edinburgh has already...
– On the inspiration of the old and new testaments
– On the power wisdom and goodness of god
– Remarks on the present position of the church of scotland
– Reply to the attempt to connect the cause of church accomodation wi...
– Report of the committee of the general assembly of the church of sc...
– Scripture references
– Selection from the correspondence of the late thomas chalmers
– A selection from the correspondence of the late thomas chalmers...
– Selection from the correspondence of the late thomas chalmers
– Series of discourses on the christian revelation viewed in connection with the modern astronomy
– Sermon delivered in the tron church, glasgow, ...
Chaloem phrakiat somdet phraborom orotsathirat chaofa maha wachiralongkon sayam makut ratchakuman – [Bangkok: Samnak Ratcha Lekhathikan 1972] [mf ed 1994] – on pt of 1r – 1 – mf#11052 r1881 n2 – us Cornell [959]
Chaloner, John Henry *see* Proces complet de [Ch]aloner-Whittaker
The chalukyan architecture of the kanarese districts / Cousens, Henry – Calcutta: Govt of India, Central Publication Branch, 1926 – (= ser Samp: indian books) – us CRL [720]
Chalutz – Gainesville, FL.Fall 1973; Apr 1974 – 1 – us AJPC [071]
Chalybaeus, Heinrich Moritz *see* Historical development of speculative philosophy from kant to hegel
Chaman Lal *see* Hindu america
The chamars / Briggs, George Weston – Calcutta: Association Press, YMCA, 1920 – (= ser Samp: indian books) – us CRL [306]
Chamba notes – 1977/78 win-1980 sum – 1r – 1 – mf#4848509 – us WHS [071]
Le chambard socialiste – Paris. n1-78. 16 dec 1893-8 juin 1895 – 1 – (mq n73) – fr ACRPP [073]
Chamber of commerce commuity building, new smyrna... – s.l, s.l? 193-? – 1r – us UF Libraries [978]
Chamber of Commerce. Constantinople *see* Journal
Chamber Of Commerce (Miami Beach, Fl) *see* Lure of miami beach, florida
Chamber of Commerce of Japan *see* Honpo shogyo kaigisho shiryo
Chamber of Commerce of the United States of America *see* Washington report
Chamber sonatas of georg muffat / Doolittle, Quenten – U of Rochester 1964 [mf ed 1978] – 1r – 1 – mf#film 2503 – us Sibley [780]
Chamber to chamber / Wisconsin Association of Manufacturers and Commerce – 1983 may-1989 oct – 1r – 1 – (cont: barter bulletin [1976]; cont by: chamber to chamber [madison wi]) – mf#1124935 – us WHS [670]
Chamberlain : the missionary – London, England. 18– – 1r – us UF Libraries [240]
Chamberlain, Alexander Francis *see* Language of the mississaga indians of skugog
Chamberlain, Basil Hall *see*
– The classical poetry of the japanese
– The invention of a new religion
– Things japanese
Chamberlain, Daniel Henry *see* Address of hon. daniel h. chamberlain to the graduating class at the commencement exercises of columbia college law school.
[Chamberlain, H L] *see* Judah and israel
Chamberlain, Heather N *see* Nutrition knowledge and self-reported eating behavior of college male athletes and non-athletes
Chamberlain, Houston Stewart *see*
– 1876-1896, die ersten zwanzig jahre der bayreuther buehnenfestspiele
– Das drama richard wagner's
– Foundations of the nineteenth century
– Grundlagen des neunzehnten jahrhunderts
– Immanuel kant
Chamberlain, Isabel Fraser *see* Abdul baha on divine philosophy
Chamberlain, J *see*
– The cobra's den
– In the tiger jungle
Chamberlain, Jacob *see*
– The cobra's den
– In the tiger jungle
– The kingdom in india
Chamberlain, Joseph. *see* The government of ireland bill
Chamberlain papers – 3 series (ongoing) – 245r – 1 – (contains the private and political papers of the chamberlain family. upon completion, coll will contain more than 61,000 items including among others, journals, correspondence, official papers, speeches of joseph chamberlain, his sons austen and neville, and other family members. a complete document bibliographic listing available both in print and digital formats) – mf#C39-28950 – us Primary [941]
The chamberlain proposals from a canadian point of view / Sutherland, John Campbell – Montreal: Montreal News, c1904 – 1mf – 9 – 0-665-73379-8 – mf#73379 – cn Canadiana [336]
Chamberlain, Tamara M *see* The development of a folk dance unit as a resource for the state of utah elementary sixth grade social studies core
Chamberlain, William Benton *see* Liturgical training as an element in the preparation for the ministry
Chamberlayne, Israel *see* Saving faith
Chamberlin, Georgia Louise *see* An introduction to the bible for teachers of children
Chambers, Arthur *see*
– Man and the spiritual world
– Our life after death
– Thoughts of the spiritual
Chambers bugle – Chambers, NE: Fern D Smith. -v28 n[39] aug 29 1917 (wkly) [mf ed 1908-17 (gaps)] – 4r – 1 – (cont: bugle. cont by: chambers sun. numbering added with v23 n24 apr 13 1911) – us NE Hist [071]
Chambers, E K *see* The mediaeval stage
Chambers, Edward Thomas Davies *see*
– The angler's guide to eastern canada
– Fur farming in the province of quebec
– Les pecheries de la province de quebec
– The philology of the ouananiche
– Quebec ancient and modern
– Quebec, lake st john and the new route to the far-famed saguenay
– The sportsman's companion
– The st louis hotel guide to quebec
Chambers, Ephraim *see* Cyclopaedia
Chambers, Ernest John *see* The origin and services of the prince of wales regiment
Chambers, George Frederick *see*
– A digest of the law relating to district councils
– A digest of the law relating to public health and local government
– The story of eclipses simply told for general readers
– The story of the comets
– The story of the solar system
– The story of the stars
Chambers handbook for judges' law clerks and secretaries – 1994 – 3mf – 9 – $4.50 – mf#llmc99-029 – us LLMC [347]
Chambers helping chambers – 1982-1987 oct – 1r – 1 – mf#1544457 – us WHS [380]
Chambers' historical newspaper – [Scotland] Edinburgh: W & R Chambers 2 nov 1832-jan 1836 (mthly) [mf ed 2003] – 39v on 1r – 1 – uk Newsplan [072]
Chambers, Jean *see* Posters prepared for the infant welfare section, department of public health, papua new guinea
Chambers, John Charles *see* The witness of the ante-nicene fathers against the claims of the roman patriarchate
Chambers, John David *see*
– Divine worship in england in the thirteenth and fourteenth centuries
– Divine worship in england in the thirteenth and fourteenth centuries contrasted with and adapted to that in the nineteenth
Chamber's journal – Edinburgh, 1832-1900 – 1 – $1242.00 – mf#0146 – us Brook [073]
Chamber's journal – London. 1832-1956 – 1 – mf#468 – us UMI ProQuest [073]
Chambers, Robert – London, 1844 – (= ser 19th c evolution & creation) – 5mf – 9 – mf#1.1.10269 – uk Chadwyck [577]
Chambers, Robert *see*
– A biographical dictionary of eminent scotsmen
– Explanations
– The picture of scotland
– Traditions of edinburgh
– Vestiges of the natural history of creation
– Walks in edinburgh
Chambers, Sue *see* Collective artistic direction
Chambers sun – Chambers, NE: A D Scott. 33v. v28 n40. sep 6 1917-v60 n35. sep 30 1948 (wkly) [mf ed with gaps] – 6r – 1 – (cont: chambers bugle. absorbed by: holt county independent (1897). suspended foll v58 n13. dec 30 1943; resumed with v58 n14 may 23 1946) – us NE Hist [071]
Chambers, Talbot W *see* The noon prayer of the north dutch church
Chambers, Talbot Wilson *see*
– The book of zechariah
– Commentary on st paul's epistle to the romans
– Companion to the revised old testament
– Essays on pentateuchal criticism by various writers
– Moses and his recent critics
Chambers, Talbot Wilson et al *see* The centennial memorial of the presbytery of carlisle
Chambers, W *see*
– A dissertation on oriental gardening...
– Plans, elevations, sections, and perspective views of the gardens and buildings at kew in surrey
Chambersburg valley spirit – Chambersburg, PA., 1849-1912 – 13 – $25.00r – us IMR [071]
Chambless quarterly – v1-5 [1969 sep-1973 jun 1] – 1r – 1 – mf#1054265 – us WHS [071]
Chambliss, J E *see* Lives and travels of livingstone and stanley
Chambord / Merle, Jean T – Paris 1832 [mf ed Hildesheim 1995-98] – 2mf [ill] – 9 – €60.00 – 3-487-29764-7 – gw Olms [914]
Chambre d'agriculture du Bas-Canada *see* Compte-rendu des travaux de la chambre d'agriculture du bas-canada
Chambre de commerce du district de Montreal *see* Memoire soumis a la commission royale
La chambre de commerce du saguenay : constitution et reglements – [Roberval, Quebec?: s.n.], 1907 – 1mf – 9 – 0-665-71832-2 – mf#71832 – cn Canadiana [360]
Chambre de Commerce, Port de Haiphong *see* Statistiques commerciales
Chambre des deputes / Dorsinville, Luc – Port-Au-Prince, Haiti. 1930 – 1r – us UF Libraries [972]
Chambres legislatives d'haiti, 1892-1894 / Mercelin, Frederic – Paris, France. 1896 – 1r – us UF Libraries [972]
Chambres Syndicales de la Ville de Paris *see* Moniteur de l'entreprise et de l'industrie
Chambrun, Charles Adolphe de Pineton *see* Le pouvoir executif aux etats-unis, etude de droit constitutionnel
Chambrun, Joseph-Dominique-Aldebert de Pineton, comte de *see*
– Wagner
– Wagner a carlsruehe
Chamerovzow, Louis Alexander *see* Letters on coolie emigration to the west indies
Chamerovzow, Louis Alexis *see* The new zealand question and the rights of aborigines
Chamier, D *see* Panstratiae catholicae
Chamier, Jacques Daniel *see* Fabulous monster
Chaminade, Cecile *see* Valse carnavalesque
Chamisso, Adelbert von *see*
– Adelbert chamisso's werke
– Adelbert von chamisso's werke
– Aus chamissos fruehzeit
– Chamissos gesammelte werke
– Fortunati glueckseckel und wuenschhuetlein
– Gedichte
– El hombre que perdio su sombra
– Leben und briefe
Chamisso de Boncourt, Louis Charles Adelaide de *see* Studien zur lyrik chamissos
Chamissos gesammelte werke / ed by Koch, Max – Stuttgart: J G Cotta [1882?] [mf ed 1993] – 4v on 1r – 1 – (incl bibl ref & ind) – mf#8536 – us Wisconsin U Libr [802]
Chamissos gesammelte werke / ed by Koch, Max – Stuttgart: J G Cotta [1882?] [mf ed 1993] – 4v on 1r – 1 – (incl bibl ref & ind) – mf#8536 – us Wisconsin U Libr [800]
Chamissos peter schlemihl / Schapler, Julius – [S.l: s.n.] 1893 (Deutsch-Krone: Druck von F Garms) [mf ed 1989] – 1r – 1 – (incl bibl ref. filmed with: die poesie, ihr wesen und ihre formen / moriz carriere) – mf#7146 – us Wisconsin U Libr [430]
Chamizo, Luis *see*
– Obras completas
– Semana santa en guarena y oracion a la virgen
– Semana santa en guarena y oracion de la virgen
Chamorro Martinez, Manuel *see* Valor militar de la "zona de defensa del noroeste peninsular"
Chamorro, Pedro Joaquin *see*
– D sofonias salvatierra y su 'comentario polemico'
– Entre dos filos
– Maximo jerez y sus contemporaneos
– El ultimo filibustero
Champ / International Union, United Automobile, Aerospace, and Agricultural Implement Workers of America – 1991 jan-1994 dec – 1r – 1 – mf#1054268 – us WHS [331]
Champ planer – Plattsburgh NY. 1962 jan 12/dec 28-1992 jan 3/dec 18 – 36r – 1 – (lacking several iss; with gaps; cont by: north country champlaner) – mf#627644 – us WHS [071]
Champa : a short sketch of her historical evolution based on architectural ruins / Sadananda, Swami – [Calcutta: SK Mitra, 1938] – (= ser Samp: indian books) – us CRL [930]
Champagne, Andre *see* Bio-bibliographie analytique de f fitz osborne
Champagne, Louis *see* Roman d'amour
Champagne, Philias *see* La guerre d'europe
Champagne-kriegszeitung – s.l, 1915 14 aug-1917 – 2r – 1 – gw Misc Inst [933]
Champaign Co. Mechanicsburg *see*
– Daily telegram
– Telegram
Champaign Co. Saint Paris *see*
– Dispatch series
– Enterprise
– Era dispatch / quiver / news / dispatch
– Examiner
– Holiday reporter
– New era
– News
– Press
Champaign Co. Urbana *see*
– Champaign republican
– Citizen and gazette
– Citizen and gazette series
– Daily citizen
– Daily democrat
– Daily times series
– Democrat series
– Informer
– Ohioan and mad river journal
– Union
Champaign republican / Champaign Co. Urbana – mar 17-sep 8 1893 (short roll) [wkly] – 1r – 1 – mf#B9525 – us Ohio Hist [071]
Champak leaves / Seshadri, P – Allahabad: Indian Press, 1923 – (= ser Samp: indian books) – us CRL [954]
Champignons du tonkin : illustrations of fungi in the farlow reference library with the mycological papers of n t patouillard / Patouillard, N T – [mf ed Chadwyck-Healey, 1985] – 3 col 4 b/w mf – 15,9 – (int by d h pfister) – uk Chadwyck [580]
Champion – London, UK. sep 1836-apr 1940 [wkly] – 3r – 1 – (aka: champion and weekly herald, nov 1836-apr 1840) – uk British Libr Newspaper [072]
Champion – Arcadia, FL. 1906-1908 – 2r – us UF Libraries [071]
Champion : containing a series of papers humorus, moral, political and critical – London. 1739-1740 (1) – mf#4224 – us UMI ProQuest [420]
Champion : the fighting voice of young canada / Young Communist League – Toronto. v1-6. jan 30 1951-jan 1957// (semimthly) – 2r – 1 – Can$175.00 – (official organ of the young communist league) – cn McLaren [321]

Champion – Norton, KS. 1884-1900 (1) – mf#68708 – us UMI ProQuest [071]
Champion / United Office and Professional Workers of America – 1950 jul-sep 7 – 1r – 1 – (cont: career [new york ny]; cont by: union voice [new york ny: 1945]) – mf#1110599 – us WHS [650]
Champion – v15 n18-20,22-23 [1988 sep 12-26, dec 5-19], v16 n3,6-16,18 [1989 feb 13, apr 12-sep 27, oct 25], v17 n1-2,4,7-8,10-17,19-20,22 [1990 jan 4-18, feb 14, apr 11-25, may 23-sep 12, oct 10-24, nov 20], v18 n1,9-12 [1991 n31, may 8-jul 17] – 1r – 1 – mf#1061358 – us WHS [331]
The champion – Lilongwe: Champion Publ [oct 25-dec 2/8 1995] – 1r – 1 – us CRL [079]
Champion and weekly herald *see* Champion
Champion city times / Clark Co. Springfield – jan 1887-jun 1888 [daily] – 3r – 1 – mf#B10817-10819 – us Ohio Hist [071]
Champion, George *see* Journal of the rev george champion
Champion, John Benjamin *see* The living atonement
Champion, L G *see* Outlook for christianity
Champion labor monthly – v1-3 n2,10. 1936-38 [all publ] – (= ser Radical periodicals in the united states, 1881-1960. series 1) – 1r – 1 – $200.00 – us UPA [331]
Champion magazine – v1 n1-8 [1916 sep-1917 apr] – 1r – 1 – mf#1054269 – us WHS [071]
Champion of fair play – 1916 apr 29-1917 jan 27, 1918 feb 23-oct 19, 1918 oct 26-1920 jul 10 – 2r – 1 – (cont: champion of freedom and right, fair play, our standard; cont by: national beverage journal) – mf#946326 – us WHS [071]
Champion or sligo news – Sligo, Ireland. 4 jun 1836-24 dec 1847, 1848-1896, 14 apr-29 dec 1923, 1926, 1930, 1983, 1986-92 – 50r – 1 – (aka: sligo champion) – uk British Libr Newspaper [072]
Champion, P *see*
– La vie du pere j rigoleuc
– La vie et la doctrine spirituelle du pere louis lallemant
Champion, Pierre *see* Histoire poetique du quinzieme siecle...
Champion post – Parkes, jan 1969-dec 1994 – 71r – at Pascoe [079]
Champion, Richard *see* Letterbooks of richard champion, 1743-1791
Champion, the... 1739-41 – 2v – 1r – 1 – mf#96355 – uk Microform Academic [073]
Champion, Thomas Edward *see* The 13th battalion of hamilton
Champion times / Baumholder Military Community – 1983 sep 9-1985 dec 31, 1986 jan 23-1988 nov 10 – 2r – 1 – (cont: community news [baumholder, germany [west]]; cont by: mainz soldier, grapevine [bad kreuznach, germany]; central rheinhold-pfalz union) – mf#1363388 – us WHS [355]
Champlain / Dawson, Samuel Edward – [Montreal?]: [s.n.], [1890?] [mf ed 1980] – 1mf – 9 – 0-665-03775-9 – mf#03775 – cn Canadiana [920]
Champlain : a drama in three acts / Harper, John Murdoch – London: F Warne; Toronto: W Briggs, c1908 – 4mf – 9 – 0-665-74429-3 – (int entitled twenty years and after) – mf#74429 – cn Canadiana [820]
Champlain : sa vie et son caractere / Casgrain, Henri Raymond – [Quebec?: s.n.], 1898 [mf ed 1980] – 1mf – 9 – 0-665-00644-6 – mf#00644 – cn Canadiana [971]
Champlain et son oeuvre : une page d'histoire – Quebec: A Talbot, 1898 [mf ed 1980] – 2mf – 9 – 0-665-02873-3 – mf#02873 – cn Canadiana [910]
Champlain, Samuel de *see*
– Oeuvres de champlain
– Les voyages de la nouvelle-france occidentale, dicte canada
– Les voyages du sieur de champlain xaintongeois
Champlain society publications : toronto, 1907-1958 – Greenwood Press – 36v on 209mf – 9 – $1295.00 – (publ related to the history & devt of all pts of canada) – us UPA [971]
Champlain's american experiences in 1613 / Harvey, Arthur – [S.l: s.n, 1886?] [mf ed 1980] – 1mf – 9 – 0-665-03987-5 – mf#03987 – cn Canadiana [917]
Champlain's tomb / Harper, John Murdoch – [S.l: s.n, 18–] [mf ed 1980] – 1mf – 9 – 0-665-05367-3 – mf#05367 – cn Canadiana [971]
Champney, Elizabeth W *see* Three vassar girls abroad
Champoeg pioneer – Champoeg OR: C F Harris, [mthly] – 1 – us Oregon Lib [071]
Champollion, Jean Francois *see*
– Lettre a m dacier
– Lettres a m le duc de blacas d'aulps relatives au musee royal egyptien deturin
– Precis du systeme hieroglyphique des anciens egyptiens
Champollion-Figeac, Jacques-Joseph *see* Charte de commune en langue romane pour la ville de grealou en quercy
Champon, E *see* Guadeloupe
Champoux, Gerard *see* Abrege d'agriculture
Chamrieng cheat niyum ou chant patriotique / Cambodia. Commission des murs et coutumes – 1.ed. Phnom-Penh: Editions de l'Institut bouddhique [mf ed 1989] – 1r with other items – 1 – (title & text in khmer; added t.p. in french) – mf#mf-10289 seam reel 122/4 [§] – us CRL [780]
El-chams – Paris. n1-35. fev-nov 1885 – 1 – fr ACRPP [073]
Chan cheng hsin wen tu fa / Lu, Yu-tung – [China]: Wu ming ch'u pan she, Min kuo 28 [1939] – (= ser P-k&k period) – us CRL [070]
Chan cheng hsing wei yu jen lei sheng huo / Wu, Nien-chung – [China: sn], Min kuo 29 [1940] – (= ser P-k&k period) – us CRL [303]
Chan cheng yu ching chi / Lu, Hsun – Shang-hai: Chung-hua shu chu, Min kuo 26 [1937] – (= ser P-k&k period) – us CRL [330]
Chan cheng yu nung ts'un : erh tz'u ta chan chung ti ko kuo nung ts'un / Vorga, E – Kuei-lin: Nung hsueh shu tien, Min kuo 31 [1942] – (= ser P-k&k period) – us CRL [630]
Chan cheng yu wen hsueh / Fan, Ch'uan, 1918- – Shang-hai: Yung hsiang yin shu kuan, Min kuo 34 [1945] – (= ser P-k&k period) – us CRL [410]
Ch'an chuan / Kan-nu – Kuei-lin: Wen hua kung ying she, Min kuo 32 [1943] – (= ser P-k&k period) – us CRL [480]
Chan hao / Cheng, Chen-to – Shang-hai: Sheng huo shu tien, Min kuo 27 [1938] – (= ser P-k&k period) – us CRL [951]
Chan hou chih jih-pen / Yang, Kung-ch'uan – Yung-an: Chung-hua ch'u pan she, Min kuo 33 [1944] – (= ser P-k&k period) – us CRL [951]
Chan hou kuo chi t'ou tzu wen t'i / Staley, Eugene – [Ch'ung-ch'ing: Chung hua shu chu, Min kuo 34 [1945]] – (= ser P-k&k period) – us CRL [330]
Chan hou shang-hai chi ch'uan kuo ko ta kung ch'ang tiao ch'a lu / ed by Hsu, Wan-ch'eng – Shang-hai: Lung wen shu tien, min kuo 29 [1940] – (= ser P-k&k period) – us CRL [670]
Chan hou shih chieh chih kai tsao wen t'i – Ch'ung-ch'ing: Tu li ch'u pan she, Min kuo 33 [1944] – (= ser P-k&k period) – us CRL [951]
Chan hou shih chieh ho p'ing i chieh shu – [China: Chung-kuo kuo min wai chiao hsieh hui, Min kuo 33 [1944]] – (= ser P-k&k period) – us CRL [951]
Chan hou shih chieh ho p'ing wen t'i / Chang, Tao-hsing – Ch'ung-ch'ing: Kuo min t'u shu ch'u pan she, Min kuo 33 [1944] – (= ser P-k&k period) – us CRL [327]
Chan hou shih chieh pi chih wen t'i / Wu, Ch'i-yuan – Ch'ung-ch'ing: Ch'ing nien shu tien, Min kuo 32 [1943] – (= ser P-k&k period) – us CRL [332]
Chan hou wen t'i lun wen chi / Chung-shan wen hua chiao yu kuan chan hou shih chieh chien she yen chiu hui – Ch'ung-ch'ing: Tu li ch'u pan she, Min kuo 32- [1943- – (= ser P-k&k period) – us CRL [951]
Chan huo jan shao ti mien tien / Hsieh, Yung-yen – Ch'eng-tu: Chin jih hsin wen she, 1942 – (= ser P-k&k period) – us CRL [951]
Chan, Leang Nio *see* Tamper moekanja sendiri
Chan Mon, U *see* Cac su kri mahosadha
Chan sheng / Kuo, Mo-jo – [Kuang-chou]: Chan shih ch'u pan she, 1938 – (= ser P-k&k period) – us CRL [951]
Chan shih an ch'uan she pei / T'ang, Ling-ko – Ch'ang-sha: Shang wu yin shu kuan, Min kuo 27 [1938] – (= ser P-k&k period) – us CRL [303]
Chan shih cheng li t'ien fu wen t'i / Kuo, Yuan – Ch'ung-ch'ing: Kuo min t'u shu ch'u pan she, Min kuo 31 [1942] – (= ser P-k&k period) – us CRL [630]
Chan shih ching ch'a yeh wu / Li, Shih-chen – Ch'ang-sha: Shang wu yin shu kuan, Min kuo 27 [1938] – (= ser P-k&k period) – us CRL [360]
Chan shih ching chi wen t'i – Ch'ang-sha: Shang wu yin shu kuan, Min kuo 29 [1940] – (= ser P-k&k period) – us CRL [951]
Chan shih ching chi wen t'i : chung-kuo ching chi hsueh she ti shih ssu chieh nien hui lun wen chi – [China]: Shang wu yin shu kuan, Min kuo 27 [1938] – (= ser P-k&k period) – us CRL [951]
Chan shih ching chi wen t'i yen chiu / Li, Hua-fei – Ch'ung-ch'ing: Hsin sheng ming shu chu, Min kuo 27 [1938] – (= ser P-k&k period) – us CRL [951]
Chan shih ching chi wen t'i yu ching chi cheng ts'e / Wang, Ya-nan & P'ing-hsin – Han-k'ou: Kuang ming shu chu, Min kuo 27 [1938] – (= ser P-k&k period) – us CRL [951]
Chan shih chung yao fa ling hui pien / China – Ch'ung-ch'ing: Shuang chiang shu wu, 1944 – (= ser P-k&k period) – us CRL [340]
Chan shih chung-kuo ching chi ti lun k'uo / Ching, Sheng – [China: sn], 1944 – (= ser P-k&k period) – us CRL [951]
Chan shih chung-kuo wu chia wen t'i / Shou, Chin-wen – Ch'ung-ch'ing: Sheng sheng ch'u pan she, Min kuo 33 [1944] – (= ser P-k&k period) – us CRL [338]
Chan shih fa kuei shu yao / Chi, Hao – Ch'ung-ch'ing: Chi Hao: Chung hsin yin shu chu, 1943 – (= ser P-k&k period) – us CRL [951]
Chan shih hsiao fei p'in chih fen p'ei t'ung chih / Wang, Po-yen – Shang-hai: Han hsueh shu tien, Min kuo 25 [1936] – (= ser P-k&k period) – us CRL [339]
Chan shih hsien cheng fang an / Liu, Ching-ch'ing – [China]: Cheng sheng ch'u pan she, Min kuo 27 [1938] – (= ser P-k&k period) – us CRL [350]
Chan shih hsin wen chi che ti chi pen hsun lien / Liu, Kuang-yen – Ch'ung-ch'ing: Tu li ch'u pan she, Min kuo 29 [1940] – (= ser P-k&k period) – us CRL [070]
Chan shih hsin wen chien ch'a ti li lun yu shih chi / Sun, I-tz'u – [China]: Chun shih wei yuan hui chan shih hsin wen chien ch'a chu, 1941 – (= ser P-k&k period) – us CRL [070]
Chan shih hsin wen kung tso ju men – Ch'ung-ch'ing: Sheng huo shu tien, Min kuo 28 [1939] – (= ser P-k&k period) – us CRL [070]
Chan shih kung chai / Wu, K'o-kang – Shang-hai: Wen hua sheng huo ch'u pan she, Min kuo 26 [1937] – (= ser P-k&k period) – us CRL [336]
Chan shih kung yeh kuan chih chien t'ao – [China]: Chung-kuo kung yeh ching chi yen chiu so, Min kuo 34 [1945] – (= ser P-k&k period) – us CRL [338]
Chan shih kuo chi hsin wen tu fa / Wu, Hao-hsiu – Shang-hai: K'ai ming shu tien, Min kuo 30 [1941] – (= ser P-k&k period) – us CRL [070]
Chan shih kuo chia tsung tung yuan / Lo, Tun-wei – [China: Kuo min cheng fu chun shih wei yuan hui cheng chih pu], Min kuo 27 [1938] – (= ser P-k&k period) – us CRL [951]
Chan shih mao i cheng ts'e / Kao, Shu-k'ang – Ch'ung-ch'ing: Tu li ch'u pan she, Min kuo 29 [1940] – (= ser P-k&k period) – us CRL [380]
Chan shih min cheng kai yao – [China]: Fu-chien sheng cheng fu mi shu ch'u kung pao shih, Min kuo 29 [1940] – (= ser P-k&k period) – us CRL [951]
Chan shih shih yu cheng ts'e / Fetzer, F & Ch'en Yun-wen – Ch'ang-sha: Shang wu yin shu kuan, Min kuo 27 [1938] – (= ser P-k&k period) – us CRL [550]
Chan shih ta ku tz'u / Chao, Ching-shen – Kuang-chou: Chan shih ch'u pan she, 1938 – (= ser P-k&k period) – us CRL [951]
Chan shih ti fang hsing cheng kung tso / Wang, Ching-wei – Ch'ung-ch'ing: Tu li ch'u pan she, Min kuo 27 [1938] – (= ser P-k&k period) – us CRL [350]
Chan shih ti jen min tzu yu / I, Shih-fang – Ch'ung-ch'ing: Tu li ch'u pan she, Min kuo 30 [1941] – (= ser P-k&k period) – us CRL [323]
Chan shih ti jih pen ching chi / P'eng, Ti-hsien – Han-k'ou: Sheng huo shu tien, Min kuo 27 [1938] – (= ser P-k&k period) – us CRL [339]
Chan shih wai chiao wen t'i / Chou, Keng-sheng – [China]: Ch'ing nien shu tien, Min kuo 29 [1940] – (= ser P-k&k period) – us CRL [327]
Chan shih wen hsueh hsuan chi : hsien tai tso chia ch'uang tso hsuan – Nan-ching: Chung yang tien hsun she, 1945 – (= ser P-k&k period) – us CRL [830]
Chan shih wen hsueh lun / Wang, P'ing-ling – Han-k'ou: Shang-hai tsa chih kung ssu, 1938 – (= ser P-k&k period) – us CRL [480]
Chan shih wu chia kuan chih / Sun, I-tz'u – Shang-hai: Chung-hua shu chu, Min kuo 32 [1943] – (= ser P-k&k period) – us CRL [380]
Chan shih wu li ts'ai li / Chu, Yuan-mao – [China]: Cheng-chung shu chu, Min kuo 29 [1940] – (= ser P-k&k period) – us CRL [951]
Chan ti chi che chiang hua / Pu, Shao-fu – Kuei-yang: Wen t'ung shu chu, 1942 – (= ser P-k&k period) – us CRL [070]
Chan ti fu wu hui i lu / Ling, Ch'ing – Han-k'ou: Kuang ming shu chu, Min kuo 27 [1938] – (= ser P-k&k period) – us CRL [951]
Chan ti i nien / Hu, Lan-ch'i et al – Ch'ung-ch'ing: Sheng huo shu tien, 1939 – (= ser P-k&k period) – us CRL [951]
Chan ti jih chi / Chou, Li-po – Han-k'ou: Shang-hai tsa chih kung ssu, 1938 – (= ser P-k&k period) – us CRL [880]
Chan ti min chung tsu chih / Chu, Yuan-mao – [China]: Cheng chung shu chu, Min kuo 29 [1940] – (= ser P-k&k period) – us CRL [951]
Chan tou ti liang nien – Shang-hai: Hsien tai Chung-kuo chou k'an she, Min kuo 28 [1939] – (= ser P-k&k period) – us CRL [951]
Chan tou ti nu hsing : ssu mu chu / Ling, Ho – Shang-hai: Shang-hai tsa chih kung ssu, Min kuo 35 [1946] – (= ser P-k&k period) – us CRL [820]
Chan tou ti su hui : k'ang chan i lai pao kao wen hsueh hsuan chi – Ch'ung-ch'ing: Tso chia shu wu, 1943 – (= ser P-k&k period) – us CRL [951]
Ch'an t'ui chi / Su, Hsueh-lin – Shang-hai: Shang wu yin shu kuan, Min kuo 35 [1946] – (= ser P-k&k period) – us CRL [480]
Chan wang / Tsou, T'ao-fen – [China: Sheng huo hsing ch'i k'an she], Min kuo 26 [1937] – (= ser P-k&k period) – us CRL [840]
Chan wang *see* Shih chieh chi-tu chiao wen chai (ccs)
Chan wang yueh k'an *see* Shih chieh chi-tu chiao wen chai (ccs)
Ch'an yu chi / Yu, Ta-fu – Shang-hai: T'ien ma shu tien, Min kuo 22 [1933] – (= ser P-k&k period) – us CRL [480]
Chan yu chu / Chiang, Po-ch'ien – Shang-hai: Shih chieh shu chu, Min kuo 36 [1947] – (= ser P-k&k period) – us CRL [480]
Chanaan / Aranha, Graca – Rio de Janeiro, Brazil. 1939 – 1r – us UF Libraries [972]
Chanakira, Elijah J *see* Shona grammar for junior secondary schools
Chanakya and chandragupta / Panchapakesa Ayyar, Aiylam Subramanier – Madras: V Ramaswamy Sastrulu & Sons, 1951 – (= ser Samp: indian books) – us CRL [830]
Chance Bros & Co *see* Designs for coloured ornamental windows
Chance, Walter W la *see* Modern schoolhouses
Chancellor – Kremlin, MT. 1922-1929 (1) – mf#64516 – us UMI ProQuest [071]
Chancellor : official organ of the retailers of omaha – Omaha, NE: Progressive Pub Co (wkly) [mf ed v7 n31. oct 3-dec 26 1912 (gaps) filmed [1979]] – 1r – 1 – (cont by: omaha nebraskan) – us NE Hist [071]
Chancellor, Edwin Beresford *see* History of the squares of london
Chancerel, Leon *see*
– Antigone
– Farce du chaudronnier
– Picrochole
Chancery chatter – 1983 aug 12-1988 dec 23, 1989 jan 6-jun 16 – 2r – 1 – (cont by: kiwi chatter) – mf#2478731 – us WHS [071]
The chancery jurisdiction and practice, according to statutes and decisions in the state of illinois, from the earliest period to 1873 / Hill, Edward Judson – Chicago: Myers, 1873. 758p. LL-780 – 1 – us L of C Photodup [348]
The chancery student's guide in the form of a didactic poem. / Carrighan, Terentius – London: Wildy, 1850. 64p. LL-1672 – 1 – us L of C Photodup [340]
Chances of success : episodes and observations in the life of a busy man / Wiman, Erastus – Toronto: F R James; New York: American News Co, 1893 – 4mf – 9 – (incl ind) – mf#27553 – cn Canadiana [307]
Chancey, R E L *see* Survey knight field
Chancy, Emmanuel *see*
– Evenements de 1902
– Faits contemporaines
– Independance nationale d'haiti
Chand, Gyan *see*
– The essentials of federal finance
– The financial system of india
– India's teeming millions
– Some aspects of fiscal reconstruction in india
Chand, Sonal *see* The history of new horizons
Chanda, Ramaprasad *see*
– The beginnings of the art in eastern india
– Exploration in orissa
– The indo-aryan races
– The indus valley in the vedic period
– Medieval indian sculpture in the british museum
– Selections from official letters and documents relating to the life of raja rammohun roy
– Survival of the prehistoric civilisation of the indus valley
Chandavarkar, G A *see* Manual of hindu ethics
Chandavarkar, Narayen Ganesh *see* The speeches and writings of sir narayen g chandavarkar
La chandelle democratique et sociale – [Paris]: Madame de Lacombe 1849. – us CRL [320]
Chandeneux, Claire de *see* La vengeance de genevieve
Chander, Jag Parvesh *see*
– Ethics of fasting
– Gandhi against fascism
– Gita the mother
– The good life
– India steps forward
– Tagore and gandhi argue
– Teachings of mahatma gandhi
– The unseen power

Chandidas : translations – Jaipur: Garg Book Co, 1941 – (= ser Samp: indian books) – us CRL [490]

Chandidas *see* Chandidas

Chandieu, A *see*
- La confirmation de la discipline ecclèsiastique
- Opera theologica

Chandieu, A de la Roche *see* Histoire des persecutions

Chandieu, A S *see* Ant sad opera theologica

Chandler arizonan – Chandler AZ. 1912 nov 15-1916 apr 14 – 1r – 1 – mf#853925 – us WHS [071]

Chandler, Arthur *see*
- Ara coeli
- The spirit of man

Chandler, David P *see* The land and people of cambodia

Chandler, George *see* Ordination services

Chandler, Henry William *see*
- Letters, lectures, and reviews
- The philosophy of mind

Chandler, Izora Chandler *see* Methodist episcopalianism

Chandler, John Scudder *see*
- History of the jesuit mission in madura south india
- Seventy-five years in the madura mission

Chandler, Peleg W *see* Thesaurus thomas a kempis

Chandler, R *see* Travels in asia minor

Chandler, Richard *see*
- The life of william waynflete
- Travels in asia minor and greece
- Voyages dans l'asie mineure et en grece

Chandler, Samuel *see* Plain reasons for being a christian

Chandler, William Eaton *see* Address before the grafton and coos bar association

Chandler, Zachariah *see*
- Papers
- Proposed annexation of winnipeg

Chandler's reports / Wisconsin. Supreme Court – v1-4. 1849-1852 (all publ) – (= ser Wisconsin supreme court reports; Pinney's reports) – 13mf – 9 – $19.50 – (for purposes of pre-nrs coverage, all of chandler's cases are included in pinney) – mf#LLMC 91-303 – us LLMC [347]

Chandonnet, Gemma *see* Bibliographie de l'oeuvre de louis-philippe audet

Chandonnet, Thomas Aime *see* L'abbe joseph aubry

Chandra, Lokesh *see* Mongolian kanjur

Chandra, Moti *see* The technique of mughal painting

Chandra, Prabodh *see* Sixty years of congress

Chandra-Natha Vasu *see* High education in india

Chandrasekhar, Sripati *see* Hungry people and empty lands

Chandrasekharan, C V *see* Political parties with special reference to india

Chandrasekharan, K *see* Sanskrit literature

Chandrasekharendra Saraswati, Jagatguru Sankaracharya of Kamakoti *see* The sanatana dharma

Chandratre, P D *see* Methodology of the major bhasyas on the brahma-sutra

Chaney, George Leonard *see* Woman's ministry, as exemplified in southern schools

Chaney, George Leonhard *see* Belief

Chaney, Theodore *see* La colonie du sacre-coeur dans les cevennes de la chine au dix-huitieme siecle

Chang, Ai-ling *see* Liu yen

Chang, Ch'ang-jen *see* Fu fu; fu, chang chung shuoo fa

Chang, Cheng-ch'uan *see* Pen shih ti ch'eng chang

Chang, Cheng-ming *see* Min chu yu t'uan chieh

Chang, Cheng-yen *see* Kuei kuo yin hsiang

Chang, Ch'i *see* Chao wen tao yun ch'u chuan chi (ccm7)

Chang, Chia-ling *see* Kuei-chou wei-pi-t'ung ning-chieh-jen huang ti ch'u yu tiao ch'a pao kao

Chang, Chih *see* T'u ti ching chi hsueh

Chang, Chih-Chiang *see* Cheng tao i chu [ccm8]

Chang, Chih-ho *see* Hsien tai chan cheng lun

Chang, Chih-i *see* Hsin-chiang chih ching chi

Chang, Chih-liang *see* So te shui chan hsing t'iao li hsiang chieh

Chang, Chi-luan *see* Chi-luan wen ts'un

Chang, Chin-chien *see*
- Hsing cheng kuan li kai lun
- Jen shih hsing cheng yuan li yu chi shu

Chang, Ch'in-fu *see* Kuo chi wen t'i chiang hua

Chang, Chin-Shih *see* Kuo nei chin shih nien lai chih tsung chiao ssu chao (ccm9)

Chang, Chin-shou *see* Lu

Chang chi-tzu chiu lu – Collected works of Chang Chien. 25v. Shanghai. 1935. 2 reels – 1 – us Chinese Res [079]

Chang, Ch'i-yun et al *see* Hsi pei wen t'i

Chang, Chu *see*
- Tsui kao fa yuan p'an li yao chih
- Wo kuo chan shih liang shih kuan li

Chang chu hsi *see* Chang chu hsi yen lun chi

Chang chu hsi yen lun chi / Chang chu hsi – [China]: Hua chung shu chu, Min kuo 27 [1938] – (= ser P-k&k period) – us CRL [951]

Chang, Chun *see*
- Hsien tai chun shih kung ch'eng hsueh
- Pa chin tai piao tso hsuan

Chang, Chun-chun *see* Shou tu ti wei yu min tsu tsai chao

Chang, Chung-fu *see* Ying jih t'ung meng

Chang, Chun-hsiang *see*
- Fu kuei fu yun
- Hsiao ch'eng ku shih
- Mei-kuo tsung t'ung hao
- Pien ch'eng ku shih
- Shan ch'eng ku shih
- Wan shih shih piao

Chang, Ch'un-i *see* Fo hua chi-tu chiao (ccm10)

Chang, Chun-mai *see*
- Li kuo chih tao, i ming, kuo chia she hui chu i
- Min tsu fu hsing chih hsueh shu chi ch'u

Chang, Chun-Yen *see* A simulation approach to crowding in outdoor recreation

Chang, Fu-liang *see* Chi-tu chiao nung ts'un yun tung [ccm252]

Chang, Han-fu *see* Mei-kuo ti tui hua cheng ts'e

Ch'ang hen ko : li shih hsiao shuo chi / T'an, Cheng-pi – Shang-hai: Shang-hai tsa chih she, Min kuo 34 [1945] – (= ser P-k&k period) – us CRL [480]

Chang, Hen-shui *see*
- Chin fen shih chia
- Chin fen shih chia hsue chi
- Mi mi ku
- Ou hsiang
- Pa shih i meng
- Shu tao nan
- Ssu shui liu nien
- T'ai p'ing hua
- Tan feng chieh

Ch'ang ho / Shen, Ts'ung-wen – Shang-hai: K'ai ming shu tien, Min kuo 38 [1949] – (= ser P-k&k period) – us CRL [830]

Chang, Ho-li *see* Man-chou-kuo chih hsien chieh tuan

Chang, Hsiao-mei *see*
- Kuei-chou ching chi
- Ssu-ch'uan sheng chih t'ung yu

Chang, Hsi-ch'ang *see* Nung ts'un she hui tiao ch'a

Chang hsing che hsueh / Yu, P'ing-k'o – Shang-hai: Hsin ling k'o hsueh shu chu, Min kuo 30 [1941] – (= ser P-k&k period) – us CRL [130]

Chang, Hsueh-ch'eng *see* Chiao ch'ou t'ung i

Chang, Hsueh-ch'eng *see* Wen shih t'ung i

Chang, Hsun-chiu *see* Shang-hai li shih yen i, yu ming, shen mi ti shang-hai

Chang, Huang *see* Pei fang ti ku shih

Chang, Huan-tou *see* I wen ch'ein

Chang, I-cheng *see* Tung-ching lao yu chung

Chang, I-ching *see*
- Kuan yu chu chih-hsin yeh-su shi shen mo tung hsi te tsa p'ing
- Yu ch'en tu-hsiu shen hsuan-lu pien tao

Chang, I-p'ing *see*
- Che shang sui pi
- Hsiao chiao niang
- Hsiu tz'u hsueh chiang hua
- Sui pi san chung

Chang, Jen-chieh *see* Hu-nan chih k'uang yeh

Chang, Jen-chien *see* K'ai fa hsi pei shih yeh chi hua

Ch'ang Jen-hsia *see* Hsien tai chung-kuo shih hsuan

Ch'ang, Jen-hsia *see* Min su i shu k'ao ku lun chi

Chang, Jen-k'an *see* Yu tai ch'u cheng k'ang ti chun jen chia shu fa kuei ch'ien shih

Chang, Jo-ku *see* Ts'ung hsiao-o tao lu hsun

Chang, Keng *see* Ta hui lao chia ch'u

Chang, Ko-nung *see* Kwan tong kie hiap

Chang, Kuang-chung *see* Hsia wan-ch'un

Chang, Kung-hui *see* Ko kuo tsung tung yuan kai k'uang

Chang, Kuo-an *see* San ta tu ts'ai cheng chih chih tu

Chang, Kuo-p'ing *see* K'ang jih ti ti pa lu chun

Chang, Kyungro *see* A systems view of quality in fitness services

Chang, Liang-jen *see* Ssu-ch'uan liang shih wen t'i

Chang, Li-sheng chu *see* Sheng tao cheng yen (ccm13)

Chang, Li-ying *see* Nu tso chia hsiao p'in hsuan

Chang, Lu-luan *see* Yin chia yu chung-kuo wu chia shui chun chih kuan hsi

Chang Marin, Carlos Francisco *see* Faragual

Chang, Min *see*
- Ch'i erh
- Hsi chu lun
- Hsueh
- Sheng lu
- Wo men ti ku hsiang
- Yeh

Chang, Ming *see* Mei-hsi shan chuang ch'ang ho chi

Chang, Ming-yang *see* Kuo chi ts'ai chun wen t'i

Chang, Nai-ch'i *see*
- Chang nai-ch'i lun wen hsuan
- Ti erh tzu ta chan yu chung-kuo
- Tzu pen chu i kuo chi yu chung-kuo

Chang nai-ch'i lun wen hsuan : [4 chuan] / Chang, Nai-ch'i – Shang-hai: Sheng-huo shu tien tsung ching shou, Min kuo 23 [1934] – (= ser P-k&k period) – us CRL [327]

Ch'ang nien tuan chi / Ou-yang, Fan-hai – Kuei-lin: Wen hsien ch'u pan she, Min kuo 31 [1942] – (= ser P-k&k period) – us CRL [840]

Chang, P'ei-fen *see* Fu nu wen t'i

Chang, P'ei-kang *see* Che-chiang sheng shih liang chih yuen hsiao

Chang, Pei-ying *see* Hsiang ts'un li pai (ccm110)

Chang, P'eng-jo *see* Nung ts'un fu hsing chih li lun yu shih chi

Chang, Pi *see* Shih chieh chih shih tu pen

Chang, P'i-chieh *see* T'u ti ching chi hsueh tao lun

Ch'ang p'ien chu pen ta ch'uan – [China]: Yu hsin shu chu, Min kuo 23 [1934] – (= ser P-k&k period) – us CRL [820]

Chang, Ping-hui *see* K'ang chan yu chiu chi shih yeh

Chang, Po-huai *see* Hsin yueh cheng ching cheng li shih (ccm15)

Chang, Sheng-chih *see* Ming jen chuan chi

Ch'ang shih i hsia / Pa, Jen – Shang-hai: To yang she ch'u pan pu, [1936] – (= ser P-k&k period) – us CRL [480]

Chang, Shih-Chang *see* Chi-tu chiao yu she hui chu i yun tung [ccm16]

Chang, Shih-chao *see* Lo chi chih yao

Chang ssu t'ai t'ai : [wu mu chu] / Ch'en, Ta-pei – Shang-hai: Hsien tai shu chu, 1931 – (= ser P-k&k period) – us CRL [820]

Chang, Su-min *see*
- K'ang chan yu ching chi t'ung chih
- Pai yin wen t'i yu chung-kuo pi chih

Chang, Tan-feng *see* Chin pai nien lai chung-kuo pao chih chih fa chan chi ch'i ch'ue shih

Ch'ang, Tao-chih *see* Tseng ting chiao yu hsing cheng ta kang

Chang, Tao-fan *see*
- Tsui hou kuan t'ou

Chang, Tao-hsing *see* Chan hou shih chieh ho p'ing wen t'i

Chang, T'ieh-chun *see* San min chu i yen chiu tao lun

Chang, T'ieh-sheng *see* Tsai hsi-pan-ya

Chang, T'ien-i *see*
- Chang t'ien-i ch'uang tso hsuan
- Ch'i jen chi
- Fan kung
- Hsiao pi-te
- Mi feng
- San hsiung ti
- Shih tai ti t'iao tung
- Su hsieh san p'ien
- T'an jen wu miao hsieh
- Tsai ch'eng shih li
- T'uan yuan
- T'ung hsiang men
- T'u-t'u ta wang; chi, hao hsiung ti
- Wan jen yueh
- Yao yuan ti hou fang

Chang t'ien-i ch'uang tso hsuan / Chang, T'ien-i – Shang-hai: Fang ku shu tien, 1936 – (= ser P-k&k period) – us CRL [480]

Chang, T'ien-i et al *see*
- Hsi-ling ti huang hun
- Yu mo hsiao shuo hsuan

Chang, Ti-fei *see* Yu mao tse-tung lun chung-kuo ko ming

Chang, Tse-yao *see* Ho tso chin jung yao i

Chang, Tso-hua *see* Yu chi chan shu chiang hua

Chang, Tsung-lin *see*
- Hsiang ts'un chiao yu ching yen t'an
- Hsiang ts'un hsiao hsueh chiao ts'ai yen chiu

Chang, Tung-sun *see* Ssu hsiang yu she hui

Chang tzu / Ou-yang, Shan – Shang-hai: Hua hsin t'u shu kung ssu, 1941 – (= ser P-k&k period) – us CRL [830]

Chang, Tzu-p'ing *see*
- Chang tzu-p'ing hsiao shuo hsuan
- Chang tzu-p'ing hsuan chi
- Hsin hung a tzu
- I tai nu yu
- Lien ai ts'o tsung
- Pu p'ing heng ti ou li
- Shang ti ti erh nu men
- She hui hsueh kang yao
- Su miao chung chung
- T'ai li
- Tzu-p'ing tzu hsuan chi

Chang tzu-p'ing hsiao shuo hsuan / Chang, Tzu-p'ing – Shang-hai: Fang ku shu tien – (= ser P-k&k period) – us CRL [480]

Chang tzu-p'ing hsuan chi / Chang, Tzu-p'ing – [Shang-hai]: Wan hsiang shu wu, Min kuo 25 [1936] – (= ser P-k&k period) – us CRL [480]

Chang, Wan-ju *see* Hong kou (ccm18)

Chang, Ya-chu *see* Hsi chua chi

Ch'ang yeh chi / Meng, Ch'ao – Kuei-lin: Wen hsien ch'u pan she, Min kuo 31 [1942] – (= ser P-k&k period) – us CRL [840]

Ch'ang yeh hsing : ssu mu chu / Yu, Ling – [China]: Hsin chih shu tien, Min kuo 35 [1946] – (= ser P-k&k period) – us CRL [820]

Chang, Yuan-jo *see* Hsien cheng chih tao

Chang, Yuan-shan *see* Hsiang ts'un chien she shih yen ti erh chi

Change – 1965 fall/winter-1966 spr/summer – 1r – 1 – mf#250809 – us WHS [071]

Change – New Rochelle. 1969+ (1) 1971+ (5) 1975+ (9) – ISSN: 0009-1383 – mf#5045 – us UMI ProQuest [370]

A change in attitude : women, war and society – 5pts. 1914-18 – (= ser Women at work) – 91r (complete) – 1 – (previous title: women at work. chronicles the involvement of women in the wartime effort. pt 1: 24r c36-28041. pt 2: 14r c36-28042. pt 3: 18r c36-28043. pt 4: 15r c36-28044. pt 5: 20r c36-28045. each pt incl printed guide) – mf#C36-28040 – us Primary [305]

The change in oxygen consumption over time in downhill versus level grade running / Pein, Wayne E – 1989 – 60p 1mf – 9 – $4.00 – us Kinesology [617]

Change of name of the protestant episcopal church : from the constitutional and legal point of view / Packard, Joseph – [s.l: s.n.] 1913 [mf ed 1992] – (= ser Prayer book papers 9) – 1mf – 9 – 0-524-03078-2 – mf#1990-4567 – us ATLA [242]

The changed cross and other religious poems – Toronto: Adam, Stevenson, 1872 – 3mf – 9 – (incl ind) – mf#26257 – cn Canadiana [810]

The changed cross and other religious poems – New ed. London: S. Low, Marston, Searle, & Rivington, 1877. 228p. Includes indexes – 1 – us Wisconsin U Libr [810]

A changed exchange broker *see* I ko shang-hai shang jen te kai p'ien (ccm213)

Changement de main / Bayard, Jean-Francois-Alfred – Paris, France. 1845? – 1r – us UF Libraries [440]

Changes in blood resistivity over a sub-maximal exercise bout / De la Cruz Napoli, Jose – Indiana University, 1994 – 1mf – 9 – $4.00 – mf#PH1457 – us Kinesology [612]

Changes in clinical students' perceptions of developmental physical education and effective teaching / Hammel, Patricia A – University of Wisconsin-La Crosse – 1mf – 9 – $4.00 – mf#PE3595 – us Kinesology [370]

Changes in clotting and fibrinolytic activity after sub-maximal exercise in males / Hegde, Sudhir S – 1999 – 2mf – 9 – $8.00 – mf#PH 1650 – us Kinesology [612]

Changes in cognitive appraisals and metabolic indices of physical exertion during at two-hour run / Acevedo, E O – 1989 – 2mf – 9 – $8.00 – us Kinesology [613]

Changes in composition of florida avocados in relation to maturity / Stahl, Arthur L – Gainesville, FL. 1933 – 1r – us UF Libraries [634]

Changes in learned motor behavior : due to the effects of various forms of augmented kinematic feedback / Hale, Trevor A – 1999 – 91p on 1mf – 9 – 5.00$ – mf#PSY 2142 – us Kinesology [150]

Changes in maternal body composition from month one to month six postpartum in 11 breastfeeding, exercising women / Kwasnicki, Sherri – 1997 – 2mf – 9 – $8.00 – mf#PH 1568 – us Kinesology [618]

Changes in power and authority relations in south african schooling since 1976 / Williams, Brian Kenneth – U of the Western Cape 1990 [mf ed S.l: s.n. 1990] – 2mf – 9 – (abstract in afrikaans & english; incl bibl) – sa Misc Inst [370]

Changes in spinal excitability preceding a voluntary movement in young and old adults / Burke, J – 1991 – 6mf – 9 – $24.00 – us Kinesology [613]

Changes in the documents of british india caused by the government of india act 1935 / Cabeen, Violet Abbott – [n.p. 1939] – us CRL [954]

Changes made by the 1951 legislature in kansas library laws / Drury, James Westbrook – Lawrence, Bureau of Government Research, University of Kansas 1952. LL-241 – 1 – us L of C Photodup [340]

Changes socialist monthly / Center for Changes [Detroit MI] – v1 n1-v5 n9 [1979 feb-1983 oct] – 1r – 1 – (cont: changes socialist monthly; cont by: against the current) – mf#586170 – us WHS [335]

Changing attitudes toward physically disabled persons using a videotape sport intervention / Bett, A – 1991 – 2mf – 9 – $8.00 – us Kinesology [362]

Changing china / Cecil, William Rupert Ernest Gascoyne, Lord & Cecil, Florence, Lady – New York: D Appleton, 1910 [mf ed 1995] – (= ser Yale coll) – xvi/342p (ill) – 1 – 0-524-10044-6 – mf#1995-1044 – us ATLA [951]

Changing creeds and social struggles / Aked, Charles Frederic – London: James Clarke, 1893 – 1mf – 9 – 0-524-07804-1 – mf#1991-3351 – us ATLA [240]

The changing east / Spender, John Alfred – London: Cassell and Co, 1926 – (= ser Samp: indian books) – us CRL [915]

Changing education – Washington. 1966-1974 (1) 1973-1974 (5) – ISSN: 0009-1413 – mf#2261 – us UMI ProQuest [370]

The changing face of human genetics in community health = Veranderende beeld van menslike genetika in gemeenskapsgesondheid / South Africa. Department of Health [Departement van Gesondheid] – Pretoria: Dept of Health [1975?] [mf ed Pretoria, RSA: State Library [199-]] – 24p on 1r with other items – 5 – (incl bibl ref) – mf#op 06458 r23 – us CRL [575]

Changing india / ed by Rao, Raja & Singh, Iqbal – London: George Allen & Unwin, 1939 – (= ser Samp: indian books) – us CRL [301]

Changing men – n8-16 [1974 nov-1974 jul] – 1r – 1 – (cont by: forum for changing men) – mf#528841 – us WHS [305]

The changing men collections : a chronicle of the modern men's movement – [mf ed 2003] – 110r in 3pts – 1 – (pt1: vertical files 37r. pt2: periodicals and newsletters 46r. pt3: archives 27r) – us Primary [305]

Changing patterns of settlement and land use / Kay, George – Hull, England. 1965 – 1r – us UF Libraries [960]

Changing perceptions of relationships of christian love and reconciliation : in small groups in the black church / Norris, Dennis Earl – Princeton, NJ, 1979. Chicago: Dep of Photodup, U of Chicago Lib, 1979 (1r); Evanston: American Theol Lib Assoc, 1984 (1r) – 1 – 0-8370-1371-2 – mf#1984-T216 – us ATLA [240]

The changing roles of women in east africa : implications for planning family-oriented programmes / FAO/SIDA Workshop for Intermediate Level Instructors in Home Economics and Rural Family-Oriented Programmes in East and Southern Africa, (1974: Njoro, Kenya) – [s.l: s.n, 1975?] – (= ser African training and research centre for women publications on microfilm) – 1mf – 9 – us CRL [305]

Changing russia / Graham, Stephen – 2nd ed. London; New York: John Lane, 1913 – 1mf – 9 – 0-7905-6749-0 – mf#1988-2749 – us ATLA [915]

The changing scene in india / Nihal Singh, Saint – [Calcutta]: sn, [1933?] – (= ser Samp: indian books) – us CRL [915]

Changing the crosses and winning the crown / Ideen, Marie A – Philadelphia: J B Lippincott, 1872 [mf ed 1984] – 2mf – 9 – 0-8370-1042-X – (incl poems) – mf#1984-4375 – us ATLA [240]

Changing the ordinance – London, England. 18-- – 1r – us UF Libraries [240]

Changing times – Washington. 1947-1991 (1) 1966-1991 (5) 1960-1991 (9) – (cont by: kiplinger's personal finance magazine) – ISSN: 0009-143X – mf#879 – us UMI ProQuest [380]

Changing woman – Portland OR: Women's Editorial Collective, 1971- [irreg] – 1 – (suspended may-aug 1973 and mar-sep 1975) – us Oregon Lib [071]

Changing work / Institute for Corporate Studies [Newton MA] – n1-7 [1984 fall-1988 fall] – 1r – 1 – (cont by: workplace democracy, grassroots economic organizing) – mf#990862 – us WHS [650]

The changing world : and, lectures to theosophical students. fifteen lectures / Besant, Annie Wood – Chicago, IL: Theosophical Book Concern, 1910 [mf ed 1991] – 1mf – 9 – 0-524-01680-1 – mf#1990-2582 – us ATLA [290]

Changnogyo yosong sinmun = The presbyterian women's news – Soul-si: Changnogyo yosong sinmunsa [biwkly] [mf ed 2004] – 1 – (mf: n134- [jan 15 2002-] lacks n144,149-150,169,171-173,175; iss in newspaper format; iss by taehan yesugyo changnohoe yo chondohoe chonguk yonhaphoe) – mf1053 – us ATLA [242]

Ch'ang-sha chung yao kung ch'ang tiao ch'a / Meng, Hsueh-ssu – [Ch'ang-sha: Hu-nan ching chi tiao ch'a so, Min kuo 23 ie 1934] – (= ser P-k&k period) – us CRL [338]

Ch'ang-sha hui chan chi shih – [China]: Chung hsing shu tien, Min kuo 29 [1940] – (= ser P-k&k period) – us CRL [951]

Chanh dao – Ho Chi Minh City, Vietnam. 1965-1969 (1) – mf#61091 – us UMI ProQuest [079]

Channel / S[/ain/]t Mary's Medical Center [Racine WI] – v10 n5-v14 n1 [1978 dec-1983 mar],v15 n1 [1984 feb] – 1r – 1 – mf#645340 – us WHS [360]

Channel / Women's Army Corps Veterans Association [US] – 1972 feb-1984 jun – 1r – 1 – mf#1007742 – us WHS [305]

Channel business : technology reselling in canada – Toronto. 1999+ (1,5,9) – ISSN: 1493-9088 – mf#18039,01 – us UMI ProQuest [000]

Channel dls – Wisconsin. 1966-1981, 1981 sep-1988 jul/aug – 2r – 1 – (cont by: channel [madison wi]) – mf#162027 – us WHS [071]

Channel islands vanguard – Channel Island Air National Guard Base CA. 1990 jun – 1r – 1 – (cont: vanguard [van nuys ca]; cont by: channel islands vanguard quarterly) – mf#1214393 – us WHS [071]

Channel one / Poets World International – 1985 – 1r – 1 – mf#5306934 – us WHS [400]

Channing, W E *see* On preaching the gospel to the poor

Channing, William E *see* The william ellery channing papers, 1791-1892

Channing, William Ellery *see*
- Memoir of william ellery channing
- Ministry for the poor
- Sermon, delivered at the ordination of the rev ezra stiles gannett, as colleague pastor of the...
- A sermon on war
- Unitarian christianity
- The works of william e. channing

Channing, William Henry *see* The life of william ellery channing, d.d

Chano / Montes Lopez, Jose – Habana, Cuba. 1938 – 1r – us UF Libraries [972]

Chanoine jean bergeron, 1868-1956 : bio-bibliographie analytique / Bergeron, Juliana – 1958 [mf ed 1978] – (= ser Bibliographies du cours...1947-66) – 2mf – 9 – (with ind; pref by felix-antoine savard) – mf#SEM105P4 – cn Bibl Nat [241]

Les chanoines reguliers de saint augustin : apercu historique / Ette, A van – Cholet, 1953 – 5mf – 8 – €12.00 – ne Slangenburg [241]

La chanson canadienne : origines, evolution, epanouissement / Morin, Victor – Toronto: University of Toronto Press, 1928 [mf ed 1987] – 1mf – 9 – mf#SEM105P771 – cn Bibl Nat [780]

Chanson de geste und hofischer roman – Heidelberg, Germany. 1963 – 1r – us UF Libraries [960]

La chanson francaise a travers les siecles : revue historique de ses auteurs et de leurs interpretes / Morin, Victor – ed de l'auteur. Toronto: University of Toronto Press, 1939 [mf ed 1987] – (= ser Operetta-dinner) – 1mf – 9 – (together with: operetta-dinner: a gastronomico-musical fantasy in two acts) – mf#SEM105P812 – cn Bibl Nat [780]

Le chansonnier cange (bibl nat paris fonds fr n846). les chansonniers des troubadours et des trouveres, no 1 facsimile-edition par jean beck / ed by Beck, Jean-B – Philadelphia-London. v1-2. 1927 – €62.00 – ne Slangenburg [780]

Le chansonnier des familles : lyre canadienne – 3e rev corr ed. Montreal: J B Rolland & Fils, editeurs, [1883?] (mf ed 1991) – 3mf – 9 – (with ind) – mf#SEM105P1339 – cn Bibl Nat [780]

Les chansonniers des troubadours et des trouveres, n1 *see* Le chansonnier cange (bibl nat paris fonds fr n846). les chansonniers des troubadours et des trouveres, no 1 facsimile-edition par jean beck

Les chansonniers des troubadours et des trouveres, n2 *see* Le manuscrit du roi (bibl nat paris fonds fr n844). les chansonniers des troubadours et des trouveres, no 2 facsimile-edition par jean beck

Chansons de beranger : ou, le tailleur et la fee / Vanderburch, Emile – Paris, France. 1839 – 1r – us UF Libraries [780]

Les chansons de colin muset / Muset, Colin; ed by Bedier, Joseph – Paris: H. Champion, 1912. xiii,44p – 1 – us Wisconsin U Libr [810]

Chansons du dodecanese / Baud-Bovy, Samuel – 2v. Athenes: J N Sideris 1935-38 [mf ed Bloomington IN: Indiana Uni Lib, Preservation Dept 1984] – 1r – 1 – us Indiana Preservation [390]

Chansons et danses de la gascogne : avec 46 airs notes et 4 hors-texte – Bordeaux: Delmas 1945 [mf ed Bloomington IN: Indiana Uni Lib, Preservation Dept 1984] – 1r – 1 – us Indiana Preservation [390]

Chansons et monologues : paroles et musique / Bruant, Aristide – Paris: H Geffroy [1896-97] [mf ed 1986] – 3v on 1r [ill] – 1 – mf#1737 – us Wisconsin U Libr [780]

Chansons et rondes enfantines / Weckerlin, Jean-Baptiste – Paris: Garnier [188-?] [mf ed Bloomington IN: Indiana Uni Lib, Preservation Dept 1984] – 1r [ill] – 1 – (incl ind) – us Indiana Preservation [780]

Chansons et rondes enfantines des provinces de la france / Weckerlin, Jean Baptiste – Paris: Garnier 1889 [mf ed Bloomington IN: Indiana Uni Lib, Preservation Dept 1984] – 1r – 1 – us Indiana Preservation [390]

Chansons populaires – [Phnom Penh] Edition de l'Institut bouddhique [mf ed 1989] – (= ser Serie de culture et civilisation khmeres 12) – 1r with other items – 1 – (extrait du bulletin edite par le departement du tourisme khmer 'connaissance du cambodge' et france-asie, no special n37-38, consacre au cambodge) – mf#mf-10289 seam reel 004/10 [§] – us CRL [780]

Chansons populaires de la france / Clairville, M – Paris, France. 1846? – 1r – us UF Libraries [780]

Chansons populaires de la france – New York: G P Putnam's Sons c1891 [mf ed Bloomington IN: Indiana Uni Lib, Preservation Dept 1984] – xxxix 282p on 1r – 1 – us Indiana Preservation [390]

Chansons populaires du canada – Quebec: "Foyer canadien", 1865 – 5mf – 9 – (ann by ernest gagnon) – mf#48455 – cn Canadiana [780]

Chansons populaires du canada : recueillies et publiees avec annotations, etc / Gagnon, Ernest – Quebec: Bureaux du "Foyer canadien", 1865 [mf ed 1974] – 1r – 5 – mf#SEM16P140 – cn Bibl Nat [780]

The chant and service book : containing the choral service for morning and evening prayer, chants for the canticles, with the pointing set forth by the general convention, music for the communion service, chants and anthems for the burial office, etc., etc / ed by Hutchins, Charles Lewis – Boston: Parish Choir, c1894 – 3mf – 9 – 0-524-07237-X – (incl ind) – mf#1991-2978 – us ATLA [780]

Chant et baiser : voix et piano / Godard, Benjamin – 1871 [mf ed 19–] – 1mf – 9 – (trans by n martin) – mf#fiche 1230 – us Sibley [780]

Chant et musique dans le culte chretien / Gelineau, J – Paris, 1962 – 4mf – 8 – €11.00 – ne Slangenburg [780]

Chant, Joseph Horatio *see* Gleams of sunshine, optimistic poems

Chant kasala des luba / Mufuta, Patrice – Paris, France. 1970, c1968 – 1r – us UF Libraries [960]

Le chant liturgique collectif a l'epoque patristique / Malherbe, G – Bruxelles, 1923 – 1mf – 8 – €3.00 – ne Slangenburg [780]

Chanta leksikon / Wright, Allen – A Choctaw in English definition. 1880 – 1 – us Southern Baptist [490]

Chante la vie... : recueil de chansons d'hier et d'aujourd'hui avec accords de guitare – [Saint-Mathieu-du-Parc: Collection Chante la vie, 1988?] (mf ed 1993) – 5mf – 9 – mf#SEM105P1776 – cn Bibl Nat [780]

Chantecler – Paris. 1er mai 1926-1er aout 1931 – 1 – fr ACRPP [073]

Chantepie de la Saussaye, Daniel *see* La crise religieuse en hollande

Chantepie de la Saussaye, Pierre Daniel *see*
- Manual of the science of religion
- The religion of the teutons
- Die vergleichende religionsforschung und der religioese glaube

Chanteuse et l'ouvriere / Xavier – Paris, France. 1832 – 1r – us UF Libraries [025]

Chanthit Krasaesin *see* Thawa thotsamat khlong dan (samnuan phra yaowarat

Chantiers cooperatifs – Paris. juin 1932-mai 1934 – 1 – (suite de: cahiers bleus. devenu: le nouvel age) – fr ACRPP [073]

Chantiers cooperatifs *see* Le nouvel age

Chan-toon *see* The nature and value of jurisprudence

Chantre, E *see*
- Mission en cappadoce – 1893-1894
- Recherches archeologiques dans l'asie centrale

Chantreau, Pierre *see*
- Voyage dans les trois royaumes d'angleterre, d'ecosse et d'irlande, fait en 1788 et 1789
- Voyage philosophique, politique et litteraire, fait en russie pendant les annees 1788 et 1789

Chantrel, Joseph *see* Annales ecclesiastiques de 1846 a 1860

Chantres et chant des psaumes a geneve – 16e siecle. [n.p. 19–] – us CRL [780]

Chantron, Jeanne *see* Le metier par l'image

Chantry certificates for cornwall *see* Documents towards a history of reformation in cornwall

Chants chretiens – Philadelphia: Presbyterian Board of Publication, [1850?] – 1mf – 9 – 0-524-06988-3 – mf#1991-2841 – us ATLA [240]

Chants de desespoir : poemes a la desolation / Lessard, Michaelena Marcon – [Cap-Rouge]: M Marcon Lessard, 1978 [mf ed 1990] – 3mf – 9 – mf#SEM105P1323 – cn Bibl Nat [780]

Les chants de la messe aux 8th et 9th siecle / Froger, L – Paris, 1950 – 1mf – 8 – €6.00 – ne Slangenburg [241]

Chants de la veillee : repertoire de romances, chansons comiques, melodies nocturnes, barcarolles, etc – Montreal: Typographie de Duvernay freres, 1855 [mf ed 1991] – 2mf – 9 – mf#SEM105P1356 – cn Bibl Nat [780]

Chants du souvenir / Dennery, Germaine – Port-Au-Prince, Haiti. 1939 – 1r – us UF Libraries [972]

Chants et chansons en pays akye : valeur expressive, valeur didactique / Aye, Agnes – 1985 – us CRL [780]

Les chants et les contes des ba-ronga de la baie de delagoa / Junod, Henri Alexandre – Lausanne: Georges Bridel & Cie [1897] – (= ser [Travel descriptions from south africa, 1711-1938]) – 4mf – 9 – mf#zah-53 – ne IDC [916]

Chants et reves / Vienx, Isnardin – Paris, France. 1896 – 1r – us UF Libraries [972]

Chants populaires d'auvergne – Aurillac: Terrisse 1910 [mf ed Bloomington IN: Indiana Uni Lib, Preservation Dept 1984] – 1r – 1 – us Indiana Preservation [390]

Chants sacres : 60 motets avec accompt. d'orgue ou piano pour messes, saluts, mariage, offices divers / Gounod, Charles – Paris: Le Beau [187-?] [mf ed 1994] – 1r – 1 – mf#pres. film 133 – us Sibley [780]

Chanute this week – v11 n25-v11 n50 [1982 jun 25-dec 17] – 1r – 1 – (cont by: pacesetter [rantoul il]) – mf#655085 – us WHS [071]

Chao, Cheng-p'ing *see* Pan pu lun yu yu cheng chih

Chao, Chia-chin *see* Ming jen chuan chi

Chao, Chia-pi chi *see* Erh shih jen so hsuan tuan p'ien chia tso chi

Chao, Ch'ing-ko *see*
- Feng
- Hua pei ti ch'iu
- Pien chu fang fa lun
- Sheng ssu lien
- T'ao li ch'un feng
- Tz'u hen mien mien
- Yu ta li hua, yu ming, huo
- Yuan yang chien

Chao, Ching-shen *see*
- Chan shih ta ku tz'u
- Hsiao mei
- Hsiao shuo hsi ch'u hsin k'ao
- Hsiao shuo hsien hua
- Hsien tai shih hsuan
- T'an tz'u k'ao cheng
- Wen hsueh kai lun
- Wen i lun chi
- Wu shih ch'i yung shih

Chao, Ching-yuan *see* Ying jih kuan hsi lun

Chao, Ch'uan-t'ien *see* Tung-pei wen t'i yu shih chieh ho p'ing

Chao, Hsiao-sung *see* Jen ho jen men

Chao, Hsia-yun *see* Chao hsia-yun tzu chuan [ccm64]

Chao hsia-yun tzu chuan [ccm64] = Life's experiences / Chao, Hsia-yun – Shanghai, 1932 [mf ed 198?] – (= ser Ccm 64) – 1 – mf#1984-b500 – us ATLA [920]

Chao, I-lin *see* K'ang jih ti ti pa lu chun

Chao, I-p'ing *see*
- Kuo chi chi t'uan ching chi
- She hui k'o hsueh yen chiu fa

Chao, Lan-p'ing *see*
- Jih-pen ching chi kai k'uang
- Jih-pen tui hua shang yeh
- Ko kuo t'ung huo cheng ts'e yu huo pi chan cheng
- T'ung huo wai hui yu wu chia

Chao shang chu san ta an / Li, Ku-fan – Shang-hai: Hsien tai shu chu, 1933 – (= ser P-k&k period) – us CRL [951]

Chao, Shu-yu *see* Hsiang ts'un chiao yu ts'ung chi

Chao thai – Bangkok, Thailand. 1967-1974 (1) – mf#67851 – us UMI ProQuest [079]

Chao, Tsu-k'ang *see* Su che wan ching hu wu sheng shih chiao t'ung wei yuan hui san nien lai kung tso kai shu

Chao, Tsung-fu *see* Chiu yueh shih (ccm65)

Chao, Tzu-ch'en *see*
- Chi-tu chiao che hsueh
- Chi-tu chiao chiao hui te i i [ccm67]
- Chi-tu chiao chin chieh [ccc68]
- Chi-tu chiao ti chung hsin hsin yang [ccm69]
- Chi-tu chiao ti lun li [ccm70]
- Chi-tu yu wo ti jen ko [ccm71]
- Chung-kuo chi-tu chiao chiao hui kai ko ti t'u ching [ccm72]
- Hsi yu chi
- Hsueh jen
- Ming chung sheng ko chi
- Pa-te ti tsung chiao ssu hsiang
- Ping hsuan chiao shih yeh ping i
- T'uan ch'i sheng ko chi
- Wo pei tai chu le

Chao, Tzu-chen *see*
- Shen hsueh 4 chiang
- Sheng pao-lo chuan
- Yeh-su chuan

Chao, Wei-jan *see* Tung hsin te hu huan (ccm83)

Chao wen tao yun ch'u chuan chi (ccm7) = God never fails san pan / Chang, Ch'i – Shanghai, 1939 [mf ed 198?] – (= ser Ccm 7) – 1 – mf#1984-b500 – us ATLA [210]

Chao yang / Pa, Chin – Shang-hai: Hsin sheng ch'u pan she, Min kuo 28 [1939] – (= ser P-k&k period) – us CRL [830]

Chao yang chi / Yu, Mu-t'ao – Shang-hai: Kuang hua shu chu, 1932 – (= ser P-k&k period) – us CRL [480]

Chao, Yu-p'ei *see* San min chu i wen i ch'uang tso lun
Ch'ao-chou wen kai / Weng, Hui-tung – Shanghai: Li kuang i yuan, Min kuo 22 [1933] – (= ser P-k&k period) – us CRL [480]
Chaos – 1,5,6,9 – us AIP [530]
Le chaos espagnol : eviternons-nous la contagion? / Bardoux, Jacques – Paris, 1937. Fiche W 745. (Blodgett Collection of Spanish Civil War Pamphlets) – 9 – us Harvard [946]
Chaos in spain / Bardoux, Jacques – London: Burns Oates & Washbourne ltd [1937?] – 9 – mf#w746 – us Harvard [335]
Chaos, solitions and fractals – Oxford. 1991-1994 (1,5,9) – ISSN: 0960-0779 – mf#49611 – us UMI ProQuest [621]
Chapado Garcia, Eusebio Maria *see* Historia general del derecho espanol
Chapais, Jean Charles *see*
– L'agriculture des regions froides de quebec
– Arbor day
– The canadian forester's illustrated guide
– Conference sur le porc et l'industrie laitiere
– La foret et le cultivateur
– Guide illustre du sylviculteur canadien
– Notes biographiques sur ed a barnard
– Notes historiques sur les ecoles d'agriculture dans quebec
– Pilote-provancher
– Un probleme d'economie sociale
– Selection of milch cows and economy in their feeding
Chapais, Jean-Charles *see*
– Arbor day
– Choix des vaches laitieres
Chapais, Thomas *see*
– Discours sur la loi de l'instruction publique
– Les hommes du jour
– Le serment du roi et les catholiques
Chap-book : semi-monthly. a miscellany and review of belles-lettres – Chicago. 1894-1898 (1) – mf#3883 – us UMI ProQuest [420]
The chap-book – Chicago. v. 1-9. May 15 1894-July 1 1898 – 1 – us NY Public [800]
The chap-book – v1-9. 1894-98 – 1 – $30.00 – us AMS Press [410]
Chapdelaine, Cecile *see* Bibliographie analytique de la delinquance juvenile
Chapeau chinois / Franc-Nohain – Paris, France. 1931, c1930 – 1r – us UF Libraries [440]
Chapeaux!! / Peyrade, Robert De La – Paris, France. 1922 – 1r – us UF Libraries [440]
Cha-pei ch'i shih san t'ien / I, Men – Hsiang-kang: Hai yen shu tien, Min kuo 29 [1940] – (= ser P-k&k period) – us CRL [951]
Chapek, Constance L *see* The effects of a ten-week step aerobic training program on aerobic capacity of college-aged females
Chapel chimes / Lake Edge Congregational Church [UCC], Madison WI – 1968 mar 13-1976 feb 1, 1976 mar-1984 dec 1 – 2r – 1 – mf#358245 – us WHS [242]
Chapel hill news – Chapel Hill, NC. 1994-2000 (1) – mf#61684 – us UMI ProQuest [071]
The chapel hymnal / ed by Benson, Louis FitzGerald – Philadelphia: Presbyterian Board of Publication and Sabbath-School Work, 1898 – 4mf – 9 – 0-524-06598-5 – mf#1991-2653 – us ATLA [780]
Chapelain, Jean *see* Lettres de jean chapelain, de l'academie francaise
Chapel-Cure *see* Heirs together of the grace of life
Le chapelet de l'amour de dieu : avec deux autres exercises portant indulgence – Quebec?: P Larose, 1888 [mf ed 1984] – 1mf – 9 – 0-665-45692-1 – mf#45692 – cn Canadiana [241]
Chapell, Frederic Leonard *see*
– Biblical and practical theology
– The great awakening of 1740
Chapell, Frederick Leonard *see* Eleventh-hour laborers
Chapelles litteraires / Lasserre, Pierre – Paris, France. 1920 – 1r – us UF Libraries [960]
Chaperon, Elisee *see* Bibliographie de monsieur l'abbe j w laverdiere
Chapin, Aaron Lucius *see* Home missions
Chapin, Anna Alice *see* Story of the rhinegold
Chapin, Edwin Hubbell *see*
– The church of the living god, and other sermons
– Discourses on the beatitudes
– Discourses on the lord's prayer
– God's requirements
– Lessons of faith and life
– Living words
Chapin, Gardner B *see* Tales of the st lawrence
Chapin, George M *see*
– Florida, 1513-1913
– Hobe sound, florida
Chapin, Howard M *see*
– List of roger williams' writings
– Trading post of roger williams with those of john wilcox and richard smith
Chapin, James Henry *see* The creation and the early developments of society
Chapin, Timothy S *see* Urban revitalization tools
Chapin, William D *see* Index to original communications in the medical journals of the united states and canada for 1877
Chaplain – Arlington. 1972-1977 (1) 1976-1977 (5) 1976-1977 (9) – ISSN: 0009-1642 – mf#8171 – us UMI ProQuest [355]
Chaplain smith and the baptists / Guild, Reuben A – or Life, journals, letters, and addresses of the Rev. Hezekiah Smith, D.D. 1737-1805 – 1 – us Southern Baptist [242]
Chaplaincy – Arlington. 1978-1981 (1,5,9) – ISSN: 0149-4236 – mf#11694 – us UMI ProQuest [355]
The chaplains and clergy of the revolution / Headley, Joel Tyler – New York: Scribner, 1864, c1861 – 1mf – 9 – 0-7905-5052-0 – mf#1988-1052 – us ATLA [975]
Chaplains' bulletin / 135 Medical Regiment [Organization] – 1st-29th [1947-1973/74] – 1r – 1 – (cont by: bull sheet [madison wi]) – mf#1054281 – us WHS [355]
Chaplains' bulletin – Camp Shelby MS. n49 [1942 feb 12] – 1r – 1 – mf#3462642 – us WHS [355]
The chaplain's narrative of the siege of delhi : from the outbreak at meerut to the capture of delhi / Rotton, John Edward Wharton – London: Smith, Elder, 1858 [mf ed 1995] – (= ser Yale coll) – vi/357p – 1 – 0-524-09925-1 – mf#1995-0925 – us ATLA [954]
Chapleau, Joseph Adolphe *see*
– Discours de l'hon j a chapleau a l'occasion de la motion censurant le ministere pour avoir permis l'execution de louis riel
– Leon 13, homme d'etat
– Speech of hon j a chapleau on the motion made before the house of commons, on the 11th march, 1886
Chapleau, Joseph-Adolphe *see*
– Le banquet donne a sir john a macdonald a quebec, le 15 octobre 1879
– Discours de l'honorable m chapleau en proposant la vente du chemin de fer quebec, nontreal, ottawa et occidental a l'assemblee legislative, seances des 27 et 28 mars 1882
– Discours de l'honorable m chapleau sur les resolutions du chemin de fer canadien du pacifique
– Du droit internationale
– Noces d'or de la saint-jean-baptiste, 1884, discours
– Report on the constitution of the dominion of canada
– The riel question
– Speech of hon mr chapleau on the canadian pacific railway resolutions, house of commons, 16th june, 1885
– Speech of hon mr chapleau...on the execution of louis riel
Chapleau sentinel – Chapleau, Ontario, CN. jan 1981-dec 1982 – 2r – 1 – cn Commonwealth Imaging [071]
Chaplin, Ada C *see* Our gold-mine
Chaplin, Dorothea *see* Matter, myth, and spirit
Chaplin, J *see* Life of henry dunster
Chaplin, Jane Dunbar *see*
– Gems of the bog
– Mother west's neighbors
– Out of the wilderness
Chaplin, Jeremiah *see* Life of henry dunster, first president of harvard college
Chapman, A T *see* The book of leviticus
Chapman, Arthur Thomas *see* An introduction to the pentateuch
Chapman, Berlin B *see*
– "How the cherokee acquired and disposed of the outlet [oklahoma]"
– Oklahoma territory and the national archives
Chapman, Charles *see*
– "Our sons and daughters"
– Pre-organic evolution and the biblical idea of god
– Revivalism and the church
– The true religion
– Why we persuade men
Chapman, Charles Edward *see* Colonial hispanic america
Chapman chatter / Alford, Gilbert K – 1983 aug 1-1988 aug – 1r – 1 – mf#1507393 – us WHS [071]
Chapman, Clowry *see* Trade-marks
Chapman, David *see* Context effects on the intrinsic dynamics of infants with spina bifida
Chapman, Edward John *see*
– Blowpipe practice
– Contributions to blowpipe-analysis
– Examples of the application of trigonometry to crystallographic calculations
– The mineral indicator
– The minerals and geology of central canada
– Note on the belmont gold veins of peterborough county, ontario
– On some deposits of titaniferous iron ore in the counties of haliburton and hastings, ontario
– On the corals and coralliform types of palaeozoic strata
– On the object of the salt condition of the sea
– On the wallbridge hematite mine
– An outline of the geology of canada
– A popular and practical exposition of the minerals and geology of canada
– Practical instructions for the determination by furnace assay of gold and silver in rocks and ores
– Preliminary report on the campbell coal area, cape breton
– Report on gatling gold and silver mines
– Report on the coal area of the medicine hat coal mining company
– Report on the copper deposit of grand manan, bay of fundy
– Report on the haycock iron location
– Report on the phosphate lands of the templeton and north ottawa mining company
– Report on the stevenson phosphate location, townships of portland and buckingham, province of quebec
– A sequel to "christabel"
– Some remarks on the classification of the trilobites
– A tabular distribution of the more commonly occurring minerals by means of which they may be easily recognized
Chapman, Edward Mortimer *see* English literature in account with religion, 1800-1900
Chapman, Frederick Spencer *see* Helvellyn to himalaya
Chapman, Gordon *see* Rex cole, junior and the grinning ghost
Chapman, Harlan P *see* Letters, ms p.p.
Chapman, [J] *see* Travels in the interior of south africa
Chapman, James *see*
– The christian character in its relation to the christian view of the world
– Jesus christ and the present age
– Proceedings of the fourth ecumenical methodist conference
Chapman, James L *see* Baptism
Chapman, John *see*
– Bishop gore and the catholic claims
– Brief outline and review of a work entitled "the principles of natu...
Chapman, John Wilbur *see*
– The life and work of dwight l. moody
– Present-day evangelism
– Records, 1880-1918
– Revivals and missions
– S H hadley of water street
Chapman, Mary Weems *see* Mother cobb, or, sixty years' walk with god
Chapman, Paul Wilbur *see* Green hand
Chapman, Robert F *see* Ventilatory influences on arterial saturation maintenance during exercise in normoxia and mild hypoxia
Chapman, William *see*
– Les aspirations
– Les fleurs de givre
– Le jour de l'an
– Les rayons du nord
Chapman's gazetteer of the province of auckland – Auckland, c1867 – 1mf – 9 – NZ$3.00 – 0-908797-12-5 – (details of places, many of which no longer exist) – mf#NZNB 1037 – nz BAB [980]
Chapmans nz advertiser – 1871 – 1r – 1 – mf#ZB 13 – nz Nat Libr [079]
Chappe d'Auteroche, abbe *see* Voyage en californie pour l'observation du passage de venus sur le disque du soleil, le 3 juin 1769
Chappe d'Auteroche, J *see* Voyage en californie
Chappell, E *see* Narrative of a voyage to hudson's bay in h m s rosamond
Chappell, Edward *see*
– Narrative of a voyage to hudson's bay in his majesty's ship rosamond
– Voyage of his majesty's ship rosamond to newfoundland and the southern coast of labrador
Chappell, Louis Watson *see* John henry
Chappell register – Chappell, NE: Yensen & Morgan. v1 n15. sep 29 1887- (wkly) [mf ed sep 29 1887-apr 4 1895 (gaps) filmed 1984] – 3r – 1 – (absorbed: big springs gazette nov 28 1895 and: big springs journal nov 16 1912. issues for jun 8 1961-jul 18 1968 called v74 n12-v80 n19. issues for jul 25 1968- called v78 n20-) – us NE Hist [071]
Chapple, Joe Mitchell *see* Heart throbs, in prose and verse dear to the american people
Chappron, E J *see* Necessaire maconique
Chappuys Tourangeau, G *see* Commentaires hieroglyphiques
Chapter 11 venue change by large public companies / Bermant, Gordon et al – 1997 – 1mf – 9 – $1.50 – mf#llmc99-045 – us LLMC [346]
A chapter from the north-west rebellion / Brooks, Geo B – Bradford. 1985-1996 (1,5,9) – 1mf – 9 – 0-665-15308-2 – mf#15308 – cn Canadiana [971]
A chapter in the history of the theological institute of connecticut or hartford theological seminary – Hartford: Wiley, Waterman & Eaton, 1879 [mf ed 1993] – 1mf – 9 – 0-524-08419-X – mf#1993-1029 – us ATLA [240]
A chapter of canadian history – S.l: s.n, 1876? – 1mf – 9 – (repr fr: mcmillan's magazine for january, 1876) – mf#02957 – cn Canadiana [971]
A chapter of mission history in modern japan : being a sketch for the period since 1869 and a report for the years since 1893... / Pettee, James Horace – [s.l: s.n, 1895?] [mf ed 1986] – 1mf – 9 – 0-8370-6302-7 – mf#1985-0302 – us ATLA [240]
A chapter on liturgies : historical sketches / Baird, Charles Washington – [authorized english ed] London: Knight, 1856 [mf ed 1992] – 1mf – 9 – 0-524-02057-4 – (expanded ed of: eutaxia, or the presbyterian liturgies. 1st ed publ in 1855) – mf#1990-4168 – us ATLA [242]
Chapters from the history of the free church of scotland / Walker, Norman Lockhart – Edinburgh: Oliphant, Anderson & Ferrier [1895?] [mf ed 1990] – (= ser The chalmers lectures [1895?]) – 1mf – 9 – 0-7905-6092-5 – (incl bibl ref) – mf#1988-2092 – us ATLA [242]
Chapters from the religious history of spain connected with the inquisition / Lea, Henry Charles – Philadelphia: Lea Bros., 1890 – 2mf – 9 – 0-7905-5771-1 – (incl bibl ref) – mf#1988-1771 – us ATLA [240]
Chapters in the early history of the church of wells *see* American expressionistic drama
Chapters in the early history of the church of wells, a d 1136-1333 : from documents in possession of the dean and chapter of wells / Church, Charles Marcus – London: E Stock 1894 [mf ed 1987] – 1r [ill] – 1 – mf#7141 – us Wisconsin U Libr [720]
Chapters of bible study : or, a popular introduction to the study of the sacred scriptures / Heuser, Herman Joseph – New York: Cathedral Library Assoc, 1895 [mf ed 1993] – 1mf – 9 – 0-524-05728-1 – mf#1992-0571 – us ATLA [220]
The chapters of coming forth by day : or, the theban recension of the book of the dead / ed by Budge, Ernest Alfred Wallis – [2nd ed] London: Kegan Paul, Trench, Truebner 1910 [mf ed 1989] – (= ser Books on egypt and chaldaea 28-30) – 3v on 2mf – 9 – 0-8370-1175-2 – (the egyptian hieroglyphic text) – mf#1987-6011 – us ATLA [390]
Chapters of early english church history / Bright, William – 3rd rev and enl ed. Oxford: Clarendon Press, 1897 – 2mf – 9 – 0-7905-5580-8 – (incl bibl ref) – mf#1988-1580 – us ATLA [240]
Chapters of the modern history of british india / Thornton, Edward – London 1840 – (= ser 19th c british colonization) – 7mf – 9 – mf#1.1.479 – uk Chadwyck [954]
Chapters on evolution / Wilson, Andrew – London, 1883 [i.e. 1882] – (= ser 19th c evolution & creation) – 5mf (ill) – 9 – mf#1.1.1536 – uk Chadwyck [575]
Chapters on jewish literature / Abrahams, Israel – Philadelphia: The Jewish Publ Society of America, 1899 – 1mf – 9 – 0-8370-2032-8 – (incl ind and bibliographies) – mf#1985-0032 – us ATLA [296]
Chapters on missions in south india / Fox, Henry Watson – London: Seeleys, 1848 [mf ed 1995] – (= ser Yale coll) – vii/213p – 1 – 0-524-09249-4 – mf#1995-0249 – us ATLA [954]
Chapters on the art of thinking : and other essays / Hinton, James; ed by Hinton, Charles Howard – London: C Kegan Paul, 1879 – 1mf – 9 – 0-7905-9965-1 – mf#1989-1690 – us ATLA [100]
Chapters on the book of mulling / Lawlor, Hugh Jackson – Edinburgh: D Douglas 1897 [mf ed 1990] – 1mf – 9 – 0-8370-1744-0 – (incl bibl ref) – mf#1987-6140 – us ATLA [225]
Chapters on the shorter catechism : a tale for the instruction of youth – 2nd amer ed fr last edinburgh ed. Philadelphia: W S Martien, [18–][mf ed 2004] – 1r – 1 – 0-524-10463-8 – mf#b00680 – us ATLA [240]
Chapters on trees : a popular account of their nature and uses / Gregg, Mary Kirby & Kirby, Elizabeth – London, Paris, New York: Cassell, Petter & Galpin 1873 [mf ed 1987] – 1r [ill] – 1 – mf#1974 – us Wisconsin U Libr [634]
Chapus, G S *see* Histoire des rois
Chapus, G-S *see* Histoire des populations de madagascar
Chaput, Donald *see* La participation de canadiens francais a la conquete de l'ouest americain
Chapuys-Montlaville, Benoit *see* Lettres sur la suisse et le pays des grisons
Chapuys-Montlaville, Benoit M de *see* Histoire du dauphine
Chaqueri, Cosroe *see* Asnad-i tarikhi-i jubnish-i kargari, susiyal-dimukrasi va kumunisti-i iran
Chara kri sakhan kuiyto mhuin / Ama, Do – Mantale: Kri Pva Re Ca Aup Tuik 1976 [mf ed 1990] – 1r with other items – 1 – (in burmese) – mf#mf-10289 seam reel 135/4 [$] – us CRL [959]
The character and call of the church of england : a charge delivered at his second visitation of the diocese of canterbury in february, 1912 / Davidson, Randall Thomas – London, New York: Macmillan, 1912 – 1mf – 9 – 0-7905-4289-7 – (incl bibl ref) – mf#1988-0289 – us ATLA [241]

Character and claims of the church of england / Marsh, W – Colchester, England. 1829 – 1r – us UF Libraries [241]

Character and composition of the turkish press, (1939-1944) – [Washington, DC]: Office of Strategic Services, Research and Analysis Branch, 1945 – us CRL [070]

Character and condition – London, England. 18– – 1r – us UF Libraries [240]

Character and death of mrs hester ann rogers / Coke, Thomas – Birmingham, England. 1796 – 1r – us UF Libraries [240]

Character and happiness of them that die in the lord / Dealtry, William – London, England. 1822 – 1r – us UF Libraries [240]

Character and office of gospel-ministers / Hill, George – Falkirk, Scotland. 1792 – 1r – us UF Libraries [240]

Character and religion / Lyttelton, Edward – London: Robert Scott; New York: S R Leland 1912 [mf ed 1991] – (= ser Library of historic theology) – 1mf – 9 – 0-7905-9321-1 – mf#1989-2546 – us ATLA [230]

Character and reward of a faithful servant of christ / Evans, Benjamin – Scarbro', England. 1831 – 11r – us UF Libraries [240]

Character and translation of enoch / James, John Angell – Shrewsbury, England. 1852 – 1r – us UF Libraries [240]

Character building : being addresses delivered on sunday evenings to the students of tuskegee institute / Washington, Booker T – Toronto: W Briggs, 1902 – 4mf – 9 – 0-665-73681-9 – mf#73681 – cn Canadiana [370]

Character building in kashmir / Tyndale-Biscoe, Cecil Earle – London: Church missionary Society, 1920 [mf ed 1995] – (= ser Yale coll) – 95p (ill) – 1 – 0-524-09465-9 – (with foreword by robert baden-powell) – mf#1995-0465 – us ATLA [954]

The character christ, fact or fiction / Lhamon, William Jefferson – New York: Fleming H Revell, c1914 – 1mf – 9 – 0-524-07019-9 – mf#1991-2872 – us ATLA [220]

The character, claims and practical workings of freemasonry / Finney, Charles Grandison – Cincinnati: Western Tract and Book Society, c1869 – 1mf – 9 – 0-7905-4732-5 – mf#1988-0732 – us ATLA [940]

Character forming in school / Ellis, F H – London; New York: Longmans, Green, 1907 – 1mf – 9 – 0-8370-7940-3 – mf#1986-1940 – us ATLA [370]

Character is everything – London, England. 18– – 1r – us UF Libraries [240]

Character of dr littledale as a controversialist / King, Owen C H – London, England. 18– – 1r – us UF Libraries [240]

Character of god / Todd, John – Northampton: Bridgman and Childs, 1867, c1856 – 1mf – 9 – 0-8370-5660-8 – mf#1985-3660 – us ATLA [210]

The character of jesus : forbidding his possible classification with men / Bushnell, Horace – New York:Charles Scribner, 1905, c1886 – 1mf – 9 – 0-8370-3063-3 – (incl bibl ref) – mf#1985-1063 – us ATLA [240]

The character of jesus / Jefferson, Charles Edward – New York: Thomas Y Crowell, c1908 – 1mf – 9 – 0-8370-3775-1 – mf#1985-1775 – us ATLA [240]

The character of jesus *see* Sheng tao kuan k'uei (ccm175)

Character of the apostle paul in some of its features delineated / Kemp, John – Edinburgh, Scotland. 1802 – 1r – us UF Libraries [240]

Character of the king / Dennis, Jonas – Exeter, England. 1800 – 1r – us UF Libraries [240]

The character of villein tenure / Ashley, William James – S.l: s.n, 18–? – 1mf – 9 – mf#44243 – cn Canadiana [333]

Character potential – Schenectady. 1962-1981 (1) 1976-1981 (5) 1976-1981 (9) – ISSN: 0009-1669 – mf#7517 – us UMI ProQuest [150]

Character, principles, and public services of the late william wilberforce / Scott, John – London, England. 1833 – 1r – us UF Libraries [240]

Character, services and reward of the faithful pastor / King, John – London, England. 1834 – 1r – us UF Libraries [240]

Character studies / Oates, James F – New York: International Comm of YMCA, 1903 – 1mf – 9 – 0-8370-4609-2 – mf#1985-2609 – us ATLA [220]

Character studies : some of the lord's "mighty men" / Vassar, Thomas Edwin – Kansas City, MO: Pearl Ptg Co, 1894 – 1mf – 9 – 0-7905-6638-9 – mf#1988-2638 – us ATLA [240]

Character studies in genesis / Blodgett, May Nellie – Chicago: American Committee Young Women's Christian Association, 1900 – 1mf – 9 – 0-8370-2380-7 – mf#1985-0380 – us ATLA [221]

Character studies in the old testament, book studies in the new testament : syllabus of a course of twelve bible lecture-studies / Young, Charles A – Indianapolis: C W B M [18–?] [mf ed 1989] – 1mf – 9 – 0-7905-1079-0 – mf#1987-1079 – us ATLA [220]

Character the christian minister should sustain and the influence h... / Williams, J C – High Wycombe, England. 1825 – 1r – us UF Libraries [240]

Character through inspiration : and other papers / Munger, Theodore Thornton – 1st ed. New York: Thos Whittaker 1897 [mf ed 1991] – (= ser Small books on great subjects 7) – 1mf – 9 – 0-7905-9419-6 – mf#1989-2644 – us ATLA [240]

Characterisation in selected sesotho novels / Zulu, Nogwaja Shadrack – Stellenbosch: U of Stellenbosch 1998 [mf ed 1998] – 6mf – 9 – mf#mf.1310 – sa Stellenbosch [470]

The characteristic costume of france – London 1819 – (= ser 19th c art & architecture) – 2mf – 9 – mf#4.2.1189 – uk Chadwyck [740]

The characteristic differences of the four gospels : considered as revealing various relations of the lord jesus christ / Jukes, Andrew John – 4th ed. London: James Nisbet, 1867 – 1mf – 9 – 0-8370-3816-2 – mf#1985-1816 – us ATLA [226]

The characteristic differences of the new testament : from the immediately preceding jewish, and the immediately succeeding christian literature, considered as evidence of the divine origin of the new testament / Sinker, Robert – Cambridge: Deighton, Bell; London: Bell and Daldy, 1865 – 1mf – 9 – 0-8370-9984-6 – mf#1986-3984 – us ATLA [225]

Characteristics and applications of resistance strain gages / United States National Bureau Of Standards – Washington, DC. 1954 – 1r – us UF Libraries [500]

The characteristics and laws of figurative language : designed for use in bible classes, schools, and colleges / Lord, David Nevins – 4th ed. New York: Franklin Knight, 1857, c1854 – 1mf – 9 – 0-7905-9022-0 – mf#1989-2247 – us ATLA [400]

Characteristics common to lawyers / O'Connor, Johnson – Hoboken, N.J.: Human Engineering Laboratories, Steven Institute of Technology, 1934. LL-2254 – 1 – us L of C Photodup [346]

Characteristics from the writings of archbishop ullathorne / Ullathorne, William Bernard – London: Burns & Oates; New York: Catholic Publ Society Co, 1889 [mf ed 1986] – 1mf – 9 – 0-8370-7514-9 – (incl ind) – mf#1986-1514 – us ATLA [241]

Characteristics of balance and posture control in development and aging / Sundermier, Lynne M – 1999 – 2mf – 9 – $8.00 – mf#PSY 2069 – us Kinesology [790]

Characteristics of christian morality : considered in eight lectures / Smith, Isaac Gregory – Oxford: James Parker, 1873 – (= ser Bampton lectures) – 1mf – 9 – 0-524-00392-0 – mf#1989-3092 – us ATLA [170]

Characteristics of christianity / Leathes, Stanley – London: James Nisbet, 1884 [mf ed 1984] – 4mf – 9 – 0-8370-0864-6 – (incl bibl ref) – mf#1984-4204 – us ATLA [240]

Characteristics of current and past participants in the university of wisconsin-la crosse cardiac rehabilitation program with a historical review of cardiac rehabilitation / Kawamura, Takayuki – 1999 – 1mf – 9 – $4.00 – mf#HE 642 – us Kinesology [616]

Characteristics of men, manners, opinions, times, etc / Shaftesbury, Anthony Ashley Cooper; ed by Robertson, John Mackinnon – London: G Richards 1900 [mf ed 1987] – 2v on 1r – 1 – (int & notes by ed) – mf#6713 – us Wisconsin U Libr [170]

Characteristics of romanism and of protestantism as developed in th... / Mcneile, Hugh – London, England. 1848? – 1r – us UF Libraries [240]

Characteristics of the gospel miracles : sermons preached before the university of cambridge / Westcott, Brooke Foss – Cambridge, Macmillan, 1859 – 1r – 1 – 0-8370-0524-8 – (with notes) – mf#1984-B297 – us ATLA [240]

Characteristics of the greek philosophers : socrates and plato / Potter, John Philips – London: JW Parker, 1845 – 1mf – 9 – 0-7905-7361-X – mf#1989-0586 – us ATLA [180]

Characteristics of thought processes and knowledge structures of novice tennis players / Oguchi-Chen, Fumiko & Sinclair, Gary D – 1990 – 2mf – 9 – $8.00 – us Kinesology [150]

Characteristics of true devotion / Grou, Jean Nicola – New York, NY. 1884 – 1r – us UF Libraries [240]

Characteristics of women, moral, poetical, and historical / Jameson, Anna – Boston, MA. 1857 – 1r – us UF Libraries [025]

Characteristics, political, philosophical, and religious / Manning, Henry Edward – London: Burns and Oates, [1885?] – 1mf – 9 – 0-524-04383-3 – mf#1991-2087 – us ATLA [240]

Characterization of glenohumeral joint laxity and stiffness using instrumented arthometry / Sauers, Eric l – 2000 – 128 on 2mf – 9 – $10.00 – mf#PE 4147 – us Kinesology [617]

Characters and characteristics of william law : nonjuror and mystic = Selections. 1893 / Law, William – London: Hodder and Stoughton, 1893 – 1mf – 9 – 0-7905-7429-2 – (incl bibl ref) – mf#1989-0654 – us ATLA [240]

Characters of modern times *see* Chin tai jen wu [ccm283]

The characters of the old testament : in a series of sermons / Williams, Isaac – London: Rivingtons, 1887 [mf ed 1985] – 1mf – 9 – 0-8370-5856-2 – mf#1985-3856 – us ATLA [221]

O charadista – Rio de Janeiro, RJ: Typ Parisiense, 10 nov-17 nov 1850 – (= ser Ps 19) – mf#P17,01,90 – bl Biblioteca [079]

Charakter und charakterisierung in den novellen von paul ernst im lichte der psychologie von ludwig klages / Semmler, Fritz – New York NY: [s.n.] 1942 [mf ed 1989] – 1r – 1 – (filmed with: manfred und beatrice / paul ernst) – mf#7222 – us Wisconsin U Libr [430]

Charakter und tendenz des johannesevangeliums / Wrede, William – Tuebingen: J C B Mohr (Paul Siebeck) 1903 [mf ed 1989] – (= ser Sammlung gemeinverstaendlicher vortraege und schriften aus dem gebiet der theologie und religionsgeschichte 37) – 1mf – 9 – 0-7905-0469-3 – mf#1987-0469 – us ATLA [225]

Das charakterbild jesu *see* Sketch of the character of jesus

Charakterbilder katholischer reformatoren des 16. jahrhunderts : ignatius von loyola, teresa de jesus, filippo neri, carlo borromeo / Pastor, Ludwig, Freiherr von – Freiburg, 1924 (mf ed 1994) – 1mf – 9 – €24.00 – 3-89349-733-1 – mf#DHS-AR 733 – gw Frankfurter [241]

Charakterisierung der viraemischen phasen und fruehe diagnostik der varizelle-zoster virus-infektion mit molekularbiologischen und immunchemischen methoden / Mainka, Claudia – (mf ed 1997) – 2mf – 9 – €40.00 – 3-8267-2470-4 – mf#DHS 2470 – gw Frankfurter [574]

Charakterisierung von endothelzell-tumorsphaeroid interaktionen bei der invasion und metastasierung / Maercker, Eva Gloria – (mf ed 2000) – 2mf – 9 – €40.00 – 3-8267-2730-4 – mf#DHS 2730 – gw Frankfurter [574]

Charakterisierung von unmodifizierten und oligomermodifizierten anorganischen metalloxidoberflaechen durch elektrokinetische messmethoden / Simon, Frank – (mf ed 1994) – 2mf – 9 – €40.00 – 3-89349-884-2 – mf#DHS 884 – gw Frankfurter [540]

Charakteristiken und kritiken von joseph goerres : aus den jahren 1804 und 1805 / ed by Schultz, Franz – Koeln: J P Bachem, 1900 [mf ed 1993] – 88p – (incl bibl ref) – mf#8219 – us Wisconsin U Libr [430]

Charakteristiken und kritiken von joseph goerres / ed by Schultz, Franz – Koeln, 1902 [mf ed 1993] – 1mf – 9 – €24.00 – 3-89349-265-8 – mf#DHS-AR 122 – gw Frankfurter [430]

Charakteristiken und kritiken von joseph goerres aus den jahren 1804 und 1805 / ed by Schultz, Franz – Koeln, 1900 [mf ed 1993] – 1mf – 9 – €24.00 – 3-89349-264-X – mf#DHS-AR 121 – gw Frankfurter [430]

Charasee press – n7-8,10-11 [1972 jan-feb, apr-may] – 1r – 1 – mf#1583020 – us WHS [071]

Charbonneau, Jeannine *see* Bibliographie de la peinture au canada

Charbonnel, Armand Francois Marie de *see*
- Copie de la correspondance echangee entre l'eveque catholique romain de toronto et le surintendant en chef des ecoles
- Copies of correspondence between the roman catholic bishop of toronto and the chief superintendent of schools
- Reponse a une adresse de l'assemblee legislative a son excellence le gouverneur-general, datee le 8 du courant...

Charbonniers / Gille, Philippe – Paris, France. 1886 – 1r – us UF Libraries [440]

Charca / Zeno Grandia, Manuel – Mexico City?, Mexico. 1958 – 1r – us UF Libraries [972]

Chard and illminster news etc – Chard, England. Jul 1875-1962; 1983- – 98+ r – 1 – uk British Libr Newspaper [072]

Chardenal, C A *see* New chardenal

Chardin, J *see* Voyages du chevalier chardin en perse et autres lieux de l'orient

Chardon democrat – Chardon, OH: J F Asper, oct 22 1850-mar 23 1852 – 1r – 1 – (aka: chardon free democrat. other titles: free democrat democrat (chardon, oh: 1850), free democrat (chardon, oh: 1849, 1852)) – mf#34 G2.1 007 – us Western Res [071]

Chardon democrat *see*
– Free democrat

Chardon, Joseph *see* Tableau historique et politique de marseille ancienne et moderne

Chardon. Ohio. Regular Baptist Church *see* Church records, ms 542

Chardon spectator and geauga gazette – Chardon, OH: A Phelps, jul 27 1833-nov 27 1835 – 1r – 1 – (wkly national republican [1833-34], later whig [1834-35] newspaper) – mf#34 G2.1 001 – us Western Res [071]

Chardzhouskaia pravda – Chardzhou, 1986-jan 1988 – 4r – 1 – us UMI ProQuest [077]

Charest, Pauline *see* Bio-bibliographie analytique de monsieur gerald godin

Charfreitag *see* Fruehlingssturm / charfreitag / der gang nach emmaus / pfingsten in weimar

A charge : delivered to the clergy of the diocese of glasgow and galloway at the visitation, sep 8 1852 / Trower, Walter John – Edinburgh: R Grant, 1852 [mf ed 1994] – 1mf – 9 – 0-524-08680-X – mf#1993-3205 – us ATLA [242]

Charge / Church Of England Archdeaconry Of Dorset – London, England. 1875 – 1r – us UF Libraries [240]

La charge – Paris. janv-sept 1870, juil 1888-89 – 1 – fr ACRPP [073]

Charge addressed to the churchwardens of the diocese of chester / Raikes, Henry – London, England. 1844 – 1r – us UF Libraries [241]

Charge addressed to the clergy of the diocese of argyll and the isl... / Ewing, Alexander – London, England. 1865 – 1r – us UF Libraries [241]

Charge addressed to the clergy of the diocese of ripon / Church Of England Diocese Of Ripon – London, England. 1850 – 1r – us UF Libraries [241]

Charge delivered / Church Of England Diocese Of Bristol – London, England. 1822 – 1r – us UF Libraries [241]

Charge delivered / Williams, Thomas – Cardiff, Wales. 1852? – 1r – us UF Libraries [241]

Charge delivered at his primary visitation / Magee, William – London, England. 1822 – 1r – us UF Libraries [241]

Charge delivered at the ordinary visitation / Wilberforce, Robert Isaac – York, England. 1843 – 1r – us UF Libraries [241]

Charge delivered at the ordinary visitation of the archdeaconry of... / Manning, Henry Edward – London, England. 1841 – 1r – us UF Libraries [241]

Charge delivered at the ordinary visitation of the archdeaconry of... / Manning, Henry Edward – London, England. 1842 – 1r – us UF Libraries [241]

Charge delivered at the ordinary visitation of the archdeaconry of... / Manning, Henry Edward – London, England. 1845 – 1r – us UF Libraries [241]

Charge delivered at the ordinary visitation of the archdeaconry of... / Manning, Henry Edward – London, England. 1848 – 1r – us UF Libraries [241]

Charge delivered at the ordinary visitation of the archdeaconry of... / Manning, Henry Edward – London, England. 1849 – 1r – us UF Libraries [241]

Charge delivered at the ordinary visitation of the archdeaconry of... / Williams, Thomas – Cardiff, Wales. 1846 – 1r – us UF Libraries [241]

Charge delivered at the ordination of the rev richard hunter / Thomson, Henry – Penrith, England. 1819? – 1r – us UF Libraries [241]

Charge delivered at the triennial visitation of john, lord bishop o... / Church Of England Diocese Of Lincoln – London, England. 1834 – 1r – us UF Libraries [241]

Charge delivered at the triennial visitation of the diocese, novemb... / Wilberforce, Samuel – Oxford, England. 1857 – 1r – us UF Libraries [241]

Charge delivered at the visitation of the archdeaconry of bristol i... / Thorp, Thomas – Bristol, England. 1842 – 1r – us UF Libraries [241]

Charge delivered in the autumn of 1834, at the visitation in hampsh... / Dealtry, William – London, England. 1835 – 1r – us UF Libraries [241]

Charge delivered on wednesday the 14th of june 1826 to the clergy o... / Low, David – Edinburgh, Scotland. 1826 – 1r – us UF Libraries [241]

A charge delivered to the clergy : at the visitation held in the cathedral church of st luke, at halifax, on the 1st day of july, 1884 / Binney, Hibbert, Lord Bishop of Nova Scotia – Halifax, NS?: s.n, 1884 – 1mf – 9 – mf#13021 – cn Canadiana [242]

Charge delivered to the clergy and churchwardens of the archdeaconry... / Crawley, William – Monmouth, England. 1852? – 1r – us UF Libraries [241]

Charge delivered to the clergy and churchwardens of the diocese... / Church Of England Diocese Of Bath And Wells – London, England. 1870? – 1r – us UF Libraries [241]

Charge delivered to the clergy and churchwardens of the diocese of... / Church Of England Diocese Of Bath And Wells – London, England. 1876 – 1r – us UF Libraries [241]

Charge delivered to the clergy and churchwardens of the diocese of... / Church Of England Diocese Of Bath And Wells – London, England. 1879 – 1r – us UF Libraries [241]

Charge delivered to the clergy and churchwardens of the diocese of... / Church Of England Diocese Of Bath And Wells – London, England. 1882 – 1r – us UF Libraries [241]

Charge delivered to the clergy and churchwardens of the diocese of ba... / Church Of England Diocese Of Bath And Wells – London, England. 1885 – 1r – us UF Libraries [241]

Charge delivered to the clergy and churchwardens of the diocese of ba... / Church Of England Diocese Of Bath And Wells – London, England. 1888 – 1r – us UF Libraries [241]

Charge delivered to the clergy and churchwardens of the diocese of ba... / Church Of England Diocese Of Bath And Wells – London, England. 1891 – 1r – us UF Libraries [241]

Charge delivered to the clergy and churchwardens of the diocese of ba... / Church Of England Diocese Of Bath And Wells – London, England. 1894 – 1r – us UF Libraries [241]

A charge delivered to the clergy at the visitation held in the cathedral church of st luke : at halifax, on the 3rd day of july, 1866 / Binney, Hibbert, Lord Bishop of Nova Scotia – Halifax, NS?: s.n, 1866 – 1mf – 9 – mf#55812 – cn Canadiana [242]

A charge delivered to the clergy at the visitation held in the cathedral church of st luke : on the 30th day of june 1874 / Binney, Hibbert, Lord Bishop of Nova Scotia – Halifax, NS?: s.n, 1874 – 1mf – 9 – mf#06144 – cn Canadiana [242]

A charge delivered to the clergy at the visitation held in the cathedral church of st luke : on the 6th day of july 1880 / Binney, Hibbert, Lord Bishop of Nova Scotia – Halifax, NS?: s.n, 1880 – 1mf – 9 – mf#06143 – cn Canadiana [242]

Charge delivered to the clergy of the archdeaconry of cleveland / Todd, Henry John – London, England. 1835 – 1r – us UF Libraries [241]

Charge delivered to the clergy of the archdeaconry of colchester / Oakeley, Herbert – London, England. 1843 – 1r – us UF Libraries [241]

Charge delivered to the clergy of the archdeaconry of derby / Shirley, Walter Augustus – London, England. 1846? – 1r – us UF Libraries [241]

Charge delivered to the clergy of the archdeaconry of essex in 1815 / Wollaston, Francis John Hyde – London, England. 1816 – 1r – us UF Libraries [241]

Charge delivered to the clergy of the archdeaconry of llandaff / Williams, Thomas – London, England. 1844 – 1r – us UF Libraries [241]

Charge delivered to the clergy of the archdeaconry of monmouth / Crawley, William – Monmouth, England. 1853? – 1r – us UF Libraries [241]

Charge delivered to the clergy of the archdeaconry of st alban's / Hale, William Hale – London, England. 1840 – 1r – us UF Libraries [241]

Charge delivered to the clergy of the archdeaconry of the east-ridi... / Wilberforce, Robert Isaac – London, England. 1844 – 1r – us UF Libraries [241]

Charge delivered to the clergy of the archdeaconry of wells at the... / Law, Henry – London, England. 1852 – 1r – us UF Libraries [241]

Charge delivered to the clergy of the archdeaconry of wilts, june... / Macdonald, William – Devizes, England. 1841? – 1r – us UF Libraries [241]

Charge delivered to the clergy of the diocese of argyll and the isl... / Chinnery-Haldane, J R Alexander – Edinburgh, Scotland. 1889? – 1r – us UF Libraries [241]

Charge delivered to the clergy of the diocese of argyll and the isl... / Chinnery-Haldane, J R Alexander – Edinburgh, Scotland. 1890? – 1r – us UF Libraries [241]

Charge delivered to the clergy of the diocese of argyll and the isl... / Chinnery-Haldane, J R Alexander – Edinburgh, Scotland. 1891? – 1r – us UF Libraries [241]

Charge delivered to the clergy of the diocese of argyll and the isl... / Chinnery-Haldane, J R Alexander – Edinburgh, Scotland. 1893 – 1r – us UF Libraries [241]

Charge delivered to the clergy of the diocese of bangor / Bethell, Christopher – London, England. 1856 – 1r – us UF Libraries [241]

Charge delivered to the clergy of the diocese of brechin in synod a... / Forbes, A P – Dundee, Scotland. 1862 – 1r – us UF Libraries [241]

Charge delivered to the clergy of the diocese of chester / Church Of England Diocese Of Chester – London, England. 1829 – 1r – us UF Libraries [241]

Charge delivered to the clergy of the diocese of chester / Church Of England Diocese Of Chester – London, England. 1832? – 1r – us UF Libraries [241]

Charge delivered to the clergy of the diocese of chester at the tri... / Church Of England Diocese Of Chester – London, England. 1835 – 1r – us UF Libraries [241]

Charge delivered to the clergy of the diocese of chester at the tri... / Church Of England Diocese Of Chester – London, England. 1838 – 1r – us UF Libraries [241]

Charge delivered to the clergy of the diocese of dublin and glandal / Church Of Ireland Diocese Of Dublin – Dublin, Ireland. 1849 – 1r – us UF Libraries [241]

Charge delivered to the clergy of the diocese of durham / Church Of England Diocese Of Durham – Dublin, Ireland. 1811 – 1r – us UF Libraries [241]

Charge delivered to the clergy of the diocese of durham / Church Of England Diocese Of Durham – London, England. 1802 – 1r – us UF Libraries [241]

Charge delivered to the clergy of the diocese of durham / Church Of England Diocese Of Durham – London, England. 1807 – 1r – us UF Libraries [241]

Charge delivered to the clergy of the diocese of durham / Van Mildert, William – Oxford, England. 1828 – 1r – us UF Libraries [241]

Charge delivered to the clergy of the diocese of exeter / Church Of England Diocese Of Exeter – London, England. 1833 – 1r – us UF Libraries [241]

Charge delivered to the clergy of the diocese of exeter / Church Of England Diocese Of Exeter – London, England. 1836 – 1r – us UF Libraries [241]

Charge delivered to the clergy of the diocese of exeter / Church Of England Diocese Of Exeter – London, England. 1839 – 1r – us UF Libraries [241]

Charge delivered to the clergy of the diocese of exeter at the trie... / Church Of England Diocese Of Exeter – London, England. 1842 – 1r – us UF Libraries [241]

Charge delivered to the clergy of the diocese of exeter at the trie... / Church Of England Diocese Of Exeter – London, England. 1848 – 1r – us UF Libraries [241]

Charge delivered to the clergy of the diocese of gloucester and bri / Church Of England Diocese Of Glousecter And Bristol – London, England. 1854 – 1r – us UF Libraries [241]

Charge delivered to the clergy of the diocese of gloucester and bri... / Church Of England Diocese Of Glousecter And Bristol – London, England. 1838 – 1r – us UF Libraries [241]

Charge delivered to the clergy of the diocese of hereford, june, 18... / Musgrave, Thomas – Hereford, England. 1842 – 1r – us UF Libraries [241]

Charge delivered to the clergy of the diocese of hereford, june, 18... / Musgrave, Thomas – London, England. 1845 – 1r – us UF Libraries [241]

Charge delivered to the clergy of the diocese of landaff in june 17... / Church Of England Diocese Of Llandaff – London, England. 1798 – 1r – us UF Libraries [241]

Charge delivered to the clergy of the diocese of landaff, june 1791 / Church Of England Diocese Of Llandaff – London, England. 1792 – 1r – us UF Libraries [241]

Charge delivered to the clergy of the diocese of lichfield and cove... / Ryder, Henry – Stafford, England. 1824 – 1r – us UF Libraries [241]

Charge delivered to the clergy of the diocese of lincoln / Church Of England Diocese Of Lincoln – Dublin, Ireland. 1812 – 1r – us UF Libraries [241]

Charge delivered to the clergy of the diocese of lincoln / Church Of England Diocese Of Lincoln – London, England. 1800 – 1r – us UF Libraries [241]

Charge delivered to the clergy of the diocese of llandaff / Ollivant, Alfred – London, England. 1872 – 1r – us UF Libraries [241]

Charge delivered to the clergy of the diocese of london / Church Of England Diocese Of London – London, England. 1834 – 1r – us UF Libraries [241]

Charge delivered to the clergy of the diocese of london / Church Of England Diocese Of London – London, England. 1842 – 1r – us UF Libraries [241]

Charge delivered to the clergy of the diocese of london / Church Of England Diocese Of London – London, England. 1850 – 1r – us UF Libraries [241]

Charge delivered to the clergy of the diocese of london in the year... / Church Of England Diocese Of London – London, England. 1804 – 1r – us UF Libraries [241]

Charge delivered to the clergy of the diocese of norwich / Bathurst, Henry – Norwich, England. 1806? – 1r – us UF Libraries [241]

Charge delivered to the clergy of the diocese of oxford / Randolph, John – Oxford, England. 1805 – 1r – us UF Libraries [241]

Charge delivered to the clergy of the diocese of oxford at his prim... / Wilberforce, Samuel – London, England. 1848 – 1r – us UF Libraries [241]

A charge delivered to the clergy of the diocese of quebec : by george j mountain...lord bishop of montreal...at his primary visitation, completed in 1838 / United Church of England and Ireland. Diocese of Quebec. Bishop (1837-1863: Mountain) – [Quebec?: s.n.] 1839 [mf ed 1983] – 1mf – 9 – 0-665-44103-7 – mf#44103 – cn Canadiana [242]

Charge delivered to the clergy of the diocese of raphoe at the prim... / Magee, William – Dublin, Ireland. 1822 – 1r – us UF Libraries [241]

Charge delivered to the clergy of the diocese of salisbury / Burgess, Thomas – Salisbury, England. 1832 – 1r – us UF Libraries [241]

Charge delivered to the clergy of the diocese of salisbury / Church Of England Diocese Of Salisbury – London, England. 1842 – 1r – us UF Libraries [241]

Charge delivered to the clergy of the diocese of salisbury / Church Of England Diocese Of Salisbury – Salisbury, England. 1839? – 1r – us UF Libraries [241]

A charge delivered to the clergy of the diocese of toronto at the visitation on wednesday, oct 12 1853 / United Church of England and Ireland. Diocese of Toronto. Bishop (1839-1867) – Toronto: H Rowsell, 1853 [mf ed 1983] – 1mf – 9 – 0-665-44106-1 – mf#44106 – cn Canadiana [242]

A charge delivered to the clergy of the diocese of winchester / Sumner, Charles Richard – London: Thomas Hatchard, 1850 [mf ed 1993] – (= ser Anglican/episcopal coll) – 2mf – 9 – 0-524-06085-1 – mf#1991-2398 – us ATLA [242]

Charge delivered to the clergy of the diocese of winchester / Church Of England Diocese Of Winchester – London, England. 1834 – 1r – us UF Libraries [241]

Charge delivered to the clergy of the diocese of winchester / Church Of England Diocese Of Winchester – London, England. 1837 – 1r – us UF Libraries [241]

Charge delivered to the clergy of the diocese of winchester / Church Of England Diocese Of Winchester – London, England. 1841 – 1r – us UF Libraries [241]

Charge delivered to the clergy of the dioceses of dublin and glande / Church Of Ireland – London, England. 1835 – 1r – us UF Libraries [241]

Charge delivered to the clergy of the dioceses of dublin, glandelag / Church Of Ireland United Diocese Of Dublin, Glendalough, And Kildare – Dublin, Ireland. 1871 – 1r – us UF Libraries [241]

Charge delivered to the clergy of the dioceses of dublin, glendelag / Church Of Ireland United Diocese Of Dublin, Glendalough, And Kildare – Dublin, Ireland. 1873 – 1r – us UF Libraries [241]

Charge delivered to the clergy of the dioceses of dublin, glendelag / Church Of Ireland United Diocese Of Dublin, Glendalough, And Kildare – Dublin, Ireland. 1875 – 1r – us UF Libraries [241]

Charge delivered to the clergy of the episcopal communion of edinbu / Walker, James – Edinburgh, Scotland. 1833 – 1r – us UF Libraries [241]

Charge delivered to the clergy of the united dioceses of ossory, fe... / O'Brien, James Thomas – London, England. 1843 – 1r – us UF Libraries [241]

Charge delivered to the clergy of the united dioceses of ossory, fe... / O'Brien, James Thomas – London, England. 1846 – 1r – us UF Libraries [241]

A charge delivered to the grand jury for the county of essex...held at ipswich. may term, 1832 / Massachusetts. Supreme Judicial Court – Boston: Steam Power Press Office, 1832. 16p. LL-976 – 1 – us L of C Photodup [340]

Charge, intended for delivery to the clergy of the diocese of cante / Longley, Charles Thomas – London, England. 1868 – 1r – us UF Libraries [240]

Charge intended to have been delivered to the clergy of norwich / Church Of England Diocese Of Norwich – Norwich, England. 1791 – 1r – us UF Libraries [241]

Charge of the bishop of london to the clergy of his diocese / Church Of England Diocese Of London – London, England. 1850 – 1r – us UF Libraries [241]

Charge of the keys / Wallis, Robert Earnes – London, England. 1873 – 1r – us UF Libraries [240]

Charge to the clergy and catechists of sierra leone / Vidal, Owen Emeric – London, England. 1854 – 1r – us UF Libraries [240]

A charge to the clergy and churchwardens of the diocese of salisbury : at his triennial visitation, in may 1867 / Hamilton, Walter Kerr – Salisbury: Brown, 1867 [mf ed 1992] – (= ser Anglican/episcopal coll) – 2mf – 9 – 0-524-05535-1 – mf#1990-5139 – us ATLA [242]

Charge to the clergy and churchwardens of the diocese of salisbury / Church Of England Diocese Of Salisbury – Salisbury, England. 1867 – 1r – us UF Libraries [241]

Charge to the clergy of the archdeaconry of durham / Thorp, Charles – Durham, England. 1838 – 1r – us UF Libraries [241]

Charge to the clergy of the archdeaconry of lewes / Church Of England Archdeaconry Of Lewes – London, England. 1855 – 1r – us UF Libraries [241]

Charge to the clergy of the archdeaconry of the east riding at the... / Wilberforce, Robert Isaac – London, England. 1845? – 1r – us UF Libraries [241]

Charge to the clergy of the archdeaconry of the east-riding at the... / Wilberforce, Robert Isaac – York, England. 1842 – 1r – us UF Libraries [241]

Charge to the clergy of the diocese of lincoln / Church Of England Diocese Of Lincoln – London, England. 1832 – 1r – us UF Libraries [241]

Charge to the clergy of the diocese of london / Church Of England Diocese Of London – London, England. 1866 – 1r – us UF Libraries [241]

Charge to the clergy of the diocese of st david's / Church Of England Diocese Of Saint David's – London, England. 1842 – 1r – us UF Libraries [241]

Charge to the clergy of the diocese of st david's / Church Of England Diocese Of Saint David's – London, England. 1872 – 1r – us UF Libraries [241]

Charge to the clergy of the dioceses of dublin and kildare / Church Of Ireland – Dublin, Ireland. 1847 – 1r – us UF Libraries [241]

Charge to the clergy of the east riding delivered at the ordinary v... / Wilberforce, Robert Isaac – London, England. 1846? – 1r – us UF Libraries [241]

Charge to the diocese of oxford / Wilberforce, Samuel – London, England. 1855 – 1r – us UF Libraries [241]

A charge to the grand jury in the district court of the united states for the district of new jersey, april 21, 1863 / U.S. District Court. New Jersey – Trenton, State Gazette and Republican Print, 1863. 24 p. LL-422 – 1 – us L of C Photodup [347]

Charges brought against the rev james morison – Kilmarnock, Scotland. 1841 – 1r – us UF Libraries [240]

Charges delivered at the ordination of the rev james m'gill, july / Symington, William – Dumfries, Scotland. 1829 – 1r – us UF Libraries [240]

Charges to the clergy of the archdeaconry of lewes : delivered at the ordinary visitations from the year 1840 to 1854... – Cambridge: Macmillan, 1856 – 3mf – 9 – 0-524-05073-2 – mf#1991-2197 – us ATLA [240]

Charikles / Becker, W A – Leipzig, Germany. v1-3. 1854 – 1r – us UF Libraries [025]

Charikles *see* Reise-briefe eines diplomaten

Charis : ein beitrag zur geschichte des aeltesten christentums / Wetter, Gilles Petersson – Leipzig: J C Hinrichs, 1913 – 1mf – 9 – 0-7905-0457-X – (incl bibl ref and indexes) – mf#1987-0457 – us ATLA [220]

Charis : leipziger mode-magazin – Leipzig DE, feb-dec 1803, 1805 – 1r – 1 – gw Misc Inst [074]

Charisma / Calvary Assembly – 1985 apr-dec, 1986, 1987 jan-may – 3r – 1 – (cont by: christian life, charisma and christian life) – mf#1238597 – us WHS [243]

Charisma : a publication of black career women, inc / Black Career Women, Inc – 1981 fall/winter, 1985 spr – 1r – 1 – mf#4841713 – us WHS [305]

Charisma – Winter Park. 1975-1987 (1) 1975-1987 (5) 1975-1987 (9) – ISSN: 0279-0424 – mf#13344 – us UMI ProQuest [240]

Charisma and christian life – 1987 jul-sep, 1987 oct-1988 jun, 1988 jul-1989 jun – 3r – 1 – (cont: charisma, christian life) – mf#1239269 – us WHS [071]

Charisma and christian life – Wheaton. 1987+ – 1,5,9 – ISSN: 0895-156X – mf#16181 – us UMI ProQuest [240]

Charisma maximum : untersuchung zu cassians vollkommenheitslehre / Kemmer, A – Loewen, 1938 – 3mf – 8 – €7.00 – ne Slangenburg [241]

La charite – [Montreal?]: L'Association, 1898[mf ed n1 15 nov 1898-n13 29 nov 1898] – 9 – mf#P04211 – cn Canadiana [360]

La charite : respectueusement dediee aux dames patronesses du bazar – [Joliette, Quebec?: s.n, 1891?] – 1mf – 9 – 0-665-91683-3 – mf#91683 – cn Canadiana [360]

La charite : souvenir du bazar a joliette, octobre 1891 – [s.l.]: [s.n.], [1891] (mf ed 1980) – 1mf – 9 – mf#SEM105P44 – cn Bibl Nat [360]

La charite et son opportunite actuelle / Lacroix, Henry – Montreal?: s.n., 1863? – 1mf – 9 – mf#23103 – cn Canadiana [360]

Charities *see* Survey

Charities and the commons *see* Survey

The charities of new york, brooklyn, and staten island / Cammann, Henry J & Camp, Hugh N – New York: Hurd and Houghton, 1868 – 2mf – 9 – 0-7905-4665-5 – mf#1988-0665 – us ATLA [362]

Charities review – New York. 1891-1901 (1) – mf#2871 – us UMI ProQuest [360]

Charities review – v1-10. 1891-1901 [all publ] – (= ser Social welfare periodicals) – 58mf – 9 – $325.00 – us UPA [360]

Chariton democrat-leader – Chariton IA. 1884 may 21 – 1r – 1 – (cont: chariton leader; cont by: chariton democrat) – mf#880433 – us WHS [071]

Charity and the clergy : being a review / Ruffner, William Henry – Philadelphia: Lippincott, Grambo, 1853 – 1mf – 9 – 0-524-00387-4 – mf#1989-3087 – us ATLA [240]

Charity, noxious and beneficent – London, England. 1853? – 1r – us UF Libraries [240]

Charity of the primitive churches / Chastel, Etienne – Philadelphia: J B Lippincott & Co 1857 [mf ed 1986] – 1r – 1 – (trans by g a matile) – mf#1764 – us Wisconsin U Libr [240]

The charity of the primitive churches : historical studies upon the influence of christian charity during the first centuries of our era, with some considerations touching its bearings upon modern society = Etudes historiques sur l'influence de la charite durant les premiers siecles chretiens / Chastel, Etienne – Philadelphia: J B Lippincott 1857 [mf ed 1990] – 1mf – 9 – 0-7905-7102-1 – (trans fr french by g a matile; incl bibl ref) – mf#1988-3102 – us ATLA [240]

The charity school movement in colonial pennsylvania, 1754-1763 : a history of the educational struggle between the colonial authorities and the german inhabitants of pennsylvania / Weber, Samuel Edwin – Philadelphia: WJ Campbell, 1905 – 1mf – 9 – 0-524-03747-7 – (incl bibl ref) – mf#1990-4852 – us ATLA [240]

Charity sermon in behalf of the gaelic episcopal society / Walker, James – Edinburgh, Scotland. 1831 – 1r – us UF Libraries [240]

The charity that covers a multitude of sins : a sermon preached on sunday, february 23rd, 1879...in the church of st alban, the martyr, ottawa / Bedford-Jones, T – Ottawa: [s.n.], 1879 – 1mf – 9 – mf#03306 – cn Canadiana [242]

Charity to the poor and afflicted, the duty and interest of the pro... / Lothian, Andrew – Edinburgh, Scotland. 1797 – 1r – us UF Libraries [240]

Charivari – Paris. 1975-1976 (1) 1975-1976 (5) 1975-1976 (9) – ISSN: 0009-1731 – mf#8572 – us UMI ProQuest [870]

Charivari – Paris, France. 1848 – 1r – 1 – uk British Libr Newspaper [870]

Le charivari – Paris. 1832-1905; 1907-juin 1908; 1909-1920; 19 juin 1926-27 fevr 1937 juil 1866-67; 25 sept 1870-1902 – 1 – fr ACRPP [870]

Char-Koosta / Confederated Salish and Kootenai Tribes of the Flathead Reservation – Dixon, Pablo MT. v2 n11-14 [1972] – 1r – 1 – (cont: char-koosta; cont by: char-koosta news) – mf#819130 – us WHS [307]

Char-koosta news / Confederated Salish and Kootenai Tribes of the Flathead Reservation – Pablo MT. 1988 dec 22-1989 dec 19 – 1r – 1 – (cont: char-koosta) – mf#3167909 – us WHS [307]

Charlan, Felix *see*
- Recherches experimentales en 1908
- Recherches experimentales en 1909

Charland, Maurice *see* Gabriel charland et sa descendance

Charland, Paul Victor *see* Questions d'histoire litteraire mises en rapport avec le programme de l'universite laval

Charland, Paul-Victor *see*
- La bonne sainte
- Madame saincte [sic] anne et son culte au moyen age, vol 1
- Les trois legendes de madame saincte anne

Charland, T-M *see* Artes praedicandi

Charlas, A *see* Tractatus de libertatibus ecclesiae gallicanae

Charlas con el presbitero jeronimo / Dangond Uribe, Alberto – Bogota, Colombia. 1963 – 1r – us UF Libraries [972]

Charlemagne / Cutts, Edward Lewes – London: SPCK; New York: E & J B Young 1882 [mf ed 1989] – (= ser The home library) – 1mf – 9 – 0-7905-4280-3 – mf#1988-0280 – us ATLA [944]

Charlemont 1719-1849 – Oxford, MA (mf ed 1995) – (= ser Massachusetts vital record transcripts to 1850) – 5mf – 9 – 0-87623-226-8 – (mf 1t-3t: family records 1719-1863. mf 2t: marriages 1795-1823. mf 3t-4t: intentions 1814-185?. mf 4t: marriages 1822-44. mf 4t-5t: births 1843-49. mf 5t: marriages; deaths 1843-49) – us Archive [978]

Charlemont 1765-1899 – Oxford, MA (mf ed 1987) – (= ser Massachusetts vital records) – 24mf – 9 – 0-87623-047-8 – (mf 1-10: town & vital records 1765-1826. mf 11-12: births & deaths 1803-43. mf 13-14: marriage intentions 1814-67. mf 15-16: militia soldiers 1840-1904. mf 17: b,m,d 1843-54. mf 18-22: b,m,d 1855-99. mf 23-34: marriage intentions 1917-68) – us Archive [978]

Charles : ou memoires historiques de m de labussiere, ex-employe au comite de salut public: servant de suite a l'histoire de la revolution francaise, avec des notes sur les evenemens extraordinaires arrives sous le regne des decemvirs / LaBussiere, Charles H de – Paris 1804 [mf ed Hildesheim 1995-98] – 4v on 12mf – 9 – €120.00 – 3-487-25823-4 – gw Olms [944]

Charles *see* Underground railroad

Charles 1 Of Spain, King *see*
- Cedulario de la monarquia espanola relativo a la i...
- Cedularios de la monarquia espanola relativos a la...

Charles 5...vie politique / Pichot, Amedee – 1854 – 9 – sp Bibl Santa Ana [320]

Charles 10 et louis 19 en exil / Villeneuve, Pons L de – Paris 1889 [mf ed Hildesheim 1995-98] – 1v on 2mf – 9 – €60.00 – 3-487-26142-1 – gw Olms [944]

[Charles a holmes] : 28 letters to him 1833-1852 – 1r – 1 – mf#B26338 – us Ohio Hist [240]

Charles abrams : papers and files – [mf ed ProQuest] – 53r – 1 – (with p/g) – us UMI ProQuest [360]

Charles bannerman papers / Bannerman, Charles – s.l, s.l? 1853-1890 – 1r – us UF Libraries [920]

Charles Borromeo, Saint *see* Testamento, o, ultima voluntad del alma

Charles bradlaugh, mp, and the irish nation – London, England. 1885 – 1r – us UF Libraries [240]

The charles bradlaugh pamphlets – ca 1875-85 – 5r – 1 – £180.00 – (main body of the pamphlets concern the advancement of radical and freethought causes) – mf#CBP – uk World [941]

The charles carroll papers / Carroll, Charles; ed by Hanley, Thomas O'Brien – 1972 – 3r – 1 – $390.00 – (with printed guide) – mf#S1612 – us Scholarly Res [920]

Charles, Cecil *see* Honduras

Charles darwin / Allen, Grant – London: Longmans, Green, 1885 [mf ed 1980] – (= ser English worthies) – 3mf – 9 – 0-665-05014-3 – (incl ind) – mf#05014 – cn Canadiana [575]

Charles darwin / Allen, Grant – London: Longmans, Green, 1888 [mf ed 1981] – (= ser English worthies) – 3mf – 9 – (incl ind) – mf#26237 – cn Canadiana [575]

Charles darwin / Allen, Grant – Paris: De Guillaumin, 1886 [mf ed 1980] – 3mf – 9 – 0-665-05013-5 – (incl ind; trans fr english by p-l le monnier) – mf#05013 – cn Canadiana [575]

Charles darwin : memorial notices reprinted from "nature" / Huxley, Thomas Henry – London, 1882 – (= ser 19th c evolution & creation) – 2mf – 9 – mf#1.1.9042 – uk Chadwyck [920]

Charles darwin and the theory of natural selection / Poulton, Edward Bagnall – [London] 1896 – (= ser 19th c evolution & creation) – 3mf – 9 – mf#1.1.11629 – uk Chadwyck [575]

Charles dickens : a sketch of his life and works / Perkins, Frederic Beecher – New York: G P Putnam & Sons 1870 [mf ed 1984] – 1r – 1 – mf#1187 – us Wisconsin U Libr [420]

Charles dickens: : the dickensian, 1905-1974 – 216mf – 1 – us Primary [420]

The charles dickens manuscripts : selected from the forster and dyce collection – 20r – 1 – (incl are dickens' personal correspondence, autograph mass drafts, page and galley proofs and other papers) – mf#C35-22401 – us Primary [420]

Charles dickens research collection : the most comprehensive dickens collection ever assembled / ed by Storey, Graham – [mf ed Chadwyck-Healey] – 102r – 1 – (pt1: the j f dexter coll at the british library. pt2: selections fr the suzannet collection. with p/g) – uk Chadwyck [420]

Charles, Elizabeth Rundle *see*
- Christian life in song
- Chronicles of the schoenberg-cotta family
- Early christian missions of ireland, scotland and england
- Ecce ancilla domini
- Martyrs and saints of the first twelve centuries
- Mary, the handmaid of the lord
- Three martyrs of the nineteenth century

Charles fenderich : lithographer of american statesmen / U.S. Library of Congress; ed by Miller, Lillian B – 1978 – 3mf – 9 – $45.00f – 0-226-69243-4 – us Chicago U Pr [760]

Charles g d roberts : and the watchers of the trails, his second book of animal life: with some mention also of his complete works / L C Page & Co – Boston: L C Page [1904] [mf ed 2000] – 1mf – 9 – 0-659-92007-7 – mf#9-92007 – cn Canadiana [800]

Charles, George *see* Last words

Charles George Douglas *see* One of those coincidences

Charles grandison finney / Wright, George Frederick – Boston: Houghton, Mifflin, 1891 [mf ed 1990] – (= ser American religious leaders) – 1mf – 9 – 0-7905-7674-0 – mf#1989-0899 – us ATLA [240]

Charles guerin : roman de moeurs canadiennes / Chauveau, Pierre J O – Montreal: Revue Canadienne, 1900 – 5mf – 9 – (int by ernest gagnon; ill by j-b lagace) – mf#11864 – cn Canadiana [830]

Charles, H *see* Le christianisme des arabes nomades sur le limes et dans le desert syro-mesopotamien aux alentours de l'hegire

Charles h spurgeon : our ally / Fulton, Justin Dewey – Montreal, Brooklyn, NY: P Propaganda, 1892 or 1893 – 5mf – 9 – mf#13153 – ne Slangenburg [242]

Charles haddon spurgeon : the puritan preacher in the nineteenth century / Lorimer, George Claude – Boston: James H Earle, 1892 – 1mf – 9 – 0-7905-9788-8 – mf#1989-1513 – us ATLA [240]

Charles johnson of zululand / Lee, Albert William – London, England. 1930 – 1r – us UF Libraries [960]

Charles kimmer – London, England. 18-- – 1r – us UF Libraries [240]

Charles kingsley : his letters and memories of his life = Correspondence / Kingsley, Charles; ed by Kingsley, Frances Eliza Grenfell – 2nd ed. London: Henry S. King, 1877 – 3mf – 9 – 0-7905-8127-2 – mf#1988-8044 – us ATLA [920]

Charles kingsley and the christian social movement / Stubbs, Charles William – Chicago: Herbert & Stone 1899 [mf ed 1991] – (= ser The victorian era series) – 1mf – 9 – 0-7905-9693-8 – mf#1989-1418 – us ATLA [230]

Charles l kades, papers / Kades, Charles L – (mf ed 2000) – 1r – 1 – $100.00 – (with guide) – University of Maryland – us UMI ProQuest [950]

Charles le temeraire : ou, le siege de nancy / Pixerecourt, Rene-Charles Guilbert De – Paris, France. 1814 – 1r – us UF Libraries [440]

Charles leckie's letter to the voluntaries of edinburgh – Edinburgh, Scotland. 1838 – 1r – us UF Libraries [240]

Charles lesieur et la fondation d'yamachiche / Desaulniers, Francois Lesieur – Montreal: Librairie Beauchemin, 1902 – 1mf – 9 – 0-665-73933-8 – mf#73933 – cn Canadiana [971]

Charles martin loeffler : his life and works / Colvin, Otis Herbert – U of Rochester 1957 [mf ed 19-] – 2v on 5mf – 9 – mf#fiche239, 533, 711 – us Sibley [780]

The charles mccarthy papers : guide to a microfilm edition / ed by Miller, Harold L – Madison: State Historical Society of Wisconsin, 1986 (mf ed 1987) – (= ser Wisconsin progressives) – 1mf – 9 – mf#FSN – 46806 – us NY Public [975]

Charles meryon : sailor, engraver, and etcher / Burty, Philippe – London 1879 – (= ser 19th c art & architecture) – 2mf – 9 – mf#4.2.1214 – uk Chadwyck [760]

Charles noble gregory / McClain, Emlin – n.p., 1911 4 p. LL-473 – 1 – us L of C Photodup [340]

Charles porterfield krauth, d.d., ll.d : norton professor of systematic theology and church polity in the lutheran theological seminary in philadelphia, professor of intellectual and moral philosophy, and vice-provost of the university of pennsylvania / Spaeth, Adolph – New York: Christian Literature Co, 1898-1909 – 1mf – 9 – 0-7905-8250-3 – (incl bibl ref) – mf#1988-8113 – us ATLA [242]

Charles, Richard *see*
- The cabinet maker
- [Three hundred] designs for window draperies fringes and mantle-board decorations

Charles, Robert Henry *see*
- The apocalypse of baruch
- Apocrypha and pseudepigrapha of the old testament
- The apocrypha and pseudepigrapha of the old testament in english
- The ascension of isaiah
- The assumption of moses
- The book of daniel
- The book of enoch
- The book of jubilees
- The book of jubilees or the little genesis
- The book of the secrets of enoch
- A critical and exigetical commentary on the book of daniel
- A critical history of the doctrine of a future life in israel, in judaism, and in christianity
- A critical history of the doctrine of a future life in israel, in judaism and in christianity
- The ethiopic version of the book of enoch
- The ethiopic version of the hebrew book of jubilees
- The greek versions of the testaments of the twelve patriarchs
- Immortality
- Masehafa kufase
- Religious development between the old and the new testaments
- Studies in the apocalypse
- The testaments of the twelve patriarchs

Charles sealsfield : ethnic elements and national problems in his works / Uhlendorf, Bernhard Alexander – Chicago IL: [s.n.] 1922 [mf ed 1991] – 1r [ill] – 1 – (incl bibl ref) – mf#2941p – us Wisconsin U Libr [410]

Charles sealsfield [carl postl] / Soffe, Emil – Bruenn: L u A Brecher [1922?] [mf ed 1991] – (= ser Veroeffentlichungen der deutschen gesellschaft fuer wissenschaft und kunst in bruenn) – 1r – 1 – (incl bibl ref; filmed with: charles sealsfield [carl postl] / albert b faust) – mf#2941p – us Wisconsin U Libr [410]

Charles sealsfield, (carl postl) der dichter beider hemisphaeren : sein leben und seine werke / Faust, Albert Bernhardt – Weimar: E Felber 1897 [mf ed 1991] – 1r – 1 – (filmed with: charles sealsfield (karl postl) / emil soffe & other titles) – mf#2941p – us Wisconsin U Libr [410]

The charles simms papers – 3r – 1 – $105.00 – Dist. us Scholarly Res – us L of C Photodup [977]

Charles starkweather clip files – Lincoln, NE. 1958-1973 [1] – mf#69092 – us UMI ProQuest [071]

Charles summer / Storey, Moorfield – Boston & NY: Houghton, Mifflin & Co, 1900 – (= ser The american statesmen series) – 6mf – 9 – $9.00 – mf#LLMC 96-033 – us LLMC [975]

Charles thain manufactures the best double mould plough : he also keeps for sale the paris straw cutter of different sizes by a whitelaw, paris... – [s.l: s.n. 187-?] [mf ed 1983] – 1mf – 9 – 0-665-39737-2 – mf#39737 – cn Canadiana [670]

Charles thomson papers / Thomson, Charles – 1729-1824 – 1 – $27.00 – us L of C Photodup [920]

Charles tomlinson griffes and his music / Davies, Katherine Currie – U of Rochester 1937 [mf ed 1983] – 3mf – 9 – mf#fiche1127 – us Sibley [780]

Charles unwin [late deputy registrar for the county of york] : begs leave to inform his friends and the public that he has commenced business as an agent for the sale and transfer of lands... – [Toronto?: s.n. 185-?] [mf ed 1983] – 1mf – 9 – 0-665-39765-8 – mf#39765 – cn Canadiana [090]

Charles w chestnutt papers *see* Chesnutt, charles w, papers, ms 3370

Charles waddell chesnutt papers, 1889-1932 / Chesnutt, Charles Waddell – [mf ed 1972] – 1r – 1 – mf#ms3370 – us Western Res [420]

Charles whittlesey papers, 1806-1909 / Whittlesey, Charles – [mf ed 1993] – 17r – 1 – (correspondence, military records, field and research notes...of this cleveland historian/ geologist/businessman, a founder of the western reserve historical society) – mf#ms3196 – us Western Res [978]

Charles wright in cuba, 1856-1867 : wright's work in cuba / Howard, Richard A – 1988 [mf ed Chadwyck-Healey] – 90p+4mf – 9 – 0-89887-059-3 – uk Chadwyck [574]

Charles wright on the boundary, 1849-1852 : or plantae wrightianae revisited / Shaw, Elizabeth A – [mf ed Chadwyck-Healey] – 42p+3mf – 9 – (monograph of wright's work in the southwest) – uk Chadwyck [574]
Charles-e harpe : president de la societe des poetes canadiens-francais, membre de la societe des ecrivains canadiens, 1909-1952: bio-bibliographie analytique / Lord, Marie-Paule – 1961 [mf ed 1979] – (= ser Bibliographies du cours...1947-66) – 3mf – 9 – (with ind; pref by roger brien) – mf#SEM105P4 – cn Bibl Nat [440]
Charles-edmond-henri de coussemaker (1805-1876) : three well-known works – Paris. 6v – 11 – $125.00 set – (l'art harmonique aux 12e et 13e siecles paris, 1865. histoire de l'harmonie au moyen-age paris, 1852. scriptorum de musica medii aevi nova series paris 4v 1864-76) – us Univ Music [780]
Charles-Liscombe, Robert S *see* The effects of accupressure therapy on exercise induced delayed onset muscle soreness and muscle function
Charles-quint : son abdication, son sejour et sa mort au monastere de yuste / Mignet, Francois-Auguste-Marie-Alexis – 2. ed. Paris: Paulin, LHeureux, 1854 – 2mf – 9 – 0-524-05157-7 – (incl bibl ref) – mf#1990-1413 – us ATLA [900]
The charleston advocate – Charleston, SC: H J Moore. feb 16 1867-1868// (mf ed 1947) – (= ser Negro Newspapers on Microfilm) – 1r – 1 – us L of C Photodup [071]
Charleston airlift dispatch – Charleston SC. v21 n7-v22 n12 [1981 may 1-1982 dec 16] – 1r – 1 – (cont: wing fact; cont by: airlift dispatch) – mf#645309 – us WHS [355]
Charleston argus – 1867 – 1r – mf#60.09 – nz Nat Libr [079]
Charleston Association. South Carolina *see* A summary of church-discipline
The charleston association to the baptist association of south carolina calling for the organization of the state baptist convention of south carolina : address – Signed by Richard Furman, John M. Roberts, and Joseph B. Cook, Nov 1820 – 1 – $5.00 – us Southern Baptist [242]
Charleston baptist church. charleston county. south carolina : church records – 1958-1971 – 4r – 1 – $170.10 – (includes financial rpts., deacon minutes, history 1923-1987) – us Southern Baptist [242]
Charleston black times / South Carolina Black Media Group – 1988 dec 1/3-19/31, 1989 jan 5/7-aug 17, 1990 apr 26-aug 23 – 2r – 1 – mf#1663915 – us WHS [305]
Charleston chronicle – Charleston SC. 1991 aug 14-dec 25, 1992 jan 1-jun 24, 1992 jul 1-1993 jan 13 – 3r – 1 – (cont by: chronicle [charleston sc: 1993]) – mf#2206300 – us WHS [071]
Charleston County, South Carolina. Democratic Party. Executive Committee *see* Minutes, 1876-1880
Charleston County. South Carolina. St Andrew's Episcopal Church *see* Transcript records, 1708-1899
Charleston daily republican – Charleston, SC: [s.n.], feb 19-jul 12 1872 – 3r – 1 – us CRL [071]
Charleston first baptist church. charleston, arkansas : church records – 1894-14 Oct 1934; 1937-56 – 1 – us Southern Baptist [242]
Charleston first baptist church. charleston, south carolina : church records – 1821-75 – 1 – us Southern Baptist [242]
Charleston first baptist church. charleston, tennessee : church records – 1901-68 – 1 – 53.73 – us Southern Baptist [242]
Charleston gospel messenger : and protestant episcopal register – Charleston. 1824-1853 (1) – mf#4557 – us UMI ProQuest [242]
Charleston industrial association minutes – [mf ed [S.l.]: Advance Access Group] – 28mf – 9 – mf#57-006 – us South Carolina Historical [366]
Charleston Library Society *see* Miscellaneous manuscripts
Charleston library society bills – 1788-1810 – 1mf – 9 – mf#51-515 – us South Carolina Historical [020]
Charleston Library Society. South Carolina *see*
– Partial manuscript catalog
– Records, 1758-1811
Charleston medical register – Charleston. 1803-1803 (1) – mf#4363 – us UMI ProQuest [610]
Charleston navel extra / Naval Electronic Systems Engineering Center, Charleston [US] – v1 n3-v10 n2 [1979 oct-1988 apr] – 1r – 1 – (cont by: navelextra) – mf#1054287 – us WHS [071]
Charleston. South Carolina. Bethel United Methodist Church *see* Records, 1845-1980, and minutes, 1845-1916
Charleston, South Carolina. Musical Art Club *see* Minute book
Charleston. South Carolina. Unitarian Church *see* Tombstone inscriptions
Charleston spectator : and ladies' literary port folio – Charleston. 1806-1806 (1) – mf#3562 – us UMI ProQuest [420]
Charlestown 1692-1874 – Oxford, MA (mf ed 1985) – (= ser Massachusetts vital records) – 258mf – 9 – 0-931248-79-5 – (mf 1-40: b,m,d 1629-1866. mf 41-80: births 1843-73. mf 81-130: marriage intentions 1725-1873. mf 131-171: marriages 1843-74. mf 172-249: deaths 1843-74. mf 250-258: church records 1632-1789, 1817-1889) – us Archive [978]
Charlestown mail / border news – Pretoria: State Library Corporate Communication, 9 nov 1909-15 jan 1910 – (= ser State library south africa newspaper microfilm project) – 1r – 1 – mf#MS00284 – sa National [079]
Charlesworth, J *see* Sermon on "doing to all men as they would do to us"
Charlesworth, J H *see* John and qumran
Charlesworth, M P *see* Trade routes and commerce of the roman empire
Charlet, Victor *see* Indicateur de dieppe
Charlevoix courier – Charlevoix, MI. 1982-2000 (1) – mf#61504 – us UMI ProQuest [071]
Charlevoix, Francois Xavier de *see* History and general description of new france
Charlevoix, P Fr X de *see* Histoire et description generale de la nouvelle france
Charlevoix, Pierre Francois Xavier de *see*
– Histoire de l'etablissement, des progres et de la decadence du christianisme dans l'empire du japon
– Histoire du christianisme au japon
Charlevoix, Pierre-Francois-Xavier de *see*
– Histoire de l'isle espagnole ou de s domingue
– Histoire du paraguay
– A voyage to north-america
Charlie-hebdo – Paris. 23 nov 1970-81 – 1 – fr ACRPP [870]
Charlier, Carl Vilhelm Ludwig *see* Untersuchung ueber die allgemeinen jupiter-stoerungen des planeten thetis
Charlier, Etienne D *see* Apercu sur la formation historique de la nation ha...
Charlotin, Marie-Joseph *see* Bibliographie analytique du reverend pere philippe deschamps
Charlotte birch-pfeiffer als dramatikerin : ein beitrag zur theatergeschichte des 19. jahrhunderts / Hes, Else – Stuttgart: J B Metzler, 1914 [mf ed 1992] – (= ser Breslauer beitraege zur literaturgeschichte. neue folge 38) – vii/227p – 1 – (incl bibl ref) – mf#8014 reel 4 – us Wisconsin U Libr [430]
Charlotte Elizabeth *see* Judaea capta
Charlotte gazette – Drakes Branch, VA. 1972-2000 (1) – mf#66683 – us UMI ProQuest [071]
Charlotte herald – Punta Gorda, FL. 1962-1963 aug – 4r – (gaps) – us UF Libraries [071]
Charlotte labor journal and dixie farm news / Central Labor Union [Charlotte, NC] – Charlotte NC. 1935 jan 24-1938, 1939-43, 1944-46, 1947-49, 1950-52, 1953 aug 21, sep 3 – 5r – 1 – mf#4946001 – us WHS [331]
Charlotte mary yonge : an appreciation / Romanes, Ethel – London: A.R. Mowbray, 1908 – 1mf – us ATLA [240]
Charlotte mary yonge : an appreciation / Romanes, Ethel Duncan – London: A.R. Mowbray, 1908 – 1mf – 9 – 0-7905-6675-3 – mf#1988-2675 – us ATLA [920]
The charlotte news – Charlotte, NC: Wade H Harris, 1901 – 4r – 1 – us CRL [071]
Charlotte post – Charlotte NC. 1987 sep 3-1998 oct/dec – 42r – 1 – mf#1573048 – us WHS [071]
Charlotte von schiller und ihre freunde : auswahl aus ihrer korrespondenz / ed by Geiger, Ludwig – Berlin. H Bondy [1908] – 1r [ill] – 1 – (incl bibl ref. filmed with: die erzahlungstechnik viktor scheffels / comp by walter grebe) – mf#2870p – us Wisconsin U Libr [860]
Charlotte von stein, goethe's freundin : ein lebensbild, mit benutzung der familienpapiere entworfen / Duentzer, Heinrich – Stuttgart: J G Cotta, 1874 [mf ed 1992] – 2v – 1 – mf#7553 – us Wisconsin U Libr [920]
Charlotte von stein und corona schroeter : eine vertheidigung / Duentzer, Heinrich – Stuttgart: J G Cotta, 1876 [mf ed 1992] – viii/301p – 1 – mf#7553 – us Wisconsin U Libr [920]
Charlottenburger tages-zeitung – Berlin DE, 1903 24 jan-30 jun, 1904 1 jan-31 mar, 1905 3 jan-31 mar, 1905 1 jul-31 dec – 4r – 1 – gw Misc Inst [074]
Charlottenburger wochenblatt – Berlin DE, 1898 7 may-1900 31 mar – 1r – 1 – gw Misc Inst [074]
Charlottesville chronicle – Charlottesville VA. 1878 apr 5-1882 apr 7/14 – 1r – 1 – (cont by: daily progress) – mf#884009 – us WHS [071]
Charlottetown herald – Canada, May 1894-Dec 1921 – 9r – 1 – uk British Libr Newspaper [071]
Charlton 1741-1849 – (= ser Massachusetts vital records Transcripts) – 0-87623-227-6 – us Archive [978]
Charlton 1742-1890 – Oxford, MA (mf ed 1986) – (= ser Massachusetts vital records) – 35mf – 9 – 0-87623-005-2 – (mf 1-4: births, deaths, int. 1742-1801. mf 5-9: b,m,d 1773-1826. mf 10-15: marriage intentions 1800-1909. mf 16-19: births, deaths 1826-45. mf 20-22: b,m,d 1844-72. mf 23: births 1871-90. mf 24-26: index to births 1844-90. mf 27-28: marriages 1862-90. mf 29-31: index to marriages 1844-90. mf 32: deaths 1870-90. mf 33-35: index to deaths 1844-90) – us Archive [978]
Charlton, John *see*
– A brief statement of objections to the policy of imposing export duties upon saw-logs, shingle bolts and stave bolts
– Speech of john charlton, mp. on unrestricted reciprocity with the united states
– Speech of mr john charlton, mp on the budget
Charlton, Lionel Evelyn Oswald *see* The military situation in spain after teruel
Charlton, Margaret Ridley *see* Louis hebert
Charlton, Robert M *see* Robert m charlton's reports
Charlton, Thomas T U P *see* Thomas t u p charlton's reports
The charm of bombay : an anthology of writings in praise of the first city of india / ed by Karkaria, R P – Bombay: DB Taraporevala Sons & Co, 1915 – (= ser Samp: indian books) – (foreword by h e lord willingdon) – us CRL [915]
The charm of indian art / Solomon, William Ewart Gladstone – London: T Fisher Unwin Ltd, 1926 – (= ser Samp: indian books) – us CRL [700]
Charm of life – London, England. no date – 1r – us UF Libraries [240]
Charm of persia / Durand, Henry Mortimer – London: Publ for the Society by J Hogg [1912?] [mf ed 1999] – 1r – 1 – mf#28582 – us Harvard [915]
The charm of sunday schools / Bliss, P P – 1871. New York – 1 – us Southern Baptist [242]
Charmant, Alcius *see*
– Notre appreciation sur le traite d'arbitrage
– Petition aux membres du corps legislatif
Charmant, Rodolphe *see* Vers les sommets par l'education et la sante
Charmetant, Felix *see*
– Extrait des annales intitulees uvres de st augustin et de sainte-monique
– Les peuplades kabyles et les tribus nomades du sahara
Charmin Paper Products Co *see* Charmin story
Charmin story / Charmin Paper Products Co – 1953 mar-dec, 1954 jan-1962 dec, 1963 feb-1969 dec – 3r – 1 – (cont: tissue topics; cont by: green bay story) – mf#1110606 – us WHS [670]
The charms of the old book : or, a study of the attractions of the bible / Huntington, George – Philadelphia: American Sunday-School Union, 1909 [mf ed 1993] – (= ser Green fund book 18b) – 1mf – 9 – 0-524-05807-5 – mf#1992-0634 – us ATLA [220]
Charnay, Desire *see* The ancient cities of the new world
Charney, Daniel *see*
– Barg aroyf
– Lider
Charnock, Joan Thomson *see* Russia
Charnock, Stephen *see* Discourse of the removal of the gospel
Charnotskii, I A *see* Edinstvo
Charns, Alexander *see* Us supreme court and federal judges subject files
Charnwood, Godfrey Rathbone Benson, Baron *see* Philosophical lectures and remains of richard lewis nettleship
Charon, Jean G *see* Plate-forme 70
Charon (klp7) : monatsschrift: dichtung, philosophie, darstellung / ed by Otto zur Linde – Berlin/Leipzig 1904-14 [mf ed 2002] – (= ser Marbacher mikrofiche-editionen (mme) 7; Kultur – literatur – politik: deutsche zeitschriften des 19./20. jahrhunderts (klp)) – 11v on 52mf – 9 – €280.00 – 3-89131-365-9 – gw Fischer [430]
Charpentier, Gustave *see* Louise
Charpentier, John *see* Goethe
Charpentier, Toussaint von *see*
– La bastille devoilee
– Bemerkungen auf einer reise von breslau ueber salzburg, durch tyrol, die suedliche schweiz nach rom, neapel und paestum
Charriere, E *see* Negociations de la france dans le levant
Charron, Kenneth C *see* Welfare of the african labourer in tanganyika
Chart analysis of the automobile liability security laws of the united states and canada. / Association of Casualty and Surety Companies. Law Dept – New York, 1947. 10 p. LL-1065 – 1 – us L of C Photodup [340]
Chart of christ's journeyings / Arnold, Charles Edward – Philadelphia, PA: John D Wattles, c1898 – 1mf – 9 – 0-524-03686-1 – mf#1990-4791 – us ATLA [220]
Chart of elocutionary drill / Browning, Thomas Blair – Toronto: Copp, Clark, 1888 [mf ed 1979] – 1mf – 9 – 0-665-00285-8 – mf#00285 – cn Canadiana [400]
Chart of the assessment life associations and friendly societies transacting business in canada : showing the business done, death claims paid, number of assessments made, income, expenses, assets, etc...1892 to 1897 inclusive – Toronto: Bulletin Pub Co, 1898 [mf ed 1979] – 1mf – 9 – 0-665-00325-0 – mf#00325 – cn Canadiana [360]
La charte coloniale. / Congo. Belgian – Bruxelles, Weissenbruch, 1910-19. 3 v. On film – v. 1 and v. 2 only. LL-12002 – 1 – us L of C Photodup [340]
La charte de 1830 – Paris. Journal du soir. oct 1836-juil 1838 – 1 – fr ACRPP [074]
Charte de commune en langue romane pour la ville de grealou en quercy : publiee avec sa traduction francaise et des recherches sur quelques points de l'histoire de la langue romane / Champollion-Figeac, Jacques-Joseph – Paris: Didot 1829 – (= ser Whsb) – 2mf – 9 – €30.00 – mf#Hu 527 – gw Fischer [410]
Charte et reglements de la cite de st-hyacinthe / Deschènes, R [comp] – St-Hyacinthe: Impr du "Courrier de St-Hyacinthe", 1895 [mf ed 1987] – 5mf – 9 – mf#SEM105P866 – cn Bibl Nat [348]
Charte et statuts de l'alliance nationale, societe de bienfaisance : fondee le 11 decembre 1892 – Montreal: C O Beauchemin, 1893 – 3mf – 9 – mf#10372 – cn Canadiana [366]
Charte et statuts de l'alliance nationale, societe de bienfaisance : fondee le 11 decembre 1892 – Montreal?: s.n, 1898 – 2mf – 9 – mf#26888 – cn Canadiana [366]
The charter – London, 27 Jan 1839-15 Mar 1840 – 4r – 1 – uk British Libr Newspaper [072]
The charter and bye laws of the shubenaccadie canal company : with the acts of the general assembly of nova scotia, relating to the canal / Shubenaccadie Canal Co – [Halifax, Nova Scotia] 1829 – (= ser 19th c british colonization) – 1mf – 9 – mf#1.1.4957 – uk Chadwyck [338]
Charter and by-laws / Bank of Montreal. Annuity and Guarantee Funds Society – Montreal?: s.n, 1861 – 1mf – 9 – mf#16942 – cn Canadiana [332]
Charter and ordinances of the city of tampa / Tampa (FL) Charters – Tampa, FL. 1918? – 1r – us UF Libraries [978]
Charter, deed of trust, by-laws and rules and regu... / Pensacola St John's Cemetery – Pensacola, FL. 1909 – 1r – us UF Libraries [978]
Charter of pass-a-grille beach / Pass-A-Grille Beach (FL) Charters – s.l, s.l? 1929? – 1r – us UF Libraries [978]
The charter of the church : six lectures on the spiritual principle of nonconformity / Forsyth, Peter Taylor – London: Alexander & Shepheard, 1896 – 1mf – 9 – 0-7905-3838-5 – mf#1989-0331 – us ATLA [240]
The charter of the church : six lectures on the spiritual principle of nonconformity / Forsyth, Peter Taylor – London: Alexander & Shepheard, 1896 – 1mf – us ATLA [240]
Charter of the city of boulder, state of colorado : official copy as framed and proposed by the charter convention elected july 24, 1917 – [Boulder?: bs.n.] (mf ed 1996) – 1r – 1 – mf#ZZ-34681 – us NY Public [323]
Charter of the city of boulder, state of colorado : official copy as framed and proposed by the charter convention elected july 24, 1917 / Boulder, CO – [Boulder?: bs.n] (mf ed 1996) – 1r – 1 – mf#MF B633c – us NY Public [323]
Charter of the city of boulder, state of colorado : official copy as framed and proposed by the charter convention elected july 24, 1917 / by authority of article 20 of the constitution of the state of colorado – [Boulder, CO?: s.n., 1917] (mf ed 1964) – 2r – 1 – (incl notes, ledgers etc) – ISSN: 0 – mf#MF B633c – us Colorado Hist [323]
Charter of the city of leesburg, florida – s.l, s.l? 1923? – 1r – us UF Libraries [978]
Charter of the city of miami, florida / Miami (FL) – Miami, FL. 1933 – 1r – us UF Libraries [978]
Charter of the city of quincy, florida / Campbell, J Baxter – Quincy, FL. 1923? – 1r – us UF Libraries [978]
Charter of the city of st petersbury, florida / Saint Petersburg (FL) Charters – St Petersburg, FL. 1931? – 1r – us UF Libraries [978]
Chartered Institute of Patent Agents *see* Cipa
The chartered institute of patent agents transactions – London. v1-50. 1882-1932 – 9 – mf#LLMC 84-341 – us LLMC [346]

Chartered Institution of Building Services *see* Journal of the chartered institution of building services
Chartered land surveyor, chartered minerals surveyor – London. 1980-1981 (1,5,9) – ISSN: 0142-520X – mf#12136 – us UMI ProQuest [520]
Chartered mechanical engineer (cme) – London. 1977-1988 (1) 1977-1988 (5) 1977-1988 (9) – (cont by: professional engineering) – ISSN: 0306-9532 – mf#11216 – us UMI ProQuest [621]
Chartered quantity surveyor – London. 1979-1993 (1,5,9) – ISSN: 0142-5196 – mf#11793 – us UMI ProQuest [690]
Chartered surveyor – London. 1960-1982 (1) 1971-1982 (5) 1974-1982 (9) – ISSN: 0009-1936 – mf#1315 – us UMI ProQuest [624]
Charteris, Archibald Hamilton *see*
- Canonicity
- The church of christ
- Historical note
- The new encyclopaedia britannica on theology
- The new testament scriptures

Charters and documents illustrating the history of the cathedral, city and diocese of salisbury (rs97) : in the twelfth and thirteenth centuries. / Salisbury (Diocese); ed by Macray, W D – 1891 – (= ser The rolls series (rs)) – €17.00 – (selected fr the capitular and diocesan registers by w rich-jones) – ne Slangenburg [241]
Charters en bescheiden over de betrekking der overijsselsche steden bijzonder van kampen op het noorden van europa gedurende de 13e eeuw – Deventer, 1861 – €5.00 – ne Slangenburg [240]
Charters of american life insurance companies. / Spectator Company. New York – New York, 1906. 408 p. LL-1416 – 1 – us L of C Photodup [346]
Charters of hereford cathedral, 1539-1900 *see* Registers of the bishops of hereford, 1275-1535/charters of hereford cathedral, 1539-1900
Charters towers daily herald etc – Queensland, Australia. 22 dec 1886 – 1/4r – 1 – uk British Libr Newspaper [072]
Les chartes coloniales et les constitutions des etats-unis de l'amerique du nord / Gourd, Alphonse – Paris: Impr nationale, 1885 [mf ed 1980] – 2v on 1mf – 9 – 0-665-07907-9 – mf#07907 – cn Canadiana [342]
Chartes de l'abbaye de saint-hubert en ardenne / Kurth, G – Bruxelles. tome premier. 1903 – €69.00 – ne Slangenburg [241]
Les chartes de l'ordre de chalais (1101-1400) (afm23) : tom 1 (1101-1200) / Romain, J Ch – 1923 – (= ser Archives de la france monastique (afm)) – €7.00 – ne Slangenburg [241]
Les chartes de l'ordre de chalais (1101-1400) (afm24) : tom 2 (1201-1300) / Romain, J Ch – (= ser Archives de la france monastique (afm)) – €11.00 – ne Slangenburg [241]
Les chartes de l'ordre de chalais (1101-1400) (afm25) : tom 3 (1301-1400) / Romain, J Ch – 1923 – (= ser Archives de la france monastique (afm)) – €7.00 – ne Slangenburg [241]
Chartes et documents de l'abbaye cistercienne de preuilly / Catel, Albert & Lecomte, Maurice – Montereau, 1927 – 11mf – 8 – €21.00 – ne Slangenburg [240]
Chartes inedites de l'abbaye d'orval / Deleseluse, A – Brussel, 1896 – 4mf – 8 – €11.00 – ne Slangenburg [241]
Chartier ancien de montmorigny / Huchet, Albert – Bourges, 1936 – 17mf – 8 – €32.00 – ne Slangenburg [241]
Chartier, Jean Baptiste *see* La colonisation dans les canton de l'est
Chartiers. Presbytery (Assoc. Pres. Ch. of No. Am.) *see* Minutes, 1805-1815
Chartism *see* Political tracts and pamphlets... 19th c
The chartist – London, 2 Feb-7 Jul 1839 – 15ft – 1 – uk British Libr Newspaper [072]
Chartist circular – Glasgow, Scotland. 28 Sept 1839-18 Sept 1841 – 38ft – 1 – uk British Libr Newspaper [072]
Chartist circular – n1-2. 1839-42 [all publ] – (= ser Radical periodicals of great britain, 1794-1914. period 1) – 7mf – 9 – $115.00 – us UPA [335]
Chartist circular, the... 1839-42 : the organ of the universal suffrage central committee for scotland – n1-146 – 1r – 1 – mf#96327 – uk Microform Academic [070]
Chartreuse de parme / Ginisty, Paul – Paris, France. 1919 – 1r – us UF Libraries [440]
Chartreuse de parme / Stendhal – Paris, France. 1839 – 1r – us UF Libraries [440]
Charts and graphs showing the conditions of afro-americans in 1900 – Drawings – 1 – us L of C Photodup [977]
Charts of south carolina baptist churches by associations / South Carolina – 1 – 5.00 – us Southern Baptist [242]
Chartularies of st mary's abbey, dublin (rs80) : and the register of its house at dunbrody; and annals of ireland / Dublin. St Mary's Abbey; ed by Gilbert, G T – (= ser The rolls series (rs)) – (v1 1884 €21. v2 1886 €23) – ne Slangenburg [241]
Chartularium universitatis parisiensis / Denifle, Heinrich – Paris. v1-4. 1889-97 – 4v on 121mf – 8 – €231.00 – ne Slangenburg [378]
O charutinho : jornal amolecado – Fortaleza, CE. 26 ago 1900 – (= ser Ps 19) – mf#P17,01,41 – bl Biblioteca [073]
Charuto : orgam do povo – Fortaleza, CE: Typ do Charuto, 21 jul, set-nov 1889; jul 1890; abr 1891; mar-abr 1896; jun 1903; 28 maio 1904 – (= ser Ps 19) – mf#P17,01,43 – bl Biblioteca [321]
Chas – Chernovtsy, U.S.S.R. -d. 10 Jan 1931-31 Dec 1939. Very imperfect. 9 reels – 1 – uk British Libr Newspaper [947]
Chas – Fuerth DE, 19 dec 1946-7 jan 1947, 27 apr 1947-1 jul 1949 – 1r – 1 – uk British Libr Newspaper [074]
Chas E Goad Co *see* Atlas of the city of montreal and vicinity
Chas g miller scrapbooks *see* Miller, chas g, scrapbooks, no 112
Chasanowitch, Leon *see* Krizis fun der idisher kolonizatsye in argentina
Chasapis belaruskae religiinae dumki – Paris, 1947-1951(6) – 34mf – 8 – (cont as: spisanie belaruskae...) – mf#R-5739 – ne IDC [243]
Chasco : queen of the calusas / Devries, Gerben M – New Port Richey, FL. 1922 – 1r – us UF Libraries [978]
Chase, Allan *see* Legacy of malthus
Chase, Alvin Wood *see*
- Dr chase's family physician, farrier, bee-keeper, and second receipt book
- Dr chase's new receipt book
- Dr chase's recipes, or, information for everybody
- Dr chase's third, last and complete receipt book and household physician

Chase, Ashton *see* Law of workmen's compensation
Chase, Chief Justice *see* Chase's reports of cases in the fourth circuit, 1865-1869
Chase county chronicle – Imperial, NE: D G Hines. v1 n1. mar 4 1886 (wkly) [mf ed -jun 10 1886 (gaps)] – 1r – 1 – us NE Hist [071]
Chase county enterprise – Imperial, NE: P W Scott, -may 1899// (wkly) [mf ed v5 n49. oct 10 1895-mar 2 1899 (gaps)] – 1r – 1 – (merged with: chase county tribune to form: chase county tribune and chase county enterprise, consolidated) – us NE Hist [071]
Chase county tribune – Imperial, NE: A C Clayburg. 2v. v1 n1. jul 30 1897-mar 1899// (wkly) [mf ed -dec 30 1898 (gaps)] – 1r – 1 – (merged with: chase county enterprise to form: chase county tribune and chase county enterprise, consolidated) – us NE Hist [071]
The chase county tribune and chase county enterprise, consolidated – Imperial, NE: A C Clayburn. 10v. mar 1899-v11 n24. dec 27 1907 (wkly) [mf ed v2 n35. mar 24 1899-dec 27 1907 (gaps)] – 3r – 1 – (formed by the union of: chase county enterprise and: chase county tribune. absorbed by: imperial republican) – us NE Hist [071]
Chase economic observer – New York. 1981-1984 (1,5,9) – ISSN: 0742-9983 – mf#13062 – us UMI ProQuest [330]
Chase, Frederic Henry *see*
- Chrysostom
- Confirmation in the apostolic age
- The credibility of the book of the acts of the apostles
- The gospels in the light of historical criticism
- The lord's prayer in the early church
- The old syriac element in the text of codex bezae
- The supernatural element in our lord's earthly life in relation to historical methods of study
- The syro-latin text of the gospels
- Thoughtful service

Chase, George *see*
- Johnson's ready legal adviser.
- Leading cases upon the law of torts

Chase, George Wingate *see* The masonic harp
Chase, Ira Joy *see* The jewish tabernacle
Chase, John Centlivres *see*
- Cape of good hope and the eastern province f algoa bay
- Cape of good hope and the eastern province of algoa bay [the]

Chase, Julia A *see* Mary a bickerdyke
Chase, Lisa A *see*
- Blood lactate responses for three competitive swimming strokes
- Perceived body image

Chase National Bank Of The City Of New York *see* Contestacion al informe de la comision especial
Chase pacesetter / Naval Air Station [Chase Field [Beeville TX]] – Beeville TX. [v1 n9,14,13,15 [1986 jun 26, sep 4, oct 30, nov 28]]-[v5 n13-14,16-19, 19-21, 24, 26, 28, 28 [1992 jan 9-23, feb 20-mar 19, apr 2,16, 30-may 14, jun 25, sep 3, nov 5, dec 3]] – 1r – 1 – (with gaps) – mf#1214440 – us WHS [355]
Chase, Philander *see* Bishop chase's reminiscences
Chase, Salmon Portland *see*
- Papers
- The salmon p chase papers

Chase, Zenas B *see* The judgment period preparatory to the establishment of the kingdom of heaven
Chase's reports of cases in the fourth circuit, 1865-1869 / Chase, Chief Justice – New York: Diossy. 1v. 1876 (all publ) – (= ser Early federal nominative reports) – 7mf – 9 – $10.50 – mf#LLMC 81-441 – us LLMC [340]
Chaset, Alan J *see* Disqualification of federal judges by peremptory challenge
Chasoslov – Krakow: Szwajpolt Fiol, 1491 – 9mf – 9 – mf#RHB-15 – ne IDC [460]
Chassay, Frederic *see* Les devoirs des femmes dans la famille
La chasse a l'heritage : comedie en quatre actes en prose / Cote, Stanislas – Montreal: Impr et litho Gebhardt-Berthiaume, 1884 [mf ed 1979] – 1mf – 9 – mf#SEM105P30 – cn Bibl Nat [830]
La chasse galerie : and other canadian stories / Beaugrand, Honore – Montreal: s.n, 1900 – 2mf – 9 – mf#03523 – cn Canadiana [830]
Chasseaud, George Washington *see* The druses of the lebanon
Le chasseur canadien – Montreal: "L'Etendard", 1885 – 7mf – 9 – mf#03018 – cn Canadiana [440]
Chasseurs canadiens / Boussenard, Louis – Paris: E Flammarion, 1892? [mf ed 1979] – 3mf – 9 – 0-665-00221-1 – mf#00221 – cn Canadiana [355]
Les chasseurs de fourrures / Bailleul, Louis – Paris: T Lefevre et E Guerin, [18–?] [mf ed 1980] – 4mf – 9 – 0-665-02487-8 – mf#02487 – cn Canadiana [636]
Chastain, J G *see* Personal diary and scrapbook
Chastel, Etienne *see*
- Charity of the primitive churches
- The charity of the primitive churches
- Christianity in the nineteenth century
- Etudes historiques sur l'influence de la charite durant les premieres siecles chretiens
- Melanges historiques et religieux

Chastelain, Pierre *see* Affectus amantis christum iesum
Chastellain, Georges *see*
- Chronique de j de lalain
- Chronique des ducs de bourgogne

Chastellux, Francois Jean, marquis de *see* Voyages de m le marquis de chastellux dans l'amerique septentrionale dans les annees 1780, 1781 et 1782
Chastenet, L *see* La vie de mgr alain de solminihac, eveque...de caors
Chastisements neglected forerunners of greater... / Pusey, E B – London, England. 1847 – 1r – us UF Libraries [240]
Chasuble, Archdeacon *see* Comedy of convocation in the english church
Le chat botte : journal humoristique hebdomadaire – Montreal: G Francq & Ls Nicolas. v1 n1 22 janv 1922- (wkly) [mf ed 1984] – 1r – 5 – (ceased 192-?) – mf#SEM16P343 – cn Bibl Nat [870]
Le chat noir – Paris. 1882-95 – 1 – fr ACRPP [440]
Chatard, Francis Silas *see*
- Christian truths
- Essays

Le chateau de montenero : comedie en trois actes et en prose / Dalayrac, Nicolas – Paris: l'auteur [c1798] [mf ed 1990] – 1r – 1 – (french words; libretto by francois benoit hoffmann; incl catalog of c dalayrac's works (on verso of t. p.)) – mf#pres. film 87 – us Sibley [780]
Le chateau des tuileries : ou recit de ce qui s'est passe dans interieur de ce palais... / Roussel, Pierre – Paris 1802 [mf ed Hildesheim 1995-98] – 2v on 6mf – 9 – €120.00 – ISBN-10: 3-487-25930-3 – ISBN-13: 978-3-487-25930-7 – gw Olms [720]
Chateaubriand et al *see* Louis-napoleon bonaparte juge
Chateaubriand, F A de *see* Buonaparte und die bourbons
Chateaubriand, [F A] de *see* Itineraire de paris...jerusalem et de jerusalem...paris
Chateaubriand, Francois R de *see*
- De la monarchie selon la charte
- Erinnerungen aus italien, england und amerika
- Itineraire de paris a jerusalem et de jerusalem a paris
- Maison de france
- Reise von paris nach jerusalem durch griechenland und kleinasien

Chateaubriand, Francois-Rene *see* The genius of christianity, or, the spirit and beauty of the christian religion
Chateaubriand, Francois-Rene, vicomte de *see* Genie du christianisme
Chateaubriand, Frandcois-Rene, Vicomte de *see* Les martyrs
Chateauguay : qui est "temoin oculaire" et sa description de la bataille est-elle correcte? / Baby, Louis Francois Georges – Montreal: Pelletier, 1900 [mf ed 1980] – 1mf – 9 – 0-665-03744-9 – (in french and english) – mf#03744 – cn Canadiana [355]
Chateauneuf, Francois de Castagneres, Abbe de *see* Dialogue sur la musique des anciens
Chatelain, Heli *see*
- Contos populares de angola
- Folk-tales of angola
- Kimbundu grammar

Chatelain, Rene *see* Lettres de sidy-mahmoud
Chatelaine : english edition – Toronto. 1972+ (1) 1977+ (5) 1977+ (9) – ISSN: 0009-1995 – mf#7201 – us UMI ProQuest [305]
Chatelaine : french edition – Toronto. 1974-1996 (1) 1976-1996 (5) 1976-1996 (9) – ISSN: 0317-2635 – mf#7209 – us UMI ProQuest [640]
Chatelet, Gabrielle-Emilie Le Tonnelier de Breteuil, Marquise du *see*
- Dissertation sur la nature et propagation de feu
- Reponse de madame a la lettre que m de mairan

Chater, James *see*
- A grammar of the cingalese language
- Grammar of the cingalese language

Chatham *see* To the right honorable charles poulett thomson, governor general of her majesty's provinces in north america
Chatham 1693-1900 – Oxford, MA (mf ed 1987) – (= ser Massachusetts vital records) – 60mf – 9 – 0-87623-002-8 – (mf 1-7: b,m,d 1693-1789. mf 8-11: vital records 1704-47. mf 12-15: meetings & marriages 1698-1748. mf 16-21: vital records 1749-87. mf 22-26: vital records & meetings 1727-1856. mf 27-33: vital & town records 1814-46. mf 34-43: vital records 1804-68. mf 44-48: marriage intentions & records 1854-74. mf 49-51: marriages 1855-1900. mf 52-53: deaths 1869-1900; births 1891-1900. mf 54-55: index to deaths 1845-1928. mf 56-57: index to births 1846-1944. mf 57-58: index to marriage intentions 1846-1944. mf 58-60: index to marriages 1846-1944) – us Archive [978]
Chatham chronicle *see* Chatham newspapers, pt 1
Chatham courier – Hudson, NY. 1999-1999 (1) – mf#69383 – us UMI ProQuest [071]
The chatham courier – Chatham, New York. Apr 4 1883-Mar 17 1897 – 1 – us NY Public [071]
Chatham growler *see* Chatham newspapers, pt 2
Chatham house series – Royal Institute of International Affairs – 9 – enquire for prices – mf#402400 – us Hein [327]
Chatham journal – Chatham, ON. 1841-44 – 1r – 1 – ISSN: 1180-6133 – cn Library Assoc [071]
Chatham news – Siler City, NC. 1999-2000 (1) – mf#68373 – us UMI ProQuest [071]
Chatham newspapers, pt 1 – Chatham, ON. 1844-55 – 2r – 1 – (incl: canadian freeman, chatham chronicle, chatham gleaner, kent advertiser, western sentinel) – cn Library Assoc [971]
Chatham newspapers, pt 2 – Chatham, ON. 1853-75 – 9r – 1 – (incl: chatham growler, chatham planet, western argus, western planet, western union) – cn Library Assoc [971]
Chatham observer – Chatham, England. Chatham & Rochester Observer – Chatham, Rochester & Brompton Observer – Chatham, Rochester & Gillingham Observer – New Observer – Kent Messenger & Observer – Kent Messenger & Chatham Observer. -w. 14 May 1870-2 Aug 1968. Lacking Jan 1896. 105 reels – 1 – uk British Libr Newspaper [072]
Chatham record – Pittsboro, NC. 1996-2000 (1) – mf#68539 – us UMI ProQuest [071]
Chatham-southeast citizen – Chicago IL. [1990 apr 26, 1991 apr 11/14-dec 26/29]-2001 jan 4/jun 28 – 19r – 1 – (with gaps; cont: chatham citizen; cont by: citizen newspaper, chatham-southeast) – mf#1886851 – us WHS [071]
Chatiee! / Gourcuff, Olivier De – s.l, s.l? 1918 – 1r – us UF Libraries [440]
Le chatiment – Ed. de Paris. [Paris: s.n. mar 25 1871 – (filmed as pt of: commune de paris newspapers; newspapers on these reels are filmed chronologically, not alphabetically.) – us CRL [074]
Chaton, Prosper *see* Avenir de la guyane francaise
Chatra Prem *see* Khmaer 500.000 chnam!
Chats on cottage and farmhouse furniture / Hayden, Arthur – Toronto: Bell & Cockburn, [1911?] – 4mf – 9 – 0-665-85919-8 – (with a chapter old english chintzes by hugh phillips) – mf#85919 – cn Canadiana [740]
Chats on english eathenware / Hayden, Arthur – Toronto: Bell & Cockburn, [1909?] – 6mf – 9 – 0-665-87646-7 – mf#87646 – cn Canadiana [730]

Chats on old coins / Burgess, Frederick William – Toronto: Bell & Cockburn, [1913?] – 5mf – 9 – 0-665-99180-0 – mf#99180 – cn Canadiana [730]

Chats on old silver / Hayden, Arthur – New York, NY. 1949 – 1r – us UF Libraries [720]

Chatskii, A *see* Partiia narodnoi svobody i demokratiia

Chatsworth...illustrated by upwards of fifty engravings / Jewitt, Llewellynn Frederick William – Buxton 1872 – (= ser 19th c art & architecture) – 1mf – 9 – mf#4.2.1299 – uk Chadwyck [760]

Chattanooga daily gazette – Chattanooga, TN. mar 5 1864-sep 2 1865 – (= ser Confederate newspapers) – 1r – 1 – us Western Res [071]

Chattanooga daily rebel – Chattanooga, TN. sep 10 1862-jul 29 1863 – (= ser Confederate newspapers) – 1 – us Western Res [071]

Chattanooga first baptist church. chattanooga, tennessee : church records – 1852-1967 – 1 – 71.51 – us Southern Baptist [242]

Chattanooga times/free press – Chattanooga, TN. 1999-2000 (1) – mf#60751 – us UMI ProQuest [071]

Chattanooga volksfreund – Chattanooga TN. 1898 jan 1-apr 30, may 14-dec 31 – 1r – 1 – mf#1225176 – us WHS [071]

Chattard, G P *see* Nuova descrizione del vaticao sia della sacrosanta basilica di s pietro

Chattel mortgage / Bahlke, William A – Detroit, Richmond & Backus, 1897. 9 p. LL-2235 – 1 – us L of C Photodup [340]

Chattel mortgages and conditional sales in the state of new york / Smith, Dix W – 3d ed. Albany: Bender, 1900. 272p. LL-1330 – 1 – us L of C Photodup [346]

Chatter from around the white tops / Circus Fans Association of America – 1927 jun-1932 nov – 1r – 1 – (title: white tops) – mf#3389605 – us WHS [790]

Chatter-box / Parents Without Partners – 1976 jul-1978 dec – 1r – 1 – (cont: newsletter [parents without partners. southern lakes chapter 730, lake geneva wi]]) – mf#619634 – us WHS [305]

Chatteris – complete set 1600-1902 – (= ser Cambridgeshire parish register transcript) – 25mf – 9 – £31.25 – (ind, briefs, quaker notes [5mf] £6.25; baptisms 1600-1902 [7mf] £8.75; marriages & banns 1600-1902 [8mf] £10; burials 1600-1902 [5mf] £6.25) – uk CambsFHS [929]

Chatterjee, Anathnath *see* The hos of seraikella

Chatterjee, Atul Chandra *see*
- The new india
- A short history of india

Chatterjee, Bijan Raj *see*
- India and java
- Indian cultural influence in cambodia

Chatterjee, Debiprasad *see* Modern bengali poems

Chatterjee, Lalitmohan *see* Representative indians

Chatterjee, Mohini Mohun *see* Indian spirituality

Chatterjee, Ramananda *see* Story of satara

Chatterjee, S C *see* The nyaya theory of knowledge

Chatterjee, Santosh *see* The art of hindu dance

Chatterjee, Satischandra *see* An introduction to indian philosophy

Chatterjee, Sris Chandra *see*
- India and new order
- Magadha architecture and culture

Chatterji, A C *see* India's struggle for freedom

Chatterji, Bankim Chandra *see*
- Indira and other stories
- Rajmohan's wife
- Sitaram
- The two rings

Chatterji, Jagadish Chandra *see* India's outlook on life

Chatterji, Jagdish Chandra *see* Hindu realism

Chatterji, Nandalal *see*
- Mir qasim, nawab of bengal, 1760-1763
- Verelst's rule in india

Chatterji, Suniti Kumar *see*
- Bengali self-taught
- Kirata-jana-krti
- The national flag
- The origin and development of the bengali language
- Scientific and technical terms in modern indian languages

Chatterton, Alfred *see* Industrial evolution of india

Chatterton, Eyre *see* The story of fifty years' mission work in chhota nagpur

Chatterton, Eyre, Bishop of Nagpur *see* The story of gondwana

Chatterton, Mason Daniel *see* Probate law

Chatting on-line : a dangerous proposition for children: hearing...house of representatives, 107th congress, 2nd session, may 13 2002 / United States. Congress. House. Committee on Energy and Commerce. Subcommittee on Telecommunications and the Internet – Washington: US GPO 2002 [mf ed 2002] – 1mf – 9 – 0-16-068725-X – (incl bibl ref) – us GPO [170]

Chattooga press – Summerville, GA. 1987-2000 (1) – mf#62471 – us UMI ProQuest [071]

Chattopaddhyaya, Nisikanta *see* The true theosophist

Chattopadhyay, Kshitis Prasad *see* Report on santals in bengal

Chattopadhyaya, Harindranath *see*
- Blood of stones
- The dark well
- Edgeways and the saint
- The feast of youth
- Five plays
- Life and myself
- Lyrics
- Perfume of earth
- The son of adam
- Strange journey

Chattopadhyaya, Kamaladevi *see*
- Japan, its weakness and strength
- Uncle sam's empire

Chatzidakis, G N *see* Einleitung in die neugriechische grammatik

Chau Seng *see* L'organisation buddhique au cambodge

Chaucer, Geoffrey *see*
- Canterbury tales
- Canterbury tales / [geoffrey chaucer] and faerie queene / [edmund spenser]
- Chaucer's legende of goode women

Chaucer review – University Park. 1988+ (1,5,9) – ISSN: 0009-2002 – mf#16967 – us UMI ProQuest [420]

Chaucer society, london. publications – v1-30 – 1 – $312.00 – (v1-21 $240 [0149]) – mf#0148 – us Brook [420]

Chaucer's legende of goode women / ed by Corson, Hiram – Philadelphia: F Leypoldt, New York: F W Christern 1864 [mf ed 1984] – 1r – 1 – (int & notes by ed) – mf#954 – us Wisconsin U Libr [810]

Chauchamayo. estudio de una region de la selva del peru. departamento...tomo 1. lima, 1969 / Ortiz, Dionisio – Madrid: Graf. Calleja, 1970 – 1 – sp Bibl Santa Ana [972]

Chaudhuri, Bhawani Prasad *see* Leftist leaders of india

Chaudhuri, Pramatha *see* Tales of four friends

Chaudhuri, Roma *see* Vedanta-parijata-saurabha of nimbarka and vedanta-kaustubha of srinivasa

Chaudhuri, S C *see* Lingua indica revealed

Chaudhury, Prabas Jivan *see* Studies in comparative aesthetics

Chaudoin, William *see* Diary

Chaughi, Rene *see* La femme esclave

Chauke, Joel *see* Kulongisela mukhongelo wa pfuxelelo

Chaula, Thomas de *see* Exposite in terentium...

Chauliac, A *see* Histoire de l'abbaye sainte-croix de bordeaux (afm9)

Chaumeton, Nigel R *see* The influence of task and ego goal orientations and perceptions of competence on affect and intrinsic motivation in competitive youth tennis

Chaumette, E J M *see* Les enfants celebres

Chaumette, Gustave *see* Documents officiels relatifs a l'avenement du gene

Chaumette, Max Gustave *see* Panamericanisme a travers l'histoire d'haiti

Chaumiere et son coeur / Scribe, Eugene – Paris, France. 1835 – 1r – us UF Libraries [440]

Chauncey w mead papers, 1862-1865 / Mead, Chauncey W – [mf ed 1981] – 1r – 1 – mf#ms3602 – us Western Res [976]

Chauncy maples...pioneer missionary in east central africa for nineteen years... : a sketch of his life...by his sister / [Maples, E] – London, 1898 – 5mf – 9 – mf#HTM-112 – ne IDC [916]

Chauncy, Maurice *see* Historia aliquot martyrum anglorum maxime octodecim cartusianorum

Chaundler, Thomas *see* The works of thomas chaundler

Chaussegros de Lery, Joseph Gaspard *see* Journal of chaussegros de lery

Chautauqua County Genealogical Society *see* Chautauqua genealogist

Chautauqua farmer – Dunkirk, NY. 1870-1894 (1) – mf#64948 – us UMI ProQuest [071]

Chautauqua genealogist / Chautauqua County Genealogical Society – 1977 oct-1993 aug – 1r – 1 – mf#2981005 – us WHS [929]

The chautauqua movement / Vincent, John Heyl – Boston: Chautauqua Press, 1886, c1885 – 1mf – 9 – 0-7905-6457-2 – mf#1988-2457 – us ATLA [970]

Chautauquan – Chautauqua. 1880-1914 – 1 – mf#5277 – us UMI ProQuest [073]

Chautauquan daily – Chautauqua, NY. 1987-1994 (1) – mf#64928 – us UMI ProQuest [071]

Chauveau, Pierre J O *see*
- Charles guerin
- Epitre a m prendergast
- Etude sur les poesies de francois-xavier garneau
- Frederic ozanam
- Relation du voyage de son altesse royale le prince de galles en amerique
- Le sacre-coeur

Chauveau, Pierre-Joseph-Olivier *see*
- Discours prononce le mercredi, 18 juillet 1855
- Noces d'or de pie 9

Chauvet, Henri *see*
- Geographie de la republique d'haiti
- Geographie de l'ile d'haiti
- Nos grandes routes nationales, inauguration
- Travers la republique d' haiti

Chauvet, Marie *see*
- Dance on the volcano
- Danse sur le volcan
- Fille d'haiti

Chauvigne, Auguste *see* Fauvette

Chauvin, Etienne *see* Nouveau journal des scavans, dresse a berlin

Chauvin, Stephanus *see* Lexicon rationale sive thesaurus philosophicus (ael1/47)

Die chauvinisten : roman / Jagow, Eugen von – Stuttgart: Deutsche Verlags-Anstalt, 1889 [mf ed 1995] – (= ser Deutsche romanbibliothek. salon-ausgabe 2 jahrg/10) – 292p – 1 – mf#8795 – us Wisconsin U Libr [830]

Chavannes *see* L'itineraire d'ou-k'ong (751-790)

Chavannes de la Giraudiere, H de *see* Les chinois pendant une periode de 4458 annees

Chavannes, Edouard *see* Le t'ai chan

Chavannes, Jules *see* Les refugies francais dans le pays de vaud

Chavannes, M E *see* Voyageurs chinois chez les khitan et les joutchen

Chavarria Flores, Manuel *see*
- Cancion de cuna
- Hacia un sistema nacional de educacion

Chavasse, A *see* Sacramentarium gelasianum

Chave, Richard Branscombe *see* Autobiographical record

O chaveco – Desterro, SC: Typ Desterrense de Jose Joaquim Lopes, 11 nov 1860-07 abr 1861 – (= ser Ps 19) – mf#UFSC/BPESC – bl Biblioteca [079]

O chaveco : jornal critico, humoristico e noticioso – Florianopolis, SC. 16 abr 1933 – (= ser Ps 19) – mf#UFSC/BPESC – bl Biblioteca [073]

Chavero, Alfredo *see* Sahagun

Chaver-Paver *see* Khaver-pavers mayselakh

Chaves, Bernabe de *see* Apuntamiento legal de la o. santiago

Chaves, Jose Maria *see* Reforma universitaria en colombia

Chaves, Manuel *see* Don bernardo marquez de la vega

Chaves Masa, Pedro *see* Llantos funebres a la sentida, lamentable,...dona ma reina de portugal

Chaves y Manso, Rafael *see*
- De la legislacion romana en las relaciones con la de los pueblos europeos
- Discurso

Chavez, A Ezequiel *see* El primero de los educadores de la nueva espana, fr pedro de gante

Chavez Alfaro, Lizandro *see* Monos de san telmo

Chavez Orozco, Luis *see* Esfuerzo de mexico por la independencia de cuba

Chavez Velasco, Waldo *see* Cuentos de hoy y de manana, cuento

Chawane, Winston Nelson *see* An evaluation of the progress of rural land reform in south africa

Chay Huk Phen *see* Pativattan dade!

Chay nhac la ra si mran ma yan kye mhu padesa – Ran Kun: Sin gi ca pe 1984 [mf ed 1990] – 1r with other items – 1 – (in burmese; with bibl) – mf#mf-10289 seam reel 146/6 [§] – us CRL [390]

Chay nhac ra si mran ma rui ra ra si pvai to mya / Ae Nuin, U – Ran Kun: Ka ba ae sa sa na re u ci tha na 1980 [mf ed 1990] – 1r with other items – 1 – mf#mf-10289 seam reel 144/1 [§] – us CRL [390]

L'chayim – La Jolla CA: UCDS Student Orgs 1974-83 [mf ed 1989] – 10v on 1r – 1 – (cont by: l'chayim quarterly) – us AJPC [378]

L'chayim quarterly – La Jolla CA: [s.n.] fall 1984- [mf ed 1989] – 1 – (cont: l'chayim) – us AJPC [378]

Chazanah : suara santri & siswa progesip / Dept Penerangan IPNU wil Djk Raya – Djakarta, 1955(1-3) – 1mf – 9 – (missing: 1955(1-2)) – mf#SE-1379 – ne IDC [950]

Chazet, M (Rene-Andre-Polydore Alissan De) *see* Philippe le savoyard

Chazet, Rene-Andre-Polydore Alissan De *see* Mademoiselle gaussin

Chazotte, Peter Stephen *see* Facts and observations on the culture of vines, olives, capers, alm

Chbab divers : texte tires de satras sur feuilles de latanier – Phnom-Penh: Edition de l'Institut bouddhique 1972 [mf ed 1990] – 1r with other items – 1 – (title in khmer on added t.p.: cpap phsen phsen; text in khmer) – mf#mf-10289 seam reel 100/03 [§] – us CRL [959]

Chbab divers – Phnom-Penh: Editions de l'Institut bouddhique 1957-59 [mf ed 1990] – 1r with other items – 1 – (title & text in khmer; added t.p. in french; v2 & 3, edited ed; contents: vithur pandit. tunmean khluon. peak chas; 2. chbab pros. chbab srey. chbab ariyasatta; 3. chbab kram. kerkal. kunchao. treinet) – mf#mf-10289 seam reel 100/04 [§] – us CRL [959]

Chbab supheasit / Nan, Okana vansasarabejn – Phnom-Penh: Editons de l'Institut bouddhique 1953 [mf ed 1990] – 1r with other items – 1 – (title in khmer on added t.p.: cpap subhasit; text in khmer) – mf#mf-10289 seam reel 111/5 [§] – us CRL [959]

Che bhi lu kri : [a novel] / Mra Cakra, Cha ra kri – Ran Kun: Tan Mon Kri Ca up Tuik [195-?] – 1r with other items – 1 – (in burmese) – mf#mf-10289 seam reel 180/2 [§] – us CRL [830]

Ch'e hsiang she hui / Feng, Tzu-k'ai – Shang-hai: Liang yu t'u shu yin shua kung ssu, Min kuo 24 [1935] – (= ser P-k&k period) – us CRL [840]

Che hsueh ta kang / Ch'u, Chu-nung – Shang-hai: Tu li ch'u pan she, [1944] – (= ser P-k&k period) – us CRL [100]

Che kan t'ieh lu lien ho kung ssu tsung pao kao : min kuo erh shih ssu nien ch'i yueh chih erh shih wu nien liu yueh – [China]: Che-chiang sheng li t'u shu kuan yin hsing so, 1935 – (= ser P-k&k period) – us CRL [380]

Che nhac rasi lu ta / Oon Kyi, U – Ran Kun: Hamsavati 1965 [mf ed 1990] – 1r with other items – 1 – (in burmese) – mf#mf-10289 seam reel 145/4 [§] – us CRL [720]

Che pu kuo shih ch'un t'ien / Li, Chien-wu – Shang-hai: Wen hua sheng huo ch'u pan she, 1940 – (= ser P-k&k period) – us CRL [820]

Che pu kuo shih ch'un t'ien / Li, Chien-wu – Shang-hai: Wen hua sheng huo ch'u pan she, Min kuo 35 [1946] – (= ser P-k&k period) – us CRL [820]

Che shang sui pi / Chang, I-p'ing – Shang-hai: Pei hsin shu chu, 1932 – (= ser P-k&k period) – us CRL [840]

Cheadle herald and general advertiser – Cheadle, Staffordshire, 8 sep 1877-28 dec 1924 [mf ed 2003] – 25r – 1 – (missing: 1882, 1896-97, 1911, 1921; cont as: cheadle herald, tean news and general advertiser [jan 1899-dec 1924]) – uk Newsplan [072]

Cheadle hulme, all saints – [North Cheshire FHS] – (= ser Cheshire monumental inscriptions) – 2mf – 9 – £3.25 – mf#119 – uk CheshireFHS [929]

Cheadle hulme, all saints: burials 1864-2000 – [North Cheshire FHS] – (= ser Cheshire church registers) – 2mf – 9 – £3.75 – mf#253 – uk CheshireFHS [929]

Cheadle, park road cemetery: burials 1903-1922 – (= ser Cheshire church registers) – 2mf – 9 – £4.00 – mf#263 – uk CheshireFHS [929]

Cheadle, park road cemetery: burials 1922-1932 – (= ser Cheshire church registers) – 1mf – 9 – £2.50 – mf#359 – uk CheshireFHS [929]

Cheadle, st mary – [North Cheshire FHS] – (= ser Cheshire monumental inscriptions) – 7mf – 9 – £8.50 – mf#148 – uk CheshireFHS [929]

Cheadle, st mary: baptisms 1813-1834 – [North Cheshire FHS] – (= ser Cheshire church registers) – 3mf – 9 – £4.00 – mf#232 – uk CheshireFHS [929]

Cheadle, st mary: baptisms 1834-1835 – [North Cheshire FHS] – (= ser Cheshire church registers) – 1mf – 9 – £2.50 – mf#233 – uk CheshireFHS [929]

Cheadle, st mary: baptisms 1836-1841 – [North Cheshire FHS] – (= ser Cheshire church registers) – 1mf – 9 – £2.50 – mf#234 – uk CheshireFHS [929]

Cheadle, st mary: baptisms 1841-1871 – [North Cheshire FHS] – (= ser Cheshire church registers) – 3mf – 9 – £4.00 – mf#235 – uk CheshireFHS [929]

Cheadle, st mary: burials 1871-1982 – [North Cheshire FHS] – (= ser Cheshire church registers) – 4mf – 9 – £5.50 – mf#252 – uk CheshireFHS [929]

Cheap repository – Philadelphia. 1800-1800 (1) – mf#4424 – us UMI ProQuest [978]

Cheap telegraph rates : address delivered at the annual meeting of the canadian press association, feb 28th, 1902 / Fleming, Sandford – [Ottawa?: s.n, 1902?] [mf ed 1995] – 1mf – 9 – 0-665-74768-3 – mf#74768 – cn Canadiana [380]

Chearful piety : or, religion without gloom – London, England. 1792 – 1r – us UF Libraries [240]

Cheatham, Tina R *see* The athletic organizational structure and administrative views of university and athletic governing personnel in the southwest conference

Cheavens, J S *see* Sobre interpretacion

Chebucto and other poems / Bell, John Allison – Halifax, NS?: s.n, 1890 – 1mf – 9 – mf#05832 – cn Canadiana [810]

Chebyshev (Tchebichef), P L *see* The theory of propability

Checa, Pedro *see*
- A un gran partido, una gran organizacion
- Tareas de organizacion y trabajo practico del partido

Chechi imwe chete yechokwadi / Mavudzi, Emmanuel – Gwelo, Zimbabwe. 1960 – 1r – us UF Libraries [960]

Che-chiang chih p'ing-shui ch'a yeh – [China: sn], 1934 – (= ser P-k&k period) – us CRL [630]

Che-chiang hsiang-shih lu – List of successful candidates in the imperial examination in Chekiang province: 1835, 1851, 1855, 1870, 1879, 1882, 1885, 1889, 1893. 1 reel – 1 – us Chinese Res [951]

Che-chiang hsing yeh yin hang pen hang erh shih liu nien chih hui ku – [China: sn], Min kuo 22 [1933] – (= ser P-k&k period) – us CRL [332]

Che-chiang hsing yeh yin hang tsung kuei ch'eng – [China: sn], 1935 – (= ser P-k&k period) – us CRL [332]

Che-chiang jih-pao – Hangchow, Chekiang. May 9, 1949-. Reel 1: Mar-May 10, 1961; Reel 2: Jan-Jul 1962; Reel 3: Aug-Dec. 1962. 3 reels – 1 – 52.50 – us Chinese Res [079]

Che-chiang lin-an hsien nung ts'un tiao ch'a – Hang-chou shih: Chien she wei yuan hui tiao ch'a Che-chiang ching chi so, min kuo 20 [1931] – (= ser P-k&k period) – us CRL [307]

Che-chiang sheng chien she nien k'an – [China]: Kai t'ing, Min kuo 22 [1933] – (= ser P-k&k period) – us CRL [339]

Che-chiang sheng nung ts'un tiao ch'a / China Nung ts'un fu hsing wei yuean hui – Shang-hai: Shang wu yin shu kuan, Min kuo 23 [1934] – (= ser P-k&k period) – us CRL [333]

Che-chiang sheng shih liang chih yuen hsiao / Chang, P'ei-kang – Shang-hai: Shang wu yin shu kuan, Min kuo 28 [1939] – (= ser P-k&k period) – us CRL [951]

Check list of the noctuidae of america, north of mexico / Grote, Augustus Radcliffe – [Buffalo?: Reinecke & Zesch], 1875-1876 – 2v on 1mf – 9 – 0-665-25030-4 – (individual vols also available separately) – mf#25030 – cn Canadiana [590]

Check to needless self-indulgence : or, an address to all whom it ma... – Whitehaven, England. 1833 – 1r – us UF Libraries [240]

Checkerboard – v1 n1-v2 n40 [1943 jan 13-1944 sep 6] – 1r – 1 – mf#1519806 – us WHS [071]

Checkered flag racing news – 1979 jun 6-1980 dec 10, 1981-86, 1987-1988 sep, 1988 oct-1989, 1990-97, 1998-2000 – 8r – 1 – (cont: checkered flag) – mf#570160 – us WHS [790]

Checklist of historical records survey publications, april 1943 / U.S. Federal Works Agency – (= ser Records Of The Federal Works Agency) – 1r – 1 – mf#T1028 – us Nat Archives [324]

A checklist of indonesian serials in the cornell university library (1945-1970) / Echols, J M & Thung, Y – Ithaca, 1973 – 3mf – 9 – mf#SE-20116 – ne IDC [959]

Checklist of inns of court holdings / Louisiana State University. Paul M. Herbert Law Center. Library – 1984 – 1mf – 9 – $1.50 – (covers add titles not offered for copyright reasons) – mf#LLMC 84-305 – us LLMC [020]

Checklist of official new jersey publications / New Jersey State Library – 1965 jul-1984 nov/dec, 1985 jan-1992 dec – 2r – 1 – mf#1054295 – us WHS [350]

Checklist of the woody cultivated plants of florida / Burch, Derek George – Gainesville, FL. 1988 – 1r – us UF Libraries [580]

Checklist of united states public documents / U.S. – 3rd ed. 1v. 1789-1909. Washington: GPO, 1909 (all publ) – 18mf – 9 – $27.00 – (v2 was never publ) – mf#LLMC 81-403 – us LLMC [324]

Checkout – [v1 n3-6,8 [1985 sep/oct-may/jun, sep/oct]]-[v5 n3-10 [1990 mar-oct]] – 1r – 1 – (cont: editor's notebook [kelly air force base [tx]]) – mf#1061823 – us WHS [071]

Cheddar valley times – England, 29 May-24 Dec 1914 – 23ft – 1 – uk British Libr Newspaper [072]

Chedwato dispatch – v1 n1-v3 n2 [1979 spr-1981 sum] – 1r – 1 – (cont by: car-del digest) – mf#633310 – us WHS [071]

Chedwato Service *see* Car-del scribe

Cheel, E *see* Results of dr e mjobergs swedish scientific expeditions to australia 1910-13

Cheer for life's pilgrimage / Meyer, Frederick Brotherton – New York: Fleming H Revell, c1897 – 1mf – 9 – 0-8370-7172-0 – mf#1986-1172 – us ATLA [240]

Cheerful ayres or ballads : first composed for one single voice and since set for three voices / Wilson, John – Oxford: printed by William Hall, for Richard Davies 1660 [mf ed 19–] – 3v on 111mf – 9 – mf#fiche 1154 – us Sibley [780]

Cheerful giver / Dick, Francis – Edinburgh, Scotland. 1832 – 1r – us UF Libraries [240]

Cheese city courier series / Lorain Co. Wellington – v1 n1. nov 1894-dec 1896 [wkly] – 1r – 1 – mf#B30729 – us Ohio Hist [071]

The cheese doll / Tagore, Abanindranath – Calcutta: Signet Press, 1945 – (= ser Samp: indian books) – us CRL [490]

Cheese reporter – 1943 oct 22/1947-1991 jan 4/dec 27 – 25r – 1 – (cont: dairy market reporter) – mf#269007 – us WHS [630]

Cheese trier / Wisconsin Swiss and Limburger Cheese Producers' Association – 1939 oct-1964 nov – 1r – 1 – mf#1054297 – us WHS [630]

Cheeseman, Lewis *see*
- Differences between old and new school presbyterians
- Ishmael and the church

Cheetham, S *see*
- A history of the christian church during the first six centuries
- A history of the christian church since the reformation
- The mysteries, pagan and christian
- A sketch of mediaeval church history

Cheetham, Samuel *see* A dictionary of christian antiquities

Cheetham, William *see* Christianity reviewed

Cheever, George Barrell *see*
- The century of preparation and the means and time of fulfillment
- God against slavery
- God's hand in america
- God's timepiece for man's eternity
- The gospel to be published and applied against all sin
- The guilt of slavery and the crime of slaveholding
- Memorabilia of george b. cheever, d.d
- The powers of the world to come, and the church's stewardship as invested with them
- Right of the bible in our public schools

Cheever, Henry Theodore *see*
- The biblical eschatology
- Correspondencies of faith and views of madame guyon

Chefs, Cooks, Pastry Cooks and Assistants Union, Local 89 [New York NY] *see* Kitchenrange

Chefs d'oeuvres classiques de l'opera francais – 38v – 1 – us L of C Photodup [780]

Chefs-d'oeuvre of the industrial arts / Burty, Philippe – London 1869 – (= ser 19th c art & architecture) – 6mf – 9 – mf#4.2.1279 – uk Chadwyck [740]

Chehalem valley news – Newberg OR: J Stamper, -1952 [wkly] – 1 – (absorbed by: newberg graphic) – us Oregon Lib [071]

Cheikh ahmed lahdcahi / Nicolas, A L M – Paris: P Geuthner 1910 [mf ed 1991] – (= ser Essai sur le cheikhisme 1) – 1mf – 9 – 0-524-01859-6 – (incl bibl ref) – mf#1990-2694 – us ATLA [260]

Le cheikhism. fascicule 3, la doctrine / Nicolas, A L M – Paris: E Leroux, 1911 – (= ser Essai sur le Cheikhisme) – 1mf – 9 – 0-524-01860-X – mf#1990-2695 – us ATLA [260]

Cheikho, L *see* Histoire de beyrouth...

Ch'ein, Kung-hsia *see*
- Hsiao p'in wen

Cheiros language of the hand : a complete practical work on the sciences of cheirognomy and cheiromancy, containing the system, rules, and experience of cheiro (comte de hamong) – 7th ed. New York: F Tennyson Neely, c1897 – us CRL [130]

O cheiroso – Rio de Janeiro, RJ. 27-28 nov 1911 – (= ser Ps 19) – mf#DIPER – bl Biblioteca [079]

Cheke, John *see* The gospel according to saint matthew

Cheke, Marcus *see* Carlota joaquina

Chekhov, Anton Pavlovich *see*
- Chorus girl and other stories
- Short stories
- Six plays of chekhov
- Stories of anton tchekov

Chekhovskoi, V I A *see* Entalpiia, teploemkost, teplota i entropiia plavleniia nekotorykh tugoplavkikh metallov

Chekoslovakian writer tells of the wave of terror sweeping over cadiz – Washington DC: American Friends of Spanish Democracy [193-?] – 9 – mf#w717 – us Harvard [946]

Chelan county courier – Spokane, WA. 1935-1938 (1) – mf#69257 – us UMI ProQuest [071]

Chelatkomplexe mit der p=o-doppelbindung : ein neues konzept fuer die asymmetrische synthese und die molekulare erkennung / Schrader, Thomas – (mf ed 2000) – 2mf – 9 – €40.00 – 3-8267-2734-7 – mf#DHS 2734 – gw Frankfurter [540]

Cheles. Ayuntamiento *see* Programa de ferias y fiestas en honor del santisimo cristo de la paz. septiembre 1976

Chelford, st john: baptisms 1673-1849 & marriages 1674-1752 – [North Cheshire FHS] – (= ser Cheshire church registers) – 2mf – 9 – £3.75 – mf#405 – uk CheshireFHS [929]

Chelford, st john: burials 1674-1894 – (= ser Cheshire church registers) – 2mf – 9 – £3.75 – mf#399 – uk CheshireFHS [929]

Chelford, st john the evangelist – [Macclesfield Ferrets] – (= ser Cheshire monumental inscriptions) – 2mf – 9 – £3.00 – mf#79 – uk CheshireFHS [929]

Cheliabinskii rabochii – Chelyabinsk, 1988 – 7r – 1 – us UMI ProQuest [077]

Chellig, Nadia *see* Pouvoirs et societe agra-pastorale dans les hautes plaines steppiques de l'algerie

Chellington – (= ser Bedfordshire parish register series) – 2mf – 9 – £5.00 – uk BedsFHS [929]

Chelmsford 1645-1849 – Oxford, MA (mf ed 1995) – (= ser Massachusetts vital record transcripts to 1850) – 23mf – 9 – 0-87623-228-4 – (mf 1t-4t: vital records 1645-1738. mf 5t-7t: births & deaths 1719-75. mf 7t: marriages & intentions 1714-39. mf 8t-9t: marriage intentions 1745-77. mf 9t: births 1704-07, 1774-75. mf 10t-12t: births & deaths 1767-1826. mf 12t-16t,18t-21t: intentions 1777-1849. mf 13t-15t: marriages 1780-1823. mf 16t-17t: births 1799-1844. mf 9t,17t-18t: marriages 1741-93, 1836-43. mf 17t-18t: out-of-town marriages 1661-1799. mf 21t: marriages 1822-36; deaths 1781-1844. mf 22t-23t: vital records 1843-49) – us Archive [978]

Chelmsford 1653-1900 – Oxford, MA (mf ed 1997) – (= ser Massachusetts vital records) – 278mf – 9 – 0-87623-388-4 – (mf 1-8: town & vital 1653-1777. mf 9-24: town & land 1653-1785. mf 12-32: vital records 1653-1826. mf 33-39: intentions 1825-1903. mf 35: marriages 1822-36. mf 40-53: vitals 1653-1843. mf 47-62: church records 1656-1901. mf 63-88: town records 1680-1767. mf 89-153: town & tax 1762-1823. mf 154-214: town record 1789-1904. mf 214-220: deeds 1837-52. mf 221-224: school 1819-69. mf 225-229: military 1861-75. mf 230-234: paupers 1817-1917. mf 235-239: voters 1877-1902. mf 240: births 1799-1844. mf 241-242: dog license 1863-72. mf 243-249: vitals 1834-70. mf 250-253: marriages 1854-87. mf 254-255: births 1871-99. mf 255-256: marriages 1887-1900. mf 253,256-257: deaths 1852-1900. mf 258-266: birth index 1843-1900. mf 267-273: marriage index 1844-1900. mf 274-278: death index 1843-1900) – us Archive [978]

Chelmsford weekly news – [London & SE] Essex Co Lib, Chelmsford Lib 1862-; BLNL 1980- – 1 – (cont as: chelmsford & south woodham weekly news [30 may 1991-]) – uk Newsplan [072]

Chelsea – New York. 1958+ (1) 1974+ (5) 1976+ (9) – ISSN: 0009-2185 – mf#9756 – us UMI ProQuest [400]

Chelsea 1655-1849 – Oxford, MA (mf ed 1995) – (= ser Massachusetts vital record transcripts to 1850) – 9mf – 9 – 0-87623-229-2 – (mf 1t-3t: births & deaths 1718-1842. mf 3t-4t: births 1842-49. mf 4t.5t: marriage intentions 1843-50. mf 5t-7t: marriages 1841-49. mf 6t-7t: deaths 1840-49. mf 7t: births & deaths 1820-49; out-of-town marriages 1655-1798. mf 7t-9t: marriage intentions 1739-1843; marriages 1738-1844) – us Archive [978]

Chelsea chronicle pimlico battersea and wandsworth gazette – London, UK. 18 may-8 jun 1860 – 1/4r – 1 – uk British Libr Newspaper [072]

Chelsea herald – London, UK. 16 feb 1884-11 dec 1886 – 3r – 1 – (aka: borough of chelsea herald; chelsea herald and west london standard; west london standard) – uk British Libr Newspaper [072]

Chelsea news and general advertiser – [London & SE] Chelsea Ref Lib 29 jul 1865-66, 1872-1975; BLNL 1978-80, 1981- – 1 – (westminster & chelsea news [2 aug 1879-25 apr 1885]; west london press, westminster & chelsea news [2 may 1885-2 mar 1962]; chelsea news, west london press, westminster & pimlico news [9 mar 1962-]) – uk Newsplan [072]

Chelsea pick and shovel – London, England. Jan-Dec 1900 – 1/4r – 1 – uk British Libr Newspaper [072]

Chelsea times – London, 24 Feb 1872; 4 Jan 1873-11 Dec 1875 – 2 1/2r – 1 – uk British Libr Newspaper [072]

Cheltenham free press – [SW England] Cheltenham nov 1834-jun 1908 [mf ed 2004] – 61r – 1 – (missing: 1873-jun 1874, 1890-91, 1897; cont by: cheltenham free press, weekly fashionable directory & gloucestershire herald [jan 1859-dec 1867]; cheltenham free press [jan 1868-jun 1908]) – uk Newsplan [072]

Cheltenham mercury – [SW England] Cheltenham jan 1890-12 sep 1903 [mf ed 2004] – 14r – 1 – uk Newsplan [072]

Chelys, minuritionum artificio exornata : sive, minuritiones ad basin, etiam ex tempore modulandi ratio / Simpson, Christopher – ed secunda, London: printed by W Godbid for H Brome 1667 [mf ed 19–] – 1mf – 9 – (1st edition, 1659, publ under title: the division-violist) – mf#fiche 182 – us Sibley [780]

O chem my sovsem ne dumaem / Mudrov, A E – Sedlets, 1888 – 1mf – 9 – mf#REF-459 – ne IDC [332]

O chem pel kolokol : stikhi / Kniazev, Vasilii – Izd 1 [mf ed 2002] Petrograd: Proletkul't, 1920 [mf ed 2002] – 1r – 1 – (filmed with: rafael' / boris zaitsev (1924)) – mf#5238 – us Wisconsin U Libr [810]

Chemawa american – 1970-76 – (= ser American indian periodicals... 1) – 16mf – 9 – $125.00 – us UPA [305]

Chemawa american / Chemawa Indian High School [Salem OR] – v52 n2 [1952 dec], v56 n1-v75 n[1955 sep, 15-1980 may], v77 n2-v78 n1 [1982 dec 10-1983 fall] – 1r – 1 – mf#514376 – us WHS [373]

Chemawa Indian High School [Salem OR] *see* Chemawa american

Chemawa indian school: register of students admitted, 1880-1928; descriptive statements of students, 1890-1914; and graduating class rolls, 1885-1921 / U.S. Bureau of Indian Affairs – (= ser Records of the bureau of indian affairs) – 1r – 1 – mf#P2008 – us Nat Archives [305]

Chemehuevi newsletter – 1968-73 – (= ser American indian periodicals... 1) – 6mf – 9 – $95.00 – us UPA [305]

Chemerinsky, Hayim *see* Ayarati motele

Chemeriskii, Aleksandr *see* Tsionistishe trayberayen

Chemical abstracts : collective indexes – 1907-86. 11 indexes – 6,9 – (1st-6th 1907-61. 7th 1962-66. 8th 1967-71. 9th 1972-76. 10th 1977-81. 11th 1982-86. 12th 1987-91. 13th 1992-96) – us Chemical [540]

Chemical age – London. 1974-1981 (1) 1974-1981 (5) 1979-1981 (9) – (cont: chemical age international) – ISSN: 0302-2900 – mf#926,01 – us UMI ProQuest [660]

Chemical age international – London. 1919-1974 (1) 1965-1974 (5) – (cont by: chemical age) – ISSN: 0009-2312 – mf#926 – us UMI ProQuest [660]

Chemical age of india – Bombay. 1968-1989 (1) 1970-1989 (5) 1973-1989 (9) – ISSN: 0009-2320 – mf#5058 – us UMI ProQuest [540]

Chemical analyses – Lake City, FL. 1889 – 1r – us UF Libraries [630]

Chemical analysis and physical tests of some florida clays / Sciutti, Walter J – s.l, s.l? 1929 – 1r – us UF Libraries [630]

Chemical and engineering news – v1- 1923- – 1,5,6,9 – us ACS [540]

Chemical and geological essays / Hunt, Thomas Sterry – Salem [MA]: S E Cassino, 1878 – 6mf – 9 – 0-665-90678-1 – (incl ind) – mf#90678 – cn Canadiana [550]

Chemical and petroleum engineering – New York. 1965-1976 (1) 1965-1976 (5) – ISSN: 0009-2355 – mf#10877 – us UMI ProQuest [550]

Chemical and process engineering – London. 1951-1972 (1) 1971-1971 (5) (9) – ISSN: 0009-2371 – mf#5723 – us UMI ProQuest [660]

Chemical business – New York. 1984-1994 (1) 1984-1994 (5) 1984-1994 (9) – ISSN: 0731-8774 – mf#13969 – us UMI ProQuest [540]

Chemical communications : chem comm – Cambridge. 1996+ (1) 1996+ (5) 1996+ (9) – (cont: journal of the chemical society chemical communications) – ISSN: 1359-7345 – mf#10060,01 – us UMI ProQuest [540]

Chemical economy and engineering review (ceer) – Tokyo. 1970-1987 [1]; 1971-1987 [5]; 1976-1987 [9] – (cont: japan chemical quarterly) – ISSN: 0009-2436 – mf#5826 – us UMI ProQuest [540]

Chemical engineer – London. 1976+ (1,5,9) – ISSN: 0302-0797 – mf#11193 – us UMI ProQuest [660]

Chemical engineering – New York. 1902+ (1) 1964+ (5) 1970+ (9) – ISSN: 0009-2460 – mf#29 – us UMI ProQuest [660]

Chemical engineering and mining review – Melbourne, Astralia. apr-nov 1918; jan 1920-dec 1921 (mthly) – 2r – 1 – (missing: sep 1918; apr 1920) – uk British Libr Newspaper [073]

Chemical engineering and processing = Genie des procedes – Lausanne. 1988-1995 (1,5,9) – ISSN: 0255-2701 – mf#42435 – us UMI ProQuest [660]

Chemical engineering journal – Lausanne. 1996+ (1,5,9) – (cont: chemical engineering journal and the biochemical engineering journal) – mf#42183,01 – us UMI ProQuest [660]

Chemical engineering journal and the biochemical engineering journal – Lausanne. 1970-1996 (1) 1970-1996 (5) 1987-1996 (9) – (cont by: chemical engineering journal) – ISSN: 0923-0467 – mf#42183 – us UMI ProQuest [660]

Chemical engineering progress – New York. 1986+ (1) 1986+ (5) 1986+ (9) – (cont: cep : chemical engineering progress) – ISSN: 0360-7275 – mf#717,01 – us UMI ProQuest [660]

Chemical engineering research and design : transactions of the institution of chemical engineers / Institution of Chemical Engineers – Rugby. 1983-1989 (1) 1983-1989 (5) 1983-1989 (9) – (cont: transactions of the institution of chemical engineers) – ISSN: 0263-8762 – mf#11192,01 – us UMI ProQuest [660]
Chemical engineering research and design : transactions of the institution of chemical engineers pt a / Institution of Chemical Engineers – Rugby. 1990-1996 (1,5,9) – ISSN: 0263-8762 – mf#17756 – us UMI ProQuest [660]
Chemical engineering science – Oxford. 1952+ [1,5,9] – ISSN: 0009-2509 – mf#49029 – us UMI ProQuest [660]
Chemical engineering science – Oxford. 1952+ [1,5,9] – ISSN: 0009-2509 – mf#49029 – us UMI ProQuest [660]
Chemical engineers' handbook / Perry, John Howard – New York, NY. 1950 – 1r – us UF Libraries [660]
Chemical equipment – Morris Plains. 1973-1981 (1) – ISSN: 0009-2525 – mf#7561 – us UMI ProQuest [660]
Chemical geology – Amsterdam. 1966+ (1) 1966+ (5) 1987+ (9) – ISSN: 0009-2541 – mf#42245 – us UMI ProQuest [550]
Chemical health and safety – v1-6. 1994-99 – 1,5,6,9 – (publ cont by elsevier science) – us ACS [540]
Chemical industry news – Bombay. 1972-1981 (1) 1972-1981 (5) 1976-1981 (9) – ISSN: 0009-2576 – mf#6741 – us UMI ProQuest [660]
Chemical market reporter – New York. 1996+ (1) 1996+ (5) 1996+ (9) – (cont: chemical marketing reporter) – mf#6631,01 – us UMI ProQuest [540]
Chemical marketing reporter – New York. 1876-1996 (1) 1979-1996 (5) 1979-1996 (9) – (cont by: chemical market reporter) – ISSN: 0090-0907 – mf#6631 – us UMI ProQuest [540]
The chemical news and journal of physical science – [London: Griffin, Bohn & Co, 1861-1921]. v14-15 1867 – 1r – 1 – us CRL [500]
Chemical physics – Amsterdam. 1973+ (1) 1973+ (5) 1987+ (9) – ISSN: 0301-0104 – mf#42145 – us UMI ProQuest [540]
Chemical physics letters – Amsterdam. 1967+ (1) 1967+ (5) 1986+ (9) – ISSN: 0009-2614 – mf#42146 – us UMI ProQuest [540]
Chemical problems associated with the control of pests in stored groundniuts in west africa / Babatunde Somade, H M – London, [1953] – us CRL [630]
Chemical processing – Chicago. 1938+ (1) 1975+ (5) 1976+ (9) – ISSN: 0009-2630 – mf#8803 – us UMI ProQuest [540]
Chemical reviews – Washington, DC: ACS. v69(1969)-v89(1989) [mthly] – 1 – mf#0009-2665 – us ACS [540]
Chemical senses – Oxford. 1988-1996 (1,5,9) – ISSN: 0379-864X – mf#16447,02 – us UMI ProQuest [612]
Chemical Society (Great Britain) *see*
– Chemical society reviews
– Faraday discussions of the chemical society
– Journal of the chemical society
– Journal of the chemical society chemical communications
Chemical Society (Great Britain). Analytical Division *see* Proceedings of the analytical division of the chemical society
Chemical Society of Japan *see* Bulletin of the chemical society of japan
Chemical society reviews / Chemical Society (Great Britain) – London. 1972+ (1) 1976+ (5) 1976+ (9) – ISSN: 0306-0012 – mf#11250 – us UMI ProQuest [540]
Chemical studies on soils from florida citrus groves / Peech, Michael – Gainesville, FL. 1939 – 1r – us UF Libraries [634]
Chemical studies on soils from florida citrus groves / Peech, Michael – Gainesville, FL. 1948 – 1r – us UF Libraries [634]
Chemical study of colloidal phospate / Purvis, E R – s.l, s.l? 1929 – 1r – us UF Libraries [630]
Chemical study of some typical soils of the florida peninsula / Persons, A A – Lake City, FL. 1897 – 1r – us UF Libraries [630]
Chemical week – New York. 1926+ (1) 1965+ (5) 1970+ (9) – ISSN: 0009-272X – mf#164 – us UMI ProQuest [660]
Chemical worker / International Chemical Workers Union – Akron OH. 1976 apr/may-1980 dec, 1981-88, v55 n1-v56 n3 [1995 jan/feb-1996 may/jun] – 3r – 1 – (cont: international chemical worker; cont by: ufcw action) – mf#569616 – us WHS [660]
Chemicke listy – Praha. 1975-1981 (1) 1975-1981 (5) 1975-1981 (9) – ISSN: 0009-2770 – mf#8944 – us UMI ProQuest [540]
Chemico-biological interactions – Amsterdam. 1969-1992 (1) 1969-1992 (5) 1987-1992 (9) – ISSN: 0009-2797 – mf#42147 – us UMI ProQuest [574]
Chemie-spiegel – Coswig DE, 1960 27 apr-1990 jul [gaps] – 5r – 1 – (title varies: chemiewerk coswig) – gw Misc Inst [074]
Chemiewerk coswig *see* Chemie-spiegel
Chemiker-zeitung – Heidelberg. 1878-1944 (1) – ISSN: 0009-2894 – mf#1141 – us UMI ProQuest [540]
Chemiker-zeitung – Koethen DE, 1880-1909 – 56r – 1 – uk British Libr Newspaper [540]
Le chemin de fer : nos communications avec l'ouest: discours / Beaubien, Louis – Quebec?: s.n, 1875? – 1mf – 9 – mf#24044 – cn Canadiana [380]
Chemin de fer de halifax et de quebec et travaux publics / Canada (Province). Gouverneur general (1847-1854: Elgin) – Montreal: Impri par Lovell & Gilbson, 1849 [mf ed 1982] – 1mf – 9 – mf#SEM105P121 – cn Bibl Nat [380]
Chemin de fer de Quebec et du Saguenay *see* Report of the chief engineer, on the survey of the line for the quebec and saguenay railway
Chemin de fer de quebec et halifax : reponse a une adresse de l'assemblee legislative...au sujet du grand tronc de chemin de fer entre halifax et quebec et de la vers l'ouest a travers le haut-canada / Canada (Province). Parlement. Assemblee legislative – Quebec: Impr par John Lovell, 1852 [mf ed 1982] – 2mf – 9 – mf#SEM105P128 – cn Bibl Nat [380]
Chemin de fer du grand tronc *see*
– Correspondence between the company and the dominion government respecting advances to the canadian pacific railway company
– Proceedings of the...annual general meeting of the shareholders of the grand trunk railway company of canada...1855-
– Proceedings of the...meeting of the shareholders of the grand trunk railway company of canada...1854
– Reponse complementaire a une adresse de l'assemblee legislative du 21 du mois dernier
– Reponse partielle a une adresse de l'assemblee legislative du 21 courant
– Statements reports and accounts of the grand trunk railway company of canada
Le chemin de fer du lac saint-jean ses origines : ses developpements passes et futur, son importance capitale, son action sur le progres et l'avenir de la province de quebec: ouvrage historique et descriptif / Buies, Arthur – Quebec: Leger Brousseau, 1895 [mf ed 1979] – 2mf – 9 – mf#SEM105P24 – cn Bibl Nat [380]
Chemin de fers de la baie des chaleurs : dossier officiel complet: correspondance officielle entre son honneur le lieutenant-gouverneur et m mercier, premier ministre / Quebec (Province) – Quebec: Belleau, 1891 – 1mf – 9 – mf#02405 – cn Canadiana [380]
Chemin de la croix / Mach, Jose – [Quebec?: s.n.] 1886 [mf ed 1984] – 1mf – 9 – 0-665-46426-6 – mf#46426 – cn Canadiana [240]
Chemin de la croix des ames du purgatoire : suivi de quelques prieres tres efficaces pour obtenir la delivrance des defunts, et de l'acte heroique de charite / Pretre de l'archidiocese de Quebec – Quebec: J A Langlais, 1886 [mf ed 1984] – 1mf – 9 – 0-665-46391-X – mf#46391 – cn Canadiana [241]
Chemin de la croix et autres prieres : a l'usage des sauvages des postes d'albany, savern, martin's falls (baie d'hudson) / Garin, Andre-Marie – Montreal: Beauchemin & Valois, 1883 [mf ed 1984] – 2mf – 9 – 0-665-04894-7 – mf#04894 – cn Canadiana [241]
Le chemin de la vie eternele compose en latin... / Sucquet, Antonii – Anvers: Henry Aertssens, 1623 – 11mf – 9 – mf#O-1920 – ne IDC [090]
Chemin Dupontes, Paul *see* Petites antilles
Chemin-Dupontes, Jean-Baptiste *see* Cours pratique de franc-maconnerie
Le cheminot algerien / Union d'Algerie – Alger. 1930-38 – 1 – fr ACRPP [380]
Le cheminot de l'etat / Union des syndicats confederes des Chemins de Fer de l'Etat francais – Paris. mai 1917-avr 1920, mai 1928, 1931-juin juil 1939 – 1 – fr ACRPP [331]
Le cheminot unifie : monats-organ des einheitsverbandes der elsass-lothr eisenbahner – Strassburg (Strasbourg F), 1935 aug-1940 may [gaps] – 1 – fr ACRPP [380]
Chemins de fer dans la province de quebec : discours prononce a l'assemblee legislative le 29e jour de decembre 1896, sur les resolutions touchant les subsides aux chemins de fer / Flynn, Edmund James – Quebec: s.n, 1897 – 2mf – 9 – mf#03113 – cn Canadiana [380]
Chemische gasphasenabscheidung von hafniumcarbid und hafniumnitrid / Wormer, Oliver Gerd – (mf ed 1993) – 2mf – 9 – €49.00 – 3-89349-644-0 – mf#DHS 644 – gw Frankfurter [540]
Chemische zeitschrift – Leipzig DE, 1901 oct-1908 – 5r – 1 – uk British Libr Newspaper [540]
Die chemisch-mechanische wurzelkanalaufbereitung : historische entwicklung der wurzelkanalinstrumente, spuelmittel und aufbereitungsmethoden / Orth, Ulrike – (mf ed 1997) – 2mf – 9 – €40.00 – 3-8267-2462-3 – mf#DHS 2462 – gw Frankfurter [617]
Chemist – Bethesda. 1929+ (1) 1967+ (5) 1970+ (9) – ISSN: 0009-3025 – mf#55 – us UMI ProQuest [540]
Chemist and druggist and pharmicist of australia *see* Chemist and druggist of australia
Chemist and druggist of australia – Melbourne, Australia. jan-1 dec 1886, 1888-1 dec 1909, 1910-9 jul 1934 – 41 1/2r – 1 – (aka: chemist and druggist and pharmicist of australia) – uk British Libr Newspaper [615]
Chemistry and chemical industry – Tokyo. 1984-1985 (1,5,9) – ISSN: 0022-7684 – mf#12595 – us UMI ProQuest [660]
Chemistry and industry – London. 1989-1992 (1,5,9) – ISSN: 0009-3068 – mf#17589,01 – us UMI ProQuest [660]
Chemistry and physics of lipids – Amsterdam. 1966+ (1) 1966+ (5) 1987+ (9) – ISSN: 0009-3084 – mf#42148 – us UMI ProQuest [574]
Chemistry and technology of fuels and oils – New York. 1965-1977 (1) 1965-1977 (5) – ISSN: 0009-3092 – mf#10878 – us UMI ProQuest [550]
Chemistry in britain – Cambridge. 1965+ [1]; 1971+ [5]; 1977+ [9] – ISSN: 0009-3106 – mf#5874 – us UMI ProQuest [540]
Chemistry in warfare / Hessel, Frederick Adam – New York, NY. 1942 – 1r – us UF Libraries [540]
Chemistry international – Oxford. 1978-1984 (1,5,9) – ISSN: 0193-6484 – mf#49306 – us UMI ProQuest [540]
Chemistry international – Oxford. 1985-1996 (1,5,9) – ISSN: 0193-6484 – mf#15590 – us UMI ProQuest [540]
Chemistry letters – Tokyo. 1975-1996 (1,5,9) – ISSN: 0366-7022 – mf#12596 – us UMI ProQuest [540]
Chemistry meteorology and the function of digestion : considered with reference to natural theology / Prout, William – [2nd ed] London, 1834 – (= ser 19th c evolution & creation) – 7mf – 9 – mf#1.1.10854 – uk Chadwyck [110]
The chemistry of common life / Johnston, James Finlay Weir – New ed., rev. and updated by Arthur Herbert Church. Edinburgh: W. Blackwood, 1879. xxvi,592p. illus. Includes index – 1 – us Wisconsin U Libr [390]
Chemistry of heterocyclic compounds – New York. 1965-1976 (1) 1965-1976 (5) – ISSN: 0009-3122 – mf#10903 – us UMI ProQuest [540]
Chemistry of natural compounds – New York. 1966-1976 (1) 1966-1976 (5) – ISSN: 0009-3130 – mf#10919 – us UMI ProQuest [540]
Chemla-Lamech, Felix *see*
– Carte topographique de la lune
– Etude monographique des plus grandes formations lunaires
Chemnitz d A, M *see*
– Apologia oder verantwortung dess christlichen concordienbuchs
– De dvabvs natvris in christo de hypostatica earvm vnione
– Die fuernemsten heupstueck der christlichen lehre
– Histori dess sacramentstreits
– Historia der passion vnsers lieben herrn vnd heilands jesu christi
– Loci theologici
– Postilla oder ausslegung der euangelien
– Repetitio sanae doctrinae de vera praesentia
Chemnitz, Martin *see*
– De dvabvs natvris in christo; de hypostatica earvm vnione, de commvincatione idiomatvm, et de aliis qaestionibvs independentibvs; libellvs ex scriptvra sententijs & ex pvrioris antiqvitatis testimonijs ... cvm praefatione nicolai selnecceri
– De incarnatione filii dei item de officio et maiestate christi tractus
– Von der ursache der suende und von der zufaelligkeit
Chemnitzer anzeiger und stadtbote – Chemnitz DE, 1883 2 sep-1900 27 mar [gaps] – 23r – 1 – (title varies: 23 may 1885: chemnitzer landes-anzeiger; 27 mar 1893: general-anzeiger fuer chemnitz und umgegend) – gw Misc Inst [074]
Chemnitzer bote – Chemnitz DE, 1919-1928 sep – 44r – 1 – (title varies: 1850: chemnitzer tageblatt und anzeiger) – gw Misc Inst [074]
Chemnitzer morgenpost *see* Dresdner morgenpost
Chemnitzer neueste nachrichten *see* Neueste nachrichten
Chemnitzer tageblatt – Chemnitz DE, 1992 jan-15 mar – 1r – 1 – (reg ed of leipziger tageblatt, leipzig) – gw Misc Inst [074]
Chemnitzer tageblatt und anzeiger *see* Chemnitzer bote
Chemosphere – Oxford. 1972+ (1,5,9) – ISSN: 0045-6535 – mf#49030 – us UMI ProQuest [333]
Chemosphere, global change science – Kidlington, 1999+ [1,5,9] – ISSN: 1465-9972 – mf#42824 – us UMI ProQuest [333]
Chemotherapy – Basel. 1966-1996 (1) 1967-1996 (5) 1970-1996 (9) – ISSN: 0009-3157 – mf#2050 – us UMI ProQuest [615]
Chemotherapy = Nihon kagaku ryoho gakkai zasshi – Tokyo. 1975-1979 (1) 1975-1979 (5) 1975-1979 (9) – ISSN: 0009-3165 – mf#9902 – us UMI ProQuest [615]
Chem-steel news / United Steelworkers of America – v37 n7-v42 n7 [1975 oct 17-1981 jul] – 1r – 1 – (cont by: chem-steel news [1981]) – mf#678895 – us WHS [660]
Chem-steel news / United Steelworkers of America – v1 n1-v10 n2 [1981 aug-1990 apr] – 1r – 1 – (cont: chem-steel news) – mf#1061988 – us WHS [660]
Chemtronics – Guildford. 1989-1991 (1) 1989-1991 (5) 1989-1991 (9) – ISSN: 0267-5900 – mf#17214 – us UMI ProQuest [540]
Chemung. Presbytery (Pres. Church in the USA) *see* Minutes, 1836-1894
Chemung valley news – Horseheads, NY. 1905-1998 (1) – mf#69197 – us UMI ProQuest [071]
Ch'en, An-jen *see*
– Ming tai hsueh shu ssu hsiang
– Sung tai ti k'ang chan wen hsueh
Ch'en, Ch'ang-heng *see*
– Min sheng chu i chih tsung ho yen chiu
– Wu ch'uan hsien fa ts'ao an ching i
Ch'en, Chao-yu *see* Kuang-tung t'ang yeh yu feng jui
Chen, Cheng-siang *see* Chen-la feng t"u chi ti yen chiu
Ch'en, Chen-hua *see*
– Huo pi yin hang yuan li
– Nung yeh hsin yung
Ch'en, Chen-lu *see*
– Hsien tai lao tung wen t'i lun ts'ung ti i chi
– Lao tung wen t'i ta kang
Ch'en chi / Hsue, Chung-nien – Shang-hai: Pei hsin shu chue, 1933 – (= ser P-k&k period) – us CRL [480]
Ch'en, Chia-ch'ing *see* Han wei liu ch'ao shih yen chiu
Ch'en chia-keng chin shih chi : kuo nei yen chung chue shih ti i mien ching tzu – [Sl]: Chin men ch'u pan she, 1941 – (= ser P-k&k period) – us CRL [951]
Ch'en, Chi-ch'eng *see* Pu p'ing teng t'iao yueh ch'ien shuo
Ch'en, Chien-hsun *see* Sheng ching chih hun yin kuan (ccm88)
Ch'en, Chih-mai *see* Chien kuo chung ti chi ko chung yao wen t'i
Ch'en, Ch'i-lu *see* Ying-kuo tui hua shang yeh
Chen ching chung wai ti wan nan shih pien mien mien kuan / Pien i ch'u pan she – [sl]: Shih chieh ch'u pan she, Min kuo 30 [1941] – (= ser P-k&k period) – us CRL [951]
Ch'en, Chin-yung *see*
– Chiang fan shang hsia p'ien
– Ch'uan tao i yu [ccm91]
Ch'en, Ch'i-su *see* Ti erh hao han chien: ssu mu chu
Ch'en, Chi-yun *see*
– Chi-tu chiao shih shen mo? [ccm85]
– Fu yin te chun pei
Chen, Chi-yun *see* Fu yin te chun pei (ccm87)
Ch'en, Chu *see*
– Chia pin li chih [ccm89]
– Shou hsuan ko wen kao hsuan
Chen, Chu *see* Ai-fan-ssu-tun hu sheng (ccm90)
Ch'en, Ch'uan *see*
– Ai meng ying
– Huang ho lou
– Hun hou
– T'ien wen
– Wen hsueh p'i p'ing ti hsin tung hsiang
– Wu ch'ing nu
Ch'en, Chu-i *see* Tu ch'i yu fang hu
Ch'en chung te ku shih (ccm140) – Shanghai, 1926 [mf ed 198?] – (= ser Ccm 140) – 1 – mf#1984-b500 – us ATLA [801]
Ch'en, Chung-fan *see* Han wei liu ch'ao wen hsueh
Ch'en, Chung-hao *see* Kuo chi hsien shih yu k'ang chan wai chiao
Ch'en, Chung-min *see* Nung yeh chien she yu ho tso
Ch'en, Fei-mo *see* Hsi chun chan
Ch'en, Heng-che *see* Heng-che san wen chi
Ch'en, Ho-k'un *see* Hsiang-pei chih chan
Ch'en hsiang chi / K'ung, Ling-ching – Shang-hai: Shih chieh shu chue, Min o 33 [1944] – (= ser P-k&k period) – us CRL [820]
Ch'en, Hsiang-ho *see* Tu shen che
Ch'en, Hsiang-po *see* Kung chiao lun
Chen hsien shih lun / T'ai-hsue – Shang-hai: Chung-hua shu chue, Min kuo 29 [1940] – (= ser P-k&k period) – us CRL [280]
Ch'en, Hsi-hao *see* Hsin she hui wen t'i
Ch'en, Hsi-meng *see* Ma-chin-na-li-ya
Ch'en, Hsueh-chao *see*
– Pai hsu chi
– Shih tai fu nu

Ch'en, Hui *see*
- Hai shang yin
- Kuang-hsi chiao t'ung wen t'i

Ch'en, Hung-chin *see* Chih min ti yue pan chih min ti
Ch'en, I *see*
- Jih-pen t'ung chih t'ai-wan ching kuo
- Tsen yang tung yuan nung min ta chung

Ch'en, I-fu *see* She hui tiao ch'a yu t'ung chi hsueh
Chen, J A *see* Selected physiological variables and distance running performance among non-elite, heterogeneous groups of male and female runners
Ch'en, Kao-yung *see* K'ang chan yu pao chia yun tung
Chen, Kevin Y *see* The effect of active recovery on the post-exercise diffusion capacity
Chen kuang [ccs] – [Kuang-tung] v24 n5-34 n5. 1925-35 [gaps] [mf ed 198?] – (= ser Chinese christian serials coll) – 5r – 1 – (began in 1902; aka: chen kuang tsa chih) – mf0298 – us ATLA [240]
Ch'en, Kuang-yao *see*
- Mi yu yen chiu
- Min chung wen i lun chi
- Tu hsing chi

Ch'en, Kung-ch'ia *see* Hsin li chien she yu hsien cheng chien she
Ch'en, Kung-po *see*
- Han feng chi
- Ko ming yu ssu hsiang

Ch'en, Kuo-chun *see* Kuei-chou miao i ko yao
Ch'en, Ku-yuan *see* T'u ti fa
Chen, Kwan *see* Personal investment in exercise and sport
Chen, Li *see*
- Lien kung tang shih
- Performances of coaches
- Timeout decisions of basketball coaches of men's and women's collegiate teams

Ch'en, Li-fu, 1899- *see* Ch'en li-fu hsien sheng yen lun chi ti i chi
Ch'en li-fu hsien sheng yen lun chi ti i chi / Ch'en, Li-fu, 1899- – [China: sn] – (= ser P-k&k period) – us CRL [951]
Ch'en, Li-t'e *see* Min chih ti ch'ien t'u
Ch'en, Li-t'ing *see* Chi-tu chiao ch'ing nien hui shih yao [ccm93]
Ch'en lo / Pa, Chin – Shang-hai: Shang wu yin shu kuan, 1948 – (= ser P-k&k period) – us CRL [480]
Ch'en, Meng-chia *see*
- Hsin yueh shih hsuan
- Meng-chia shih chi
- Pu k'ai hua ti ch'un t'ien

Ch'en, Mien *see* Pan yeh
Chen, Ming *see* Mei li fu jao ti kuo chia chien-pu-chai
Ch'en, Ming-chung *see*
- Hsi chu yu chiao yu
- Sheng huo hsien shang

Chen, Moon S *see* Pressure sore prevention self-efficacy and outcome expectations in the spinal cord-injured
Ch'en, Mu et al *see* Fei ch'ang shih ch'i chih chun shih chih shih
Ch'en, Nien-chung *see* Hsien ko chi min i chi kuan
Ch'en, Pai-ch'en *see*
- Ch'a yeh pang tzu
- Chieh hun chin hsing ch'ue
- Feng yu chih yeh
- Hou fang hsiao hsi chu
- Hsi chu ch'uang tso chiang hua
- K'ai ko kuei
- Luan shih nan nu
- Min tsu wan sui
- Ni t'ui tzu
- Shih ta-k'ai ti mo lu
- Sui han t'u
- Ta ti huang chin, i ming, ch'iu shou

Ch'en, Pai-ch'en et al *see* Sheng li hao
Ch'en, Pei-ou *see*
- Hsien cheng chi ch'u chih shih

Ch'en, Ping-po *see* Chin jih chih hsien cheng
Ch'en, Ping-yuan *see* Ko kuo ping i hsing cheng kai lun
Ch'en, Po-hsin *see* Ti fang tzu chih yu hsin hsien chih
Chen shang chi / Fang, Hsi – Ch'ang-sha: Shang wu yin shu kuan, 1938 – (= ser P-k&k period) – us CRL [480]
Ch'en, Shao-yu *see*
- K'ang jih chiu kuo cheng ts'e
- Lun fan ti t'ung i chan hsien wen t'i
- Mu ch'ien kuo nei wai hsing shih yu ts'an cheng hui ti ssu tz'u ta hui ti ch'eng chi
- T'o p'ai tsai chung-kuo

Ch'en, Shih *see* Jen chien tsa chi
Chen shih chih ko : huang yeh tuan shu shang chuean / Feng, Hsueeh-feng – Ch'ung-ch'ing: Tso chia shu wu, 1943 – (= ser P-k&k period) – us CRL [810]
Ch'en, Shih-hung *see* Ts'an kuan ch'uan ch'ien kuei yueh hsiang kan ko sheng hsin hsien chih pao kao
Chen, Shing-Jye *see* Effects of arch support on changes in arch height, vertical ground reaction force and center of pressure under different foot positions while loading and demonstrated by contact bone-on-bone forces
Ch'en, Shou-chu *see* Nai ho t'ien
Ch'en, Shu-i *see* I-se-lieh te ku shih (ccm6)
Ch'en, Shu-shih *see* Fang kung chou hsin lun
Ch'en Ta *see* Chung-kuo lao-kung wen-t'i
Ch'en, Ta *see* Jen k'ou wen t'i
Chen tao ch'ang shih [ccc202] = Abiding knowledge of christian truth / Huang, To – Shanghai 1934 [mf ed 198?] – (= ser Ccm 202) – 1 – mf#1984-b500 – us ATLA [220]
Ch'en, Ta-pei *see* Chang ssu t'ai t'ai
Chen, Ta-san *see* Hsin chiu yueh wen ta (ccm266)
Ch'en, Ta-tz'u *see* Hua chiao
Ch'en, Te-cheng *see* T'ien ts'ai erh t'ung chiao yu
Ch'en, T'ing-t'ing *see* Jen shih t'ai-wan
Chen tsai chi yao : ch'an-an pien chi – Shang-hai: kuang yi shu tien, Min kuo 25 [1936] – (= ser P-k&k period) – us CRL [951]
Ch'en, Tso-liang *see* Pi chiao chiao yu
Ch'en, Tsui-yun *see* Nung ts'un ching chi kai lun
Ch'en, Tsu-jun *see* I chiu ssu erh nien ti t'ai-p'ing yang
Ch'en, Tsu-lien *see* Shan-hsi tiao ch'a chi
Ch'en, Tuan-chih *see* K'ang chan yu she hui wen t'i
Ch'en, Tzu-chan *see* T'ang tai wen hsueh shih
Ch'en, Tzu-mi *see* Shih yung kung chai k'u ch'uan hui pien
Ch'en, Wang-tao *see*
- Hsiao p'in wen ho man hua
- Hsiu tz'u hsueh fa fan

Ch'en, Wei *see* Jih-pen fu nu yun tung k'ao ch'a chi lueh
Ch'en, Wei-sung *see* Fu jen chi
Ch'en, Wen-chien *see* Hsueh sha hsing ts'ao
Ch'en, Wen-yuan *see* Tsung chiao yu jen ke (ccm94)
Chen, Xinhua *see*
- Die bedeutung der wirtschaftlichen kooperation fuer die wirtschaftsentwicklung chinas am beispiel joint ventures
- Die chinesische wirtschaftsreform und inflationsproblematik seit 1979

Ch'en, Yen-lin *see* Shang-hai ti ch'an ta ch'uan
Ch'en, Ying *see* Nu hsing
Ch'en, Yin-k'o *see* T'ang tai cheng chih shih shu lun kao
Ch'en, Yuan *see* Ming chi tien ch'ien fo chiao k'ao
Ch'en yuean-yuean / Chiang, Ch'i – Shang-hai: Kuo min shu tien, 1940 – (= ser P-k&k period) – us CRL [820]
Ch'en, Yueh *see* Kuan-tung lei
Ch'en Yun-wen *see* Chan shih shih yu cheng ts'e
Ch'en, Yu-to *see* Yu-to pi chi
Chenango american and whitney point reporter – Greene, NY. 1855-1997 (1) – mf#64988 – us UMI ProQuest [071]
Chenango telegraph – Norwich, NY. 1835-1876 (1) – mf#65129 – us UMI ProQuest [071]
Chenaye-Desbois, F A de la *see* Dictionnaire de la noblesse de la france
Cheney bulletin – n35-40 [1974 may-1976 jan] – 1r – 1 – (cont: great cheney clan's bulletin) – mf#351524 – us WHS [071]
Cheney Cowles Memorial Museum. Eastern Washington State Historical Society *see* Cornhusk bags of the plateau indians
Cheney, Mary Bushnell *see* Life and letters of horace bushnell
Chen-fu kung-pao *see* Seifu koho
Cheng, Ch'en-chih *see* Han chien ch'ou shih
Cheng, Chen-to *see*
- Chan hao
- Cheng chen-to chieh tso hsuean
- Chin pai nien ku ch'eng ku mu fa chueeh shih, cheng chen-to chu
- Hsi hsing shu chien
- K'un hsueh chi
- Tuan chien chi
- Wen i chien shang yu p'i p'ing
- Wen t'an

Cheng chen-to chieh tso hsuean / Cheng, Chen-to – Shang-hai: Hsin hsiang shu tien, Min kuo 30 [1941] – (= ser P-k&k period) – us CRL [480]
Cheng, Chen-wen *see* Min hsien shih lueh ch'u kao
Cheng, Chen-yu *see* Hsien ko chi tsu chih chung chi hsiang chung yao wen t'i
Cheng chi kai k'uang – [China]: Chuen shih wei yuean hui ch'uean kuo chih shih ch'ing nien chih yuean ts'ung chuen pien lien tsung chien pu, Min kuo 34 [1945] – (= ser P-k&k period) – us CRL [951]
Cheng ch'i ko / Wu, Tsu-kuang – Shang-hai: K'ai ming shu tien – (= ser P-k&k period) – us CRL [820]
Cheng chiao an wei (ccm132) / Gutzlaff, Karl Friedrich August – [s.l: s.n, 1837] [mf ed 198?] – (= ser Ccm 132) – 4v – 1 – mf#1984-b500 – us ATLA [240]
Cheng chiao chen ch'uean / Liang, I-chuen – Ch'ung-ch'ing: T'ieh hua feng, Min kuo 33 [1944] – (= ser P-k&k period) – us CRL [210]
Cheng chih ch'ang shih – [China]: Che tung T'ao-fen shu tien, Min kuo 34 [1945] – (= ser P-k&k period) – us CRL [951]
Cheng chih chien she yue chih tu ching shen / Liu, Nai-ch'eng – Ch'ung-ch'ing: Kuo min t'u shu ch'u pan she, Min kuo 30 [1941] – (= ser P-k&k period) – us CRL [951]
Cheng chih hsueeh kai lun / Li, Chien-nung – Ch'ang-sha: Shang wu yin shu kuan, [Min kuo 23 [1934]] – (= ser P-k&k period) – us CRL [951]
Cheng chih k'o hsueeh ta kang / Teng, Ch'u-min – Shang-hai: K'un lun shu tien, 1932 – (= ser P-k&k period) – us CRL [951]
Ch'eng, Chih-i *see* Hsin-yueh yen chiu chih nan (ccm96)
Cheng, Chih-i *see* Chu chi te chiao hui [ccm95]
Ch'eng, Ch'ing-fang *see* Fei ch'ang shih ch'i chih kuo fang chien she
Ch'eng, Ch'i-p'an *see* Tiao-ku lou i i
Ch'eng, Chu-hsi *see* Tzu se cha yao
Cheng fa yen chiu = Studies in political science and law – 1954-1961 (1) – mf#2594 – us UMI ProQuest [320]
Ch'eng, Fang-wu *see* Liu lang
Ch'eng hsia chi / Chien, Hsien-ai – Shang-hai: K'ai ming shu tien, 1936 – (= ser P-k&k period) – us CRL [840]
Ch'eng, Hsiao-kang *see* San min chu i chih chi hua ching chi
Cheng hsin lu – [China]: Yue Shan Kan chen tsai wei yuean hui – (= ser P-k&k period) – us CRL [951]
Cheng, Hsueeh-chia *see* Chia-li-po-ti chuan
Cheng, Hsueh-chia *see* Ti tsu lun
Cheng, Hua *see* Fu-chien hsi nan lu k'uang chi hua
Cheng, I-hung *see* Kuang yuan lun
Cheng, I-mei *see*
- I-mei hsiao p'in hsu chi
- Ku fang chi

Ch'eng jen chih lu [ccm209] = Road to mature manhood / Kuo, Ching-i – Shanghai, 1930 [mf ed 198?] – (= ser Ccm 209) – 1 – mf#1984-b500 – us ATLA [240]
Cheng, Jui-mei *see* Fei yueh yun tung shih mo
Ch'eng kung jih pao – Ho Chi Minh City, Vietnam. 1966-1967 (1) – mf#67822 – us UMI ProQuest [079]
Cheng li chiang-hsi kung lu ying yuen kuan li chi hua / Hsiung, Ta-hui – [China: sn], Min kuo 25 [1936] – (= ser P-k&k period) – us CRL [380]
Cheng, Lien-te *see* T'ai-wan chi-tu chang lao chiao hui pei pu chiao hui chiu shih chou nien chien shih (ccm188)
Ch'eng, Lu-ting *see* Hsin sheng
Cheng, Pao-chao *see* T'ieh tao shih chi yun chuan
Cheng, Pi-jen *see* Ti fang tzu chih li lun yu shih shih
Cheng, Po-ch'i *see* Ta huo chi
Cheng, Shih-hsu *see* T'ung ku k'ao lueh
Cheng, Shih-hsue *see* Ch'i ch'i k'ao
Ch'eng, Shou-chung *see* Kung yeh an ch'uan yu kuan li
Cheng t'ai t'ieh lu chieh shou chi nien k'an – [China]: Cheng T'ai t'ieh lu kuan li chue, Min kuo 22 [1933] – (= ser P-k&k period) – us CRL [951]
Cheng t'ai t'ieh lu chieh shou chou nien chi nien k'an – [China]: Cheng T'ai t'ieh lu kuan li chue, Min kuo 23 [1934] – (= ser P-k&k period) – us CRL [380]
Cheng tang kai lun / Yang, Kung-ta – Shang-hai: Shen chou kuo kuang she, Min kuo 22 [1933] – (= ser P-k&k period) – us CRL [951]
Cheng, T'ao *see* Hsiu cheng ping i fa chung mien huan i wen t'i
Cheng tao chi [ccm172] = Reasons for christian faith / Hsieh, Hung-lai – 2nd ed. Shanghai, 1918 [mf ed 198?] – (= ser Ccm 172) – 1 – mf#1984-b500 – us ATLA [210]
Cheng tao chih lun (ccm131) / Gutzlaff, Karl Friedrich August – [s.l: s.n, 1837] [mf ed 198?] – (= ser Ccm 131) – 1 – mf#1984-b500 – us ATLA [240]
Cheng tao i chu [ccm8] = Art of preaching / Chang, Chih-Chiang – Shanghai, 1929 [mf ed 198?] – (= ser Ccm 8) – 1 – mf#1984-b500 – us ATLA [240]
Ch'eng, Ting-sheng *see* Shih ti yuan li
Cheng tsai hsiang: / Ts'ao, Yue – Shang-hai: Wen hua sheng huo ch'u pan she, Min kuo 30 [1941] – (= ser P-k&k period) – us CRL [820]
Cheng tun san feng ts'an k'ao ts'ai liao – [China: Su chung ch'ue tang wei], 1940 – (= ser P-k&k period) – us CRL [951]
Cheng, Yuan-tsou *see* Ta li yuan chieh shih li ch'uan chi
Ch'eng, Yu-shu *see* Hsin hsien chih chih li lun yu shih chi
Chengeta mari neungwaru / Preston, Hilary – Gwelo, Zimbabwe. 1965 – 1r – us UF Libraries [960]
Cheng-Heng *see* Declaration made by cheng heng, vice president of the high political council at the press conference held on september 5 1973
Ch'eng-tu shih hsin wen chi che kung hui ch'eng li ta hui chi nien k'an – Ch'eng-tu: Chung yang jih pao she, Min kuo 33 [1944] – (= ser P-k&k period) – us CRL [070]
Ch'eng-tu shih lin shih ts'an i hui ti san tz'u kung tso pao kao shu / Wang, Li-chung – [China: sn], Min kuo 33 [1944] – (= ser P-k&k period) – us CRL [350]
Ch'eng-tu shih lin shih ts'an i hui ti ssu tz'u kung tso pao kao shu / Wang, Li-chung – [China: sn], Min kuo 34 [1945] – (= ser P-k&k period) – us CRL [350]
Ch'eng-tu shih shih cheng t'ung chi – [China: Ch'eng-tu shih cheng fu mi shu ch'u], Min kuo 29 [1940] – (= ser P-k&k period) – us CRL [315]
Chenier, Marie-Joseph *see* Henri 8
Chenjera ninga yo moto – Gwelo, Zimbabwe. 1959 – 1r – us UF Libraries [960]
Chen-la feng t"u chi ti yen chiu = Zhenla feng tu ji di yan jiu / Chen, Cheng-siang – Xianggang, Di li yan jiu zhong xin 1969 [mf ed 1989] – (= ser Xianggang Zhongwen da xue yan jiu yuan di li yan jiu zhong xin yan jiu bao gao di 22 hao) – 1r with other items – 1 – (added cover title: a study of chin la feng tu chi) – mf#mf-10289 seam reel 030/08 [§] – us CRL [951]
Chennechot, L E *see* Histoire de la vie politique, militaire et privee de napoleon bonaparte
Chenone, E *see* L'heresie a la charite-sur-loire
Chenowith, Edith *see* Scrapbooks
Chen-tan jen yue chou-k'ou-tien wen hua / Yeh, Wei-tan – Shang-hai: Shang wu yin shu kuan, Min kuo 25 [1936] – (= ser P-k&k period) – us CRL [951]
Chenu, Adolphe *see* Die verschwoerer
Chen-ya lang mo / Hsue, Chen-ya – Shang-hai: Ch'ing hua shu chue, 1933 – (= ser P-k&k period) – us CRL [840]
Cheo alvanez / Alvarez, C – Santa Clara, Cuba. 1962 – 1r – us UF Libraries [972]
Chepstow and caldicot news – Chepstow, Wales 11 jul 1991-25 feb 1993 – 1 – (cont: chepstow news [5 oct 1989-4 jul 1991]; cont by: chepstow & caldicot news & weekly argus [4 mar 1993-]) – uk British Libr Newspaper [072]
Chepstow chronicle – [Wales] LLGC 11 apr 1884-5 sep 1888 [mf ed 2004] – 4r – 1 – (filmed with: the chepstow chronicle & advocate of progress) – uk Newsplan [072]
Chepstow express – [Wales] LLGC 20 may 1865-dec 1866 [mf ed 2004] – 2r – 1 – uk Newsplan [072]
Chepstow mercury – [Wales] LLGC 30 may 1863-5 nov 1870 [mf ed 2004] – 6r – 1 – (missing: 1871-74; cont by: chepstow & county mercury, volunteers gazette etc [jan 1868-nov 1879]) – uk Newsplan [072]
Chequamegon sun – Washburn WI. 1977 dec 1-15 – 1r – 1 – mf#956116 – us WHS [071]
Cheraw baptist church. south carolina : church records – 1822-1934 – 1 – us Southern Baptist [242]
Cherbury, Edward, Lord Herbert of *see* Autobiography 1764
Le chercheur – [Quebec: J Dussault], 1888-1889 – 9 – mf#P04129 – cn Canadiana [440]
Le chercheur de tresors : ou l'influence d'un livre / Aubert de Gaspe, Philippe – Quebec?: L Brousseau, 1878 – 2mf – 9 – mf#32891 – cn Canadiana [440]
Chercheuse / Janvier, Louis Joseph – Paris, France. 1889 – 1r – us UF Libraries [972]
Chercheuse d'esprit / Gersin, M – Bruxelles, Belgium. 1827 – 1r – us UF Libraries [440]
Cheremnykh, K M *see* Zveno
Cheremshanova, O *see* Sklep
Cherepnin, V A *see* Ialtinskii golos
Cherepovetskii kooperator – Cherepovets, 1922-1923(14) – 22mf – 9 – (missing:1922(1-12)) – mf#COR-704 – ne IDC [335]
Cherevanin, N *see* Sovremennoe polozhenie i vozmozhnoe budushchee
Cherevkov, S P *see* Donskoi krai
Cherez kooperatsiiu k elektrifikatsii : kak kooperatsiia-staruiu ladu na novyi lad peredelala / Erokhin, N V – 1925 – 76p 1mf – 9 – mf#COR-474 – ne IDC [335]
Cheriton, glamorgan, parish church of st cadoc : baptisms 1672-1925, burials 1672-1991, marriages 1671-1950 – [Glamorgan]: GFHS [mf ed [2001?] – 1mf – 9 – £1.25 – uk Glamorgan FHS [929]
Cheriton, st cadoc, monumental inscriptions – 1mf – 9 – £1.25 – uk Glamorgan FHS [929]
Cherkeshenka – Vladikavkaz, 1906 – 1 – (reel contains short runs of multiple titles. for complete listing of titles on a reel, please inquire) – us UMI ProQuest [077]
Chermenskii, E D *see* Burzhuaziia i tsarizm v pervoi russkoi revoliutsii

Chernaia sotnia / ed by Gorbunova, L & Shakhovskii, S – 1905 – 15p 1mf – 9 – mf#RPP-175 – ne IDC [325]
Cherneeva, L I et al *see* Entalpiia plavleniia solevykh evtektik
Chernenkov, N N *see* Agrarnaia programma partii narodnoi svobody i ee posleduiushchaia razrabotka
Chernev, Irving *see*
– Chessboard magic
– Fireside book of chess
Chernevskii, P O *see*
– Ukazatel materialov dlia istorii torgovli promyshlennosti i finansov v predelakh rossiiskoi imperii
– Ukazatel materialov dlia istorii torgovli, promyshlennosti i finansov v predelakh rossiiskoi imperii
Cherniaev, N I *see*
– Iz zapisnoi knizhki russkogo monarkhista
– Neobkhodimost samoderzhaviia dlia rossii, priroda i znachenie monarkhicheskikh nachal
Chernigovskie gubernskie vedomosti – Chernigov, 1838-1917 – 80r – 1 – us UMI ProQuest [077]
Chernigovskii listok – N.p., 1861-1862 – 13mf – 9 – (missing: 1862, v30(p 233-240)) – mf#R-11329 – ne IDC [077]
Chernikov, S S *see* Zagadka zolotogo kurgana
Cherniss, Harold Frederick *see* Aristotle's criticism of plato and the academy
Chernomordik, S *see* Esery
Chernomorskaia zdravnitsa – Sochi, 1986-88 – 4r – 1 – us UMI ProQuest [077]
Chernomorski front – Burgas, Bulgaria. -d. 1 Jan 1951-31 Dec 1955. 10 reels – 1 – uk British Libr Newspaper [949]
Chernomorski front – Burgas, Bulgaria. Jun 1951-1982; 1985-1992 – 39r – 1 – us L of C Photodup [077]
Chernomorskii maiak / ed by Petrov, P G – Novorossiisk [Chernomor gub]: [s n] 1918 [1918 [?]-1920 [mart]] – (= ser Asn 1-3) – n38-89 [1918] [gaps] item 441, on reel n86 – 1 – mf#asn-1 441 – ne IDC [077]
Chernov, V *see*
– Agrarnyi vopros i sovremennyi moment
– K obosnovaniiu programmy partii sotsialistov-revoliutsionerov
– Proletariat, trudovoe krestianstvo i revoliutsiia
– Trekhmesiachnoe literaturno-politicheskoe obozrenie
– Zemliia i pravo
Chernov, V M *see* Zemelnyi vopros
Chernov, Viktor Mikhailovich *see* Great russian revolution
Chernovskii, A *see* Soiuz russkogo naroda
Chernyi peredel – St Petersburg, 1880-81 – 1 – us UMI ProQuest [077]
Chernyi peredel *see* Organ sotsialistov-federalistov
Chernykh, Pavel Iakovlevich *see* Istoricheskaia grammatika russkogo iazyka
Chernyshev, Illarion *see* Klassovyia osnovy izbiratel'nago prava
Chernyshev, V R *see* Narodnoe khoziaistvo sssr i zheleznodorozhnye perevozki
Cherokee advocate – Tahlequah OK. 1977 feb-1981 fall, 1982 aug-1987 dec, 1988-89, 1990-93 – 4r – 1 – (cont: cherokee nation news, cherokee voices; cont by: cherokee phoenix and indian advocate) – mf#470304 – us WHS [071]
Cherokee advocate – Tahlequah OK. [1870 oct 22-1896 jul 25] – 1r – 1 – (cont: cherokee phoenix) – mf#633782 – us WHS [071]
Cherokee banner – Jacksonville, TX. 1986+ (1) – mf#68258 – us UMI ProQuest [071]
Cherokee bible – 1860 – 1 – us Southern Baptist [242]
Cherokee Center for Family Services *see* Cherokee voice
Cherokee county daily times – Cherokee, IA. 1992-1999 (1) – mf#61422 – us UMI ProQuest [071]
Cherokee County Genealogical Society of Southeast Kansas *see* Relatively seeking
Cherokee county herald – Centre, AL. 1988-2000 (1) – mf#68370 – us UMI ProQuest [071]
Cherokee County, KS *see* Cemetery records
Cherokee Examiner *see* Rainbow people
Cherokee hymn book – 1 – $5.00 – us Southern Baptist [780]
Cherokee hymn book – n.d – 1 – $5.00 – us Southern Baptist [780]
Cherokee nation news – 1967 aug 24-1974 dec 27, 1975 jan 3-1977 jan 7 – 2r – 1 – mf#273427 – us WHS [071]
Cherokee one feather – Cherokee. 1973-1980 (1) 1980-1980 (5) 1980-1980 (9) – ISSN: 0045-6543 – mf#7582 – us UMI ProQuest [305]
Cherokee one feather / Eastern Band of Cherokee Indians – Cherokee NC. v20 n4-5 [1987 jan 28-feb 4] – 1r – 1 – mf#765744 – us WHS [305]
Cherokee one feather / Eastern Board of Cherokee Indians – 1966-82 – (= ser American indian periodicals... 2) – 98mf – 9 – $625.00 – us UPA [305]
Cherokee phoenix – New Echota, Feb 1828-May 1834 – 3r – 1 – uk British Libr Newspaper [071]
Cherokee phoenix – Norman, OK. 1828-1834 (1) – mf#65785 – us UMI ProQuest [071]
Cherokee phoenix / Perry Wheeler for Chief Committee – v1 n1-2 [1983 feb 18-apr 10] – 1r – 1 – mf#1002634 – us WHS [305]
Cherokee rose / Youngblood, Alice P – s.l, s.l? 193-? – 1r – us UF Libraries [978]
Cherokee times – Cherokee IA. 1918 jul 4-1919 jul 1 – 1r – 1 – (cont: cherokee times-herald; cont by: cherokee daily times) – mf#851080 – us WHS [071]
Cherokee voice : the quarterly newsletter of the cherokee children's home / Cherokee Center for Family Services – v1 n1-v3 n4 [1981 dec-1983/84 win], v4 n1 [1984 spr], v5 n5-v6 n4 [1985 win-1986 fall] – 1r – 1 – mf#1597987 – us WHS [366]
Cherokees of Georgia Intertribal Council *see*
– Newsletter
– Tribal scroll
Cheron, Louis Claude *see*
– Homme a sentimens
– Tartufe de meurs
Cherrier, A Benjamin *see* History of the quebec directory
Cherrier, Andre-Romuald *see* Proces de joseph n cardinal et autres
Cherrier, Come Seraphin *see*
– Discours de c s cherrier, ecr, cr
– Memoire contenant un resume du plaidoyer de c s cherrier, ecuier, cr
Cherrier, Come Seraphin et al *see* Discours sur la confederation prononces
Cherriman, John Bradford *see*
– An elementary treatise on mechanics, pt 1
– An elementary treatise on mechanics, pt 2
– Plane trigonometry as far as the solution of triangles
Cherrington, Ernest H *see* Papers
Cherrington, Kristy Y *see* Relationship between the marital satisfaction of the wife and the viewing of televised sports by the husband
Cherry county independent – Valentine, NE: Farris & Hawkes. -v9 n10. apr 2 1896 (wkly) [mf ed 1892, 1895-96 (gaps) filmed [1974]] – 1r – 1 – (cont by: valentine democrat) – us NE Hist [071]
Cherry county news – Valentine, NE: D W Reed. 25v. v46 n1. feb 13 1930-72nd yr n4. jan 31 1957 (wkly) [mf ed lacks apr 10 1941 filmed [1974]] – 10r – 1 – (cont: valentine democrat (1900). merged with: valentine republican to form: valentine republican and cherry county news) – us NE Hist [071]
[Cherry creek-] miner – NV. apr 1903 [wkly] – 1r – 1 – $60.00 – mf#U04478 – us Library Micro [071]
Cherry hinton 1538-1950 – (= ser Cambridgeshire parish register transcript) – 10mf – 9 – £12.50 – uk CambsFHS [929]
Cherry lake farms / Atkinson, Dorothy – s.l, s.l? 1936 – 1r – us UF Libraries [978]
Chertablon, J de *see*
– Christlicher krancken-spiegel
– La maniere de se bien preparer a la mort
– Sterben und erben
Chertkov, V G *see* Dvukhnedelnoe obozrenie, posviashchennoe voprosam bratskoi zhizni, kak ikh obiasnial liudiam khristos i kak napominaet teper l n tolstoi
Chertkov, Vladimir *see*
– Christian martyrdom in russia
O cherubim – Rio de Janeiro, RJ: Typ Montenegro, 13 set 1885-25 dez 1887 – (= ser Ps 19) – mf#P17,01,107 – bl Biblioteca [079]
The cherubim / the ordering of human life : being the 3rd annual lecture and sermon... 1880 / Jeffers, W & Ross, William Wilson – Toronto: Methodist Book & Pub House, 1880 – 1mf – 9 – mf#09335 – cn Canadiana [210]
Cherubini, Luigi Carlo Zenobio Salvatore Maria *see* Favorite songs from the opera il marchese tulipano
Cherubinischer wandersmann : geistreiche sinn-und schlussreime / Silesius, Angelus; ed by Ellinger, Georg – Halle a. S: Max Niemeyer, 1895 [mf ed 1993] – (= ser Neudrucke deutscher literaturwerke des 16. und 17. jahrhunderts n135-138) – lxxix/[1]/174p (ill) – 1 – (incl bibl ref) – mf#8413 reel 6 – us Wisconsin U Libr [430]
Cherveno zname – Vidin, Bulgaria. Aug 1950-1970 – 14r – 1 – us L of C Photodup [077]
Chervin, Arthur *see* Anthropologie bolivienne
Chervonii skalat – Skalat, 1940-41 – 1 – us UMI ProQuest [934]
Chervonii zhovten' – Borschev, 1941 – 1 – us UMI ProQuest [934]
Cherwell – Oxford. 1973-1973 – 1 – ISSN: 0308-731X – mf#9052 – us UMI ProQuest [373]
Chery *see* Saint epvre
Chesapeake and Ohio Canal Association *see* C and o canaller
Chesapeake and ohio historical magazine – Clifton Forge. 1986-1995 (1) – (cont: chesapeake and ohio historical newsletter) – ISSN: 0886-6287 – mf#8327,01 – us UMI ProQuest [978]
Chesapeake and ohio historical newsletter – Alderson. 1969-1985 (1) 1977-1980 (5) 1977-1980 (9) – (cont by: chesapeake and ohio historical magazine) – ISSN: 0883-587X – mf#8327 – us UMI ProQuest [978]
Chesapeake bay banner – Easton, MD. 1973-1991 (1) – mf#63606 – us UMI ProQuest [071]
Chesapeake cousins / Upper Shore Genealogical Society of Maryland – v1 n1 [1974 aug], v2 n1 [1975 oct 6], v3 n1 – 1r – 1 – mf#1659962 – us WHS [929]
Chesapeake & Ohio Canal Association *see* Level walker
Chesapeake science – Lawrence. 1960-1977 (1) 1972-1977 (5) 1976-1977 (9) – ISSN: 0009-3262 – mf#6824 – us UMI ProQuest [574]
Cheshikhin, V E *see* Kak dumaet partiia narodnoi svobody reshit zemelnyi vopros
Cheshire 1793-1892 – Oxford, MA (mf ed 1983) – (= ser Massachusetts vital records) – 12mf – 9 – 0-931248-53-1 – (mf 1: vital records 1764-1812. mf 2: vital records 1802-37. mf 3: marriages & intent 1837-64. mf 4: births & deaths 1782-1835. mf 4: military pension 1815. mf 5: vital records 1845-59. mf 6: deaths 1860-68. mf 7: births 1868-83. mf 8: births 1883-92. mf 9: marriages 1860-83. mf 10: marriages 1884-92. mf 11: deaths 1860-78. mf 12: deaths 1878-92) – us Archive [978]
Cheshire county news, stockport & lancashire, derbyshire, and north western counties advertiser – [NW England] Stockport 1873, 1896-97, 1910 [mf ed 2004] – 4r – 1 – uk Newsplan [072]
Cheshire, F J *see* Account of the proceedings and doings of the government commissioners
Cheshire first baptist church (formerly new providence). cheshire, massachusetts : church records – 1769-1848 (the entire life of the church); Werden Church. Manuscript Copy. 1769-1841 – 1 – us Southern Baptist [242]
Cheshire, Joseph Blount *see* The church in the conferderate states
Cheshire oaths of allegiance 1723 – (= ser Cheshire church registers) – 4mf – 9 – £5.00 – mf#355 – uk CheshireFHS [929]
Cheshire observer – [North West] Chester 4 feb 1871-6 jul 1979 [mf 1978-89] – 1 – (cont: cheshire observer & chester, birkenhead & north wales times [4 jul 1863-28 jan 1871]; cont by: chester observer (main ed) 13 jul 1979-18 mar 1987) – uk MLA; uk Newsplan [072]
Cheshire quarter sessions index 1700-1730 – (= ser Cheshire church registers) – 4mf – 9 – £5.00 – mf#392 – uk CheshireFHS [929]
Cheshire wills beneficiaries index, vol 1-3 – (= ser Cheshire church registers) – 9 – (v1 [3mf] £4.50; v2 [2mf] £4.00; v3 [4mf] £5.00) – mf#75, 244, 360 – uk CheshireFHS [929]
Cheshskiia glossy v mater verborum / Patera, Adolf – Sanktpeterburg: Tip Imp akademii nauk, 1878 [mf ed 2002] – (= ser Prilozhenie k 31-mu tomu zapisok imp akademii nauk 4) – 1r – 1 – (filmed with: gaagskaia konferentsiia, iiun'-iiul' 1922 g / ed by g n lashkevicha (1922) and: na perevale / andrei belyi (1923)) – mf#5256 – us Wisconsin U Libr [460]
Cheshunt and waltham cross mercury *see* Lea valley mercury
Cheshunt and waltham weekly telegraph – Cheshunt, Waltham, England. 1876-77, 1879, 1884-85, 1888-90, 1893-94, 1900, 1970-63+ r – 1 – uk British Libr Newspaper [072]
Chesneau, A *see* Orpheus eucharisticus sive deus absconditus humanitatis...
Chesneau, Ernest Alfred *see*
– The education of the artist
– The english school of painting
Chesnee first baptist church. chesnee, south carolina : church records – 1910-21 – 1 – us Southern Baptist [242]
Chesnel, Adolphe de *see* Voyage dans les cevennes et la lozere
Chesney, F R *see*
– The expedition for the survey of the rivers euphrates and tigris, carried on by order of the british government, in the years 1835, 1836 and 1837
– Das tuerkische reich in historisch-statistischen schilderungen
Chesney, George Tomkyns *see* Indian polity
Chesnut, James et al *see* James chesnut papers, 1815-1900
Chesnutt, charles w, papers, ms 3370 – 1891-1932 – 1r – 1 – (correspondence, speeches, other writings) – us Western Res [920]
Chesnutt, Charles Waddell *see* Charles waddell chesnutt papers, 1889-1932
Chess life – New Windsor. 1980+ (1) 1980+ (5) 1980+ (9) – (cont: chess life and review) – ISSN: 0197-260X – mf#6663,01 – us UMI ProQuest [790]
Chess life and review – New Windsor. 1966-1979 [1]; 1972-1979 [5]; 1974-1979 [9] – (cont by: chess life) – ISSN: 0009-3351 – mf#6663 – us UMI ProQuest [790]
Chessboard magic : a collection of 160 brilliant chess endings / Chernev, Irving – 1st ed. New York: N Y Chess Review 1943 [mf ed 1987] – 1r [ill] – 1 – mf#1979 – us Wisconsin U Libr [790]
Chesshyre, W J *see* Messenger of christ
Chessman – 1969 jul, 1969 jul – 2r – 1 – mf#720847 – us WHS [071]
Chesson, Frederick William *see*
– The atlantic cables
– The dutch boers and slavery in the trans-vaal republic
– The dutch republics of south africa
– Mr chesson on manitoba
– The opium trade between india and china in some of its present aspects
Chest – Chicago. 1935+ (1) 1966+ (5) 1970+ (9) – ISSN: 0012-3692 – mf#418 – us UMI ProQuest [616]
Chester 1762-1849 – Oxford, MA (mf ed 1995) – (= ser Massachusetts vital record transcripts to 1850) – 11mf – 9 – 0-87623-230-6 – (mf 1t-2t: marriages & intentions 1770-91. mf 1t-5t: births & deaths 1762-1858. mf 5t-8t: marriages & intentions 1791-1848. mf 8t-10t: births & deaths 1802-58. mf 10t: marriages 1831-49. mf 11t: births, marriages 1843-49; deaths 1846-49) – us Archive [978]
Chester 1793-1892 – Oxford, MA (mf ed 1988) – (= ser Massachusetts vital records) – 26mf – 9 – 0-87623-058-3 – (mf 1-7: town & vital records 1766-87. mf 8-11: births, marriages, intentions, deaths 1829-64. mf 12: index to births: 1842-57. mf 13: index to marriages 1844-57. mf 14: index to deaths 1846-56. mf 15: b,m,d 1844-57. mf 16-17: index to births 1857-1970. mf 18-22: b,m,d 1857-93. mf 23: index to deaths 1893-1900. mf 24-26: deaths 1893-1940) – us Archive [978]
Chester a arthur papers – 1843-1926 (mf ed 1959) – (= ser Presidential papers microfilm) – 3r – 1 – us L of C Photodup [975]
[Chester-] bantam bugle – CA. nov 13 1963 – 1r – 1 – $60.00 – mf#B02105 – us Library Micro [071]
Chester chronicle : or commercial intelligencer – [NW England] Chester Record Off 1775-77, 1782-83 – 1 – (title change: chester chronicle & general advertiser [1783-84, 1789]; chester chronicle [1802, 1806-14]; chester chronicle & cheshire & north wales general advertiser [1830-74]; chester chronicle & cheshire & north wales advertiser [1874-1964]; chester chronicle [1964-77, 1985-86]) – uk MLA; uk Newsplan [072]
Chester chronicle (city ed) – 1986-90; Jul-Dec 1991; 1992-96 – 93 1/2r – 1 – uk British Libr Newspaper [072]
Chester county democrat – West Chester, PA. -w 1889-1898 – 13 – $25.00r – us IMR [071]
Chester County Genealogical Society *see* Bulletin of the chester county...
Chester County, PA *see* Chester county, pennsylvania, estate papers, 1700-1820
Chester county, pennsylvania, estate papers, 1700-1820 / Chester County, PA – 1978 – 1 – (dorothy lapp collection, chester county estate papers 1700-1810 55r $7150 s1821.p1. chester county orphans court, minors' estate papers 1700-1820 11r $1430 s1821.p2. chester county orphans court, decedents' papers 1700-1810 19r $2470 s1821.p3) – us Scholarly Res [978]
Chester county village record – West Chester, PA. -w 1889-1895 – 13 – $25.00r – us IMR [071]
Chester daily times – Chester PA. 1879 feb 4 – 1r – 1 – (cont by: delaware county daily times) – mf#888527 – us WHS [071]
Chester, Deon D *see* University presidential involvement in intercollegiate athletics
Chester first baptist church. chester, south carolina : church records – 1873-1926, 1939-57, 1960-61, 1963-72. Deacons' Minutes. 1921-47 – 1 – $36.59 – us Southern Baptist [242]
Chester guardian and cambrian intelligencer – [NW England] Chester Record Off 1820 – 1 – uk MLA; uk Newsplan [072]
Chester herald – Chester, NE: C F Bedell (wkly) [mf ed v4 n48. aug 30 1889- (gaps) filmed 1957] – 1 – (absorbed: byron herald. suspended in 1895; resumed with v11 n6 aug 28 1896. suspended foll feb 25 1943; resumed nov 1945. issues for aug 30 1889-dec 9 1909 called also whole n204-whole n127[4]) – us NE Hist [071]
Chester, Joseph Lemuel *see*
– John rogers
Chester, New Hampshire. Chester Baptist Church *see* Records

Chester. Ohio. Free Will Baptist Church *see* Church records, ms 669
Chester, Pennsylvania. Crozer Theological Seminary *see* Trustees' minutes
Chester. Presbytery (Pres. Church in the USA) *see* Minutes, 1870-1926
[Chester-] progressive – CA. jan 7 1955-sep 27 1957; 1972; 1977 (wkly) – 3r – 1 – $180.00 – mf#B02106 – us Library Micro [071]
Chester Record – [NW England] Chester Record Off jan-mar 1857 – 1 – (title change: chester record and general advertiser for cheshire, flintshire and denbighshire [apr 1857-1866, sep 1867-jun 1868]; chester guardian, record & news [jan-24 jul 1869]) – uk MLA; uk Newsplan [072]
Chester, Samuel Hall *see* Lights and shadows of mission work in the far east
Chester, st giles cemetery – (= ser Cheshire monumental inscriptions) – 1mf – 9 – £2.50 – mf#11 – uk CheshireFHS [929]
Chester, st mary on the hill: baptisms 1547-1627 – (= ser Cheshire church registers) – 1mf – 9 – £2.50 – mf#60 – uk CheshireFHS [929]
Chester, st mary on the hill: baptisms 1628-1646 – (= ser Cheshire church registers) – 1mf – 9 – £2.50 – mf#61 – uk CheshireFHS [929]
Chester, st mary on the hill: baptisms 1646-1667 – (= ser Cheshire church registers) – 1mf – 9 – £2.50 – mf#62 – uk CheshireFHS [929]
Chester, st mary on the hill: baptisms 1668-1701 – (= ser Cheshire church registers) – 1mf – 9 – £2.50 – mf#63 – uk CheshireFHS [929]
Chester, st mary on the hill: baptisms 1813-1814 & 1820-1824 – (= ser Cheshire church registers) – 1mf – 9 – £2.50 – mf#64 – uk CheshireFHS [929]
Chester, st mary on the hill: baptisms 1825-1831 – (= ser Cheshire church registers) – 1mf – 9 – £2.50 – mf#65 – uk CheshireFHS [929]
Chester, st mary on the hill: burials 1547-1628 (partial) – (= ser Cheshire church registers) – 1mf – 9 – £2.50 – mf#192 – uk CheshireFHS [929]
Chester, st mary on the hill: burials 1628-1651 – (= ser Cheshire church registers) – 2mf – 9 – £4.00 – mf#193 – uk CheshireFHS [929]
Chester, st mary on the hill: burials 1651-1665 – (= ser Cheshire church registers) – 1mf – 9 – £2.50 – mf#194 – uk CheshireFHS [929]
Chester, st mary on the hill: burials 1665-1679 – (= ser Cheshire church registers) – 1mf – 9 – £2.50 – mf#195 – uk CheshireFHS [929]
Chester, st mary on the hill: burials 1679-1706 – (= ser Cheshire church registers) – 1mf – 9 – £2.50 – mf#196 – uk CheshireFHS [929]
Chester, st mary on the hill: burials 1823-1829 – (= ser Cheshire church registers) – 1mf – 9 – £2.50 – mf#197 – uk CheshireFHS [929]
Chester, st mary on the hill: burials 1837-1844 – (= ser Cheshire church registers) – 1mf – 9 – £2.50 – mf#198 – uk CheshireFHS [929]
Chester, st mary on the hill: burials 1851-1854 – (= ser Cheshire church registers) – 1mf – 9 – £2.50 – mf#199 – uk CheshireFHS [929]
Chester, st mary on the hill: marriages 1547-1683 (pt), 1754-1766 – (= ser Cheshire church registers) – 1mf – 9 – £2.50 – mf#66 – uk CheshireFHS [929]
Chester wright's labor letter – 1946 aug 31-1950 feb 11 – 1r – 1 – (cont: wright's washington labor letter; cont by: john herling's labor letter) – mf#1546332 – us WHS [331]
Chesterfield 1762-1892 – Oxford, MA (mf ed 1983) – (= ser Massachusetts vital records) – 30mf – 9 – 0-931248-26-4 – (mf 1-5: birth index cards 1762-1963. mf 6-8: death index cards 1762-1962. mf 9-11: marriage index cards 1762-1962. mf 12-15: marriage intention index cards 1762-1962. mf 16: town records 1762-1783: b,m,i,d, vol 1. mf 17-18: town records 1782-97: b,m,i,d, vol 2. mf 19-21: b,m,d 1802-45. mf 22-25: marriage intentions 1803-1909 bk 2. mf 26-28: b,m,d 1844-75. mf 29-30: b,m,d 1858-92) – us Archive [978]
Chesterfield, Massachusetts. Chesterfield Baptist Church *see* Records
Chesterfield, staveley, bolsover, clay cross and district gazette – Chesterfield, Derbyshire, n1 [16 nov 1929-14 nov 1930]; 13-27 mar 1931 [mf ed 2004] – 1r – 1 – (missing: jan-feb 1931) – uk Newsplan [072]
The chesterian – 1915-61 – 1 – us AMS Press [073]
Chester-le-street and district advertiser – England. 30 Oct 1891-20 May 1892.-w. men reel – 1 – uk British Libr Newspaper [072]
Chester-le-street chronicle – England. Jan 1913-Dec 1940.-w – 1 – uk British Libr Newspaper [072]
Chester-le-street observer – England. 22 March 1894-9 May 1895.-w. 1 reel – 1 – uk British Libr Newspaper [072]
Chester-le-street times – Chester-le-Street, England. -w. 26 Nov 1870-3 Sept 1887. 5 1 2 reels – 1 – uk British Libr Newspaper [072]
Chesterton, G K *see* Appreciations and criticisms of the works of charles dickens
Chesterton, GK *see* La supersticion del divorcio
Chesterton, st andrew 1564-1812 – (= ser Cambridgeshire parish register transcript) – 5mf – 9 – £6.25 – uk CambsFHS [929]
Chesterton st lukes [church founded 1869] 1869-1940 – 8mf – 9 – £10.00 – uk CambsFHS [929]
Chestnut hill baptist church. saluda county. sorth carolina : church records – 1832-91 – 1 – 7.56 – us Southern Baptist [242]
Chestnut Knob District Primitive Baptist Association *see* Minutes of the...annual session of the chestnut knob district primitive baptist association
Chestnut ridge baptist church. laurens county. south carolina : church records – 1816-90, 1920-55, 1959, 1964, 1971-74 – 1 – us Southern Baptist [242]
Chestnut tree : official organ / Pierre Chastain Family Association – [v1 n1] 1976 jul-1984 apr – 1r – 1 – mf#711533 – us WHS [929]
Chestnyi slon – Paris, 1945 – 1 – us UMI ProQuest [934]
Chests, chairs, cabinets and old english woodwork / Andre, J Lewis – Horsham, 1879 – (= ser 19th c art & architecture) – 1mf – 9 – mf#4.2.426 – uk Chadwyck [740]
The chet rami sect : paper / Griswold, Hervey De Witt – Cawnpore: Christ Church Mission Press, [1904?] – 1mf – 9 – 0-524-01547-3 – mf#1990-2501 – us ATLA [280]
Chetek alert – Chetek WI. 1882 sep 15/1884 mar 22-2001 sep/dec – 95r – 1 – (with gaps) – mf#1138825 – us WHS [071]
Chetham Society *see*
– Remains historical and literary connected with the palatine counties of lancaster and chester, 3rd series
– Remains historical and literary connected with the palatine counties of lancaster and chester, new series
– Remains historical and literary connected with the palatine counties of lancaster and chester, old series
Chetopa. Kansas *see* Ordinances
Chetty, D Gopaul *see* New light upon indian philosophy
Chetty, V Venugopaul *see* A collection of the inscriptions on copper-plates and stones in the nellore district
[Chetvertaia sonata dlia fortepiano] : [iz starykh tetradei [no 4 op 29, c minor]] = Quatrieme sonate pour piano: [d'apres des vieux cahiers]: op 29: c-moll : 1917 [1908] / Prokofiev, Sergey; ed by Schneider, F H – Moscou: A Gutheil; Berlin: Breitkopf & Haertel c1921 [mf ed 19–] – 1r – 1 – (russian title transliterated) – mf#film 1745, 854 – us Sibley [780]
Chetyrkina, V I *see* V pomoshch uchiteliu – stroiteliu derevenskoi kooperatsii
Cheung, Vanessa S *see* Reflective conversation in the choreographic process
Cheval de troie : l'accord du 7 aout 1933 / Thezan, Emmanuel – Port-Au-Prince, Haiti. 1933? – 1r – us UF Libraries [972]
Chevalier, A *see* Exploration botanique de l'Afrique occidentale francaise...
Chevalier, Alexis *see* Les freres des ecoles chretiennes et l'enseignement primaire
Un chevalier apotre : celestin-godefroy chicard, missionnaire du yun-nan / Drochon, Jean-Emmanuel B – nouv ed. Paris: Typographie Augustinienne [1891] [mf ed 1995] – (= ser Yale coll) – iii/432p (ill) – 1 – 0-524-10077-2 – (in french) – mf#1995-1077 – us ATLA [920]
Chevalier, Auguste *see* L'afrique centrale francaise
Chevalier de canolle / Souque, Joseph Francois – Paris, France. 1816 – 1r – us UF Libraries [440]
Chevalier du guet / Lockroy, M – Paris, France. 1840 – 1r – us UF Libraries [440]
Chevalier, Henri Emile *see* L'heroine de chateauguay
Chevalier, Henri-Emile *see* 39 men for one woman
Chevalier, N *see* Histoire de guillaume 3
Le chevalier noel brulart de sillery / Bois, Louis-Edouard – Quebec: A Cote, 1871 – 1mf – 9 – mf#04791 – cn Canadiana [920]
Chevalier, Omer *see*
– De la necessite de l'assolement dans la culture du tabac
– Recherches experimentales en 1908
Chevalier, Thomas Wm *see* Defence of the athanasian creed
Chevalier, U *see*
– Ordinaires de l'eglise de laon (12 et 13s)
– Sacramentaire et martyrologe de l'abbaye de saint-remy. martyrologe, calendrier...de la metropole de reims
Chevalier, Ulysse *see*
– Poesie liturgique des eglises de france aux 17e et 18e siecles, ou, recueil d'hymnes et de proses
– Poesie liturgique du maoyen age
– Repertoire des sources historiques du moyen-age
– Repertorium hymnologicum
Chevallard, P *see* Saint agobard, archeveque de lyon
Chevalley, L *see* La declaration du droit...
Chevallier, Andre Fontanges Felicite *see* Bakoulou
Chevallier, Temple *see* Rich and the poor meet together
Chevallier, Temple et al *see* Results of astronomical observations made at the observatory of the university, durham, from ...
Cheveley 1559-1902 – (= ser Cambridgeshire parish register transcript) – 7mf – 9 – £8.75 – uk CambsFHS [929]
Cheves' cases in equity / South Carolina. Supreme Court – 1v. 1839-1840 (all publ) – (= ser Pre-nrs nominative equity reports) – 3mf – 9 – $4.50 – mf#LLMC 94-033 – us LLMC [342]
Cheves, Langdon *see* Langdon cheves papers, 1777-1864
Cheves' law reports / South Carolina. Supreme Court – 1v. 1839-1840 (all publ) – (= ser Pre-nrs nominative law reports) – 4mf – 9 – $6.00 – mf#LLMC 94-021 – us LLMC [340]
Cheveu sur la langue / Bastien, Rene – Paris, France. 19–? – 1r – us UF Libraries [440]
Chevillard, Valbert *see* Tirelire
Chevilles de maitre adam : menuisier de nevers / Francis, M – Paris, France. 1810 – 1r – us UF Libraries [440]
Chevrette, John M *see* The effect of oral smokeless tobacco on the cardiovascular and metabolic responses in humans during rest and exercise
Chevreul, Michel Eugene *see*
– The laws of contrast of colour
– The principles of harmony and contrast of colours
Chevrier, Francois Antoine *see* Les ridicules du siecle
Chevrier, O *see* Mosotho oa khale oa mohedene
Chevron – San Diego CA. 1942 jan 10/1943 dec-1990 – 27r – 1 – (cont: whatsmyname) – mf#703191 – us WHS [071]
Chew on : newsletter of the common market food buying co-op / Common Market, Ltd – 1971 apr-1977 oct – 1r – 1 – mf#1110619 – us WHS [334]
Chewett, J H *see* Pocket manual of mining
Cheyenne and arapaho bulletin / Cheyenne-Arapaho Tribal Office – 1978 aug/sep-1980 jul/iss 1, 1980 jul/iss 1-2 – 2r – 1 – (cont: cheyenne arapaho bulletin; cont by: southern cheyenne and arapaho nation news) – mf#633683 – us WHS [307]
Cheyenne and arapaho messenger – Canton OK: Workers Conference of the Missionaries in Oklahoma of the General Conference of the Mennonites of North America 1930-39 [mf ed 2007] – (= ser Religious periodical literature of the hispanic and indigenous peoples of the americas, 1850-1985) – 10v on 1r – 1 – (imprint varies) – mf#2007i-s009 – us ATLA [071]
Cheyenne and arapahoe sword – Cantonment OK: S K Mosiman 1900- (1900-dec 1901) [mf ed 2007] – (= ser Religious periodical literature of the hispanic and indigenous peoples of the americas, 1850-1985) – 1r – 1 – (lacks: v1 n3 [apr 1900], v1 n6-? [jul-sep 1900?], v2 n2,4,6-7 [nov 1900], jan, mar-apr 1901, v2 n11-v3 n1 [aug-oct 1901?]) – mf#2007i-s026 – us ATLA [071]
Cheyenne arapaho bulletin / Cheyenne-Arapaho Tribes of Oklahoma – v8 n7, 9, 10 [1977 feb, apr, may/ jun], v8 n16 – 1r – 1 – (cont by: cheyenne and arapaho bulletin) – mf#633282 – us WHS [307]
Cheyenne county citizen – Gurley, NE: Citizen's Pub Co, 1921-25// (wkly) [mf ed with gaps] – 1r – 1 – us NE Hist [071]
The cheyenne county record – Sidney, NE: Pindell & Long. 3v. 4th yr n45. aug 9 1934-v6 n45. aug 6 1936 (wkly) [mf ed with gaps] – 1r – 1 – (cont: gurley gazette. absorbed by: telegraph) – us NE Hist [071]
Cheyenne Indians *see* Collection
The cheyenne news and ivywild times *see* El paso county miscellaneous newspapers, reel 2
Cheyenne transporter – Darlington, Indian Territories: W A Eaton, – apr 1882; George West Maffett, may 1882- (oct 1885- with Lafe Merritt) – 2r – 1 – (reel 1: dec 5 1879-jun 25 1880. reel 2: aug 25 1880-aug 12 1886) – ISSN: 0 – mf#MF C429t – us Colorado Hist [071]
Cheyenne-Arapaho Tribal Office *see* Cheyenne and arapaho bulletin
Cheyenne-Arapaho Tribes of Oklahoma *see*
– Cheyenne arapaho bulletin
– Southern cheyenne and arapaho nation news
Cheyne, T K *see* The decline and fall of the kingdom of judah
Cheyne, Thomas Kelly *see*
– Aids to the devout study of criticism
– Bible problems and the new material for their solution
– The book of psalms
– The christian use of the psalms
– Encyclopaedia biblica
– Founders of old testament criticism
– Fresh voyages on unfrequented waters
– The hallowing of criticism
– Hosea
– Introduction to the book of isaiah
– Jeremiah, his life and times
– Jewish religious life after the exile
– Job and solomon
– The mines of isaiah re-explored
– Notes and criticisms on the hebrew text of isaiah
– The origin and religious contents of the psalter in the light of old testament criticism and the history of religions
– The reconciliation of races and religions
– Traditions and beliefs of ancient israel
Cheyne, Thomas Kelly et al *see* The holy bible
Cheyney Training School for Teachers *see* Annual report of the cheyney training school for teachers (institute for colored youth)
Cheyney University *see* Record: the student newspaper of cheyney university of pennsylvania
Chez les civils / Herment-Grenie – Paris, France. 1917 – 1r – us UF Libraries [440]
Chez les fang au quinze annees de sejour au congo francais / Trilles, H – Lille: Soc St Augustan, Desclee, de Brouwer, [1912] – 1 – us CRL [960]
Chez les femmes a crinieres du sud-angola / Balsan, Francois – Paris, France. 1963 – 1r – us UF Libraries [960]
Chez nos freres les acadiens : notes d'histoire et impressions de voyage / Dubois, Emile – Montreal: Bibliotheque de l'Action francaise, 1920 – 3mf – 9 – 0-665-72811-5 – mf#72811 – cn Canadiana [390]
Chez nous – New York. 1969-1981 (1) 1970-1981 (5) 1976-1981 (9) – ISSN: 0009-3424 – mf#5840 – us UMI ProQuest [370]
Chezy, Antoine Leonard de *see* Theorie du sloka ou metre heroique sanskrit
Chezy, Helmina von *see* Leben und kunst in paris seit napoleon dem ersten
Chezy, Wilhelm von *see*
– Hildebrand pfeiffer
– Zehn geschichten aus meister haemmerlings leben und denkwuerdigkeiten
Chhak, Sarin *see* Les frontieres du cambodge...
Chhim-Soum *see* Jataka dhammapada
Ch'i / Sun, Tien – Shang-hai: Hsi wang she, 1947 – (= ser P-k&k period) – us CRL [810]
Ch'i / Teng, Chao-hui – Shang-hai: Shih chieh shu chue, Min kuo 33 [1944] – (= ser P-k&k period) – us CRL [820]
Lo chi / Chin, Yueh-lin – Ch'ung-ch'ing: Shang wu yin shu kuan, Min kuo 31 [1942] – (= ser P-k&k period) – us CRL [160]
Chi che tao / Yuean, Shu – Shang-hai: Ch'uen li shu tien, 1936 – (= ser P-k&k period) – us CRL [070]
Ch'i ch'i chi nien jih tsung ts'ai wen kao hui pien – [China]: Chung-kuo kuo min tang chung yang chih hsing wei yuean hui hsuean ch'uan pu, Min kuo 31 [1942] – (= ser P-k&k period) – us CRL [951]
Ch'i ch'i k'ao / Cheng, Shih-hsue – Shang-hai: Chung-hua shu chue, Min kuo 26 [1937] – (= ser P-k&k period) – us CRL [951]
Ch'i chien ch'u yuen hao / Sung, Chih-ti – Han-k'ou: Shang-hai tsa chih kung ssu, Min kuo 27 [1938] – (= ser P-k&k period) – us CRL [951]
Lo chi chih yao / Chang, Shih-chao – Ch'ung-ch'ing: Shih tai ching shen she, Min kuo 32 [1943] – (= ser P-k&k period) – us CRL [160]
Chi chin hsiao shuo : wen t'an chieh ching / Yen, Tu-ho – Shang-hai: Ta Chung-hua shu chue, Min kuo 22 [1933] – (= ser P-k&k period) – us CRL [480]
Ch'i ch'ing / Wang, Jen-shu – Shang-hai: Ta kuang shu chue, Min kuo 24 [1935] – (= ser P-k&k period) – us CRL [830]
Ch'i erh / Chang, Min – [Shang-hai?]: Hsin yen chue she, Min kuo 26 [1937] – (= ser P-k&k period) – us CRL [820]
Chi feng / Lo, Chia-lun, 1897-1969 – Ch'ung-ch'ing: Shang wu yin shu kuan, Min kuo 32 [1943] – (= ser P-k&k period) – us CRL [810]
Ch'i feng shu hsin ti tzu chuan / Wei, Chin-chih – Shang-hai: Hu feng shu chue, 1931 – (= ser P-k&k period) – us CRL [480]
Chi, Hao *see* Chan shih fa kuei shu yao
Chi hen ch'u ch'u / Yue, Ta-fu – Shang-hai, Fu hsing shu chue, Min kuo 25 [1936] – (= ser P-k&k period) – us CRL [920]

Chi hua ching chi hsueeh ta kang / Shen, Chih-yuean – Shang-hai: Shen pao kuan, Min kuo 21 [1932] – (= ser P-k&k period) – us CRL [330]

Chi hua ching chi lun, i ming, t'ung chih ching chi lun – Pei-p'ing: Min yu shu chue, min kuo 22 [1933] – (= ser P-k&k period) – us CRL [330]

Chi hua ti min chu cheng chih / Holcombe, Arthur Norman – [Shang-hai]: Cheng chung shu chue, Min kuo 29 [1940] – (= ser P-k&k period) – us CRL [951]

Ch'i i ti lue ch'eng / Sha, T'ing – Ch'ung-ch'ing: Tang chin ch'u pan she, 1944 – (= ser P-k&k period) – us CRL [830]

Ch'i jen chi / Chang, T'ien-i – Shang-hai: Liang yu t'u shu yin shua kung ssu, Min kuo 34 [1945] – (= ser P-k&k period) – us CRL [480]

Ch'i jen chih yue / Sha, Ch'ien-li – Shang-hai: Sheng huo shu tien, Min kuo 27 [1938] – (= ser P-k&k period) – us CRL [920]

Chi kuan kuan li / Hsiao, Ming-hsin – [China]: Chung yang hsuen lien wei yuean hui, Min kuo 30 [1941] – (= ser P-k&k period) – us CRL [350]

Chi kuan kuan li i te / Huang, Yen-p'ei – Ch'ung-ch'ing: Shang wu yin shu kuan, Min kuo 32 [1943] – (= ser P-k&k period) – us CRL [350]

Chi kuan kuan li shu yao – [China]: Chung-kuo kuo min tang chung yang chih hsing wei yuean hui hsuen lien wei yuean hui, Min kuo 31 [1942] – (= ser P-k&k period) – us CRL [350]

Chi, Kuo-an *see* Liang tu chi

Ch'i lin chai / Shao, Ch'uean-lin – Fu-chien Yung-an: Kai chin ch'u pan she, Min kuo 32 [1943] – (= ser P-k&k period) – us CRL [820]

Chi ming tsao k'an t'ien : san mu nao chue / Hung, Shen – Han-k'ou: Hua chung t'u shu kung ssu, Min kuo 34 [1945] – (= ser P-k&k period) – us CRL [820]

Chi mo / Lo, Sun – ch'ung-ch'ing: Mei hsueeh ch'u pan she, 1944 – (= ser P-k&k period) – us CRL [480]

Chi mo ti kuo / Wang, Ching-chih – Shang-hai: K'ai ming shu tien, Min kuo 20 [1931] – (= ser P-k&k period) – us CRL [810]

Ch'i nien chi / Ou-yang, Shan – Shang-hai: Sheng huo shu tien, Min kuo 24 [1935] – (= ser P-k&k period) – us CRL [830]

Ch'i nue shu / Yue-ch'ieh – Shang-hai: T'ai p'ing shu chue, 1945 – (= ser P-k&k period) – us CRL [830]

Ch'i shih lu, i ming, ssu ch'i shih: san mu chue / Ting, Po-liu – Ch'ung-ch'ing: Hsi chue kung tso she, 1943 – (= ser P-k&k period) – us CRL [480]

Chi shih wen fan – [China]: Chung-hua shu chue, [1937] – (= ser P-k&k period) – us CRL [480]

Chi, Ta *see* Hsuan ch'uan hsueh yu hsin wen chi che

Chi tao te tao shih [ccm242] = Lord, teach us to pray / McNeur, George Hunter – 1st ed. Hong Kong, 1953 [mf ed 198?] – (= ser Ccm 242) – 1 – mf#1984-b500 – us ATLA [240]

Chi t'i an ch'uean yue kuo chi hsin chuen shih : k'ang chan pan nien chung ti kuo chi hsing shih / T'ao, Hsi-sheng – Han-k'ou: Chan shih wen hua ch'u pan she, Min kuo 27 [1938] – (= ser P-k&k period) – us CRL [327]

Chi t'ing chang pai jih kung tso pao kao – Shan-hsi: Shan-hsi sheng chiaoyue t'ing, 1932 – (= ser P-k&k period) – us CRL [370]

Ch'i t'u / Hsue, Kung-mei – [Shang-hai: Shang wu yin shu kuan], 1932 – (= ser P-k&k period) – us CRL [820]

Chi tu tu chih fo hsueh yen chiu [ccc305] = Christian study of buddhism / Wang, Chih-hsin – 3rd ed. Shanghai, 1941 [mf ed 198?] – (= ser Ccm 305) – 1 – mf#1984-b500 – us ATLA [230]

Ch'i, T'ung *see*
- Hsin sheng tai
- Lien

Chi wai chi / Lu, Hsuen – Shang-hai: Lu Hsuen ch'uean chi ch'u pan she, Min kuo 30 [1941] – (= ser P-k&k period) – us CRL [480]

Chi wai chi shih i : i chiu ling chiu nien shih / Lu, Hsuen – Shang-hai: Lu Hsuen ch'uean chi ch'u pan she, Min kuo 30 [1941] – (= ser P-k&k period) – us CRL [480]

Ch'i wang t'ien heng / Hsiao, Cho-lin – Ch'ung-ch'ing: Ching wei ch'u pan she, 1943 – (= ser P-k&k period) – us CRL [951]

Ch'i yeh hui i lu / T'ung, Shih-heng – [Sl]: Kuang hua yin shu kuan, [1941] – (= ser P-k&k period) – us CRL [951]

Ch'i yeh tsu chih : wang tan-ju pien – Shang-hai: Chung-hua shu chue, Min kuo 25 [1936] – (= ser P-k&k period) – us CRL [951]

Chi yuean ts'ao / Hu, Feng – Shang-hai: Hsi wang she, 1947 – (= ser P-k&k period) – us CRL [840]

Chia hsue kuang-te han tsai ta shih chi – [China: sn, 1935] – (= ser P-k&k period) – us CRL [951]

Chia pao-yue ti ch'u chia – Fu-chien Yung-an: Tung nan ch'u pan she, 1945 – (= ser P-k&k period) – us CRL [951]

Chia pin li chih [ccm89] = Honourable guest / Ch'en, Chu – 1st ed. Hong Kong, 1953 [mf ed 198?] – (= ser Ccm 89) – 1 – mf#1984-b500 – us ATLA [820]

Chia, Shih-i *see* Hua hui chien wen lu

Chia t'ing chiao yu te yen chiu (ccm179) = Short study of education in the christian home – Shanghai, 1930 [mf ed 198?] – (= ser Ccm 179) – 1 – mf#1984-b500 – us ATLA [640]

Chia tsu ssu yu ts'ai ch'an chi kuo chia chih ch'i yuean / Engels, Friedrich – Shang-hai: Ming hua shu tien, Min kuo 27 [1938] – (= ser P-k&k period) – us CRL [920]

Chia, Tsu-chang *see* Niao yu wen hsueh

Chia: wu mu chue / Wu, T'ien – Shang-hai: Kuang ming shu chue, Min kuo 31 [1942] – (= ser P-k&k period) – us CRL [820]

Chia yu kung pao [ccs] = Educational bulletin – Shanghai. n1-10. dec 1925-feb 1928 [complete] [mf ed 198?] – (= ser Chinese christian serials coll) – 1r – 1 – (filmed with later titles: hui wu ts'ung k'an [bulletin of the east china christian education association] n11-22 1928-31 [s0707e] and: hua tung chiao yu [bulletin of the east china christian education] n23-30 1931-32 [s0707f]) – mf0707d – us ATLA [240]

Chia-li-po-ti chuan / Cheng, Hsueeh-chia – [China]: Ch'ing nien ch'u pan she, Min kuo 31 [1942] – (= ser P-k&k period) – us CRL [920]

Chiang che wan shih t'ai tiao ch'a hui chi – [China]: Wei hsin hsueeh yuean, Min kuo 28 [1939] – (= ser P-k&k period) – us CRL [951]

Chiang, Chen *see* Nung ts'un ching chi chi ho tso

Chiang, Ch'i *see*
- Ch'en yuean-yuean
- Chiao yue shih
- Shang-hai hsiao ching

Chiang, Chien-ts'e *see* Shih cheng yu hsin chung-kuo

Chiang chuen / Pa, Chin – Shang-hai: Sheng huo shu tien, Min kuo 26 [1937] – (= ser P-k&k period) – us CRL [480]

Chiang, Chun-chang *see*
- Hsi nan ching chi ti li kang yao
- Hsin-chiang ching ying lun

Chiang, Chung-cheng *see*
- Tang ch'ien shih chu ti chung hsin wen t'i
- Tsung ts'ai chiang shu ta hsueh chung yung ching i

Chiang fan shang hsia p'ien (ccm92) = Addresses and sermons for preachers / Ch'en, Chin-yung – Shanghai, 1927 [mf ed 198?] – (= ser Ccm 92) – 1 – mf#1984-b500 – us ATLA [240]

Chiang, Fang-chen *see* Hsin ping chih yu hsin ping fa

Chiang, Feng-ch'en *see* Lu han ch'iao pao pei nan chi

Chiang han yue ko : hsin ko chue / T'ien, Han – Shang-hai: Shang-hai tsa chih kung ssu, Min kuo 29 [1940] – (= ser P-k&k period) – us CRL [820]

Chiang, Heng-yuan *see* Nung ts'un kai chin ti lun yu shih chi

Chiang hu ching yen mi chueeh – Shang-hai: Ming ming shu chue, Min kuo 25 [1936] – (= ser P-k&k period) – us CRL [480]

Chiang, Hung-chiao *see* Hui se yen ching

Chiang, I-chen *see* Chien k'u chung ch'eng chang te chia (ccm97)

Chiang, J *see* Cardiorespiratory responses to circuit weight training as measured by a biokinetic swim-bench test and a treadmill run test

Chiang, Kai-shek *see*
- K'ang chan i nien
- Ling hsiu shih nien lai k'ang chan yen lun chi
- Tsung ts'ai tui ch'ing nien ti chiao hsun
- Tsung ts'ai yen lun hsuan chi

Chiang, Kuang-tz'u *see* Kuang-tz'u i chi

Chiang, K'uei-wu *see* Kuo min chun shih ch'ang shih

Chiang, Kung-ku *see* Hsien ching san yueh chi

Chiang lai chih hua yuean / Hsue, Yue-no – Shang-hai: Shang wu yin shu kuan, Min kuo 20 [1931] – (= ser P-k&k period) – us CRL [810]

Chiang, Liang-fu *see* Wen hsueh kai lun chiang shu

Chiang, Liu *see* Wei tsou chih ch'ien

Chiang, Meng-lin *see* Kuo tu shih tai chih ssu hsiang yu chiao yu

Chiang, Nai-yung *see* Kai tsao shih chieh hsin lun

Chiang nan ch'ien hsien / Chu, Min-wei – Ch'ung-ch'ing: I wen yen chiu hui, 1938 – (= ser P-k&k period) – us CRL [951]

Chiang nan chih ch'un / Ma, Yen-hsiang – Ch'ung-ch'ing: Cheng chung shu chue, 1943 – (= ser P-k&k period) – us CRL [820]

Chiang nan min chien ch'ing ko chi : li pai-ying pien – Shang-hai: Ta kuang shu chue, 1935 – (= ser P-k&k period) – us CRL [390]

Chiang, Pai-li *see* Kuo fang lun

Chiang, Po-ch'ien *see*
- Chan yu chu
- T'i ts'ai yu feng ko shang, hsia ts'e

Chiang shang / Hsiao, Chuen – Shang-hai: Wen hua sheng huo ch'u pan she, Min kuo 25 [1936] – (= ser P-k&k period) – us CRL [480]

Chiang, Shan-kuo *see* San pai p'ien yen lun

Chiang, Shih-chieh *see* Li chia chih tu k'ao lueh

Chiang, Shu-ko *see* T'ung-ch'eng wen p'ai p'ing shu

Chiang, Tieh-lu *see* Hu tieh pei

Chiang, Tsu-i *see* T'i ts'ai yu feng ko shang, hsia ts'e

Chiang, Tung-pai *see* Hsin ssu-ch'uan

Chiang, Wei-ch'iao *see* Fo chiao kai lun

Chiang, Wen-han *see*
- Chi-tu chiao yu ma lieh chu i [ccm205]
- Shen me shih chi-tu chiao hsin yang

Chiang, Wen-hsin *see* Lieh ch'iang chun pei

Chiang, Yin-en *see* Kuo chi wen t'i tz'u hui

Chiang, Yin-hsiang *see* Ts'ung chu lei

Chiang, Yin-sung *see* K'en chih ch'ien shuo

Chiang, Yu-ching *see*
- Li tai hsiao shuo pi chi hsuan

Chiang, Yung-hung *see* Shih yeh chiang yen chi

Chiang-hsi chih mi mai wen t'i – Chang-ch'ang: Chiang-hsi sheng cheng fu ching chi wei yuean hui, Min kuo 22 [1933] – (= ser P-k&k period) – us CRL [307]

Chiang-hsi hsiang-shih lu – List of successful candidates in the imperial examination in Kiangsi province: 1873, 1875, 1876, 1879, 1885, 1888, 1889, 1891, 1893, 1894, 1897, 1900, 1901. 1 reel – 1 – us Chinese Res [951]

Chiang-hsi min-cheng kung-pao / Kiangsi. China. (Province) – Gazette of Civil Affairs...16 Jan 1928-1 Jul 1930. Incomplete. 5 reels – 1 – us Chinese Res [951]

Chiang-hsi sheng k'en wu kai k'uang – [China]: Chiang-hsi sheng k'en wu ch'u, Min kuo 30 [1941] – (= ser P-k&k period) – us CRL [304]

Chiang-hsi sheng kung lu kai k'uang – [Nan-ch'ang]: Chiang-hsi sheng cheng fu mi shu ch'u, min kuo 24 [1935] – (= ser P-k&k period) – us CRL [380]

Chiang-hsi sheng-cheng fu kung-pao / Kiangsi. China. (Province) – Kiangsi Provincial government Gazette. 1st series. Oct 1927-Dec 1928. 45 issues. 2nd series. Jan 1929-Apr 1931. 87 issues. 3rd series. May 1931-Dec 1931. 32 issues. 4th series. Jan 1932-Sep 1934. 99 issues. 5th series. Oct 1934-Dec 1948. 1680 issues – 7r – 1 – $185.00 – us Chinese Res [951]

Chiang-nan hsiang-shih lu – List of successful candidates in the imperial examination in Kiangsu province: 1879, 1885, 1888, 1893, 1901, 1903. 1 reel – 1 – us Chinese Res [951]

Chiang-ning hsien cheng kai k'uang – [Nan-ching shih?]: Chiang-ning tzu chih shih yen hsien hsien cheng fu mi shu shih, min kuo 23 [1934] – (= ser P-k&k period) – us CRL [350]

Chiang-ning tzu chih hsien cheng shih yen / Wu, Ch'un – [China: sn], 1936 – (= ser P-k&k period) – us CRL [350]

Chiang-pei hsien chien she t'e k'an – Chiang-pei hsien: Hsien cheng fu chien she k'o, Min kuo 23 [1934] – (= ser P-k&k period) – us CRL [330]

Chiang-su hsiang hsien chuan lueeh ch'u kao – Shang-hai: Cheng chung shu chue, Min kuo 25 [1936] – (= ser P-k&k period) – us CRL [920]

Chiang-su. (kiangsu) – No.1-35, 37-40, 42, 45. 1 Sept 1928-11 Dec 1929. Propaganda of the Chinese Revolution. 2 reels – 1 – 60.00 – us Chinese Res [951]

Chiang-su sheng chin yen kai k'uang – Chiang-su sheng: Min cheng t'ing, Min kuo 25 [1936] – (= ser P-k&k period) – us CRL [360]

Chiang-su sheng hsien hsing chiao yue fa ling hui pien – [China]: Chiang-su sheng chiao yue t'ing mi shu shih, 1933 – (= ser P-k&k period) – us CRL [370]

Chiang-su sheng li Hsue-chou min chung chiao yue kuan *see* Chiang-su sheng li hsue-chou min chung chiao yue kuan chou nien chi nien t'e k'an

Chiang-su sheng li hsue-chou min chung chiao yue kuan chou nien chi nien t'e k'an / Chiang-su sheng li Hsue-chou min chung chiao yue kuan – [China]: Chiang-su sheng li Hsue-chou min chung chiao yue kuan, 1933 – (= ser P-k&k period) – us CRL [370]

Chiang-su sheng nung min yin hang li nien fang k'uan chih hui ku chi kai chin chi hua – [China]: Chiang-su sheng nung min yin hang tsung hang, Min kuo 21 [1932] – (= ser P-k&k period) – us CRL [951]

Chiang-su sheng nung min yin hang pan li nung yeh ts'ang k'u chi ho tso shih yeh kai k'uang – [China]: Chiang-su sheng nung min yin hang tsung hang, Min kuo 23 [1934] – (= ser P-k&k period) – us CRL [951]

Chiang-su sheng nung min yin hang wu nien lai chih hui ku – [China]: Chiang-su sheng nung min yin hang tsung hang, Min kuo 22 [1933] – (= ser P-k&k period) – us CRL [951]

Chiang-su sheng nung ts'un tiao ch'a = Rural survey in kiangsu province / China Nung ts'un fu hsing wei yuean hui – Shang-hai: Shang wu yin shu kuan, Min kuo 23 [1934] – (= ser P-k&k period) – us CRL [307]

Chiang-su sheng shang-hai shih kai chin yue yeh hsuean ch'uan hui chi nien ts'e – [Shang-hai: Chiang-su sheng Shang-hai shih kai chin yue yeh hsuean ch'uan hui, 1931] – (= ser P-k&k period) – us CRL [951]

Chiang-su sheng t'ien fu cheng fu shui t'ung chi piao – [Kiangsu Province (China): sn, Min kuo 22 [1933]] – (= ser P-k&k period) – us CRL [630]

Chiang-su wu-chin nan-t'ung t'ien fu tiao ch'a pao kao / Wan, Kuo-ting – Nan-ching: Ts'an mou pen pu kuo fang she chi wei yuean hui, Min kuo 23 [1934] – (= ser P-k&k period) – us CRL [630]

Ch'i-ao *see* Chih tan yue ch'iang tan, i ming, han-k'ou meng: san mu chue

Ch'iao, Ch'i-min *see* Nung hui hui wu yu yeh wu

Ch'iao, Ch'i-ming *see*
- Nung hui tsu chih hsu chih
- Nung yeh chin jung

Chiao ch'ou t'ung i / Chang, Hsueeh-ch'eng – [China]: Shih chieh shu chue, Min kuo 32 [1943] – (= ser P-k&k period) – us CRL [480]

Chiao hui li shih [ccm145] = Church history / Hayes, Watson M – Shanghai. 2v. 1929-31 [mf ed 198?] – (= ser Ccm 145) – 1 – (v1 6th ed 1931. v2 5th ed 1929) – mf#1984-b500 – us ATLA [240]

Chiao hui li wen [ccm248] = Liturgies of christian churches – Chiu-chiang. 1v. 1891 [mf ed 198?] – (= ser Ccm 248) – 1 – mf#1984-b500 – us ATLA [240]

Chiao hui shih chi ju men [ccm271] = Introduction to church history: the apostolic and post-apostolic ages / Ridgely, Laurenc Butler – Hankow, 1915 [mf ed 198?] – (= ser Ccm) – 1 – mf#1984-b500 – us ATLA [240]

Chiao hui shih kung ping i [ccm282] = Success and failure of the church / Shih, Tao-hung – 1st ed. Hong Kong, 1964 [mf ed 198?] – (= ser Ccm 282) – 1 – mf#1984-b500 – us ATLA [230]

Chiao hui te cheng t'ung (ccm98) – Taipei, 1955 [mf ed 198?] – (= ser Ccm 98) – 1 – mf#1984-b500 – us ATLA [220]

Chiao i shen hsueh [ccm147] = Systematic theology / Hayes, Watson M – Shanghai. 2v. 1930-33 [mf ed 198?] – (= ser Ccm 147) – 1 – (v1 1st ed 1930. v2 2nd ed 1931) – mf#1984-b500 – us ATLA [240]

Chiao lai yuen ho / Li, Hsiu-chieh – Ch'ang-sha: Shang wu yin shu kuan, Min kuo 27 [1938] – (= ser P-k&k period) – us CRL [951]

Ch'iao le ts'un / China Ch'iao wu wei yuean hui – Nan-ching: [Ch'iao wu wei yuean hui], Min kuo 24 [1935] – (= ser P-k&k period) – us CRL [951]

Ch'iao shan tsa chu / P'an, Ching – [China]: P'an Ching, 1931 – (= ser P-k&k period) – us CRL [840]

Chiao t'ung cheng ts'e / Liu, Kuang-hua – Shang-hai: Nan-ching shu tien, Min kuo 21 [1932] – (= ser P-k&k period) – us CRL [380]

Chiao t'ung ching chi hsueeh / Yue, Sung-yuen – Shang-hai: Shang wu yin shu kuan, Min kuo 26 [1937] – (= ser P-k&k period) – us CRL [380]

Chiao t'ung nien chien – [China]: Chiao t'ung pu tsung wu ssu, 1935 – (= ser P-k&k period) – us CRL [380]

Chiao t'ung pu kung tso pao kao / China Chiao t'ung pu – [China: Chiao t'ung pu, 1932] – (= ser P-k&k period) – us CRL [380]

Chiao t'ung pu kung tso pao kao / China Chiao t'ung pu – [China: Chiao t'ung pu, 1933] – (= ser P-k&k period) – us CRL [380]

Chiao t'ung pu kung tso pao kao : erh shih nien tu / China Chiao t'ung pu – [China: Chiao t'ung pu, 1931] – (= ser P-k&k period) – us CRL [380]

Chiao t'ung pu kung tso pao kao : erh shih san nien tu / China Chiao t'ung pu – [China: Chiao t'ung pu, 1934] – (= ser P-k&k period) – us CRL [380]

Chiao t'ung pu kung tso pao kao : erh shih ssu nien tu / China Chiao t'ung pu – [China: Chiao t'ung pu, 1935] – (= ser P-k&k period) – us CRL [380]

Chiao t'ung pu kung tso pao kao : min kuo shih pa nien / China Chiao t'ung pu – [China: Chiao t'ung pu, 1929] – (= ser P-k&k period) – us CRL [380]

Chiao t'ung yin hang *see* Chiao t'ung yin hang ch'eng li san shih nien chi nien ts'e

Chiao t'ung yin hang ch'eng li san shih nien chi nien ts'e / Chiao t'ung yin hang – [Shang-hai: Chiao t'ung yin hang tsung hang], Min kuo 26 [1937] – (= ser P-k&k period) – us CRL [951]

Chiao t'ung yin hang pao kao : chiao t'ung yin hang tsung hang pien – [China]: Chiao t'ung yin hang tsung hang, [1933] – (= ser P-k&k period) – us CRL [951]

Chiao t'ung yin hang pao kao : chiao t'ung yin hang tsung hang pien – [China]: Chiao t'ung yin hang tsung hang, 1934 – (= ser P-k&k period) – us CRL [951]

Chiao t'ung yin hang pao kao : chiao t'ung yin hang tsung hang pien – [China]: Chiao t'ung yin hang tsung hang, 1935 – (= ser P-k&k period) – us CRL [951]

Chiao yu [ccm116] = Church member / Ch'uan, shao-wu – Shanghai, 1924 [mf ed 198?] – (= ser Ccm 116) – 1 – mf#1984-b500 – us ATLA [240]

Chiao yu chi k'an *see* Chung-hua chi-tu-chiao chiao-yu chi-k'an [ccs25]

Chiao yu chien i tu pen [ccm186] = Short term readers for illiterate church members / Hsieh, Sung-kao – Hong Kong. 6v. 1953 [mf ed 198?] – (= ser Ccm 186) – 1 – mf#1984-b500 – us ATLA [230]

Chiao yu kung pao / China. Ministry of Education – Gazette. Peking. v.3, n.5-v.12, n.2. May 1916-Mar 1925, Very incomplete. 14 reels – 1 – us Chinese Res [324]

Chiao, yu pu *see* Min chung hsueh hsiao k'o pen chiao hsueh fa ti i, erh ts'e

Chiao yu pu kung pao / China. Ministry of Education – Gazette. Nanking, Chungking. 1929-48. v.1-v.20, n.6. Lacking: v.2, n.25, 28-31,34,36,37; v.9, n.17-40; v.10,n.9-12; v.11-17. 18 reels – 1 – us Chinese Res [324]

Chiao yu wen ta [ccm158] = Catechism / Ho, Shou-liang & Lin, Chih-shih – 1st ed. Kowloon, 1967 [mf ed 198?] – (= ser Ccm 158) – 1 – mf#1984-b500 – us ATLA [242]

Chiao yue che hsueeh ta kang = Ueber philosophie als die grundwissenschaft der paedagogik oder paedagogische philosophie / Fan, Shou-k'ang – Shang-hai: Chung-hua hsueeh i she: Shang wu yin shu kuan, 1933 – (= ser P-k&k period) – us CRL [370]

Chiao yue chih k'o hsueeh yen chiu fa / Chung, Lu-chai – Shang-hai: Shang-wu yin shu kuan, min kuo 24 [1935] – (= ser P-k&k period) – us CRL [951]

Chiao yue fa ling hui pien ti 1-5 chi – [China]: Chiao yue pu, Min kuo 25-29 [1936-1940] – (= ser P-k&k period) – us CRL [370]

Chiao yue hsing cheng chi hua – [China]: Ch'ing-tao shih chiao yue chue, Min kuo 22 [1933] – (= ser P-k&k period) – us CRL [370]

Chiao yue hsing cheng chih li lun yue shih chi – [China]: Chiao yue pien i kuan, [1935] – (= ser P-k&k period) – us CRL [370]

Chiao yue hsing cheng pao kao : ch'ing-tao shih chiao yue chue pien – [China]: Ch'ing-tao shih chiao yue chue, Min kuo 21 [1932] – (= ser P-k&k period) – us CRL [370]

Chiao yue kai lun / Lo, T'ing-kuang – Shang-hai: Shih chieh shu chue, Min kuo 22 [1933] – (= ser P-k&k period) – us CRL [370]

Chiao yue po yin chiang yen chi ti i chi, min chung chiao yue p'ien – Shang-hai: Shang wu yin shu kuan, Min kuo 25 [1936] – (= ser P-k&k period) – us CRL [370]

Chiao yue pu tu hsueeh shih ch'a hu-pei sheng chiao yue tsung pao kao – Han-k'ou: Hu-pei chiao yue t'ing pien shen wei yuean hui, Min kuo 23 [1934] – (= ser P-k&k period) – us CRL [370]

Chiao yue shih / Chiang, Ch'i – Shang-hai: Shang wu yin shu kuan, Min kuo 21 [1932] – (= ser P-k&k period) – us CRL [370]

Chiao yue ta tz'u shu : t'ang yueeh teng pien – Shang-hai: Shang wu yin shu kuan, Min kuo 22 [1933] – (= ser P-k&k period) – us CRL [370]

Chiao yue ts'e yen chi t'ung chi : p'u i-jen, huang ming-tsung ho pien – Shang-hai: Li ming shu chue, Min kuo 26 [1937] – (= ser P-k&k period) – us CRL [370]

Chiao yue yen chiu fa : chu chih-hsien chu / Chu, Chih-hsien – Nan-ching: Cheng chung shu chue, Min kuo 23 [1934] – (= ser P-k&k period) – us CRL [370]

Chiao yue yue hsueeh hsiao hsing cheng yuean li / Tu, Tso-chou – Shang-hai: Shang wu yin shu kuan, Min kuo 22 [1933] – (= ser P-k&k period) – us CRL [370]

Chiao, Yu-t'ing *see* Ho sso shih yeh

Chiapas. Mexico (State) *see* Periodico oficial del gobierno constitucional del estado de chiapas

Chiaramonte, M *see* Le simpatie della citta di messina coll'aquila augusta rinfiammate nelle solenne acclamazione dell'imperator carlo 6...

Chiaretti, Giuseppe *see* Archivo leonessano

Chiari, Pietro *see* Favorite songs from the opera il marchese tulipano

Chiavacci, Vincenz *see* Letzte dorfgaenge

Chiavelloni, Vincenzo *see* Discorsi della musica

Chibuku chenzanga yavana vamaria – Gweru?, Zimbabwe. 1959 – 1r – us UF Libraries [960]

Chica, Luis Alonso *see* Deberes de centroamerica con guatemala ante el cas...

Chica moderna / Rivas Bonilla, Alberto – San Salvador, El Salvador. 1945 – 1r – us UF Libraries [972]

Chicago american – Chicago IL. 1839 aug 30, sep 27, 1840 jan 3, 24, feb 7, aug 7, dec 11, 1841 jul 30, 1842 apr 6, jul 13, aug 10,17 – 1r – 1 – mf#854917 – us WHS [071]

Chicago and North Western Railway Co *see*
– Annual report of the...
– Annual report of the...for the fiscal year ending may 31st...
– Report of the chicago and north western railway company

Chicago and the baptists / Stackhouse, Perry J – 1782-1933 – 1 – 9.87 – us Southern Baptist [242]

Chicago and the old northwest, 1673-1835 : a study of the evolution of the northwestern frontier, together with a history of fort dearborn / Quaife, Milo Milton – Chicago: University of Chicago press, c1913 [mf ed 1970] – (= ser Library of american civilization 16750) – vii/480p on 1mf – 9 – (with bibl) – us Chicago U Pr [978]

Chicago Area Committee on Occupational Safety and Health *see* Cacosh health and safety news

Chicago Area Military Project *see* Camp news

Chicago Bar Association. Board of Managers *see* Report...concerning an inquiry conducted by it in reference to the activities of judges in partisan politics

Chicago bar record – Chicago. 1969-1986 (1) 1971-1986 (5) 1976-1986 (9) – ISSN: 0009-3505 – mf#5881 – us UMI ProQuest [340]

Chicago bar record – v1-67. 1910-1986 [all publ] – 9 – $798.00 set – (cont as: cba record; suspended april 1932-oct 1934) – mf#101481 – us Hein [340]

Chicago bee – Chicago IL. 1943 jan-dec, 1944 jan-dec, 1945 jan-dec, 1946 jan-dec, 1947 jan-aug 17 – 9r – 1 – mf#3453864 – us WHS [071]

The chicago bee – Chicago: Bee Publ Co, jan 1943-aug 17 1947 – 9r – us CRL [071]

[Chicago-] bitalian news – IL. 1977-78 – 1r – 1 – $60.00 – mf#R04300 – us Library Micro [071]

Chicago breeze – 1983 apr-1993 dec – 1r – 1 – mf#1110623 – us WHS [071]

Chicago Brown Berets Organization *see* Mi sangre

Chicago Building Trades Council et al *see* Federation news

Chicago christian events guide – 1992 oct/nov, 1993 apr-jul – 1r – 1 – mf#4027758 – us WHS [071]

Chicago chronicle – Chicago IL 21 nov 1904 – 1 – uk British Libr Newspaper [071]

Chicago citizen – Chicago IL 6 feb-25 dec 1897 – 1 – (cont: citizen [4 jan 1890-30 jan 1897]) – uk British Libr Newspaper [071]

Chicago Commission on Human Relations *see* Human relations news of chicago

Chicago commons through forty years / Taylor, Graham – Chicago, IL: Chicago Commons Association, [1936] [mf ed 1970] – (= ser Library of american civilization 16837) – xiv/322p on 1mf – 9 – us Chicago U Pr [360]

Chicago Conference on Trusts (1899) *see* Speeches, debates, resolutions, list of the delegates, committees, etc

Chicago courier – Chicago IL. 1932 oct 22 – 1r – 1 – mf#5012826 – us WHS [071]

Chicago courier – Chicago IL. 1974 apr 13-20, 1975 oct 25, nov 15 – 1r – 1 – mf#4164316 – us WHS [071]

Chicago daily commercial advertiser – Chicago IL. 1852 sep 7,10, oct 18, dec 7 – 1r – 1 – mf#845766 – us WHS [071]

Chicago daily journal – Chicago, IL: R L Wilson, jun 1847-dec 1849 – 1 – us CRL [071]

Chicago daily news – Chicago IL 2 jan 1941-31 dec 1945 – 1 – uk British Libr Newspaper [071]

Chicago daily news – Chicago IL. 1903 jun 22-jul 24 – 1r – 1 – (cont: chicago journal; chicago daily journal; chicago evening post) – mf#765746 – us WHS [071]

Chicago daily news – Chicago IL. 1886 dec 18, 1889 apr 30, 1893 jun 24 – 1r – 1 – (cont: chicago morning news; cont by: chicago news-record; chicago daily news-record) – mf#977107 – us WHS [071]

Chicago daily socialist / Chicago Socialist Party – Chicago, 1906-12 – 1 – us NY Public [335]

Chicago daily socialist – Chicago IL. all mutilated: v2 n76,99,111-112,129,177,266 [1908 jan 24, feb 21, mar 16-13,27, may 22, sep 5] v3 n50 [1908 dec 23] – 1r – 1 – (cont by: evening world; chicago daily socialist and evening world) – mf#611219 – us WHS [071]

Chicago daily sun-times – Chicago IL. 1949 mar 1-10 – 1r – 1 – (cont: chicago sun; daily times) – mf#846095 – us WHS [071]

Chicago daily tribune – Chicago IL apr 1849, 28 dec 1850, 4 jan, 4,28 oct 1851, 6 may, 1 dec 1852-30 jun, 2 sep 1853-28 jun 1858 – 1 – (1853-72 imperfect; cont by: chicago daily press and tribune [1 jul 1858-16 mar 1859]; press & tribune [17 mar 1859-24 oct 1860]; chicago daily tribune [25 oct 1860-20 aug 1864]; chicago tribune [21 aug 1864-8 oct 1872]; chicago daily tribune [9 oct 1872-18 mar 1886]; chicago tribune [19 mar 1886-18 jul 1890]; chicago daily tribune [19 jul 1890-16 feb 1963]; chicago tribune [17 feb 1963-]) – uk British Libr Newspaper [072]

Chicago daily tribune – Chicago IL, Milwaukee WI. 1941 dec 4,8, 1945 aug 11-23 – 1r – 1 – mf#765747 – us WHS [071]

Chicago daily tribune – Chicago: Tribune Co 1872-1963, jul 1873-75, may-dec 1881, jun-nov 1892, may 16-31 1923, oct 16-31 1933 – us CRL [071]

Chicago daily tribune [paris ed] – Chicago IL 9 jul 1917-9 jan 1929 – 1 – (cont by: chicago daily tribune (paris ed) [10 jan 1929-30 nov 1934]) – uk British Libr Newspaper [071]

Chicago defender – Chicago IL. 1931 may 16,30 – 1r – 1 – mf#5012913 – us WHS [071]

Chicago defender – Chicago IL. 1941 jan 4-jun 28, 1941 jul 5-dec 27, 1942 jan 3-jun 27, 1942 jul 4-dec 26, 1943 jan 2-jun 26, 1943 jul 3-dec 25, 1944 jan 1-jun 25, 1944 jul 1-dec 30, 1945 jan 6-dec 29 – 9r – 1 – (cont by: chicago daily defender [chicago il : weekly]) – mf#153129 – us WHS [071]

Chicago defender [city ed] – Chicago IL 31 jul 1909-7 jul 1945 (imperfect) – 1 – (numeration irreg; cont by: chicago defender [city ed] [14 jul 1945-29 apr 1966]; chicago daily defender [30 apr 1966-15 feb 1973]; chicago defender [daily ed] 19 feb 1973-30 dec 1993) – uk British Libr Newspaper [071]

Chicago Dental Society *see* Fortnightly review of the chicago dental society

Chicago dollar tribune – Chicago IL. 1884 jan 9,16 – 1r – 1 – mf#1010717 – us WHS [071]

The chicago eagle – Chicago IL, 1889-1936// (mf ed 1947) – (= ser Negro Newspapers on Microfilm) – 1 – us L of C Photodup [071]

Chicago. Educational Commission *see* Report...of the commission...appointed by the mayor, hon. carter h. harrison, jan 19th, 1898

Chicago enterprise – Chicago IL. 1926 mar 27, aug 7 – 1r – 1 – (cont by: chicago world [chicago il: 1929]) – mf#5012955 – us WHS [071]

Chicago evening journal – Chicago IL. 1855 dec 10-1917 jun 15 – 1r – 1 – (cont: chicago daily journal [chicago il: 1844]; cont by: chicago journal [chicago il: daily]) – mf#851277 – us WHS [071]

Chicago evening post – Chicago IL. 1871 oct 10,12,16-19,21,23,28, 1916 sep 21 – 1r – 1 – (cont: daily chicago post; cont by: chicago post and mail; chicago evening mail [chicago il: 1870]) – mf#851232 – us WHS [071]

Chicago evening post – Chicago IL. 1903 nov 11/1904 jan 23-1905 may 25/aug 3 – 10r – 1 – (with gaps; cont by: chicago daily news [chicago il: 1875]) – mf#846098 – us WHS [071]

Chicago express – 1972 jun 21-1974 jul – 1r – 1 – mf#1110624 – us WHS [071]

Chicago fed letter – Chicago. 1987-1996 (1,5,9) – ISSN: 0895-0164 – mf#16320 – us UMI ProQuest [332]

Chicago Federation of Labor and Industrial Union Council *see*
– Daily labor bulletin
– New majority
– Union labor advocate

Chicago field – Chicago. v1-15. feb 1874-jun 1881 [all publ] – (= ser Sports periodicals, 1822-1922) – 5r – 1 – $855.00 – us UPA [790]

Chicago fire fighter / Chicago Firefighters' Union Local 2 – v30 n4-v38 n5 [1975 win-1983 hol ed], v32 n3 [1977 fall], v33 n3-4 [1978 fall-spring], v35 n3-v36 n3 [1979 fall-1980 fall], v38 n3 [1983:winter/spring] – 1r – 1 – (cont by: local 2 news) – mf#1048154 – us WHS [360]

Chicago Fire Fighters' Union *see* Local two news

Chicago Firefighters' Union Local 2 *see* Chicago fire fighter

Chicago Hebrew Mission *see* Annual report of the chicago hebrew mission

The chicago herald – Chicago: [s.n.], jan-jun 1891; sep-oct 1891; may-jun 1893 – us CRL [071]

Chicago herald american – Chicago IL. 1940 jan 2-feb 10, 1940 feb 12-mar 21, 1940 mar 22-apr 30, 1940 may 1-jun 7, 1940 jun 8-jul 20, 1940 jul 22-aug 31, 1940 sep 3-oct 9, 1940 oct 9-nov 13, 1940 nov 14-dec 31, 1951 mar 12, apr 8 – 1r – 1 – (cont: chicago herald and examiner; chicago evening american [chicago il: 1914]; cont by: chicago american [chicago il: 1953]) – mf#849355 – us WHS [071]

Chicago herald and examiner – Chicago IL. 1919 feb 2,9, jun 7, jul 14 – 1r – 1 – (cont: chicago examiner; chicago herald [chicago il: 1914]; chicago examiner; chicago herald; cont by: chicago american [chicago il: 19uu : daily]; chicago herald american; chicago american; chicago herald-american) – mf#853342 – us WHS [071]

Chicago hilltop = chicago area alumni newsletter / Howard University Alumni Association – 1982 sep – 1r – 1 – mf#4990652 – us WHS [378]

Chicago Historical Society *see* News review

Chicago history – Chicago. 1978+ (1,5,9) – ISSN: 0272-8540 – mf#11914 – us UMI ProQuest [978]

Chicago IL *see* Lever and new voice series

Chicago illustrated press and the women's press – v3 n3-v3 n21 [1919 sep 20-1920 jan 31] – 1r – 1 – (cont: chicago sunday press and the women's press) – mf#939559 – us WHS [070]

[Chicago-] in these times – IL. 1977-86 – 12r – 1 – $720.00 – mf#R04301 – us Library Micro [071]

Chicago independent bulletin – Chicago IL. 1973 nov 15/1974 may 30-2000 jul 6/dec 28 – 27r – 1 – (with gaps) – mf#874825 – us WHS [071]

The chicago israelite : the jewish society paper of chicago 1908-1909 – Chicago, IL: Chicago Israelite Pub Co (wkly) [mf ed 197-?] – 1 – mf#*ZAN-*P918 – us NY Public [071]

Chicago jewish forum – Chicago. 1942-1969 (1) – mf#3309 – us UMI ProQuest [939]

Chicago journal of international law – Chicago. 2000+ (1,5,9) – ISSN: 1529-0816 – mf#32207 – us UMI ProQuest [341]

Chicago journal of international law – v1-3. 2000-2002 – 9 – $100.00 set $55.00v – ISSN: 1529-0816 – mf#118220 – us Hein [341]

Chicago journalism review – Chicago. 1973-1975 (1) 1975-1975 (5) (9) – ISSN: 0009-3580 – mf#7818 – us UMI ProQuest [070]

The chicago juvenile court / Jeter, Helen Rankin – Washington, Govt. Print. Off., 1922. 119 p. LL-1294 – 1 – us L of C Photodup [347]

Chicago kaleidoscope – Chicago IL. v1 n1-14 [1969 mar 28/apr 10-may 31/jun 13] – 1r – 1 – (cont: kaleidoscope chicago; cont by: chicago seed) – mf#1110804 – us WHS [071]

Chicago kent law review – v1-75. 1923-2000 – 5,6,9 – $1070.00 set – (v1-60 1923-84 in reel $635. v61-75 1985-2000 in mf $435) – ISSN: 0009-3599 – mf#101491 – us Hein [340]

Chicago law journal / ed by Barber, G L – E B Meyers Co. v1-2 n1. 1876-78 (all publ) – 8mf – 9 – $12.00 – mf#LLMC 91-095 – us LLMC [340]

Chicago law journal – v1-2. 1876-77 (all publ) – (= ser Historical legal periodical series) – 1 – $45.00 – mf#408900 – us Hein [340]

Chicago law journal / ed by Wachob, I S – Chicago Law Book Co. os: v4-10 1883-89. ns: v1-7 1890-96 [all publ] – 94mf – 9 – $141.00 – (cont: the central law monthly in 1896; absorbed by: the chicago law journal weekly) – mf#llmc82-913 – us LLMC [340]

Chicago law journal *see* Loyola university chicago law journal

The chicago law journal – Chicago. v1, n.9-v9,n.12, 1880-88. Title varies. Incomplete. LL-048 – 1 – us L of C Photodup [340]

Chicago law journal weekly – v1-24. 1896-1907 (all publ) – (= ser Chicago law journal) – 378mf – 9 – $328.00 – (absorbs: chicago law journal in 1896. the set is complete as provided by llmc, but its vol numeration is very confused, with some vol numbers, 7-16 & 20, not being used at all. v6 ends in 1901 and the next vol, v17, starts in 1902) – mf#LLMC 91-096 – us LLMC [340]

Chicago law times – v1-3. 1886-89 (all publ) – (= ser Historical legal periodical series) – 1 – $45.00 set – mf#101681 – us Hein [340]

The chicago law times – v1-3. 1886-89 (all publ) – 3mf – 9 – $16.50 – (lacking: v2) – mf#LLMC 84-436 – us LLMC [340]

Chicago lawyer – Chicago. 1986+ (1,5,9) – ISSN: 0199-8374 – mf#15052 – us UMI ProQuest [340]

Chicago ledger – Chicago IL. v4 n19 [1876 may 6], v20 n29 [1892 jul 20], v22 n1-v23 n52 [1894 jan 3-1895 dec 25], 1910 jan-nov – 3r – 1 – mf#851194 – us WHS [071]

Chicago legal news – v1-57 1868-1925 (all publ) – (= ser Historical legal periodical series) – 1 – $825.00 set – mf#408730 – us Hein [340]
Chicago mahogany – v1 n1 [1980], v2 n2-4 [1981 feb-may/jun] – 1r – 1 – mf#4717681 – us WHS [071]
Chicago medical journal and examiner – Chicago. 1844-1889 (1) – mf#4802 – us UMI ProQuest [610]
Chicago medical school quarterly – Chicago. 1940-1973 (1) 1970-1973 (5) – ISSN: 0009-3629 – mf#2467 – us UMI ProQuest [610]
Chicago Medical Society *see* Official proceedings
Chicago Men's Gathering *see* Newsletter
Chicago men's gathering – iss 32-47 [1982 aug-1984 jun] – 1r – 1 – (cont: newsletter [chicago men's gathering]) – mf#1050819 – us WHS [071]
Chicago merchant – 1930 sep – 1r – 1 – mf#4364626 – us WHS [071]
Chicago metro news – Chicago IL. 1973 jan 20/oct 27-1990 jan 13/oct 6 – 32r – 1 – mf#870839 – us WHS [071]
Chicago, Milwaukee, and St Paul Railway Co *see* Milwaukee employes' magazine
Chicago, Milwaukee, and St. Paul Railway Co *see* Milwaukee railway system employes' magazine
Chicago, Milwaukee, St Paul and Pacific Railroad Co *see* Milwaukee magazine
Chicago morning news – Chicago IL. 1881 sep 20 – 1r – 1 – (cont by: chicago daily news [chicago il: 1882]; chicago daily news [1882]) – mf#851727 – us WHS [071]
[Chicago-] muhammad speaks – IL. 1971-75 – 7r – 1 – $420.00 – mf#R04302 – us Library Micro [071]
Chicago Newspaper Guild, Local 71 *see* Memo
Chicago Office, Theater, and Amusement Building Janitors' Union, Local 25 et al *see* Local 25 voice
Chicago outlines : the voice of the gay and lesbian community – Chicago. v1-10. jun 4 1987-may 1997 (wkly, mthly) – (= ser Chicago Outlines) – 11r – 1 – Can$1275.00 – (title varies: chicago outlines) – cn McLaren [305]
The chicago packer – Kansas City, MO: The Packer. [v17 n4-v46 n39]. jan 9 1915-sep 7 1946 – 32r – 1 – us CRL [071]
Chicago pnyx – v39 n645-v44 n740 [1978 sep 15-1983 jun], v44 n741-v50 n853 [1983 jul 1-1989 aug 15] – 2r – 1 – mf#679628 – us WHS [071]
Chicago police officer – v1 n1-v3 n4 [1976 aug/sep-1978 aug/sep] – 1r – 1 – mf#641256 – us WHS [360]
A chicago princess / Barr, Robert – Toronto: McLeod & Allen, c1904 – 4mf – 9 – 0-665-73559-6 – (ill by francis p wightman) – mf#73559 – cn Canadiana [830]
Chicago record-herald – Chicago IL. 1903 aug 16 – 1r – 1 – (cont: chicago record; chicago times-herald; chicago record; chicago times-herald; cont by: chicago record-herald and the inter ocean; inter ocean [chicago il: daily]; chicago record-herald and the inter ocean; inter ocean) – mf#856599 – us WHS [071]
The chicago republican – Chicago: A W Mack, 1865- [1869-70] – us CRL [071]
Chicago review – Chicago. 1957+ (1) 1970+ (5) 1976+ (9) – ISSN: 0009-3696 – mf#1035 – us UMI ProQuest [400]
Chicago saturday record – Chicago IL. 1893 apr-nov 18, 1893 nov 25-1895 aug 17, 1895 aug 24-dec 28 – 3r – 1 – (cont: chicago weekly news record) – mf#871365 – us WHS [071]
Chicago sentinel – Chicago. Ill. 1911-23 – 1 – us AJPC [071]
Chicago shoreland news – Chicago IL. 1995 aug 5-12, sep 23-oct 7, 1996 feb 17-mar 2, sep 21, oct 5-dec 28, 1997 jan 4-dec 20, 1998 jan 3-dec 25 – 3r – 1 – (cont: chicago shoreland) – mf#3363870 – us WHS [071]
Chicago socialist – Chicago. Daily. Oct 25 1906-Dec 4 1912. Incomplete – 1 – us NY Public [071]
Chicago socialist – 1902-1907 apr 6 – 1r – 1 – (cont: workers' call; cont by: chicago socialist [daily ed]) – mf#945537 – us WHS [335]
Chicago socialist – n389-429 [1920 apr 24-1921 jan 29] – 1r – 1 – mf#945537 – us WHS [335]
Chicago Socialist Party *see* Chicago daily socialist
Chicago standard news – Chicago Heights IL. 1984 feb 18-1986 jan 18, 1990 jan 4-dec 27, 1992 jan 2-dec 31, [1992 jul 9-1998 jan 8], 1999 feb 25-dec 30, 2000 jan 6-jun 29, 2000 jul 6-dec 28, 2001 jan 4-jun 28 – 8r – 1 – mf#2697111 – us WHS [071]
Chicago State University *see* Csu magazine
Chicago studies – Mundelein. 1962-1994 [1]; 1971-1994 [5]; 1976-1994 [9] – ISSN: 0009-3718 – mf#2444 – us UMI ProQuest [240]
Chicago sunday press and the women's press – v1 n1-2 [1919 jul 13-20], v3 n2 [1919 sep 14] – 1r – 1 – (cont by: chicago illustrated press and the women's press) – mf#939550 – us WHS [071]
Chicago sun-times – Chicago, IL. 1948+ (1) – mf#60458 – us UMI ProQuest [071]
Chicago teacher – Chicago. 1874-1875 – 1 – mf#4637 – us UMI ProQuest [370]
Chicago teacher – v40 n1, 3-4,5,7,9 [1975 sep, nov-dec, 1976 jan, mar, may] – 1r – 1 – (cont: chicago union teacher [1931]; cont by: chicago union teacher [1976]) – mf#666008 – us WHS [370]
Chicago Teachers Union *see* Ctu newsletter
[Chicago-] the call – IL. 1974-1982 – 7r – 1 – $420.00 – mf#R04304 – us Library Micro [071]
Chicago theological seminary : register – v1-80. mar 1908-1990 [complete] – 12r – 1 – mf#ATLA S0258 – us ATLA [200]
Chicago times – Chicago Junction, OH. 1916-1919 (1) – mf#65408 – us UMI ProQuest [071]
Chicago times – Chicago IL. 1888 jul 3 – 1r – 1 – (cont: chicago times [1888: sunday]; times [chicago il: 1881: daily]; cont by: times [chicago il: 1894]) – mf#874084 – us WHS [071]
Chicago times – Chicago IL. 1864 dec 27, 1869 oct 10, 1872 jul 28, 1879 jun 4,25 – 4r – 1 – (cont: daily chicago times; cont by: times [chicago il: 1881: daily]) – mf#872736 – us WHS [071]
Chicago times – Willard, OH. 1883-1915 (1) – mf#65715 – us UMI ProQuest [071]
The chicago times – Chicago: Wilbur F Storey, [jul 22 1862;, ar 22 1863; sep 22 1863-jun 10 1864; apr-sep 1869; apr-dec 1870] – us CRL [071]
Chicago tribune – Chicago IL 21 aug 1864-8 oct 1872 – 1 – (1853-72 imperfect; cont: chicago daily tribune [25 oct 1860-20 aug 1864; cont by: chicago daily tribune [9 oct 1872-18 mar 1886]) – uk British Libr Newspaper [071]
[Chicago-] tricontinental news service – IL. 1972-74 – 2r – 1 – $120.00 – mf#R04303 – us Library Micro [071]
Chicago union label bulletin – 1896 nov 14, 1899 apr 18, 1902 apr 22, jun 10, aug 12-28, dec 24 – 1r – 1 – (cont: union label bulletin) – mf#3442552 – us WHS [331]
Chicago union teacher – 1976 jun-1977 jun, 1977 dec-1981, 1982 jan-1989 dec, 1990 jan-1993 nov/dec – 4r – 1 – (cont: chicago teacher) – mf#599397 – us WHS [071]
Chicago volunteer : a t r, no 2611 – Woodstock, Ont?: s.n, 18–? – 1mf – 9 – mf#53868 – cn Canadiana [636]
Chicago weekend – Chicago IL. [1991 apr 11/14-jun 27/30]-2001 jan 4/jun 28 – 21r – 1 – (cont by: citizen [chicago weekend ed]) – mf#1884212 – us WHS [071]
Chicago weekly news – Chicago IL. 1887 aug 11, dec 15, 1888 jan 12, mar 8, 1892 jan 7-dec 29, 1893 jan 5-mar 16 – 2r – 1 – (cont by: chicago weekly news record) – mf#851114 – us WHS [071]
Chicago weekly news record – Chicago IL. 1893 mar 23-30 – 1r – 1 – (cont: chicago weekly news; cont by: chicago saturday record) – mf#871367 – us WHS [071]
Chicago weekly times – Chicago IL. 1854 nov 2 – 1r – 1 – (cont by: weekly chicago times) – mf#1010882 – us WHS [071]
Chicago Women's Liberation Union *see* Womankind
Chicago world – Chicago IL. 1925 oct 29, nov 28 – 1r – 1 – mf#5013017 – us WHS [071]
Chicago world – Chicago IL. 1929 jun 29, nov 2, 1932 oct 8, 1935 jun 15 – 1r – 1 – (cont: chicago enterprise [chicago il: 1926]) – mf#5012971 – us WHS [071]
The chicago world – Chicago, IL: B F Harris & Co. jan 20 1900 (wkly) [mf ed 1947] – (= ser Negro Newspapers on Microfilm) – 1r – 1 – us L of C Photodup [071]
The chicago world – Chicago. Ill. Sept. 2, 9, 30; Oct. 7, 1950; Dec. 29, 1951 – 1 – us NY Public [071]
Chicagoan – Chicago. 1973-1974 – 1 – mf#8575 – us UMI ProQuest [073]
Chicago-bladet / Evangelical Free Church of Amnerica – Chicago IL. [1877 apr 13/1891 mar 10]-[1938 nov 22/1942 jun 30] – 25r – 1 – (with gaps; cont: zion's banner) – mf#765745 – us WHS [243]
Chicagoer arbeiter-zeitung – [1887 oct 24/1890 feb 24]-1919 nov 2/1920 sep 12 – 45r – 1 – mf#851912 – us WHS [331]
Chicagoer arbeiter-zeitung – Chicago, IL: Socialistic Publ Soc, feb 13,18,28, mar 16,25, apr 11, may 25, nov 12 1885; may 19, jun 22,24,26,28, jul 1, oct 2-dec 1886; 1887-jun 1894; jul 1897-1900; may 1910-apr 1920 – 1 – us CRL [071]
Chicagoer arbeiter-zeitung – Chicago IL. 46 jahrg n18-37 [1924 mai 11-sep 28], 1919 oct 19-1920 jan 25, 1919 nov 2-1920 sep 12, 1920 feb 1-1921 may 29, 1921 jun 5-1924 oct 12 – 5r – 1 – (cont: fackel [chicago il], vorbote) – mf#967590 – us WHS [331]
Chicagoer arbeiter-zeitung – Chicago IL (USA), 1894-1910 – 30r – 1 – (filmed by other misc inst: 1920 22 feb-1924 12 oct, 1931 feb-dec) – gw Misc Inst [331]
Chicagoer deutsche zeitung – Chicago IL (USA), 1924 20 dec-1926 22 may [gaps] – 1r – 1 – gw Misc Inst [071]
Chicagoer deutsche zeitung : wochen-ausgabe – Chicago: Chicago-German Gazette Publ Co, dec 20 1924; jan 10 1925; aug 1925-may 22 1926 – 1r – 1 – us CRL [071]
Chicagoer frauen-zeitung – Chicago: [German-American Publ Co, [dec 10 1893-jun 2 1901] – us CRL [071]
Chicagoer freie presse – Chicago: Richard Michaelis, [1891-1901] – 49r – 1 – us CRL [071]
Chicagoer weckruf – Chicago IL (USA), dec 27 1933-nov 1 1935 – 1r – 1 – gw Misc Inst [071]
Chicago-posten – Chicago IL. 1892 aug 17/dec 28-1916 nov 16/1918 sep 19 – 14r – 1 – (with gaps) – mf#869045 – us WHS [071]
Chicago-south suburban news – Chicago, Harvey IL. 1870 jan 17-oct 3, 1968 sep 28-1869 may 24, 1969 may 31-1970 jan 10, 1970 oct 17-1971 jul 24, 1971 jul 31-1972 oct 14 – 5r – 1 – (cont: south suburban news) – mf#874821 – us WHS [071]
Chicano – Colton, San Bernardino CA. 1969 apr-1971 apr 5, 1972 apr 12-1973 jul 26, 1973 aug 3-1975 dec 25, 1976 jan 1-1978 may 25, 1978 jun-dec, 1979-81, 1987 jul 17-1988 jun – 9r – 1 – mf#267432 – us WHS [071]
Chicano – San Bernadino. 1975+ (1) 1979-1984 (5) 1979-1984 (9) – ISSN: 0009-3777 – mf#9482 – us UMI ProQuest [305]
Chicano Cultural Corporation [Houston TX] et al *see* Papel chicano
Chicano history newspaper clipping file : union city, alameda county library branch clippings from regional and leading papers on articles of import to the chicano community in the san francisco bay area – jun 1968-jan 1976 – 400r – 9 – $400.00 – mf#B63007 – us Library Micro [071]
Chicano scrapbook : clippings from california newspapers of happenings in santa clara valley – San Jose, CA: Mexican-American Services Agency, 1968-73 – 1r – 1 – $50.00 – (also avail on mf $100.00) – mf#B63020 – us Library Micro [305]
Chicano studies library serial collection : complete collection, sections 1-13 – 426r – 1 – $21,485.00 set – (sect 1 + 2: 5r $375 b62001. sect 3: 3r $225 b62002. sect 4 39r $2925 b62003. sect 5: 89r $6675 b62004. sect 6: 64r $3520 b62005. sect 7: 75r $5625 b62006. sect 8: 16r $1200 b62007. sect 9: 32r $2400 b62008. sect 10: 27r $2025 b62009. sect 11: 32r $2400 b62010. sect 12: 44r $3300 b62011. sect 13: 40r $3000 b62012) – mf#B62000- – us Library Micro [305]
Chicano studies newsletter / University of California, Los Angeles – v13 n1 [1985 nov] – 1r – 1 – (cont: mirlo; cont by: noticias de aztlan) – mf#1041596 – us WHS [972]
Chicco's big decision – Pretoria: Dept of National Health & Population Development [1993] [mf ed Pretoria, RSA: State Library [199-]] – (= ser Teen talk) – 16p [ill] on 1r with other items – 5 – mf#op 11254 r24 – us CRL [362]
Chicherin, Boris N *see* Feelosofia prava
Chichester, Charles Raleigh *see* Amalgamation of unions and proposed modifications in the poor-law (ireland)
Chichester courier – England. -w. 21 Oct-16 Nov 1868, 6-20 Jan 1869. (14 ft) – 1 – uk British Libr Newspaper [072]
Chichester customary : the rites of the church as observed throughout the year in chichester cathedral with an introductory essay / Duncan-Jones, Arthur Stuart – London: SPCK 1948 [mf ed (S.l): Micro Methods c1971] – (= ser Alcuin club colls 36) – 1r – 1 – (on reel 6 with on the epiclesis of the eucharistic liturgy.../ atchley) – sa Misc Inst [242]
Chichester, Edward *see*
– Documents illustrative of the oppressions and cruelties of irish revenue officers
– A second letter to a british member of parliament
Chichester express and west sussex journal – England. -w. 6 Jan 1863-30 Dec 1902. (13 reels) – 1 – uk British Libr Newspaper [072]
Chichester journal – England. 28 nov 1860-27 apr 1864 [wkly] – 2r – 1 – (aka: southern star) – uk British Libr Newspaper [072]
Chichester, New Hampshire. Chichester Baptist Church and Society *see* Records
Chick, Susan A *see* Gender classifications of female athletes and nonathletes at women's and coeducational colleges
Chickasaw Historical and Genealogical Society *see* Chickasaw times past
Chickasaw newsletter – 1972 jan-1976 jan/mar – 1r – 1 – (cont by: chickasaw times) – mf#1555135 – us WHS [071]
Chickasaw times – 1976 apr/jun-1987 dec – 1r – 1 – (cont: chickasaw newsletter) – mf#1555133 – us WHS [071]
Chickasaw times past / Chickasaw Historical and Genealogical Society – 1982 apr/jun-1987 jan/mar – 1r – 1 – mf#1054341 – us WHS [929]
[Chico-] chico enterprise record – CA. 1907- – 439+ r – 1 – $26,340.00 (subs $800/y) – (aka: chico record) – mf#BC02017 – us Library Micro [071]
[Chico-] chico news and review – CA. 1994- – 1+ r – 1 – $60.00 (subs $50/y) – mf#B06016 – us Library Micro [071]
Chico record *see* [Chico-] chico enterprise record
Chico rising – v1 n11-14 [1972 feb 3-Apr] – 1r – 1 – mf#714268 – us WHS [071]
Chicopee 1848-1890 – Oxford, MA (mf ed 1984) – (= ser Massachusetts vital records) – 88mf – 9 – 0-931248-66-3 – (mf 1-7: births 1848-79 vol a 1-192. mf 8-11: births 1880-90 vol b 193-283. mf 12-16: index to births 1848-91. mf 17-27: marriage intentions 1848-61 vol 1. mf 28-38: marriage intentions 1861-81 vol 2. mf 39-48: marriage intentions 1881-93. mf 49-53: index to marriage intention 1848-90. mf 54-60: marriages 1848-62 vol a. mf 61-66: marriages 1862-82 vol b. mf 67-70: marriages 1883-90 vol 3. mf 71-75: index to marriages 1848-90. mf 76-82: deaths 1848-82 vol a. mf 83-84: deaths 1883-90 vol b. mf 85-88: index to deaths 1848-90) – us Archive [978]
Chidvilas Foundation *see* Rajneesh times of india
Chidyausiku, Paul *see*
– Karumekangu
– Nhoroondo dzokuwanana
– Nyadzi dzinokunda rufu
– Pfungwa dzasekuru mafusire
Chidzero, Bernard T *see* Nzvengamutsvairo
Chidzero, Bernard T G *see*
– Nzvengamutsvairo
– Tanganyika and international trusteeship
Chiec la : truyen dai gia dinh xa hoi / To Dien Lang – [s. l: s. n.] 1974 (Saigon: an-Quan Hanh) [mf ed 1992] – on pt of 1r – 1 – mf#11052 r222 n5 – us Cornell [830]
Chiec ma t na da nguoi / Nguyen, Xuan Quang – [Saigon]: Tri Dang 1974 [mf ed 1992] – on pt of 1r – 1 – mf#11052 r382 n5 – us Cornell [959]
Chiec vong dinh hon : truyen dai tinh cam tam-ly xa-hoi / Lan Phuong – Saigon: Ba Lan Phuong 1975 [mf ed 1992] – on pt of 1r – 1 – mf#11052 r152 n1 – us Cornell [830]
Chief – Bridgeport, WA. 1952-1956 (1) – mf#68901 – us UMI ProQuest [071]
Chief – Oshkosh WI. 1889 dec 14-1890 nov 1 – 1r – 1 – mf#959522 – us WHS [071]
The chief actors in the puritan revolution / Bayne, Peter – 2nd ed. London: James Clarke, 1879 – 2mf – 9 – 0-7905-5922-6 – mf#1988-1922 – us ATLA [243]
The chief corner-stone : essays towards an exposition of the christian faith for to-day / Davison, William Theoplilus et al; ed by Davison, William Theophilus – 1st ed. London: Charles H Kelly, 1914 – 1mf – 9 – 0-7905-0570-3 – (incl bibl ref) – mf#1987-0570 – us ATLA [240]
Chief, council and commissioner / Holleman, J F – Assen, Netherlands. 1969 – 1r – us UF Libraries [960]
The chief currents of contemporary philosophy / Datta, Dhirendra Mohan – Calcutta: University of Calcutta, 1950 – (= ser Samp: indian books) – us CRL [180]
The chief end of revelation / Bruce, Alexander Balmain – London: Hodder & Stoughton, 1890 – 1mf – 9 – 0-8370-2482-X – mf#1985-0482 – us ATLA [221]
Chief executive – London. 1978-1988 (1) 1978-1988 (5) 1978-1988 (9) – (cont: chief executive monthly) – ISSN: 0262-5865 – mf#6020,02 – us UMI ProQuest [650]
Chief executive – New York. 1980+ (1,5,9) – ISSN: 0160-4724 – mf#14421 – us UMI ProQuest [650]
Chief executive monthly – London. 1977-1978 (1) 1977-1978 (5) 1977-1978 (9) – (cont: business administration. cont by: chief executive) – ISSN: 0140-8453 – mf#6020,01 – us UMI ProQuest [650]
Chief information officer journal – New York. 1992-1993 (1,5,9) – ISSN: 0899-0182 – mf#18363 – us UMI ProQuest [650]
Chief joseph herald – Joseph OR: C E Heard, -1959 [wkly] – 1 – (cont: joseph herald (joseph, or). absorbed by: wallowa county chieftain) – us Oregon Lib [071]
Chief justice marshall's decisions *see* Brockenbrough's reports of cases in the fourth circuit, 1802-1833
Chief of sinners in heaven / Stock, John – London, England. 18– – 1r – us UF Libraries [240]
The chief of the herd / Mukerji, Dhan Gopal – New York: EP Dutton & Co, 1938 – (= ser Samp: indian books) – (ill by mahlon blaine) – us CRL [490]

The chief periods of european history : six lectures read in the university of oxford in trinity term, 1885 / Freeman, Edward Augustus – London; New York: Macmillan, 1886 – 1mf – 9 – 0-7905-5393-7 – mf#1988-1393 – us ATLA [940]

The chief periods of european history : six lectures read in the university of oxford in trinity term, 1885: with an essay on greek cities under roman rule / Freeman, Edward A – London; New York: Macmillan, 1886 – 1mf – us ATLA [940]

Chief points of difference betwixt the established and the free chu... / Dods, Selby Ord – Edinburgh, Scotland. 1847 – 1r – us UF Libraries [240]

The chief superintendent's report on education in upper canada for the year 1856 : omitting the statistical tables and appendix / Ryerson, Egerton – Toronto: Printed by Lovell & Gibson, 1857 – 1mf – 9 – mf#47554 – cn Canadiana [370]

The chief works of benedict de spinoza / Spinoza, Benedictus de – 2nd rev ed. London: G Bell 1887 [mf ed 1993] – (= ser Bohn's philosophical library) – 2v on 3mf – 9 – 0-524-08650-8 – (trans fr latin with int by r h m elves) – mf#1993-2110 – us ATLA [190]

Chiefly among women – [New York: The Catholic Publ House, 1875] [mf ed 1984] – (= ser Women & the church in america 7) – 1mf – 9 – 0-8370-1606-1 – mf#1984-2007 – us ATLA [305]

Chiefs and families of note in the punjab – Lahore, [India]: [sn], 190 (Punjab: Supt, Govt Print) – (= ser Samp: indian books) – us CRL [954]

Chiefs of state and cabinet members of foreign governments / U.S. Central Intelligence Agency – Aug 1962-Dec 1975 – 1 $262.00 $18.00y 8 $58.00y – us L of C Photodup [324]

Chieftain – Pueblo, CO. 1946-2001 (1) – mf#61242 – us UMI ProQuest [071]

The chieftains of ceylon / Sanden, J C van – Colombo: Plate, 1936 – 1 – us CRL [954]

The chieftains of champlain : a story of adventure in the new world – New York: Hickey, 1883 – (= ser The vatican library 9) – 2mf – 9 – mf#00607 – cn Canadiana [830]

Chief-union / Wyandot Co. Upper Sandusk – jan 1983-dec 1984 [daily] – 8r – 1 – mf#B27873-27880 – us Ohio Hist [071]

Chieh hou / Ch'en, Keng-hsin – Shang-hai: K'ai ming shu tien, Min kuo 20 [1931] – (= ser P-k&k period) – us CRL [840]

Chieh hou shih i / Mao, Tun – [Kuei-lin]: Hsueeh i ch'u pan she, [Min kuo 31 ie 1942] – (= ser P-k&k period) – us CRL [830]

Chieh hou shih i / Mao, Tun – [Kuei-lin]: Hsueeh i ch'u pan she, [Min kuo 31 ie 1942] – (= ser P-k&k period) – us CRL [830]

Chieh hun chin hsing ch'ue / Ch'en, Pai-ch'en – Ch'ung-ch'ing: Tso chia shu wu, Min kuo 33 [1944] – (= ser P-k&k period) – us CRL [820]

Chieh kuan chao shang chue liang chou nien chi nien k'an – [China: sn] – (= ser P-k&k period) – us CRL [951]

Chieh lou sui pi / Lin, Keng-pai – Shang-hai: Ch'en pao she ch'u pan pu, 1934 – (= ser P-k&k period) – us CRL [840]

Chieh p'ou hsueh pao = Journal of anatomy – 1963-1964 (1) – mf#2596 – us UMI ProQuest [611]

Chieh shih hui tsuan – Ch'eng-tu: Ch'eng ch'eng ch'u pan she, Min kuo 32 [1943] – (= ser P-k&k period) – us CRL [951]

Chieh teng hsia – Shang-hai: Hsin ti shu tien: Kuo feng shu tien, Min kuo 29 [1940] – (= ser P-k&k period) – us CRL [820]

Chieh t'ou chiang hua / Liu, Shih – Shang-hai: Sheng huo shu tien, 1939 – (= ser P-k&k period) – us CRL [840]

Chieh t'ou chue / Shen, Hsi-ling – Han-k'ou: Hsing hsing ch'u pan she, 1938 – (= ser P-k&k period) – us CRL [951]

Chieh t'ou wen t'an / Hsue, Mao-yung – Shang-hai: Kuang ming shu chue, Min kuo 26 [1937] – (= ser P-k&k period) – us CRL [951]

Chieh-fang jih-pao – Yenan. China. 1941-47 – 1 – us Chinese Res [079]

Chiel : an illustrated, national, social, musical, & dramatic journal – [Scotland] Glasgow: R Graham 17 feb 1883-25 jan 1890 (wkly) [mf ed 2003] – 15v on 15r – 1 – uk Newsplan [072]

Chiemgau-bote – Traunstein DE, 1929 2 nov-1933 – 4r – 1 – gw Misc Inst [074]

Chien ch'ai / Kao, Shen – Pei-ching: Hsin min yin shu kuan, Min kuo 33 [1944] – (= ser P-k&k period) – us CRL [480]

Ch'ien, Chao-hsiung *see* T'ui kuang chiao yu

Ch'ien, Chia-chu *see*
– Nung ts'un yu tu shih
– Wu chia wen t'i

Chien ch'ih ho-pei k'ang chan yue kung ku t'uan chieh – Ch'ung-ch'ing: Hsin hua jih pao kuan, 1939 – (= ser P-k&k period) – us CRL [951]

Ch'ien, Chi-po *see*
– Hsien tai chung-kuo wen hsueh shih
– Ming tai wen hsueh

Chien chu hsueh pao = Journal of agriculture – 1963-1964 (1) – mf#2597 – us UMI ProQuest [630]

Ch'ien chuang hsueeh / Shih, Po-heng – Shang-hai: Shang-hai shang yeh chu suan hsueeh she, min kuo 23 [1934] – (= ser P-k&k period) – us CRL [332]

Ch'ien ch'ue kuo yue lo-ma tzu piao chun kuo yue chiao pen – Hsia-men: Hsia-men ta hsueeh wen hsueh yuean yue yen hsueeh hsi, 1935 – (= ser P-k&k period) – us CRL [480]

Ch'ien, Chun-jui *see*
– Kei chiu wang t'ung chih ti kung k'ai hsin
– Lun chan cheng
– Tsen yang yen chiu chung-kuo ching chi
– Wang ching-wei mai kuo ti li lun yu shih chien

Ch'ien, Chun-t'ao *see* Su miao

Le chien d'or : legende canadienne / Kirby, William – [Montreal?: s.n.] 1884 [mf ed 1986] – 2v on 1mf – 9 – 0-665-07985-0 – mf#07985 – cn Canadiana [830]

Le chien d'or *see* La terre paternelle

Ch'ien, Ho *see* Jen ko chiao yu hsueh kai

Ch'ien hou fang / Hsue, Ying – Ch'ung-ch'ing: Chien kuo shu tien, 1943 – (= ser P-k&k period) – us CRL [951]

Ch'ien hsi / Chin, I – Ch'ung-ch'ing: Wen hua sheng huo ch'u pan she, Min kuo 32 [1943] – (= ser P-k&k period) – us CRL [830]

Ch'ien, Hsiao-yue *see* Chien wen i pan: t'ao nan pi chi

Ch'ien hsien kuei lai – [China]: Min kuang shu tien, Min kuo 26 [1937] – (= ser P-k&k period) – us CRL [951]

Ch'ien hsien te chi-tu t'u ch'ing nien [ccm224] = Christian youth at the front / Liu, Liang-mo – Shanghai, 1940 [mf ed 198?] – (= ser Ccm 224) – 1 – mf#1984-b500 – us ATLA [951]

Chien, Hsien-ai *see* Ch'eng hsia chi

Ch'ien, Hsing-ts'un *see*
– Hai shih chi
– Hsien tai chung-kuo wen hsueh lun

Chien i hsiang ts'un shih fan hsueeh hsiao k'o ch'eng piao chun – Shang-hai: Chung-hua shu chue, Min kuo 24 [1935] – (= ser P-k&k period) – us CRL [370]

Chien i shih fan hsueeh hsiao k'o ch'eng piao chuen – [China]: Shang wu yin shu kuan, Min kuo 24 [1935] – (= ser P-k&k period) – us CRL [951]

Ch'ien, I-shih *see* Pai lang t'ao t'ien ti t'ai-p'ing yang wen t'i

Chien, I-ts'ung *see* Pi nan jih chi

Ch'ien, Keng-hsin *see* Chieh hou

Chien k'u chung ch'eng chang te chia (ccm97) / Chiang, I-chen – Shanghai, 1950 [mf ed 198?] – (= ser Ccm 97) – 1 – mf#1984-b500 – us ATLA [230]

Chien, Kuan-san *see*
– Li lun she hui hsueh
– Ta shih tai chung ti ch'ing nien wen t'i

Ch'ien Kung-hsia *see* Hsi chu

Chien kuo chi lien ho pao – Ho Chi Minh City, Vietnam. 1966-1967 (1) – mf#67824 – us UMI ProQuest [079]

Chien kuo chung ti chi ko chung yao wen t'i / Ch'en, Chih-mai – Nan-wen-ch'uean: Hsin p'ing lun pan yueeh k'an she, Min kuo 30 [1941] – (= ser P-k&k period) – us CRL [951]

Chien kuo chung yen / Li, Tz'u-shan – [Shang-hai: sn, Min kuo 22 ie 1933] – (= ser P-k&k period) – us CRL [951]

Chien kuo fang lueeh / Sun, Yat-sen – [China]: Chung-kuo kuo min tang chung yang hsuean ch'uan pu: Chung-kuo wen hua fu wu she, [1941?] – (= ser P-k&k period) – us CRL [951]

Chien kuo ta kang ch'ien shih / Chung-kuo kuo min tang Hsuean ch'uan pu – [Ch'ung-ch'ing], Min kuo 29 [1940] – (= ser P-k&k period) – us CRL [951]

Chien kuo t'u ching / Ch'ien, Tuan-sheng – Ch'ung-ch'ing: Kuo min t'u shu ch'u pan she, Min kuo 31 [1942] – (= ser P-k&k period) – us CRL [951]

Ch'ien lu / Ping-ying – Shang-hai: Kuang ming shu chue, Min kuo 24 [1935] – (= ser P-k&k period) – us CRL [480]

Chien lu fan ko ming fen tzu wang ming-tao te fan tung yen lun (ccm293) / T'ien feng chou kan tzu liao shih pien – Shanghai, 1955 [mf ed 198?] – (= ser Ccm 293) – 1 – mf#1984-b500 – us ATLA [240]

Ch'ien, Mu *see* Kuo shih ta kang

Chien pei p'ien / Lao, She – Ch'ung-ch'ing: Wen i chiang chu chin kuan li wei yuean hui ch'u pan pu, 1942 – (= ser P-k&k period) – us CRL [951]

Ch'ien pen / Wang, Hsi-p'eng – Shang-hai: Liang yu t'u shu yin shua kung ssu, Min kuo 21 [1932] – (= ser P-k&k period) – us CRL [480]

Chien pi ch'ing yeh / Hsue, Ch'ang-lin – Ch'ung-ch'ing: Kuo min t'u shu ch'u pan she, 1945 – (= ser P-k&k period) – us CRL [820]

Ch'ien pi ko ming shih hsing fang an hui lan – [China]: Chung-kuo ch'ien pi ko ming hsieh chin hui, Min kuo 22 [1933] – (= ser P-k&k period) – us CRL [332]

Chien she ti "ta chung yue" wen hsueeh / Li, Chin-hi – Shang-hai: Shang wu yin shu kuan, 1936 – (= ser P-k&k period) – us CRL [480]

Ch'ien, Shih-yun *see* Nu tzu shih yung hsin ch'ih tu

Chien tang yue chien kuo / T'ao, Pai-ch'uan – Han-k'ou: Tu li ch'u pan she, Min kuo 27 [1938] – (= ser P-k&k period) – us CRL [951]

Ch'ien t'u: ssu mu hua chue / Liu, Tzu-ch'ing – [China: Chuen shih wei yuean hui cheng chih pu], Min kuo 32 [1943] – (= ser P-k&k period) – us CRL [820]

Chien tu wen t'i lun chi – Ch'ung-ch'ing: Tu li ch'u pan she, Min kuo 33 [1944] – (= ser P-k&k period) – us CRL [951]

Ch'ien, Tuan-sheng *see*
– Chien kuo t'u ching
– Min kuo cheng chih shih

Ch'ien tzu wen [ccm264] = Ploughman's song / Price, Philip Francis – Hankow, 1938 [mf ed 198?] – (= ser Ccm 264) – 1 – mf#1984-b500 – us ATLA [780]

Chien wen i pan: t'ao nan pi chi / Ch'ien, Hsiao-yue – Shang-hai: Chien Hsiao-yue: Hsin ya shu tien tsung ching shou, Min kuo 29 [1940] – (= ser P-k&k period) – us CRL [951]

Ch'ien yeh / Tai, Wan-yeh – Shang-hai: Ya tung t'u shu kuan, Min kuo 29 [1940] – (= ser P-k&k period) – us CRL [830]

Ch'ien yeh / Yang, Han-sheng – [China]: Hsi chue shu tien, Min kuo 27 [1938] – (= ser P-k&k period) – us CRL [820]

Chien ying chi / Yao, P'eng-tzu – Shang-hai: Liang yu t'u shu yin shua kung ssu, 1933 – (= ser P-k&k period) – us CRL [830]

Chien yue chih tu lun / Jui, Chia-jui – Shang-hai: Shang wu yin shu kuan, 1934 – (= ser P-k&k period) – us CRL [951]

Chien yue kung ch'ang kuan li fa / Jui, Chia-jui – Shang-hai: Shang wu yin shu kuan, Min kuo 23 [1934] – (= ser P-k&k period) – us CRL [951]

Ch'ien Yun-chan *see* Hung hsing ch'u ch'iang t'an tz'u

Chien Yu-wen *see* Chuan chiao wei jen ma-li-hsun (ccm251)

Chien, Yu-wen *see*
– Chung-kuo chi-tu chiao te kai shan shih yeh [ccm100]
– Hsin tsung chiao kuan
– Ku yu-tai ko ming shih yen i
– Tsung chiao yu jen sheng
– Tsung chiao yu k'o hsueh

Chiengmai khonmuang – Chiengmai, Thailand. 1953-54; 1975-76 – 3r – 1 – us L of C Photodup [079]

Chienne / Mouezy-Eon, Andre – Paris, France. 1931 – 1r – us UF Libraries [440]

Chien-pu-chai = Jianpuzhai / I, Chun – Pei-ching [China] Shih-chieh chih shih chu pan she 1957 [mf ed 1989] – 1r with other items – 1 – mf#mf-10289 seam reel 030/07 [§] – us CRL [959]

Chien-pu-chai feng wu = Jianpuzhai feng wu / Shao, Min-chih – [Hsiang-kang] Hsiang-kang i ch'un ch'u pan she [1970] [mf ed 1989] – 1r with other items – 1 – mf#mf-10289 seam reel 030/11 [§] – us CRL [915]

Ch'ien-t'u : [the future] – Shanghai, feb 1933-feb 1939 – 7r – 1 – (scattered issues missing) – us Chinese Res [073]

Chiera, E *see*
– Joint expedition with the iraq museum at nuzi
– Sumerian religious texts

Chiera, Edward *see* Inscriptions from adab

Chiesa De Perez, Carmen *see* Proyecciones del modernismo

La chiesa nuova in aissi... / Terzi, Arduino – Madrid: Arch. Ibero Americano, 1964 – 1 – sp Bibl Santa Ana [240]

La chiesa russa : le sue odierne condizioni e il suo riformismo dottrinale / Palmieri, Aurelio – Firenze: Libreria editrice fiorentina, 1908 – 2mf – 9 – 0-8370-7727-3 – (incl bibl ref) – mf#1986-1727 – us ATLA [947]

La chiesa russa, le sue odierne condizioni / Palmieri, A – Firenze, 1908 – €25.00 – ne Slangenburg [241]

Chiesi, Danielle *see* Personal characteristics of beginning, intermediate, and advanced sport performers

Chiesi, Gustavo *see* La colonizzazione europea nell'est africa

Chifamba, Jane *see* Ngano dzepasi chigare

Ch'i-feng i pien / Ch'i-feng-ch'iao-tao-jen – Shang-hai: Feng yue shu wu: Chung-hua ta hsueeh t'u shu yu hsien kung ssu, Min kuo 27 [1938] – (= ser P-k&k period) – us CRL [951]

Ch'i-feng-ch'iao-tao-jen *see* Ch'i-feng i pien

Chiffre und kabbala in goethe's faust : neue beitraege zur neuen faustforschung / Louvier, Ferdinand August – Dresden: Henkler 1897 [mf ed 1990] – 1r – 1 – (incl bibl ref. filmed with: vorlesungen über goethe's faust / fr kreyssig) – mf#7354 – us Wisconsin U Libr [430]

The chignecto ship railway : the substitute for the baie verte canal / Ketchum, Henry George Clopper – [Fredericton, NB?: s.n, 1892?] [mf ed 1982] – 1mf – 9 – mf#07830 – cn Canadiana [380]

Chignell, Arthur Kent *see*
– An outpost in papua
– Twenty-one years in papua

Chignell, Robert *see* The life and paintings of vicat cole, r a

Chih hsien yue k'ang jih / Li, Tsung-wu – Ch'eng-tu: Li Tsung-wu, 1937 – (= ser P-k&k period) – us CRL [951]

Ch'ih i pu chueeh – Shang-hai: Sheng huo shu tien, Min kuo 22 [1933] – (= ser P-k&k period) – us CRL [390]

Chih min ti yue pan chih min ti / Ch'en, Hung-chin – Shang-hai: Hei pei ts'ung shu she, Min kuo 27 [1938] – (= ser P-k&k period) – us CRL [951]

Ch'ih pei ou t'an / Wang, Shih-chen – Shang-hai: Ta ta t'u shu kung ying she, 1935 – (= ser P-k&k period) – us CRL [480]

Chih ping yue chih hsin / Liu, Chien-hsue – Chin-hua: T'ien hsing tsa chih she, Min kuo 31 [1942] – (= ser P-k&k period) – us CRL [355]

Chih shih – Kuang-chou: Kuang-tung sheng cheng fu mi shu ch'u pien i shih, 1942 – (= ser P-k&k period) – us CRL [951]

Chih shih ti ying yung: tu shu wen ta ti erh chi / Ai, Ssu-ch'i – Shang-hai: Tu shu sheng huo ch'u pan she, Min kuo 25 [1936] – (= ser P-k&k period) – us CRL [951]

Chih tan yue ch'iang tan, i ming, han-k'ou meng: san mu chue / Ch'i-ao – [China: sn, 1944] – (= ser P-k&k period) – us CRL [951]

Chih t'ang kung yeh pao kao shu – Shang-hai: Ch'uean kuo ching chi wei yuean hui, Min kuo 25 [1936] – (= ser P-k&k period) – us CRL [951]

Chih t'ang wen chi / Chou, Tso-jen – Shang-hai: Tien ma shu tien, Min kuo 22 [1933] – (= ser P-k&k period) – us CRL [840]

Ch'ih tu chue chieh: wen yen tui chao / Hsiung, Shih-seng – Shang-hai: Hua chung shu chue, Min kuo 30 [1941] – (= ser P-k&k period) – us CRL [951]

Chih tu yue jen ts'ai – [China]: Pei tou shu tien, Min kuo 33 [1944] – (= ser P-k&k period) – us CRL [650]

Chih tzu yue kuei / Yao, Su-feng – Ch'ung-ch'ing: Hsin sheng t'u shu wen chue kung ssu, 1943 – (= ser P-k&k period) – us CRL [820]

Chih we hsueh pao = Journal of botany – 1960-1964 (1) – mf#2598 – us UMI ProQuest [580]

Chih yeh chiao yue chih li lun yue shih chi – [China]: Chung-hua chih yeh chiao yue she, 1933 – (= ser P-k&k period) – us CRL [370]

Chih yeh hsueeh hsiao ko k'o chiao ts'ai ta kang, k'o ch'eng piao, she pei kai yao hui pien ti 1 ts'e – [China]: Chiao yue pu, 1934 – (= ser P-k&k period) – us CRL [370]

Chih-hao shih chi / Ho, Chih-hao – Shang-hai: Nan hua shu tien, Min kuo 23 [1934] – (= ser P-k&k period) – us CRL [810]

Chih-hsing *see* Hu fen

Chih-hsing shih ko hsue chi / T'ao, Hsing-chih – Shang-hai: Erh t'ung shu chue, Min kuo 24 [1935] – (= ser P-k&k period) – us CRL [810]

Chih-hsing shih ko pieh chi, i ming, ch'ing feng ming yueeh chi / T'ao, Hsing-chih – Shang-hai: Erh t'ung shu chue, Min kuo 24 [1935] – (= ser P-k&k period) – us CRL [810]

Chih-sheng hsiang-shih lu – List of successful candidates in the imperial examination in Chih-li province: 1831, 1834, 1843, 1851, 1904, 1907. 2 reels – 1 – 62.00 – us Chinese Res [951]

Chih-shih *see* Nung min lei

Chihuahua de mis amores y otros despachos de mexic... / Rembao, Alberto – Mexico City?, Mexico. 1949 – 1r – us UF Libraries [972]

Chihuahua. Mexico. (State) *see*
– Periodico oficial
– Periodico oficial del gobierno del estado

Chiiko chinonzi 'communism' / Reich, J – Gwelo, Zimbabwe. 1959 – 1r – us UF Libraries [960]

Chi-jomvu / Lambert, H E – Kampala, Uganda. 1958 – 1r – us UF Libraries [960]

Child – Washington. 1936-1953 – 1 – mf#5134 – us UMI ProQuest [370]

The child / Dupanloup, Felix – Boston: P Donahoe, 1875 – 1r – 1 – us Wisconsin U Libr [920]

The child / Tagore, Rabindranath – London: George Allen & Unwin, 1931 – (= ser Samp: indian books) – us CRL [490]

Child abuse and neglect – New York. 1977+ (1,5,9) – ISSN: 0145-2134 – mf#49253 – us UMI ProQuest [360]

Child abuse review – Croydon. 1992+ (1,5,9) – ISSN: 0952-9136 – mf#19117 – us UMI ProQuest [360]

Child, Alfred Thurston *see* Our virgin islands

Child and adolescent social work journal : c and a – New York. 1984+ (1,5,9) – ISSN: 0738-0151 – mf#14129 – us UMI ProQuest [640]

Child and family behavior therapy / ed by Franks, Cyril M – v1- 1989- – 1,9 ($325.00 in US $455.00 outside hardcopy subsc) – us Haworth [306]

The child and religion : eleven essays / Jones, Henry, Sir et al; ed by Stephens, Thomas – London: Williams and Norgate; New York: Putnam, 1905 – (= ser Crown Theological Library) – 1mf – 9 – 0-7905-9978-3 – mf#1989-1703 – us ATLA [240]

The child and the curriculum / Dewey, John – Chicago: The University of Chicago Press, c1902 [mf ed 1970] – (= ser University of chicago. contributions to education 5; Library of american civilization 40014) – 40p on 1mf – 9 – us Chicago U Pr [370]

Child and youth care forum – New York. 1991+ (1,5,9) – (cont: child and youth care quarterly) – ISSN: 1053-1890 – mf#11173,02 – us UMI ProQuest [150]

Child and youth care quarterly – New York. 1987-1990 (1,5,9) – (cont: child care quarterly. cont by: child and youth care forum) – ISSN: 0893-0848 – mf#11173,01 – us UMI ProQuest [150]

Child and youth services / ed by Beker, Jerome – v1- 1977- – 1,9 ($200.00 in US $280.00 outside hardcopy subsc) – us Haworth [305]

The child as god's child / Rishell, Charles Wesley – New York: Eaton & Mains; Cincinnati: Jennings & Graham, c1904 – 1mf – 9 – 0-7905-9613-X – mf#1989-1338 – us ATLA [240]

Child care, health and development – Oxford. 1980+ (1,5,9) – ISSN: 0305-1862 – mf#15511 – us UMI ProQuest [640]

Child care information exchange – Redmond. 1989+ (1,5,9) – ISSN: 0164-8527 – mf#17464 – us UMI ProQuest [360]

Child care quarterly – New York. 1971-1986 (1) 1971-1986 (5) 1971-1986 (9) – (cont by: child and youth care quarterly) – ISSN: 0045-6632 – mf#11173 – us UMI ProQuest [150]

Child, Daphne *see* Yesterday's children

Child, David Lee *see* The culture of the best

Child day care planning project – 1983-90 – 9r – 1 – (minutes, correspondence, and publs of this cooperative project of cuyahoga county, united way, and federation for community planning, including both personnel and committee files) – us Western Res [360]

Child development – Malden. 1930+ (1) 1930+ (5) 1930+ (9) – ISSN: 0009-3920 – mf#5648 – us UMI ProQuest [640]

Child development abstracts and bibliography – Malden. 1959+ (1) 1971+ (5) 1977+ (9) – ISSN: 0009-3939 – mf#5649 – us UMI ProQuest [150]

Child education – London. 1963+ (1) 1976+ (5) 1976+ (9) – ISSN: 0009-3947 – mf#7629 – us UMI ProQuest [370]

Child education quarterly – London. 1974-1978 (1) 1975-1978 (5) 1975-1978 (9) – ISSN: 0045-6640 – mf#7628 – us UMI ProQuest [370]

Child, Gilbert William *see* Church and state under the tudors

Child, Harold *see*
- History and extent of recognition of tribal law in rhodesia
- History of the amandebele

The child in india : a symposium commemorating the coming of age of the society for the protection of children in western india / ed by Manshardt, Clifford – Bombay: DB Taraporevala Sons & Co, [1937] – (= ser Samp: indian books) – (int by lord brabourne) – us CRL [362]

Child labor and street trades permits issued in wisconsin / Wisconsin. Division of Labor Standards – 1915-68. 14 fiches. (Harvard Law School Library Collection.) – 9 – us Harvard Law [331]

Child labor bulletin *see* American child / child labor bulletin

Child labor legislation : handbook / National Consumers' League – 1904-06 – 1r – 1 – mf#3187742 – us WHS [344]

Child language survey, transcripts of the... – 105mf – 9 – mf#86915 – uk Microform Academic [370]

Child life – Indianapolis. 1922+ (1) 1971+ (5) 1976+ (9) – ISSN: 0009-3971 – mf#2195 – us UMI ProQuest [370]

Child, Lydia Maria Francis *see*
- Isaac t. hopper
- The right way the safe way

Child, Marcus *see* An address, delivered to the inhabitants of the county of stanstead at a public meeting of that county

Child obscenity and pornography prevention act of 2002 and the sex tourism prohibition improvement act of 2002 : hearing...house of representatives, 107th congress, 2nd session, on h.r. 4623 and h.r. 4477, may 9 2002 / United States. Congress. House. Committee on the Judiciary. Subcommittee on Crime, Terrorism, and Homeland Security – Washington: US GPO 2002 [mf ed 2002] – 1mf – 9 – 0-16-068720-9 – us GPO [345]

Child of destiny / Fischer, William Joseph – Toronto: W Briggs, 1909 [mf ed 1995] – 4mf – 9 – 0-665-74218-5 – (ill by carlo cattapani & george a loughbridge) – mf#74218 – cn Canadiana [830]

Child of pallas – Baltimore. 1800-1801 – 1 – mf#3563 – us UMI ProQuest [420]

The child of the kingdom / Barbour, Margaret Frazer – 2nd ed. London: James Nisbet & Co, 1862 – (= ser 19th c women writers) – 3mf – 9 – mf#5.1.138 – uk Chadwyck [420]

Child psychiatry and human development – New York. 1970+ (1) 1973+ (5) 1973+ (9) – ISSN: 0009-398X – mf#11174 – us UMI ProQuest [616]

Child psychiatry quarterly – Hyderabad. 1972-1989 [1]; 1975-1989 [5,9] – ISSN: 0009-3998 – mf#7660 – us UMI ProQuest [616]

Child study : a journal of parent education – New York. 1952-1960 (1) – mf#817 – us UMI ProQuest [150]

Child study and child training / Forbush, William Byron – Toronto: McClelland, Goodchild & Stewart, c1915 [mf ed 1995] – 4mf – 9 – 0-665-73640-1 – mf#73640 – cn Canadiana [150]

Child study journal – Buffalo. 1970+ (1) 1970+ (5) 1977+ (9) – ISSN: 0009-4005 – mf#6629 – us UMI ProQuest [370]

Child, Theodore *see* Praise of paris

Child welfare – New York. 1922+ (1) 1971+ (5) 1977+ (9) – ISSN: 0009-4021 – mf#1615 – us UMI ProQuest [360]

Child Welfare Association of British Columbia *see* Constitution and by-laws

Child welfare statistics – 1951 apr/jun-1960 oct/dec – 1r – 1 – (cont by: child welfare and juvenile court statistics) – mf#629025 – us WHS [360]

Childe, Vere Gordon *see* Man makes himself

Childhood : the text-book of the age, for parents, pastors and teachers, and all lovers of childhood / Crafts, Wilbur Fisk – Toronto: A Miller, 1877 – 3mf – 9 – mf#26254 – cn Canadiana [150]

Childhood education – Olney. 1924+ (1) 1967+ (5) 1973+ (9) – ISSN: 0009-4056 – mf#821 – us UMI ProQuest [370]

Childhood leukaemia : family patterns overtime / Pradhuman, Rheola Gillian – Uni of South Africa 2000 [mf ed Johannesburg 2000] – 4mf – 9 – (incl bibl ref) – mf#mfm14772 – sa Unisa [618]

The childhood of jesus / Gannett, William Channing – English ed. London: Sunday School Association, 1885 – 1mf – 9 – 0-8370-3228-8 – mf#1985-1228 – us ATLA [240]

Children – Washington. 1954-1971 [1]; 1968-1971 [5] – ISSN: 0009-4064 – mf#1434 – us UMI ProQuest [150]

Children and art in the ussr / Marshak, Samuil – Moscow: Foreign Languages Pub House 1939 [mf ed 2000] – 1r – 1 – mf#29077 – us Harvard [700]

Children and schools – Washington, 2000+ [1,5,9] – (cont: social work in education) – ISSN: 1532-8759 – mf#11632,01 – us UMI ProQuest [360]

Children and their primary schools : report of the central advisory council for england (plowden report), 1967 – 14mf – 9 – mf#86966 – uk Microform Academic [324]

Children and youth services review – New York. 1979+ (1,5,9) – ISSN: 0190-7409 – mf#49292 – us UMI ProQuest [360]

The children for christ : thoughts for christian parents on the consecration of the home life / Murray, Andrew – Toronto: S. R. Briggs, 1887. Beltsville, Md: NCR Corp, 1978 (5mf); Evanston: American Theol Lib Assoc, 1984 (5mf) – 9 – 0-8370-0994-4 – mf#1984-4350 – us ATLA [240]

Children in concentration camps / North American Committee to Aid Spanish Democracy – [New York: Medical Bureau...1939?] [mf ed 1977] – (= ser Blodgett coll) – 1mf – 9 – mf#w1035 – us Harvard [946]

Children in court / Puner, Helen Walker – 1st ed. New York: Public Affairs Committee, 1954. 28p – 1 – us L of C Photodup [347]

Children of central africa / Wareham, J M – London: Longmans, Green, [1957] – us CRL [305]

The children of god *see* Shang ti ti erh nu men (ccm17)

Children of god and union with christ / Schieffelin, Samuel Bradhurst – NY: Board of Publication of the Reformed Church in America, 1896 – 1mf – 9 – 0-8370-5418-4 – mf#1985-3418 – us ATLA [240]

Children of God [Movement] *see* Mo letters

Children of loneliness / Yezierska, Anzia – New York, NY. 1923 – 1r – us UF Libraries [939]

The children of madagascar / Standing, Herbert F – [London]: Religious Tract Society, 1887 – 1mf – 9 – 0-8370-6413-9 – mf#1986-0413 – us ATLA [960]

Children of peace, the history of a novel sect in york co : established in the early part of this century: their ceremonies and how they conducted their services... – S.l: s.n, 1898? – 1mf – 9 – mf#09369 – cn Canadiana [243]

The children of the church : thoughts on the relation of baptized children to the church, and the duty and responsibility which it involves / Dewart, Edward Hartley – Toronto: Printed for the author, Guardian Office, 1861 – 1mf – 9 – 0-665-89044-3 – (incl bibl ref) – mf#89044 – cn Canadiana [240]

Children of their fathers / Read, Margaret – New Haven, CT. 1960 – 1r – us UF Libraries [960]

Children of their fathers / Read, Margaret – New York, NY. 1968 – 1r – us UF Libraries [960]

Children of this world and the children of light / Macfie, Daniel – Newcastle, England. 1867 – 1r – us UF Libraries [240]

Children of wrath / Buchet, Edmond Edouard – London, England. 1947 – 1r – us UF Libraries [025]

Children of yayoute / Des Pres, Francois Marcel-Turenne – Port-Au-Prince, Haiti. 1949 – 1r – us UF Libraries [972]

Children receiving ssi by state / United States. General Accounting Office. Health, Education, and Human Services Div – Washington DC: The Office [mf ed 1996] – 1mf – 9 – us US Gen Account [360]

Children today – Washington. 1972-1996 (1) 1972-1996 (5) 1975-1996 (9) – ISSN: 0361-4336 – mf#6658 – us UMI ProQuest [640]

Children with specific reading difficulties : report of the advisory committee on handicapped children, 1972 – 1mf – 9 – mf#87020 – uk Microform Academic [324]

Children's Aid Society of Wisconsin *see*
- Quarterly bulletin from the children's aid society of wisconsin
- Quarterly bulletin from the children's service society of wisconsin

Children's business – 2r/yr – 1 – $200.00/yr – us Fairchild Micro [680]

Children's choices in science books / Williams, Alice Marietta – New York, NY. 1939 – 1r – us UF Libraries [500]

The children's crusade : an episode of the thirteenth century / Gray, George Zabriskie – Boston: Houghton, 1900 – 1mf – 9 – 0-524-02795-1 – (incl bibl ref) – mf#1990-0699 – us ATLA [940]

Children's digest – New York. 1950-1980 (1) 1971-1980 (5) 1975-1980 (9) – (cont by: children's digest and children's playcraft) – ISSN: 0009-4099 – mf#5902 – us UMI ProQuest [640]

Children's digest – Indianapolis. 1980+ (1) 1980+ (5) 1980+ (9) – (cont: children's digest and children's playcraft) – ISSN: 0272-7145 – mf#5902,02 – us UMI ProQuest [640]

Children's digest and children's playcraft – Bergenfield. 1980-1980 (1) 1980-1980 (5) 1980-1980 (9) – (cont: children's digest. cont by: children's digest) – ISSN: 0273-7582 – mf#5902,01 – us UMI ProQuest [640]

The children's garland from the best poets / Patmore, Coventry Kersey Dighton – London, Cambridge: Macmillan & Co, 1862 – (= ser 19th c children's literature) – 4mf – 9 – mf#6.1.53 – uk Chadwyck [810]

Children's health care – Thorofare. 1980+ (1,5,9) – ISSN: 0273-9615 – mf#12161,01 – us UMI ProQuest [360]

Children's Home Society of Wisconsin *see* Financial report of the children's home society of wisconsin, 1897-1898

Children's Hospital National Medical Center *see* Clinical proceedings

Children's hour / Milwaukee Children's Hospital – 1982 jul-1986 nov – 1r – 1 – mf#1131008 – us WHS [360]

Children's legal rights journal – v1-21. 1979-2002 – 9 – $329.00 set – ISSN: 0278-7210 – mf#108521 – us Hein [340]

Children's literature in education – New York. 1970+ – 1,5,9 – ISSN: 0045-6713 – mf#11146 – us UMI ProQuest [370]

Children's magazine – Hartford. 1789-1789 – 1 – mf#3514 – us UMI ProQuest [305]

The children's missionary and sabbath school record – Montreal: J C Becket, 1844-[1845?] – 9 – (incl ind) – mf#P04233 – cn Canadiana [240]

The children's missionary magazine – London: J Nisbet [mthly] [mf ed 2006] – v10 (1847);ns: v2-3 (1849-50); n10-30 (oct 1851-jul 1853); jan 1855-jul 1858 [gaps] on 2r – 1 – (began in 1838? vols for 1847 called v10; for 1849-50 called ns v2-3; vol numbering ceased with 1851(?), but iss are numbered consecutively fr oct 1851-jul 1853 & are called new ser; iss numbering ceased in 1855(?), but iss are identified by date; 1859-also called new ser; some iss lacking; some pgs damaged) – mf#2005c-s095 – us ATLA [242]

Children's playcraft – New York. 1976-1978 (1) 1976-1978 (5) 1976-1978 (9) – mf#9905 – us UMI ProQuest [370]

Children's playmate magazine – Indianapolis. 1965+ [1]; 1970+ [5]; 1974+ [9] – ISSN: 0009-4161 – mf#1902 – us UMI ProQuest [370]

Children's record – New Glasgow, NS: E Scott, [1886?-1899] – 9 – (cont by: king's own) – ISSN: 1190-6472 – mf#P04614 – cn Canadiana [240]

Children's robinson crusoe : or, the remarkable adv... / Farrar, John (Mrs) – Boston, MA. 1830 – 1r – us UF Libraries [830]

Children's service news : from the children's service society of wisconsin – v1 n1-v6 n2 [1961 apr-1966 oct] – 1r – 1 – (cont by: children's service society reports to you) – mf#599353 – us WHS [366]

Children's service newsletter – v1 n1-v5 n2 [1975 jan/feb-1979 win] – 1r – 1 – (cont: children's service society reports to you; cont by: cssw newsletter) – mf#599358 – us WHS [360]

Children's Service Society of Wisconsin *see* Newsletter

Children's service society reports to you – v7 n1-v10 n2 [1968 spr-1971 christmas], 1968 spr-1971 christmas, 1972 sum, 1973 spr – 1r – 1 – (cont: children's service news; cont by: children's service newsletter) – mf#599355 – us WHS [366]

Children's services – Mahwah. 1998+ (1,5,9) – ISSN: 1093-9644 – mf#31728 – us UMI ProQuest [305]

Children's sunday : its history and methods of observing it / Dunning, Albert Elijah – Boston: Congregational Sunday-School and Pub Society, c1887 – 1mf – 9 – 0-524-03004-9 – mf#1990-4526 – us ATLA [240]

Children's theatre review : [the journal of the children's theatre association of america] – Washington. 1980-1986 (1) 1980-1986 (5) 1980-1986 (9) – ISSN: 0009-4196 – mf#12054,02 – us UMI ProQuest [790]

Children's times – 1993 dec, 1994 nov/dec – 1r – 1 – mf#4841829 – us WHS [305]

Children's treasury of bible stories : part 2: new testament / Gaskoin, Herman (Mrs); ed by Maclear, George Frederick – London: Macmillan, 1879 – 1mf – 9 – 0-8370-7382-0 – mf#1986-1382 – us ATLA [220]

The child's bible expositor : or, lessons and records of the sunday school – Toronto: H.Rowsell, [1840?-18–] – 9 – mf#P04343 – cn Canadiana [220]

Child's bible history / Knecht, Friedrich Justus – Mariannhill, South Africa. 1915 – 1r – us UF Libraries [220]

The child's book of ballads / Leeson, Jane Eliza – London: Joseph Masters, 1849 – (= ser 19th c children's literature) – 3mf – 9 – mf#6.1.58 – uk Chadwyck [810]

The child's book of homilies / Taylor, Helen – London: Edwards & Hughes, 1844 – (= ser 19th c children's literature) – 2mf – 9 – mf#6.1.19 – uk Chadwyck [240]

Child's friend and family magazine – Boston. 1843-1858 (1) – mf#3958 – us UMI ProQuest [640]

Child's gem – 1900-29 – 1 – us Southern Baptist [242]

Child's index – v. 1, no. 1-V. 3, no. 4. Sep 1862-Apr 1865 – 1 – 5.63 – us Southern Baptist [242]

Childs, Nancy M *see* Journal of nutraceuticals, functional and medical foods

Child's nervous system: chns – Heidelberg. 1990-1996 (1) – ISSN: 0256-7040 – mf#16150 – us UMI ProQuest [618]

Child's newspaper – Cincinnati. 1834-1834 (1) – mf#3959 – us UMI ProQuest [305]

Child's paper – New York, NY. -w. Jan 1852-Nov 1880. 2 reels – 1 – uk British Libr Newspaper [071]

Chile – National Library of Chile – 198r – 1 – (coll incl: el araucana; sala barros arana; el ferrocarril; periodicos various; el mercurio de valparaiso; sala toribio medina; archivo ramon freire (1820-1850)) – Pan-American Institute of Geography and History (IPGH) – us UMI ProQuest [972]

Chile *see*
- Diario oficial
- Diario oficial de la republica de chile
- Gaceta de los tribunales

Chile and the chilians / Aldana, Abelardo – London, England. 1910 – 1r – us UF Libraries [972]

Chile. Consejo de Defensa Fiscal *see* Memoria...
Chile. Direccion General de Correos y Telegrafos *see* Boletin oficial
Chile. direccion general de estadistica. anuario – v1-17. 1858-75 – 1 – $240.00 – (in spanish) – mf#0150 – us Brook [318]
Chile. Direccion Nacional de Estadística y Censos *see* Sinopsis estadística 1882-1969/70
Chile. Empresa de los Ferrocarriles del Estado *see* Memoria...
Chile. Inspeccion Jeneral de Tierras i Colonizacion *see* Memoria
Chile Laws, Statutes, etc *see* Boletin de leyes y decretos sobre ferrocarriles dictados
Chile. Ministerio de Bienestar Social *see* Memoria...
Chile. Ministerio de Culto *see* Memoria...
Chile. Ministerio de Hacienda *see*
– Boletin
– Memoria...
Chile. Ministerio de Justica *see* Memoria
Chile. Ministerio de Justicia *see*
– Memoria...
– Memoria i anuario...
Chile. Ministerio de Relaciones Exteriores *see*
– Memoria...
– Memoria
– Memoria...
Chile. Ministerio de Relaciones Exteriores, Culto y Colonizacion *see*
– Memoria...
Chile. Ministerio de Relaciones Exteriores i Colonizacion *see* Memoria...
Chile. Ministerio de Relaciones Exteriores y Comercio *see* Memoria...
Chile. Ministerio de Relaciones Exteriores y Culto *see* Memoria...
Chile. Ministerio del Ferrocarriles *see* Memoria...
Chile. Ministerio del Interior *see* Memoria...
Chile missions: our baptist work in chile – By Mary P. Moore and "Reminiscenses of Baptist Missions in Chile" by W. E. Davidson. (Unpubl. mss.). 168p – 1 – 5.88 – us Southern Baptist [242]
Chile. Oficina Central de Estadistica *see* Estadistica comercial de la republica de chile
Chile. Servicio Nacional de Estadistica y Censos *see* Anuario estadistico 1848/1858-1937
Chilenos en la antartica / Vila Labra, Oscar – Santiago, Chile. 1947 – 1r – us UF Libraries [972]
Chilgoopie the glad : a story of korea and her children / Perry, Jean – London: S W Partridge, [1906] [mf ed 1986] – 1mf – 9 – 0-8370-6594-1 – mf#1986-0594 – us ATLA [240]
Chili, New York. Chili Baptist Church *see* Records
Der chiliasmus : seiner neuesten bekaempfung gegenueber / Volck, Wilhelm – Dorpat: W Glaeser, 1869 – 1mf – 9 – 0-7905-2199-7 – (incl bibl ref) – mf#1987-2199 – us ATLA [240]
Chilidugu sive res chilenses : vel descriptio status tum naturalis, tum civilis, cum moralis regni populique chilensis, inserta suis locis perfectae ad chilensem linguam manuductioni / Havestadt, Bernhard – Monasterii Westphaliae: Aschendorf 1777 – (= ser Whsb) – 12mf – 9 – €100.00 – mf#Hu 386 – gw Fischer [440]
Ch'i-lien-shan pei lu tiao ch'a pao kao – [SI]: Meng Tsang wei yuean hui, Min kuo 31 [1942] – (= ser P-k&k period) – us CRL [951]
Chi-lin jih pao – Ch'ang-ch'un, China. 1958-1960 – 6r – 1 – us L of C Photodup [079]
Chi-lin jih-pao – Kirin, Kirin. Oct 10, 1945-. Reel 1: Jan, Apr-Jun 1962; Reel 2: Jul-Dec 1962. 2 reels – 1 – us Chinese Res [079]
Chillicothe. Ohio. First New Jerusalem Society *see* First new jerusalem society [chillicothe, ohio] records, 1838-1879
Chilliwack progress – Chilliwack, British Columbia, CN. apr 1891- – 6r/y – 1 – Can$93.00r – cn Commonwealth Imaging [071]
Chillombo, A *see*
– Iminshoni ya umca
Chilmark 1673-1849 – Oxford, MA (mf ed 1995) – (= ser Massachusetts vital record transcripts to 1850) – 4mf – 9 – 0-87623-231-4 – (mf 1t-2t: vital records 1673-1881. mf 2t-3t: marriages & intentions 1699-1849. mf 3t: family records 1752-1859. mf 4t: births 1810-49; marriages, deaths 1844-49; births & deaths 1718-1806) – us Archive [978]
Chilmark 1674-1900 – Oxford, MA (mf ed 1994) – (= ser Massachusetts vital records) – 43mf – 9 – 0-87623-193-8 – (mf 1-6: town & vitals 1688-1871. mf 7-13: town records 1801-67. mf 8-13: marriges 1824-38. mf 13: out-of-town marriages. mf 14: vitals. mf 14-16: marriages 1838-90; intentions 1844-1905. mf 17: marriages 1674-1837. mf 17-18: births 1690-1807. mf 18: intentions 1800-38; deaths 1691-1829+. mf 19-20: west family 1606-1944. mf 21: baptisms 1852-1908; marriages 1819-1905. mf 21-23: churchgoers 1810-1921. mf 24-27: church records 1837-88. mf 28-30: treasurer 1820-55. mf 31: valuation list 1862. mf 32-33: valuation list 1891. mf 34: voters 1892-1915. mf 35-36: birth index 1844-1975. mf 37-38: marriage index 1844-1975. mf 39-40: death index 1845-1975. mf 41,43: births 1810-1900. mf 41-43: marriages 1844-1905. mf 42-43: deaths 1845-1910) – us Archive [978]
Chilocco Indian Agricultural School et al *see* Indian school journal
Chiloquin review – Chiloquin, Klamath County, OR: A W Priaulx. v1 n1-v2 n6 . sep 4 1925-oct 8 1926 – 1 – us Oregon Hist [071]
Chiloquin review – Chiloquin OR: A W Priaulx, 1925- [wkly] – 1 – us Oregon Lib [071]
Chilton, C *see* The subantarctic islands of new zealand
Chilton spirit – Chilton WI. 1993 aug 3/dec 28-2000 jul/dec – 16r – 1 – (cont by: kiel tri county record; new holstein reporter; tri-county news) – mf#2813483 – us WHS [071]
Chilton times – Chilton WI. 1857 sep/1860-1932 apr 7/1933 feb 23 – 40r – 1 – (with gaps; cont by: independent journal [chilton wi]; chilton times-journal) – mf#986440 – us WHS [071]
Chilton times-journal – Chilton WI. 1933 mar 2/nov 2-2002 sep/dec – 90r – 1 – (with gaps; cont: chilton times; independent times [chilton wi]) – mf#1010192 – us WHS [071]
Chilton's automotive industries – Radnor. 1976-1994 (1,5,9) – (cont by: automotive industries) – ISSN: 0273-656X – mf#45,01 – us UMI ProQuest [629]
Chilton's automotive marketing – Radnor. 1979-1998 (1,5,9) – (cont by: automotive marketing) – ISSN: 0193-3264 – mf#11779,03 – us UMI ProQuest [380]
Chilton's ccj – Radnor. 1911-1982 (1) 1972-1982 (5) 1973-1982 (9) – (cont by: chilton's commercial carrier journal) – ISSN: 0193-628X – mf#942 – us UMI ProQuest [380]
Chilton's commercial carrier journal – Radnor. 1982-1984 (1) 1982-1984 (5) 1982-1984 (9) – (cont: chilton's ccj) – ISSN: 0734-1423 – mf#942,01 – us UMI ProQuest [380]
Chilton's commercial carrier journal for professional fleet managers (ccj) – Radnor. 1984-1997 (1) 1984-1997 (5) 1984-1997 (9) – (cont by: commercial carrier journal for professional fleet managers : ccj) – ISSN: 1062-0060 – mf#942,02 – us UMI ProQuest [380]
Chilton's distribution – Radnor. 1986-1992 (1) 1986-1992 (5) 1986-1992 (9) – (cont: chilton's distribution for traffic and transportation decision makers. cont by: distribution) – ISSN: 1057-9710 – mf#944,03 – us UMI ProQuest [380]
Chilton's distribution – Radnor. 1979-1980 (1,5,9) – (cont: chilton's distribution worldwide. cont by: chilton's distribution for traffic and transportation decision makers) – ISSN: 0195-7244 – mf#944,01 – us UMI ProQuest [380]
Chilton's distribution for traffic and transportation decision makers – Radnor. 1980-1985 (1) 1980-1985 (5) 1980-1985 (9) – (cont: chilton's distribution. cont by: chilton's distribution) – ISSN: 0273-6721 – mf#944,02 – us UMI ProQuest [380]
Chilton's distribution worldwide – Radnor. 1909-1979 (1) 1970-1979 (5) 1976-1979 (9) – (cont by: chilton's distribution) – ISSN: 0193-3248 – mf#944 – us UMI ProQuest [380]
Chilton's electronic component news – Radnor. 1957-1997 [1]; 1979-1997 [5,9] – ISSN: 0193-614X – mf#1418 – us UMI ProQuest [621]
Chilton's food engineering – Radnor. 1928-1998 (1) 1971-1998 (5) 1976-1998 (9) – (cont by: food engineering) – ISSN: 0193-323X – mf#23 – us UMI ProQuest [660]
Chilton's food engineering international – Radnor. 1978-1996 (1,5,9) – ISSN: 0148-4478 – mf#11781 – us UMI ProQuest [660]
Chilton's hardware age – Radnor. 1894-1981 (1) 1971-1981 (5) 1977-1981 (9) – (cont by: hardware age) – ISSN: 0162-5896 – mf#946 – us UMI ProQuest [680]
Chilton's hardware age – Radnor. 1984-1994 (1) 1984-1994 (5) 1984-1994 (9) – (cont: hardware age. cont by: hardware age home improvement marketplace) – ISSN: 8755-254X – mf#946,02 – us UMI ProQuest [680]
Chilton's i and cs – Radnor. 1983-1992 (1) 1983-1992 (5) 1983-1992 (9) – (cont: chilton's instruments and control systems. cont by: instrumentation and control systems : i&cs) – ISSN: 0746-2395 – mf#403,01 – us UMI ProQuest [621]
Chilton's iami – Radnor. 1962-1985 (1) 1978-1985 (5) 1978-1985 (9) – ISSN: 0195-2323 – mf#1627 – us UMI ProQuest [660]
Chilton's ian – Radnor. 1964-1993 (1) 1979-1993 (5) 1979-1993 (9) – (cont by: instrumentation and automation news: ian) – ISSN: 0193-6174 – mf#1662 – us UMI ProQuest [621]
Chilton's industrial safety and hygiene news – Radnor. 1982-1991 (1,5,9) – (cont: industrial safety and hygiene news) – ISSN: 8755-2566 – mf#12611,03 – us UMI ProQuest [360]
Chilton's instruments and control systems – Radnor. 1928-1983 [1]; 1965-1983 [5]; 1976-1983 [9] – (cont by: chilton's i and cs) – ISSN: 0164-0089 – mf#403 – us UMI ProQuest [621]
Chilton's iron age – Radnor. 1977-1983 (1) 1977-1983 (5) 1977-1983 (9) – (cont: iron age) – ISSN: 0164-5137 – mf#919,01 – us UMI ProQuest [660]
Chilton's iron age manufacturing management – New York. 1984-1986 (1,5,9) – (cont by: iron age manufacturing management) – ISSN: 0747-6310 – mf#13869 – us UMI ProQuest [660]
Chilton's iron age metals producer – Radnor. 1984-1986 (1,5,9) – (cont by: iron age metals producer) – ISSN: 0747-6329 – mf#13991 – us UMI ProQuest [660]
Chilton's jewelers' circular/keystone – Radnor. 1892-1989 (1) 1970-1989 (5) 1976-1989 (9) – (cont by: jewelers' circular-keystone: jck) – ISSN: 0194-2905 – mf#948 – us UMI ProQuest [730]
Chilton's motor/age – Radnor. 1899-1997 (1) 1971-1997 (5) 1976-1997 (9) – (cont by: motor age) – ISSN: 0193-7022 – mf#941 – us UMI ProQuest [629]
Chilton's oil and gas energy – Radnor. 1975-1976 (1) 1975-1975 (5) 1975-1975 (9) – mf#10376 – us UMI ProQuest [550]
Chilton's product design and development – Radnor. 1978-1996 (1,5,9) – (cont by: product design and development) – ISSN: 0193-6182 – mf#11783 – us UMI ProQuest [620]
Chilton's review of optometry – Radnor. 1977-1997 (1) 1977-1997 (5) 1977-1997 (9) – (cont: optical journal and review of optometry. cont by: review of optometry) – ISSN: 0147-7633 – mf#952,01 – us UMI ProQuest [617]
Chilton's truck and off-highway industries – Radnor. 1979-1983 (1,5,9) – ISSN: 0194-1410 – mf#11784 – us UMI ProQuest [380]
Chi-luan wen ts'un / Chang, Chi-luan – Ch'ung-ch'ing: Ta kung pao kuan, 1944 (1945 printing) – (= ser P-k&k period) – us CRL [840]
Chilvers, Hedley Arthur *see*
– Out of the crucible
– Seven lost trails of africa
Chimani, Leopold *see* Meine ferien-reise, von wien durch das land unter und ob der enns
Chimba, Barnabas *see* A history of the baushi
Chim-chim / Smith, Pamela Colman – London, England. 1905 – 1r – us UF Libraries [972]
Chimere – Montpellier. n1-6, 8-9, 19. aout 1891-avr 1893 – 1 – fr ACRPP [073]
Chimes *see* Kuranty
La chimica industriale : (l'industria chimica) – Turin, Italy. oct 1899-dec 1904. -f – 3r – 1 – uk British Libr Newspaper [660]
La chimie agricole mise a la portee de tout le monde : ouvrage tres simplifie a l'usage des agriculteurs canadiens et particulierement des ecoles elementaires / Aubin, Napoleon – Quebec: de l'impr de J B Frechette, pere, 1847 [mf ed 1976] – 1r – 5 – mf#SEM16P264 – cn Bibl Nat [630]
La chimie appliquee aux arts et metiers : a l'usage de toutes les familles – [Quebec?: s.n.] 1859 [mf ed 1984] – 2mf – 9 – 0-665-45054-0 – mf#45054 – cn Canadiana [640]
Chimney rock transcript – Baynard, NE: E M Totten, dec 1888 (wkly) [mf ed dec 13 [1888]-jul 3 1891 (gaps) filmed 1999] – 1r – 1 – (cont by: baynard transcript) – us NE Hist [071]
Chimolula, A *see* Akatanshi takalisha
Chin ch'a chi pien ch'ue yin hsiang chi / Chou, Li-po – Han-k'ou: Tu shu sheng huo ch'u pan she, Min kuo 27 [1938] – (= ser P-k&k period) – us CRL [951]
Chin, Ch'ang-yu *see* Jih-pen cheng fu
Chin ch'eng yin hang *see* Shih pien hou chih shang-hai kung yeh
Chin, Chung-hua *see*
– Fu nu t'an sou
– Fu nu wen t'i ti ko fang mien
– T'ai-p'ing yang hsun li
Chin erh shih nien lai chih chung jih mao i chi ch'i chu yao shang p'in / Ts'ai, Ch'ien – Shang-hai: Shang wu yin shu kaun, min kuo 25 [1936] – (= ser P-k&k period) – us CRL [380]
Chin fen shih chia / Chang, Hen-shui – Shang-hai: Shih chieh shu chue, Min kuo 21 [1932] – (= ser P-k&k period) – us CRL [830]
Chin fen shih chia hsue chi / Chang, Hen-shui – Shang-hai: Shih chieh shu chue, Min kuo 22 [1933] – (= ser P-k&k period) – us CRL [830]
Chin hsi chi / Kuo, Mo-jo – Ch'ung-ch'ing: Tung fang shu she, 1943 – (= ser P-k&k period) – us CRL [480]
Chin hsing (forward march) in china / Hart, Edith & Sturgis, Lucy C – New York: Domestic and Foreign missionary Society, [1913] [mf ed 1995] – (= ser Yale coll) – 98p (ill) – 1 – 0-524-09469-1 – mf#1995-0469 – us ATLA [951]
Chin hsiu ho shan – Shang-hai: Sheng huo shu tien, Min kuo25 [1936] – (= ser P-k&k period) – us CRL [915]
Ch'in huai shih chia / Fan, Yen-ch'iao – Shang-hai: Ta chung ying hsuen she, 1940 – (= ser P-k&k period) – us CRL [480]
Chin, Hui *see* Hsin chung-kuo chih hsien cheng chien she
Chin, I *see*
– Ch'ien hsi
– Chin i tuan p'ien hsiao shuo i chi
– Huang sha
– Hung chu
– Hung liu
– Huo hua
– Jen shih pai t'u
– Mao yu tuan chien
– Niao shu hsiao chi
– Sheng hsing
– Ts'an yang
– Wo men ti hsieh
– Wu chi ch'i t'a
– Yao yuan ti ch'eng
– Yuan t'ien ti ping hsueh
Chin i tuan p'ien hsiao shuo i chi / Chin, I – Shang-hai: K'ai ming shu tien, 1940 – (= ser P-k&k period) – us CRL [480]
Chin jen pai hua wen hsuean – Shang-hai: Shang wu yin shu kuan, Min kuo 21 [1932] – (= ser P-k&k period) – us CRL [480]
Chin jih chih hsien cheng / Ch'en, Ping-po – Shang-hai: T'ung wen t'u shu yin shua kung ssu, Min kuo 22 [1933] – (= ser P-k&k period) – us CRL [350]
Chin jih chih nei meng / Li, Sheng-lun – Ch'ung-ch'ing: Tu li ch'u pan she, 1941 – (= ser P-k&k period) – us CRL [951]
Chin jih chih shang-hai / Hsia, Yen – Han-k'ou: Hsien shih ch'u pan she, 1938 – (= ser P-k&k period) – us CRL [951]
Chin jih chung-kuo chih lao kung wen t'i / Chu, Hsueeh-fan – [China: sn], 1936 – (= ser P-k&k period) – us CRL [951]
Chin jih hsin wen – Beijing, China. 1962-1969 (1) – mf#67675 – us UMI ProQuest [079]
Chin jih ti hsin hsi nan / Pai, Shui – [China]: Yen hsing ch'u pan she, 1939 – (= ser P-k&k period) – us CRL [951]
Chin, John *see* Good shepherd
Chin jung chieh fu wu chi pen chih shih / Li, Ch'uean-shih – Shang-hai: Shih chieh shu chue, Min kuo 23 [1934] – (= ser P-k&k period) – us CRL [332]
Chin jung yeh – [Shang-hai: sn, 1934] – (= ser P-k&k period) – us CRL [332]
Chin, Kuo-pao *see*
– T'ung chi hsin lun
– T'ung chi hsueh ta kang
Ch'in lueeh wen t'i chih kuo chi fa ti yen chiu / Chu, Chien-min – Ch'ang-sha: Shang wu yin shu kuan, Min kuo 29 [1940] – (= ser P-k&k period) – us CRL [951]
Chin pai nien ku ch'eng ku mu fa chueeh shih, cheng chen-to chu / Cheng, Chen-to – Shang-hai, Shang wu yin shu kuan, Min kuo 24 [1935] – (= ser P-k&k period) – us CRL [951]
Chin pai nien lai chung-kuo hsin chiao yue chih fa chan / Tu, Tso-chou – [China: sn] – (= ser P-k&k period) – us CRL [370]
Chin pai nien lai chung-kuo pao chih chih fa chan chi ch'i ch'ue shih / Chang, Tan-feng – Shang-hai: K'ai ming shu tien, Min kuo 31 [1942] – (= ser P-k&k period) – us CRL [070]
Chin pei ta chan yue ti pa lu chuen – Shang-hai: Shih tai shih liao pao ts'un she, 1937 – (= ser P-k&k period) – us CRL [951]
Chin pei yu chi chan cheng chi shih / Lin, Piao – [Ch'ang-sha]: Chan shih ch'u pan she, Min kuo 27 [1938] – (= ser P-k&k period) – us CRL [951]
Chin, Po-ming *see* Yin hang shih chien
Chin pu ch'u pan she *see* Hsiang ch'ih chieh tuan chung ti hsing shih yu jen wu
Chin p'u t'ieh lu liang nien lai chih kung tso kai yao – [China: sn, Min kuo 23 [1934]] – (= ser P-k&k period) – us CRL [625]
Chin shu wen ta [ccm265] = Abc catechism / Price, Philip Francis – Hankow, 1938 [mf ed 198?] – (= ser Ccm 265) – 1 – mf#1984-b500 – us ATLA [480]
Chin shih chung-kuo kuo wai mao i – Nan-ching: Li fa yuean mi shu ch'u t'ung chi k'o, Min kuo 22 [1933] – (= ser P-k&k period) – us CRL [380]
Chin shih tung-pei kuo chi kuan hsi jih chi / Yang, Chia-lo – [China]: Tung-pei wen t'i yen chiu she, Min kuo 30 [1941] – (= ser P-k&k period) – us CRL [951]
Chin, Shih-hsuan *see* T'ieh lu yun shu yeh wu
Ch'in, Shou-ou *see* Erh chiu
Chin shu hsueh pao = Journal of metallurgy – 1959-1960 [1] – mf#2600 – us UMI ProQuest [660]
Chin ssu ch'ueeh / Chou, I-pai – Shang-hai: Shih chieh shu chue, Min o 33 [1944] – (= ser P-k&k period) – us CRL [820]
Chin ssu niao / Mu, Ni – Shang-hai: Shuo feng shu wu, 1942 – (= ser P-k&k period) – us CRL [480]

Chin sui chi hsing / Shih, Chang-ju – Ch'ung-ch'ing: Tu li ch'u pan she, Min kuo 31 [1942] – (= ser P-k&k period) – us CRL [951]
Chin tai ch'ih tu hsuean chu / T'an, Cheng-pi – Shang-hai: Kuang ming shu chue, Min kuo 24 [1935] – (= ser P-k&k period) – us CRL [860]
Chin tai chung-kuo chiao yue ssu hsiang shih – Shang-hai: Chung-hua shu chue, Min kuo 21 [1932] – (= ser P-k&k period) – us CRL [370]
Chin tai chung-kuo nue tzu chiao yue / Liang, Ou-ti – Nan-ching: Cheng chung shu chue, Min kuo 25 [1936] – (= ser P-k&k period) – us CRL [305]
Chin tai chung-kuo shih yeh t'ung chih / Yang, Ta-chin – Nan-ching: Shou ch'ang, Min kuo 22 [1933] – (= ser P-k&k period) – us CRL [951]
Chin tai hsi chue hsuean / Ou-yang, Yue-ch'ien – Shang-hai: I liu shu tien, Min kuo 31 [1942] – (= ser P-k&k period) – us CRL [820]
Chin tai jen wu [ccm283] = Characters of modern times: for middle schools and general reading – Hong Kong, 1952 [mf ed 198?] – (= ser Ccm 283) – 1 – mf#1984-b500 – us ATLA [920]
Chin tai k'o hsueh chia te tsung chiao kuan [ccm180] = Science and religion / Thomson, John Arthur; ed by Hsieh, Sung-kao – Shanghai, 1927 [mf ed 198?] – (= ser Ccm 180) – 1 – (chinese trans of the english) – mf#1984-b500 – us ATLA [210]
Chin, Ting-i *see* Hsiang ts'un hsiao hsueh shih chi wen t'i
Chin yen kung pao / China. National Committee for Opium Suppression – Gazette. Nanking, Jan 1930, n.12. Also: CHIN YEN WEI YUAN HUI KUNG PAO. Gazette, Nanking. 1931. n.1-12. 2 reels – 1 – us Chinese Res [360]
Chin, Yueh-Lin *see* Lun tao (ccm104)
Chin, Yueh-lin *see* Lo chi
Chin, Yung *see* Pi hsueh chien
China – Boston: The American Board, 1867 [mf ed 1995] – (= ser Yale coll) – 16p – 1 – 0-524-09357-1 – mf#1995-0357 – us ATLA [951]
China : containing illustrations of the manners, customs, character, and costumes of the people of that empire / ed by Shoberl, Frederick – London [1823] [mf ed Hildesheim 1995-98] – 2v on 4mf – 9 – €120.00 – 3-487-27563-5 – gw Olms [951]
China : ergebnisse eigener reisen und darauf gegruendeter studien / Richthofen, F von – Berlin, 1877-1911. 5v – 44mf – 9 – mf#H-6147 – ne IDC [915]
China : a geographical, statistical and political sketch / Hippisley, Alfred Edward – [Shanghai], 1876 – (= ser 19th c books on china) – 1mf – 9 – mf#7.1.12 – uk Chadwyck [951]
China / Gorst, Harold Edward – London: Sands, 1899 [mf ed 1995] – (= ser Yale coll; Imperial interest library) – 1 – 0-524-09230-3 – mf#1995-0230 – us ATLA [915]
China : her claims and call / John, Griffith – London: Hodder & Stoughton, 1882 [mf ed 1995] – (= ser Yale coll) – 62p – 1 – 0-524-10057-8 – mf#1995-1057 – us ATLA [951]
China : historisch, romantisch, malerisch – Carlsruhe [1843] [mf ed Hildesheim 1995-98] – 1v on 4mf [ill] – 9 – €120.00 – 3-487-27581-3 – gw Olms [951]
China : in a series of views, displaying the scenery, architecture, and social habits, of that ancient empire... / Allom, T – London, Paris. 4v. [1843] – 11mf – 9 – mf#HT-708 – ne IDC [915]
China : in a series of views, displaying the scenery, architecture, and social habits, of that ancient empire / Wright, George Newenham – [London]: Fisher, Son & Co; Paris: Rue St Honore. 4v. [1843] – (= ser 19th c books on china) – 9mf – 9 – mf#7.1.4 – uk Chadwyck [951]
China : internal affairs and foreign affairs, 1930-jan 1963 / U.S. State Dept – (= ser Confidential u s state department central files) – 1 – $62,215.00 coll – (china: internal affairs: 1930-39 105r isbn 0-89093-638-2 $14,980. 1940-44 51r isbn 0-89093-639-0 $5820. 1945-49 75r isbn 0-89093-640-4 $8280. us-china relations, 1940-49 7r isbn 0-89093-713-3 $1010. china: internal affairs, 1950-54 47r isbn 0-89093-771-0 $6695. foreign affairs, 1950-54 6r isbn 0-89093-772-9 $885. internal affairs, 1955-59: pt1: political, governmental, & national defense affairs 31r isbn 0-89093-901-2 $4695; pt2: social, economic, & industrial affairs 17r isbn 0-89093-902-0 $2580. foreign affairs, 1955-59 10r isbn 0-89093-973-X $1575. internal affairs, 1960-jan 1963 27r isbn 1-55655-706-X $5225. foreign affairs, 1960-jan 1963 5r isbn 1-55655-705-1 $970. subject-numeric files, feb 1963-66: pt1: political, governmental, & national defense affairs 41r isbn 1-55655-838-4 $7940. subject-numeric files, 1967-69: pt1: political, governmental, & national defense affairs 30r isbn 1-55655-974-7 $5810. with p/g) – us UPA [951]
China : an interpretation / Bashford, James Whitford – New York, Cincinnati: Abingdon Press, [1916] [mf ed 1995] – (= ser Yale coll) – 630p (ill) – 1 – 0-524-09290-7 – mf#1995-0290 – us ATLA [951]
China : its costume, arts, manufactures, etc; edited principally from the originals in the cabinet of the late M Bertin; with observations explanatory, historical, and literary / Breton de LaMartiniere, Jean – London 1812 [mf ed Hildesheim 1995-98] – 4v on 7mf – 9 – €140.00 – 3-487-27580-5 – (trans fr french) – gw Olms [951]
China : its past history and future hopes / Rhind, William Graeme – London, 1850 – (= ser 19th c books on china) – 3mf – 9 – mf#7.1.14 – uk Chadwyck [951]
China : its state and prospects: with especial reference to the spread of the gospel: containing allusions to the antiquity, extent, population, civilization, literature and religion of the chinese / Medhurst, Walter Henry – London: John Snow, 1842 [mf ed 1995] – (= ser Yale coll) – xv/592p (ill) – 1 – 0-524-10147-7 – mf#1995-1147 – us ATLA [951]
China : its state and prospects, with special reference to the spread of the gospel / Medhurst, W H – London: John Snow, 1838 – 7mf – 9 – mf#HT-541 – ne IDC [915]
China / Norris, Francis Lushington – London: A R Mowbray; New York: Thomas Whittaker, [1908] [mf ed 1995] – (= ser Yale coll; Handbooks of english church expansion 3) – xii/219p (ill) – 1 – 0-524-09999-5 – mf#1995-0999 – us ATLA [951]
China : an outline of its government, laws, and policy: and of the british and foreign embassies to, and intercourse with, that empire / Auber, Peter – London 1834 [mf ed Hildesheim 1995-98] – 1v on 5mf – 9 – €100.00 – 3-487-27498-1 – gw Olms [951]
China : an outline of its government, laws and policy: and of the british and foreign embassies to, and intercourse with, that empire / Auber, Peter – London: Parbury, Allen & Co, 1834 – (= ser 19th c books on china) – 5mf – 9 – mf#7.1.3 – uk Chadwyck [951]
China : political, commercial, and social / Martin, Robert Montgomery – London: James Madden. 2v. 1847 – (= ser 19th c books on china) – 11mf – 9 – mf#7.1.7 – uk Chadwyck [951]
China – 3ser – (= ser Confidential british foreign office political correspondence files) – 1 – (ser1: 1906-11 $78,260.00 for 4pt coll. pt1: 1906-08 105r $20,895 isbn 1-55655-519-9; pt2: 1909-11 106r $21,105 isbn 1-55655-559-8; pt3: 1912-14 99r $19,695 isbn 1-55655-647-0; pt4: 1915-19 127r isbn 1-55655-681-0 $25,265; printed ind for pt1+2 $735 ea. ser2: 1920-31: pt1: 1920-23 109r ISBN 1-55655-801-5 $21,700; pt2: 1924-26 81r ISBN 1-55655-864-3 $15,720. ser3: 1932-45: pt1: 1932-33 78r ISBN 1-55655-662-4 $15,520; pt2: 1934-35 80r ISBN 1-55655-700-0 $15,905; pt3: 1936-38 103r ISBN 1-55655-772-8 $20,500; pt4: 1939-41 97r ISBN 1-55655-798-1 $18,820. with printed ind) – us UPA [327]
China *see*
- Chan shih chung yao fa ling hui pien
- Chung-kuo chan-shih ching-chi fa-kui hui-pien
- Hsien ko chi tsu chih kang yao chi ti fang tzu chih ts'an k'ao ts'ai liao
- Hsien tzu chih fa ts'ao an, hsien tzu chih fa shih hsing fa ts'ao an, shih tzu chih fa ts'ao an, shih tzu chih fa shih hsing fa ts'ao an
- Jen li tung yuan fa kuei hui pien
- Kuan hsia tsai hua wai kuo jen shih shih t'iao li an
- Kuo min ta hui tai piao hsuan chu chih nan
- Liu fa ch'uan shu
- Nei cheng fa kuei hui pien
- Tairiku nenkau
- Wai chiao kung pao

China (1927-) Ts'ai cheng pu T'ung chi ch'u *see* Shih nien lai chih ts'ai wu t'ung chi
China, 1911-1941 – (= ser U s military intelligence reports) – 15r – 1 – $2615.00 – 0-89093-425-8 – (with p/g) – us UPA [355]
China, 1946-1976 / U.S. Central Intelligence Agency – (= ser Cia research reports) – 6r – 1 – $920.00 – 0-89093-424-X – (with p/g) – us UPA [951]
China and christianity / Michie, Alexander – Boston: Knight & Millet, 1900 [mf ed 1995] – (= ser Yale coll) – xv/232p – 1 – 0-524-09362-8 – mf#1995-0362 – us ATLA [951]
China and formosa : the story of the mission of the presbyterian church of england / Johnston, James – New York: Fleming H Revell, [1897?] – 1mf – 9 – 0-524-04074-5 – mf#1991-2019 – us ATLA [242]
China and india : 1941-1949 / U.S. Office of Strategic Services & U.S. State Dept – (= ser Oss/state department intelligence and research reports 3) – 6r – 1 – $920.00 – 0-89093-119-4 – (with p/g) – us UPA [327]
China and india : 1950-1961 supplement / U.S. Office of Strategic Services & U.S. State Dept – (= ser Oss/state department intelligence and research reports 9) – 5r – 1 – $770.00 – 0-89093-249-2 – (with p/g) – us UPA [327]
China and its people / Cornaby, William Arthur – London: Christian Literature Society for India, 1910 [mf ed 1995] – (= ser Yale coll) – 79p (ill) – 1 – 0-524-09276-1 – mf#1995-0276 – us ATLA [951]
China and japan : a record of observations made during a residence of several years in china, and a tour of official visitation to the missions of both countries in 1877-78 / Wiley, Isaac William – Cincinnati: Hitchcock & Walden; New York: Phillips & Hunt, 1879 [mf ed 1995] – (= ser Yale coll) – 548p (ill) – 1 – 0-524-09239-7 – mf#1995-0239 – us ATLA [950]
China and methodism / Bashford, James Whitford – Cincinnati: Jennings and Graham; New York: Eaton and Mains, c1906 – 1mf – 9 – 0-8370-7201-8 – (includes bibliography) – mf#1986-1201 – us ATLA [240]
China and religion / Parker, Edward Harper – New York: EP Dutton, 1905 – 1mf – 9 – 0-524-00951-1 – mf#1990-2174 – us ATLA [290]
China and the boxers : a short history on the boxer outbreak... / Beals, Zephaniah Charles – Toronto: W Briggs, 1901 – 2mf – 9 – 0-665-71880-2 – mf#71880 – cn Canadiana [951]
China and the chinese : a general description of the country and its inhabitants, its civilization and form of government, its religious and social institutions, its intercourse with other nations, and its present condition and prospects / Nevius, John Livingston – Rev. ed. Philadelphia: Presbyterian Board of Publication, c1882 – 2mf – 9 – 0-7905-6309-6 – mf#1988-2309 – us ATLA [915]
China and the chinese / Giles, Herbert Allen – New York: Columbia UP, 1912 [mf ed 1995] – (= ser Yale coll) – ix/229p – 1 – 0-524-09548-5 – (lectures, mar 1902 at columbia university, new york, to inaugurate the foundation by general horace w. carpentier of the dean lung chair of chinese) – mf#1995-0548 – us ATLA [480]
China and the chinese / Nevius, John L – 1869 – 1 – $50.00 – us Presbyterian [951]
China and the chinese : their religion, character, customs, and manufactures / Sirr, H C – London: Wm S Orr & Co, 1849. 2v – 10mf – 9 – mf#HT-555 – ne IDC [915]
China and the chinese : their religion, character, customs, and manufactures: the evils arising from the opium trade: with a glance at our religious, moral, political, and commercial intercourse with that country / Sirr, Henry Charles – London: Wm S Orr. 2v. 1849 – (= ser 19th c books on china) – 10mf – 9 – mf#7.1.8 – uk Chadwyck [951]
China and the gospel : an illustrated report of the china inland mission: being the story of time redeemed amid the evil days of 1936 / China Inland Mission – London: China Inland Mission 1906- (annual) [mf ed 2003] – 23v on 2r – 1 – (each iss has also distinctive title fr 1923-37) – mf#2003-s062 – us ATLA [240]
China and the gospel / Muirhead, William – London: James Nisbet, 1870 [mf ed 1995] – (= ser Yale coll) – vii/305p – 1 – 0-524-09682-1 – mf#1995-0682 – us ATLA [951]
China and the roman orient : researches into their ancient and mediaeval relations as represented in old chinese records / Hirth, Friedrich – Leipsic [Leipzig]: G Hirth, 1885 – 1mf – 9 – 0-524-03305-6 – mf#1990-3190 – us ATLA [915]
China and the united states : from hostility to engagement, 1960-1998 – [mf ed Chadwyck-Healey] – (= ser National security archive, washington dc: the making of us policy) – 374mf – 9 – (with p/g & ind) – uk Chadwyck [327]
China and the west : the maritime customs service archive: from the second historical archives, nanjing, china – [mf ed 2003] – ca 350r in 7 units – 1 – (incl electronic catalogue) – us Primary [951]
China as a mission field / Moule, Arthur Evans – 2nd rev ed. London: Church missionary House, 1891 [mf ed 1995] – (= ser Yale coll) – 80p – 1 – 0-524-09823-9 – mf#1995-0823 – us ATLA [951]
China as a mission field / Moule, Arthur Evans – London: Church Missionary House, [1881] – (= ser 19th c books on china) – 1mf – 9 – mf#7.1.28 – uk Chadwyck [240]
China Assessor to the Commission of Enquiry into Sino-Japanese Dispute *see* Ts'an yu kuo chi lien ho hui tiao ch'a wei yuan hui chung-kuo tai piao ch'u shuo t'ieh
The china bookman *see* Chi-tu chiao ch'u pan chieh [ccs22]
China, burmah, ceylon, etc / Missionary records – London: the Religious Tract Society, 1799 – 4mf – 9 – mf#HT-954 – ne IDC [915]
China business review – Washington. 1983+ (1) 1983+ (5) 1974+ (9) – ISSN: 0163-7169 – mf#13392 – us UMI ProQuest [338]
China centenary missionary conference records : report of the great conference held at shanghai, april 5th [read 25th] to may 8th, 1907 – New York: American Tract Society, [1907?] Chicago: Department of Photodup, U of Chicago Lib, 1969 (1r); Evanston: American Theol Lib Assoc, 1984 (1r) – 1 – 0-8370-0156-0 – mf#1984-B105 – us ATLA [240]
China. Central Bank *see* Chung yang yin hang yueh pao
China Chiao t'ung pu *see*
- Chiao t'ung pu kung tso pao kao
- erh shih wu nien tu

China chiao t'ung pu : shih chiu nien tu – [China: Chiao t'ung pu, 1930] – (= ser P-k&k period) – us CRL [380]
China Chiao t'ung pu K'ao ch'a t'uan *see* K'ao ch'a ou mei chiao t'ung pao kao
China Ch'iao wu wei yuean hui *see* Ch'iao le ts'un
China Chiao yu pu Chung teng chiao yu ssu *see* Ko sheng shih fan chiao yu she shih chih yen chin
China Ching chi pu *see* Ssu nien lai ti ching chi chien she
China christian advocate – Shanghai: Methodist Episcopal Church, South, 1914-39; China Central Conference of the Methodist Church, 1940-41. v1-29 n10/11. feb 1914-oct/nov 1941 (frequency varies) [all publ?] – (= ser Missionary periodicals from the china mainland) – 4r – 1 – $605.00 – us UPA [242]
China christian educational association newsletter *see* Hui hsun (ccs)
China christian yearbook / the china mission handbook / a century of protestant missions in china – (= ser Missionary periodicals from the china mainland) – 3 titles in 6r – 1 – $920.00 set – (the china christian yearbook (title varies), shanghai: christian literature soc for china v1-21 1910-1938/39 (frequency varies) [all publ]. the china mission handbook, shanghai, 1896. a century of protestant missions in china, shanghai, 1907) – us UPA [242]
China church yearbook *see* Chung-hua chi-tu-chiao-hui nien chien [ccs]
The china collector : a guide to the porcelain of the english factories / Lewer, William – Toronto: Bell & Cockburn, 1914 [mf ed 1996] – 5mf – 9 – 0-665-81002-4 – mf#81002 – cn Canadiana [730]
China. Commission of Mongolian and Tibetan Affairs *see* Meng tsang wei yuan hui kung pao
China continuation committee *see* The art of using the china missionary survey
China daily – 1982- – 2r per y – 1 – (in english) – us UMI ProQuest [079]
China daily news – New York. July 8 1940-Dec 1947 – 1 – us NY Public [071]
China daily news – New York, NY. 1962-1989 (1) – mf#65061 – us UMI ProQuest [071]
China economic review – Greenwich, 1998+ [1,5,9] – ISSN: 1043-951X – mf#19769 – us UMI ProQuest [330]
China from within : impressions and experiences – New York: Fleming H Revell [1917] [mf ed 1995] – (= ser Yale coll; Students' lectures on missions, princeton theological seminary, 1914-1915) – 327p (ill) – 1 – 0-524-09456-X – (int by j ross stevenson) – mf#1995-0456 – us ATLA [951]
China gleanings – Honan, China: Lutheran United Mission. v13-15 (1936-38) [mf ed 2005 – 1r [ill] – 1 – (cont: gleanings (hunan sheng, china)) – mf#2005C-S017 – us ATLA [242]
China Hai chun pu *see* Hai chun t'ung chi
China Hai kuan tsung shui wu ssu shu *see* Shih nien lai chih hai kuan
China. Han-k'ou *see* Han-k'ou shih cheng kai k'uang
China, historical and descriptive / Eden, Charles Henry – London: Marcus Ward, 1877 [mf ed 1995] – (= ser Yale coll) – 334p (ill) – 1 – 0-524-09721-6 – (with app on korea) – mf#1995-0721 – us ATLA [951]
China. Hsing cheng yuan nung ts'un fu hsing wei yuan hui *see* Ho-nan sheng nung ts'un tiao ch'a
China in convulsion / Smith, Arthur Henderson – New York: F H Revell, 1901 [mf ed 1995] – (= ser Yale coll) – 2v (ill) – 1 – 0-524-09639-2 – mf#1995-0639 – us ATLA [951]
China in historischer beleuchtung : eine denkschrift zu seinem 30 jaehrigen dienstjubilaeum als missionaer in china / Faber, Ernst – Berlin: A Haack, 1900 [mf ed 1995] – (= ser Yale coll; Sechste (doppel-)flugschrift des allgemeinen evangelisch-protestantischen missionsvereins) – 66p (ill) – 1 – 0-524-10042-X – (in german) – mf#1995-1042 – us ATLA [951]
China in legend and story / Brown, Colin Campbell – Edinburgh; London: Oliphant Anderson & Ferrier, 1907 [mf ed 1995] – (= ser Yale coll) – 253p (ill) – 1 – 0-524-09359-8 – mf#1995-0359 – us ATLA [390]

China in transformation / Colquhoun, Archibald Ross – New York, NY. 1912 – 1r – us UF Libraries [951]

China Inland Mission *see*
- China and the gospel
- He purposeth a crop
- The land of sinim
- Modern pentecost
- Part of the story of the china inland mission in...
- Recent survey of the work of the china inland mission
- The story of...
- The story of the year

China inland mission, 1865-1951 : from the school of oriental and african studies, london – 1 – (pt1: james hudson taylor papers: correspondence & journals [25r] £2400; pt2: james hudson taylor papers: subject files [14r] £1350; pt3: minutes & papers of china inland mission [22r] £2100; with d/g) – uk Matthew [951]

China inside out / Muller, George Amos – New York; Cincinnati: Abingdon Press, [1917] [mf ed 1995] – (= ser Yale coll) – 180p (ill) – 1 – 0-524-09526-4 – (ill by alice and a w best fr photos by aut) – mf#1995-0526 – us ATLA [951]

China. Inspectorate General of Customs *see*
- Returns of trade at the treaty ports
- The trade of china

China, its state and prospects : with special reference to the spread of the gospel: containing allusions to the antiquity, extent, population, civilization, literature, and religion of the chinese / Medhurst, Walter Henry – Boston: Crocker & Brewster, 1838 [mf ed 1995] – (= ser Yale coll) – xv/472p – 1 – 0-524-09507-8 – mf#1995-0507 – us ATLA [951]

China journal = Chung kuo yen chiu – Canberra. 1995-1995 (1) – (cont: australian journal of chinese affairs=ao chung) – mf#17530,01 – us UMI ProQuest [951]

China journal of science and arts – Shanghai 1923-40 – v1-33 on 446mf – 9 – mf#ch-146/2 – ne IDC [079]

China K'ao shih yuan *see* K'ao shih yuan kung tso pao kao

China Kuang-tung ts'ai cheng t'e p'ai yuan kung shu *see* Kuang-tung sheng ts'ai cheng chi shih

China. Kwangtung Provincial Government *see* Kwang-tung ching-wu chuang-k'uang

China law reporter (aba) – v1-8. 1980-1999 – 9 – $121.00 set – ISSN: 0891-6829 – mf#108531 – us Hein [340]

China. Laws, Statutes, etc *see* Nung tsun ching chi chin jung fa kuei huipien

China looking west : a missionary study text-book on China / Hughes-Hallett, F – London: Church missionAry Society, 1919 [mf ed 1995] – (= ser Yale coll) – 60p (ill) – 1 – 0-524-09504-3 – mf#1995-0504 – us ATLA [951]

China Lu chun Ti 18 chi t'uan chun *see* K'ang chan pa nien lai ti pa lu chun yu hsin ssu chun

China mail – Hong Kong. -w. 1856-58, 1863-64; 1866-73. (25mqn reels) – 1 – uk British Libr Newspaper [072]

The china mail – Hong Kong: Andrew Shortrede, 1861-65; 1874-1961 – 1 – us CRL [079]

The china martyrs of 1900 : a complete roll of the christian heroes martyred in china in 1900 / ed by Forsyth, Robert Coventry – New York: Fleming H Revell, [1904?] – 2mf – 9 – 0-8370-6493-7 – (incl ind) – mf#1986-0493 – us ATLA [951]

The china medical journal – Peiping, Shanghai, etc: China Medical Missionary Assoc. v1-35 n6. mar 1887-nov 1921 (frequency varies) – (= ser Missionary periodicals from the china mainland) – 6r – 1 – $920.00 – (title varies) – us UPA [242]

China. Ministry of Education *see*
- Chiao yu kung pao
- Chiao yu pu kung pao
- Ta hsueh yuan kung pao

China. Ministry of Foreign Affairs *see* Wai chiao pu kung pao

China. Ministry of Public Health *see* Wei sheng kung pao

The china mission : embracing a history of the various missions of all denominations among the chinese: with biographical sketches of deceased missionaries / Dean, William – New York: Sheldon, 1859 – 1mf – 9 – 0-8370-6487-2 – mf#1986-0487 – us ATLA [240]

The china mission advocate – Louisville KY. v1 n1-12. jan-dec 1839 [all publ] – (= ser Missionary periodicals from the china mainland) – 1r – 1 – $165.00 – us UPA [242]

China mission handbook *see* China christian yearbook / the china mission handbook / a century of protestant missions in china

China mission society records, 1901-1919 / Evangelical Lutheran Augustana Synod of North America. China Mission Society – 1r – 1 – (Forms pt of: Subgroup AUG 24/4 China Mission Society; with finding aid) – mf#xa0113r – us ATLA [242]

China mission year book – Shanghai. v1-21. 1910-39 – 1 – $240.00 – mf#0151 – us Brook [951]

China monthly review – Shanghai: Millard Pub Co, jun 9 1917- – 1 – mf#02214 – us L of C Photodup [951]

China. National Committee for Opium Suppression *see* Chin yen kung pao

China. National Financial Conference *see* Proceedings

China Nei cheng pu Nien chien pien tsuan wei yuan hui *see* Nei cheng nien chien

The china news *see* English-language newspapers published in china

China news-letter – Shanghai: Lutheran World Federation, 1946-59 [mf ed 2001] – (= ser Christianity's encounter with world religions, 1850-1950) – 2r – 1 – (missing: 1 iss) – mf#2001-s199 – us ATLA [242]

China Nung ts'un fu hsing wei yuean hui *see*
- Che-chiang sheng nung ts'un tiao ch'a
- Chiang-su sheng nung ts'un tiao ch'a

China opened : or, a display of the topography, history, customs...etc. of the chinese empire / Gutzlaff, K – London: Smith, Elder and Co, 1838. 2v – 13mf – 9 – mf#HT-523 – ne IDC [915]

China. Parliament. Lower House *see*
- Tsung i yuan kung pao
- Tsung i yuan kung pao fu lu, ti i tzu hui i su chi lu

China. Parliament. Upper House *see* Ts'an i yuan kung pao

China. People's Republic of China *see* Chuan kuo chin jung chi kou i lan

China pictorial – v1-26. 1951-76 – 1 – us AMS Press [073]

China pictorial, descriptive and historical : with some account of ava and the burmese, siam, and anam / Corner, Julia – London: Henry G Bohn, 1853 [mf ed 1995] – (= ser Yale coll) – xx7521p (ill) – 1 – 0-524-10154-X – mf#1995-1154 – us ATLA [915]

China Ping i pu I cheng ssu *see* Hsueh sheng ts'ung chun chi shih

China post *see* English-language newspapers published in china

China, present and past : foreign intercourse, progress and resources, the missionary question, etc / Gundry, Richard Simpson – London: Chapman & Hall, 1895 [mf ed 1995] – (= ser Yale coll) – xxxi/414p – 1 – 0-524-09406-3 – mf#1995-0406 – us ATLA [951]

China press – Shanghai, China. 1925-1938 (1) – mf#67676 – us UMI ProQuest [079]

China press *see* English-language newspapers published in china

The china press – Shanghai. Jun 1946-11 May 1949. Incomplete issues. 7 reels – 1 – us Chinese Res [951]

The china press = Ta-lu pao – Shanghai: China Press [oct 1 1938-mar 31 1949] – 55r – 1 – (daily ex hols, oct 1 1938-) – us CRL [079]

China. Provisional Legislative Assembly *see* Ts'an i yuan i shih lu

China quarterly – Oxford. 1960+ [1,5,9] – ISSN: 0305-7410 – mf#6097 – us UMI ProQuest [951]

China reconstructs – v1-25. 1952-76 – 1 – us AMS Press [073]

China relief notes / Mennonite Central Committee – Kaifeng, Honan, China: [Mennonite Central Committee] 1946- (qrtly, mthly) [v1-4 (1946-48)][mf ed 2005] – 1r – 1 – (publ in akron pa feb. 1947-may 1948; merged with: philippine relief notes; and java-pakistan news, to form: far eastern relief notes; cont : china relief sheet) – mf#2005c-s031 – us ATLA [243]

China relief sheet / Mennonite Central Committee – Chengchow, Honan, China: [Mennonite Central Committee] 1946 [mf ed 2005] – 1v on 1r with other item – 1 – (cont: china sheet; cont by: china relief notes; filmed with: china sheet [2005c-s029]) – mf#2005c-s030 – us ATLA [243]

China (Republic: 1949-) Tsui kao fa yuan *see* Tsui kao fa yuan p'an li yao chih

China. Republic of China *see* Taiwan sotokufu tokeisho

China review : or notes and queries on the far east – Hong Kong. v1-25. 1872-1901 – 184mf – 9 – (missing: 1896/1897 v22(4); 1897/1898 v23(1-3); 1899/1900 v24(1)) – mf#CH-205 – ne IDC [951]

China revolutionized / Thomson, John Stuart – Indianapolis: Bobbs-Merrill [1913] [mf ed 1995] – (= ser Yale coll) – 590p (ill) – 1 – 0-524-10251-1 – mf#1996-1251 – us ATLA [951]

China. Second National Financial Conference *see* Proceedings

China securities *see* Zhongguo zheng quan bao

China sheet / Mennonite Central Commitee – Chungking, Ichang, Hupeh, China: [s n] 1945 [mf ed 2005] – 1v on 1r with other item – 1 – (cont by: china relief sheet; Filmed with: china relief sheet [2005c-s030]) – mf#2005c-s029 – us ATLA [243]

China Shih yeh pu *see* Shih yeh ssu nien chi hua ts'ao an

China Shih yeh pu Kuo chi mao i chu *see* Tsui chin san shih ssu nien lai chung-kuo t'ung shang k'ou an tui wai mao i t'ung chi

China Ssu fa yuan *see* Ssu fa yuan chieh shih hui pien 1

China. Third National Financial Conference *see* Proceedings

China through western eyes : manuscript records of traders, travellers, missionaries and diplomats, 1792-1942 – [mf ed Marlborough 1996] – 8pt – 1 – (pt1: sources fr the william r perkins library [15r] £1400; pt2:... [16r] £1500; pt3: papers of j a thomas, c1905-23 fr william r perkins library [17r] £1600; pt4: mss diaries & papers fr china records project at yale divinity library [23r] £2150 [mf ed 2000]; pt5:... [18r] £1700 [mf ed 2000]; pt6: correspondence & papers of sir ernest satow [1843-1929] relating to china from public record office class pro 30/33 [15r] £1400; pt7: diaries of g e morrison [1862-1920], peking correspondent of the times from 1897, & political advisor to the president of china, 1912-20, from the mitchell library, state library of nsw [20r] £1850; pt8: diaries, notebooks & writings of rewi alley [1897-1988] fr national library of nz [15r] £1400; pt9: addis & geller collections fr school of oriental & african studies, london [14r] £1350; with d/g) – uk Matthew [951]

China T'ieh tao pu Ts'ai wu ssu *see* Pao ning hsien pao lin tuan ching chi tiao ch'a pao kao shu

The china times : journal international = I wen hsi pao – Peking: China Times [jun 3 1901] (daily ex sun and hols) – 1r – 1 – us CRL [079]

China today / American Friends of the Chinese People – ser1: n1-8 1934 [all publ]. ser 2: v1-8 1934-42 [all publ] – (= ser Radical periodicals in the united states, 1881-1960. series 1) – 23mf – 9 – $210.00 – us UPA [335]

China today – 1977- – 9 – enquire for prices – (in english) – us UMI ProQuest [079]

China today – v1-12. 1958-70 – 1 – us AMS Press [073]

"China traveler" / Lu-hsing tga-chih – Shanghai, 1927-1954. 28v – 492mf – 9 – (Missing: several vols) – mf#CH-312 – ne IDC [915]

China Ts'ai cheng pu Ch'i ssu shu *see* Shih nien lai chih ch'i ssu

China Ts'ai cheng pu Ch'ien pi ssu *see* Shih nien lai chih chin jung

China Ts'ai cheng pu Kuan shui shu *see* Shih nien lai chih kuan shui

China Ts'ai cheng pu Kung chai ssu *see* Shih nien lai chih kung chai

China Ts'ai cheng pu Shui wu shu *see* Shih nien lai chih huo wu shui

China Ts'ai cheng pu Ti fang ts'ai cheng ssu *see* Shih nien lai chih ti fang ts'ai cheng

China Ts'ai cheng pu Ts'an shih t'ing *see* Shih nien lai chih ts'ai wu fa chih

China Ts'ai cheng pu Yen cheng ssu *see* Shih nien lai chih yen cheng

China Ts'ai cheng pu Yen wu tsung chu *see* Yen chuan mai fa kuei hui pien ti i chi

China Tsui kao fa yuan *see* Tsui kao fa yuan ts'ai p'an yao chih hui pien

China weekly herald – China. -w. 28 Jan-27 Feb 1932. 3 ft – 1 – uk British Libr Newspaper [072]

China weekly mail *see* Month of reign of terror in shanghai

China weekly review – Shanghai, China. 1917-1953 (1) – mf#67677 – us UMI ProQuest [079]

The china weekly review = Mi-le shih p'ing lun pao – Shanghai: Millard Publ House, 1923-50 (wkly) [mf ed 1930-aug 5 1950] – 1 – (publ suspended dec 13 1941-oct 13 1945. v25 n4 (jun 23 1923)-v118 n10 (aug 5 1950)) – mf#02214 – us L of C Photodup [951]

The china weekly review / ed by Powell, John William – 1917-1947 – 50r – 1 – $3,000.00 – us UMI ProQuest [320]

China-Burma-India Veterans Association *see* Ex-cbi roundup

China-dienst : halbmonatsschrift fuer die foerderung der deutsch-chinesischen beziehungen – Schanghai (VR), 1932 oct-1933 – 1r – 1 – gw Misc Inst [327]

Chinakal, N A *see* Voprosy obogashcheniia poleznykh iskopaemykh sibiri

The chinaman as we see him : and fifty years of work for him / Condit, Ira M – Chicago: F.H. Revell, c1900 – 1mf – us ATLA [240]

The chinaman as we see him : and fifty years of work for him / Condit, Ira Miller – Chicago: F.H. Revell, c1900 – 1mf – 9 – 0-7905-4669-8 – mf#1988-0669 – us ATLA [240]

Chinamato chamauro – Gwelo, Zimbabwe. 1960 – 1r – us UF Libraries [960]

Chinamen at home / Selby, T G – London, 1900 – 4mf – 9 – mf#HT-131 – ne IDC [915]

Chinamen at home / Selby, Thomas Gunn – London: Hodder & Stoughton, 1900 [mf ed 1995] – (= ser Yale coll) – viii/295p – 1 – 0-524-09498-5 – mf#1995-0498 – us ATLA [306]

China's book of martyrs : a record of heroic martyrdoms and marvelous deliverances of chinese christians during the summer of 1900 / Miner, Luella – Philadelphia: Westminster Press, 1903 [mf ed 1995] – (= ser Yale coll) – 512p (ill) – 1 – 0-524-09134-X – mf#1995-0134 – us ATLA [240]

China's challenge in manchuria : anti-japanese activities in manchuria prior to the mukden incident / Ito, Takeo – [Darien?]: South Manchuria Railway Co [1932] [mf ed 1986] – 1r – 1 – mf#7083 – us Wisconsin U Libr [951]

China's "diamonds in the rough" *see* T'a shan shih yu (ccm123)

China's millions – London, 1875-1892 v1-17; n.s. 1893-1899 v1-7 – 120mf – 8 – mf#CH-152 – ne IDC [951]

China's millions – [London: Morgan & Scott], 1876-1952 [mthly, bimthly] [mf ed 2004] – 77v on 8r – 1 – (mf: n7-114 [1876-84]; v10-78 [1885-1952] lacks: 1888 n10 p129-130. iss for 1893-1952 also called new ser v1-58. iss by china inland mission. iss also in an australasian, & north america ed) – mf#2003-s096 – us ATLA [240]

China's millions and our work among them – London: Morgan & Scott. n1 jul-6 dec 1875 [mthly] [mf ed 2003] – 1v on 1r – 1 – (iss by china inland mission) – mf#2003-s095 – us ATLA [240]

China's new literature and art : essays and addresses / Yang, Chou – Peking: Foreign Language Press, 1954 – us CRL [480]

China's place in philology : an attempt to show that the languages of europe and asia have a common origin / Edkins, Joseph – London: Truebner, 1871 – 2mf – 9 – 0-8370-8094-0 – mf#1986-2094 – us ATLA [400]

China's revolution, 1911-1912 : a historical and political record of the civil war / Dingle, Edwin John – New York: McBride, Nast, 1912 [mf ed 1995] – (= ser Yale coll) – 304p (ill) – 1 – 0-524-09402-0 – mf#1995-0402 – us ATLA [951]

China's spiritual need and claims : [conspectus of protestant missions in china, march, 1884] / Taylor, James Hudson – 5th ed. London: Morgan & Scott, 1884 [mf ed 1995] – (= ser Yale coll) – iv, 91p (ill) – 1 – 0-524-10090-X – mf#1995-1090 – us ATLA [242]

Chinatown news – v2 n2-v3 n3 [1975 jun/jul-1976 apr] – 1r – 1 – mf#359366 – us WHS [071]

Chinatown news – 1975 mar 18/1978-1988/1989 jun – 9r – 1 – (with gaps; cont: chinatown [vancouver bc]) – mf#555788 – us WHS [071]

Chinchilla Aguilar, Ernesto *see*
- Ayuntamiento colonial de la ciudad de guatemala
- Danza del sacrificio, y otros estudios

Chinchilla, Anastasio *see* Anales...medicina...y biografico

Chinchon, Condesa de *see* Esposocion(sic) que dirige a las cortes constintuyentes en defensa de su padre don manuel godoy...

Chincilla Aguilar, Ernesto *see* Historia del arte en guatemala 1524-1962

La chine : huit ans au yun-nan, recit d'un missionnaire / Pourias, Emile Rene – 3rd ed. [Bruges]: Societe de Saint-Augustin, 1892 [mf ed 1995] – (= ser Yale coll) – viii/188p (ill) – 1 – 0-524-09743-7 – (in french) – mf#1995-0743 – us ATLA [241]

La chine / Maspero, Georges – Paris: Delagrave [1918] [mf ed 1995] – (= ser Yale coll; Bibliotheque d'histoire et de politique) – 453p – 1 – 0-524-09374-1 – (in french) – mf#1995-0374 – us ATLA [951]

La chine : sa religion, ses moeurs, ses missions / Piton, Charles – Lausanne: Georges Bridel; Paris: Librairie Fischbacher, 1902 [mf ed 1995] – (= ser Yale coll) – 286p (ill) – 1 – 0-524-10152-3 – (in french) – mf#1995-1152 – us ATLA [951]

La chine avec ses beautes et ses singularites : ou lettres d'un jeune voyageur a sa famille, sur les moeurs, les usages, l'education des chinois... – Paris 1823 [mf ed Hildesheim 1995-98] – 2v on 4mf – 9 – €120.00 – 3-487-27575-9 – gw Olms [951]

La chine catholique : tableau des progres du christianisme dans cet empire / Condurier – Paris: chez l'auteur, 1829 [mf ed 1995] – (= ser Yale coll) – 51p (ill) – 1 – 0-524-09844-1 – mf#1995-0844 – us ATLA [241]

La chine en miniature : ou choix de costumes, arts et metiers de cet empire, representes par...gravures, la plupart d'apres les originaux inedits du cabinet de feu m bertin, ministre / Breton de LaMartiniere, Jean – Paris 1811-12 [mf ed Hildesheim 1995-98] – 5v on 10mf – 9 – €100.00 – 3-487-27579-1 – gw Olms [951]

La chine et les puissances chretiennes / Mas y Sans, Sinibaldo de – Paris: L Hachette, 1861 [mf ed 1995] – (= ser Yale coll) – 2v (ill) – 1 – 0-524-09940-5 – (in french) – mf#1995-0940 – us ATLA [951]

La chine et les religions etrangeres : kiao-ou ki-lio..."resume des affaires religieuses" / Li, Kang-chi – Chang-hai: Impr de la Mission Catholique, 1917 [mf ed 1995] – (= ser Yale coll; Variete sinologiques 47) – 1 – 0-524-09660-0 – (in french and chinese in parallel clms. trans, comm and app by p jerome tobar) – mf#1995-0660 – us ATLA [230]

La chine, ou description generale des moeurs et des coutumes, du gouvernment, des lois, des religions, des sciences, de la literature, des productions naturelles, des arts, des manufactures et du commerce de l'empire chinois / Davis, J F – Paris: Libraire de Paulin, 1837. 2v – 10mf – 9 – mf#HT-505 – ne IDC [915]

La chine ouverte : aventures d'un fan-kouei dans le pays de tsin / Forgues, Emile – Paris 1845 [mf ed Hildesheim 1995-98] – 1v on 3mf – 9 – €90.00 – 3-487-27558-9 – gw Olms [915]

La chine ouverte : aventures d'un fan-kouei dans le pays de tsin / [Forgues, P E D] – Paris: H Fournier, 1845 – 6mf – 9 – mf#HT-656 – ne IDC [915]

The chinese : a general description of the empire of china and its inhabitants / Davis, J F – London: Charles Knight, 1836. 2v – 11mf – 9 – mf#HT-506 – ne IDC [915]

The chinese : a general description of the empire of china and its inhabitants / Davis, John – London 1836 [mf ed Hildesheim 1995-98] – 2v on 10mf – 9 – €100.00 – 3-487-27573-2 – gw Olms [951]

The chinese : their present and future: medical, political, and social / Coltman, Robert – Philadelphia: FA Davis, 1891 – 1mf – 9 – 0-524-02941-5 – mf#1990-3153 – us ATLA [390]

Chinese account of the opium war / Wei, Yuan & Parker, Edward Harper – Shanghai: Kelly & Walsh, 1888 [mf ed 1995] – (= ser Yale coll; Pagoda library 1) – ii/82p – 1 – 0-524-09003-3 – mf#1995-0003 – us ATLA [951]

The chinese and the ministry : an inquiry into the origin and progress of our present difficulties with china, and into the expediency, justice, and necessity of the war / Murray, John Fisher – London: printed for T Cadell; Edinburgh: W Blackwood & Sons, 1840 – (= ser 19th c books on china) – 1mf – 9 – mf#7.1.33 – uk Chadwyck [951]

Chinese art motives interpreted / Tredwell, Winifred Reed – New York, London: G P Putnam's Sons, 1915 [mf ed 1995] – (= ser Yale coll) – xiii/110p (ill) – 1 – 0-524-09335-0 – mf#1995-0335 – us ATLA [700]

The chinese as they are : their moral and social character, manners, customs, language; with remarks on their arts and sciences, medical skill, the extent of missionary enterprise, etc / Lay, George Tradescant – Albany: George Jones; New York: Burgess & Stringer, 1843 [mf ed 1995] – (= ser Yale coll) – iv/116p – 1 – 0-524-09101-3 – mf#1995-0101 – us ATLA [951]

The chinese as they are : their moral, social, and literary character... / Lay, G T – London: William Ball & Co., 1841 – 4mf – 9 – mf#HT-76 – ne IDC [915]

The chinese as they are [...] by g. tradescent lay... : containing also, illustrative and corroborative notes, additional chapters on the ancient and modern history...compiled from authentic sources. by e.g. squier / Lay, George Tradescant & Squier, Ephraim George – Albany: George Jones; New York...Boston...Philadelphia...Baltimore...1843 – (= ser 19th c books on china) – 2mf – 9 – mf#7.1.57 – uk Chadwyck [306]

Chinese astronomy and astrophysics – Oxford. 1977-1994 (1,5,9) – ISSN: 0275-1062 – mf#49293 – us UMI ProQuest [520]

Chinese biographical archive (cba) = Chinesisches biographisches archiv (cba) / Minden, Stephan von [comp] – [mf ed 1996-99] – 453mf (1:24) – 9 – diazo €10,060.00 (silver €11,080 isbn: 978-3-598-33911-0) – ISBN-10: 3-598-33910-0 – ISBN-13: 978-3-598-33910-3 – (with printed ind) – gw Saur [951]

A chinese boy's personal problem *see* Chung-kuo shao nien te ke jen wen ti [ccm210]

Chinese buddhism : a volume of sketches, historical, descriptive, and critical / Edkins, Joseph – 2nd rev ed. London: Kegan Paul, Trench, Treubner, 1893 [mf ed 1995] – (= ser Yale coll; Treubner's oriental series) – xxxiii/453p – 1 – 0-524-09079-3 – mf#1995-0079 – us ATLA [280]

Chinese central asia : a ride to little tibet / Lansdell, H – London, 1893. 2v – 12mf – 9 – mf#HT-74 – ne IDC [915]

Chinese central asia : a ride to little tibet / Lansdell, Henry – New York: Charles Scribner's Sons, 1894 [mf ed 1996] – (= ser Yale coll) – 2v (ill) – 1 – 0-524-10256-2 – (with aut's pref, app and bibl) – mf#1996-1256 – us ATLA [915]

Chinese characteristics / Smith, Arthur Henderson – New York, Chicago: Fleming H Revell [1894] [mf ed 1995] – (= ser Yale coll) – 342p (ill) – 1 – 0-524-09379-2 – mf#1995-0379 – us ATLA [306]

Chinese christian collection : monographs – S.l.: s.n., 19-- – 2r – 1 – 0-8370-1700-9 – (in chinese) – mf#1984-B500 – us ATLA [240]

Chinese christian collection : serials in chinese – 44 serial titles 307 monographs – 55r – 1 – (individual titles also listed separately) – mf#ATLA S0296 (A-S)-0321 – us ATLA [240]

Chinese christian general : feng yu hsiang / Goforth, Jonathan – [Chefoo: printed by J McMullan, 1919] [mf ed 1995] – (= ser Yale coll) – 12p – 1 – 0-524-09645-7 – mf#1995-0645 – us ATLA [951]

The chinese christian intelligencer *see* T'ung wen pao (ccs)

A chinese chronicle : by abdalla of beyza / Weston, S – London: William Clarke, 1820 – 1mf – 9 – mf#HT-694 – ne IDC [915]

Chinese churchman *see* Sheng kung hui pao (ccs34)

Chinese classical stories *see* Ku wen chin i chung-kuo ku shih (ccm308)

A chinese commercial guide : consisting of a collection of details respecting foreign trade in china / Morrison, John Robert – Canton: printed at the Albion Press, 1834 – (= ser 19th c books on china) – 2mf – 9 – mf#7.1.44 – uk Chadwyck [380]

Chinese courier and canton gazette – Canton: [Markwick & Lane], 1831-32 – (= ser Chinese courier) – 1r – 1 – (filmed with: chinese courier jul 28 1831-apr 5 1832) – us CRL [079]

Chinese courier etc – Canton, China. 28 jul 1831-23 sep 1833 – 1/2r – 1 – uk British Libr Newspaper [072]

Chinese culture – Taiwan. 1957+ (1) 1971+ (5) 1975+ (9) – ISSN: 0009-4544 – mf#3273 – us UMI ProQuest [480]

Chinese daily news – feb 1946-dec 1996 – 269r – 1 – ch Transmission [079]

Chinese dialogues, questions, and familiar sentences : literally rendered into english, with a view to promote commercial intercourse... / Medhurst, Walter Henry – Shanghae: London Mission Press, 1863 [mf ed 1995] – (= ser Yale coll) – 225p – 1 – 0-524-09375-X – (rev by aut's son) – mf#1995-0375 – us ATLA [480]

Chinese diamonds for the king of kings / Goforth, Rosalind – Toronto: Evangelical Publ [1920] [mf ed 1995] – (= ser Yale coll) – 117p – 1 – 0-524-09446-2 – mf#1995-0446 – us ATLA [951]

Chinese directory, vancouver island = Chung hua ch'u pan she (vancouver) – Vancouver: Chinese Publicity Bureau, Ltd – 9 – (in chinese or english; title also in chinese) – mf#*XLM-5 – us NY Public [971]

Chinese economic journal and bulletin – Peking, January 1927-June 1937 – 12r – 1 – $720.00 – us UMI ProQuest [330]

Chinese economic studies – Armonk. 1989-1996 (1) – (cont by: chinese economy) – ISSN: 0009-4552 – mf#16879 – us UMI ProQuest [330]

Chinese economy – Armonk. 1997+ (1) – (cont: chinese economic studies) – ISSN: 1097-1475 – mf#16879,01 – us UMI ProQuest [330]

Chinese education – Armonk. 1982-1992 (1) 1982-1992 (5) 1982-1992 (9) – (cont by: chinese education and society) – ISSN: 0009-4560 – mf#13346 – us UMI ProQuest [370]

Chinese education and society – Armonk. 1993+ – 1,5,9 – (cont: chinese education) – ISSN: 1061-1932 – mf#13346,01 – us UMI ProQuest [370]

The chinese empire : forming a sequel to the work entitled "recollections of a journey through tartary and thibet" = L'empire chinois, faisant suite a l'ouvrage intitule Souvenirs d'un voyage dans la tartarie et le thibet / Huc, Evariste Regis – London: Longman, Brown, Green & Longmans, 1855 [mf ed 1995] – (= ser Yale coll) – 2v – 1 – 0-524-09381-4 – mf#1995-0381 – us ATLA [915]

The chinese empire : a general and missionary survey: with portraits and illustrations / ed by Broomhall, Marshall – New York: Fleming H Revell; Philadelphia: China Inland Mission, [ca 1907] – 2mf – 9 – 0-8370-6806-1 – (incl indes) – mf#1986-0806 – us ATLA [951]

Chinese fiction / Candlin, George T – Chicago: Open Court 1898 [mf ed 1992] – (= ser The religion of science library 33) – 1mf – 9 – 0-524-02346-8 – mf#1990-2957 – us ATLA [480]

Chinese foreign policy / Ross, John – Shanghai, 1877 – (= ser 19th c books on china) – 1mf – 9 – mf#7.1.37 – uk Chadwyck [327]

Chinese geography and environment – Armonk. 1988-1988 (1,5,9) – ISSN: 0896-2979 – mf#16880 – us UMI ProQuest [915]

"Chinese" gordon, r e, c b / Allen, Charles H – London, England: Abraham Kingdon & Co 1884 – 1r – 1 – us UF Libraries [355]

Chinese horrors and persecutions of the christians : containing a full account of the great insurrection in china, atrocities of the "boxers,"...together with the complete history of china down to the present time... / Northrop, Henry Davenport – Philadelphia: World Bible House [1900] [mf ed 1995] – (= ser Yale coll) – 420p (ill) – 1 – 0-524-10284-8 – mf#1996-1284 – us ATLA [951]

The chinese in cambodia / Willmott, William E – Vancouver [Canada]: Publ Centre, University of British Columbia 1967 [mf ed 1989] – 1r with other items – 1 – (with bibl) – mf#mf-10289 seam reel 019/08 [§] – us CRL [305]

Chinese in the mother lode / Minke, Pauline – 1r – 1 – $50.00 – mf#C63001 – us Library Micro [305]

Chinese journal – New York, NY. 1962-1976 (1) – mf#65062 – us UMI ProQuest [071]

Chinese Journalists Delegation to Cambodia *see* Le cambodge en lutte

Chinese labour in transvaal mines, pamphlets relating to... 1904-07 : from the john burns library, trades union congress and the royal commonwealth society library – 3r – 1 – (with guide; int by baruch hirson; contains 1r of uk government command papers relating to the iss) – mf#97240 – uk Microform Academic [305]

Chinese Language Teachers Association *see* Journal of the chinese language teachers association

Chinese law : journals and serials – 15r – 1 – $525.00 in US $40.00r outside – (cheng-fa yen-chiu (political and legal research): peking, 1954-1963 (some incomplete years) 3r l9300129. chung-yang jen-min cheng-fu fa-ling hui-pien (compendium of laws and decrees of the central people's government): peiping, 1949-54 2r l9300130. chung-hua jen-min kung-ho-kuo ch'uean-kuo jen-min tai-piao ta-hui ch'ang-wu wei-yuean-hui kung-pao (official gazette of standing committee of the national people's congress of the people's republic of china): peking, 1959-1963 3r l9300131. chung-hua jen-min kung-ho-kuo kuo-wu-yuean kung-pao (official gazette of the state council of the people's republic of china): peking, 1957, 1958 and index, 1959 and index (some incomplete years) 3r l9300132. chung-hua jen-min kung-ho-kuo fa-kuei hui-pien (compendium of laws and regulations of the people's republic of china): jul-dec 1956, 1959-jun 1960, jul-dec 1961, jan-dec 1963 3r l9300133, fa hsueeh (jurisprudence): shanghai, 1956-1958 1r l9300134. in chinese) – mf#L9300129-L9300134 – Dist. us Scholarly Res – us L of C Photodup [340]

Chinese law and government – v1-33. 1968-2000 – 5,6,9 – $1436.00 set – (v1-17 1968-85 in reel $330.00. v18-33 1985-2000 in mf $1106.00) – ISSN: 0009-4609 – mf#101701 – us Hein [340]

Chinese literature : essays, articles, reviews – Madison. 1979+ (1,5,9) – ISSN: 0161-9705 – mf#11795 – us UMI ProQuest [480]

Chinese literature – Peking: Cultural Press, 1951-54 – 1 – us CRL [480]

Chinese literature – v1-26. 1951-76 – 1 – us AMS Press [480]

Chinese mathematics – Providence. 1962-1967 (1) 1962-1967 (5) 1962-1967 (9) – ISSN: 0577-909X – mf#13413 – us UMI ProQuest [510]

Chinese mechanical engineering abstracts : english ed – Elmsford. 1988-1990 (1,5,9) – ISSN: 1001-0378 – mf#49565 – us UMI ProQuest [621]

Chinese military studies and materials in english translation – 10r – 1 – $350.00 in US $40.00r outside – (collection can only be purchased as a set. in english) – mf#L9300070-9 – Dist. us Scholarly Res – us L of C Photodup [355]

Chinese moral maxims : with a free and verbal translation / Davis, John Francis, 1st Bart. – London, 1823 – (= ser 19th c books on linguistics) – 3mf – 9 – (transliteration of chinese characters at head of title: hsien-wen-shu hien wun shoo) – mf#2.1.51 – uk Chadwyck [480]

Chinese moral maxims : with a free and verbal translation; affording examples of the grammatical structure of the language / Davis, John Francis – London, Macao: Murray, Thomas 1823 – (= ser Whsb) – 3mf – 9 – €40.00 – mf#Hu 300 – gw Fischer [480]

Chinese moral sentiments before confucius : a study in the origin of ethical valuation / Rudd, Herbert Finley – 1914 [mf ed 1991] – 1mf – 9 – 0-524-01379-9 – (incl bibl ref) – mf#1990-2391 – us ATLA [170]

Chinese muslim = Hung kuo mu suu lin – n25. 10 jan 1960* – (= ser Chinese christian coll 51) – 1r – 1 – (in chinese) – mf#ATLA S0296G – us ATLA [260]

The chinese nation through the eyes of jesus *see* Yeh-su yen li ti chung-hua min tsu (ccm196)

Chinese nationalist daily – New York. Jan 21 1927-Mar 1958 – 1 – us NY Public [071]

Chinese newspapers, 1 – 471r including section 2 – 1 – (ch'ang chiang jih pao, hankow: nov 1949-59 15r l9400090. che-chiang jih pao, hangchow: dec 1949-oct 1954, 1955-nov 1959 12r l9400091. chieh fang jih pao, shanghai: may 1949-jul 1955, feb 1956-dec 1959, may 1960-nov 1962 16r l9300092. chiang hsi jih pao, nanchang: nov 1950-dec 1960 9r l9300093. chin jih hsin wen, peking: sep 1959-1961 4r l9300094. ch'ing-tao jih pao, tsingtoo: apr 1950-aug 1959 9r l9300095. ch'un chung jih pao, shansi: mar 1950-oct 1954 9r l9300096. ch'ung ch'ing jih pao, chungking: sep 1952-59 3r l9300097. chung kuo hsin wen, canton: sep 1954-56, 1957-90 96r l9300098. chung kuo shih pao, taipei: 1970-88 85r l9300099. chung pao, nanking: apr 1940-jun 1945 10r l9300100. ho-nan jih pao, kaifeng: nov 12 1950-apr 1962 9r l9300101. ho-pei jih pao, pao-ting: aug 1950-jan 1958, oct-nov 1959 11r l9300102. hsin chiang jih pao, urmuchi: 1943-sep 1945, 1948-60 9r l9300103. hsin-chien jih pao, kewichow: jul 1951-nov 1956 5r l9300104. hsin hua jih pao: chungking: 1942-46, oct 1950-54 14r l9300105. nanking, sep 1949-aug 1961 11r l9300106. hsin min pao, peking: feb 1941-apr 1944 5r l9300107. hsin shen pao, shanghai: oct 1937-may 1945 20r l9300108. hsin wen pao, shanghai: aug-dec 1928, jul 1944-may 1946, may 1947-dec 1959 23r l9300109;) – Dist. us Scholarly Res – us L of C Photodup [079]

Chinese newspapers, 2 – 471r including section 1 – 1 – (kan su- jih pao lanchow. july 1951-60. 10r. l9300110; kuang chou jih pao, canton. 1952-54; 1956-60. 4r. l9300111; kuang hsi jih pao, nanning. dec 1942-43; dec 1951-nov 1954. 9r. l9300112; lao-tung pao, shanghai. july 1949-may 1958. 5r. l9300113; li pao, hengyang. july 1940-42; apr-oct 1943. 5r. l9300114; lu ta jih pao, liaoning. apr 1956-oct 1958. 4r. l9300115; nang-fang jih pao, canton, feb 1950-64. 20r. l9300116; shan-hsi jih pao, taiyuan. 1950-59. 13r l9300117; shih chia chuang jih pao, hopeh. aug 1950-nov 1957. 4r l9300118; ssu-ch'uan jih pao, chengtu. 1952-3; 1955-59. 4r. l9300119; su nan jih pao, wusih. dec 1949-52. 5r. l9300120; tung pei jih pao, mukden. mar 1949-dec 1950; aug 1951-july 1954. 7r. l9300121; yun-nan jih pao, kunming. 1950; 1952-55; 1957; sept 1958-oct 1960. 6r. l9300122; comes in chinese) – Dist. us Scholarly Res – us L of C Photodup [079]

Chinese peasant cults / Day, Clarence B – 1940 – 1 – us Southern Baptist [290]

The chinese people : a handbook on china (with maps and illustrations) / Moule, Arthur Evans – London: SPCK; New York: E S Gorham, 1914 [mf ed 1995] – (= ser Yale coll) – xiv/469p (ill) – 1 – 0-524-09294-X – mf#1995-0294 – us ATLA [951]

Chinese philosophy : an exposition of the main characteristic features of chinese thought / Carus, Paul – Chicago: Open Court, 1898 – 1mf – 9 – 0-524-02347-6 – mf#1990-2958 – us ATLA [180]

Chinese philosophy in classical times / Hughes, Ernest Richard – London, England. 1954 – 1r – us UF Libraries [180]

Chinese press review / U.S. Consulate General.Canton, China – Chungking, China: US Consulate General, 1945-Oct 21 1946 – 3r – 1 – us L of C Photodup [073]

Chinese press review / U.S. Consulate. Peiping-Tientsin – Peiping-Tientsin: US Consulate, 1946-Jul 1948 – 3r – 1 – us L of C Photodup [073]

The chinese press review / U.S. Consulate General.Chungking, China – Mukden, China: US Consulate General, Jun 6 1947-Jul 26 1948 – 2r – 1 – us L of C Photodup [073]

Chinese press summaries and related publications, 1944-1950 – 33r – 1 – $1,155.00 in US $40.00r outside – (us consulate general (& us information service), canton, china, chinese press review, 1946-48 l9300045 3r. us consulate general (& us office of war information), chungking, china, chinese press review, 1945-46 l9300046 3r. us information service, chungking, china, domei news, 1944-45 l9300047 1r. us consulate general, kunming, china, chinese press review, 1945-48 l9300048 2r. us consulate general, mukden, china, chinese press review, 1947-48 l9300049 2r. us embassy, nanking, china, chinese magazine review, 1948-49 l9300050 1r. us embassy, nanking, china, chinese press review, 1946-48 l9300051 3r. us consulate general (& us information service), peiping, china, chinese press review, 1946-48 l9300052 3r. us consulate general, peiping, china, translations radio broadcasts of communist hsin hua station, north shensi, 1947-49 l9300053 1r. us consulate general (& us information service), shanghai, china, chinese press review, 1945-50 l9300054 10r. us information service, shanghai, china, for your informations: yenan broadcasts, 1946-47 l9300055 1r. us consulate general (& us information service), tientsin, china, tientsin chinese review, 1945-48 l9300056 2r. us consulate general, hong kong, china, summary of new china news agency chinese news dispatches, jun- oct 1950 l9300057 1r. in english) – mf#L9300045-L9300057 – Dist. us Scholarly Res – us L of C Photodup [322]

The chinese public opinion = Ying-wen pei-ching jih-pao – Peking: Chinese Public Opinion, may 5 1908-apr 1909 – (= ser Pekinger deutsche zeitung) – (filmed with: pekinger deutsche zeitung) – us CRL [079]

The chinese reader's manual : a handbook of biographical, historical, mythological, and general literary reference / Mayers, William Frederick – Shanghai: American Presbyterian Mission Press, 1910 [mf ed 1995] – (= ser Yale coll) – xvi/444p – 1 – 0-524-09367-9 – mf#1995-0367 – us ATLA [951]

Chinese recorder – Fu-Chou, China. 1868-1940 (1) – mf#67678 – us UMI ProQuest [079]

The chinese recorder, 1867-1941 – 1986 – 18r – 1 – $2340.00 – (the chinese recorder index: a guide to the christian missions in asia 1867-1941, 1986 2v $15000 isbn:0-8420-2250-3) – us Scholarly Res [951]

Chinese recorder and missionary journal – Shanghai. 1867-1906 (1) – mf#5737 – us UMI ProQuest [240]

Chinese religion through hindu eyes : a study in the tendencies of asiatic mentality / Sarkar, Benoy Kumar – Shanghai: Commercial Press, 1916 – 1mf – 9 – 0-524-04351-5 – (incl bibl ref) – mf#1990-3335 – us ATLA [280]

Chinese repository – Kuang-Chou, China. 1832-1851 (1) – mf#67665 – us UMI ProQuest [079]

Chinese repository – v1-20. may 1832-dec 1851 [complete] – 5r – 1 – mf#ATLA S0014 – us ATLA [073]

The chinese repository – Canton, 1832-1851. 20v – 142mf – 9 – mf#HT-563 – ne IDC [915]

Chinese researches / Wylie, A – Shanghai, 1897 – 6mf – 9 – mf#HT-161 – ne IDC [915]

Chinese scenes and people : with notices of christian missions and missionary life in a series of letters from various parts of china / Edkins, J R – London, 1863 – 4mf – 9 – mf#HTM-188 – ne IDC [915]

Chinese scenes and people : with notices of christian missions and missionary life in a series of letters from various parts of china / Edkins, Jane R – London: J Nisbet, 1863 [mf ed 1995] – (= ser Yale coll) – vi/307p – 1 – 0-524-09466-7 – (with narrative of a visit to nanking by her husband, joseph edkins; also a memoir by her father, william stobbs) – mf#1995-0466 – us ATLA [951]

Chinese scenes and people : with notices of christian missions and missionary life in a series of letters from various parts of china / Edkins, Jane Rowbotham (Stobbs) – London: James Nisbet & Co, 1863 – (= ser 19th c books on china) – 4mf – 9 – mf#7.1.27 – uk Chadwyck [240]

Chinese self-taught : by the natural method with phonetic pronunciation, thimm's system / Darroch, John – 2nd ed. London: E Marlborough, 1916 [mf ed 1995] – (= ser Yale coll) – vi/154p – 1 – 0-524-09492-6 – mf#1995-0492 – us ATLA [480]

The chinese social and political science review – Peking, April 1916-March 1941. v1, No1-v24, No 4 – 13r – 1 – $780.00 – us UMI ProQuest [320]

Chinese society in cambodia : with special reference to the system of congregations in phnom-penh / Willmott, William E – [London? 1964] – 1r with other items – 1 – (with bibl) – mf#mf-10289 seam reel 019/10 [§] – us CRL [305]

Chinese sociology and anthropology – Armonk. 1991-1996 (1) – ISSN: 0009-4625 – mf#16881 – us UMI ProQuest [301]

The chinese speaker or extracts from works written in the mandarin language, as spoken in peking, pt 1 : compiled for the use of students / Thom, Robert – Ningpo: Presbyterian Mission Press, 1846 – (= ser 19th c books on linguistics) – 3mf – 9 – mf#2.1.28 – uk Chadwyck [480]

A chinese st francis : or, the life of brother mao / Brown, Colin Campbell – London, New York: Hodder & Stoughton [1911?] [mf ed 1990] – 1mf – 9 – 0-7905-4663-9 – mf#1988-0663 – us ATLA [920]

Chinese studies in history – Armonk. 1989-1996 (1) – ISSN: 0009-4633 – mf#16883 – us UMI ProQuest [951]

Chinese studies in philosophy – Armonk. 1989-1996 (1) – (cont by: contemporary chinese thought) – ISSN: 0023-8627 – mf#16884 – us UMI ProQuest [100]

Chinese thought : an exposition of the main characteristic features of the chinese world-conception / Carus, Paul – Chicago: Open Court 1907 [mf ed 1992] – 1mf [ill] – 9 – 0-524-02420-0 – (continuation of aut's essay: chinese philosophy) – mf#1990-3004 – us ATLA [180]

Chinese times – British Columbia, CN. jan 1914-dec 1970 – 148r – 1 – (in chinese) – cn Commonwealth Imaging [071]

Chinese times – Tientsin, China. -w. 6 nov 1886-28 mar 1891 – 4r – 1 – uk British Libr Newspaper [072]

The chinese times – Tientsin: Printed & publ for the proprietors by Tientsing Printing Co, nov 6 1886-mar 28 1891 – 4r – 1 – us CRL [079]

Chinese topography : being an alphabetical list of the provinces, departments and districts in the chinese empire, with their latitudes and longitudes / Williams, S Wells – n.p, 1844 – 5mf – 9 – mf#HT-615 – ne IDC [915]

The chinese trade unions – v1-20. 1951-1970 – 1 – us AMS Press [331]

The chinese traveller : containing a geographical, commercial, and political history of china – London: E and C Dilly in the Poultry. 2v. 1775 – 7mf – 9 – mf#HT-504 – ne IDC [915]

Chinese voice – British Columbia, CN. jan 1954-dec 1970 – 37r – 1 – (in chinese) – cn Commonwealth Imaging [071]

The chinese war : an account of all the operations of the british forces from the commencement to the treaty of nanking / Ouchterlony, John – London: Saunders & Otley, 1844 – (= ser 19th c books on china) – 8mf – 9 – mf#7.1.10 – uk Chadwyck [951]

Chinese-american bulletin = T'ien hsia i chia – Maryknoll NY: Chinese Catholic Foreign Mission Society of America [v1-5 n4 (1942-may 1946)] [bimthly] [mf ed 2005] – 1r – 1 – (v4 n2-3 not publ?) – mf#2005c-s050 – us ATLA [241]

Die chinesen und die christliche mission = Kineserne og den kristne mission / Coucheron-Aamot, William – Leipzig: Robert Baum, [n.d.] [mf ed 1995] – (= ser Yale coll) – 77p – 1 – 0-524-09398-9 – (in german. trans fr norwegian by friedrich von kaenel) – mf#1995-0398 – us ATLA [951]

Die chinesische fremden- und christenverfolgung vom sommer 1900 : ein bild aus der neuesten missionsgeschichte / Schlatter, Wilhelm – Basel: Missionsbuchhandlung, 1901 [mf ed 1995] – (= ser Yale coll; Basler missionsstudien 7) – 77p – 1 – 0-524-09144-7 – (in german) – mf#1995-0144 – us ATLA [240]

Die chinesische mission im gerichte der deutschen zeitungspresse / Warneck, Gustav – 7th ed. Berlin: Martin Warneck, 1900 [mf ed 1995] – (= ser Yale coll) – 45p – 1 – 0-524-09363-6 – (in german) – mf#1995-0363 – us ATLA [951]

Der chinesische prediger / Voskamp, C I – Berlin: Berliner Evangelischen Missionsgesellschaft, 1919 [mf ed 1995] – (= ser Yale coll) – 95p – 1 – 0-524-10198-1 – (in german) – mf#1995-1198 – us ATLA [242]

Das chinesische reich / Huc, Evariste R – Leipzig 1856 [mf ed Hildesheim 1995-98] – 2v on 4mf – 9 – €120.00 – 3-487-27568-6 – gw Olms [951]

Der chinesische ritenstreit / Huonder, Anton – Aachen [um 1903] (mf ed 1993) – 1mf – 9 – €19.00 – 3-89349-320-4 – mf#DHS-AR 176 – gw Frankfurter [951]

Chinesische texte : zu dr. joh. heinrich plath's abh. 2. der cultus der alten chinesen – Muenchen: K Akademie, 1864 – 1mf – 9 – 0-524-07218-3 – mf#1991-0080 – us ATLA [951]

Die chinesische wirtschaftsreform und inflationsproblematik seit 1979 / Chen, Xinhua – (mf ed 1995) – 4mf – 9 – €56.00 – 3-8267-2188-8 – mf#DHS 2188 – gw Frankfurter [330]

Der chinesische zopf / Rohrbach, Paul – Heidelberg: Evangelischer Verlag, 1910 [mf ed 1995] – (= ser Yale coll; Volksschriften des allgemeinen evangelischen-protestantischen missionsvereins 10) – 1 – 0-524-09445-4 – (in german) – mf#1995-0445 – us ATLA [242]

Ching chiao pei kao [ccm129] = Notes on the nestorian inscription of si-an / Feng, Ch'eng-chun – 1st ed. Shanghai, 1931 [mf ed 198?] – (= ser Ccm 129) – 1 – mf#1984-b500 – us ATLA [240]

Ching feng : english edition – Hong Kong. 1985+ (1,5,9) – ISSN: 0009-4668 – mf#15377,01 – us UMI ProQuest [290]

Ch'ing nien tu shu yun tung (ccc199) – Shanghai, 1927 [mf ed 198?] – (= ser Ccm 199) – 1 – mf#1984-b500 – us ATLA [400]

Ching, Sheng *see* Chan shih chung-kuo ching chi ti lun k'u

Ching-chi pan-yueh k'an : semi-monthly economic journal – Peking and Shanghai, China. nov 1927-nov 1928. – 1 – us Chinese Res [330]

Ch'ing-chih *see* Lun hsin chung-kuo

Chingford herald and post – London, UK. 1990-2 oct 1991 – 2r – 1 – uk British Libr Newspaper [072]

Chingford midweek observer and epping forest herald *see* Chingford observer and epping forest herald

Chingford observer and epping forest herald – London, UK. 27 apr 1946-apr 1950 – 2 1/2r – 1 – (aka: chingford midweek observer and epping forest herald, chingford observer midweek) – uk British Libr Newspaper [072]

Chingford observer midweek *see* Chingford observer and epping forest herald

Chingford, walthamstow, leyton and leytonstone independent *see* Waltham forest guardian and independent extra

Ch'ing-hua hsueh-pao – Peking. 1-15, no.1, 1924-Oct 1948, n.s.1, no.1-2, 1956-57 – 1 – us L of C Photodup [951]

Ching-kuo ch'ing-nien chih-yeh wen-t'i [ccm149] = Youth and vocation / Ho, Ch'ing-ju – Shanghai, 1934 [mf ed 198?] – (= ser Ccm 149) – 1 – mf#1984-b500 – us ATLA [331]

Ch'ing-nien chin-pu [ccs23] = Association progress – Shanghai. n14-150. 1918-32 [gaps] [mf ed 198?] – (= ser Chinese christian serials coll 23) – 3r – 1 – (began with n1 in may 1917) – mf0300 – us ATLA [240]

Ch'ing-shan-hsien-nung *see* Hung lou meng kuang i

Chinh luan – Ho Chi Minh City, Vietnam. 1964-1975 (1) – mf#67825 – us UMI ProQuest [079]

Chiniquy, Charles *see*
– L'eglise de rome
– The priest, the woman, and the confessional

Chiniquy, Charles Paschal Telesphore *see*
– Adresse des associes de la temperance de longueuil au rev pere chiniquy
– Fifty years in the church of rome
– Le pretre, la femme et le confessionnal
– Rome and education

Chin-ling hsieh ho shen hsueh chih [ccs45] = Chin-ling theological seminary journal – n6-7. feb-aug 1957 [complete] [mf ed 198?] – (= ser Chinese christian serials coll 45) – 1r – 1 – mf0296a – us ATLA [240]

Chin-ling shen hsueh chih *see* Shen hsueh chin (ccs)

Chin-ling shen hsueh yuan hua hsi t'e k'an *see* Chin-ling shen hsueh yuan hua hsi t'e k'an [ccs46]

Chin-ling shen hsueh yuan hua hsi t'e k'an [ccs46] = Chin-ling theological seminary. west china issue – Ch'eng-tu, aug 1945 [complete] [mf ed 198?] – (= ser Chinese christian serials coll 46) – 1 – (only no publ?) – mf0296b – us ATLA [240]

Chin-ling theological seminary journal *see* Chin-ling hsieh ho shen hsueh chih [ccs45]

Chinnery, E W P *see* Anthropological reports

Chinnery-Haldane, J R Alexander *see*
– Charge delivered to the clergy of the diocese of argyll and the isl...

[Chino-] chino champion – CA. nov 1887- – 93+ r – 1 – $5580.00 (subs $150/y) – (aka: chino hills news) – mf#R02108 – us Library Micro [071]

Chino hills news *see* [Chino-] chino champion

[Chino-] south ontario news – CA. nov 1978- – 15+ r – 1 – $900.00 (subs $50/y) – mf#R02109 – us Library Micro [071]

Les chinois chez eux / Aubry, Jean-Baptiste – Lille: Societe Saint-Augustin, Desclee, de Brouwer, 1889 [mf ed 1996] – (= ser Yale coll) – 300p (ill) – 1 – 0-524-10283-X – (in french) – mf#1996-1283 – us ATLA [241]

Chinois et missionnaires : une persecution dans la province de ning-ko-fou chinois and missionnaires / Bizeul, Severe Jacques – Limoges: Marc Barbou [1896?] [mf ed 1995] – (= ser Yale coll) – 335p (ill) – 1 – 0-524-10087-X – (in french) – mf#1995-1087 – us ATLA [951]

Les chinois pendant une periode de 4458 annees : histoire, gouvernement, sciences, arts, commerce, industrie, navigation, moeurs et usages / Chavannes de la Giraudiere, H de – 2nd ed. Tours: A Mame, 1854 [mf ed 1995] – (= ser Yale coll; Bibliotheque de la jeunesse chretienne) – 380p (ill) – 1 – 0-524-09641-4 – (in french) – mf#1995-0641 – us ATLA [951]

Chinono, Richard *see* Mashoko e wanu

Chinook advance – Alberta, CN. jan 1915-dec 1945 – 6r – 1 – cn Commonwealth Imaging [071]

Chinook observer – Long Beach, WA. 1963-1979 (1) – mf#69572 – us UMI ProQuest [071]

Chinook texts / Boas, Franz – Washington: GPO, 1894 [mf ed 1980] – 4mf – 9 – 0-665-00155-X – mf#00155 – cn Canadiana [490]

Chinos llegaron antes que colon / Loayza, Francisco A – Lima, Peru. 1948 – 1r – us UF Libraries [972]

Chin-shan shih-pao = Chinese times – San Francisco: Chinese Times Publ Co, [oct 4 1928-jun 1929; jun 1930-aug 1931] – us CRL [071]

Chin-shen lu – List of Chinese gentry. Government posts (Ching Dynasty) and lists of those who filled them. 1757-1917. Many years missing – 1 – us Chinese Res [951]

Chintamani, C Y *see* Speeches and writings of the honourable sir pherozeshah m mehta

Chintamani, Chirravoori Yajneswara *see* Indian politics since the mutiny

Chi-nyanja comprehension by lusaka schoolchildren / Serpell, Robert – Lusaka, Zambia. 1970 – 1r – us UF Libraries [960]

Chinyanja exercise book / Woodward, M E – London, England. 1898 – 1r – us UF Libraries [470]

Chi-nyanja simplified / Caldwell, Robert – 2nd ed. London: Zambesi Industrial Mission, [1897] – 1 – us CRL [490]

Chiossone, Tulio *see* Temas sociales venezolanos

Chiovenda, E *see* Le collezioni botaniche della missione stefanini-paoli nella somalia italiana

Chipere / Barnard, T H – Fort Victoria, Zimbabwe. 1960 – 1r – us UF Libraries [960]

Chipiez, Charles *see* History of art in sardinia, judaea, syria, and asia minor

Chipley banner – Chipley, FL. 1893 jul 15-1929 – 8r – (gaps) – us UF Libraries [071]

Chipman, D *see* D chipman's reports

Chipman, N *see* N chipman's reports

Chipmunk press – Stoddard WI. 1977 apr 6-1978 dec 28, 1979 jan 3-dec 26, 1980-81 – 3r – 1 – (cont: last stoddard tuesday advertiser) – mf#1079942 – us WHS [071]

Chippendale, sheraton and hepplewhite furniture designs / Bell, J Munro [comp] – London 1900 – (= ser 19th c art & architecture) – 5mf – 9 – mf#4.2.855 – uk Chadwyck [740]

Chippenham 1559-1950 – (= ser Cambridgeshire parish register transcript) – 7mf – 9 – £8.75 – uk CambsFHS [929]

Chippewa and Munsee Indians *see* Journal of proceedings of council

Chippewa anzeiger – Eau Claire WI. 1876 aug 3, 1886 jul 16 – 1r – 1 – mf#875210 – us WHS [071]

Chippewa county independent – Chippewa Falls WI. 1881 dec 29-1883 jul 29, 1883 aug 2-1884 apr 17, 1885 dec 31, 1887 feb 9-dec 27, 1888 jan-jul 16 – 4r – 1 – (cont by: chippewa valley independent) – mf#921681 – us WHS [071]

Chippewa current – Chippewa Falls WI. 1895 jan 12,14,18, feb 1,4-6,8-9, mar 25 – 1r – 1 – mf#1269584 – us WHS [071]

Chippewa daily gazette – Chippewa Falls WI. 1921 nov 1/dec 31-1924 oct 7/nov 8 – 10r – 1 – (with gaps; cont: wisconsin daily press; cont by: chippewa telegram) – mf#923577 – us WHS [071]

Chippewa daily press – Chippewa WI. 1919 dec 12-1920 feb 2, 1920 aug 10-sep 30, 1920 feb 3-aug 9 – 3r – 1 – (cont: evening independent; cont by: wisconsin daily press) – mf#923787 – us WHS [071]

Chippewa falls [city directory] : listing – 1885, 1889-90, 1893-94, 1901 – 4r – 1 – mf#3188887 – us WHS [917]

Chippewa falls democrat – Chippewa Falls WI. 1869 jun 3-1872 nov 14 – 1r – 1 – (cont: chippewa union and times) – mf#921510 – us WHS [071]

Chippewa falls workman – Chippewa Falls WI. 1886 dec 18 – 1r – 1 – (cont by: chippewa times and independent) – mf#960678 – us WHS [331]

Chippewa herald – 1870 jan 29-1873 jul 12, 1873 jul 18-1875 dec 24, 1875 dec 31-1878 aug 2, 1878 aug 9-1881 aug 5, 1881 aug 12-1884 aug 29, 1884 sep 5-1887 aug 26, 1887 sep 2-1890 sep 19 – 9r – 1 – (cont by: weekly herald [chippewa falls wi]) – mf#922586 – us WHS [071]

Chippewa herald – 1995 mar-1996 dec – 22r – 1 – (cont: chippewa herald-telegram) – mf#4290412 – us WHS [071]

Chippewa herald – 1894 jun 28/dec 31-1926 sep/dec 4 – 1r – 1 – (with gaps; cont by: chippewa telegram; chippewa herald-telegram) – mf#922958 – us WHS [071]

Chippewa herald-telegram – Chippewa Falls WI. 1926 dec 6/31-1995 feb – 438r – 1 – (cont: chippewa herald [1894]; chippewa telegram; cont by: chippewa herald [chippewa falls wi: 1995]) – mf#845956 – us WHS [071]

Chippewa hills courier – Barryton, MI. 1973-1980 (1) – mf#63682 – us UMI ProQuest [071]

Chippewa hills courier shopper – Big Rapids, MI. 1969-1972 (1) – mf#63701 – us UMI ProQuest [071]

Chippewa journal-tribune – Chippewa WI. 1912 oct 24-dec 31, 1913 jan 1-jun 30 – 2r – 1 – (cont: menominee journal) – mf#923524 – us WHS [071]

Chippewa Lake. Ohio. First Presbyterian Church of Lafayette *see* Church record, ms 1217

Chippewa observer – Chippewa Falls WI. 1898 jan 5-1899 oct 4 – 1r – 1 – mf#922610 – us WHS [071]

Chippewa sun – Hayward WI. 1999 feb 12-jun 25 – 1r – 1 – (cont: lco times; cont by: times [hayward wi]) – mf#4337769 – us WHS [071]
Chippewa telegram – Chippewa WI. 1924 nov 10-1925 jan 31, 1925 feb 2-may 28, 1925 may 29-sep 25, 1925 sep 25-dec 31 – 4r – 1 – (cont: chippewa daily gazette; cont by: chippewa herald [1894]; chippewa herald-telegram) – mf#923515 – us WHS [071]
Chippewa times – Chippewa Falls WI. 1892 apr 12/may 10-1915 apr 20/1916 aug 15 – 18r – 1 – (cont: chippewa times and independent; chippewa county independent) – mf#921678 – us WHS [071]
Chippewa times – Chippewa Falls WI. 1875 nov 3/1876 dec 31-1892 may 17/1893 oct 31 – 8r – 1 – (with gaps; cont by: chippewa valley independent; chippewa times and independent) – mf#921529 – us WHS [071]
Chippewa times and independent – Chippewa Falls WI. 1889 dec 18-1890 nov 5, 1890 nov 12-apr 5 – 2r – 1 – (cont: chippewa times [1875]; chippewa valley independent; chippewa falls workman; cont by: chippewa times [1892]) – mf#921675 – us WHS [071]
Chippewa union and times – Chippewa Falls WI. 1867 jan 12-1869 nov 27 – 1r – 1 – (cont: chippewa valley union; cont by: chippewa falls democrat) – mf#922615 – us WHS [071]
Chippewa valley commonwealth advocate – Eau Claire WI. 1937 mar 19-sep 23 – 1r – 1 – (cont by: eau claire advocate and the chippewa valley commonwealth advocate) – mf#962628 – us WHS [071]
Chippewa valley courier – Cornell, Holcombe, Jim Falls WI. 1918 mar 22/1919 sep 24-1958 jan/oct – 18r – 1 – (cont: cornell courier [cornell wi: 1912]; cont by: cornell courier [cornell wi: 1958]) – mf#1047273 – us WHS [071]
Chippewa valley independent – Chippewa Falls WI. 1888 jul 23-1889 apr 1 – 1r – 1 – (cont: chippewa county independent; cont by: chippewa times [1875]; chippewa times and independent) – mf#921690 – us WHS [071]
Chippewa valley news – Eau Claire, West Eau Claire WI. 1870 jul 23-1874 feb 19 – 1r – 1 – mf#938299 – us WHS [071]
Chips – London, Ont: London Collegiate Institute, [1890-189- or 19–] [mf ed dec 1890-mar 1891] – 9 – mf#P04800 – cn Canadiana [370]
Chips and ships / Bay County Genealogical Society [MI] – v5 n3-v10 n1 [1974 spr-1978 fall] – 1r – 1 – mf#361196 – us WHS [929]
Chips from many blocks / Burritt, Elihu – Toronto: Rose-Belford, 1878 [mf ed 1979] – 4mf – 9 – 0-665-00353-6 – mf#00353 – cn Canadiana [890]
Chiquilinga / Dominquez Alba, Bernardo – Panama, Panama. 1961 – 1r – us UF Libraries [972]
El chiquitin charlatan – Villafranca de los Barros, 1896 – 5 – sp Bibl Santa Ana [073]
Chiragh – shumarah-'i 1-5. pa'iz 1360-mihr 1363 [fall 1981-sep 1984] – 18mf – 9 – $290.00 – us MEDOC [079]
Chirambo, G R *see* Kugomezgeka
Chirikov, Evgenii Nikolaevich *see* Izbrannye razskazy
Chirkov, A *see* Rabochii den' [orenburg: 1918]
Chirogram: the chiropractic physician – Glendale. 1972-1977 (1) 1972-1977 (5) 1975-1977 (9) – ISSN: 0009-4692 – mf#7418 – us UMI ProQuest [615]
Chirol, Valentine *see*
– India
– India, old and new
– Indian unrest
Chiropodist – London. 1973-1980 (1) 1976-1979 (5) 1976-1980 (9) – ISSN: 0009-4706 – mf#8652 – us UMI ProQuest [617]
Chiropractic journal of australia – Castlemaine. 1991+ (1,5,9) – (cont: journal of the australian chiropractors' association) – ISSN: 1036-0913 – mf#15997,01 – us UMI ProQuest [615]
Chiropractic sports medicine – v1-10. 1987-1996 – 1,5,6,9 – $65.00r – us Lippincott [617]
Chiropractic technique – v1-8. 1989-1996 – 1,5,6,9 – $65.00r – us Lippincott [615]
Chirurg – Heidelberg. 1982-1983 (1) 1982-1983 (5) 1982-1983 (9) – ISSN: 0009-4722 – mf#13153 – us UMI ProQuest [617]
Chirurgie – Paris. 1968-1980 (1) 1971-1980 (5) 1975-1980 (9) – ISSN: 0001-4001 – mf#3406 – us UMI ProQuest [617]
Chirurgie pediatrique – Paris. 1978-1979 (1) 1978-1980 (5,9) – (cont: annales de chirurgie infantile) – ISSN: 0180-5738 – mf#3409,01 – us UMI ProQuest [617]
Chirurgische behandlungsergebnisse gutartiger schilddruesenerkrankungen / Sengupta, Rahul – (mf ed 1995) – 1mf – 9 – €30.00 – 3-8267-2220-5 – mf#DHS 2220 – gw Frankfurter [617]
Chirurgische praxis – Munich. 1973-1980 (1) 1979-1980 (5) 1979-1980 (9) – ISSN: 0009-4846 – mf#8202 – us UMI ProQuest [617]
Chisaka, Bornface Chenjerai *see* Ability grouping in harare secondary schools
Chiselberti, D F Godfr *see* Opera
Chisholm, Alexander *see* The bible in the light of nature, of man, and of god
Chisholm, Caroline *see*
– The abc of colonization
– Comfort for the poor!
Chisholm, Murdoch *see*
– Cape breton intererts sic sacrificed
– Glimpses of destiny from the book
– Napoleon 1
Chisholm trail / Williamson County Genealogical Society – 1982 jan-1988 spr – 1r – 1 – mf#1054391 – us WHS [071]
Chisholme, David *see* Catalogue of the household furniture, books and other effects and property, belonging to david chisholme, esq
Chisholm's all round route and panoramic guide of the St Lawrence : the hudson river, trenton falls, niagara, toronto... – Montreal: Chisholm 1870 [mf ed 1986] – 3mf [ill] – 9 – 0-665-43621-1 – mf#43621 – cn Canadiana [917]
Chishti, Unwan *see* Tanqidi pairai
Chisla – Paris. v. 1-10. 1930-1934 – 1 – us NY Public [073]
Chislehurst and district times – 1911, 1950-51, 1984-20 oct 1989, 13 jun 1990-1996, 9 jan 1997-jun 1998, nov-dec 1998 67 1/2r – 1 – (aka: chislehurst & kentish times, chislehurst times; not publ between 27 oct 1989 and 6 jun 1990 during which it was amalgamated with beckenham times and bromley times and publ as beckenham bromley chislehurst times) – uk British Libr Newspaper [072]
Chislehurst & kentish times *see* Chislehurst and district times
Chislehurst times *see* Chislehurst and district times
Chislennye metody mekhaniki sploshnoi sredy / Akademiia nauk SSSR, Sibirskoe otdelenie & Otdelenie mekhaniki i protsessov upravleniia Novosibirsk: VTS SO AN SSSR 1970-1986. v1 n1-5 (1970) & v2 (1971) & v3 n1,4-5 1972) – us CRL [947]
Chism, J W *see* Campbellism, what is it?
Chispa – Coral Gables, FL. 1971 jan 15-1975 jan 30 – 1r – (1972 oct 30) – us UF Libraries [071]
Chispazos (aqui-dulces de tiempos pasados) / Sanchez Arjona, Vicente – Sevilla: Imprenta Carlos Acuna, Tomo 1-3. 1942, 1943, 1945 – 1 – sp Bibl Santa Ana [810]
Chistiakov, P S *see* Rechi oktiabrista, 1905-1907 gg
Chistovich, I A *see* Feofan prokopovich i ego vremia
Chisum's pilgrimage, and others : by popular request, resurrected and republished from the union – Cincinnati: [s.n], 1927 (mf ed 1978) – mf#ZZ-16001 – us NY Public [830]
Chisungu / Richards, Audrey Isabel – New York, NY. 1956 – 1r – us UF Libraries [960]
Chiswick & brentford gazette – Brent, England 6 apr 1972- [mf 1986-] – 1 – (variant ed of: ealing gazette) – uk British Libr Newspaper [072]
Chiswick fulham and hammersmith recorder *see* West london recorder
Chiswick guardian – London, UK. 8 jun-21 dec 1990; 4 jan-8 mar 1991 – 3/4r – 1 – uk British Libr Newspaper [072]
Chiswick mail – [London & SE] Hounslow 21 feb-20 mar 1924 [mf ed 2003] – 1r – 1 – uk Newsplan [072]
Chiswick review and sales advertiser – [London & SE] Hounslow jul-dec 1913; mar-jun 1914 [mf ed 2003] – 1r – 1 – (missing: np jan & feb 1914) – uk Newsplan [072]
Chiswick review shopping and amusement guide – [London & SE] Hounslow feb 1924-jun 1926 [mf ed 2003] – 1r – 1 – uk Newsplan [380]
Chit 'n chatter / Amalgamated Meat Cutters and Butcher Workmen of North America – v4 n10-v7 n1 [1974 oct-1977 jan] – 1r – 1 – (cont by: local 593's allied progress report) – mf#622469 – us WHS [660]
Chitambo, Beritha Ruth *see* The expectations of mothers regarding community participation in antenatal care
Chitatel i kniga : sbornik nauchnykh trudov / Nauchnyi sovet po istorii mirovoi kultury Akademii nauk SSSR, Gosudarstvennaia ordena Lenina biblioteka SSSR imeni V I Lenina [redaktsionnaia 111=Markushevich, A I et al – Moskva: Gos. bib-ka SSSR im. V I Lenina 1978 – us CRL [947]
Chitatele – M., 1896-1897, 1901 – 179mf – 9 – (missing: 1896(1-50); 1897(25); 1901(4-50)) – mf#R-1576 – ne IDC [077]
Chitenderano chitsva / Bible NT Shona – Gwelo, Zimbabwe. 1966 – 1r – us UF Libraries [960]
Chitepo, H W *see* Soko risina musoro
Chitevedzero cha kriste – Chishawasha, Zimbabwe. 1935 – 1r – us UF Libraries [960]
Chitevedzero cha kriste – Chishawasha, Zimbabwe. 1936 – 1r – us UF Libraries [960]
Chitevedzero cha kriste – Gwelo, Zimbabwe. 19–? – 1r – us UF Libraries [960]
Chitevedzero cha kriste : rugwaro rwe china / Imitatio Christi Book 4 Shona 1936 – Chishawasha, Zimbabwe. 1936 – 1r – us UF Libraries [960]
Chitevedzero cha kriste : rugwaro rwe chitatu / Imitatio Christi Book 3 Shona 1937 – Chishawasha, Zimbabwe. 1937 – 1r – us UF Libraries [960]
Chitinskaia regional'naia organizatsii Liberal'no-Demokraticheskoi partii Rossii *see* Golos naroda
Chitonga vocabulary of the zambesi valley / Griffin, A W – London, England. 1915 – 1r – us UF Libraries [470]
Chittenden, Hiram Martin *see* Life, letters and travels of father pierre-jean de smet, s.j., 1801-1873
Chittenden, Newton H *see*
– Health seekers', tourists' and sportsmen's guide to the sea-side, lake-side, foothill, mountain and mineral spring health and pleasure resorts of the pacific coast
– Official report of the exploration of the queen charlotte islands
– Settlers, prospectors and tourists guide
– Travels in british columbia and alaska
Chittick, H Neville *see* Guide to the ruins of kilwa
Chitty, Joseph *see*
– A practical treatise on the criminal law
– A treatise on pleading, and parties to actions
– A treatise on pleading, and parties to actions.
– A treatise on the law of bills of exchange, checks on bankers, promissory notes, bankers' cash notes, and banknotes
– A treatise on the laws of commerce and manufactures, and the contracts relating thereto
Chitty's law journal : and family law review – v1-48. 1950-2000 – 1,5,6 – $437.00 set – (v1-29 1950-81 in reel $357; v30-48 1982-2000 in mf $80; title varies: v1-29 1950-81 as chitty's law journal) – ISSN: 0009-4889 – mf#101711 – us Hein [340]
Chitty's law journal *see* Chitty's law journal
Chi-tu chiao che hsueh (ccm66) / Chao, Tzu-ch'en – [s.l: s.n, 1925] [mf ed 198?] – (= ser Ccm 66) – 1 – mf#1984-b500 – us ATLA [240]
Chi-tu chiao chiang hua [ccm339] = Talks on christian faith / Wu, Yao-tsung – Shanghai, 1950 [mf ed 198?] – (= ser Ccm 339) – 1 – mf#1984-b500 – us ATLA [240]
Chi-tu chiao chiao hui te i i [ccm67] = Meaning of the church / Chao, Tzu-ch'en – Shanghai, 1948 [mf ed 198?] – (= ser Ccm 67) – 1 – mf#1984-b500 – us ATLA [240]
Chi-tu chiao chin chieh [ccc68] = Interpretation of christianity / Chao, Tzu-ch'en – Shanghai, 1948 [mf ed 198?] – (= ser Ccm) – 1 – mf#1984-b500 – us ATLA [240]
Chi-tu chiao ch'ing nien hui shih yao [ccm93] = Essentials of ymca / Ch'en, Li-t'ing – Shanghai, 1927 [mf ed 198?] – (= ser Ccm 93) – 1 – mf#1984-b500 – us ATLA [360]
Chi-tu chiao ch'ing nien hui yuan li (ccm167) = Principles of the young men's christian association / Hsieh, Fu-ya – Shanghai, 1923 [mf ed 198?] – (= ser Ccm 167) – 1 – mf#1984-b500 – us ATLA [366]
Chi-tu chiao ching shen [ccm231] = Spirit of christianity / Lo, Ren Yen – 1st ed. Hong Kong, 1958 [mf ed 198?] – (= ser Ccm 231) – 1 – mf#1984-b500 – us ATLA [240]
Chi-tu chiao chiu kuo chu i k'an hsing chih san (ccm189) / Hsu, Ch'ien – [s.l: s.n, 1920] [mf ed 198?] – (= ser Ccm 189) – 1 – mf#1984-b500 – us ATLA [230]
Chi-tu chiao ch'u pan chieh [ccs22] = China bookman – Shanghai. v1-28 n108. 1918-51 [gaps] [mf ed – (= ser Chinese christian serials coll 22) – 3r – 1 – (in english & chinese) – mf0299 – us ATLA [240]
Chi-tu chiao hua ti chiao ting chiao yu [ccm2] = Christian home education / Barbour, Dorothy Dickinson – 5th ed Shanghai, 1933 [mf ed 198?] – (= ser Ccm) – 1 – mf#1984-b500 – us ATLA [240]
Chi-tu chiao nung ts'un yun tung [ccm252] = Christian rural movement / Chang, Fu-liang – Shanghai, 1930 [mf ed 198?] – (= ser Ccm 252) – 1 – mf#1984-b500 – us ATLA [240]
Chi-tu chiao she hui chu i (ccm281) / Ishikawa, Sanshiro & Li, Po – [s.l: s.n, 1929] [mf ed 198?] – (= ser Ccm 281) – 1 – mf#1984-b500 – us ATLA [240]
Chi-tu chiao shih erh chiang [ccm156] = Twelve talks on christianity / Ho, Shih-ming – 1st ed. Hong Kong, 1955 [mf ed 198?] – (= ser Ccm 156) – 1 – mf#1984-b500 – us ATLA [240]
Chi-tu chiao shih shen mo? [ccm85] = What is the christian church? / Ch'en, Chi-yun – Shanghai, 1949 [mf ed 198?] – (= ser Ccm 85) – 1 – mf#1984-b500 – us ATLA [240]
Chi-tu chiao shih shen mo [ccm279] = What is christianity? / Shen, Ch'ing-lai – Shanghai, 1923 [mf ed 198?] – (= ser Ccm 279) – 1 – mf#1984-b500 – us ATLA [240]
Chi-tu chiao ssu hsiang shih [ccm261] = History of christian thought / P'eng, Peter – Hong Kong, 1953 [mf ed 198?] – (= ser Ccm 261) – 1 – (incl bibl ref) – mf#1984-b500 – us ATLA [240]
Chi-tu chiao te chi pen hsin yang [ccm321] = Basic elements of the christian faith / Wei, Chuo-ming – Shanghai, 1950 [mf ed 198?] – (= ser Ccm 321) – 1 – mf#1984-b500 – us ATLA [240]
Chi-tu chiao ti chung hsin hsin yang [ccm69] / Chao, Tzu-ch'en – 2nd ed. Shanghai, 1934 [mf ed 198?] – (= ser Ccm 69) – 1 – mf#1984-b500 – us ATLA [210]
Chi-tu chiao ti li shih kuan [ccm217] : Christian interpretation of history / Liang, Hsiao-ch'u – 1st ed. Hong Kong, 1968 [mf ed 198?] – (= ser Ccm 217) – 1 – mf#1984-b500 – us ATLA [240]
Chi-tu chiao ti lun li [ccm70] = Christian ethics / Chao, Tzu-ch'en – Shanghai, 1948 [mf ed 198?] – (= ser Ccm 70) – 1 – mf#1984-b500 – us ATLA [230]
Chi-tu chiao ts'ung k'an [ccs] = Christian omnibook – n4 nov 1 1943 [complete] [mf ed 198?] – (= ser Chinese christian serials coll) – 1 – mf0296d – us ATLA [240]
Chi-tu chiao yu chung-kuo [ccm106] = Christianity and china / Chou, I-fu – 1st ed. Hong Kong, 1965 [mf ed 198?] – (= ser Ccm 106) – 1 – mf#1984-b500 – us ATLA [240]
Chi-tu chiao yu chung-kuo (ccm168) / Hsieh, Fu-ya – Chiu-lung, 1965 [mf ed 198?] – (= ser Ccm 168) – 1 – mf#1984-b500 – us ATLA [240]
Chi-tu chiao yu chung-kuo hsiang tsun chien she yun tung [ccm351] = Christianity and the rural reconstruction movement in china / Yu, Mu-jen – 3rd ed. Shanghai, 1948 [mf ed 198?] – (= ser Ccm 351) – 1 – (incl bibl ref) – mf#1984-b500 – us ATLA [240]
Chi-tu chiao yu chung-kuo wei hua [ccm331] = Christianity and chinese culture / Wu, Lei-ch'uan – Shanghai, 1936 [mf ed 198?] – (= ser Ccm 331) – 1 – mf#1984-b500 – us ATLA [230]
Chi-tu chiao yu chung-kuo wen hua [ccm192] = Christianity and chinese culture / Hsu, Sung-shih – Hong Kong: Baptist Press, 1965 [mf ed 198?] – (= ser Ccm 192) – 1 – mf#1984-b500 – us ATLA [230]
Chi-tu chiao yu hsien tai ssu hsiang [ccm166] = Christianity and modern thought / Hsieh, Fu-ya – Shanghai, 1941 [mf ed 198?] – (= ser Ccm 166) – 1 – (incl bibl ref) – mf#1984-b500 – us ATLA [230]
Chi-tu chiao yu hsin chung-kuo [ccm229] = Christianity and new china / Lo, Ren Yen – 2nd ed. Shanghai, 1923 [mf ed 198?] – (= ser Ccm 229) – 1 – mf#1984-b500 – us ATLA [240]
Chi-tu chiao yu hsin chung-kuo [ccm333] : a symposium = Christianity and the new china / ed by Wu, Yao-tsung – Shanghai, 1940 [mf ed 198?] – (= ser Ccm 333) – 1 – (incl bibl ref) – mf#1984-b500 – us ATLA [240]
Chi-tu chiao yu hsin wu li hsueh [ccm299] = Christianity and the new physics / Tu, Yu-ching – Shanghai, 1939 [mf ed 198?] – (= ser Ccm 299) – 1 – mf#1984-b500 – us ATLA [210]
Chi-tu chiao yu k'o hsueh [ccm173] = Christianity and science / Hsieh, Hung-lai – 2nd ed. Shanghai, 1921 [mf ed 198?] – (= ser Ccm 173) – 1 – mf#1984-b500 – us ATLA [210]
Chi-tu chiao yu kung ch'an chu i [ccm4] = Christianity and communism / Bates, Miner Searle – Hong Kong, 1939 [mf ed 198?] – (= ser Ccm 4) – 1 – mf#1984-b500 – us ATLA [230]
Chi-tu chiao yu ma lieh chu i [ccm205] = Christianity and marx-leninism / Chiang, Wen-han – Shanghai, 1950 [mf ed 198?] – (= ser Ccm 205) – 1 – mf#1984-b500 – us ATLA [230]
Chi-tu chiao yu she hui chu i yun tung [ccm16] / Chang, Shih-Chang – Hong Kong, 1939 [mf ed 198?] – (= ser Ccm 16) – 1 – mf#1984-b500 – us ATLA [230]
Chi-tu chiao yu shih chieh ho p'ing (ccm223) : hsueh hsi shou ts'e / ed by Liu, Liang-mo – Shanghai, 1951 [mf ed 198?] – (= ser Ccm 223) – 1 – mf#1984-b500 – us ATLA [230]
Chi-tu chiao yu wen hsueh [ccm114] = Christianity and literature / Chu, Wei-chih – Shanghai, 1948 [mf ed 198?] – (= ser Ccm 114) – 1 – mf#1984-b500 – us ATLA [230]
Chi-tu hua ching chi kuan hsi ch'uan kuo ta hui pao kao (ccm253) / National Christian Council of China – Shanghai, 1927 [mf ed 198?] – (= ser Ccm) – 1 – mf#1984-b500 – us ATLA [230]
Chi-tu hua te hun yin [ccm117] : chung-kuo chia t'ing ch'uan p'an chi-tu hua = Christian marriage: christianizing the home week pamphlet... – Shanghai, 1948 [mf ed 198?] – (= ser Ccm 117) – 1 – mf#1984-b500 – us ATLA [230]

Chi-tu t'u chia t'ing [ccm225] = Christian family / Liu, Mei-li – 1st ed. Hong Kong, 1955 [mf ed 198?] – (= ser Ccm 225) – 1 – mf#1984-b500 – us ATLA [240]

Chi-tu t'u chin pu wen ta [ccm344] = Introductory catechism / Yang, Tzu-hung; ed by Price, Philip Francis – [12th ed] Hankou. v2. 1939 [mf ed 198?] – (= ser Ccm 344) – 1 – mf#1984-b500 – us ATLA [240]

Chi-tu tu sheng huo te pei yang (ccm218) / Liang, Hsiao-ch'u – Hong Kong, 1963 [mf ed 198?] – (= ser Ccm 218) – 1 – mf#1984-b500 – us ATLA [240]

Chi-tu tu te hsi wang [ccm330] = Christian hope / Wu, Chen-chun – Shanghai, 1940 [mf ed 198?] – (= ser Ccm 330) – 1 – mf#1984-b500 – us ATLA [240]

Chi-tu tu te hsin yang yu sheng huo [ccm150] = Christian faith and life / Ho, Shih-ming – 1st ed. Hong Kong, 1956 [mf ed 198?] – (= ser Ccm 150) – 1 – (incl bibl ref) – mf#1984-b500 – us ATLA [210]

Chi-tu tu te lien ko wen t'i (ccm203) / Hughes, Ernest Richard – [s.l: s.n, 19–?] [mf ed 198?] – (= ser Ccm 203) – 1 – mf#1984-b500 – us ATLA [240]

Chi-tu t'u ti yen yu [ccm314] = Works [i e words] of my mouth / Wang, Ming-tao – 2nd ed. Singapore, 1957 [mf ed 198?] – (= ser Ccm 314) – 1 – mf#1984-b500 – us ATLA [240]

Chi-tu t'u yu chan shih fu wu [ccm311] = Christians and wartime service / Wang, Hsiang-hsien & Ying, Yuan-t'ao – Shanghai, 1940 [mf ed 198?] – (= ser Ccm 311) – 1 – mf#1984-b500 – us ATLA [951]

Chi-tu tu yu chiu kuo yun tung [ccm278] = Christians and the national salvation movement / Shen, T'i-lan – Shanghai, 1938 [mf ed 198?] – (= ser Ccm 278) – 1 – mf#1984-b500 – us ATLA [240]

Chi-tu yen hsing lu hsin pien [ccm323] = Good news / Wickings, H F – 1st ed. Hong Kong, 1952 [mf ed 198?] – (= ser Ccm 323) – 1 – (chinese trans of the english) – mf#1984-b500 – us ATLA [240]

Chi-tu yu wo ti jen ko [ccm71] = Jesus and my character / Chao, Tzu-ch'en – Shanghai, 1925 [mf ed 198?] – (= ser Ccm 71) – 1 – mf#1984-b500 – us ATLA [230]

Chi-tu-chiao sheng-ho chou-k'an [ccs47] = Christian life weekly – v1 nr17-34. 7 may 1955-10 sep 1955 [complete] [mf ed 198?] – (= ser Chinese christian serials coll 47) – 1 – (no iss publ on jul 2) – mf0296c – us ATLA [240]

Chi-tu-chiao tao-te-kuan yu chung-kuo lun li [ccm329] = Christian morality and traditional chinese ethics / Huang, Hua-chieh – 1st ed. Hong Kong, 1962 [mf ed 198?] – (= ser Ccm 329) – 1 – (incl bibl ref) – mf#1984-b500 – us ATLA [230]

Chi-tu-chiao tsai tai-wan te fa chan [ccm301] = Development of christianity in taiwan / Tong, Hollington Kong – [Taipei: s.n, 1970] [mf ed 198?] – (= ser Ccm 301) – 1 – mf#1984-b500 – us ATLA [240]

Ch'iu, Chih-chung *see* Tu shih she hui shih

Ch'iu, Han-p'ing *see* Hua ch'iao wen t'i

Ch'iu, Jen-hao et al *see* Hu-nan sheng chin jung kai k'uang

Ch'iu, Jih-ch'ing *see* P'iao chu fa yao lun

Ch'iu, Pin-ts'un *see* Kuang-tung pi chih yu chin jung

Chiu shih chu yeh-su chih sheng hsun (ccm133) / Gutzlaff, Karl Friedrich August – Hsin-chia-p'o, 1837 [mf ed 198?] – (= ser Ccm 133) – 1 – mf#1984-b500 – us ATLA [240]

Chiu yueh jen wu [ccm256] = Old testament characters / ed by Nieh, E C – Hong Kong, 1952 [mf ed 198?] – (= ser Ccm 256) – 1 – mf#1984-b500 – us ATLA [221]

Chiu yueh shih (ccm65) = History of the old testament / Chao, Tsung-fu – Che-chiang, 1938 [mf ed 198?] – (= ser Ccm 65) – 1 – mf#1984-b500 – us ATLA [221]

Chiu yueh yen chiu chih nan [ccm216] = Guide to the study of the old testament / Li, Jung-fang – Hong Kong, 1954 [mf ed 198?] – (= ser Ccm 216) – 1 – mf#1984-b500 – us ATLA [221]

Ch'iu,Han-p'ing *see* Ti fang yin hang kai lun

Ch'iu-lang *see* Ma jen ti i shu

Chiume, M W Kanyama *see* Caro ncinonono

Chiurazzi, Luigi *see* Scelta di canzoni popolari in dialetto napolitano

Ch'iu-shih *see* Hen

Ch'iu-yang *see* Tsen yang cheng ch'u tsui hou sheng li

Chivot, Henri *see* Grand mogol

Chiwororo chavakuru / Jackson, S K – Fort Victoria, Zimbabwe. 1962 – 1r – us UF Libraries [960]

Chizhikov, O L *see* Proizvodstvennaia kooperatsiia i elektrifikatsiia selskogo khoziaistva

Chkalovskaia kommuna – Orenburg, 1973 – 4r – 1 – us UMI ProQuest [077]

Chladni, Ernst Florens Friedrich *see*
– Entdeckungen ueber die theorie des klanges
– Ueber feuer-meteore

Chlawson, James William *see* Effect of footing shape on foundation vibrations

Chleni 1-i gosudarstvennoi dumy : biografii, kharakteristiki, politicheskie vzgliady, obshchestvennaia deiatelnost, vybory i prochee – 1906 – 79p 2mf – 9 – mf#RPP-53 – ne IDC [325]

Chlodovech : historischer roman aus der voelkerwanderung / Dahn, Felix – 2. aufl. Leipzig: Breitkopf & Haertel 1895 [mf ed 1979] – (= ser Kleine romane aus der voelkerwanderung 8) – 2r – 1 – (filmed with: bissula, attila, vom chiemgau & stilicho) – mf#film mas c494 – us Harvard [830]

Chlopska droga – Warsaw, Poland. 1981-92 – 12r – 1 – us L of C Photodup [077]

Chloroform exposure and dose determination associated with competitive swimmers during a two-hour swim practice / Berkoff, David C – University of Montana, 1995 – 1mf – 9 – mf#PE 3629 – us Kinesology [617]

Chlumberg, Hans *see* Miracle at verdun

Chm *see* Regeneration

Chmel, Joseph *see* Urkunden, briefe und actenstuecke zur geschichte maximilians 1. und seiner zeit

Chnay tac qalay : pralom lok phnaek manosancetana / Lamn Pen Siak – Bhnam Ben: Qariyadharm [mf ed 1990] – 1r with other items – 1 – (in khmer) – mf#mf-10289 seam reel 110/3 [§] – us CRL [830]

Cho, Ho S *see* Factors related to fasting behavior among american adults

Cho, Kwang M et al *see* Attitudes of korean national athletes and coaches toward athletics participation

Cho lon [Vietnam] *see* Budget...

Cho, Won-Kyung *see* Dances of korea

Chobham, Thomas de *see*
– Sermones
– Summa de arte praedicandi

Choboy, Jon A *see* Influence of mental imagery on tennis service accuracy of intermediate level tennis players

Choc / Laleau, Leon – Port-Au-Prince, Haiti. 1932 – 1r – us UF Libraries [972]

Chocarne, pere (Bernard) *see* The inner life of the very reverend pere lacordaire of the order of preachers

Choco en la independencia de colombia / Velasquez Rogerio – Bogota, Colombia. 1965 – 1r – us UF Libraries [972]

Choctaw baptist hymnal – Orig. and trans. hymns – 1 – 5.00 – us Southern Baptist [242]

Choctaw Community Action Agency *see* Choctaw community news

Choctaw community news / Choctaw Community Action Agency – 1980 apr 30-1984 dec – 1r – 1 – mf#801115 – us WHS [360]

Choctaw Nation *see* Bishinik

Chodera, Jan *see* Die deutsche polenliteratur 1918-193?

Choderlos de Laclos, P A F *see* De l'education des femmes

Chodorov, Edward *see* Oh, men! oh, women!

Choffletti, Caryn E *see* The effects of exercise on weight loss, fat loss and circumference changes

Choh lin : the chinese boy who became a preacher / Davis, John A – Philadelphia: Presbyterian Board of Publ & Sabbath-School Work, 1901, c1884 [mf ed 1986] – 1mf – 9 – 0-8370-7209-3 – mf#1986-1209 – us ATLA [920]

Chohatsu bukken ichiranhyo : requisitioned materials list in the meiji era / Special Staff Office of the Japanese Army [comp] – 1875-1911 – 75v on 24r – 1 – Y220,000 – (in japanese) – ja Yushodo [315]

Choi, Monica W *see* Choosing whether or not to use hormone replacement therapy during the menopausal transition

Choice – Middletown. 1964+ (1) 1972+ (5) 1975+ (9) – ISSN: 0009-4978 – mf#6770 – us UMI ProQuest [070]

Choice / Sojourner Truth House [Milwaukee WI] – 1979 apr-1984 dec – 1r – 1 – (cont by: shelter [milwaukee wi]) – mf#940357 – us WHS [243]

Choice before south africa / Sachs, Emil Solomon – New York, NY. 1952 – 1r – us UF Libraries [960]

Choice examples of art workmanship : selected from the exhibition of mediaeval art / Delamotte, Philip Henry – London 1851 – (= ser 19th c art & architecture) – 2mf – 9 – mf#4.2.1662 – uk Chadwyck [700]

Choice examples of wedgewood art : a selection of plaques, cameos, medallions, vases, etc, from the designs of flaxman and others / Meteyard, Eliza – London: George Bell & Sons, 1878 – (= ser 19th c art & architecture) – 2mf – 9 – mf#4.1.136 – uk Chadwyck [730]

"Choice fragments" : being a collection of wise and witty sayings of celebrated men; anecdotes, conundrums, poetry etc – [Montreal?: s.n.] 1866? [mf ed 1984]) – 2mf – 9 – 0-665-41451-X – mf#41451 – cn Canadiana [880]

Choice recipes : how to use fleischmann's compressed yeast / Kirk, Eleanor [comp] – New York: C Jourgensen, 1889 [mf ed 1984] – 1mf – 9 – 0-665-01164-4 – mf#01164 – cn Canadiana [640]

Choice selections in prose and poetry – Quebec: J Walsh, 1891 [mf ed 1980] – 1mf – 9 – 0-665-00630-6 – mf#00630 – cn Canadiana [800]

Choir of the future / Fraser, Duncan – Edinburgh, Scotland. 1896 – 1r – us UF Libraries [240]

Choir office-book : the daily and occasional offices and the order of holy communion set to anglican and plain-song music, as used in trinity church, new york / ed by Messiter, Arthur Henry – New York: E & JB Young, 1894 – 3mf – 9 – 0-524-08803-9 – mf#1993-3295 – us ATLA [780]

Choiseul, Claude Antoine Gabriel de *see* Relation du depart de louis 16, le 20 juin 1791

Choiseul et la france d'outre-mer apres le traite de paris : etude sur la politique coloniale au 18e siecle, avec un appendice sur les origines de la question de terre-neuve / Daubigny, Eugene – Paris: Hachette, 1892 [mf ed 1980] – 4mf – 9 – 0-665-02508-4 – mf#02508 – cn Canadiana [944]

Choiseul, Etienne F de *see* Maemoire historique sur la naegociation de la france & de l'angleterre

Choiseul, Etienne Francois, duc de *see* Memoire historique sur la negociation de la france et de l'angleterre

Choiseul, G F A *see* Voyage pittoresque de la grece

Choisy, Eugene *see*
– L'etat chretien calviniste a geneve
– La theocratie a geneve au temps de calvin

Choix de chansons / Marius-Anselme, frere [comp] – 3e ed, 14e mille. Montreal: [les freres des ecoles chretiennes], 1916 [mf ed 1992] – 3mf – 9 – (with ind) – mf#SEM105P1679 – cn Bibl Nat [780]

Choix de chansons / Marius-Anselme, frere [comp] – 5e ed, 23e mille. Montreal: les freres des escoles chretiennes, 1936 [mf ed 1992] – 3mf – 9 – (with ind) – mf#SEM105P1680 – cn Bibl Nat [780]

Choix de chansons et poesies wallonnes – Liege: F Oudart 1844 [mf ed Bloomington IN: Indiana Uni Lib, Preservation Dept 1984] – xxii 220p on 1r – 1 – us Indiana Preservation [440]

Choix de contes populaires de la haut-bretagne / Sebillot, Paul – New York, NY. 1909 – 1r – us UF Libraries [944]

Choix de rapports, opinions et discourssupplement au premier volume / ed by Lallement, Guillaume – Paris 1818-25 [mf ed Hildesheim 1995-98] – 73mf – 9 – €730.00 – 3-487-26173-1 – gw Olms [944]

Choix de textes religieux assyro-babyloniens – Paris: J Gabalda 1907 [mf ed 1989] – (= ser Etudes bibliques) – 2mf – 9 – 0-7905-1931-3 – (transcr, trans & comm by paul dhorme; texts in french & akkadian; comm in french, akkadian & greek; incl bibl ref & ind) – mf#1987-1931 – us ATLA [470]

Choix d'eglises bysantines en grece / Couchaud, A – Paris, 1842 – 2mf – 9 – mf#OA-129 – ne IDC [720]

Choix des vaches laitieres : economie dans leur alimentation / Chapais, Jean-Charles – Montreal: Herald Pub Co, 1898 [mf ed 1985] – 1mf – 9 – 0-665-10432-4 – mf#10432 – cn Canadiana [630]

Choix des vaches laitieres d'apres le systeme guenon / Couture, Joseph Alphonse – Quebec: impr Leger Brousseau, 1884 [mf ed 1980] – 2mf – 9 – 0-665-05271-5 – mf#05271 – cn Canadiana [636]

Le choix, le monde, l'existence / Wahl, J A et al – Paris, 1948 – 4mf – 8 – €11.00 – ne Slangenburg [120]

Chojecki, Charles *see* Voyage dans les mers du nord a bord de la corvette la reine hortense

Die chokma (sophia) in der juedischen hypostasenspekulation : ein beitrag zur geschichte der religioesen ideen im zeitalter des hellenismus / Schencke, Wilhelm – Kristiania: In Kommission bei J Dybwad, 1913 – (= ser Videnskapsselskapets Skrifter) – 1mf – 9 – 0-524-08055-0 – mf#1991-0271 – us ATLA [270]

Choksey, R D *see* The last phase

Choksey, Rustom Dinshaw *see*
– The aftermath
– Economic history of the bombay, deccan, and karnatak, 1818-1868 170=foreword by dr gadgil
– A history of british diplomacy at the court of the peshwas, 1786-1818

Chokwe grammar / White, C M N – s.l, s.l? 19–? – 1r – us UF Libraries [470]

Cholera : and its consequences – London, England. 1832? – 1r – us UF Libraries [614]

Cholera : le cholera, le regime sanitaire du pays, mesures d'hygiene individuelle destines a preserver du cholera sic... / Desroches, Joseph Israel – S.l: s.n, 1885? – 1mf – 9 – mf#63036 – cn Canadiana [614]

Le cholera : son historique, son origine, sa nature, les causes qui le produisent, ainsi que celles des autres maladies epidémiques ou contagieuses produites par les microbes... / Crevier, Joseph Alexandre – Montreal: impr generale...1885 [mf ed 1980] – 1mf – 9 – mf#SEM105P48 – cn Bibl Nat [616]

Le cholera, comment le prevenir et le combattre : conseils pratiques aux familles publies par le conseil provincial d'hygiene (province de quebec) – Montreal: Conseil provincial d'hygiene, [18–?] [mf ed 1985] – 1mf – 9 – 0-665-01775-8 – mf#01775 – cn Canadiana [614]

Le cholera electoral : de profundis proudhonien – Paris [1848?] – us CRL [944]

Cholera epidemics in east africa. : an account of the several diffusions of the disease in that country from 1821 till 1872 / Christie, James – London: Macmillan, 1876 – 1 – us CRL [610]

Cholet, [A P] de *see* Voyage en turquie d'asie

Cholet, Louis F de *see* Madame, nantes, blaye, paris

Cholmondeley, Charles *see* The protestant doctrine of justification and scheme of salvation

Chome, Jules *see* Moise tshombe et l'escroquerie katangaise

Chomedey de maisonneuve : drame chretien en trois actes: samuel de champlain, pages oratoires: trois aureoles / Corbeil, Sylvio – Montreal: Cadieux & Derome, 1899 [mf ed 1979] – 2mf – 9 – 0-665-00233-5 – mf#00233 – cn Canadiana [820]

Chomel, Auguste *see* Histoire du sergent flavigny

Chomsky, Noam *see* Two essays on cambodia

Chomton, Werner *see* Heinrich der loewe

Chon, K S (Kaye) *see* Journal of travel and tourism marketing

Chonbilal ch'ulelal / Gomez Takiwah, Mariano – Chicago: University of Chicago Library, 1977 (mf ed) – (= ser Microfilm coll of manuscripts on cultural anthropology) – (ein gebetstext auf maya-tzotzil, p44-96, uebersetzt und erlaeutert von ulrich koehler) – us Chicago U Pr [490]

Chones, Isaac Bear *see* Sefer 'orekh ha-milim veha-pitronim

Chones, Simon Moses *see*
– Toldot (ha-geonim) ha-poskim
– Toldot ha-poskim

Chong, Kwong Y R *see* [Ready], set, go!

Choosing whether or not to use hormone replacement therapy during the menopausal transition : a qualitative study / Choi, Monica W & Ruzek, Sheryl – 1993 – 3mf – 9 – $12.00 – us Kinesology [613]

Chop, Max *see* Vademecum fuer wagnerfreunde

Chopra, Gulshan Lall *see* The panjab as a sovereign state, 1799-1839

Chopra, I C *see* A review of work on indian medicinal plants

Chopra, R N *see* A review of work on indian medicinal plants

Chopra, Ram Nath *see* Indigenous drugs inquiry

Choquette, Charles Philippe *see*
– 12e congres international de geologie
– Premier congres des universites de l'empire anglais (juillet 1912)

Choquette, Ernest *see*
– Carabinades
– La terre

Der choral bei johann sebastian bach / Meyer, Juergen – (mf ed 1995) – 1mf – 9 – €30.00 – 3-8267-2224-8 – mf#DHS-AR 2224 – gw Frankfurter [780]

Choral book of the ephrata cloister – Manuscript, ca.1745. MUSIC 1154, Reel 1 – 1 – us L of C Photodup [780]

Choral harmonie. enthaltend kirchen-melodien / Gerhart, Isaac – 1822. with Pennsylvanische Sammlung von Kirchen-Musik. 1840 – 1 – $50.00 – us Presbyterian [780]

Choral journal – Lawton. 1959+ (1,5,9) – ISSN: 0009-5028 – mf#11450 – us UMI ProQuest [780]

The choral journal / American Choral Directors Association – v1-10. 1959-70 – 1 – us AMS Press [780]

Choral-system / Vogler, Georg Joseph – Offenbach a M: J Andre [ca 1800] [mf ed 1982] – 2v in 1 on 1r – 1 – mf#10121 – us Wisconsin U Libr [780]

Chord : a quarterly devoted to music – London. 1899-1900 (1) – mf#5278 – us UMI ProQuest [780]

Choreographer's journey into the world of dr seuss [giesel, theodor seuss] / Stoddard, Lisa – 1996 – 1mf – 9 – $4.00 – mf#PE 3843 – us Kinesology [790]

Choreographing as teaching/teaching as choreographing : dancing and dialoguing with mark taylor / Batemann, Joylyn – 2000 – 26p on 1mf – 9 – $5.00 – mf#PE 4124 – us Kinesology [790]

The choreography and performance of a japanese folk tale / Ito, Sayuri – Brigham Young University, 1993 – 1mf – 9 – $4.00 – mf#PE3602 – us Kinesology [790]

Die chorfuge in haendels werken / Wieber, Georg-Friedrich – Frankfurt a.M., 1958 – 3mf – 3-89349-847-8 – gw Frankfurter [780]

Chorister / Boys Choir of Harlem, Inc – 1980 fall – 1r – 1 – mf#5296675 – us WHS [780]
Chorley and district times – [NW England] Chorley Lib 1942-29 may 1959 – 1 – uk MLA; uk Newsplan [072]
Chorley echo – [NW England] Chorley Lib 24 mar 1883-12 jan 1884* – 1 – (missing: 5 may, 4 aug, 18 aug 1883) – uk MLA; uk Newsplan [072]
Chorley guardian and leyland hundred advertiser – [NW England] Chorley Lib 4 nov 1871-12 oct 1935 – 1 – (title change: Chorley Guardian & Leyland Advertiser [19 oct 1935-25 feb 1966]; Chorley Guardian [4 mar 1966]) – uk MLA; uk Newsplan [072]
Chorley standard and district advertiser – [NW England] Chorley 3 sep 1864-1 feb 1908* [wkly] – 1 – (title change: chorley & district weekly news [8 feb 1908-16 feb 1918]; chorley weekly news [23 feb 1918-1931; chorley news [2 mar-26 oct 1948]) – uk MLA; uk Newsplan [072]
Chorley weasel – [NW England] Chorley Lib 26 mar 1881-17 mar 1883 – 1 – uk MLA; uk Newsplan [072]
Chorografia / Barrientos, Gaspar – 1561 – 9 – sp Bibl Santa Ana [946]
Chorokodza / Basset, Bernard – Gwelo, Zimbabwe. 1964 – 1r – us UF Libraries [960]
Chorti indians of guatemala / Wisdom, Charles – Chicago, IL. 1940 – 1r – us UF Libraries [307]
Chortovo koleso / Bukhov, Arkadii Sergeevich – [Petrograd]: Novyi Satirikon 19– [mf ed 1980] – 1r – 1 – us UW Libraries [460]
Chorus girl and other stories / Chekhov, Anton Pavlovich – New York, NY. 1921 (c1920) – 1r – us UF Libraries [830]
Chorus lady / Forbes, James – New York, NY. 1908 – 1r – us UF Libraries [025]
The chosen friend – Toronto: Chosen Friend Print and Pub Co, [1892-189- or 19–] – 9 – mf#P04601 – cn Canadiana [350]
Chosen sotokufu / Korea. (Government-General of Chosen, 1910-45) – Official Gazette. Seoul. On film: v1-5521; 1910-45 (incomplete). LL-02008 – 1 – us L of C Photodup [340]
Chosenshi genpon collection : imanishi collection of original texts on korean history. in the holdings of the tenri central library, nara prefecture – 1532pts on 154r – 1 – Y950,000 – (in korean) – ja Yushodo [950]
Choses d'autrefois : feuilles eparses / Gagnon, Ernest – [Quebec?: Dussault & Proulx], 1905 – 4mf – 9 – 0-665-71191-3 – mf#71191 – cn Canadiana [880]
Choses d'autrefois : feuilles eparses / Gagnon, Ernest – Quebec: Typ Dussault & Proulx, 1905 [mf ed 1992] – 4mf – 9 – (with ind) – mf#SEM105P1610 – cn Bibl Nat [971]
Choses d'haiti / Audain, Leon – Port-Au-Prince, Haiti. 1916 – 1r – us UF Libraries [972]
Choses du pays : melanges, litterature, histoire – Montreal: Bibliotheque de l'Action francaise, [1921?] (mf ed 1990) – 2mf – 9 – (ill by j mcisaac; with ind) – mf#SEM105P1290 – cn Bibl Nat [440]
Choses passees / Loisy, Alfred Firmin – Paris: Emile Nourry, 1913 – 1mf – 9 – 0-7905-1224-6 – mf#1987-1224 – us ATLA [220]
Choses vues / Danache, Berthomieux – Port-Au-Prince, Haiti. 1939 – 1r – us UF Libraries [972]
Chossat, Ed de *see* Repertoire sumerien (accadien)
Chosun ilbo – 1920- – 1 – enquire for prices – (yrly reel count varies) – us UMI ProQuest [070]
Chota nagpore : a little-known province of the empire / Bradley-Birt, Francis Bradley – London: Smith, Elder, 1903 [mf ed 1995] – (= ser Yale coll) – xiv/310p (ill) – 1 – 0-524-09749-6 – (int by earl of northbrook) – mf#1995-0749 – us ATLA [915]
Choteau county independent – Fort Benton, MT. 1910-1916 (1) – mf#64387 – us UMI ProQuest [071]
Chotscho : facsimile-wiedergaben der wichtigeren funde der ersten koenigl preuss expedition nach turfan in ost-turkistan / Le Coq, A von – Berlin, 1913 – 25mf – 9 – mf#U-624 – ne IDC [915]
Chou, An-kuo *see* Pei ya p'o min tsu chan cheng lun
Chou, Chen-fu *see* Yen fu ssu hsiang shu p'ing
Chou, Chien-ch'en *see* Hsi-t'ai-hou
Chou, Chih-ying *see* Hsin ch'eng-tu
Chou, Ching-wen *see* Min chu chu i ti tou cheng
Chou, Ch'ing-yun *see* Tuan ch'i i wu hsiao hsueh shih shih fa
Chou, Ch'uan-p'ing *see* Lou t'ou ti fan nao
Chou, Chung-jen *see* Chai kung ch'ien shuo
Chou, En-lai *see* K'ang chan cheng chih kung tso kang ling
Chou, Erh-fu *see*
– Ti shih san li tzu tan
– Tzu ti ping ti i chi
– Yang ko chu ch'u chi
– Yeh hsing chi
Chou, Fan et al *see* Tsui ch'u ti mi
Chou, Fo-hai *see* Wang i chi
Chou, Hsien-wen *see* Hsin nung pen chu i p'i p'an
Chou, Hsin-ming *see* T'ai-p'ing yang ti liang an
Chou, I-fu *see* Chi-tu chiao yu chung-kuo [ccm106]
Chou, I-pai *see*
– Chin ssu ch'ueeh
– Li hsiang-chun
– Lien huan chi
– Lu ch'uang hung lei
Chou, I-wu *see*
– Jih su kuan hsi lun
– Tsui chin chih ying jih wai chiao
Chou, Keng-sheng *see*
– Chan shih wai chiao wen t'i
– Hsien tai kuo chi fa wen t'i
– Kuo chi fa ta kang
Chou king : texte chinois avec traduction / Couvreur, Seraphin – 2. ed. Hien Hien: Imprimerie de la Mission Catholique, 1916 – 1mf – 9 – 0-524-07726-6 – mf#1991-0148 – us ATLA [930]
Chou, Leng-ch'ieh *see*
– Feng feng yu yu
– Han tsai, chou leng-ch'ieh chu
– T'ien yuan chi
– Yu lin
– Yueh ch'iu lu hsing chi
Chou, Li-an *see* Hua fa chi
Chou, Liang-ts'ai *see* Mao i fa ling chang tse hui pien
Chou, Li-po *see*
– Chan ti jih chi
– Chin ch'a chi pien ch'ue yin hsiang chi
Chou, Li-Shan *see* Measurement and predictions of obstructed und unobstructed gait
Chou, Meng-tieh *see* Hsiao hsiao lu
Chou, Mu-chai *see* Hsiao ch'ang chi
Chou nien chiang tan [ccm303] = Sermons for the year / Walter, E – 1st ed. Shanghai, 1931 [mf ed 198?] – (= ser Ccm 303) – 1 – (chinese trans of the english) – mf#1984-b500 – us ATLA [240]
Chou, Pai-ch'in *see* K'ung hsiang chu i (ccm105)
Chou, Po-ti *see* Huo pi yu chin jung 2
Chou, Ta-kuan *see* Memoires sur les coutumes du cambodge
Chou, Ti-ch'in *see* Wo men ti ch'ih ju
Chou Tso-jen *see* K'u ch'a an hsiao hua hsuan
Chou, Tso-jen *see*
– Chih t'ang wen chi
– Erh t'ung wen hsueh hsiao lun
– I shu yu sheng huo
– K'an yun chi
– K'u ch'a sui pi
– K'u chu tsa chi
– K'u k'ou kan k'ou
– K'u yu chai hsu pa wen
– Kua tou chi
– Kuo ch'u ti sheng ming
– Ping chu hou t'an
– Ping chu t'an
– Shu fang i chiao
– Yao t'ang tsa wen
– Yeh tu ch'ao
– Yu t'ien ti shu
Chou, Tzu-ya *see* Wai chiao wen shu yu wai chiao li chieh
Chou, Wei *see* Ya-chou ku ping ch'i yu wen hua i shu chih kuan hsi
Chou, Wen *see*
– Tsai pai-sen chen
– Yen miao chi
– Yen miao chi hou pu
Chou, Yang-wen *see* Kuo nei hui tui chi ya hui yeh wu
Chou, Yeh-sun *see* Nan feng
Chou, Yen *see*
– T'ao hua shan, chi, mo ling feng yu
– Wo men shih hsi chu ti t'ieh chun
Chou, Ying *see* Tsung ts'ai chiang shu ta hsueh chung yung ching i
Chou, Yin-hsin et al *see* She shen ch'u i
Chou, Yueh-jan *see* Liu shih hui i
Chou, Yu-t'ung *see* Wu shih nien lai chung-kuo chih hsin shih hsueh
Choublier, M *see* La question d'orient depuis le traite de berlin
Chouinard, Edouard Pierre *see* Histoire de la paroisse de saint-joseph de carleton (baie des chaleurs) 1755-1906
Chouinard, Francois-Xavier *see* Quebec, the historic city
Chouinard, Honore Julien Jean Baptiste *see* Troisieme centenaire de la fondation de quebec, berceau du canada, par champlain, 1608-1908
Chouinard, Honore Julien Jean Baptiste [comp] *see* Annales de la societe st-jean-baptiste de quebec
Chouinard, Honore-Julien-Jean-Baptiste *see* Paul de chomedey, sieur de maisonneuve, fondateur de montreal
Chouinard, Mathias *see* Code de l'instruction publique de la province de quebec
Chouman, Joumana *see* Les relations publiques au liban, entre le secteur public et prive, de 1982 a 1988
Choun-Nath, Preas Krou Samsattha *see* Vinaya patimokkha samvarasila samkhepa et khandhaka vinaya samkhepa
Choun-Nath, Preas Krou Samsattha et al *see* Samanera-vinaya
Chourbaji, Ranja *see* Der palaestina-konflikt
Choussy, Felix *see* Actual panorama economico agricola de el salvador
Chow, B C *see* Comparisons of domain score and reliability estimates using trials-tocriterion, sequential probability ratio, and pre-set trial length tests
Chow, Franklin Hon-Ching *see* Colonization of neoaplectana dutkyi jackson
Chow, Sergio et al *see* Die taetigkeit des kindes in der spieltherapie
Chowan baptist church. chowan association. north carolina : church records – 1884-90 – 1 – 6.48 – us Southern Baptist [242]
[Chowchilla-] chowchilla news – CA. 1913-81; 1987- – 29+ r – 1 – $1740.00 (subs $50/y) – mf#RC02110 – us Library Micro [071]
[Chowchilla-] chowchilla today – CA. apr-dec 1980; 1985-86 – 4r – 1 – $240.00 – mf#B02111 – us Library Micro [071]
Chowdhury, R *see* Mahatma gandhi and india's struggle for swaraj
Chowdowski, Salomo *see* Kritik des midrash schir-haschirim
Choy, Valerie E *see* The effect of exogenous recombinant porcine somatotropin on pig common calcanean tendon biochemistry
Cho-ying *see* Ying ch'un
Chr. jac. bostroems foerelaesningar i religionsfilosofi = Foerelaesningar i religionsfilosofi / Bostroem, Christopher Jacob; ed by Ribbing, Sigurd – Stockholm: Norstedt, 1885 – 1mf – 9 – 0-7905-3641-2 – mf#1989-0134 – us ATLA [200]
Chranitel = Guardian – [Johnstown, PA: St Mary's Greek Catholic Congregation. v1 n1-v1 n9/10. jul 1920-mar/apr 1921 – 1 – us CRL [241]
Chrehan, J *see* Early christian baptism and the creed
Chreitzberg, Abel McKee *see* Early methodism in the carolinas
Chrestiennes meditations... / Beza, Theodor de – [Geneve, Laimarie], 1583 – 2mf – 9 – mf#PFA-107 – ne IDC [240]
Chrestomatheia ekklesiastikes mousikes : periechousa pan oti anagkaion to ieropsalte, kai egcheiridion pros didaskalian / Sakellarides, Ioannes Th – En Athenais: Ek ... Philadelpheos, 1880. Chicago: Dep of Photodup, U of Chicago Lib, 1978 (1r); Evanston: American Theol Lib Assoc, 1984 (1r) – 1 – 0-8370-0699-6 – mf#1984-T091 – us ATLA [780]
Chrestomathia aethiopica edita et glossario explanata / Dillmann, A – Lipsiae, 1866 – 4mf – 9 – mf#NE-20245 – ne IDC [956]
Chrestomathia aethiopica edita et glossario explanata / Dillmann, August – Lipsiae [Leipzig]: T O Weigel, 1866 – 1mf – 9 – 0-8370-2915-5 – (includes an ethiopic-latin glossary) – mf#1985-0915 – us ATLA [470]
Chrestomathia syriaca : quam glossario et tabulis grammaticis – Halis Saxonum: Sumptibus Orphanotrophei, 1868 – 1mf – 9 – 0-8370-8614-0 – (texts and glossary in syriac; discussion in latin) – mf#1986-2614 – us ATLA [470]
Chrestomathia targumica : quam collatis libris manu scriptis antiquissimis tiberiensibus editionibusque impressis celeberrimis / ed by Merx, Adalbert – Berlin: H Reuther; New York: B Westermann 1888 [mf ed 1986] – (= ser Porta linguarum orientalium 8) – 1mf – 9 – 0-8370-7170-4 – mf#1986-1170 – us ATLA [470]
Chrestomathia targumico-chaldaica : addito lexico / Kaerle, Joseph – Viennae: Typis Caes Reg Aulae et Imperii Typographiae, 1852 – 1mf – 9 – 0-8370-7550-5 – mf#1986-1550 – us ATLA [470]
Chrestomathie en turk oriental : contenant plusieurs ouvrages de l'emir ali-schir, des extraits des memoires du sultan baber, du traite du miradje, du tezkiret-el-avlia et du bakhtiar-nameh / Quatremere, M – Paris, 1841 – 3mf – 8 – mf#U-369 – ne IDC [956]
Chrestomathie mandchou : ou recueil de textes mandchou: destine aux personnes qui veulent s'occuper de l'etude de cette langue / Klaproth, Heinrich Julius – Paris: Impr royale 1828 – (= ser Whsb) – 3mf – 9 – €40.00 – mf#Hu 303 – gw Fischer [480]
Chrestos, a religious epithet : its import and influence / Mitchell, James Barr – London: Williams & Norgate, 1880 [mf ed 1992] – 1mf – 9 – 0-524-05233-6 – mf#1992-0366 – us ATLA [450]
Chretien, De Troyes *see* Arthurian romances
Les chretiens dans l'empire romain : de la fin des antonins au milieu du 3e siecle (180-249) / Aube, Benjamin – 2e ed. Paris: Didier, 1881 [mf ed 1990] – 2mf – 9 – 0-7905-7101-3 – (in french. incl bibl ref) – mf#1988-3101 – us ATLA [240]
Les chretiens et l'empire romain a l'epoque du nouveau testament / Goguel, Maurice – Paris: Fischbacher, 1908 [mf ed 1990] – 1mf – 9 – 0-7905-6290-1 – (in french. incl bibl ref) – mf#1988-2290 – us ATLA [230]
Chretiens et musulmans : voyages et etudes / Contenson, L de – Paris, 1901 – 4mf – 9 – mf#AR-1587 – ne IDC [910]
La chretiente africaine de dakar : partie descriptive et statistique / Martin, V – Dakar, Fraternite Saint-Dominique 1964 – us CRL [960]
La chretiente romaine, 1198-1274 (he10) – Paris, 1950 – (= ser Histoire de l'eglise (he)) – €25.00 – ne Slangenburg [241]
Les chretientes celtiques / Gougaud, Louis – Paris: Lecoffre, 1911 – (= ser Bibliotheque de l'enseignement de l'histoire ecclesiastique) – 2mf – 9 – 0-7905-5887-4 – (incl bibl ref) – mf#1988-1887 – us ATLA [240]
Chrichton, David *see* Family
Chrischona blaettchen – 1934-60 [complete] – 1r – 1 – (filmed with: chrischona gruesse) – mf#atla s0718a – us ATLA [242]
Chrischona blaettchen *see* Chrischona gruesse
Chrischona gruesse – 1961-72 [complete] – 1r – 1 – (filmed with chrischona blaettchen) – mf#atla s0718b – us ATLA [242]
Chrischona gruesse *see* Chrischona blaettchen
Chriscoe, Stephen B *see* The validity of the tecumseh self-administered occupational activity questionnaire
Chrisphonte, Prosper *see* Deuxieme these de doctorat
Christ : the bread of life – London, England. no date – 1r – us UF Libraries [240]
Christ : the glory of israel / Goode, F – London, England. 1835? – 1r – us UF Libraries [240]
Christ : the light of the world / Melvill, Henry – London, England. 1852 – 1r – us UF Libraries [240]
Christ : the savior of society / Kaufmann, M – London, England. 1895 – 1r – us UF Libraries [240]
The christ / Brookes, James Hall – New York: Fleming H Revell, c1893 – 1mf – 9 – 0-524-04792-8 – mf#1992-0212 – us ATLA [220]
Christ, a home missionary / Williams, William R – (A discourse). 1836 – 1 – 5.00 – us Southern Baptist [242]
Christ, Ad *see* William carey und seine mitarbeiter
Christ and antichrist : or, jesus of nazareth proved to be the messiah and the papacy proved to be the antichrist predicted in the holy scriptures / Cassels, Samuel Jones – Philadelphia: Presbyterian Board of Publ, c1846 [mf ed 1991] – 1mf – 9 – 0-524-01103-6 – mf#1990-0317 – us ATLA [240]
Christ and buddha / Cushing, Josiah Nelson – Philadelphia: American Baptist Publ Society, 1907 – 1mf – 9 – 0-524-00826-4 – mf#1990-2072 – us ATLA [230]
Christ and casar : or, the cardross case viewed in the light of god'... / Buchanan, Robert – Glasgow, Scotland. 1860? – 1r – us UF Libraries [240]
Christ and christendom : the boyle lectures for the year 1866 / Plumptre, Edward Hayes – London: Alexander Strahan, 1867 – 1mf – 9 – 0-8370-5480-X – mf#1985-3480 – us ATLA [240]
Christ and christian life : sermons preached in zion church, brantford, 1875 / Cochrane, William – Toronto: Adam, Stevenson; Brantford Ont: J Sutherland, 1876 – 4mf – 9 – mf#08595 – cn Canadiana [240]
Christ and christianity : studies on christology, creeds and confessions, protestantism and romanism, reformation principles, sunday observance, religious freedom, and christian union / Schaff, Philip – New York: Charles Scribner's Sons, 1885. Chicago: Dep of Photodup, U of Chicago Lib, 1979 (1r); Evanston: American Theol Lib Assoc, 1984 (1r) – 1 – 0-8370-1336-4 – mf#1984-T183 – us ATLA [240]
Christ and christianity : a vindication of the divine authority of the christian religion, grounded on the historical verity of the life of christ / Alexander, William Lindsay – Edinburgh: Adam & Charles Black, 1854 [mf ed 1989] – 1mf – 9 – 0-7905-0900-8 – (incl bibl ref & ind) – mf#1987-0900 – us ATLA [225]
Christ and civilization : a survey of the influence of the christian religion upon the course of civilization / Bennett, William Henry et al; ed by Paton, John Brown et al – London: National Council of Evangelical Free Churches, 1910 – 2mf – 9 – 0-524-05819-9 – mf#1992-0646 – us ATLA [240]
Christ and economics : in the light of the sermon on the mount / Stubbs, Charles William – London: Isbister, 1893 – 1mf – 9 – 0-7905-9694-6 – (incl bibl ref) – mf#1989-1419 – us ATLA [240]
Christ and his apostles : or the critic, which? / Bull, Bartle E – [Toronto?: s.n, 1910?] – 1mf – 9 – 0-665-88001-4 – mf#88001 – cn Canadiana [230]

The christ and his church : some occasional, special, and other sermons / Seiss, Joseph Augustus – Philadelphia: Board of Publ of the General Council, 1902 – 1mf – 9 – 0-7905-2068-0 – mf#1987-2068 – us ATLA [220]
Christ and his church in the book of psalms / Bonar, Andrew Alexander – New York: Robert Carter 1860 [mf ed 1989] – 2mf – 9 – 0-7905-3011-2 – mf#1987-3011 – us ATLA [221]
Christ and his critics : studies in the person and problems of jesus / Hitchcock, Francis Ryan Montgomery – London: Robert Scott, 1910 – 1mf – 9 – 0-7905-1105-3 – mf#1987-1105 – us ATLA [240]
Christ and his religion / Reid, John – New York: Wilbur B Ketcham [c1880] [mf ed 1984] – 4mf – 9 – 0-8370-0813-1 – mf#1984-4175 – us ATLA [240]
Christ and his times / Benson, Edward White – London; New York: Macmillan, 1889 – 1mf – 9 – 0-7905-0862-1 – (incl bibl ref) – mf#1987-0862 – us ATLA [306]
Christ and human life : lectures...jan 1901 / Stone, Darwell – London: Longmans, Green, 1901 [mf ed 1991] – 1mf – 9 – 0-7905-8593-6 – (incl bibl ref) – mf#1989-1818 – us ATLA [240]
Christ and humanity : with a review, historical and critical, of the doctrine of christ's person / Goodwin, Henry Martyn – New York: Harper, 1875 – 1mf – 9 – 0-8370-4910-5 – (incl bibl ref) – mf#1985-2910 – us ATLA [240]
Christ and life / Speer, Robert Elliott – New York: Fleming H Revell, c1901 [mf ed 1988] – 1mf – 9 – 0-7905-0346-8 – mf#1987-0346 – us ATLA [240]
Christ and man : sermons / Dods, Marcus – New York: Hodder & Stoughton, [1909?] – 1mf – 9 – 0-7905-1513-X – mf#1987-1513 – us ATLA [240]
Christ and modern thought : with a preliminary lecture, on the methods of meeting modern unbelief / Cook, Joseph et al – Boston: Roberts 1881 [mf ed 1985] – (= ser Boston monday lectures) – 1mf – 9 – 0-8370-2325-4 – mf#1985-0325 – us ATLA [240]
Christ and modern unbelief / McKim, Randolph Harrison – New York: Thomas Whittaker, 1893 [mf ed 1985] – 1mf – 9 – 0-8370-4359-X – (incl bibl ref) – mf#1985-2359 – us ATLA [240]
Christ and other masters : an historical inquiry into some of the chief parallelisms and contrasts between christianity and the religious systems of the ancient world / Hardwick, Charles; ed by Procter, Francis – 3rd ed. London: Macmillan, 1874 – 2mf – 9 – 0-7905-4967-0 – (incl bibl ref) – mf#1988-0967 – us ATLA [230]
Christ and peace : a discussion of some fundamental issues raised by the war / Heath, J St George et al; ed by Fry, Joan Mary – London: Headley Bros, [1915?] – 1mf – 9 – 0-524-03899-6 – (incl bibl ref) – mf#1990-1158 – us ATLA [240]
Christ and society / Macleod, Donald – London: Isbister, 1892 – 1mf – 9 – 0-8370-9965-X – mf#1986-3965 – us ATLA [240]
Christ and the bible : four lectures / Leathes, Stanley – London: SW Partridge [c1885] [mf ed 1985] – 1mf – 9 – 0-8370-4068-X – mf#1985-2068 – us ATLA [220]
Christ and the cherubim : or the ark of the covenant a type of christ our saviour / Otts, John Martin Philip – Richmond, VA: Presbyterian Cttee of Publ, c1896 [mf ed 1985] – 1mf – 9 – 0-8370-4650-5 – mf#1985-2650 – us ATLA [240]
Christ and the church : essays concerning the church and the unification of christendom / Bradford, Amory Howe – New York: Fleming H Revell c1895 [mf ed 1985] – 1mf – 9 – 0-8370-3315-2 – (int by aut) – mf#1985-1315 – us ATLA [240]
Christ and the controversies of christendom / Dale, Robert William – New-York: T Whittaker, [1869?] – 1mf – 9 – 0-524-02521-5 – mf#1990-0621 – us ATLA [240]
Christ and the covenant : francis turretin's federal theology as a defense of the doctrine of grace / Beach, James Mark – [mf ed 2006] – 1r [complete] – 1 – 0-524-10552-9 – (incl bibl ref) – mf#d00013 – us ATLA [240]
Christ and the dramas of doubt : studies in the problem of evil / Flewelling, Ralph Tyler – New York: Eaton & Mains, c1913 – 1mf – 9 – 0-7905-3836-9 – (incl bibl ref) – mf#1989-0329 – us ATLA [210]
Christ and the eastern soul : the witness of the oriental consciousness to jesus christ / Hall, Charles Cuthbert – Chicago: University of Chicago Press, 1909 – (= ser Barrows lectures) – 1mf – 9 – 0-8370-3451-5 – (incl bibl ref) – mf#1985-1451 – us ATLA [240]
Christ and the eternal order / Buckham, John Wright – Boston:Pilgrim Press, 1906 – 1mf – 9 – 0-8370-3034-X – (incl bibl ref) – mf#1985-1034 – us ATLA [240]
Christ and the human race, or, the attitude of jesus christ toward foreign races and religions : being the william belden nobel lectures for 1906 / Hall, Charles Cuthbert – Boston: Houghton, Mifflin, 1906 – 1mf – 9 – 0-8370-9785-1 – (incl bibl ref) – mf#1986-3785 – us ATLA [240]
Christ and the nations : an examination of old and new testament teaching / Tait, Arthur James – London: Hodder and Stoughton, 1910 – 1mf – 9 – 0-7905-0161-9 – (incl bibl ref and indexes) – mf#1987-0161 – us ATLA [220]
Christ and war : the reasonableness of disarmament on christian, humanitarian and economic grounds / Wilson, William Ernest – London: James Clarke, 1913 – 1mf – 9 – 0-524-00217-7 – (incl bibl ref) – mf#1989-2917 – us ATLA [240]
Christ bearing witness to himself : being the donnellan lectures for the year 1878-9 / Chadwick, George Alexander – New York: ADF Randolph [1879?] [mf ed 1985] – (= ser The donnellan lectures 1878-79) – 1mf – 9 – 0-8370-2620-2 – mf#1985-0620 – us ATLA [240]
Christ came again : the parousia of christ a past event, the kingdom of christ a present fact, with a consistent eschatology / Urmy, William Smith – New York: Eaton & Mains, c1900 [mf ed 1993] – 1mf – 9 – 0-524-05640-4 – mf#1992-0495 – us ATLA [240]
Christ church, montreal : as parish church and cathedral, a report to the rector of the parish, with appendices, containing opinions of canadian counsel and evidence of the chief cathedral authorities in england... – Montreal: Lovell, 1875 [mf ed 1980] – 2mf – 9 – 0-665-00634-9 – mf#00634 – cn Canadiana [240]
Christ Church. Oxford *see* The early printed music collection
Christ crucified : a sermon preached...on february 7, 1837, by previous appointment of the presbytery, and published at their request / George, James – Toronto?: W J Coates, 1837 – 1mf – 9 – mf#21609 – cn Canadiana [240]
Christ dying for the helpless and ungodly / Ferguson, Archibald – Aberdeen, Scotland. 1861 – 1r – us UF Libraries [240]
Christ enough / Smith, Hannah Whitall – New York: Ketcham [1895] [mf ed 1984] – (= ser Women & the church in america 131) – 1mf – 9 – 0-8370-1407-7 – mf#1984-2131 – us ATLA [240]
Christ et le siecle / Bungener, Felix – Paris: J Cherbuliez, 1856 – 1mf – 9 – 0-7905-9909-0 – mf#1989-1634 – us ATLA [240]
Christ for all / Assemblies of God – v2 n5-v6 n2 [1974 nov/dec-1978 mar/apr] – 1r – 1 – (cont by: reach out [springfield mo]; prayer and praise; assemblies of god home missions) – mf#403987 – us WHS [243]
Christ for india : being a presentation of the christian message to the religious thought of india / Lucas, Bernard – London: Macmillan, 1910 – 2mf – 9 – 0-8370-6583-6 – mf#1986-0583 – us ATLA [230]
The christ from without and within : a study of the gospel by st john / Clark, Henry William – New York: Fleming H Revell, 1907 – 1mf – 9 – 0-8370-2666-0 – mf#1985-0666 – us ATLA [220]
Christ glorified! / Gribble, Charles Bessly – London, England. 1841? – 1r – us UF Libraries [240]
Christ, Grace H *see* Journal of psychosocial oncology
The christ has come : the second advent an event of the past / Hampden-Cook, Ernest – 3rd ed. London: Simpkin, Marshall, Hamilton, Kent, 1905 [mf ed 1990] – 1mf – 9 – 0-7905-3894-6 – (incl bibl ref) – mf#1989-0387 – us ATLA [240]
Christ, Hermann *see* Eine fruehlingsfahrt nach den canarischen inseln
Christ imitable : or, the religious value of the doctrine of christ'... / Higginson, Edward – London, England. 1837 – 1r – us UF Libraries [240]
Christ in creation; and, ethical monism = Selections. 1899 / Strong, Augustus Hopkins – Philadelphia: Roger Williams Press, 1899 – 2mf – 9 – 0-7905-7472-1 – mf#1989-0697 – us ATLA [240]
Christ in everyday life / Bosworth, Edward Increase – New York: Association Press, c1910 – 1mf – 9 – 0-7905-3311-1 – mf#1987-3311 – us ATLA [220]
Christ in his humiliation – Bath, England. 18– – 1r – us UF Libraries [240]
Christ in his ministry – Bath, England. 18– – 1r – us UF Libraries [240]
Christ in his temptation – Bath, England. 18– – 1r – us UF Libraries [240]
Christ in history / Turnbull, Robert – new and rev ed. Boston: Gould and Lincoln, 1860 – 2mf – 9 – 0-524-07662-6 – mf#1992-1103 – us ATLA [240]
Christ in isaiah : expositions of isaiah 40-55 / Meyer, Frederick Brotherton – New York: Fleming H Revell, c1895 – 1mf – 9 – 0-8370-4405-7 – mf#1985-2405 – us ATLA [221]
Christ in modern life : sermons / Brooke, Stopford Augustus – New York: D Appleton, 1872 – 1mf – 9 – 0-7905-7560-4 – mf#1989-0785 – us ATLA [240]
Christ in song : hymns of immanuel / ed by Schaff, Philip – new rev enl ed. New York: Anson DF Randolph, c1895 – 8mf – 9 – 0-7905-7076-9 – (incl bibl ref) – mf#1988-3076 – us ATLA [780]
Christ in the christian year and in the life of man : sermons for laymen's reading; trinity to advent / Huntington, Frederic Dan – New York: E P Dutton, 1881 [mf ed 1984] – 4mf – 9 – 0-8370-0791-7 – mf#1984-4162 – us ATLA [242]
Christ in the gospels : or, the life of our lord in the words of the evangelists, american revision, a d 1881... / Cadman, James Piper – 6th ed. Chicago: American Publ Society of Hebrew, 1886 [mf ed 1986] – 1mf – 9 – 0-8370-9132-2 – (incl ind) – mf#1986-3132 – us ATLA [226]
Christ in the midst of us / Barry, A – Leeds, England. 1862 – 1r – us UF Libraries [240]
Christ in the social order / Clow, William Maccallum – New York: Hodder & Stoughton, [1913?] – 1mf – 9 – 0-524-03040-5 – mf#1990-0797 – us ATLA [240]
Christ in the tabernacle / Simpson, Albert B – New York: Christian Alliance Pub Co, c1896 [mf ed 1992] – (= ser Christian & missionary alliance coll) – 1mf – 9 – 0-524-02142-2 – mf#1990-4208 – us ATLA [240]
Christ in theology / Bushnell, Horace – 1851 – 1 – $50.00 – us Presbyterian [240]
Christ is all / Henson, G M – [s.l: s.n] c1919 [mf ed 2005] – 1r – 1 – 0-524-10529-4 – mf#b00741 – us ATLA [240]
Christ, Lena *see* Mathias bichler
Christ lore : being the legends, traditions, myths, symbols, customs, and superstitions of the christian church / Hackwood, Frederick William – London: Elliot Stock, 1902 – 1mf – 9 – 0-8370-3444-2 – mf#1985-1444 – us ATLA [230]
Christ Methodist Episcopal Church (Pittsburgh PA) *see* Year book
Christ mystical : or, the blessed union of christ and his members: from general gordon's copy / Hall, Joseph – London: Hodder & Stoughton 1908 [mf ed 1985] – 1mf – 9 – 0-8370-4745-5 – (int on theology of general gordon by h carruthers wilson) – mf#1985-2745 – us ATLA [240]
The christ myth : a study / Evans, Elizabeth Edison – New York:Truth Seeker Co., c1900 – 1mf – 9 – 0-8370-3078-1 – mf#1985-1078 – us ATLA [240]
The christ of english poetry / Stubbs, Charles William – London: JM Dent, 1906 – (= ser Hulsean Lectures) – 1mf – 9 – 0-7905-9695-4 – mf#1989-1420 – us ATLA [420]
The christ of history : an argument grounded in the facts of his life on earth / Young, John – New York: Robert Carter, 1857 – 1mf – 9 – 0-8370-5737-X – mf#1985-3737 – us ATLA [240]
The christ of history and of experience / Forrest, David William – 3rd ed. New York: Scribner, 1901 – (= ser Kerr Lectures) – 2mf – 9 – 0-7905-9925-2 – mf#1989-1650 – us ATLA [240]
The christ of nineteen centuries / Behrends, Adolphus Julius Frederick – Brooklyn, NY: TB Ventres, I904 – 5mf – 9 – 0-524-07849-1 – mf#1991-3394 – us ATLA [240]
The christ of paul, or, the enigmas of christianity : st john never in asia minor, irenaeus the author of the fourth gospel, the frauds of the churchmen of the second century exposed / Reber, George – New York: Charles P. Somerby, 1876, c1875 – 1mf – 9 – 0-7905-3211-5 – mf#1987-3211 – us ATLA [240]
The christ of the forty days / Simpson, Albert B – New York: Christian Alliance Pub Co, [1890?] [mf ed 1992] – (= ser Christian & missionary alliance coll) – 1mf – 9 – 0-524-02498-7 – mf#1990-4357 – us ATLA [240]
The christ of the gospels / Holdsworth, W W – 1st ed. London: Charles H Kelly, 1911 – 1mf – 9 – 0-7905-1106-1 – (incl bibl ref) – mf#1987-1106 – us ATLA [226]
The christ of the gospels and the christ of modern criticism : lectures on m. renan's "vie de jesus" / Tulloch, John – Cincinnati: Poe & Hitchcock, 1865 – 1mf – 9 – 0-8370-5585-7 – mf#1985-3585 – us ATLA [240]
The christ of to-day / Gordon, George Angier – Boston: Houghton, Mifflin, 1895 – 1mf – 9 – 0-8370-4843-5 – (incl bibl ref) – mf#1985-2843 – us ATLA [240]
Christ on parnassus : lectures on art, ethic, and theology / Forsyth, Peter Taylor – New York: Hodder and Stoughton, [1911?] – 1mf – 9 – 0-7905-3677-3 – mf#1989-0170 – us ATLA [700]
Christ on the throne of power and antichrist : a treatise on the book of revelation, to st john the divine / Brown, Fortune Charles – Rochester, NY: Union & Advertiser, 1885 [mf ed 1985] – 1mf – 9 – 0-8370-2477-3 – (incl ind) – mf#1985-0477 – us ATLA [225]
Christ or confucius, which? : or, the story of the amoy mission / Macgowan, John – London: London Missionary Society, 1889 [mf ed 1995] – (= ser Yale coll; Missionary manuals) – 208p (ill) – 1 – 0-524-09616-3 – mf#1995-0616 – us ATLA [951]
Christ or napoleon – which? : a study of the cure for world militarism and the church's scandal of division / Ainslie, Peter – New York: Fleming H. Revell, c1915 – 1mf – 9 – 0-7905-3361-8 – mf#1987-3361 – us ATLA [320]
Christ our example as a teacher of religious truth / Trafford, J – s.l, s.l? no date – 1r – us UF Libraries [240]
Christ our life : in its origin, law, and end / Angus, Joseph – Philadelphia: American Baptist Publication Society, c1853 – 1mf – 9 – 0-7905-3067-8 – mf#1987-3067 – us ATLA [240]
Christ our life : the scriptural argument for immortality through christ alone / Hudson, Charles Frederic – Boston: JP Jewett, 1860 – 1mf – 9 – 0-524-07181-0 – (incl bibl ref) – mf#1992-1051 – us ATLA [240]
Christ our life : sermons chiefly preached in oxford / Moberly, Robert Campbell – New York: Longmans, Green, 1902 – 1mf – 9 – 0-7905-2183-0 – mf#1987-2183 – us ATLA [240]
Christ our passover – London, England. 18– – 1r – us UF Libraries [240]
Christ our passover : or, thoughts on the atonement / Cumming, John – London: Arthur Hall, Virtue 1854 [mf ed 1991] – 1mf – 9 – 0-7905-8775-0 – mf#1989-2000 – us ATLA [240]
Christ outreach magazine – 1982 apr/may – 1r – 1 – mf#4717835 – us WHS [240]
Le christ paien au 3e siecle *see* Apollonius of tyana
Christ, Paul *see* Die lehre vom gebet nach dem neuen testament
Christ preaching to spirits in prison : or, christ's preaching to the dead explained by the change from the inferior to the celestial paradise / Love, William De Loss – Boston: publ...by Congregational Sunday- School & Publ Soc, 1883 [mf ed 1984] – 2mf – 9 – 0-8370-0969-3 – (incl bibl ref & ind) – mf#1984-4347 – us ATLA [240]
Christ pre-eminent / Guinness, H Grattan – Dublin, Ireland. 1858 – 1r – us UF Libraries [240]
Le christ republicain – Paris: Bonaventure et Ducessois, jun 8-22/25 1848 – us CRL [944]
Le christ republicain-democrate-socialiste – Paris: Beaule et Maignand, jan-mar 1849 – us CRL [944]
Christ, Richard et al *see* Dabeisein – mitgestalten
Christ satisfying the instincts of humanity : eight lectures / Vaughan, Charles John – 2nd ed London: Macmillan, 1873 – 1mf – 9 – 2-524-00524-0 – mf#1992-0373 – us ATLA [240]
The christ story / Tappan, Eva March – Boston: Houghton, Mifflin, 1903 – 2mf – 9 – 0-524-06054-1 – mf#1992-0767 – us ATLA [220]
Christ tempted in all points like as we are : yet without sin / Carlile, Warrand – London, England. 1831 – 1r – us UF Libraries [240]
The christ that is to be / Dougall, Lily – New York: Macmillan, 1907 – 1mf – 9 – 0-8370-3634-8 – (incl bibl ref) – mf#1985-1634 – us ATLA [240]
Christ the author and end of civil government / Dudley, W M – Poole, England. 1836 – 1r – us UF Libraries [240]
Christ, the book, and the church / Allon, Henry – London, England. 1864 – 1r – us UF Libraries [240]
Christ the bread of life : an attempt to give a profitable direction to the present occupation of thought with romanism / Campbell, John McLeod – 2nd ed London: Macmillan, 1869 – 1mf – 9 – 0-7905-3652-8 – mf#1989-0145 – us ATLA [240]
Christ the central evidence of christianity / Cairns, John – New York:American Tract Society, [188-?] [mf ed 1985] – (= ser Book for the times 3) – 1mf – 9 – 0-8370-2567-2 – mf#1985-0567 – us ATLA [240]
Christ, the church, and man : an essay on new methods in ecclesiastical studies and workshop: with some remarks on a new apologia for christianity in relation to the social question / Capecelatro, Alfonso – London: Burns & Oates; St Louis, MO: B Herder, 1909 – 1mf – 9 – 0-8370-6970-X – mf#1986-0970 – us ATLA [240]
Christ the creative ideal : studies in colossians and ephesians / Walker, W L – Edinburgh: T & T Clark; New York: Charles Scribner [distributor], 1913 – 1mf – 9 – 0-7905-2443-0 – (incl ind) – mf#1987-2443 – us ATLA [240]

Christ the interpreter of scripture : a series of discourses showing how to read the bible wisely and profitably, with a preliminary essay on the sources and guarantees of the gospel history / Beard, John Relly – London: Whitfield, Green, [1865?] – 1mf – 9 – 0-524-04445-7 – mf#1992-0114 – us ATLA [220]
Christ the king / Foster, James Mitchell – Boston: James H Earle, c1894 – 2mf – 9 – 0-524-04190-3 – mf#1990-1229 – us ATLA [240]
Christ the lord / Thompson, Henry – London, England. 1870 – 1r – us UF Libraries [240]
Christ, the morning star : and other sermons / Cairns, John; ed by Cairns, William & Cairns, David – London: Hodder and Stoughton, 1892 – 1mf – 9 – 0-7905-3707-9 – mf#1989-0200 – us ATLA [240]
Christ the only sacrificing priest under the gospel / Price, Thomas C – London, England. 1854 – 1r – us UF Libraries [240]
Christ the orator : or, never man spake like this man: ecce orator / Hyde, Thomas Alexander – Boston: Arena 1893 [mf ed 1985] – 1mf – 9 – 0-8370-3713-1 – mf#1985-1713 – us ATLA [240]
Christ the sole master – Edinburgh, Scotland. 1854 – 1r – us UF Libraries [240]
Christ, the son of god : a discourse in review of the rev dr wilkes' sermon, entitled "who is christ?":...jan 19, 1851 / Cordner, John – Montreal?: J Potts, 1851 – 1mf – 9 – mf#26225 – cn Canadiana [240]
The christ, the son of god : a life of our lord and saviour jesus christ = Vie de n-s jesus-christ / Fouard, Constant – New York: Longmans, Green 1892 c1890 [mf ed 1989] – 2v on 2mf [ill] – 9 – 0-7905-2653-0 – (trans fr french by george francis xavier griffith; int by cardinal manning; incl bibl ref) – mf#1987-2653 – us ATLA [240]
Christ the substitute for his people / Smith, James – London, England. 18-- – 1r – us UF Libraries [240]
Christ the true essiah – London, England. 1796 – 1r – us UF Libraries [240]
Christ the truth : an essay towards the organization of christian thinking / Medley, William – London, New York: Macmillan 1900 [mf ed 1991] – (= ser The angus lectureship 3) – 1mf – 9 – 0-7905-8850-1 – mf#1989-2075 – us ATLA [240]
Christ the way : four addresses / Paget, Francis – London; New York: Longmans, Green, 1902 – 1mf – 9 – 0-7905-8546-4 – mf#1989-1771 – us ATLA [240]
Christ the world's peace / Garnier, Thomas – London, England. 1856 – 1r – us UF Libraries [240]
Der christ und die suende bei paulus / Wernle, Paul – Freiburg i.B: J C B Mohr (Paul Siebeck), 1897 – 1mf – 9 – 0-8370-9345-7 – (incl bibl ref) – mf#1986-3345 – us ATLA [220]
Christ und welt – Stuttgart, Duesseldorf DE, 1975 10 jan-1979 21 dec – 9r – 1 – (title varies: 2 apr 1971: deutsche zeitung, christ und welt, 1 jan 1980 merged with: rheinischer merkur, koblenz/bonn; cont: see stuttgart, fr 2 apr 1971 in duesseldorf; filmed by mikropress: 1948 6 jun-1979 21 dec [2r/yr] order#7259, filmed by misc inst: 1953-56 [10r], 1946 10 may-1969, 1952-1957 jun) – gw Mikrofilm; gw Mikropress; gw Misc Inst [230]
Christ und welt *see* Rheinischer merkur 1946
Christ upon the waters / Newman, John Henry – Birmingham, England. 1850? – 1r – us UF Libraries [240]
Christ versus christianity : the christian church cross-examined by a modern lawyer / Hale, William Pillsbury – Boston: American Elzevir, 1892 – 1mf – 9 – 0-8370-3450-7 – mf#1985-1450 – us ATLA [240]
Christ, W *see* Anthologia graeca carminum christianorum
The christ we forget : a life of our lord for men of to-day / Wilson, Philip Whitwell – New York: Fleming H Revell, c1917 – 1mf – 9 – 0-524-03999-2 – mf#1992-0042 – us ATLA [220]
Christ won by faith – London, England. 1812 – 1r – us UF Libraries [240]
Christa : ein kinderroman / Flake, Otto – Berlin: S Fischer c1931 [mf ed 1989] – 1r [ill] – 1 – (filmed with: der tod vor dem spiegel / edmund finke) – mf#7242 – us Wisconsin U Libr [830]
Christaller, Johann Gottlieb *see*
– A collection of three thousand and six hundred tshi proverbs in use among the negroes of the gold coast speaking the asante and fante language
– Twi manuscripts, 1855-1907
Christaller, Th. *see* Handbuch der duala-sprache
Christchurch mid-week mail – 24 sep 1986-aug 1989; oct-dec 1989 – 10r – 1 – mf#70.29 – nz Nat Libr [079]
Christchurch press – jan 1979-oct 1982 – 92r – 1 – mf#70.16 – nz Nat Libr [079]
Christchurch star – 1 – (previously known as: the star may 1868-dec 1925, jan 1979-oct 1982, the star [dunedin ed] jul 1980-sep 16 1980, jul-dec 1981) – mf#70.19 – nz Nat Libr [079]
Christchurch star *see* The star
Christchurch star sports and magzine edition – jan 1975-aug 1976 – 48r – 1 – (title changes to: weekend star fr sep 1976-dec 1988) – mf#70.21 – nz Nat Libr [079]
Christchurch star-sun – New Zealand. Christchurch Star. -d. 21 March 1941-16 Dec 1944; 7-25 May, 20 July 1945-10 Sept 1960; 7, 11 March, 30 Nov, 1, 30 Dec 1961; 3 Jan-19 Feb, 14 Sept 1962; 3 April 1963-Oct 1971. 292 reels – 1 – uk British Libr Newspaper [079]
Christchurch, upwell 1863-1950 – (= ser Cambridgeshire parish register transcript) – 3mf – 9 – £3.75 – uk CambsFHS [929]
Christ-comoedia : ein weihnachtsspiel / Huebner, Johann; ed by Brachmann, Friedrich – Berlin: B Behr (E Bock), 1899 [mf ed 1993] – (= ser Deutsche litteraturdenkmale des 18. und 19. jahrhunderts 82, n f n32) – xxvii/39p – 1 – mf#8676 reel 5 – us Wisconsin U Libr [820]
Christe, Pierre *see* En avant...marche!
Christelige ethik *see*
– Christian ethics
Christelige taler / Kierkegaard, Soeren – Kobenhavn: C A Reitzel 1848 [mf ed 1990] – 2pt on 1mf – 9 – 0-7905-7413-6 – mf#1989-0638 – us ATLA [240]
Den christelige vished : apologetiske undersoegelser med saerligt hensyn til franks "system der christlichen gewissheit" / Ussing, Henry – Kobenhavn: G E C Gad 1883 [mf ed 1991] – 1mf – 9 – 0-7905-8614-2 – mf#1989-1839 – us ATLA [240]
Christelijcken waerseggher : de principale stucken van t'christen geloof en leven int cort begrijpende / David, J – t'Antwerpen: Plantijnsche Druckerije, Jan Moerentorf, 1603 – 7mf – 9 – mf#O-3060 – ne IDC [090]
Christelijcke kercken-ordeninge der stadt, steden ende landen van vtrecht / Wtenbogaert, J – Vtrecht, 1612 – 1mf – 9 – mf#H-2500 – ne IDC [240]
Het christelijk geloof van schleiermacher in verband tot het rationalismus beschouwd / Reddingius, Jodocus Henricus – Groningen: W Zuidema, 1836 – 1mf – 9 – 0-7905-9451-X – (incl bibl ref) – mf#1989-2676 – us ATLA [240]
Christelijk of heidensch / Henzel, J – [Rotterdam: J M Bredee, 1906] [mf ed 1995] – (= ser Yale coll; Lichtstralen op den akker der wereld [12. jaarg 1906] 1) – 29p (ill) – 1 – 0-524-09938-3 – (in dutch) – mf#1995-0938 – us ATLA [951]
Christelijke gereformeerde kerken : jaarboek – 1968-92 [complete] – Inquire – 1 – mf#ATLA S0506 – us ATLA [242]
Christelyke aandachten of vlammende ziel-zuchten : eener godvreesende ziele... / Hesman, Gerrit – Deventer: H W van Welbergen, 1728 – 2mf – 9 – mf#O-3080 – ne IDC [090]
Christelyke bedenkingen en voorbeeldlyke zeedelessen afgeleid uit 's werelds eerste toestand / Graauwhart, Hendrik – Amsterdam: J. ter Beek, 1756 – 4mf – 9 – mf#O-3072 – ne IDC [090]
Christelyke staets-vorst in hondert sin-spreuken... / Saavedra Faxardo, Didaco de – t'Amsterdam: Jan Jacobsz. Schipper en Borrit Jansz. Smit, 1662 – 11mf – 9 – mf#O-3160 – ne IDC [090]
Christen, Ada *see*
– Aus dem leben
– Unsere nachbarn
Der christen sabath... / Wolf, J – Zuerych, Christoffel Froschouer, 1563 – 1mf – 9 – mf#PBU-656 – ne IDC [240]
The christen state of matrimony... / Bullinger, Heinrich – London, 1546 [i.e. 1543] – 3mf – 9 – mf#PBU-678 – ne IDC [240]
The christen state of matrymonye / Bullinger, Heinrich – [London, 1543] – 3mf – 9 – mf#PBU-138 – ne IDC [240]
Christendom : the christian churches constitutional, forms and ways / Molland, E – London, 1959 – 9mf – 8 – €18.00 – ne Slangenburg [240]
Christendom : a journal of christian sociology – v1-16. 1931-50 [complete] – 3r – 1 – mf#ATLA S0006 – us ATLA [240]
Christendom en leven : toespraken over opvoeding en vereenigingsleven / Berkhof, Louis – Grand-Rapids: Eerdmans-Sevensma, 1912 – 1mf – 9 – 0-524-06478-4 – mf#1991-2578 – us ATLA [240]
Christendom's divisions : being a philosophical sketch of the divisions of the christian family in east and west / Ffoulkes, Edmund S – London: Longman, Green, Longman, Roberts, & Green, 1865 – 1mf – 9 – 0-7905-4514-4 – (incl bibl ref) – mf#1988-0514 – us ATLA [240]
Christenfragen / Zorn, Carl Manthey – [2. Aufl.] Milwaukee, Wis: Northwestern Pub House, 1915 – 1mf – 9 – 0-524-05273-5 – mf#1991-2265 – us ATLA [240]
Der christenheit rechte vollkommenheit / Bullinger, Heinrich – [Zuerych, Andrea Geszner d. j., Ruodolff Wyssenbach, 1551] – 2mf – 9 – mf#PBU-170 – ne IDC [240]
Die christenkatastrophe unter nero : nach ihren quellen, insbesondere nach tac. ann. 15, 44 / Klette, Emil Theodor – Tuebingen: J.C.B. Mohr, 1907 – 1mf – 9 – 0-7905-6196-4 – (incl bibl ref) – mf#1988-2196 – us ATLA [240]
Christenkreuz und hakenkreuz – Dresden DE, 1935 n1-12 – 1 – gw Misc Inst [074]
Christenlehre : eine handreichung fuer den konfirmandenunterricht und den religionsunterricht hoeherer stufe / Rohde, Adolf – 2., verb. und verm. Aufl. Leipzig: Friedrich Fleischer, 1895 – 1mf – 9 – 0-8370-7733-8 – (incl ind) – mf#1986-1733 – us ATLA [240]
Christennlich ordnung vnd bruch der kirchen zuerich / Bullinger, Heinrich – [Zuerich, Christoffel Froschouwer], 1535 – 1mf – 9 – mf#PBU-260 – ne IDC [240]
Christensen, Anders *see* Ungdom der vaagner
Christensen, C F A *see* Index of flora aegyptiaco-arabica and herbarium forskilii
Christensen, Karen *see* Implementation of religious symbols in a choreographic work
Christensen, Kimberly M *see* Effects of an interval training dance class on select cardiovascular variables
Das christentum justins des maertyrers : eine untersuchung ueber die anfaenge der katholischen glaubenslehre / Engelhardt, Moritz von – Erlangen: A Deichert, 1878 – 2mf – 9 – 0-524-04959-9 – (incl bibl ref) – mf#1990-1362 – us ATLA [241]
Das christenthum und die christliche kirche der drei ersten jahrhunderte / Baur, Ferdinand Christian – Tuebingen: Fues, 1853 – 5mf – 9 – 0-7905-5921-8 – (incl bibl ref) – mf#1988-1921 – us ATLA [240]
Das christenthum und die einsprueche seiner gegner : eine apologetik fuer jeden gebildeten / Vosen, Christian Hermann; ed by Rheinstaedter, Ferdinand – 4. aufl. Freiburg i.B; St Louis, MO: Herder, 1881 [mf ed 1986] – 2mf – 9 – 0-8370-6954-8 – mf#1986-0954 – us ATLA [240]
Christenthum und kirche im einklange mit der culturentwicklung : zwanzig betrachtungen / Schenkel, Daniel – Wiesbaden: CW Kreidel, 1867 – 2mf – 9 – 0-524-00785-3 – mf#1990-0217 – us ATLA [240]
Christenthum und lutherthum / Kahnis, Karl Friedrich August – Leipzig: Doerffling und Franke, 1871 – 1mf – 9 – 0-524-06425-3 – mf#1991-2547 – us ATLA [242]
Christenthum und moderne cultur : studien, kritiken und charakterbilder / Hamberger, Julius – Erlangen: Theodor Blaesing, 1863 – 1mf – 9 – 0-8370-5113-4 – (incl bibl ref) – mf#1985-3113 – us ATLA [240]
Christenthum und socialismus / Hohoff, Kaplan – Leipzig, 1878 – 1 – gw Mikropress [240]
Das christentum : fuenf einzeldarstellungen / Cornill, Carl Heinrich et al – Leipzig:Quelle & Meyer, 1908 – 1mf – 9 – 0-8370-2656-3 – mf#1985-0656 – us ATLA [240]
Das christentum / Hammerstein, Ludwig von – Trier: Paulus-Druckerei, 1893 [mf ed 1986] – 1mf – 9 – 0-8370-7064-3 – (incl ind) – mf#1986-1064 – us ATLA [240]
Das christentum des neuen testaments / Hartmann, Eduard von – 2. umgearb aufl. Sachsa (Harz): Hermann Haacke, 1905 – 1mf – 9 – 0-7905-0952-0 – (incl bibl ref) – mf#1987-0952 – us ATLA [225]
Das christentum in den ersten drei jahrhunderten / Achelis, Hans – Leipzig: Quelle & Meyer, 1912 – 2mf – 9 – 0-7905-4421-0 – (incl bibl ref) – mf#1988-0421 – us ATLA [240]
Das christentum justins des maertyrers / Engelhardt, M von – Erlangen, 1878 – €18.00 – ne Slangenburg [180]
Christentum und buddhismus : eine studie zur geisteskultur des ostens und des westens / Luettge, Willy – Goettingen: Vandenhoeck and Ruprecht, 1916 – 1mf – 9 – 0-524-01620-8 – mf#1990-2559 – us ATLA [230]
Christentum und buddhismus : ein vortrag / Falke, Robert – Berlin: F Ruehe, 1898 – 1mf – 9 – 0-524-01439-6 – mf#1990-2434 – us ATLA [230]
Christentum und die geschichte / Harnack, Adolf von – London: Adam & Charles Black, 1896 – 1mf – 9 – 0-8370-3478-7 – mf#1985-1478 – us ATLA [240]
Das christentum und die heutige vergleichende religionsgeschichte / Happel, Julius – Leipzig: Otto Schulze, 1882 – 1mf – 9 – 0-524-01496-5 – (incl bibl ref) – mf#1990-2472 – us ATLA [230]
Das christentum und die monistische religion / Werner, Max – Berlin: Karl Curtius, 1908, c1905 – 1mf – 9 – 0-8370-5791-4 – (incl bibl ref) – mf#1985-3791 – us ATLA [240]
Das christentum und die philosophie : ein vortrag / Kaftan, Julius – Leipzig: J C Hinrichs, 1895 – 1mf – 9 – 0-8370-3828-6 – mf#1985-1828 – us ATLA [240]
Christentum und geschichte bei schleiermacher : die geschichtsphilosophischen grundlagen der schleiermacherschen theologie / Sueskind, Hermann – Tuebingen: JCB Mohr, 1911 – 1mf – 9 – 0-524-00114-6 – (incl bibl ref) – mf#1989-2814 – us ATLA [240]
Christentum und geschichte bei wilhelm herrmann : mit besonderer beruecksichtigung der erkenntnis-theoretischen seite des problems / Hermann, Rudolf – Leipzig: A Deichert, 1914 – 1mf – 9 – 0-524-00758-6 – (incl bibl ref) – mf#1990-0190 – us ATLA [240]
Christentum und geschichtlichkeit : untersuchungen zur entstehung des christentums und zu augustins buergerschaft gottes / Kamlah, Wilhelm – 2., neubearb und erg Aufl. Stuttgart: W Kohlhammer, 1951 – 1mf – 9 – 0-524-08112-3 – mf#1993-9018 – us ATLA [240]
Christentum und kirche in russland und dem orient / Mulert, Hermann – Tuebingen: J C B Mohr 1916 [mf ed 1991] – (= ser Religionsgeschichtliche volksbuecher fuer die deutsche christliche gegenwart 4/22-23) – 1mf – 9 – 0-7905-9418-8 – mf#1989-2643 – us ATLA [243]
Christentum und kultur : gedanken und anmerkungen zur modernen theologie / Overbeck, Franz Camille; ed by Bernoulli, C A – Basel, 1919 – €14.00 – (aus dem nachlass) – ne Slangenburg [230]
Christentum und kultur : ein orientierender vortrag / Haack, Ernst – Schwerin i. M.: Fr Bahn, 1897 – 1mf – 9 – 0-8370-3438-8 – mf#1985-1438 – us ATLA [306]
Christentum und wissenschaft in schleiermachers glaubenslehre : ein beitrag zum verstaendnis der schleiermacherschen theologie / Scholz, Heinrich – Berlin: A Glaue, 1909 – 1mf – 9 – 0-524-00101-4 – (incl bibl ref) – mf#1989-2801 – us ATLA [240]
Die christenverfolgung in nord-schansi (china) im jahre 1900 / Volling, Arsenius – Trier: Paulinus-Druckerei, 1911 [mf ed 1995] – (= ser Yale coll; Aus allen zonen 6) – 128p (ill) – 1 – 0-524-09949-9 – (in german) – mf#1995-0949 – us ATLA [241]
Die christenverfolgungen im roemischen reiche : vom standpunkte des juristen / Cohn, Max Conrat – Leipzig: J.C. Hinrichs, 1897 – 1mf – 9 – 0-7905-5518-2 – (incl bibl ref) – mf#1988-1518 – us ATLA [340]
Christherre-chronik (cima29) : farbmikrofiche-edition der handschrift linz, bundesstaatliche studienbibliothek, cod 472 – (mf ed 1994) – (= ser Codices illuminati medii aevi (cima) 29) – 49p on 12 color mf – 15 – €390.00 – 3-89219-029-1 – (int & description by ralf plate) – gw Lengenfelder [090]
Christi lehrthaetigkeit : nach den evangelienberichten fuer schulzwecke / Schaarschmidt, U – Chemnitz: J C F Pickenhahn, 1895 – 1mf – 9 – 0-8370-7665-X – mf#1986-1665 – us ATLA [240]
Christi person und werk : mit bezug auf die christologie ritschl's und dessen schule / Lamm, Karl – Frankfurt a M: Heyder & Zimmer, 1896 [mf ed 1991] – 1mf – 9 – 0-7905-7890-5 – mf#1989-1115 – us ATLA [240]
Christi predigt an die geister (1 petr. 3, 19 ff : ein beitrag zur neutestamentlichen theologie / Spitta, Friedrich – Goettingen: Vandenhoeck & Ruprecht, 1890 – 1mf – 9 – 0-7905-2137-7 – (incl bibl ref) – mf#1987-2137 – us ATLA [220]
Christi worte ueber vollendung der wege gottes : mit seiner kirche, dem volke israel und der ganzen menschheit und schoepfung / Caird, William Renny & Lutz, Johann Georg – Augsburg:Richard Preyss, 1879 – 1mf – 9 – 0-8370-2565-6 – mf#1985-0565 – us ATLA [240]
Christi zeugniss von seiner person und seinem werk : nach seiner geschichtlichen entwicklung / Gess, Wolfgang Friedrich – Basel: Bahnmaier, 1870 – 1mf – 9 – 0-7905-0838-9 – (incl bibl ref) – mf#1987-0838 – us ATLA [240]
Christian – St. Louis. 1959-1973 (1) 1972-1973 (5) (9) – ISSN: 0009-5206 – mf#6351 – us UMI ProQuest [240]
The christian : a monthly periodical devoted to the faith and practice of primitive christianity – St John, NB: W W Eaton, 1839-1848 – 9 – (incl ind) – mf#P04161 – cn Canadiana [240]
The christian : a weekly record of christian life – London, 1882-1925 – 34r – 1 – (lacks 1890-93. 1908-12) – mf#ATLA S0445 – us ATLA [240]
Christian A Herter *see* Papers of john foster dulles and of christian a herter, 1953-1961
Christian accent / American Council of Christian Churches – v1 n1-v5 n4 [1972 spr-1977? dec] – 1r – 1 – mf#645610 – us WHS [240]

Christian achievement in america / Beardsley, Frank Grenville – Chicago, Ill.: Winona, c1907 – 1mf – 9 – 0-7905-5504-2 – mf#1988-1504 – us ATLA [240]

Christian advance / Knox-Little, William John – Manchester, England. 1877? – 1r – us UF Libraries [240]

Christian adventures in south africa / Taylor, William – New York: Nelson & Phillips 1877 [mf ed 1990] – 1r [ill] – 1 – (int by william b boyce. filmed with: the imitation of christ / kempis, t) – mf#1761p – us Wisconsin U Libr [240]

Christian adventures in south africa / Taylor, William MacKergo – London: Jackson, Walford & Hodder; New York: Carlton & Porter [1867?] [mf ed 1986] – 2mf – 9 – 0-8370-6529-1 – mf#1986-0529 – us ATLA [240]

Christian advocate – Belfast Ireland 1886-1896 – 11r – 1 – (aka: irish christian advocate) – uk British Libr Newspaper [072]

Christian advocate – Birmingham. 1978-1978 (1,5,9) – ISSN: 0305-3652 – mf#11522 – us UMI ProQuest [240]

Christian advocate – London. -w. Jan 7 1830-Sep 2 1839 – 4r – 1 – uk British Libr Newspaper [072]

Christian advocate / Methodist Episcopal Church – 1929 oct 24-1930 aug 28, 1930 sep 4-1931 aug 27, 1931 sep 3-1932 aug 25, 1932 sep 1-1933 feb 23, 1938 jan 6-1938 dec 29, 1939 jan 5-1939 dec 28, 1940 jan 4-1940 dec 26 – 7r – 1 – (cont: northwestern christian advocate; cont by: methodist recorder [baltimore md]; christian advocate [chicago il: 1941]) – mf#964321 – us WHS [242]

Christian advocate – Nashville. 1956-1973 (1) 1971-1973 (5) – ISSN: 0577-9936 – mf#2103 – us UMI ProQuest [240]

Christian advocate – Philadelphia. 1823-1834 (1) – mf#4059 – us UMI ProQuest [240]

Christian advocate – Sydney – 2r – A$77.00 vesicular A$88.00 silver – at Pascoe [079]

Christian advocate *see* Hua mei chiao pao (ccs29)

Christian agnosticism as related to christian knowledge : the critical principle in theology / Johnson, Elias Henry; ed by Vedder, Henry Clay – Philadelphia: Griffith & Rowland Press, 1907 – 1mf – 9 – 0-7905-7793-3 – mf#1989-1018 – us ATLA [240]

The christian alliance and missionary weekly – New York: Christian Alliance Pub Co. v3-11. 1889-93 [wkly] [mf ed 2003] – (= ser Christian and missionary alliance) – 9v on 3r – 1 – (lacks: ind for v3-5; v4 n2; v5 n78 p279-280; v5 n24?) – mf1030 – us ATLA [240]

The christian alliance and missionary weekly – [New York: Christian Alliance Pub Co] v12-17. 1894-96 [wkly] [mf ed 2003] – (= ser Christian and missionary alliance) – 7v on 2r – 1 – mf1031 – us ATLA [240]

The christian alliance for fellowship, prayer and service in the four-fold gospel – New York: Word, Work & World Pub Co. v1-2 n6. 1888-jun 1889 [mthly] [mf ed 2003] – (= ser Christian and missionary alliance) – 2v on 1r – 1 – (lacks: ind for v2?) – mf1029 – us ATLA [240]

The christian alliance year book 1888 / ed by Simpson, Albert B – New York: Word, Work & World, [1888?] [mf ed 1993] – (= ser Christian & missionary alliance coll) – 1mf – 9 – 0-524-06073-8 – mf#1990-5187 – us ATLA [240]

Christian and church (a sermon) / Clarke, W Newton – 1882 – 1 – 5.00 – us Southern Baptist [242]

Christian and Missionary Alliance *see*
- Annual report of the...
- Indian christian
- Report and retrospect of the work of the...

Christian and missionary alliance – New York: Christian Alliance, 1897-1911 [wkly] [mf v18-36 1897-1911 filmed 2003] – (= ser Christian and missionary alliance) – 31v on 9r – 1 – (lacks: ind & regular iss for several vols) – mf1032 – us ATLA [240]

Christian and Missionary Alliance. General Council *see*
- Annual report for...and minutes of the general council
- Annual report to the general council
- Minutes of the general council...and annual report...

Christian and mohammedan : a plea for bridging the chasm / Herrick, George Frederick – New York: FH Revell, c1912 – 1mf – 9 – 0-7905-5153-5 – mf#1988-1153 – us ATLA [230]

Christian anti-communism crusade – Long Beach. 1967-1996 (1) 1967-1996 (5) 1977-1996 (9) – (cont by: schwarz report) – ISSN: 0195-9387 – mf#3254 – us UMI ProQuest [240]

Christian apologetics : or, a rational exposition of the foundations of faith = Cours d'apologetique chretienne / Devivier, Walter; ed by Sasia, Joseph Casimir – San Jose CA: Popp & Hogan 1903 [mf ed 1986] – 2v on 3mf – 9 – 0-8370-9056-3 – (trans fr 16th ed of original french. with int and treatise by l peeters. incl bibl ref and ind) – mf#1986-3056 – us ATLA [241]

Christian apologetics : a series of addresses / Henslow, George et al; ed by Seton, Walter Warren – New York: E P Dutton, 1903 [mf ed 1985] – 1mf – 9 – 0-8370-5232-7 – (int by w d mclaren) – mf#1985-3232 – us ATLA [240]

Christian apology = Apologie des christenthums / Schanz, Paul – 2nd rev ed. New York: Pustet 1896, c1891 [mf ed 1993] – 4mf – 9 – 0-524-06318-4 – (english trans fr german by michael f glancey & victor j schobel; incl bibl ref) – mf#1991-2491 – us ATLA [241]

The christian apostolate : its principles, methods and promise in evangelism, missions, and in social progress / Everts, William Wallace – Chicago: FH Revell, c1890 – 2mf – 9 – 0-524-08235-9 – mf#1993-2010 – us ATLA [240]

Christian archaeology / Bennett, Charles Wesley – New York: Hunt & Eaton; Cincinnati: Cranston & Jennings c1888 [mf ed 1989] – (= ser Library of biblical and theological literature 4) – 2mf – 9 – 0-7905-4080-0 – (incl bibl ref) – mf#1988-0080 – us ATLA [930]

Christian art 1 / National Art Library – (= ser The art periodicals coll at the v and a museum, 1750-1920, pt 4) – 39r – 1 – £1950.00 – (christian art 1 [39r] £1950; revue de l'art chretien 1857-1914 [32r] £1650; organ fuer christliche kunst 1857-1914 [7r] £360) – mf#VAU – uk World [700]

Christian art 2 / National Art Library – (= ser The art periodicals coll at the v and a museum, 1750-1920, pt 5) – 25r – 1 – £1200.00 – (christian art 2 [26r] £1200; christliches kunstblatt 1859-1905 [8r] £390; zeitschrift fuer christliche kunst 1888-1921 [8r] £390; bulletino di archaeologica cristiana [10r] £520) – mf#VAV – uk World [700]

Christian aspects of life / Westcott, Brooke Foss – London; New York: Macmillan, 1897 – 1mf – 9 – 0-7905-8972-9 – mf#1989-2197 – us ATLA [240]

Christian association for psychological studies bulletin – Farmington. 1975-1981 [1,5,9] – (cont by: journal of psychology and christianity) – ISSN: 0147-7978 – mf#12179 – us UMI ProQuest [240]

Christian association for psychological studies proceedings of the annual convention – Farmington. 1963-1967 (1) – (cont: christian association for psychological studies proceedings of the annual convention) – mf#12334,02 – us UMI ProQuest [240]

Christian association for psychological studies proceedings of the annual convention – Grand Rapids. 1957-1962 (1) – (cont: proceedings of the calvinistic conference on psychology and psychiatry. cont by: christian association for psychological studies proceedings of the annual convention) – ISSN: 0092-072X – mf#12334,01 – us UMI ProQuest [240]

The christian at work : evangelical always! sectarian never! / ed by Spurgeon, Bonar – [s.l: s.n. 187-?] [mf ed 1987] – 1mf – 9 – 0-665-37361-9 – mf#37361 – cn Canadiana [240]

The christian atonement, its basis, nature, and bearings : or, the principle of substitution illustrated as applied in the redemption of man / Gilbert, Joseph – 3rd ed. London: Jackson & Walford 1854 [mf ed 1993] – (= ser The congregational lecture 3 ser) – 1mf – 9 – 0-524-07099-7 – (incl bibl ref & ind) – mf#1991-2922 – us ATLA [240]

Christian attire : our personal responsibility. address. delivered at the annual conference of the church of the brethren at winona lake, ind... / Taylor, Lydia E – Elgin IL: Brethren Pub House 1916 [mf ed 1992] – 1mf – 9 – 0-524-02756-0 – mf#1990-4431 – us ATLA [230]

The christian awakening of faith (ccc195) = Sheng tao ch'i hsin lun / Hsu, Sung-shih – Shanghai: Christian Literature Society, 1940 – (= ser Chinese christian coll) – 1r – 1 – (in chinese) – mf#1984-B500 – us ATLA [240]

Christian banner – Cobourg [Ont]: D Oliphant, [1852?-1858] – 9 – (cont by: the banner of the faith) – mf#P04344 – cn Canadiana [240]

Christian banner – Fredericksburg, VA. 1850-1862 (1) – mf#68503 – us UMI ProQuest [071]

Christian banner – Fredericksburg VA. 1862 may 20, jun 7-11,18-26 – 1r – 1 – mf#881677 – us WHS [240]

Christian banner – Philadelphia. New England Missionary and Educational Baptist Convention/ South Jersey Association/National Baptist Convention of America. 1900-01, 1915-19 – 1 – us ABHS [240]

The christian banner – Philadelphia, PA: Christian Banner Pub Co, 1888 (wkly) [mf ed 1947] – (= ser Negro Newspapers on Microfilm) – 1r – 1 – us L of C Photodup [071]

Christian baptism : action and subject / Fee, John Gregg – Cincinnati: J G Fee 1878 [mf ed 1992] – 1mf – 9 – 0-524-02465-0 – mf#1990-4324 – us ATLA [240]

Christian baptism : the duty, design, subjects, and act / Ball, George Harvey – 5th rev enl ed. Boston MA: Free Baptist Print Est 1889 [mf ed 1993] – 1mf – 9 – 0-524-07669-3 – mf#1991-3254 – us ATLA [242]

Christian baptism : in two parts / Hibbard, Freeborn Garretson – New York: Nelson & Philipps c1841 [mf ed 1992] – 2mf – 9 – 0-524-05479-7 – (incl bibl ref) – mf#1990-5126 – us ATLA [242]

Christian baptism : its subjects / Ingham, Richard – London: E Stock 1871 [mf ed 1994] – (= ser A hand-book on christian baptism 2) – 7mf – 9 – 0-524-08795-4 – mf#1993-3287 – us ATLA [240]

Christian baptism / Kershner, Frederick Doyle – Cincinnati: Standard Pub c1917 [mf ed 1992] – 1mf – 9 – 0-524-02476-6 – mf#1990-4335 – us ATLA [240]

Christian baptism : a lecture in reply to a sermon by rev d m welton / Annand, Edward – Halifax, [NS: s.n.], 1870 [mf ed 1980] – 1mf – 9 – 0-665-04110-1 – mf#04110 – cn Canadiana [242]

Christian baptism / Scott, Walter – Hamilton, Scotland. 18– – 1r – us UF Libraries [242]

Christian baptism : with its antecedents and consequents / Campbell, Alexander – Bethany VA: printed & publ by A Campbell 1852 [mf ed 1991] – 1mf – 9 – 0-7905-8769-6 – mf#1989-1994 – us ATLA [240]

Christian baptism (illustrated) : its proper subjects and proper act, with a brief history of baptist principles and practices, from the planting of the apostolic church to the present time... / Carey, T H – S.l: s.n.], 1891 [mf ed 1980] – 2mf – 9 – 0-665-00478-8 – mf#00478 – cn Canadiana [242]

Christian baptism illustrated and greatly simplified by means of a number of ingenious charts and diagrams / Wilkinson, Thomas Lottridge – Toronto: W Briggs; Montreal: C W Coates, [1890?] [mf ed 1981] – 2mf – 9 – mf#25786 – cn Canadiana [242]

Christian baptist – Buffalo Creek. 1823-1829 (1) – mf#4425 – us UMI ProQuest [242]

Christian battle / Knox-Little, W J – Manchester, England. 1877? – 1r – us UF Libraries [240]

Christian beacon – 1965 jun 24-1967 jun 15, 1968 jan 4-1971 jul 1, 1971 jul 8-1974 apr 25, 1974 may 2-1977 mar 31, 1977 apr-1979 sep 27, 1979 oct-1981, 1982-84, 1985-87 – 8r – 1 – mf#1494434 – us WHS [071]

Christian beacon – Collingswood NJ: Christian Beacon. v15-18 [1950-64] [mf ed 195?-1961] – 5r [ill] – 1 – (lacks: v15 n2,6; v16 n39,50; v18 n10 [p1]) – ISSN: 0009-5265 – mf#S0089 – us ATLA [230]

Christian belief and life : [discourses delivered in the chapel of harvard university] / Peabody, Andrew Preston – Boston: Roberts, 1876 [mf ed 1984] – 4mf – 9 – 0-8370-0868-9 – mf#1984-4199 – us ATLA [240]

Christian belief interpreted by christian experience : lectures delivered in india, ceylon, and japan on the barrows foundation / Hall, Charles Cuthbert – Chicago: University of Chicago Press, 1905 – (= ser Barrows lectures) – 1mf – 9 – 0-8370-3452-3 – mf#1985-1452 – us ATLA [240]

Christian beliefs reconsidered in the light of modern thought / Henslow, George – London: Frederic Norgate, 1884 – 1mf – 9 – 0-8370-3562-7 – (incl bibl ref) – mf#1985-1562 – us ATLA [240]

Christian biographies (ccm270) / Richard, Timothy [Mrs] – Shanghai. 10v. 1900-01 [mf ed 198?] – (= ser Ccm 270) – 1 – mf#1984-b500 – us ATLA [920]

The christian book of concord : or, symbolical books of the evangelical lutheran church. comprising the three chief symbols, the unaltered augsburg confession, the apology, the smalcald articles, luther's smaller and larger catechisms, the formula of concord, and an appendix – 2nd rev ed. Newmarket: SD Henkel, 1854 – 8mf – 9 – 0-524-08799-7 – mf#1993-3291 – us ATLA [242]

Christian bookseller – Wheaton. 1972-1976 (1) 1972-1976 (5) 1976-1976 (9) – ISSN: 0009-5273 – mf#7367 – us UMI ProQuest [070]

Christian brotherhood : a letter to the hon heman lincoln / Stow, Baron – Boston: Gould & Lincoln 1859 [mf ed 1992] – 1mf – 9 – 0-524-05447-9 – mf#1990-1479 – us ATLA [240]

Christian brotherhoods / Leete, Frederick DeLand – Cincinnati: Jennings & Graham; New York: Eaton & Mains c1912 [mf ed 1990] – 1mf – 9 – 0-7905-5418-6 – (incl bibl ref) – mf#1988-1418 – us ATLA [240]

Christian Brothers *see* Compendium of the history of canada and of the british north american provinces

The christian brothers : their origin and work / Wilson, R F – London: Kegan Paul, Trench, 1883 – 1mf – 9 – 0-8370-7998-5 – mf#1986-1998 – us ATLA [240]

Christian cabinet – Philadelphia. 1802-1802 (1) – mf#3564 – us UMI ProQuest [240]

The christian calling / Davies, John Llewelyn – London: Macmillan, 1875 – 1mf – 9 – 0-7905-3823-7 – mf#1989-0316 – us ATLA [240]

Christian catechism for the use of schoolchildren and young people *see* Khrystyianskyy katekhyzm dlia uzhytku shkilnykh ditei i molodezhy

Christian Catholic Apostolic Church in Zion *see* Theocrat

Christian century – Chicago. 1900+ (1) 1968+ (5) 1960+ (9) – ISSN: 0009-5281 – mf#760 – us UMI ProQuest [240]

The christian certainties : discourses and addresses in exposition and defence of the christian faith / Clifford, John – London: Isbister, 1894 [mf ed 1993] – 1mf – 9 – 0-524-08270-7 – mf#1993-3025 – us ATLA [240]

The christian certainty amid the modern perplexity : essays, constructive and critical, towards the solution of some current theological problems / Garvie, Alfred Ernest – London: Hodder and Stoughton, 1910 – 2mf – 9 – 0-7905-3935-7 – mf#1989-0428 – us ATLA [240]

Christian certitude : its intellectual basis / Touche, Everard Digges la – London: James Clarke 1910 [mf ed 1989] – 1mf – 9 – 0-7905-1278-5 – (incl bibl & ind. pref by h c g moule) – mf#1987-1278 – us ATLA [240]

Christian character : being some lectures on the elements of christian ethics / Illingworth, John Richardson – London, New York: Macmillan 1904 [mf ed 1991] – 1mf – 9 – 0-7905-9969-4 – mf#1989-1694 – us ATLA [240]

Christian character exemplified / Althens, Margaret Magdalen – London, England. 1791 – 1r – us UF Libraries [240]

The christian character in its relation to the christian view of the world / Chapman, James – Nashville, TN: Pub House ME Church, South, 1905, c1904 – (= ser The Cole Lectures) – 1mf – 9 – 0-8370-3201-6 – (incl bibl ref) – mf#1985-1201 – us ATLA [240]

Christian charity, exerting itself by means of missionary incitement for the correction of hindoo immorality : or, cursory remarks on a pamphlet, entitled "missionary incitement, and hindoo demoralization, etc" / Biddulph, Thomas Tregenna – Bristol, 1821 – (= ser 19th c books on british colonization) – 1mf – 9 – mf#1.1.2980 – uk Chadwyck [230]

Christian charity in the ancient church / Uhlhorn, Gerhard – New York: C Scribner's sons 1883 [mf ed 1988] – 1r – 1 – (filmed with: a selection from the correspondence of the late thomaschalmers) – mf#2133 – us Wisconsin U Libr [240]

Christian charity in the ancient church / Uhlhorn, Gerhard – New York: Charles Scribner's, 1883. Chicago: Dep of Photodup, U of Chicago Lib, 1972 (1r); Evanston: American Theol Lib Assoc, 1984 (1r) – 1 – 0-8370-0296-6 – (incl ind) – mf#1984-B318 – us ATLA [240]

Christian chronicle – Bennington. 1818-1818 (1) – mf#3687 – us UMI ProQuest [240]

The christian church / Stone, Darwell – London: Rivingtons, 1905 – 2mf – 9 – 0-524-00789-6 – mf#1990-0221 – us ATLA [240]

The christian church, comprising the reigns of hadrian and antoninus pius (a.d. 117-161) / Renan, Ernest – London: Mathieson, [1888?] – 1mf – 9 – 0-524-06525-X – mf#1992-0909 – us ATLA [240]

The christian church during the first three centuries / Blunt, John James – 7th ed. London: J Murray, 1888 – 1mf – 9 – 0-524-05429-0 – (incl bibl ref) – mf#1990-1461 – us ATLA [240]

The christian church in its foundation, essence, appearance, and work / Loy, Matthias – Columbus: Lutheran Book Concern, 1896 [mf ed 1991] – 1mf – 9 – 0-7905-7907-3 – mf#1989-1132 – us ATLA [240]

The christian church in these islands before the coming of augustine : three lectures delivered at st paul's... / Browne, George Forrest – 4th (rev) ed. London: SPCK; New York: E & J B Young 1899 [mf ed 1990] – 1mf – 9 – 0-7905-5926-9 – (incl ind; original ed publ in 1894) – mf#1988-1926 – us ATLA [240]

Christian circumspection / Craig, Edward – Edinburgh, Scotland. 1826 – 1r – us UF Libraries [240]

Christian citizen / Harris, John – London, England. 1837 – 1r – us UF Libraries [240]

The christian citizen – Omaha, NE: R L Wheeler. v27 n7. feb 10 1925 (wkly) [mf ed 1988] – 1r – 1 – us NE Hist [071]

Christian citizen in life and in death / James, John Angell – London, England. 1852 – 1r – us UF Libraries [240]

Christian Citizens Crusade *see* Independent voice

Christian civilisation : with special reference to india / Cunningham, William – London: Macmillan 1880 [mf ed 1990] – 1mf – 9 – 0-7905-5933-1 – mf#1988-1933 – us ATLA [240]

Christian communications – Ottawa. 1973-1973 (1) 1972-1972 (5) (9) – ISSN: 0009-5303 – mf#7703 – us UMI ProQuest [380]

Christian compassion / Sulivan, Henry William – London, England. 1847 – 1r – us UF Libraries [240]

Christian conception and experience / Gill, William Icrin – New York: Authors' Publ Co 1877 [mf ed 1985] – 1mf – 9 – 0-8370-3286-5 – mf#1985-1286 – us ATLA [240]

The christian conception of god / Adeney, Walter Frederic – New York: Fleming H Revell c1912 [mf ed 1989] – (= ser Christian faith and doctrine series) – 1mf – 9 – 0-7905-0840-0 – (incl bibl ref) – mf#1987-0840 – us ATLA [210]

The christian conception of holiness / Askwith, Edward Harrison – London, New York: Macmillan, 1900 – 1mf – 9 – 0-8370-2515-X – mf#1985-0515 – us ATLA [230]

The christian conquest of asia : studies and personal observations of oriental religions / Barrows, John Henry – New York: Scribner, 1899 – (= ser Morse Lectures) – 1mf – 9 – 0-7905-6402-5 – mf#1988-2402 – us ATLA [240]

The christian conquest of asia : studies and personal observations of oriental religions: being the morse lectures of 1898 / Barrows, John Henry – New York: Scribner, 1899 – 1mf – us ATLA [240]

The christian conquest of india / Thoburn, James Mills – 1st ed. New York: Young People's Missionary Movt c1906 [mf ed 1986] – 1mf – 9 – 0-8370-6420-1 – (incl ind) – mf#1986-0420 – us ATLA [240]

The christian conscience : a contribution to christian ethics: being the fernley lecture for 1888 / Davison, William Theophilus – 2nd ed. London: T Woolmer [1888?] [mf ed 1991] – 1mf – 9 – 0-7905-9261-4 – mf#1989-2486 – us ATLA [240]

The christian consciousness : its relation to evolution in morals and in doctrine / Black, John Sutherland – Boston: Lee and Shepard, 1895 – 1mf – 9 – 0-7905-0864-8 – mf#1987-0864 – us ATLA [240]

Christian consolations : sermons designed to furnish comfort and strength to the afflicted / Peabody, Andrew Preston – 9th ed. Boston: American Unitarian Assoc 1890, c1857 [mf ed 1991] – 1mf – 9 – 0-524-00301-7 – mf#1989-3001 – us ATLA [243]

Christian contributor and free missionary – 1845 jan-1847 aug 18, 1847 aug 25-1849 feb 21 – 2r – 1 – mf#1054402 – us WHS [240]

Christian council quarterly – n23-88. 1949-68 – 1r – 1 – (lacks n24,44,45; cont by: kairos) – ISSN: 0022-7765 – mf#S0704b – us ATLA [240]

Christian courtesy : addressed particularly to young members of the... – London, England. 1841 – 1r – us UF Libraries [240]

The christian creed : its theory and practice, with a preface on some present dangers of the english church / Leathes, Stanley – New York: E. P. Dutton, 1878. Beltsville, Md: NCR Corp, 1978 (5mf); Evanston: American Theol Lib Assoc, 1984 (5mf) – 9 – 0-8370-0863-8 – mf#1984-4205 – us ATLA [240]

The christian creed and the creeds of christendom : seven lectures...1898 / Green, Samuel Gosnell – London, New York: Macmillan 1898 [mf ed 1990] – (= ser The angus lectureship 2) – 1mf – 9 – 0-7905-7395-4 – mf#1989-0620 – us ATLA [240]

Christian creeds and confessions : a short account of the symbolical books of the churches and sects of christendom and of the doctrines dependent on them = Kurzgefasste christliche symbolik / Gumlich, Gotthold Albertus – [3rd ed] New York: Funk & Wagnalls 1894 [mf ed 1992] – 1mf – 9 – 0-524-04310-8 – (incl bibl ref; trans fr german by l a wheatley) – mf#1990-1236 – us ATLA [240]

Christian crose family newsletter – v1 n1-v4 n2 [1976 jul-1980 apr] – 1r – 1 – mf#675639 – us WHS [071]

Christian crusade – 1978 nov-1984 may – 1r – 1 – (cont: christian crusade weekly) – mf#710308 – us WHS [240]

Christian crusade – Tulsa. 1978+ (1) 1979-1989 (5) 1979-1989 (9) – (cont: christian crusade weekly) – ISSN: 0195-265X – mf#3243,01 – us UMI ProQuest [240]

Christian crusade – Tulsa. 1968-1969 (1) – mf#3265 – us UMI ProQuest [240]

Christian crusade for a warless world / Gulick, Sidney Lewis – New York, NY. 1922 – 1r – us UF Libraries [025]

Christian crusade weekly – Tulsa. 1960-1978 (1) – (cont by: christian crusade) – mf#3243 – us UMI ProQuest [240]

Christian crusade weekly : a national christian newspaper / Christian Echoes National Ministry, Inc – 1969 oct 19-1972 dec 31, 1973 jan 7-1975 mar 30, 1975 apr 6-1978 oct 1 – 3r – 1 – (cont: weekly crusader; christian crusade [1961]; cont by: christian crusade [1978]) – mf#714105 – us WHS [240]

Christian Defense League [US] *see* Cdl report

The christian demand for social justice / Wedel, Theodore O et al; ed by Scarlett, William – New York: New American Library, c1949 – (= ser Signet Special) – 1mf – 9 – 0-524-08138-7 – mf#1993-9044 – us ATLA [240]

The christian democracy : a history of its suppression and revival / Leavitt, John McDowell – New York: Eaton and Mains; Cincinnati: Curts and Jennings, 1896 – 1mf – 9 – 0-524-01464-7 – mf#1990-0413 – us ATLA [240]

Christian dietrich grabbe / Gottschall, Rudolf von – Leipzig: P Reclam [1901] [mf ed 1990] – (= ser Dichter-biographien 7) – 1r [ill] – 1 – (incl ind. filmed with: rudolf von gottschall / moritz brasch) – mf#2685p – us Wisconsin U Libr [430]

Christian dietrich grabbe in der nachschillerischen entwickelung / Gieben, Joseph – [Luedinghausen]: Selbstverlag [19–?] [mf ed 1990] – 1r – 1 – (incl bibl ref. filmed with: shakespeare's influence upon grabbe / horace lind hoch) – mf#2687p – us Wisconsin U Libr [430]

Christian dietrich grabbe's saemmtliche werke und handschriftlicher nachlass / ed by Blumenthal, Oskar – Berlin: G Grote, 1875 [mf ed 1993] – 4v in 2 – 1 – (incl bibl ref) – mf#8698 – us Wisconsin U Libr [802]

Christian difficulties in the second and twentieth centuries : a study of marcion and his relation to modern thought / Foakes-Jackson, Frederick John – Cambridge: W Heffer 1903 [mf ed 1990] – (= ser Hulsean lectures) – 1mf – 9 – 0-7905-7583-3 – mf#1989-0808 – us ATLA [240]

Christian diligence / Rawnsley, R Drummond – London, England. 1855 – 1r – us UF Libraries [240]

Christian disciple and theological review – Boston. 1813-1823 (1) – mf#3688 – us UMI ProQuest [240]

Christian dispensation miraculous / Boys, Thomas – London, England. 1831 – 1r – us UF Libraries [240]

Christian doctrine / Dale, Robert William – New York: A C Armstrong 1895 [mf ed 1985] – 1mf – 9 – 0-8370-3440-X – (incl bibl ref) – mf#1985-1440 – us ATLA [240]

Christian doctrine / Greene, William Brenton – Philadelphia: Westminster Press 1906, c1905 [mf ed 1985] – 1mf – 9 – 0-8370-4803-6 – mf#1985-2803 – us ATLA [240]

Christian doctrine and morals viewed in their connexion : being the 24th fernley lecture / Findlay, George Gillanders – London: C H Kelly 1894 [mf ed 1990] – 1mf – 9 – 0-7905-3741-9 – mf#1989-0234 – us ATLA [230]

Christian doctrine and practice in the 12th century / ed by [Cornwallis, Caroline Frances] – London: William Pickering 1850 [mf ed 1991] – (= ser Small books on great subjects 17) – 1mf – 9 – 0-524-01105-2 – mf#1990-0319 – us ATLA [931]

Christian doctrine and systematic theology / Schultze, Augustus – 2nd rev ed. Bethlehem PA: Bethlehem Print Co 1914 [mf ed 1991] – 1mf – 9 – 0-524-00340-8 – mf#1989-3040 – us ATLA [242]

Christian doctrine harmonized and its rationality vindicated / Kedney, John Steinfort – New York: G P Putnam 1889, c1888-89 – 2v on 2mf – 9 – 0-7905-7862-X – mf#1989-1087 – us ATLA [240]

Christian doctrine in contrast with hinduism and islam : intended for young missionaries in north india / Hooper, William – [2nd ed] [London] 1896 – (= ser 19th c british colonization) – 2mf – 9 – mf#1.1.5273 – uk Chadwyck [230]

The christian doctrine of god / Clarke, William Newton – New York: C Scribner, 1910, c1909 – (= ser The International Theological Library) – 2mf – 9 – 0-7905-3656-0 – mf#1989-0149 – us ATLA [240]

The christian doctrine of god : lectures / Sparrow-Simpson, William John – London: Published for the S Paul's Lecture Society by R Flint, 1906 – 1mf – 9 – 0-7905-9675-X – (incl bibl ref) – mf#1989-1400 – us ATLA [240]

The christian doctrine of immortality / Salmond, Stewart Dingwall Fordyce – 4th rev ed. Edinburgh: T & T Clark, 1901 – (= ser Cunningham Lectures) – 2mf – 9 – 0-7905-8575-8 – (incl bibl ref) – mf#1989-1800 – us ATLA [240]

The christian doctrine of justification and reconciliation = Die positive entwickelung der lehre / Ritschl, Albrecht; ed by Mackintosh, Hugh Ross & Macaulay, Alexander Beith – [3rd ed] Edinburgh: T & T Clark, 1900 [mf ed 1991] – 2mf – 9 – 0-524-00084-0 – (in english) – mf#1989-2784 – us ATLA [240]

The christian doctrine of prayer : an essay / Clarke, James Freeman – 8th ed. Boston: American Unitarian Association, 1874, c1854 – 1mf – 9 – 0-7905-1581-4 – mf#1987-1581 – us ATLA [240]

The christian doctrine of prayer / ed by Hastings, James – New York: Scribner, 1915 – (= ser The Great Christian Doctrines) – 2mf – 9 – 0-7905-3947-0 – (incl bibl ref) – mf#1989-0440 – us ATLA [240]

The christian doctrine of prayer for the departed / Lee, Frederick George – London: Strahan, 1872 – 1mf – 9 – 0-7905-9780-2 – (incl bibl ref) – mf#1989-1505 – us ATLA [240]

The christian doctrine of salvation / Stevens, George Barker – New York: C. Scribner, 1905 – (= ser The International Theological Library) – 2mf – 9 – 0-7905-3290-5 – (incl bibl ref) – mf#1987-3290 – us ATLA [240]

The christian doctrine of sin = Die christliche lehre von der suende / Mueller, Julius – Edinburgh: T & T Clark 1868 [mf ed 1991] – (= ser Clark's foreign theological library) – 2v on 2mf – 9 – 0-7905-8861-7 – (trans fr 5th german ed by william urwick) – mf#1989-2086 – us ATLA [240]

The christian doctrine of sin / Tulloch, John – New York: Scribner, Armstrong, [1876?] – (= ser Croall Lectures) – 1mf – 9 – 0-7905-7613-9 – mf#1989-0838 – us ATLA [240]

The christian doctrine of the lord's supper / Adamson, Robert M – Edinburgh: T & T Clark; New York: Charles Scribner [distributor], 1905 – 1mf – 9 – 0-7905-1622-5 – (incl bibl ref and ind) – mf#1987-1622 – us ATLA [240]

The christian doctrine of the soul : an essay / Estes, Hiram Cushman – Boston: Noyes, Holmes, 1873 – 1mf – 9 – 0-524-08234-0 – mf#1993-2009 – us ATLA [240]

Christian doctrines : a compendium of theology / Pendleton, James Madison – Philadelphia: American Baptist Pub Soc [c1878] [mf ed 1984] – 5mf – 9 – 0-8370-1066-7 – (incl bibl ref & ind) – mf#1984-4406 – us ATLA [242]

Christian doctrines and modern thought / Bonney, Thomas George – London, New York: Longmans, Green 1892 [mf ed 1990] – (= ser The boyle lectures 1891) – 1mf – 9 – 0-7905-3593-9 – mf#1989-0086 – us ATLA [240]

Christian dogmatics : a compendium of the doctrines of christianity = Christelige dogmatik / Martensen, Hans – Edinburgh: T & T Clark 1866 [mf ed 1986] – (= ser Clark's foreign theological library. 4th series 12) – 2mf – 9 – 0-8370-9717-7 – (incl ind. trans fr german by william urwick) – mf#1986-3717 – us ATLA [240]

Christian duty of feeding the poor of the flock / Bickersteth, Edward Henry – London, England. 1845? – 1r – us UF Libraries [240]

Christian duty of granting the claims of the roman catholics / Arnold, Thomas – Oxford, England. 1829 – 1r – us UF Libraries [240]

The christian ecclesia : a course of lectures on the early history and early conceptions of the ecclesia, and four sermons / Hort, Fenton John Anthony – London: Macmillan, 1897 – 1mf – 9 – 0-524-07655-3 – mf#1992-1096 – us ATLA [240]

Christian Echoes National Ministry, Inc *see* Christian crusade weekly

Christian economics – 1950-57, 1958-1971 apr, 1971 may-1972 dec – 3r – 1 – mf#1054403 – us WHS [240]

Christian economics / Richmond, Wilfrid – New York: E P Dutton 1888 [mf ed 1992] – 1mf – 9 – 0-524-02799-4 – mf#1990-0703 – us ATLA [230]

Christian economics with reference to the land question / Spicer, Albert – London, England. 1891 – 1r – us UF Libraries [240]

Christian education / O'Connell, Cornelius Joseph – New York: Benziger 1906 [mf ed 1986] – 1mf – 9 – 0-8370-6925-4 – mf#1986-0925 – us ATLA [377]

Christian education : a sermon by maurice s baldwin...preached in christ church cathedral, sunday, january 22, 1871 – Montreal: J Lovell, 1871 [mf ed 1980] – 1mf – 9 – 0-665-00839-2 – mf#00839 – cn Canadiana [240]

Christian education : a sermon preached in christ church cathedral, sunday, january 22, 1871 / Baldwin, Maurice Scollard – Montreal: J Lovell, 1871 – 1mf – 9 – mf#00839 – cn Canadiana [240]

Christian education *see* Methodist education

Christian education and the national consciousness in china *see* Tsung chiao chiao yu yu kuo hun (ccm320)

Christian education in america : a lecture / Keane, John J – Washington DC: Church News Pub Co 1892 [mf ed 1986] – 1mf – 9 – 0-8370-7551-3 – mf#1986-1551 – us ATLA [241]

Christian education in the first centuries : a d 33-a d 476 / Magevney, Eugene – New York: Cathedral Library Assoc 1900 [mf ed 1986] – 1mf – 9 – 0-8370-7960-8 – (incl bibl ref) – mf#1986-1960 – us ATLA [240]

Christian education journal – Glen Ellyn. 1985+ – 1,5,9 – ISSN: 0739-8913 – mf#15299,01 – us UMI ProQuest [377]

Christian education newsletter / Urban Outreach [Organization] – 1994 spr – 1r – 1 – mf#4023934 – us WHS [240]

Christian education the remedy for the growing ungodliness of the times / Dix, Morgan – Boston: publ...by E P Dutton [1866?] [mf ed 1986] – 1mf – 9 – 0-8370-7785-0 – mf#1986-1785 – us ATLA [230]

Christian Educational Association *see* Common sense

Christian egypt : past, present, and future / Fowler, Montague – London: Church Newspaper Co, 1901 – 1mf – 9 – 0-8370-7632-3 – (incl bibl ref and index) – mf#1986-1632 – us ATLA [240]

Christian endeavor / Wisconsin Christian Endeavor Union – 1895 feb-1898 mar – 1r – 1 – (cont: wisconsin christian endeavorer) – mf#1054404 – us WHS [071]

Christian endeavor world / Franklin Co. Columbus – nov 1946-win 77, (aug 78-1985) [irreg] – 5r – 1 – mf#B27900-27904 – us Ohio Hist [071]

Christian endeavor world / International Society of Christian Endeavor – v16 [1901 oct-1902 sep] – 1r – 1 – (cont: golden rule [boston ma: 1886]; cont by: christian endeavor world quarterly) – mf#1430809 – us WHS [240]

The christian endeavour manual for india, burma, and ceylon – Agra: India Christian Endeavour Union, 1909 [mf ed 1995] – (= ser Yale coll) – ii/177p (ill) – 1 – 0-524-09488-8 – mf#1995-0488 – us ATLA [240]

Christian endeavour world / Boston, MA – oct 1886-sep 1900 [wkly] – 15r – 1 – (incl precedessors) – mf#B8263-8277 – us Ohio Hist [240]

Christian endeavour world / Boston, MA – oct 1900-sep 1911 [wkly] – 11r – 1 – mf#B8278-8288 – us Ohio Hist [240]

Christian endeavour world / Boston, MA – oct 1911-oct 1946 [irreg] – 19r – 1 – mf#B27881-27899 – us Ohio Hist [240]

Christian epigraphy : an elementary treatise = Epigrafia cristiana / Marucchi, Orazio – Cambridge: University Press 1912 [mf ed 1991] – 2mf [ill] – 9 – 0-524-00573-7 – (incl bibl ref: trans by j armine willis) – mf#1990-0073 – us ATLA [240]

Christian epoch-makers : the story of the great missionary eras in the history of christianity / Vedder, Henry Clay – Philadelphia: Griffith & Rowland 1908 [mf ed 1986] – 1mf – 9 – 0-8370-6438-4 – (incl bibl & ind) – mf#1986-0438 – us ATLA [240]

Christian equality – Edinburgh, Scotland. 1849 – 1r – us UF Libraries [240]

Christian equality / Kyle, Robert Wood – London, England. 1837? – 1r – us UF Libraries [240]

Christian ernst von brandenburg-baireuth : die aufnahme reformirter fluechtlingsgemeinden in ein lutherisches land, 1686-1712 / Ebrard, Johannes Heinrich August – Guetersloh: C Bertelsmann 1885 [mf ed 1992] – 1mf [ill] – 9 – 0-524-03897-X – (incl bibl ref) – mf#1990-1156 – us ATLA [242]

Christian essentials : a re-statement for the people of to-day / Ballard, Frank – London: Robert Culley [1907?] [mf ed 1991] – 1mf – 9 – 0-7905-8758-0 – (incl bibl ref) – mf#1989-1983 – us ATLA [240]

The christian ethic / Knight, William Angus – London: John Murray, 1893 – 1mf – 9 – 0-7905-8818-8 – mf#1989-2043 – us ATLA [170]

The christian ethic of war / Forsyth, Peter Taylor – London; New York: Longmans, Green, 1916 – 1mf – 9 – 0-7905-7734-8 – mf#1989-0959 – us ATLA [240]

Christian ethics = Christelige ethik / Martensen, Hans – Edinburgh: T & T Clark [187-?] [mf ed 1986] – (= ser Clark's foreign theological library. 3rd series 39) – 2mf – 9 – 0-8370-9560-3 – (incl bibl ref & ind; trans fr danish by c spence) – mf#1986-3560 – us ATLA [230]

Christian ethics = Christelige ethik / Martensen, Hans – Edinburgh: T & T Clark 1899 [mf ed 1986] – (= ser Clark's foreign theological library. new series 11) – 1mf – 9 – 0-8370-9562-X – (incl bibl ref & ind; trans fr german by sophia taylor) – mf#1986-3562 – us ATLA [230]

Christian ethics = Christelige ethik / Martensen, Hans – 4th ed. Edinburgh: T & T Clark [189-?] [mf ed 1986] – (= ser Clark's foreign theological library. new series 7) – 2mf – 9 – 0-8370-9561-1 – (trans fr german by william affleck) – mf#1986-3561 – us ATLA [230]

Christian ethics : eight lectures...oxford in the year 1895... / Strong, Thomas Banks – London, New York: Longmans, Green 1896 [mf ed 1986] – (= ser Bampton lectures 1895) – 1mf – 9 – 0-8370-6417-1 – (incl bibl ref & ind) – mf#1986-0417 – us ATLA [230]

Christian ethics = Handbuch der christlichen sittenlehre / Wuttke, Adolf – New York: Nelson & Phillips 1873 [mf ed 1991] – 2v on 2mf – 9 – 0-7905-8751-3 – (incl bibl ref; trans by john p lacroix; pref by dr riehm) – mf#1989-1976 – us ATLA [230]

Christian ethics / Smyth, Newman – New York: Scribner 1892 [mf ed 1990] – (= ser The international theological library) – 2mf – 9 – 0-7905-7470-5 – mf#1989-0695 – us ATLA [230]

Christian ethics : a system based on martensen and harless / Weidner, Revere Franklin – 3rd rev ed. Chicago, IL: Wartburg, c1897 [mf ed 1991] – 2mf – 9 – 0-7905-9653-9 – mf#1989-1378 – us ATLA [230]

Christian ethics *see* Chi-tu chiao ti lun li [ccm70]

Christian ethics and modern thought / d'Arcy, Charles Frederick – London, New York: Longmans, Green 1912 [mf ed 1990] – (= ser Anglican church handbooks) – 1mf – 9 – 0-7905-3778-8 – mf#1989-0271 – us ATLA [230]

Christian ethics and social progress / Harper, James Wilson – London: J Nisbet 1912 [mf ed 1990] – 1mf – 9 – 0-7905-3855-5 – mf#1989-0348 – us ATLA [230]

Christian ethics and wise sayings – London: James Nisbet 1883 [mf ed 1986] – 1mf – 9 – 0-8370-6078-8 – (incl ind) – mf#1986-0078 – us ATLA [880]

The christian eucharist and the pagan cults / Groton, William Mansfield – New York: Longmans, Green, 1914 – 1mf – 9 – 0-7905-1408-7 – (incl bibl ref and index) – mf#1987-1408 – us ATLA [240]

The christian eucharist as it might be celebrated in unitarian congregational churches / Silliman, Vincent B – [Chicago, 1929] Chicago: Dep of Photodup, U of Chicago Lib, 1971 (1r); Evanston: American Theol Lib Assoc, 1984 (1r) – 1 – 0-8370-0283-4 – mf#1984-B154 – us ATLA [243]

The christian evangel – Chicago, IL. v1-23. 1910-33 [complete] – (= ser Mennonite serials coll) – 4r – 1 – mf#ATLA 1993-S012 – us ATLA [242]

The christian evangel – Scottdale, PA. v24-45. 1936-57 [complete] – (= ser Mennonite serials coll) – 3r – 1 – mf#ATLA 1993-S013 – us ATLA [242]

Christian evidences / Robinson, Ezekiel Gilman – New York: Silver, Burdett, 1895 [mf ed 1985] – 1mf – 9 – 0-8370-4935-0 – (incl bibl ref & ind) – mf#1985-2935 – us ATLA [240]

Christian evidences viewed in relation to modern thought : eight lectures / Row, Charles Adolphus – 3rd ed. London: F. Norgate; Edinburgh: Williams & Norgate, 1881 [mf ed 1989] – (= ser Bampton lectures 1877) – 2mf – 9 – 0-7905-3052-X – mf#1987-3052 – us ATLA [240]

Christian examiner – Lexington. Kentucky. v. 1. 1830 – 1 – us Southern Baptist [242]

Christian examiner – London. Jan 1841-June 1848. -m. 1mqn reels – 1 – uk British Libr Newspaper [072]

Christian examiner – New York. 1824-1869 (1) – mf#3836 – us UMI ProQuest [240]

Christian examiner and church of ireland magazine – Dublin, Ireland. Feb 1855-jun 1868 – 5r – 1 – uk British Libr Newspaper [072]

Christian exertion : or, the duty of private members of the church of christ to labor for the souls of men, explained and enforced / ed by Peck, George – New York:...for the Methodist Episcopal Church, 1845 [mf ed 1984] – 2mf – 9 – 0-8370-0785-2 – mf#1984-4117 – us ATLA [240]

Christian experience / Cox, John – London, England. 1849? – 1r – us UF Libraries [240]

Christian experience / Franklin, B – London, England. 1855 – 1r – us UF Libraries [240]

Christian experience : or, sincerity seeking the way to heaven / Franklin, Benjamin – Cincinnati, Ohio: Standard Pub Co, [18–?] – 1mf – 9 – 0-524-07817-3 – mf#1991-3364 – us ATLA [240]

Christian experience : or, the spiritual exercises of eminent christ – Edinburgh, Scotland. 1825 – 1r – us UF Libraries [240]

The christian experience : an inquiry into its character and its contents / Faunce, Daniel Worcester – Philadelphia: American Baptist Publication Society, c1881 – 1mf – 9 – 0-7905-7730-5 – mf#1989-0955 – us ATLA [240]

The christian exponent – v1-5. 1924-28 [complete] – (= ser Mennonite serials coll) – Inquire – 1 – mf#ATLA 1993-S010 – us ATLA [242]

Christian express *see* South african outlook

Christian facts and forces / Smyth, Newman – New York: Charles Scribner, 1887 – 1mf – 9 – 0-8370-9824-6 – mf#1986-3824 – us ATLA [240]

Christian faith : five sermons / Smith, William Saumarez – London: Macmillan, 1869 – 1mf – 9 – 0-8370-5166-5 – mf#1985-3166 – us ATLA [240]

Christian faith : its nature, object, causes, and effects / Godwin, John Henry – London: Jackson, Walford, & Hodder, 1862 – 1mf – 9 – 0-8370-4918-0 – mf#1985-2918 – us ATLA [240]

The christian faith : a system of dogmatics = Christliche glaube / Haering, Theodor – London; New York: Hodder and Stoughton, 1913 – 3mf – 9 – 0-7905-7827-1 – (incl bibl ref. in english) – mf#1989-1052 – us ATLA [240]

The christian faith / Curtis, Olin Alfred – New York: Eaton & Mains; Cincinnati: Jennings & Graham, c1905 – 1mf – 9 – 0-7905-3664-1 – (incl bibl ref) – mf#1989-0157 – us ATLA [240]

Christian faith and life – v37-45. 1931-39 [complete] – Inquire – 1 – mf#ATLA 1993-S518 – us ATLA [240]

Christian faith and life *see* Chi-tu tu te hsin yang yu sheng huo [ccm150]

The christian faith and the old testament / Thomas, John M – New York: Thomas Y. Crowell, c1908 – 1mf – 9 – 0-8370-5515-6 – (incl bibl ref) – mf#1985-3515 – us ATLA [221]

Christian faith, comprehensive, not partial; definite, not uncertain : eight sermons / Jelf, William Edward – Oxford: J H & Jas Parker [distributor], 1857 – 1mf – 9 – 0-7905-0955-5 – mf#1987-0955 – us ATLA [240]

Christian faith in an age of science / Rice, William North – [3rd ed.] New York: Hodder & Stoughton: George H Doran c1903 [mf ed 1985] – (= ser Library of standard religious authors) – 1mf – 9 – 0-8370-4892-3 – (incl ind) – mf#1985-2892 – us ATLA [210]

The christian faith in japan / Moore, Herbert – 2nd ed. Westminster: Society for the Propagation of the Gospel in Foreign Parts, 1904 – 1mf – 9 – 0-524-01005-6 – mf#1990-0282 – us ATLA [240]

The christian faith under modern searchlights / Johnson, William Hallock – New York: FH Revell, c1916 – (= ser Stone Lectures) – 1mf – 9 – 0-7905-7845-X – (incl bibl ref) – mf#1989-1070 – us ATLA [240]

The christian family *see* Chi-tu t'u chia t'ing [ccm225]

Christian family advocate – [Scotland] Edinburgh: W Whyte 15 may 1852-1 jun 1854, jan 1855-1 jan 1856 (mthly) [mf ed 2003] – 4v on 3r – 1 – (cont by: christian family advocate and literary review) – uk Newsplan [242]

Christian family chronicles – 1979 jan-1983 jul, 1988 jan-jul – 1r – 1 – mf#816986 – us WHS [929]

Christian family companion – Tyrone City PA: H R Holsinger 1865-73 [wkly] [mf ed 2005] – 9v on 4r – 1 – (preceded by 2 int iss dated may 10 1864 & oct 4 1864; with ind; merged with: gospel-visitor to form: christian family companion and gospel visitor) – mf1073 – us ATLA [242]

Christian family companion *see* The gospel-visitor

Christian family companion and gospel visitor / ed by Quinter, James – Dale City PA: J Quinter 1874-75 [wkly] [mf ed 2005] – v1-2 (1874-1875) on 1r – 1 – (publ in meyersdale, pa beginning with v1 n29 (jul 21 1874); iss for 1874-75 also carry vol numbering of previous titles: christian family companion, v10-11; and gospel visitor, v24-25; with ind; formed by the union of: christian family companion and: gospel-visitor; cont by: primitive christian (meyersdale, pa); some pgs damaged) – mf1071 – us ATLA [242]

Christian family companion and gospel visitor *see* The gospel-visitor

The christian family in changing east asia *see* Yen pien chung te tung-ya chi-tu hua chia t'ing sheng huo (ccm246)

Christian farmer *see* T'ien chia (ccs38)

The christian fathers / Perry, George Gresley – London: SPCK, [1893?] – 1mf – 9 – 0-524-04847-9 – mf#1990-1339 – us ATLA [240]

Christian felix weisse und seine beziehungen zur deutschen literatur des achtzehnten jahrhunderts / Minor, Jacob – Innsbruck: Wagner 1880 [mf ed 1991] – 1r – 1 – (incl bibl ref. filmed with: "und alles ist zerstoben" / werner weisbach) – mf#3038p – us Wisconsin U Libr [430]

Christian fellowship : a letter to the right rev l meurin...bishop, vicar-apostolic of bombay / Rivington, Luke – Bombay: Education Society's Press, 1883 [mf ed 1995] – (= ser Yale coll) – 11p – 1 – 0-524-09868-9 – mf#1995-0868 – us ATLA [241]

Christian fellowship hymns *see* T'uan ch'i sheng ko chi (ccm80)

Christian fellowship monthly *see* En yu (ccs)

Christian focus : a series of college sermons / McConnell, Francis John – Cincinnati: Jennings & Graham, c1911 – 1mf – 9 – 0-7905-9797-7 – mf#1989-1522 – us ATLA [240]

Christian Foodship Committee *see* Christians and spain

Christian fraternity – Edinburgh, Scotland. 1849 – 1r – us UF Libraries [240]

Christian free schools : the subject discussed at rochester, n.y. / McQuaid, Bernard John – [Rochester, NY?: s.n., 1872?] – 1mf – 9 – 0-8370-7718-4 – mf#1986-1718 – us ATLA [377]

Christian freedom / Macgregor, William Malcolm – New York: Hodder & Stoughton 1914 [mf ed 1990] – (= ser The baird lecture 1913) – 1mf – 9 – 0-7905-7913-8 – (incl bibl ref) – mf#1989-1138 – us ATLA [225]

Christian freeman – Hartford CT. 1843 jan 6-1845 dec 25 – 1r – 1 – (cont by: charter oak) – mf#875325 – us WHS [071]

Christian freeman : the monthly organ of the scottish unitarian association – [Scotland] Glasgow: A MacLaren & Sons oct 1919-feb 1926 (mtly) [mf ed 2004] – 7v on 2r – 1 – uk Newsplan [242]

Christian friedrich hunold [menantes] : [1681-1721]: sein leben und seine werke: eine monographie / Vogel, Hermann – Leipzig: E Graefe [1898] [mf ed 1990] – 1r – 1 – (incl bibl ref; filmed with: ricarda huch / gertrud baumer) – mf#2734p – us Wisconsin U Libr [430]

Christian friedrich junii kurzgefasste reformations- geschichte : aus des hrn. veit ludwigs von seckendorf historia lutheranismi = Commentarius historicus et apologeticus de lutheranismo. selections / Seckendorf, Veit Ludwig von; ed by Lindner, Benjamin – Stereotyp-Ausg. Baltimore: A Schlitt, 1865 – 10mf – 9 – 0-524-07973-0 – (in german) – mf#1990-5418 – us ATLA [242]

Christian friedrich scherenberg : und das literarische berlin von 1840 bis 1860 / Fontane, Theodor – Berlin: W Hertz, 1885 [mf ed 1995] – 260p – 1 – mf#8864 – us Wisconsin U Libr [430]

The christian front – v1-4. 1936-39 [complete] – Inquire – 1 – mf#ATLA 1994-S500 – us ATLA [240]

Christian frontiers : a journal of baptist life and thought – North Carolina: Baptist Book Club. v1-2. 736p – 1r – 1 – mf#7056 – us Southern Baptist [242]

Christian fuerchtegott gellert : dichter und erzieher / Durach, Moritz – Dresden: Verlag Heimatwerk Sachsen 1938 [mf ed 1989] – 1r – 1 – (filmed with: emanuel geibel / ed by arno holz) – mf#7287 – us Wisconsin U Libr [430]

The christian fulfilments and uses of levitical sin-offering / Batchelor, Henry – London: James Nisbet 1887 [mf ed 1989] – (= ser Nisbet's theological library) – 1mf – 9 – 0-7905-1023-5 – (incl bibl ref) – mf#1987-1023 – us ATLA [240]

The christian fundamentalist – Glendale CA: World's Christian Fundamentals Assoc. v1-2 (1941-1942), v1-8 (1943-1951) [mf ed 2005] – 2r [ill] – 1 – mf#2005c-s026 – us ATLA [240]

The christian fundamentalist – v1-6. 1927-32 [complete] – Inquire – 1 – mf#ATLA 1993-S513 – us ATLA [240]

The christian gem – Halifax, NS: W Cunnabell & J Belcher, [1845] – 9 – mf#P04613 – cn Canadiana [240]

Christian giving / Mackay, W P – London, England. 18– – 1r – us UF Libraries [240]

Christian giving illustrated and enforced by ancient tithing : a discourse preached in st paul's church, montreal, on sunday morning, feb. 13, 1881 / Jenkins, John – Montreal?: Mitchell & Wilson, 1881? – 1mf – 9 – mf#08443 – cn Canadiana [240]

The christian gleaner – [Halifax, NS?: s.n, 1833-1838?] – 9 – (incl ind) – mf#P04296 – cn Canadiana [240]

The christian graces : a series of discourses on faith and its fruits / Trail, William – 2nd ed. Glasgow: William Collins; London: James Nisbet, 1859 – 1mf – 9 – 0-7905-2389-2 – mf#1987-2389 – us ATLA [240]

Christian guide for plain people / Miller, John – London, England. 1821 – 1r – us UF Libraries [240]

Christian health and happiness – [Scotland] Glasgow: Maclaren & Sons 1 jan-mar 1892 (mthly) [mf ed 2003] – 1r – 1 – (cont: health and happiness) – uk Newsplan [240]

The christian helper : a baptist monthly journal for christian workers – Toronto: Dudley & Burns, [1877-18–?] – 9 – mf#P05056 – cn Canadiana [242]

Christian herald – Chappaqua. 1941-1992 (1) 1970-1992 (5) 1973-1992 (9) – ISSN: 0009-5354 – mf#2169 – us UMI ProQuest [240]

Christian herald – 1928 dec 8-1929 aug 31, 1929 sep 7-1930 may 31 – 2r – 1 – (cont: american messenger; christian herald and signs of our times; world outlook) – mf#765752 – us WHS [071]

Christian herald – Portland OR: D T Stanley, [wkly] – 1 – (cont: pacific christian messenger [1870-77]) – us Oregon Lib [071]

Christian herald – Portsmouth. 1818-1825 (1) – mf#4426 – us UMI ProQuest [975]

Christian herald *see*
– The herald and torchlight
– Michigan christian herald

The christian herald : organ of the micgigan baptist convention – Detroit MI: Rev L H Trowbridge [wkly] [mf v5-33 1877-1902 filmed 1981] – 10r – 1 – (with gaps. name changed fr: herald and torchlight to relfect the revival of the michigan christian herald. incl: some iss of michigan christian centre [detroit 1842]; v2-4 of herald and torchlight and some iss of: michigan christian herald [detroit 1902]. some iss called: a wide-awake baptist newspaper) – mf#r0132c – us ATLA [242]

Christian herald and seaman's magazine – New York. 1816-1824 (1) – mf#4427 – us UMI ProQuest [240]

Christian heritage – Hackensack. 1883-1978 [1]; 1971-1978 [5]; 1977-1978 [9] – ISSN: 0009-5362 – mf#1919 – us UMI ProQuest [240]

A christian hero : life of rev william cassidy / Simpson, Albert Benjamin – [New York?: s.n, 1888?] [mf ed 1992] – (= ser Christian & missionary alliance coll) – 1mf – 9 – 0-524-04718-9 – mf#1990-5070 – us ATLA [920]

Christian heroism in heathen lands / Royer, Galen Brown – Elgin, IL: Brethren Pub House, 1914 – 1mf – 9 – 0-524-03563-6 – mf#1990-4758 – us ATLA [240]

Christian higher education in china : a study for the year 1925-26 / Cressy, Earl Herbert – Shanghai: China Christian Educational Association, [1928]. Chicago: Dep of Photodup, U of Chicago Lib, 1975 (1r); Evanston: American Theol Lib Assoc, 1984 (1r) – 1 – 0-8370-0547-7 – (incl ind) – mf#1984-B432 – us ATLA [377]

Christian history : containing accounts of the revival and propagation of religion in great britain, america, etc – Boston. 1743-1745 (1) – mf#3516 – us UMI ProQuest [240]

Christian history – Worcester. 1989+ (1,5,9) – ISSN: 0891-9666 – mf#17630 – us UMI ProQuest [240]

Christian history in its three great periods / Allen, Joseph Henry – Boston: Roberts Bros., 1884-1890, c1882-1883 – 3mf – 9 – 0-7905-4424-5 – (incl bibl ref) – mf#1988-0424 – us ATLA [240]

Christian hofmann von hofmannswaldau : ein beitrag zur literaturgeschichte des siebzehnten jahrhunderts / Ettlinger, Josef – Halle a.d.S: M Niemeyer 1891 [mf ed 1990] – 1r – 1 – (incl bibl ref. filmed with: e t a hoffmann / werner bergengruen) – mf#2728p – us Wisconsin U Libr [430]

Christian hofmann von hofmannswaldaus grabschriften / Friebe, Karl – Greifswald: F W Kunike 1893 [mf ed 1990] – (= ser Jahresbericht ueber das staedtische gymnasium und das mit demselben verbundene realgymnasium zu greifswald 1892-93) – 1r – 1 – (filmed with: e t a hoffmann / werner bergengruen) – mf#2728p – us Wisconsin U Libr [430]

Christian holiness almanac and year book / ed by Hughes, George – New York: Palmer & Hughes 1884-88 [mf ed 2004] – 1884-88 [complete] 5v on 1r – 1 – (cont by: illustrated holiness year book for...) – mf1055a – us ATLA [240]

Christian home education *see* Chi-tu chiao hua ti chiao ting chiao yu [ccm2]

Christian hope *see* Hsi wang yueh k'an (ccs26)

The christian hope : a study in the doctrine of immortality / Brown, William Adams – London: Duckworth 1912 [mf ed 1990] – (= ser Studies in theology) – 1mf – 9 – 0-7905-3700-1 – mf#1989-0193 – us ATLA [240]

The christian hope *see* Chi-tu tu te hsi wang [ccm330]

The christian hope in the apocalypse / Mozley, John Kenneth – London: R Scott, 1915 – 1mf – 9 – 0-7905-9527-3 – mf#1989-1232 – us ATLA [221]

Christian horizons – v1-9. 1938-46 (complete) – 1r – 1 – mf#ATLA 1994-S516 – us ATLA [240]

Christian hymns : or, a collection of spiritual songs – 2nd corr enl ed. Nashville TN: printed by Thomas G Bradford 18[13?] [mf ed 2005] – 1r – 1 – 0-524-10537-5 – (without music; incl ind; lacks: p77-78, some pgs damaged) – us ATLA [240]

Christian iconography : or, the history of christian art in the middle ages = Iconographie chretienne / Didron, Adolphe Napoleon – London: Henry G Bohn, 1851-91 [mf ed 1990] – 2v on 3mf – 9 – 0-7905-8025-X – (english trans by e j millington; incl bibl ref) – mf#1988-6006 – us ATLA [700]

Christian iconography : or, the history of christian art in the middle ages / Didron, Adolphe Napoleon – London 1851-86 – (= ser 19th c art & architecture) – 11mf – 9 – mf#4.2.1188 – uk Chadwyck [700]

The christian idea of atonement : lectures / Tymms, Thomas Vincent – London, New York: Macmillan 1904 [mf ed 1991] – (= ser Angus lectureship) – 2mf – 9 – 0-7905-8608-8 – mf#1989-1833 – us ATLA [240]

The christian idea of education as distinguished from the secular idea of education / Robins, Henry Ephraim – Philadelphia: American Baptist Publication Society, 1895 – 1mf – 9 – 0-524-00085-9 – mf#1989-2785 – us ATLA [377]

The christian idea of sacrifice : a discourse preached...on sunday, 12th september, 1858 / Cordner, John – Montreal: H Rose, 1858 – 1mf – 9 – mf#67008 – cn Canadiana [240]

Christian in his trade and profession / Woodford, James Russell – London, England. 1851 – 1r – us UF Libraries [240]

The christian in hungarian romance : a study of dr. maurus jokai's novel, there is a god, or, the people who love but once / Fretwell, John – Boston, USA: JH West Co, c1901 – 1mf – 9 – 0-524-07682-0 – mf#1991-3267 – us ATLA [490]

The christian in the world / Faunce, Daniel Worcester – Boston: Roberts, 1875. Beltsville, Md: NCR Corp, 1978 (3mf); Evanston: American Theol Lib Assoc, 1984 (3mf) – (= ser The Fletcher prize essay) – 9 – 0-8370-0924-3 – mf#1984-4248 – us ATLA [240]

The christian in war time / Lynch, Frederick et al – New York: Fleming H Revell, c1917 – 1mf – 9 – 0-524-03823-6 – mf#1990-1139 – us ATLA [240]

The christian index – Atlanta, GA. 222p. 1822-1999 – 1 – mf#0334 – us Southern Baptist [242]

The christian index – v69-78. 1938-46 [complete] – 9r – 1 – (v75 not publ) – mf#ATLA S0883 – us ATLA [240]

The christian inheritance / Hedley, John Cuthbert – London: Burns & Oates; New York: Benziger, [19–?] – 1mf – 9 – 0-8370-7155-0 – mf#1986-1155 – us ATLA [240]

Christian initiation and first communion / Lavis, Allan Albert – [Princeton, NJ: s.n.], 1978. Chicago: Dep of Photodup, U of Chicago Lib, 1979 (1r); Evanston: American Theol Lib Assoc, 1984 (1r) – 1 – 0-8370-1351-8 – mf#1984-T198 – us ATLA [240]

Christian inquirer *see* The inquirer

Christian instincts and modern doubt : essays and addresses in aid of a reasonable, satisfying, and consolatory religion / Craufurd, Alexander Henry – New York: T Whittaker, 1897 [mf ed 1985] – 1mf – 9 – 0-8370-2770-5 – mf#1985-0770 – us ATLA [230]

Christian institutions / Allen, Alexander Viets Griswold – New York: Scribner, 1897 – 2mf – 9 – 0-7905-4008-8 – (incl bibl ref) – mf#1988-0008 – us ATLA [240]

Christian institutions : essays on ecclesiastical subjects / Stanley, Arthur Penrhyn – 4th ed. London: John Murray, 1884 – 2mf – 9 – 0-524-00144-8 – mf#1989-2844 – us ATLA [240]

Christian institutions / Stanley, Arthur Penrhyn – London, England. 1881 – 1r – us UF Libraries [240]

The christian instructed in the nature and use of indulgences = Chretien eclaire sur la nature et l'usage des indulgences / Maurel, F Antoine – 6th rev enl ed. Dublin: M H Gill, 1901 – 1mf – 9 – 0-8370-7410-X – (in english. incl ind) – mf#1986-1410 – us ATLA [240]

Christian instructor and missionary register of the presbyterian church of nova scotia – [Halifax, N.S.?: s.n, 1856?-1860] – 9 – (cont by: the home and foreign record of the presbyterian church of the lower provinces of british north america) – mf#P05017 – cn Canadiana [242]

Christian intelligencer – Gardiner. 1821-1836 (1) – mf#4428 – us UMI ProQuest [240]

The christian intelligencer – [S.l: s.n. v1 1829] [mf ed 2004] – 1r – 1 – (cont by: christian intelligencer, and evangelical guardian; damaged: v1 n4 [apr 1829] p127) – mf#2004-S082 – us ATLA [240]

The christian intelligencer, and evangelical guardian – [S.l: s.n. v2-13. 1830-43] [mf ed 2004] – 3r – 1 – (suspended 1832-feb 1833; lacks: v2 p342-352; cont: christian intelligencer [1829]; cont by: evangelical guardian) – mf#2004-S083 – us ATLA [240]

Christian interpretation of history *see* Chi-tu chiao ti li shih kuan [ccm217]

The christian interpretation of life and other essays / Davison, William Theophilus – London: CH Kelly, 1898 – 1mf – 9 – 0-7905-9180-4 – mf#1989-2405 – us ATLA [240]

Christian jensen : ein lebensbild / Evers, Ernst – Breklum: Christlichen Buchh, 1908 – 1mf – 9 – 0-524-02912-1 – mf#1990-0728 – us ATLA [240]

Christian, John *see*
– Behar proverbs
– The oxford union murals

Christian, John T *see* History of baptists of louisiana

Christian, John Tyler *see*
– America or rome, which?
– Baptist history vindicated
– Did they dip?

Christian journal : and literary register – New York. 1817-1830 (1) – mf#3710 – us UMI ProQuest [240]

Christian journal – Dublin, Ireland. Feb-dec 1846 – 1/4r – 1 – uk British Libr Newspaper [072]

Christian journal [relief synod (scotland)] : by members of the relief church – [Scotland] Glasgow: J Reid jan 1833-dec 1845 [mf ed 2003] – 25r – 1 – (cont by: christian journal, or, relief magazine [jan 1838-jun 1845]) – uk Newsplan [240]

Christian journal [united presbyterian church (scotland)] – [Scotland] Glasgow: R Jackson jan 1850-jun 1855 (mthly) [mf ed 2004] – 6v on 8r – 1 – (separated fr: united presbyterian magazine) – uk Newsplan [240]

Christian joy / Knox-Little, William John – Manchester, England. 1877? – 1r – us UF Libraries [240]

Christian, Juan T *see* Inmersion; el acto del bautismo cristiano

Christian kingdom / Coleridge, Henry James – London, England. 1870 – 1r – us UF Libraries [240]

Christian labor herald – v24 n3-v41 n2 [1963 jun-1979 spr] – 1r – 1 – (cont: christian journal) – mf#665289 – us WHS [240]

Christian Labour Association of Canada *see* Guide

The christian lady's friend and family repository (london) – sep 1831-sep1833 – (= ser 19th c british periodicals) – r25 – 1 – us Primary [073]

Christian leader / United States Conference of Mennonite Brethren Churches – 1977-87 – 6r – 1 – mf#500710 – us WHS [243]

Christian leader [Glasgow, Scotland : 1882] – [Scotland] Glasgow: Aird & Coghill 4 jan 1882-6 jul 1905 (wkly) [mf ed 2003] – 62r – 1 – (cont by: weekly leader [jul 1902-jun 1904]; christian leader [jul 1904-jul 1905]) – uk Newsplan [240]

Christian Legal Society *see* Quarterly

Christian legends = Zur nachfolge christi / Buelow, Eduard von – London: W Swan Sonnenschein, [1884?] – 1mf – 9 – 0-524-04008-7 – (in english) – mf#1990-1180 – us ATLA [240]

Christian liberty – Edinburgh, Scotland. 1849 – 1r – us UF Libraries [240]

Christian liberty in its relation to the usages of the evangelical lutheran church : the substance of two sermons. delivered in st. mark's lutheran church, philadelphia... / Krauth, Charles Porterfield – Philadelphia: Henry B Ashmead, 1860 – 1mf – 9 – 0-524-08475-0 – mf#1993-3120 – us ATLA [242]

Christian librarian – Three Hills. 1982+ [1,5,9] – ISSN: 0412-3131 – mf#12525 – us UMI ProQuest [020]

Christian life – Wheaton. 1973-1988 (1) 1973-1988 (5) 1977-1988 (9) – ISSN: 0009-5427 – mf#8714 – us UMI ProQuest [240]

The christian life : its course, its hindrances, and its helps / Arnold, Thomas – from the 5th London ed. Philadelphia: Lindsay & Blakiston, 1856 – 1mf – 9 – 0-524-08331-2 – mf#1993-2021 – us ATLA [240]

The christian life / Russell, Elbert – Philadelphia: WH Jenkins, 1916 – 1mf – 9 – 0-524-06663-9 – mf#1991-2718 – us ATLA [240]

The christian life : a study / Bowne, Borden Parker – Cincinnati: Jennings & Pye; New York: Eaton & Mains, c1899 – 1mf – 9 – 0-8370-2813-2 – mf#1985-0813 – us ATLA [240]

Christian life and character of the civil institutions of the united states : developed in the official and historical annals of the republic / Morris, Benjamin Franklin – Philadelphia: George W Childs; Cincinnati: Rickey & Carroll, 1864 – 2mf – 9 – 0-7905-6247-2 – mf#1988-2247 – us ATLA [240]

[Christian life and doctrine pamphlets] / Simpson, Albert B – New York City: Christian Alliance Pub Co [1885?]-1915 [mf ed 1992] – (= ser Christian alliance tracts) – 1v on 3mf – 9 – 0-524-04238-1 – mf#1990-5029 – us ATLA [240]

Christian life and theology : or, the contribution of christian experience to the system of evangelical doctrine / Foster, Frank Hugh – New York: Fleming H Revell c1900 [mf ed 1985] – (= ser Stone lectures 1900) – 1mf – 9 – 0-8370-4956-3 – (incl bibl ref & ind) – mf#1985-2956 – us ATLA [240]

Christian life bulletin – 1955-58. Reel also includes Light, 1948-79. 406p – 1 – us Southern Baptist [242]

Christian Life Commission. Advisory Council of Southern Baptist Work with Negroes *see* Minutes, reports and correspondence

Christian life in germany : as seen in the state and the church / Williams, Edward Franklin – New York: Fleming H Revell, c1896 – 1mf – 9 – 0-524-01031-5 – mf#1990-0308 – us ATLA [240]

Christian life in song : or, hymns and hymn-writers of many lands and ages / Charles, Elizabeth Rundle – 4th ed. London: T Nelson, 1888 [mf ed 1992] – 1mf – 9 – 0-524-03812-0 – (earlier eds publ as: the voice of christian life in song; later eds as: te deum laudamus) – mf#1990-1128 – us ATLA [780]

The christian life in the modern world / Peabody, Francis Greenwood – New York: Macmillan, 1914 – (= ser John Calvin McNair Lectures) – 1mf – 9 – 0-7905-9836-1 – mf#1989-1561 – us ATLA [240]

Christian life in the primitive church / Dobschuetz, Ernst von; ed by Morrison, William Douglas – New York: G P Putnam; London: Williams and Norgate, 1904 – 2mf – 9 – 0-8370-9615-4 – (incl indes) – mf#1986-3615 – us ATLA [240]

[Christian life pamphlets] / Montgomery, Carrie Judd – Oakland CA: Triumphs of Faith, [19–?] [mf ed 1992] – (= ser Christian & missionary alliance coll) – 1v on 1mf – 9 – 0-524-04225-X – mf#1990-5016 – us ATLA [240]

[Christian life pamphlets] / Pardington, George Palmer – [s.l: s.n, 1898-1915?] – (= ser Christian & missionary alliance coll) – 1mf – 9 – 0-524-03731-0 – mf#1990-4836 – us ATLA [240]

Christian life weekly *see* Chi-tu-chiao sheng-ho chou-k'an [ccs47]

Christian literature – New York. 1889-1897 (1) – mf#2872 – us UMI ProQuest [240]

Christian literature in the mission field : a survey of the present situation... / Ritson, John Holland – Edinburgh: Continuation Committee of the World Missionary Conference, 1910 – 1mf – 9 – 0-7905-8067-5 – mf#1988-6048 – us ATLA [240]

Christian Literature Society for China *see* Christian literature society for china

Christian literature society for china : annual reports / Christian Literature Society for China – v1-60. 1887-1947 – 2r – 1 – (lacks v54-58 1941-45) – mf#ATLA S0111 – us ATLA [240]

Christian living / Meyer, Frederick Brotherton – New York: Fleming H Revell, c1892 – 1mf – 9 – 0-8370-7173-9 – mf#1986-1173 – us ATLA [240]

Christian looks at the jewish question / Maritain, Jacques – New York, NY. 1939 – 1r – us UF Libraries [025]

Christian love : or, charity an essential element of true christian character / Wise, Daniel – New-York: Lane & Scott, 1850 – 2mf – 9 – 0-524-07774-6 – mf#1991-3342 – us ATLA [240]

Christian loyalty / Mortimer, Thomas – Wycombe, England. 1820 – 1r – us UF Libraries [240]

Christian lunds relation til kong frederik 3 om david danells tre rejser til gronland 1652-1654 / Lund, C; ed by Bobe, L – Kobenhavn, 1916. v2 – 2mf – 9 – mf#N-298 – ne IDC [919]

The christian lyre / Leavitt, Joshua – Collections of hymns and tunes. Boston. 1832 – 1 – us Southern Baptist [242]

Christian lyrics for public and social worship / ed by Webb, E – 6th ed. Nagercoil: Madras Tract and Book Society, 1878 [mf ed 1995] – (= ser Yale coll) – 465p – 1 – 0-524-10132-9 – (in tamil. parallel title in tamil characters. bound with the tamil hymn book) – mf#1995-1132 – us ATLA [780]

Christian lyrics for public and social worship *see* The tamil hymn book

Christian magazine – Providence. 1824-1827 (1) – mf#4364 – us UMI ProQuest [240]

Christian magistrate / Houston, Thomas – Belfast, Northern Ireland. 1832 – 1r – us UF Libraries [240]

The christian man, the church and the war / Speer, Robert Elliott – New York: Macmillan, 1918 – 1mf – 9 – 0-524-06500-4 – mf#1991-2600 – us ATLA [240]

Christian mandate – 1986 apr-1987 apr – 1r – 1 – (cont: christian action digest; cont by: aids protection) – mf#1289067 – us WHS [240]

Christian manifesting his lord's glory / Lear, Francis – Salisbury, England. 1859 – 1r – us UF Libraries [240]

Christian marriage *see* Chi-tu hua te hun yin [ccm117]

Christian martyrdom in russia : an account of the members of the universal brotherhood or doukhobortsi, now migrating from the caucasus to canada / ed by Chertkov, Vladimir – Chicago. 1997+ (1,5,9) – 2mf – 9 – (containing a concluding chap and letter by leo tolstoy; int by james mavor) – mf#26918 – cn Canadiana [243]

Christian martyrdom in russia : persecution of the doukhobors / ed by Tchertkoff, Vladimir – Maldon England; London: Free Age Press, 1900 – 2mf – 9 – (containing a concluding chap and letter by leo tolstoy) – mf#26934 – cn Canadiana [243]

Christian martyrdom in russia : persecution of the spirit-wrestlers (or doukhobortsi) in the caucasus / ed by Chertkov, Vladimir – London: Brotherhood Pub Co, 1897 – 2mf – 9 – (containing a concluding chap and letter by leo tolstoy) – mf#00600 – cn Canadiana [243]

Christian matured for heaven / Smith, George – London, England. 1849 – 1r – us UF Libraries [240]

Christian Medical Association of India, Pakistan, Burma and Ceylon *see* Journal of the christian medical association of india, pakistan, burma and ceylon

Christian medical dental society journal – Richardson. 1988-1995 (1,5,9) – (cont: christian medical society journal. cont by: today's christian doctor) – mf#15359,01 – us UMI ProQuest [610]

Christian medical society journal – Richardson. 1986-1988 (1,5,9) – (cont by: christian medical dental society journal) – ISSN: 0009-546X – mf#15359 – us UMI ProQuest [610]

Christian memento – London, England. 1823 – 1r – us UF Libraries [240]

Christian memorials of the war : or, scenes and incidents illustrative of religious faith and principle, patriotism and bravery in our army / Hackett, Horatio Balch – Boston: Gould & Lincoln, 1864 [mf ed 1990] – 1mf – 9 – 0-7905-4965-4 – mf#1988-0965 – us ATLA [355]

A christian merchant : a memoir of james c crane / Burrows, J L – 1858 – 1 – $5.00 – us Southern Baptist [920]

Christian messenger – Baltimore. 1817-1819 (1) – mf#4365 – us UMI ProQuest [240]

Christian messenger – Monmouth OR: T F Campbell, 1870-77 [wkly] – 1 – (cont by: pacific christian messenger [1870-77]) – us Oregon Lib [240]

Christian messenger : devoted to doctrine, religion and morality – Philadelphia. 1819-1821 (1) – mf#3690 – us UMI ProQuest [240]

Christian messenger – Pittsford. 1815-1816 (1) – mf#3689 – us UMI ProQuest [240]

The christian messenger – London, 1884-1889. v19-24 – 30mf – 9 – mf#H-2740 – ne IDC [240]

The christian method of ethics / Clark, Henry William – New York: Fleming H Revell, c1908 – 1mf – 9 – 0-524-00012-3 – mf#1989-2712 – us ATLA [170]

Christian minister / Jones, Thomas – London, England. 1861 – 1r – us UF Libraries [240]

Christian minister's duty and encouragement / Walker, Thomas Horatio – Plymouth, England. 1834 – 1r – us UF Libraries [240]

The christian ministers' manual : for the use of church officers in the various relations of evangelists, pastors, bishops and deacons / Green, Francis Marion – St Louis: Christian Pub Co, [1883?] – 1mf – 9 – 0-524-07566-2 – mf#1991-3186 – us ATLA [240]

Christian ministry – Chicago. 1969-1999 (1) 1969-1999 (5) 1969-1999 (9) – ISSN: 0033-4138 – mf#6272 – us UMI ProQuest [240]

Christian ministry / Pinder, John Hothersall – London, England. 1840 – 1r – us UF Libraries [240]

The christian ministry : its origin, constitution, nature, and work / Lefroy, William – London: Hodder and Stoughton, 1890 – (= ser Donnellan Lectures) – 2mf – 9 – 0-7905-5537-9 – (incl bibl ref) – mf#1988-1537 – us ATLA [240]

The christian ministry / Lightfoot, Joseph Barber – New York, T. Whittaker, 1879 – 1r – 1 – 0-8370-1523-5 – mf#1984-B231 – us ATLA [240]

Christian ministry and its requirements / Irvine, W F – Edinburgh, Scotland. 1867 – 1r – us UF Libraries [240]

The christian ministry and the social order : lectures delivered in the course in pastoral functions at yale divinity school, 1908-1909 / ed by Macfarland, Charles S – New Haven, Conn.: Yale University Press; London: Henry Frowde, 1909 – 1mf – us ATLA [240]

The christian ministry and the social order : lectures delivered in the course in pastoral functions at yale divinity school, 1908-1909 / ed by Macfarland, Charles Stedman – New Haven, Conn.: Yale University Press; London: Henry Frowde, 1909 – 1mf – 9 – 0-7905-4949-2 – mf#1988-0949 – us ATLA [240]

The christian ministry at the close of the 19th century / Littlejohn, Abram Newkirk – New York: T Whittaker 1884 [mf ed 1990] – (= ser The bishop paddock lectures 1884) – 1mf – 9 – 0-7905-6536-6 – mf#1988-2536 – us ATLA [240]

Christian mirror – Charleston. 1814-1814 (1) – mf#3691 – us UMI ProQuest [240]

Christian mirror – Portland ME: Arthur Shirley 1822-99 [mf ed 1974] – 77v on 29r – 1 – (various imprints; some iss accomp by suppls; lacks: v44 n52; several pp damaged; absorbed: new hampshire observer on apr 15 1830, separated again on aug 5 1830; cont by: congregationalist and christian mirror) – mf0257 – us ATLA [242]

The christian mission magazine, 1870-78... – 3r – 1 – (filmed with: the east london evangelist 1868-69; the salvationist 1879) – mf#97094 – uk Microform Academic [240]

Christian missions : and historical sketches of missionary societies among the disciples of christ / Green, Francis Marion – St Louis: John Burns Pub Co, 1884 [mf ed 1993] – (= ser Christian church (disciples of christ) coll) – 1mf – 9 – 0-524-06415-6 – mf#1991-2537 – us ATLA [240]

Christian missions / Seelye, Julius Hawley – New York: Dodd, Mead, c1875 – 1mf – 9 – 0-8370-6514-3 – (incl bibl ref) – mf#1986-0514 – us ATLA [240]

Christian missions : their agents, and their results / Marshall, Thomas William M – 2nd ed. London: Longman, Green, Longman, Roberts, & Green, 1863 – 3mf – 9 – 0-7905-7117-X – (incl bibl ref) – mf#1988-3117 – us ATLA [240]

Christian missions and foreign relations in china : an historical study / Drury, Clifford Merrill – 1932 – 1r – 1 – 0-8370-0584-1 – mf#1984-B324 – us ATLA [951]

Christian missions and government education in India. *see* Review of a letter...

Christian missions and government education in india : review of a letter addressed to the court of directors of the east-india company by the earl of ellenborough... – London, 1858 – (= ser 19th c books on british colonization) – 1mf – 9 – mf#1.1.510 – uk Chadwyck [954]

Christian missions and social progress : a sociological study of foreign missions / Dennis, James Shepard – New York: Fleming H Revell 1899-1906 [mf ed 1986] – (= ser Students' lectures on missions 1896) – 3v on 7mf – 9 – 0-8370-6108-3 – (incl ind) – mf#1986-0108 – us ATLA [240]

Christian missions before the reformation / Walrond, Francis Frederick – London: SPCK [1873?] [mf ed 1992] – 1mf – 9 – 0-524-03828-7 – mf#1990-1144 – us ATLA [242]

Christian missions in burma / Purser, William Charles Bertrand – Westminster: Society for the Propagation of the Gospel in Foreign Parts, 1911 – 1mf – 9 – 0-7905-5676-6 – (incl bibl ref) – mf#1988-1676 – us ATLA [240]

Christian missions in china / Estes, Charles Sumner – 1mf – 9 – 0-524-07872-6 – (incl bibl ref) – mf#1991-3417 – us ATLA [240]

Christian missions in japan / Kinnosuke, Adachi – Boston: American Board of Commissioners for Foreign Missions, 1911 [mf ed 1995] – (= ser Yale coll) – 29p (ill) – 1 – 0-524-09609-0 – (repr fr the century magazine for sep 1911) – mf#1995-0609 – us ATLA [950]

Christian missions in the east and west : in connection with the baptist missionary society, 1792-1872 – London: Yates & Alexander, 1873 – 2mf – 9 – 0-524-07467-4 – mf#1991-3127 – us ATLA [242]

Christian missions in the far east : addresses on the subject / Montgomery, Henry Hutchinson & Stock, Eugene – 2nd ed. London: SPCK, 1906 [mf ed 1995] – (= ser Yale coll) – 96p – 1 – 0-524-09859-X – mf#1995-0859 – us ATLA [240]

Christian missions in the telugu country / Hibbert-Ware, George – Westminster: Society for the Propagation of the Gospel in Foreign Parts, 1912 – 1mf – 9 – 0-7905-6809-8 – mf#1988-2809 – us ATLA [240]

Christian missions of the middle ages : or, a thousand years / Lyndon, John W – London: SPCK [1872?] [mf ed 1992] – 1mf – 9 – 0-524-03904-6 – mf#1990-1163 – us ATLA [240]

Christian moderation / Hall, Joseph – London, England. no date – 1r – us UF Libraries [240]

Christian monasticism : from the fourth to the ninth centuries of the christian era / Smith, Isaac Gregory – London: A D Innes, 1892 – 1mf – 9 – 0-7905-5965-X – (incl bibl ref) – mf#1988-1965 – us ATLA [240]

Christian monasticism in egypt to the close of the fourth century / Mackean, W H – London, 1920 – 3mf – 8 – €7.00 – ne Slangenburg [240]

Christian monitor – Hallowell. 1814-1818 (1) – mf#3692 – us UMI ProQuest [240]

Christian monitor / Rawlet, John – London, England. 1797 – 1r – us UF Libraries [240]

Christian monitor : a religious periodical work – Boston. 1806-1811 (1) – mf#3565 – us UMI ProQuest [240]

Christian monitor – Richmond. 1815-1817 (1) – mf#4429 – us UMI ProQuest [240]

Christian monitor – v1-45. 1909-53 [complete] – (= ser Mennonite serials coll) – 15r – 1 – mf#ATLA 1991-S001 – us ATLA [240]

Christian monitor and religious intelligencer : designed to promote experimental and practical religion – New York. 1812-1813 (1) – mf#4430 – us UMI ProQuest [240]

Christian monitor and weekly register – Providence, Rhode Island. May 22-Nov 13 1824 – 1r – 1 – us L of C Photodup [071]

Christian monthly / Apostolic Lutheran Church of America – 1977 jun-1980 dec, 1981-87 – 2r – 1 – mf#573066 – us WHS [242]

Christian monthly history – 1743-46 – 1 – $50.00 – us Presbyterian [240]

Christian monuments in england and wales / Boutell, Charles – London 1854 – (= ser 19th c art & architecture) – 2mf – 9 – mf#4.2.1328 – uk Chadwyck [720]

Christian morality and traditional chinese ethics *see*
– Chi-tu-chiao tao-te-kuan yu chung-kuo lun li [ccm329]

Christian morgensterns dichtungen von "ich und du" / Klemm, Guenther – Bonn a. Rh: L Roehrscheid 1933 [mf ed 1992] – (= ser Mnemosyne 12) – 1r – 1 – (incl bibl ref. filmed with: das gotterlebnis des germanischen menschen / lisel etscheid) – mf#3114p – us Wisconsin U Libr [430]

Christian mourning : a sermon, occasioned by the death of mrs isabella graham / Mason, John Mitchell – New York: Whiting & Watson, 1814 [mf ed 1984] – (= ser Women & the church in america 125) – 1mf – 9 – 0-8370-1390-9 – mf#1984-2125 – us ATLA [240]

The christian movement in japan : fifth annual issue / ed by Greene, Daniel Crosby & Clement, Ernest Wilson – Tokyo: Published for the Standing Committee of Co-operating Christian Missions [by the] Methodist Pub House, 1907 – 1mf – 9 – 0-524-05435-5 – mf#1990-1467 – us ATLA [240]

Christian mysteries – London, England. no date – 1r – us UF Libraries [240]

Christian mysticism : considered in eight lectures / Inge, William Ralph – New York: Scribners, 1899 – (= ser Bampton lectures) – 1mf – 9 – 0-524-00560-5 – mf#1990-0060 – us ATLA [240]

Christian name / Malcolm, James – London, England. 1853 – 1r – us UF Libraries [240]

Christian news – Bethany, NE: News Print & Pub Co, 1894-v12 n42. jan 27 1906 (wkly) [mf ed with gaps filmed 1975?] – 3r – 1 – (cont by: nebraska christian news) – us NE Hist [071]

Christian news – Glasgow, Scotland. -w. 1859-1870. Lacking 1868. 11 reels – 1 – uk British Libr Newspaper [072]

Christian news – 1968/1970 apr 6-1989 oct 9/1990 dec 31 – 15r – 1 – (with gaps; cont: lutheran news) – mf#1330936 – us WHS [242]

Christian news from israel – Jerusalem. 1972-1982 (1) 1972-1982 (5) 1972-1982 (9) – ISSN: 0009-5532 – mf#7529 – us UMI ProQuest [240]

Christian news [glasgow, scotland] : the advocate of moral and religious progress – [Scotland] Glasgow: H Nisbet 5 aug 1846-27 dec 1849, jan 1871-27 jul 1906 (wkly) [mf ed 2004] – 73r – 1 – (publ varies) – uk Newsplan [240]

Christian non-resistance : in all its important bearings / Ballou, Adin – Philadelphia: J Miller M'Kim, 1846 – 1mf – 9 – 0-524-00981-3 – mf#1990-0258 – us ATLA [240]

Christian nurture / Bushnell, Horace – New York: Scribner, 1883 – 1mf – 9 – 0-524-07308-2 – mf#1991-3023 – us ATLA [240]

Christian observatory : a religious and literary magazine – Boston. 1847-1850 (1) – mf#4783 – us UMI ProQuest [240]

Christian observer – Catlettsburg, KY. 1867-1872 (1) – mf#63455 – us UMI ProQuest [071]

Christian observer – Toronto: A T McCord & J Pyper, [1851?-1852] [mf ed v1 n1 jan 1851-v2 n12 dec 1852] – 9 – (cont by: toronto christian observer) – mf#P04888 – cn Canadiana [242]

Christian observer – 1865 jun 22, 1865 nov 9-1873 jul 30, 1873 sep 17-1886 jan 27, 1901 sep 4-1902 dec 3, 1903 mar 4 – 4r – 1 – (cont: religious telegraph and observer) – mf#683830 – us WHS [071]

Christian observer : from the london ed – Boston. 1802-1825 (1) – mf#4431 – us UMI ProQuest [240]

Christian observer – Louisville. 1965-1976 (1) 1975-1976 (5) 1975-1976 (9) – mf#1718 – us UMI ProQuest [240]

Christian observer – Louisville. v. 83-97. 1895-1909 – 1 – us NY Public [240]

Christian occupation of china : a translation = chung hua kuei chu – n36-216. nov 1923-may 1941* – (= ser Chinese christian coll 50) – 1r – 1 – mf#ATLA S0296F – us ATLA [240]

The christian occupation of china : a general survey of the numerical strength and geographical distribution of the christian forces in china...1918-1921 / ed by Stauffer, Milton Theobald – Shanghai: China Continuation Committee, 1922. Chicago: Dep of Photodup, U of Chicago Lib, 1971 (1r); Evanston: American Theol Lib Assoc, 1984 (1r) – 1 – 0-8370-0521-3 – (incl ind) – mf#1984-6293 – us ATLA [240]

Christian oesers geschichte der deutschen poesie in umrissen und schilderungen : nebst charakteristischen proben: fuer gebildete leser / Oeser, Christian [pseud of: Tobias Gottfried Schroeer]; ed by Schaefer, Johann Wilhelm – 3rd rev ed. Leipzig: F Brandstetter 1871 [mf ed 1992] – 2v in 1 on 1r – 1 – (incl ind; filmed with: im urteil der dichter / ed by arno mulot) – mf#3320p – us Wisconsin U Libr [430]

The christian of to-day : a brief description of his thought and life / Veitch, Robert – London: J Clarke, 1909 [mf ed 1991] – 1mf – 9 – 0-7905-9729-2 – mf#1989-1454 – us ATLA [240]

Christian omnibook *see* Chi-tu chiao ts'ung k'an [ccs]

The christian opportunity : being sermons and speeches / Davidson, Randall Thomas – New York: Macmillan; London: Macmillan, 1904 – 1mf – 9 – 0-7905-4456-3 – mf#1988-0456 – us ATLA [240]

The christian opportunity : being sermons and speeches / Davidson, Randall Thomas – New York: Macmillan; London: Macmillan, 1904 – 1mf – us ATLA [240]

Christian organizer – Lynchburg, VA. Virginia Baptist State Convention. 1899, 1902-03 – 1 – us ABHS [240]

Christian orthodoxy reconciled with the conclusions of modern biblical learning : a theological essay, with critical and controversial supplements / Donaldson, Joh William – London: Williams & Norgate, 1857 [mf ed 1989] – 2mf – 9 – 0-7905-1146-0 – (incl bibl ref) – mf#1987-1146 – us ATLA [240]

Christian painter of the nineteenth century / Lear, H L Sidney – New York, NY. 1875 – 1r – us UF Libraries [720]

Christian panoply / Mavor, William – London, England. 1803 – 1r – us UF Libraries [240]

Christian parlor magazine – New York. 1844-1854 (1) – mf#5282 – us UMI ProQuest [240]

The christian pastor : his work and the needful preparation. a discourse in favor of theological education. delivered before the n.b.e. society at north adams... / Hovey, Alvah – Boston: Gould and Lincoln, 1857 – 1mf – 9 – 0-524-07886-6 – mf#1991-3431 – us ATLA [240]

The christian pastor and the working church / Gladden, Washington – New York: Charles Scribner, 1898 – 2mf – 9 – 0-8370-9866-1 – (incl ind) – mf#1986-3866 – us ATLA [240]

Christian patience : the strength and discipline of the soul: a course of lectures / Ullathorne, William Bernard – 4th ed. London: Burns and Oates; New York: Catholic Publication Society, 1890 – 1mf – 9 – 0-8370-7110-0 – (incl bibl ref) – mf#1986-1110 – us ATLA [240]

Christian patriot – Belfast. Ireland. -w. 13 Apr 1838-13 Mar 1840. (1 reel) – 1 – uk British Libr Newspaper [072]

Christian patriotism / Fuller, Andrew – Dunstable, England. 1803 – 1r – us UF Libraries [240]

Christian peace conference – Prague. 1985-1990 (1,5,9) – ISSN: 0009-5567 – mf#15382 – us UMI ProQuest [327]

Christian peace officer : an official publication of fellowship of christian peace officers – 1978 oct/nov, 1979 feb, 1980 apr/may, 1981 may/jun, sep/oct-nov/dec, 1982 jan/feb, may/jun, sep/oct-nov-dec, 1983 jan/feb, spr, fall,1984 win, sum, 1985 win, apr, 1986 1-2, 1987 jan, jul, oct [n1-3] – 1r – 1 – (cont by: christian peace officer [1989]) – mf#1533926 – us WHS [240]

Christian peaceableness / Kennedy, Benjamin Hall – Shrewsbury, England. 1840 – 1r – us UF Libraries [240]

Christian pedagogy : or, the instruction and moral training of youth / Halpin, Patrick Albert – New York: JF Wagner, 1909 – 1mf – 9 – 0-524-06184-X – mf#1991-2440 – us ATLA [377]

Christian perfection / Forsyth, Peter Taylor – London: Hodder & Stoughton [1899?] [mf ed 1989] – (= ser Little books on religion) – 1mf – 9 – 0-7905-2588-7 – mf#1987-2588 – us ATLA [240]

Christian philanthropist : devoted to literature and religion – New Bedford. 1822-1823 (1) – mf#3693 – us UMI ProQuest [240]

Christian philosophy / Frothingham, Ephraim Langdon & Frothingham, Arthur Lincoln – Baltimore: AL Frothingham, 1888-1890 – 1mf – 9 – 0-7905-3684-6 – mf#1989-0177 – us ATLA [240]

The christian philosophy of life : reflections on the truths of religion = Christliche lebensphilosophie / Pesch, Tilmann – London: Sands, 1909 – 2mf – 9 – 0-524-08553-6 – (in english) – mf#1993-2078 – us ATLA [240]

The christian platonists of alexandria / Bigg, Charles; ed by Brightman, Frank Edward – Reprinted with some additions and corrections. Oxford: Clarendon Press, 1913 – (= ser Bampton lectures) – 1mf – 9 – 0-7905-5510-7 – (incl bibl ref) – mf#1988-1510 – us ATLA [240]

The christian point of view : three addresses / Knox, George William et al – New York:Charles Scribner, 1902 – 1mf – 9 – 0-8370-3326-8 – mf#1985-1326 – us ATLA [240]

The christian policy of life / Brown, James Baldwin – London: E Stock, 1870 – 1mf – 9 – 0-7905-3620-X – mf#1989-0113 – us ATLA [240]

Christian prayer and general laws : being the burney prize essay for the year 1873 / Romanes, George John – London: Macmillan, 1874 – 1mf – 9 – 0-7905-9093-X – mf#1989-2318 – us ATLA [240]

The christian preacher's companion : or, the gospel facts sustained by the testimony of unbelieving jews and pagans / Campbell, Alexander – Centreville, Ky: Published for RB Neal, 1891 – 1mf – 9 – 0-524-06392-3 – mf#1991-2514 – us ATLA [220]

Christian preaching as exemplified in the conduct of st paul / Davies, J – London, England. 1827 – 1r – us UF Libraries [240]

Christian preaching considered / Benson, Christopher – Worcester, England. 1833 – 1r – us UF Libraries [240]

Christian predestination : or, the predetermined providential appointment of them that love god to suffer with jesus, that with him they may be glorified / Evans, John Swanton – Quebec?: Middleton and Dawson, 1862 – 1mf – 9 – mf#51208 – cn Canadiana [225]

Christian press : an evangelical, independent, and non-sectarian journal – [Scotland] Glasgow: J Cossar 15 sep 1877-1 may 1880 (wkly) [mf ed 2004] – 2v on 4r – 1 – uk Newsplan [240]

Christian priesthood and the church of england vindicated from the... / Perceval, Arthur Philip – London, England. 1838 – 1r – us UF Libraries [241]

Christian Prison Volunteers et al *see* Hollywood free paper

The christian profession of the society of friends : commended to its members / Ash, Edward – London: John and Arthur Arch, 1837 – 1mf – 9 – 0-524-06759-7 – mf#1991-2766 – us ATLA [240]

Christian progress in china : gleanings from the writings and speeches of many workers / Foster, Arnold – London: Religious Tract Society, 1889 – 1mf – 9 – 0-8370-6044-3 – (incl bibl ref) – mf#1986-0044 – us ATLA [240]

The christian prophets and the prophetic apocalypse / Selwyn, Edward Carus – London, New York: Macmillan, 1900 [mf ed 1985] – 1mf – 9 – 0-8370-5229-7 – (incl bibl ref and ind) – mf#1985-3229 – us ATLA [225]

The christian psalmist / Leonard, SW – Louisville. 1850 – 1 – us Southern Baptist [242]

The christian psalter : a manual of devotion containing responsive readings for public worship / ed by Dowling, William Worth – 2nd ed. St Louis, MO: Christian Pub Co, c1890 – 1mf – 9 – 0-524-02445-6 – mf#1990-4304 – us ATLA [220]

Christian pulpit / Henry, J – Belfast, Northern Ireland. 1892 – 1r – us UF Libraries [240]

Christian quarterly – Cincinnati. 1869-1876 (1) – mf#3158 – us UMI ProQuest [240]

Christian race / Peake, Arthur Samuel – London: Hodder & Stoughton 1908 [mf ed 1989] – (= ser Aids to the devotional study of scripture 1) – 1mf – 9 – 0-7905-1775-2 – mf#1987-1775 – us ATLA [225]

Christian reality in modern light / Ballard, Frank – 1st ed. London: Charles H Kelly 1916 [mf ed 1991] – 2mf – 9 – 0-7905-7681-3 – mf#1989-0906 – us ATLA [240]

Christian reconstruction in the south / Douglass, Harlan Paul – Boston: Pilgrim Press, c1909 [mf ed 1986] – 1mf – 9 – 0-8370-6488-0 – (incl bibl ref & ind) – mf#1986-0488 – us ATLA [240]

The christian record : a religious magazine / ed by Dunbar, Hugh – Pictou, NS: Publ...by Stiles & Fraser, 1843 – 9 – ISSN: 1190-6898 – mf#P04164 – cn Canadiana [240]

Christian recorder / African Methodist Episcopal Church – 1994-1995 dec 18, 1996-98 – 2r – 1 – (cont: a m e christian recorder) – mf#1095139 – us WHS [242]

The christian recorder – [Toronto: U C Gazette], 1819-[1821] – 9 – (incl ind) – mf#P04132 – cn Canadiana [240]

Christian reflector – Worcester. 1838-1848 (1) – mf#3697 – us UMI ProQuest [240]

Christian reformer : or evangelical miscellany – Harrisburg. 1828-1829 (1) – mf#3960 – us UMI ProQuest [240]

Christian register – Boston. 1821-1850 (1) – mf#4533 – us UMI ProQuest [240]

Christian register – Boston MA, Chicago IL. v61 n28-30,33-37,39-40,42,45-46,49-52 [1882 jul 13-dec 28]-v62 [1883], v63-64 [1884-85] – 2r – 1 – (cont: christian register and boston observer; unitarian [chicago il]; unitarian advance; unitarian word and work; cont by: christian register; unitarian) – mf#1110654 – us WHS [071]

Christian register – Lexington. 1822-1823 (1) – mf#4778 – us UMI ProQuest [240]

Christian register and moral theological review – New York. 1816-1817 (1) – mf#3694 – us UMI ProQuest [240]

Christian relations of the east and the west : a sermon in behalf of the american home missionary society. preached in the broadway tabernacle church, new york... / Bartlett, Samuel Colcord – New York: American Home Missionary Society, 1871 – 1mf – 9 – 0-524-06704-X – mf#1991-2734 – us ATLA [240]

The christian religion / Fisher, George Park – New York: Chautauqua Press, 1886 [mf ed 1985] – 1mf – 9 – 0-8370-4977-6 – (incl bibl ref) – mf#1985-2977 – us ATLA [240]

The christian religion : its meaning and proof / Lidgett, John Scott – New York: Eaton & Mains, c1907 [mf ed 1991] – 1mf – 9 – 0-7905-8504-9 – mf#1989-1729 – us ATLA [240]

The christian religion as a healing power : a defense and exposition of the emmanuel movement / Worcester, Elwood & McComb, Samuel – New York: Moffat, Yard, 1909 – 1mf – 9 – 0-7905-8980-X – mf#1989-2205 – us ATLA [240]

The christian religion as profes'd by a daughter of the church of england / Astell, Mary – London: Printed by SH for R Wilkin, 1705 – 2mf – 9 – 0-524-01100-1 – mf#1990-0314 – us ATLA [241]

Christian remembrancer : a quarterly review – London. 1819-1868 (1) – mf#4225 – us UMI ProQuest [240]

The christian remembrancer for... – Montreal: printed and publ by T.A. Starke for the Canada Young Men's Society, 1831?-18– or 19– – 9 – mf#A01761 – cn Canadiana [030]

Christian reporter – Bethany, NE: DeForest Austin, dec 1906-v30 n29. jul 17 1936 (wkly) [mf ed with gaps filmed 1987] – 5r – 1 – (absorbed: nebraska christian news. publ in lincoln ne, jul 1 1927-jul 17 1936) – us NE Hist [071]

The christian reporter : an unsectarian record of christian thought and labour – Toronto: Bengough, Moore, [1880?-18– or 19–] – 9 – mf#P04143 – cn Canadiana [240]

Christian researches in india... : to which are prefixed, a memoir of the author, and an introductory sketch of protestant missions in india... / Buchanan, C – London, 1840 – 2mf – 9 – mf#HTM-25 – ne IDC [242]

Christian researches in syria and the holy land, in 1823 and 1824 : in furtherance of the objects of the church missionary society / Greaves, Joseph – London 1825 [mf ed Hildesheim 1995-98] – 1v on 4mf [ill] – 9 – €120.00 – 3-487-27683-6 – gw Olms [240]

Christian responsibility : or, the duty of individual effort for the... / Thornton, John – Belfast, Northern Ireland. 1837 – 1r – us UF Libraries [240]

Christian responsibility in the matter of popular amusements / Cochrane, William – [Stratford, Ont?: Beacon Steam Print], 1874 – 1mf – 9 – 0-665-89705-7 – (incl bibl ref) – mf#89705 – cn Canadiana [230]

Christian Restoration Association *see* Restoration herald

Christian re-union / Johnson, P B – Dublin?, Ireland. v1. 1895? – 1r – us UF Libraries [240]

Christian re-union / Johnson, P B – Dublin?, Ireland. v2. 1895? – 1r – us UF Libraries [240]

Christian re-union / Johnson, P B – Dublin?, Ireland. v3. 1895? – 1r – us UF Libraries [240]

The christian revelation / Bowne, Borden Parker – 2nd ed. Cincinnati: Curts & Jennings, 1898 – 1mf – 9 – 0-8370-2428-5 – mf#1985-0428 – us ATLA [240]

Christian review – Boston. 1836-1863 (1) – mf#3961 – us UMI ProQuest [240]

Christian review – Philadelphia. Afro-American Baptist. 1949-51 – 1 – us ABHS [240]

Christian review *see* Baptist review

The christian review – v1-10. 1932-41 [complete] – 3r – 1 – mf#ATLA 1993-S515 – us ATLA [240]

Christian rewards : or, 1: the everlasting rewards for christian workers; supperadded to everlasting salvation by faith only; 2: the antecedent millennial reward for christian martyrs / Evans, John Swanton – Toronto: W Briggs, 1880 – 2mf – 9 – mf#12839 – cn Canadiana [242]

The christian rural movement *see* Chi-tu chiao nung ts'un yun tung [ccm252]

Christian Rural Overseas Program *see* Crop news

Christian sabbath – Dublin, Ireland. 1859 – 1r – us UF Libraries [240]

Christian sabbath / Gibson, James – Edinburgh, Scotland. 18– – 1r – us UF Libraries [240]

The christian sabbath : its history, authority, duties, benefits, and civil relations: a series of discourses / Rice, Nathan Lewis – New York: Robert Carter, 1862 – 1mf – 9 – 0-7905-0112-0 – mf#1987-0112 – us ATLA [240]

The christian sabbath : its nature, design, and proper observance / Dabney, Robert Lewis – Philadelphia: Presbyterian Board of Publ, c1882 – 1mf – 9 – 0-524-05097-X – mf#1991-2221 – us ATLA [240]

The christian sacraments / Candlish, James S – Edinburgh: T & T Clark, [1879?] – (= ser Handbooks for Bible Classes) – 1mf – 9 – 0-7905-7503-5 – mf#1989-0728 – us ATLA [240]

Christian sacrifice / Ferrer, William Hugh – Dublin, Ireland. 1866 – 1r – us UF Libraries [240]

The christian sanctified by the lord's prayer = Chretien sanctifie par l'oraison dominicale / Grou, Jean Nicolas – New York: T Whittaker, 1885 – 1mf – 9 – 0-524-04574-7 – (in english) – mf#1992-0162 – us ATLA [240]

Christian scholar / Kennet, White – London, England. 1797 – 1r – us UF Libraries [240]

Christian scholar's review – Wenham. 1970+ (1) 1976+ (5) 1976+ (9) – ISSN: 0017-2251 – mf#8529 – us UMI ProQuest [240]

Christian school monthly *see* Hsueh shu yueh pao (ccs)

Christian science : as a religious belief and a therapeutic agent / Flower, Benjamin Orange – Boston: Twentieth Century Co, 1909 – 1mf – 9 – 0-524-03817-1 – mf#1990-1133 – us ATLA [240]

Christian science : the faith and its founder / Powell, Lyman Pierson – New York: Putnam, 1907 – 1mf – 9 – 0-7905-5735-5 – (incl bibl ref) – mf#1988-1735 – us ATLA [240]

Christian science and its problems / Bates, J H – New York: Eaton & Mains, 1898 – 1mf – 9 – 0-524-04828-2 – (incl bibl ref) – mf#1990-1320 – us ATLA [240]

Christian science and legislation : together with testimonies, editorial comments and appendix / Kimball, Edward Ancel – Boston MA: Christian Science Pub Soc 1906 [mf ed 1993] – 1mf – 9 – 0-524-08431-9 – mf#1993-1041 – us ATLA [240]

Christian science before the bar of reason / Lambert, Louis Aloisius; ed by Quinlan, Aloysius Stanislaus – New York: Christian Press Assoc Publ 1908 [mf ed 1989] – 1mf – 9 – 0-8370-6991-2 – mf#1986-0991 – us ATLA [210]

Christian science monitor – Boston MA, Chicago IL. 1912 jun 17,18 – 1r – 1 – mf#1010707 – us WHS [240]

Christian science monitor [london ed] – Boston MA 25 oct 1960-28 mar 1975 [mf oct 1960-aug 1961] – 1 – (wanting: aug 1974) – uk British Libr Newspaper [240]

Christian science monitor [weekly international ed] – Boston MA 2 jun 1975-10 nov 2000 [mf apr 1987-dec 1991] – 1 – (cont by: monitor world: international ed of the christian science monitor [20/26 nov 2000-20 aug 2005]) – uk British Libr Newspaper [240]

Christian science so-called : an exposition and an estimate / Sheldon, Henry Clay – New York: Abingdon, c1913 – 1mf – 9 – 0-7905-6437-8 – mf#1988-2437 – us ATLA [210]

Christian science unchristian / Simpson, Albert B – New York: Alliance Press Co, [1907?] [mf ed 1992] – (= ser Christian & missionary alliance coll) – 1mf – 9 – 0-524-03739-6 – mf#1990-4844 – us ATLA [240]

Christian science unmasked / Hogg, Wilson Thomas – 3rd ed. Syracuse, NY: AW Hall, 1892 – 1mf – 9 – 0-524-05146-1 – mf#1990-1402 – us ATLA [240]

Christian scotsman / ed by Robertson, John – [Scotland] Glasgow: C Glass & Co apr 1892-dec 1901 (mthly) [mf ed 2004] – 10v on 8r – 1 – uk Newsplan [240]

Christian secretary – Hartford. 1822-1851 (1) – mf#4432 – us UMI ProQuest [240]

Christian self-dedication and "departure" / Sheppard, John – London, England. 1833 – 1r – us UF Libraries [240]

Christian self-denial / Beecher, Henry Ward – London, England. 1886 – 1r – us UF Libraries [240]

The christian sentinel, vols 1-19 (1883-1901) – Pittsburgh, PA – 1r – 1 – $50.00 – (incomplete) – us Presbyterian [240]

Christian series = Chi-tu-chiao ts'ung k'an – n4 1943* – (= ser Chinese christian coll 48) – 1r – 1 – (in chinese) – mf#ATLA S0296D – us ATLA [240]

Christian service among educated bengalese / Wilder, Robert Parmelee – Lahore: Civil & Military Gazette Press, 1895 [mf ed 1995] – (= ser Yale coll) – vi/76p – 1 – 0-524-09092-0 – mf#1995-0092 – us ATLA [240]

Christian service and the modern world / Macfarland, Charles Stedman – New York: F.H. Revell, c1915 – 1mf – 9 – 0-7905-4892-5 – mf#1988-0892 – us ATLA [240]

Christian sincerity / Churton, Edward – London, England. 1851 – 1r – us UF Libraries [240]

Christian singers of germany / Winkworth, Catherine – [London]: Macmillan, [1869?] – 1mf – 9 – 0-7905-6971-X – mf#1988-2971 – us ATLA [240]

Christian social action – Washington. 1988+ (1,5,9) – (cont: esa engage/social action) – ISSN: 0897-0459 – mf#16417 – us UMI ProQuest [301]

Christian social action – v4-7. 1939-42 [complete] – Inquire – 1 – mf#ATLA 1994-S501 – us ATLA [240]

Christian social association : information bulletin – 1965-79; 1985-89 – Inquire – 1 – mf#ATLA S0356 – us ATLA [240]

Christian social economist – Dublin, Ireland. 22 nov-27 dec 1851 – 1/4r – 1 – uk British Libr Newspaper [072]

Christian social union – apr 1895-apr 1908 [complete] – 2r – 1 – (cont: church social union) – mf#atla s0583 – us ATLA [240]

Christian socialism / Kaufmann, Moritz – London: K Paul, Trench, 1888 – 1mf – 9 – 0-7905-9985-6 – (incl bibl ref) – mf#1989-1710 – us ATLA [240]

Christian socialism : what and why... / Sprague, Philo Woodruff – New York: EP Dutton, 1891 [mf ed 1991] – 1mf – 9 – 0-7905-8905-2 – mf#1989-2130 – us ATLA [335]

Christian socialism in england / Woodworth, Arthur V – London: S Sonnenschein 1903 [mf ed 1992] – 1mf – 9 – 0-524-05834-2 – (incl bibl ref) – mf#1990-1529 – us ATLA [335]

Christian socialist – Chicago. v. 2-18. 1905-1921. Incomplete – 1 – us NY Public [240]

Christian socialist – Chicago, Danville IL. 1907 jan-1911 dec, 1908 jan-1909 dec 15, 1918 jun-aug, 1912 jan-1918 feb, 1918 mar-aug – 4r – 1 – (cont by: real democracy) – mf#403978 – us WHS [071]

Christian socialist – London. v. 1 no. 1-v. 9 no. 103. June 1883-Dec 1891. Incomplete – 1 – us NY Public [240]

Christian sociology / Stuckenberg, John Henry Wilbrandt – New York: I.K. Funk, 1880 – 1mf – 9 – 0-7905-6684-2 – mf#1988-2684 – us ATLA [240]

Christian soldier / Broughton, Thomas – London, England. 1795 – 1r – us UF Libraries [240]

Christian soldier / Dale, Rev Canon – London, England. 1862 – 1r – us UF Libraries [240]

Christian soldier / Gosse, Philip Henry (Mrs) – London, England. no date – 1r – us UF Libraries [240]

Christian soldier – Providence, RI. 1842-1850 (1) – mf#66273 – us UMI ProQuest [071]

The christian soldiers penny bible : london, printed by r. smith for sam. wade, 1693 – London: Willis and Sotheran, 1862 – 1mf – 9 – 0-7905-0023-X – mf#1987-0023 – us ATLA [220]

Christian standard – Cincinnati. 1988+ (1,5,9) – ISSN: 0009-5656 – mf#16352 – us UMI ProQuest [240]

Christian standard / Hamilton Co. Cincinnati – feb 1870-aug 1871 [wkly] – 1r – 1 – mf#B1315 – us Ohio Hist [240]

Christian standard – 1977-80, 1981/1982 apr-1987 jul/1988 – 11r – 1 – (with small gaps) – mf#470998 – us WHS [071]

The christian state : the state, democracy and christianity / Batten, Samuel Zane – Philadelphia: Griffith & Rowland, c1909 – 2mf – 9 – 0-524-08255-3 – mf#1993-3010 – us ATLA [321]

The christian state of life : or, sermons on the principal duties of christians in general, and of different states in particular = Christliche sittenlehre ueber die evangelischen wahrheiten / Hunolt, Franz – 2nd ed. New York: Benziger, 1886 [mf ed 1986] – 2v on 4mf – 9 – 0-8370-7295-6 – (english trans fr german ed of 1740 by j allen. incl ind) – mf#1986-1295 – us ATLA [241]

Christian statesman – Beaver Falls. 1989-1994 (1) – ISSN: 0009-5664 – mf#15386 – us UMI ProQuest [240]

Christian statesman / National Reform Association [United States] – v22-24 n14 [1888 sep 6-1890 dec 4] – 1r – 1 – mf#765758 – us WHS [360]

The christian statesman – 12r – 1 – $600.00 – (v1-31, 1867-1897) – us Presbyterian [240]

The christian statesman, 1867-1897 – 12r – 1 – $1,020.00 – mf#D3331 – Scholarly Resources – us Presbyterian [302]

Christian stewardship, morristown, tenn / Cox, E K – 1887 – 1 – 5.00 – us Southern Baptist [242]

Christian Student Foundation *see* Koinonia kaller

Christian study of buddhism *see* Chi tu tu chih fo hsueh yen chiu [ccc305]

Christian suffering / Knox-Little, William John – Manchester, England. 1877? – 1r – us UF Libraries [240]

Christian Tabernacle Baptist Church (Chicago, IL) *see* Gospel times

Christian teacher – London, England. 1835-44 [mf ed 2001] – (= ser Christianity's encounter with world religions, 1850-1950) – 5r – 1 – (filmed with: christian teacher and chronicle; christian teacher and chronicle of beneficence; christian teacher [london 1838]) – mf#2001-s046-049 – us ATLA [242]

Christian teacher – Wheaton. 1974-1979 (1) 1974-1979 (5) 1976-1979 (9) – ISSN: 0009-5672 – mf#8945 – us UMI ProQuest [377]

Christian teacher and chronicle *see* Christian teacher

Christian teacher and chronicle of beneficence *see* Christian teacher

Christian teaching and life / Hovey, Alvah – Philadelphia: American Baptist Publication Society, 1895 – 1mf – 9 – 0-7905-7766-6 – mf#1989-0991 – us ATLA [220]

Christian telescope and universalist miscellany – Providence. 1824-1828 (1) – mf#4433 – us UMI ProQuest [420]

Christian temper / Clowes, John – Manchester, England. 1822 – 1r – us UF Libraries [240]

Christian thankfulness – London, England. 1856 – 1r – us UF Libraries [240]

Christian theism : a brief and popular survey of the evidence upon which it rests / Row, Charles Adolphus – 2nd ed. New York: Thomas Whittaker, 1890 – 1mf – 9 – 0-8370-4984-9 – mf#1985-2984 – us ATLA [210]

Christian theism : its claims and sanctions / Purinton, Daniel Boardman – New York: G P Putnam, 1889 – 1mf – 9 – 0-8370-4812-5 – (includes bibliographies & index) – mf#1985-2812 – us ATLA [240]

Christian theism and a spiritual monism : god, freedom, and immortality in view of monistic evolution / Walker, William Lowe – Edinburgh: T & T Clark, 1906 [mf ed 1991] – 2mf – 9 – 0-7905-8744-0 – mf#1989-1969 – us ATLA [230]

Christian theology / Clarke, Adam – London: Thomas Tegg, 1835 [mf ed 1993] – (= ser Methodist coll) – 2mf – 9 – 0-524-06173-4 – mf#1991-2429 – us ATLA [240]

Christian theology : a concise and practical view of the cardinal doctrines and institutions of christianity / Weaver, Jonathan – Memorial ed. Dayton, Ohio: United Brethren Publ House, 1900 – 1mf – 9 – 0-8370-5665-9 – (incl ind) – mf#1985-3665 – us ATLA [240]

Christian theology / Robinson, Ezekiel Gilman – Rochester, NY: E R Andrews, c1894 – 1mf – 9 – 0-8370-6353-1 – (incl bibl ref and index) – mf#1986-0353 – us ATLA [240]

Christian theology / Valentine, Milton – Philadelphia: United Lutheran Publication House, c1906 – 3mf – 9 – 0-524-00188-X – mf#1989-2888 – us ATLA [240]

Christian theology and social progress / Bussell, Frederick William – London: Methuen, 1907 – (= ser Bampton lectures) – 1mf – 9 – 0-7905-3765-6 – mf#1989-0258 – us ATLA [240]

Christian theology in outline / Brown, William Adams – New York: C Scribner, 1906 – 2mf – 9 – 0-7905-8641-X – mf#1989-1866 – us ATLA [240]

Christian, Thomas *see* Campaign of 1813 on the ohio frontier

Christian thought and hindu philosophy : a treatise / Bowman, Arthur Herbert – London: Religious Tract Society, 1917 – 2mf – 9 – 0-524-02292-5 – (incl bibl ref) – mf#1990-2915 – us ATLA [230]

Christian thought on present-day questions : sermons on special occasions / Whitworth, William Allen – London; New York: Macmillan, 1906 – 1mf – 9 – 0-524-00212-6 – mf#1989-2912 – us ATLA [240]

Christian thought to the reformation / Workman, Herbert Brook – New York: Scribner 1911 [mf ed 1990] – (= ser Studies in theology) – 1mf – 9 – 0-7905-6276-6 – mf#1988-2276 – us ATLA [240]

Christian times and witness – Illinois. 1853-67 – 1 – 87.69 – us Southern Baptist [242]

Christian times and witness *see* Michigan christian herald

The christian tradition / Pullan, Leighton – London: Longmans, Green 1902 [mf ed 1992] – (= ser The oxford library of practical theology) – 1mf – 9 – 0-524-04849-5 – mf#1990-1341 – us ATLA [240]

The christian tradition and its verification / Glover, Terrot Reaveley – London: Methuen 1913 [mf ed 1989] – (= ser The angus lectureship 8) – 1mf – 9 – 0-7905-1405-2 – (incl ind) – mf#1987-1405 – us ATLA [240]

Christian treasury : containing contributions from ministers and members of various evangelical denominations – [Scotland] Edinburgh: Johnstone & Hunter 1848-96 (wkly) [mf ed 2004] – 52r – 1 – (began with v1 feb 28 1845; ceased with new ser: v13 1896? imprint varies) – uk Newsplan [240]

Christian truth and life : sermons / Valentine, Milton – Philadelphia, PA: Lutheran Pub Soc, c1898 [mf ed 1991] – 1mf – 9 – 0-7905-9650-4 – mf#1989-1375 – us ATLA [242]

Christian truth and modern opinion : seven sermons preached in new york by clergymen of the protestant episcopal church – 4th ed. New York: Thomas Whittaker, 1885, c1884 – 1mf – 9 – 0-8370-3488-4 – mf#1985-1488 – us ATLA [210]

Christian truth and other intellectual forces : speeches and discussions together with the papers published for the consideration of the congress / Pan-Anglican Congress (1908: London, England). Section B – London: Society for Promoting Christian Knowledge; New York: ES Gorham, 1908 – 1mf – 9 – 0-8370-9092-X – (includes bibliographies) – mf#1986-3092 – us ATLA [240]

Christian truth viewed in relation to plymouthism / Mearns, Peter – Edinburgh: William Oliphant, 1874 – 1mf – 9 – 0-7905-5662-6 – mf#1988-1662 – us ATLA [243]

Christian truths : lectures / Chatard, Francis Silas – New York: Catholic Publ Society, 1881 – 1mf – 9 – 0-8370-7050-3 – (includes appendix) – mf#1986-1050 – us ATLA [230]

Christian types of heroism : a study of the heroic spirit under christianity / Adams, John Coleman – Boston: Universalist Pub House, 1891, c1890 – 1mf – 9 – 0-7905-4306-0 – mf#1988-0306 – us ATLA [240]

Christian und die kataloge : eine erzaehlung fuer buecherfreunde / Hoinkes, Carl – Berlin: Verlag Die Heimbuecherei 1942 [mf ed 1990] – 1r [ill] – 1 – (ill by helmuth von geyer. filmed with: der tod des empedokles / holderlin & other titles) – mf#2727p – us Wisconsin U Libr [830]

Christian union / Chalmers, Thomas – Edinburgh, Scotland. 1843 – 1r – us UF Libraries [240]

Christian union – Edinburgh, Scotland. 1863 – 1r – us UF Libraries [240]

Christian union – Edinburgh, Scotland. 1863 – 1r – us UF Libraries [240]

Christian union : a historical study / Garrison, James Harvey – St. Louis, MO: Christian Pub Co, c1906 – 1mf – 9 – 0-7905-5044-X – mf#1988-1044 – us ATLA [240]

Christian union : real and unreal / Armitage, Thomas – London, England. no date – 1r – us UF Libraries [240]

Christian union and the protestant episcopal church in its relations to church unity / Lewis, William Henry – New York: Published for the author by F.D. Harriman, 1858 – 1mf – 9 – 0-7905-5422-4 – mf#1988-1422 – us ATLA [242]

Christian union quarterly : interdenominational and international – v1-24. jul 1911-apr 1935 [complete] – 4r – 1 – mf#ATLA S0024 – us ATLA [240]

Christian union witness – 1977-83 – 1r – 1 – mf#669148 – us WHS [243]

Christian unity : a sermon preached before the congregational union of eastern canada, at stanstead, september 21, 1842... / Atkinson, T – Quebec?: s.n, 1842 (Quebec: T Cary) – 1mf – 9 – mf#67004 – cn Canadiana [240]

Christian unity and the bishops' declaration : lectures...1895 / Gailor, Thomas Frank et al – New York: E & J B Young 1895 [mf ed 1992] – (= ser The church club lectures 1895) – 1mf – 9 – 0-524-02913-X – (incl bibl ref) – mf#1990-0729 – us ATLA [240]

Christian unity and the historic episcopate / Forrester, Henry – New York: T. Whittaker, 1889 – 1mf – 9 – 0-7905-6060-7 – mf#1988-2060 – us ATLA [240]

Christian unity at work : the federal council of the churches of christ in america, in quadrennial session at chicago, illinois, 1912 / ed by Macfarland, Charles Stedman – New York: The Council, c1913 – 1mf – 9 – 0-7905-6058-5 – mf#1988-2058 – us ATLA [240]

The christian unity of capital and labor / Cadman, Harry W – Philadelphia: American Sunday-School Union, c1888 – (= ser Green fund book) – 1mf – 9 – 0-7905-4613-2 – (incl bibl ref) – mf#1988-0613 – us ATLA [240]

The christian use of the psalms : with essays on the proper psalms in the anglican prayer book / Cheyne, Thomas Kelly – New York: E P Dutton, 1900 – 1mf – 9 – 0-8370-2644-X – mf#1985-0644 – us ATLA [220]

Christian vanguard / New Christian Crusade Church – n50 [1976 feb], n69 [1977 sep], n83-157/159, [1978 nov-1985 win] – 1r – 1 – (cont by: cdl report) – mf#957515 – us WHS [071]

Christian vernacular education society for india. annual report – 1861-1922 [mf ed 2001] – (= ser Christianity's encounter with world religions, 1850-1950) – 3r – 1 – (filmed with: Christian literature society for india. annual report) – mf#2001-s190-191 – us ATLA [377]

Christian vestiges of creation / Sewell, William – Oxford: JH & Jas Parker, 1861 [mf ed 1991] – 1mf – 9 – 0-7905-8891-9 – mf#1989-2116 – us ATLA [210]

The christian view of god and the world as centring in the incarnation / Orr, James – 3rd ed. New York City: Charles Scribner, 1897 – 2mf – 9 – 0-8370-9893-9 – (incl bibl ref and index) – mf#1986-3893 – us ATLA [240]

Christian view of moral evil / Martineau, James – Liverpool, England. 1839 – 1r – us UF Libraries [240]

Christian view of retribution hereafter / Giles, Henry – Liverpool, England. 1839 – 1r – us UF Libraries [240]

The christian view of the old testament / Eiselen, F C – New York: Eaton & Mains; Cincinnati: Jennings & Graham, c1912 – 1mf – 9 – 0-7905-1597-0 – (includes bibliographies and index) – mf#1987-1597 – us ATLA [221]

The christian view of the world : lectures 1910-1911 / Blewett, George John – New Haven: Yale UP 1912 [mf ed 1990] – (= ser Nathaniel william taylor lectures 1910-11) – 1mf – 9 – 0-7905-3586-6 – mf#1989-0079 – us ATLA [240]

Christian virtues and the means for obtaining them : containing the practice of the love of our lord jesus christ, treatise on prayer as the great means of obtaining salvation, directions for acquiring the christian virtues, rule of life for a christian, etc = Selections. 1855 / Liguori, Alfonso Maria de, Saint; ed by Coffin, Robert A – New York: P.J. Kenedy, c1855 – 2mf – 9 – 0-8370-7303-0 – (in english) – mf#1986-1303 – us ATLA [240]

Christian visitant – Albany. 1815-1816 (1) – mf#3695 – us UMI ProQuest [240]

Christian visitor – Providence. 1823-1823 (1) – mf#3962 – us UMI ProQuest [240]

Christian wahnschaffe : roman / Wassermann, Jakob – Berlin: S Fischer 1928 [mf ed 1991] – 2v on 1r – 1 – (filmed with: der aufruhr um den junker ernst / jakob wassermann) – mf#3025p – us Wisconsin U Libr [830]

Christian warfare / M'caig, Charles Neilson – Glasgow, Scotland. 1873 – 1r – us UF Libraries [240]

Christian warfare : or, the character of a gospel minister / Fisher, Samuel – Liverpool, England. 1791 – 1r – us UF Libraries [240]

The christian warfare against the devill world and flesh / Downame, J – Ed 4. London: William Stansby, 1634 – 22mf – 9 – mf#PW-11 – ne IDC [240]

Christian watching / Knox-Little, W J – Manchester, England. 1877 – 1r – us UF Libraries [240]

Christian watchman / Cawood, John – Worcester, England. 1821 – 1r – us UF Libraries [240]

The christian way : for advanced scholars in sunday schools and bible classes / Smith, Benton – Boston: Universalist Pub House, 1868 – 1mf – 9 – 0-524-06448-2 – mf#1991-2570 – us ATLA [220]

The christian way : whither it leads and how to go on / Gladden, Washington – New York: Dodd, Mead, 1877 – 1mf – 9 – 0-7905-1601-2 – (cont.: being a christian) – mf#1987-1601 – us ATLA [240]

Christian weekly *see* Tien feng (ccs)

Christian weise : ein saechsischer gymnasialrektor aus der reformzeit des 17. jahrhunderts / Kaemmel, Otto – Leipzig: B G Teubner 1897 [mf ed 1991] – 1r – 1 – (a commissioned festschrift. incl bibl ref. filmed with: weckherlin's eclogues of the seasons / elizabeth frienchjohnson) – mf#2948p – us Wisconsin U Libr [370]

Christian weise und moliere : eine studie zur entwicklungsgeschichte des deutschen lustspiels / Levinstein, Kurt – Berlin, 1899 (mf ed 1995) – 1mf – 9 – €24.00 – 3-8267-3136-0 – mf#DHS-AR 3136 – gw Frankfurter [410]

Christian weise und moliere : eine studie zur entwicklungsgeschichte des deutschen lustspiels / Levinstein, Kurt – Berlin: G Schade (O Francke), [1899?] [mf ed 1992] – 45p – 1 – (incl bibl ref) – mf#7935 – us Wisconsin U Libr [430]

Christian weise's bauern-komoedie von tobias und der schwalbe : aufgefuehrt im jahre 1682 – Berlin: U Hofmann 1882 [mf ed 1991] – (= ser Bibliothek deutscher curiosa 5) – 1r – 1 – (int by rudolph genee; filmed with: weckherlin's eclogues of the seasons / elizabeth frienchjohnson) – mf#2948p – us Wisconsin U Libr [820]

Christian weises historische dramen und ihre quellen / Hess, Adolf [comp] – Rostock: Adler's Erben 1893 [mf ed 1991] – 1r – 1 – (incl bibl ref. filmed with: "und alles ist zerstoben" / werner weisbach) – mf#3038p – us Wisconsin U Libr [430]

Christian weises romane und ihre nachwirkung / Becker, Rudolf – 1910 – 135p – 1 – mf#7935 – us Wisconsin U Libr [430]

Christian witness / Franklin Co. Columbus – jun 1866-nov 1867 [wkly] – 1r – 1 – mf#B307 – us Ohio Hist [240]

Christian witness and church advocate – v2 n46-v3 n45, v6 n51, v17 n41, v37 [1837, 1841 jan 12, 1851 nov 21, 1871] – 1r – 1 – mf#360422 – us WHS [240]

Christian witness in the resistance : experiences of some members of european student christian movements, 1939-1945 / ed by Maury, Philippe & Schanke, Andreas – Geneva: World's Student Christian Federation, [1947?] [mf ed 1993] – 1mf – 9 – 0-524-08120-4 – mf#1993-9026 – us ATLA [933]

Christian womanhood / Hack, Mary Pryor – London: Hodder & Stoughton, 1883 [mf ed 1984] – (= ser Women & the church in america 115) – 1mf – 9 – 0-8370-1382-8 – mf#1984-2115 – us ATLA [305]

Christian work / Knox-Little, W J – Manchester, England. 1877 – 1r – us UF Libraries [240]

Christian work in florence / M Dougall, John R – Stirling, Scotland. 1870 – 1r – us UF Libraries [240]

Christian work in rural china *see* Hsiang tsun chuan tao kung tso ching yen tan (ccm350)

Christian worker – Meaford, Ont: H R Sherman, [1881-1886] – 9 – (cont by: the ontario evangelist) – mf#P04387 – cn Canadiana [240]

Christian world – Boston: George G Channing 1843-48 [mf ed 1969] – 6v on 2r – 1 – mf0159 – us ATLA [243]

Christian world / Methodist Protestant Church – 1844 p81-1845 – 1r – 1 – $35.00 – mf#um-40 – us Commission [242]

Christian world : news of the week – London: [s.n] 1894-1961 [mf ed 1972] – 58r – 1 – (lacks some iss & scattered pgs; merged with: british weekly (london, england: 1886) to form: british weekly & christian world (edinburgh, scotland)) – mf0204 – us ATLA [240]

Christian world – v3 n33 1843; v3 jan 1843-dec; v4-pt1 1844, pt2 1844 – 1r – 1 – $35.00 – mf#um-39 – us Commission [242]

Christian world *see* The messenger of the evangelical and reformed church

Christian worship : its principles and forms / Richard, James William & Painter, Franklin Verzelius Newton – 2nd rev ed. Philadelphia: Lutheran Publication Society, c1908 – 1mf – 9 – 0-7905-6008-9 – mf#1988-2008 – us ATLA [240]

Christian worship / O'beirne, Thomas Lewis – London, England. 1819 – 1r – us UF Libraries [240]

Christian worship : ten lectures. delivered in the union theological seminary, new york... / Hall, Charles Cuthbert et al – New York: Scribner, 1897 – 1mf – 9 – 0-524-02914-8 – (incl bibl ref) – mf#1990-0730 – us ATLA [240]

The christian writers of the inner emigration / Klieneberger, H R – The Hague; Paris: Mouton, c1968 – 1r – 1 – (incl bibl ref and index) – us Wisconsin U Libr [430]

Christian year / Keble, John – New York, NY. 1905 – 1r – us UF Libraries [960]

The christian year / Horn, Edward Trail – Philadelphia: Lutheran Book Store, 1876 – 1mf – 9 – 0-524-04662-X – (incl bibl ref) – mf#1990-5058 – us ATLA [240]

The christian year : its purpose and its history / Gwynne, Walker – New York: Longmans, Green, c1915 – 1mf – us ATLA [240]

The christian year : its purpose and its history / Gwynne, Walker – New York: Longmans, Green, c1915 – 1mf – 9 – 0-7905-5231-0 – mf#1988-1231 – us ATLA [240]

Christian youth / Williams Temple Church of God in Christ [Gainesville FL] – 1994 oct, 1995 apr – 1r – 1 – mf#4025061 – us WHS [243]

Christian youth at the front *see* Ch'ien hsien te chi-tu t'u ch'ing nien [ccm224]

Christiana et catholica doctrina, fides, opera, ecclesia diui petri apostoli...per theodorvm bibliandrvm collecta / Bibliander, T – Basileae, [Iacobvs Parcvs], 1550 – 2mf – 9 – mf#PBU-577 – ne IDC [241]

La christiana vittoria maritima / Bolognetti, F – Bologna, 1572 – 2mf – 9 – mf#H-8323 – ne IDC [956]

Christianae isagoges ad locos communes, libri 2... / Daneau, Lambert – [Geneve], E Vignon, 1583 – 4mf – 9 – mf#PFA-127 – ne IDC [240]

Christianae religionis institutio, totam fere pietatis summam, et quicquid est in doctrina salutis cognitu necessarium, complectens : omnibus pietatis studiosis lectu dignissimum opus, ac recens editum... / Calvin, J – Basileae: Per Thomam Platterum et Balthasarem Lasium, 1536 – 6mf – 9 – mf#CL-3 – ne IDC [240]

Christianae theologiae medulla didactico-elenctica / Marck, J – Amstelaedami, 1690 – 4mf – 9 – mf#PBA-244 – ne IDC [240]

Christian-erlangischer zeitungs-extract – Erlangen DE, 1741-98, 1813-1829 30 may – 42r – 1 – (several title changes: 1742/43: auszug der neuesten weltgeschichte; 1763: real-zeitung; 1804: erlanger real-zeitung; dez 1821: erlanger zeitung) – gw Misc Inst [074]

Christiani weisii nucleus ethicae : sub uno conspectu singula breviter ac fideliter referens quaecunque tum de hominis felicitate tum de mediis virtutum & ipso prudentiae officio disputari solent / Weise, Christian – Lipsiae: Sumptibus Jacob Gerthesii; Typis Krügerianis 1694 – (= ser Ethics in the early modern period) – 1mf – 9 – mf#pl-324 – ne IDC [170]

Die christianisierung der fuerstentuemer reuss / Priegel, F – [Koenigsbrueck, um 1905] (mf ed 1993) – 1mf – 9 – €19.00 – 3-89349-342-5 – mf#DHS-AR 195 – gw Frankfurter [240]

Le christianisme dans l'empire perse : sous la dynastie sassanide (224-632) / Labourt, J – Paris: V Lecoffre, 1904 [mf ed 1990] – (= ser Bibliotheque de l'enseignement de l'histoire ecclesiastique) – 1mf – 9 – 0-7905-6484-X – (in french. incl bibl ref) – mf#1988-2484 – us ATLA [956]

Le christianisme dans l'empire perse sous la dynastie sassanide (224-632) / Labourt, J – Paris, 1904 – €15.00 – ne Slangenburg [243]

Le christianisme de luther / Kuhn, Felix – Paris: Fischbacher, 1900 – 1mf – 9 – 0-7905-6757-1 – mf#1988-2757 – us ATLA [242]

Le christianisme des arabes nomades sur le limes et dans le desert syro-mesopotamien aux alentours de l'hegire / Charles, H – Paris, 1936 – 2mf – 9 – mf#H-3073 – ne IDC [956]

Le christianisme en chine : en tartarie et au thibet / Huc, Evariste Regis – Paris: Gaume Freres, 1857-58 [mf ed 1995] – (= ser Yale coll) – 4v – 1 – 0-524-09653-8 – (in french) – mf#1995-0653 – us ATLA [241]

Le christianisme en koree / Delpech, Jacques – Paris: Societe generale d'impression, 1913 [mf ed 1995] – (= ser Yale coll) – 110p – 1 – 0-524-09697-X – (in french) – mf#1995-0697 – us ATLA [241]

Christianisme et bouddhisme / Thomas, M l'abbe – Paris: Librairie Bloud 1909 [mf ed 1991] – (= ser Histoire des religions) – 1mf – 9 – 0-524-01312-8 – mf#1990-2348 – us ATLA [280]

Le christianisme et le progres / Perraud, Charles-Alexis – 2e ed. Paris: Jules Gervais: A Sauton, 1883 – 1mf – 9 – 0-8370-8928-X – (incl bibl ref) – mf#1986-2928 – us ATLA [240]

Le christianisme et l'extreme orient / Joly, Leon – Paris: P Lethielleux, [1907] [mf ed 1995] – (= ser Yale coll) – 2v – 1 – 0-524-09166-8 – (in french) – mf#1995-0166 – us ATLA [241]

Christianisme et liberte / Giran, Etienne – Saint-Blaise, Suisse: Foyer Solidariste, 1909 – 1mf – 9 – 0-524-03520-2 – mf#1990-1025 – us ATLA [240]

Le christianisme moderne : etude sur lessing / Fontanes, Ernest – Paris: G Balliere; New York: Bailliere brothers, 1867 – 1r – 1 – us Wisconsin U Libr [240]

O christianismo : semanario religioso – Sao Luis, MA: Typ de J L M da Cunha Torres, 1854-30 abr 1855 – (= ser Ps 19) – mf#P17,02,54 – bl Biblioteca [240]

Christianismvs sempiternvs, vervs, certvs et immvtabilis... / Bibliander, T – Tigvri, [Christoph] Froschover, 1556 – 1mf – 9 – mf#PBU-588 – ne IDC [240]

Christianissimi martini lutheri et annemundi cocti pro sequentibus commentariis epistolae : evangelici in minoritarum regulam commentarii / Lambert, F – Strasbourg, 1523 – 2mf – 9 – mf#PPE-110 – ne IDC [242]

Christianity : the deliverance of the soul and its life / Mountford, William – Boston: W Crosby & H P Nichols, 1847 [mf ed 1984] – 2mf – 9 – 0-8370-1004-7 – (int by f d huntington) – mf#1984-4360 – us ATLA [240]

Christianity : a divine revelation / Stock, John – London, England. 18– – 1r – us UF Libraries [240]

Christianity : the fortress of great britain / Steele, Robert – London, England. 18– – 1r – us UF Libraries [240]

Christianity : an intellectual and individual religion / Grundy, John – Liverpool, England. 1811 – 1r – us UF Libraries [240]

Christianity : an interpretation / McConnell, Samuel David – New York: Longmans, Green, 1912 – 1mf – 9 – 0-7905-9505-2 – mf#1989-1210 – us ATLA [240]

Christianity : its nature and its truth / Peake, Arthur S – London: Duckworth, 1908 – 1mf – 9 – 0-8370-5226-2 – mf#1985-3226 – us ATLA [240]

Christianity / Schweitzer, Albert – New York, NY. 1951 – 1r – us UF Libraries [240]

Christianity according to christ : a series of papers / Gibson, John Monro – New York: Robert Carter 1888 [mf ed 1985] – (= ser Nisbet's theological library) – 1mf – 9 – 0-8370-2882-5 – mf#1985-0882 – us ATLA [240]

Christianity against coercion / Redford, George – London, England. 1840? – 1r – us UF Libraries [240]

Christianity against infidelity : or, the truth of the gospel history / Thayer, Thomas Baldwin – new rev enl ed. Cincinnati: JA Gurley, 1849 [mf ed 1991] – 1mf – 9 – 0-7905-9711-X – mf#1989-1436 – us ATLA [240]

Christianity agreeable to reason in its evidence, its doctrine of the atonement, and its commemorative sacrament : to which is added, baptism from the bible / Mortlock, Edmund – Cambridge: Macmillan 1862 [mf ed 1985] – 1mf – 9 – 0-8370-4497-9 – (incl bibl ref) – mf#1985-2497 – us ATLA [240]

Christianity and agnosticism : a controversy / Wace, Henry et al – New York: D Appleton, 1889 [mf ed 1985] – 1mf – 9 – 0-8370-2658-X – (incl bibl ref) – mf#1985-0658 – us ATLA [210]

Christianity and agnosticism : reviews of some recent attacks on the christian faith / Wace, Henry – [Edinburgh] 1895 – (= ser 19th c evolution & creation) – 4mf – 9 – mf#1.1.11596 – uk Chadwyck [140]

Christianity and agnosticism : reviews of some recent attacks on the christian faith / Wace, Henry – London: SPCK; New York: E S Gorham, 1905 [mf ed 1985] – 1mf – 9 – 0-8370-5677-2 – mf#1985-3677 – us ATLA [240]

Christianity and anti-christianity in their final conflict / Andrews, Samuel James – 2nd rev ed. New York: G P Putnam, 1899, c1898 [mf ed 1989] – 1mf – 9 – 0-7905-0850-8 – (incl bibl ref & ind) – mf#1987-0850 – us ATLA [230]

Christianity and buddhism : comparison and a contrast: being the donellan lectures...1889-90 / Berry, Thomas Sterling – London: SPCK; New York: E & J B Young [1891?] [mf ed 1990] – (= ser Non-christian religious systems) – 1mf – 9 – 0-7905-3581-5 – mf#1989-0074 – us ATLA [230]

Christianity and buddhism compared / Hardy, Robert Spence – Colombo: Wesleyan Mission Press 1874 [mf ed 1991] – 1mf – 9 – 0-524-01905-3 – mf#1990-2718 – us ATLA [230]

Christianity and china *see* Chi-tu chiao yu chung-kuo [ccm106]

Christianity and chinese culture *see*
– Chi-tu chiao yu chung-kuo wei hua [ccm331]
– Chi-tu chiao yu chung-kuo wen hua [ccm192]

Christianity and common sense / Jones, Willoughby, Sir – London: Longman, Green, Longman, Roberts, & Green, 1863 – 1mf – 9 – 0-7905-9981-3 – (incl bibl ref) – mf#1989-1706 – us ATLA [240]

Christianity and communism *see* Chi-tu chiao yu kung ch'an chu i [ccm4]

Christianity and confucianism / Kozaki, Hiromichi – 3rd ed. Tokyo: [s.n.] 1892 [mf ed 1995] – (= ser Yale coll) – 136p – 1 – 0-524-09646-5 – (in japanese) – mf#1995-0646 – us ATLA [230]

Christianity and crisis – New York. 1941-1993 (1) 1970-1993 (5) 1976-1993 (9) – ISSN: 0009-5745 – mf#1565 – us UMI ProQuest [240]

Christianity and economic science / Cunningham, William – London: John Murray, 1914 – 1mf – 9 – 0-7905-4275-7 – mf#1988-0275 – us ATLA [330]

Christianity and emancipation : or, the teachings and the influence of the bible against slavery / Thompson, Joseph Parrish – New York: A D F Randolph, 1863 – 1mf – 9 – 0-7905-6894-2 – (incl bibl ref) – mf#1988-2894 – us ATLA [230]

Christianity and ethics : a handbook of christian ethics / Alexander, Archibald Browning Drysdale – New York: Scribner 1914 [mf ed 1990] – (= ser Studies in theology) – 1mf – 9 – 0-7905-3512-2 – mf#1989-0005 – us ATLA [230]

Christianity and evolution : modern problems of the faith / Matheson, George et al – London: James Nisbet 1887 [mf ed 1985] – (= ser Nisbet's theological library) – 1mf – 9 – 0-8370-2659-8 – mf#1985-0659 – us ATLA [210]

Christianity and filial piety *see* Ts'ung chi-tu chiao kan chung-kuo hsiao tao (ccm151)

Christianity and greek philosophy : or, the relation between spontaneous and reflective thought in greece and the positive teaching of christ and his apostles / Cocker, Benjamin Franklin – New York: Harper, 1870 [mf ed 1990] – 2mf – 9 – 0-7905-3822-9 – (incl bibl ref) – mf#1989-0315 – us ATLA [180]

Christianity and history / Figgis, John Neville – London: James Finch, 1905 – 1mf – 9 – 0-7905-4677-9 – mf#1988-0677 – us ATLA [240]

Christianity and infallibility : both or neither / Lyons, Daniel – 2nd ed, 3rd impr. New York: Longmans, Green, 1916, c1891 – 1mf – 9 – 0-7905-9320-3 – (incl bibl ref) – mf#1989-2545 – us ATLA [240]

Christianity and international peace : six lectures at grinnell college...feb 1915, on the george a gates memorial foundation / Jefferson, Charles Edward – New York: Thomas Y Crowell c1915 [mf ed 1990] – 1mf – 9 – 0-7905-7441-1 – mf#1989-0666 – us ATLA [230]

Christianity and islam : the bible and the koran / Stephens, William Richard Wood – New York: Scribner, Armstrong, 1877 – 1mf – 9 – 0-524-01302-0 – (incl bibl ref) – mf#1990-2338 – us ATLA [230]

Christianity and islam = Christentum und islam / Becker, Carl Heinrich – London; New York: Harper, 1909 – (= ser Harper's library of living thought) – 1mf – 9 – 0-524-01041-2 – (in english) – mf#1990-2189 – us ATLA [230]

Christianity and islam in spain, a.d. 756-1031 / Haines, Charles Reginald – London: K. Paul, Trench, 1889 – 1mf – 9 – 0-7905-5467-4 – (incl bibl ref) – mf#1988-1467 – us ATLA [946]

Christianity and its evidence no 1 : a request to dr lee for furth – London, England. 1874 – 1r – us UF Libraries [240]

Christianity and judaism : an essay = Christentum und judentum / Dalman, Gustaf – Oxford: Williams & Norgate, 1901 – 1mf – 9 – 0-7905-0935-0 – (incl bibl ref) – mf#1987-0935 – us ATLA [240]

Christianity and literature *see* Chi-tu chiao yu wen hsueh [ccm114]

Christianity and marx-leninism *see* Chi-tu chiao yu ma lieh chu i [ccm205]

Christianity and modern civilization : being some chapters in european history, with an introductory dialogue on the philosophy of history / Lilly, William Samuel – London: Chapman & Hall, 1903 – 1mf – 9 – 0-524-03764-7 – mf#1990-1111 – us ATLA [940]

Christianity and modern thought – Boston: American Unitarian Association, 1873, c1872 – 1mf – 9 – 0-8370-8724-4 – mf#1986-2724 – us ATLA [240]

Christianity and modern thought *see* Chi-tu chiao yu hsien tai ssu hsiang [ccm166]

Christianity and morality : or, the correspondence of the gospel with the moral nature of man / Wace, Henry – 3rd ed. London: Basil Montagu Pickering, 1877 [mf ed 1988] – (= ser The boyle lectures 1874-75) – 1mf – 9 – 0-7905-0410-3 – (incl bibl ref) – mf#1987-0410 – us ATLA [240]

Christianity and mythology / Robertson, John Mackinnon – 2nd rev enl ed. London: Watts, 1910 – 2mf – 9 – 0-524-07062-8 – (incl bibl ref) – mf#1992-1025 – us ATLA [230]

Christianity and natural science / Guthrie, David – Edinburgh, Scotland. 1866 – 1r – us UF Libraries [210]

Christianity and naturalism / Shafer, Robert – New Haven, CT. 1926 – 1r – us UF Libraries [210]

Christianity and new china *see* Chi-tu chiao yu hsin chung-kuo [ccm229]

Christianity and non-christian religions compared : containing 800 library references to facilitate further study / Marshall, Edward A – Chicago: Bible Institute Colportage Association, c1910 – 1mf – 9 – 0-7905-7971-5 – (incl bibl ref) – mf#1989-1196 – us ATLA [230]

Christianity and other faiths : an essay in comparative religion / Tisdall, William St Clair – London: Robert Scott 1912 [mf ed 1991] – (= ser Library of historic theology) – 1mf – 9 – 0-7905-8741-6 – mf#1989-1966 – us ATLA [230]

Christianity and other religions : three short sermons / Driver, Samuel Rolles – London; New York: Longmans, Green, 1908 – 1mf – 9 – 0-7905-1593-8 – mf#1987-1593 – us ATLA [240]

Christianity and politics / Cunningham, W – Boston: Houghton Mifflin, 1915 – 1mf – 9 – 0-7905-4337-0 – mf#1988-0337 – us ATLA [240]

Christianity and positivism : a series of lectures to the times on natural theology and apologetics / McCosh, James – New York: Robert Carter, 1871 [mf ed 1985] – 1mf – 9 – 0-8370-2418-8 – (incl bibl ref) – mf#1985-0418 – us ATLA [210]

Christianity and scepticism : comprising a treatment of questions in biblical criticism / Mead, Charles Marsh et al – Boston: Congregational Pub Soc, c1871 [mf ed 1990] – (= ser Boston lectures 1871) – 2mf – 9 – 0-7905-3388-X – mf#1987-3388 – us ATLA [220]

Christianity and science : the introductory lecture delivered 13th nov 1860, in the class of natural science, new college, edinburgh / Duns, John – Edinburgh, 1860 – (= ser 19th c evolution & creation) – 1mf – 9 – mf#1.1.11602 – uk Chadwyck [210]

Christianity and science : a series of lectures...new york in 1874 / Peabody, Andrew Preston – New York: Robert Carter, 1875, c1874 [mf ed 1985] – 1mf – 9 – 0-8370-4681-5 – (incl app & ind) – mf#1985-2681 – us ATLA [210]

Christianity and science *see* Chi-tu chiao yu k'o hsueh [ccm173]

Christianity and sin / Mackintosh, Robert – London: Duckworth 1913 [mf ed 1991] – (= ser Studies in theology) – 1mf – 9 – 0-7905-7963-4 – (incl bibl ref) – mf#1989-1188 – us ATLA [230]

Christianity and social problems / Abbott, Lyman – Boston: Houghton, Mifflin, 1896 – 1mf – 9 – 0-7905-4001-0 – (incl bibl ref) – mf#1988-0001 – us ATLA [240]

Christianity and social questions / Cunningham, William – London: Duckworth 1910 [mf ed 1989] – (= ser Studies in theology) – 1mf – 9 – 0-7905-4276-5 – (incl bibl ref) – mf#1988-0276 – us ATLA [230]

Christianity and socialism : the 23rd fernley lecture...cardiff / Nicholas, William – London: R Culley [1908?] – 1mf – 9 – 0-7905-9826-4 – (first printed in 1893, with new int) – mf#1989-1551 – us ATLA [240]

Christianity and socialism / Gladden, Washington – New York: Eaton & Mains; Cincinnati: Jennings & Graham, c1905 – 1mf – 9 – 0-7905-0030-2 – mf#1987-0030 – us ATLA [240]

Christianity and society : a guide to the thought of reinhold niebuhr / Allen, Edgar Leonard – London: Hodder & Stoughton, [1950] [mf ed 2004] – 1r – 1 – 0-524-10490-5 – mf#b00705 – us ATLA [230]

Christianity, and some of its evidences : an address / Mowat, Oliver – Toronto: Williamson, 1890 [mf ed 1985] – 1mf – 9 – 0-8370-4514-2 – mf#1985-2514 – us ATLA [240]

Christianity and spain / Negrin, Juan – [New York: Spanish Information Bureau 1939?] [mf ed 1977] – (= ser Blodgett coll) – 1r – 9 – (trans of radio address made by premier juan negrin of the spanish republic in barcelona on christmas eve, 1938) – mf#w1070 – us Harvard [946]

Christianity and the american commonwealth : or, the influence of christianity in making this nation / Galloway, Charles Betts – Nashville, TN: Pub House, ME Church, South, 1898 [mf ed 1990] – (= ser The quillian lectures 1898) – 1mf – 9 – 0-7905-3843-1 – mf#1989-0336 – us ATLA [240]

Christianity and the christ : a study of christian evidences / Raymond, Bradford Paul – Cincinnati: Cranston & Curtis, 1894 [mf ed 1984] – 3mf – 9 – 0-8370-1012-8 – (incl bibl ref) – mf#1984-4368 – us ATLA [240]

Christianity and the labor movement / Balch, William Monroe – Boston: Sherman, French, 1912 – 1mf – 9 – 0-7905-4064-9 – (incl bibl ref) – mf#1988-0064 – us ATLA [240]

Christianity and the modern mind / McComb, Samuel – New York: Dodd, Mead, 1910 – 1mf – 9 – 0-7905-9796-9 – mf#1989-1521 – us ATLA [240]

Christianity and the nations / Speer, Robert Elliott – New York: Fleming H Revell, c1910 – 1mf – 9 – 0-8370-6390-6 – mf#1986-0390 – us ATLA [240]

Christianity and the new china *see* Chi-tu chiao yu hsin chung-kuo [ccm333]

Christianity and the new idealism : a study in the religious philosophy of to-day = Hauptprobleme der religionsphilosophie der gegenwart / Eucken, Rudolf – London, New York: Harper, 1909 – (= ser Harper's library of living thought) – 1mf – 9 – 0-7905-3729-X – (in english) – mf#1989-0222 – us ATLA [100]

Christianity and the new physics *see* Chi-tu chiao yu hsin wu li hsueh [ccm299]

Christianity and the progess of man : as illustrated by modern missions / Mackenzie, William Douglas – Chicago: Fleming H Revell, 1897 – 1mf – 9 – 0-8370-6147-4 – (incl bibl ref) – mf#1986-0147 – us ATLA [240]

Christianity and the religions : being three lectures delivered...harvard university in july 1908 / Lloyd, Arthur Selden – New York: EP Dutton, c1909 [mf ed 1985] – 1mf – 9 – 0-8370-4153-8 – mf#1985-2153 – us ATLA [230]

Christianity and the religions of india : essays / Kennedy, James – Mirzapore: Orphan School Press, 1874 [mf ed 1992] – 1mf – 9 – 0-524-03120-7 – mf#1990-3173 – us ATLA [230]

Christianity and the roman government : a study in imperial administration / Hardy, Ernest George – London: Longmans, Green 1894 [mf ed 1986] – 1r – 1 – (incl bibl ref. filmed with: with: hume / huxley, t h & other titles) – mf#1736 – us Wisconsin U Libr [240]

Christianity and the rural reconstruction movement in china *see* Chi-tu chiao yu chung-kuo hsiang tsun chien she yun tung [ccm351]

Christianity and the science of religion : a discourse... london, aug 2nd, 1880...10th [fernley] lecture... / Banks, John Shaw – London: Wesleyan Conference Off 1880 [mf ed 1991] – 1mf – 9 – 0-7905-9132-4 – mf#1989-2357 – us ATLA [240]

Christianity and the shona / Murphree, Marshall W – London, England. 1969 – 1r – us UF Libraries [960]

Christianity and the social crisis / Rauschenbusch, Walter – New York: Macmillan, 1908, c1907 – 1mf – 9 – 0-7905-9600-8 – mf#1989-1325 – us ATLA [240]

Christianity and the social order / Campbell, Reginald John – New York: Macmillan, 1907 – 1mf – 9 – 0-7905-7923-5 – mf#1989-1148 – us ATLA [240]

Christianity and the social rage / Berle, Adolf Augustus – New York: McBride, Nast, 1914 – 1mf – 9 – 0-7905-9136-7 – mf#1989-2361 – us ATLA [240]

Christianity and the social state / Lorimer, George Claude – Philadelphia: AF Rowland, c1898 – 2mf – 9 – 0-7905-7906-5 – mf#1989-1131 – us ATLA [240]

Christianity and the socialist movement *see*
– Chi-tu chiao yu she hui chu i yun tung [ccm16]

Christianity and the supernatural / D'Arcy, Charles Frederick – London, New York: Longmans, Green 1909 [mf ed 1985] – (= ser Anglican church handbooks) – 1mf – 9 – 0-8370-2824-8 – mf#1985-0824 – us ATLA [230]

Christianity and the united states / Goucher, John Franklin – New York: Eaton & Mains, c1908 – 1mf – 9 – 0-524-05319-7 – mf#1990-1437 – us ATLA [240]

Christianity as mystical fact : and the mysteries of antiquity = Christentum als mystische thatsache / Steiner, Rudolf; ed by Collison, Harry – 3rd rev and enl ed. New York: GP Putnam, 1914 – 1mf – 9 – 0-524-03374-9 – (in english) – mf#1990-3208 – us ATLA [240]

Christianity as taught by s paul / Irons, William Josiah – Oxford: J Parker, 1870 – 2mf – 9 – 0-7905-1414-1 – mf#1987-1414 – us ATLA [225]

Christianity at the cross-roads / Tyrrell, George – London, New York: Longmans, Green, 1910 [mf ed 1986] – xxii/282p on 1mf – 9 – 0-8370-8873-9 – mf#1986-2873 – us ATLA [240]

Christianity at the fountain / Hays, Daniel – Elgin, Ill: Brethren Pub House, 1916 – 1mf – 9 – 0-524-02826-5 – mf#1990-4447 – us ATLA [240]

Christianity contrasted with hindu philosophy : an essay, in five books, sanskrit and english / Ballantyne, James Robert – London: J Madden, 1859 – 1mf – 9 – 0-524-01038-2 – mf#1990-2186 – us ATLA [230]

Christianity established by jewish and pagan testimony / Bradlaugh, William Robert – London, England. 18– – 1r – us UF Libraries [240]

Christianity in a new light – London, England. 1875 – 1r – us UF Libraries [240]

Christianity in africa / Northcott, Cecil – Philadelphia, PA. 1963 – 1r – us UF Libraries [960]

Christianity in celtic lands / Gougaud, L – London, 1932 – €18.00 – ne Slangenburg [240]

Christianity in ceylon : its introduction and progress under the portuguese, the dutch, the british, and american missions / Tennent, James Emerson – London: John Murray, 1850 – 1mf – 9 – 0-524-01071-4 – mf#1990-2219 – us ATLA [240]

Christianity in china : a fragment / Marshall, Thomas William M – London: Longman, Brown, Green, Longmans & Roberts, 1858 [mf ed 1995] – (= ser Yale coll) – 188p – 1 – 0-524-09496-9 – mf#1995-0496 – us ATLA [240]

Christianity in china – London, England. 1850 – 1r – us UF Libraries [240]

Christianity in china, tartary, and thibet / Huc, Evariste Regis – London Longman, Brown, Green, Longmans & Roberts, 1857-58 [mf ed 1995] – (= ser Yale coll) – 3v – 1 – 0-524-09776-3 – mf#1995-0776 – us ATLA [241]

Christianity in early britain / Williams, Hugh – Oxford: Clarendon Press 1912 [mf ed 1990] – (= ser Davies lecture 1905) – 2mf – 9 – 0-7905-6151-4 – (incl bibl ref) – mf#1988-2151 – us ATLA [240]

Christianity in history : a study of religious development / Bartlet, James Vernon & Carlyle, Alexander James – London: Macmillan, 1917 – 2mf – 9 – 0-524-01942-8 – mf#1990-0531 – us ATLA [240]

Christianity in india : an historical narrative / Kaye, John William – London: Smith, Elder, 1859 [mf ed 1995] – (= ser Yale coll) – xvi/522p – 1 – 0-524-09842-5 – mf#1995-0842 – us ATLA [240]

Christianity in india / Smith, George – Edinburgh, Scotland. 1864 – 1r – us UF Libraries [240]

Christianity in its relation to the state and the church : two sermons preached in st andrew's church, ottawa, on april 7th and april 14th, 1889 / Herridge, William Thomas – [Ottawa?: s.n.], 1889 [mf ed 1980] – 1mf – 9 – 0-665-05557-9 – mf#05557 – cn Canadiana [230]

Christianity in japan / Harris, Merriman Colbert – Cincinnati: Jennings and Graham; New York: Eaton and Mains [1907] [mf ed 1995] – (= ser Yale coll; Little books on missions) – 88p – 1 – 0-524-09567-1 – mf#1995-0567 – us ATLA [240]

Christianity in modern japan / Clement, Ernest Wilson – Philadelphia: American Baptist Publ Soc, 1905 – 1mf – 9 – 0-8370-6096-6 – (includes appendix and index) – mf#1986-0096 – us ATLA [240]

Christianity in polynesia : a study and a defence / King, Joseph Hillery – Sydney: William Brooks, 1899 [mf ed 1995] – (= ser Yale coll) – 184p – 1 – 0-524-09597-3 – mf#1995-0597 – us ATLA [240]

Christianity in relation to science and morals / MacColl, Malcolm – 3rd ed. New York: James Pott, 1890 – 1mf – 9 – 0-8370-4219-4 – (includes an appendix containing the author's review of the unseen universe, or, physical speculations on a future state, and an index) – mf#1985-2219 – us ATLA [240]

Christianity in talmud and midrash / Herford, Robert Travers – London: Williams & Norgate, 1903 – 2mf – 9 – 0-7905-1154-1 – (incl ind) – mf#1987-1154 – us ATLA [240]

Christianity in the cartoons : referred to artistic treatment and historic fact / Lloyd, William Watkiss – [London], Edinburgh: Williams and Norgate, 1865 – (= ser 19th c art & architecture) – 5mf – 9 – mf#4.1.5 – uk Chadwyck [740]

Christianity in the first century / Fairbairn, Andrew Martin – London, England. 1883 – 1r – us UF Libraries [240]

Christianity in the light of today *see* Hsien tai ssu hsiang chung te chi-tu chiao (ccm200)

Christianity in the modern world / Cairns, David Smith – New York: AC Armstrong; London: Hodder & Stoughton [1906?] [mf ed 1985] – 1mf – 9 – 0-8370-3122-2 – mf#1985-1122 – us ATLA [240]

Christianity in the nineteenth century : a religious and philosophical survey of the immediate past, according to the spirit of jesus = Christianisme au dix-neuvieme siecle / Chastel, Etienne – London: Williams and Norgate, 1874 – 1mf – 9 – 0-7905-4500-4 – mf#1988-0500 – us ATLA [240]

Christianity in the nineteenth century / Lorimer, George Claude – Philadelphia: Griffith & Rowland, 1900 – (= ser Lowell institute lectures) – 2mf – 9 – 0-524-00765-9 – mf#1990-0197 – us ATLA [240]

Christianity in the nineteenth century : a sermon. delivered in the universalist church, charlestown, mass... / Townley, Robert – Boston: Bazin & Chandler, 1852 – 1mf – 9 – 0-524-04745-6 – mf#1991-2150 – us ATLA [240]

Christianity in the t'ang dynasty *see* Tang ch'ao chi-tu chiao chih yen chiu (ccm208)

Christianity in the united states : from the first settlement down to the present time / Dorchester, Daniel – rev ed. New York: Hunt & Eaton, c1895 – 2mf – 9 – 0-524-03635-7 – (incl bibl ref) – mf#1990-1063 – us ATLA [240]

Christianity is a life / unitarianism and original congregationalism in new england / the unitarians / Hale, Edward Everett – Boston: American Unitarian Assoc [191-?] [mf ed 1993] – 1v on 1mf – 9 – 0-524-08682-6 – mf#1993-3207 – us ATLA [243]

Christianity is christ / Thomas, William Henry Griffith – London, New York: Longmans, Green 1909 – (= ser Anglican church handbooks) – 1mf – 9 – 0-7905-2747-2 – mf#1987-2747 – us ATLA [240]

Christianity judged by its fruits / Croslegh, Charles – London: SPCK, 1884 [mf ed 1985] – 1mf – 9 – 0-8370-2783-7 – (incl bibl ref) – mf#1985-0783 – us ATLA [240]

Christianity not the property of critics and scholars... / Thom, John Hamilton – Liverpool, England. 1839 – 1r – us UF Libraries [240]

Christianity not the religion either of the bible only – Oxford, England. 1830 – 1r – us UF Libraries [240]

The christianity of jesus christ : is it ours? / Pearse, Mark Guy – Cincinnati: Jennings & Pye, [18–]Beltsville, Md: NCR Corp, 1978 (3mf); Evanston: American Theol Lib Assoc, 1984 (3mf) – (= ser Little Books on Devotion) – 9 – 0-8370-0840-9 – mf#1984-4228 – us ATLA [240]

The christianity of jesus christ : is it ours? / Pearse, Mark Guy – Cincinnati: Jennings & Pye; New York: Eaton & Mains, [1901] – (= ser Little Books on Devotion) – 1mf – 9 – 0-8370-5222-X – mf#1985-3222 – us ATLA [240]

The christianity of st paul / Alexander, Sidney Arthur – London, NY: Longmans, Green, 1899 – 1mf – 9 – 0-8370-2063-8 – mf#1985-0063 – us ATLA [225]

Christianity old and new : lectures / Bacon, Benjamin Wisner – New Haven: Yale UP 1914 [mf ed 1990] – (= ser Earl lectures [1911?]) – 1mf – 9 – 0-7905-3526-2 – mf#1989-0019 – us ATLA [240]

Christianity reviewed / Cheetham, William – [Brockville, Ont?: Recorder Print Co], 1896 – 4mf – 9 – 0-665-00595-4 – mf#00595 – cn Canadiana [240]

Christianity revived in the east : or, a narrative of the work of god among the armenians of turkey / Dwight, H G O – New York, 1850 – 4mf – 9 – mf#HT-165 – ne IDC [910]

Christianity revived in the east : or, a narrative of the work of god among the armenians of turkey / Dwight, Harrison Gray Otis – New York: Baker & Scribner, 1850 [mf ed 1989] – 1mf – 9 – 0-7905-4462-8 – mf#1988-0462l – us ATLA [240]

Christianity supernatural : a brief essay on christian evidence / Minton, Henry Collin – Philadelphia: Westminster Press, 1900 [mf ed 1991] – (= ser Westminster handbooks) – 1mf – 9 – 0-7905-9520-6 – mf#1989-1225 – us ATLA [240]

Christianity, the logic of creation / James, Henry – New York: D Appleton, 1857 [mf ed 1990] – 1mf – 9 – 0-7905-7783-6 – mf#1989-1008 – us ATLA [240]

Christianity, the religion of nature : lectures delivered before the lowell institute / Peabody, Andrew P – Boston: Gould and Lincoln, 1864, c1863 – (= ser Lowell institute lectures) – 1mf – 9 – 0-8370-4682-3 – mf#1985-2682 – us ATLA [240]

Christianity the science of manhood / Savage, Minot Judson – London: Simkin, Marshall, [1883?] [mf ed 1985] – (= ser Modern handbooks of religion) – 1mf – 9 – 0-8370-5138-X – mf#1985-3138 – us ATLA [240]

Christianity, the world-religion : lectures delivered in india by john henry barrows / Barrows, John Henry – 1st ed. Madras: Christian Lit Society for India, 1897 – 1mf – 9 – 0-8370-2189-8 – (includes appendix) – mf#1985-0189 – us ATLA [230]

Christianity today – Carol Stream. 1956+ (1) 1970+ (5) 1970+ (9) – ISSN: 0009-5753 – mf#1647 – us UMI ProQuest [240]

Christianity today : a presbyterian journal – v1-11. may 1930-may 1949 – 5r – 1 – (lacks some iss) – mf#ATLA S0189 – us ATLA [242]

Christianity today – v25 n5,10,14,16 [1981 mar 13, may 29, aug 7, sep 18, 1983 sep 2-dec 31], 1985 jan 18/jun 14-1988 oct 7-1989 jun 16 – 9r – 1 – mf#26952 – us WHS [240]

Christianity triumphant over infidelity! / Woffendale, Z B – London, England. 1888 – 1r – us UF Libraries [240]

Christianity vindicated by its enemies / Dorchester, Daniel – New York: Hunt & Eaton, c1896 [mf ed 1985] – 1mf – 9 – 0-8370-2951-1 – mf#1985-0951 – us ATLA [240]

Christianity, what is it? : five lectures on dr harnack's wesen des christentums / Mason, Arthur James – London: SPCK 1902 [mf ed 1992] – (= ser Church historical society (series) 66) – 1mf – 9 – 0-524-05047-3 – (incl bibl ref) – mf#1992-0300 – us ATLA [240]

Christianity, what is it? and what has it done? / Tayler, John James – London: Williams and Norgate, 1868 – 1mf – 9 – 0-524-00348-3 – mf#1989-3048 – us ATLA [240]

Christianity without judaism : a second series of essays... / Powell, Baden – London: Longman, Brown, Green, Longmans & Roberts, 1857 [mf ed 1989] – 1mf – 9 – 0-7905-1841-4 – mf#1987-1841 – us ATLA [242]

Christianity without priest : and without ritual / Martineau, James – Liverpool, England. 1839 – 1r – us UF Libraries [240]

Christianity's challenge : and some phases of christianity / Johnson, Herrick – Chicago: Cushing, Thomas, 1881 [mf ed 1985] – 1mf – 9 – 0-8370-4574-6 – mf#1985-2574 – us ATLA [240]

Christianity's encounter with world religions, 1850-1950 – [mf ed 2002] – 239 titles on 577r (phases 1+2) – 1 – $75,010.00 coll $130.00r – (coll is representative of non-christian, missionary, and syncretistic religious journals, documenting three areas: (1) the dramatic commitment to missions that north american churches demonstrated at the turn of the nineteenth century; (2) the initial journals available in north america representing the theological viewpoint of non-western religions; and (3) titles that represent experimental and syncretistic religious movements, incorporating elements of both western and non-western religions. phase 1+2 completed. phase 3 scheduled completion dec 2003. titles in coll listed individually) – us ATLA [230]

Christianity's storm centre : a study of the modern city / Stelzle, Charles – New York: FH Revell, c1907 – 1mf – 9 – 0-7905-6019-4 – mf#1988-2019 – us ATLA [301]

Christianized rationalism and the higher criticism : a reply to professor harnack's "what is christianity" / Anderson, Robert – Chicago: Winona Pub Co, 1903 [mf ed 1985] – 1mf – 9 – 0-8370-2093-X – mf#1985-0093 – us ATLA [240]

The christianizing of china / Pratt, Edwin A – London: SPCK; New York: E S Gorham, 1915 [mf ed 1995] – (= ser Yale coll) – 109p – 1 – 0-524-09518-3 – mf#1995-0518 – us ATLA [951]

Christianizing the social order / Rauschenbusch, Walter – New York: Macmillan, 1912 – 2mf – 9 – 0-7905-9450-1 – mf#1989-2675 – us ATLA [301]

Il christiano – 1913-17 [complete] – 1r – 1 – (title varies: the new aurora) – mf#ATLA RS0144 – us ATLA [240]

Christians : the temple of god / Hall, Newman – London, England. 1863 – 1r – us UF Libraries [240]

Christians! : seek the rest of god in his millennial kingdom / Govett, Robert – Norwich, England. no date – 1r – us UF Libraries [240]

O christians! : why do ye believe not on christ? / Kheiralla, Ibrahim George – [Chicago?: s.n.], c1917 [mf ed 1992] – 1mf – 9 – 0-524-01966-5 – mf#1990-2757 – us ATLA [290]

Christians and infidels – s.l, s.l? no date – 1r – us UF Libraries [240]

Christians and spain : a statement / Christian Foodship Committee – Watford [England]: The Committee [1937] – 9 – mf#w799 – us Harvard [946]

Christians and the national salvation movement *see* Chi-tu tu yu chiu kuo yun tung [ccm278]

Christians and wartime service *see* Chi-tu t'u yu chan shih fu wu [ccm311]

Christians at chen-chiang fu / Moule, Arthur Christopher & Giles, Lionel – [s.l: s.n, s.n, 1915?] [mf ed 1995] – (= ser Yale coll) – p[627]-686 – 1 – 0-524-09278-8 – mf#1995-0278 – us ATLA [240]

The christian's companion / Whitefield, George – or Sermons on several subjects. 1738 – 1 – us Southern Baptist [242]

Christian's death, life, prospects, and duty : and an apostle's grou... / Punshon, William Morley – London, England. 1861 – 1r – us UF Libraries [240]

Christian's desire to depart / Beckett, William – Edinburgh, Scotland. 1869 – 1r – us UF Libraries [240]

Christian's directory : or, sentiments of christian piety – London, England. 1825 – 1r – us UF Libraries [240]

Christian's duty : arising out of the christian's privilege / Menzies, John – Farnham, England. 1838 – 1r – us UF Libraries [240]

Christian's golden chain : or, the divine human titles of the lord / Sibly, Manoah – London, England. 1796 – 1r – us UF Libraries [240]

The christian's guide to heaven : a manual of spiritual exercises for catholics with the evening office of the church, in latin and english, and a selection of pious hymns – New York: Catholic Publication Society, [1870?] – 4mf – 9 – 0-524-08672-9 – mf#1993-3197 – us ATLA [241]

A christian's habits / Speer, Robert Elliott – Philadelphia: Westminster Press, 1911 [mf ed 1989] – 1mf – 9 – 0-7905-3231-X – mf#1987-3231 – us ATLA [240]

Christian's hope / Dick, Thomas – Glasgow, Scotland. no date – 1r – us UF Libraries [240]

The christian's hope / Webb, Robert Alexander – Jackson, Miss.: Presbyterian School for Christian Workers, 1914 – (= ser Smyth Lectures) – 1mf – 9 – 0-7905-7489-6 – mf#1989-0714 – us ATLA [240]

Christians in india / Money, Robert Cotton – Bombay: sold by Narayan Shankar, at the Scottish mission-house, 1834 [mf ed 1995] – (= ser Yale coll) – 72p – 1 – 0-524-09987-1 – mf#1995-0987 – us ATLA [954]

The christian's lady magazine (london) – jan 1834-dec 1834 – (= ser 19th c british periodicals) – r26 – 1 – us Primary [073]

The christian's lady magazine (london) – jan 1835-dec 1835 – (= ser 19th c british periodicals) – r27 – 1 – us Primary [073]

The christian's lady magazine (london) – jan 1836-dec 1836 – (= ser 19th c british periodicals) – r28 – 1 – us Primary [073]

The christian's lady magazine (london) – jan 1837-dec 1837 – (= ser 19th c british periodicals) – r29 – 1 – us Primary [073]

The christian's lady magazine (london) – jan 1838-dec 1838 – (= ser 19th c british periodicals) – r30 – 1 – us Primary [073]

The christian's lady magazine (london) – jan 1839-dec 1839 – (= ser 19th c british periodicals) – r31 – 1 – us Primary [073]

The christian's lady magazine (london) – jan 1840-dec 1840 – (= ser 19th c british periodicals) – r32 – 1 – us Primary [073]

Christian's magazine : designed to promote the knowledge and influence of evangelical truth and order – New York. 1806-1811 (1) – mf#3696 – us UMI ProQuest [240]

Christian's magazine, reviewer and religious intelligencer – Portsmouth. 1805-1808 (1) – mf#3566 – us UMI ProQuest [240]

The christian's manual : a treatise on christian perfection, with directions for obtaining that state / ed by Merritt, Timothy – Cincinnati: Swormstedt & Poe, 1854 – 2mf – 9 – 0-524-08698-2 – mf#1993-3223 – us ATLA [240]

Christian's manual of faith and practice – Bath, England. 1816 – 1r – us UF Libraries [240]

Christian's monitor : adapted to the present alarming crisis / Fisher, Samuel – Wisbech, England. 1798 – 1r – us UF Libraries [240]

Christian's monitor – Portland. 1799-1799 (1) – mf#3517 – us UMI ProQuest [240]

The christians of assyria commonly called "nestorians" / Badger, George Percy – London: WH Bartlett, 1869 – 1mf – 9 – 0-524-03331-5 – mf#1990-0912 – us ATLA [240]

The christians of st thomas and their liturgies : comprising the anaphorae of st. james, st. peter, the twelve apostles, mar dionysius, mar xystus, and mar evannis, together with the ordo communis / Howard, George Broadley – Oxford: J. Henry and J. Parker, 1864 – 1mf – us ATLA [240]

The christians of st. thomas and their liturgies : comprising the anaphorae of st. james, st. peter, the twelve apostles, mar dionysius, mar xystus, and mar evannis, together with the ordo communis / Howard, George Broadley – Oxford: J. Henry and J. Parker, 1864 – 1mf – 9 – 0-7905-5233-7 – (incl bibl ref) – mf#1988-1233 – us ATLA [240]

Christians of the copperbelt / Taylor, John Vernon – London, England. 1961 – 1r – us UF Libraries [960]

Christians on earth and in heaven : the substance of a discourse, delivered in the adelaide street wesleyan-methodist church, toronto, on sabbath evening, october 29th, 1848... / Ryerson, Egerton – Toronto: W M Book Room, 1848 – 1mf – 9 – 0-665-88939-9 – mf#88939 – cn Canadiana [240]

Christian's pathway to power – London, 1874-78 [mf ed 2001] – (= ser Christianity's encounter with world religions, 1850-1950) – 1r – 1 – mf#2001-s180 – us ATLA [240]

The christian's plea against modern unbelief : a handbook of christian evidence / Redford, Robert Ainslie – London: Hodder & Stoughton, 1883 [mf ed 1988] – 2mf – 9 – 0-7905-0209-7 – (incl ind) – mf#1987-0209 – us ATLA [240]

The christian's present for all seasons : containing devotional thoughts of eminent divines, from joseph hall to william jay / ed by Harsha, David Addison – New York: American Tract Society, [c1866] Beltsville, Md: NCR Corp, 1977 (7mf); Evanston: American Theol Lib Assoc, 1984 (7mf) – 9 – 0-8370-0141-2 – (incl ind) – mf#1984-0028 – us ATLA [240]

Christian's race / Moore, Daniel – London, England. 1860 – 1r – us UF Libraries [240]

The christian's relation to evolution : a question of gain or loss / Johnson, Franklin – Chicago: Fleming H Revell, 1904 – 1mf – 9 – 0-8370-3787-5 – (incl bibl ref) – mf#1985-1787 – us ATLA [210]

The christians rescue from the grand error of the heathen : touching the fatal necessity of all events... / Pierce, T – London, Richard Royston, 1658 – 9 – mf#ZWI-108 – ne IDC [240]

The christian's sacrifice and service of praise, or, the two great commandments : being an exposition of the twelth chapter of the epistle to the romans / Candlish, Robert Smith – Edinburgh: Adam and Charles Black, 1867 – 1mf – 9 – 0-8370-2579-6 – mf#1985-0579 – us ATLA [220]

Christian's, scholar's, and farmer's magazine : calculated, in an eminent degree, to promote religion – Elizabeth Town. 1789-1791 (1) – mf#3518 – us UMI ProQuest [240]

The christian's secret of a happy life / Smith, Hannah Whitall – New and enl. ed. New York: Revell, 1888. El Segundo, Ca: Micro Publication Systems, 1980 (1mf); Evanston: American Theol Lib Assoc, 1984 (1mf) – (= ser Women & the church in america) – 9 – 0-8370-1394-1 – mf#1984-2132 – us ATLA [240]

Christian's thank-offering / Hessey, Francis – Huddersfield, England. 1841 – 1r – us UF Libraries [240]

Christians under the crescent in asia / Cutts, Edward Lewes – London: SPCK; New York: Pott, Young, [1877?] – 1mf – 9 – 0-7905-5934-X – mf#1988-1934 – us ATLA [240]

Christian's view of the cause and remedy of the present national di... / Elliott, Edward B – London, England. 1830 – 1r – us UF Libraries [240]

Christian's way to heaven / Divine Of The Church Of England – London, England. 1804 – 1r – us UF Libraries [240]

Christian's weekly monitor – Sangerfield. 1815-1818 (1) – mf#4434 – us UMI ProQuest [240]

CHRISTIANSBURG

Christiansburg Industrial Institute *see*
- Annual catalogue
- Annual report of the christiansburg industrial institute
- Catalogue of the christiansburg industrial institute

Christiansen, Carl *see* Index filicum
Christiansen, Monty L *see* An exploration of the opinions of recreation and parks/leisure studies faculty and public sector practitioners concerning the computer competency skills of recreation and parks/leisure studies baccalaureate students
Christianstads annonsblad – Kristianstad, 1888-89 – 1r – 1 – sw Kungliga [078]
Christianstads annonsblad – Kristianstad, Sweden. 1888-89 – 1r – 1 – sw Kungliga [078]
Christianstads weckoblad – Kristianstad, Sweden. 1810-40 – 4r – 1 – sw Kungliga [078]
Christianstatesman – 1873 jan 2-1875 nov 25, 1875 dec 2-1878 nov 7, 1878 nov 14-1881 sep 8, 1881 sep 15-1884 jan 31, 1884 oct 30-1889 sep 26 – 5r – 1 – mf#360424 – us WHS [071]
Christiany, Ludwig *see* Eva von buttler, die messaline und muckerin, als prototype der "seelenbraeute"
Christic Institute *see* Convergence
Christie, Agatha *see* Crooked house
Christie, Dugald *see*
- Ten years in manchuria
- Thirty years in the manchu capital, in and around moukden in peace and war

Christie, James *see*
- Cholera epidemics in east africa.
- Disquisitions upon the painted greek vases
- The records of the commissions of the general assemblies of the church of scotland holden in edinburgh in 1650, in st. andrews and dundee in 1651 and in edinburgh in 1652
- The records of the commissions of the general assemblies of the church of scotland holden in edinburgh in the years 1646 and 1647
- The records of the commissions of the general assemblies of the church of scotland holden in edinburgh in the years 1648 and 1649

Christie, Jas *see* For king and kingdom
Christie, John F *see* Ministry a call to endure hardness
Christie, Manson and Woods, Ltd, London *see*
- Addenda of the remainder of the furniture...of ralph bernal
- Catalogue of the celebrated collection of...ralph bernal
- Catalogue of the celebrated collection...british india
- Catalogue of the celebrated fontaine collection
- Catalogue of the choice collection...from blenheim palace
- Catalogue of the collection of...his grace the duke of hamilton
- Catalogue of the highly important collection of...james price
- Catalogue of the magnificent contents of alton towers...
- Catalogue of the marlborough gems
- Catalogue of the renowned collection of... hollingworth magniac
- Catalogue of the...collection...by the late adrian hope, esq

Christie, Michael John *see* Simonstown agreements
Christie, Robert *see* The dignity of labor
Christie, S J *see* International symposium on the ecological effects of arctic airborne contaminants
Christie, Thomas William *see* Fall of rome
Christie, William *see* Alliance work in western china and tibet
Christie's impressionist and modern art – 1950-83 – 158mf – 9 – $1225.00 – 0-907006-57-4 – (suppl to christie's pictorial archive; 9400 captioned photos of items sold by christie's in 4 sub-sections: british & foreign, paintings & sculpture; with printed ind of artists) – uk Mindata [700]
Christie's pictorial archive : decorative and applied art – 710mf – 9 – $4950.00 coll – 0-907006-47-7 – (available in 5 binders; individual titles also listed separately) – uk Mindata [700]
Christie's pictorial archive : painting and graphic art – 498mf – 9 – $3450.00 coll – 0-907006-17-5 – (available in 4 binders; individual titles also listed separately) – uk Mindata [700]
Christie's pictorial archive : painting and graphic art – decorative and applied art – 1900-79 – 1208mf – 9 – $8140.00 set – 0-907006-52-3 – (complete set consists of 9 sect; may be purchased separately; individual titles also listed) – uk Mindata [700]
Christie's pictorial archive new york – 599mf – 9 – $4600.00 coll – 0-907006-92-2 – (over 35,000 captioned photos of items sold by new york christie's with full catalogue details, prices realized & date of sale; in alphabetical order, by artist; 6 sects with binder for each; also listed individually) – uk Mindata [700]
Christie's pictorial sales review – 341mf – 9 – $2760.00 coll – 0-907006-18-3 – (comprehensive cumulative update of christie's pictorial archive; full catalogue details, prices realized, date of sale, arr in alphabetical order by artist; over 40,000 reproductions in 4 sects, with binder for each) – uk Mindata [700]
Christina de Wonderbare *see* Gedenkboek 1150-1940
Christina mortens ehe : [a novel] / Boger, Margot – Berlin: W Limpert, 1943 [mf ed 1989] – 425p – 1 – mf#7050 – us Wisconsin U Libr [830]
Christinehamnsbladet – Kristinehamn, Sweden. 1850-54 – 2r – 1 – sw Kungliga [078]
Christine-Marie, soeur *see* Bibliographie analytique de l'oeuvre du docteur pierre jobin...
Christkatholische dogmatik / Hermes, Georg; ed by Achterfeld, Johann Heinrich – Muenster: Coppenrath, 1834 – 1mf – 9 – 0-524-04072-9 – (incl bibl ref) – mf#1991-2017 – us ATLA [241]
The christless nations : a series of addresses on christless nations and kindred subjects / Thoburn, James Mills – New York: Hunt and Eaton; Cincinnati: Cranston and Curts, 1895 – 1mf – 9 – 0-524-01024-2 – mf#1990-0301 – us ATLA [240]
Der christlich eestand / Bullinger, Heinrich – [Zuerich, Christoffel Froschouer], 1540 – 3mf – 9 – mf#PBU-137 – ne IDC [240]
Das christlich kinderlied d martini lutheri : erhalt vns herr etc itziger zeit nuetz. vnd noetig zu singen. in sechs stimmen gesetzt vnd gedruckt zu witteberg – Wittenberg 1565 [mf ed 1989] – 6pt on 1r – 1 – mf#pres. film 51 – us Sibley [780]
Ein christlich psalter gebett : der betrangten kirchen gottes zu trost gestellet und auss den cl. psalmen davids zusamen gezogen / Mylius, Georg – Ulm: Ulhart 1585 – (= ser Hqab. literatur des 16. jahrh.) – 1mf – 9 – €20.00 – mf#1585c – gw Fischer [780]
Die christlich-arabische litteratur bis zur fraenkischen zeit... / Graf, G – (= ser Strassburger Theologische Studien) – 1mf – 9 – (strassburger theologische studien. freiburg, 1905. v7(1)) – mf#H-2979 – ne IDC [470]
Der christliche altar / Braun, J – Muenchen. v1-2. 1924 – 2v on 50mf – 8 – €95.00 – ne Slangenburg [240]
Der christliche altar und seine schmuck : archaeologisch-liturgisch dargestellt / Schmid, A – Regensburg, 1871 – €12.00 – ne Slangenburg [930]
Das christliche alterthum : oder, die katholische kirche in ihrem kampfe mit den verfolgungen und irrlehren: ein vollstaendiges leben der heiligen des christlichen alterthums im anschluss an die kirchengeschichte / Bayerle, Bernard Gustav – New York: S Zickel [distributor], 1862 – 2mf – 9 – 0-8370-6883-5 – (incl ind of saints) – mf#1986-0883 – us ATLA [241]
Christliche apokryphen / Geffcken, Johannes – Tuebingen: J C B Mohr 1908 [mf ed 1989] – (= ser Religionsgeschichtliche volksbuecher fuer die deutsche christliche gegenwart 1/15) – 1mf – 9 – 0-7905-1661-6 – mf#1987-1661 – us ATLA [225]
Der christliche apologete – Cincinnati OH (USA) 1839-1941 3 dec [gaps] [mf ed 2004] – 46r – 1 – gw Mikrofilm [240]
Christliche apologetik : versuch eines handbuchs / Sack, Karl Heinrich – Hamburg: Friedrich Perthes, 1829 [mf ed 1991] – 2mf – 9 – 0-524-00315-7 – mf#1989-3015 – us ATLA [240]
Die christliche apologetik im neunzehnten jahrhundert : lebensbilder und charakteristiken deutscher evangelischer glaubenszeugen aus der juengsten vergangenheit / Zoeckler, Otto – Guetersloh: C Bertelsmann, 1904 [mf ed 1986] – 1mf – 9 – 0-8370-8879-8 – (incl bibl ref) – mf#1986-2879 – us ATLA [240]
Christliche ausslegung in das erste capitel des euangelisten s johannis / Mathesius, J – Leipzig, 1589 – 4mf – 9 – mf#TH-1 mf 970-973 – ne IDC [242]
Das christliche des platonismus, oder, sokrates und christus : eine religionsphilosophische untersuchung / Baur, Ferdinand Christian – Tuebingen: LF Fues, 1837 – 1mf – 9 – 0-7905-5681-2 – mf#1988-1681 – us ATLA [240]
Christliche dogmatik / Biedermann, Alois Emanuel – 2. erw Aufl. Berlin: G Reimer, 1884-1885 – 3mf – 9 – 0-7905-3582-3 – (incl bibl ref) – mf#1989-0075 – us ATLA [240]
Christliche dogmatik / Ebrard, Johannes Heinrich August – 2. Aufl. Koenigsberg: AW Unzer, 1862-1863 – 4mf – 9 – 0-7905-9375-0 – mf#1989-2600 – us ATLA [240]
Christliche dogmatik / Schmidt, Wilhelm – Bonn: E Weber 1895-98 [mf ed 1991] – (= ser Sammlung theologischer handbuecher 4/1) – 2v on 3mf – 9 – 0-7905-8729-7 – (imprint of v2: bonn : a marcus & e weber 1898) – mf#1989-1954 – us ATLA [240]
Die christliche dogmatik vom standpunkte des gewissens aus dargestellt / Schenkel, Daniel – Wiesbaden: Kreidel und Niedner, 1858-1859 – 5mf – 9 – 0-524-00326-2 – (incl bibl ref) – mf#1989-3026 – us ATLA [240]
Christliche dogmengeschichte *see* Lectures on the history of christian dogmas
Die christliche dogmengeschichte : nach ihrem organischen entwickelungsgange, uebersichtlich dargestellt / Noack, Ludwig – 2. Aufl. Erlangen: F Enke, 1856 – 2mf – 9 – 0-7905-9427-7 – mf#1989-2652 – us ATLA [240]
Die christliche erfahrung : ihre entstehung und entwickelung: luthers katechismus artikel 3 / Scholz, Hermann – Berlin: Julius Springer, 1902 – 1mf – 9 – 0-8370-5147-9 – mf#1985-3147 – us ATLA [242]
Christliche eschatologie / Kliefoth, Theodor – Leipzig: Doerffling und Franke, 1886 – 1mf – 9 – 0-7905-9775-6 – (incl bibl ref) – mf#1989-1500 – us ATLA [240]
Die christliche eschatologie in den stadien ihrer offenbarung im alten und neuen testaments : mit besonderer beruecksichtigung der juedischen eschatologie im zeitalter christi / Atzberger, Leonhard – Freiburg i B, St Louis MO: Herder, 1890 – 4mf – 9 – 0-7905-9118-9 – (incl bibl ref) – mf#1989-2343 – us ATLA [220]
Die christliche familie : im kampfe gegen feindliche maechte: vortraege ueber christliche ehe und erziehung / Hug, Gall Joseph – 2., vielfach verm Aufl. Freiburg (Schweiz): Universitaetsbuchhandlung (B Veith), 1896 – 1mf – 9 – 0-8370-6985-8 – mf#1986-0985 – us ATLA [240]
Die christliche frau – 1902-41 [mf ed 2001] – (= ser Hq 45) – 39v on 195mf – 9 – €900.00 – 3-89131-364-0 – gw Fischer [305]
Der christliche freiheitsbegriff / Ziegelmeier, Otto W – (mf ed 1994) – 1mf – 9 – €30.00 – 3-89349-909-1 – mf#DHS 909 – gw Frankfurter [240]
Christliche gebet vnd gesaeng auff die heilige zeit vnd fayertage vber das gantze jar – Prag: Peterle 1581 – (= ser Hqab. literatur des 16. jahrh.) – 2mf – 9 – €30.00 – mf#1581b – gw Fischer [780]
Der christliche gemeindegottesdienst im apostolischen und altkatholischen zeitalter / Harnack, Theodosius – Erlangen: Theodor Blaesing, 1854 – 2mf – 9 – 0-7905-1329-3 – (in german, greek and latin. incl bibl ref) – mf#1987-1329 – us ATLA [241]
Das christliche gesang-buch : eine zusammenstellung der besten lieder der alten und neuen dichter zum gottesdienstlichen gebrauch aller gottsuchenden und heilsbegierigen seelen – Lancaster, Pa: Johann Baer, 1879 – 2mf – 9 – 0-524-04206-3 – mf#1990-4997 – us ATLA [240]
Der christliche glaube : nach dem bekenntniss der lutherischen kirche: vortraege / Langbein, Bernhard Adolf – Leipzig: Justus Naumann, 1873 – 1mf – 9 – 0-8370-4390-5 – mf#1985-2390 – us ATLA [242]
Der christliche glaube im kampfe mit dem modernen aufklaerungschriftenthum und der widerspruch des letztern mit der vernunft / Hanne, Johann Wilhelm – Jena: Fr Frommann, 1850 [mf ed 1991] – (= ser Vorhoefe zum glauben 1) – 1mf – 9 – 0-7905-9384-X – mf#1989-2609 – us ATLA [240]
Der christliche glaube in acht buechern / Hackenschmidt, Karl; ed by Calwer Verlagsverein – Calw: Vereinsbuchh., 1901 – 1mf – 9 – 0-8370-4760-9 – (incl ind) – mf#1985-2760 – us ATLA [240]
Der christliche glaube nach den grundsaezen der evangelischen kirche im zusammenhange dargestellt / Schleiermacher, Friedrich [Ernst Daniel] – 2. umgearb ausg. Berlin: G Reimer 1830-31 [mf ed 1991] – 2v on 3mf – 9 – 0-524-00464-1 – mf#1989-3164 – us ATLA [242]
Christliche glaubens- und sittenlehre : leitfaden fuer den religionsunterricht hauptsaechlich an hoeheren klassen von realanstalten und realgymnasien / Wurster, Paul – 2. Aufl. Heilbronn: E Salzer, 1906 – 1mf – 9 – 0-7905-8983-4 – mf#1989-2208 – us ATLA [240]
Das christliche glaubensbekenntnis : protestantismus gegen orthodoxismus / Richter, F – Berlin: Franz Lobeck, 1868 – 1mf – 9 – 0-8370-8707-4 – mf#1986-2707 – us ATLA [242]
Christliche glaubenslehre / Sulzberger, Arnold – 2. Aufl. Bremen: Verlag des Tractathauses, [1886?] – 2mf – 9 – 0-524-05022-8 – mf#1991-2192 – us ATLA [240]
Die christliche glaubenslehre / Luthardt, Christoph Ernst – 2. aufl. Leipzig: Doerffling & Franke, 1906 [mf ed 1984] – 7mf – 9 – 0-8370-0855-7 – (1st ed publ 1898) – mf#1984-4213 – us ATLA [240]
Die christliche glaubenslehre des herrn dr. david friedrich strauss : erster band, tuebingen und stuttgart, 1840 / Koester, Friedrich – Hannover: Hahn, 1841 – 1mf – 9 – 0-524-08085-2 – mf#1992-1145 – us ATLA [240]
Die christliche glaubenslehre im gegensatze der modernen gewissenslaxheit : ein beitrage zur wissenschaftlichen beurtheilung der straussschen dogmatik / Sartorius, Ernst – Koenigsberg: JH Bon, 1842 – 1mf – 9 – 0-524-08090-9 – mf#1992-1150 – us ATLA [240]
Christliche glaubenslehre in leitsaetzen : fuer eine akademische vorlesung / Reischle, Max – 2. Aufl. Halle a.S: Max Niemeyer, 1902 – 1mf – 9 – 0-8370-5150-9 – mf#1985-3150 – us ATLA [240]
Die christliche gnosis, oder, die christliche religions-philosophie in ihrer geschichtlichen entwiklung / Baur, Ferdinand Christian – Tuebingen: CF Osiander, 1835 – 2mf – 9 – 0-7905-7376-8 – (incl bibl ref) – mf#1989-0601 – us ATLA [240]
Der christliche gottesbegriff : beitrag zur speculativen theologie / Rocholl, Rudolf – Goettingen: Vandenhoeck & Ruprecht, 1900 – 1mf – 9 – 0-8370-5275-0 – (includes authors index) – mf#1985-3275 – us ATLA [210]
Der christliche gottesglaube : in seinem verhaeltnis zur heutigen philosophie und naturwissenschaft / Wobbermin, Georg – 2. umgearb Aufl. Berlin: Alexander Duncker, 1907 – 1mf – 9 – 0-8370-6469-4 – mf#1986-0469 – us ATLA [240]
Das christliche gottvertrauen und der glaube an christus : eine dogmatische untersuchung auf biblisch-theologischer grundlage und unter beruecksichtigung der symbolischen litteratur / Mayer, E W – Goettingen: Vandenhoeck und Ruprecht, 1899 – 1mf – 9 – 0-8370-4124-4 – (incl bibl ref) – mf#1985-2124 – us ATLA [220]
Die christliche heilsgewissheit : eine systematische darstellung des mittelpunkts evangelischen heilsverstaendnisses / Clasen, L – Halle a. S: Eugen Strien, 1897 – 1mf – 9 – 0-8370-3458-2 – mf#1985-1458 – us ATLA [240]
Die christliche heilslehre : auf grund der heiligen schrift fuer die evangelische gemeinde / Buttmann, Philipp – Leipzig: Zangenberg & Himly, 1872 – 1mf – 9 – 0-8370-3049-8 – mf#1985-1049 – us ATLA [240]
Der christliche jugend-freund – v1-74. 1878-1951 [gaps] – (= ser Mennonite serials coll) – Inquire – 1 – mf#ATLA 1994-S025 – us ATLA [242]
Die christliche kinder-zeitung – Augsburg DE, 1841 n1-12 – 1 – gw Misc Inst [240]
Die christliche kinder-zeitung – Duesseldorf DE, 1833-36 – 1r – 1 – (filmed by other misc inst: 1850 apr, 1851-62, 1863 nov-1867 feb, 1868 jan-jun [3r]) – gw Misc Inst [240]
Die christliche kirche des mittelalters in den hauptmomenten ihrer entwicklung / Baur, Ferdinand Christian; ed by Baur, Ferdinand Friedrich – 2. aufl. Leipzig: Fues, 1869 [mf ed 1989] – (= ser Geschichte der christlichen kirche 3) – 2mf – 9 – 0-7905-4185-8 – (incl bibl ref) – mf#1988-0185 – us ATLA [240]
Die christliche kirche vom anfang des vierten bis zum ende des sechsten jahrhunderts in den hauptmomenten ihrer entwicklung / Baur, Ferdinand Christian – 2. Ausg. Tuebingen: L Fr Fues, 1863 – (= ser Geschichte der christlichen kirche) – 1mf – 9 – 0-7905-7435-7 – (incl bibl ref) – mf#1989-0660 – us ATLA [240]
Die christliche legende des abendlandes / Guenter, Heinrich – Heidelberg: C Winter, 1910 – (= ser Religionswissenschaftliche Bibliothek) – 1mf – 9 – 0-524-02081-7 – (incl bibl ref) – mf#1990-2845 – us ATLA [240]
Die christliche lehre : nach dem gegenwaertigen stande der theologischen wissenschaft und ihre vermittlung an die gemeinde / Dorner, Isaak August – Berlin: Schwetschke, 1904 – 1mf – 9 – 0-8370-2952-X – mf#1985-0952 – us ATLA [240]
Die christliche lehre auf heilsgeschichtlichem grunde : dem deutsch-evangelischen volke / Buchholtz, Ludwig – Hoerter: Otto Buchholtz, 1879 – 1mf – 9 – 0-8370-3040-4 – mf#1985-1040 – us ATLA [240]
Die christliche lehre von der dreieinigkeit und menschwerdung gottes in ihrer geschichtlichen entwicklung / Baur, Ferdinand Christian – Tuebingen: CF Osiander, 1841-1843 – 28mf – 9 – 0-524-03390-0 – (incl bibl ref) – mf#1990-0944 – us ATLA [240]
Die christliche lehre von der gnade : apologie des biblischen christentums: insbesondere gegenueber der ritschlschen rechtfertigungslehre / Dieckmann, August – Berlin: C A Schwetschke, 1901 [mf ed 1985] – 1mf – 9 – 0-8370-3561-9 – mf#1985-1561 – us ATLA [240]

Christliche lehre von der rechtfertigung und versoehnung *see* Critical history of the christian doctrine of justification and reconciliation

Die christliche lehre von der rechtfertigung und versoehnung / Ritschl, Albrecht – Bonn: A Marcus, 1870-74 [mf ed 1990] – 3v on 4mf – 9 – 0-7905-9614-8 – (incl bibl ref) – mf#1989-1339 – us ATLA [240]

Die christliche lehre von der suende : erster teil, die biblische lehre / Clemen, Carl – Goettingen: Vandenhoeck und Ruprecht, 1897 – 1mf – 9 – 0-8370-2682-2 – mf#1985-0682 – us ATLA [220]

Die christliche lehre von der suende *see* The christian doctrine of sin

Die christliche lehre von der versoehnung : in ihrer geschichtlichen entwicklung von der aeltesten zeit bis auf die neuste / Baur, Ferdinand Christian – Tuebingen: C F Osiander, 1838 – 2mf – 9 – 0-7905-4067-3 – (incl bibl ref) – mf#1988-0067 – us ATLA [240]

Die christliche lehre von der versoehnung in ihrer geschichtlichen entwicklung / Bauer, F C – Tuebingen, 1838 – €27.00 – ne Slangenburg [240]

Die christliche liebesthaetigkeit / Uhlhorn, Gerhard – 2. verb Aufl. Stuttgart: D Gundert, 1895 – 2mf – 9 – 0-7905-7154-4 – mf#1988-3154 – us ATLA [240]

Christliche mission in sudwestafrika / Loth, Heinrich – Berlin, Germany. 1963 – 1r – us UF Libraries [960]

Die christliche mystik des mittelalters / Noack, Ludwig – Koenigsberg: Gebrueder Borntraeger, 1853 [mf ed 1991] – (= ser Die christliche mystik 1) – 1mf – 9 – 0-524-00293-2 – mf#1989-2993 – us ATLA [240]

Die christliche mystik seit dem reformationszeitalter / Noack, Ludwig – Koenigsberg: Gebrueder Borntraeger, 1853 – (= ser Christliche Mystik) – 1mf – 9 – 0-524-00294-0 – mf#1989-2994 – us ATLA [240]

Der christliche orient / ed by Lepsius, Johannes – Berlin: W Faber 1897 [mthly] [mf ed 2006] – n1-12 (jan-dec 1897) [complete] on 1r – 1 – (cont by: christliche orient (berlin, germany: 1900)) – mf#2005c-s093 – us ATLA [240]

Die christliche philosophie : nach ihrem begriff, ihren aeussern verhaeltnissen und in ihrer geschichte bis auf die neuesten zeiten / Ritter, Heinrich – Goettingen: Dieterich, 1858-1859 – 4mf – 9 – 0-7905-8568-5 – mf#1989-1793 – us ATLA [100]

Christliche polemik / Sack, Karl Heinrich – Hamburg: F Perthes, 1838 – 1mf – 9 – 0-524-04275-6 – (incl bibl ref and ind) – mf#1991-2059 – us ATLA [240]

Christliche predigt aus dem 48 capittel esaiae von den grossen wolthaten gottes / Hoffmann, D – Helmstadt, 1589 – 1mf – 9 – mf#TH-1 mf 699 – ne IDC [242]

Eine christliche predigt vber der leiche des herrn m: matthiae flacij jllyrici gestellet / Flacius Illyricus d A, M – np, 1575 – 3mf – 9 – mf#TH-1 mf 467-469 – ne IDC [242]

Ein christliche predigt, von christlicher einigkeit der theologen augspuergischer confession / Andreae d A, J – Wolffenbuettel, 1570 – 1mf – 9 – mf#TH-1 mf 42 – ne IDC [242]

Christliche predigt von der vocal und instrumentalischen music : [als den 4. septemb. anno 1605. in der newen euangelischen kirchen zu kauffbeuren ein orgel auffgerichtet] gehalten vnd in margine mit dem artificio rhetorico... / Anwander, Georg – Tuebingen: G Gruppenbach 1606 [mf ed 1992] – 1r – 1 – mf#pres. film 111 – us Sibley [780]

Christliche predigten vber den 129 psalm dauids darinne angezeiget wird wie die caluinische schwermer der kirch zu wittemberg vnnd im gantzen churkreiss seyen mit jhrem heilloses pflug / Huber, S – Wittemberg, 1594 – 1mf – 9 – mf#TH-1 mf 710 – ne IDC [242]

Die christliche religion im urteil ihrer gegner : die kritische bewegung gegen das christentum in neuerer zeit / Foerster, Erich – Tuebingen: J C B Mohr 1916 [mf ed 1992] – (= ser Lebensfragen 27) – 1mf – 9 – 0-524-04489-9 – mf#1990-1251 – us ATLA [240]

Der christliche religions-unterricht : auf grundlage der heiligen schrift und nach paedagogischen grundsaetzen in der oberklasse der volksschule: ein handbuch fuer lehrer / Kehr, C – 3. Aufl. Gotha: C F Thienemann. 2v. 1875 – 2mf – 9 – 0-8370-8583-7 – (incl bibl ref) – mf#1986-2583 – us ATLA [240]

Christliche reuter lieder / Winnenberg und Beichelsteyn, Philip zu – Strassburg: Jobin 1586 – (= ser Hqab. literatur des 16. jahrh.) – 3mf – 9 – €40.00 – mf#1586c – gw Fischer [780]

Die christliche sitte / Schleiermacher, Friedrich [Ernst Daniel]; ed by Jonas, Ludwig – 2. aufl. Berlin: G Reimer 1884 [mf ed 1991] – 3mf – 9 – 0-524-00331-9 – mf#1989-3031 – us ATLA [240]

Christliche sitten-lehre ueber die evangelischen wahrheiten *see* The christian state of life

Der christliche staendestaat – Wien (A), 1934-38 [gaps] – 1 – gw Misc Inst [240]

Christliche und juedische ostertafeln / Schwartz, Eduard – Berlin, 1905 – 6mf – 8 – €14.00 – (agwg.pg bd 8 (1904-1905)) – ne Slangenburg [230]

Christliche und juedische ostertafeln / Schwartz, Eduard – Berlin: Weidmann 1905 [mf ed 1990] – (= ser Abhandlungen der koeniglichen gesellschaft der wissenschaften zu goettingen 8/6) – 1mf – 9 – 0-7905-3413-4 – (in german, greek & latin) – mf#1987-3413 – us ATLA [270]

Ein christliche vnd ernstlich antwurt der prediger des euangelij zuo basel, warumb sy die mess einen greuwel gescholten habind / Oecolampadius, J – [Zuerich, Christoph Froschauer, 1527] – 1mf – 9 – mf#PBU-379 – ne IDC [240]

Das christliche volks-blatt – v1-10. 1856-66 [gaps] – (= ser Mennonite serials coll) – 1r – 1 – mf#ATLA 1994-S024 – us ATLA [242]

Christliche volkszeitung – Osijek (Esseg HR), 1938 6 jan-28 jul, 1940 4 mar-19 dec – 2r – 1 – gw Misc Inst [240]

Christliche volkszeitung – Osijek, Yugoslavia. Sept 1925-Aug 1935; Jun-Dec 1937; Aug-Dec 1938; 1941 – 5r – 1 – us L of C Photodup [949]

Die christliche wahrheitsgewissheit : ihr letzter grund und ihre entstehung / Ihmels, Ludwig – Leipzig: A Deichert, 1901 [mf ed 1985] – 1mf – 9 – 0-8370-3719-0 – (incl bibl ref) – mf#1985-1719 – us ATLA [240]

Die christliche weltanschauung und kant's sittlicher glaube : eine religioese untersuchung / Schrempf, Christoph – Goettingen: Vandenhoeck & Ruprecht, 1891 – 1mf – 9 – 0-7905-8881-1 – mf#1989-2106 – us ATLA [170]

Die christlichen grundwahrheiten : oder, die allgemeinen principien der christlichen dogmatik / Goltz, Hermann von der – Gotha: F A Perthes, 1873 – 1mf – 9 – 0-8370-3259-8 – mf#1985-1259 – us ATLA [240]

Die christlichen literaturen des orients / Baumstark, A – Leipzig, 1911. 3 pts – 4mf – 9 – mf#AR-1855 – ne IDC [956]

Die christlichen literaturen des orients / Baumstark, Anton – Leipzig: G J Goeschen 1911 [mf ed 1989] – (= ser Sammlung goeschen) – 1mf – 9 – 0-7905-4132-7 – (incl bibl ref) – mf#1988-0132 – us ATLA [240]

Ein christlicher bericht vo dem brot vnd weyn desz herren / Hoen, K H & Zwingli, H – [Augsburg, Philipp Ulhart], 1526 – 1mf – 9 – mf#PBU-522 – ne IDC [240]

Christlicher bundes-bote – v1-66. 1882-1947 [complete] – (= ser Mennonite serials coll) – 21r – 1 – mf#ATLA 1993-S022 – us ATLA [242]

Christlicher familien-kalender – 1884-85 [complete] – (= ser Mennonite serials coll) – 1r – 1 – mf#ATLA 1993-S023 – us ATLA [242]

Christlicher hochtheurer helden tugend-lauff : in sinnbildern vorgestellt / Hagelganss, J H – Nuernberg: Zufinden bey Paulus Fuersten, Kunsthaendlern, 1651 – 2mf – 9 – mf#O-1598 – ne IDC [090]

Christlicher krancken-spiegel : in welchem so wohl denen augen, als dem gemueth eines krankcken gantz klar und lehr-reich vergestellet wird / Chertablon, J de – Wienn: Schwendimann – 4mf – 9 – mf#O-96 – ne IDC [090]

Christlicher textilarbeiter – Krefeld DE, 1901 28 sep-1933 8 jul – 7r – 1 – (1906: textilarbeiter-zeitung) – mf#2668 – gw Mikropress [074]

Christlicher underricht und warhafftige erweiszung das jhesus christus durch d personl vereinigung d gottl u vleueschl naturen in alle goettliche herrlichkeyt gesetzt seye / Marbach, J – Straszburg, 1567 – 13mf – 9 – mf#TH-1 mf 945-957 – ne IDC [242]

Christlicher vnd warhafftiger underricht von den worten der einsatzung des heyligen abendtmals jhesu christi / Marbach, J – Straszburg, 1566 – 7mf – 9 – mf#TH-1 mf 938-944 – ne IDC [242]

Christlicher volksdienst – Duesseldorf DE, 1930-1933 31 oct – 1r – 1 – gw Misc Inst [240]

Christliches baettbuechlin / Bullinger, Heinrich – Zuerich, Joh. Rudolff Wolff, 1623 – 3mf – 9 – mf#PBU-269 – ne IDC [240]

Christliches bedencken des ministerij der kirchen zu brunschwig auff d maiors repitition vnd endtliche erklerung gelangend den streit : ob gute wercke zur seligkeit noetig sind, also, das es vnmueglich sey, ohne gute wercke selig zuwerden / Major, G – np, 1568 – 1mf – 9 – mf#TH-1 mf 917 – ne IDC [242]

Christliches erbe und lyrische gestaltung : eine kritische bestandsaufnahme der christlichen lyrik der gegenwart / Giesecke, Hans Heinrich – Leipzig: Koehler & Amelang 1961 [mf ed 1993] – 1r – 1 – (incl bibl ref & ind. filmed with: geschichte der deutschen ode / karl vieetor & other titles) – mf#8297 – us Wisconsin U Libr [430]

Christliches gesangbuch; christian song book or collection of psalms and hymns – 1 – us Southern Baptist [242]

Christliches kunstblatt *see* Christian art 2

Christliches magazin / ed by Pfenninger, J K – Zuerich, Winterthur 1779-80 – (= ser Dz. abt theologie) – 4v on 17mf – 9 – €170.00 – mf#k/n2144 – gw Olms [240]

"Christliches volk" *see* Der bote aus kurpfalz

Die christlich-germanische baukunst und ihr verhaeltniss zur gegenwart / Reichensperger, A – Trier, 1852 – 2mf – 9 – mf#OA-139 – ne IDC [720]

Christlich-palaestinische fragmente aus der omajjaden-moschee zu damaskus / Schulthess, Friedrich – Berlin, 1905 – 3mf – 8 – €7.00 – ne Slangenburg [240]

Christlich-palaestinische fragmente aus der omajjaden-moschee zu damaskus / ed by Schulthess, Friedrich – Berlin: Weidmann 1905 [mf ed 1993] – (= ser Abhandlungen der koeniglichen gesellschaft der wissenschaft zu goettingen. philologisch-historische klasse. neue folge 8/3) – 1mf [ill] – 9 – 0-524-06512-8 – mf#1992-0896 – us ATLA; ne Slangenburg [470]

Christlieb, Emily *see* The odor christlieb

Christlieb, Marie Luise *see* A struggle for a soul

Christlieb, Max *see* History of protestant missions in japan

Christlieb, Th *see* Leben und lehre des johannes scotus erigena

Christlieb, Theodor *see*
- Aerztliche missionen
- The best methods of counteracting modern infidelity
- The odor christlieb
- Protestant foreign missions: their present state

Christlieb, Theodor et al *see* The bremen lectures on great religious questions of to-day

Christmann, Friedrich *see* Australien

Christmas : its origin and associations, together with its historical events and festive celebrations during nineteen centuries / Dawson, William Francis – London: E Stock 1902 [mf ed 1986] – 1r [ill] – 1 – (filmed with: lietuviskai-vokiskas zodynas / paskevicius, j & other titles) – mf#6687 – us Wisconsin U Libr [390]

Christmas box – 1828-29 – (= ser English gift books and literary annuals, 1823-1857) – 6mf – 9 – uk Chadwyck [800]

Christmas bulletin / 135 Medical Regiment [Organization] – 38th ed-46th ed [1983-91] – 1r – 1 – (cont: bull sheet [madison wi]) – mf#2802249 – us WHS [355]

Christmas day, and other sermons / Maurice, Frederick Denison – London: John W Parker, 1843 – 1mf – 9 – 0-7905-9339-4 – mf#1989-2564 – us ATLA [240]

Christmas evans : the preacher of wild wales / Hood, Edwin Paxton – London: Hodder and Stoughton, 1881 – 1mf – 9 – 0-7905-4755-4 – mf#1988-0755 – us ATLA [240]

Christmas eve : a choral / Carman, Bliss – [S.l: s.n, 1913?] – 1mf – 9 – 0-665-78117-2 – mf#78117 – cn Canadiana [810]

Christmas eve at s kavin's / Carman, Bliss – New York: I Kimball, 1901 – 1mf – 9 – 0-665-77794-9 – mf#77794 – cn Canadiana [810]

Christmas eve entertainment for poor children under the auspices of the children's fresh air fund : over twelve hundred children will be given a free entertainment... – [s.l: s.n. 1889?] [mf ed 1987] – 1mf – 9 – 0-665-34737-5 – mf#34737 – cn Canadiana [362]

Christmas festival – Dundee, Scotland. no date – 1r – us UF Libraries [390]

Christmas, Henry *see*
- Pictures of canadian life, vol 1
- Pictures of canadian life, vol 2
- Pictures of canadian life, vols 1-2

The christmas holydays in rome / Kip, William Ingraham – New-York: D Appleton, 1846 – 1mf – 9 – 0-524-03585-7 – mf#1990-1045 – us ATLA [914]

Christmas in french canada / Frechette, Louis – New York: C Scribner's Sons, 1899 – 4mf – 9 – 0-665-91196-3 – (ill by frederick simpson coburn. also available in french) – mf#91196 – cn Canadiana [390]

Christmas in ritual and tradition, christian and pagan / Miles, Clement A – London: T. Fisher Unwin, 1912 – 1mf – 9 – 0-7905-6307-X – (incl bibl ref) – mf#1988-2307 – us ATLA [390]

Christmas nights' entertainments / Palafox Y Mendoza, Juan De – Dublin, Ireland. 1840 – 1r – us UF Libraries [306]

Christmas Seal and Charity Stamp Society *see* Seal news

Christmas sermons / McConnell, Francis John – Cincinnati: Jennings & Graham, c1909 [mf ed 1991] – 1mf – 9 – 0-7905-9798-5 – mf#1989-1523 – us ATLA [242]

Christmas valley gazette – Christmas Valley OR: A Santana Publ, 1962- [mthly] [mf ed 1970] – 1r – 1 – us Oregon Lib [071]

Christmas, Walter *see* Amazonfloden

Christner endeavors – v1 n1-v2 n2 [1980 aug-1982 spr] – 1r – 1 – mf#651591 – us WHS [071]

Christoamerica / Escobar Velado, Oswaldo – s.l, s.l? 1959 – 1r – us UF Libraries [972]

Christoffel, Karl *see* Rebe und wein in goethes weltbild

Christoffel, R *see*
- Heinrich bullinger und seine gattin
- Huldreich zwingli

Christoffel, Raget *see* Huldreich zwingli

Christoforo, Armeno *see* Die reise der soehne giaffers

Christological theology : an address / Harbaugh, H – Philadelphia: S R Fisher, [1865?] – 1mf – 9 – 0-7905-3254-9 – mf#1987-3254 – us ATLA [240]

Die christologie der apokalypse des johannes / Holtz, T – Berlin, 1962 – (= ser Tugal 5-85) – 5mf – 9 – €12.00 – ne Slangenburg [226]

Die christologie der bekenntnisse und die moderne theologie – atheistische methoden in der theologie / Schaeder, Erich & Schlatter, Adolf von – Guetersloh: C Bertelsmann, 1905 – (= ser Beitraege zur foerderung christlicher theologie) – 1mf – 9 – 0-7905-8577-4 – mf#1989-1802 – us ATLA [240]

Christologie des alten testamentes : oder, auslegung der wichtigsten messianischen weissagungen / Buehl, Eduard – Wien: Wilhelm Braumueller 1882 [mf ed 1985] – 1mf – 9 – 0-8370-2393-9 – mf#1985-0393 – us ATLA [221]

Christologie des alten testamentes und commentar ueber die messianischen weissagungen *see* Christology of the old testament and a commentary on the messianic predictions

Die christologie des h ignatius von antiochien / Rackl, M – Freiburg Brsg, 1914 – 8mf – 9 – €17.00 – ne Slangenburg [240]

Die christologie des neuen testaments : ein biblisch-theologischer versuch / Beyschlag, Willibald – Berlin: Ludwig Rauh, 1866 – 1mf – 9 – 0-8370-2328-9 – mf#1985-0328 – us ATLA [225]

Die christologie seit schleiermacher : ihre geschichte und ihre begruendung / Faut, S – Tuebingen: J C B Mohr (Paul Siebeck), 1907 – 1mf – 9 – 0-8370-4595-9 – mf#1985-2595 – us ATLA [240]

Christologie traditionnelle et la foi protestante *see* Collected works

La christologie traditionnelle et la foi protestante / Lobstein, Paul – Paris: Librairie Fischbacher, 1894. Chicago: Dep of Photodup, U of Chicago Lib, 1975 (1r); Evanston: American Theol Lib Assoc, 1984 (1r) – (= ser HIS etudes christologiques) – 1 – 0-8370-0558-2 – (incl bibl ref) – mf#1984-6060 – us ATLA [242]

Christology : or, the doctrine of the person of christ / Weidner, Revere Franklin – Chicago: Wartburg Publ House, [1913?] [mf ed 1991] – (= ser A system of dogmatics 4) – 1mf – 9 – 0-7905-9748-9 – (incl bibl ref) – mf#1989-1473 – us ATLA [230]

Christology and personality : containing 1. christologies ancient and modern, 2. personality in christ and in ourselves / Sanday, William – New York: Oxford UP, American Branch, c1911 [mf ed 1988] – 1mf – 9 – 0-7905-0227-5 – (incl bibl ref & ind) – mf#1987-0227 – us ATLA [240]

Christology of p t forsyth / Thompson, Douglas Brian – Princeton: Princeton Theological Seminary, [1950] – 1r – 1 – 0-8370-1118-3 – mf#1984-B522 – us ATLA [240]

The christology of paul's opponents in second corinthians and its relationship to their concept of apostleship / Howell, David B – 1982 – 1 – 5.04 – us Southern Baptist [242]

The christology of the epistle to the hebrews : including its relation to the developing christology of the primitive church / MacNeill, Harris Lachlan – Chicago, IL: University of Chicago Press, 1914 – 1mf – 9 – 0-7905-1431-1 – (incl ind) – mf#1987-1431 – us ATLA [227]

Christology of the old testament and a commentary on the messianic predictions = Christologie des alten testamentes und commentar ueber die messianischen weissagungen / Hengstenberg, Ernst Wilhelm – 2nd ed. Edinburgh: T & T Clark, 1856-58 [mf ed 1989] – (= ser Clark's foreign theological library) – 4v on 5mf – 9 – 0-7905-2042-7 – (english trans fr german by theodore and james meyer; incl bibl ref & ind) – mf#1987-2042 – us ATLA [221]

Christoph columbus : der don quichote des ozeans: ein portraet / Wassermann, Jakob – 11.-20. aufl. Berlin: S Fischer 1929 [mf ed 1991] – 1r [ill] – 1 – (incl bibl ref. filmed with: der aufruhr um den junker ernst / jakob wassermann) – mf#7855 – us Wisconsin U Libr [910]

Christoph ernst freiherr von houwald als dramatiker / Schmidtborn, Otto – Marburg a.L: N G Elwert 1909 [mf ed 1992] – 1r – 1 – (incl bibl ref. filmed with: gustav freytag und das junge deutschland / otto mayrhofer) – mf#3091p – us Wisconsin U Libr [430]

Christoph marlow : trauerspiel in vier akten / Wildenbruch, Ernst von – 2. aufl. Berlin: G Grote 1902 [mf ed 1991] – 1r – 1 – (filmed with: kinderthranen & other titles) – mf#2963p – us Wisconsin U Libr [420]

Christoph martin wieland's leben und wirken in schwaben und in der schweiz / Offterdinger, Ludwig Felix – Heilbronn: Gebr Henninger 1877 [mf ed 1992] – 1r [ill] – 1 – (filmed with: wieland und martin und regula künzli / ludwig hirzel) – mf#3050p – us Wisconsin U Libr [430]

Christoph pankratius mieserich unter den seligen / Wellems, Hugo – Berlin: Nordland Verlag [1943] [mf ed 1991] – (= ser Nordland buecherei) – 1r [ill] – 1 – (Filmed with: Josef Weinheber / Franz Koch) – mf#2957p – us Wisconsin U Libr [890]

Christoph panzer : roman / Reichelt, Johannes – Dresden: Wodni & Lindecke [194-?] [mf ed 1991] – 1r – 1 – (filmed with: die kinder und die frauen & other titles) – mf#2845p – us Wisconsin U Libr [830]

Christoph pechlin : eine internationale lebensgeschichte / Raabe, Wilhelm Karl – Leipzig: E J Guenther 1873 [mf ed 1995] – (= ser Bibliothek der deutschen literatur) – 2v in 1 on 1r – 1 – (filmed with:gertrud von loden / c quandt) – mf#8834 – us Wisconsin U Libr [830]

Christoph von schallenberg : ein oesterreichischer lyriker des 16. jahrhunderts / ed by Hurch, Hans – Stuttgart: Litterarischer Verein, 1910 (Tuebingen: H Laupp, Jr) [mf ed 1993] – (= ser Blvs 253) – xxxix/230p – 1 – (early modern german and latin text. int in german. incl bibl ref and ind) – mf#8470 reel 52 – us Wisconsin U Libr [430]

Christoph von schallenberg : ein oesterreichischer lyriker des 16. jahrhunderts / Schallenberg, Christoph von; ed by Hurch, Hans – Stuttgart: Litterarischer Verein 1910 (Tuebingen: H Laupp, Jr) [mf ed 1993] – (= ser Blvs 253) – 58r – 1 – (incl bibl ref & ind. early modern german & latin text, int in german) – mf#3420p – us Wisconsin U Libr [810]

Christophaneia : the doctrine of the manifestations of the son of god under the economy of the old testament / Kidd, George Balderston; ed by Dobbin, Orlando Thomas – London: Ward, 1852 [mf ed 1990] – 2mf – 9 – 0-7905-3454-1 – (incl bibl ref) – mf#1987-3454 – us ATLA [225]

Christophe, Eduard Curt *see* Vom leichten laecheln

Christophe, J-B *see* Histoire de la papaute pendant le 14e siecle

Christopher and gay : a partisan's view of the greenwich village homosexual scene / Hamilton, Wallace – New York: Saturday Review Press [1973] [mf ed 1986] – 1r – 1 – mf#1738 – us Wisconsin U Libr [305]

Christopher columbus collection of the library of congress / ed by Larson, Everette E – 9 – $10,515.00 coll – (pt1: works in english 634mf $5925 isbn 1-55655-391-9. pt2: non-english works 740mf $5165 isbn 1-55655-392-7. with p/g) – us UPA [910]

Christopher cowan papers, ms 1328 / Cowan, Christopher – 1784, 1787, 1798, 1815-26 – 1r – 1 – (letters, agreements relating to lands in virginia military district of ohio) – us Western Res [920]

Christopher, Hiram *see*
– The relations of god to the world
– The remedial system

Christopher, Luella S *see* Palau's evolving relationship with the u.s.

Christopher schultz (1718-1789) : [memorial issue] – Norristown PA: Board of Pub of the Schwenckfelder Church, 1940 [mf ed 2003] – (= ser Schwenckfeldiana 1/1) – 1r – 1 – (in english. incl trans fr german sources) – mf#2003-s008a – us ATLA [242]

Christopher street – New York. 1990-1995 – 1,5,9 – ISSN: 0146-7921 – mf#18240 – us UMI ProQuest [305]

Christopher, Tara L *see* Application of lma principles in ethnic dance training

Christophers, Joseph S *see* Cape of good hope and the eastern province of algoa bay [the]

Christophers, Samuel Woolcock *see* The epworth singers and other poets of methodism

Christophilus *see* Vindiciae britannicae

Christophilus, A *see* Die lage der christen in der tuerkei und das russische protectorat

Christophori clavii bambergensis e societate les operum mathematicorum tomus primus-quintus... / Clavius, Christoph – Moguntiae: sumptibus A Hierat. 5v. 1611-12 [mf ed 1972] – 1r – 1 – mf#SEM35P61 – cn Bibl Nat [510]

Christophori clavii...in sphaeram ioannis de sacro bosco commentarius – Venetiis: apud Bernardum Basam 1596 – 1 – (lacking: p106-107) – mf#1512p – us Wisconsin U Libr [520]

Christophori moralis hyspaniensis missarum liber primus (-secundus) / Morales, Cristobal de – Lugduni [i.e.Lyon]: Jacobus Moder [Moderne] 1546 [mf ed 19–] – 2bk on 1r – 1 – (incl bk2 [lugduni [i.e. lyon], jacobus moder [moderne] 1551]) – mf#film 67 – us Sibley [780]

Christophori wittichii annotationes ad renati des-chartes meditationes – Dordrechti, 1688 – 2mf – 9 – mf#PBA-413 – ne IDC [240]

Christophoro d'Avalos, Felice A de *see* Tableau historique, politique, physique et moral de malte et de ses habitans

Christrosen im mariengarten : oder, die geheimnisse des heiligen rosenkranzes / Hattler, Franz – 3., verm Aufl. Innsbruck: Fel Rauch, 1894 – 1mf – 9 – 0-8370-8347-8 – mf#1986-2347 – us ATLA [240]

Christ's atonement / Marsh, Frederick Edward – London: Marshall Bros [1898?] [mf ed 1991] – 1mf – 9 – 0-7905-8843-9 – mf#1989-2068 – us ATLA [240]

Christ's blueprint for the south : a social action bulletin of the new orleans province institute of social order / Loyola University [New Orleans LA] – v1 n1-v16 n8 [1948 nov 15-1964 may] – 1r – 1 – (cont by: blueprint for the christian reshaping of society) – mf#1826652 – us WHS [230]

Christ's covenant the best defence of christ's crown / White, William – Edinburgh, Scotland. 1844 – 1r – us UF Libraries [240]

Christ's cure for care / Pearse, Mark Guy – New York: Eaton & Mains, [190-] [mf ed 1984] – 2mf – 9 – 0-8370-0839-5 – mf#1984-4229 – us ATLA [240]

Christ's finished work / Claughton, Thomas Legh – London, England. 1870 – 1r – us UF Libraries [240]

Christ's Hospital, Topeka, KS *see* Patient registers

Christ's kingdom : and its antagonist / Longmuir, John – Edinburgh, Scotland. 1843 – 1r – us UF Libraries [240]

Christ's kingdom / Ranken, Arthur – Aberdeen, Scotland. 1883? – 1r – us UF Libraries [240]

Christ's kingdom on earth : or, the church and her divine constitution, organization, and framework: explained for the people / Meagher, James Luke – New York: Russell, 1892 – 2mf – 9 – 0-8370-6920-3 – (incl ind) – mf#1986-0920 – us ATLA [240]

Christ's kingdom upon earth : a series of discourses / Flint, Robert – Edinburgh: W Blackwood, 1865 – 1mf – 9 – 0-7905-3674-9 – mf#1989-0167 – us ATLA [240]

Christ's message of the kingdom : a course of daily study for private students and for bible circles / Hogg, Alfred George – Edinburgh: T & T Clark 1912 [mf ed 1989] – 1mf – 9 – 0-7905-1997-6 – (incl ind) – mf#1987-1997 – us ATLA [225]

Christ's object in preaching to the spirits in prison / Welch, Adam – Edinburgh, Scotland. 1871 – 1r – us UF Libraries [240]

Christ's object lessons / White, Ellen Gould Harmon – Oakland CA: Pacific Press c1900 [mf ed 1985] – 1mf – 9 – 0-8370-5820-1 – (incl ind) – mf#1985-3820 – us ATLA [240]

Christ's "own house" / Martin, Hugh – London, England. 1859 – 1r – us UF Libraries [240]

Christ's people : imitators of him / Spurgeon, C H – Finsbury, England. 1855 – 1r – us UF Libraries [240]

Christ's presence in the gospel history / Martin, Hugh – London, New York: T Nelson 1860 [mf ed 1985] – 1mf – 9 – 0-8370-4080-9 – mf#1985-2080 – us ATLA [240]

Christ's second coming / Mathias, Benjamin William – Dublin, Ireland. 1821 – 1r – us UF Libraries [240]

Christ's second coming : will it be premillennial? / Brown, David – 5th ed. Edinburgh: T & T Clark, 1859 [mf ed 1989] – 2mf – 9 – 0-7905-1028-6 – (incl bibl ref & ind) – mf#1987-1028 – us ATLA [240]

Christ's secret of happiness / Abbott, Lyman – New York: Thomas P Crowell 1907 [mf ed 1989] – 1mf – 9 – 0-7905-1565-2 – mf#1987-1565 – us ATLA [230]

Christ's sermon on the mount and orientalism *see* Pa fuh chen ching (ccm112)

Christ's social remedies / Montgomery, Harry Earl – New York: Putnam, 1911 – 1mf – 9 – 0-524-04844-4 – (incl bibl ref) – mf#1990-1336 – us ATLA [360]

Christ's teaching concerning divorce in the new testament : an exegetical study / Gigot, Francis Ernest – New York: Benziger Bros, 1912 – 1mf – 9 – 0-524-04796-0 – (incl bibl ref) – mf#1992-0216 – us ATLA [225]

Christ's teaching concerning the last things : and other papers / Caven, William – London: Hodder and Stoughton; Toronto: Westminster, [1908] – 4mf – 9 – 0-665-76981-4 – mf#76981 – cn Canadiana [240]

Christ's temptation and ours / Hall, Arthur Crawshay Alliston – London, New York: Longmans, Green 1897, c1896 [mf ed 1989] – (= ser The baldwin lectures 1896) – 1mf – 9 – 0-7905-1526-1 – (incl bibl ref) – mf#1987-1526 – us ATLA [220]

Christ's tenderness towards the fallen / Longley, Charles Thomas – London, England. 1865 – 1r – us UF Libraries [240]

Christ's testimony to the doctrine of everlasting punishment / Kerr, James – Edinburgh, Scotland. 18– – 1r – us UF Libraries [240]

Christ's work of reform : a bible view – Boston: Crocker & Brewster, 1862 [mf ed 1985] – 1mf – 9 – 0-8370-3335-7 – mf#1985-1335 – us ATLA [240]

Christus : das evangelium und seine weltgeschichtliche bedeutung / Schell, Herman – Mainz: Kirchheim, 1906 – 1mf – 9 – 0-8370-5081-2 – (incl ind) – mf#1985-3081 – us ATLA [240]

Christus – Mexico: Centro de Reflexion teologica, [ano 48 n562-ano 63 n709 (feb 1983-nov/dic 1998)] (bimthly) – 7r – 1 – us CRL [240]

Christus – no. 1-76. Paris. 1954-72. tb: 1954-63 – 5 – fr ACRPP [240]

Christus consummator : some aspects of the work and person of christ in relation to modern thought / Westcott, Brooke Foss – London, New York: Macmillan, 1886 [mf ed 1984] – (= ser Biblical crit – us & gb 56) – 3mf – 9 – 0-8370-0255-9 – mf#1984-1056 – us ATLA [220]

Der christus der geschichte und sein christenthum / Laengin, Georg – Leipzig: Otto Wigand, 1897-1898 – 1mf – 9 – 0-7905-2008-7 – (incl bibl ref) – mf#1987-2008 – us ATLA [240]

Christus der herr : erlaeuterungen zu philipper 2, 5-11 / Koegel, Julius – Guetersloh: C Bertelsmann, 1908 – (= ser Beitraege zur foerderung christlicher theologie) – 1mf – 9 – 0-524-05916-0 – mf#1992-0673 – us ATLA [220]

Christus, der zweite adam : das suehnopfer fuer den angehorsam des ersten adam und fuer die suenden seiner nachkommen: zwanzig conferenzen = Second adam / Coret, Jacques – Regensburg: Georg Joseph Manz, 1870 [mf ed 1986] – 2mf – 9 – 0-8370-6894-0 – (german trans fr french by h scheid; incl bibl ref) – mf#1986-0894 – us ATLA [240]

Christus – ein inder? : versuch einer entstehungsgeschichte des christentums unter benutzung der indischen studien louis jacolliots / Plange, Theodor J – 2. Aufl. Stuttgart: H Schmidt, [1906?] – 1mf – 9 – 0-524-02030-2 – mf#1990-2805 – us ATLA [240]

Christus en de heidenwereld : opwekkende rede / Oosterzee, Johannes Jacobus van – Rotterdam: Van der Meer & Verbruggen, 1851 – 1mf – 9 – 0-7905-3088-0 – mf#1987-3088 – us ATLA [240]

Christus fuer uns : passionspredigten / Rueling, Joseph – Leipzig: Friedrich Jansa, 1906 – 1mf – 9 – 0-8370-9738-X – mf#1986-3738 – us ATLA [240]

Christus im modernen geistesleben : christliche einfuehrung in die geisteswelt der gegenwart / Pfennigsdorf, Emil – 4. verm verb aufl. Schwerin i M: Fr Bahn, 1901 [mf ed 1991] – 1mf – 9 – 0-7905-9439-0 – mf#1989-2664 – us ATLA [240]

Christus imperator : a series of lecture-sermons on the universal empire of christianity / ed by Stubbs, Charles William – London, New York: Macmillan, 1894 [mf ed 1988] – 1mf – 9 – 0-7905-0396-4 – mf#1987-0396 – us ATLA [242]

Christus in der modernen sozialen bewegung / Dausch, Petrus – 1.& 2. aufl. Muenster in Westfalen: Aschendorff 1920 [mf ed 2005] – (= ser Biblische zeitfragen; 9. folge 5/6) – 1r with other items – 1 – 0-524-10545-6 – (Incl bibl ref) – us ATLA [230]

Christus in ecclesia : sermons on the church and its institutions / Rashdall, Hastings – Edinburgh: T & T Clark, 1904 – 1mf – 9 – 0-7905-9597-4 – mf#1989-1322 – us ATLA [240]

Christus in seiner kirche *see* History of the catholic church

Christus liberator : an outline study of africa / Parsons, Ellen C – New York: Macmillan, 1905 – 1mf – 9 – 0-8370-6295-0 – (incl ind) – mf#1985-0295 – us ATLA [240]

Christus magister : some teachings from the sermon on the mount / Pearson, Alfred – London: James Nisbet, 1892 [mf ed 1984] – 4mf – 9 – 0-8370-0795-X – mf#1984-4140 – us ATLA [220]

Christus mediator / Elliott, Charles – New York: A.C. Armstrong, 1891, c1890 – 1mf – 9 – 0-7905-1598-9 – (incl bibl ref and index) – mf#1987-1598 – us ATLA [240]

Christus oder buddha? : vortrag / Haack, Ernst – Schwerin i M: Fr Bahn, 1898 – 1mf – 9 – 0-524-01442-6 – mf#1990-2437 – us ATLA [230]

Christus redemptor : an outline study of the island world of the pacific / Montgomery, Helen Barrett – new rev ed. New York: Macmillan 1909 [mf ed 1990] – (= ser United study of missions) – 1mf – 9 – 0-7905-6660-5 – (incl bibl ref) – mf#1988-2660 – us ATLA [240]

Christus redemptor : an outline study of the island world of the pacific / Montgomery, Helen Barrett – New York: Macmillan, 1906 [mf ed 1995] – (= ser Yale coll; United study of missions 6) – viii/282p – 1 – 0-524-09117-X – mf#1995-0117 – us ATLA [240]

Christus und buddha / Wecker, Otto – 1. & 2. aufl. Muenster i W: Aschendorff 1908 [mf ed 1992] – (= ser Biblische zeitfragen 1/9) – 1mf – 9 – 0-524-05469-X – (incl bibl ref) – mf#1990-3495 – us ATLA [230]

Christus und buddha in ihrem himmlischen vorleben / Englert, Winfried Philipp – Wien: Mayer 1898 [mf ed 1991] – (= ser Apologetische studien 1/1) – 1mf – 9 – 0-524-01543-0 – (incl bibl ref) – mf#1990-2497 – us ATLA [230]

Christus und christentum; j.t. becks theologische arbeit : zwei reden / Schlatter, Adolf von – Guetersloh: C Bertelsmann, 1904 – (= ser Beitraege zur foerderung christlicher theologie) – 1mf – 9 – 0-7905-2425-2 – mf#1987-2425 – us ATLA [240]

Das christusbild der apostel und der nachapostolischen zeit / Schenkel, Daniel – Leipzig: F A Brockhaus, 1879 – 1mf – 9 – 0-8370-5143-6 – mf#1985-3143 – us ATLA [225]

Das christusbild der geschichte und das christusbild der dogmatik : ein vortrag / Holtzmann, Oskar – Darmstadt: J Waitz, 1890 – 1mf – 9 – 0-524-07965-X – mf#1992-1120 – us ATLA [240]

Das christusbild des paulus / Juncker, Alfred – Halle a. S: Max Niemeyer, 1906 – 1mf – 9 – 0-7905-2004-4 – mf#1987-2004 – us ATLA [225]

Das christusbild des urchristlichen glaubens in religionsgeschichtlicher beleuchtung : its significance and value in the history of religion = The early christian conception of christ / Pfleiderer, Otto – 1mf – 9 – 0-8370-5449-4 – (in english) – mf#1985-3449 – us ATLA [240]

Das christusbild des urchristlichen glaubens in religionsgeschichtlicher beleuchtung : vortrag / Pfleiderer, Otto – Berlin: Georg Reimer, 1903 – 1mf – 9 – 0-8370-5153-3 – (incl bibl ref) – mf#1985-3153 – us ATLA [240]

Christusbilder : untersuchungen zur christlichen legende / Dobschuetz, Ernst von – Leipzig, 1899 – (= ser Tugal 2-18) – 15mf – 9 – €29.00 – ne Slangenburg [240]

Christusfrommigkeit in ihrer historischen entfaltung / Richstaetter, C – Koeln, 1949 – 9mf – 8 – €18.00 – ne Slangenburg [240]

Das christus-problem : grundlinien zu einer sozial-theologie / Kalthoff, Albert – 2. Aufl. Leipzig: Eugen Diederichs, 1903 – 1mf – 9 – 0-8370-3838-3 – mf#1985-1838 – us ATLA [240]

Christy, David *see* Cotton is king, and pro-slavery arguments

Christy, David et al *see* Cotton is king, and pro-slavery arguments

Christy's nigga songster : containing songs as are sung by christy's, pierce's, white's, sable brothers, and dumbleton's band of minstrels – New York: T W Strong [184-?] (mf ed 1974) – 1r – 1 – (without the music) – mf#Sc Micro R-0791.1-C – us NY Public [780]

Chromatics : or, an essay on the analogy and harmony of colours / Field, George – London: printed by A J Valpy, 1817 – (= ser 19th c art & architecture) – 1mf – 9 – mf#4.1.19 – uk Chadwyck [700]

Chromatographia – Braunschweig. 1968-1986 (1,5,9) – ISSN: 0009-5893 – mf#49031 – us UMI ProQuest [540]

Chromatographic reviews – Amsterdam. 1959-1971 (1) 1959-1971 (5) (9) – ISSN: 0009-5907 – mf#42246 – us UMI ProQuest [540]

Chromatography – Newton. 1987-1987 (1,5,9) – (cont: chromatography forum) – ISSN: 0892-8797 – mf#16063,01 – us UMI ProQuest [540]

Chromatography : or, a treatise on colours and pigments / Field, George – new ed. London 1841 – (= ser 19th c art & architecture) – 5mf – 9 – mf#4.2.1035 – uk Chadwyck [770]

Chromatography forum – Barrington. 1986-1986 (1,5,9) – (cont by: chromatography) – ISSN: 0892-8800 – mf#16063 – us UMI ProQuest [540]

Chrome dust – Nye, MT. 1954-1960 (1) – mf#64590 – us UMI ProQuest [071]

The chromolithograph *see* The artist 1880-82 – l'artist et courier de l'art

Chromosoma – Heidelberg. 1939-1996 (1) 1939-1996 (5) 1939-1996 (9) – ISSN: 0009-5915 – mf#13155 – us UMI ProQuest [575]

Chrona1my przyrode ojczysta – Cracow. v1-7. 1945-1951 (incomplete) – 1 – us NY Public [073]

Chroni, Stiliani *see* Incentive motivation, competitive orientation and gender in collegiate alpine skiers

Chronia monastereii s albani 1 [rs28] : thomas walsingham: historia anglicana / S[/ain/]t Albans Abbey; ed by Riley, H T – v1-2. 1863 – (= ser The rolls series [rs]) – €35.00 – ne Slangenburg [241]

Chronia monastereii s albani 2 [rs28] : william rishanger: chronica et annales 1259-1307 / S[/ain/]t Albans Abbey; ed by Riley, H T – 1865 – (= ser The rolls series [rs]) – €21.00 – ne Slangenburg [241]

Chronia monastereii s albani 3 [rs28] : johannis de trokelowe et henrici de blaneforde: chronica et annales 1259-1296, 1307-1324, 1392-1406 / S[/ain/]t Albans Abbey; ed by Riley, H T – 1866 – (= ser The rolls series [rs]) – €19.00 – ne Slangenburg [241]

Chronia monastereii s albani 4 [rs28] : thomas walsingham: gesta abbatum monasterii s albani, a thomas walsingham, regnante ricardo secundo, ejusdem ecclesiae praecentore, compilata / S[/ain/]t Albans Abbey; ed by Riley, H T – (= ser The rolls series [rs]) – (v1 1867 €19 v2 1867 €19 v3 1869 €25) – ne Slangenburg [241]

Chronia monastereii s albani 5 [rs28] : johannes amundesham: annales monasterii s albani, a johanne amundesham, monacho, ut videtur, conscripta [ad 1421-1440], quibus praefigitur chronicon rerum gest in mon s albani [ad 1422-1431], a quodam auctore ign comp / S[/ain/]t Albans Abbey; ed by Riley, H T – (= ser The rolls series [rs]) – (v1 1870 €19 v2 1871 €21) – ne Slangenburg [241]

Chronia monastereii s albani 6 [rs28] : st albans abbey: registra quorundam abbatum monasterii s albani qui saec 15 floruere [joh wethamstede, will albon, etc] / S[/ain/]t Albans Abbey; ed by Riley, H T – (= ser The rolls series [rs] 28) – (v1 1872 €19 v2 1873 €21) – ne Slangenburg [241]

Chronia monastereii s albani 7 [rs28] : thomas walsingham: ypodigma neustrie / S[/ain/]t Albans Abbey; ed by Riley, H T – 1876 – (= ser The rolls series [rs]) – €25 – ne Slangenburg [241]

Chronic cannabis use in costa rica / University Of Florida Center For Latin American Studies – Gainesville, FL. 1976 – 1r – us UF Libraries [362]

Chronic disease : advances in diagnosis and treatment – Darien CT 1974-75 – 1,5,9 – (cont: chronic disease management) – ISSN: 0095-0270 – mf#6666,01 – us UMI ProQuest [616]

Chronic disease management – Darien CT 1972 – 1 – (cont by: chronic disease: advances in diagnosis and treatment) – ISSN: 0016-8661 – mf#6666.01 – us UMI ProQuest [616]

Chronic effects of exercise / Steinhaus, Arthur H – 1933 – 2mf – 9 – $6.00 – us Kinesology [790]

Chronic exercise and the effects on the immune response / Kenton, Mark A – 1996 – 2mf – 9 – $8.00 – mf#PH 1554 – us Kinesology [612]

Chronica aevi suevici (mgh5:23.bd) – 1874 – (= ser Monumenta germaniae historica 5: scriptores in folio (mgh5)) – €52.00 – ne Slangenburg [240]

Chronica apostolica : y seraphica de todos los colegios de Propaganda fide de esta Nueva-Espana, de missioneros franciscanos observantes: erigidos con autoridad pontificia... / Espinosa, Isidro Felix de – en Mexico:...D Joseph Bernardo de Hogal...1746-1792 – (= ser Books on religion...1543/44-c1800: ordenes, etc: congregacion de propaganda fide) – 2v on 4mf – 9 – mf#crl-181 – ne IDC [241]

Chronica austriae (mgh6:13.bd) / Ebendorfer, Thomas; ed by Lhotsky, A – 1967 – (= ser Monumenta germaniae historica 6: scriptores rerum germanicarum, nova series (mgh6)) – €27.00 – ne Slangenburg [240]

Chronica de 25 anos / Badajoz. Espana en Paz – Madrid: Publicaciones Espanolas, 1964 – 1 – sp Bibl Santa Ana [946]

Chronica de a santa provincia de san joseph... / Santa Rosa o Alcala, Marcos de – Madrid: Imp. y Lib. de Manuel Fernandez, s.a. – sp Bibl Santa Ana [240]

Chronica de la provincia de n p s francisco de zacatecas : compuesta por el m r p fr joseph arlegui, lector jubilado, calificador del sto officio... / Arlegui, Jose de – en Mexico: Por Joseph Bernardo de Hogal...ano de 1737 – (= ser Books on religion...1543/44-c1800: franciscanos) – 5mf – 9 – mf#crl-207 – ne IDC [241]

Chronica de la provincia del santissimo no[m]bre de jesus de guatemala : de el orden de n seraphico padre san francisco en el reyno de la nueva espana: dividida en dos tomos / Vazquez, Francisco – en Guatemala: impr S Francisco, ano de 1714-16 – (= ser Books on religion...1543/44-c1800: franciscanos) – 2v on 8mf – 9 – mf#crl-204 – ne IDC [241]

Chronica de la santa provincia de san diego de mexico : de religiosos descalcos de n s p s francisco en la nueva-espana... / Medina, Baltasar de – en Mexico: Por Juan de Ribera, ano de 1682 – (= ser Books on religion...1543/44-c1800: franciscanos) – 7mf – 9 – mf#crl-200 – ne IDC [241]

Chronica de susenyos, rei de ethiopia... / Pereira, F M E – Lisboa, 1892-1900. 2v – 8mf – 9 – (missing: v1) – mf#SEP-86 – ne IDC [960]

Chronica de veinticinco anos / Caceres. Espana en Paz – Madrid: Publicaciones Espanolas, 1964 – 1 – sp Bibl Santa Ana [946]

Chronica del esforcado principe y capitan iorge castrioto rey de epiro, o albania... / George, C – Lisboa, 1588 – 7mf – 9 – mf#H-8370 – ne IDC [956]

Chronica del muy esclarecido principe, y rey don alfonso – Valladolid: [s.n.], 1554 [mf ed 1980] – 77lea – 1 – mf#1806 – us Wisconsin U Libr [946]

Chronica do imperio : revista quinzenal – Rio de Janeiro, RJ: Typ de Domingos Luiz dos Santos, 1876 – (= ser Ps 19) – mf#P17,01,99 – bl Biblioteca [073]

Chronica et annales 1259-1296, 1307-1324, 1392-1406 *see* Chronia monastereii s albani 3 [rs28]

Chronica et annales 1259-1307 *see* Chronia monastereii s albani 2 [rs28]

Chronica et annales aevi salici (mgh5:6.bd) – 1844 – (= ser Monumenta germaniae historica 5: scriptores in folio (mgh5)) – €42.00 – ne Slangenburg [220]

Chronica et annales aevi salici (mgh5:9.bd) – 1851 – (= ser Monumenta germaniae historica 5: scriptores in folio (mgh5)) – €46.00 – ne Slangenburg [240]

Chronica et gesta aevi salici (mgh5:7.bd) – 1846 – (= ser Monumenta germaniae historica 5: scriptores in folio (mgh5)) – €46.00 – ne Slangenburg [240]

Chronica et gesta aevi salici (mgh5:8.bd) – 1848 – (= ser Monumenta germaniae historica 5: scriptores in folio (mgh5)) – €35.00 – ne Slangenburg [220]

Chronica fratris jordani / Jordanus de Yano; ed by Boehmer, Heinrich – Paris: Librairie Fischbacher 1908 [mf ed 1990] – (= ser Coll d'etudes et de documents sur l'histoire religieuse et litteraire du moyen age 6) – 1mf – 9 – 0-7905-8109-4 – (notes & comm by ed; text in latin, int in french) – mf#1988-6071 – us ATLA [241]

Chronica heinrici surdi de selbach (mgh6:1.bd) / ed by Bresslau, H – 1922 – (= ser Monumenta germaniae historica 6: scriptores rerum germanicarum, nova series (mgh6)) – €12.00 – ne Slangenburg [240]

Chronica hispana saeculi 12 pars 2 chronica naierensis : formae tplila 87 – 1995 – (= ser ILL – ser a; Cccm 71a) – 4mf+49p – 9 – €40.00 – 2-503-63714-0 – be Brepols [400]

Chronica hispana saeculi 13 : formae tplila 95 – [mf ed 1997] – (= ser ILL – ser a; Cccm 73) – 4mf+93p – 9 – €40.00 – 2-503-63732-9 – be Brepols [400]

Chronica litteraria : jornal de instruccao e recreio – Rio de Janeiro, RJ: Typ Guanabarense de L A F de Menezes, 02 jan-12 nov 1848 – (= ser Ps 19) – mf#P01B,05,18 – bl Biblioteca [440]

Chronica magistri rogeri de houedene [rs51] / Roger of Hoveden; ed by Stubbs, W – (= ser The rolls series [rs]) – (v1 1868 €15. v2 1869 €17. v3 1871 €17. v4 1870 €19) – ne Slangenburg [931]

Chronica majora [rs57] / Paris, Matthew; ed by Luard, H R – (= ser The rolls series [rs]) – (v1: the creation-1066 1872 €21. v2: 1067-1216 1874 €25. v3: 1216-1239 1876 €23. v4: 1240-1247 1878 €23. v5: 1248-1259 1880 €25. v6: additamenta 1882 €19. v7: index, glossary 1884 €21) – ne Slangenburg [241]

Chronica mathiae de nuwenburg (mgh6:4.bd) / ed by Hofmeister, A – 1924-1940 – (= ser Monumenta germaniae historica 6: scriptores rerum germanicarum, nova series (mgh6)) – €25.00 – ne Slangenburg [240]

Chronica minora saec 4, 5, 6, 7 (mgh1:9.bd) : vol 1 / ed by Mommsen, Theodor – 1892 – (= ser Monumenta germaniae historica 1: scriptores – auctores antiquissimi) – €38.00 – ne Slangenburg [240]

Chronica minora saec 4, 5, 6, 7 (mgh1:11.bd) : vol 2 / ed by Mommsen, Theodor – 1894 – (= ser Monumenta germaniae historica 1: scriptores – auctores antiquissimi) – €25.00 – ne Slangenburg [240]

Chronica minora saec 4, 5, 6, 7 (mgh1:13.bd) : vol 3 / ed by Mommsen, Theodor – 1898 – (= ser Monumenta germaniae historica scriptores 1: scriptores – auctores antiquissimi) – €37.00 – ne Slangenburg [240]

Chronica monasterii de melsa, a fundatione usque ad annum 1396 (rs43) : accedit continuatio ad annum 1406 a monacho quodam ipsius domus / Thomas de Burton; ed by Bond, E A – (= ser The rolls series (rs)) – (v1 1866 €19. v2 1867 €17. v3 1868 €19) – ne Slangenburg [241]

Chronica ordinis carthusiensis ab anno 1084 ad annum 1510 / Bohic, O Carth – Tornaci-Parkmonasterii. v1-4. 1911-1954 – €267.00 – ne Slangenburg [241]

Chronica regia coloniensis (annales maximi colonienses) (mgh7:18.bd) : cum continuationibus in monasterio s pantaleonis scriptis aliisque historiae coloniensis monumentis – 1880 – (= ser Monumenta germaniae historica 7: scriptores rerum germanicarum in usum scholarum (mgh7)) – €18.00 – ne Slangenburg [240]

Chronica (rs13) / Johannes de Oxenedes; ed by Ellis, H – 1859 – (= ser The rolls series (rs)) – €18.00 – ne Slangenburg [931]

Chronicae bavaricae saec 14 (mgh7:19.bd) – 1918 – (= ser Monumenta germaniae historica 7: scriptores rerum germanicarum in usum scholarum (mgh7)) – €12.00 – ne Slangenburg [240]

Chronicals and memorials of the reign of richard 1 (rs38) / ed by Stubbs, W – (= ser The rolls series (rs)) – (v1: itinerarium peregrinorum et gesta regis ricardi (1187-1199) auctore ut videtur can.s. trinitatis londoniensis 1864 €23. v2: epistolae cantuarienses, 1187-1199 1865 €27) – ne Slangenburg [931]

Chronicals and memorials of the reign of the reign of richard 1 / ed by Stubbs, W – (pt1: itinerarium peregrinorum et gesta regis ricardi (1187-1199) auctore ut videtur can s trinitatis, londoniensis 1864 €23. pt2: epistolae cantuarienses (1187-1199) 1865 €27) – ne Slangenburg [941]

Chronicas / Machado De Assis – Rio de Janeiro, Brazil. v1-4. 1938 – 1r – us UF Libraries [972]

Chronica...san gabriel de francisco descalzos / San Francisco Membrio, Andres – 1753 – 9 – sp Bibl Santa Ana [240]

Chronica...san miguel / Santa Criz, Fr. Jose – 1671 – 9 – sp Bibl Santa Ana [240]

Chronicle / The American Baptist Historical Society – 1938-57 – 1 – $149.31 – us Southern Baptist [242]

Chronicle – Battle Creek, Lansing MI. 1995 mar 23/31-1996 dec 19/28, 1997 jan 3/10-dec 31/1998 jan 6, 1998 jan 8/18-dec 16/30, 1999 jan 6-dec 30, 2000 – 5r – 1 – mf#3281357 – us WHS [071]

Chronicle / Central Labor Council of Cincinnati and Vicinity – Cincinnati OH. 1903 dec 5, 26, 1907 jul 13/1908 dec 26-1967 jan 5/1968 jul 25 – 31r – 1 – mf#685554 – us WHS [331]

Chronicle – Beloit WI. 1983 jan 11-aug, 1983-1988 jun, 1990 jan-1992 jul 22, 1993, 1994, 1995 jan 6-1996 mar 15/22 – 6r – 1 – (cont: beloit chronicle) – mf#955604 – us WHS [071]

Chronicle – Charleston SC. 1994 jul 20-dec 28, 1995 jan 4-jun 28, 1995 jul 5-dec 27, 1996 jan 3-jun 26, 1996 jul 3-dec 25 – 5r – 1 – (cont: charleston chronicle [charleston sc: 1971]) – mf#2902509 – us WHS [071]

Chronicle – Deshler, NE: C B Langley. v7 n37. may 4 1906-1906// (wkly) [mf ed -jun 15 1906 (gaps)] – 1r – 1 – (cont: deshler chronicle. cont by: deshler rustler) – us NE Hist [071]

Chronicle – La Crosse WI. 1879 feb 14-1880 jan 1, 1880 sep 9-1882 may 25, 1882 jun 1-sep 14 – 3r – 1 – (cont: liberal democrat; cont by: weekly chronicle [la crosse wi: 1882]) – mf#933782 – us WHS [071]

Chronicle – La Crosse WI. 1878 aug 1-1879 jul 31, 1879 aug 1-1880 jul 30, 1880 jul 31-oct 31 – 3r – 1 – (cont: morning liberal democrat; cont by: morning chronicle [la crosse wi]) – mf#933720 – us WHS [071]

Chronicle – Adelaide, Australia 2 jan 1904-24 jun 1922, 2 jan 1930-30 jul 1936, 1 nov 1951-1 jan 1953 (imperfect) – 1 – (cont: south australian chronicle & weekly mail [4 jan 1868-31 dec 1870]) – uk British Libr Newspaper [079]

Chronicle – Montreal: R. Wilson-Smith, [1898-19–] – 9 – (cont: the insurance and finance chronicle) – mf#P04460 – cn Canadiana [360]

Chronicle – Ansley, NE: J H Chapman (wkly) – 1r – 1 – (cont: western echo; cont by: ansley chronicle) – us Bell [071]

Chronicle – Whitewater WI. 1879 jun 18-1879 dec 17, 1882 sep 20-1884 dec 25, 1885 mar 14 – 3r – 1 – (cont: whitewater chronicle) – mf#951631 – us WHS [071]

Chronicle / Early American Industries Association, Inc – v1-2 [1939 sep-1944 apr] – 1r – 1 – (cont: chronicle of early american industries) – mf#847889 – us WHS [338]

Chronicle – Jacksonville, FL. 1949 may-1968 – 18r – (gaps) – us UF Libraries [071]

Chronicle – Jacksonville, FL. 1969-1971 – 3r – (gaps) – us UF Libraries [071]

Chronicle – Sarasota, FL. v14 n21-v25 n21. 1984 dec 24-1996 – 4r – (gaps) – us UF Libraries [071]

Chronicle / Hamilton Co. Cincinnati – 1,1958-12,1959/1,1961-7,1968 – 6r – 1 – mf#B36228-36233 – us Ohio Hist [071]

Chronicle / Hamilton Co. Cincinnati – feb 1892-jan 1910, 1916-18 [mthly, wkly] – 7r – 1 – mf#B11097-11013 – us Ohio Hist [331]

Chronicle – Katoomba, Aug 15-Oct 10 1929 – 9 – at Pascoe [079]

Chronicle – Kenosha WI. 1877 dec 23, 30, 1878 jan 13, 20, feb 10 – 1r – 1 – mf#929567 – us WHS [071]

Chronicle – Fort Pierce FL. 1991 mar 28-dec 19, 1992 jan 16-nov 19 – 2r – 1 – (lacks: 1991 jun 13, 1992 feb 20, jul 9, aug 6-sep 10) – mf#1886884 – us WHS [071]

Chronicle : london missionary society / London Missionary Society – 1892-1966 [complete?] – 13r – 1 – (title varies) – mf#ATLA T0005 – us ATLA [240]

Chronicle / Mississippi River Commission – v1 n1-9 [1987 may-1988 jan], v2 n3 [1988 jul] – 1r – 1 – mf#1546698 – us WHS [380]

Chronicle / Richland Co. Shelby – v1 n2. mar 1867-aug 1868 [wkly] – 1r – 1 – mf#B8456 – us Ohio Hist [071]

Chronicle – Superior WI. 1984 may 18-1985 jun 24, 1985 jun 25-1986 jun 30, 1986 jul 8-1987 jul 6 – 3r – 1 – mf#1095529 – us WHS [071]

Chronicle / Tuscarawas Co. Uhrichsville – jan 1902-dec 1918 [wkly] – 7r – 1 – mf#B33767-33773 – us Ohio Hist [071]

Chronicle / Tuscarawas Co. Uhrichsville – (jan 1918-jun 1929) very damaged [irreg] – 14r – 1 – mf#B31172-31185 – us Ohio Hist [071]

Chronicle / Tuscarawas Co. Uhrichsville – jan-aug 1986// [daily] – 3r – 1 – mf#B4394-4396 – us Ohio Hist [071]

Chronicle / Warren Co. Franklin – jan-dec 1972 – 1r – 1 – mf#B4050 – us Ohio Hist [071]

Chronicle – Two Rivers WI. 1899 jul 4/dec 26-1926 jul 28/1927 apr 13 – 20r – 1 – (with gaps; cont: manitowoc county chronicle; cont by: two rivers reporter; reporter-chronicle) – mf#955578 – us WHS [071]

Chronicle – Melrose WI. 1980 mar/dec-1999 jul/dec – 24r – 1 – (with gaps; cont: melrose chronicle; cont by: melrose chronicle [melrose wi: 2001]) – mf#1130812 – us WHS [071]

Chronicle – Winston-Salem NC. 1997 may 1/jun 26-1999 apr 1/jun 24 – 9r – 1 – (with gaps; cont: winston-salem chronicle) – mf#3989443 – us WHS [071]

Chronicle *see*
- Coleraine chronicle
- Lakeside news and pleasantimes

The chronicle – Davenport, NE: C C Snowden, 1898-v2 n8. jun 2 1899 (wkly) [mf ed with gaps filmed 1958] – 1r – 1 – us NE Hist [071]

The chronicle – Featherston, NZ. 1975-88 – 14r – 1 – mf#48.15 – nz Nat Libr [079]

The chronicle – Journal of the American Baptist Historical Society, Chester, P. 1938-57. 4,266p – 1 – us Southern Baptist [242]

The chronicle – Lilongwe: Lilongwe Publ [dec 7 1993-apr 19/25 1994, may 3/9-16, jun 2/6-20/21 1994] (3 times/wk) – 1r – 1 – us CRL [079]

The chronicle *see*
- The chronicle
- Horowhenua daily chronicle

Un chronicle – New York. 1975+ (1,5,9) – (cont: un monthly chronicle) – ISSN: 0251-7329 – mf#2167,01 – us UMI ProQuest [327]

Un chronicle / United Nations – v1-32 n1. 1964-mar 1995 – 410mf – 9 – $615.00 – (v1-12 n3 entitled: un monthly chronicle. now a qrterly. updates planned) – mf#LLMC 81-907 – us LLMC [341]

Chronicle and echo – 1950-51, 1986-jun 30 1997 – 206 1/4r – 1 – (aka: northampton chronicle and echo) – uk British Libr Newspaper [072]

Chronicle and munster advertiser – Waterford, Ireland 30 mar 1844-8 mar 1848 [mf 1844-75, 1896-99] – 1 – (cont by: chronicle [11 mar 1848-5 may 1849]) – uk British Libr Newspaper [072]

Chronicle and munster advertiser – Waterford. Ireland. -w. 30 Mar 1844-5 May 1849. (6 reels) – 1 – uk British Libr Newspaper [072]

Chronicle [banbridge] – Banbridge, Ireland 19 sep 1985-27 oct 1988 – 1 – (cont: banbridge chronicle [15 mar 1968-5 sep 1985]) – uk British Libr Newspaper [072]

Chronicle [creswell, or] – Creswell OR: D Hunt, 1971- [wkly] – 1 – (cont: creswell chronicle [creswell, or]; absorbed: lakeside news and pleasantimes [1973-74]) – us Oregon Lib [071]

Chronicle [edgbaston & harborne ed] – Birmingham, England 30 mar 1990-29 dec 1995 – 1 – (cont: birmingham chronicle [8 sep 1989-23 mar 1990]) – uk British Libr Newspaper [072]
Chronicle farmer series / Medina Co. Seville – (dec 1948-mar 1984) [wkly, biwkly] – 18r – 1 – mf#B33163-33180 – us Ohio Hist [071]
Chronicle for south and mid glamorgan – [Wales] Cardiff jan 1895-14 aug 1914 [mf ed 2003] – 13r – 1 – (missing: 1896, 1897) – uk Newsplan [072]
Chronicle [kings heath & moseley ed] – Birmingham, England 30 mar 1990-18 sep 1992 – 1 – (cont: birmingham chronicle [kings heath & moseley ed] 12 jan-23 mar 1990; discontinued) – uk British Libr Newspaper [072]
The chronicle (levin) – 1 – (first title for this paper was: horowhenua daily chronicle jan-jun 1915, jan-dec 1916. title change to: levin daily chronicle on jan 20 1917, jan 1917-dec 1921, jan 1923-dec 1939, mar 1973-feb 1976. title change to: the chronicle mar 1976-aug 1977, oct 1977-apr 1994, jun 1994-oct 1998) – mf#46.1 – nz Nat Libr [079]
Chronicle [northfield & kings norton ed] – Birmingham, England 30 mar 1990-18 sep 1992 – 1 – (cont: birmingham chronicle [northfield & kings norton ed] 12 jan-23 mar 1990; discontinued) – uk British Libr Newspaper [072]
The chronicle of dino compagni / Compagni, Dino – Trans. by Else C.M. Benecke and A.G. Ferrers Howell. London: Dent, 1906. vii,284p. illus – 1 – us Wisconsin U Libr [945]
Chronicle of england (rs1) / Capgrave, John; ed by Hingeston, C – 1958 – (= ser The rolls series (rs)) – €18.00 – ne Slangenburg [941]
Chronicle of higher education – Washington DC 1966- – 1,5,9 – ISSN: 0009-5982 – mf#3363 – us UMI ProQuest [378]
Chronicle of jeremiah goldswain [the] : albany settler of 1820: vol 1 1819-1836 / ed by Long, Una – Cape Town: The Van Riebeeck Society 1946 – (= ser [Travel descriptions from south africa, 1711-1938]) – 3mf – 9 – mf#zah-28 – ne IDC [920]
The chronicle of king theodore / Littmann, E – Princeton, 1902 – 1mf – 9 – mf#NE-20314 – ne IDC [956]
Chronicle of philanthropy – Washington DC 1988- – 1,5,9 – ISSN: 1040-676X – mf#16729 – us UMI ProQuest [360]
Chronicle of pierre de langtoft [rs47] : in french verse from the earliest period to the death of edward 1 / ed by Wright, T – v1 1866 v2 1868 – (= ser The rolls series [rs]) – €18.00v – ne Slangenburg [931]
A chronicle of st john's cemetery on the humber / Denison, George Taylor – Toronto: printed for the use of the members of the Denison family, 1868 – 1mf – 9 – mf#03767 – cn Canadiana [929]
Chronicle of the augsburg confession / Krauth, Charles Porterfield – Philadelphia: J Fred'k Smith, 1879 [mf ed 1986] – 1mf – 9 – 0-8370-8755-4 – (filmed with: a question of latinity by henry eyster jacobs; incl bibl ref) – mf#1986-2755 – us ATLA [240]
Chronicle of the diocese of fredericton / Church of England. Diocese of Fredericton – St John, [NB]: Diocesan Church Society of New Brunswick, [1886-188-?] [mf ed v1 n1 jan 1886-v1 n2 feb 1886; v1 n4 apr 1886-v1 n11 nov 1886] – 9 – ISSN: 1190-6650 – mf#P04526 – cn Canadiana [242]
The chronicle of the discovery and conquest of guinea / Azurara, G E de – London, 1896. 2v – 13mf – 9 – mf#A-272 – ne IDC [916]
Chronicle of the early american industries association, inc / Early American Industries Association, Inc – Dartmouth MA 1933- – 1,5,9 – ISSN: 0012-8147 – mf#9632 – us UMI ProQuest [338]
Chronicle of the haynes family association – 1982 dec-1987 jun – 1r – 1 – mf#1533213 – us WHS [929]
Chronicle of the horse – Middleburg VA 1937+ (1) 1970+ (5) 1974+ (9) – ISSN: 0009-5990 – mf#5856 – us UMI ProQuest [636]
The chronicle of the reigns of henry 2 and richard 1, 1169-1192 (rs49) : known commonly under the name of benedict of peterborough = Gesta regis henrici secundi benedicti abbatis / ed by Stubbs, W – (= ser The rolls series (rs)) – (v1 1867 €17. v2 1867 €19) – ne Slangenburg [931]
Chronicle record – Chico, CA. 1896-1897 (1) – mf#62122 – us UMI ProQuest [071]
Chronicle review – Montreal, Quebec, Canada. 1968-76 – 1 – us AJPC [071]
Chronicle [rs82/4] / Robert of Torigni – 1890 – (= ser The rolls series [rs]) – €18.00 – ne Slangenburg [931]
Chronicle (sandbach ed) – [NW England] Sandbach, Cheshire Record Off 1947-56, 1958-64 – 1 – uk MLA; uk Newsplan [072]
Chronicle series / Hamilton Co. Cincinnati – v1 n1. (1/1827-3/1835, 10/1838-9/1839) [wkly] – 2r – 1 – mf#B14143-14144 – us Ohio Hist [071]
Chronicle series / Montgomery Co. Vandalia – dec 1955-dec 1972 [wkly, semiwkly] – 12r – 1 – mf#B5291-5302 – us Ohio Hist [071]
Chronicle series / Montgomery Co. Vandalia – jan 1973-jan 1984 [wkly] – 10r – 1 – mf#B33238-33247 – us Ohio Hist [071]
Chronicle series / Tuscarawas Co. Uhrichsville – 1/1932-6/33,(5-12/80),1/81-12/1985 [daily] – 23r – 1 – mf#B27905-27927 – us Ohio Hist [071]
Chronicle: sf, fantasy & horror's monthly trade journal – Brooklyn NY 1979- – 1,5,9 – mf#12542.02 – us UMI ProQuest [700]
Chronicle telegram – Elyria, OH. 1995-2000 (1) – mf#61706 – us UMI ProQuest [071]
Chronicle telegram-morning edition – Elvira, OH. 1999-2000 (1) – mf#69530 – us UMI ProQuest [071]
Chronicle telegraph – Quebec, Canada. 1971-1973 (1) – mf#67658 – us UMI ProQuest [071]
Chronicle telegraph and pred – Pittsburgh, PA. 1842-1927 (1) – mf#66028 – us UMI ProQuest [071]
Chronicle tribune – Marion, IN. 1968-2000 (1) – mf#61393 – us UMI ProQuest [071]
Chronicle [waterford] – Waterford, Ireland 11 mar 1848-5 may 1849 [mf 1844-75, 1896-99] – 1 – (cont: chronicle, & munster advertiser [30 mar 1844-8 mar 1848]; cont by: waterford chronicle [3 aug 1850-4 sep 1866]) – uk British Libr Newspaper [072]
Chronicle-citizen – Ansley, NE: A H Barks. 1v. v26 n28-31. oct 5-26 1909 (wkly) – 1r – 1 – (cont: argosy and the chronicle-citizen; merged with: argosy (1909) to form: argosy and the chronicle-citizen (1909); cont numbering of: argosy and the chronicle-citizen) – us Bell [071]
Chronicle-citizen – Ansley, NE: Barles & Wright. 6v. v19 n25. oct 3 1902-v24 n26. sep 27 1907 (wkly) – 3r – 1 – (formed by the union of: citizen (ansley ne) and: the ansley chronicle; merged with: argosy (ansley ne) to form: argosy and the chronicle-citizen) – us Bell [071]
Chronicle-citizen *see* The citizen
The chronicle-herald – Halifax, Nova Scotia, CN. 1880- – 24r /y – 1 – Can$2130.00 silver Can$1975.00 vesicular – cn Commonwealth Imaging [071]
Chronicler / Cincinnati AFL-CIO Labor Council – v6 n13 [1976 aug 13]-v15 n3 [1985 sep/oct] – 1r – 1 – (cont: chronicle [cincinnati oh]) – mf#1239527 – us WHS [331]
Chronicles : introduction, revised version / ed by Harvey-Jellie, Wallace Raymond – New York: Henry Frowde, 1906 – (= ser The new-century bible) – 1mf – 9 – 0-8370-3515-5 – (incl ind and notes) – mf#1985-1515 – us ATLA [220]
Chronicles / Rockford Institute – 1986 mar-1987 jun, 1987 jul-1988 dec, 1989 jan-1990 jun, 1990 jul-1991 dec, 1992 jan-1993 mar – 5r – 1 – (cont: chronicles of culture) – mf#1350173 – us WHS [071]
Chronicles and documents of medieval england, c1150-c1500 – 39r coll – 1 – (the most important vols from the mss holdings of cambridge university library. pt 1: mss dd-gg 18r c39-16501. pt 2: mss hh-oo and additional 21r c39-16502) – mf#C39-16500 – us Primary [941]
Chronicles and stories of old bingley / Speight, Harry – London, England. 1898 – 1r – us UF Libraries [914]
Chronicles and the mosaic legislation / Terry, Milton Spenser – New York: Funk & Wagnalls, 1888, c1887 – 1mf – 9 – 0-8370-5788-4 – (incl bibl ref) – mf#1985-3788 – us ATLA [220]
Chronicles concerning early babylonian kings : including records of the early history of the kassites and the country of the sea / ed by King, Leonard William – London: Luzac 1907 [mf ed 1986] – 2v on 2mf [ill] – 9 – 0-8370-8266-8 – (discussion in english; texts in akkadian & english. incl bibl ref & ind) – mf#1986-2266 – us ATLA [470]
Chronicles of an old inn : or, a few words on gray's inn / Harvey, Annie J – London: Chapman & Hall, 1887 – 3mf – 9 – $4.50 – mf#LLMC 84-292 – us LLMC [941]
Chronicles of convocation of canterbury, 1854-1914 : from lambeth palace library – 20r – 1 – mf#96045 – uk Microform Academic [240]
Chronicles of culture / Rockford College – 1977 sep-1981 dec, 1982 jan-1983 dec, 1984 jan-1985 dec, 1986 jan-feb – 4r – 1 – (cont by: chronicles [rockford il]) – mf#2745620 – us WHS [071]
Chronicles of enguerrand de monstrelet : a history of fair example, and of great profit to the french, beginning at the year 1400, where that of sir john froissart finishes... / Monstrelet, Enguerrand de – London 1810 [mf ed Hildesheim 1995-98] – 13v on 37mf – 9 – €370 – ISBN-10: 3-487-26300-9 – ISBN-13: 978-3-487-26300-7 – gw Olms [944]
Chronicles of florida / Athanase – Norfolk, VA. 1886 – 1r – us UF Libraries [630]
The chronicles of jerahmeel : or, the hebrew bible historial. being a collection of apocryphal and pseudo-epigraphical books... – London: Printed and published under the patronage of the Royal Asiatic Society, 1899 – (= ser Oriental Translation Fund (Series)) – 2mf – 9 – 0-7905-3431-2 – mf#1987-3431 – us ATLA [270]
The chronicles of kartdale : our jeames / ed by Harper, John Murdoch – Montreal: W Drysdale, 1896 – 4mf – 9 – mf#08714 – cn Canadiana [830]
Chronicles of the ancient british church – London, England. 1840 – 1r – us UF Libraries [240]
Chronicles of the builders of the commonwealth : historical character study / Bancroft, Hubert Howe – San Francisco: History Co, 1891-1892 – 7v on 1mf – 9 – mf#14085 – cn Canadiana [975]
Chronicles of the builders of the commonwealth : historical character study / Bancroft, Hubert Howe – San Francisco: History Co, 1891 – v1 on 8mf – 9 – mf#14086 – cn Canadiana [975]
Chronicles of the builders of the commonwealth : historical character study / Bancroft, Hubert Howe – San Francisco: History Co, 1892 – v2 on 8mf – 9 – mf#14087 – cn Canadiana [975]
Chronicles of the builders of the commonwealth : historical character study / Bancroft, Hubert Howe – San Francisco: History Co, 1892 – v3 on 8mf – 9 – mf#14088 – cn Canadiana [975]
Chronicles of the builders of the commonwealth : historical character study / Bancroft, Hubert Howe – San Francisco: History Co, 1892 – v4 on 8mf – 9 – mf#14089 – cn Canadiana [975]
Chronicles of the builders of the commonwealth : historical character study / Bancroft, Hubert Howe – San Francisco: History Co, 1891 – v5 on 8mf – 9 – mf#14090 – cn Canadiana [975]
Chronicles of the builders of the commonwealth : historical character study / Bancroft, Hubert Howe – San Francisco: History Co, 1892 – v6 on 8mf – 9 – mf#14091 – cn Canadiana [975]
Chronicles of the builders of the commonwealth : historical character study / Bancroft, Hubert Howe – San Francisco: History Co, 1892 – v7 on 8mf – 9 – mf#14092 – cn Canadiana [975]
Chronicles of the crusades : contemporary narratives of the crusade of richard coeur de lion, by richard of devizes and geoffrey de vinsauf, and of the crusade of saint louis, by lord john de joinville – London: G Bell 1903 [mf ed 1986] – 1r – 1 – (filmed with: a brief sketch of the zoroastrian religion and customs / bharucha, s c) – mf#6903 – us Wisconsin U Libr [931]
The chronicles of the follies of the children of montreal – [s.l: s.n. 1887?] [mf ed 1987] – 1mf – 9 – 0-665-40986-9 – mf#40986 – cn Canadiana [971]
Chronicles of the north american savages – La Salle IL 1835 – 1 – mf#3963 – us UMI ProQuest [305]
Chronicles of the nzef 1916-1919 – [mf ed 2002] – 5v on 21mf – 9 – NZ$93.00 – 0-908989-53-9 – nz BAB [355]
Chronicles of the reigns of edward 1 and edward 2 (rs76) / ed by Stubbs, W – v1 1882 v2 1883 – (= ser The rolls series (rs)) – ne Slangenburg [931]
Chronicles of the reigns of stephen, henry 2 and richard 1 *see*
- Chronicle [rs82/4]
- Gesta stephani regis anglorum
- Historia rerum anglicarum, bk 5
- Historia rerum anglicarum, bks 1-4

Chronicles of the reigns of stephen, henry 2 and richard 1 (rs82) / ed by Howlett, R – (= ser The rolls series (rs)) – 4v – (individual vols listed separately) – ne Slangenburg [931]
Chronicles of the schoenberg-cotta family / Charles, Elizabeth Rundle – London; New York: T Nelson, 1871 – 2mf – 9 – 0-524-00524-9 – mf#1990-0024 – us ATLA [240]
Chronicles of uganda / Ashe, R P – London, 1894 – 6mf – 9 – mf#HT-2 – ne IDC [916]
Chronicle-telegram / Lorain Co. Lorain – may 1921 (damaged) [daily] – 1r – 1 – mf#B33259 – us Ohio Hist [071]
Chronicle-telegraph – Quebec: Chronicle-Telegraph Pub Co. v1 n1 jul 2 1925-v9 n29 feb 3 1934 (daily) [mf ed 1987] – 14r – 1 – (merged of: the quebec chronicle and quebec gazette, and the: quebec daily telegraph (1922) to become: the quebec chronicle-telegraph) – mf#SEM35P239 – cn Bibl Nat [071]
Chronicon : cum reliquiis ex consularibus caesaraugustanis. formae tplila 136 / Tunnunensis, Victor – [mf ed 2003] – (= ser ILL – ser a; Ccsl 173a) – 2mf+vii/23p – 9 – €27.00 – 2-503-61734-4 – be Brepols [400]
Chronicon / Tyrensis, Willelmus – 1986 – (= ser ILL – ser a; Cccm 63-63a) – 24mf+149p – 9 – €60.00 – 2-503-60632-6 – be Brepols [400]
Chronicon abbatiae de evesham ad annum 1418 (rs29) / ed by Macray, W D – 1863 – (= ser The rolls series (rs)) – €17.00 – ne Slangenburg [241]
Chronicon abbatiae ramesiensis (rs83) / Ramsey Abbey; ed by Macray, W D – 1886 – (= ser The rolls series (rs)) – €19.00 – ne Slangenburg [241]
Chronicon angliae (rs64) : ab anno domini 1328 usque ad annum 1388 / ed by Thompson, E M – 1874 – (= ser The rolls series (rs)) – €19.00 – ne Slangenburg [931]
Chronicon anglicanum [rs66] : de expugnatione terrae sanctae libellus / Ralph of Coggeshall; ed by Stevenson – 1875 – (= ser The rolls series [rs]) – €18.00 – (thomas agnellus: de morte et sepultura henrici regis angliae juniores, gesta fulconis filii warini, excerpta ex otiis imperialibus gervasii tileburiensis) – ne Slangenburg [242]
Chronicon benedictoburanum : opera et studio car meichelbeck – Benedictoburani. v1-2. 1753 – €71.00 – ne Slangenburg [241]
Chronicon cartusiense / Diestensis, Petri Dorlandi & Prioris, Cartusiae – Col Agrippinae, 1608 – 6mf – 8 – €15.00 – ne Slangenburg [241]
Chronicon (cbh30,2) : versio latina / Georgii Phranzae – Venetiis, 1733 – (= ser Corpus byzantinae historiae (cbh)) – €14.00 – ne Slangenburg [240]
Chronicon cisterciensis / Miraeus, A – Colonia Agrippina, 1604 – 6mf – 8 – €14.00 – ne Slangenburg [241]
Chronicon ephratense : a history of the community of seventh day baptists at ephrata, lancaster county, penn'a / Lamech, Brother – Lancaster, PA: S H Zahm, 1889 – 1mf – 9 – 0-7905-4996-4 – (in english) – mf#1988-0996 – us ATLA [240]
Chronicon gotwicense : seu annales liberi et exempti monasterii gotwicensis o s b – Tegernsensis. v1-2. 1753 – €90.00 – ne Slangenburg [241]
Chronicon henrici knighton [rs92] : vel cnitthon, monachi leycestrensis / Knighton, Henry; ed by Lumby, J R – (= ser The rolls series [rs]) – (v1 1899 €18. v2 1895 €17) – ne Slangenburg [931]
Chronicon hispaniae (siecle 15) / Rada, Rodrigo Jimenez de – Copenhague – 1r – 5,6 – sp Cultura [946]
Chronicon mellicense / Schramb, A – Viennae Austriae, 1702 – €69.00 – ne Slangenburg [240]
Chronicon moguntinum (mgh7:20.bd) – 1885 – (= ser Monumenta germaniae historica 7: scriptores rerum germanicarum in usum scholarum (mgh7)) – €7.00 – ne Slangenburg [240]
Chronicon monasterii aldenburgensis / ed by Malou, J-B – Brugis, 1840 – €12.00 – ne Slangenburg [241]
Chronicon monasterii de abington (rs2) / Abington Abbey; ed by Stevenson, J – (= ser The rolls series (rs)) – (v1 1858 €19. v2 1858 €25) – ne Slangenburg [241]
Chronicon monasterii evershamensis : consriptum per gerardum de meestere winnoci-bergensen – Brugis, 1852 – €11.00 – ne Slangenburg [241]
Chronicon mundi : formae tplila 143 / Lucas Tudensis – [mf ed 2003 – (= ser ILL – ser a; Cccm 74) – 8mf+vii/108 – 9 – €60.00 – 2-503-63742-6 – be Brepols [400]
Chronicon novaliciense (mgh7:21.bd) – 1846 – (= ser Monumenta germaniae historica 7: scriptores rerum germanicarum in usum scholarum (mgh7)) – €7.00 – ne Slangenburg [240]
Chronicon orientale (cbh20) : latinitate donatum a abr ecchellensi – Parisiis, 1685 – (= ser Corpus byzantinae historiae (cbh)) – €23.00 – ne Slangenburg [240]
Chronicon paschale ad exemplar vaticanum (cshb16,17) / ed by Dindorfius, L – Bonnae. v1-2. 1832 – (= ser Corpus scriptorum historiae byzantinae (cshb)) – €44.00 – ne Slangenburg [240]
Chronicon paschale (cbh21) / ed by Cange, C du – Parisiis, 1688 – (= ser Corpus byzantinae historiae (cbh)) – €57.00 – ne Slangenburg [240]
Chronicon samaritanum : arabice conscriptum, cui titulus est liber josuae / ed by Juynboll, Th J – Lugduni Batavorum, 1848 – 14mf – 8 – €27.00 – ne Slangenburg [220]
Chronicon scotorum (rs46) : a chronicle of irish affairs from the earliest times to 1135, and a supplement containing the events from 1141-1150 / ed by Hennessy, W M – 1866 – (= ser The rolls series (rs)) – €17.00 – ne Slangenburg [931]
Chronicon vormelense – Brugis, 1847 – €11.00 – ne Slangenburg [241]
Chronicon warnestoniensis : ordinis canonicorum regularium s augustini – Brugis, 1852 – €5.00 – ne Slangenburg [241]

CHRONOLOGICAL

Chronicon windeshemense und liber de reformatione monastica des augustinerpropstes joh busch / ed by Grube, K – Halle, 1887 – €29.00 – ne Slangenburg [241]
Chronicorum turcicorum / Lonicerus, P – Francofvrti ad Moenvm, 1584. 2v – 8mf – 9 – mf#H-8365 – ne IDC [956]
Chronicorum turcicorum... / Lonicerus, P – Francoforti ad Moenvm. 3v. 1578 – 20mf – 9 – mf#H-8417 – ne IDC [956]
Chronicvm... : continens historiam rervm memorabilivm, a nino assyriorvm rege ad tempora friderici 2... / Burchardus, U – Argentorati, [1540] – 9mf – 9 – mf#H-8249 – ne IDC [956]
Chronik : auf das jahr 1790 [-91] / ed by Schubart, Christian Friedrich Daniel – Stuttgart 1790-91 – (= ser Dz. historisch-politische abt) – 2v on 12mf – 9 – €120.00 – mf#k/n1235 – gw Olms [900]
Chronik der evangelischen gemeinde zu krakau von ihren anfaengen bis 1657 = Kronika zboru ewangelichiego krakowskiego / Wdegierski, Wojciech – Breslau [Wroclaw]: Max Schlesinger, 1880 – 1mf – 9 – 0-524-08696-6 – (in german) – mf#1993-3221 – us ATLA [242]
Chronik der francken [...] – s.l, 1800 29 sep – 1 – fr ACRPP [943]
Chronik der teutschen *see* National-chronik der teutschen
Chronik der ukrainischen sevcenko-gesellschaft der wissenschaften in lemberg – Lemberg, 1900-1914. 59 nos – 42mf – 9 – mf#R-1708 – ne IDC [077]
Chronik des abenteuerlichen, wundervollen und seltsamen in den schicksalen beruehmter reisenden / Ehrenstein, August – Pesth 1816-17 [mf ed Hildesheim 1995-98] – 3v on 6mf – 9 – €120.00 – 3-487-29913-5 – gw Olms [910]
Die chronik des bernhard wyss 1519-1530 / Wyss, B; ed by Finsler, G – Basel, Basler Buch- und Antiquariatshandlung, 1901 – 3mf – 9 – mf#PBU-450 – ne IDC [240]
Chronik des bickenklosters zu villingen 1238 bis 1614 / ed by Glatz, Karl Jordan – Stuttgart: Litterarischer Verein 1881 (Tuebingen: L F Fues) [mf ed 1993] – (= ser Blvs 151) – 58r – 1 – (incl bibl ref. filmed with: tristrant und isalde / ed by fridrich pfaff) – mf#3420p – us Wisconsin U Libr [241]
Chronik des edeln en ramon muntaner / ed by Lanz, Karl – Stuttgart: Literarischer Verein, 1844 [mf ed 1993] – (= ser Blvs 8) – xxxvi/550p – 1 – (catalan text; int in german) – mf#8470 reel 2 – us Wisconsin U Libr [880]
Die chronik des hippolytos im matritensis graecus 121 / Bauer, Adolf – Leipzig: J C Hinrichs, 1905 [mf ed 1989] – (= ser Tugal 29/1) – 1mf – 9 – 0-7905-4184-X – (in german, greek & latin. with: stadiasmus maris magni by otto cuntz) – mf#1988-0184 – us ATLA [930]
Die chronik des hippolytos in matritensis graecus 121 / Bauer, Adolf – Leipzig, 1905 – (= ser Tugal 2-29/1) – 5mf – 9 – €7.00 – ne Slangenburg [240]
Chronik des johan oldecop / ed by Euling, Karl – Stuttgart: Litterarischer Verein 1891 [Tuebingen: H Laupp] [mf ed 1993] – (= ser Blvs 190) – 58r – 1 – (annalistic account of events 1500-73, esp of reformation in hildesheim; incl bibl ref & ind) – mf#3420p – us Wisconsin U Libr [943]
Die chronik des klosters kaisheim / Knebel, Johannes; ed by Huettner, Franz – Stuttgart: Litterarischer Verein, 1902 (Tuebingen: H Laupp, Jr) – us Wisconsin U Libr [914]
Die chronik des klosters kaisheim / Knebel, Johannes; ed by Huettner, Franz – Stuttgart: Litterarischer Verein, 1902 (Tuebingen: H Laupp, Jr) [mf ed 1993] – (= ser Blvs 226) – 625p – 1 – mf#8470 reel 47 – us Wisconsin U Libr [241]
Die chronik des laurencius bosshart von winterthur 1485-1532 / ed by Hauser, K – Basel, Basler Buch- und Antiquariatshandlung, 1905 – 5mf – 9 – mf#PBU-451 – ne IDC [240]
Chronik einer deutschen wandlung : 1925-1935 / Euringer, Richard – Hamburg: Hanseatische Verlagsanstalt, c1936 [mf ed 1989] – 1r – 1 – (filmed with: die arbeitslosen) – mf#7226 – us Wisconsin U Libr [943]
Chronik und stamm der pfalzgrafen bei rhein und herzoge in bayern 1501 : die aelteste gedruckte bayerische chronik, zugleich der aelteste druck der stadt landshut in bayern: in faksimiledruck / ed by Leidinger, George – Strassburg: J H Ed Heitz 1901 [mf ed 1993] – 1r – (incl bibl ref. filmed with: kleines deutsches sagenbuch / ed by will-erich peuckert) – mf#3367p – us Wisconsin U Libr [943]
Die chronik von barlete : kulturgeschichte eines niedersaechsischen dorfes / Frenssen, Gustav – Berlin: G Grote 1928 [mf ed 1990] – 1r [ill] – 1 – (filmed with: ferdinand freiligrath / schmidt-weissenfels) – mf#7263 – us Wisconsin U Libr [943]
Chronik von goethes leben / Biedermann, Flodoard, Freiherr von [comp] – Leipzig: Insel-Verlag [1931] [mf ed 1999] – 1r – 1 – (incl bibl ref. filmed with: goethe-forschungen / woldemar freiherr von biedermann) – mf#4652p – us Wisconsin U Libr [430]
Die chronika des fahrenden schuelers : urfassung / Brentano, Clemens – Leipzig: Wolkenwanderer-Verlag, 1923 [mf ed 1989] – xv/94p (ill) – 1 – mf#7082 – us Wisconsin U Libr [830]
Chronika eines fahrenden schuelers / Brentano, Clemens – 9. aufl. Heidelberg: C Winter, 1901 [mf ed 1989] – 262p (ill) – 1 – (cont & completed by a von der elbe) – mf#7082 – us Wisconsin U Libr [830]
Chronikalische nachrichten von nossen und umgebung – Nossen DE, 1886-90 – 1r – 1 – gw Misc Inst [943]
Chroniken *see* Die chroniken des karthaeuser klosters in klein-basel
Die chroniken der oberrheinischen staedte – Leipzig 1961-1871 [mf ed Hildesheim 1995-98] – 4v on 8mf – 9 – €160.00 – 3-487-25959-1 – gw Olms [943]
Die chroniken des karthaeuser klosters in klein-basel : 1401-1532 – Leipzig, S. Hirzel: 1872. v1 (p 231-591) – 5mf – 9 – mf#PBU-463 – ne IDC [240]
Das chronikon des konrad pellikan / ed by Riggenbach, B – Basel, 1877 – 3mf – 9 – mf#PBU-472 – ne IDC [240]
La chronique artistique et litteraire – Paris. 1857 – 1 – fr ACRPP [073]
Chronique concernant le prochain concile – Quebec: P G Delisle, 1869 [mf ed 1980] – 2v on 1mf – 9 – 0-665-05506-4 – mf#05506 – cn Canadiana [241]
Chronique de galawdewos [claudius] : roi d'ethiopie / Conzelman, W E – Paris, 1895 – 3mf – 9 – mf#NE-20311 – ne IDC [956]
Chronique de j de lalain / Chastellain, Georges – Paris 1825 [mf ed Hildesheim 1995-98] – 1v on 3mf – 9 – €90.00 – 3-487-26220-7 – gw Olms [880]
Chronique de jersey – St. Helier, England. -w. Jan 1814-Dec 1815; Jan 1820-Dec 1821; Jan 1825-Dec 1828; Jan 1830-Dec 1840; April 1849-Dec 1850. 4 1 2 reels – 1 – uk British Libr Newspaper [072]
Chronique de juillet 1830 / Rozet, Louis – Paris 1832 [mf ed Hildesheim 1995-98] – 2v on 6mf – 9 – €120.00 – 3-487-26059-X – gw Olms [944]
Chronique de la colonie reformee francaise de friedrichsdorf : suivie de documents et pieces explicatives – Hombourg-es-Monts: Imprimerie J. G. Steinhaeusser, 1887. Chicago: Dep of Photodup, U of Chicago Lib, 1978 (1r); Evanston: American Theol Lib Assoc, 1984 (1r) – 1 – 0-8370-0762-3 – mf#1984-T124 – us ATLA [240]
Chronique de la conquete de constantinople et de l'etablissement des francais en moree : ecrite en vers politiques par un auteur anonyme dans les premieres annees du 14e siecle / Buchon, Jean Alexandre C – Paris 1825 [mf ed Hildesheim 1995-98] – 1v on 3mf – 9 – €90.00 – 3-487-26314-9 – gw Olms [931]
Chronique de la dynastie alaouie du maroc / [Esslaoui, Ahmed Ennasiri] – Paris: E Leroux 1906-1907 – us CRL [960]
Chronique de la prise de constantinople par les francs : suivie de la continuation de henri de valenciennes, et de plusieurs autres morceaux en prose et en vers, relatifs a l'occupation de l'empire grec par les francais au treizieme siecle / Villehardouin, Geoffroi de – Paris 1828 [mf ed Hildesheim 1995-98] – 1v on 3mf – 9 – €90.00 – 3-487-26316-5 – gw Olms [931]
Chronique de londres – London, UK. 25 Mar 1899-12 Jul 1924 – 1 – uk British Libr Newspaper [072]
Chronique de paris – Paris. aout. 1789-aout 1793 – 1 – fr ACRPP [073]
La chronique de paris *see* Les evenements de paris
Chronique de ramon muntaner – Paris 1827 [mf ed Hildesheim 1995-98] – 2v on 11mf – 9 – €110.00 – 3-487-26313-0 – gw Olms [440]
Chronique de Rimouski / Guay, Charles – [Quebec?: s.n.], 1873-74 [mf ed 1980] – 2v on 1mf – 9 – 0-665-06497-7 – mf#06497 – cn Canadiana [971]
La chronique des arts *see* Art and decoration
Chronique des ducs de bourgogne / Chastellain, Georges – Paris 1827 [mf ed Hildesheim 1995-98] – 2v on 5mf – 9 – €100.00 – 3-487-26219-3 – gw Olms [944]
Chronique du mois : ou les cahiers patriotiques de e claviere, c condorcet, l mercier, etc – Paris. nov 1791-juil 1793 – 1 – (puis ou les cahiers patriotiques des amis de la verite) – fr ACRPP [073]
Chronique du monastere d'oudenbourg de l'ordre de s benoit / ed by Putte, F van de – Gand, 1843 – €15.00 – ne Slangenburg [241]
Chronique d'un anonyme de bethune, histoire des ducs de normandie et des rois d'angleterre jusqu'en 1220 / Mace de Gastines, Edith – 2mf – 9 – (10339) – fr Atelier National [440]
Chronique europeene – London, UK. 20 Jan-8 Jun 1872 – 1 – uk British Libr Newspaper [072]
La chronique guyanaise – Cayenne. no1-8. oct 1884-janv 1885 – 1 – fr ACRPP [073]
La chronique illustree – Paris. 14 aout 1868-72 – 1 – fr ACRPP [073]
Chronique metrique de godefroy de paris – Paris 1827 [mf ed Hildesheim 1995-98] – 1v on 3mf – 9 – €90.00 – 3-487-26307-6 – gw Olms [440]
La chronique musicale – Paris: [s.n.] v1 n1-v11 n66. jul 1873-jun 15 1876 – 4r – us CRL [780]
Chronique parisienne et departementale – Paris: Impr de Poussielgue, aug 13 1848 – 1r – us CRL [944]
Chronique religieuse – Paris. 1819-21 (I-VI) – 1 – fr ACRPP [200]
Chronique royale du cambodge / Garnier, Francis – [Paris Impr nationale 1871-72] [mf ed 1989] – 1r with other items – 1 – (from journal asiatique, 1871, 1872) – mf#mf-10289 seam reel 004/04 [§] – us CRL [915]
Chronique sur le concile du vatican, vol 2 – Quebec: P G Delisle, 1870 [mf ed 1980] – 6mf – 9 – 0-665-05508-0 – mf#05508 – cn Canadiana [241]
Chroniques / Fabre, Hector – Quebec: Impr de l'Evenement, 1877 [mf ed 1979] – 3mf – 9 – mf#SEM105P21 – cn Bibl Nat [920]
Chroniques canadiennes / Buies, Arthur – new ed. Montreal?: s.n, 1884-1875 – 2v on 1mf – 9 – mf#11854 – cn Canadiana [917]
Chroniques de fouta senegalais de sire-abbas-soh / Delafosse, Maurice – Paris: E Leroux, 1913 – 1 – us CRL [960]
Chroniques de jean froissart – Paris 1824-26 [mf ed Hildesheim 1995-98] – 15v on 45mf – 9 – €450.00 – 3-487-26305-X – gw Olms [931]
Chroniques de jean molinet – Paris 1827-28 [mf ed Hildesheim 1995-98] – 5v on 13mf – 9 – €130.00 – 3-487-26218-5 – gw Olms [440]
Chroniques de la mauritanie senegalaise / Hamet, Ismael – Nacer edd. Paris: E Leroux, 1911 – 1 – (text in arabic) – us CRL [960]
Chroniques de lundi de francoise / Francoise – [S.l: s.n, 1896?] [mf ed 1980] – 4mf – 9 – 0-665-03171-8 – mf#03171 – cn Canadiana [440]
Les chroniques de oualata et de nema (soudan francais) – Paris: P Geuthner, 1927 – 1 – us CRL [960]
Les chroniques de zar'a ya'eqob et de ba'eda maryam, rois d'aethiopie de 1434 a 1478 / Perruchon, J – Paris, 1893 – 3mf – 9 – mf#NE-20232 – ne IDC [960]
Chroniques d'enguerrand de monstrelet – Paris 1826-27 [mf ed Hildesheim 1995-98] – 15v on 47mf – 9 – €470.00 – 3-487-26301-7 – gw Olms [944]
Chroniques des comtes de flandres – Ghent, 1477 – (= ser Holkham library manuscript books 659) – 1r – 1 – (1 col reel [ill only] c506. ill in grisaille by the "master of mary of burgundy. notes by w o hassall) – mf#2113 – uk Microform Academic [931]
Chroniques des comtes de hainault / Guise, Jacques de – late 15th c – (= ser Holkham library manuscript books 658) – 1r – 1 – (french trans. notes by w o hassall. 6 col slides [ill only] s7049) – mf#2114 – uk Microform Academic [900]
Chroniques laurentiennes / Lesage, Jules Simeon – Quebec: L. Brousseau, 1901 [mf ed 1998] – 2mf – 9 – 0-665-97448-5 – mf#97448 – cn Canadiana [840]
Chroniques litteraires publiees dans "l'union liberale" de quebec / DeGuise, Charles et al – [Quebec: s.n, 1912?] – 3mf – 9 – 0-659-91750-5 – mf#9-91750 – cn Canadiana [870]
Chroniques neustriennes : ou precis de l'histoire de normandie, ses ducs, ses heros, ses grands hommes: influence des normands sur la civilisation, la litterature, les sciences et les arts...depuis le 9e. siecle jusqu'a nos jours... / Marie DuMesnil, Ange – Paris 1825 [mf ed Hildesheim 1995-98] – 1v on 3mf [ill] – 9 – €90.00 – 3-487-25949-4 – gw Olms [931]
Chroniques pittoresques et critiques de l'oeil de boeuf, des petits appartemens de la cour et des salons de paris, sous louis 14, la regence, louis 15 et louis 16 / Touchard-Lafosse, Georges – Paris 1830-32 [mf ed Hildesheim 1995-98] – 8v on 25mf – 9 – €250.00 – 3-487-26090-5 – gw Olms [944]
Chroniques, vol 1 : humeurs et caprices / Buies, Arthur – nouv ed. [Quebec: s.n.] 1873 [mf ed 1980] – 5mf – 9 – 0-665-07700-9 – mf#07700 – cn Canadiana [880]
Chroniques, vol 2 : voyages, etc etc / Buies, Arthur – nouv ed. [Quebec: s.n.] 1875 [mf ed 1980] – 4mf – 9 – 0-665-07701-7 – mf#07701 – cn Canadiana [880]
O chronista (1836-1839) *see* O brasil
Chronobiologia – Milano. 1979-1979 (1,5,9) – ISSN: 0390-0037 – mf#11545 – us UMI ProQuest [574]
Chronograms 5000 and more in number excerpted out of various authors and collected at many places / Hilton, James – London. 3v. 1882-95 – (= ser 19th c publishing...) – 19mf – 9 – (v2 dated 1885, v3 1895) – mf#3.1.7 – uk Chadwyck [070]
Der chronograph aus dem zehnten jahre antonins / Schlatter, Adolf von – Leipzig, 1894 – (= ser Tugal 1-12/1a) – 2mf – 9 – €5.00 – ne Slangenburg [240]
Der chronograph aus dem zehnten jahre antonins / Schlatter, Adolf von – Leipzig: J C Hinrichs, 1894 [mf ed 1989] – (= ser Tugal 12/1) – 1mf – 9 – 0-7905-4051-7 – (together with: zur ueberlieferungsgeschichte der altchristlichen literatur by adolf von harnack) – mf#1988-0051 – us ATLA [221]
Chronographia / Theophanis; ed by Boor, C de – Lipsiae. v1-2. 1883-85 – 2v on 23mf – 8 – €44.00 – ne Slangenburg [240]
Chronographia [cbh5] / Georgii Syncelli; ed by Goar, J – Parisiis, 1652 – (= ser Corpus byzantinae historiae [cbh]) – €56.00 – (filmed with: nicephori patriarchae: breviarium chronographicum) – ne Slangenburg [240]
Chronographia [cbh6] / Theophanis; ed by Goar, J – Parisiis, 1655 – (= ser Corpus byzantinae historiae [cbh]) – €61.00 – (filmed with: leonis grammatici: vitae recentiorum imperatorrum) – ne Slangenburg [240]
Chronographia compendiaria *see* Historia (cbh14)
Chronographia (cshb15) / Joannis Antiocheni Malalae; ed by Dindorfius, L – Bonnae, 1831 – (= ser Corpus scriptorum historiae byzantinae (cshb)) – €29.00 – (accedunt chilmeadi hodiique annotationes et ric bentlei epistola ad io millium) – ne Slangenburg [240]
Chronographia (cshb44) / Leonis Grammatici; ed by Bekkeri, Imm – Bonnae, 1842 – (= ser Corpus scriptorum historiae byzantinae (cshb)) – €19.00 – (accedit eusthatii de capta thessalonica liber) – ne Slangenburg [240]
Chronographia [cshb39,40] / Theophanis; ed by Classeni, Io – Bonnae. v1-2. 1839-1840 – (= ser Corpus scriptorum historiae byzantinae [cshb]) – €29.00 – (v2 cont anastasii bibliothecarii: historiam ecclesiasticam ex rec imm bekkeri €25) – ne Slangenburg [240]
Chronographiae libri quatuor / Genebrardus, Gilb. – Parisiis, 1585 – 43mf – 8 – €82.00 – ne Slangenburg [220]
A chronographical description of england, scotland, ireland and islands adjacent / Camden, W – v1-4. 1906 – 2r – 1 – mf#924 – uk Microform Academic [914]
Chronologen : ein periodisches werk / ed by Wekhrlin, Wilhelm Ludwig – Frankfurt, Leipzig 1779-81 – (= ser Dz) – 12v on 35mf – 9 – €350.00 – mf#k/n5520 – gw Olms [900]
Chronologia sacra / Usser, J; ed by Barlow, T – Genevae, 1722 – 4mf – 8 – mf#1087 – ne IDC [700]
Chronologia siue de tempore et eivs mvtationibvs ecclesiasticis tractatio theologica libris duobus comprehensa... / Wolf, H – Tigvri, in ofeicina [!] Froschoviana, 1585 – 2mf – 9 – mf#PBU-666 – ne IDC [240]
Chronological account of india : showing the principal events connected with the mahomedan and european governments in india... / Burgoyne, John Charles – London. 2pt. 1859 – (= ser 19th c books on british colonization) – 2mf – 9 – mf#1.1.8449 – uk Chadwyck [954]
Chronological and alphabetical tables of the principal facts of the history of canada, 1492-1887 / Gosselin, David – Quebec: J A Langlais, 1887 – 2mf – 9 – mf#06584 – cn Canadiana [971]
Chronological and geographical introduction to the life of christ = Chronologisch-geographische einleitung in das leben jesu christi / Caspari, Chretian Edouard – Edinburgh: T & T Clark, 1876 [mf ed 1985] – 1mf – 9 – 0-8370-2606-7 – (fr original german rev by aut, trans with additional notes by maurice j evans; incl ind & app) – mf#1985-0606 – us ATLA [240]
Chronological annals of the war from its beginning to the present time : in 2 parts: pt1. containing from apr 2 1755 to the end of 1760; pt 2: from the beginning of 1761 / Dobson, John – Oxford [England]: At the Clarendon Press: 1763 [mf ed 1983] – 4mf – 9 – 0-665-44376-5 – (with int pref, conclusion and ind) – mf#44376 – cn Canadiana [940]

Chronological correspondence series *see* Papers of john foster dulles and of christian a herter, 1953-1961

Chronological files of the alaskan governor, 1884-1913 / Alaska. Governor's Office – 44r – 1 – mf#T1200 – us Nat Archives [324]

Chronological handbook of the history of china : a manuscript left by ernst faber / ed by Kranz, P – Shanghai: General Evangelical Protestant Missionary Society of Germany, 1902 [mf ed 1995] – (= ser Yale coll) – xvi/250p (ill) – 1 – 0-524-09078-5 – mf#1995-0078 – us ATLA [951]

Chronological history / Frost, Jules A – s.l, s.l? 193-? – 1r – us UF Libraries [978]

A chronological history of the north-eastern voyages of discovery : and of the early eastern navigations of the russians / Burney, James – London 1819 [mf ed Hildesheim 1995-98] – (= ser Fbc) – 2mf – 9 – €60.00 – 3-487-28974-1 – gw Olms [910]

A chronological history of the people called methodists : of the connexion of the late rev john wesley, from their rise in the year 1729, to their last conference, in 1812 / Myles, William – 4th enl ed. London: Thomas Cordeux, 1813 [mf ed 1992] – (= ser Methodist coll) – 2mf – 9 – 0-524-06962-X – mf#1990-5326 – us ATLA [242]

A chronological history of voyages into the arctic regions undertaken chiefly for the purpose of discovering a north-east, north-west, or polar passage between the atlantic and pacific : from the earliest periods of scandinavian navigation, to the departure of the recent expedition / Barrow, John – London 1818 [mf ed Hildesheim 1995-98] – 1v on 3mf – 9 – €90.00 – 3-487-27078-1 – gw Olms [919]

Chronological list of members / Lawrence. Kansas. First Christian Church – 1884-1980 – 1 – us Kansas [920]

Chronological list of the practicing members of the philadelphia bar / Campbell, William J – 2nd ed. Philadelphia: Gallagher, 1890. 48p. With an appendix containing the bar of Camden, N.J. LL-1361 – 1 – us L of C Photodup [340]

Chronological outlines of jamaica history, 1492-19... / Cundall, Frank – Kingston, Jamaica. 1927 – 1r – us UF Libraries [972]

Chronological synopsis of the four gospels = Chronologische synopse der vier evangelien / Wieseler, Karl – 2nd rev corr ed. London: G Bell, 1877 [mf ed 1990] – (= ser Bohn's theological library) – 2mf – 9 – 0-7905-3498-3 – (trans fr german by edmund venables) – mf#1987-3498 – us ATLA [226]

Chronological tables of the chinese dynasties : (from the chow dynasty to the ch'ing dynasty) / Wong, Theodore; ed by Lyman, E R – [Shanghai]: Shanghai printing Co, 1902 [mf ed 1995] – (= ser Yale coll) – iii/103p – 1 – 0-524-09480-2 – mf#1995-0480 – us ATLA [951]

Chronologie de l'histoire des etats-unis d'amerique / Begin, Louis Nazaire – Quebec: A Cote, 1895 [mf ed 1980] – 1mf – 9 – mf#03547 – cn Canadiana [975]

Chronologie de l'histoire des etats-unis d'amerique / Begin, Louis-Nazaire – [Quebec?: s.n.], 1895 [mf ed 1986] – 1mf – 9 – 0-665-60947-7 – mf#60947 – cn Canadiana [975]

Chronologie de l'histoire des etats-unis d'amerique / Gagnon, Charles-Octave – Quebec: A Cote, 1895 [mf ed 1980] – 1mf – 9 – 0-665-03269-2 – mf#03269 – cn Canadiana [975]

Chronologie de l'histoire du canada / Begin, Louis Nazaire – 3e ed. Quebec: [s.n.], 1899 [mf ed 1982] – 1mf – 9 – mf#04782 – cn Canadiana [971]

Chronologie de l'histoire du canada / Begin, Louis-Nazaire – Quebec: C Darveau, 1886 [mf ed 1980] – 1mf – 9 – 0-665-02335-9 – mf#02335 – cn Canadiana [971]

Die chronologie der bibel : im einklange mit der zeitrechnung der egypter und assyrier / Raska, Johann – Wien: Wilhelm Braumueller, 1878 – 1mf – 9 – 0-8370-4842-7 – (incl bibl ref) – mf#1985-2842 – us ATLA [220]

Die chronologie der bibel des manetho und beros / Floigl, Victor – Leipzig: W. Friedrich, 1880 – 1mf – 9 – 0-7905-3195-X – mf#1987-3195 – us ATLA [220]

Die chronologie der biblischen urgeschichte (gen 5 und 11) / Euringer, Sebastian – 1. & 2. aufl. Muenster i W: Aschendorff 1909 [mf ed 1993] – (= ser Biblische zeitfragen 2/11) – 1mf – 9 – 0-524-06127-0 – mf#1992-0794 – us ATLA [221]

Die chronologie der buecher der koenige und paralipomenon : im einklang mit der chronologie der aegypter, assyrer, babylonier, phoenizier, meder... / Alker, Emmerich – Leobschuetz: George Schnurpfeil, 1889 – 1mf – 9 – 0-8370-2075-1 – (includes chronological tables. includes appendix) – mf#1985-0075 – us ATLA [221]

Die chronologie der geschichte israels, aegyptens, babyloniens und assyriens von 2000-700 v chr / Niebuhr, Carl – Leipzig: Eduard Pfeiffer 1896 [mf ed 1985] – 1mf – 9 – 0-8370-4006-X – (incl bibl ref) – mf#1985-2006 – us ATLA [221]

Die chronologie der hebraeischen koenige : eine geschichtliche untersuchung / Kamphausen, Adolf – Bonn: Max Cohen (Fr. Cohen), 1883 – 1mf – 9 – 0-8370-3844-8 – mf#1985-1844 – us ATLA [221]

Die chronologie der paulinischen briefe / Clemen, Carl – Halle a S: Max Niemeyer, 1893 – 1mf – 9 – 0-7905-0684-X – (incl bibl ref) – mf#1987-0684 – us ATLA [227]

Chronologie der roemischen bischoefe : bis zur mitte des vierten jahrhunderts / Lipsius, Richard Adelbert – Kiel: Schwer, 1869 – 1mf – 9 – 0-8370-8838-0 – (incl bibl ref) – mf#1986-2838 – us ATLA [241]

Chronologie des apostolischen zeitalters : bis zum tode der apostel paulus und petrus... / Wieseler, Paul – Goettingen: Vandenhoeck & Ruprecht, 1848 [mf ed 1989] – 2mf – 9 – 0-7905-2630-1 – (incl bibl ref & ind) – mf#1987-2630 – us ATLA [240]

Chronologie des evenements du cambodge [1954-1970] – [s.l: s.n. 1971?] [mf ed 1989] – 1r with other items – 1 – mf#mf-10289 seam reel 022/04 [§] – us CRL [959]

Die chronologie des josephus / Destinon, Justus Von – Kiel: Schmidt & Klaunig, 1880 – 1mf – 9 – 0-8370-2895-7 – (in english. incl bibl ref) – mf#1985-0895 – us ATLA [270]

Die chronologie des lebens des apostels paulus / Hoennicke, Gustav – Leipzig: A Deichert, 1903 – 1mf – 9 – 0-8370-9549-2 – (incl bibl ref) – mf#1986-3549 – us ATLA [920]

Chronologie des lebens jesu von hermann sevin / Sevin, Hermann – 2. umgearb. Aufl. Tuebingen:H Laupp, 1874 – 1mf – 9 – 0-8370-5234-3 – (incl bibl ref) – mf#1985-3234 – us ATLA [240]

La chronologie des temps chretiens de l'egypte et de l'aethiopie / Chaine, M – Paris, 1925 – 4mf – 9 – mf#NE-20240 – ne IDC [960]

La chronologie du canzoniere de petrarque / Cochin, Henry – Paris: E Bouillon, 1898 – 1 – us Wisconsin U Libr [440]

Chronologie septenaire de l'histoire de la paix entre les roys de france et d'espagne : contenant les choses plus memorables advenues en france, espagne, allemagne... / Cayet, Pierre-Victor – 2e ed. A Paris: par Jean Richer...1605 [mf ed 1982] – 11mf – 9 – 0-665-32433-2 – mf#32433 – cn Canadiana [944]

Chronologie septenaire de l'histoire de la paix entre les roys de france et d'espagne : contenant les choses plus memorables advenues en france, espagne, allemagne... / Cayet, Pierre-Victor – 3e ed. A Paris: par Jean Richer...1607 [mf ed 1983] – 11mf – 9 – 0-665-32475-8 – mf#32475 – cn Canadiana [944]

Chronologie septenaire de l'histoire de la paix entre les roys de france et d'espagne : contenant les choses plus memorables advenues en france, espagne, allemagne... / Cayet, Pierre-Victor – A Paris: par Jean Richer...1605 [mf ed 1982] – 11mf – 9 – 0-665-33072-3 – mf#33072 – cn Canadiana [944]

Die chronologische reihenfolge : in welcher die briefe des neuen testaments verfasst sind insofern diese abzuleiten ist / Brueckner, Wilhelm – Haarlem: De Erven F Bohn, 1890 – 1mf – 9 – 0-8370-9532-8 – (incl bibl ref) – mf#1986-3532 – us ATLA [225]

Chronologische synopse der vier evangelien *see* Chronological synopsis of the four gospels

Chronologisch-geographische einleitung in das leben jesu christi *see* Chronological and geographical introduction to the life of christ

Chronology : gainesville – s.l, s.l? 193-? – 1r – us UF Libraries [978]

Chronology of american case law covering all reported cases, state and federal, from the earliest period to 1897 / Phelps, W W – St. Paul: West, 1897. 506p. LL-1433 – 1 – us L of C Photodup [340]

A chronology of army administration, 1858-1938 – 1r – 1 – mf#96705 – uk Microform Academic [355]

The chronology of bible history : and how to remember it / Munger, C – New York: Nelson & Phillips; Cincinnati: Hitchcock & Walden c1876 [mf ed 1985] – 1mf [ill] – 9 – 0-8370-4539-8 – mf#1985-2539 – us ATLA [221]

The chronology of effects of caffeine during prolonged cycle ergometry / Dennis, D – 1991 – 1mf – 9 – $4.00 – us Kinesology [612]

Chronology of events pertaining to u.s. involvement in the war in vietnam and southeast asia / U.S. Military Assistance Command. Vietnam – 1v. 1972 – 1 – us L of C Photodup [977]

A chronology of important military events in republican china, 1924-1950, part 1, 1924-1928 / U.S. Office of the Chief of Military History – 1971 – 1 – us L of C Photodup [951]

Chronology of international events – London. 1945-1955 (1) – mf#540 – us UMI ProQuest [900]

The chronology of modern india : for four hundred years, from the close of the 15th century, a d 1494-1894 / Burgess, James – Edinburgh: John Grant, 1913 [mf ed 1995] – (= ser Yale coll) – vi/483p – 1 – 0-524-09054-8 – mf#1995-0054 – us ATLA [954]

A chronology of the baptists in nigeria, west africa, and related subjects, 1782-1968 / Roberson, Cecil F – rev ed 1978 – 1 – $15.30 – us Southern Baptist [242]

The chronology of the bible : connected with contemporaneous events in the history of babylonians, assyrians and egyptians / Bunsen, Ernest de – London: Longmans, Green 1874 [mf ed 1985] – 1mf – 9 – 0-8370-2530-3 – (pref by archibald henry sayce; incl notes) – mf#1985-0530 – us ATLA [220]

The chronology of the bible / Sharpe, Samuel – London: J Russell Smith, 1868 – 1mf – 9 – 0-8370-5243-2 – mf#1985-3243 – us ATLA [220]

The chronology of the bible, connected with contemporaneous events in the history of babylonians, assyrians, and egyptians / De Bunsen, Ernest – Preface by A.H. Sayce. London: Longmans, Green and Co., 1874. xiv, 138p – 1 – us Wisconsin U Libr [520]

The chronology of the early tamils : based on the synchronistic tables of their kings, chieftains, and poets appearing in the sangam literature / Sivaraja Pillai, K Narayanan – [Madras]: University of Madras, 1932 – (= ser Samp: indian books) – us CRL [954]

Chronology of the most important events connected... / Ranson, Robert – St Augustine, FL. 1928 – 1r – us UF Libraries [978]

La chronophotographie / Marey, Etienne-Jules – Paris: Gauthier-Villars, 1899 [mf ed 1975] – (= ser Conferences du Conservatoire national des arts et metiers) – 1r – 5 – mf#SEM16P236 – cn Bibl Nat [770]

Chronos athenon – Athens, Greece. -d. 9 March 1885-27 Jan 1887. Imperfect. 4 reels – 1 – uk British Libr Newspaper [949]

Chronotype – Rice Lake WI. 1890 aug 7-1893 feb 2 – 1r – 1 – (cont: barron county chronotype; cont by: rice lake chronotype [rice lake wi: 1893]) – mf#1012046 – us WHS [071]

Chronotype – Rice Lake WI. 1894 jul 5-1896 may 1 – 1r – 1 – (cont: rice lake chronotype [rice lake wi: 1893]; cont by: rice lake chronotype [rice lake wi: 1896]) – mf#1012048 – us WHS [071]

Chronotype – Rice Lake WI. 2002 mar 13/apr-2004 mar-apr – 13r – 1 – (cont: rice lake chronotype [rice lake wi: 1896]) – mf#5519284 – us WHS [071]

Chroust, A *see* Historia de expeditione friderici imperatoris et quidam alii rerum gestarum fontes eiusdem expeditionis (mgh6:5.bd)

Chruch and its endowments / Dealtry, William – London, England. 1831 – 1r – us UF Libraries [240]

Chruch question and the approaching election / Bathurst, W A – London, England. 1885 – 1r – us UF Libraries [240]

Chruch reform on christian principles considered in a letter to the... / Robinson, Hastings – London, England. 1833 – 1r – us UF Libraries [240]

Chruch's lamentation / Noel, Baptist Wriothesley – London, England. 1867 – 1r – us UF Libraries [240]

Chrusalida : jornal scientifico, litterario e critico – Rio de Janeiro, RJ: Typ de Domingos Luiz dos Santos, 05 jul-set 1867; maio-16 set 1869 – (= ser Ps 19) – mf#P05,04,137 – bl Biblioteca [073]

Chrut und uchrut im seelegaertli / Abbondio-Kuenzle, Christine – 1. ufl. Fryburg: Schwyzerluet-Verlag (G Schmidt), 1952 [mf ed 1995] – (= ser Schwyzerluet 14. jahrg n4) – 64p – 1 – mf#8917 – us Wisconsin U Libr [810]

Chrysalida : folha litteraria, critica e recreativa – Rio de Janeiro, RJ. 15 maio 1884 – (= ser Ps 19) – mf#P17,01,103 – bl Biblioteca [079]

Chrysalida : folha litteraria, critica e theatral – Rio de Janeiro, RJ: Typ Fluminense, 12 jul 1873 – (= ser Ps 19) – mf#P05,04,136 – bl Biblioteca [079]

Chrysalida : jornal do collegio santa thereza – Rio de Janeiro, RJ: Typ da Chrysalida, 01 nov-dez 1887; jan,fev,maio-ago,out,dez 1888; 01 fev 1889 – (= ser Ps 19) – mf#P17,01,104 – bl Biblioteca [073]

Chrysalida : jornal scientifico, litterario e critico – Rio de Janeiro, RJ: Typ de Domingos Luiz dos Santos, 05 jul-set 1867; maio-16 set 1869 – (= ser Ps 19) – mf#P05,04,137 – bl Biblioteca [073]

La chrysalide – Port-au-Prince, Haiti: [s.n.], 1re annee n1-n5. 22 avril 1911-22 dec 1911 – 2 sheets – us CRL [972]

Chrysalis – Des Moines IA. v1 n1 [1969] – 1r – 1 – mf#1583022 – us WHS [071]

Chrysalis – Los Angeles. 1979-1980 (1,5,9) – ISSN: 0197-1867 – mf#12017 – us UMI ProQuest [305]

Chrysander, Friedrich *see* Georg friedrich handel's (1685-1759) works

Chrysanthemum and the sword / Benedict, Ruth – Boston, MA. 1946 – 1r – us UF Libraries [830]

Chrysologus, Petrus *see* Collectio sermonum

Chrysostom – Granville, New York. v1-4 n3 oct 1935-mar 1938 – 1 – us CRL [073]

Chrysostom – London. 1974-1980 (1) 1974-1980 (5) 1974-1980 (9) – ISSN: 0529-5025 – mf#10103 – us UMI ProQuest [240]

Chrysostom : a study in the history of biblical interpretation / Chase, Frederic Henry – Cambridge: Deighton, Bell; London: George Bell, 1887 – 1mf – 9 – 0-7905-3077-5 – mf#1987-3077 – us ATLA [221]

Chrysostom, John, Saint, Archbishop of Constantinople *see* Opera omnia, opera et studio d bern

Chrysostomus (Chrysostom, John, Saint) *see*
- Kommentar zu den briefen des hl paulus an die galater und epheser, 8. bd (bdk15 2.reihe)
- Kommentar zu den briefen des hl paulus an die philipper und kolosser, 7. bd (bdk45 1.reihe)
- Kommentar zum briefe des hl paulus an die roemer, 5. bd 1. teil (bdk39 1.reihe)
- Kommentar zum briefe des hl paulus an die roemer, 6. bd 2. teil (bdk42 1.reihe)
- Kommentar zum evangelium des hl matthaeus, 1. bd (bdk23 1.reihe)
- Kommentar zum evangelium des hl matthaeus, 2. bd (bdk25 1.reihe)
- Kommentar zum evangelium des hl matthaeus, 3. bd (bdk26 1.reihe)
- Kommentar zum evangelium des hl matthaeus, 4. bd (bdk27 1.reihe)

Chrystal, George *see*
- Lectures and essays of william robertson smith
- The life of william robertson smith

Chrystal, James *see*
- Authoritative christianity
- A history of the modes of christian baptism

Chtenie dlia detei ot 5 do 8 let / Zadushevnoe slovo – Spb., 1877. nos 1-12 – 18mf – 9 – mf#R-8193 – ne IDC [077]

Chtenie dlia iunoshestva ot 12-16 let / Zadushevnoe slovo; ed by Lapin, V – M., Spb., 1877 – 24mf – 9 – mf#R-8194 – ne IDC [077]

Chtenie dlia malchikov i devochek vsekh soslovii – Spb., 1864-1866 – 95mf – 9 – (missing: 1864, v7-9) – mf#R-2349 – ne IDC [077]

Chtenie dlia mladshego vozrasta / Zadushevnoe slovo.; ed by Makarova, S – M., Spb., 1879. Pt 1(1-12) – 18mf – 9 – mf#R-8195 – ne IDC [077]

Chtenie dlia starshego vozrasta / Zadushevnoe slovo; ed by Makarova, S M – M., Spb., 1879. v1-12 – 27mf – 9 – mf#R-8196 – ne IDC [077]

Chtenie dlia vkusa, razuma i chuvstvovanii – M., 1791-1793. 12 pts – 96mf – 9 – mf#R-18579 – ne IDC [077]

Chtenie v besede liubitelei russkogo slova – Edmonton. 1965-1976 (1) 1971-1976 (5) – 36mf – 9 – mf#1699 – ne IDC [077]

Chteniia v istoricheskom obshchestve nestora-letopistsa – Kiev, 1879, 1888-1914. 24 v – 132mf – 9 – mf#1217 – ne IDC [077]

Chteniia v moskovskom obshchestve liubitelei dukhovnogo prosveshcheniia – M., 1863-1894. v1-31 – 657mf – 9 – (missing: 1889 v27(1), 1891 v29(1) title pp & contents; v29(2)) – mf#R-18495 – ne IDC [077]

Chteniia v moskovskom obshchestvie liubitelei dukhovnago prosvieshcheniia – n3,10. 1868; 1874; n1-4,7,10. 1894 (complete) – (= ser Corpus of russian orthodox periodicals) – 2r – 1 – (incl ind1863-80. 1880-94. 1910-12) – mf#ATLA S0193B – us ATLA [243]

Chteniia v tserkovno-arkheologicheskom obshchestve – Kiev, 1883-1916. v1-13 – 42mf – 9 – (missing:1914, v12) – mf#R-1578 – ne IDC [077]

Chto chitat po bogosloviiu? : sistematicheskii ukazatel apologeticheskii literatury na russkom, nemetskom, frantsuzskom i angliiskom iazykakh (248-1906 gg / Svetlov, P – Kiev, 1907 – 273p 5mf – 8 – mf#R-7282 – ne IDC [243]

Chto chitat po promyslovoi kooperatsii / Merkulov, A V – 1930 – 124p 2mf – 9 – mf#COR-538 – ne IDC [335]

Chto dielat? : nabolievshie voprosy nashego dvizheniia / Lenin, Vladimir Ilyich – Stuttgart: Verlag von J H W Dietz 1902 [mf ed 1987] – 1r – 1 – mf#1831 – us Wisconsin U Libr [335]

Chto mozhet dat kooperatsiia rabochim / Kheisin, M L – Ekaterinburg, 1916 – 39p 1mf – 9 – mf#COR-136 – ne IDC [335]

Chto takoe anarkhiia? / Tiukhanov, A – 1917 – 16p 1mf – 9 – mf#RPP-90 – ne IDC [325]

Chto takoe obshchestvo potrebitelei, kak ego osnovat i vesti / Ozerov, I K – 1909 – 136p 2mf – 9 – mf#COR-84 – ne IDC [335]

Chto takoe proizvoditelno-trudovaia artel i kak ee organizovat / Golikov, P I – 1920 – 15p 1mf – 9 – mf#COR-419 – ne IDC [335]

Chto takoe promyslovaia artel i kak ee ustroit / Simakov, V – 1929 – 116p 2mf – 9 – mf#COR-445 – ne IDC [335]

Chto takoe trudoviki? / Vasilev, N P – 1907 – 76p 1mf – 9 – mf#RPP-194 – ne IDC [325]

Chto takoe tsentrosoiuz? / Merkulov, A V – 1919 – 40p 3mf – 9 – mf#COR-335 – ne IDC [335]

Chu, Chao-ts'ui *see* Erh t'ung sheng huo

Chu chi te chiao hui [ccm95] = Life of the early church / Cheng, Chih-i – Shanghai, 1939 [mf ed 198?] – (= ser Ccm 95) – 1 – mf#1984-b500 – us ATLA [240]

Chu, Chia-hua *see* Pien chiang wen t'i yu pien chiang kung tso

Chu chiao te yen chiu [ccm183] = Short study of religions / Hsieh, Sung-kao & Yu, Mu-jen – 13th ed. Shanghai, 1946 [mf ed 198?] – (= ser Ccm 183) – 1 – (incl bibl ref) – mf#1984-b500 – us ATLA [200]

Chu, Ch'ien-chih *see* Wen hua che hsueh

Chu, Chien-min *see* Ch'in lueeh wen t'i chih kuo chi fa ti yen chiu

Chu, Chih-hsien *see* Chiao yue yen chiu fa

Chu, Chih-hsin *see* Yeh-su shih shen mo tung hsi (ccm107)

Ch'u, Chih-sheng *see* Hua pei min chung shih liao ti i ko ch'u pu yen chiu

Chu, Ching-i *see*
- Chung-kuo hsiang ts'un chiao hui chih hsin chien she
- Hsiang ts'un li pai
- I ko shih yen te hsiang tsun chiao hui

Chu, Ching-nung *see* Shih chieh ho p'ing yun tung

Chu, Ch'ing-yuan *see* T'ang sung kuan ssu kung yeh

Ch'u, Chu-nung *see* Che hsueh ta kang

Chu, Hao *see* Mai yin wen t'o

Chu, Hsi *see*
- Chu-tzu hsiao hsueh [ccm111]
- Confucian cosmogony

Chu, Hsiang *see* Shih men chi

Chu, Hsiao-ch'un *see*
- Jen li tung yuan lun
- Kuo chia tsung tung yuan ti shih chi wen t'i

Chu, Hsueeh-fan *see* Chin jih chung-kuo chih lao kung wen t'i

Chu, Hung-ta *see*
- Ssu fa yuan chieh shih li yao chih hui lan
- Ta li yuan chieh shih li ch'uan chi

Chu, I-ts'ai *see* Tang tai ch'uang tso hsiao shuo hsuan

Chu, James A *see* Journal of trauma & dissociation

Chu jih yuan liu (ccm272) / Schramm, George & Li, Jui-fang – [s.l: s.n, 1920] [mf ed 198?] – (= ser Ccm 272) – 1 – mf#1984-b500 – us ATLA [240]

Chu, Jo-hsi *see* Nung ts'un ching chi chi ho tso

Chu, Kuang-ch'ien *see*
- T'an hsiu yang
- T'an mei
- Wen i hsin li hsueh
- Wo yu wen hsueh chi ch'i t'a

Chu, Kung-yen *see* I ch'an shui yuan li chi shih wu

Chu, Lei *see* Wei tsu kuo fei hsing

Ch'u, Min-i *see* Hsing cheng yuan wen wu pao kuan wei yuan hui nien k'an

Chu, Min-wei *see* Chiang nan ch'ien hsien

Chu, Mo *see* Sheng ti i chih

Chu nghia mac va van hoa viet-nam / Truong Chinh – Ha-Noi: Su That 1974 [mf ed 1993] – on pt of 1r – 1 – mf#11052 r436 n6 – us Cornell [959]

Chu nhat uyen uong : truyen dai / Dinh, Tien Luyen – Saigon: Tuoi Ngoc 1974 [mf ed 1992] – on pt of 1r – 1 – mf#11052 r386 n8 – us Cornell [830]

Chu, Pai-ying *see* She hui k'o hsueh chiang hua

Chu pan lay mha kattipa kyvan / Kyo Kyo nay – Ran kun Mrui: Ca pe Biman 1974 [mf ed 1995] – (= ser Ca pe biman thut prannsu lak cvai ca can) – on pt of 1r – 1 – (in burmese) – mf#11052 r1951 n3 – us Cornell [830]

Chu, Pang-hsing et al *see* Shang-hai ch'an yeh yu shang-hai chih kung

Ch'u, P'ing *see* T'u ti cheng ts'e yao lun

Chu, Po-k'ang *see* Ts'ung ching chih ti chung-kuo tao tung tang ti chung-kuo

Chu, Pu-ch'uan *see* Man yun chi

Chu Rhan, Mon *see* Mrat bhu ra rvhe ti gum

Chu, Sha-lang *see* Ko nen-niang

Chu, Shih-ming *see* Ming chi ai yin lu

Chu, T'an *see* Ming chi she tang yen chiu

Chu tao wen mo hsiang lu [ccm162] = Meditations on the lord's prayer / Holth, Sverre – 1st ed. Hong Kong, 1955 [mf ed 198?] – (= ser Ccm 162) – 1 – mf#1984-b500 – us ATLA [240]

Chu tao wen tu pen [ccm273] = Reader on the lord's prayer – Wu-ch'ang, 1932 [mf ed 198?] – (= ser Ccm 273) – 1 – mf#1984-b500 – us ATLA [240]

Chu, Te *see* K'ang jih yu chi chan cheng

Chu, Tuan-chun *see* Yuan huang chi

Chu, T'ui-yu *see* Pa fuh chen ching (ccm112)

Chu, Tui-yu *see* T'ien fu chen ching (ccm113)

Chu, T'ung *see*
- Feng nu
- Yu lei, i ming, pao-yu yu tai-yu

Chu, Tzu-ch'ing *see*
- Lun-tun tsa chi
- Ni wo
- Ou yu tsa chi

Chu Van *see* Dat man

Chu, Wei-chih *see* Chi-tu chiao yu wen hsueh [ccm114]

Chu, Wei-yu *see* P'u t'ao yuan (ccm115)

Chu, Wen *see*
- Pai hua chou p'an
- Pu yuan tso nu li ti jen men
- Ts'ung wen hsueh tao lien ai
- Yu yueh chieh

Chu, Yu-an *see* Shih yung kung chai k'u ch'uan hui pien

Chu, Yuan-mao *see*
- Chan shih wu li ts'ai li
- Chan ti min chung tsu chih

Ch'u, Yun *see* Fu nu wen t'i

Chu, Yun-ying *see* Kuo ying shih yeh lun

Chu, Yu-ts'ang *see* Hsu tzu yung fa

Chuan chiao wei jen ma-li-hsun (ccm251) = Robert morrison: a master-builder / Broomhall, Marshall & Chien Yu-wen – 1st ed. Hong Kong, 1956 [mf ed 198?] – (= ser Ccm 251; The modern series of missionary biographies) – 1 – (chinese trans fr the english. also filmed: London 1924 edition [mf ed 2002] ser: the modern series of missionary biographies [with ind]) – mf#1984-b500 – us ATLA [240]

Ch'uan jen chu yueh (ccm135) / Gutzlaff, Karl Friedrich August – [s.l: s.n, 1837] – (= ser Ccm 135) – 1 – mf#1984-b500 – us ATLA [240]

Chuan kuo chin jung chi kou i lan / China. People's Republic of China – Directory of financial organizations in China. Shanghai, 1947. Rev. ed. 1 reel – 1 – 8.80 – us Chinese Res [324]

Ch'uan kuo hsin shu mu = Bibliography of new publications of the entire country – Beijing, China 1951-63 – 1 – mf#2601 – us UMI ProQuest [020]

Ch'uan kuo shang p'in chien yen hui i (2nd: 1933 Nanking, China) *see* Ti 2 tz'u ch'uan kuo shang p'in chien yen hui i hui pien

Ch'uan kuo tsung shu-mu – (Cumulative national bibliography). Peking, 1958, 1960, 1962-65, 1970, 1972-77. 7 reels – 1 – us Chinese Res [920]

Ch'uan kuo yun tung ta hui (1933: Nanking, China) *see* 22 nien ch'uan kuo yun tung ta hui tsung pao kao

Ch'uan min k'ang chan she pien *see*
- Hsien cheng yun tung ts'an k'ao ts'ai liao
- Hsien cheng yun tung ts'an k'ao tzu liao ti 2 chi

Chuan, S Peter *see* Church member (ccc 116)

Ch'uan, shao-wu *see* Chiao yu [ccm116]

Ch'uan tao i yu [ccm91] = Hints on preaching / Ch'en, Chin-yung – Shanghai, 1930 [mf ed 198?] – (= ser Ccm 91) – 1 – mf#1984-b500 – us ATLA [240]

Ch'uan tao wei jen chi [ccm263] : lessons of the women's missionary service league, 1925 = Pioneers of the chung hua shen kung hui / Pott, Francis Lister Hawks – Shanghai, 1926 [mf ed 198?] – (= ser Ccm 263) – 1 – (pref in english) – mf#1984-b500 – us ATLA [240]

Chuang, Ch'ing-kuang *see* Sheng ming ti ch'an tung

Chuang kan hao *see*
- Jie fang ri bao

Chuang, Tse-hsuan *see* Wo ti chiao yu ssu hsiang

Chuang, Tsu-t'ung *see* Hua ch'iao wen t'i

Chuang tz'u : mystic, moralist, and social reformer = Nan-hua ching / Chuang-tzu – London: Bernard Quaritch, 1889 – 2mf – 9 – 0-524-07993-5 – (in english) – mf#1991-0215 – us ATLA [180]

Chuang-tzu *see*
- Chuang tz'u
- The divine classic of nan-hua
- Das wahre buch vom suedlichen bluetenland, nan hua dschen ging

Chuang-tzu (ccm118) / ed by Yeh, Yu-lin – Shanghai, 1938 [mf ed 198?] – (= ser Ccm 118) – 1 – mf#1984-b500 – us ATLA [240]

Ch'uan-kuo hsiang hui shih t'i-ming lu – National lists of successful candidates in the imperial examinations. Scattered years from 1673-1909. 12 reels – 1 – us Chinese Res [951]

Chuchin, F G *see* Katalog bon i denznakov rossii, rsfsr, sssr, okrain i obrazovanii, (1769-1927)

Chudleigh, Daniel W *see* Muscle temperature change during ultrasound treatments of 2 and 6 era

Chudnovtsev, M *see*
- Politcheskaya rol' tserkovnikov i sektantov v sssr
- Tserkovniki i sektanti v bor'be protiv kul'turnoi revolyutsee

Chueh wu [ccs] = Consciousness – Shanghai. n1-23. 1924-25 [mf ed 198?] – (= ser Chinese christian serials coll) – 1 – mf0296e – us ATLA [240]

Chufas in florida / Killinger, G B – Gainesville, FL. 1946 – 1r – us UF Libraries [630]

Chug kreis der buecherfreunde – Tel Aviv, Haifa, Jerusalem [IL], 1943-1945 jun – 1r – 1 – (title varies: nov 1944: heute und morgen, jan 1945: heute und morgen / antifaschistische revue) – gw Misc Inst [800]

Chugach Natives, Inc *see* Newsletter

Chugoku chosaryoko hokokusho : reports of research travel in china, 1916-1935 – 10th-29th reports. 1916-35 – 136r – 1 – Y2040,000 – (in japanese) – ja Yushodo [915]

Chugoku kindai-shi shiryo : series 1: (political history) – 161bks on 51r – 1 – Y396,000 – (coll of the government publications compiled by the nationalist government of china; in chinese) – ja Yushodo [951]

Chugoku kindai-shi shiryo : series 2: (economical materials) – (= ser Govt publ) – 264bks on 64r – 1 – Y528,000 – (in chinese) – ja Yushodo [951]

Chugoku nenkan – China yearbook. In Japanese. Shanghai, 1934-37, 1939. 3 reels – 1 – 49.50 – us Chinese Res [951]

Chugoku seiji keizai kankei shinbun kirinukishu : matsumoto collection of the chinese press cuttings of politics and the economy in the 20th century – 1908-23 – 10 classifications on 10r – 1 – Y103,000 – (in chinese) – ja Yushodo [951]

Chui ray kui to ma me nuin : [a novel] / Nnui Van – Ran Kun: Ca pe bi man 1975 [mf ed 1990] – 1r with other items – 1 – (in burmese) – mf#mf-10289 seam reel 165/7 [§] – us CRL [830]

Chui rhay lac bhan lup nan / Maung Maung, U – Ran Kun: Ca pe bi man a phvai 1973 [mf ed 1990] – 1r with other items – 1 – (in burmese) – mf#mf-10289 seam reel 147/7 [§] – us CRL [332]

Chuirhaylac lay ya cuik pyui re samavayama / Mrat Su, U – Ran kun Mrui: Sa khan Lvan, Prannsu Ca up tuik 1975 [mf ed 1995] – on pt of 1r – 1 – mf#11052 r1951 n5 – us Cornell [959]

Chujoy, Anatole *see* Civic ballet

[Chula vista-] chula vista star news – CA. 1930-1934; feb 1935-37; 1970- – 99+ r – 1 – $5940.00 (subs $165/y) – mf#H03182 – us Library Micro [071]

Chum Na Bangchang *see* Khun phon kalasing

Chumacero, Ali *see* Poesia romantica

Chumacero y Carrillo, Juan *see* Memorial...ano de 1633

Chumachenko, K D *see* Taganrogskii vestnik

Chun, C *see* Wissenschaftliche ergebnisse der deutschen tiefsee-expedition auf dem dampfer valdivia 1898-1899

The chun tsew : with the tso chuen = Chun chiu – Hongkong: Lane, Crawford 1872 [mf ed 1993] – (= ser The chinese classics 5) – 2v on 12mf – 9 – 0-524-08775-X – (in english & chinese) – mf#1993-4015 – us ATLA [180]

Ch'un-chung jih-pao – Sian, China. Ch'un-chung Daily. Starting 16 Oct 1954, title changed to Shensi Daily. May 1952-Oct 1954. 6 reels – 126.00 – 1 – us Chinese Res [079]

Chundra lela : the converted fakir / Lee, Ada – Cincinnati: Curtis & Jennings, 1899 [mf ed 1995] – (= ser Yale coll) – 125p (ill) – 1 – 0-524-09762-3 – mf#1995-0762 – us ATLA [240]

Chundra lela : the story of a hindu devotee and christian missionary / Griffin, Zebina Flavius – Philadelphia: Griffith & Rowland Press [1911] [mf ed 1995] – (= ser Yale coll) – 84p (ill) – 1 – 0-524-09893-X – mf#1995-0893 – us ATLA [240]

Chung, Ch'ung-min *see* Ssu-ch'uan ts'an ssu ch'an hsiao tiao ch'a pao kao

Chung, Ho Lee *see* The archeology of the white buffalo robe site

Chung hsi jih pao = Chung sai yat po – San Francisco [CA: Chung Sai Yat Po Publ Co], jul 1947-1950 – us CRL [071]

Chung hsi jih pao fu chang – San Francisco CA. 1904 jul 5 – 1r – 1 – mf#881879 – us WHS [071]

Chung hua jen min kung ho kuo jen min pai paio to hui = Official gazette of the permanent committee, people's republic of china – 1957-1963 (1) – mf#2603 – us UMI ProQuest [338]

Chung hua kuei chu [ccs] = China for christ – Shanghai. n36-218. 1923-24 [gaps] [mf ed 198?] – (= ser Chinese christian serials coll) – 1 – (later title: hsieh chin) – mf0296f – us ATLA [240]

Chung hua tsung hui : berita tionghoa – Bandoeng, 1949. v1(1-2) – 1mf – 9 – (missing: 1949 v1(1)) – mf#SE-351 – ne IDC [950]

Chung hua wai k'o tsa chih = Chinese journal of surgery – 1963-1964 (1) – mf#2604 – us UMI ProQuest [617]

Chung kuo k'o hsueh = Journal of science – 1963-1964 (1) – mf#2607 – us UMI ProQuest [500]

Chung kuo nung yeh k'o hsueh = Chinese agricultural science – 1962-1964 [1] – mf#2609 – us UMI ProQuest [630]

Chung kuo yu wen = Chinese language – 1952-1963 (1) – mf#2610 – us UMI ProQuest [480]

Chung kuoshih pao = Chinese daily times – Toronto, Chinese free Mason society (varies), 1928-56// (wkly (to 1929?), daily) – 6r – 1 – Can$875.00 – (a window on toronto's chinese community in the mid-50's; title varies: hung chung she po. on film: jan 18 1954-dec 14 1956) – cn McLaren [071]

Chung, Lu-chai *see* Chiao yue chih k'o hsueeh yen chiu fa

Chung, Nai-k'o *see* Tien ch'uan chih tu lun

Chung nung ching-chi t'ung-chi – (Economic and Statistical review). Chungking, Nanking, China. 31 Jul 1941-31 Dec 1947. v.1, no.2-v.7, no.2. 3 reels – 1 – 62.50 – us Chinese Res [330]

Chung nung yueh kan – Farmers' Bank Monthly. Chungking, Shanghai. May 1940-Oct 1948. v.1, n.5-v.9, no.10. Incomplete. 6 reels – 1 – us Chinese Res [332]

Chung, Pak K *see* Self-esteem and health related physical fitness of male college students in hong kong

Chung ta mat het, chi con nhau : tho / Vu, Hoang Chuong – Paris: Rung Truc 1974 [mf ed 1992] – on pt of 1r – 1 – mf#11052 r88 n1 – us Cornell [810]

Chung, Tao-tsan *see* Hsien tai chung-kuo chih yeh chiao yu chih ch'an sheng yu ch'i fa chan

Chung wai ching-chi chou-k'an – Peking, China. Chinese weekly economic bulletin. March 1923-Oct 1927. Incomplete – 1 – us Chinese Res [330]

Chung yang ho tso t'ung hsun = Central cooperative bulletin – 1952-1958 (1) – mf#2611 – us UMI ProQuest [334]

Chung yang hsuan ch'uan pu pien *see* Hsien cheng yu ti fang tzu chih

Chung yang yin hang yueh pao / China. Central Bank – Shanghai. -m. v1-6, n.6. Aug 1932-Jun 1937. Lacking: v1, n.1-3. 14 reels – 1 – $237.00 – (n.s. v1-4. n.3. lacking: v2, n.8. 5 reels. $93.50) – us Chinese Res [332]

Chung, Yuan-chao *see* Ti kuo chu i lun

Chung yung *see* The conduct of life

Ch'ung-ching jih-pao – Chungking, China. Chungking Daily. Sept 1952-Jun 1953; Oct 1954-Dec 1960 – 14r – 1 – us Chinese Res [079]

Chung-hsi chiao-hui pao [ccs] = Missionary review – Shanghai: SPCK. v2 n12 1896, v4 n37-48 1898 [complete] [mf ed 198?] – (= ser Chinese christian serials coll) – 1r – 1 – mf0302 – us ATLA [240]

Chung-hua chi-tu chiao ch'ing nien hui er shih wu nien hsiao shih [ccm237] = Brief history of the first 25 years' history of the ymca's in china / Lyon, David Willard – Shanghai, 1920 [mf ed 198?] – (= ser Ccm 237) – 1 – mf#1984-b500 – us ATLA [360]

Chung-hua chi-tu chiao ching nien hui shi lueh [ccm349] = Indigenization of the ymca in china / Yu, Rih-chang – Shanghai, 1927 [mf ed 198?] – (= ser Ccm 349) – 1 – mf#1984-b500 – us ATLA [360]

Chung-hua chi-tu-chiao chiao-yu chi-k'an [ccs25] = China christian education quarterly – Shanghai. v1-6. 1925-30 [gaps] – (= ser Chinese christian serials coll 25) – 1r – 1 – (also incl some iss of: chiao yu chi k'an [china christian educational quarterly] v7-12 n3 1931-sep 1936 2r [so303b]) – mf0303a – us ATLA [240]

Chung-hua chi-tu-chiao-hui nien chien [ccs] = China church yearbook – v1-13. 1914-36 [complete] [mf ed 1974-77] – (= ser Chinese christian serials coll) – 4r – 1 – mf0269 – us ATLA [240]

Chung-hua hsueh tso chih = Chinese medical journal – Peking. 1963-1964 (1) – mf#2602 – us UMI ProQuest [610]

Chung-hua min tsu yen li ti yeh-su [ccm193] = Jesus through the eyes of the chinese nation / Hsu, Sung-shih – Shanghai, 1934 [mf ed 198?] – (= ser Ccm 193) – 1 – mf#1984-b500 – us ATLA [230]

Chung-kuo chan-shih ching-chi fa-kui hui-pien / China – (Collection of Laws and Regulations Governing China's Wartime Economy). Shanghai, 1940. 1 reel – 1 – us Chinese Res [951]

Chung-kuo ch'ing kung yeh = Light industry of china – 1953-1960 (1) – mf#2605 – us UMI ProQuest [338]

Chung-kuo ching-chi kai-tsao / Ma Yin-Ch'u – (The Economic Reform of China). Shanghai. 1935. 1 reel – 1 – us Chinese Res [330]

Chung-kuo ching-chi (the chinese economy) – Nanking, Apr 1933-Apr 1937. Scattered issues missing. 7 reels – 1 – us Chinese Res [339]

Chung-kuo chi-tu chiao chiao hui kai ko ti t'u ching [ccm72] = Reconstruction of christian church in china / Chao, Tzu-ch'en – Shanghai, 1950 [mf ed 198?] – (= ser Ccm 72) – 1 – mf#1984-b500 – us ATLA [240]

Chung-kuo chi-tu chiao shih (ccm343) / Yang, Sen-fu – Taipei, 1968 [mf ed 198?] – (= ser Ccm 343) – 1 – mf#1984-b500 – us ATLA [240]

Chung-kuo chi-tu chiao shih kang [ccm307] = History of christianity in china / Wang, Chih-hsin – Shanghai, 1940 [mf ed 198?] – (= ser Ccm 307) – 1 – (incl bibl ref) – mf#1984-b500 – us ATLA [240]

Chung-kuo chi-tu chiao te kai shan shih yeh [ccm100] = Pioneers of the protestant church in china / Chien, Yu-wen – 1st ed. Hong Kong, 1956 [mf ed 198?] – (= ser Ccm 100) – 1 – (incl bibl ref) – mf#1984-b500 – us ATLA [242]

Chung-kuo chuan tung wen-hua yu tien-chu ku chiao [ccm285] = The old testament and the chinese classical books / Su, Hsueh-Lin – 2nd ed. Hong Kong, 1957 [mf ed 198?] – (= ser Ccm 285) – 1 – mf#1984-b500 – us ATLA [221]

Chung-kuo fang chih = Chinese textiles – 1951-1960 (1) – mf#2606 – us UMI ProQuest [670]

Chung-kuo hsiang ts'un chiao hui chih hsin chien she (ccm108) / Chu, Ching-i – Shanghai, 1927 [mf ed 198?] – (= ser Ccm 108) – 1 – mf#1984-b500 – us ATLA [240]

Chung-kuo hui chiao shih chien (ccm239) / Ma, I-yu – [Shanghai, 1941] [mf ed 198?] – (= ser Ccm 239) – 1 – (incl bibl ref) – mf#1984-b500 – us ATLA [260]

Chung-kuo kuan shen fan chiao ti yuan yin [1860-1874] [ccm233] = =Origin and cause of the anti-christian movement by chinese officials and gentry, 1860-1874 / Lu, Shih-chiang – Nanking, Taipei, 1966 [mf ed 198?] – (= ser Ccm 233) – 1 – (incl bibl ref & ind) – mf#1984-b500 – us ATLA [230]

Chung-kuo kung yeh = China industry – 1949-1958 (1) – mf#2608 – us UMI ProQuest [338]

Chung-kuo kuo min tang Chung yang t'ung chi ch'u *see* Min kuo erh shih san nien chih chien she

Chung-kuo kuo min tang Hsuan ch'uan pu *see*
– Fu nu wen t'i chung yao yen lun chi
– K'ang chan ying hsiung chuan chi

Chung-kuo kuo min tang Hsuean ch'uan pu *see* Chien kuo ta kang ch'ien shih

Chung-kuo lao-kung wen-t'i / Ch'en Ta – (Labor Problems in China). Shanghai, 1929. 1 reel – 1 – us Chinese Res [951]

Chung-kuo lao-tung fa ling hui-pien / Ku Ping-Yuan – (A Collection of Chinese Labor Laws and Decrees). Shanghai, 1937. 1 reel – 1 – $11.00 – us Chinese Res [951]

Chung-kuo li shih te shang ti kuan (ccm306) / Wang, Chih-hsin – Shanghai, 1926 [mf ed 198?] – (= ser Ccm 306) – 1 – mf#1984-b500 – us ATLA [290]

Chung-kuo mu-ssu-lin [ccs] = Muslims in china – Pei-ching. n25 jan 10 1960 [complete] [mf ed 198?] – (= ser Chinese christian serials coll) – 1 – mf0296g – us ATLA [260]

Chung-kuo nung-ts'un : (chinese villages) – Shanghai, oct 1934-may 1943. – 4r – 1 – (scattered issues missing.) – us Chinese Res [951]

Chung-kuo san-chiao ti kung t'ung pen chih [ccm169] = Common ground of confucianism, taoism and chinese buddhism / Hsieh, Fu-ya – Hong Kong, 1966 [mf ed 198?] – (= ser Ccm 169) – 1 – mf#1984-b500 – us ATLA [290]

Chung-kuo shao nien te ke jen wen ti [ccm210] = Chinese boy's personal problem / Rounds, H J – Shanghai, 1923 [mf ed 198?] – (= ser Ccm 210) – 1 – mf#1984-b500 – us ATLA [305]

Chung-kuo sheng hsien yao tao lei pien [ccm300] = Collection of the teachings of famous chinese / ed by Tung, Ching-an – [China: s.n, 19–?] [mf ed 198?] – (= ser Ccm 300) – 1 – mf#1984-b500 – us ATLA [170]

Chung-kuo ti cheng yen chiu so *see* P'ing chun ti ch'uan yu t'u ti kai ko

Chung-kuo tien chu chiao chuan chiao shih (ccm121) / Elia, Paschal d' – Shanghai, 1934 [mf ed 198?] – (= ser Ccm 121) – 1 – mf#1984-b500 – us ATLA [241]

Chung-kuo tsung chiao ssu hsiang shih ta kang (ccm309) / Wang, Chih-hsin – Shanghai, 1933 [mf ed 198?] – (= ser Ccm 309) – 1 – mf#1984-b500 – us ATLA [951]

Chung-kuo wei hsin pao = Chinese reform news – New York City: Chinese Reform News Publ Co, [aug 22 1917-1928] – 7r – us CRL [071]

Chung-nung yueh-k'an – (The Farmer's Bank Monthly). Chungking Shanghai. v.1, no.5-v.9, no.10. Incomplete. 6 reels – 1 – us Chinese Res [332]

Chung-qang jih-pao, 1928-1990 – 198r – 1 – $30.00r in U.S. $6,930.00r outside U.S. (L94A0001-L94A0016) – (in chinese) – us L of C Photodup [321]

Chung-shan ta hsueh (Canton, China) Hua hsueh kung yeh yen chiu so *see* Kang hu hua hsueh kung yeh k'ao ch'a chi

Chung-shan wen hua chiao yu kuan chan hou shih chieh chien she yen chiu hui *see* Chan hou wen t'i lun wen chi

Chung-wai-jim-pao universal gazette – Shanghai. Sep-dec 1908 – 2r – 1 – uk British Libr Newspaper [072]

Chung-yang jih-pao : changsha (hunan) edition – Apr-Nov 1938 – 1r – 1 – mf#L94A0001 – Dist. us Scholarly Res – us L of C Photodup [320]

Chung-yang jih-pao : chengtu (szechwan) edition – Apr 1943-Jun 1946 – 4r – 1 – mf#L94A0002 – Dist. us Scholarly Res – us L of C Photodup [320]

Chung-yang jih-pao : chikiang (hunan) edition – Aug-Dec 1943 – 1r – 1 – mf#L94A0003 – Dist. us Scholarly Res – us L of C Photodup [320]

Chung-yang jih-pao : chungking (szechwan) edition – Sept 1938-Dec 1941; Apr 1942-Jul 1948 – 8r – 1 – mf#L94A0004 – Dist. us Scholarly Res – us L of C Photodup [320]

Chung-yang jih-pao : kunming (junan) edition – May 1943-Aug 1945 – 1r – 1 – mf#L94A0005 – Dist. us Scholarly Res – us L of C Photodup [320]

Chung-yang jih-pao : kweiyang (kweichow) edition – Apr 1943-48 – 1r – 1 – mf#L94A0006 – Dist. us Scholarly Res – us L of C Photodup [320]

Chung-yang jih-pao : nanking (kiangsu) edition – Nov 1932-Dec 1937; Mar 1945-Apr 1949 – 22r – 1 – mf#L94A0007 – Dist. us Scholarly Res – us L of C Photodup [320]

Chung-yang jih-pao : nanking (kiangsu) edition – (Wang Ching-wei regime). Mar-Jul 1945 – 1r – 1 – mf#L94A0008 – Dist. us Scholarly Res – us L of C Photodup [320]

Chung-yang jih-pao : pai-se (kwangsi) edition – Jan-Jul 1945 – 1r – 1 – mf#L94A0009 – Dist. us Scholarly Res – us L of C Photodup [320]

Chung-yang jih-pao : shanghai edition – Feb-Sept 1928; Aug-Dec 1945; 1946; 1947; 1948; 1949 – 12r – 1 – mf#L94A0010 – Dist. us Scholarly Res – us L of C Photodup [320]

Chung-yang jih-pao : shaoyang (hunan) edition – Aug 1942-Sept 1945 – 2r – 1 – mf#L94A0011 – Dist. us Scholarly Res – us L of C Photodup [320]

Chung-yang jih-pao : shenyang (liaoning) edition – Jul 1948 – 1r – 1 – mf#L94A0012 – Dist. us Scholarly Res – us L of C Photodup [320]

Chung-yang jih-pao : taipei (taiwan) edition – Oct-Dec 1950; 1951-1990; 1950-1959: 20r; 1960-1969: 30r; 1970-1979: 35r; 1980-1989: 49r; 1990: 6r – 140r – 1 – mf#L94A0013 – Dist. us Scholarly Res – us L of C Photodup [320]

Chung-yang jih-pao : t'un-hsi (anhwei) edition – Jul 1944-Jun 1945 – 1r – 1 – mf#L94A0014 – Dist. us Scholarly Res – us L of C Photodup [320]

Chung-yang jih-pao : wu-chou (kwangsi) edition – Oct 1943-Jul 1944 – 1r – 1 – mf#L94A0015 – Dist. us Scholarly Res – us L of C Photodup [320]

Chung-yang jih-pao : yung-an (fukien) edition – Jul 1943-Jun 1944 – 1r – 1 – mf#L94A0016 – Dist. us Scholarly Res – us L of C Photodup [320]

Chung-yang jih-pao, 1928-1990 = Central daily news – 198r – 1 – (in chinese. individual titles also listed) – mf#L94A0001-L94A00016 – Dist. us Scholarly Res – us L of C Photodup [079]

Chun-sheng *see*
– Hsien tai jih chi wen hsuan
– Hsien tai nu tso chia hsiao p'in hsuan
– Hsien tai nu tso chia shih ko hsuan
– Hsien tai nu tso chia shu hsin hsuan
– Hsien tai nu tso chia sui pi hsuan

Chuo bijutsu : a monthly magazine, 1915-1936 – 1st period v1 n1-v15 n6(1915-29). 2nd period n1-40(1933-36) – 36r – 1 – Y480,000 – (in japanese; with 140p guide) – ja Yushodo [700]

Chuo kikuu cha dar es salaam : history dept maji maji research project, 1968: collected papers – [Dar es Salaam: s.n., 1969?] – us CRL [960]

Chuquet, Arthur *see* Les guerres de la revolution

Church : a banqueting-house, and christ's banner love – Edinburgh, Scotland. 1848 – 1r – us UF Libraries [240]

Church : the guardian of her children, her guide, the oracles of god / Church Of England – London, England. 1850 – 1r – us UF Libraries [240]

Church : her dangers and duties / Wright, Charles – Dublin, Ireland. 1857 – 1r – us UF Libraries [240]

Church – New York. 1990-1996 (1) – ISSN: 0883-5667 – mf#16157 – us UMI ProQuest [240]

Church – s.l, s.l? 18– – 1r – us UF Libraries [240]

Church : the teacher of her children / Denison, Edward – London, England. 1839 – 1r – us UF Libraries [240]

The church / Binnie, William – Edinburgh: T & T Clark; New York: Scribner and Welford [distributor], 1882 – (= ser Handbooks for bible classes and private students) – 1mf – 9 – 0-7905-3309-X – mf#1987-3309 – us ATLA [240]

The church = De ecclesia / Hus, Jan – New York: Scribner, 1915 – 1mf – 9 – 0-7905-4821-6 – (incl bibl ref. in english) – mf#1988-0821 – us ATLA [240]

The church = Ecclesia / Boardman, George Dana – New York: Scribner, 1901 – 1mf – 9 – 0-7905-7379-2 – mf#1989-0604 – us ATLA [240]

The church : her ministry and sacraments / Van Dyke, Henry Jackson – Philadelphia: Presbyterian Board of Publication and Sabbath-School Work, 1903, c1890 – (= ser Stone Lectures) – 1mf – 9 – 0-7905-9726-8 – mf#1989-1451 – us ATLA [240]

The church : its origin, its history, its present position / Luthardt, Christoph Ernst et al – Edinburgh: T & T Clark, 1867. Beltsville, Md: NCR Corp, 1978 (4mf); Evanston: American Theol Lib Assoc, 1984 (4mf) – 9 – 0-8370-0854-9 – mf#1984-4214 – us ATLA [240]

The church : its polity and ordinances / Harvey, Hezekiah – Philadelphia: American Baptist Publication Society, c1879 – 1mf – 9 – 0-524-08379-7 – mf#1993-3079 – us ATLA [240]

The church : a sermon. preached at the opening of the synod of the german reformed church at carlisle... / Nevin, John Williamson – Chambersburg, Pa: Printed at the Publication Office of the German Ref Church, 1847 – 1mf – 9 – 0-524-08766-0 – mf#1993-3271 – us ATLA [240]

The church a composite life / Prestridge, John Newton – Louisville, KY: World Press, 1911 – 1mf – 9 – 0-524-07708-8 – mf#1991-3293 – us ATLA [240]

Church, A M *see* Picturesque cuba, porto rico, hawaii, and the phil...

Church administration – 1927-31. N.S. Oct 1959-61 – 1 – 81.97 – us Southern Baptist [242]

Church administration – Nashville. 1962+ (1) 1970+ (5) 1970+ (9) – ISSN: 0412-4553 – mf#2178 – us UMI ProQuest [240]

Church advocate – Dublin, Ireland [ns] aug 1879-15 nov 1891 – 1 – (cont: irish church advocate [ns] jan 1876-1 jul 1879) – uk British Libr Newspaper [240]

Church, Alfred John *see*
– Pliny's letters
– To the lions

The church and country life : report of conference held by the commission on church and country life under the authority of the federal council of churches of christ in america, columbus, ohio, december 8-10, 1915 / ed by Vogt, Paul Leroy – New York: Missionary Education Movement of the United States and Canada, 1916 – 1mf – 9 – 0-524-07768-1 – mf#1991-3336 – us ATLA [240]

Church and creed / Newton, Richard Heber – New York: G P Putnam, 1891 – 1mf – 9 – 0-8370-3916-9 – mf#1985-1916 – us ATLA [240]

Church and dissent in wales – Liverpool, England. 1886 – 1r – us UF Libraries [240]

The church and her children / Hulbert, Henry Woodward – New York: F H Revell, c1912 – 1mf – 9 – 0-7905-5234-5 – (incl bibl ref) – mf#1988-1234 – us ATLA [240]

The church and her children : a sermon / Wilson, James Patriot – New-York: John A Gray, 1856 – 1mf – 9 – 0-524-00119-7 – mf#1989-2819 – us ATLA [240]

The church and her teaching : addresses delivered in cornwall / Robinson, Charles Henry – London; New York: Longmans, Green, 1893 – 1mf – 9 – 0-8370-8613-2 – mf#1986-2613 – us ATLA [240]

Church and home – Saint John, NB: [s.n, 1896-189-?] [mf ed v1 n2 feb 1896-v2 n6 jun 1897; v2 n8 aug 1897-v2 n12 dec 1897] – 9 – ISSN: 1190-6766 – mf#P04297 – cn Canadiana [242]

Church and home *see* The telescope-messenger

The church and human society : speeches and discussions together with the papers published for the consideration of the congress / Pan-Anglican Congress 1908, Section A – London: Society for Promoting Christian Knowledge; New York: E S Gorham, 1908 – 1mf – 9 – 0-8370-9091-1 – (includes bibliographies) – mf#1986-3091 – us ATLA [240]

The church and its polity / Hodge, Charles; ed by Durant, William & Hodge, Archibald Alexander – London: T Nelson, 1879 – 2mf – 9 – 0-524-05011-2 – mf#1991-2181 – us ATLA [240]

The church and its social mission / Lang, John Marshall – New York: T Whittaker 1902 [mf ed 1990] – (= ser The baird lecture 1901) – 1mf – 9 – 0-7905-7893-X – mf#1989-1118 – us ATLA [303]

The church and labor / Stelzle, Charles – Boston: Houghton Mifflin, 1910 – (= ser Modern Religious Problems) – 1mf – 9 – 0-7905-6086-0 – (incl bibl ref) – mf#1988-2086 – us ATLA [240]

Church and life – v28-41. 1979-92 [complete] – Inquire – 1 – (cont: kirche og folk) – mf#atla s0743 – us ATLA [240]

The church and life of to-day / Browne, George Forrest et al – London: Hodder and Stoughton, 1910 – 1mf – 9 – 0-7905-8767-X – mf#1989-1992 – us ATLA [240]

Church and manor : a study in english economic history / Addy, Sidney Oldall – London: George Allen, 1913 – 2mf – 9 – 0-7905-4363-X – (incl bibl ref) – mf#1988-0363 – us ATLA [330]

The church and modern life / Gladden, Washington – Boston: Houghton, Mifflin, 1908 – 1mf – 9 – 0-7905-4530-6 – mf#1988-0530 – us ATLA [240]

The church and modern problems in the light of the teachings of paul in first corinthians / Fitzwater, Perry Braxton – Chicago: Bible Institute Colportage Ass'n, c1914 – 1mf – 9 – 0-524-01814-6 – mf#1990-4152 – us ATLA [227]

The church and modern society : lectures and addresses / Ireland, John – Chicago: D H McBride. 2v. 1897-1905 – 2mf – 9 – 0-8370-6668-9 – (includes indexes) – mf#1986-0668 – us ATLA [240]

Church and nation / Temple, William – London: Macmillan 1915 [mf ed 1990] – (= ser The bishop paddock lectures 1914-15) – 1mf – 9 – 0-7905-7478-0 – mf#1989-0703 – us ATLA [230]

The church and popular education / Adams, Herbert Baxter – Baltimore: Johns Hopkins Press 1900 [mf ed 1986] – (= ser Johns hopkins university studies in historical and political science 18/8-9) – 1mf – 9 – 0-8370-7520-3 – mf#1986-1520 – us ATLA [377]

Church and realm in the stuart times : a course of ten illustrated lectures / Lane, Charles Arthur – London: Edward Arnold, [1898?] – 1mf – 9 – 0-524-03235-1 – mf#1990-0863 – us ATLA [240]

Church and reform in scotland : a history from 1797 to 1843 / Mathieson, William Law – Glasgow: James Maclehose 1916 – 1 – (this work with aut's "politics and religion", "scotland and the union" and "the awakening of scotland", forms a continuous history of scotland fr 1550-1843. filmed with: etudes sur l'ancien poeme francais du voyage / goulet, j) – mf#2171 – us Wisconsin U Libr [941]

The church and religious unity / Kelly, Herbert – London; New York: Longmans, Green, 1913 – 1mf – 9 – 0-7905-5292-2 – mf#1988-1292 – us ATLA [240]

The church and social problems / Husslein, Joseph – New York: American Press, 1912 – 1mf – 9 – 0-524-02980-6 – mf#1990-0767 – us ATLA [240]

Church and society – Philadelphia, PA. v1-81. sep 1908-aug 1991 – 19r – 1 – (lacks v1-4 p5-6. title varies) – mf#atla s0147 – us ATLA [240]

Church and state / Americans United for Separation of Church and State – 1979-1983 jun – 1r – 1 – (cont: church and state newsletter) – mf#153208 – us WHS [230]

Church and state / Beeching, H C – London, England. 1887 – 1r – us UF Libraries [230]

Church and state / Candlish, Robert Smith – London, England. no date – 1r – us UF Libraries [230]

Church and state / Galt, Alexander Tilloch – Montreal: Dawson, 1876 – 1mf – 9 – mf#24094 – cn Canadiana [230]

Church and state : a historical handbook / Innes, Alexander Taylor – Edinburgh: T & T Clark; New York: Scribner and Welford, [1890?] – (= ser Handbooks for bible classes and private students) – 1mf – 9 – 0-7905-4872-0 – mf#1988-0872 – us ATLA [240]

Church and state – London, England. 1850 – 1r – us UF Libraries [230]

Church and state / Potter, S G – London, England. 1874 – 1r – us UF Libraries [230]

Church and state : their relations historically developed = Staat und kirche / Geffcken, Friedrich Heinrich; ed by Taylor, Edward Fairfax – London: Longmans, Green, 1877 – 3mf – 9 – 0-7905-5696-0 – (in english) – mf#1988-1696 – us ATLA [230]

Church and state : thoughts applicable to present conditions / Ridding, George; ed by Ridding, Laura – London: AR Mowbray, [1912?] – 1mf – 9 – 0-524-06290-0 – mf#1990-5219 – us ATLA [240]

Church and state – Washington. 1948+ [1]; 1971+ [5]; 1976+ [9] – ISSN: 0009-6334 – mf#1540 – us UMI ProQuest [320]

Church and state in america, pt 2 : review of the bishop of london / Colotn, Calvin – London, England. 1834 – 1r – us UF Libraries [230]

Church and state in early maryland / Petrie, George – Baltimore: Johns Hopkins Press 1892 [mf ed 1990] – (= ser Johns hopkins university studies in historical and political science 10/4) – 1mf – 9 – 0-7905-5258-2 – (incl bibl ref) – mf#1988-1258 – us ATLA [230]

Church and state in england before the conquest / Collins, William Edward – London: SPCK 1903 [mf ed 1993] – (= ser Church historical society (series) 79) – 1mf – 9 – 0-524-05493-2 – mf#1990-1488 – us ATLA [230]

Church and state in france, 1300-1907 / Galton, Arthur – London: Edward Arnold, 1907 – 1mf – 9 – 0-7905-4526-8 – (incl bibl ref) – mf#1988-0526 – us ATLA [240]

Church and state in new england / Lauer, Paul Erasmus – Baltimore: Johns Hopkins Press 1892 [mf ed 1990] – (= ser Johns hopkins university studies in historical and political science 10/2-3) – 1mf – 9 – 0-7905-5247-7 – (incl bibl ref) – mf#1988-1247 – us ATLA [230]

Church and state in north carolina / Weeks, Stephen Beauregard – Baltimore: Johns Hopkins Press 1893 [mf ed 1990] – (= ser Johns hopkins university studies in historical and political science 11/5-6) – 1mf – 9 – 0-7905-5259-0 – (incl bibl ref) – mf#1988-1259 – us ATLA [230]

Church and state in scotland : a narrative of the struggle for independence from 1560 to 1843 / Brown, Thomas – new ed. Edinburgh: MacNiven & Wallace 1892 [mf ed 1989] – (= ser The chalmers lectures [1891?]) – 1mf – 9 – 0-7905-4160-2 – (incl bibl ref) – mf#1988-0160 – us ATLA [230]

Church and state in the united states : or, the american idea of religious liberty and its practical effects: with official documents / Schaff, Philip – New York: Charles Scribner, 1888 [mf ed 1986] – 1mf – 9 – 0-8370-9983-8 – (incl bibl ref) – mf#1986-3983 – us ATLA [230]

Church and state in the united states / Schaff, Philip – 1889 – 1 – $50.00 – us Presbyterian [240]

Church and state in the united states : with an appendix on the german population / Thompson, Joseph Parrish – Boston: J R Osgood, 1873 – 1mf – 9 – 0-7905-6369-X – mf#1988-2369 – us ATLA [230]

The church and state responsible to christ / Irving, Edward – 1829 – 1 – $50.00 – us Presbyterian [240]

Church and state two hundred years ago : a history of ecclesiastical affairs in england from 1660 to 1663 / Stoughton, John – London: Jackson, Walford, and Hodder, 1862 – 2mf – 9 – 0-7905-6089-5 – mf#1988-2089 – us ATLA [240]

Church and state under the tudors / Child, Gilbert William – London; New York: Longmans, Green, 1890 – 2mf – 9 – 0-7905-4203-X – (incl bibl ref) – mf#1988-0203 – us ATLA [941]

Church and synagogue libraries – Bryn Mawr. 1986-1996 (1) 1986-1996 (5) 1986-1996 (9) – ISSN: 0009-6342 – mf#15210 – us UMI ProQuest [020]

Church and the age / Garbett, James – Brighton, England. 1851 – 1r – us UF Libraries [230]

The church and the age : an exposition of the catholic church in view of the needs and aspirations of the present age / Hecker, Isaac Thomas – New York: Catholic Book Exchange, 1896, c1887 [mf ed 1986] – 1mf – 9 – 0-8370-8348-6 – (incl bibl ref) – mf#1986-2348 – us ATLA [241]

The church and the age / Inge, William Ralph – London, New York: Longmans, Green, 1912 [mf ed 1990] – 1mf – 9 – 0-7905-7346-6 – mf#1989-0571 – us ATLA [240]

The church and the barbarians : being an outline of the history of the church from a d 461 to a d 1003 / Hutton, William Holden – London: Rivingtons, 1906 [mf ed 1990] – (= ser The church universal 3) – 1mf – 9 – 0-7905-4868-2 – (incl bibl ref) – mf#1988-0868 – us ATLA [240]

The church and the bible / Sparrow-Simpson, William John – London, New York: Longmans, Green, 1897 – 1mf – 9 – 0-7905-0344-1 – mf#1987-0344 – us ATLA [220]

The church and the changing order / Mathews, Shailer – New York: Macmillan, 1907 – 1mf – 9 – 0-7905-8516-2 – mf#1989-1741 – us ATLA [240]

The church and the churches : or, the papacy and the temporal power: an historical and political review = Kirche und kirchen / Doellinger, Johann Joseph Ignaz von – London: Hurst & Blackett 1862 [mf ed 1990] – 2mf – 9 – 0-7905-4625-6 – (trans fr german by william bernard mac cabe; incl bibl ref) – mf#1988-0625 – us ATLA [241]

Church and the citizen / Maitland, Edward – Ramsgate, England. 1872 – 1r – us UF Libraries [230]

The church and the civil law / Howell, Charles Boynton – Detroit, 1886. 58p. LL-712 – 1 – us L of C Photodup [346]

The church and the college : a discourse delivered at the thirteenth anniversary of the society for the promotion of collegiate and theological education at the west, in the first congregational church, bridgeport, ct, nov 11 1856 / Kirk, Edward Norris – Boston: T. R. Marvin, 1856. Beltsville, Md: NCR Corp, 1978 (1mf); Evanston: American Theol Lib Assoc, 1984 (1mf) – 9 – 0-8370-1010-1 – mf#1984-4366 – us ATLA [242]

The church and the divine order / Oman, John – London; New York: Hodder and Stoughton, [1911?] – 1mf – 9 – 0-7905-7537-X – mf#1989-0762 – us ATLA [240]

The church and the eastern empire / Tozer, Henry Fanshawe – New York: A.D.F. Randolph, [1888?] – (= ser Epochs Of Church History) – 1mf – 9 – 0-7905-6132-8 – mf#1988-2132 – us ATLA [240]

The church and the empire : being an outline of the history of the church from a d 1003 to a d 1304 / Medley, D J – New York: Macmillan, 1910 [mf ed 1986] – (= ser The church universal 4) – 1mf – 9 – 0-8370-7719-2 – (incl ind) – mf#1986-1719 – us ATLA [240]

The church and the empires : historical periods / Wilberforce, Henry William – London: Henry S King, 1874 – 1mf – 9 – 0-8370-8237-4 – mf#1986-2237 – us ATLA [240]

The church and the future = L'eglise et l'avenir / Bourdon, Hilaire – Abridged and rearranged. [S.l.: s.n.], 1903 (Edinburgh: Turnbull and Spears) – 1mf – 9 – 0-8370-9446-1 – mf#1986-3446 – us ATLA [240]

The church and the future / Tyrrell, George – London: Priory Press, 1910 [mf ed 1986] – 192p on 1mf – 9 – 0-8370-8951-4 – mf#1986-2951 – us ATLA [241]

The church and the hour : reflections of a socialist churchwoman / Scudder, Vida Dutton – New York: E. P. Dutton, [c1917] Beltsville, Md: NCR Corp, 1978 (2mf); Evanston: American Theol Lib Assoc, 1984 (2mf) – (= ser Women & the church in america) – 9 – 0-8370-0738-0 – mf#1984-2093 – us ATLA [240]

The church and the ideal / Lawrence, William – New York: Church Peace Union, [1916?] – (= ser The Church and International Peace) – 1mf – 9 – 0-7905-9297-5 – mf#1989-2522 – us ATLA [240]

The church and the jew / Gruenstein, Bernard – Sewanee, TN: University Press at the University of the South, [1907] (mf ed 1995) – 1r – 1 – mf#*ZP-1487 – us NY Public [230]

The church and the kingdom / Denney, James – London: Hodder and Stoughton, [19–?] – (= ser Little books on religion) – 1mf – 9 – 0-8370-7211-5 – mf#1986-1211 – us ATLA [240]

The church and the kingdom / Gladden, Washington – New York: Fleming H Revell, c1894 [mf ed 1986] – 1mf – 9 – 0-8370-9782-7 – mf#1986-3782 – us ATLA [210]

The church and the kingdom : a new testament study / Thomas, Jesse Burgess – Louisville: Baptist Book Concern, c1914 – 1mf – 9 – 0-7905-2333-7 – mf#1987-2333 – us ATLA [240]

The church and the labor conflict / Womer, Parley Paul – New York: Macmillan, 1913 – 1mf – 9 – 0-7905-6153-0 – (incl bibl ref) – mf#1988-2153 – us ATLA [240]

Church and the labor movement / Stelzle, Charles – Philadelphia: American Baptist Publ Society, 1910 – 1mf – 9 – 0-7905-6506-4 – mf#1988-2506 – us ATLA [240]

Church and the meeting house – London, England. 1846 – 1r – us UF Libraries [240]

The church and the ministry : a review of the rev. e. hatch's bampton lectures / Gore, Charles – 2nd ed. London: Rivingtons, 1882 – 1mf – 9 – 0-7905-3876-8 – (incl bibl ref) – mf#1989-0369 – us ATLA [240]

The church and the ministry in the early centuries / Lindsay, Thomas Martin – 2nd ed. London: Hodder and Stoughton, 1903 – (= ser Cunningham Lectures) – 1mf – 9 – 0-524-01001-3 – mf#1990-0278 – us ATLA [240]

The church and the nation : charges and addresses / Creighton, Mandell; ed by Creighton, Louise – London, New York: Longmans, Green, 1901 – 1mf – 9 – 0-7905-4552-7 – mf#1988-0552 – us ATLA [240]

The church and the people's play / Atkinson, Henry A – Boston: Pilgrim Press, c1915 – 1mf – 9 – 0-7905-4314-1 – (incl bibl ref) – mf#1988-0314 – us ATLA [240]

The church and the puritans, 1570-1660 / Wakeman, Henry Offley – New York: A.D.F. Randolph, [1894?] – (= ser Epochs Of Church History) – 1mf – 9 – 0-7905-6143-3 – mf#1988-2143 – us ATLA [243]

The church and the rebellion : a consideration of the rebellion against the government of the united states and the agency of the church, north and south, in relation thereto / Stanton, Robert Livingston – New York: Derby & Miller, 1864 – 2mf – 9 – 0-7905-6571-4 – mf#1988-2571 – us ATLA [975]

The church and the social problem : a study in applied christianity / Plantz, Samuel – Cincinnati: Jennings and Graham; New York: Eaton and Mains, c1906 – 1mf – 9 – 0-8370-9728-2 – mf#1986-3728 – us ATLA [240]

The church and the social question / Backus, Edwin Burdette – [Meadville, Pa.], 1912. Chicago: Dep of Photodup, U of Chicago Lib, 1971 (1r); Evanston: American Theol Lib Assoc, 1984 (1r) – 1 – 0-8370-0274-5 – mf#1984-B147 – us ATLA [240]

Church and the world / Noel, Baptist Wriothesley – London, England. 1849? – 1r – us UF Libraries [230]

The church and the world in idea and in history : eight lectures / Hobhouse, Walter – London: Macmillan, 1910 – (= ser Bampton lectures) – 1mf – 9 – 0-7905-3957-8 – (incl bibl ref) – mf#1989-0450 – us ATLA [240]

Church architecture considered / Mant, Richard – Belfast, Northern Ireland. 1843 – 1r – us UF Libraries [720]

Church army gazette, 1888-1914 – 9r – 1 – mf#97166 – uk Microform Academic [240]

Church, Arthur Herbert *see* English earthenware

Church as a national establishment unimpaired as a christian church – London, England. 1836 – 1r – us UF Libraries [240]

Church as an apostle to the heathen : and, a modern christ – London, England. 1873 – 1r – us UF Libraries [240]

The church – as it was, as it is, as it ought to be : a discourse / Clarke, James Freeman – Boston: Greene 1848 [mf ed 1989] – 1mf – 9 – 0-7905-4450-4 – mf#1988-0450 – us ATLA [243]

Church as shareholder : an occasional bulletin / United Church Board for World Ministries – v1 n1-v8 n2 [1974 jan-1981 sum] – 1r – 1 – mf#601817 – us WHS [240]

Church Association for the Advancement of the Interests of Labor *see* Hammer and pen

The church association of the diocese of toronto : instituted 1873, to uphold the principles and doctrines of the protestant church of england, and to counteract the efforts now being made to pervert her teaching / Church of England Diocese of Toronto. Church Association – Toronto: [s.n.], 1875 [mf ed 1983] – 1mf – 9 – 0-665-25151-3 – mf#25151 – cn Canadiana [242]

The church association of the diocese of toronto : instituted in 1873, to uphold the principles and doctrines of the protestant church of england, and to counteract the efforts now being made to pervert her teaching / Church of England Diocese of Toronto. Church Association – Toronto: [s.n.], 1874 [mf ed 1983] – 1mf – 9 – mf#24353 – cn Canadiana [242]

The church at home and abroad – Philadelphia, PA. v1-24. 1887-1898 – 6r – 1 – $300.00 – us Presbyterian [240]

The church at the center / Wilson, Warren Hugh – New York: Missionary Education Movement of the United States and Canada, 1914 – (= ser Library Of Christian Progress) – 1mf – 9 – 0-7905-6218-9 – mf#1988-2218 – us ATLA [240]

Church authority and power in medieval and early modern england : the episcopal registers – 8pt-coll – 111r – 1 – (pt 1: registers of the archbishops of york, 1215-1650 22r c39-24801. pt 2: registers of the bishops of lincoln, 1209-1663 20r c39-24802. pt 3: registers of the bishops of coventry & lichfield, 1295-1632; carlisle, 1292-1656; chester, 1502-1686; and durham, 1311-1683 12r c39-24803. pt 4: registers of the bishops of salisbury, 1297-1689 10r c39-24804. pt 5: registers of the bishops of london, 1304-1660 9r c39-24805. pt 6: registers of christ church cathedral, prior canterbury, 1284-1661 22r c39-24806. pt 7: registers of the bishops of ely, 1337-1619; oxford, 1592-1663; wales, 1389-1705 8r c39-24807. pt 8: registers of the bishops of chichester, 1396-1675; gloucester, 1541-1681; and rochester, 1319-1683 8r c39-24808) – us Primary [240]

The church bell – Round Hill, NB: [s.n, 1899?-19–] – 9 – ISSN: 1190-6642 – mf#P04529 – cn Canadiana [242]

Church bells of england / Walters, H B – London, 1912 – 8mf – 8 – mf#H-1248 – ne IDC [700]

Church Bible and Prayer Book Society *see* Annual report...for the year ending october 31st...

Church book / Evangelical and Reformed Church, Hoisington, KS – 1912-1946 – 1 – us Kansas [240]

Church book – Greenbottom Church of Jesus Christ (Cabell Co), West Virginia, Jan 1836-Dec 1858 – $20.00 – us ABHS [240]

Church book / Trinity Lutheran Church, Lehigh, KS – 1900-1953 – 1 – us Kansas [240]

Church books of ford or cuddington and amersham in the county of... / Ford, Eng (Buckinghamshire) Baptists – London, England. 1912 – 1r – us UF Libraries [240]

Church building : a study of the principles of architecture in their relation to the church / Cram, Ralph Adams – Boston: Small, Maynard, 1901 – 1mf – 9 – 0-7905-4451-2 – mf#1988-0451 – us ATLA [720]

Church building : a study of the principles of architecture in their relation to the church / Cram, Ralph Adams – Boston: Small, Maynard, 1901 – 1mf – us ATLA [720]

Church calendar, clergy list, and general almanack for the diocese of worcester – Birmingham, 1862 [mf ed 1987] – 1r – 1 – (filmed with: das verhaltnis des staates... / maurer, w) – mf#2047 – us Wisconsin U Libr [240]

The church catechism : the christian's manual / Newbolt, William Charles Edmund – London: Longmans, Green 1903 [mf ed 1992] – (= ser The oxford library of practical theology) – 1mf – 9 – 0-524-04774-X – (incl bibl ref) – mf#1991-2160 – us ATLA [242]

The church catechism : its history and contents / Allen, Andrew James Campbell – London, New York: Longmans, Green, 1892 – 1mf – 9 – 0-7905-3628-5 – mf#1989-0121 – us ATLA [240]

The church catechism : with explanations, notes, and proofs from scripture / Stowell, Thomas Alfred – London: James Nisbet, 1894 – (= ser Nisbet's Scripture Hand-Books) – 1mf – 9 – 0-524-07112-8 – mf#1991-2935 – us ATLA [240]

The church cevenant idea : its origin and its development / Burrage, Champlin – Philadelphia: American Baptist Publication Society, 1904 – 1mf – us ATLA [240]

The church chant book : a series of chants adapted to the daily psalter from the book of common prayer / ed by Davies, Charles F – New York: Wm A Pond, c1880 – 2mf – 9 – 0-524-08748-2 – mf#1993-3253 – us ATLA [780]

Church & chapel register indexes – [Cardiff?]: GFHS [1996-] – 9 – uk Glamorgan FHS [929]

Church, Charles Marcus *see* Chapters in the early history of the church of wells, a d 1136-1333

Church chimes – Toronto: [s.n, 1874-18– or 19–] [mf ed v1 n4 dec 1874; v1 n8 apr 1875] – 9 – mf#P04425 – cn Canadiana [241]

Church choirs, and church music : their origin, and usefulness / Pinnock, W H – Cambridge, England. 1866 – 1r – us UF Libraries [780]

The church chronicle – Toronto: H Rowsell, [1863-187-?] – 9 – mf#P06031 – cn Canadiana [242]

The church chronicle extra, toronto, september, 1865 : the clergy commutation fund / Church of England. Church Society of the Diocese of Toronto – Toronto?: H Rowsell, 1865? – 1mf – 9 – mf#32800 – cn Canadiana [242]

Church chronicle for the diocese of montreal – Montreal: J Lovell, 1860-1862 [mf ed v1 n1 may 1860-v2 n13 may 1862] – 9 – ISSN: 1190-6839 – mf#P04210 – cn Canadiana [242]

Church Coalition for Human Rights in the Philippines *see* Philippine witness

Church considered in its relation to the social nature of man / Sadleir, William Digby – Dublin, Ireland. 1852 – 1r – us UF Libraries [230]

Church constitution of the bohemian and moravian brethren : the original latin, with a translation, notes, and introduction = Ratio disciplinae ordinisque ecclesiastici in unitate fratrum bohemorum / Seifferth, Benjamin – London: W Mallalieu, 1866 – 1mf – 9 – 0-524-08221-9 – (in english and latin) – mf#1993-1006 – us ATLA [242]

The church covenant committing : a study in the social dynamics of local church membership committing in times of intrachurch group conflict with documentation from a pastoral experiment / Brownfield, Richard Charles – Princeton, New Jersey, 1976. Chicago: Dep of Photodup, U of Chicago Lib, 1976 (1r); Evanston: American Theol Lib Assoc, 1984 (1r) – 1 – 0-8370-1284-8 – mf#1984-T012 – us ATLA [240]

The church covenant idea / Burrage, Champlin – 1904 – 1 – 8.19 – us Southern Baptist [242]

The church covenant idea : its origin and its development / Burrage, Champlin – Philadelphia: American Baptist Publication Society, 1904 – 1mf – 9 – 0-7905-4107-6 – (incl bibl ref) – mf#1988-0107 – us ATLA [240]

Church defended : in her principle, constitution, and effects / Garbett, John – London, England. 1833 – 1r – us UF Libraries [240]

Church defense : report of a conference on the present dangers of the church / Marshall, Thomas William M – New York: Catholic Publ Society, 1873 – 1mf – 9 – 0-8370-6756-1 – mf#1986-0756 – us ATLA [230]

Church design for congregations : its development and possibilities / Cubitt, James – London: Smith, Elder & Co, 1870 – (= ser 19th c art & architecture) – 2mf – 9 – mf#4.1.78 – uk Chadwyck [720]

A church dictionary / Hook, Walter Farquhar – London: John Murray, 1877 [mf ed 1992] – (= ser Anglican/episcopal coll) – 2mf – 9 – 0-524-05359-6 – mf#1990-5110 – us ATLA [052]

Church difficulties of 1851 / Church Of England – London, England. 1851 – 1r – us UF Libraries [240]

Church difficulties of 1851 / Church Of England – London, England. 1851 – 1r – us UF Libraries [240]

Church discipline / Dover Baptist Association. Virginia – Summary. 1824. 30p – 1 – 5.00 – us Southern Baptist [242]

Church discipline : an ethical study of the church of rome / McCabe, Joseph – London: Duckworth, 1903 – 1mf – us ATLA [240]

Church discipline : an ethical study of the church of rome / McCabe, Joseph – London: Duckworth, 1903 – 1mf – 9 – 0-7905-5486-0 – mf#1988-1486 – us ATLA [240]

Church discipline : in two parts, formative and corrective, in which is developed the true philosophy of religious education / Savage, Eleazer – New York: Sheldon, 1863 – 3mf – 9 – 0-524-07915-3 – mf#1991-3460 – us ATLA [240]

Church discussion, baptists and disciples : the ray and lucas debate / Ray, David Burcham – Cincinnati: Geo E Stevens, 1873 – 2mf – 9 – 0-524-06563-2 – mf#1991-2647 – us ATLA [242]

Church divisions and christianity / Grane, William Leighton – London: Macmillan, 1916 – 1mf – 9 – 0-7905-7744-5 – mf#1989-0969 – us ATLA [240]

Church echoes – Janesville WI. 1989 mar-1905 jan – 1r – 1 – mf#1054486 – us WHS [240]

Church eclectic – v1-41. 1873-1908 [complete] – 19r – 1 – mf#ATLA S0882 – us ATLA [240]

Church, Elihu Dwight *see* A catalogue of books relating to the discovery and early history of north and south america

Church embroidery ancient and modern practically illustrated / Dolby, Anastasia – London 1867 – (= ser 19th c art & architecture) – 3mf – 9 – mf#4.2.326 – uk Chadwyck [740]

Church endowments / Fagan, George Hickson – Leeds, England. 1873 – 1r – us UF Libraries [240]

Church enlargement and church arrangement / Cambridge Camden Society – Cambridge 1843 – (= ser 19th c art & architecture) – 1mf – 9 – mf#4.2.1007 – uk Chadwyck [720]

The church essential to the republic : a sermon in behalf of the american home missionary society. preached in the cities of new-york and brooklyn... / Kirk, Edward Norris – New-York: Printed for the American Home Missionary Society, by Leavitt, Trow, 1848 – 1mf – 9 – 0-524-06633-7 – mf#1991-2688 – us ATLA [240]

Church established in scotland – Edinburgh, Scotland. 1879 – 1r – us UF Libraries [242]

Church establishment / Mcneile, Hugh – London, England. 1837 – 1r – us UF Libraries [240]

Church establishment anti-christian : the house of bondage / Baker, Franklin – London, England. 1832 – 1r – us UF Libraries [240]

Church establishment considered in its relation to the state and th... – London, England. 1837 – 1r – us UF Libraries [240]

Church establishment inconsistent with the spirit of christianity a... / Fox, William Johnson – London, England. 1834? – 1r – us UF Libraries [240]

Church establishments – London, England. 1864 – 1r – us UF Libraries [240]

Church establishments : viewed in relation to their political effect / Stuart, J G – Cupar, Scotland. 1843 – 1r – us UF Libraries [240]

Church establishments considered : especially in reference to the church of england / Ingham, Richard; ed by Green, Samuel Gosnell – London: Elliot Stock, 1875 – 2mf – 9 – 0-524-07315-5 – mf#1991-3030 – us ATLA [241]

Church establishments defended / Brown, Charles J – Glasgow, Scotland. 1833 – 1r – us UF Libraries [240]

Church evangelist – v17 n1 [1895 jun 6]-v17 n44 [1896 apr 2] – 1r – 1 – (cont: church guardian) – mf#1166745 – us WHS [240]

The church evangelist – Toronto: Church of England Pub Co, [1895-189-] – 9 – mf#P04994 – cn Canadiana [242]

Church extension : two sermons / Davis, Emerson – Westfield, MA: Day & Davis, 1856 [mf ed 1992] – (= ser Congregational coll) – 1mf – 9 – 0-524-03259-9 – mf#1990-4662 – us ATLA [242]

A "church farm" in assiniboia, north-west, canada / Anson, Adelbert – S.l: s.n, 1885? – 1mf – 9 – mf#15125 – cn Canadiana [630]

Church federation : inter-church conference on federation, new york, november 15-21, 1905 / ed by Sanford, Elias B – New York: Fleming H. Revell, c1906 – 2mf – us ATLA [240]

Church federation : inter-church conference on federation, new york, november 15-21, 1905 / ed by Sanford, Elias Benjamin – New York: Fleming H. Revell, c1906 – 2mf – 9 – 0-7905-4825-9 – mf#1988-0825 – us ATLA [240]

Church finance / Wood, James – London, England. 1873 – 1r – us UF Libraries [240]

Church finances / Stevens, L C – or God's law. 1849 – 1 – 5.00 – us Southern Baptist [242]

The church for americans / Brown, William Montgomery – 4th rev enl ed. New York: T Whittaker, 1896 [mf ed 1993] – (= ser Anglican/episcopal coll) – 2mf – 9 – 0-524-06943-3 – mf#1990-5307 – us ATLA [242]

Church formation in india : address at foreign missions conference of north america, garden city, new york, january 13 1915 / Fleming, Daniel Johnson – [s.l: s.n, s.n, 1915?] [mf ed 1995] – (= ser Yale coll) – 14p (ill) – 1 – 0-524-10184-1 – mf#1995-1184 – us ATLA [240]

Church glorious before its lord / Wilson, John – Oxford, England. 1844 – 1r – us UF Libraries [240]

Church going : an address at the bi-centennial of the first parish in framingham / Hoar, George Frisbie – Boston: American Unitarian Association, [19–?] – 1mf – 9 – 0-524-08683-4 – mf#1993-3208 – us ATLA [240]

Church guardian – v15 n1-44 [1893 aug 30-1894 jun 27] – 1r – 1 – (cont by: church evangelist) – mf#1166743 – us WHS [071]

The church handbook for teacher training classes / Caley, Llewellyn Neville & Burk, William Herbert – rev and enl ed. Philadelphia: George W Jacobs, c1915 – 2mf – 9 – 0-524-05647-1 – (incl bibl ref) – mf#1991-2316 – us ATLA [240]

Church herald – Grand Rapids. 1973+ (1) 1974+ (5) 1974+ (9) – ISSN: 0009-6393 – mf#8636 – us UMI ProQuest [240]

The church herald – Toronto: Church Print and Pub Co, [1869-1875] – 9 – mf#P04457 – cn Canadiana [242]

A church history : continuation from the council of constantinople, a d 381 / Wordsworth, Christopher – 2nd ed. New York: J Pott, 1892 [mf ed 1992] – 1mf – 9 – 0-524-02937-7 – (incl bibl ref) – mf#1990-0753 – us ATLA [240]

A church history : continuation to the council of chalcedon, a d 451 / Wordsworth, Christopher – 2nd ed. New York: J Pott, 1892 [mf ed 1992] – 1mf – 9 – 0-524-02936-9 – (incl bibl ref and ind) – mf#1990-0752 – us ATLA [240]

A church history : from the council of nicaea to that of constantinople, a d 381 / Wordsworth, Christopher – 3rd ed. New York: J Pott, 1892 [mf ed 1992] – 1mf – 9 – 0-524-02938-5 – (incl bibl ref) – mf#1990-0754 – us ATLA [240]

A church history : to the council of nicaea a d 325 / Wordsworth, Christopher – 4th rev ed. New York: J Pott, 1892 – 2mf – 9 – 0-524-03438-9 – (incl bibl ref) – mf#1990-0992 – us ATLA [240]

Church history / Besse, Henry True – San Jose, CA: HT Besse, c1908 – 1mf – 9 – 0-524-03572-5 – mf#1990-1032 – us ATLA [240]

Church history – Chicago. 1932+ (1) 1970+ (5) 1976+ (9) – ISSN: 0009-6407 – mf#223 – us UMI ProQuest [240]

Church history / Learned, Dwight Whitney & Hayami, T – [Kyoto], 1889 [mf ed 1995] – (= ser Yale coll) – (30)/799p – 1 – 0-524-10097-7 – (in japanese) – mf#1995-1097 – us ATLA [240]

Church history association of india : bulletin – n7-10. 1965-67* – 1r – 1 – mf#ATLA S0714A – us ATLA [240]

Church history for busy people / Klingman, George Adam – Cincinnati, O[hio]: FL Rowe, 1909 – 2mf – 9 – 0-524-02259-3 – mf#1990-4266 – us ATLA [240]

A church history for the use of schools and colleges / Loevgren, Nils – Rock Island, IL: Augustana Book Concern, c1906 [mf ed 1992] – 1mf – 9 – 0-524-03045-6 – (with a ser of biogr by august edman. trans by m wahlstroem and c w foss) – mf#1990-0802 – us ATLA [240]

Church history handbooks / Vedder, Henry Clay – Philadelphia: American Baptist Publication Society, 1909 – 2mf – 9 – 0-7905-8279-1 – mf#1988-6157 – us ATLA [240]

Church history in brief / Moffat, James Clement – Philadelphia: Presbyterian Board of Publication, c1885 – 2mf – 9 – 0-7905-5310-4 – mf#1988-1310 – us ATLA [240]

Church history in queen victoria's reign / Fowler, Montague – London: S.P.C.K.; New York: E. & J.B. Young, 1896 – 1mf – 9 – 0-7905-5037-7 – (incl bibl ref) – mf#1988-1037 – us ATLA [240]

Church history in the modern sunday school / Coleman, Christopher Bush – St Louis, MO: Christian Board of Pub, c1911 [mf ed 1992] – (= ser Front rank teacher training series; Advanced teacher-training course 6; Christian church (disciples of christ) coll) – 1mf – 9 – 0-524-04255-1 – mf#1991-2039 – us ATLA [240]

The church history of ethiopia / Geddes, M – London, 1696 – 6mf – 9 – mf#NE-20231 – ne IDC [960]

The church history of scotland : from the commencement of the christian era to the present time / Cunningham, John – 2nd ed. Edinburgh: J. Thin, 1882 – 3mf – 9 – 0-7905-5527-1 – (incl bibl ref) – mf#1988-1527 – us ATLA [240]

A church history of the first three centuries : from the 30th to the three hundred and twenty-third year of the christian era / Mahan, Milo – New York: D Dana, 1860 [mf ed 1992] – 2mf – 9 – 0-524-03407-9 – (incl bibl ref) – mf#1990-0961 – us ATLA [240]

The church history of the first three centuries = Kirchengeschichte der drei ersten jahrhunderte / Baur, Ferdinand Christian – 3rd ed. London: Williams and Norgate, 1878-1879 – 2mf – 9 – 0-7905-4021-5 – (incl bibl ref. in english) – mf#1988-0021 – us ATLA [240]

Church honesty / Ker, William T – Aberdeen, Scotland. 1861 – 1r – us UF Libraries [240]

The church hymnal / ed by Hutchins, Charles Lewis – rev enl ed. Boston: Parish Choir, 1911 – 10mf – 9 – 0-524-06816-X – mf#1991-2803 – us ATLA [780]

The church hymnary : authorized for use in public worship by the church of scotland, the free church of scotland, the united presbyterian church, the presbyterian church in ireland / ed by Stainer, John – Edinburgh: Henry Frowde, 1898 – 10mf – 9 – 0-524-06667-1 – mf#1991-2722 – us ATLA [780]

Church ideals in education : a pre-convention statement, 1916. a description of the work and aims of the general board of religious education of the protestant episcopal church – [New York?: s.n., 1916?] – 1mf – 9 – 0-524-05250-6 – mf#1991-2242 – us ATLA [242]

The church identified by a reference to the history of its origin, perpetuation, and extension into the united states / Wilson, William Dexter – new rev ed. New York: James Pott, 1889 – 1mf – 9 – 0-524-02239-9 – mf#1990-4250 – us ATLA [240]

The church in a workhouse : a record of effort / Henning, James – London: SPCK, 1897 – 1mf – 9 – 0-524-07245-0 – mf#1991-2986 – us ATLA [240]

Church in america / Coleman, Leighton – New York: J Pott, [1895?] – 1mf – 9 – 0-524-02566-5 – (incl bibl ref) – mf#1990-4378 – us ATLA [240]

The church in america and its baptisms of fire : being an account of the progress of religion in america in the eighteenth and nineteenth centuries as seen in the great revivals in the christian church, and in the growth and work of various religious bodies / Halliday, Samuel Byram & Gregory, Daniel Seeley – New York: Funk & Wagnalls, 1896, c1895 – 2mf – 9 – 0-7905-8105-1 – mf#1988-6067 – us ATLA [242]

The church in corea / Trollope, Mark Napier, Bishop of Korea – London: A R Mowbray, 1915 [mf ed 1995] – (= ser Yale coll) – 132p (ill) – 1 – 0-524-09610-4 – mf#1995-0610 – us ATLA [240]

Church, in england : the pillar and ground of the truth / Snow, Thomas – London, England. 1834 – 1r – us UF Libraries [241]

The church in england : from william 3 to victoria / Hore, Alexander Hugh – Oxford: Parker, 1886 – 3mf – 9 – 0-7905-5156-X – (incl bibl ref) – mf#1988-1156 – us ATLA [240]

Church in fetters / Tillett, Jacob Henry – London, England. 1848 – 1r – us UF Libraries [240]

Church in france / Smith, Richard Travers – London: Wells Gardner, Darton, 1894 – 2mf – 9 – 0-7905-6081-X – mf#1988-2081 – us ATLA [240]

Church in germany / Baring-Gould, S – London: Wells Gardner, Darton, 1891 – 1mf – 9 – 0-7905-4125-4 – mf#1988-0125 – us ATLA [240]

Church in germany / Baring-Gould, Sabine – New York, NY. 1891 – 1r – us UF Libraries [025]

Church in italy / Pennington, Arthur Robert – London: Wells Gardner, Darton, [1893?] – 2mf – 9 – 0-7905-6873-X – mf#1988-2873 – us ATLA [240]

Church in puerto ricos dilemma / International Missionary Council Dept Of Social... – New York, NY. 1942 – 1r – us UF Libraries [972]

The church in relation to sceptics : a conversational guide to evidential work / Harrison, Alexander James – London, New York: Longmans, Green, 1892 [mf ed 1985] – 1mf – 9 – 0-8370-3500-7 – (incl ind) – mf#1985-1500 – us ATLA [210]

The church in roman gaul / Smith, Richard Travers – London: SPCK; New York: E & J B Young [1882?] [mf ed 1990] – (= ser The home library) – 2mf [ill] – 9 – 0-7905-6448-3 – mf#1988-2448 – us ATLA [240]

The church in rome in the first century : an examination of various controverted questions relating to its history, chronology, literature and traditions / Edmundson, George – London; New York: Longmans, Green, 1913 – (= ser Bampton lectures) – 1mf – 9 – 0-7905-1940-2 – (incl bibl ref and indexes) – mf#1987-1940 – us ATLA [240]

Church in scotland – London, England. 1845 – 1r – us UF Libraries [242]

Church in scotland – London, England. 1845 – 1r – us UF Libraries [242]

Church in scotland / Luckock, Herbert Mortimer – London: Wells Gardner, Darton, [1893?] – 1mf – 9 – 0-7905-6932-9 – (incl bibl ref) – mf#1988-2932 – us ATLA [240]

The church in scotland : a history of its antecedents, its conflicts, and its advocates, from the earliest recorded times to the first assembly of the reformed church / Moffat, James Clement – Philadelphia: Presbyterian Board of Publication, c1882 – 2mf – 9 – 0-7905-5666-9 – mf#1988-1666 – us ATLA [240]

Church in spain / Meyrick, Frederick – London: Wells Gardner, Darton, 1892 – 2mf – 9 – 0-524-00771-3 – mf#1990-0203 – us ATLA [240]

The church in spain / Bayle, Constantino & Allison Peers, E – Burgos: Razon y Fe, 1939 – 1 – sp Bibl Santa Ana [240]

The church in the british isles : sketches of its continuous history from the earliest times to the restoration / Doane, William Croswell et al – 4th ed. New York: E & JB Young, 1894 – (= ser Church Club Lectures) – 1mf – 9 – 0-524-05474-6 – mf#1990-5121 – us ATLA [240]

The church in the cherubim : or, the glory of the saints / Tanner, James Gosset – London: Hatchards, Piccadilly, 1875 [mf ed 1993] – 1mf – 9 – 0-524-06344-3 – (incl bibl ref) – mf#1992-0882 – us ATLA [220]

The church in the city / Leete, Frederick DeLand – New York: Abingdon, c1915 – (= ser Constructive Church Series) – 1mf – 9 – 0-7905-5170-5 – (incl bibl ref) – mf#1988-1170 – us ATLA [240]

The church in the confederate states : a history of the protestant episcopal church in the confederate states / Cheshire, Joseph Blount – New York: Longmans, Green, 1912, c1911 – 1mf – 9 – 0-7905-4501-2 – mf#1988-0501 – us ATLA [242]

Church in the country town / Bemies, Charles Otto – Philadelphia: Published for the Social Service Commission of the Northern Baptist Convention [by] American Baptist Publ. Society, 1912 – 1mf – 9 – 0-7905-5923-4 – mf#1988-1923 – us ATLA [240]

The church in the highlands : or, the progress of evangelical religion in gaelic scotland, 563-1843 / Mackay, John – London, New York: Hodder & Stoughton [1914?] [mf ed 1990] – (= ser The chalmers lectures [1914?]) – 1mf – 9 – 0-7905-5253-1 – (incl bibl ref) – mf#1988-1253 – us ATLA [242]

The church in the mirror of history : studies on the progress of christianity = Aus der geschichte des christentums / Sell, Karl – New York: Scribner & Welford [1890?] [mf ed 1990] – 1mf – 9 – 0-7905-6434-3 – (trans fr german by elizabeth stirling) – mf#1988-2434 – us ATLA [240]

The church in the mission field : with supplement, presentation and discussion of the report in the conference on 16th june 1910 – Edinburgh: Publ for the World Missionary Conference by Oliphant, Anderson & Ferrier; New York: Fleming H Revell, [1910?] – 1mf – 9 – 0-8370-6473-2 – (incl indes) – mf#1986-0473 – us ATLA [240]

The church in the nation : pure and apostolical, god's authorized representative / Lay, Henry Champlin – New York: E P Dutton 1885 [mf ed 1991] – (= ser The bishop paddock lectures 1885) – 1mf – 9 – 0-7905-9779-9 – mf#1989-1504 – us ATLA [242]

Church in the netherlands / Ditchfield, P H – London: Wells Gardner, Darton, 1893 – 1mf – 9 – 0-7905-4507-1 – (incl bibl ref) – mf#1988-0507 – us ATLA [240]
Church in the smaller cities / Patterson, Frederic William – Philadelphia: American Baptist Publ. Society, 1911 – 1mf – 9 – 0-7905-5959-5 – mf#1988-1959 – us ATLA [240]
The church in the south american republics 2nd edic. westmenster, 1943 / Ryan, D D Edwin – Madrid: Missionalia Hispanica, 1946 – 1 – sp Bibl Santa Ana [240]
The church in victoria during the episcopate of the right reverend charles perry : first bishop of melbourne, prelate of the order of st michael and st george / Goodman, George – Melbourne: Melville, Mullen and Slade; London: Seeley, 1892 – 2mf – 9 – 0-7905-5886-6 – mf#1988-1886 – us ATLA [240]
Church in wales / Gladstone, William Ewart – London, England. 1871 – 1r – us UF Libraries [242]
Church in wales / Great Britain, Parliament, House Of Commons – London, England. 1892 – 1r – us UF Libraries [242]
Church in wales / Wells, Ca – London, England. 1891 – 1r – us UF Libraries [242]
Church its own enemy / Black, Adam – Edinburgh, Scotland. 1835 – 1r – us UF Libraries [240]
Church League of America *see* News and views digest
Church leases / Grey, William Henry – London, England. 1851 – 1r – us UF Libraries [240]
Church library – 1960-61 – 1 – 9.59 – us Southern Baptist [242]
Church library magazine – Nashville. 1962-1970 (1) – ISSN: 0578-2279 – mf#2179 – us UMI ProQuest [240]
Church life and thought in north africa a.d. 200 / Donaldson, Stuart Alexander – Cambridge: University Press, 1909 – 1mf – 9 – 0-8370-7623-4 – (incl indes) – mf#1986-1623 – us ATLA [240]
Church life in colonial maryland / Gambrall, Theodore Charles – Baltimore: G. Lycett, 1885 – 1mf – 9 – 0-7905-5142-X – (incl bibl ref) – mf#1988-1142 – us ATLA [240]
Church life in scotland / Boyd, Andrew Kennedy Hutchion – Edinburgh, Scotland. 1890 – 1r – us UF Libraries [242]
The church magazine – Montreal: J D Borthwick, [1868] – 9 – ISSN: 1190-6863 – mf#P04201 – cn Canadiana [242]
The church magazine – St. John, NB: W M Wright, [1865-1868] – 9 – mf#P04126 – cn Canadiana [242]
Church management : the clergy journal – Austin. 1924-1992 (1) 1971-1992 (5) 1975-1992 (9) – (cont by: clergy journal) – ISSN: 0009-6431 – mf#210 – us UMI ProQuest [240]
Church manual : designed for the use of baptist churches / Pendleton, James Madison – Philadelphia: American Baptist Pub Soc [1867] [mf ed 1984] – 2mf – 9 – 0-8370-1069-1 – (incl ind) – mf#1984-4403 – us ATLA [242]
The church manual : containing the declaration of faith, rules of order, how to conduct religious meetings, etc / Brumbaugh, Henry Boyer – rev ed. Elgin, Ill: Brethren's Publ House, 1901 – 1mf – 9 – 0-524-02814-1 – mf#1990-4435 – us ATLA [240]
Church, Mary C *see*
– Life and letters of dean church
– Occasional papers
Church matters – London, England. 1842 – 1r – us UF Libraries [240]
Church member *see* Chiao yu [ccm116]
Church member (ccc 116) = Chiao-yu / Chuan, S Peter – Shanghai: Mission Book Co., 1922 – (= ser Chinese christian coll) – 1r – 1 – (in chinese) – mf#1984-B500 – us ATLA [240]
The church member's hand-book : a guide to the doctrines and practice of baptist churches / Crowell, William – Boston: Gould and Lincoln, 1853 – 1mf – 9 – 0-524-02686-6 – mf#1990-4393 – us ATLA [240]
The church member's manual : for the church, the home and the...closet: prepared for the churches of the american madura mission – Madras: American Madura Mission, 1872 [mf ed 1995] – (= ser Yale coll) – xii/126p – 1 – 0-524-09557-4 – (in hindi) – mf#1995-0557 – us ATLA [240]
The church member's manual of ecclesiastical principles, doctrine, and discipline : presenting a systematic view of the structure, polity, doctrines, and practices of christian churches, as taught in the scriptures / Crowell, William – new rev ed. Boston: Gould and Lincoln, 1854 – 1mf – 9 – 0-524-03789-2 – mf#1990-4861 – us ATLA [240]
Church membership : or, the conditions of new testament and methodist church membership examined and compared / Bond, S – Toronto: W Briggs; Montreal: C Coates; Halifax: S Huestis, 1882 [mf ed 1980] – 1mf – 9 – 0-665-07201-5 – mf#07201 – cn Canadiana [242]
Church membership and what it involves : a lecture delivered under the auspices of the auxiliary home mission board of hants county, ns, april 20th, 1880 / Manning, James William – Windsor, NS: publ by members of the Board, 1880 [mf ed 1983] – 1mf – 9 – mf#24487 – cn Canadiana [240]
Church membership in the past and the future / Jellett, John Hewitt – Dublin, Ireland. 1869 – 1r – us UF Libraries [240]
The church memorial : containing important historical facts and reminiscences connected with the associate and associate reformed churches... / Beveridge, Thomas Hanna et al; ed by Harper, Robert D – Columbus, OH: Follett, Foster; Xenia, OH: Fleming and Crawford, 1858 – 1mf – 9 – 0-524-01727-1 – mf#1990-4119 – us ATLA [240]
Church messenger = Tserkovnyi viestnik = cerkovnyj / American Carpatho-Russian Orthodox Greek Catholic Diocese in USA – 1977 sep 25-1981, 1982 jan 10-1984 dec 23, 1985 jan-1988 dec, 1989-90, 1991-93, 1994-97 – 6r – 1 – mf#4019733 – us WHS [243]
The church messenger – v1-21 n9. oct 1875-sep 1896 [complete] – 2r – 1 – (title varies) – mf#ATLA S0097 – us ATLA [242]
The church messenger = Tserkovnyi viestnik – Pemberton, NJ: [American Carpatho-Russian Orthodox Greek Catholic Diocese in USA. nov 10 1946; aug 15 1947-1975 – us CRL [071]
Church messenger for the diocese of qu'apelle – Qu'Apelle Station [Sask]: S John's College, [188- or 189–189- or 19–] – 9 – (cont: our messenger) – mf#P04393 – cn Canadiana [242]
Church minshull, st bartholomew – (= ser Cheshire monumental inscriptions) – 1mf – 9 – £2.50 – mf#387 – uk CheshireFHS [929]
Church minutes / Cullom Association. North Carolina. Warrenton Baptist Church – 1849-68 – 1 – $14.67 – us Southern Baptist [242]
The church miscellany – [Kingston, Ont?: First Congregational Church, Kingston?, 18–?] – 9 – ISSN: 1190-674X – mf#P04327 – cn Canadiana [242]
The church missionary atlas : containing an account of the various countries in which the church missionary / Society Labours, And Of Its Missionary Operations – London: Church Missionary Society, 1896 – 1mf – us ATLA [240]
The church missionary atlas : containing an account of the various countries in which the church missionary society labours, and of its missionary operations – 8th ed. London: Church Missionary Soc 1896 [mf ed 1989] – 1mf [ill] – 9 – 0-7905-4173-4 – mf#1988-0173 – us ATLA [240]
The church missionary gleaner *see* The canadian church missionary gleaner
Church missionary intelligencer, the... 1849-1906 / church missionary society review, 1907-27 : from the archives of the church missionary society, london – v1-78 – 38r – 1 – mf#96567 – uk Microform Academic [240]
Church missionary sociaety jubilee address no 2 / Fox, Henry Watson – London, England. 1848? – 1r – us UF Libraries [240]
Church Missionary Society *see*
– Instructions of the committee to missionaries proceeding to the west africa, india, ceylon, china, and the mediterranean missions
– Mengo-uganda notes, 1900-21
– Proceedings of the church missionary society for africa and the east, 1801-1921
– Taveta chronicle, the... 1895-1901
– Uganda mission, 1915-1934
– West indies mission records of the church missionary society, 1819-61
– Yoruba and northern nigeria missions, 1915-1925, archives
Church missionary society archive, section 1 : east asia missions – 20pt – 1 – (pt1: japan 1869-1949 (incl loochoo naval mission 1843-61) [21r] £1950; pt2: japan 1869-1949 [21r] £1950; pt3: japan 1869-1949 [21r] £1950; pt4: church of england zenana missionary society 1880-1957 [21r] £1950; pts5,6,7,8,9: church of england zenana missionary society [20r £1850, 20r £1850, 21r £1950, 32r £3000, 33r £3100]; pt10: china mission 1834-1914 [30r] £2800; pt11,12: south china mission 1885-1934 [20r £1850, 17r £1600]; pt13,14: chekiang mission 1888-1934 [21r £1950, 19r £1775]; pt15: western china mission 1897-1934 [20r] £1850; pt16: western china mission 1898-1934, and fukien mission 1900-34 [21r] £1950; pt17: fukien mission 1911-34 [18r] £1675; pt18: fukien mission 1900-34, kwangsi-hunan mission 1911-34, china general 1935-51, & south china 1935-51 [24r] £2250; pt19: south china mission 1935-51, chekiang mission 1935-51, west china mission 1935-51, fukien mission 1935-51, & kwangsi-hunan mission 1935-51 [26r] £2450; pt20: east asia general 1935-49, & annual letters for japan, china & canada 1917-49 [26r] £2450; pt21: periodicals for south, central & west china, 1899-1970, & japan, 1905-41, incl papers of fukien conferences, 1906-37 [20r] £1850; with d/g) – uk Matthew [240]
Church missionary society archive, section 2 : missions to women – 5pt – 1 – (pt1: society for promoting female education [fes] in china, india & the east 1834-99 [10r] £925; pt2: india's women & china's daughters 1880-1939, & looking east at india's women & china's daughters 1940-57 [19r] £1775; pt3: homes of the east 1910-48 [incl torchbearer fr 1914], daybreak 1889, 1893-94 & 1906-09, & the indian female evangelist 1872-80 [6r] £565; pt4: indian female evangelist & successors 1881-1956 [covering the indian female evangelist 1881-93, the zenana: or, woman's work in india 1893-1935, the zenana, women's work in india and pakistan 1936-56] from interserve, london [10r] £925; pt5: minutes of the zenana, medical and bible mission 1865-1937, & the annual reports of the indian female normal school & instruction society 1863-79, from interserve, london [13r] £1225; with d/g) – uk Matthew [240]
Church missionary society archive, section 3 : central records – 19pt – 1 – (pt1: annotated register of cms missionaries, history of the cms by eugene stock, & catalogues to the overseas archive, cezms & fes archives [10r] £925; pt2: cms gleaner 1841-1921 (also cms gleaner pictorial album 1888 & cms missionary atlas 1879) [12r] £1125; pt3: cms outlook, 1922-1972 (a cont fr cms gleaner) [10r] £925; pt4: annual letters 1886-1912 [13r] £1225; pt5: cms medical journals: mercy & truth 1897-1921, mission hospital 1922-39, way of healing 1940; medical mission quarterly 1892-96, & preaching and healing 1900-06 [17r] £1600; pt6: cms circular books & letters 1799-1921 [12r] £1125; pt7: cms minutes 1799-1875 [10r] £925; pt8: cms minutes 1837-53 [10r] £925; pt9: cms minutes 1854-76, & indexes to minutes 1799-1878 [12r] £1125; pt10: missionary papers 1816-84, cms monthly paper 1828-29, a quarterly token for juvenile subscribers, 1856-78 & 1888-1917, home gazette, 1905-06, & the cms gazette, 1907-34 [10r] £925; pt11: general review of missions 1919, cont as annual reports 1922-44, & cms historical record 1944-86 [21r] £1950; pt12: cms juvenile instructor 1842-90, children's world 1891-1900, & the round world 1901-58 [25r] £2350; pts13,14: cms collection of lives of missionaries held at the church mission society library [15r £1400, 14r £1400]; pt15: church missionary society record 1830-1875 held at the church mission society library [23r] £2150; pt16: cms awake! – a missionary magazine for general readers 1891-1921 cont as eastward ho! 1922-1940 held at the church mission society library [13r] £1225 [mf ed 2003]; pt17: cms minutes 1876-98 &, indexes to minutes 1875-1907 [24r] £2250 [mf ed 2003/4]; pt18: cms minutes 1898-1949 [27r] £2550; pt19: papers of henry venn (secretary of cms, 1841-1873) & family [19r] £1775; with d/g) – uk Matthew [240]
Church missionary society archive, section 4 : africa missions – 26pt – 1 – (pt1: west africa (sierra leone) 1803-80 [27r] £2550; pt2: west africa (sierra leone) 1820-80 [18r] £1675; pt3: nigeria – yoruba 1844-80 [17r] £1600; pt4: nigeria – yoruba 1844-80 [17r] £1600; pt5: west africa (sierra leone) 1820-80 [17r] £1600; pt6: nigeria – niger 1857-82 [12r] £1125; pt7: sudan 1905-49 [21r] £1950; pt8: nigeria – yoruba 1880-1934 [23r] £2150; pt9: nigeria – yoruba 1880-1934 [23r] £2150; pt10: nigeria – niger 1881-1934 [36r] £3350; pt11: nigeria – niger 1880-1934, & nigeria – northern nigeria 1900-34 [20r] £1850 [mf ed jun 2000]; pt12: west africa [sierra leone] 1881-1934 [26r] £2450 [mf ed sep 2000]; pt13: west africa [sierra leone] 1935-49 & nigeria missions 1935-49 [32r] £3000 [mf ed nov 2000]; pt14: egypt 1889-1934 [17r] £1600 [mf ed may 2001]; pt15: egypt 1889-1949 [14r] £1300; pt16: south africa 1836-1843, kenya 1841-88, & nyanza 1876-82 [16r] £1500; pt17: kenya 1880-1934 [29r] £2700; pt18: kenya 1880-1934 [29r] £2700; pt19: tanganyika 1900-34, nyanza 1880-86, & rwanda 1933-34 [14r] £1300 [mf ed spring 2003]; pt20: uganda 1898-1934 [17r] £1600 [mf ed summer 2003]; pt21: kenya 1935-49 [19r] £1775 [mf ed fall 2003]; pt22: uganda 1898-1934 [22r] £2050 [mf ed spring 2004]; pt23: uganda, tanganyika & rwanda 1935-49 [25r] £2350 [mf ed summer 2004]; pt24: mauritius, madagascar & the seychelles 1856-1929 [20r] £1850; pt25: africa general, 1935-49 [18r] £1675; pt26: africa general, 1935-49 [18r] £1675; with d/g) – uk Matthew [240]
Church missionary society archive, section 5 : missions to the americas – [mf ed Marlborough 1998-99] – 4pt – 1 – (pt1: west indies 1819-61 [20r] £1850; pt2: north west canada 1821-80 [30r] £2800; pt3: north west canada 1881-1930 [33r] £3100; pt4: british columbia 1856-1925 [12r] £1125; with d/g) – uk Matthew [240]
Church missionary society archive, section 6 : missions to india – 4pt – 1 – (pt1: india general 1811-15, and north india mission 1815-81 [21r] £1950; pt2: north india mission 1844-86 [23r] £2150; pt3: india general 1811-15, & south india mission 1815-84 [24r] £2250; pt4: south india mission 1834-80 [24r] £2250; with d/g) – uk Matthew [240]
Church missionary society archive, section 7 : general secretary's papers – 3pt – 1 – (pt1: papers relating to africa, 1847-1950 [18r] £1675; pt2:...1873-1949 [22r] £2050; pt3: papers relating to japan & china, 1874-1952 [13r] £1850; with d/g) – uk Matthew [240]
Church missionary society archive, section 8 : home papers – 1pt – 1 – (pt1: papers of the cms education secretaries, 1910-59 [10r] £825) – uk Matthew [240]
Church Missionary Society. London *see* [Records, 1803?-1914]
Church missionary society review, 1907-27 *see* Church missionary intelligencer, the... 1849-1906 / church missionary society review, 1907-27
Church Missionary Society. South India Mission. Madras *see*
– Correspondence, 1830-1865
– Records, 1834-1860
The church monthly *see* The cathedral monthly
The church monthly and the haldimand deanery magazine – [Dunnville, Ont.?: s.n, 1900] – 9 – mf#P04380 – cn Canadiana [242]
Church music – St. Louis. 1966-1980 (1) 1977-1980 (5) 1977-1980 (9) – ISSN: 0009-6458 – mf#8533 – us UMI ProQuest [780]
Church music, a manuscript of an essay-sermon 14 mar 1875 / Smith, W S D – 1 – 5.00 – us Southern Baptist [242]
Church music in theory and practice in selected baptist churches, an exploratory study / Benson, David P – 1961 – 1 – 5.00 – us Southern Baptist [242]
The church music problem : six essays / Pratt, Waldo Selden – New-York: Century Co, c1887 – 1mf – 9 – 0-524-03047-2 – mf#1990-0804 – us ATLA [780]
Church musician – Nashville. 1962-1997 (1) 1970-1997 (5) 1976-1997 (9) – (cont by: church musician today) – ISSN: 0009-6466 – mf#2176 – us UMI ProQuest [780]
Church musician – v. 1. 1950-61 – 1 – 64.96 – us Southern Baptist [242]
Church musician today – Nashville. 1997+ (1,5,9) – (cont: church musician) – ISSN: 0009-6466 – mf#26780 – us UMI ProQuest [780]
Church never forsaken / O'Sullivan, Mortimer – Dublin, Ireland. 1838 – 1r – us UF Libraries [240]
The church news *see* Miscellaneous newspapers of mesa county
Church nursery guide – 1957-61 – 1 – us Southern Baptist [242]
Church observer – Montreal: printed for P Wilson, [1868?-18– or 19–] – 9 – mf#P04363 – cn Canadiana [242]
The church observer – Springhill, NS: [s.n, 1896?-189- or 19–] [mf ed v3 n38 sep 1897; v3 n41 dec 1897-v4 n2 feb 1898; v4 n4 apr 1898-v4 n12 dec 1898] – 9 – mf#P04277 – cn Canadiana [242]
The church of armenia : her history, doctrine, rule, discipline, liturgy, literature, and existing condition / Ormanean, Maghakia – London: A.R. Mowbray, [1912?] – 1mf – 9 – 0-7905-6005-4 – mf#1988-2005 – us ATLA [240]
Church of christ : what is it? – London, England. 1845 – 1r – us UF Libraries [240]
The church of christ : its life and work: an attempt to trace the work of the church in some of its departments... / Charteris, Archibald Hamilton – London, New York: Macmillan 1905 [mf ed 1989] – (= ser The baird lecture 1887) – 1mf – 9 – 0-7905-1578-4 – (incl bibl ref & ind) – mf#1987-1578 – us ATLA [240]
The church of christ / Phillips, Thomas Wharton – 6th rev ed. New York: Funk & Wagnalls, 1907, c1906 [mf ed 1986] – 1mf – 9 – 0-8370-6077-X – (incl ind) – mf#1986-0077 – us ATLA [240]
The church of christ : a treatise on the nature, powers, ordinances, discipline, and government of the christian chuch / Bannerman, James; ed by Bannerman, David Douglas – Edinburgh: T. & T. Clark, 1868. Beltsville, Md: NCR Corp, 1978 (11mf); Evanston: American Theol Lib Assoc, 1984 (11mf) – 9 – 0-8370-0984-7 – (incl bibl ref and index) – mf#1984-4332 – us ATLA [240]
Church of christ and sunday school extension / Alexander, Disney – London, England. 1873 – 1r – us UF Libraries [240]
The church of christ in corea / Fenwick, Malcolm C – New York: Hodder & Stoughton; George H Doran, [1911] [mf ed 1995] – (= ser Yale coll) – vi/134p (ill) – 1 – 0-524-10191-4 – mf#1995-1191 – us ATLA [240]

The church of christ, in its idea, attributes, and ministry : with a particular reference to the controversy on the subject between romanists and protestants / Litton, Edward Arthur – London: Longman, Brown, Green, and Longmans, 1851 – 2mf – 9 – 0-7905-9410-2 – mf#1989-2635 – us ATLA [240]
The church of christ in japan : a course of lectures / Imbrie, William – Philadelphia: Westminster Press, c1906 – 1mf – 9 – 0-8370-6578-X – mf#1986-0578 – us ATLA [240]
Church of Christ [Temple Lot] *see* Zion's advocate
The church of christ the same forever / McErlane, Daniel – St Louis, MO: B Herder, 1900 – 1mf – 9 – 0-8370-6684-0 – (incl ind) – mf#1986-0684 – us ATLA [240]
The church of cyprus / Duckworth, Henry Thomas Forbes – London: S.P.C.K.; New York: E. & J.B. Young, 1900 – 1mf – 9 – 0-7905-4415-6 – mf#1988-0415 – us ATLA [240]
Church Of England *see*
– Church
– Church difficulties of 1851
– Collects for sundays and holydays throughout the year
Church of England *see* Records of the church of england during the commonwealth period
Church of england : the nursing mother of her people / Sutcliffe, W – Blackburn, England. 1844 – 1r – us UF Libraries [241]
Church of england : a witness and keeper of the catholic tradition / Churton, Edward – Durham, England. 1836 – 1r – us UF Libraries [241]
The church of england : an appeal to facts and principles / Newbolt, William Charles Edmund & Stone, Drawell – London: Longmans, Green, 1903 [mf ed 1992] – (= ser [Oxford library of practical theology]; Anglican/episcopal coll) – 1mf – 9 – 0-524-03265-3 – mf#1990-4668 – us ATLA [241]
The church of england : a history for the people / Spence-Jones, Henry Donald Maurice – London: Cassell, 1897-1898 – 5mf – 9 – 0-7905-8050-0 – mf#1988-6031 – us ATLA [241]
The church of england / Watson, Edward William – London: Williams and Norgate; New York: H Holt, 1914 – (= ser Home university library of modern knowledge) – 1mf – 9 – 0-7905-6908-6 – mf#1988-2908 – us ATLA [241]
Church of england a blessing in the land / Langley, Thomas – London, England. 1840 – 1r – us UF Libraries [241]
Church of england admonished by the examples of former times / Goodwin, Harvey – Cambridge, England. 1854 – 1r – us UF Libraries [241]
Church of england and dissent / Cawood, John – London, England. 1831 – 1r – us UF Libraries [241]
The church of england and episcopacy / Mason, Arthur James – Cambridge: University Press; New York: Putnam [distributor], 1914 – 2mf – 9 – 0-7905-4894-1 – mf#1988-0894 – us ATLA [240]
The church of england and recent religious thought / Whittuck, Charles Augustus – London; New York: Macmillan, 1893 – 1mf – 9 – 0-7905-8629-0 – mf#1989-1854 – us ATLA [241]
Church of england and ritualism / Gladstone, W E – London, England. 1875? – 1r – us UF Libraries [241]
Church of england and the church of rome / Garbett, James – London, England. 1851 – 1r – us UF Libraries [230]
Church of england and the education of the people from the earliest... / Wells, Charles Arthur – London, England. 1891 – 1r – us UF Libraries [241]
Church Of England Archdeaconry Of Dorset *see* Charge
Church Of England Archdeaconry Of Lewes *see* Charge to the clergy of the archdeaconry of lewes
Church of england chronicle – Sydney – 1r – A$74.93 vesicular A$80.43 silver – at Pascoe [079]
Church of England. Church Society of the Diocese of Toronto *see* The church chronicle extra, toronto, september, 1865
Church of england common prayer book / Smedley, John – Lea Mills, England. 1858 – 1r – us UF Libraries [241]
Church of England Diocese of Algoma. Bishop (1882-1897: Sullivan) *see* "Restoration of church unity"
Church of England. Diocese of Athabasca. Synod *see* Journal of proceedings of the... meeting of the synod of the diocese of athabasca
Church Of England Diocese Of Bath And Wells *see*
– Charge delivered to the clergy and churchwardens of the diocese...
– Charge delivered to the clergy and churchwardens of the diocese of...
– Charge delivered to the clergy and churchwardens of the diocese of ba...
Church Of England Diocese Of Bristol *see* Charge delivered
Church Of England Diocese Of Canterbury *see* Primate and church defence
Church Of England Diocese Of Chester *see*
– Charge delivered to the clergy of the diocese of chester
– Charge delivered to the clergy of the diocese of chester at the tri...
Church Of England. Diocese Of Chester *see* Letter to the clergy of the diocese of chester
Church Of England Diocese Of Durham *see*
– Charge delivered to the clergy of the diocese of durham
– Grounds of union between the churches of england and of rome consid...
– Grounds on which the church of england separated from the church of...
Church Of England Diocese Of Exeter *see*
– Charge delivered to the clergy of the diocese of exeter
– Charge delivered to the clergy of the diocese of exeter at the trie...
Church Of England. Diocese Of Exeter *see*
– Letter to the churchwardens of the parish of brampford speke
– Letter to the clergy of the diocese of exeter
Church of England. Diocese of Fredericton *see* Chronicle of the diocese of fredericton
Church Of England Diocese Of Glousecter And Bristol *see*
– Charge delivered to the clergy of the diocese of gloucester and bri
– Charge delivered to the clergy of the diocese of gloucester and bri...
Church Of England Diocese Of Lincoln *see*
– Charge delivered at the triennial visitation of john, lord bishop o...
– Charge delivered to the clergy of the diocese of lincoln
– Charge to the clergy of the diocese of lincoln
Church Of England Diocese Of Llandaff *see*
– Charge delivered to the clergy of the diocese of landaff in june 17...
– Charge delivered to the clergy of the diocese of landaff, june 1791
Church Of England Diocese Of London *see*
– Charge delivered to the clergy of the diocese of london
– Charge delivered to the clergy of the diocese of london in the year...
– Charge of the bishop of london to the clergy of his diocese
– Charge to the clergy of the diocese of london
Church of England Diocese of Montreal. Synod *see* Constitution, rules and regulations and canons of the synod of the diocese of montreal
Church of England. Diocese of Niagara *see* Act of incorporation, declaration, constitution, rules, canons and by-laws of the synod of the diocese of niagara
Church Of England Diocese Of Norwich *see* Charge intended to have been delivered to the clergy of norwich
Church of England. Diocese of Nova Scotia *see*
– A charge delivered to the clergy
– A charge delivered to the clergy at the visitation held in the cathedral church of st luke
– Constitution, canons, rules and regulations of the diocesan synod of nova scotia
Church of England. Diocese of Ontario *see*
– Canons of the synod of the diocese of ontario
– Draft of the revised canons of the diocese of ontario
Church Of England Diocese Of Ripon *see* Charge addressed to the clergy of the diocese of ripon
Church Of England Diocese Of Saint David's *see*
– Charge to the clergy of the diocese of st david's
Church Of England Diocese Of Salisbury *see*
– Charge delivered to the clergy of the diocese of salisbury
– Charge to the clergy and churchwardens of the diocese of salisbury
Church of England. Diocese of Toronto *see*
– Canons, by-laws and resolutions adopted by the synod of the diocese of toronto
– Churchwardens' manual
Church of England Diocese of Toronto. Church Association *see*
– The church association of the diocese of toronto
Church of England Diocese of Toronto. Synod *see* Constitution, canons, by-laws and resolutions of the incorporated synod of the diocese of toronto
Church Of England Diocese Of Winchester *see*
– Charge delivered to the clergy of the diocese of winchester
Church of England in Canada *see* Shall we change the communion service?
Church of england in her liturgy or prayer book – London, England. 1865 – 1r – us UF Libraries [241]
The church of england in nova scotia and the tory clergy of the revolution / Eaton, Arthur Wentworth Hamilton – New York: T. Whittaker, 1891 – 1mf – 9 – 0-7905-4417-2 – mf#1988-0417 – us ATLA [241]
The church of england in the eighteenth century / Plummer, Alfred – London: Methuen, 1910 – (= ser Handbooks of english church history) – 1mf – 9 – 0-7905-5552-2 – mf#1988-1552 – us ATLA [241]
Church of england leaves her children free to whom to open their gr... / Pusey, E B – Oxford, England. 1850 – 1r – us UF Libraries [241]
Church of england not of roman catholic origin / Wells, Charles Arthur – Windermere, England. 1884? – 1r – us UF Libraries [241]
Church of england past and present / Goodwin, Harvey – London, England. 1881 – 1r – us UF Libraries [241]
Church of england right / Bardsley, Joseph – London, England. 1864 – 1r – us UF Libraries [241]
Church of england schoolmaster / Freeman, John – Lynn, England. 1856? – 1r – us UF Libraries [241]
Church of england teaching / Carmichael, James – Montreal: W Drysdale, 1890 [mf ed 1980] – 1mf – 9 – 0-665-00494-X – mf#00494 – cn Canadiana [242]
Church of england's commission to her priests considered / Haddon, T C – Cambridge, England. 1846 – 1r – us UF Libraries [241]
Church of england's portrait / Cole, Henry – Cambridge, England. 1847 – 1r – us UF Libraries [241]
Church of God *see*
– Church of god
– Pctii
Church of god – 1944-1967 dec 1, 1968 jan-1976 dec – 2r – 1 – mf#630977 – us WHS [240]
Church of god : its constitution, government, and laws – London, England. 1814 – 1r – us UF Libraries [240]
Church of god : yearbook / Church of God – Anderson, IN. 1902-90 [complete] – 9r – 1 – mf#ATLA S0198 – us ATLA [240]
The church of god : a catechism for families, sunday schools, and churches / Ross, Abel Hastings – Boston: Congregational Pub Society, c1881 – 1mf – 9 – 0-524-06777-5 – mf#1991-2784 – us ATLA [240]
Church of God [7th Day] *see* Wand
The church of god and the bishops : an essay suggested by the convocation of the vatican council = Kirche gottes und die bischoefe / Liaeno, Heinrich St A von – London: Rivingtons; New York: Pott and Amery, 1870 – 1mf – 9 – 0-8370-8444-X – (in english) – mf#1986-2444 – us ATLA [240]
The church of god and what and whence is it? / Peebles, Isaac Lockhart – Nashville, Tenn: Publishing House of the ME Church, South, 1914, c1913 – 1mf – 9 – 0-524-00078-6 – mf#1989-2778 – us ATLA [240]
Church of god confessing her guilt and depravity / Milner, Joseph – London, England. 18– – 1r – us UF Libraries [240]
Church of god evangel – Cleveland. 1910-1996 (1) 1969-1989 (5) 1977-1989 (9) – ISSN: 0745-6778 – mf#3208 – us UMI ProQuest [240]
Church of god evangel – 1981 mar 9-1982 jan, 1983-1984 feb 27, 1984 mar 12-dec 24, 1985 jan 14-dec 23, 1986-1987 mar, 1987 apr-1989 jun – 6r – 1 – (cont: evening light and church of god evangel) – mf#699887 – us WHS [240]
Church of God in Christ *see*
– Bible band topics for weekly meetings
– Cogic challenger
– Evangelist speaks
– Good news
– Good shepherd
– National informer
– Pentecostal interpreter
– Power for living
– Sunshine band topics
– Voice of missions
– Whole truth
– Y p w w quarterly topics for weekly meetings
Church of God in Christ Jesus *see* Rehoboth beacon
Church of God in Christ [Los Angeles CA] *see* International outlook
Church of God in Christ of Western New York *see* In touch
Church of God of the Mountain Assembly *see* Gospel herald
Church Of Ireland *see*
– Charge delivered to the clergy of the dioceses of dublin and glande
– Charge to the clergy of the dioceses of dublin and kildare
Church of ireland / Olden, Thomas – 2nd ed. London: Wells Gardner, Darton, 1895 – 2mf – 9 – 0-524-00774-8 – mf#1990-0206 – us ATLA [240]
Church of ireland : reasons for dissenting from the legislation of the general synod / Meade, Joseph Fulton – Dublin, 1875 – (= ser 19th c ireland) – 1mf – 9 – mf#1.1.2631 – uk Chadwyck [241]
Church of ireland and the reformation / Olden, Thomas – Limerick, Ireland. 1895 – 1r – us UF Libraries [241]
Church of ireland defended / Massingham, John Derren – London, England. 1868 – 1r – us UF Libraries [241]
Church Of Ireland Diocese Of Dublin *see* Charge delivered to the clergy of the diocese of dublin and glandal
Church Of Ireland Diocese Of Dublin Archbishop *see* Infant-baptism considered
Church of ireland gazette *see* Irish ecclesiastical gazette
Church Of Ireland General Convention *see* Statutes passed at the first session of the general convention, 187...
Church Of Ireland United Diocese Of Dublin, Glendalough, And Kildare *see*
– Charge delivered to the clergy of the dioceses of dublin, glandelag
– Charge delivered to the clergy of the dioceses of dublin, glendelag
Church of Jesus Christ *see* Torch
Church of Jesus Christ Christian, Aryan Nations *see* Way
Church of Jesus Christ of Latter Day Saints *see*
– Items on pacific islands from reports of annual and semi-annual conferences
– Legends, journals, diaries, correspondence etc re various pacific island missions
– Manuscript mission histories
Church of Jesus Christ of Latter-Day Saints *see*
– Deseret news
– Success quarterly
Church of Jesus Christ of Latter-Day Saints et al *see* Improvement era
Church of Nature *see* Exegesis
Church of old england / Breen, John Dunstan – London, England. 1886 – 1r – us UF Libraries [240]
The church of old england : devoted to the interests of the church in canada, the advancement of education and temperance – Montreal: J P McMillin, 1866-[1867?] – 9 – mf#P04134 – cn Canadiana [242]
The church of our fathers : as seen in st. osmund's rite for the cathedral of salisbury / Rock, Daniel; ed by Hart, George Waldegrave & Frere, Walter Howard – new ed. London: J Hodges. 4v. 1903-04 – 4mf – 9 – 0-7905-8069-1 – (incl bibl ref) – mf#1988-6050 – us ATLA [240]
Church of Our Lord Jesus Christ of the Apostolic Faith *see* Educational leader
Church of rome / Alford, Charles Richard – London, England. 1851 – 1r – us UF Libraries [241]
The church of rome : her present moral theology, scriptural instruction, and canon law / M'Ghee, R J – London: Partridge and Oakey, 1852 – 1mf – 9 – 0-8370-8036-3 – (incl ind) – mf#1986-2036 – us ATLA [240]
Church of rome brought to the test of the epistle to the romans / Brown, Charles J – Edinburgh, Scotland. no date – 1r – us UF Libraries [241]
Church of rome guilty of idolatry : in the worship offered to the vi... – London, England. no date – 1r – us UF Libraries [241]
The church of sancta sophia constantinople : a study of byzantine building / Lethaby, W R & Swainson, H – London, 1894 – €17.00 – ne Slangenburg [720]
Church of Scientology International *see* Freedom
Church Of Scotland *see*
– Report of the committee of the general assembly for increasing the...
– Report of the debate in the general assembly of the church of scotl...
Church of scotland / Eden, Robert – Edinburgh, Scotland. 1876 – 1r – us UF Libraries [242]
Church of scotland : the poor man's church / Collins, William – Glasgow?, Scotland. 18– – 1r – us UF Libraries [242]
Church of scotland / Scott-Moncrieff, W – London, England. 1880 – 1r – us UF Libraries [242]
The church of scotland : her divisions and her re-unions / McCrie, Charles Greig – Edinburgh: Macniven & Wallace, 1901 – 1mf – 9 – 0-7905-5067-9 – mf#1988-1067 – us ATLA [242]
Church of scotland and its assailants / Fraser, William R – Montrose, Scotland. 1885 – 1r – us UF Libraries [242]
Church of scotland and the clerical scandals in old greyfriars' chu... / Free Lance – Edinburgh, Scotland. 1871 – 1r – us UF Libraries [242]
Church of scotland and the free church / Macgeorge, Andrew – Glasgow, Scotland. 1870 – 1r – us UF Libraries [242]

Church of Scotland. Committee on Public Worship and Aids to Devotion *see* Ordinal and service book

Church of scotland crisis 1843 and 1874, and the duke of argyll / Innes, Alexander Taylor – Edinburgh, Scotland. 1874 – 1r – us UF Libraries [242]

Church of scotland, endowment scheme – Edinburgh, Scotland. 1860 – 1r – us UF Libraries [242]

Church Of Scotland General Assembly *see*
- Pastoral admonition by the general assembly, to the people of scotl...
- Second report of the general assembly's committee on the endowment of... chapels of ease

Church of Scotland. General Assembly. Committee on Colonial Churches *see*
- Report...28th may 1838, and deliverance of the assembly
- Report...30th may 1836

Church Of Scotland General Assembly Commmittee On Church Extension *see* Fourth report of the committee of the general assembly

Church Of Scotland General Assembly Special Commission *see* Minute of the general assembly's special commission

The church of scotland in the thirteenth century : the life and times of david de berham of st. andrews, bishop a.d. 1239 to 1253: with list of churches dedicated by him, and dates / Lockhart, William – Edinburgh: W. Blackwood, 1892 – 1mf – us ATLA [242]

The church of scotland in the thirteenth century : the life and times of david de bernham of st. andrews, bishop a.d. 1239 to 1253 / Lockhart, William – 2nd ed. Edinburgh: W Blackwood, 1892 – 1mf – 9 – 0-7905-6538-2 – mf#1988-2538 – us ATLA [242]

Church of Scotland. Jewish Mission Committee *see* Report on jewish missions

Church Of Scotland Ladies' Association For Foreign Missions, Incl *see* Questions submitted by norman macleod

Church of scotland missionary archive : from the national library of scotland – (= ser Missions to india and china, 1829-1933) – 1pt – 1 – (pt1: missions to india & china, 1829-1933 [7r] £675; with d/g) – uk Matthew [242]

Church of scotland not erastian – Glasgow, Scotland. 1874 – 1r – us UF Libraries [242]

The church of scotland, past and present : its history, its relation to the law and the state, its doctrine, ritual, discipline, and patrimony / Campbell, James et al; ed by Story, Robert Herbert – London: William Mackenzie, [1890?] – 4mf – 9 – 0-524-02454-5 – mf#1990-4313 – us ATLA [242]

Church of scotland's endowment / McLean, T A & Brymner, Douglas – Ottawa?: s.n, 188-? [mf ed 1980] – 1mf – 9 – 0-665-09637-2 – mf#09637 – cn Canadiana [242]

Church of scotland's india mission / Duff, Alexander – Edinburgh, Scotland. 1835 – 1r – us UF Libraries [242]

Church of scripture and the church of the disruption / Buchanan, Robert – Glasgow, Scotland. 1859 – 1r – us UF Libraries [240]

The church of st mary the virgin oxford / Jackson, Thomas Graham. baronet – Oxford 1897 – (= ser 19th c art & architecture) – 4mf – 9 – mf#4.2.1558 – uk Chadwyck [720]

The church of sweden and the anglican communion / Williams, Gershom Mott – Milwaukee: Young Churchman, 1910 – 1mf – 9 – 0-524-00806-X – mf#1990-0238 – us ATLA [240]

Church of the Advocate [Philadelphia PA] *see* North philly free press

Church of the Apocalypse *see* Eye of the beast

Church of the apostles / Kip, William Ingraham – New York, NY. 1877 – 1r – us UF Libraries [025]

The church of the apostles : being an outline of the history of the church of the apostolic age / Ragg, Lonsdale – New York: Macmillan, 1909 – (= ser The Church Universal) – 1mf – 9 – 0-8370-4824-9 – (incl ind) – mf#1985-2824 – us ATLA [220]

The church of the apostles : an historical inquiry / Capes, John Moore – London: Kegan Paul, Trench, 1886 – 1mf – 9 – 0-8370-6569-0 – mf#1986-0569 – us ATLA [240]

The church of the apostles / Kip, William Ingraham – New York: D. Appleton, 1877 – 1mf – 9 – 0-7905-4985-9 – mf#1988-0985 – us ATLA [240]

The church of the bible, or, scripture testimonies to catholic doctrines & catholic principles / Oakeley, Frederick – London: Charles Dolman, 1857 – 1mf – 9 – 0-8370-8285-4 – mf#1986-2285 – us ATLA [241]

Church of the brethren : directory – 1973-81 [complete] – 2r – 1 – mf#ATLA S0886 – us ATLA [242]

Church of the brethren : statistics – 1973-80 [complete] – 1r – 1 – mf#ATLA S0891 – us ATLA [242]

Church of the brethren : yearbook [1918] – 1918-72 [complete] – 4r – 1 – mf#ATLA S0901 – us ATLA [242]

Church of the brethren : yearbook [1982] – 1982-1989 [complete] – 2r – 1 – mf#ATLA S0902 – us ATLA [242]

Church of the brethren meeting. full report of the proceedings of the brethren's annual meeting – Elgin IL: Brethren Publ House 1908 [mf ed 2007] – 1v on 1r – 1 – (cont by: church of the brethren. meeting. full report of the proceedings of the annual meeting of the church of the brethren [1909-18] 1r [1095]; church of the brethren. meeting. full report of the proceedings of the annual meeting of the church of the brethren [1919-27] 1r [1096]; church of the brethren. conference. report of the proceedings of the...annual conference, church of the brethren [1928-30] 1r [1097]) – mf1094 – us ATLA [242]

The church of the disciples in boston : a sermon on the principles and methods of the church of the disciples in boston / Clarke, James Freeman – Boston: G H Ellis, 1909. Beltsville, Md: NCR Corp, 1978 (1mf); Evanston: American Theol Lib Assoc, 1984 (1mf) – 9 – 0-8370-1081-0 – mf#1984-4440 – us ATLA [240]

The church of the early fathers : external history / Plummer, Alfred – London: Longmans, Green, 1903 – (= ser Epochs of Church History) – 1mf – 9 – 0-524-02706-4 – (incl bibl ref) – mf#1990-0687 – us ATLA [240]

Church of the epiphany parish magazine – Parkdale [Ont: s.n, 189-?-19–] [mf ed v5 n10 jan 1895; v7 n4 jul 1897] – 9 – mf#P06036 – cn Canadiana [242]

The church of the fathers : being an outline of the history of the church a.d. 98 to a.d. 461 / Pullan, Leighton – 3rd ed. London: Rivingtons, 1909 – (= ser The Church Universal) – 2mf – 9 – 0-524-03418-4 – (incl bibl ref) – mf#1990-0972 – us ATLA [240]

The church of the first three centuries : or, notices of the lives and opinions of the early fathers. with special reference to the doctrine of the trinity... / Lamson, Alvan – London: British and Foreign Unitarian Association, 1875 – 2mf – 9 – 0-524-04379-5 – (incl bibl ref) – mf#1991-2083 – us ATLA [240]

Church of the future / Allon, Henry – London, England. 1881 – 1r – us UF Libraries [240]

The church of the future : its catholicity, its conflict with the atheist, its conflict with the deist, its conflict with the rationalist, its dogmatic teaching, practical counsels for its work, its cathedrals, appendices / Tait, Archibald Campbell – New York: Macmillan, 1881 – 1mf – 9 – 0-7905-7211-7 – mf#1988-3211 – us ATLA [241]

Church of the Heavenly Rest [New York NY] *see* Newsletter

Church of the Holy Apostles [Oneida WI] *see* Oneida

The church of the living god; also, the swiss and belgian confessions and expositions of the faith : containing a distinct delineation of each and of all the veritable doctrines of the glorious gospel of the blessed god / Jones, Owen – London: Caryl Book Society, 1865 – 1mf – 9 – 0-8370-9631-6 – mf#1986-3631 – us ATLA [240]

The church of the living god, and other sermons / Chapin, Edwin Hubbell – New York: J Miller, 1881 – 1mf – 9 – 0-524-08355-X – mf#1993-3055 – us ATLA [240]

Church of the Lord Jesus Christ of the Apostolic Faith *see* Guiding light

Church of the Lutheran Brethren et al *see* Faith and fellowship

Church of the Lutheran Confession *see*
- Journal of theology
- Lutheran spokesman

Church of the Nazarene *see*
- Herald of holiness
- Journal of the general assembly

The church of the open country : a study of the church for the working farmer / Wilson, Warren Hugh – New York: Literature Dept, Presbyterian Home Missions c1911 – (= ser Forward mission study courses) – 1mf – 9 – 0-7905-6097-6 – (incl bibl ref) – mf#1988-2097 – us ATLA [240]

Church of the People *see* Sea-turtle and the shark

Church Of The Province Of Central Africa *see* Buka re munamato wevese

Church of the redeemer parish magazine – Toronto: [s.n, 1891?-18– or 19–] [mf ed v1 n3 jan 1892-v1 n4 feb 1892; v1 n10 aug 1892; v3 n4 feb 1894; v4 n12 oct 1895; v6 n4 apr 1897] – 9 – mf#P04397 – cn Canadiana [242]

The church of the sixth century : six chapters in ecclesiastical history / Hutton, William Holden – London; New York: Longmans, Green, 1897 – (= ser Birkbeck lectures) – 1mf – 9 – 0-7905-6233-2 – (incl bibl ref) – mf#1988-2233 – us ATLA [240]

The church of the sub-apostolic age : its life, worship, and organization, in the light of "the teaching of the twelve apostles" / Heron, James – London: Hodder and Stoughton, 1888 – 1mf – 9 – 0-7905-4754-6 – mf#1988-0754 – us ATLA [240]

Church of the United Brethren in Christ *see* United brethren

The church of the west in the middle ages / Workman, Herbert Brook – [2nd ed] London: Charles H Kelly [mf ed 1986] – (= ser Books for bible students) – 2v on 2mf – 9 – 0-8370-7758-3 – (incl bibl & ind) – mf#1986-1758 – us ATLA [240]

Church on a rock / Macfarlan, D – Paisley, Scotland. 1843 – 1r – us UF Libraries [240]

Church order / Murray, John Walton – Dublin, Ireland. 1869 – 1r – us UF Libraries [240]

Church organization and methods – 1917 – 1 – 5.00 – us Southern Baptist [242]

Church parties – London, England. 1854 – 1r – us UF Libraries [240]

Church past and present / Woodford, James Russell – London, England. 1852 – 1r – us UF Libraries [240]

The church, past and present : a review of its history / ed by Gwatkin, Henry Melvill – London: J Nisbet, 1900 – 1mf – 9 – 0-7905-5335-X – mf#1988-1335 – us ATLA [240]

Church pastorals : hymns and tunes for public and social worship / ed by Adams, Nehemiah – Boston: Ticknor & Fields, 1864 [mf ed 1984] – 6mf – 9 – 0-8370-0746-1 – (incl ind) – mf#1984-6242 – us ATLA [780]

Church patient in her mode of dealing with controversies / Haddan, Arthur West – Oxford, England. 1851 – 1r – us UF Libraries [240]

Church, Pharcellus *see*
- Religious dissensions
- Seed-truths

The church polity of the pilgrims : a sermon / Wellman, Joshua Wyman – Boston: Congregational Board of Publication, [1857?] – 1mf – 9 – 0-524-01539-2 – mf#1990-0445 – us ATLA [243]

Church praise / Fraser, Duncan – Edinburgh, Scotland. 1898 – 1r – us UF Libraries [240]

"Church principles" of nice, rome, and oxford... : compared with the christian principles of the new testament, on baptismal regeneration, lay-baptism, the ancient mode of baptism, &c &c – 2nd rev enl ed. London, England: Francis Baisler 1842 – 1r – 1 – us UF Libraries [242]

Church principles of the new testament / Godkin, James – London, England. 1845 – 1r – us UF Libraries [225]

Church problems : a view of modern anglicanism / ed by Henson, Hensley – London: J Murray, 1900 – 2mf – 9 – 0-7905-5997-8 – (incl bibl ref) – mf#1988-1997 – us ATLA [241]

Church psalmody : or, hymns for public worship: selected from dr watt's psalms and hymns and the congregational hymn book / Atkinson, T [comp] – Quebec?: s.n, 1845 (Quebec: G Stanley) – 6mf – 9 – (incl ind) – mf#33280 – cn Canadiana [242]

Church quarterly review – London. 1875-1907 (1) – ISSN: 0269-4034 – mf#2874 – us UMI ProQuest [240]

Church quarterly review – London: SPCK, 1907-68 [mf ed 2001] – (= ser Christianity's encounter with world religions, 1850-1950) – 22r – 1 – mf#2001-s181 – us ATLA [242]

Church question no 3 : the headship – Ayr?, Scotland. 18– – 1r – us UF Libraries [240]

Church, R W *see* On some influences of christianity upon national character

Church rate opposition / Harvey, Frederick Burn – London, England. 1867 – 1r – us UF Libraries [240]

Church rates – London, England. 1837 – 1r – us UF Libraries [240]

Church rates : neither antiscriptural nor unjust / Pretyman, John Radclyffe – Aylesbury, England. no date – 1r – us UF Libraries [240]

Church rates / Nicholl, John Iltid – London, England. 1837 – 1r – us UF Libraries [240]

Church record : the monthly organ of the anglican church in british columbia – New Westminster, BC: H Morey, [1897-1897 or 1898] – 9 – (cont by: the church record for diocese of new westminster) – mf#P04656 – cn Canadiana [242]

Church record – Philadelphia. 1822-1823 (1) – mf#4435 – us UMI ProQuest [240]

The church record for diocese of new westminster – New Westminster [BC]: H Morey, [1897 or 1898-189- or 19–] – 9 – mf#P04657 – cn Canadiana [242]

Church record, ms 1217 / Chippewa Lake. Ohio. First Presbyterian Church of Lafayette – 1853-83 – 1r – 1 – us Western Res [240]

Church records / First Presbyterian Church, Topeka, KS – 1962-1984, Records of the church and the Synod of Kansas – 1 – us Kansas [240]

Church records / Garfield Heights. Ohio. St. John Evangelical Lutheran Church – 1854-1953 – 1r – 1 – us Western Res [240]

Church records / Grand River. Ohio. Grand River Baptist Association – 1817-42, 1853-71 – 1r – 1 – us Western Res [240]

Church records / St. Paul's Church, Geary County, KS – 1870-1915 – 1 – us Kansas [240]

Church records / United Emmanuel Lutheran Church (ALC), Milberger, KS – 1883-1973, Records of the church and its predecessors – 1 – us Kansas [240]

Church records, ms 40 / Randolph. Ohio. First Congregational Church – 1846-99 – 1r – 1 – us Western Res [240]

Church records, ms 241 / Cleveland. Ohio. Presbytery. The Women's Foreign Mission Society – 1872-1914 – 1r – 1 – us Western Res [240]

Church records, ms 384 / Muskingham. Ohio. Presbytery – 1861-87 – 1r – 1 – us Western Res [240]

Church records, ms 415 / Cleveland. Ohio. Woodland Ave. Methodist-Episcopal Church – 1874-86 – 1r – 1 – us Western Res [240]

Church records, ms 421 / Franklin. Ohio. First Congregational Church – 1819-98 – 1r – 1 – us Western Res [240]

Church records, ms 542 / Chardon. Ohio. Regular Baptist Church – 1831-1905 – 1r – 1 – us Western Res [240]

Church records, ms 642 / Jefferson. Ohio. First Baptist Church – 1811-1921 – 1r – 1 – us Western Res [240]

Church records, ms 647 / Cleveland. Ohio. First Regular Baptist Church – 1820-92; 1901-16 – 1r – 1 – us Western Res [240]

Church records, ms 654 / Mentor Willoughby. Ohio. Baptist Church. (Ohio Plains Conference) – 1836-48 – 1r – 1 – us Western Res [240]

Church records, ms 668 / Dover. Ohio. Baptist Church – 1836-56 – 1r – 1 – us Western Res [240]

Church records, ms 669 / Chester. Ohio. Free Will Baptist Church – 1863-1904 – 1r – 1 – us Western Res [240]

Church records, ms 788 / Elyria. Ohio. First Baptist Church, Women's Home Mission Society – 1887-1900 – 1r – 1 – us Western Res [240]

Church records, ms 1204 / Montville. Connecticut. Baptist Church – 1749-79; 1779-1801; 1807-27 – 1r – 1 – us Western Res [240]

Church records, ms 1468 / Parma. Ohio. First Congregational Church – 1835-74 – 1r – 1 – us Western Res [240]

Church records, ms 1509 / New Hope. Ohio. Second Creek New Hope Baptist Church – 1836-81 – 1r – 1 – us Western Res [240]

Church records, ms 1512 / Warren County. Ohio. Providence Baptist Church – 1820-46 – 1r – 1 – us Western Res [240]

Church records, ms 1528 / East Cleveland. Ohio. First Presbyterian Church – 1807-1911 – 1r – 1 – us Western Res [240]

Church records, ms 1559 / Upper Sandusky. Ohio. First Universalist Church – 1870-1912 – 1r – 1 – us Western Res [240]

Church records, ms 1585 / Jefferson. Ohio. Bethel Union Baptist Church – 1829-87 – 1r – 1 – us Western Res [240]

Church records, ms 1797 / Bradford. New Hampshire. Christian Church – 1829-45 – 1r – 1 – us Western Res [240]

Church records, ms 2041 / Bellevue. Ohio. Methodist Episcopal Church – 1852-86 – 1r – 1 – us Western Res [240]

Church records, ms 2087 / Garrettsville. Ohio. Baptist Church – 1808-60 – 1r – 1 – us Western Res [240]

Church records, ms 2125 / Farmington. Ohio. United Presbyterian and Congregational Church – 1817-66 – 1r – 1 – us Western Res [240]

Church records, ms 2335 / Cleveland. Ohio. Third Baptist Church – 1852-67; 1880-1900 – 1r – 1 – us Western Res [240]

Church records, ms 2822 / Canton. OH. (Elkton Circuit). Methodist-Episcopal Church – 1864-1940 – 1r – 1 – us Western Res [240]

Church records, ms 2843 / Garrettsville. Ohio. Church of Christ – 1889-1902 – 1r – 1 – us Western Res [240]

Church records, ms 3066 / Cleveland. Ohio. St. John's Episcopal Church – 1835-71 – 1r – 1 – us Western Res [240]

Church records, ms 3168 / Brecksville. Ohio. Brecksville Congregational Church – 1816-1947 – 1r – 1 – us Western Res [240]

Church records, ms 3190 / Streetsboro. Ohio. Congregational Church – 1833-81 – 1r – 1 – us Western Res [240]

Church records, ms 3324 / Cleveland. Ohio. Bethany Presbyterian Church – 1889-1917 – 1r – 1 – us Western Res [240]

Church recreation – 1960-61 – 1 – 9.45 – us Southern Baptist [242]

Church recreation magazine – Nashville. 1962-1995 (1) 1971-1995 (5) 1977-1995 (9) – ISSN: 0162-4652 – mf#2177 – us UMI ProQuest [240]

Church reform in spain and portugal : a short history of the reformed episcopal churches of spain and portugal, from 1868 to the present time / Noyes, Henry Edward – London: Cassell, 1897 – 1mf – 9 – 0-7905-6713-X – mf#1988-2713 – us ATLA [241]
Church register / Ogallah. Kansas. Swedish Evangelical Lutheran Church – 1900-45 – 1 – us Kansas [978]
Church register / St. Mark's Episcopal Church, Medicine Lodge, KS – 1890-1981, The canonical church register (ledger) (1890-1981), 1904-1981 – 1 – us Kansas [240]
Church restoration / Grimthorpe, Edmund Beckett – London, England. 1880 – 1r – us UF Libraries [240]
Church restoration / Grimthorpe, Edmund Beckett. 1st Baron – London 1880 – (= ser 19th c art & architecture) – 1mf – 9 – mf#4.2.1318 – uk Chadwyck [720]
Church reunion : discussed on the basis of the lambeth propositions of 1888 – New York: Church Review, 1890 – 1mf – 9 – 0-524-03270-X – mf#1990-0881 – us ATLA [240]
Church review – New York. 1848-1891 (1) – mf#5283 – us UMI ProQuest [240]
The church review – Lunenburg, NS: G Haslam, [1891?-1893?] – 9 – mf#P04868 – cn Canadiana [242]
The church revival : thoughts thereon and reminiscences / Baring-Gould, Sabine – London: Methuen, 1914 – 2mf – 9 – 0-7905-5448-8 – mf#1988-1448 – us ATLA [240]
Church, Richard William *see*
– Essays and reviews
– Life and letters of dean church
– Occasional papers
– On some influences of christianity upon national character
– The oxford movement
– The sacred poetry of early religions
Church scholiast / Bishop Welles Brotherhood – 1885 oct-1888 feb – 1r – 1 – (cont: nashotah scholiast; church militant) – mf#1054489 – us WHS [230]
The church school / Athearn, Walter Scott – Boston: Pilgrim Press, c1914 – 1mf – 9 – 0-524-06007-X – (incl bibl ref) – mf#1991-2367 – us ATLA [240]
The church school hymnal : being one hundred hymns chosen from the new hymnal... – New York: HW Gray...c1916 [mf ed 1993] – (= ser Anglican/episcopal coll) – 2mf – 9 – 0-524-06615-9 – mf#1991-2670 – us ATLA [242]
The church seasons / Grant, Alexander Henley – London: J Hogg, [1869?] – 2mf – 9 – 0-524-08367-3 – mf#1993-3067 – us ATLA [240]
Church sentinel – Sydney – 1r – A$33.40 vesicular A$38.90 silver – at Pascoe [079]
Church services and service-books before the reformation / Swete, Henry Barclay – London: SPCK; New York: E & J B Young 1896 [mf ed 1990] – 1mf [ill] – 9 – 0-7905-6734-2 – mf#1988-2734 – us ATLA [241]
Church Society of the Archdeaconry of New Brunswick *see*
– Abstract of the proceedings of the church society of the archdeaconry of new brunswick
– Fifth report of the proceedings of the church...
– Fourth report of the proceedings of the church...
– Second report of the proceedings of the church society of the archdeaconry of new brunswick
– Third report of the proceedings of the church...
Church song : for the uses of the house of god / Stryker, Melancthon Woolsey – New York: Biglow & Main 1889 [mf ed 1993] – 5mf – 9 – 0-524-06668-X – mf#1991-2723 – us ATLA [242]
Church song : a repertory of music for the use of english evangelical lutheran congregations / Seiss, Joseph Augustus et al – new rev enl ed. Philadelphia: General Council Publ Bd 1908 [mf ed 1992] – 4mf – 9 – 0-524-05572-6 – mf#1991-2306 – us ATLA [242]
Church standard – Sydney – 1r – A$92.49 vesicular A$97.99 silver – at Pascoe [079]
The church standard – Toronto: [s.n, 1868?-18–] – 9 – mf#P04443 – cn Canadiana [241]
Church, state, and dissent – London, England. 18– – 1r – us UF Libraries [240]
Church, state and politics in sixteenth and seventeenth century england : the most important volumes selected from the tanner manuscripts in the bodleian library, oxford – 85r coll – 1 – (pt 1: ...after the civil war 1648-99 17r c39-16801. pt 2:...in 16th and 17th century england 1570-1647 25r c39-16802. pt 3:...in 17th century england 1600-1700 22r c39-16803. pt 4:..in england 1550-1700 21r c39-16804 with printed guide) – mf#C39-16800 – us Primary [941]
Church stretton advertiser and visitors' list – [West Midlands] Shropshire 7 jul 1898-4 nov 1938 [mf ed 2002] – 33r – 1 – (missing: 1913) – uk Newsplan [072]
The church teacher's manual of christian instruction : being the church catechism expanded and explained in question and answer for the use of clergymen, parents, and teachers / Sadler, Michael Ferrebee – 12th ed. London: G Bell, 1890 – 1mf – 9 – 0-524-05711-7 – mf#1991-2325 – us ATLA [240]
The church, the churches, and the sacraments / Beet, Joseph Agar – London: Hodder & Stoughton 1907 [mf ed 1989] – 1mf – 9 – 0-7905-0796-X – (incl ind) – mf#1987-0796 – us ATLA [240]
The church, the people, and the age / Anderson, Robert et al; ed by Scott, Robert & Gilmore, George William – New York: Funk & Wagnalls, 1914 [mf ed 1990] – 2mf – 9 – 0-7905-7024-6 – mf#1988-3024 – us ATLA [240]
The church, the state, and the poor : a series of historical sketches / Chadwick, William Edward – London: R Scott, 1914 – 1mf – 9 – 0-7905-4198-X – (incl bibl ref) – mf#1988-0198 – us ATLA [360]
Church theological review – Pelham Manor. 1961-1970 [1,5,9] – mf#2022 – us UMI ProQuest [240]
Church times / Episcopal Church – v1-6 n12 [1890 sep-1896 aug] – 1r – 1 – (cont by: milwaukee churchman) – mf#1224686 – us WHS [242]
Church times – London. 1975+ (1) – ISSN: 0009-658X – mf#10182 – us UMI ProQuest [240]
Church times – London. -w. 1863-92. (21 reels) – 1 – uk British Libr Newspaper [072]
Church times : a weekly journal of religious news – New York, Utica, NY. v1-6. 1940-46 [complete] – 2r – 1 – mf#ATLA P0003 – us ATLA [240]
The church times – Halifax, NS: W Gossip, [1848-1858] – 9 – mf#P04967 – cn Canadiana [242]
Church training – Nashville. 1962-1989 (1) 1971-1989 (5) 1974-1989 (9) – (cont by: discipleship training) – ISSN: 0162-4601 – mf#2456 – us UMI ProQuest [240]
The church treasury of history, custom, folk-lore, etc / Tyack, George Smith et al; ed by Andrews, William – London: W Andrews, 1898 – 1mf – 9 – 0-524-03032-4 – mf#1990-0789 – us ATLA [941]
The church under queen elizabeth : an historical sketch / Lee, Frederick George – New and rev. ed. London: W.H. Allen, 1892 – 1mf – 9 – 0-7905-5609-X – (incl bibl ref) – mf#1988-1609 – us ATLA [240]
Church union as affected by the question of valid orders : from a presbyterian point of view / Fotheringham, Thomas Francis – St John, NB: E G Nelson, [1908?] – 1mf – 9 – 0-665-86542-2 – mf#86542 – cn Canadiana [242]
Church union in scotland – Edinburgh, Scotland. 18– – 1r – us UF Libraries [242]
Church unity : an address...philadelphia, pa, jan 26th 1893... / Brown, Francis – New York: s.n. [1893?] [mf ed 1992] – 1mf – 9 – 0-524-02974-1 – mf#1990-0761 – us ATLA [240]
Church unity and a new name / Slattery, Charles Lewis – [s.l: s.n.] 1913 [mf ed 1992] – (= ser Prayer book papers 10) – 1mf – 9 – 0-524-03023-5 – mf#1990-4545 – us ATLA [242]
The church universal / Adams, Clayton (Mrs) – London: E W Allen, 1886 – (= ser 19th c women writers) – 1mf – 9 – mf#5.1.10 – uk Chadwyck [420]
The church universal : a series of discourses on the true comprehension of the church, as exhibited mainly in the holy scriptures and subordinately in the standards of the protestant episcopal church / Stone, John Seely – New York: Houel & Macoy, Printers, 1846 – 1mf – 9 – 0-7905-6629-X – mf#1988-2629 – us ATLA [242]
Church university of upper canada : pastoral letter from the lord bishop of toronto: proceedings of the church university board: list of subscribers etc / United Church of England and Ireland. Diocese of Toronto, Bishop (1839-67: Strachan) – Toronto: printed by A F Plees, 1851 [mf ed 1984] – 1mf – 9 – 0-665-22269-6 – mf#22269 – cn Canadiana [377]
Church vestments / Darby, William Arthur – London, England. 1866 – 1r – us UF Libraries [240]
Church vestments : their origin, use, and ornament...illustrated / Dolby, Anastasia – London 1868 – (= ser 19th c art & architecture) – 3mf – 9 – mf#4.2.327 – uk Chadwyck [740]
Church village, salem welsh baptist; & upper church village, bryntirion calvinistic methodist, monumental inscriptions – 1mf – 9 – £1.25 – uk Glamorgan FHS [929]
Church visible, and the church invisible / Carpenter, Henry – London, England. 1845? – 1r – us UF Libraries [240]
Church, W S *see* Church's digest of cases in vols 25-48 of the american state reports
Church watchman – Springfield/Ashtabula, OH. Ohio Free Communion Baptist Association (Free Will Baptist General Conference). 1892-93, 1895, 1897-98. Incomplete – 1 – us ABHS [242]
Church windows : a series of designs / Evans, Sebastian – Birmingham 1862 – (= ser 19th c art & architecture) – 2mf – 9 – mf#4.2.1050 – uk Chadwyck [740]
Church work – Dorchester, NB: J D H Browne, [1876?-19–?] – 9 – mf#P04286 – cn Canadiana [242]
Church work in british columbia : being a memoir of the episcopate of acton windeyer sillitoe...first bishop of new westminster / Gowen, Herbert Henry – London; New York: Longmans, Green, 1899 [mf ed 1981] – 4mf – 9 – mf#15031 – cn Canadiana [242]
Church world – 1987 jan-sep, 1987 oct-1988 jun, 1988 jul 21-1989 mar 30, 1989 apr-dec, 1990 jan-sep – 5r – 1 – mf#1311644 – us WHS [071]
Church year and kalendar / Dowden, John – Cambridge: University Press; New York: Putnam [distributor], 1910 – (= ser Cambridge handbooks of liturgical study) – 1mf – 9 – 0-7905-4459-8 – (incl bibl ref) – mf#1988-0459 – us ATLA [240]
Churches : a blessing or a curse / Wilberforce, Samuel – London, England. 1844 – 1r – us UF Libraries [240]
Churches : historical (manatee county) / Liddle, Carl – s.l, s.l? 1936 – 1r – us UF Libraries [978]
Churches and chapels / Chalmers, Thomas – Glasgow, Scotland. 1834 – 1r – us UF Libraries [720]
Churches and church workers in fiji / Ross, Charles Stuart – Geelong: H Thacker 1909 [mf ed 1991] – 1mf [ill] – 9 – 0-524-00648-2 – mf#1990-0148 – us ATLA [240]
The churches and educated men : a study of the relation of the church to makers and leaders of public opinion / Hardy, Edwin Noah – Boston: Pilgrim Press c1904 [mf ed 1990] – 1mf – 9 – 0-7905-5336-8 – mf#1988-1336 – us ATLA [230]
Churches and education – Glasgow, Scotland. 1870 – 1r – us UF Libraries [240]
The churches and modern thought : an inquiry into the grounds of unbelief and an appeal for candour / Vivian, Philip – [3rd ed] London: Watts 1911 [mf ed 1991] – 1mf – 9 – 0-7905-8618-5 – (incl bibl ref) – mf#1989-1843 – us ATLA [240]
The churches and monasteries of egypt and some neighboring countries / Abu Salih – Oxford: Clarendon Press 1895 [mf ed 1986] – 1mf [ill] – 9 – 0-8370-8000-2 – (trans fr arabic by b t a evetts, added notes by alfred j butler; incl bibl ref & ind) – mf#1986-2000 – us ATLA [243]
The churches and monasteries of egypt and some neighbouring countries : attributed to ab- salih, the armenian / Evetts, B T A – Oxford, 1895 – 6mf – 9 – (anecdota oxoniensia. text,documents and extracts chiefly fr mss in the bodleian and other oxford libraries. semitic series. pt7)) – mf#AR-1877 – ne IDC [243]
The churches and monasteries of egypt and some neighbouring countries : attributed to abu salih, the armenian / Abu Salih, the Armenian; ed by Evetts, B T A & Butler, A J – Oxford. v I-P.7. 1895 – (= ser Anecdota oxoniensia. semitic series) – €35.00 – ne Slangenburg [720]
The churches and sects of the united states : containing a brief account of the origin, history, doctrines... / Gorrie, Peter Douglass – New York: L Colby 1850 [mf ed 1990] – 1mf – 9 – 0-7905-4912-3 – mf#1988-0912 – us ATLA [243]
The churches and the wage earners : a study of the cause and cure of their separation / Thompson, Clarence Bertrand – New York: Scribner 1909 [mf ed 1990] – 1mf – 9 – 0-7905-6124-7 – (incl bibl ref) – mf#1988-2124 – us ATLA [301]
Churches at bosra and samaria-sebaste / Crowfoot, J W – BSA, 1937 – 9 – $10.00 – us IRC [240]
Churches at jerash / Crowfoot, J W – BSA, 1931 – 9 – $10.00 – us IRC [240]
The churches in britain before a d 1000 / Plummer, Alfred – London: R Scott 1911-12 [mf ed 1990] – (= ser Library of historic theology) – 2v on 2mf – 9 – 0-7905-5731-2 – (incl bibl ref) – mf#1988-1731 – us ATLA [240]
Churches in the modern state / Figgis, John Neville – 2nd ed. London; New York: Longmans, Green 1914 [mf ed 1990] – 1mf – 9 – 0-7905-4516-0 – mf#1988-0516 – us ATLA [230]
The churches of asia : a methodical sketch of the second century / Cunningham, William – London: Macmillan 1880 [mf ed 1992] – 1mf [ill] – 9 – 0-524-02851-6 – (incl bibl ref) – mf#1990-0708 – us ATLA [240]
Churches of cambridgeshire and the isle of ely / Cambridge Camden Society – Cambridge: T Stevenson; London...Oxford...7pt. 1843, 1844 – (= ser 19th c art & architecture) – 2mf – 9 – mf#4.1.171 – uk Chadwyck [720]
Churches of christ : a historical, biographical, and pictorial history of churches of christ in the united states, australasia, england and canada / Briney, John Benton et al; ed by Brown, John Thomas – Louisville KY: J P Morton 1904 [mf ed 1990] – 2mf – 9 – 0-7905-5451-8 – mf#1988-1451 – us ATLA [240]
Churches of Christ Mission. New Hebrides *see* Record of births at nduindui hospital, aoba, new hebrides
Churches of christendom : lectures, critical and historical / Bray, Alfred James – Montreal: Milton League, 1877 [mf ed 1979] – 2mf – 9 – 0-665-00252-1 – mf#00252 – cn Canadiana [240]
The churches of london / Godwin, George – London 1839 – (= ser 19th c art & architecture) – 8mf – 9 – mf#4.1.239 – uk Chadwyck [720]
The churches outside the church / Coleman, George William – Philadelphia: publ...by American Baptist Publ Soc 1910 [mf ed 1990] – 1mf – 9 – 0-7905-5930-7 – mf#1988-1930 – us ATLA [230]
The churches separated from rome = Eglises separees / Duchesne, Louis – London: Kegan Paul, Trench, Truebner 1907 [mf ed 1989] – (= ser The international catholic library 9) – 1mf – 9 – 0-7905-4412-1 – (trans fr french by arnold harris mathew) – mf#1988-0412 – us ATLA [241]
Churchill, A *see* A collection of voyages and travels...
Churchill, Asa Gildersleeve *see* Poetical directory of the town of lindsay and business men of the surrounding country
Churchill at war : the prime minister's office papers : prem 3 and prem 4 form the public record office (pro), london – 1940-45 (mf ed 1998) – 291r in 11 units – 1 – us Primary [940]
Churchill, Charles Henry *see*
– The druzes and the maronites under the turkish rule from 1840 to 1860
– Mount lebanon
Churchill county standard *see* [Fallon-] churchill standard
Churchill drive baptist church. shelby, north carolina : church records – 1950-63 – 1 – 8.37 – us Southern Baptist [242]
Churchill, Edward Perry *see* Oyster and the oyster industry of the atlantic and...
Churchill, J *see* A collection of voyages and travels...
Churchill, James *see* Royal tomb re-opened
Churchill, Lord Randolph Henry Spencer *see* The irish land purchase bill
[Churchill-] news – NV. 31 mar 1888 [wkly] – 1r – 1 – $60.00 – mf#U04479 – us Library Micro [071]
Churchill, Randolph Henry Spencer *see* Men, mines and animals in south africa
Churchill, William *see* Weather words of polynesia
Churching of women – London, England. no date – 1r – us UF Libraries [305]
The church-kingdom : lectures on congregationalism / Ross, A Hastings – Boston: Congregational Sunday-school and Publ. Society, c1887 – 1mf – us ATLA [242]
The church-kingdom : lectures on congregationalism. delivered on the southworth foundation in the andover theological seminary, 1882-86 / Ross, Abel Hastings – Boston: Congregational Sunday-school and Pub Society, c1887 – 1mf – 9 – 0-7905-6492-0 – mf#1988-2492 – us ATLA [242]
Church-life? : or sect-life? / Martineau, James – London, England. 1859 – 1r – us UF Libraries [240]
Churchman – St Petersburg. 1969-1985 (1) 1971-1985 (5) 1976-1985 (9) – (cont by: churchman's human quest) – ISSN: 0009-6628 – mf#3364 – us UMI ProQuest [240]
Churchman – London, UK. v62-64. 1948-50 [complete] – Inquire – 1 – ISSN: 0009-661X – mf#ATLA 1993-S507 – us ATLA [240]
Churchman – Watford. 1975+ (1,5,9) – ISSN: 0009-661X – mf#10183 – us UMI ProQuest [240]
The churchman – New York. jan 1893-dec 1909; jul 1910-jun 1911; jul-dec 1914 [wkly] – 37r – 1 – uk British Libr Newspaper [240]
Churchman, Philip H *see* Byron and espronceda
The churchman's friend : for the diffusion of information relative to the united church of england and ireland her doctrine and her ordinances – Windsor, CW [Ont: s.n, 1855-1857?] – 9 – mf#P04423 – cn Canadiana [242]
Churchman's human quest – 1985 dec/1986 jan-1989 sep/oct – 1r – 1 – (cont: churchman; cont by: human quest) – mf#1586182 – us WHS [071]

Churchman's human quest – St Petersburg. 1985-1989 (1,5,9) – (cont: churchman. cont by: human quest) – ISSN: 0897-8786 – mf#3364,01 – us UMI ProQuest [240]

Churchman's human quest – St Petersburg. 1995-1996 (1,5,9) – (cont: human quest. cont by: human quest) – ISSN: 1089-5035 – mf#3364,03 – us UMI ProQuest [240]

The churchman's life of wesley / Urlin, Richard Denny – new rev corr ed. London: SPCK [188-?] [mf ed 1992] – (= ser The home library) – 1mf – 9 – 0-524-04544-5 – (incl bibl ref) – mf#1990-5051 – us ATLA [242]

Churchman's magazine – Middletown. 1804-1827 (1) – mf#4436 – us UMI ProQuest [240]

The churchman's magazine and monthly review – Hamilton, Ont: T and R White, 1869-[1871?] – 9 – ISSN: 1182-736X – mf#P04131 – cn Canadiana [242]

Churchman's manual – London, England. 1838 – 1r – us UF Libraries [240]

The churchman's reasons for his faith and practice / Richardson, Nathaniel Smith – 2nd ed. New York: Pott & Amery, 1863 – 1mf – 9 – 0-8370-8612-4 – (incl bibl ref) – mf#1986-2612 – us ATLA [240]

The churchman's record of the colored council of the protestant episcopal church of the diocese of georgia – Savannah GA: [s.n.] [mf ed 2004] – 1r – 1 – (mf: oct-nov 1919, apr 1920) – mf#2004-s005 – us ATLA [242]

Churchman's repository for the eastern diocese – Newburyport. 1820-1820 (1) – mf#4437 – us UMI ProQuest [240]

Churchmanship and character : three years' teaching in birmingham cathedral / Carnegie, William Hartley – New York: E P Dutton, 1909 – 1mf – 9 – 0-8370-3189-3 – (incl ind) – mf#1985-1189 – us ATLA [240]

The churchmanship of john wesley and the relations of wesleyan methodism to the church of england / Rigg, James Harrison – new rev ed. London: Wesleyan-Methodist Book-Room, [1886?] – 1mf – 9 – 0-7905-6721-0 – (incl bibl ref) – mf#1988-2721 – us ATLA [242]

The churchmember's guide and complete church manual / Essig, Montgomery Ford – Nashville, TN: Southwestern Co, c1907 [mf ed 1989] – 1mf – 9 – 0-7905-4470-9 – mf#1988-0470 – us ATLA [242]

Church-members' handbook of theology / Robertson, Norvell – 1874. 328p – 1 – us Southern Baptist [242]

Churchmen and dissenters / Gray, John Hamilton – Chesterfield, England. 1831 – 1r – us UF Libraries [240]

The church's attitude towards truth / Usher, Edward Preston – Grafton, Mass, USA: EP Usher, 1907 – 1mf – 9 – 0-7905-8948-6 – mf#1989-2173 – us ATLA [240]

The church's best state : or, constant revivals of religion / Harkey, Simeon Walcher – 2nd ed Baltimore: Publication Rooms, 1843 – 1mf – 9 – 0-7905-7235-4 – mf#1988-3235 – us ATLA [240]

The church's certain faith gray / Gray, George Zabriskie – Boston: Houghton, Mifflin; Cambridge: Riverside Press, 1890 – 1mf – 9 – 0-8370-4805-2 – mf#1985-2805 – us ATLA [240]

Church's creed or the crown's creed? / Ffoulkes, Edmund Salisbury – London, England. 1868? – 1r – us UF Libraries [240]

The church's creed, or the crown's creed? : a letter to the most rev. archbishop manning, etc / ed by Foulkes, Edmund Salisbury – New York: Pott & Amery, 1869 – 1mf – 9 – 0-8370-8422-9 – (incl bibl ref) – mf#1986-2422 – us ATLA [241]

Church's digest of cases in vols 25-48 of the american state reports / Church, W S – San Francisco: Bancroft-Whitney, 1896 (all publ) – (= ser American state reports. trinity series, pt 3) – 14mf – 9 – $21.00 – mf#LLMC 78-038H – us LLMC [348]

Church's duty at the present time / Nicolson, J – Dundee, Scotland. 1885 – 1r – us UF Libraries [240]

Church's hope / Dow, William – Edinburgh, Scotland. 1869 – 1r – us UF Libraries [240]

The church's ministry : speeches and discussions together with the papers published for the consideration of the congress / Pan-Anglican Congress 1908, Section C – London: Society for Promoting Christian Knowledge; New York: E S Gorham, 1908 – 2mf – 9 – 0-8370-9093-8 – mf#1986-3093 – us ATLA [240]

The church's ministry of grace : lectures... 1892 / Clark, William et al – New York: E & J B Young 1893, c1892 [mf ed 1991] – (= ser The church club lectures 1892) – 1mf – 9 – 0-7905-9167-7 – mf#1989-2392 – us ATLA [242]

The church's mission as to war and peace / Remensnyder, Junius Benjamin – New York: Church Peace Union, [1916?] – (= ser The Church and International Peace) – 1mf – 9 – 0-7905-9454-4 – mf#1989-2679 – us ATLA [240]

The church's missions in christendom : speeches and discussions together with the papers published for the consideration of the congress / Pan-Anglican Congress 1908, Section E – London: Society for Promoting Christian Knowledge; New York: E S Gorham, 1908 – 2mf – 9 – 0-8370-9095-4 – (includes bibliographies) – mf#1986-3095 – us ATLA [240]

The church's missions in non-christian lands : speeches and discussions together with the papers published for the consideration of the congress / Pan-Anglican Congress 1908, Section D – London: Society for Promoting Christian Knowledge; New York: E S Gorham, 1908 – 2mf – 9 – 0-8370-9094-6 – mf#1986-3094 – us ATLA [240]

Church's musical visitor, 1878-1883 – With which is incorporated Root's Song Messenger. 1862 – 1 – us Southern Baptist [242]

Church's office towards the young / Armstrong, John – Oxford, England. 1853 – 1r – us UF Libraries [240]

The church's one foundation : christ and recent criticism / Nicoll, William Robertson – New York: A.C. Armstrong, 1902 – 1mf – 9 – 0-8370-6234-9 – (incl bibl ref) – mf#1986-0234 – us ATLA [240]

Church's quarrel expoused – 1715. and A vindication of the government of the New England churches. 1717 – 1 – 5.00 – us Southern Baptist [242]

The church's task under the roman empire : four lectures, with preface, notes, and an excursus / Bigg, Charles – Oxford: Clarendon Press, 1905 – 1mf – 9 – 0-7905-4089-4 – (incl bibl ref) – mf#1988-0089 – us ATLA [240]

Church's unity in diversity / Candlish, Robert Smith – London, England. 1862 – 1r – us UF Libraries [240]

Church's war with national intemperance? / Clifford, John – London, England. 1874? – 1r – us UF Libraries [240]

The church's work in our large towns / Huntington, George – 2nd ed., rev. and enl. Oxford: J. Parker, 1871 – 1mf – 9 – 0-7905-6108-5 – mf#1988-2108 – us ATLA [240]

Churchwardens' accounts : from the 14th century to the close of the 17th century / Cox, John Charles – London: Methuen, 1913 [mf ed 1990] – (= ser The antiquary's books) – 1mf – 9 – 0-7905-4788-0 – mf#1988-0788 – us ATLA [240]

Churchwardens' manual : concise memorandum of laws, canons, rules and regulations respecting churchwardens and sidesmen in the diocese of toronto... / Church of England. Diocese of Toronto – [Toronto?: s.n.], 1886 [mf ed 1983] – 1mf – 9 – mf#01078 – cn Canadiana [242]

Der churfuerstlich-saechsische privilegierte postilion *see* Der privilegirte churfuerstlich saechsische postilion

Churgin, Yaakov *see* Kana'im ha-tse'irim

Churman, Philip H *see* Blanca de borbon

Der chursaechsische land-physicus – Naumburg DE, 1771-73 – 1r – 1 – gw Misc Inst [610]

Churton, Edward *see*
- Christian sincerity
- Church of england
- Memoir of joshua watson

Churton, Edward Townson *see* Foreign missions

Churton, Ralph *see* Constitution and example of the seven apocalyptic churches

Churton, William Ralph *see* The influence of the septuagint version of the old testament upon the progress of christianity

Chushingura *see* Sandiwara chusingura

Chut tinh xin lang quen / Nguyen, Thi Hoang – Saigon: Truong-Vi-nh-Ky 1974 [mf ed 1993] – on pt of 1r – 1 – mf#11052 r453 n6 – us Cornell [959]

The chutch in the catacombs : a description of the primitive church of rome illustrated by its sepulchral remains / Maitland, Charles – 2nd rev ed. London: Longman, Brown, Green, and Longmans, 1847 – 1mf – 9 – 0-7905-5008-3 – mf#1988-1008 – us ATLA [240]

Chute, Arthur Crawley *see* William carey

Chute, Chaloner William *see* A history of the vyne in hampshire

La chute de l'empire de rabah / Gentil, Emile – Paris: Hatchette, 1902 – 1 – us CRL [960]

Chute, Marchette Gaylord *see* Geoffrey chaucer of england

Chute's western herald – (The Western Herald). Tralee. Ireland. -w. 27 Aug 1812, 2 Jan 1828-4 May 1835. (7 reels) – 1 – uk British Libr Newspaper [072]

Chutes western herald : or kerry advertiser – Tralee, Ireland. 27 aug 1812, 2 jan 1828-4 may 1835 – 7r – 1 – (aka: western herald or tralee and killarney advertiser, western herald or kerry advertiser) – uk British Libr Newspaper [072]

Chutzpah – Chicago, IL. 1972-79 – 1 – us AJPC [071]

Chutzpah – nl-17 [1972 feb-1980] – 1r – 1 – mf#676202 – us WHS [071]

Chu-tzu hsiao hsueh [ccm111] = Ethical teachings for the young / Chu, Hsi – Shanghai, 1926 [mf ed 198?] – (= ser Ccm 111) – 1 – mf#1984-b500 – us ATLA [230]

Chuuk – truk district charter, 1977 : as approved by the seventh congress of micronesia / Federated States of Micronesia – 1st spec sess. Saipan: the Congress, 29 aug 1977 – 2mf – 9 – $3.00 – mf#LLMC 82-100H Title 2 – us LLMC [324]

[Chuuk] truk district code – Seattle: Book Publ Co 1970 – 12mf – 9 – $18.00 – mf#llmc82-100f, title 26 – us LLMC [348]

Chuvinskii, P P *see* Trudy etnograficheskostatisticheskoi ekspeditsii v zapadno-russkii krai, snariazhennoi imperatorskim russkim geograficheskim obshchestvom

Chuyen hay su cu – Ha Noi: Thanh Nien 1974-[mf ed 1993] – on pt of 1r – 1 – mf#11052 r436 n1 – us Cornell [959]

Chvam ma ra na ga pum pran mya / Lu Thu U Lha – Mantale: Kri Pva re pum nhip tuik 1966 [mf ed 1990] – (= ser Prann thon su tuin ran sa lu mhyi mya pum pran 19) – [ill] 1r with other items – 1 – (in burmese) – mf#mf-10289 seam reel 143/6 [§] – us CRL [390]

Chwa, Daudi *see* Why sir apolo kaggwa, kcmg, mbe, prime minister of buganda

Y chwarelwr cymreig – Bangor, Wales jun 1893-mar 1902 – 1 – (incorp with: y clorianydd; wanting 1897) – uk British Libr Newspaper [072]

Y chwarelwr cymreig – [Wales] LLGC 8 meh 1893-maw 1902 [mf ed 2004] – 5r – 1 – (missing: 1897) – uk Newsplan [072]

Chwolson, D A *see* Syrisch-nestorianische grabinschriften aus semirjetschie

Chynoweth, Tracy L *see* Medical services available for the participants of intramural sports at the schools of the mid-american conference

Chytraeus, D *see*
- Catechesis davidis chytraei postremo recognita
- Commentarivs in matthaevm evangelistam
- Explicatio malachiae prophetae
- Historia avgvstanae confessionis
- In devteronomivm mosis enarratio
- In exodvm enarratio
- In genesin en arratio, tradita
- In genesin enarratio
- In historiam iudicum populi israel commentarius
- In leviticvm, complecten
- In nvmeros enarratio
- In psalmvm 118 praelectiones
- ...Oratio de statv ecclesiarvm hoc tempore in graecia, asia, africa, vngaria...
- Oratio de statv ecclesiarvm hoc tempore in graecia, asia, boemia...
- Tertius liber moysis, qvi inscribitvr leviticvs addita enarratione

Ci pva re lam pra aim rhan ma mya lak cvai khyak nann prut nann 500 / Kri Pu, U – Ran Kun: Ce Ta Na ca pe tuik 1972 [mf ed 1990] – 1r with other items – 1 – (in burmese) – mf#mf-10289 seam reel 177/5 [§] – us CRL [640]

Ci pva re samavayama = Economics of cooperation / Bhui Van, U – Ran Kun: Bhasa pran Ca pe A san 1952 [mf ed 1990] – 1r with other items – 1 – (in burmese) – mf#mf-10289 seam reel 202/8 [§] – us CRL [337]

Cia research reports / U.S. Central Intelligence Agency – (complete asia regional set $3520) – us UPA [327]

Ciacono, A *see* Historia utriusque belli dacici a traiacaesare gesti...quae in columna eiusdem romae visuntur...

Ciadoncha, Marques de *see*
- Alonso fernandez de barrantes. su testamento (1390) apuntes genealogicos de su casa
- Marinos extremnos

Ciamarra, Guglielmo *see* La giustizia nella somalia, guglielmo ciamarra. raccolta di giurisprudenza coloniale.

Ciampini, J *see*
- De sacris aedificiis a constanti mag constructis...
- Vetera monumenta in quibus praecipus musiva opera sacrarum, profanarumque aedium structura, nonnulli antiqui ritus, dissertationibus, iconibusque illustrantur

Ciampino, J *see* Sacro-historica disquisitio

Ciano, Galeazzo, Count *see* Papers of count ciano (lisbon papers) received from the department of state

Cias / Buenos Aires. Centro de Investigacion y Accion Social – Buenos Aires: El Centro. v10 n105 july 1961. v11 n113, 119 may, nov 1962. v12 n128 oct 1963. v13 n136, n138 aug, oct 1964. v18 n181/182 apr/may 1969 – us CRL [350]

Cias – Buenos Aires: El Centro, [ano 10 n105-ano 18 nr181-182 (jul 1961-abr/mayo 1969)] (mthly) – 1r – 1 – us CRL [300]

Ciasca, Augustino *see*
- Examen critico-apologeticum super constitutionem dogmaticam de fide catholica editam in sessione tertia SS. oecumenici Concilii Vaticani
- Sacrorum bibliorum fragmenta copto-sahidica

Ciau athletes' use and intentions to use performance enhancing drugs : a study utilizing the theory of planned behaviour / Allemeier, Meredith Frances – University of British Columbia, 1996 – 3mf – 9 – $12.00 – mf#PSY 1878 – us Kinesology [150]

Ciba journal – Basel. 1967-1970 (1) – ISSN: 0007-8395 – mf#2369 – us UMI ProQuest [540]

Ciba journal – Basel. 1992-1993 (1) 1992-1993 (5) 1992-1993 (9) – (cont: ciba-geigy journal) – ISSN: 0007-8395 – mf#6269,01 – us UMI ProQuest [540]

Ciba-geigy journal – Basel. 1971-1992 (1) 1971-1992 (5) 1976-1992 (9) – (cont by: ciba journal) – ISSN: 0366-5380 – mf#6269 – us UMI ProQuest [540]

Cibao / Hernandez Franco, Tomas Rafael – Ciudad Trujillo, Dominican Republic. 1951 – 1r – us UF Libraries [972]

Cibat, A *see*
- Elementos de matematicas o...introduccion a la fisica experimental
- Memoria sobre la calentura amarilla...que invadio a cadiz y sevilla
- Memoria sobre la naturaleza del contagio de la fiebre amarilla...
- Memorias fisicas sobre el influjo delges indiogeno en la anestitucion del nombre sobre el opigeno... del aire atmosferico
- Por que motivos o causas las tercianas se han hecho tan comunes y graves...

Cibles : l'ami du campeur: manuel des techniques scoutes / Scouts catholiques du Canada – 8e ed. Montreal: Quartier general: [edite pour les Scouts catholiques – Canada par la Cordee], 1965 [mf ed 2001] – 5mf – 9 – (with ind) – mf#SEM105P3291 – cn Bibl Nat [360]

Cibles : manuel des techniques scoutes / Federation des scouts catholiques, Canada – 4e rev augm ed. Montreal: Quartier general, 1958 [mf ed 1999] – 5mf – 9 – (with ind and bibl ref) – mf#SEM105P3210 – cn Bibl Nat [360]

Cibot, P M *see* Notices du royaume de ha-mi

Cicero *see* Fifteenth century italian manuscripts

Cicero, Marcus Tullius *see*
- Ciceros rede fuer t annius milo
- De amicitia...
- De finibus bonorum et malorum
- Divinae institutiones...
- Epistola ad quintum fratrem...
- In 100 verrem actionis secundae libri 4, 5
- M T ciceronis tusculanarum quaestionum libri quinque
- M tulli ciceronis scripta quae manserunt omnia vol vi, pt2
- Rhetorica ad herennium...
- Select orations of cicero

Ciceron orateur : analyse et critique des discours de ciceron / Cucheval, Victor – Paris: E Belin 1901 [mf ed 1987] – 2v on 1r – 1 – (incl bibl ref. filmed with other titles) – mf#8662 – us Wisconsin U Libr [850]

El cicerone / Pascual Ayago, Julian – Guia kilometrica de la provincia de badajoz, ano 1946. badajoz, graficas iberia – 1 – sp Bibl Santa Ana [946]

El cicerone / Pascual, Julian – Puerta de Palma: Guia-Callejero de Badajoz. Badajoz, Imp. Clasica – 1 – sp Bibl Santa Ana [946]

The cicerone : an art guide to painting in italy / Burckhardt, Jakob Christoph – London 1879 – (= ser 19th c art & architecture) – 4mf – 9 – mf#4.2.1137 – uk Chadwyck [750]

El cicerone del pueblo / Gutierrez Macias, Valeriano – Badajoz: Dip. Provincial, 1969. Sep. REE – 1 – sp Bibl Santa Ana [946]

Ciceronis amor: tullies love / Greene, Robert – 1589. A quip for an upstart courtier. 1592 – 9 – us Scholars Facs [840]

Cicero's 'de inventione rhetorica', twelfth century commentary on... : york minster library, ms. 16/m7 – 1r – 1 – mf#4581 – uk Microform Academic [760]

Ciceros rede fuer t annius milo : mit dem kommentar des asconius und den bobienser scholien / Cicero, Marcus Tullius; ed by Wessner, Paul – Bonn: A Marcus & E Weber 1911 [mf ed 1992] – (= ser Kleine texte fuer theologische vorlesungen und uebungen 71) – 1mf – 9 – 0-524-04695-6 – mf#1990-3404 – us ATLA [450]

Cicily : ou, le lion amoureux / Scribe, Eugene – Paris, France. 1840? – 1r – us UF Libraries [440]

Ciclo : coloquios empresariales / Delegacion Provincial de Sindicatos – Caceres: Imp. T. Rodriguez, 1972 – 1 – sp Bibl Santa Ana [946]

Ciclo de lo ausente / Arrivi, Francisco – San Juan, Puerto Rico. 1962 – 1r – us UF Libraries [972]

El ciclo del cerdo en espana : investigaciones sobre las fluctuaciones de la produccion y de los precios desde 1939 a 1956... / Wienberg, Dieter & Sobrino, Francisco – Madrid: consejo s.i.c. y diput. prov. de badajoz, 1958 – 1 – sp Bibl Santa Ana [946]

El ciclon – Badajoz, 1921. 1 numero – 5 – sp Bibl Santa Ana [073]

Ciclon de 1926 sobre la habana – Havana, Cuba. 1926 – 1r – us UF Libraries [972]
Ciclopes / Lamarque, Nydia – Buenos Aires, Argentina. 1930 – 1r – us UF Libraries [972]
Ciclos de lirismo colombiano / Caparraso, Carlos Arturo – Bogota, Colombia. 1961 – 1r – us UF Libraries [972]
Cicognani, H J *see* La caridad en los primeros siglos del cristianismo
Cicognara, L *see* Storia della scultura
Cid – Chicago, IL. 1980 dec 14-1984 dec 01 – 1r – us UF Libraries [071]
Cid *see* Tale of the warrior lord
Der cid : nach spanischen romanzen besungen durch johann gottfried von herder: mit einer einleitung ueber herder und seine bedeutung fuer die deutsche literatur / ed by Schmidt, Julian – Leipzig: F A Brockhaus 1868 [mf ed 1990] – (= ser Bibliothek der deutschen nationalliteratur des achtzehnten und neunzehnten jahrhunderts) – 1r – 1 – (clarifications by karoline michaelis. filmed with: eck segge man bloss / wilhelm henze) – mf#2722p – us Wisconsin U Libr [410]
Cid Fernandez, Enrique Del *see* Don gabino de gainza y otros estudios
O cidadao : jornal satyrico e litterario – Rio de Janeiro, RJ: Typ Particular de S T Aquino, 21 maio-16 jun 1877 – (= ser Ps 19) – mf#DIPER – bl Biblioteca [870]
O cidadao : orgao noticioso, commercial, litterario e industrial – Braganca, PA. 20 mar 1890 – (= ser Ps 19) – bl Biblioteca [079]
O cidadao – Vitoria, ES: Typ do Cidadao, 15 mar-30 jul 1868 – (= ser Ps 19) – mf#DIPER – bl Biblioteca [079]
Cidadao do mundo / Costa, Licurgo – Rio de Janeiro, Brazil. 1943 – 1r – us UF Libraries [972]
A cidade – Ouro Preto, MG: Typ A Cidade, out 1901-dez 1902; 20 out 1904 – (= ser Ps 19) – 1,5,6 – mf#P11B,03,79 – bl Biblioteca [079]
Cidade antiga do brasil, ouro preto / Krull, Germaine – Lisboa, Portugal. 1943 – 1r – us UF Libraries [972]
Cidade da empreza : orgam official da prefeitura do alto acre – Empreza, AC. 15 jul-22 out 1910 – (= ser Ps 19) – mf#P25,01,19 – bl Biblioteca [350]
Cidade da vigia – Cidade da Vigia, PA: Typ da Cidade da Vigia, 06 jul 1890; 13 ago 1893 – (= ser Ps 19) – bl Biblioteca [079]
Cidade de barbacena : orgam dos interesses do municipio e do povo – Barbacena, MG. 23 jan 1898-dez 1906; 06 mar 1949 – (= ser Ps 19) – bl Biblioteca [079]
Cidade de caldas : folha popular – Caldas, MG: [s.n.] 01 maio, 28 ago 1892 – (= ser Ps 19) – mf#P11B,03,78 – bl Biblioteca [079]
A cidade do rio de janeiro – Rio de Janeiro, RJ: Typ do Diario do Rio de Janeiro de N L Vianna, 23 mar 1850 – (= ser Ps 19) – 1,5,6 – mf#P15,01,48 – bl Biblioteca [079]
Cidade do sacramento : orgao dos interesses municipaes – Sacramento, MG. fev 1903 – (= ser Ps 19) – mf#P31,03,08 – bl Biblioteca [079]
Cidade enferma / Dantas, Paulo – Sao Paulo, Brazil. 1950 – 1r – us UF Libraries [972]
Cidade pouso alegre – Pouso Alegre, MG: Typ Sul Mineira, 30 abr 1906 – (= ser Ps 19) – mf#P11B,03,76 – bl Biblioteca [079]
Cidade sitiada / Lispector, Clarice – Rio de Janeiro, Brazil. 1948 – 1r – us UF Libraries [972]
Ci-devant jeune homme / Merle, Jean Toussaint – Paris, France. 1835 – 1r – us UF Libraries [440]
Cidoncha, Marques de *see* Dona mencia de los nidos
Ciel – Zilina, Czechoslovakia. Nov 1955-Mar 1960 – 2r – 1 – us L of C Photodup [077]
Le ciel bleu – Brussels, 1945 [mf ed Chadwyck-Healey] – (= ser Art periodicals on microform) – 1r – 1 – uk Chadwyck [750]
Le ciel, sejour des elus / Frederic, de Ghyvelde, pere – Montreal: Revue du Tiers Ordre et de la Terre sainte, 1912 [mf ed 1985] – 5mf – 9 – mf#SEM105P536 – cn Bibl Nat [241]
Cielo negro / Caro, Nestor – Ciudad Trujillo, Dominican Republic. 1949 – 1r – us UF Libraries [972]
Cielslewicz, Lindsy S *see* Dance and doctrine
Ciemny, Melech *see* Uzbekistan
Cien anos de poesia en panama (1852-1952) / Miro, Rodrigo – Panama, Panama. 1953 – 1r – us UF Libraries [972]
Cien anos de vida universitaria / Pacheco, Juan Rafael – Ciudad Trujillo, Dominican Repuplic. 1944 – 1r – us UF Libraries [378]
Cien de las mejores poesias cubanas / Estenger, Rafael – Habana, Cuba. 1948 – 1r – us UF Libraries [972]
Cien de las mejores poesias liricas salvadorenas / Espinosa, Francisco – San Salvador, El Salvador. 1951 – 1r – us UF Libraries [972]
Cien letrillas / Sanchez Arjona, Vicente – Sevilla: Graficas Sevillanas, 1951 – 1 – sp Bibl Santa Ana [810]
Las cien mejores poesias (liricas) de la lengua castellana / Menendez y Pelayo, Marcelino; ed by Artigas, Miguel – rev ed. Madrid, 1932 – 1 – sp Bibl Santa Ana [810]
Cien mejores poesias liricas de panama / Rubinos, Jose – New York, NY. 1964 – 1r – us UF Libraries [972]
Cien razones (aunque sea tarde) / Galan, Leocadio – Caceres: Tip. El Noticiero, 1977 – 1 – sp Bibl Santa Ana [946]
Cien sinrazones / Cadilla Ruibal, Carmen – San Juan, Puerto Rico. 1962 – 1r – us UF Libraries [972]
La cienaga / Reyes Huertas, Antonio – Madrid: Edic. Hispano-Americanas, s.a. – 1 – sp Bibl Santa Ana [946]
La cienaga / Reyes Huertas, Antonio – Madrid: Ediciones Hispano Americanas, 1921 – 1 – sp Bibl Santa Ana [946]
Ciencia de la hacienda publica / Vasquez, Juan Ernesto – San Salvador, El Salvador. 1943 – 1r – us UF Libraries [972]
La ciencia de las mugeres / Sanchez Arjona y Sanchez Arjona, Jose – 1874 – 9 – sp Bibl Santa Ana [810]
Ciencia e investigacion – Buenos Aires. 1950-1954 (1) – ISSN: 0009-6733 – mf#566 – us UMI ProQuest [500]
La ciencia, la naturaleza y el milagro / Caba, Pedro – Madrid: Diana, Artes Graficas, 1965. Sep. Rev. Filosofia, vol XXIV, no 92-93, Enero-Junio 1965 – sp Bibl Santa Ana [240]
Ciencia politica – 1-9, 1941-44. Incomplete – 1 – us L of C Photodup [320]
Ciencia y fe / San Miguel FCP. Facultades de Filosofia y Teologia – Buenos Aires; Montivideo: Editorial Verbum. v1-20. 1944-1964 – 7r – us CRL [972]
Ciencias administrativas – La Plata. 1975-1979 (1) 1975-1979 (5) 1975-1979 (9) – ISSN: 0009-6784 – mf#7721 – us UMI ProQuest [350]
Ciencias economicas e a vida nacional / Pinto, Jose Gomes Pereira – Rio de Janeiro, Brazil. 1950 – 1r – us UF Libraries [330]
Ciencias medicas en guatemala / Martinez Duran, Carlos – Guatemala, . 1945 – 1r – us UF Libraries [610]
Cienfuegos Linares, Julio *see* Pregon de la fiestas patronales de l'erena, 11 de agos to de 1968
Ciento cincuenta anos de periodismo en caceres y salamanca / Colegio Universitario de Caceres – Caceres: Imprenta Diputacion Provincial, 1973 – sp Bibl Santa Ana [946]
Cieza Leon, Pedro de *see*
– La cronica del peru
– Guerras civiles del peru
– Segunda parte de la cronica...incas yupanquis..
– Tercero libro de las guerras civiles...de quito
Cifra antologica de fabio baudrit gonzalez – San Jose, Costa Rica. 1956 – 1r – us UF Libraries [972]
Cifras e notas : economia e financas do brasil / Tavares, Joao De Lyra – Rio de Janeiro, Brazil. 1925 – 1r – us UF Libraries [972]
Cig – cryogenics and industrial gases – Cleveland. 1965-1976 (1) 1971-1976 (5) – ISSN: 0011-2283 – mf#2346 – us UMI ProQuest [620]
Cigala, C Albin de *see* Vie intime de pie 10
Ciganek, David S *see* A study of multi-dimensional evaluation processes in the professional preparation of athletic trainers
Cigar industry of tampa, florida / Campbell, Archer Stuart – Gainesville, FL. 1939 – 1r – us UF Libraries [338]
Cigar makers' official journal – 1876 mar-1885 sep, 1885 oct-1890 dec, 1891 jan-1895 jun – 1r – 1 – mf#780648 – us WHS [670]
Cigar Makers Progressive Union of America *see* Progress
Cigar workers official journal – 1876-1972 – (= ser Labor union periodicals, pt 3: food and agricultural industries) – 15r – 1 – $3115.00 – 1-55655-624-1 – us UPA [660]
Cigarmakers' union dispute in tampa / Bryan, Lindsay M – s.l, s.l? 1938-1939 – 1r – us UF Libraries [331]
A cigarra – Maranhao: Typ Nacional e Imperial, 12 out 1829-17 abr 1830 – (= ser Ps 19) – mf#P17,02,51 – bl Biblioteca [079]
A cigarra – Rio de Janeiro, RJ: Officinas Graphicas de J Bevilacqua & C. 1895, v1(1-34); 1896, v2(35-37) – (= ser Ps 19) – mf#P03,02,23 – bl Biblioteca [079]
Cigoi, Alois *see* Historisch-chronologische schwierigkeiten im zweiten makkabaeerbuche
Cikagas Latviesu Organizaciju Apvienibas *see* Cikagas zinas
Cikagas zinas = Chicago news / Cikagas Latviesu Organizaciju Apvienibas – 1976 jan-1981 dec, 1982-1987 feb – 2r – 1 – mf#1160162 – us WHS [071]
Cikar, Jutta [comp] *see*
– Arab-islamic biographical archive. series 2
– German books on islam from the 16th century to 1900, pt 1
– German books on islam from the 16th century to 1900, pt 2
Ciles alucinada, y otras poesias / Herrera Y Reissig, Julio – San Jose, Costa Rica. 1916 – 1r – us UF Libraries [972]
Cilleros. Ayuntamiento *see*
– Ferias y fiestas 1971
– Tradicionales fiestas en honor de la santisima virgen de navelonga
– Tradicionales fiestas en honor de la santisima virgen de navelonga, 1978
Cillier zeitung – Celje, Yugoslavia. 1921; 1923-Feb 1928 – 5r – 1 – us L of C Photodup [079]
Cillier zeitung – Cilli [Celje SLO], 1922, 1930 5 jan-29 jun, 1932 7 jan-1936 7 may – 4r – 1 – (later: deutsche zeitung; 1934: some iss missing) – gw Misc Inst [079]
Cilliers, Andries Charl *see* The state and the universities
Cilybebyll, glamorgan, parish church of st john evangelist : baptisms 1721-1911, burials 1721-1986, marriages 1654-1837 – [Glamorgan]: GFHS [mf ed c1999] – 1mf – 9 – £1.25 – uk Glamorgan FHS [929]
Cilybebyll, st john evangelist, monumental inscriptions – 1mf – 9 – £1.25 – uk Glamorgan FHS [929]
Cim bulletin / Canadian Institute of Mining and Metallurgy – Montreal. 1984+ (1,5,9) – ISSN: 0317-0926 – mf#15107 – us UMI ProQuest [622]
Cim ne u mann kyvan to mre : [a novel] / Bhun Nuin, Takkasuil – Ran Kun: Mra Ca pe tuik 1965 [mf ed 1990] – 1r with other items – 1 – (in burmese) – mf#mf-10289 seam reel 192/5 [§] – us CRL [830]
Cim, Panna re *see* Rajadhatukalya
Cim review – Pennsauken. 1989-1990 (1) (5) 1990-1990 (9) – ISSN: 0748-0474 – mf#14373 – us UMI ProQuest [000]
Cim technology : casa, sme's magazine of computers in design and manufacturing / Computer and Automated Systems Association of SME – Dearborn. 1984-1986 (1,5,9) – (cont: cad/cam technology) – mf#13437,01 – us UMI ProQuest [000]
Cima, Andrea *see* Il secondo libro delli concerti a due tre, & quattro, voci...opera seconda
Cimaise : Revue de l'art actuel – n1-50. Paris. nov 1953-60 – 1 – fr ACRPP [700]
Cimambwe grammar / London Missionary Society – Lusaka, Zambia. 1962 – 1r – us UF Libraries [470]
Cimarosa, Domenico *see*
– L'artemisia
– Duo del matrimonio per raggiro...
– Ouverture [des horaces musique de cimarosa]
– [Recueil d'airs]
– Le sacrifice d'abraham
– Stravaganze di(!) amore
Cimarron review – Stillwater. 1967+ (1) 1972+ (5) 1976+ (9) – ISSN: 0009-6849 – mf#6708 – us UMI ProQuest [300]
La cimbarra / Martinez de Carnero y Diaz, Rafael – 1846 – 9 – sp Bibl Santa Ana [830]
Cimbri, emblemata...christiano 4...dicata / Westhovius, W – Hafniae: Impensis Ioachimi Moltkenii, 1640 – 1mf – 9 – mf#O-802 – ne IDC [090]
Cimbrishamnsbladet – Simrishamn, Sweden. 1857-1944 – 1 – sw Kungliga [078]
Cimetiere Notre-Dame des Neiges (Montreal, Quebec) *see* Reglement du cimetiere de notre-dame-des-neiges
Cimon, Constance *see* Bibliographie analytique des ecrits canadiens sur l'oeuvre et la personnalite de cornelius krieghoff
Cin – Brno, Czechoslovakia. Aug 1945-Oct 1946 – 2r – 1 – us L of C Photodup [077]
Cin Aon Man, Vanna Kyo Than *see* Lo ba mhon khvan a ron lan
Cin Chan, Cha ra *see* Bedan pon khyup kyam
Cin Cin *see* Lvam sacca tuin
Cin phran pru lup tha so nhalum sa puin rhan mya nhan a khra vatthu tui mya : [short stories] / Khyui Phru Khan – Ran Kun: Ca pe Biman pum nhip tuik 1985 [mf ed 1990] – (= ser Ca pe biman thut prann su lak cvai ca can) – 1r with other items – 1 – (in burmese) – mf#mf-10289 seam reel 149/7 [§] – us CRL [830]
Cin sa muin / Tan Cin Nay – Ran Kun: Ca pe Bi Man A Phvai 1973 [mf ed 1990] – 1r with other items – 1 – (in burmese) – mf#mf-10289 seam reel 177/9 [§] – us CRL [550]
Cin Tan *see* Mran ma a kron nhan kambha a kron
Cin Tan, Takkasuil *see*
– Ekari cu phura
– Kyvan to tui a me
– Mran ma nuin nam phakchac to lhan re samuin
Cin Thve, U *see* Mon cin thve e kuiy tuin re atthuppatti
Cina, latvia – Riga, 1904-45 – 8r – 1 – (missing: 1916) – us UMI ProQuest [077]
Cinagli, Ang *see* Le monete dei papi descritte in tavolle sinottiche
Cincinnati abend=post / Hamilton Co. Cincinnati – oct 22 1878-oct 24 1880 – 1r – 1 – mf#B37467 – us Ohio Hist [071]
Cincinnati AFL-CIO Labor Council *see* Chronicler
Cincinnati american / Hamilton Co. Cincinnati – oct 1913-feb 1914 – 1r – 1 – mf#B36625 – us Ohio Hist [071]
Cincinnati Area Teacher Center *see* Catc helpline
Cincinnati art museum bulletin – Cincinnati. 1950-1987 (1) 1971-1987 (5) 1977-1987 (9) – ISSN: 0069-4061 – mf#2483 – us UMI ProQuest [060]
Cincinnati business courier – Cincinnati. 1989-1997 (1) – (cont by: business courier) – ISSN: 0882-8881 – mf#16673 – us UMI ProQuest [650]
Cincinnati chronicle / Hamilton Co. Cincinnati – apr-sep 1842 – 1r – 1 – mf#B37466 – us Ohio Hist [071]
Cincinnati chronicle and literary gazette / Hamilton Co. Cincinnati – jan 1830-sep 1837 – 1r – 1 – mf#B37469 – us Ohio Hist [071]
Cincinnati commercial – Cincinnati, Ohio. Daily. Apr 12 1861-Dec 30 1865. Incomplete – 1 – us NY Public [071]
Cincinnati daily chronicle / Hamilton Co. Cincinnati – dec 1869-jun 1871 – 4r – 1 – mf#B37454-37457 – us Ohio Hist [071]
Cincinnati daily columbian / Hamilton Co. Cincinnati – 6/19/1854-6/55.7/21-9/10/56 – 3r – 1 – mf#B36964-36966 – us Ohio Hist [071]
Cincinnati daily commercial – Cincinnati, OH: M D Potter & Co, [nov 13 1854-may 25 1861] – 1 – us CRL [071]
Cincinnati daily nonpareil / Hamilton Co. Cincinnati – nov 1851-may 1852 – 1r – 1 – mf#B36962 – us Ohio Hist [071]
Cincinnati democrat – Cincinnati, dec 8 1845 – 1r – us CRL [071]
Cincinnati emporium / Hamilton Co. Cincinnati – aug 10,17& 31, 1823 – 1r – 1 – mf#B37533 – us Ohio Hist [071]
Cincinnati enquirer – Cincinnati, OH. 1921+ (1) – mf#60554 – us UMI ProQuest [071]
Cincinnati evening chronicle / Hamilton Co. Cincinnati – jul 1 1869-nov 30 1869 – 1r – 1 – mf#B37532 – us Ohio Hist [071]
Cincinnati fed / American Postal Workers Union – v8 n2-v34 [1949 feb-1972 aug] – 1r – 1 – mf#63381 – us WHS [380]
Cincinnati Federation of Teachers *see* Federation teacher
Cincinnati gazette – Cincinnati, Ohio. Daily. June 26 1828-Dec 31 1881. Incomplete – 1 – us NY Public [071]
Cincinnati, hamilton, and dayton railroad, 1850-1862 – 1r – 1 – mf#B26389 – us Ohio Hist [380]
Cincinnati herald – Cincinnati OH. 1978-1998 jul-dec – 29r – 1 – mf#907909 – us WHS [071]
Cincinnati herald / Hamilton Co. Cincinnati – jan 1979-dec 1994 – 13r – 1 – mf#B37415-37427 – us Ohio Hist [071]
Cincinnati jewish world – Cincinnati, OH. 4 Apr 1952 – 1 – us AJPC [071]
Cincinnati kurier / Hamilton Co. Cincinnati – apr 1964-may 1982 – 21r – 1 – (in german) – mf#B37428-37448 – us Ohio Hist [071]
Cincinnati literary gazette – Cincinnati. 1824-1825 [1] – mf#4438 – us UMI ProQuest [420]
Cincinnati mirror, and chronicle – 1835 apr 18-oct 24 – 1r – 1 – (cont: cincinnati mirror; and western gazette of literature and science; cont by: buckeye and cincinnati mirror) – mf#3094490 – us WHS [071]
Cincinnati mirror, and ladies' parterre – v1 n2-25 [1831 oct 15-1832 sep 1], v2 n9 [1833 jan 19] – 1r – 1 – (cont by: cincinnati mirror, and western gazette of literature and science) – mf#1054495 – us WHS [071]
Cincinnati mirror, and western gazette of literature and science – 1833 oct 5-1835 apr 11 – 1r – 1 – (cont: cincinnati mirror, and ladies' parterre; cont by: cincinnati mirror, and chronicle) – mf#2739068 – us WHS [071]
Cincinnati mirror and western gazette of literature, science and the arts – Cincinnati. 1831-1836 – 1 – mf#3762 – us UMI ProQuest [073]
Cincinnati mirror, and western gazette of literature, science, and the arts – 1836 jan 30-sep 17 – 1r – 1 – (cont: buckeye and cincinnati mirror; cont by: cincinnati chronicle and literary gazette) – mf#768562 – us WHS [071]
Cincinnati morgan=post / Hamilton Co. Cincinnati – mar 12-oct 15, 1878 – 1r – 1 – mf#B37533 – us Ohio Hist [071]
Cincinnati morning herald / Hamilton Co. Cincinnati – oct 1843-nov 1845 – 2r – 1 – mf#B37452-37453 – us Ohio Hist [071]

Cincinnati news journal / Hamilton Co. Cincinnati – apr 1 1884-jun 7 1884 – 1r – 1 – mf#B37530 – us Ohio Hist [071]
Cincinnati Observatory [Ohio] *see* Publications of the cincinnati observatory
Cincinnati, OH *see*
– Day star
– Selections
Cincinnati post – Cincinnati, OH. 1882+ (1) – mf#60143 – us UMI ProQuest [071]
Cincinnati. Presbytery (Pres. Church in the USA) *see* Minutes, 1822-1911
Cincinnati reporter – Cincinnati OH. v1 n1-41 [1977 jun 22-1978 jun 16] – 1r – 1 – mf#630446 – us WHS [071]
Cincinnati republikaner = Cincinnati republican – Cincinnati, OH: Carl Hiller & Wm Ed Becht, [jan 4 1858-mar 23 1861] – 1r – 1 – us CRL [071]
The cincinnati star – Cincinnati, Ohio. Weekly. Mar 14 1872-June 23 1880. Incomplete – 1 – us NY Public [071]
Cincinnati superior court decisions / Ohio. Superior Court – 1v. 1903-07 (all publ) – (= ser Ohio appellate decisions) – 6mf – 9 – $9.00 – mf#llmc 84-181 – us LLMC [347]
Cincinnati superior court decisions / Ohio. Superior Court – v1-2. 1854-1855 (all publ) – (= ser Ohio appellate decisions) – 8mf – 9 – $12.00 – mf#llmc 91-036 – us LLMC [347]
Cincinnati superior court decisions / Ohio. Superior Court – v1-2. 1870-1873 (all publ) – (= ser Ohio appellate decisions) – 14mf – 9 – $21.00 – mf#llmc 91-035 – us LLMC [347]
Cincinnati tageblatt / Hamilton Co. Cincinnati – 7/22/1895-10/17/1896 – 2r – 1 – mf#B37399-37400 – us Ohio Hist [071]
Cincinnati tagliche morgan=post / Hamilton Co. Cincinnati – dec 1877-oct 122, 1878 – 3r – 1 – mf#B37458-37460 – us Ohio Hist [071]
Cincinnati taglicher abend=post / Hamilton Co. Cincinnati – mar 3 1877-oct 20 1880 – 7r – 1 – mf#B37501-37507 – us Ohio Hist [071]
Cincinnati telegram / Hamilton Co. Cincinnati – nov 11-dec 30 1888 – 1r – 1 – mf#B37533 – us Ohio Hist [071]
Cincinnati telephone books (1915-1989) – 79r – 1 – mf#B31910-31988 – us Ohio Hist [978]
Cincinnati times – Cincinnati, Ohio. Weekly. July 15 1880-Dec 27 1888 – 1 – us NY Public [071]
Cincinnati times star – Covington, KY. 1902-1958 (1) – mf#63456 – us UMI ProQuest [071]
Cincinnati volksfreund – Cincinnati OH. 1863 feb 18/1864 feb 10-1895 jan 2/1896 jan 1 – 10r – 1 – mf#912636 – us WHS [071]
Cincinnati weekly chronicle / Hamilton Co. Cincinnati – jan 1869-dec 1869 – 1r – 1 – mf#B37401 – us Ohio Hist [071]
Cincinnati weekly enquirer / Hamilton Co. Cincinnati – 7,1868-2,1921 (scattered) – 29r – 1 – mf#B36967-36995 – us Ohio Hist [071]
Cincinnati weekly gazette / Hamilton Co. Cincinnati – jan 1878-dec 1881 – 3r – 1 – mf#B37408-37410 – us Ohio Hist [071]
Cincinnati weekly herald – Cincinnati: Sperry & Brewster, dec 16 1846-feb 7 1847 – 1r – (filmed with: philanthropist (new richmond, oh), and: cincinnati weekly herald and philanthropist) – us CRL [071]
Cincinnati weekly herald *see* Cincinnati weekly herald and philanthropist
Cincinnati weekly herald and philanthropist – Cincinnati. 1836-1846 (1) – mf#5284 – us UMI ProQuest [360]
Cincinnati weekly herald and philanthropist – Cincinnati OH: Gamaliel Bailey Jr 1843-46 – 1r – 1 – (filmed with: philanthropist (new richmond, ohio) and cincinnati weekly herald oct 18 1843-dec 9 1846) – us CRL [071]
Cincinnati weekly news / Hamilton Co. Cincinnati – jan 1883-jun 11 1884 – 2r – 1 – mf#B36234-36235 – us Ohio Hist [071]
Cincinnati weekly times / Hamilton Co. Cincinnati – jan 1874-dec 1889 (scattered) – 4r – 1 – mf#B37411-37414 – us Ohio Hist [071]
Cincinnatier freie presse – Cincinnati OH (USA), 1922 1 apr-1927 30 sep [gaps], 1928-29 [gaps], 1930 1 apr-30 sep, 1931 1 apr-20 oct, 1932 12 feb-31 dec, 1933 1 apr-1937, 1938 1 apr-31 may, 1939 1 feb-24 aug – 42r – 1 – gw Misc Inst [071]
Cincinnatier zeitung – Cincinnati, OH: Cincinnatier Zeitung Pub Co, jul 1887-oct 20 1901 – 59r – 1 – us CRL [071]
Cincinnati-kurier – Omaha NE [USA], 1972-82 – 1 – (cont by: amerika-woche, chicago) – gw Misc Inst [071]
Cinco de janeiro : orgao do partido liberal do amazonas – Manaus, AM: Typ do Amazonas de J Carneiro dos Santos, 29 maio 1879 – (= ser Ps 19) – mf#P11B,06,15 – bl Biblioteca [325]
Cinco discursos del general marcos perez jimenez – Caracas, Venezuela. 1955 – 1r – us UF Libraries [972]
Cinco libros de arquitectura / Serlio, Sebastian – SL, 1563 – 8mf – 9 – sp Cultura [720]
Cinco poetas universitarios / Universidad De Costa Rica – San Jose, Costa Rica. 1952 – 1r – us UF Libraries [378]
Cinco reporteros y el personaje de la semana – Bogota, Colombia. 1964? – 1r – us UF Libraries [972]
Cinco sentidos / Blanco, Tomas – San Juan, Puerto Rico. 1955 – 1r – us UF Libraries [972]
Cinco tesis sobre las pasiones / Puerta Flores, Ismael – Caracas, Venezuela. 1949 – 1r – us UF Libraries [972]
Cincpacflt interim evaluation reports, 1950-1953 – Korean conflict. 1978 – 6r – 1 – $780.00 – mf#S1653 – us Scholarly Res [951]
Cincuenta anos de literatura puertorriquena / Alegria, Jose S – San Juan, Puerto Rico. 1955 – 1r – us UF Libraries [972]
Cincuenta y dos anos de politica, oriente / Riera Hernandez, Mario – Habana, Cuba. 1953 – 1r – us UF Libraries [972]
Cincuenta y seis anos de historia patria / Santana Calzada, Luis – Trinidad de Cuba, Cuba. 1948 – 1r – us UF Libraries [972]
Cincuentenario del 95 – Habana, Cuba. v1-2. 1945- – 1r – us UF Libraries [972]
Cinderella : three hundred and forty-five variants of cinderella, catskin, and cap o'rushes / Cox, Marian Roalfe [comp] – London: publ for the Folk-lore Society by David Nutt 1893 [mf ed 1993] – (= ser Publications of the folk-lore society 31) – 2mf – 9 – 0-524-05839-3 – (incl bibl ref; discussion & notes by comp; int by andrew lang) – mf#1990-3503 – us ATLA [390]
Cineaste – New York. 1967+ (1) 1967+ (5) 1967+ (9) – ISSN: 0009-7004 – mf#11704 – us UMI ProQuest [790]
Cinegram magazine – Ann Arbor. 1976-1977 [1,5,9] – mf#11153 – us UMI ProQuest [790]
Cine-liberte – no. 1-5. Paris. mai-nov 1936. Collection privee – 1 – fr ACRPP [790]
Cinelli, G *see* Le bellezze della citta di firenze
Cinema – 17 no. Paris. oct 1952-janv 1955 – 1 – fr ACRPP [790]
Cinema – Beverly Hills. 1962-1976 (1) 1971-1972 (5) (9) – ISSN: 0009-7047 – mf#7024 – us UMI ProQuest [790]
Cinema canada – Montreal, Quebec, CN. 1972-nov 1989 – 9r – 1 – cn Commonwealth Imaging [790]
Le cinema canadien : valse chantee / Dandurand, J L – [Montreal]: Star of the Canadian Moving Picture Service, 1921 [mf ed 1988] – 1mf – 9 – mf#SEM105P927 – cn Bibl Nat [790]
Le cinema et l'echo du cinema reunis – Paris. mars 1912-mars 1914, mai 1916-mai 1923 – 1 – fr ACRPP [790]
Cinema journal – Lawrence. 1983+ (1,5,9) – ISSN: 0009-7101 – mf#13916,01 – us UMI ProQuest [790]
Cinema news – San Francisco. 1976-1980 (1) 1976-1980 (5) 1976-1980 (9) – ISSN: 0198-7305 – mf#7615,01 – us UMI ProQuest [790]
Cinema nuovo – Rome. 1972-1991 (1) 1972-1980 (5) 1975-1980 (9) – ISSN: 0009-711X – mf#7194 – us UMI ProQuest [790]
Cinema pressbooks of the major hollywood studios : from the original studio collections: united artists, 1919-1949; warner brothers, 1922-1949; monogram pictures, 1937-1946 – 38r – 1 – (previous title: mass communications and the twentieth century. sect a: pressbooks for united artists 1919-49; warner bros 1922-49; monogram pictures 1937-46 19r c39-10901. sect b: pressbooks for united artists 1919-49; warner bros 1922-49; and monogram pictures 1937-46 19r c39-10902. title listing available) – mf#C39-10900 – us Primary [790]
A cinematographical analysis and force measure of three styles of the karate back punch and side kick / Powell, Steven W – 1989 – 215p 3mf – 9 – $12.00 – us Kinesology [790]
A cinematographical and biomechanical analysis of the approach run phase for the pole vault / Hsu, Hung-Yi – 1997 – 1mf – 9 – $4.00 – mf#PE 3883 – us Kinesology [612]
La cinematographie / Bull, Lucien – Paris: A Colin 1928 [mf ed 1986] – 1r – 1 – (with: international conference on unemployment) – mf#6805 – us Wisconsin U Libr [790]
La cinematographie francaise – Paris. nov 1918-29 nov 1928 – 1 – fr ACRPP [790]
O cinematographo : jornal semanal de propaganda commercial, humoristico e noticioso – Rio de Janeiro, RJ. 19-? – (= ser Ps 19) – bl Biblioteca [790]
Cine-theatro : orgao de propaganda da empreza cinematographica m cruz – Tijucas, SC: Typ Santa Cruz, 22 ago-29 out 1929; 01 nov 1931 – (= ser Ps 19) – mf#UFSC/BPESC – bl Biblioteca [790]
Cinethique – Paris. 1974-1983 (1) 1974-1983 (5) 1974-1983 (9) – mf#8748 – us UMI ProQuest [790]
Cing han wen hai bithe *see* Qing han wen hai
Cing wen diyen yoo bithe *see* Qing wen dian yao
Cing wen ki meng bithe = Qing wen qi meng / Shou, Ping – [China]: San huai tang, [1730?] [mf ed 1966] – (= ser Tenri coll of manchu-books in manchu-characters. series 1, linguistics 51; Mango bunkenshu. 1, gogaku hen) – 4v – 1 – (in manchu and chinese) – ja Yushodo [480]
Cingirakli tatar – n1-29. 24 mart-6 temmuz 1289 [all publ] – (= ser O & t journals) – 3mf – 9 – $55.00 – us MEDOC [956]
Cingoez – Istanbul: A Asaduryan Matbaasi, 1908. Sahib-i Imtiyaz: Seyyid Hasan. n1-7. 26 agustos-26 eyluel 1324 [9 sep-19 oct 1908] – (= ser O & t journals) – 1mf – 9 – $25.00 – us MEDOC [956]
Cinotti, Mia *see* Femme nue dans la sculpture
Cinq annees d'administration reformiste : la ruine a l'interieur quand la fortune est a la porte: choisissez! – [Montreal?: s.n, 1878?] [mf ed 1981] – 2mf – 9 – mf#11941 – cn Canadiana [320]
Cinq annees d'administration reformiste : la ruine a l'interieur quand la fortune est a la porte: choisissez! – [S.l: s.n, 1878?] [mf ed 1980] – 2mf – 9 – 0-665-03189-0 – mf#03189 – cn Canadiana [320]
Cinq annees de residence au canada / Talbot, Edward Allen – Paris: Boulland. 2v. 1825 [mf ed 1984] – 2v on 1mf – 9 – mf#47848 – cn Canadiana [917]
Cinq ans de sejour au soudan francais / Bechet, Eugene – Paris: E Plon, 1889 – 9 – us CRL [916]
Les cinq decades / Bullinger, Heinrich – [Geneve], Thomas Courteau, 1565 – 1mf0mf – 9 – mf#PBU-162 – ne IDC [240]
Cinq mars / Vigny, Alfred De – New York, NY. v1-2. 1923 – 1r – us UF Libraries [960]
Cinq mois de l'histoire de france : ou fin de la vie politique de napoleon / Regnault-Warin, Jean Baptiste Joseph Innocent Philadelphe – Paris 1815 [mf ed Hildesheim 1995-98] – 1v on 3mf [ill] – 9 – €90.00 – 3-487-26379-3 – gw Olms [944]
Cinq mois de l'histoire de paris en mil huit cent trente / Lamothe-Langon, Etienne L de – Paris 1831 [mf ed Hildesheim 1995-98] – 1v on 3mf – 9 – €90.00 – 3-487-26061-1 – gw Olms [944]
Cinq-mars : or, a conspiracy under louis 13 / Vigny, Alfred de – Boston: Little, Brown & Co 1889 – 2v [ill] – 1 – (trans by william hazlitt; ill by a dawant & by gaujean) – mf#1519 – us Wisconsin U Libr [830]
Les cinquante-deux serviteurs de dieu : francais – annamites – chinois, mis a mort pour la foi en extreme-orient de 1815 a 1856, dont la cause de beatification a ete introduite en 1840, 1843, 1857; biographies / Launay, Adrien – Paris: Tequi, 1893 [mf ed 1995] – (= ser Yale coll) – 2v (ill) – 1 – 0-524-10276-7 – (in french) – mf#1996-1276 – us ATLA [951]
Cinquantenaire de la banque d'epargne de la cite et du district de montreal – [Montreal?]: [s.n.], [1896?] [mf ed 1980] – 1mf – 9 – 0-665-04080-6 – mf#04080 – cn Canadiana [332]
Cinquantenaire de la fondation de l'asile du bon pasteur de quebec : celebre les 3, 4 et 5 janvier: 1850-1900 – [Quebec?: s.n, 1900?] [mf ed 1980] – 3mf – 9 – 0-665-03868-2 – mf#03868 – cn Canadiana [360]
Cinquantenaire de la fondation de l'asile du bon pasteur de quebec : celebre les 3, 4 et 5 janvier 1900 – [Quebec?: s.n, 1900?] [mf ed 1980] – 3mf – 9 – 0-665-04102-0 – mf#04102 – cn Canadiana [360]
Le cinquantenaire de la mission de calcutta : 28 nov 1859-28 nov 1909 – [Roulers: Jules de Meester, 1909?] [mf ed 1995] – (= ser Yale coll) – 14p – 1 – 0-524-10065-9 – (in french) – mf#1995-1065 – us ATLA [241]
Cinquantenaire de l'arrivee des peres oblats a montreal – Montreal: s.n, 1891?] [mf ed 1980] – 1mf – 9 – 0-665-03867-4 – mf#03867 – cn Canadiana [241]
Cinquantenaire d'enseignement de mm f-x toussaint et n lacasse, professeurs a l'ecole normale laval : soiree donnee a la salle de promotions de l'universite laval, le 19 mai 1893, programme – [Quebec?: s.n, 1893?] [mf ed 1986] – 1mf – 9 – 0-665-57885-7 – mf#57885 – cn Canadiana [378]
Cinquantenaire des oblats de marie immaculee en canada : fetes jubilaires les 7, 8 et 9 decembre 1891 / Guillet, Didace – Montreal: C O Beauchemin, 1891? [mf ed 1980] – 2mf – 9 – 0-665-05131-X – mf#05131 – cn Canadiana [241]
Cinquantenaire des oblats de marie immaculee en canada : fetes jubilaires les 7, 8 et 9 decembre 1891 – Montreal: C O Beauchemin, [1891?] [mf ed 1980] – 2mf – 9 – 0-665-03046-0 – mf#03046 – cn Canadiana [241]
Cinquantenaire des religieuses de notre-dame de charite du bon pasteur d'angers a montreal : fetes jubilaires les 23, 24 et 25 juin 1894 – [Montreal: s.n, 1894?] [mf ed 1981] – 1mf – 9 – mf#13805 – cn Canadiana [360]
Cinquantenaire du college de l'assomption : fetes jubilaires celebrees les 12, 13 et 14 juin 1883 – Montreal: Beauchemin, 1893 [mf ed 1980] – 2mf – 9 – 0-665-03041-X – (in french and english) – mf#03041 – cn Canadiana [378]
Cinquantieme anniversaire de la charte des travailleurs : album souvenir publie par les syndicats catholiques de montreal a l'occasion de la fete du travail 1941 – Montreal: bureau de Jean Nolin, conseil en publicite, [1941?] (mf ed 1992) – 1mf – 9 – (incl english text) – mf#SEM105P1699 – cn Bibl Nat [241]
Cinquantieme anniversaire de la fondation de l'universite laval : programme officiel complet des fetes artistiques, lundi, mardi et mercredi, 23, 24 et 25 juin 1902 – [Quebec?: s.n, 1902?] – 1mf – 9 – 0-665-78161-X – mf#78161 – cn Canadiana [378]
Cinquantieme anniversaire de la fondation du seminaire de ste therese : souvenir des fetes du 22 et 23 juin 1875 – Montreal?: s.n, 1875 – 1mf – 9 – mf#00652 – cn Canadiana [378]
Cinque ports *see* The white and black books of the cinque ports from 1433
Cinque ports chronicle and east sussex observer – [London & SE] East Sussex, Hastings Ref Lib sep 1838-dec 1839 [wkly] – 1 – uk Newsplan [072]
Cinquentenario de belo horizonte / Senna, Caio Nelson De – Belo Horizonte, Brazil. 1948 – 1r – us UF Libraries [972]
Cinquieme conference economique nationale / Toure, Ahmed Sekou – Conakry: Imprimerie nationale "Patrice Lumumba" 1976 – us CRL [330]
Cint Uttam *see* Sneh pancras dis
Cintra, Francisco De Assis *see* Homen da independencia
El cinturon de afrodita / Hurtado de Mendoza, Publio – Caceres: luciano jimenez merino, impresor, 1922 – 1 – sp Bibl Santa Ana [946]
Cinvanja hulpboekie / Ferreira, M – Mkhoma, Malawi. 1937 – 1r – us UF Libraries [960]
Cio – Framingham. 1989-1993 (1) 1989-1993 (5) 1989-1993 (9) – ISSN: 0894-9301 – mf#16395 – us UMI ProQuest [650]
Cio files of john l lewis – 2pt – (= ser Cio and industrial unionism in america; Research colls in labor studies) – 9 – (pt1: correspondence with cio unions, 1929-62 25r isbn 1-55655-048-0 $4465. pt2: cio general files, 1929-55 20r isbn 1-55655-049-9 $3560. with p/g) – us UPA [331]
Cio industrial worker – Portland OR: Labor Newdealer Publ Assoc, 1941 [wkly] – 1 – (cont: labor newdealer) – us Oregon Lib [331]
Cio insight – New York. 2001+ (1,5,9) – ISSN: 1535-0096 – mf#32111 – us UMI ProQuest [000]
Cio news : aluminium workers edition, 1938-1943 : aluminium workers news digest, 1943-1944 / Aluminium Workers of America – (= ser Labor union periodicals, pt 1: the metal trades) – 2r – 1 – $430.00 – 1-55655-232-7 – us UPA [680]
Cio news / American Federation of Labor – 1937 dec 7-1943, 1944-55 – 7r – 1 – (cont by: afl news-reporter; afl-cio news) – mf#1110417 – us WHS [331]
Cio news / American Federation of Labor – v1 n31-v6 n52 [1938 jul 16-1943 dec 27] – 1r – 1 – mf#1426630 – us WHS [331]
Cio news : die casters edition, 1938-1942 / National Association of Die Casting Workers – (= ser Labor union periodicals, pt 1: the metal trades) – 1r – 1 – $210.00 – 1-55655-233-5 – us UPA [680]
Cio news / Congress of Industrial Organizations [US] – 1948-49, 1952-53 – 1r – 1 – (cont by: michigan cio news) – mf#1110417 – us WHS [331]
Cio news / Congress of Industrial Organizations [US] – 1945 aug 6-1955 nov 14 – 1r – 1 – (cont: paper worker [cincinnati oh]; cont by: paperworkers news) – mf#3633132 – us WHS [670]
Cio news / Farm Equipment and Metal Workers of America – v1 n52-v6 n9 [1938 dec 5-1943 mar 1] – 1r – 1 – (cont by: fe [chicago il]) – mf#1053773 – us WHS [630]
Cio news / Federation of Glass, Ceramic and Silica Sand Workers of America – v3 n36-v7 n35 [1940 sep 2-1944 aug 28] – 1r – 1 – (cont: cio news [glass workers edition]; cont by: cio news [glass workers ed: 1944]) – mf#659723 – us WHS [680]
Cio news / Franklin Co. Columbus – jun 1940-42, feb 1943-55 [irreg] – 4r – 1 – mf#B9757-9760 – us Ohio Hist [331]

Cio news / Insurance Workers of America – v3 n1-11/12 [1953 feb 2-dec 28] – 1r – 1 – (cont: cio insurance news letter; cont by: insurance worker [washington [dc]: 1954]) – mf#1125534 – us WHS [360]

Cio news / International Union, Aluminum Workers of America [CIO] – 1938 jul 16-1939 dec 11, 1940 jun 8-1943 jun 21 – 2r – 1 – (cont by: aluminum workers news digest) – mf#1053770 – us WHS [660]

Cio news / International Union of Mine, Mill, and Smelter Workers – v1 n53-v5 n13 [1938 dec 12-1942 mar 30] – 1r – 1 – (cont by: cio news [die casters ed]; union [denver co: 1942]) – mf#1110419 – us WHS [331]

Cio news / International Union, United Automobile, Aircraft, and Agricultural Implement Workers of America – 1943 jan 18-1944, 1945 jan-jul 2 – 2r – 1 – (cont by: wisconsin cio news [local 248 edition]) – mf#3575178 – us WHS [331]

Cio news / Milwaukee County Industrial Union Council – 1938 mar 26-1941, 1942-43, 1944-1945 jul 2 – 3r – 1 – (cont by: wisconsin cio news) – mf#1410322 – us WHS [331]

Cio news : mine, mill and smelter workers international edition, 1938-1942 / International Union of Mine, Mill and Smelter Workers – (= ser Labor union periodicals, pt 1: the metal trades) – 1r – 1 – $210.00 – 1-55655-234-3 – us UPA [680]

Cio news / Montgomery Co. Dayton – mar 1944-sep 1958 [biwkly] – 5r – 1 – mf#B5435-5439 – us Ohio Hist [331]

Cio news / Oil Workers International Union – v1 n22-v6 n15 [1938 may 7-1943 apr 12] – 1r – 1 – (cont: international oil worker [1937]; cont by: cio oil facts) – mf#1008298 – us WHS [622]

Cio news : packinghouse workers edition, 1938-1942 / United Packinghouse Workers of America – (= ser Labor union periodicals, pt 3: food and agricultural industries) – 2r – 1 – $405.00 – 1-55655-622-5 – us UPA [660]

Cio news / Packinghouse Workers Organizing Committee – 1938 oct 1-1940 may 13, 1940 may 27-1942 jan 5 – 2r – 1 – (cont by: packinghouse worker) – mf#1110420 – us WHS [660]

Cio news / Tennessee Industrial Union Council – v7 n9 [1944 feb 28], v8 n52 [1945 dec 24], v9 n4,13,17,21,25,33,43,48,53 [1946 jan 21, mar 25, apr 22, may 20, jun 17, aug 12, oct 21, nov 25, dec 30], v10 n4,16,29,35,43 [1947 jan 27, apr 21, jul 21, sep 1, oct 27] – 1r – 1 – mf#862094 – us WHS [331]

Cio news : united farm equipment and metal workers edition, 1938-1943 / United Farm Equipment and Metal Workers of America – (= ser Labor union periodicals, pt 1: the metal trades) – 1r – 1 – $210.00 – 1-55655-235-1 – us UPA [680]

Cio news / United Gas, Coke, and Chemical Workers of America – 1947 jun 13-1950, 1951 jan-mar – 2r – 1 – (cont: cio news [victory edition]; cont by: international oil worker; united chemical worker) – mf#1053771 – us WHS [660]

Cio news / United Gas, Coke, and Chemical Workers of America – 1944 nov 13-1946 dec 9 – 1r – 1 – (cont: victory [washington dc: 1942]; cont by: cio news [united chemical worker edition]) – mf#1079874 – us WHS [660]

Cio news / United Glass and Ceramic Workers of North America – 1944 sep-1947, 1948-51, 1952-54, 1955 jan-dec 5 – 4r – 1 – (cont: cio news [glass, ceramic and silica sand ed]; cont by: glass workers news) – mf#659721 – us WHS [680]

Cio news / United Railroad Workers of America – v7 n26-v11 n5 [1944 jun 26-1948 feb 23] – 1r – 1 – (cont by: cio railroad news) – mf#1053776 – us WHS [380]

Cio news / United Shoe Workers of America – 1938 dec 12-1950 sep 18 – 1r – 1 – mf#1053775 – us WHS [380]

Cio news / United Stone and Allied Products Workers of America – 1954 nov-1955 dec – 1r – 1 – mf#1052747 – us WHS [690]

Cio news / United Transport Service Employees of America – 1943 may 10-1945 jul 2 – 1r – 1 – (cont: bags and baggage) – mf#1053778 – us WHS [380]

Cio news cannery workers edition *see* Cannery and field union news, 1937 / cio news cannery workers edition, 1938-1939 / ucapawa news, 1939-1944 / fta news, 1945-1950

Cio news/ district 50 edition / United Mine Workers of America – v4 n6, 12-40 [1941 feb 10/mar 24-oct 6] – 1r – 1 – (cont by: district 50 news) – mf#1110418 – us WHS [622]

Cio news for railroad workers / Congress of Industrial Organizations [US] – v1 n1-9 [1951 apr-dec] – 1r – 1 – (cont: cio railroad news; cont by: railroad news, cio) – mf#1053779 – us WHS [380]

Cio news/ retail and wholesale edition / Congress of Industrial Organizations [US] – 1938 may 28-1940 jun 29 – 1r – 1 – (cont: retail employee [new york ny]; cont by: retail and wholesale employee) – mf#3564728 – us WHS [380]

Cio oil facts / Congress of Industrial Organizations [US] – v1 n1-v3 n4 [1943 jun e 9-1945 aug 21] – 1r – 1 – (cont: cio news [oil workers ed]; cont by: international oil worker [1945]) – mf#1008394 – us WHS [622]

Cio railroad news / United Railroad Workers of America – v1 n1-9 [1951 apr-dec] – 1r – 1 – (cont: cio news. railroad labor ed; cont by: cio news for railroad workers) – mf#3633944 – us WHS [380]

Cio world affairs bulletin / Congress of Industrial Organizations – v1-v3 n1 [1951 oct-1954 may] – 1r – 1 – mf#1053783 – us WHS [337]

Cio-pac news service / Congress of Industrial Organizations – v1 n1-23 [1945 dec 22-1946 nov 4] – 1r – 1 – mf#1053781 – us WHS [338]

Cipa : the journal of the chartered institute of patent agents / Chartered Institute of Patent Agents – London. 1971-1974 (1) – mf#10100 – us UMI ProQuest [343]

Cipere / Barnard, T H – Fort Victoria, Zimbabwe. 1953 – 1r – us UF Libraries [960]

The cipher in the plays, and on the tombstone / Donnelly, Ignatius – Minneapolis: The Verulam Publ. Co., 1899. 5,(9),372p. With: Shakspere Not Shakespeare by W.H. Edwards. 1 reel. 1295 – 1 – us Wisconsin U Libr [420]

Cipolletta, Eugenio *see* Memorie politiche sui conclavi da pio 7 a pio 9

Cipreses creen en dios / Gironella, Jose Maria – Barcelona, Spain. 1963 – 1r – us UF Libraries [025]

Cipriani motectorum liber primus – 1544 – (= ser Mssa) – 3mf – 9 – €50.00 – mfchl 89 – gw Fischer [780]

Cipriano *see* Heredia

Cipriano, De Utrera *see* Ntra sra de altagracia

Ciquard, Francois *see* Portrait d'un missionnaire apostolique

Circa sacra / Bryce, James Bryce, Viscount – Edinburgh, Scotland. 1843 – 1r – us UF Libraries [240]

Circannuale schwankungen in der inzidenz des morbus basedow : retrospektive analysen der daten einer endokrinologischen fachpraxis von 1985-1995 / Nordmann, Thomas – (mf ed 1999) – 1mf – 9 – €30.00 – 3-8267-2630-8 – mf#DHS 2630 – gw Frankfurter [616]

Circe o el amor / Belaval, Emilio S – Barcelona, Spain. 1963 – 1r – us UF Libraries [972]

Circle / Minneapolis American Indian Center – 1979 aug-1987 dec – 1r – 1 – mf#1277521 – us WHS [307]

Circle : a publication / Boston Indian Council – v1 n3-6 [1976 jun-nov], v1 n20 [1977 mar], v2 n1-12, [1977 apr-1978 jun], v3 n1-12 [1978 jul/aug-1981 apr] v4 1,9-10 [1981 may/jun, dec-1982 jan], v5 n11-12 [1982 feb-mar/apr], v6 n1-9 [1982 may/jun-1984 aug], 1983 nov-dec – 1r – 1 – (cont by: knowledge of the circle) – mf#1363371 – us WHS [307]

Circle f / Woodworkers' Industrial Union of Canada – n2-3 [1949 apr 4-18] – 1r – 1 – mf#681696 – us WHS [680]

Circle news – iss/v1-v3 n4 [1978 jul-1980 apr,jul?] – 1r – 1 – (cont by: four elements) – mf#639191 – us WHS [071]

The circle of christian doctrine : a handbook of faith framed out of a layman's experience / Kinloch, William Penney – 2nd ed. Edinburgh: Edmonston & Douglas, 1861 [mf ed 1985] – 1mf – 9 – 0-8370-4470-7 – mf#1985-2470 – us ATLA [240]

A circle of the arts and sciences, for the use of school and young persons : containing a clear yet brief explanation of the principles and objects of the most important branches of human knowledge / Mavor, William Fordyce – London: printed for Richard Phillips, 1808 – (= ser 19th c children's literature) – 6mf – 9 – mf#6.1.43 – uk Chadwyck [000]

The circle of theology : an introduction to theological study / Clarke, William Newton – Cambridge: University Press, 1897 – 1mf – 9 – 0-8370-2673-3 – (incl bibl ref) – mf#1985-0673 – us ATLA [200]

Circle tour of pinellas county / Phillips, Roland – s.l, s.l? 1936 – 1r – us UF Libraries [978]

Circlet / Union of National Defence Employees – 1978 jan-1985, v20 n1-2,19-v27 n4 [1986 jan 10-24, may 23-1990 may 25, oct-1993 nov/dec – 2r – 1 – mf#1046172 – us WHS [355]

Circling the caribbean / Marvel, Tom – New York, NY. 1937 – 1r – us UF Libraries [972]

El circo romano de merida : memoria de las excavaciones practicadas de 1920 a 1925 / Melida, Jose Ramon – Madrid: rev. arch. bibl, 1925 – 1 – sp Bibl Santa Ana [946]

Circolo Matematico di Palermo *see* Rendiconti

Circonscriptions indigenes / Magotte – Dison-Verviers, Belgium. 1934 – 1r – us UF Libraries [960]

Circuit design – Alpharetta. 1990-1990 (1) 1990-1990 (5) 1990-1990 (9) – (cont by: printed circuit design) – ISSN: 1047-5567 – mf#16433,02 – us UMI ProQuest [621]

Circuit litteraire sur voltaire et rousseau : ou, paysages d'un songe a la derive / Fortin-Roussel, Robert – [Trois-Rivieres: Module de lettres et de linguistique, UQTR, 1979] (mf ed 2001) – 2mf – 9 – (pref by paul langlois; [ill by pierre jeanson]) – mf#SEM105P3310 – cn Bibl Nat [914]

A circuit of the globe / Galloway, Charles Betts – Nashville, TN: Pub House ME Church, South, 1897 [mf ed 1993] – (= ser Methodist coll) – 2mf – 9 – 0-524-06902-6 – mf#1991-2815 – us ATLA [242]

A circuit of the globe : a series of letters of travel across the american continent... / McLean, Archibald – St Louis: Christian Pub Co, 1897 [mf ed 1992] – (= ser Christian church (disciples of christ) coll) – 2mf – 9 – 0-524-05017-1 – mf#1991-2187 – us ATLA [917]

Circuit reports, 1835-1898, and the swanston collection on the ra and ba military campaigns, 1873 / Fiji. Methodist Church – 5r – 1 – (restricted access) – mf#PMB1093 – at Pacific Mss [355]

Circuit rider / Washington State American Revolution Bicentennial Commission – 1975 oct-1976 oct – 1r – 1 – (cont: newsletter) – mf#361197 – us WHS [975]

Circuit rider / Washington State Historical Society – v8 n4-v10 n2 [1978-1980 dec] – 1r – 1 – (cont: news notes [washington state historical society]; cont by: history highlights) – mf#614663 – us WHS [978]

The circuit rider : a tale of the heroic age / Eggleston, Edward – New York: J.B. Ford, 1874 – 1mf – 9 – 0-7905-5986-2 – mf#1988-1986 – us ATLA [420]

Circuit world – Bradford. 2001+ (1,5,9) – ISSN: 0305-6120 – mf#21874 – us UMI ProQuest [621]

Circuits assembly – Manhasset. 1990+ (1,5,9) – ISSN: 1054-0407 – mf#18750 – us UMI ProQuest [621]

Circuits manufacturing – San Francisco. 1961-1990 (1) 1971-1990 (5) 1976-1990 (9) – ISSN: 0009-7306 – mf#1528 – us UMI ProQuest [621]

Circulaire : aux abonnes du moniteur canadien, en terminant la septieme annee du moniteur canadien... – S.l: s.n, 1855? – 1mf – 9 – mf#63699 – cn Canadiana [331]

Circulaire : j'ai l'honneur de vous transmettre un certain nombre d'exemplaires des nos 88, 89 (le 90e n'est pas encore arrive ici) et 91 des annales de la propagation de la foi... – [S.l: s.n, 1844?] [mf ed 1986] – 1mf – 9 – 0-665-53809-X – mf#53809 – cn Canadiana [241]

Circulaire : je vous informe que la retraite de mm les cures s'ouvrira, au seminaire, vendredi, le 28 aout prochain... / Baillargeon, Charles-Francois – [s.l: s.n, 1868?] [mf ed 1986] – 1mf – 9 – mf#56165 – cn Canadiana [241]

Circulaire : une uvre sainte et destinee a produire de grands fruits s'est etablie, depuis quelques annees, au milieu de notre peuple religieux... / Catholic Church. Archidiocese de Quebec – Quebec?: s.n, 1855? – 1mf – 9 – mf#25795 – cn Canadiana [360]

Circulaire a messieurs les cures, missionnaires et autres pretres du diocese de montreal : en vertu d'un indult ad decennium, que j'ai recu du st. siege, en date du 31 mai dernier / Bourget, Ignace – [Montreal?: I Bourget? 1840?] [mf ed 1985] – 1mf – 9 – 0-665-14487-3 – mf#14487 – cn Canadiana [241]

Circulaire a messieurs les pretres et autres ecclesiastiques du diocese de montreal / Catholic Church. Diocese de Montreal. Eveque (1836-1840: Lartigue) – [Montreal?: s.n, 1839?] [mf ed 1985] – 1mf – 9 – 0-665-18926-5 – mf#18926 – cn Canadiana [241]

Circulaire a mm les cures et missionnaires du diocese de montreal : en vous adressant la lettre pastorale ci-jointe / Bourget, Ignace – [Montreal?: I Bourget? 1856?] [mf ed 1985] – 1mf – 9 – 0-665-04230-2 – mf#04230 – cn Canadiana [241]

Circulaire a mrs les cures du diocese de montreal / Catholic Church. Diocese de Montreal. Eveque (1836-1840: Lartigue) – [s.l: s.n, 1839?] [mf ed 1985] – 1mf – 9 – 0-665-18925-7 – mf#18925 – cn Canadiana [241]

Circulaire annoncant au clerge la retraite pastorale et le second synode diocesain / Bourget, Ignace – [Montreal?: s.n, 1864?] [mf ed 1985] – 1mf – 9 – 0-665-18693-2 – mf#18693 – cn Canadiana [241]

Circulaire annoncant la celebration du troisieme concile provincial de quebec... / Catholic Church. Diocese de Montreal. Eveque (1840-1876 : Bourget) – [s.l: s.n.] 1863 [mf ed 1985] – 1mf – 9 – 0-665-01698-0 – mf#01698 – cn Canadiana [241]

Circulaire au clerge : 1: caisse de st joseph; 2: documents officiels appartenant a la fabrique... / Taschereau, Elzear-Alexandre – S.l: s.n, 1877? – 1mf – 9 – mf#56597 – cn Canadiana [241]

Circulaire au clerge : 1. cinquantieme anniversaire de l'ordination de mgr cazeau... / Archdiocese of Quebec. Catholic Church – S.l: s.n, 1879? – 1mf – 9 – mf#56598 – cn Canadiana [241]

Circulaire au clerge : 1. union spirituelle du clerge. 2. cinquantieme anniversaire de l'episcopat de pie 9... / Archdiocese of Quebec. Catholic Church – S.l: s.n, 1876? – 1mf – 9 – mf#56582 – cn Canadiana [241]

Circulaire au clerge / Bourget, Ignace – [Montreal?: s.n, 1853?] [mf ed 1985] – 1mf – 9 – 0-665-27113-1 – mf#27113 – cn Canadiana [241]

Circulaire au clerge : dans quelques heures, je me mettrai en route... – S.l: s.n, 1871? – 1mf – 9 – mf#57383 – cn Canadiana [241]

Circulaire au clerge : evidemment le seigneur est irrite contre son peuple, puisque les pluies continuelles menacent serieusement le succes de la recolte... – S.l: s.n, 1888? – 1mf – 9 – mf#60652 – cn Canadiana [630]

Circulaire au clerge : les lettres que je recois de rome font instance pour que nous vous opposions de toutes nos forces aux mariages entre cousins germains... – S.l: s.n, 1860? – 1mf – 9 – mf#51639 – cn Canadiana [230]

Circulaire au clerge : les souffrances de nsp le pape sont, a nos yeux, une mine precieuse qu'il faut exploiter au profit de la loi de notre bon peuple... / Catholic Church. Diocese de Montreal. Eveque – S.l: s.n, 1849? – 1mf – 9 – mf#51640 – cn Canadiana [241]

Circulaire au clerge : la st jean-baptiste a coutume de resserrer les biens qui unissent ici la religion et la patrie... – S.l: s.n, 1868? – 1mf – 9 – mf#55254 – cn Canadiana [241]

Circulaire au clerge : vous gemissez comme moi, de l'etrange disposition du peuple par rapport au st pere, et c'est vraiment a n'y rien comprendre que d'en voir un si grand nombre livre a un tel vertige... / Catholic Church. Diocese de Montreal. Eveque – S.l: s.n, 1860? – 1mf – 9 – mf#49402 – cn Canadiana [241]

Circulaire au clerge, accompagnant le mandement de visite pour 1861 et 1862 : pour etre cependant envoyee des maintenant a chaque cure du diocese / Bourget, Ignace – [Montreal?: I Bourget?, 1861?] [mf ed 1985] – 1mf – 9 – 0-665-01699-9 – mf#01699 – cn Canadiana [241]

Circulaire au clerge concernant les 40 heures, l'ordo, l'indulgence des chapelets, l'annee religieuse, etc / Bourget, Ignace – [Montreal?: I Bourget?, 1861?] [mf ed 1985] – 1mf – 9 – 0-665-01700-6 – mf#01700 – cn Canadiana [241]

Circulaire au clerge de montreal : je m'empresse de vous adresser ci-jointe copie d'une lettre que je viens de recevoir de s em le card barnabo au sujet du vin de messe / Bourget, Ignace – [Montreal?: I. Bourget?, 1861?] [mf ed 1985] – 1mf – 9 – 0-665-18912-5 – mf#18912 – cn Canadiana [240]

Circulaire au clerge de montreal accompagnant le mandement du 8 dec 1862 / Bourget, Ignace – [Montreal?: I Bourget?, 1862?] [mf ed 1985] – 1mf – 9 – 0-665-01701-4 – mf#01701 – cn Canadiana [241]

Circulaire au clerge du diocese de Montreal : j'accompagne le mandement du jubile de cette circulaire qui mettra... / Bourget, Ignace – [s.l: I Bourget?, 1854?] [mf ed 1985] – 1mf – 9 – 0-665-18485-9 – mf#18485 – cn Canadiana [241]

Circulaire au clerge du diocese de montreal : au sortir de notre retraite, imitons par nos dispositions, comme par notre nombre... / Bourget, Ignace – [Montreal?: I Bourget? 1852?] [mf ed 1985] – 1mf – 9 – 0-665-01703-0 – mf#01703 – cn Canadiana [241]

Circulaire au clerge du diocese de montreal / Catholic Church. Diocese de Montreal. Eveque (1836-1840: Lartigue) – [Montreal?: s.n, 1837?] [mf ed 1985] – 1mf – 9 – 0-665-18928-1 – mf#18928 – cn Canadiana [241]

Circulaire au clerge du diocese de montreal : comme rien n'est plus important que l'uniformite dans le clerge d'un meme diocese... / Bourget, Ignace – [Montreal?: I Bourget?, 1845?] [mf ed 1985] – 1mf – 9 – 0-665-01704-9 – (in french and latin) – mf#01704 – cn Canadiana [241]

Circulaire au clerge du diocese de montreal : dans sa cirulaire du 19 oct dernier, mgr l'eveque de montreal... / Larocque, Joseph, – [Montreal?: s.n.] 1855 [mf ed 1985] – 1mf – 9 – 0-665-18103-5 – (in french and latin) – mf#18103 – cn Canadiana [241]

Circulaire au clerge du diocese de montreal : en lui adressant le mandement du 18 avril 1838 / Catholic Church. Diocese de Montreal. Eveque (1836-1840: Lartigue) – [s.l: s.n, 1838?] [mf ed 1985] – 1mf – 9 – 0-665-18927-3 – mf#18927 – cn Canadiana [241]

Circulaire au clerge du diocese de montreal : en vous envoyant le rapport ci-contre de l'assemblee du clerge, tenue le jour de la st jacques... / Bourget, Ignace – [Montreal?: I Bourget?, 1848?] [mf ed 1985] – 1mf – 9 – 0-665-01705-7 – mf#01705 – cn Canadiana [241]

Circulaire au clerge du diocese de montreal : j'ajoute au mandement et au precis ci- joints concernant la pieuse association de l'immaculee conception... / Bourget, Ignace – [Montreal?: I Bourget?, 1854?] [mf ed 1985] – 1mf – 9 – 0-665-18909-5 – mf#18909 – cn Canadiana [241]

Circulaire au clerge du diocese de montreal : je crois devoir, par la pastorale ci- jointe, informer le diocese du resultat de mon voyage en europe... / Bourget, Ignace – [Montreal?: I Bourget?, 1841?] [mf ed 1985] – 1mf – 9 – 0-665-18910-9 – mf#18910 – cn Canadiana [241]

Circulaire au clerge du diocese de montreal : je me borne en ce moment, (ecrivait, le 9 oct dernier... / Bourget, Ignace – [Montreal?: I Bourget? 1860?] [mf ed 1985] – 1mf – 9 – 0-665-18911-7 – mf#18911 – cn Canadiana [241]

Circulaire au clerge du diocese de montreal accompagnant le mandement du 1 janvier 1865 / Bourget, Ignace – [s.l: I Bourget?, 1865?] [mf ed 1985] – 1mf – 9 – 0-665-01702-2 – mf#01702 – cn Canadiana [241]

Circulaire au clerge du diocese de montreal, sur le cholera / Bourget, Ignace – [Montreal?: s.n, 1854?] [mf ed 1985] – 1mf – 9 – 0-665-07273-2 – mf#07273 – cn Canadiana [241]

Circulaire au clerge du diocese de montreal sur le grand incendie du huit juillet : j'accompagne la lettre pastorale de ce jour de quelques observations... / Bourget, Ignace – [Montreal?: I. Bourget?, 1852?] [mf ed 1985] – 1mf – 9 – 0-665-10836-2 – mf#10836 – cn Canadiana [241]

Circulaire au clerge et au peuple pour demander du beau temps : vous direz desormais, jusqu'a nouvel ordre, apres celle deja prescrite pour le pape, la collecte... – S.l: s.n, 1867? – 1mf – 9 – mf#57903 – cn Canadiana [240]

Circulaire de l'association d'annexion de montreal / Association d'annexion de Montreal – [Montreal?: s.n.], 1850 [mf ed 1984] – 1mf – 9 – 0-665-32235-6 – mf#32235 – cn Canadiana [971]

Circulaire du comite de l'association d'annexion de montreal / Association d'annexion de Montreal – [S.l: s.n, 1849?] [mf ed 1984] – 1mf – 9 – 0-665-22157-6 – mf#22157 – cn Canadiana [971]

Circulando el cuadrado / Lopez, Cesar – Habana, Cuba. 1963 – 1r – us UF Libraries [972]

Circular : as you are aware the heart of our holy father pope leo 13, profoundly touched by the miseries which at present afflict human society... – [Saint John, NB?: s.n, 1901?] – 1mf – 9 – 0-665-98516-9 – mf#98516 – cn Canadiana [241]

Circular – La Plata: El Observatorio 1949-51 (irreg) [mf ed 2001] – (= ser Publicaciones del observatorio astronomico de la universidad nacional de la plata) – n1-9[1949-51] on 1r [ill] – 1 – (cont by: serie circular (universidad nacional de la plata. observatorio astronomico); n5-6 publ out of chronological order (in 1949)) – mf#film mas c5057 – us Harvard [520]

Circular – 1851 nov 6-1855 oct 11, 1855 oct 18-1862 jul 10, 1862 jul 17-1868 jul 13, 1868 jul 20-1870 dec 26 – 4r – 1 – (cont: free church circular; cont by: oneida circular) – mf#3070771 – us WHS [071]

Circular : an election of great importance to mcgill university will take place on thursday the 21st, viz, the election of an attending in-door physician to the general hospital... / Howard, Robert Palmer – S.l: s.n, 1885? – 1mf – 9 – mf#53490 – cn Canadiana [360]

Circular / Harvard College Observatory – Cambridge [MA]: The Observatory 1900- (irreg) [1-457(1895-(1951)] [mf ed 1999] – 5r [ill] – 1 – (1st vol incl circulars 1 to 50, iss oct 30 1895-may 9 1900) – mf#film mas c4138 – us Harvard [520]

Circular : in response to the circular which i had the honor to address to your municipality... on the subject of municipal tax exemptions... – S.l: s.n, 1889? – 1mf – 9 – mf#54270 – cn Canadiana [336]

Circular – [Scotland] Alloa: printed & publ by J Waddell & D Melville jan 1869-dec 1950 (wkly) [mf ed 2003] – 40r – 1 – (missing: jul 1869-73, 1894; cont by: alloa circular [may 1874-dec 1879]; alloa circular and clackmannanshire herald [jan 1880-dec 1938]; alloa circular and hillsfoot record [jan 1939-dec 1950]) – uk Newsplan [072]

Circular : office of "the favorite", 319 st antoine street, montreal... / Bosse, C L – S.l: s.n, 187-? – 1mf – 9 – mf#46263 – cn Canadiana [420]

Circular : to be read by the pastor to the faithful / O'Brien, Cornelius, Archbishop – Halifax, NS?: s.n, 1887? – 1mf – 9 – mf#06308 – cn Canadiana [241]

Circular : to the shareholders of the commercial bank of canada – S.l: s.n, 1865? – 1mf – 9 – mf#62434 – cn Canadiana [332]

Circular : to the shareholders of the saint john gas light company – S.l: s.n, 1891? – 1mf – 9 – mf#59270 – cn Canadiana [650]

Circular : to the shareholders of the toronto, grey and bruce railway company... – S.l: s.n, 1881? – 1mf – 9 – mf#28853 – cn Canadiana [380]

Circular : we have purchased from messrs i and f burpee and co all their stock of hardware... – [Saint John, NB?: s.n, 1877?] [mf ed 1986] – 1mf – 9 – 0-665-54023-X – mf#54023 – cn Canadiana [680]

Circular – Wilmington. 1821-1825 (1) – mf#4439 – us UMI ProQuest [240]

[Circular] : association of mechanics' institutes of ontario, i beg to forward you, in accordance with a resolution passed at the last annual meeting, the modified scheme for awarding prizes to mechanics' institutes – [s.l: s.n. 1873?] [mf ed 1987] – 1mf – 9 – 0-665-28840-9 – mf#28840 – cn Canadiana [374]

[Circular] : i take the earliest opportunity of presenting you with the following extracts from a law of this province, regulating its intercourse with the united states ... / Colt, Jabez – [Montreal?: s.n, 1819?] [mf ed 1993] – 1mf – 9 – 0-665-91313-3 – mf#91313 – cn Canadiana [343]

[Circular] : order for western advertiser enclosed please find one dollar, being the subscription of mpo to the western advertiser for one year – [s.l: s.n. 1873?] [mf ed 1987] – 1mf – 9 – 0-665-28848-4 – mf#28848 – cn Canadiana [071]

[Circular] : to subscription to banner [1] year from [7 march] 184[5] to [27 feb] 184[6]...i take the liberty of annexing the above account for the banner newspaper / Brown, George – [s.l: s.n. 1846?] [mf ed 1987] – 1mf – 9 – 0-665-18704-1 – mf#18704 – cn Canadiana [071]

[Circular] : to the shareholders of the toronto, grey & bruce railway company on the 25th of may 1...on the subject of the proposed lease of your railway to the grand trunk railway company – [s.l: s.n. 1881?] [mf ed 1987] – 1mf – 9 – 0-665-28853-0 – mf#28853 – cn Canadiana [380]

[Circular] : we ask your attention to the enclosed documents which should be sufficient to give you confidence in the superior and reliable character of our undertaking – [Toronto?: s.n. 1874?] [mf ed 1987] – 1mf – 9 – 0-665-28854-9 – mf#28854 – cn Canadiana [912]

Circular 1 cost of living index for foreign family / Commercial Advisory Foundation in Indonesia – Djakarta, 1967-1971 – 1mf – 9 – (missing: 1967-1969) – mf#SE-1380 – ne IDC [959]

Circular address / The Mississippi Society for Baptist Missions – 1817 – 1 – $5.00 – us Southern Baptist [242]

Circular address on botany and zoology / Rafinesque-Schmaltz, C S – Court House Station. 1967-1982 (1) 1976-1982 (5) 1976-1982 (9) – 1mf – 9 – mf#8102 – ne IDC [500]

Circular b / Commercial Advisory Foundation in Indonesia – Djakarta, [1956]-1963 – 2mf – 9 – mf#SE-677 – ne IDC [959]

Circular de la agrupacion socialista de valencia y reglamento de la academia / Academia Socialista Preparatoria...Escuelas Populares de Guerra – Valencia: Artes Graficas 1937 [mf ed 1977] – (= ser Blodgett coll) – 1mf – 9 – mf#w801 – us Harvard [946]

Circular en la que el director.. / Canseco, Manuel – 1824 – 9 – sp Bibl Santa Ana [946]

Circular letter / Association of Research Libraries. Foreign Newspaper Microfilm Project – v1-16 1955-63 – 1 – us CRL [020]

Circular letter / Cone, Spencer H – 1824 – 1 – 5.00 – us Southern Baptist [242]

Circular letter from the president, pontiac pacific junction railway co... : ottawa, dec 15th, 1893 / Beemer, J J – Ottawa: s.n, 1893 – 1mf – 9 – mf#11392 – cn Canadiana [380]

A circular letter suggesting a petition to parliament for the protection of the temporalities fund / Barclay, John – Toronto: J Barclay, 1882 – 1mf – 9 – mf#02973 – cn Canadiana [242]

Circular letters / Nicolas, Charles Joseph – 30 dec 1918-20 jun 1941 – 1r – 1 – mf#pmb doc209 – at Pacific Mss [240]

Circular letters of ministers and messengers / Northamptonshire Baptist Association, England – (ms.). 1765-1820 – 1 – us Southern Baptist [242]

Circular letters of the secretary of the treasury ("t" series), 1789-1878 / U.S. Treasury Dept. Office of the Secretary – (= ser General records of the department of the treasury) – 5r – 1 – (with printed guide) – mf#M735 – us Nat Archives [336]

Circular n_o 11 de organizacion encuadramiento de profesiones por sindicatos y secciones / Servicio Nacional de Sindicatos – Caceres: Garcia Floriano, 1938 – 1 – sp Bibl Santa Ana [946]

Circular of the anti-persecution union *see* Investigator, 1843

Circular of the astronomical observatory of the warsaw university = Okolnik obserwatorjum astronomicznego uniwersytetu warszawskiego / Uniwersytet Warszawski. Obserwatorium Astronomiczne – Warsaw: The Observatory [1945- [irreg] [mf ed 2002] – 1r – 1 – (cont: okolnik obserwatorjum astronomicznego uniwersytetu jozefa pilsudskiego w warzawie; in english#) – mf#film mas c5403 – us Harvard [520]

Circular of the committee of the annexation association of montreal / Association d'annexion de Montreal – [Montreal?: s.n, 1849?] [mf ed 1984] – 1mf – 9 – 0-665-22156-8 – mf#22156 – cn Canadiana [971]

Circular of the committee of the annexation association of montreal / Association d'annexion de Montreal – [S.l: s.n, 1849?] [mf ed 1984] – 1mf – 9 – 0-665-22185-1 – mf#22185 – cn Canadiana [971]

Circular of the department of agriculture containing "the copyright act of 1875" = Circulaire du department de l'agriculture contenant "l'acte de 1875 sur la propriete litteraire et artistique" – Ottawa: B Chamberlin, 1875 [mf ed 1986] – 1mf – 9 – 0-665-57488-6 – (in english and french) – mf#57488 – cn Canadiana [346]

Circular of the gammon school of theology / Gammon Theological Seminary – Atlanta GA: Clark UP [annual] [mf ed 2006] – 1885/1886 [complete] on 1r – 1 – (filmed with suppl: address of the rev atticus g hargood...at the fourth annual opening of the gammon school of theology; cont by: gammon theological seminary. catalogue of gammon school of theology) – mf#2006-s006 – us ATLA [242]

Circular recomendando la puntualidad del ayuntamiento.. / Canseco, Manuel – 1828 – 9 – sp Bibl Santa Ana [946]

Circular to bankers, 1827-60 – 17r – 1 – mf#97092 – uk Microform Academic [336]

Circulars / Republic Observatory, Johannesburg – Pretoria: Govt Printer. v7 n121 [1962]-v8 n131 [1971] [irreg] [mf ed 2001] – 2v on 1r – 1 – (cont: circular...of the union observatory; cont by: south african astronomical observatory. circulars [issn 0376-7884]; imprint varies) – mf#film mas c5078 – us Harvard [520]

Circulars and regulations.. / U.S. Dept of the Interior. General Land Office – With reference tables and index.Comp. by C.G. Fisher. 1696p. Washington: GPO, 1930.85-303 – 9 – us LLMC [324]

Circulation – New York. 1950+ (1) 1966+ (5) 1970+ (9) – ISSN: 0009-7322 – mf#2255 – us UMI ProQuest [612]

La circulation dans le sud cameroun / Billard, Pierre – Lyons, Impr des beaux-arts 1961 – us CRL [960]

Circulation of roman catholic versions of the bible by the british – London, England. 1868 – 1r – us UF Libraries [220]

The circulation of the blood : and, andrea cesalpino of arezzo / Arcieri, Giovanni P – New York: S F Vanni, 1945 [mf ed 1995] – 193p (ill) – 1 – (incl ind) – mf#1282 – us Wisconsin U Libr [612]

Circulation research – Dallas. 1953+ (1) 1966+ (5) 1970+ (9) – ISSN: 0009-7330 – mf#2278 – us UMI ProQuest [612]

Circulatory system of the cow's udder / Becker, R B – Gainesville, FL. 1942 – 1r – us UF Libraries [636]

Circulo de Artesanos *see*
- Memoria, ano 1937
- Memoria, ano 1958

Circulo de Artesanos. *see* Memoria, ano 1938

Circunscripciones electorales y division politico-... – Bogota, Colombia. 1964 – 1r – us UF Libraries [972]

Circus – New York. 1966-1978 (1) 1975-1978 (5) 1975-1978 (9) – (cont by: circus weekly) – ISSN: 0009-7365 – mf#10944 – us UMI ProQuest [790]

Circus – New York. 1979-1993 (1,5,9) – (cont: circus weekly) – ISSN: 0009-7365 – mf#10944,02 – us UMI ProQuest [790]

Circus Fans Association of America *see*
- Chatter from around the white tops
- White tops

Circus Historical Society *see*
- C h s bandwagon
- Hobby-bandwagon

Circus raves – New York. 1974-1975 (1) – mf#11201 – us UMI ProQuest [790]

Circus weekly – New York. 1978-1979 (1,5,9) – (cont: circus. cont by: circus) – ISSN: 0164-9248 – mf#10944,01 – us UMI ProQuest [790]

Circus World Museum *see*
- Monthly report
- Newsletter

Circus World Museum [Baraboo WI] *see* Bulletin of the circus...

Circus world museum news release – Baraboo WI. 1960 mar 9-1971 apr 29 – 1r – 1 – mf#1054501 – us WHS [060]

Cirenaica sconosciuta – [Firenze] Sansoni, [1952] – 9 – us CRL [960]

Ciriaco perez bustamante, la fundacion de un imperio... / Bayle, Constantino – Madrid: Razon y Fe, 1941 – 1 – sp Bibl Santa Ana [946]

Cirilli, Rene *see* Les praetres danseurs de rome

Cirni, A F *see*
- Svccessi dell' armata della mta cca destinata all' impresa di tripoli di barberia, della presa delle gerbe, e progressi dell'armata turchesca...
- Svccessi della armata della maesta' catolica destinata all' impresa di tripoli di barberia, della presa delle gerbe, e progressi dell' armata turchesca...

Cirp : annals of the international institution for production engineering research – Berne; Stuttgart: Technische Rundschau, v23. 1974 – 1r – us CRL [620]

Cirp annals ... manufacturing technology – Berne: Technische Rundschau. v24 1975; v29 1980 – 2r – us CRL [670]

Cirri, Giovanni Battista *see* Sei trij per violino, viola e violoncello concertanti, opera 18

Ciruelo de yuan pei fu / Pedroso, Regino – Habana, Cuba. 1955 – 1r – us UF Libraries [972]

Ciruelo, P *see*
- Curso de geometria y matematicas
- Cursus quattor mathematicorum artium liberalium

Cirugia, osteologia, miologia y vasos, operaciones, medicamentos, supuracion... – SL, SA – 10mf – 9 – sp Cultura [617]

Cirurgia y cirujanos – Mexico City. 1949-1954 (1) – ISSN: 0009-7411 – mf#436 – us UMI ProQuest [617]

Cis congressional bills, resolutions and laws on microfiche : retrospective collection / U.S. Congress – 73rd congress (1933-34)-104th congress (1995-96) – 9 – apply for price – us CIS [324]

Cis congressional member organizations and caucuses : guide to publications and policy materials / U.S. Congress – 1991- – ca 325mf per yr – 9 – apply for price – us CIS [324]

Cis presidential executive orders and proclamations on microfiche – 2pts. 1789-1983 – 9117mf – 9 – $29,990.00 set – (index and mf set $34,800.00. index pts 1 + 2 $8,675.00) – us CIS [324]

Cis unpublished u.s. house of representatives committee hearings on microfiche : retrospective index coverage and microfiche collection / U.S. Congress. House of Representatives – Pts 1-5. 1833-1964 – (= ser Cis Unpublished U.S. Congressional Committee Hearings) – 9 – (1833-1936 1691mf $6,925 index $820. 1937-46 2491mf $10,200 index $1,345. 1947-54 5211mf $19,900 index $1,495. 1955-58 2975mf $13,355 index $845. 1959-64 ca 3500mf $14,995 index $995) – us CIS [324]

Cis unpublished us senate committee hearings on microfiche : retrospective index coverage and microfiche collection / U.S. Congress. Senate – Pts 1-3. 1823-1972 – (= ser Cis Unpublished U.S. Congressional Committee Hearings) – 9925+ mf – 9 – (1823-1964 9072mf $29,815 index $2,585. 1965-68 853mf $3,945 index $425. 1969-72 ca 725mf $3,355 index $345) – us CIS [324]

Cis us congressional committee hearings on microfiche library catalog records / U.S. Congress – Pts 1-8. 1833-1969 – 84,067mf – 9 – apply for prices – (print index accompanies 8pt coll. available individually or as a complete coll (except pts1-3, sold only as a combined coll). pts 1-3: 1833-1934 23rd-74th congress. pt 4: 1935-94 74th-78th congress. pt 5: 1945-52 79th-82nd congress. pt 6: 1953-58 83rd-85th congress. pt 7: 1959-64 86th-88th congress. pt 8: 1965-69 89th-91st congress, 1st session.) – us CIS [324]

Cis us congressional committee prints on microfiche : retrospective index coverage and microfiche collection / U.S. Congress – Pts 1-3. 1830-1969 – 17,743mf – 9 – (pt1: 1911-1969 $22,585. pt2: 1917-1969 $5,095. pt3: 1830-1969 $29,475. combined coll: pts1-3 1830-1969 $51,440. complete ind set $2170) – us CIS [324]

Cis u.s. congressional journals on microfiche : retrospective collection / U.S. Congress – 1789-1978 – 4177mf – 9 – $8380.00 – us CIS [073]

Cis us executive branch documents, 1789-1909 on microfiche – 1995 – pt1-6 + suppl – 9 – apply for prices – (provides precise access to documents from federal executive agencies in existence during the time covered by the us govt's "checklist of united states public documents, 1789-1909". printed ind) – us CIS [324]

Cis u.s. executive branch documents, 1910-1932 – Pts 1-7. 1996- – ca 8500mf – 9 – apply for price – (7pt-series to be publ over 7-yr period between 1996-2002) – us CIS [324]

Cis u.s. senate executive documents and reports on microfiche : retrospective index coverage and microfice collection / U.S. Senate – 1817-1969 – 1153mf + index – 9 – $6,125.00 – (index available $820.00. access to facts behind treaties and nominations) – us CIS [324]

Cis u.s. serial set on microfiche / U.S. Congress – pt1-14. 1789-1969 – 116,000mf – 9 – apply for price – (provides access to pre-1970 congressional reports and documents. group 1: american state papers and the 15th-34th congresses (1789-1857). groups 2-12: 35th-91st congresses (1857-1969). group 13: index by reported bill numbers (1819-1969). group 14: index and carto-bibliography of maps (1789- 1897, 1897-1925 and 1925-1969)) – us CIS [324]

Cisalpinische blaetter / Lessmann, Daniel – Berlin 1828 [mf ed Hildesheim 1995-98] – 2v on 4mf – 9 – €120.00 – 3-487-29293-9 – gw Olms [880]

Cis/index and cis/microfiche library – 1970-1997 – 9 – apply for prices – (microfiche and ind coll incl complete, hearings, ltd ed, and serial set) – us CIS [020]

Cisne / Branly, Roberto – Habana, Cuba. 1956 – 1r – us UF Libraries [972]

Cisne de apolo, de las excelencias, y dignidad y todo que al arte poetica y versificatoria pertenece : los metodos y estylos que en sus obras deue seguir el poeta – Medina del Campo: Godinez de Millis, 1602 – 9 – us CRL [810]

Cisneros, D *see* Sitio, naturaleza y propiedades de la ciudad de mejico, aguas y vientos. a que esta sujeta y si tiempos del ano. necesidad de su cono cimiento para el exercicio de la medicina...

Cisneros, Juan *see*
– Un caso de extirpacion de la laringe...de esta operacion
– Congreso internacional de medicina verificado en berlin del 4 al 9 de agosto de 1890
– Contribucion al estudio del colera
– Papiloma comeo de la laringe, laringotomia curacion

Cisneros, Luys de *see* Historia de el principio

Cisneros Y Betancourt, Salvador *see* Appeal to the american people on behalf of cuba

Cissell, William B *see* Dental health attitudes and knowledge levels of rural and suburban texas

Cissoko, Sekene-Mody *see* Recueil des traditions orales des mandingue de gambie et de casamance

Cistercian studies quarterly – Sonoita. 1981+ (1,5,9) – ISSN: 1062-6549 – mf#13039 – us UMI ProQuest [240]

Cistercienser-chronik – 1(1889)-38(1926) – 387mf – 9 – €738.00 – ne Slangenburg [241]

Cistercii reflorescentis / Morotius, C – Augustae Taurinorum, 1690 – 13mf – 8 – €25.00 – ne Slangenburg [241]

Der cisterzienser : eine erzaehlung aus der zeit des markgrafen otto 1. von brandenburg / Schmidt, Ferdinand – Duesseldorf: Felix Bagel [18–?] [mf ed 1995] – 1r [ill] – 1 – (filmed with: sueden und norden / hermann schmid) – mf#3738p – us Wisconsin U Libr [880]

Cist's weekly advertiser – Cincinnati: Charles Cist, mar 22 1847-apr 29 1853 – (bound with: western general advertiser) – us CRL [071]

Cistula entomologica – London 1869-85 – v1-3 on 27mf – 9 – mf#z-932/2 – ne IDC [590]

Cit : [a novel] / Ma Ma Le – Ran Kun: Gya nay kyo Ma Ma Le 1951 [mf ed 1990] – 1r with other items – 1 – (in burmese) – mf#mf-10289 seam reel 187/1 [§] – us CRL [830]

Cit panna rhu thon a thve thve / Kyo Cin, Dok Ta – Ran Kun: Mrui To Ca pe 1968 [mf ed 1990] – 1r with other items – 1 – (in burmese) – mf#mf-10289 seam reel 132/5 [§] – us CRL [150]

Cit tim mon : [a novel] / Bhui San, Cac kuin U – Mantale: Rai Ran Ca pe thana Rum to kri 1959 [mf ed 1990] – 1r with other items – 1 – (in burmese) – mf#mf-10289 seam reel 203/9 [§] – us CRL [830]

Cita de prensa / Seminario Venezolano-Norteamericano De Periodismo – Caracas, Venezuela. 1961 – 1r – us UF Libraries [972]

The citadel of ethiopia / Gruehl, M – London, 1932 – 5mf – 9 – mf#NE-20292 – ne IDC [916]

Citadel square baptist church. charleston county. south carolina : church records – 1868-Apr 1969 – 1 – us Southern Baptist [242]

Citadel square baptist church. charleston county. south carolina : church records – April 17, 1955-1969; Ladies Benevolent Society, 1867-1904; Burial Records and List of Members. 1828-1830; Register of Names of Colored Members. 1826 – 1 – us Southern Baptist [242]

Citanias extremenas / Monsalud, Marques de – Caceres: Tip. Enc. y Lib. de Jimenez, 1901. Rev. Extremadura, 1901 – 1 – sp Bibl Santa Ana [946]

Citation / American Medical Association – v1-24. 1958-71 – 1 – us AMS Press [616]

Citation – dec 1964-jun 1968 – 1r – at Pascoe [079]

Citations of the statutes of alberta and saskatchewan (1886-1912) / Miall, Edward – 2nd.ed. Toronto: Carswell, 1912 – 1 – (supplement. (1912-17). toronto, 1917. 105p. II-2330) – us L of C Photodup [348]

Cite – London, UK. 26 Jul 1880-9 Jun 1882 – 1 – uk British Libr Newspaper [072]

Cite de saint-jerome, la porte des laurentides : histoire, industrie, commerce, statistique – [Quebec (Province): s.n.], 1950 [mf ed 2001] – 2mf – 9 – mf#SEM105P3319 – cn Bibl Nat [971]

La cite des trois-rivieres = The city of three rivers / Trois-Rivieres. Chambre de commerce – [Trois-Rivieres?]: Chambre de commerce de Trois-Rivieres, [1923?] (mf ed 1995) – 1mf – 9 – (in french and english) – mf#SEM105P2371 – cn Bibl Nat [971]

Cite des voix / Descaves, Pierre – Paris, France. 1938 – 1r – us UF Libraries [440]

Cite educative / Association generale des etudiants de la Faculte de l'education permanente de l'Universite de Montreal – v2 n1 sep 1986- (bimthly) [mf ed 1988] – 9 – (cont: revue de l'ageefep) – mf#SEM105P1076 – cn Bibl Nat [378]

Cite nouvelle – Brussels Belgium. 20 sep 1944-jul 1945 – 1r – 1 – uk British Libr Newspaper [074]

Citeaux in de nederlanden – 1(1950)-25(1974) – 161mf – 9 – €94.00 – ne Slangenburg [241]

Cithara – St. Bonaventure. 1961+ (1) 1972+ (5) 1976+ (9) – ISSN: 0009-7527 – mf#7174 – us UMI ProQuest [000]

Cithara christiana : psalmodiarum sacrarum libri septem = Christliche harpfen / Lauterbach, Johann – Lipsiae: Steinman 1586 – (= ser Hqab. literatur des 16. jahrh.) – 9mf – 9 – €85.00 – mf#1586a – gw Fischer [780]

Cithara christiana : psalmodiarum sacrarum libri 7 = Christliche harpffen / Lauterbach, Johann – Leipzig: Steinman 1585 – (= ser Hqab. literatur des 16. jahrh.) – 9mf – 9 – €85.00 – mf#1585b – gw Fischer [780]

Cithera melica, vel opus musicum plane novum... : modulos aloquot sacros praecipuis festicitatibus anniversarijs inservientes continens...vocibus 12, 10 and 8 / Duling, Anton – Magdaeburgi: typis J Joachimi Boelij...1620 [mf ed 19–] – 8pt on 1r – 1 – mf#film 1250 – us Sibley [780]

Citibank monthly economic letter – New York. 1976-1981 (1) 1976-1981 (5) 1976-1981 (9) – (cont: monthly economic letter) – ISSN: 0015-279X – mf#6203,01 – us UMI ProQuest [332]

Cities – Kidlington. 1983+ (1,5,9) – ISSN: 0264-2751 – mf#17215 – us UMI ProQuest [350]

The cities and bishoprics of phrygia : being an essay of the local history of phrygia from the earliest times to the turkish conquest / Ramsay, William Mitchell – Oxford: Clarendon Press. 2v. 1895-97 – 3mf – 9 – 0-7905-0196-1 – (incl bibl ref) – mf#1987-0196 – us ATLA [930]

The cities and bishoprics of phrygia, vol 1 pts 1 and 2 : pt 1: the lycos valley and south-western phrygia; pt 2: west and west-central phrygia / Ramsay, William Mitchell – Oxford 1895, 1897 – 15mf – 8 – €29.00 – ne Slangenburg [930]

The cities and cemeteries of etruria / Dennis, George – London 1848 – (= ser 19th c art & architecture) – 14mf – 9 – mf#4.1.271 – uk Chadwyck [930]

The cities and principal towns of the world – London 1830 [mf ed Hildesheim 1995-98] – 3mf – 9 – €90.00 – 3-487-29947-X – gw Olms [910]

The cities and towns of china : a geographical dictionary / Playfair, George Macdonald Home – Shanghai, Hong Kong: Kelly & Walsh Ltd, 1910 [mf ed 1996] – (= ser Yale coll) – xii/582p/lxxvi – 1 – 0-524-10281-3 – (with pref) – mf#1996-1281 – us ATLA [915]

Cities of belgium / Allen, Grant – London: G Richards, 1897 – (= ser Grant allen's historical guides) – 3mf – 9 – (incl ind) – mf#05015 – cn Canadiana [914]

Cities of our faith : and other discourses and addresses / Caldwell, Samuel Lunt – Boston: Houghton, Mifflin, 1890 [mf ed 1993] – 1mf – 9 – 0-524-08265-0 – (biogr sketch of dr caldwell by oakman sprague stearns) – mf#1993-3020 – us ATLA [242]

Cities of paul : beacons of the past rekindled for the present / Wright, William Burnet – Boston: Houghton, Mifflin, 1905 [mf ed 1986] – 1mf – 9 – 0-8370-7438-X – mf#1986-1438 – us ATLA [930]

Cities of southern italy ând sicily / Hare, Augustus John Cuthbert – New York: G Routledge & Sons [188-] [mf ed 1986] – 1r [ill] – 1 – (filmed with: with: essays on milton / thompson, e n s) – mf#1554 – us Wisconsin U Libr [914]

The cities of st paul : their influence on his life and thought / Ramsay, William Mitchell – New York: A C Armstrong; London: Hodder and Stoughton, 1908 – (= ser The Dale Memorial Lectures) – 2mf – 9 – 0-8370-9574-3 – mf#1986-3574 – us ATLA [240]

Cities of the eastern roman provinces / Jones, A H M – 1937 – 9 – $21.00 – us IRC [930]

The cities of the eastern roman provinces / Jones, A H M – Oxford, 1937 – 12mf – 9 – mf#NE-107 – ne IDC [956]

Citimart news and tempo – Providence, RI. 1977-1981 (1) – mf#68461 – us UMI ProQuest [071]

Citizen – 1894 mar 31-apr 2 – 1r – 1 – mf#3094081 – us WHS [071]

Citizen – Halifax, Nova Scotia. 20 dec 1864, 8 feb 1871-31 jan 1888 – 37r – 1 – (aka: citizen and evening chronicle) – uk British Libr Newspaper [071]

Citizen / Ashtabula Co. Andover – aug 1977-jan 1982 [wkly] – 2r – 1 – mf#B13020-13021 – us Ohio Hist [071]

Citizen / Butler Co. Oxford – 8 iss 1857, 1890 [wkly] – 1r – 1 – mf#B9579 – us Ohio Hist [071]

Citizen / Colorado Association of Public Employees – 1976 dec-2-1982, 1983-88 – 2r – 1 – mf#655042 – us WHS [350]

Citizen – Chicago IL 4 jan 1890-30 jan 1897 – 1 – (cont by: chicago citizen [6 feb-25 dec 1897]) – uk British Libr Newspaper [071]

Citizen – Cedarburg, Grafton WI. 1968 jun 20-1969 mar 13, 1969 mar 20-1970 jan 9 – 2r – 1 – (cont by: squire [thiensville wi]) – mf#960666 – us WHS [071]

Citizen – South Omaha, NE: Citizen Print Co. -v4 n37. aug 6 1909 (wkly) [mf ed v3 n18. mar 27 1908-aug 6 1909 filmed 1980] – 1r – 1 – (cont: independent. cont by: globe=citizen) – us NE Hist [071]

Citizen – Los Angeles CA. 1907 mar 1/1908 jan 31-1916 jan 7/apr 28 – 9r – 1 – (cont: union labor news [los angeles ca]; cont by: los angeles citizen) – mf#702049 – us WHS [071]

Citizen / Crawford Co. Crestline – v1 n1. oct 1903-dec 1906 [wkly] – 2r – 1 – mf#B10571-10572 – us Ohio Hist [071]

Citizen / Cuyahoga Co. Cleveland – feb 1891-jan 1893 [wkly] – 1r – 1 – mf#B10570 – us Ohio Hist [331]

Citizen / Cuyahoga Co. East Cleveland – v1 n1. dec 1970-dec 1975 [wkly, mthly, biwkly] – 1r – 1 – mf#B30885 – us Ohio Hist [071]

Citizen / Franklin Co. Columbus – 1910, 1917, 1919, 1936 (gap filler) [daily] – 6r – 1 – mf#B1431-1436 – us Ohio Hist [071]

Citizen / Franklin Co. Columbus – jan 1947-apr 1951 [daily] – 38r – 1 – mf#B1393-1430 – us Ohio Hist [071]

Citizen / Franklin Co. Columbus – sep-oct 1942 (gap filler) [daily] – 2r – 1 – mf#B308-309 – us Ohio Hist [071]

Citizen – Harriman TN. v2 n35-44 [1904 aug 31-nov 2] – 1r – 1 – mf#868695 – us WHS [071]

Citizen / Highland Co. Leesburg – (1928-33, 41, 7/44-1/70, 87-1993) [wkly] – 23r – 1 – mf#B34059-34081 – us Ohio Hist [071]

Citizen / Highland Co. Leesburg – jan 1914-dec 1926 [wkly] – 5r – 1 – mf#B12373-12377 – us Ohio Hist [071]

Citizen / Highland Co. Leesburg – jan 1971-dec 1986 [wkly] – 14r – 1 – mf#B29562-29575 – us Ohio Hist [071]

Citizen – Jackson MS 1955-89 – 1,5,9 – ISSN: 0578-3283 – mf#3266 – us UMI ProQuest [322]

Citizen – Johannesburg, South Africa – 122r – 1 – sa State Libr [079]

Citizen – Kaukauna WI. 1904 jan – 1r – 1 – mf#876771 – us WHS [071]

Citizen / Knox Co. Fredericktown – v1 n1. (aug 1922-dec 1936) [wkly] – 5r – 1 – mf#B34728-34732 – us Ohio Hist [071]

Citizen – London, UK. 1913-Jun 1922. -w.4 reels – 1 – uk British Libr Newspaper [072]

Citizen / Montgomery Co. Dayton – jan 1950-jun 1951 [wkly] – 1r – 1 – mf#B5466 – us Ohio Hist [071]

Citizen / Morgan Co. McConnelsvill – (aug 1905-jun 1906) [daily] – 1r – 1 – mf#B11570 – us Ohio Hist [071]

Citizen / National Wallace for President Committee – 1948 apr-oct – 1r – 1 – mf#1393559 – us WHS [320]

Citizen – New York NY. 7 jan 1854-3 oct 1857 – 4r – 1 – ie National [071]

Citizen : [north dayton-northridge edition] / Montgomery Co. Dayton – nov 1946-nov 1948 [biwkly] – 1r – 1 – mf#B5462 – us Ohio Hist [071]

Citizen – North Vancouver, British Columbia, CN. apr 1938-dec 1973 – 22r – 1 – cn Commonwealth Imaging [071]

Citizen – Ottawa, Canada. Jul 1908-9 aug 1909; 18 may 1939 – 7r – 1 – uk British Libr Newspaper [071]

Citizen *see*
– The collinwood citizen
– Ottawa citizen (daily ed)
– Review / citizen / torch / news

The citizen – Cape Town: SA Library, 15 dec 1897-9 jul 1898 – 1r – 1 – (cont: kimberley elector; cont by: south african citizen) – mf#ms00269 – sa National [079]

The citizen – Greeley, NE: Edward P Curran. 1v. v1 n1-n7. jul 1-aug 12 1918 (wkly) – 1r – 1 – us NE Hist [071]

The citizen – Ansley, NE: A H Barks, may 17 1901-sep 1902// (wkly) – 1r – 1 – (merged with: ansley chronicle to form: chronicle-citizen) – us Bell [071]

The citizen – Collinwood, OH: [Frank A Bowman]. n5 feb 1 1901-sept 15 1905 – 2r – 1 – (weekly republican newspaper publ in a cleveland suburb later annexed to the city; issues are intermixed on the film with those of the nottingham citizen and the collinwood citizen) – us Western Res [071]

The citizen *see* Miscellaneous newspapers of las animas county, reel 2

Citizen Action for Lasting Security et al *see* Nuclear news bureau

Citizen airman: the official magazine of the air national guard and air force reserve – Washington. 1984+ (1) 1984+ (5) 1984+ (9) – (cont: air reservist) – ISSN: 0887-9680 – mf#7421,01 – us UMI ProQuest [629]

Citizen and artisan *see* Citizen and irish artisan

Citizen and evening chronicle *see* Citizen

Citizen and gazette / Champaign Co. Urbana – (apr 1852-54,56-61,76,86-mar 1887) [wkly] – 4r – 1 – mf#B33404-33407 – us Ohio Hist [071]

Citizen and gazette – Urbana, OH. 1848-1891 (1) – mf#65696 – us UMI ProQuest [071]

Citizen and gazette series / Champaign Co. Urbana – (mar 1890-oct 1914), 1915-16 [wkly, daily] – 20r – 1 – mf#B9543-9562 – us Ohio Hist [071]

Citizen and irish artisan – Dublin, Ireland. 31 may 1879-17 apr 1880 – 1/2r – 1 – (aka: citizen and artisan) – uk British Libr Newspaper [072]

Citizen and pred – Keller, TX. 1978-1984 [1] – mf#69075 – us UMI ProQuest [071]

Citizen and waterford commercial record – Waterford, Ireland. 9 sep 1859-1888, 1890-24 dec 1896 – 31r – 1 – (missing 1889, aka: waterford citizen and commercial record; waterford citizen and bi weekly advertiser; waterford citizen county news and bi weekly advertiser; waterford citizen new ross news and weekly advertiser) – uk British Libr Newspaper [072]

Citizen [banbury ed] – Banbury, England 24 aug 1989-10 jan 2003 [mf ed 1986-] – 1 – (cont: banbury citizen [2 mar-17 aug 1989]; cont by: banbury & district citizen [17 jan 2003-24 jun 2005]) – uk British Libr Newspaper [072]

Citizen [blackburn ed] – [NW England] Blackburn Lib 1978-79 – 1 – uk MLA; uk Newsplan [072]

The citizen. (chicago citizen) – Chicago. 4 Jan 1890-25 Dec 1897.-w. 8 reels – 1 – uk British Libr Newspaper [071]

Citizen cio / Congress of Industrial Organization – v1 n1-11 [1945 nov-1946 dec] – 1r – 1 – mf#1054506 – us WHS [331]

Citizen (darwen ed) – [NW England] Darwen, Blackburn Lib 1978-79 – 1 – (title change: darwen citizen [24 may 1979-dec 1980*]) – uk MLA; uk Newsplan [072]

Citizen democrat / Carroll Co. Carrollton – jan 1856-dec 1857 [wkly] – 1r – 1 – mf#B3982 – us Ohio Hist [071]

Citizen [dundee, scotland] : incorporating dundee labour bulletin – [Scotland] Dundee: Dundee Trades & Labour Council nov 1946-mar 1949 (mthly) [mf ed 2003] – 5v on 1r – 1 – (cont by: dundee citizen) – uk Newsplan [072]

Citizen garden city record and advertising journal – sep 22-nov 24, dec 1 1906, 1908-10, jan 14-dec 29 1911, 1912-jun 1925, jul 3-dec 25 1925, 1926-dec 25 1931, 1932-38, jan 6-jun 30 1939, jul 7-dec 29 1939, 1940-aug 1978, sep 21 1978-89, jul 1990-92, jan 8-jun 25 1993, jul 1993-sep 1994, oct 7-dec 23 1994, 1995-jun 1996, jul 5-dec 1996 – 153 1/2r – 1 – (aka: the gazette [letchworth and baldock]; letchworth and baldock citizen; letchworth and baldock citizen gazette) – uk British Libr Newspaper [072]

Citizen intelligencer – v1 n12-v2 n3 [1976 oct-1977 fall] – 1r – 1 – mf#363410 – us WHS [071]

Citizen [inverkeithing, scotland] – [Scotland] Fife, Inverkeithing: Inverkeithing Labour Party 20 oct 1919 (wkly) [mf ed 2004] – 1r – 1 – uk Newsplan [335]

Citizen [johannesburg] – Pretoria: State Library [mf ed 29 dec 1993-15 jan 1994-] – 122r – 1 – sa State Libr [079]

Citizen journal – Columbus, OH. 1959-1985 (1) – mf#60555 – us UMI ProQuest [071]

Citizen news / Citizens for Social Responsibility – 1983 sum-1986 feb – 1r – 1 – (cont: newsletter [citizens for social responsibility]; cont by: citizen's news [arcata ca]) – mf#1005333 – us WHS [360]

The citizen newspaper *see* Miscellaneous newspapers of mesa county

The citizen of england : his rights and duties / Armitage-Smith, G – London/Edinburgh, 1895 – 2mf – 9 – $3.00 – mf#LLMC 92-163 – us LLMC [322]

Citizen participation / Lincoln Filene Center for Citizenship and Public Affairs – v1 n1-v7 n3 [1979 sep/oct-1986 sum] – 1r – 1 – mf#456289 – us WHS [322]

Citizen patriot – Jackson, MI. 1918-2000 (1) – mf#60163 – us UMI ProQuest [071]

Citizen power / Citizen/Labor Energy Coalition – n1-16 [1980 dec-1985:fall] – 1r – 1 – mf#1131023 – us WHS [360]

Citizen register – Ossining, NY. 1991-1998 (1) – mf#61952 – us UMI ProQuest [071]

Citizen series / Trumbull Co. Youngstown – jul 1915-jun 1922, jun 1924-jun 1925 [wkly] – 6r – 1 – (klu klux klan) – mf#B3218-3223 – us Ohio Hist [071]

Citizen series (klu klux klan) / Mahoning Co. Youngstown – jul 1915-jun 1922,jun 1924-jun 1925 [wkly] – 6r – 1 – mf#B3218-3223 – us Ohio Hist [071]

Citizen times – Asheville, NC. 1991-2000 (1) – mf#60683 – us UMI ProQuest [071]

Citizen times journal – premiere iss-4th iss [1982 jan 14-feb 11] – 1r – 1 – mf#597184 – us WHS [071]

Citizen toussaint / Korngold, Ralph – New York, NY. 1965, 1944 – 1r – us UF Libraries [972]

Citizen / town / news / Butler Co. Oxford – (1885-95,1926-30) scattered [semiwkly] – 3r – 1 – (title changes) – mf#B29889-29891 – us Ohio Hist [071]

Citizen/Labor Energy Coalition *see* Citizen power

Citizens action news – 1975 jun-1978 feb, 1980 aug/sep-1986 sum – 2r – 1 – (cont by: united states of acorn) – mf#1219011 – us WHS [360]

Citizens Alert [Organization : Chicago IL] *see* Bridge

Citizens Alliance of Minneapolis *see* Records

Citizens' Association of Montreal *see* [Rules and regulations of the...]

Citizens Budget Commission [New York NY] *see* Cbc quarterly/

Citizens call – Philipsburg, MT. 1894-1901 (1) – mf#64595 – us UMI ProQuest [071]

Citizen's Choice, Inc *see* Citizen's voice

Citizen's claw / Citizens for Constitutional Law [US] – v2 n1 [1984 jan], 1984 mar-1987 dec, 1988 jan, may-jun/jul, sep/oct-dec, 1989 jan/feb – 1r – 1 – mf#1054510 – us WHS [322]

Citizen's Constitutional Committee *see* Newsletter

Citizens' Council of Louisiana *see* Councilor

Citizens Energy Council *see*
- Bulletin of the citizens...
- Energy news digest
- Energy news digest of nuclear hazards versus alternative energies
- Nuclear opponents
- People and power
- Watch on the aec

Citizens for Constitutional Law [US] *see* Citizen's claw

Citizens for Dave Obey Committee *see* News from the citizens for dave obey committee

Citizens for Decency through Law *see* National decency reporter

Citizens for Decency through Law [US] *see* Cdl reporter

Citizens for Social Responsibility *see*
- Citizen news
- Newsletter

Citizens for the republic newsletter – 1977 feb 1-1985 nov – 1r – 1 – (cont by: closed circuit) – mf#622872 – us WHS [323]

Citizens' governmental research bureau – v51 n1-v57 n11 [1963 jan 12-1969 dec 20] – 1r – 1 – (cont by: bulletin [citizens' governmental research bureau [wi]]) – mf#601947 – us WHS [322]

Citizens' Governmental Research Bureau [WI] *see* Bulletin of the citizens'...

Citizen's guard – New Orleans LA. 1871 aug 20 – 1r – 1 – mf#861365 – us WHS [360]

Citizens Historical Association IN *see* Ohio biographical sketches, 1938-1951

Citizens in Defense of Civil Liberties [US] et al *see* Public eye

Citizens Insurance Company of Canada *see* Report of the directors to the shareholders of the citizens insurance company

Citizens Natural Resources Association of Wisconsin *see* Wisconsin conservationist

Citizens of to-morrow : a study of childhood and youth from the standpoint of home mission work / Guernsey, Alice Margaret – New York: Fleming H Revell c1907 [mf ed 1986] – (= ser Home mission study course) – 1mf – 9 – 0-8370-6574-7 – (incl app) – mf#1986-0574 – us ATLA [240]

Citizens paper – Limerick, Ireland. 5 oct-26 oct 1867 – 1/4r – 1 – uk British Libr Newspaper [072]

Citizens party of minnesota news – [v2 n1]-v4 n8 [1981 aug-1984 aug, 1985 apr-aug] – 1r – 1 – (cont: minnesota citizens party newsletter) – mf#1345126 – us WHS [325]

Citizen's press / Noble Co. Caldwell – sep 1880-apr 1884 [wkly] – 2r – 1 – mf#B8562-8563 – us Ohio Hist [071]

Citizens' report / South Louisiana Citizens' Council – v10 n3 [1968 apr]-1982 jun – 1r – 1 – mf#615832 – us WHS [360]

Citizens' reporter – River Falls WI. 1863 jan 24, mar 14, jun 27 – 1r – 1 – mf#935582 – us WHS [071]

Citizens Utility Board [WI] *see* Cub prints

Citizen's voice / Citizen's Choice, Inc – v8 n1-v11 n6 [1983/84 dec/jan-1986 oct/nov] – 1r – 1 – mf#1519137 – us WHS [322]

Citizens voice – Kohima, India. May 1969-1975 – 5r – 1 – us L of C Photodup [079]

Citizens weekly – St Georges, Australia. 21 dec 1959; 4 jan-19 dec 1960; 10 apr-4 sep 1961 – 1/4r – 1 – uk British Libr Newspaper [072]

Citizenship and salvation, or, greek and jew : a study in the philosophy of history / Lloyd, Alfred Henry – Boston: Little, Brown, 1897 – 1mf – 9 – 0-8370-4296-8 – mf#1985-2296 – us ATLA [100]

Citonga grammar and vocabulary for the use of the settlers / Casset, A – n.p., Zambia. 19–? – 1r – us UF Libraries [470]

Citonga reading and writing / Torrend, J – Chikuni, Zambia. 1934 – 1r – us UF Libraries [470]

Le citoyen / Defoy, Henri – Paris: C Amat, 1912 – 5mf – 9 – 0-665-97864-2 – mf#97864 – cn Canadiana [322]

Le citoyen : journal quotidien politique, industriel et commercial – Paris. no. 1-54. 4 mars-29 avr 1870 – 1 – fr ACRPP [074]

Le citoyen – Paris. no. 1-602. 1er oct 1881-27, mai 1883. mq no. 8-12, 83 – 1 – fr ACRPP [074]

Citoyen Pinto *see* Association libertiste

La citoyenne – Paris. no. 1-165. 13 fevr 1881-90. mq no. 78 – 1 – fr ACRPP [074]

Citrus / Edwards – s.l, s.l? 193-? – 1r – us UF Libraries [634]

Citrus – Tampa, FL. v1-4. 1938/39-1941 – 2r – us UF Libraries [634]

Citrus and vegetable magazine – Tampa, FL. v22 n11-v54 n12. 1960-1991 – 18r – us UF Libraries [634]

Citrus and vegetable world – Winter Haven. 1972-1973 (1) 1972-1973 (5) (9) – ISSN: 0009-7608 – mf#7076 – us UMI ProQuest [634]

Citrus canker – Gainesville, FL. 1914 – 1r – us UF Libraries [634]

Citrus canker : a preliminary report / Stevens, H E – Gainesville, FL. 1914 – 1r – us UF Libraries [634]

Citrus canker, iii / Stevens, H E – Gainesville, FL. 1915 – 1r – us UF Libraries [634]

Citrus center, glades county, florida / Huss, Veronica E – s.l, s.l? 193-? – 1r – us UF Libraries [634]

Citrus county / Coll, Aloyisus – s.l, s.l? 1936 – 1r – us UF Libraries [634]

Citrus county – s.l, s.l? 1939 – 1r – us UF Libraries [634]

Citrus county chronicle – Inverness, FL. 1932 jan 07-1999 nov – 242r – (gaps) – us UF Libraries [071]

Citrus county star – Mannfield (Manville?), FL. v1 n2-v2 n35. 1888 jan 21-1889 sep 05 – 1r – (missing: 1888 jan 28-feb 18, mar 03, 17, 21, apr 14-21, may 05-12, 26-jun 02, 16-jul 07, 21-1989 aug) – us UF Libraries [071]

Citrus culture in florida / Wheeler, H J – Jacksonville, FL. 1923 – 1r – us UF Libraries [634]

Citrus fertilizer experiments / Collison, S E – Gainesville, FL. 1919 – 1r – us UF Libraries [634]

Citrus fruit laws / Florida – Tallahassee, FL. 1939 – 1r – us UF Libraries [634]

Citrus fruits and health – Lakeland, FL. 1940 – 1r – us UF Libraries [634]

Citrus fruits and their culture / Hume, H Harold – New York, NY. 1915 – 1r – us UF Libraries [634]

Citrus industry – Bartow. 1920+ (1) 1971+ (5) 1977+ (9) – mf#2441 – us UMI ProQuest [634]

Citrus industry – Tampa, FL. v1-v53 n1-9. 1920-1972:jan-sep – 16r – (missing: v1 n2-4) – us UF Libraries [634]

Citrus industry magazine – Bartow, FL. v53 n10-v62 n9. 1972 oct-dec – 4r – us UF Libraries [634]

Citrus industry of florida – Tallahassee, FL. 1947 – 1r – us UF Libraries [634]

Citrus insects and their control / Watson, J R – Gainesville, FL. 1926 – 1r – us UF Libraries [634]

Citrus magazine – Tampa, FL. v4 no7-v22 n 10. 1942-1959/60 – 19r – us UF Libraries [634]

Citrus profits – Lakeland, FL. 1931 – 1r – us UF Libraries [634]

Citrus propagation / Camp, A F – Gainesville, FL. 1931 – 1r – us UF Libraries [634]

Citrus pulp silage – Gainesville, FL. 1946 – 1r – us UF Libraries [634]

Citrus scab / Fawcett, H S – Gainesville, FL. 1912 – 1r – us UF Libraries [634]

Citrus-grove cooperative caretaking / Brooke, Donald Lloyd – s.l, s.l? 1942 – 1r – us UF Libraries [634]

Citt kramum / Ken Vac Sak – Bhnam Ben: Ron Bumb Deb [mf ed 1989] – 1r with other items – 1 – (title on added t.p.: "coeur vierge" poesies; in khmer) – mf#mf-10289 seam reel 130/7 [§] – us CRL [480]

Citt muay thloem muay : pralom lok knun manosancetana / Kuy Yak Hu – Pat Tam Pan: Pannagar Q'yn H'uen 2502 [1959] [mf ed 1989] – 1r with other items – 1 – (in khmer) – mf#mf-10289 seam reel 095/05 [§] – us CRL [480]

Citta *see* Gati sacca

Il cittadino – New York. Weekly, Italian Jan 14 1915-Aug 28 1919. Incomplete. Not collated – 1 – us NY Public [071]

Cittadino italo-americano / Mahoning Co. Youngstown – jan 1920-dec 1932 [wkly] – 5r – 1 – (in italian) – mf#B3977-3981 – us Ohio Hist [071]

Cittadino italo-americano / Trumbull Co. Youngstown – jan 1920-dec 1932 [wkly] – 5r – 1 – (in italian) – mf#B3977-3981 – us Ohio Hist [071]

Cittavisuddhiprakarana : sanskrit and tibetan texts / ed by Patel, Prabhubhai Bhikhabhai – Santiniketan: Visva-Bharati 1939 [mf ed 1982] – (= ser Visva-bharati studies 8) – 1r – 1 – (incl bibl; pref by vidhushekhara bhattacharya) – mf#238 – us Wisconsin U Libr [280]

La citt...d'iddio incarnato : descritta per don vincenzo giliberto... / Giliberto, Vincenzo – Modona: Appresso Giulian Cassiani, 1608-15. 3v – 13mf – 9 – mf#O-1584 – ne IDC [090]

City – nos. 1-4. 1967-68 – 1 – us AMS Press [800]

City – Washington. 1967-1972 (1) 1971-1972 (5) – ISSN: 0009-7675 – mf#2466 – us UMI ProQuest [710]

City advertiser – [Northern Ireland] Belfast jul 1894-may 1895 [mf ed 2002] – 1r – 1 – uk Newsplan [072]

The city advertiser and monthly visitor – Montreal: Wilsons and Nolan, [1852-185-?] – 9 – ISSN: 1190-7592 – mf#P04216 – cn Canadiana [420]

City advocate – Lyons, IA. 1856-1873 (1) – mf#63296 – us UMI ProQuest [071]

City and country homes – 1990 oct 6/20-1991 mar 9, 1991 mar 9/23-sep 7/21, sep 21-dec 14 – 3r – 1 – mf#1825891 – us WHS [071]

City and county cork general advertiser *see* Scraggs cork general advertiser

City and society – Washington. 1989-1992 (1) – ISSN: 0893-0465 – mf#16332 – us UMI ProQuest [301]

City and state – Chicago. 1985-1994 (1,5,9) – ISSN: 0885-940X – mf#14394 – us UMI ProQuest [338]

City and west end news – Auckland, NZ. sep 1973-dec 1974, jul-dec 1975, jan 1976-dec 1977 – 1 – (title changes to: city news fr jan 1976-dec 1977) – mf#11.19 – nz Nat Libr [079]

"The city below the hill" : a sociological study of a portion of the city of montreal, canada / Ames, Herbert Brown – Montreal?: s.n, 1897 – 2mf – 9 – mf#17943 – cn Canadiana [305]

City centres of early christianity / Aytoun, Robert Alexander – London; New York: Hodder and Stoughton, 1915 – 1mf – 9 – 0-7905-4317-6 – mf#1988-0317 – us ATLA [240]

The city church and its social mission : a series of studies in the social extension of the city church / Trawick, Arcadius McSwain – New York: Association Press, 1913 – 1mf – 9 – 0-7905-6133-6 – (incl bibl ref) – mf#1988-2133 – us ATLA [240]

City club bulletin – v1 n1-v6 n5 [1915 nov-1921 jun 15] – 1r – 1 – (cont by: city club news [milwaukee wi: 1921]) – mf#1108266 – us WHS [071]

City club news – 1921 nov 11-1927 jun 10, 1927 sep 16-1930 jun 13, 1930 sep 19-1937 oct 11 – 3r – 1 – (cont: city club bulletin [milwaukee wi]; cont by: city club news [milwaukee, wi: 1938]) – mf#1108267 – us WHS [071]

City club news – v23-26 [1937 sep/1938 may-1941], v27-30 [1941 may/1942 may-1944 jun/1945 may], v31-34 [1945 jun-1948/49], v35-40 [1949 jun-1955 may], v41-45 [1956-1970 may] – 5r – 1 – (cont: city club news [milwaukee [wi]]) – mf#1124817 – us WHS [071]

City club of cleveland speeches, ms 3517 – Cleveland, Cuyahoga, OH. 6 Nov 1917-25 May 197? – 1 – us Western Res [978]

City code of the city of fort pierce, florida – s.l, s.l? 1929? – 1r – us UF Libraries [978]

City commission agrees to deed beach tract to fort – s.l, s.l? 1939 – 1r – us UF Libraries [978]

City council journal / Bruce Publishing Co – v1 n4-8 [1892 jun 15-1894 sep] – 1r – 1 – mf#1054522 – us WHS [071]

City directories, 1843-1862 / Columbus, OH – 2r – 1 – mf#B26232-26233 – us Ohio Hist [978]

City directories, (1889-1894, 1899-1923) / Warren, OH – 4r – 1 – mf#B29162-29165 – us Ohio Hist [978]

City directories of the united states – 1,9 – (segment 1 through 1860 6292mf. segment 2: 1861-81 372r. segment 2 suppl 80r. segment 3: 1882-1901 746r. segment 3 suppl 26 units 25r ea. segment 4: 1902-35 87 units (ongoing) 50r ea. segment 5: 1936-60 32 units (ongoing) 50r ea) – us Primary [978]

City directory, 1928 / Barberton, OH – 1r – 1 – mf#B27451 – us Ohio Hist [978]

City echo – [Scotland] Dundee: G Montgomery mar 1908-feb 1909 (mthly) [mf ed 2004] – 1r – 1 – (absorbed by: piper o' dundee (dundee, scotland : 1886)) – uk Newsplan [072]

City edition – Chicago IL. v1 n2 [1980 dec], v2 n7-v3 n9 91981 oct-1982 sep] – 1r – 1 – mf#656919 – us WHS [071]

City enterprise – McComb, MS. 1889-1931 (1) – mf#64051 – us UMI ProQuest [071]

The city enterprise – Hamilton [Ont]: R Theophilus, [1864] – 9 – (some iss have title: the daily enterprise) – mf#P04908 – cn Canadiana [071]

City gazette – Charleston. South Carolina. Jan 3-Dec 31 1789 – 1 – us NY Public [071]

City gazette – Providence, RI. 1833-1833 (1) – mf#66364 – us UMI ProQuest [071]

City gazette and mercantile register – Belfast Ireland, 9 jan-27 feb 1891 – 1/4r – 1 – uk British Libr Newspaper [072]

City gazette, and the daily advertiser – Charleston SC. 1788, 1791 jul 6-aug 16, aug 17-oct 21, 1792 feb 8-sep 27, 1794 sep 15-nov 6, 1795 jan 1-1796 jan 22, 1796 jan 23-dec 23, dec 24-1797 aug 14, 1797 aug 15-1798 jul 4, 1806 jul-dec – 6r – 1 – (cont: charleston morning post and daily advertiser; cont by: city gazette [charleston sc: 1804]) – mf#2731346 – us WHS [071]

City goeteborg – Goeteborg, Sweden. 2006- – 1 – sw Kungliga [078]

City government of daytona beach / Goebel, Rubye K – s.l, s.l? 1936 – 1r – us UF Libraries [978]

The city hall clock : addressed to the citizens of fredericton, june-1878 / Fenety, George Edward – [Fredericton, NB?: s.n, 1878?] – 1mf – 9 – 0-665-94514-0 – mf#94514 – cn Canadiana [336]

City hall, london : tuesday, wednesday and thursday evenings, october 5th, 6th and 7th, 1880...cantata of the flower queen!... – [London, Ont?: s.n, 1880?] [mf ed 1986] – 1mf – 9 – 0-665-54273-9 – mf#54273 – cn Canadiana [780]

City hall recorder – New York. 1816-1822 (1) – mf#3964 – us UMI ProQuest [323]

City hospital worker : the voice of local 420... / American Federation of State, County, and Municipal Employees – 1979 jul-1987 jan – 1r – 1 – mf#1477010 – us WHS [350]

The city, its sins and sorrows : being a series of sermons from luke 29, 41 / Guthrie, Thomas – Edinburgh: Adam and Charles Black, 1857 – 1mf – 9 – 0-524-00756-X – mf#1990-0188 – us ATLA [240]

City jackdaw – Manchester. 1875-1880 (1) – mf#4714 – us UMI ProQuest [420]

City journal – Canyon City OR: Typographical Soc [irreg] – 1 – (began with nov 9 1868 iss; cont by: grant county news) – us Oregon Lib [071]
City leader – Hastings, NZ. feb-jun 1985 – 1r – 1 – mf#35.9 – nz Nat Libr [079]
City life – n68-87 [1983 oct/nov-1989 mar/apr], special ed – 1r – 1 – (cont: community news [jamaica plain ma]) – mf#1110691 – us WHS [071]
The city life – Montreal: The City Life Pub. Co., [1879?-187- or 188-] – 9 – ISSN: 1190-7231 – mf#P04215 – cn Canadiana [870]
City Life [Organization] *see* Labor page
City lights : canada's metropolitan magazine – Toronto. v1-2 n2. nov 1934-dec 1935 (mthly) – 1r – 1 – Can$95.00 – (no more publ?) – cn McLaren [790]
City lights – 1980 aug 27-1983 apr 8/12 – 1r – 1 – (cont: back porch radio pilot) – mf#595162 – us WHS [071]
City limits / Association of Neighborhood Housing Developers – 1983-86 – 1r – 1 – mf#1277550 – us WHS [360]
City link – Ft. Lauderdale, FL. 1999-2000 (1) – mf#69379 – us UMI ProQuest [071]
City magazine – New York, NY. 1965-1967 (1) – mf#65089 – us UMI ProQuest [071]
The city magazine : devoted to the interests of young men engaged in commercial pursuits – Montreal: Printed...by J W Harrison [1847-18–?] – 9 – mf#P05971 – cn Canadiana [650]
City malmoe lund – Malmoe, Sweden. 2006- – 1 – sw Kungliga [078]
City miner – v1 n1-v5 n1 [1976 spr-1980 final iss] – 1r – 1 – mf#669074 – us WHS [071]
City missions / McVickar, William Augustus – 2nd ed. New-York: Pott & Amery, 1868 – 2mf – 9 – 0-524-06642-6 – mf#1991-2697 – us ATLA [240]
City news – Holland, MI. 1872-1977 (1) – mf#63771 – us UMI ProQuest [071]
City news – London UK, 30 jan-31 dec 1864 – 1/2r – 1 – uk British Libr Newspaper [072]
City news *see* City and west end news
City newspaper – Rochester, NY. 1982-1995 (1) – mf#68084 – us UMI ProQuest [071]
The city of akhenaten : pt 1: excavations of 1921 and 1922 at el-'amarneh / Peet, T E & Woolley, C L – London, 1921 – (= ser Mees 38) – 14mf – 8 – €27.00 – ne Slangenburg [930]
The city of benin / Egharevba, Jacob U – [Benin City: s.n, 1952?] – 1 – us CRL [960]
City of birmingham museum and art gallery catalogue...of...modern english animal painters / Birmingham. Museum and Art Gallery – Birmingham 1892 – (= ser 19th c art & architecture) – 1mf – 9 – mf#4.2.1617 – uk Chadwyck [750]
City of bristol, newport and welch towns directory / Edward Hunt & Co. Bristol, England – London, 1848 [mf ed 1986] – 1v on 1r – 1 – (with: american institute of homoeopathy) – mf#8548 – us Wisconsin U Libr [914]
City of chartres : its cathedral and churches / Masse, Henri Jean Louis Joseph – London: George Bell 1905 [mf ed 1992] – (= ser Bell's handbooks to continental churches) – 1mf [ill] – 9 – 0-524-03907-0 – mf#1990-1166 – us ATLA [914]
City of Edinburgh Charity Organisation Society *see* Report on the physical condition of fourteen hundred school children in the city, together with some account of their homes and surroundings
City of gainesville – s.l, s.l? 193-? – 1r – us UF Libraries [978]
The city of gloucester parliamentary register *see* Ward lists and other records of the city of gloucester, 1843-86
The city of god = De civitate dei / Augustine, Saint, Bishop of Hippo – Edinburgh: T & T Clark, 1871-72 [mf ed 1985] – (= ser The works of aurelius augustine, bishop of hippo) – 2v on 4mf – 9 – 0-8370-5886-4 – (trans by marcus dode) – mf#1985-3886 – us ATLA [240]
City of god and the city of man in africa / Brookes, Edgar Harry – Lexington, KY. 1964 – 1r – us UF Libraries [960]
City of hope national medical center pilot – (Los Angeles, CA) Winter 1962-Winter 1985 – (missing: spr.-sum. 1962, spr. 1963, 1964-1969, wint. 1969/70, spr. 1970, fall 1973, wint. 1973/4, fall 1974-fall 1975, wint. 1979-1980, sum. 1980, spr.-sum. 1981, fall 1982-sum. 1984, spr.-sum. 1985) – us AJPC [610]
City of hope reporter – Los Angeles, CA.Jan-Feb 1959; Jan 1960 – 1 – us AJPC [071]
The city of jerusalem / Conder, Claude Reignier – London: John Murray, 1909 – 1mf – 9 – 0-7905-1640-3 – (incl ind) – mf#1987-1640 – us ATLA [956]
City of london illustrated *see* North london illustrated and ratepayers' guardian
City of london, ontario, canada : the pioneer period and the london of today / Bremner, Archie – London, Ont: London Print & Lithographing Co, 1900 – 3mf – 9 – mf#26696 – cn Canadiana [720]
City of london post – London UK, 11 nov 1977-78, 11 jan-15 feb, 4 apr-19 dec 1980, 1981-jun 1982, 9 jul-24 dec 1982, 1983-21 dec 1984, 11 jan-20 dec 1985, 1986-14 dec 1989, jan-20 dec 1990, jan-19 dec 1991, jan-16 dec 1992, jan-23 dec 1993 – 30r – 1 – (aka: city post) – uk British Libr Newspaper [072]
City of london recorder – London UK, 11 mar 1976-29 dec 1978, 4 jan-20 dec 1990, 1991, 10 jan-20 jun 1992, 3 jul-18 dec 1992, 2 jan-24 dec 1993 – 12 1/2r – 1 – (aka: city recorder of london) – uk British Libr Newspaper [072]
City of ottawa : capital of the dominion of canada – Ottawa: Ottawa Free Press, 1899 [mf ed 1980] – 2mf – 9 – 0-665-03044-4 – mf#03044 – cn Canadiana [321]
City of refuge / Beaumont, Joseph – London, England. 1851 – 1r – us UF Libraries [240]
City of refuge – Kelso, Scotland. 18– – 1r – us UF Libraries [240]
City of refuge – London, England. 1841 – 1r – us UF Libraries [240]
City of san francisco : financial records – San Francisco, CA. 1849-50 – 1r – 1 – $50.00 – mf#B40304 – us Library Micro [978]
City of san francisco municipal employee – San Francisco, CA. v1-12. 1927-37 – 2r – 1 – $100.00 – mf#B40305 – us Library Micro [350]
The city of springs : or, mission work in chinchew / Duncan, Annie N – Edinburgh, London: Oliphant Anderson & Ferrier, 1902 [mf ed 1995] – (= ser Yale coll) – 110p (ill) – 1 – 0-524-09539-6 – mf#1995-0539 – us ATLA [240]
The city of two gateways : the autobiography of an indian girl / Nanda, Savitri Devi – London: George Allen & Unwin Ltd, 1950 – (= ser Samp: indian books) – us CRL [390]
City of washington gazette – Washington DC. 1816 oct 19, 1818 dec 3 – 1r – 1 – (cont: washington city weekly gazette; cont by: washington gazette) – mf#851641 – us WHS [071]
City pages – v4 n130, 136-v5 n150, 176 [1983 jun 1, jul 6-oct 19, 1984 may 2] – 1r – 1 – (cont: sweet potato) – mf#1023350 – us WHS [071]
City paper – Baltimore, MD. 1977-1985 (1) – mf#61012 – us UMI ProQuest [071]
City paper / Washington Free Weekly – Washington DC. [1983 feb 4/11-dec 30, 1984 jan 5]-[1991 nov/dec] – 21r – 1 – mf#1041692 – us WHS [071]
City post *see* City of london post
City press – [London & SE] City of London 1857-1976 – 1 – uk Newsplan [072]
City press – Newburgh, NY. 1866-1866 (1) – mf#65107 – us UMI ProQuest [071]
City recorder of london *see* City of london recorder
City, rice-swamp, and hill / Johnson, William – London: London Missionary Society, 1893 [mf ed 1995] – (= ser Yale coll; Missionary manuals) – 224p (ill) – 1 – 0-524-10074-8 – mf#1995-1074 – us ATLA [954]
City smoke signals – 1968-79 – (= ser American indian periodicals... 1) – 8mf – 9 – $95.00 – us UPA [305]
City star – Blantyre, [Malawi]: G E Publ, apr 16/23 1993 – 1r – us CRL [079]
City star – v1-v4 n4 [1973 may 1-1977 mar 20] – 1r – 1 – (cont: liberated guardian) – mf#293155 – us WHS [071]
City sun – Brooklyn, NY. 1984-1989 (1) – mf#68405 – us UMI ProQuest [071]
City temple sermons / Campbell, Reginald John – New York: Fleming H Revell c1903 [mf ed 1990] – (= ser The international pulpit) – 1mf – 9 – 0-524-00850-7 – mf#1990-4010 – us ATLA [242]
City times – Akron, OH. 1869-1892 (1) – mf#65361 – us UMI ProQuest [071]
City times – Auburn, RI. 1895-1927 (1) – mf#66175 – us UMI ProQuest [071]
City times / Muskingum Co. Zanesville – v1 n1. sep 1852-aug 1854,feb 1860-oct 1863 [wkly] – 3r – 1 – mf#B5560-5562 – us Ohio Hist [071]
City times – premiere iss-v1 n8 [1980 sum-1982 dec] – 1r – 1 – mf#516487 – us WHS [071]
City times / Summit Co. Akron – jan 1886-dec 1886 [wkly] – 1r – 1 – mf#B30884 – us Ohio Hist [071]
City times / Summit Co. Akron – jan-dec 1885, jan 1887-dec 1889 [wkly] – 2r – 1 – mf#B34610-34611 – us Ohio Hist [071]
City times – Warwick, RI. 1932-1933 (1) – mf#66421 – us UMI ProQuest [071]
City tribune – New York, NY. 1984-1991 (1) – mf#60536 – us UMI ProQuest [071]
City University of New York *see*
- C c n y black alumni news
- Caribbean Research Center focus
- Continuities
- Paper
- Tieline

City University of New York. City College *see* Notes from workshop center for open education
City weekly – Washington. 1978-1978 (1,5,9) – ISSN: 0164-5595 – mf#11922 – us UMI ProQuest [350]
The city with foundations / McFadyen, John Edgar – New York: Hodder & Stoughton: George H. Doran, [1909?] – 1mf – 9 – 0-7905-1240-8 – mf#1987-1240 – us ATLA [220]
City worker : the voice of local 400 / Service Employees International Union – v6 n1-v7 n11 [1981 feb-1982 nov] – 1r – 1 – (cont: san francisco city worker; cont by: united action; united worker [san francisco ca]) – mf#647465 – us WHS [331]
City-anzeiger – Dortmund DE, 1988 3 jun-2002 [gaps] – 1 – gw Mikrofilm [074]
City-post – Dortmund DE, 1985 16 nov-1988 20 may – 1 – gw Mikrofilm [074]
The city-state of the greeks and romans : a survey introductory to the study of ancient history / Fowler, William Warde – London, New York: Macmillan, 1893 – 1mf – 9 – 0-7905-5326-0 – (incl bibl ref) – mf#1988-1326 – us ATLA [930]
Ciudad cerrada / Menendez, Aldo – Cienfuegos, Cuba. 1955 – 1r – us UF Libraries [972]
La ciudad de dios : revista quincenal religiosa, cientifica y literaria dedicada al gran padre san agustin – v1-153. Jun 1881-Aug 1936 – 1 – (v. 6, no. 1; v. 9, no. 3; v. 18; v. 117 wanting) – us L of C Photodup [073]
Ciudad de marta y marta de la ciudad / Marquina, Rafael – Habana, Cuba. 1950 – 1r – us UF Libraries [972]
Ciudad inefable / Mieses Burgos, Franklin – s.l, s.l? 1949 – 1r – us UF Libraries [972]
Ciudad portatil / Briceno Valero, Americo – Caracas, Venezuela. 1939 – 1r – us UF Libraries [972]
Ciudad romantica / Cestero, Tulio Manuel – Paris, France. 1911? – 1r – us UF Libraries [972]
Ciudad trujillo : the oldest city in the new world / Inchaustegui Cabral, Joaquin Marino – s.l, s.l? 1948 – 1r – us UF Libraries [972]
Ciudad Trujillo Universidad De Santo Domingo *see*
- Catalogo de libros y revistas donados por el gobie
- Concursos y premios para los estudiantes
- Homenaje a pedro henriquez urena
- Primera exposicion de arte indigena autoctono
- Trabajos premiados en distintas facultades

Ciudad vencida / Devis Echandia, Julian – Bucaramanga, Colombia. 1937 – 1r – us UF Libraries [972]
Ciudadela baptisms – Minorca, Spain. v1-18. 1566-1803 – 8r – us UF Libraries [324]
Ciudadela cadastre del manifest – Minorca, Spain. 1768-1772 – 1r – us UF Libraries [324]
Ciudadela census – Minorca, Spain. 1758 – 1r – us UF Libraries [324]
Ciudadela deaths – Minorca, Spain. v1-13. 1600?-1666 – 6r – us UF Libraries [324]
Ciudadela index of deaths and baptisms – Minorca, Spain. no date – 1r – us UF Libraries [324]
Ciudadela marriages – Minorca, Spain. v2-8. 1640-1814 – 4r – us UF Libraries [324]
Ciudades arqueologicas de mexico / Pina Chan, Roman – Mexico City?, Mexico. 1963 – 1r – us UF Libraries [930]
Ciudades y rutas de colombia / Mendoza Velez, Jorge – Bogota, Colombia. 1940 – 1r – us UF Libraries [972]
Civananacittiyar *see* Sivajnana siddhiyar of arunandi sivacharya
Civic ballet / Chujoy, Anatole – New York: Publ for Central Service for Regional Ballets by Dance News, inc, 1958 – 1 – mf#*ZBD-*MGO pv15 – Located: NYPL – us Misc Inst [790]
Civic public works – Toronto. 1979-1994 (1) 1979-1979 (5) 1979-1979 (9) – ISSN: 0829-772X – mf#12040,01 – us UMI ProQuest [350]
Civic training in soviet russia / Harper, Samuel Northrup – Chicago IL: The University of Chicago Press [1929] [mf ed 1987] – 1r – 1 – (filmed with: phaenomenologie des sittlichen bewusstseins / hartmann, edward von) – mf#1826 – us Wisconsin U Libr [947]
Civic Understudies [Group: Two Rivers WI] *see* Sojourner
Civil aeronautics administration annual reports / U.S. Civil Aeronautics Board – 1940-42 (all publ) – 3mf – 9 – $4.50 – (cont by: civil aeronautics board annual reports) – mf#LLMC 81-202 – us LLMC [350]
Civil aeronautics authority annual reports / U.S. Civil Aeronautics Board – 1939-40 (all publ) – 2mf – 9 – $3.00 – (cont by: civil aeronautics administration annual reports) – mf#LLMC 81-203 – us LLMC [350]
Civil aeronautics board annual reports / U.S. Civil Aeronautics Board – 1941-83 (all publ) – (= ser Civil aeronautics board reports) – 67mf – 9 – $100.00 – (lacking: 1979-80) – mf#LLMC 81-204 – us LLMC [324]
Civil aeronautics board reports : index/digest of the c a b reports – v1-2. 1938-60 [all publ] – 27mf – 9 – $40.50 – mf#llmc78-205 – us LLMC [380]
Civil aeronautics board reports – v1-75 – 838mf – 9 – $1257.00 – mf#LLMC 78-204 – us LLMC [380]
Civil aeronautics journal – v1-13 no 7 – 32mf – 9 – $48.00 – (cont: civil aeronautics authority's air commerce bulletin) – mf#LLMC 81-201 – us LLMC [380]
Civil affairs and military government in the mediterranean theater / U.S. Army. Office of Military History – v. 1. 1948? – 5 – us L of C Photodup [945]
Civil affairs journal and newsletter : a civil affairs association publication – v24 n5/6-v32 n5/7 [1971 may/jun-1979 may/jun] – 1r – 1 – (cont: military government journal and news letter) – mf#667587 – us WHS [366]
Civil and criminal codes of practice of kentucky; rev. and cor. to july 1, 1908 / Kentucky. Laws, Statutes, etc – Pocket ed. Lexington: Hughes 1908. 481p. LL-936 – 1 – us L of C Photodup [345]
Civil and mechanical engineering designs of john smeaton : from the royal society, london – 11v – 2r – 1 – mf#96885 – uk Microform Academic [621]
Civil and military gazette – Lahore, Pakistan. may 1876-dec 1914 – 458r – 1 – €48,090.00 – mfm48 – ne IDC [355]
The civil and natural history of jamaica / Browne, P – London, 1756 – 29mf – 9 – mf#1272 – ne IDC [918]
The civil and natural history of jamaica in three parts / Browne, P – 1959-1960 (1) – 28mf – 9 – mf#2600 – ne IDC [918]
The civil and penal codes of the territory of guam, 1953 / Bohn, John A – Agana: Office of the Sec of the Gov of Guam, 1953 – 10mf – 9 – $15.00 – mf#LLMC 82-100B Title 3 – us LLMC [324]
Civil and religious forces / Halstead, William Riley – Cincinnati: Jennings and Pye; New York: Eaton and Mains, c1890 – 1mf – 9 – 0-7905-9280-0 – mf#1989-2505 – us ATLA [240]
Civil and religious institutions necessarily and inseparably connec... / Esdaile, James – Perth, Australia. 1833 – 1r – us UF Libraries [240]
Civil and religious intelligencer : or the gleanor and monitor – Sangerfield. 1816-1817 (1) – mf#4584 – us UMI ProQuest [240]
The civil architecture of vitruvius – London 1812 – (= ser 19th c art & architecture) – 12mf – 9 – mf#4.2.1702 – uk Chadwyck [720]
Civil aviation in el salvador / Gilbert, Glen Alexander – Montreal, Quebec. 1952 – 1r – us UF Libraries [380]
Civil case files of the u(nitedstates district court for the district and territory of alaska, second division (nome), 1908-1955 / United.States. District Court – (= ser Records Of U.S. District Courts) – 87r – 1 – mf#M1968 – us Nat Archives [347]
The civil code / Field, David Dudley – New York, 1886. 10p. LL-676 – 1 – us L of C Photodup [348]
The civil code in force in cuba, porto rico, and the philippines, 1899 – Washington: GPO, 1899 – 4mf – 9 – $6.00 – mf#LLMC 92-305 – us LLMC [324]
The civil code of guam, 1933 : approved by the naval government of guam may 1 1933, effective february 1 1934 in 4 divisions / Robinson, Stephen B – Govt House, 28 Dec 1933 – 8mf – 9 – $12.00 – mf#LLMC 82-100B Title 1 – us LLMC [324]
The civil code of guam, 1947 : in 4 divisions / Guam. (Commonwealth). Laws, Statutes, etc – Washington: GPO, 1947 – 4mf – 9 – $6.00 – mf#LLMC 82-100B Title 9 – us LLMC [324]
The civil code of louisiana as a democratic institution / Fenner, Charles Erasmus – New Orleans? n.d. 19p. LL-656 – 1 – us L of C Photodup [348]
The civil code of the german empire : as enacted on august 18, 1896 = Buergerliche gesetzbuch – Boston: Boston Book Co, 1909 – (= ser Civil law 3 coll) – 8mf – 9 – (incl bibl ref) – mf#LLMC 96-530 – us LLMC [348]
The civil code of the republic of panama and amendatory laws continued in force in the canal zone, isthmus of panama, by executive order of may 9 1904 / Panama Canal Zone – Washington: Isthmanian Canal Commission, 1905 – 8mf – 9 – $12.00 – (also contains decree no 4 nov 4 1903 of the junta of the provisional government of the republic, and a historical introduction to the development of the panamanian civil law) – mf#LLMC 82-100D Title 1 – us LLMC [348]

The civil code of the territory of guam, 1970 / Bohn, John A – Agana: n.p. 2v. 1970 – 13mf – 9 – $19.50 – (with 1974 suppl) – mf#LLMC 82-100B Title 4 – us LLMC [324]

Civil defence for home and flat dwellers = Burgerlike beskerming vir huis- en woonstelbewoners / South Africa. Department of National Health and Population Development [Departement van Nasionale Gesondheid en Bevolkingsontwikkeling – Pretoria: Dept of National Health & Population Development 1991 [mf ed Pretoria, RSA: State Library [199-]] – 14p [ill] on 1r with other items – 5 – (In english & afrikaans) – mf#op 10002 r26 – us CRL [360]

Civil defense – Bethlehem PA. v1 n4 [1971 aug] – 1r – 1 – mf#1583046 – us WHS [360]

Civil Defense Forum *see* Survive

Civil disturbance, chartism and riots in nineteenth century england : pro class 45, home office registered papers – 37r – 1 – (pt 1: 1841-44 16r c39-16901. pt 2: 847-76 21r c39-16902) – mf#C39-16900 – us Primary [941]

Civil disturbances / Mullik, B N – Delhi: Manager of Publ, 1966 – (filmed with: dutt, surendra nath. the life of benoyendra nath sen) – us CRL [303]

Civil duties of christians / Thomas, Thomas – London, England. 1839? – 1r – us UF Libraries [240]

Civil engineer in south africa = Die siviele ingenieur in suid-afrika – Johannesburg. 1980-1980 (1,5,9) – ISSN: 0009-7845 – mf#10759 – us UMI ProQuest [624]

Civil engineering – London, 1976-1988 [1,5,9] – (cont by: construction weekly) – ISSN: 0305-6473 – mf#10982,01 – us UMI ProQuest [624]

Civil engineering – New York. 1930+ (1) 1965+ (5) 1973+ (9) – ISSN: 0885-7024 – mf#616 – us UMI ProQuest [624]

Civil engineering for practicing and design engineers – Elmsford. 1982-1986 (1) 1982-1986 (5) 1984-1986 (9) – ISSN: 0277-3775 – mf#49396 – us UMI ProQuest [624]

Civil engineers convene : annual meeting of the canadian society yesterday: the president's address – [Montreal?: s.n, 1889?] – 1mf – 9 – 0-665-90501-7 – mf#90501 – cn Canadiana [624]

Civil establishments of christianity tried by their only authoritat... / Wardlaw, Ralph – Glasgow, Scotland. 1833 – 1r – us UF Libraries [240]

Civil establishments of religion unjust in their principle... / Heugh, Hugh – Glasgow, Scotland. 1835 – 1r – us UF Libraries [200]

Civil government : an exposition of romans 13, 1-7 / Willson, James McLeod – Philadelphia: W S Young, 1853 – 2mf – 9 – 0-524-07929-3 – mf#1991-3474 – us ATLA [220]

Civil government in the united states : considered with some reference to its origins / Fiske, John – Boston/New York/Chicago: Houghton, Mifflin, 1890 – 5mf – 9 – $7.50 – mf#LLMC 95-079 – us LLMC [323]

Civil government – the late conspiracy : a discourse delivered in kingston, u c december 31, 1837 / Ryerson, Egerton – Toronto?: s.n, 1838 – 1mf – 9 – mf#21653 – cn Canadiana [971]

The civil justice reform act expense and delay reduction plans : a sourcebook / Rauma, David et al – 1995 – 8mf – 9 – $12.00 – mf#llmc99-022 – us LLMC [340]

Civil law 1 : france: a basic collection – permanent ed. Kaneohe, 1996 – (= ser The llmc civil law 1 catalogs on fiche) – 1mf – 9 – mf#LLMC 86-000B – us LLMC [346]

Civil law 2 : italy, spain, portugal and the low countries: a basic collection – permanent ed. [mf ed Kaneohe 1996] – (= ser The llmc civil law 2 catalogs on fiche) – 1mf – 9 – mf#LLMC 86-001B – us LLMC [346]

Civil law 3 : germany, austria and switzerland: a basic collection / ed by Reynolds, Thomas H – permanent ed 1999 – 9 – $4,330.00 coll – (civil law development in germany, austria and switzerland. from the coll of the u.c. berkeley law library) – mf#LLMC 82-300C – us LLMC [346]

The civil law and the church / Lincoln, Charles Zebina – New York: Abingdon Press, c1916 – 3mf – 9 – 0-524-03765-5 – mf#1990-1112 – us ATLA [240]

Civil liberties / American Civil Liberties Union – 1949 sep-1965, 1966-69, 1970 feb-1980 nov – 3r – 1 – (cont: civil liberties quarterly) – mf#765770 – us WHS [322]

Civil liberties – London, UK: The National Council for Civil Liberties. v1 1937-1970 (all publ) – 27mf – 9 – $40.50 – mf#LLMC 84-437 – us LLMC [322]

Civil liberties docket / Meiklejohn Civil Liberties Library – v. 1-12. 1955-66 – 1 – us AMS Press [321]

Civil liberties in arizona / Arizona Civil Liberties Union – v3 n3-v7 n4 [1974 jan/Feb-1979 nov/dec] – 1r – 1 – mf#500718 – us WHS [322]

Civil liberties in new york – New York. 1953-1974 (1) – (cont by: ny civil liberties) – ISSN: 0009-7926 – mf#3235 – us UMI ProQuest [323]

Civil liberties news / Wisconsin Civil Liberties Union – v1 n1-v18 n4 [1964 nov-1981] – 1r – 1 – (cont by: civil liberties [milwaukee wi]) – mf#579296 – us WHS [322]

Civil liberties publications / Meiklejohn Civil Liberties Library – 1 – (incl: bulletin of the american committee for the protection of the foreign born 1953-59; lamp 1944-59; review of the year 1947-50; proceedings of the national conference for protection of foreign born 1947; civil rights law letter of the civil rights congress 1955-56; civil liberties reporter 1950-52) – us AMS Press [321]

Civil liberties quarterly / American Civil Liberties Union – n1-n73 [1931 jun-1949 jun] – 1r – 1 – (cont by: monthly bulletin, civil liberties) – mf#630421 – us WHS [322]

Civil liberties reporter *see* Civil liberties publications

Civil liberty in lower canada / Galt, Alexander Tilloch – Montreal: D Bentley, 1876 – 1mf – 9 – mf#24095 – cn Canadiana [322]

Civil regulations with the force and effect of law in guam, 1947 / Guam. (Commonwealth) – Washington: GPO, 1947 – 2mf – 9 – $3.00 – mf#LLMC 82-100B Title 10 – us LLMC [324]

Civil report of major john r brooke – Havana, Cuba. 1899 – 1r – us UF Libraries [972]

Civil report of major-general john r brooke – Washington, DC. 1900 – 1r – us UF Libraries [972]

Civil rights – Cape Town, Civil Rights League. [4-15] 1957-1968; v16 n5 jun 1969; v17 n8 sep 1970 – us CRL [322]

Civil rights and present wrongs / Paton, Alan – Johannesburg, South Africa. 1968 – 1r – us UF Libraries [322]

Civil rights bulletin / Connecticut Commission on Civil Rights – v1 n1-v3 n4 [1954 jan-1961 jun], 1961 oct-1967 may/jun – 1r – 1 – (cont by: rights, opportunities, action reporter) – mf#642952 – us WHS [322]

Civil rights digest – Washington. 1972-1979 (1) 1972-1979 (5) 1975-1979 (9) – (cont by: perspectives) – ISSN: 0009-7969 – mf#7373 – us UMI ProQuest [323]

Civil rights digest / National Endowment for the Arts – 1984 sum-fall – 1r – 1 – mf#5296650 – us WHS [322]

Civil rights digest / U[/nited/] S[/tates/] Commission on Civil Rights – v1-18 n1. 1968-86 – 56mf – 9 – $84.00 – (v1-11 titled: civil rights digest; after v11: new perspectives: civil rights quarterly) – mf#llmc81-206 – us LLMC [348]

Civil rights digest *see* New perspectives

Civil rights division of the u s department of justice : hearing...house of representatives, 107th congress, 2nd session, june 25 2002 / United States. Congress. House. Committee on the Judiciary. Subcommittee on the Constitution – Washington: US GPO 2002 [mf ed 2002] – 1mf – 9 – 0-16-068785-3 – us GPO [322]

Civil rights during the johnson administration, 1963-1969 / ed by Lawson, Steven F – 5pt – (= ser Black studies research sources) – 1 – $11,730.00 coll – (pt1: white house central files & aides files [15r] isbn 0-89093-690-0 $2685; pt2: equal employment opportunity commission administrative history [3r] isbn 0-89093-691-9 $525; pt3: oral histories [3r] isbn 0-89093-692-7 $525. pt4: papers of the white house conference on civil rights [20r] isbn 0-89093-693-5 $3560; pt5: records of the national advisory commission on civil disorders [kerner commission] [28r] isbn 0-89093-903-9 $4990; with p/g) – us UPA [322]

Civil rights during the kennedy administration / ed by Brauer, Carl M – 2pt – (= ser Black studies research sources) – 1 – $7985.00 coll – (pt1: the white house central files & staff files & the president's office files [19r] isbn 0-89093-900-4 $3395; pt2: the papers of burke marshall, assistant attorney general for civil rights [28r] isbn 0-89093-364-2 $4990; with p/g) – us UPA [322]

Civil rights during the nixon administration, 1969-1974 : pt 1: the white house central files / ed by Graham, Hugh Davis – (= ser Black studies research sources) – 46r – 1 – $8485.00 – 1-55655-133-9 – (with p/g) – us UPA [322]

Civil rights handbook / National Association for the Advancement of Colored People – New York: The Association, 1953 (mf ed 1976) – 1r – 1 – mf#ZZ-14168 – us NY Public [323]

Civil rights issues of euro-ethnic americans in the u.s. : opportunities and challenges / U.S. Commission on Civil Rights – Washington: GPO, 1980 – 7mf – 9 – $10.50 – mf#LLMC 94-332 – us LLMC [322]

Civil rights journal : commentary / United Church of Christ – n1-266 [1962 jan 1-1986 dec 31], n267-418 [1987 jan 6-1989 dec 25], 1992 jan-1993 may 3, 1994 jan 10-1995 dec, 1996 jan 8-1999 jun 26 – 4r – 1 – (cont by: witness for justice) – mf#2898927 – us WHS [322]

Civil rights journal – Washington. 1995+ (1,5,9) – (cont: new perspectives) – mf#25677 – us UMI ProQuest [323]

Civil Rights League. Cape Town *see* Annual report

Civil Rules Advisory Committee *see* Report on mass tort litigation

The civil sabbath restored – New York: EO Jenkins, [1861?] – (= ser Document of the New York Sabbath Committee) – 1mf – 9 – 0-524-02918-0 – mf#1990-0734 – us ATLA [240]

Civil Service Association of Canada *see* Csac journal

Civil Service Association of Ontario *see* C s a o news

Civil service current employment opportunities – Wisconsin. 1978 jan 9-dec 17 – 1r – 1 – (cont: wisconsin career candidate vacancy bulletin; cont by: state service current employment opportunities bulletin) – mf#1173801 – us WHS [350]

Civil Service Employees Association *see*
– Public sector
– Suffolk area c s e a regional reporter

Civil Service Employees Association [NY] *see* Long island c s e a regional reporter

Civil service journal – Perth, Australia. -m. May 1914-Dec 1921. 2 reels – 1 – uk British Libr Newspaper [360]

Civil service journal / U.S. Civil Service Commission – v. 1-8. 1960-68 – 1 – us AMS Press [324]

Civil service journal – Washington. 1960-1979 (1) 1972-1979 (5) 1975-1979 (9) – ISSN: 0009-7985 – mf#6387 – us UMI ProQuest [350]

Civil service news / Montgomery Co. Dayton – jan 1945-dec 1972 [mthly] – 4r – 1 – mf#B5198-5201 – us Ohio Hist [350]

Civil service news; the public employes' weekly – v. 6, no. 10-v. 11, no. 36. 8 Jan 1914-23 Oct 1919. (Wanting scattered numbers) – 1 – us L of C Photodup [350]

Civil service reporter – v1 n1-3 [1973 jul-sep] – 1r – 1 – (cont by: long island c s e a regional reporter) – mf#1132046 – us WHS [350]

The civil service review : a journal devoted to the interests of the services in canada – Ottawa: Paynter, [1893-189- or 19–] – 9 – mf#P04221 – cn Canadiana [350]

Civil service servant / Reissman, Leonard – Madison, WI. 1947 – 1r – us UF Libraries [025]

Civil service standard / State, County and Municipal Workers of America – 1937 oct 11-1939 dec 29, 1940 jan 5-1942 aug 3 – 2r – 1 – mf#1054533 – us WHS [350]

The civil service system of the royal khmer government of cambodia / Ouellette, James L et al – Phnom Penh: Public Admin Div, US Agency for International Devt [mf ed 1989] – (= ser United States. Agency for International Development. Public Administration Division. Public administration report 2) – 1r with other items – 1 – mf#mf-10289 seam reel 026/27 [§] – us CRL [350]

Civil Service Technical Guild *see*
– Cstg press
– Guild newsletter
– Pstg press

Civil war : 31 ovvi recollections – 1r – 1 – mf#B31207 – us Ohio Hist [976]

The civil war : the papers of the white army / Russian State Military Archive (RGVA) – 1917-21 – ca 100r – 1 – us Primary [947]

Civil war and reconstruction in florida / Davis, William Watson – New York, NY. 1913 – 1r – us UF Libraries [978]

Civil war and reconstruction: the making of modern america, series 1 : the papers of jay cooke [1821-1905] from the historical society of pennsylvania – 5pt – 1 – (pt1: general correspondence 1843-apr 1865 [20r] £1850; pt2: general correspondence, may 1865-dec 1867 [20r] £1850; pt3: general correspondence, jan 1868-apr 1870 [20r] £1850; pt4: general correspondence, may 1870-dec 1871 [20r] £1850 [mf ed oct 2001]; pt5: general correspondence, jan 1872-jun 1874 & n d [20r] £1850 [mf ed 2001]; with d/g) – uk Matthew [976]

Civil war and the confederacy : the business records of fraser, trenholm and company of liverpool and charleston, south carolina, 1860-1877, from the merseyside maritime museum, liverpool – 13r – 1 – £1225.00 – (with d/g) – uk Matthew [976]

Civil war army nurses' scrapbook / North, Mary – 1925 – 1 – us L of C Photodup [976]

Civil war band books / United States Army Corps, 15th Division, 3rd Brigade, 1st Band – 1864-65 [mf ed 198-?] – 1r – 1 – mf#319p – us Wisconsin U Libr [780]

Civil war battles and campaigns – 3pt – (= ser Civil war research colls) – 9 – $23,765.00 coll – (pt1: eastern theater 801mf isbn 1-55655-589-x $7675. pt2: western theater 513mf isbn 1-55655-590-3 $4925. pt3: general references & coll works 1165mf isbn 1-55655-591-1 $11,170. with p/g) – us UPA [976]

Civil war collection : coll 11, mic-17 – 22r – 1 – (write for details) – us Ohio Hist [976]

Civil war correspondence, diaries, and journals : at the massachusetts historical society – [mf ed 1986] – 29r – 1 – (with p/g. coll contains correspondence, diaries, and journals written by young men as they served in the conflict) – us MA Hist [976]

Civil war diary / Beck, Aaron N – 1861-64 – 1 – us Kansas [920]

Civil war diary, 1864-1865 / Rogall, Albert – 1r – 1 – mf#B34921 – us Ohio Hist [355]

Civil war direct tax assessment lists : tennessee / U.S. Treasury Dept – (= ser Records of the accounting officers of the department of the treasury) – 6r – 1 – mf#T227 – us Nat Archives [336]

Civil war drawings by edwin forbes / U.S. Library of Congress. Prints and Photographs Division – 322 pencil drawings and 15 oil paintings. 1 reel. P&P11950 – 1 – $23.00r – us L of C Photodup [976]

Civil war history – Kent. 1955+ (1) 1971+ (5) 1975+ (9) – ISSN: 0009-8078 – mf#2353 – us UMI ProQuest [976]

Civil war in china, 1945-1950 / U.S. Office of the Chief of Military History – v. 1-2 – 1 – us L of C Photodup [951]

The civil war in spain / Morrow, Felix – N.Y., 1936. Fiche W 1064. (Blodgett Collection of Spanish Civil War Pamphlets) – 9 – us Harvard [946]

Civil war letters / Flanders, George E – 1861-64 – 1 – us Kansas [976]

Civil war letters / Fleischer, George W – 1861-65 – 1 – us Kansas [976]

Civil war letters / Wakefield, Amor William – 20 May 1861-5 Aug 1864 – 1 – us Kansas [978]

Civil war material / Johnson's Island, OH – 2r – 1 – mf#B26007-26008 – us Ohio Hist [976]

Civil war material / Robinson, James S – 1r – 1 – mf#B32659 – us Ohio Hist [355]

Civil war memories : 8th ovi / Sexton, Samuel – 1r – 1 – mf#B27940 – us Ohio Hist [355]

Civil war muster in records – 179v. 5r – 1 – mf#B30549-30553 – us Ohio Hist [976]

Civil war railroad album / U.S. Library of Congress. Prints and Photographs Division – Produced by U.S. Military Railroad Department in 1862-63. 82 Photoprints. 1 reel. P&P9209 – 1 – us L of C Photodup [080]

Civil war regimental histories and reunions – 19r – 1 – mf#B29783-29801 – us Ohio Hist [355]

Civil war times illustrated – Harrisburg. 1962+ (1,5,9) – ISSN: 0009-8094 – mf#12074 – us UMI ProQuest [976]

The civilian conservation corps : what it is and what it does – Washington, DC: Civilian Conservation Corps, Office of the Director 1939 – us CRL [060]

The civilian conservation corps and colored youth / Brown, Edgar G – Washington, DC: Office of the Director, 1939 – us CRL [060]

Civilian conservation corps bibliography : a list of references on the united states civilian conservation corps 102=civilian conservation corps. office of the director – Washington, DC: [GPO], 1939 – us CRL [060]

Civilian conservation corps camp papers – Chicago, IL: Filmed by Mid-Atlantic Preservation Service for The Center for Research Libraries, 1989-1991 – 306r 7694mf – 1,9 – us CRL [060]

Civilian conservation corps camp papers – Chicago, IL: filmed by Mid-Atlantic Preservation Service for The Center for Research Libraries, 1989-1991 – 1,9 – (ind publ separately) – us CRL [360]

Civilian conservation corps educational activities – [Spokane, Washington?: s.n.] 1937 – us CRL [060]

Civilian conservation corps newspaper, "happy days," 1933-1940 / U.S. Civilian Conservation Corps – (= ser Records Of The Civilian Conservation Corps) – 6r – 1 – mf#M1783 – us Nat Archives [071]

Civilian Conservation Corps (US) *see*
– Happy days
– Headquarters star

Civilian Conservation Corps [US] *see*
– District digest
– District review
– Nu-wud-nus
– Rib mountaineer
– Spartan
– Wekantakit
– Woodchopper

Civilian defense news : official bulletin / Milwaukee Council of Defense – v1 n1-v4 n7 [1942 aug-1945 jun] – 1r – 1 – mf#479288 – us WHS [350]
Der civilingenieur – Freiberg: Verlag von J G Engelhardt, 1854-1896 – us CRL [624]
Le civilisateur : ou les hommes illustres – Paris. oct 1852-54 – 1 – fr ACRPP [073]
Civilisation : its cause and cure, and other essays / Carpenter, Edward – 2nd ed. London: S Sonnenschein 1891 [mf ed 1987] – (= ser Social science series 2) – 1r – 1 – mf#2081 – us Wisconsin U Libr [900]
Civilisation at the cross roads : four lectures / Figgis, John Neville – London, New York: Longmans, Green, 1913, c1912 [mf ed 1990] – (= ser William belden noble lectures) – 1mf – 9 – 0-7905-4517-9 – (incl bibl ref) – mf#1988-0517 – us ATLA [230]
La civilisation du tchad...: suivi d'une etude sur les bronzes sao, par raymond lantier / Lebeuf, Jean Paul – Paris: Payot, 1950. 198p. illus. maps – 1 – us Wisconsin U Libr [306]
Civilismo y militarismo / Mancera Galletti, Angel – Caracas, Venezuela. 1960 – 1r – us UF Libraries [972]
La civilite des petites filles / Juranville, Clarisse – Paris: Larousse, 1900 – (= ser Les femmes [coll]) – 2mf – 9 – mf#12751 – fr Bibl Nationale [390]
Civilizacao do couro / Castelo Branco, Renato – Teresina, Brazil. 1942 – 1r – us UF Libraries [972]
Civilizacao holandesa no brasil / Rodrigues, Jose Honorio – Sao Paulo, Brazil. 1940 – 1r – us UF Libraries [972]
Civilizacion chibcha / Triana, Miguel – Bogota, Colombia. 1951 – 1r – us UF Libraries [972]
La civilizacion cristiana del choco (1554-1810) / Tommasini, Gabriel – Buenos Aires, 1937; Madrid: Razon y Fe, 1941.-2v – 1 – sp Bibl Santa Ana [306]
Civilizacion de los antiguos mayas / Ruz Lhuillier, Alberto – Santiago, Cuba. 1957 – 1r – us UF Libraries [972]
Civilization during the middle ages : especially in relation to modern civilization / Adams, George Burton – rev ed. New York: C Scribner, c1914 [mf ed 1991] – 2mf – 9 – 0-524-01340-3 – (original ed c1894. incl bibl ref) – mf#1990-0386 – us ATLA [931]
The civilization of babylonia and assyria : its remains, language, history, religion, commerce, law, art, and literature / Jastrow, Morris – 2nd ed. Philadelphia: JB Lippincott, c1915 – (= ser Richard W. Westbrook Lectures) – 2mf – 9 – 0-524-02210-0 – mf#1990-2884 – us ATLA [930]
The civilization of christendom : and other studies / Bosanquet, Bernard – London: S Sonnenschein; New York: Macmillan, 1893 – (= ser The Ethical Library) – 1mf – 9 – 0-7905-3596-3 – mf#1989-0089 – us ATLA [240]
The civilization of the east / Hommel, Fritz – London: JM Dent, 1900 – (= ser The Temple Primers) – 1mf – 9 – 0-7905-1159-2 – mf#1987-1159 – us ATLA [930]
Die civilprocess-ordnung nach mosaisch-rabbinischem rechte / Bloch, Moses – Budapest: Universitaets-Buchdruckerei 1882 [mf ed 1985] – 1mf – 9 – 0-8370-5982-8 – mf#1985-3982 – us ATLA [270]
Civilta cattolica on father faber's spiritual works – London, England. 1872 – 1r – us UF Libraries [241]
Civilta khmer : testo di donatella mazzeo e chiara silvi antonini. presentazione di han suyin / Mazzeo, Donatella – Milano [Italy]: A Mondadori 1972 [mf ed 1989] – (= ser Grandi monumenti [9]) – 1r with other items – 1 – (with bibl) – mf#mf-10289 seam reel 021/07 [$] – us CRL [930]
Civinini, Guelfo *see* Recordi di carovana
Civitates orbis terrarvm / Braun, G & Hohenberg, F – [Cologne, 1575-1618]. 6v – 44mf – 9 – mf#H-8199 – ne IDC [956]
Ciwororo cavakuru / Jackson, S K – Fort Victoria, Zimbabwe. 1950 – 1r – us UF Libraries [960]
Cj international – Chicago. 1989-1996 (1,5,9) – (cont by: crime and justice international) – ISSN: 0882-0244 – mf#14964 – us UMI ProQuest [360]
Cj management and training digest – Fairfax. 1995-1998 (1,5,9) – ISSN: 1079-1574 – mf#21213 – us UMI ProQuest [360]
Cjem : journal of the canadian association of emergency physicians – Ottawa. 1999+ (1,5,9) – ISSN: 1481-8035 – mf#30978 – us UMI ProQuest [610]
The cl psalms of david, in scottish meter.. – 1615 – 1 – $50.00 – us Presbyterian [240]
Claassen, Antoon *see* Cambodja
Claassen, Johannes *see*
– Die falschmuenzerische theologie albrecht ritschls und die christliche wahrheit
– Das licht und die farben
Clackamas county banner – Oswego OR: Clackamas County Banner Pub Co, 1918-19 [wkly] – 1 – (merged with: oregon city courier [oregon city, or: 1902], to form: banner-courier; cont: oswego times; ceased in 1919) – us Oregon Lib [071]
Clackamas county banner *see* Banner-courier
Clackamas county independent – Canby OR: H L Gill, [wkly] [1968] – 1r – 1 – us Oregon Lib [071]
Clackamas county news – Canby OR: B E Lee, [wkly] [mf ed 1968] – 1r – 1 – (merged with: canby herald [canby or: 1914] to form: canby herald and clackamas county news) – us Oregon Lib [071]
Clackamas county news [estacada, or] – Estacada OR: G E Parks, 1928- [wkly] – 1 – (cont: eastern clackamas news; cont by: estacada's clackamas county news [estacada, or]) – us Oregon Lib [071]
Clackamas county news [estacada, or: 1976] – Estacada OR: R C Horn, 1976-91 [wkly] – 1 – (cont: estacada's clackamas county news [estacada, or]; cont by: estacada's clackamas county news [estacada, or: 1991]) – us Oregon Lib [071]
Clackamas county record – Oregon City OR: Record Pub Co, 1903- [wkly] – 1 – us Oregon Lib [071]
Clackamas county review – Milwaukie OR: B T Casey, 1983-88 [wkly] – 1 – (cont: new review [milwaukie, or]; cont by: review [clackamas, or]. place of publ moves to clackamas, or, jul 3 1986) – us Oregon Lib [071]
Clackamas county review – Clackamas OR: Eastside Pub Co, 1992-95 [wkly] – 1 – (cont: review [clackamas, or]; cont by: clackamas review) – us Oregon Lib [071]
Clackamas democrat – Oregon City OR: B Fithian, [wkly] – 1 – us Oregon Lib [071]
Clackamas review – Clackamas OR: Eastside Pub Inc, 1995- [wkly] – 1 – (cont: clackamas county review [1983-88]) – us Oregon Lib [071]
Clackmannanshire advertiser : and alloa journal of news – Alloa, Scotland 1 feb 1851-19 feb 1859 (wkly) – 1 – (cont by: alloa journal & clackmannanshire advertiser [26 feb 1859-12 feb 1916]) – uk British Libr Newspaper [072]
Clackmannanshire advertiser, and alloa journal of news – Alloa, Scotland 18 jan 1851-19 feb 1859 – 1 – (wanting: no.21,23,30,31,45; cont: clackmannanshire advertiser, & monthly journal of general literature [n1-48, ns: n1-36] 20 jan 1844-18 dec 1847, 15 jan 1848-21 dec 1850) – uk British Libr Newspaper [072]
Clackmannanshire advertiser, and monthly journal of general literature. – Alloa, Scotland 20 jan 1844-18 dec 1847, 15 jan 1848-21 dec 1850 – 1 – (wanting: n21,23,30,31,45; cont by: clackmannanshire advertiser, & alloa journal of news [18 jan 1851-19 feb 1859]) – uk British Libr Newspaper [072]
Cladel, Leon *see* Revanche
Een claer bewijs : van het recht gebruyck des nachtmaels christi... / Micronius, M – London, 1554 – 4mf – 9 – mf#PBA-270 – ne IDC [240]
Claeys, Hector *see* Leven van sinte godelieve
Clagett, Charles *see* A review relative to the court of appeals of maryland
Claiborne County Historical Society [TN] *see* Reflections
Claim in the hills / Wickenden, James – New York, NY. 1957 – 1r – us UF Libraries [972]
Claim of ireland / Thom, John Hamilton – London, England. 1847 – 1r – us UF Libraries [241]
The claims and opportunities of the christian ministry / Gordon, George Angier et al; ed by Mott, John Raleigh – New York: Young Men's Christian Association Press, 1911 – 1mf – 9 – 0-524-08872-1 – mf#1993-3336 – us ATLA [240]
Claims between shippers and carriers: a digest of the american decisions / Merriam, Ralph – Chicago: La Salle Extension University, 1916. 1815p. LL-773 – 1 – us L of C Photodup [344]
Claims for georgia militia campaigns against indians on the frontier, 1792-1827 / U.S. Treasury Dept – Washington, DC: R J Taylor, Jr (mf ed 1993) – (= ser Records of the accounting officers of the department of the treasury) – 5r – 1 – (with printed guide) – mf#M1745 – us Nat Archives [975]
The claims of christian science as so styled : and its peculiar philosophy / Jewell, Frederick Swartz – 2nd ed. Milwaukee, Wis: Young Churchman, 1897 – 1mf – 9 – 0-524-04617-4 – mf#1990-1277 – us ATLA [240]
Claims of christianity examined from a rationalist standpoint / Watts, Charles – London, England. 18– – 1r – us UF Libraries [240]
The claims of decorative art / Crane, Walter – London: Lawrence & Bullen, 1853 – (= ser 19th c art & architecture) – 3mf – 9 – mf#4.1.204 – uk Chadwyck [740]
Claims of edward quinn, lumberer : for indemnity against the government, for losses sustained in the st maurice territory / Quinn, Edward – [Quebec?: s.n.] 1858 [mf ed 1994] – 1mf – 9 – 0-665-94712-7 – mf#94712 – cn Canadiana [346]
The claims of japan and malaysia upon christendom : exhibited in notes of voyages made in 1837, from canton, in the ship morrison and brig himmaleh... – New York: E French, 1839 [mf ed 1995] – (= ser Yale coll) – 2v (ill) – 1 – 0-524-09703-8 – mf#1995-0703 – us ATLA [950]
The claims of jesus christ : lent lectures / Sparrow-Simpson, William John – London: Longmans, Green, 1899 – 1mf – 9 – 0-524-05635-8 – mf#1992-0490 – us ATLA [220]
Claims of jesus of nazareth examined / Raffles, Thomas – London, England. 1810 – 1r – us UF Libraries [240]
Claims of rome / Smith, Samuel – London, England. 1896 – 1r – us UF Libraries [240]
Claims of swedenborg / Mill, John – London, England. 1856 – 1r – us UF Libraries [240]
Claims of the additional curates' fund society / Butcher, Samuel – Dublin, Ireland. 1857 – 1r – us UF Libraries [240]
Claims of the catholic church / Sibthorpe, Richard Waldo – Oxford, England. 1841 – 1r – us UF Libraries [241]
The claims of the catholic church : a letter to the parishoners of saint paul's, halifax, nova scotia / Maturin, Edmund – [Halifax, NS?: s.n.] 1859 [mf ed 1983] – 2mf – 9 – 0-665-38049-6 – mf#38049 – cn Canadiana [241]
Claims of the church of england upon her members / Symons, Benjamin Parsons – Oxford, England. 1842 – 1r – us UF Libraries [241]
Claims of the churchmen and dissenters of upper canada : brought to the test in a controversy between several members of the church of england and a methodist preacher – Kingston, Ont?: s.n, 1828 (Kingston Ont: Herald) – 3mf – 9 – (incl bibl ref) – mf#54883 – cn Canadiana [242]
Claims of the established church – London, England. 1817 – 1r – us UF Libraries [240]
Claims of the free church of scotland – Princeton, NJ. 1844 – 1r – us UF Libraries [243]
Claims of the missionary enterprise on the medical profession / Macgowan, Daniel Jerome – Edinburgh, Scotland. 1847 – 1r – us UF Libraries [240]
The claims of the old testament : lectures delivered in connection with the sesquicentennial celebration of princeton university / Leathes, Stanley – New York: Charles Scribner's Sons, 1897. Beltsville, Md: NCR Corp, 1978 (1mf); Evanston: American Theol Lib Assoc, 1984 (1mf) – 9 – 0-8370-0862-X – mf#1984-4206 – us ATLA [221]
Claims of the protestant association on public support / Woodward, George Henry – London, England. 1836 – 1r – us UF Libraries [242]
The claims of the roman catholic church examined and tested by scripture / Spochynski, Stephen – Paterson, NJ: T Warren, 1853 – 1mf – 9 – 0-8370-8067-3 – mf#1986-2067 – us ATLA [241]
Le clairon – Paris. no. 1-1153. Red. en chef J. Cornely. 7 mars 1881-, 1er mai 1884. mq no. 730. 4 mars 1889-91, janv-24 juin 1893, 22 mars-5 juin 1902 – 1 – fr ACRPP [074]
Le clairon – Port-au-Prince: P Errie. 1re annee. n2-n30. 28 juin 1902-17 dec 1902 – 2 sheets – us CRL [079]
Clair-Tisdall, W St *see* Manual of the leading muhammadan objections to christianity
Clairvaux, Bernard of, Saint *see* Predigten des h bernhard in altfranzoesischer uebertragung
Clairville, M *see*
– Ah! enfin!
– Amour dans tous les quartiers
– Avenir dans le passe
– Breda-street
– Chansons populaires de la france
– Congres de la paix
– Daphnis et chloe
– Jeune et la vieille garde
– Ma niece et mon ours
– Madame marneffe, ou, le pere prodigue
– Moulin joli
– Propriete c'est le vol
– Rhum
– Roger bontemps
– Semaine a londres, ou, les trains de plaisirs
– Trois loges
– Vie a bon marche
Clairvoyance / Leadbeater, Charles Webster – 2nd ed. London: Theosophical Publ Soc, 1903 [mf ed 1999] – 1r – 1 – (incl ind) – mf#28693 – us Harvard [290]
Clam, Ernst *see* Lord cohn
Clamor africano : lourenco marques: "o brado africano", [dec 10 1932-feb 25 1933] – reel 5 – us CRL [960]
Clamor africano – Quelimane: Tip. Progresso, apr 21-27, aug 30 1892; aug 30 1893 – us CRL [960]
O clamor publico : jornal politico, industrial e litterario – Rio de Janeiro, RJ: Typ Guanabarense de L A F de Menezes, 18 ago 1860-29 jan 1861 – (= ser Ps 19) – mf#P25,03,08 n02 – bl Biblioteca [073]
Clan conrey – v1 n1-v7 n4 [1982 dec-1988 dec] – 1r – 1 – mf#1054539 – us WHS [071]
Clan Douglas Society of North America *see* Newsletter
Clan macbean in north america register – 1983 jan-1985 dec – 1r – 1 – (cont: clan macbean register) – mf#1312385 – us WHS [929]
Clan macbean register – v2 n7 [1970 mar], v2 n11-12 [1971 jun-sep], v2 n14-15 [1972 mar-jun], v3 n1 [1972 dec], v3 n2-5, [1973 mar-dec], v3 n7 [1974 sep], v3 n11-12, [1975 sep-dec], v3 n14-16 [1976 jul-dec], v3 n17 [1977 mar], v4 n3 [1977 dec], v4 n4 [1978 mar], v4 n14 [1980 dec], v4 v15-18 [1981 mar-dec] v5 n2 [1982 jun] – 1r – 1 – (cont: macbean register; cont by: clan macbean in north america register) – mf#615161 – us WHS [929]
Clan mccullough/mcculloch newsletter – 1977 nov-1984 nov – 1r – 1 – mf#965681 – us WHS [929]
Clanchy, T J *see* Ireland in the twentieth century
Clancy, James J *see* Ireland
Clandestinos / Candanedo, Cesar A – Panama, Panama. 1957 – 1r – us UF Libraries [972]
Clangores tubae adversvs theodorvm bezam vt priusquam in alterum exspirat secvlvm, hvc respiciat et perpendat / Huber, S – Vrsellis, 1598 – 1mf – 9 – mf#TH-1 mf 714 – ne IDC [242]
Clanin, Douglas E *see* The papers of william henry harrison 1800-1815
The clans of the baganda / Kagwa, Apolo – Mengo, Uganda 1949 – (filmed with: rowe, j a: selected articles on the bataka controversy ... bukalasa 1922-1923. gray, j m: mutesa of buganda 1934. katate, a g: abagabe b'ankole, kampala 1955) – us CRL [307]
Clao journal : official publication of the contact lens association of ophthalmologists, inc / Contact Lens Association of Ophthalmologists – St. Louis. 1983+ (1,5,9) – ISSN: 0733-8902 – mf#13373,01 – us UMI ProQuest [617]
Claparede, R *see* La reforme en bourgogne
Claparede, Theodore *see* Histoire des eglises reformees du pays de gex
Claparede, Theodore et al *see* Pour michel servet
Clapham – (= ser Bedfordshire parish register series) – 1mf – 9 – £3.00 – uk BedsFHS [929]
Clapham gazette and local advertiser – London, UK. Feb-may 1854; jul 1854-jul 1885 – 1/2r – 1 – uk British Libr Newspaper [072]
Clapham mercury or clapham wandsworth battersea streatham tooting putney and south western general advertiser – London, UK. may, 30 jun, 7 jul-21 jul – 1/4r – 1 – (incl specimen issue dated may 1855) – uk British Libr Newspaper [072]
Clapham observer tooting = [And balham times and surrey advertiser] – London. 1 may 1869-dec 1871, 7, 21, 28 sep 1872, 31 may 1873-dec 1965 [wkly] – 114 1/2r – 1 – (aka: balham times; surrey advertiser; clapham news observer; clapham news and observer; clapham & lambeth news) – uk British Libr Newspaper [072]
Clapham, Samuel *see* Duty of the clergy to enforce the frequent receiving of the sacrame...
Clapham, st thomas a becket monumental inscriptions – Bedfordshire Family HS 1978 – (= ser Bedfordshire parish register series) – 1mf – 9 – £1.25 – uk BedsFHS [929]
Clapin, Sylva *see*
– Londres et paris
– Ne pas dire mais dire
– Sensations de nouvelle-france
Clapin, Sylvia *see* Le canada
Clapp, Cephas F et al *see* Mrs. abbie walker staver
Clapp, Henry *see* With raleigh to british guiana
Clapp, Jacob Crawford et al *see* Historic sketch of the reformed church in north carolina
Clapp, Theodore *see*
– Autobiographical sketches and recollections
– Theological views
Clappe, Arthur A *see*
– Masque entitled "canadas sic welcome"
– Wind-band and its instruments
Clapper, David K *see*
– History of the clappers
– The selection of a church
Clapper, Valiant Abel *see* Organisational culture and transformation
Clapperton, H *see*
– Journal of a second expedition into the interior of africa
– Narrative of travels and discoveries in northern and central africa in the years 1822, 1823 and 1824
– Narrative of travels and discoveries in northern and central africa in the years 1822, 1823, and 1824...

Clapperton, Hugh *see*
- Journal of a second expedition to africa
- Travels and discoveries in africa

Clapperton, John Alexander *see*
- First steps in new testament greek
- Pitfalls in bible english

Clapton, Edward *see* The precious stones of the bible – descriptive and symbolical

La claque – [Montreal]: [s.n.] n1 [1970]- (irreg) [mf ed 1978] – 1r – 1 – (ceased 1970?) – mf#SEM35P161 – cn Bibl Nat [073]

The clara barton papers – 123r – 1 – $4,305.00 – Dist. us Scholarly Res – us L of C Photodup [362]

Clara viebig und der frauenroman des deutschen naturalismus / Wingenroth, Sascha – Endingen/Baden: E Wild, 1936 [mf ed 1989] – 109p – 1 – (incl bibl) – mf#7156 – us Wisconsin U Libr [430]

Clare advertiser – Kilrush, Ireland. 17 Jul 1869-29 Nov 1873; 1875-76; 1878-82; 1887. -w. 15 1/2 reels – 1 – uk British Libr Newspaper [072]

Clare advertiser and kilrush gazette – Kilrush, Ireland. 17 jul 1869-1887 – 18 1/2r – 1 – uk British Libr Newspaper [072]

Clare champion – Ennis, Clare. 1923-50 – 24r – 1 – (cont as: clare champion [28 mar 1903-; not publ 30 mar-28 sep 1918]) – ie National [072]

Clare champion – Ennis, Ireland. 1930; 1986-92 – 20r – 1 – uk British Libr Newspaper [072]

Clare examiner and limerick advertiser – Ennis, Ireland. may 1879-nov 1887 – 3 1/2r – 1 – uk British Libr Newspaper [072]

Clare freeman and ennis gazette – Ennis, Ireland. 14 feb 1853-26 jan 1884 – 25 1/2r – 1 – uk British Libr Newspaper [072]

Clare independent and tipperary catholic times – Ennis, Ireland. 24 jan 1877-11 jun 1881 – v2 n11- – 1 – (cont as: independent and munster advertiser) – uk British Libr Newspaper [072]

Clare independent and tipperary catholic times – Ennis, Clare. aug 1876-jan 1877 – 1r – 1 – (cont as: independent and munster advertiser [18 jun 1881-19 dec 1885]) – ie National [072]

Clare, John *see*
- Original manuscripts and papers from the collection in northampton central library
- Original manuscripts from the collection in peterborough museum and art gallery

Clare journal – Ennis 1808, 1876, 1899-1909, 1913-apr 1917 – 11r – 1 – ie National [072]

Clare journal and ennis advertiser – Ennis, Ireland. 1828-nov 1874; 1875-aug 1877; may 1878-1896 – 67 1/2r – 1 – uk British Libr Newspaper [072]

Een clare uitlegginghe vanden apocalypsis... / Taffin, J – Middelburgh, 1611 – 11mf – 9 – mf#PBA-309 – ne IDC [240]

Clare weekly news – Ennis, Ireland. jun 1879-may 1880 – 1/2r – 1 – uk British Libr Newspaper [072]

[Claremont-] circuit west – CA. feb 1975-1976 – 1r – 1 – $60.00 – mf#R03183 – us Library Micro [071]

[Claremont-] claremont courier – CA. sep 1912-mar 1913; aug 27 1969- – 53+ r – 1 – $3180.00 (subs $85/y) – mf#CR02112 – us Library Micro [071]

Claremont courier – Claremont, CA. 1994-2000 (1) – mf#68926 – us UMI ProQuest [071]

[Claremont-] courier (laverne, montclair and upland comb) – CA. 1983-1989 – 7r – 1 – $420.00 – mf#R04014 – us Library Micro [071]

[Claremont-] inland valley times – CA. 1991 – 1r – 1 – $60.00 – mf#R04015 – us Library Micro [071]

Claremont, New Hampshire. Claremont Baptist Church and Society *see* Records

[Claremont-] news pulse – CA. apr 1973-feb 7 1975 – 1r – 1 – $60.00 – mf#R02113 – us Library Micro [071]

Claremont reading conference yearbook – Claremont. 1936+ (1) 1971+ (5) 1976+ (9) – ISSN: 0886-6880 – mf#2475 – us UMI ProQuest [370]

Clarence and richmond examiner – Grafton, Australia. 10 Sep 1892-31 Dec 1901; 8 Mar 1902.-d. 15 reels – 1 – uk British Libr Newspaper [079]

Clarence and richmond examiner – Grafton, jul 1859-jun 1915 – 32r – 9 – A$1976.90 vesicular A$2152.90 silver – at Pascoe [079]

Clarence cameron white papers : from the holdings of the schomburg center for research in black culture, manuscripts, archives and rare books division: the new york public library, astor, lenox and tilden foundations – 1995 – 10r – 1 – $850.00 – (guide which covers all coll under "literature and the arts" sold separately for $20.00 d3305.g6) – mf#D3305P20 – Dist. us Scholarly Res – us L of C Photodup [780]

Clarence darrow's two great trials : reports of the scopes and the dr. sweet negro trial / Halderman-Julius, M – Girard, Kansas: H-J Company, 1927 – 1mf – 9 – $1.50 – mf#LLMC 91-059 – us LLMC [340]

Clarence, frere *see* Bibliographie analytique des oeuvres de m jean-marie laurence

Clarence jordan: a prophet in blue jeans / Barnette, Henlee H – 1982. Lecture delivered at Southern Baptist Theological Seminary, Louisville, KY – 1 – $5.00 – us Southern Baptist [242]

Clarence muse chicago fan club : [newsletter] – 1937 mar – 1r – 1 – mf#4862671 – us WHS [790]

Clarence river advocate – Maclean, jan 1898-sep 1909 – 5r – A$325.65 vesicular A$353.15 silver – at Pascoe [079]

Clarendon, Edward Hyde, Earl of *see* The history of the rebellion and civil wars in england

The clarendon papers, 1867-1870 / Villiers, George William Frederick – State Papers Foreign Collection 361. Orig. publ. by Michael Glazier Inc – 1r – 1 – $130.00 – mf#D3264 – us Scholarly Res [941]

Clarendon papers, the american material in the... 1853-1870 : from the bodleian library, oxford – (= ser British records relating to america in microform) – 15r – 1 – (with guide. int by colin bonwick) – mf#97403 – uk Microform Academic [975]

Clarendon second baptist church. clarendon, vermont : church records – 1798-1832 – 1 – 5.58 – us Southern Baptist [242]

Claresholm advertiser – Alberta, CN. 1914-16 – 1r – 1 – cn Commonwealth Imaging [071]

Claresholm local press – Alberta, CN. jan1926-dec 1996 – 33r – 1 – cn Commonwealth Imaging [071]

Claresholm review – Alberta, CN. jan 1907-dec 1916 – 3r – 1 – cn Commonwealth Imaging [071]

Clareson, Thomas D *see* Early science fiction novels

Claretian Fathers and Brothers *see* Salt

Clarici, P B *see* Istoria e coltura delle piante...

Claridad – 1972 mar 19-1973 dec 23, 1974 jan 13-1975 apr 27, 1975 may 4-1976 dec, 1976 dec 28-1977 jul 7 – 4r – 1 – (cont by: claridad [new york ny: 1977]) – mf#359365 – us WHS [071]

Claridad – v4 n249-n263 [1977 apr 1/7-jul 1/7] – 1r – 1 – (cont: claridad [new york ny: 1972]; cont by: claridad [new york ny: 1979]) – mf#599399 – us WHS [071]

Claridad – 1979 jun 22-1980 jan 3, 1980 jan-1981 jan 3, 1981, 1982-1983 mar, 1983 feb 25-1984 mar, 1984 mar-1985 jan 24 – 6r – 1 – (cont: claridad [new york ny: 1977]) – mf#599400 – us WHS [071]

Claridad / Movimiento Pro Independencia de Puerto Rico – San Juan. v17 n884-v19 n1310 [1975 oct 3-1978 mar 2], 1984 dec 27/1985jun-1990 jan-sep – 12r – 1 – mf#486363 – us WHS [972]

La claridad – [Habana: s.n. v1 n5,6,?,? (nov 1,8, [no date], 29 1890); v2 n12-22,?,?-? (mar 21-may 22, may 30, jun 20-27 1891) – 3mf – us CRL [079]

Claridge, R *see* An answer to richard allen's essay

Claridge, Richard *see* Extracts from the writings of william penn and richard claridge

O clarim da fama : periodico satyrico – Recife, PE: Typ Popular, 04 dez 1863-[?] – (= ser Ps 19) – mf#P17,02,145 – bl Biblioteca [870]

O clarim da monarchia : folha politica e litteraria – Maranhao: Typ A Conservadora, 30 out 1861-27 mar 1862 – (= ser Ps 19) – bl Biblioteca [079]

O clarim dos bastidores : jornal theatral, critico e recreativo – Rio de Janeiro, RJ: Typ de Domingos Luiz dos Santos, 07 nov 1861 – (= ser Ps 19) – mf#P17,1,91 – bl Biblioteca [790]

O clarim dos theatros : publicacao critica – Rio de Janeiro, RJ: Typ Franceza, 17 maio 1851 – (= ser Ps 19) – mf#P15,01,77 – bl Biblioteca [790]

Clarin – Miami, FL. 1976 may 27-1988 aug 01 – 1r – us UF Libraries [071]

Clarin de la Rive, Abel *see*
- La femme et l'enfant dans la francmaconnerie universelle
- Le juif dans la franc-maconnerie

Clarinadas / Cebrian De Quesada, Arnaldo – Miami, FL. 1963 – 1r – us UF Libraries [972]

Clarines de feria – Merida, 1964, 1965 y 1969 – 5 – sp Bibl Santa Ana [073]

Clarinet choir : a means of teaching and performing music / Danfelt, Edwin Douglas – U of Rochester 1964 [mf ed 19–] – 9mf / 1r – 9,1 – (contents: appendices [pt1];; pt2: transcr of works by claudio monteverdi, carlo gesualdo, thomas weelkes, henry purcell, j s bach, w a mozart, johannes brahms & igor stravinsky) – mf#fiche 414 / film 2557 – us Sibley [780]

Clarinet choir: a means of teaching and performing music / Danfelt, Edwin Douglas – U of Rochester 1964 [mf ed 19–] – 9mf / 1r – 9,1 – mf#fiche414 / film 2557 – us Sibley [780]

Clarington news / Monroe Co. Clarington – 1933-46 (scattered) – 1r – 1 – mf#B36221 – us Ohio Hist [071]

Clarion – Belize, 19 dec 1897-10 jul 1919, 1 sep-18 dec 1919, 12 feb 1920-31 may 1940, 1 feb-30 nov 1943, 1 oct 1951-6 nov 1954, 1957-20 oct 1961 – 87 1/2r – 1 – (aka: daily clarion) – uk British Libr Newspaper [079]

Clarion / Bethune-Cookman College [Daytona Beach FL] – 1992 spr, 1993 win-spring/summer, 1994 spr-1996 win, 1997 win, 1998 jan – 1r – 1 – mf#2907015 – us WHS [071]

Clarion – Camp Cook, CA. 1942-1946 (1) – mf#61967 – us UMI ProQuest [071]

Clarion – Medford OR: Clarion Pub Co, 1922-24 [wkly] – 1 – (cont by: jackson county news [1924-26]; merged with: ashland square deal) – us Oregon Lib; us Oregon Hist [071]

Clarion – Superior WI. 1901 apr 6-jul 20 – 1r – 1 – (cont by: superior citizen; clarion-citizen) – mf#933596 – us WHS [071]

Clarion – Taylor, NE: E Andrews. 4v. v13 n28. may 7 1896-v16 n12. jan 12 1899 (wkly) [mf ed with gaps] – 2r – 1 – (cont: loup county clarion. cont by: taylor clarion) – us NE Hist [071]

Clarion / Lorain Co. Columbia Stat – v1 n1. feb 1952-sep 1954// [mthly] – 1r – 1 – mf#B33544 – us Ohio Hist [071]

Clarion / Mahoning Co. Poland – mar 1971-jan 1973, apr 1973-76 [wkly] – 5r – 1 – mf#B6891-6895 – us Ohio Hist [071]

Clarion – Poland, OH. 1974-1984 (1) – mf#65638 – us UMI ProQuest [071]

Clarion / Professional Staff Congress/City University of New York – v1 n1-v4 n5 [1972 may 5-1975 feb 3] – 1r – 1 – (cont by: psccuny clarion) – mf#656941 – us WHS [378]

Clarion – v1 n1-v2 n20 [1941 apr 12-dec 27] – 1r – 1 – mf#919871 – us WHS [071]

Clarion – v1-2 n2. 1932-34 [all publ] – (= ser Radical periodicals in the united states, 1881-1960. series 2) – 1r – 1 – $115.00 – us UPA [335]

Clarion *see*
- Denver county miscellaneous newspapers, reel 3
- Guardian

The clarion – Belize. British Honduras. (Daily Clarion). -d. 19 Nov 1897-31 May 1940; 1 Feb-30 Nov 1943, 1 Oct 1951-6 Nov 1954, 2 Jan 1957-20 Oct 1961. (90 reels) – 1 – uk British Libr Newspaper [079]

The clarion – Cape Town [South Africa]: Stewart Printing Co Ltd, may 29-aug 14 1952 – us CRL [079]

The clarion – Manchester and London. England. -w. 12 Dec 1891-22 Apr 1927. (31mqn reels) – 1 – uk British Libr Newspaper [072]

The clarion : official organ of the communist party of canada / Communist Party of Canada – Toronto: Communist Party of Canada. v16 n1851-70. mar 23 1940-apr 5 1941//? – 1r – 1 – Can$60.00 – (issues publ illegally after the clarion was suspended under the wartime measures act) – cn McLaren [335]

The clarion *see* Huerfano county miscellaneous newspapers

Clarion call – 1979 jan 3-1981 sep 30 – 1r – 1 – (cont by: clarion call [national city ca: 1982]) – mf#347418 – us WHS [071]

Clarion call – n1-151 [1982 jan-1985 aug] – 1r – 1 – (cont: clarion call [national city ca]) – mf#577182 – us WHS [071]

Clarion defender – Portland OR: [s.n.] 1965-[wkly] – 1 – us Oregon Lib [071]

Clarion democrat – Princeton, IN. 1958-1960 (1) – mf#62940 – us UMI ProQuest [071]

The clarion democrat – Clarion, PA. -w 1868-69; 1884-1941; 1945-46. 22 rolls – 13 – $25.00r – us IMR [071]

Clarion leader – Princeton, IN. 1897-1901 (1) – mf#62941 – us UMI ProQuest [071]

Clarion ledger – Jackson, MS. 1947+ (1) – ISSN: 0744-9526 – mf#60504 – us UMI ProQuest [071]

Clarion news – Clarion, PA., 1972-1975 – 13 – $25.00r – us IMR [071]

Clarion of freedom / Guernsey Co. Cambridge – apr-sep 1847 [wkly] – 1r – 1 – mf#B6739 – us Ohio Hist [071]

Clarion of freedom / Muskingum Co. New Concord – oct 1847-sep 1848 [wkly] – 1r – 1 – mf#B6739 – us Ohio Hist [071]

Clarion of skye and highland trader : onward skye! queen of the west – Struan, Isle of Skye: Cristobel Nicolson 1953-57 (mthly) – 1 – (cont: clarion of skye & inter-island advertiser; title varies) – uk Scotland NatLib [072]

Clarion of skye and inter-island advertiser : onward skye! queen of the west – Struan, Isle of Skye: Cristobel Nicolson 1951-53 (mthly) – 31v – 1 – (cont by: clarion of skye and highland trader) – uk Scotland NatLib [072]

Clarion (preston) – [NW England] Lancashire 10 dec 1891-15 may 1897 [mf ed 2002] – 6r – 1 – uk Newsplan [072]

The clarion republican – Clarion, PA., 1899-1958 – 13 – $25.00r – us IMR [071]

Clarion republican/gazette – Clarion, PA. -w 1884-1899. 5 rolls – 13 – $25.00r – us IMR [071]

Clarion-citizen – Superior WI. 1901 jul 27-1902 dec 21 – 1r – 1 – (cont: clarion [superior wi]; superior citizen) – mf#933599 – us WHS [071]

Clarissa / Verissimo, Erico – Porto Alegre, Brazil. 1943 – 1r – us UF Libraries [972]

Clarissimi viri d. andreae alciati emblematum libellus, vigilanter recognitus... / Alciato, Andrea – Parisiis: Apud Christianum Wechelum, 1542 – 3mf – 9 – mf#O-1473 – ne IDC [090]

Clarissimi viri d. andreae alciati emblematum libri duo/ [and] in d andreae alciati emblemata succincta commentariola... / Alciato, Andrea – Lugduni: Apud loan Tornaesium, & Gul. Gazeium, 1554, 1556 – 4mf – 9 – mf#O-1824 – ne IDC [090]

Clarity / Young Communist League – v1-4 n2,1. 1940-43 [all publ] – (= ser Radical periodicals in the united states, 1881-1960. series 1) – 10mf – 9 – $115.00 – us UPA [335]

Clark Air Base [Philippines] *see* Philippines flyer

Clark, Albert Curtis *see* Recent developments in textual criticism

Clark and finnelly's reports : reports of cases heard and decided in the house of lords on appeals and writs of error / Clark, Charles & Finnelly, W – v1-12. 1831-46. London: J & W T Clark, 1835-47 (all publ) – 103mf – 9 – $154.00 – mf#LLMC 95-248 – us LLMC [324]

Clark and scully's ontario drainage cases / Ontario. Canada – v1-2. 1898-1903 (all publ) – 13mf – 9 – $19.50 – mf#LLMC 81-059 – us LLMC [340]

Clark, Andrew *see* First world war: the home front

Clark, Benjamin C *see*
- Geographical sketch of st domingo, cuba and nicar...
- Remarks upon united states intervention in hayti

Clark, Calvin Montague *see* History of bangor theological seminary

Clark, Charles *see*
- Clark and finnelly's reports
- House of lords cases (clark and finnelly)

Clark, Charles Allen [comp] *see* Digest of the presbyterian church of korea (chosen)

Clark clan newsletter – v1 n1-v4 n1 [1978 fall-1981 sum] – 1r – 1 – mf#671683 – us WHS [929]

Clark clarion – 1977 jan-1985 oct – 1r – 1 – mf#1018991 – us WHS [071]

Clark Co. New Carlisle *see* Sun

Clark Co. South Charles *see*
- Sentinel
- Sentinel series

Clark Co. Springfield *see*
- Champion city times
- Daily democrat
- Farm and fireside
- Gazette
- Gazette series
- Journal und adler series
- Press-republic
- Republic
- Republic times
- Springfield tribune series
- Springfielder journal
- Times
- Times series
- Tribune
- Weekly gazette
- Weekly news

Clark county advocate – Neillsville WI. 1864 mar 21, 1866 feb 8, 1866 apr 19, 1863 aug 22 – 1r – 1 – mf#957363 – us WHS [071]

Clark county courier – Neillsville WI. 1880 jan 27-1881 feb 1 – 1r – 1 – mf#956603 – us WHS [071]

Clark county herald – Dorchester WI. 1906 jan 5-1907 aug 30, 1907 sep 6-1909 apr 30, 1909 may 7-1910 nov 25, 1910 dec 2-1912 jun 28, 1912 jul 5-1913 dec 26 – 5r – 1 – (cont by: dorchester herald) – mf#965407 – us WHS [071]

Clark county journal – Withee WI. 1912 oct 4, 1915-17, 1923-25 – 3r – 1 – (cont by: withee journal) – mf#952485 – us WHS [071]

Clark county news – Vancouver, WA. 1951-1952 (1) – mf#69366 – us UMI ProQuest [071]

Clark County press – Granton, Neillsville WI. 1938 oct 6/1940-2002 jul/dec – 74r – 1 – (cont: neillsville press) – mf#1005759 – us WHS [071]

Clark county press – Neillsville WI. 1873 jun 27-1876 apr 8 – 1r – 1 – (cont: clark county republican [neillsville wi]; cont by: clark county republican and press) – mf#1005740 – us WHS [071]

CLARK

Clark County republican – Neillsville WI. [1870 jul 6-1870 nov 30], 1871 sep-1872, 1873-1876 apr 7 – 3r – 1 – (cont by: clark county press [neillsville wi: 1873]) – mf#987039 – us WHS [071]
Clark county republican and press – Neillsville WI. 1876 apr-dec, 1877-1878 jun 14 – 2r – 1 – (cont: clark county press [neillsville wi: 1873]; cont by: republican and press) – mf#1005744 – us WHS [071]
Clark, Daniel *see*
– Address of the retiring president of "the association of medical superintendents of american institutions for the insane"
– An animated molecule and its nearest relatives
– Brain lesions and functional results
– Brain stuffing and forcing
– Ghosts and their relations
– Heredity, worry and intemperance as causes of insanity
– Medical evidence in courts of law
– Neurasthenia
– A newly discovered system of electrical medication
– Physiology in thought, conduct and belief
– A psycho-medical history of louis riel
– A report on cerebro-spinal pathology
– Wrinkles in ancient asylum reports
Clark, Davis Wasgatt *see*
– Asbury and his coadjutors
– Death-bed scenes
– Essays, moral and religious
– Experience of german methodist preachers
– Importance of doctrinal truth in religion
– Man all immortal
– The methodist episcopal pulpit
– Our friends in heaven
– Sermons
Clark, Dawn *see* An interpretive inquiry of the professional life histories of selected women dance/physical educators
Clark, Dougan *see*
– The holy ghost dispensation
– Instructions to christian converts
– The offices of the holy spirit
– The theology of holiness
Clark, E C *see* History of roman private law
Clark, Edson Lyman *see*
– The arabs and the turks
– Fundamental questions
Clark Electric Cooperative *see* Annual report
Clark, Francis Barnard *see* Clark's form book, containing legal and business forms useful to the private citizen, as well as to judges, attorneys.
Clark, Francis Edward *see*
– Danger signals
– Drawing the net
– The everlasting arms
– Fellow travellers
– The gospel in latin lands
– Junior societies of christian endeavor
– Looking out on life
– The lookout committee and its work
– The mossback correspondence
– Old lanterns for present paths
– Our business boys
– Our vacations
– Reorganization
– The united society of christian endeavor
Clark, Francis Edward [comp] *see* The work of the committees in the young people's society of christian endeavor
Clark, Fred *see* Plant beds for flue-cured tobacco
Clark, G J *see* Great sayings by great lawyers
Clark, Gabriel Penn *see* Business records
Clark, Gavin Brown *see* The transvaal and bechuanaland
Clark, George Little *see* Use of blackstrap molasses in a ration for the growing and fattening...
Clark, George Thomas *see* Mediaeval military architecture in england
Clark, George W *see*
– Notes on the gospel of matthew
– Romans and 1. and 2. corinthians
Clark, George Whitefield *see*
– The acts of the apostles
– Galatians, ephesians, philippians, colossians, 1 and 2 thessalonians, 1 and 2 timothy, titus and philemon
– Harmony of the acts of the apostles
– A new harmony of the four gospels in english
Clark, Gilbert J *see* Life sketches of eminent lawyers
Clark, Grenville *see* Microfiche inventory of the papers of grenville clark as preserved within the library of dartmouth college
[Clark Guernsey] / Guernsey, Clark – 1r – 1 – mf#B27418 – us Ohio Hist [910]
Clark, Hannah Belle *see* Public schools of chicago
Clark, Harriet Elizabeth *see* The gospel in latin lands
Clark, Henry Martyn *see* Robert clark of the punjab
Clark, Henry W *see* Liberal orthodoxy
Clark, Henry William *see*
– The christ from without and within
– The christian method of ethics
– The gospel according to st john
– History of english nonconformity from wiclif to the close of the nineteenth century
– Laws of the inner kingdom
– Liberal orthodoxy
– The philosophy of christian experience
Clark, Horace F et al *see* Clark's mineral law
Clark, Horace Fletcher et al *see* Miners' manual, united states, alaska, the klondike
Clark, Hugh *see* An introduction to heraldry
Clark, J M *see* Computer input microfilm (cim) feasibility study
Clark, J Reuben, Jr *see* Emergency legislation of the us 1775-1918
Clark, Janet M *see* Women and politics
Clark, Jeremiah Simpson *see*
– The acadian exile and sea shell essays
– Rand and the micmacs
Clark, JK *see* Two curricular settings of a hiv education unit
Clark, John *see*
– The amateur's assistant
– Brief comments on unusual happenings in early jack...
– Historical personal interview
Clark, John King *see* Systematic moral education: with daily lessons in ethics
Clark, John Murray *see*
– Canadian mining law
– The future of canada
– The law of mines in canada
Clark, John Ruskin *see* William bentley and his place in the development of unitarian theology
Clark, Joseph B *see* Leavening the nation; the story of american (protestant) home missions
Clark, Joseph Bourne *see* Blue sky
Clark kinsey collection : checklist and photographs / University of Washington. Libraries. Special Collections Division – [Seattle] WA: Special Coll Div, University of Washington Libraries [1982] – 95mf – 9 – (incl ind) – us UW Libraries [770]
Clark, Lindley Daniel *see* The law of the employment of labor
Clark, Lucien *see* Religion for the times
Clark, Marona M (Still) *see* Autobiographical notes
Clark, Mary *see* Biographical sketches of the fathers of new england
Clark, Mary Mead *see* Corner in india
Clark, Nathaniel George *see* Discourse commemorative of rev. rufus anderson, d.d., ll.d
Clark, Robert *see* The missions of the church missionary society
Clark, Rufus Wheelwright *see*
– A memoir of the rev john edwards emerson
– The question of the hour
– A review of the rev moses stuart's pamphlet on slavery entitled conscience and the constitution
– Romanism in america
– The work of god in great britain
Clark, S *see* The marrow of ecclesiastical historie
Clark, S H *see* Practical public speaking
Clark, Salter S *see* The government class book
Clark, Samuel *see* The bible atlas of maps and plans
Clark, Sean *see* Task and support surface constraints on the coordination and control of posture in older adults
Clark, Sereno Dickenson *see* Utility and glory of god's immutable purposes
Clark, Sidney J W *see* The art of using the china missionary survey
Clark, Sidney James Wells *see*
– The art of using the china missionary survey
– The indigenous church
Clark, Susan *see* History of coconut grove
Clark, Susan D *see* Quality ranking and evaluation of accredited undergraduate athletic training programs
Clark, Sybil G de *see* The evangelical missionaries and the basotho, 1833-1933
Clark, Sydney *see*
– All the best in bermuda, the bahamas, puerto rico
– All the best in central america
– All the best in cuba...
– All the best in south america
– All the best in south america west coast
– All the best in the caribbean
– Cuban tapestry
Clark, Tara J *see* The relationship between physical self-perceptions and functional muscular strength in young adult females
Clark, Thomas G *see* Field
Clark, Thomas March *see* Primary truths of religion
Clark, Virginia M *see* What women wrote
Clark, Walter H *see* History of platte presbytery
Clark, Wilfrid E Le Gros *see* Report to the committee on vaccination on an anatomical investigation into the routes by which infections may pass from the nasal cavities into the brain
Clark, William *see*
– The anglican reformation
– The comforter
– History of the christian councils
– Pascal and the port royalists
– Savonarola, his life and times
– Witnesses to christ
Clark, William Bullock *see* The promise of the spirit
Clark, William et al *see* The church's ministry of grace
Clark, William George *see* Macbeth
Clark, William Jared *see* Commercial cuba
Clark, William Lawrence *see* A treatise on the law of crimes
Clark, William Philo *see* The indian sign language
Clark, William R *see* Saint augustine
Clark, William Robinson *see*
– The anglican reformation
– Looking at the things of others
– The paraclete
– Savonarola
– Witnesses to christ
Clark-Bekederemo, J P (John Pepper) *see* Ozidi
Clarke, A M *see* The life of st. francis borgia of the society of jesus
Clarke, Adam *see*
– Christian theology
– Discourses on various subjects relative to the being and attributes of god, and his works in creation, providence, and grace
– Love of god to a lost world demonstrated by the incarnation and dea...
Clarke and hall's cases in contested elections in congress, 1789-1834 / Clarke, M St. Clair & Hall, David A – Washington: Gales & Seaton. 1v. 1834 (all publ) – 11mf – 9 – $16.50 – mf#LLMC 95-122 – us LLMC [340]
Clarke, Arthur Charles *see* Exploration of space
Clarke, Charles *see*
– Architectura ecclesiastica londini
– Examination of objections made to unitarianism by the rev j c miller, m a
Clarke, Charles Baron *see* A letter to the right hon w e gladstone
Clarke, Charles Cowley *see* Handbook of the divine liturgy
Clarke, Comer *see* Eichmann
Clarke county atlas, 1870 – 1r – 1 – mf#B7070 – us Ohio Hist [978]
Clarke courier – Berryville, VA. 1869-2000 (1) – mf#61898 – us UMI ProQuest [071]
Clarke, Dorus *see* Orthodox congregationalism and the sects
Clarke, E D *see* Travels in various countries of europe asia and africa
Clarke, Edward *see* Voyages en russie, en tartarie et en turquie
Clarke, Emily Smith *see* William newton clarke
Clarke, Geoffrey *see* The post office of india and its story
Clarke, George *see*
– Pompeii
– Tuvalu physical development plans, reports and related papers
Clarke, George Herbert *see* The essays or counsels civil and moral of francis bacon
Clarke, H H *see* The shipping ring and the south african trade
Clarke, Henry Green *see*
– The art-union exhibition, for 1843
– A critical examination and complete catalogue of the works of art now exhibiting in westminster hall
– A critical examination of the cartoons, frescos, and sculpture, exhibited in westminster hall
Clarke, Henry Harrison *see*
– Biographies of fellows, american academy of physical education
– Oregon cable-tension strength test batteries for boys and girls from fourth grade through college
– Reflections
Clarke, Henry J O C *see* A short sketch of the life of the hon thomas d'arcy mcgee
Clarke, Henry Lowther *see* Studies in the english reformation
Clarke, Hyde *see*
– Colonization, defence, and railways in our indian empire
– Serpent and siva worship and mythology, in central america, africa, and asia
– Serpent and siva worship and mythology in central america, africa, and asia – the origin of serpent worship
Clarke, J I *see* American leading cases
Clarke, James *see* Naufragia
Clarke, James Freeman *see*
– The christian doctrine of prayer
– The church – as it was, as it is, as it ought to be
– The church of the disciples in boston
– Common-sense in religion
– Essentials and non-essentials in religion
– Events and epochs in religious history
– Every-day
– False witnesses answered
– The fourth gospel
– Go up higher
– The ideas of the apostle paul
– The introduction to the gospel of john
– James freeman clarke
– The legend of thomas didymus, the jewish sceptic
– Manual of unitarian belief
– Memorial and biographical sketches
– Nineteenth century questions
– Orthodoxy, its truths and errors
– Self-culture
– Self-culture, physical, intellectual, moral and spiritual
– Sermon on channing
– Steps of belief
– Ten great religions
– The transfiguration of life
Clarke, James Freeman et al *see* Modern unitarianism
Clarke, James Langton *see* The eternal saviour-judge
Clarke, John *see*
– Memorials of baptist missionaries in jamaica
– Specimens of dialects
Clarke, John Caldwell Calhoun *see*
– The origin and varieties of the semitic alphabet
– The revelation rediscovered
Clarke, John H T *see* Transportes interiores de el salvador
Clarke, Joseph *see*
– Infinite benevolence
– Schools and school houses
Clarke, kerr and thorne : general hardware merchants and dealers in silver ware and fancy goods, at moore's nail factory building, portland bridge, and n44 city market building, germain st – [S.l: s.n, 18–?] [mf ed 1986] – 1mf – 9 – 0-665-53056-0 – mf#53056 – cn Canadiana [680]
Clarke, M St. Clair *see* Clarke and hall's cases in contested elections in congress, 1789-1834
Clarke press (greater enterprise news north ed) *see* Clarke press [north clackamas news ed]
Clarke press [greater enterprise news north ed] – Portland OR: Clarke Pub Co Inc, 1967 [wkly] – 1 – (related to: clarke press [portland, or: north clackamas news ed]; cont: greater enterprise news north; cont by: press [portland, or]) – us Oregon Lib [071]
Clarke press (north clackamas news ed) *see* Clarke press [greater enterprise news north ed]
Clarke press [north clackamas news ed] – Portland OR: Clarke Pub Co, 1967- [wkly] – 1 – (related to: clarke press [portland, or: greater enterprise news north ed], press [portland, or]; cont: north clackamas news [-1967]) – us Oregon Lib [071]
Clarke press (portland, or: north clackamas news ed) *see* Press (portland, or)
Clarke, R Floyd *see* The science of law and lawmaking
Clarke, Richard Frederick *see*
– Cardinal lavigerie; and, the african slave trade
– The catechism explained
– Lourdes
– A pilgrimage to the holy coat of treves
Clarke, Richard Henry *see* Lives of the deceased bishops of the catholic church in the united states
Clarke, Samuel Robinson *see*
– The constables' manual: being a summary of the law relating to the rights, powers, and duties of constables
– The law of lis pendens and in part of mechanics' liens.
– The magistrates' manual
– A new light on annexation
– A treatise on the criminal law of canada
Clarke, Thomas Hutchings *see*
– The domestic architecture of the reign of queen elizabeth and james the first
– Eastbury illustrated, by elevations, plans, sections, views
Clarke, Tom *see* Word of an englishman
Clarke, W K Lowther *see* Liturgy and worship
Clarke, W Newton *see* Christian and church (a sermon)
Clarke, Walter *see* Half century discourse
Clarke, William *see*
– The boy's own book
Clarke, William Fletcher *see*
– Baptism
– Canadian bicentenary papers
– "In memoriam"
– In memoriam
– Lord tennyson's pessimism
– A mother in israel
– My farm of lindenbank
– The nobility of agriculture
– Review of a discourse preached by the rev t s ellerby in zion church, toronto, oct 30, 1864
Clarke, William Kemp Lowther *see* St basil the great
Clarke, William Newton *see*
– The christian doctrine of god
– The circle of theology
– Commentary on the gospel of mark
– An outline of christian theology
– Sixty years with the bible
– A study of christian missions
– The use of the scriptures in theology
– What shall we think of christianity?

Clarke, Wm. H *see* Travels and explorations of
Clarke's chancery appeals reports / New York. (State) – 1v. 1839-41 (all publ) – 5mf – 9 – $7.50 – (a pre-nrs title) – mf#LLMC 80-200 – us LLMC [340]
Clarke's critical catalogue of the works of art sent in / Great Britain. Royal Commission of Fine Arts, Westminster Hall – London 1847 – (= ser 19th c art & architecture) – 1mf – 9 – mf#4.2.670 – uk Chadwyck [700]
The clarks enterprise – Clarks, NE: Geo W Cornell. 51v. old ser: v7 n26. jan 7 1898-57th yr n48. apr 29 1949 (wkly) [mf ed 1898-1903,1914-49 (gaps)] – 12r – 1 – (cont: clarks leader. absorbed by: central city nonpareil (1903). jan 7-14 1898 called also new ser v 1 n1-v1 n2) – us NE Hist [071]
Clark's form book, containing legal and business forms useful to the private citizen, as well as to judges, attorneys. / Clark, Francis Barnard – Montgomery, Ala.: Hold & Crawford, 1882. 460p. LL-12 – 1 – us L of C Photodup [346]
Clarks leader – Clarks, NE: Walrath Bros. - v7 n25. dec 31 1897 (wkly) [mf ed 1892, 1896-97 (gaps)] – 1r – 1 – (cont by: clarks enterprise) – us NE Hist [071]
Clark's mineral law : digest of decisions of the courts and land department under the public mineral laws / Clark, Horace F et al – Chicago: Callaghan. 1v. 1897 (all publ) – 6mf – 9 – $9.00 – mf#LLMC 95-134 – us LLMC [348]
Clarks news – Clarks, NE: John B Carter. 16v. v1 n1. sep 7 1950-v16 n19. dec 31 1964 (wkly) [mf ed with gaps filmed -1977] – 4r – 1 – (absorbed by: osceola record) – us NE Hist [071]
Clarks weekly messenger – Clarks, NE: Jno C Hartwell (wkly) [mf ed 1884, 1889 (gaps) filmed 1980] – 1r – 1 – (cont: clarksville messenger) – us NE Hist [071]
Clarksburg 1798-1915 – Oxford, MA (mf ed 1988) – (= ser Massachusetts vital records) – 9mf – 9 – 0-87623-072-9 – (mf 1-5: town records 1798-1846. mf 6: births 1846-1900. mf 8: marriages 1847-1916; deaths 1846-72. mf 9: deaths 1873-1915) – us Archive [978]
The clarkson herald – Clarkson, NE: H E Phelps. -may 30 1916// (wkly) [mf ed 1909-14 (gaps)] – 2r – 1 – (cont by: colfax county press and the clarkson herald consolidated) – us NE Hist [071]
Clarkson, Thomas *see*
– Abolition and emancipation
– Letters on the slave-trade and the state of the natives in those parts of africa which are contiguous to fort st louis and goree.
– A portraiture of quakerism
Clarkson, William *see*
– India and the gospel
– Missionary encouragements in india
Clarksville messenger – Clarksville, NE: Jas G Kreider, 1878 (wkly) [mf ed v1 n41. feb 8-mar 8 1879 (gaps) filmed 1980] – 1r – 1 – (cont by: clarks weekly messenger) – us NE Hist [071]
Claros, Jose Ma de *see* Discursos de...sobre cuestiones de caracter politico...legislatura de 1864-65
Clarte – Paris. n1-35, no. special de dec 1938. aout 1936-aout 1939 – 1 – (lacking: n3-5; n18, 20, 29) – fr ACRPP [073]
Clarte – Paris. 25 oct 1919-dec 1927 janv 1928 – 1 – fr ACRPP [074]
Clarte *see* La lutte de classes
Clary, Dexter *see* History of the churches and ministers connected with the presbyterian and congregational convention of wisconsin
Clary institute news bulletin for indian leaders – v1 n1-v3 n8 [1979 aug 15-1981 apr 30] – 1r – 1 – mf#639935 – us WHS [307]
Clary, J M *see* Eating disorders among athletes
Clasen, L *see* Die christliche heilsgewissheit
Clasey, Jody Lee *see* The relationship between finger flexion force production and selected hand, forearm and body physique measurements
Class – 1986-87, 1988 sep-1989 nov, 1990 jan-nov, 1991 dec/1992 jan-nov, 1991 jan-nov, 1993 jan-nov, 1994-96 – 7r – 1 – (cont by: black diaspora) – mf#1353105 – us WHS [305]
Class 3 a : folk festivals, pageants, celebrations / Ramsdell, Nellie B – s.l, s.l? 1936 – 1r – us UF Libraries [978]
Class and colour in south africa, 1850-1950 / Simons, Harold Jack – Harmondsworth, England. 1969 – 1r – us UF Libraries [960]
Class book, 1843 / Smith, Whitefoorde – [mf ed Duke University Library Repr Services] – 1v – 1 – mf#45-339 – us South Carolina Historical [242]
Class meetings : their origin, and advantages / Barrass, Edward – Sherbrooke, Quebec?: s.n, 1865 – 1mf – 9 – mf#48792 – cn Canadiana [242]
Class struggle – 1973 jan-1975 apr/may – 1r – 1 – mf#203705 – us WHS [335]
Class struggle / Communist League of Struggle – v1-7 n2,9. 1931-37 [all publ] – (= ser Radical periodicals in the united states, 1881-1960. series 1) – 17mf – 9 – $155.00 – us UPA [335]
Class struggle : devoted to international socialism – v1-3 n4. 1917-19 [all publ] – (= ser Radical periodicals in the united states, 1881-1960. series 1) – 18mf – 9 – $155.00 – us UPA [335]
Class struggle – New York. v. 1-3. May 1917-Nov 1919 – 1 – us NY Public [335]
Class struggle / Spark [Organization: US] – 1980 mar-1986 sum – 1r – 1 – (cont: class struggle [paris, france: 1972]; cont by: lutte de classe [paris, france: 1986]) – mf#1098851 – us WHS [335]
A class-book of biblical history and geography / Osborn, Henry Stafford – New York: American Tract Society, c1890 [mf ed 1985] – 1mf – 9 – 0-8370-4639-4 – (incl bibl ref) – mf#1985-2639 – us ATLA [220]
A class-book of old testament history / Maclear, George Frederick – London: Macmillan, 1879 [mf ed 1992] – 2mf – 9 – 0-524-04581-X – (incl bibl ref) – mf#1992-0169 – us ATLA [221]
Classbook of old testament history / Hodges, George – New York: Macmillan, 1914 – 1mf – 9 – 0-524-04461-9 – mf#1992-0130 – us ATLA [221]
Classen, Walther *see* Suchen wir einen neuen gott?
Classeni, Io *see* Chronographia [cshb39,40]
Classic american homes – New York. 2000+ (1) – (cont: colonial homes) – ISSN: 1528-2864 – mf#12241,02 – us UMI ProQuest [640]
Classic baptism : an inquiry into the meaning of the word baptizo, as determined by the usage of classical greek writers / Dale, James Wilkinson – Boston: Draper & Halliday, 1867 – 1mf – 9 – 0-524-03459-1 – mf#1990-1002 – us ATLA [240]
Classic commentary / Wisconsin Rescue Mission and Halfway House [Madison WI] – v1-v5 n3 [1971 aug ?-1973 may?] – 1r – 1 – (cont: mission classic) – mf#1054556 – us WHS [360]
Classic film collector – Davenport. 1973-1978 (1) – (cont by: classic film/video images) – ISSN: 0009-8329 – mf#7482 – us UMI ProQuest [790]
Classic film/video images – Muscatine. 1978-1979 [1] – (cont by: classic images) – ISSN: 0164-5560 – mf#7482,01 – us UMI ProQuest [790]
Classic film/video images – Muscatine. 1978-1979 (1) – (cont: classic film collector) – ISSN: 0164-5560 – mf#7482,01 – us UMI ProQuest [790]
Classic images – Muscatine. 1980+ (1) 1980+ (5) 1980+ (9) – (cont: classic film/video images) – ISSN: 0275-8423 – mf#7482,02 – us UMI ProQuest [790]
Classic myth and legend / Moncrieff, A R Hope – New York, NY. 1934 – 1r – us UF Libraries [390]
The classic myths : in english literature and in art / Gayley, Charles Mills – new rev and enl ed. Boston: Ginn, c1911 – 2mf – 9 – 0-524-06688-4 – (incl bibl ref) – mf#1990-3549 – us ATLA [250]
The classic test of authorship, authenticity and authority : founded on jurists' rules of interpreting records, applied to supposed inaccuracies in the text of the old and new testament scriptures / Samson, George Whitefield – New York: F Scott c1893 [mf ed 1985] – 1mf – 9 – 0-8370-5031-6 – mf#1985-3031 – us ATLA [220]
The classical age of german literature, 1748-1805 / Willoughby, Leonard Ashley – London: Oxford University Press, H Milford, 1926 – 1 – (incl bibl ref and index) – us Wisconsin U Libr [430]
The classical age of german literature, 1748-1805 / Willoughby, Leonard Ashley – London: Oxford University Press, H Milford, 1926 – 1 – (incl bibl ref and index) – us Wisconsin U Libr [430]
Classical antiquity – Berkeley. 1988+ (1,5,9) – ISSN: 0278-6656 – mf#15674 – us UMI ProQuest [450]
Classical association proceedings – Cardiff. 1976-1980 (1) 1977-1980 (5) 1977-1980 (9) – mf#10184 – us UMI ProQuest [450]
Classical bulletin – Wilmore. 1925+ (1) 1971+ (5) 1976+ (9) – ISSN: 0009-8337 – mf#3365 – us UMI ProQuest [450]
A classical dictionary of hindu mythology and religion, geography, history, and literature / Dowson, John – 3rd ed. London: Kegan Paul, Trench, Truebner, 1891 [mf ed 1991] – (= ser Truebner's oriental series) – 1mf – 9 – 0-524-00876-0 – mf#1990-2099 – us ATLA [280]
A classical dictionary of hindu mythology and religion, geography, history, and literature / Dowson, John – London: Kegan Paul, Trench, Truebner & Co, 1903 – (= ser Samp: indian books) – us CRL [390]
A classical dictionary of india : illustrative of the mythology, philosophy, literature, antiquities, arts, manners, customs, etc of the hindus / Garrett, John – Madras: Higginbotham & Co, 1871 [mf ed 1986] – x/11/793p – 1 – mf#8260 – us Wisconsin U Libr [390]
The classical element in the new testament : considered as a proof of its genuineness / Hoole, Charles H – London, New York: Macmillan, 1888 – 1mf – 9 – 0-7905-1109-6 – (incl bibl ref) – mf#1987-1109 – us ATLA [225]
Classical english poetry : for the use of schools, and young persons in general / Mavor, William Fordyce – new rev ed. London: Longman, Hurst, Rees, Orme, and Brown, 1823 – (= ser 19th c children's literature) – 6mf – 9 – mf#6.1.6 – uk Chadwyck [810]
The classical english spelling-book : in which the hitherto difficult art of orthography is rendered easy and pleasant, and speedily acquired / Vasey, George G – Montreal: Printed & publ by J Lovell; Toronto: R & A Miller, 1860 – (= ser Lovell's series of school books) – 3mf – 9 – mf#42442 – cn Canadiana [420]
Classical excursion from rome to arpino / Kelsall, Charles – Geneva 1820 [mf ed Hildesheim 1995-98] – (= ser Fbc) – 2mf [ill] – 9 – €60.00 – 3-487-29239-4 – gw Olms [914]
The classical heritage of the middle ages / Taylor, Henry Osborn – 3rd ed. New York: Macmillan, 1911 [mf ed 1990] – 1mf – 9 – 0-7905-6695-8 – (1st ed 1901. later eds publ under title: the emergence of christian culture in the west. incl bibl ref) – mf#1988-2695 – us ATLA [931]
Classical journal – Gainesville. 1905+ (1) 1968+ (5) 1976+ (9) – ISSN: 0009-8353 – mf#979 – us UMI ProQuest [450]
The classical moralists : selections illustrating ethics from socrates to martineau / Rand, Benjamin – Boston: Houghton, Mifflin, c1909 – 2mf – 9 – 0-8370-6309-4 – (incl ind) – mf#1986-0309 – us ATLA [170]
Classical museum : a journal of philology, and of ancient history and literature – London. 1844-1850 (1) – mf#4715 – us UMI ProQuest [450]
Classical outlook – Oxford. 1923+ (1) 1971+ (5) 1977+ (9) – ISSN: 0009-8361 – mf#969 – us UMI ProQuest [450]
Classical philology – Chicago. 1906+ (1) 1969+ (5) 1978+ (9) – ISSN: 0009-837X – mf#479 – us UMI ProQuest [450]
Classical philology – Chicago. v18-41. 1923-1946 – 150mf – 8 – mf#58c – ne IDC [450]
The classical poetry of the japanese / ed by Chamberlain, Basil Hall – London: Truebner, 1880 – (= ser Truebner's Oriental Series) – 1mf – 9 – 0-524-01259-8 – (incl bibl ref) – mf#1990-2295 – us ATLA [480]
The classical psychologists : selections illustrating psychology from anaxagoras to wundt / Aristotle et al – Boston: Houghton Mifflin, c1912 – 1mf – 9 – 0-524-00081-6 – mf#1989-2781 – us ATLA [150]
Classical quarterly – Oxford. 1907+ (1) 1971+ (5) 1975+ (9) – ISSN: 0009-8388 – mf#1215 – us UMI ProQuest [450]
Classical review – Oxford. 1887+ [1]; 1971+ [5]; 1975+ [9] – ISSN: 0009-840X – mf#1216 – us UMI ProQuest [450]
Classical revision of the greek new testament : tested and applied on uniform principles with suggested alterations of the english version / Nicolson, W Millar – London: Williams and Norgate, 1878 – 1mf – 9 – 0-8370-4590-8 – (includes an appendix and index) – mf#1985-2590 – us ATLA [450]
A classical tour through italy anno 1802 [eighteen hundred and two] / Eustace, John C – London 1815 [mf ed Hildesheim 1995-98] – (= ser Fbc) – 4v on 12mf – 9 – €120.00 – 3-487-29303-X – gw Olms [914]
Classical world – Pittsburgh. 1907+ (1) 1970+ (5) 1977+ (9) – ISSN: 0009-8418 – mf#581 – us UMI ProQuest [930]
Classics of international law. carnegie institution. micro-mini-prints edition... / ed by Scott, James Brown – Washington: Carnegie Institution, 22 tit in 40bks. 1911- – 9 – $575.00 set – 0-89941-203-3 – mf#400040 – us Hein [341]
Classification nominale dans les langues negro-africaines / Colloque International Sur La Classification Nominale Dans... – Paris, France. 1967 – 1r – us UF Libraries [960]
The classification of religions : different methods, their advantages and disadvantages / Ward, Duren James Henderson – Chicago: Open Court, 1909 – 1mf – 9 – 0-524-02056-6 – mf#1990-2831 – us ATLA [200]
The classification of the sciences : to which are added reasons for dissenting from the philosophy of m comte / Spencer, Herbert – London, 1864 – (= ser 19th c evolution & creation) – 1mf – 9 – mf#1.1.4121 – uk Chadwyck [100]
Classification outline with topical index for decisions of the nlrb and related court decisions / U.S. National Labor Relations Board – Washington: GPO, 1988 – (= ser National labor relations board decisions) – 12mf – 9 – $18.00 – mf#LLMC 95-033 – us LLMC [344]
Classified directory of wisconsin manufacturers / Wisconsin Manufacturers and Commerce – 1939, 1941 – 2r – 1 – (cont by: wisconsin manufacturers directory) – mf#35151 – us WHS [670]
Classified index of rate cases, years 1925, 1926, 1927 / American Telephone and Telegraph Co – New York 1928 75 p. LL-1066 – 1 – us L of C Photodup [348]
Classified index, regional directors' decisions in representation proceedings / U.S. National Labor Relations Board – 1977-89 (all publ) – (= ser National labor relations board decisions) – 13mf – 9 – $19.50 – mf#LLMC 95-034 – us LLMC [344]
A classified index to the leonine, gelasian and gregorian sacramentaries : according to the text of muratori's liturgia romana vetus / Wilson, Henry Austin – Cambridge: University Press 1892 [mf ed 1992] – 1mf – 9 – 0-524-03868-6 – mf#1990-4915 – us ATLA [241]
A classified list of photographs of drawings, paintings, and sculpture, precious metals and enamels / Arundel Society, London – London 1867 – (= ser 19th c art & architecture) – 4mf – 9 – mf#4.2.1656 – uk Chadwyck [700]
Classified list of published bibliographies in physics, 1910-1922 / National Research Council (Us) Research Information Service – Washington, DC. 1924 – 1r – us UF Libraries [025]
The classified mail – Limbe, Malawi: [s.n, [dec1/14 1993] – 1r – 1 – us CRL [079]
Classified minutes of the annual meetings of the brethren : a history of the general councils of the church from 1778-1885 – Mt Morris IL: Brethren's Pub Co 1886 [mf ed 1992] – 1mf – 9 – 0-524-02820-6 – mf#1990-4441 – us ATLA [242]
Classified psalter arranged by subjects – New York, NY. 1899 – 1r – us UF Libraries [939]
Classified table of the public general statutes of canada, wholly or partly in force at the end of the session of 1882 : with notices of those repealed or expired, or effete by the accomplishment of their purpose / Wicksteed, Gustavus William – [Ottawa?: s.n.] 1883 [mf ed 1992] – 1mf – 9 – 0-665-94694-5 – mf#94694 – cn Canadiana [348]
Classified (waltham forest ed) *see* Waltham forest classified
Das classische heidenthum und die christliche religion / Arneth, Franz Hektor, Ritter von – Wien: C Konegen, 1895 – 2mf – 9 – 0-524-05835-0 – mf#1990-3499 – us ATLA [230]
Classroom computer learning – Belmont. 1983-1990 (1) 1983-1990 (5) 1983-1990 (9) – (cont: classroom computer news. cont by: technology and learning) – ISSN: 0746-4223 – mf#13044,01 – us UMI ProQuest [370]
Classroom computer news – Watertown. 1980-1983 – 1,5,9 – (cont by: classroom computer learning) – ISSN: 0731-9398 – mf#13044 – us UMI ProQuest [370]
Classroom interaction newsletter – Washington. 1965-1976 (1) 1975-1976 (5) 1975-1976 (9) – (cont by: journal of classroom interaction) – ISSN: 0009-8485 – mf#10394 – us UMI ProQuest [370]
Clastic huronian rocks of western ontario / Coleman, Arthur Philemon – Rochester NY: publ by the society, 1898 – 1mf – 9 – (incl bibl ref) – mf#59509 – cn Canadiana [550]
The clatonia leader – Clatonia, NE: Chris Baker. v1 n1. dec 5 1935- (wkly) [mf ed 1935-38,1944 (gaps)] – 1r – 1 – (publ in cortland dec 5 1935-may 14 1936; in clatonia and cortland may 28 1936-aug 17 1944) – us NE Hist [071]
The clatonia observer – Clatonia, NE: H L Gardner. v1 n1. jun 13 1907- (wkly) [mf ed 1907-09 (gaps)] – 1r – 1 – (publ in clatonia jun 13-jul 18 1907; in cortland jul 25 1907-apr 2 1909) – us NE Hist [071]
Clatskanie chief – Clatskanie OR: E C Blackford, [wkly] – 1 – (began in 1891. 1925-69 incl newspaper publ during school terms by clatskanie high school students) – us Oregon Lib [071]
Clatsop county argus – Warrenton OR: J H Walker, [wkly] – 1 – (began in 1925; ceased in 1925; merged with: warrenton news to form: clatsop county argus the warrenton news) – us Oregon Lib [071]
Clatsop county argus *see*
– Clatsop county argus the warrenton news
– Warrenton news (warrenton, or)
Clatsop county argus the warrenton news – Warrenton OR: G C Barlow, [wkly] – 1 – (merger of: clatsop county argus [1925, warrenton news [-1925]) – us Oregon Lib [071]

Clatsop county argus the warrenton news *see*
– Clatsop county argus
– Warrenton news (warrenton, or)
Clatsop County Historical Society *see* Cumtux
Claude *see* Historia da missao dos padres capuchinhos na ilha
The claude a barnett papers : pt 1: associated negro press news releases, 1928-64 / The Associated Negro Press – 3ser – 1 – $14,110.00 set – (ser a: 1928-44 29r $5185 p/g isbn: 0-89093-698-6; ser b: 1945-55 29r $5185 p/g isbn: 0-89093-699-4; ser c: 1956-64 25r $4465 p/g isbn: 0-89093-697-8) – us UPA [380]
The claude a barnett papers : pt 2: associated negro press organizational files, 1920-66 / The Associated Negro Press – 24r – 1 – $4290.00 – 0-89093-739-7 – (with p/g) – us UPA [380]
The claude a barnett papers : pt 3: subject files on black americans, 1918-67 / The Associated Negro Press – 11ser – 1 – $15,470.00 set – (ser a: agriculture, 1923-66 11r $1945 isbn 0-89093-759-1; ser b: colleges & universities, 1918-66 16r $2870 isbn 0-89093-760-5; ser c: economic conditions, 1918-66 13r $2330 isbn 0-89093-761-3; ser d: entertainers, artists and authors, 1928-65 7r $1260 isbn 0-89093-762-1; ser e: medicine, 1927-65 7r $1260 isbn 0-89093-763-X; ser f: the military, 1925-65 3r $525 isbn 0-89093-764-8; ser g: philanthropic and social organizations, 1925-66 5r $885 isbn 0-89093-765-6; ser h: politics and law, 1920-66 9r $1,595 isbn 0-89093-766-4; ser i: race relations, 1923-65 8r $1435 isbn 0-89093-767-2; ser j: religion, 1924-66 9r $1595 isbn 0-89093-768-0; ser k: claude a barnett papers, personal and financial, 1920-67 3r $525 isbn 0-89093-769-9. with p/g) – us UPA [380]
Claude, J *see*
– Defence de la reformation
– Reponse au livre de m l'evesque de meaux
– Reponse au livre de mr arnaud
– Reponse aux deux traites
[Claude, J] *see* Les plaintes des protestants
Claude, Jean *see* Cruel persecutions of the protestants in the kingdom of france
Claude lorrain / Friedlaender, Walter F – Berlin, Germany. 1921 – 1r – us UF Libraries [750]
Claude McKay Secondary School [May Pen, Jamaica] *see* Flame heart
Claudel, Paul *see*
– Pain dur
– Pere humilie
Claudette *see* Les aventures du prince romanic
Claudian as an historical authority / Crees, James Harold Edward – Cambridge, England. 1908 – 1r – us UF Libraries [930]
Claudians gedicht vom gotenkrieg – Berlin, Germany. 1927 – 1r – us UF Libraries [450]
Claudianus, Claudius *see* Claudian as an historical authority
Claudii claudiani carmina (mgh1:10.bd) / ed by Birt, Th – 1892 – (= ser Monumenta germaniae historica 1: scriptores – auctores antiquissimi) – €42.00 – ne Slangenburg [240]
Claudin, Fernando *see* La juventud espanola continua su lucha
Claudio jose domingo brindis de salas / Guillen, Nicolas – Habana, Cuba. 1935 – 1r – us UF Libraries [972]
Claudius, Hermann *see*
– Hoerst du nicht den eisenschritt
– Lieder der unruh
– Mank muern
– Matthias claudius
Claudius, Matthias *see*
– Briefe an freunde
– Matthias claudius werke
– Der wandsbecker bote
Claudon, F *see* Abbayes et prieures de l'ancienne france (afm45)
Claughton, T L *see* Our present duties in regard of holy baptism
Claughton, Thomas Legh *see* Christ's finished work
Claus, Guenther *see* Unsere nationale volksarmee
The clause compromissoire : its validity in quebec / Johnson, Walter Seely – Montreal, Johnson, 1945. 166 p. LL-2319 – 1 – us L of C Photodup [340]
Clausen, Carl Christian *see* Under palmer
Clausen, Carl Jon *see* A-356 site and the florida archaic
Clausen, Ernst Alexander *see* Der heiligen kind
Clausen, J *see* Papst honorius 3, 1216-27
Clausen, Julius *see*
– Dagboger fra 1792
– Jens baggesen
Clauses generales du contrat de louage ou affermage et de la mise en exploitation du chemin de fer – [s.l: s.n.] 1879. [mf ed 1984] – 1mf – 9 – 0-665-04130-6 – mf#04130 – cn Canadiana [380]
Clausewitz, Karl von *see* Principles of war
Clauss, Gertrud *see* Die frau in der dichtung conrad ferdinand meyers
Clauss, Walter *see* Deutsche literatur
Claussen, Martin P *see* The state-war-navy coordinating committee (swncc) and state-army-navy-air force coordinating committee (sanacc) case files, 1944-49
Claussen, Sophus *see* Byen : junker friklover : nutidsroman
Claustro y tres maestros / Romero Lozano, Armando – Cali, Colombia. 1958 – 1r – us UF Libraries [972]
Claveau, Antoine G *see* De la police de paris, de ses abus, et des reformes dont elle est susceptible
Le clavecin bien tempere : ou preludes et fugues dans tous les tons et demintons du mode majeur et mineur / Bach, Johann Sebastian – Vienne: Hoffmeister & Comp; Leipsic: Bureau de Musique [18–?] [mf ed 19–] – 2v on 1r – 1 – mf#pres. film 116 – us Sibley [780]
Clavego, P *see* El trabajo de los comisarios politicos
Claverite / Knights of Peter Claver – v75 n1-2 [1994 sum-winter], v77 n2 [1996 win], v78 n1 [1997 sum] – 1r – 1 – mf#347180 – us WHS [241]
Claves de marti y el plan de alzamiento para cuba / Rosell Planas, Rebeca – Habana, Cuba. 1948 – 1r – us UF Libraries [972]
Clavicula Salomonis *see* Clavicula salomonis
Clavicula salomonis = The key of solomon the king / Clavicula Salomonis – London: Kegan Paul, Trench, Truebner, 1909 – 1mf – 9 – 0-8370-4179-1 – (in english. incl bibl ref) – mf#1985-2179 – us ATLA [920]
Clavicula salomonis *see* Mafteah shelomoh
[Claviere, E de] *see* Figure emblematique en trois langues
Claviere, Etienne *see* De la france et des etats-unis
Clavieruebung bestehend in einer aria : mit verschiedenen veraenderungen vors clavicimbal mit 2 manualen. denen liebhabern zur gemuethsergetzung verfertigt von... / Bach, Johann Sebastian – Nuernberg: im Verlagung Balthasar Schmids [1742] [mf ed 19–] – 1r – 1 – mf#film 795 – us Sibley [780]
Clavigo : eine studie zur sprache des jungen goethe / Schmidt, Georg – Gotha: Druck von F A Perthes 1893 [mf ed 1990] – 1r – 1 – (filmed with: das volkslied und sein einfluss auf goethe's lyrik / j suter) – mf#7320 – us Wisconsin U Libr [430]
Clavijo : drama / Goethe, Johann Wolfgang von – Madrid: Calpe 1920 [mf ed 1990] – 1r – 1 – (trans fr german into spanish by r m tenreiro. filmed with: das volkslied und sein einfluss auf goethe's lyrik / j suter) – mf#7320 – us Wisconsin U Libr [820]
Clavijo Tisseur, Arturo *see*
– Estampas martianas
– Poemas para el alma
Clavijo y Clavijo, Salvador *see* La trayectoria hospitalaria de la armada espanola. madrid, 1944
Clavijos / Alvarez Garzon, Juan – Pasto, Colombia. 1964 – 1r – us UF Libraries [972]
Clavis librorum veteris testamenti apocryphorum philologica / Wahl, Christian Abraham – Lipsiae [Leipzig]: J A Barth, 1853 – 5mf – 9 – 0-8370-1999-0 – mf#1987-6386 – us ATLA [221]
Clavis orientalis, pt 1 : or, lecture card of the london oriental institution / Arnot, Sandford – [London]: Oriental Institution, 1827 – (= ser 19th c books on linguistics) – 1mf – 9 – (containing an easy int to the principles of oriental writing...) – mf#2.1.59 – uk Chadwyck [400]
Clavis orientalis, pt 2 : or, lecture card of the london oriental institution / Arnot, Sandford & Forbes, Duncan – London, 1827 – (= ser 19th c books on linguistics) – 1mf – 9 – (consisting of a brief int to the reading and writing of the most useful and important of the oriental characters, called nuskhee and taliq) – mf#2.1.60 – uk Chadwyck [400]
Clavis syriaca : a key to the ancient syriac version, called "peshito," of the four holy gospels / Whish, Henry F – London: George Bell; Cambridge: Deighton, Bell, 1883 – 2mf – 9 – 0-8370-8634-5 – (incl ind) – mf#1986-2634 – us ATLA [221]
Clavius, Christoph *see* Christophori clavii bambergensis e societate les operum mathematicorum tomus primus-quintus...
Clawson, Cindy A *see* The effects of toys, prompts, and flotation devices on the learning of water orientation skills
Claxton's music store : thos claxton, importer, wholesale and retail dealer in english and french band instruments, violins, accordeons, german and anglo-german concertinas, and all kinds of musical merchandise – [Toronto?: s.n. 187-?] [mf ed 1983] – 1mf – 9 – 0-665-39738-0 – mf#39738 – cn Canadiana [780]
Clay, A T *see* Babylonian records in the library of j pierpont morgan
Clay, Albert Tobias *see*
– Amurru
– Light on the old testament from babel
Clay center dispatch – Clay Center, KS. 1956-2000 (1) – mf#68154 – us UMI ProQuest [071]
Clay county crescent – Green Cove Springs, FL. 1946 oct-1996 – 49r – (gaps) – us UF Libraries [071]
Clay county free press – Clay, WV. 1934+ (1) – mf#67257 – us UMI ProQuest [071]
Clay county leader – Clay Center, NE: Ostdiek Pub. -v15 n49 dec 30 1976 (wkly) [mf ed v2 n32. sep 11 1963-74 (gaps) filmed 1977] – 5r – 1 – (absorbed by: clay county sun) – us NE Hist [071]
Clay county news – Sutton, NE: [Howard C King, Burlin B King, Roy M King] v66 n32. aug 3 1950- (wkly) [mf ed lacks sep 30 1954, feb 29 1968 filmed [1974?]] – 1 – (cont: sutton news (1942). absorbed: harvard courier 1977 and: clay county sun (1979)) – us NE Hist [071]
Clay county patriot – Clay Center, NE: Henry B Funk, jun 1894-v27 n30. oct 28 1920 (wkly) [mf ed 1895-1920 (gaps)] – 9r – 1 – (cont: clay county progress. cont by: clay county republican) – us NE Hist [071]
Clay county progress – Clay Center, NE: Eric Johnson, 1892-94// (wkly) [[mf ed v1 n7. may 27 1892 filmed [1973]] – 1r – 1 – (cont by: clay county patriot) – us NE Hist [071]
Clay county register – Edgar, NE: E M Burr, 1891-93// (wkly) [mf ed v1 n36. jun 17 1892 filmed [1979]] – 1r – 1 – us NE Hist [071]
Clay county republican – Clay Center, NE: Chas H Epperson Jr. 2v. v27 n31. nov 4 1920-v28 n15. aug 25 1921 (wkly) [mf ed lacks jun 9 1921] – 2v – 1 – (cont: clay county patriot. absorbed: ong sentinel. absorbed by harvard courier) – us NE Hist [071]
Clay county republican – Clay Center, NE: Chas H Epperson Jr. 2v. v27 n31.. nov 4 1920-v28 n15. aug 25 1921 (wkly) [mf ed lacks jun 9 1921] – 1r – 1 – (cont: clay patriot republican. absorbed: ong sentinel. absorbed by: harvard courier) – us NE Hist [071]
Clay county sun – Clay Center, NE: Howard & Ojers. 67v. v28 n46. jun 7 1912-v94 n52 dec 28 1978 (wkly) [mf ed 1917-18,1920-78 (gaps) filmed -1979] – 28r – 1 – (cont: sun. absorbed: fairfield auxiliary oct15 1965 and: edgar sun (1914), 1977 and: clay county leader 1977. absorbed by: clay county news. some irregularities in numbering) – us NE Hist [071]
Clay county times – Green Cove Springs, FL. 1918-1920; 1926 [scattered] – 1r – us UF Libraries [071]
Clay cross chronicle – Clay Cross, England, 11 may 1900-20 dec 1910 (wkly) – 10r – 1 – (ceased publ) – uk Newsplan [072]
Clay, Dawn E *see* Comparing kilocalorie expenditure between a stair-stepper, a treadmill, and an elliptical trainer
Clay, Gervas *see* Your friend, lewanika
Clay, Henrietta *see* Bits of family history
Clay, Henry *see*
– Papers
– Works
Clay, Jehu Curtis *see* Annals of the swedes on the delaware
Clay, Rotha Mary *see*
– The hermits and anchorites of england
– The mediaeval hospitals of england
Clay today – Orange Park, FL. v25 n1-v27 n51. 1997-1998 jun – 7r – (gaps) – us UF Libraries [071]
Clayden, Arthur *see*
– British colonisation
– The england of the pacific
Claymore – [Scotland] Edinburgh: Claymore Press 12 dec 1933-10 jul 1934 (wkly) [mf ed 2003] – 31v on 1r – 1 – ("only weekly for scottish boys") – uk Newsplan [072]
Clays and clay minerals – Long Island City. 1968-1977 (1) 1959-1977 (5) 1968-1977 (9) – ISSN: 0009-8604 – mf#49032 – us UMI ProQuest [550]
Clayton, Albert Charles *see* The rig-veda and vedic religion
Clayton, Anna *see* Les colonies francaises
Clayton, B S *see* Water control in the peat and muck soils of the florida everglades
Clayton clarion – 1986 jan – 1r – 1 – mf#4863848 – us WHS [071]
Clayton commercial – Plainfield, IN. 1930-1952 (1) – mf#62936 – us UMI ProQuest [071]
Clayton, George *see* Coming of christ desired
Clayton, H G *see* Study of some varieties of japanese cane...
Clayton, Henry James *see* Our national church
Clayton, Henry R *see* Anglo-canadian copyright
Clayton, John *see*
– A collection of the ancient timber edifices of england
– The works of sir christopher wren
Clayton, Joseph *see* St hugh of lincoln, a biography...
Clayton, Richard *see* Oratorios unsuited to the house of prayer
Clayton, Robert *see* A journey from aleppo to jerusalem, at easter, a d 1696
Clayton, William *see* Ritualism in high places
Clc today / Canadian Labour Congress – 1990 aug-1993 apr/may – 1r – 1 – (cont: canadian labour) – mf#1829241 – us WHS [331]
Cle – Paris. n1-2. janv-fevr 1939 – 1 – fr ACRPP [073]
Cle journal (ali-aba) – Philadelphia. v1-3. 1998-2001 – $50.00 – mf#118891 – us Hein [340]
Cle journal and register [ali-aba] – v1-44. 1965-98 – 9 – $907.00 set – (cont by: cle journal) – ISSN: 0193-693X – mf#113121 – us Hein [340]
Cleal, Edward E *see* The story of congregationalism in surrey
Clean air – Mount Waverly. 1978-1981 (1,5,9) – ISSN: 0009-8647 – mf#11068 – us UMI ProQuest [333]
Clean politics / Prohibition Party [US] – n36,46,53,60 [1910 mar 24, jun 2, jul 21, sep 8], n134,141,146-188,191 [1912 feb 8, mar 28, may 2-1913 feb 20, mar 13], n192,199-204,207,211 [1913 apr 10, may 29, jul 3, 24, aug 21], n240 [1914 mar 12] – 1r – 1 – mf#944684 – us WHS [325]
Clear creek – San Francisco. 1971-1972 (1) – ISSN: 0045-7124 – mf#6122 – us UMI ProQuest [333]
Clear creek – v2 n1,8 [1971 apr, nov], n2,16,18 [1972 mar, oct, dec] – 1r – 1 – mf#1583048 – us WHS [071]
Clear creek baptist church. adams county. mississippi : church records – 1835-1873 – 1 – 5.13 – us Southern Baptist [242]
Clear Creek Baptist School. Pineville, Kentucky *see* Catalogs and college records
Clear creek county miscellaneous newspapers – Clear Creek County, Empire, Idaho Springs, CO [mf ed 1991] – (empire true fissure [jul 3 1901-oct 4 1901]; the arbitrator [feb 1 1887]; the centennial [jan-feb 1876]; clear creek democrat [may 11 1904-may 18 1904]; clear creek topics [jan 25 1902-oct 1 1903]; colorado miner [sep 29 1870-jan 5 1884]; gold rush gazette [jul 1951]; idaho springs advance [dec 29 1881-dec 21 1882]; idaho springs iris [jan 27 1892-mar 16 1892]; idaho springs reporter [aug 31 1872], silver plume mining news [aug 12 1881]) – ISSN: 0 – mf#MF Z99 C58 – us Colorado Hist [071]
Clear creek democrat *see* Clear creek county miscellaneous newspapers
Clear cut : the deforestation of america / Wood, Nancy – (= ser Sierra club books out of print) – 1r – 5,9 – $50.00 – mf#B70008 – us Library Micro [574]
Clear fork baptist church – Albany, Clinton Co, KY. 740p. dec 1960-92 – 1 – $33.30 – mf#6708 – us Southern Baptist [242]
Clear hills standard – Clear Hills, jun 20 1914 – 1r – 9 – A$28.07 vesicular A$33.57 silver – at Pascoe [079]
Clear Lake star – Clear Lake WI. 1912 jan 18/1915-1988 jul-nov 17 – 50r – 1 – (with small gaps) – mf#983632 – us WHS [071]
Clearing house – Washington. 1920+ (1) 1968+ (5) 1969+ (9) – ISSN: 0009-8655 – mf#445 – us UMI ProQuest [370]
Clearing the air / Center for Public Representation. United States – n1,2 [1979 jan, feb 14], n3 [1980 feb], n2 [1979 feb 14] – 2r – 1 – mf#646925 – us WHS [350]
Clearinghouse for Citizen Participation *see* Neighborhood exchange
Clearinghouse on Women's Issues in Congress [US] *see* Cwic
Clearinghouse review – Chicago. 1979+ (1,5,9) – ISSN: 0009-868X – mf#12252 – us UMI ProQuest [340]
[Clearlake highlands-] clearlake observer-american – CA. jan 1976- – 31+ r – 1 – $1860.00 (subs $140/y) – mf#B02114 – us Library Micro [071]
Clearmont baptist church. oconee county. westminster, south carolina : church records – 1890-1909, 1922-53 – 1 – 9.45 – us Southern Baptist [242]
Clearwater : florida west coast on the gulf – Clearwater, FL. 1926? – 1r – us UF Libraries [978]
Clearwater / Phillips, Roland – s.l, s.l? 1936 – 1r – us UF Libraries [978]
Clearwater / Phillips, Roland – s.l, s.l? 1936 – 1r – us UF Libraries [978]
Clearwater : supplementary history and color / Coll, Aloyisus – s.l, s.l? 1936 – 1r – us UF Libraries [978]
Clearwater headlight – Clearwater, NE: C E Fields. v1 n1. sep 30 1886- (wkly) [mf ed -1887 (gaps)] – 1r – 1 – us NE Hist [071]
Clearwater message – Clearwater, NE: Fred E Seeley. 1st yr. jul 8 1887- (wkly) [mf ed 1887-96 (gaps)] – 1r – 1 – us NE Hist [071]
Clearwater record – Clearwater, NE: F S Delanoy. 68v. v1 n1. apr 23 1897-v68 n29. jun 22 1967 (wkly) [mf ed 1953-67 (lacks jan 10 1963)] – 6r – 1 – (merged with: ewing news to form: clearwater record-ewing news) – us NE Hist [071]

Clearwater record – Clearwater, NE: F S Delanoy. 68v. v1 n1. apr 23 1897-v68 n29. jun 22 1967 (wkly) – 13r – 1 – (merged with: ewing news to form: clearwater record-ewing news; cont by: ewing news) – us Bell [071]
Clearwater record-ewing news – Clearwater, NE: Clearwater Pub Co. v68 n[30] jun 29 1967- (wkly) [mf ed with gaps filmed 1977-] – 1 – (formed by the union of: clearwater record and: ewing news. publ in neligh jul 23 1987- . cont the numbering of clearwater record) – us NE Hist [071]
Clearwater rewrite / Phillips, Roland – s.l, s.l? 1936 – 1r – us UF Libraries [978]
Cleary, James Vincent *see*
– A doctrinal instruction on the indulgences and masses for the dead
– Sermon of the right rev james vincent cleary... bishop of kingston
Cleary, Michelle A *see* The time course of the repeated bout effect of eccentric exercise on delayed onset muscle soreness
Cleary, Reuben *see* Manuscript of his chronicos lageanas
Cleary, Thomas *see* A bond to save from bondage
Cleaveland gazette and commercial register – Cleveland, OH, jul 31 1818-mar 7 1820 – 1r – 1 – (weekly general newspaper, the first publ in cleveland) – us Western Res [071]
Cleburne county historical society journal – 1974 fall-1979 win, 1980 spr-1986 win – 2r – 1 – mf#693303 – us WHS [978]
Cledat, J *see* Le monastere et la necropole de baouit
Cledat, Leon *see* Le nouveau testament
Clef – Santa Monica CA. v1 n1-7. mar-sep 1946 [all publ] – (= ser Jazz periodicals, 1914-1977) – 1r – 1 – $115.00 – us UPA [780]
La clef des principales difficultes de la grammaire francaise ou cours raisonne sur la grammaire francaise : le meme qui a ete donne avec succes durant plusieurs annees en soixante lecons / Lassiseraye, Charles Hubert – Montreal: J B Rolland, 1850 [mf ed 1984] – 1mf – 9 – 0-665-45244-6 – mf#45244 – cn Canadiana [440]
La clef du cabinet des princes de l'europe : or recueil historique et politique sur les matieres du tems / Jordan, Claude – Luxembourg. juil 1704-06 (v1-5) – 1 – fr ACRPP [073]
La clef du cabinet des souverains – Paris, 1797-sept 1805 (1-32) – 1 – fr ACRPP [944]
La clef du mystere : vicit leo de tribu juda / Leroy, Pierre – [Nantes, France?: s.n.] 1885 [mf ed 1985] – 4mf – 9 – 0-665-08975-9 – mf#08975 – cn Canadiana [370]
Clef du nouveau systeme de toiser tous les corps-segments, troncs et onglets de ces corps par une seule et meme regle... / Baillairge, Charles P Florent – Quebec: C Darveau, 1875 – 1mf – 9 – mf#02372 – cn Canadiana [510]
Clef du tableau stereometrique baillairge : nouveau systeme de toiser tous les corps-segments, troncs et onglets de ces corps par une seule et meme regle... / Baillairge, Charles P Florent – Quebec: C Darveau, 1874 – 3mf – 9 – mf#02498 – cn Canadiana [510]
Clef synoptique : ou, abregee du tableau stereometrique baillairge: nouveau systeme de toiser tous les corps-segments, troncs et onglets de ces corps par une seule et meme regle... / Baillairge, Charles P Florent – Quebec: C Darveau, 1874 – 1mf – 9 – mf#02497 – cn Canadiana [624]
Cleft lip and palate = Gesplete lip en verhemelte / South Africa. Department of National Health and Population Development [Departement van Nasionale Gesondheid en Bevolkingsontwikkeling – 2nd ed. Pretoria: Dept of National Health & Population Development 1987 [mf ed Pretoria, RSA: State Library [199-]] – 18p [ill] on 1r with other items – 5 – mf#op 08512 r23 – us CRL [616]
Clegg, Alan G *see* The relationship between selected health risk factors and health care costs and utilization
Clegg, Samuel *see* Architecture of machinery
Cleghorn, George *see* Remarks on the intended restoration of the parthenon of athens as the national monument of scotland
Cleghorn, Robert *see* A short history of baptist missionary work in british honduras
Cleirac, Estienne *see*
– Us et coutumes de la mer
– Les us
Cleisz, Augustin *see* Etude sur les missions nestoriennes en chine au 7 et au 8 siecles d'apres l'inscription syro-chinoise de si-ngan-fou
Cleland, James *see* The institution of a young noble man
Cleland, Sharon M *see* The mediating effect of goal setting on exercise efficacy of efficacious older adults
Cleland, Thomas *see* The trial and acquittal of john the baptist, the apostles, and evangelists
Cleland, W I et al *see* History of all the religious denominations in the united states
Cleland, William *see* History of the presbyterian church in ireland
Clelland, Thomas *see* Marriage register and account book
Clemen, August *see*
– Der gebrauch des alten testaments in den neutestamentlichen schriften
– Die wunderberichte ueber elia und elisa in den buechern der koenige
Clemen, Carl *see*
– Die apostelgeschichte im lichte der neueren text-, quellen- und historisch-kritischen forschungen
– Die christliche lehre von der suende
– Die chronologie der paulinischen briefe
– Der einfluss der mysterienreligionen auf das aelteste christentum
– Die einheitlichkeit der paulinischen briefe
– Die entstehung des johannesevangeliums
– Die entstehung des neuen testaments
– Die entwicklung der christlichen religion
– Der geschichtliche jesus
– Die himmelfahrt des mose
– Niedergefahren zu den toten
– Paulus
– Primitive christianity and its non-jewish sources
– Die religionsgeschichtliche methode in der theologie
– Die religionsphilosophische bedeutung des stoisch-christlichen eudaemonismus in justins apologie
– Die reste der primitiven religion im aeltesten christentum
– Schleiermachers glaubenslehre
– Der ursprung des heiligen abendmahls
Clemen, Carl Christian *see* Die religionen der erde
Clemen, Otto *see*
– Alte einblattdrucke
– Geschichte der reformation
Clemen, Paul *see* Gedenkrede auf stefan george
Clemence et waldemar / Pelletier-Volmeranges, Benoit – Paris, France. 1803 – 1r – us UF Libraries [440]
Clemenceau, Georges *see* South america to-day
Clemens alexandrinus (gcsej3) / ed by Staehlin, O – (= ser Griechische christlichen schriftsteller der ersten jahr- hunderte (gcsej)) – (bd1: 1909 €18. bd2: 1906 €21. bd3: 1909 €15. bd4: 1936 €37) – ne Slangenburg [240]
Clemens alexandrinus in seiner abhaengigkeit von der griechischen philosophie / Merk, C – Leipzig, 1879 – €5.00 – ne Slangenburg [180]
Clemens alexandrinus und das neue testament / Kutter, H – Giessen, 1897 – €7.00 – ne Slangenburg [240]
Clemens alexandrinus und das neue testament : eine untersuchung / Kutter, Hermann – Giessen: J Ricker, 1897 – 1mf – 9 – 0-8370-9636-7 – (incl bibl ref) – mf#1986-3636 – us ATLA [225]
Clemens brentano : irrtum des herzens, einkehr bei gott / Michels, Josef – Muenster: Regensberg, 1948 [mf ed 1989] – 1r – 1 – (incl bibl ref. filmed with: clemens brentanos liebesleben / lujo brentano) – mf#7085 – us Wisconsin U Libr [430]
Clemens brentano : ein romantisches dichterleben / Pfeiffer-Belli, Wolfgang – Freiburg (Breisgau): Herder 1947 [mf ed 1990] – 1r – 1 – (incl bibl ref & ind. filmed with: ausgewahlte werke) – mf#7071 – us Wisconsin U Libr [430]
Clemens brentano / Seidel, Ina – Stuttgart: J G Cotta, c1944 [mf ed 1989] – (= ser Die dichter der deutschen) – 1r – 1 – (filmed with: clemens brentanos religioser werdegang / ernst koethke) – mf#7086 – us Wisconsin U Libr [430]
Clemens brentano und apollonia diepenbrock : eine seelenfreundschaft in briefen: 25 brentanobriefe / ed by Reinhard, Ewald – Muenchen: Parcus [1914?] [mf ed 1989] – (= ser Romantische buecherei 51-52) – 1r – 1 – (int & ann by ed. filmed with: godwi / alfred kerr) – mf#7084 – us Wisconsin U Libr [860]
Clemens brentano und die landschaft der romantik : mit besonderer beruecksichtigung seiner beziehungen zur romantischen malerei / Harms, Susanne – Wuerzburg: C J Becker 1932 [mf ed 1989] – 1r – 1 – (filmed with: clemens brentanos liebesleben / lujo brentano) – mf#7085 – us Wisconsin U Libr [430]
Clemens brentano und minna reichenbach : ungedruckte briefe des dichters / ed by Limburger, W – Leipzig: Insel-Verlag 1921 [mf ed 1989] – 1r – 1 – (filmed with: godwi / alfred kerr) – mf#7084 – us Wisconsin U Libr [860]
Clemens brentanos fruehlingskranz : aus jugend, briefen ihm geflochten wie er selbst schriftlich verlangte / Arnim, Bettina von – Berlin: im Propylaen-Verlag c1920 [mf ed 1993] – 1r – 1 – (filmed with: saemtliche werke / ed by waldemar oehlke) – mf#3247p – us Wisconsin U Libr [860]
Clemens brentanos fruehlyrik : chronologie und entwicklung / Jaeger, Hans – Frankfurt am Main: M Diesterweg 1926 [mf ed 1993] – 1r – 1 – (incl bibl ref) – mf#8023 reel 3 – us Wisconsin U Libr [430]
Clemens brentanos jugenddichtungen : abschnitt 1, der ideengehalt des godwi / Kerr, Alfred – Halle: [s.n.] 1894 [mf ed 1989] – 1r – 1 – (incl bibl ref. filmed with: clemens brentanos liebesleben / lujo brentano) – mf#7085 – us Wisconsin U Libr [430]
Clemens brentanos liebesleben : eine ansicht / Brentano, Lujo – Frankfurt/M: Frankfurter Verlags-Anstalt 1921 [mf ed 1989] – 1r – 1 – (filmed with: un poete romantique allemand / rene guignard) – mf#7085 – us Wisconsin U Libr [920]
Clemens brentanos religioeser werdegang / Koethke, Ernst – Hamburg: [s.n.] 1927 [mf ed 1989] – 1r – 1 – (filmed with: tiecks einfluss auf brentano / erich nippold & other titles) – mf#7086 – us Wisconsin U Libr [430]
Clemens brentanos weltliche lyrik / Schubert, Kurt – Breslau: F Hirt 1910 [mf ed 1992] – (= ser Breslauer beitraege zur literaturgeschichte. neue folge 10) – 1r – 1 – (incl bibl ref. filmed with: das gasel in der deutschen dichtung und das gasel bei platen / hubert tschersig) – mf#3102p – us Wisconsin U Libr [430]
Clemens, Bruno *see* Bruno brehm zum fuenfzigsten geburtstag
Clemens, Franz Jakob *see*
– De scholasticorum sententia philosophiam esse theologiae ancillam commentatio
– Giordano bruno und nicolaus von cusa
Clemens Romanus *see* Epsitolae binae de virginitate, syriace
Clemens, Samuel Langhorne *see*
– Boys' life of mark twain
– Contributions to the galaxy 1868-1871
Clement 1, Pope *see*
– Bruchstuecke des ersten clemensbriefes
– Epistles of ss clement of rome and barnabas and the shepherd of hermas
– Der erste clemensbrief
– Der erste clemensbrief in altkopischer uebersetzung
– First epistle of clemens romanus to the church at corinth
– Die homilien und recognitionen des clemens romanus
Clement 7, Pope *see*
– Bulla erectionis sanctae metropolitanae ecclesiae mexiceae
– Clementis 7. epistolae per sadoletum scriptae
Clement 12, Pope *see* Ad futuram rei memoriam
Clement 14, Pope *see*
– Breve de n m s p clemente 14 en respuesta a la carta de el ill[ustrissi]mo s[eno]r arzobispo metropolitano de mexico
– Clementis 14 pont. max. epistolae et brevia
Clement, A John *see* Kalahari and its lost city
Clement, Alex *see* Aux jeunes gens qui veulent reussir
Clement, Charles Francois *see* Essai sur l'accompagnement du clavecin
Clement, Clara Erskine *see* Egypt
Clement d'alexandrie : etude sur les rapports de christianisme et de philosophie grecque au 2e siecle / Faye, E de – Paris, 1898 – €14.00 – ne Slangenburg [240]
Clement d'alexandrie : etude sur les rapports du christianisme et de la philosophie grecque au 2e siecle / Faye, Eugene de – 2e ed. Paris: Ernest Leroux, 1906 [mf ed 1989] – (= ser Bibliotheque de l'ecole des hautes etudes. sciences religieuses 12) – 1mf – 9 – 0-7905-4296-X – (incl bibl ref) – mf#1988-0296 – us ATLA [240]
Clement d'alexandrie / Freppel, Charles – 3. ed. Paris: Retaux-Bray, [1873?] [mf ed 1990] – 1mf – 9 – 0-7905-4525-X – (incl bibl ref) – mf#1988-0525 – us ATLA [240]
Clement, David *see* Bibliotheque curieuse historique et critique
Clement, Ernest Wilson *see*
– The christian movement in japan
– Christianity in modern japan
– Handbook of modern japan
– Short history of japan
Clement, James A *see* An exposition of the pretensions of baptists to antiquity
Clement marot et le psautier huguenot : etude historique, litteraire, musicale et bibliographique / Douen, O – Paris: Imprimerie Nationale, 1878-1879 – 1mf – 9 – 0-7905-1982-8 – mf#1987-1982 – us ATLA [240]
Clement of alexandria / Patrick, John – Edinburgh: W Blackwood 1914 [mf ed 1991] – (= ser The croall lectures 1899-1900) – 1mf – 9 – 0-7905-9565-6 – mf#1989-1290 – us ATLA [240]
Clement of alexandria : a study in christian liberalism / Tollinton, Richard Bartram – London: Williams & Norgate 1914 – 2v on 1r [ill] – 1 – (filmed with: life and administration of edward, first earl of clarendon / lister, t h) – mf#1344p – us Wisconsin U Libr [240]
Clement of alexandria : a study in christian liberalism / Tollinton, Richard Bartram – London: Williams & Norgate 1914 [mf ed 1990] – 2v on 2mf – 9 – 0-7905-8095-0 – (incl bibl ref) – mf#1988-8031 – us ATLA [240]
Clement of alexandria, quis dives salvetur (ts5/2) : with an introduction on the mss of clement's works / ed by Barnard, P M – 1897 – (= ser Texts and studies (ts)) – 2mf – 9 – €5.00 – ne Slangenburg [240]
Clement of Alexandria, Saint *see* Quis dives salvetur
Clement, Richard Gray *see* Dosis-wirkungsbeziehung von unretadiertem isosorbiddinitrat in kleinen dosen bei patienten mit koronarer herzkrankheit
Clement, William Henry Pope *see* The law of the canadian constitution
Clementi, Cecil *see* Constitutional history of british guiana
Clementi, Marie Penelope Rose *see* Through british guiana to the summit of roraima
Clementi, Muzio *see*
– 3 sonates pour le clavecin ou pianoforte avec flute et basse
– Sonata per clavicembalo o piano forte con un violino o flauto e violoncello, opera
– Sonata per clavicembalo o piano-forte con un violino o flauto e violoncello...opera [31]
– Two capriccios for the piano forte...dedicated to mrs clementi...op 47
Die clementinischen recognitionen und homilien / Hilgenfeld, Adolf – Jena: JG Schreiber in Commission bei C Hochhausen, 1848 – 1mf – 9 – 0-7905-7647-3 – (incl bibl ref) – mf#1989-0872 – us ATLA [240]
Clementis 7. epistolae per sadoletum scriptae : quibus accedunt variorum ad papam et ad alios epistolae = Epistolae per sadoletum scriptae / Clement 7, Pope; ed by Balan, Pietro – Oeniponte [Innsbruck]: Libraria Academica Wagneriana, 1885 – (= ser Monumenta saeculi) – 2mf – 9 – 0-8370-9049-0 – mf#1986-3049 – us ATLA [240]
Clementis 14 pont. max. epistolae et brevia : selectiora ac nonnulla alia aacta pontificatum ejus = Epistolae et brevia / Clement 14, Pope; ed by Theiner, Augustin – Parisiis: F Didot, 1852 – 1mf – 9 – 0-524-03542-3 – mf#1990-4737 – us ATLA [240]
Clementis alexandrini de logoi doctrina / Laemmer, Hugo – Lipsiae: FA Brockhaus, 1855 – 1mf – 9 – 0-7905-7891-3 – (incl bibl ref) – mf#1989-1116 – us ATLA [180]
Clements, Ernest *see* Introduction to the study of indian music
Clements, Frank *see*
– Rhodesia
– This is our land
Clements, James I *see* The klondyke
Clements, Rex S *see* Sermons
Clements, S *see* An itinerant ministry
Clements, W H *see* The glamour and tragedy of the zulu war
Clemmer exchange – v1 n1-v3 n2 [1988 mar-1990 sum] – 1r – 1 – mf#1789262 – us WHS [071]
Clemmer, Myrtle M *see* United presbyterian historical directory
Clemson spectator – 1992 apr, nov, 1993 mar, sep, oct, 1994 jan, feb, mar, oct – 1r – 1 – mf#2896995 – us WHS [071]
Clendenen, Frank Leslie *see* Cake walks
Clendenin, William Ritchie *see* Use of the french chanson in some polyphonic masses by french and netherlands composers, 1450-1550
Clennell, Walter James *see* The historical development of religion in china
Cleobury mortimer journal & ludlow advertiser – [West Midlands] Shropshire 17 mar 1933-4 nov 1938 [mf ed 2004] – 6r – 1 – uk Newsplan [072]
Cleopatra's needle : a history of the london obelisk, with an exposition of the hieroglyphics / King, James – [London]: Religious Tract Soc [1893?] [mf ed 1989] – (= ser By-paths of bible knowledge 1) – 1mf – 9 – 0-7905-2240-3 – mf#1987-2240 – us ATLA [730]
Cleopatra's needles and other egyptian obelisks / Budge, Ernest Alfred Wallis – London, 1926 – 4mf – 9 – mf#NE-20017 – ne IDC [930]
Cleopatre / Soumet, Alexandre – Paris, France. 1825 – 1r – us UF Libraries [440]
Cleophas lachance : son crime, son proces, son execution – Levis Quebec: A G Routhier, 1881 – 1mf – 9 – mf#03014 – cn Canadiana [360]
Cleophas lachance pendu le 28 janvier 1881 pour meurtre d'odelide desilets – Arthabaskaville, Quebec?: s.n, 1881? – 1mf – 9 – mf#03871 – cn Canadiana [360]
Clephane, Walter Collins *see* The organization and management of business corporations
Cler, Jean Joseph Gustave *see* Reminiscences of an officer of zouaves
Clerambault, Louis Nicolas *see* Cantates francoises a 1 et 2 voix

Clerc, Jean le *see* Le grand dictionnaire historique (ael1/44.2)

Clercq, V C de *see* Ossius of cordova (sca13)

The clerc's book of 1549 (hbs25) / Wickham Legg, J – 1901 – (= ser Henry bradshaw society (hbs)) – 4mf – 8 – €11.00 – ne Slangenburg [240]

Clerfeyt, Joseph Maximilian Louis *see* L'affaire clerfeyt.

Le clerge canadien : sa mission, son oeuvre / David, Laurent Olivier – Montreal: s.n, 1896 – 2mf – 9 – mf#02510 – cn Canadiana [241]

Clerge indigene / Gayot, Gerard G – Port-Au-Prince, Haiti. 1956 – 1r – us UF Libraries [972]

Le clerge protestant du bas-canada de 1760 a 1800 / Audet, Francis-Joseph – Ottawa: J Hope, 1901 – 1mf – 9 – 0-665-73116-7 – (incl bibl ref) – mf#73116 – cn Canadiana [242]

The clergy a source of danger to the american republic / Jamieson, William F – 2nd ed. Chicago: WF Jamieson, 1873, c1871 – 1mf – 9 – 0-524-00994-5 – mf#1990-0271 – us ATLA [240]

Clergy and laity concerned minnesota report – 1983 sum-1987 spr – 1r – 1 – (cont: clergy and laity concerned report) – mf#1477008 – us WHS [240]

Clergy and Laity Concerned [US] *see*
– Calc report
– Wisconsin calc newsletter

The clergy and popular education / Fowler, William Chauncey – [S.l.: s.n., 1868?] – 1mf – 9 – 0-8370-7791-5 – mf#1986-1791 – us ATLA [240]

The clergy and the creeds : a sermon / Gore, Charles – London: Rivingtons, 1887 – 1mf – 9 – 0-7905-3846-6 – mf#1989-0339 – us ATLA [240]

The clergy and the pulpit in their relations to the people / Mullois, Isidore – 1st American ed. New-York: Catholic Publication Society, 1867 – 1mf – 9 – 0-524-03954-2 – mf#1991-2008 – us ATLA [240]

Clergy bulletin / Evangelical Lutheran Synod – 1960 sep-1961 mar – 1r – 1 – (cont by: lutheran synod quarterly) – mf#5199299 – us WHS [242]

The clergy in american life and letters / Addison, Daniel Dulany – New York: Macmillan, 1900 – (= ser National Studies In American Letters) – 1mf – 9 – 0-7905-4307-9 – mf#1988-0307 – us ATLA [240]

Clergy in maryland of the protestant episcopal church since the independence of 1783 / Allen, Ethan – Baltimore: James S Waters 1860 [mf ed 1992] – 1mf – 9 – 0-524-04032-X – mf#1990-4940 – us ATLA [242]

Clergy journal – Inver Grove Heights. 1992+ (1) 1992+ (5) 1992+ (9) – (cont: church management: the clergy journal) – mf#210,01 – us UMI ProQuest [240]

The clergy list : with which is incorporated the clerical guide and ecclesiastical directory, 1895 – London: Kelly, 1895 – 12mf – 9 – 0-524-08860-8 – mf#1993-3324 – us ATLA [240]

The clergy monthly : [catholic church in india] – Madras, India. 1938-74 [mf ed 2001] – (= ser Christianity's encounter with world religions, 1850-1950) – 9r – 1 – mf#2001-s156 – us ATLA [241]

Clergy not a priesthood / Marsden, J B – Birmingham, England. 18– – 1r – us UF Libraries [240]

The clergy of america : anecdotes illustrative of the character of ministers of religion in the united states / Belcher, Joseph – Philadelphia: JB Lippincott, 1849, c1848 [mf ed 1991] – 2mf – 9 – 0-524-00506-0 – mf#1990-0006 – us ATLA [880]

Clergy of the church in ireland weighed in the balance / Hamilton, George Alexander – London, England. 1868 – 1r – us UF Libraries [240]

The clergy reserve question : as a matter of history, a question of law, a subject of legislation / Ryerson, Egerton – Toronto?: s.n, 1839 (Toronto: L H Lawrence) – 2mf – 9 – mf#24231 – cn Canadiana [240]

Clergy review – London. 1931-1987 (1) 1971-1987 (5) 1974-1987 (9) – (cont by: priests and people) – ISSN: 0009-8736 – mf#2170 – us UMI ProQuest [240]

Clergyman Of The Church Of England *see* Rational piety and prayers for fair weather

Clergyman's remonstrance with a dissenting minister / Harding, W – Chelmsford, England. 1840 – 1r – us UF Libraries [240]

Clerical declaration on the athanasian creed – London, England. 1872 – 1r – us UF Libraries [240]

Clerical education / Perry, Charles – London, England. 1841 – 1r – us UF Libraries [240]

Clerical intemperance / Kirkman, Thomas Penyngton – Ramsgate, England. 1871 – 1r – us UF Libraries [240]

Clerical politics in the methodist episcopal church / Townsend, Luther Tracy – Boston: McDonald, Gill c1892 [mf ed 1990] – 1mf – 9 – 0-7905-6328-2 – mf#1988-2328 – us ATLA [242]

Clerical "pooh, pooh!" rhetoric – London, England. 1875 – 1r – us UF Libraries [240]

Clerical sketches : or, pulpit preaching in 1840-1-2 / Anthroposophus – Glasgow, Scotland. 1842 – 1r – us UF Libraries [240]

Clerical studies / Hogan, John Baptist – 2nd ed. Boston: Marlier, c1898 – 2mf – 9 – 0-524-04378-7 – mf#1991-2082 – us ATLA [240]

Le clericalisme au canada – Montreal: [en vente chez J Grant], 1896 [mf ed 1977] – 1r – 5 – (fasc 1: cures et bedeaux. fasc 2: saintes comedies) – mf#SEM16P282 – cn Bibl Nat [241]

Clerici, J *see* Opera omnia

Clericus *see* Letters to liberationists

Los clerigos y la extirpacion de la idolatria entre los neofitos americanos / Bayle, Constantino – Madrid: Missionalia Hispanica, 1946 – 1 – sp Bibl Santa Ana [240]

Clerke, Agnes Mary *see* The system of the stars

Clerke, Francis *see* Praxis francisci clerke, tam jus dicentibus quam aliis omnibus qui in foro ecclesiastico versantur apprime utilis

The clerkenwell chronicle, st luke's examiner, holborn reporter and north london observer – Islington, England. 19 jul 1884-25 sep 1886 [mf 1895] – 1 – (incorp with: hackney gazette. aka: weekly news and clerkenwell chronicle; weekly news and chronicle; finsbury weekly news and chronicle; finsbury weekly news and clerkenwell chronicle and st lukes examiner) – uk British Libr Newspaper [072]

Clerkenwell press and general advertiser – London, UK. 14 jul 1877-17 mar 1886 – 5 1/2r – 1 – (aka: clerkenwell press; st lukes guardian and holborn news) – uk British Libr Newspaper [072]

Clerkenwell times and north london gazette – London, UK. 26 oct-20 dec 1856 – 1/4r – 1 – uk British Libr Newspaper [072]

Clerkenwell watchman and reformers gazette and general advertiser – London, UK. 10 jan-16 may 1857 – 1/4r – 1 – uk British Libr Newspaper [072]

The clerk's assistant, containing a large variety of legal forms and instruments. / McCall, Henry Strong – 4th ed. Albany, Gould, 1884. 618 p. LL-874 – 1 – (5th ed. new york, banks, 1898. 1136p. ll-930. 1. 6th ed. new york, banks, 1902. 1216p. ll-1412. 1) – us L of C Photodup [340]

Clerk's bulletin – Shorewood Hills WI. n429-560 [1979 feb 2-1988 dec], v62 n1-12 [1989 jan-dec], v63 n1-9 [1990 jan-sep] – 1r – 1 – (cont by: village bulletin) – mf#1222520 – us WHS [071]

Clerk's record / Shawnee County. Kansas. School District 26 – 1927-43 – 1 – us Kansas [978]

Clermont Co. Batavia *see*
– Clermont courier
– Clermont courier-press series
– Clermont sun
– Ohio sun
– Spirit / chronicle of times

Clermont Co. Bethel *see* Journal

Clermont Co. Cincinnati *see* Clermont county review

Clermont Co. Loveland *see* Herald

Clermont Co. Milford *see* Valley enterprise

Clermont Co. New Richmond *see*
– Ohio star
– Philanthropist

Clermont co, ohio, records, mss 1086 – Clermont, OH. 1801-63 – 1r – 1 – (scattered census records, tax accounts, treasurers' reports, loose papers of the sheriff, justices of peace, and court of common pleas, vital statistics, and poll books) – us Western Res [350]

Clermont Co. Williamsburg *see* Times

Clermont County, OH *see* Miscellaneous records, ms 1086

Clermont county review / Clermont Co. Cincinnati – jan 1971-may 1994 [wkly] – 20r – 1 – mf#B34971-34990 – us Ohio Hist [071]

Clermont county review/w / Hamilton Co. Cincinnati – jan 1971-may 1994 – 20r – 1 – mf#B34971-34990 – us Ohio Hist [071]

Clermont courier / Clermont Co. Batavia – 1909-, 1928-29, 31-33, 36-1959 [wkly, semiwkly] – 20r – 1 – mf#B8952-8971 – us Ohio Hist [071]

Clermont courier / Clermont Co. Batavia – 1919-31, 1933-36, 1960 [wkly] – 8r – 1 – mf#B10504-511 – us Ohio Hist [071]

Clermont courier / Clermont Co. Batavia – 1961-82 [wkly,semiwkly, wkly] – 21r – 1 – mf#B13074-13094 – us Ohio Hist [071]

Clermont courier / Clermont Co. Batavia – apr 2-aug 13 1847 (gap fillers) [wkly] – 1r – 1 – mf#B30907 – us Ohio Hist [071]

Clermont courier / Clermont Co. Batavia – v1 n1. (1836-52, 75-99, 1902-04,06-1907) [wkly] – 16r – 1 – mf#B11070-11085 – us Ohio Hist [071]

Clermont courier-press series / Clermont Co. Batavia – jan 1983-mar 1989 [wkly] – 7r – 1 – mf#B31165-31171 – us Ohio Hist [071]

Clermont press – Clermont, FL. 1928-1930 – 2r – us UF Libraries [071]

Clermont sun / Clermont Co. Batavia – 1926, 1929-52, 1954-68 [wkly] – 29r – 1 – mf#B10706-734 – us Ohio Hist [071]

Clermont sun / Clermont Co. Batavia – 1928, 1953, 1969-71, 1983-85 [wkly] – 8r – 1 – mf#B4400-4407 – us Ohio Hist [071]

Clermont sun / Clermont Co. Batavia – aug 1852-dec 1857 [wkly] – 1r – 1 – mf#B28799 – us Ohio Hist [071]

Clermont sun / Clermont Co. Batavia – feb 1971-dec 1982 [wkly] – 12r – 1 – mf#B13041-13052 – us Ohio Hist [071]

Clermont sun / Clermont Co. Batavia – jan-dec 1986 [wkly] – 1r – 1 – mf#B6616 – us Ohio Hist [071]

Clermont sun / Clermont Co. Batavia – sep 1876-1925, 1927 [wkly] – 22r – 1 – mf#B9280-9301 – us Ohio Hist [071]

Clermont sun – Batavia, OH, feb 22 1854-dec 26 1860 – 1r – 1 – (weekly democratic newspaper) – us Western Res [071]

Clermont-Ganneau, Charles *see* Archaeological researches in palestine during the years 1873-1874

Clero : la milicia y las revoluciones / Mora, Jose Maria Luis – Mexico City? Mexico. 1951 – 1r – us UF Libraries [972]

El clero secular y la evangelizacion de america / Bayle, Constantino – Madrid: csic instituto santo toribio de mogroviejo, 1950 – 1 – sp Bibl Santa Ana [240]

El clero vasco, fiel al gobierno de la republica, se dirige al sumo pontifice / Vitoria. Spain (Diocese) – Madrid, 1937 – (= ser Blodgett coll) – 9 – mf#fiche w804 – us Harvard [946]

El clero y los catolicos vasco-separatistas y el movimiento nacional – Madrid: imp y enc de los sobrinos de la sucesoera de m minuesa, 1940 – sp Bibl Santa Ana [240]

Clerq, L De *see* Grammaire du kiyombe

Clery, Jean-Baptiste Cant Hanet *see* Journal de ce qui s'est passe a la tour du temple

Cleveland bay herald – Townsville, 3 mar 1866 – 1r – A$27.98 vesicular A$33.48 silver – at Pascoe [079]

Cleve, Karl *see*
– Goethes verhaeltnis zu hans sachs
– Nicolais feyner kleyner almanach

Clevedon mercury – [SW England] North Somerset 24 jan 1863-dec 1950 [mf ed 2003] – 77r – 1 – (incorp: clevedon courier to become: clevedon mercury and courier [jan 1865-dec 1869]; cont as: clevedon mercury and courier and somersetshire weekly advertiser [jan 1870-dec 1937]; clevedon mercury and courier and somerset weekly advertiser [jan 1938-dec 1950]) – uk Newsplan [072]

Cleveland advocate – Cleveland, OH, jun 15 1918-dec 18 1920 – 2r – 1 – (weekly african-american republican newspaper) – us Western Res [071]

Cleveland, akron and columbus railway "observation car," 1892 – 1r – 1 – mf#B27451 – us Ohio Hist [380]

Cleveland anzeiger – Cleveland, OH, sep 29 1878-sep 13 1891 – (= ser Ethnic newspapers) – 7r – 1 – (sunday ed of this german language republican newspaper) – us Western Res [071]

Cleveland anzeiger and deutsche presse – Cleveland, OH, sep 19 1891-sep 30 1893 – 3r – 1 – (daily ed of this german language republican newspaper) – us Western Res [071]

Cleveland anzeiger and deutsche presse – Cleveland, OH, sep 20 1891-oct 1 1893 – 3r – 1 – (sunday ed of this german language republican newspaper) – us Western Res [071]

Cleveland Area Genealogical Enterprises [TX] *see* Cleveland area pioneer

Cleveland area pioneer / Cleveland Area Genealogical Enterprises [TX] – v1 n1-v2 n4 [1978 jun-1980 mar] – 1r – 1 – mf#669243 – us WHS [929]

Cleveland artisan – 1899 jul 25-aug 17 – 1r – 1 – mf#3238285 – us WHS [071]

Cleveland bar journal – v1-55. 1927-84 – 198mf – 9 – $297.00 – (cont: cleveland bar association journal. lacking: v1 n5,11. v2 no 2. v5 n1-4. v32 n9. v47 n 8. updates planned) – mf#LLMC 84-438 – us LLMC [340]

Cleveland bar journal – v1-72. 1927-2001 – 9 – $945.00 set – (title varies: v1-40 1927-68 as journal of the cleveland bar association; suspended mar 1933-sep 1936) – ISSN: 0160-1598 – mf#101771 – us Hein [340]

Cleveland beacon : a publication / New American Movement [Organization] – v1 n1-v3 n3 [1981 feb-1983 sum] – 1r – 1 – mf#951644 – us WHS [320]

Cleveland Building and Construction Trades Council *see* Cleveland citizen

Cleveland, Catharine Caroline *see* The great revival in the west, 1797-1805

Cleveland Centennial Commission. Women's Dept *see* Genealogical data relating to women in the western reserve before 1840 (1850)

Cleveland citizen / Cleveland Building and Construction Trades Council – Cleveland OH. 1891 jun 5/1899-1995/97 – 34r – 1 – (with gaps) – mf#1425438 – us WHS [690]

Cleveland clinic journal of medicine – Cleveland. 1987+ (1) 1987+ (5) 1987+ (9) – (cont: cleveland clinic quarterly) – ISSN: 0891-1150 – mf#3270,01 – us UMI ProQuest [610]

Cleveland clinic quarterly – Cleveland. 1932-1986 (1) 1971-1986 (5) 1975-1986 (9) – (cont by: cleveland clinic journal of medicine) – ISSN: 0009-8787 – mf#3270 – us UMI ProQuest [610]

Cleveland correspondent – Cleveland, OH, sep 14 1908-mar 4 1911 – 2r – 1 – (weekly german language newspaper) – us Western Res [071]

Cleveland daily express – Cleveland, OH, jun 21-dec 30 1854 – 1r – 1 – (daily know nothing/american party newspaper) – us Western Res [071]

Cleveland daily herald – Cleveland, OH, aug 6 1839-sep 20 1843 – 6r – 1 – (daily whig newspaper) – us Western Res [071]

Cleveland daily herald – Cleveland, OH, jul 21 1856-jun 22 1857 – 2r – 1 – (daily whig newspaper) – us Western Res [071]

Cleveland daily herald – Cleveland, OH, jun 1-dec 31 1874 – 2r – 1 – (evening ed of this daily republican newspaper) – us Western Res [071]

Cleveland daily herald – Cleveland, OH, jul 1-dec 2 1874 – 1r – 1 – (morning ed of this daily republican newspaper) – us Western Res [071]

Cleveland federationist – Cleveland, OH, apr 14 1910-jul 6 1933 – (= ser 20th-century united states newspapers) – 8r – 1 – (weekly labor paper, publ by the cleveland branch, american federation of labor) – us Western Res [071]

Cleveland gazette – Cleveland, OH, jun 1 1836-mar 21 1837 – 1r – 1 – (daily whig newspaper) – us Western Res [071]

Cleveland germania – Cleveland, OH, jan 2-jun 30 1889 – 2r – 1 – (daily german language, republican newspaper) – us Western Res [071]

Cleveland, Grover *see* Papers

Cleveland, Harold Irwin *see* Massacres of christians by heathen chinese and horrors of the boxers

Cleveland herald – Cleveland, OH, jun 5 1880-jun 30 1884 – 6r – 1 – (daily republican newspaper, morning ed) – us Western Res [071]

Cleveland herald – Cleveland, OH. oct 19 1819-apr 12 1832 – 4r – 1 – (weekly whig newspaper, the second newspaper publ in cleveland) – us Western Res [071]

Cleveland herald and gazette – Cleveland, OH, mar 25 1837-sep 20 1843 – 2r – 1 – (weekly whig newspaper) – us Western Res [071]

Cleveland jewish miscellany, ms 3669 / Nebel, Abraham Lincoln – 1831-1971 – 1 – (correspondence, clippings, notes and legal documents relating to jewish american genealogy and jewish community in cleveland, ohio) – us Western Res [920]

Cleveland jewish news *see* The jewish review and observer

The cleveland jewish news – Cleveland. Ohio. 1964-67 – 1 – us AJPC [071]

Cleveland journal – Cleveland, OH, mar 28 1903-apr 9 1910 – 2r – 1 – (weekly african-american republican newspaper) – us Western Res [073]

Cleveland journal and south durham advertiser – Cleveland, South Durham, England. 13 Feb-23 Oct 1873. 45 ft – 1 – uk British Libr Newspaper [072]

Cleveland law record – v1. 1855-56 (all publ) – 1mf – 9 – $1.50 – mf#LLMC 84-441 – us LLMC [348]

Cleveland law reporter – v1-2. 1878-79 (all publ) – 9mf – 9 – $13.50 – mf#LLMC 84-439 – us LLMC [340]

Cleveland law school journal – v1. 1916 (all publ) – 1mf – 9 – $1.50 – mf#LLMC 84-440 – us LLMC [340]

Cleveland magazine – Cleveland. 1973+ (1) 1973+ (5) 1976+ (9) – ISSN: 0160-8533 – mf#8589 – us UMI ProQuest [978]

Cleveland marshall law review *see* Cleveland state law review

Cleveland methodist messenger – Cleveland, England. -m. Oct 1874-Jan 1875. 5 ft – 1 – uk British Libr Newspaper [072]

Cleveland morning daily herald – Cleveland, OH, dec 2 1871-may 30 1874 – 5r – 1 – (morning ed of this daily republican newspaper) – us Western Res [071]

Cleveland morning herald – Cleveland, OH, jan 2-nov 30 1871 – 3r – 1 – (morning ed of this daily republican newspaper) – us Western Res [071]

Cleveland Museum of Art *see* Bulletin of the cleveland museum of art
Cleveland news – England. 7-28 Jun 1873; 6 Aug 1875-29 Oct 1887.-w. 12 reels – 1 – uk British Libr Newspaper [072]
Cleveland observer – Cleveland, OH, sep 28 1837-apr 1 1840 – 1r – 1 – (weekly presbyterian newspaper; originally publ in hudson, oh, as the ohio observer, the title returned to hudson in 1840) – us Western Res [071]
Cleveland, OH *see*
– Selections
– Telephone directories, (1914-1916, 1924-1979)
Cleveland. Ohio. Bethany Presbyterian Church *see* Church records, ms 3324
Cleveland. Ohio. Fifth Circuit Court *see* Tappan's common pleas reports
Cleveland. Ohio. First Regular Baptist Church *see* Church records, ms 647
Cleveland. Ohio. Presbytery. The Women's Foreign Mission Society *see* Church records, ms 241
Cleveland. Ohio. St. John's Episcopal Church *see* Church records, ms 3066
Cleveland, ohio, taxes, ms v.f. o – 1802 – 1r – 1 – (appraisement of cleveland giving proprietors, owners or occupiers names) – us Western Res [978]
Cleveland, ohio, taxes, ms v.f. o / Ayres, Ebenezer – 1802 – 1r – 1 – (return of list of inhabitants of the district of cleveland) – us Western Res [978]
Cleveland, ohio, taxes, ms v.f. o / Gilbert, Stephen & Spafford, Amos – 1801 – 1r – 1 – (lists all properties liable to be appraised for raising county levies in the town of cleveland in county trumbull) – us Western Res [350]
Cleveland, ohio, taxes, ms v.f.o. – 1801 – 1r – 1 – (list of all the polls and taxable property found in cleveland town) – us Western Res [978]
Cleveland. Ohio. Third Baptist Church *see* Church records, ms 2335
Cleveland. Ohio. Woodland Ave. Methodist-Episcopal Church *see* Church records, ms 415
Cleveland post – Cleveland, OH, oct 4 1879-jan 17 1880 – 1r – 1 – (daily german language newspaper) – us Western Res [071]
Cleveland. Presbytery (Pres. Church in the USA) *see* Minutes, 1830-1883
Cleveland, President *see*
– American rights in samoa
– Samoan affairs
Cleveland shopping news – Cleveland, OH, dec 26 1945-jul 1 1954 – 23r – 1 – (bi-weekly advertising newspaper) – us Western Res [380]
Cleveland shopping news – Cleveland, OH. oct 15 1921-jun 27 1929 – 9r – 1 – (weekly, later bi-weekly advertising paper) – us Western Res [380]
Cleveland standard – Cleveland, England. -w. 1 Feb, 21 Aug 1908-1 May 1953. 28 1 2 reels – 1 – uk British Libr Newspaper [072]
Cleveland star – Shelby, NC. 1902-1936 (1) – mf#65338 – us UMI ProQuest [071]
Cleveland state law review – v1-47. 1952-99 – 5,6,9 – $565.00 set – (v1-33 1972-85 in reel $229; v34-47 1985-99 in mf $336; title varies: v1-18 1952-69 as cleveland-marshall law review) – ISSN: 0009-8876 – mf#101801 – us Hein [342]
Cleveland Teachers Union *see* Critique
Cleveland und sein deutschthum – [Cleveland], Cuyahoga, OH. 1907 – 1r – 1 – (a german language history and biographical dictionary of cleveland) – us Western Res [920]
Cleveland union leader – 1939 sep 21-1941, 1942-55, 1956-1959 apr 24 – 8r – 1 – (cont by: cleveland citizen) – mf#1425783 – us WHS [331]
Cleveland union leader – Cleveland, OH, sep 3 1937-sep 14 1939 – 1r – 1 – (weekly labor newspaper, publ by the cleveland congress of industrial organizations) – us Western Res [331]
Cleveland volksfreund – 1903 jul 4-18 – 1r – 1 – mf#3253655 – us WHS [071]
Cleveland weekly gazette – Cleveland, OH, jan 4-mar 22 1837 – 1r – 1 – (weekly whig newspaper) – us Western Res [071]
Cleveland weekly herald – Cleveland, OH. mar 28 1868-dec 30 1876 – 4r – 1 – (weekly republican newspaper) – us Western Res [071]
Cleveland whig – Cleveland, OH, aug 20 1834-dec 28 1836 – 1r – 1 – (weekly whig newspaper) – us Western Res [071]
Cleveland Women's Counsel *see* What she wants
Cleveland worker / Revolutionary Union – v1 n2-v3 n4 [1973 apr/may-1975 sum] – 1r – 1 – mf#1110702 – us WHS [331]
The cleveland workhouse and house of refuge and correction records, 1855-1950 – 3 series – 24r – 1 – (guide available separately: d3494.g $15. records broken into 3 series: series 1: minutes of the workhouse board of directors. series 2: inmate records. series 3: operational records) – mf#D3494 – us Western Res [978]
Cleveland world – Cleveland, OH. apr 1 1890-jun 30 1905 – 64r – 1 – (daily republican newspaper) – us Western Res [071]
Clevelander anzeiger und deutsche presse : cleveland advertiser and german press – Cleveland, OH: German- American Pub Co, sep 23 1891-sep 27 1893 – 1r – 1 – (weekly ed of this german language republican newspaper. formed by the union of: cleveland anzeiger (1876: weekly), and: woechentliche deutsche presse. merged with: waechter am erie, to form: waechter und anzeiger (cleveland, ohio: 1893: weekly). daily and sunday eds issued concurrently) – us Western Res [071]
Clevelander herold – Cleveland, OH, mar 2 1901-aug 24 1908 – (= ser Ethnic newspapers) – 13r – 1 – (weekly, later daily, german language newspaper) – us Western Res [071]
Clevelandska amerika – Cleveland, Oh. 1918 – 1 – us CRL [071]
Clevelandska amerika – Cleveland OH, 1909, 1914* – 1r – 1 – (slovenian newspaper) – us IHRC [071]
Clevenger, Shobal Vail *see* Medical jurisprudence of insanity; or, forensic psychiatry
Clevenger, William May *see* The courts of new jersey; their origin, composition and jurisdiction
Clewiston news – Clewiston, FL. 1928 feb-1997 – 58r – (gaps) – us UF Libraries [071]
Clews to holy writ : or, the chronological scripture cycle / Carus-Wilson, Ashley, Mrs – 8th ed. New York: American Tract Society, [1893] – 1mf – 9 – 0-8370-2600-8 – (incl ind of biblical books and an ind to the psalms) – mf#1985-0600 – us ATLA [220]
De cleyn werelt : daer in claerlijcken door seer schoone poetische, moralische en historische exempelen betoont wort... / [Moerman, J] – Amstelredam: Dirck Pietersz, 1608 – 2mf – 9 – mf#O-827 – ne IDC [090]
De cleyne catechismus... / Micronius, M – [Londen], 1559 – 3mf – 9 – mf#PBA-269 – ne IDC [240]
Clichtove, J *see*
– Antilutherus...tres libros complectens
– Compendium veritatum ad fidem pertinentium...
– De sacramento eucharistiae contra oecolampadium opusculum...
– De veneratione sanctorum, opusculum duos libros coplectens
– Elucidatorium ecclesiasticum ad officium ecclesiae pertinentia planius exponens in quattuor libros completens
– Homiliae seu sermones
– Improbatio quorundam articulorum martini lutheri...
– Propugnaculum ecclesiae adversus lutheranos...
[Clichtove, J] *see*
– De vita et moribus sacerdotum opusculum...
– Elucidatorium ecclesiasticum ad officium ecclesiae pertinentia planius exponens in quatuor libros completens
Clichtove, Josse *see* Dogma moralium philosophorum, compendiose & studiose collectum
Click, Barry Clinton *see* Toward a theology of marriage and family ministry: a practicum development in a pastoral counseling center. 1982
Client motivation for rehabilitation / Barry, John R – Gainesville, FL. 1965 – 1r – us UF Libraries [025]
The client princes of the roman empire under the republic / Sands, Percy Cooper – Cambridge: University Press, 1908 – (= ser Cambridge Historical Essays) – 1mf – 9 – 0-7905-6559-5 – (incl bibl ref) – mf#1988-2559 – us ATLA [930]
Cliffe, Charles *see* The resources of british columbia in minerals, agriculture, lumber, and the fisheries
Clifford, H *see* Studies in brown humanity being scrawls and smudges in sepia white, and yellow
Clifford henderson papers *see* Henderson, clifford, papers, ms 4309
Clifford, Henry *see* Reflections on the appointment of a catholic bishop to the london d...
Clifford, John *see*
– The christian certainties
– Church's war with national intemperance?
– Daily strength for daily living
– The dawn of manhood
– The english baptists
– The gospel of gladness
– The inspiration and authority of the bible
– Secret of jesus
– Social worship
– Typical christian leaders
– The ultimate problems of christianity
Clifford, John Henry *see* John h. clifford, esq., attorney-general, &c
Clifford, John Herbert *see* The odore parker
Clifford, N *see* Clifford's reports of cases in the first circuit, 1858-1878
Clifford, W H *see* Clifford's reports of cases in the first circuit, 1858-1878
The clifford w henderson national air races collection, 1928-1939 – 14r – 1 – $1,610.00 – (guide sold separately: d3495.g $15) – mf#D3495 – us Western Res [790]
Clifford, W K *see* Bearing of morals on religion
Clifford, William *see* Development of the christian life
Clifford, William G *see* Cape florida lighthouse
Clifford, William Kingdon *see* Lectures and essays
Clifford's reports of cases in the first circuit, 1858-1878 / Clifford, N & Clifford, W H – Boston: Little-Brown. v1-4. 1869-80 (all publ) – (= ser Early federal nominative reports) – 32mf – 9 – $48.00 – mf#LLMC 81-442 – us LLMC [340]
Clift, C Winifred Lechmere *see* Very far east
Clifton – (= ser Bedfordshire parish register series) – 3mf – 9 – £7.50 – uk BedsFHS [929]
Clifton and redland free press – Bristol, England. 9 May 1890-15 Jan 1931. -w. -irr – 14r – 1 – uk British Libr Newspaper [072]
Clifton directory – [SW England] Bristol 13 nov 1850-sep 1928 [mf ed 2004] – 80r – 1 – (missing: 1873, 1899-1900; cont by: clifton chronicle etc [jan 1854-sep 1928]) – uk Newsplan [072]
Clifton Park, New York. Clifton Park Baptist Church *see* Records
Clifton, Robert T *see* Gender differences in the relationships among self-confidence, gender-appropriateness, and value
Clifton society – Bristol, England. Nov 1890-Mar 1916. Missing: 1893, 1898. -w. 23 1/4 reels – 1 – uk British Libr Newspaper [072]
Cliftonville and hove mercury, brighton, preston and sussex county news – [London & SE] East Sussex, Hove Ref Lib 13 sep 1878-16 jan 1880 (wkly) – 1 – (discontinued) – uk Newsplan [072]
Clima e saude / Peixoto, Afranio – Sao Paulo, Brazil. 1938 – 1r – us UF Libraries [972]
Clima, paisaje y naturaleza en la obra de gabriel y galan / Lopez Bustos, Carlos – Badajoz: Dip. Provincial, 1970. Sep. REE – 1 – sp Bibl Santa Ana [440]
Climacus, Johannes *see* Afsluttende uvidenskabelig efterskrift til de philosophiske smuler
Climacus, John *see* The ladder of divine ascent
Climate – Paris, France. 25 jul 1946-1 oct 1947 – 1r – 1 – uk British Libr Newspaper [072]
Climate control – New Delhi. 1973-1974 (1) – ISSN: 0009-8930 – mf#9073 – us UMI ProQuest [690]
Climate of florida / Mitchell, A J – Gainesville, FL. 1928 – 1r – us UF Libraries [630]
Climate policy – Amsterdam, 2001+ [1,5,9] – ISSN: 1469-3062 – mf#42831 – us UMI ProQuest [550]
Climatic change – Dordrecht. 1984+ (1,5,9) – ISSN: 0165-0009 – mf#14743 – us UMI ProQuest [550]
Climatic data of the east coast of florida / Florida East Coast Railway. Land Dept – St Augustine, FL. 1912 – 1r – us UF Libraries [550]
Climatological records of the weather bureau, 1819-1892 / U.S. Weather Bureau – (= ser Records Of The Weather Bureau) – 564r – 1 – mf#T907 – us Nat Archives [550]
Climatologie, pedologie et ecologie forestieres au canada, 1937-1956 : bibliographie / Deslandes, Germain – 1958 [mf ed 1978] – (= ser Bibliographies du cours...1947-66) – 2mf – 9 – (with ind; pref by l-z rousseau) – mf#SEM105P4 – cn Bibl Nat [550]
Climatology of jacksonville : florida and vicinity / Davis, Thomas Frederick – Jacksonville, FL. 1908 – 1r – us UF Libraries [550]
The climax of protection and free trade, capped by annexation – [Montreal?: s.n.], 1849 [mf ed 1984] – 1mf – 9 – 0-665-22159-2 – mf#22159 – cn Canadiana [337]
Climbing and exploration in the karakoram-himalayas / Conway, W M – London, 1894 – 8mf – 9 – mf#HT-34 – ne IDC [915]
Climbing plants / Watson, William – New York, NY. 1920? – 1r – us UF Libraries [580]
Climo, V C *see* Precis of information concerning the colony of the gold coast and ashanti
Climstein, M *see* Myocardial structure and function differences between steroid using and non-steroid using elite powerlifters and endurance athletes
Clinch valley news – Tazewell, VA. 1886-2000 (1) – mf#66892 – us UMI ProQuest [071]
Clinch valley times – Saint Paul, VA. 1988-2000 (1) – mf#68309 – us UMI ProQuest [071]
Clinica chimica acta – Amsterdam. 1956+ (1) 1956+ (5) 1987+ (9) – ISSN: 0009-8981 – mf#42044 – us UMI ProQuest [610]
Clinical allergy – Oxford. 1980-1988 (1) 1980-1988 (5) 1980-1988 (9) – (cont by: clinical and experimental allergy) – ISSN: 0009-9090 – mf#15512 – us UMI ProQuest [616]
Clinical and experimental allergy – Oxford. 1989+ (1,5,9) – (cont: clinical allergy) – ISSN: 0954-7894 – mf#15512,01 – us UMI ProQuest [616]
Clinical and experimental dermatology – Oxford. 1980-1996 (1,5,9) – ISSN: 0307-6938 – mf#15514 – us UMI ProQuest [616]
Clinical and experimental immunology – Oxford. 1980-1996 (1,5,9) – ISSN: 0009-9104 – mf#15515 – us UMI ProQuest [616]
Clinical and experimental pharmacology and physiology – Oxford. 1980-1993 (1) 1980-1993 (5) 1980-1993 (9) – ISSN: 0305-1870 – mf#15516 – us UMI ProQuest [615]
Clinical and genetic investigations into tuberous sclerosis and recklinghausen's neurofibromatosis : contribution to elucidation of interrelationship and eugenics of the syndromes / Borberg, Allan – Copenhagen: Munksgaard 1951 – (transl fr the danish by elisabeth aagesen) – us CRL [616]
Clinical and laboratory haematology – Oxford. 1979-1996 (1,5,9) – ISSN: 0141-9854 – mf#15517 – us UMI ProQuest [616]
Clinical biofeedback and health – Toronto. 1985-1987 (1,5,9) – (cont: american journal of clinical biofeedback. cont by: medical psychotherapy) – mf#11712,01 – us UMI ProQuest [616]
Clinical biomechanics – Kidlington. 1986+ (1,5,9) – ISSN: 0268-0033 – mf#17216 – us UMI ProQuest [615]
Clinical bulletin / Memorial Sloan-Kettering Cancer Center – New York. 1979-1981 (1) 1979-1981 (5) 1979-1981 (9) – ISSN: 0047-6706 – mf#12337 – us UMI ProQuest [616]
Clinical bulletin of myofascial therapy : the practical journal for the soft-tissue practitioner / ed by Lowe, John C – v2 n1. 1997- – 1,9 – $105.00 in US $147.00 outside hardcopy subsc – us Haworth [615]
Clinical chemistry – Washington. 1955+ (1) 1955+ (5) 1955+ (9) – ISSN: 0009-9147 – mf#12411 – us UMI ProQuest [540]
Clinical chemistry news – Washington. 1991-1994 (1) – (cont by: clinical laboratory news) – ISSN: 0161-9640 – mf#12412 – us UMI ProQuest [574]
Clinical diabetes – New York. 1989-1996 (1,5,9) – ISSN: 0891-8929 – mf#16048 – us UMI ProQuest [616]
Clinical electroencephalography – Wheaton. 1973+ (1,5,9) – ISSN: 0009-9155 – mf#10201 – us UMI ProQuest [616]
Clinical endocrinology – Oxford. 1980+ (1,5,9) – ISSN: 0300-0664 – mf#15513 – us UMI ProQuest [616]
Clinical eye and vision care – New York. 1988-1995 (1,5,9) – ISSN: 0953-4431 – mf#17130 – us UMI ProQuest [617]
Clinical, functional, and radiographic assessment of the conventional and modified boyd-anderson surgical procedures for repair of distal biceps tendon ruptures : a 3-year follow-up study / d'Arco, Patrick H – Temple University, 1996 – 2mf – 9 – $8.00 – mf#PE 3635 – us Kinesology [617]
Clinical gerontologist / ed by Brink, T L – v1-1982- – 1, 9 ($300.00 in US $420.00 outside hardcopy subsc) – us Haworth [618]
Clinical hemorheology – New York. 1981-1994 (1) 1981-1994 (5) 1984-1994 (9) – ISSN: 0271-5198 – mf#49391 – us UMI ProQuest [616]
Clinical imaging – New York. 1989+ (1,5,9) – (cont: journal of computed tomography) – ISSN: 0899-7071 – mf#42411,01 – us UMI ProQuest [616]
Clinical infectious diseases – Chicago. 1992+(1,5,9) – (cont: reviews of infectious diseases) – ISSN: 1058-4838 – mf#12197,01 – us UMI ProQuest [616]
Clinical instruction in athletic training / Gardner, Gregory A – University of Southern Mississippi, 1995 – 2mf – 9 – $8.00 – mf#PE3590 – us Kinesology [617]
Clinical journal of pain – New York. 1993+ (1,5,9) – ISSN: 0749-8047 – mf#18701 – us UMI ProQuest [616]
Clinical journal of sport medicine – New York. 1993+ (1,5,9) – ISSN: 1050-642X – mf#18702 – us UMI ProQuest [617]
Clinical kinesiology – San Diego. 1988+ (1,5,9) – (cont: american corrective therapy journal) – ISSN: 0896-9620 – mf#3189,01 – us UMI ProQuest [616]
Clinical laboratory news – Washington. 1994-1996 (1) – (cont: clinical chemistry news) – mf#12412,01 – us UMI ProQuest [574]
Clinical laboratory science – Bethesda. 1988+ (1,5,9) – (cont: journal of medical technology: official publication of american medical technologists and american society for medical technology) – ISSN: 0894-959X – mf#16303 – us UMI ProQuest [619]
Clinical lecture on the surgical treatment of perforated gastric ulcer : delivered at the montreal general hospital on the 6th of november, 1895 / Armstrong, George E – S.l: s.n, 1896? – 1mf – 9 – mf#39093 – cn Canadiana [617]

Clinical linguistics and phonetics – London. 1993-1996 (1) – ISSN: 0269-9206 – mf#17296 – us UMI ProQuest [616]
Clinical management – Alexandria. 1990-1992 (1,5,9) – (cont: clinical management in physical therapy) – mf#14143,01 – us UMI ProQuest [610]
Clinical management in physical therapy – Alexandria. 1989-1989 (1) – (cont by: clinical management) – ISSN: 0276-8038 – mf#14143 – us UMI ProQuest [610]
Clinical materials – London. 1990-1994 (1,5,9) – ISSN: 0267-6605 – mf#42464 – us UMI ProQuest [610]
Clinical medicine – Northfield. 1895-1978 (1) 1975-1978 (5) 1975-1978 (9) – ISSN: 0412-7994 – mf#2423 – us UMI ProQuest [610]
Clinical microbiology reviews – Washington. 1988+ (1,5,9) – ISSN: 0893-8512 – mf#16471 – us UMI ProQuest [576]
Clinical neurophysiology – Limerick. 1999+ (1,5,9) – (cont: electroencephalography and clinical neurophysiology) – ISSN: 1388-2457 – mf#42253,01 – us UMI ProQuest [616]
Clinical neuropsychology – Stoughton. 1979-1983 (1,5,9) – (cont by: international journal of clinical neuropsychology) – ISSN: 0197-3681 – mf#11955 – us UMI ProQuest [616]
Clinical neuroscience research – Oxford, 2001+ [1,5,9] – ISSN: 1566-2772 – mf#42832 – us UMI ProQuest [612]
Clinical notes on respiratory diseases – New York. 1962-1983 (1) 1970-1983 (5) 1977-1983 (9) – ISSN: 0009-9198 – mf#2426 – us UMI ProQuest [616]
Clinical nuclear medicine – Philadelphia. 1976+ (1,5,9) – ISSN: 0363-9762 – mf#11383 – us UMI ProQuest [616]
Clinical nurse specialist : the journal for advanced nursing practice – v1-10. 1987-1996 – 1,5,6,9 – $65.00r – us Lippincott [610]
Clinical nursing research – Thousand Oaks. 1996+ (1,5,9) – ISSN: 1054-7738 – mf#21664 – us UMI ProQuest [610]
Clinical nutrition : official journal of the european society of parenteral and enteral nutrition / European Society of Parenteral and Enteral Nutrition – Edinburgh. 1982+ (1,5,9) – ISSN: 0261-5614 – mf#13425 – us UMI ProQuest [613]
Clinical obstetrics and gynecology – Philadelphia. 1958+ (1) 1973+ (5) 1975+ (9) – ISSN: 0009-9201 – mf#8779 – us UMI ProQuest [618]
Clinical orthopaedics and related research – Philadelphia. 1971+ (1) 1971+ (5) 1975+ (9) – ISSN: 0009-921X – mf#6890 – us UMI ProQuest [617]
Clinical otolaryngology – Oxford. 1980-1994 (1) 1980-1994 (5) 1980-1994 (9) – ISSN: 0307-7772 – mf#15518 – us UMI ProQuest [617]
Clinical pediatrics – Glen Head. 1962+ (1) 1966+ (5) 1970+ (9) – ISSN: 0009-9228 – mf#1569 – us UMI ProQuest [618]
Clinical pharmacology and therapeutics – St. Louis. 1960+ (1) 1965+ (5) 1970+ (9) – ISSN: 0009-9236 – mf#1885 – us UMI ProQuest [615]
Clinical pharmacy – Bethesda. 1990-1993 (1,5,9) – ISSN: 0278-2677 – mf#18313 – us UMI ProQuest [615]
Clinical physiology – Oxford. 1981-1996 (1,5,9) – ISSN: 0144-5979 – mf#15519 – us UMI ProQuest [612]
Clinical physiology and functional imaging – Oxford. 2002+ (1,5,9) – ISSN: 1475-0961 – mf#15519,01 – us UMI ProQuest [610]
Clinical preventive dentistry – Philadelphia. 1979-1992 (1) 1979-1992 (5) 1979-1992 (9) – ISSN: 0163-9633 – mf#11971 – us UMI ProQuest [617]
Clinical proceedings / Children's Hospital National Medical Center – Washington. 1944-1984 (1) 1971-1984 (5) 1974-1984 (9) – ISSN: 0092-7813 – mf#2322 – us UMI ProQuest [618]
Clinical psychiatry news – New York. 1989-1996 (1) – ISSN: 0270-6644 – mf#12975 – us UMI ProQuest [616]
Clinical psychologist – Philadelphia. 1947-1987 (1) 1971-1987 (5) 1975-1987 (9) – ISSN: 0009-9244 – mf#6388 – us UMI ProQuest [616]
The clinical psychologist in the department of health = Kliniese sielkundige in die Departement van Gesondheid / South Africa. Department of Health [Departement van Gesondheid] – Pretoria: Dept of Health [1976?] [mf ed Pretoria, RSA: State Library [199-]] – 8p [ill] on 1r with other items – 5 – mf#op 06641 r24 – us CRL [362]
Clinical psychology review – New York. 1981+ (1,5,9) – ISSN: 0272-7358 – mf#49382 – us UMI ProQuest [150]
Clinical psychology: science and practice – New York. 1994-1996 (1,5,9) – ISSN: 0969-5893 – mf#20907 – us UMI ProQuest [150]
Clinical radiology – Oxford. 1970+ [1]; 1949+ [5]; 1975+ [9] – ISSN: 0009-9260 – mf#5959 – us UMI ProQuest [616]
Clinical reproduction and fertility – Oxford. 1982-1987 (1) 1982-1987 (5) 1982-1987 (9) – (cont by: reproduction, fertility, and development) – ISSN: 0725-556X – mf#15520 – us UMI ProQuest [618]
Clinical respiratory physiology – Oxford. 1965-1982 (1,5,9) – ISSN: 0272-7587 – mf#49288 – us UMI ProQuest [612]
Clinical social work journal – New York. 1973+ (1) 1973+ (5) 1975+ (9) – ISSN: 0091-1674 – mf#11175 – us UMI ProQuest [360]
The clinical supervisor : a journal of supervision in psychotherapy and mental health / ed by Munson, Carlton – v1- 1983- – 1, 9 ($275.00 in US $385.00 outside hardcopy subsc) – us Haworth [610]
Clinical techniques in small animal practice – Orlando, 1998+ [1,5,9] – ISSN: 1096-2867 – mf#21099,01 – us UMI ProQuest [636]
Clinical toxicology – New York. 1968-1981 (1) 1968-1981 (5) 1968-1981 (9) – (cont by: journal of toxicology clinical toxicology) – ISSN: 0009-9309 – mf#12924 – us UMI ProQuest [615]
Clinical vision sciences – Oxford. 1987-1992 (1,5,9) – ISSN: 0887-6169 – mf#49498 – us UMI ProQuest [617]
Clinician's letter / California Urban Indian Health Council – 1981 jan 15, apr, aug, dec, 1982 feb, jun, sep, dec, 1983 mar, jun, sep, 1984 mar, jun, nov, 1985:winter, 1986 feb 28, May 21, aug 20, 1987 mar, sep 3 attch a-g, nov 19, 1988 may – 1r – 1 – mf#1054567 – us WHS [360]
Clinics in anaesthesiology – London. 1983-1986 (1) 1983-1986 (5) 1983-1986 (9) – ISSN: 0261-9881 – mf#13377 – us UMI ProQuest [617]
Clinics in chest medicine – Philadelphia. 1980+ (1,5,9) – ISSN: 0272-5231 – mf#12717 – us UMI ProQuest [616]
Clinics in communication disorders – Reading. 1991-1994 (1,5,9) – ISSN: 1054-8505 – mf#18969 – us UMI ProQuest [616]
Clinics in dermatology – Philadelphia. 1992-1994 (1,5,9) – ISSN: 0738-081X – mf#42666 – us UMI ProQuest [616]
Clinics in endocrinology and metabolism – Philadelphia. 1972-1986 (1) 1972-1986 (5) 1972-1986 (9) – ISSN: 0300-595X – mf#12718 – us UMI ProQuest [616]
Clinics in gastroenterology – Philadelphia. 1972-1986 (1) 1972-1986 (5) 1972-1986 (9) – ISSN: 0300-5089 – mf#12719 – us UMI ProQuest [616]
Clinics in geriatric medicine – Philadelphia. 1985+ (1,5,9) – ISSN: 0749-0690 – mf#14730 – us UMI ProQuest [618]
Clinics in haematology – London. 1972-1986 (1) 1972-1986 (5) 1972-1986 (9) – ISSN: 0308-2261 – mf#12720 – us UMI ProQuest [616]
Clinics in immunology and allergy – Philadelphia. 1981-1986 (1) 1981-1986 (5) 1981-1986 (9) – (cont by: immunology and allergy clinics of north america) – ISSN: 0260-4639 – mf#12721 – us UMI ProQuest [616]
Clinics in laboratory medicine – Philadelphia. 1981+ (1,5,9) – ISSN: 0272-2712 – mf#13378 – us UMI ProQuest [619]
Clinics in obstetrics and gynaecology – Philadelphia. 1980-1986 (1) 1980-1986 (5) 1980-1986 (9) – ISSN: 0306-3356 – mf#12722 – us UMI ProQuest [618]
Clinics in oncology – Philadelphia. 1982-1986 (1) 1982-1986 (5) 1982-1986 (9) – ISSN: 0261-9873 – mf#13520 – us UMI ProQuest [616]
Clinics in perinatology – Philadelphia. 1977+ (1,5,9) – ISSN: 0095-5108 – mf#11460 – us UMI ProQuest [618]
Clinics in plastic surgery – Philadelphia. 1974+ (1,5,9) – ISSN: 0094-1298 – mf#11461 – us UMI ProQuest [617]
Clinics in podiatric medicine and surgery – Philadelphia. 1986+ (1,5,9) – ISSN: 0891-8422 – mf#13521,01 – us UMI ProQuest [617]
Clinics in rheumatic diseases – Philadelphia. 1975-1986 (1,5,9) – ISSN: 0307-742X – mf#12723 – us UMI ProQuest [616]
Clinics in sports medicine – Philadelphia. 1982+ (1,5,9) – ISSN: 0278-5919 – mf#13379 – us UMI ProQuest [617]
Clint, Harold Cuthbert *see* Colonel william wood, soldier, historian, archivist
Clinton 1850-1900 – Oxford, MA (mf ed 1991) – (= ser Massachusetts vital records) – 54mf – 9 – 0-87623-138-5 – (mf 1-5: births 1850-81. mf 6-10: births 1882-98. mf 11-15: marriages 1850-83. mf 16-20: marriages 1884-1900. mf 21-25: deaths 1850-92. mf 26-30: death 1893-1903. mf 31-39: birth index 1850-1990. mf 40-47: marriage index 1850-1990. mf 48-54: death index 1850-1990) – us Archive [978]
Clinton Co. Blanchester *see* Star-republican
Clinton Co. New Vienna *see* Reporter
Clinton Co. Wilmington *see*
– Clinton county democrat
– Clinton republican
– Herald of freedom
– Independent
– Journal series
– Journal-republican
– Watchman
– Weekly empyrean
Clinton county advertiser – Lyons, IA. 1874-1881 (1) – mf#63297 – us UMI ProQuest [071]
Clinton county democrat / Clinton Co. Wilmington – (may 1880-1905, 1919-dec 1928) [wkly] – 11r – 1 – mf#B31214-31224 – us Ohio Hist [071]
Clinton county democrat – Wilmington, OH. 1888-1950 (1) – mf#65716 – us UMI ProQuest [071]
Clinton County Farmers Union [IA] *see* County farmer
Clinton County Grammar School (Ont). Board of Trustees *see* The grammar school system of ontario
Clinton County Historical Association [NY] *see* North country notes
Clinton county historical society quarterly – 1978-87 – 1r – 1 – mf#1277589 – us WHS [978]
Clinton county times – Lock Haven, PA. 1903-1968 (1) – mf#68958 – us UMI ProQuest [071]
Clinton democrat – Lock Haven, PA. 1868-1923 (1) – mf#68929 – us UMI ProQuest [071]
The clinton democrat – Lock Haven, PA. -w 1889-1912; 1918-1923 – 13 – $25.00r – us IMR [071]
Clinton first baptist church. clinton, south carolina : church records – 1881-1966 – 1 – us Southern Baptist [242]
Clinton, Henry Lauren *see* Extraordinary cases
Clinton herald – Clinton WI. 1942 may 7-dec 31 – 1r – 1 – mf#963550 – us WHS [071]
Clinton herald – Clinton WI. 1880 aug 12/1881 jan 26-1905 sep 19/1907 oct 1 – 13r – 1 – (with gaps; cont: weekly herald [clinton wi]) – mf#963565 – us WHS [071]
Clinton independent – Clinton WI. 1875 may 19-1877 jun 27, 1877 jul 4-1880 feb 25 – 2r – 1 – (cont: independent [clinton wi]; cont by: rock county republican [clinton wi]) – mf#963573 – us WHS [071]
Clinton, Iris *see* Hope foundation story
Clinton labor congress – 1944 sep 4 – 1r – 1 – (cont by: clinton labor review) – mf#3925600 – us WHS [331]
Clinton labor day news / Tri-City Labor Congress [Clinton IA] – 1938 sep 5 – 1r – 1 – mf#3925623 – us WHS [331]
Clinton labor review – 1945 sep 3 – 1r – 1 – (cont: clinton labor congress) – mf#3925605 – us WHS [331]
Clinton line railroad company records, 1852-1859 – [mf ed 194?] – 1r – 1 – mf#ms463 – us Western Res [380]
Clinton new era – Ontario, CN. apr 1894-dec 1895 – 13r – 1 – cn Commonwealth Imaging [071]
Clinton republican / Clinton Co. Wilmington – (jan 1882-dec 1906) [wkly] – 10r – 1 – mf#B30954-30963 – us Ohio Hist [071]
Clinton republican / Clinton Co. Wilmington – jan 1907-dec 1912 [wkly] – 3r – 1 – mf#B31211-31213 – us Ohio Hist [071]
Clinton republican / Clinton Co. Wilmington – jan 3-dec 26, 1862 [wkly] – 1r – 1 – mf#B31405 – us Ohio Hist [071]
Clinton republican – Lock Haven, PA. 1863-1923 (1) – mf#68957 – us UMI ProQuest [071]
Clinton republican – Wilmington, OH. 1846-1892 (1) – mf#65721 – us UMI ProQuest [071]
The clinton republican – Lock Haven, PA. -w 1889-1912 – 13 – $25.00r – us IMR [071]
Clinton street quarterly – [Portland/Eugene ed]: CSQ, [qrterly] – 1 – (1st publ spring 1979. regional qrterly featuring humor, culture, fiction, artwork and comm by northwest authors & artists) – us Oregon Lib [071]
Clinton times observer – Clinton WI. 1923 sep 7/1925 aug 28-1938 feb 2/1941 jan 16 – 10r – 1 – (with gaps) – mf#965738 – us WHS [071]
Clinton topper – Clinton WI. 1938 apr 28/1942-1997 – 37r – 1 – mf#1001664 – us WHS [071]
Clinton/huron news – Ontario, CN. jan 1874-dec 1911 – 29r – 1 – cn Commonwealth Imaging [071]
Clintonville gazette – Clintonville WI. 1919 jun 5-1920 sep 3, 1920 sep 9-1921 nov 10, 1921 nov 17-1923 feb 1, 1923 feb 8-22 – 4r – 1 – (cont by: dairyman-gazette) – mf#966866 – us WHS [071]
Clintonville herald – Clintonville WI. 1879 mar 14-oct 10 – 1r – 1 – mf#963576 – us WHS [071]
Clintonville tribune – Clintonville, Marion WI. 1885 jul 11-dec 26, 1886 jan-1888 apr 26 – 2r – 1 – (cont: tribune [clintonville wi]; cont by: dual-city tribune) – mf#1009203 – us WHS [071]
Clintonville tribune – Clintonville WI. 1891 mar 20/dec-1938/1940 jul 21 – 20r – 1 – (with gaps; cont: dual-city tribune; cont by: dairyman-gazette; clintonville tribune-gazette) – mf#1009207 – us WHS [071]
Clintonville tribune-gazette – Clintonville WI. 1940 jul 18/dec-2002 jan/jun – 81r – 1 – (with gaps; cont: clintonville tribune [clintonville wi: 1891]; dairyman-gazette) – mf#983634 – us WHS [071]
Clio – Fort Wayne. 1971+ (1) 1971+ (5) 1972+ (9) – ISSN: 0884-2043 – mf#6865 – us UMI ProQuest [400]
Clio *see* Canadian poets in miniature
Clionian Debating Society. Charleston, South Carolina *see* Proceedings of the clionian debating society
Clip sheet / American Game Protective Association – 1926 aug 1, 1927 feb 1, jul 1, sep 1-nov 1, 1928 jan 1, jun 1-aug 1, oct 1-nov 1, 1929 jan 1-aug 1, nov 1-dec 1, 1930 jan 1-nov 1 – 1r – 1 – mf#715023 – us WHS [360]
Clipboard / American Appraisal Co – 1962 mar-1970 mar/apr – 1r – 1 – (cont: clipboard [milwaukee wi: 1956]; cont by: american appraisal news briefs) – mf#1337552 – us WHS [071]
Clipper – v1-2 n2,9. 1940-41 [all publ] – (= ser Radical periodicals in the united states, 1881-1960. series 1) – 5mf – 9 – $85.00 – us UPA [073]
The clipper – New York. v1-72 n3. may 7 1853-july 12 1924 – 1 – (lack: jan 1855-apr 19 1856 and few scattered issues) – us NY Public [073]
Clipper-citizen – Lexington, NE: Holmes & Wickizer. -v20 n526. mar 31 1922 (wkly) [mf ed 1893-1922 (gaps) filmed 1972-79] – 12r – 1 – (formed by the union of: lexington clipper and: cozad citizen. absorbed: lexington gazette (1893) dawson county enterprise (1897) and: lexington news. cont by: lexington clipper (1922)) – us NE Hist [071]
Clipper-herald – Lexington, NE: Western Pub Co. 102nd yr n15. dec 4 1991- (semiwkly) [mf ed filmed 1992-] – 1 – (formed by the union of: lexington clipper (1983) and: dawson county herald (1938)) – us NE Hist [071]
[Clippings and handbills by or about isabelo de los reyes...] : from his personal notebook... university of the philippines library – [mf ed 1985] – 1r – 1 – mf#6583 – us Wisconsin U Libr [959]
[Clippings from the union's spanish newspaper...] : los obreros, from the issues of feb. 20, 24, 26 and may 2-3,9, 1903 / Union Obrera Democratica – Manila: Union Obrera Democratica de Filipinas, 1903 [mf ed 1985] – (= ser Philippine labor publications 1/7) – 47p – 1 – mf#6580 reel 1 n7 – us Wisconsin U Libr [079]
Clippings scrapbooks / Northern Pacific Railway Company – 1866-1904. 10 rolls including filmed inventory – 1 – $300.00; $30.00r – us Minn Hist [380]
Clitheroe times – [NW England] Clitheroe, Whalley Lib 7 dec 1888-1895, 1898-1920 – 1 – (title change: clitheroe advertiser & times [1924, 1927-30, 1935, 1961-62, 1974-75, 1980-jun 1985]) – uk MLA; uk Newsplan [072]
Clive forrester's gold / Kenyon, Charles Richard – Toronto: Musson [1904?] [mf ed 1996] – 3mf – 9 – 0-665-80963-8 – mf#80963 – cn Canadiana [830]
The cloak room and weekly letter / Lambertson, William Purnell – 1930-1950, Lambertson communicated with his constituents through a weekly newspaper column called "The Cloak Room" or "The Crossroads" (used between sessions) and a weekly letter to certain supporters. The microfilm consists of typed drafts but not the acutal newspaper columns – 1 – us Kansas [978]
Cloarec-Heiss, France *see* Banda-linda de ippy
La cloche – Paris. 19 dec 1869-17 dec 1870, 6 fevr 1871-21 dec 1872 – 1 – (devenu: l' etat. 22 dec 1872-13 mai 1873.) – fr ACRPP [073]
La cloche – Paris: Imp. Dubuisson et Ce, mar 26-apr 9, apr 11-16, 18-19 1871-1872] – (issues at mf-7496 and neg. mf-1751 filmed as part of: commune de paris newspapers; newspapers on these reels are filmed chronologically, not alphabetically.) – us CRL [074]
La cloche – Paris. n1-52. avr 1892-nov 1894 – 1 – (journal litteraire, artistique, satirique et humoristique. mq n21, 27, 40-43, 50) – fr ACRPP [073]
La cloche – Tamatave. oct 1880-juil 1892 – 1 – (replaced by: the bee. d'oct a dec 1880) – fr ACRPP [073]
La cloche felee – Saigon. dec 1923-mai 1926 – 1 – fr ACRPP [073]
Le clocher – [Paris]: A Desrez, apr 8 1849- – us CRL [074]
Les cloches de plurs = Die glocken von plurs / Pasque, Ernst – Geneve: H Robert, 1901 – 1r – 1 – us Wisconsin U Libr [830]

The clockmaker : or, the sayings and doings of samuel slick of slickville / Haliburton, Thomas Chandler – New York: W H Colyer, 1840 – 2mf – 9 – mf#48209 – cn Canadiana [880]
Clockmakers Company *see* Papers of the clockmakers' company
Clodd, Edward *see*
– Animism
– Myths and dreams
– Pioneers of evolution from thales to huxley
– The story of creation
– The story of the alphabet
Clodius, Christian August *see* Versuche aus der literatur und moral
Clodore, A M de *see* Souvenirs d'un voyageur en asie, depuis 1802 jusqu'en 1815 inclusivement
Cloet, Jean J de *see* Geographie historique, physique et statistique du royaume des pays-bas et de ses colonies
Cloete, Henry *see* The history of the great boer trek and the origins of south african republics
Cloete, Rehna *see* Nylon safari
Cloete, Stuart *see*
– African giant
– African portraits
– Against these three
– Fiercest heart
– Mask
– Rags of glory
– Soldiers' peaches
– Watch for the dawn
Cloister life in the days of coeur de lion / Spence-Jones, Henry Donald Maurice – London: Isbister; Philadelphia: JB Lippincott, 1892 – 1mf – 9 – 0-7905-7199-4 – mf#1988-3199 – us ATLA [240]
The cloisters of monreale in sicily : thirty photographs / Arundel Society, London – London 1870 – (= ser 19th c art & architecture) – 2mf – 9 – mf#4.2.1433 – uk Chadwyck [720]
Clokey, Joseph Waddell *see* David's harp in song and story
Cloninger, Karl W *see* Off season resident camp utilization in the contignous united states
Clonmel advertiser – Clonmel, Ireland. -w. Jan 1828-7 apr 1838 – 10r – 1 – uk British Libr Newspaper [072]
Clonmel advertiser and literary journal – Clonmel, Ireland. 8 jul-16 sep 1843 – 1/4r – 1 – uk British Libr Newspaper [072]
Clonmel chronicle – Tipperary 1897-1913, 1922-23, 1926-mar 1935 – 29r – 1 – ie National [072]
Clonmel chronicle etc – Clonmel, Ireland. 21 jul 1848-1896 – 48r – 1 – uk British Libr Newspaper [072]
Clonmel gazette : or, hibernian advertiser – Tipperary 1788-95 – 2.5r – 1 – ie National [072]
Clonmel gazette and munster mercury – Tipperary 1802-04 – 1r – 1 – ie National [072]
Clonmel herald – Clonmel. Ireland. 1828-1840 – 13r – 1 – uk British Libr Newspaper [072]
Clonmel people – Clonmel 25 nov 1944 – 0.25r – 1 – ie National [072]
Clophill – (= ser Bedfordshire parish register series) – 2mf – 9 – £5.00 – uk BedsFHS [929]
Cloran, Henry Joseph *see* Report on the jury system
Y clorianydd – [Wales] Ynys Mon ion 1899-16 meh 1904, ion 1909-rhag 1950 [mf ed 2004] – 37r – 1 – (cont by: y clorianydd a'r gwalia) – uk Newsplan [072]
Y clorianydd – Bangor, Wales 13 aug 1891-30 apr 1969 [mf 1897, 1912] – 1 – (wanting: 1911; discontinued) – uk British Libr Newspaper [072]
"Close communion" : the english and greek of it / Adams, Henry – Yarmouth, NS?: C Carey, 1888 – 1mf – 9 – mf#07193 – cn Canadiana [240]
Close, F *see*
– Justification of the charges brought against the british and foreig...
– Mystery of iniquity
– Restoration of churches is the restoration of popery
Close, Francis *see*
– Female chartists' visit to the parish church
– The restoration of churches is the restoration of popery
Close of sermon preached in free north church, stirling,30th september... / Buchanan, Robert – Stirling? Scotland. 1866? – 1r – us UF Libraries [240]
The close of the middle ages, 1273-1494 / Lodge, Richard – 4th ed. London: Rivingtons, 1915 [mf ed 1992] – (= ser Periods of european history 3) – 2mf – 9 – 0-524-02915-6 – (incl bibl ref) – mf#1990-0731 – us ATLA [931]
Close of the tenth century of the christian era / Dixon, Richard Watson – Oxford: T & G Shrimpton, 1858 – 1mf – 9 – 0-7905-7224-9 – (incl bibl ref) – mf#1988-3224 – us ATLA [240]
Close, Percy L *see* A prisoner of the germans in south-west africa
Close up – London. v. 1-10. july 1927-dec 1933 – 1 – us NY Public [073]
Closener, Fritsche *see* Strassburgische chronik
Closer association of the british west indian colonies / Great Britain Colonial Office – London, England. 1947 – 1r – us UF Libraries [972]
The closer walk : or, the believer's sanctification / Darling, Henry – Philadelphia: J B Lippincott, 1863, c1862 – 1mf – 9 – 0-8370-3249-0 – mf#1985-1249 – us ATLA [240]
Closer-ups / Marah, Inc – 1964 jul-1966 mar – 1r – 1 – mf#1054570 – us WHS [071]
Close-up – Cambridge. 1970-1977 (1) 1975-1977 (5) 1975-1977 (9) – mf#10548 – us UMI ProQuest [370]
Closeup : the howard university magazine – 1970 sum – 1r – 1 – (cont: howard university magazine) – mf#4848518 – us WHS [378]
Closing address / Gloag, Paton James – Edinburgh, Scotland. 1889? – 1r – us UF Libraries [240]
Closing century / Macleod, Norman – Inverness, Scotland. 1900 – 1r – us UF Libraries [240]
Closing scenes of the present dispensation and the manifestation of... – London, England. 1846 – 1r – us UF Libraries [240]
Closing the gaps in florida's wildlife habitat conservation / Florida Game And Fresh Water Fish Commission – Tallahassee, FL. 1994 – 1r – us UF Libraries [639]
Closs, August *see*
– Die freien rhythmen in der deutschen lyrik
– Medusa's mirror
Closure as reflected in northern sotho narratives / Makgopa, Mokgale Albert – Uni of South Africa 2000 [mf ed Johannesburg 2000] – 4mf – 9 – (incl bibl ref) – mf#mfm14774 – sa Unisa [470]
Clothes / Hunter, C M – s.l, s.l? 1936 – 1r – us UF Libraries [680]
Clotilde, duchessa di salerno : ballo tragicomico del fu salvatore vigano, posto in iscena dal di lui fratello signor giulio. da rappresentarsi nell' i r teatro alla canobbiana l'autunno dell'anno 1825 / Vigano, Salvatore – Milano: N Bettoni, 1825 – 1 – mf#*ZBD-*MGTZ pv1-Res – Located: NYPL – us Misc Inst [790]
Clotilde tejidor / Macau, Miguel Angel – Habana, Cuba. 1958 – 1r – us UF Libraries [972]
Clotilde-Angele de Jesus, mere *see* Bibliographie analytique de l'oeuvre de beraud de saint maurice
Clotten, Francis Egon *see* England and south africa
Cloture de l'annee academique 1879-1880 / Universite Laval a Montreal – Montreal: Chapleau & Lavigne, 1880 – 1mf – 9 – mf#35790 – cn Canadiana [378]
Cloture de l'annee academique 1880-1881 : seance de cloture du 30 juin 1881 / Universite Laval a Montreal – Montreal: E Senecal, 1881 – 1mf – 9 – mf#35792 – cn Canadiana [378]
Cloture de l'annee academique 1881-1882 / Universite Laval a Montreal – Montreal: E Senecal, 1883 – 1mf – 9 – mf#35793 – cn Canadiana [378]
Clou / Royer, Alphonse – Paris, France. 1835 – 1r – us UF Libraries [440]
Cloud family journal [cfj] – v7-v11 [1984/85-1988/89] – 1r – 1 – (cont: cloud family newsletter) – mf#1619286 – us WHS [929]
Cloud family newsletter – v1 n1/2 [1978 oct 1], v1 n3-4 [1979 jan 1-apr 1]v2 n1-v4 n4 [1979/1980-1981/82], v5-6 [1982/83-1983/84], v1-2 [1978/79-1979/80], v5-6 [1982/83-1983/84] – 1r – 1 – (cont by: cloud family journal) – mf#1619277 – us WHS [929]
Cloud, Frederick D *see* Hangchow, the "city of heaven"
A cloud of faithfull witnesses : leading to the heavenly canaan... / Perkins, W – London: Humphrey Townes, 1608 – 11mf – 9 – mf#PW-77 – ne IDC [240]
Cloud of unknowing and book of privy counselling / ed by Hodgson, Ph – London, 1944 – €18.00 – ne Slangenburg [240]
A cloud of witnesses : containing selections from the writings of poets and other literary and celebrated persons, expressive of the universal triumph of good over evil / ed by Hanson, John Wesley – Chicago: Star & Covenant Office, 1880 [mf ed 1992] – (= ser Unitarian/universalist coll) – 1mf – 9 – 0-524-03950-X – mf#1991-2004 – us ATLA [243]
A cloud of witnesses for the royal prerogatives of jesus christ : or, the last speeches and testimonies of those who have suffered for the truth in scotland since the year 1680 / M'Main, John – [ill ed] Edinburgh: Schenck & McFarlane, [1871?] [mf ed 1992] – 2mf – 9 – 0-524-03352-8 – (with int. pref dated 1871. originally publ in 1714) – mf#1990-0933 – us ATLA [242]
The cloud-messenger : an indian love lyric / Kalidasa – London: John Murray, 1930 – (= ser Samp: indian books) – (trans fr original sanskrit of kalidasa by charles king) – us CRL [810]
Clouds on the horizon : an essay on the various forms of belief, which stand in the way of the acceptance of real christian faith by the educated natives of asia, africa, america, and oceania / Cust, Robert Needham – 3rd enl ed. Hertford: S Austin, 1904 – 2mf – 9 – 0-7905-7216-8 – (incl bibl ref) – mf#1988-3216 – us ATLA [230]
Clough, Arthur Hugh *see* Nineteenth century literary manuscripts
Clough, Benjamin *see* A dictionary of the english and singhalese, and singhalese and english languages
Clough, James Cresswell *see* On the existence of mixed languages
Clough, John Everett *see*
– From darkness to light: the story of a telugu convert
– Social christianity in the orient
Cloughly, Alfred *see* Mechanics' lien law of the state of new jersey and builders' guide.
Clouston, T S *see* Female education from a medical point of view
Clouston, Thomas Smith *see* Unsoundness of mind
Clouston, William Alexander *see* Arabian poetry for english readers
"Cloven" hoof – v10 n6 iss 76 [1978 nov/dec] a.s. 13; v11 n4-v21 n3; iss 80-125 [1979 jul/aug-88 may/jun] a.s. 14-23 – 1 – mf#1609442 – us WHS [071]
Clover creek baptist church. medon, tennessee : church records – Sep 1848-99, 1937-67 – 1 – 49.32 – us Southern Baptist [242]
[Cloverdale-] cloverdale reveille – CA. sep 9 1880-sep 12 1896 (broken series); 1897-feb 16 1907; 1908-24 (broken series); 1925-26; 1928-– 50+ r – 1 – $3000.00 (subs $50/y) – (detailed ind available) – mf#B02115 – us Library Micro [071]
Cloverdale courier – Cloverdale OR: C E Trombley, [wkly] – 1r – 1 – us Oregon Lib [071]
Cloverdale sentinel – CA. jun 30 1965-mar 23 1966 (wkly) – 1r – 1 – $60.00 – mf#B02116 – us Library Micro [071]
Cloverland star – Cumberland Wl. 1909 apr 22 – 1r – 1 – mf#963563 – us WHS [071]
[Clovis-] clovis independent – CA. mar 12 1933-mar 1966; apr 1967-mar 1976 – 27r – 1 – $1620.00 – mf#BC02117 – us Library Micro [071]
[Clovis-] clovis tribune – CA. jan 1915-dec 1925 – 6r – 1 – $360.00 – mf#B02119 – us Library Micro [071]
[Clovis-] independent and tribune – CA. apr 2 1976- – 15+ r – 1 – $900.00 (subs $50/y) – mf#B02118 – us Library Micro [071]
Clow, William Maccallum *see*
– The bible reader's encyclopaedia and concordance
– Christ in the social order
Clowes, Francis *see*
– Importance of right views on baptism
Clowes, J *see*
– Combined duties of the citizen and of the christian considered
– Few plain answers to the question, why do you receive the testimony
Clowes, John *see* Christian temper
Clowes, William *see* Profitable and necessarie booke of observations
Clozel, Francois Joseph *see* Dix ans a la cote d'ivoire
Clu forum report / American Society of Chartered Life Underwriters – Bryn Mawr. 1980-1980 (1,5,9) – ISSN: 0066-0590 – mf#12509 – us UMI ProQuest [366]
Clu journal / American Society of Chartered Life Underwriters – Bryn Mawr. 1972-1983 (1) 1972-1983 (5) 1975-1983 (9) – (cont by: journal of the american society of clu) – ISSN: 0007-8573 – mf#7377 – us UMI ProQuest [366]
Club Alfaya *see* Estatutos
Club canadien de montreal : catalogue de la bibliotheque – [Montreal?: s.n.] 1883 [mf ed 1984] – 9 – 0-665-01117-2 – mf#01117 – cn Canadiana [020]
Club de raquettes de Levis *see* Constitution et reglements
Club de raquettes "L'etoile" de Sainte-Cunegonde *see* Reglements et constitution...fonde le 2 decembre 1885
Club de solteros : una guinolada en tres espantos / Arrivi, Francisco – Barcelona, Spain. 1962 – 1r – us UF Libraries [972]
Club de Tenis Cabeza-Rubia *see* Reglamento de regimen interior
Club Deportivo Cacereno *see* Reglamento dela sociedad...
Club der Filmindustrie *see* Geschaeftsbericht
Le club des 21 en 1879 *see* 20 ans apres, le club des 21 en 1879
Club Lacrosse Fraserville *see* Constitution, regles et reglements
Club life – London. jan 1899-dec 1902 [wkly] – 4r – 1 – uk British Libr Newspaper [073]
Club management – St. Louis. 1958+ (1) 1971+ (5) 1975+ (9) – ISSN: 0009-9589 – mf#2520 – us UMI ProQuest [366]
Club Nautico Lago Gabriel y Galan. Plasencia *see* 11th campeonato de espana de la clase de...y copa nacional juvenil 1974
Club Polideportivo de Tiro Puerto de los Castanos (Caceres) *see* Tiradas extraordinarias puntuables para el primer campeonato ruta de la plata. 1971
Club revolucionario juan bruno zayas / Lubian, Silvia – Santa Clara, Cuba. 1961 – 1r – us UF Libraries [972]
Club Taurino Emeritense *see* Memoria de actividades 1975-76
The club woman – v1-12 n2. 1897-1904 [all publ] – (= ser Periodicals on women and women's rights, series 2) – 1r – 1 – $210.00 – us UPA [305]
Club world – 1957 late win-1960 win – 1r – 1 – (cont by: club world newsmagazine) – mf#4852987 – us WHS [071]
Club-Cartier (Montreal, Quebec) *see* Constitution et reglements du club cartier
Clubdate – 1980 win-nov/dec – 1r – 1 – mf#5266243 – us WHS [071]
Club-room – Boston. 1820-1820 (1) – mf#3705 – us UMI ProQuest [420]
Clubs – s.l, s.l? 1936 – 1r – us UF Libraries [978]
Clue : a guide through greek to hebrew scripture / Abbott, Edwin Abbott – London: Adam and Charles Black; New York: Macmillan [distributor], 1900 – 1mf – 9 – 0-8370-9680-4 – (incl bibl ref) – mf#1986-3680 – us ATLA [220]
Clugston, W G *see* "Cat-wagon trails."
Clum, John Philip *see* A trip to the klondike through the stereoscope
Clune, Frank *see* To the isles of spice with frank clune
Cluniacensis, Petrus *see* De miraculis libri duo
Das cluniazensische totengedaechtniswesen (910-954) / Jorden, W – Muenster Westf, 1930 – 3mf – 8 – €7.00 – ne Slangenburg [241]
The cluster of spiritual songs, divine hymns and sacred poems / Mercer, Jesse – 3rd ed. Philadelphia. 1823 – 1 – us Southern Baptist [242]
Cluster trends : a report on nab college/industry relations cluster activities / National Alliance of Business – 1985 win – 1r – 1 – mf#4888694 – us WHS [338]
Clustering techniques / Breytenbach, Mariette – Uni of South Africa 2000 [mf ed Johannesburg 2000] – 4mf – 9 – (incl bibl ref) – mf#mfm15093 – sa Unisa [510]
Clute, O *see*
– Cassava, the velvet bean, prickly comfrey, taro, chinese yam, canaigre, alfalfa, flat pea, sachaline
– Pineapple at myers
Clutha leader – Balclutha, NZ. jul 1874-dec 1971 – 1 – mf#84.3 – nz Nat Libr [079]
Clutter, Marcia Oral *see* Me and pa in florida
Clutterbuck, George W *see* In india (the land of famine and of plague)
Clutton, Henry *see* Remarks...on the domestic architecture of france
Clutton-Brock, Alan Francis *see* Introduction to french painting
Clydach, glamorgan, parish churches of st john & st mary : baptisms 1845-1925, burials 1847-1927 – [Glamorgan]: GFHS [mf ed c2004] – 1mf – 9 – £1.25 – uk Glamorgan FHS [929]
Clyde bill of entry and shipping list : including grangemouth – [Scotland] Glasgow: J Harper dec 1841-sep 1939 (3times/wk) [mf ed 2003-04] – 50r – 1 – (cont as: clyde and forth bill of entry and shipping list [jul 1919-sep 1939]) – uk Newsplan; uk British Libr Newspaper [380]
Clyde commercial advertiser – [Scotland] Glasgow: printed by S Hunter & Co 3 dec 1806-26 dec 1810 (wkly) [mf ed 2003] – 279v on 2r – 1 – uk Newsplan [072]
Clyde commercial register and shipping list – [Scotland] Glasgow: S & T Dunn 9 oct 1845-7 feb 1846 (3iss/wk) [mf ed 2003] – 1r – 1 – uk Newsplan [380]
Clyde & forth bill of entry – Glasgow, Scotland 5 apr 1919-1 sep 1939 [mf 1915-39] – 1 – (cont: clyde bill of entry & shipping list [7 jul 1874-3 apr 1919]) – uk British Libr Newspaper [380]
Clyde Mitchell, J *see* The kalela dance
Clyde shipping advertiser and glasgow mercantile and shipping list – [Scotland] Glasgow: W C Cunningham 2 jan-31 dec 1844 (2iss/wk) [mf ed 2003] – 1r – 1 – uk Newsplan [380]
Clyde, Thomas *see* Doctrines of personal election

Clydebank leader – [Scotland] Clydebank: Clydebank Printing Works 11 nov 1905-3 feb 1911, jan 1914-25 jun 1915 (wkly) [mf ed 2003] – 6r – 1 – (cont by: clydebank & district leader) – uk Newsplan [072]
Clydebank post – Clydebank: Clydebank Post 1983- (wkly) [mf ed 5 jan 1995-] – 1 – (cont: clydebank press) – ISSN: 1356-8655 – uk Scotland NatLib [072]
Clydebank & refrew press – Scotland, UK. 15 Aug 1891-1913.-w. 11 reels – 1 – uk British Libr Newspaper [072]
Clydesdale catholic herald – [Scotland] Greenock: Scottish Catholic Printing Press...1 jan 1910-6 jan 1939 (wkly) [mf ed 2004] – 54r – 1 – (merged with: glasgow observer and catholic herald and: lanarkshire catholic herald to form: glasgow observer and scottish catholic herald; began in 1892) – uk Newsplan [241]
Clymer, Joseph Floyd *see* Indianapolis 500-mile race history
Clypeus theologiae thomisticae contra novos eius impugnatorus / Gonet, J B – Parisiis, 1669. 5v – 57mf – 9 – mf#CA-51 – ne IDC [240]
Clytemnestre et saul / Soumet, Alexandre – Gand, Belgium. 1822 – 1r – us UF Libraries [440]
Cm : a reviewing journal of canadian materials for young people 1981-90 / Canadian Library Association – Ottawa, ON: Canadian Library Association, 1981-94 – 6r – 1 – cn Library Assoc [020]
Cma : the management accounting magazine – Hamilton. 1985-1998 (1) 1985-1998 (5) 1985-1998 (9) – (cont: cost and management. cont by: cma management) – ISSN: 0831-3881 – mf#5716,01 – us UMI ProQuest [650]
Cma management – Hamilton. 1999+ (1) 1999+ (5) 1999+ (9) – (cont: cma: the management accounting magazine) – mf#5716,02 – us UMI ProQuest [650]
The cmba herald – Kingston [Ont]: The Association, [1891] – 9 – mf#P04041 – cn Canadiana [360]
Cmba journal and catholic society news – Montreal: Catholic Societies Pub Co, [1891-189- or 19–] – 9 – (incl some text in french) – ISSN: 1190-688X – mf#P04168 – cn Canadiana [241]
Cme news / University of Oregon – 1987:winter, spr-1989 spr [v10 n3-v12 n3] – 1r – 1 – (cont by: multicultural affairs issues) – mf#1551041 – us WHS [321]
Cmlea cmlea journal – Concord. 1979-1995 (1) 1979-1995 (5) 1979-1995 (9) – (cont by: journal – california school library association) – ISSN: 0196-3309 – mf#11980 – us UMI ProQuest [020]
Cmm – confectionery manufacture and marketing – London. 1964-1967 (1) – ISSN: 0007-8654 – mf#1342 – us UMI ProQuest [660]
Cms, new name or same old game? : hearing...house of representatives, 107th congress, 2nd session, washington dc, may 16 2002 / United States. Congress. House. Committee on Small Business – Washington: US GPO 2002 [mf ed 2002] – 2mf – 9 – (incl bibl ref) – us GPO [338]
Cn and v : collegiate news and views – Cincinnati. 1972-1984 (1) 1972-1984 (5) 1972-1984 (9) – ISSN: 0010-1222 – mf#6763 – us UMI ProQuest [378]
CNS Vicesecretaria Provincial de O Sindicales de E D *see* 3rd exposicion filatelica cacerna
Cnutonis regis gesta sive encomium emmae reginae auctore monacho s bertini (mgh7:22.bd) – 1865 – (= ser Monumenta germaniae historica 7: scriptores rerum germanicarum in usum scholarum (mgh7)) – €3.00 – ne Slangenburg [240]
Co, A Kui *see* Le la mi sa myha ka tu ka nan de sa
Co entarue academiae scientiarum imperialis petropolitanae – Petropoli 1726-46 – v1-14 on 157mf – 9 – mf#r-5814/2 – ne IDC [590]
Co gai xa nieng : truyen dai / Vu, Hanh – [Saigon]: Anh Vu 1974 [mf ed 1992] – on pt of 1r – 1 – mf#11052 r87 n6 – us Cornell [830]
Co khyay *see* Sasana pru ca tam
Coach : women's athletics – Wallingford. 1976-1976 (1,5,9) – (cont by: coaching: women's athletics) – ISSN: 0145-9570 – mf#11839,01 – us UMI ProQuest [790]
Coach and athlete – New York. 1938-1982 (1) 1938-1982 (5) 1938-1982 (9) – ISSN: 0009-9872 – mf#5819 – us UMI ProQuest [790]
Coach and athletic director – Jefferson City. 1995+ (1) 1995+ (5) 1995+ (9) – (cont: scholastic coach and athletic director) – ISSN: 1087-2000 – mf#345,02 – us UMI ProQuest [790]
Coach perceptions of psychological characteristics and behaviors of male and female athletes and their impact on coach behaviors / Tuffey, Suzanne L – 1995 – 3mf – 9 – $12.00 – mf#PSY 1978 – us Kinesology [150]
[Coachella-] coachella valley submarine – CA. jul 27 1917-nov 13 1942 – 13r – 1 – $780.00 – mf#R02121 – us Library Micro [071]
[Coachella-] coachella valley sun news – CA. 1911-89 – 43r – 1 – $2580.00 – mf#R02123 – us Library Micro [071]
[Coachella-] desert barnacle – CA. dec 17 1945-mar 6 1952 – 3r – 1 – $180.00 – mf#R02120 – us Library Micro [071]
[Coachella-] desert rancher – CA. 1969-1986 – 4r – 1 – $240.00 – mf#R03184 – us Library Micro [071]
[Coachella-] sun shopper – CA. 1971-1977 – 2r – 1 – $120.00 – mf#R03185 – us Library Micro [071]
Coaches association quarterly / Wisconsin High School Coaches Association – fall sports ed-spring sports ed [1962 sep-1965 mar] – 1r – 1 – mf#683714 – us WHS [790]
A coaches' intervention to enhance practice, motor time, and skill in youth basketball players / Rukavina. Paul B – 1997 – 342p on 4mf – 9 – $20.00 – mf#PE 4198 – us Kinesology [790]
Coaching : women's athletics – Madison. 1977-1981 (1,5,9) – (cont: coach: women's athletics) – ISSN: 0160-2624 – mf#11839,02 – us UMI ProQuest [790]
Coaching motivation and efficiency / Gentry, Grier B – 1998 – 1mf – 9 – $4.00 – mf#PE 3851 – us Kinesology [790]
The coaching philosophy of dr. don shondell / Tiernan, Mark – 1997 – 1mf – 9 – $4.00 – mf#PE 3778 – us Kinesology [370]
Coaching staff cohesion and success of intercollegiate field hockey teams / Martin, Kathleen A – 1997 – 2mf – 9 – $8.00 – mf#PE 3762 – us Kinesology [790]
Coahuila. Mexico *see* Periodico oficial-gobierno constitucional del estado independiente, libre y soberano de coahuila de zaragoza
Coahuila. Mexico. (State) *see* Periodico oficial
Coal : an argument for reciprocity in coal between united states and canada / Milner, William Cochrane – [Nova Scotia?: s.n, 1906?] – 1mf – 9 – 0-665-75195-8 – mf#75195 – cn Canadiana [622]
Coal – Chicago. 1988-1996 (1,5,9) – (cont by: coal age) – ISSN: 1040-7820 – mf#16467 – us UMI ProQuest [622]
Coal age – Overland Park. 1996+ (1,5,9) – (cont: coal) – ISSN: 1091-0646 – mf#16467,01 – us UMI ProQuest [622]
Coal age – New York. 1911-1988 (1) 1967-1988 (5) 1975-1988 (9) – ISSN: 0009-9910 – mf#28 – us UMI ProQuest [622]
Coal city chronicle – New Castle, PA. -sw 1860 – 13 – $25.00r – us IMR [071]
Coal city item – New Castle, PA. -w 1856-1858 – 13 – $25.00r – us IMR [071]
Coal field defender / United Mine Workers of America – v1 n1-v6 n1 [1978 jul-1983 jan] – 1r – 1 – mf#998829 – us WHS [622]
The coal fields and coal trade of the island of cape breton / Brown, Richard – Stellarton, NS: Maritime Mining Record Office, 1899 – 2mf – 9 – mf#26704 – cn Canadiana [622]
Coal miner republican – Madison, WV. 1911-1913 (1) – mf#67346 – us UMI ProQuest [071]
Coal mining – Chicago. 1984-1988 (1) 1984-1988 (5) 1984-1988 (9) – (cont: coal mining and processing) – ISSN: 0749-1948 – mf#6181,01 – us UMI ProQuest [622]
Coal mining – Pittsburgh. 1949-1961 (1) – mf#107 – us UMI ProQuest [622]
Coal mining and processing – Chicago. 1964-1984 (1) 1971-1984 (5) 1976-1984 (9) – (cont by: coal mining) – ISSN: 0009-9961 – mf#6181 – us UMI ProQuest [622]
Coal mining women's support team news! – 1978 jun e-1984 nov, 1984 dec/1985 jan-1990 win/spring – 2r – 1 – (cont by: cep news) – mf#929393 – us WHS [622]
Coal outlook : [incorporating coal week] – New York. 2001+ [1,5,9] – mf#22815,01 – us UMI ProQuest [622]
Coal trade bulletin : a journal devoted to the coal industry – v6-39. Dec 1901-Nov 1918 – 1 – $336.00 – us L of C Photodup [380]
The coal trade of the new dominion / Haliburton, Robert Grant – Halifax, NS?: T Chamberlain, 1868 – 1mf – 9 – mf#05328 – cn Canadiana [622]
Coal week *see* Coal outlook
Coalfield progress – Norton, VA. 1935-2000 (1) – mf#66785 – us UMI ProQuest [071]
Coalfield times – Dhanbad, India. 1962-9 Apr 1980; Apr 1967-1970; 1972-1980 – 18r – 1 – us L of C Photodup [079]
La coaliciob – Badajoz, 1895 y 1895-1900 – 5 – sp Bibl Santa Ana [073]
La coalicion – Badajoz, 1902-1904, 1906-1908 y 1911 – 5 – sp Bibl Santa Ana [073]
La coalicion boletin *see* [Santa cruz-] miscellaneous titles
[Coalinga-] coalinga record – CA. 1983-1985 – 3r – 1 – $180.00 – mf#B03186 – us Library Micro [071]
Coalition – New York, NY. v2 n1-v12 n1. 1986 sep-1996 nov – 1r – (missing: sep 1988; mar,.nov1989) – us UF Libraries [071]
Coalition close-up : newsletter / Coalition for a New Foreign and Military Policy [US] – v1 n2-v10 n4 [1979 fall-1988 win] – 1r – 1 – mf#837997 – us WHS [071]
Coalition for a New Foreign and Military Policy [US] *see* Coalition close-up
Coalition for Alternatives in Jewish Education *see* News
Coalition for Black Unity *see* Cbu drumbeat
Coalition for Peaceful Schools *see* Newsletter: boletin de noticias
Coalition for Student Awareness *see* Uwm times
Coalition for the Medical Rights of Women [California] *see* Second opinion
Coalition insider : report / American Security Council. Coalition for Peace through Strength – v1 n1-v4 [i.e. 5] n4 [1978 dec-1982 may] – 1r – 1 – mf#422504 – us WHS [071]
The coalition of indian controlled school boards – 1973-76 – (= ser American indian periodicals... 1) – $95.00 – us UPA [370]
Coalition of the thermal and mineral waters of france against the s... – London, England. 1873? – 1r – us UF Libraries [944]
Coalition Opposed to Medical and Biological Attack *see* Combat ethnic weapons
Coalsmouth journal / Saint Albans Historical Society [W VA] – v2 n1 [1976 spr], 1978 sep-1979 sep, 1981 mar-1984 mar, 1985 spr-1987 spr – 1r – 1 – mf#932022 – us WHS [978]
Coalville times – Coalville, Leicestershire, 29 sep 29 1893-1895; 1897-to date – 1 – (has 2 localised ed: ashby times and swadlincote times) – uk Newsplan [072]
Coan, L B *see* Titus coan
Coan, Lydia Bingham *see* Titus coan
Coan, Titus *see*
– Adventures in patagonia
– Life in hawaii
Coar, John Firman *see* Studies in german literature in the nineteenth century
Coaracy, Vivaldo *see* Rio de janeiro no seculo 17...
Coast burgh's reporter – Anstruther, Scotland, UK. 8 Oct 1896-22 Dec 1898. -w. 1 reel – 1 – uk British Libr Newspaper [072]
Coast chronicle for leven, methil, buckhaven & wemyss, etc – [Scotland] Fife, Cupar: J & G Innes jul 1907-jun 1910 (wkly) [mf ed 2004] – 157v on 4r – 1 – (merged with: fife news (cupar, scotland : 1870) to form: fife news and coast chronicle [9 jul 1910-4 jun 1921]) – uk Newsplan [072]
Coast gazette – Ormond Beach, FL. v2 n6-v5 n22. 1891 apr-oct – 1r – us UF Libraries [071]
Coast guard – Taft OR: R E Collins, 1931- [irreg] – 1 – (cont by: north lincoln coast guard) – us Oregon Lib [071]
Coast guard bulletin – Washington. 1942-1953 (1) – mf#5761 – us UMI ProQuest [355]
Coast guard [lincoln city, or: 1937] – Nelscott OR: H A Veatch, 1937 [wkly] – 1 – (cont: north lincoln coast guard [1932-37]; cont by: lincoln coast guard [1937-39]) – us Oregon Lib [071]
Coast mail – Marshfield OR: Webster, Hacker & Lockhart, -1902 [wkly] – 1 – (related to daily ed: daily coast mail, 1902; cont by: weekly coast mail) – us Oregon Lib [071]
Coast mail *see* Daily coast mail
Coast seamen's journal – 1887 nov/1891 oct-1916 sep/1918 apr – 18r – 1 – (cont by: seamen's journal) – mf#1413224 – us WHS [360]
Coastal bantu of the cameroons : [the kpe-mboko, duala-limba and tanga-yasa groups of the british and french trusteeship territories of the cameroons] / Ardener, Edwin – London: International African Institute 1956 [mf ed Wooster OH: Bell & Howell Micro...1974] – (= ser Ethnographic survey of africa. western africa 11) – 1r – 1 – us UMI ProQuest [305]
Coastal chronicle – Georgetown, SC. 1922-1926 (1) – mf#66486 – us UMI ProQuest [071]
Coastal engineering – Amsterdam. 1976+ (1) 1976+ (5) 1987+ (9) – ISSN: 0378-3839 – mf#42149 – us UMI ProQuest [627]
Coastal line : official publication of southwest coastal area local / American Postal Workers Union – 1995 mar/apr-1996 apr/may – 1r – 1 – mf#1054579 – us WHS [380]
Coastal management – New York. 1987+ (1,5,9) – (cont: coastal zone management journal) – ISSN: 0892-0753 – mf#11080,01 – us UMI ProQuest [550]
Coastal management – v1-28. 1973-2000 – 9 – $1122.00 set – (title varies: v1-14 1973-86 as coastal zone management journal) – ISSN: 0045-723X – mf#101811 – us Hein [333]
Coastal observer – Pawley's Island, SC. 1983-1999 (1) – mf#68319 – us UMI ProQuest [071]
Coastal views – at Pascoe [079]
Coastal zone management journal – New York. 1973-1986 (1,5,9) – (cont by: coastal management) – ISSN: 0090-8339 – mf#11080 – us UMI ProQuest [550]
Coastal zone management journal *see* Coastal management
Coaster – sep 1981-dec 1987 – 12r – 1 – (previous title: the hibiscus coaster) – mf#12.14 – nz Nat Libr [079]
The coaster *see* Hibiscus coaster
Coastline – Lyttelton, NZ. 1983-84 – 1r – 1 – mf#70.32 – nz Nat Libr [079]
Coast-valley journal – Greenleaf OR: [H Kantor], 1972- [wkly] – 1 – us Oregon Lib [071]
Coat cutting / Holding, Thomas Hiram – 3rd ed. London 1893 – (= ser 19th c art & architecture) – 2mf – 9 – mf#4.2.167 – uk Chadwyck [740]
Coatbridge and airdrie standard and lanarkshire and linlithgowshire observer – [Scotland] North Lanarkshire, Coatbridge: W Craig 3 oct-26 dec 1868 (wkly) [mf ed 2004] – 1r – 1 – (cont by: weekly standard for coatbridge, airdrie, wishaw, lanark, and carluke and linlithgowshire advertiser) – uk Newsplan [072]
Coatbridge express – Coatbridge: Baird & Hamilton – 1 – (absorbed by: airdrie & coatbridge advertiser; iss 1885-1951 on microfilm; place of publ varies: coatbridge and airdrie) – uk Scotland NatLib [072]
Coatbridge reader – Scotland, UK. 1905-28 Dec 1957.-w. 26 reels – 1 – uk British Libr Newspaper [072]
Coates, Austin *see* Basutoland
Coates, E *see* Journal of the siege of quebec, 1759
Coates, James *see* Photographing the invisible
Coates, Peter Ralph *see* The cape town english press index 1871-75
Coatesville weekly times – Coatesville, PA. -w 1889-1923 – 13 – $25.00r – us IMR [071]
Coatings in canada – Toronto. 1978-1978 (1,5,9) – (cont: canadian paint and finishing) – ISSN: 0706-5124 – mf#10773,01 – us UMI ProQuest [660]
Coats, W *see* The geography of hudson's bay
Coats, Walter William *see* Glory of young men
Co-authoring spiritual ways of being : a narrative group approach to christian spirituality / Hudson, Trevor Allan – Uni of South Africa 2000 [mf ed Pretoria: UNISA 2000] – 3mf – 9 – (incl bibl ref) – mf#mfm15121 – sa Unisa [240]
Coaybay / Ramos, Jose Antonio – Habana, Cuba. 1926 – 1r – us UF Libraries [972]
Cobar age – Cobar, jan 1969-jul 1975 – 7r – A$323.14 vesicular A$361.64 silver – at Pascoe [079]
Cobar herald – Cobar, jan 1899-oct 1914 – 5r – A$357.02 vesicular A$384.52 silver – at Pascoe [079]
Cobar leader – Cobar, oct 1897-dec 1907 – 3r – A$214.90 vesicular A$231.40 silver – at Pascoe [079]
Cobargo chronicle – Cobargo, nov 1898-sep 1944 (misc periods) – 6r – A$367.53 vesicular A$400.53 silver – at Pascoe [079]
Cobb, Cyril S *see* Rationale of ceremonial, 1540-1543
Cobb, David *see* The david cobb papers, 1708-1833
Cobb, Edward M *see* Bible institute series, no 2
Cobb, Eunice Hale *see* Memoir of james arthur cobb
Cobb, Irvin S *see* Old judge priest
Cobb, Ivo Geikie *see* The glands of destiny (a study of the personality)
Cobb, L E *see* The influence of goal setting on exercise adherence of apparently healthy adults
Cobb, Louise S *see* A study of the functions of physical education in higher education
Cobb, Sanford H *see* The rise of religious liberty in america
Cobb, Sanford Hoadley *see* The rise of religious liberty in america
Cobb, Sylvanus *see*
– Autobiography of the first forty-one years of the life of sylvanus bobb
– A compend of christian divinity
– Discussion of the scripturalness of future endless punishment
– Human destiny, a discussion
– Memoir of james arthur cobb
– The new testament of our lord and saviour jesus christ
– Review of the conflict of ages by edward beecher
Cobb, William Frederick *see*
– Commentarius in primam epistolam ad corinthios
– Mysticism and the creed
– Origines judaicae
– Spiritual healing
– Theology old and new
Cobb, William Henry *see* A criticism of systems of hebrew metre
Cobbe, Frances Power *see*
– Darwinism in morals
– Dawning lights
– Essays on the pursuits of women
– The hopes of the human race, hereafter and here
– The peak in darien

Cobbe, Francis Power *see* Lessons from the world of matter and the world of man
Cobbett, William *see*
– Advice to young men, and (incidentally) to young women, in the middle and higher ranks of life
– A history of the protestant reformation in england and ireland
– Parliamentary history of england from the norman conquest in 1066 to the year 1803
– A year's residence, in the united states of america
Cobbett, William et al *see* English state trials, 1163-1858
Cobbett's complete collection of state trials – Microcard Editions – 288mf (24:1) – 9 – $1495.00 – us UPA [345]
Cobbett's evening post – London, UK. 1820. -d. 1 1/2 reels – 1 – uk British Libr Newspaper [072]
Cobbett's magazine : a monthly review of politics, history, science, domestic pursuits – London. 1833-1834 (1) – mf#4226 – us UMI ProQuest [320]
Cobbett's political register – n1-89. 1802-35 [all publ] – (= ser Radical periodicals of great britain, 1794-1914. period 1) – 501mf – 9 – $2750.00 – us UPA [325]
Cobbett's political register – New York. 1816-1818 (1) – mf#3706 – us UMI ProQuest [320]
Cobbett's state trials / howell's state trials / Great Britain. England – London: Bagshaw/ Longman. v1-33+ind. 1809-28 [all publ] – 268mf – 9 – $402.00 – (cobbett's complete coll of state trials and proceedings for high treason and other crimes and misdemenors: fr the earliest to the present time 5th ed by thomas b howell; first 10v of ed known as "cobbett's state trials" and the remainder as "howell's state trials") – mf#llmc84-762 – us LLMC [345]
Cobbett's weekly political register – London. 1802-1835 (1) – mf#4227 – us UMI ProQuest [941]
Cobbin, Ingram *see* The book of popery
Cobbold, George Augustus *see* Religion in japan
Cobden first baptist church. cobden, illinois : church records – 1876-1973 – 1 – 96.03 – us Southern Baptist [242]
Cobden, Richard *see*
– Correspondence between mr jonas and mr cobden
– Richard cobden papers
Cobern, Camden M *see* The new archaeological discoveries
Cobern, Camden McCormack *see*
– Ezekiel and daniel
– Recent explorations in the holy land and kadesh-barnea, the "lost oasis" of the sinaitic peninsula
Cobernadores y capitanes generales de venezuela / Sucre, Luis Alberto – Caracas, 1928; Madrid: Razon y Fe, 1932 – 1 – sp Bibl Santa Ana [350]
Cobett's magazine shilling magazine – London, UK. 1833-34. -w – 1r – 1 – uk British Libr Newspaper [072]
Cobham, Claude Delaval *see*
– Excerpta cypria
– The patriarchs of constantinople
Cobham, H W *see* The effects of cardiac rehabilitation on coronary heart disease risk factors in post myocardial infarction patients
Coblenz, Felix *see* Ueber das betende ich in den psalmen
Coblenzer tageblatt 1847 – Koblenz DE, 1848 1 apr-1849 – 2r – 1 – gw Misc Inst [074]
Coblenzer volkszeitung – Koblenz DE, 1916 3 jul-30 sep, 1924 1 oct-31 dec – 2r – 1 – (title varies: 1 jul 1926: koblenzer volkszeitung) – gw Mikrofilm [074]
Coblenzer zeitung – Koblenz DE, 1870 2 jul-31 dec, 1900 2 jan-31 jul – 3r – 1 – gw Mikrofilm [074]
Cobley, Leslie S *see* Introduction to the botany of tropical crops
Cobo Sampedro, Ramon *see* Sermon...a maria santisma...el beneficio de la lluvia
Cobos de Belchite, Baron de *see* Marquesado de aguilafuente, el
Cobos de Villalobos, Amantina *see* Romances caballerescos
Cobourg star – Coburg, ON. v24-27. jan 2 1856-dec 28 1859 (wkly) – 2r – 1 – Can$325.00 – (prepared from original issues in the metropolitan toronto reference library. the 4v on microfilm fill gap in otherwise complete run filmed by the archives of ontario) – cn McLaren [071]
Cobra de vidro / Holanda, Sergio Buarque De – Sao Paulo, Brazil. 1944 – 1r – us UF Libraries [972]
The cobra's den : and other stories of missionary work among the telugus of india / Chamberlain, J – Edinburgh, London, 1900 – 4mf – 9 – mf#HTM-35 – ne IDC [915]
The cobra's den : and other stories of missionary work among the telugus of india / Chamberlain, Jacob – New York: Fleming H Revell, c1900 – 1mf – 9 – 0-8370-6170-9 – mf#1986-0170 – us ATLA [240]
Coburg countryman – Coburg OR: Coburg Lions Club, [irreg] [mf ed 1974] – 1r – 1 – (cont by: countryman [1971-79]) – us Oregon Lib [071]
Coburger tageblatt 1848 – Coburg DE, apr 29 1848-oct 17 1851 – 2r – 1 – gw Misc Inst [074]
Coburn, Foster Dwight *see* Coburn's manual
Coburn's manual / Coburn, Foster Dwight – Garden City, NY. 1915 – 1r – us UF Libraries [500]
Coc ron khrann : [short stories] / Khan Khan Le, Dagun – Ran Kun: Naga Ni pum nhip tuik [195-?] [mf ed 1990] – 1r with other items – 1 – (in burmese) – mf#mf-10289 seam reel 192/6 [S] – us CRL [830]
Cocaine and excercise : temporal changes in the plasma concentrations of catecholamines, lactate, glucose, and cocaine / Han, Dong H – 1994 – 2mf – $8.00 – us Kinesology [612]
Cocaine and exercise : alteration in carbohydrate metabolism in adrenodemedullated rats / Ojuka, Edward O – 1994 – 2mf – $8.00 – us Kinesology [619]
Cocarde – Paris, France. 13 mar 1888-1897 – 28r – 1 – uk British Libr Newspaper [072]
La cocarde – Paris, 13 Mar 1888-31 Dec 1897 – 28r – 1 – uk British Libr Newspaper [074]
La cocarde – Paris. 17 janv 1888, 13 mars 1888-15 oct 1897, 3 mai 1898-13 sept 1906, 12 oct 1907-25, 1928-mars 1932. contient egalement: La Cocarde du lundi. 8, 14 oct 1907, 13 mai 1908, 9 juin, 2 oct 1913 et 1 no. de mai juin 1906 de La Cocarde – 1 – fr ACRPP [944]
Cocceji, Samuel, Freiherr von *see* Novum systema iustitiae naturalis et romanae
Coccejus, J *see*
– Opera anekdota theologica et philologica
– Opera omnia theologica
– Opera omnia theologica, exegetica, didactica, polemica, philologica...
Coccejus, [Johann] Heinrich *see* Tractatio ethica de summo bono morali
Coccius, M A *see* Le historie vinitiane di marco antonio sabellico...
Cochelet, Charles *see* Naufrage du brick francais la sophie, perdu, le 30 mai 1819
Cochemer anzeiger – Cochem DE, 1851-1867 27 dec 27 – 1 – (title varies: 3 jan 1861: cochemer kreis-anzeiger) – gw Misc Inst [074]
Cochemer kreis-anzeiger *see* Cochemer anzeiger
Cochenhausen, Friedrich von *see* Gedanken
Cocheril, Maurice *see* Etudes sur le monachisme en espagne...paris, 1966
Cocheris, J *see* Situation internationale de l'egypte et du soudan
Cochin : administrative volume – Ernakulam: Printed at the Cochin Govt Press, 1911-21 – (= ser Census of india) – 9 – us CRL [315]
Cochin China. Conseil colonial *see* Proces-verbaux du conseil colonial...cochinchine francaise
Cochin, Henry *see*
– Le bienheureux fra giovanni angelico de fiesole
– La chronologie du canzoniere de petrarque
The cochin tribes and castes / Anantha Krishna Iyer, L Krishna, Diwan Bahadur – Madras: publ for the govt of Cochin by Higginbotham, 1909-12 [mf ed 1995] – (= ser Yale coll) – 2v (ill) – 1 – 0-524-09840-9 – mf#1995-0840 – us ATLA [305]
Cochinchine, Direction de l'interiur *see* Budget local pour l'exercice...
La cochinchine religieuse / Louvet, Louis Eugene – Paris: Challamel Aine, Libraire et Commisionnaire, 1885. Chicago: Dep of Photodup, U of Chicago Lib, 1972 (1r); Evanston: American Theol Lib Assoc, 1984 (1r) – 1 – 0-8370-0302-4 – mf#1984-B291 – us ATLA [240]
Cochiti lake sun – v14 n1-2 [1981 mar-oct] – 1r – 1 – mf#618274 – us WHS [071]
Cochlaeus, J *see*
– De canonicae scripturae et catholicae ecclesiae autoritate...
– De sanctorum invocatione et intercessione... adversus henricum bullingerum helvetium
– Historiae husitarum
– In primum musculi anticochlaeum replica brevis...
– Ein kurtze replica
– Replica brevis...adversus prolixam responsionem henrici bullingeri...
Cochlaeus, Johannes *see* Ein heimlich gespraech von der tragedia johannis hussen
Cochran and Company – v1 n1-v3 n3 [1983 mar-1985 fall] – 1r – 1 – mf#1497420 – us WHS [071]
Cochran, Hamilton *see* Buccaneer islands
Cochran, John *see* The revelation of john
Cochran, Samuel Davies *see* The moral system and the atonement
Cochran, Thomas Childs *see*
– Hombre de negocios puertorriqueno
– Puerto rican businessman
Cochran, Thomas Everette *see* History of public school education in florida
Cochran, William Cox *see* The students' law lexicon.
Cochrane, Charles *see*
– Journal of a residence and travels in colombia
– Journal of a tour made by senor juan de vega, the spanish minstrel of 1828-9
Cochrane, H P *see* Among the burmans
Cochrane, Henry Park *see* Among the burmans
Cochrane, Herndone *see* Stories of florida
Cochrane, Hugh *see* Roundels
Cochrane, J D *see* Narrative of a pedestrian journey through russia and sibirian tartary, from the frontiers of china to the frozen sea and kamchatka; performed during the years 1820, 1821, 1822, and 1823
Cochrane, James *see* Discourses on some of the most difficult texts of scripture
Cochrane, John *see*
– Fussreise durch russland und die sibirische tartarei
– Narrative of a pedestrian journey through russia and siberian tartary
Cochrane northern post – Cochrane, Canada. 28 feb 1914-19 dec 1919 – 2r – 1 – uk British Libr Newspaper [071]
Cochrane recorder – Cochrane WI. 1920 may 20/1922 jan 19-1959 jan/sep 3 – 27r – 1 – (cont by: buffalo county republican; cochrane-fountain city recorder) – mf#1139311 – us WHS [071]
Cochrane, Thomas *see*
– The quest of cathay
– Survey of the missionary occupation of china
Cochrane, William *see*
– Christ and christian life
– Christian responsibility in the matter of popular amusements
– General grant, the lessons of his life and death
– The heavenly vision and other sermons
– Memoirs and remains of the reverend walter inglis
– The negative theology and the larger hope
– The old paths and the new
– A quiet and gentle life
– Warning and welcome
Cochrane, William S *see* Conflict and victory
Cochrane-fountain city recorder – Cochrane, Fountain City WI. 1959 sep 10/1961-1997 – 31r – 1 – (with gaps; cont: buffalo county republican; cochrane recorder) – mf#1139341 – us WHS [071]
Cock, Alfons de *see* Kinderspel und kinderlust in zuid-nederland
Cock Arango, Alfredo *see* Tratado de derecho internacional privado
Cock fighting / Phillips, Roland – s.l, s.l? 1936 – 1r – us UF Libraries [790]
Cockatoo and north queensland figaro – Townsville, Australia. 1 sep-15 sep 1883 – 1/4r – 1 – uk British Libr Newspaper [072]
The cockatoo and north queensland figaro – Townsville, Australia. 1-15 Sept 1883 – 4ft – 1 – uk British Libr Newspaper [079]
Cockayne hatley – (= ser Bedfordshire parish register series) – 1mf – 9 – £3.00 – uk BedsFHS [929]
Cockayne, T O *see* Leechdoms, wortcunning and starcraft of early england (rs35)
Cockburn, Alexander James Edmund *see* Letter to the rt hon lord penzance
Cockburn, Alexander Peter *see* Political annals of canada
Cockburn, Francis *see* Return to an address of the honourable the house of commons, dated 4th march 1828
Cockburn, G F *see* Report on proposed docks and extension of the lachine canal through the city of montreal
Cockburn, George *see* A voyage to cadiz and gibraltar
Cockburn, George Ralph Richardson *see*
– Speech of mr cockburn, mp, on the tariff
– Speech of mr g r r cockburn, mp on the tariff and free trade
– Statement of geo r r cockburn, esq, ma
Cockburn, James *see* A review of the general and particular causes which have produced the late disorders and divisions in the yearly meeting of friends
Cockburn, James Seton *see* Canada for gentlemen
Cockburn, William *see* The creation of the world
Cockburn, William Sarsfield Rossiter *see* An address to the citizens of bath
Cockburne, William *see* Authentic account of the late unfortunate death of lord camelford
Cocke, Alonzo Rice *see* Studies in ephesians
Cocker, Benjamin Franklin *see*
– Christianity and greek philosophy
– Lectures on the truth of the christian religion
– The theistic conception of the world
Cockerell, Charles Robert *see*
– Antiquities of athens
– Iconography of the west front of wells cathedral
– The temples of jupiter panhellenius at aegina
Cockerell, Douglas *see* Bookbinding
Cockerell, Sydney Carlyle *see*
– Some german woodcuts of the fifteenth century
– The work of w de brailes
Cockin, Hereward Kirby *see* Gentleman dick of the greys
Cockle, Mary *see*
– An explanation of dr watt's hymns for children, in question and answer
– Moral truths, and studies from natural history
Cockney : past and present / Matthews, William – New York, NY. 1938 – 1r – us UF Libraries [420]
Cocks, Norman F *see*
– Struts and frets his hour
– Struts and frets his hour, 1987
Cockshott, H M *see* The statutes of new south wales of ractical utility passed prior to 1894 and still in force
Coclico, Adrianus Petit *see* Compendivm mvsices descriptvm ab adriano petit coclico, discipvlo iosqvinide pres
Cocoa beach current – Cocoa Beach, FL. 1989-1999 (1) – mf#68527 – us UMI ProQuest [071]
Coconnier, Marie Thomas *see* L'ame humaine
Coconut bud rot in florida / Seal, J L – Gainesville, FL. 1928 – 1r – us UF Libraries [630]
Cocotologia / Bustamente, Coton – Habana, Cuba. 1958 – 1r – us UF Libraries [972]
Cocu magnifique / Crommelynck, Fernand – Paris, France. 1930? – 1r – us UF Libraries [440]
Cocuk duenyasi – Istanbul: Ikdam Matbaasi, Kader Matbaasi, Ayyildiz Matbaasi, Ahmediye Matbaasi, 1913-19. Mesuel Mueduerue: Tevfik Nureddin, Muallim Ahmed Halid. n12,41,83-91,93-94. 30 mayis 1329 [1913]-9 kanunisani 1919 – (= ser O & t journals) – 4mf – 9 – $90.00 – us MEDOC [956]
Cocuk duenyasi – Istanbul: Milli Matbaa, 1926-27. Sahibi ve Mueduerue: Ahmed Halid [Yasaroglu]. n1-30 (2 Kanunievvel 1926-22 Haziran 1927) – (= ser O & t journals) – 9mf – 9 – $150.00 – us MEDOC [079]
COD Cattle Ranch *see* Record book
Coda : poets and writers newsletter – New York. 1973-1986 (1,5,9) – (cont by: poets and writers) – ISSN: 0091-5645 – mf#11428 – us UMI ProQuest [400]
Coda magazine – Toronto. 1973+ (1) 1977+ (5) 1977+ (9) – ISSN: 0820-926X – mf#6976 – us UMI ProQuest [780]
Coddington, Henry *see* A few remarks on the "new library" question
Le code bolchevik du mariage / Prouvost, Leon – Conflans-Ste-Honorine: L'Idee libre, 1921 – (= ser Les femmes [coll]) – 1mf – 9 – mf#8738 – fr Bibl Nationale [346]
Le code catholique : ou commentaire du catechisme des provinces ecclesiastiques de quebec, montreal et ottawa / Gosselin, David – Quebec: H Chasse, 1898 – 3mf – 9 – mf#29072 – cn Canadiana [241]
Le code civil annote etant le code civil du bas-canada : en force depuis le premier aout 1866... – 2nd rev corr enl ed. Montreal: C O Beauchemin, 1889 [mf ed 1984] – 10mf – 9 – 0-665-45520-8 – (incl ind and bibl ref) – mf#45520 – cn Canadiana [348]
Code civil d'haiti / Haiti Laws, Etc – Port-Au-Prince, Haiti. 1931 – 1r – us UF Libraries [323]
Code civil du bas canada : d'apres le role amende depose dans le bureau du greffier du conseil legislatif, tel que prescrit par l'acte 29 vict chap 41, 1865 = Civil code of lower canada: from the amended roll deposited... / Canada. Bas-Canada – Ottawa: printed by Malcom Cameron, 1866 – 9mf – 9 – (in french and english) – mf#SEM105P430 – cn Bibl Nat [348]
Code civil du bas canada : d'apres le role amende depose dans le bureau du greffier du conseil legislatif, tel que prescrit par l'acte 29 vict chap 41, 1865 = Civil code of lower canada: from the amended roll deposited in the office of the clerk of the legislative council as directed by the act 29 vict chap 41, 1865 / Canada (Province) – Ottawa: printed by Malcom Cameron, 1866 [mf ed 1984] – 9mf – 9 – (in english and french) – mf#SEM105P430 – cn Bibl Nat [350]
Code civil du bas canada : [rapports / des commissaires pour la codification des lois du bas canada qui se rapportent aux matieres civiles] = Civil code of lower canada: [reports / of the commissioners for the codification of the laws of lower canada relating to civil matters] / Canada (Province) – Quebec: Impr Georges E Desbarats. 3v. 1865 [mf ed 1984] – 16mf – 9 – (in french and english) – mf#SEM105P426 – cn Bibl Nat [348]
Code civil du bas canada : titre des obligations...nommes en vertu du statut 20 vic chap 43 = Report / of the commissioners for the codification of the laws of lower canada relating to civil matters, appointed under the statute 20 vic cap 43 / Canada (Province) – Quebec: Impr Stewart Derbishire & Georges Desbarats, [1862?] (mf ed 1984) – 19mf – 9 – (in french and english) – mf#SEM105P380 – cn Bibl Nat [348]

Code civil du bas canada. / Quebec. (Province). Laws, Statutes, etc – Ottawa: Cameron, 1866. 4, 747p. LL-2387 – 1 – (table analytique du code civil du bas-canada...ottawa, 1867. 98p. ll-2387. analytical index to the civil code of lower-canada. ottawa, 1867. 100p. ll-2387) – us L of C Photodup [348]

Code civil du bas-canada : contenant sous chaque article les amendements et autres dispositions legislatives qui affectent le texte jusqu'au 1er janvier 1888... – Montreal: A Periard, 1888 [mf ed 1984] – 8mf – 9 – 0-665-10822-2 – mf#10822 – cn Canadiana [348]

Le code civil du bas-canada : contenant sous chaque article, les amendements et autres dispositions legislatives qui affectent le texte...avec le code napoleon et le code de commerce francais / Lareau, Edmond – Montreal: A Periard, 1885 [mf ed 1984] – 8mf – 9 – 0-665-10894-X – mf#10894 – cn Canadiana [340]

Le code civil du bas-canada (en force depuis le 1er aout 1866) : tel qu'il a ete amende... au 1er janvier 1885 / Bellefeuille, Edouard Lefebvre de [comp] – Montreal: Beauchemin & Valois, 1885 [mf ed 1984] – 7mf – 9 – 0-665-10769-2 – (incl ind) – mf#10769 – cn Canadiana [348]

Code civil et de prodedure civile cambodgiens : d'apres les travaux des commissions instituees par arretes des 5 juillet et 3 septembre 1912, 9 avril 1913, 19 juillet 1918 et 29 janvier 1919 – Phnom-Penh: Impr du Protectorat 1920 [mf ed 1989] – 1r with other items – 1 – mf#mf-10289 seam reel 026/14 [§] – us CRL [340]

Der code civil franzoesisch und deutsch : verbesserte cramer'sche uebersetzung, nebst den ihn ab abaendernden und ergaenzenden reichs- und preussischen gesetzes und den noch geltenden artikeln des code de procedure civile und des code de commerce / ed by Loersch, Hugo – 3. verb u verm aufl. Leipzig: K Baedeker, 1887 – (= ser Civil law 3 coll) – 8mf – 9 – (the code itself is in french and german on opposite pages, which are paged in duplicate. the code of procedure follows in both languages in double columns. incl ind) – mf#LLMC 96-500 – us LLMC [348]

Code constitutionel de la belgique, ou commentaire sur la constitution, la loi electorale, la loi communale et la loi provinciale / Bivort, Jean Baptiste – Nouv. ed. rev. Bruxelles, Decq, 1859-62 4 pt. in 1 v. LL-4017 – 1 – us L of C Photodup [348]

Code criminel : ou commentaire sur l'ordonnance de 1670 / Serpillon, Francois – nouv ed. Lyon: chez les Freres Perisse. 2v. 1788 [mf ed 1971] – 1r – 1 – mf#SEM35P80 – cn Bibl Nat [345]

Code criminel de l'empereur charles, vulgairement appelle la caroline : contenant les loix qui sont suivies dans les jurisdictions criminelles de l'empire; et a l'ufage des conseils de guerre des troupes suisses – A Maestricht: Chez Jean-Edme Dufour & Phil Poux, 1779 – 4mf – 9 – $6.00 – mf#LLMC 89-018 – us LLMC [348]

Code de commerce / Haiti (Republic) Laws, Statutes, Etc – Port-Au-Prince, Haiti. 1945 – 1r – us UF Libraries [380]

Code de commerce d'haiti : contenant la conference / Haiti Laws, Statutes, Etc – Berlin, Germany. 1910 – 1r – us UF Libraries [380]

Code de commerce haitien / Haiti Laws, Statutes, Etc – Port-Au-Prince, Haiti. 1910 – 1r – us UF Libraries [380]

Le code de droit canonique : ses canons les plus pratiques pour le ministere avec references a la discipline locale / Emard, Joseph-Medard – Valleyfield [Quebec]: Bureaux de la chancellerie, 1918 [mf ed 1994] – 4mf – 9 – 0-665-73176-0 – (incl latin text) – mf#73176 – cn Canadiana [240]

Code de l'instruction publique de la province de quebec : comprenant les lois scolaires et un grand nombre de decisions judiciaires s'y rapportant et les reglements du comite catholique du conseil de l'instruction publique... / Cazes, Paul de [comp] – Quebec?: s.n, 1890 – 4mf – 9 – (incl ind) – mf#54641 – cn Canadiana [348]

Code de l'instruction publique de la province de quebec : etant une compilation des divers statuts sur cette matiere / Chouinard, Mathias – Quebec: J O Filteau, 1888 – 5mf – 9 – 0-665-54640-8 – (incl ind) – mf#54640 – cn Canadiana [370]

Code de musique pratique : ou methodes pour apprendre la musique, meme a des aveugles, pour former la voix & l'oreille... / Rameau, Jean Philippe – Paris: De l'Impr royale 1760 [mf ed 19–] – 5mf – 9 – mf#fiche 505 – us Sibley [780]

Code de musique sacree et liste de pieces recommandees : pour le culte divin (messes, motets, cantiques, morceaux d'orgues) / Comite interdiocesain de musique sacree (Quebec) – Quebec: Presses universitaires Laval, 1952 [mf ed 1994] – 1mf – 9 – (with ind) – mf#SEM105P2106 – cn Bibl Nat [780]

Code de procedure civile : avec les dernieres modif / Haiti (Republic) Laws, Statutes, Etc – Port-Au-Prince, Haiti. 1943 – 1r – us UF Libraries [350]

Code de procedure civile : [acte concernant le code de procedure civile du bas canada] = Code of civil procedure: [an act respecting the code of civil procedure of lower canada] / Canada (Province) – [Ottawa?: s.n, 1866?] (mf ed 1998) – 1mf – 9 – (in english and french) – mf#SEM105P2902 – cn Bibl Nat [348]

Code de procedure civile / Haiti (Republic).Laws, Statutes, Etc – Roche-sur-Yon? France. 1959 – 1r – us UF Libraries [350]

Code de procedure civile du bas canada : [huitieme rapport] = Code of civil procedure of lower canada: [eighth report] / Canada (Province). Commissaires charges de codifier les lois du Bas Canada, en matieres civiles – Ottawa: impr par George E Desbarats, 1866 [mf ed 1998] – 4mf – 9 – (in french and english) – mf#SEM105P2903 – cn Bibl Nat [348]

Code de procedure civile du bas canada : [dixieme rapport] = Code of civil procedure of lower canada: [tenth report] / Canada (Province). Commissaires charges de codifier les lois du Bas Canada, en matieres civiles – Ottawa: impr par G E Desbarats, 1866 [mf ed 1998] – 4mf – 9 – (in french and english) – mf#SEM105P2898 – cn Bibl Nat [348]

Code de reforme et de discipline formant la troisieme partie du systeme de lois penales prepare pour l'etat de la louisiane / Livingston, Edward – [Quebec?: s.n.] 1831 [mf ed 1984] – 1mf – 9 – 0-665-45573-9 – mf#45573 – cn Canadiana [345]

Code des cures, marguilliers et paroissiens : accompagne de notes historiques et critiques / Beaudry, Joseph Alphonse Ubalde – Montreal: La Minerve, 1870 – 4mf – 9 – (incl ind) – mf#03034 – cn Canadiana [241]

Code des institutions politiques du rwanda precolonial / Kagame, Alexis – Bruxelles, Belgium. 1952 – 1r – us UF Libraries [350]

Code des lois usuelles, recueil des lois et de jur... / Haiti Laws, Statutes, Etc – Port-Au-Prince, Haiti. 1954 – 1r – us UF Libraries [323]

Code d'identification pour la classification des recettes et des depensed du budget national / United States. Agency for International Development – [Phnom-Penh? 1961?] [mf ed 1989] – 1r with other items – 1 – (in french & english) – mf#mf-10289 seam reel 024/05 [§] – us CRL [336]

Code d'instruction criminelle et code penal / Haiti – Paris, France. 1909 – 1r – us UF Libraries [360]

Le code du mahaayaana en chine : son influence sur la vie monacale et sur le monde laique / Groot, Jan Jakob Maria de – Amsterdam: Johannes Mueller, 1893 – (= ser Verhandelingen der Koninklijke Akademie van Wetenschappen te Amsterdam) – 1mf – 9 – 0-524-02422-7 – (incl text of the fan wang ching in french and chinese) – mf#1990-3006 – us ATLA [280]

Code du travail / Haiti Laws, Statutes, Etc – Port-Au-Prince, Haiti. 1961 – 1r – us UF Libraries [323]

Le code du travail malgache : son application pratique / Goyat, Michel & Mouric, Rene – Tananarive [1962] – (filmed with garlick, peter: african traders in kumasi, legon 1959; and nypan, astrid: market trade, legon 1960) – us CRL [960]

Code et guide de l'etat civil a l'usage des minist... / Bistoury, Andre F – Port-Au-Prince, Haiti. 1956 – 1r – us UF Libraries [350]

Code fiscal haitien / Haiti Laws, Statutes, Etc – Port-Au-Prince, Haiti. 1953 – 1r – us UF Libraries [332]

Code, Joseph Bernard *see* Spanish war and lying propaganda

Code militaire : ou compilation des ordonnances des rois de france, concernant les gens de guerre / Briquet, Pierre de – nouv ed aumg. Paris: chez Durand...1761 [mf ed 1984] – 1r – 5 – mf#SEM16P344 – cn Bibl Nat [355]

Code militaire / Suzor, Louis Timothee [comp] – Quebec: G & G E Desbarats, 1864 [mf ed 1983] – 3mf – 9 – 0-665-44374-9 – (trans by comp. incl ind) – mf#44374 – cn Canadiana [355]

Code militaire / Suzor, Louis-Timothee [comp] – Quebec: G & G E Desbarats...1864 [mf ed 1983] – 3mf – 9 – mf#SEM105P317 – cn Bibl Nat [355]

Code municipal de la province de quebec annote 1898-1902 : suivi d'un supplement qui le met au courant de la legislation et de la jurisprudence jusqu'au 1er juillet 1902... / Bedard, Joseph-Edouard – Montreal: C Theoret, 1902 [mf ed 1993] – 7mf – 9 – (in french and english; with ind) – mf#SEM105P1865 – cn Bibl Nat [348]

Code municipal de la province de quebec, annote; comprenant tous les amendements jusqu'au 1 janvier 1888. / Quebec. (Province). Laws, Statutes, etc – 1er. ed. Quebec: Filteau, 1888. 494p. LL-1654 – 1 – us L of C Photodup [348]

Code municipal de la province de quebec annote mis au courant de la legislation et de la jurisprudence : et suivi des dispositions statutaires concernant les officiers municipaux quant aux elections parlementaires, licences, jures, etc / Bedard, Joseph-Edouard – 2e ed. Montreal: C Theoret, 1905 – 8mf – 9 – (in french and english) – mf#SEM105P1864 – cn Bibl Nat [348]

Code municipal de la province de quebec mis au courant de la legislation et de la jurisprudence : suivi d'un appendice comprenant des extraits des statuts concernant les corporations municipales et leurs officiers... / Mathieu, Michel – Montreal: A Periard, 1887 [mf ed 1993] – 5mf – 9 – mf#SEM105P1863 – cn Bibl Nat [348]

Code municipal de la province de quebec tel qu'en force le 1er janvier 1881 : auquel on a ajoute la jurisprudence des arrets s'y rapportant, l'acte des licences de quebec de 1878... – 2e ed. Montreal: E Senecal, 1881 [mf ed 1984] – 6mf – 9 – 0-665-45567-4 – (incl ind. also available in english) – mf#45567 – cn Canadiana [342]

Code municipal de la province de quebec tel qu'en force le 1er juillet 1882 : auquel on a ajoute la jurisprudence des arrets s'y rapportant, l'acte des licences de quebec de 1878... – 3e ed. [Montreal?: s.n.] 1882 [mf ed 1984] – 6mf – 9 – 0-665-45568-2 – (incl ind) – mf#45568 – cn Canadiana [342]

Code of american samoa, 1946 edition / American Samoa – Office of the Governor, n.d. – 4mf – 9 – $6.00 – mf#LLMC 82-100C Title 1 – us LLMC [324]

Code of american samoa, 1973 edition / American Samoa – Equity Publ Co. 2v. 1973 – 22mf – 9 – $33.00 – (with 1979 pocket pts) – mf#LLMC 82-100C Title 3 – us LLMC [324]

Code of canons of the episcopal church in scotland / Episcopal Church In Scotland General Synod (1838) – Edinburgh, Scotland. 1838 – 1r – us UF Libraries [242]

Code of canons of the episcopal church in scotland / Episcopal Church In Scotland General Synod (1862-1863) – Edinburgh, Scotland. 1863 – 1r – us UF Libraries [242]

Code of census procedure / Punjab. India. Superintendent of Census Operations – 1911, pt. 1 – 1 – us CRL [315]

The code of civil procedure and probate code of guam, 1953 / Bohn, John A – Agana: Off of the Sec of the Gov of Guam, 5 Nov 1953 – 10mf – 9 – $15.00 – (incl 1964 suppl) – mf#LLMC 82-100B Title 7 – us LLMC [324]

The code of civil procedure and probate code of guam, 1970 / Bohn, John A – 2v – 12mf – 9 – $18.99 – (with 1972 and 1974 suppls) – mf#LLMC 82-100B Title 13 – us LLMC [324]

The code of civil procedure of guam, 1947 / Guam. (Commonwealth). Laws, Statutes, etc – Washington: GPO. 4pts. 1947 – 3mf – 9 – $4.50 – mf#LLMC 82-100B Title 11 – us LLMC [324]

The code of civil procedure of lower canada : together with the amendments thereto made since its promulgation... / Foran, Thomas Patrick – 2nd ed. Toronto: T Moore; Edinburgh: Carswell, 1886 [mf ed 1984] – 11mf – 9 – 0-665-45523-2 – mf#45523 – cn Canadiana [347]

The code of civil procedure of the canal zone : enacted by executive order of president theodore roosevelt – 1 may 1907. Washington: GPO, 1907 – 3mf – 9 – $4.50 – mf#LLMC 82-100D Title 17 – us LLMC [348]

Code of commerce in force in cuba, porto rico, and the philippines, 1897 : including the commercial registry regulations, exchange regulations,and other provisions of a similar character, the code of 1885 as amended by the law of june 10, 1897 – Washington: GPO, 1899 – 4mf – 9 – $6.00 – mf#LLMC 92-306 – us LLMC [348]

Code of criminal procedure: preliminary draft / American Law Institute – Philadelphia The Institute, 1927. 88 p. LL-2341 – 1 – us L of C Photodup [345]

The code of evidence... / New York. (State). Commissioners to Report a Code of Evidence – Albany? 1889. LL-326 – 1 – us L of C Photodup [348]

Code of federal regulations – Backfile 1938-2001 – 9 – $30,720.00 set (1996 subs $795.00 set) – 0-89941-200-9 – (1975-94 $775 per yr. 1938-74 price varies. 1995-98 $795 per yr. 1999 $825v. 2000 $860v. 2001 $895v. 2002 subs $930v) – mf#400010 – us Hein [324]

Code of federal regulations / U.S. National Archives and Records Administration, Office of the Federal Register – 1994 – 9 – $264.00y in US; $330.00y outside – mf#S-N 869-029-00000-9. Sub-list ID-CFRM4 – us GPO [324]

Code of federal regulations *see* Public land statutes and regulations in force june 1, 1938

Code of federal regulations: 1939-1982 / U.S. Laws, Statutes, etc – 577 reels. Includes printed guide – 1 – us Trans-Media [324]

Code of federal regulations on microfiche / U.S. – 1938- – ca 36,000mf – 9 – apply for price – (printed index available separately) – us CIS [348]

The code of hammurabi, king of babylon about 2250 b.c : autographed text, transliteration, translation, glossary... / Harper, Robert Francis – Chicago: University of Chicago Press, 1904 [mf ed 1986] – 1mf – 9 – 0-8370-7218-2 – (incl ind) – mf#1986-1218 – us ATLA [470]

Code of law, practice and forms for justices' and other inferior courts in the western states / Hillyer, Curtis – San Francisco: Bender-Moss Co., 1912. 2v. LL-1251 – 1 – us L of C Photodup [347]

Code of massachusetts regulations – Boston: Office of the Massachusetts Secretary of State, 1992 revision with 2001 service – 9 – $795.00 set – (2002 subs $820) – mf#400920 – us Hein [324]

Code of nature – London, England. 1832 – 1r – us UF Libraries [240]

Code of ordinances of the city of apalachicola, fl... / Apalachicola (Fla) Ordinances, Etc – Tallahassee, FL. 1913 – 1r – us UF Libraries [350]

Code of ordinances of the city of pensacola / Pensacola, Fla Ordinance, Etc – Pensacola, FL. 1920 – 1r – us UF Libraries [350]

Code of practice for industrial radiography / South Africa. Directorate Electromedical Devices and Radiological Health [comp] – rev ed. [Pretoria]: Dept of Health 1994 [mf ed Pretoria, RSA: State Library [199-]] – 16p [ill] on 1r with other items – 5 – (incl bibl ref) – mf#op 12118 r26 – us CRL [360]

Code of practice for the safe use of soil moisture and density gauges containing radioactive sources / South Africa. Directorate Electromedical Devices and Radiological Health [comp] – rev ed. [Pretoria]: Dept of National Health & Population Development 1994 [mf ed Pretoria, RSA: State Library [199-]] – 10p on 1r with other items – 5 – (incl bibl ref) – mf#op 12121 r26 – us CRL [360]

Code of public instruction of the province of quebec : comprising the school law, with notes of numerous decisions thereon and the regulations of the protestant committee of the council of public instruction / Cazes, Paul de [comp] – Montreal?: s.n, 1891 – 4mf – 9 – (incl ind; trans by john ahern) – mf#54643 – cn Canadiana [370]

A code of reform and prison discipline / Livingston, Edward – New York, National Prison Association of the United States 1872 140 p. LL-4077 – 1 – us L of C Photodup [348]

Code of regulations of the ttpi / Trust Territory of the Pacific Islands. (US) – Saipan: TTPI Gov, n.d. – 4mf – 9 – $6.00 – mf#LLMC 82-100F Title 5 – us LLMC [324]

Code of the city of starke, 1932 / Starke (Fla) Ordinances, Etc – Starke, FL. no date – 1r – us UF Libraries [350]

Code of the federated states of micronesia, 1982 – Seattle: for the Govt by Book Publ Co. 2v. 1982 & 1987 – 19mf – 9 – $28.50 – (incl 1987 suppl) – mf#llmc82-100h, title 20 – us LLMC [342]

Code of the trust territory of the pacific, 1952 – Honolulu: Office of the High Commissioner, 22 dec 1952 – 5mf – 9 – $7.50 – (with an appendix of executive orders) – mf#LLMC 82-100F Title 20 – us LLMC [324]

Code of the trust territory of the pacific, 1959 revision – Agana, Guam: Office of the High Commissioner, 31 dec 1959 – 5mf – 9 – $7.50 – mf#LLMC 82-100F Title 21 – us LLMC [324]

Code of the trust territory of the pacific, 1966 rev – Saipan: Off of the High Commissioner, oct 10 1966 – 6mf – 9 – $9.00 – mf#LLMC 82-100F Title 22 – us LLMC [323]

Code of the trust territory of the pacific, 1970 rev / ed by Steincipher, John – Seattle: Book Publ Co. 2v. 1970 – 21mf – 9 – $31.50 – (includes 1973 & 1975 suppl) – mf#LLMC 82-100F Title 23 – us LLMC [323]

Code of the trust territory of the pacific, 1980 – Charlottesville: Michie. 2v. 1980 – 14mf – 9 – $21.00 – mf#LLMC 82-100F Title 24 – us LLMC [323]

Code penal / Congo. Free State. Laws, Statutes, etc – Bruxelles, Hayez, 1888. 21 p. LL-12004 – 1 – us L of C Photodup [348]

Code penal / Guinea. French. Laws, Statutes, etc – Conakry, Imprimerie Lumumba, 1966. 122 p. LL-12035 – 1 – us L of C Photodup [345]

Code penal avec les dernieres modifications / Haiti (Republic) Laws, Statutes, Etc – Port-Au-Prince, Haiti. 1938 – 1r – us UF Libraries [360]

Code penal du royaume de siam promulgue le ler juin 1908, entre en vigeur le 22 septembre 1908 / Thailand. Laws, Statutes, etc – Paris, Imprimerie Nationale, 1909. 110 p. LL-10015 – 1 – us L of C Photodup [348]

Code recreatif des francs-macons : poesies, cantiques et discours a leur usage / Grenier, F – Paris: Caillot 5807 [1807] – 4mf – 9 – mf#vrl-49 – ne IDC [366]

Code remedies and remedial rights by the civil action according to the reformed american procedure. / Pomeroy, John Norton – 4th ed. Boston: Little, Brown, 1904. 983p. LL-1399 – 1 – us L of C Photodup [348]

Code rural a l'usage des habitants tant anciens que nouveaux du bas-canada : concernant leurs devoirs religieux et civils, d'apres les loix en force dans le pays / Perrault, Joseph-Francois – [Quebec?: s.n.] 1832 [mf ed 1984] – 1mf – 9 – 0-665-21360-3 – mf#21360 – cn Canadiana [340]

Code scolaire de la province de quebec : contenant la loi de l'instruction publique et un grand nombre de decisions judiciaires s'y rapportant, les reglements scolaires du comite catholique du conseil de l'instruction publique... / Cazes, Paul de – Montreal: C Theoret, 1899 – 5mf – 9 – mf#10549 – cn Canadiana [348]

Codebooks...1867-1876 / U.S. Dept of State – (= ser General records of the department of state) – 1r – 1 – mf#T1171 – us Nat Archives [975]

Code-formulaire de l'etat civil d'haiti – Port-Au-Prince, Haiti. 1888 – 1r – us UF Libraries [350]

Codera, Francisco *see*
– Los beniverman en merida y badajoz
– Inscripcion arabe en trujillo
– Nueva lapida romana de montanchez, capital de partido en la provincia de caceres

Les codes cambodgiens / Leclere, Adhemard – Paris (I-II). 1898 – 1 – fr ACRPP [959]

Codes congolais et lois usuelles en vigueur au congo, collationnes d'apres les textes officiels et annotes / Congo. Free State. Laws, Statutes, etc – Bruxelles, Larcier, 1900. 604 p. LL-12005 – 1 – us L of C Photodup [348]

Les codes du congo, suivis des decrets, ordonnances et arretes complementaires / Congo. Free State. Laws, Statutes, etc – 2. ed. Bruxelles, Larcier, 1892. 360 p. LL-12001 – 1 – us L of C Photodup [348]

Codes et lois du burundi – Bruxelles, Belgium. 1970 – 1r – us UF Libraries [323]

Codes of census procedures for the hyderabad assigned districts / Hyderabad. India. (State). Superintendent of Census Operations – pt. 11901 – 1 – us CRL [315]

The codes of hammurabi and moses : with copious comments, index, and bible references / Davies, William Walter – Cincinnati: Jennings and Graham; New York: Eaton and Mains, c1905 – 1mf – 9 – 0-8370-7290-5 – (incl ind) – mf#1986-1290 – us ATLA [348]

Codex 1 of the gospels and its allies / Lake, Kirsopp – Cambridge: University Press 1902 [mf ed 1991] – (= ser Texts and studies (cambridge, england) 7/3) – 3mf – 9 – 0-7905-8329-1 – (in greek; crit app in english) – mf#1987-6428 – us ATLA [226]

Codex 1 of the gospels and its allies (ts7/3) / Lake, Kirsopp – 1902 – (= ser Texts and studies (ts)) – 5mf – 9 – €12.00 – ne Slangenburg [226]

Codex alexandrinus (royal ms 1 d 5-8) – British Museum – (old testament pt 1 genesis-ruth, 1915 €114. old testament pt 2 1 samuel- 2 chronicles, 1930 €84. old testament pt 3 hosea-judith, 1936 €114. old testament pt 4 1 esdras-ecclesiasticus, 1957 €138. new testament and clementine epistles, 1909 €102) – ne Slangenburg [221]

Codex apocryphis novi testamenti – Lipsiae: F C G Vogel, 1832 – 1r – 1 – 0-8370-1105-1 – mf#1984-6240 – us ATLA [225]

Codex apocryphus novi testamenti – Hamburg: Benjamin Schiller, 1703 – 1r – 1 – 0-8370-1096-9 – mf#1984-B525 – us ATLA [225]

Codex bezae : a study of the so-called western text of the new testament / Harris, James Rendel – Cambridge: University Press; New York: Macmillan [dist] 1891 [mf ed 1989] – (= ser Texts and studies: contributions to biblical and patristic literature 2/1) – 1mf – 9 – 0-7905-1330-7 – (in english, greek & latin) – mf#1987-1330 – us ATLA [225]

Codex bezae cantabrigiensis : quattuor evangelia et actus apostolorum complectens graece et latine – Londini: Venevnt apud CJ Clay et Filios in emporio preli Academici Catabrigiensis, 1899 – 1r – 1 – 0-7905-8324-0 – mf#1987-B002 – us ATLA [090]

Der codex boernerianus : der briefe des apostels paulus (msc. dresd. a 145b) in lichtdruck nachgebildet / ed by Koenigliche Oeffentliche Bibliothek zu Dresden – Leipzig: Karl W Hiersemann, 1909 – 1mf – 9 – 0-7905-8283-X – mf#1987-6388 – us ATLA [220]

Codex claromontanus : sive, epistulae pauli omnes graece et latine / ed by Tischendorf, Constantin von – Lipsiae: F A Brockhaus, 1852 [mf ed 1989] – 6mf – 9 – 0-8370-1297-X – mf#1987-6036 – us ATLA [090]

Der codex d in der apostelgeschichte : textkritische untersuchung / Weiss, Bernhard – Leipzig: J C Hinrichs, 1897 – (= ser Tugal) – 1mf – 9 – 0-7905-1857-0 – mf#1987-1857 – us ATLA [220]

Der codex d in der apostelgeschichte / Weiss, Bernhard – Leipzig, 1897 – (= ser Tugal 2-17/1) – 2mf – 9 – €5.00 – ne Slangenburg [225]

Codex diplomaticus brandenburgensis : sammlung der urkunden, chroniken und sonstigen quellen fuer die geschichte der mark brandenburg und ihrer regesten / ed by Riedel, Adolf Friedrich – Berlin 1838-69 [mf ed Hildesheim 1997] – 41v on 237mf – 9 – diazo €998.00 silver €1248.00 – gw Olms [943]

Codex diplomaticus rheno-mosellanus – Koblenz DE. v1-3. 1822-25 – 25mf – 9 – gw Mikropress [943]

Codex diplomaticus silesiae / ed by Verein fuer Geschichte und Altertum Schlesien – Breslau 1857-1933 [mf ed Hildesheim 1997] – 36v on 120mf – 9 – diazo €528.00 silver €648.00 – gw Olms [943]

Codex dunensis / ed by Lettenhove, J Kervijn de – Brussel, 1875 – 20mf – 8 – €38.00 – ne Slangenburg [241]

Codex el musical de las huelgas / Angles, Higini – Barcelona, 1931 – (= ser Musica a veus dels segles 13-14) – €131.00 – (v1: introduccio (bcat dm 6). v2: facsimil (bcat dm 6). v3: transcripcio (bcat dm 6)) – ne Slangenburg [780]

Codex fuldensis : novum testamentum latine interprete hieronymo / ed by Ranke, Ernst – Marburgi: Sum[p]tibus N G Elwerti Bibliopolae Academici, 1868 – 2mf – 9 – 0-8370-9499-2 – mf#1986-3499 – us ATLA [220]

Codex graecus quatuor evangeliorum : e bibliotheca universitatis pestinensis / ed by Markfi, Samuele – Pestini: Typis Gustavi Emich, 1860 – 5mf – 9 – 0-8370-1854-4 – mf#1987-6241 – us ATLA [220]

Codex hirsaugiensis – Stuttgardiae: Sumtibus Societatis litterariae stuttgardiensis, 1843 [mf ed 1993] – (= ser Blvs 1/5) – viii/131p – 1 – mf#8470 reel 1 – us Wisconsin U Libr [240]

Codex laudianus : sive, actus apostolorum graece et latine: ex codice olim laudiano iam bodleiano sexto fere saeculi... / ed by Tischendorf, Constantin von – Lipsiae: JC Hinrichs, 1870 [mf ed 1986] – 3mf – 9 – 0-8370-9428-3 – mf#1986-3428 – us ATLA [220]

Codex liturgicus ecclesiae lutheranae : in epitomen redactus / ed by Daniel, Hermann Adalbert – Lipsiae [Leipzig]: T O Weigel 1848 [mf ed 1993] – 2mf – 9 – 0-524-08279-0 – mf#1993-3034 – us ATLA [242]

Codex liturgicus ecclesiae universa / Daniel, H A – Lipsiae, 1847, 1848, 1851, 1853 – 8 – (v1 ecclesiae romano-catholicae, lipsiae, 1847 8mf €17. v2 ecclesiae lutheranae, lipsiae, 1848 11mf €21. v3 ecclesiae reformatae atque anglicanae, lipsiae, 1851 11mf €21. v4 ecclesiae orientalis, lipsiae, 1853 13mf €25) – ne Slangenburg [240]

Codex liturgicus ecclesiae universae / Assemanus, J A – Roma. v1-13. 1749-66 – 106mf – 8 – €202.00 – ne Slangenburg [240]

Codex marianus glagoliticus : quattuor evangeliorum versionis palaeoslovenicae. mariiskoe chetveroevangelie. pamiatnik glagolicheskoi pismennosti / ed by Iagich, V – Berolini, 1883 – 12mf – 8 – mf#770 – ne IDC [243]

Der codex mf 2016 des musikalischen instituts bei der universitaet breslau : eine palaeographische und stilistische beschreibung / Feldmann, Fritz – Breslau: Priebatsch's Buchhandlung 1932 [mf ed 19–] – (= ser Schriften des musikalischen instituts bei der universitaet breslau bd2) – 2v on 1r – 1 – (pref by arnold schmitz; contents: 1. teil. darstellung: palaeographischer teil. stilisticher teil: messen. motetten; 2. teil. verzeichnisse. uebertragungen: missa anonyma 1; missa anonyma 2; [25] motetten) – mf#film 702 – us Sibley [780]

Codex napoleon : uebersetzt nach der neuen offiziellen ausgabe von einer gesellschaft rechtsgelehrter und durch noten erlaeutert von I spielmann = Code napoleon – Strassburg, Paris, 1808 [mf ed 1995] – 8mf – 9 – 3-8267-3155-7 – mf#DHS-AR 3155 – gw Frankfurter [944]

Codex purpureus petropolitanus : the text of codex n of the gospels / ed by Cronin, Harry Stovell – Cambridge: University Press 1899 [mf ed 1989] – (= ser Texts and studies (cambridge, england) 5/4) – 1mf – 9 – 0-7905-1866-X – (in greek; int in english; incl bibl ref) – mf#1987-1866 – us ATLA [226]

Codex purpureus petropolitanus (codex n of the gospels) (ts5/4) / ed by Cronin, H S – 1899 – (= ser Texts and studies (ts)) – 3mf – 9 – €7.00 – ne Slangenburg [226]

Codex regularum : amplificatus a m brockie / Holstenius, L – Aug Vindelicorum. v1-6. 1759 – €294.00 – ne Slangenburg [241]

Codex regularum et constitutionum clericalium / Miraeus, A – Antverpiae, 1638 – €12.00 – ne Slangenburg [240]

Codex rehdigeranus : die vier evangelien nach der lateinischen handschrift r 169 der stadtbibliothek breslau / ed by Vogels, Heinrich Joseph – Rom: F Pustet, 1913 – (= ser Collectanea biblica latina) – 1mf – 9 – 0-7905-0598-3 – (incl bibl ref) – mf#1987-0598 – us ATLA [220]

The codex rescriptus dublinensis of st matthew's gospel (z) : first published by dr. barrett in 1801. also, fragments of the book of isaiah... together with a newly discovered fragment of the codex palatinus / Abbott, Thomas Kingsmill – new rev and augm ed. Dublin: Hodges, Foster, and Figgis; London: Longmans, Green, 1880 – 1mf – 9 – 0-8370-9360-0 – mf#1986-3360 – us ATLA [226]

Codex saeculi 11 : continens tractatus various ad historiam musicae medii aevi pertinentes... – [10–] [mf ed 19–] – 6mf / 1r – 9,1 – mf#fiche 1043 / pres. film 222 – us Sibley [780]

The codex sangallensis : a study in the text of the old latin gospels / Harris, James Rendel – London: C J Clay, 1891 – 1mf – 9 – mf#1986-3243 – us ATLA [220]

Codex schlierbach 1, 31 (12 saec.) : kalendarium antiphonale missae, sacramentarium, lectionarium monasterii s. agapiti cremifani (kremsmuenster) – 14mf – 8 – €27.00 – ne Slangenburg [241]

Codex sinaiticus petropolitanus et friderico-augustanus lipsiensis : the old testament / ed by Lake, H & Lake, K – Oxford, 1922 – €180.00 – ne Slangenburg [221]

Codex syro-hexaplaris ambrosianus / ed by Ceriani, Antonio Maria – Mediolani [Milan]: Impensis Bibliothecae Ambrosianae, 1874 – (= ser Monumenta sacra et profana) – 1r – 1 – 0-7905-8325-9 – mf#1987-B003 – us ATLA [090]

Der codex theresianus und seine umarbeitungen / ed by Harras, Philipp, Ritter von Harrasowsky – Wien: C Gerold's Sohn. v1-5. 1883-86 – (= ser Civil law 3 coll) – 26mf – 9 – (incl bibl ref and indexes) – mf#LLMC 96-623 – us LLMC [348]

Codex vercellensis / ed by Gasquet, Francis Aidan, Cardinal – Romae; Neo-Eboraci [New York]: F. Pustet, 1914 – (= ser Collectanea biblica latina) – 2mf – 9 – 0-7905-0835-4 – mf#1987-0835 – us ATLA [220]

Codex vercellensis : quatuor evangelia ante hieronymum latine translata ex reliquiis codicis vercellensis saeculo ut videtur quarto script et ex editione iriciana principe / ed by Belsheim, J – Christianiae: Libraria Mallingiana, 1894 – 1mf – 9 – 0-8370-1300-3 – mf#1987-6037 – us ATLA [220]

Codice franciscano s. 16. informe de la provincia del santo evangelio al visitador lc. juan de ovando – 1869 – 9 – sp Bibl Santa Ana [240]

Codice mendieta / Mendieta, Geronimo – Tomo I. 1892 – 9 – (tomo 21892) – sp Bibl Santa Ana [890]

Codice siete partidas de alfonso x / Lopez de Tovar, Gregorio – 1877. Tomo I, primera partida – 9 – (tomo 2 1887) – sp Bibl Santa Ana [946]

Codices mayas... / Villacorte, J Antonio & Villacorta, Carlos A – Guatemala, 1933; Madrid: Razon y Fe, 1934 – 1 – sp Bibl Santa Ana [348]

Codices mejicanos de fr. bernardino de sahagun / Ramirez, J F – Madrid: Fortanet, 1885. B.R.A.H. vi/pp. 85-124 – 1 – sp Bibl Santa Ana [240]

Codificacion de las leyes y disposiciones ejecutiv... / Republic Of Colombia, 1886- – Bogota, Colombia. 1937 – 1r – us UF Libraries [323]

Codificacion del trabajo / Perez Hernandez, Ramon – Bogota, Colombia. 1936 – 1r – us UF Libraries [972]

Codification of african music and textbook project / Tracey, Hugh – Roodepoort, South Africa. 1969 – 1r – us UF Libraries [780]

Codification of presidential proclamations and executive orders : apr 13 1954 to jan 20 1989 / Office of the Federal Register, National Archives and Records Administration – Washington: GPO n.d. [all publ] – 12mf – 9 – $18.00 – mf#llmc81-238 – us LLMC [324]

Codification of the regulations and orders for the government of american samoa : 1921, 1931 and 1937 editions / American Samoa. Executive Branch – 1921-37 – 13mf – 9 – $19.50 – mf#LLMC 82-100C Title 9 – us LLMC [324]

Codignola, Arturo *see* Anna giustiniani

Codigo civil / Cuba – Habana, Cuba. 1910 – 1r – us UF Libraries [350]

Codigo civil / Cuba – Habana, Cuba. 1916 – 1r – us UF Libraries [350]

Codigo civil / Cuba – Habana, Cuba. 1924 – 1r – us UF Libraries [350]

Codigo civil brasileiro / Brazil – Lisboa, Portugal. 1917 – 1r – us UF Libraries [350]

Codigo civil colombiano / Colombia Laws, Statutes, Etc – Bogota, Colombia. 1962 – 1r – us UF Libraries [350]

Codigo civil de costa rica – Madrid, Spain. 1962 – 1r – us UF Libraries [350]

Codigo civil de la republic del salvador en centro / El Salvador Laws,Statutes, Etc – San Salvador, El Salvador. 1960 – 1r – us UF Libraries [323]

Codigo civil de la republica de el salvador / El Salvador Laws, Statutes, Etc – San Salvador, El Salvador. 1913 – 1r – us UF Libraries [323]

Codigo civil de la republica de panama – Panama, 1960 – 1r – us UF Libraries [323]

Codigo civil de la republica dominicana. / Dominican Republic. Laws, Statutes, etc – Edicion autorizada. Santo Domingo, Casanova N., 1930. 279 p. LL-8006 – 1 – us L of C Photodup [348]

Codigo civil de la republica oriental del uruguay – Ed oficial. Montevideo: Impr de la Nacion 1893 [mf ed 1980] – 1r – 1 – mf#41 – us Wisconsin U Libr [348]

Codigo civil del ano de 1860 / El Salvador Laws, Statutes, Etc – San Salvador, El Salvador. v1-3. 1911 – 2r – us UF Libraries [323]

Codigo civil interpretado por el tribunal supremo... / Cuba Laws, Statutes, Etc – Habana, Cuba. v1-2. 1926 – 1r – us UF Libraries [350]

Codigo contencioso administrativo, ley 167 de 1941 / Colombia Laws, Statutes, Etc – Bogota, Colombia. 1964 – 1r – us UF Libraries [350]

Codigo de aduanas 1938 / Colombia Laws, Etc – Bogota, Colombia. 1938 – 1r – us UF Libraries [323]

Codigo de agricultura de la republica del salvador / El Salvador Laws, Statutes, Etc – San Salvador, El Salvador. 1893 – 1r – us UF Libraries [630]

Codigo de comercio de la republica dominicana – Ciudad Trujillo, Dominican Republic. 1956 – 1r – us UF Libraries [380]

Codigo de comercio terrestre / Colombia Laws, Statutes, Etc – Bogota, Colombia. 1963 – 1r – us UF Libraries [333]

Codigo de comercio vigente en la republica de cuba – Habana, Cuba. 1917 – 1r – us UF Libraries [380]

Codigo de comercio vigentes en la republica de cuba / Cuba Laws, Statutes, Etc – Habana, Cuba. 1909 – 1r – us UF Libraries [380]

Codigo de comercio y sus reformas / Costa Rica Laws, Statutes, Etc – San Jose, Costa Rica. 1965 – 1r – us UF Libraries [380]

Codigo de educacion / Costa Rica – San Jose, Costa Rica. 1965 – 1r – us UF Libraries [370]

Codigo de instruccion criminal de la republica de... / El Salvador Laws, Statutes, Etc – San Salvador, El Salvador. 1917 – 1r – us UF Libraries [360]

Codigo de justicia militar de la republica de el s... / El Salvador Laws, Statutes, Etc – Salvador, El Salvador. 1918 – 1r – us UF Libraries [355]

Codigo de la circulacion / Espana. Leyes, decretos, etc – Caceres: Imprenta Moderna, 1959 – 1 – sp Bibl Santa Ana [340]

Codigo de las siete partidas del rey d. alfonso el sabio glosadas por el lic. gregorio lopez de tovar / Alfonso 10 el Sabio, Rey de Castilla – Partida 1st. Madrid: Imp. de la Nueva Prensa, 1877 – 1 – sp Bibl Santa Ana [348]

Codigo de leyes y decretos del estado s del cauca / Cauca (Colombia) – Popayan, Colombia. 1871 – 1r – us UF Libraries [323]

Codigo de minas y codigo de petroleos / Colombia Laws, Statutes, Etc – Bogota, Colombia. 1961 – 1r – us UF Libraries [622]

Codigo de minas y leyes del petroleo / Colombia Laws, Statutes, Etc – Bogota, Colombia. 1950 – 1r – us UF Libraries [622]

Codigo de minas y petroleos / Colombia – Bogota, Colombia. 1939 – 1r – us UF Libraries [622]

Codigo de procedimiento civil / Colombia Laws, Statutes, Etc – Bogota, Colombia. 1960 – 1r – us UF Libraries [350]

Codigo de procedimiento civil y legislacion comple... / Dominican Republic – Ciudad Trujillo, Dominican Republic. 1956 – 1r – us UF Libraries [350]

Codigo de procedimiento criminal de la republica d... / Dominican Republic – Ciudad Trujillo, Dominican Republic. 1953 – 1r – us UF Libraries [360]
Codigo de procedimientos administrativos / Honduras. Laws, Statutes, etc – Tegucigalpa: Nacional, 1930. 22p. LL-8031 – 1 – us L of C Photodup [340]
Codigo de procedimientos civiles / Costa Rica Laws, Statutes, Etc – San Jose, Costa Rica. 1945 – 1r – us UF Libraries [350]
Codigo de processo penal : decreto-lei n 3689, 3- / Brazil – Rio de Janeiro, Brazil. 1941 – 1r – us UF Libraries [360]
Codigo de trabajo / Guatemala Laws, Statutes, Etc – Guatemala, 1947? – 1r – us UF Libraries [323]
Codigo de trabajo / El Salvador Laws, Statutes, Etc – San Salvador, El Salvador. 1963 – 1r – us UF Libraries [323]
Codigo de trabajo, 26 de agosto de 1943 / Costa Rica Laws, Statutes, Etc – San Jose, Costa Rica. 1943 – 1r – us UF Libraries [323]
Codigo de trabajo (decreto numero 330 del congreso / Guatemala Laws, Statutes, Etc – Guatemala, 1956 – 1r – us UF Libraries [323]
Codigo dos uzos e costumes dos habitantes nao-christaos de damao / Nery Xavier, Filippe – Nova Goa: Imprensa Nacional, 1854 [mf ed 1995] – (= ser Yale coll) – 16p – 1 – 0-524-10103-5 – (in portuguese) – mf#1995-1103 – us ATLA [241]
Codigo electoral / Panama – Panama, 1964 – 1r – us UF Libraries [320]
Codigo fiscal / Panama Laws, Statutes, Etc – Panama, 1962 – 1r – us UF Libraries [332]
Codigo fiscal de los estado unidos de colombia / Colombia (United States Of Colombia, 1863-1885) – Bogota, Colombia. 1882 – 1r – us UF Libraries [332]
Codigo judicial / Panama Laws, Statutes, Etc – Panama, 1961 – 1r – us UF Libraries [348]
Codigo militar de la republica de guatemala – Guatemala, 1958 – 1r – us UF Libraries [355]
Codigo militar expedido por el congreso de los est... / Colombia – Bogota, Colombia. 1881? – 1r – us UF Libraries [355]
Codigo militar expedido por el congreso de los est... / Colombia – Bogota, Colombia. suppl. 1881? – 1r – us UF Libraries [355]
Codigo penal / Brazil – Rio de Janeiro, Brazil. 1941 – 1r – us UF Libraries [360]
Codigo penal / Panama – Barcelona, Spain. 1917 – 1r – us UF Libraries [360]
Codigo penal boliviano / Bolivia. Laws, Statutes, etc – La Paz, Gamarra, 1902. 267 p. LL-8003 – 1 – us L of C Photodup [348]
Codigo penal brasileiro (decreto-lei n2848... / Brazil – Rio de Janeiro, Brazil. 1950? – 1r – us UF Libraries [360]
Codigo penal brazileiro / Brazil Laws, Statutes, Etc – Sao Paulo, Brazil. 1918 – 1r – us UF Libraries [360]
Codigo penal de 1879 : para las islas de cuba y puerto rico, y ley provisional para la aplicacion de sus disposiciones, concordado con las legislaciones romana, patria y extranjeras... / Orozco y Arascot, Andres de – Habana: Imprenta de G Montiel, 1879 – 2mf – 9 – $3.00 – mf#LLMC 92-314 – us LLMC [348]
Codigo penal para las islas de cuba y puerto rico / ed by Ochotorena, Manuel D – 2nd ed. Madrid: Centro Editorial de Gongora, 1891 – 7mf – 9 – $10.50 – (with catalogue) – mf#LLMC 92-302 – us LLMC [345]
Codigo penal y codigo de policia / Costa Rica – San Jose, Costa Rica. 1965 – 1r – us UF Libraries [323]
Codigos de cuba / Cuba Laws, Statutes, Etc – Barcelona, Spain. 1922 – 1r – us UF Libraries [323]
Codigos de procedimientos civiles y criminales y d... / El Salvador Laws, Statutes, Etc – Guatemala, 1858 – 1r – us UF Libraries [350]
Los codigos espanoles / Lopez de Tovar, Gregorio – Tomo V. 1872 – 9 – sp Bibl Santa Ana [946]
Codina, Luis *see* Cartas a floro
Coding and development of movement recall in laboratory and field settings / Quek, Jin-Jong – University of Queensland, 1990 – 6mf – 9 – $24.00 – mf#PSY 1899 – us Kinesology [150]
Codini Curopalatae *see* De officialibus palatii constantinopolitani (cshb37)
Codka macallinka = Teacher's voice – Mogadishu, Somalia, feb, may, dec 1973; jun 1975; jul 1976; apr 1978; feb, jun, sep 1989; feb 1990 – 1r – us CRL [370]
Codman, John *see* Ten months in brazil: with incidents of voyages and travels, descriptions of scenery and character, notices of commerce.
Codman, Ogden *see* The decoration of houses
Codorus chronicles / Southwest Pennsylvania Genealogical Services – v1 n1-v6 n4 [1983 may-1990 feb] – 1r – 1 – mf#1110707 – us WHS [929]
Codrington family *see* Records relating to the codrington family estates, barbados
Codrington, Graeme Trevor *see* Multi-generational ministries in the context of a local church
Codrington, Kenneth de Burgh *see* An introduction to the study of mediaeval indian sculpture
Codrington, Robert Henry *see* The melanesians
Cody cow boy – Cody, NE: El L Heath. 28v. v1 n1. dec 6 1900-v28 n31. jun 10 1927 (wkly) [mf ed 1900-20,1922-27 (gaps)] – 11r – 1 – (absorbed by: valentine democrat (1900)) – us NE Hist [071]
Cody, Hiram Alfred *see* An apostle of the north
Cody, Jennie L *see* Letters to betsey
Cody round-up – Cody, NE: Glen Stinson. -v16 n31. oct 28 1943 (wkly) [mf ed 1936-41 (gaps)] – 2r – 1 – (cont: cody booster) – us NE Hist [071]
Cody, SM *see* The effects of the strength shoe on vertical jump performance in male collegiate basketball players
Coe, Charles H *see* Debunking the so-called spanish mission near new smyrna beach, volusia county, florida
Coe College *see* Mwendo
Coe, Curtis P *see* Moqui mission messenger
Coe, George Albert *see* The psychology of religion
Coe, Joseph *see* The true american
Co-ed – New York. 1969-1985 (1) 1971-1985 (5) 1974-1985 (9) – (cont by: scholastic choices) – ISSN: 0009-9724 – mf#5843 – us UMI ProQuest [370]
Coedes, George *see*
– Angkor, an introduction
– Note sur l'apotheose au cambodge
Coedes, George et al *see* Un grand roi du cambodge, jayavarman 7
Coeducacion / Aradillas Agudo, Antonio – Madrid: Ediciones Studium, 1970 – sp Bibl Santa Ana [370]
Coeffeteau, N *see* Response au livre intitule le mystere d'iniquite
Coehoorn, M *see*
– Nieuwe vestingbouw, op een natte of lage horisont
– Uvelle fortification, tant pour un terrain bas...
– Versterckinge des vijf-hoecks
– Wederlegginge der architectura militaris...
Coelestina : eine maerchenlegende / Binding, Rudolf Georg – Hamburg: H Dulk, [194-?] [mf ed 1989] – 63p – 1 – mf#7024 – us Wisconsin U Libr [390]
Coelestis urbs ierusalem : aphorismen: nebst einer beilage / Laemmer, Hugo – Freiburg i.B.: Herder, 1866 – 1mf – 9 – 0-8370-7300-6 – (incl bibl ref) – mf#1986-1300 – us ATLA [780]
Coelho, Angelica *see* Ritmos humanos
Coelho De Souza, Jose Pereira *see* Pensamento politico de assis brasil
Coelho, Jose Saldanha *see* Deputado no exilio
Coelho, Jose Simoes *see* Brasil contemporaneo
Coelho Netto, Henrique *see*
– Agua de juventa
– Contos da vida e da morte
– Paginas escolhidas
– Rei negro
Coelina : ou l'enfant du mystere / Pixerecourt, Rene-Charles Guilbert De – Paris, France. 1801 – 1r – us UF Libraries [440]
Coelln, Freiherr von *see* Neue feuerbraende
Coello, Francisco *see*
– Boletin de la real academia de la historia. informes
– Vias romanas entre toledo y merida. informes
Coelner vereins-zeitung – Koeln DE, 1902-1903 – 1 – gw Misc Inst [360]
Coelum empyreum : non vanis et fictis constellationum monstris belluatum... / Engelgrave, H – Coloniae: Apud Gabrielem...Roy; Amstelodami, 1669 – 17mf – 9 – mf#O-3066 – ne IDC [090]
Coelum empyreum : non variis et fictis constellationum monstris belluatum... / Engelgrave, H – Coloniae Agrippinae: Sumptibus Haered. Thomae von Coellen et Josephi Huisch, 1727 pt1 – 11mf – 9 – mf#O-1566 – ne IDC [090]
Coen, Edwidg *see* En marge d'une confederation economique inter-anti...
Coen, Jan Pieterszoon *see* Bescheiden omtrent zijn bedrijf in indie
Coens eerherstel / Gerretson, Frederik Carel – Amsterdam: P N van Kampen 1944 [mf ed 1987] – 1r – 1 – (incl bibl footnotes. filmed with: numbers in history / delbruck, h & other titles) – mf#6748 – us Wisconsin U Libr [959]
Coepenicker dampfboot – Berlin DE, 1890 11 nov & 1893 24 apr, 1896 1 jul-1897, 1898 1 jul-1900 30 jun, 1901 2 jan-29 jun, 1902 1 mar-30 jun, 1903 2 jan-30 jun, 1904 2 jan-30 jun, 1905 1 jul-1906 30 jun, 1907 1 jul-31 dec, 1909, 1911 2 jan-30 jun, 1912-13, 1915 1917, 1921, 1925 2 jan-31 mar, 1925 2 jun-1928 sep, 1929 jan-mar, 1929 jul-1943 31 jul, 1944 1 jul-31 aug – 1 – (title varies: 2 jan 1923: das dampfboot; 2 jan 1934: berliner neueste nachrichten; 15 mar 1943: das dampfboot-berliner ostzeitung; 1890 11 nov & 1893 24 apr) – gw Misc Inst [074]
Coepenicker tageblatt – Berlin DE, 1922, 1 jul-1923, 1925-1928 31 mar, 1928 2 jul-31 aug, 1928 1 oct-1937 30 sep, 1938 1 jul-1939, 1940 11 jan-29 nov, 1941 3 jan-30 aug & oct, 1942-1943 27 feb – 1 – gw Misc Inst [074]
"Coercive agency" : james henderson's lovedale, 1916-1930 / Duncan, Graham Alexander – Uni of South Africa 2000 [mf ed Pretoria: UNISA 2000] – 7mf – 9 – (incl bibl ref) – mf#mfm14725 – sa Unisa [377]
Coerper, F *see* Jeremia
Coerver, Hubert *see* Kalunga
Coester, Alfred Lester *see*
– Literary history of spanish america
Coetlogon, C De *see* Seasonable caution against the abominations of the church of rome
Coetser, Paulus Petrus Johannes *see* Gebeurtenisse uit di kaffer-oorloge fan 1834
Coetzee, Gerrit Abraham *see* The republic
Coetzee, Pieter Hendrik *see* Onverwagse uitdienstrede
Le coeur – Paris. no. 1-10. avr 1893-juin 1895. [mnthly] – 1 – (Esoterisme, litterature, sciences, arts. mq n8) – fr ACRPP [073]
Coeur d'Alene Tribal Council *see* Council fires
Coeur ebloui / Descaves, Lucien – Paris, France. 1927 – 1r – us UF Libraries [440]
Le coeurs embellis : 1914-1915: 2. milleth / Hepp, Alexandre – Paris: E Fasquelle, 1916 [mf ed: Bethlehem, PA: Mid-Atlantic Preservation Service for NYPL, 1987] – 1r – 1 – mf#*Z-4872 – Located: NYPL – us Misc Inst [940]
Coevolution quarterly – Sausalito. 1974-1984 (1) 1974-1984 (5) 1974-1984 (9) – ISSN: 0095-134X – mf#12706 – us UMI ProQuest [073]
Co-existence – Dordrecht. 1991-1995 (1,5,9) – (cont by: international politics) – ISSN: 0587-5994 – mf#16774 – us UMI ProQuest [300]
Coexistence – v1-31. 1964-94 – 9 – $655.00 set – ISSN: 0587-5994 – mf#114571 – us Hein [073]
Coffee and tea industries, spices and flavors – New York. 1950-1963 (1) – mf#209 – us UMI ProQuest [630]
Coffee break / Office and Professional Employees International Union – v12 n3-v15 n2 [1977 jun-1980 fall] – 1r – 1 – mf#679079 – us WHS [331]
Coffee break / United Steelworkers of America – v2 n2-v5 n5 [1977 jun, special iss-1980 mar] – 1r – 1 – mf#669862 – us WHS [331]
Coffee room journal : illustrated monthly reader – [Scotland] Glasgow: Wilson & Co jun 1886-dec 1888 (mthly) [mf ed 2003] – 2r – 1 – (cont by: glasgow literary journal and coffee-room advertiser; glasgow literary journal and family reader [jan 1889-may 1891]) – uk Newsplan [400]
Coffee Tavern Co Ltd *see* Practical hints for the management of coffee taverns
Coffey, Achilles *see* History of the regular baptists
Coffey, Aeneas *see*
– Observations on the rev edward chichester's pamphlet
Coffey, Peter *see* Ontology, or, the theory of being
Coffey's probate reports / California – v1-6. 1883-1908 (all publ) – (= ser California appellate reports) – 41mf – 9 – $61.00 – mf#LLMC 91-037 – us LLMC [340]
Coffin, Fulton Johnson *see* The third commandment
Coffin, G *see* Histoire veritable et naturelle des moeurs et productions du pays de la nouvelle france
Coffin, Henry Sloane *see*
– Social aspects of the cross
– The ten commandments
Coffin, Rhoda Moorman *see* Rhoda m. coffin
Coffin, Robert A *see* Christian virtues and the means for obtaining them
Coffin, Victor *see* The province of quebec and the early american revolution
Le coffret ou le tresor enfoui : maniere de decouvrir un tresor: histoire merveilleusement veritable et veritablement merveilleuse / Bois, Louis-Edouard – [Montreal: s.n, 1872] – 2mf – 9 – 0-665-01163-6 – mf#01163 – cn Canadiana [440]
Coffs harbour advocate – Coffs Harbour, 1907, 1929, 1946-1968 – at Pascoe [079]
Coffs harbour advocate – Coffs Harbour, apr 1907-apr 1942 – 14r – A$815.89 vesicular A$892.89 silver – at Pascoe [079]
Coffs harbour advocate – Coffs Harbour, jan 1969-jun 1997 – at Pascoe [079]
La cofradia cacerena de nuestra senora de la paz / Munoz de San Pedro, Miguel – Badajoz: Diput.Prov. Badajoz, 1949 – 1 – sp Bibl Santa Ana [946]
Cofradia de la Santisima Virgen del Pilar de Casas de Don Antonio *see* Estatutos para el regimen y administracion de la...
Cofradia de los Ramos, Cristo de la Buena Muerte y Virgen de la Esperanza *see* Estatutos de la cofradia de los ramos, cristo de la buena muerte y virgen de la esperanza
Cofradia de Ntra. Sra. de la Montana, Real. Caceres *see*
– Memoria y cuenta general 1947-48
– Memoria y cuenta general...ano 1942
– Memoria y cuenta general...ano 1949
Cofradia de Nuestra Senora De la Santisima Virgen de la Montana *see* Ordenanzas para su regimen y administracion. estatutos a actualizar
Cofradia de Nuestra Senora de la Soledad y del Santo Entierro *see* Estatutos de la real...de caceres
Cofradia de San Cristobal *see*
– Fiestas de san cristobal 1974
– Fiestas de san cristobal 1976
Cofradia Santisimo Sacramento *see* Estatutos de la...en la iglesia parroquial de almendralejo
Cofrancesco, Lisa *see* Hostility and coronary risk factors among native americans and caucasians
Cofresi : novela / Tapia Y Rivera, Alejandro – San Juan, Puerto Rico. 1944 – 1r – us UF Libraries [830]
Cogan, E *see* Sermon on the purity and perfection of christian morality
Cogan, glamorgan, parish church of st peter : baptisms 1724-1857, burials 1724-1823, marriages 1728-1821; & lavernock, st lawrence, baptisms 1724-1904, burials 1724-1902, marriages 1725-1836 – 1mf – 9 – £1.25 – uk Glamorgan FHS [929]
Cogan, st peter; and lavernock, st lawrence; and sully, st john baptist, monumental inscriptions – 1mf – 9 – £1.25 – uk Glamorgan FHS [929]
Coggeshall, George *see* History of the american privateers, and letters-of-marque during our war with england in the years 1812, '13 and '14
Coggin, Frederick Ernest *see* Man's estate
Coggins memorial baptist church's involvement in a meaningful world hunger ministry / Bunce, Dearl Linwood I – 1982 – 1 – 5.84 – us Southern Baptist [242]
Cogic challenger / Church of God in Christ – v1 n8 [[197-?] jan/mar] – 1r – 1 – mf#4114889 – us WHS [240]
Cogitata physico-mathematica *see* F marini mersenni minimi cogitata physico-mathematica
Cogitationes et dissertationes theologicae / Turrettini, J A – Geneve, Barillot, 1737. 2 v – 10mf – 9 – mf#PFA-187 – ne IDC [240]
Cogitationum rationalium de deo, anima et malo, libri quatuor / Poiret, P – Amsterdam, 1677 – 4mf – 9 – mf#PPE-199 – ne IDC [240]
Cogitationum rationalium libri quatuor : accesit dissertatio ubi de duplici descendi methodo... / Poiret, P – Amsterdam, 1715 – 11mf – 9 – mf#PPE-201 – ne IDC [240]
Cogitationum rationalium libri quatuor / Poiret, P – Ed 2. Amsterdam, 1685 – 10mf – 9 – (cum animadversionibus p. bayle) – mf#PPE-200 – ne IDC [240]
La cognee : organe du front de liberation du quebec – [Quebec]: FLQ. n1 oct 1963-n66 15 avril 1967 (mthly) [mf ed 1982] – 1r – 1 – mf#SEM35P164 – cn Bibl Nat [325]
Cogniard, Hippolyte *see* De schim van maria
Cogniard, Theodore *see*
– 1841 et 1941
– Biche au bois
– Diners a trente-deux sous
– Trois quenouilles
Cognition – Lausanne. 1972+ (1) 1972+ (5) 1987+ (9) – ISSN: 0010-0277 – mf#42150 – us UMI ProQuest [150]
Cognition and athletic behavior : an investigation of the nlp principle of congruence / Ingalls, Joan S – 1987 – 2mf – 9 – $8.00 – mf#PSY 2072 – us Kinesology [150]
Cognition and instruction – Hillsdale, 1996+ – 1,5,9 – ISSN: 0737-0008 – mf#25212 – us UMI ProQuest [370]
The cognitive, affective, and behavioral characteristics of students enrolled in physical education activity classes at brigham young university / Jorgenson, Shane M – 1998 – 1mf – 9 – $4.00 – mf#PE 4011 – us Kinesology [370]
Cognitive, affective and behavioral neuroscience – Austin. 2001+ (1,5,9) – ISSN: 1530-7026 – mf#32352 – us UMI ProQuest [150]
Cognitive brain research – Amsterdam. 1992+ (1,5,9) – ISSN: 0926-6410 – mf#42683 – us UMI ProQuest [612]
Cognitive development – Norwood. 1998+ (1,5,9) – ISSN: 0885-2014 – mf#26005 – us UMI ProQuest [150]
Cognitive science – Norwood, 1999+ [1,5,9] – ISSN: 0364-0213 – mf#19204 – us UMI ProQuest [150]
Cognitive therapy and rational education : a theory based program using adventure, challenge, and recreation / Lundberg, Neil R – 1997 – 1mf – 9 – $4.00 – mf#PSY 1991 – us Kinesology [150]
Cognitive therapy and research – New York. 1989+ (1,5,9) – ISSN: 0147-5916 – mf#11499 – us UMI ProQuest [150]

The cognitive validity of the pedagogy of the critical cross-field outcomes / Majolo, Dina Matseliso – Pretoria: Vista University 2000 [mf ed 2000] – 2mf – 9 – mf#mfm15323 – sa Unisa [370]

Cogolludo, Francisco de *see* Historia de las vidas y milagros de nuestro beato p fr pedro de alcantara

Cogswell, William *see* Letters to young men preparing for the christian ministry

Cogumelos / Accioly, Breno – Rio de Janeiro, Brazil. 1944 – 1r – us UF Libraries [972]

Cohasset 1717-1900 – Oxford, MA (mf ed 1992) – (= ser Massachusetts vital records) – 38mf – 9 – 0-87623-120-2 – (mf 1-2: births & deaths 1732-1864. mf 2: marriages 1798-1835. mf 3-4: vital & town records 1739-1864. mf 5-10: town meetings 1717-1813. mf 11-18: town records 1814-50. mf 19-21: vital records 1844-66. mf 22-24: marriage certificates 1848-76. mf 25-26: index to births 1844-1900. mf 27-28: index to marriages 1844-1900. mf 29-30: index to deaths 1844-1900. mf 31-32: births 1844-66; marriages 1844-53. mf 32-34: deaths 1844-1900. mf 35-36: births 1867-1900. mf 36-38: marriages 1862-19004) – us Archive [978]

Cohasset 1737-1849 – Oxford, MA (mf ed 1995) – (= ser Massachusetts vital record transcripts to 1850) – 6mf – 9 – 0-87623-232-2 – (mf 1t-2t: births & deaths 1737-1843. mf 2t: marriages 1798-1849. mf 2t-4t: births & deaths 1769-1881. mf 4t: marriages 1837-1844. mf 4t,6t: marriages & intentions 1846-49. mf 4t-5t: hingham marriages 1747-91. mf 5t: out-of-town marriages 1770-96; births, deaths 1844-49. mf 6t: deaths 1844-49) – us Archive [978]

Coheleth, commonly called the book of ecclesiastes / Ginsburg, Christian David – London: Longman, Green, Longman, and Roberts, 1861 – 2mf – 9 – 0-7905-1399-4 – (in english and hebrew. incl bibl ref) – mf#1987-1399 – us ATLA [221]

Cohen, A *see* Le talmud

Cohen, Andrew *see* British policy in changing africa

Cohen, Chapman *see* What is freethought?

Cohen, David William *see* Selected texts

Cohen, E G W *see* Air of mozart

Cohen, Gustave *see* Un grand romancier d'amour et d'aventure au 13e siecle. chretien de troyes.

Cohen, Hermann *see*
– Festschrift zum siebzigsten geburtstage jakob guttmanns
– Hermann cohens juedische schriften

Cohen, Isidor *see* Historical sketches and sidelights of miami, florida

Cohen, J M *see* Pengiun book of spanish verse

Cohen, Jenna S *see* A case study of a multiple-joint resistance exercise for an individual with cerebral palsy

Cohen, Joseph *see*
– Les deicides
– The deicides

Cohen, Liber *see* Hidushe haviva

Cohen, M *see* Traite de langue amharique

Cohen, Marx Edwin *see* Plantation records

Cohen, Morris L *see* The yale law library blackstone collection

Cohen, Ronald *see* Dominance and defiance

Cohen, Selma Jeanne *see* Dance notation conversation; facts on the only successful system of recording human movement

Cohen, Tobias *see* Ma'aseh toviyah

Cohen's gazette and lottery register – Baltimore. 1814-1830 (1) – mf#4440 – us UMI ProQuest [978]

The cohensive elemnts of british imperialism / Ireland, Alleyne – S.l: s.n, 1899? – 1mf – 9 – (in dble clms) – mf#17807 – cn Canadiana [320]

Cohesion and perceived parental purposes of sport / Parrow, Darlene M – 1999 – 2mf – 9 – $8.00 – mf#PSY 2088 – us Kinesology [150]

Cohesion of coacting and interacting female intercollegiate athletes / Matheson, H E – 1991 – 2mf – 9 – $8.00 – us Kinesology [150]

Cohn, Alfons Fedor *see* Briefe an brinkmann, henriette v finckenstein, wilhelm v humboldt, rahel, friedrich tieck, ludwig tieck und wiesel

Cohn, Emil *see* Juedischer bote vom rhein

Cohn, Georg *see* Die gesetze hammurabis

Cohn, Gustav *see* Gemeindeblatt der israelitischen religionsgemeinde zu leipzig

Cohn, Julius *see* Des samuel al-magrebi abhandlungen ueber die pflichten der priester und richter bei den karaeern

Cohn, Max Conrat *see* Die christenverfolgungen im roemischen reiche

Cohn, Michael A *see* Some questions and an appeal

Cohocton valley times and index – Cohocton, NY. 1893-1963 (1) – mf#64931 – us UMI ProQuest [071]

Cohrs, Ferdinand *see* Philipp melanchthon

Cohu, J R *see*
– The old testament in the light of modern research
– Vital problems of religion

Cohu, John Rougier *see*
– Our father
– Through evolution to the living god

Coicou, Massillon *see* Genie francais et l'ame haitienne

Coiffier de Moret, Simon *see* Histoire du bourbonnais et des bourbons qui l'ont possede

Coignet, Clarisse *see* La reforme francaise avant les guerres civiles, 1512-1559

Coignet, Clarisse Gauthier *see* Francis the first and his times

Coile, Nancy C *see*
– Common plants of florida's aquatic plant industry
– Notes on nomenclature of citrus and some related genera...

Coillard, Francois *see*
– On the threshold of central africa
– Sur le haut-zambeze

Coillard of the zambesi : the lives of francois and christina coillard, of the paris missionary society, in south and central africa, 1858-1904 / Mackintosh, Catharine Winkworth – 2nd ed. London: T F Unwin, 1907 – 2mf – 9 – 0-7905-4833-X – (incl bibl ref) – mf#1988-0833 – us ATLA [240]

Coin and stamp – Toronto: Greenslade Bros, [1882] – 9 – ISSN: 1190-7509 – mf#P04567 – cn Canadiana [730]

The coin collector / Hazlitt, William Carew – London 1896 – (= ser 19th c art & architecture) – 4mf – 9 – mf#4.2.1529 – uk Chadwyck [730]

Coin collector and shopper – v33 n462-v34 n479 [1968 jul-1969 dec] – 1r – 1 – mf#1054593 – us WHS [730]

Coin de rue : ou, le rempailleur de chaises / Brazier, Nicholas – Paris, France. 1820 – 1r – us UF Libraries [440]

Le coin du feu – Basse-Ville [Quebec]: Frechette, 1840-1841 – 9 – mf#P04133 – cn Canadiana [870]

Le coin du feu – [Montreal: s.n, 1829] – 9 – mf#P04140 – cn Canadiana [370]

Le coin du feu – Montreal: [s.n.] ([s.l.]: impr Desaulniers) (mthly) [mf ed 1984] – 1r – 5 – mf#SEM16P342 – cn Bibl Nat [305]

Coin prices : the standard guide to current values for all united states coins – 1975 mar-1977 may, 1977 jul-1979 nov, 1980-81, 1982-1983 jul, 1983 sep-1985 may, 1985 jul-1986 nov, 1986 jan-1988 may – 7r – 1 – mf#515932 – us WHS [730]

Coin world – Sidney. 1960+ (1) 1988+ (5) 1988+ (9) – ISSN: 0010-0447 – mf#7901 – us UMI ProQuest [929]

Coin world – 1970 may 13/1970 sep 2-1981 aug/oct 14 – 44r – 1 – (with gaps) – mf#1054596 – us WHS [730]

Coinage – Ventura. 1972-1996 [1,5]; 1976-1996 [9] – ISSN: 0010-0455 – mf#6665 – us UMI ProQuest [929]

The coinage of the british empire... : from the earliest period to the present time / Humphreys, Henry Noel – London 1854 – (= ser 19th c art & architecture) – 3mf – 9 – mf#4.2.1728 – uk Chadwyck [730]

Coin-op – New York. 1960-1973 (1) 1971-1973 (5) – (cont by: american coin-op) – ISSN: 0010-0404 – mf#1675 – us UMI ProQuest [660]

Coins – Iola. 1955+ (1) 1971+ (5) 1977+ (9) – ISSN: 0010-0471 – mf#5823 – us UMI ProQuest [929]

Coins and medals – London. 1967-1972 (1) 1971-1972 (5) – ISSN: 0010-048X – mf#2553 – us UMI ProQuest [929]

Coin's financial series – n7 [1895] – 1r – 1 – mf#2699051 – us WHS [332]

Coins of bible days / Banks, Florence Aiken – 1959 – 9 – $10.00 – us IRC [730]

Coins of the jews / Madden, Frederic William – London: Truebner, 1903 – 4mf – 9 – 0-8370-1672-X – (incl bibl ref) – mf#1987-6102 – us ATLA [930]

Cointelpro : the counterintelligence program of the fbi / U.S. Federal Bureau of Investigation – 1978 – 30r – 1 – $3900.00 – mf#S1753 – us Scholarly Res [360]

O coio : hebdomadario illustrado e humoristico – Rio de Janeiro, RJ. 01 abr-jun 1901; jan-13 mar 1902 – (= ser Ps 19) – mf#P05,04,151 – bl Biblioteca [079]

Coisas que o povo diaz / Cascudo, Luis Da Camara – Rio de Janeiro, Brazil. 1968 – 1r – us UF Libraries [972]

Coiscou-Weber, Rodolfo Juan *see* Velero del regreso

Coit, Stanton *see* National idealism and the book of common prayer

Coit, Stanton et al *see* Ethical world series, 1898-1916

Coit, Thomas Winthrop *see*
– Lectures on the early history of christianity in england

Coittant, Philippe *see* Tableau des prisons de paris, sous le regne de robespierre

Coity, glamorgan, parish church of st mary : baptisms 1713-1920, burials 1720-1916, marriages 1728-1833 – [Glamorgan]: GFHS [mf ed [2001?] – 3mf – 9 – £3.75 – uk Glamorgan FHS [929]

Coity, st mary, monumental inscriptions – 2mf – 9 – £2.50 – uk Glamorgan FHS [929]

Coke, Edward *see*
– Axiomata ex commentariis ejus
– Legal and state memoranda
– Legal, political and personal papers

Coke family *see* Correspondence of the coke family

Coke, Henry John *see* The domain of belief

Coke, Thomas *see* Character and death of mrs hester ann rogers

Coke, Thomas William *see*
– Holkham office cash accounts, 1808-1844
– Letters to t w coke

Coker, Ernest George *see* An investigation into the elastic constants of rocks

Coker kin newsletter – v1 n1-2 [1977 nov-1978 jun] – 1r – 1 – mf#403988 – us WHS [071]

Co-kim van-tap = Gu jin wen ji – Hanoi: [s.n.] n1 [jan 1930-] – 1r – 1 – mf#mf-12548 seam – us CRL [079]

Cokwe expansion, 1850-1900 / Miller, Joseph Calder – Madison, WI. 1969 [1967] – 1r – us UF Libraries [960]

Col david fanning's narrative of his exploits and adventures as a loyalist of north carolina in the american revolution : supplying important omissions in the copy published in the united states / Fanning, David – Toronto: [s.n.] 1908 [mf ed 1998] – 1mf – 9 – 0-665-98061-2 – (int & notes by alfred william savary; repr fr: canadian magazine) – mf#98061 – cn Canadiana [975]

Col. h. c. hart's new and improved instructor for the drum / Hart, H C – New York: the Author, 1862. MUSIC 3081 – 1 – us L of C Photodup [780]

Cola call : monthly newsletter of the cola clan – 1984 jan-1986 jun – 1r – 1 – (cont by: coca cola collectors news) – mf#1098870 – us WHS [929]

La colaboracion de la caja extremena de prevision social en el fomento de las construcciones escolares / Leal Ramos, Leon – Caceres: Imprenta Moderna, 1929 – sp Bibl Santa Ana [946]

Colaboradores de santander en la organizacion de l... / Acevedo Latorre, Eduardo – Bogota, Colombia. 1944 – 1r – us UF Libraries [972]

Colaco, P A *see* Select writings of the most reverend dr leo meurin

Colam, Lance *see* The death treasure of the khmers

Colani, Timothee *see* Jesus christ et les croyances messianiques de son temps

Colas et colinette ou le bailli dupe : comedie en trois actes, et en prose, melee [sic] d'ariettes / Quesnel, Joseph – Quebec: chez John Neilson, 1808 [mf ed 1988] – 1mf – 9 – mf#SEM105P889 – cn Bibl Nat [780]

Colbeck, George H *see* Letters from mandalay

Colbeck, James Alfred *see* Letters from mandalay

Colbert county banner – Tuscumbia, AL. 1895-1896 (1) – mf#62046 – us UMI ProQuest [071]

Colbert county reporter – Tuscumbia, AL. 1911-1969 (1) – mf#62047 – us UMI ProQuest [071]

Colbert, William *see* A journal of the travels of william colbert, methodist preacher, thro' parts of maryland, pennsylvania, new york, delaware and virginia in 1790, 1, 2, 3, 4, 5, 6, 7, 8

The colbertine breviary, vol 1-2 (hbs43-44) / Gambier-Parry, T R – 1912-1913 – (= ser Henry bradshaw society (hbs)) – 2v on 10mf – 8 – €19.00 – ne Slangenburg [241]

Colby, Charles Carroll *see*
– Canada's national policy
– An open letter from mr c c colby mp to mr c h mackintosh, editor, ottawa citizen
– Tariff re-adjustment

Colby community caller – Colby WI. 1950 apr-1959 nov – 1r – 1 – mf#1224727 – us WHS [071]

Colby, John H *see* The statute railroad laws of the state of new york.

Colby. Kansas. Public Library. Board of Directors *see* Pioneer memorial library records

Colby library quarterly – Waterville. 1943-1989 [1] 1970-1989 [5] 1973-1989 [9] – (cont by: colby quarterly) – ISSN: 0010-0552 – mf#1602 – us UMI ProQuest [020]

Colby, Merle Estes *see* Virgin islands

Colby phonograph – Abbotsford, Colby WI. 1918 mar 21/jul 18-1959/1963 jan 31 – 21r – 1 – (with gaps; cont: phonograph; cont by: tribune-phonograph) – mf#966856 – us WHS [071]

Colby quarterly – Waterville. 1990+ [1] 1990+ [5] 1990+ [9] – (cont: colby library quarterly) – ISSN: 1050-5873 – mf#1602,01 – us UMI ProQuest [020]

Colchester baptist church. colchester, connecticut : church records – 1780-1939. (Scott Hill). 2v. 276p – 1 – us Southern Baptist [242]

Colchester, Elizabeth Susan (Law) Abbot, Baroness *see* Fitz-edward

Cold regions science and technology – Amsterdam. 1979+ (1) 1979+ (5) 1986+ (9) – ISSN: 0165-232X – mf#42151 – us UMI ProQuest [600]

Cold storage studies of florida citrus fruits : effect of temperature and maturity / Stahl, Arthur L – Gainesville, FL. 1936 – 1r – us UF Libraries [634]

Cold storage studies of florida citrus fruits / Stahl, Arthur L – Gainesville, FL. 1937 – 1r – us UF Libraries [634]

Cold storage studies of florida citrus fruits n ii : effect of various wrappers and temperatures / Stahl, Arthur L – Gainesville, FL. 1936 – 1r – us UF Libraries [634]

Colden, C *see* The history of the five indian nations of canada

Coldstream, John Phillips *see* The development of the teaching of law in the university of edinburgh

Coldwater baptist church. collierville, tennessee : church records – 1867-1979. 1276p – 1 – 51.04 – us Southern Baptist [242]

Coldwater. Kansas. Police Court *see* Docket

Coldwell, William *see* The maitland distillery case

Cole, Alan Summerly *see*
– Cantor lectures on the art of lace-making
– A renascence of the irish art of lace-making

Cole, Alfred W *see* Das kap und die kaffern

Cole, Arthur Augustus *see* Mineral wealth along the temiskaming and northern ontario railway

Cole, C A *see* Memorials of henry the fifth (rs11)

Cole, C W *see* Report on land tenure

Cole county democrat – Jefferson City, MO. 1884-1909 (1) – mf#64168 – us UMI ProQuest [071]

Cole, David *see* The teaching of our lord

Cole, Desmond T *see*
– Course in tswana
– Some features of ganda linguistic structure

Cole, E L *see* An application of item response theory to the rest of gross motor development

Cole, Edna Earle *see* The good samaritan

Cole, Emma L Taylor *see* Guide to the mushrooms

Cole, Ernest *see* House of bondage

Cole, F G *see* Mother of all churches

Cole Family *see* Papers

Cole, Frederick Minden *see*
– The canadian artillery team at shoeburyness, 1896
– Canadian boy scouts

Cole, George *see*
– How can a man be born when he is old?
– How shall i put thee among the children?
– How then can man be justified with god?
– Lord, what wilt thou have me to do?

Cole, George Percy *see* The conservation of natural resources through the electrification of railways

Cole, Grenville Arthur James *see* The growth of europe

Cole, H *see* Documents illustrative of english history in the 13th and 14th centuries

Cole, Harriet *see* Songs from the valley

Cole, Henry *see*
– Church of england's portrait
– Journal of design and manufactures, 1849-1862
– Letter most respectfully addressed to the lord bishop of london
– Notes for a universal art inventory of works of fine art which may be found throughout europe, for the most part in ecclesiastical buildings and in connexion with architecture

Cole, Henry Hardy *see*
– The architecture of ancient delhi
– Fifty-one photographic illustrations
– First exercises for children in light, shade, and colour
– A hand-book for the architecture, tapestries, paintings...and grounds of hampton court
– Illustrations of ancient buildings in kashmir
– Illustrations of buildings near muttra and agra
– Notes for a universal art inventory of works of fine art

Cole, Henry [pseud. Felix Summerly] *see* A hand-book for the architecture, sculptures, tombs, and decorations of westminster abbey

Cole, Isaac *see* Diary

Cole, Jim *see* The serials librarian

Cole, John Y *see* The library of congress

Cole, Joseph *see* Memoir of miss hannah ball, of high-wycomb, in buckinghamshire

Cole, Kelly J *see*
– The effect of altering speed of backward movement of the trunk on anticipatory postural adjustments
– The effects of imaginary maximal muscle contraction training on the voluntary neural dirve to muscle

Cole, Monica M *see* South africa

Cole, Robert Henry *see* The anglican church

Cole, Sydney William *see* Practical physiological chemistry

Cole, Timothy *see* Old dutch and flemish masters

Cole, William *see* Life in the niger

COLEBROOKE

Colebrooke, Henry Thomas see Essays on the religion and philosophy of the hindus
Colecao da casa dos contos de ouro preto / Mathias, Herculano Gomes – Rio de Janeiro, Brazil. 1966 – 1r – us UF Libraries [972]
Colecao de portugal / Arquivo Nacional (Brazil) – Rio de Janeiro, Brazil. 1959 – 1r – us UF Libraries [972]
Coleccion arqueologica antillana / Hostos, Adolfo De – San Juan, Puerto Rico. 1955 – 1r – us UF Libraries [930]
La coleccion canonica hispana 1. estudio por...madrid, 1966 / Martinez Diez, Gonzalo – Madrid: Graf. Calleja, 1966 – 1 – sp Bibl Santa Ana [946]
Coleccion completa de las disposiciones legislativas expedidas desde la independencia de la republica / Mexico. Laws, Statutes, etc – v1-42. 1876-1912 – 9 – $588.00 – mf#0360 – us Brook [348]
Coleccion completa de leyes nacionales sancionadas por el honorable congreso / Argentine Republic. Laws, Statutes, etc – 1852-1934 – 1 – us L of C Photodup [324]
Coleccion Contemporaneos see Poetas de guatemala
Coleccion de articulos de anselmo suarez y romero / Suarez Y Romera, Anselmo – Habana, Cuba. 1859 – 1r – us UF Libraries [972]
Coleccion de articulos satiricos y de costumbres / Cardenas Y Rodriguez, Jose Maria De – Havana, Cuba. 1963 – 1r – us UF Libraries [972]
Coleccion de bulas, breves y otros documentos relativos a la iglesia de america y filipinas – Bruselas: A Vromant, 1879 – 5mf – 9 – 0-524-05000-7 – mf#1990-5088 – us ATLA [240]
Coleccion de cuadrillas : segunda edicion, que comprende doce figurados – Mexico: Oficina de la Testamentaria de Valdes, a Cargo de J M Gallegos, 1835 – 1 – (without music) – mf#*ZBD-*MGO pv11 – Located: NYPL – us Misc Inst [790]
Coleccion de "cuadros sinopticos" de los pueblos, haciendad y ranchos del estado libre y soberano de oaxaca : anexo numero 50 a la memoria administrativa presentada al h. congreso del mismo el 17 de setiembre de 1883 / Martinez Gracida, Manuel – Oaxaca: Impr del Estado, a cargo de I Candiani 1883 [mf ed 1980] – 1v on 1r – 1 – mf#102 – us Wisconsin U Libr [972]
Coleccion de cuentos / Febres Cordero, Julio – Caracas, 2nd ed 1930; Madrid: Razon y Fe, 1931 – 1 – sp Bibl Santa Ana [390]
Coleccion de decretos del rey fernando 7th – Madrid, 1814-1833 – 178mf – 9 – sp Cultura [340]
Coleccion de decretos y ordenes expedidas por las cortes generales yextraordinarias – 1810-1814; 1820-1823 – 63mf – 9 – sp Cultura [340]
Coleccion de diarios y relaciones para la historia de los viajes y descubrimientos. vol 1...vol 2...madrid, 1943 / Bayle, Constantino – Madrid: Razon y Fe, 1946 – 1 – sp Bibl Santa Ana [946]
Coleccion de discursos y poesias leidos en el acto de la inauguracion del monumento que se ha de erigir en badajoz a la memoria de d. jose moreno nieto – Fregenal: Imprenta El Eco, 1883 – 1 – sp Bibl Santa Ana [810]
Coleccion de documentos historicos – Coruna: Academia 1915-70 [mf ed 1985] – 2r – 1 – (filmed with: boletin de la real academia gallega) – mf#1528 supp – us Wisconsin U Libr [360]
Coleccion de documentos importantes relativos / El Salvador Ministerio De Relaciones Exteriores – San Salvador, El Salvador. 1921 – 1r – us UF Libraries [972]
Coleccion de documentos ineditos del... / Spain. Archivo General de la Corona de Aragon, Barcelona – v1-41. 1847-1910 – 1 – $498.00 – mf#0565 – us Brook [946]
Coleccion de documentos ineditos para la historia de chile / Medina, Jose Toribio – Santiago de Chile. v. 1-30. 1888-1902. (Wanting v. 19) – 1 – us L of C Photodup [972]
Coleccion de documentos ineditos para la historia de espana – Madrid. v1-113. 1842-95 – 1 – $720.00 – mf#0156 – us Brook [946]
Coleccion de documentos ineditos para la historia de espana / Spain – v. 44-112. 1864-95 – 1 – us L of C Photodup [946]
Coleccion de documentos ineditos para la historia de iberoamerica / Montoto, Santiago; ed by Bayle, Constantino – Madrid: Razon y Fe, 1928 – 9 – sp Bibl Santa Ana [972]
Coleccion de documentos ineditos, relativos al descubrimiento, conquista y organizacion de las antiguas posesiones espanolas de america y oceania / America – v. 1-42. 1864-84 – 1 – us L of C Photodup [946]
Coleccion de documentos ineditos, relativos al descubrimiento, conquista y organizacion de las antiguas posesiones espanolas de america y oceania – Madrid. v1-42. 1864-84 – 1 – $276.00 – mf#0157 – us Brook [900]
Coleccion de documentos ineditos relativos al descubrimiento, conquista y organizacion de las antiguas posesiones espanolas de ultramar – Madrid. v1-25. 1885-1932 – 1 – $276.00 – mf#0158 – us Brook [900]
Coleccion de documentos para la historia de costa-... / Fernandez, Leon – San Jose, Costa Rica. v1-10. 1881-1907 – 3r – us UF Libraries [972]
Coleccion de documentos relativos al adelantado capitan don sebastian de belalcazar 1535-1560 / Bayle, Constantino – Quito, 1936; Burgos: Razon y Fe, 1938 – 1 – sp Bibl Santa Ana [946]
Coleccion de epigrames / Salas, Francisco Gregorio de – 1816 – 9 – (1827) – sp Bibl Santa Ana [880]
Coleccion de escritores castellanos / Lopez de Ayala, Adelardo – 1885 – 9 – sp Bibl Santa Ana [800]
Coleccion de historiadores de chile / Olivares, Miguel de, y otro – 1864 – 9 – sp Bibl Santa Ana [972]
Coleccion de historiadores de chile...: historia de gongora marmolejo – 1536-1575. 1862 – 9 – sp Bibl Santa Ana [972]
Coleccion de historiadores de chile y documentos relativos a la historia nacional / Marino de Lovera, Pedro – Santiago de Chile: Imprenta del Ferrocarril, 1865 – 1 – sp Bibl Santa Ana [972]
Coleccion de inscripciones y antiguedades de extremadura / D F de Viu – Caceres: Imp. Concha y Cia, 1846 – 1 – sp Bibl Santa Ana [946]
La coleccion de lapidas de d. claudio constanzo / Jimenez Navarro, Ernesto – Badajoz, 1949 – 1 – sp Bibl Santa Ana [946]
Coleccion de las memorias o relaciones que escribieron los virreyes del peru acerca del estado en que dejaban las cosas generales del reino, tomo 2 / Altolaguirre, Angel de – Madrid, 1931 – 1 – sp Bibl Santa Ana [972]
Coleccion de las obras / Las Casas, Bartolome de; ed by Llorente, J A – Paris. v1-2. 1822 – 18mf – 8 – €35.00 – ne Slangenburg [240]
Coleccion de las obras sueltas, assi en prosa, como en verso / Vega Carpio, Lope de – Madrid. v1-21. 1776 – 1 – $216.00 – (in spanish) – mf#0664 – us Brook [802]
Coleccion de leyes...circulares...de la mesta desde el ano 1729 al de 1827 / Brieva, Matias – Madrid: Imp. de Repulles, 1828 – 1 – sp Bibl Santa Ana [946]
Coleccion de leyes...ramo de la mesta / Brieva, Matias – 1828 – 9 – sp Bibl Santa Ana [946]
Coleccion de libros y documentos referentes a la historia de america / America – Madrid. v. 1-21. 1904-29 – 1 – us L of C Photodup [972]
Coleccion de los discursos mas notables... congreso de los diputados... en la legislatura 1818-19 – 1849 – 9 – sp Bibl Santa Ana [323]
Coleccion de los epigramas y otras poesias criticas, satiricas y jocosas / Salas, Francisco Gregorio de – Madrid: Repulles, 1827 – 1 – sp Bibl Santa Ana [800]
Coleccion de los mas preciosos adelantamientos de la medicina... / Ellerker, R – Malaga, S.A. – 11mf – 9 – sp Cultura [610]
Coleccion de noticias de muchas de las indulgencias plenarias y perpetuas que pueden ganar todos los fieles de christo : que con la debida disposicion visitaren en sus respectivos dias las iglesias que se iran nombrando en ellos... / Avila, Jose de – en Mexico: Felipe de Zuniga y Ontiveros 1787 – (= ser Books on religion...1543/44-c1800: iglesias, catedrales) – 2mf – 9 – mf#crl-355 – ne IDC [241]
Coleccion de obras y documentos relativos a la historia antigua y moderna de las provincias del rio de la plata / Angelis, Pedro de – 5v. 1910 – 1 – us L of C Photodup [972]
Coleccion de poesias / Chacon Villareso, Carmen – Madrid: Hijos de Rens. Impresores, 1914 – 1 – sp Bibl Santa Ana [810]
Coleccion de poesias / Malendez Valdes, Juan – 1798 – 9 – sp Bibl Santa Ana [810]
Coleccion de poesias latinas y castellanas / Santa Lucia y Amaya, Jose – 1883 – 9 – sp Bibl Santa Ana [450]
Coleccion de providencias dadas a fin de establecer la santa vida comun : a que se dio principio en el dia tres de diciembre domingo primero de adviento del ano proximo pasado de mil setecientos sesenta y nueve... / Fabian y Fuero, Francisco – [Puebla: s.n. 1770] – (= ser Books on religion...1543/44-c1800: obispos: obispos de puebla) – 3mf – 9 – mf#crl-399 – ne IDC [241]
Coleccion de providencias diocesanas del obispado de la puebla de los angeles / Fabian y Fuero, Francisco – [Puebla]: impr del Real Seminario Palafoxiano, ano de 1770 – (= ser Books on religion...1543/44-c1800: obispos: obispos de puebla) – 9mf – 9 – mf#crl-400 – ne IDC [241]
Coleccion de silogismos (por via inductiva) (promanuscripto) / Colegio de San Antonio – Caceres: Imp. La Minerva, 1963 – 1 – sp Bibl Santa Ana [946]
Coleccion de silogismos (por via inductiva) t. 2. el universal "in re"... / Colegio de San Antonio – Caceres: Imp. La Minerva, 1963 – 1 – sp Bibl Santa Ana [946]
Coleccion de todas las leyes / Spain. Laws, Statutes, etc – 1792 – 9 – sp Bibl Santa Ana [324]
Coleccion de...del peru / Odriozola, Manuel de – 1863 – 9 – sp Bibl Santa Ana [800]
Coleccion diplomatica de san andres de faulo (958-1270). zaragoza, 1964 / Canellas, Angel – Madrid: Graf. Calleja, 1966 – 1 – sp Bibl Santa Ana [321]
Coleccion documentos...chile / Medina, Jose Toribio – Tomo 8. 1896 – 9 – (tomo 9 1896. tomo 10 1896. tomo 11 1897. tomo 13 1897. tomo 15. tomo 17 1899. tomo 1818. tomo 20) – sp Bibl Santa Ana [972]
Coleccion general de documentos relativos a las islas filipinas / Bayle, Constantino – Madrid: Razon y Fe, 1924.-v5 – 1 – sp Bibl Santa Ana [959]
Coleccion historiadores de chile-documento / Marino de Lovera, Pedro – 1865 – 9 – sp Bibl Santa Ana [972]
Coleccion legislativa de espana : tomos 19 a 100 – Madrid, 1833-1808 – 9 – sp Cultura [348]
Coleccion legislativa de espana, continuacion de la coleccion de decretos – 1869-1874 – 275mf – 9 – sp Cultura [340]
Coleccion legislativa de espana, continuacion de la coleccion de decretos – 1874-1885 – 388mf – 9 – sp Cultura [342]
Coleccion legislativa de instruccion publica... 1888-1935 – Madrid, 1889-1940 – 293mf – 9 – sp Cultura [340]
Coleccion legislativa de primera ensenanza / Pimentel y Donaire, Miguel – Badajoz: Tip La Economica. Rodriguez y Cia, 1894 – 1 – sp Bibl Santa Ana [350]
Coleccion legislativa...ensenanza / Pimentel y Donaire, Miguel – 1894 – 9 – sp Bibl Santa Ana [340]
Coleccion oficial de leyes, decretos, ordenes, resoluciones. / Bolivia. Laws, Statutes, etc – Paz de Ayacucho. v1-16; 1825-54; 2nd series, v1-6; 1857-63. LL-065 – 1 – 92.00 – us L of C Photodup [348]
O coleccionador de sellos : revista mensal – Sorocaba, SP: Typ Durski, 01 jul-ago, dez 1896; jan, mar-maio, jul, set-dez 1897; mar-abr, out 1898; abr-31 ago 1899 – (= ser Ps 19) – mf#P17,02,242 – bl Biblioteca [760]
Coleccion...antiguedades de estremadura (sic) / Viu, Jose de – 1846 – 9 – sp Bibl Santa Ana [930]
Coleccion...chile y documentos.. / Ovalle, Alonso de – v12. 1888 – 9 – sp Bibl Santa Ana [972]
Coleccionistas / Martinez Herrera, Alberto – Habana, Cuba. 1957 – 1r – us UF Libraries [972]
Coleccion...la diplomacia / Donoso Cortes, Juan Francisco – 1848 – 9 – sp Bibl Santa Ana [946]
Coleccion...mejores poetas espanoles / Blanco, Indalecio – 1883 – 9 – sp Bibl Santa Ana [810]
Coleccion...proyecto de ley / Donoso Cortes, Juan Francisco – 1848 – 9 – sp Bibl Santa Ana [946]
Colegio de Abogados de Badajoz see Lista de abogados del ilustre colegio de badajoz. ano de 1930
Colegio de Abogados de Caceres see Lista de los abogados del ilustre colegio de caderes en este ano de 1873
Colegio De Abogados De La Habana see Informes y discursos
El colegio de la constancia : robo de cuatro ruillones / Plasencia, Juan de – Plasencia: tip e pinto sanchez, 1900 – sp Bibl Santa Ana [946]
Colegio de la Inmaculada Concepcion see
– Jmj reglamento para el colegio de senoritas... sagrada familia... plasencia
– Reglamento para el colegio de senoritas dirigido por las hermanas de la sagrada familia bajo el titulo de la inmaculada concepcion establecido en plasencia
Colegio de San Antonio see
– Coleccion de silogismos (por via inductiva)
– Coleccion de silogismos (por via inductiva) t. 2. el universal "in re"...
– Mesa revuelta. redacciones de no 6 y pren. t. 4-b. dia de la asuncion 1961
Colegio de San Antonio. Caceres see
– Mesa revuelta
– Mesa revuelta. redacciones de 6th y preu. dia de san pedro de alcantara de 1961
Colegio de San Jose de Villafranca (Badajoz) see
– Memoria del ano escolar 1936-37
– Memoria del ano escolar 1938-39
– Memoria del ano escolar de 1927 a 1928
– Memoria del ano escolar de 1928 a 1929
– Memoria del ano escolar de 1929 a 1930
– Memoria del ano escolar de 1931-1932
Colegio de San Jose de Villafranca de los Barros (Badajoz) see
– Laudate pueri. oraciones y cantos
– Memoria del a.esc.37-38
Colegio de san jose de villafranca. distribucion de premios 1922-1923 – Villafranca: Imp Rodriguez, 1923 – 1 – sp Bibl Santa Ana [946]
Colegio de san jose de villafranca. distribucion de premios. ano 1921-1922 – Villafranca: Imp Rodriguez, (1922) – 1 – sp Bibl Santa Ana [946]
Colegio de san jose. distribucion de premios. curso 1920-1921 – Zafra: Morera, 1921 – 1 – sp Bibl Santa Ana [946]
Colegio de San Jose. Estremoz (Portugal) see
– Memoria del ano escolar 1932-1933
– Memoria del ano escolar 1933-34
– Memoria del ano escolar 1934-35
– Memoria del ano escolar 1935-36
Colegio de san jose. memoria del curso academico 1917-18 – Villafranca de los Barros, Madrid: Blass y Cia., 1918 – 1 – sp Bibl Santa Ana [946]
Colegio de Santa Cecilia de Caceres see
– Dirigido por hermanas carmelitas de la caridad legalmente reconocido para ensenanza media
– Reglamento del...para la educacion de senoritas dirigido por las religiosas carmelitas de la caridad
Colegio Extremeno de Arbitros de Futbol... see Reglamento de futbol
Colegio Maristas San Calixto see
– Proyecto educativo 1977-78
– Proyecto educativo curo 1980-1981
– Proyecto educativo. curso 1976-77
Colegio Mayor del Conde Duque see Aprobacion y confirmacion que dio el...al parecer y adicion que hizo. fray juan de los reyes...
Colegio Oficial de Abogados de Badajoz see
– Lista oficial de los senores abogados y procuradores y guia judicial de la provincia, ano 1958
– Lista oficial de los sres. abogados y procuradores y guia judicial de la provincia. ano de 1965
Colegio Oficial de Agentes Comerciales see
– Censo oficial de los senores agentes comerciales en 1st de junio de 1961
– Memoria de los trabajos realizados durante el ano 1959...
– Memoria de los trabajos realizados durante el ano 1960 leida y aprobada en sesion del 25 de marzo de 1961
Colegio Oficial de Ayudantes Tecnicos Sanitarios see Tarifa de honorarios minimos aprobada por junta de gobierno
Colegio Oficial de Farmaceuticos. Caceres see
– Anales, 1959
– Lista de farmaceuticos colegiados ano 1943
– Lista de farmaceuticos colegiados, ano 1945
Colegio Oficial de Medicos de la Provincia de Badajoz see Lista de sres. medicos colegiados 1969
Colegio Oficial de Medicos de la Provincia de Caceres see
– La escuela de medicina de guadalupe
– Lista de senores colegiados por partidos judiciales y por orden de incorporacion a este colegio provincial
Colegio oficial de peritos agricolas de badajoz. aranceles 1960 – Badajoz: Graficas Jimenez, 1960 – sp Bibl Santa Ana [630]
Colegio oficial de secretarios, interventores y deportivos de administracion local de la provincia de caceres. memoria que formula el secretario...del ejercicio 1941... / Rubio, Julio – Caceres: Tip. El Noticiero – sp Bibl Santa Ana [350]
Colegio oficial veterinario de badajoz. circular no 144 – Badajoz: La Minerva Extremena, 1945 – sp Bibl Santa Ana [946]
Colegio Provincial de Abogados see
– Lista oficial de los senores abogados y procuradores y guia judicial
– Lista oficial de los senores abogados y procuradores y guia judicial. ano 1967
Colegio Provincial de Abogados de Badajoz see
– Bases orientadoras para la fijacion de honorarios aprobados por la junta general de este ilustre colegio, el 31 de marzo de 1961
– Lista oficial de los senores abogados y procuradores y guia judicial de la provincia
– Lista oficial de los senores abogados y procuradores y guia judicial de la provincia. ano de 1957
– Lista oficial de los sres. abogados y procuradores y guia judicial de la provincia. ano de 1959
– Lista oficial de los sres. abogados y procuradores y guia judicial de la provincia. ano de 1960

– Lista oficial de los sres. abogados y procuradores y guia judicial de la provincia. ano de 1961
– Lista oficial de los sres. abogados y procuradores y guia judicial de la provincia. ano de 1962
– Lista oficial de los sres. colegiados, de los sres. procuradores y guia judicial de la provincia. ano de 1968
– Lista oficial de los sres. colegiados, de los sres. procuradores y guia judicial de la provincia. ano de 1971

Colegio Provincial de Abogados de Caceres see
– Lista oficial de los senores abogados del ilustre colegio provincial de caceres y guia judicial de los tribunales de dicha ciudad para el ano 1956
– Lista oficial de los senores abogados del ilustre colegio provincial de caceres y guia judicial de los tribunales de dicha ciudad para el ano 1959
– Lista oficial de los senores abogados del ilustre colegio provincial de caceres y guia judicial de los tribunales de dicha ciudad para el ano 1960
– Lista oficial de los senores colegiados y guia judicial de los tribunales de la ciudad, ano 1961
– Normas de honorarios minimos

Colegio Provincial de Procuradores de Badajoz see Estatutos del...y el general de procuradores de los tribunales de espana

Colegio provincial de veterinarios de badajoz. circular no 1 – Badajoz: Imp. Barrena, 1944 – sp Bibl Santa Ana [946]

Colegio San Antonio see Ensayos de etice. duns escoto en extremadura 2

Colegio San Francisco Javier. Fuente de Cantos (Badajoz) see Cantos liturgicos

Colegio San Jose see
– Catalogo
– Catalogo de los alumnos del colegio de san jose. 1904-1905
– Congregaciones marianas del...bajo la advocacion de la inmaculada concepcion, san luis gorzaga y san estanislao de kostka

Colegio San Jose de Azuaga see Procultura. trabajos de extension pedagogica

Colegio Santiago y Santa Margarita. Sociedad Cooperativa see Ideario de la colegio de egb y formacion profesional

Colegio Universitario de Caceres see Ciento cincuenta anos de periodismo en caceres y salamanca

Cole-King, Pa see Cape maclear

Coleman, Arthur Philemon see
– Canadian rockies
– The canadian rockies
– Canoeing on the columbia
– Clastic huronian rocks of western ontario
– Glacial and inter-glacial deposits near toronto
– Interglacial fossils from the don valley, toronto
– The iroquois beach
– The michipicoten iron ranges
– Microscopic petrography of the drift of central ontario
– Mount brown
– The nickel industry
– The sudbury nickel field

Coleman, Christopher Bush see
– Church history in the modern sunday school
– Constantine the great and christianity

Coleman, Eli see Journal of psychology and human sexuality

Coleman, Eliphalet Beecher see The sabbath school catechism

Coleman, George William see The churches outside the church

Coleman, Gina M see Time out

Coleman, James M see Coleman's general index to printed pedigrees

Coleman, John Noble see Ecclesiastes

Coleman, Julius Archer see
– The mechanic's lien law of the state of illinois
– A treatise on the mechanic's lien law of the state of illinois

Coleman, Leighton see
– Church in america
– A history of the american church to the close of the 19th century

Coleman lite – 1981 aug-1989 feb – 1r – 1 – mf#1054602 – us WHS [071]

Coleman, Lyman see
– Ancient christianity exemplified
– The apostolical and primitive church
– Genealogy of the lyman family
– An historical geography of the bible
– An historical text book and atlas of biblical geography

Coleman, Robert H see Publications of music

Coleman, Thomas see Memorials of the independent churches of northamptonshire

Coleman world – v2 n1-6 [1981 jan-nov], v3 n1-4 [1982 win-fall], v4 n1-4 [1983-1984], v5 n1 [1985], v6 n1 [1986 win] – 1r – 1 – mf#1611758 – us WHS [071]

Coleman's general index to printed pedigrees : which are to be found in all the principal county and local histories, and in...genealogies / Coleman, James M – London: J Coleman 1866 [mf ed 1987] – 1r – 1 – (incl bibl ref. filmed with: lectures on the comparative grammar of the semitic languages) – mf#7599 – us Wisconsin U Libr [929]

Colemokee baptist church. early county. georgia : church records – 1889-1908 – 1 – us Southern Baptist [242]

Colenso, Frances Ellen see History of the zulu war and its origin

Colenso, J W see Ten weeks in natal

Colenso, John William see
– First steps in zulu
– Lectures on the pentateuch and the moabite stone
– Letter to his grace the archbishop of canterbury
– The pentateuch and book of joshua critically examined
– St paul's epistle to the romans
– Trial of the bishop of natal for erroneous teaching
– Zulu-english dictionary

Colenso, John William, bishop of Natal see Foreign missions, and mosaic traditions

Colephous de manuscrito / Bounere, Benedictius du – Madrid: Graf. Calleja, 1967 – 1 – sp Bibl Santa Ana [946]

Coler, Bird S see Two and two make four

Coler, Bird Sim see Deux et deux font quatre

Coler, J see
– Bericht von dem exorcismo bey der tauffe
– Historia dispvtationis sev potivs colloqvii inter iacobvm colervm et mathiam flacivm illyricvm de peccato originis

Coler, William Nichols see A practical treatise on the law of municipal bonds.

El colera / Oliveres, Luis – 1884 – 9 – sp Bibl Santa Ana [610]

El colera morbo en badajoz en 1883 / Guerra Camacho, Mercedes – Badajoz: dip provincial, 1970 – 1 – sp Bibl Santa Ana [946]

Coleraine chronicle – Coleraine, Ireland. 14 apr 1844-1929, 28 jun 1930-1988, may 1989-1998 [wanting jan-jun 1930] – 225 1/2r – 1 – (aka: the chronicle) – uk British Libr Newspaper [072]

Coleraine constitution – Coleraine, Ireland. 21 apr 1877-1984, 1986-mar 1991, 30 apr 1991-dec 1998, jan-jun 1999 – 185r – 1 – (aka: coleraine constitution and northern counties advertiser; northern constitution) – uk British Libr Newspaper [072]

Coleraine times – Coleraine, Ireland. 7 mar 1990-1998 – 24 1/2r – 1 – uk British Libr Newspaper [072]

Coleraine tribune – Coleraine, Ireland. jan-2 jul 1986 – 1r – 1 – uk British Libr Newspaper [072]

La colere et le desespoir d'un vieux republicain – [Paris]: Bonaventure et Ducessoi. n1-2. 1848 – reel 1, item 27 – us CRL [944]

Coleridge and his followers / Hetherington, William Maxwell – London, England. 1853 – 1r – us UF Libraries [420]

Coleridge and literary society, 1790-1834 : the papers of samuel taylor coleridge (1772-1834) from the british library, london – 15r – 1 – £1400.00 – (with d/g) – uk Matthew [420]

The coleridge blade – Coleridge, NE: A J Watson (wkly) [mf ed v1 n27. jun 16 1892,1896-1902,1906-35- (gaps)] – 1 – (vol numbering irregular oct 24-nov 14 1912. issues for nov 21 1912- called v26 n7-) – us NE Hist [071]

Coleridge, Derwent see The scriptural character of the english church

Coleridge, Henry see
– Six months in the west indies, in 1825

Coleridge, Henry James see
– Christian kingdom
– The life and letters of st. francis xavier
– The life of mother frances mary teresa ball
– The story of st stanislaus kostka

Coleridge, Samuel Taylor see
– Aids to reflection in the formation of a manly character
– Friend
– On the constitution of the church and state according to the idea...
– Poetical and dramatic works of s t coleridge

Coleridge sentinel – Coleridge, NE: J L Kroesen (wkly) [mf ed v3 n2. sep 9 1886-jan 27 1887 (lacks oct 7 1886) filmed 1976] – 1r – 1 – us NE Hist [071]

Colerus, Egmont see Archimedes in alexandrien

Coles, Charles see Observations on the claims of the west-india colonists to a protecting duty on east india sugar

Coles, V S see Salvation

Coles, William see The art of simpling

The colesberg advertiser – Colesberg SA, 1 jan 1861-1902 – 22r – 1 – (1899 incomplete) – sa National [079]

The colesberg herald – 1869-76 (wkly) [mf ed Cape Town: SA library 1986] – 1 – (suspended sep 13 1873-jun 8 1875) – mf#MS00427 – sa National [079]

Coleshill chronicle and nuneaton standard – [West Midlands] Warwickshire jan 1911-dec 1950 [mf ed 2004] – 23r – 1 – (cont by: coleshill chronicle and kenilworth advertiser [jan 1925-dec 1934]; coleshill chronicle and castle bromwich advertiser [jan 1935-dec 1950]) – uk Newsplan [072]

Colet, John see Ioannis coleti enarratio in primam epistolam s pauli ad corinthios

Coletanea de poetas pernambucanos / Silva, Francisco De Oliveira E – Rio de Janeiro, Brazil. 1951 – 1r – us UF Libraries [810]

Coletanea de poetas sul-riograndenses, 1834-1951 / Machado, Antonio Carlos – Rio de Janeiro, Brazil. 1952 – 1r – us UF Libraries [810]

Colette, A see Histoire du breviaire de rouen

Coley, J see Last lent lecture

[Colfax-] colfax record – CA. 1944-58; 1987 – 9r – 1 – $540.00 – mf#C02124 – us Library Micro [071]

Colfax county call – Schuyler, NE: H A McCormick & Son. v1 n1. jul 27 1933-v22 n44. may 19 1955 (wkly) [mf ed lacks may 12 1938 and feb 20 1947 filmed 1970] – 14r – 1 – (merged with: schuyler sun to form: schuyler sun and colfax county call) – us NE Hist [071]

The colfax county press – Clarkson, NE: Odvarka Bros. v50 n1. jul 7 1954- (wkly) – 1 – (cont: colfax county press and the clarkson herald consolidated) – us NE Hist [071]

Colfax county press and the clarkston herald consolidated – Clarkson, NE: Odvarka Press. 41v. v12 n40. jun 8 1916-v49 n52. jun 30 1954 (wkly) [mf ed with gaps] – 14r – 1 – (cont: clarkson herald. cont by: colfax county press) – us NE Hist [071]

Colfax messenger – Colfax WI. 1897 apr 30/1917 dec 27-1995 – 66r – 1 – (with gaps) – mf#1007253 – us WHS [071]

Colfax, Schuyler see Papers

Colfax sentinel – CA. 1893-1907 – 8r – 1 – $480.00 – mf#B02125 – us Library Micro [071]

Colgate baptist church. baltimore, maryland : church records – 1945-69 – 1 – us Southern Baptist [242]

Colgate, Robert see The immigrant

Colgate rochester divinity school : bexley hall bulletin – v41-42 n4. 1968-70 [gaps] – Inquire – 1 – mf#ATLA 1994-S526 – us ATLA [200]

Colgate rochester divinity school bulletin – v1-40. 1928-68 [complete] – Inquire – 1 – mf#ATLA 1994-S525 – us ATLA [200]

O colibri : orgao dedicado ao bello sexo – Manaus, AM: Typ do Corneta, 24 fev 1888 – (= ser Ps 19) – mf#P11B,06,16 – bl Biblioteca [079]

Coligny, Gustav Adolf see Wallenstein

Coligny – gustav adolf – wallenstein : drei zeitgenoessische lateinische dramen von rhodius, narssius, vernulaeus / Rhode, Theodor & Noorssen, Johann van & Vernulz, Nicolas de; ed by Bolte, Johannes – Leipzig: K W Hiersemann 1933 [mf ed 1993] – (= ser Blvs 280) – 58r – 1 – (latin texts, int in german. incl bibl ref. filmed with: das rheinische marienlob / adolf bach & other titles) – mf#3420p – us Wisconsin U Libr [820]

Colima. Mexico. (State) see
– El estado de colima
– Periodico oficial

Colin clout's calendar : the record of a summer, april-october / Allen, Grant – London: Chattos & Windus, 1883 – 3mf – 9 – (originally appeared in st james's gazette) – mf#05016 – cn Canadiana [580]

Colin, Frederic Louis de Gonzague see Discours sur l'ouvrier prononce par le reverend m colin...devant l'institut des artisans canadiens le 2 avril 1869

Colin, Frederic-Louis-de-Gonzague see
– Discours sur l'ouvrier
– Le pape honorius

Colin legum's african collection : material from the personal library of colin legum, one of the world's most eminent international press correspondents and writers on africa – 1936-89 [mf ed Microform Academic Publ [Altair]] – (= ser African coll) – 312mf+10r – 9,1 – mf#97534 – uk Microform Academic [070]

Colin legum's writings from the 1940's to the 1980's : material from the personal library of colin legum / Legum, Colin – (= ser African coll) – 40mf – 9 – (with guide) – uk Microform Academic [960]

Los colinas people – Irving, TX. 1987-1991 (1) – mf#68543 – us UMI ProQuest [071]

Colizza, G see Lingua 'afar nel nord-est dell'africa

Colizzi, Johannes see First lessons for the harpsichord or spinnet

Coll, Aloysius see
– Citrus county
– Clearwater
– Indian mounds
– Pasco county history

Coll, Blanche D see Minutes of trustees of the poor

Coll, Cornelius van see Beknopte geschiedenis der katholieke missie in suriname

Coll Y Toste, Cayetano see
– Narraciones historicas
– Seleccion de leyendas puertorriquenas

Collaboration : a dancer's phenomenological study of combined vision in art-making / Smith, Colleen A – 1997 – 2mf – 9 – $8.00 – mf#PE 3775 – us Kinesology [790]

Collaborative adult learning principles in baccalaureate community health education curricula / Lucas, Martha M – 1987 – 114p 2mf – 9 – $8.00 – us Kinesology [362]

Collaborators / Ainslie, Rosalynde – London, England. 1963 – 1r – us UF Libraries [960]

Collado, Diego see
– Additiones ad dictionarium japonicum
– Ars grammaticae japonicae linguae

Collado, Fr. Diego see
– Ars gramatica japonicae linguae
– Dictionarum sive thesauri linguae japonicae

Collado, L see
– Ex hippocratis et galeni monumentis isagoge...
– Platica manual de artilleria

Collado, Martell A see Cuentos absurdos

Collage – v1 n1-3 [1970 may-nov] – 1r – 1 – mf#1110709 – us WHS [071]

El collar de lescot / Hurtado, Antonio – 1868 – 9 – sp Bibl Santa Ana [830]

Collar, Jimmie Oliver see Bibliography, history of duval county

Collard, Edgar Andrew see Old montreal

Collard, Frederick J M see
– Letter addressed to his excellency, lord aylmer, governor general of lower canada
– Letter addressed to louis joseph papineau, esquire, speaker of the house of assembly

Collard, Paul Maria Alexandre see Cambodge et cambodgiens

Collarenebri gazette – Collarenebri, aug 1935-1967 (misc issues) – 1r – 9 – A$33.97 vesicular A$39.47 silver – at Pascoe [079]

Collas, J P L see Extrait d'une lettre, crite de pekin

Collasse, Pascal see Achille et polixene

Collateral consanguinity – Philadelphia 1860. 11 p. LL-14 – 1 – us L of C Photodup [340]

Collateral guide – Pittsburgh, PA. 1908-1935 (1) – mf#66029 – us UMI ProQuest [071]

Collatio : oder gegeneinanderhaltung vnd vergleichung der augspurgischen confession vnd der zwinglischen oder calvinischen lehr vnd glaubens: / Mentzer, B – Giessen, Hampel, 1607 – 2mf – 9 – mf#TH-1 mf 1169-1170 – ne IDC [242]

Collatio codicis lewisiani rescripti evangeliorum sacrorum syriacorum : cum codice curetoniano / Bonus, Albert – Oxford, 1896 – €11.00 – ne Slangenburg [221]

Collatio codicis lewisiani rescripti evangeliorum sacrorum syriacorum cum codice curetoniano (mus. brit. add. 14,451) : cui adiectae sunt lectiones e peshitto desumptae / Bonus, Albert – Oxonii [Oxford]: E prelo Clarendoniano, 1896 – 1mf – 9 – 0-524-08214-6 – mf#1993-0009 – us ATLA [090]

A collation of four important manuscripts of the gospels : with a view to prove their common origin and to restore the text of their archetype – Dublin: Hodges, Foster, and Figgis; London: Macmillan, 1877 [mf ed 1989] – 2mf – 9 – 0-7905-1659-4 – (text in greek, int in english) – mf#1987-1659 – us ATLA [226]

Collation of the sacred scriptures : the old testament from the translations of john rogers... – Dundee: M'Cosh, Park, and Dewars, 1847 [mf ed 1992] – 1mf – 9 – 0-524-02763-3 – mf#1987-6457 – us ATLA [220]

Collbran journal see Miscellaneous newspapers of mesa county

Colle, Pierre see Societes secretes en uruwa

Colle, R P see Les baluba (congo belge)

Collecao de memorias aos fazendeiros see O auxiliador da industria nacional

Colleccao de observacoes grammaticaes sobre a lingua bunda : ou angolense / Cannecattim, Bernardo Maria de – Lisboa: Impr regia – (= ser Whsb) – 3mf – 9 – €40.00 – mf#Hu 350 – gw Fischer [470]

Colleccao dos fac-similes das assignaturas : e rubricas dos arcebispos primazes do oriente e dos vigarios capitulares do arcebispado, coordenada, por detfrminacao sic] / Nery Xavier, Filippe – Nova-Goa: Imprensa Nacional, 1853 [mf ed 1995] – (= ser Yale coll) – 37lea – 1 – 0-524-10031-4 – (in portuguese) – mf#1995-1031 – us ATLA [241]

Collectanea : [first series with a preface by c graham botha] – Capetown: The Van Riebeeck society 1924 – (= ser [Travel descriptions from south africa, 1711-1938]) – 2mf – 9 – mf#zah-65 – ne IDC [916]

Collectanea anglo-premonstratensia : documents drawn from the original register of the order... / ed by Gasquet, Francis Aidan – London: Offices of the Royal Hist Soc 1904-06 [mf ed 1991] – (= ser Camden third series 6,10,12) – 3v on 2mf – 9 – 0-524-01889-8 – (in latin) – mf#1990-0516 – us ATLA [241]

Collectanea chemica : being certain select treatises on alchemy and hermetic medicine / Philalethes, Eirenaeus [pseud] et al – London: J Elliott & Co 1893 [mf ed 1984] – 1r – 1 – (filmed with: sogno d'una notte d'estate / shakespeare, william) – mf#6716 – us Wisconsin U Libr [540]
Collectanea commissionis synodalis : digest of the synodal commission of the catholic church in china – 1928-47 – (= ser Religious history research colls) – 229mf (24:1) – 9 – $2180.00 – (with p/g. filmed fr holdings of maryknoll missioners) – us UPA [241]
Collectanea graea et latina : selections from the greek and latin fathers; with notes biographical and illustrative / Willis, Michael – [Toronto?: s.n.], 1865 [mf ed 1982] – 3mf – 9 – (text in greek and latin; int and biogr notes in english) – mf#35211 – cn Canadiana [200]
Collectanea monumentorum veterum ecclesiae graecae et latinae / ed by Zacagnio, A – Romae, 1698 – €44.00 – ne Slangenburg [240]
Collectaneum miscellaneum : formae tplila 47 / Scottus, Sedulius – 1990 – (= ser ILL – ser a; Cccm 67) – 8mf+105p – 9 – €60.00 – 2-503-63472-9 – be Brepols [400]
Collecte pour payer la dette de l'eglise st jean-baptiste de quebec – Quebec?: Paroisse de St Jean-Baptiste de Quebec. 2e annee, n1 1er oct 1900-1909 [mf ed 1989] – 4mf – 9 – (cont: bulletin de collecte (paroisse de saint-jean-baptiste (quebec); ceased 1909) – mf#P04493 – cn Canadiana [336]
Collected diplomatic documents relating to the outbreak of the european war / Great Britain. Foreign Office – London: H M Stationery Off 1915 [mf ed 1987] – 1r – 1 – (incl ind) – mf#9955 – us Wisconsin U Libr [933]
Collected essays of rudolf eucken : professor of philosophy in the university of jena, nobel prizeman, 1908 / ed by Booth, Meyrick – London: T F Unwin, 1914 – 1mf – 9 – 0-7905-3730-3 – (in english) – mf#1989-0223 – us ATLA [190]
Collected field reports on the phonology and grammar of chakosi / Stanford, Ronald – Legon, Ghana. 1970 – 1r – us UF Libraries [470]
Collected field reports on the phonology of basari / Abbott, Mary – Legon, Ghana. 1966 – 1r – us UF Libraries [470]
Collected field reports on the phonology of dagaari / Kennedy, Jack – Legon, Ghana. 1966 – 1r – us UF Libraries [470]
Collected field reports on the phonology of konkomba / Steele, Mary – Legon, Ghana. 1966 – 1r – us UF Libraries [470]
Collected field reports on the phonology of kusal / Spratt, David – Legon, Ghana. 1968 – 1r – us UF Libraries [470]
Collected field reports on the phonology of tampulma / Bergman, Richard – Legon, Ghana. 1969 – 1r – us UF Libraries [470]
Collected field reports on the phonology of vagala / Crouch, Marjorie – Legon, Ghana. 1966 – 1r – us UF Libraries [470]
Collected hymns, sequences and carols of john mason neale / Neale, John Mason – London; New York: Hodder and Stoughton, 1914 – 2mf – 9 – 0-7905-8268-6 – mf#1988-6146 – us ATLA [240]
Collected legal papers / Holmes, Oliver Wendell, Jr – New York: Harcourt, Brace and Howe, 1920. 316p. LL-347 – 1 – us L of C Photodup [340]
Collected pamphlets / Lenin, N – 1917-24 – 1 – us L of C Photodup [943]
Collected papers : zoology: a collection of 15 articles published in various journals / Rafinesque-Schmaltz, C S – 2mf – 9 – mf#Z-2226 – ne IDC [590]
Collected papers, ms 3031 / Lincoln, Abraham – 1841-70 – 1r – 1 – us Western Res [920]
The collected papers of john packer, 1616-40 : from the hartley-russell collection in the berkshire record office, ref. d/etty 01 – 1r – 1 – mf#96828 – uk Microform Academic [920]
Collected papers of srinivasa ramanujan – Cambridge, England. 1927 – 1r – us UF Libraries [510]
Collected papers on analytical psychology = Selections. 1916 / Jung, Carl Gustav; ed by Long, Constance E – London: Bailliere, Tindall and Cox, 1916 – 1mf – 9 – 0-7905-7853-0 – (in english) – mf#1989-1078 – us ATLA [150]
Collected papers on aspects of png history : transcripts of interviews with png defence force personnel; png chinese; arthur duna on the japanese landing at buna; michael mell, phillip kamen and anton parao on the highlands liberation front; and others, / Denoon, Donald – 1972-75 – 1r – 1 – mf#pmb1250 – at Pacific Mss [980]
Collected papers on bantu linguistics / Guthrie, Malcolm – Farnborough, England. 1970 – 1r – us UF Libraries [470]
The collected poems of isabella valancy crawford / ed by Garvin, John William – Toronto: W Briggs, 1905 – 4mf – 9 – 0-665-71230-8 – (int by ethelwyn wetherald) – mf#71230 – cn Canadiana [810]
Collected records / Afro-American Clubwoman's Project – undated – 1 – us Kansas [305]
Collected reports on land and related matters in papua new guinea / Sack, Peter – 1960-79 – r1-2 – 1 – (available for ref) – mf#pmb1167 – at Pacific Mss [980]
Collected tongan papers / Latukefu, Sione – 1884-1965 – 2r – 1 – mf#pmb1124 – at Pacific Mss [980]
Collected works / Lobstein, Paul – A corpus of the monographic publications of Paul Lobstein [1850-1922] – 1 – us ATLA [240]
The collected works of philipp melanchthon – Microcard Editions – (= ser Religious history research colls) – 28v on 231mf (21:1) – 9 – $1105.00 – (with ind. in latin & german) – us UPA [242]
The collected works of the late dastur darab peshotan sanjana – Bombay: British India Press, 1932 – (= ser Samp: indian books) – us CRL [290]
Collected works on the bible and theology / Schlatter, Adolf von – Princeton: Princeton Theo. Sem., [1975] – 1r – 1 – 0-8370-1087-X – mf#1984-B528 – us ATLA [220]
Collected writings of caroline norton *see* Norton
The collected writings of edward irving / Irving, Edward; ed by Carlyle, Rev. G – Alexander & Straham & Co., 1864. 5v – 9 – $96.00 – us IRC [920]
The collected writings of james henley thornwell, d.d., ll. d : late professor of theology in the theological seminary at columbia, south carolina = Works. 1871 / Thornwell, James Henley; ed by Adger, John Bailey – Richmond: Presbyterian Committee of Publication, 1871-1873 – 7mf – 9 – 0-524-05963-2 – (incl ind) – mf#1991-2363 – us ATLA [240]
Collected writings of margaret oliphant *see* Oliphant
Collected writings of thomas de quincey – London, England. v1-2. 1896 – 1r – us UF Libraries [420]
Collectio confessionum in ecclesiis reformatis publicatarum / ed by Niemeyer, Hermann Agathon – Lipsiae: Sumptibus Iulii Klinkhardti, 1840 – 3mf – 9 – 0-7905-8221-X – mf#1988-6121 – us ATLA [240]
Collectio nova patrum et scriptorum graecorum / Montfaucon, B de – Parisiis, 1707. 2v – 28mf – 9 – mf#H-3112 – ne IDC [956]
Collectio nova patrum et scriptorum graecorum : studio et opera / De Montfaucon, Bern – Parisiis. v1-2. 1707 – 9 – €126.00 – (v1 34mf v2 32mf) – ne Slangenburg [220]
Collectio scriptorum : rerum historico-monastico-ecclesiasticarum variorum religiosorum ordinum / Kuen, M – Ulmae. v1-6. 1755-1768 – 6v on 75mf – 9 – €143.00 – ne Slangenburg [220]
Collectio sermonum : formae tplila 3 / Chrysologus, Petrus – 1982 – (= ser ILL – ser a; Cccm) – 15mf+116p – 9 – €40.00 – 2-503-60242-8 – be Brepols [400]
Collectio sermonum : lemmata tplilb 3 / Petrus Chrysologus – 1982 – (= ser ILL – ser b; Ccsl 24-24a-24b) – 10mf+38p – 9 – €30.00 – 2-503-70242-2 – be Brepols [400]
Collection / Adair, Samuel Lyle and Florella (Brown) Family – undated, Correspondence between S.L. and Florella Adair and with other members of their family. The Adairs were missionaries at Osawatomie, KS, and were involved with the territorial conflict in eastern Kansas. The Adairs were relatives of John Brown – 1 – us Kansas [920]
Collection / Bailey, Lawrence D – 1855-1881, A one volume scrapbook of newspaper clippings, correspondence, and personal manuscripts. Much of the material relates to the monetary question of the 1870's – 1 – us Kansas [920]
Collection / Carver, W O – Handwritten notes by John A. Broadus and E. C. Dargan while students at Southern Baptist Theological Seminary. 1522p – 1 – 53.27 – us Southern Baptist [242]
Collection / Cheyenne Indians – undated, Material relating to the Cheyenne Indians in Kansas – 1 – us Kansas [305]
Collection / Cooley, Anna M – 1891-1950 – 1 – us Kansas [920]
Collection / Custer, Elizabeth B – undated, Letters and related papers of Mrs. Custer; mostly about her experiences and her husband – 1 – us Kansas [920]
Collection / Force, Peter – Series 8 and 9 – 1 – us L of C Photodup [920]
Collection / Jackson, Sheldon – Alaska selected documents – 1 – $50.00 – us Presbyterian [978]
Collection / Monday, Henry A – 1822-1932 – 1 – 897.00 – us L of C Photodup [972]
Collection / Morgan, Dale L – undated, Correspondence as an advisor to the Kansas State Historical Society on various historical research projects – 1 – us Kansas [978]
Collection / Nelson, Erik Alfred – 1 – us Southern Baptist [242]
Collection / Newton, Louis Devotie – Correspondence, pamphlets, addresses and manuscripts. Papers prepared while he was President of the Southern Baptist Convention. Also materials relating to naming and dedication of Louis D. Newton Hall, Mercer University. 388p – 1 – us Southern Baptist [242]
Collection / Roche, William Lundy – 1873-1976, Papers of a prominent Washington, KS, family containing much local history – 1 – us Kansas [920]
Collection / Spilman, B W – 1871-1950. 2558p – 1 – 89.53 – us Southern Baptist [242]
Collection / Woodson, Carter Godwin – 1804-1936 – 1 – us L of C Photodup [920]
Collection, 1578-1787 *see* Huguenot records, 1578-1787
Collection, 1781-1906 / Courtenay, William Ashmead – [mf ed 1981] [Spartanburg SC: Reprint Co, dist] – 134mf – 9 – mf#51-550/560 – us South Carolina Historical [978]
Collection, 1925-52 / Norris, J Frank – 46,360p – 1 – us Southern Baptist [242]
Collection building – Bradford. 2001+ (1,5,9) – ISSN: 0160-4953 – mf#29047 – us UMI ProQuest [020]
Collection complete des decrets de la convention nationale / France. Convention Nationale – v1-8. 1792-95 – 1 – $90.00 – mf#0219 – us Brook [944]
Collection complete des lois, decrets, ordonnances, reglemens avis du conseil d'etat / France. Conseil d'Etat – 12th ed. Paris. v1-30. 1788-1930 – 9 – $300.00 – mf#0218 – us Brook [324]
Collection complete des lois et decrets nationale de leur / France. Convention Nationale. Comite de Salut Public – Paris. v1-12. 1793-95 – 1 – $120.00 – mf#0220 – us Brook [324]
Collection de documents concernant madagascar et les pays voisions – Tananarive, Academie malgache. v4. 1953-58 – us CRL [960]
Collection de documents inedits sur le canada et l'amerique – [Quebec: s.n.] 1888 [mf ed 1980] – 3v on 1mf – 9 – 0-665-05321-5 – mf#05321 – cn Canadiana [971]
Collection de documents relatifs a l'histoire de paris pendant la revolution francaise et l'epoque contemporaine, publiee sous le patronage du conseil municipal – Paris. v1-53. 1888-1942 – 16 titles – 1 – $810.00 – mf#0159 – us Brook [944]
Collection de documents relatifs...l'histoire de paris pendant la revolution francaise – Paris, 1883-1923 – 693mf – 8 – mf#690 – ne IDC [700]
Collection de manuscrits de marechal de levis / Levis, Francois Gaston – Quebec. v1-12. 1889-95 – 1 – $108.00 – mf#0328 – us Brook [978]
Collection de musique canadienne : fonds jean-chatillon / Bibliotheque nationale du Quebec. Departement des manuscrits – [mf ed 1971] – 3r – 1 – mf#SEM35P112 – cn Bibl Nat [780]
Collection de planches pour servir au voyage aux indes orientales et a la chine / Sonnerat, P – Paris: Dentu, 1806 – 3mf – 9 – mf#HT-705 – ne IDC [915]
Collection de plusieurs des actes et ordonnances les plus utiles en force dans le bas-canada : concernant la loi criminelle et les devoirs des magistrats = A collection of some of the most useful acts and ordinances in force in lower canada, relating to criminal law and to the duties of magistrates / Canada (Province) – Quebec: impr par Stewart Derbishire & George Desbarats, 1854 [mf ed 1983] – 2mf – 9 – mf#SEM105P190 – cn Bibl Nat [345]
Collection de resumes geographiques : ou bibliotheque portative de geographie physique, historique, ancienne et moderne – Paris o J [mf ed Hildesheim 1995-98] – 4mf – 9 – €120.00 – 3-487-29817-1 – gw Olms [910]
Collection des bons romans – Montreal: Lamarre. v1 n1 (25 mai 1887), v1 n3 (25 juin 1887) – 9 – mf#P04064 – cn Canadiana [440]
Collection des chroniques nationales francaises / ed by Buchon, Jean A – Paris. v1-47. 1826-38 – 1 – $480.00 – mf#0118 – us Brook [944]
Collection des memoires relatifs a l'histoire de france, pt 2 / ed by Petitot, Claude-Bernard – Paris 1820-29 [mf ed Hildesheim 1995-98] – 85v on 253mf – 9 – €1518.00 – 3-487-25854-4 – gw Olms [944]
Collection des monographies ethnographiques / Overbergh, Cyr. van – Bruxelles: Albert de Wit. v1-11. 1907-13 – 1 – us CRL [305]
Collection des ouvrages anciens concernant madagascar / Grandidier, Alfred et al – Paris: Comite de Madagascar. 5v. 1903-20 – 1 – us CRL [080]
Collection: diaries, 1832-85; notebooks, manuscripts and correspondence with contemporaries / Crane, William Carey – 40,629p – 1 – us Southern Baptist [242]
Collection e z massicotte – [mf ed 1979] – 2r – 1 – (contains papers and clippings, dated ca 1886-1947) – mf#SEM35P162 – cn Bibl Nat [971]
Collection, five notebooks. / Dawson, J M – 594p – 1 – us Southern Baptist [242]
A collection in the making : works from the phillips collection / Grogan, Kevin [comp] – 1976 – (= ser Chicago visual library) – 15 – $60.00f – 0-226-69538-7 – (foreword by milton w brown) – us Chicago U Pr [700]
The collection laws, special, exemption, property, banking, and interest laws, of illinois, indiana, michigan, iowa, wisconsin and minnesota. / Hoyt, James T – Chicago: Cooke, 1859. 406p. LL-734 – 1 – us L of C Photodup [346]
Collection management / ed by Lee, Sul H – v1- 1976- – 1, 9 ($125.00 in US $175.00 outside hardcopy subsc) – us Haworth [020]
Collection, manuscripts and books / Howell, Robert Boyte C – 1820. 6190p – 1 – us Southern Baptist [242]
Collection marie-paule caty-lacroix – [mf ed 1987] – 2mf – 9 – (incl: album constitue by perpetue girouard; with ind) – mf#SEM105P559 – cn Bibl Nat [780]
Collection, ms 3947 p268 / Palmer, William P – 1r – 1 – (johnson's island [prison] lists and narratives; lists of prisoners held at johnson's island, ohio, and a mss history entitled "confederates abroad, or leisure activities at johnson's island") – us Western Res [355]
Collection, ms 3947 p575 / Palmer, William P – 1r – 1 – (coll of letters written to messrs. bradlee & sears of boston, mass., by correspondents in charleston, sc, and mobile, alabama, regarding lincoln's election and the southern reaction) – us Western Res [976]
Collection, ms 3947 p1060 / Palmer, William P – 1r – 1 – (diary of john m. butler, acting master of the uss new ironsides, may 6 1862-april 12 1864, during the blockade of charleston, sc, during the american civil war. butler notes the boredom of blockade duty) – us Western Res [355]
Collection, mss 3947 p746 / Palmer, William P – 1r – 1 – (excerpted from the palmer collection, the report of confederate general preston smith regarding the battle of perryville, kentucky, 8 oct 1862) – us Western Res [355]
Collection of 18th and 19th-century songs and piano pieces – [180-] [mf ed 1989] – 1r – 1 – mf#pres. film 25 – us Sibley [780]
Collection of african newspaper issues mainly in africa, 1959 to may 1962 – [s.l, s.n, 19–?] – 1 – us CRL [079]
[Collection of africana] / St Clair Drake – Chicago, IL: Uni of Chicago Photodup Dept, [19–?] – 43r – 1 – us CRL [080]
Collection of akan (twi-fante) materials / Warren, Denise M – 1 – us CRL [960]
Collection of ancient tunes : step shuffling & quick. from various authers [!] and churches of believers, both far and near, and herein transcribed for the purpose of retaining them / Haskins, Orren N – 1852 [mf ed 19–] – 1r – 1 – mf#pres. film 8 – us Sibley [780]
Collection of arrangements and transcriptions for the organ from the works of celebrated masters / Parker, Horatio William – New York: G Schirmer [1895] [mf ed 199-] – 1r – 1 – mf#pres. film 138 – us Sibley [780]
A collection of articles, injunctions, canons, orders, ordinances, and canons ecclesiastical, with other publick records of the church of england : chiefly in the times of king edward 6, queen elizabeth, king james, and king charles 1 – London: [s.n.], 1846 [mf ed 1993] – 1mf – 9 – 0-524-08267-7 – (in english and latin) – mf#1993-3022 – us ATLA [242]
Collection of booklets and pamphlets obtained in 1965-66 in the lower congo / Janson, John M – [n.p.], 1959-65 – 1r – 1 – (items are primarily official publs of several religious organizations) – us CRL [960]
A collection of chronicles and ancient histories of great britain (rs40) : now called england / Jehan de Wavrin, seigneur du Forestel; ed by Hardy, W & Hardy, E L C P – (= ser The rolls series (rs)) – (v1 1864 €19. v2 1887 €18. v3 1891 €15; trans by eds) – ne Slangenburg [931]
[Collection of clippings about auguste bournonville from danish periodicals] – [Copenhagen, 1865-1918] – 1r – 1 – us Misc Inst [920]
Collection of correspondence and documents of the civil war period – s.l, s.l? no date – 1r – us UF Libraries [976]

Collection of correspondence of herbert von bismarck, 1881-1883 / Bismarck, Herbert von – (= ser National archives coll of foreign records seized, 1941-) – 1r – 1 – mf#T972 – us Nat Archives [943]

A collection of curious travels and voyages / Ray, J – Bangkok. 1968-1972 (1) – 13mf – 9 – mf#7731 – ne IDC [910]

A collection of curious travels and voyages : vol 2: observations made in many parts of greece, egypt, asia minor... / Ray, J – London, 1693 – 4mf – 9 – mf#7095 – ne IDC [910]

Collection of data, 1784-1840 / Lane, Tidence – 166p – 1 – 5.81 – us Southern Baptist [242]

[Collection of documents from amnesty international's research archives] : amnesty's country dossiers [1975-] and publications [1962-] – 15 titles on 2518mf – 9 – ne IDC [327]

Collection of documents on ruanda-urundi and rwanda – s.l, s.l? no date – 1r – us UF Libraries [025]

Collection of early census listings – 5mf – 9 – £6.25 – (balsham 1811; cambridge st benet 1749, 1815, 1821; cambridge st botolph 1854; cambridge st edward 1801, 1811, 1821; cambridge st giles c 1760; cambridge st peter c 1760; cambridge st mary the great 1801, 1811; downham 1821; duxford 1821; ely st mary 1801, 1811; gamlingay 1798; girton 1801; great abington 1686; hildersham 1810; hinxton 1802; kirtling 1860; landbeach 1781, 1798; little wilbraham 1801; melbourn 1831; parson drove 1821; swaffham bulbeck 1854; trumpington 1786, 1811, c1838) – uk CambsFHS [929]

A collection of emblemes, ancient and moderne : quickened with metricall illustrations, both morall and divine... / Wither, G – London: A M for Robert Milbourne, 1635 – 7mf – 9 – mf#O-825 – ne IDC [090]

[Collection of ephemera related to the african national congress and dr s m molema – Johannesburg: Microfile, [19–] – 1r – 1 – (presumably coll by dr s m molema. contains minutes and programs of the congress, photos, newspaper clippings, correspondence, etc) – us CRL [960]

Collection of esoteric writings of t subba row – Bombay: printed at the "Tatva-Vivechaka" Press, 1895 [mf ed 1991] – 1mf – 9 – 0-524-01515-5 – mf#1990-2491 – us ATLA [290]

Collection of evidences for the divinity of our lord jesus christ / Freston, Anthony – London, England. 1807 – 1r – us UF Libraries [240]

A collection of facts and observations made relative to the state of morals and religion in holland in 1814 / Romeyn, John B – 1 – $50.00 – us Presbyterian [949]

A collection of favorite songs arranged for the voice and piano forte / Reinagle, Alexander – Philadelphia: Printed for A. Reinagle 1789?. music 1456 – 1 – us L of C Photodup [780]

The collection of finnish literature 1810-1944 – Helsinki – ca 80,000mf – 9 – (collection increases annually with 5,500 mf) – fi Helsinki [490]

A collection of gaelic proverbs and familiar phrases – Edinburgh: Maclachlan and Stewart; London: Simpkin, Marshall 1882 [mf ed Bloomington IN: Indiana Uni Lib, Preservation Dept 1984] – 1r – 1 – us Indiana Preservation [390]

Collection of government documents, mainly economic and statistical 150=ouagadougou 1960-[67?] – 3r – us CRL [324]

Collection of hausa manuscripts in arabic script – Chicago, Uni of Chicago, Photodup Dept, [19–?] – 20r – us CRL [470]

Collection of histories of gaul and france – Microcard Editions – 356mf (20:1) – 9 – $1535.00 – us UPA [944]

Collection of hungarian political and military records, 1909-45 – (= ser National archives coll of foreign records seized, 1941-) – 21r – 1 – $483.00 – mf#T973 – us Nat Archives [947]

A collection of hymns and sacred songs : suited to both private and public devotions... of the brethren of the old german baptist church – 1st ed. Kinsey's Station OH: Office of the Vindicator 1882 [mf ed 1993] – 1mf – 9 – 0-524-06869-0 – mf#1990-5288 – us ATLA [242]

Collection of important acts of parliament and assembly – Edinburgh, Scotland. 1840 – 1r – us UF Libraries [323]

Collection of italian military records, 1935-1943 – (= ser National archives coll of foreign records seized, 1941-) – 506r – 1 – mf#T821 – us Nat Archives [355]

Collection of mediaeval and renaissance manuscripts / Trinity College. Dublin – 4 sects – 1 – (sect 1: the roman inquisition 34r £1600 tce. sect 2: secular studies – 4pts 61r £2700 tcf. sect 3: waldensian and icelandic mss 12r £620 tcg. sect 4: literature 22r £990 tch) – mf#TCH – uk World [090]

Collection of negro spirituals / Work, John – 1874 – 1 – us Southern Baptist [242]

Collection of organ music by pupils of johann sebastian bach / Mulbury, David G [comp] – U of Rochester 1969 [mf ed 19–] – 2v on 4mf – 9 – (incl bibl ref) – mf#fiche 624 – us Sibley [780]

A collection of ornamental designs : after the manner of the antique / Smith, George – London [1812] – (= ser 19th c art & architecture) – 2mf – 9 – mf#4.2.1280 – uk Chadwyck [740]

Collection of ornaments at austin's artificial stone works / Austin, F – London 1838 – (= ser 19th c art & architecture) – 1mf – 9 – mf#4.2.1398 – uk Chadwyck [730]

[Collection of pamphlets on the churches and missions in the former federation of Rhodesia and Nyasaland] – Chicago, IL: Uni of Chicago Photodup Dept, 1967 (mf ed) – 1r – 1 – us CRL [080]

Collection of pamphlets relating to the methodist church in fiji, 1878-1970 / Gribble, Cecil F – Suva, Fiji – 1r – 1 – mf#PMB Doc408 – at Pacific Mss [242]

A collection of papers and facts relative to the dismission of william sandford oliver, esq : from the office of sheriff of the city and county of st john, in the province of new brunswick – [s.l: s.n.] 1791. [mf ed 1987] – 1mf – 9 – 0-665-61371-7 – mf#61371 – cn Canadiana [360]

A collection of papers connected with the theological movement of 1833 / Perceval, Arthur Philip [comp] – 2nd ed. London: Rivington, 1843 [mf ed 1991] – 1mf – 9 – 0-524-00304-1 – mf#1989-3004 – us ATLA [242]

Collection of papers read before the... / Bucks County Historical Society – v1, v3 – 2r – 1 – (cont by: papers read before the society and other historical papers) – mf#2800184 – us WHS [978]

[Collection of philippine literature in the bikol language] – [1895?-1913?] [mf ed 1984] – 1r – 1 – us Indiana Preservation [490]

[Collection of philippine literature in the bisaya language] – 1894?-1913? [mf ed 1984] – 2r – 1 – us Indiana Preservation [490]

[Collection of philippine literature in the pampanga language] – 1902?-05? [mf ed 1984] – 1r – 1 – us Indiana Preservation [490]

[Collection of philippine literature in the tagalog language] – [1904?-23?] [mf ed 1984] – 1r – 1 – us Indiana Preservation [490]

Collection of photographs and theatre, dance and sports programmes : relating to banaba (ocean isand) / Miller, Frank – 1909-39 – 1r – 1 – (available for reference) – mf#pmb1157 – at Pacific Mss [790]

[Collection of piano and vocal music, and piano duets] : handwritten during last decade of 1700s or first of 1800s – [1796?] [mf ed 19–] – 1v on 1r – 1 – (most composers are not identified but are identifiable from the titles...hook, reinagle, shuster, shield et al) – mf#pres. film 60 – us Sibley [780]

Collection of pioneer stories / Monroe, Lilla Day – undated, Copyright protected. No copies without permission – 1 – us Kansas [920]

Collection of popular tales from the norse and north german / Dasent, George Webbe – London, England. 1907 – 1r – us UF Libraries [430]

Collection of popular tales from the norse and north german / Dasent, George Webbe – London: Norroena Society, 1907 [mf ed 1993] – (= ser Anglo saxon classics) – 4mf – 9 – 0-524-08189-1 – mf#1991-0302 – us ATLA [390]

A collection of private acts of practical utility in force in new south wales : embracing the local private legislation from the year 1832 to the year 1885 / Tarleton, W W – Sydney: Thomas Richards, 1886 – 14mf – 9 – $21.00 – mf#LLMC 96-005 – us LLMC [323]

Collection of programs / Daly's Fifth Avenue Theatre. New York – (Ada Rehan Collection). New York. 1879-1899 – 1 – us NY Public [790]

A collection of psalms, hymns, and spiritual songs : suited to the various kinds of christian worship... – Covington, Miami Co, O[hio]: James Quinter 1875 [mf ed 1992] – 2mf – 9 – 0-524-03700-0 – mf#1990-4805 – us ATLA [242]

[A collection of publications relating to travel and description in north america between 1697 and 1775] / Masterson, James Raymond [comp] – [v.p, 195-?] – 14r – 1 – us Misc Inst [917]

A collection of published music including 17 piano pieces and 3 vocal selections / Bethune, Thomas Greene – Music-3087 – 1 – us L of C Photodup [780]

The collection of ralph waldo emerson, 1822-1903 (brram) : from the alexander ireland collection, manchester central library – 2r – 1 – (int by brian harding) – mf#97571 – uk Microform Academic [080]

Collection of reports of meetings held in favour of mr. james peiris' nomination as representative of the low country sinhalese in the legislative council – [Colombo]: Printed at the "Ceylon Examiner" Press, [1895?] – 1r – 1 – us CRL [954]

A collection of right merrie garlands for north country anglers – Newcastle-on-Tyne: G Rutland 1864 [mf ed Bloomington IN: Indiana Uni Lib, Preservation Dept 1984] – xv 312p on 1r [ill] – 1 – us Indiana Preservation [810]

[Collection of russian literary avant-garde, 1904-1946] – 778 titles on 1361mf – 9 – (individual titles also listed separately) – ne IDC [460]

A collection of several commissions : and other public instruments, proceeding on his majesty's royal authority... – London: printed by W & J Richardson...1772 [mf ed 1983] – 4mf – 9 – 0-665-44019-7 – mf#44019 – cn Canadiana [971]

A collection of some of the most useful acts and ordinances in force in lower canada : relating to criminal law and to the duties of magistrates / Canada (Province) – Quebec: printed by Stewart Derbishire & George Desbarats, 1854 [mf ed 1983] – 2mf – 9 – mf#SEM105P191 – cn Bibl Nat [345]

[Collection of songs and piano music in manuscript] – Bethlehem PA [181-?] [mf ed 1989] – 7v on 1r – 1 – mf#pres. film 59 – us Sibley [780]

Collection of southern rhodesia archives, manuscripts and documents / Haddon, Eileen – 1950s and 1960s – 1 – us CRL [960]

[Collection of spanish devotional literature] – [1882?]-1907 [mf ed 1984] – 1r – 1 – us Indiana Preservation [241]

Collection of spirituals (john w. work, fisk university) – 1 – us Southern Baptist [242]

Collection of state documents / Harvard University. Law School Library – Banking, insurance, labor, public utilities and taxation. 199 serial titles from 13 states: California, Connecticut, Illinois, Indiana, Maryland, Massachusetts, Michigan, New Jersey, New York, Ohio, Pennsylvania, Virginia and Wisconsin. 35,269mf. See individual states for further information – 9 – ca $44,087.00 – (all 32 banking titles. 6359mf. $1.50f. all 18 insurance titles. 12,655mf. $1.50f. all 60 labor titles. 3115mf. $1.50f. all 4 public utilities titles. 10,294mf. $1.50f. all 40 taxation titles. 2846mf. $1.50f. all 15 california titles. 2685mf. $1.50f. all 14 connecticut titles. 3357mf. $1.50f. all 14 illinois titles. 2154mf. $1.50f. all 11 indiana titles. 642mf. $1.50f. all 9 maryland titles. 1119mf. $1.50f. all 28 massachusetts titles. 4271mf. $1.50f. all 13 michigan titles. 1693mf. $1.50f. all 17 new jersey titles. 1674mf. $1.50f. all 26 new york titles. 8273mf. $1.50f. all 13 ohio titles. 2261mf. $1.50f. all 10 pennsylvania titles. 2805mf. $1.50f. all 7 virginia titles. 1386mf. $1.50f. all 22 wisconsin titles. 2949mf. $1.50f) – us Harvard Law [336]

A collection of statutes affecting new south wales : containing all the statutes of practical utility to the present time / ed by Cary, Henry – Sydney/Melbourne: Sands & Kenny. v1-2. 1841 – 24mf – 9 – $36.00 – mf#LLMC 96-003 – us LLMC [323]

A collection of telugu literature – India: [s.n.], [n.d.] [mf ed 1984] – 1v – 1 – mf#7217 – us Wisconsin U Libr [800]

A collection of temperance dialogues : for divisions of sons, good templar lodges, sections of cadets, bands of hope, and other temperance societies / Hammond, S T [comp] – Ottawa: S T Hammond, 1869 [mf ed 1981] – 2mf – 9 – mf#09887 – cn Canadiana [360]

A collection of the acts, deliverances, and testimonies of the supreme judicatory of the presbyterian church : from its origin in america to the present time, with notes and documents explanatory and historical, constituting a complete illustration of her polity, faith, and history / Baird, Samuel John [comp] – [2nd ed] Philadelphia: Presbyterian Board of Publ [1858?], c1855 [mf ed 1992] – (= ser Presbyterian coll) – 2mf – 9 – 0-524-03925-9 – mf#1990-4919 – us ATLA [242]

A collection of the ancient timber edifices of england / Clayton, John – London 1846 – (= ser 19th c art & architecture) – 2mf – 9 – mf#4.2.1342 – uk Chadwyck [720]

A collection of the charges, opinions and sentences of general courts-martial : as published by authority, from the year 1795 to the present time / James, Charles – 10mf – 9 – $15.00 – (intended to serve as an appendix tytler's "essay on military law...", and forming a book of cases and references, with a copious index) – mf#LLMC 88-119 – us LLMC [347]

A collection of the inscriptions on copper-plates and stones in the nellore district / Butterworth, Alan & Chetty, V Venugopaul – Madras: Supt, Govt Press, 1905 – (= ser Samp: indian books) – us CRL [730]

A collection of the judgments of the judicial committee of the privy council : in ecclesiastical cases relating to doctrine and discipline / ed by Brodrick, George Charles & Fremantle, William H – London: John Murray, 1865 [mf ed 1993] – (= ser Anglican/episcopal coll) – 2mf – 9 – 0-524-06872-0 – mf#1990-5291 – us ATLA [240]

Collection of the national conference of industry and commerce in china – Nan King, 1931. Chinese text. 1 reel – 1 – us Chinese Res [951]

A collection of the statutes of practical utility, colonial and imperial, in force in new south wales : embracing the local legislation from the year 1824 to 1879 / Oliver, Alexander – Sydney: Thomas Richards. v1-2 + app. 1879 – 28mf – 9 – $42.00 – mf#LLMC 96-004 – us LLMC [323]

Collection of the teachings of famous chinese *see* Chung-kuo sheng hsien yao tao lei pien [ccm300]

A collection of three thousand and six hundred tshi proverbs in use among the negroes of the gold coast speaking the asante and fante language / Christaller, Johann Gottlieb – Basel: Basel German Evangelical Missionary Society, 1879 – 1 – us CRL [390]

[Collection of travel accounts and travelogues] : [dating from the 16th to 19th centuries] – 1246 titles on 31,279mf – 9 – (individual titles also listed separately) – ne IDC [910]

Collection of trials on microfiche – Pts 1-8 – 9 – $6165.00 set – mf#409140 – us Hein [340]

Collection of upper volta political and other ephemera / Martens, George F – 1 – us CRL [080]

A collection of voyages and travels : consisting of authentic writes in our tongue, which have not before been collected in english... / [Osborne, T] – London: Thomas Osborne, 1745. 2v – 31mf – 9 – mf#HT-676 – ne IDC [910]

A collection of voyages and travels... / Churchill, A & Churchill, J – London: Awnsham, John Churchill, 1704. 4v – 51mf – 9 – (missing: v3) – mf#HT-672 – ne IDC [910]

A collection of voyages in four volumes, vol 1 : a new voyage round the world / Dampier, William – London 1729 [mf ed Hildesheim 1995-98] – 4mf – 9 – €120.00 – 3-487-29958-5 – gw Olms [910]

A collection of voyages in four volumes, vol 2 : a supplement to the voyage round the world / Dampier, William – London 1729 [mf ed Hildesheim 1995-98] – 4mf – 9 – €120.00 – 3-487-29957-7 – gw Olms [910]

A collection of voyages in four volumes, vol 3 : a voyage to new-holland / Dampier, William – London 1729 [mf ed Hildesheim 1995-98] – 4mf – 9 – €120.00 – 3-487-29956-9 – gw Olms [910]

A collection of voyages in four volumes, vol 4 : being an account of capt william dampier's expedition into the south-seas, in the ship st george... / Funnell, William – London 1729 [mf ed Hildesheim 1995-98] – 4mf – 9 – €120.00 – 3-487-29942-9 – gw Olms [910]

A collection of wills / Grutze, Albert Lewis – Portland, Ore., Title and Trust Co. 1927 99 p. LL-334 – 1 – us L of C Photodup [340]

Collection of works : [Works] / Billington, Thomas – London: printed for aut [1780-90] [mf ed 1990] – 1r – 1 – mf#pres. film 94 – us Sibley [780]

A collection of ye old-fashioned dances of 1850 : containing over 50 contra dances, cotillons, quadrilles / Carville, F E – [Lewiston, ME: F E Carville, 1926] – 1 – mf#*ZBD-*MGO pv20 – Located: NYPL – us Misc Inst [790]

[Collection on european freemasonry] : [early sources, 1717-1870, from the grand lodge library in the hague] – 651 titles on 2345mf – 9 – ne IDC [366]

[Collection on richard wagner] : [wagner literature of the 19th century as held by the zentralbibliothek zuerich] – 145 titles on 366mf – 9 – (individual titles also listed separately) – ne IDC [780]

[Collection on rural and regional development] – 2410mf – 9 – (sect: africa [329 titles on 881mf]; east asia [11 titles on 49mf]; eastern europe [2 titles on 5mf]; europe [2 titles on 9mf]; latin america [233 titles on 637mf]; middle east/ north africa [54 titles on 210mf]; north america [1 title on 3mf]; pacific [18 titles on 30mf]; south asia [93 titles on 216mf]; south east asia [112 titles on 370mf]; with printed catalogue) – ne IDC [338]

Collection on the exclusion of dr. calvin and caroline cutter / Nashua, New Hampshire. First Baptist Church – c1838 – 1 – us ABHS [242]

Collection "question libanaise" – [Beirut?: s.n.] 1975-76 – 1r – us CRL [956]

Collection(mexican and peruvian documents) / Harkness, Edward S – 1 – us L of C Photodup [972]

Le collectionneur illustre des monnaies canadiennes : premier supplement annuel, juin 1892 = Breton's illustrated canadian coin collector. first annual supplement, june 1892 / Breton, Pierre Napoleon – Montreal: P Breton, 1892 [mf ed 1980] – 1mf – 9 – 0-665-03721-X – (in french and english) – mf#03721 – cn Canadiana [730]

Le collectionneur illustre des monnaies canadiennes : le seul livre donnant la valeur approximative des monnaies du canada = Breton's illustrated canadian coin collector: the only work giving the approximate value of canadian coins / Breton, Pierre Napoleon – Montreal: P Breton, [1890] [mf ed 1980] – 1mf – 9 – 0-665-03720-1 – (in french and english) – mf#03720 – cn Canadiana [730]

Collections / Holland Society of New York – v1 pt1 [1891] – 1r – 1 – mf#4187008 – us WHS [978]

Collections / Massachusetts Historical Society – ser 7 v6 – 1r – 1 – mf#1063575 – us WHS [978]

Collections / Minnesota Historical Society – 1872-1920. 9 rolls. Printed guide 2210.00, available separately – 1 – $270.00; $30.00r – us Minn Hist [025]

Collections / Minnesota Historical Society – v8 – 1r – 1 – mf#3075801 – us WHS [978]

Collections / South Carolina Historical Society – 1857-97. 5v – 34mf – 9 – us South Carolina Historical [978]

Collections / State Historical Society of Wisconsin – v13 – 1r – 1 – (cont: report and collections of the state historical society of wisconsin, for the years...) – mf#167342 – us WHS [978]

Collections / Worcester Society of Antiquity – v9 – 1r – 1 – (cont by: proceedings of the worcester society of antiquity) – mf#3820875 – us WHS [071]

Collections de documents inedits sur l'histoire de france (guizot collection) / France. Ministere de l'Instruction Publique – v1-322. 1835-1951 – (= ser Guizot coll) – 1 – us AMS Press [944]

Les collections de nouvelles de l'empereur justinien / Noailles, P – Paris, 1912. 2v – 7mf – 9 – mf#H-2897 – ne IDC [956]

Collections for a handbook of the makua language / Maples, Chauncy – London, England. 1879 – 1r – us UF Libraries [470]

Collections from the library of the jewish theological seminary of america : valuable books and manuscripts from international centers of judaism – [mf ed UMI] – 383r – 1 – (with p/g for all but 3 of the coll, separate ind for these 3 units) – us UMI ProQuest [270]

Collections from the royal society / The Royal Society – $16,875.00 coll – (book catalogue of the library of the royal society 5v $1240; council minutes, 1660-1800 [3r] isbn 0-89093-876-8 $490; the early letters & classified papers, 1660-1740 [23r] isbn 1-55655-171-1 $4115; journal books of scientific meetings, 1660-1800 [18r] isbn 0-89093-875-X $2820; letters & papers of robert boyle [16r] isbn 1-55655-170-3 $2870; letters & papers of robert boyle: a guide to the mss & microfilm isbn 1-55655-2-17-3 $125; letters & papers of sir john herschel [28r] isbn 1-55655-169-X $4990; letters & papers of sir john herschel: a guide to the mss & microfilm isbn 1-55655-218-1 $125; misc mss [10r] isbn 0-89093-877-6 $1575; with p/g) – us UPA [500]

Collections, historical and miscellaneous, and monthly literary journal – Concord, 1822-1824 – 1,5,9 – mf#3720 – us UMI ProQuest [073]

Collections of the alcuin club – London, 1899-1901. v1-4 – 16mf – 8 – mf#H-762 – ne IDC [700]

Collections of the massachusetts historical society, 1792-1941 / Massachusetts Historical Society – 7 series of 10v ea (mf Massachusetts Hist Soc 1990) – 24r – 1 – $2105.00y – (vol 10 of each series contains a general table of contents and ind) – us MA Hist [978]

Collections of the state historical society of wisconsin / Wisconsin. State Historical Society – v1-31. 1854-1931 – 1 – $378.00 – mf#0678 – us Brook [978]

Collection(spanish-american documents) / Kraus, Hans P – 1500-1819 – 1 – 87.00 – us L of C Photodup [972]

Collective artistic direction : the dynamics of a modern dance company / Chambers, Sue – Texas Woman's University, 1994 – 1mf – 9 – mf#PE 3634 – us Kinesology [790]

Collective bargaining report / AFL-CIO [American Federation of Labor and Congress of Industrial Organizations] – v1-6 n4 [1956 jan-1961 apr] – 1r – 1 – mf#1110711 – us WHS [331]

Collective bargaining settlements in new york state / New York. Dept. of Labor. Division of Research and Statistics – 1948-79. 174 fiches. (Harvard Law School Library Collection.) – 9 – us Harvard Law [331]

Collective Black Artists *see* Expansions

The collective catalogue of hebrew manuscripts : two major catalogues at the jewish national and university library in jerusalem / The Jewish National and University Library. Jerusalem – [mf ed Chadwyck-Healey] – 812mf + update on 50 COM – 9 – (with ind, p/g and int) – uk Chadwyck [090]

Collective of Black Students [Temple University of Delaware] *see* Pamoja!

Collective [South Portland, Maine] *see* Common scold

Collector / American Flyer Collectors Club – v1 n1-v2 n4 [1978-79] – 1r – 1 – mf#638917 – us WHS [790]

Collector – 1975 apr-1976 sep, 1976 sep-1978 sep, 1978 oct-1979 dec – 3r – 1 – (cont: collector's weekly; antiques today; cont by: american collector) – mf#464414 – us WHS [740]

The collector – London, 1902-07 – 1r – 1 – (selections from 'the queen') – uk British Libr Newspaper [072]

The collector / Tawfik Al-Hakim – Np, 1974 – 1mf – 9 – mf#NE-382 – ne IDC [956]

The collector and art critic – 1899-1907 [mf ed Chadwyck-Healey] – (= ser Rare 19th century american art journals) – 17mf – 9 – uk Chadwyck [700]

Collector editions – New York. 1989-1996 [1] – (cont: collector editions quarterly) – ISSN: 0733-2130 – mf#8740,02 – us UMI ProQuest [790]

Collector editions quarterly – New York. 1977-1981 (1,5,9) – (cont: acquire: the magazine for collectors; cont by: collector editions) – ISSN: 0199-929X – mf#8740,01 – us UMI ProQuest [790]

Collector investor – Chicago. 1980-1982 (1,5,9) – ISSN: 0197-2367 – mf#12606 – us UMI ProQuest [700]

Collectorium in 4 libros sententiarum : cum oratione soluta wendelin steinbach / Biel, Gabriel – Tuebingen. v1-2. 1501 – €101.00 – ne Slangenburg [241]

Collector's exchange – v1 n1-10 [1971 jul/aug-1976 aug] – 1r – 1 – mf#359364 – us WHS [790]

Collector's ledger – v2 n23-v4 n5 [1887 oct-1888 aug] – 1r – 1 – mf#219516 – us WHS [790]

Collectors' marks / Fagan, Louis Alexander – London 1883 – (= ser 19th c art & architecture) – 3mf – 9 – mf#4.2.1531 – uk Chadwyck [700]

Collectors motor news [cmn] – 1984 jan-1986 feb – 1r – 1 – (cont: antique motor news; cars for sale; cont by: coast car collector; collector car news) – mf#1028724 – us WHS [629]

Collectors' network news – 1977 jan/feb-1978 jan/dec – 1r – 1 – (cont: top secret) – mf#505596 – us WHS [790]

Collector's weekly – v5 n259-v6 n288 [1974 sep 3-1975 mar 25] – 1r – 1 – (cont by: collector [kermit tx]) – mf#464579 – us WHS [790]

Collects for sundays and holydays throughout the year / Church Of England – London, England. 1820 – 1r – us UF Libraries [240]

Colledge, T R *see* The medical missionary society in china

Os collegas – Rio de Janeiro, RJ. 13 nov-dez 1881 – (= ser Ps 19) – mf#P17,01,97 – bl Biblioteca [440]

College addresses and sermons / Lindsay, Thomas Martin – Glasgow: J. Maclehose, 1915 [mf ed 1990] – 1mf – 9 – 0-7905-5173-X – mf#1988-1173 – us ATLA [242]

College and career – 1956-70 – 1 – us Southern Baptist [242]

College and research libraries – Chicago. 1939+ [1]; 1969+ [5]; 1975+ [9] – ISSN: 0010-0870 – mf#412 – us UMI ProQuest [020]

College and research libraries news – Chicago. 1980+ (1,5,9) – ISSN: 0099-0086 – mf#12476 – us UMI ProQuest [020]

College and undergraduate libraries / ed by Bahr, Alice Harrison – v3 n1. 1996- – 1,9 – $60.00 in US $84.00 outside hardcopy subsc – us Haworth [020]

College and university – Washington. 1925+ (1) 1968+ (5) 1976+ (9) – ISSN: 0010-0889 – mf#1459 – us UMI ProQuest [378]

College and university bulletin / American Association for Higher Education – Washington. 1972-1978 (1) 1972-1978 (5) 1973-1978 (9) – (cont by: aahe bulletin) – ISSN: 0010-0897 – mf#6980 – us UMI ProQuest [378]

College and university business – Chicago. 1946-1974 (1) 1969-1974 (5) – ISSN: 0010-0900 – mf#322 – us UMI ProQuest [378]

College and university journal – Washington. 1962-1974 [1]; 1971-1974 [5,9] – ISSN: 0010-0927 – mf#2005 – us UMI ProQuest [378]

College and University Personnel Association *see* Journal of the college and university personnel association

College and university sermons / Lyttelton, Arthur Temple – London; New York: Macmillan, 1894 – 1mf – 9 – 0-7905-8509-X – mf#1989-1734 – us ATLA [240]

Le college anglais de douai : son histoire heroique / Fabre, F – 1930 – (= ser Revue de litterature comparee 10) – €3.00 – ne Slangenburg [378]

College architecture in america and its part in the development... / Klauder, Charles Zeller – New York, NY. 1929 – 1r – us UF Libraries [720]

College board review – New York. 1947+ (1) 1974+ (5) 1975+ (9) – ISSN: 0010-0951 – mf#9757 – us UMI ProQuest [378]

College catalog collections: national / Career Guidance Foundation – 1500mf – 9 – $698.00 – (more than 2900 accredited postsecondary schools represented by approx. 3600 catalogs. Annual subscription includes supplements and indexes and most recent catalogs) – us Career [378]

College catalog collections: regional / Career Guidance Foundation – 9 – $298.00 each region – (More than 700 regional schools and 900 or more catalogs. eastern, 350mf; western, 256mf; north central, 354mf; southern, 355mf) – us Career [370]

College chapel sermons / Nevin, John Williamson; ed by Kieffer, Henry Martyn – Philadelphia: Reformed Church Publ House, 1891 [mf ed 1991] – 1mf – 9 – 0-7905-9824-8 – mf#1989-1549 – us ATLA [242]

College composition and communication – Urbana. 1950+ [1]; 1971+ [5]; 1976+ [9] – ISSN: 0010-096X – mf#1502 – us UMI ProQuest [378]

College corner news / Butler Co. Col.Corner – 2,1902-12,1915/1,1917-6,1991 – 26r – 1 – mf#B35308-35333 – us Ohio Hist [071]

College crampton : komoedie in 5 akten / Hauptmann, Gerhart – 3. aufl. Berlin: S Fischer 1896 [mf ed 1990] – 1r – 1 – (filmed with: die armseligen besenbinder / carl hauptmann) – mf#2700p – us Wisconsin U Libr [820]

College days / Ripon College – 1868 may, v1 n1-v2 n9 [1868 may-1870 jul], v4 n5-v6 n2, [1872 feb-1873 nov], v6 n3-6 [1873 dec-1874 mar] – 1r – 1 – mf#1725057 – us WHS [378]

College days / Ripon College – v1 n1-v3 n9 [1882 jan-1884: jun], v2 n5-7 [1883 feb-apr], v4 n2 [1884 nov], v9 n8-9, [1890 may-jun], filmed out of sequence, at end of reel, v9 n1-v10 n9 [1889 oct-1891 jun], v12 n9 [1893 may], v14 n7-12 [1895 feb 14-jun 19], 29th:1-31st 2 [1895 sep 21-oct 19]=n179-n205, 1897 nov 9-1903 mar 30, 1903 apr 22-1906 dec, 1907-1909 jun – 5r – 1 – (cont: ripon college news-letter; cont by: ripon college days) – mf#1725310 – us WHS [378]

The college days of calvin / Blackburn, William Maxwell – Philadelphia: Presbyterian Board of Publ, c1865 – 1mf – 9 – 0-7905-4487-3 – mf#1988-0487 – us ATLA [242]

College de pataphysique. dossiers *see* Cahiers du college de pataphysique

College des medecins et chirurgiens de la province de Quebec. Bureau provincial de medecine. Assemblee *see* Proces-verbaux des assemblees

College english – Urbana. 1939+ [1]; 1968+ [5]; 1975+ [9] – ISSN: 0010-0994 – mf#480 – us UMI ProQuest [378]

College English Association *see*
- Cea critic
- Cea forum

College Entrance Examination Board *see* Examination questions in latin and greek

College l'assomption, hommage d'un medaillon presente par m maximilien bibaud : doyen de l'ecole de droit du college ste marie, montreal – Montreal: impr de la Minerve, 1865 – 1mf – 9 – mf#33335 – cn Canadiana [378]

College lectures on democracy of religion / McWhinney, Thomas Martin – Dayton, Ohio: Christian Publishing Association, 1907 – 1mf – 9 – 0-7905-8523-5 – mf#1989-1748 – us ATLA [240]

College library: catalogue of early indian imprints, 1714-1850 / Carey, William – 1,296p – 1 – us Southern Baptist [242]

College literature – West Chester. 1974+ (1) 1974+ (5) 1976+ (9) – ISSN: 0093-3139 – mf#10295 – us UMI ProQuest [400]

College management – Stamford. 1966-1974 [1]; 1971-1974 [5,9] – ISSN: 0010-1036 – mf#2080 – us UMI ProQuest [378]

College mathematics journal : an official publication of the mathematical association of america / Mathematical Association of America – Washington. 1984+ (1,5,9) – (cont: two-year college mathematics journal) – ISSN: 0746-8342 – mf#10929,01 – us UMI ProQuest [510]

College mercury / Racine College [WI] – v1-9 [1867 jun 15-1871 jul 6], v10 n1-4,6-8 [1871 sep 30-1872 feb 15], v11-13 [1872 mar 1-1873 jul 7], v14 n1 [1873 sep 27] – 1r – 1 – mf#2976207 – us WHS [378]

College music symposium – Binghampton. 1989+ (1,5,9) – ISSN: 0069-5696 – mf#15695 – us UMI ProQuest [780]

College news / Ripon College – v1 n1-2 [1879 oct 18-nov 1] – 1r – 1 – (cont: ripon college quarterly; cont by: ripon college news-letter) – mf#2547969 – us WHS [378]

A college of colleges / Moody, Dwight Lyman & Drummond, Henry; ed by Shanks, T J – Chicago: Fleming H Revell, c1887 [mf ed 1986] – 1mf – 9 – 0-8370-6614-X – mf#1986-0614 – us ATLA [240]

College of Ganado *see* Ganado today

College of medicine news / Howard University – 1977 oct – 1r – 1 – mf#4989508 – us WHS [610]

College of physicians. transactions – Philadelphia. v1-40. 1875-1918 – 1 – $432.00 – mf#0160 – us Brook [610]

The college of st. francis xavier : a memorial and a retrospect, 1847-1897 – New York: Meany, c1897 – 1mf – 9 – 0-8370-8708-2 – mf#1986-2708 – us ATLA [240]

College of William and Mary et al *see* News letter from the institute of early american history and culture

College Physical Education Association Committee on Terminology *see* Glossary of physical education terms, part 1

College planning and management – Dayton. 1998+ – 1 – mf#27962 – us UMI ProQuest [370]

College press service – Denver. 1980-1981 (1) 1980-1981 (5) 1980-1981 (9) – ISSN: 0010-1125 – mf#7936 – us UMI ProQuest [378]

College Saint-Denis *see* Retrospective generale et projet de nouvelles conventions memoire du college saint-denis

College station first baptist church. college station, texas : church records – 1920-63. 1688p – 1 – 67.52 – us Southern Baptist [242]

College store journal – Oberlin. 1977-1981 – 1,5,9 – ISSN: 0010-115X – mf#10522 – us UMI ProQuest [378]

College student journal – Mobile. 1967+ (1) 1970+ (5) 1976+ (9) – ISSN: 0146-3934 – mf#5830 – us UMI ProQuest [378]

College student personnel abstracts – Claremont. 1973-1984 (1) 1973-1984 (5) 1973-1984 (9) – (cont by: higher education abstracts) – ISSN: 0010-1168 – mf#9170 – us UMI ProQuest [378]

College teaching – Washington. 1985+ (1) 1985+ (5) 1985+ (9) – (cont: improving college and university teaching) – ISSN: 8756-7555 – mf#6044,01 – us UMI ProQuest [378]

College times – [Toronto?]: Upper Canada Literary Society, [1871?-19–] – 9 – mf#P05018 – cn Canadiana [378]

College times *see* [San jose-] the spartan daily

College topics : devoted to the interests of the students in the universities and colleges of toronto – Toronto. v1-5 n10. nov 11 1897-jan 14 1902// – 1r – 1 – Can$90.00 – cn McLaren [378]

College topics – Toronto: [s.n, 1897-1902?]; (Toronto: [C B Robinson]) – 9 – mf#P06026 – cn Canadiana [378]

College transcript / Delaware Co. Delaware – 6,1898-6,1899 – 1r – 1 – mf#B36315 – us Ohio Hist [378]

College transcript / Delaware Co. Delaware – oct 1874-may 1903 [semiwkly, wkly] – 6r – 1 – mf#B30130-30135 – us Ohio Hist [378]

College union 1600 voice / Cook County College Teachers Union, Local 1600 – v13 n8-v20 n3 [1977 mar-1984 may], v20 n4-v21 n3 [1984 jul-1984 oct] – 2r – 1 – mf#5123505 – us WHS [378]

College union voice / Cook County College Teachers Union, Local 1600 – v21 n4-v31 n3 [1984 dec-1994 nov] – 1r – 1 – mf#5123556 – us WHS [331]

College view enterprise – College View, NE: Enterprise Pub Co (wkly) [mf ed v2 n15. apr 12 1893 filmed [1973]] – 1r – 1 – us NE Hist [071]

College view gazette – College View, NE: Gazette Pub Co. 8v. v1 n1. nov 10 1910-v8 n1. dec 27 1917 [wkly] [mf ed with gaps filmed 1958] – 3r – 1 – (merged with: college view advocate to form: college view gazette-advocate) – us NE Hist [071]

College view gazette-advocate – College View, NE: Norman Ott. v8 n2. jan 3 1918-1922// [wkly] [mf ed with gaps] – 2r – 1 – (formed by the union of: college view gazette to form: college view advocate; cont by: college view herald) – us NE Hist [071]

College view herald – College View, NE: L W Evans, 1922 [wkly] [mf ed -1925 [gaps]] – 1r – 1 – (cont: college view gazette-advocate) – us NE Hist [071]

College women, alcohol consumption, and negative sexual outcomes / Good, DL – 1992 – 2mf – 9 – $8.00 – us Kinesology [613]

College women athletes' knowledge and perceptions of title 9 / Jacob, Michael P – Iowa State University, 1993 – 2mf – 9 – $8.00 – mf#PE3603 – us Kinesology [790]
The colleges and theological institutions of america : a lecture / Blaikie, William Garden – Edinburgh: A. Elliot, 1870 – 1mf – 9 – 0-7905-5570-0 – mf#1988-1570 – us ATLA [240]
Colleges of education, report of the study group on the government of... 1966 – 1mf – 9 – mf#87026 – uk Microform Academic [324]
Collegia theologica quae extant omnia / Maccovius, J – Franekerae, 1641 – 11mf – 9 – mf#PBA-241 – ne IDC [240]
Collegian : or, american students' magazine – New York. 1819-1819 (1) – mf#3721 – us UMI ProQuest [378]
Collegiate and other ancient manchester / Smith, James Hicks – London, England. 1877 – 1r – us UF Libraries [240]
Collegiate baseball – Tucson. 1958+ (1) 1982-1982 (5) 1982-1982 (9) – ISSN: 0530-9751 – mf#10488 – us UMI ProQuest [790]
Collegiate church : yearbook of the reformed protestant dutch church of the city of new york – New York, NY. 1950-78 [complete] – 3r – 1 – mf#ATLA S0387 – us ATLA [242]
Collegiate coaches' knowledge of eating disorders / Turk, Joanne C – University of North Carolina at Chapel Hill, 1995 – 2mf – 9 – $8.00 – mf#HE559 – us Kinesology [616]
Collegiate football performance as a predictor of wonderlic personnel test scores / Travis, Kelly C – 2000 – 86p on 1mf – 9 – $5.00 – mf#PSY 2135 – us Kinesology [790]
Collegiate microcomputer – West Point. 1983-1993 (1) 1983-1993 (5) 1983-1993 (9) – ISSN: 0731-4213 – mf#13360 – us UMI ProQuest [378]
Collegiate soccer players' perceptions of sport psychology, sport psychologists and sport psychological services / Francis, Nicholas C & Gould, Daniel – 1991 – 2mf – 9 – $8.00 – us Kinesology [150]
Collegiate times – Blacksburg, VA. 1971-2001 (1) – mf#66673 – us UMI ProQuest [071]
Collegiate trends / Inter-Varsity Christian Fellowship – iss 14-41 [1986 jan-1988 jun] – 1r – 1 – mf#1699027 – us WHS [378]
Collegium – Villafranca de los Barros: Colegio de San Jose, 1939. 1 numero – 5 – sp Bibl Santa Ana [073]
Collegium ethicum : in quo triginta quatuor disputationibus in ill amstelodamensium gymnasio, publice ventilatis, ethica proponitur / Senguerdius, Arnoldus – Amstelaedami: Sumptibus Joannis Ravesteinii 1654 – (= ser Ethics in the early modern period) – 3mf – 9 – mf#pl-473 – ne IDC [170]
Collegium S. Bonaventurae *see*
– Opuscula sancti patris francisci assisiensis
– Speculum beatae mariae virgins
Collegium theologicum / Desmarets, S – Groningae, 1659 – 8mf – 9 – mf#PFA-140 – ne IDC [240]
Collet, Sophia Dobson *see*
– The brahmo somaj
– Keshub chunder sen's english visit
– Lectures and tracts
– The life and letters of raja rammohun roy
Collett, Sidney *see* The scripture of truth
Collette, Charles Hastings *see*
– Chair of peter
– Dr mccave (a roman priest in kidderminster) on the reformation
– Dr wiseman's popish literary blunders exposed first series
– Is the honour or veneration given to images and relics by roman catho...
– The novelties of romanism
– Rev s baring-gould on "luther and justification"
– Sacramental confession
Colletter, Charles Hastings *see* Roman priests as described by themselves
Le collezioni botaniche della missione stefanini-paoli nella somalia italiana : appendice: le raccolte di mangano, scassellati, mazzocchie provenzale in somalia / Chiovenda, E – Carbondale. 1971+ (1) 1927+ (5) 1976+ (9) – 10mf – 9 – mf#6125 – ne IDC [914]
Collichio, Gary S *see* Peer group support and propensity for violence against womem
Collie, James *see* Plans, elevations, sections, details and views of the cathedral of glasgow
Collie, James H *see* Old disciple
Collier County Historical Society *see* Timepiece
Collier county news – Naples, FL. 1928 apr 26-1967 mar – 53r – (gaps) – us UF Libraries [071]
Le collier de coquillages / Ouane, Ibrahima Mamadou – [Andrezieux, France, Impr moderne 1958] – us CRL [944]
Collier, George *see* West-african sketches
Collier, J Payne *see* Complete works of william shakespeare
Collier on bankruptcy – 1st-14th ed. 1898-1982 (all publ) – 9 – $795.00 set – mf#402450 – us Hein [332]
Collier, Price *see* The west in the east from an american point of view
Collier, Richard *see* Pay-off in calcutta
Collier, Robert Laird *see* Meditations on the essence of christianity
Collier, William Miller *see* The law and practice in bankruptcy under the national bankruptcy act of 1898
Collier's : the national weekly – New York. 1891-1915 (1) – mf#5709 – us UMI ProQuest [305]
Collier's quarterly – 1976 jan-1977 dec, 1981-85 – 2r – 1 – mf#363409 – us WHS [071]
Colliery guardian – Redhill. 1973-1989 (1) 1974-1989 (5) 1974-1989 (9) – ISSN: 0010-1281 – mf#7030 – us UMI ProQuest [622]
Colliery workman's times – Manchester, 2 Dec 1893-13 Jan 1894 – 7ft – 1 – uk British Libr Newspaper [072]
Colligan, James Hay *see*
– The arian movement in england
– Eighteenth century nonconformity
Colligere fragmenta *see* Das irische palimpsest-sakramentar in clm 14429 [tab53-54]
Collignon, Maxime *see* Manual of mythology in relation to greek art
Collin, August Zacharias *see* Sur les conjonctions gothiques
Collin de Plancy, J A S *see* Dictionnaire infernal; repertoire universel des etres, des personnages, des livres, des faits et des choses qui tiennent aux esprits.
Collin de Plancy, Jacques-Albin-Simon *see*
– La botte de paille
– Jacquemin le franc-macon
Collin D'harleville, Jean Francois *see* Moeurs du jour
Collin, Eric *see* Barron g collier
Collin, Jean-Pierre *see* L'evolution du marche foncier en peripherie du centre-ville de montreal au cours des annees soixante
Collin, Joseph *see* Goethes faust in seiner aeltesten gestalt
Colling, James Kellaway *see*
– Art foliage
– Gothic ornaments drawn from existing authorities
Collingridge, Ignatius *see* Correspondence respecting the spiritual condition of catholics in t...
Collingwood, Robin George *see* Religion and philosophy
Collingwood, William Gershom *see*
– The art teaching of john ruskin
– The philosophy of ornament
Collins – An appendix concerning the ordinance of singing. London. 1680 – 1 – $5.00 – us Southern Baptist [242]
Collins, A Frederick *see* Design and construction of induction coils
Collins, A Jefferies *see* Manuale ad usum percelebris ecclesiae sarisburiensis (hbs91)
Collins, Almer M et al *see* The contradictions of orthodoxy
Collins, B *see*
– Elementary tonga grammar
– Tonga grammar
Collins, Cornelius Francis *see* The municipal court practice act, annotated.
Collins, D *see* An account of the english colony in new south wales
Collins, David *see* An account of the english colony in new south wales
Collins, Francis *see* Voyages to portugal, spain, sicily, malta, asia minor, egypt, &c &c
Collins, Fred Chrysler *see* Study of the use of rigid circular bearing plates of small...
Collins, Fred K *see* An abstract of the statutory law of corporations as respects their formation, officers, meetings, liability of members, etc
Collins, G Colleen *see* Effects of individual leisure counseling on perceived freedom in leisure, perceived self-efficacy, depression, and abstinence of adults in a residential program for substance
Collins, George *see*
– The apocalypse explained
– An explanation of the eleventh chapter of the book of daniel
– An explanation of the visions of the four beasts, daniel 7
Collins, John H *see* Propaganda, ethics and psychological assumptions in caesar's writings
Collins, John Owen *see* Panama guide
Collins, Joseph Edmund *see*
– Annette, the metis spy
– Canada under the administration of lord lorne
– Canada's patriot statesman
– The four canadian highwaymen
– The future of the dominion of canada
– The story of louis riel, the rebel chief
Collins, Mabel *see* Light on the path
Collins, Michael G *see* Effects of three different hyperhydration strategies on cardiovascular and thermoregulatory resonses, blood volume and running performance
Collins, Perry McDonough *see* A voyage down the amoor
Collins, Richard *see*
– Missionary enterprise in the east
– The philosophy of jesus christ as unfolded in the physical aspect of his miracles
Collins, Sherry L *see* Impressionism in the arts and its influence on selected dance works
Collins, William *see*
– Church of scotland
– On the harmony between the gospel and temperance societies
Collins, William A *see* The divorce question
Collins, William Edward *see*
– The beginnings of english christianity
– Church and state in england before the conquest
– The english reformation and its consequences
– Four recent pronouncements
– Internal evidence of the letter "apostolicae curae" as to its own...
– The internal evidence of the letter "apostolicae curae" as to its own origin and value
– Lectures on archbishop laud
– The nature and force of the canon law
– Queen elizabeth's defence of her proceedings in church and state
– The rights of a particular church in matters of practice
– The study of ecclesiastical history
– Thomas becket
– Typical english churchmen from parker to maurice
Collins, William W *see* Free statia
Collins, William Wilkie *see* Memoirs of the life of william collins...
Collins, Winfield Hazlitt *see* The domestic slave trade of the southern states
Collinson and Lock *see* Sketches of artistic furniture
Collinson, R *see*
– Account of the proceedings of h m s enterprise from behring strait to cambridge bay
– Journal of h m s enterprise 1850-1855
– The three voyages of...in search of a passage to cathaia and india by the north-west, a d 1576-1578
Collinsworth, James Ragan *see* The pseudo church doctrine of anti-pedo-baptists defined and refuted
Collinus, R *see* Vita rodolphi collini...ab ipso collino descripta...
The collinwood citizen : official republican newspaper of the village – Cleveland, OH: Frank A. Bowman, sep 22 1905-jun 27 1918 – 6r – 1 – (wkly republican newspaper) – mf#(M) 34 C9.3 182 – us Western Res [071]
Collis, James *see* The builders' portfolio
Collis, Maurice *see*
– Ba ma' a tvan sa bho
– Cortes and montezume
Collison, Harry *see* Christianity as mystical fact
Collison, S E *see*
– Citrus fertilizer experiments
– Loss of fertilizers by leaching
– Prussic acid in sorghum
– Sugar and acid in oranges and grapefruit
Collison, W H *see* In the wake of the war canoe
Collison-Morley, Lacy *see* Greek and roman ghost stories
Collocott, E E V *see*
– Correspondence
– 'King taufa"
Colloid journal of the ussr – New York. 1965-1977 (1) 1965-1976 (5) – ISSN: 0010-1303 – mf#10817 – us UMI ProQuest [540]
Colloids and surfaces – Amsterdam. 1980+ (1) 1980+ (5) 1987+ (9) – ISSN: 0166-6622 – mf#42152 – us UMI ProQuest [540]
Colloids and surfaces a : physicochemical and engineering aspects – Amsterdam. 1993+ (1,5,9) – ISSN: 0927-7757 – mf#42714 – us UMI ProQuest [540]
Colloids and surfaces b : biointerfaces – Amsterdam. 1993+ (1,5,9) – ISSN: 0927-7765 – mf#42715 – us UMI ProQuest [540]
Collom, John *see* The prophetic numbers of daniel and the revelation
Colloque de bande dessinee de Montreal *see* Actes
Colloque International Sur La Classification Nominale Dans... *see* Classification nominale dans les langues negro-africaines
Colloque kino-quebec : gaspe 1983, 14, 15, 16, 17, 18, 19, aout – Quebec: [Ministere du loisir, de la chasse et de la peche], [1983] (mf ed 1985) – 3mf – 9 – mf#SEM105P485 – cn Bibl Nat [790]
Colloque sur le multilinguisme = Symposium on multilingualism (1962 : brazzaville, congo) – London, England. 1964 – 1r – us UF Libraries [400]
Colloquia peripatetica : deep-sea soundings: being notes of conversations with the late john duncan, ll. d., professor of hebrew in the new college, edinburgh / Knight, William Angus – 5th ed, enl. Edinburgh: David Douglas, 1879 – 1mf – 9 – 0-8370-4403-0 – (incl ind) – mf#1985-2403 – us ATLA [240]
Colloquial arabic; shuwa dialect of bornu, nigeria, and of the region of the chad / Lethem, Gordan James et al – London: Publ for the Govt of Nigeria by the Crown Agents for the Colonies, 1920 – 1 – (grammar & vocabulary, with some proverbs & songs) – us CRL [470]
Colloquiorum scholasticorum libri 4 / Cordier, M – Lipsiae, 1588 – 4mf – 9 – mf#PPE-106 – ne IDC [240]
Colloquium – v1-18. 1964-86 [complete] – 3r – 1 – (cont: new zealand theological review) – mf#atla s0843 – us ATLA [240]
Colloquium de peccato originis inter d iacobum andreae, et m matthiam flaccivm illyricum / Andreae d A, J – Tvbingae, 1574 – 2mf – 9 – mf#TH-1 mf 43-44 – ne IDC [242]
Colloquium metricum... / Faber, Benedikt – 1609 – (= ser Mssa) – 1mf – 9 – €20.00 – mfchl 213 – gw Fischer [780]
Colloquium on regulatory design in theory and practice : 1982-83 and 1984-85, in 3 vols / Administrative Conference of the US (ACUS) – Acus: np. nd. (all publ) – 3mf – 9 – $4.50 – mf#LLMC 94-342 – us LLMC [340]
Colloqvivm so den 9 vnd septembris des 1577 jars zu sangerhausen / Spangenberg, C – np, 1578 – 1mf – 9 – mf#TH-1 mf 1400 – ne IDC [242]
Collot, Victor *see* Voyage dans l'amerique septentrionale
Collver, John *see* The upper canada hymn book, for all christian denominations
Collyer, James N *see* An historical record of the light horse volunteers of london and westminster
Collyer, John *see* A practical treatise on the law of partnership; with an appendix of forms
Collyer, Robert *see*
– Father taylor
– Some memories
Collyer, William Bengo *see*
– Aspect of prophecy respecting the present and future state of the j...
– Joy turned into mourning
Collyer, William Bengo' *see* Invisible church
Collymore, Frank A *see* Notes for a glossary of words and phrases of barba...
Colm, Gerhard *see* Beitrag zur geschichte und soziologie des ruhraufstandes vom marz-april 1920
[Colma-] record – CA. 1910-11 [wkly] – 1r – 1 – $60.00 – mf#B02126 – us Library Micro [071]
Colman, Benjamin *see* The papers of benjamin colman, 1641-1763
Colman, George *see* Inkle and yarico
Colman's rural world – St. Louis. 1849-1916 (1) – mf#4441 – us UMI ProQuest [720]
Colman-schwarzenberg family scrapbook – 1840-1972 – 1r – 1 – (manuscripts, documents, photographs, artifacts, and memorabilia chronicling the colman and schwarzenberg families, early jewish settlers in cleveland ca. 1840, and their careers) – us Western Res [920]
Colmarer liederhandschrift *see* Meisterlieder der kolmarer handschrift
Colmarer zeitung – Colmar / Elsass (F), 1869-1870 n11 – 1r – 1 – gw Misc Inst [074]
Colmarer zeitung – Colmar / Elsass (F), 1888-1893 6 aug, 1895 30 oct-1910, 1914 jan-jul – 1 – fr ACRPP [074]
Colmenares Fernandez de Cordoba, Felipe Urbano, marques de Zelada de la Fuente *see* El dia deseado, relacion de la solemnidad
Colmenero, J *see* Reprobacion del...abuso de los polvos del quarango...
Colmenero Ledesma, A *see* Curioso tratado de la naturaleza y calidad...del chocolate...
Colmer, Joseph Grose *see*
– The canadian census
– Some canadian railway and commercial statistics
Colmworth – (= ser Bedfordshire parish register series) – 2mf – 9 – £5.00 – uk BedsFHS [929]
Colnago, Bernard *see*
– Exercice tres devot envers s antoine de padoue le thaumaturge
Colne and nelson guardian, barrowfield, brierfield and general advertiser – [NW England] Colne, Nelson Lib 1863-sep 1864, aug 1867-jul 1869 [mf ed 2004] – 1 – uk Newsplan; uk MLA [072]
Colne and nelson times – [NW England] Colne Lib 1875-30 aug 1935 – 1 – (title change: colne times (a) [1974]) – uk MLA; uk Newsplan [072]
Colne miscellany – [NW England] Colne Lib 14 oct 1854-1857 – 1 – uk MLA; uk Newsplan [072]
Colne observer – [NW England] Colne Lib 1899-1903 – 1 – uk MLA; uk Newsplan [072]
Colne times (b) – [NW England] Colne 6 sep 1935, 1977 – 1 – uk MLA; uk Newsplan [072]

COLNE

Colne valley labour party records, 1891-1951 – (= ser Labour party in britain, origins and development at local level. series 1) – 9r – 1 – (int by david clark) – mf#97065 – uk Microform Academic [325]
Colnet DuRaval, Charles *see* L'hermite du faubourg saint-germain
Coloana printre ruini / Relgis, Eugen – Bucuresti, Romania. 1921 – 1r – us UF Libraries [025]
Cologne. Museum fuer Ostasiatische Kunst *see* Veroeffentlichungen
Cologne post : upper silesian edition – Oppeln, Germany. 17 June-6 Aug 1921 – 15ft – 1 – uk British Libr Newspaper [072]
The cologne post : eine tageszeitung veroeffentlicht von der armee am rhein – Koeln DE, 1919 31 mar-1920 13 feb – 1r – 1 – mf#3971 – gw Mikropress [355]
The cologne post / oberschlesien – Koeln DE, 1921 17 jun-6 aug – 1 – (wkly as: cologne post / oppeln) – uk British Libr Newspaper [355]
Cologne. Regierungsbezirk *see* Amtsblatt der regierung zu coeln
Cologny, L *see* L'antitrinitarisme a geneve au temps de calvin
Colomb dans les fers, a ferdinand et isabelle, apres la decouverte de l'amerique : epitre qui a remporte le prix de l'academie de marseille... / Langeac, Egide Louis Edme Joseph de Lespinesse, chevalier de – Londres, Paris: Chez Alexandre Jombert...et Jacques Esprit...1782 [mf ed 1984] – 2mf – 9 – 0-665-45225-X – mf#45225 – cn Canadiana [880]
Colomb, John Charles Ready *see* The protection of our commerce and distribution of our naval forces considered
Colomb, Romain *see* Journal d'un voyage en italie et en suisse
Colombani Bey, E *see* Les questions de nationalite en egypte
Colombia : being a geographical, statistical, agricultural, commercial, and political account of that country; adapted for the general reader, the merchant, and the colonist / Walker, Alexander – London 1822 [mf ed Hildesheim 1995-98] – 2v on 10mf – 9 – €100.00 – 3-487-26905-8 – gw Olms [972]
Colombia / Conder, Josiah – London 1825 [mf ed Hildesheim 1995-98] – 1v on 3mf [ill] – 9 – €90.00 – 3-487-26899-X – gw Olms [918]
Colombia / Duque Gomez, Luis – Mexico City? Mexico. v1-2. 1955 – 1r – us UF Libraries [972]
Colombia : estados unidos y el canal interoceanico / Leduc, Alberto – Mexico City? Mexico. 1904 – 1r – us UF Libraries [972]
Colombia / Franco R, Ramon – Bogota, Colombia. 1952 – 1r – us UF Libraries [972]
Colombia / Galbraith, W O – London, England. 1953 – 1r – us UF Libraries [972]
Colombia : gateway to south america / Henion, Doris Volz – New York, NY. 1963 – 1r – us UF Libraries [972]
Colombia : gateway to south america / Romoli, Kathleen – Garden City, NY. 1941 – 1r – us UF Libraries [972]
Colombia : its present state, in respect of climate, soil, productions, population, government, commerce... / Hall, Francis – 2nd ed. London: Baldwin, Cradock & Joy, 1827 [mf ed 1984] – 3mf – 9 – 0-665-45061-3 – (incl bibl ref) – mf#45061 – cn Canadiana [918]
Colombia : its present state, in respect of climate, soil, productions, population, government, commerce...and itineraries, partly from spanish surveys... / Hall, Francis – London 1824 [mf ed Hildesheim 1995-98] – 1v on 1mf [ill] – 9 – €40.00 – 3-487-26903-1 – gw Olms [972]
Colombia : land of miracles / Niles, Blair – New York, NY. 1924 – 1r – us UF Libraries [972]
Colombia / Levine, V – New York, NY. 1914 – 1r – us UF Libraries [972]
Colombia : pais formal y pais real / Montana Cuellar, Diego – Buenos Aires, Argentina. 1963 – 1r – us UF Libraries [972]
Colombia : posesiones presidenciales / Monsalve Martinez, Manuel – Bogota, Colombia. 1954 – 1r – us UF Libraries [972]
Colombia / Ramirez, Plutarco Elias – Habana, Cuba. 1964 – 1r – us UF Libraries [972]
Colombia / Reclus, Elisee – Bogota, Colombia. 1958 – 1r – us UF Libraries [972]
Colombia / Reichel-Dolmatoff, Gerardo – New York, NY. 1965 – 1r – us UF Libraries [972]
Colombia / Unesco Siscence Cooperation Office For Latin America – Montevideo, Uruguay. 1965 – 1r – us UF Libraries [972]
Colombia *see*
– Codigo de minas y petroleos
– Codigo militar expedido por el congreso de los est...
– Compilacion cafetera, 1939-1951
– Compilacion electoral
– Constitucion nacional
– Contrato chaux-folsom y documentos relacionados co...
– Decretos del libertador
– Diario oficial
– Jurisprudencia de los tribunales of colombia
– Legislacion de aguas de uso publico
– Proyectos de ley presentados al congreso de 1946
– Tratado sobre limites y libre navegacion y conveni...
– Tratados y convenios de colombia
Colombia a la mano / Echeverri, Elio Fabio – Bogota, Colombia. 1955 – 1r – us UF Libraries [972]
Colombia al borde de la guerra / Puentes, Milton – Bogota, Colombia. 1938 – 1r – us UF Libraries [972]
Colombia and venezuela and the guianas / Maceoin, Gary – New York, NY. 1965 – 1r – us UF Libraries [972]
Colombia Comision Corografica *see*
– Jeografia fisica i politica de las provincias de l...
Colombia Comision De Estduios Constitucionales *see* Estudios constitucionales
Colombia Comision De Estudios Economicos Y Social *see* Conclusiones, 1965
Colombia Congreso Camara De Prepresentantes *see* Comisiones especiales para estudiar con caracte in...
Colombia Congreso Camara De Representantes Comi... *see* Supia y marmato ante la camara
Colombia Congreso Senado *see* Congreso de 1825
Colombia. Congreso Senado *see* Derecho internacional privado
Colombia Congreso Senado Comision Quinta Consti... *see* Informe de la comision quinta constitucional perma
Colombia constitucional / Moreno Jaramillo, Miguel – Medellin, Colombia. 1915 – 1r – us UF Libraries [323]
Colombia Contraloria General De La Republica *see* Indice universal de inventarios
Colombia coruscante que yo conoci / Ydigoras Fuentes, Miguel – Guatemala, 1963 – 1r – us UF Libraries [972]
Colombia de norte a sur. 2 vol. madrid, 1943 / Perez de Barradas, Jose – Madrid: Razon y Fe, 1946 – 1 – sp Bibl Santa Ana [946]
Colombia Departamento Administrativo Nacional De... *see*
– Division politico-administrativa de colombia
– Reglamentacion de las estadisticas continuas
Colombia. Departamento Administrativo Nacional de Estadistica *see* Anuario general de estadistica 1905-1969/1970
Colombia Direccion De Informacion Y Propaganda *see*
– Colombia trabaja
– Seis meses de gobierno
Colombia Direccion Nacional De Estadistica *see* Sintesis estadistica de colombia, 1939-1943
Colombia donde los andes se disuelven / Osorio Lizarazo, Jose Antonio – Santiago, Chile. 1955 – 1r – us UF Libraries [972]
Colombia Ejercito 8 Brigada *see* De laviolencia a la paz
Colombia Ejercito Estado Mayor General *see* Participacion de colombia en la libertad del peru
Colombia en cifras – Bogota, Colombia. 1963 – 1r – us UF Libraries [972]
Colombia en el sur / Davalos, Pedro Maria – Pasto, Colombia. 1941 – 1r – us UF Libraries [972]
Colombia en korea / Hernandez B, Ernesto – Bogota, Colombia. 1953 – 1r – us UF Libraries [327]
Colombia en la guerra de corea / Torres Almeyda, Pablo E – Bogota, Colombia. 1960? – 1r – us UF Libraries [972]
Colombia en la hora cero / Lopez Michelsen, Alfonso – Bogota, Colombia. v1-2. 1963 – 1r – us UF Libraries [972]
Colombia Fuerzas De Policia *see* Reglamento de uniformes
Colombia Junta Militar De Gobierno *see* Itinerario historico
Colombia Laws, Etc *see*
– Codigo de aduanas 1938
– Decretos de caracter extradorinario
Colombia Laws, Statutes, Etc *see*
– Codigo civil colombiano
– Codigo contencioso administrativo, ley 167 de 1941
– Codigo de comercio terrestre
– Codigo de minas y codigo de petroleos
– Codigo de minas y leyes del petroleo
– Codigo de procedimiento civil
– Compilacion electoral
– Compilacion legal
– Conductas antisociales
– Ley de reforma social agraria
– Leyes de 1948 y ley 91 de 1947
– Proyecto de codigo administrativo
– Proyectos de ley presentados por el gobierno nacio...
– Reorganica de la carrera de oficiales de las fuerz
Colombia. Laws, Statutes, etc *see*
– Historia de las leyes
– Leyes y decretos
Colombia Ministerio De Gobierno *see*
– Ano de gobierno, 1950-1951
– Crisis politica
– Limites entre santander y boyaca
– Teoria y practica de una politica colombianista
Colombia Ministerio De Guerra *see*
– En el darien
– Operaciones contra las fuerzas irregulares
Colombia. Ministerio de Guerra *see* Informe del secretario de guerra de la nueva granada al congreso constitucional de...
Colombia Ministerio de Relaciones Exteriores *see* Anales diplomaticos y consulares de colombia
Colombia. Ministerio de Relaciones Exteriores *see*
– Esposicion...
– Esposicion que el secretario de estado en el despacho de relaciones esteriores de la republica de Colombia hace al congreso de...sobre los negocios de su departamento
– Informe...
– Informe del secretario de relaciones esteriores de la confederacion granadina al congreso nacional de...
– Memoria...
– Memoria de la secretaria de estado y relaciones esteriores de la republica de colombia leida al primer congreso constitucional...
– Memoria del secretario de relaciones esteriores de la confederacion granadina al congreso nacional de...
– Memoria...al congreso de...
Colombia Oficina De Comercio Exterior *see* Concesiones arancelarias y cambiarias otorgadas
Colombia President Lopez *see* Documentos relacionados con la recuncia del presid...
Colombia (Republic Of Colombia, 1886-) *see* Compilacion parlamentaria y administrativa
Colombia (Republic Of Colombia, 1886-) Laws... *see* Reglamento general de sanidad del ministerio de ob...
Colombia (Republic Of Colombia 1886-) Ministe... *see* Sentido y realizacion de una politica social
Colombia (Republic Of Colombia, 1819-1831) *see* Congreso de 1823
Colombia. Secretaria de Hacienda *see* Exposicion...
Colombia. Secretaria de lo Interior i Relaciones Exteriores *see*
– Esposicion...
– Exposicion...
– Memoria...
– Memoria...del gobierno de la nueva granada, dirije al congreso constitucional de...
Colombia Secretaria De Organizacion E Inspeccion *see* Manual de organizacion de la rama ejecutiva del po...
Colombia. Secretaria del Interior *see*
– Esposicion que el secretario de estado en el despacho de lo interior de la nueva granada presenta al congreso constitucional de...
– Esposicion que el secretario de estado en el despacho del interior de la republica de Colombia hizo al congreso de...sobre los negocios de su departamento
– Memoria...presento al congreso de colombia
Colombia. Senado *see* Anales
Colombia Superintendencia De Sociedades Anonimas *see* Doctrinas
Colombia today : and tomorrow / Holt, Pat M – New York, NY. 1964 – 1r – us UF Libraries [972]
Colombia trabaja / Colombia Direccion De Informacion Y Propaganda – Bogota, Colombia. 1954 – 1r – us UF Libraries [972]
Colombia Treaties, Etc *see* Tratados y convenios de colombia, 1938-1948, compi...
Colombia (United States Of Colombia, 1863-1885) *see* Codigo fiscal de los estado unidos de colombia
Colombia y cuba / Merchan, Rafael Maria – Bogota, Colombia. 1897 – 1r – us UF Libraries [327]
Colombia y los estados unidos de america / Uribe, Antonio Jose – Bogota, Colombia. 1931 – 1r – us UF Libraries [972]
Colombia y su pueblo / Arias Ramirez, Fernando – Manizales, Colombia. 1948 – 1r – us UF Libraries [972]
Colombian and venezuelan republics / Scruggs, William Lindsay – Boston, MA. 1900 – 1r – us UF Libraries [972]
Colombo – Belem, PA: Typ do Jornal do Amazonas, 25 abr 1869 – (= ser Ps 19) – bl Biblioteca [079]
Colombo : periodico critico e litterario – Desterro, SC: Typ Commercial, 14-28 maio; 07 jul 1881 – (= ser Ps 19) – mf#P17,03,90 – bl Biblioteca [440]
Colomen columbia newyddiadur wythnosol at wasanaeth y cymry yn america – [Wales] LLGC gorff 1888-1 chw 1894 (incomplete) [mf ed 2002] – 3r – 1 – uk Newsplan [072]
Colomer, B M *see* Idylls et caprices
Un colomniateur demasque par lui-meme / Frechette, Louis – [S.l: s.n, 1862?] – 1mf – 9 – 0-665-51324-0 – mf#51324 – cn Canadiana [346]
Colon : precursor literario / Balaquer, Joaquin – Buenos Aires, Argentina. 1958 – 1r – us UF Libraries [972]
Colon, Edmundo Dimas *see* Gestion agricola despues de 1898
Colon, Eduardo *see* Ley organica del poder ejecutivo y reglamento para...
Colon en barcelona : sevilla, 1944 / Rumeu de Armas, Antonio – Madrid: Razon y Fe, 1947 – 1 – sp Bibl Santa Ana [946]
Colon estremena / Fita, Fidel & Fernandez D, Cesareo – Madrid: Tip. de Fortanet, 1903 – 1 – sp Bibl Santa Ana [946]
Colon extremeno? de vicente paredes / Fita, Fidel & Fernandez D, Cesareo – Madrid: Fortanet, 1903. B.R.A.H. 42, 1903, pp. 237-238 – sp Bibl Santa Ana [946]
Colon, Hernando *see* Historia del almirante don cristobal colon. tomo 1
Colon italiano? colon espanol? / Bayle, Constantino – Madrid: Razon y Fe, 1923 – 1 – sp Bibl Santa Ana [440]
Colon starlet – Colon, Panama. 1904; 1905-08; misc dates – 3r – 1 – us L of C Photodup [079]
Colon sur sa plantation / La Barre, Gaspard Alexis – Dakar, Senegal. 1959 – 1r – us UF Libraries [972]
Colon telegram – Colon, Panama. 1902-11 (incomplete) – 4r – 1 – us L of C Photodup [079]
Colonel charles l decker's collection of records relating to military justice and the revision of military law, 1948-1956 / Decker, Charles L – (= ser Records of the office of the judge advocate general (army)) – 31r – 1 – (with printed guide) – mf#M1739 – us Nat Archives [355]
Le colonel dambourges / Bois, Louis-Edouard – Quebec?: A Cote, 1877 – 2mf – 9 – mf#26522 – cn Canadiana [971]
Colonel gardner – Edinburgh, Scotland. 18-- – 1r – us UF Libraries [240]
Colonel Kalpak's Latvian School (Wauwatosa, WI) *see* Gudra puce
Colonel mahlon burwell : land surveyor / Blue, Archibald – Toronto?: s.n, 18-- – 1mf – 9 – mf#03706 – cn Canadiana [920]
Colonel russell's baby / Adams, Ellinor Davenport – London: Walter Smith & Innes, (late Mozley), 1889 – (= ser 19th c women writers) – 5mf – 9 – mf#5.1.40 – uk Chadwyck [830]
Colonel william wood, soldier, historian, archivist : an analytical bibliography of the writings of colonel william wood / Clint, Harold Cuthbert – 1951 [mf ed 1979] – (= ser Bibliographies du cours...1947-66) – 1mf – 9 – (with ind; pref by george cartwright; int by aut) – mf#SEM105P4 – cn Bibl Nat [355]
A colonia – S Tome: C Lopes Alpoim, sep 29 1923-apr 10 1924; apr 24-oct 5 1924 – us CRL [079]
Colonia de mocambique : territoire de manica et sofala – Lisboa, Portugal. 1931 – 1r – us UF Libraries [960]
[Colonia del valle-] informacions sistematica – MX. 1976-82 – 8r – 1 – $400.00 – mf#R04213 – us Library Micro [079]
La colonia eritrea dalle sur origini fino al io. marzo 1899... – Parma: L Battei 1899 – us CRL [960]
Colonia hacia la nacion / Congreso Nacional De Historia, 3d – Habana, Cuba. 1946 – 1r – us UF Libraries [972]
A colonia portuguesa – Belem, PA: Typ d'A Colonia Portuguesa, 13 set 1885 – (= ser Ps 19) – bl Biblioteca [079]
Colonia portuguesa de mocambique – Lourenco Marques, Mozambique. 1929 – 1r – us UF Libraries [960]
La colonia svizzera – San Francisco, CA: Swiss Pub Co, [dec 7 1917-1933] – 32r – us CRL [071]
El coloniaje y sus detractores : el paso, 1927 / Planchet, Regis; ed by Bayle, Constantino – Madrid: Razon y Fe, 1928 – 9 – sp Bibl Santa Ana [972]
Colonial / American society of colonial families, Boston – 1913 mar-1917 mar – 1r – 1 – mf#1054620 – us WHS [929]
Colonial advocate – Hobart, 1828 – 1r – 1 – A$27.50 vesicular A$33.00 silver – at Pascoe [079]
The colonial advocate, no 6 : published sept 27th, 1824, containing an essay on canals and inland navigation... / Mackenzie, William Lyon – Queenston, Ont?: W L Mackenzie, 1824 – 1mf – 9 – mf#21168 – cn Canadiana [380]
The colonial and coloured peoples : a programme for their freedom and progress / Ranga, N G – Bombay: Hind Kitabs, 1946 – (= ser Samp: indian books) – us CRL [322]
Colonial application for extended land area *see* Northern territory land applications – various
Colonial application for land – 160 acres *see* Northern territory land applications – various

The colonial church in virginia / Goodwin, Edward Lewis – Milwaukee. 1927 – 1 – us CRL [975]
Colonial church legislation / Venn, Henry – London, England. 1850 – 1r – us UF Libraries [240]
The colonial churchman – Lunenburg, NS: E A Moody, [1835?-1841?] – 9 – (incl ind) – mf#P04181 – cn Canadiana [242]
Colonial constitutions : an outline of the existing forms of government in the british dependencies / Mills, Arthur – London 1891 – (= ser 19th c british colonization) – 1mf – 9 – mf#1.1.3807 – uk Chadwyck [323]
Colonial defence commission under lord carnarvon : 1881-82 – 2r – 1 – £110.00 – (unpubl records) – mf#CDC – uk World [324]
Colonial discourses, series 1 : women, travel and empire, 1660-1914 – 4pt – 1 – (pt1: early travel accounts by women, & women's experiences in india, africa, australasia & canada [25r] £2350; pt2,3: women & 'the orient' [25r £2350, 26r £2450]; pt4: women, the americas & world travel [25r] £2350; with d/g) – uk Matthew [305]
Colonial discourses, series 2 : imperial adventurers and explorers – 2pt – 1 – (pt1: papers of richard burton [1821-90] fr wiltshire & swindon record office [14r] £1300; pt2: papers of james augustus grant [1827-92] & john hanning speke [1827-64] fr national library of scotland [17r] £1600; with d/g) – uk Matthew [910]
Colonial discourses, series 3 : colonial fiction, 1650-1914 – 3pt – 1 – (pts1,2,3: general works & fiction fr india fr the british library, london [29r £2700, 25r £2350, 30r £2700 [mf ed summer 2004]; with d/g) – uk Matthew [800]
Colonial economy in the netherlands indies, pt 1 : the commission to investigate the sugar manufactories on java, 1854-1857 – [mf ed 2008] – ca 260mf – 9 – €1995.00 – (printed p/guide & concordance based on: inventaris van het archief van de commissie voor de opname van de verschillende suikerfabrieken op java (commissie umbgrove) [1853-58], 1854-57, by h b n b adam (nationaal archief, den haag, 1979)) – mf#mmp135 – ne Moran [330]
Colonial economy in the netherlands indies, pt 2 : the commission for industrial development in the netherlands indies, 1915-1926 – [mf ed 2008] – ca 250-270mf – 9 – €2295.00 – (printed p/guide & concordance based on: inventaris van het archief van de commissie voor de opname van de verschillende suikerfabrieken op java (commissie umbgrove) [1853-58], 1854-57, by h b n b adam (nationaal archief, den haag, 1979)) – mf#mmp136 – ne Moran [330]
The colonial empire of great britain : considered chiefly with reference to its physical geography and industrial productions / Rowe, George – London [1864,1865] – (= ser 19th c british colonization) – 9 – mf#1.1.4821 (11 mf) – uk Chadwyck [910]
The colonial empire of great britain, especially in its religious aspect : a lecture, addressed...on dec 3, 1849 / Lyttelton, George William, 4th Baron – London, [1850?] – (= ser 19th c british colonization) – 1mf – 9 – mf#1.1.532 – uk Chadwyck [330]
Colonial enterprise : review of the mines, manufacturers and industries of greater britain, 1894-1899 – [mf ed Marlborough 1996] – (= ser Business and financial papers series) – 3r – 1 – £285.00 – (r14: 14 jun 1894-17 jan 1895 [v1 n1-v2 n32], 1 jan 1895-24 dec 1895 [v2 n34-v3 n81], missing iss: 24 jan 1895 [v2 n33]; r15: 2 jan 1896-24 jun 1897 [v4 n82-v6 n159]; r16: 1 jul 1897-15 dec 1898 [v7 n160-v9 n236], 29 dec 1898-2 mar 1899 [v9 n238-v10 n247], 30 mar 1899 [v10 n251]; missing iss: 22 dec 1898 [v9 n237], 9 mar 1899 [v10 n248], 16 mar 1899 [v10 n249], 23 mar 1899 [v10 n250]) – uk Matthew [338]
Colonial era / Fisher, George Park – New York: Scribner, 1892 – 1mf – 9 – 0-7905-5391-0 – (incl bibl ref) – mf#1988-1391 – us ATLA [975]
Colonial experiences : or, incidents and reminiscences of thirty four years in new zealand / Pratt, William Tidd – London 1877 – (= ser 19th c british colonization) – 4mf – 9 – mf#1.1.9599 – uk Chadwyck [880]
Colonial families of the united states / MacKenzie, George N – New York. v1-7. 1907-20 – 1 – $120.00 – mf#0341 – us Brook [929]
Colonial farmer – Fredericton, NB. 1863-73 – 3r – 1 – ISSN: 1483-0175 – cn Library Assoc [630]
The colonial farmer : devoted to the agricultural interests of nova-scotia, new brunswick, and prince edward island – [Halifax, NS?]: R Nugent, [1841-1843?] – 9 – mf#P04769 – cn Canadiana [630]
Colonial florida / Mendelis, Louis J – s.l, s.l? 193-? – 1r – us UF Libraries [978]
Colonial gazette – London. -w. Dec 1838-Jan 1847. (8 reels) – 1 – uk British Libr Newspaper [072]
Colonial gazette, 1838-1847 – [mf ed Marlborough 1996] – (= ser Business and financial papers series) – 8r – 1 – £750.00 – uk Matthew [073]
Colonial guardian – Belize, British Honduras, 1882-19 dec 1885; 1886-16 dec 1905; 1906-17 may 1913 – 16r – 1 – (imperfect) – uk British Libr Newspaper [079]
Colonial heights baptist church. fairfield county. columbia, south carolina : church records – 1971-76 – 1 – us Southern Baptist [242]
Colonial herald – Charlottetown, PEI. 1840-44 – 2r – 1 – cn Library Assoc [071]
Colonial heritage – 1970 sum-1972 nov, 1973 oct-1976 nov – 2r – 1 – mf#1054623 – us WHS [975]
Colonial heritage – York. 1970-1976 (1) – mf#9508 – us UMI ProQuest [978]
Colonial hispanic america / Chapman, Charles Edward – New York, NY. 1933 – 1r – us UF Libraries [972]
Colonial hispanic america / George Washington University Seminar Conference... – Washington, DC. 1936 – 1r – us UF Libraries [972]
Colonial homes – New York. 1979-1999 (1) 1979-1999 (5) 1979-1999 (9) – (cont: house beautiful's colonial homes. cont by: classic american homes) – ISSN: 0195-1416 – mf#12241,01 – us UMI ProQuest [640]
Colonial inquiry : speech of the honorable francis scott, mp, on moving the appointment of a select committee, on the 16th apr 1849 / Scott, Francis – London 1849 – (= ser 19th c british colonization) – 1mf – 9 – mf#1.1.530 – uk Chadwyck [330]
Colonial latin american manuscripts and transcripts in the obadiah rich collection : from the holdings of the new york public library, astor, lennox and tilden foundations – 33r – 1 – (coll of documents on spanish discovery, conquest and administration of the americas from 1492 to the early 19th century. based on papers from juan bautista muñoz and obadiah rich. includes printed guide) – mf#C39-27950 – us Primary [972]
Colonial law journal : reports of cases argued and determined in the supreme court of new zealand, and on appeal to the court of appeal of new zealand – n.p. 1v. n.d. – 2mf – 9 – $3.00 – (covers selected cases from 1865-75 and includes some articles, a 14-pg biography of wilson gray, and other editorial matter. missing title-pg) – mf#LLMC 96-016 – us LLMC [347]
Colonial lawyer – College of William and Mary. v1-20. 1967-91 – 9 – $113.00 set – (cont by: william and mary bill of rights journal; v7 never publ) – mf#111571 – us Hein [340]
The colonial life assurance company : capital, £1,000,000 sterling...canada, head office, montreal...manager, a davidson parker – [S.l: s.n, 1854?] [mf ed 1986] – 1mf – 9 – 0-665-50577-9 – mf#50577 – cn Canadiana [368]
Colonial office 824 / 1-2. brunei. sessional papers / Great Britain. Public Record Office – 1r – 1 – (on reel with co648) – mf_3r66h – us CRL [324]
Colonial office records : class 5 files (c o 5) – 5pt – 1 – (pt1: westward expansion 1700-83 12r isbn 0-89093-410-X $1865. pt2: the board of trade 1689-1775 6r isbn 0-89093-411-8 $935. pt3: the french & indian war 1754-63 8r isbn 0-89093-412-6 $1260. pt4: royal instructions & commissions to colonial officials 1702-71 12r isbn 0-89093-415-0 $1865. pt5: the american revolution 1772-84 15r isbn 0-89093-416-9 $2345. with p/g) – us UPA [975]
Colonial orange court hotel – s.l, s.l? 193-? – 1r – us UF Libraries [978]
Colonial parliamentary bulletin – London, African Press Agency. v1 n3-v3 n6 mar/apr 1946-sep 1948 – us CRL [320]
Colonial patriot – Pictou, NS: William Milne, 1827-34 – 2r – 1 – cn Library Assoc [071]
Colonial phrenological journal and repository of science, literature, and general intelligence – Pictou, NS: A B Parker, [1860] (mf ed v1 n1 may 1860-v1 n2 jun 1860) – 9 – ISSN: 1190-6480 – mf#P04612 – cn Canadiana [130]
Colonial poems / Anderson, Frances – London: E Marlborough & Co, 1869 – (= ser 19th c women writers) – 2mf – 9 – mf#5.1.145 – uk Chadwyck [810]
Colonial policies in africa / Wieschhoff, H A – Philadelphia, PA. 1944 – 1r – us UF Libraries [960]
The colonial policy of england examined by ebenezer telltruth : owing to those philanthropic labors, many facts, not generally known... – St Catharines Ont: J M'Mullen, 1849 – 1mf – 9 – (incl bibl ref) – mf#48189 – cn Canadiana [941]
The colonial policy of lord john russell's administration / Grey, Henry George Grey, 3rd earl – London, 1853 – (= ser 19th c books on british colonization) – 10mf – 9 – mf#1.1.9758 – uk Chadwyck [941]
The colonial protestant and journal of literature and science / ed by Taylor, W & Cramp, J M – Montreal: Publ by R Campbell, 1848-[1849] – 9 – mf#P04200 – cn Canadiana [242]
The colonial question : being essays on imperial federalism / Jenkins, Edward – Montreal: Dawson Bros, 1871 – 2mf – 9 – mf#57062 – cn Canadiana [320]
The colonial question : a brief consideration of colonial emancipation, imperial federalism and colonial conservation / Fuller, William Henry – Kingston, Ont?: s.n, 1875 – 1mf – 9 – mf#07133 – cn Canadiana [971]
Colonial questions pressing for immediate solution : in the interest of the nation and the empire. papers and letters... / MacFie, Robert Andrew – London 1871 – (= ser 19th c british colonization) – 2mf – 9 – mf#1.1.3787 – uk Chadwyck [330]
Colonial reckoning / Perham, Margery Freda – London, England. 1961 – 1r – us UF Libraries [960]
Colonial records of spanish florida / Connor, Jeannette M Thurber – Deland, FL. v1-2. 1925 and 1930 – 1r – us UF Libraries [978]
The colonial review : a weekly journal of politics, literature and society – St John, NB: J & A McMillan, [1862-186-?] – 9 – mf#P04288 – cn Canadiana [071]
Colonial secretary's papers, 1788-1825 *see* Australia: colonial life and settlement
Colonial session laws – 9 – $5450.00 set – mf#400981 – us Hein [348]
Colonial Society [London, England] *see* Resolutions and rules...
Colonial society of massachusetts. publications – Boston. v1-25+ind. 1895-1924 – 1 – $270.00 – (v26-41 1927-61 $144 [0161]) – mf#0161 – us Brook [978]
Colonial standard – Pictou, NS. 1862-73 – 4r – 1 – ISSN: 0844-4366 – cn Library Assoc [079]
Colonial times *see* Cape frontier times
Colonial times etc – Hobart, Australia. 6 jan 1826-28 dec 1827; 10 jan, 7 mar, 1 aug 1837; 5 nov (supp) 1939 – 1/2r – 1 – uk British Libr Newspaper [072]
Colonial Williamsburg Foundation *see* Early american history research reports from the colonial williamsburg foundation library
Colonial williamsburg research collections in microform : a guide – 122mf – 9 – $1135.00 – 1-55655-284-X – (with p/g) – us UPA [975]
Colonial-hispanic legal documents (18th-19th centuries) / U.S. Library of Congress. Law Library – 820 items on 13 reels. Jurisdictions: Colombia, Mexico, Peru, Portugal, Puerto Rico – 1 – us L of C Photodup [340]
Une colonie de commerce francaise : etude sur le protectorat de la cote somali / Bacquart, A – Paris, 1907 – 3mf – 9 – mf#ILM-331 – ne IDC [960]
La colonie du sacre-coeur dans les cevennes de la chine au dix-huitieme siecle / Chaney, Theodore – Paris: Librairie Retaux-Bray, 1889 [mf ed 1996] – (= ser Yale coll) – 96p – 1 – 0-524-10265-1 – (in french) – mf#1996-1265 – us ATLA [241]
Une colonie feodale en amerique : l'acadie (1604-1881) / Rameau, Edme – nouv ed. Paris: E Plon, Nourrit & Cie, 1889 [mf ed 1981] – 2v on 1mf – 9 – 0-665-12373-6 – (incl bibl ref) – mf#12373 – cn Canadiana [971]
Colonies – (= ser The Goldsmiths'-Kress Library of Economic Literature) – 121r – 1 – us Primary [330]
Les colonies – Saint Pierre, Martinique. 1881-1902 (1) – mf#67950 – us UMI ProQuest [079]
The colonies – London. The Colonies and India. 17 Jan 1872-20 May 1898.-f. 40mqn reels – 1 – uk British Libr Newspaper [072]
The colonies : treating of their value generally of the ionian islands in particular; the importance of the latter in war and commerce ..., strictures on the administration of sir frederick adam / Napier, Charles – London 1833 [mf ed Hildesheim 1995-98] – 4mf – 9 – €120.00 – 3-487-29058-8 – gw Olms [949]
The colonies, 1492-1950 / Thwaites, Reuben Gold – 4th ed. New York: Longmans, Green, 1893 c1890. xviii,301p. maps – 1 – us Wisconsin U Libr [975]
Les colonies agricoles – Port-au-Prince: [s.n. v1 n1-4 june-dec 1938; v2-v3 n5/1 aug 1939-apr/jun 1940 – 2 sheets – us CRL [630]
The colonies and imperial unity or the "barrel without the hoops" : inaugural address...at westminster palace hotel, in london, july 19, 20, and 21, 1871 / Jenkins, Edward – [London], 1871 – (= ser 19th c british colonization) – 1mf – 9 – mf#1.1.4939 – uk Chadwyck [320]
The colonies and the century / Robinson, John – London 1899 – (= ser 19th c british colonization) – 2mf – 9 – mf#1.1.7286 – uk Chadwyck [337]
Colonies des vallees de la rouge, de la kiamika et de la lievre – [Quebec (Province)]: [s.n.], [1898?] (mf ed 1988) – 1mf – 9 – mf#SEM105P959 – cn Bibl Nat [971]
Les colonies francaises / Bert, Paul & Clayton, Anna – Paris: C Bayle, 1889 – 3mf – 9 – mf#08935 – cn Canadiana [944]
Colonisacao de angola / Pereira Do Nascimento, Jose – Lisboa, Portugal. 1912 – 1r – us UF Libraries [960]
La colonisation dans les canton de l'est / Chartier, Jean Baptiste – St Hyacinthe, Quebec: Courrier de St Hyacinthe, 1871 – 2mf – 9 – mf#02980 – cn Canadiana [333]
La colonisation du canada envisagee au point de vue national / Drapeau, Stanislas – [Quebec?: s.n.], 1858 [mf ed 1985] – 1mf – 9 – 0-665-44172-X – mf#44172 – cn Canadiana [971]
La colonisation et le gouvernement mercier – Montreal: s.n, 1892? – 1mf – 9 – mf#11944 – cn Canadiana [320]
The colonisation of indo-china / Chailley-Bert, Joseph – London, 1894 – (= ser 19th c books on british colonization) – 5mf – 9 – (trans fr french of j chailly-bert by arthur baring brabant) – mf#1.1.7940 – uk Chadwyck [951]
Colonisation Society *see* Competence in a colony contrasted with poverty at home
Le colonise – Paris: Isaac Becon, nov 15-dec 20 1936 – (filmed with les continents and 11 other titles) – us CRL [944]
Colonist – Georgetown British Guiana, 11 dec 1863-9 feb 1884 – 38r – 1 – uk British Libr Newspaper [079]
Colonist – Georgetown, Guyana. 11 Dec 1863-9 Feb 1884.-d. 38 reels – 1 – uk British Libr Newspaper [079]
Colonist – Maryborough, Australia. 3 Jan 1885-15 Jul 1916; 7 Apr 1917-30 Mar 1940.-w. 106mqn reels – 1 – uk British Libr Newspaper [072]
Colonist – Maryborough, Australia. 3 jan-26 dec 1885; 1886-15 jul 1916; 7 apr 1917-1930; 31 jan-28 mar 1931; 22 aug 1931-30 mar 1940; 1 feb 1941-11 apr 1942 – 107 1/2r – 1 – uk British Libr Newspaper [072]
Colonist : a monthly magazine devoted to the interests of manitoba and the territories – Winnipeg: The Colonist, [1890-1898] – 9 – (cont: manitoba colonist) – mf#P05023 – cn Canadiana [971]
Colonist – Sydney, 1835-40 – 3r – 1 – A$115.50 vesicular A$132.00 silver – at Pascoe [079]
Colonist – Sydney, Australia. 1835-28 dec 1837 – 1 1/2r – 1 – uk British Libr Newspaper [072]
Colonist – Sydney, Australia. -w. 1836-37. 1 reel – 1 – uk British Libr Newspaper [072]
Colonist – Winnipeg, Canada. jun 1892-nov 1897 – 2r – 1 – uk British Libr Newspaper [072]
Colonist – Winnipeg, Canada. -m. June 1892-Nov 1897. 2 reels – 1 – uk British Libr Newspaper [071]
A colonist on the colonial question / Mathews, John – London 1872 – (= ser 19th c british colonization) – 3mf – 9 – mf#1.1.3716 – uk Chadwyck [327]
The colonist's and emigrant's handbook of the mechanical arts / Burn, Robert Scott – [Edinburgh], 1854 – (= ser 19th c books on british colonization) – 2mf – 9 – mf#1.1.7855 – uk Chadwyck [600]
Colonizacao dos planaltos de angola / Dias, Manuel Da Costa – Lisboa, Portugal. 1913 – 1r – us UF Libraries [960]
Colonizacion agricola de costa rica / Sandner, Gerhard – San Jose, Costa Rica. v1-2. 1962 – 1r – us UF Libraries [972]
Colonizacion antioquena el el occidente de colombia / Parsons, James Jerome – Bogota, Colombia. 1961 – 1r – us UF Libraries [972]
Colonizacion y parcelaciones / Merchan Merchan, Felipe & Lopez Santamaria, Francisco – Badajoz: Graficas Iberia, 1950 – 1 – sp Bibl Santa Ana [946]
Colonization : or a project for rendering our colonial territories accessible to the population of the united kingdom / Brown, David Stevens – London, 1852 – (= ser 19th c books on british colonization) – 1mf – 9 – mf#1.1.533 – uk Chadwyck [330]
Colonization and abolition : an address delivered...at the anniversary meeting of the new york state colonization society, held in metropolitan hall, may 13th, 1852 / Latrobe, John Hazlehurst Boneval – Baltimore: J D Toy, 1852 (mf ed 1976) – 1r – 1 – mf#ZZ-14667 – us NY Public [320]
Colonization and church work in victoria / Ross, Charles Stuart – Melbourne: Melville, Mullen & Slade; Dunedin: Wise, Caffin 1891 [mf ed 1991] – 1mf – 9 – 0-524-00596-6 – mf#1990-0096 – us ATLA [980]

COLONIZATION

Colonization and missions : a historical examination of the state of society in western africa, as formed by paganism and muhammedanism, slavery, the slave trade and piracy / Tracy, Joseph – Boston: T R Marvin, 1844 – 1mf – 9 – 0-7905-6895-0 – (incl bibl ref) – mf#1988-2895 – us ATLA [960]
Colonization, defence, and railways in our indian empire / Clarke, Hyde – London, 1857 – (= ser 19th c books on british colonization) – 3mf – 9 – mf#1.1.8797 – uk Chadwyck [330]
Colonization herald – Philadelphia: Pennsylvania Colonization Soc [mf ed 1968-] – 1r [incomplete] – 1 – (sometimes publ as: colonization herald and general register; with: liberia christian advocate, and new york colonization journal) – mf#21 – us Wisconsin U Libr [073]
Colonization of neoaplectana dutkyi jackson / Chow, Franklin Hon-Ching – s.l, s.l? 1972 – 1r – us UF Libraries [500]
Colonizationist and journal of freedom – Boston. 1833-1834 (1) – mf#4366 – us UMI ProQuest [320]
The colonizer : an imperial medium of colonization, emigration, exploration, and travel – v. 1-37, no. 1-439. Jan 1896-Jul 1932 – 1 – us L of C Photodup [910]
The colonizer – Toronto: Temperance Colonization Society, [1881?-18– or 19–] – 9 – mf#P04223 – cn Canadiana [971]
La colonizzazione della cirenaica nell'antichita e nel presente – Bengasi: Stabilimento tipografico Fratelli Pavone, 1934 – 1 – us CRL [960]
La colonizzazione europea nell'est africa : italia, inghilterra, germania / Chiesi, Gustavo – Torino: Unione tipografico-editrice torinese 1909 – us CRL [960]
[Colonna, F] *see*
– Discours du songe de poliphile...
– La hypnerotomachia di poliphilo
– Le tableau des riches inventions couvertes du voile des feintes amoureuses
La colonna italiana in spagna – [Paris: s.n. 1936] [mf ed 1977] – (= ser Blodgett coll) – 1mf – 9 – mf#w806 – us Harvard [934]
Colonna, M A *see* I trionfi feste, et livree fatte dalli signori conservatori, and popolo romano, and da tutte le arti di roma...
Colonna, Maria Elisabetta *see* Gli storici bizantini dal 4 al 4 secolo
Colons de saint-domingue et la revolution / Debien, Gabriel – Paris, France. 1953 – 1r – us UF Libraries [972]
The colony and provincial reporter – (Colonial and Provincial Reporter West Africa Mail and Trade Gazette). Freetown. Sierra Leone. -w. 25 May 1911-Dec 1932. (Imperfect). (19 reels) – 1 – uk British Libr Newspaper [072]
Colony newsletter / Socialist Party [US] – v4 [1977 mar-dec] – 1r – 1 – mf#361201 – us WHS [325]
Colony of british guyana and its labouring population / Bronkhurst, H V P – London, England. 1883 – 1r – us UF Libraries [972]
The colony of british honduras, its resources and prospects / Morris, Daniel – London. 1883 – 1 – us CRL [972]
The colony of british honduras, its resources and prospects : with particular reference to its indigenous plants and economic productions / Morris, Daniel – London 1883 – (= ser 19th c british colonization) – 2mf – 9 – mf#1.1.5496 – uk Chadwyck [972]
A colony of mercy : or, social christianity at work / Sutter, Julie – 3rd ed. London: R Brimley Johnson, 1904 [mf ed 1986] – 1mf – 9 – 0-8370-6419-8 – mf#1986-0419 – us ATLA [230]
The colony of natal : an account of the characteristics and capabilities of this british dependency – London: Jarold, [1860] – 1 – (publ under the authority of the govt immigration board for the guidance and info of emigrants) – us CRL [960]
The colony of new zealand : its history, vicissitudes and progress / Gisborne, William – London, 1888 – (= ser 19th c books on british colonization) – 4mf – 9 – mf#1.1.6853 – uk Chadwyck [980]
Colophons de manuscrits... / Bonaeret, Benedictius de – Madrid: Graf. Calleja, 1966 – 1 – sp Bibl Santa Ana [240]
Coloquio : publicacion periodica del congreso judio latinoamericano, rama del congreso judio mundial – Buenos Aires: Congreso Judio Latinoamericano. n1-24. 1979-92 – 2r – us CRL [972]
Coloquio de la paz y tranquilidad christiana en lengua mexicana: con licencia, y priuilegio / Gaona, Juan de – en Mexico: E[n] casa de Pedro Ocharte 1582 – (= ser Books on religion...1543/44-c1800: doctrina cristiana, obras de devocion) – 122lea on 3mf – 9 – mf#crl-39 – ne IDC [241]
Coloquio de las damas / Aretino, Pietro – Madrid, Spain. 1900 – 1r – us UF Libraries [025]
Color blindness in its relation to railway employees and the public / Ryerson, George Sterling – Toronto: J E Bryant, 1889? – 1mf – 9 – mf#12800 – cn Canadiana [616]
Color line : a monthly round-up of the facts of negro american progress and of the growth of american democracy – Mt Vernon NY. v1-2 n6. 1946-47 [all publ] – (= ser Black journals, series 1) – 1mf – 9 – $20.00 – us UPA [305]
Color research and application – New York. 1976+ (1,5,9) – ISSN: 0361-2317 – mf#11050 – us UMI ProQuest [660]
Colorado : colorado revised statutes – Bradford-Robinson, 1973-mar 2002 update – 9 – $4289.00 set – mf#402240 – us Hein [348]
Colorado : session laws of american states and territories – 1859-1999 – 9 – $1,729.00 set – mf#402540 – us Hein [348]
Colorado *see*
– Colorado law reporter
– Colorado nisi prius decisions
– Reports and opinions
– Reports, pre-nrs
Colorado addendum to green's pleading and practice / Green, Thomas Andre – St. Louis: Gilbert, 1880. 123p. LL-601 – 1 – us L of C Photodup [340]
Colorado AFL-CIO [American Federation of Labor and Congress of Industrial Organizations] *see* Colorado labor advocate
Colorado and the west – Denver. 1965-1978 [1] 1974-1978 [5] 1974-1978 [9] – (cont by: colorado/rocky mountain west) – ISSN: 0161-7168 – mf#9121 – us UMI ProQuest [071]
Colorado appellate reports / Colorado. Court of Appeals – v1-27. 1891-1915 – 59mf (1:42) 21mf (1:24) – 9 – $297.00 – (no pre-nrs vols. updates planned) – mf#LLMC 84-126 – us LLMC [340]
Colorado Association of Public Employees *see* Citizen
Colorado attorney general reports and opinions – 1887-2001 – 6,9 – $848.00 set – (1887-1966 on reel $105. 1967-2001 on mf $743) – mf#408150 – us Hein [340]
Colorado baptist bulletin – Denver. Colorado and Wyoming Baptist Bulletin/Colorado Baptist. Colorado Baptist State Convention, Wyoming Baptist Convention. 1904, 1911-12, 1915, 1917, 1929-30, 1931-72. Incomplete. Single reels available – 1 – $120.10 – us ABHS [242]
Colorado bar association annual reports – v1-43. 1884-1940 [all publ] – 50mf – 9 – $225.00 – (cont by: colorado looseleaf service, 1941-46) – mf#llmc84-442 – us LLMC [340]
Colorado Centennial-Bicentennial Commission *see* Directions 76
Colorado chronicle – Denver CO. 1901 oct 16-1903 jul 22 – 1r – 1 – mf#854417 – us WHS [071]
Colorado chronicle *see* Miscellaneous newspapers of las animas county, reel 2
Colorado city iris *see* El paso county miscellaneous newspapers, reel 2
Colorado city journal *see* El paso county miscellaneous newspapers, reel 2
Colorado clipper *see* Miscellaneous newspapers of washington county
Colorado. Committee on Child Welfare Legislation *see* First and second reports of governor shoup's committee on child welfare legislation for colorado
Colorado courier – Denver CO. 1884 mar 16 – 1r – 1 – mf#854374 – us WHS [071]
Colorado. Court of Appeals *see* Colorado appellate reports
Colorado democrat *see* Jefferson county miscellaneous newspapers
Colorado. Dept of Natural Resources *see* 1970 colorado comprehensive outdoor recreation plan
Colorado. Div of Parks & Outdoor Recreation *see* 1976 colorado comprehensive outdoor recreation plan
Colorado Division of Parks and Outdoor recreation *see* Interim colorado comprehensive outdoor recreation plan, 1974
Colorado exchange journal *see* Denver city and county miscellaneous newspapers
Colorado farmer *see* Miscellaneous newspapers of weld county
Colorado field ornithologist – Fort Collins. 1967-1973 (1) – (cont by: cfo journal) – ISSN: 0010-1591 – mf#8388 – us UMI ProQuest [590]
Colorado Field Ornithologists *see* Cfo journal
Colorado health profile / United States. Centers for Disease Control and Prevention – [Atlanta, GA?]: US Dept of Health and Human Services... 1996 [mf ed 19–] – 9 – us GPO [614]
Colorado herald *see* Gilpin county miscellaneous newspapers
Colorado heritage news – Denver, CO: State Historical Society of Colorado, 1981 (mf ed 1991) – 1r – 1 – ISSN: 0 – mf#MF C714he – us Colorado Hist [978]
Colorado herold – Denver, CO: Harburg & Co, sep 2 1917-jun 27 1918 – 4r – us CRL [071]
Colorado history news – Denver, CO: Colorado Historical Society, 1987- – 1r – 1 – ISSN: 0 – mf#MF C714his – us Colorado Hist [978]
Colorado independent *see* Alamosa county miscellaneous newspapers
Colorado journal of international environmental law and policy – v1-12. 1990-2001 – 9 – $320.00 set – ISSN: 1050-0391 – mf#112971 – us Hein [333]
Colorado labor advocate / Colorado AFL-CIO [American Federation of Labor and Congress of Industrial Organizations] – Denver CO. 1926 mar 11/1928-1989 jul 7/1994 dec 23 – 23r – 1 – mf#3363169 – us WHS [331]
Colorado law reporter / Colorado – v1-24. 1880-1884 (all publ) – (= ser Colorado appellate reports) – 31mf – 9 – $46.50 – mf#LLMC 91-026 – us LLMC [340]
Colorado law reporter – Denver. v1-4. 1880-84 (all publ) – (= ser Historical legal periodical series) – 1 – $50.00 set – mf#500731 – us Hein [340]
Colorado law reporter – v1-4. 1880-84 (all publ) – 9 – mf#LLMC 82-914 – us LLMC [340]
Colorado. Laws, Statutes, etc *see* The juvenile court laws of the state of colorado
Colorado lawyer – v1-30. 1971-2001 – 9 – $1847.00 set – ISSN: 0363-7867 – mf#101851 – us Hein [340]
Colorado legion magazine / American Legion – Telluride CO. v1 n1 [1920 aug] – 1r – 1 – (cont by: colorado service star) – mf#1054635 – us WHS [355]
Colorado lifestyle – Denver, CO: Colorado Lifestyle Magazine, 1982-83 (mf ed 1991) – 1r – 1 – ISSN: 0 – mf#MF C714lif – us Colorado Hist [640]
Colorado magazine – Denver. 1975-1980 (1) 1976-1980 (5) 1976-1980 (9) – ISSN: 0010-1648 – mf#10320 – us UMI ProQuest [978]
The colorado magazine – Denver, CO: State Historical Society of Colorado, 1923 – 1r – 1 – (publ suspended may 1925 to feb 1926, inclusive. reel 1: v1-v 42. reel 2: v7, 10, 25 v4 n4. reel 3: v13 n5 v1 n4, v25 n4. reel 4: v26 n3. reel 5: v40 n1-v51 n4 (some issues missing). reel 6: v45 n4. reel 7: v46 n1-2. reel 8: v57 n1-4) – ISSN: 0 – mf#MF C714ma – us Colorado Hist [978]
Colorado Media Project *see* Unsatisfied man
Colorado medicine – Denver. 1980+ (1,5,9) – (cont: rocky mountain medical journal) – ISSN: 0199-7343 – mf#610,01 – us UMI ProQuest [610]
Colorado miner *see*
– Clear creek county miscellaneous newspapers
– Gilpin county miscellaneous newspapers
Colorado mountain history photo collection – Leadville, CO: Lake County Civic Center Assoc, 1982 (mf ed 1982) – 2r – 1 – ISSN: 0 – mf#MF Pho1 – us Colorado Hist [770]
Colorado mountaineer *see* El paso county miscellaneous newspapers, reel 2
Colorado nisi prius decisions / Colorado – 1v. 1900-1902 (all publ) – (= ser Colorado appellate reports) – 6mf – 9 – $9.00 – mf#LLMC 91-027 – us LLMC [340]
Colorado pioneer *see* Miscellaneous newspapers of las animas county, reel 2
Colorado place names / Rogers, James Grafton – [S.l: s.n, 19–] (mf ed 1967) – 6r – 1 – ISSN: 0 – mf#MF R632c – us Colorado Hist [978]
Colorado portrait and biography index / Bromwell, Henrietta Elizabeth – [Denver, CO?: Bromwell. 5v. 1930] (mf ed 1973) – 2r – 1 – mf#MF B788 – us Colorado Hist [978]
Colorado quarterly – Boulder. 1952-1980 (1) 1971-1980 (5) 1976-1980 (9) – ISSN: 0010-1710 – mf#5969 – us UMI ProQuest [978]
Colorado River Indian Tribes *see* Manataba messenger
Colorado school of mines quarterly – Golden. (1) 1961-1961 (5) 1961-1961 (9) – ISSN: 0010-1753 – mf#6985 – us UMI ProQuest [622]
Colorado service star / American Legion – Montrose CO. v2 n2-7 [1922 oct 15-1923 mar 15] – 1r – 1 – (cont: colorado legion magazine) – mf#854406 – us WHS [355]
Colorado springs farm news *see* El paso county miscellaneous newspapers, reel 2
Colorado springs independent *see* El paso county miscellaneous newspapers, reel 2
Colorado springs minority press *see* El paso county miscellaneous newspapers, reel 2
Colorado springs observer *see* El paso county miscellaneous newspapers, reel 2
Colorado springs sentinel *see* El paso county miscellaneous newspapers, reel 2
Colorado. State Bar Association *see* Proceedings, 1882-1940
Colorado State Federation of Labor *see*
– Denver labor bulletin
– Official proceedings of the...annual convention of the Colorado State Federation of Labor
– Official year book
– Report of proceedings of the...annual convention
– United labor bulletin
Colorado State Federation of Labor et al *see* Labor news
Colorado State Grange *see* Journal of proceedings of the...annual session of the colorado state grange of the patrons of husbandry
Colorado. State Historical and Natural History Society *see* Biennial report of the state historical and natural history society of colorado
Colorado state republic *see* El paso county miscellaneous newspapers, reel 2
The colorado statesman – Denver, CO: J D D Rivers, 1895-1961// (wkly) [mf ed 1947] – (= ser Negro Newspapers on Microfilm) – 1r – 1 – us L of C Photodup [071]
The colorado statesman – Denver. Colo. aug. 16, 1940; Oct. 30, 1948 – 1 – us NY Public [071]
Colorado sun *see* Miscellaneous newspapers of pueblo county
Colorado. Supreme Court *see* Colorado supreme court reports
Colorado supreme court reports / Colorado. Supreme Court – v1-78. 1864-1926 – 152mf (1:42) 165mf (1:24) – 9 – $931.00 – (pre-nrs: v1-6 1864-83 15mf $67.00. updates planned) – mf#LLMC 84-125 – us LLMC [347]
Colorado. Synod (Cum. Pres. Ch.) *see* Minutes, 1854-1887
Colorado telegraph semi-weekly edition *see* El paso county miscellaneous newspapers, reel 3
Colorado telegraph weekly edition *see* El paso county miscellaneous newspapers, reel 3
Colorado, Vicente *see* Fundamentos de sociologia
The colorado voice *see* El paso county miscellaneous newspapers, reel 2
Colorado worker *see* Denver county miscellaneous newspapers, reel 2
Coloradobiz – Englewood. 1999+ (1,5,9) – ISSN: 1523-6366 – mf#17858,02 – us UMI ProQuest [332]
Colorado/rocky mountain west – Denver. 1978-1979 [1,5,9] – (cont: colorado and the west) – ISSN: 0194-052X – mf#9121,01 – us UMI ProQuest [917]
Colorado-Wyoming Council of Teamsters *see* Rocky mountain teamster
Coloradske novice – Pueblo CO, 1905* – 1r – 1 – (slovenian newspaper) – us IHRC [071]
Coloradske solnce – Denver CO, 1908* – 1r – 1 – (slovenian newspaper) – us IHRC [071]
Coloration technology – Bradford. 2001+ (1,5,9) – mf#703,01 – us UMI ProQuest [660]
The colored american – New York. Mar 14 1840-Mar 13 1841 – 1 – us NY Public [071]
Colored american magazine – 1900 may-1902jun, 1902 jul-1904 nov, 1904 dec-1907 mar, 1907 apr-1909 nov – 4r – 1 – mf#1054639 – us WHS [305]
Colored american magazine – Boston; New York. 1900-1909 (1) – mf#3086 – us UMI ProQuest [305]
Colored american magazine – Boston, New York. v1-17 n5. 1900-09 [all publ] – (= ser Black journals, series 1) – 89mf – 9 – $855.00 – us UPA [305]
The colored american magazine – v1-17. may 1900-nov 1909 – 1 – 74.00 – us L of C Photodup [305]
Colored and white : unite and fight for a workers world – v1 n1-v3 n6 [1959 mar-1963 mar 24] – 1r – 1 – (cont by: black and white; unite and fight for a workers world) – mf#1044264 – us WHS [331]
Colored baptists family tree: a compendium of organized negro baptists church history / Moses, W H – Nashville, TN: The Sunday School Publishing Board of the National Baptist Convention, U.S.A., Inc., c1925 – 1r – 1 – $5.00 – us Southern Baptist [242]
Colored citizen – Pensacola, FL. 1914 aug 28-1948 jul 23 – 1r – (filmed scattered issues only) – us UF Libraries [071]
Colored citizen – Helena MT. 1894 sep 3-nov 5 – 1r – 1 – mf#900331 – us WHS [305]
The colored citizen – Cincinnati, [OH]: Colored Citizen Co. v3 n29. may 19 1866 (wkly) [mf ed 1947] – (= ser Negro Newspapers on Microfilm) – 1r – 1 – us L of C Photodup [071]
Colored Cumberland Presbyterian Church *see* Cumberland flag
Colored Disciples of Christ. General Assembly *see* Minutes of the proceedings of the...general assembly of the colored disciples of christ in eastern north carolina, virginia, pennsylvania, new york and new jersey
Colored Disciples of Christ. Washington and Norfolk District *see*
– Minutes of the washington and norfolk district of the colored disciples of christ in eastern north carolina and virginia
– Minutes of the...annual assembly of the colored disciples of christ of the washington and norfolk district, eastern north carolina and virginia
Colored Home and Hospital (New York NY) *see* Annual report of the colored home and hospital

The colored lady evangelist : being the life, labors and experiences of mrs harriet a baker / Acornley, John Holmes – Brooklyn, N.Y.: [s.n.], 1892. El Segundo, Ca: Micro Publication Systems, 1981 (1mf); Evanston: American Theol Lib Assoc, 1984 (1mf) – (= ser Women & the church in america) – 9 – 0-8370-1469-7 – mf#1984-2142 – us ATLA [240]

The colored man in the methodist episcopal church / Hagood, Lewis Marshall – Cincinnati: Cranston & Stowe; New York: Hunt & Eaton, 1890 – 1mf – 9 – 0-7905-5938-2 – mf#1988-1938 – us ATLA [242]

The colored messenger : a magazine exclusively devoted to the cause of the colored missions – Techny IL: Mission Press. v1-2 1916-17 [qrterly] [mf ed 2004] – 2v on 1r – 1 – (publ by: negro missions of the society of the divine word) – mf#2004-s011 – us ATLA [241]

Colored Methodist Episcopal Church. General Conference *see* Journal of the...general conference and the...quadrennial session of the colored methodist episcopal church

The colored patriot – Topeka, KS. v1 n1. apr 20 1882- [mf ed 1947] – (= ser Negro Newspapers on Microfilm) – 1r – 1 – us L of C Photodup [071]

The colored patriots of the american revolution / Nell, William C – New York: Arno Press 1968 – (filmed with: [baird, h c] washington, u. jackson uber die neger als soldaten) – us CRL [975]

Colored presbyterian : [in the interest of presbyterian and the general welfare of the colored race in america] – Nashville TN: S Jackson, H A Cameron, v1 n40 (feb 24 1904) [mf ed 2005] – 1r – 1 – (reel also contains 2 other titles) – mf#2005-s016 – us ATLA [242]

Colored Teachers State Association of Texas *see* Proceedings of the...annual session of the colored teachers' state association of texas and the principals' division

Colored Teachers State Association of Texas. Meeting (1925: Dallas TX) *see* Minutes of the colored teachers' state association of texas

Colored Teachers State Association of Texas. Meeting (1926: Waco TX) *see* Minutes of the colored teachers state association of texas

The colored tennessean – Nashville, TN, 1865-1866? or 1867?// [mf ed 1947] – (= ser Negro Newspapers on Microfilm) – 1 – us L of C Photodup [071]

The colored visitor – Logansport, IN: [S M Raines, jul 1879 (semimthly) [mf ed 1947] – (= ser Negro newspapers on microfilm) – 1r – 1 – us L of C Photodup [071]

Colored Western Disciples of Christ *see* Proceedings of the...annual session of the western disciples of christ, colored

Colored Woman's League of Washington, DC *see* Annual report of the colored woman's league of washington, dc

Colored Young Women's Christian Association (Washington DC) *see* Years' report of the colored y w christian association

La colorina. badajoz, arqueros, 1928 / Reyes Huertas, Antonio; ed by Zurbitu, D – Madrid: Razon y Fe, 1929 – 9 – sp Bibl Santa Ana [946]

Colosi, Christopher B *see*
- La mediacion
- Mediation

Colosi, Thomas R *see*
- La mediacion
- Mediation

Colossian studies : lessons in faith and holiness from st. paul's epistles to the colossians and philemon / Moule, Handley Carr Glyn – New York: A C Armstrong, 1898 – 1mf – 9 – 0-8370-4504-5 – mf#1985-2504 – us ATLA [227]

Colosso, N A *see* ...Sev tvrcarvm expeditio in siculum fretum

Colotn, Calvin *see* Church and state in america, pt 2

Colour and colour theories / Franklin, Christine Ladd – London, England. 1929 – 1r – us UF Libraries [025]

Colour bar in the copper belt / Lewisn, Julius – Johannesburg, South Africa. 1941 – 1r – us UF Libraries [320]

Colour problem / Richmond, Anthony H – Baltimore, MD. 1961 – 1r – us UF Libraries [305]

Colour question in imperial policy *see* Empire and commonwealth

Colour sergeant no 1 company / Laffan, Bertha Jane (Grundy) – London: Jarrold & Sons. 2v. 1894 – (= ser 19th c women writers) – 8mf – 9 – mf#5.1.33 – uk Chadwyck [830]

Colourage – Bombay. 1976-1981 (1) 1976-1981 (5) 1976-1981 (9) – ISSN: 0010-1826 – mf#6610 – us UMI ProQuest [660]

Colourful burma : her infinite variety: a collection of stories and essays / Khin Myo Chit, Daw – Rangoon, Burma: KMCT Sazin 1976 [mf ed 1995] – on pt of 1r – 1 – mf#11052 r1973 n1 – us Cornell [800]

The colour-sense : its origin and development: an essay in comparative psychology / Allen, Grant – London: Kegan Paul, Trench, Trubner, 1892 – (= ser The english and foreign philosophical library) – 2mf – 9 – mf#28976 – cn Canadiana [150]

Colpitts, W W *see* Baptism

Le colporteur parisien – [Paris]: H Vrayet de Surcy, jul 2 1848 – us CRL [074]

Colquhoun, A R *see* The 'overland' to china

Colquhoun, Archibald Ross *see*
- China in transformation
- Dan to beersheba
- Key of the pacific
- Matabeleland
- The renascence of south africa
- Russia against india

Colquhoun, J C *see*
- On the object and uses of protestant associations
- Uses of the established church to the protestantism and civilization...

Colquhoun, John *see* Catechism for the instruction and direction of young communicants

Colquhoun, John C *see* Ireland

Colquhoun, John Campbell *see* Progress of the church of rome towards ascendancy in england

Colrain 1740-1849 – Oxford, MA (mf ed 1995) – (= ser Massachusetts vital record transcripts to 1850) – 7mf – 9 – 0-87623-233-0 – (mf 1t: baptisms 1828-38; church marriages 1829-35; church deaths 1829-62. mf 1t-2t: births 1792-1849. mf 2t: marriages 1844-49. mf 2t-3t: births & deaths 1740-1822. mf 3t: marriages 1789-92. mf 3t-4t: deaths 1843-49. mf 4t,7t: marriages 1842-49. mf 4t: marriage intentions 1773-88; out-of-town marriges 1747-95. mf 4t-7t: marriages & intentions 1804-49) – us Archive [978]

Colrain 1741-1895 – Oxford, MA (mf ed 1987) – (= ser Massachusetts vital records) – 36mf – 9 – 0-87623-052-4 – (mf 1-6: town & vital records 1741-96. mf 7-15: town & vital records 1803-39. mf 16-17: marriages & publishments 1803-25. mf 18-20: marriage intentions 1828-1908. mf 21-22: b,m,d 1843-53. mf 23-25: index to births 1854-1986. mf 26-28: index to marriages: 1854-1986. mf 29-30: index to deaths 1854-1986. mf 31-32: births 1854-95 bk 3. mf 33-34: marriages 1854-95. mf 35-36: deaths 1854-95) – us Archive [978]

Colson, E *see*
- Plateau tonga of northern rhodesia: studies
- Studies on the plateau tonga of northern rhodesia

Colson, Elizabeth *see*
- Marriage and the family among the plateau tonga of northern rhodesia
- Plateau tonga of northern rhodesia
- Seven tribes of british central africa

Colson, Jaime *see* Maestro del valle

Colson, Leon Clement *see* Transports & tarifs; regime administratif des voies de communication, conditions techniques et commerciales des transports.

Colston, Marianne *see* Journal of a tour in france, switzerland, and italy, during the years 1819, 20, and 21

Colt collector – v1-2 n6 [1974 aug-1976 jun] – 1r – 1 – mf#359832 – us WHS [730]

Colt, Jabez *see* [Circular]

Coltman, Robert *see* The chinese

Colton – 1909-34 – (= ser California telephone directory coll) – 25r – 1 – $1250.00 – mf#P00018 – us Library Micro [917]

Colton, Asa Smith *see* Successful missions

Colton, Calvin *see*
- The genius and mission of the protestant episcopal church in the united states
- History and character of american revivals of religion
- Tour of the american lakes and among the indians of the north-west territory, in 1830
- Tour of the american lakes109=and among the indians of the north-west territory, in 1830; disclosing the character and prospects of the indian race

[Colton-] colton courier – CA. 1971- – 23 – 1 – $1380.00 (subs $50/y) – mf#R02127 – us Library Micro [071]

[Colton-] colton page (the sun) – CA. 1971-1973 – 2r – 1 – $120.00 – mf#R03188 – us Library Micro [071]

[Colton-] colton semitropic – CA. 1877-1878 – 1r – 1 – $60.00 – mf#R04016 – us Library Micro [071]

Colucci, M *see* Principle di diritto consuetudinario della somalia, italiana meriodinale

Columban *see* Poems of saint columban

Columbia / Summit Co. Akron – jul 1920-dec 1925 [twice wkly, semiwkly] – 6r – 1 – (in german) – mf#B3936-3941 – us Ohio Hist [071]

Columbia basin herald – Moses Lake, WA. 1941-2000 (1) – mf#67040 – us UMI ProQuest [071]

Columbia beacon (umatilla, or) – Umatilla OR: E E Johnson, 1978- [wkly] – 1 – us Oregon Lib [071]

Columbia beacon (warrenton, or) – Warrenton OR: L J Anderson, 1954- [wkly] – 1 – us Oregon Lib [071]

Columbia business law review – 1986-2001 – 9 – $332.00 set – ISSN: 0898-0721 – mf#110801 – us Hein [346]

Columbia College *see* CBMR digest

The columbia college mss of meghilla (babylonian talmud) : with an autotype facsimile / Margolis, Max Leopold – New York: s.n. 1892 (A Ginsberg) [mf ed 1985] – 1mf [ill] – 9 – 0-8370-4277-1 – mf#1985-2277 – us ATLA [270]

Columbia College [New York, NY]. Observatory *see* Contributions from the observatory of columbia college, new york

Columbia collegian *see* Milton eagle

Columbia county and lake city / Williamson – s.l, s.l? 1939 – 1r – us UF Libraries [978]

Columbia county dispatch – Dayton, WA. 1917-1926 (1) – mf#68417 – us UMI ProQuest [071]

Columbia county, florida – Jacksonville, FL. 1883 – 1r – us UF Libraries [630]

Columbia county herald – Scappoose OR: Sel-Mor inc 1974 [wkly] – 1 – (merger of: veronia eagle [1922-74]; scappoose spotlight [1961-74]; cont by: columbia herald [1974-78]) – us Oregon Lib [071]

Columbia county herald *see*
- Scappoose spotlight
- Veronia eagle

Columbia county reporter – Rio WI. 1886 sep 10/1887 apr 29-1905 aug 25/1906 dec 26 – 14r – 1 – (with gaps) – mf#966915 – us WHS [071]

Columbia county wecker – Portage WI. 1874 sep 7-1880 sep 3 – 1r – 1 – mf#1097569 – us WHS [071]

Columbia courier – Marcus, WA. 1925-1925 (1) – mf#67032 – us UMI ProQuest [071]

Columbia empire – Umatilla OR: Empire Print Co, -1949 [wkly] – 1 – us Oregon Lib [071]

Columbia first baptist church. columbia, south carolina : church records – 1809-40, 1870-1949, 1984-1988 – 1 – $528.57 – us Southern Baptist [242]

Columbia forum – New York. 1957-1975 (1) 1971-1975 (5) – ISSN: 0010-1907 – mf#1509 – us UMI ProQuest [320]

Columbia Free Press Association *see* New morning

Columbia gateway – Meyers Falls, WA. 1908-1908 (1) – mf#67038 – us UMI ProQuest [071]

Columbia herald – Scappoose OR: Sel-Mor inc, 1974-78 [wkly] – 1 – (cont: columbia county herald [1974]; cont by: scappoose spotlight [1978-84]; 1974 incl newspaper publ by scappoose high school students) – us Oregon Lib [071]

Columbia human rights law review – New York. 1967+ (1) 1967+ (5) 1967+ (9) – ISSN: 0090-7944 – mf#10628 – us UMI ProQuest [322]

Columbia human rights law review – v1-31. 1967-2000 – 5,6,9 – $620.00 set – (v1-16 1967-85 on reel or mf $220; v17-31 1985-2000 on mf $400; title varies: v1-3 1967-71 as columbia survey of human rights law review) – ISSN: 0090-7944 – mf#101861 – us Hein [341]

Columbia independent – Columbia, PA. -w 1891-1912 – 13 – $25.00r – us IMR [071]

Columbia journal of asian law – (title varies: v1-9 1986-95 as journal of chinese law) – ISSN: 1094-8449 – mf#111551 – us Hein [342]

Columbia journal of environmental law – v1-25. 1974-2000 – 5,6 – $559.00 set – (v1-9 1974-84 on reel or mf $121. v10-25 1985-2000 on mf $438) – ISSN: 0098-4582 – mf#101871 – us Hein [344]

Columbia journal of law and social problems – v1-34. 1965-2001 – 5,6,9 – $734.00 set – (v1-18 1965-85 on reel or mf $264. v19-34 1985-2001 on mf $470) – ISSN: 0010-1923 – mf#101881 – us Hein [360]

Columbia journal of transnational law – v1-39. 1961-2001 – 5,6,9 – $818.00 set – (v1-23 1961-85 on reel or mf $314; v24-39 1985-2001 on mf $504; title varies: v1-2 1961, 1963-63 as columbia society of international law bulletin; v2 1963 as international law bulletin) – ISSN: 0010-1931 – mf#101891 – us Hein [340]

Columbia journal of world business – New York. 1971-1996 (1) 1965-1996 (5) 1975-1996 (9) – (cont by: journal of world business) – ISSN: 0022-5428 – mf#6182 – us UMI ProQuest [338]

Columbia journalism review – New York. 1962+ (1) 1971+ (5) 1975+ (9) – ISSN: 0010-194X – mf#5955 – us UMI ProQuest [070]

Columbia jurist – New York. v1-3. 1885-87 (all publ) – 1 – $47.00 set – mf#101931 – us Hein [340]

The columbia jurist – v1-3. 1885-87 (all publ) – 4mf – 9 – $18.00 – mf#LLMC 82-915 – us LLMC [340]

Columbia law review – v1-26. 1901-26 – 222mf – 9 – $333.00 – (lacking: p179-80 v13. updates planned) – mf#LLMC 84-443 – us LLMC [340]

Columbia law review – v1-100. 1901-2000 – 1,5,6,9 – $3522.00 set – (v1-96 1901-96 in reel or mf $3316. v97-100 1997-2000 in mf $206.) – ISSN: 0010-1958 – mf#101941 – us Hein [340]

Columbia law times – v1-6. 1887-93 (all publ) – 10mf – 9 – $45.00 – (lacking: v3,4,6) – mf#LLMC 82-916 – us LLMC [340]

Columbia law times – New York. v1-6. 1887-93 (all publ) – 1 – $60.00 set – mf#101951 – us Hein [340]

Columbia magazine – Hudson. 1814-1815 (1) – mf#3722 – us UMI ProQuest [240]

Columbia midlands black pages – 1994/95, 1995-96, 1997-98 – 1r – 1 – mf#3958863 – us WHS [305]

Columbia Mission *see* Twelfth annual report of the columbia mission for the year 1870

Columbia news *see*
- Stanfield standard (stanfield, or)
- Umatilla spokesman (umatilla, or)

Columbia, newsletter – Oregon. v1 n4-v3 n4 [[1973 apr?]-1976 mar?]] – 1r – 1 – mf#359973 – us WHS [071]

Columbia press [astoria, or] – Astoria OR: Columbia Press, 1949- [wkly] – 1 – (absorbed: lannen uutiset [1946-51]; place of publ moves to warrenton, or, jun 22 1978; text is english until mar 1951 and after dec 4 1958, in interim predominantly finnish, with occasional english articles) – us Oregon Lib [071]

Columbia press (umatilla, or) – Umatilla OR: J M Moore, [wkly] – 1 – us Oregon Lib [071]

Columbia register – Houlton OR: R H Mictmell [Mitchell] 1904-06 [wkly] – 1 – us Oregon Lib [071]

Columbia reporter – Columbus WI. 1854 nov 18 – 1r – 1 – mf#960669 – us WHS [071]

The columbia river : or, scenes and adventures during a residence of six years on the western side of the rocky mountains, among various tribes of indians hitherto unknown; together with a journey across the american continent / Cox, Ross – London 1832 [mf ed Hildesheim 1995-98] – 2v on 6mf – 9 – €120.00 – 3-487-27156-7 – gw Olms [917]

Columbia river sun – Cathlamet, WA. 1905-1938 (1) – mf#66955 – us UMI ProQuest [071]

Columbia second baptist church. columbia, south carolina : church records – 1890-1915 – 1 – 7.16 – us Southern Baptist [242]

Columbia society of international law bulletin *see* Columbia journal of transnational law

Columbia spectator – New York, NY. 1989-1992 (1) – mf#65063 – us UMI ProQuest [071]

Columbia spy – Columbia, PA. -w 1889-1912 – 13 – $25.00r – us IMR [071]

Columbia Students for a Democratic Society *see* Hard core

Columbia survey of human rights law review *see* Columbia human rights law review

[Columbia-] topics – NV. 1908-09 – 1r – 1 – $60.00 – mf#U04480 – us Library Micro [071]

Columbia University *see*
- Lectures on science, philosophy and art, 1907-8
- Reports of the bureau of applied social research

Columbia university. fact finding commission. proceedings : (cox commission) – New York, 1968 – 1 – $180.00 – mf#0163 – us Brook [301]

Columbia University. Observatory *see* Contributions from the observatory of columbia university

Columbia university quarterly – New York. 1898-1907 – 1 – mf#2876 – us UMI ProQuest [378]

Columbia University. Rutherfurd Observatory *see* Contributions from the rutherfurd observatory of columbia university

Columbia University. Teachers College *see* Contributions to education

Columbia University Teachers College Institute... *see* Report of the survey of the schools of tampa, flor...

Columbia valley advocate – Washougal, WA. 1953-1959 (1) – mf#67175 – us UMI ProQuest [071]

Columbian – Columbia Falls, MT. 1891-1897 (1) – mf#64333 – us UMI ProQuest [071]

Columbian – New Westminster, British Columbia, CN. sept 1899-nov 1983 – 401r – 1 – cn Commonwealth Imaging [071]

Columbian – Vancouver, WA. 1890-1921 (1) – mf#69365 – us UMI ProQuest [071]

Columbian – Vancouver, WA. 1973+ (1) – ISSN: 1043-4151 – mf#61907 – us UMI ProQuest [071]

Columbian chemical society of philadelphia memoirs – Philadelphia. 1813-1813 (1) – mf#3723 – us UMI ProQuest [540]

Columbian chronicle – 1795 mar 13, 1795 mar 13 – 2r – 1 – mf#850808 – us WHS [071]

COLUMBIAN

Columbian College *see* Correspondence
Columbian college correspondence – Baptist records in Library of Congress, 1822-1936. 144p – 1 – us Southern Baptist [242]
Columbian gazette – Ithaca, NY. 1813-15 – 1r – 1 – us Western Res [071]
The columbian harmonist, or songster's repository: being a selection of the most approved sentimental, patriotic, and other songs – New York: Smith & Forman, 1814. Text only. MUSIC 519 – 1 – us L of C Photodup [780]
Columbian herald – Charleston, SC. 1787-1790 (1) – mf#66469 – us UMI ProQuest [071]
Columbian historian – New Richmond. 1824-1825 (1) – mf#3724 – us UMI ProQuest [975]
Columbian lady's and gentleman's magazine : embracing literature in every department – New York. 1844-1849 [1] – mf#3965 – us UMI ProQuest [780]
Columbian magazine – Danbury. 1806-1806 (1,5,9) – mf#3600,01 – us UMI ProQuest [230]
Columbian magazine *see* Columbian lady's and gentleman's magazine
Columbian methodist recorder – Victoria, BC: Province Pub Co, [1899] (mf ed v1 n1 apr 1899) – 9 – (cont by: methodist recorder) – mf#P04489 – cn Canadiana [242]
Columbian mirror and alexandria gazette – Alexandria VA. 1795 jul 9, aug 8,27-29, sep 15 – 1r – 1 – mf#881612 – us WHS [071]
Columbian museum : or, universal asylum – Philadelphia. 1793-1793 (1) – mf#3519 – us UMI ProQuest [630]
Columbian museum and savannah commercial advertiser – Savannah GA. 1817 jan 24 – 1r – 1 – (cont by: savannah gazette; columbian museum and savannah daily gazette) – mf#780654 – us WHS [071]
Columbian museum and savannah daily gazette – Savannah GA. 1818 dec 10 – 1r – 1 – (cont: columbian museum and savannah advertiser; savannah gazette) – mf#845989 – us WHS [071]
Columbian observer – Philadelphia. 1822-1825 (1) – mf#5286 – us UMI ProQuest [978]
Columbian phenix and boston review – Boston. 1800-1800 (1) – mf#3567 – us UMI ProQuest [420]
Columbian post-boy – Warren RI. 1813 jan 9, feb 13 – 1r – 1 – mf#858857 – us WHS [071]
Columbian star – Washington. 1822-1829 (1) – mf#4442 – us UMI ProQuest [240]
Columbian (weekly ed) *see* Weekly columbian
Columbiana american and new-lisbon free press – New Lisbon, OH, feb 9-mar 1 1828 – 1r – 1 – (weekly national republican newspaper) – us Western Res [071]
Columbiana baptist church. columbiana, alabama : church records – 1909-52 – 1 – us Southern Baptist [242]
Columbiana Co. Columbiana *see*
– Columbiana heritage
– Independent
– Independent register
– Ledger
– Ledger series
– True press
Columbiana Co. East Liverpool *see*
– Crisis series
– Daily crisis
– Gazette
– Mercury
– Morning tribune
– News
– Potter's gazette
– Potter's herald
– Potters herald
– Tribune
Columbiana Co. East Liverpooll *see* Tribune
Columbiana Co. East Palestine *see*
– Daily leader
– Reveille echo
– Reveille series
– Valley echo
Columbiana Co. Leetonia *see*
– Courier
– Reporter
Columbiana Co. Lisbon *see*
– Buckeye state
– Daily patriot
– Evening / morning journal
– Morning journal
– Morning journal (east palestine edition)
– Ohio patriot
Columbiana Co. New Lisbon *see*
– Aurora
– Journal
– Ohio patriot
– Western palladium
Columbiana Co. Salem *see*
– Anti-slavery bugle
– Columbiana heritage
– Daily herald
– Era series
– Journal
– News
– Record
Columbiana Co. Salineville *see* Record
Columbiana Co. Wellsville *see*
– Daily union
– Daily union series
– Evening record
– Local
– Press
– Union series
– Weekly union
Columbiana county records, ms 1134 – 1803-54 – 2r – 1 – (general records of county incl court cases, ohio laws, township voting abstracts, justices of the peace...) – us Western Res [978]
Columbiana heritage / Columbiana Co. Columbiana – 1988-96 – 6r – 1 – mf#B36929-36934 – us Ohio Hist [071]
Columbiana heritage / Columbiana Co. Salem – v1 n1. apr 1988-dec 1989 [wkly] – 2r – 1 – mf#B30882-30883 – us Ohio Hist [071]
Columbia's war for cuba / Tupper, Henry Allen – New York, NY. 1898 – 1r – us UF Libraries [972]
Columbia-vla journal of law and the arts – v1-23. 1974-2000 – 5,6,9 – $545.00 set – (v1-9 1974-85 in reel $125; v10-23 1985-2000 on mf $420; title varies: v1-6 1974-81 as art and the law; v7-9 1982-85 as columbia-vla art and the law) – ISSN: 0888-4226 – mf#101661 – us Hein [340]
The columbine herald *see* El paso county miscellaneous newspapers, reel 2
Columbus : amerikanische miscellen – Hamburg 1825-32 [mf ed Hildesheim 1995-98] – 15v on 44mf – 9 – €440.00 – 3-487-27219-9 – gw Olms [975]
Columbus : don quixote of the seas / Wassermann, Jakob – Boston: Little, Brown, & Co 1930 [mf ed 1991] – 1r [ill] – 1 – (trans fr german by eric sutton; incl ind. filmed with: der aufruhr um den junker ernst / jakob wassermann) – mf#3025p – us Wisconsin U Libr [910]
Columbus Air Force Base (MS) *see* Blueprint
Columbus Air Force Base [MS] *see* Silver wings
The columbus booster – Columbus, NE: Art Printery. v1 n1. aug 12 1932-v1 n41. jun 2 1933 (wkly) [mf ed filmmed 1999] – 1r – 1 – (cont by: columbus shopping news) – us NE Hist [071]
[Columbus-] borax miner – NV. oct 1873; 1875-77 [wkly] – 2r – 1 – $120.00 – mf#U04481 – us Library Micro [071]
Columbus, Christopher *see*
– Select letters and other original documents relating to the new world
– Select letters of christopher columbus
Columbus chronicle – Columbus, GA: J T Coleman, 1895-1900 (wkly) [mf ed 1947] – (= ser Negro Newspapers on Microfilm) – 1r – 1 – us L of C Photodup [071]
Columbus communicator – Columbus OH. 1998 apr 30-jun 25/28 – 1r – 1 – (cont: columbus minority communicator) – mf#4133451 – us WHS [071]
Columbus daily telegram – Columbus, NE: D F Davis. 4v. apr 22 1889-v4 n1026. jul 20 1892 [daily ex sun & mon] [mf ed v1 n218. jan 1 1890-jul 20 1892 filmed 2000] – 2r – 1 – (absorbed by: columbus weekly telegram; numbering incl sun ed) – us NE Hist [071]
Columbus daily telegram – Columbus, NE: Telegram Co. 9v. 82nd yr n127. may 31 1961-90th yr n267. nov 12 1969 (daily ex sun) – 33r – 1 – (cont: daily telegram (columbus ne); cont by: columbus telegram (1969)) – us Bell [071]
The columbus daily telegram – Columbus, NE: Telegram Co. 18v. 43rd yr n7. apr 3 1922-60th yr n11. jan 14 1939 (daily ex sun) [mf ed with gaps] – 43r – 1 – (formed by the union of: columbus telegram and: columbus daily news. cont by: daily telegram) – us NE Hist [071]
Columbus democrat – Columbus, NE: John G Higgins. 5v. v6 n51. jul 24 1885-v10 n3. apr 19 1889 (wkly) [mf ed with gaps] – 2r – 1 – (cont: democrat. cont by: columbus weekly telegram) – us NE Hist [071]
Columbus democrat – Columbus, Fall River WI. 1868 sep 10/1870 aug-1936 nov/1937 dec – 40r – 1 – (with gaps; cont by: columbia journal [columbus wi]) – mf#986407 – us WHS [071]
Columbus dispatch – Columbus, OH. 1871+ (1) – mf#60556 – us UMI ProQuest [071]
Columbus free press – Columbus OH. v1 n4-7 [1971 jan 4/17-feb 16/mar 1], v3 n5-10, 12 [1971 aug 16/29-1982 nov 8/21, dec 6/19] – 1r – 1 – (cont by: columbus freepress and cowtowntimes) – mf#709063 – us WHS [071]
Columbus free press series / Franklin Co. Columbus – jan 1969-oct 1995 – 6r – 1 – mf#B36326-36331 – us Ohio Hist [071]
Columbus freepress and cowtowntimes – Columbus OH. v4 n4-5,15 [1974 jan 16/19-30/feb 1, oct 23/nov 12] – 1r – 1 – (cont: columbus free press [columbus oh: 1970]; cont by: columbus freepress [columbus oh: 1976]) – mf#606018 – us WHS [071]
Columbus herold – Columbus OH (USA), 1932 16 sep-1934, 1936-1939 4 oct [gaps] – 3r – 1 – gw Misc Inst [071]
Columbus journal – Columbus WI. 1938 jan 7-dec, 1939 jan-oct 6 – 2r – 1 – (cont: columbus democrat; cont by: journal-republican [columbus wi]) – mf#1009960 – us WHS [071]
Columbus journal – Columbus, NE: M K Turner & Co. 39v. v4 n45. mar 18 1874-42nd yr n7. may 17 1911=whole n201-n2060 (wkly) [mf ed with gaps filmed 1958] – 15r – 1 – (cont: platte journal. absorbed: columbus times apr 1904 and: platte county argus jan 1906. merged with: columbus tribune to form: columbus tribune-journal) – us NE Hist [071]
The columbus journal – Columbus, NE: Journal Pub Co. v44 n37. dec 10 1913- (wkly) [mf ed -1917] – 2r – 1 – (cont: columbus tribune-journal) – us NE Hist [071]
Columbus journal-republican – Columbus Wi: 1939 oct 20/1941-2001 – 16r – 1 – (with gaps; cont: journal-republican [columbus wi: 1939]; cont by: journal-republican [columbus wi: 1968]) – mf#1009998 – us WHS [071]
Columbus Memorial Library *see*
– Selected list of books (in english) on latin america
– Selected list of recent books (in english) on latin america
Columbus minority communicator – Columbus OH. 1996 sep 26/oct 3-dec 26/jan 1 1997, 1997 jan 2/8-jun 26, 1997 jul 3-dec 25, 1998 jan 1-apr 16 – 4r – 1 – (cont: communicator news [columbus, ohio]; cont by: columbus communicator) – mf#3673376 – us WHS [071]
Columbus monthly – Columbus. 1984+ – 1,5,9 – mf#14617 – us UMI ProQuest [073]
The columbus news – Columbus, NE: Art Printery. 29v. v1 n12. aug 25 1933-v29 n14. aug 2 1961 (wkly) [mf ed with gaps filmed -1979] – 13r – 1 – (cont by: columbus shopping news) – us NE Hist [071]
Columbus, OH *see*
– City directories, 1843-1862
– Selections
Columbus record – Portland OR: Columbus Record Pub Co, [wkly] – 1 – ("northwest's greatest italian newspaper") – us Oregon Lib [071]
Columbus record monthly – Portland OR: Columbus Record Pub Co, [wkly] – 1 – ("the northwest's leading italian-american newspaper") – us Oregon Lib [071]
Columbus republican – Columbus, NE: Frank P Burgess. v1 n1. may 13 1875-77// (wkly) – 1r – 1 – us NE Hist [071]
Columbus republican – Columbus WI. 1868 oct 14/1871-1938/1939 oct 6 – 23r – 1 – mf#965737 – us WHS [071]
Columbus shopping news – Columbus, NE: Art Printery. 1v. v1 n1-11. jun 9-aug 18 1933 [wkly] [mf ed filmed 1999] – 1r – 1 – (cont: columbus booster; cont by: columbus news) – us NE Hist [071]
Columbus standard – Columbus, OH: P W Chavers, 1898-1901// (wkly) [mf ed 1947] – (= ser Negro Newspapers on Microfilm) – 1r – 1 – us L of C Photodup [071]
Columbus sunday telegram – Columbus, NE: D F Davis. 4v. v1 n282. mar 16 1890-v4 n1026. jul 20 1892 [wkly] [mf ed filmed 2000] – 1r – 1 – (absorbed by: columbus weekly telegram; numbering foll columbus daily telegram) – us NE Hist [071]
Columbus telegram – Columbus, NE: Freedom Newspapers. 90th yr n268. nov 13 1969- (daily ex sat & hols) – 97r – 1 – (cont: columbus daily telegram (1961)) – us Bell [071]
Columbus telegram – Columbus, NE: Freedom Newspapers. 90th yr n268 nov 13 1969)- (daily ex sat and hols) – 1 – (cont: columbus daily telegram (columbus, ne: 1961)) – us Crest [071]
Columbus telegram – Columbus, NE: N H Parks. 29v. v15 n16. jul 12 1894-43rd yr n6. mar 31 1922 [wkly] [mf ed with gaps] – 12r – 1 – (cont: columbus weekly telegram; merged with: columbus daily news to form: columbus daily telegram [1922]) – us NE Hist [071]
The columbus telegram – Columbus, NE: Freedom Newspapers. 90th yr n268. nov 13 1969- (daily ex sat & holidays) [mf ed 1986-88 filmed 1987-88] – 11r – 1 – (cont: columbus daily telegram (1961)) – us NE Hist [071]
Columbus theological magazine : devoted to the interests of the evangelical lutheran church – [Columbus OH: Printing House of the Ohio Synod 1881-1910] [mf ed 1974] – 30v on 6r – 1 – (merged with: theologische zeitblaetter [columbus, oh: 1882] to form: theologische zeitblaetter [columbus, oh: 1911]) – mf0259 – us ATLA [242]
Columbus times – Columbus, NE: Times Print Co. v4 n1. apr 8 1899-apr 1904 (wkly) [mf ed 1899-1902,1904 (gaps)] – 1r – 1 – (cont: platte county times. absorbed by: columbus journal) – us NE Hist [071]
Columbus times – [1979 feb 21/dec]-[1997 jul 2/8-dec 31/jan 6] – 29r – 1 – (with small gaps) – mf#571873 – us WHS [071]
The columbus tribune – Columbus, NE: Frederick H Abbott, oct 3 1906-v5 n34. may 17 1911 (wkly) [mf ed with gaps] – 2r – 1 – (merged with: columbus journal to form: columbus tribune-journal) – us NE Hist [071]
Columbus tribune-journal – Columbus, NE: Tribune Print Co. 3v. v42 n8. may 24 1911-v44 n36. dec 3 1913 [wkly] – 2r – 1 – (formed by the union of: columbus tribune and: columbus journal; cont by: columbus journal [1913]) – us NE Hist [071]
Columbus union banner – Columbus WI. 1862 jun 26-aug 7 – 1r – 1 – mf#963536 – us WHS [071]
Columbus weekly journal – Columbus WI. 1861 jun 6-1864 jun 29 – 1r – 1 – mf#965419 – us WHS [071]
Columbus weekly telegram – Columbus, NE: D F Davis. 6v. v10 n4. apr 26 1889-v15 n15. jul 5 1894 [wkly] [mf ed 1891-94 [gaps]] – 4r – 1 – (cont: columbus democrat; absorbed: columbus daily telegram [1890] and: columbus sunday telegram; cont by: columbus telegram) – us NE Hist [071]
Columbus wochenblatt – Columbus, NE: J R Kilian (wkly) [mf ed jahrg 25 n50. apr 22-may 20 1892 (lacks apr 29-may 13) filmed 1996] – 1r – 1 – (in german. cont by: nebraska biene) – us NE Hist [071]
Columella, Lucius Junius Moderatus *see* L junius moderatus columella de re rustica
Column left / Veterans Club [University of New York at Buffalo] – v2 n3 [1972 feb], v2 n3 [1972 feb] – 2r – 1 – mf#720846 – us WHS [305]
Column of comment – 1975 may 22-1980 nov 14 – 1r – 1 – mf#641481 – us WHS [071]
Columna antoniniana marci aurelii antonini augusti... : cum notis excerptis ex declarationibus j p bellorii / Bartolo, P S – Romae, [1672] – 4mf – 9 – mf#O-1083 – ne IDC [700]
Columnas volantes – Lipa: Club democratico independista, jun 18-jul 2 1899 – (filmed with other miscellaneous titles from duke university) – us CRL [079]
Columnis, Guido de *see* Historia destructionis troiae (cima3)
Colusa – 1913-50 – (= ser California telephone directory coll) – 18r – 1 – $900.00 – mf#P00019 – us Library Micro [917]
[Colusa-] colusa county sun herald – CA. 1934- – 107 – 1 – $6420.00 (subs $105/y) – mf#C02131 – us Library Micro [071]
[Colusa-] colusa sun – CA. 1911 – 1r – 1 – $60.00 – mf#C01228 – us Library Micro [071]
[Colusa county-] butte, colusa, glenn, nevada, placer, shasta, sutter, tehama and yuba counties – CA. 1892-1894 – 1r – 1 – $50.00 – mf#D008 – us Library Micro [978]
[Colusa county-] butte, colusa, sutter, tehama and yuba counties – CA. 1881; 1884-1885 – 4r – 1 – $200.00 – mf#D007 – us Library Micro [978]
[Colusa county-] colusa county – CA. 1878 – 1r – 1 – $50.00 – mf#D011 – us Library Micro [978]
[Colusa-] daily sun – CA. Jul 1919-1932 – 27r – 1 – $1620.00 – mf#C02130 – us Library Micro [071]
[Colusa-] daily times – CA. 1916-17; Jan-Mar 1933; Jul 1933-52 – 28r – 1 – $1680.00 – mf#B02132 – us Library Micro [071]
[Colusa-] herald – CA. 1916-32 [wkly] – 18r – 1 – $1080.00 – mf#B02133 – us Library Micro [071]
[Colusa-] tri-weekly colusa sun – CA. 1916-19 – 3r – 1 – $180.00 – mf#C02129 – us Library Micro [071]
Colusi County Historical Society *see* Wagon wheels
Colver/culver crossings – v6-v8 [1987 jan-1988 mar] – 1r – 1 – (cont: culver crossings; cont by: culver/colver crossings) – mf#1703967 – us WHS [071]
Colville, William Juvenal *see* Mystic light library bulletin
Colvin, Otis Herbert *see* Charles martin loeffler
Colvin, Sidney *see*
– Notes on the exhibitions at the royal academy and old water-colour society
– Papers literary, scientific etc
Colvin's weekly register – Washington. 1808-1808 (1) – mf#3966 – us UMI ProQuest [320]
Colwell, Gregory B *see* Interrelationships among stress, social support, health behaviors and self-assessed health status
Colwell, James *see* A century in the pacific
Colwell, Raymond G *see* Southern colorado communities
Colwell, Stephen *see*
– New themes for the protestant clergy
– New themes for the protestant clergy : creeds without charity, theology without humanity, and protestantism without christianity
– The position of christianity in the united states

Colwinston, glamorgan, parish church of st michael : baptisms 1696-1951, burials 1696-1992, marriages 1725-1836 – 1mf – 9 – £1.25 – uk Glamorgan FHS [929]

Colwinston, st michael, monumental inscriptions – 1mf – 9 – £1.25 – uk Glamorgan FHS [929]

Colwyn and conway bay pioneer and general advertiser for the north wales coast – [Wales] Conway 19 nov 1898-dec 1950 [mf ed 2003] – 80r – 1 – (cont as: colwyn bay and welsh coast pioneer and reporter for the town and vale of conway [1899]; colwyn bay and welsh coast pioneer and review for north cambria [jan 1900-dec 1901]; welsh coast pioneer and review for north cambria [jan 1902-jun 1908]; welsh coast pioneer [jul 1908-dec 1917]; north wales pioneer [jan 1918-dec 1950]) – uk Newsplan [072]

Colzel, M *see* Bibliographie des ouvrages relatifs a la senegambie et au soudan occidental

Com mon sense / Concerned Officers Movement – v2 n1-v2 n2 [1970 dec-1971 mar] – 1r – 1 – (cont: newsletter) – mf#720843 – us WHS [360]

Com systems in libraries : current british practice / ed by Teague, S J – Guildford, Surrey: Microfilm Assoc of Great Britain, 1978 (mf ed 1982) – 1mf – 9 – (incl bibl ref) – mf#FSN 37,956 – us NY Public [020]

Les comalis / Ferrand, Gabriel – Paris: E Leroux, 1903 – 1 – us CRL [960]

Comanche County. Kansas. Union Church Ediface Society *see* History

Comanche newsletter – v2 n13 [1976 oct] – 1r – 1 – mf#626718 – us WHS [307]

Comandante cazimajou / Porteil Vila, Herminio – Habana, Cuba. 1950 – 1r – us UF Libraries [972]

Comarcas naturales de la alta extremadura. la jara cacerena / Gutierrez Macias, Valeriano – Badajoz: Dip. Provincial, 1975. Sep. REE – 1 – sp Bibl Santa Ana [946]

Comaroff, Irene *see* Social workers and their work situation

Comas, Jose *see* Mundo pintoresco

Comas Roca, Jose M *see* La ruta de lo desconocido...

Comba, Emilio *see*
- Enrico arnaud
- Histoire des vaudois
- Histoire des vaudois. introduction
- History of the waldenses of italy
- I nostri protestanti
- Lezioni di storia della chiesa
- Visita ai grigioni riformati italiani
- Waldo and the waldensians before the reformation

Combalot, Abbe *see* Le culte de la b. vierge marie, mere de dieu

Combat – Algiers. 27 feb 1943-21 jul 1945 – 1r – 1 – uk British Libr Newspaper [072]

Combat – New York. 1968-1971 (1) 1971-1971 (5) – ISSN: 0010-2113 – mf#3328 – us UMI ProQuest [355]

Combat – Paris. 1936-juin 1939 [mnthly] – 1 – fr ACRPP [073]

Combat – Paris. Le journal de Paris. Quot. aout 1944-30 aout 1974 – 1 – fr ACRPP [074]

Le combat – aout 1897-sept 1898 – 1 – fr ACRPP [073]

Le combat – Cayenne, French Guiana. 1897-1899 (1) – mf#67703 – us UMI ProQuest [079]

Le combat – Federation socialiste de l'Allier. Montlucon. oct 1903-aout 1911 – 1 – fr ACRPP [073]

Le combat – Paris. n1-131. 6 sept 1870-23 janv 1871 – 1 – fr ACRPP [073]

Le combat – Paris. Organe republicain quotidien. 20 janv-25 26 juin 1893 – 1 – fr ACRPP [944]

Combat crew / United States. Air Force. Strategic Air Command – Offutt AFB, NE: The Command, Washington, DC: Supt of Docs, USGPO, [distributor] v1 n1 nov 1950-v42 n5 may 1992 (mthly) – 9 – (cont: professional pilot; related to: tac attack and: combat edge; merged with: tac attack to form: combat edge) – us GPO [355]

Le combat des deux armees : regrets de nos pauvres affliges dans l'emeute du 15 aout – Quebec: [s.n.] 1879. [mf ed 1985] – 1mf – 9 – 0-665-03106-8 – mf#03106 – cn Canadiana [971]

Combat des montagnes : ou, la folie beaujon / Scribe, Eugene – Paris, France. 1817 – 1r – us UF Libraries [440]

Combat estimates : europe, 1920-1943 – (= ser U s military intelligence reports) – 4r – 1 – $645.00 – 0-89093-665-X – (with p/g) – us UPA [355]

Combat estimates : the western hemisphere, 1920-1943 – (= ser U s military intelligence reports) – 2r – 1 – $335.00 – 0-89093-658-7 – (with p/g) – us UPA [355]

Combat ethnic weapons / Coalition Opposed to Medical and Biological Attack – v1 n1, v1 n1 – 1r – 1 – mf#720844 – us WHS [170]

Combat le journal de paris – Paris: Combat, 1951-57. 1953-55; 1956-62 – us CRL [073]

Le combat marxiste – Paris. n1-30. oct 1933-avr 1936 – 1 – (mq n12-14) – fr ACRPP [325]

Le combat marxiste *see* Idee et action

Le combat social – Alger. n1-5. mars-juin 1927 – 1 – fr ACRPP [073]

Combat socialiste – n1-41. Paris. 15 nov 1971-nov 1980 – 1 – fr ACRPP [325]

Combat socialiste / Organisation socialiste – Montreal: OCS, 1980-81 (mthly) [mf ed 1984] – 1r – 1 – (cont: lutte ouvriere) – mf#SEM35P201 – cn Bibl Nat [335]

Combat socialiste – v1 n1-10 [1980 oct-1981 nov] – 1r – 1 – mf#605594 – us WHS [071]

Combat socialiste pour la republique des travailleurs du quebec – Montreal: [Groupe marxiste revolutionnaire] n1 15 oct 1975-n32/33 juil/aout 1977 [mf ed 1984] – 1r – 1 – (cont: la taupe rouge) – mf#SEM35P199 – cn Bibl Nat [335]

Le combat syndicaliste – Limoges. Confederation generale du travail syndicaliste revolutionnaire. n140-200. 1936-19 mars 1937 – 1 – fr ACRPP [320]

Combat weapons – 1986 win, sum, fall – 1r – 1 – (cont: sof's combat weapons) – mf#1104606 – us WHS [355]

O combate – S Tome: Joas Carragoso, mar 21-apr 25 1925 – us CRL [079]

O combate : semanario politico, litterario e noticioso – Dois Corregos, SP. 09 out 1898 – (= ser Ps 19) – mf#P46,06,45 – bl Biblioteca [079]

Combate de el obrajuelo / Castaneda S, Gustavo A – Tegucigalpa, Mexico. 1944 – 1r – us UF Libraries [972]

Combates y capitulacion de santiago de cuba / Muller Y Tejeiro, Jose – Madrid, Spain. 1898 – 1r – us UF Libraries [972]

Combatiendo la fabula / Reyes Testa, Benito – Panama, 1943 – 1r – us UF Libraries [972]

Le combattant europeen – Paris. juin 1943-juil 1944 – 1 – fr ACRPP [073]

Combe, E *see* Histoire du culte de sin en babylonie et en assyrie

Combe, Ernest *see* Grammaire grecque du nouveau testament

Combe, George *see*
- Moral philosophy
- Outlines of phrenology
- Phrenology applied to painting and sculpture

Combe, Taylor *see* A description of the collection of ancient terracottas in the british museum

Combe, William *see* Doctor syntax, his three tours in search of the picturesque, of consolation, of a wife

Combefis, Fr *see* Scriptores post theophanem (cbh7)

Comberton 1564-1950 – (= ser Cambridgeshire parish register transcript) – 5mf – 9 – £6.25 – uk CambsFHS [929]

Combes, E *see* Voyage en abyssinie, dans le pays des galla, de choa et d'ifat

Combes, Ernest *see* Profils et types de la litterature allemande

Combes, Le, sieur *see* Brest on the quebec labrador

Combes, Louis de *see* The finding of the cross

Combes, Pierre de *see* Recueil tire des procedures civiles faites en l'officialite de paris

Combination and social progress : an address / Emery, James Augustin – Boston, MA: Associated Industries, [1919?] (mf ed 19–) – 31p – mf#ZT-TN pv84 n2 – us Harvard [303]

Combination laws, select committee on the... : minutes of evidence / Great Britain. Laws, Statutes, etc – London, 1825 – 1r – 1 – mf#95651 – uk Microform Academic [324]

Combination of unity with progressiveness of thought in the books o... / Titcomb, Jonathan Holt – London, England. 1874 – 1r – us UF Libraries [240]

Combinatorica – Budapest. 1989-1996 (1) – ISSN: 0209-9683 – mf#16983 – us UMI ProQuest [510]

Combine edition – Cambridge, NE: C Don Harpst, Mrs C Don Harpst. 1v. v1 n1-n8. oct 3-nov 21 1952 (wkly) – 1r – 1 – (absorbed by: cambridge clarion) – us NE Hist [071]

Combined duties of the citizen and of the christian considered / Clowes, J – Birmingham, England. 1807 – 1r – us UF Libraries [240]

Combines investigation act : investigation into the amalgamated builders' council and related organizations... – Ottawa: F A Acland, 1930 (mf ed 19–) – 38p – mf#ZT-TN pv96 n1 – us Harvard [690]

Combines investigation act, 1923 : investigation into the proprietary articles trade association, an alleged combine of wholesale and retail druggists and manufacturers... – Ottawa: F A Acland, 1926 (mf ed 19–) – 37p – mf#ZT-TN pv83 n7 – us Harvard [615]

Combing the caribbees / Foster, Harry La Tourette – New York, NY. 1929 – 1r – us UF Libraries [972]

Combining walking, jogging, and running into a single vo 2 max prediction test / Larsen, Gary E – 2000 – 67p on 1mf – 9 – $5.00 – mf#PH 1706 – us Kinesology [612]

Combs, George Hamilton *see* Some latter-day religions

Combustion – Battleboro. Engineering & Finance. -m. Sep 1924-Jul 1931 – 5r – 1 – uk British Libr Newspaper [621]

Combustion – Stamford. 1929-1981 (1) 1970-1981 (5) 1975-1981 (9) – ISSN: 0010-2172 – mf#930 – us UMI ProQuest [690]

Combustion and flame – New York. 1957+ (1) 1957+ (5) 1987+ (9) – ISSN: 0010-2180 – mf#42153 – us UMI ProQuest [620]

Combustion, explosion, and shock waves – New York. 1965-1977 (1) 1965-1977 (5) – ISSN: 0010-5082 – mf#10902 – us UMI ProQuest [660]

Combustion toxicology – Westport. 1974-1975 (1) 1974-1975 (5) 1974-1975 (9) – (cont by: journal of combustion toxicology) – ISSN: 0094-8128 – mf#10458 – us UMI ProQuest [360]

Come home : an appeal on behalf of reunion / Langtry, John – Toronto: Church of England Pub Co, 1900 – 1mf – 9 – 0-7905-4942-5 – mf#1988-0942 – us ATLA [240]

Come, let us celebrate : meeting god in christian and non-christian feasts / ed by Puthiadam, I – Bangalore: Asian Trading Corp, 1976 – us CRL [230]

Come now – London, England. 18– – 1r – us UF Libraries [240]

Come out fighting – Hollywood CA: Lavender & Red Union 1975- [mf ed 1984] – 1r – 1 – mf#881 – us Wisconsin U Libr [305]

Come out of the kitchen! / Miller, Alice Duer – New York, NY. 1916 – 1r – us UF Libraries [640]

Come unity – 1975 may-1988 sep – 1r – 1 – mf#1712346 – us WHS [071]

Comeau-Stender, Susan M *see* The effect of orthotic correction on walking and running efficiency in subjects with excessive pronation

Come-back / Walter Reed General Hospital – 1918 dec 4-1919 nov 12, 1919 nov-1921 mar 19 – 2r – 1 – (cont: fort sheridan recall) – mf#919865 – us WHS [360]

Comedia can panahon nin paghadit / Ariate, Nicolas – Nueva Caceres: Libreria Mariana [190-?] [mf ed Bloomington IN: Indiana Uni Lib, Preservation Dept 1984] – (= ser Coll...in the bikol language) – 1r [ill] – 1 – us Indiana Preservation [490]

La comedia de la vida / Hurtado, Antonio – 1871 – 9 – sp Bibl Santa Ana [870]

Comedia literaria / Borba, Jose Osorio De Morais – Rio de Janeiro, Brazil. 1959 – 1r – us UF Libraries [440]

Comedia na dapit sa dios, o, magna cahayagan can magna misterios nin navidad, 2nd pt / Inciso, Flaviano – Nueva Caceres: La Sagrada Familia 1896 [mf ed Bloomington IN: Indiana Uni Lib, Preservation Dept 1984] – (= ser Coll...in the bikol language) – 1r – 1 – us Indiana Preservation [490]

Comedia na dapit sa dios, o, magna cayayagan can magna misterios nin navidad, 3rd pt / Inciso, Flaviano – N a Caceres: La Sagrada Familia 1896 [mf ed Bloomington IN: Indiana Uni Lib, Preservation Dept 1984] – (= ser Coll...in the bikol language) – 1r – 1 – us Indiana Preservation [490]

Comedia prodiga / Miranda, Luis de – 1868 – 9 – sp Bibl Santa Ana [820]

Comedia prodiga / Miranda, Luis De – Sevilla, Spain. 1868 – 1r – us UF Libraries [440]

Comedia sin titulo. ms. / Garcia Lorca, Federico – MS. DE 1935. Fragmento – 1mf – 9 – sp Cultura [820]

Comedia trofea / Torres Naharro, Bartolome – Reimp. prefaciada por Fidelino de Figueiredo. Sao Paulo. Oficina Jose Magalhaes, 1943 – 1 – sp Bibl Santa Ana [946]

La comedianta. ms. / Garcia Lorca, Federico – MS DE 1923 – 2mf – 9 – sp Cultura [820]

Comedias nuevas escogidas de los mejores ingenios de espana – Madrid, 1652-1704 – 447mf – 9 – sp Cultura [820]

La comedie infernale : pieces justificatives – [Montreal?: s.n.], 1872 [mf ed 1980] – 3v on 1mf – 9 – 0-665-06187-0 – mf#06187 – cn Canadiana [241]

Comediens / Delavigne, Casimir – Paris, France. 1820 – 1r – us UF Libraries [440]

The comedies of aristophanes – London: G Bell & Sons, 1905 [mf ed 1984] – (= ser Bohn's classical library) – 2v on 1r – 1 – (new and literal trans fr rev text of dindorf. notes and extracts fr best metrical versions by william james hickie) – mf#8120 – us Wisconsin U Libr [920]

The comedy of catherine the great / Gribble, Francis Henry – London: E. Nash, 1912. xix,367p. Front., ports – 1 – us Wisconsin U Libr [920]

Comedy of convocation in the english church / Chasuble, Archdeacon – London, England. 18– – 1r – us UF Libraries [240]

A comedy of terrors / De Mille, James – Boston: J R Osgood, 1872 – 2mf – 9 – mf#06018 – cn Canadiana [830]

Comendadores...alcantara / Manera de Rezar – 1663 – 9 – sp Bibl Santa Ana [946]

Comenius, Johann Amos *see* The orbis pictus of john amos comenius

Comentaire...poeme...ibn abdoun / Ibn-Badroun – 1846 – 9 – sp Bibl Santa Ana [440]

Comentarii in quator evangelistas / Maldonado, Juan – Tomo I. 1840 – 9 – (tomo1 1862. tomo 2-4 1841. tomo 2 1874. 1601, 1611) – sp Bibl Santa Ana [240]

Comentarii..3 librum...duns scoti / Ovando, Juan de – 1597 – 9 – sp Bibl Santa Ana [240]

Comentario al articulo 1361 del codigo civil / Carrasco Alvarez, Antonio – 1898 – 9 – sp Bibl Santa Ana [347]

Comentario in profetas mimos / Arias Montano, Benito – Amberes: Plantino, 1571 – 1 – sp Bibl Santa Ana [946]

Comentarios a la embajada de persia / Silva y Figueroa, Garcia de – Madrid: Fortanet, 1904. B.R.A.H. 44, 1904, p. 196 – sp Bibl Santa Ana [946]

Comentarios a la vida de pedro de valdivia : escrita por a. miguel-romero y gil de zuniga / Mena, Vicente – Madrid: Impresos Madrid, 1929 – 1 – sp Bibl Santa Ana [920]

Comentarios a un regimen / Gomez, Laureano – Bogota, Colombia. 1935 – 1r – us UF Libraries [972]

Comentarios al codigo civil venezuolano (reformado... / Venezuela – Caracas, Venezuela. v1-4. 1962 – 2r – us UF Libraries [350]

Comentarios al codigo de procedimiento penal colom... / Moncada R, Timoleon – Bogota, Colombia. 1940 – 1r – us UF Libraries [360]

Comentarios de don garcia de silva y figueroa de la embajada que de parte del rey de espana don felipe 3rd hizo al rey xa abas de persia / Asin Palacios, Miguel – de Serrano Sanz. Madrid: Tip. de la Rev. de Archivos. Bibliotecas y Museos, 1928 – 1 – sp Bibl Santa Ana [946]

Comentarios de francisco zarco sobre la intervenci... / Zarco, Francisco – Mexico City? Mexico. 1929 – 1r – us UF Libraries [972]

Comentarios y juicios de la prensa nacional sobre / Venezuela Ministerio De La Defensa Nacional – Caracas, Venezuela. 1952 – 1r – us UF Libraries [972]

Comento – London, UK. 8 Jul 1922; 10 May, 10 Sept-26 Nov 1924 – 1 – uk British Libr Newspaper [072]

Comento – London, UK. Monthly Bulletin of the Italian Antifascist Federation. Nov 1943 – 1 – uk British Libr Newspaper [072]

Comento de los majoretes, 1500 / Meseguer Fernandez, Juan – Madrid: Graf. Calleja, 1500 – 1 – sp Bibl Santa Ana [946]

Comento filologico-esegetico sul primo salmo / Garzia, Gabriele – Napoli: Tipografia della Reale accademia delle scienze fis e mat, 1886 – 1mf – 9 – 0-524-06132-7 – mf#1992-0799 – us ATLA [220]

Comentos criticos sobre la fundacion de cartagena de indias. bogota / Otero D'Acosta, Enrique – Madrid: Razon y Fe, 1935 – 1 – sp Bibl Santa Ana [970]

Comer, G *see* A geographical description of southampton island and notes on the eskimo

Comer, John *see* Diary

El comercio – Lima: jose ayarza, 1938- – 1 – us CRL [079]

El comercio – Quito, Ecuador: Carlos Mantilla, jan 4 1940- – 1 – us CRL [079]

El comercio – Quito, Ecuador. jan 1948-dec 1955 – 44r – 1 – us L of C Photodup [079]

Comercio anglo-latino – London, UK. Apr-Aug 1912 – 1 – uk British Libr Newspaper [072]

Comercio argentino-britanico – London, UK. mar/may 1940-sep/nov 1942 – 1 – (comercio britanico dec 1942/feb 1943-jul 1948; ingenieria britanico aug 1948-mar/apr 1958) – uk British Libr Newspaper [072]

Comercio britanico *see* Comercio argentino-britanico

Comercio colombiano y la economia nacional / Federacion Nacional De Comerciantes (Colombia) – Bogota, Colombia. 1952 – 1r – us UF Libraries [330]

Comercio exterior / Honduras. Direccion General de Estadistica y Censos – 1957-65 – 1 – us L of C Photodup [380]

Comercio hispano-britanico – London, UK. Aug 1939-Oct/Dec 1945 – 1 – uk British Libr Newspaper [072]

Comercio hispano-britanico – London, UK. Jan/Mar 1921-Apr/Jun 1936; Jan/Mar 1948 – 1 – uk British Libr Newspaper [072]

Comercio internacional / Lleras Restrepo, Carlos – Medellin, Colombia. 1965 – 1r – us UF Libraries [337]

Comercio y comerciantes, y sus proyecciones / Alvarez, F & Mercedes, M – Caracas, Venezuela. 1964 – 1r – us UF Libraries [380]

El comercio y la banca / Martinez Perez, Eloy – 1892 – 9 – sp Bibl Santa Ana [380]

Comerford, Michael *see* Naas

Comet / Bedford-Stuyvesant Youth in Action Community Corporation – 1972 jul – 1r – 1 – mf#4848345 – us WHS [360]

Comet : a belfast journal of fun, frolic, and literature – Belfast, Ireland 9 aug 1849-14 jun 1850 – 1 – uk British Libr Newspaper [072]

Comet – Boston. 1811-1812 (1) – mf#3746 – us UMI ProQuest [390]

Comet – Charlestown IN. 1835 may 9 – 1r – 1 – (cont by: western farmer) – mf#856280 – us WHS [071]

Comet – Dublin, Ireland. 1 may 1831-1 dec 1833 – 1r – 1 – uk British Libr Newspaper [072]

Comet – Hamilton, OH. nov 1865 – 1r – 1 – us Western Res [073]

Comet – New York. v1 n1-5. dec 1940-jul 1941 [all publ] – (= ser Science fiction periodicals, 1926-1978. series 1) – 1r – 1 – $105.00 – us UPA [830]

Comet – [NW England] Wigan jan 1889-apr 1894 – 1 – uk MLA; uk Newsplan [072]

Comet : the organ of the aberdeen social-democrats – [Scotland] Aberdeen City 25 jun 1898-mar 1908 [mf ed 2003] – 1r – 1 – uk Newsplan [325]

The comet – Dublin. Ireland. -w. 1 May 1831-1 Dec 1833 – 1r – 1 – uk British Libr Newspaper [072]

The comet – Lagos, Nigeria. Daily Comet. -d. 16 May 1944-1 Dec 1945; 23 April-2 Nov 1946. Imperfect. 2 reels – 1 – uk British Libr Newspaper [079]

The comet – Lagos, Nigeria. -w. 22 July 1933-13 May 1944. 6 reels – 1 – uk British Libr Newspaper [072]

The comet : a weekly news-magazine of west africa – Lagos, Nigeria. june 12, 1937-may 26, 1943 – 1 – us NY Public [073]

The comet and local advertiser – Johannesburg SA, 22 may 1897-31 jul 1897 – 1r – 1 – sa National [960]

The comet (nanaimo, bc) – Kamloops [BC: s.n, 1875] – 9 – mf#P06109 – cn Canadiana [320]

The comet (ottawa, ont) – Ottawa: Thoburn, [1894-189-?] – 9 – mf#P04232 – cn Canadiana [073]

The comet (quebec) – Quebec: [s.n, 1866-1868?] – 9 – mf#P04157 – cn Canadiana [870]

Cometa / Hermoso, Eugenio – Madrid: Chulilla y Angel, 1932 – 1 – sp Bibl Santa Ana [946]

Cometarios a refranes, modismos, locuciones de con... / Llorens, Washington – San Juan, Puerto Rico. 1962 – 1r – us UF Libraries [972]

Cometarvm omnivm fere catalogvs qvi ab avgvsto qvo imperante christus natus est usque ad hunc 1556. annum apparuerunt... / Lavater, L – Tigvri, Andreas Gesner & Jacob Gesner, [1556] – 1mf – 9 – mf#PBU-600 – ne IDC [240]

Cometographie : ou, traite historique et theorique des cometes / Pingre, Alexandre Guy – Paris: Imprimerie royale 1783-84 – 2v on 1r [ill] – 1 – mf#film mas 28230 – us Harvard [520]

Comets / Olivier, Charles Pollard – Baltimore: Williams & Wilkins c1930 [mf ed 1998] – 1r [pl/ill] – 1 – (incl bibl ref & ind) – mf#film mas 28228 – us Harvard [520]

Comets : their constitution and phases: being an attempt to explain the phenomena on known principles of physical laws / Kemplay, Christopher – London: Longman, Brown, Green, Longmans & Roberts 1859 [mf ed 1998] – 1r – 1 – mf#film mas 28415 – us Harvard [520]

The comets : a descriptive treatise upon those bodies: with a condensed account of the numerous modern discoveries respecting them: and a table of all the calculated comets... / Hind, John Russell – London: J W Parker & Son 1852 [mf ed 1998] – 1r [ill] – 1 – (incl ind) – mf#film mas 28228 – us Harvard [520]

Comets and meteors : their phenomena in all ages: their mutual relations: and the theory of their origin / Kirkwood, Daniel – Philadelphia: J B Lippincott & Co 1873 [mf ed 1998] – 1r [pl/ill] – 1 – (incl bibl ref) – mf#film mas 28407 – us Harvard [520]

Comets and their tails : and the gegenschein light / Shaw, Frederick G – London: Bailliere, Tindall & Cox 1903 [mf ed 1998] – 1r [pl/ill] – 1 – mf#film mas 28406 – us Harvard [520]

[Comets pamphlets] – 1772-1857 [mf ed 1998] – 14 pcs on 1r – 1 – (in english, dutch, german & italian; incl various repr concerning description of comets attested fr 10th c; misc papers, 18th & 19th c works bound together) – mf#film mas 28417 – us Harvard [520]

Comfort and counsel under affliction – London, England. 1823 – 1r – us UF Libraries [240]

Comfort and economy in clothes / Holding, Thomas Hiram – London, [1891] – (= ser 19th c art & architecture) – 1mf – 9 – mf#4.1.142 – uk Chadwyck [740]

Comfort for the bereaved / Pirie, James – Edinburgh, Scotland. 1866 – 1r – us UF Libraries [240]

Comfort for the feeble-minded / Bailey, John – London, England. 1812 – 1r – us UF Libraries [240]

Comfort for the jews / Rutherford, J F – Brooklyn, NY. 1925 – 1r – us UF Libraries [939]

Comfort for the poor! : meat three times a day!! voluntary information from the people of new south wales...in 1845-46 / Chisholm, Caroline – London, 1847 – (= ser 19th c books on british colonization) – 1mf – 9 – mf#1.1.3581 – uk Chadwyck [360]

Comfort, George F *see* Woman's education, and woman's health

Comforter : even the spirit of trugh, who dwelleth in us, and teache / Thom, John Hamilton – Liverpool, England. 1839 – 1r – us UF Libraries [240]

The comforter : or, the pastor's friend / Bartholomew, John Glass – Boston: Tompkins, 1863 – 1mf – 9 – 0-524-02948-2 – mf#1990-4500 – us ATLA [240]

The comforter : a series of sermons on certain aspects of the work of the holy ghost / Clark, William – London: Rivingtons, 1864 – 1mf – 9 – 0-7905-7278-8 – mf#1989-0503 – us ATLA [240]

Comhaire Sylvain, Suzanne *see* Contes du pays d'haiti

Comhaire-Sylvain, Suzanne *see*
- Creole haitien
- Roman de bouqui

Comic australian – Sydney, oct 1911-jun 1913 – 1r – A$77.04 vesicular A$82.54 silver – at Pascoe [079]

The comic blackstone : (parodies on the commentaries) – (= ser The yale law library blackstone coll) – ca 22mf – 9 – mf#LLMC 82-800 titles 185-197 – us LLMC [340]

The comic crisis *see* Miscellaneous newspapers of pueblo county

The comic guide to the royal academy, for 1864 / A Beckett, Arthur William & A Beckett, Gilbert Abbott [Joint pseud: Gemini] – [London] 1864 – (= ser 19th c art & architecture) – 1mf – 9 – mf#4.2.1387 – uk Chadwyck [060]

The comic magazine / funny pages – iss n1-5. may-sep 1936 – (= ser Funny Pages) – 15 – mf#001CM – us MicroColour [740]

Comic offering – 1831-35 – (= ser English gift books and literary annuals, 1823-1857) – 21mf – 9 – uk Chadwyck [800]

The comic offering : or ladies' melange of literary mirth (london) – 1831-35 – (= ser 19th c british periodicals) – reel 33 – 1 – us Primary [870]

Comic tunes in the celebrated entertainment call'd harlequin sorcerer as they are perform'd at the theatre-royal in covent-garden : for the harpsicord, violin, &c / Arne, Thomas Augustine – London: I Walsh in Catherine Street in the Strand...[1752] [mf ed 1989] – 1r – 1 – (incl: "comic tunes in [harlequin] dr faustus" by samuel arnold; incl bibl notes by alfred moffat) – mf#pres. film 46 – us Sibley [780]

Comic world – New York, NY. 1876 – 1r – 1 – us Western Res [073]

Comics and scrapbooks : collection no 396 / Grey, Zane – 1r – 1 – mf#B29873 – us Ohio Hist [080]

Comics buyer's guide – 1983 feb 11/jun-1993 sep/oct – 28r – 1 – (cont: comic buyer's guide price list; buyer's guide for comic fandom) – mf#693405 – us WHS [740]

Comics collector – 1983 spr, 1984 win-1986 win – 1r – 1 – mf#932809 – us WHS [740]

Comics on microfilm – 1 – (titles include: barney google and snuffy smith. 1933-72. blondie. 1933-72. bringing up father. 1933-72. cap'n and the kids. 1931-72. cisco kid. 1951-68. felix the cat. 1933-67. ferd'nand. 1947-72. flash gordon. 1934-72. the good old days. 1946-72. jungle jim. 1935-54. katzenjammer kids. 1937-72. king of the royal mounted. 1935-54. krazy kat. 1933-44. l'il abner. 1934-72. little annie rooney. 1933-66. the little king. 1937-72. little orphan annie. 1927-72. mandrake the magician. 1934-72. nancy. 1922-72. the phantom. 1937-72. popeye. 1933-72. prince valiant. 1937-72. tarzan. 1929-72. tillie the toiler. 1933-59. tim tyler's luck. 1933-72. toots and casper. 1933-56) – us AMS Press [790]

Die comicsprache in der ddr : eine untersuchung der verstaendlichkeit / Schmidt, Dana – (mf ed 1999) – (= ser Leipziger arbeiten zur fachsprachenforschung) – 1mf – 9 – €30.00 – 3-8267-2661-8 – mf#DHS 2661 – gw Frankfurter [430]

Cominciamento e progresso dell'arte dell'intagliare in rame, colle vite di molti de'pi- eccellenti maestri della stessa professione / Baldinucci, F – Firenze, 1686 – 4mf – 9 – mf#O-137 – ne IDC [700]

The coming and kingdom of christ / Nangle, Edward – Dublin: Dublin Tract Repository [1862?] [mf ed 1991] – (= ser Select series of christian tracts and books 31) – 1mf – 9 – 0-7905-9823-X – mf#1989-1548 – us ATLA [240]

The coming and reign of christ / Lord, David Nevins – New York: F Knight, 1858 [mf ed 1991] – 1mf – 9 – 0-7905-8506-5 – mf#1989-1731 – us ATLA [240]

Coming back / YMCA of the USA – New York NY. n1-37 [1919 jan 1-sep 12] – 1r – 1 – mf#1110750 – us WHS [071]

The coming china / Goodrich, Joseph King – Chicago: A C McClurg, 1911 [mf ed 1995] – (= ser Yale coll) – xx/298p (ill) – 1 – 0-524-09219-2 – mf#1995-0219 – us ATLA [951]

The coming commonwealth : an australian handbook of federal government / Garran, Robert Randolph – Sydney, 1897 – (= ser 19th c books on british colonization) – 3mf – 9 – mf#1.1.1013 – uk Chadwyck [980]

The coming creed / Womer, Parley Paul – Boston: Sherman, French, 1911 – 1mf – 9 – 0-7905-8978-8 – mf#1989-2203 – us ATLA [240]

The coming event! : or freedom and independence for the seven united provinces of australia / Lang, John Dunmore – London, 1870 – (= ser 19th c british colonization) – 6mf – 9 – mf#1.1.3490 – uk Chadwyck [320]

Coming events – Derby, Derbyshire n1 [7 mar 1896]-n1046 [25 mar 1916] – 18r – 1 – uk Newsplan [072]

Coming king / Antipas, F D – London, England. 18-- – 1r – us UF Libraries [240]

Coming nation – 1893 sep 23/1901 jul 27, 1902 mar 22-1903 dec 26, 1910 sep 10-1911, 1912 jan 6-1913 jun – 4r – 1 – mf#1054659 – us WHS [071]

Coming nation / Socialist Party [WI] – v1 n1-41 [1916 jun 17-1917 mar 31] – 1r – 1 – (cont: wisconsin comrade) – mf#3499478 – us WHS [335]

The coming nation / the progressive woman – ns: v1 n1-8. 1913-14 [all publ] – (= ser Periodicals on women and women's rights, series 2) – 1r – 1 – $155.00 – us UPA [305]

Coming new era – Wahoo, NE: Eric Johnson & Sons. v10 n23. jan 3 1900-01// (wkly) [mf ed with gaps filmed [1967]] – 2r – 1 – (in english and swedish. merged with: pilen to form: lincoln svenska tribun. companion to: saunders county new era) – us NE Hist [071]

The coming of christ : both pre-millennial and imminent / Haldeman, Isaac Massey – Los Angeles, CA: Bible House; New York: Charles C Cook, c1906 [mf ed 1989] – 1mf – 9 – 0-7905-2358-2 – mf#1987-2358 – us ATLA [240]

Coming of christ, ad 1947 / Fysy, Frederic – Bath, England. 1839 – 1r – us UF Libraries [240]

Coming of christ desired / Clayton, George – London, England. 1814 – 1r – us UF Libraries [240]

The coming of messiah in glory and majesty / Irving, Edward – Preliminary discourse. Also includes Irving's Ordination Charge and an introductory essay to Bishop Harnes' Commentary on the Psalms. Bosworth & Harrison, 1859 – 9 – $10.00 – us IRC [240]

The coming of peace : a family catastrophe / Hauptmann, Gerhart – Chicago: C H Sergel, 1900 – 1r – 1 – us Wisconsin U Libr [830]

The coming of the friars and other historic essays / Jessopp, Augustus – London: T F Unwin; New York: Putnam, 1913 – 1mf – 9 – mf#1988-1344 – us ATLA [941]

The coming of the great king : or, an examination and discussion of the subject of the second coming of christ, and of questions thereto related / Houliston, Wm – Minneapolis, MN: Great Western, 1897 [mf ed 1988] – 1mf – 9 – 0-7905-0092-2 – mf#1987-0092 – us ATLA [240]

Coming of the lord / Kapff, Sixt Karl – London, England. 1837 – 1r – us UF Libraries [240]

The coming of the lord / Pierson, Arthur Tappan – London: Passmore & Alabaster, 1896 [mf ed 1990] – 1mf – 9 – 0-7905-3399-5 – mf#1987-3399 – us ATLA [240]

Coming of the loyalists / Haight, Canniff – Toronto: Haight, 1899 – 1mf – 9 – mf#05146 – cn Canadiana [971]

The coming of the world-teacher; and, death, war and evolution : a book of extracts from lectures and writings / Leadbeater, Charles Webster et al – London: George Allen & Unwin, 1917 – 1mf – 9 – 0-524-07730-4 – mf#1991-0152 – us ATLA [210]

The coming one / Simpson, Albert B – New York: Christian Alliance, c1912 [mf ed 1992] – (= ser Christian & missionary alliance coll) – 1mf – 9 – 0-524-02499-5 – mf#1990-4358 – us ATLA [240]

The coming one *see* Eternal punishment / the coming one

Coming pentecost / Winslow, Octavius – London, England. 1861 – 1r – us UF Libraries [240]

The coming religion / Dole, Charles Fletcher – Boston: Small, Maynard, c1910 – 1mf – 9 – 0-8370-8808-9 – mf#1986-2808 – us ATLA [230]

The coming religion / Van Ness, Thomas – Boston: Roberts Bros, 1893, c1892 – 1mf – 9 – 0-524-03054-5 – mf#1990-0811 – us ATLA [210]

Coming struggle for south africa / Sandor – London, England. 1963 – 1r – us UF Libraries [960]

Coming struggle with rome : not religious but political / Connelly, Pierce – London, England. 1853 – 1r – us UF Libraries [240]

Coming up / Lysistrata Restaurant and Bar – Madison WI. 1978 apr/may-1980 dec – 1r – 1 – (cont by: lysistrata letter) – mf#676328 – us WHS [071]

Coming up from the wilderness / Philpot, J C – Stamford, England. 1857? – 1r – us UF Libraries [240]

Coming wars : and other momentous prophetic events at hand / Baxter, Michael Paget – London, England. 1876 – 1r – us UF Libraries [240]

Coming whirlwind among the nations of the earth – London, England. 1868 – 1r – us UF Libraries [240]

Comings, Albert Gallatin *see* Jesus in his offices

Comision catequistica de zaragoza. religion y cultura, un grafico y ejemplos zaragoza, 1940 / Marquez, Gabino – Madrid: Razon y Fe, 1945 – 1 – sp Bibl Santa Ana [240]

Comision catolica de habla hispana / Spanish Speaking Catholic Commission – v3 n1-2 [1976 jan/feb-mar] – 1r – 1 – (cont: newsletter [spanish speaking catholic commission]; cont by: cara a cara) – mf#620790 – us WHS [241]

Comision Cubana Pro Centenario De Hostos *see* Hostos y cuba

Una comision de carlos 5th al...p. francisco de los angeles quinones (despues obispo de coria) / Gracia Villacampa, Carlos – Archivo Ibero-Americano, 1916 – 1 – sp Bibl Santa Ana [240]

Comision de Estadistica *see*
- Censo de poblacion de espana segun el recuento de 21 de mayo de 1857
- Nomenclator de los pueblos de espana

Comision de Monumentos *see*
- Acta de 22 de octubre de 1918
- Acta de la sesion de 1 de octubre de 1920
- Acta de la sesion de 22 de marzo de 1920
- Acta de la sesion del 14 de marzo de 1920
- Cargos

Comision de monumentos, antiguedades romanas / Alcuescar – 1900 – 9 – sp Bibl Santa Ana [930]

Comision de Monumentos de Caceres *see* Antiguedades romanas de alcuescar

Comision de Semana Santa *see*
- Semana santa cacerena 1960
- Semana santa. caceres 1974

Comision Diocesana de Apostolada Rural *see* La familia rural hacia la conquista de un mejor nivel de cultura. encuesta campana experimental 1963-1964

Comision especial de aguas. dictamen... / Caceres – Caceres: Imp. Moderna, 1936 – sp Bibl Santa Ana [946]

Comision especial de aguas. dictamen...6.11.1934 / Caceres – Caceres: Imprenta Moderna, 1935 – sp Bibl Santa Ana [628]

Comision militar ejecutiva y permanente / Llaverias Y Martinez, Joaquin – Habana, Cuba. 1929 – 1r – us UF Libraries [355]

Comision Mixta De Limites Entre Guatemala Y El Sal... *see* Informe de la comision mixta

Comision no 2 : transformacio en regadio / Consejo Economico Sindical Provincial – Badajoz: Imprenta Inca, 1965 – sp Bibl Santa Ana [330]

Comision no 4 : industrias basicas del hierro / Consejo Economico Sindical Provincial – Badajoz: Imp. Inca, 1965 – sp Bibl Santa Ana [338]

Comision no 5: productos quimicos / Consejo Economico Sindical Provincial – Badajoz: Imp. Inca, 1965 – sp Bibl Santa Ana [338]

Comision no 6: industrias relacionadas – con la alimentacion / Consejo Economico Sindical Provincial – Badajoz: Imprenta Inca, 1965 – sp Bibl Santa Ana [338]

Comision no 8: transportes. comunicaciones y servicios de informacion / Consejo Economico Sindical Provincial – Badajoz: Imp. Inca, 1965 – sp Bibl Santa Ana [380]

Comision no 9 : turismo / Consejo Economico Sindical Provincial – Badajoz: Imp. Inca, 1965 – sp Bibl Santa Ana [338]

Comision no 13: financiacion / Consejo Economico Sindical Provincial – Badajoz: Imprenta Inca, 1965 – sp Bibl Santa Ana [332]

Comision no 14: trabajo / Consejo Economico Sindical Provincial – Badajoz: Imp. Inca, 1965 – sp Bibl Santa Ana [331]

Comision no 15: factores humanos y sociales. productividad / Consejo Economico Sindical Provincial – Badajoz: Imprenta Inca, 1965 – sp Bibl Santa Ana [338]

Comision organizadora de festejos. memoria de 1920 / Fomento de Caceres – Caceres: Tip. de Santos Floriano Gonzalez, y 1921-22, 1924 – sp Bibl Santa Ana [946]

Comision Organizadora Del Homenaje El Dr Emeterio S Santovenia *see* Libro jubilar de emeterio s santovenias en su cincuentenario

Comision Pro Celebracion Del Centenario... *see* America y hostos

Comision Provincial de Monumentos *see*
- Constitucion en 1925. noticia de vicente castaneda
- Constitucion en 1928, noticia de vicen te castaneda

Comision provincial de monumentos historicos y artisticos de badajoz / Solar, Antonio del – Madrid: Fortanet, 1918. B.R.A.H 73, pp.383-384 – sp Bibl Santa Ana [946]

Comision Provincial de subsidio el Combatiente *see* Decreto de 25 de abril de 1938 y reglamento del mis mo mes, reorganizando el servicio del subsidio al combatiente

Comision Semana Santa. Caceres *see*
- Semana santa cacerena, 1958
- Semana santa cacerena, 1959
- Semana santa cacerena 1961
- Semana santa cacerena 1976
- Semana santa cacerena 1978
- Semana santa. caceres, 1963
- Semana santa caceres 1977
- Semana santa de plasencia. marzo 1959

Comision Tecnica De Demarcacion De La Frontera Ent... *see* Informe detallado de la comision...

Comisiones especiales para estudiar con caracte in... / Colombia Congreso Camara De Prepresentantes – Bogota, Colombia. v1-2. 1943 – 1r – us UF Libraries [972]

Comiso columns – Comiso, Sicily. 1984 aug 3-1986 oct 10, 1986 oct 17-1988 jul – 2r – 1 – mf#1508434 – us WHS [071]

Comite Central des Artistes *see* La voix des artistes

Comite Cubano Pro Libertad De Patriotas Puertorriq *see* Por la independencia de puerto rico, por la libert...

Comite d'Action Antifasciste et de Vigilance *see* Vigilance

Comite De Estudiantes Universitarios Anticomunista *see* Calvario de guatemala

Comite de Homenaje a "El Diario Israelita", Buenos Aires *see* Antologie fun der yidisher literatur in argentina

Comite de la Troisieme Internationale *see* Bulletin communiste

Comite de l'Afrique francaise *see* Bulletin

Comite de l'Asie francaise *see* Asie francaise

Comite de recrutement canadien-francais (Montreal, Quebec) *see* Album de la grande guerre

Comite de Vigilance des Intellectuels Antifascistes *see* La presse de franco

Comite der Chicago Schiller Gedenkfeier, Mai 1905 *see* Zur wuerdigung schiller's in amerika

Comite d'Etudes Berberes de Rabat *see* Les archives berberes

Comite d'Etudes Historiques et Scientifiques de l'Afrique Occidentale Francaise *see* Bulletin

Comite Ejecutivo Pro-Celebracion Del 4 Centenario *see* San salvador

Comite electoral des medecins : la prochaine election du bureau provincial et la circulaire beausoleil – S.l: s.n, 1898? – 1mf – 9 – mf#54403 – cn Canadiana [610]

Comite executif canadien de l'Exposition universelle a Paris (1855) *see*
- Canada at the universal exhibition of 1855
- Le canada et l'exposition universelle de 1855

Comite, Farel *see* Guillaume farel, 1489-1565

Comite federal-provincial sur la publicite destinee aux enfants (Canada) *see* Les effets de la loi quebecoise interdisant la publicite destinee aux enfants rapport

Comite Franco-Espagnole *see* Durango, ville martyre; ce que furent les bombardements de la ville de durango par les avions allemands

Comite intellectuels pour le soutien du Gouvernement de sauvetage national *see* Manifeste du comite des intellectuels pour le soutien du gouvernement de sauvetage national

Comite interdiocesain de musique sacree (Quebec) *see* Code de musique sacree et liste de pieces recommandees

Comite kota pki surabaja / Madjalah PKI – Surabaja, 1962. 1-6 – 4mf – 9 – (missing: 1962(1)) – mf#SE-383 – ne IDC [950]

El Comite – Movimiento de Izquierda Nacional Puertorriqueno *see* Obreros en marcha

Comite National Du Kivu *see* Vingt ans d'activite en matiere de colonisation europeenne

Comite national du kivu, 1928-1953 – Bruxelles, Belgium. 1953? – 1r – us UF Libraries [960]

Comite Regional de l'Oranie du Parti Communiste *see* L'enchaine

Comite Revolucionario de Mocambique *see* Newsletter

Comite sectoriel d'adaptation de la main-d'oeuvre industrie du meuble et des articles d'ameublement : bilan situationnel / Turcotte, Andre – [Montreal]: le Comite, 1990 [mf ed 1992] – 4mf – 9 – mf#SEM105P1622 – cn Bibl Nat [680]

Comite Tecnico de Ayuda a los Espanoles en Mexico *see*
- Cultural creations.
- Spanish professors and artists in the emigration

Comm net / Naval Electronic Systems Engineering Center, Portsmouth [US] – v4 n2,6,8 [1990 mar, aug/sep, dec], v5 n2-4, [1991 feb-jul], v6 n2 [1992 apr] – 1r – 1 – mf#1789394 – us WHS [623]

Command magazine – Arlington. 1978-1979 (1,5,9) – (cont by: command policy) – ISSN: 0198-7313 – mf#11637 – us UMI ProQuest [355]

Command policy – Arlington. 1979-1979 (1,5,9) – (cont: command magazine. cont by: defense) – ISSN: 0270-9015 – mf#11637,01 – us UMI ProQuest [355]

Command post – Clacutta. 1944 jul 14 – 1r – 1 – mf#2892888 – us WHS [071]

Command post – O'Fallon, Scott Air Force Base IL. 1981 may/1982 mar-1989 mar/jun – 14r – 1 – (with small gaps; cont: broadcaster [scott air force base il]) – mf#627544 – us WHS [071]

Le commandant marchand et ses compagnons d'armes a travers l'afrique : histoire complete et anecdotique de la mission – Paris: E Geffroy, [1892] – 1 – us CRL [960]

Commandant's bulletin – 1982 oct 11-1983 oct 28, 1983 nov 11-1984 dec 21, 1985 jan 4-1986 jun 20, 1986 jul-1987 dec, 1988 jan-jul – 5r – 1 – (cont: bulletin [united states. coast guard: 1988]) – mf#717291 – us WHS [355]

Commander in chief, u.s. fleet, battle experiences, dec. 1941-aug. 1945 / U.S. Navy – 1989 – 2r – 5 – $260.00 – (with printed guide) – mf#S3178 – us Scholarly Res [355]

Commander kentucky volunteers, 1794 / Scott, Charles – 1r – 1 – mf#B26328 – us Ohio Hist [355]

Commanders digest – Washington. 1965-1978 (1) 1972-1978 (5) 1975-1978 (9) – ISSN: 0010-2482 – mf#6389 – us UMI ProQuest [355]

Commandment of god made of none effect by the traditions of men / Gilbert, Ashurst Turner – London, England. 1845 – 1r – us UF Libraries [240]

The commandments considered as instruments of national reformation / Maurice, Frederick Denison – London: Macmillan, 1866 – 1mf – 9 – 0-7905-9030-1 – mf#1989-2255 – us ATLA [242]

Commando / Hurlburt Field [FL]. United States – Fort Walton Beach FL. 1981 may-1984, 1985-1986 sep, 1986 oct-1987, 1988-1989 jun, 1992 jan 10-1993 dec 17 – 5r – 1 – mf#1042975 – us WHS [355]

Commando / Reitz, Deneys – London, England. 1929 – 1r – us UF Libraries [960]

Commando / Reitz, Deneys – New York, NY. 1930 – 1r – us UF Libraries [960]

Commando ses sor / Bouvier, Roger – Paris: La Pensee universelle c1975 [mf ed 1989] – 1r with other items – 1 – (recollections of a french officer of khmer troops during world war 2 & the indochinese war) – mf#mf-10289 seam reel 021/03 [§] – us CRL [934]

Comme il vous plaira – Mensuel d'art et de litterature. no. 1-2. Bruxelles. oct-nov 1897 – 1 – fr ACRPP [800]

Comme les autres / Tery, Simone – Paris, France. 1932 – 1r – us UF Libraries [025]

Commemorating the 100th anniversary of joseph pulitzer – St Louis, MO. 1947 – 1r – us UF Libraries [420]

Commemoration sermon preached in the chapel of trinity college / Whewell, William – Cambridge, England. 1828 – 1r – us UF Libraries [240]

Commemorative discourses : preached in the beneficent congregational church, providence, r.i., october 18, 1868 / Vose, James Gardiner – Providence: Beneficent Congregational Church, 1869 – 1mf – 9 – 0-524-08604-4 – mf#1993-3189 – us ATLA [242]

The commemorative services of the first baptist church of boston, massachusetts (250th anniversary) / Wells, Edwin P – 1915 – 1 – 9.17 – us Southern Baptist [242]

Commemorazione del centenario della nascita di simone corleo / Merenda, Pietro – Palermo: Boccone del Povero, 1925-1926 – 1mf – 9 – 0-524-08123-9 – mf#1993-9029 – us ATLA [100]

Comment – Johannesburg: Broadcasting Centre, [1981- – (issues for mar 3 1981-jun 21 1982 filmed with: current affairs (south african broadcasting corporation), may 31 1978-mar 2 1981) – us CRL [070]

Comment : a new zealand quarterly review – 1959-aug 1964; oct 1964-nov 1970 – 2r – mf#ZB 21 – nz Nat Libr [079]

Comm/ent: a journal of communications and entertainment *see* Hastings communications and entertainment law journal (comm/ent)

Comment and review / Community Training and Development, Inc – 1972 jan-1973 win – 1r – 1 – mf#1054664 – us WHS [360]

Comm/ent: hastings journal of communications and entertainment law *see* Hastings communications and entertainment law journal (comm/ent)

Comment je conjcois une constitution d'haiti / Doret, Frederic – Port-Au-Prince, Haiti. 1916 – 1r – us UF Libraries [323]

Comment je suis entre dans la francmaconnerie et comment j'en suis sorti / Copin-Albancelli, Paul – Paris: Perrin et Cie 1912 – 1mf – 9 – mf#vrl-129 – ne IDC [366]

Comment la province de quebec s'appauvrit : etude politique / Vallee, Roch Pamphile – Quebec?: L Brousseau, 1876 – 1mf – 9 – mf#24135 – cn Canadiana [336]

Comment l'eglise romaine n'est plus l'eglise catholique / Michaud, Eugene – Paris: Sandoz & Fischbacher, 1872 [mf ed 1986] – 1mf – 9 – 0-8370-9014-8 – (in french. incl bibl ref) – mf#1986-3014 – us ATLA [241]

Comment [nairobi, kenya] – Nairobi: East Africa News Review Ltd 1949-56 [wkly] [mf ed [Nairobi]: Kenya National Archives Photographic Service 1970] – n1-342 [1949 sep 17-1956 29th mar] [gaps] on 9r – 1 – (publ: nairobi: comment -1956; n326-342 [1955 8th dec-1956 29th mar] on reel with: new comment [n1-18 [1956 6th apr-3rd aug]]; and other items) – mf#mf-1550 camp r30-38 – us CRL [079]

Comment on the arts / WMTV [Television station : Madison WI] – 1979 jan 12-1982 apr 16 – 1r – 1 – mf#636806 – us WHS [700]

Comment se sont formes les evangiles : la question synoptique, l'evangile de saint jean / Calmes, Th – Paris: Bloud, 1904 – 1mf – 9 – 0-7905-0915-6 – (incl bibl ref) – mf#1987-0915 – us ATLA [220]

Comment vivre longtemps / Fisher, Irving – [Ottawa?]: Metropolitan Life Insurance Co, 1916 [mf ed 1998] – 1mf – 9 – 0-665-65368-9 – (also available in english 65368) – mf#65368 – cn Canadiana [613]

Commentaar op de brieven van paulus aan de thessalonikers, efeziers, kolossers in aan filemon / Baljon, Johannes Marinus Simon – Utrecht: J van Boekhoven, 1907 – 1mf – 9 – 0-8370-2168-5 – mf#1985-0168 – us ATLA [227]

Commentaar op de katholieke brieven / Baljon, Johannes Marinus Simon – Utrecht: J Van Boekhoven, 1904 – 1mf – 9 – 0-524-05653-6 – mf#1992-0503 – us ATLA [227]

Commentaar op de openbaring van johannes / Baljon, Johannes Marinus Simon – Utrecht: J van Boekhoven, 1908 – 1mf – 9 – 0-524-05788-5 – mf#1992-0615 – us ATLA [220]

Le commentaire d'origene sur rom 3, 5-v, 7 : d'apres les extraits du papyrus no 88748 du musee du caire / Origenes (Origen); ed by Scherer, Jean – Le Caire, 1957 – 10mf – 9 – €19.00 – ne Slangenburg [240]

Commentaire historique sur le poeme d'ibn-abdoen / Ibn-Badroun; ed by Dozy, R – Leyde, 1846 – €17.00 – ne Slangenburg [470]

Commentaire litteral, historique et moral : sur la regle de saint benoit / Calmet, Aug – Paris. v1-2. 1734 – €75.00 – ne Slangenburg [241]

Commentaire sur la genese / Mestral, Armand de – Lausanne: Georges Bridel, 1863 – 1mf – 9 – 0-7905-1467-2 – mf#1987-1467 – us ATLA [221]

Commentaire sur la regle de s benoit / Mege, Joseph – Paris, 1687 – 22mf – 8 – €42.00 – ne Slangenburg [241]

Commentaire sur le yacna, l'un des livres liturgiques des parses : ouvrage contenant le texte zend explique pour la premiere fois le variantes des quatre manuscrits de la bibliotheque royale et la version sancrite inedite de neriosengh / Burnouf, Eugene – Paris: Impr royale 1833 – (= ser Whsb) – 10mf – 9 – €90.00 – mf#Hu 224 – gw Fischer [410]

Commentaire sur l'evangile de saint jean *see* Commentary on the gospel of john

Commentaire sur l'evangile de saint luc *see* Commentary on the gospel of st luke

Commentaire sur l'exode / Mestral, Armand de – Lausanne: Georges Bridel, 1864 – 1mf – 9 – 0-7905-1362-5 – mf#1987-1362 – us ATLA [220]

Commentaires... / Montluc, B – Paris, [1821]. 7v – 16mf – 9 – mf#OA-194 – ne IDC [720]

Les commentaires de cesar / Caesar, G J – Amsterdam, 1678 – 7mf – 9 – mf#OA-186 – ne IDC [720]

Commentaires de m jean calvin, sur les cinq livres de moyse... / Calvin, Jean – Geneve: Imprime par François Estienne, 1564 – 20mf – 9 – mf#CL-70 – ne IDC [242]

Commentaires hieroglyphiques : ou images des choses de ian pierius valerian... / Valeriano Bolzani, G P; ed by Chappuys Tourangeau, G – Lyons: Par Barthelemy Honorat, 1576 – 22mf – 9 – mf#O-53 – ne IDC [090]

Commentaires memorables de don bernardin de mendoce / Mendoza, B – Paris, 1591 – 9mf – 9 – mf#OA-158 – ne IDC [720]

Commentaires, o- sont decrits tous les combats, rencontres, escarmouches, batailles...prises...de villes [et] places fortes... / Montluc, B – Paris, 1746. 4v – 21mf – 9 – mf#OA-275 – ne IDC [720]

Commentaires sur les lois du bas-canada : ou conferences de l'ecole de droit liee au college des rr pp jesuites: suivis d'une notice historique / Bibaud, Maximilien – Montreal: Cerat & Bourguignon. 2v. 1859 [mf ed 1985] – 2v on 1mf – 9 – mf#46876 – cn Canadiana [340]

Commentar ueber das allgemeine buergerliche gesetzbuch : fuer die gesammten deutschen erblaender der oesterreichischen monarchie / Zeiller, Franz von – Wien: Geistinger. v1-4+index vol. 1811-13 – (= ser Civil law 3 coll) – 31mf – 9 – (incl bibl ref and index) – mf#LLMC 96-621 – us LLMC [346]

Commentar ueber das avesta / Spiegel, Friedrich – Wien: K K Hof- und Staatsdruckerei 1864-68 [mf ed 1992] – 2v on 3mf – 9 – 0-524-04655-7 – mf#1990-3398 – us ATLA [290]

Commentar ueber das buch "esther" mit seinen "zusaetzen" und ueber "susanna" / Scholz, Anton – Wuerzburg: Leo Woerl, 1892 – 1mf – 9 – 0-7905-3412-6 – mf#1987-3412 – us ATLA [221]

Commentar ueber das buch judith und ueber bel und drache / Scholz, Anton – Wuerzburg: Leo Woerl, 1896 – 1mf – 9 – 0-8370-5140-1 – (incl app) – mf#1985-3140 – us ATLA [221]

Commentar ueber das evangelium des heiligen marcus / Schanz, Paul – Freiburg im Breisgau; St Louis, MO: Herder, 1881 – 2mf – 9 – 0-7905-2058-3 – (incl bibl ref and index) – mf#1987-2058 – us ATLA [226]

Commentar ueber den brief an die hebraeer / Keil, Carl Friedrich – Leipzig: Doerffling und Franke, 1885 – 1mf – 9 – 0-8370-3865-0 – mf#1985-1865 – us ATLA [227]

Commentar ueber den brief pauli an die ephesier / Harless, Gottlieb Christoph Adolf von – 2. unveraend aufl. Stuttgart: S G Liesching, 1858 – 2mf – 9 – 0-524-04402-3 – (incl bibl ref) – mf#1992-0095 – us ATLA [227]

Commentar ueber den brief pauli an die galater : mit besonderer ruecksicht auf die lehre und geschichte des apostels / Wieseler, Karl – Goett7 – 2mf – 9 – 0-8370-9671-5 – (incl indes) – mf#1986-3671 – us ATLA [220]

Commentar ueber den brief pauli an die roemer / Stoeckhardt, G – St Louis, MO: Concordia Pub. House, 1907 – 2mf – 9 – 0-7905-2152-0 – mf#1987-2152 – us ATLA [227]

Commentar ueber den ersten brief pauli an die korinther / Maier, Adalbert – Freiburg i.B: Friedrich Wagner, 1857 – 1mf – 9 – 0-8370-9638-3 – (incl bibl ref) – mf#1986-3638 – us ATLA [227]

Commentar ueber den ersten brief pauli an die korinthier / Osiander, Johann Ernst – Stuttgart: Chr Belser, 1847 – 2mf – 9 – 0-8370-9644-8 – (incl bibl ref) – mf#1986-3644 – us ATLA [227]

Commentar ueber den propheten jesaia / Stoeckhardt, G – St Louis, MO: Concordia Pub. House, 1902 – 1mf – 9 – 0-7905-2153-9 – mf#1987-2153 – us ATLA [221]

Commentar ueber die apostelgeschichte des lukas / Nosgen, C F – Leipzig, 1882 – 9mf – 8 – €18.00 – ne Slangenburg [226]

Commentar ueber die briefe des petrus und judas / Keil, Carl Friedrich – Leipzig: Doerffling & Franke, 1883 – 1mf – 9 – 0-8370-3866-9 – (incl bibl ref) – mf#1985-1866 – us ATLA [227]

Commentar ueber die buecher der makkaber / Keil, Carl Friedrich – Leipzig: Doerffling & Franke, 1875 [mf ed 1989] – (= ser Supplement zu dem biblischen commentare ueber das alte testament) – 1mf – 9 – 0-7905-1272-6 – (incl bibl ref) – mf#1987-1272 – us ATLA [221]

Commentar ueber die psalmen / De Wette, Wilhelm Martin Leberecht – Heidelberg: Mohr und Zimmer, 1811 – 5mf – 9 – 0-8370-1831-5 – (incl bibl ref) – mf#1987-6219 – us ATLA [220]

Commentar zu den briefen des paulus an die corinther *see* Commentary on the epistles of paul to the corinthians

Commentar zu der weissagung des propheten obadja / Johannes, Adolf – Wuerzburg: Thein, 1885 – 1mf – 9 – 0-8370-3784-0 – mf#1985-1784 – us ATLA [221]

Commentar zu kants kritik der reinen vernunft / ed by Vaihinger, Hans – Stuttgart: W Spemann, 1881-1892 – 3mf – 9 – 0-524-00414-5 – (incl bibl ref) – mf#1989-3114 – us ATLA [120]

Commentar zum briefe an die hebraeer : mit archaeologischen und dogmatischen excursen ueber das opfer und die versoehnung / Delitzsch, Franz – Leipzig: Doerffling & Franke, 1857 – 2mf – 9 – 0-8370-9612-X – mf#1986-3612 – us ATLA [227]

Commentar zum briefe an die hebraer *see* Commentary on the epistle to the hebrews

Commentar zum buche des propheten hoseas / Scholz, Anton – Wuerzburg, Leo Woerl, 1882 – 1mf – 9 – 0-8370-5141-X – mf#1985-3141 – us ATLA [221]

Commentar zum buche des propheten joel / Scholz, Anton – Wuerzburg:Leo Woerl, 1885 – 1mf – 9 – 0-8370-5142-8 – mf#1985-3142 – us ATLA [221]

Commentar zum buche tobias / Scholz, Anton – Wuerzburg: Leo Woerl, 1889 – 1mf – 9 – 0-7905-3225-5 – mf#1987-3225 – us ATLA [221]

Commentar zum oesterreichischen allgemeinen buergerlichen gesetzbuche / Pfaff, Leopold & Hofmann, Franz – Wien, Manz. 2v in 1. 1877- – (= ser Civil law 3 coll) – 13mf – 9 – (issued in pts, with v1 incomplete, and no more publ. incl bibl ref) – mf#LLMC 96-616 – us LLMC [346]

Commentar zur rigveda-uebersetzung / Ludwig, Alfred – Prag: F Tempsky, 1881-1883 – 3mf – 9 – 0-524-07610-3 – mf#1991-0136 – us ATLA [280]

Commentari in prophetas. jeremian ezechielem / Maldonado, Juan – Turnoni: Horatij Candon, 1611 – 1 – sp Bibl Santa Ana [240]

Commentaria bibliorvm / Pellican, C – Zuerich, Christoph Froschauer, 1532-1535. 5 v – 54mf – 9 – mf#PBU-471 – ne IDC [240]

Commentaria in 1. p. summae theologicae s. thomae aquinatis, o.p., a q. 1. ad q. 23 (de deo uno) / Buonpensiere, Enrico – Romae: F Pustet, 1902 – 3mf – 9 – 0-524-07514-X – mf#1991-3144 – us ATLA [210]

Commentaria in aristotelem graeca – Berolini. v1-23. 1882-1909 – 1 – $360.00 – mf#0164 – us Brook [180]

Commentaria in concordiam et historiam evangelicam / Barradas, S – Antverpiae. v1-4. 1622 – 4v on 94mf – 8 – €179.00 – ne Slangenburg [220]

Commentaria in decretales (siecle 14) – Barcelona – 4r – 5,6 – sp Cultura [240]

Commentaria in duodecim prophetas / Arias Montano, Benito – 1583 – 9 – sp Bibl Santa Ana [240]

Commentaria in hossean prophetam / Guadelupe, Andres – 1581 – 9 – sp Bibl Santa Ana [240]

Commentaria in isiae prophetae / Arias Montano, Benito – 1599 – 9 – sp Bibl Santa Ana [240]

Commentaria in ruth : formae tplila 57 – 1990 – (= ser ILL – ser a; Cccm 81) – 9mf+82p – 9 – €50.00 – 2-503-63812-0 – be Brepols [400]

Commentaria in ruth. tractatus de tabernaculo : formae tplila 11 / Cellensis, Petrus – 1983 – (= ser ILL – ser a; Cccm 54) – 8mf+85p – 9 – €30.00 – 2-503-60542-7 – be Brepols [400]

Commentaria in ruth. tractatus de tabernaculo : lemmata tplilb 11 / Petrus Cellensis – 1987 – (= ser ILL – ser b; Cccm 54) – 6mf+42p – 9 – €30.00 – 2-503-73542-8 – be Brepols [400]

Commentaria in scripturam sacram = The great commentary of cornelius a lapide / Lapide, Cornelius a – 3rd ed. London: John Hodges. 6v. 1891-96 – 9mf – 9 – 0-8370-6992-0 – mf#1986-0992 – us ATLA [220]

Commentaria, libri 13 = Commentaries / Aeneas Sylvius Piccolomini [Pius 2, Pope] – 18th c – (= ser Holkham library manuscript books 482) – 1r – 1 – mf#97256 – uk Microform Academic [920]

Commentaria quaedam in cantica canticorum / Alfonso de Orozco, Beato – 1581 – 9 – sp Bibl Santa Ana [780]

Commentaria symbolica in duos tomos distributa... / Ricciardi, A – Venetiis: Apud Franciscum de Franciscis Senensem, 1591 – 16mf – 9 – mf#O-2017 – ne IDC [090]

Commentaria una cum quaestionibus, in universam aristotelis logicam / Toletus, F – Coloniae Agrippinae, 1583 – 5mf – 9 – mf#CA-39 – ne IDC [180]

Commentaries on american law / Kent, James – 4th ed. New York, 1840. 4 v. LL-861 – 1 – (12th ed. boston, little, brown, 1873. 4 v. II-1665. new and thoroughly rev. ed. philadelphia, blackstone, 1889. 4 v. II-775.. new and thoroughly revised ed., by william m. lacy. new york, banks, 1891-92. 4 v. II-852. 12th ed. ed. by o. w. holmes, jr. 14th ed. ed. by john m. gould. boston, little, brown, 1896. 4 v. II-903) – us L of C Photodup [340]

Commentaries on equity jurisprudence, as administered in england and america. 11th ed / Story, Joseph – Boston, Little, Brown, 1873. 2 v. LL-1011 – 1 – (12th ed. boston, little, brown, 1877. 2v. II-1068. 1) – us L of C Photodup [342]

Commentaries on equity pleadings, and the incidents thereof, according to the practice of the courts of equity, of england and america. 6th ed / Story, Joseph – Boston, Little, Brown, 1857. 871 p. LL-1114 – 1 – (8th ed. boston, little, brown, 1870. 816 p. II-532. 1. 10th ed. boston, little, brown, 1892. 823 p. II-1640 1) – us L of C Photodup [347]

Commentaries on the conflict of laws, foreign and domestic, in regard to contracts, rights, and remedies, and especially in regard to marriages, divorce, wills, successions, and judgments. 8th ed / Story, Joseph – Boston Little, Brown, 1883. 901 p. LL-1013 – 1 – us L of C Photodup [346]

Commentaries on the constitution of the united states / Story, Joseph – 4th ed. Boston: Little, Brown and Co. 2v. 1873 – 17mf – 9 – $25.50 – mf#LLMC 90-367 – us LLMC [323]

Commentaries on the history, constitution, and chartered franchises of the city of london / Norton, George – 3rd rev ed. London: Longmans, Green & Co 1869 [mf ed 1986] – 1r – 1 – (filmed with: the sermons of the right rev jeremy taylor) – mf#1710 – us Wisconsin U Libr [941]

Commentaries on the jurisdiction of courts / Brown, Timothy – Chicago, Callaghan, 1891. 624 p. LL-1465 – 1 – us L of C Photodup [347]

Commentaries on the law of agency as a branch of commercial and maritime jurisprudence...4th ed / Story, Joseph – Boston, Little and Brown, 1851. 719 p. LL-1151 – 1 – (7th ed., boston, little, brown, 1869. 674 p. II-1156. 1. 8th ed., boston, little, brown, 1874. 706 p. II-1143. 1) – us L of C Photodup [346]

Commentaries on the law of bailments, with illustrations from the civil and foreign law. 8th ed / Story, Joseph – Boston, Little, Brown, 1870. 653 p. LL-1413 – 1 – us L of C Photodup [340]

Commentaries on the law of municipal corporations / Dillon, John Forrest – 4th ed. Boston, Little, Brown, 1890. 2 v. LL-803 – 1 – us L of C Photodup [346]

Commentaries on the law of negligence in all relations. / Thompson, Seymour Dwight – Indianapolis, Bowen-Merrill, 1901-05. 6 v. LL-1250 – 1 – (a supplement. being volume vii of the series. indianapolis, 1907. 1148 p. a supplement. being volume viii of the series. indianapolis, 1914. 1192 p) – us L of C Photodup [340]

Commentaries on the law of receivers. / Beach, Charles Fisk – New York, Strouse, 1888. 796 p. LL-899 – 1 – us L of C Photodup [340]

Commentaries on the laws of england : (american editions of the commentaries) – (= ser The yale law library blackstone coll) – 814mf – 9 – mf#LLMC 82-800 titles 80-146 – us LLMC [343]

Commentaries on the laws of england / Blackstone, William – 8th ed. Oxford: Clarendon Press. v1-4. 1778 – 16mf – 9 – $24.00 – (this ed was missing from the yale law library blackstone collection) – mf#LLMC 90-463 – us LLMC [340]

Commentaries on the laws of england : (english and irish editions of the commentaries) – 1st-23rd ed + new ed – (= ser The yale law library blackstone coll) – ca 774mf – 9 – mf#LLMC 82-800 titles 1-45 – us LLMC [343]

Commentaries on the laws of england : (foreign editions, abridgements and extracts): french – (= ser The yale law library blackstone coll) – 62mf – 9 – mf#LLMC 82-800 titles 179-182 – us LLMC [343]

Commentaries on the laws of england : (foreign editions, abridgements and extracts): german – (= ser The yale law library blackstone coll) – 10mf – 9 – mf#LLMC 82-800 Title 183 – us LLMC [342]

Commentaries on the laws of england : (foreign editions, abridgements and extracts): italian – (= ser The yale law library blackstone coll) – 6mf – 9 – mf#LLMC 82-800 Title 184 – us LLMC [342]

Commentaries on the laws of england : (works founded on blackstone's commentaries) – (= ser The yale law library blackstone coll) – 395mf – 9 – mf#LLMC 82-800 titles 198-216 – us LLMC [343]

Commentaries on the laws of england applicable to real property / Blackstone, William – Toronto: W C Chewett, 1864 [mf ed 1985] – 5mf – 9 – 0-665-30007-7 – (incl ind) – mf#30007 – cn Canadiana [346]

Commentaries on the laws of england; in four books / Blackstone, William – 4th ed. Chicago, Callaghan, 1899. 2 v. LL-840 – 1 – us L of C Photodup [340]

Commentaries on the laws of moses = Mosaisches recht / Michaelis, Johann David – London: Rivington 1814 [mf ed 1987] – 1r – 1 – (trans fr german by alexander smith) – mf#2080 – us Wisconsin U Libr [221]

Commentaries on the modern law of municipal corporations...being a rev., re-written and enl. ed. of beach on public corporations. / Smith, John Wilson – Indianapolis: Bowen-Merrill, 1903. 2v. LL-1304 – 1 – us L of C Photodup [346]

Commentaries on the the law of marriage and divorce. / Bishop, Joel Prentiss – 6th ed. Boston, Little, Brown, 1881. 2 v. LL-902 – 1 – us L of C Photodup [346]

Commentaries on the written laws and their interpretation / Bishop, Joel Prentiss – Boston, Little, Brown, 1882. 354 p. LL-378 – 1 – us L of C Photodup [340]

Commentaries upon topics demanding them – Halifax, N.S: W S Hall, [1878-18–?] – 9 – mf#P04965 – cn Canadiana [071]

Commentarii academiae scientiarum imperialis petropolitanae – Petropoli, 1726-1746. v1-14 – 157mf – 9 – mf#R-5814 – ne IDC [077]

Commentarii cistercienses – 10(1959)-38(1987) – 9 – €309.00 – ne Slangenburg [241]

Commentarii collegii conimbricensis e societatis iesu in universam dialecticam aristotelis / Conimbricenses. – Lugduni, 1605. 2v – 14mf – 9 – mf#CA-11 – ne IDC [240]

Commentarii collegii conimbricensis societatis iesu, in duos libros de generatione et corruptione aristotelis – Lugduni, 1606 – 6mf – 9 – mf#CA-12 – ne IDC [180]

Commentarii collegii conimbriencis sociatatis iesu in aristotelis logicam – Venetiis, 1607. 2v – 7mf – 9 – mf#CA-180 – ne IDC [180]

Commentarii de causis excaecationis multorum saeculorum / Lambert, F – [Strasbourg, 1524] – 2mf – 9 – mf#PPE-111 – ne IDC [240]

Commentarii de origine et progressu legum iuriumque germanicorum / ed by Biener, Christian Gottlob – Lipsiae, Apud: G E Beer. 3v in 2. 1787-95 – (= ser Civil law 3 coll) – 13mf – 9 – (incl bibl ref) – mf#LLMC 96-628 – us LLMC [340]

Commentarii de prophetia : deque litera et spiritu / Lambert, F – Strasbourg, 1526 – 4mf – 9 – mf#PPE-119 – ne IDC [240]

Commentarii de rebus assyro-babylonicis, arabicis, aegyptiacis etc editi a pontificio instituto biblico – Roma, S.1, 1920-1930, v1-55; n.s., 1932-1936, v1-5 – 104mf – 9 – (missing: 1927 v25; 1930 v50) – mf#NE-20062c – ne IDC [956]

Commentarii de rebus byzantinis (cbh3,4) / Nicephori Caesaris Bryennii; ed by Possinus, P – Parisiis, 1661 – (= ser Corpus byzantinae historiae (cbh)) – €147.00 – ne Slangenburg [240]

Commentarii de religione revelata ejusque fontibus ac de ecclesia christi / MacGuinness, Joannes – Parisiis: Des Irlandais; Turonibus: Bousrez, 1900 [mf ed 1986] – 1mf – 9 – 0-8370-6915-7 – (incl bibl ref) – mf#1986-0915 – us ATLA [241]

Commentarii et annotationes in epistolam b pauli apostoli ad romanos / Toledo, F – Romae, 1602 – 10mf – 9 – mf#CA-71 – ne IDC [241]

Commentarii hieronymi comitis alexandrini de acerrimo, ac omnium difficillimo turcarum bello, in insulam melitam gesto, anno 1565 / Conti, N – Venetiis, 1566 – 2mf – 9 – mf#H-8304 – ne IDC [956]

Commentarii illustres...in quinque mosaicos libros / Cajetan, Tommaso de Vio Gaetani – Parisiis, 1539 – 24mf – 8 – €46.00 – ne Slangenburg [241]

Commentarii in epistolas catholicas / Lefevre (D'Etaples), J – Bale, 1527 – 3mf – 9 – mf#PRS-159 – ne IDC [241]

Commentarii in evangelicam historiam, et in acta apostolorum... / ed by Salmeron, A – Coloniae Agrippinae, 1602-1604. 16v – 154mf – 9 – mf#CA-63 – ne IDC [240]

Commentarii in libros aristotelis stagiritae philosophorum principis de anima... / Rubio, A – [Alcala], 1611 – 7mf – 9 – mf#CA-36 – ne IDC [180]

Commentarii in libros aristotelis stagyritae philosophorum principis, de anima... / Rubio, A – Lugduni, 1620 – 9mf – 9 – mf#CA-37 – ne IDC [180]

Commentarii in libros veteris et novi testamenti / Piscator, J – Herbornae, Sigenae, 1591-1622 – 126mf – 9 – mf#PBA-292 – ne IDC [240]

Commentarii in micheam, naum et abacuc / Lambert, F – Strasbourg, 1525 – 4mf – 9 – mf#PPE-115 – ne IDC [240]

Commentarii in octo libros aristotelis de physico auditu, seu auscultatione continentur / Rubio, A – Lugduni, 1620 – 9mf – 9 – (missing: title p) – mf#CA-35 – ne IDC [180]

Commentarii in omnes pauli epist et epist catholicas. de testamento..dei...- de utraque in christo natura / Bullinger, Heinrich – Tiguri, 1539 – 41mf – 8 – €79.00 – ne Slangenburg [227]

Commentarii in omnes pauli epistolas : in epistolam ad hebraeos et in epistolas canonicas / Bullinger, Heinrich – Tiguri, 1603 – 30mf – 8 – €57.00 – ne Slangenburg [227]

Commentarii in prima 12 capita...evangelii secundam lucam / Toledo, F – Romae, 1600 – 15mf – 9 – mf#CA-70 – ne IDC [241]

Commentarii in prophetas minores / Daneau, Lambert – Geneve, Vignon, 1586 – 12mf – 9 – mf#PFA-132 – ne IDC [240]

Commentarii in quattuor evangelistas / Maldonatus, Ioan. – Mussiponti. v1-2. 1596-97 – 49mf – 8 – €94.00 – ne Slangenburg [226]

Commentarii in quatuor ultimos prophetas nempe sophoniam / Lambert, F – Strasbourg, 1526 – 6mf – 9 – mf#PPE-117 – ne IDC [240]

Commentarii in sacrosanctum iesu christi d n : evangelium secundum lucam / Toletus, Franciscus – Parisiis, 1600 – 37mf – 8 – €71.00 – ne Slangenburg [220]

Commentarii in summam theologicam thomae aquinatis / Cajetan, Tommaso de Vio Gaetani – Lier, 1892 – 27mf – 8 – €52.00 – ne Slangenburg [241]

Commentarii initiatorii in quatuor evangelia / Lefevre (D'Etaples), J – n.p, 1523 – 15mf – 9 – mf#PRS-158 – ne IDC [240]

Commentarii juris civilis. / Acevedo, Alfonso de – 1737. v. I-VI – 9 – sp Bibl Santa Ana [346]

Commentarii mathematici helvetici – Basel. 1991-1996 (1) – ISSN: 0010-2571 – mf#13941 – us UMI ProQuest [510]

Die commentarii zu den zwoelf kleinen propheten / Rahmer, Moritz – Berlin: M Poppelauer, 1902 – (= ser Hebraeische Traditionen in den Werken des Hieronymus) – 1mf – 9 – 0-7905-3049-X – (incl bibl ref) – mf#1987-3049 – us ATLA [221]

Commentarii...in omnes d. pauli apostoli epistolas / Hyperius, A – Tijuri, 1582-84 – 8mf – 9 – mf#PBA-208 – ne IDC [240]

Commentario de le cose de tvrchi, di pavlo iovio, vescovo di nocera, a carlo qvinto imperadore avgvsto / Giovio, P – [Venice], 1538 – 1mf – 9 – mf#H-8252 – ne IDC [956]

Commentario de le cose de tvrchi, di pavlo iovio, vescovo di nocere, a carlo qvinto imperadore avgvsto / Giovio, P – Venetia, 1540 – 1mf – 9 – mf#H-8263 – ne IDC [956]

Commentario de le cose de tvrchi, et del s georgio scanderbeg, principe di epyrro / Giovio, P – [Venice], 1539 – 2mf – 9 – mf#H-8257 – ne IDC [956]

Commentario de le cose de tvrchi, et del s georgio scanderbeg, principe di epyrro / Giovio, P – [Venice], 1541 – 2mf – 9 – mf#H-8268 – ne IDC [956]

Commentario...della origine de tvrchi, et imperio della casa ottomanna / Cambini, A – n.p, 1538 – 2mf – 9 – mf#H-8251 – ne IDC [956]

Commentario...della origine de tvrchi, et imperio della casa ottomanna / Cambini, A – [Venice], 1540 – 2mf – 9 – mf#H-8261 – ne IDC [956]

Commentariorum continuatio ad leges regias. / Acevedo, Alfonso de – 1600 – 9 – sp Bibl Santa Ana [340]

Commentariorum de religione christiana / [Ramus, P] – Francforti, 1576 – 5mf – 9 – mf#PRS-168 – ne IDC [240]

Commentariorum et disputationum in genesim / Peperius, Ben – Moguntiae. v1-4. 1612 – 46mf – 8 – €88.00 – ne Slangenburg [221]

Commentariorum in evangelistam ioannem, heptas altera, item tertia et postrema in eundem / Musculus, W – Basilea, Johann Herwagen, 1548 – 9mf – 9 – mf#PBU-335 – ne IDC [242]

Commentariorum in evangelistam ioannem, heptas prima / Musculus, W – Basilea, Bartholomaeus Westhemer, 1545 – 6mf – 9 – mf#PBU-334 – ne IDC [242]

Commentariorum iuris civilis / Acevedo, Alfonso de – 1591 – 9 – sp Bibl Santa Ana [347]

Commentariorum iuris civilis / Acevedo, Alfonso de – 1612. v. I-VI – 9 – sp Bibl Santa Ana [347]

Commentariorum iuris civilis, tomus quintus / Acevedo, Alfonso de – 1596 – 9 – sp Bibl Santa Ana [347]

Commentariorum iuris civilis, tomus secundus / Acevedo, Alfonso de – 1595 – 9 – sp Bibl Santa Ana [340]

Commentariorum iuris civilis, tomus sextus / Acevedo, Alfonso de – 1598 – 9 – sp Bibl Santa Ana [347]

Commentariorum joannis calvini in acta apostolorum : liber 1 ad sereniss. daniae regem / Calvin, J – Genevae: Ex officina Joannis Crispini, 1552 – 4mf – 9 – mf#CL-71 – ne IDC [242]

Commentariorum joannis calvini in acta apostolorum, liber posterior : additus est utriusque libri index rerum et sententiarum / Calvin, J – Genevae: Ex officina Joannis Crispini, 1554 – 3mf – 9 – mf#CL-72 – ne IDC [242]

Commentariorum libri 10 in...evangelium secundum ioannem / Bullinger, Heinrich – Tiguri, 1543 – 20mf – 8 – €38.00 – ne Slangenburg [226]

Commentariorum libri 12 in...evangelium secundum matthaeum / Bullinger, Heinrich – Tiguri, 1542 – 25mf – 8 – €48.00 – ne Slangenburg [240]

Commentariorum memorabilium multiplicis hystoriae tarvisinae locuples promptuarium libris quatuor distributum... / Burchelati, B – Tarvisii: Apud Angelum Righetinum, 1616 – 8mf – 9 – mf#O-1532 – ne IDC [090]

Commentariorum theologicorum... / Gregorius de Valencia – Lugduni, 1603-1609. 4v – 59mf – 9 – mf#CA-76 – ne IDC [240]

Commentariorum theologicorum tomi quatuor / Valentia, Greg de – Inglostadii, 1592-1597 – 178mf – 8 – €339.00 – ne Slangenburg [240]

Commentariorvm in aratvm reliqviae / Maass, Ernst – Berolini, Germany. 1898 – 1r – us UF Libraries [200]

Commentarios in carmina sacra melodorum cosmae hierosolymitani et ioannis damasceni / Prodromus, Theodore [Ptochoprodromus]; ed by Stevenson, H M – Romae, 1888 – €13.00 – ne Slangenburg [240]

Commentarios ineditos a la tercera parte de santo tomae / Banes, dom – Matriti. v1-2. 1951-1953 – 17mf – 8 – €32.00 – ne Slangenburg [241]

Commentarium et disputationum in genesim, tomi quatuor / Pererius, B – Coloniae Agrippinae, 1601. 4v – 39mf – 9 – mf#CA-29 – ne IDC [240]

Commentarium in quartum sententiarum / Soto, D de – [Salamanca], 1581. 2v – 32mf – 9 – mf#CA-73 – ne IDC [240]

Commentarium juris civilis...salmantical, didacuscusio, 1591 / Acevedo, Alfonso de – 1 – sp Bibl Santa Ana [347]

Commentarius captae vrbis, dvctorecarolo borbonio, ad exquisitum modum confectus... / Giovio, P – Parisiis, 1539 – 1mf – 9 – mf#H-8258 – ne IDC [956]

Commentarius criticus in n t : quo loca graviora et difficiliora lectionis dubiae accurate recensentur et explicantur / Reiche, Johann Georg – Gottingae: Vandenhoeck et Ruprecht, 1853-1862. Chicago: Dep of Photodup, U of Chicago Lib, 1975 (1r); Evanston: American Theol Lib Assoc, 1984 (1r) – 1 – 0-8370-0021-1 – mf#1984-B405 – us ATLA [225]

Commentarius de praerogativis beati petri, apostolorum principis / Passaglia, Carlo – Ratisbonae: J Manz, 1850 – 2mf – 9 – 0-7905-8716-5 – (incl bibl ref) – mf#1989-1941 – us ATLA [240]

Commentarius de sacris ecclesiae ordinationibus / Morin, J – Parisiis, 1686 – 37mf – 8 – €71.00 – ne Slangenburg [240]

Commentarius exegetico-philologicus in hebraismos novi testamenti, seu, de dictione hebraica novi testamenti graeci / Schilling, David – Mechliniae [Malines]: H Dessain, 1886 – 1mf – 9 – 0-524-06860-7 – mf#1992-1002 – us ATLA [225]

Commentarius in actus apostolorum / Camerlynck, Achille – ed 6. Brugis: Car Beyaert, 1910 [mf ed 1986] – (= ser Commentarii brugenses in s scripturam) – 2mf – 9 – 0-8370-7131-3 – (incl ind) – mf#1986-1131 – us ATLA [226]

Commentarius in apocalypsin : formae tplila 26 / Primasius – 1985 – (= ser ILL – ser a; Cccm 92) – 7mf+70p – 9 – €40.00 – 2-503-60922-8 – be Brepols [400]

Commentarius in catechesin paltino-belgicam / Lubbertus, S – Franicae, 1618 – 10mf – 9 – mf#PBA-238 – ne IDC [240]

Commentarius in danielem prophetam, lamentationes et baruch / Knabenbauer, Joseph – Parisiis: P Lethielleux 1891 [mf ed 1991] – (= ser Cursus scripturae sacrae 3/4) – 6mf – 9 – 0-8370-1936-2 – mf#1987-6323 – us ATLA [221]

Commentarius in deuteronomium / Hummelauer, Franz von – Parisiis: P Lethielleux 1901 [mf ed 1991] – (= ser Cursus scripturae sacrae 1/3/2) – 6mf – 9 – 0-8370-1930-3 – (incl bibl ref) – mf#1987-6317 – us ATLA [221]

Commentarius in duos libros machabaeorum / Knabenbauer, Joseph – Parisiis: P Lethielleux 1907 [mf ed 1991] – (= ser Cursus scripturae sacrae 1/11) – 5mf – 9 – 0-8370-1937-0 – (incl bibl ref; in latin & greek; comm in latin) – mf#1987-6324 – us ATLA [221]

Commentarius in ecclesiasten et canticum canticorum / Gietmann, Gerhard – Parisiis: P Lethielleux 1890 [mf ed 1991] – (= ser Cursus scripturae sacrae 2/[4?]) – 6mf – 9 – 0-8370-1929-X – mf#1987-6316 – us ATLA [221]

Commentarius in ecclesiasticum : cum appendice, textus "ecclesiastici" hebraeus, descriptus secundum fragmenta nuper reperta, cum notis et versione litterali latina / Knabenbauer, Joseph – Parisiis: P Lethielleux 1902 [mf ed 1991] – (= ser Cursus scripturae sacrae 2/6) – 6mf – 9 – 0-8370-1938-9 – (incl bibl ref) – mf#1987-6325 – us ATLA [221]

Commentarius in epistolam b pauli apostoli ad hebraeos / Paenek, Joannes – Oeniponte [Innsbruck]: Libraria Academica Wagneriana, 1882 – 1mf – 9 – 0-8370-4662-9 – (incl bibl ref) – mf#1985-2662 – us ATLA [227]

Commentarius in epistolas ad thessalonicenses / Voste, Jacques-Marie – Romae: F Ferrari, 1917 – 1mf – 9 – 0-524-05704-4 – (incl bibl ref) – mf#1992-0554 – us ATLA [227]

Commentarius in exodum et leviticum / Hummelauer, Franz von – Parisiis: P Lethielleux 1897 [mf ed 1991] – (= ser Cursus scripturae sacrae 1/2) – 6mf – 9 – 0-8370-1931-1 – (incl bibl ref) – mf#1987-6318 – us ATLA [221]

Commentarius in ezechielem prophetam / Knabenbauer, Joseph – Parisiis: P Lethielleux 1890 [mf ed 1991] – (= ser Cursus scripturae sacrae 3/3) – 6mf – 9 – 0-8370-1939-7 – (incl bibl ref) – mf#1987-6326 – us ATLA [221]

Commentarius in genesim / Hummelauer, Franz von – Parisiis: P Lethielleux 1895 [mf ed 1991] – (= ser Cursus scripturae sacrae 1/1) – 6mf – 9 – 0-8370-1932-X – (incl bibl ref) – mf#1987-6319 – us ATLA [221]

Commentarius in ieremiam prophetam / Knabenbauer, Joseph – Parisiis: P Lethielleux 1889 [mf ed 1991] – (= ser Cursus scripturae sacrae 3/2) – 6mf – 9 – 0-8370-1940-0 – (incl bibl ref) – mf#1987-6327 – us ATLA [221]

Commentarius in libros iudicum et ruth / Hummelauer, Franz von – Parisiis: Sumptibus P Lethielleux 1888 [mf ed 1991] – (= ser Cursus scripturae sacrae 1/4) – 4mf – 9 – 0-8370-1995-8 – mf#1987-6382 – us ATLA [221]

Commentarius in libros samuelis : seu, 1 et 2 regum / Hummelauer, Franz von – Parisiis: P Lethielleux 1886 [mf ed 1991] – (= ser Cursus scripturae sacrae 1/5) – 5mf – 9 – 0-8370-1933-8 – mf#1987-6320 – us ATLA [221]

Commentarius in librum iob / Knabenbauer, Joseph – Parisiis: P Lethielleux 1886 [mf ed 1991] – (= ser Cursus scripturae sacrae 2/1) – 5mf – 9 – 0-8370-1941-9 – mf#1987-6328 – us ATLA [221]

Commentarius in librum iosue / Hummelauer, Franz von – Parisiis: P Lethielleux 1903 [mf ed 1991] – (= ser Cursus scripturae sacrae 2/3/3) – 6mf – 9 – 0-8370-1934-6 – (incl bibl ref) – mf#1987-6321 – us ATLA [221]

Commentarius in librum primum paralipomenon / Hummelauer, Franz von – Parisiis: Sumptibus P Lethielleux 1905 [mf ed 1992] – (= ser Cursus scripturae sacrae 1/7/1) – 1mf – 9 – 0-524-02781-1 – mf#1987-6475 – us ATLA [221]

Commentarius in librum sapientiae / Cornely, Rudolph; ed by Zorell, Franz – Parisiis: P Lethielleux 1910 [mf ed 1991] – (= ser Cursus scripturae sacrae 2/5) – 6mf – 9 – 0-8370-1923-0 – (incl bibl ref; in latin & greek; comm in latin) – mf#1987-6310 – us ATLA [221]

Commentarius in numeros / Hummelauer, Franz von – Parisiis: P Lethielleux 1899 [mf ed 1991] – (= ser Cursus scripturae sacrae 1/3/1) – 4mf – 9 – 0-8370-1935-4 – (incl bibl ref) – mf#1987-6322 – us ATLA [221]

Commentarius in omnes s pauli epistolas : ad usum seminariorum et cleri / Steenkiste, J-A van – 6. ed. Brugis [Brugge]: Apud C Beyaert, 1899 – 3mf – 9 – 0-524-04595-X – mf#1992-0183 – us ATLA [227]

Commentarius in postremos tres prophetas, nempe haggaeum, zachariam & malachiam / Oecolampadius, J – Basileae, Andreas Cratander, 1527 – 3mf – 9 – mf#PBU-374 – ne IDC [240]

Commentarius in primam epistolam ad corinthios / Lapide, Cornelius; ed by Cobb, William Frederick – London: John Hodges, 1896 – 1mf – 9 – 0-7905-0136-8 – mf#1987-0136 – us ATLA [227]

Commentarius in prophetas minores / Knabenbauer, Joseph – Parisiis: Sumptibus P Lethielleux 1886 [mf ed 1992] – (= ser Cursus scripturae sacrae 3/1) – 10mf – 9 – 0-524-03883-X – (latin, greek & hebrew; incl bibl ref) – mf#1987-6496 – us ATLA [221]

Commentarius in proverbia : cum appendice, de arte rhythmica hebraeorum / Knabenbauer, Joseph – Parisiis: P Lethielleux 1910 [mf ed 1991] – (= ser Cursus scripturae sacrae 2/3) – 3mf – 9 – 0-8370-1942-7 – (with app: de arte rhythmica hebraeorum by francisco zorell) – mf#1987-6329 – us ATLA [221]

Commentarius in quatuor s evangelia domini n iesu christi. 2 : evangelium secundum s marcum / Knabenbauer, Joseph – Paaisiis [sic]: P Lethielleux 1894 [mf ed 1991] – (= ser Cursus scripturae sacrae 1/2) – 5mf – 9 – 0-8370-1944-3 – (biblical text in latin & greek in parallel clm) – mf#1987-6331 – us ATLA [225]

Commentarius in quatuor s evangelia domini n iesu christi. 3 : evangelium secundum lucam / Knabenbauer, Joseph – Parisiis: P Lethielleux 1896 [mf ed 1991] – (= ser Cursus scripturae sacrae 1/3) – 7mf – 9 – 0-8370-1970-2 – mf#1987-6357 – us ATLA [225]

Commentarius in quatuor s evangelia domini n iesu christi. 4 : evangelium secundum ioannem / Knabenbauer, Joseph – Parisiis: P Lethielleux 1898 [mf ed 1991] – (= ser Cursus scripturae sacrae 1/4) – 6mf – 9 – 0-8370-1971-0 – (in latin & greek; comm in latin) – mf#1987-6358 – us ATLA [225]

Commentarius in quatuor s evangelia domini n jesu christi. 1 : evangelium secundum s matthaeum / Knabenbauer, Joseph – Parisiis: P Lethielleux 1892 [mf ed 1991] – (= ser Cursus scripturae sacrae 1/1) – 2v on 11mf – 9 – 0-8370-1943-5 – (in latin & greek; comm in latin) – mf#1987-6330 – us ATLA [225]

Commentarius in s pauli apostoli epistolas. 1 : epistola ad romanos / Cornely, Rudolph – Parisiis: P Lethielleux 1896 [mf ed 1991] – (= ser Cursus scripturae sacrae 2/1) – 8mf – 9 – 0-8370-1924-9 – (in latin & greek; comm in latin) – mf#1987-6311 – us ATLA [227]

Commentarius in s pauli apostoli epistolas. 2 : prior epistola ad corinthios / Cornely, Rudolph – Parisiis: P Lethielleux 1909 [mf ed 1991] – (= ser Cursus scripturae sacrae 2/2) – 5mf – 9 – 0-8370-1965-6 – (in latin & greek; comm in latin) – mf#1987-6352 – us ATLA [227]

Commentarius in s pauli apostoli epistolas. 3 : epistolae ad corinthios altera et ad galatas / Cornely, Rudolph – ed 2a emendata. Parisiis: P Lethielleux 1909 [mf ed 1991] – (= ser Cursus scripturae sacrae 2/3) – 6mf – 9 – 0-8370-1991-5 – (in latin & greek; comm in latin) – mf#1987-6378 – us ATLA [227]

Commentarius in s pauli apostoli epistolas. 4 : epistolae ad ephesios ad philippenses et ad colossenses / Knabenbauer, Joseph – Parisiis: Sumptibus P Lethielleux 1912 [mf ed 1991] – (= ser Cursus scripturae sacrae 2/4) – 4mf – 9 – 0-8370-1997-4 – (in latin & greek; comm in latin) – mf#1987-6384 – us ATLA [227]

Commentarius in s pauli apostoli epistolas. 5 : epistolae ad thessalonicenses, ad timotheum, ad titum et ad philemonem / Knabenbauer, Joseph – Parisiis: P Lethielleux 1913 [mf ed 1991] – (= ser Cursus scripturae sacrae 2/5) – 4mf – 9 – 0-8370-1972-9 – (in latin & greek; comm in latin) – mf#1987-6359 – us ATLA [227]

Commentarius in vaticinium michae / Roorda, Taco – Lugduni Batavorum [Leiden]: P Engels; Lipsiae [Leipzig]: T O Weigel, 1869 – 1mf – 9 – 0-8370-4962-8 – (incl ind) – mf#1985-2962 – us ATLA [220]

Commentarius literalis, historico-moralis in regulam s p benedicti / Calmet, Aug – Lincii. v1-2. 1750 – €35.00 – ne Slangenburg [241]

Commentarius perpetuus in johannis marckii compendium theologiae christianae didactico-elencticum / Moor, B de – Lugduni Batavorum, 1761-71. 6v – 65mf – 9 – mf#PBA-276 – ne IDC [240]

Commentarius porretanus in primam epistolam ad corinthios / Landgraf, Arthur M – Romae, 1945 – (= ser Studi e testi) – 7mf – 8 – €16.00 – ne Slangenburg [227]

Commentarius super genesin : in quo textus declaratur, quaestiones dubiae solvuntur, observationes eruuntur / Gerhard, J – Jenae, 1637 – 10mf – 9 – mf#TH-1 mf 542-551 – ne IDC [242]

Commentarius theologico-canonico-criticus de ecclesiis / Assemani, J A – Romae, 1766 – €46.00 – (filmed with: j de bonis: tractatus de oratoriis publicis and: f a brixia: tractatus de oratoriis domesticis) – ne Slangenburg [240]

Commentarivs de regibvs persicis, sev familia artaxerxis magvsaei... / Reineck, R – Helmaestadii, 1588 – 1mf – 9 – mf#H-8373 – ne IDC [956]

Commentarivs in matthaevm evangelistam / Chytraeus, D – [Strassburg], 1556 – 5mf – 9 – mf#TH-1 mf 331-335 – ne IDC [242]

Commentary – New York. 1945+ (1) 1969+ (5) 1970+ (9) – ISSN: 0010-2601 – mf#925 – us UMI ProQuest [073]

A commentary by writers of the first five centuries on the place of st peter in the new testament : and that of st peter's successors in the church / Waterworth, James – London: Thomas Richardson; New York: Henry H Richardson, 1871 [mf ed 1986] – 2mf – 9 – 0-8370-6712-X – (incl bibl ref & ind) – mf#1986-0712 – us ATLA [225]

A commentary, critical, exegetical, and doctrinal, on st paul's epistle to the galatians : with a revised translation / Gwynne, George John – Dublin: University Press, 1863 [mf ed 1985] – 1mf – 9 – 0-8370-3437-X – (incl app) – mf#1985-1437 – us ATLA [227]

A commentary, critical, expository and practical, on the gospel of john : for the use of ministers, theological students... / Owen, John Jason – New York: Leavitt & Allen, 1861, c1860 [mf ed 1989] – 2mf – 9 – 0-7905-2934-3 – mf#1987-2934 – us ATLA [226]

A commentary, critical, expository and practical, on the gospel of luke : for the use of ministers, theological students... / Owen, John Jason – New York: Leavitt & Allen, 1861, c1859 [mf ed 1989] – 1mf – 9 – 0-7905-2729-4 – mf#1987-2729 – us ATLA [226]

A commentary, critical, expository and practical, on the gospels of matthew and mark : for the use of ministers, theological students... / Owen, John Jason – New York: Leavitt & Allen, 1864, c1857 [mf ed 1989] – 2mf – 9 – 0-7905-2859-2 – mf#1987-2859 – us ATLA [226]

A commentary, grammatical and exegetical, on the book of job : with a translation / Davidson, Andrew Bruce – [London]: Williams & Norgate, 1862 [mf ed 1985] – 1mf – 9 – 0-8370-2833-7 – (no more publ) – mf#1985-0833 – us ATLA [221]

The commentary of origen on s john's gospel : the text = Origenous ton eis to kata ioannen euangelion exegetikon / Origen – Cambridge: University Press, 1896 – 2mf – 9 – 0-7905-9429-3 – mf#1989-2654 – us ATLA [226]

A commentary on ecclesiastes / Stuart, Moses; ed by Robbins, Rensselaer David Chanceford – Boston: Draper & Halliday, 1880, c1862 [mf ed 1986] – 1mf – 9 – 0-8370-9422-4 – mf#1986-3422 – us ATLA [221]

Commentary on ecclesiastes : with treatises on song of solomon, job, isaiah, sacrifices, etc / Hengstenberg, Ernst Wilhelm – Edinburgh: T & T Clark, 1876 [mf ed 1984] – (= ser Clark's foreign theological library 6) – 6mf – 9 – 0-8370-1070-5 – (incl bibl ref & ind) – mf#1984-4416 – us ATLA [221]

Commentary on ezra and nehemiah / Saadiah, Rabbi; ed by Mathews, H J – Oxford. v1-pt 1. 1882 – (= ser Anecdota oxoniensia. semitic series) – 3mf – 8 – €17.00 – ne Slangenburg [221]

A commentary on hegel's logic / McTaggart, John McTaggart Ellis – Cambridge: University Press, 1910 [mf ed 1986] – 1mf – 9 – 0-7905-9513-3 – mf#1989-1218 – us ATLA [160]

Commentary on mechanic's lien law for the state of new york; chapter xlix of the general laws. / Heydecker, Edward Le Moyne – Albany, NY: Bender, 1897. 251p. LL-790 – 1 – us L of C Photodup [340]

Commentary on pan american problems / Alfaro, Ricardo J – Cambridge, MA. 1938 – 1r – us UF Libraries [972]

A commentary on st paul's epistle to the galatians / Beet, Joseph Agar – 3rd ed. New York: Thomas Whittaker, 1891 [mf ed 1986] – 1mf – 9 – 0-8370-6014-1 – (incl ind) – mf#1986-0014 – us ATLA [227]

A commentary on st paul's epistle to the romans / Beet, Joseph Agar – 2nd ed. London: Hodder & Stoughton, 1881 [mf ed 1985] – 1mf – 9 – 0-8370-2246-0 – (incl ind) – mf#1985-0246 – us ATLA [227]

Commentary on st paul's epistle to the romans = Commentar ueber den brief pauli an die roemer / Philippi, Friedrich Adolph – Edinburgh: T & T Clark. 2v. 1878-79 – 2mf – 9 – 0-8370-9894-7 – (in english and greek) – mf#1986-3894 – us ATLA [227]

Commentary on st paul's epistle to the romans / Godet, Frederic Louis; ed by Chambers, Talbot Wilson – New York: Funk & Wagnalls, 1883 [mf ed 1984] – 6mf – 9 – 0-8370-0124-2 – (trans fr french by alexander cusin; rev & ed with int & app by talbot w chambers. incl bibl ref & app) – mf#1984-0011 – us ATLA [227]

A commentary on st paul's epistles to the corinthians / Beet, Joseph Agar – 2nd ed. London: Hodder & Stoughton, 1883 [mf ed 1992] – 2mf – 9 – 0-524-03962-3 – mf#1992-0005 – us ATLA [227]

A commentary on st paul's epistles to the ephesians, philippians, colossians, and to philemon / Beet, Joseph Agar – NY: Thomas Whittaker, 1891 [mf ed 1985] – 1mf – 9 – 0-8370-2247-9 – (incl diss on paul's thought and letters) – mf#1985-0247 – us ATLA [227]

A commentary on the acts of the apostles / Hackett, Horatio Balch; ed by Hovey, Alvah – rev enl ed. Philadelphia: American Baptist Publ Society, c1882 [mf ed – 1mf – 9 – 0-8370-3442-6 – (incl ind) – mf#1985-1442 – us ATLA [226]

Commentary on the acts of the apostles / Summers, Thomas Osmond – Nashville, TN: Methodist Episcopal Church, South, 1882, c1874 [mf ed 1985] – 1mf – 9 – 0-8370-5466-4 – mf#1985-3466 – us ATLA [226]

Commentary on the acts of the apostles / Veil, Charles Marie de – London: J. Haddon, 1846-1854 – 1r – 1 – 0-8370-1698-3 – mf#1984-6073 – us ATLA [240]

A commentary on the apocalypse / Stuart, Moses – Andover: Allen, Morrill & Wardwell; New York: M H Newman, 1845 [mf ed 1989] – 3mf – 9 – 0-7905-2433-3 – mf#1987-2433 – us ATLA [225]

A commentary on the book of daniel / Jephet Inb Ali the Karaite; ed by Margoliouth, David Samuel – Oxford. v1-pt 3. 1889 – (= ser Anecdota oxoniensia. semitic series) – 9mf – 8 – €18.00 – ne Slangenburg [221]

A commentary on the book of daniel / Jepheth ben Eli; ed by Margoliouth, David Samuel – Oxford: Clarendon Press, 1889 [mf ed 1985] – (= ser Anecdota oxoniensia) – 1mf – 9 – 0-8370-3770-0 – (text in arabic & hebrew with trans in english) – mf#1985-1770 – us ATLA [221]

A commentary on the book of daniel / Stuart, Moses – Boston: Crocker & Brewster, 1850 [mf ed 1986] – 2mf – 9 – 0-8370-9583-2 – mf#1986-3583 – us ATLA [221]

Commentary on the book of deuteronomy / Jordan, W G – New York: Macmillan, 1911 – (= ser The bible for home and school) – 1mf – 9 – 0-7905-2120-2 – (incl ind) – mf#1987-2120 – us ATLA [221]

Commentary on the book of deuteronomy / Jordan, William George – New York: Macmillan, 1911 – (= ser The bible for home and school) – 4mf – 9 – 0-665-78245-4 – (incl ind) – mf#78245 – cn Canadiana [221]

A commentary on the book of ecclesiastes / Young, Loyal – Philadelphia: Presbyterian Board of Publ, c1865 [mf ed 1985] – 1mf – 9 – 0-8370-5938-0 – (int by alexander taggart mcgill & melancthon william jacobus) – mf#1985-3938 – us ATLA [221]

A commentary on the book of job : from a hebrew manuscript in the university library, cambridge / ed by Wright, William Aldis – London: publ for the Text and Translation Society by Williams & Norgate, 1905 [mf ed 1993] – 1mf – 9 – 0-524-07648-0 – (authorship attr to berechiah ben natronai) – mf#1992-1089 – us ATLA [221]

Commentary on the book of job : with translation = Buch 1 job / Ewald, Heinrich – London: Williams and Norgate, 1882 – 1mf – 9 – 0-8370-3085-4 – mf#1985-1085 – us ATLA [221]

Commentary on the book of joshua / Keil, Carl Friedrich – Edinburgh: T & T Clark, 1857 [mf ed 2004] – (= ser Clark's foreign theological library. new series 14) – 1r – 1 – 0-524-10499-9 – (trans by james martin) – mf#b00714 – us ATLA [221]

A commentary on the book of leviticus / Bonar, Andrew Alexander – 9 – $18.00 – us IRC [221]

A commentary on the book of leviticus : expository and practical, with critical notes / Bonar, Andrew Alexander – New York: R Carter, 1877 [mf ed 1992] – 2mf – 9 – 0-524-04119-9 – mf#1992-0077 – us ATLA [221]

A commentary on the book of proverbs / Stuart, Moses – Andover: Warren F Draper, 1870, c1852 [mf ed 1986] – 1mf – 9 – 0-8370-9423-2 – mf#1986-3423 – us ATLA [221]

A commentary on the book of psalms : in which their literal or historical sense, as they relate to king david and the people of israel, is illustrated / Horne, George – new ed. London; New York: Ward, Lock, [1824?] [mf ed 1989] – 2mf – 9 – 0-8370-1196-5 – mf#1987-6026 – us ATLA [221]

Commentary on the book of psalms = Biblischer commentar ueber die psalmen / Delitzsch, Franz – New York: Funk and Wagnalls [ca 1883] – (= ser The foreign biblical library) – 3v on 6mf – 9 – 0-7905-0367-0 – (english trans fr german by david eaton & james e duguid; incl bibl ref & ind) – mf#1987-0367 – us ATLA [221]

A commentary on the book of the acts of the apostles / Humphry, William Gilson – 2nd rev ed. London: John W Parker, 1854 [mf ed 1989] – 1mf – 9 – 0-7905-2107-5 – (in english and greek) – mf#1987-2107 – us ATLA [226]

A commentary on the books of amos, hosea, and micah / Smith, John Merlin Powis – New York: Macmillan, 1914 [mf ed 1985] – 1mf – 9 – 0-8370-4630-0 – (incl bibl and ind) – mf#1985-2630 – us ATLA [221]

Commentary on the books of kings / Keil, Carl Friedrich – Edinburgh: T & T Clark, 1857 – 1r – 1 – 0-8370-0473-X – (bibliographical footnotes) – mf#1984-B246 – us ATLA [221]

A commentary on the catholic epistles / Demarest, John Terhune – New York: Board of Publ of the Reformed Church in America, 1879 [mf ed 1989] – 2mf – 9 – 0-7905-1929-1 – mf#1987-1929 – us ATLA [227]

A commentary on the confession of faith : with questions for theological students and bible classes / Hodge, Archibald Alexander – Philadelphia: Presbyterian Board of Publ [c1869] [mf ed 1984] – 7mf – 9 – 0-8370-0932-4 – (incl ind) – mf#1984-4293 – us ATLA [242]

Commentary on the de consolatione philosophiae by boethius, c1258-1328 / Trivet, Nicholas – 1r – 1 – mf#95671 – uk Microform Academic [180]

Commentary on the epistle of james / Winkler, Edwin Theodore – Philadelphia: American Baptist Pub Soc, c1888 [mf ed 1989] – (= ser American commentary on the new testament) – 1mf – 9 – 0-7905-2279-9 – mf#1987-2279 – us ATLA [227]

Commentary on the epistle of paul to the galatians / Bacon, Benjamin Wisner – New York: Macmillan, 1909 – (= ser The bible for home and school) – 1mf – 9 – 0-8370-2140-5 – (incl ind) – mf#1985-0140 – us ATLA [227]

Commentary on the epistle to the colossians / Dargan, Edwin Charles – Philadelphia: American Baptist Pub Soc, c1890 [mf ed 1988] – (= ser American commentary on the new testament) – 1mf – 9 – 0-7905-0076-0 – mf#1987-0076 – us ATLA [227]

A commentary on the epistle to the ephesians / Hodge, Charles – New York: R Carter, 1857 [mf ed 1984] – 5mf – 9 – 0-8370-0898-0 – mf#1984-4273 – us ATLA [227]

Commentary on the epistle to the ephesians / Smith, Justin Almerin – Philadelphia: American Baptist Pub Soc, c1890 [mf ed 1988] – (= ser American commentary on the new testament) – 1mf – 9 – 0-7905-0115-5 – mf#1987-0115 – us ATLA [227]

Commentary on the epistle to the galatians / Hovey, Alvah – Philadelphia: American Baptist Pub Soc, c1890 [mf ed 1988] – (= ser American commentary on the new testament) – 1mf – 9 – 0-7905-0093-0 – mf#1987-0093 – us ATLA [227]

A commentary on the epistle to the hebrews / Stuart, Moses – 2nd corr enl ed. Andover: Flagg, Gould & Newman; New York: J Leavitt, 1833 [mf ed 1990] – 6mf – 9 – 0-8370-1677-0 – mf#1987-6105 – us ATLA [227]

Commentary on the epistle to the hebrews = Commentar zum briefe an die hebraer / Delitzsch, Franz – Edinburgh: T & T Clark 1868-70 [mf ed 1988] – (= ser Clark's foreign theological library. 4th series 20,25) – 2v on 3mf – 9 – 0-7905-0007-8 – (trans fr german by thomas l kingsbury) – mf#1987-0007 – us ATLA [227]

Commentary on the epistle to the hebrews / Kendrick, Asahel Clark – Philadelphia: American Baptist Pub Soc, c1889 [mf ed 1989] – (= ser American commentary on the new testament) – 1mf – 9 – 0-7905-1124-X – mf#1987-1124 – us ATLA [227]

Commentary on the epistle to the hebrews = Kommentar zum briefe an die hebrer / Tholuck, August – [2nd ed] Edinburgh: Thomas Clark, 1842 [mf ed 1989] – (= ser The biblical cabinet 29,38) – 2v on 2mf – 9 – 0-7905-2498-8 – (english trans by james hamilton; incl app trans by j e ryland) – mf#1987-2498 – us ATLA [225]

Commentary on the epistle to the philippians / Pidge, John Bartholomew Gough – Philadelphia: American Baptist Pub Soc, c1896 [mf ed 1988] – (= ser American commentary on the new testament) – 1mf – 9 – 0-7905-0110-4 – mf#1987-0110 – us ATLA [227]

A commentary on the epistle to the romans / Foster, Robert Verrell – Nashville, TN: Cumberland Presbyterian Pub House, 1891 [mf ed 1985] – 1mf – 9 – 0-8370-3169-9 – mf#1985-1169 – us ATLA [227]

A commentary on the epistle to the romans / Stuart, Moses; ed by Robbins, Rensselaer David Chanceford – 4th ed. Andover: Warren F Draper, 1868, c1859 [mf ed 1986] – 2mf – 9 – 0-8370-9831-9 – (incl bibl ref) – mf#1986-3831 – us ATLA [227]

Commentary on the epistle to the romans / Arnold, Albert Nicholas – Philadelphia: American Baptist Pub Soc c1889 [mf ed 1988] – (= ser American commentary on the new testament) – 1mf – 9 – 0-7905-0062-0 – (incl bibl ref) – mf#1987-0062 – us ATLA [227]

Commentary on the epistle to the romans : embracing the latest results of criticism / Brown, David – Glasgow: William Collins, 1860 – 1mf – 9 – 0-8370-2475-7 – mf#1985-0475 – us ATLA [227]

Commentary on the epistle to the romans / Hodge, Charles – new rev ed and in great measure rewritten. Philadelphia: J S Claxton, 1866. Beltsville, Md: NCR Corp, 1978 (8mf); Evanston: American Theol Lib Assoc, 1984 (8mf) – 9 – 0-8370-0897-2 – mf#1984-4274 – us ATLA [227]

Commentary on the epistles of john / Sawtelle, Henry A – Philadelphia: American Baptist Pub Soc, c1888 [mf ed 1988] – (= ser American commentary on the new testament) – 1mf – 9 – 0-7905-0228-3 – mf#1987-0228 – us ATLA [227]

Commentary on the epistles of jude / Williams, Nathaniel Marshman – Philadelphia: American Baptist Pub Soc, c1888 [mf ed 1988] – (= ser American commentary on the new testament) – 1mf – 9 – 0-7905-0239-9 – mf#1987-0239 – us ATLA [227]

Commentary on the epistles of paul to the corinthians = Commentar zu den briefen des paulus an die corinther / Billroth, Gustav – Edinburgh: T Clark, 1837-38 [mf ed 1989] – (= ser The biblical cabinet 21,23) – 2v on 2mf – 9 – 0-7905-1503-2 – (english trans fr german with additional notes by w lindsay alexander) – mf#1987-1503 – us ATLA [227]

Commentary on the epistles of peter / Williams, Nathaniel Marshman – Philadelphia: American Baptist Pub Soc, c1888 [mf ed 1988] – (= ser American commentary on the new testament) – 1mf – 9 – 0-7905-0299-2 – mf#1987-0299 – us ATLA [227]

Commentary on the epistles of st john = Einleitende untersuchungen und commentar ueber die briefe / Luecke, Friedrich – Edinburgh: Thomas Clark, 1837 [mf ed 1989] – (= ser The biblical cabinet 15) – 1mf – 9 – 0-7905-2024-9 – (english trans fr german with additional notes by thorleif gudmundson repp) – mf#1987-2024 – us ATLA [227]

Commentary on the epistles to the corinthians / Gould, Ezra Palmer – Philadelphia: American Baptist Pub Soc, c1887 [mf ed 1988] – (= ser American commentary on the new testament) – 1mf – 9 – 0-7905-0087-6 – mf#1987-0087 – us ATLA [227]

Commentary on the epistles to the seven churches in asia : revelation 2, 3 / Trench, Richard Chenevix – New York:Charles Scribner 1872 [mf ed 1985] – 1mf – 9 – 0-8370-5569-5 – (incl also revelation 1, 4-20) – mf#1985-3569 – us ATLA [227]

Commentary on the epistles to the thessalonians / Stevens, William Arnold – Philadelphia: American Baptist Pub Soc, c1890 [mf ed 1985] – (= ser American commentary on the new testament) – 1mf – 9 – 0-8370-5411-7 – mf#1985-3411 – us ATLA [227]

Commentary on the first epistle of st john : in the form of addresses = Erste brief johannis in predigten / Dryander, Ernst von; ed by Oesterley, William Oscar Emil – London: Elliot Stock, 1899 [mf ed 1986] – 1mf – 9 – 0-8370-9937-4 – (trans by ed) – mf#1986-3937 – us ATLA [227]

A commentary on the first epistle to the corinthians / Edwards, Thomas Charles – 2nd ed. New York: A C Armstrong, 1886 [mf ed 1986] – 2mf – 9 – 0-8370-9618-9 – (incl bibl ref and ind) – mf#1986-3618 – us ATLA [227]

Commentary on the five classics : adapted to modern times, for use in christian schools and colleges / Woods, Henry M – Shanghai: Christian Literature Society, 1917 [mf ed 1995] – (= ser Yale coll) – 278p – 1 – 0-524-09524-8 – (in chinese) – mf#1995-0524 – us ATLA [951]

Commentary on the four books : adapted to modern times / Woods, Henry McKee – 3rd rev enl ed. Shanghai: Christian Literature Society for China, 1914 [mf ed 1995] – (= ser Yale coll) – 120/116p – 1 – 0-524-10003-9 – (in chinese. only 1v publ?) – mf#1995-1003 – us ATLA [480]

Commentary on the gospel according to luke : giving critical, exegetical and applicative notes, and illustrations drawn from life and thought in the east / Rice, Edwin Wilbur – new enl ed. Philadelphia: Union Press, 1898 – (= ser Green fund book) – 1mf – 9 – 0-8370-4889-3 – (incl ind) – mf#1985-2889 – us ATLA [226]

Commentary on the gospel according to matthew : giving critical and exegetical notes / Rice, Edwin Wilbur – 6th newly rev ed. Philadelphia: American Sunday-School Union, 1909 – (= ser Green fund book) – 1mf – 9 – 0-8370-4888-5 – (incl ind) – mf#1985-2888 – us ATLA [226]

Commentary on the gospel according to s john / Cyril, Saint, Patriarch of Alexandria – London: Rivingtons; Oxford: sold by J Parker, 1874-1885. Chicago: Dep of Photodup, U of Chicago Lib, 1969 (1r); Evanston: American Theological Library Association, 1984 (1r) – (= ser A library of fathers of the holy catholic church) – 1 – 0-8370-0267-2 – mf#1984-B112 – us ATLA [226]

A commentary on the gospel by st luke / McLaughlin, George Asbury – Boston: McDonald, Gill, c1889 [mf ed 1992] – 1mf – 9 – 0-524-04804-5 – mf#1992-0224 – us ATLA [226]

Commentary on the gospel of john : with an historical and critical introduction = Commentaire sur l'evangile de saint jean / Godet, Frederic Louis – New York: Funk & Wagnalls, 1886 [mf ed 1989] – 2v on 3mf – 9 – 0-7905-3021-X – (trans fr 3rd french ed into english by timothy dwight; with pref, int & notes by trans; incl bibl ref) – mf#1987-3021 – us ATLA [226]

Commentary on the gospel of john / Hovey, Alvah – Philadelphia: American Baptist Pub Soc, c1885 [mf ed 1988] – (= ser American commentary on the new testament) – 1mf – 9 – 0-7905-0192-9 – mf#1987-0192 – us ATLA [226]

Commentary on the gospel of luke / Bliss, George Ripley – Philadelphia: American Baptist Pub Soc, c1884 [mf ed 1988] – (= ser American commentary on the new testament) – 1mf – 9 – 0-7905-0066-3 – (incl bibl ref) – mf#1987-0066 – us ATLA [226]

Commentary on the gospel of mark / Clarke, William Newton – Philadelphia: American Baptist Pub Soc, c1881 [mf ed 1988] – (= ser American commentary on the new testament) – 1mf – 9 – 0-7905-0073-6 – (incl bibl ref) – mf#1987-0073 – us ATLA [226]

Commentary on the gospel of mark / Weidner, Revere Franklin – 2nd ed. Allentown, PA: TH Diehl, 1888 – 1mf – 9 – 0-524-04417-1 – (incl bibl ref) – mf#1992-0110 – us ATLA [226]

Commentary on the gospel of matthew / Broadus, John Albert – Philadelphia: American Baptist Pub Soc, c1886 – (= ser American commentary on the new testament) – 2mf – 9 – 0-7905-0068-X – (incl ind) – mf#1987-0068 – us ATLA [226]

A commentary on the gospel of s matthew / Goodwin, Harvey – Cambridge: Deighton, Bell; London: Bell & Daldy, 1857 [mf ed 1990] – 2mf – 9 – 0-8370-1724-6 – mf#1987-6120 – us ATLA [226]

Commentary on the gospel of st john = Das evangelium des heiligen johannes / Hengstenberg, Ernst Wilhelm – Edinburgh: T & T Clark 1865 [mf ed 1989] – (= ser Clark's foreign theological library. 4th series 5,7) – 2v on 3mf – 9 – 0-7905-2779-0 – (trans fr german) – mf#1987-2779 – us ATLA [226]

Commentary on the gospel of st luke = Commentaire sur l'evangile de saint luc / Godet, Fredric Louis – New York: I K Funk, 1881 [mf ed 1989] – (= ser The standard series) – 2mf – 9 – 0-7905-2906-8 – (english trans by e w shalders & m d cusin; pref & notes to amer ed by john hall) – mf#1987-2906 – us ATLA [226]

A commentary on the gospels of matthew and mark : intended for popular use / Whedon, Daniel Denison – New York: Carlton & Porter, 1860 [mf ed 1990] – 1mf – 9 – 0-8370-1659-2 – mf#1987-6089 – us ATLA [226]

Commentary on the gospels of matthew and mark : critical, doctrinal, and homiletical = Evangelien von matthaeus und markus / Nast, Wilhelm – Cincinnati: Poe & Hitchcock, 1864 [mf ed 1989] – 2mf – 9 – 0-7905-1540-7 – (trans fr german into english) – mf#1987-1540 – us ATLA [226]

A commentary on the greek text of the epistle of paul to the colossians / Eadie, John; ed by Young, W – 2nd ed. Edinburgh: T & T Clark, 1884 [mf ed 1985] – 1mf – 9 – 0-8370-3015-3 – (incl ind) – mf#1985-1015 – us ATLA [227]

A commentary on the greek text of the epistle of paul to the ephesians / Eadie, John – 3rd ed. Edinburgh: T & T Clark, 1883 [mf ed 1989] – 2mf – 9 – 0-8370-1316-X – (in english and greek. incl bibl ref) – mf#1987-6049 – us ATLA [227]

A commentary on the greek text of the epistle of paul to the galatians / Eadie, John – Edinburgh: T & T Clark, 1869 [mf ed 1989] – 2mf – 9 – 0-8370-1320-8 – (in english and greek. incl bibl ref) – mf#1987-6050 – us ATLA [227]

A commentary on the greek text of the epistle of paul to the philippians / Eadie, John – London: Richard Griffin, 1859 [mf ed 1985] – 1mf – 9 – 0-8370-3016-1 – (incl ind of subjects and greek terms) – mf#1985-1016 – us ATLA [227]

A commentary on the greek text of the epistles of paul to the thessalonians / Eadie, John; ed by Young, William – London: Macmillan, 1877 [mf ed 1985] – 1mf – 9 – 0-8370-3017-X – mf#1985-1017 – us ATLA [227]

A commentary on the law of agency and agents / Wharton, Francis A – Philadelphia, Kay, 1876. 620 p. LL-1532 – 1 – us L of C Photodup [346]

A commentary on the law of evidence in civil issues / Wharton, Francis A – Philadelphia, Kay, 1877. 2 v. LL-1478 – 1 – us L of C Photodup [346]

A commentary on the mining legislation of congress with a preliminary review of the repealed sections of the mining act of 1866...2nd ed / Weeks, Edward P – San Francisco, Whitney, 1880. 580 p. LL-1645 – 1 – us L of C Photodup [343]

A commentary on the new testament / Weiss, Bernhard – New York: Funk & Wagnalls, 1906 [mf ed 1986] – 1mf – 9 – 0-8370-9663-4 – (english trans fr german by george h schodde & epiphanius wilson. int by james s riggs) – mf#1986-3663 – us ATLA [225]

Commentary on the new testament / Godbey, William Baxter – Cincinnati OH: Revivalist Office c1896-1900 [mf ed 1993] – 7v on 8mf – 9 – 0-524-05726-5 – mf#1992-0569 – us ATLA [225]

Commentary on the new testament : luke – the acts = Das neue testament. 1. haelfte / Weiss, Bernhard – New York: Funk & Wagnalls, 1906 [mf ed 1986] – 2mf – 9 – 0-8370-9664-2 – (trans fr german by george h schodde & epiphanius wilson; int by james s riggs) – mf#1986-3664 – us ATLA [225]

Commentary on the new testament : romans – colossians = Das neue testament. 2. haelfte / Weiss, Bernhard – New York: Funk & Wagnalls, 1906 [mf ed 1986] – 2mf – 9 – 0-8370-9665-0 – (trans fr german by george h schodde and epiphanius wilson; int by james s riggs) – mf#1986-3665 – us ATLA [227]

Commentary on the new testament : thessalonians – revelation = Das neue testament. 2. haelfte / Weiss, Bernhard – New York: Funk & Wagnalls, 1906 [mf ed 1986] – 2mf – 9 – 0-8370-9666-9 – (trans fr german by george h schodde and epiphanius wilson; int by james s riggs) – mf#1986-3666 – us ATLA [225]

Commentary on the new testament, vol 3 : acts-romans / Whedon, Daniel Denison – New York: Carlton & Lanahan 1871 [mf ed 1990] – 1mf – 9 – 0-8370-1679-7 – mf#1987-6107 – us ATLA [226]

Commentary on the new testament, vol 4 : 1: corinthians 2: timothy / Whedon, Daniel Denison – New York: Eaton & Mains; Cincinnati: Curts & Jennings c1875 [mf ed 1990] – 2mf – 9 – 0-8370-1680-0 – mf#1987-6108 – us ATLA [227]

Commentary on the new testament, vol 5 : titus-revelation / Whedon, Daniel Denison – New York: Phillips & Hunt; Cincinnati: Walden & Stowe c1880 [mf ed 1990] – 2mf – 9 – 0-8370-1682-7 – mf#1987-6109 – us ATLA [227]

A commentary on the original text of the acts of the apostles / Hackett, Horatio Balch – new rev enl ed. Boston: Gould and Lincoln, 1858 [mf ed 1989] – 2mf – 9 – 0-7905-0572-X – (incl bibl ref and ind) – mf#1987-0572 – us ATLA [226]

Commentary on the pastoral epistles, first and second timothy and titus, and the epistle to philemon / Hervey, Hezekiah – Philadelphia: American Baptist Pub Soc, c1890 [mf ed 1988] – (= ser American commentary on the new testament) – 1mf – 9 – 0-7905-0189-9 – mf#1987-0189 – us ATLA [227]

Commentary on the pentateuch / Gerlach, Otto von – Edinburgh: T & T Clark, 1860 – 2mf – 9 – 0-8370-1665-7 – mf#1987-6095 – us ATLA [221]

A commentary on the proverbs / Miller, John – New York: Anson D F Randolph, c1872 [mf ed 1989] – 2mf – 9 – 0-7905-1364-1 – (incl ind) – mf#1987-1364 – us ATLA [221]

A commentary on the psalms : designed chiefly for the use of hebrew students and of clergymen / Phillips, George – London: Williams & Norgate, 1872 [mf ed 1989] – 2v on 2mf – 9 – 0-7905-2788-X – (in english & hebrew) – mf#1987-2788 – us ATLA [221]

Commentary on the psalms = Commentar ueber die psalmen / Hengstenberg, Ernst Wilhelm – Edinburgh: T & T Clark 1846-51 [mf ed 1989] – (= ser Clark's foreign theological library 1-2,12) – 3v on 4mf – 9 – 0-8370-1544-8 – (trans fr german by p fairbairn & j thomson) – mf#1987-6070 – us ATLA [221]

Commentary on the revelation / Smith, Justin Almerin – Philadelphia: American Baptist Pub Soc, c1884 [mf ed 1988] – (= ser American commentary on the new testament) – 1mf – 9 – 0-7905-0231-3 – mf#1987-0231 – us ATLA [225]

Commentary on the ritual of the methodist episcopal church, south / Summers, Thomas Osmond – Nashville TN: publ by A H Redford...1874 [mf ed 1992] – 1mf – 9 – 0-524-03194-0 – (incl ind) – mf#1990-4643 – us ATLA [242]

A commentary on the second epistle of the apostle peter / Demarest, John Terhune – New York: Sheldon, 1862 [mf ed 1985] – 1mf – 9 – 0-8370-2884-1 – mf#1985-0884 – us ATLA [227]

Commentary on the sermon of the mount = Die bergpredigt / Tholuck, August – Edinburgh: T & T Clark 1869 [mf ed 1986] – 2mf – 9 – 0-8370-9750-9 – (trans fr 4th rev enl ed by r lundin brown; incl ind) – mf#1986-3750 – us ATLA [225]

Commentary on the song of songs and ecclesiastes / Delitzsch, Franz – Edinburgh: T & T Clark 1877 [mf ed 1984] – (= ser Clark's foreign theological library. 4th series 54) – 1r – 1 – 0-8370-0413-6 – (incl bibl ref; trans fr german by matthew george easton) – mf#1984-b352 – us ATLA [221]

A commentary on the whole epistle to the hebrews : being the substance of thirty years' wednesday's lectures at blackfriars, london / Gouge, William – Edinburgh: James Nichol, 1866-67 [mf ed 1993] – (= ser Nichol's series of commentaries) – 3v on 10mf – 9 – 0-524-08707-5 – mf#1993-0052 – us ATLA [227]

Commentary on water development in the jordan valley region / The Arab Palestine Office – Beirut, Lebanon: The Arab Palestine Office 1954 – us CRL [333]

A commentary upon the epistle of st paul written to the colossians / Cartwright, Thomas – Edinburgh: James Nichol, 1864 [mf ed 1985] – (= ser Nichol's series of commentaries) – 1mf – 9 – 0-8370-5984-4 – mf#1985-3984 – us ATLA [227]

A commentary upon the prophecy of malachi / Stock, Richard – Edinburgh: James Nichol, 1865 [mf ed 1985] – (= ser Nichol's series of commentaries) – 1mf – 9 – 0-8370-5422-2 – (incl ind) – mf#1985-3422 – us ATLA [221]

Commentatio de personis vvlgo larvis sev mascheris, von der carnavals-lvst : critico historico morali atqve ivridico modo diligenter conscripta / Berger, Christoph Heinrich von – Francofvrti, Lipsiae: Knoch 1723 – (= ser Whsb) – 7mf – 9 – €75.00 – mf#Hu 466 – gw Fischer [450]

Commentatio seu declaratio adillud geneseos / Correas, Gonzalo – 1622 – 9 – sp Bibl Santa Ana [400]

Commentationes physico-mathematicae – Helsinki. 1924-1973 (1) 1972-1972 (5) (9) – ISSN: 0069-6609 – mf#7122 – us UMI ProQuest [510]

Commentator – Marietta OH. 1810 apr 3 – 1r – 1 – (cont: commentator and marietta recorder) – mf#890572 – us WHS [071]

Commentator – Jacksonville, FL. 1943-1950 – 2r – us UF Libraries [071]

Commentator / Wappingers Congress of Teachers – 1972 dec 22-1976 dec 17, v6 n1-v10 n12 [1977 jan 14-1981 feb 6] – 1r – 1 – mf#676206 – us WHS [370]

Commentator of the community council – Jacksonville, FL. 28 Oct, 8 Dec 1960; 10 Mar, 6 May, 25 May 1961; 12 Jan, 16 Apr 1962; 28 Oct 1964.Ceased publication. In English – 1 – us AJPC [071]

Un commento a giobbe di giuliano di eclana / Vaccari, Alberto – Roma: Pontificio Istituto Biblico, 1915 – (= ser Scripta Pontificii Instituti Biblici) – 1mf – 9 – 0-524-06583-7 – mf#1992-0926 – us ATLA [220]

Commento en defensa del libro 4...nebrija / Lopez, Diego – Salamanca: Casa de Autoria. Ramirez Vda., 1610 – 1 – sp Bibl Santa Ana [440]

Comments / Cooperative Services, Inc – 1978 fall-1980 fall, 1982 fall – 1r – 1 – (cont: cooperative comment) – mf#665550 – us WHS [071]

Comments of the federal military government on the report of the tribunal of inquiry into the affairs of waac (nigeria) limited : otherwise known as nigeria airways, for the period 1st march 1961 to 31st december 1965 – Lagos, Federal Ministry of Information, Printing Division 1968 – us CRL [380]

Comments of the lagos state government on the report of the tribunal of inquiry into the affairs of the lagos executive development board for the period 1st october 1960 to 31st december 1965 – Lagos, Information Division, Lagos State Government 1968 – us CRL [350]

Comments on the dharmapada / Hack, Wilton – Madras: Oriental Pub Co, 1911 [mf ed 1991] – (= ser Sadharana dharma series 5) – 1mf – 9 – 0-524-01443-4 – mf#1990-2438 – us ATLA [280]

Commerce – (= ser The Goldsmiths'-Kress Library of Economic Literature) – 327r – 1 – us Primary [380]

Commerce – Montreal: Chambre de commerce. v52 n1 janv 1950- [mthly] [mf ed 1978-] – 9 – (cont: bulletin [chambre de commerce du district de montreal]) – mf#SEM105P1 – cn Bibl Nat [380]

Commerce : official journal / Sydney Chamber of Commerce – Sydney. v. 29-46 n1. jan 1940-feb 1957 – 1 – us NY Public [380]

Le commerce : journal commercial, politique, scientifique et litteraire – Paris. n1-17. 6 juin-26 sept 1869 – 1 – fr ACRPP [073]

Le commerce – Port-au-Prince: J J Andain, [6e annee n8-n42]. (12 fevr. 1876-31 mars 1877) – 2 sheets – us CRL [079]

Commerce america – Washington. 1976-1978 (1,5,9) – ISSN: 0361-0438 – mf#10724 – us UMI ProQuest [380]

The commerce between the roman empire and india / Warmington, Eric Herbert – Cambridge: University Press, 1928 – (= ser Samp: indian books) – us CRL [380]

Commerce Clearing House *see* Business law case method.

Commerce Clearing House Canadian Limited *see*
- Canadian estate tax and succession duties acts; including all amendments to december 1, 1966
- Expenses under canadian income tax act

Commerce first baptist church : church records – Commerce, GA. 2030p. 1874-1989 – 1 – us Southern Baptist [242]

Commerce hot-line / Metropolitan Milwaukee Association of Commerce – v1 n1-v5 n10 [1980 jun 24-1984 apr 2] – 1r – 1 – (cont: metropolitan milwaukee economic trends; cont by: milwaukee commerce hot-line) – mf#596851 – us WHS [380]

Commerce today – Washington. 1970-1975 (1) 1970-1975 (5) (9) – ISSN: 0020-6385 – mf#5954 – us UMI ProQuest [380]

Commercial – Bangor, ME. may 4-aug 4 1956 – 3r – us CRL [070]

Commercial / Denver Civic and Commercial Association – Denver CO. 1918 jan 3-1920 sep 30, 1920 oct 10-1922 may 18 – 2r – 1 – (cont by: denver commercial) – mf#854390 – us WHS [071]

Commercial – Florianopolis, SC. 10 nov 1894 – (= ser Ps 19) – mf#UFSC/BPESC – bl Biblioteca [079]

Commercial : a journal devoted to the financial, mercantile and manufacturing interests of the canadian north-west – Winnipeg: Steen & Boyce, [1882?-1922] – 9 – (cont by: commercial and retail merchants review) – mf#P04957 – cn Canadiana [380]

Commercial : periodico semanal – Desterro, SC: Typ do Commercial, 20 ago 1885; 28 fev 1886 – (= ser Ps 19) – mf#UFSC/BPESC – bl Biblioteca [079]

Commercial / Williams Co. Edon – mar 1971-dec 1975 [wkly] – 1r – 1 – mf#B29343 – us Ohio Hist [071]

Commercial – Winnipeg, Canada. 5 jan 1886-1907 – 39 1/2r – 1 – uk British Libr Newspaper [072]

The commercial – Meyersdale, PA., 1894-1897 – 13 – $25.00r – us IMR [071]

The commercial – Winnipeg, Canada. -w. 1886-1907. 41 reels – 1 – uk British Libr Newspaper [072]

Commercial accountant, the.../accountants review, 1947-77 – 8r 6mf – 1,9 – mf#96473 – uk Microform Academic [650]

Commercial advertiser – Milwaukee WI. [1850 apr 16-1850 nov 16], [1851 apr 4-1852 may 22], 1851 apr 8 – 3r – 1 – (cont by: milwaukee morning news) – mf#1138088 – us WHS [071]

Commercial advertiser – Red Cloud, NE: A C Hosmer. 49v. 10th yr n1349. mar 27 1911-v56 n31. apr 9 1959 (wkly) [mf ed with gaps filmed 1969] – 57r – 1 – (cont: commercial advertiser semi-weekly (1910). absorbed: webster county argus and: guide rock signal (1908). cont by: commercial advertiser and the red cloud chief. sect called guide rock signal retained separate numbering through nov 21 1957. shares numbering with wkly fri eds. drops yr and whole numbering with 26th yr n3922. feb 20 1929; resumes numbering with v46 n34 nov 9 1948) – us NE Hist [071]

Commercial advertiser – Red Cloud, NE: Red Cloud Print. -7th yr n955. sep 23 1908 (triwkly) [mf ed 7th yr n894-955. may 1 1908 (gaps) filmed 1969 – 1r – 1 – (cont: tri-weekly commercial advertiser (1902). cont by: commercial advertiser tri-weekly (1908)) – us NE Hist [071]

Commercial advertiser – Huron, OH. 1837-1842 (1) – mf#65540 – us UMI ProQuest [071]

Commercial advertiser – Apalachicola, FL. v2-7. 1844 jan-1849 mar 15 – 1r – (missing: 1844 jul 20; 1845; 1847 feb 20, mar 13-27, apr 03,17-24, may 01-08,22-29, jun 07-14, jul 06-13,24-31, oct 09, nov 25; 1848 may 25; jun 24; aug 5,26; sep 2,23-dec 1,16; 1849 feb 1) – us UF Libraries [071]

Commercial advertiser – New York. Oct 2 1797-Jan 31 1904. Incomplete – 1 – us NY Public [071]

Commercial advertiser – Waterford, Ireland. -w. 15 jan-4 mar 1848 – 1/4r – 1 – uk British Libr Newspaper [072]

Commercial advertiser *see* Transvaal advocate / commercial advertiser

Commercial advertiser and the red cloud chief – Red Cloud, NE: D Furse. 8v. v56 n32. apr 16 1959-v63 n24. feb 23 1967 (wkly) [mf ed with gaps filmed 1969] – 6r – 1 – (cont: commercial advertiser. cont by: red cloud chief and the commercial advertiser. publ as: commercial advertiser jan 31-feb 21 1963) – us NE Hist [071]

Commercial advertiser semi-weekly – Red Cloud, NE: A C Hosmer. 2v. 9th yr n1284. oct 26 1910-10th yr n1347. mar 22 1911 (semiwkly) [mf ed lacks 10th yr n1316 filmed 1969] – 1r – 1 – (cont: commercial advertiser tri-weekly (1908). cont by: commercial advertiser. shares numbering with wkly fri ed) – us NE Hist [071]

Commercial advertiser tri-weekly – Red Cloud, NE: A C Hosmer. 3v. 7th yr n956. sep 25 1908-9th yr n1283. oct 24 1910 (triwkly) [mf ed with gaps filmed 1969] – 2r – 1 – (cont: commercial advertiser (1908). cont by: commercial advertiser semi-weekly (1910)) – us NE Hist [071]

Commercial advertiser weekly edition – Red Cloud, NE: A C Hosmer. 3v. 9th yr n1285. oct 28 1910-11th yr n1476. jan 19 1912 (wkly) [mf ed with gaps filmed 1969] – 2r – 1 – (cont by: weekly advertiser (1912). shares numbering with semiwkly ed) – us NE Hist [071]

Commercial adviser – Canton, NY. 1952-1958 (1) – mf#64926 – us UMI ProQuest [071]

Commercial Advisory Foundation in Indonesia *see*
- Agriculture section circular tan
- Banking sector circular bnk
- Capital investment sector circular mod
- Circular 1 cost of living index for foreign family
- Circular b
- Economic section circular
- Economic section circular cr 1-4
- Economic section circular e
- Economic section circular fr
- Economic section circular ta
- Finance sector circular keu
- General section circular u
- Industries sector circular ind
- Labour and social section berita
- Labour and social section circular cw
- Labour and social section report
- Laporan-bulanan
- Legal section circular h
- Sector of mining circular tam
- Statistics and price survey section circular k
- Statistics and price survey section circular ph
- Summaries laws
- Surat edaran
- Trade sector circular dag
- Warta cafi

Commercial age – Olympia, WA. 1868-1870 (1) – mf#67050 – us UMI ProQuest [071]

Commercial and financial chronicle – New York. Nov 1875-Dec 1876; Jan 1897-Dec 1908.-w. 63 reels – 1 – uk British Libr Newspaper [071]

The commercial and financial chronicle – New York. v. 1-19 no. 496. July 1865-Dec 1874 – 1 – us NY Public [071]

Commercial and notarial precedents. / Montefiore, Joshua – 2d Amer. ed. from the last London ed. Philadelphia: Carey & Lea, 1822. 480p. LL-867 – 1 – us L of C Photodup [346]

Commercial appeal – Memphis TN. 1919 may 19-20 – 1r – 1 – (cont: memphis evening appeal; memphis appeal-avalanche; memphis commercial) – mf#850356 – us WHS [071]

Commercial appeal – Danville, VA. 1947+ (1) – mf#66699 – us UMI ProQuest [071]

Commercial appeal – Memphis, TN. 1894+ (1) – ISSN: 0745-4856 – mf#60585 – us UMI ProQuest [071]

Commercial appeal (3 star mississippi edition) – Memphis, TN. 1983-1986 (1) – mf#68004 – us UMI ProQuest [071]

Commercial appeal (vietnam edition) – Memphis, TN. 1966-1971 (1) – mf#66551 – us UMI ProQuest [071]

Commercial bank of canada : special general meeting of shareholders – S.l: s.n, 1865? – 1mf – 9 – mf#42609 – cn Canadiana [332]

Commercial bank vs wilmot and others : judgment of his honor chief justice ritchie, delivered 9th march, 1866 – St John, NB?: Barnes, 1866 – 1mf – 9 – mf#54527 – cn Canadiana [345]

Commercial bulletin – Marshfield & Coos Co OR: Bulletin Pub Co – 1 – us Oregon Lib [071]

Commercial bulletin – New Orleans, LA. 1836-1871 (1) – mf#63501 – us UMI ProQuest [071]

Commercial bulletin – Richmond VA. 1865 jul 12 – 1r – 1 – mf#881819 – us WHS [071]

Commercial carrier journal for professional fleet managers (ccj) – Radnor. 1997+ (1) 1997+ (5) 1997+ (9) – (cont: chilton's commercial carrier journal for professional fleet managers: ccj) – ISSN: 1099-4173 – mf#942,03 – us UMI ProQuest [380]

Commercial chronicle and daily marylander – Baltimore, Maryland. Apr 14 1829-Dec 27 1833 – 1r – 1 – us L of C Photodup [071]

Commercial citizen – Dazey, Barnes Co, ND: Leo Ratcliff. v8 n21 aug 16 1918-v11 n34 nov 11 1921 (wkly) – (= ser Dazey commercial; Rogers citizen) – 1 – (missing: 1920 apr 16; 1921 aug 26; formed by the union of: dazey commercial (1911) and: rogers citizen (1913)) – mf#11030 – us North Dakota [071]

Commercial construction news – Morris Plains. 1973-1973 (1) – mf#8070 – us UMI ProQuest [720]

Commercial control of citrus scab in florida / Ruehle, George D – Gainesville, FL. 1939 – 1r – us UF Libraries [634]

Commercial cuba : a book for business men / Clark, William Jared – Toronto: G N Morang, 1898 – 7mf – 9 – 0-665-93671-0 – (incl ind. int by e sherman gould) – mf#93671 – cn Canadiana [972]

Commercial daily list – London. 2 March 1838-26 March 1870.-d. 39mqn reels – 1 – uk British Libr Newspaper [072]

Commercial directory of latin america / International Bureau Of American Republics – Washington, DC. 1892 – 1r – us UF Libraries [380]

Commercial directory of manila – [Manila: s.n.], 1901 – 1 – us CRL [380]

Commercial dispatch – Columbus, MS. 1972-2000 (1) – mf#61546 – us UMI ProQuest [071]

The commercial exhibit – Omaha, NE: Exhibit Pub Co. v2 n50. aug 15 1895 (wkly) [mf ed 1895-96] – 1r – 1 – (lacks: dec 26 1896) – us NE Hist [071]

Commercial feeds of wisconsin – 65th-68th [1965-1968] – 1r – 1 – (cont: feed inspection for...) – mf#535581 – us WHS [071]

Commercial fertilizer and plant food industry – Atlanta. 1950-1970 (1) – ISSN: 0097-2738 – mf#295 – us UMI ProQuest [630]

Commercial fisheries abstracts – Seattle. 1972-1973 (1) 1972-1973 (5) 1972-1973 (9) – (cont by: marine fisheries abstracts) – ISSN: 0010-2970 – mf#7374 – us UMI ProQuest [639]

Commercial gazette *see* Transvaal argus / commercial gazette

A commercial geography of the british empire / Lyde, Lionel William – London, 1894 – (= ser 19th c british colonization) – 3mf – 9 – mf#1.1.8370 – uk Chadwyck [338]

Commercial herald – Milwaukee WI. 1843 jul 3,14 – 1r – 1 – (cont by: milwaukie commercial herald) – mf#1166156 – us WHS [071]

Commercial herald – Milwaukee WI. 1844 nov 25-dec 27 – 1r – 1 – (cont: milwaukie commercial herald; cont by: milwaukie sentinel) – mf#1166157 – us WHS [071]

Commercial herald – Oswego, NY. 1836-1838 (1) – mf#65143 – us UMI ProQuest [071]

Commercial index – Eau Claire WI. 1872 oct 19 [v1 n11] – 1r – 1 – mf#962642 – us WHS [071]

Commercial ireland – Dublin, Ireland. 1890-2 jul 1892 – 3r – 1 – (aka: irish manufacturers journal etc) – uk British Libr Newspaper [338]

Commercial ireland *see* Irish manufacturers journal

Commercial journal – Pittsburgh, PA. 1846-1852 (1) – mf#66031 – us UMI ProQuest [071]

Commercial journal – Sydney, jul 1835-jun 1842, apr-nov 1845 – 4r – A$202.88 vesicular A$224.88 silver – at Pascoe [079]

Commercial journal *see*
– Commercial journal and family herald
– Dublin shipping and mercantile gazette

Commercial journal and family herald – Dublin, Ireland. 23 dec 1848-26 dec 1874 – 15r – 1 – (incorp with: dublin advertising gazette fr 1875; aka: commercial journal) – uk British Libr Newspaper [072]

Commercial law / Gano, Darwin Curtis – New York: American Book Co. 1904. 399p. LL-247 – 1 – (commercial law. new york: american book co. 1913. 399p. ll-205) – us L of C Photodup [346]

Commercial law / Lyon, Newell – New York, 1901-04. 148l. LL-1238 – 1 – us L of C Photodup [346]

Commercial law bulletin – Chicago. 1986+ (1,5,9) – ISSN: 0888-8000 – mf#16459 – us UMI ProQuest [346]

Commercial law journal – Chicago. 1973+ (1) 1977+ (5) 1977+ (9) – ISSN: 0010-3055 – mf#9172 – us UMI ProQuest [346]

Commercial law journal – v6-105. 1902-2000 – 9 – $1760.00 set – (title varies: v6-27 n7 1902-1922 as bulletin of commercial law league of america; v27 n8-v28 n8 1922-23 as commercial law legal bulletin; v28-35 1923-30 as commercial law league journal) – ISSN: 0010-3055 – mf#101181 – us Hein [346]

Commercial law league bulletin *see* Commercial law journal

Commercial law questions / Cathcart, Wallace Daniel – San Francisco: Pacioli, 1946. 74p. LL-210 – 1 – us L of C Photodup [346]

The commercial laws of the states: a summary of the laws relating to arrest assignments attachment...&c / Homans, Isaac Smith – New York: Bankers', 1870. 328p. LL-900 – 1 – us L of C Photodup [346]

Commercial lending review – New York. 1987+ (1,5,9) – ISSN: 0886-8204 – mf#15758 – us UMI ProQuest [332]

Commercial mail – Columbia City, IN. 1879-1919 (1) – mf#62741 – us UMI ProQuest [071]

Commercial master file-restricted / U.S. National Technical Information Service – Commercial operator and restricted operator licensees in operator name sequence. Cross-reference by serial number included. Covers last 7 years – 9 – us NTIS [000]

Commercial motor – Sutton. 1958-1985 (1) 1973-1985 (5) 1973-1985 (9) – ISSN: 0010-3063 – mf#1236 – us UMI ProQuest [380]

Commercial news – Danville, IL. 1989-2000 (1) – mf#61326 – us UMI ProQuest [071]

Commercial news – Lynwood. 1977-1980 (1) 1978-1980 (5) 1978-1980 (9) – ISSN: 0047-5068 – mf#7769 – us UMI ProQuest [338]

Commercial operations in space [aasms34] – 1981 – (= ser Aasms 1968) – 2papers on 1mf – 9 – $10.00 – 0-87703-165-7 – (suppl to vol 51, science and technology) – us Univelt [338]

The commercial policy of the moguls / Pant, D – Bombay: DB Taraporevala Sons, 1930 – (= ser Samp: indian books) – (foreword by lord meston) – us CRL [380]

The commercial power of great britain : exhibiting a complete view of the public works of this country, under the several heads of streets, roads, canals, aqueducts, bridges, coasts, and maritime ports / Dupin, Charles – London 1825 [mf ed Hildesheim 1995-98] – 2v on 7mf – 9 – €140.00 – 3-487-28834-6 – (trans fr french...with a quarto atlas of plans, elevations, etc) – gw Olms [380]

Commercial precedents : selected from the column of replies and decisions of the new york journal of commerce, also selected decisions from other sources / Putzel, Charles & Baehr, H A – Hartford, CO: American Publ Co, 1897 – 8mf – 9 – $12.00 – mf#LLMC 92-122 – us LLMC [346]

Commercial price current – Canton, China. 12 sep 1835-3 sep 1836 – 1/4r – 1 – uk British Libr Newspaper [072]

Commercial record – Bridgeport, CT. 1981-1997 [1] – mf#69076 – us UMI ProQuest [071]

Commercial record of pensacola, florida / Waterman, G A – Pensacola, FL. 1909 – 1r – us UF Libraries [380]

The commercial review of the south and the west *see* Debow's review

Commercial rose culture, under glass and outdoors / Holmes, Eber – New York, NY. 1926 – 1r – us UF Libraries [630]

Commercial series / Lucas Co. Toledo – 1853,(6/1875-2/1877),7/1898-4/1900 [daily] – 7r – 1 – mf#B34594-34600 – us Ohio Hist [071]

Commercial series / Lucas Co. Toledo – jul 1892-jun 1898 (fire damaged) [daily] – 19r – 1 – mf#B30934-30952 – us Ohio Hist [071]

Commercial shipping and general advertiser for west cornwall – Penryn, England. 8 Jun 1867-27 Sep 1912.-w. 10 reels – 1 – uk British Libr Newspaper [072]

Commercial space – New York. 1985-1986 (1) 1985-1986 (5) 1985-1986 (9) – ISSN: 8756-4831 – mf#15654 – us UMI ProQuest [629]

Commercial sponges and the sponge fisheries / Moore, Henry Frank – Washington, DC. 1910 – 1r – us UF Libraries [639]

Commercial telegraphers' journal : the official organ of the commercial telegraphers union of america – 1903 aug-1905, 1906-66, 1967 feb-1968 jul – 10r – 1 – (cont: journal [commercial telegraphers' union of america]; cont by: telegraph workers journal) – mf#864635 – us WHS [380]

Commercial Telegraphers' Union *see* Key and sounder

Commercial Telegrapher's Union of America *see* Journal

Commercial times – New Orleans, LA. 1846-1849 (1) – mf#68753 – us UMI ProQuest [071]

The commercial tourist – Omaha, NE: Redfield Bros. v1 n2. may 1879 (mthly) [mf ed 1981] – 1r – 1 – us NE Hist [071]

Commercial travelers' guide to latin america / Filsinger, Ernst B – Washington, DC. 1920 – 1r – us UF Libraries [918]

Commercial travelers' guide to latin america / Filsinger, Ernst B – Washington, DC. 1922 – 1r – us UF Libraries [918]

Commercial travelers' guide to latin america / Filsinger, Ernst B – Washington, DC. 1926 – 1r – us UF Libraries [918]

Commercial tribune – Cincinnati, OH. 1901-1930 (1) – mf#65411 – us UMI ProQuest [071]

The commercial tribune – Cincinnati, Ohio. Jan 1 1928-Dec 3 1930. Incomplete – 1 – us NY Public [071]

Commercial union : a study – S.l: s.n, 1886? – 1mf – 9 – mf#00724 – cn Canadiana [337]

Commercial union between the united states and canada : some letters, papers and speeches – Toronto: Printed by the Toronto New Company for Erastus Wiman, New York, 1887? – 1mf – 9 – mf#25992 – cn Canadiana [337]

Commercial union between the united states and canada : speech of erastus wiman at lake dufferin, ontario, july 1 1887 – Toronto: Printed by the Toronto News Co...for Erastus Wiman, 1887 – (= ser Commercial union document) – 1mf – 9 – mf#25964 – cn Canadiana [337]

Commercial union between the united states and canada : speech...delivered in the house of assembly of nova scotia, may 2 1887 / Longley, James Wilberforce – S.l: s.n, 1887? – 1mf – 9 – mf#51511 – cn Canadiana [337]

Commercial utilization of space [aasms3] – 1968 – (= ser Aasms 1968) – 72papers on 24mf – 9 – $35.00 – 0-87703-216-5 – (suppl to vol 23, advances) – us Univelt [380]

Commercial vegetable varieties for florida – Gainesville, FL. 1944 – 1r – us UF Libraries [634]

Commercialist & weekly advertiser : a weekly supplement – London, England 1 jan 1832-28 dec 1834 – 1 – uk British Libr Newspaper [072]

Commercial-review – Portland, IN. 1993-2000 (1) – mf#61402 – us UMI ProQuest [071]

Commercio – Quito, Ecuador. 23 jun 1955-21 may 1962; jan-feb 1964; 25 may-28 dec 1969 – 98 3/4r – 1 – uk British Libr Newspaper [079]

O commercio : orgao commercial, litterario e noticioso – Goias: Typ do Commercio, 06 abr 1879-jan 1881; 31 jan 1882 – (= ser Ps 19) – mf#P11B,06,02 – bl Biblioteca [380]

O commercio : orgao dos interesses do commercio – Barra do Pirai, RJ. 07-21 jan 1909 – (= ser Ps 19) – mf#DIPER – bl Biblioteca [079]

O commercio de lourenco marques – Lourenco Marques: Baptista de Carvalho & Irmao, sep 3-17 1892 – us CRL [380]

Commercio do amazonas – Manaus, AM: [s.n.] 01-27 jul, dez 1870; abr 1872; jan, maio-jun 1874; jan-jul 1875; dez 1876; dez 1877; jan 1878; maio 1879; maio-out 1880; ago,dez 1881; mar 1884; out 1891; ago 1897; maio 1898-dez 1899; abr-jun,out-dez 1900; ago 1903; 29 mar 1912 – (= ser Ps 19) – mf#P11,01,36 – bl Biblioteca [079]

Commercio do espirito santo – Vitoria, ES. 31 mar, maio 1892; jan-nov 1893; jan 1894-dez 1897; jan-nov 1900; jan-nov 1901; jan-jun, set-nov 1902; jun-ago, out-dez 1908; jan, mar, maio-dez 1909; jan-jun, ago-30 dez 1910 – (= ser Ps 19) – mf#P11B,05,15 – bl Biblioteca [380]

Commercio do madeira : orgao especial do commercio – Manicore, AM: Typ Canto do Largo da Matriz, 11 maio 1884 – (= ser Ps 19) – mf#P11B,06,17 – bl Biblioteca [380]

O commercio do para – Belem, PA. 10 dez 1887 – (= ser Ps 19) – bl Biblioteca [380]

O commercio do porto – Oporto, Portugal. -d. 1 Aug 1916-9 Aug 1919. Imperfect. 6 reels – 1 – uk British Libr Newspaper [072]

Commercio mineiro : orgam dos interesses da classe commercial do estado – Ouro Preto, MG. 05 fev 1905 – (= ser Ps 19) – bl Biblioteca [079]

Commercio suburbano – Rio de Janeiro (Piedade), RJ. 15 maio-15 jul 1902 – (= ser Ps 19) – mf#DIPER – bl Biblioteca [079]

Commercium litterarium ad rei medicae et scientiae naturalis incrementum institutum quo quicquid novissime observatum agitatum scriptum vel peractum est succinte dilucideque exponitur – Norimbergae: Societas Litteris Joh Ernesti Adelbulneri 1731-45 [mf ed 2005] – 89mf [pl] – 9 – €550.00 – 3-89131-462-0 – gw Fischer [610]

Commerford, Sophia Elizabeth Jacoba *see* 'N konstruktivistiese beskrywing van veranderende persepsies in 'n welsynsorganisasie

Commissaire extraordinaire / Duvert, Felix-Auguste – Paris, France. 1840 – 1r – us UF Libraries [440]

Commissaires du havre de Montreal *see* Amendements...

Commissariat, Manekshah Sorabshah *see* A history of gujarat

Commissie Hulp aan Spanje *see* Spanje helpen is werken voor de vrede

Commission *see* Periodicals

Commission de colonisation de la province de Quebec *see*
– Rapport
– Rapport de la commission de colonisation de la province de quebec

Commission de la Fondation Piot *see* Monuments et memoires publies par l'academie des inscriptions et belles-lettres

Commission des ecoles catholiques de Montreal *see* Memoire presente au gouvernement de la province de quebec

Commission d'etude sur l'integrite du territoire du Quebec *see* Rapport

Commission du chemin de fer de Quebec, Montreal, Ottawa et Occidental *see*
– Rapport...sur les operations de la commission, et l'etendue et la nature des travaux executes jusqu'au 1er decembre 1877

Commission for Catholic Missions *see* Our negro and indian missions

Commission for the Elimination of Racism et al *see* Martin luther king community newsletter

Commission geologique du Canada *see*
– Rapport de progres depuis son commencement jusqu'a 1863
– Report of progress from its commencement to 1863

Commission Internationale pour l'Aide aux Refugies Espanols *see* Listes de souscription

Commission minutes, committee minutes and proposals, january 1966-june 1966 / Florida. Constitution Revision Commission – Tallahassee 1966. LL-2248 – 1 – us L of C Photodup [340]

Commission of assembly : and professor smith's reply to the committee's report / Innes, James – Edinburgh: John Maclaren, 1881. Princeton: Speer Lib, and Dep of Photodup, U of Chicago Lib, 1978 (1r); Evanston: American Theol Lib Assoc, 1984 (1r) – (= ser Case of william robertson smith in the free church of scotland) – 1 – 0-8370-0636-8 – mf#1984-6271 – us ATLA [242]

Commission of enquiry into the 1986 unrest and alleged mismanagement in kwandebele – Hamden, CT: Micrographic Systems of Connecticut 1992 – us CRL [350]

Commission of Inquiry into the Rehabilitation of the Worked-out Phosphate Lands in Nauru *see* Transcript of proceedings

Commission of the European Communities *see* The historical records of the high authority of the european coal and steel community, part 1

Commission on Interracial Cooperation *see* Southern frontier

Commissioner of Indian Affairs *see* Birdseye view of indian policy

Commissioner of indian affairs annual reports / U.S. Dept of the Interior – 1837-1968 [all publ] – (= ser Native american coll) – 504mf – 9 – $756.00 – (lacking: 1899 pt 2. 1901 pt2. 1902 pt2. 1961-62. 1964) – mf#llmc 88-002 – us LLMC [340]

The commissioners of the alms-house, vs. alexander whistelo, a black man : being a remarkable case of bastardy... – New York, NY: David Longworth, 1808 – 1r – 1 – us Western Res [340]

Commissioner's sic journal / Leavenworth County. Kansas. Board of Commissioners – 1855-62 – 1 – us Kansas [978]

Commitment / National Catholic Conference for Interracial Justice – 1969 jul-1974 sum – 1r – 1 – mf#1071178 – us WHS [230]

Commitment : a publication of the coca-cola company highlighting activities which further its commitment to responsible corporate citizenship – 1984 aug – 1r – 1 – mf#4881474 – us WHS [071]

Commitment to physical activity and body-image distortion in college students / Triola, Danielle P – 1996 – 2mf – 9 – $8.00 – mf#PSY 1964 – us Kinesology [150]

Committee Against Racism *see* Car wheel

The committee appointed by the honourable the legislative council and house of assembly : to consider and report upon the subject matter of certain resolutions of the house of assembly...respecting the financial concerns of this province with lower canada... / Haut-Canada. Parliament. Legislative Council – [York, Haut-Canada: s.n, 1821?] (mf ed 1996) – 1mf – 9 – mf#SEM105P1978 – cn Bibl Nat [330]

Committee for a Labor Party [US] *see* Workers' action

Committee for an Independent Canada *see* Independencer

Committee for Anglophone Social Action *see* Spec (new carlisle, quebec)

Committee for Free Press *see* Media monitor

The committee for moslem religious affairs – [Jerusalem, 1921] – 1mf – 9 – mf#J-28-64 – ne IDC [956]

Committee for Non-violent Revolution *see* Alternative

Committee for Original Peoples Entitlement *see* Inuvialuit

Committee for Puerto Rican Decolonization *see* Puerto rico libre!

Committee for the Preservation of Socialist Policies Within the Socialist Party *see* Socialist voice

Committee for the Re-election of the President *see* Re-elector

Committee of Enquiry into Breaches of International Law Relating to Intervention in Spain *see* Report and findings.

Committee of fifteen records, 1900-1901 : from the holdings of the rare books and manuscripts division – Center for the Humanities, the New York Public Library; Astor, Lenox, Tilden Foundations, 1997 – 17r – 1 – $2210.00 – (with guide) – mf#S3358 – us Scholarly Res [300]

Committee of Returned Volunteers *see* Crv

Committee of returned volunteers newsletter – v2 n?-v3 n5 [1968 mar-1969 jul] – 1r – 1 – (cont by: crv [committee of returned volunteers]) – mf#684918 – us WHS [360]

Committee of the Sturbridge Association *see* History of the baptist churches composing the sturbridge association from their origin to 1843

Committee of University Industrial Relations Librarians *see*
– Exchange bibliographies

Committee On Africa, The War, And Peace Aims *see* Atlantic charter and africa from an american standpoint

Committee on Canadian Labour History *see* Bulletin of the committee...

Committee on Convention Procedure and Jurisdiction, Marshall Islands Constitutional Convention *see* Draft constitution of ralik ratak, 1977

Committee on disarmament, 1962-1984 : meetings and documents – 30r – 1 – $5225.00 – 0-89093-517-3 – (with p/g) – us UPA [327]

Committee on organization of the ninth international medical congress : to be held in washington, d c in 1887 – [s.l: s.n, 1884?] [mf ed 1985] – 1mf – 9 – 0-665-01778-2 – mf#01778 – cn Canadiana [610]

Committee on public information : official bulletin – v1-3 in 8bks n1-575 may 10 1917 to mar 31 1919 [daily] [all publ] – 103mf – 9 – (title changed with v2 n389 aug 17 1918 to: official u s bulletin; cont by: the united states bulletin [apr 3 1919-] wh is not offered here) – mf#llmc97-210 – us LLMC [350]

Committee on Secondary School Examinations other than G C E *see* Report of the committee on secondary schools examinations other than g c e, 1958

Committee on the Religious Needs of Anglo-American Communities in Asia, Africa and South America *see* Tourist directory of christian work in the chief cities of the far east, india and egypt

Committee reports n1-67 : and outline of constitution and articles 1-16 – Saipan, n.p, 1975 – (= ser Micronesian constitutional convention, 1975) – 12mf – 9 – $18.00 – (various pagination) – mf#LLMC 82-100F, Title 91 – us LLMC [323]

Committee to Defend the Panthers *see* Panther 21 trial news

Committee to End the War in Vietnam *see* Crisis

Committee to Fight Repression [US] *see* Insurgent

Committee to Frame a World Constitution *see* Document

Committee to Rebuild the National Welfare Rights Organization et al *see* Organizer

Committee to secure justice for morton sobell. papers, 1950-68 – 1950-68 – 1 – $1440.00 – (court transcripts, correspondence, press releases, circulars, reports, and cttee records. incl transcript of the rosenberg-sobell trial & documentation relating to appeals by sobell) – mf#0165 – us Brook [347]

The committees of the continental congress, chosen to hear and determine appeals from courts of admiralty / Davis, John Chandler Bancroft – New York, Banks, 1888. 24 p. LL-438 – 1 – us L of C Photodup [976]

Commlaw conspectus : journal of communications law and policy – Catholic University: v1-7. 1993-99 – 9 – $118.00 set – ISSN: 1068-5871 – mf#115791 – us Hein [380]

Commodities – Cedar Falls. 1972-1983 (1) 1972-1983 (5) 1972-1983 (9) – (cont by: futures) – ISSN: 0279-5590 – mf#8097 – us UMI ProQuest [332]

Commoediae... / Varro, Marcus Terentius – 15th c – (= ser Holkham library manuscript books 300,339) – 1r – 1 – (filmed with: catilina et jugurtha by sallustius, orationes homeri et eulogium othonis by l aretinus) – mf#96515 – uk Microform Academic [450]

A common apologie of the church of england : against the...brownists... / Hall, J – London: Samuel Macham, 1610 – 2mf – 9 – mf#PW-47 – ne IDC [241]

Common bench reports, new series : cases argued and determined in the court of common pleas, the exchequer chamber and in the courts of error / Great Britain. England. Court of Common Pleas – v1-20. 1856-65. London: W G Benning & Co/Wm W Gearing, 1857-66 (all publ) – 197mf – 9 – $295.00 – (incl ind to series bound into v20) – mf#LLMC 95-270 – us LLMC [324]

Common bench reports, old series : cases argued and determined in the court of common pleas and in the exchequer chamber / Great Britain. England. Court of Common Pleas – v1-18 + index. 1845-56. London: Wm Benning & Co, 1846-56; 1858 (all publ) – 195mf – 9 – $292.00 – mf#LLMC 95-269 – us LLMC [324]

Common bond – v1 n1-v2 n8 [1974?-1978 nov] – 1r – 1 – mf#3899757 – us WHS [071]

The common bond – Powell River. B.C. Canada. jul-oct 1946 – 1 – cn Commonwealth Imaging [071]

Common carrier land mobile base station cumulative staff study listing / U.S. National Technical Information Service – Weekly – 9 – us NTIS [000]

Common carrier land mobile base stations data base (suppliers) / U.S. National Technical Information Service – Monthly.Lists all telephone companies in the Common Carrier Land Mobil Base Station System that provide service to individual consumers–in call sign sequence – 9 – us NTIS [000]

Common carrier microwave construction permit file / U.S. National Technical Information Service – Every two months. Permit data such as licensee name, address, transmitter, receiver, path link and antenna information. Data is in State, county latitude and longitude sequence – 9 – us NTIS [000]

Common carrier microwave pending application dump / U.S. National Technical Information Service – Every two months – 9 – us NTIS [621]

Common cause – 1980 oct-1983 – 1r – 1 – (cont: frontline [washington dc]; in common; cont by: common cause magazine) – mf#701276 – us WHS [071]

Common cause – Sydney. jul 1921-dec 1924, nov 1935-dec 1970; Labour daily supp jan 1925-mar 1931 – 14r – A$925.98 vesicular A$1002.98 silver – at Pascoe [079]

Common cause magazine – Washington. 1985-1996 (1,5,9) – ISSN: 0884-6537 – mf#15666,02 – us UMI ProQuest [320]

Common cause report from washington – v1 n3-v5 n4 [1971 feb-1975 mar] – 1r – 1 – (cont by: in common [washington dc: 1975]) – mf#646729 – us WHS [071]

Common Cause [US] *see*
– Frontline
– In common

Common cause, wisconsin – v2 n1-v8 n1 [1975 win-1982 spr] – 1r – 1 – (cont: wisconsin common cause news; cont by: common cause in wisconsin) – mf#659257 – us WHS [071]

Common colics of the horse / Reeks, Harry Caulton – Chicago, IL. 1912 – 1r – us UF Libraries [636]

Common council proceedings – 1983 apr 19-1987 apr 21, 1987 may 5-1989 apr 18, 1989 may 28-1991 apr 23, 1991 may 9-1993 apr 20 – 4r – 1 – mf#3046466 – us WHS [071]

Common crier – Raf Greeham Common, Raf Welford England. 1983 mar 17-1986 apr 25, 1986 aug 15-1987 aug 28, 1987 sep 4-1988 nov 11 – 3r – 1 – mf#1047594 – us WHS [071]

Common defense / American Defense Preparedness Association – n452-470 [1978 jun-1980 sep] – 1r – 1 – (cont: industrial preparedness bulletin; cont by: national defense) – mf#399329 – us WHS [355]

Common errors / Sherwood, Mrs – London, England. 18– – 1r – us UF Libraries [240]

Common errors respecting christian experience / Buckinghamshire Association Of Baptist Churches – London, England. 1832 – 1r – us UF Libraries [240]

Common forest trees of florida / Mattoon, Wilbur R – Jacksonville, FL. 1925 – 1r – us UF Libraries [634]

Common good – London, UK. 1880-81. -irr. 21 feet – 1 – uk British Libr Newspaper [072]

Common ground – London. 1946-1991 (1) 1971-1981 (5) 1976-1981 (9) – ISSN: 0010-325X – mf#961 – us UMI ProQuest [200]

Common ground – n5-7 [1982 spr-fall], v2 n1 [1982/83 win], 1983 apr-dec, 1984 jan/feb-1988 apr/may – 1r – 1 – mf#1110772 – us WHS [071]

Common ground / New Vocations Project – n1-7 [1974 spr-1976 win/spring] – 1r – 1 – mf#156310 – us WHS [331]

Common ground / Women's Coalition – 1977 sum, v3 n1-v9 n2 [1978 jan/feb-1984 sum] – 1r – 1 – mf#812641 – us WHS [071]

The common ground of confucianism, taoism and chinese buddhism *see*
– Chung-kuo san-chiao ti kung t'ung pen chih [ccm169]

Common knowledge – New York. 1992-1998 (1,5,9) – ISSN: 0961-754X – mf#18227 – us UMI ProQuest [301]

The common law; its origin, sources, nature and development, and what the state of new york has done to improve upon it / Daly, Charles Patrick – New York, Banks, 1894. 71 p. LL-539 – 1 – us L of C Photodup [346]

The common law jurisdiction and practice, according to statutes and decisions in the state of illinois, from the earliest period to 1872 / Hill, Edward Judson – Chicago: Myers, 1872. 2v. LL-779 – 1 – us L of C Photodup [348]

Common law practice in civil actions / Cox, Walter Smith – Washington, D.C., Morrison, 1877. 362 p. LL-521 – 1 – us L of C Photodup [346]

Common law practice in civil actions / Cox, Walter Smith – Washington, D.C., Morrison, 1880. 362 p. LL-530 – 1 – us L of C Photodup [346]

Common man series / Montgomery Co. Englewood – v1 n1. nov 1971-jan 1975 [mthly] – 1r – 1 – mf#B33693 – us Ohio Hist [071]

Common market law review – Dordrecht. 1988-1991 (1,5,9) – ISSN: 0165-0750 – mf#16775 – us UMI ProQuest [346]

Common market law review – v1-38. 1963-2001 – 9 – $2992.00 set – ISSN: 0165-0750 – mf#101921 – us Hein [346]

Common Market, Ltd *see* Chew on

Common people – Stillwater OK. 1904 aug 25, sep 1, nov 17 – 1r – 1 – mf#868987 – us WHS [071]

Common plants of florida's aquatic plant industry / Coile, Nancy C – Tallahassee, FL. 1995 – 1r – us UF Libraries [580]

Common pleas act, 1911 / Greaves, William Herbert – s.l, s.l? 1911 – 1r – us UF Libraries [323]

Common school advocate – Madison; Cincinnati. 1837-1841 – 1 – mf#3967 – us UMI ProQuest [370]

Common school assistant – Albany. 1836-1840 – 1 – mf#3698 – us UMI ProQuest [370]

Common school journal – Boston. 1838-1852 – 1 – mf#4443 – us UMI ProQuest [370]

The common school system : its principle, operation and results / Dallas, Angus – Toronto?: Thompson, 1855 – 1mf – 9 – mf#47458 – cn Canadiana [370]

Common scold / Collective [South Portland, Maine] – 1983 mar, jun-jul, sep-dec – 1r – 1 – mf#922811 – us WHS [334]

Common sense / Christian Educational Association – n268 [1957 jan 15]-1961, 1962 jan 1-1971 may 15, 1962 jan 1-1972 may 15 – 3r – 1 – (cont: think weekly) – mf#1054685 – us WHS [240]

Common sense – Johannesburg: Soc of Jews & Christians. v1-12. jul 1939-51 – 1 – us CRL [073]

Common sense – London. -w. 1 Aug 1824-5 Feb 1826; 7 Oct 1916-Dec 1919. 3 reels – 1 – uk British Libr Newspaper [072]

Common sense / Peoples Bicentennial Commission – v1 n7 [1972 oct?]-v4 n2 [1976 oct] – 1r – 1 – mf#203236 – us WHS [071]

Common sense / Socialist Party [CA] – Los Angeles CA. 1904 aug 20-1905 jul 15, 1905 jul 22-1909 aug 7 – 2r – 1 – (cont: los angeles socialist) – mf#709796 – us WHS [335]

Common sense – Springfield MA. v1 n2-15 [1969 mar 15-oct 15] – 1r – 1 – mf#1110779 – us WHS [071]

Common sense – Union. 1946-1972 (1) – ISSN: 0010-3306 – mf#7583 – us UMI ProQuest [320]

Common sense – v1 n1-v2 n10 [1971 apr 15-1972 mar 27], v3 n5 [1972 dec 1/15], v5 n6-v6 n3 [1973 dec-1974 apr] – 1r – 1 – mf#823891 – us WHS [071]

Common sense – v1 n5 [1895 jan 12] – 1r – 1 – mf#1336298 – us WHS [071]

Common sense – v1-15 n2,1. 1932-46 [all publ] – (= ser Radical periodicals in the united states, 1881-1960. series 1) – 63mf – 9 – $585.00 – us UPA [303]

Common sense and the disestablishment question – London, England. 1885 – 1r – us UF Libraries [240]

Common sense coalition newsletter – 1978 oct-1982 feb/mar – 1r – 1 – mf#648872 – us WHS [071]

Common sense, common good – 1975 oct-1976 nov – 1r – 1 – mf#361199 – us WHS [071]

The common sense of the constitution of the united states / Southworth, A T – Boston, etc: Allyn & Bacon 1924 – 2mf – 9 – $3.00 – mf#llmc92-231 – us LLMC [323]

Common sense versus judicial legislation. / Sargent, John Osborne – New York: Putnam, 1871. 34p. LL-580 – 1 – us L of C Photodup [340]

Commonefactio historica de statv eivs temporis, cui inserta est breuiter confessio de doctrina iustificationis et bonorum operum / Major, G – Witebergae, 1567 – 1mf – 9 – mf#TH-1 mf 919 – ne IDC [242]

Commoner – Colfax, WA. 1888-1932 (1) – mf#66973 – us UMI ProQuest [071]

Commoner / Hamilton Co. Cincinnati – sep 1868-aug 1869 [wkly] – 1r – 1 – mf#B5604 – us Ohio Hist [071]

The commoner / ed by Bryan, William Jennings – Lincoln, NE: William J Bryan. v1 n1. jan 23 1901-v23 n4. apr 1923 (mthly) [mf ed 1949] – 9r – 1 – (publ in periodical format, aug 13 1913- . publ varies: charles w bryan. v2 n5-v23 n4 called also whole n57-768) – us NY Public [071]

The commoner / ed by Bryan, William Jennings – Lincoln, NE: William J Bryan. 23v. v1 n1. jan 23 1901-v23 n4. apr 1923 (mthly) [mf ed 1968] – 10r – 1 – (publ in periodical format aug 13 1913- . publ varies. v2 n5-v23 n4 called also whole n57-768) – us NE Hist [071]

The commoner – Lincoln, Nebraska. v1-23 n4. jan 23 1901-apr 1923 – 1 – us NY Public [320]

Commonism *see* K'ung hsiang chu i (ccm105)

Commonitorium *see* Schriften ueber den hl martinus (bdk20 1.reihe)

The commonitorium of vincentius of lerins / ed by Moxon, Reginald Stewart – Cambridge: University Press, 1915 [mf ed 1991] – (= ser Cambridge patristic texts) – 1mf – 9 – 0-524-00194-4 – mf#1989-2894 – us ATLA [240]

The commonly received version of the new testament of our lord and savior jesus christ : with several hundred emendations / ed by Cone, Spencer Houghton & Wyckoff, William Henry – New-York: Lewis Colby, 1850 – 1mf – 9 – 0-8370-1301-1 – mf#1987-6038 – us ATLA [225]

Commonplace book : from the library of the royal college of surgeons, down house, kent / Darwin, Erasmus – 1r – 1 – mf#96764 – uk Microform Academic [920]

Commonplace book / Lowndes, William Thomas – 1803 [mf ed 1981] – 1mf – 9 – mf#51-544 – us South Carolina Historical [025]

Commonplace book *see* State papers and family documents / commonplace book

Commons : for industrial justice, efficient philanthropy, educational freedom and the people's control of public utilities – Chicago. 1896-1905 (1) – mf#5148 – us UMI ProQuest [331]

Commons *see* Survey

Commons, John R *see* Wisconsin progressives

Commons, John Rogers *see* Races and immigrants in america

Commons, open spaces and footpaths : preservation society journal / Preservation Society – London. 1978-1981 (1) 1978-1981 (5) 1978-1981 (9) – ISSN: 0010-3322 – mf#7289 – us UMI ProQuest [790]

Commonsense / Hoopa Valley Indian Reservation – v2 iss 2 [1979 mar], v3 iss 24-v4 iss 21 [1980 dec 22-1981 dec 28] – 1r – 1 – mf#944698 – us WHS [307]

Commonsense : newsmonthly of the san francisco socialist coalition / Northern California Alliance – 1973 oct-1978 apr – 1r – 1 – mf#513617 – us WHS [335]

Commonsense : philadelphia action report / Philadelphia Resistance [PA] – v3 n8-v5 n8 [1974 may 1-1976 sum] – 1r – 1 – mf#359835 – us WHS [071]

Commonsense *see* Political pamphlets... 19th c

Common-sense clothing / Barnett, Edith A – [London], New York: Ward, Lock, & Co [1882] – (= ser 19th c art & architecture) – 2mf – 9 – mf#4.1.70 – uk Chadwyck [640]

A commonsense digest of american negligence cases – Chicago: Callaghan. 1v. 1914 (all publ) – (= ser American negligence cases) – 13mf – 9 – $19.50 – (covers mainly american negligence cases but recommended that it be used also for american negligence reports) – mf#LLMC 84-699C – us LLMC [340]

Common-sense in religion : a series of essays / Clarke, James Freeman – 13th ed. Boston: Houghton, Mifflin, 1890, c1873 – 2mf – 9 – 0-8370-9851-3 – mf#1986-3851 – us ATLA [240]

A commonsense index to the notes of american negligence cases – Chicago: Callaghan. 1v. 1914 (all publ) – 2mf – 9 – $3.00 – mf#LLMC 84-699D – us LLMC [348]

Common-sense theology : a second series of tracts for the times – London: British and Foreign Unitarian Association, 1893 – 2mf – 9 – 0-524-07858-0 – mf#1991-3403 – us ATLA [240]

Commonweal – [London & SE] Hammersmith & Fulham Archives feb 1885-12 may 1894 [mf ed 2003] – 1 – uk Newsplan [072]

Commonweal – New York. 1924+ (1) 1965+ (5) 1960+ (9) – ISSN: 0010-3330 – mf#330 – us UMI ProQuest [073]

Commonweal : organ of the socialist league – (= ser Radical periodicals of great britain, 1794-1914. period 2) – 2r – 1 – $230.00 – us UPA [335]

The commonweal / Socialist League – London. feb. 1885-sep. 1892; may 1893-may 1894 [wkly] – 2r – 1 – uk British Libr Newspaper [320]

Commonweal, 1885-94 : the official organ of the london socialist league – 2r – 1 – mf#5217 – uk Microform Academic [073]

Commonweal/peak hill golden age – Peak hill – 1r – A$53.68 vesicular A$59.18 silver – at Pascoe [079]

Commonwealth – A Monthly Magazine and Library of Sociology. v. 1-9. 1893-1902 – 1 – 85.00 – us L of C Photodup [335]

Commonwealth – Richmond VA. 1880 mar 4,26 – 1r – 1 – (cont by: state [richmond va: 1876]) – mf#882384 – us WHS [071]

Commonwealth – Fond Du Lac WI. 1885 apr 3/sep 25-1901 jan 2-aug 2 – 14r – 1 – (cont: fond du lac commonwealth; cont by: commonwealth and the commercial) – mf#941400 – us WHS [071]

Commonwealth – Ripon WI. 1882 oct 20 – 1r – 1 – (cont: ripon commonwealth [ripon wi: 1864]; cont by: ripon commonwealth [ripon wi: 1887]) – mf#1012812 – us WHS [071]

Commonwealth – Duluth, MN. 1893-1896 (1) – mf#63909 – us UMI ProQuest [071]

Commonwealth – Gary, IN. 1930-1932 (1) – mf#62787 – us UMI ProQuest [071]

Commonwealth – Mineral Point WI. 1840 aug 25 – 1r – 1 – mf#1109177 – us WHS [071]

Commonwealth : official journal of the commonwealth club of california / Commonwealth Club of California – San Francisco. 1972+ (1) 1972+ (5) 1975+ (9) – ISSN: 0010-3349 – mf#7080 – us UMI ProQuest [350]

Commonwealth – [Scotland] Glasgow oct 1853-2 feb 1861 [mf ed 2003] – 10r – 1 – uk Newsplan [072]

Commonwealth / Socialist Party [Washington] – Everett WA. 1912 nov 12, dec 27, 1913 feb 27, mar 7, apr 11, 18, oct 9,30, 1914 jan 22, apr 9 – 1r – 1 – mf#869013 – us WHS [325]

Commonwealth – Sydney – 1r – A$31.46 vesicular A$36.96 silver – at Pascoe [079]

Commonwealth / Virginia State Chamber of Commerce – v51 n3,10-v52 n3 [1984 mar, 1984 oct-1985 mar] – 1r – 1 – (cont: metro; richmond lifestyle) – mf#970710 – us WHS [380]

Commonwealth *see* Harrisburg bulletin (harrisburg, or: 1901)

The commonwealth – Freetown. Sierra Leone. -f. 4 Aug-1 Sep 1888. (4 ft) – 1 – uk British Libr Newspaper [072]

Commonwealth advocate – Manila. v1 n10-v. 7 n9. nov 1935-oct 1941 – 1 – us NY Public [073]

Commonwealth and the commercial – Fond Du Lac WI. 1901 aug 6-1902 feb 28 – 1r – 1 – (cont: commonwealth [fond du lac wi]) – mf#941403 – us WHS [071]

Commonwealth Club of California *see* Commonwealth

The commonwealth covenant : and related u s documents – Saipan: the Commonwealth Govt, n.d. – 1mf – 9 – $1.50 – mf#llmc82-100j, title 14 – us LLMC [348]

The commonwealth covenant : hearing before the senate committee on armed services, 94th cong, 1st sess, nov 17 1975 / United States. Congress – Washington: GPO 1976 – 2mf – 9 – $3.00 – mf#llmc82-100j, title 7 – us LLMC [342]

The commonwealth covenant : hearing before the subcommittee on territorial and insular affairs, house comm on interior and insular affairs, 94th cong, 1st sess, jul 14 1975 / United States. Congress – Washington: GPO 1975 – 8mf – 9 – $12.00 – mf#llmc82-100j, title 8 – us LLMC [342]

Commonwealth covenant, 1975 : with technical agreement and history: n.a. – n.p., n.d. – 2mf – 9 – $3.00 – mf#llmc82-100j, title 10 – us LLMC [348]

Commonwealth digest – London. 1964-1964 (1) – ISSN: 0588-7607 – mf#5923 – us UMI ProQuest [332]

Commonwealth engineer – Melbourne, Australia. -m. 1 Mar 1920; 1 Jan-1 Dec 1921. Lacking Dec 1920 – 2r – 1 – uk British Libr Newspaper [620]

Commonwealth england / Brown, John – London: National Council of Evangelical Free Churches 1904 [mf ed 1990] – (= ser Eras of nonconformity 6) – 1mf – 9 – 0-7905-4607-8 – mf#1988-0607 – us ATLA [941]

Commonwealth [everett, wa] – Everett WA: Commonwealth Pub Co, [wkly] – 1 – (began in 1911; cont by: washington socialist) – us Oregon Lib [335]

Commonwealth [harrisburg, or] – Harrisburg OR: C A Dimond, -1916 [wkly] [mf ed 1964] – 1r – 1 – (merged with: harrisburg bulletin [harrisburg, or: 1901] to form: harrisburg bulletin and commonwealth [1916-25]) – us Oregon Lib [071]

Commonwealth home – Sydney, Australia. -w. 14 Aug 1925-1 Nov 1927. 1 1 2 reels – 1 – uk British Libr Newspaper [072]

Commonwealth Institute *see* The arts of the hausa

Commonwealth law review – Australia. v1-6. 1903-09 (all publ) – 6mf – 9 – $27.00 – mf#LLMC 84-444 – us LLMC [340]

Commonwealth of australia gazette – Canberra, etc, 1950-jun 1973 – 59r – us CRL [980]

Commonwealth of australia gazette – Canberra. General ed: jul 1977-apr 28 1987 (24r); Public Service ed: jul 1977-dec 20 1984 (20r) – us CRL [980]

Commonwealth of massachusetts publications – ongoing – 5,6 – mf#C39-28870 – us Primary [324]

Commonwealth producer – London. 1950-1952 (1) – ISSN: 0010-342X – mf#561 – us UMI ProQuest [338]

Commonwealth register – adopted regulations : reprint of regulations published in the register between oct 1978 and feb 1 1985 – Saipan: Office of the Attorney General 1985 – 2v on 18mf – 9 – $27.00 – (with ind) – mf#llmc82-100j, title 13 – us LLMC [348]

Commonwealth scientific council biological diversity project / Walls, Geoff – 1986-87 – 1r – 1 – mf#pmb doc395 – at Pacific Mss [574]

Commonwealth Solar Observatory [Australia] *see* Memoirs of the commonwealth solar observatory, mount stromlo, canberra, australia

Commonwealth spy – Carlisle, PA., 1823 – 13 – $25.00r – us IMR [071]

The commonwealths and the kingdom : a study of the missionary work of state conventions / Padelford, Frank William – Philadelphia: Griffith & Rowland, 1913 – 1mf – 9 – 0-7905-6939-6 – mf#1988-2939 – us ATLA [240]

Commonwoman – v1 n1-v6 n4 [i.e. v5 n5] [1978 aug-1982 sep] – 1r – 1 – (cont by: commonwomon) – mf#1042507 – us WHS [071]

Commonwomon – v6 n5-v8 n4 [1982 oct-1984 sep/oct], 1985 jan – 1r – 1 – (cont: commonwoman) – mf#1042512 – us WHS [071]

The communal triangle in india / Mehta, Asoka & Patwardhan, Achyut – Allahabad: Kitabistan, 1942 – (= ser Samp: indian books) – us CRL [280]

Communalanzeiger fuer die staedte im regierungsbezirk breslau, liegnitz und oppeln – Oels (Olesnica PL), 1854-66 – 1 – (title varies: n2 1855: communalanzeiger fuer die staedte im regierungsbezirk breslau, bromberg, liegnitz, oppeln und posen) – gw Misc Inst [350]

Communal-blatt – Koenigsberg [Kaliningrad RUS], 1914 may, 1915 mar/apr-1944 jun – 92r – 1 – (with gaps. title varies: 7 may 1878: koenigsberger allgemeine zeitung; filmed by other misc inst: 1877 24 may & 1890 16 mar, 1914 may & 1921 [gaps], 1925 1 nov [jub-no], 1929 feb-apr, 1939 oct-1944 jul, 1944 oct-dec [27r]) – gw Misc Inst [077]

Communale *see* Heidelberger rundschau

Communate des chretiens – 1960-85 (complete) – 3r – 1 – mf#ATLA S0737 – us ATLA [242]

La communaute : senate. debats – 1959-1960 – 1 – us NY Public [320]

La communaute noire au quebec : bilan de la tournee organisee par le mouvement quebecois pour combattre le racisme – Montreal: le Mouvement, [1980] (mf ed 1994) – 1mf – 9 – mf#SEM105P2061 – cn Bibl Nat [321]

Communaute ou secession? / Ehrhard, Jean – Paris, Calmann-Levy [1959] – us CRL [944]

Communautes syriaques en iran et irak des origines a 1552 / Fiey, J M – London, 1979 – 7mf – 8 – €15.00 – ne Slangenburg [240]

Commune / ed by Aldred, Guy A – [Scotland] Glasgow: Bakunin Press may 1923-dec 1929 [mf ed 2003] – 2r – 1 – (incl special anti-parliamentary communist gazette) – uk Newsplan [335]

Commune – puis Revue litteraire. pour la defense de la culture. Paris. juil 1933-sept 1939 – 1 – fr ACRPP [800]

La commune – Paris. Dir. politique Felix Pyat. no. spec. du 17 aout, no. 1-45. 21 sept-4 nov 1880 – (= ser Journal quotidien, politique et socialiste) – 1 – (aka: journal quotidien, politique et socialiste) – fr ACRPP [074]

La commune – Paris: Impr Dubuisson et Ce, mar 26-31, apr 1-8,10-16,20-30, may 1,3-4,6-15,17-19 1871 – (filmed as pt of: commune de paris newspapers; newspapers on these reels are filmed chronologically, not alphabetically.) – us CRL [074]

La commune – Geneve. Almanach socialiste pour l'annee 1877 – 1 – fr ACRPP [325]

La commune – no. 1-6. Paris. avr-sept 1874 – 1 – fr ACRPP [335]

La commune / Parti Communiste Internationaliste Bolchevik-Leniniste pour la Construction de la IVe Internationale – Paris. dec 1935-38 – 1 – fr ACRPP [335]

The commune *see* The works of guy aldred

La commune de paris – Paris. n1-86 . 9 mars-7 juin 1848; n1-2. fevr-mars 1849 – 1 – fr ACRPP [073]

La commune devoilee par un ami des travailleurs – Paris: [Auguste Petit, 1871] – (filmed as pt of: commune de paris newspapers; newspapers on these reels are filmed chronologically, not alphabetically) – us CRL [074]

Commune sigilli secret (anno 1412-1416) / Fernando 1 – Barcelona – 1r – 5,6 – sp Cultura [946]

Communicant's remembrancer / Jackson, Miles – London, England. 1817 – 1r – us UF Libraries [240]

Communicate / Northwest Territories Teacher's Association – 1981 jan-1989 jan – 1r – 1 – (cont: northwest territories teachers' association's newsletter) – mf#1054700 – us WHS [370]

Communication abstracts – Thousand Oaks. 1989+ (1,5,9) – ISSN: 0162-2811 – mf#17944 – us UMI ProQuest [380]

Communication between adolescents and parents in restructured families / Sibiya, Tembisa Bellinda – Uni of South Africa 2000 [mf ed Johannesburg 2000] – 4mf [ill] – 9 – (incl bibl ref) – mf#mfm15065 – sa Unisa [302]

Communication between cultures and the importance of english as an international language : an interdisciplinary approach / Rink, Thomas – (mf ed 2002) – 70p 1mf – 9 – €30.00 – 3-8267-2776-2 – mf#DHS 2776 – gw Frankfurter [420]

Communication booknotes quarterly: cbq – Mahwah. 1998+ (1) – ISSN: 1094-8007 – mf#22365,03 – us UMI ProQuest [380]

Communication bulletin / Wisconsin Board of Vocational, Technical and Adult Education – 1971-75, 1975-78, 1978-81 – 3r – 1 – (cont: informational letters) – mf#362940 – us WHS [374]

Communication disorders quarterly – Austin. 1999+ – 1,5,9 – (cont: journal of children's communication development: jccd) – ISSN: 1525-7401 – mf#17529,02 – us UMI ProQuest [370]

Communication du greffier de la couronne en chancellerie : transmettant le rapport du nombre de votes donnes durant la derniere election, etc / Canada (Province) – Toronto: impr par John Lovell...1858 [mf ed 1983] – 1mf – 9 – mf#SEM105P169 – cn Bibl Nat [325]

Communication education – Annandale. 1976+ (1) 1976+ (5) 1976+ (9) – (cont: speech teacher) – ISSN: 0363-4523 – mf#1590,01 – us UMI ProQuest [370]

Communication et langage – Paris. mars 1969-75; 1985-1993 – 1 – fr ACRPP [302]

Communication monographs – Annandale. 1976+ (1) 1976+ (5) 1976+ (9) – (cont: speech monographs) – ISSN: 0363-7751 – mf#1591,01 – us UMI ProQuest [302]

The communication of genderization in sport : a content analysis of women's national basketball association and national basketball association media guides / Luif, Jennifer – 1999 – 1mf – 9 – $4.00 – mf#PSY 2087 – us Kinesology [306]

Communication par h jeannotte, ecr, mp, a ses electeurs du comte de l'assomption : tarte contre laurier / Jeannotte, Hormidas – Montreal: [s.n.], 1894 [mf ed 1980] – 1mf – 9 – 0-665-07531-6 – mf#07531 – cn Canadiana [320]

The communication process between a coach and an athlete as a predictor of success of failure / Poitras, John D – 1997 – 1mf – 9 – $4.00 – mf#PSY 1955 – us Kinesology [150]

Communication quarterly – University Park. 1976+ (1) 1976+ (5) 1977+ (9) – (cont: today's speech) – ISSN: 0146-3373 – mf#1576,01 – us UMI ProQuest [380]

Communication reports. – Pullman. 1988+ (1,5,9) – ISSN: 0893-4215 – mf#17461 – us UMI ProQuest [370]

Communication research – Beverly Hills. 1974+ (1,5,9) – ISSN: 0093-6502 – mf#12642 – us UMI ProQuest [380]

Communication strategies of women principals of secondary schools / Thakhathi, Tshilidzi – Uni of South Africa 2001 [mf ed Johannesburg 2001] – 5mf – 9 – (incl bibl ref) – mf#mfm14783 – sa Unisa [302]

Communication studies – West Lafayette. 1989+ – 1,5,9 – (cont: central states speech journal) – ISSN: 1051-0974 – mf#14901,01 – us UMI ProQuest [370]

Communication theory: ct – New York. 1991+ (1,5,9) – ISSN: 1050-3293 – mf#18385 – us UMI ProQuest [380]

Communication world – San Francisco. 1983+ (1,5,9) – ISSN: 0744-7612 – mf#14450 – us UMI ProQuest [650]

Communications – Englewood. 1965-1996 (1) 1973-1996 (5) 1973-1996 (9) – ISSN: 0010-356X – mf#8074 – us UMI ProQuest [380]

Communications / Professional Institute of the Public Service of Canada – 1984 jan 27-1994 dec – 1r – 1 – (cont: communications) – mf#1110793 – us WHS [380]

Communications, 1862-1990 – 474mf – 9 – $4580.00 – 1-55655-475-3 – (p/g only $500) – us UPA [380]

Communications and the law – v1-23. 1979-2001 – 5,6,9 – $696.00 set – (v1-6 1979-84 on reel $143. v7-23 1985-2001 on mf $553) – ISSN: 0162-9093 – mf#108721 – us Hein [340]

Communications between the colonial office and the governors of upper and lower canada : on the subject of the civil government of canada, as established by the act of 31 geo 3... / Grande-Bretagne. Colonial Office – [S.l.]: [s.n.], [1830] (mf ed 1982) – 1mf – 9 – mf#SEM105P103 – cn Bibl Nat [323]

Communications convergence – San Francisco, 2001+ [1,5,9] – ISSN: 1534-2840 – mf#25832,01 – us UMI ProQuest [380]

Les communications de mercator : sur la conteste entre le comte de selkirk, et la compagnie de la baye d'hudson d'une part, et la compagnie du nord-ouest... / Ellice, Edward – Montreal: C B Pasteur & H Meziere, 1817 [mf ed 1971] – 1r – 5 – mf#SEM16P37 – cn Bibl Nat [380]

Les communications des indigenes du kasai avec les ames des morts / Tiarko Fourche, J A & Morlighem, H – Bruxelles: G van Campenhout, 1939 – 1 – us CRL [290]

Communications in applied numerical methods – Chichester. 1985-1992 (1,5,9) – ISSN: 0748-8025 – mf#14804 – us UMI ProQuest [510]

Communications in mathematical physics – Heidelberg. 1965-1995 (1) 1965-1995 (5) 1965-1995 (9) – ISSN: 0010-3616 – mf#13156 – us UMI ProQuest [510]

Communications in psychopharmacology – New York. 1977-1980 (5,9) – ISSN: 0145-5699 – mf#49254 – us UMI ProQuest [615]

Communications journal / American Radio Telegraphists Association – v1 n10-12 [1937 jun-aug] – 1r – 1 – mf#1816617 – us WHS [380]

Communications news – Nokomis. 1964+ [1]; 1991+ [5,9] – ISSN: 0010-3632 – mf#1749 – us UMI ProQuest [380]

Communications of the association for computing machinery – Baltimore. 1958-1959 (1) 1958-1959 (5) 1958-1959 (9) – (cont by: communications of the acm) – mf#12688 – us UMI ProQuest [000]

Communications on pure and applied mathematics – New York. 1948+ (1) 1948+ (5) 1948+ (9) – ISSN: 0010-3640 – mf#11051 – us UMI ProQuest [510]

Communications on the case of professor robertson smith : in the general assembly of the free church of scotland, held at glasgow in 1878 / Moncreiff, Henry Wellwood, Sir – Edinburgh: John Maclaren, 1879. Chicago: Dep of Photodup, U of Chicago Lib, 1978 (1r); Evanston: American Theol Lib Assoc, 1984 (1r) – (= ser Case of william robertson smith in the free church of scotland) – 1 – 0-8370-0619-8 – mf#1984-6292 – us ATLA [242]

Communications to the national academy of sciences / Carnegie Institution of Washington, Mount Wilson Solar Observatory – [Washington DC: s.n.] 1918- (irreg) [v1-3(1915-38)] [mf ed 1999] – 3v on 1r – 1 – (repr fr: proceedings of the national academy of sciences; ceased with n122; n57-122 iss by the observatory under its later name: mount wilson observatory) – ISSN: 1063-3022 – mf#film mas c4199 – us Harvard [520]

Communications Workers of America *see*
- Cable
- Communicator
- Contactor
- Cwa news
- Cwa voice
- Cwa weekly news letter
- Cwa wire tap
- Cwa-cio coast coordinator
- Hot line
- Labor news
- Local 3603 news
- Local 4400 newsletter
- Local 5502 communique
- Local 13000 news
- Minutes, executive board
- Minutes, executive board report
- News-report
- Orange empire news
- Tele time
- Tele-a-fact
- Typographical news
- Unioneer

Communications Workers of America et al *see* No 6 bulletin

Communications Workers of Canada *see* Cwc news

Communicationsweek – Manhasset. 1991-1995 (1,5,9) – (cont by: internetweek) – ISSN: 0746-8121 – mf#19185 – us UMI ProQuest [380]

The communicative approach for effective english second language teaching and learning in grade eight / Thobedi, Alfred Motsamai – Pretoria: Vista University 2002 [mf ed 2002] – 3mf – 9 – (incl bibl ref) – mf#mfm15213 – sa Unisa [420]

Communicative speaking / Gray, John Stanley – Boston, MA. 1928 – 1r – us UF Libraries [400]

Communicative speaking / Gray, John Stanley – Boston, MA. 1928 – 1r – us UF Libraries [400]
Communicator – Cincinnati, Ohio. Aug 1976-Aug 1984 – 1 – us AJPC [071]
Communicator / Communications Workers of America – 1941 may-1951 feb – 1r – 1 – mf#599518 – us WHS [380]
Communicator / Communications Workers of America – 1981 apr-1987 sep – 1r – 1 – mf#1313295 – us WHS [380]
Communicator / Communications Workers of America – 1984 dec-1989 jul – 1r – 1 – mf#1508926 – us WHS [380]
Communicator / Communications Workers of America – 1984-87 – 1r – 1 – mf#1313293 – us WHS [380]
Communicator – v1 n6 [v8 n5 [1970 dec-1978 feb] – 1r – 1 – (cont by: great lakes communicator) – mf#176326 – us WHS [071]
Communicator – Albany. 1978-1980 – 1,5,9 – (cont by: outdoor communicator) – mf#11710 – us UMI ProQuest [370]
Communicator – 1980 dec-1984 apr – 1r – 1 – (cont: simsoc) – us WHS [071]
Communicator – Hawaii. 1984 dec-1993 dec – 1r – 1 – mf#1071342 – us WHS [071]
Communicator / Madison Area Community of Churches – 1973 mar-1985 dec – 1r – 1 – (cont: malc communicator) – mf#1047937 – us WHS [242]
Communicator / New York State Public Employees Federation, AFL-CIO – 1979 may-1988 – 1r – 1 – mf#1614490 – us WHS [350]
Communicator / Western Conference of Teamsters – 1977 mar-1978 jun – 1r – 1 – mf#398304 – us WHS [302]
Communicator / Wisconsin Association of Homes and Services for the Aging – 1974 jul-1979 dec 19, 1980 jan 16-1987 dec 29 – 2r – 1 – (cont: wcha communicator) – mf#599518 – us WHS [360]
Communicator news – Cleveland, Columbus OH. 1992 dec 17/23, 1992 jan 8-aug 26, 1994 jan 6/12-jun 30/jul 6, 1994 jul 7/13-dec 29/jan 4,1995, 1995 jan 5/11-jun 29/jul 5, 1995 jul 6/12-nov 23/29 – 5r – 1 – (cont by: columbus minority communicator) – mf#2901375 – us WHS [302]
Communicator news – Kenosha, Racine WI. 1985 nov-1988 dec 5, 1987 jan 2-dec 16, 1989 jun 2-1990 nov 14, 1991 jan 23-dec 31 – 5r – 1 – mf#1239690 – us WHS [380]
Communio : american edition – Washington. 1980+ (1,5,9) – ISSN: 0094-2065 – mf#12713 – us UMI ProQuest [240]
Communio – Santiago, Chile: Communio. v1-7 n25. jun/jul 1982-1991 – 2r – us CRL [079]
Communion : french edition – Taize. 1947-1981 [1]; 1971-1981 [5]; 1977-1981 [9] – ISSN: 0042-370X – mf#1904 – us UMI ProQuest [240]
Communion – 11 miscellaneous pamphlets – (= ser The Terms Of Communion At The Lord's Table) – 1 – $13.30 – (includes: cone, spencer h. the terms of communion at the lord's table. n.d. haldane, james alex. the foundation of the observance of the lord's day and of the lord's supper. 1807) – us Southern Baptist [242]
The communion of life / Fry, Joan Mary – London: Publ for the Woodbrooke Extension Committee by Headley, 1910 – 1mf – 9 – 0-8370-8901-8 – mf#1986-2901 – us ATLA [240]
Communion of saints / Garbett, James – Brighton, England. 1842 – 1r – us UF Libraries [240]
Communion of saints / Moore, Daniel – London, England. 1859 – 1r – us UF Libraries [240]
Communion of saints / Sutcliffe, Joseph – London, England. 1815 – 1r – us UF Libraries [240]
The communion of saints : an attempt to illustrate the true principles of christian union / Wilson, Henry Bristow – Oxford: W Graham; London: Hatchard, 1851 – (= ser Bampton lectures) – 1mf – 9 – 0-7905-9765-9 – mf#1989-1490 – us ATLA [240]
The communion of the christian with god = Verkehr des christen mit gott / Herrmann, Wilhelm – 2nd English ed, enl and altered in accordance with the 4th German ed of 1903. New York: Putnam, 1906 – (= ser Crown Theological Library) – 1mf – 9 – 0-7905-3952-7 – (in english) – mf#1989-0445 – us ATLA [240]
Communion sermon manuscripts / Tennent, William – 1r – 1 – $50.00 – us Presbyterian [240]
The communion table : the approach, the service, the retrospect / Boyd, James Robert – Philadelphia: Presbyterian Board of Publ, c1866 – 2mf – 9 – 0-524-02387-5 – mf#1990-4289 – us ATLA [240]
Communion wine and bible temperance : being a review of dr. thos. laurie's article in the bibliotheca sacra, of january 1869 / Thayer, William Makepeace – [2nd ed]. New York: National Temperance Society and Publ House, 1870, c1869 – 1mf – 9 – 0-7905-0236-4 – mf#1987-0236 – us ATLA [240]
Communion wine question / Reid, William – Glasgow, Scotland. 1873 – 1r – us UF Libraries [240]
Communique : los angeles county employees seiu, local 660, consumer publication / Service Employees International Union – v5 n8, 10-11, 13-v15 n8 [1977 jul 1, sep 1-oct 1, dec, 1-1987 may/jun-oct] – 1r – 1 – mf#1672259 – us WHS [331]
Communique / Association for Preservation Technology – v4 n4-6 [1975 apr-dec], v5 n1, 3, 5-6 [1976 feb, jun [with suppl], oct-dec] – 1r – 1 – (cont: newsletter; cont by: apt communique) – mf#328102 – us WHS [770]
Communique / Engineers' Association and Scientists' Association [Marconi] – 1980 fall-1985 dec – 1r – 1 – mf#1476945 – us WHS [621]
Communique / Gouvernement revolutionnaire de l'Angola en exil – [Alger]: Gouvernement. n45 feb 4 1965; n51-52 mar 26-29 1965; n54 apr 15 1965; jul 12 1972 – us CRL [320]
Communique / Partido Africano da Independencia da Guine e Cabo Verde – Conakry: O Partido, jun 9 1961-sep 25 1972 – us CRL [320]
Communique : quarterly newsletter / New Jersey Coalition of One Hundred Black Women – 1983 spr – 1r – 1 – mf#4881507 – us WHS [305]
Communique de la chambre de commerce francaise de londres – London, UK. jun 1913-jan/feb 1916 – 1 – (aka: bulletin de la chambre de commerce francaise de grande-bretagne, oct 1943-sep 1956; bulletin de la chambre de commerce francaise de londres, mar/apr-jun 1916, 1917-jun 1924, 1930-mar 1934; deux cotes du detroit, apr 1934-sept 1943; revue du commerce franco-britannique, oct 1956-73) – uk British Libr Newspaper [380]
Communique de presse du deuxieme congres national du kampuchea [24 et 25 fevrier 1975] / Congres national du Kampuchea, 2d, 1975 – Moscou [Russia]: Ambassade royale du Cambodge en URSS [1975] [mf ed 1989] – 1r with other items – 1 – mf#mf-10289 seam reel 017/19 [§] – us CRL [959]
Communiques of military operations – dec 11 1967-may 2 1969 – us CRL [355]
Communiques of military operations – [mid-sep 1968-mid-dec 1969] – us CRL [355]
Communism and socialism : minutes of the first german evangelical lutheran congregation u.a.c. at st. louis, mo / Walther, Carl Ferdinand Wilhelm – St Louis, MO: Luth. Concordia Pub House, 1879 – 1mf – 9 – 0-524-06672-8 – mf#1991-2727 – us ATLA [335]
Communism and the theologians : study of an encounter / West, Charles C – New York: Macmillan, c1958 – 1mf – 9 – 0-524-08151-4 – (incl bibl ref) – mf#1993-9057 – us ATLA [240]
Communism in spain, 1931-1936 / Godden, Gertrude M – New York: America Press [1937] – 9 – (repr fr dublin review, oct 1936) – mf#w918 – us Harvard [335]
Der communismus der maehrischen wiedertaeufer im 16. und 17. jahrhundert : beitraege zu ihrer geschichte, lehre und verfassung / Loserth, Johann – [Wien: F Tempsky, 1895] – (= ser Archiv fuer oesterreichische geschichte) – 1mf – 9 – 0-524-05439-8 – mf#1990-1471 – us ATLA [943]
Communist / Communist Party of America – ser1: n1-7 1919 [all publ]. ser2: v1-3. 1919-21 [all publ] – (= ser Radical periodicals in the united states, 1881-1960. series 1) – 1r – 1 – $200.00 – us UPA [335]
Communist – 1974 aug 15-1975 apr – 1r – 1 – (cont by: movin' on!) – mf#361200 – us WHS [335]
Communist / Friendship Community – Cincinnati OH, Saint Louis MS. 1868 jan-1885 feb – 1r – 1 – (cont by: altruist [saint louis mo]) – mf#1440867 – us WHS [335]
Communist : official organ / United Communist Party of America – v1 n3 [1920 jul 17] – 1r – 1 – mf#464949 – us WHS [335]
Communist : official organ of the communist party of america [section of the communist international] – v1 n2 [1921 aug] – 1r – 1 – (cont: communist [chicago il: 1919]) – mf#464948 – us WHS [335]
Communist : official paper – v1 n1 [1919 jul 19] – 1r – 1 – (cont by: communist [communist party of america: 1919]) – mf#700124 – us WHS [335]
Communist *see* Australian communist / communist / workers weekly / tribune
The communist : a journal for the theory and practice of marxism – Sydney, 1925-26// (mthly) [mf ed n2-12. feb 1925-mar 1926 filmed 1961] – 1r – 1 – (journal of the communist party of australia) – mf#*ZAN-T1981 – us NY Public [335]
Communist activity in the entertainment industry : fbi surveillance files on hollywood, 1942-1958 / ed by Leab, Daniel – (= ser Federal bureau of investigation confidential files) – 14r – 1 – $2510.00 – 1-55655-414-1 – (with p/g) – us UPA [320]
Communist and post-communist studies – Kidlington. 1993+ (1,5,9) – (cont: studies in comparative communism) – ISSN: 0967-067X – mf#17260,01 – us UMI ProQuest [335]
Communist attack in great britain, london 1938 / Bayle, Constantino & Godden, G M – Burgos: Razon y Fe, 1938 – 1 – sp Bibl Santa Ana [335]
Communist infiltration in guatemala / Martz, John D – New York, NY. 1956 – 1r – us UF Libraries [972]
Communist infiltration of the southern christian leadership conference : fbi investigation file, 1958-1980 / U.S. Federal Bureau of Investigation – 1984 – 9r – 1 – $1170.00 – mf#S1754 – us Scholarly Res [360]
Communist International *see*
- Pamphlets
- Unity for spain; correspondence between the communist international and the labor and socialist international, jun-july, 1937

Communist international / Communist International. Library – Hamburg. no.1-40 – (= ser Kommunistische Internationale. Bibliothek) – 1 – us NY Public [335]
Communist international – Earl Browder, Editor. New York. no. 1-12. Jan-Dec 1940 – 1 – us NY Public [335]
Communist international : english edition – London etc. v. 6-12. May 1929-Aug 1935. Incomplete – 1 – us NY Public [335]
Communist international – New York. v. 11 no. 2-v. 16 no 12. Jan 15 1934-Dec 1939 – 1 – us NY Public [335]
Communist international – ser1: n1-30 1919-24 [all publ]. ser2: v1-17 1924-40 [all publ] – (= ser Radical periodicals in the united states, 1881-1960. series 1) – 9r – 1 – $1680.00 – us UPA [335]
Communist international : special american edition – New York. v. 9 no. 8-12. May 15-July 1 1932 – 1 – us NY Public [335]
The communist international / Roy, Manabendra Nath – [Bombay]: Radical Democratic Party Publication, 1943 – (= ser Samp: indian books) – us CRL [335]
Communist International. Comite Executif *see* L'internationale syndicale rouge
Communist International. Library *see* Communist international
Communist international periodicals from the feltrinelli archives – 1329mf (24:1) – 9 – $6310.00 coll – (individual titles listed separately) – us UPA [335]
Communist Labor Party *see* Revolutionary age
Communist Labor Party of the United States of North America et al *see* People's tribune
Communist League of America *see*
- Militant
- Young spartacus

Communist League of Struggle *see* Class struggle
Communist manifesto / Marx, Karl – Chicago, IL. 1954 – 1r – us UF Libraries [335]
The communist network / Hinkel, John Vincent – N.Y., 1939. Fiche W944. (Blodgett Collection of Spanish Civil War Pamphlets) – 9 – us Harvard [946]
Communist operations in spain / Godden, Gertrude M – London: Burns Oates & Washbourne 1936 – 9 – mf#w919 – us Harvard [335]
Communist pamphlets, 1907-1982 : from the labadie collection, university of michigan – [mf ed Chadwyck-Healey] – 1429mf – 9 – (with free printed list of titles) – uk Chadwyck [335]
Communist party *see* Department of justice investigative files
Communist Party. France *see*
- L'araldo.
- Bulletin colonial
- Comptes rendus et rapports
- France-nouvelle
- L'humanite-dimanche
- La terre
- La voix des travailleurs
- La voix du peuple

Communist Party. Germany *see* Bericht ueber den parteitag
Communist Party. Great Britain *see*
- The link
- Worker's weekly

Communist Party In South Africa *see* Road to south african freedom
Communist Party [Marxist-Leninist] *see* Call
Communist Party of America *see* Communist
Communist Party of Canada *see*
- The clarion
- Marxiste-leniniste
- Marxist-leninist
- Worker

Communist Party of Canada [Marxist-Leninist] *see*
- People's canada daily news release
- People's canada news service
- Quotidien du canada populaire

Communist party of great britain : complete archives – 4 series. 1916-92 – 1 – £7500.00 coll – (series 1: journals 1921-92 – titles: imprecor 1921-37; world news and views 1938-62; comment 1963-82; focus 1982-92 57r £3200 cpb. series 2: newspapers 1916-29 – titles: the call 1916-20; the communist 1920-23; the workers' weekly 1923-27; workers' life 1927-29 9r £450 cpc. series 3: theoretical journals 1921-92 – titles: the communist review 1921-35; the communist 1927-28; discussion 1936-38; modern quarterly 1938-53; marxist quarterly 1854-57; marxism today 1957-92 40r £1950 cpd. series 4: pamphlets, 1920-92 50r £2350 cpe. after 1992: journals of the democratic left (successor to the cpgb) 1993- £300) – uk World [325]
Communist Party of the United States of America *see*
- Daily worker
- Daily world
- Ford worker
- Negro affairs quarterly
- Sunday worker
- Unity
- Viewpoint
- Western worker
- Wisconsin people's voice
- Worker

Communist Party of the United States of America et al *see* Lavoratore
Communist Party of the United States of America [WI] *see* Nothing but the truth
Communist Party of the USA *see* Harlem pointer
Communist Party. USA *see*
- Party organizer
- Revolutionary age
- Western worker

The communist party usa and radical organizations, 1953-1960 : fbi reports from the eisenhower library / ed by Naison, Mark & Isserman, Maurice – 7r – 1 – $1260.00 – 1-55655-195-9 – (with p/g) – us UPA [325]
Communist Party. USA International Labor Defense *see* Equal justice
Communist Party. USA National Committee *see* National issues
Communist Party USA/Marxist-Leninist *see* Unite!
The communist plot in spain / Azcarate y Florez, Pablo de – n.p. 1938? Fiche W 743. (Blodgett Collection of Spanish Civil War Pamphlets) – 9 – us Harvard [946]
Communist Political Association of Wisconsin *see* Wisconsin cpa guide
Communist vietnamese publications – 1 – 59.00 – us L of C Photodup [959]
Communist Workers Party [US] *see*
- Punto de vista obrero
- Workers viewpoint

Communist world – v1 n1-8 [1919 nov 1-dec 20] – 1r – 1 – mf#465135 – us WHS [335]
Communist Youth League of USA *see* Young worker
Le communiste – Paris. n1-13 14. nov 1931-sept 1933 – 1 – (mq no. 2-3, 5. n.s., n1-2) – fr ACRPP [325]
Le communiste – n1. Paris. mars 1849 – 1 – fr ACRPP [325]
Le communiste / Parti Communiste et des Soviets – Paris. oct-dec 1919, juil-aout 1920 – 1 – fr ACRPP [335]
Les communistes et la lutte armee – (city unknown) 1944 – 1 – (in french) – us UMI ProQuest [934]
The communistic societies of the united states : from personal visit and observation / Nordhoff, Charles – New York: Harper, 1875, c1874 – 2mf – 9 – 0-7905-5852-1 – (incl bibl ref) – mf#1988-1852 – us ATLA [300]
Die "communitarians" / Beiwaltes, Andreas – (mf ed 1995) – 1mf – 9 – €30.00 – 3-8267-2137-3 – mf#DHS 2137 – gw Frankfurter [320]
Communities – 1972 dec-1976 jul, 1976 may, 1980 jun-dec, 1981 jan, 1981 feb-1984/1985 win – 3r – 1 – mf#361206 – us WHS [307]
Communities – Louisa. 1972+ (1) 1974+ (5) 1974+ (9) – ISSN: 0199-9346 – mf#8530 – us UMI ProQuest [307]
Communities in action – Washington. 1966-1969 (1) – mf#3200 – us UMI ProQuest [350]
Community / AFL-CIO [American Federation of Labor and Congress of Industrial Organizations] – 1969 feb-1980 jan – 1r – 1 – us WHS [307]
Community – Chicago. 1941-1983 (1) 1974-1983 (5) 1975-1983 (9) – ISSN: 0010-3772 – mf#10271 – us UMI ProQuest [307]
Community – 1988 aug 5-1988 dec 30 – 1r – 1 – (cont: butler special) – mf#3475161 – us WHS [360]
Community – v1 [n1]-v3 n12 [[1978] oct 19-1981 jun 29/jul 20 – 1r – 1 – (cont by: philadelphia's community) – mf#647476 – us WHS [307]
Community / Friendship House – Chigaco IL. [1961 jun-jul, oct-nov, 1962 feb-1983 spr] – 1r – 1 – (cont: catholic interracialist) – mf#825790 – us WHS [241]

COMMUNITY

Community access news – Albany GA. 1995 feb-mar, aug-sep, 1996 feb, 1997 feb-apr, jun – 1r – 1 – mf#3430263 – us WHS [360]
Community Action on Latin America *see* Cala newsletter
Community Action Program *see*
– Moccasin tracks
– War cry
Community advertising / American Newspaper Publishers Association. Bureau of Advertising – New York: The Association, 1927 (mf ed 19–) – 35p (ill) – mf#ZT-TB+ pv458 n13 – us Harvard [650]
Community advocate – 1982 may – 1r – 1 – mf#5305268 – us WHS [360]
Community analysis reports and community analysis trend reports of the war relocation authority, 1942-1946 / U.S. War Relocation Authority – (= ser Records Of The War Relocation Authority) – 29r – 1 – (with printed guide) – mf#M1342 – us Nat Archives [324]
Community and junior college journal – Washington. 1930-1985 (1) 1968-1985 (5) 1969-1985 (9) – (cont by: community, technical, and junior college journal) – ISSN: 0190-3160 – mf#829 – us UMI ProQuest [378]
Community and junior college libraries / ed by Todaro, Julie Beth – v1- 1982- – 1, 9 ($60.00 in US $84.00 outside hardcopy subsc) – us Haworth [020]
Community and Social Agency Employees Union *see* Voice of 1707
Community banker – Washington. 2002+ (1) – (cont: america's community banker) – ISSN: 1529-1332 – mf#19538,02 – us UMI ProQuest [332]
Community care – Sutton. 1974-1992 (1) 1974-1992 (5) 1974-1992 (9) – ISSN: 0307-5508 – mf#10643 – us UMI ProQuest [360]
Community change, inc newsletter – 1983 jan-1991 sep – 1r – 1 – (cont by: community changing) – mf#4862506 – us WHS [071]
Community changing : the newsletter of community change, inc – 1991 nov, 1992 mar-sep, 1993 feb-may, sep-dec, 1995 dec-1997 oct – 1r – 1 – (cont: community change, inc newsletter) – mf#4862474 – us WHS [071]
Community civics / Hughes, R O – Boston etc: Allyn & Bacon, 1917 – 6mf – 9 – $9.00 – mf#LLMC 96-062 – us LLMC [360]
Community college enterprise – Livonia. 2002+ (1,5,9) – mf#31303,01 – us UMI ProQuest [378]
Community college frontiers – Springfield. 1972-1981 (1) 1975-1981 (5) 1975-1981 (9) – mf#10523 – us UMI ProQuest [378]
Community college journal – Washington. 1992+ (1) 1992+ (5) 1992+ (9) – (cont: community, technical, and junior college journal) – ISSN: 1067-1803 – mf#829,02 – us UMI ProQuest [378]
Community college journal of research and practice – Washington. 1993+ – 1,5,9 – (cont: community/junior college) – ISSN: 1066-8926 – mf#11133,02 – us UMI ProQuest [378]
Community college journalist – Midland. 1976-1996 (1,5,9) – mf#11418,01 – us UMI ProQuest [070]
Community college review – Raleigh. 1973+ (1) 1975+ (5) 1975+ (9) – ISSN: 0091-5521 – mf#10692 – us UMI ProQuest [378]
Community college social science journal – El Cajon. 1977-1982 (1,5,9) – mf#10974 – us UMI ProQuest [300]
Community comment – Hamilton, NZ. 1981-83 – 1r – 1 – mf#15.39 – nz Nat Libr [079]
Community contact – Montrbeal QC. 1995 feb-1996 apr – 1r – 1 – (cont by: montreal community contact) – mf#3401721 – us WHS [307]
Community corrections in penological perspective / Ntuli, Ronald Mpuru – Uni of South Africa 2000 [mf ed Johannesburg 2000] – 4mf – 9 – (incl bibl ref) – mf#mfm14835 – sa Unisa [365]
Community Council for Public Television *see* Programs 21
Community development journal – Oxford. 1966+ (1) 1975+ (5) 1975+ (9) – ISSN: 0010-3802 – mf#9853 – us UMI ProQuest [360]
Community Development Society *see* Journal of the community development society
Community Economics and Cultural Exchange *see* Cece guide
Community education journal – Fairfax. 1971+ (1) 1978+ (5) 1978+ (9) – ISSN: 0045-7736 – mf#10521 – us UMI ProQuest [370]
Community empowerment through participation : a south african small town case study / Tamasane, Tsiliso A – Stellenbosch: U of Stellenbosch 1998 [mf ed 1998] – 5mf – 9 – mf#mf.1266 – sa Stellenbosch [307]
Community for Creative Non-violence (Washington, DC) *see* Gamaliel
Community herald – Monona WI. 1983 mar 9/aug-1998 jul/dec – 20r – 1 – (cont: monona community herald; cont by: independent [deerfield wi: cottage grove ed]; herald-independent [monona wi]) – mf#851860 – us WHS [071]
Community jobs / The Youth Project [US] – v2 n5-v3 n9 [1979 may-1980 nov] – 1r – 1 – mf#665975 – us WHS [360]
Community journal press (northern edition) / Hamilton Co. Cincinnati – jun 1986-dec 1989/w – 6r – 1 – mf#B35946-35951 – us Ohio Hist [071]
Community journal press (southern edition) / Hamilton Co. Cincinnati – oct 1986-mar 1991/ – 7r – 1 – mf#B35449-35455 – us Ohio Hist [071]
Community journal series (southern edition) / Hamilton Co. Cincinnati – jan 1971- sep 1986 – 18r – 1 – mf#B35928-35945 – us Ohio Hist [071]
Community leader – Baton Rouge LA. 1985 jun 13/15 – 1r – 1 – (cont: baton rouge community leader) – mf#3912661 – us WHS [071]
Community leader – Rushville, NY. 1928-1930 (1) – mf#65218 – us UMI ProQuest [071]
Community Liberation Movement [St Petersburg FL] *see* Community liberator
Community liberator / Community Liberation Movement [St Petersburg FL] – v1 n10 [1970 aug 6] – 1r – 1 – mf#1583053 – us WHS [320]
Community mental health journal – New York. 1965+ (1) 1974+ (5) 1975+ (9) – ISSN: 0010-3853 – mf#11176 – us UMI ProQuest [360]
Community Mercantile Food Cooperative *see* Public notice
Community news / Baumholder Military Community – 1981 oct 5-1982 sep 24 – 1r – 1 – (cont: baumholder community news; cont by: champion times) – mf#1363386 – us WHS [355]
Community news / Carroll Co. Malvern – jan 1978-dec 1987 [wkly] – 7r – 1 – mf#B29603-29609 – us Ohio Hist [071]
Community news – n55-67 [1981 aug/sep-1983 aug/sep] – 1r – 1 – (cont by: city life) – mf#698123 – us WHS [307]
Community news – Flint, MI. 1945-1948 (1) – mf#63735 – us UMI ProQuest [071]
Community news – Marcus, WA. 1922-1923 (1) – mf#67033 – us UMI ProQuest [071]
Community news – Mt. Clemens, MI. 1977-1986 (1) – mf#63820 – us UMI ProQuest [071]
Community news – Saratoga Springs, NY. 1996-1999 (1) – mf#65221 – us UMI ProQuest [071]
Community news – Upper Arlington, OH. 1922-1929 (1) – mf#65694 – us UMI ProQuest [071]
Community news reporter / Jewish Telegraphic Agency – New York. N.Y. 1962-67 – 1 – us AJPC [939]
Community post / Auglize Co. Minster – jan-dec 1965 [wkly] – 1r – 1 – mf#B248 – us Ohio Hist [071]
Community post – Camas, WA. 1977-1978 (1) – mf#66951 – us UMI ProQuest [071]
Community practitioner – London. 1998+ (1) 1998+ (5) 1998+ (9) – (cont: health visitor) – mf#8320,01 – us UMI ProQuest [360]
Community press – Portland OR: Clarke Pub Co, 1973 [wkly] – 1 – (began in 1973; cont: suburban press [1970-1973]; cont by: suburban community press [1973-]; publ in several regional ed) – us Oregon Lib [071]
Community press – Milwaukee WI. 1935 dec 5-19 – 1r – 1 – (cont by: north milwaukee community press) – mf#1166137 – us WHS [071]
Community press – Dancy, Junction City etc WI. 1949 oct 27-1952, 1953-57, 1958-59, 1961 feb 2-1962, 1963-1966 jan 27 – 4r – 1 – mf#1012707 – us WHS [071]
Community press – Milwaukee WI. 1958 jan 9-1961 apr 13, 1961 apr 20-1964, 1965 jan 7-1966 dec 15 – 3r – 1 – mf#1166142 – us WHS [071]
Community press features / Urban Planning Aid, Inc – 1973 dec-1976 oct, 1976 nov-1983 dec – 2r – 1 – (cont: community press service) – mf#361205 – us WHS [307]
Community press service – Cambridge MA. n2-25 [1971 sep-1973 nov] – 1r – 1 – (cont by: community press features) – mf#700804 – us WHS [071]
Community review – New Brunswick. 1980-1996 – 1,5,9 – ISSN: 0163-8475 – mf#12221 – us UMI ProQuest [370]
Community School District 12 [New York NY] et al *see* Newsletter
Community schools : the magazine about schools for people – Toronto: Community School Workshop. v1-4 n4. jun 1971-jul 1974// – 18 sheets – 9 – Can$200.00 – cn McLaren [370]
Community service news / Council of Social Agencies [Milwaukee WI] et al – v1 n1,3-4 [1944], v2 n1-2 [1945], v3-6 [1946-49], v7 n1-3 [1949-1950 feb] – 1r – 1 – mf#2697639 – us WHS [360]
Community service newsletter – 1976 jan/feb-1982 – 1r – 1 – (cont: community comments; cont by: community journal [yellow sprs oh]) – mf#1393255 – us WHS [360]
Community services catalyst – Blacksburg. 1985-1994 – 1,5,9 – (cont by: catalyst) – ISSN: 0739-9227 – mf#14618 – us UMI ProQuest [374]
Community stew / Langdon Area Grocery Collective [Madison WI] – v2-3 n2 [1977 jan-[1978] may/jun] – 1r – 1 – (cont: sum salad) – mf#499195 – us WHS [334]
Community, technical, and junior college journal – Washington. 1985-1992 (1) 1985-1992 (5) 1985-1992 (9) – (cont: community and junior college journal. cont by: community college journal) – ISSN: 0884-7169 – mf#829,01 – us UMI ProQuest [378]
Community times – Florence SC. 1997 feb 6/12-jun, 1997 jul-dec – 2r – 1 – (cont: times [florence sc]) – mf#3835587 – us WHS [071]
Community times / United States Military Community Activity, Pirmasens – v9 n9-12,14 [1983 aug 15-nov 1, dec 1], v9 n18-22, [1984 feb 15-apr 15]-v18 n2-4 [1992 feb 1-mar 1] – 1r – 1 – (with gaps) – mf#1054710 – us WHS [355]
Community times – Westminster, MD. 1992-1994 (1) – mf#68931 – us UMI ProQuest [071]
Community times (mad river) / Montgomery Co. Dayton – mar 1963-dec 1967 [wkly] – 3r – 1 – mf#B5304-5306 – us Ohio Hist [071]
Community to community gazette / Xpressions Journal – v1 n1-v2 n21, 22, 23-24 [1995 nov 15/dec 15-1998 feb 15] – 1r – 1 – mf#3465714 – us WHS [307]
Community Training and Development, Inc *see* Comment and review
Community unit cable full record / U.S. National Technical Information Service – Every two years – 9 – us NTIS [000]
Community voice – Fort Myers FL. 1993 dec 9 – 1r – 1 – mf#2733586 – us WHS [307]
Community voice – Providence, RI. 1974-1976 (1) – mf#66276 – us UMI ProQuest [071]
Community/junior college – Washington. 1981-1992 – 1,5,9 – (cont: community/junior college research quarterly. cont by: community college journal of research and practice) – ISSN: 0277-6774 – mf#11133,01 – us UMI ProQuest [378]
Community/junior college research quarterly – New York. 1976-1981 – 1,5,9 – (cont by: community/junior college) – ISSN: 0361-6975 – mf#11133 – us UMI ProQuest [378]
Communiviews / Black Teacher, Parent, Student Coalition – 1970 may, 1972 jan/feb-mar/apr – 1r – 1 – mf#4990731 – us WHS [305]
Commutation / Salmon, George – Dublin, Ireland. 1871 – 1r – us UF Libraries [240]
Como cantan alla / Espino, Miguel Angel – San Salvador, El Salvador. 1960 – 1r – us UF Libraries [972]
Como el fascismo defiende a los trabajadores – Bilbao: Impr Moderna 1939 [mf ed 1977] – (= ser Blodgett coll) – 1mf – 9 – mf#w812 – us Harvard [946]
Como ensenar a los hijos a vivir con alegria / Aradillas Agudo, Antonio – Madrid: Alameda, 1968 – sp Bibl Santa Ana [946]
Como es la guajira / Felix Maria – Caracas, Venezuela. 1951? – 1r – us UF Libraries [972]
Como es la vida / Fernandez Abelehira, Maria Isabel – Sevilla: Ed. Catolica Espanola, S.A., s.a. – sp Bibl Santa Ana [240]
Como fizeram os portugueses em mocambique / Costa, Mario Augusto Da – Lisboa, Portugal. 1928 – 1r – us UF Libraries [960]
Como nace un monasterio y muere un cesar / Jimenz Vasco, Felipe – Jaraiz de la Vera: Imp. Romero, 1969 – 1 – sp Bibl Santa Ana [240]
Como nasceu goiania / Monteiro, Ofelia Socrates Do Nascimento – Sao Paulo, Brazil. 1938 – 1r – us UF Libraries [972]
Como nos ven los de fuera : anecdotario de cosas extremenas / Vera Camacho, Juan Pedro – Caceres: Diputacion Provincial de Caceres, 1953 – 1 – sp Bibl Santa Ana [946]
Como piensa el partido sindicalista en este momento historico de la vida espanola – [Valencia?: s.n. 1937] [mf ed 1977] – (= ser Blodgett coll) – 1mf – 9 – mf#w1099 – us Harvard [946]
Como record *see* Miscellaneous newspapers of park county
Como se alhajaban casas e iglesias en maynas / Bayle, Constantino – Madrid: Missionalia Hispanica, 1948 – 1 – sp Bibl Santa Ana [240]
Como se enfrento al fascismo en toda espana / Montseny, Federica – Buenos Aires: Ed del Servicio de Propaganda Espana 1938 – 9 – mf#w813 – us Harvard [946]
Como se evapora un ejercito / Cuervo, Angel – Bogota, Colombia. 1953 – 1r – us UF Libraries [972]
Como se inicio el glorioso movimiento nacional en valladolid y la gesta heroica del alto del leon / Raymundo, Francisco J de – Valladolid [Spain]: Imp Catolica 1936 – 9 – mf#w1129 – us Harvard [946]
Como se va el amor / Nieto De Herrera, Carmela – Habana, Cuba. 1926 – 1r – us UF Libraries [972]
Como ser fieles a varona / Perez, Emma – Habana, Cuba. 1949 – 1r – us UF Libraries [972]
Como vio antonio j valdes la toma de la habana po... / Valdes, Antonio Jose – Habana, Cuba. 1962 – 1r – us UF Libraries [972]
Como vio jacobo de la pezuela la toma de la habana / Pezuela Y Lobo, Jacobo De La – Habana, Cuba. 1962 – 1r – us UF Libraries [972]
Comoedia – Paris. 1er oct 1907-6 aout 1914, 1er oct 1919-1er janv 1937, 21 juin 1941-5 aout 1944 – 1 – fr ACRPP [073]
Comoedia – Paris, France. 27 sep 1941; 27 jun-24 dec 1942; 9 jan 1943-24 jun 1944 – 2r – 1 – uk British Libr Newspaper [072]
Comoedia vom studentenleben *see* Joh georg schoch's comoedia vom studentenleben
Comoediae 18 / Plautus – 15th c – (= ser Holkham library manuscript books 298) – 1r – 1 – mf#96545 – uk Microform Academic [870]
Comoediae 18 / Plautus – 15th c – (= ser Holkham manuscripts 298) – 1r – 1 – mf#96545 – uk Microform Academic [760]
Comox district free press – Comox, British Columbia, CN. aug 1931-dec 1973 – 37r – 1 – cn Commonwealth Imaging [071]
Compact – Denver. 1978-1978 – 1,5,9 – (cont by: interstate compact for education) – ISSN: 0010-3934 – mf#11829 – us UMI ProQuest [370]
Compact of free association : redraft 1983 – Koror: Off of the President of Palau, oct 1983 – (= ser Republic Of Belau (Palau) – Status Negotiations With The U.S.) – 2mf – 9 – $3.00 – (unpag) – mf#LLMC 82-100G, Title 26 – us LLMC [323]
The compact of free association : history and materials / Schwalbenberg, Henry M – Henry M. Schwalbenberg, 29 jan 1983 – (= ser Micronesia: evolution to separate political entities) – 2mf – 9 – $3.00 – (covers all micronesian jurisdictions) – mf#LLMC 82-100F Title 13 – us LLMC [324]
Compact of free association act of 1985 (fsm and marshall islands) : public law 99-239, jan 14, 1986 / U.S. Congress – Washington: GPO, 1986 – (= ser Micronesia: evolution to separate political entities) – 1mf – 9 – $1.50 – (unpag) – mf#LLMC 82-100F, Title 106 – us LLMC [323]
Compact of free association and related agreements : between the governments of the marshall islands and the united states – Majuro: The Committee on Political Education 1982? – 4mf – 9 – $6.00 – mf#llmc82-100i, title 12 – us LLMC [327]
Compact of free association and related agreements between the federated states of micronesia and the united states, 1 october 1982 / Federated States of Micronesia – Kolonia, Ponape: Plebiscite Commission, 1982 – 3mf – 9 – $4.50 – mf#LLMC 82-100H Title 7 – us LLMC [324]
Compact of free association and subsidiary agreements – Koror: Off of the President, mar 1981 – (= ser Republic Of Belau (Palau) – Status Negotiations With The U.S.) – 8mf – 9 – $12.00 – (various pagination) – mf#LLMC 82-100G, Title 21 – us LLMC [323]
Compact of free association between the united states and the governments of palau, the marshall islands and the federated states of micronesia / Micronesia. (U.S.) – 31 Oct 1980; 11 Nov 1980 – (= ser Micronesia: evolution to separate political entities) – 2mf – 9 – $3.00 – (mimeo of initialed copies, with add agreement between the us and the fsm regarding aspects of marine sovereignty and jurisdiction) – mf#LLMC 82-100F Title 10 – us LLMC [324]
Compact of free association between the u.s. and palau : hearings and markup before the house comm on foreign affairs and its subcomm on asian and pacific affairs, 99th cong, 2nd sess, may-jun 1986 / U.S. Congress – Washington: GPO, 1986 – (= ser Republic Of Belau (Palau) – Status Negotiations With The U.S.) – 3mf – 9 – $4.50 – mf#LLMC 82-100G, Title 9 – us LLMC [327]
The compact of free association, foreign political provisions : a section by section legal analysis / Zafren, Daniel H – Washington: The Library of Congress Congressional Research Service, jul 19, 1984 – (= ser Micronesia: evolution to separate political entities) – 1mf – 9 – $1.50 – mf#LLMC 82-100F, Title 63 – us LLMC [323]

Compact of free association with palau : (as signed on may 23, 1984) / U.S. Congress. Hse.Comm on Interior and Insular Affairs – Washington: GPO, 1985 – (= ser Republic Of Belau (Palau) – Status Negotiations With The U.S.) – 1mf – 9 – $1.50 – mf#LLMC 82-100G, Title 27 – us LLMC [327]

Compact of free association with palau : hearing before the senate comm on energy and natural resources, 99th cong, 2nd sess, may 9, 1986 / U.S. Congress – Washington: GPO, 1986 – (= ser Republic Of Belau (Palau) – Status Negotiations With The U.S.) – 3mf – 9 – $4.50 – mf#LLMC 82-100G, Title 30 – us LLMC [327]

Compact zulu dictionary / Dent, George Robinson – Pietermaritzburg, South Africa. 1964 – 1r – us UF Libraries [470]

Compadecido bosque / Hernandez Rivera, Sergio Enrique – Habana, Cuba. 1964 – 1r – us UF Libraries [972]

Compadre mon / Cabral, Manuel Del – Bogota, Colombia. 1948 – 1r – us UF Libraries [972]

Compagni, Dino *see* The chronicle of dino compagni

Compagnie d'assurance agricole du Canada *see* Almanach agricole des cultivateurs pour l'annee...

Compagnie d'assurance agricole du canada : procedes de l'assemblee annuelle des actionnaires tenue le 22 janvier 1878 – Montreal?: s.n, 1878 (Montreal: Impr canadienne) – 1mf – 9 – mf#55990 – cn Canadiana [368]

Compagnie d'assurance de Montreal contre les accidents du feu *see* Articles d'association de la compagnie d'assurance de montreal contre les accidents du feu

Compagnie d'assurance de Quebec contre les accidens du feu *see*
- Articles d'association etablissant une compagnie d'assurance contre les accidens du feu dans la cite de quebec
- Introduction explicative concernant les statuts, regles et reglemens...
- Statuts, regles et reglemens de la compagnie d'assurance de quebec contre les accidens du feu

Compagnie d'assurance des citoyens du Canada *see* Rapport des directeurs pour l'annee...

Compagnie d'assurance du Canada *see* Reglements pour le gouvernement de la corporation...

Compagnie d'assurance du feu de Quebec *see* Extraits des minutes du comite nomme le 2e mars, 1816

La compagnie d'assurance mutuelle contre le feu des comtes de rimouski, temiscouata et kamouraska : fondee en 1876 – Rimouski, Quebec?: s.n, 1894 (Rimouski Quebec: A G Dion) – 1mf – 9 – mf#51941 – cn Canadiana [368]

Compagnie d'assurance mutuelle contre le feu du comte de Montreal *see* Regles et reglements de la...tel qu'apprrouves [sic] le 4 fevrier 1836...

Compagnie d'assurance mutuelle de montmagny : contre le feu et la foudre, fondee en 1872 – St Hyacinthe, Quebec?: s.n, 1888 (St Hyacinthe Quebec: Impr de l'Union) – 1mf – 9 – mf#01247 – cn Canadiana [368]

Compagnie d'assurance provinciale ecossaise : etablie en 1825 – S.l: s.n, 1863? – 1mf – 9 – mf#60179 – cn Canadiana [368]

Compagnie d'assurance Stadacona *see* Liste des actionnaires enregistres jusqu'au 31 decembre 1874

Compagnie de colonisation et de credit des cantons de l'est *see* La compagnie de colonisation et de credit des cantons de l'est

La compagnie de colonisation et de credit des cantons de l'est : notice sur son but et son organisation / Compagnie de colonisation et de credit des cantons de l'est – Sherbrooke: "Pionnier", 1884 – 1mf – 9 – mf#52082 – cn Canadiana [338]

Compagnie des Cent-Associes *see* Factum

Compagnie du chemin de fer canadien du Pacifique *see*
- Canadian pacific railway annotated time table
- Contract between the government of the dominion of canada and the canadian pacific railway company
- From emory's bar at the west end of contract 60 to port moody (burrard inlet), british columbia

Compagnie du chemin de fer de Credit Valley *see* Application of the credit valley railway for right of way and crossings at the city of toronto

Compagnie du chemin de fer de la rive Nord *see* Reports of chief engineer on the survey of the north shore railway

La compagnie du haras national : vente et affermage de chevaux percherons, arabes, et carossiers normands: catalogue pour 1889-1890 – Montreal?: s.n, 1889 – 1mf – 9 – mf#04235 – cn Canadiana [636]

Compagnons – Lyons, France. 4 jan 1941-1942; 9 jan-11 dec 1943 – 2 1/2r – 1 – uk British Libr Newspaper [072]

Compagnons du devoir : ou, le tour de france / Lafontaine, W – Paris, France. 1827 – 1r – us UF Libraries [440]

Companeras! : y otros poemas / Winer, Sonia – Habana, Cuba. 1938 – 1r – us UF Libraries [810]

El companero – v1-4, n.d; 1962 n1-5 – 374p – 1 – us Southern Baptist [242]

Un companero de hernando cortes, juan cano de saavedra, verno de moctezuma (caceres, ?1501? sevilla, 2 de septiembre de 1572) / Lopez de Meneses, Amada – Badajoz: Dip. Provincial, 1965. Sep. REE – sp Bibl Santa Ana [946]

Companhia Do Caminho De Ferro De Benguela *see* Benguela railway

Companhias de colonizacao / Moraes Carvalho, Arthur De – Coimbra, Portugal. 1903 – 1r – us UF Libraries [972]

[Compania colombiana de tabaco, Bogota] *see* Ade, antologia del tabaco

Compania De Caminos De Hierro De La Habana *see* Alegato presentado a nombre de la....

Compania fundidora de fierro y acero de monterrey, s.a. ... / ed by Bayle, Constantino – Madrid: Razon y Fe, 1926 – 1 – sp Bibl Santa Ana [946]

The companies ordinance of hongkong : being no 1 of 1865 / Hong Kong. Laws, Statutes, etc – Hongkong: Kelly & Walsh, 1907. 121p. LL-10005 – 1 – us L of C Photodup [348]

Companion – London. 1828-1828 (1) – mf#4228 – us UMI ProQuest [941]

The companion / Cates, J M D – Second ed. contains 239 hymns published in 1846 – 1 – $8.82 – us Southern Baptist [242]

Companion and teacher – London, Ont: Companion Pub. Co., [1875?-18– or 19–] – 9 – mf#P04402 – cn Canadiana [370]

Companion and weekly miscellany – Baltimore. 1804-1806 (1) – mf#3568 – us UMI ProQuest [975]

Companion animal practice – Santa Barbara. 1987-1989 (1) 1987-1989 (5) 1987-1989 (9) – ISSN: 0894-9794 – mf#16552 – us UMI ProQuest [636]

Companion for the afflicted – London, England. 1834 – 1r – us UF Libraries [240]

Companion for the penitent : and for persons troubled in mind / Kettlewell, John – London, England. 1794 – 1r – us UF Libraries [240]

Companion for the working classes – London, England. 18– – 1r – us UF Libraries [240]

Companion to a chart / Denny, Edward – London, England. 18– – 1r – us UF Libraries [240]

A companion to biblical studies : being a revised and re-written edition of the cambridge companion to the bible / Ryle, Herbert Edward et al; ed by Barnes, William Emery – Cambridge: University Press, 1916 [mf ed 1991] – 7mf – 9 – 0-8370-1976-1 – mf#1987-6363 – us ATLA [220]

A companion to latin studies / ed by Sandys, John Edwin – Cambridge: University Press, 1910 [mf ed 1990] – 3mf – 9 – 0-7905-7075-0 – (incl bibl ref) – mf#1988-3075 – us ATLA [930]

Companion to roman history / Jones, Henry Stuart – Oxford: Clarendon Press, 1912 – 2mf – 9 – 0-524-02214-3 – (incl bibl ref) – mf#1990-2888 – us ATLA [930]

Companion to the bible / Barrows, Elijah Porter – New York: American Tract Society, c1867 – 2mf – 9 – 0-8370-2188-X – (includes appendixes. incl indof subjects and texts cited) – mf#1985-0188 – us ATLA [220]

Companion to the encyclical "satis cognitum" : with a reply to the bishop of stepney / Smith, Sydney Fenn – London: Catholic Truth Society, 1896 – 1mf – 9 – 0-8370-6778-2 – (includes the text of the encyclical and bibliographical references) – mf#1986-0778 – us ATLA [240]

Companion to the grapes of wrath / French, Warren G – New York, NY. 1963 – 1r – us UF Libraries [410]

A companion to the greek testament and the english version / Schaff, Philip – 4th rev ed. New York: Harper, 1892, c1887 [mf ed 1986] – 2mf – 9 – 0-8370-9898-X – (incl bibl and ind) – mf#1986-3898 – us ATLA [225]

Companion to the most celebrated private galleries of art / Jameson, Anna Brownell (Murphy) – London 1844 – (= ser 19th c art & architecture) – 5mf – 9 – mf#4.2.26 – uk Chadwyck [700]

A companion to the new testament : being a plain commentary on scripture history from the birth of our lord to the end of the apostolic age / Blunt, John Henry – London, New York: Longmans, Green, 1900 [mf ed 1989] – 1mf – 9 – 0-7905-0740-4 – (incl ind) – mf#1987-0740 – us ATLA [225]

The companion to the newspaper; and journal of facts : in politics, statistics and public economy (london) – mar 1833-jan 1837 – (= ser 19th c british periodicals) – reel 50 – 1 – (filmed with: the dublin family magazine (dublin), apr-sep 1829) – us Primary [073]

A companion to the old testament / Blunt, John Henry – London: Rivingtons, 1872 [mf ed 1989] – 2mf – 9 – 0-7905-2835-5 – mf#1987-1835 – us ATLA [221]

Companion to the revised old testament / Chambers, Talbot Wilson – New York: Funk & Wagnalls, 1885 – 1mf – 9 – 0-8370-2624-5 – (incl ind) – mf#1985-0624 – us ATLA [221]

Companions for the devout life : lectures. delivered in st. james church, london, in 1875-6 / Farrar, Frederic William – new ed. New York: EP Dutton, 1877 – 1mf – 9 – 0-524-08236-7 – mf#1993-2011 – us ATLA [240]

The companions of st paul / Howson, John Saul – London: Strahan, 1871 – 1mf – 9 – 0-8370-3680-1 – mf#1985-1680 – us ATLA [920]

The company of the haras national : importers and breeders of french coach, percheron and arabian horses: catalogue for 1889-90 – Montreal: Gazette, 1889 – 1mf – 9 – mf#04236 – cn Canadiana [636]

Companys, Lluis *see* Discurso pronunciado por luis companys el dia 27 de diciembre de 1936 en el palacio de bellas artes de barcelona

Companys y Jover, Luis *see* Address of president lluis companys to the parliament of catalunya, march 1st, 1938

Comparable worth project newsletter – Oakland CA. v1 n1-v6 n1 [1981 jan-1986 win/spring] – 1r – 1 – mf#1161323 – us WHS [071]

Comparaciones historicas / Cervi, Emilio – Madrid: Razon y Fe, 1940 – sp Bibl Santa Ana [240]

Comparaison du basque avec les idiomes asiatiques, et principalement avec ceux qu'on apelle semitiques / Klaproth, Heinrich Julius – Paris: Dondey-Dupre 1823 – (= ser Whsb) – 1mf – 9 – €20.00 – (fr: journal asiatique; ser 1, t3 1823) – mf#Hu 052 – gw Fischer [410]

Comparative analysis [and translation] of vincent d'indy's cours de composition musicale / Montgomery, Merle – U of Rochester 1946 [mf ed 19–] – 7v on 1r – 1 – mf#film 171 – us Sibley [780]

A comparative analysis of equivalent submaximal treadmill and bicycle ergometer exercise / Black, John G – 1980 – 1mf – 9 – $4.00 – us Kinesology [790]

Comparative analysis of factors influencing participation in an employee health promotion program, including characterizations of participants and nonparticipants / Teschner, Pamela J S & Donatelle, Rebecca J – 1992 – 2mf – 9 – $8.00 – us Kinesology [613]

A comparative analysis of outcomes based education in australia and south africa / Williamson, Merryl Cheryne – Uni of South Africa 2001 [mf ed Johannesburg 2001] – 5mf – 9 – (incl bibl ref) – mf#mfm15035 – sa Unisa [370]

A comparative analysis of the leadership styles of head football coaches from taditionally black and white universities and colleges / Thomas, Johnny D & Waigandt, Alex – 1992 – 3mf – 9 – $12.00 – us Kinesology [150]

Comparative and international law journal of southern africa – v1-34. 1968-2001 – 5,6,9 – $716.00 set – (v1-17 1968-84 on reel $204. v18-34 1985-2001 on mf $512) – ISSN: 0010-4051 – mf#108731 – us Hein [341]

The comparative archeology of early mesopotamia / Perkins, Ann L – 1948 – 9 – $10.00 – us IRC [930]

Comparative biochemistry and physiology – London. 1960-1970 (1,5,9) – ISSN: 0010-406X – mf#49033 – us UMI ProQuest [612]

Comparative biochemistry and physiology pt a : molecular and integrative physiology – New York. 1971+ (1,5,9) – ISSN: 1095-6433 – mf#49249 – us UMI ProQuest [612]

Comparative biochemistry and physiology pt b : biochemistry and molecular biology – New York. 1971+ (1,5,9) – ISSN: 1096-4959 – mf#49250 – us UMI ProQuest [574]

Comparative biochemistry and physiology pt c : pharmacology, toxicology and endocrinology – New York. 1975-1999 (1,5,9) – ISSN: 1096-4932 – mf#49251 – us UMI ProQuest [615]

Comparative biochemistry and physiology. toxicology and pharmacology – New York, 2000+ [1,5,9] – (cont: comparative biochemistry and physiology, pt c, pharmacology, toxicology and endocrinology) – ISSN: 1532-0456 – mf#49251,01 – us UMI ProQuest [615]

A comparative case study of the differential effects of animal-assisted therapy, plush toy-facilitated therapy, and one-on-one therapy for a child with autism / Nakanishi, Akiko – 1999 – 2mf – 9 – $8.00 – mf#RC 528 – us Kinesology [615]

Comparative criticism – Cambridge. 1989-1993 (1) – ISSN: 0144-7564 – mf#16525 – us UMI ProQuest [410]

Comparative darstellung des lehrbegriffs der verschiedenen christlichen kirchenparteihen *see* Comparative view of the doctrines and confessions of the various communities of christendom

Comparative darstellung des religionsbegriffes in den verschiedenen auflagen der schleiermacher'schen "reden" / Braasch, E F – Kiel: Lipsius & Tischer, 1883 – 1mf – 9 – 0-7905-9906-6 – (incl bibl ref) – mf#1989-1631 – us ATLA [200]

Comparative drama – Kalamazoo. 1967+ (1) 1975+ (5) 1975+ (9) – ISSN: 0010-4078 – mf#10736 – us UMI ProQuest [410]

Comparative economic studies – New Brunswick. 1987+ (1,5,9) – ISSN: 0888-7233 – mf#14599,03 – us UMI ProQuest [330]

Comparative education review – Chicago. 1957+ (1) 1971+ (5) 1976+ (9) – ISSN: 0010-4086 – mf#5804 – us UMI ProQuest [370]

The comparative effectiveness of static stretching and proprioceptive neuromuscular facilitation stretching techniques in increasing hip flexion range of motion / Sundquist, Robert D – Oregon State University, 1996 – 1mf – 9 – mf#PH 1511 – us Kinesology [612]

A comparative evaluation of stenographic and audiotape methods for us district court reporting / Greenwood, Michael et al – Washington: FJC, July 1983 – 3mf – 9 – $4.50 – mf#LLMC 95-353 – us LLMC [347]

Comparative feeding value of silages made from napier grass, sorghum and sugarcane / Shealy, A L – Gainesville, FL. 1941 – 1r – us UF Libraries [630]

Comparative free government / Macy, Jesse – New York, NY. 1915 – 1r – us UF Libraries [320]

The comparative geography of palestine and the sinaitic peninsula = Vergleichende erkunde der sinai-halbinsel, von palaestina und syrien / Ritter, Karl – New York: D Appleton, 1866 – 4mf – 9 – 0-8370-1203-1 – (incl bibl ref. in english) – mf#1987-6033 – us ATLA [916]

A comparative glossary of the gothic language : with especial reference to english and german / Balg, Gerhard Hubert – Mayville, WI: GH Balg, 1887-89 [mf ed 1990] – 2mf – 9 – 0-8370-1848-X – mf#1987-6235 – us ATLA [430]

Comparative graduation rates and grade point averages among regular admit, non-competitive admit, and admissions exception student-athletes of the university of north carolina at chapel hill / Durand, Jennings F – 1999 – 2mf – 9 – $8.00 – mf#PE 3952 – us Kinesology [790]

A comparative grammar of the dravidian : or, south-indian family of languages / Caldwell, Robert, 1814-1891 – London: Harrison, 1856 – (= ser Samp: indian books) – us CRL [490]

A comparative grammar of the hebrew language : for the use of classical and philological students / Donaldson, John William – London: John W Parker, 1853 [mf ed 1986] – 1mf – 9 – 0-8370-9227-2 – mf#1986-3227 – us ATLA [470]

A comparative grammar of the hittite language / Sturtevant, Edgar H – 1933 – 9 – $12.00 – us IRC [470]

Comparative group studies – Beverly Hills. 1970-1972 (1) 1970-1972 (5) 1970-1972 (9) – (cont by: small group behavior) – ISSN: 0010-4108 – mf#11976 – us UMI ProQuest [150]

A comparative history of religions / Moffat, James Clement – New York: Dodd & Mead, 1871-c1873 [mf ed 1992] – 2mf – 9 – 0-524-06930-1 – (incl bibl ref) – mf#1990-3556 – us ATLA [230]

Comparative immunology, microbiology and infectious diseases – Oxford. 1978+ (1,5,9) – ISSN: 0147-9571 – mf#49295 – us UMI ProQuest [616]

A comparative investigation on the efficacy of integrated and segregated physical education settings for students with disabilities / Perkins, Jennifer L – 1998 – 2mf – 9 – $8.00 – mf#PSY 2048 – us Kinesology [362]

Comparative labor law *see* Comparative labor law journal

Comparative labor law and policy journal *see* Comparative labor law journal

Comparative labor law journal – University of Pennsylvania. v1-18. 1976-97 + cum ind 1-15 – 9 – $358.00 set – (title varies: v1-7 1976-1986 as comparative labor law) – ISSN: 0147-9202 – mf#101971 – us Hein [344]

Comparative law yearbook *see* Comparative law yearbook of international business

Comparative law yearbook of international business – v1-23. 1977-2001 – 9 – $972.00 set – (title varies: v1-11 1977-87 as comparative law yearbook) – ISSN: 0962-0435 – mf#111021 – us Hein [341]

Comparative literature – Eugene. 1949+ (1) 1949+ (5) 1949+ (9) – ISSN: 0010-4124 – mf#1024 – us UMI ProQuest [410]

Comparative literature studies – University Park. 1971+ (1) 1963+ (5) 1976+ (9) – ISSN: 0010-4132 – mf#6101 – us UMI ProQuest [410]
Comparative medicine east and west – New York. 1977-1978 (1"5,9) – (cont: american journal of chinese medicine. cont by: american journal of chinese medicine) – ISSN: 0147-2917 – mf#10055,01 – us UMI ProQuest [615]
Comparative mythology : an essay / Mueller, Friedrich Max; ed by Palmer, Abram Smythe – London: G Routledge; New York: EP Dutton, [1909?] – 1mf – 9 – 0-524-01288-1 – mf#1990-2324 – us ATLA [230]
The comparative number of the saved and lost : a study / Walsh, Nicholas – Dublin: MH Gill, 1899 – 1mf – 9 – 0-7905-8961-3 – mf#1989-2186 – us ATLA [240]
Comparative phonetics of the suto-chuana group of bantu languages / Tucker, Archibald Norman – London, England. 1929 – 1r – us UF Libraries [470]
Comparative political studies – Beverly Hills. 1968+ (1) 1971+ (5) 1975+ (9) – ISSN: 0010-4140 – mf#5065 – us UMI ProQuest [320]
Comparative politics – New Brunswick. 1968+ (1) 1971+ (5) 1975+ (9) – ISSN: 0010-4159 – mf#6213 – us UMI ProQuest [320]
Comparative print of the texts of the immigration and naturalization act and the immigration and naturalization laws existing prior to the enactment of p.l. 414, june 27, 1952, 66 stat. 163 / U.S. Dept of Immigration. Senate Committee on the Judiciary – Washington: GPO, 1952 (all publ) – 3mf – 9 – $4.50 – mf#LLMC 81-500 – us LLMC [340]
Comparative psychology : or, the growth and grades of intelligence / Bascom, John – New York: G P Putnam's Sons, 1878 [mf ed 1987] – 297p – 1 – mf#1956 – us Wisconsin U Libr [150]
Comparative religion / Carpenter, Joseph Estlin – New York: H Holt; London: Williams and Norgate, [1913?] – (= ser Home university library of modern knowledge) – 1mf – 9 – 0-524-00704-7 – mf#1990-2032 – us ATLA [230]
Comparative religion / Geden, Alfred Shenington – New York: Macmillan, 1917 – 1mf – 9 – 0-524-01961-4 – (incl bibl ref) – mf#1990-2752 – us ATLA [230]
Comparative religion / Jevons, Frank Byron – Cambridge: University Press 1913 [mf ed 1990] – (= ser The cambridge manuals of science and literature) – 1mf – 9 – 0-7905-7950-2 – (incl bibl ref) – mf#1989-1175 – us ATLA [230]
Comparative religion / St Clair Tisdall, William – London, New York: Longmans, Green 1909 [mf ed 1991] – (= ser Anglican church handbooks) – 1mf – 9 – 0-524-01072-2 – mf#1990-2220 – us ATLA [230]
Comparative religion, its adjuncts and allies / Jordan, Louis Henry – London; New York: Oxford University Press, 1915 – 2mf – 9 – 0-7905-7850-6 – (incl bibl ref) – mf#1989-1075 – us ATLA [012]
Comparative religion, its genesis and growth / Jordan, Louis Henry – Edinburgh: T & T Clark, 1905 – 2mf – 9 – 0-7905-7851-4 – (incl bibl ref) – mf#1989-1076 – us ATLA [230]
Comparative religion, its method and scope : a paper read (in part) at the third international congress of the history of religions, oxford, september 17, 1908 / Jordan, Louis Henry – London; New York: Oxford University Press, 1908 – 1mf – 9 – 0-7905-9982-1 – mf#1989-1707 – us ATLA [230]
Comparative religion, its origin and outlook : a lecture / Jordan, Louis Henry – London; New York: Oxford University Press, 1913 – 1mf – 9 – 0-7905-7852-2 – mf#1989-1077 – us ATLA [230]
Comparative religion, its range and limitations : a lecture / Jordan, Louis Henry – London; New York: Oxford University Press, 1916 – 1mf – 9 – 0-7905-9983-X – (incl bibl ref) – mf#1989-1708 – us ATLA [230]
Comparative strategy – New York. 1978+ (1,5,9) – ISSN: 0149-5933 – mf#11625 – us UMI ProQuest [320]
Comparative studies in religion : an introduction to unitarianism / Secrist, Henry Thomas – Teachers' ed. Boston: Unitarian Sunday-School Society, c1909 – (= ser Beacon series (boston, mass.)) – 1mf – 9 – 0-524-05961-6 – mf#1991-2361 – us ATLA [243]
Comparative studies in society and history – Cambridge. 1958+ [1,5,9] – ISSN: 0010-4175 – mf#13016 – us UMI ProQuest [900]
Comparative study of cultural, morphological, and histological characteristics of species / Hocking, George Macdonald – s.l, s.l? 1942 – 1r – us UF Libraries [630]
A comparative study of experiential learning utilizing indoor-centered training and outdoor-centered training / Moorefield, David L – Texas Woman's University, 1994 – 1mf – 9 – mf#PSY 1895 – us Kinesology [150]
A comparative study of injuries in division i and division iii men's lacrosse / Ferraro, Joseph A – 1999 – 2mf – 9 – $8.00 – mf#PE 4054 – us Kinesology [617]
Comparative study of kgalagadi, kwena, and other sotho dialects / Merwe, D F Van Der – Cape Town, South Africa. 1943 – 1r – us UF Libraries [470]
A comparative study of leisure lifestyles : three developmentally disabled and three non-disabled older adults / Kerr, S R – 1991 – 2mf – 9 – $8.00 – us Kinesology [301]
A comparative study of recruiting techniques : between proposition 48 and proposition 16 student-athletes in division 1a football / Parenteau, Robert A – 1997 – 1mf – 9 – $4.00 – mf#PE 3768 – us Kinesology [790]
A comparative study of selected sports management programs at the master's degree level / Tungjaroenchai, Amnart – 2000 – 250p on 3mf – 9 – $15.00 – mf#PE 4188 – us Kinesology [378]
Comparative study of sex discrimination at the workplace / Nyman, Roseline Mary – U of the Western Cape 1994 [mf ed [S.l: s.n. 1994] – 2mf – 9 – (incl bibl ref & bibl) – sa Misc Inst [305]
Comparative study of some aspects of the supernatural / Fleming, Sarah Hollis – s.l, s.l? 1944 – 1r – us UF Libraries [130]
Comparative study of some central african gong-languages / Carrington, John F – Bruxelles, Belgium. 1949 – 1r – us UF Libraries [470]
A comparative study of the literatures of egypt, palestine and mesopotamia / Peet, T E – 1929 – 9 – $10.00 – us IRC [470]
A comparative study of the social background of adult education problems : in africa and india / Banerji, Santi Kumar – London: [s.n.], 1950 – 1 – us CRL [374]
A comparative study of the social background of adult education problems : in africa and india / Banerji, Santi Kumar – London: [s.n.] 1950 – us CRL [374]
A comparative study of the student-athlete academic support programs at the schools in the mid-american conference / Lambertson, Amy J – 1998 – 1mf – 9 – $4.00 – mf#PE 3853 – us Kinesology [790]
Comparative study of the variations on a theme of paganini by paganini, liszt, brahms and rachmaninoff / Scott, Margaret Lois – U of Rochester 1950 [mf ed 19–] – 3mf – 9 – (with bibl) – mf#fiche 85, 941 – us Sibley [780]
A comparative study on the effectiveness of preventive knee braces / Szczodrowski, Daniel F – 1988 – 91p 1mf – 9 – $4.00 – us Kinesology [617]
Comparative urban and community research – New Brunswick. 1988+ (1,5,9) – (cont: comparative urban research) – ISSN: 0892-5569 – mf#16699 – us UMI ProQuest [307]
Comparative urban research – New York. 1976-1985 (1,5,9) – (cont by: comparative urban and community research) – ISSN: 0090-3892 – mf#11102 – us UMI ProQuest [307]
Comparative value of grazing crops for fattening feeder pigs / Kirk, W Gordon – Gainesville, FL. 1943 – 1r – us UF Libraries [636]
A comparative view of church organizations, primitive and protestant : with a supplement on methodist secessions and methodist union / Rigg, James Harrison – 3rd rev enl ed. London: Charles H Kelly, 1897 [mf ed 1991] – 1mf – 9 – 0-7905-9463-3 – (1st publ in 1887) – mf#1989-2688 – us ATLA [242]
Comparative view of the doctrines and confessions of the various communities of christendom : with illustrations from their original standards = Comparative darstellung des lehrbegriffs der verschiedenen christlichen kirchenpartheien / Winer, Georg Benedict; ed by Pope, William Burt – Edinburgh: T & T Clark 1873 [mf ed 1989] – (= ser Clark's foreign theological library. 4th series 35) – 2mf – 9 – 0-7905-7554-X – (english trans fr german; int by ed; incl bibl ref) – mf#1989-0779 – us ATLA [230]
A comparative view of the gospels : with an introduction intended to further aid their illustration... / Platt, Isaac L – New York: Thomas Holman, 1860 [mf ed 1985] – 1mf – 9 – 0-8370-4761-7 – mf#1985-2761 – us ATLA [226]
A comparative view of the words bathe, wash, dip, sprinkle and pour of the english bible : and of their originals in the hebrew and septuagint copies / Barclay, E D – Cincinnati: Standard, 1881 [mf ed 1992] – (= ser Christian church (disciples of christ) coll) – 1mf – 9 – 0-524-02244-5 – mf#1990-4251 – us ATLA [220]
Compare – New York NY, 1976 – 1r – 1 – (italian periodical) – us IHRC [073]
Comparing kilocalorie expenditure between a stair-stepper, a treadmill, and an elliptical trainer / Clay, Dawn E – 2000 – 53p on 1mf – 9 – $5.00 – mf#PH 1704 – us Kinesology [613]
Comparing the effectiveness of constant task and task variation teaching methods when working with adolescents who have behavioral disorders / Schauer, Aaron T – 1998 – 1mf – 9 – $4.00 – mf#PE 3906 – us Kinesology [790]
Comparing tort liability knowledge of future teacher coaches and current practicing teacher coaches / Kautz, Robert E & Adams, Samuel Houston – 1993 – 2mf – $8.00 – us Kinesology [340]
A comparison among the track and grab starts in swimming and a stand-up response task / Stone, Randall A – 1988 – 68p 1mf – 9 – $4.00 – us Kinesology [612]
A comparison and cross-analysis of the constitution of the fsm : and the draft compact of free association / White, Michael A [comp] – n.p, n.d. – 1mf – 9 – $1.50 – mf#llmc82-100h, title 17 – us LLMC [348]
The comparison between an aquatic running program versus a hard surface running program on aerobic capacity and body composition / Hartman, H – 1988 – 1mf – 9 – $4.00 – us Kinesology [612]
A comparison between anthropometric regression equations and hydrostatic weighing for predicting percent body fat of adult males with down syndrome / Ovalle, S E – 1992 – 2mf – 9 – $8.00 – us Kinesology [790]
A comparison between professional and non-professional football players using selected anthropometric and performance variables / Woodfork, David R – 1998 – 1mf – 9 – $4.00 – mf#PE 3890 – us Kinesology [790]
A comparison between stairclimbing and graded walking in peripheral vascular occlusive disease / Vaughan, N R – 1991 – 2mf – 9 – $8.00 – us Kinesology [612]
A comparison between the number of european americans, african americans, and other minority races who enter phase 1 and phase 2 cardiac rehabilitation programs / Shields, Lionell – University of Wisconsin-La Crosse, 1995 – 1mf – 9 – mf#HE 571 – us Kinesology [613]
A comparison between the rate of wages in some of the british colonies and in the united states; with observations thereupon / Everest, Robert – London, 1861 – (= ser 19th c books on british colonization) – 1mf – 9 – mf#1.1.4688 – uk Chadwyck [331]
The comparison of active plantarflexor muscle stiffness between young and elderly human females / Blanpied, Peter R & Smidt, Gary L – 1989 – 3mf – 9 – $12.00 – us Kinesology [617]
Comparison of acute heart rate and blood pressure responses among isometric, isotonic, and isokinetic exercise / Hui, Sai C & Mahar, Matthew T – 1992 – 2mf – $8.00 – us Kinesology [612]
The comparison of aerobic fitness levels in repect to visual-perceptual performance at rest and during exercise / Pate, John G & Pleasants, Frank – 1993 – 1mf – $4.00 – us Kinesology [150]
Comparison of androgyny levels in team and individual sport female athletes / Thibodeau, Laura – 1990 – 67p 1mf – 9 – $4.00 – us Kinesology [150]
A comparison of anterior tibial-femoral laxity in female intercollegiate gymnasts to a normal population / Brannan, Tori L – 1994 – 1mf – $4.00 – us Kinesology [617]
A comparison of athletic ankle taping techniques with respect to ankle inversion / Plummer, P E – 1991 – 1mf – 9 – $4.00 – us Kinesology [790]
Comparison of attitudes : toward physical activity and physical activity levels of sixth grade boys and girls of various ethnic origins / Parkhurst, Diana L – 2000 – 169p on 2mf – 9 – $10.00 – mf#HE 680 – us Kinesology [305]
Comparison of attitudes towards sportmanship in intramural basketball participants / Vogt, Geoffrey D – 1999 – 1mf – 9 – $4.00 – mf#PSY 2061 – us Kinesology [790]
Comparison of balance and maximal oxygen consumption among hearing, congenital non-hearing and acquired non-hearing female intercollegiate athletes / Ellis, Marjorie K – 1991 – 1mf – $4.00 – us Kinesology [612]
A comparison of balance/stability assessment systems / Stanchina, Carol J – 1mf – 9 – $4.00 – mf#PE 3776 – us Kinesology [612]
Comparison of behavior modification techniques used in physical education with institutionalized profoundly mentally retarded students at different ages / Culver, April D – 1989 – 120p 2mf – 9 – $8.00 – us Kinesology [150]
Comparison of bilateral normal tibial rotation in adult males / Carver, T A – 1990 – 1mf – 9 – $4.00 – us Kinesology [613]
A comparison of blood flow of the vastus lateralis during exercise in trained and untrained cyclists as measured by 133 xenon clearance / Sharpe, Glen P – 1998 – 1mf – 9 – $4.00 – mf#PH 1611 – us Kinesology [612]
Comparison of body composition between german and american adults with mental retardation / Frey, Bernd – 1994 – 1mf – $4.00 – us Kinesology [612]
A comparison of body composition changes in moderately obese and extremely obese women who experience the same caloric deficit / Stowe, Stephen R & Fisher, A Garth – 1991 – 1mf – $4.00 – us Kinesology [617]
A comparison of body density and percent body fat using functional residual capacity and residual volume and development of immersed functional residual capacity and residual volume prediction formulas / Koenig, Joseph M & Floyd, William – 1990 – 1mf – $4.00 – us Kinesology [612]
A comparison of bone mineral density between active and nonactive men with spinal cord injuries / Eddins, William C, Jr – Oregon State University, 1995 – 2mf – 9 – $8.00 – mf#PH 1491 – us Kinesology [612]
Comparison of bone mineral density in 10- to 13-year-old female gymnasts and swimmers / Kearney, Kristi D – 2000 – 126 on 2mf – 9 – $10.00 – mf#PE 4141 – us Kinesology [612]
A comparison of cardiopulmonary variables in male smokers and non-smokers during endurance exercise / Yakey-Ault, Jennifer L – 1998 – 2mf – 9 – $8.00 – mf#PH 1657 – us Kinesology [612]
A comparison of cardiorespiratory fitness of trained and untrained middle-aged women / Upton, S Jill – 1981 – 2mf – 9 – $8.00 – us Kinesology [790]
A comparison of career advancement for male and female head athletic trainers at the ncaa division 1, 2 and 3 levels / Rudd, Lorraine L – 1997 – 1mf – 9 – $4.00 – mf#PE 3771 – us Kinesology [790]
Comparison of citrus fruit grown on various rootstocks / Brooks, Richard L – s.l, s.l? 1934 – 1r – us UF Libraries [634]
A comparison of coaches' and athletic administrators' perceptions on the desirability of hosting non-revenue, postseason athletic events / Smith, Linda J – 1994 – 1mf – $4.00 – us Kinesology [790]
A comparison of different durations of static stretch of the hamstring muscle group in an elderly population / Feland, Jeffrey B – 1999 – 2mf – 9 – $8.00 – mf#PE 4063 – us Kinesology [617]
A comparison of direct instruction and computer-assisted instruction on learning a motor skill by fourth grade students / Ross, James R & Robertson, James B – 1992 – 2mf – $8.00 – us Kinesology [150]
A comparison of division ia football players' grades in season and out of season / Hickey, Kathleen P – 1994 – 1mf – $4.00 – us Kinesology [790]
A comparison of energy cost during forward and backward exercise on the precor c544 transport / Bakken, Angela J – 1997 – 1mf – 9 – $4.00 – mf#PH 1608 – us Kinesology [612]
Comparison of factors affecting the career paths of male and female directors of intercollegiate athletics / Sweany, Lisa – Ball State University, 1996 – 1mf – 9 – mf#PE 3674 – us Kinesology [790]
A comparison of functional ability and physical activity in a community dwelling elderly population / Weisner, Kristie K – 1996 – 1mf – 9 – $4.00 – mf#PSY 1968 – us Kinesology [618]
A comparison of grip strength in young athletes and non-athletes / Garcia, R – 1991 – 1mf – 9 – $4.00 – us Kinesology [790]
A comparison of ground reaction forces during running and form skipping / Johnson, Samuel K – 2000 – 72p on 1mf – 9 – $5.00 – mf#PE 4128 – us Kinesology [612]
Comparison of hard and soft surfaces during maximal vertical jumps in a depth jump plyometric exercise / Olson, Michael W – 1999 – 1mf – 9 – $4.00 – mf#PE 3991 – us Kinesology [611]
A comparison of heart rates among fourth grade students while jumping rope and hula hooping using heart rate monitors / Moris, William D – 1999 – 1mf – 9 – $4.00 – mf#HE 643 – us Kinesology [612]
A comparison of hemodynamic responses to arm and leg exercise of the same intensities / Holmes, Richard J – 1983 – 9 – $4.00 – us Kinesology [790]
A comparison of hydrostatic weighing and displacement plethysmography for determining body density of young elite female gymnasts / Graham, Bruce J & Klug, Gary A – 1993 – 1mf – 9 – $4.00 – us Kinesology [617]
Comparison of hydrostatic weighing to bioelectrical impedance analysis in women greater than thirty percent body fat / Bauer, Shari R & Porcari, John P – 1991 – 1mf – $4.00 – us Kinesology [617]

Comparison of ingesting various forms of carbohydrate on glucose and insulin levels / Rasmussen, Christopher J – 2000 – 1mf – 9 – $4.00 – mf#PE 4075 – us Kinesology [612]
A comparison of intermittent exercise and relaxation versus steady state exercise on fitness levels and attitudes in seventh grade girls / Jordan, Mary C – 1993 – 1mf – $4.00 – us Kinesology [612]
A comparison of isometric strength test results between low back injured patients and normals / Kendrick, R J – 1991 – 1mf – 9 – $4.00 – us Kinesology [612]
A comparison of lower back pain and injury in competitive and non-competitive gymnasts / Parks, Laura M – 2000 – 1mf – 9 – $4.00 – mf#PE 4090 – us Kinesology [617]
A comparison of maladaptive behaviors of athletes and non-athletes / Weiss, Steven M – Springfield College – 2mf – 9 – $8.00 – mf#PSY1871 – us Kinesology [150]
A comparison of manual and machine assisted proprioceptive neuromuscular facilitation flexibility techniques / Burke, Darren – Dalhousie University, 1995 – 2mf – 9 – $8.00 – mf#PH1453 – us Kinesology [612]
Comparison of muscle force production for the smith machine and free weight modes using similar exercises / Cotterman, Michael L – 1998 – 2mf – 9 – $8.00 – mf#PE 3867 – us Kinesology [612]
Comparison of nutritional supplements during a six-week resistance training program / Moroney, Daniel R – 1998 – 2mf – 9 – $8.00 – mf#PH 1613 – us Kinesology [612]
A comparison of occupational stress and related variables among salespersons, clerical staff, service technicians, and managers of the mid-ohio district of the xerox corporation / Gardner, J K – 1991 – 3mf – 9 – $12.00 – us Kinesology [150]
A comparison of offensive efficiency between three types of basketball offense / Papachatzis, Thanasis – 1994 – 3mf – $12.00 – us Kinesology [301]
Comparison of one continous bout versus a split bout of aerobic exercise on 13-hour ambulatory blood pressure in hypertensive females / Fietkau, Rebecca – 2000 – 1mf – 9 – $4.00 – mf#PH 1693 – us Kinesology [612]
A comparison of orthostatic tolerance and fluid loading in trained and non-trained subjects during simulated weightlessness / Mauro, B A – 1992 – 2mf – 9 – $8.00 – us Kinesology [612]
Comparison of pcr-based profiling techniques for the identification of tetraploid potato (solanum tuberosum l.) cultivars / Lambert, Carol-Ann – Stellenbosch: U of Stellenbosch 1998 [mf ed Gauteng: MGX 1998] – 4mf – 9 – (incl bibl ref; abstract in english & afrikaans) – mf#mf.1348 – sa Misc Inst [580]
A comparison of perceived health and quality of life between cardiac rehabilitation participants and nonparticipants / Briner, Megan A – 1996 – 1mf – 9 – $4.00 – mf#PSY 1979 – us Kinesology [617]
A comparison of perceptions of the importance between physical education graduate teaching assistants and graduate program coordinators : on selection of graduate teaching assistants and their teaching performance criteria / Wei, Bing & Cooper, Walter E – 1992 – 2mf – 9 – $8.00 – us Kinesology [378]
Comparison of phosphate carriers for establishment and maintenance / Willson, George Cralle – s.l, s.l? 1940 – 1r – us UF Libraries [630]
A comparison of physiologic responses to forward and retrograde simulated stair stepping on the stairmaster / Ryan, Patrick T – 1993 – 1mf – $4.00 – us Kinesology [612]
A comparison of physiologic responses when exercising on five exercise modalities at a self-selected exercise intensity / Allaback, Nicole J – 1998 – 1mf – 9 – $4.00 – mf#PH 1646 – us Kinesology [612]
A comparison of preferred coaching leadership behaviors : of college athletes in individual and team sports / Lindauer, Jeffrey R – 2000 – 55p on 1mf – 9 – $5.00 – mf#PE 4138 – us Kinesology [150]
Comparison of preferred coaching leadership behaviors of basketball players at the ncaa division 3 level / Peng, Hsiao-hwei – 1997 – 1mf – 9 – $4.00 – mf#PSY 2001 – us Kinesology [150]
Comparison of purebred and cross bred cockerels with respect to fattening and dressing qualities / Mehrhof, N R – Gainesville, FL. 1945 – 1r – us UF Libraries [636]
A comparison of rating of perceived exertion in treadmill vs track walking and running / Schroeder, Lisa M & Butts, Nancy Kay – 1991 – 1mf – $4.00 – us Kinesology [150]
A comparison of ratings of perceived exertion during stairmaster and treadmill exercise / Budzinski, Karen M – 1996 – 2mf – 9 – $8.00 – mf#PSY 1980 – us Kinesology [612]
Comparison of resting metabolic rate and excess post-exercise oxygen consumption in normal and low calorie dieting females / Hilbert, Carey A – Oregon State University, 1995 – 2mf – 9 – $8.00 – mf#PH 1496 – us Kinesology [612]
The comparison of resting metabolic rate in trained vs. untrained females / Allen, David M – 1991 – 1mf – 9 – $4.00 – us Kinesology [613]
Comparison of risk factors for coronary heart disease in sedentary and physically active college students / Jensen, Marian & Heiner, Steven W – 1992 – 2mf – $8.00 – us Kinesology [613]
A comparison of safety belt public service announcement strategies on the attitudes of driver education students / Moss, Brian J – 1998 – 1mf – 9 – $4.00 – mf#HE 636 – us Kinesology [360]
A comparison of selected coronary heart disease risk factors in weight trained males / Lambert, Christopher M & Floyd, William – 1991 – 1mf – $4.00 – us Kinesology [612]
A comparison of selected neuromuscular and kinematic variables before and after learning an aiming task / Steciak, David M & Brush, Florence C – 1992 – 1mf – 9 – $4.00 – us Kinesology [150]
A comparison of selected physiological responses to indoor rock climbing in beginner and advanced sport climbers / Sanders, Benjamin V – 1999 – 1mf – 9 – $4.00 – mf#PH 1683 – us Kinesology [612]
Comparison of serratus anterior emg activity during stable and unstable closed kinetic chain activity / Mauger, Curtis – 2000 – 43p on 1mf – 9 – $5.00 – mf#PE 4171 – us Kinesology [617]
A comparison of sideline versus clinical cognitive test performance in collegiate athletes / Onate, James A – 1997 – 2mf – 9 – $8.00 – mf#PE 3767 – us Kinesology [790]
A comparison of six personality factors between professional, college, and high school basketball players / Bowe, William G Jr – State University of New York College at Brockport, 1994 – 1mf – 9 – $4.00 – mf#PSY1839 – us Kinesology [150]
A comparison of skeletal muscle responsiveness to exercise in male and female sprague-dawley rats treated with an anabolic steroid / Brown, Gordon D et al – 1992 – 2mf – 9 – $8.00 – us Kinesology [613]
Comparison of skinfold measurements under normally hydrated and dehydrated conditions in females ages to 54 / Botenhagen, Kim A & Alejandro-De Leon, Daniel – 1992 – 1mf – 9 – $4.00 – us Kinesology [617]
Comparison of sorghum silage,peanut hay and cottonseed hulls as roughages for fattening steers / Shealy, A L – Gainesville, FL. 1938 – 1r – us UF Libraries [636]
Comparison of soy bean silage and alfalfa hay for milk production / Dawson, C R – s.l, s.l? 1931 – 1r – us UF Libraries [630]
Comparison of sport achievement orientation between wheelchair basketball athletes and able-body basketball athletes / Skordilis, Emmanouil K – Springfield College, 1995 – 2mf – 9 – $8.00 – mf#PSY1864 – us Kinesology [150]
A comparison of standing posture between female junior high school students in hong kong with and without ballet training / Lee, B W – 1991 – 2mf – 9 – $8.00 – us Kinesology [790]
A comparison of student development outcomes among male revenue athletes, non-revenue athletes, and club sport athletes at an ncaa divison 1 university : a case study / Jackovic, Terence J – 1999 – 231p on 3mf – 9 – $15.00 – mf#PE 4183 – us Kinesology [150]
Comparison of substrate utilization patterns in males and amenorrheic females during submaximal exercise / Zderic, Theodore W – 1997 – 1mf – 9 – $4.00 – mf#PH 1580 – us Kinesology [612]
Comparison of tagalog and iloko : dissertation for the obtention of the doctor's degree in the faculty of philosophy of the university of hamburg / Lopez, Cecilio – 1928 – us CRL [490]
A comparison of tagalog and malay lexicographies (on phonetic-semantic basis) / Lopez, Cecilio – Manila: Bureau of Print, 1939 – us CRL [490]
Comparison of television viewing, arcade game play, and resting metabolic rates in youth / Cox, Lori M – 1999 – 1mf – 9 – $4.00 – mf#HE 650 – us Kinesology [613]
A comparison of the american red cross and young men's christian association teaching methods : for beginning swimming / Roan, Sarah A – 1987 – 117p on 2mf – 9 – $8.00 – us Kinesology [370]
Comparison of the association of cervical spinal canal stenosis and intervertebral foraming canal stenosis and transient upper extremity parasthesias / Aliquo, David – Temple University, 1996 – 1mf – 9 – mf#PE 3626 – us Kinesology [617]
A comparison of the cardio glide, cross walk, and treadmill walking in the development of cardiovascular endurance dynamic strength and flexibility / Taylor, Julie E – 1995 – 1mf – 9 – $4.00 – mf#PH 1591 – us Kinesology [612]
A comparison of the cardioglide, crosswalk, and treadmill walking in body composition and blood lipids in middle-aged men and women / McAlpine, Christine M – 1996 – 1mf – 9 – $4.00 – mf#PH 1603 – us Kinesology [612]
A comparison of the coach leadership behavior preferred by male and female track and field athletes / Wang, Yao T – 1996 – 2mf – 9 – $8.00 – mf#PE 3781 – us Kinesology [790]
A comparison of the effectiveness of interactive video in teaching the ability to analyze two motor skills in swimming / Mathias, KE – 1990 – 2mf – 9 – $8.00 – us Kinesology [790]
A comparison of the effects of a wrestling practice and a weightlifting workout on the body fat percent of wrestlers / Bergerson, Mark – 1994 – 1mf – $4.00 – us Kinesology [612]
Comparison of the efficacy of calcium hydroxide, iodine potassium iodide and chlorhexidine on actinomyces israelii contaminated dentinal tubules using an in vitro model / Tait, Carol Margaret Elizabeth – Stellenbosch: U of Stellenbosch 1998 [mf ed 1998] – 3mf – 9 – mf#mf.1267 – sa Stellenbosch [617]
Comparison of the lactate and ventilatory thresholds during prolonged work / Loat, Christopher E R & Rhodes, E C – 1991 – 1mf – 9 – $4.00 – us Kinesology [613]
A comparison of the leadership behavior of winning and losing high school basketball coaches / Lam, Tak C – Springfield College, 1995 – 2mf – 9 – $8.00 – mf#PSY1849 – us Kinesology [150]
Comparison of the mid-ameriacn conference athletic department regarding compliance with title 9 / Fiscus, Douglas L – Ball State University, 1996 – 1mf – 9 – mf#PE 3639 – us Kinesology [790]
Comparison of the number of classification trials for various criterion-referenced testing methods / McManis, BG – 1991 – 2mf – 9 – $8.00 – us Kinesology [613]
A comparison of the pharmaceutical practices of head athletic trainers in the treatment of athletic injuries at the national collegiate athletic association division 1 level / Mackey, Theresa R – 1998 – 1mf – 9 – $4.00 – mf#PE 3886 – us Kinesology [615]
A comparison of the physiological and psychological responses to exercise on a virtual reality recumbent cycle versus a non-virtual reality recumbent cycle / Maldari, Monica M – 1997 – 1mf – 9 – $4.00 – mf#PSY 1950 – us Kinesology [612]
A comparison of the physiological responses to exercise on five different upper and lower body ergometers / Spranger, Lance L – 1998 – 1mf – 9 – $4.00 – mf#PH 1643 – us Kinesology [612]
A comparison of the political attitudes of female and male athletes / Martindell-Jensen, Jane – 1980 – 1mf – 9 – $4.00 – us Kinesology [790]
A comparison of the ratio of shoulder concentric internal rotation to eccentric external rotation of male swimmers vs. nonswimmers / Werner-Ferrel, Sally D – 1989 – 40p 1mf – 9 – $4.00 – us Kinesology [612]
A comparison of the sportsmanship attitudes and/or moral reasoning of interscholastic coaches / Gillentine, John A – University of Southern Mississippi, 1995 – 1mf – 9 – $4.00 – mf#PSY1845 – us Kinesology [150]
A comparison of the submaximal and maximal responses to upright versus semi-recumbent cycling in females / Pauly, Marsha A – 1999 – 1mf – 9 – $4.00 – mf#PH 1672 – us Kinesology [612]
A comparison of the submaximal and maximal responses to upright verus semi-recumbent cycling in males / Johnson, Charles C – 1999 – 1mf – 9 – $4.00 – mf#PH 1671 – us Kinesology [612]
A comparison of the tensile strength of the umbilical cord of babies of smoking and non-smoking mothers / England, Joyce & McGuire, Don – 1988 – 2mf – 9 – $8.00 – us Kinesology [613]
A comparison of the yellow springs instruments : and accusports tm portable lactate analyzer for measuring blood lactate in cold environments / Franklin, Jodi L – 2000 – 28p on 1mf – 9 – $5.00 – mf#PE4135 – us Kinesology [612]
A comparison of three hydrostatic weighing techniques : without head submersion, total lung capacity, and residual volume / Wa, K M – 1991 – 2mf – 9 – $8.00 – us Kinesology [790]
A comparison of three warm-up protocols on power output in the wingate anaerobic test / Thomas, R S – 1990 – 1mf – 9 – $4.00 – us Kinesology [613]
A comparison of time management behaviors : between the physical education setting and the coaching setting of five middle school teachers / Vosylius, Gaile – 2000 – 73p on 1mf – 9 – $5.00 – mf#PE 4099 – us Kinesology [650]
Comparison of total peripheral resistance and blood velocity as obtained from doppler ultrasound waveforms during rest, exercise and recovery / Socha, M L – 1991 – 1mf – 9 – $4.00 – us Kinesology [612]
Comparison of two instruments for needs assessment and evaluation of an employee health promotion program / Hoffman, Marilyn A – 1989 – 78p 1mf – 9 – $4.00 – us Kinesology [613]
A comparison of two intermittent external compression devices on pitting ankle edema / Lemley, T R – 1991 – 1mf – 9 – $4.00 – us Kinesology [790]
A comparison of two methods for teaching three-ball juggling / Catanzariti, Jason C – 1998 – 1mf – 9 – $4.00 – mf#PE 3868 – us Kinesology [790]
A comparison of two methods of teaching volleyball : skill teaching, with game and equipment modifications, and mastery learning / Preece, Lisa A – 1996 – 1mf – 9 – $4.00 – mf#PE 3808 – us Kinesology [370]
Comparison of two methods of training special olympics volunteers to teach and coach bowling / Albright, Cindy W – 1989 – 191p on 2mf – 9 – $8.00 – us Kinesology [370]
Comparison of unstructured feedback to structured feedback on initial learning of cpr / Jankovich, Gina – 1997 – 1mf – 9 – $4.00 – mf#HE 607 – us Kinesology [612]
A comparison of valued outcomes of youth sport programs : participants, parents, coaches, and administrators / Millard, Linda – 1990 – 2mf – 9 – $8.00 – us Kinesology [150]
A comparison of various fitness parameters and their relationship to cholesterol levels / Nowotny, David C – 1998 – 1mf – 9 – $4.00 – mf#PH 1630 – us Kinesology [612]
A comparison of vertical jump height measurements on the vertec and biotran using three jumping techniques / McCollum, Jennifer D – University of North Carolina at Chapel Hill, 1995 – 1mf – 9 – $4.00 – mf#PE3611 – us Kinesology [612]
A comparison of village and staged versions of selected hungarian dance styles / Geslison, Jeannette E K – Brigham Young University, 1995 – 2mf – 9 – $8.00 – mf#PE 3645 – us Kinesology [790]
Comparison of waist to hip ratio measurements in relation to cardiovascular risk factors in healthy menopausal women / Hurtubise, Cheryl L – University of Carolina at Greensboro, 1995 – 1mf – 9 – $4.00 – mf#HE555 – us Kinesology [616]
Comparisons of domain score and reliability estimates using trials-tocriterion, sequential probability ratio, and pre-set trial length tests / Chow, B C – 1991 – 2mf – 9 – $8.00 – us Kinesology [370]
Comparsas populares del carnaval habanero, cuestio – Habana, Cuba. 1937 – 1r – us UF Libraries [972]
Compass – 1987 jun/jul-1993 aug/sep – 1r – 1 – mf#1071430 – us WHS [071]
Compassion for prisoners recommended / Young, George – Edinburgh, Scotland. 1809 – 1r – us UF Libraries [240]
Compayre, Gabriel *see*
– Abelard and the origin and early history of universities
– L'education
– Elements d'instruction morale et civique
– Herbert spencer and scientific education
– Jean jacques rousseau and education from nature
– Pestalozzi and elementary education
Compel – Bradford. 2001+ (1,5,9) – ISSN: 0332-1649 – mf#18248 – us UMI ProQuest [510]
A compend of christian divinity / Cobb, Sylvanus – 4th ed. Boston: S Cobb, 1849 [mf ed 1992] – (= ser Unitarian/universalist coll) – 1mf – 9 – 0-524-06990-5 – mf#1991-2843 – us ATLA [243]

Compend of lutheran theology : a summary of christian doctrine = Compendium locorum theologicorum ex scripturis sacris et libro concordiae / Hutter, Leonhard – Philadelphia: Lutheran Book Store, 1868 – 1mf – 9 – 0-524-04770-7 – (in english) – mf#1991-2156 – us ATLA [242]

Compendaria in graecam via / Gonzalez, Castro – Tip. Regia, 1792 – sp Bibl Santa Ana [946]

Compendia of awards / Connecticut. Compensation Commissioners – v.1-12, 1914-48. 102 fiches. (Harvard Law School Library Collection.) – 9 – us Harvard Law [336]

Compendiaria expositio doctrinae de essentia orig : peccati / Flacius Illyricus d A, M – Vrsellis, 1572 – 1mf – 9 – mf#TH-1 mf 434 – ne IDC [242]

Compendiaria praecipvarvm rervm tvrcicarvm... / Risebergius, L – Helmaestadii, 1596 – 4mf – 9 – mf#H-8390 – ne IDC [956]

Compendio das eras da provincia do para... / Monteiro Baena, Antonio Ladislau – Belem, Brazil. 1969 – 1r – us UF Libraries [972]

Compendio de aritmetica / Munoz de Rivera, Juan – 1868 – 9 – sp Bibl Santa Ana [510]

Compendio de aritmetica.. / Botello del Castillo, Carlos – 1878 – 9 – sp Bibl Santa Ana [510]

Compendio de elementos de matematicas / Morales Fernandez, Francisco – 1884 – 9 – sp Bibl Santa Ana [510]

Compendio de filosofia escolastica. tomo 1 : logica y psicologia / Marquez, Gabino – Jerez: Tipolitografia de Salido Hermanos, 2nd ed 1917 – 1 – sp Bibl Santa Ana [170]

Compendio de filosofia escolastica. tomo 2. filosofia moral / Marquez, Gabino – Jerez: Tipolitografia de Salido Hermanos, 3rd ed 1921 – 1 – sp Bibl Santa Ana [170]

Compendio de geografia de colombia para uso de las... / Martinez Silva, Carlos – Bogota, Colombia. 1923 – 1r – us UF Libraries [972]

Compendio de geometria y trigonometria.. / Botello del Castillo, Carlos – 1879 – 9 – sp Bibl Santa Ana [510]

Compendio de gramatica castellana / Diez Olivares, J Ma – 1874 – 9 – sp Bibl Santa Ana [440]

Compendio de historia da literatura brasileira / Romero, Silvio – Rio de Janeiro, Brazil. 1906 – 1r – us UF Libraries [972]

Compendio de historia de centriamerica / Salvatierra, Sofonias – Managua, Nicaragua. 1946 – 1r – us UF Libraries [972]

Compendio de historia de cuba / Fonseca, Miguel Angel – Habana, Cuba. 1943 – 1r – us UF Libraries [972]

Compendio de historia economica y hacendaria de co... / Soley Guell, Tomas – San Jose, Costa Rica. 1940 – 1r – us UF Libraries [330]

Compendio de historia general de mexico / Zarate, Julio – Paris, France. 1913 – 1r – us UF Libraries [972]

Compendio de indulgencias concedidas a los ministros, y demas personas, que se emplean en el servicio del santo oficio de la inquisicion / Tribunal de la Inquisicion en Mexico – [Mexico: s.n. 17-] – (= ser Books on religion...1543/44-c1800: inquisicion) – 1mf – 9 – mf#crl-258 – ne IDC [241]

Compendio de la arte de la lengua tagala : (ano 1703) / Gaspar de San Augustin – 2. impr. Sampaloc: Impr de N S de Loreto del pueblo 1787 – (= ser Whsb) – 3mf – 9 – €40.00 – (incl: confessionario copioso en lengua espanola, y tangela) – mf#Hu 323 – gw Fischer [490]

Compendio de la historia / Garcia, Jose Gabriel – Santo Domingo, Dominican Republic. v1-4. 1893-1906 – 1r – us UF Libraries [972]

Compendio de la historia de centro-america / Saravia, Miguel G – Guatemala, 1881 – 1r – us UF Libraries [972]

Compendio de la historia de colombia / Bermudez, Jose Alejandro – Bogota, Colombia. 1934 – 1r – us UF Libraries [972]

Compendio de la historia de el bienaventurado felipe de jesus, y devocion : consagrada a celebrar su memoria el dia cinco de cada mes – en Mexico: Por D Felipe de Zuniga y Ontiveros, ano de 1781 – (= ser Books on religion...1543/44-c1800: biografias de religiosos) – 1mf – 9 – mf#crl-158 – ne IDC [241]

Compendio de la historia de la literatura de guate... / Ydigoras Fuentes, Carmen – Guatemala, 1959 – 1r – us UF Libraries [972]

Compendio de la historia de venezuela / Yanes, Francisco Javier – Caracas, Venezuela. 1944 – 1r – us UF Libraries [972]

Compendio de la prodigiosa vida del apostol de valencia, san vicente ferrer, con su novena / Vidal, Francisco – Manila: E Balbas 1887 [mf ed Bloomington IN: Indiana Uni Lib, Preservation Dept 1984] – (= ser Coll of spanish devotional literature) – 1r – 1 – us Indiana Preservation [241]

Compendio de la salud humana, zaragoza, 1494 / Kethan, J – 9 – sp Cultura [610]

Compendio de la vida de s andres avelino, confessor / San Felix, Juan de – en Mexico: En la imprenta nueva Antuerpiana de D Phelipe de Zuniga y Ontiveros, ano de 1767 – (= ser Books on religion...1543/44-c1800: vidas y cultos de santos) – 2mf – 9 – mf#crl-126 – ne IDC [241]

Compendio de la vida del v j gabriel perboyre : presbitero de la congregacion de la mision, fundada por san vicente de paul – Madrid: Tipografia de Los Huerfanos, 1890 [mf ed 1995] – (= ser Yale coll) – 120p (ill) – 1 – 0-524-09707-0 – (in spanish) – mf#1995-0707 – us ATLA [241]

Compendio de las constituciones y reglas de la m[uy] il[lus]tre 5. apostolica congegacion de n g p s s pedro : canonicamente fundada por el exemplar clero secular de esta ciudad de mexico, en su colegio, hospital e iglesia de la ss trinidad, el ano de 1577... – en Mexico: impr dona Maria de Rivera, ano de 1747 – (= ser Books on religion...1543/44-c1800: ordenes, etc: congregacion de san pedro) – 1mf – 9 – mf#crl-183 – ne IDC [241]

Compendio de las excelencias de la bulla de la sancta cruzada en lengua mexicana / Elias, de San Juan Baptista – en Mexico: impr de Enrico Martinez, ano de 1599 – (= ser Books on religion...1543/44-c1800: miscelanea) – 1mf – 9 – mf#crl-410 – ne IDC [241]

Compendio de las instituciones de derecho canonigo segun el metodo de domingo cavallario / Cervantes Bermudez de Cana, Tomas – 1870 – 9 – sp Bibl Santa Ana [340]

Compendio de literatura y artes de guatemala / Ydigoras Fuentes, Carmen – Guatemala, 1956 – 1r – us UF Libraries [972]

Compendio de los comentarios extendidos por el maestro antonio gomez a las ochenta y tres leyes de toro... / Nolasco de Llano, Pedro – Madrid: Imp. y Lib. de D. Manuel Martin, 1777 – 1 – sp Bibl Santa Ana [946]

Compendio de los comentarios...antonio gomez...toro / Nolasco de Llano, Pedro – 1785 – 9 – sp Bibl Santa Ana [636]

Compendio de teologia cristiana / Pendleton, J M – 1910. 292p – 1 – us Southern Baptist [242]

Compendio de un curso de historia universal para uso de los alumnos de segunda ensenanza – 1875 – 9 – sp Bibl Santa Ana [370]

Compendio del'historia romana... = Romanae historiae compendium / Leto, Giulio Pomponio – Venice: Aperesso Gabriel di Ferrarii 1549 [mf ed 1983] – 1r – 1 – (trans by francesco baldelli) – mf#7158 – us Wisconsin U Libr [930]

Il compendio della mvsica, nel qvale brevemente si tratta dell'arte del contrapunto, diviso in qvatro libri / Tigrini, Orazio – Venetia: appresso Ricciardo Amadino 1588 [mf ed 19–] – 3mf – 9 – mf#fiche 854 – us Sibley [780]

Compendio della vita e delle gesta di giuseppe balsamo : denominato il conte cagliostro, che si e estratto dal processo contro di lui formato in roma l'anno 1790... / Barberi, Giovanni – In Roma: Nella stamperia della Rev Camera Apostolica 1791 – 3mf – 9 – mf#vrl-32 – ne IDC [366]

Compendio estadistico de cuba 1965-1966, 1968, 1976 / Cuba. Direccion Central de Estadistica – (= ser Latin american & caribbean...1821-1982) – 4mf – 9 – uk Chadwyck [318]

Compendio estadistico descriptivo de la republica de panama : con los datos sinopticos de comercio internacional de 1909 a 1916 / Panama. Direccion General de Estadistica – (= ser Latin american & caribbean...1821-1982) – 3mf – 9 – uk Chadwyck [318]

Compendio gramatical para la inteligencia del idioma tarahumar : oraciones, doctrina cristiana, platicas, y otras cosas necesarias para la recta administracion de los santos sacramentos en el mismo idioma / Tellechea, Miguel – Mexico: Impr de la Federacion en Palacio 1826 – (= ser Whsb) – 3mf – 9 – €40.00 – mf#Hu 414 – gw Fischer [440]

Compendio historia eclesiastica / Pedrajas y Nunez-Moreno, Eloy – 1889 – 9 – sp Bibl Santa Ana [240]

Compendio historial de la vida de la gloriosa s[eno]rita de casia monja en el obletvantissimo monasterio de s maria de n p s agustin / Lisperguer y Solis, Mathias – en Lima: Por Joseph de Contreras, ano de 1699 – (= ser Books on religion...1543/44-c1800: vidas y cultos de santos) – 3mf – 9 – mf#crl-118 – ne IDC [241]

Compendio historico / Acosta Y Albear, Francisco De – Madrid, Spain. 1875 – 1r – us UF Libraries [972]

Compendio historico, delle gverre vltimamente successe tra christiani, and turchi, and tra turchi, and persiani... / Campana, C – Vinegia, 1597 – 1mf – 9 – mf#H-8391 – ne IDC [956]

Compendio historico...di ferrara... / Guarini, M A – Ferrara, 1621 – 5mf – 9 – mf#O-1052 – ne IDC [720]

Compendio historicos de las vidas de los santos...del orden de predicadores / Amado, Manuel – 1829 – 9 – sp Bibl Santa Ana [920]

Compendio instructivo sobre el mejor metodo de curar las tercianas y quartanas... / Puig, S – Madrid, 1786 – 3mf – 9 – sp Cultura [615]

Compendio narrativo do peregrino da america / Marques Pereira, Nuno – Rio de Janeiro, Brazil. v1-2. 1939 – 1r – us UF Libraries [972]

Compendio quirurgico... / Robledo, D – Barcelona, 1702 – 8mf – 9 – sp Cultura [617]

Compendio...antiguedad...ntr. sra. guadalupe / Vicente Yanez, Juan – 1889 – 9 – sp Bibl Santa Ana [240]

Compendio...antonio gomez...toro / Llano, Pedro N – 1785 – 9 – sp Bibl Santa Ana [790]

Compendio...cristiana / Rodrigo de la Cerda, Jose – 1891 – 9 – (1888 ed) – sp Bibl Santa Ana [240]

Compendio...de fortificacion... / Rojas, C – Madrid, 1613 – 3mf – 9 – sp Cultura [355]

Compendiolo di molti dvbbi : segreti et sentenze intorno al canto fermo, et figurato, da molti eccellenti & consumati musici dichiarate / Aron, Pietro – Milano: per I A da Castelliono stampatore [1550?] [mf ed Watertown MA: General Microfilm Co [19–]] – 1r – 1 – us UMI ProQuest [780]

Compendio...ntra. sra. guadalupe / Yanez, Juan Vicente – 1899 – 9 – sp Bibl Santa Ana [240]

Compendiosa et diserta ad annotationes papistae cuiusdem anonymi... / Dathenus, P – n.p, 1558 – 1mf – 9 – mf#PBA-160 – ne IDC [240]

Compendiosa narratio : de statu missionis chinnensis. ab anno 1581, usque ad annum 1669 / Intorcetta, P – Romae, 1671 – 1mf – 9 – mf#HT-895 – ne IDC [915]

Compendioso...tratado de la oracion / Pedro de Alcantara, Saint – 1731 – 9 – sp Bibl Santa Ana [410]

Compendious anglo-saxon and english dictionary / Bosworth, Joseph – London, England. 1888 – 1r – us UF Libraries [420]

A compendious grammar of the egyptian language as contained in the coptic and sahidic dialects : with observations on the bashmuric / Tattam, Henry – London: John & Arthur Arch, 1830 – (= ser 19th c books on linguistics) – 4mf – 9 – mf#2.1.35 – uk Chadwyck [470]

Compendious grammar of the egyptian language as contained in the coptic, sahidic, and bashmuric dialects : together with alphabets and numerals in the hieroglyphic and enchorial characters and a few explanatory observations / Tattam, Henry – London: Arch 1830 – (= ser Whsb) – 2mf – 9 – €30.00 – (app by thomas young; incl: rudiments of an egyptian dictionary in the ancient enchorial character: containing all the words of which the sense has been ascertained / thomas young) – mf#Hu 382 – gw Fischer [470]

A compendious history of american methodism : abridged from the author's history of the methodist episcopal church / Stevens, Abel – New York: Carlton & Porter, 1868, c1867 [mf ed 1990] – 2mf – 9 – 0-7905-7027-0 – mf#1988-3027 – us ATLA [242]

A compendious history of new england : from the discovery by europeans to the first general congress of the anglo-american colonies / Palfrey, John Gorham – Boston: Houghton Mifflin, c1883 [mf ed 1992] – (= ser Congregational coll) – 5mf – 9 – 0-524-03266-1 – mf#1990-4669 – us ATLA [978]

Compendious history of new england / Palfrey, John Gorham – Boston, MA. v1-4. 1883 – 1r – us UF Libraries [978]

A compendious history of the rise and progress of the methodist church both in europe and america : consisting principally of selections from various approved and authentic documents, arranged in proper order / Meacham, Albert Gallatin – [Picton, Ont?: s.n.] 1832 [mf ed 1983] – 9 – 0-665-38233-2 – mf#38233 – cn Canadiana [242]

A compendious introduction to the study of the bible / Horne, Thomas Hartwell; ed by Ayre, John – 11th ed. London: Longmans, Green, 1868 [mf ed 1991] – 2mf – 9 – 0-7905-8307-0 – (rev by ed) – mf#1987-6412 – us ATLA [220]

A compendious law dictionary : containing both an explanation of the terms and the law itself; intended for the use of the country gentleman, the merchant, and the professional man / Potts, Thomas – new rev corr enl ed. London: B & R Crosby, 1813 – 8mf – 9 – $12.00 – mf#LLMC 95-424 – us LLMC [340]

Compendious syriac grammar / Noeldeke, Theodor – 9 – $12.00 – us IRC [470]

Compendious syriac grammar / Noeldeke, Theodor – London: Williams & Norgate 1904 [mf ed 1989] – 1mf – 9 – 0-7905-1257-2 – (incl ind; trans fr 2nd german ed by james a crichton) – mf#1987-1257 – us ATLA [470]

Compendiu musices confectu ad faciliore instructione cantum chorale discentiu : necno ad introductione huius libelli... / Machold, Johann – Venetijs...[i.e. 1513] [mf ed 19–] – 4mf – 9 – mf#fiche 395 – us Sibley [780]

Compendium / Harris County Heritage Society [TX] – v4 n1 [1977/78 win/spring] – 1r – 1 – mf#502071 – us WHS [978]

Compendium. / American Bar Association. Committee on Unauthorized Practice of the Law – Chicago. 1942. 122, 14 p. LL-1097 – 1 – us L of C Photodup [340]

A compendium and digest of the laws of massachusetts / White, William Charles – Boston, Munroe Francis and Parker, 1809-11. 4 v. in 2. LL-465 – 1 – us L of C Photodup [348]

Compendium cantionum ecclesiasticarum / Holthusius, Joannes – 1579 – (= ser Mssa) – 5mf – 9 – €70.00 – mfchl 79 – gw Fischer [780]

Compendium christianae religionis / Bullinger, Heinrich – Tigvri, [Christoph] Frosch[auer], 1556 – 4mf – 9 – mf#PBU-186 – ne IDC [240]

Compendium der judischen gesetzeskunde aus dem vierzehnten / Rosin, David – Breslau, Germany. 1871 – 1r – us UF Libraries [939]

Compendium der kirchengeschichte / Groene, Valentin – Regensburg: G.J. Manz, 1869 – 2mf – 9 – 0-7905-5229-9 – (incl bibl ref) – mf#1988-1229 – us ATLA [240]

Compendium errorum johannis 22 papae / Occam, Guillelmus de (Ockham, William of) – Lugduni apud Jo Trechsel, 1495 – €3.00 – ne Slangenburg [241]

Compendium ethicae aristotelicae ad normam veritatis christiana revocatum / Walaeus, Antonius – editio postrema, prioribus auctior & emendatior. Lugduni Batavorum: Ex officina Bonaventurae & Abrahami Elzevir 1627 – (= ser Ethics in the early modern period) – 3mf – 9 – mf#pl-467 – ne IDC [170]

Compendium germanicolatinum musices practicae quaestionibus : facilibus & perspicuis expolitum, omissis omnibus non admodum necessarijs conscriptum = Musica deutsch und lateinisch in kurtze und leichte regulen verfasset / Machold, Johann – Impressum Erphordiae: in Georgij Baumanni senioris Officina 1596 [mf ed 19–] – 1mf / 1r – 9,1 – mf#fiche 790, 39 / film 39 – us Sibley [780]

Compendium harmonicum : oder kurzer begrif der lehre von der harmonie... / Sorge, Georg Andreas – Lobenstein: Im Verlag des Verfassers, und in Commission in Hof bey Ludwig [1760] [mf ed 19–] – 1r – 1 – mf#film 1391 – us Sibley [780]

Compendium historiae literariae novissimae – 1746-69 [mf ed 1993] – 24v on 247mf – 9 – €2510.00 – 3-89131-102-8 – (incl: erlangische gelehrte anmerkungen und nachrichten v25-42 1770-87; annalen der gesammten litteratur 1788-1789; erlangische gelehrte zeitung 1790-1798) – gw Fischer [430]

Compendium historiarum (cbh8) : ex versione guillelmi xylandri / Georgii Cedreni – Parisiis, 1647 – (= ser Corpus byzantinae historiae (cbh)) – €88.00 – ne Slangenburg [240]

Compendium moralis theologiae / Laymann, P – Lugduni, 1631 – 8mf – 9 – mf#CA-67 – ne IDC [240]

Compendium musicae latino-germanicum / Gumpelzhaimer, Adam – 1675 – (= ser Mssa) – 2mf – 9 – €35.00 – mfchl 57 – gw Fischer [780]

Compendium musicae latino-germanicum : studio & opera adami gumpelzhaimeri / Gumpeltzhaimer, Adam – nunc corr ed. Augustae: typis & impensis J V Schoenigij 1616 [mf ed 1983?] – 5mf – 9 – mf#fiche 546 – us Sibley [780]

A compendium of all the instructions and indictments which are approved and criticised in the missouri supreme and appellate court reports / West, James Columbus – Ozark, Mo., 1899. 275 11 p. LL-570 – 1 – us L of C Photodup [347]

A compendium of american criminal law / Desty, Robert – San Francisco, Whitney, 1882. 713 p. LL-550 – 1 – us L of C Photodup [345]

Compendium of baptist history : showing the origin and history of the baptists, from the days of the apostles to the present time, with an original chart, giving a comparative view of some of the denominations of christians with which they have come in contact / Shackelford, James Alfred – Louisville, Ky: Press Baptist Book Concern, 1892 – 4mf – 9 – 0-524-08812-8 – (incl bibliographic references and ind) – mf#1993-3304 – us ATLA [242]

A compendium of christian theology : being analytical outlines of a course of theological study / Pope, William Burt – 2nd rev enl ed. New York: Phillips & Hunt, 1881 [mf ed 1992] – (= ser Methodist coll) – 4mf – 9 – 0-524-04388-4 – mf#1991-2092 – us ATLA [242]

Compendium of church history / Zenos, Andrew Constantinides – Philadelphia PA: Presbyterian Board of Publ & Sabbath-School Work 1896 [mf ed 1986] – 1mf – 9 – 0-8370-9519-0 – (incl ind) – mf#1986-3519 – us ATLA [240]

A compendium of commercial law / Townsend, Calvin – New York/Chicago: Ivison, Blakeman, Taylor & Co, 1877 – 7mf – 9 – $10.50 – mf#LLMC 92-138 – us LLMC [346]

Compendium of evidence / Straker, David Augustus – Detroit, Mich., Richmond & Backus, 1899. 255 p. LL-1189 – 1 – us L of C Photodup [347]

Compendium of kafir laws and customs / Maclean, John – London, England. 1968 – 1r – us UF Libraries [960]

A compendium of methodism : embracing the history and present condition of its various branches in all countries. with a defence of its doctrinal, governmental, and prudential peculiarities / Porter, James – New York: Carlton & Porter, c1851 [mf ed 1992] – (= ser Methodist coll) – 2mf – 9 – 0-524-03080-4 – mf#1990-4569 – us ATLA [242]

A compendium of official documents relative to native affairs in the south island / Mackay, Alexander [comp] – Wellington, NZ: Govt Printer, 1872-3 [mf ed 1991] – 14mf – 9 – NZ$49.95 – 0-477-07433-2 – (with a new chronology and ind comp by rohi williams and josie laing in the canterbury museum) – nz Nat Libr [342]

Compendium of papers on broadening the tax base / U.S. Congress. House Committee on Ways and MeansTax Revision Study – Washington: GPO. 3v. 1959 – 26mf – 9 – $43.50 – mf#LLMC 82-701 – us LLMC [336]

A compendium of religious faith and practice : designed for young persons of the society of friends / Murray, Lindley – New-York: Samuel Wood, 1817 [mf ed 1993] – (= ser Society of friends (quakers) coll) – 1mf – 9 – 0-524-07445-3 – mf#1991-3105 – us ATLA [243]

Compendium of statistics 1965-1970 / Malawi (formerly Nyasaland). National Statistical Office – (= ser African official statistical serials, 1867-1982) – 4mf – 9 – (1967-69 not publ) – uk Chadwyck [316]

Compendium of the art of always rejoicing / Sarasa, Alphonso Antonio de – Boston: Henry A Young, [1872?] – 1mf – 9 – 0-8370-5404-4 – mf#1985-3404 – us ATLA [240]

A compendium of the common law in force in kentucky / Humphreys, Charles – Lexington: Hunt, 1822. 594p. LL-868 – 1 – us L of C Photodup [346]

Compendium of the history of canada and of the british north american provinces / Christian Brothers – Quebec: C Darveau, 1873 – 2mf – 9 – mf#61466 – cn Canadiana [971]

A compendium of the history of the catholic church : from the commencement of the christian era to the ecumenical council of the vatican – 22nd rev enl ed. Baltimore: John Murphy, 1882, c1870 [mf ed 1986] – 2mf – 9 – 0-8370-6924-6 – (incl chronological table and ind) – mf#1986-0924 – us ATLA [241]

A compendium of the law and practice of injunctions, and of interlocutory orders in the nature of injunctions / Henley, Robert Henley Eden – 3d ed. New York: Banks, Gould, 1852. 2v. LL-740 – 1 – us L of C Photodup [340]

A compendium of the law of marine insurance / Annesley, Alexander – Middletown, Conn. Alsop 1808. 258 p. LL-102 – 1 – us L of C Photodup [346]

Compendium of the law on prisoners' rights / Sensenich, Ila Jeanne – Washington: FJC, Apr 1979 – 7mf – 9 – $11.50 – (plus suppl feb 2 1981) – mf#LLMC 95-391 – us LLMC [345]

A compendium of the laws and decisions relating to mobs, riots, invasion, civil commotion, insurrection, &c., as affecting fire insurance companies in the united states / Ecclesine, Joseph B – New York, Grierson & Ecclesine, 1863. 112 p. LL-584 – 1 – us L of C Photodup [346]

Compendium of the speeches presented by educators, olympic champions, administrators, and avery brundage at the international olympic academy 1961-1985 – Colorado Springs, 1986 – (= ser United states olympic academy 10/2) – 3mf – 9 – $12.00 – us Kinesology [790]

Compendium of the summa theologica of st thomas aquinas, pars prima / Bonjoannes, Berardus – London: Thomas Baker; New York: Benziger Bros, 1906 [mf ed 1991] – 1mf – 9 – 0-7905-9713-6 – (in english. rev by wilfrid lescher. int & app by carlo falcini) – mf#1989-1438 – us ATLA [241]

Compendium on continuing education for the practicing veterinarian – Princeton. 1982+ (1,5,9) – ISSN: 0193-1903 – mf#13397,01 – us UMI ProQuest [636]

A compendium on the soul / Abau-Aly al-Husayn ibn Abdallah ibn Sainaa – Verona, Italy: [s.n.] 1906 [mf ed 1991] – 1mf – 9 – 0-7905-9122-7 – (trans fr arabic original by edward abbott van dyck) – mf#1989-2347 – us ATLA [260]

Compendium sacrae theologiae... / Daneau, Lambert – Monspelii, Gilet, 1595 – 4mf – 9 – mf#PFA-135 – ne IDC [240]

Compendium theologiae christianae didactico-elencticum / Marck, J – Amstelaedami, 1690 – 8mf – 9 – mf#PBA-246 – ne IDC [240]

Compendium theologiae dogmaticae et moralis : una cum praecipuis notionibus theologiae canonicae, liturgicae, pastoralis et mysticae, ac philosophiae christianae / Berthier, Jean-Baptiste – 4. ed, aucta et emendata. La Salette: Corps in Gallia; Paris: Haton, 1898 – 2mf – 9 – 0-7905-9236-3 – mf#1989-2461 – us ATLA [240]

Compendium theologiae moralis / Gury, Jean Pierre – ed 15. Romae: Ex Officina libraria instituendorum opificum S Iosephi, 1907 – 5mf – 9 – 0-524-07003-2 – (incl bibl ref) – mf#1991-2856 – us ATLA [240]

Compendium theologiae moralis / Lehmkuhl, Augustinus – Editio quinta emendata et aucta. Friburgi Brisgoviae [Freiburg im Breisgau]: Herder, 1907 – 2mf – 9 – 0-8370-7079-1 – (incl ind) – mf#1986-1079 – us ATLA [240]

Compendium theologiae naturalis... / Lampe, F A – Trajecti ad Rhenum, 1734 – 3mf – 9 – mf#PBA-216 – ne IDC [240]

Compendium theologie, methodi qvaestionibvs tractatvm / Heerbrand, J – Tvbingae, 1575 – 7mf – 9 – mf#TH-1 mf 605-611 – ne IDC [242]

Compendium veritatum ad fidem pertinentium... / Clichtove, J – Parisiis, 1529 – 4mf – 9 – mf#CA-83 – ne IDC [241]

Compendivm mvsices descriptvm ab adriano petit coclico, discipvlo iosqvinide pres : in quo praeter caetera tractantur haec: de modo ornate canendi, de regula contrapuncti, de compositione... / Coclico, Adrianus Petit – Norimbergae: I Montanus & V Neuber 1552 [mf ed 19–] – 2mf – 9 – mf#fiche 394 – us Sibley [780]

Compensation and benefits management – Greenvale. 1996-1996 (1,5,9) – ISSN: 0748-061X – mf#16396 – us UMI ProQuest [650]

Compensation and benefits management – New York: Aspen Law & Business. v1 – (inquire for info) – mf#118831 – us Hein [346]

Compensation and benefits review – Saranac Lake. 1985+ (1) 1985+ (5) 1985+ (9) – (cont: compensation review) – ISSN: 0886-3687 – mf#6284,01 – us UMI ProQuest [650]

Compensation and working conditions – Washington. 1991+ (1) 1991+ (5) 1991+ (9) – (cont: current wage developments) – ISSN: 1059-0722 – mf#7346,01 – us UMI ProQuest [331]

Compensation review – New York. 1969-1985 (1) 1972-1985 (5) 1975-1985 (9) – (cont by: compensation and benefits review) – ISSN: 0010-4248 – mf#6284 – us UMI ProQuest [650]

The compensator / Ohio. Bureau of Unemployment Compensation – v.1, n.1-v.13, n.4, 1938-49. 15 fiches. (Harvard Law School Library Collectio.) – 9 – us Harvard Law [336]

Comper, J *see* On the restoration of the office of metropolitan in the scottish ch...

Competence in a colony contrasted with poverty at home : or, relief to landlords and labourers held out by australian colonization and emigration. a memorial addressed to the right hon lord john russell... / Colonisation Society – London, 1848 – (= ser 19th c books on british colonization) – 1mf – 9 – mf#1.1.525 – uk Chadwyck [339]

Competencies for adapted physical educators in thailand / Suphawibul, M – 1992 – 2mf – 9 – $8.00 – us Kinesology [790]

Competencies of sport event managers in the united states / Peng, Hsiao-hwei – 2000 – 221p on 3mf – 9 – $15.00 – mf#PE 4172 – us Kinesology [650]

A competency analysis of ncaa athletic administrators / Nielsen, FE – 1990 – 2mf – 9 – $8.00 – us Kinesology [790]

The competency of witnesses in civil causes in pennsylvania. / Miller, Nicholas Dubois – Philadelphia: Welsh, 1881. 181p. LL-730 – 1 – us L of C Photodup [347]

Competition, exercise, provocation, and verbal comments: influences on attributions and aggression? / Miller, David L – 1981 – 1mf – 9 – $4.00 – us Kinesology [616]

Competition in the pharmaceutical marketplace : antitrust implications of patent settlements : hearing before the committee on the judiciary, united states senate, 107th congress, 1st session, may 24 2001 / United States. Congress. Senate. Committee on the Judiciary – Washington: US GPO 2002 [mf ed 2002] – 1mf – 9 – (incl bibl ref) – us GPO [346]

Competitive orientation of basketball starters and nonstarters across different playing levels / Schoen, Christopher H – 1993 – 2mf – $8.00 – us Kinesology [150]

Competitive position of the port of durban / Shaffer, N Manfred – Evanston, IL. 1965 – 1r – us UF Libraries [960]

Competitive shop organizer / International Union, United Automobile, Aircraft and Agricultural Implement Workers of America – v1-v2 n6 [1946 nov-1947 sep] – 1r – 1 – mf#1110814 – us WHS [331]

Competitor – Pittsburgh. v1-3 n4. 1920-21 [all publ] – (= ser Black journals, series 1) – 11mf – 9 – $115.00 – us UPA [305]

Compflash – New York. 1989-1995 (1) – ISSN: 0147-1570 – mf#12449 – us UMI ProQuest [650]

Compilacion cafetera, 1939-1951 / Colombia – Bogota, Colombia. 1951 – 1r – us UF Libraries [972]

Compilacion de algunos de sus poemas y cuentos / Solano Blanco, Hector – Alajuela, Costa Rica. 1954 – 1r – us UF Libraries [800]

Compilacion de articulos referentes a las ordenes... / Inchaurre Aldape, Diego – Madrid: Archivo Ibero Americano, 1963 – 1 – sp Bibl Santa Ana [240]

Compilacion de decretos del sr presidente / Cuba – Habana, Cuba. 1904 – 1r – us UF Libraries [972]

Compilacion electoral / Colombia – Bogota, Colombia. 1924 – 1r – us UF Libraries [325]

Compilacion electoral / Colombia Laws, Statutes, Etc – Bogota, Colombia. 1949 – 1r – us UF Libraries [325]

Compilacion legal / Colombia Laws, Statutes, Etc – Medellin, Colombia. 1961 – 1r – us UF Libraries [323]

Compilacion ordenada y completa de la legislacion / Borges, Milo Adrian – Habana, Cuba. 1935 – 1r – us UF Libraries [323]

Compilacion ordenada y completa de la legislacion / Borges, Milo Adrian – Habana, Cuba. v1-3. 1952 – 2r – us UF Libraries [972]

Compilacion parlamentaria y administrativa / Colombia (Republic Of Colombia, 1886-) – Bogota, Colombia. 1925 – 1r – us UF Libraries [323]

O compilador : jornal dos jornaes – Rio de Janeiro, RJ: Typ de Vianna & Companhia, 03 maio 1852-24 jul 1853 – (= ser Ps 19) – bl Biblioteca [910]

Compilador constitucional politico e literario brasiliense – Rio de Janeiro, RJ: Typ Nacional, 05 jan-26 abr 1822 – (= ser Ps 19) – mf#P01,04,02 – bl Biblioteca [079]

Compilador mineiro – Ouro Preto, MG: Officina Patrícia Barboza e C, 22 out-19 nov 1823 – (= ser Ps 19) – mf#P17,02,64 – bl Biblioteca [972]

Compilateur – New Orleans LA. 1862 sep 2 – 1r – 1 – mf#861289 – us WHS [071]

Compilatio praesens (cccm 51) : compilatio praesens / Petrus Pictaviensis; ed by Longere, J – 1980 – (= ser Corpus christianorum continuatio mediaevalis (cccm)) – 258p+3mf – 9 – €80.00 – 2-503-03511-6 – be Brepols [180]

Compilation of agreements and related documents : between the united states and the freely associated state of the republic of the marshall islands – Honolulu: Hawaiian/Pacific Coll of the Univ of Hawaii Library, n.d. – 5mf – 9 – $7.50 – (contains a wide assortment of agreements, presidential proclamations, & portions of us statutes implementing bilateral compacts between the two govts) – mf#llmc82-100i, title 25 – us LLMC [327]

Compilation of court martial orders, 1916-1937 / U.S. Naval Dept – Washington: GPO. 2v + cum index. 1940 – 11mf – 9 – $16.50 – mf#LLMC 84-213 – us LLMC [347]

The compilation of items and calibration for a survery of infant motor behavior / Bayer, Emily C – 1999 – 4mf – 9 – $16.00 – mf#PSY 2122 – us Kinesology [612]

Compilation of laws with 1937 inserts; digest of state, federal and canadian laws, relating to foods, drugs, pharmacy. / Standard Remedies Publishing Company, Inc – Washington, D.C., Proprietary 1937? 1 v. LL-1044 – 1 – us L of C Photodup [348]

A compilation of massachusetts law relating to privacy and personal data / Massachusetts. Laws, Statutes, etc – 2nd ed. Boston. Governor's Special Commission on Privacy and Personal Data, 1974. LL-2343 – 1 – us L of C Photodup [346]

Compilation of reports of the foreign relations committee, 1789-1901 / U.S. Senate Committee on Foreign Relations – Washington: GPO. 1st session. 1st Congress-56th Congress. 8v. 1902 – 80mf – 9 – $120.00 – mf#LLMC 81-106 – us LLMC [323]

Compilation of tennessee census reports, 1820 / U.S. Bureau of Census – (= ser Records Of The Bureau Of The Census) – 1r – 1 – mf#T911 – us Nat Archives [317]

Compilation of the laws and amendments thereto relating to building societies, loan companies, joint stock companies, and interest on mortgages and other acts pertaining to monetary institutions : as passed by the dominion parliament and the several provincial legislatures... / Garland, Nicholas Surrey – Ottawa: A S Woodburn, 1882 – 5mf – 9 – (incl: the laws relating to banks and banking comp by william wilson) – mf#03293 – cn Canadiana [332]

Compilation of the laws of the united states applicable to the duties of the governor, attorney, judge, clerk, marshal, and commissioners of the district of alaska / U.S. Laws, Statutes, etc – Washington, Govt. Print. Off., 1884. 60 p LL-916 – 1 – us L of C Photodup [348]

Compilation of the messages and papers of the presidents, 1789-1922 / U.S. President – v1-20. 1907-24 – 9 – $168.00 – mf#0653 – us Brook [324]

Compilation of the organic provisions of the administration of justice in force in the spanish colonial provinces : and appendices relating thereto – 1891. Washington: GPO, 1899 (mf ed) – 2mf – 9 – $3.00 – (a description of the organization of the courts in puerto rico, cuba & the philippines) – mf#LLMC 92-304 – us LLMC [340]

A compilation of the statutes passed since confederation relating to banks and banking : government and other savings banks, promissory notes and bills of exchange, and interest and usury in new brunswick and nova scotia / Davidson, Charles Peers – Montreal?: s.n, 1876 – 2mf – 9 – (incl ind) – mf#02589 – cn Canadiana [348]

The compiled and revised laws of the territory of idaho. / Idaho. (Territory). Laws, Statutes, etc – Boise: Kelly, 1875. 877,22p. LL-2350 – 1 – us L of C Photodup [348]

Compiled military service record, company f, 15th kansas cavalry, relating to orren a. curtis / U.S. Army. Adjutant General's Office – 1 – us Kansas [324]

Compiled military service records of major uriah blue's detachment of chickasaw indians in the war of 1812 / U.S. War Dept. Adjutant General's Office – (= ser Records of the adjutant general's office, 1780's-1917) – 1r – 1 – mf#M1829 – us Nat Archives [355]

Compiled military service records of michigan and illinois volunteers who served during the winnebago indian disturbances of 1827 / U.S. War Dept. Adjutant General's Office – (= ser Records of the adjutant general's office, 1780's-1917) – 3r – 1 – (with printed guide) – mf#M1505 – us Nat Archives [355]

Compiled military service records of volunteer union soldiers who served with the united states colored troops : 54th massachusetts infantry regiment (colored) / U.S. War Dept. Adjutant General's Office – (= ser Records of the adjutant general's office, 1780's-1917) – 20r – 1 – (with printed guide) – mf#M1898 – us Nat Archives [355]

Compiled minutes and history of the church of the brethren of the southern district of illinois – [S.l: s.n., 1907?] – 1mf – 9 – 0-524-03139-8 – mf#1990-4588 – us ATLA [242]

Compiled ordinances of the city of council bluffs, iowa / Hazelton, A S & Capell, Frank J [comp] – [Council Bluffs: Monarch Print Co, 1920 (mf ed 1996)] – 1r – 1 – (incl ind) – mf#ZZ-34745 – us NY Public [350]

Compiled records showing service of military units in confederate organizations / U.S. War Dept. – (= ser Military service records in the war department coll of confederate records) – 74r – 5 – (with printed guide) – mf#M861 – us Nat Archives [355]

Compiled records showing service of military units in volunteer union organizations / U.S. War Dept. Adjutant General's Office – (= ser Records Of Movements And Activities Of Volunteer Union Organizations) – 225r – 5 – (with printed guide) – mf#M594 – us Nat Archives [355]

Compiled service records of american naval personnel and members of the departments of the quartermaster general and the commissary general of military stores who served during the revolutionary war / U.S. War Dept – (= ser War department coll of revolutionary war records) – 4r – 1 – (with printed guide) – mf#M880 – us Nat Archives [355]

Compiled service records of confederate generals and staff officers, and non-regimental enlisted men / U.S. War Dept. – 275r – 5 – (with printed guide) – mf#M331 – us Nat Archives [355]

Compiled service records of confederate soldiers who served in organizations from the state/territory of [...] / U.S. War Dept. – (= ser War department coll of confederate records) – 5 – (alabama 508r m311. arizona 1r m318. arkansas 256r m317. florida 104r m251. georgia 607r m266. kentucky 136r m319. louisiana 414r m320. maryland 22r m321. mississippi 427r m269. missouri 193r m322. north carolina 580r m270. south carolina 392r m267. tennessee 359r m268. texas 445r m323. virginia 1075r m324. with printed guides) – us Nat Archives [355]

Compiled service records of confederate soldiers who served in organizations raised directly by the confederate government / U.S. War Dept. – (= ser Military service records in the war department coll of confederate records) – 123r – 5 – (with printed guide) – mf#M258 – us Nat Archives [355]

Compiled service records of former confederate soldiers who served in the 1st through 6th u.s. volunteer infantry regiments, 1864-1866 / U.S. War Dept. Adjutant General's Office – (= ser Records of the adjutant general's office, 1780's-1917) – 65r – 1 – (with printed guide) – mf#M1017 – us Nat Archives [355]

Compiled service records of soldiers who served in the american army during the revolutionary war / U.S. War Dept. Adjutant General's Office – (= ser War department coll of revolutionary war records) – 1096r – 1 – (with printed guide) – mf#M881 – us Nat Archives [355]

Compiled service records of volunteer soldiers who served during the mexican war in organizations from the state of [...] / U.S. War Dept. Adjutant General's Office – (= ser Records of volunteer soldiers who served during the mexican war) – 1 – (mormon organizations 3r m351. mississippi 9r m863. pennsylvania 13r m1028. tennessee 15r m638. texas 19r m278. with printed guides) – us Nat Archives [355]

Compiled service records of volunteer soldiers who served during the war of 1812 in organizations from the territory of mississippi / U.S. War Dept. Adjutant General's Office – (= ser Records Of Volunteer Soldiers Who Served In The War Of 1812) – 22r – 5 – (with printed guide) – mf#M678 – us Nat Archives [355]

Compiled service records of volunteer soldiers who served from 1784 to 1811 / U.S. War Dept. Adjutant General's Office – (= ser Records Of Volunteer Soldiers Who Served From 1784 Until 1811) – 32r – 5 – (with printed guide) – mf#M905 – us Nat Archives [355]

Compiled service records of volunteer soldiers who served in organizations from the state of florida during the florida indian wars, 1835-1858 / U.S. War Dept. Adjutant General's Office – (= ser Records of volunteer soldiers who served during the florida wars) – 63r – 1 – (with printed guide) – mf#M1086 – us Nat Archives [355]

Compiled service records of volunteer soldiers who served in the florida infantry during the war with spain / U.S. War Dept. Adjutant General's Office – (= ser Records of volunteer soldiers who served during the war with spain) – 13r – 1 – (with printed guide) – mf#M1087 – us Nat Archives [355]

Compiled service records of volunteer union soldiers who served in organizations from the state of [...] / U.S. War Dept. Adjutant General's Office – (= ser Records of the adjutant general's office, 1780's-1917) – 5 – (alabama 10r m276. arkansas 60r m399. florida 11r m400. georgia 1r m403. kentucky 515r m397. louisiana 50r m396. maryland 238r m384. mississippi 4r m404. missouri 854r m405. territory of nebraska 43r m1787. territory and state of nevada 16r m1789. new mexico 46r m427. north carolina 25r m401. oregon 34r m1816. tennessee 220r m395. texas 13r m402. utah 1r m692. virginia 7r m398. west virginia 261r m508. with printed guides) – us Nat Archives [355]

Compiled statutes : from the kingdom (1827) to 1994 / Hawaii Legislature – 1827-1994 – (= ser Hawaii appellate reports) – 942mf – 9 – $1413.00 – (add vols after 1993 planned) – mf#LLMC 77-103 – us LLMC [348]

Compiler – Gettysburg, PA. 1857-1953 (1) – mf#65904 – us UMI ProQuest [071]

Compiler of Laws, Office of the Attorney General *see* Guam-federal digest, 1950-1987

Compitum : or, the meeting of the ways at the catholic church / Digby, Kenelm Henry – London: C Dolman, 1851 – 7mf – 9 – 0-8370-8501-2 – (incl bibl ref) – mf#1986-2501 – us ATLA [241]

The complaint and the answer : being allama sir muhammad iqbal's shikwah and jawab-i-shikwah done into english verse / Husain, Altaf – Lahore: Shaikh Muhammad Ashraf, 1943 – (= ser Samp: indian books) – us CRL [320]

The complaint of nature / Lille, Alain de – New York, 1908 – 2mf – 8 – €5.00 – (trans fr latin by d m moffat) – ne Slangenburg [110]

The complaint of peace / Erasmus, Desiderius – 1559 – 9 – us Scholars Facs [323]

A compleat history of the empire of china : being the observations of above ten years travels through that country / Le Comte, L D – London: James Hodges, 1739 – 6mf – 9 – mf#HT-534 – ne IDC [915]

A compleat history of the piratical states of barbary : viz algiers, tunis, tripoli and morocco: containing the origins, revolutions, and present state of these kingdoms, their forces, revenues, policy, and commerce – London: R Griffiths, 1750 – 1 – us CRL [960]

Compleat library : or, news for the ingenious – London. 1692-1694 (1) – mf#4229 – us UMI ProQuest [941]

Compleat linguist – London. 1719-1722 (1) – mf#4230 – us UMI ProQuest [420]

Compleete uitgave van de officieele stukken betreffende den uitgang uit het nederl. herv. kerkgenootschap / Scholte, Hendrik Peter et al – 2de druk. Kampen: Zalsman, 1884 – 6mf – 9 – 0-524-07916-1 – mf#1991-3461 – us ATLA [240]

Complement des memoires et revelations d'un page de la cour imperiale : pour servir a l'histoire de l'interieur des cours de france, de naples, de madrid, de hollande, de westphalie et de suede; sous le consulat et l'empire / Deville, V J de – Paris 1832 [mf ed Hildesheim 1995-98] – 2v on 6mf – 9 – €120.00 – 3-487-26400-5 – gw Olms [914]

Complete aas microfiche series collection : papers / American Astronautical Society; ed by Jacobs, H – v1-73. 1968-96 – 868mf – 9 – $915.00 – (vols available separately) – ISSN: 0 – us Univelt [629]

A complete analysis of the holy bible : containing the whole of the old and new testaments...in thirty books, based on the work of the learned talbot / West, Nathaniel – New York: Charles Scribner, 1853 [mf ed 1989] – 3mf – 9 – 0-7905-3057-0 – (incl ind) – mf#1987-3057 – us ATLA [220]

Complete archive program / Human Relations Area Files – 1958. 38 installments – 9 – us HRAF [300]

Complete author catalogue of the national art library, 1843-1986 : from the victoria and albert museum, london – 706mf – 9 – (bibliography of the fine and applied arts: complete author catalogue) – us Primary [700]

Complete book of engines – Los Angeles. 1965-1973 (1) 1971-1972 (5) – ISSN: 0069-7974 – mf#3142 – us UMI ProQuest [621]

The complete cabinet-maker : and upholsterer's guide / Stokes, J – London [1838] – (= ser 19th c art & architecture) – 2mf – 9 – mf#4.1.451 – uk Chadwyck [740]

The complete catholic directory, almanack and registry... / Battersby, W J – Dublin, 1840-1845 – us CRL [030]

A complete catholic registry, directory and almanack... / Battersby, W J – Dublin, Printed by W Powell [etc], 1836; 1838-39 – us CRL [030]

The complete commentary of oecumenius on the apocalypse / Hoskier, H C – 1928 – 9 – $10.00 – us IRC [240]

Complete concordance to the holy scriptures of the old and new / Cruden, Alexander – New York, NY. 1854 – 1r – us UF Libraries [220]

The complete conspiracy trial book / Hayden, Samuel Augustus – 1907. 356p – 1 – us Southern Baptist [242]

Complete course in massage, swedish movement and mechanical therapeutics : based on the teachings of peter henrik ling, founder of the royal central gymnastic insitute, stockholm, sweden and supplemented by original discoveries and teachings / Evans, James Gwallia – Toronto: School of Massage, Royal College of Science, c1911 [mf ed 1996] – 1mf – 9 – 0-665-80834-8 – mf#80834 – cn Canadiana [615]

The complete dressmaker for the million / Whiteley, T (Mrs) – Manchester: John Heywood; London: Simkin, Marshall & Co [1875] – (= ser 19th c art & architecture) – 1mf – 9 – mf#4.1.144 – uk Chadwyck [740]

Complete fifer's museum : a collection of marches of all kinds, now in use in the military line / Hulbert, James – Northhampton, MA: Simeon Butler [1807] [mf ed 19–] – 1r – 1 – mf#film 492 – us Sibley [780]

Complete forms of proceedings in civil and criminal cases before justices of the peace, in the state of ohio / Palm, Jefferson – Warren, Ohio: Ritezel, 1855. 55p. LL-1392 – 1 – us L of C Photodup [345]

Complete greek drama : all the extant tragedies of aeschylus, sophocles and euripides, and the comedies of aristophanes and menander, in a variety of translations / ed by Oates, Whitney Jennings & O'Neill, Eugene – New York NY: Random house [c1938] [mf ed 1994] – 2v on 1r – 1 – us UF Libraries [450]

Complete handbook of the virgin islands / Murray, Stuart – New York, NY. 1951 – 1r – us UF Libraries [972]

Complete hanoverian state papers domestic, 1714-1782 – 164r coll – 1 – (coll of papers on the politics and society of 18th century england. george 1 (1714-27) 53r c39-27520. george 2 (1727-60) 91r c39-27530. george 3 (1760-82) 20r c39-27531) – mf#C39-27510 – us Primary [941]

A complete harmony of daniel and the apocalypse / Litch, Josiah – Philadelphia: Claxton, Remsen & Haffelfinger, 1873, c1872 [mf ed 1985] – 1mf – 9 – 0-8370-4146-5 – mf#1985-2146 – us ATLA [221]

Complete henry crabb robinson diaries, travel journals and reminiscences, 1790-1867 : insight into the lives of hundreds of famous persons – 11r – 1 – uk Academic [880]

A complete hermeneutical manual on the book of ecclesiastes / Strong, James – New York: Hunt & Eaton; Cincinnati: Cranston & Curts, c1893 [mf ed 1985] – (= ser The student's commentary) – 1mf – 9 – 0-8370-5454-0 – (incl bibl ref & ind) – mf#1985-3454 – us ATLA [221]

Complete history of methodism as connected with... / Jones, John Griffing – Nashville, TN. v1-2. 1908 – 1r – us UF Libraries [242]

Complete history of the present war : from its commencement in 1756, to the end of the campaign, 1760... – London: printed for W Owen & others, 1761 [mf ed 1971] – 1r – 5 – mf#SEM16P16 – cn Bibl Nat [971]

The complete home : an encyclopaedia of domestic life and affairs / Wright, Julia McNair – Philadelphia, Brantford, Ont: Bradley, Garretson, 1879 – 7mf – 9 – 0-665-92074-1 – (incl ind) – mf#92074 – cn Canadiana [640]

The complete indian housekeeper and cook : giving the duties of mistress and servants the general management of the house and practical recipes for cooking in all its branches / Steel, Flora Annie (Webster) & Gardiner, Grace Anne Marie Louise (Napier) – new ed. London 1898 – (= ser 19th c british colonization) – 4mf – 9 – mf#1.1.2505 – uk Chadwyck [640]

The complete law quizzer...the ohio supreme court examination questions for admission to the bar / Kinkead, Edgar Benton – 5th ed. Cincinnati, Anderson, 1915. 1025z-10 (i.e. 1731) p. LL-1426 – 1 – us L of C Photodup [347]

Complete lectures of col. r.g. ingersoll / Ingersoll, Robert Green – Chicago: J Regan, [191-?] – 1mf – 9 – 0-8370-6266-7 – mf#1986-0266 – us ATLA [080]

A complete legal advertising form book. / Monaghan, James Patrick – Cincinnati?: Legal Advertising, 1907. 335p. LL-848 – 1 – us L of C Photodup [340]

Complete life of william mckinley / Neil, Henry – Chicago, IL. 1901 – 1r – us UF Libraries [975]

A complete list of the descendants of joseph smith and deliverance lane : antecedents of joseph smith of rowley, ma... / Simpson, John K – Arlington Heights, MA: J K Simpson, [19–] (mf ed 1986) – 1r – 1 – mf#*ZI-474 n20 – us NY Public [978]

The complete l.r.a. digest – New York: Lawyers' Co-op. 10v. 1921-24 (all publ) – (= ser Lawyers' Reports Annotated, 2nd Series) – 150mf – 9 – $225.00 – (covers both series) – mf#LLMC 78-046C – us LLMC [348]

The complete man – v2. 1875 [complete] – (= ser Mennonite serials coll) – 1r – 1 – mf#ATLA 1993-S020 – us ATLA [242]

The complete martindale-hubbell, 1868-1980 *see* Lawyers' directories, 1868-1980

The complete martindale-hubbell's law directory, 1868-1980 – 11,980mf – 9 – $11,900.00 – (directories providing a unique, comprehensive and continuous record of the american bar for the past 114 years) – mf#LLMC 92-001 – us LLMC [340]

A complete narrative of the celebration of the nuptials of her most gracious majesty queen victoria with his royal highness prince albert of saxe coburg and gotha : by the nova scotia philanthropic society... / Crosskill, John H – Halifax: publ by the Nova Scotia Philanthropic Society...[1840?] [mf ed 1983] – 1mf – 9 – 0-665-44036-7 – mf#44036 – cn Canadiana [941]

Complete novels and selected tales of nathaniel hawthorne – New York, NY. 1937 – 1r – us UF Libraries [802]

Complete paradigms of the samaritan verbs : regular and irregular / Young, Robert – Edinburgh: Robert Young, [18–?] – 1mf – 9 – 0-8370-7116-X – mf#1986-1116 – us ATLA [470]

Complete poems / Dunbar, Paul Laurence – New York, NY. 1922 – 1r – us UF Libraries [810]

The complete poems.. / Dickinson, Emily – With introd. by her niece, Martha Dickinson Bianchi. Boston: Little, Brown, 1925. 330p. Illus – 1 – us Wisconsin U Libr [810]

The complete poetical works of dante gabriel rossetti / Rossetti, Dante Gabriel; ed by Rossetti, William Michael – Boston: Roberts Bros 1894 [c1887] [mf ed 1987] – 1r [ill] – 1 – (filmed with: across central america / boddan-whetham, j w & other titles) – mf#1990 – us Wisconsin U Libr [810]

A complete popular encyclopedia of virginia law and forms and business guide or how-book for the businessman and citizen. / Hurst, Samuel Need – Pulaski, Va.: Hurst, 1922. 2148p. LL-963 – 1 – us L of C Photodup [346]

The complete preacher : sermons preached by some of the most prominent clergymen in this and other countries and in the various denominations / ed by Funk, Isaac Kaufman – New York: Funk & Wagnalls, c1877-1878 – 3mf – 9 – 0-524-08342-8 – mf#1993-2032 – us ATLA [240]

A complete preceptor for the clarionet, containing the most approved instructions relative to that instrument – New York: W. Dubois 1818. Includes: "General Washingtons March," "Rule Britannia," and "Yankee Doodle." MUSIC 778, Item 4 – 1 – us L of C Photodup [780]

A complete record of unity talks / ed by Singh, Durlab – Lahore: Hero Publ, [1945] – (= ser Samp: indian books) – us CRL [954]

Complete records of the mission of general george c. marshall to china, dec 1945-jan 1947 / Marshall, George C – 1987 – 50r – 1 – $6500.00 – (with printed guide) – mf#S3048 – us Scholarly Res [951]

The complete ski guide. / ed by Elkins, Frank – New York: Doubleday, 1940. xvii,286p. illus., maps, tables, plates – 1 – us Wisconsin U Libr [790]

The complete state papers domestic, 1547-1702 – 3 series – 903r coll – 1 – (series 1: units 1-5: edward 6, mary 1, elizabeth 1, 1547-1603 123r c39-27461. units 6-9: james 1, 1603-25 86r c39-27462. unit 10: addenda, 1547-1625 18r c39-27468. series 2: units 11-18: charles 1, 1625-48 206r c39-27463. units 19-24: interregnum, 1649-60 108r c39-27464. units 25-34: charles 2 227r c39-27465. unit 35: william 3 and mary – king william's chest 11r c39-27466. series 3: units 36-41: henry 8, 1509-47 124r c39-27467) – mf#C39-27460 – us Primary [941]

Complete state papers regencies, 1716-1755 – 16r – 1 – mf#C39-17000 – us Primary [941]

A complete system of christian theology : or, a concise, comprehensive, and systematic view of the evidences, doctrines, morals and institutions of christianity / Wakefield, Samuel – New York: Carlton & Porter, 1862 [mf ed 1993] – (= ser Methodist coll) – 1mf – 9 – 0-524-07050-4 – (incl bibl ref) – mf#1991-2903 – us ATLA [242]

Complete system of harmony : or, a regular and easy method to attain a fundamental knowledge and practice of thorough bass / Heck, Johann Caspar – [London]: printed for & sold by aut [1768] [mf ed 19–] – 1mf – 9 – mf#fiche 550, 729 – us Sibley [780]

The complete text of the pahlavi dinkard under the supervision of d m madan – Bombay, 1911 – 6mf – 9 – mf#NE-20148 – ne IDC [490]

A complete treatise on the art of retouching photographic negatives : and clear directions how to finish and colour photographs / Johnson, Robert – [London]: Marion & Co, 1889 – (= ser 19th c art & architecture) – 2mf – 9 – mf#4.1.101 – uk Chadwyck [770]

Complete triumph of moral good over evil / Barnes, John – London: Longmans, Green, 1870 [mf ed 1991] – 2mf – 9 – 0-7905-8760-2 – mf#1989-1985 – us ATLA [230]

Complete works / Kauder, Hugo – 9 – $210.00 – mf#0309 – us Brook [780]

Complete works of abraham lincoln / ed by Nicolay, John G – v1-12. 1905 – 9 – $267.00 – mf#0332 – us Brook [975]

The complete works of flavius josephus / Whiston, W – 1859 – 1r – 1 – mf#2172 – uk Microform Academic [939]

Complete works of rev thomas smyth / ed by Flinn, John William – new ed. Columbia, SC: repr by the RL Bryan Co, 1908-1912 [mf ed 1992] – (= ser Presbyterian coll) – 10v on 16mf – 9 – 0-524-02503-7 – mf#1990-4362 – us ATLA [242]

The complete works of saint john of the cross, doctor of the church / ed by Peers, E Allison – London: Burns Oates. v1-3. 1934-35 [mf ed 2003] – 3v on 1r – 1 – (trans fr spanish by p silverio de santa teresa. with ind and bibl) – mf#b00663 – us ATLA [241]

The complete works of swami vivekananda – Almora: Advaita Ashrama, 1926-1936 – (= ser Samp: indian books) – us CRL [280]

Complete works of the most rev john hughes, d d, archbishop of new york : comprising his sermons, letters, lectures, speeches, etc = Works. 1865 / Hughes, John; ed by Kehoe, Lawrence – 2nd rev corr ed. New York: American News Co 1865 [mf ed 1993] – 2v on 16mf – 9 – 0-524-07432-1 – mf#1991-3092 – us ATLA [241]

The complete works of the rev. andrew fuller = Works. 1845 / Fuller, Andrew – Philadelphia: American Baptist Publication Society, c1845 – 26mf – 9 – 0-524-08759-8 – mf#1993-3264 – us ATLA [240]

The complete works of the rev. andrew fuller, with a memoir of his life by andrew gunton fuller / Fuller, Andrew – London: G. & J. Dyer, 1846. xciii,1012p – 1 – us Wisconsin U Libr [920]

Complete works of william shakespeare : comprising his plays and poems, with a history of the stage... / ed by Collier, J Payne – New York: F M Lupton Publ Co [1855?] [mf ed c1995] – (text of plays corr by mss emendations contained in the recently discovered folio of 1632 by j payne collier; to which are added, glossarial & explanatory notes...; ann by malcolm lowry) – cn UBC Preservation [802]

Complete writings / Musset, Alfred de – v1-10. 1908 – 9 – $132.00 – mf#0386 – us Brook [802]

Completed research in health, physical education, and recreation – Washington. 1978-1979 (1,5,9) – ISSN: 0516-916X – mf#11673 – us UMI ProQuest [613]

Complex south africa / Macmillan, William Miller – London, England. 1930 – 1r – us UF Libraries [960]

Complex training compared to a combined weight training and plyometric training program / Burger, Troy – 1999 – 1mf – 9 – $4.00 – mf#PE 4009 – us Kinesology [790]

Compliance and cardiac rehabilitation / Harris-Sund, Valarie – 1995 – 2mf – 9 – $8.00 – mf#HE 591 – us Kinesology [616]

Complimentary banquet tendered to rt hon sir wilfred sic laurier...by the board of trade of the city of toronto, oct 6th 1897 – S.l: s.n, 1897? – 1mf – 9 – mf#28850 – cn Canadiana [920]

Complimentary banquet to the hon john rose – Montreal?: s.n, 1869? – 1mf – 9 – mf#23567 – cn Canadiana [920]

A complimentary epistle to james bruce, esq, the abyssinian traveller / Pindar, Peter [pseud] – London: H D Symonds, 1792 – 1 – us CRL [916]

Complimentary farewell dinner to prof william osler : by the medical profession of montreal, windsor hotel, oct 9th 1884 – [Montreal?: s.n, 1884?] [mf ed 1985] – 1mf – 9 – 0-665-01779-0 – mf#01779 – cn Canadiana [610]

Compliments of company 670, civilian conservation corps, camp bitely f-22, bitely, michigan, april 4th 1937 – [Bitely, MI: s.n.] 1937 – us CRL [060]

Compliments of lieut col tyrwhitt, government candidate, north riding york : the national policy, canadian pacific ry, prosperity – [s.l: s.n. 1882?] [mf ed 1983] – 1mf – 9 – 0-665-39740-2 – mf#39740 – cn Canadiana [325]

Compliments of the season : the owl club assembly – S.l: s.n, 18–? – 1mf – 9 – mf#53886 – cn Canadiana [366]

Compliments of the titusville civic league – Titusville, FL. 19–? – 1r – us UF Libraries [978]

Le complot franc-maconnique et le droit d'accroissement / Francais, Gabriel – 4e ed. Paris: Maison de la Bonne Presse [189-?] – 2mf – 9 – mf#vrl-187 – ne IDC [366]

Un complot terrorista en el siglo 15th. madrid, 1927 / Castellano, Conde de; ed by Bayle, Constantino – Madrid: Razon y Fe, 1928 – 9 – sp Bibl Santa Ana [946]

Compos, Humberto De *see* Contrastes (cronicas)

Le compose verbal en ge- et ses fonctions grammaticales en moyen haut allemand : etude fondee sur l'iwein de hartman von aue et sur les sermons de berthold von regensburg / Marache, Maurice – Paris: Didier, 1960 [mf ed 1996] – (= ser Germanica 1) – 445p – 1 – mf#9575 – us Wisconsin U Libr [430]

Composer – Hamilton. 1969-1981 (1) 1977-1981 (5) 1977-1981 (9) – ISSN: 0010-4345 – mf#10856 – us UMI ProQuest [780]

Composicion del 4 pleno del consejo economico sindical / Organizacion Sindical – Badajoz: Imprenta Inca, 1963 – sp Bibl Santa Ana [330]

Composiciones...dona isabel 2 / Martinez Abeytia, Mateo – 1852 – 9 – sp Bibl Santa Ana [810]

Composite structures – Barking. 1983-1996 (1) 1983-1996 (5) 1987-1996 (9) – ISSN: 0263-8223 – mf#42409 – us UMI ProQuest [624]

Composites – Guildford. 1969-1995 (1) 1969-1995 (5) 1969-1995 (9) – ISSN: 0010-4361 – mf#13323 – us UMI ProQuest [620]

Composites engineering – Elmsford. 1991-1994 (1,5,9) – ISSN: 0961-9526 – mf#49613 – us UMI ProQuest [620]

Composites manufacturing – Guildford. 1990-1991 (1,5,9) – ISSN: 0956-7143 – mf#17217 – us UMI ProQuest [670]

Composites pt a : applied science and manufacturing – Oxford. 1996+ (1,5,9) – ISSN: 1359-835X – mf#42773 – us UMI ProQuest [620]

Composites science and technology – Barking. 1968-1992 (1) 1968-1992 (5) 1987-1992 (9) – ISSN: 0266-3538 – mf#42016 – us UMI ProQuest [670]

Compositio mathematica – Leyden. 1991-1996 (1,5,9) – ISSN: 0010-437X – mf#16776 – us UMI ProQuest [510]

Die composition des buches daniel / Meinhold, Johannes – Greifswald: Julius Abel, 1884 – 1mf – 9 – 0-8370-4378-6 – (incl bibl ref) – mf#1985-2378 – us ATLA [221]

Die composition des deuteronomischen richterbuches (richter 2, 6-16) : nebst einer kritik von richter 17-24 / Frankenberg, Wilhelm – Marburg: N G Elwert, 1895 – 1mf – 9 – 0-524-06131-9 – (incl bibl ref) – mf#1992-0798 – us ATLA [221]

Die composition des hexateuchs und der historischen buecher des alten testaments / Wellhausen, Julius – Berlin: Georg Reimer, 1899 – 1mf – 9 – 0-8370-5769-8 – (incl bibl ref) – mf#1985-3769 – us ATLA [221]

Die composition des matthaeus-evangeliums von paul schanz / Schanz, Paul – Tuebingen:Ludwig Friedrich Fues, 1877 – 1mf – 9 – 0-8370-5075-8 – (incl bibl ref) – mf#1985-3075 – us ATLA [225]

Die composition des pseudopetrinischen evangelien-fragments / Schubert, Hans von – Berlin: Reuther & Reichard, 1893 – 1mf – 9 – 0-7905-0380-8 – mf#1987-0380 – us ATLA [221]

La composition du livre d'habacuc / Nicolardot, Firmin – Paris: Librairie Fischbacher, 1908 – 1mf – 9 – 0-8370-4587-8 – (incl ind of authors) – mf#1985-2587 – us ATLA [220]

Composition in bankruptcy / Bump, Orlando Franklin – St. Louis, Jones, 1877. 48 p. LL-583 – 1 – us L of C Photodup [346]

Composition of florida-grown vegetables mineral composition of commercially grown vegetables / Sims, Guilford Trice – Gainesville, FL. 1947 – 1r – us UF Libraries [634]

Composition of miscellaneous tropical and sub-tropical florida fruits / Stahl, Arthur L – Gainesville, FL. 1935 – 1r – us UF Libraries [630]

Composition of some of the concentrated feeding stuffs on sale in florida / Blair, A W – Lake City, FL. 1905 – 1r – us UF Libraries [630]

The composition of the book of isaiah in the light of history and archaeology / Kennett, R H – London: publ for the British Academy by Henry Frowde, Oxford University Press, 1910 – (= ser The Schweich Lectures) – 1mf – 9 – 0-7905-2234-9 – (incl ind) – mf#1987-2234 – us ATLA [221]

The composition of the four gospels : a critical inquiry / Wright, Arthur – London, New York: Macmillan, 1890 – 1mf – 9 – 0-8370-6548-8 – (incl indes) – mf#1986-0548 – us ATLA [226]

The composition of the hexateuch : an introduction with select lists of words and phrases / Carpenter, Joseph Estlin – London, New York: Longmans, Green, 1902 – 2mf – 9 – 0-7905-0565-7 – mf#1987-0565 – us ATLA [221]

Composition studies – Chicago, 1999+ – 1,5,9 – (cont: composition studies: freshman english news) – mf#10315,02 – us UMI ProQuest [370]

Composition studies/freshman english news – Chicago. 1992-1998 (1) 1992-1998 (5) 1992-1998 (9) – (cont: freshman english news) – mf#10315,01 – us UMI ProQuest [420]

Compositionen fur pianoforte solo / Mendelssohn-Bartholdy, Felix – Leipzig, Germany. cA.1885 – 1r – us UF Libraries [780]

Compost science – Emmaus. 1960-1977 (1) 1970-1977 (5) 1977-1977 (9) – (cont by: compost science/land utilization) – ISSN: 0010-4388 – mf#2712 – us UMI ProQuest [333]

Compost science/land utilization – Emmaus. 1978-1980 (1) 1978-1980 (5) 1978-1980 (9) – (cont: compost science. cont by: biocycle) – ISSN: 0160-7413 – mf#2712,01 – us UMI ProQuest [333]

Comprehensive advocacy program / National Federation of Republican Women – v3 n7-v8 [i.e. 7] n2 [1980 aug 15-1984 jun 15], [1984 aug, 25-dec 29] – 1r – 1 – mf#999835 – us WHS [325]

The comprehensive church, or, christian unity and ecclesiastical union / Vail, Thomas Hubbard – Hartford: Huntington, 1841 – 1mf – 9 – 0-7905-6901-9 – mf#1988-2901 – us ATLA [240]

A comprehensive history of methodism : in one volume, embracing origin, progress, and present spiritual, educational, and benevolent status in all lands / Porter, James – Cincinnati: Hitchcock & Walden, 1876 [mf ed 1992] – (= ser Methodist coll) – 2mf – 9 – 0-524-02897-4 – mf#1990-4488 – us ATLA [242]

A comprehensive history of the disciples of christ : being an account of a century's effort to restore primitive christianity in its faith, doctrine, and life / Moore, William Thomas – New York: Fleming H Revell, 1909 [mf ed 1992] – (= ser Christian church (disciples of christ) coll) – 3mf – 9 – 0-524-02485-5 – (incl bibl ref) – mf#1990-4344 – us ATLA [240]

Comprehensive index to architectural periodicals / British Architectural Library – 21r – 5 – £780.00 – (incl printed guide) – mf#RND – uk World [720]

Comprehensive psychiatry – Orlando, 1997+ [1,5,9] – ISSN: 0010-440X – mf#6902 – us UMI ProQuest [612]

A comprehensive system of book-keeping by single and double entry : simplified by detailed explanations of the phrases and books in general use, and by numerous examples... / Johnson, Thomas Richard – new rev ed. Montreal: J Lovell, 1868 [mf ed 1984] – 2mf – 9 – 0-665-27335-5 – mf#27335 – cn Canadiana [650]

Comprehensive treatise on inorganic and theoretical chemistry / Mellor, Joseph W – v1-16. 1922-37 – 9 – $330.00 – mf#0355 – us Brook [540]

Comprendio de historia de la america central / Gomez Carrillo, Augustin – Guatemala, 1900 – 1r – us UF Libraries [972]

Comprendio historial de coria para uso del colegio de ninas del sagrado corazon de dicha ciudad escrito por... / Prebendado de la Catedral, un sudonimo de Eugenio Escobar Prieto – Madrid: Imp. Viudad e hija de Gomez Fuentenebro, 1897 – 1 – sp Bibl Santa Ana [946]

Comprendre : serie blanche – Paris. [n1-48]. may 3 1956-dec 15 1963 – us CRL [944]

Comprendre : serie bleue – Paris. n2-28,30-37. may 16 1956-jan 1 1964 – us CRL [944]

Comprendre : serie jaune – Paris. n1-22,24-25,27-31,33. may 16 1956-feb 1 1964 – us CRL [944]

Comprension de venezuela / Picon-Salas, Mariano – Caracas, Venezuela. 1949 – 1r – us UF Libraries [972]

Compressed air – Washington. 1896-1999 (1) 1971-1999 (5) 1977-1999 (9) – ISSN: 0010-4426 – mf#929 – us UMI ProQuest [621]

Compromiso y responsabilidad. campamento nacional "francisco franco". cavaleda (soria) 74 / Delegacion Nacional de la Juventud – Caceres: Imp. M. Sergio Dorado, 1974 – 1 – sp Bibl Santa Ana [350]

Compstall methodist church, george st – (= ser Cheshire monumental inscriptions) – 1mf – 9 – £2.50 – mf#12a – uk CheshireFHS [929]

Compstall, st paul – [North Cheshire FHS] – (= ser Cheshire monumental inscriptions) – 1mf – 9 – £2.50 – mf#149 – uk CheshireFHS [929]

Comptabilite des affaires a enzel et leurs consequences juridiques pour les commercants / Bonan, Jules – Tunis, Imprimerie commerciale (L. Rombi) 1913. 30 p. LL-12023 – 1 – us L of C Photodup [340]

Compte administratif / Republique francaise, gouvernement general de l'Indo-chine, Directorat du Cambodge – Phnom-Penh: Impr du protectorat 1906, 1908-10, 1913-19, 1921, 1923-39 – 6r – 1 – mf#mf-12623 seam – us CRL [350]

Compte administratif... / Haut-commissariat de France en indochine – [Saigon?: Le Commissariat 1946 – reel 6 – 1 – (filmed with: haut-commissariat de france en indochine. budget federal 1935-36) – mf#mf-12275 seam – us CRL [350]

Compte administratif... / Republique francaise, gouvernement general de l'Indochine, Laos – Saigon: Conseil superieur 1901-02, 1905-32, 1934-42 – 5r – 1 – mf#mf-12621 seam – us CRL [350]

Compte administratif du budget general de l'indochine / Republique francaise, gouvernement general de l'indo-chine francaise – Saigon: Impr Coloniale 1899, 1900, 1914-15, 1918-19, 1922, 1924-26, 1932-36, 1946 – 6r – 1 – mf#mf-12275 seam – us CRL [350]

Compte rendu analytique des seances / Belgium. Conseil Colonial – 1908-48 – 1 – $702.00 – (in french) – mf#0099 – us Brook [949]

Compte rendu annuel de la federation des syndicats des travailleurs de la terre *see* Le travailleur rural

Compte rendu des debats / Confederation Generale du Travail. Congres – III-XI, XIII-XV, XVII-XXXVIII Congres. 1897-1975 – 1 – fr ACRPP [331]

Compte rendu des seances / Congres International des Traditions Populaires – Paris: Bibliotheque des Annales economiques, Societe d'editions scientifiques 1891 [mf ed Bloomington IN: Indiana Uni Lib, Preservation Dept 1984] – 168p on 1r [ill] – 1 – us Indiana Preservation [390]

Compte rendu des seances / France. Assemblee Nationale Constituante – v1-10. 1848-49 – 1 – $216.00 – mf#0216 – us Brook [324]

Compte rendu des seances / France. Assemblee nationale legislative – v1-17. 28 may 1849-dec 1851 – 9 – $534.00 – mf#0213 – us Brook [323]

Compte rendu des travaux de la quatrieme session tenue a geneve du 24 au 29 aout 1896 : sous le haut patronage du conseil federal suisse et du gouvernement de la republique et canton de geneve – Geneve: Georg & Co, 1897 – us CRL [949]

Compte-rendu / Congres National des Droits Civils et du Suffrage des Femmes – Paris. 1908 – 1 – us CRL [323]

[Compte-rendu] – Toulouse: Impr regionale, 1950 – 1r – 1 – us CRL [616]

Compte-rendu de la conference ... – Tananarive, 6-12 sep 1961 [Tananarive, Impr. nationale 1962] – (filmed with three conference reports of the union africaine et malgache: compte-rendu ... de la conference de bangui ... cotonou [1962?].–compte-rendu ... de la conference de libreville ... cotonou [1962?].-compte–rendu... de la conference de ouagadougou ... cotonou [1963?]) – us CRL [960]

Compte-rendu de la conference de bangui, 25-27 mars 1962 – Cotonou: [s.n. 1962?] – (filmed with conference des chefs d'etat et de gouvernement africains et malgaches, tananarive 1961 [tananarive 1962]) – us CRL [960]

Compte-rendu des travaux / Depince, M Ch – Paris: Comite d'organization du Congres, 1909 – 1 – us CRL [960]

Compte-rendu des travaux de la chambre d'agriculture du bas-canada : annee 1859 / Chambre d'agriculture du Bas-Canada – Montreal: De Montigny & compagnie...[s.d.] (mf ed 1983) – 5mf – 9 – mf#SEM105P155 – cn Bibl Nat [630]

Compte-rendu des travaux de la conference de libreville 10-13 sep 1962 / ed by Le Secretariat general de l'UAM – Cotonou [1962?] – (filmed with conference des chefs d'etat et de gouvernement africains et malgaches, tananarive 1961 [tananarive 1962]) – us CRL [960]

Compte-rendu des travaux de la conference de ouagadougou 10-14 mars 1963 / ed by Le Secretariat general de l'UAM – Cotonou [1963?] – (filmed with conference des chefs d'etat et de gouvernement africains et malgaches, tananarive 1961 [tananarive 1962]) – us CRL [960]

Compte-rendu des travaux du seminaire d'anthropologie sociale / Seminaire D'anthropologie Sociale (1951 : Astrida) – Astrida, Rwanda. 1952 – 1r – us UF Libraries [301]

Compte-rendu du grand concert donne au benefice des zouaves pontificaux a levis, le 22 fevrier 1868 : suivi du discours de m chapleau – Levis Quebec: J N Doucet, 1868 – 1mf – 9 – mf#04162 – cn Canadiana [241]

Compte-rendu du meeting democratique de patignies – Bruxelles: Desire Brismee 1864 – 2mf – 9 – mf#vrl-127 – ne IDC [366]

[Compte-rendu] instituee / Maginot, M A – Paris: E Larose, 1917 – 1 – us CRL [944]

Comptes rendus / Academie des Sciences d'Outre-Mer – Paris. 1941-67 – 1 – fr ACRPP [500]

Comptes rendus / Academie des Sciences. Russie – Petrograd-Leningrad. Serie A: Section physico-mathematique et sciences naturelles. 1922-33 – 1 – (serie b: section historico-philologique 1924-31. nouvelle serie: 1933-43) – fr ACRPP [500]

Comptes rendus / Parti Republicain Radical et Radical-Socialiste. Congres du Parti – 1933-38 – 1 – fr ACRPP [335]

Comptes rendus et rapports / Communist Party. France – 1929-59 – 1 – fr ACRPP [335]

Comptes rendus et rapports des 30-44 congres / Parti Socialiste. Section Francaise de l'Internationale Ouvriere – 1933-52 – 1 – fr ACRPP [335]

Comptes-rendus des seances de l'academie des inscriptions et belles-lettres – Paris. v1-8. 1857-1864; ns: v1-7. 1865-1871; S3 v1 1872; S4 v1-48 1873-1920 – 450mf – 9 – mf#0-1203 – ne IDC [400]

Le comptoir francais de juda (ouidah) au 18e siecle – Paris: E Larose, 1942 – 1 – us CRL [336]

Compton bulletin – California. 1991 jun 19/dec 25-1998 jul/dec 30 – 15r – 1 – (with gaps) – mf#2178523 – us WHS [071]

[Compton-] the herald american – CA. 1960-72 – 67r – 1 – $4020.00 – mf#C02134 – us Library Micro [071]

Compton-Rickett, Joseph *see* Origins and faith

Comptroller general annual reports – 1937-91 – (= ser Comptroller general decisions) – 141mf – 9 – $217.00 – (missing: 1938-39; 1942; 1986-89. updates planned) – mf#llmc 90-375 – us LLMC [336]

Comptroller general decisions / U.S. General Accounting Office – v1-73. 1921-94 – 932mf – 9 – $1398.00 – (with index/digest for 1921-81. updates planned) – mf#LLMC 79-412 – us LLMC [324]

Comptroller of the treasury decisions : 1894-1921 / U.S. Treasury Dept. Comptroller – v1-27 (all publ) – 283mf – 9 – $424.00 – (cont by: decisions of the comptroller general of the us llmc 79-412) – mf#LLMC 78-212 – us LLMC [336]

Compulsion of the gospel / Beith, Alexander – Edinburgh, Scotland. 1837 – 1r – us UF Libraries [220]

Compulsory drinking usages / Dunlop, John – London, England. 1840 – 1r – us UF Libraries [080]

Compulsory licensing of patents: a legislative history / U.S. Library of Congress. Legislative Reference Service – Washington, Govt. Print. Off., 1958. 70 p. LL-2310 – 1 – us L of C Photodup [346]

Computational economics – Dordrecht. 1993+ (1,5,9) – (cont: computer science in economics and management) – ISSN: 0927-7099 – mf#16777,01 – us UMI ProQuest [330]

Computational intelligence = Intelligence informatique – Cambridge. 1993-1996 (1) – ISSN: 0824-7935 – mf#17208 – us UMI ProQuest [000]

Computational linguistics – Cambridge. 1988+ (1,5,9) – ISSN: 0891-2017 – mf#16440,01 – us UMI ProQuest [400]

Computational mechanics – Berlin. 1986-1993 (1,5,9) – ISSN: 0178-7675 – mf#16984 – us UMI ProQuest [621]

Computational statistics and data analysis – Amsterdam. 1983-1996 – 1,5,9 – ISSN: 0167-9473 – mf#42482 – us UMI ProQuest [310]

Compute – New York. 1979-1994 (1,5,9) – ISSN: 0194-357X – mf#12857 – us UMI ProQuest [000]

Compute-a-tree – v1 n3-v3 n6 [1983 mar-1986 sep], v2 n1 added [1984 oct] – 1r – 1 – mf#1130728 – us WHS [000]

Computer age edp weekly – Annandale. 1982-1995 (1,5,9) – (cont: edp weekly. cont by: edp weekly) – ISSN: 0884-206X – mf#12413,01 – us UMI ProQuest [000]

Computer aided design – Kidlington. 1970+ (1,5,9) – ISSN: 0010-4485 – mf#13324 – us UMI ProQuest [000]

Computer aided geometric design – Amsterdam. 1989-1995 (1,5,9) – ISSN: 0167-8396 – mf#42557 – us UMI ProQuest [000]

Computer aided surgery – New York. 1997+ (1) – ISSN: 1092-9088 – mf#21843,01 – us UMI ProQuest [617]

Computer and Automated Systems Association of SME *see* Cim technology

Computer and communications decisions – Hasbrouck Heights. 1987-1988 (1) 1987-1988 (5) 1987-1988 (9) – (cont: computer decisions. cont by: computer decisions) – ISSN: 0894-1246 – mf#6558,01 – us UMI ProQuest [000]

Computer and internet lawyer – Frederick. 2000+ (1) – mf#25000,01 – us UMI ProQuest [000]

Computer applications in the biosciences: cabios – Oxford. 1989-1996 (1,5,9) – (cont by: bioinformatics) – ISSN: 0266-7061 – mf#16457 – us UMI ProQuest [500]

Computer bulletin – London. 1982+ (1,5,9) – ISSN: 0010-4531 – mf#13302,01 – us UMI ProQuest [000]

Computer business news – Framingham. 1978-1982 (1,5,9) – (cont by: iso world) – ISSN: 0162-5853 – mf#11663 – us UMI ProQuest [000]

Computer communications – Amsterdam. 1978+ (1,5,9) – ISSN: 0140-3664 – mf#13325 – us UMI ProQuest [000]

Computer contributions – Lawrence. 1966-1970 (1) – mf#8808 – us UMI ProQuest [550]

Computer crime digest – Annandale. 1982-87 (1,5,9) – ISSN: 0889-5694 – mf#13374 – us UMI ProQuest [365]

Computer data – Downsview. 1983-1989 (1) 1989-1989 (5) 1989-1989 (9) – (cont by: info canada) – ISSN: 0383-7319 – mf#15027 – us UMI ProQuest [000]

Computer decisions – Hasbrouck Heights. 1969-1987 [1]; 1972-1987 [5]; 1975-1987 [9] – (cont by: computer and communications decisions) – ISSN: 0010-4558 – mf#6558 – us UMI ProQuest [000]

Computer decisions – Teaneck. 1988-1989 (1,5,9) – (cont: computer and communications decisions) – ISSN: 0898-1825 – mf#6558,02 – us UMI ProQuest [000]

Computer design – Tulsa. 1962-1998 (1) 1973-1998 (5) 1976-1998 (9) – ISSN: 0010-4566 – mf#8508 – us UMI ProQuest [000]

Computer enhanced spectroscopy: ces – Chichester. 1983-1986 (1) 1983-1986 (5) 1983-1986 (9) – ISSN: 0734-3051 – mf#13303 – us UMI ProQuest [530]

Computer graphics today – New York. 1984-1989 (1,5,9) – ISSN: 0747-9670 – mf#15221 – us UMI ProQuest [000]

Computer graphics world – San Francisco. 1984+ (1,5,9) – ISSN: 0271-4159 – mf#14648,01 – us UMI ProQuest [000]

Computer input microfilm (cim) feasibility study / Burford, J B & Clark, J M – [Beltsville, MD]: Agricultural Research Service, US Dept of Agriculture, 1974 (mf ed 1987) – 1mf – 9 – (incl bibl ref) – mf#FSN 41,670 – us NY Public [000]

Computer journal – London. 1982+ (1,5,9) – ISSN: 0010-4620 – mf#13304 – us UMI ProQuest [000]

Computer language – San Francisco. 1986-1993 (1,5,9) – ISSN: 0749-2839 – mf#15990 – us UMI ProQuest [000]

Computer languages – New York. 1975+ (1,5,9) – ISSN: 0096-0551 – mf#49035 – us UMI ProQuest [000]

Computer law and tax report – v1-26. 1974-2000 – 9 – $397.00 set – ISSN: 0361-7203 – mf#117061 – us Hein [343]

Computer law review and technology law journal – 1997- – mf#119051 – us Hein [346]

Computer law review and technology law journal – (1997-99 $150 set. 1997, 1999 $45v. 1998 $60v) – mf#119050 – us Hein [346]

Computer lawyer – v1-16. 1984-2000 – 9 – $449.00 set – ISSN: 0742-1192 – mf#116361 – us Hein [340]

Computer lawyer *see* International computer lawyer

Computer literature index : annual cumulation – Phoenix. 1989-1996 (1) – mf#17138,01 – us UMI ProQuest [000]

Computer marketing newsletter – Orange. 1991-1993 (1) – ISSN: 0886-7194 – mf#15081 – us UMI ProQuest [000]

Computer methods and programs in biomedicine – Amsterdam. 1970-1991 (1) 1970-1991 (5) 1987-1991 (9) – ISSN: 0169-2607 – mf#42156 – us UMI ProQuest [610]

Computer methods in applied mechanics and engineering – Amsterdam. 1972+ (1) 1972+ (5) 1987+ (9) – ISSN: 0045-7825 – mf#42154 – us UMI ProQuest [621]

Computer music journal – Cambridge. 1977+ (1,5,9) – ISSN: 0148-9267 – mf#11761 – us UMI ProQuest [780]

Computer networks – Amsterdam. 1999+ (1) – (cont: computer networks and isdn systems) – ISSN: 1389-1286 – mf#42155,01 – us UMI ProQuest [000]

Computer networks and isdn systems – Amsterdam. 1976-1998 (1) 1976-1998 (5) 1987-1998 (9) – ISSN: 0169-7552 – mf#42155 – us UMI ProQuest [000]

Computer optics – Oxford. 1989-1990 (1,5,9) – ISSN: 0955-355X – mf#49599 – us UMI ProQuest [530]

Computer output microfilm (com) hardware and software: the state of the art / American Library Association – RLMS Micro-File Series, v. 5. 1977 – 9 – 8.00 – us L of C Photodup [770]

Computer performance – Guildford. 1982-1984 (1,5,9) – ISSN: 0143-9642 – mf#13326 – us UMI ProQuest [000]

Computer physics communications – Amsterdam. 1969+ (1) 1969+ (5) 1979+ (9) – ISSN: 0010-4655 – mf#42159 – us UMI ProQuest [500]

Computer physics reports – Amsterdam. 1988-1990 (1,5,9) – ISSN: 0167-7977 – mf#42555 – us UMI ProQuest [530]

Computer publishing magazine – Malibu. 1990-1991 (1,5,9) – (cont: electronic publishing and printing) – ISSN: 1054-0415 – mf#15283,02 – us UMI ProQuest [070]

Computer reseller news – Manhasset. 1993-1998 (1,5,9) – ISSN: 0893-8377 – mf#19186,01 – us UMI ProQuest [000]

Computer science in economics and management – Dordrecht. 1991-1992 (1,5,9) – (cont by: computational economics) – ISSN: 0921-2736 – mf#16777 – us UMI ProQuest [000]

Computer services journal – Los Angeles. 1969-1971 (1) – ISSN: 0025-1763 – mf#7334 – us UMI ProQuest [000]

Computer standards and interfaces – Amsterdam. 1982-1995 (1) 1982-1995 (5) 1987-1995 (9) – ISSN: 0920-5489 – mf#42410 – us UMI ProQuest [000]

Computer supported cooperative work: cscw – Dordrecht. 1992-1994 (1,5,9) – ISSN: 0925-9724 – mf#18641 – us UMI ProQuest [334]

Computer sweden – Stockholm. 1989-1991 (1) – ISSN: 0280-9982 – mf#15235 – us UMI ProQuest [000]

Computer systems science and engineering – London. 1985-1992 (1,5,9) – ISSN: 0267-6192 – mf#15166 – us UMI ProQuest [000]

Computer technology review – Los Angeles. 1985+ (1,5,9) – ISSN: 0278-9647 – mf#15053 – us UMI ProQuest [000]

Computer weekly – Sutton. 1978-1982 (1) – ISSN: 0010-4787 – mf#11144 – us UMI ProQuest [000]

Computer-aided civil and infrastructure engineering – Malden. 1998+ (1) – ISSN: 1093-9687 – mf#20645,01 – us UMI ProQuest [624]

Computer-aided engineering: cae – Cleveland. 1982+ (1,5,9) – ISSN: 0733-3536 – mf#13833 – us UMI ProQuest [620]

Computer-aided recording and mathematical analysis of team performance in volleyball / Eom, Han J & Schutz, Robert W – 1989 – 2mf – 9 – $8.00 – us Kinesology [790]

Computer-aided transcription : a survey of federal court-reporters' perceptions / Greenwood, J Michael – Washington: FJC, 1981 – 1mf – 9 – $1.50 – mf#LLMC 95-306 – us LLMC [347]

Computer-integrated manufacturing systems – Kidlington. 1988-1998 (1,5,9) – ISSN: 0951-5240 – mf#17218 – us UMI ProQuest [000]

Computerized medical imaging and graphics – New York. 1988+ (1,5,9) – (cont: computerized radiology) – ISSN: 0895-6111 – mf#49255,01 – us UMI ProQuest [610]

Computerized radiology – Elmsford. 1977-1987 (1,5,9) – (cont by: computerized medical imaging and graphics) – ISSN: 0730-4862 – mf#49255 – us UMI ProQuest [610]

Computer-processed tabulations of data from seamen's protective certificate applications to the collector of customs for the port of philadelphia, 1812-1815 / U.S. Bureau of the Customs – (= ser Records Of The U.S. Customs Service) – 1r – 1 – (with printed guide) – mf#M972 – us Nat Archives [336]

Computers and automation – Newtonville. 1952-1969 (1) 1952-1969 (5) 1952-1969 (9) – ISSN: 0010-4795 – mf#12583 – us UMI ProQuest [000]

Computers and chemical engineering – Oxford. 1977+ (1,5,9) – ISSN: 0098-1354 – mf#49036 – us UMI ProQuest [621]

Computers and chemistry – Oxford. 1976+ (1,5,9) – ISSN: 0097-8485 – mf#49037 – us UMI ProQuest [500]

Computers and composition – Norwood. 1999+ (1,5,9) – ISSN: 8755-4615 – mf#25447 – us UMI ProQuest [000]

Computers and education – New York. 1976+ – 1,5,9 – ISSN: 0360-1315 – mf#49039 – us UMI ProQuest [370]

Computers and electrical engineering – New York. 1974+ (1,5,9) – ISSN: 0045-7906 – mf#49040 – us UMI ProQuest [621]

Computers and electronics in agriculture – Amsterdam. 1994-1996 (1,5,9) – ISSN: 0168-1699 – mf#42488 – us UMI ProQuest [630]

Computers and fluids – New York. 1973+ (1,5,9) – ISSN: 0045-7930 – mf#49041 – us UMI ProQuest [621]

Computers and geosciences – Elmsford. 1975+ (1,5,9) – ISSN: 0098-3004 – mf#49042 – us UMI ProQuest [500]

Computers and geotechnics – New York. 1989+ (1,5,9) – ISSN: 0266-352X – mf#42449 – us UMI ProQuest [621]

Computers and graphics – New York. 1975+ (1,5,9) – ISSN: 0097-8493 – mf#49043 – us UMI ProQuest [000]

Computers and industrial engineering – New York. 1977+ (1,5,9) – ISSN: 0360-8352 – mf#49045 – us UMI ProQuest [621]

Computers and mathematics with applications – Oxford. 1975+ (1) 1975+ (5) 1976+ (9) – ISSN: 0898-1221 – mf#49047 – us UMI ProQuest [510]

Computers and operations research – New York. 1974+ (1,5,9) – ISSN: 0305-0548 – mf#49049 – us UMI ProQuest [000]

Computers and programming – New York. 1981-1981 (1) 1981-1981 (5) 1981-1981 (9) – (cont: science and electronics) – ISSN: 0279-070X – mf#8371,02 – us UMI ProQuest [000]

Computers and security – Amsterdam. 1982+ (1,5,9) – ISSN: 0167-4048 – mf#42543 – us UMI ProQuest [000]

Computers and structures – New York. 1971+ (1,5,9) – ISSN: 0045-7949 – mf#49051 – us UMI ProQuest [621]

Computers and the humanities – New York. 1985+ (1,5,9) – ISSN: 0010-4817 – mf#15301 – us UMI ProQuest [000]

Computers and the social sciences – Osprey. 1985-1986 (1) 1985-1986 (5) 1985-1986 (9) – ISSN: 0748-9269 – mf#15302 – us UMI ProQuest [500]

Computers, environment and urban systems – New York. 1975+ (1,5,9) – ISSN: 0198-9715 – mf#49053 – us UMI ProQuest [333]

Computers in accounting – New York. 1987-1993 (1,5,9) – (cont by: accounting technology) – ISSN: 0883-1866 – mf#16340 – us UMI ProQuest [000]

Computers in biology and medicine – New York. 1970+ (1,5,9) – ISSN: 0010-4825 – mf#49054 – us UMI ProQuest [610]

Computers in healthcare – Englewood. 1982-1993 (1) 1982-1993 (5) 1982-1993 (9) – (cont: computers in hospitals. cont by: health management technology) – ISSN: 0745-1075 – mf#12663,01 – us UMI ProQuest [610]

Computers in hospitals – Englewood. 1980-1982 (1,5,9) – (cont by: computers in healthcare) – ISSN: 0274-631X – mf#12663 – us UMI ProQuest [610]

Computers in human behavior – Elmsford. 1985+ (1,5,9) – ISSN: 0747-5632 – mf#49471 – us UMI ProQuest [150]

Computers in human services / ed by Schoech, Dick – v1- 1985- – 1, 9 ($200.00 in US $280.00 outside hardcopy subsc) – us Haworth [000]

Computers in industry – Amsterdam. 1979+ (1) 1979+ (5) 1982+ (9) – ISSN: 0166-3615 – mf#42247 – us UMI ProQuest [000]

Computers in libraries – Medford. 1989+ (1,5,9) – (cont: small computers in libraries) – ISSN: 1041-7915 – mf#14910,01 – us UMI ProQuest [000]

Computers in mechanical engineering: cime – New York. 1982-1988 (1) 1982-1988 (5) 1982-1988 (9) – ISSN: 0745-9726 – mf#14249 – us UMI ProQuest [621]

Computers in nursing – Philadelphia. 1983+ (1,5,9) – ISSN: 0736-8593 – mf#13826 – us UMI ProQuest [610]

Computers in physics / American Institute of Physics – v1- 1987- – 1,5,6 – us AIP [530]

Computers in the schools / ed by Johnson, D LaMont – v1- 1984- – 1, 9 ($200.00 in US $280.00 outside hardcopy subsc) – us Haworth [000]

Computers, informatics, nursing (cin) – Philadelphia. 2002+ (1,5,9) – ISSN: 1538-2931 – mf#13826,01 – us UMI ProQuest [610]

Computers, topic work and young children / Galpin, Barrie – BLBNS [mf ed Wakefield: Microform [1988]] – (= ser Library & information research report 68) – 2mf – 9 – 0-7123-3161-1 – uk Microform Academic [370]

Computertomographische kieferschnittbilder zur evaluation von knochenneubildung und regenerierten implantatlokalisationen / Buerger, Mark Claus – (mf ed 1999) – 2mf – 9 – €40.00 – 3-8267-2603-0 – mf#DHS 2603 – gw Frankfurter [617]

Computeruser. twin cities – Minneapolis, 2000+ [1,5,9] – ISSN: 1533-5585 – mf#20120,02 – us UMI ProQuest [621]

Computerworld – Framingham. 1967+ (1) 1979+ (5) 1979+ (9) – ISSN: 0010-4841 – mf#6206 – us UMI ProQuest [000]

Compute!'s gazette – Greensboro. 1983-1990 (1) 1983-1990 (5) 1983-1990 (9) – ISSN: 0737-3716 – mf#13544 – us UMI ProQuest [000]

Computing and communications : law and protection report – Madison. 1993-1996 (1,5,9) – (cont: computing and communications protection) – mf#11931,03 – us UMI ProQuest [360]

Computing and communications protection – Madison. 1992-1992 (1) 1992-1992 (5) 1992-1992 (9) – (cont: data processing and communications security. cont by: computing and communications: law and protection report) – mf#11931,02 – us UMI ProQuest [360]

Computing archiv fuer informatik und numerik = Archives for informatics and numerical computing – Wien. 1983-1996 (1,5,9) – ISSN: 0010-485X – mf#13264 – us UMI ProQuest [000]

Computing for business – Paramount. 1985-1985 (1,5,9) – (cont: interface age: computing for business) – ISSN: 0883-4350 – mf#11898,02 – us UMI ProQuest [000]

Computing report – Yorktown Heights. 1973-1974 (1) – ISSN: 0010-4876 – mf#8222 – us UMI ProQuest [000]

Computing reviews – New York. 1960+ (1,5,9) – ISSN: 0010-4884 – mf#12689 – us UMI ProQuest [000]

Computing surveys – New York. 1969-1970 (1,5,9) – (cont by: acm computing surveys) – ISSN: 0010-4892 – mf#12685 – us UMI ProQuest [000]

Computing systems in engineering – Elmsford. 1990-1993 (1,5,9) – ISSN: 0956-0521 – mf#49580 – us UMI ProQuest [621]

Computing teacher – Eugene. 1983-1994 – 1,5,9 – (cont by: learning and leading with technology) – ISSN: 0278-9175 – mf#15055 – us UMI ProQuest [370]

The comrade – New York. -m. Oct 1901-Dec 1904. (1 reel) – 1 – uk British Libr Newspaper [071]

The comrade, 1913-14 – Delhi. India – 2r – 1 – mf#4896 – uk Microform Academic [079]

Comrade bill / Cope, Robert K – [Cape Town: Stewart Printing, 1944?] – 1 – us CRL [960]

Comrades in service / Burton, Margaret Ernestine – New York: Missionary Education Movement of the United States and Canada, 1915 – 1mf – 9 – 0-524-00518-4 – mf#1990-0018 – us ATLA [920]

Comrades of the road / Andrews, Matthew T – 1 – 5.00 – us Southern Baptist [242]

Comstock, Bertha *see* Capron trail

Comstock, Bertha A *see*
- Big spring
- First highway along the southeast coast of florida

Comstock, John adams *see* Butterflies of california

Comstock, John Moore *see* The congregational churches of vermont and their ministry, 1762-1914

Comstock news – Comstock, NE: [E E Wimmer] -v87 n18 sep 24 1992 (wkly) 1909-17,1920-92 (gaps) [mf ed -1992] – 1 – (absorbed by: sargent leader. publ in comstock ne aug 27 1909-apr 23 1987; in burwell ne aug 30 1987-sep 24 1992) – us NE Hist [071]

Comtat, Paulin *see* Notre frontiere

Comte and mill / Whittaker, Thomas – New York: Dodge, 1908 [mf ed 1993] – (= ser Philosophies ancient and modern (new york, ny)) – 1mf – 9 – 0-524-08662-1 – (incl bibl ref) – mf#1993-2122 – us ATLA [100]

Comte, Auguste *see* The catechism of positive religion

Le comte de frontenac : etude sur le canada francais a la fin du 17e siecle / Lorin, Henri – Paris: Armand Colin & cie, editeurs...1895 – 6mf – 9 – mf#SEM105P31 – cn Bibl Nat [971]

Comte, F le *see* Cabinet des singularitez d'architecture, peinture, sculpture, et graveure...

Comte julien : ou, l'expiation / Guiraud, Pierre Marie Theresa Alexandre – Paris, France. 1823 – 1r – us UF Libraries [440]

Comte, mill, and spencer : an outline of philosophy / Watson, John – Glasgow, 1895 – (= ser 19th c evolution & creation) – 4mf – 9 – mf#1.1.6089 – uk Chadwyck [100]

Comte ory / Scribe, Eugene – Paris, France. 1816 – 1r – us UF Libraries [440]

Comtesse d'altenberg / Royer, Alphonse – Paris, France. 1844 – 1r – us UF Libraries [440]

La comune – Philadelphia PA, 1911-15 – 1r – 1 – (italian periodical) – us IHRC [073]

Comuneros / Cardenas Acosta, Pablo Enrique – Bogota, Colombia. 1945 – 1r – us UF Libraries [972]

Comuneros / Posada, Eduardo – Bogota, Colombia. 1905 – 1r – us UF Libraries [972]

Comunicacion sobre la escultura de juan martinez montanes, "san jeronimo penitente" existente en el convento de clarisas de llerena / Lepe de la Camara, Jose Maria – Badajoz: Dip. Provincial, 1970. Sep. REE – 1 – sp Bibl Santa Ana [240]

Comunicacion y culturas de masas / Pasquali, Antonio – Caracas, Venezuela. 1964 – 1r – us UF Libraries [972]

Comunicaciones *see* Lineas de autobuses

Comunicaciones en la administracion / Redfield, Charles E – San Jose, Costa Rica. 1958 – 1r – us UF Libraries [350]

Comunicacoes do iser – Rio de Janeiro: Comunicacoes do ISER. v1 n1-9 n39. may 1982-1990 – 2r – us CRL [972]

Comunicador – 1983 jun-1984 dec – 1r – 1 – (cont by: comunicador [united states. naval computer and telecommunications station, spain]) – mf#1054718 – us WHS [071]

Comunidad de labradores. ordenanzas de la comunidad de labradores de villafranca de los barros / Villafranca de los Barros – Villafranca de los Barros: Impresor Bolanos, 1937 – 1 – sp Bibl Santa Ana [946]

Comunidad de Regantes de Badajoz. Canal Montijo *see* Ordenanzas y reglamentos

Comunidad de regantes de Badajoz por el Canal de Lobon *see* Ordenanzas y reglamentos para el sindicato y jurado e riego

Comunidad de regantes de la margen derecha del rio Salor *see* Ordenanzas y reglamentos...

Comunidad de regantes de Merida *see* Ordenanzas y reglamentos

Comunidad de Regantes de Montijo *see* Ordenanzas y reglamentos

Comunidad de regantes de montijo. canal de montijo (badajoz). ordenanzas y reglamentos – Badajoz: Imp. Barrena, 1964 – sp Bibl Santa Ana [340]

Comunidad de regantes. ordenanzas / Navasfrias (Salamanca) – Caceres: Edit. Extremadura, 1970 – 1 – sp Bibl Santa Ana [060]

Comunidade a metropole / Morse, Richard M – Sao Paulo, Brazil. 1954 – 1r – us UF Libraries [972]

Comunidade amazonica / Wagley, Charles – Sao Paulo, Brazil. 1957 – 1r – us UF Libraries [972]

Comunidade luso-brasileira / Almeida, Lourival Nobre De – Rio de Janeiro, Brazil. 1969 – 1r – us UF Libraries [972]

La comunion entre los indios americanos / Bayle, Constantino – Madrid: Missionalia Hispanica, 1944 – 1 – sp Bibl Santa Ana [240]

Las comuniones en la espana roja / Bayle, Constantino – Burgos: Razon y Fe, 1939 – 1 – sp Bibl Santa Ana [240]

El comunismo en espana. cinco anos en el partido : su organizacion y sus misterios / Bayle, Constantino & Karl, Mauricio – Madrid, 2nd ed 1931; Madrid: Razon y Fe, 1932 – 1 – sp Bibl Santa Ana [335]

Comyn, T de *see* State of the philippine islands

Comyns, John *see* A digest of the laws of england

Con be toi yeu / Vo, Ha Anh – [Gia Dinh, Vietnam]: Thuy Duong [1974] [mf ed 1992] – on pt of 1r – 1 – mf#11052 r350 n9 – us Cornell [959]

Con krui tan sann si khyan sam : vatthu mya / Than Lan – Ran Kun: Po Than Lan Ca pe 1977 [mf ed 1990] – 1r with other items – 1 – (in burmese) – mf#mf-10289 seam reel 174/3 [§] – us CRL [830]

Con la voz del corazon / Quintero Carrasco, J – Fregenal: Imprenta Angel Verde, 1962 – sp Bibl Santa Ana [810]

Con lo que tengo a bordo (mecanografiado) / Belloso Rodriguez, Pedro – 1975 – 1 – sp Bibl Santa Ana [810]

Con los brazos abiertos / Bauza, Guillermo – Barcelona, Spain. 1963 – 1r – us UF Libraries [972]

Con maceo en la invasion / Llorens Y Maceo, Jose Silvino – Habana, Cuba. 1928 – 1r – us UF Libraries [972]

Los con razon olvidados / Sanchez Arjona, Vicente – Sevilla: Imprenta Zamb, Tomo 1. 1955 – 1 – sp Bibl Santa Ana [810]

Con sandino en nicaragua / Belausteguidoitia, Ramon De – Madrid, Spain. 1934 – 1r – us UF Libraries [972]

Con suoi mat guong : tho / Xuan Sach – Ha-noi: Quan Doi Nhan Dan 1974 [mf ed 1992] – on pt of 1r – 1 – mf#11052 r15 n12 – us Cornell [810]

Conan, Laure *see*
- Un amour vrai
- Angeline de montbrun
- Elisabeth seton
- L'oublie
- Silhouettes canadiennes

Conangla Fontanilles, Jose *see*
- Cuba y pi y margall
- Tomas gener

Conant, Hannah Chaplin *see* The popular history of the translation of the holy scriptures into the english tongue

Conant, Thomas *see* Life in canada

Conant, Thomas J *see*
- Defence of the hebrew grammar of gesenius against prof. stuart's translation
- The meaning and use of baptizein

Conant, Thomas Jefferson *see*
- Bible word-book
- The psalms

Conant, William C *see* Narratives of remarkable conversions and revival incidents

Conard, Elizabeth Laetitia Moon *see* Les idees des indiens algonquins relatives a la vie d'outre-tombe

Concanen, E *see* Gems of art from the great exhibition

Conceicao, Antonio Pereira Da *see* Roteiro de cabo frio ate ao porto de santos

Conceito de civilisacao brasileira / Franco, Afonso Arinos De Melo – Sao Paulo, Brazil. 1936 – 1r – us UF Libraries [972]

Conceito e a imagem na poesia brasileira / Campos, Humberto De – Rio de Janeiro, Brazil. 1929 – 1r – us UF Libraries [972]

Concenti musicali a 8, 12 et 16 voci, lib. sec. / Stivori, Francesco – 1601 – (= ser Mssa) – 3mf – 9 – €50.00 – mfchl 411 – gw Fischer [780]

Concenti musicali con le sue sinfonie a otto voci / Gastoldi, Giovanni Giacomo – 1604 – (= ser Mssa) – 2mf – 9 – €35.00 – mfchl 238 – gw Fischer [780]

Concentracion nacional de las falanges femeninas en honor del caudillo y del ejercito espanol – Bilbao: [s.n. 1939] [mf ed 1977] – (= ser Blodgett coll) – 1mf – 9 – mf#w816 – us Harvard [946]

Concentracion publica / Gonzalez De Cascorro, Raul – Habana, Cuba. 1964 – 1r – us UF Libraries [972]

Concentration of energy : bruce grit uses plain language in emphasizing the power of organization / Bruce, John Edward – [New York: Edgar Print & Stationery Co, 1899?] (mf ed 1969) – 1r – 1 – mf#Sc Micro R-1163 – us NY Public [305]

Concentric and eccentric strength differences in the lead and back legs of division 1 college level fencers / Casey, Kevin M – 1994 – 1mf – $4.00 – us Kinesology [612]

Concentus octo, sex, quinque & quatuor vocum – 1545 – (= ser Mssa) – 4mf – 9 – €60.00 – mfchl 92 – gw Fischer [780]

The concept of consciousness / Holt, Edwin Bissell – London: G Allen, 1914 – 1mf – 9 – 0-7905-7303-2 – mf#1989-0528 – us ATLA [100]

The concept of mission in the roman catholic church in light of vatican 2 / Lindell, Charles G. – [Chicago], 1967. Chicago: Dep of Photodup, U of Chicago Lib, 1968 (1r); Evanston: American Theol Lib Assoc, 1984 (1r) – 1 – 0-8370-0486-1 – mf#1984-B090 – us ATLA [241]

Concept of sarasvati (in vedic literature) / Airi, Raghunath – 1st ed. Rohtak: Rohtak Co-operative Print & Pub Society, [New Delhi]: exclusively distributed by Munshiram Manoharlal Publ, c1977 – us CRL [490]

The concept of sin / Tennant, Frederick Robert – Cambridge: University Press, 1912 – 1mf – 9 – 0-7905-0167-8 – (incl bibl ref and index) – mf#1987-0167 – us ATLA [290]

Conception and development of poetry in zulu / Vilakazi, B W – [Johannesburg] 1937 – us CRL [490]

La conception du mariage d'apres les juristes romaines / Volterra, Edoardo – Padova, 'La Garangola,' 1940. 66 p. LL-4204 – 1 – us L of C Photodup [340]

The conception of god : a philosophical discussion concerning the nature of the divine idea as a demonstrable reality / Royce, Josiah et al – New York: Macmillan, 1897 – 1mf – 9 – 0-8370-4991-1 – mf#1985-2991 – us ATLA [210]

The conception of immortality / Royce, Josiah – Boston, New York: Houghton Mifflin, 1900. (The Ingersoll Lecture, 1899) – 1 – us Wisconsin U Libr [210]

The conception of immortality / Royce, Josiah – Boston: Houghton, Mifflin, c1900 – (= ser The Ingersoll Lecture) – 1mf – 9 – 0-7905-9860-4 – (incl bibl ref) – mf#1989-1585 – us ATLA [210]

The conception of priesthood in the early church and in the church of england : four sermons / Sanday, William – 2nd ed. London; New York: Longmans, Green, 1899 – 1mf – 9 – 0-7905-0278-X – (incl bibl ref) – mf#1987-0278 – us ATLA [241]

The conception of surplus in theoretical economics / Dasgupta, Amiya Kumar – Calcutta: Dasgupta & Co, 1942 – (= ser Samp: indian books) – us CRL [330]

The conception of the infinite and the solution of the mathematical antinomies : a study in psychological analysis / Fullerton, George Stuart – Philadelphia: JB Lippincott, 1887 – 1mf – 9 – 0-7905-8791-2 – mf#1989-2016 – us ATLA [510]

The conceptions of existence and essence in anthropology : a critical study of the major alternatives in the doctrine of man as revealed in plato, aristotle, augustine, hume and pascal / Hayward, John Frank – Chicago, 1943. Chicago: Dep of Photodup, U of Chicago Lib, 1971 (1r); Evanston: American Theol Lib Assoc, 1984 (1r) – 1 – 0-8370-0326-1 – mf#1984-B183 – us ATLA [100]

Concepto del federalismo en la guerra y en la revolucion; conferencia pronunciada en el cine coliseum de barcelona, el dia 7 de febrero de 1937 / Lopez, Juan – Barcelona, 1937? Fiche W1006. (Blodgett Collection of Spanish Civil War Pamphlets) – 9 – us Harvard [946]

Conceptos fundamentales de literatura comparada / Gicovate, Bernard – San Juan, Puerto Rico. 1962 – 1r – us UF Libraries [972]

Concepts : the journal of defense systems acquisition management – Washington. 1980-1982 (1,5,9) – ISSN: 0279-6759 – mf#12112,01 – us UMI ProQuest [324]

Conceptualising school-based management development : priorities, alternatives, strategies and future directions for school management / Rasool, Mohamed Hoosen Abbas – Uni of South Africa 2000 [mf ed Johannesburg 2000] – 7mf – 9 – (incl bibl ref) – mf#mfm15060 – sa Unisa [370]

Conceptualization and development of the sources of enjoyment in youth sport questionnaire / Wiersma, Lenny D – 2000 – 238p on 3mf – 9 – $15.00 – mf#PE 4143 – us Kinesology [790]

Concern – v3 1961-v10 1968 – 2r – $70.00 – mf#um-81 – us Commission [242]

Concern for dying (association) – New York. 1978-1985 (1) 1978-1985 (5) 1978-1985 (9) – (cont: euthanasia news. cont by: concern for dying newsletter) – ISSN: 0192-1096 – mf#6771,01 – us UMI ProQuest [170]

Concern for dying newsletter – New York. 1985-1989 (1) 1985-1989 (5) 1985-1989 (9) – (cont: concern for dying) – mf#6771,02 – us UMI ProQuest [170]

Concern magazine/newsfold – New York. 1959-1988 (1) 1959-1988 (5) 1959-1988 (9) – (cont by: horizons) – ISSN: 0010-5163 – mf#9625 – us UMI ProQuest [305]

Concern magazine/newsfold / United Presbyterian Women – v19 n9 [1977 aug], v21 n5-v24 n4 [1979 apr-1982 apr] – 1r – 1 – (cont: concern; cont by: horizons [new york ny]) – mf#1054718 – us WHS [242]

Concerned Citizens for Choice on Abortion [Vancouver BC] *see* Ccca newsletter

Concerned citizens of palau : a petition to the hon adrian p winkel, high commissioner of the trust territory of the pacific – feb 19, 1978 – (= ser Republic Of Belau (Palau) – Status Negotiations With The U.S.) – 1mf – 9 – $1.50 – mf#LLMC 82-100G, Title 15 – us LLMC [323]

Concerned Educators Against Forced Unionism [US] *see* Recaps

Concerned Military [San Diego CA] *see* Liberty call

Concerned Officers Movement *see*
- Com mon sense
- Newsletter
- Puget sound sound off

Concerned Officers Movement [San Diego CA] *see* Liberty call

Concerned Servicemen's Movement *see* Fid amchitka

Concerning dade county / Garcia, Helen M – s.l, s.l? 1937 – 1r – us UF Libraries [978]

Concerning jesus christ the son of god / Wilkinson, William Cleaver – Philadelphia: Griffith and Rowland, 1916 – 1mf – 9 – 0-524-07599-9 – mf#1991-3219 – us ATLA [220]

Concerning jesus christ the son of man / Wilkinson, William Cleaver – Philadelphia: Griffith and Rowland Press, c1918 – 1mf – 9 – 0-524-07600-6 – mf#1991-3220 – us ATLA [220]

Concerning life : sermons / Latimer, George Dimmick – Boston: American Unitarian Assoc, 1907 [mf ed 1993] – (= ser Unitarian/universalist coll) – 1mf – 9 – 0-524-07754-1 – mf#1991-3322 – us ATLA [243]

Concerning poetry – Bellingham. 1968-1987 (1) 1973-1987 (5) 1975-1987 (9) – ISSN: 0010-5201 – mf#7013 – us UMI ProQuest [400]

Concerning sea power / Jordan, David Starr – Boston: World Peace Foundation 1912 [mf ed 1992] – (= ser World peace foundation pamphlet series 2/4/1) – 1mf – 9 – 0-524-03233-5 – (originally printed in new york independent, july 6) – mf#1990-0861 – us ATLA [327]

Concerning the date of the bohairic version : covering a detailed examination of the text of the apocalypse and a review of some of the writings of the egyptian monks / Hoskier, H C – London: Bernard Quaritch, 1911 – 1mf – 9 – 0-7905-2001-X – (incl ind) – mf#1987-2001 – us ATLA [220]

Concerning the disciples of christ / Tyler, Benjamin Bushrod – Cleveland, Ohio: Bethany CE Co, c1897 – (= ser Bethany c.e. hand-book series; Bethany c.e. reading courses) – 2mf – 9 – 0-524-02274-7 – (incl bibl ref) – mf#1990-4281 – us ATLA [240]

Concerning the genesis of the versions of the new testament : remarks suggested by the study of p and the allied questions as regards the gospels / Hoskier, Herman Charles – London: Bernard Quaritch, 1910-11 [mf ed 1988] – 2v on 3mf – 9 – 0-7905-0041-8 – (in english, greek and latin. incl ind) – mf#1987-0041 – us ATLA [226]

Concerning the import and export of funds and holdings by travellers / Cambodia. Office of Exchange Control – [Phnom-Penh 1955] [mf ed 1989] – 1r with other items – 1 – mf#mf-10289 seam reel 026/16 [§] – us CRL [346]

Concerning them that are asleep / Gorham, Barlow Weed – Boston: publ by aut, 1885 [mf ed 1984] – 1mf – 9 – 0-8370-0948-0 – mf#1984-4297 – us ATLA [210]

Concert aux champs-elysees / Lafortelle, A M – Paris, France. 1802 – 1r – us UF Libraries [440]

Le concert interrompu : opera comique en un acte de mm marsollier et favieres / Berton, Henri – Paris: Freres Gaveaux [1802?] [mf ed 1990] – 1r – 1 – mf#pres. film 78 – us Sibley [780]

Concert pour le violoncell [et orchestre] : edition pour violoncelle et piano / Ritter, Peter – [London (?) 1904?] [mf ed 1988] – 1r – 1 – mf#pres. film 7 – us Sibley [780]

Concerted piano music of stravinsky / Stevens, Willis Alvin – U of Rochester 1961 [mf ed 1982] – 3mf – 9 – (with bibl) – mf#fiche 1093 – us Sibley [780]
Concerti di andrea et gio. gabrieli... / Gabrieli, Andrea – 1587 – (= ser Mssa) – 8mf – 9 – €100.00 – mfchl 230 – gw Fischer [780]
Concerti ecclesiastici..., libro secondo / Viadana, Lodovico da – 1607 – (= ser Mssa) – 2mf – 9 – €35.00 – mfchl 437 – gw Fischer [780]
Concerti ecclesiatici a due et a quatro voci / Molinaro, Simone – 1605 – (= ser Mssa) – 3mf – 9 – €50.00 – mfchl 349 – gw Fischer [780]
Concerti grossi con due violini, viola e violoncello di concertino obligati, e due altri violini e basso di concerto grosso... : composti della seconda parte del opera quinta d'arcangelo corelli per francesco geminiani / Geminiani, Francesco – London: N Prevost [1735?] [mf ed 1992] – 1r – 1 – mf#pres. film 117 – us Sibley [780]
Concerti grossi con due violini, viola e violoncello di concertino obligati, e due altri violini e basso di concerto grosso... : composti delli sei soli della prima parte del opera quinta d'arcangelo corelli / Geminiani, Francesco – Amsterdam: M C Le Cene [1735?] [mf ed 1992] – 1r – 1 – mf#pres. film 117 – us Sibley [780]
Concerti grossi con due violini, viola e violoncello di concertino obligati, e due altri violini e basso di concerto grosso...no 1-6 / Geminiani, Francesco – Amsterdam: M C Le Cene [1735?] [mf ed 1992] – 7pt on 1r – 1 – mf#pres. film 94 – us Sibley [780]
Concerti grossi con due violini, viola e violoncello di concertino, obligati, e due altri violini e basso di concerto grosso...opera terza / Geminiani, Francesco – Amsterdam: M C Le Cene [1735?] [mf ed 1992] – 7pt on 1r – 1 – mf#pres. film 117 – us Sibley [780]
Concerti grossi con due violini, violoncello e viola di concertino obligati : e due altri violini, e basso di concerto grosso ad arbitrio... opera seconda / Geminiani, Francesco – London: printed for aut & sold by I Walsh [1732] [mf ed 19–] – 7pt on 1r – 1 – mf#pres. film 91 – us Sibley [780]
Concerti grossi con due violini, violoncello e viola di concertino obligati, e due altri violini, e basso di concerto grosso ad arbitrio...opera seconda : il 4. 5. e 6. si potranno suonare con due flauti traversieri, o due violini con violoncello / Geminiani, Francesco – Amsterdam: M C Le Cene [1730?] [mf ed 1992] – 1r – 1 – mf#pres. film 117 – us Sibley [780]
Concerti grossi con duoi violini e violoncello di concertino obligati e duoi altri violini, viola e basso di concerto grosso ad arbitrio, che si potranno radoppiare, parte prima-[seconda, per camera : preludii, allemande, correnti, gighe, sarabande, gavotte e minuetti] op 6 / Corelli, Arcangelo – Amsterdam: E Roger & M C Le Cene [1712] [mf ed 1992] – 1r – 1 – mf#pres. film 117 – us Sibley [780]
Concerti musicali a dve, tre e qvattro voci : con vna messa a quattro concertata, et introiti, pange lingua a quattro da capella. opera terza / Beria, Giovanni Battista – Milano: C Camagno 1650 [mf ed 19–] – 5pt on 1r – 1 – mf#film 2536 – us Sibley [780]
Concertino per il cembalo : accompagnato da due violini, viola e violoncello / Benda, Georg – Lipsia: Schwickert [1779] [mf ed 1995] – 5pt on 1r – 1 – mf#pres. film 140 – us Sibley [780]
Concerto for clarinet and string orchestra by aaron copland : a stylistic analysis / Fleisher, Gerald – U of Rochester 1961 [mf ed 19–] – 2mf – 9 – (incl abstract) – mf#fiche 994 – us Sibley [780]
Concerto for the grand or small piano forte with accompaniments : in which is introduced, the favorite air of the plough boy, op 15 / Dussek, Johann Ladislaus – London: Broderip & Wilkinson...[179-?] [mf ed 19–] – 1mf – 9 – mf#fiche 1150 – us Sibley [780]
Concerto for the violincello (!) : with accompanyments for two violins, tenor, bass, flutes and horns / Paxton, Stephen – London: printed for G Goulding [179-?] [mf ed 1989] – 1r – 1 – mf#pres. film 42 – us Sibley [780]
Concerto (no 1) for piano forte, op 1 / Griffin, George Eugene – London: printed for aut by Clementi & Co...[between 1801 and 1830] [mf ed 199-] – 1r – 1 – mf#pres. film 134 – us Sibley [780]
Concerto, no 2, in b minor for violin and orchestra [op 36] : violino principale part / Nachez, Tivadar – [1908] [mf ed 1988] – 1r – 1 – mf#pres. film 7 – us Sibley [780]
Concerto per il organo o cembalo / Hook, James – 1799. Composer's holograph score for organ and orchestra. MUSIC 1873, Item 3 – 1 – us L of C Photodup [780]
Concesiones arancelarias y cambiarias otorgadas – Bogota, Colombia. 1963 – 1r – us UF Libraries [972]
Concesiones arancelarias y cambiarias otorgadas / Colombia Oficina De Comercio Exterior – Bogota, Colombia. 1964 – 1r – us UF Libraries [972]
Concessions, documents and opinions of the attorney – s.l, s.l? 1899 – 1r – us UF Libraries [972]
Concessions to america the bane of britain : or the cause of the present distressed situation of the british colonial and shipping interests explained, and the proper remedy suggested / Marryat, Joseph – London: W J & J Richardson, & J Hatchard, 1807 – (= ser 19th c economics) – 1mf – 9 – mf#1.1.414 – uk Chadwyck [337]
Concha, Antonio *see* Contestacion al manifiesto publicado...mauricio ceresoles...pasadas eleciones..
Concha Castaneda, Juan de la *see* Discursos leidos...ciencias morales y politicas
Concherias, romances, epigramas y otras poemas / Echeverria, Aquileo J – San Jose, Costa Rica. 1953 – 1r – us UF Libraries [440]
Conchis, Guillelmus de *see* Dragmaticon philosophiae (cccm 152)
Conchologia iconica : or, illustrations of the shells of molluscous animals / ed by Reeve, L A – London 1843-78 – 20v on 351mf – 9 – mf#2168/2 – ne IDC [590]
Conchological magazine – Kyoto 1907-09 – v1-3 on 41mf – 9 – mf#z-895/2 – ne IDC [590]
Conchologist – London 1891-93. v1-2 – 27mf – 9 – (cont as: journal of malacology [london 1894-1902] v3-9) – mf#5623/2 – ne IDC [590]
Conchudo, J *see* Carta al autor de la oracion apologetica por la espena y su merito literario
Conciencia al espejo / Ortiz, Jose Antonio – Barcelona, Spain. 1960 – 1r – us UF Libraries [972]
Conciencia espanola / Menendez Y Pelayo, Marcelino – Madrid, Spain. 1948 – 1r – us UF Libraries [025]
Le concil de turin / Babut, Ernest Ch – Paris, 1904 – 6mf – 8 – €14.00 – ne Slangenburg [241]
Concil und jesuitismus : brennende fragen zur orientierung fuer das deutsche volk – Stuttgart: Vogler & Beinhauer, 1870 – 1mf – 9 – 0-8370-8549-7 – mf#1986-2549 – us ATLA [241]
Das concil zu konstanz : ein protestantenvereins-vortrag / Trautz, Th – Karlsruhe: G Braun, 1874 – 1mf – 9 – 0-524-00178-2 – mf#1989-2878 – us ATLA [242]
Le concile de turin : essai sur l'histoire des eglises provencales au 5e siecle et sur les origines de la monarchie ecclesiastique romaine (417-450) / Babut, Ernest Ch – Paris: A Picard, 1904 – 1mf – 9 – 0-7905-5446-1 – mf#1988-1446 – us ATLA [240]
Le concile du vatican : son histoire et ses consequences politiques et religieuses / Pressense, Edmond de – Paris: Sandoz et Fischbacher, 1872 – 2mf – 9 – 0-8370-8933-6 – mf#1986-2933 – us ATLA [241]
Concilia aevi karolini (mgh leges 2c:2.bd) – 1906-1908 – (= ser Monumenta germaniae historica leges 2. leges in quarto. c legum sectio 3: concilia (mgh leges 2)) – €52.00 – ne Slangenburg [240]
Concilia aevi merivingici (mgh leges 2c:1.bd) – 1893 – (= ser Monumenta germaniae historica leges 2. leges in quarto. c legum sectio 3: concilia (mgh leges 2)) – €17.00 – ne Slangenburg [240]
Concilia antiqua galliae : opera et studio / Sirmondi, J – Lutetiae Parisiorum. v1-3. 1629 – 3v on 91mf – 9 – €174.00 – ne Slangenburg [241]
Concilia galliae narbonensis / Baluzius, Stephanus – Parisiis, 1668 – 72mf – 8 – €137.00 – (filmed with: miscellaneorum libri 7, hoc est collectio veterum monumentorum quae hactenus latuerant in variis codicibus ac bibliothecis, parisiis, 1678-1715) – ne Slangenburg [240]
Concilia magnae britanniae et hiberniae a synodo verolamiensi a.d. 446 ad londinensem a.d. 1717 a davide wilkins collecta – Londini. v1-4. 1737 – €273.00 – ne Slangenburg [240]
Concilia omnia, tum generalia, tum provincialia atque particularia / Surius, L et al – Coloniae Agrippinae. v1-4. 1567 – 4v on 101mf – 8 – €193.00 – ne Slangenburg [240]
Concilia provincialia, baltimori habita ab anno 1829 usque ad annum 1849 – ed altera. Baltimori: J Murphy 1851 [mf ed 1992] – 1mf – 9 – 0-524-03138-X – (incl: statuta synodi baltimorensis anno 1791...incl ind) – mf#1990-4587 – us ATLA [241]
Conciliabulo de basilea / Caceres, Diego de – S.l., s.i., s.a. Hacia 1641 – 1 – sp Bibl Santa Ana [946]
A conciliacao : orgao do partido republicano – Santarem, PA, 11 jan 1890 – (= ser Ps 19) – bl Biblioteca [079]
O conciliador : jornal politico e noticioso da provincia de santa catharina – Desterro, SC: Typ de Jose Joaquim Lopes, 07 mar 1872; 06 fev-30 out 1873 – (= ser Ps 19) – bl Biblioteca [320]
O conciliador pernambucano – Olinda, PE: Typ de Pinheiro Faria, 25 jan-16 abr 1832 – (= ser Ps 19) – mf#P19,02,25 – bl Biblioteca [320]
Conciliateur : ou, l'homme aimable / Demoustier, Charles Albert – Paris, France. 1802 – 1r – us UF Libraries [440]
Le conciliateur : nouvelles du jour – Paris: Bureau du journal. v1 n18-39. jul 7-28 1848 – us CRL [073]
Conciliationis ecclesiae armenae cum romana : ex ipsis armenorum patrum, et doctorum testimoniis, in duas partes: historialem et controversialem divisae / Galanus, Clemens – Romae: Typis Sacrae Congregationis de Propaganda Fide, 1658-1690 – 2r – 1 – 0-8370-1515-4 – mf#1984-B293 – us ATLA [240]
Conciliatory suggestions on the subject of regeneration / Cunningham, J W – London, England. 1816 – 1r – us UF Libraries [240]
Conciliengeschichte *see* History of the christian councils
Concilii plenarii baltimorensis 2 : in ecclesia metropolitana baltimorensi, a die 7. ad diem 21. octobris, a.d., 1866, habiti, et a sede apostolica recogniti: decreta – Baltimorae: excudebat Joannes Murphy, 1868 – 1mf – 9 – 0-8370-9849-1 – (incl bibl ref and ind) – mf#1986-3849 – us ATLA [240]
Il concilio 2 di leone... / Franchi, Antonino – Madrid: Graf. Calleja, 1966 – 1 – sp Bibl Santa Ana [946]
El concilio 3 emeritense / Garcia de la Fuente, P Arturo – Badajoz, 1932 – 1 – sp Bibl Santa Ana [946]
Concilio for the Spanish Speaking *see* Voz
Concilio for the Spanish Speaking [Seattle WA] *see* Concilio newsletter
Concilio, Januarius de *see*
– Catholicity and pantheism
– The knowledge of mary
Concilio newsletter / Concilio for the Spanish Speaking [Seattle WA] – n1-4 [1979 feb-may] – 1r – 1 – mf#1047565 – us WHS [071]
Concilio y vida cristiana / Aradillas Agudo, Antonio – Madrid: Sociedad de Educacion. Atenas, 1966 – sp Bibl Santa Ana [240]
Concilios provinciales primero, y segundo : celebrados en la muy noble, y muy real ciudad de mexico, presidiendo el illmo y rmo senor d fr alonso de montufar, en los anos de 1555, y 1565 / Catholic Church Province of Mexico City [Mexico]. Concilio Provincial [1st: 1555] – en Mexico: impr D Joseph Antonio de Hogal, ano de 1769 – (= ser Books on religion...1543/44-c1800: concilios y sinodos) – 2v on 5mf – 9 – mf#crl-403 – ne IDC [241]
Concilium basileense scriptores – Vindobonae v1-4 1857; Basileae 1935 – €243.00 – ne Slangenburg [220]
Concilium mexicanum provinciale 3. celebratum mexici anno 1585 : praeside d d petro moya, et contreras archiepiscopo ejusdem urbis / Catholic Church Province of Mexico City [Mexico]. Concilio Provincial [3rd: 1585] – Mexici: Ex Typographia Bac Josephi Antonij de Hogal 1770 – (= ser Books on religion... 1543/44-c1800: concilios y sinodos) – 6mf – 9 – mf#crl-404 – ne IDC [241]
Concilium sacrosanctvm domini nostri iesu christi, angelorum, apostolorum...decreta sacrosancti concilij... regis sapientissime solomonis sermo de sapientia uera... / Bibliander, T – [Basileae, Ioannes Oporinus], 1552 – 5mf – 9 – mf#PBU-482 – ne IDC [240]
Concilium tridentinum non institutum esse ad inquirendam...veritatem...demonstratio / Bullinger, Heinrich – 2mf – 9 – mf#PBU-164 – ne IDC [240]
Concio ad clerum : a sermon delivered in the chapel of yale college, sep 10, 1828 / Taylor, Nathaniel William – New Haven: A H Maltby & Homan Hallock, 1842 [mf ed 1978] – (= ser Revivalism and revival preachers in america 6) – 1mf – 9 – 0-8370-0247-8 – (incl bibl ref) – mf#1984-3006 – us ATLA [242]
Concise account of the late rev. noah davis / Davis, Noah – 38p – 1 – 5.00 – us Southern Baptist [242]
A concise account of the principal works in stained glass / Willement, Thomas – [London?] 1840 – (= ser 19th c art & architecture) – 1mf – 9 – mf#4.2.1067 – uk Chadwyck [740]
A concise account of the religious society of friends, commonly called quakers : embracing a sketch of their christian doctrines and practices / Evans, Thomas – Philadelphia: for sale at Friends' Book Store, 1872 [mf ed 1986] – 1mf – 9 – 0-8370-8896-8 – mf#1986-2896 – us ATLA [243]
Concise and familiar exposition of the leading prophecies regarding... – Edinburgh, Scotland. 1834 – 1r – us UF Libraries [240]
A concise and practical treatise of the law of vendors and purchasers of estates / Sugden, Edward Burtneshaw – 14th ed.; 8th American ed., by J.C. Perkins. Philadelphia: Kay, 1873. 2v – 1 – us L of C Photodup [340]
Concise dictionary of middle english from ad 1150 to 1580 / Mayhew, Anthony Lawson – Oxford, England. 1938 – 1r – us UF Libraries [420]
The concise dictionary of religious knowledge and gazetteer / ed by Jackson, Samuel Macauley et al – 3rd ed thoroughly rev. New York: Christian Literature Co, 1898 – 3mf – 9 – 0-7905-8265-1 – mf#1988-6143 – us ATLA [052]
Concise english-kafir dictionary / Mclaren, James – London, England. 1923 – 1r – us UF Libraries [040]
A concise history of kehukee baptist association, nc, pts 1 and 2 / Biggs, Joseph et al – 1803, 1834 – 1 – $10.64 – us Southern Baptist [242]
A concise history of missions / Bliss, Edwin Munsell – New York: Fleming H Revell, c1897 [mf ed 1986] – 1mf – 9 – 0-8370-6646-8 – (incl a chronology of principal foreign societies and ind) – mf#1986-0646 – us ATLA [240]
A concise history of painting / Heaton, Mary Margaret [Keymer] et al – London 1888 – (= ser 19th c art & architecture) – 6mf – 9 – mf#4.2.420 – uk Chadwyck [750]
A concise history of religion / Gould, Frederick James – London: Watts, [1893?]-1897 [mf ed 1992] – 2mf – 9 – 0-524-02206-2 – (incl bibl ref) – mf#1990-2880 – us ATLA [200]
A concise history of the colony and natives of new south wales / Kittle, Samuel – Edinburgh [1815] [mf ed Hildesheim 1995-98] – 1v on 3mf – 9 – €90.00 – 3-487-26759-4 – gw Olms [980]
A concise history of the foreign christian missionary society – Cincinnati: Foreign Christian Missionary Society, 1910 [mf ed 1992] – (= ser Christian church (disciples of christ) coll) – 1mf – 9 – 0-524-04372-8 – mf#1991-2076 – us ATLA [240]
A concise history of the introduction of protestantism into mississippi and the southwest / Jones, John G – 1772-1817 – 1 – $10.29 – us Southern Baptist [242]
A concise history of the kehukee baptist association : from its original rise down to 1803 / Burkitt, Lemuel & Read, Jesse – Philadelphia: Lippincott, Grambo, 1850 [mf ed 1992] – (= ser Baptist coll) – 1mf – 9 – 0-524-04356-6 – mf#1990-5039 – us ATLA [242]
A concise history of the ketocton baptist association / Fristoe, William – 1808 – 1 – $5.81 – us Southern Baptist [242]
A concise history of the methodist protestant church : from its origin, with biographical sketches of several leading ministers of the denomination, and also a sketch of the author's life / Bassett, Ancel Henry – 3rd rev nl ed. Pittsburgh: Wm McCracken, 1887 [mf ed 1992] – (= ser Methodist coll) – 2mf – 9 – 0-524-05355-3 – (originally publ in 1877) – mf#1990-5106 – us ATLA [242]
Concise history of tithes / Fry, Joseph Storrs – London, England. 1820 – 1r – us UF Libraries [941]
Concise lives of famous iyases of benin / Egharevba, Jacob U – 2nd ed. [Lagos: s.n., 1947] – 1 – us CRL [920]
A concise statement of the law of partnership / Brown, Benjamin F – Indianapolis, 1871. 32 p. LL-1456 – 1 – us L of C Photodup [346]
A concise summary of the law of libel as it affects the press / Henderson, William Graham – Rutherford, NJ: Chemical Bank Note Co., 1915. 120p. LL-242 – 1 – us L of C Photodup [340]
Concise swahili and english dictionary / Perrott, Daisy Valerie – London, England. 1965 – 1r – us UF Libraries [040]
A concise treatise on the law of landlord and tenant adapted to the province of ontario : with an appendix of statutes and forms / Sinclair, James Shaw – Toronto: Goodwin & Wingfield, 1890 – 3mf – 9 – (incl ind) – mf#10728 – cn Canadiana [346]
A concise treatise on the law of wills / Theobald, Henry Studdy – London: Stevens & Sons; Toronto: Canada Law Book Co, 1908 – 15mf – 9 – 0-665-77092-8 – (1st publ london: stevens, 1876; with notes of canadian statutes and cases, by e d armour; incl app) – mf#77092 – cn Canadiana [340]
A concise treatise on the principles of equity pleading / Heard, Franklin Fiske – Boston: Boston Book Co., 1889. 217p. LL-535 – 1 – us L of C Photodup [340]
A concise treatise on the principles of equity pleading. / Heard, Franklin Fiske – Boston: Soule and Bugbee, 1882. 217p. LL-1300 – 1 – us L of C Photodup [340]
Concise xhosa-english dictionary / Mclaren, James – London, England. 1936 – 1r – us UF Libraries [040]

El conciso – Ano 1810 (Diciembre)-1814 (Mayo) – 132mf – 9 – sp Cultura [946]

El conciso – Cadiz, Spain. 1 apr-27 jun 1811 [wkly] – 32ft – 1 – uk British Libr Newspaper [074]

The conclave of clement 10 (1670) / Bildt, Carl Nils Daniel Bildt, Freiherr von – London: Publ for the British Academy by Henry Frowde, [1903?] – 1mf – 9 – 0-8370-7845-8 – mf#1986-1845 – us ATLA [920]

Conclave thesauri magnae artis musicae... / Vogt, Moritz Johann – Vetero-Pragae: typis Georgij Labaun 1719 [mf ed 19–] – 9mf – 9 – mf#fiche 907 – us Sibley [780]

Conclusion des observations d'anti-banque sur les banques du canada – [s.l: s.n, 1831?] [mf ed 1984] – 1mf – 9 – 0-665-21346-8 – mf#21346 – cn Canadiana [332]

Conclusiones, 1965 / Colombia Comision De Estudios Economicos Y Social – Bogota, Colombia. 1965 – 1r – us UF Libraries [972]

Conclusions / Berliner, Emile – New York:Kaufman, 1902 – 1mf – 9 – 0-8370-2292-4 – mf#1985-0292 – us ATLA [360]

Concord = Baptist associations. tennessee – 1866, 1900, 1957-79 – 1 – 58.32 – us Southern Baptist [242]

Concord – London. 1980-1980 (1) – ISSN: 0300-4384 – mf#10185 – us UMI ProQuest [320]

Concord baptist church – New Windsor. 1966-1979 (1) 1972-1979 (5) 1974-1979 (9) – 1r – 1 – $16.47 – mf#6663 – us Southern Baptist [242]

Concord baptist church. anderson county. south carolina : church records – 1963-72 – 1 – 6.17 – us Southern Baptist [242]

Concord baptist church. chattanooga, tennessee : church records – 1848-72 – 1 – 5.00 – us Southern Baptist [242]

Concord baptist church. madison, florida : church records – 1841-68 – 1 – 6.84 – us Southern Baptist [242]

Concord baptist church. rose hill association. duplin county. north carolina : church records – 1813-80 – 1 – us Southern Baptist [242]

Concord baptist church. st louis, missouri : church records – 1955-60 – 1 – us Southern Baptist [242]

[Concord-] diablo beacon – CA. 1949; 1951-1967 – 5r – 1 – $300.00 – mf#B03588 – us Library Micro [071]

Concord lectures on philosophy : comprising outlines of all the lectures at the concord summer school of philosophy in 1882, with an historical sketch / Harris, William Torrey et al – Cambridge, MA: Moses King, c1883 – 1mf – 9 – 0-7905-3785-0 – mf#1989-0278 – us ATLA [100]

Concord, New Hampshire. Concord Baptist Church *see* Records

The concord of ages : or, the individual and organic harmony of god and man / Beecher, Edward – New York: Derby & Jackson, 1860, c1859 [mf ed 1989] – 2mf – 9 – 0-7905-0905-9 – mf#1987-0905 – us ATLA [210]

[Concord-] transcript – CA. 1905-07; 1909-10; 1910-23; 1925-26; 1928-30; 1936-37; 1951-82 [daily] – 156r – 1 – $9360.00 – mf#B02135 – us Library Micro [071]

[Concord-] transcript shopper – CA. 1933-51 (broken file) – 6r – 1 – $360.00 – mf#B02136 – us Library Micro [071]

Concordance latine des pseudepigraphes d'ancien testament / ed by Denis, A-M – 1993 – (= ser Thesauri – supplementum (formae)) – 5mf+664p – 9 – €260.00 – 2-503-50343-8 – be Brepols [400]

A concordance of parallels : collected from bibles and commentaries... / Cruttwell, Clement – [s.l: s.n.] 1790 [mf ed 1992] – 6mf – 9 – 0-524-03879-1 – mf#1987-6492 – us ATLA [220]

Concordance of the armenian bible – Jerusalem, 1895 – €402.00 – ne Slangenburg [220]

Concordance to gregory of nyssa / Fabricus, Cajus – Goeteborg: Acta Universitatis Gothoburgensis 1989 – (= ser Studia graeca et latina gothoburgensia 50) – 31mf – 9 – mf#mf93 – sw Gothenburg University [450]

A concordance to the canonical books of the old and new testaments : to which are added a concordance to the books called the apocrypha; and a concordance to the psalter, contained in the book of common prayer – London: SPCK [1859?] [mf ed 2004] – 1r – 1 – 0-524-10480-8 – mf#b00696 – us ATLA [220]

A concordance to the greek testament : according to the texts of westcott and hort, tischendorf and the english revisers / ed by Moulton, William Fiddian & Geden, Alfred Shenington – New York: Scribner, 1897 [mf ed 1990] – 10mf – 9 – 0-8370-1910-9 – mf#1987-6297 – us ATLA [225]

Concordance to the septuagint : and the other greek versions of the old testament / Hatch, E & Redpath, H A – Oxford. v1-3. 1897 – 102mf – 8 – €195.00 – ne Slangenburg [221]

Concordances to the major writings of william hazlitt : keys to the work of england's first critical literary historian – [mf ed UMI] – 9 – (mf in envelopes bound within guides. completed concordances: round table; characters of shakespear's [sic] plays; table talk) – us UMI ProQuest [420]

Concordantiae augustinianae : labore davidis lenfant o.p. – Lutetiae Parisiorum. v1-2. 1656-1665 – (tomus primus 42mf tomus alter 41mf) – 8 – €159.00 – ne Slangenburg [241]

Concordantiae augustinianae / Lenfant, Dav – Lutetiae Parisiorum. v1-2. 1656-1665 – €158.00 – ne Slangenburg [240]

Concordantiae corani arabicae / Fluegel, Gustav – Editio stereotypa Caroli Tauchnitii. Lipsiae [Leipzig]: Sumtibus Ernesti Bredtii, 1898 – 1mf – 9 – 0-524-03115-0 – mf#1990-3168 – us ATLA [260]

Concordantiae iuris canonici cum legibus partitarum... / Jimenez, Sebastian – Madrid: Juan de la Cuesta, 1611. 2nd parte – 1 – sp Bibl Santa Ana [946]

Concordantiae librorum novi testamenti domini nostri jesu christi juxta vulgatam editionem : jussu sixti 5, pontificis max., recognitam / Legrand, C – Bruges: Ch Beyaert-Storie, 1889 – 2mf – 9 – 0-8370-1946-X – mf#1987-6333 – us ATLA [225]

Concordantiae veteris testamenti graecae : ebraeis vocibvs respondentes ... simvl enim et lexicon ebraicolatinum, ebraicograecum, graecoebraicum / Kircher, Konrad – Francofurti: Apud C. Marnium & heredes I. Aubrii, 1607. Chicago: Dep of Photodup, U of Chicago Lib, 1972 (2r); Evanston: American Theol Lib Assoc, 1984 (2r) – 1 – 0-8370-0099-8 – mf#1984-B289 – us ATLA [220]

Concordanz der deutschen national-literatur / ed by Berlepsch, Hermann Alexander von – Leipzig: A Lehmann 1859 [mf ed 1993] – 1r – 1 – (filmed with: das buch deutscher briefe / ed by walter heynen & other title) – mf#8503 – us Wisconsin U Libr [430]

Concordat between his holiness pope pius 7 and bonaparte – Dublin, Ireland. 1802 – 1r – us UF Libraries [240]

La concorde : journal historique, politique et litteraire. – Haiti. mai 1821-juil 1822 – 1 – fr ACRPP [073]

Concordia – Green Bay WI. 1875 aug 5-1877 aug 16, 1877 aug 23-1879 may 15, 1879 may 22-1880 dec 9, 1880 dec 16-1881 dec 1 – 4r – 1 – (cont: wisconsin staats-zeitung [green bay wi: 1874]; cont by: green bay courier) – mf#944351 – us WHS [071]

Concordia : eine deutsche kaisergeschichte aus baiern / Schmid, Herman – 2.aufl. Leipzig: Keil [18–?] – (= ser Gesammelte schriften. volks- und familien-ausgabe 42-46 nf10-14) – 5v in 2 – 1 – mf#film mas c438 – us Harvard [830]

Concordia – Paris. v. 3, 5-21. aug sept 1897, dec 1898-july 1914 – 1 – us NY Public [073]

Concordia 1849 – Berlin, Leipzig, Bremen DE, 1849-50 – 1 – gw Misc Inst [074]

Concordia baptist church. havana, florida : church records – 1858-1917 – 1 – us Southern Baptist [242]

Concordia discors et antichristus revelatus / Desmarets, S – Amsterdam, Janssonius, 1642. 2 v – 16mf – 9 – mf#PFA-139 – ne IDC [240]

The concordia eagle – Vidalia, LA: David Young, 1873-90// (wkly) [mf ed 1947] – (= ser Negro Newspapers on Microfilm) – 1r – 1 – us L of C Photodup [071]

Concordia historical institute quarterly – St. Louis. 1986+ (1,5,9) – ISSN: 0010-5260 – mf#15394 – us UMI ProQuest [240]

Concordia journal – St. Louis. 1989+ (1,5,9) – ISSN: 0145-7233 – mf#17531 – us UMI ProQuest [240]

Concordia Luthern Church, Hoisington, KS *see* Records

Concordia, oder, die bekenntnisschriften der evangelisch-lutherischen kirche – 11. Aufl. Basel: P Kober, 1898 – 2mf – 9 – 0-8370-8859-3 – (incl indes) – mf#1986-2859 – us ATLA [242]

Concordia pia : evangeliskt-lutherska kyrkans symboliska boecker – Rock Island IL: Lutheran Augustana Book Concern [1878?] [mf ed 1993] – 2mf – 9 – 0-524-06428-8 – (in swedish) – mf#1991-2550 – us ATLA [242]

Concordia theological monthly – St. Louis. 1930-1972 (1) 1971-1972 (5) – (cont by: ctm) – ISSN: 0010-5279 – mf#1536 – us UMI ProQuest [240]

Concordia theological quarterly – Fort Wayne. 1980+ (1,5,9) – ISSN: 0038-8610 – mf#12650,01 – us UMI ProQuest [240]

Concordiantae...glossematibusque gregorii lopez / Jimenez, Sebastian – 1611 – 9 – sp Bibl Santa Ana [440]

Concordienbuch : das ist, die symbolischen buecher der ev. luth. kirche – 2. Aufl. St Louis, MO: Lutherischer Concordia-Verlag, 1881 – 2mf – 9 – 0-524-06429-6 – mf#1991-2551 – us ATLA [242]

Der concordienformel, kern und stern : mit einer geschichtlichen einleitung und mit kurzen erklaerenden anmerkungen versehen / Walther, Carl Ferdinand Wilhelm – 3. Aufl. St Louis, MO: Lutherischer Concordia-Verlag, 1887 – 1mf – 9 – 0-524-07097-0 – (incl ind) – mf#1991-2920 – us ATLA [242]

Concours d'eloquence de 1876 : seance de la proclamation du laureat, 13 octobre, 1876 – Quebec?: A Cote, 1876 – 1mf – 9 – mf#35528 – cn Canadiana [910]

Les concours publics d'architecture – Paris, 1895-1914 – 4r – 1 – $525.00 – us UPA [720]

Concrete – Addison. 1956-1961 (1) – mf#1014 – us UMI ProQuest [690]

Concrete – London. 1967+ (1) 1972+ (5) 1974+ (9) – ISSN: 0010-5317 – mf#7257 – us UMI ProQuest [690]

Concrete and constructional engineering – London. 1950-1966 (1) – mf#538 – us UMI ProQuest [690]

Concrete construction – Addison. 1999+ (1,5,9) – (cont: aberdeen's concrete construction) – mf#6545,02 – us UMI ProQuest [690]

Concrete construction – Addison. 1956-1990 (1) 1972-1990 (5) 1973-1990 (9) – (cont by: aberdeen's concrete construction) – ISSN: 0010-5333 – mf#6545 – us UMI ProQuest [690]

Concrete products – Chicago. 1989-1996 (1) – ISSN: 0010-5368 – mf#12384,01 – us UMI ProQuest [690]

Concurrence – 1969 [complete] – Inquire – 1 – mf#ATLA S0884 – us ATLA [073]

Concurrence and dissent; some recent supreme court cases / Prenner, Manuel – New York: Merrill, 1933. 243p. LL-1037 – 1 – us L of C Photodup [347]

Concurrence, consommation et repression des fraudes – 1981- – 9 – €34.00 – (1941-80 eur121.96) – fr Journal Officiel [350]

Concurrency and computation : practice and experience – Chichester, 2001+ [1,5,9] – (cont: concurrency, practice and experience) – ISSN: 1532-0626 – mf#17044,01 – us UMI ProQuest [000]

Concurrency, practice and experience – Chichester. 1989-2000 (1,5,9) – ISSN: 1040-3108 – mf#17044 – us UMI ProQuest [000]

Die concurrenz fuer entwuerfe zum neuen reichstagsgebaeude / Eggert, H – Berlin, 1882 – 2mf – 9 – mf#O-242 – ne IDC [720]

Concurso hipico internacional (No oficial) *see* Programa general del concurso hipico internacional

Concurso literario de trujillo organizado por la comision de fiestas para solucionar la inauguracion del monumento a pizarro, levan tado...vda. de d. carlos rumsey / Trujillo – Trujillo: Tip. de Sobrino de Benito Pena, 1928 – sp Bibl Santa Ana [946]

Concurso regional de ganados que organiza la junta provincial de badajoz en representacion de la asociacion general de ganaderos del reino que se celebrara en don benito...en 1925 – Don Benito: Tip. de Trejo, 1925 – sp Bibl Santa Ana [946]

Concursos y premios para los estudiantes / Ciudad Trujillo Universidad De Santo Domingo – Ciudad Trujillo, Dominican Republic. 1941 – 1r – us UF Libraries [972]

Condado de la gomera / Regulo Perez, Juan & Siete Iglesias, Marques de – Madrid, 1955 – 1 – sp Bibl Santa Ana [946]

Condado news – Hialeah, FL. 1985 apr 25-1990 oct 01 – 1r – us UF Libraries [071]

La condanna del modernismo : appunti polemici / Deho, Ettore – Roma: Desclee, 1908 – 1mf – 9 – 0-8370-9772-X – (incl bibl ref) – mf#1986-3772 – us ATLA [240]

Conde Berron, Felix *see* Libro primero. metodo simultaneo de lectura y dibujo

Conde, Carmen *see* Acompanando a francisca sanchez resumen de una vida junto a ruben dario

Conde de Canilieros *see* Brozas. la encomienda mayor

El conde de canilleros / Valgoma y Diaz-Varela, Dalmiro de la – Madrid: imp y editorial maestre, 1972 – 1 – sp Bibl Santa Ana [946]

Conde de castralla / Lopez de Ayala, Adelardo – 1856 – 9 – sp Bibl Santa Ana [820]

El conde de gandomar y su intervencion en el proceso, prision y muerte de sir walter raleigh *see* Espanoles e ingleses en america durante el siglo 17. el conde de gandomar y su intervencion en el proceso, prision y muerte de sir walter raleigh

O conde de linhares / Funchal, Agostinho de Sousa Coutinho – Lisboa, Typ: Bayard, 1908 – 1 – us Wisconsin U Libr [946]

El conde de montecristo / Dumas, Alexandre – 1867 – 9 – (trans by vicente barrantes) – sp Bibl Santa Ana [830]

Conde d'eu / Cascudo, Luis Da Camara – Sao Paulo, Brazil. 1933 – 1r – us UF Libraries [972]

Conde dos arcos e a revolucao de 1817 / Pagano, Sebastiao – Sao Paulo, Brazil. 1938 – 1r – us UF Libraries [972]

Conde irlos *see* Corrido qng bierang [sic] delanan ning conde irlos ila ning condesang asauana qng cayarian francia

Conde, Jean de [Jehan de Condet] *see* Gedichte

Conde, Jose *see*
- Ramo para luisa
- Terra de caruaru

Conde, Louis 1 de Bourbon *see*
- Protestation de moseigneur le prince de conde
- Requeste presentee au roy par, monsieur le prince de conde: ac compagne d'un grand nombre de seigneurs gentils-hommes, & autres qui font profession de la religion reformee en ce royaume

Conde, Louis 1st Prince of Bourbon *see* Discours veritable des propos tenus par monsieur le prince de conde, auec les seigneurs deputez par le roy: contenant les causes gui ont contraint ledict seigneur prince & autres de sa copagnie a prendre les arms

Conde nast's traveler – New York. 1989+ (1,5,9) – ISSN: 0893-9683 – mf#11371,01 – us UMI ProQuest [910]

Conde, Prudencio *see* Etica general

Conde y Corral, Bernardo *see* Carta pastoral al inaugurar su pontificado

Condemned to devil's island / Niles, Blair – New York, NY. 1928 – 1r – us UF Libraries [972]

A condensed index to the minor liens filed in the office of the clerk of kings county, ny. / Sparrow, George – New York, Lawyers' Real Estate Agency, 1897. 254 1 p. LL-1350 – 1 – us L of C Photodup [348]

Condensed lectures on eschatology : or, the doctrine of final things / Myers, Tobias T – Mt Morris IL: [s.n. 1903?] [mf ed 1992] – (= ser The word for the worker series) – 1mf – 9 – 0-524-03623-3 – mf#1990-4783 – us ATLA [240]

Condensed proceedings of the southern chemurgic conference / Southern Chemurgic Conference (1936 : Lafayette, La) – Dearborn, MI. 1936 – 1r – us UF Libraries [630]

Condensed report of the proceedings of the... day's session / Progressive Mine Workers of America – 1st-6th [1940 aug 26-31] – 1r – 1 – mf#3314587 – us WHS [622]

Condensed reports of general executive officers...cio, national convention / United Cannery, Agricultural, Packing, and Allied Workers of America – 5th [1944] – 1r – 1 – mf#3153374 – us WHS [660]

Conder, C R *see*
- The bible and the east
- The hebrew tragedy
- Judas maccabaeus and the jewish war of independence
- The rise of man
- The survey of western palestine
- Tent work in palestine

Conder, Claude Reignier *see*
- The city of jerusalem
- A handbook to the bible
- Heth and moab
- The latin kingdom of jerusalem, 1099 to 1291 a.d
- The survey of eastern palestine
- Syrian stone-lore
- Tent work in palestine

Conder, Claude Reignier et al *see* The survey of western palestine

Conder, Francis Roubiliac *see* A handbook to the bible

Conder, Josiah *see*
- Africa
- Arabia
- Birmah, siam, and anam
- Brazil and buenos ayres
- Colombia
- Egypt, nubia, and abyssinia
- Greece
- India
- Italy
- Landscape gardening in japan
- Law of the sabbath
- Mexico and guatimala
- North america
- Palestine
- Persia and china
- Peru-chile
- Russia
- Spain and portugal
- Syria and asia minor

Condet, Jehan de *see* Gedichte

Condicion resolutoria tacita por incumplimiento / Santiso Galvez, Gustavo – Guatemala, 1946 – 1r – us UF Libraries [972]

Condicion social de la mujer en espana, la. su estado actual : su posible desarrollo / Nelken, Margarita – Barcelona: Editorial Minerva, S.A. – 1 – sp Bibl Santa Ana [946]

Condict, Alice Byram *see* Old glory and the gospel in the philippines

Condit, Blackford *see*
- The history of the english bible
- Short titles of familiar bible texts, mistranslated, misinterpreted, and misquoted

Condit, Ira M *see* The chinaman as we see him

Condit, Ira Miller *see*
- The chinaman as we see him
- The language, literature, religions, and evolution of china

Condition and extent of the natural oyster beds an... / Danglade, Ernest – Washington, DC. 1917 – 1r – us UF Libraries [639]

Condition and prospects of the greek or oriental church / Waddington, George – London, England. 1854 – 1r – us UF Libraries [240]

The condition and prospectus of architectural art / Beresford-Hope, Alexander James Beresford – London 1863 – (= ser 19th c art & architecture) – 1mf – 9 – mf#4.2.959 – uk Chadwyck [720]

Condition humaine – Dakar. fevr 1948-aout 1956 – 1 – fr ACRPP [073]

La condition internationale de l'egypte – Montauban, 1904 – 2mf – 9 – mf#ILM-1946 – ne IDC [960]

The condition of the english working class : from the british library of political and economic science, london – Papers of Rev. Henry Solly, 1813-1903. British Library of Political and Economic Science – 6r – 1 – us Primary [320]

The condition of the financial markets : hearing...united states senate, 107th congress, 1st session, on the condition of the financial markets and regulatory responses following the september 11 terrorist attacks, sep 20 2001 / United States. Congress. Senate. Committee on Banking, Housing, and Urban Affairs – Washington: US GPO 2002 [mf ed 2002] – 1mf – 9 – us GPO [332]

Conditional and future interests and illegal conditions and restraints in illinois / Kales, Albert Martin – Chicago,Callaghan, 1905. 453 p. LL-599 – 1 – us L of C Photodup [340]

Conditional immortality : a help to sceptics / Stokes, George Gabriel, Sir – London: J Nisbet, 1897 – 1mf – 9 – 0-7905-8592-8 – mf#1989-1817 – us ATLA [240]

Conditional immortality : a help to sceptics: a series of letters addressed...to james marchant / Stokes, George Gabriel – London, 1897 – (= ser 19th c evolution & creation) – 1mf – 9 – mf#1.1.4780 – uk Chadwyck [120]

Conditional immortality : plain sermons on a topic of present interest / Huntington, William Reed – New York: EP Dutton, 1878 – 1mf – 9 – 0-7905-3919-5 – mf#1989-0412 – us ATLA [240]

Conditional reflex – Philadelphia. 1966-1973 (1) 1971-1973 (5) (9) – (cont by: pavlovian journal of biological science) – ISSN: 0010-5392 – mf#6891 – us UMI ProQuest [150]

Conditions and politics in occupied western europe 1940-1945, pt 1 : 1940: belgium, france, norway, sweden, spain – 12r – 1 – mf#C39-27823 – us Primary [940]

Conditions and politics in occupied western europe 1940-1945, pt 2 : 1941: belgium, co-ordination files, denmark, france, general files, italy, netherlands, norway, spain, sweden, vatican – 27r – 1 – mf#C39-27824 – us Primary [940]

Conditions and politics in occupied western europe 1940-1945, pt 3 : 1942: belgium, co-ordination files, denmark, france, general files, italy, netherlands, norway, portugal, spain, sweden, switzerland, vatican – 24r – 1 – mf#C39-27825 – us Primary [940]

Conditions and politics in occupied western europe 1940-1945, pt 4 : 1943: belgium, co-ordination files, denmark, france, general files, italy, netherlands, norway, portugal, spain, sweden, switzerland, vatican – 32r – 1 – mf#C39-27826 – us Primary [940]

Conditions and politics in occupied western europe 1940-1945, pt 5a : 1944: belgium and luxembourg, denmark, economic and reconstruction files, france and general files – 28r – 1 – mf#C39-27827 – us Primary [940]

Conditions and politics in occupied western europe 1940-1945, pt 5b : 1944: italy, netherlands, norway, portugal, spain, sweden, switzerland, vatican – 20r – 1 – mf#C39-27828 – us Primary [940]

Conditions and politics in occupied western europe 1940-45 – 184r coll – 1 – (coll includes detailed information collected from papers received in the foreign office documenting conditions in europe during ww2. coll indexed by year and country) – mf#C39-27820 – us Primary [940]

Conditions and politics in occupied western europe 1940-45, pt 6a : 1945: belgium, denmark, european general files, france – 20r – 1 – mf#C39-27829 – us Primary [940]

Conditions and politics in occupied western europe 1940-45, pt 6b : 1945: italy, netherlands, norway, portugal, spain, sweden, switzerland, vatican – 21r – 1 – mf#C39-27830 – us Primary [940]

Les conditions de l'senseignement religieux dans les eglises nationales de la suisse romande : rapport / Dumont, Emile – Lausanne: Georges Bridel, 1898 – 1mf – 9 – 0-8370-7626-9 – mf#1986-1626 – us ATLA [377]

Les conditions du retour au catholicisme : enquaete philosophique et religieuse / Rifaux, Marcel – 3e ed. Paris: Librairie Plon, 1907 – 1mf – 9 – 0-8370-8856-9 – mf#1986-2856 – us ATLA [241]

The conditions of church life in the first six centuries : a paper / Stone, Darwell et al – London: SPCK 1905 [mf ed 1992] – (= ser Church historical society (series) 92) – 1mf – 9 – 0-524-05516-5 – (incl bibl ref) – mf#1990-1511 – us ATLA [240]

Conditions of life in the sea / Johnstone, James – Cambridge, MA. 1908 – 1r – us UF Libraries [574]

Conditions of obtaining salvation by jesus christ – London, England. 18– – 1r – us UF Libraries [240]

The conditions of our lord's life on earth : being five lectures...new york 1896 / Mason, Arthur James – New York: Longmans, Green 1896 [mf ed 1985] – (= ser The bishop paddock lectures 1896) – 1mf – 9 – 0-8370-4148-1 – (incl bibl ref) – mf#1985-2148 – us ATLA [240]

The conditions on which local societies will be received into membership with the "working men's club and institute union," : and the advantages to be thereby obtained / Working Men's Club and Institute Union – London, 1863 – (= ser 19th c economics) – 1mf – 9 – mf#1.1.264 – uk Chadwyck [331]

Conditions rurales en haiti / Dartigue, Maurice – Port-Au-Prince, Haiti. 1938 – 1r – us UF Libraries [972]

Condivi, A *see* Das leben des michelangelo buonarroti

Condobolin argus – Condobolin, jul 1896-oct 1900 – 2r – A$125.09 vesicular A$136.09 silver – at Pascoe [079]

Condon globe – Condon OR: S P Shutt, [wkly] [mf ed 1968-76] – 5r – 1 – (merged with: condon times [1905-19] to form: condon globe-times [1919-75]; ceased in 1919) – us Oregon Lib [071]

Condon globe *see* Condon globe-times

Condon globe-times – Condon OR: G H Flagg, 1919-75 [wkly] [mf ed 1963-75] – 18r – 1 – (merger of: condon globe [189?-1919], condon times [1905-19]; merged with: fossil journal [1886-1998] to form: times-journal [1975-]) – us Oregon Lib [071]

Condon globe-times *see*
- Condon globe
- Fossil journal
- Times-journal

Condon times – Condon OR: E Curran, 1905-19 [wkly] [mf ed 1968-77] – 4r – 1 – (merged with: condon globe [189?-1919] to form: condon globe-times [1919-75]; cont: weekly times [condon, or]) – us Oregon Lib [071]

Condon times *see*
- Condon globe
- Condon globe-times

Condor – Lawrence. 1900+ (1) 1900+ (5) 1900+ (9) – (cont: bulletin of the cooper ornithological club of california) – ISSN: 0010-5422 – mf#12347,01 – us UMI ProQuest [590]

Condor – Santiago de Chile (RCH), 1972- – 1 – gw Misc Inst [079]

O condor – Madureira, RJ. 20 maio 1908 – (= ser Ps 19) – mf#DIPER – bl Biblioteca [079]

O condor : revista litteraria – Rio de Janeiro, RJ: Typ Moraes, 08 jun-20 jul 1901 – (= ser Ps 19) – mf#DIPER – bl Biblioteca [440]

Condorcets "esquisse d'un tableau historique" und seine stellung in der geschichtsphilosophie / Niedlich, Joachim Kurd – [S.l.]: J K Niedlich, 1907 (Sorau N-L [Germany]: Rauert & Pittius) [mf ed 19–] – 1 – (incl bibl) – mf#*Z-72 – us NY Public [944]

The conduct of life : or, the universal order of confucius. a translation of one of the four confucian books hitherto known as the doctrine of the mean = Chung yung / Ku, Hung Ming – New York: E P Dutton 1912 [mf ed 1993] – (= ser Wisdom of the east series (new york, ny)) – 1mf – 9 – 0-524-07994-3 – mf#1991-0216 – us ATLA [180]

The conduct of life / Emerson, Ralph Waldo – Boston: Ticknor and Fields, 1860 – 1mf – 9 – 0-7905-7565-5 – mf#1989-0790 – us ATLA [170]

Conduct of monetary policy : report of the federal reserve board pursuant to section 2b of the federal reserve act, and the state of the economy: hearing...house of representatives, 107th congress, 2nd session, feb 27 2002 / United States. Congress. House. Committee on Financial Services – Washington: US GPO 2002 [mf ed 2002] – 2mf – 9 – us GPO [332]

The conduct of public worship / Greenhough, John Gersham – London: Baptist Union Publication Dept., [1901?] – 1mf – 9 – 0-7905-4797-X – mf#1988-0797 – us ATLA [210]

Conduct of the elder brother on account of the father's treatment o... / Wright, R – London, England. 1826 – 1r – us UF Libraries [240]

Conduct of the understanding / Locke, John – [s.l: s.n. 18–?] [mf ed 1986] – 1r – 1 – (filmed with: friends, society of / lower, t) – mf#1669 – us Wisconsin U Libr [120]

Conduct of the voir dire examination : practices and opinions of federal district judges / Bermant, Gordon – Washington: FJC, Sept 1977 – 1mf – 9 – $1.50 – mf#LLMC 95-812 – us LLMC [340]

Conductas antisociales / Colombia Laws, Statutes, Etc – Bogota, Colombia. 1964 – 1r – us UF Libraries [350]

Conducting job interviews : a guide for federal judges / Hendrickson, David K – 1999 – 1mf – 9 – $1.50 – mf#llmc99-016 – us LLMC [347]

Conductor and brakeman : the roadman's magazine / Order of Railway Conductors and Brakemen – 1965 jan-1967 dec 16, 1968 jan 1-1969 jan 25 – 2r – 1 – (cont: railway conductor) – mf#866040 – us WHS [380]

Condurier *see* La chine catholique

Co-ne bstan-'gyur – A.D. 1753-1773 repr 1928 – 209v on 2227mf – 9 – $1560.00 – (co-ne is located in the wu-t'ai-shan area of shansi, china. contains commentaries and other works transl primarily from sanskrit) – us IASWR [090]

Cone, Orello *see*
- The epistles to the hebrews, colossians, ephesians, and philemon, the pastoral epistles, the epistles of james, peter, and jude
- Gospel-criticism and historical christianity
- Rich and poor in the new testament
- Salvation

Cone, Spencer H *see* Circular letter

Cone, Spencer Houghton *see* The commonly received version of the new testament of our lord and savior jesus christ

Conejo Valley Genealogical Society *see* Rabbit tracks

Conemaugh country / Southwest Pennsylvania Genealogical Services – v1 n1-v1 n4 [1981 jun-1982 mar] – 1r – 1 – mf#626149 – us WHS [929]

Coneross baptist church. oconee county. south carolina : church records – West Union included. 1833-91, 1907-20, 1934-45 – 1 – us Southern Baptist [242]

Conexion de yuste con los puntos historicos de la provincia de caceres / Pablos Abril, Juan – (Plasencia): Imprenta La Victoria S.A., 1976? – sp Bibl Santa Ana [946]

Coney, John *see* Engravings of ancient cathedrals, hotels de ville, and other public buildings of celebrity, in france, holland, germany, and italy

Con-fab / Convalescent Facility of Percy Jones G&C Hospital [Fort Custer MI] – v1 n15 [1945 feb 2] – 1r – 1 – mf#3394943 – us WHS [360]

A confederacao artistica : orgao das classes operarias – Belem, PA, 20 set 1888 – (= ser Ps 19) – bl Biblioteca [079]

Confederacao do trabalho : orgam das classes laboriosas – Manaus, AM: Typ da Confederacao do Trabalho, 14 nov-25 dez 1909 – (= ser Ps 19) – mf#P11B,06,18 – bl Biblioteca [079]

Confederacao Nacional da Industria Departamento E *see* Analise do intercambio comercial, brasil-reino-uni

Confederacao ou separacao / Ellis Junior, Alfredo – Sao Paulo, Brazil. 1934 – 1r – us UF Libraries [972]

La confederacion de las clases. el programa de un nuevo partido / Cascales Munoz, Jose – 1894 – 9 – sp Bibl Santa Ana [360]

Confederacion Nacional de Trabajo. Spain *see*
- La obra de los trabajadores unidos
- Los sucesos de barcelona: relacion documental de las tragicas jornadas de la semana de mayo de 1937
- Los sucesos de mayo en barcelona. relato autentico
- La verdad sobre la tragedia de casas viejas, por el comite nacionale.

Confederate baptist – South Carolina. 1862-65 – 1 – us Southern Baptist [242]

Confederate blockade running through bermuda / Vandiver, Frank Everson – Austin, TX. 1947 – 1r – us UF Libraries [972]

Confederate court records, eastern district of north carolina, 1861-1864 / U.S. District Court – (= ser Records of district courts of the united states) – 1r – 1 – (with printed guide) – mf#M1430 – us Nat Archives [347]

Confederate imprints – 144r – 1 – (coll contains 6188 publ of the confederate states of america. coll is based on marjorie lyle crandall's confederate imprints: a check list, and richard b. harwell's more confederate imprints. with printed guide) – mf#C39-23000 – us Primary [976]

Confederate Memorial Literary Society, Richmond *see* Catalogue of the florida department of the confede...

Confederate military manuscripts / ed by Glatthaar, Joseph T – 4ser – (= ser Civil war research colls) – 1 – (ser a: holdings of the virginia historical society 42r isbn 1-55655-632-2$8130. ser b: holdings of louisiana state university 22r isbn 1-55655-659-4 $4270. ser c: holdings of the center for american history, university of texas at austin, pt1: the trans-mississippi west 22r isbn 1-55655-714-0 $4245. ser d: holdings of the university of virginia library, pt1: albemarle county historical soc papers – sergeant h b johnston confederate furlough papers 17r isbn 1-55655-775-2 $3290. with p/g) – us UPA [976]

Confederate newspapers – 1861-65 – 10r – 1 – (coll is a mixture of issues and papers from florida, georgia, tennessee, virginia and alabama. titles are also listed separately) – mf#D3490P01 – us Western Res [071]

Confederate papers of the u.s. district court for the eastern district of north carolina, 1861-1865 / U.S. District Court – (= ser Records of district courts of the united states) – 1r – 1 – mf#M436 – us Nat Archives [348]

Confederate papers relating to citizens or business firms / U.S. War Dept. Confederate Records – (= ser War department coll of confederate records) – 1158r – 5 – (with printed guide) – mf#M346 – us Nat Archives [324]

Confederate records of the state of georgia / Candler, Allen D – Atlanta. 6v. 1909 – 1 – $162.00 – (v5 never publ) – mf#0140 – us Brook [976]

Confederate states army casualties : lists and narrative reports, 1861-1865 / U.S. War Dept. Confederate Records – (= ser War department coll of confederate records) – 7r – 1 – (with printed guide) – mf#M836 – us Nat Archives [355]

Confederate states medical and surgical journal / New York Academy of Medicine – 1864 mar – 1r – 1 – mf#1117914 – us WHS [610]

Confederate States of America *see*
- Presidential messages and papers of the confederacy
- Records

The confederate states of america and border states – (= ser Civil war unit histories: regimental histories and personal narratives 1; Civil war research colls) – 1937mf – 9 – $18,585.00 – 1-55655-216-5 – (alabama 84mf $1075. arkansas 13mf $165. florida 16mf $210. georgia 150mf $1915. kentucky 155mf $1970. louisiana 94mf $1210. maryland 83mf $1060. mississippi 55mf $720. missouri 133mf $1705. north carolina 142mf $1820. south carolina 123mf $1575. tennessee 126mf. $1595. texas 92mf $1165. virginia 333mf $4270. higher & independent commands & naval forces 338mf $4330. p/g only isbn 1-55655-257-2 $270) – us UPA [976]

Confederate States of America. Army. Dept of South Carolina, Georgia, and Florida *see* Records, 1863

Confederate States of America. Secretary of the Treasury *see*
- Letters received by the confederate secretary of the treasury, 1861-1865
- Letters sent by the confederate secretary of the treasury, 1861, 1864-1865

Confederate States of America. War Dept *see* Records of the cotton bureau of the trans-mississippi department, 1862-1865

Confederate veteran / Confederated Southern Memorial Association [US] et al – v1, n3-4,7-12 [1893], v2 n3-5 [1894] – 1r – 1 – mf#1217453 – us WHS [305]

Confederate veteran – v. 1-40. 1893-1932 – 1 – us L of C Photodup [976]

Confederate war journal – 1893 apr-1895 mar – 1r – 1 – mf#695436 – us WHS [976]

Confederated Salish and Kootenai Tribes of the Flathead Reservation *see*
- Char-Koosta
- Char-koosta news

Confederated Southern Memorial Association [US] et al *see* Confederate veteran

Confederated Tribes of Siletz *see* Siletz news

Confederated Tribes of the Colville Reservation *see*
- Bulletin: a monthly newsletter
- Tribal tribune

Confederated Tribes of the Umatilla Reservation in Oregon *see* Confederated umatilla journal

Confederated Tribes of the Warm Springs Reservation of Oregon *see* Spilyay tymoo

Confederated umatilla journal / Confederated Tribes of the Umatilla Reservation in Oregon – [1976 jan-1980 1 may] scattered iss – 1r – 1 – mf#626876 – us WHS [307]

Csa manuscripts, 1861-1865 – [mf ed 1981] [Spartanburg SC: Reprint Co, dist] – 32mf – 9 – mf#51-033 – us South Carolina Historical [976]

Confederation : a letter to the right honourable the earl of carnarvon, principal secretary of state for the colonies / Annand, William – London: E Stanford, 1866 – 1mf – 9 – mf#33327 – cn Canadiana [971]

La confederation : couronnement de dix annees de mauvaise administration / Lusignan, Alphonse – Montreal?: s.n, 1867 – 1mf – 9 – mf#23425 – cn Canadiana [971]

Confederation des loisirs du Quebec *see* Loisir-plus

Confederation des syndicats nationaux *see* Memoire commun sur le projet de loi 45 presente

Confederation des syndicats nationaux. Quebec *see* Les accidents du travail

Confederation Generale du Travail *see*
– L'atelier pour le plan
– Le peuple

Confederation Generale du Travail. Congres *see* Compte rendu des debats

Confederation Generale du Travail Unitaire *see* La vie syndicale

Confederation, independence, annexion : conference faite a l'institut canadien de quebec, le 15 mars 1871 / Fabre, Hector – [Quebec?: s.n.], 1871 [mf ed 1981] – 1mf – 9 – mf#23672 – cn Canadiana [971]

Confederation of Indians of Quebec *see* Indians of quebec

Confederation of Iranian Students [US] *see* Struggle

Conference africaine sur la tse-tse et la trypanomiase (1948 brazzaville) *see* [Compte-rendu]

La conference africaine-francaise, brazzaville, 1944 – Alger: Commissariat aux colonies, 1944 – 1 – us CRL [960]

Conference avec m claude...sur la matiere de l'eglise / Bossuet, J-B – Paris, 1682 – 6mf – 9 – mf#CA-116 – ne IDC [241]

Conference between two men that had doubts about infant-baptism / Wall, William – London, England. 1795 – 1r – us UF Libraries [242]

Conference bulletin / National Conference on Social Welfare – Washington. 1873-1984 [1]; 1973-1984 [5,9] – mf#1193 – us UMI ProQuest [301]

Conference coloniale (1917 paris) *see* [Compte-rendu] instituee

Conference de chretiens evangeliques de toute nation a paris, 1855 / ed by Monod, Guillaume – Paris: Ch Meyrueis, 1856 [mf ed 1992] – 2mf – 9 – 0-524-03762-0 – mf#1990-1109 – us ATLA [240]

Conference de l'honorable monsieur l s beaubien : prononce a l'assomption lors de la reunion annuelle des membres de la societe laitiere de la province de quebec... – Montreal?: s.n, 1889? – 1mf – 9 – mf#55879 – cn Canadiana [630]

Conference de Quebec (1864) *see* Resolutions relatives a l'union proposee des provinces de l'amerique britannique du nord

Conference de son altesse royale samdech preah norodom sihanouk, chef de l'etat du cambodge a la faculte de droit de l'universite de paris [29 juin 1964 / Norodom Sihanouk, Prince – Phnom-Penh: Impr du Ministere de l'information 1964] [mf ed 1989] – 1r with other items – 1 – mf#mf-10289 seam reel 015/30 [§] – us CRL [959]

Conference donnee a la cathedrale d'ottawa par le r p damen...1871 : la bible ne suffit pas pour enseigner [1]es verites necessaires au salut – [Ottawa?: Impr du Canada], 1880 – 1mf – 9 – 0-665-90899-7 – mf#90899 – cn Canadiana [210]

Conference et lettres de p. savorgnan de brazza sur ses trois explorations dans l'ouest africain de 1875 a 1886 / Ney, Napoleon – Paris: M Dreyfous, 1887 – 1 – us CRL [960]

Conference Group in Women's History *see*
– Ccwhp newsletter
– Newsletter

Conference On Health Education Of The Methodist Church *see* Zvekudya zvakanaka

Conference in the interest of physical training, boston, 1889 / ed by Barrows, Isabel C – 1899 – 3mf – 9 – $9.00 – us Kinesology [790]

Conference. minutes... / Wesleyan methodist church – London, Bristol, 1791-1850 – 88mf – 9 – (missing:1812,1815,1819,1836,1840,1844) – mf#MP-345 – ne IDC [242]

Conference newsletter = Hui hsun – v3-4. 1949-50; v5 jan 1951* – (= ser Chinese christian coll 57) – 1r – 1 – (in chinese) – mf#ATLA S0296M – us ATLA [073]

Conference Of Anglo-Jewish Ministers *see* Proceedings

Conference of Bar Association Delegates. Committee on Cooperation of the Press and the Bar *see* Report

Conference of Missionary Societies in Great Britain and Ireland *see* Bibliography of african christian literature

Conference of missionary societies in great britain and ireland. reports and minutes of the annual conference / handbook – 1912-78 [mf ed 2001] – (= ser Christianity's encounter with world religions, 1850-1950) – 5r – 1 – (filmed with: british council of churches. conference for world mission. handbook) – mf#2001-s148-151/152 – us ATLA [240]

Conference of representatives of commissions in the united states appointed to arrange for a world conference on faith and order : hotel astor, new york, thursday, may 8th, 1913 – [S.l.: s.n., 1913?] (New York: Chas P Young) – 1mf – 9 – 0-524-02793-5 – mf#1990-0697 – us ATLA [240]

Conference of Southern Mountain Workers *see* Mountain life and work

Conference of the foreign missions board in the united states and canada *see* Interdenominational conference of foreign missionary boards

Conference of the officers and representatives of foreign mission board and societies in the united states and canada *see* Interdenominational conference of foreign missionary boards

Conference on African Land Tenure in East and Central Africa (1956: Arusha, Tanganyika) *see* Report on the conference on african land tenure in east and central africa [in] arusha, feb 1956

Conference on christian union : (1845 : liverpool, england) – London, England. 1845 – 1r – us UF Libraries [240]

Conference on disarmament, 1985-1989 : meetings and documents – 20r – 1 – $3475.00 – 1-55655-224-6 – (with p/g) – us UPA [327]

Conference On Economic Coordination In The Caribbe... *see* Official records

Conference On Education And Economic And Social De... *see* Cuba y la conferencia de educacion y desarrollo ec

Conference On Education And Small Scale Farming *see* Small scale farming in the caribbean

Conference on florida everglades reclamation, july – Baltimore, MD. 1927 – 1r – us UF Libraries [630]

Conference on Foreign Missions...London, 1886 *see* Proceedings...mildmay park, london, october 5th to 7th 1886

Conference on great lakes research proceedings / International Association for Great Lakes Research – Ann Arbor. 1953-1974 (1) 1972-1974 (5) (9) – ISSN: 0045-8058 – mf#6335 – us UMI ProQuest [574]

Conference on missions held in 1860 at liverpool : including the papers read, the deliberations, and the conclusions reached / Mullens, Joseph et al – Rev. London: James Nisbet, 1860 – 2mf – 9 – 0-8370-6097-4 – (incl ind) – mf#1986-0097 – us ATLA [240]

Conference on Russia, (1922 : Hague, Netherlands) *see* Gaagskaia konferentsiia, iiun'-iiul' 1922 g

Conference on tax planning for 501(c)(3) organizations (new york university) : proceedings – v1-27. 1953-99 – 1,5,6 – $380.00 set – (v1-22 1953-94 in reel $275. v23-27 1995-99 in fiche $105) – mf#101981 – us Hein [343]

Conference On The Caribbean (12th : 1961) *see* Caribbean

Conference On The Caribbean (13th : 1962) *see* Caribbean

Conference on the copyright question – Ottawa: S E Dawson, 1896 – 1mf – 9 – 0-665-93913-2 – mf#93913 – cn Canadiana [346]

Conference On The Teaching Of African Languages In Schools *see* African languages in school

Conference papers / African Literature Association – [Chicago: The Association, 1976-[1976-1986] (annual) – 11r – 1 – us CRL [490]

Conference papers, or, analyses of discourses, doctrinal and practical = Selections. 1879 / Hodge, Charles – New York: C Scribner, c1879 – 1mf – 9 – 0-7905-7302-4 – mf#1989-0527 – us ATLA [240]

Conference Pleniere Des Ordinaires Des Missions Du Congo Belge *see* Deuxieme conference pleniere des ordinaires des missions

Conference proceedings / National Conference on Law and Poverty. Washington, DC. 1965 – Washington: GPO, 1966 200p. LL-2281 – 1 – us L of C Photodup [340]

Conference proceedings / Nigerian Institute of Social and Economic Research – Ibadan, 6th-8th. 1958-1962 – us CRL [330]

Conference proceedings / West African Institute of Social and Economic Research. Ibadan – v1-5. 1952-56 – 1 – us CRL [960]

Conference sur chenier / Ethier, Joseph Arthur Calixte – [Montreal?: s.n, 1905? – 1mf – 9 – 0-665-75121-4 – mf#75121 – cn Canadiana [971]

Conference sur chenier / Ethier, Joseph Arthur Calixte – [Montreal?: s.n, 1905? [mf ed 1995] – 1mf – 9 – 0-665-75121-4 – mf#75121 – cn Canadiana [971]

Conference sur haiti / Bowler, Arthur – Paris, France. 1888 – 1r – us UF Libraries [972]

Conference sur haiti / Justin, Joseph – Paris, France. 1894 – 1r – us UF Libraries [972]

Conference sur la litterature canadienne / Lesage, Jules Simeon – [Quebec: s.n.] 1901 [mf ed 1998] – 1mf – 9 – 0-665-97449-3 – mf#97449 – cn Canadiana [840]

Conference sur le porc et l'industrie laitiere / Chapais, Jean Charles – Montreal: Herald Pub Co, 1899 – 1mf – 9 – mf#02101 – cn Canadiana [630]

Conference syndicale mondiale de solidarite avec les travailleurs et les peuples d'indochine en lutte contre l'agression des etats unis : rapport de thiounn mumm / Thiounn, Mumm – [n.p. 1970?] [mf ed 1989] – 1r with other items – 1 – mf#mf-10289 seam reel 016/03 [§] – us CRL [959]

Conference with certain ministers and elders of the church of scotland / Chalmers, Thomas – Glasgow, Scotland. 1837 – 1r – us UF Libraries [240]

Conferences : la presse canadienne-francaise et les ameliorations de quebec / Buies, Arthur – Quebec?: s.n, 1875 [mf ed 1981] – 1mf – 9 – mf#24047 – cn Canadiana [070]

Conferences : la presse canadienne-francaise et les ameliorations de quebec, 20 septembre 1875 / Buies, Arthur – Quebec: Typographie de C Darveau, 1875 [mf ed 1979] – 2mf – 9 – mf#SEM105P25 – cn Bibl Nat [070]

Conferences : la presse canadienne-francaise et les ameliorations de quebec / Buies, Arthur – [Quebec?: s.n.], 1875 [mf ed 1980] – 1mf – 9 – 0-665-02436-3 – mf#02436 – cn Canadiana [070]

Conferences apologetiques *see* Lectures in defence of the christian faith

Conferences de notre-dame de paris : conferences delivered at notre dame in paris = God and man / Lacordaire, Henri-Dominique – New York: Scribner, Welford and Armstrong, 1872 – 1mf – 9 – 0-8370-4028-0 – (in english) – mf#1985-2028 – us ATLA [240]

Conferences de notre-dame de paris *see* Jesus christ

Conferences de notre-dame de quebec / Holmes, Jean – Quebec: A Cote, 1850 [mf ed 1980 – 2mf – 9 – 0-665-45119-9 – mf#45119 – cn Canadiana [241]

Conferences de saint-etienne : (ecole pratique d'etudes bibliques) 1909-1910 / Dhorme, Edouard et al – Paris: V Lecoffre 1910 [mf ed 1993] – (= ser Etudes palestiniennes et orientales) – 1mf – 9 – 0-524-06199-8 – mf#1992-0837 – us ATLA [933]

Conferences de saint-etienne : ecole pratique d'etudes bibliques, 1910-1911 / Lagrange, Marie-Joseph et al – Paris: Victor Lecoffre 1911 [mf ed 1993] – (= ser Etudes palestiniennes et orientales) – 1mf – 9 – 0-524-06336-2 – mf#1992-0874 – us ATLA [933]

Conferences du rev pere damen, sj – Quebec?: A Cote, 1891? – 1mf – 9 – (trans fr english by r p gladu) – mf#56015 – cn Canadiana [241]

Conferences economiques nationales – 2. ed. [Conakry]: Imprimerie National "Patrice Lumumba" 1973 – us CRL [330]

Conferences et discours / Bourassa, Gustave – Montreal: C Beauchemin, 1899 – 4mf – 9 – mf#00200 – cn Canadiana [241]

Conferences historiques : dessalines devant l'histo... / Vaval, Duracine – Port-Au-Prince, Haiti. 1906 – 1r – us UF Libraries [972]

Conferences of the rev. pere lacordaire / Lacordaire, Henri-Dominique – New York: P O'Shea, c1870 – 2mf – 9 – 0-8370-7557-2 – mf#1986-1557 – us ATLA [240]

Conferences on palestine : united kingdom – jewish agency delegation. secretary's notes of the ninth meeting held at st james palace – London, 1939 – 1mf – 9 – mf#J-28-86 – ne IDC [956]

Conferences on the spiritual life / Ravignan, Pere de – London: R Washbourne, 1873 – 1mf – 9 – 0-8370-7417-7 – mf#1986-1417 – us ATLA [240]

Conferences sur la question ouvriere : donnees a l'eglise saint-saveur de quebec / Gohiet, Francois – Quebec: Leclerc & Roy, 1892 – 3mf – 9 – (pref by joseph jules fillatre) – mf#03467 – cn Canadiana [331]

Conferences sur l'instruction obligatoire : faites au cercle catholique de quebec / Paquin, Louis Philibert – Quebec: J A Langlais, 1881 – 2mf – 9 – mf#29927 – cn Canadiana [377]

Conferencia : semana agricola de badajoz / Fernandez Santana, Ezequiel – Badajoz: Imp. de Vicente Rguez., 1912 – 1 – sp Bibl Santa Ana [630]

Conferencia das Organizacoes Nacionalistas das Colonias Portuguesas (CONCP) *see* Boletim de informacao

Conferencia de las naciones unidas sobre alimentac... / United Nations Conference On Food And Agriculture – Habana, Cuba. 1943 – 1r – us UF Libraries [630]

La conferencia de lausana sobre fe y disciplina / ed by Bayle, Constantino – Madrid: Razon y Fe, 1927 – 1 – sp Bibl Santa Ana [210]

Conferencia de ministros de hacienda, seccion asun... / Argentina – Montevideo, Uruguay. 1939 – 1r – us UF Libraries [630]

Conferencia De Organismos De Fomento De La Producc... *see* Memoria

La conferencia nacional de juventudes / Serrano Poncela, Segundo – Guia de divulgacion y de trabajo. Valencia, 1937. Fiche W1159. (Blodgett Collection of Spanish Civil War Pamphlets) – 9 – us Harvard [946]

Conferencia recital sobre el romancero gitano : manuscrito / Garcia Lorca, Federico – 1mf – 9 – sp Cultura [850]

Conferencia sobre el seguro de maternidad dedicada especialmente a patronos agricolas / Leal Ramos, Leon – Madrid: Imp. y Enc. de los sobrinos de la sucesora de M. Minussa de los Rios, 2nd ed 1932 – sp Bibl Santa Ana [946]

Conferencias de historia habanera – Habana, Cuba. v1-4. 1937- – 1r – us UF Libraries [972]

Conferencias del niagara falls / Guerrero Yoacham, Cristian – Santiago, Chile. 1966 – 1r – us UF Libraries [972]

Conferencias del shoreham (el cesarismo en cuba) / Marquez Sterling, Manuel – Mexico City? Mexico. 1933 – 1r – us UF Libraries [972]

Conferencias dictadas en la universidad de puerto – San Juan, Puerto Rico. 1949 – 1r – us UF Libraries [378]

Conferencias militares / Leon Gutierrez, Florencio – 1894 – 9 – sp Bibl Santa Ana [355]

Conferencias no prata / Rego, Jose Lins Do – Rio de Janeiro, Brazil. 1946 – 1r – us UF Libraries [972]

Conferencias pronunciadas en la liga contra la blasfemia de oliva de jerez / Serrano Serrano, Ildefonso – Segura de Leon: Imp. de Ntra. Sra. de Gracia, 1924 – sp Bibl Santa Ana [240]

Conferencias sobre estetica y literatura / Tapia Y Rivera, Alejandro – San Juan, Puerto Rico. 1945 – 1r – us UF Libraries [972]

Conferencias sobre la plurotonia / Huertas y Barrero, Francisco – 1885 – 9 – sp Bibl Santa Ana [946]

Conferencias teosoficas en america del sur / Roso de Luna, Mario – Madrid: Gregorio Pueyo, editor, S.A. – 1 – sp Bibl Santa Ana [240]

Conferencias teosoficas en america del sur / Roso de Luna, Mario – Madrid: Libreria de Pueyo, Tomo 2, 1911 – 1 – sp Bibl Santa Ana [290]

Conferenze religiose e sociale *see* Die wahrheit

Confesion de fe de la iglesia espanola reformada : aprobada por la asamblea general de sevilla en el ano de 1869 = Confesion de fe – Sevilla: Hijos de Fe, [1869?] – 1mf – 9 – 0-8370-8750-3 – mf#1986-2750 – us ATLA [240]

Confessio augustana *see* The unaltered augsburg confesssion

Confessio et apologia pastorum / [Amsdorff, N von] – Magdeburg, 1550 – 1mf – 9 – mf#TH-1 mf 16 – ne IDC [242]

Confessio et expositio simplex orthodoxae fidei / Bullinger, Heinrich – Tigvri, Christoph Froschouer, 1566 – 2mf – 9 – mf#PBU-226 – ne IDC [240]

Confessio fidei de evcharistiae sacramento / Westphal, J aus Hamburg – Magdeburgae, 1557 – 4mf – 9 – mf#TH-1 mf 1476-1479 – ne IDC [242]

Confessio joannis...lasco : de nostra cum christo domino communione – [Emden, 1554] – 1mf – 9 – mf#PBA-219 – ne IDC [240]

Confession / Peppin, S F B – Wells, England. 1873 – 1r – us UF Libraries [240]

La confession / Normand, Victor – Paris: F. Rieder, 1926. 172p. (Christianisme) – 1 – us Wisconsin U Libr [240]

Confession, absolution, and holy communion / Denison, George Anthony – Oxford, England. 1873? – 1r – us UF Libraries [240]

Confession and absolution / Neale, J M – London, England. 1854 – 1r – us UF Libraries [240]

Confession and absolution / Phillpotts, Henry – London, England. 1852 – 1r – us UF Libraries [240]

Confession as taught by the church of england / Gray, C N – Manchester, England. 18– – 1r – us UF Libraries [241]

La confession aux laiques dans l'eglise latine depuis le ville / Teetaert, Amedee – Wetteren: J de Meester et fils; Paris: J Gabalda, [etc., etc.], 1926 – 1 – us Wisconsin U Libr [240]

Confession, covenants, and secession testimony, vindicated and defe... / Thomson, George – Glasgow, Scotland. 1799 – 1r – us UF Libraries [240]

Confession de foy : ...auec vne remonstrance aux magistrats, de flandres, braban... / Bres, G de – [Lyon, S Barbier pour J Frellon], 1561 – 1mf – 9 – mf#PBA-150 – ne IDC [240]

Confession de foy : faicte d'vn commun accord par les fideles qui coeuersent... / Bres, G de – [Rouen, Abel Clemence], 1561 – 1mf – 9 – mf#PBA-149 – ne IDC [240]

Confession de foy... / Bres, G de – n.p, 1562 – 1mf – 9 – mf#PBA-433 – ne IDC [240]

Confession de foy... / Bres, G de – n.p, 1562 – 1mf – 9 – mf#PBA-434 – ne IDC [240]

Confession de la foy chrestienne / Beza, Theodor de – [Geneve], J Crespin, 1561 – 5mf – 9 – mf#PFA-102 – ne IDC [240]

Confession de la foy chrestienne... / Beza, Theodor de – [Geneve], Badius, 1559 – 4mf – 9 – mf#PFA-101 – ne IDC [240]

La confession de la sentinelle *see* La sentinelle du peuple

Confession et simple exposition de la vraye foy / Bullinger, Heinrich – Geneve, Francois Perrin, pour Iean Durant, 1566 – 4mf – 9 – mf#PBU-228 – ne IDC [240]

Confession in the church of rome / Morin, Andre Saturnin – London, England. 18– – 1r – us UF Libraries [240]

A confession of faith – Charleston, SC. 153p. 1774 – 1 – $5.35 – us Southern Baptist [242]

Confession of faith – Edinburgh, Scotland. 18– – 1r – us UF Libraries [240]

Confession of faith – Edinburgh, Scotland. 1884 – 1r – us UF Libraries [240]

Confession of faith adopted by the baptist association met at philadelphia, a, 25 sep 1742 / Baptist Association – 1 – 5.00 – us Southern Baptist [242]

The confession of faith with the scripture proofs – Montreal: W Drysdale, 187-? – 1mf – 9 – (int pref by robert campbell; incl bibl ref) – mf#08657 – cn Canadiana [242]

A confession of fayth / Bullinger, Heinrich – London, Henry Wykes, for Lucas Harrison, [1568] – 4mf – 9 – mf#PBU-229 – ne IDC [240]

Confession of the patriarchs, that they were strangers and pilgrims / Crombie, John – London, England. 1829 – 1r – us UF Libraries [240]

Confession sociale, commentaire critique tant de la declaration faite au peuple – Paris, 1848 – us CRL [300]

Confession vnd bekentnis johanns agricole eisslebens vom gesetze gottes / Johannes Agricola aus Eisleben – Berlin, 1541 – 1mf – 9 – mf#TH-1 mf 802 – ne IDC [242]

The confessional : an appeal to the primitive and catholic forms in the east and in the west / Seymour, Michael Hobart – London: Seeley, Jackson, & Halliday, 1870 – 1mf – 9 – 0-524-04022-2 – mf#1990-1194 – us ATLA [241]

The confessional history of the lutheran church / Richard, James William – Philadelphia, Pa: Lutheran Publication Society, c1909 – 2mf – 9 – 0-8370-8706-6 – (incl bibl ref and ind) – mf#1986-2706 – us ATLA [242]

The confessional principle and the confessions of the lutheran church : as embodying the evangelical confession of the christian church / Schmauk, Theodore Emanuel & Benze, Charles Theodore – Philadelphia: General Council Publication Board, 1911 – 3mf – 9 – 0-7905-8876-5 – (incl bibl ref) – mf#1989-2101 – us ATLA [242]

Confessional revision : being a collection of 395 articles that have appeared in the religious press between september, 1887 and october, 1890, on the subject of revising the westminster confession of faith – Pittsburg: s.n., 1890 – 1r – 1 – 0-8370-1293-7 – mf#1984-B369 – us ATLA [240]

Confessionario breue, en lengua mexicana y castellana / Molina, Alonso de – en Mexico: En casa de Antonio de Espinosa, imp[re]ssor 1565 – (= ser Books on religion...1543/44-c1800: confesionarios) – 20lea on 1mf – 9 – mf#crl-17 – ne IDC [241]

Confessionario copioso en lengua espanola, y tangela : para direccion de los confessores, y instruccions de los penitentes. (ano 1713) / Gaspar de San Augustin – 2. impr. Sampaloc: Impr de N S de Loreto del pueblo 1787 – (= ser Whsb) – 2mf – 9 – €30.00 – mf#Hu 323a – gw Fischer [490]

Confessionario en lengua mexicana y castellana : con muchas aduertencias muy necessarias para los confessores / Baptista, Juan – en Sanctiago Tlatilulco: Por Melchior Ocharte, ano de 1599 – (= ser Books on religion...1543/44-c1800: confesionarios) – 3mf – 9 – mf#crl-20 – ne IDC [241]

Confessionario mayor, instruction y doctrina, para el que se quiere bien confessar / Molina, Alonso de – En la muy insigne y gran ciudad de Mexico: En casa de Antonio de Espinosa impressor de libros...ano de 1565 – (= ser Books on religion...1543/44-c1800: confesionarios) – 3mf – 9 – mf#crl-18 – ne IDC [241]

Confessionario mayor y menor en la lengua mexicana : y platicas contra las supresticiones [sic] de idolatria, que el dia de oy an quedado a los naturales desta nueua espana, e instrucion de los santos sacramentos &c... / Alva, Bartolome de – en Mexico: Por Francisco Salbago...Por Pedro de Quinones, ano de 1634 – (= ser Books on religion...1543/44-c1800: confesionarios) – 2mf – 9 – mf#crl-22 – ne IDC [241]

Confessionario para los curas de indios : con la instrucion contra sus ritos, y exhortacion para ayudar a bien morir, y summa de sus priuilegios, y forma de impedimentos del matrimonio / Catholic Church. Province of Peru. Concilio Provincial – en la Ciudad de los Reyes: Por Antonio Ricardo...ano de 1585 – (= ser Books on religion...1543/44-c1800: confesionarios) – 2mf – 9 – mf#crl-19 – ne IDC [241]

La confessione elvetica... / Bullinger, Heinrich – Coira, 1777 – 2mf – 9 – mf#PBU-688 – ne IDC [240]

La confessione vocale de'peccati : praticata nella sinagoga antica e innalzata a sacramento da ges u cristo nella chiesa cristiana / Vincenzi, Luigi – Roma: Tipografia Paterno, 1850 – 1mf – 9 – 0-524-00659-8 – mf#1990-0159 – us ATLA [240]

Die confessionelle entwicklung der altprotestantischen kirche deutschlands : die altprotestantische union und die gegenwaertige confessionelle lage und aufgabe des deutschen protestantismus / Heppe, Heinrich – Marburg: N.G. Elwert, 1854 – 1mf – 9 – 0-7905-5998-6 – (incl bibl ref) – mf#1988-1998 – us ATLA [242]

Confessiones / Augustinus, St – 1983 – (= ser ILL – ser a; Cccm 27) – 8mf+82p – 9 – €30.00 – 2-503-60272-X – be Brepols [400]

Confessions / Balbaith, M J A – Toulouse, France. 1837 – 1r – us UF Libraries [240]

Les confessions de theoroigne de mericourt / Ravelsberg, Ferdinand Strobl von – Paris, 1892 – (= ser Les femmes [coll]) – 3mf – 9 – mf#7266 – fr Bibl Nationale [360]

Confessions of a medium : with five illustrations – London: Griffith & Farran; New York: E P Dutton & Co [1882] [mf ed 1986] – 1r [ill] – 1 – (with: der flohhaz von johann fischart und mathiss holtzwart... / koch, p) – mf#1677 – us Wisconsin U Libr [130]

The confessions of augustine = Confessiones / Augustine, Saint, Bishop of Hippo; ed by Gibb, John & Montgomery, William – Cambridge: University Press, 1908 – (= ser Cambridge Patristic Texts) – 2mf – 9 – 0-524-08332-0 – (incl ind) – mf#1993-2022 – us ATLA [240]

The confessions of augustine = Confessiones / Augustine, Saint, Bishop of Hippo; ed by Shedd, William Greenough Thayer – Andover: Warren F Draper, 1860 – 2mf – 9 – 0-8370-9841-6 – (in english) – mf#1986-3841 – us ATLA [240]

Confessions of faith : and other public documents, illustrative of the history of the baptist churches of england in the 17th century / ed by Underhill, Edward B – Hanserd Knolleys Society – 1 reel – 1 – $15.28 – (382p) – us Southern Baptist [242]

Confessions of faith and formulas of subscription : in the reformed churches of great britain and ireland especially in the church of scotland / Cooper, James – Glasgow: James Maclehose, 1907 – 1mf – 9 – 0-8370-8804-6 – (incl bibl ref) – mf#1986-2804 – us ATLA [240]

Confessions of faith, and other public documents illustrative of the history of the baptist churches of england in the 17th century / Underhill, Edward Bean – London: J. Haddon, 1846-1854 – 1r – 1 – 0-8370-1695-9 – mf#1984-6072 – us ATLA [242]

Confessions of st augustine / Augustine – New York, NY. 1943 – 1r – us UF Libraries [240]

The confessions of st augustine / Augustine, Saint, Bishop of Hippo – New York: New American Library, 1963 [mf ed 2004] – (= ser A mentor-omega book) – 1r – 1 – 0-524-10495-6 – (new trans by rex warner) – mf#b00710 – us ATLA [240]

The confessions of the church of scotland : their evolution in history / McCrie, Charles Greig – Edinburgh: Macniven & Wallace, 1907 – (= ser Chalmers Lectures) – 1mf – 9 – 0-7905-5301-5 – (incl bibl ref) – mf#1988-1301 – us ATLA [242]

Confessionschrifft: etlicher predicanten in den herrschafften / Musaeus, S – np, 1567 – 4mf – 9 – mf#TH-1 mf 1196-1199 – ne IDC [242]

Confessiun da la vera cardienscha... / Bullinger, Heinrich – Cuera, 1776 – 2mf – 9 – mf#PBU-687 – ne IDC [240]

Confessoins ie, confessions of an infidel / P, Phillip – London, England. no date – 1r – us UF Libraries [240]

Confessonario breve activo, y passivo, en lengua mexicana : con el qual los que comienzan [sabiendolo bien de memoria] parece, que qualquiera estara suficiente mientras aprende mas / Saavedra, Marcos de – en Mexico: impr Dona Maria de Rivera, en la Empedradillo, ano de 1746 – (= ser Books on religion...1543/44-c1800: confesionarios) – 1mf – 9 – mf#crl-25 – ne IDC [241]

Confessore, R J *see* Quantification factors describing physical activity involvement and their relationsship to current criterion reference standards for aerobic capacity in children and youth

Confessors of florence, 1852-1853 – London, England. 1853? – 1r – us UF Libraries [240]

Confidence of the church : a letter to sir henry w moncreiff, bart., d.d / Innes, Alexander Taylor – Edinburgh: John Maclaren, [1881] Princeton: Speer Lib, and Dep of Photodup, U of Chicago Lib, 1978 (1r); Evanston: American Theol Lib Assoc, 1984 (1r) – (= ser Case of william robertson smith in the free church of scotland) – 1 – 0-8370-0631-7 – mf#1984-6269 – us ATLA [242]

Confident / Scribe, Eugene – Paris, France. 1828 – 1r – us UF Libraries [440]

Confident par hasard / Faur, Louis Francois – Paris, France. 1801 – 1r – us UF Libraries [440]

Confidential and unofficial letters sent by the secretary of war, 1814-1847 / U.S. War Dept. Office of the Secretary – (= ser Records of the office of the secretary of war) – 2r – 1 – (with printed guide) – mf#M7 – us Nat Archives [324]

Confidential bulletin / Kappa Alpha Psi Fraternity – 1986 jun-nov, 1989 fall, 1990 may, nov, 1991 sep, 1993 apr+suppl – 1r – 1 – mf#4862729 – us WHS [378]

Confidential bulletin / Wisconsin Newspaper Association – 1965 jan 7-1969 dec 10, 1970 jan 7-oct 20, 1971 oct-1975 jun, 1979-80, 1981-82, 1983-1985 apr, 1985 may-1987 mar 11 – 7r – 1 – mf#1278894 – us WHS [070]

Confidential circular to agents : i herewith send you prospectuses, containing particulars of my great premium sale of engravings – [s.l: s.n. 1871?] [mf ed 1987] – 1mf – 9 – 0-665-28855-7 – mf#28855 – cn Canadiana [760]

The confidential file of the johnson white house, 1963-1969 – 2pt – (= ser Research colls in american politics) – 1 – (pt1: confidential subject & name files 79r isbn 1-55655-668-3 $15,275. pt2: confidential reports file 24r isbn 1-55655-669-1 $4640. with p/g in prep) – us UPA [977]

Confidential files of the eisenhower white house : pt 1: subject files – 25r – 1 – $4840.00 – 1-55655-959-3 – (with p/g) – us UPA [324]

Confidential intelligence report / Herald of Freedom – 1973 jan-1980 oct – 1r – 1 – mf#665399 – us WHS [071]

Confidential publications and home political files / India. National Archives – Reports, etc. on Indian publications, ca. 1900-45 – 1 – us CRL [954]

Confidential publications and home political files / India. National Archives – Two major series and two publications. ca. 1920-45 – 1 – us CRL [954]

Confidential report on the northern rhodesia police : pts 3, 4 and 9 / Dowbiggin, Herbert – [s.l: s.n.], 1937 (Lusaka Govt Printer) – 1 – (filmed with: report on the northern rhodesia police, pts 1-2, 5-9 & summary by herbert dowbiggin) – us CRL [960]

Confinado en las hurdes (una victima de la inquisicion republicana) / Albinana, Jose Maria – Madrid: Imp. El Financiero, 1933 – 1 – sp Bibl Santa Ana [946]

Confinia neurologica = Borderlands of neurology – Basel. 1967-1973 (1) 1970-1973 (5) 1970-1973 (9) – ISSN: 0010-5678 – mf#2051 – us UMI ProQuest [616]

Confinia psychiatrica – Basel. 1966-1973 (1) 1970-1972 (5) – ISSN: 0010-5686 – mf#2052 – us UMI ProQuest [616]

Confirmatio doctrinae orthodoxae contra genebardi scriptum / Daneau, Lambert – Geneve, E Vignon, 1585 – 1mf – 9 – mf#PFA-129 – ne IDC [240]

Confirmation / Bickersteth, H V – London: Longmans, Green 1916 [mf ed 1992] – 2mf – 9 – 0-524-05564-5 – mf#1991-2298 – us ATLA [242]

Confirmation / Hall, Arthur Crawshay Alliston – London, New York: Longmans, Green [1900?] [mf ed 1989] – (= ser The oxford library of practical theology) – 1mf – 9 – 0-7905-0495-2 – (incl bibl ref & ind) – mf#1987-0495 – us ATLA [242]

Confirmation / Holland, C – London, England. 18– – 1r – us UF Libraries [240]

Confirmation : its authority and benefits plainly stated – London, England. 1842 – 1r – us UF Libraries [240]

Confirmation – London, England. 18– – 1r – us UF Libraries [240]

Confirmation : or the laying on of hands, catechetically explained / Blunt, Walter – London, England. 1846 – 1r – us UF Libraries [240]

Confirmation : or, what is your motive? – London, England. 18– – 1r – us UF Libraries [240]

Die confirmation / Kliefoth, Theodor – Schwerin: Stiller, 1856 – 1mf – 9 – 0-524-04216-0 – mf#1990-5007 – us ATLA [242]

La confirmation de la discipline ecclèsiastique / Chandieu, A – [Geneve, H Estienne], 1566 – 3mf – 9 – mf#PFA-124 – ne IDC [240]

Confirmation in the apostolic age / Chase, Frederic Henry – London; New York: Macmillan, 1909 – 1mf – 9 – 0-7905-1579-2 – (incl ind) – mf#1987-1579 – us ATLA [240]

Confirmation lectures delivered to a village congregation in the di... / Pott, Alfred – London, England. 1860 – 1r – us UF Libraries [240]

The confirmation rubric and christian fellowship / Hodges, George – New York City: Prayer Book Papers Joint Cttee 1915 [mf ed 1992] – (= ser Prayer book papers 14) – 1mf – 9 – 0-524-03009-X – mf#1990-4531 – us ATLA [242]

Confirmation sermon : preached in the church of s michael / Gibbs, Joseph – Clifton Hampden? England. 1864? – 1r – us UF Libraries [240]

A confirmatory factor analysis of the carolina sport confidence inventory / Mink, Randi S – University of North Carolina at Chapel Hill, 1995 – 2mf – 9 – $8.00 – mf#PSY1855 – us Kinesology [150]

Confissoes de minas / Andrade, Carlos Drummond De – Rio de Janeiro, Brazil. 1944 – 1r – us UF Libraries [972]

Confitebor a 5 : voix avec des instruments / Pergolesi, Giovanni Battista – Roma: Gli Amici della Musica da Camera 1942 [mf ed 19–] – 1r – 1 – mf#pres. film 31, 62 – us Sibley [780]

Confitemini et confitebor : motets a grand choeur / Lalande, M R – [170-?] [mf ed 19–] – 1r – 1 – (confitemini domino [psalm 104] composed 1705 [paris: philidor 1705]; confitebor tibi domine in consilio [psalm 110] composed 1699 [paris: boivin 1729]) – mf#film 589, pres. film 62 – us Sibley [780]

Confitures de ma tante / Goby, Emile – Paris, France. 1863 – 1r – us UF Libraries [440]

Conflict – New York. 1978-1991 (1) 1978-1991 (5) 1978-1991 (9) – ISSN: 0149-5941 – mf#11688 – us UMI ProQuest [327]

Conflict and consensus in british industrial relations, 1916-1948 – 51r – 1 – us Primary [331]

Conflict and solidarity in a guianese plantation / Jayawardena, Chandra – London, England. 1963 – 1r – us UF Libraries [306]

Conflict and victory / Cochrane, William S – Cincinnati: Jennings and Graham, c1907 – 3mf – 9 – 0-659-90021-1 – mf#9-90021 – cn Canadiana [230]

The conflict between despotism and liberty, or the right of trial by jury abolished in civil suits, and sustained by the courts of the state of new york. – Rochester, Curtis, Butts, 1860. 15 p. LL-15 – 1 – us L of C Photodup [347]

The conflict between paganism and christianity in the fourth century / Momigliano, A – Oxford, 1963 – €12.00 – ne Slangenburg [230]

Conflict in indo-china : a reader on the widening war in laos and cambodia / ed by Gettleman; Marvin et al – 1st ed. New York: Random House [1970] [mf ed 1989] – 1r with other items – 1 – (with bibl) – mf#mf-10289 seam reel 020/11 [§] – us CRL [959]

Conflict in spain 1920-1937 / Gooden, G M – London, 1938; Burgos: Razon y Fe, 1938 – 1 – sp Bibl Santa Ana [946]

Conflict management : a case study of selected secondary schools in the kwamhlanga district of mpumalanga department of education / Seloane, Rasani Aubrey – Pretoria: Vista University 2001 [mf ed 2001] – 2mf – 9 – (incl bibl ref) – mf#mfm15201 – sa Unisa [303]

The conflict of ages : or, the great debate on the moral relations of god and man / Beecher, Edward – 3rd ed. Boston: Phillips, Sampson; London: Sampson Low, 1853 [mf ed 1989] – 2mf – 9 – 0-7905-0906-7 – mf#1987-0906 – us ATLA [240]

The conflict of ages : or, the great debate on the moral relations of god and man / Beecher, Edward – 7th ed. Boston: Phillips, Sampson, 1855. c1853 [mf ed 1987] – xii/552p – 1 – mf#9954 – us Wisconsin U Libr [210]

The conflict of christianity with heathenism = Kampf des christenthums mit dem heidenthum / Uhlhorn, Gerhard; ed by Smyth, Egbert C & Ropes, CJH – New York: Scribner, 1879 – 2mf – 9 – 0-524-01139-7 – (in english) – mf#1990-0353 – us ATLA [240]
The conflict of duties and other essays / Gardner, Alice – London: T. Fisher Unwin, 1903 – 1mf – 9 – 0-7905-4646-9 – mf#1988-0646 – us ATLA [200]
The conflict of good and evil in our day : twelve letters to a missionary / Maurice, Frederick Denison – London: Smith, Elder, 1865 – 1mf – 9 – 0-524-00062-X – mf#1989-2762 – us ATLA [210]
The conflict of ideals in the church of england / Little, William John Knox – London: Isaac Pitman, 1905 – 1mf – 9 – 0-7905-8831-5 – mf#1989-2056 – us ATLA [241]
Conflict of races in south africa / Aiyar, P S – Durban: African Chronicle Print Works, [1946?] – 1 – us CRL [960]
The conflict of severus patriarch of antioch by athanasius / Goodspeed, E J & Crum, W E – 3mf – 9 – mf#H-2821 mf48-50 – ne IDC [243]
The conflict of truth / Capron, Frederick Hugh – 3rd ed. Cincinnati: Jennings & Graham 1903 [mf ed 1991] – 2mf – 9 – 0-7905-7806-9 – mf#1989-1031 – us ATLA [210]
Conflict resolution quarterly – San Francisco, 2001+ [1,5,9] – (cont: mediation quarterly) – ISSN: 1536-5581 – mf#17617,01 – us UMI ProQuest [150]
El conflicto de espana ante el mundo cristiano republica, monarquia, fascismo y justicia de una causa / Thompson, Emmanuel – San Jose, Costa Rica, 1937 – (= ser Blodgett coll) – 9 – mf#fiche w1225 – us Harvard [946]
Conflicto religiosa de 1926 / Moctezuma, Aquiles P – Mexico City? Mexico. 1929 – 1r – us UF Libraries [303]
Conflictos / Salazar, Ramon A – Guatemala, 1898 – 1r – us UF Libraries [303]
Los conflictos del proletariado. el movimiento social contemporaneo: por que, cuando y como ha nacido el problema obrero / Cascales Munoz, Jose – Madrid: Imp. Alrededor del mundo, 1912 – 1 – sp Bibl Santa Ana [946]
Conflictos familiares y problemas humanos / Calle Restrepo, Arturo – Madrid, Spain. 1964 – 1r – us UF Libraries [303]
The conflicts of the age – New York: Charles Scribner, 1881 – 1mf – 9 – 0-8370-2725-X – mf#1985-0725 – us ATLA [240]
Conflits de lois relatifs aux successions ab intestat en tunisie / Slama, R – Paris, 1935 – 2mf – 9 – mf#ILM-2963 – ne IDC [956]
Confluences – Lyon, Paris. juil 1941-47 – 1 – fr ACRPP [073]
Confluencia – Albuquerque. 1978-1980 (1,5,9) – mf#11316 – us UMI ProQuest [900]
The conformation of the material by the spiritual : to imperfection by the spirit of error, to perfection and beauty by the spirit of truth: christian idealism / Thomas, W Cave – London: Strangeways and Walden, 1862 – 1mf – 9 – 0-8370-7432-0 – mf#1986-1432 – us ATLA [210]
Conformite de la conduite de l'eglise de france...avec celle de l'eglise d'affrique... / [Dubois-Goibaud, P] – Paris, 1685 – 4mf – 9 – mf#CA-127 – ne IDC [240]
La confrerie musulmane de saidi mohammed ben ali es-senoausai et son domaine geographique : en l'annee 1300 de l'hegire = 1883 de notre ere / Duveyrier, Henri – Paris: Societe de geographie, 1886 – 1mf – 9 – 0-524-01434-5 – (incl bibl ref) – mf#1990-2429 – us ATLA [260]
La confrerie musulmane de sidi mohammed ben 'ali es senousi et son domaine geograohique en l'annee 1300 de l'hegire : 1883 de notre ere / Duveyrier, Henri – Paris: Societe de geographie, 1886 – 1 – us CRL [260]
Les confreries musulmanes / Petit, Louis – Paris: B Bloud, 1902 – (= ser Science et Religion) – 1mf – 9 – 0-524-02357-3 – mf#1990-2968 – us ATLA [260]
Les confreries religieuses musulmanes / Depont, O & Coppolani, X – Alger, 1897 – 8mf – 9 – mf#NE-303 – ne IDC [956]
Les confreries religieuses musulmanes / Depont, Octave & Coppolani, Xavier – Alger: A Jourdan, 1897 – 2mf – 9 – 0-524-04979-3 – (incl bibl ref) – mf#1990-3437 – us ATLA [260]
Confrontation – Greenvale. 1972+ (1) 1972+ (5) 1973+ (9) – ISSN: 0010-5716 – mf#7795 – us UMI ProQuest [810]
Confrontation / Heath, Alan – v1 n1-7 [1970 nov-1971 may], v2 n1 [1971 sep], v3 n1,2,3 [1972 sum, sep, dec] – 1r – 1 – mf#690240 – us WHS [071]
Confronti – v16-20. 1989-93 [complete] – 3r – 1 – (cont: nuovi tempi) – mf#atla s0635 – us ATLA [240]
Confucian analects / ed by Kamei, Nanmei – [Osaka, 1880] [mf ed 1995] – (= ser Yale coll) – 10v – 1 – 0-524-10258-9 – (in chinese) – mf#1996-1258 – us ATLA [290]
Confucian analects : the great learning, and the doctrine of the mean = Lun yue / Confucius – Hongkong: J Legge 1861 [mf ed 1993] – (= ser The chinese classics 1) – 6mf – 9 – 0-524-08186-7 – (in english & chinese; trans by james legge) – mf#1991-0299 – us ATLA [180]
Confucian cosmogony : a translation of section forty-nine of the complete works of the philosopher choo-foo-tze / Chu, Hsi – Shanghai: American Presbyterian Mission Press, 1874 – 1mf – 9 – 0-524-08041-0 – mf#1991-0257 – us ATLA [180]
Confucianism and its rivals : lectures...london, oct-dec 1914 / Giles, Herbert Allen – London: Williams & Norgate 1915 [mf ed 1990] – (= ser Hibbert lectures (london, england) 1914) – 1mf – 9 – 0-7905-7166-8 – mf#1988-3166 – us ATLA [290]
Confucianism and taouism : with a map / Douglas, Robert K – London: SPCK; New York: E & J B Young 1889 [mf ed 1991] – (= ser Non-christian religious systems) – 1mf [ill] – 9 – 0-524-00875-2 – (incl bibl ref) – mf#1990-2098 – us ATLA [290]
Confucianism in relation to christianity : a paper / Legge, James – Shanghai: Kelly and Walsh, 1877 – 1mf – 9 – 0-524-01839-1 – mf#1990-2674 – us ATLA [230]
Confucianism in relation to christianity : a paper read before the missionary conference in shanghai / Legge, James – Shanghai: Kelly & Walsh; London: Truebner & Co, 1877 – (= ser 19th c books on china) – 1mf – 9 – mf#7.1.22 – uk Chadwyck [230]
Confucius / Stuebe, Rudolf – Tuebingen: J C B Mohr 1913 [mf ed 1992] – (= ser Religionsgeschichtliche volksbuecher fuer die deutsche christliche gegenwart 3/15) – 1mf – 9 – 0-524-01990-8 – (also filmed & available on 35mm – enquire) – mf#1990-2781 – us ATLA [290]
Confucius *see*
- The analects of confucius
- Confucian analects
- The discourses and sayings of confucius
- The ethics of confucius
- The sayings of confucius
- Tchrouen tsriou et tso tchouan

Confucius and confucianism : four lectures [james long lectures] / Walshe, W Gilbert – Shanghai: Kelly & Walsh 1911 [mf ed 1993] – 1mf – 9 – 0-524-08066-6 – mf#1991-0282 – us ATLA [180]
Confucius and mencius / Wu, Ting-fang – Philadelphia: S Burns Weston, 1901 [mf ed 1992] – (= ser Ethical addresses 8/2) – 1mf – 9 – 0-524-02682-3 – mf#1990-3112 – us ATLA [180]
Confucius and new china : confucius' idea of the state and its relation to the constitutional government = Staatsidee des konfuzius und ihre beziehung zur konstitutionellen verfassung / Wang, Ching-tao – Shanghai: Printed at the Commercial Press, 1912 – 1mf – 9 – 0-524-02675-0 – (incl bibl ref. in english) – mf#1990-3105 – us ATLA [320]
Confucius and the chinese classics : or, readings in chinese literature / ed by Loomis, Augustus Ward – San Francisco: A Roman, 1867 – 5mf – 9 – 0-524-08830-6 – mf#1993-4022 – us ATLA [480]
Confucius and the chinese classics : or, readings in chinese literature / Loomis, Augustus Ward [comp] – San Francisco, New York: A Roman & Co 1867 [mf ed 1985] – 1r – 1 – (selections fr legge's trans of the four books, & fr various other sources. with: records of later life / kemble, f) – mf#1302 – us Wisconsin U Libr [480]
Confucius, der weise china's / Haug, Martin – Berlin: C Habel 1880 [mf ed 1991] – (= ser Sammlung gemeinverstaendlicher wissenschaftlicher vortraege 15/338) – 1mf – 9 – 0-524-01604-6 – mf#1990-2543 – us ATLA [180]
Confucius und seine lehre / Dvorak, Rudolf – Muenster i W: Aschendorff 1895 [mf ed 1992] – (= ser Darstellungen aus dem gebiete der nichtchristlichen religionsgeschichte 12) – 1mf – 9 – 0-524-02298-4 – (incl bibl ref) – mf#1990-2921 – us ATLA [290]
Confucius und seine lehre / Gabelentz, Georg von der – Leipzig: FA Brockhaus, 1888 – 1mf – 9 – 0-524-01357-8 – mf#1990-2369 – us ATLA [290]
A confutation of monstrous and horrible heresies : taught by h.n. and embraced of a number, who call themselves the family of love / Knewstub, J – London: Thomas Dawson, 1579 – 3mf – 9 – mf#PW-50 – ne IDC [240]
Confvsion de la secte de mvhamed / Andres, J – Paris, 1573. – 3mf – 9 – mf#H-8343 – ne IDC [956]
Confvtatio brevis libri, sub alieno nomine editi, de controversia inter theologos vvittebergenses / Huber, S – np, 1595 – 1mf – 9 – mf#TH-1 mf 715 – ne IDC [242]
Cong bao : bulletin officiel en langue indigene / Indochina. French – Bac-ky Bao ho, Quoc nqu Cong bao, Saigon. 1936-40 – 1 – fr ACRPP [073]
Cong bao viet-nam : Journal officiel / Vietnam – juin 1948-50 – 1 – fr ACRPP [073]
Cong giao dong thinh – Saigon: Impr A Portail n583-627 [jan 7-dec 1932], n772-860 [jan 3 1936-sep 17 1937] – 1r – 1 – mf#mf-12542 seam – us CRL [079]
Cong of Micronesia. Joint Committee on Future Status *see* Proceedings of the 6th round of future political status negotiations
Congar, Y M J *see* The catholic church and the race question
Congaree baptist church. richland county. south carolina : church records – 1849-1953 – 1 – 47.84 – us Southern Baptist [242]
Cong-bao – n3-16. Hanoi. 18 juil-2 aout 1930 – 1 – fr ACRPP [073]
Cong-bao viet-nam cong-hoa / Vietnam – An-ban Quoc-hoi; (Ha-Ngi-Vien). Saigon. May 26-July 22, 1968, Jan.-Aug. 29, 1969 – 1 – us NY Public [959]
Cong-bao viet-nam cong-hoa / Vietnam – Oct 1955-April 1975 – 1 – us L of C Photodup [959]
Congdon, Frederick Tennyson *see* A digest of the nova scotia common law, equity, vice-admiralty and election reports
Conger, Arthur Bloomfield *see* Religion for the time
Cong-giao dong-thin – Saigon. 16 sept 1927-nov 1937 – 1 – fr ACRPP [073]
Congleton chronicle – [NW England] Congleton 1893, 1912-90 – 1 – uk MLA; uk Newsplan [072]
Congleton, Henry Brooke Parnell, 1st Baron *see* A history of the penal laws against the irish catholics
Congleton rd (rg 12/2843-2848) – (= ser 1891 census surname/location indexes [cheshire]) – 2mf – 9 – £4.00 – mf#58 – uk CheshireFHS [929]
Congleton, st peter – (= ser Cheshire monumental inscriptions) – 1mf – 9 – £2.50 – mf#12 – uk CheshireFHS [929]
Congleton, st stephen (& grave register) – (= ser Cheshire monumental inscriptions) – 3mf – 9 – £4.50 – mf#274 – uk CheshireFHS [929]
Cong-luan bao : (xuat ban thu'o'ng ngay) – Saigon. 15 mars 1917-18-oct 1939 – 1 – fr ACRPP [073]
Congo : internal affairs and foreign affairs, 1960-jan 1963 / U.S. State Dept – (= ser Confidential u s state department central files) – 40r – 1 – $7745.00 – 1-55655-809-0 – (with p/g) – us UPA [327]
Congo – Leopoldville: [P Kanza, aug 27-sep 2 1960] – (issues filmed as pt of: r e bartlett collection of african newspapers) – us CRL [079]
Congo / Latouche, John – New York, NY. 1945 – 1r – us UF Libraries [960]
Congo – Leopoldville: [P Kanza, apr 13-20, jun 1,15, jul 13-20 1957; nov 21-dec 26 1959; jan 2-jul 2, aug 3-sep 13 1960 – us CRL [079]
Congo / Merlier, Michel – Paris, France. 1962 – 1r – us UF Libraries [960]
The congo / Campbell, Henry D – New York: Christian & Missionary Alliance, [19-?] [mf ed 1992] – (= ser Missionary series; Christian and missionary alliance coll) – 1mf – 9 – 0-524-03693-4 – mf#1990-4798 – us ATLA [240]
The congo : a report of the commission of enquiry...; a complete and accurate translation / Congo Free State. Commission to Investigate the State Territories – New York and London: G.P. Putnam's Sons, 1906. iii,171p. On cover: Questions of the Day – 1 – us Wisconsin U Libr [324]
The congo and the founding of its free state : a story of work and exploration / Stanley, Henry Morton – London: S Low, Marston, Searle, & Rivington, 1885 – 3mf – 9 – 0-7905-7081-5 – mf#1988-3081 – us ATLA [916]
Congo belge / Bertrand, Jean – Bruxelles, Belgium. 1909 – 1r – us UF Libraries [960]
Congo belge / Steenackers, E M – Bruxelles, Belgium. 1909 – 1r – us UF Libraries [960]
Congo belge, 1944 / Belgian Congo Service De L'information – Leopoldville, Congo. 1944 – 1r – us UF Libraries [960]
Congo belge en images – Bruxelles, Belgium. 191-? – 1r – us UF Libraries [960]
Congo belge et la weltpolitik, 1894-1914 / Willequet, Jacques – Bruxelles, Belgium. 1962 – 1r – us UF Libraries [327]
Le congo belge illustre : ou, l'etat independent du congo (afrique centrale) sous la souverainete de s m leopold 2... / Alexis, M G – 2. augm. ed. Liege: H Dessain, 1888 – 1 – us CRL [960]
Congo. Belgian *see*
- Bulletin administratif
- Bulletin officiel
- La charte coloniale.

Congo. (Brazzaville) *see* Journal officiel
Congo crisis / Bayly, Joseph – Grand Rapids, MI. 1966 – 1r – us UF Libraries [960]
Congo (Democratic Republic) *see* From leopoldville to lagos
Congo Democratic Republic *see* Moniteur congolaise de la republique democratique du congo
Congo et afrique noire : bulletin hebdomadaire interafricain "belga n1, 323, 25-96 – aug 30/sep 6 1958-jun 21/30 1960 – us CRL [960]
Congo et angola : h / Daye, Pierre – Bruxelles: Renaissance du livre, 1929 (mf ed 1977) – 3mf – 9 – (incl bibl footnotes) – mf#Sc Micro F-72 – us NY Public [960]
The congo for christ : the story of the congo mission / Myers, John Brown – New York: Fleming H Revell, [1895?] – 1mf – 9 – 0-8370-6229-2 – (incl ind) – mf#1986-0229 – us ATLA [240]
The congo for christ : the story of the congo mission / Myers, John Brown – New York: Fleming H Revell, [pref. 1895] – 1r – 1 – 0-8370-0462-4 – (incl ind) – mf#1984-B216 – us ATLA [240]
Congo (formerly French Congo). Service National de la Statistique, des Etudes Demographiques et Economiques *see* Annuaire statistique 1958-1963, 1969
Le congo francais du gabon a brazzaville / Guiral, Leon – Paris: Plon, Nourrit, 1889 – 1 – us CRL [960]
Congo Free State. Commission to Investigate the State Territories *see* The congo
Congo. Free State. Laws, Statutes, etc *see*
- Code penal
- Codes congolais et lois usuelles en vigueur au congo, collationnes d'apres les textes officiels et annotes
- Les codes du congo, suivis des decrets, ordonnances et arretes complementaires

Congo. (French) *see* Journal officiel des possessions du congo francais
Congo independant – Stanleyville: Philppe Elebe, mar 19 1960 – us CRL [079]
The congo independent state : a report on a voyage of inquiry / Mountmorres, William Geoffrey Bouchard de Montmorency – London: Williams & Norgate, 1906 – 1 – us CRL [960]
Congo kitabu / Hallet, Jean Pierre – New York, NY. 1965 – 1r – us UF Libraries [960]
Congo mission archives: the papers of bishop ridsdale (1916-2000) : missionary to the eastern congo, from the henry martyn centre, cambridge – 6r – 1 – £570.00 – (with d/g) – uk Matthew [240]
Congo mission news, 1912-1960 – Leopoldville. nos1-190 – 61mf – 9 – (the joint imc/ cbms missionary archives africa 1910-1945.soas,london. missing: 1913(2), 1914(8), 1917(17), 1918(22), 1922(38, 41), 1923(42), 1927(62), 1928(63)1930(70), 1933(82)) – mf#H-2009 – ne IDC [240]
Congo news letter – 1919-52 [complete] – 2r – 1 – mf#ATLA S0497 – us ATLA [073]
Congo political ephemera – Chicago, IL: [mf for Coop Africana Microfilm Project by] Uni of Chicago Photodup Dept, 1965 (mf ed) – 4r – 1 – us CRL [080]
Congo. Republic *see* Journal officiel de la republique du congo
Congo since independence, january 1960-december, 1961 / Hoskyns, Catherine – London, England. 1965 – 1r – us UF Libraries [960]
Congopresse – Leopoldville: Secretariat general du Congo belge, Section information. n4-9,11-18,21-95. nov 1947-jan 1948; feb 15-jun 1 1948; jul 15 1948-sep 1951 – 2r – us CRL [079]
Congratulation a venerable prestre messire gabriel de saconnay precenteur de l'eglise de lyon, touchant la belle preface & mignonne, dont il a rempare le livre du roy d'angleterre / Calvin, J – [Geneva], 1561 – 1mf – 9 – mf#CL-11 – ne IDC [240]
Congratulatory letter to the rev herbert marsh... / Gandolphy, Peter – London, England. 1812 – 1r – us UF Libraries [240]
Congregacion de la Inmaculada San Luis y San Estanislao *see* Memoria 1948-49
Congregacion de Nuestra Senora de los Dolores de el Colegio de S Pedro y S Pablo *see* Motivos piadosos para adelantar la devocion tierna de los dolores de la ss virgen
Congregaciones marianas del...bajo la advocacion de la inmaculada concepcion, san luis gorzaga y san estanislao de kostka (1900-1901) / Colegio San Jose – Madrid: Imp. Aurial, 1901 – 1 – sp Bibl Santa Ana [240]
Congregation Anshai Lebowitz [Milwaukee WI] *see* Bulletin of the congregation...
Congregation de Notre-Dame *see* Petit questionnaire pour faciliter l'etude de l'arbre historique du canada

Congregation des hommes de la paroisse saint-roch de quebec : erigee sous le titre de l'immaculee conception de marie par diplome episcopal du 24 decembre 1839 – Quebec?: L Brousseau, 1883 – 3mf – 9 – mf#57048 – cn Canadiana [241]
Congregation des jeunes gens de la haute-ville de quebec erigee sous le titre de l'immaculee-conception – Quebec: Leger Brousseau imprimeur, 1902 [mf ed 2000] – 9 – (in latin and french) – cn Bibl Nat [241]
Congregation des oblats de marie immaculee : missions du canada et des sauvages de l'amerique septentrionale – [Marseille, France?: s.n, 1846?] [mf ed 1984] – 1mf – 9 – 0-665-46500-9 – mf#46500 – cn Canadiana [241]
Congregation des Sacres-Coeurs et de l'Adoration *see* Annales des sacres-coeurs
La congregation du grand-orient et les congregations – Paris: Maison de la Bonne Presse [1902?] – 4mf – 9 – mf#vrl-170 – ne IDC [366]
Congregation Ezra Bessaroth [Seattle WA] *see* Records, 1914-1983 [bulk 1945-1957]
Congregation lubavitch times – South Brookline, Mass., Winter 1990/91-Spring 1991 – us AJPC [240]
Congregation of Notre-Dame *see* Souvenir of the golden jubilee of sister st aloysia, st patrick's academy, may, 1915
Congregation of the daughters of our lady of the sacred heart, kiribati : house diaries and accounts of the mission – 1921-1967 – 1r – 1 – mf#pmb1155 – at Pacific Mss [241]
Congregation of the daughters of our lady of the sacred heart, kiribati : house diaries and historical accounts of the mission – 1921-67 – 1r – 1 – (available for reference) – mf#pmb1155 – at Pacific Mss [241]
Congregation Sons of Israel and David *see* Organ of congregation sons of israel and david
Congregational – (= ser Transcripts of nonconformist registers (pre 1837)) – 3mf – 9 – £5.00 – uk BedsFHS [242]
Congregational administration : the carew lectures...1908-09 / Nash, Charles Sumner – Boston: Pilgrim Press 1909 [mf ed 1993] – 1mf – 9 – 0-524-07027-X – mf#1991-2880 – us ATLA [242]
Congregational Christian Churches *see*
– Handbook
– The yearbook
Congregational christian historical society : news – v5-10. 1973-78* – 1r – 1 – mf#ATLA S0403 – us ATLA [242]
The congregational churches of vermont and their ministry, 1762-1914 : historical and statistical / Comstock, John Moore – St Johnsbury VT: Caledonian Co 1915 [mf ed 1991] – 1mf [ill] – 9 – 0-524-00849-3 – mf#1990-4009 – us ATLA [242]
Congregational convention of vermont : minutes – 1899-1907 [complete] – 1r – 1 – mf#ATLA S0607 – us ATLA [242]
Congregational Convention of Wisconsin *see* Minutes of the...annual meeting of the congregational convention of wisconsin
Congregational creeds and covenants / Barton, William Eleazar – Chicago: Advance Publishing, 1917 – 1mf – 9 – 0-524-03256-4 – mf#1990-4659 – us ATLA [242]
Congregational Historical Society *see* Transactions
Congregational history, 1200-1567 / Waddington, John – London: J Snow, 1869 [mf ed 1992] – (= ser Congregational coll) – 2mf – 9 – 0-524-03201-7 – mf#1990-4650 – us ATLA [242]
Congregational history, 1567-1700 : in relation to contemporaneous events, and the conflict for freedom, purity, and independence / Waddington, John – London: Longmans, Green, 1874 – 2mf – 9 – 0-524-02905-9 – mf#1990-4496 – us ATLA [242]
Congregational history, 1700-1800 : in relation to contemporaneous events, education, the eclipse of faith, revivals and christian missions / Waddington, John – London: Longmans, Green, 1876 – 2mf – 9 – 0-524-02906-7 – mf#1990-4497 – us ATLA [242]
Congregational history, 1800-1850 : with special reference to the rise, growth, and influence of institutions, representative men, and the inner life of the churches / Waddington, John – London: Longmans, Green, 1878 – 2mf – 9 – 0-524-06965-4 – mf#1990-5329 – us ATLA [242]
Congregational history, 1850-1880 / Waddington, John – London: Longmans, Green, and Co, 1880 – 2mf – 9 – 0-524-06966-2 – mf#1990-5330 – us ATLA [242]
Congregational history, continuation to 1850 : with special reference to the rise, growth, and influence of institutions, representative men, and the inner life of the churches / Waddington, John – London: Longmans, Green, 1878 – 2mf – 9 – 0-524-02904-0 – mf#1990-4495 – us ATLA [242]
Congregational Home Missionary Society *see* Home missionary and american pastor's journal
Congregational home missionary society : reports – n1-110. 1827-1936 [complete] – 11r – 1 – mf#ATLA S0184 – us ATLA [242]
Congregational magazine – London. 1818-1845 (1) – mf#5289 – us UMI ProQuest [242]
Congregational Ministers and Churches of Vermont *see* General convention of congregational ministers and churches of vermont
Congregational missionary work in porto rico : conducted by the american missionary association / Douglass, Harlan Paul – New York: American Missionary Assoc [1910?] [mf ed 1991] – (= ser Congregational coll) – 1mf – 9 – 0-524-00731-4 – mf#1990-4000 – us ATLA [242]
Congregational music and some of its hindrances : a paper read at the church conference in kingston, october 19th, 1887... / Bedford-Jones, T – [Napanee, Ont?: s.n.], 1887 – 1mf – 9 – 0-665-89252-7 – mf#89252 – cn Canadiana [780]
Congregational quarterly – Boston. 1859-1878 (1) – mf#3154 – us UMI ProQuest [242]
The congregational quarterly – v1-28. 1932-50 [complete] – Inquire – 1 – mf#ATLA 1993-S505 – us ATLA [242]
Congregational review – Boston. 1861-1871 (1) – mf#5079 – us UMI ProQuest [242]
Congregational review – London. 1887-1891 – 1 – mf#4628 – us UMI ProQuest [073]
Congregational Union Of England And Wales *see* Report of special committee on intemperance
Congregational union of scotland : manual – 1978-82 [complete] – Inquire – 1 – mf#ATLA S0922 – us ATLA [242]
Congregational union of scotland : year book [1962] – 1962/1963-77 [complete] – Inquire – 1 – mf#ATLA S0921 – us ATLA [242]
Congregational union of scotland : year book [1984] – 1984-93 [complete] – Inquire – 1 – mf#ATLA S0923 – us ATLA [242]
The congregational way : a hand-book of congregational principles and practices / Boynton, George Mills – rev ed. New York: Pilgrim Press, c1903 – 1mf – 9 – 0-524-01077-3 – mf#1990-4042 – us ATLA [242]
Congregational work *see* The american missionary
Congregationalism / Allon, Henry – London, England. 1881 – 1r – us UF Libraries [242]
Congregationalism / Goodwin, T S – [S.l.: s.n., 1868?] – 1mf – 9 – 0-524-02636-X – mf#1990-4391 – us ATLA [242]
Congregationalism / Jefferson, Charles Edward – Boston: Pilgrim Press, c1910 – 1mf – 9 – 0-524-02391-3 – mf#1990-4293 – us ATLA [242]
Congregationalism : a premium tract / Pond, Enoch – Boston: Congregational Board of Publication, [18–?] – 1mf – 9 – 0-524-06654-X – mf#1991-2709 – us ATLA [242]
Congregationalism : the scripturalness of its polity, with a sketch of its history and methods of work / Boardman, George Nye – Chicago: Advance Publ Co [18–] [mf ed 1984] – 1mf – 9 – 0-8370-0934-0 – mf#1984-4291 – us ATLA [242]
Congregationalism : what it is, whence it is, how it works, why it is better than any other form of church government, and its consequent demands / Dexter, Henry Martyn – 3rd rev enl ed. Boston: Noyes, Holmes, 1871 – 1mf – 9 – 0-524-01329-2 – (incl bibl ref) – mf#1990-4078 – us ATLA [242]
Congregationalism in kansas / Cordley, Richard – Boston: Alfred Mudge [printers] 1876 [mf ed 1992] – 1mf – 9 – 0-524-02735-8 – mf#1990-4410 – us ATLA [242]
Congregationalism in maine – Maine: Congregational Conference & Missionary Society, 1914-30 [mf ed 2001] – (= ser Christianity's encounter with world religions, 1850-1950) – 2r – 1 – mf#2001-s211 – us ATLA [242]
Congregationalism in new hampshire during the nineteenth century : including a sketch of the general association for one hundred years / Thayer, Lucius Harrison – [New Hampshire?: s.n., 1909?] – 1mf – 9 – 0-524-06758-9 – mf#1990-5284 – us ATLA [242]
The congregationalism of the last three hundred years, as seen in its literature : with special reference to certain recondite, neglected, or disputed passages / Dexter, Henry Martyn – New York: Harper, 1880 – 3mf – 9 – 0-8370-8980-8 – (incl indes) – mf#1986-2980 – us ATLA [242]
Congregationalist – 1856 jan 11-1859 dec 23, 1864 mar 4-1867 may 17 – 2r – 1 – (cont: boston reporter, christian times; cont by: boston recorder [1858]; congregationalist and recorder) – mf#577314 – us WHS [242]
Congregationalist – 1870 nov 3/1871 dec 21-1891 feb 26/1892 oct 27 – 16r – 1 – (cont: congregationalist and boston recorder) – mf#577317 – us WHS [242]
Congregationalist – v1-152. 1957-92 [complete] – Inquire – 1 – (cont: the free churches) – mf#atla s0839 – us ATLA [242]
Congregationalist – London. 1872-1886 (1) – mf#2788 – us UMI ProQuest [242]
Congregationalist *see* The american missionary
Congregationalist and boston recorder – 1867 sep 5-1869 apr 22, 1869 apr 29-1870 oct 27 – 2r – 1 – (cont: Congregationalist and recorder) – mf#577316 – us WHS [071]
Congregationalist and herald of gospel liberty – Boston. 1816-1906 (1) – mf#4563 – us UMI ProQuest [242]
Congregationalist and herald of gospel liberty *see* The american missionary
Congregationalist and recorder – 1867 aug 23-30, 1867 may 24-aug 16 – 2r – 1 – (cont: boston recorder [1858]; congregationalist [boston ma: 1849]; cont by: congregationalist and boston recorder) – mf#577315 – us WHS [071]
Congregationalists : who they are and what they do / Prudden, Theodore Philander – Boston: Pilgrim Press, 1906 [mf ed 1992] – 1mf – 9 – 0-524-02694-7 – mf#1990-4401 – us ATLA [242]
Congregationalists in america : a popular history of their origin, belief, polity, growth and work / Dunning, Albert Elijah et al – New York: J A Hill, c1894 – 2mf – 9 – 0-7905-4632-9 – (incl bibl ref) – mf#1988-0632 – us ATLA [242]
Congregations religieuses a saint-domingue, 1681-1 / Jan, Jean Marie – Port-Au-Prince, Haiti. 1951 – 1r – us UF Libraries [972]
La congregazione dell'indice ed il cardinale zigliara : studii critici / Passaglia, Carlo – Roma: GB Paravia, 1882 – 1mf – 9 – 0-524-00299-1 – mf#1989-2999 – us ATLA [240]
Le congres : poemes / Auclair, Joseph – Quebec: C Darveau, 1875 – 1mf – 9 – mf#04091 – cn Canadiana [810]
Congres catholique canadien francais (1er : 1880 : Quebec) *see* Actes et deliberations du premier congres catholique canadien francais tenu a quebec les 25, 26, et 27 juin 1880
Congres colonial national *see* La question sociale au congo
Le congres de la baie saint paul / Auclair, Joseph – Quebec: Darveau, 1882 – 1mf – 9 – mf#02451 – cn Canadiana [971]
Congres de la paix / Clairville, M – Paris, France. 1849? – 1r – us UF Libraries [440]
Congres des inspecteurs d'ecoles : tenu a st-hyacinthe le 21 et le 22 aout 1895 – s.l: s.n, 1895? – 1mf – 9 – mf#04244 – cn Canadiana [370]
Le congres des religions a chicago en 1893 / Bonet-Maury, Gaston – Paris: Hachette, 1895 – 1mf – 9 – 0-7905-5808-4 – mf#1988-1808 – us ATLA [200]
Congres du trente-cinquieme anniversaire de la societe genealogique canadienne-francaise : cegep du vieux-montreal, 7 et 8 octobre 1978 [programme souvenir] / Societe genealogique canadienne-francaise – Montreal: la Societe, [1978] (mf ed 1998) – 2mf – 9 – mf#SEM105P3022 – cn Bibl Nat [929]
Le congres eucharistique de montreal / Emard, Joseph-Medard – Valleyfield [Quebec: s.n, 1910?] – 1mf – 9 – 0-665-74237-1 – mf#74237 – cn Canadiana [241]
Congres interamericain de philosophie (7e : 1967 : Quebec, Quebec) *see* Proceedings of the congress
Le congres international de droit commercial a anvers. l'unification de la legislation en matiere de lettres de change / Nobele, Edouard de – Gand: Annoot-Braeckman, 1886. 48p. LL-4079 – 1 – us L of C Photodup [343]
Congres International de Folklore *see* Travaux du 1er congres...
Congres international des raquetteurs : montreal 26, 27, 28, 29, 30 janv 1938 = International convention of snowshoers – [Quebec (Province): s.n, 1938 ?] (mf ed 1992) – 1mf – 9 – (incl english text) – mf#SEM105P1695 – cn Bibl Nat [790]
Congres International des Traditions Populaires *see* Compte rendu des seances
Congres National des Chambres Syndicales et Groupes Corporatifs *see* La greve generale
Congres National des Droits Civils et du Suffrage des Femmes *see* Compte-rendu
Congres National Des Femmes Haitiennes *see* Voeux du premier congres national des femmes haiti
Congres national du Kampuchea, 2d, 1975 *see* Communique de presse du deuxieme congres national du kampuchea [24 et 25 fevrier 1975]
Congres Ouvrier de France (3rd: 1879: Marseille) *see* Seances du congres ouvrier socialiste de france.20-31 octobre 1879
Congreso Catolico Latinoamericano Sobre Los Proble *see* Primer congreso catolico latinoamericano sobre...
Congreso Continental Anticomunista 1st, Mexico *see* Libro negro del comunismo en guatemala
Congreso Contral El Racismo Y El Antisemitismo *see* Primer congreso contra el racismo y el antisemitismo
Congreso de 1823 / Colombia (Republic Of Colombia, 1819-1831) – Bogota, Colombia. 1926 – 1r – us UF Libraries [972]
Congreso de 1825 / Colombia Congreso Senado – Bogota, Colombia. 1952 – 1r – us UF Libraries [972]
Congreso de barcelona / Federation of European National Societies of the Theosophical Society. Congress – Madrid: Sociedad Teosofica Espanola, 1934 [irreg] [mf ed 2003] – 1r – 1 – (mf: 12th [1934]. no official transactions publ at 13th-18th congress. film incl earlier & later titles: transactions of the...annual congress of the federation of european sections of the theosophical society; transactions of the...congress of the federation of european national societies of the theosophical society; federation of european national societies of the theosophical society, congress, emlekkoenyve; and: history of the efts summary) – mf#2003-s124e – us ATLA [290]
Congreso De Escritores Martianos *see* Memoria
Congreso de espiritualidad franciscana / Barrado Manzano, Arcangel – Madrid: Archivo Ibero Americano, 1963 – 1 – sp Bibl Santa Ana [240]
El congreso de guinea / Vinajeras, Antonio – Matanzas, Ferro-Carril 1879 – us CRL [972]
Congreso De Historia Centro America-Panama *see* Memoria
Congreso de historia y geografia hispano-americanas en commemoracion del descubrimiento del mar pacifico por vasco nunez de balboa / Nunez de Balboa, Vasco – Madrid: Est. Tip. Fortanet, 1914. B.R.A.H. Tomo 64, 1914, pp. 556 – 1 – sp Bibl Santa Ana [946]
Congreso de las provincias unidas / New Granada (United Provinces, 1811-1816) – Bogota, Colombia. 1924 – 1r – us UF Libraries [323]
Congreso De Municipios Dominicanos *see* Al pais
Congreso de panama (1826) / Peru Ministerio De Relaciones Exteriores – Lima, Peru. 1930 – 1r – us UF Libraries [972]
Congreso de panama en 1826 / Velarde, Fabian – Panama, 1922 – 1r – us UF Libraries [972]
Congreso De Poesia Puertorriquena *see* Critica y antologia de la poesia puertorriquena
Congreso eucaristico de dublin / Bayle, Constantino – Madrid: Razon y Fe, 1932 – 1 – sp Bibl Santa Ana [241]
Congreso hispano-americano de historia, cartagena... – Cartagena, Colombia. 1935 – 1r – us UF Libraries [972]
Congreso Internacional De Catedraticos De Literatu... *see*
– Memoria del primer congreso internacional
– Memoria del tercer congreso internacional
Congreso Internacional De Folklore *see* Trabajo de base
Congreso internacional de medicina verificado en berlin del 4 al 9 de agosto de 1890 / Cisneros, Juan – Madrid: Revista Clinica de los Hospitales, 1890 – 1 – sp Bibl Santa Ana [610]
Congreso internacional de panama en 1826 / O'leary, Daniel Florencio – Madrid, Spain. 1920 – 1r – us UF Libraries [972]
El congreso mariano de sevilla / Bayle, Constantino – Madrid: Razon y Fe, 1929 – 1 – sp Bibl Santa Ana [946]
Congreso medico nacional cubano (2nd : 1911 : Havana, Cuba) *see* Actas y trabajos del segundo congreso medico nacional, habana, febrero 24-28 de 1911
Congreso Nacional De Escritores Y Artistas De Cuba *see* Memoria
Congreso Nacional De Historia, 3d *see* Colonia hacia la nacion
Congreso Nacional De Historia, 4th *see* Historia y americanidad
Congreso Nacional De Historia, 5th, Havana, 1946 *see* Lustro de revaloracion historica
Congreso Nacional De Historia, 6th, Trinidad, Cuba *see* Historia y patria
Congreso Nacional De Historia, 7th, Santiago De Cu... *see* Reivindicaciones historicas
Congreso Nacional De Historia, 8th *see* Conmemoraciones historica
Congreso Nacional De Sociologia (Colombia), 1st, B... *see* Memoria
El congreso y exposicion : misionales de barcelona / ed by Bayle, Constantino – Madrid: Razon y Fe, 1929 – 1 – sp Bibl Santa Ana [240]
Congresos y conferencias internacionales / Peru Ministerio De Relaciones Exteriores – Lima, Peru. v1-5. 1909 – 2r – us UF Libraries [327]
Congress and the courts : a legislative history 1787-1984 / ed by Reams, Bernard D Jr & Haworth, Charles B – Series 1 and 2 pt 1 in 52bks (complete) – 9 – $995.00 set – 0-89941-812-0 – (v1-6 1781-1978 index and suppl 1 in 30 books (series 1) $600 set. v1-22 1978-84 (series 2) $395 set) – mf#301301 – us Hein [350]

Congress and the presidency – Washington. 1981+ (1) 1981+ (5) 1981+ (9) – (cont: congressional studies) – ISSN: 0734-3469 – mf#7498,02 – us UMI ProQuest [320]

Congress beacon – Chicago, IL. 2 Jan 1942, 2 Mar-6 Apr 1945-15 Jun 1945. Many missing – 1 – us AJPC [071]

Congress monthly / American Jewish Congress – 1973 jan 12-1976 aug, 1976 sep-1981, 1982-88 – 3r – 1 – (cont: congress bi-weekly; cont by: american jewish congress monthly) – mf#310388 – us WHS [939]

Congress monthly – New York. 1989+ (1,5,9) – ISSN: 0887-0764 – mf#12842,03 – us UMI ProQuest [305]

Congress of Democrats *see* Correspondence, minutes, circulars, clippings, etc

Congress of Industrial Organization *see*
- Citizen cio
- Memo from pac

Congress of Industrial Organizations *see*
- Cio world affairs bulletin
- Cio-pac news service
- Minnesota labor
- Ohio cio council monthly review
- Union news service

Congress of Industrial Organizations [CIO] [US] *see* Dayton union news

Congress of Industrial Organizations. Industrial Union Councils *see* New jersey cio news

Congress of Industrial Organizations [US] *see*
- Cio news
- Cio news for railroad workers
- Cio news/ retail and wholesale edition
- Cio oil facts
- Political action of the week

Congress of Liberal Religion *see* Unity

Congress of Micronesia *see*
- House journals 1966-79
- Journal of the general assembly
- Laws and resolutions
- Political status digest
- Senate journals 1966-79

Congress of Racial Equality *see*
- Corelator
- Northwest area core-lator

Congress of racial equality papers, 1959-1976 – 3pts – 1 – $13,595 coll – (pt1: western regional office 1962-65 5r isbn 0-89093-582 3 $885. pt2: southern regional office 1959-66 15r isbn 0-89093-598-X $2685. pt3: sedfre: ser a: administrative files 1960-76 13r isbn 0-89093-599-8 $2330 ser b: leadership development files 1960-76 10r isbn 0-89093-578-5 $1795 ser c: legal department files 1960-76 37r isbn 0-89093-579-3 $6620. with p/g) – us UPA [324]

Congress probe – v1 iss 1-v2 n23 [1979 mar 12-1980 jun 23] – 1r – 1 – mf#639509 – us WHS [071]

Congress today : a publication / National Republican Congressional Committee – v1 n1-2 [1981 jul-aug] – 1r – 1 – (cont: congress today) – mf#625993 – us WHS [323]

Congress watcher / Public Citizen, Inc – 1980 mar/apr-1985 feb/mar – 1r – 1 – (cont by: congress watcher action alert) – mf#1043433 – us WHS [323]

Congressional bills and resolutions, 1st-72d and 92d congresses – 595r – 1 – $20,825.00 – Dist. us Scholarly Res – us L of C Photodup [323]

Congressional Black Caucus *see*
- For the people
- Report to the people

Congressional black caucus reports for the people – 1980 nov – 1r – 1 – (cont: for the people [washington dc]; cont by: report to the people [washington dc]) – mf#2171805 – us WHS [324]

Congressional budget office publications – 1975-2002 update 1 – 9 – $3250.00 set – (incl chronological and tit ind) – mf#402180 – In japan: far eastern book-sellers – us Hein [324]

Congressional clearinghouse on women's rights : [newsletter] – v3 n1-v5 n9 [1977 jan 11-1979 jun 11] – 1r – 1 – (cont by: clearinghouse on women's issues in congress) – mf#1212327 – us WHS [322]

Congressional debates, 1824-37 / U[nited/] S[tates/] Congress – 18th, 2nd sess. to 25th congress. gales and seaton, publishers. 29 bk [all publ] – 216mf – 9 – $324.00 – (followed by congressional globe) – mf#llmc80-034 – us LLMC [323]

Congressional digest – Washington. 1921+ (1) 1921+ (5) 1921+ (9) – ISSN: 0010-5899 – mf#16420 – us UMI ProQuest [323]

Congressional digest, v11 – 1932. Foreign debts; Tribute to George Washington; Tariff, politics, and the 72nd Congress; Congress considers economic planning; New tax bill; Bankruptcy problem; National campaign of 1932; Five-day work week; Veterans legislation; Investigation of "short" selling – 1,5,9 – us CIS [324]

Congressional digest, v22 – 1943. Problems of the new Congress; The Ruml pay-as-you-go tax plan; Balancing civilian and armed forces; Equal rights for women amendment; Reciprocal trade act extension; Outlawing labor racketeering; Reconstitution of League of Nations; Treaty-making authority of Senate; The New Deal food subsidy program; Proposed federal retail sales tax – 1,5,9 – us CIS [324]

Congressional digest, v48 – 1969. Nuclear nonproliferation treaty; School desegregation guidelines; Federal postal corporation; Controversy over bail reform; Tax status of foundations; Radio-TV cigarette advertising; U.S. foreign military commitments; Federal revenue sharing; Extension of Voting Rights Act; Federal role in urban mass transit – 1,5,9 – us CIS [324]

Congressional digest, v50 – 1971. Women's "Equal Rights Amendment"; Federal consumer protection agency; Federally-subsidized jobs proposals; the Nixon revenue-sharing plan; An all-volunteer U.S. armed force; "No-fault"' automobile insurance; Revising the U.S. jury system; Federal information-gathering on private individuals; Strengthening equal employment opportunity laws; Adoption of metric system – 1,5,9 – us CIS [324]

Congressional digest, vol 1 – 1921-22. Sheppard-Towner maternity bill; public Welfare Dept. bill; Beer bill; Dept. of Educ. bill; Dog bill; Congressional reapportionment; Fed. empl. reclassif.; Peyote bill; Anti-lynching; Cooperative marketing; New naturalization bill; Fordney-McCumber tariff bill; Soldier bonus controversy; Ship subsidy controversy; Political parties and issues-1922; St. Lawrence Seaway – 1,5,9 – us CIS [324]

Congressional digest, vol 2 – 1922-23. Muscle shoals development; U. S. budget system; Cancellation of inter-allied war debt; Rural credits legislation; Child labor amendment; Constitution and the maternity act; Federal civil service; World Court proposal; Curbing Supreme Court powers; Immigration problem; Federal taxation – 1,5,9 – us CIS [324]

Congressional digest, vol 3 – 1923-24. Congress and the railroads; Organization of the new Congress; Problems before the 68th Congress; Foreign service reorganization; Sterling-Reed education bill; Woman's "equal rights" amendment; Immediate Philippine independence; McNary-Haugen agricultural bill; Peace proposals and the 68th Congress; Political parties and issues-1924; Developing our inland waterways – 1,5,9 – us CIS [324]

Congressional digest, vol 4 – 1924-25. Enforcement of prohibition; Dawes plan for reparations payments; Repeal of income tax publicity law; Naval armaments limitation; Postal pay and rate issue; Review of the 68th Congress; Federal Dept. of Aeronautics; Congress and cooperative marketing; Congress and the coal problem; Repeal of the federal estate tax – 1,5,9 – us CIS [324]

Congressional digest, vol 5 – 1926. Govt. aid to U. S. shipping; U. S. and the World Court; McFadden banking bill; Adoption of the metric system; A Federal Department of Education; Congress and prohibition enforcement; Changing inauguration day; Direct primary system; Cloture in the U. S. Senate; Settlement of U. S. war claims – 1,5,9 – us CIS [324]

Congressional digest, vol 6 – 1927. St. Lawrence Seaway; Boulder Dam project; Problem of railroad consolidations; U. S.-Nicaragua controversy; America and the Chinese conflict; A uniform marriage and divorce law; Question of capital punishment; Problems of copyright revision; Issues involved in seating a senator; Capital of the United States – 1,5,9 – us CIS [324]

Congressional digest, vol 7 – (1928). New Congress and tax question; Congress and flood control; Question of outlawing war; Third-term controversy; Immigration problems; Merchant Marine Act; 1928 Presidential election; Problem of radio reallocation; Uncle Sam and the movies; Effect of a "pocket veto" – 1,5,9 – us CIS [324]

Congressional digest, vol 8 – 1929. Question of more U. S. cruisers; Congressional reapportionment; New administration begins; Question of a thirteen-month year; 1929 farm relief problem; Making a tariff law; German reparations; Limitation of naval armaments; Modification of the jury system; Congress and the lobby problem – 1,5,9 – us CIS [324]

Congressional digest, vol 9 – 1930. Freedom of seas; Censorship of foreign books; 71st Congress and prohibition; Communications problem; Federal operation of Muscle Shoals; London naval treaty; Congress and chain stores; 1930 Congressional elections; Equal nationality rights for women: Outlook for 2nd session, 71st Congress – 1,5,9 – us CIS [324]

Congressional digest, vol 10 – 1931. Congress and the employment problem; New maternity and infancy bill; Motor bus controversy; Federal birth control legislation; Philippine independence; Congress and the oil problem; Compulsory unemployment insurance; America's war debt policy; Congress and the silver question; Congress and the banking situation – 1,5,9 – us CIS [324]

Congressional digest, vol 12 – 1933. Congress and the beer problem; Domestic allotment plan; Congress and currency expansion; Congress faces the banking problem; America and tariff reciprocity; Suspension of anti-trust laws; U.S. vs. British radio control systems; U.S. recognition of the U.S.S.R.; Increasing power of the President; Supreme Court and the N.I.R.A – 1,5,9 – us CIS [324]

Congressional digest, vol 13 – 1934. Gold policy and commodity prices; U.S. public school crisis; New food and drug bill; National defense problem; Federal Securities Act of 1933; Congress and the New Deal; Federal monies for public schools; Federal ownership of power utilities; Investigation of munitions industries; Crop control experiment – 1,5,9 – us CIS [324]

Congressional digest, vol 14 – 1935. Congress and the New Deal; Compulsory unemployment insurance; Old-age pensions; Congress and the labor problem; Abolition of utility holding cos.; Lynching as a federal crime; Socialized medicine; Federal sedition legislation; Alien deportation controversy; Congress and Supreme Court powers – 1,5,9 – us CIS [324]

Congressional digest, vol 15 – 1936. American neutrality; Federal aid to shipping; Substitute for AAA; New Deal's housing activities; New administration tax proposal; Proposed state control of relief; National campaign of 1936; Federal ownership of electric utilities; Minimum wage law controversy; Federal all-risk crop insurance – 1,5,9 – us CIS [324]

Congressional digest, vol 16 – 1937. Digest teaching plan; New congress; Aid to farm tenants; Roosevelt Supreme Court plan; Proposed nationalization of munitions; Congress and the sit-down strikes; Congress and the budget; Unicameral legislative system; Roosevelt's reorganization plan; Civil Service reform; Labor and compulsory arbitration – 1,5,9 – us CIS [324]

Congressional digest, vol 17 – 1938. Question of regional planning; War referendum proposal; Roosevelt's national defense program; Proposed Japanese boycott; Question of presidential tenure; Pump-priming and recovery; Alliance with Great Britain; Proposed federal sales tax; Roosevelt "Purge"; Monopoly in American radio – 1,5,9 – us CIS [324]

Congressional digest, vol 18 – 1939. Profit sharing and increased employment; Moves to control relief; Extension of U.S. defense frontier; 150th anniversary of U.S. Congress; Expansion of social security program; Labor Relations act; Federal ownership of the railroads; Amendment of neutrality law; Investigation of un-American activities; Reciprocal tariff-making powers – 1,5,9 – us CIS [324]

Congressional digest, vol 19 – 1940. Extending New Deal borrowing power; Joint committee to balance the budget; Curbing powers of the NLRB; Hatch Act and state officials; Court review of U.S. rulings; War widows and orphans pensions; Increasing federal powers; Prohibition of presidential third term; Conscription of American industry; Financial aid to Latin America – 1,5,9 – us CIS [324]

Congressional digest, vol 20 – 1941. Military defense of the Americas; Union of American Republics; Abolition of electoral college; National defense labor-control laws; St. Lawrence Seaway; Federal union of world democracies; Perm. compulsory military training; Federal price controls and inflation; Federal regulation of labor unions; Congress and state poll tax laws – 1,5,9 – us CIS [324]

Congressional digest, vol 21 – 1942. Question of a separate air force; Help to small business; Abolition of tax-exempt securities; Fed. unemployment compensation; Abolition of 40-hour week; Abolition of the CCC and NYA; Administration and inflation; Post-war world organization; Poll tax legislation; War manpower problem – 1,5,9 – us CIS [324]

Congressional digest, vol 23 – 1944. Absentee soldier voting; Federal aid to public schools; Post-war education of veterans; Civilian conscription for war; Presidential fourth term; Federal vs. states rights in voting; Reduction of voting age to 18; Federal control of insurance; Limitation of federal taxing powers; Demobilization of war workers – 1,5,9 – us CIS [324]

Congressional digest, vol 24 – 1945. Comp. peacetime military training; Little Steel formula; Montgomery Ward seizure; Constitutional treaty process; Bretton Woods proposals; Employment discrimination; Reorganizing congressional procedures; Murray "full employment" bill; Proposed increase in minimum wage; A single armed forces department – 1,5,9 – us CIS [324]

Congressional digest, vol 25 – 1946. Truman's labor fact-finding boards; Federal funds for public schools; Presidental Succession Act; Control of employment services; Atomic energy control; Prospectus for the 1946 elections; Compulsory prepaid medical care; St. Lawrence Seaway; Administration housing policy; Woman's equal rights amendment – 1,5,9 – us CIS [324]

Congressional digest, vol 26 – 1947. Powers of the U. S. President; Outlawing the closed shop; Curbing "unfair" labor practices; Budget, debt, and taxes; Knutson income tax reduction bill; Progress of new Republican Congress; Arbitration of labor disputes; Universal military training; Hawaii and Alaska statehood; Universal "split-income" tax law – 1,5,9 – us CIS [324]

Congressional digest, vol 27 – 1948. Admission of displaced persons; Voice of America; Marshall Plan controversy; Electoral College revision; Reciprocal trade program extension; Long-range housing problem; Federal world government; Tideland oil dispute; International Trade Organization; National Science Foundation – 1,5,9 – us CIS [324]

Congressional digest, vol 28 – 1949. National aviation policy; Administration's legislative program; Federal compulsory health insurance; Repeal of Taft-Hartley Act; Oleomargarine tax fight; North Atlantic Treaty; Presidential election reform; Curbing communism in the U.S.; Federal aid to education; Extension of Social Security – 1,5,9 – us CIS [324]

Congressional digest, vol 29 – 1950. New valley power authorities; Federal civil rights program; Brannan Plan; D. P. program; Big unions and anti-trust laws; Point 4 program; Aid to Great Britain; Welfare state; A non-communist world org.; Question of new states for the U.S.; U.N. genocide treaty – 1,5,9 – us CIS [324]

Congressional digest, vol 30 – 1951. 81st Congress and outlook for 82nd; Congress and U.S. foreign relations; Great federal budget; Reciprocal trade program; Government propaganda; Basic principles of American govt.; Civilian conscription in wartime; Federal tax revision; Foreign economic aid; Military factors in U.S. foreign policy – 1,5,9 – us CIS [324]

Congressional digest, vol 31 – 1952. Point 4 and world politics; U.S. conflict with the U.N.; U.S. primary system; Great federal payroll; Congressional investigation; St. Lawrence Seaway; U.S. and world federation; 1952 national election; Curbing the treaty power; Home rule for District of Columbia – 1,5,9 – us CIS [324]

Congressional digest, vol 32 – 1953. Curbing federal power to tax; Limiting debate in the Senate; U.S.-Spanish relations; Abolition of the RFC; Federal lobbying act; "Stand-by" controls bill; Changing the electoral system; Raising the federal debt limit; Revision of the Taft-Hartley Act; Public lands problem – 1,5,9 – us CIS [324]

Congressional digest, vol 33 – 1954. U.S. foreign trade policy; Social Security revision; Lowering the voting age to 18; "Flexible" farm price supports; Internal security measures; Hawaii and Alaska statehood; Foreign economic policy; 1954 Biennial election; Federal funds for public schools; Congress studies juvenile delinquency – 1,5,9 – us CIS [324]

Congressional digest, vol 34 – 1955. "Dixon-Yates" controversy; Organization of 84th congress; Federal health reinsurance plan; New national military reserve plan; Ten-year highway program; Federal control of gas producers; Federal subsidies to higher education; Bricker treaty power amendment; U.N. charter review conference; President's partnership power policy – 1,5,9 – us CIS [324]

Congressional digest, vol 35 – 1956. U.S. immigration policy; "Right to work" laws; Farm aid proposals; Electoral revision; Organization for Trade Cooperation; Postal rate increase; Federal farm policy; 1956 federal elections; U.S. foreign aid; Federal small business policy – 1,5,9 – us CIS [324]

Congressional digest, vol 36 – 1957. Atomic energy policy; Senate cloture rule 22; Mid-East doctrine; Civil rights legislation; Federal budget reduction; Minimum wage legislation; U.S. foreign aid policy; Right-to-work legislation; Federal aid for school construction; Foreign jurisdiction over U.S. forces – 1,5,9 – us CIS [324]

Congressional digest, vol 37 – 1958. Provision for presidential disability; Congress and the job ahead; 1958 farm aid proposals; Renewal of reciprocal trade act; Curbing the Supreme Court; Tax cut proposals; Challenge of today's public schools; Nuclear disarmament; Labor reform legislation; Senate cloture rule 22 – 1,5,9 – us CIS [324]

Congressional digest, vol 38 – 1959. Hawaii statehood; Federal aid for depressed areas; Extending federal housing aid; Future financing of REA programs; FY 1960 foreign aid program; Federal aid to public education; Curbing labor union abuses; 86th Congress and the Supreme Court; Raising rate ceiling on U.S. bonds; Unemployment insurance changes – 1,5,9 – us CIS [324]

Congressional digest, vol 39 – 1960. New civil rights legislation; Federal minimum wage law; Adding medical aid to OASI program; NDEA loyalty provisions; Congress and the wheat problem; Curtailing U.S. mutual security aid; Strengthening the U. N.; Guide to the 1960 federal elections; U.S. policy and the Caribbean situation; Proposed youth conservation corps – 1,5,9 – us CIS [324]

Congressional digest, vol 40 – 1961. U. S. and World Court jurisdiction; Economic aid for Latin America; Kennedy program and the new Congress; Aid to depressed areas; Congress and the railroad situation; New Frontier's housing program; Federal aid to education; Labor unions and anti-trust laws; Dept. of Urban Affairs and Housing; A wilderness area system – 1,5,9 – us CIS [324]

Congressional digest, vol 41 – 1962. New medicare plan for the aged; Administered prices in drug industry; Proposed U. N. bond issue; Administration's new farm proposals; Literacy test and poll tax; FY 1963 foreign aid proposals; Foreign Trade Expansion Act; Non-Communist econ. community; U. S. and the situation in Cuba, Record of the 87th Congress – 1,5,9 – us CIS [324]

Congressional digest, vol 42 – 1963. Federal aid for urban mass transit; Aid to higher education; Alliance for Progress; Administration's proposed tax revision; Proposed 1964 budget; The FY 1964 foreign aid proposals; Proposed medicare for the aged; Omnibus higher education bill; Civil rights and "public accommodations"; Youth employment legislation – 1,5,9 – us CIS [324]

Congressional digest, vol 43 – 1964. A domestic "peace corps"; Communist bloc trade; Civil rights and FEPC; Federal urban renewal; Presidential succession and inability; Federal food stamp program; Disarmament and intl. arms control; Fed. public works and unemployment; Supreme Court and "school prayer"; "Appalachia" program – 1,5,9 – us CIS [324]

Congressional digest, vol 44 – 1965. State legislative apportionment; U.S. space program; Administration "medicare" program; U.S. and situation in Vietnam; Amendment of U.S. immigration laws; Fiscal 1966 foreign aid program; Labor-management relations policy; Crime and law enforcement in U.S.; U.S. Latin American policy; Proposed U.S. Soviet consular treaty – 1,5,9 – us CIS [324]

Congressional digest, vol 45 – 1966. Administration rent subsidy program; Federal minimum wage revision; Federal anti-poverty program; Review of U.S. Vietnam policy; Proposals for a longer House term; "Truth in packaging" legislation; Controversy over U.S. foreign aid; U. S. foreign policy commitments; "Open housing" controversy; Proposed "gun control" legislation – 1,5,9 – us CIS [324]

Congressional digest, vol 46 – 1967. Unemployment compensation revision; "Demonstration cities" program; U.S. policy toward Rhodesia; Controversy over the FY68 budget; Proposed selective service changes; Expansion of east-west trade; Congress and national crime problem; Federally-guaranteed minimum income; Changing the electoral system; Question of "pay television" – 1,5,9 – us CIS [324]

Congressional digest, vol 47 – 1968. Federal Job Corps; Community Action Program; Consumer protection proposals; Federal anti-riot legislation; Congress and U.S. travel deficit; Public welfare revision; Revision of the military draft; Congress foreign policy role; Anti-ballistic missile defense; Congress and airport development – 1,5,9 – us CIS [324]

Congressional digest, vol 49 – 1970. Changing the U.S. electoral system; Casualty insurance regulation; The "Philadelphia Plan"; Federal school racial policy; 18-year voting age; Public welfare revision; The federal anti-pollution role; Standby wage and price controls; Broadening U.S. import controls; Debate over the supersonic transport – 1,5,9 – us CIS [324]

Congressional digest, vol 51 – 1972. Federal voter registration by mail; National health insurance proposals; A six-year presidential term; Minimum wage law expansion; Federal day-care proposals; Future U.S. space program; Financing elementary and secondary education; Presidential election issues; Panama Canal treaty revision; Federal spending ceiling controversy – 1,5,9 – us CIS [324]

Congressional digest, vol 52 – 1973. Capital punishment controversy; U.S. Rhodesia policy; Future of foreign aid program; Presidential impoundment of funds; Newsmen's "shield" legislation; Reconstruction Aid for North Vietnam; Federal role in public assistance; Federal Petroleum policy; Expansion of U.S.-Soviet trade; Federal-State Land-Use policy – 1,5,9 – us CIS [324]

Congressional digest, vol 53 – 1974. A "School Prayer" Amendment; Federal Campaign Financing Reform; Automobile Emission Controls; "School Busing" Controversy; Federal Regulation of Surface Mining; National Health Insurance Program; Revising the U.S. Electoral System; Controversy Over the Amnesty Question; a U.S. Consumer Protection Agency; U.S. Food Export Policy – 1,5,9 – us CIS [324]

Congressional digest, vol 54 – 1975. The Problem of Illegal Aliens; Controversy Over Sugar Controls; U.S. Financial Role in the United Nations; U.S. Military Aid to Turkey; The Food Stamp Controversy; Extension of Federal Voting Rights Act; Allocation of Scarce World Resources; Congress & U.S.-Soviet Detente; Safeguarding Classified Information; Controversy over Handgun Controls – 1,5,9 – us CIS [324]

Congressional digest, vol 55 – 1976. Basic Principles of American Government; Public Employee Strikes; Mandatory Budget Balancing; Panama Canal Treaty Revision; Breakup of Major Oil Companies; Humphrey-Hawkins Full Employment Bill; Mandatory Sentencing & Crime Control; 1976 Election Campaign Issues: Postal Service Reorganization; B-1 Bomber Controversy – 1,5,9 – us CIS [324]

Congressional digest, vol 56 – 1977. U.S. Southern Africa Policy; Nuclear Energy Safety; Congressional Campaign Financing; Executive Reorganization Powers; Minimum Wage Revision; Women's Equal Rights Amendment; National Health Insurance; Amnesty for Illegal Aliens; Mandatory Retirement Age – 1,5,9 – us CIS [324]

Congressional digest, vol 57 – 1978. Labor Reform Act; U.S. Cuban Policy; Beverage Container Problem; U.S. Foreign Arms Sales; Welfare Reform; Airline Deregulation; U.S. Energy Policy; Congress Congress seats for D.C.; New U.S. Dept of Education; Alaska Lands Controversy – 1,5,9 – us CIS [324]

Congressional digest, vol 58 – 1979. Student Aid Policy; Marihuana Use Policy; Direct Presidential Election; "Real Wage Insurance"; Balanced Budget Amendments; Treaty Termination; SALT Treaty; Trucking Regulation; National Defense Spending; House Campaign Financing – 1,5,9 – us CIS [324]

Congressional digest, vol 59 – 1980. Welfare Reform Amendments; The Consumer and Antitrust Law; "Sunset" Legislation; Draft Registration; Limitations on Covert Intelligence; Federal Revenue Sharing; Occupational Safety and Health Act; 1980 Election Campaign Issues; MX Missile Controversy; Congress & "School Prayer" – 1,5,9 – us CIS [324]

Congressional digest, vol 60 – 1981. Food Stamp Program; School Busing; "Fair Housing" Proposals; CETA Program; Federal Legal Aid Program; Reagan Tax-Reduction Program; Social Security Financing; Illegal Aliens; International Lending Agencies; Extension of Voting Rights Act – 1,5,9 – us CIS [324]

Congressional digest, vol 61 – 1982. Clean Air Act; Freedom of Information Act; Urban Enterprise Zones; Curtailing Federal Student Aid; Curbing Federal Courts; A Gold-Based U.S. Currency; Nuclear Freeze Proposal; Mandatory Balanced Federal Budget; Abolition of Mandatory Retirement; Wilderness Protection Act – 1,5,9 – us CIS [324]

Congressional digest, vol 62 – 1983. Law of the Sea Treaty; Auto "Domestic Content"; Caribbean Basin Initiative; Social Security Reform; American Conservation Corps; Export Controls over U.S. Technology; Immigration Reform; Central American Policy; War Powers Act; Legislative Veto – 1,5,9 – us CIS [324]

Congressional digest, vol 63 – 1984. Tuition Tax Credit; Federal Budget Deficit; Space Weapons Policy; Broadcast Deregulation; School Prayer Controversy; Federal Criminal Sentencing; NATO Cost Sharing Controversy; 1984 Election Campaign Issues; U.S. Policy towards Nicaragua; The Genocide Treaty – 1,5,9 – us CIS [324]

Congressional digest, vol 64 – 1985. Civil Rights Legislation; Acid Rain Controversy; Star Wars Controversy; Subminimum Wage for Youth; Enterprise Zones; The MX Missile; Federal Subsidy of AMTRAK; Policy toward South Africa; Presidential Line-Item Veto; The Clean Water Act – 1,5,9 – us CIS [324]

Congressional digest, vol 65 – 1986. Farm Credit System; The Tax Reform Controversy; Immigration Legislation; Policy toward Angola; Gun Control Controversy; Environmental Superfund; Foreign Trade Legislation; SALT II Agreements; Omnibus Drug Legislation; National 55 MPH Speed Limit – 1,5,9 – us CIS [324]

Congressional digest, vol 66 – 1987. Product Liability; Limiting PACs; Bilingual Education; Catastrophic Health Insurance; Drug Testing; U.S. Foreign Trade Policy; Minimum Wage; Broadcasting Fairness Doctrine; ABM Treaty; War Powers Act; Persian Gulf – 1,5,9 – us CIS [324]

Congressional digest, vol 67 – 1988. Fed. Emp. Polit. Act. Legislation; Welfare Reform; Contra Aid; Missiles Treaty; Med. Leave Policy; Fair Housing Act; Election Camp. Financing; 1988 Elections; Child Care Leg.; Intelligence Oversight Act – 1,5,9 – us CIS [324]

Congressional digest, vol 68 – 1989 – 1,5,9 – (text. and apparel trade act; clean air act; corporate merger leg.; occ. health controversy; minimum wage leg.; savings and loan leg.; flag desecration; immigration reform; balanced budget amend.; disabilities act) – us CIS [324]

Congressional digest, vol 69 – 1990 – 1,5,9 – (capital gains; child care; clean air; voter reg.; national service; line item veto; civil rights; campaign finance reform; 1990 omnibus crime bill; budget reconciliation act) – us CIS [324]

Congressional digest, vol 70 – 1991 – 1,5,9 – (funding of the arts; cable tv reregulation; persian gulf policy; family and medical leave; energy policy; brady handgun act; title x pregnancy counseling; u.s.-china act; striker replacement; choice in schools) – us CIS [324]

Congressional digest, vol 71 – 1992 – 1,5,9 – (banking reform; u.s.-mexico free trade agreement; campaign finance; omnibus crime control; defense spending and budget; capping fed. entitlement programs; aid to the former soviet union; election issues; product liability; economic competitiveness) – us CIS [324]

Congressional digest, vol 72 – 1993 – 1,5,9 – (family and medical leave; line item veto & rescission authority; voter reg.; space station funding; 1994 budget; replacement of striking workers; fed. employees' political activities; nat. and comm. service; n.a. free trade agreement; violence on tv) – us CIS [324]

Congressional digest, vol 73 – 1994 – 1,5,9 – (education reform; proposed dept. of the environment; mining and public lands; campaign finance; second-hand tobacco smoke; crime control; u.s. policy toward haiti; health care; gen. agreement on tariffs and trade) – us CIS [324]

Congressional directory / United States Congress – Washington. 1971-1979 (1) 1969-1979 (5) 1975-1979 (9) – mf#2576 – us UMI ProQuest [920]

The congressional directory, 1909-1997/98 / U.S. Congress – Washington: GPO. 60th Congress 2nd session-105rd Congress – 1391mf – 9 – $2086.00 – (updates planned) – mf#llmc 81-109 – us LLMC [323]

Congressional globe *see* Congressional debates, 1824-37

Congressional globe, 1833-73 / U[/nited/] S[/tates/] Congress – 23rd to 42nd Congress. 109 bk [all publ] – 1190mf – 9 – $1785.00 – (followed by congressional record) – mf#llmc80-035 – us LLMC [323]

Congressional investigating committees / Dimock, Marshall Edward – Baltimore, Johns Hopkins Press, 1929. 182 p. LL-1609 – 1 – us L of C Photodup [340]

The congressional manual : compiled from the congressional record and other public documents – N.Y./Chicago: International Survey Co, 1901 – 3mf – 9 – $4.50 – mf#LLMC 92-132 – us LLMC [324]

Congressional oversight : the general accounting office: statement of charles a bowsher, comptroller general of the united states... – [Washington DC]: US General Accounting Office [2000] – 1mf – 9 – (incl bibl ref) – us US Gen Account [336]

Congressional quarterly almanac – Washington. 1945+ (1) 1981+ (5) 1981+ (9) – ISSN: 0095-6007 – mf#6390 – us UMI ProQuest [320]

Congressional quarterly weekly report – Washington. 1953-1998 (1) 1966-1998 (5) 1970-1998 (9) – (cont by: cq weekly) – ISSN: 0010-5910 – mf#858 – us UMI ProQuest [320]

Congressional quarterly's editorial research reports – Washington. 1987-1991 (1) 1987-1991 (5) 1987-1991 (9) – (cont: editorial research reports. cont by: cq researcher) – ISSN: 1057-0926 – mf#971,01 – us UMI ProQuest [320]

Congressional record – Law Library Microfilm Corporation, 1873-1975 – 9 – $26,450.00 set – mf#402030 – us Hein [340]

Congressional record / Philippines. Congress. House of Representatives – 1945-61 – 1 – 690.00 – us L of C Photodup [959]

Congressional record / Philippines. Congress. Senate – 1945-61 – 1 – us L of C Photodup [959]

Congressional record / U.S. Congress – 1873-1996. 43rd to 99th Congress. 142v in approx. 2000 bks – 6914mf (1:42) 5078mf (1:24) – 9 – $38,730.00 – mf#LLMC 85-200 – us LLMC [323]

Congressional record / U.S. Congress – Daily periodical with biweekly indexes when Congress is in session. Subscriptions for six-month or one-year period only – 9 – $118.00y in US; $147.50 outside – mf#S-N 752-029-00000-8. Sub-list ID-CRM – us GPO [324]

Congressional record / U.S. Congress – Backfile 1976-2001 – 9 – $11,785.00 set (2002 subs $810.00) – 0-89941-201-7 – (per yr $495. updated as released by gpo) – us Hein [324]

Congressional record *see* Congressional globe, 1833-73

Congressional record on microfiche : retrospective collection / U.S. Congress – 43rd-100th congresses. v1-134. 1873-1988 – 9 – $31,680.00 coll – (provides a complete transcript of debates and other floor proceedings of the us congress. available as a complete coll or by individual congress) – us CIS [324]

The congressional records 1789-1990 – 4 titles – 9164mf – 9 – $31,200.00 set – (the annals of congress 1789-1824 343mf $515.00; the congressional debates 1824-1837 216mf $324.00; the congressional globe 1833-1873 1190mf $1785.00; the congressional record 1873-1996 5078mf $38,730.00; this title still growing) – us LLMC [323]

Congressional research service digest of public general bills and resolutions / United States Library of Congress – Washington. 1936-1988 (1) 1972-1988 (5) 1972-1988 (9) – ISSN: 0012-2785 – mf#6830 – us UMI ProQuest [348]

Congressional studies – Washington. 1979-1981 (1) 1979-1981 (5) 1979-1981 (9) – (cont by: congress and the presidency) – ISSN: 0194-4053 – mf#7498,01 – us UMI ProQuest [320]

Congressional studies – Washington. 1979-1981 [1]; 1979-1981 [5]; 1979-1981 [9] – (cont: capitol studies) – ISSN: 0194-4053 – mf#7498,01 – us UMI ProQuest [320]

Congressman al baldus reports from washington – v1 n1-2 [1975 spr-sum] – 1r – 1 – (cont by: your congressman al baldus reports from washington) – mf#641478 – us WHS [323]

Congressman bob kastenmeier reports – v23 n2-27 [1981 apr-1985 dec] – 1r – 1 – (cont: bob kastenmeier reports from the u s house of representatives) – mf#1048499 – us WHS [323]

Congressman glen davis on the line from washington – n1-8 [1967 feb 6-1974 aug 5] – 1r – 1 – mf#1054809 – us WHS [323]

Congressman henry reuss reports – 1961 feb 6-1982 feb – 1r – 1 – mf#361202 – us WHS [071]

Congressman Jim Sensenbrenner reports – 1979 apr 2-1984 dec 24 – 1r – 1 – (cont by: weekly column) – mf#1125331 – us WHS [323]

Congressman robert j cornell [newsletter] – 1976 jun-[1978 jul] – 1r – 1 – mf#401088 – us WHS [323]

Congresso De Historia da Revolucao de 1894 (1st) *see* Anais do primeiro congresso de historia de revolucao

Congresso nacional e o programa de integracao soci... / Brazil. Congresso Nacional Senado Federal Direto... – Brasilia, Brazil. 1970 – 1r – us UF Libraries [972]

Congreve, William *see* William congreve

Conheca a pre-historia brasileira / Mendes, Josue Camargo – Sao Paulo, Brazil. 1970 – 1r – us UF Libraries [972]

Conheca a vegetacao brasileira / Joly, Aylthon Brandao – Sao Paulo, Brazil. 1970 – 1r – us UF Libraries [972]

Coni, Emilio Angel *see* Gaucho

Conimbricenses. *see* Commentarii collegii conimbricensis e societatis iesu in universam dialecticam aristotelis

Coninck, Mathilde Courant De *see* Nouveau

Coningham, William *see* The national gallery in 1856

Conington 1538-1950 – (= ser Cambridgeshire parish register transcript) – 4mf – 9 – £5.00 – uk CambsFHS [929]

Conjectures on original composition *see* Edward youngs gedanken ueber die originalwerke

Conjonction – Port-Au-Prince, Haiti. n96-99. 1946 jan – 1r – us UF Libraries [972]

Conjugal and parental duties stated and enforced / Fisher, Samuel – Wisbech, England. 1802 – 1r – us UF Libraries [640]

Conjura / Castellanos, Jesus – Madrid, Spain. 1916 – 1r – us UF Libraries [972]

A conjuracao : orgam republicano – Campanha, MG, 22 maio 1888 – (= ser Ps 19) – 1,5,6 – bl Biblioteca [079]

Conjuration du general malet contre napoleon / Aubignosc, L P d' – Paris 1824 [mf ed Hildesheim 1995-98] – 1v on 2mf – 9 – €60.00 – 3-487-26386-6 – gw Olms [944]

Conklin, Charles H *see* A treatise on the powers and duties of justices of the peace in civil, criminal and special cases in the state of iowa.

Conkling, Alfred *see*
- The powers of the executive department of the government of the united states, and the political institutions and constitutional law of the united states
- A treatise on the organization and jurisdiction of the supreme, circuit and district courts of the united states.

- A treatise on the organization, jurisdiction and practice of the courts of the united states
- The young citizen's manual

Conkling, Roscoe *see* Papers

Conkwright, S J *see* History of the churches of boone's creek baptist association of kentucky, 1780-1923

Conlatio critica codicis sinaitici cum textu elzeviriano : vaticani quoque codicis ratione habita / ed by Tischendorf, Constantin von – Lipsiae: Hermann Mendelssohn, 1869 [mf ed 1985] – 1mf – 9 – 0-8370-5536-9 – mf#1985-3536 – us ATLA [225]

Conley, John Wesley *see*
- The bible in modern light
- History of the baptist young people's union of america
- The young christian and the early church

Conmemoracion del cuarto centenario de los francis – Bogota, Colombia. 1953 – 1r – us UF Libraries [972]

Conmemoraciones historica / Congreso Nacional De Historia, 8th – Habana, Cuba. 1950 – 1r – us UF Libraries [972]

Conn, Herbert William *see*
- Method of evolution
- The story of life's mechanism

Connacht champion *see* Connaught champion

Connacht sentinel – Galway, Ireland. 1950; jan-16 dec 1986; 1987-22 dec 1992; 1993 – 9r – 1 – uk British Libr Newspaper [072]

Connacht tribune and tuam news – Galway, Ireland. 1930; 1986-92; 1993 – 26r – 1 – uk British Libr Newspaper [072]

La connaissance – Paris. 1920-nov 1922 – 1 – fr ACRPP [073]

Connaught champion – Galway, Ireland. 15 jul 1904-13 jan 1911 [wkly] – 6r – 1 – (aka: connacht champion) – uk British Libr Newspaper [072]

Connaught journal – Galway, Ireland. 24 may 1813, 1823-1836, 12 aug 1839, 6 feb-31 dec 1840 [wkly] – 13r – 1 – (aka: burkes connaught journal) – uk British Libr Newspaper [072]

Connaught leader – Ballinasloe 1905-07 – 3r – 1 – ie National [072]

Connaught patriot – Tuam, Ireland. 27 aug 1859-1864; 7 oct 1865-27 mar 1869 – 3 3/4r – 1 – uk British Libr Newspaper [072]

Connaught people and ballinasloe independent – Ballinasloe, Ireland. 19 apr 1884-21 nov 1885; 2 jan-25 sep 1886 (imperfect) – 3/4r – 1 – uk British Libr Newspaper [072]

Connaught telegraph *see* Telegraph or connaught ranger

Connaught watchman *see* Ballina chronicle

Connaught watchman etc – Ballina, Ireland. Aug 1851-13 jul 1860; 22 feb 1862-3 oct 1863 – 3 3/4r – 1 – uk British Libr Newspaper [072]

Conneautville courier – Conneautville, PA. -w 1876-1955 – 13 – $25.00r – us IMR [071]

Connecticut : session laws of american states and territories – 1776-2001 – 9 – $2433.00 set – mf#402550 – us Hein [348]

Connecticut *see*
- The new joint stock act of connecticut, passed january session, 1880.
- Reports and opinions
- Reports, post-nrs
- Reports, pre-nrs

Connecticut 1670 census – Oxford MA (mf ed 1977) – 84p on 1mf – 9 – 0-931248-04-3 – (census lists names of all household heads residing in connecticut between 1667-73. over 2,400 names indexed alphabetically by surname and by town. for each individual there is a date, place of residence, biographical notation & reference to source document. first-time publ of names fr obscure listings discovered in town records of guildford, killingworth, milford, new london, norwich, saybrook and stamford) – us Archive [978]

Connecticut academy of arts and science memoirs – New Haven. 1810-1816 (1) – mf#3726 – us UMI ProQuest [700]

Connecticut Academy of Arts and Sciences *see* Transactions

Connecticut afl-cio news – 1978 mar, 1979 jan-1986 dec/1987 jan – 1r – 1 – (cont by: labor news [hartford ct]) – mf#1222773 – us WHS [331]

Connecticut Afro-American Historical Society *see* Newsletter

Connecticut attorney general reports and opinions – 1899-2001 – 6,9 – $402.00 set – (1899-1968 on reel $175. 1967-68, 1983-2001 on mf $227.) – mf#408160 – us Hein [340]

Connecticut baptist – Rockville/Hartford, CT. Conn. Baptist State Convention/Conn. Convention of American Baptist Churches. 1922-70. Incomplete. Single reels available – 1 – (single reels available) – us ABHS [242]

Connecticut baptist materials: miscellaneous booklets, sermons and addresses – 1632p – 1 – 65.28 – us Southern Baptist [242]

Connecticut bar journal – v1-37. 1927-63 – 173mf – 9 – $259.00 – mf#LLMC 84-445 – us LLMC [340]

Connecticut bar journal – v1-74. 1927-2000 – 9 – $1091.00 set – ISSN: 0010-6070 – mf#102001 – us Hein [340]

Connecticut bicentennial gazette / American Revolution Bicentennial Commission of Connecticut – v1-v5 n9, v6 n4 [1971 fall-1976 dec, 1978 jun] – 1r – 1 – mf#384132 – us WHS [975]

Connecticut. Board of Compensation Commissioners *see* Reports

Connecticut. Bureau of Labor Statistics *see* Reports

Connecticut business and industry – Hartford. 1923-1973 (1) – mf#9810 – us UMI ProQuest [338]

Connecticut Business and Industry Association *see* Cbia news

Connecticut catholic – Hartford, CT. 1876-1897 (1) – mf#62347 – us UMI ProQuest [071]

Connecticut colonists 1635-1703 – Oxford MA (mf ed 1986) – 317p on 1mf – 9 – 0-931248-40-X – (compiles all available public docs for windsor fr 1635-1703; incl 3 different sets of vital records; church, census, & probate records; and lists of freemen, ratables, & petitioners) – us Archive [978]

Connecticut Commission on Civil Rights *see* Civil rights bulletin

Connecticut common school journal and annals of education – Hartford. 1838-1866 – 1 – mf#3763 – us UMI ProQuest [370]

Connecticut. Compensation Commissioners *see* Compendia of awards

Connecticut. Comptroller *see* Report in relation to the criminal business of the courts..

Connecticut Co-ops *see* Co-op times

Connecticut courant – Hartford CT. [1823-66] – 1r – 1 – (cont: connecticut courant, and the weekly intelligencer; hartford weekly journal; connecticut whig [hartford ct]; connecticut press) – mf#888449 – us WHS [071]

Connecticut courant – Hartford, CT. 1764-1914 (1) – mf#60156 – us UMI ProQuest [071]

Connecticut cpa quarterly – Hartford. 1987-1993 (1) – ISSN: 0884-2817 – mf#12900,01 – us UMI ProQuest [650]

Connecticut. Dept. of Factory Inspection *see* Reports

Connecticut. Dept. of Labor and Factory Inspection *see* Reports

Connecticut. Dept. of Revenue Services *see* Information relative to the assessment and collection of taxes

Connecticut Employees Union Independent *see* Independent union

Connecticut. Employment Security Division *see*
- Bulletin: connecticut labor department
- Reports of the administrator of the unemployment compensation law

Connecticut evangelical magazine and religious intelligencer – Hartford. 1800-1815 (1) – mf#3804 – us UMI ProQuest [242]

Connecticut Federation of Labor *see*
- Official proceedings...annual convention of the connecticut federation of labor
- Proceedings of the...annual convention of the connecticut state branch of the american federation of labor

Connecticut general statutes annotated – St Paul: West Pub Co, 1958-jun 2002 update – 9 – $2978.00 set – mf#401450 – us Hein [348]

Connecticut herald – New Haven CT. 1808 may 31, 1818 jan 27, 1825 jun 28, dec 6, 1828 sep 23 – 1r – 1 – (cont: connecticut post and new haven visitor; cont by: connecticut herald and general advertiser) – mf#875329 – us WHS [071]

Connecticut herald and general advertiser – New Haven CT. 1818 dec 8 – 1r – 1 – (cont: connecticut herald [new haven ct: 1803]; cont by: connecticut herald [new haven ct: 1821]) – mf#850492 – us WHS [071]

Connecticut historical society. hartford. collections – v1-31. 1860-1967 – 1 – $356.00 – mf#0166 – us Brook [978]

Connecticut history : a publication / Association for Study of Connecticut History – n14-21 [1974 jun-1980 jan] – 1r – 1 – (cont by: connecticut history newsletter of the association for the study of connecticut history) – mf#1004995 – us WHS [978]

Connecticut history newsletter / Association for Study of Connecticut History – 1967 oct-1970 jun – 1r – 1 – (cont by: connecticut history newsletter...) – mf#976791 – us WHS [978]

Connecticut history newsletter of the... / Association for Study of Connecticut History – 1970 nov-1974 jan – 1r – 1 – (cont: connecticut history newsletter; cont by: connecticut history) – mf#1004834 – us WHS [071]

Connecticut. Insurance Dept *see* Reports of the insurance commissioner

Connecticut journal of international law – v1-16. 1985-2001 – 9 – $295.00 set – ISSN: 0897-1218 – mf#110181 – us Hein [341]

Connecticut law journal – v1-19. 1935-55 – 120mf – 9 – $180.00 – mf#LLMC 84-446 – us LLMC [340]

Connecticut law, public utilities and carriers / Connecticut. Laws, Statutes, etc – Hartford, Public Utilities Commission, 1944. 264 p. LL-1567 – 1 – us L of C Photodup [348]

Connecticut law review – v1-33. 1968-2001 – $632.00 – ISSN: 0010-6151 – mf#102011 – us Hein [340]

Connecticut law review – West Hartford. 1968+ (1) 1972+ (5) 1972+ (9) – ISSN: 0010-6151 – mf#6550 – us UMI ProQuest [340]

Connecticut. Laws, Statutes, etc *see*
- Connecticut law, public utilities and carriers
- Index to general statutes of connecticut, and public acts, from 1875 to 1882
- The joint stock act of connecticut, from the revised statutes.

Connecticut lawyer – v1-11. 1990-2001 – 9 – $223.00 set – ISSN: 1057-2384 – mf#116941 – us Hein [340]

Connecticut libraries – Hamden. 1970+ (1) 1972+ (5) 1975+ (9) – ISSN: 0010-616X – mf#6488 – us UMI ProQuest [020]

Connecticut loan office records relating to the loan of 1790 / U.S. Treasury Dept. Bureau of the Public Debt – (= ser Records of the bureau of the public debt) – 10r – 1 – mf#T654 – us Nat Archives [336]

Connecticut magazine : or gentleman's and lady's monthly museum of knowledge and rational entertainment – Bridgeport. 1801-1801 (1) – mf#3569 – us UMI ProQuest [640]

Connecticut maple leaf / French-Canadian Genealogical Society of Connecticut, Inc – v1 n1-v3 n4 [1983 jun-1988 win] – 1r – 1 – mf#1054827 – us WHS [929]

Connecticut news : news letter of the archeological society of connecticut / Archaeological Society of Connecticut – n98-n142 [1966 oct-1980 win] – 1r – 1 – (cont by: newsletter of the archeological society of connecticut, inc) – mf#679164 – us WHS [978]

Connecticut nursing news – Meriden. 1979+ (1) 1979+ (5) 1979+ (9) – (cont: nursing news) – ISSN: 0278-4092 – mf#9630,01 – us UMI ProQuest [610]

Connecticut. Office of Bank Commissioner *see*
- Reports
- Reports of bank commissioner relative to licensed credit unions

Connecticut probate law journal *see* Quinnipiac probate law journal

Connecticut. Public Utilities Commission *see* Reports

Connecticut. Railroad Commissioners *see* Reports

Connecticut republican magazine – Suffield. 1802-1803 (1) – mf#3570 – us UMI ProQuest [320]

Connecticut school journal – New Haven. 1871-1874 – 1 – mf#4766 – us UMI ProQuest [370]

Connecticut spectator – Middletown CT. 1814 sep 21 – 1r – 1 – mf#851742 – us WHS [071]

Connecticut staats-zeitung – Holyoke, MA: [s.n.], [mar 1918-jun 1942] – us CRL [071]

Connecticut staatszeitung – Hartford CT (USA), 1921 17 mar, 1921 1 dec-1923 8 nov [gaps], 1924 1 may-1939 30 nov [gaps] – 6r – 1 – gw Misc Inst [071]

Connecticut State AFL-CIO [American Federation of Labor and Congress of Industrial Organizations] *see* Labor news

Connecticut. State Bar Association *see* Proceedings

Connecticut. State Board of Mediation and Arbitration *see* Reports

Connecticut State Employees Association *see*
- Csea news
- Government news

Connecticut State Federation of Teachers *see*
- Union teacher
- Unionist: the journal of the connecticut state federation of teachers, aft, afl-cio

Connecticut. Supreme Court *see*
- Connecticut supreme court reports
- Day's reports
- Kirby's reports
- Root's reports

Connecticut supreme court reports / Connecticut. Supreme Court – v1-104. 1814-1926 – 795mf – 9 – $1192.00 – (pre-nrs: v1-52 1814-85 356mf $534.00. updates planned) – mf#LLMC 80-801 – us LLMC [347]

Connecticut Trust for Historic Preservation *see* News from the connecticut trust for historic preservation

Connecticut Union of Telephone Workers *see* Cutw voice

Connecticut vanguard – Waterbury CT. 1958 – 1r – 1 – (cont: connecticut cio vanguard [waterbury ct: 1947]) – mf#3632811 – us WHS [071]

Connecticut workers and technological change *see* The worker and technological change, 1930-80

Connection – 1970 – 1r – 1 – mf#4852924 – us WHS [071]

Connection – Fairfax, VA. 1987-2000 (1) – mf#68395 – us UMI ProQuest [071]

Connection – Teaneck NJ. 1993 jun 5-dec 18, 1994 jan 8-dec 17, 1995 jan 7-dec 16, 1996 feb 3-dec 14, 1997 jan 11, 25-feb 1-8, 22-mar 8, jun 14, dec 13 – 5r – 1 – mf#2748519 – us WHS [071]

Connection between famine and pestilence and the great apostasy / Nagnatus – Dublin, Ireland. 1847 – 1r – us UF Libraries [240]

The connection between literature and commerce : in two essays... / Burn, William Scott – Toronto: H & W Rowsell, 1845 – 1mf – 9 – mf#67299 – cn Canadiana [380]

Connection between ministerial character and success / Hyatt, Charles – London, England. 1826 – 1r – us UF Libraries [240]

Connection between old and new testaments / Rae, George Milne – London: J M Dent [19–?] [mf ed 1992] – (= ser The temple series of bible handbooks) – 1mf – 9 – 0-524-04807-X – mf#1992-0227 – us ATLA [220]

Connection of the holy sacraments with the spiritual life and their... / Bartholomew, Charles Charles – Exeter, England. 1842 – 1r – us UF Libraries [240]

Connection of the redeemer's heavenly state with the advancement of... / Smith, John Pye – London, England. 1820 – 1r – us UF Libraries [240]

Connections – 1967 mar 1-1969 may 6, 1967 mar 1-1969 may 6 – 2r – 1 – mf#1110911 – us WHS [071]

Connections / Connections Guidance Center – 1970 mar-1973 dec – 1r – 1 – mf#1110914 – us WHS [360]

Connections / Societe de l'histoire des familles du Quebec – 1978 sep-1989 jun – 1r – 1 – mf#1573051 – us WHS [929]

Connections Guidance Center *see* Connections

The connections of the geste des loherains with other french epics and mediaeval genres / Bowman, Russell Keith – New York: [s.n.] 1940 [mf ed Bloomington IN: Indiana Uni Lib, Preservation Dept 1984] – xv 168p on 1r – 1 – us Indiana Preservation [390]

Connector of the... / Hamilton National Genealogical Society – 1979 mar-1985 – 1r – 1 – mf#1027216 – us WHS [929]

Connell, Arthur Knatchbull *see*
- Discontent and danger in india
- The economic revolution of india and the public works policy

Connell, Charles James *see* Our land revenue policy in northern india

Connelley, William Elsey *see* Papers

Connellsville daily news – Connellsville PA. 1928 jul 2-sep 29, 1929 apr-jun 29, jul-aug, sep 3-nov 1, nov 2-dec, 1930 jan-feb, mar – 7r – 1 – mf#1269188 – us WHS [071]

Connelly, James Henderson *see* The yoga aphorisms of patanjali

Connelly, Marc *see* Verts paturages

Connelly, Pierce *see*
- Coming struggle with rome
- Reasons for abjuring allegiance to the see of rome

Connelly, Teresa A *see*
- Motivation and goal orientation of male and female golfers
- The relationship of golf orientation to motivation among women golfers

Conner, C M *see*
- Forage crops
- Pig feeding with cassava and sweet potatoes
- Preliminary report on growing irish potatoes

Conner, Charles M *see* Feeding horses and mules on home-grown feed-stuffs

Conner, Christopher P *see* The incidence of post-traumatic stress disorder symptoms in certified athletic trainers

Conner, James *see* The history of a suit at law, according to the practice of this state

Conner, Jeremiah Frederick *see* Employers' liability, workmen's compensation and liability insurance.

Connexion between christian benevolence and spiritual propserity / Eastman, Theophilus – Portsea, England. 1825 – 1r – us UF Libraries [240]

Connexion between faith in christ, and eternal life – Blackburn, England. 1818 – 1r – us UF Libraries [240]

Connexion between the headship of christ and revival in the church / Martin, Hugh – Edinburgh, Scotland. 1859 – 1r – us UF Libraries [240]

Connexion of morality with religion / Fitzgerald, William – London, England. 1851 – 1r – us UF Libraries [230]

Connexion of religion with national character / Rice, Edward – London, England. 1833 – 1r – us UF Libraries [230]

Connexion of the divine dispensation with the divine glory / Winter, Robert – London, England. 1830 – 1r – us UF Libraries [240]

Connexion of the east-india company's government with the superstitious and idolatrous customs etc – London, 1838 [mf ed 1995] – (= ser Yale coll) – 152p – 1 – 0-524-09129-3 – mf#1995-0129 – us ATLA [306]

CONNOISSEUR

Connoisseur : a journal of music and the fine arts – London. 1845-1846 (1) – mf#5291 – us UMI ProQuest [780]
Connoisseur – New York. 1979-1992 (1) 1979-1992 (5) 1979-1992 (9) – ISSN: 0010-6275 – mf#12248 – us UMI ProQuest [700]
The connoisseur – 1886-89 [mf ed Chadwyck-Healey] – (= ser Rare 19th century american art journals) – 10mf – 9 – uk Chadwyck [700]
The connoisseur, 1901-79 – 94r – 1 – mf#95701 – uk Microform Academic [073]
Connoisseur by mr town : critic and censor general – London. 1754-1756 – 1 – mf#4702 – us UMI ProQuest [073]
Connolly, Declan A *see* The effects of an application of sunscreen on selected physiological variables during exercise in the heat
Connolly memorial baptist church. delbarton, west virginia : church records – 1940-81 – 1 – us Southern Baptist [242]
Connolly, R H *see*
– The explanatio symboli ad initiandos
– The liturgical homilies of narsai
– The so-called egyptian church order and derived documents
Connolly, Richard Hugh *see* The so-called egyptian church order and derived documents
Connoquenessing valley news – Zelienople, PA. 1901-1932 (1) – mf#66172 – us UMI ProQuest [071]
Connor, Jeanette Thurber *see* History of turtle mound and the indian river from...
Connor, Jeannette M Thurber *see* Colonial records of spanish florida
Connor, Ralph *see* The life of james robertson, missionary superintendent in the northwest territories
Connotton valley times / Carroll Co. Leesville – jan 1884-dec 1886 [wkly] – 1r – 1 – mf#B11636 – us Ohio Hist [071]
Conny de LaFay, Felix J de *see*
– Les bourbons
– La france sous le regne de la convention
Conocimiento, curacion y preservacion de la peste / Frailas, A – Jaen, 1606 – 10mf – 9 – sp Cultura [614]
Conolly, Arthur *see* Journey to the north of india, overland from england, through russia, persia, and affghaunistaun
Conover, Milton *see* Working manual of original sources in american government
Conquering anxiety in grade school aged swimmers through the use of imaginative play / Gup, Marc L & Jensen, Barbara E – 1992 – 2mf – $8.00 – us Kinesology [150]
The conquering cross (the church) / Haweis, Hugh Reginald – New York: T. Y. Crowell, 1887. Beltsville, Md: NCR Corp, 1978 (4mf); Evanston: American Theol Lib Assoc, 1984 (4mf) – (= ser HIS Christ and christianity) – 9 – 0-8370-0926-X – mf#1984-4246 – us ATLA [240]
The conqueror – Lahore, shradhe mohan for the deva dharma mission. v1 n10. feb/mar 1892 – us CRL [070]
The conquest of canaan : lectures on the first ttwelve chapters of the book of joshua / Mackay, Alexander Bisset – London: Hodder and Stoughton, 1884 – 1mf – 9 – 0-7905-2921-1 – mf#1987-2921 – us ATLA [221]
The conquest of death / Kinney, Abbot – New York, 1893. ix,259p. illus – 1 – us Wisconsin U Libr [306]
Conquest of devil's island / Pean, Charles – London, England. 1953 – 1r – us UF Libraries [972]
The conquest of florida...hernando de soto / Irving, Theodore – 1877 – 9 – sp Bibl Santa Ana [975]
The conquest of new granada / Markham, Clements Robert – London: Smith, Elder & Co, 1912 (mf ed 2000) – 1r – 1 – (transl of the duquesne memoir on the chibcha calendar) – mf#*Z-8638 – us NY Public [972]
Conquest of power / Weisbord, Albert – New York, NY. v1-2. 1937 – 1r – us UF Libraries [025]
The conquest of quebec : an epic poem in eight books / Murphy, Henry – Dublin: W Porter, 1790 [mf ed 1972] – 1r – 5 – mf#SEM16P72 – cn Bibl Nat [810]
The conquest of red spain / Fuller, John Frederick Charles – London, 1937. Fiche W 901. (Blodgett Collection of Spanish Civil War Pamphlets) – 9 – us Harvard [946]
Conquest of siberia : and the history of the transactions, wars, commerce etc carried on between russia and china, from the earliest period / Mueller, G F & Pallas, P S – London: Smith, Elder, and Co, 1842 – 2mf – 9 – mf#HT-630 – ne IDC [915]
The conquest of the cross in china / Speicher, Jacob – New York: Fleming H Revell, c1907 – 1mf – 9 – 0-8370-6525-9 – (incl ind) – mf#1986-0525 – us ATLA [240]
Conquest of the maya / Mitchell, J Leslie – London, England. 1934 – 1r – us UF Libraries [972]
The conquest of the sioux / Gilman, Samuel C – new rev ed. Indianapolis: Carlon & Hollenbeck, 1897 [mf ed 1986] – 1mf – 9 – 0-8370-7143-7 – mf#1986-1143 – us ATLA [240]
The conquest of the weast india / Gomara, Francisco Lopez de – 1578 – 9 – us Scholars Facs [972]
The conquest of trouble; and, the peace of god : musings / Brent, Charles Henry – Philadelphia: George W Jacobs, c1916 – 1mf – 9 – 0-7905-9147-2 – mf#1989-2372 – us ATLA [220]
Les conquestes et les trophees des normand-francois aux royaumes de naples et de sicile : aux duchez de calabre, d'antioche, de galilee, et autres principautez d'italie et d'orient / Du Moulin, Gabriel – Rouen: Chez David du Petit Val...et Jean et David Berthelin, 1658 [mf ed 1984] – 6mf – 9 – (with ind) – mf#SEM105P384 – cn Bibl Nat [940]
Conquests of the cross : a record of missionary work throughout the world / ed by Hodder, Edwin – London, New York: Cassell, 1890 – 5mf – 9 – 0-7905-7113-7 – mf#1988-3113 – us ATLA [240]
A la conquete de la liberte en france et au canada / DeCelles, Alfred Duclos – Levis, Quebec: P-G Roy, 1898 – 1mf – 9 – mf#06207 – cn Canadiana [971]
Conquete de l'andalousie / Lapene, Edouard – 1823 – 9 – sp Bibl Santa Ana [946]
La conquete pacifique du maroc et du tafilalet / Cruchet, Rene – Paris: Berger-Levrault, 1934 – 1 – us CRL [960]
La conquete spirituelle...du mexique...nouvelle espagne 1523-1524 a 1572 / Ricard, Robert – Paris, 1933; Madrid: Razon y Fe, 1934 – 1 – sp Bibl Santa Ana [240]
Conquista de las islas malucas al rey felipe 3 / Argensola, B L de – Madrid, 1609 – 5mf – 9 – mf#SEP-19 – ne IDC [910]
Conquista de mejico / Lopez de Gomara, Francisco – Tomo I. 1887 – 9 – (tomo 2, 1888) – sp Bibl Santa Ana [972]
La conquista de mejico / Diaz del Castillo, Bernal – Madrid: Atlas, 1943 – 1 – sp Bibl Santa Ana [972]
La conquista de mejico / Gomez de Arteche, Jose – 1892 – 9 – sp Bibl Santa Ana [972]
A conquista do brasil = Conquest of Brazil / Nash, Roy – Sao Paulo [etc]: Companhia editora nacional, 1939 (mf ed 1977) – 6mf – 9 – (transl into portuguese. incl bibl footnotes) – mf#Sc Micro F-1234 – us NY Public [972]
Conquista do deserto ocidental / Craveiro Costa, Joao – Sao Paulo, Brazil. 1940 – 1r – us UF Libraries [972]
Conquista espiritual de mexico / Ricard, Robert – Mexico City? Mexico. 1947 – 1r – us UF Libraries [972]
Conquista espiritual do oriente...introducao e notas de f. felix lopes, 3 parte. lisboa, 1967 / Trindade, Paula da – Madrid: Graf. Calleja, 1968 – 1 – sp Bibl Santa Ana [240]
Conquista y colonizacion de mejico / Garcia Icazbalceta, Joaquin – Madrid: Tip. Fortanet, 1894 – 1 – sp Bibl Santa Ana [972]
Conquistadors / Fuentes, Patricia De – New York, NY. 1963 – 1r – us UF Libraries [972]
Conquistadors : the struggle for colonial power in latin america, 1492-1825: from the british library, department of manuscripts / ed by Scammell, Geoffrey – [mf ed 2003] – ca 100r in 3pts – 1 – us Primary [972]
Las conquistas de caceres por fernando 2 y alfonso 9 de leon y su fuero latino anotado / Orti Belmonte, Miguel Angel – Badajoz: Imp de la Diputacion Provincial, 1947 – (sep de la rev de estudios extremenos) – sp Bibl Santa Ana [946]
Conrad, Arcturus Zodiac *see* Boston's awakening
Conrad, Christine *see* Berechtigung und praktische anwendung des verbots der werbung mit sonderangeboten bei mengenmaessig beschraenkter abgabe
Conrad ferdinand meyer / Holzamer, Wilhelm – Berlin: Schuster & Loeffler [1904?] [mf ed 1990] – (= ser Die dichtung 23) – 1r [ill] – 1 – (filmed with: der anti-necker j h mercks und der minister fr k v moser / richard loebell) – mf#2834p – us Wisconsin U Libr [430]
Conrad ferdinand meyer : leben und werke / Nussberger, Max – Frauenfeld: Huber 1919 [mf ed 1995] – 1r – 1 – (incl bibl ref & ind. filmed with: georg enach / k f meier) – mf#3697 – us Wisconsin U Libr [430]
Conrad ferdinand meyer : saggio psicologico estetico / Pensa, Mario – Bologna: R Patron c1963 [mf ed 1995] – 1r – 1 – (incl bibl ref. filmed with: georg enach / k f meier) – mf#3697 – us Wisconsin U Libr [920]
Conrad ferdinand meyer : sein religioeses und sittliches vermaechtnis / Fischer, Richard – Stuttgart: Calwer Verlag 1949 [mf ed 1995] – 1r – 1 – (filmed with: georg enach / k f meier) – mf#3697 – us Wisconsin U Libr [170]
Conrad ferdinand meyer und julius rodenberg : ein briefwechsel = Correspondence. selections / ed by Langmesser, August – Berlin: Paetel, 1918 [mf ed 1995] – 322p – 1 – (incl bibl ref) – mf#8824 – us Wisconsin U Libr [920]
Conrad, Frederick William *see* The lutheran doctrine of baptism
Conrad, Johannes *see* The german universities for the last fifty years
Conrad, Joseph *see*
– The arrow of gold
– The nigger of the "narcissus"
– A personal record
– Shorter tales of joseph conrad
– Tales of unrest
Conrad, Michael Georg *see*
– Der freimaurer!
– Von emile zola bis gerhart hauptmann
Conrad, Sven *see*
– Grab- und weihdenkmaeler aus den territorien von augusta traiana und kabyle vom 1. bis 3. jahrhundert n. chr
– Grab- und weihdenkmaeler aus den territorien von augusta traiana und kabyle vom 1. bis 3. jahrhundert n chr
Conradi, Hermann *see*
– Brutalitaeten
– Hermann conradis gesammelte schriften
– Phrasen
Conradi, Johannes *see* Schleiermachers arbeit auf dem gebiete der neutestamentlichen einleitungswissenschaft
Conradi, Ludwig Richard *see* History of the sabbath and first day of the week
Conradi, Matthias *see*
– Praktische deutsch-romanische grammatik
– Taschenwoerterbuch der romanisch-deutschen sprache
Conradiana – Lubbock. 1968+ (1) 1970+ (5) 1977+ (9) – ISSN: 0010-6356 – mf#3069 – us UMI ProQuest [400]
Conradie, Catharina Maria *see* Khayelitsha
Conrads von weinsberg, des reichs-erbkaemmerers, einnahmen- und ausgaben-register von 1437 und 1438 / ed by Albrecht, Joseph – Tuebingen: Literarischer Verein, 1850 [mf ed 1993] – (= ser Blvs 18) – viii/95p – 1 – mf#8470 reel 5 – us Wisconsin U Libr [943]
Conradus de Brunopoli, OFM *see* Speculum beatae mariae virgins
Conradus gemnicensis : konrads von haimburg und seiner nachahmer, alberts von prag und ulrichs von wessobrunn / ed by Dreves, Guido Maria – Leipzig: Fues, 1888 [mf ed 1986] – (= ser Analecta hymnica medii aevi 3) – 1mf – 9 – 0-8370-7455-X – (hymns in latin. int in german. incl ind) – mf#1986-1455 – us ATLA [450]
Conrady, Alexander Hubert Alphons *see* Rheinlande in der franzosenzeit (1750 bis 1815)
Conrady, L *see* Vier rheinische palaestina-pilgerschriften des 14, 15 und 16 jahrhunderts
Conrath, Annemarie *see* Adalbert stifters "witiko"
Conroy, Ellen *see* The symbolism of colour
Conroy, George *see* Occasional sermons, addresses, and essays
Conroy, Theresa R *see* Plyometric training and its effects on speed, strength, and power of intercollegiate athletes
Cons pratiques de lingala / Rubben, E – Dison, Belgium. 1928 – 1r – us UF Libraries [960]
Consaca / Quintero, Jaime – Cali, Colombia. 1944 – 1r – us UF Libraries [972]
La consacrazione eucharistica nella chiesa copta / Giamberandini, G – Cairo, 1957 – 3mf – 8 – €7.00 – ne Slangenburg [243]
La consagracion de la republica del ecuador al sagrado corazon de jesus. quito, 1935 / Heredia, Jose Felix – Madrid: Razon y Fe, 1936 – 1 – sp Bibl Santa Ana [972]
Conscience : with preludes on current events / Cook, Joseph – Boston: Houghton, Osgood 1879 [mf ed 1986] – (= ser Boston monday lectures) – 1mf – 9 – 0-8370-6172-5 – mf#1986-0172 – us ATLA [170]
The conscience : lectures on casuistry, delivered in the university of cambridge / Maurice, Frederick Denison – London: Macmillan, 1868. Chicago: Dep of Photodup, U of Chicago Lib, 1973 (1r); Evanston: American Theol Lib Assoc, 1984 (1r) – 1 – 0-8370-0411-X – mf#1984-B341 – us ATLA [100]
Conscience and christ : six lectures on christian ethics [haskell lectures...1913] / Rashdall, Hastings – New York: C Scribner 1916 [mf ed 1991] – (= ser Haskell lectures) – 1mf – 9 – 0-7905-9449-8 – mf#1989-2674 – us ATLA [230]
Conscience and faith : five lectures = Conscience et la foi / Coquerel, Athanase – London: British and Foreign Unitarian Association, 1878 – 1mf – 9 – 0-524-08232-4 – (in english) – mf#1993-2007 – us ATLA [240]
Conscience and law : or, principles of human conduct / Humphrey, William – London: Thomas Baker, [1896] – 1mf – 9 – 0-8370-6127-X – mf#1986-0127 – us ATLA [170]
Conscience and the constitution : with remarks on the recent speech of the hon daniel webster in the senate of the united states on the subject of slavery / Stuart, Moses – Boston: Crocker & Brewster, 1850 [mf ed 1989] – 1mf – 9 – 0-7905-3296-4 – mf#1987-3296 – us ATLA [976]
Conscience guyanais – Cayenne, French Guiana. 1959-1962 (1) – mf#67704 – us UMI ProQuest [079]
La conscience populaire : son education par une eglise d'etat et par une eglise de professants: conference prononcee a lausanne le 18 fevrier 1895 / Rittmeyer, Charles – Lausanne: Georges Bridel, 1895 – 1mf – 9 – 0-8370-7663-3 – mf#1986-1663 – us ATLA [230]
Conscience with the power and cases there of : divided into 5 bookes / Ames, William – [London], 1639 – 5mf – 9 – mf#PW-33 – ne IDC [240]
Conscientious objector – v1-8 n2,6. 1939-46 [all publ] – (= ser Radical periodicals in the united states, 1881-1960. series 1) – 1r – 1 – $200.00 – us UPA [303]
Conscientious objectors in the civil war / Wright, Edward Needles – New York: A S Barnes 1961, c1931 [mf ed 1993] – 1mf – 9 – 0-524-08152-2 – (incl bibl ref) – mf#1993-9058 – us ATLA [976]
Consciousness *see* Chueh wu [ccs]
Consciousness, being, immortality; divine healing and christian science / Burgess, O O – [S.l.: s.n., 1899?] – 1mf – 9 – 0-524-08438-6 – mf#1993-2043 – us ATLA [240]
Conscription and true liberalism / Fisher, Sydney – [Montreal?: s.n.] 1917 [mf ed 1994] – (= ser The case against conscription 6) – 1mf – 9 – 0-665-72973-1 – mf#72973 – cn Canadiana [971]
Conscription news – nos 1-262. 1944-59 (all publ) – (= ser Military law coll) – 16mf – 9 – $24.00 – (offered as part of llmc's military law collection) – us LLMC [343]
Conscrit / Merle, M (Jean Toussaint) – Berlin, Germany. 1830 – 1r – us UF Libraries [440]
Le conscrit : organe antimilitariste revolutionaire – Paris. n1-6. 1900-05 – 1 – (mq no. 4-5) – fr ACRPP [320]
Le conscrit – Paris. 1924-39 – 1 – fr ACRPP [073]
Le conscrit ou le retour de crimee : drame comique en deux actes / Doin, Ernest – Montreal: Librairie Beauchemin, 1902 [mf ed 1985] – 1mf – 9 – mf#SEM105P461 – cn Bibl Nat [820]
Le conscrit, ou, le retour de crimee : drame comique en deux actes / Doin, Ernest – Montreal: Beauchemin & Valois, 1878 – 1mf – 9 – mf#27035 – cn Canadiana [820]
Consculluela Y Barreras, Juan Antonio *see* Cuatro anos en la cienaga de zapata
A consecrated life : a sketch of the life and labors of rev ransom dunn...1818-1900 / Gates, Helen Dunn – Boston, MA: Morning Star, 1901 [mf ed 1990] – 1mf – 9 – 0-7905-5210-8 – mf#1988-1210 – us ATLA [920]
Consecrated womanhood : a sermon preached in the first congregational church, portland, oregon / Marvin, Frederic Rowland – New York: J O Wright, 1903 [mf ed 1984] – (= ser Women & the church in america 28) – 1mf – 9 – 0-8370-0727-5 – mf#1984-2028 – us ATLA [242]
Consecrated women / Hack, Mary Pryor – Philadelphia: Longstreth [1882?] [mf ed 1984] – (= ser Women & the church in america 116) – 1mf – 9 – 0-8370-1383-6 – mf#1984-2116 – us ATLA [305]
Consecration : encouragements of temple building; peace in the sanct... / Weeks, S – Newcastle upon Tyne, England. 1859 – 1r – us UF Libraries [240]
Consecration and purity : or, the will of god concerning me / Wheeler, Mary Sparkes – Ocean Grove, NJ: MS Wheeler, c1913 – 1mf – 9 – 0-524-00419-6 – mf#1989-3119 – us ATLA [240]
La consecration de l'eglise de st-lin des laurentides : d'apres les rapports de la presse: 29 avril 1891 – Montreal: C O Beauchemin, 1891? – 1mf – 9 – mf#04246 – cn Canadiana [241]
Conseil a mon pays / Pommayrac, A – Paris, France. 1894 – 1r – us UF Libraries [972]
Conseil Attikamek-Montagnis *see* Tepatshimuwin
Conseil economique et social – 1981- – 9 – €27.59y – (1947-1980 €381.12) – fr Journal Officiel [330]
Conseil International Provisoire des Syndicats Ouvriers *see* Mouvement ouvrier international
Conseil special pour les affaires du bas-Canada *see* Ordinances passed by the governor and special council of lower canada
Le conseiller du peuple : ou reflexions adressees aux canadiens-francais / Beaudry, David-Hercule – Quebec?: s.n, 1877 (Quebec: J A Langlais) – 3mf – 9 – mf#29586 – cn Canadiana [305]
Le conseiller du peuple – Paris. avr 1849-nov 1851 [mnthly] – 1 – fr ACRPP [073]

Conseils aux mauvais poetes : poeme / Taki, Mir – Paris: Dondey-Dupre 1826 – us CRL [810]

Les conseils du pere jean : ou du chiffonier de paris, a ses amis des faubourgs – [Paris]: Bureau central, [1848?] – us CRL [944]

Conseils d'une maitresse de penision a ses eleves – Lyon, France. 1865 – 1r – us UF Libraries [025]

Consejero del lobo / Hinostroza, Rodolfo – Habana, Cuba. 1964 – 1r – us UF Libraries [972]

Consejo de castilla : sala de alcaldes de casa y corte... / Archivo Historico Nacional – Madrid: Razon y Fe, 1926 – 1 – sp Bibl Santa Ana [946]

Consejo de Hacienda *see* Defensas omitidas en el memorial ajustado hecho por el relator en la causa de don juan de ovando...

Consejo Diocesano de los hombres de Accion Catolica *see*
- Plan de formacion-accion (reuniones de equipo). primer ciclo-guiones
- Revision de vida

Consejo Economico Sincial *see* 4th asamblea plenaria del consejo economico sindical

Consejo Economico Sindical *see*
- 4 pleno
- Cesiex en cifras
- Normas de constitucion y funcionamiento de los consejos economicos sindicales comarcales e intercomarcales o de zona

Consejo economico sindical – Badajoz: Graficas Iberia, 1948 – sp Bibl Santa Ana [330]

Consejo economico sindical interprovincial de extremadura y huelva. evolucion socio-economica – Badajoz: Graf. N. Jimenez, 1970 – sp Bibl Santa Ana [330]

Consejo Economico Sindical Provincial *see*
- 4. comision no 1
- 4. comision no 1 a
- 4. comision no 1 b
- 4. comision no 7
- 11th comision
- Comision no 2
- Comision no 4
- Comision no 5: productos quimicos
- Comision no 6: industrias relacionadas – con la alimentacion
- Comision no 8: transportes. comunicaciones y servicios de informacion
- Comision no 9
- Comision no 13: financiacion
- Comision no 14: trabajo
- Comision no 15: factores humanos y sociales. productividad

Consejo Episcopal Latinoamericano, Secretariado General *see* Documentacion celam

Consejo Nacional De Planificacion Economica (Guate... *see* Plan de desarrollo economico de guatemala, 1955-19

Consejo provincial de agricultura, industria y energia de caceres / Caceres. Spain – 1884 – 9 – sp Bibl Santa Ana [946]

Consejo Provincial de Fomento. Badajoz *see* Cartilla redactada para dar a conocer los trabajos que se realizan en la granja escuela practica de agricultura de badajoz

El consejo real y supremo de las indias : su historia, organizacion y labor administrativa hasta la terminacion de la casa de austria / Schaefer, Ernst – Sevilla: imp m carmona, 1935-47 – 2v – 1 – us Wisconsin U Libr [970]

Consejo Superior de Investigaciones Cientificas *see* Memoria, ano 1967, 1968

Consejo Superior Universitario Centroamericano *see*
- Sistema educativo en costa rica
- Sistema educativo en honduras

Consejos aun joven / Moreno Torrado, Luis – Villafranca de los Barros: Imp. Ventura Rodriguez, 1909 – 1 – sp Bibl Santa Ana [946]

Conselheiro francisco jose furtado / Franco De Almeida, Tito – Rio de Janeiro, Brazil. 1867 – 1r – us UF Libraries [972]

Conselheiro francisco jose furtado / Franco De Almeida, Tito – Sao Paulo, Brazil. 1944 – 1r – us UF Libraries [972]

O conselho da boa amizade – Rio de Janeiro, RJ: Typ de Silva Porto e Cia, 1823 – (= ser Ps 19) – mf#P17,01,87 – bl Biblioteca [321]

CONSELHO NACIONAL DE ESTAT ISTICA *see* Bresil d'aujourd'hui

Conselho Nacional De Estat Istica *see* Bresil d'aujourd'hui

Conselho Nacional De Estatistica *see* Divisao territorial do brasil

Conselho Nacional De Estatistica (Brazil) *see* Populacao

Conselho Nacional De Protecao Aos Indios (Brazil) *see* Catalogo geral das publicacoes da comissao rondon

Consensio mvtva in re sacramentaria / Bullinger, Heinrich – Tigvri, Rodolph Vuissenbach, [1549] – 1mf – 9 – mf#PBU-261 – ne IDC [240]

Consensus / National Citizens' Coalition – 1975 oct-1985 aug – 1r – 1 – mf#792895 – us WHS [303]

Consensus of opinions against instrumental music in worship – Greenock, Scotland. no date – 1r – us UF Libraries [780]

Consensus of opinions on behalf of the psalms alone in praise – Glasgow, Scotland. no date – 1r – us UF Libraries [780]

The consensus tigurinus and john calvin / Bunting, Ian David – 1r – 1 – 0-8370-0718-6 – mf#1984-6101 – us ATLA [242]

Consensvs orthodoxvs sacrae scriptvrae et veteris ecclesiae : de sententiae et veritate verborvm coenae dominicae... / [Hardesheim, C] – Tigvri, [Christoph] Froschover, 1578 – 8mf – 9 – mf#PBU-594 – ne IDC [240]

Consentius, Ernst *see*
- Buergers gedichte in zwei teilen
- "Freygeister, naturalisten, atheisten"

Consequences / Conway, Moncure Daniel – London, England. no date – 1r – us UF Libraries [240]

The consequences of high population growth in developing countries : a case study of south africa / Lekganyane-Maleka, Mokgadi Rahab – Pretoria: Vista University 2004 [mf ed 2004] – 4mf – 9 – (incl bibl ref) – mf#mfm15259 – sa Unisa [304]

The consequences of mandatory minimum prison terms : a summary of recent findings / Vincent, Barbara S & Hofer, Paul J – Washington: FJC, 1994 – 1mf – 9 – $1.50 – mf#LLMC 95-831 – us LLMC [345]

O conservador : folha politica e industrial – Maranhao: Typ da Temperanca, 13 dez 1858; jan-dez 1859; jan-fev, abr, jun-dez 1860; jan-ago 1861; jan-out, 06 dez 1862 – (= ser Ps 19) – mf#P19B,02,15 – bl Biblioteca [079]

O conservador : jornal politico e commercial – Manaus, AM. 27 out 1912-abr 1917; 25 set 1921 – (= ser Ps 19) – mf#P11B,06,25 – bl Biblioteca [073]

O conservador : jornal politico e industrial – Sao Paulo, SP: Typ do Governo, 26 abr-07 ago 1850 – (= ser Ps 19) – mf#P18,01,84 – bl Biblioteca [320]

O conservador : jornal politico noticioso e commercial da provincia de santa catharina – Desterro, SC: Typ Jornal do Commercio, 13 dez 1873-jun 1879; jan-11 fev 1880 – (= ser Ps 19) – mf#UFSC/BPESC – bl Biblioteca [073]

O conservador : jornal politico, noticioso e mercantil – Desterro, SC: Typ do Conservador, 17 jan 1854-28 dez 1855 – (= ser Ps 19) – bl Biblioteca [073]

O conservador : orgao do partido – Aracaju, SE: Typ do Conservador, 11-30 dez 1871; 18 fev 1873 – (= ser Ps 19) – mf#DIPER – bl Biblioteca [325]

O conservador : orgao do partido – Desterro, SC. 02 out 1884-14 nov 1889 – (= ser Ps 19) – mf#UFSC/BPESC – bl Biblioteca [325]

O conservador : orgao do partido republicano catharinense – Tijucas, SC: Typ de Joao Barthem Jr, 25 maio 1913 – (= ser Ps 19) – mf#UFSC/BPESC – bl Biblioteca [325]

O conservador : orgao político – Estancia, SE. 11 jun 1881 – (= ser Ps 19) – mf#DIPER – bl Biblioteca [320]

Le conservateur : ou collection de morceaux rares et d'ouvrages anciens et modernes – Paris. 1756-58, janv-nov 1760 – 1 – fr ACRPP [073]

Le conservateur – Paris. oct 1818-mars 1820 (I-VI) – 1 – fr ACRPP [073]

Le conservateur litteraire – Par Abel et Victor Hugo. no. 1-30. Paris. dec 1819-mars 1821 (I-III) – 1 – fr ACRPP [073]

Le conservateur litteraire *see* Societe des textes francais modernes

Les conservateurs et la politique nationale de 1878 a 1882 – St Hyacinthe Quebec: Des Presses a moteur hydraulique du Courrier, 1882 – 2mf – 9 – mf#11945 – cn Canadiana [325]

Conservation – Ottawa: Commission of Conservation, [1912-1921] – 9 – mf#P04995 – cn Canadiana [333]

Conservation and recycling – Oxford. 1976-1987 (1,5,9) – ISSN: 0361-3658 – mf#49256 – us UMI ProQuest [639]

Conservation biology – Boston. 1987+ (1,5,9) – ISSN: 0888-8892 – mf#15615 – us UMI ProQuest [639]

Conservation in the eastern caribbean / Eastern Caribbean Conservation Conference, 1st – Charlotte Amalie, St Thomas. 1965? – 1r – us UF Libraries [639]

Conservation news – Vienna. 1975-1979 (1) 1975-1979 (5) 1975-1979 (9) – ISSN: 0010-647X – mf#2376 – us UMI ProQuest [639]

Conservation of national ideals / Wells, Delphine Bartholomew et al – New York: F H Revell c1911 [mf ed 1990] – (= ser Home mission study course 8) – 1mf – 9 – 0-7905-3866-0 – mf#1989-0359 – us ATLA [975]

The conservation of natural resources through the electrification of railways / Cole, George Percy – [S.l: s.n, 1913] – 1mf – 9 – 0-659-90037-8 – mf#9-90037 – cn Canadiana [380]

The conservation of pictures / Holyoake, Manfred – London 1870 – (= ser 19th c art & architecture) – 1mf – 9 – mf#4.2.1143 – uk Chadwyck [700]

Conservationist – Albany. 1946-1994 (1) 1972-1994 (5) 1975-1994 (9) – (cont by: new york state conservationist) – ISSN: 0010-650X – mf#8071 – us UMI ProQuest [639]

Conservatismo una ideologia cristiana / Jurado E, Gerardo A – s.l, s.l? 196-? – 1r – us UF Libraries [972]

Conservative : and general advertiser for drogheda, meath, louth, monaghan and cavan – Drogheda, Ireland 17 sep 1853-20 feb 1864 [mf 1849-96] – 1 – (cont: conservative; and drogheda, louth, meath, monaghan & cavan advertiser [16 jun 1849-10 sep 1853]; cont by: drogheda conservative, & general advertiser for the counties of meath, louth, dublin, monaghan & cavan [27 feb 1864-3 oct 1908]) – uk British Libr Newspaper [072]

Conservative – Drogheda, 7 apr 1855, aug 1887-17 oct 1891 – 1r – 1 – ie National [072]

Conservative : a journal devoted to the discussion of political, economic, and sociological questions / ed by Morton, J Sterling – Nebraska City, NE: Morton Print Co. 4v. v1 n1. jul 14 1898-v4 n47. may 29 1902 (wkly) [mf ed filmed 1988] – 2r – 1 – (cont by: nebraska city weekly) – us NE Hist [071]

Conservative – Linden WI. 1909 apr 22-1911 feb 9 – 1r – 1 – mf#932824 – us WHS [071]

Conservative / Morgan Co. McConnelsvill – v1 n1. aug 1866-mar 1871 [wkly] – 2r – 1 – mf#B5528-5529 – us Ohio Hist [071]

Conservative Baptist Association of America *see* Cba builder

The Conservative Baptist Foreign Mission Society *see* News and views

Conservative Baptist Foreign Mission Society *see* Impact

Conservative baptist foreign mission society : annual report – 1959-65; 1971-83 [complete] – 1r – 1 – mf#ATLA S0450 – us ATLA [242]

Conservative baptists – Sep 1960-Nov 1965 – 1 – us Southern Baptist [242]

Conservative Caucus Research, Analysis and Education Foundation *see* Senate report

Conservative character of the english reformation... / Butcher, Samuel – Dublin, Ireland. no date – 1r – us UF Libraries [242]

Conservative correspondenz – Berlin DE, 1887-89, 1891-95, 1897, 1902 – 1 – gw Misc Inst [074]

Conservative digest – Washington. 1987-1989 (1) 1987-1989 (5) 1987-1989 (9) – ISSN: 0146-0978 – mf#16731 – us UMI ProQuest [320]

Conservative journal including wallace for president news – v1 n1-v2 n6 [1967 jun-aug/sep] – 1r – 1 – mf#1054846 – us WHS [325]

Conservative judaism – New York. 1980+ [1,5,9] – ISSN: 0010-6542 – mf#12691 – us UMI ProQuest [270]

The conservative reformation and its theology : as represented in the augsburg confession, and in the history and literature of the evangelical lutheran church / Krauth, Charles Porterfield – Philadelphia: J B Lippincott, 1871 – 2mf – 9 – 0-8370-8686-8 – (incl bibl ref and index) – mf#1986-2686 – us ATLA [242]

Conservative review – Washington. 1899-1901 (1) – mf#2877 – us UMI ProQuest [320]

Conservative review – Washington. 1990-1997 (1,5,9) – ISSN: 1047-5990 – mf#18388 – us UMI ProQuest [338]

The conservative review – v1-5. 1899-1901 – 1 – us L of C Photodup [073]

Conservator – Menasha, Neenah WI. 1856 may 14-1859 nov 28 – 1r – 1 – (cont by: menasha conservator) – mf#1108824 – us WHS [071]

Conservator – Elkhorn WI. 1859 jun 28 – 1r – 1 – mf#875424 – us WHS [071]

The conservator – Chicago, IL: A F Bradley. v5 n42. nov 18 1882 (wkly) [mf ed 1947] – (= ser Negro Newspapers on Microfilm) – 1r – 1 – us L of C Photodup [071]

The conservator – Philadelphia. v1-9. mar 1890-feb 1899 – 1 – us NY Public [073]

Conservatorio di Musica Luigi Cherubini. Florence *see* Annuario

Conserve neighborhoods – n1-61 [1978 jun/jul-1986 sep] – 1r – 1 – mf#1507224 – us WHS [071]

Conseso Provincial de Fomento *see* Folleto de las conferencias dadas durante la semana agricola de badajoz del 12-18 nov. 1912

Consider your ways / Ryle, J C – Ipswich, England. no date – 1r – us UF Libraries [240]

Consideracines sobre el credito en el salvador hac... / Ortiz Mancia, Alfredo – San Salvador, El Salvador. 1938 – 1r – us UF Libraries [972]

Consideraciones acerca de la segunda paradoja de el brocense / Gonzalez de la Calle, Pedro Urbano – Caceres: Tip. Extremadura – sp Bibl Santa Ana [946]

Consideraciones demograficas sobre el censo de buenos aires / Instituto Geografico – Buenos Aires, 1883 – 1mf – 9 – sp Cultura [972]

Consideraciones en torno al regimen colonial / Padin, Jose – Rio Piedras? Puerto Rico. 1945 – 1r – us UF Libraries [972]

Consideraciones sobre diversas categorias de fuerzas... / Nieto y Serrano, M – Madrid, 1886 – 9 – sp Cultura [610]

Consideraciones sobre la diplomacia / Donoso Cortes, Juan Francisco – Madrid: Tip. R. Rodriguez de Rivera, Tomo 1. 1848 – 1 – sp Bibl Santa Ana [320]

Consideraciones sobre la historia sismica / Martinez Barrio, Domingo – Ciudad Trujillo, Dominican Republic. 1946 – 1r – us UF Libraries [972]

Consideraciones sobre lo que significa... cristiano / Hernandez Cardenal, Garcia – 1570 – 9 – sp Bibl Santa Ana [240]

Consideraciones sobre los dolores de la santisima virgen maria. primer dolor / Tena Fernandez, Juan – Trujillo: Tip. sobrino de B. Pena, 1928 – 1 – sp Bibl Santa Ana [240]

Consideraciones...ferrocarriles...caceres / Godinez de la Paz, Carlos – 9 – sp Bibl Santa Ana [380]

Consideracoes sobre o problema das transferencias de angola / Associacao Do Comercio E Industria De Luanda – Lisboa, Portugal. 1932 – 1r – us UF Libraries [960]

Consideratien over de toestand in makasar, 1699, 1708 – 2mf – 8 – mf#SD-102 mf 10-11 – ne IDC [959]

Consideration / Doucet, Camille – Paris, France. 1860 – 1r – us UF Libraries [440]

Considerations concerning the sacrament of our lord's body and blood / Hall, Arthur Crawshay Alliston – New York: Longmans, Green, 1917 – 1mf – 9 – 0-524-02688-2 – mf#1990-4395 – us ATLA [240]

Considerations, explanatory and recommendatory... / Buchanan, Dr – Edinburgh, Scotland. 1850 – 1r – us UF Libraries [240]

Considerations on civil establishments of religion / Heugh, Hugh – Glasgow, Scotland. 1833 – 1r – us UF Libraries [230]

Considerations on modern theories of geology / Gisborne, Thomas – London, England. 1837 – 1r – us UF Libraries [550]

Considerations on some of the more popular mistakes and misrepresen... – London, England. 1830 – 1r – us UF Libraries [240]

Considerations on the divine authority of the lord's day / Cameron, Charles Richard – Oxford, England. 1831 – 1r – us UF Libraries [240]

Considerations on the economics and platform of the free church of... / Chalmers, Thomas – Glasgow, Scotland. 1843 – 1r – us UF Libraries [240]

Considerations on the expediency of procuring an act of parliament : for the settlement of the province of quebec / Maseres, Francis – London: Printed by Robert Wilks...sold by John White...1809? – 2mf – 9 – mf#61352 – cn Canadiana [241]

Considerations on the expediency of revising the liturgy and artic... – London, England. 1790 – 1r – us UF Libraries [240]

Considerations on the expediency of the congregation of st paul's – Aberdeen, Scotland. 1831 – 1r – us UF Libraries [240]

Considerations on the general distribution of the bible – London, England. 1819 – 1r – us UF Libraries [220]

Considerations on the importance of canada and the bay and river of st lawrence : and of the american fisheries dependant on the islands of cape breton, st john's, newfoundland, and the seas adjacent... – London: printed for W Owen, 1759 [mf ed 1983] – 1mf – 9 – 0-665-44017-0 – mf#44017 – cn Canadiana [971]

Considerations on the pentateuch / Taylor, Isaac – 3rd ed. London: Jackson, Walford, and Hodder, 1863 – 1mf – 9 – 0-8370-5483-4 – mf#1985-3483 – us ATLA [221]

Considerations on the practicability, policy, and obligation of communicating to the natives of india the knowledge of christianity : with observations on the "prefatory remarks" to a pamphlet published by major scott waring / Teignmouth, John Shore, Baron – London 1808 – (= ser 19th c british colonization) – 2mf – 9 – mf#1.1.490 – uk Chadwyck [240]

Considerations on the present political state of india : embracing observations on the character of the natives, on the civil and criminal courts, the administration of justice, the state of the land-tenure, the condition of the peasantry / Tytler, Alexander Fraser – London 1815 – (= ser 19th c british colonization) – 10mf – 9 – mf#1.1.8339 – uk Chadwyck [320]

Considerations on the proceedings of a general court-martial, upon the trial of lieutenant-general sir john moraunt sic : (as published by authority), with an answer to the expedition against rochefort... – London: Printed for S Hooper and A Morley...1758 – 1mf – 9 – mf#20261 – cn Canadiana [355]

Considerations on the proposed re-union of the canadian and english wesleyan conferences – Picton, CW [Ont]: printed for the aut, 1847 – 1mf – 9 – 0-665-91078-9 – mf#91078 – cn Canadiana [242]

Considerations on the question...whether the proceedings of commanders in chief of fleets and armies...are subject to the review of the civil courts of law... / Pulteney, William – London: J Stockdale, 1786 – 1mf – 9 – $1.50 – mf#LLMC 89-019 – us LLMC [347]

Considerations on the state of british india : embracing the subjects of colonization; missionaries; the state of the press; the nepaul and mahrattah wars; the civil government; and indian army / White, Adam – Edinburgh 1822 [mf ed Hildesheim 1995-98] – 1v on 3mf – 9 – €90.00 – 3-487-27526-0 – gw Olms [954]

Considerations on the universality and uniformity of the theocracy – London, England. 1796 – 1r – us UF Libraries [240]

Considerations on the uses of paper money and the effects of the banking system of canada : together with the project of a law to regulate banks of issue – [Montreal?: s.n.] 1836 [mf ed 1984] – 1mf – 9 – 0-665-44016-2 – mf#44016 – cn Canadiana [332]

Considerations on two fundamental principles of the anglican reformation : with preliminary observations on certain characteristic failures of different forms of christianity / Waterman, Lucius – New York: Edwin S Gorham, 1916 – 1mf – 9 – 0-524-03027-8 – mf#1990-4549 – us ATLA [241]

Considerations on volcanos : the probable causes of their phenomena, the laws which determine their march, the disposition of their products, and their connexion with the present state and past history of the globe / Scrope, George Julius Duncombe Poulett – London, 1825 – (= ser 19th c evolution & creation) – 4mf – 9 – mf#1.1.1190 – uk Chadwyck [550]

Considerations proposees aux eveques du concile sur la question de... – Ratisbonne, Germany. 1869 – 1r – us UF Libraries [240]

Considerations suited to the present crisis – Edinburgh, Scotland. 1831 – 1r – us UF Libraries [240]

Considerations sur la conscience nationale / Baguidy, Joseph D – Port-Au-Prince, Haiti. 1945? – 1r – us UF Libraries [972]

Considerations sur l'agriculture canadienne : au point de vue religieux, national et du bien-etre materiel... / Pelletier, Thomas Benjamin – Quebec?: s.n, 1860 – 1mf – 9 – mf#37161 – cn Canadiana [630]

Considerations sur l'annexion / Desjardins, Louis-Georges – Quebec: [s.n.], 1891 [mf ed 1980] – 1mf – 9 – 0-665-02684-6 – mf#02684 – cn Canadiana [327]

Considerations sur les cercles agricoles et les societes d'agriculture avec instructions aux juges dans les concours et les expositions : echelles de points (officielles) pour juger les cheveaux, le betail, les moutons, porcs... / Dallaire, O-E – Quebec: [s.n.], 1902 (mf ed 1995) – 9 – cn Bibl Nat [630]

Considerations sur les effets qu'ont produit en canada : la conservation des etablissemens du pays, les moeurs, l'education, etc... / Viger, Denis-Benjamin – Montreal: J Brown, 1809 [mf ed 1971] – 1r – 5 – mf#SEM17P91 – cn Bibl Nat [350]

Considerations sur les lois civiles du mariage / Girouard, Desire – Montreal?: s.n, 1868 – 1mf – 9 – mf#03437 – cn Canadiana [346]

Considerations sur l'etat present du canada : d'apres un manuscrit aux archives du bureau de la marine a paris / Tremais, Querdisien – [s.l: s.n, 1840?] [mf ed 1984] – 1mf – 9 – 0-665-20307-1 – mf#20307 – cn Canadiana [330]

Considine, William J *see* The effect of two types of plyometric training in improving vertical jump ability in female college soccer players

Consilia / Gutierrez, Juan – 1595 – 9 – sp Bibl Santa Ana [240]

Consilia...perfecta...per johannem de acevedo / Acevedo, Alfonso de – 1737 – 9 – sp Bibl Santa Ana [340]

Consiliorum sive responsorum / Gutierrez, Juan – vl. 1611 – 9 – (v1 1618. v1 1730) – sp Bibl Santa Ana [240]

Consistencia de el jubileo maximo de el ano santo : y de la suspension de indulgencias dentro de el: respeuesta a las proposiciones de vn anonymo... / Robles, Antonio de – en Mexico: Por Dona Maria de Benavides, viuda de Juan de Ribera, ano de 1700 – (= ser Books on religion...1543/44-c1800: miscelanea) – 1mf – 9 – mf#crl-411 – ne IDC [241]

The consistency of the divine conduct in revealing the doctrines of redemption : being the hulsean lectures for the year 1841 / Alford, Henry – Cambridge: Printed for J & J J Deighton; London: J G F & J Rivington. 2v. 1842-43 – 2mf – 9 – 0-7905-0901-6 – (incl bibl ref) – mf#1987-0901 – us ATLA [220]

The consistent – Odell, NE: Consistent Pub Co. v4 n24. dec 2 1892 (wkly) [mf ed filmed 1979] – 1r – 1 – us NE Hist [071]

Consistoire Central Israelite De France *see* Memorial en souvenir de nos rabbins et ministres officiants vic...

Consolacion de la filosofia = [Consolation of philosophy] / Boethius, Anicius Manlius Severinus – Buenos Aires, Argentina. 1943 – 1r – 1 – us UF Libraries [180]

Consolatio ad senatum argentinensem de morte... / Sturm, J – Argentorati, 1553 – 1mf – 9 – mf#PPE-140 – ne IDC [240]

Consolation : in discourses on select topics, addressed to the suffering people of god / Alexander, James Waddel – 6th ed. New York: Scribner, Armstrong, 1873 [mf ed 1984] – 5mf – 9 – 0-8370-1017-9 – mf#1984-4400 – us ATLA [240]

Consolation for mourners / Hill, John – London, England. 1818 – 1r – us UF Libraries [240]

A consolation for our grammar schooles / Brinsley, John – 1622 – 9 – us Scholars Facs [370]

Consolation to the church / Culbertson, Robert – Edinburgh, Scotland. 1807 – 1r – us UF Libraries [240]

Consolations a ceux qui pleurent ou tresor des malades / Picard, Eustache – Montreal: E Senecal, 1872 – 4mf – 9 – mf#00424 – cn Canadiana [240]

Les consolations de l'ame fidele, contre les frayeurs de la mort / Drelincourt, C – Amsterdam, 1660 – 9mf – 9 – mf#PRS-138 – ne IDC [240]

Consolations in travel : or the last days of a philosopher / Davy, Humphry – London, 1830 – (= ser 19th c evolution & creation) – 4mf – 9 – mf#1.1.10847 – uk Chadwyck [920]

The consolations of the cross : addresses on the seven words of the dying lord / Brent, Charles Henry – New York: Longmans, Green, 1904 – 1mf – us ATLA [240]

The consolations of the cross : addressses on the seven words of the dying lord / Brent, Charles Henry – New York: Longmans, Green, 1904 – 1mf – 9 – 0-7905-3550-5 – mf#1989-0043 – us ATLA [240]

La consolatrice / Ferland, Albert – [Montreal?: s.n, 1898?] – 1mf – 9 – 0-665-94593-0 – mf#94593 – cn Canadiana [810]

Consoletter / Consolidated Papers, Inc – 1958 oct-1973 jun 13 – 1r – 1 – mf#1110917 – us WHS [670]

Consolidacao da republica / Magalhaes, Joao Baptista – Rio de Janeiro, Brazil. 1947 – 1r – us UF Libraries [972]

Consolidated Edison Company of New York, Inc *see* Observations of the total solar eclipse of january 24, 1925

Consolidated edition as at the 31st dec 1955 of the penal code and the criminal procedure code – Zomba, Govt print, [1956] – (prepared by the attorney general under section 44 of the interpretation and general clauses ordinance n2 of 1953) – us CRL [348]

Consolidated history of the churches of the oxford baptist association, state of maine : and a historical sketch of the association / Crockett, George B – Bryant's Pond ME: AM Chase 1905 [mf ed 1993] – 2mf – 9 – 0-524-08747-4 – mf#1993-3252 – us ATLA [242]

Consolidated Independent Union et al *see* Labor review

Consolidated index to compiled service records of confederate soldiers / U.S. War Dept – (= ser Records of confederate soldiers who served during the civil war) – 535r – 5 – (with printed guide) – mf#M253 – us Nat Archives [355]

Consolidated list of debarred, suspended, and ineligible contractors as of ... – 1982-86 – 1r – 1 – (cont: consolidated list of current administrative debarments by executive agencies; gsa debarred bidders list) – mf#2045484 – us WHS [690]

The consolidated municipal act, 1883 / Bell, George – Toronto: Hart, 1883? – 4mf – 9 – (incl ind) – mf#59400 – cn Canadiana [350]

Consolidated news / Consolidated Papers, Inc – 1963 sep-1975 dec, 1973 nov-1983 dec – 2r – 1 – mf#950449 – us WHS [670]

Consolidated news / Consolidated Vultee Aircraft Corporation – v2 n1-v3 n48 [1943 jan 7-1944 dec 28] – 1r – 1 – mf#716296 – us WHS [629]

Consolidated Papers, Inc *see*
- Consoletter
- Consolidated news

The consolidated statutes for lower canada : proclaimed and published under the authority of the act 23 vict cap 56, ad 1860 = Les statuts refondus pour le bas-canada:... L'acte 23 vic cap 56, ad 1860 / Canada (Province) – Quebec: printed by Stewart Derbishire & George Desbarats, 1861 [mf ed 1982] – 13mf – 9 – (with ind) – mf#SEM105P139 – cn Bibl Nat [348]

The consolidated statutes for upper canada : proclaimed and published under the authority of the act 22 vict cap 30, ad, 1859 = Acte relatif aux statuts refondus pour le haut canada / Canada (Province) – Toronto: printed by Stewart Derbishire & George Desbarats, 1859 [mf ed 1982] – 13mf – 9 – (with ind) – mf#SEM105P138 – cn Bibl Nat [348]

The consolidated statutes of canada : proclaimed and published under the authority of the act 22 vict cap 29, ad 1859 / Canada (Province) – Toronto: printed by Stewart Derbishire & George Desbarats, 1859 [mf ed 1982] – 15mf – 9 – (with ind) – mf#SEM105P126 – cn Bibl Nat [348]

Consolidated statutes respecting the militia : approved by proclamation of his excellency the governor general / Canada (Province) – Toronto: printed by Stewart Derbishire & George Desbarats, 1859 [mf ed 1983] – 1mf – 9 – mf#SEM105P179 – cn Bibl Nat [355]

Consolidated Vultee Aircraft Corporation *see* Consolidated news

The consolidation of the christian power in india / Basu, Baman Das – Calcutta: R Chatterjee, 1927 – (= ser Samp: indian books) – us CRL [954]

The consolidation of the church in canada : a plea for a general synod with legislative powers, from the north-west / Anson, Adelbert – Toronto: J P Clougher, 1892? – 1mf – 9 – mf#05872 – cn Canadiana [242]

Consommation : annales du c.r.e.d.o.c. – Paris. 1958-72 – 1 – fr ACRPP [073]

Consonanzen und dissonanzen : gesamte schriften aus alterer und neurer zeit / Lobe, Johann Christian – Leipzig: Baumgaertner's Buchhandlung 1869 [mf ed 19–] – 1r – 1 – mf#pres. film 101 – us Sibley [780]

Consortium on Peace Research, Education and Development [US] *see* Copred peace chronicle

Consort-journal of the dolmetsch foundation – Godalming. 1929-1990 (1) 1973-1989 (5) 1973-1989 (9) – ISSN: 0268-9111 – mf#8905 – us UMI ProQuest [780]

Conspiracies unlimited – v1 n1-v4 n4 [1978-86] – 1r – 1 – mf#1219907 – us WHS [071]

Conspiracion de los alcarrizos / Henriquez Urena, Max – Lisboa, Portugal. 1941 – 1r – us UF Libraries [972]

Conspiracy / National Lawyers Guild – 1970 sep-1975 jun, 1975 jul-1988 jul – 2r – 1 – mf#361203 – us WHS [340]

La conspiration de cellamare : episode de la regence / Vatout, Jean – Paris 1832 [mf ed Hildesheim 1995-98] – 2v on 6mf – 9 – €120.00 – ISBN-10: 3-487-26207-X – ISBN-13: 978-3-487-26207-9 – gw Olms [944]

La conspiration des poudres – [Paris]: Impr Populaire de J Dupont, jun 4-8 1848 – us CRL [944]

Conspiration pour l'egalite dite de babeuf : suivie du proces auquel elle donna lieu, et des pieces justificatives, etc, etc / Buonarroti, Filippo M – Bruxelles 1828 [mf ed Hildesheim 1995-98] – 2v on 4mf – 9 – €120.00 – 3-487-26259-2 – gw Olms [944]

Conspirators in conflict / Rhoodie, Denys O – Cape Town, South Africa. 1967 – 1r – us UF Libraries [960]

Constable and old suffolk artists / Ipswich. Fine Arts Club – 1887 – 9 – $4.90 – uk Chadwyck [750]

Constable, Archibald *see* Nineteenth century literary manuscripts

Constable, Henry *see*
- The duration and nature of future punishment
- Duration and nature of future punishment

The constable's guide and form book / Marsh, Howard Franklin – Wellsboro, 167p. LL-675 – 1 – us L of C Photodup [340]

The constables' manual: being a summary of the law relating to the rights, powers, and duties of constables / Clarke, Samuel Robinson – Toronto: Ure, 1881. 120p. LL-2293 – 1 – us L of C Photodup [340]

Constance fenimore woolson papers, 1875-1894 / Woolson, Constance Fenimore – [mf ed 1961] – 1r – 1 – us Western Res [860]

Constance, Lady Arbuthnot *see* Memories of rugby and india

Constancy, and other poems : in which may be found poems of home life, of school life, and of the christian life / Baker, Naaman Rimnon – Mt Morris, Ill: Brethren's Pub Co, 1894 – 1mf – 9 – 0-524-04035-4 – mf#1990-4943 – us ATLA [420]

Constant, Benjamin *see* De la religion

Constant captain, gonzalo de sandoval / Gardiner, Clinton Harvey – Carbondale, IL. 1961 – 1r – us UF Libraries [972]

Constant de Rebecque, Benjamin de *see* Memoires sur les cent jours

Constant, G *see*
- Rapport sur une mission scientifique aux archives d'autriche et d'espagne
- The reformation in england 1: the english schism henry 8th (1509-1547). london, 1934
- The reformation in england. the english schim. henry 7th 1509-1547, london 1934

Constant, Victor Nevers *see* Vie a port-au-prince par l'image

Constantes fondamentales de l'astronomie : paris, 27 mars-1er avril 1950. / Centre national de la recherche scientifique (France) – Paris: Service des publications du Centre national de la recherche scientifique 1950 [mf ed 2006] – (= ser Colloques internationaux du centre national de la recherche scientifique 25) – 1r – 1 – (in french & english; "le present volume...est constitue par la reunion des articles parus dans le bulletin astronomique, tome 15, fasc 3 et 4"; incl bibl ref) – mf#film mas 37490 – us Harvard [520]

Constantin der grosse als religionspolitiker : kirchengeschichtlicher essay / Brieger, Theodor – Gotha: FA Perthes, 1880 – 1mf – 9 – 0-524-03272-6 – (incl bibl ref) – mf#1990-0883 – us ATLA [240]

Constantin tischendorf in seiner fuenfundzwanzigjaehrigen schriftstellerischen wirksamkeit : literar-historische skizze / Volbeding, Johann Ernst – Leipzig: C Fr. Fleischer, 1862 – 1mf – 9 – 0-7905-6210-3 – (incl bibl ref) – mf#1988-2210 – us ATLA [430]

Constantine, Learie Nicholas *see* Cricket and i

Constantine the great : the reorganisation of the empire and the triumph of the church / Firth, John B – New York: G P Putnam, 1905 – (= ser Heroes of the nations) – 1mf – 9 – 0-7905-4568-3 – mf#1988-0568 – us ATLA [930]

Constantine the great : the union of the state and the church / Cutts, Edward Lewes – London: SPCK; New York: Pott, Young 1881 [mf ed 1990] – (= ser The home library) – 1mf – 9 – 0-7905-5596-4 – mf#1988-1596 – us ATLA [240]

Constantine the great and christianity : three phases: the historical, the legendary, and the spurious / Coleman, Christopher Bush – New York: Columbia University Press; London: PS King, 1914 – 1mf – 9 – 0-7905-4212-9 – (incl bibl ref) – mf#1988-0212 – us ATLA [930]

Constantineau, Philippe *see* Parmenides und hegel ueber das sein

Constantini Manassis *see*
- Breviarium historiae metricum
- Breviarium historicum

Constantini Porphyrogenneti *see* Libri duo de ceremoniis aulae byzantinae (cbh33)

Die constantinische schenkung / Friedrich, Johann – Noerdlingen: Beck, 1889 – 1mf – 9 – 0-7905-6172-7 – (includes text of the constitutum constantini, and bibliographical references) – mf#1988-2172 – us ATLA [930]

Constantinople : a sketch of its history from its foundation to its conquest by the turks in 1453 / Brodribb, William Jackson & Besant, Walter – London: Seeley, Jackson, & Halliday 1879 [mf ed 1990] – 1r – 1 – (filmed with: roman emperor worship / sweet, l m) – mf#1770p – us Wisconsin U Libr [931]

Constantinople and its problems, its peoples, customs, religions and progress / Dwight, H G O – London, 1901 – 4mf – 9 – mf#HT-174 – ne IDC [915]

Constantinople and the scenery of the seven churches of asia minor / Allom, Thomas – London [1838] – (= ser 19th c art & architecture) – 7mf – 9 – mf#4.2.1576 – uk Chadwyck [720]

Constantinople et le bosphore de thrace : pendant les annees 1812, 1813 et 1814, et pendant l'annee 1826 / Andreossy, Antoine F – Paris 1828 [mf ed Hildesheim 1995-98] – (= ser Fbc) – 4mf – 9 – €120.00 – 3-487-29124-X – gw Olms [915]

Constantinople in 1828 : a residence of sixteen months in the turkish capital and provinces: with an account of the present state of the naval and military power, and of the resources of the ottoman empire / Mac Farlane, C – London, 1829 – 5mf – 9 – mf#AR-2026 – ne IDC [956]

Constantinopoleos *see* Historia politica et patriarchica (cshb47)

Constantinus porphyrogenitus imperator, vols 1-3 (cshb9-11) – (= ser Corpus scriptorum historiae byzantinae (cshb)) – (v1+2: de cerimoniis aulae byzantinae libri duo, graece et latine, e recensione io iac reiskii, cum eiusdem commentariis integris v1 bonnae 1829 €29. v2 bonnae 1830 €31. v3: de thematibus et de administrando imperio. accedit hieroclis synecdemus cum bandurii et wesselingii commentariis. rec imm bekkerus, bonnae 1840 €19) – ne Slangenburg [240]

Constantin-Weyer, Maurice *see* Stratageme des roues

Constantius *see* The patriarchate of antioch

Constantius, Constantin *see* Gjentagelsen

Constanze – Hamburg. 1968-1969 (1) – mf#5099 – us UMI ProQuest [305]

Constelacion de celebres terciarios / Arcila Robledo, Gregorio – Bogota, Colombia. 1950? – 1r – us UF Libraries [972]

Constellation – Paris. 1830-1833 – 1 – ISSN: 0010-6615 – mf#10618 – us UMI ProQuest [073]

Constellation [Aircraft carrier] *see* Slep times

The constellation (no. 4 of mr. francis's ballroom assistant) / Reinagle, Alexander – n.p., n.d.. MUSIC 3082, Item 10 – 1 – us L of C Photodup [780]

Constellations – Oxford. 1994+ (1,5,9) – (cont: praxis international=praxis) – ISSN: 1351-0487 – mf#20775 – us UMI ProQuest [320]

Le constitionnel *see* Mechroutiette

Constitucion de 1886 y las reformas proyectadas – Bogota, Colombia. 1936 – 1r – us UF Libraries [323]

Constitucion de comision nacional de cooperacion economica : una etapa mas en la ejecucion de la doctrina peronista en el orden economico – Buenos Aires. 1950 – 1 – us CRL [339]

Constitucion de guatemala como obra / Vidaurre, Adrian – Guatemala, 1935 – 1r – us UF Libraries [323]

Constitucion de la corporacion municipal el 6 de febrero de 1949 / Caceres – Caceres: Tip El Noticiero, 1949 – sp Bibl Santa Ana [628]

Constitucion de la corporacion municipal el 31 de julio de 1948 / Caceres – Caceres: Tip. El Noticiero, 1948 – sp Bibl Santa Ana [338]

Constitucion de la republica / Honduras Constitution – Tegucigalpa, Mexico. 1965 – 1r – us UF Libraries [323]

Constitucion de la republica de colombia y sus ant... – Bogota, Colombia. 1950 – 1r – us UF Libraries [323]

Constitucion de la republica de guatemala – Guatemala, 1956 – 1r – us UF Libraries [323]

Constitucion de la republica de guatemala / Guatemala Constitution – Guatemala, 1928 – 1r – us UF Libraries [323]

Constitucion de la republica dominicana / Dominican Republic Constitution – Ciudad Trujillo, Dominican Republic. 1947 – 1r – us UF Libraries [323]

Constitucion de la republica dominicana / Dominican Republic Constitution – Santo Domingo, Dominican Republic. 1934 – 1r – us UF Libraries [323]

Constitucion de los estados-unidos / U.S. Constitution – Mayaguez, Tip. Comercial, 1898. 52 p. LL-1678 – 1 – us L of C Photodup [340]

Constitucion en 1925. noticia de vicente castaneda / Comision Provincial de Monumentos – Madrid: Tip. Rev. Arch. y Museos, 1927. Boletin R.A. Historia, 1927 – 1 – sp Bibl Santa Ana [324]

Constitucion en 1928, noticia de vicen te castaneda / Comision Provincial de Monumentos – Madrid: Tip. R.B. Arch. y Museos, 1928 – 1 – sp Bibl Santa Ana [324]

Constitucion nacional / Colombia – Bogota, Colombia. 1910 – 1r – us UF Libraries [323]

Constitucion politica / Arboleda, Sergio – Bogota, Colombia. 1952 – 1r – us UF Libraries [321]

Constitucion politica de la republica de el salvador – San Salvador, El Salvador. 1965 – 1r – us UF Libraries [323]

Constitucion politica y leyes constitutivas de la... / Honduras Constitution – Tegucigalpa, Mexico. 1915 – 1r – us UF Libraries [323]

Constitucion politica y reformas constitucionales / Dominican Republic – Santiago, Dominican Republic. v1-2. 1944 – 1r – us UF Libraries [323]

Constitucion y codigos de la republica / El Salvador Laws, Statutes, Etc – San Salvador, El Salvador. 1947 – 1r – us UF Libraries [323]

Constitucion y codigos de la republica de guatemala / Guatemala Laws, Statutes, Etc – Guatemala, 1957 – 1r – us UF Libraries [323]

O constitucional – Bahia: Typ da Viuva Serva & Carvalho, 10 abr-21 ago 1822 – (= ser Ps 19) – mf#P19,01,13 – bl Biblioteca [323]

O constitucional – Bahia: Typ da Viuva Serva & Carvalho, 10 abr-21 ago 1822 – (= ser Ps 19) – mf#P19,01,13 – bl Biblioteca [323]

O constitucional : diario mercantil, politico, e litterario – Rio de Janeiro, RJ: Typ de Gueffier & C, 04 maio-10 set 1831 – (= ser Ps 19) – mf#P02,04,21 – bl Biblioteca [323]

O constitucional : jornal politico – Sao Paulo, SP: Typ do Constitucional, 13 abr 1861-30 abr 1863 – (= ser Ps 19) – mf#P06,01,16 – bl Biblioteca [320]

O constitucional : jornal politico e literario – Pernambuco: Typ do Diario, 06 jul-dez 1829; jan-abr,ago-dez 1830; abr, 06-09 jun 1831 – (= ser Ps 19) – mf#P19,03,01 – bl Biblioteca [073]

O constitucional : jornal politico e noticioso – Desterro, SC: Typ de Jose Joaquim Lopes, 17 jul 1867-01 abr 1869 – (= ser Ps 19) – mf#UFSC/BPESC – bl Biblioteca [320]

O constitucional : jornal politico, litterario, industrial e noticioso – Desterro, SC: Typ Brasilea de F P M de Carvalho, 25 mar 1870-28 abr 1871 – (= ser Ps 19) – bl Biblioteca [073]

O constitucional – Macae, RJ: Typ do Constitucional, 20 jul 1881-abr 1882; jul,out 1884; dez 1885; 02 out 1886 – (= ser Ps 19) – mf#P18A,04,13 – bl Biblioteca [323]

O constitucional – Ouro Preto, MG. 21 jan 1846 – (= ser Ps 19) – mf#P19B,01,19 – bl Biblioteca [323]

O constitucional – Ouro Preto, MG: Typ de Paula Castro, 25 abr-07 ago 1878 – (= ser Ps 19) – mf#P11B,03,70 – bl Biblioteca [323]

O constitucional – Rio de Janeiro, RJ: Typ Franceza, 18 set 1841-12 mar 1842 – (= ser Ps 19) – mf#P14,01,19 – bl Biblioteca [323]

O constitucional aiuruoca, mg : orgao da ideia republicana – 02 jan 1898 – (= ser Ps 19) – mf#P31,03,10 – bl Biblioteca [323]

O constitucional mineiro – Sao Joao del Rei, MG: [s.n.] 18 set 1832-16 abr 1833 – (= ser Ps 19) – mf#P19B,01,29 – bl Biblioteca [320]

Constitucionalismo colombiano / Sachica, Luis Carlos – Bogota, Colombia. 1962 – 1r – us UF Libraries [323]

Constituciones / Escuelas de Maria Santisima – 1832 – 9 – sp Bibl Santa Ana [370]

Constituciones de cataluna (impresas en pergamino, siecle 15) – Barcelona – 1r – 5,6 – sp Cultura [348]

Constituciones de colombia – Bogota, Colombia. v1-4. 1951 – 1r – us UF Libraries [323]

Constituciones de costa rica / Costa Rica Constitucion Politica (1949) – Madrid, Spain. 1962 – 1r – us UF Libraries [323]

Constituciones de la congregacion de nuestra senora con el titulo de covadonga : defensora, y restauradora de la libertad espanola, funda baxo la real proteccion por los naturales y originarios del principado de asturias, y obispado de oviedo... – en Mexico: impr D Joseph de Jauregui...ano de 1785 – (= ser Books on religion...1543/44-c1800: ordenes, etc: congregacion de nuestra senora de covadonga) – 1mf – 9 – mf#crl-180 – ne IDC [241]

Constituciones de la congregacion y escuela de cristo – 1788 – 9 – sp Bibl Santa Ana [240]

Constituciones de la provincia de san diego de mexico / de los menores descalcos de la mas estrecha observancia regular de n s p s francisco en esta nueva-espana / Franciscans. Provincia de San Diego de Mexico – Mexico: ...Francisco Rodriguez Lupercio 1698 – (= ser Books on religion... 1543/44-c1800: franciscanos) – 7mf – 9 – mf#crl-202 – ne IDC [241]

Constituciones de la provincia de san gabriel (1580) / Barrado Manzano, Arcangel – Madrid: Graf. Calleja, 1967 – 1 – sp Bibl Santa Ana [340]

Constituciones de panama / Panama Constitution – Madrid, Spain. 1954 – 1r – us UF Libraries [323]

Constituciones de puerto rico / Puerto Rico Constitution – Madrid, Spain. 1953 – 1r – us UF Libraries [323]

Constituciones eclesiasticas (siecle 14) – Zamora – 1r – 5,6 – sp Cultura [240]

Constituciones generales de la orden de los frailes menores = Constitutiones generales ordinis minorum – [Mexico: s.n., 1891] (Queretaro: Luciano Frias y Soto) – 1mf – 9 – 0-524-08863-2 – (incl ind . in spanish) – mf#1993-3327 – us ATLA [240]

Constituciones que el ilustrisimo senor doctor don alonso nunez de haro y peralta : del consejo de su magestad, y arzobispo de esta santa iglesia metropolitana de mexico... / Nunez de Haro y Peralta, Alonso – en Mexico: impr D Felipe de Zuniga y Ontiveros, ano de 1777 – (= ser Books on religion...1543/44-c1800: arzobispos: arzobispos de mexico) – 2mf – 9 – mf#crl-386 – ne IDC [241]

Constituciones, que para el mejor govierno : y direccion de la real casa del senor s joseph de ninos expositos de esta ciudad de mexico formo el ilmo sr dr d alonso nunez de haro, y peralta... / Nunez de Haro y Peralta, Alonso – en Mexico: impr Lic Joseph de Jauregui [1774] – (= ser Books on religion...1543/44-c1800: arzobispos: arzobispos de mexico) – 1mf – 9 – mf#crl-385 – ne IDC [241]

Constituciones sinodales del obispado de plasencia / Casas Souto, Pedro – 1892 – 9 – sp Bibl Santa Ana [240]

Constituciones synodales / Roys y Mendoza, Francisco – 1673 – 9 – sp Bibl Santa Ana [240]

Constituciones synodales del obispado de coria / Carvajal, Pedro de – 1608 – 9 – sp Bibl Santa Ana [240]

Constituciones synodales...granada / Guerrero, Pedro – 1573 – 9 – sp Bibl Santa Ana [240]

Constituciones y exercicios de la venerable madre sor maria de la antigua...iglesia parroquial santa maria la real – Ecija: Benito Daza. S. XVIII – 1 – sp Bibl Santa Ana [240]

Constituciones y regla de la minima congregacion de los hermanos enfermeros / Obregon, B – Madrid, 1634 – 9 – sp Cultura [610]

Constituent – 1993 jan/mar-jul/sep – 1r – 1 – mf#4851595 – us WHS [960]

Constituents of climate / Lente, Frederick D – Louisville, KY. 1878 – 1r – us UF Libraries [550]

Constituicao : do brasil ao alcance de todos / Sarasate, Paulo – Rio de Janeiro, Brazil. 1968 – 1r – us UF Libraries [323]

Constituicao de dez de novembro / Brazil – Rio de Janeiro, Brazil. 1940 – 1r – us UF Libraries [323]

Constituicoes federal e estaduais / Brazil. Ministerio da Justica e Negocios Interiore – Rio de Janeiro, Brazil. 1952 – 1r – us UF Libraries [323]

Constituido...2 de febrero de 1958 / Caceres – Caceres: Imprenta Moderna, s.a. – sp Bibl Santa Ana [946]

O constituinte – Niteroi, RJ: Typ Nictheroyense de M G de S Rego, 21 mar-16 jun 1855 – (= ser Ps 19) – mf#P17,02,20 – bl Biblioteca [320]

Constituinte republicana / Roure, Agenor De – Rio de Janeiro, Brazil. v1-2. 1920 – 1r – us UF Libraries [321]

Constituintes de 46 / Pereira De Silva, Gastao – Rio de Janeiro, Brazil. 1947 – 1r – us UF Libraries [323]

Constituiton of ngonde / Wilson, Godfrey – Livingstone, Zambia. 1939 – 1r – us UF Libraries [323]

Constitution – Cork, Ireland. 18 aug, 15 sep 1823, 1826-37, 1839-96, jul-sep 1921 – 163 1/4r – 1 – (aka: constitution or cork morning post; constitution or cork advertiser; cork constitution) – uk British Libr Newspaper [072]

Constitution – Baltimore, MD, Washington DC. 1844 oct 22-25, nov 1-5 – 1r – 1 – (cont: spectator [washington dc: 1843]) – mf#846132 – us WHS [071]

Constitution – Dublin, Ireland. may 1854-oct 1855 – 1/2r – 1 – uk British Libr Newspaper [072]

Constitution / Haiti – Port-Au-Prince, Haiti. 1932 – 1r – us UF Libraries [323]

Constitution – Keokuk, IA. 1879-1888 (1) – mf#63274 – us UMI ProQuest [071]

Constitution : or anti union evening post – Dublin, Ireland. 17 dec 1799-7 jan 1800; 11, 14, 21 jan; 8, 15, 18 feb; 22 mar; 7 jun 1800 – 1/4r – 1 – uk British Libr Newspaper [072]

Constitution : or, anti-union evening post – Dublin 1799/1800 – 1r – 1 – ie National [072]

Constitution – Robinson, IL. 1865-1919 (1) – mf#62686 – us UMI ProQuest [071]

Constitution – St Johns Newfoundland, Canada. 3 nov-17 nov 1883 – 1/4r – 1 – uk British Libr Newspaper [071]

Constitution – Toronto, ON: W L Mackenzie, 1836-37 – 1r – 1 – ISSN: 1181-3830 – cn Library Assoc [971]

Constitution / Trumbull Co. Warren – 1877-78, 1883-84 [wkly] – 2r – 1 – mf#B10500-10501 – us Ohio Hist [071]

Constitution / Trumbull Co. Warren – v1 n1. jul 1862-jun 1870, 1872-dec 1875 [wkly] – 4r – 1 – mf#B5569-5572 – us Ohio Hist [071]

Constitution – Washington DC. 1845 jan 25, apr 26 – 1r – 1 – mf#953739 – us WHS [071]

Constitution – Erie, PA. july 16, 1856-oct 20, 1858 – 1r – 1 – (weekly republican newspaper) – us Western Res [071]

The constitution – London. -w. 3 Apr 1831-22 Jan 1832. (1 reel) – 1 – uk British Libr Newspaper [072]

The constitution : or anti-union evening post – Dublin. Ireland. -sw. 17 Dec 1799-7 Jan 1800, 11, 14, 21 Jan, 8, 15, 18 Feb, 22 Mar, 7 Jun 1800. (16 ft) – 1 – uk British Libr Newspaper [072]

Constitution, 1786-1956 / Presbyterian Church in the U.S.A. General Assembly – 1 – $800.00 – us Presbyterian [240]

Constitution, 1958-1983 / United Presbyterian Church in the U.S.A – 1 – $200.00 – us Presbyterian [240]

Constitution and address to the christian public / Southern Aid Society – 1854. Annual report. v1-7. 1854-60 – 1 – $50.00 – us Presbyterian [240]

Constitution and by-laws : adopted september 24th, 1918 / Child Welfare Association of British Columbia – [British Columbia?: s.n, 1918?] – 1mf – 9 – 0-665-99272-6 – mf#99272 – cn Canadiana [360]

Constitution and by-laws : as adopted at the convention held at the queen's hotel, toronto, on thursday, march 21st, a d, 1889 / Ontario Association of Architects – Toronto?: s.n, 1889? – 1mf – 9 – mf#11445 – cn Canadiana [720]

Constitution and by-laws / British Columbia Mountaineering Club – [Vancouver?: Western Speciality], 1914) – 1mf – 9 – 0-659-91017-9 – mf#9-91017 – cn Canadiana [790]

Constitution and by-laws / British Columbia Rifle Association (Victoria, BC) – S.l: s.n, 1901? – 1mf – 9 – mf#25012 – cn Canadiana [790]

Constitution and by-laws : revised april, 1888 / Ottawa Lawn Tennis Club – Ottawa?: Mason & Reynolds, 1888 – 1mf – 9 – mf#11579 – cn Canadiana [790]

Constitution and by-laws / Schubert Club (Ottawa, Ont) – Ottawa: s.n, 1894 – 1mf – 9 – mf#13257 – cn Canadiana [780]

Constitution and by-laws / Toronto Produce Exchange – [Toronto?: s.n, 191-?] – 1mf – 9 – 0-659-91119-1 – mf#9-91119 – cn Canadiana [630]

Constitution and by-laws : with the...annual report / Hamilton Horticultural Society – Hamilton, Ont?: The Society, 1858-18– or 19– – 9 – mf#A01733 – cn Canadiana [635]

Constitution and by-laws... : revised and adopted dec 14, 1901: instituted a d 1875, incorporated by act of dominion parliament 1880 / Dominion Commercial Travellers' Association – Montreal: [s.n.], 1901 – 1mf – 9 – 0-665-86003-X – mf#86003 – cn Canadiana [360]

Constitution and by-laws of eureka council, no 13 / Canadian Order of Chosen Friends. Eureka Council – Hamilton, Ont?: E Barker, 1889 – 1mf – 9 – mf#56928 – cn Canadiana [360]

Constitution and by-laws of the... / Hamilton Mercantile Library Association and General News Room – [Hamilton, Ont?: s.n.] 1845 [mf ed 1994] – 1mf – 9 – 0-665-94598-1 – mf#94598 – cn Canadiana [360]

Constitution and by-laws of the aurora snow shoe club / Aurora Snow Shoe Club – Quebec: C Darveau, 1879 – 1mf – 9 – mf#04089 – cn Canadiana [360]

Constitution and by-laws of the chamber of commerce – Pensacola, FL. 1903 – 1r – us UF Libraries [380]

Constitution and by-laws of the chateauguay literary and historical society, ormstown, pq : organized october 26th, 1888 – Montreal: W Drysdale, 1889 [mf ed 1987] – 1mf – 9 – 0-665-56362-0 – mf#56362 – cn Canadiana [366]

Constitution and by-laws of the fruit growers' association of upper canada / Fruit Growers' Association of Upper Canada – Hamilton, Ont?: Spectator, 1859 [mf ed 1983] – 1mf – 9 – 0-665-44525-3 – mf#44525 – cn Canadiana [634]

Constitution and by-laws of the hamilton co-operative association : constituted december, 1864 / Hamilton Co-Operative Association (Ont) – [Hamilton, Ont?: s.n.] 1864 [mf ed 1983] – 1mf – 9 – 0-665-44948-8 – mf#44948 – cn Canadiana [366]

Constitution and by-laws of the natural history society of montreal : with the amending act, 20th vict, ch 118... / Natural History Society of Montreal – Montreal?: J Lovell, 1859 – 1mf – 9 – mf#36758 – cn Canadiana [366]

Constitution and by-laws of the ottawa bicycle club / Ottawa Bicycle Club – Ottawa?: A Bureau, 1890 – 1mf – 9 – mf#11587 – cn Canadiana [366]

The constitution and by-laws of the st george's society : established in the city of montreal in 1834, for the purpose of relieving their brethren in distress / St George's Society of Montreal – [Montreal?: s.n.] 1855 [mf ed 1983] – 1mf – 9 – mf#44673 – cn Canadiana [366]

Constitution and canons of the diocese of qu'appelle, assiniboia, n w canada : as formulated at a meeting of the diocese, june 3rd, 1885 – Winnipeg?: Manitoba Free Press Print, 1885 – 1mf – 9 – mf#30100 – cn Canadiana [242]

Constitution and canons of the synod of the diocese of toronto : with explanatory notes and comments / Bovell, James – [Toronto?: s.n.], 1858 [mf ed 1987] – 1mf – 9 – 0-665-67036-2 – mf#67036 – cn Canadiana [242]

Constitution and church sentinel – Dublin, Ireland. 14 apr 1849-15 dec 1852 – 1r – 1 – uk British Libr Newspaper [072]

Constitution and example of the seven apocalyptic churches / Churton, Ralph – Oxford, England. 1803 – 1r – us UF Libraries [240]

Constitution and general laws of the farmers' and mechanics' institute of streetsville, in the county of peel : incorporated, apr 3rd 1854 / Farmers' and Mechanics' Institute of Streetsville (Ont) – [Streetsville, Ont?: s.n.] 1858 [mf ed 1983] – 1mf – 9 – 0-665-44461-3 – mf#44461 – cn Canadiana [630]

The constitution and law of the church in the first two centuries = Entstehung und entwickelung der kirchenverfassung und des kirchenrechts in den zwei ersten jahrhunderten / Harnack, Adolf von; ed by Major, Henry Dewsbury Alves – London: Williams & Norgate; New York: G P Putnam, 1910 – 1mf – 9 – 0-7905-1060-X – (incl bibl ref) – mf#1987-1060 – us ATLA [240]

Constitution and laws : as revised and amended, july 1881 / Royal Canadian Academy of Arts – Toronto?: Globe Print Co, 1881 – 1mf – 9 – mf#12893 – cn Canadiana [700]

Constitution and laws of the... / Hamilton Literary Society – [Hamilton, Ont?: s.n.] 1837 [mf ed 1994] – 1mf – 9 – 0-665-94606-6 – mf#94606 – cn Canadiana [366]

Constitution and laws of the knights of the ku klux klan (incorporated) – Atlanta, GA: The Klan, c1921 – us CRL [366]

Constitution and playing rules / National League and American Association of Professional Baseball Clubs – New York etc. 1876-1890; 1895-1900 – 1 – us NY Public [790]

Constitution and polity of the new testament church / Weston, Henry Griggs – Philadelphia: American Baptist Publ Soc 1895 [mf ed 1988] – 1mf – 9 – 0-7905-0175-9 – (incl bibl ref & ind) – mf#1987-0175 – us ATLA [240]

Constitution and proceedings / Iowa State Federation of Labor – 10th-11th [1902-1903], 14th [1906], 27th-28th [1919-1920], 34th – 1r – 1 – (cont by: proceedings of the iowa state federation of labor as enacted at the... annual convention) – mf#3145919 – us WHS [350]

Constitution and rules of the marshlands club, amherst, ns / Marshlands Club (Amherst, NS) – [Amherst, NS?: s.n, 1907?] – 1mf – 9 – 0-665-65256-9 – mf#65256 – cn Canadiana [366]

Constitution and statutes of the grand lodge, knights of pythias of the province of quebec / Knights of Pythias. Grand Lodge of Quebec – [Quebec (Province)?: s.n, 1905?] [mf ed 1995] – 2mf – 9 – mf#74739 – cn Canadiana [366]

The constitution and what it means today / Corwin, Edward A – Princeton, London: Princeton UP, Oxford UP, 1920 – 2mf – 9 – $3.00 – mf#LLMC 95-093 – us LLMC [323]

Constitution, by-laws, and list of members : revised oct 1886 / University College (Toronto, Ont). Literary and Scientific Society – Toronto: printed for the Society, [1886?] [mf ed 1984] – 1mf – 9 – 0-665-32315-8 – mf#32315 – cn Canadiana [378]

Constitution, by-laws and rules, 1920 / North Pacific Association of Amateur Oarsmen – [S.l: s.n, 1903?] – 1mf – 9 – 0-665-99267-X – mf#99267 – cn Canadiana [360]

Constitution, by-laws and rules of order... : incorporated june 20th, 1881 / Montreal Amateur Athletic Association – S.l: s.n, 1881? – 2mf – 9 – mf#27592 – cn Canadiana [790]

Constitution, by-laws and rules of order of alpha lodge : no 1, knights of jericho, canada west / Knights of Jericho. Alpha Lodge, No 1 (Brockville, Ont) – [Brockville, Ont?: s.n.] 1853 [mf ed 1984] – 1mf – 9 – 0-665-45206-3 – mf#45206 – cn Canadiana [366]

Constitution, by-laws and rules of the home for aged and infirm colored persons : also the proceedings of the...annual meeting / Home for Aged and Infirm Colored Persons (Philadelphia PA) – Philadelphia: Merrihew & Son, printers 1865-67 [annual] [mf ed 2006] – 1st-3rd (1865-67) [complete] on 1r – 1 – (cont by: home for aged and infirm colored persons (philadelphia, pa). proceedings of the...annual meeting of the home for aged and infirm colored persons, held...) – mf#2006-s015 – us ATLA [362]

Constitution, by-laws, rules of order etc... : instituted at dundas, july 10th 1873 / Independent Order of Odd Fellows. Valley City Lodge, No 117 (Dundas, Ont) – [Dundas, Ont?: s.n.] 1885 [mf ed 1994] – 1mf – 9 – 0-665-94614-7 – mf#94614 – cn Canadiana [366]

Constitution, by-laws, rules of order, etc... : instituted at lynden, jul 20th 1887 / Independent Order of Odd Fellows. Lynden Lodge, No 259 (Ont) – [Dundas, Ont?: s.n, 1888 [mf ed 1994] – 1mf – 9 – 0-665-94613-9 – mf#94613 – cn Canadiana [366]

Constitution, by-laws, rules of order, etc, no 30 : instituted at london, january 30th, 1854... / Independent Order of Odd Fellows. Eureka Lodge – London, Ont?: "Free Press", 1861 – 1mf – 9 – mf#50027 – cn Canadiana [366]

Constitution, by-laws, sailing regulations, yacht routine, list of members, list of yachts, signal code, etc : station, st john, nb, 1899 / Royal Kennebeccasis Yacht Club (Saint John, NB) – St John, NB: Barnes, 1899 – 2mf – 9 – mf#25566 – cn Canadiana [790]

Constitution, canons, by-laws and resolutions of the incorporated synod of the diocese of toronto : together with the constitution and canons of the provincial synod, forms, documents of importance and statutes affecting the diocese of toronto / Church of England Diocese of Toronto. Synod – [Toronto?: s.n.], 1886 [mf ed 1983] – 4mf – 9 – (incl ind) – mf#29862 – cn Canadiana [242]

Constitution, canons, rules and regulations of the diocesan synod of nova scotia / Church of England. Diocese of Nova Scotia – [Halifax, NS?: s.n.], 1892 [mf ed 1980] – 1mf – 9 – mf#08275 – cn Canadiana [242]

Constitution, canons, rules and regulations of the diocesan synod of nova scotia / Eglise d'Angleterre en Canada Diocese of Nova Scotia. Diocesan Synod – [S.l: s.n, 1896?] [mf ed 1983] – 1mf – 9 – (incl ind) – mf#34948 – cn Canadiana [242]

La constitution comme je la voudrais, avec des debats imaginaires – [Paris]: Impr centrale de Napoleon Chaix, oct 1848 – us CRL [944]

Constitution, correspondence with bishop c j nicolas, sm / Catholic Young Mens' Society of Fiji – 1923-1929 – 1r – 1 – mf#pmb450 – at Pacific Mss [241]

Constitution de la republique francaise : votee par l'assemblee nationale dans sa seance du 4 novembre 1848 – Stuttgart 1848 [mf ed Hildesheim 1995-98] – 1v on 1mf – 9 – €40.00 – 3-487-26021-2 – gw Olms [323]

Constitution democrat – Keokuk, IA. 1889-1902 (1) – mf#63275 – us UMI ProQuest [071]

Constitution democrat – Keokuk, IA. 1902-1912 (1) – mf#63276 – us UMI ProQuest [071]

Constitution des societes anonymes en france, dans l'empire allemand et en grande-bretagne / Ostrorog, Leon – Paris: Larose & Forcel, 1893. 178p. LL-4080 – 1 – us L of C Photodup [342]

Constitution d'haiti en face de la convocation / Sejourne, Georges – Port-Au-Prince, Haiti. 1933 – 1r – us UF Libraries [972]

Constitution du club choquette – [Quebec?: s.n, 1905?] – 1mf – 0-665-77062-6 – mf#77062 – cn Canadiana [366]

Constitution du royaume du cambodge – [Pnom-penh? 1969?] [mf ed 1989] – 1r with other items – 1 – (in french; title also in cambodian) – mf#mf-10289 seam reel 027/07 [§] – us CRL [342]

Constitution du royaume du cambodge – Phnom Penh: Hem-Chiam-Reun [1947?] [mf ed 1989] – 1r with other items – 1 – mf#mf-10289 seam reel 026/05 [§] – us CRL [342]

Constitution du royaume du cambodge – [Phnom-Penh] Ministere de l'information 1963 [mf ed 1989] – 1r with other items – 1 – (title & text also in cambodian) – mf#mf-10289 seam reel 027/13 [§] – us CRL [342]

Constitution du royaume du cambodge – [Phnom Penh?: s.n. 1959] [mf ed 1989] – 1r with other items – 1 – (title & text also in khmer) – mf#mf-10289 seam reel 027/12 [§] – us CRL [342]

Constitution du royaume du cambodge – Phnom Penh: Impr du Palais Royal 1966 [mf ed 1989] – 1r with other items – 1 – (title & text in french & khmer) – mf#mf-10289 seam reel 027/06 [§] – us CRL [342]

Constitution establishing self-government in the islands of cuba and porto rico : promulgated by royal decree of november 25, 1897 – Washington: GPO, 1899 – 1mf – 9 – $1.50 – mf#LLMC 92-303 – us LLMC [324]

Constitution et reglements / Club de raquettes de Levis – Levis [Quebec]: Mercier, 1886 – 1mf – 9 – 0-665-91904-2 – mf#91904 – cn Canadiana [790]

Constitution et reglements... : fondee le 28 janvier, 1861 par pierre imbleau, fondeur; medecin, dr m s boulet / Societe de bienfaisance et de secours mutuels de l'Industrie et du comte de Joliette – [Joliette, Quebec?: s.n.] 1863 [mf ed 1983] – 1mf – 9 – 0-665-41766-7 – mf#41766 – cn Canadiana [331]

Constitution et reglements de... / Association St. Antoine de Montreal – [s.l: s.n.] 1857 [mf ed 1983] – 1mf – 9 – 0-665-43086-8 – mf#43086 – cn Canadiana [366]

Constitution et reglements de la ligue du sacre-coeur : forme speciale de l'apostolat de la priere parmi les hommes: livret d'admission / Ligue du Sacre-Coeur – Montreal: Bureau central du Sacre-Coeur, [1901?] [mf ed 1995] – 1mf – 9 – 0-665-74912-0 – mf#74912 – cn Canadiana [241]

Constitution et reglements de l'association bienveillante des pompiers de montreal / Association bienveillante des pompiers de Montreal – Montreal: impr par Owler & Stevenson, 1854 [mf ed 1994] – 9 – cn Bibl Nat [366]

Constitution et reglements de l'association des instituteurs en rapport avec l'ecole normale jacques-cartier / Association des instituteurs de la circonscription de l'Ecole normale Jacques-Cartier – Montreal?: s.n, 1858 (Montreal: Senecal, Daniel) – 1mf – 9 – mf#42551 – cn Canadiana [366]

Constitution et reglements de l'association saint antoine de montreal / Association Saint-Antoine de Montreal – [Montreal]: Senecal & Daniel, impr, 1857 [mf ed 1985] – 1mf – 9 – mf#SEM105P524 – cn Bibl Nat [366]

Constitution et reglements de l'institut canadien : societe fondee par de jeunes canadiens-francais de montreal, le dix-sept decembre 1844 – [Montreal?: s.n.] 1845 [mf ed 1984] – 1mf – 9 – 0-665-45128-8 – mf#45128 – cn Canadiana [366]

Constitution et reglements du club cartier : franc et sans dol / Club-Cartier (Montreal, Quebec) – Montreal: s.n, 1874 – 1mf – 9 – mf#23970 – cn Canadiana [366]

The constitution explained / Atwood, Harry – New York/Chicago: Harcourt, Brace & Co, 1927 – 3mf – 9 – $4.50 – mf#llmc92-228 – us LLMC [323]

Constitution haitienne de 1889 et sa revision / Dube, Charles – Paris, France. 1897 – 1r – us UF Libraries [323]

The constitution in the year 2000 : choices ahead in constitutional interpretation, 11 oct 1988 – n.p., n.d. – (= ser Office of legal policy, reports to the attorney general) – 3mf – 9 – $4.50 – mf#llmc 94-367 – us LLMC [342]

Constitution of american samoa, 1960 / Samoa – Washington: GPO, 1961 – 1mf – 9 – $1.50 – (effective oct 17 1960. contained in committe print no 1, us hse. comm. on interior and insular affairs, 87th congress 1st session) – mf#LLMC 82-100C Title 20 – us LLMC [324]

Constitution of botswana – Gaberone, Botswana. 1966 – 1r – us UF Libraries [323]

Constitution of canada the british north america act, 1867; its interpretation. / Doutre, Joseph – Montreal, Lovell, 1880. 414 p. LL-2376 – 1 – us L of C Photodup [342]

The constitution of mcgill university : being the annual university lecture in the session of 1888-89 / Dawson, John William – Montreal?: s.n, 1888 – 1mf – 9 – mf#02137 – cn Canadiana [378]

The constitution of ngonde / Wilson, G – (= ser Institute for social research, university of zambia. papers 3) – 2mf – 7 – mf#363/2 – uk Microform Academic [960]

The constitution of northern nigeria law, 1953 / Nigeria. Northern Region – Kanduna: Gov't. Printer 1963 35p. LL-12039 – 1 – us L of C Photodup [342]

Constitution of the... / Wisconsin Interscholastic Athletic Association – 1933/34-1965/65 – 1r – 1 – mf#683660 – us WHS [790]

Constitution of the african civilization society / African Civilization Society – [s.l: s.n., 1861?] (mf ed 1982) – 1r – 1 – mf#Sc Micro R-1447 – us NY Public [060]

Constitution of the american federation of labor and congress of industrial organizations : as amended by the 13th constitutional convention of the afl-cio, nov 15-20 1979 – n.p., n.d. (AFL-CIO publ no 1) – 1mf – 9 – $1.50 – mf#LLMC 96-045 – us LLMC [323]

Constitution of the central canada chamber of mines, winnipeg, canada / Central Canada Chamber of Mines – Winnipeg: Stovel, 1899 – 1mf – 9 – mf#02560 – cn Canadiana [333]

Constitution of the commonwealth of the northern mariana islands, 1976 / Government of the Northern Mariana Islands – n.d. – 1mf – 9 – $1.50 – (contains also president carter's "certification" and proclamation dated oct 24 1977) – mf#LLMC 82-100J Title 1 – us LLMC [324]

Constitution of the federated states of micronesia – Micronesian Constitutional Convention, 1975 – 6mf – 9 – $9.00 – (multi-lingual ed) – mf#LLMC 82-100F Title 103 – us LLMC [323]

The constitution of the federation of rhodesia and nyasaland... st 109=incorporating relevant papers and indices – Salisbury, Government Printer 1959 – us CRL [323]

The constitution of the free christian union – [London?: s.n., 1868?] (London: Woodfall and Kinder) – 1mf – 9 – 0-524-00260-6 – mf#1989-2960 – us ATLA [240]

Constitution of the grand and subordinate temples of the independent order of good templars of canada : revised nov 1 1864 / Independent Order of Good Templars of Canada. Grand Temple – Hamilton, CW [Ont]: printed for the Grand Temple by A Lawson, 1866 [mf ed 1984] – 1mf – 9 – 0-665-45123-7 – mf#45123 – cn Canadiana [366]

The constitution of the human soul : six lectures / Storrs, Richard S – New York: Robert Carter, 1857, c1856 – (= ser Graham Lectures) – 1mf – 9 – 0-7905-3234-4 – mf#1987-3234 – us ATLA [210]

Constitution of the jewish agency for palestine – London, 1929 – 1mf – 9 – mf#J-28-34 – ne IDC [956]

The constitution of the later roman empire / Bury, John Bagnell – Cambridge: University Press; New York: Putnam [distributor], 1910 – (= ser Creighton Memorial Lecture) – 1mf – 9 – 0-7905-5452-6 – (incl bibl ref) – mf#1988-1452 – us ATLA [930]

Constitution of the liberal-conservative association of the town and township of cornwall / Liberal-Conservative Association of Cornwall (Ont) – [Cornwall, Ont?: s.n.] 1878 [mf ed 1993] – 1mf – 9 – 0-665-91411-3 – mf#91411 – cn Canadiana [325]

Constitution of the marshall islands, 1979 / Marshall Islands – Micronitor News and Printing Co, may 1983 – 1mf – 9 – $1.50 – mf#LLMC 82-100I Title 5 – us LLMC [324]

Constitution of the marshall islands, december 1 1978 / Republic of the Marshall Islands – Majuro: Constitutional Convention, 4 jan 1979 – 2mf – 9 – $3.00 – mf#LLMC 82-100I Title 1 – us LLMC [324]

Constitution of the republic of guatemala, 1956 – Washington, DC. 1960 – 1r – us UF Libraries [323]

The constitution of the republic of palau, 1979 / Palau Constitutional Convention, Koror – Koror: Palau Constitutional Convention, 1979 – (= ser Republic Of Belau (Palau) – Constitution And Laws) – 1mf – 9 – $1.50 – (palauan and english versions. various pagination) – mf#LLMC 82-100G Title19 – us LLMC [323]

The constitution of the republic of palau, 1979 / Palau. (The Republic of Belau) – Koror, Palau. 1979 – 1mf – 9 – $1.50 – (mimeo copy of the original signed version from the palau constitution convention, jan 28-apr 2 1979) – mf#LLMC 82-100G Title 1 – us LLMC [324]

Constitution of the st george's society of quebec : established 1835 / St George's Society of Quebec – [Quebec?: s.n, 1837?] [mf ed 1994] – 1mf – 9 – 0-665-94574-4 – mf#94574 – cn Canadiana [366]

Constitution of the state of chuuk, 1988 – n.p, n.d. – 1mf – 9 – $1.50 – mf#llmc82-100h, title 16 – us LLMC [342]

Constitution of the state of florida – Jacksonville, FL. 1887 – 1r – us UF Libraries [350]

Constitution of the state of truk, 1985 – n.p. – 1mf – 9 – $1.50 – (incl call to the general election of mar 11 1986) – mf#llmc82-100h, title 15 – us LLMC [342]

Constitution of the state of yap, 1982 / Federated States of Micronesia – n.p, n.d. – 1mf – 9 – $1.50 – mf#LLMC 82-100H Title 14 – us LLMC [324]

Constitution of the union of south africa / Hutton, James – Cape Town, South Africa. 1946 – 1r – us UF Libraries [323]

The constitution of the united states : a critical discussion of its genesis, development and interpretation / Tucker, John Randolph; ed by Tucker, Henry ST George – Chicago: Callaghan. 2v. 1899 – 12mf – 9 – $18.00 – mf#LLMC 95-066 – us LLMC [323]

The constitution of the united states / Hickey, W – 7th ed. Philadelphia: T K & F G Collins, 1854 – 6mf – 9 – $9.00 – mf#LLMC 91-084 – us LLMC [323]

The constitution of the united states / Tucker, John Randolph – Chicago, Callaghan, 1899. 2 v. LL-1603 – 1 – us L of C Photodup [348]

The constitution of the united states at the end of the first century / Boutwell, George S – Boston: D C Heath, 1896 – 5mf – 9 – $7.50 – mf#LLMC 95-096 – us LLMC [323]

The constitution of the united states defined and carefully annotated / Paschal, George Washington – 2nd ed. Washington, D.C.: Morrison, 1876. 644p LL-1039 – 1 – (3d ed. washington, d.c.: morrison, 1882. 644p ll-1155) – us L of C Photodup [348]

The constitution of the united states with notes of the decisions of the supreme court thereon. / Bryant, Edwin Eustace – Madison, Wis., Democrat, 1901. 418 p. LL-503 – 1 – us L of C Photodup [348]

Constitution of the zionist organisation – Jerusalem, 1938 – 1mf – 9 – mf#J-28-51 – ne IDC [956]

Constitution of the...good templars of canada : revised, apr 3 1862; by-laws of phoenix temple, n275, organized, oct 5th 1864 / Independent Order of Good Templars of Canada. Grand Temple – [Picton, Ont?: s.n.] 1865 [mf ed 1984] – 1mf – 9 – 0-665-45096-6 – mf#45096 – cn Canadiana [366]

La constitution politique et sociale – Paris: Impr Ch Schiller, may 18, 20 1871 – (filmed as pt of: commune de paris newspapers; newspapers on these reels are filmed chronologically, not alphabetically. le constitutionnel port-au-prince: libr. bouchereau & cie., (2 sheets) [1ere annee n1-2e annee n37/38/39] (29 avril 1876-23 mars 1878)) – us CRL [074]

Constitution, regles et reglements / Club Lacrosse Fraserville – [Fraserville (Quebec): J E Mercier, 1885?] – 1mf – 9 – 0-665-91890-9 – mf#91890 – cn Canadiana [790]

Constitution rules and canons of the incorporated synod of the diocese of huron : as well as those of the provincial synod and statutes of parliament affecting ecclesiastical rights – [London, Ont?: s.n.], 1879 [mf ed 1981] – 3mf – 9 – (incl ind) – mf#08837 – cn Canadiana [242]

Constitution, rules and canons of the incorporated synod of the diocese of huron / Eglise d'Angleterre en Canada Diocese of Huron. Synod – [London, Ont?: s.n.], 1893 [mf ed 1981] – 1mf – 9 – mf#08838 – cn Canadiana [242]

Constitution, rules and regulations and canons of the synod of the diocese of montreal / Church of England Diocese of Montreal. Synod – [Montreal?: s.n.], 1872 [mf ed 1983] – 1mf – 9 – mf#29978 – cn Canadiana [242]

Constitution, rules of order, by-laws, rules and canons of the synod of the diocese of montreal / Eglise d'Angleterre en Canada Diocese of Montreal. Synod – [Montreal?: s.n.], 1893 [mf ed 1986] – 1mf – 9 – 0-665-62919-2 – (incl ind) – mf#62919 – cn Canadiana [242]

Constitution, rules of order, by-laws, rules and canons of the synod of the diocese of montreal / Eglise d'Angleterre en Canada Diocese of Montreal. Synod – [Montreal?: s.n.], 1899 [mf ed 1986] – 1mf – 9 – 0-665-62920-6 – mf#62920 – cn Canadiana [242]

Constitution, rules of order, canons etc of the synod of the province of "canada" : revised to 16th session, 1895, inclusive... / Eglise d'Angleterre en Canada Province of Canada – [Montreal?: s.n.], 1896 [mf ed 1980] – 1mf – 9 – mf#07533 – cn Canadiana [242]

Constitution, rules of order, canons, etc of the synod of the province of "canada" : revised to 16th session, 1895... / Eglise d'Angleterre en Canada. Province du Canada – Montreal: printed by John Lovell & Son, 1896 [mf ed 1994] – 9 – cn Bibl Nat [241]

Constitution, statutes, rules and order of business, rules of procedure of grand forum, incorporation of the order, and form of complaint of the benevolent and protective order of elks of the united states of america – Ed 1970-71. Chicago, 1970 – 2mf – 9 – $3.00 – mf#LLMC 92-243 – us LLMC [366]

Die constitution unigenitus : ihre veranlassung und ihre folgen: ein beitrag zur geschichte des jansenismus / Schill, Andreas – Freiburg i B; St Louis, MO: Herder, 1876 [mf ed 1986] – 1mf – 9 – 0-8370-8711-2 – (incl bibl ref) – mf#1986-2711 – us ATLA [241]

Constitutional *see* Niagara peninsula newspapers, pt 2

Constitutional Alliance, Inc *see* Politics

Constitutional and governmental rights of the mormons, as defined by congress and the supreme court of the united states. – Salt Lake City, Parry, 1890. 116 p. LL-162 – 1 – us L of C Photodup [342]

Constitutional and parliamentary history of the methodist episcopal church / Buckley, James Monroe – New York: Methodist Book Concern, c1912 – 1mf – 9 – 0-7905-4442-3 – (incl bibl ref) – mf#1988-0442 – us ATLA [242]

The constitutional authority of bishops in the catholic church : illustrated by the history and canon law of the undivided church, from the apostolic age to the council of chalcedon, a.d. 451 / Wirgman, Augustus Theodore – London, New York: Longmans, Green 1899 [mf ed 1990] – 1mf – 9 – 0-7905-6973-6 – (in english, greek & latin; incl bibl ref) – mf#1988-2973 – us ATLA [241]

Constitutional commentary – University of Minnesota. v1-17. 1984-2000 – 9 – $256.00 set – ISSN: 0742-7115 – mf#109701 – us Hein [323]

Constitutional convention... / Marine Cooks and Stewards Association of the Pacific Coast – 1st [1945] – 1r – 1 – (cont by: proceedings of the...biennial convention of the national union...) – mf#3269479 – us WHS [640]

Constitutional convention, 1976 – briefing papers for delegates : prepared by the law firm of wilmer, cutler and pickering – Saipan: Office of Transition Studies & Planning 1976 – 2v on 14mf – 9 – $21.00 – mf#llmc82-100j, title 19 – us LLMC [342]

Constitutional Convention, Koror *see* Draft constitutions. alternatives 1 and 2

Constitutional Convention of the Congress of Industrial Organizations *see* Proceedings of the constitutional convention...

Constitutional convention. report / Philippines – Washington, 1934-1935 [1,5,9] – mf#2571 – us UMI ProQuest [323]

Constitutional decisions of john marshall / Cotton, Joseph Jr – New York, London: Putnam's Sons. 2v. 1905 – 11mf – 9 – $16.50 – mf#LLMC 84-248 – us LLMC [342]

Constitutional development of palau : fact sheet – Washington: The State Department?, jul 1980 – (= ser Republic Of Belau (Palau) – Constitution And Laws) – 1mf – 9 – $1.50 – mf#LLMC 82-100G, Title 20 – us LLMC [323]

The constitutional documents of the puritan revolution, 1625-1660 / ed by Gardiner, Samuel Rawson – 3d ed., rev. Oxford: Clarendon Press, 1906 – 2mf – 9 – 0-7905-4679-5 – (incl bibl ref) – mf#1988-0679 – us ATLA [323]

Constitutional government in the united states / Wilson, Woodrow – New York: Columbia UP, 1908 – 3mf – 9 – $4.50 – mf#LLMC 95-069 – us LLMC [323]

The constitutional guarantees of the right of property as affected by recent decisions. / Hoadly, George – New York: Evening Post Job Printing Office, 1889. 53p. LL-427 – 1 – us L of C Photodup [342]

The constitutional history and constitution of the church of england = Verfassung der kirche von england / Makower, Felix – London: Swan Sonnenschein; New York: Macmillan, 1895 – 2mf – 9 – 0-7905-5193-4 – (incl bibl ref. in english) – mf#1988-1193 – us ATLA [241]

A constitutional history of american episcopal methodism / Tigert, John James – 6th rev enl ed. Nashville, TN: Pub House of the ME Church, South, 1916, c1904 [mf ed 1992] – (= ser Methodist coll) – 2mf – 9 – 0-524-02902-4 – (incl bibl ref) – mf#1990-4493 – us ATLA [242]

Constitutional history of british guiana / Clementi, Cecil – London, England. 1937 – 1r – us UF Libraries [323]

The constitutional history of england in its origin and development / Stubbs, W – Oxford. v1-3. 1874-1878 – €67.00 – ne Slangenburg [323]

Constitutional history of the american people / Thorpe, Francis F – New York, London: Harper. 2v. 1898 – 23mf – 9 – $34.50 – mf#LLMC 84-255 – us LLMC [323]

The constitutional history of the presbyterian church in the united states of america / Hodge, Charles – Philadelphia: W.S. Martien, 1839-1840 – 2mf – 9 – 0-7905-5155-1 – mf#1988-1155 – us ATLA [242]

Constitutional history of the united states : as seen in the development of american law. a course of lectures before the political science association of the university of michigan / Cooley, Thomas M et al – New York, London: G P Putnam's Sons, 1890 – 4mf – 9 – $6.00 – mf#LLMC 95-075 – us LLMC [323]

Constitutional history of the united states : from their declaration of independence to the close of their civil war / Curtis, George Ticknor – New York: Harper & Bros. 2v. 1897 – 18mf – 9 – $27.00 – mf#LLMC 95-061 – us LLMC [323]

Constitutional history of the united states / Holst, Herman E von – Chicago: Callaghan. v1-7. 1889-92 – 45mf – 9 – $67.50 – (incl ind and list of authorities) – mf#LLMC 84-251 – us LLMC [323]

Constitutional history of the united states, 1765-1895 / Thorpe, Francis Newton – New York: Harper. 3v. 1901 – 23mf – 9 – $34.50 – mf#LLMC 82-714 – us LLMC [323]

Constitutional history of the united states as seen in the development of american law. / Michigan. University. Political Science Association – New York: Putnam's, 1889. 296p. LL-192 – 1 – us L of C Photodup [342]

Constitutional law. / Chadman, Charles Erehart – Chicago: American School of Law, 1906. 316p. LL-1322 – 1 – us L of C Photodup [342]

Constitutional law; from the article on this subject in the encyclopedia of united states supreme court reports. / Richey, Homer – Charlottesville, Va.: Michie, 1928. 4v. LL-1033 – 1 – us L of C Photodup [342]

The constitutional law of the united states / Willoughby, Westel Woodbury – New York, Baker, Voorhis, 1910. 2 v. LL-1543 – 1 – us L of C Photodup [348]

Constitutional law questions now pending in the methodist episcopal church : with a suggestion on the future of the episcopacy / Warren, William Fairfield – Cincinnati: Cranston & Curts, 1894 – (= ser Women & the church in america) – 1mf – 9 – 0-8370-1575-8 – mf#1984-2215 – us ATLA [242]

Constitutional & perthshire agricultural & general advertiser – [Scotland] Perth 24 apr 1835-28 dec 1836, jan 1844-dec 1850 [mf ed 2004] – 5r – 1 – uk Newsplan [072]

Constitutional power and world affairs / Sutherland, George – New York: Columbia UP, 1919 – 3mf – 9 – $4.50 – mf#LLMC 95-082 – us LLMC [323]

The constitutional powers of the general conference : with a special application to the subject of slaveholding / Harris, William Logan – Cincinnati: Methodist Book Concern, 1860 – 1mf – 9 – 0-524-05358-8 – (incl bibl ref) – mf#1990-5109 – us ATLA [240]

Constitutional rights of military personnel : summary-report of hearings by the subcommittee on constitutional rights of the committee on the judiciary – Senate. 88th Congress, 1st session. Washington: GPO, 1963 – 1mf – 9 – $1.50 – mf#LLMC 96-087 – us LLMC [342]

Constitutional rights of military personnel : summary-report of hearings by the subcommittee on constitutional rights of the committee on the judiciary – US Senate Pursuant to Senate Resolution 58. 88th Congress, 1st session. Washington: GPO, 1963 – 1mf – 9 – $1.50 – mf#LLMC 96-080 – us LLMC [355]

Constitutional whig – Richmond VA. 1824 jun 22, dec 21, 1826 aug 18-22, sep 15, 19-29, oct 17, 24-27, 1827 apr 27 – 1r – 1 – (cont by: richmond whig and public advertiser) – mf#851283 – us WHS [071]

Constitutionalist – Augusta GA. 1825 dec 16 – 1r – 1 – mf#869559 – us WHS [071]

Constitutionalist – Bath, NY. 1837-1842 (1) – mf#64896 – us UMI ProQuest [071]

Constitutionalist – Plainfield, NJ. 1868-1911 (1) – mf#64842 – us UMI ProQuest [071]

The constitutionality of the lawyers test oath; argument of w. d. porter, in the district court of the united states / Porter, William Dennison – Charleston: Walker, 1866. 26p. LL-2357 – 1 – us L of C Photodup [347]

Constitution...and by-laws of the national council / Sovereigns of Industry – 1874 jan 14, mar 5/7, 1875 jan 12/15 – 1r – 1 – mf#3400989 – us WHS [338]

Constitution...as revised : resolutions adopted and proceedings of the annual convention... / Ohio Farmers' Alliance – 1891-92 – 1r – 1 – mf#5324894 – us WHS [630]

Constitutionelle zeitung – Berlin DE, 1851-1851 30 jun – 3r – 1 – gw Misc Inst [074]

Constitutionelle zeitung *see* Neues dresdner journal

Constitutionelles blatt aus boehmen – Prag (CZ), 1848 apr-1849 – 3r – 1 – gw Misc Inst [323]

Constitutiones canonicorum regularium : ordinis s p augustini ep congregationis windesemensis – Lovanii, 1639 – 7mf – 8 – €15.00 – ne Slangenburg [241]

Constitutiones ecclesiasticae disputandae propugnandaeque in aula reg et antiq ss aa petri et pauli s q ildefonsi collegii... / San Juan Hermoso y Rio de Loza, Faustino de – Mexici: Ex typographia Philippi de Zunniga et Ontiveros [1792] – (= ser Books on religion... 1543/44-c1800: colegios religiosos) – 1mf – 9 – mf#crl-362 – ne IDC [241]

Constitutiones et acta generalium / Jimenez Samaniego, Jose – 1704 – 9 – sp Bibl Santa Ana [342]

Constitutiones et acta regum germanicorum (mgh leges. 1: 2.bd. pars 1a) : tomi primi supplementa – 1837 – (= ser Monumenta germaniae historica leges 1. leges in folio (mgh leges 1)) – €40.00 – (pars 2a: capitularia spuria. canones ecclesiastici. bullae pontificum) – ne Slangenburg [240]

Constitutiones ordinis frartum eremitraum [sic] sancti augustini : nuper recognitae, & in ampliorem formam ac ordinem redactae – Mexici: Excudebat Petrus Ocharte cum licencia, anno 1587 – (= ser Books on religion... 1543/44-c1800: ordenes, etc: agostinos) – 6mf – 9 – mf#crl-167 – ne IDC [241]

Constitution-making for a democracy / Cowen, Denis Victor – Johannesburg, South Africa. 1960 – 1r – us UF Libraries [321]

Constitutionnel – Paris, France. 15 oct 1909, 10 jan 1910, 1912-may 1914 – 3 1/4r – 1 – (aka: mechroutiette) – uk British Libr Newspaper [074]

Le constitutionnel – Paris. 1817 a mai 1819. – 1 – (a paru sous le titre: Journal du commerce de juil. oct 1815-21 juil 1914, 1907-juil 1914 lac) – fr ACRPP [073]

Constitutions and canons ecclesiasticall : treated upon by the bishop of london... – London: A Warren, 1662 – 2mf – 9 – mf#PW-10 – ne IDC [240]

Constitutions and laws of the knights of the maccabees of the world : governing the supreme tent, gt camps and subordinate tents... / Knights of the Maccabees of the World – London, Ont: Southam and Brierly, 1881 – 1mf – 9 – mf#61584 – cn Canadiana [366]

Constitutions de l'institut des petites filles de saint-joseph / Petites filles de Saint-Joseph – Montreal: [s.n.] 1901 [mf ed 1999] – 9 – cn Bibl Nat [366]

Les constitutions du canada / DeCelles, Alfred Duclos – Montreal: Librairie Beauchemin ltee, 1918 [mf ed 1985] – 1mf – 9 – mf#SEM105P517 – cn Bibl Nat [323]

Constitutions et reglements de la ligue du sacre-cur de jesus : apostolat de la priere (section des hommes) – S.l: s.n, 1892? – 1mf – 9 – mf#64594 – cn Canadiana [360]

Constitutions of the americas : as of january 1, 19... / Fitzgibbon, Russell Humke – Chicago, IL. 1948 – 1r – us UF Libraries [972]

The constitutions of the several states of the union and united states in the year 1859 – New York, Barnes 1879? 602, 17 p. LL-1131 – 1 – us L of C Photodup [348]

Constitutions of the world 1850 to the present, pt 1 : europe = Verfassungen der welt 1850 bis zur gegenwart, teil 1: europa / Wispelwey, Berend [comp]; ed by Dippel, Horst – [mf ed 2002-05] – 1040mf (1:24) in 7 installments + ind – 9 – diazo €12,600.00 (silver €15,750 isbn: 978-3-598-35205-8) – ISBN-10: 3-598-35204-2 – ISBN-13: 978-3-598-35204-1 – (incl: index of european constitutions 1850 to 2003) – gw Saur [323]

Constitutions of the world 1850 to the present, pt 2 : north and south america = Verfassungen der welt 1850 bis zur gegenwart, teil 2: nord- und suedamerika / Wispelwey, Berend [comp]; ed by Dippel, Horst – [mf ed 2004-07] – 1750mf (1:24) in 9 installments + suppl – 9 – diazo €16,200.00 (silver €19,800 isbn: 978-3-598-35501-1) – ISBN-10: 3-598-35500-9 – ISBN-13: 978-3-598-35500-4 – (with: index of north and south american constitutions 1850 to 2007) – gw Saur [323]

Constitutions, regles et reglements de l'assemblee legislative du canada : adoptes par la chambre dans la 3e session du 6e parlement, et revises dans les sessions subsequents = [rules, orders, and forms of proceedings of the legislative assembly of Canada] / Canada (Province). Parlement. Assemblee legislative – Quebec: impr pour les entrepreneurs, par Hunter, Rose et cie, 1863 [mf ed 1999] – 2mf – 9 – (in english and french; with ind) – mf#SEM105P3146 – cn Bibl Nat [348]

Constitutions, regles et reglements de l'assemblee legislative du canada : deposes sur le bureau de la chambre par m l'orateur, le 4 mai 1860 = [rules, orders, and forms of proceedings of the legislative assembly of Canada: ...4th may, 1860] / Canada (Province). Parlement. Assemblee legislative – Quebec: impr par Stewart Derbishire & George Desbarats, 1860 [mf ed 1999] – 2mf – 9 – mf#SEM105P3145 – cn Bibl Nat [348]

Constitutions, regles et reglements du conseil legislatif du canada / Canada (Province). Parlement. Conseil legislatif – Quebec: impr pour les entrepreneurs, par Hunter, Rose & Lemieux, [1864?] (mf ed 1999) – 1mf – 9 – (with ind) – mf#SEM105P3143 – cn Bibl Nat [348]

Das constitutum constantini (konstantinische schenkung) text (mgh leges 4:10.bd) – 1968 – (= ser Monumenta germaniae historica leges 4. fontes iuris germanici antiqui in usum scholarum separatim editi (mgh leges 4)) – €7.00 – ne Slangenburg [342]

Constitvtiones regvm regni vtrivsqve siciliae mandante friderico 2 imperatore per petrvm de vinea. / Naples. (Kingdom). Laws, Statutes, etc – Neapoli, ex Roma Typographia, 1786. 459p. LL-4015 – 1 – us L of C Photodup [340]

Constrained by jesus' love / Berg, J van den – Kampen, 1956 – 5mf – 8 – €12.00 – ne Slangenburg [240]

Constrained by jesus' love : an inquiry into the motives of the missionary awakening in great britain in the period between 1698 and 1815 / Berg, Johannes van den – Kampen: J. H. Kok, 1956. Chicago: Dep of Photodup, U of Chicago Lib, 1958 (1r); Evanston: American Theol Lib Assoc, 1984 (1r) – 1 – 0-8370-0104-8 – (incl ind) – mf#1984-B001 – us ATLA [240]

Constraining power of the love of christ / Jackson, Miles – Edinburgh, Scotland. 1806 – 1r – us UF Libraries [240]

Constraints on functional competence in persons with multiple sclerosis / Kasser, Susan L – 1998 – 2mf – 9 – $8.00 – mf#HE 622 – us Kinesology [616]

Construccion grammatical de los hymnos ecclesiasticos : dividida an siete libros, por el orden de el breviario romano: explicacion y medida de sus versos... / Rivas, Manuel Jose de la – en Mexico: impr Dona Maria de Rivera 1747 – (= ser Books on religion... 1543/44-c1800: himnos, villancicos) – 3mf – 9 – mf#crl-62 – ne IDC [241]

Construccion y explicacion de las reglas de los generos y preteritos, conforme al arte de antonio, muy util y provechosa para los que comienzan a estudiar / Lopez, Diego – Sevilla: Imprenta Castellana y Latina de los Herederos de Tomas Lopez de Haro, s.a. – sp Bibl Santa Ana [946]

The construct validity of a scale to measure teacher enthusiasm in secondary physical education / Fischer, Joseph C & Brockmeyer, Gretchen A – 1992 – 3mf – $12.00 – us Kinesology [790]

Constructeur d'usines a gaz – Paris, France. 15 jul 1894-15 dec 1905 – 1/2r – 1 – uk British Libr Newspaper [072]

Le constructeur d'usines a gaz – Paris, France. july 1894-dec 1905 [mnthly] – 1 2 r – 1 – uk British Libr Newspaper [680]

Constructing and validating competencies of sport managers (cosm) instrument : a model development / Toh, Kian L – 1997 – 2mf – 9 – $8.00 – mf#PE 3863 – us Kinesology [790]

Construction : a journal for the architectural, engineering and contracting interests of canada – Toronto: H Gagnier [etc]. v1-27 n5. oct 1907-oct/Nnov 1934// – 17r – 1 – Can$925.00 – cn McLaren [690]

Construction – Port-au-Prince, Haiti: Imp Construction, [1ere annee n10-2eme annee n253]. 1951-1953 – 11 sheets – us CRL [079]

The construction and initial validation of the physique anxiety scale / Lutter, Candice D – 1993 – 2mf – $8.00 – us Kinesology [150]

Construction contracting – Torrance. 1979-1983 (1) 1979-1983 (5) 1979-1983 (9) – (cont: mcgraw-hill, inc mcgraw-hill's construction contracting) – ISSN: 0270-1588 – mf#27,02 – us UMI ProQuest [690]

La construction de l'opinion publique dans le sondage de la question au discours de reformulation / Zappella, Jeannine – 2mf – 9 – (10345) – fr Atelier National [380]

La construction des navires a quebec et ses environs : greves et naufrages / Rosa, Narcisse – Ouvrage inedit. Quebec: impr Leger Brousseau, 1897 [mf ed 1976] – 1r – 5 – mf#SEM16P275 – cn Bibl Nat [623]

La construction du chemin de fer / Neering, Rosemary – Longueuil: editions Julienne, 1978 [mf ed 1994] – (= ser Coll une nation en marche) – 1mf – 9 – mf#SEM105P2069 – cn Bibl Nat [380]

Construction equipment – Boston. 1985+ (1) 1985+ (5) 1985+ (9) – (cont: construction equipment and materials) – ISSN: 0192-3978 – mf#1704,01 – us UMI ProQuest [690]

Construction equipment and materials – New York. 1964-1968 [1,5,9] – (cont by: construction equipment) – mf#1704 – us UMI ProQuest [690]

Construction in the south – Baltimore. 1949-1953 (1) – mf#62 – us UMI ProQuest [690]

Construction labor news – 1978 jun 30-1981 dec, 1982 jan 8-1985 jan 18, 1988 jan 1-1990 dec 21, 1991 jan-1994 dec – 4r – 1 – mf#572100 – us WHS [331]

Construction management and economics – London. 1983+ (1,5,9) – ISSN: 0144-6193 – mf#14399 – us UMI ProQuest [690]

Construction methods and equipment – New York. 1919-1978 (1) 1970-1978 (5) 1975-1978 (9) – (cont by: mcgraw-hill, inc mcgraw-hill's construction contracting) – ISSN: 0010-6844 – mf#27 – us UMI ProQuest [690]

Construction news – London. 1979-1979 (1) – ISSN: 0010-6860 – mf#11486 – us UMI ProQuest [690]

The construction of the bible / Adeney, Walter Frederic – New York: Thomas Whittaker, 1898 – 1mf – 9 – 0-7905-0300-X – mf#1987-0300 – us ATLA [220]

Construction of the great victoria bridge in canada / Hodges, James – London: John Weale, 1860 [mf ed 1988] – 1r – 1 – mf#SEM35P301 – cn Bibl Nat [624]

Construction products – Newton. 1993-1996 (1) 1993-1996 (5) 1993-1996 (9) – (cont: highway and heavy construction products) – ISSN: 1070-4531 – mf#273,03 – us UMI ProQuest [690]

Construction review – Washington. 1955-1997 (1) 1973-1997 (5) 1976-1997 (9) – ISSN: 0010-6917 – mf#6293 – us UMI ProQuest [690]

Construction weekly – London. 1989-1994 (1,5,9) – (cont: civil engineering) – ISSN: 0956-9189 – mf#17136 – us UMI ProQuest [624]

Constructive action newsletter / American Conference of Therapeutic Selfhelp/Selfhealth/Social Action Club – 28th yr n240, 243, 29th yr n244-255 [1988 sep, dec, 1989 jan-dec] – 1r – 1 – mf#1580453 – us WHS [150]

A constructive basis for theology / Ten Broeke, James – London: Macmillan, 1914 [mf ed 1993] – (= ser Baptist coll) – 1mf – 9 – 0-524-07765-7 – mf#1991-3333 – us ATLA [240]

Constructive engagement and south africa, 1981-1986 : development of a diplomatic strategy / September, Peter E – U of the Western Cape 1989 [mf ed S.l: s.n. 1989] – 2mf – 9 – (incl bibl) – sa Misc Inst [327]

Constructive ethics : a review of modern moral philosophy in its three stages of interpretation, criticism, and reconstruction / Courtney, William Leonard – London: Chapman & Hall, 1886 – 1mf – 9 – 0-7905-7331-8 – mf#1989-0556 – us ATLA [170]

Constructive natural theology : lectures for [1913?] / Smyth, Newman – New York: Scribner 1913 [mf ed 1990] – (= ser Nathaniel william taylor lectures [1913?]) – 1mf – 9 – 0-7905-7374-1 – mf#1989-0599 – us ATLA [210]

Constructive principles of the bahai movement : a summary of the history, object and institutions of the bahai religious teachings / Remey, Charles Mason – Chicago: Bahai Pub Soc 1917 [mf ed 1992] – 1mf [ill] – 9 – 0-524-01981-9 – mf#1990-2772 – us ATLA [290]

A constructive survey of upanishadic philosophy : being a systematic introduction to indian metaphysics / Ranade, Ramchandra Dattatraya – Poona: Oriental Book Agency, 1926 – (= ser Samp: indian books) – us CRL [180]

Constructive work of the christian ministry / Westcott, Brooke Foss – London, England. 1870 – 1r – us UF Libraries [240]

Constructor – Washington. 1950-1954 (1) – mf#450 – us UMI ProQuest [690]

Consuegra, Walfredo I *see* Estudio acerca de la guerra de guerrillas en cuba

Consuelo / Lopez de Ayala, Adelardo – 1878 – 9 – sp Bibl Santa Ana [820]

Consuetudines monasticae / Albers, B – Stuttgardiae et Vindobonae. v1-5. 1900-12 – 5v on 30mf – 8 – €57.00 – ne Slangenburg [241]

Consuetudines sci victoris parisiensis : sted archief. 15e s / Zutphen – 3mf – 8 – €13.00 – ne Slangenburg [240]

The consul : a sketch of emma booth tucker / Booth-Tucker, Frederick de Latour – New York: Salvation Army Pub Dept, 1903 – 1mf – 9 – 0-524-01646-1 – mf#1990-0467 – us ATLA [240]

El consulado de buenos aires y sus proyecciones en la historia del rio de la plata : buenos aires, 1962 / Barrado Manzano, Arcangel – Madrid: graf calleja, 1969 – 1 – sp Bibl Santa Ana [972]

Consular despatches from united states consuls in [...] / U.S. Consuls – (= ser Records of miscellaneous civilian agencies) – 1 – (mexico: guaymas 1832-96 5r m284; guerrero 1871-68 1r m292; matamoras 1826-1906 12r m281; mexico city 1822-1906 15r m296; mier 1870-78 1r m297; monterey, upper california 1834-48 1r m138; monterrey 1849-1906 7r m165; nogales 1889-1906 4r m283; nuevo laredo 1871-1906 4r m280; piedras negras 1868-1906 5r m299; santa fe 1830-46 1r m199. texas: galveston 1832-46 2r t151. canada: fort erie 1865-1906 3r t465; moncton 1885-1905 2r t636; sherbrooke 1879-1906 3r t680; sorel 1882-98 1r t684; toronto 1864-1906 9r t491; victoria 1862-1906 16r t130; windsor 1864-1906 4r t492; winnipeg 1869-1906 10r t24. not all with printed guides) – us Nat Archives [327]

Consular despatches...in asuncion, paraguay, 1844-1906 / U.S. Dept of State – (= ser General records of the department of state, 1910-1929 decimal file) – 6r – 1 – mf#T329 – us Nat Archives [324]

Consular despatches...in bahia, brazil, 1850-1906 / U.S. Dept of State – (= ser General records of the department of state, 1910-1929 decimal file) – 8r – 1 – mf#T331 – us Nat Archives [324]

Consular despatches...in grand bassa, liberia, 1868-1882 / U.S. Dept of State – (= ser General records of the department of state, 1910-1929 decimal file) – 1r – 1 – mf#M171 – us Nat Archives [324]

Consular despatches...in la rochelle, france, 1794-1906 / U.S. Dept of State – (= ser General records of the department of state, 1910-1929 decimal file) – 8r – 1 – mf#T394 – us Nat Archives [324]

Consular despatches...in monrovia, liberia, 1852-1906 / U.S. Dept of State – (= ser General records of the department of state, 1910-1929 decimal file) – 7r – 1 – (with printed guide) – mf#M169 – us Nat Archives [324]

Consular instructions of the department of state, 1801-1834 / U.S. Dept of State – (= ser General records of the department of state) – 7r – 1 – mf#M78 – us Nat Archives [327]

Consular trade reports, 1943-1950 – (= ser General records of the department of state) – 680 – 5 – mf#M238 – us Nat Archives [380]

Consulta canonica, eclesiastica, regular y ceremonial por los conventos... / Zambrano, Juan – Sevilla: s.i., s.a. – 9 – sp Bibl Santa Ana [240]

Consultant – Greenwich. 1961+ [1]; 1970+ [5]; 1975+ [9] – ISSN: 0010-7069 – mf#1950 – us UMI ProQuest [610]

Consulta...sobre asunto p. caceres y respuesta de andres barbosa / Davila, Andres – S.l., s.i., 1641 – 1 – sp Bibl Santa Ana [946]

Consultation – Los Angeles. 1984-1990 (1,5,9) – ISSN: 8756-6508 – mf#14479 – us UMI ProQuest [380]

Consultation de m dupin : avocat a la cour royale de paris, pour le seminaire de montreal, en canada / Dupin, Andre Marie Jean Jacques – Paris: de l'impr d'Everat, 1826 [mf ed 1988] – 1mf – 9 – mf#SEM105P921 – cn Bibl Nat [971]

Consultation on church union...the fourth meeting, april 5-8 1965, at lexington, kentucky : papers, reports, preliminary studies, etc – [s.l.: s.n., 1965?] Lexington: U of Kentucky, 1965 (1r); Evanston: American Theol Lib Assoc, 1984 (1r) – 1 – 0-8370-0454-3 – mf#1984-B034 – us ATLA [240]

Consulta...universidad de salamanca...censuras impuestas...p. caceres / Toledo, Jose de – 1641 – 9 – sp Bibl Santa Ana [946]

The consulting architect : practical notes on administrative difficulties and disputes / Kerr, Robert – London: John Murray, 1886 – (= ser 19th c art & architecture) – 4mf – 9 – mf#4.1.155 – uk Chadwyck [720]

Consulting engineer – Barrington. 1952-1986 (1) 1971-1986 (5) 1975-1986 (9) – ISSN: 0010-7107 – mf#1149 – us UMI ProQuest [620]

Consulting engineer – London. 1978-1984 (1,5,9) – ISSN: 0010-7093 – mf#11485 – us UMI ProQuest [620]

Consulting to management – Burlingame. 2000+ (1) – (cont: journal of management consulting) – ISSN: 1530-0153 – mf#15818,01 – us UMI ProQuest [650]

Consulting-specifying engineer – Denver. 1987+ (1,5,9) – ISSN: 0892-5046 – mf#16060 – us UMI ProQuest [620]

Consumer and marketing reserves: hearings...march 21 and 22, 1974 / U.S. Congress. Senate. Committee on Agriculture and Forestry. Subcommittee on Agricultural Production, Marketing and Stabilization of Prices – Washington, Govt. Print. Off., 1974. 222 p. LL-2256 – 1 – us L of C Photodup [340]

Consumer and responsive government news / Center for Public Representation – v1 n3-v2 n4 [1978 jun-1979 dec] – 1r – 1 – mf#639500 – us WHS [350]

Consumer bulletin – Washington. 1931-1973 (1) 1967-1973 (5) 1960-1973 (9) – (cont by: consumers' research magazine) – ISSN: 0010-7123 – mf#440 – us UMI ProQuest [380]

Consumer credit leader – Washington. 1949-1952 (1) – (cont by: credit) – ISSN: 0010-7166 – mf#426 – us UMI ProQuest [332]

Consumer information for employees – New York. 1973-1973 (1) – mf#8110 – us UMI ProQuest [380]

Consumer information: hearings...february 27, march 13, and april 15, 1975 / U.S. Congress. House. Committee on Banking, Currency and Housing Subcommittee on Consumer Affairs – Washington, Govt. Print. Off., 1975. 156 p. LL-2393 – 1 – us L of C Photodup [346]

Consumer legislative monthly report – Washington. 1972-1975 (1) 1972-1975 (5) (9) – mf#7345 – us UMI ProQuest [380]

Consumer markets / Standard Rate & Data Service – [v30 n22] 1948-1949 county & city data listings complete 1950, & 1951 thru p448, 1950-51 p449-end, 1952-53, 1953 consumer income data, suppl [complete], 1954 & 1955 – 3r – 1 – (cont: standard rate and data service) – mf#772795 – us WHS [380]

Consumer news / Marion Co. Marion – jul 22-nov 6, 1977 [daily] – 1r – 1 – mf#B29896 – us Ohio Hist [071]

Consumer news – Washington. 1971-1979 (1) 1971-1979 (5) 1971-1979 (9) – ISSN: 0045-8260 – mf#7908 – us UMI ProQuest [380]

Consumer news and reviews – Columbia. 1996-2000 (1,5,9) – (cont: american council on consumer interests newsletter) – ISSN: 1086-9107 – mf#9141,01 – us UMI ProQuest [380]

Consumer product safety commission annual reports – 1980-83 (each year in 2 pts) – 14mf – 9 – $21.00 – mf#LLMC 95-022 – us LLMC [344]

Consumer reports – Mount Vernon. 1936+ (1) 1967+ (5) 1960+ (9) – ISSN: 0010-7174 – mf#762 – us UMI ProQuest [380]

Consumer reports news digest – Yonkers. 1990-1992 (1,5,9) – (cont: consumers union news digest) – ISSN: 1047-4048 – mf#11510,01 – us UMI ProQuest [380]

The consumer's commercial cyclone – Yutan, NE: R W Parmenter (wkly) [mf ed v4 n5. mar 25 1905-1910 filmed 1978] – 3r – 1 – us NE Hist [071]

Consumers digest – Chicago. 1972-2000 (1) 1975-2000 (5) 1976-2000 (9) – ISSN: 0010-7182 – mf#9751 – us UMI ProQuest [380]

Consumers Education and Protection Association *see* Consumers voice

Consumers guide – Washington. 1933-1947 (1) – mf#5762 – us UMI ProQuest [338]

Consumers' league bulletin – Cincinnati. 1932 mar – 1r – 1 – (cont: bulletin [consumers' league of cincinnati]) – mf#3144576 – us WHS [350]

Consumers' League of Cincinnati *see* Bulletin of the consumers'...

Consumers' League of Massachusetts *see* Bulletin of the consumers'...

Consumers' League of New York City *see* Bulletin of the consumers'...

Consumer's League of Ohio *see* Records, ms 3546

Consumers' League of Pennsylvania *see* Annual report of the council, philadelphia branch...

Consumers' League of Philadelphia *see* Annual report for the year ending...

Consumers' news – Ottawa: Consumer Branch, Wartime Prices & Trade Board. n2-60. may 5 1942-apr 1947//? – 7mf – 9 – $65.00 – (directed to housewives, this paper's main objective was the enlistment of women in "the battle against inflation") – cn McLaren [640]

Consumers' research magazine – Washington. 1973+ (1) 1973+ (5) 1973+ (9) – (cont: consumer bulletin) – ISSN: 0095-2222 – mf#440,01 – us UMI ProQuest [380]

Consumers union news digest – Mount Vernon. 1976-1989 (1) 1976-1989 (5) 1976-1989 (9) – (cont by: consumer reports news digest) – ISSN: 0279-5353 – mf#11510 – us UMI ProQuest [380]

Consumers voice / Consumers Education and Protection Association – 1966 sep 1-1982 n3 – 1r – 1 – mf#621105 – us WHS [350]

Consumption – Seattle. 1967-1970 (1) – ISSN: 0010-7204 – mf#3463 – us UMI ProQuest [810]

Consumptive death-bed / Gosse, P H, Mrs – London, England. 18– – 1r – us UF Libraries [240]

cont by: sports collectors news *see* Sport fan

cont: international worker *see* Struggle

cont: la crosse citireport; cont by: la crosse-winona cityreport *see* La crosse-winona citireport

Conta : que a sua magestade o imperador da' o ministro e secretario d'estado dos negocios da justica, do tempo da sua administracao / Brazil. Ministerio da Justica – Rio de Janeiro: Typo. Imperial e Nacional, [1826] (annual) – 1r – 1 – us CRL [340]

Contact – Cape Town, Selemela Publications. [v1-10 n1. feb 8 1958-jan 1967] – us CRL [079]

Contact – Fiji, nov 1974-dec 1984 – 4r – at Pascoe [079]

Contact : fiji's catholic newspaper – 10 nov 1974-30 dec 1984 – 4r – 1 – mf#pmb doc392 – at Pacific Mss [241]

Contact – nos. 1-10. 1952-54 – 1 – us AMS Press [800]

Contact : official publication / National Association of Free Will Baptists [US] – v22 n11 [1975 nov], v24 n7-8 [1977 jul-aug], v26:n4-v28 [1979 apr-1981], v29-30 [1982-83], v31-34 [1984-87] – 3r – 1 – mf#602091 – us WHS [242]

Contact – Sausalito. 1959-1965 – 1 – ISSN: 0589-5049 – mf#1194 – us UMI ProQuest [073]

Contact – v1 1959 – 3r – 1 – $105.00 – mf#um-81 – us Commission [242]

Contact – Wellington, NZ. jan 1983-dec 1986 – 8r – 1 – mf#41.24 – nz Nat Libr [079]

Contact / Wisconsin Rural Electric Cooperative Association – 1937 may-1938 dec – 1r – 1 – mf#963652 – us WHS [360]

[Contact-] contact miner – NV. 1915 – 1r – 1 – $60.00 – mf#U03704 – us Library Micro [622]

Contact lens and anterior eye – Houndmills. 2002+ (1,5,9) – ISSN: 1367-0484 – mf#42880,01 – us UMI ProQuest [617]

Contact Lens Association of Ophthalmologists *see* Clao journal

[Contact-] the contact news – NV. 1909 – 1r – 1 – $60.00 – mf#U04837 – us Library Micro [071]

Contact-concern – v2 1960 – 2r – 1 – $70.00 – mf#um-81 – us Commission [242]

Contacto lipocratico y entrada de nuevos conceptos en el saber medico de dos epocas muy ligadas : fracastoro-luis de toro-sydenham / Sayans Castanos, Marcelo – Plasencia: Graf. Sandoval, 1973 – 1 – sp Bibl Santa Ana [610]

Contactor / Communications Workers of America – 1981 jul-1992 aug – 1r – 1 – (cont: 10 20 news; cont by: st louis/southern illinois labor tribune) – mf#1064233 – us WHS [331]

Contactos y cambios culturales en la sierra nevada / Reichel-Dolmatoff, Gerardo – Bogota, Colombia. 1953 – 1r – us UF Libraries [972]

Contacts : revue francaise de l'orthodoxie – 12(1960)-29(1977) – 114mf – 9 – €217.00 – ne Slangenburg [073]

Contacts : revue francaise de l'orthodoxie – n1-10. Paris. avr 1949-janv 1950; n.s., n1-80. 1955-72 – 1 – fr ACRPP [243]

Contaduria. memoria del ano 1910 / Caceres – Tip. El noticiero – 1 – sp Bibl Santa Ana [946]

Container news – Atlanta. 1966-1991 (1) 1975-1991 (5) 1975-1991 (9) – (cont by: intermodal container news) – ISSN: 0010-7360 – mf#8781 – us UMI ProQuest [380]

Containerisation international – London. 1977-1993 (1,5,9) – ISSN: 0010-7379 – mf#11457 – us UMI ProQuest [380]

Contant, Alexis *see* Vive laurier

Contant Dorville, Andre Guillaume *see* Histoire des differens peuples du monde

Contant d'Orville, Andre Guillaume *see*
- Histoire de l'opera bouffon
- Histoire des differens peuples du monde

Contarini, A *see* Travels to tana and persia

Contarini, G P *see* ...Historiae de bello nvper venetis a selimo 2 tvrcarvm imperatore illato, liber vnvs, ex italico sermone in latinum conuersus...

Contarsy, Steven A *see* Physiological comparison of high performance swimmers and runners

Conte Aguero, Luis *see*
- Betancourt y el comunismo
- Ideario de un combatiente
- Jose marti y la oratoria cubana

Conte de noel / Bouchor, Maurice – Paris, France. 1896 – 1r – us UF Libraries [440]
Conte, Joseph Le *see* Religion and science
Conte, Joseph le *see* Evolution and its relation to religious thought
Contemplacion / Villaronga, Luis – San Juan, Puerto Rico. 1947 – 1r – us UF Libraries [972]
Contemplaciones europeas / Mejia Sanchez, Ernesto – San Salvador, El Salvador. 1957 – 1r – us UF Libraries [972]
Contemplation of heathen idolatry : an excitement to missionary zeal / Wardlaw, Ralph – London, England. 1818? – 1r – us UF Libraries [240]
Contemplations scientifiques / Flammarion, Camille – 2nd ser. Paris: Hachette 1887 [mf ed 1998] – 1r – 1 – mf#film mas 28229 – us Harvard [520]
Contempora – Atlanta. 1970-1971 (1) 1970-1971 (5) (9) – ISSN: 0010-7433 – mf#7487 – us UMI ProQuest [400]
La contemporaine en egypte : pour faire suite aux souvenirs d'une femme, sur les principaux personnages de la republique, du consulat, de l'empire et de la restauration / Saint-Elme, Ida – Paris 1831 [mf ed Hildesheim 1995-98] – 6v on 18mf – 9 – €180.00 – ISBN-10: 3-487-25837-4 – ISBN-13: 978-3-487-25837-9 – gw Olms [960]
La contemporaine en miniature : ou abrege de ses memoires / Sevelinges, Charles L de – Paris 1828 [mf ed Hildesheim 1995-98] – 1v on 3mf – 9 – €90.00 – ISBN-10: 3-487-25838-2 – ISBN-13: 978-3-487-25838-6 – gw Olms [944]
Les contemporains – Paris. oct 1892-1914 [wkly] – 1 – fr ACRPP [073]
Contemporaneo : jornal litterario e scientifico – Sao Paulo, SP: Typ de F Gerbach & Comp, 12 out-17 nov 1896 – (= ser Ps 19) – mf#P18,02,27 – bl Biblioteca [073]
O contemporaneo – Rio de Janeiro, RJ: Typ G Leuzinger & Filhos, out 1882; jan-dez 1883 – (= ser Ps 19) – mf#P19A,04,163 – bl Biblioteca [870]
O contemporaneo – Sabara, MG. 29 jun-ago, out-dez 1890; jan, abr-dez 1891; jan, mar-abr, out-dez 1892; jan 1893-dez 1894; jan-fev, abr-jun 1895; jun, set-out 1896; out-dez 1897; jan, mar-maio, jul, nov 1898; abr 1899; maio-jul, set-dez 1902 – (= ser Ps 19) – mf#P11B,03,69 – bl Biblioteca [079]
Contemporaneous reputation of james russell lowell / Mcfadyen, Alvan Robbins – s.l, s.l? 1955 – 1r – us UF Libraries [420]
Contemporanul – Bucuresti, 1976-79 – 4r – 1 – gw Mikropress [949]
Contemporaries / Higginson, Thomas Wentworth – Boston: Houghton, Mifflin, 1900, c1899 – 1mf – 9 – 0-7905-5840-8 – mf#1988-1840 – us ATLA [920]
Contemporary accounting research – Mississauga. 1991+ (1,5,9) – ISSN: 0823-9150 – mf#18674 – us UMI ProQuest [650]
Contemporary africa : continent in transition / Wallbank, T Walter – Princeton, NJ. 1964 – 1r – us UF Libraries [321]
Contemporary africa / Prasad, Bisheshwar – Bombay, India. 1960 – 1r – us UF Libraries [960]
Contemporary black styles – 1998 win – 1r – 1 – mf#4261869 – us WHS [321]
Contemporary chinese thought – Armonk. 1997+ (1) – (cont: chinese studies in philosophy) – ISSN: 1097-1467 – mf#16884,01 – us UMI ProQuest [180]
Contemporary crises – Amsterdam. 1986-1990 (1) 1986-1990 (5) 1986-1990 (9) – (cont by: crime, law and social change) – ISSN: 0378-1100 – mf#16037 – us UMI ProQuest [360]
Contemporary drug problems – New York. 1971+ (1) 1971+ (5) 1975+ (9) – ISSN: 0091-4509 – mf#6680 – us UMI ProQuest [360]
Contemporary economic policy – Huntington Beach. 1994+ (1,5,9) – (cont: contemporary policy issues) – ISSN: 1074-3529 – mf#14015,01 – us UMI ProQuest [338]
Contemporary education – Terre Haute. 1929-2000 (1) 1969-2000 (5) 1975-2000 (9) – ISSN: 0010-7476 – mf#953 – us UMI ProQuest [370]
Contemporary essays in theology / Hunt, John – London: Strahan, 1873 – 2mf – 9 – 0-7905-7406-3 – mf#1989-0631 – us ATLA [240]
Contemporary estimates of his life and character / Gladstone, William Ewart – v. 1-19. 1898. W ind – 1 – 81.00 – us L of C Photodup [920]
Contemporary european affairs – Oxford. 1989-1991 (1,5,9) – ISSN: 0955-3843 – mf#49589 – us UMI ProQuest [934]
Contemporary european history – Cambridge. 1994-1996 (1,5,9) – ISSN: 0960-7773 – mf#21013 – us UMI ProQuest [940]
Contemporary evolution : an essay on some recent social changes / Mivart, Saint George Jackson – [London], 1876 – (= ser 19th c evolution & creation) – 3mf – 9 – mf#1.1.2924 – uk Chadwyck [120]
Contemporary evolution of religious thought in england, america / Goblet D'alviella, Eugene – London, England. 1885 – 1r – us UF Libraries [200]
The contemporary evolution of religious thought in england, america and india = L'evolution religieuse contemporaine chez les anglais, les americains et les hindous / Goblet d'Alviella, Eugene, comte – New York: Putnam, 1886 [mf ed 1991] – 1mf – 9 – 0-7905-9937-6 – (trans by j moden fr french. incl bibl ref) – mf#1989-1662 – us ATLA [190]
Contemporary family therapy – New York. 1986+ (1,5,9) – (cont: international journal of family therapy) – ISSN: 0892-2764 – mf#11640,01 – us UMI ProQuest [306]
Contemporary french painters / Hamerton, Philip Gilbert – London 1868 – (= ser 19th c art & architecture) – 2mf – 9 – mf#4.2.741 – uk Chadwyck [750]
Contemporary german art at the centenary festival of the royal academy of arts, berlin / Pietsch, Ludwig – London: George Bell & Sons, 1888 – (= ser 19th c art & architecture) – 2v on 7mf – 9 – mf#4.1.13 – uk Chadwyck [700]
Contemporary issues – New York. 1948-1966 [1]; 1966-1966 [5,9] – mf#1731 – us UMI ProQuest [321]
Contemporary jewish record – New York: American Jewish Cttee, 1938-45 [mf ed 2001] – (= ser Christianity's encounter with world religions, 1850-1950) – 2r – 1 – mf#2001-s064 – us ATLA [939]
Contemporary jewry – New Brunswick. 1976-1987 (1,5,9) – ISSN: 0147-1694 – mf#11103,02 – us UMI ProQuest [939]
Contemporary keyboard – Cupertino. 1975-1981 (1) 1976-1981 (5) 1976-1981 (9) – (cont by: keyboard) – ISSN: 0361-5820 – mf#10650 – us UMI ProQuest [780]
Contemporary literature – Madison. 1960+ (1) 1971+ (5) 1975+ (9) – ISSN: 0010-7484 – mf#2370 – us UMI ProQuest [400]
Contemporary longterm care – New York. 1988+ (1,5,9) – ISSN: 8750-9652 – mf#16483,02 – us UMI ProQuest [360]
Contemporary manchuria : a bi-monthly magazine – South Manchuria Railway Company, April 1937-January 1941 – 2r – 1 – $125.00 – us UMI ProQuest [380]
Contemporary ob/gyn – Montvale. 1973+ (1) 1975+ (5) 1976+ (9) – ISSN: 0090-3159 – mf#8560 – us UMI ProQuest [618]
Contemporary opinions on current topics – Translated from Japanese Magazines, Books, and Government Bulletins, Pamphlets, and Reports from Various Sources. nos. 117-378. 1936-41. 95 scattered issues wanting – 1 – us L of C Photodup [950]
Contemporary paintings, drawings and sculpture – (= ser Christie's pictorial archive new york) – 61mf – 9 – $550.00 – 0-907006-87-6 – (extensive coverage of post war & living artists; 3600 reproductions) – uk Mindata [700]
Contemporary pediatrics – Montvale. 1998+ (1,5,9) – ISSN: 8750-0507 – mf#20429 – us UMI ProQuest [618]
Contemporary philosophies of music education / Marple, Hugo Dixon – U of Rochester 1949 [mf ed 19–] – 3v on 1r / 8mf – 1,9 – mf#film 169 / fiche 1020 – us Sibley [780]
Contemporary physics – London. 1988+ (1,5,9) – ISSN: 0010-7514 – mf#17305 – us UMI ProQuest [530]
Contemporary poetry – Bryn Mawr. 1973-1982 (1) 1975-1982 (5) 1975-1982 (9) – (cont by: poesis) – ISSN: 0193-8339 – mf#9128 – us UMI ProQuest [810]
Contemporary policy issues – Huntington Beach. 1982-1993 (1) 1982-1993 (5) 1982-1993 (9) – (cont by: contemporary economic policy) – ISSN: 0735-0007 – mf#14015 – us UMI ProQuest [338]
Contemporary ponapean land tenure / Fischer, John L – n.d. – 6mf – 9 – $9.00 – mf#llmc82-100f, title 71 – us LLMC [333]
Contemporary portraits : thiers, strauss compared with voltaire, arnaud de l'ariege, dupanloup, adolphe monod, vinet, verny, robertson = Etudes contemporaines / Pressense, Edmond de – New York: ADF Randolph, 1880 – 1mf – 9 – 0-7905-5671-5 – (in english) – mf#1988-1671 – us ATLA [240]
Contemporary psychoanalysis – New York. 1985+ (1,5,9) – ISSN: 0010-7530 – mf#14622 – us UMI ProQuest [150]
Contemporary psychology – Washington. 1956+ [1]; 1965+ [5]; 1970+ [9] – ISSN: 0010-7549 – mf#1151 – us UMI ProQuest [150]
Contemporary review – Cheam. 1953+ (1) 1971+ (5) 1976+ (9) – ISSN: 0010-7565 – mf#865 – us UMI ProQuest [321]
Contemporary reviews in obstetrics and gynaecology – Guildford. 1988-1991 (1,5,9) – ISSN: 0955-9182 – mf#17219 – us UMI ProQuest [618]
Contemporary rhythmic devices / Fennell, Dorothy Codner – U of Rochester 1939 [mf ed 19–] – 4mf – 9 – mf#fiche 721 – us Sibley [780]
Contemporary security policy – London. 1994-1995 (1,5,9) – ISSN: 1352-3260 – mf#18546,01 – us UMI ProQuest [320]
Contemporary socialism / Rae, John – 3rd ed. New York: C Scribner's Sons 1910 [mf ed 1986] – 1r – 1 – (filmed with: la religion de l'empereur julien / farney, r) – mf#1725 – us Wisconsin U Libr [335]
Contemporary sociology – Washington. 1972+ (1,5,9) – ISSN: 0094-3061 – mf#11116 – us UMI ProQuest [301]
The contemporary south african short story in english with special references to the work of nadine gordimer, doris lessing, alan paton, jack cope, uys krige and dan jacobson / Millar, Clive – Cape Town 1962 – us CRL [420]
Contemporary southeast asia – Singapore. 1994+ (1,5,9) – ISSN: 0129-797X – mf#19218 – us UMI ProQuest [959]
Contemporary spanish dramatists / Turrell, Charles Alfred – Boston, MA. 1919 – 1r – us UF Libraries [440]
Contemporary stone and tile design – Troy. 1999+ (1,5,9) – ISSN: 1527-7690 – mf#31548,01 – us UMI ProQuest [690]
Contemporary surgery – Redondo Beach. 1972+ (1) 1974+ (5) 1974+ (9) – ISSN: 0045-8341 – mf#8561 – us UMI ProQuest [617]
Contemporary trends in the musical settings of the liturgical mass / Rosner, Mary Christian – U of Rochester 1957 [mf ed 19–] – 2v on 1r / 6mf – 9 – (with bibl & app) – mf#film 1113 / fiche 1044 – us Sibley [780]
Contenant la 1,2,3,4 livraison des monuments des regnes de saint louis, de philippe le hardi, de philippe le bel, de louis 10, de philippe 5 et de charles 4 (rgfs20-23) : depuis l'an 1226-1328 – 1893-94 – (= ser Rerum gallicarum et francicarum scriptores (rgfs)) – €201.00 – ne Slangenburg [241]
Contenant une relation de son voyage de canton...pe-king / Lettres d'un missionaire...M l'Abbe G*** – Paris, 1776-1783. v8 – 1mf – 9 – mf#CH-1003 – ne IDC [915]
The contender for the faith – New York City: Church of Christ Pub Co. v5 n9-v6 (dec 1946-1950), may-jun 1951 [mf ed 2005] – 1r [ill] – 1 – (lacks: v5 n10, v6 n2 p15, v7 n1,7,9-10, v6 n6, mar-dec 1950]) – mf#2005-s028 – us ATLA [242]
Contending for the faith / Anderdon, William Henry – London, England. 1850 – 1r – us UF Libraries [240]
Contending for the faith / Barr, James – Glasgow, Scotland. 1845 – 1r – us UF Libraries [240]
The contendings of the apostles : being the histories of the lives and martyrdoms and deaths of the twelve apostles and evangelists / Budge, Ernest Alfred Wallis – London, 1899-1901. 2pts – 15mf – 9 – mf#NE-20319 – ne IDC [240]
The contendings of the apostles, vol 2, the english translation : being the histories of the lives and martyrdoms and deaths of the twelve apostles and evangelists / ed by Budge, Ernest Alfred Wallis – London: Oxford University Press, 1901 – 2mf – 9 – 0-524-08306-1 – mf#1993-0011 – us ATLA [225]
Contenson, L de *see* Chretiens et musulmans
Content / Humber College of Applied Arts and Technology – n50-65, 96-h11 [1975 apr-1976 apr, 1979 may-1984:winter] – 1r – 1 – (cont: content for canadian journalists; cont by: content for canadian journalists [1984]; sources [b zwicker: publisher]) – mf#181511 – us WHS [700]
Content and form of yoruba ijala / Babalola, S A – Oxford, England. 1966 – 1r – us UF Libraries [960]
The content of broadcasting in nigeria / Ekwelie, Sylvanus Ajani – Madison, 1968 – us CRL [070]
The content of indian and iranian studies : an inaugural lecture delivered on 2 may 1938 / Bailey, Harold Walter – Cambridge: University Press, 1938 – (= ser Samp: indian books) – us CRL [490]
Contenta : ricoldi ordinis praedicatorum contra sectam ma-humeticam... / Montecroce, Riccoldo de – Parisiis, 1509 – 2mf – 9 – mf#H-8222 – ne IDC [956]
Contentio veritatis : essays in constructive theology / Rashdall, Hastings et al – London:John Murray, 1902 – 1mf – 9 – 0-8370-2728-4 – mf#1985-0728 – us ATLA [240]
The contents and origin of the acts of the apostles : critically investigated / Zeller, Eduard – London: Williams and Norgate, 1875. Chicago: Dep of Photodup, U of Chicago Lib, 1973 (1r); Evanston: American Theol Lib Assoc, 1984 (1r) – 1 – 0-8370-0289-3 – (incl bibl ref) – mf#1984-6008 – us ATLA [226]
Contents of 1913 cornerstone / Nashville. Tennessee. Grace Baptist Church – 1 – 5.00 – us Southern Baptist [242]
Contents of current legal periodicals – Wilmington. 1972-1976 (1) 1972-1976 (5) 1972-1976 (9) – (cont by: cclp: contents of current legal periodicals) – ISSN: 0300-7391 – mf#9481 – us UMI ProQuest [340]
O conterraneo : periodico imparcial, litterario e noticioso – Valenca, RJ. 11 jul 1885 – (= ser Ps 19) – bl Biblioteca [079]
Contertulios de la gruta simbolica / Mora, Luis Maria – Bogota, Colombia. 1936 – 1r – us UF Libraries [972]
Contes africains : livre de lectures africaines / Beaumont, Pierre de – [Abidjan]: Editions Africaines 1964 – us CRL [960]
Contes bizarres : [short stories] / Arnim, Ludwig Achim, Freiherr von – Paris: Cahiers libres, 1933 [mf ed 1993] – 199p/3pl (ill) – 1 – (ill by valentine hugo. int by andre breton. pref by theophile gautier) – mf#8464 – us Wisconsin U Libr [830]
Contes canadiens – Montreal: Librairie Beauchemin ltee, [1919?] [mf ed 1992] – 2mf – 9 – 0-665-99788-4 – mf#99788 – cn Canadiana [800]
Les contes d'amadou-koumba / Diop, Birago – Paris: Fasquelle, [1947] – 1 – us CRL [490]
Contes de la reine de navarre : ou, la revanche de... / Scribe, Eugene – Paris, France. 1850 – 1r – us UF Libraries [440]
Les contes de m mercier – S.l: s.n, 1883? – 2mf – 9 – mf#03431 – cn Canadiana [320]
Contes de noel / Dandurand, Josephine – Montreal: J Lovell, 1889 – 2mf – 9 – 0-665-06536-1 – (pref by louis frechette) – mf#06536 – cn Canadiana [830]
Les contes de perrault et les recits paralleles / Saintyves, Pierre – Paris: E Nourry 1923 [mf ed Bloomington IN: Indiana Uni Lib, Preservation Dept 1984] – xxiii 646p on 1r – 1 – us Indiana Preservation [390]
Contes des fees = Fairy tales / Aulnoy, Marie-Catherine d' – London: G Routledge 1855 [mf ed Bloomington IN: Indiana Uni Lib, Preservation Dept 1984] – xx 619p on 1r [ill] – 1 – us Indiana Preservation [390]
Contes du chevalier de la morliere / Lamorliere, Jacques Rochette De – Paris, France. 1879 – 1r – us UF Libraries [944]
Contes du larhalle, suivis d'un recueil de proverbes et de devises du pays mossi / Tiendrebeogo, Yamba – Ouagadougou: Tiendrebeogo, 1963 [cover 1964] – us CRL [960]
Contes du pays d'haiti / Comhaire Sylvain, Suzanne – Port-Au-Prince, Haiti. 1938 – 1r – us UF Libraries [972]
Contes et legendes / Lacerte, Adele Bourgeois – [Ottawa?: s.n.] 1915 [mf ed 1995] – 3mf – 9 – 0-665-73893-5 – mf#73893 – cn Canadiana [390]
Contes maconniques dedies aux soeurs et aux freres / Bazot, Etienne-Francois – Paris: Teissier 1845 – 2mf – 9 – mf#vrl-12 – ne IDC [366]
Contes persans – Liege: Vaillant-Carmanne; Paris: H Champion 1910 [mf ed Bloomington IN: Indiana Uni Lib, Preservation Dept 1984] – xv 526p on 1r – 1 – us Indiana Preservation [830]
Contes populaires : recueillis a bournois / Roussey, Charles – Paris: C Roussey: H Welter 1894 [mf ed Bloomington IN: Indiana Uni Lib, Preservation Dept 1984] – xi 303p on 1r – 1 – us Indiana Preservation [390]
Contes populaires des sakalava et des tsimihety de la region d'analalava / Dandouau, Andre – Algiers: J Carbonel, 1922 – 1 – us CRL [390]
Contes pour enfants canadiens / Marjolaine – Montreal: Librairie d'Action canadienne-francaise, ltee, 1931 [mf ed 1991] – 2mf – 9 – (ill by james mcisaac) – mf#SEM105P1316 – cn Bibl Nat [971]
Contes vrais / Lemay, Pamphile – Montreal: Beauchemin, 1907 [mf ed 1995] – 7mf – 9 – 0-665-74852-3 – mf#74852 – cn Canadiana [830]
Contest courier – [Scotland] Glasgow: J Cossar 18 may-18 aug 1908 (mthly) [mf ed 2003] – 4v on 1r – 1 – (a comprehensive and up-to-date record of the contesting movement) – uk Newsplan [072]
The contest for liberty of conscience in england / St John, Wallace – Chicago: University of Chicago Press, 1900 – (= ser Divinity Studies of the University of Chicago) – 1mf – 9 – 0-524-07719-3 – (incl bibl ref) – mf#1991-3304 – us ATLA [100]
The contest with rome : a charge to the clergy of the archdeaconry of lewes / Hare, Julius Charles – 2nd ed. Cambridge: Macmillan, 1856 – 1mf – 9 – 0-8370-8025-8 – mf#1986-2025 – us ATLA [240]
Contestacion al informe de la comision especial / Chase National Bank Of The City Of New York – New York, NY. 1934 – 1r – us UF Libraries [332]

Contestacion al informe...joaquin munoz bueno... linea ferrea...caceres / Tamarit de Plaza – 1867 – 9 – sp Bibl Santa Ana [946]

Contestacion al manifiesto publicado...mauricio ceresoles...pasadas elecciones.. / Concha, Antonio – 1839 – 9 – sp Bibl Santa Ana [946]

Contestacion de los fabricantes de jabon de la hab... – Habana, Cuba. 1891 – 1r – us UF Libraries [972]

Contestacion de rocafuerte al articulo...en la concordia 5 de febrero de 1844 / Rocafuerte, Vicente – 1 – us CRL [972]

Contestacion de teoria al programa de musica de magisterio segundo – Caceres: Imprenta Moderna, 1960 – 1 – sp Bibl Santa Ana [780]

Contestacion que da la senora dona agustina orellana – Caceres, 1849. Imp. de D.A. Concha y Cia – 1 – sp Bibl Santa Ana [946]

Contestaciones al derecho hipotecario / Munoz Casillas, Juan & Munoz Casillas, Joaquin – Caceres: Tip.El Noticiero, Tomo 1. 1921 – 1 – sp Bibl Santa Ana [946]

Contestaciones oficiales de doctrina del movimento, para oposiciones en las que se exige titulo elemental – Madrid: Editora Nacional 1945 [mf ed 1977] – (= ser Blodgett coll) – 1mf – 9 – mf#w818 – us Harvard [946]

Contestacion...extremadura por el aviso / Hore, Rafael – 1811 – 9 – sp Bibl Santa Ana [946]

Contestacion...pleito...falsedad y nulidad / Quintanilla, Condessa de – 1849 – 9 – sp Bibl Santa Ana [340]

Contestacion...vecino de burguillos...venta de arbolada / Martinez de Santa Maria, Juan – 1863 – 9 – sp Bibl Santa Ana [830]

Conteur : ou, les deux postes / Picard, Louis-Benoit – Paris, France. 1813 – 1r – us UF Libraries [440]

Conteurs canadiens-francais du 19e siecle : 2e se rie – Montreal: Librairie Beauchemin, 1913 – 2mf – 9 – 0-665-65358-1 – mf#65358 – cn Canadiana [830]

Context : a commentary on the interaction of religion and culture – Chicago. 1992-1993 (1) – ISSN: 0361-8854 – mf#15467 – us UMI ProQuest [230]

Context – Venice, CA. Spring 1982-Winter 1985 – 1 – us AJPC [071]

Context effects on the intrinsic dynamics of infants with spina bifida / Chapman, David – 1997 – 2mf – 9 – $8.00 – mf#PSY 2073 – us Kinesology [612]

Context Foundation [Sequim WA] et al *see* In context

The contextual interference effect in learning a soccer passing skill / Lima, Rogerio P – 2001 – 44p on 1mf – 9 – $5.00 – mf#PSY – us Kinesology [150]

The contextual interference effect on the memory system : motoric or perceptual? / Whitman, Shawn P – 2000 – 1mf – 9 – $4.00 – mf#PSY 2116 – us Kinesology [612]

Contextual interference in the motor domain : the effects of related and unrelated task practice / Dollar, John E – 1995 – 220p on 3mf – 9 – $15.00 – mf#PSY 2154 – us Kinesology [612]

Conti, Francesco *see* Most celebrated aires & duets in the opera of clotilda

Conti, Giovanni *see* La giustizia fra intrighi e tradimenti

Conti, Luigi de *see*
– Mercury's caducean rod: or, the great and wonderful office of the universal mercury, or god's vicegerent, displayed..
– Trifertes sagani, or, immortal dissolvent: being a brief but candid discource of the matter and manner of preparing the liquor alkahest of helmont, the great hilech of parocelsus..

Conti, N *see* Commentarii hieronymi comitis alexandrini de acerrimo, ac omnium difficillimo turcarum bello, in insulam melitam gesto, anno 1565

Conti Rossini, Carlo *see* La langue des kemant en abyssinie

Contient ce qui s'est passe dans les gaules (rgfs2,3) : et ce que les francois ont fait sous les rois de la premiere race – 1869 – (= ser Rerum gallicarum et francicarum scriptores (rgfs)) – 2v – (v2 €46 v3 €44) – ne Slangenburg [241]

Contient ce qui s'est passe depuis le commencement du regne de louis le begue (rgfs8) : fils de charles le chauve, jusqu'a la fin du regne de louis 5, derniere roi de la seconde race, c'est-a-dire depuis l'an 877-987 – 1871 – (= ser Rerum gallicarum et francicarum scriptores (rgfs)) – €46.00 – ne Slangenburg [944]

Contient ce qui s'est passe sous les regnes de pepin et de charlemagne (rgfs5) : c'est a dire depuis l'an 752-814, avec les lois, etc de ces deux rois – 1869 – (= ser Rerum gallicarum et francicarum scriptores (rgfs)) – €46.00 – ne Slangenburg [931]

Contient la suite des monuments des trois regnes de philippe 1, de louis 6 et de louis 7 depuis l'an 1060-1180 (rgfs13) – 1869 – (= ser Rerum gallicarum et francicarum scriptores (rgfs)) – €48.00 – ne Slangenburg [241]

Contient la suite des monuments des trois regnes...depuis l'an 1060-1180 (rgfs14-16) – J J Brial: 1877-78 – (= ser Rerum gallicarum et francicarum scriptores (rgfs)) – €143.00 – ne Slangenburg [241]

Contient le 1,2,3 (rgfs17-19) : livraison des monuments des regnes de philippe-auguste et de louis 8, depuis l'an 1180-1226 – 1818-33 – (= ser Rerum gallicarum et francicarum scriptores (rgfs)) – €136.00 – ne Slangenburg [241]

Contient les enquetes administratives du regne de saint louis et la chronique de l'anonyme de bethune (rgfs24) – 1904 – (= ser Rerum gallicarum et francicarum scriptores (rgfs)) – €65.00 – ne Slangenburg [241]

Contient les gestes de louis le debonnaire, d'abord roi d'aquitaine, et ensuite empereur (rgfs6) : depuis l'an 781-840, avec les lois, etc de ce prince, etc – 1870 – (= ser Rerum gallicarum et francicarum scriptores (rgfs)) – €42.00 – ne Slangenburg [241]

Contient les lettres historiques (rgfs4) : les loix...qui concernent les gaules et la france sous les rois de la premiere race – 1869 – (= ser Rerum gallicarum et francicarum scriptores (rgfs)) – €40.00 – ne Slangenburg [944]

Contient principalement ce qui s'est passe sous le regne de henri 1 (rgfs11) : fils de robert le pieux, c a d depuis l'an 1031-1060 – 1876 – (= ser Rerum gallicarum et francicarum scriptores (rgfs)) – €56.00 – ne Slangenburg [944]

Contient surtout ce qui s'est passe depuis le commencement du regne de hugues capet jusqu'a celui du roi henri 1 (rgfs10) : fils de robert le pieux – 1874 – (= ser Rerum gallicarum et francicarum scriptores (rgfs)) – €46.00 – ne Slangenburg [944]

Contient tout ce qui a ete fait par les gaulois, et qui s'est passe dans les gaulois avant l'arrive des francois (rgfs1) : et plusieurs autres choses qui regardent les francois depuis leur origine jusqu'a clovis – 1869 – (= ser Rerum gallicarum et francicarum scriptores (rgfs)) – €52.00 – ne Slangenburg [241]

Contient une partie de ce qui s'est passe sous les trois regnes de philippe 1, de louis 6 dit le gros (rgfs12) : et de louis 6 surnomme le jeune, depuis l'an 1060-1180 – 1877 – (= ser Rerum gallicarum et francicarum scriptores (rgfs)) – €52.00 – ne Slangenburg [944]

Contigo pan y cebolla / Quintero, Hector – Habana, Cuba. 1965 – 1r – us UF Libraries [972]

Contin Aybar, Pedro Rene *see*
– Antologia poetica dominicana
– Notas acerca de la poesia dominicana

Continent – v1-57. mar 1870-apr 1926 – 90r – 1 – (lacks some iss. title varies; cont: the interior) – mf#ATLA S0160 – us ATLA [073]

Continent – Philadelphia. 1882-1884 (1) – mf#3871 – us UMI ProQuest [420]

The continent – 6v. 1882-84 – 130mf – 9 – (an illustrated weekly magazine) – mf#C35-22100 – us Primary [073]

A continent decides / Birdwood, Christopher Bromhead, Baron – London: Robert Hale, 1953 – (= ser Samp: indian books) – us CRL [954]

Continental and island life, a review of wallace : with reference to the bearing of geological facts and theories of evolution on the distribution of life / Dawson, John William – S.l: s.n, 1881? – 1mf – 9 – mf#02104 – cn Canadiana [550]

Continental bank journal of applied corporate finance – New York. 1988-1994 (1,5,9) – (cont by: bank of america journal of applied corporate finance) – ISSN: 0898-4484 – mf#17450 – us UMI ProQuest [332]

Continental drift emphasizing the history of the south atlantic area / ed by Wilson, J T – 1972 – ca 1000p – 1,5,6,9 – $10.00 – us AGU [550]

Continental europe, 500 to 1980 *see* History of glass

Continental india : travelling sketches and historical recollections, illustrating the antiquity, religion, and manners of the hindoos, the extent of british conquests, and the progress of missionary operations / Massie, James William – London: Thomas Ward, 1840 [mf ed 1995] – (= ser Yale coll) – 2v (ill) – 1 – 0-524-09932-4 – mf#1995-0932 – us ATLA [915]

Continental marine – v1 n1-v2 n10 [1976 feb 19-1977 oct] – 1r – 1 – (cont by: continental marine and digest) – mf#703093 – us WHS [355]

Continental marine – 1979-90 – 7r – 1 – (cont: continental marine and digest) – mf#703082 – us WHS [355]

Continental marine and digest – 1977 nov-dec, 1978 – 2r – 1 – (cont: continental marine; cont by: continental marine [1979]) – mf#703086 – us WHS [355]

Continental monthly : devoted to literature and national policy – New York. 1862-1864 – 1 – mf#4588 – us UMI ProQuest [073]

Continental philosophy review – Dordrecht. 2000+ (1,5,9) – (cont: man and world) – ISSN: 1387-2842 – mf#16826,01 – us UMI ProQuest [100]

The continental reformation / Kidd, Beresford James – New York: Edwin S Gorham, 1902 – (= ser Oxford church text books) – 1mf – 9 – 0-524-00868-X – mf#1990-0253 – us ATLA [242]

The continental reformation in germany, france and switzerland : from the birth of luther to the death of calvin / Plummer, Alfred – London: R. Scott, 1912 – 1mf – 9 – 0-7905-5553-0 – (incl bibl ref) – mf#1988-1553 – us ATLA [242]

Continental review – brussels, belgium 3 oct 1908-21 oct 1909 – 1 – (cont: belgian times & news & european express [12 jan 1907-26 sep 1908]) – uk British Libr Newspaper [072]

Continental shelf research – Oxford. 1982+ (1,5,9) – ISSN: 0278-4343 – mf#49397 – us UMI ProQuest [550]

The continental teutons / Merivale, Charles – London: SPCK; New York: Pott, Young, [1878?] – (= ser Conversion Of The West) – 1mf – 9 – 0-7905-6305-3 – mf#1988-2305 – us ATLA [240]

Continental times – Toronto, dec 3 1948-mar 30 1982// (semiwkly) – 35r – 1 – Can$1925.00 – (in japanese & english, feb 1 1949-82. cont by: canada times, toronto) – cn McLaren [071]

The continental times – Berlin DE, 1914 18 sep-1919 14 jul [gaps] – 2r – 1 – (filmed by misc inst: 1915 10 nov-1 dec [gaps], 1916 3 jul-1918. 1 jan-30 jun 1916 not publ) – uk British Libr Newspaper; gw Misc Inst [074]

Continental Union Association of Ontario *see* Our best policy

Continental union versus reciprocity : erastus wiman answered by an ex-member of the canadian parliament / Glen, Francis Wayland – [S.l: s.n, 1893?] [mf ed 1981] – 1mf – 9 – mf#01466 – cn Canadiana [337]

Continental unity : an address delivered in music hall, boston, by invitation of prominent citizens, december 13, 1888 / Murray, William Henry Harrison – 1st ed. [Boston?: s.n.], 1888 [mf ed 1981] – 1mf – 9 – 0-665-11185-1 – mf#11185 – cn Canadiana [971]

Continente – Mexico City. 1953-1953 – 1 – mf#472 – us UMI ProQuest [073]

Continente de la esperanza / Henriquez Urena, Max – Bruselas, Belgium. 1939 – 1r – us UF Libraries [972]

Les continents – Paris: M Boivent, may 15, jun 15, jul 15, sep 1-dec 15 1924 – (filmed with 12 other titles) – us CRL [074]

The contingency of the laws of nature = de la contingence des lois de la nature / Boutroux, Emile – Chicago: Open Court, 1916 – 1mf – 9 – 0-7905-7690-2 – (in english) – mf#1989-0195 – us ATLA [100]

Contingent accounts of the sheriff of the district of montreal : from the 11th april to the 10th october 1833 – [Quebec]: the Standing Committee on Public Accounts, [1834 ?] (mf ed 1992) – 3mf – 9 – mf#SEM105P1393 – cn Bibl Nat [336]

Contino, Giovanni *see*
– Modulationum quinque vocum. liber primus
– Modulationum quinque vocum. liber secundus

Continua : newsletter of the ecumenical partnership on peace and justice / Wisconsin Conference of Churches – 1981 mar-1983 nov – 1r – 1 – (cont by: metanoia [dodgeville wi]) – mf#958116 – us WHS [240]

Continuacion de las nuevas poesias de... / Salas, Francisco Gregorio de – Madrid: Andres Ramirez, 1776 – 1 – sp Bibl Santa Ana [946]

Continuacion del juicio critico del ano de 1788 / Salas, Francisco Gregorio de – Madrid: Joseph Otero, 2ª ed 1788 – 1 – sp Bibl Santa Ana [946]

Continuacion...poesias / Salas, Francisco Gregorio de – 9 – sp Bibl Santa Ana [810]

Continuatio cantionum sacrarum quatuor...et plurium vocum – 1588 – (= ser Mssa) – 8mf – 9 – €100.00 – mfchl 140 – gw Fischer [780]

Continuatio chronicarum [rs93] / Murimuth, Adam; ed by Thompson, E M – 1889 – (= ser The rolls series [rs] 93) – €19.00 – (filmed with: robertus de avesbury: de gestis mirabilibus regis edwardii tertii) – ne Slangenburg [931]

Continuatio des abenteuerlichen simplicissimi *see* Der abenteuerliche simplicissimus

Continuatio des abentheurlichen simplicissimi : oder der schluss desselben / Grimmelshausen, Hans Jakob Christoph von; ed by Scholte, Jan Hendrik – Halle/Saale: M Niemeyer 1939 [mf ed 1993] – (= ser Neudrucke deutscher literaturwerke des 16. und 17. jahrhunderts 310-314) – 1r – 1 – (incl bibl ref) – mf#3387p – us Wisconsin U Libr [830]

Continuatio missarum sacrarum / Regnart, Jacob – 1603 – (= ser Mssa) – 6mf – 9 – €80.00 – mfchl 386 – gw Fischer [780]

The continuation committee conferences in asia, 1912-1913 : a brief account of the conferences together with their findings and lists of members – New York: Chairman of the Continuation Committee, 1913 – 2mf – 9 – 0-7905-4179-3 – mf#1988-0179 – us ATLA [240]

Continuation de l'histoire generale des voyages ou collection nouvelle : 1e des relations de voyages par mer, decouvertes, observations...de feu m l'abbe provost... – Paris: Chez Rozet... 1768 [mf ed 1985] – 8mf – 9 – 0-665-50078-5 – mf#50078 – cn Canadiana [910]

Continuation des copies de communications officielles : rapports et autres documens qui ont rapport aux evenemens qui ont eu lieu, a montreal, le 21 mai, 1832... – [s.l: s.n, 1832?] [mf ed 1984] – 1mf – 9 – 0-665-21368-9 – (also available in english) – mf#21368 – cn Canadiana [303]

Continuation of essays towards the history of painting / Callcott, Maria (Dundas) Graham, lady – London 1838 – (= ser 19th c art & architecture) – 1mf – 9 – mf#4.2.941 – uk Chadwyck [750]

Continuation of henry's journal : covering adventures and experiences in the fur trade on the red river, 1799-1801 / Bell, Charles Napier – Winnipeg: Manitoba Free Press, 1889 – 1mf – 9 – mf#30237 – cn Canadiana [380]

Continuation of the ampleforth discussion / Keary, Wiliam – York, England. 18– – 1r – us UF Libraries [240]

Continuation of the copies of official communications, reports and other documents : having reference to the occurrences which took place in montreal, on the 21st may, 1832... – [s.l: s.n, 1832?] [mf ed 1985] – 1mf – 9 – 0-665-21369-7 – (in dble clms. also available in french) – mf#21369 – cn Canadiana [303]

Continued professional learning and the experienced elementary school physical education specialist / Pissanos, B W – 1989 – 4mf – 9 – $16.00 – us Kinesology [790]

Continuing education – Bensalem. 1968-1974 – 1,5 – ISSN: 0010-7751 – mf#6489 – us UMI ProQuest [374]

A continuing education programme for family nurse practitioners in swaziland / Mathunjwa, Murmly D – Uni of South Africa 2000 [mf ed Johannesburg 2000] – 5mf – 9 – (incl bibl ref) – mf#mfm14959 – sa Unisa [610]

Continuing inquiry – 1982 jan 22-1984 apr 22, 1986 aug 22-1981 – 2r – 1 – mf#68751 – us WHS [120]

Continuities : journal from the black studies department / City University of New York – 1974, 1975 spr – 1r – 1 – mf#4851568 – us WHS [305]

Continuity and change – Cambridge. 1989-1996 (1) – ISSN: 0268-4160 – mf#16526 – us UMI ProQuest [303]

The continuity of christian thought : a study of modern theology in the light of its history / Allen, Alexander Viets Griswold – Boston: Houghton, Mifflin, 1884 – (= ser The bohlen lectures) – 2mf – 9 – 0-524-08846-2 – mf#1993-2131 – us ATLA [240]

Continuity of possession at the reformation / Browne, George Forrest – London, England. 1897 – 1r – us UF Libraries [240]

The continuity of possession at the reformation / Browne, George Forrest – rev ed. London: SPCK 1897 [mf ed 1993] – (= ser Church historical society (series) 24) – 1mf – 9 – 0-524-05489-4 – (incl ind) – mf#1990-1484 – us ATLA [242]

The continuity of scripture : as declared by the testimony of our lord and of the evangelists and apostles / Wood, William Page – rev ed. London: SPCK, [c1869] – 1mf – 9 – 0-8370-3522-8 – mf#1985-1522 – us ATLA [220]

The continuity of the church of england : before & after its reformation in the sixteenth century, with some account of its present condition / Puller, Frederick William – London: Longmans, Green, 1912 – 1mf – 9 – 0-524-02395-6 – mf#1990-4297 – us ATLA [241]

The continuity of the church of england in the sixteenth century : two discourses, with an appendix and notes / Seabury, Samuel – New-York: Pudney & Russell, 1853 – 1mf – 9 – 0-524-02495-2 – mf#1990-4354 – us ATLA [241]

Continuity of the english church / Croft, Aloysius – London, England. 1886 – 1r – us UF Libraries [241]

CONTRIBUCIONES

Continuity of the holy catholic church in england / Browne, George Forrest – London, England. 1896 – 1r – us UF Libraries [241]

The continuity of the holy catholic church in england : a lecture...haggerston, in 1896 / Browne, George Forrest – London: SPCK 1903 [mf ed 1993] – (= ser Church historical society (series) 8) – 1mf – 9 – 0-524-05531-9 – (incl ind) – mf#1990-5135 – us ATLA [242]

Continuity of the platonic tradition during the middle ages / Klibansky, Raymond – London, England. 1939 – 1r – us UF Libraries [180]

Continuity or collapse? : the question of church defence / McCave, Canon James & Breen, J D; ed by Mackinlay, James Boniface – New ed. London: Art and Book; New York: Benziger, 1891 – 1mf – 9 – 0-8370-7086-4 – (incl ind) – mf#1986-1086 – us ATLA [240]

Continuous learning – Toronto. 1962-1971 (1) 1971-1971 (5) – ISSN: 0010-7778 – mf#1594 – us UMI ProQuest [374]

Continuum – Chicago. 1963-1970 [1] – ISSN: 0010-7786 – mf#5871 – us UMI ProQuest [240]

Con-tob retailing – Ilford. 1963-1971 [1] – ISSN: 0041-3089 – mf#1341 – us UMI ProQuest [660]

Contos da vida e da morte / Coelho Netto, Henrique – Porto, Portugal. 1927 – 1r – us UF Libraries [972]

Contos do brasil / Hamilton, Daniel Lee – New York, NY. 1944 – 1r – us UF Libraries [972]

Contos e cronicas / Andrade, Nuno Ferreira De – Rio de Janeiro, Brazil. 1941 – 1r – us UF Libraries [972]

Contos e lendas do brasil / Orico, Osvaldo – Sao Paulo, Brazil. 1939 – 1r – us UF Libraries [972]

Contos fluminenses / Machado De Assis – Rio de Janeiro, Brazil. v1-2. 1937 – 1r – us UF Libraries [972]

Contos leves / Lobato, Jose Bento Monteiro – Sao Paulo, Brazil. 1941 – 1r – us UF Libraries [972]

Contos populares brasileiros / Gomes, Lindolfo – Sao Paulo, Brazil. 1948 – 1r – us UF Libraries [972]

Contos populares de angola / Chatelain, Heli – Lisboa, Portugal. 1964 – 1r – us UF Libraries [960]

Contos reunidos / Cruls, Gastao – Rio de Janeiro, Brazil. 1951 – 1r – us UF Libraries [972]

Contos tradicionais do brasil / Cascudo, Luis Da Camara – Rio de Janeiro, Brazil. 1967 – 1r – us UF Libraries [972]

Contour notes / Wisconsin Regional Artists Association – 1970 win-1983 fall – 1r – 1 – (cont: contour [tripoli wi]) – mf#1659965 – us WHS [700]

Contra adversarium legis et prophetarum. contra priscillianistas et orienistas. de errore priscillianistarum et origenistarum : formae tplila 29 / Augustinus, Orosius – 1985 – (= ser ILL – ser a; Ccsl 49) – 4mf+43p – 9 – €30.00 – 2-503-60492-7 – be Brepols [400]

...Contra alchoranum and sectam machometicam libri quinque / Leuwis, D de – Coloniae, 1533 – 7mf – 9 – mf#H-8242 – ne IDC [956]

Contra arianos; de laude sanctorum; libellus emendationis; epistulae; commonitorium. excerptis ex operibus s. augistini; altercatio legis inter simonem iudaeum et theophilum christianum / Foebadius Aginnensis et al – 1985 – (= ser ILL – ser a; Ccsl 64) – 7mf+104p – 9 – €40.00 – 2-503-60642-3 – be Brepols [400]

Contra costa county – 1906-38 – (= ser California telephone directory coll) – 32r – 1 – $1,600.00 – (east 1992- 3r $150. west 1992- 3r $150. north 1992- 4r $200. south 1992- 4r $200. north and south 1994- 2r $100) – mf#P00020 – us Library Micro [917]

[Contra costa county-] alameda, contra costa, monterey, san benito, san mateo, santa clara and santa cruz counties – CA. 1879 – 1r – 1 – $50.00 – mf#D002 – us Library Micro [978]

[Contra costa county-] antioch and pittsburg city directories – CA. 1931; 1947-1950 – 3r – 1 – $150.00 – mf#D014 – us Library Micro [917]

[Contra costa county-] history of contra costa county / Fraser, J P Munro – CA. 1882 – 1r – 1 – $50.00 – mf#B40213 – us Library Micro [978]

Contra costa county labor journal – 1968 jan 5-1970 jul 31, 1970 aug 7-1972 dec 8 – 2r – 1 – mf#643881 – us WHS [331]

[Contra costa county-] richmond and martinez city directories – CA. 1921-1950 – 22r – 1 – $1100.00 – mf#D013 – us Library Micro [917]

Contra Costa Labor Health and Welfare Council *see* News letter

Contra costa labor news – 1961 jun-1978 may – 1r – 1 – (cont: news letter [contra costa labor health and welfare council]; cont by: contra costa; napa, solano labor news) – mf#643864 – us WHS [331]

Contra costa, napa, solano, labor news – 1978 jun-1983 mar, 1983 jan-1988 dec, 1989-94 – 3r – 1 – (cont: contra costa labor news; cont by: labor community news) – mf#643866 – us WHS [331]

Contra fatalitatis errorem : formae tplila 96 / Bartholomaeus Exoniensis – [mf ed 1999] – (= ser ILL – ser a; Cccm 157) – 5mf+42p – 9 – €40.00 – 2-503-64572-0 – be Brepols [400]

Contra felicem : formae tplila 59 / Aquileiensis, Paulinus – 1990 – (= ser ILL – ser a; Cccm 95) – 4mf+44p – 9 – €30.00 – 2-503-63952-6 – be Brepols [400]

Contra impugnantes dei cultum et religionem *see* Apology for the religious orders

Contra iohannem (sl 79a). altercatio luciferiani et orthodoxi (sl 79b) / Hieronymus – (mf ed 2001) – 4mf+54p – 9 – €30.00 – 2-503-60794-2 – be Brepols [400]

Contra komintern – Berlin DE, 1937 apr-1939 aug – 1 – gw Misc Inst [074]

Contra la anexion / Saco, Jose Antonio – Habana, Cuba. v1-2. 1928 – 1r – us UF Libraries [972]

Contra rufinum / Hieronymus – 1982 – (= ser ILL – ser a; Ccsl 79) – 4mf+48p – 9 – €20.00 – 2-503-60792-6 – be Brepols [400]

Contra sacramentarios dispvtationes dvae, prima de coena domini, altera de communicatione idiomatum item declarationes duae : et vera sententia avgvstane confessionis, in articulo, de coena domini quibus subscripserunt inferiores saxoniae theologi, qui fuerunt in proximo conuentu brunschuicensi, anno domini 1561 / Moerlin, J – [Eisleben, 1561] – 1mf – 9 – mf#TH-1 mf 1171 – ne IDC [242]

Contra turrim traiectensem *see* Ornatus spiritualis desponsationis / Contra turrim traiectensem

Contraception – New York. 1970+ (1) 1973+ (5) 1976+ (9) – ISSN: 0010-7824 – mf#9674 – us UMI ProQuest [610]

The contraceptive injection = Kontraseptiewe inspuiting / South Africa. Department of Health [Departement van Gesondheid] – [Pretoria: Dept of Health 1976?] [mf ed Pretoria, RSA: State Library [199-]] – 4p on 1r with other items – 5 – mf#op 06605 r24 – us CRL [360]

Contraceptive practices among division 1 collegiate women swimmers / Hinton, Stephanie A – 1998 – 2mf – 9 – $8.00 – mf#PE 3932 – us Kinesology [615]

Contraceptive technology update – Atlanta. 1989-1991 (1) – ISSN: 0274-726X – mf#12277 – us UMI ProQuest [680]

Contract – New York. 1969-1990 (1) 1972-1990 (5) 1976-1990 (9) – (cont by: contract design) – ISSN: 0010-7832 – mf#3489 – us UMI ProQuest [640]

Contract – San Francisco. 2000+ (1) – ISSN: 1530-6224 – mf#3489,02 – us UMI ProQuest [640]

Contract and captive electronic manufacturing and printed circuit production – Libertyville. 1989-1990 (1,5,9) – (cont: electronic manufacturing) – ISSN: 1053-1017 – mf#16957,01 – us UMI ProQuest [621]

Contract between the government of the dominion of canada and the canadian pacific railway company : also, the consolidated railway act (1879), and the act of 1881 amending it / Compagnie du chemin de fer canadien du Pacifique – [Ottawa?: s.n.], 1882 [mf ed 1980] – 2mf – 9 – (incl ind) – mf#02046 – cn Canadiana [380]

Contract design – New York. 1990-1996 (1) 1990-1996 (5) 1990-1996 (9) – (cont: contract. cont by: contract) – ISSN: 1053-5632 – mf#3489,01 – us UMI ProQuest [640]

Contract interiors – New York. 1976-1978 (1) 1976-1978 (5) 1976-1978 (9) – (cont: interiors. cont by: interiors) – ISSN: 0148-012X – mf#1037,01 – us UMI ProQuest [740]

Contract journal – Surrey. 1960-1994 (1) 1977-1979 (5) 1977-1979 (9) – ISSN: 0010-7859 – mf#2984 – us UMI ProQuest [690]

Contracting business – Cleveland. 1981+ (1,5,9) – (cont: airconditioning and refrigeration business) – ISSN: 0279-4071 – mf#797,01 – us UMI ProQuest [690]

Contractor – Newton. 1969-1996 (1) 1975-1981 (5) 1975-1981 (9) – ISSN: 0897-7135 – mf#5034 – us UMI ProQuest [690]

Contractor news – Radnor. 1962-1970 (1) – ISSN: 0589-5693 – mf#1937 – us UMI ProQuest [690]

Contractors and engineers magazine – New York. 1920-1975 (1) – ISSN: 0010-7905 – mf#3249 – us UMI ProQuest [690]

Contracts and combinations in restraint of trade / Kales, Albert Martin – Chicago, Callaghan, 1918. 169 p. LL-1623 – 1 – us L of C Photodup [343]

The contradictions of orthodoxy : or, "what shall i do to be saved?" as answered by several representative clergymen of chicago... / Collins, Almer M et al – Chicago IL: Central Book Concern, 1880 [mf ed 1990] – 1mf – 9 – 0-7905-3774-5 – mf#1989-0267 – us ATLA [210]

Contrails – Wichita KS. 1981 may-1983 jun, 1983 jul-1985 dec, 1986 jan 10-1987 dec 18 – 3r – 1 – mf#679271 – us WHS [071]

Contra-mao / Pereira, Antonio Olavo – Rio de Janeiro, Brazil. 1950 – 1r – us UF Libraries [972]

Contrapuntal practices of victoria / Young, Edward – U of Rochester 1942 [mf ed 19–] – 1r – 1 – (with bibl) – mf#film 193 – us Sibley [780]

Contrapunteo cubano del tabaco y el azucar / Ortiz, Fernando – Habana, Cuba. 1940 – 1r – us UF Libraries [972]

Contrast – 1969 oct-1972 nov 24; 1972 dec 1-1974 jun 28; 1974 jul 5-1975 dec 19; 1976 jan 9-1977 jun 30; 1977 jul 7-1978 dec 22; 1979 jan 11-dec 21; 1980; 1981; 1982-1983 jun; 1983 jul-1984 jun; 1984 jul-1985 sep; 1985 oct-1987 dec; 1988 jan 1988 oct-1989 jun; 1989 dec 6-1990 jun 28; 1989 jul-nov; 1990 jul 5-1991 jan 31 – 1 – mf#2464455 – us WHS [071]

Contrast – London, England. 18– – 1r – us UF Libraries [240]

Contrast : or a prophet and a forger / Abbott, Edwin Abbott – London: Adam and Charles Black, 1903. Chicago: Dep of Photodup, U of Chicago Lib, 1974 (1r); Evanston: American Theol Lib Assoc, 1984 (1r) – 1 – 0-8370-0055-6 – (includes bibliographical footnotes) – mf#1984-6022 – us ATLA [220]

Contrast : or, one missing – London, England. 18– – 1r – us UF Libraries [240]

Contrast : serving canada's black community – Toronto. v1-23 n3. feb 1969-jan 31 1991// – 23r – 1 – Can$1960.00 – (weekly since feb 1972) – cn McLaren [305]

Contrast *see* Monthly bulletin 1934-91

The contrast : or, the evangelical and tractarian systems compared in their structure and tendencies / Stone, John Seely – New-York: Protestant Episcopal Society for the Promotion of Evangelical Knowledge, 1853 – 3mf – 9 – 0-7905-6836-5 – mf#1988-2836 – us ATLA [242]

The contrast between good and bad men : illustrated by the biography and truths of the bible / Spring, Gardiner – New York: M W Dodd, 1855 – 2mf – 9 – 0-7905-2429-5 – mf#1987-2429 – us ATLA [220]

A contrast of planning skills between expert and novice college tennis coaches / Lubbers, Paul A – 1998 – 2mf – 9 – $8.00 – mf#PE 3847 – us Kinesology [790]

Contraste : libro de poemas / Morales Cardenas, Cirilo – Guanabacoa, Cuba. 1956 – 1r – us UF Libraries [972]

Contrastes (cronicas) / Compos, Humberto De – Rio de Janeiro, Brazil. 1938 – 1r – us UF Libraries [972]

Contrastes, cuentos, aguafuertes, cronicas / Diaz Nadal, Roberto – San Juan, Puerto Rico. 1965 – 1r – us UF Libraries [972]

Contrastes e confrontos / Cunha, Euclydes Da – Porto, Portugal. 1941 – 1r – us UF Libraries [972]

The contrasts of christianity with heathen and jewish systems : or, nine sermons / Rawlinson, George – London: Longman, Green, Longman, and Roberts, 1861 – 1mf – 9 – 0-7905-0204-6 – mf#1987-0204 – us ATLA [230]

Contratacion colectiva en la industria azucarera d... / Nogueras Rivera, Nicolas – San Juan, Puerto Rico. 1955 – 1r – us UF Libraries [972]

Contrato chaux-folsom y documentos relacionados co... / Colombia – Bogota, Colombia. 1931 – 1r – us UF Libraries [972]

Contrato de opcion / Nunez Y Nunez, Eduardo Rafael – Habana, Cuba. 1940 – 1r – us UF Libraries [972]

Contratos civiles / Aguilar, Leopoldo – Mexico City? Mexico. 1964 – 1r – us UF Libraries [972]

Contratos de la united fruit company / Leon Aragon, Oscar De – Guatemala, 1950 – 1r – us UF Libraries [338]

Contratos diversos / Martinez Escobar, Manuel – Habana, Cuba. 1939 – 1r – us UF Libraries [972]

Contratos y actuaciones de la companias del ferroc... / Saenz, Alfredo – San Jose, Costa Rica. 1929 – 1r – us UF Libraries [972]

Contratos y cuasicontratos mineros en las legislaciones sudamericanas / Alvarez Madariaga, Luz – Santiago de Chile, Talleres graficos "Simiente" 1944. 67 p. LL-8001 – 1 – $23.00r – us L of C Photodup [340]

Contrats de banque et emprunt... / Pouget, Louis Edouard – Port-Au-Prince, Haiti. no date – 1r – us UF Libraries [972]

Contre la guerre – 1-4. Paris. 23 nov 1912-5 mars 1913 [bimnthly] – 1 – fr ACRPP [325]

Contre la secte phantastique et furieuse des libertins : qui se nomment spirituelz / Calvin, J – Geneva: Jean Girard, 1545 – 3mf – 9 – mf#CL-49 – ne IDC [240]

Contre la secte phantastique et furieuse des libertins : qui se nomment spirituelz / Calvin, Jean – [Geneva: Jean Girard], 1547 – 3mf – 9 – mf#CL-25 – ne IDC [240]

Contre le courant : Organe de l'opposition communiste – n1-38. Paris. nov 1927-oct 1929 – 1 – fr ACRPP [325]

Contre le courant : organe de l'opposition communiste – Paris. v1 n1-5-v2 n38. nov 1927-oct 1929 – (= ser Communist international periodicals from the feltrinelli archives) – 7mf – 9 – $95.00 – us UPA [355]

Contre les barbares de l'orient : etudes sur la turquie, ses felonies et ses crimes sur la marche des allies dans l'asie anterieure sur la solution de la question d'orient / Morgan, J de – Paris, 1918 – 3mf – 9 – mf#AR-1611 – ne IDC [956]

Contre les doutes de plethon sur aristote / Georges Scholarus [Gennadius]; ed by Mynas, M – Paris, 1858 – €12.00 – ne Slangenburg [180]

Le contre-poison : ou preservatif contre les motions insidieuses, cabales, erreurs, mensonges, calomnies et faux principes repandus dans les feuilles de la semaine – Paris. n1-36. janv-avr 1791 – 1 – fr ACRPP [944]

Contreras Gallardo, Pedro de *see* Don lope diaz de azmendariz marquez de carreyta de su magestad su mayrdomo

Contreras Labarca, Carlos *see* Hacia donde va chile

Contreras R, J Daniel *see* Breve historia de guatemala

Contreras, Raul *see* Presencia de humo

Contribucao para o estudo da antropologia de mocambique / Santos Junior, Joaquim Rodrigues Dos – Porto, Portugal. 1944 – 1r – us UF Libraries [960]

Contribucao para o estudo da antropologia de mocambique / Santos Junior, Joaquim Rodrigues Dos – Porto, Portugal. 1945 – 1r – us UF Libraries [960]

Contribucion a la historia de la colonia del sacramento. la epopeya de manuel lobo... / Arazola Gil, Luis Enrique – Madrid, 1931; Madrid: Razon y Fe, 1932 – 1 – sp Bibl Santa Ana [946]

Contribucion a la historia de las instituciones co... / Zavala, Silvio Arturo – Guatemala, 1953 – 1r – us UF Libraries [972]

Contribucion a la prehistoria y protohistoria de rio muni / Perramon Marti, R – Santa Isabel de Fernando Poo, Guinea. 1968 – 1r – us UF Libraries [960]

Contribucion a una antologia de la realidad historica / Frutos Cortes, Eugenio – Madrid, 1943. Rev. Filosofia (Tomo II. Num. 4) del Instituto Luis Vives – sp Bibl Santa Ana [100]

Contribucion a una campana (cuatro... / Pena Batlle, Manuel Arturo – Santiago, Dominican Republic. 1942 – 1r – us UF Libraries [972]

Contribucion al estudio de la composicion acomatica de los vinos y derivados vinicos de tierra de barros / Maynar Marino, Juan – Badajoz: Universidad de Extremadura, 1981 – 1 – sp Bibl Santa Ana [370]

Contribucion al estudio de la guerra federal en ve... / Rodriguez, Jose Santiago – Caracas, Venezuela. v1-2. 1933 – 1r – us UF Libraries [972]

Contribucion al estudio de la osteosintesis en caracas / Parra Leon, Jose Antonio – Caracas, 1930; Madrid: Razon y Fe, 1931 – 1 – sp Bibl Santa Ana [610]

Contribucion al estudio de los pastos extremenos / Moreno Marquez, Victor – Madrid: Graficas Agma, 1952 – 1 – sp Bibl Santa Ana [946]

Contribucion al estudio de los vinagre de vino de "tierra de barros" y estudio de algunos componentes volatiles de los mismos / Olivares del Valle, Francisco Javier – Badajoz: Universidad de Extremadura, Facultad de Ciencias. Departamento de Quimica-fisica, 1975 – 1 – sp Bibl Santa Ana [550]

Contribucion al estudio de nuestra toponimia / Deletang, Luis F – Buenos Aires, 1931; Madrid: Razon y Fe, 1933 – 1 – sp Bibl Santa Ana [972]

Contribucion al estudio del bogotano / Gonzalez De La Calle, Pedro Urbano – Bogota, Colombia. 1963 – 1r – us UF Libraries [972]

Contribucion al estudio del colera / Cisneros, Juan – 1886 – 9 – sp Bibl Santa Ana [610]

Contribucion al proceso de concientizacion en america latina – Montevideo, Uruguay: Junta Latino Americana de Iglesia y Sociedad, 1968 – us CRL [972]

Contribucion de la republica dominicana / Henriquez, Gustavo Julio – Mexico City? Mexico. 1943 – 1r – us UF Libraries [972]

Contribuciones. impuestos, aranceles y gravamenes / Bardaji y Buitrago, A – Madrid: Mag, 1954 – 1 – sp Bibl Santa Ana [946]

Contribuicao a historia administrativa do brasil / Andrade, Almir De – Rio de Janeiro, Brazil. v1-2. 1950 – 1r – us UF Libraries [350]

Contribuicao a historia da imprensa brasileira, 18... / Vianna, Helio – Rio de Janeiro, Brazil. 1945 – 1r – us UF Libraries [972]

Contribuicao a historia das ideias no brasil / Cruz Costa, Joao – Rio de Janeiro, Brazil. 1956 – 1r – us UF Libraries [972]

Contribuicao a historia das ideias no brasil / Cruz Costa, Joao – Rio de Janeiro, Brazil. 1967 – 1r – us UF Libraries [972]

Contribuicoes para a archeologia paulista / Lofgren, Alberto – Sao Paulo, Brazil. 1893 – 1r – us UF Libraries [930]

Contribuicoes para a etnologia do brasil / Ehrenreich, Paul Max Alexander – Sao Paulo, Brazil. 1948 – 1r – us UF Libraries [305]

Contribuindo / Ribeiro De Andrada, Martim Francisco – Sao Paulo, Brazil. 1921 – 1r – us UF Libraries [972]

Contributi del r osservatorio astronomico di milano-merate / Reale Osservatorio di Brera in Milan – Pavia: Tip-legatoria Mario Ponzio 1938-43 [irreg] [mf ed 2003] – 14v on 1r – 1 – (cont: contributi della r. specola di brera-milano; cont by: contributi dell'osservatorio astronomico di milano-merate; imprint varies) – mf#film mas c5517 – us Harvard [520]

Contributi dell'osservatorio astronomico di milano-merate / Osservatorio astronomico di Brera – Pavia: Tip Mario Ponzio 1944- [irreg] [mf ed 2003] – 1r – 1 – (cont: contributi del r osservatorio astronomico di milano-merate; iss by the osservatorio astronomico di brera under a variant form of name; title varies slightly; imprint varies) – mf#film mas c5518 – us Harvard [520]

Contributi dell'osservatorio astronomico di torino [pino torinese] / Osservatorio astronomico di Torin – Torino: Tip v Bona 1944 [irreg] [mf ed 2003] – 1r – 1 – (some iss publ in combined form; some iss publ out of chronologial order) – mf#film mas c5587 – us Harvard [520]

Contributing variables to depth jump performance / Vaczi, Mark – 2000 – 1mf – 9 – $4.00 – mf#PE 4072 – us Kinesology [611]

Contribution a l'etude de la flore du maroc / Pitard, C J M – Paris, 1931 – 1mf – 9 – mf#11912 – ne IDC [956]

Contribution a l'etude de l'histoire de l'ancien royaume de porto-nova / Akindele, Adolphe & Aguessy, C – Dakar: IFAN, 1953 – 1 – us CRL [960]

Contribution a l'etude de l'homme haitien / Jacob, Kleber Georges – Port-Au-Prince, Haiti. 1946 – 1r – us UF Libraries [972]

Une contribution a l'etude des partis politiques nigeriens: le temoignage de adamou mayaki / Talba, Aly – Centre d'Etude d'Afrique noire (Institut d'Etudes politiques de Bordeaux). 1984 – 1 – us Wisconsin U Libr [960]

Contribution a l'etude du droit coutumier berbere marocain : etude sur les coutumes des tribus zayanes / Aspinion, Robert – Casablanca: A Moynier, 1937 – 1 – us CRL [960]

Contribution a l'histoire diplomatique et contempo... / Benjamin, George J – Port-Au-Prince, Haiti. 1951 – 1r – us UF Libraries [972]

Contribution de la guadeloupe a la pensee francais / Lara, H Adolphe – Paris, France. 1936 – 1r – us UF Libraries [972]

Contribution du nord-ouest a l'independance nation / Alcindor, Fernand – Port-de-Paix? Haiti. 1954 – 1r – us UF Libraries [972]

Contribution of c van riet lowe to prehistory in southern africa / Malan, B D – Claremont, South Africa. 1962? – 1r – us UF Libraries [960]

The contribution of force and velocity in the development of peak power output / McLario, David J – 1981 – 1mf – 9 – $4.00 – us Kinesology [790]

The contribution of recreation and sport participation to the quality of life of children with disabilities / Kinkade, Kristen M – 1998 – 1mf – 9 – $4.00 – mf#RC 525 – us Kinesology [790]

Contribution semiotique a l'etude du conte africain : applications pratiques et theorisation a partir de quelques contes de birago diop – Saer Dione, 1983 – us CRL [960]

Contribution to a knowledge of certain freshwater turtles / Marchand, Lewis Jelfs – Gainesville, FL. 1942 – 1r – us UF Libraries [590]

Contribution to an historical sketch of the roman catholic church at macao : and the domestic and foreign relations of macao / Ljungstedt, Andrew – Canton, China, [s.n], 1834 [mf ed 1995] – (= ser Yale coll) – 53p – 1 – 0-524-10058-6 – mf#1995-1058 – us ATLA [241]

Contribution to the biology and control of the green citrus aphid, aphis spiraecola patch / Miller, R L – Gainesville, FL. 1929 – 1r – us UF Libraries [630]

A contribution to the cause of christianity unity : or, the thoughts of an indian missionary on the controversies of the day / O'Neill, Simeon Wilberforce – London: J T Hayes [1879?] [mf ed 1990] – 1mf – 9 – 0-7905-6662-1 – mf#1988-2662 – us ATLA [240]

A contribution to the dynamics of racial diet in british india / Johnston, John Wilson – Edinburgh, 1876 – (= ser 19th c british colonization) – 1mf – 9 – mf#1.1.3293 – uk Chadwyck [640]

Contribution to the knowledge of florida odonata / Byers, Charles Francis – Gainesville, FL. 1930 – 1r – us UF Libraries [550]

Contribution to the palaontology sic of the post-pliocene deposits of the ottawa valley / Ami, Henry Marc – S.l: s.n, 1897? – 1mf – 9 – mf#08080 – cn Canadiana [560]

A contribution towards a bibliography of hosiery and lace, etc / Briscoe, John Potter & Kirk, S J – Nottingham: Free Public Ref Lib, 1896 – (= ser 19th c publishing...) – 1mf – 9 – mf#3.1.62 – uk Chadwyck [016]

Contribution towards an argument for the plenary inspiration of scr... / White, Adam – London, England. 1851 – 1r – us UF Libraries [240]

Contributions / Boyce Thompson Institute for Plant Research – v1-24. 1925-71 – 1 – us AMS Press [980]

Contributions a la flore d'egypte... / Sickenberger, E – Le Caire, 1901 – 2mf – 9 – mf#13206 – ne IDC [956]

Contributions a l'etude de la vegetation du senegal / Trochain, Jean-Louis – Paris: Larose, 1940 – 1 – us CRL [580]

Contributions a l'etude du determinisme fonctionnel de l'industrie dans l'indigene congolais / Mottoulle, Leopold – [Bruxelles: G van Campenhout, 1934] – 1 – us CRL [960]

Contributions chiefly to the early history of the late cardinal newman : with comments / Newman, Francis William – London: K Paul, Trench, Truebner, 1891 – 1mf – 9 – 0-7905-7015-7 – mf#1988-3015 – us ATLA [240]

Contributions from the astronomical institute of the charles university, prague / Univerzita Karlova. Astronomicky ustav – Praha: [s.n.] 1949-52 [irreg] [mf ed 2001] – 4v on 1r – 1 – (n1-3 repr fr: bulletin of the astronomical institutes of czechoslovakia v1 n5 & v2 n1-2; n4 extracted fr: acta mathematica v88) – mf#film mas c5083 – us Harvard [520]

Contributions from the herbarium of the geological survey of canada / Macoun, James Melville – S.l: s.n, 1897? – 1mf – 9 – mf#26637 – cn Canadiana [550]

Contributions from the institute of astrophysics, university of kyoto / Kyoto Daigaku. Uchu Butsurigaku Kankyushitsu – [Kyoto, Japan: Institute of Astrophysics, College of Science, University of Kyoto 1949?- [irreg] [mf ed 2002] – 1r – 1 – (cont by: contributions from the institute of astrophysics and kwasan observatory, university of kyoto [issn 0451-1514]; n1 erroneously dated nov 1948 on vol t.p; some iss publ in combined form; iss consist of repr fr various scientific journals) – mf#film mas c5303 – us Harvard [520]

Contributions from the lick observatory – Sacramento: printed by authority of the Regents of the University of California 1889-1991 (irreg) [mf ed 1999] – 1r [ill] – 1 – (began in 1889; publ suspended foll n5 iss in 1895; resumed with ser 2 v1 n1 (jan 1942); publ: mount hamilton ca 1951?-; santa cruz ca 1973-; papers repr fr astrophysical journal and similar publ) – ISSN: 0457-7833 – mf#film mas c4110 – us Harvard [520]

Contributions from the mcdonald observatory, fort davis, texas / McDonald Observatory – [s.l: The University 1936- [irreg] [mf ed 2003] – 3r – 1 – (consists mainly of repr) – ISSN: 1062-8045 – mf#film mas c5520 – us Harvard [520]

Contributions from the metropolitan museum of natural history academia sinica / Sinensia – New York. 1973-1978 (1) 1973-1978 (5) 1976-1978 (9) – 55mf – 9 – mf#8149 – ne IDC [500]

Contributions from the mount wilson observatory / Carnegie Institution of Washington – [Washington DC?]: The Institution 1905- [v1-29 (1905-49)] [mf ed 1999] – 10r [ill] – 1 – (majority of contribtuions repr fr: astrophysical journal) – ISSN: 0898-1892 – mf#film mas c4144 – us Harvard [520]

Contributions from the observatory of columbia college, new york / Columbia College [New York, NY]. Observatory – New York NY: [The College] 1892-95 (irreg) [mf ed 1999] – n1/2-8(1892-1906) on 1r – 1 – (cont by: contributions from the observatory of columbia university [issn 1063-3030]; iss out of chronological sequence, n3 in 1892, n1 & 2 in 1906 with later title: contributions from the observatory of columbia university) – mf#film mas c4356 – us Harvard [520]

Contributions from the observatory of columbia university / Columbia University. Observatory – New York: [s.n.] 1898-1915 (irreg) [mf ed 1999] – n9-29(1896-15) on 1r – 1 – (cont: contributions from the observatory of columbia college, new york; cont by: contributions from the rutherfurd observatory of columbia university [issn 1063-3049]; iss out of chronological sequence: n1 & 2 in 1906 and carry this title, n9 in 1906, n10 in 1896 with earlier title; n12-14 publ together) – ISSN: 1063-3030 – mf#film mas c4357 – us Harvard [520]

Contributions from the perkins observatory, ohio wesleyan university and ohio state university / Perkins Observatory – [Delaware OH]: Ohio Wesleyan U & Ohio State U [irreg] [mf ed 2002] – 37v on 1r – 1 – (cont in pt by: contributions from the perkins observatory, ser 1 [issn 0096-6975] and in pt by: contributions from the perkins observatory, ser 2 [issn 0099-3085]; with ind) – ISSN: 0195-7465 – mf#film mas c5413 – us Harvard [520]

Contributions from the princeton university observatory – Princeton NJ: The Observatory 1911- (irreg) [n1-26(1911-53)] [mf ed 1999-2004] – 2r [ill] – 1 – (n24 publ out of chronological order; n26 publ in 1952) – ISSN: 1063-5807 – mf#film mas c4147 – us Harvard [520]

Contributions from the rutherfurd observatory of columbia university / Columbia University. Rutherfurd Observatory – New York: [s.n.] (irreg) [mf ed 1999] – n30-33(1937-56) on 1r – 1 – (cont: contributions from the observatory of columbia university [issn 1063-3030]) – ISSN: 1063-3049 – mf#film mas c4358 – us Harvard [520]

Contributions of the visual and somatosensory perceptual systems to the development of postural control in infants / Sveistrup, Heidi – 1993 – 2mf – $8.00 – us Kinesology [150]

Contributions of vision in aerial performances / Hondzinski, Jan M – 1998 – 3mf – 9 – $12.00 – mf#PSY 2059 – us Kinesology [790]

Contributions to a new revision : or, a critical companion to the new testament / Young, Robert – Edinburgh: G A Young, 1881 [mf ed 1989] – 1mf – 9 – 0-7905-0477-4 – mf#1987-0477 – us ATLA [225]

Contributions to assyriology, semitic languages and philology *see* Beitraege zur assyriologie und vergleichenden semitischen sprachwissenschaft / Beitraege zur assyriologie und semitischen sprachwissenschaft

Contributions to biblical and patristic literature : texts and studies / ed by Robinson, D – 1891-1922. v1-9+ind – 71mf – 9 – mf#H-2914 – ne IDC [240]

Contributions to blowpipe-analysis / Chapman, Edward John – [Toronto?: Lovell & Gibson], 1865 – 1mf – 9 – mf#93885 – cn Canadiana [540]

Contributions to education / Columbia University. Teachers College – nos. 1-974. 1905-51 – 1,9 – us AMS Press [370]

Contributions to mineralogy and petrology – Heidelberg. 1966-1993 (1) 1966-1993 (5) 1980-1993 (9) – (cont: beitraege zur mineralogie und petrographie) – ISSN: 0010-7999 – mf#13157,02 – us UMI ProQuest [550]

Contributions to the criticism of the greek new testament : being the introduction to an edition of the codex augiensis and fifty other manuscripts / Scrivener, Frederick Henry Ambrose – Cambridge: Deighton, Bell, 1859 – 1mf – 9 – 0-524-08091-7 – mf#1992-1151 – us ATLA [225]

Contributions to the galaxy 1868-1871 / Clemens, Samuel Langhorne – 9 – $10.00 – us Scholars Facs [830]

Contributions to the history of bantu linguistics / Doke, Clement Martyn – Johannesburg, South Africa. 1961 – 1r – us UF Libraries [470]

Contributions to the history of the eastern townships : a work containing an account of the early settlement of st armand, dunham, sutton, brome, polton, and bolton. with a history of the principal events that have transpired in each of these townships up to the present time / Thomas, Cyrus – [Montreal?: s.n.] 1866 [mf ed 1983 – 5mf – 9 – 0-665-41414-5 – mf#41414 – cn Canadiana [971]

Contributions to the knowledge of south african marine mollusca, pt 1 : gastropoda, prosobranchiata, toxoglossa / Barnard, K H – (= ser Annals of the South African Museum) – 2mf – 9 – (annals of the south african museum. 1958. v44 pt.4(p73-163))) – mf#2841 – ne IDC [590]

Contributions to the literature of the fine arts / Eastlake, Charles Lock – 2nd ed. London 1870 – (= ser 19th c art & architecture) – 9mf – 9 – mf#4.2.1008 – uk Chadwyck [700]

Contributions to the science of mythology / Mueller, Friedrich Max – London: Longmans, Green, 1897 – 3mf – 9 – 0-524-02220-8 – mf#1990-2894 – us ATLA [200]

Contributions to the science of mythology / Muller, F Max – London, England. v1-2. 1897 – 1r – us UF Libraries [390]

Contributions to the technique of violin playing made by nicolo paganini / Deegan, Mabel Alice – U of Rochester 1941 [mf ed 1987] – 2mf – 9 – mf#fiche1207 – us Sibley [780]

Contributions towards a glossary of the assyrian language / Talbot, William Henry Fox – [s.l: s.n, 18–?] [mf ed 1986] – (= ser Journal of the royal asiatic society. new series 3) – 1mf – 9 – 0-8370-8622-1 – (in english and akkadian. incl bibl ref) – mf#1986-2622 – us ATLA [470]

Contributions towards the exposition of the book of genesis / Candlish, Robert Smith – Edinburgh: J Johnstone, 1843-52 [mf ed 2003] – 2v on 1r – 1 – 0-524-10459-X – mf#b00678 – us ATLA [221]

Contributo alla storia della riforma del messale promulgato da san pio 5th nel 1570 / Frutaz, A P – 1960 – 1mf – 8 – €6.00 – ne Slangenburg [241]

Contributory negligence. a comparative study of the law of torts (u.s.a., england and denmark) / Kragh, Karsten – Nykobing, Forfatteren, 1965. 5, 89 p. LL-2323 – 1 – us L of C Photodup [340]

Control – London. 1958-1969 (1) – mf#2995 – us UMI ProQuest [629]

Control as an instrument of effective education management : in six selected secondary schools in the brits district [north west province] / Moloko, Nkobo Jacob – Pretoria: Vista University 2001 [mf ed 2001] – 2mf – 9 – (incl bibl ref) – mf#mfm15241 – sa Unisa [370]

Control Council for Germany *see* The statutory criminal law of germany

Control de la constitucionalidad en panama / Pedreschi, Carlos Bolivar – Panama? Panama. 1965 – 1r – us UF Libraries [972]

Control de natalidad. informe para expertos, llos documentos de roma / Javierre, Jose Maria et al – Madrid: Editorial Alameda, 1967 – 1 – sp Bibl Santa Ana [946]

Control engineering – Barrington. 1954+ (1) 1965+ (5) 1970+ (9) – ISSN: 0010-8049 – mf#874 – us UMI ProQuest [629]

Control engineering practice – Oxford. 1993-1994 (1,5,9) – ISSN: 0967-0661 – mf#49627 – us UMI ProQuest [620]

The control of movements which vary in accuracy and complexity / Lajoie, Jennifer M – University of British Columbia, 1996 – 2mf – 9 – $8.00 – mf#PSY 1892 – us Kinesology [150]

Control of root-knot, 2 / Watson, J R – Gainesville, FL. 1921 – 1r – us UF Libraries [630]

Control of root-knot by calsium cyanamide and other means / Watson, J R – Gainesville, FL. 1917 – 1r – us UF Libraries [630]

Control of root-knot in florida / Watson, J R – Gainesville, FL. 1937 – 1r – us UF Libraries [630]

Control of serial pointing movements / Pizzimenti, Marco A – 1999 – 2mf – 9 – $8.00 – mf#PSY 2096 – us Kinesology [612]

Control of the celery leaf-tier in florida – Gainesville, FL. 1932 – 1r – us UF Libraries [634]

Control of the velvet bean caterpillar / Watson, J R – Gainesville, FL. 1916 – 1r – us UF Libraries [634]

Controle financier du gouvernement des etats-unis / Beauvoir, Vilfort – Paris, France. 1930 – 1r – us UF Libraries [336]

Controlled clinical trials – New York. 1980+ (1) 1980+ (5) 1987+ (9) – ISSN: 0197-2456 – mf#42248 – us UMI ProQuest [619]

Controlling damping-off and other losses in celery seedbeds / Townsend, G R – Gainesville, FL. 1944 – 1r – us UF Libraries [634]

Controlling the citrus aphis (aphis spiraecola patch) / Watson, J R – Gainesville, FL. 1925 – 1r – us UF Libraries [634]

Controlling tobacco downy mildew (blue mold) with paradichlorobenzene / Tisdale, W B – Gainesville, FL. 1939 – 1r – us UF Libraries [630]

Il contro-pelo – Barre VT, 1911-12* – 1r – 1 – (italian periodical) – us IHRC [073]

Controversia de limites entre venezuela / Rodriguez, Jose Santiago – Caracas, Venezuela. 1944 – 1r – us UF Libraries [972]

Controversia sobre belice durante el ano de 1946 / Guatemala Secretaria De Relaciones Exteriores – Guatemala, 1948 – 1r – us UF Libraries [972]

Controversia sobre el territorio de belice / Garcia Bauer, Carlos – Guatemala, 1958 – 1r – us UF Libraries [972]

Controversiae inter theologos vvittenbergenses de regeneratione et electione dilvcida explicatio dd egidii hvnnii, polycarpi leyseri, salomonis gesneri / Gesner, S – [Francoforti ad Moenvm, 1594] – 1mf – 9 – (missing title pg) – mf#TH-1 mf 598 – ne IDC [242]

The controversial methods of romanism / Brinckman, Arthur – London: Swan Sonnenschein, Lowrey, 1888 – 1mf – 9 – 0-524-03759-0 – mf#1990-1106 – us ATLA [240]

Controversial statistics of romanism / Brinckman, Arthur – London, England. 1898 – 1r – us UF Libraries [241]

The controversial statistics of romanism / Brinckman, Arthur – 2nd ed. London: SPCK 1898 [mf ed 1993] – (= ser Church historical society (series) 43) – 1mf – 9 – 0-524-05487-8 – mf#1990-1482 – us ATLA [241]

Controversial tracts on christianity and mohammedanism / Martyn, Henry et al – Cambridge: J Smith, 1824 – 2mf – 9 – 0-524-06929-8 – (incl bibl ref) – mf#1990-3555 – us ATLA [230]

Controversiarum de divinae gratiae liberique arbitrii concordia : initia et progressus / Schneemann, Gerhard – Friburgi Brisgoviae [Freiburg i B]: Herder, 1881 – 6mf – 9 – 0-524-08723-7 – (incl bibl ref) – mf#1993-2128 – us ATLA [240]

Controversiarum epitomes... / Gordon, J – Coloniae Agrippinae, 1620. 3v – 10mf – 9 – mf#CA-52 – ne IDC [240]

Controversiarum roberti bellarmi responsio / Daneau, Lambert – [Geneve], Le Preux, 1598. 2 pts – 9mf – 9 – mf#PFA-137 – ne IDC [240]

Controversie of singing brought to an end / Marlow, Isaac – London. 1696 – 1 – 5.00 – us Southern Baptist [242]

Controversy – 1976 jan 1-1977 jun – 1r – 1 – mf#363416 – us WHS [071]

Controversy – Carmel, CA. 1934-1934 (1) – mf#62113 – us UMI ProQuest [071]

Controversy about prayer / Newman, Francis William – London, England. 1873 – 1r – us UF Libraries [240]

Controversy between dr ryerson, chief superintendent of education in upper canada, and rev j m bruyere, rector of st michael's cathedral, toronto : on the appropriation of the clergy reserves funds – [Toronto: Leader and Patriot], 1857 – 2mf – 9 – 0-665-22644-6 – mf#22644 – cn Canadiana [230]

Controversy between rev. messrs. hughes and breckenridge : on the subject "is the protestant religion the religion of christ?" / Hughes, John – 6th ed. Philadelphia: Eugene Cummiskey, 1885 – 2mf – 9 – 0-8370-7546-7 – mf#1986-1546 – us ATLA [230]

Controversy between the puritans and the stage / Thompson, Elbert Nevius Sebring – New York: H Holt, 1903 – (= ser Yale studies in english) – 1mf – 9 – 0-524-04975-0 – (incl bibl ref) – mf#1990-1378 – us ATLA [790]

The controversy between true and pretended christianity : an essay / Townsend, Luther Tracy – Boston: James P Magee, 1869 [mf ed 1991] – (= ser Methodist coll) – 1mf – 9 – 0-524-00977-5 – mf#1990-4035 – us ATLA [210]

Controversy of faith / Dodgson, Charles – London, England. 1850 – 1r – us UF Libraries [240]

Controversy on the constitutions of the jesuits between dr littledale and fr drummond / Littledale, Richard Frederick – Winnipeg?: Manitoba Free Press Print, 1889 – 1mf – 9 – mf#09180 – cn Canadiana [241]

Controversy on the subjects and mode of baptism : between mr john torrance...baptist preacher, and the editor of the "theological instructor" – Toronto?: Lovelock, Stovel, 1875 – 1mf – 9 – (in dble clms) – mf#41471 – cn Canadiana [240]

The controversy over a new canal treaty between the u.s. and panama : a selective, annotated bibliography of u.s., panamanian, columbian, french and international organization titles / Bray, Wayne D – Washington: Library of Congress, 1976 – 1mf – 9 – $1.50 – mf#LLMC 82-100D Title 15 – us LLMC [341]

The controversy over the proposition for an american episcopate, 1767-1774 : a bibliography of the subject / Nelson, William – Paterson, NJ: Paterson History Club, 1909 – 1mf – 9 – 0-7905-8220-1 – mf#1988-6120 – us ATLA [240]

[Contry costa county-] contra costa county – 1871-1872 – 1r – 1 – $50.00 – mf#D012 – us Library Micro [978]

Contuzzi, F P *see* La neutralizzazione del canale di suez...

Convalescent Facility of Percy Jones G&C Hospital [Fort Custer MI] *see* Con-fab

Convalescente / Melesville, M – Bruxelles, Belgium. 1830 – 1r – us UF Libraries [440]

Convenant ministers' quarterly / convenant quarterly : theological quarterly of the evangelical covenant church – 1941-99 [mf ed 2001] – (= ser Christianity's encounter with world religions, 1850-1950) – 6r – 1 – mf#2001-s125-126 – us ATLA [242]

Convenanting struggle – London, England. 1880 – 1r – us UF Libraries [240]

Convencao preliminar de paz de 1828 / Docca, Emilio Fernandes De Souza – Sao Paulo, Brazil. 1929 – 1r – us UF Libraries [972]

Convencion de ibague, 1922 / Paz, Felipe Santiago – Bogota, Colombia. 1926? – 1r – us UF Libraries [972]

Convencion Latinoamericana De Solidaridad Con Israel (1956 : Monte... *see* Por la paz en el medio oriente

Convencion sobre administracion provisional de col... / Meeting Of Consultation Of Ministers Of Foreign Affairs – Habana, Cuba. 1940 – 1r – us UF Libraries [972]

Convenciones internacionales de nicaragua, 1913 / Nicaragua Treaties, Etc – Managua, Nicaragua. 1913 – 1r – us UF Libraries [972]

Convenio colectivo sindical de sidermetalurgia (n_o 188 de 18.8.76) / Sindicato Provincial del Metal – Caceres: Tip El Noticiero, 1976 – 1 – sp Bibl Santa Ana [946]

Convenio colectivo sindical de siderometalurgica. numero 159 de 13-7-74 / Sindicato Provincial del Metal – Caceres: Rdit. Extremadura, 1974 – 1 – sp Bibl Santa Ana [946]

Convenio colectivo sindical de trabajo agricola, aprobado y suscrito por los representantes...de coria / Delegacon Provincial de Sindicatos – Caceres: Imprenta Moderna, 1961 – 1 – sp Bibl Santa Ana [630]

Convenio colectivo sindical de trabajo agricola, aprobado y suscrito por los representantes... de villamiel (caceres) / Delegacion Provincial de Sindicatos – Caceres: Imp. Moderna, 1963 – 1 – sp Bibl Santa Ana [630]

Convenio colectivo sindical de trabajo agricola de alcantara (caceres) y sus agregados de estorninos y piedras albas / Delegacion Provincial de Sindicatos – Caceres: Imp. Moderna, 1962 – 1 – sp Bibl Santa Ana [630]

Convenio colectivo sindical de trabajo agricola...de casas de millan caceres / Delegacion Provincial de Sindicatos – Caceres: Imprenta Moderna, 1962 – 1 – sp Bibl Santa Ana [630]

Convenio colectivo sindical de trabajo agricola...de guadalupe / Delegacion Provincial de Sindicatos – (Caceres): Imprenta Moderna, 1962 – 1 – sp Bibl Santa Ana [630]

Convenio colectivo sindical de trabajo, de ambito provincial para el comercio de la alimentacion (mayor y menor) aprobado por el ilmo. sr. delegado prov. de trabajo / Sindicato Prov. de la Alimentacion y Productos Coloniales, Badajoz – Badajoz: Imp. Inca, 1969 – sp Bibl Santa Ana [331]

Convenio colectivo sindical de trabajo...de la empresa textil lanera sobrino de benito matas en hervas (caceres) / Delegacion Provincial de Sindicatos – Caceres: Imp. Moderna, 1962 – 1 – sp Bibl Santa Ana [331]

Convenio colectivo sindical provincial para las industrias del aceite y sus derivados de badajoz / Delegacion Prov. de la Org. Sindical – Badajoz: Tipografia Clasica, 1961 – sp Bibl Santa Ana [338]

Convenios y tratados celebrados / Cuba Treaties, Etc – Habana, Cuba. 1912 – 1r – us UF Libraries [972]

Convent experiences – London, England. 1875 – 1r – us UF Libraries [240]

The convent horror : story of barbara ubryk, twenty-one years in the dungeon, eight feet long six feet wide – Aurora, Mo: Menace Pub, [ca. 1890] – 1mf – 9 – 0-8370-8468-7 – mf#1986-2468 – us ATLA [240]

Convent jubilee memorial / Abbott, S J – London, England. 1898 – 1r – us UF Libraries [240]

Convent life unveiled / O'gorman, Edith – London, England. no date – 1r – us UF Libraries [240]

La convention anti-seigneuriale de montreal au peuple / Anti-Seigniorial Convention (1854 : Montreal, Quebec) – Montreal?: s.n, 1854 – 1mf – 9 – mf#22442 – cn Canadiana [333]

Convention baptist press release materials – SBC, 1954-60. 2818p – 1 – us Southern Baptist [242]

La convention collective de travail (loi du 24 juin 1936) et l'arbitrage obligatoire / Capeau, Charles – Nice? 1937. 81p. LL-4091 – 1 – us L of C Photodup [340]

La convention de la baie james et du nord quebecois : convention entre le gouvernement du quebec et la societe d'energie de la baie james, la societe de developpement de la baie james... / Quebec (Province). Conseil executif et al – [Quebec]: editeur officiel du Quebec, c1976 [mf ed 1984] – (= ser Convention de la baie james et du nord quebecois (1975)) – 6mf – 9 – mf#SEM105P445 – cn Bibl Nat [333]

Convention de la Baie James et du Nord quebecois (1975) *see* Agreement between

Convention de la baie james et du nord quebecois (1975) *see* Agreement between: the government of quebec and the societe d'energie de la baie james and the societe de developpement de la baie james and the commission hydroelectrique de quebec (hydro-quebec) and...

Convention de la baie james et du nord quebecois [1975] *see* Convention entre

La convention de la baie james et du nord quebecois et les conventions complementaires nos 1, 2, 3, 4, 5 et 6 : convention entre le gouvernement du quebec, la societe d'energie de la baie james, la societe de developpement de la baie james... / Quebec (Province). Conseil executif et al – rev corr ed. Quebec: editeur officiel, c1980 [mf ed 1985] – (= ser Convention de la Baie James et du Nord quebecois (1975)) – 8mf – 9 – mf#SEM105P449 – cn Bibl Nat [333]

La convention de reciprocite laurier, 1911 – [Quebec (Province)]: [s.n.], [1911?] (mf ed 1990) – 2mf – 9 – mf#SEM105P1244 – cn Bibl Nat [337]

Convention du 15 septembre et l'encyclique du 8 decembre = Remarks on the encyclical of the 8th of december, a.d. 1864 / Dupanloup, Felix – 2nd ed. London:Burns, Lambert, & Oates, 1865 – 1mf – 9 – 0-8370-3651-8 – (in english) – mf#1985-1651 – us ATLA [240]

Convention entre : le gouvernement du quebec et la societe d'energie de la baie james et la societe du developpement de la baie james... / Quebec [Province]. Conseil executif et al – [Montreal: Conseil executif: Negociations indiens Inuit de la Baie James, 1974] [mf ed 1984] – 11mf – 9 – mf#SEM105P447 – cn Bibl Nat [333]

Convention Evangelica Bautista. Argentine Baptist convention *see* Minutes

Convention forestiere canadienne tenue a montreal, les 11 et 12 mars 1908 : discours prononces par mgr j-c k-laflamme, m g-c piche / Laflamme, Joseph Clovis Kemler – Quebec: Departement des terres et forets, 1908 [mf ed 1997] – 1mf – 9 – 0-665-83379-2 – mf#83379 – cn Canadiana [634]

Convention journal – Saipan, n.p, 1975 – (= ser Micronesian constitutional convention, 1975) – 42mf – 9 – $63.00 – mf#LLMC 82-100F, Title 93 – us LLMC [323]

Convention nationale des Canadiens-francais des Etats-Unis *see* Manifeste aux societes nationales et autres des etats-unis et du canada

Convention of American Instructors of the Deaf *see* Report of the proceedings of the meeting

Convention of baptist social unions at the athenaeum, brooklyn, n.y : december 9th and 10th, 1874 – New York: LH Biglow, 1875 – 1mf – 9 – 0-524-04042-7 – mf#1990-4950 – us ATLA [242]

Convention of Bible Societies of New Jersey, 1880 *see* The wycliffe semi-millennial bible celebration

Convention of the... / United Hatters of North America – 1911, 1919, 1927 – 1r – 1 – (cont by: proceedings of the convention of the united hatters of north america; proceedings of the...convention of the united hatters, cap and millinery workers international union) – mf#3144959 – us WHS [366]

Convention proceedings of the...annual session / Oklahoma State Federation of Labor – 9th [1912] – 1r – 1 – (cont by: official proceedings of the...annual convention) – mf#3147242 – us WHS [331]

Convention souvenir – 1890, 1892, 1893 – 1r – 1 – mf#2837891 – us WHS [060]

Convention souvenir / Federated Association of Letter Carriers (Canada) – Hamilton, Ont: Branch n3, Federated Association of Letter Carriers, 1918 – 2mf – 9 – 0-659-91760-2 – mf#9-91760 – cn Canadiana [360]

The convention system of teacher training / Burroughs, P E – 1914 – 1 – 5.00 – us Southern Baptist [242]

El convento de santa maria de jesus... / Marti Mayor, Jose – Madrid: archivo ibero americano, 1963 – 1 – sp Bibl Santa Ana [240]

El convento de tepotzotlan...1924 / Eliodoro Valle, Rafael; ed by Bayle, Constantino – Madrid: Razon y Fe, 1928 – 9 – sp Bibl Santa Ana [946]

El convento dominicano de nuestra sra : del rosario en santa fe y su universidad / Mesanza, Andres – Burgos: Razon y Fe, 1939 – 1 – sp Bibl Santa Ana [241]

Conventos de monjas en la nueva espana. mexico, 1946 / Mauriel, Josefina – Madrid: Razon y Fe, 1947 – 1 – sp Bibl Santa Ana [946]

Convents – London, England. no date – 1r – us UF Libraries [240]

Convents / Wiseman, Nicholas Patrick – London, England. 1852 – 1r – us UF Libraries [240]

Convents and monasteries – London, England. 1870? – 1r – us UF Libraries [240]

Conventual and monastic inquiry bill – London, England. 1874? – 1r – us UF Libraries [240]

Conventual Franciscan Friars of Marytown *see* Mission of the immaculata

Convergence / Christic Institute – 1982 win-1991 fall – 1r – 1 – mf#1163087 – us WHS [071]

Convergence – Toronto. 1968+ (1) 1972+ (5) 1976+ (9) – ISSN: 0010-8146 – mf#7520 – us UMI ProQuest [370]

Convergencia – Rio de Janeiro: CRB. v6 n55-v8 n79 mar 1973-mar 1975; v8 n82-v10 n99 jun 1975-jan/feb 1977; v10 n102 may 1977; v10 n104-v26 n258 jul/aug 1977-dec 1992 – 7r – us CRL [079]

Conversaciones con goethe en los ultimos anos de su vida / Eckermann, Johann Peter – Madrid: Calpe 1920 [mf ed 1990] – 2v in 1 on 1r – 1 – (trans by j perez bances. filmed with: goethe im gesprach / ed by franz deibel & friedrich gundelfinger) – mf#2794p – us Wisconsin U Libr [880]

Conversaciones familiares entre el censor / Forner Segarra, Juan Pablo – 1787 – 9 – sp Bibl Santa Ana [840]

Conversao en el batey / Fonfrias, Ernesto Juan – San Juan, Puerto Rico. 1958 – 1r – us UF Libraries [972]

Conversatio politico-christiana ad leges ethico-politico-morales, moralium philosophorum doctoris... / Senftleben, J – Prague: Typis Univ. Carolo-Ferd., 1681 – 2mf – 9 – mf#O-1898 – ne IDC [090]

Conversation – 1977 spr-1978 win – 1r – 1 – (cont: heritage conversation, conversation autour du patrimoine; cont by: conversation) – mf#681529 – us WHS [929]

Conversation between a minister of the gospel and one who doubts his... – London, England. no date – 1r – us UF Libraries [240]

Conversation between pm (modified protestant) and pp (pronounce... / Long, Harry Alfred – Glasgow? Scotland. 1895 – 1r – us UF Libraries [242]

Conversations khmero-franco-viet-namiennes – [n.p. 196-?] [mf ed 1989] – 1r with other items – 1 – (in khmer & french) – mf#mf-10289 seam reel 029/07 [§] – us CRL [480]

Conversations nouvelles sur divers sujets, dediees au roy / Scudery, Madeleine de – La Haye: Abraham Arondeus. 2v. 1685 [mf ed 1974] – 1r – 5 – mf#SEM16P81 – cn Bibl Nat [840]

Conversations of dr. doellinger / Doellinger, Johann Joseph Ignaz von – London: R. Bentley, 1892 – 1mf – 9 – 0-7905-5030-X – mf#1988-1030 – us ATLA [920]

Conversations of goethe with eckermann and soret – rev ed. London: G Bell & Sons 1883 [mf ed 1970?] – 1r – 1 – (trans fr german by john oxenford; also filmed by uw madison: london 1874 [mf ed 1993] order#8551) – mf#film mas 2143 – us Harvard; us Wisconsin U Libr [080]

Conversations of goethe with eckermann and soret *see* Briefe des dichters ludwig zacharias werner

Conversations of james northcote / Hazlitt, William – London: Henry Colburn & Richard Bentley, 1830 – (= ser 19th c art & architecture) – 4mf – 9 – mf#4.1.127 – uk Chadwyck [750]

Conversations of james northcote / Hazlitt, William – rev ed. London 1894 – (= ser 19th c art & architecture) – 4mf – 9 – mf#4.2.142 – uk Chadwyck [750]

Conversations of jesus christ with representative men / Adams, William – New York: American Tract Society, c1868 – 1mf – 9 – 0-8370-2045-X – mf#1985-0045 – us ATLA [240]

Conversations on liberalism and the church / Brownson, Orestes Augustus – New York: D & J Sadlier, 1870, c1869 – 1mf – 9 – 0-8370-7369-3 – mf#1986-1369 – us ATLA [240]

Conversations on political economy : or, a series of dialogues supposed to take place between a minister of state and representatives of the agricultural, manufacturing...and monied interests, as well as of the labouring classes of society / Pinsent, Joseph – London: Printed for J M Richardson...and Hatchard & son...1821 – 2mf – 9 – mf#21115 – cn Canadiana [336]

Conversations on scripture – Plymouth, England. no date – 1r – us UF Libraries [220]

Conversations on scripture no 1 : on romans 8 18-23 – Plymouth, England. no date – 1r – us UF Libraries [220]

Conversations on scripture no 2 : on zechariah 14 1 – Plymouth, England. no date – 1r – us UF Libraries [220]

Conversations on sin and salvation / Neff, Felix – London, England. 1839 – 1r – us UF Libraries [240]

Conversations on the bible : its statements harmonized and mysteries explained / Pond, Enoch – Springfield, MA: CA Nichols, 1881 – 2mf – 9 – 0-524-05997-7 – mf#1992-0734 – us ATLA [220]

Conversations on the creation : chapters on genesis and evolution – London [1881] – (= ser 19th c evolution & creation) – 2mf – 9 – mf#1.1.9052 – uk Chadwyck [210]
Conversations on the mass – London, England. no date – 1r – us UF Libraries [240]
Conversations on the office of sponsors for infants : and the use of the sign of the cross in baptism / Belt, William – Toronto: s.n, 1870 – 1mf – 9 – mf#10314 – cn Canadiana [240]
Conversations with a ranter / Campbel, Charles – London, England. 1835 – 1r – us UF Libraries [240]
Conversations with christ : a biographical study / Lucas, Bernard – London, New York: Macmillan, 1905 – 1mf – 9 – 0-8370-2729-2 – mf#1985-0729 – us ATLA [240]
Conversations with eckermann : being appreciations and criticism on many subjects / Goethe, Johann Wolfgang von & Eckermann, Johann Peter – New York: M W Dunne, c1901 [mf ed 1993] – (= ser Universal classics library) – xii/397p – 1 – (pref by eckermann. special int by wallace wood) – mf#8613 – us Wisconsin U Libr [080]
Das conversationsblatt – Berlin DE, 1836-39 – 1 – gw Misc Inst [074]
Conversations-lexikon (ael1/35.2) : oder kurzgefasstes handwoerterbuch fuer die in der gesellschaftlichen unterhaltung aus den wissenschaften und kuensten vorkommenden gegenstaende mit bestaendiger ruecksicht auf die ereignisse der aelteren und neueren zeit – [1st ed]. Amsterdam 1809 [mf ed 1997] – (= ser Das brockhaus conversations-lexikon 1796-1898 (ael1/35)) – 6v on 18mf – 9 – €390.00 – 3-89131-252-0 – gw Fischer [030]
Conversations-lexikon (ael1/35.3) : oder kurzgefasstes handwoerterbuch...nachtraege – Amsterdam 1809; Leipzig 1811 [mf ed 1997] – (= ser Das brockhaus conversations-lexikon 1796-1898 (ael1/35)) – 2v on 8mf – 9 – €90.00 – 3-89131-253-9 – gw Fischer [030]
Conversations-lexikon (ael1/35.4) : oder hand-woerterbuch fuer die gebildeten staende ueber die in der gesellschaftlichen unterhaltung und bei der lectuere vorkommenden gegenstaende, namen und begriffe... – 2nd ed. Leipzig 1812-19 [mf ed 1997] – (= ser Das brockhaus conversations-lexikon 1796-1898 (ael1/35)) – 10v on 55mf – 9 – €510.00 – 3-89131-254-7 – gw Fischer [030]
Conversations-lexikon (ael1/35.5) : oder encyclopaedisches handwoerterbuch fuer gebildete staende – 3rd ed. Leipzig 1814-19 [mf ed 1997] – (= ser Das brockhaus conversations-lexikon 1796-1898 (ael1/35)) – 10v on 57mf – 9 – €540.00 – 3-89131-255-5 – gw Fischer [030]
Conversations-lexikon (ael1/35.6) : oder encyclopaedisches handwoerterbuch fuer gebildete staende – 4th ed. Leipzig 1817-19 [mf ed 1997] – (= ser Das brockhaus conversations-lexikon 1796-1898 (ael1/35)) – 10v on 59mf – 9 – €570.00 – 3-89131-256-3 – gw Fischer [030]
Conversations-lexikon (ael1/35.10) – new series. Leipzig 1822-26 [mf ed 1997] – (= ser Das brockhaus conversations-lexikon 1796-1898 (ael1/35)) – 2v on 23mf – 9 – €220.00 – 3-89131-260-1 – gw Fischer [030]
Conversations-lexikon (ael1/35.21) : allgemeine deutsche real-encyklopaedie – 12th ed. Leipzig 1875-79 [mf ed 1997] – (= ser Das brockhaus conversations-lexikon 1796-1898 (ael1/35)) – 15v on 82mf – 9 – €720.00 – 3-89131-270-9 – gw Fischer [030]
Conversationslexikon mit vorzueglicher ruecksicht auf die gegenwaertigen zeiten (ael1/35.1) : [loebel-ausgabe] – Leipzig 1796-1808 [mf ed 1997] – (= ser Das brockhaus conversations-lexikon 1796-1898 (ael1/35)) – 6pt on 18mf – 9 – €200.00 – 3-89131-251-2 – gw Fischer [030]
Conversazioni della domenica – Milan, Italy. -w. 3 Jan 1886-28 Dec 1890. 2 reels – 1 – uk British Libr Newspaper [074]
Converse, John Melvin *see*
- Diary
- Scrapbook

Converse, Mildred *see* Historical data
Conversi, Girolamo *see* Il primo libro delle canzoni a cinque voci
Conversion – 19 aug 1924-25 feb 1925* – (= ser Chinese christian coll 49) – 1r – 1 – (in chinese) – mf#ATLA S0296E – us ATLA [242]
Conversion : or, the new birth / Marshall, Newton Herbert – London: National Council of Evangelical Free Churches 1909 [mf ed 1991] – (= ser Christian faith and doctrine series 3) – 1mf – 9 – 0-7905-7972-3 – (incl bibl ref) – mf#1989-1197 – us ATLA [240]
Conversion : a revelation in the soul / Handford, T W – London, England. 1872? – 1r – us UF Libraries [240]
La conversion – Paris. 29 nov 1881-25 dec 1882 [wkly] – 1 – fr ACRPP [073]
Conversion and election : a plea for a united lutheranism in america = Die grunddifferenz in der lehre von der bekehrung und gnadenwahl / Pieper, Franz – St Louis MO: Concordia Publ House 1913 [mf ed 1991] – 1mf – 9 – 0-7905-9581-8 – (trans fr german; incl bibl ref) – mf#1989-1306 – us ATLA [242]
Conversion and restoration of the jews / M'caul, Alexander – London, England. 1838 – 1r – us UF Libraries [270]
Conversion de figaro / Brousson, Jean-Jacques – Paris, France. 1928? – 1r – us UF Libraries [440]
Conversion in american unitarianism / Cunningham, Michael Frank – Chicago, 1968. Chicago: Dep of Photoduplic, U of Chicago Lib, 1971 (1r); Evanston: American Theol Lib Assoc, 1984 (1r) – 1 – 0-8370-0273-7 – mf#1984-B200 – us ATLA [243]
Conversion monetaria de la republica de nicaragua / Ruiz Y Ruiz, Frutos – Granada, Nicaragua. 1918 – 1r – us UF Libraries [972]
The conversion of armenia to the christian faith / Tisdall, William St Clair – London: Religious Tract Society, 1897 – 1mf – 9 – 0-524-00402-1 – mf#1989-3102 – us ATLA [240]
The conversion of children : can it be effected? how young? will they remain steadfast?... / Hammond, Edward Payson – Chicago: Fleming H Revell, [1877?] – 1mf – 9 – 0-8370-6065-6 – mf#1986-0065 – us ATLA [240]
The conversion of children : can it be effected? how young? will they remain steadfast? what means to be used? when to be received and how trained in the church? / Hammond, Edward Payson – Chicago: Fleming H. Revell, [189-]Beltsville, Md: NCR Corp, 1978 (3mf); Evanston: American Theol Lib Assoc, 1984 (3mf) – (= ser Revivalism and revival preachers in america) – 9 – $1.50 – 0-8370-0198-6 – mf#1984-3009 – us ATLA [240]
The conversion of india : from pantaenus to the present time, a.d. 193-1893 / Smith, George Adam – New York: Young People's Missionary Movement, [1894] – 1mf – 9 – 0-8370-6518-6 – (incl bibl ref and index) – mf#1986-0518 – us ATLA [240]
The conversion of india : or, reconciliation between christianity and hinduism. being studies in indian missions / Berg, Emil P – London: Arthur H Stockwell, 1911 – 1mf – 9 – 0-524-01679-8 – mf#1990-2581 – us ATLA [240]
Conversion of st paul: three discourses / Geer, George Jarvis – New York: Samuel R. Wells, 1871.1 fiche – 9 – us ATLA [240]
Conversion of the ethiopian / Holloway, James Thomas – London, England. 1818? – 1r – us UF Libraries [240]
The conversion of the heptarchy : seven lectures. given at st. paul's / Browne, George Forrest – rev ed. London: SPCK, 1914 – 1mf – 9 – 0-524-00515-X – mf#1990-0015 – us ATLA [240]
The conversion of the maoris / MacDougall, Donald – Philadelphia, PA: Presbyterian Board of Publication and Sabbath-Schoolwork, 1899 – 1mf – 9 – 0-8370-6584-4 – mf#1986-0584 – us ATLA [240]
The conversion of the northern nations / Merivale, Charles – London: Longmans, Green, 1866 – (= ser The boyle lectures) – 1mf – 9 – 0-7905-5435-6 – mf#1988-1435 – us ATLA [240]
The conversion of the roman empire / Merivale, Charles – London: Longman, Green, Longman, Roberts, & Green, 1864 – (= ser The boyle lectures) – 1mf – 9 – 0-7905-4897-6 – mf#1988-0897 – us ATLA [240]
Conversion of the world consequent upon the improvement of the church / Wright, George – Edinburgh, Scotland. 1820 – 1r – us UF Libraries [240]
Conversion planner / National Committee for a Sane Nuclear Policy – v2 n1-v5 n4 [1979 jan/feb-1982 jun/aug] – 1r – 1 – mf#652342 – us WHS [350]
The conversion policy of the jesuit in india. bombay, 1933 / Heras, H – Madrid: Razon y Fe, 1934 – 1 – sp Bibl Santa Ana [241]
Conversion to the roman catholic faith – Dublin, Ireland. 1830 – 1r – us UF Libraries [241]
The convert : or, leaves from my experience / Brownson, Orestes Augustus; ed by Brownson, Henry Francis – new ed. New York: D & J Sadlier, 1877, c1876 – 1mf – 9 – 0-7905-8001-2 – mf#1988-8001 – us ATLA [240]
Converted dealer / Bayne, R – London, England. 18– – 1r – us UF Libraries [240]
Converted deist's profession of faith / P, W – London, England. 1868? – 1r – us UF Libraries [240]
Converted sailor – London, England. 18– – 1r – us UF Libraries [240]
Converter – Croydon. 1977-1991 (1) 1977-1981 (5) 1977-1981 (9) – ISSN: 0010-8189 – mf#11094 – us UMI ProQuest [670]
Converting a business into a private company / Jordan, Herbert William – London: Jordan & Sons, 1922 (mf ed 19–) – 48p – mf#ZT-TN pv74 n5 – us Harvard [346]
Die convertiten seit der reformation : nach ihrem leben und aus ihren schriften / Raess, Andreas – Freiburg i B: Herder, 1866-1880 – 27mf – 9 – 0-8370-8293-5 – (incl bibl ref and ind) – mf#1986-2293 – us ATLA [242]
Conveyance news – v5 n4-v6 n3 [1982 jul/aug-1983 sep/oct] – 1r – 1 – (cont: ancsa news; cont by: alaska conveyance news) – mf#855386 – us WHS [071]
The conveyancer's and notary's manual...in the state of minnesota / Booth, Walter Sherman – 1st ed. Minneapolis, Booth, 1892. 135 p. LL-665 – 1 – us L of C Photodup [348]
The conveyancer's and notary's manual...in the state of south dakota / Booth, Walter Sherman – 1st ed. Minneapolis, Booth, 1892. 135 p. LL-1677 – 1 – us L of C Photodup [348]
Conveyor / International Union, United Automobile, Aerospace, and Agricultural Implement Workers of America – 1965-1982 nov, 1982 oct/nov-1991 jan – 2r – 1 – (cont: solidarity [conveyor edition]) – mf#2910333 – us WHS [331]
Conveyor / Milwaukee Coke and Gas Co – v5 n9,12 [1918 sep, dec], v6 n1-7,9-12 [1919 jan-jul, sep-dec], v7 n1,3-6,9,11-12 [1920 jan, mar-jun, sep, nov-dec], v8 n1-3,5-6 [1921 jan-mar, may-jun] – 1r – 1 – mf#1819176 – us WHS [660]
Conveyor / Woodworkers' Industrial Union of Canada – v1 n1 [1948 dec 1] – 1r – 1 – mf#681701 – us WHS [331]
The convict ship : and other poems / Peace, M S [Mrs] – Greenock [Scotland]: R A Baird 1850 [mf ed 1987] – 1mf – 9 – 0-665-33824-4 – mf#33824 – cn Canadiana [810]
Convict transportation and the metropolis : the letterbooks and papers of duncan campbell (1726-1803) from the state library of new south wales – 4r – 1 – £375.00 – (with d/g) – uk Matthew [980]
Convicting the innocent; errors of criminal justice / Borchard, Edwin Montefiore – New Haven, Yale, 1932. 421 p. LL-398 – 1 – us L of C Photodup [345]
Convictions and expectation of the patriarch job / Way, Lewis – London, England. 1827 – 1r – us UF Libraries [240]
Convictions of agrippa / Bickersteth, Robert – Oxford, England. 1858 – 1r – us UF Libraries [240]
Convictions of balaam / Bickersteth, Edward Henry – Oxford, England. 1858 – 1r – us UF Libraries [240]
Convicts.. / Kansas. State Penitentiary – 1864-84 – 1 – us Kansas [360]
Convicts and colonies : thoughts on transportation and colonization, with reference to the islands and mainland of northern australia / Morris, George Sculthorpe – London 1853 – (= ser 19th c british colonization) – 1mf – 9 – mf#1.1.6077 – uk Chadwyck [980]
Convito musicale... / Vecchi, Orazio – 1597 – (= ser Mssa) – 4mf – 9 – €60.00 – mfchl 430 – gw Fischer [780]
Il convivio: the banquet of dante alighieri / Dante, Alighieri – Trans. by Elizabeth Price Sayer.London, New York: G. Routledge and Sons, 1887. 286p – 1 – us Wisconsin U Libr [810]
Convocation / Blakeney, Richard Paul – London, England. 1852? – 1r – us UF Libraries [240]
Convocation / Herbert, Samuel Asher – Newcastle upon Tyne, England. 1868 – 1r – us UF Libraries [240]
Convocation / Maitland, Samuel Roffey – London, England. 1855 – 1r – us UF Libraries [240]
Convocation / Williams, Thomas – Cardiff, Wales. 1853 – 1r – us UF Libraries [240]
Convocations and synods – London, England. 1850 – 1r – us UF Libraries [240]
Convocatoria y reglamento de la 1st asamblea provincial de cultura popular / Badajoz, Delegacion Provincial de Informacion y Turismo – Badajoz: Direccion General de Cultura del Ministerio de Informacion y Turismo, 1970 – sp Bibl Santa Ana [338]
Convoy / Teamsters for a Democratic Union – 1976 jan 23/feb 6-1979 oct – 1r – 1 – (cont by: prod dispatch; convoy dispatch) – mf#499210 – us WHS [071]
Convoy (and successor) / Cuyahoga Co. Cleveland – jan 1976-dec 1984 [mthly, bimthly] – 1r – 1 – mf#B3279 – us Ohio Hist [331]
Convoy dispatch / Detroit MI – jan 1976-dec 1984 [mthly, bimthly] – 1r – 1 – mf#B3279 – us Ohio Hist [331]
Convoy dispatch / Teamsters for a Democratic Union – Detroit MI. 1979 nov/dec-1987 nov/dec – 1r – 1 – (cont: convoy; cont by: teamster convoy dispatch) – mf#1576562 – us WHS [331]
Conway 1749-1892 – Oxford, MA (mf ed 1989) – (= ser Massachusetts vital records) – 43mf – 9 – 0-87623-100-8 – (mf 1-3: births & deaths 1750-1849. mf 4-7: town & vital records 1752-91. mf 8-12: town & vital records 1770-1806. mf 13-20: town & vital records 1806-36. mf 21-28: town & vital records 1829-51. mf 29-33: index: b,m,d 1843-1986. mf 34-35: b,m,d 1843-58. mf 36-38: births 1859-92. mf 38-40: marriages 1856-92. mf 40-43: deaths 1855-92) – us Archive [978]
Conway 1750-1849 – Oxford, MA (mf ed 1995) – (= ser Massachusetts vital record transcripts to 1850) – 10mf – 9 – 0-87623-234-9 – (mf 1t: out-of-town marriages 1784-90; births & deaths 1752-81. mf 1t-2t: marriage intentions 1769-91. mf 1t-5t: births 1750-1849. mf 4t,6t,8t: deaths 1768-1845. mf 4t-7t: marriage intentions 1791-1849. mf 5t-6t: births & deaths 1758-1801. mf 6t-9t: marriages 1791-1801, 1828-48. mf 7t-8t: births 1787-1808. mf 9t-10t: vital records 1843-49) – us Archive [978]
Conway, Bertrand Louis *see*
- The question-box answers
- Studies in church history

Conway first baptist church. conway, south carolina : church records – 1899-1981 – 1 – (deacon minutes and financial reports, 1973-1981) – us Southern Baptist [242]
Conway, James *see*
- The rights of our little ones
- The state last

Conway, James Joseph *see* The fundamental principles of christian ethics
Conway, Jim *see* Marx and jesus
Conway library – 6pt+5-yr update – 9 – (pt1: architecture: france & italy [1543mf] £6200; pt2: architecture: britain, germany & rest of the world [1602mf] £6200; pt3: architectural drawings [408mf] £2100; pt4: sculpture [1427mf] £6200; pt5: medieval arts [1100mf] £5150; pt6: illuminated mss [967mf] £5150; 5-yr update: containing c75,000 additional images to update all categories [860mf] £5150; with d/g) – uk Matthew [720]
Conway log cabin – Conway AK. 1888 aug 11 – 1r – 1 – (cont by: conway democrat; log cabin democrat) – mf#853688 – us WHS [071]
Conway, Moncure Daniel *see*
- Autobiography, memories and experiences of Moncure Daniel Conway
- Consequences
- Demonology and devil-lore
- Mazzini
- My pilgrimage to the wise men of the east
- The sacred anthology
- Solomon and solomonic literature
- Travels in south kensington
- The true and the false in prevalent theories of divine dispensations

Conway, W M *see* Climbing and exploration in the karakoram-himalayas
Conway, William Martin *see* Literary remains of albrecht durer
Conway, William Martin, Baron *see* The woodcutters of the netherlands in the fifteenth century
Conway, William Martin Conway, Baron *see* Early flemish artists and their predecessors on the lower rhine
Conway, william martin conway, Baron *see* Literary remains of Albrecht duerer
Conwell, Effie (Wood) *see* Reminiscences
Conwell, Russell Herman *see*
- Acres of diamonds
- The life, speeches, and public services of james a. garfield, twentieth president of the united states

Conwray, G R G *see* Postrera voluntad y testamento de hernando cortes marques del valle
Cony, Carlos Heitor *see* Ato e o fato
Conybeare, Frederick C *see* Rituale armenorum
Conybeare, Frederick Cornwallis *see*
- The armenian apology and acts of apollonius
- The dialogues of athanasius and zacchaeus and of timothy and aquila
- The historical christ
- History of new testament criticism
- The key of truth
- Roman catholicism as a factor in european politics
- Selections from the septuagint

Conybeare, Frederick Cornwallis et al *see* The story of ahikar
Conybeare, William John *see*
- Essays ecclesiastical and social
- The life and epistles of st paul

Conze, Edward *see* Buddhism
Conzelman, W E *see* Chronique de galawdewos [claudius]
Cook, Albert Stanburrough *see*
- The authorized version of the bible and its influence
- Biblical quotations in old english prose writers
- Extracts from the anglo-saxon laws

Cook, Alden Stoddard *see* The development of unitarian thought in america from arminianism to transcendentalism

Cook, Arthur Bernard *see*
– The metaphysical basis of plato's ethics
– Zeus, god of the bright sky
Cook, Ben T *see* An investigation of north carolina high school football coaches
Cook, Charles Augustus *see*
– Stewardship
– Stewardship and missions
Cook, Charles Henry *see* The curiosities of ale and beer: an entertaining history
The cook chronicle – Cook, NE: S W McCoy. v1 n1. dec 11 1947- (wkly) [mf ed -1949 (gaps)] – 1r – 1 – us NE Hist [071]
Cook County College Teachers Union, Local 1600 *see*
– College union 1600 voice
– College union voice
Cook, Dane B *see* A description of leg muscle pain and the effect of acetylsalicylic acid on the perception of pain and effort during and after cycle ergometry
Cook, E Wake *see* The endless future
Cook, Edmund Francis *see* Young j allen...1859-1907
Cook, Edward Tyas *see*
– The irish land act, 1881
– A popular handbook to the tate gallery
– Studies in ruskin
Cook, Edward Tyas [comp] *see* A popular handbook to the national gallery
Cook, Eveline Bosworth *see* Recollections, ms 3143
Cook, Forrest *see* Equity
Cook, Frederic Charles *see*
– The holy bible
– The origins of religion and language
– The revised version of the first three gospels
Cook, George *see*
– Few plain observations on the enactment of the general assembly, 18...
– An illustration of the general evidence establishing the reality of christ's resurrection
Cook, George Cram *see* Greek coins; poems... with memorabilia by floyd dell, edna kenton and susan glaspell
Cook islands annual reports [a3s] / Wellington. Dept of Island Territories – 1894-1945 [incomplete] – 1r – (lacking: 1896/97/98, 1903, 1918, 1920, 1921 and 1929) – mf#PMB Doc403 – at Pacific Mss [350]
Cook islands betela dance troupe : performances in japan – 1971-76 – 1r – 14 – mf#pmb doc394 – at Pacific Mss [790]
Cook Islands Collector of Customs *see* Records of arrivals and departures
Cook islands legislative assembly : proceedings, 1st session – 1959 – 1r – 1 – mf#pmb doc7 – at Pacific Mss [323]
Cook islands legislative assembly : proceedings, 2nd session – 1960 – 1r – 1 – mf#pmb doc8 – at Pacific Mss [323]
Cook islands legislative assembly : proceedings, 3rd session – 1961 – 1r – 1 – mf#pmb doc9 – at Pacific Mss [323]
Cook islands legislative assembly : proceedings, 4th session – 1962 – 1r – 1 – mf#pmb doc10 – at Pacific Mss [323]
Cook islands legislative assembly: proceedings, 5th to 8th sessions – 1r – 1 – mf#pmb doc6 – at Pacific Mss [323]
Cook islands legislative assembly : proceedings, 8th session – v1. 1954 – 1r – 1 – mf#pmb doc3 – at Pacific Mss [323]
Cook islands legislative council : proceedings – 1947-49 – 1r – 1 – mf#pmb doc23 – at Pacific Mss [323]
Cook islands legislative council : proceedings – 1950 – 1r – 1 – mf#pmb doc24 – at Pacific Mss [323]
Cook islands legislative council : proceedings, 10th session – 1956 – 1r – 1 – mf#pmb doc5 – at Pacific Mss [323]
Cook islands legislative council : proceedings, 11th session – 1957 – 1r – 1 – mf#pmb doc5a – at Pacific Mss [323]
Cook islands legislative council : proceedings, 5th to 8th sessions – n2-3, 3a. 1951-53 – 2r – 1 – mf#pmb doc2 – at Pacific Mss [323]
Cook islands legislative council : proceedings, 8th session – v2. 1954 – 1r – 1 – mf#pmb doc3a – at Pacific Mss [323]
Cook Islands legislative council : proceedings, 9th session – 1955 – 1r – 1 – mf#pmb doc4 – at Pacific Mss [323]
Cook Islands Library and Museum Society *see*
– Miscellaneous manuscripts, 1847-1977
– Miscellaneous manuscripts, 1891-1973
– Miscellaneous manuscripts, 1903-1939
– Miscellaneous manuscripts, 1933-1970
Cook islands news – 1961 – 2r – 1 – mf#pmb doc371 – at Pacific Mss [079]
Cook islands news – 1961-71 – 1r – 1 – mf#pmb doc385 – at Pacific Mss [079]
Cook islands news – 1962 – 4r – 1 – mf#pmb doc372 – at Pacific Mss [079]
Cook islands news – 1963 – 4r – 1 – mf#pmb doc373 – at Pacific Mss [079]
Cook islands news – 1965 – 4r – 1 – mf#pmb doc375 – at Pacific Mss [079]
Cook islands news – 1966 – 4r – 1 – mf#pmb doc376 – at Pacific Mss [079]
Cook islands news – 1967 – 4r – 1 – mf#pmb doc377 – at Pacific Mss [079]
Cook islands news – 1968 – 5r – 1 – mf#pmb doc378 – at Pacific Mss [079]
Cook islands news – 1969 – 5r – 1 – mf#pmb doc379 – at Pacific Mss [079]
Cook islands news – 1970 – 5r – 1 – mf#pmb doc380 – at Pacific Mss [079]
Cook islands news – 1971 – 4r – 1 – mf#pmb doc381 – at Pacific Mss [079]
Cook islands news – 27 may-30 dec 1960 – 2r – 1 – mf#pmb doc370 – at Pacific Mss [079]
Cook islands news – 4 jan 1972-21 jul 1972 – 2r – 1 – mf#pmb doc382 – at Pacific Mss [079]
Cook, J *see*
– A voyage to the pacific ocean
– A voyage towards the south pole, and round the world
Cook, J A Bethune *see* Sunny singapore
Cook, James *see*
– Journal of his voyage round the world in h.m.s. "endeavour", 1768-71
– Journal of his voyage round the world in hms "resolution", 1772-1775
– Log and journal of his journey round the world in the bark "endeavour", 1768-71
– Three voyages of captain james cook round the world
– Voyage au paele austral et autour du monde
Cook, Joel *see*
– America picturesque and descriptive, vol 1
– America picturesque and descriptive, vol 2
– America picturesque and descriptive, vol 3
– America picturesque and descriptive, vols 1-3
Cook, John *see*
– The advantages of life assurance to the working classes
– Early moral and religious education
– Evidence on church patronage
– Review of the proceedings of the general assembly
– A sermon preached on occasion of the general thanksgiving on the proclamation of peace
– A sermon preached on the occasion of the death of the rev robert mcgill, dd, minister of st paul's church, montreal
– A voice from the tomb of the late east india company
Cook, John Angus Bethune *see* Sunny singapore
Cook, John Wilson *see* On the history of canada
Cook, Joseph *see*
– Biology
– Conscience
– Current religious perils
– Heredity, with preludes on current events
– Labor, with preludes on current events
– Marriage
– Occident, with preludes on current events
– Orient
– Orthodoxy, with preludes on current events
– Rev, joseph cook's monday lectures on emerson's view of immortality
– Socialism
– Transcendentalism, with preludes on current events
Cook, Joseph et al *see* Christ and modern thought
Cook, Keningale *see* The fathers of jesus
Cook, Mercer *see*
– Education in haiti
– Haitian-american anthology
– Introduction to haiti
Cook, Millicent Whiteside *see*
– How to dress on £15 a year
– Tables and chairs
Cook Observatory *see* Publications
Cook, Richard Briscoe *see* The story of the baptists in all ages and countries
Cook, Robert S *see* Ecological issues on reintroducing wolves into yellowstone national park
Cook, Samuel *see* Sketches in spain during the years 1829, 30, 31, and 32
Cook, Stanley Arthur *see*
– Critical notes on old testament history
– The foundations of religion
– A glossary of the aramaic inscriptions
– Kinship and marriage in early arabia
– The laws of moses and the code of hammurabi
– Laws of moses and the code of hammurabi
– The religion of ancient palestine in the second millennium b c
Cook, T *see* Days of god's right hand
Cook weekly courier – Cook, NE: [James W Hammond] feb 26 1892-v53 n13. dec 28 1944 (wkly) [mf ed 1896-1944 (gaps) filmed [1974]] – 16r – 1 – (cont by: johnson county courier (1945)) – us NE Hist [071]
Cook, William Azel *see* By horse, canoe and float through the wilderness...
Cook, William Wilson *see* "Trusts"
Cook, Yvonne H *see* The relative effects of a live and videotaped instructor o ratings of perceived exertion and subjective feelings of students in an aerobic exercise class
Cooke, Alfred Fuller *see* Growing of easer lily bulbs under florida conditions
Cooke, Britton Bertrand *see* The first traveler
Cooke, Frances E *see*
– The story of dorothea lynde dix
– The story of john greenleaf whittier
– The story of theodore parker
Cooke, G A *see*
– The history and song of deborah
– Progress of revelation
– A text book of north-semitic inscriptions
– A text-book of north-semitic inscriptions
Cooke, George *see*
– Topographical and statistical description of the british isles
– Topographical and statistical description of the county of bedford
– Topographical and statistical description of the county of berks
– Topographical and statistical description of the county of buckingham
– Topographical and statistical description of the county of cambridge
– Topographical and statistical description of the county of chester
– Topographical and statistical description of the county of cumberland
– Topographical and statistical description of the county of derby
– Topographical and statistical description of the county of dorset
– Topographical and statistical description of the county of durham
– Topographical and statistical description of the county of essex
– Topographical and statistical description of the county of gloucester
– Topographical and statistical description of the county of hants
– Topographical and statistical description of the county of hereford
– Topographical and statistical description of the county of hertford
– Topographical and statistical description of the county of huntingdon
– Topographical and statistical description of the county of kent
– Topographical and statistical description of the county of lancaster
– Topographical and statistical description of the county of leicester
– Topographical and statistical description of the county of lincoln
– Topographical and statistical description of the county of middlesex
– Topographical and statistical description of the county of monmouth
– Topographical and statistical description of the county of norfolk
– Topographical and statistical description of the county of northampton
– Topographical and statistical description of the county of northumberland
– Topographical and statistical description of the county of nottingham
– Topographical and statistical description of the county of oxford
– Topographical and statistical description of the county of rutland
– Topographical and statistical description of the county of salop
– Topographical and statistical description of the county of somerset
– Topographical and statistical description of the county of stafford
– Topographical and statistical description of the county of suffolk
– Topographical and statistical description of the county of surrey
– Topographical and statistical description of the county of sussex
– Topographical and statistical description of the county of warwick
– Topographical and statistical description of the county of westmoreland
– Topographical and statistical description of the county of wilts
– Topographical and statistical description of the county of worcester
– Topographical and statistical description of the county of york
– Topographical and statistical description of the principality of wales
– Topographical survey of the county of cornwall
– Topographical survey of the county of devon
– Walks through kent
Cooke, George A *see*
– A general description of scotland
– A topographical description of the northern division of scotland
– A topographical description of the southern division of scotland
Cooke, George Albert *see*
– The book of amos
– The book of joshua
– The book of judges
– The book of ruth
Cooke, George Alexander *see* Topographical library of great britain. the british travellers' guide; or, pocket county directory
Cooke, George Willis *see*
– The american scholar
– A bibliography of ralph waldo emerson
– A guide-book to the poetic and dramatic works of robert browning
– The poets of transcendentalism
– Ralph waldo emerson
– The transient and permanent in christianity
– Unitarianism in america
– The world of matter and the spirit of man
Cooke, H *see*
– Letter from the rev dr cooke, belfast
– Papal aggresion
Cooke, Harriette J *see* Mildmay
Cooke, Henry *see* Second letter from the rev henry cooke, dd to the rev, john ritchie, dd in reply to his
Cooke, Henry et al *see* The true psalmody
Cooke, J P *see* Dhammapada
Cooke, Jay *see* Civil war and reconstruction: the making of modern america, series 1
Cooke, John H *see* Johann gerhard oncken, his life and work
Cooke, John T *see* Religious legislation
Cooke, Josiah Parsons *see*
– The credentials of science the warrant of faith
– Religion and chemistry
Cooke, Morris Llewellyn *see* Brazil on the march
Cooke, Parsons *see*
– The baptismal question
– Modern universalism exposed
Cooke, Richard Joseph *see*
– History of the ritual of the methodist episcopal church
– The incarnation and recent criticism
Cooke, Sarah A *see* The handmaiden of the lord
Cooke, William *see*
– Appeal to british protestants
– The fallacies of the alleged antiquity of man proved
– The shekinah
– Six views at rome, milan, and pisa
Cookeville first baptist church. cookeville, tennessee : church records – 1873-1956 – 1 – 58.59 – us Southern Baptist [242]
Cooking for profit – Fond du Lac. 1969+ (1) 1971+ (5) 1975+ (9) – ISSN: 0091-861X – mf#3329 – us UMI ProQuest [640]
Cooking light – Birmingham. 1989+ (1,5,9) – ISSN: 0886-4446 – mf#18206 – us UMI ProQuest [640]
Cooking without cans / Wason, Elizabeth – New York, NY. 1943 – 1r – us UF Libraries [500]
Cooks australasian travellers gazette – Melbourne, Australia. may, jun, aug 1911, jul 1914-dec 1915, 1918-21 – 1 3/4r – 1 – (aka: australasian travellers gazette) – uk British Libr Newspaper [919]
Cook's australasian traveller's gazette (the australasian traveller's gazette) – Melbourne, Australia. -m. 1911-15; 1918-21 – 2r – 1 – uk British Libr Newspaper [919]
The cook's favorite – Bruce Mines, Ont: W A of St George's Church, 1916 – 1mf – 9 – 0-659-92257-6 – mf#9-92257 – cn Canadiana [640]
Cook's lower canada admiralty court cases / Canada. Quebec. (Province) – 1v. 1873-84 (all publ) – 5mf – 9 – $7.50 – (contains decisions by the hon. george o smart) – mf#LLMC 81-071 – us LLMC [347]
Cookstown news, and ulster advertiser – [Northern Ireland] Belfast oct 1904-dec 1916 [mf ed 2002] – 10r – 1 – uk Newsplan [072]
Cool, Amanda *see* Diary
Cool, J *see* Diary
Cool, M F J *see* Struktuurveranderingen in nederlandsch-indie in de laatste 25 jaar
Cool springs primitive baptist church – Greenville Co, SC. 394p. 1834-40, 1886-1948, 1964-80 – 1 – $17.73 – mf#6557 – us Southern Baptist [242]
Cool, W *see* With the dutch in the east
Coolamon echo – Coolamon, sep 1898-dec 1905 – 1r – A$89.19 vesicular A$94.69 silver – at Pascoe [079]
Coolamon farmers review – Coolamon, nov 1906-dec 1951 – 10r – A$659.65 vesicular A$714.65 silver – at Pascoe [079]
Cooledge, Charles Edwin *see* The religious life of goethe
Cooley, Anna M *see* Collection
Cooley law review *see* Thomas m. cooley law review
Cooley, Roger William *see* Briefs on the law of insurance
Cooley, Thomas M et al *see* Constitutional history of the united states
Cooley, Thomas McIntyre *see*
– The elements of torts
– The general principles of constitutional law in the united states
– The law of taxation
– Liability of public officers to private actions for neglect of official duty
– A treatise on the constitutional limitations
– A treatise on the law of taxation, including the law of local assessments
– A treatise on the law of torts, or the wrongs which arise independent of contract

Cooley, William *see* The history of maritime and inland discovery
Coolgardie miner – Australia. Sep 1902-Oct 1910 (1910 imperfect). -w – 17r – 1 – uk British Libr Newspaper [622]
Coolgardie pioneer – Australia. 22 Apr 1899-30 Mar 1901.-w. 4mqn reels – 1 – uk British Libr Newspaper [072]
Coolidge, Calvin *see* Papers
Coolidge, James Ivers Trecothick *see* [Unitarian interpretations of jesus christ]
Coolombia en la encrucijada / Restrepo, Felix – Bogota, Colombia. 1951 – 1r – us UF Libraries [972]
Coolus, Romain *see*
– Amour
– Fifille a sa memere
Cooma express – Cooma, apr 1882-dec 1931 – 17r – A$1065.26 vesicular A$1158.76 silver – at Pascoe [079]
Cooma express – Cooma, jan 1932-dec 1968 – 21r – A$1454.64 vesicular A$1570.14 silver – at Pascoe [079]
Cooma monaro express – Cooma. jan 1969-dec 1990, apr 1992-jun 1993 – at Pascoe [079]
Coomaraswamy, Ananda Kentish *see*
– Buddha and the gospel of buddhism
– The dance of shiva
– Elements of buddhist iconography
– Essays in national idealism
– Figures of speech
– Hinduism and buddhism
– The indian craftsman
– The living thoughts of gotama the buddha
– The message of the east
– Myths of the hindus and buddhists
– A new approach to the vedas
– The new orient asiatic art
– The rg veda as land nama-bok
– Spiritual authority and temporal power in the indian theory of government
– The transformation of nature in art
– Why exhibit works of art?
Coomassie and magdala / Stanley, H M – London, 1874 – 6mf – 9 – mf#NE-20236 – ne IDC [916]
Coombs, James Vincent *see* Religious delusions
Coon, D Burdett et al *see* Seventh day baptists in europe and america
Coon dissector / Montgomery Co. Dayton – v1 n1. may-nov 1844// [wkly] – 1r – 1 – mf#5000 or 7195 – us Ohio Hist [320]
Coonabarabran times – Coonabarabran, 1946-68 – at Pascoe [079]
Coonabarabran times – Coonabarabran, jan 1969-dec 1996 – at Pascoe [079]
Coonamble independent – Coonamble, jan 1898-dec 1909 – 6r – A$436.44 vesicular A$469.44 silver – at Pascoe [079]
Coonamble independent – Jan 7 1898-dec 31 1909 – 6r – 9 – A$436.44 vesicular A$469.44 silver – at Pascoe [079]
Coonamble times – Coonamble. 1899-1905, 1956-59, 1962, 1965-68 – 7r – A$384.38 vesicular A$422.88 silver – at Pascoe [079]
Coonamble times – Coonamble, jan 1969-dec 1996 – at Pascoe [079]
Cooney, Rian *see* Icarus
Co-op : the harbinger of economic democracy – Ann Arbor. 1979-1979 (1,5,9) – (cont: new harbinger. cont by: co-op magazine) – ISSN: 0190-2741 – mf#11086,02 – us UMI ProQuest [338]
Co-op / North American Students of Cooperation – v6 n1-7 [1979 mar/apr-1979 nov/dec] – 1r – 1 – (cont: new harbinger; cont by: co-op magazine [ann arbor [mi]]) – mf#676069 – us WHS [302]
Co-op banknotes / National Consumer Cooperative Bank [US] – v1 n1-v4 n2 [1980 oct/dec-1984 may] – 1r – 1 – (cont by: bank notes [washington dc]) – mf#841756 – us WHS [332]
Co-op country news / Farmers Union Central Exchange [Saint Paul MN] – 1974 oct 7-1975 jun 6, 1975 jul 7-1978, 1979-81, 1982-1987 mar – 4r – 1 – (cont: farmers union herald; cont by: land o'lakes mirror; cooperative partners) – mf#1383299 – us WHS [334]
Co-op magazine – Ann Arbor. 1980-1980 (1,5,9) – (cont: co-op: the harbinger of economic democracy) – ISSN: 0199-459X – mf#11086,03 – us UMI ProQuest [338]
Co-op magazine – v7 n1-v8 n2 [1980 jan/Feb-1981 spr] – 1r – 1 – mf#676070 – us WHS [071]
Co-op news – v2 n11-v9 n12 [1943 nov 4-1945 aug 14] – 1r – 1 – mf#568905 – us WHS [334]
Co-op times / Connecticut Co-ops – v1 n1-v2 n4 [1982 jul/aug-1983 fall] – 1r – 1 – (cont by: northeast co-op times) – mf#709586 – us WHS [334]
Coope, William Jesser *see* Swazieland as an imperial factor
Cooper, Charles Henry *see* Memoir of margaret, countess of richmond and derby
Cooper, Charles William *see*
– Digest of reports of cases decided in the court of chancery, in the court of error & appeal, on appeal from the court of chancery, and in chancery chambers.
– The mechanics' lien law of illinois, as amended by act of 1887.
Cooper, Clayton Sedgwick *see* Brazilians and their country
Cooper collection – v1 n1-v6 n4 [1977 jan-1982 oct] – 1r – 1 – mf#697364 – us WHS [080]
Cooper, Edith *see* Michael field and fin-de-siecle culture and society
Cooper, Elizabeth *see* My lady of the chinese courtyard
Cooper, Florence Kendrick *see* Martin b. anderson, ll. d
Cooper, George *see* Designs for the decoration of rooms
Cooper, H J of South Hampstead *see* The art of furnishing on rational and aesthetic principles
Cooper, Jal *see* Stamps of india
Cooper, James *see*
– Confessions of faith and formulas of subscription
– Journal of geriatric drug therapy
– Notions of the americans
– The testament of our lord
Cooper, James Fenimore *see*
– Deerslayer
– Early critical essays 1820-1822
– Tales for fifteen
Cooper, John *see* Miscellanies in verse and prose
Cooper, John Francis *see* Study of the cost of growing beans in florida, 1927-28
Cooper, Morris *see* The law and practice of referees and references under the code of civil procedure and statutes of the state of new york
Cooper Ornithological Club *see* Bulletin of the cooper ornithological club of california
Cooper, Page *see* Sambumbia
Cooper, R Bransby *see* Letter to a clergyman on the peculiar tenets of the present day
Cooper, R F *see* Course of study in vocational agriculture for individualized instruction
Cooper, Reginald Davey *see* Hunting and hunted in the belgian congo
Cooper river baptist church. charleston county. south carolina : church records – 1955-72 – 1 – us Southern Baptist [242]
Cooper, Robert *see*
– Brewin grant refuted
– Holy scriptures analyzed
– The infidel's text-book
Cooper, Thomas *see*
– The bridge of history over the gulf of time
– Evolution, the stone book
– Some information respecting america
Cooper, Thomas Sidney *see* My life
Cooper, Trevor K *see* Peripheral chemoresponsiveness and exercise induced arterial hypoxemia in highly trained endurance athletes
Cooper Union Museum for the Arts of Decoration. New York *see* Italian drawings for jewelry 1700-1875
Cooper, Walter E *see* A comparison of perceptions of the importance between physical education graduate teaching assistants and graduate program coordinators
Cooper, William B *see* Lecture on the manners and customs of the japanese
Cooper, William Frierson *see* Removal of causes from state to federal courts
Cooper, William Henry *see* The book of mormon proved to be a fraud
Cooper, William M *see* Flagellation and the flagellants
Cooper, William Ricketts *see*
– Egypt and the pentateuch
– The serpent myths of ancient egypt
Cooperacion de roman gomez villafranca a la bibliografiade arias montano, nº 1 : la biblia regia / Gomez Villafranca, Ramon – Badajoz: Imp. Provincial, 1928. Rev. Centro de Estudios Extremenos Tomo 2, no 1-2 – 1 – sp Bibl Santa Ana [946]
Cooperaction – v1 n1-v2 n2 [1976 jul-1979 spr] – 1r – 1 – mf#641278 – us WHS [334]
Co-operation *see* Miscellaneous newspapers of the colorado historical society
Cooperation : its essence and background / Durell, Fletcher – Cape May, NJ. 1936 – 1r – us UF Libraries [025]
Cooperation – Alger. mars 1963-65 – 1 – (journal francais paraissant a alger) – fr ACRPP [073]
Cooperation and conflict – London. 1985-1995 (1,5,9) – ISSN: 0010-8367 – mf#13025 – us UMI ProQuest [327]
Co-operation and the promotion of unity : with supplement, presentation and discussion of the report in the conference on 21st june 1910 – Edinburgh: Publ for the World Missionary Conference by Oliphant, Anderson & Ferrier; New York: Fleming H Revell, [1910?] – 1mf – 9 – 0-8370-6474-0 – (incl indes) – mf#1986-0474 – us ATLA [240]
La cooperation audiovisuelle franco-algerienne de 1975 a nos jours / Rabia, Ali – 2mf – 9 – (10271) – fr Atelier National [790]
Cooperation between jews and arabs – [London, 1931]. 2pts – 1mf – 9 – mf#J-28-134 – ne IDC [956]
Cooperation, principles and practices : the application of cooperation to the assembling, processing, and marketing of farm products, to the purchase of farm supplies and consumers' goods and to credit and insurance / University of Wisconsin. College of Agriculture. Dept of Agricultural Economics – Madison WI: Extension Service of the College of Agriculture, The University of Wisconsin, [1937] [mf ed 1998] – (= ser Usain state and local literature preservation project: wisconsin) – 1r – 1 – (filmed with: Land economic inventory of the state of Wisconsin. suppl readings at end of each chapter) – mf#9835 n4 – us Wisconsin U Libr [334]
Co-operation with employees : a study / Forster, Hans Walter – Philadelphia: Independence Bureau, c1919 (mf ed 19–) – 15p – mf#ZT-TB pv161 n3 – us Harvard [331]
COOPERATIVA BANANERA COSTARRICENSE *see* Certamen del patriotismo
Cooperativa Bananera Costarricense *see* Certamen del patriotismo
Cooperativa de pequenos campesinos de castuera – Castuera: Tip. Republica, 1937 – sp Bibl Santa Ana [946]
Cooperativa de Suministros y Consumo de Nuestra Senora de Guadalupe de Caceres *see* Estatutos de la...
Cooperativa de viviendas de proteccion oficial *see* San francisco de asis
Cooperativa de Viviendas "San Antonio" *see* Estatutos y reglamento de la...
Cooperativa del Campo *see* Estatutos...
Cooperativa del campo "Arrago" *see* Estatutos de la cooperativa del campo "arrago"
Cooperativa del campo Nuestra Senora de Botoa y de la Caja Rural de Badajoz *see*
– Memoria, 1970
– Memoria 1970 y cincuenta aniversario de su precursora la caja rural de ahorros y prestamos del sindicato catolico agrario
– Memoria 1972
Cooperativa del Campo Union de Cultivadores de tabaco de Jaraiz de la Vera *see* Estatutos de la...
Cooperativa Farmaceutica Extremena "Cofex" *see* Estatutos de la...
Cooperativa Industrial Cacerena *see* Estatutos por los que ha de regirsela...
Cooperativa Local de Consumo *see* Estatutos de la...denominada de santiago y santa margarita
Cooperativa Local del Campo y Ganaderos "San Isidro". Miajadas *see* Estatutos
Cooperativa se Casas Baratas *see* Estatutos de la cooperativa de casas baratas nuestra senora de la asuncion de caceres
Cooperativa y Caja Rural Comarcal del Campo Nuestra Senora de Piedraescrita. Espinaso *see* Memoria y balances 1975
Co-operative builder / Central Co-operative Wholesale [US] – 1933 jan 7/1937 apr 3-1980 – 21r – 1 – (cont: co-operative pyramid builder; cont by: cooperative world) – mf#701204 – us WHS [334]
Cooperative builder – Superior WI, 1933, 1977-82 – 3r – 1 – (finnish newspaper) – us IHRC [071]
Co-operative Central Exchange [US] *see* Keskusosuuskunnan tiedonantaja
Cooperative Children's Book Center circular / Wisconsin Free Library Commission – v1 n1-v16 n1 [1964 mar-1979 mar], 1980 aug – 1r – 1 – mf#390113 – us WHS [020]
The cooperative commonwealth in its outlines : an exposition of modern socialism / Gronlund, Laurence – Boston: Lee and Shepard; New York: C.T. Dillingham, 1884. 278p – 1 – us Wisconsin U Libr [335]
Cooperative competition : a discussion of the acute legal and economic perplexities confronting trade associations / New York evening post – New York: The Post, [1922?] (mf ed 19–) – 56p – mf#ZT-TB+pv491 n5 – us Harvard [338]
Cooperative economic insect report – Washington. 1975-1975 (1) 1975-1975 (5) 1975-1975 (9) – ISSN: 0045-8465 – mf#7928 – us UMI ProQuest [630]
Cooperative Education and Internship Program [WI] *see* Co-oportunity knocks
Co-operative educator / Farmers' Co-operative Packing Company [Madison WI] – 1917 jun-1918 jan – 1r – 1 – mf#1054876 – us WHS [334]
Cooperative learning : as a didactic strategy to improve the english proficiency of second language speakers / Matlou, Makoma Piet – Pretoria: Vista University 2002 [mf ed 2002] – 5mf – 9 – (incl bibl) – mf#mfm15165 – sa Unisa [370]
Cooperative lutheranism : the helen m knubel archives – 1920-87 – 90r – 1 – $130.00r – mf#xa0001r-xa0034r – us ATLA [242]
Cooperative marketing of citrus fruits in florida / O'byrne, Frank Mccord – s.l, s.l? 1913 – 1r – us UF Libraries [634]
Co-operative news – [NW England] Manchester ALS jan 1966-dec 1996 – 1 – uk MLA; uk Newsplan [334]
Co-operative news – Sydney, jul 1921-dec 1947 – 3r – A$189.00 vesicular A$205.50 silver – at Pascoe [079]
Cooperative news – Freewater OR: New Era Pub Co, [semimthly] [mf ed 1967] – 1r – 1 – us Oregon Lib [334]
Cooperative news-budget – Augusta, Fairchild, Fall Creek WI. 1918 oct 18-1919 aug 15, aug 22-nov 28 – 2r – 1 – (cont: augusta eagle [augusta wi: 1915]; fairchild observer [fairchild wi: 1897]; fall creek journal [fall creek wi: 1916]; cont by: eau claire county union) – mf#1044327 – us WHS [334]
Co-operative revision of the new testament : notes of the method and progress of the work, and of the share of the american committee therein / Lee, Alfred – New York: A D F Randolph, [1881?] – 1mf – 9 – 0-524-05987-X – mf#1992-0724 – us ATLA [225]
Co-operative school governance : from policy to practice / Looyen, Roger – Uni of South Africa 2000 [mf ed Johannesburg 2000] – 4mf – 9 – (incl bibl ref) – mf#mfm14773 – sa Unisa [370]
Cooperative Services, Inc *see* Comments
Cooperative Union of Canada *see* Canadianco-operator
Cooperative world / Land O'Lakes, Inc – 1982 oct-1983 dec – 1r – 1 – (cont: cooperative builder; cont by: land o'lakes mirror) – mf#1418716 – us WHS [334]
Cooperativismo e comunitarismo / Chacon, Vamireh – Belo Horizonte, Brazil. 1959 – 1r – us UF Libraries [972]
Co-operator – 1898 dec-1902 may, 1899 dec-1900 dec, 1902 jul-1903 dec, 1905-06 – 4r – 1 – mf#3437499 – us WHS [071]
Cooperator – Canoga Park. 1971-1973 (1) – ISSN: 0045-849X – mf#7888 – us UMI ProQuest [240]
La cooperazione – Barre VT, 1911 – 1r – 1 – (italian newspaper) – us IHRC [071]
Cooper-Chadwick, J *see* Three years with lobengula and experiences in south africa
Cooper-Marsdin, Arthur Cooper *see* The school of lerins
Cooperrider, George Trout *see*
– Be true!
– The last things
Coopers and lybrand journal – New York. 1973-1973 (1) – (cont: lybrand journal) – ISSN: 0190-2237 – mf#6283,01 – us UMI ProQuest [340]
Coopers and lybrand newsletter – New York. 1975-1980 (1) 1976-1980 (5) 1976-1980 (9) – (cont: lybrand newsletter) – mf#6423,01 – us UMI ProQuest [338]
Cooper's chancery reports / Tennessee. Supreme Court – v1-3. 1872-1878 (all publ) – (= ser Tennessee Supreme Court Reports) – 25mf – 9 – $37.50 – mf#LLMC 91-040 – us LLMC [347]
Cooper's clarksburg register – Clarksburg, WV. 1851-1858 (1) – mf#67247 – us UMI ProQuest [071]
Coopers International Union of North America *see* Proceedings of the...annual session of the coopers' international union of north america
Coopers' International Union of North America *see* Official handbook
Cooper's journal : or unfettered thinker and plain speaker for truth, freedom and progress – v1-30. 1850 [all publ] – (= ser Radical periodicals of great britain, 1794-1914. period 1) – 5mf – 9 – $95.00 – us UPA [303]
Coopers journal : devoted to the interests of the coopers of north america – v1, n1 [1870 jul] – 1r – 1 – (cont by: coopers' monthly journal) – mf#3253900 – us WHS [331]
Coopers' monthly journal : devoted to the interests of the coopers of north america – in english: v1 n1 [1870 jul], v2 n10 [1871 nov], v3 n7,9-10 [1872 jul, sep-oct], v6 n1-2,5 [1875 jan, mar, jun], in german: bd2 n10 [1871 nov], bd3 n7-9,11 [1872 jul-sep, nov], bd4 n3,7 [1873 mar, jul] – 1r – 1 – (cont: coopers journal; cont by: coopers' international journal) – mf#3253929 – us WHS [331]
Coopers' ritual – 1870, undated german ed of 24p – 1r – 1 – mf#3260464 – us WHS [331]
Coopersamy, Ibrahim *see* Misconceptions of bsc students at vista university concerning newton's laws
Co-oportunity knocks / Cooperative Education and Internship Program [WI] – v2 n3-v8 n2 [1977 sum-1982 sum] – 1r – 1 – mf#670414 – us WHS [370]
Coordinated collective bargaining quarterly [cbq] / AFL-CIO [American Federation of Labor and Congress of Industrial Organizations] – v13 n3-v14 n4 [1985 3rd qtr-1986 4th qtr] – 1r – 1 – (cont: iud coordinated bargaining quarterly) – mf#1362002 – us WHS [331]

Co-ordinated community service news – New York NY. 1963 nov/dec – 1r – 1 – mf#4881937 – us WHS [360]
Coordinating Committee on Women in the Historical Profession *see*
– Ccwhp
– Ccwhp newsletter
– Newsletter
Coordination / Schmidt, Mary Brainerd – [Chicago, 1955] – 1 – mf#*ZBD-*MGO pv17 – Located: NYPL – us Misc Inst [611]
Coordination chemistry reviews – Amsterdam. 1966+ (1) 1966+ (5) 1988+ (9) – ISSN: 0010-8545 – mf#42161 – us UMI ProQuest [540]
Coornhert, D V *see*
– Recht ghebruyck ende misbruyck
– Recht ghebruyck ende misbruyck van tydlicke have
[Coornhert, D V] *see* Recht ghebruyck ende misbruyck van tydlycke have
Coornhert, D van *see* Zedekunst
Coors courier – Golden, CO : Adolph Coors Co, Corporate Communications Dept, 1973- – 3r – 1 – ISSN: 0 – mf#MF C788 – us Colorado Hist [660]
Coos bay empire builder – Coos Bay OR: W N & M E Grannell, 1976-77 [wkly] – 1 – (cont: builder [1975-76]; cont by: empire builder [1977-78]) – us Oregon Lib [071]
Coos bay harbor – North Bend OR: [s.n.] -1950 [wkly] – 1 – (began in 1905; cont by: north bend news and coos bay harbor [1951-56]; 1925-40 incl newspaper publ by north bend high school students) – us Oregon Lib [071]
Coos bay news – Empire OR: T G Owen & J M Siglin, 1873-1917 [wkly] – 1 – us Oregon Lib [071]
Coos bay news – Empire City, OR: T G Owen and J M Siglin. v1 n1-v46 n15. mar 20 1873-oct 30 1917 – 1 – (place of publ moved to marshfield, or, dec 5 1877-oct 30 1917) – us Oregon Hist [071]
Coos bay times – Marshfield OR: Coos Bay Times Pub Co, -1957 [daily] – 1 – (began in 1906; formed by the union of: daily coast mail, and: weekly coast mail, and: advertiser; cont by: world) – us Oregon Lib [071]
Coos bay times *see* Daily coast mail
Coos bay world *see* World (coos bay, or)
Coos country courier – Coquille OR: W E Hassler, -1931 [wkly] – 1 – (cont: powers courier; cont by: oregon coos district courier [1931-34]) – us Oregon Lib [071]
Coos County Education Service District [OR] *see* Indian education
Coos Genealogical Forum *see* Bulletin of the coos...
Coosawhatchie baptist church – Jasper Co, SC. 395p. 1941-sep 1973 – 1 – $17.78 – mf#5003-29b – us Southern Baptist [242]
Cootamundra herald – Cootamundra – 27r – 9 – A$1552.89 vesicular A$1701.39 silver – at Pascoe [079]
Cootamundra herald – Cootamundra, jan 1969-dec 1992 – 46r – at Pascoe [079]
Cootamundra liberal – Cootamundra, jan 1899-dec 1906 – 3r – A$199.98 vesicular A$216.48 silver – at Pascoe [079]
Cootie courier / Military Order of the Cootie – 1976 jan-1989 aug – 1r – 1 – mf#1064386 – us WHS [355]
Copanti : jardin maya 'la concordia' / Morales Y Sanchez, Augusto – Tegucigalpa, Mexico. 1947 – 1r – us UF Libraries [972]
Copas, J V *see* Diary
Cope, Charles Henry *see* Reminiscences of charles west cope r a
Cope, Edward Drinker *see* The origin of the fittest essays on evolution
Cope, Henry Frederick *see*
– The evolution of the sunday school
– Religious education in the family
Cope, Jack *see*
– Penguin book of south african verse
– Seismograph
Cope, John Patrick *see*
– King of the hottentots
– South africa
Cope, Robert K *see* Comrade bill
Copeia – Carbondale. 1913+ (1) 1913+ (5) 1913+ (9) – ISSN: 0045-8511 – mf#12593 – us UMI ProQuest [590]
Copeland, David Graham *see* Policy
Copeland, Edward Brent *see* The development of long-range goals for the first baptist church of fairfield, ohio
Copeland, Glenda [et al] *see* A flashlight and compass
Copeland, R P *see* Sport sponsorship in canada
Copeland, Ralph *see* Annals of the royal observatory, edinburgh
Coperario, John *see* M coperario[s 3 pts] ayres
Copernicus atau rahasia-rahasia langit : bersama riwajat giordano bruno, galilei, kepler dan newton / Djakarta. Balai Poestaka – Djakarta: Balai Poestaka, 2602 – 39p 1mf – 9 – mf#SE-2002 mf22 – ne IDC [520]
Copete Lizarralda, Alvaro *see* Lecciones de derecho constitucional colombiano, ap
Copia augmentada de la carta de edificacion del v p sebastian de estrada : professor de la compania de jesus / Ansaldo, Matheo – Mexico: Impr real del superior gobierno, etc 1743 – (= ser Books on religion...1543/44-c1800: jesuitas) – 1mf – 9 – mf#crl-224 – ne IDC [241]
Copia certificada del privilegio real de la villa de acehuche. ano 1573 – 1 – sp Bibl Santa Ana [946]
Copia de la protesta...congreso...eleccion de coria / Gutierrez Utrera, Benigno – 1886 – 9 – sp Bibl Santa Ana [946]
Copia de la relacion...viage...cadiz a cartagena de indias / Soto y Marne, Francisco de – 1753 – 9 – sp Bibl Santa Ana [946]
Copia poetica del cuadro de la anunciacion... / Salas, Francisco Gregorio de – Madrid, 1781 – 1 – sp Bibl Santa Ana [810]
Copia...anunciacion...antonio r. mengs / Salas, Francisco Gregorio de – 1781 – 9 – sp Bibl Santa Ana [810]
Copiador de ordenes del regimiento de milicias de... / Cundinamarca Regimiento De Milicias De Infanteria – Bogota, Colombia. 1963 – 1r – us UF Libraries [355]
Copie d'avtres novvelles de rome... – Paris, 1561 – 1mf – 9 – mf#H-8158 – ne IDC [956]
Copie de la correspondance echangee entre les membres du gouvernement et le surintendant en chef des ecoles : au sujet de la loi des ecoles pour le haut canada, et de l'education en general... – Toronto: Lovell & Gibson, 1850 – 1mf – 9 – mf#28559 – cn Canadiana [370]
Copie de la correspondance echangee entre l'eveque catholique romain de toronto et le surintendant en chef des ecoles : au sujet des ecoles separees, dans le haut-canada – Quebec?: J Lovell, 1852 – 1mf – 9 – (with app) – mf#22321 – cn Canadiana [370]
Copie de la correspondance echangee entre l'eveque catholique romain de toronto et le surintendant en chef des ecoles : au sujet des ecoles separees, dans le haut-canada = Copies of correspondence between the roman catholic bishop of toronto and the chief superintendent of schools, on the subject of separate common schools in upper canada / Charbonnel, Armand Francois Marie de – Quebec: impr par John Lovell, 1852 [mf ed 1983] – 1mf – 9 – (with app) – mf#SEM105P330 – cn Bibl Nat [370]
Copie de la lettre de mr de montcalm / Montcalm, Louis-Joseph, marquis de – [s.l: s.n, 1758?] [mf ed 1984] – 1mf – 9 – 0-665-44073-1 – mf#44073 – cn Canadiana [971]
Copie de la lettre enuoiee par selim empereur des turqz, au seigneur domp iouan d'austrie, capitaine general de la ligue saincte – Paris, 1572 – 1mf – 9 – mf#H-8190 – ne IDC [956]
Copie de la petition adressee au gouverneur en conseil : par les honorables messieurs chapleau, church et angers demandant la destitution de son honneur luc letellier, lieutenant-gouverneur de la province de quebec – Ottawa: Maclean, Roger & Cie, 1879 [mf ed 1987] – 2mf – 9 – mf#SEM105P833 – cn Bibl Nat [971]
Copie des lettres de son exce : envoiees a monseigneur le baron de billy gouverneur de lille, douay and orchies, touchant la grande heureuse and memorable victoire de l'armee de sa majeste imperialle contre les turcs and hongres – Dovay, 1594 – 1mf – 9 – mf#H-8210 – ne IDC [956]
Copie d'une lettre escrite par le pere jacques bigot de la compagnie de jesus, l'an 1684 : pour accompagner un collier de pourcelaine... – [Manate, New York?: s.n.] 1858 [mf ed 1984] – 1mf – 9 – 0-665-20032-3 – mf#20032 – cn Canadiana [241]
Copies de correspondances entre le surintendant-en-chef des ecoles pour le haut-canada et autres personnes : au sujet des ecoles separees; (etant une continuation du rapport mis devant le parlement et imprime le 17 septembre 1852) = Copies of correspondence...on the subject of separate schools...17th september, 1852 / Canada (Province) – Toronto: Lovell et Gibson, 1855 [mf ed 1983] – 3mf – 9 – mf#SEM105P338 – cn Bibl Nat [370]
Copies de correspondances entre le surintendant-en-chef des ecoles pour le haut-canada, et autres personnes : au sujet des ecoles separees: (etant une continuation du rapport mis devant le parlement, et imprime le 17 septembre 1852) – Toronto: Lovell et Gibson, 1855 – 3mf – 9 – mf#28561 – cn Canadiana [370]
Copies of address of the house of assembly to the governor-general respecting the civil list : of report of seigniorial tenures etc – [London, England: s.n, 1844] (mf ed 1992) – 1mf – 9 – mf#SEM105P1737 – cn Bibl Nat [336]
Copies of all ordinances...passed by the special council and governor of lower canada : since the 24th day of november 1838 – [London, England: s.n, 1839] (mf ed 1992) – 1mf – 9 – mf#SEM105P1372 – cn Bibl Nat [323]
Copies of any despatches from the governor-general of canada to her majesty's secretary of state for the colonies in regard to the commercial changes now under the consideration of the imperial legislature : (in continuation of parliamentary paper, no 321, of the present session) / Grande-Bretagne. Colonial Office – [London, England: s.n, 1846] (mf ed 1996) – 1mf – 9 – mf#SEM105P2765 – cn Bibl Nat [324]
Copies of correspondence between members of the government and the chief superintendent of schools : on the subject of the school law for upper canada and education generally, with appendices = Copie de la correspondance echangee entre les membres du gouvernement et le surintendant en chef des ecoles au sujet de la loi des ecoles pour le haut canada, et de l'education en general, avec appendices... / Ryerson, Egerton – Toronto: printed by Lovell & Gibson, 1850 [mf ed 1983] – 1mf – 9 – (incl correspondence on the subject from march 3 1846 to april 25 1850) – mf#SEM105P339 – cn Bibl Nat [370]
Copies of correspondence between members of the government and the chief superintendent of schools : on the subject of the school law for upper canada and education generally... – Toronto?: Lovell & Gibson, 1850 – 1mf – 9 – (with app) – mf#28562 – cn Canadiana [350]
Copies of correspondence between the chief superintendent of schools for upper canada and other persons : on the subject of separate schools (being a continuation of the return laid before the house, and printed on the 17th september, 1852) – Toronto: Lovell & Gibson, 1855 – 3mf – 9 – mf#44178 – cn Canadiana [370]
Copies of correspondence between the roman catholic bishop of toronto and the chief superintendent of schools : on the subject of separate common schools, in upper canada / Charbonnel, Armand Francois Marie de – Quebec: Printed by John Lovell, 1852 [mf ed 1983] – 1mf – 9 – (with app) – mf#SEM105P331 – cn Bibl Nat [370]
Copies of correspondence between the roman catholic bishop of toronto and the chief superintendent of schools : on the subject of separate common schools in upper canada – Quebec?: J Lovell, 1852 – 1mf – 9 – (with app) – mf#22368 – cn Canadiana [370]
Copies of correspondence relative to the affairs of canada / Grande-Bretagne. Parliament – [London, England] [s.n.], 1839 [mf ed 1984] – 1mf – 9 – mf#SEM105P402 – cn Bibl Nat [323]
Copies of judgments of the international military tribunal for the far east, 1948 – (= ser Records of allied operational and occupation headquarters, world war 2) – 7r – 1 – mf#M1660 – us Nat Archives [355]
Copies of letters and telegrams received and sent by governor zebulon b. vance of north carolina, 1862-1865 / U.S. War Dept. Confederate Records – (= ser War department coll of confederate records) – 1r – 1 – mf#T731 – us Nat Archives [324]
Copies of letters received, 1858-1928, for africa / United Society for the Propagation of the Gospel. Archives – 12r – 1 – mf#97373 – uk Microform Academic [220]
Copies of letters sent, 1836-1931, for africa / United Society for the Propagation of the Gospel. Archives – 7r – 1 – mf#97374 – uk Microform Academic [220]
Copies of letters sent by the outfactors of the royal african company of st england to the chief agents at cape coast castle from january 1681-1699 – [Cape Coast Castle, 1699] – us CRL [960]
Copies of lists of passengers arriving at miscellaneous ports on the atlantic and gulf coasts and at ports on the great lakes, 1820-1873 – (= ser Records of the united states customs service, 1820-c1891) – 16r – 1 – mf#M575 – us Nat Archives [975]
Copies of papers connected with the branch of railway constructed by the municipalities of port hope and peterborough : from millbrook to peterborough – [Peterborough, Ont?: s.n.] 1862 [mf ed 1983] – 1mf – 9 – 0-665-44012-X – mf#44012 – cn Canadiana [380]
Copies of speeches / Case, Nelson – 1867-1920 – 1 – us Kansas [978]
Copies of the petition addressed to the governor in council by the honorable messieurs chapleau, church and angers : praying for the dismissal of his honor luc letellier, lieutenant-governor of the province of quebec – Ottawa?: Maclean, Roger, 1879 – 2mf – 9 – mf#64077 – cn Canadiana [324]
Copies or extracts of any despatches from the governor-general of canada to the secretary of state for the colonies : and of his replies, respecting the conduct of the returning officer of montreal during the late election – [London, England: s.n, 1845?] (mf ed 1997) – 1mf – 9 – mf#SEM105P2801 – cn Bibl Nat [325]
Copies or extracts of correspondence alluded to in lord glenelg's despatch to sir francis head, 7th september 1837 : between himself and persons communicating with him on behalf of the churches of england and scotland / Grande-Bretagne. Colonial Office – [London, England: s.n, 1840?] (mf ed 1996) – 1mf – 9 – mf#SEM105P2778 – cn Bibl Nat [241]
Copies or extracts of correspondence relative to the affairs of british north america – [London, England]: [s.n.], [1839] (mf ed 1982) – 5mf – 9 – mf#SEM105P111 – cn Bibl Nat [971]
Copies or extracts of correspondence relative to the affairs of canada / Grande-Bretagne. Parliament – [London, England]: [s.n.], 1839 [mf ed 1984] – 1mf – 9 – mf#SEM105P403 – cn Bibl Nat [323]
Copies or extracts of correspondence relative to the affairs of canada / Grande-Bretagne. Parliament – [London, England]: [s.n.], [1839] (mf ed 1984) – 1mf – 9 – mf#SEM105P404 – cn Bibl Nat [323]
Copies or extracts of despatches from sir f b head : on the subject of canada, with copies or extracts of the answers from the secretary of state – [S.l: s.n, 1839] [mf ed 1998] – 6mf – 9 – mf#SEM105P2896 – cn Bibl Nat [971]
Copilacion...orden...santiago del espada / Tapia, Gregorio de – 1605 – 9 – sp Bibl Santa Ana [946]
Copin-Albancelli, Paul *see*
– Comment je suis entre dans la francmaconnerie et comment j'en suis sorti
– Le drame maconnique
Coping with tension = Hoe om spanning te klop / ed by Ackermann, Manie – Pinelands: Medical Ass of South Africa [1982] [mf ed Pretoria, RSA: State Library [199-]] – 1r with other items – 5 – (also in afrikaans) – mf#82-2632 r25 – us CRL [150]
Copinger, Walter Arthur *see*
– The bible and its transmission
– Catalogue of the copinger collection of editions of the latin bible
– The manors of suffolk; notes on their history and devolution, with some illustrations of the old manor houses
– On the english translations of the "imitatio christi"
– Supplement to hain's repertorium bibliographicum
– A treatise on predestination, election, and grace
Copland, James *see* A dictionary of practical medicine (ael3/14)
Copland, S *see* A history of the island of madagascar...
Copland, Samuel *see* A history of the island of madagascar
Coplas colombianas – Bogota, Colombia. 1951 – 1r – us UF Libraries [972]
Coplas del baile del pandero / Gutierrez Macias, Valeriano – Madrid: C. Bermejo Impresor, 1961 – 1 – (sep rev dialectologia y tradiciones populares tomo 17, 1961 cuaderno n3 p401-415) – sp Bibl Santa Ana [946]
Cople – (= ser Bedfordshire parish register series) – 1mf – 9 – £3.00 – uk BedsFHS [929]
Cople, all saints monumental inscriptions monumental inscriptions – Arthur Weight Matthews 1914 – (= ser Bedfordshire parish register series) – 1mf – 9 – £1.25 – uk BedsFHS [929]
Copleston, Edward *see*
– False liberality, and the power of the keys
– Remains of the late edward copleston, d.d., bishop of llandaff
Copleston, Reginald Stephen *see* Buddhism, primitive and present
Copp, Henry N [comp] *see* Us mineral lands
Copp, Henry Norris *see*
– American mining code
– "Copp's mining decisions"
– Mining decisions of the secretary of the interior and the commissioner of the general land office
– Public land laws of the united states, 1869-1882
– Public land laws of the united states, 1882-1890
– Public laws passed by congress from april 1, 1882 to january 1, 1890.
– United states mineral lands.
Coppee, Francois *see*
– Fais ce que dois
– Luthier de cremone
– Naufrage
– Passant
– Pater
– Rendez-vous
Coppell, William G *see* Development and education in the cook islands

Coppell, William George *see* Miscellaneous papers concerning education in the cook islands

Coppenhall (crewe), st michael & all angels – (= ser Cheshire monumental inscriptions) – 4mf – 9 – £5.00 – mf#411 – uk CheshireFHS [929]

Coppens, Charles *see*
- A brief text-book of moral philosophy
- Moral principles and medical practice
- The mystic treasures of the holy mass
- The protestant reformation
- A systematic study of the catholic religion
- Who are the jesuits?

Coppens, J *see* L'imposition des mains et les rites connexes dans le nouveau testament et dans l'eglise ancienne

Coppens, Urbain *see* Der palast des kaiphas und der neue st. petersgarten der p.p. assumptionisten auf dem berge sion

Coppenstein, I A *see* Ex bellarmino epitome controversiarum omnium huius aevi luthero-calvinisticarum

Coppenstein, Ioan *see* Homiliae

Copper deficiency of tung in florida / Dickey, R D – Gainesville, FL. 1948 – 1r – us UF Libraries [630]

Copper state bulletin / Arizona State Genealogical Society – 1979 spr-1987 win, 1987 spr-1992 win – 2r – 1 – (cont: bulletin [southern arizona genealogical society]; cont by: copper state journal) – mf#202369 – us WHS [929]

Copper town / Powdermaker, Hortense – New York, NY. 1962 – 1r – us UF Libraries [960]

Copernicus-Institut *see* Veroeffentlichungen des kopernikus-instituts [astronomisches rechen-institut] zu berlin-dahlem

Coppieters, Honoratus *see* De historia textus actorum apostolorum

Coppin State College *see* Ronald e mcnair post-baccalaureate achievement program at coppin state college newsletter

Coppola, Raffaele *see* Dei concilii ecumenici

Coppolani, X *see* Les confreries religieuses musulmanes

Coppolani, Xavier *see* Les confreries religieuses musulmanes

Copp's land owner – v1-18+ind, 1874-92 [all publ] – 68mf – 9 – $102.00 – (lacking: v8 p45-46. v9 p181-82; index vol v18 p40; suppl to: the western landowner which is not offered by llmc) – mf#llmc84-448 – us LLMC [333]

"Copp's mining decisions" : decisions of the commissioner of the general land office and the secretary of the interior under the u s mining statutes / Copp, Henry Norris – Bancroft & Co, 1874 – 9 – $8.00 – mf#84-113 – us LLMC [343]

Copred peace chronicle / Consortium on Peace Research, Education and Development [US] – 1979 oct, 1980 jun-1981 feb – 1r – 1 – mf#653670 – us WHS [327]

Coptic aopcryphal gospels (ts4/2) – 1986 – (= ser Texts and studies (ts)) – 6mf – 9 – €14.00 – (trans by f robinson) – ne Slangenburg [226]

Coptic apocryphal gospels : translations together with the texts of some of them – Cambridge: University Press 1896 [mf ed 1989] – (= ser Texts and studies (cambridge, england) 4/2) – 1mf – 9 – 0-7905-1843-0 – (in english & coptic; incl ind) – mf#1987-1843 – us ATLA [226]

Coptic biblical texts in the dialect of upper egypt / ed by Budge, Ernest Alfred Wallis, Sir – London: Printed by order of the Trustees, sold at the British Museum, 1912 – 2mf – 9 – 0-8370-1792-0 – mf#1987-6180 – us ATLA [090]

Coptic church review – Lebanon. 1989-1996 (1) – ISSN: 0273-3269 – mf#16049 – us UMI ProQuest [240]

The coptic element in languages of the indo-european family / Campbell, John – Toronto: Copp, Clark, 1872 – 1mf – 9 – mf#00369 – cn Canadiana [410]

The coptic morning service for the lord's day – London, 1908 – 4mf – 8 – €7.00 – (trans by john, marquis of butet) – ne Slangenburg [243]

A coptic palimpsest : containing joshua, judges, ruth, judith and esther in the sahidic dialect / ed by Thompson, Herbert – London, New York: Oxford UP, 1911 [mf ed 1991] – 1mf – 9 – 0-7905-8284-8 – (text in coptic, int in english) – mf#1987-6389 – us ATLA [221]

The coptic (sahidic) version of certain books of the old testament : from a papyrus in the british museum / ed by Thompson, Herbert, Sir – London, New York: Oxford University Press, 1908 – 1mf – 9 – 0-8370-1793-9 – mf#1987-6181 – us ATLA [221]

Coptic time / Ethiopian Zion Coptic Church – 1978 nov ?-1980 sep – 1r – 1 – mf#573062 – us WHS [243]

The coptic version of the new testament in the northern dialect : otherwise called memphitic and bohairic, with introduction, critical apparatus, and literal english translation – Oxford: Clarendon Press, 1898-1905 – 6mf – 9 – 0-8370-1161-2 – mf#1987-6002 – us ATLA [225]

Coptius, F *see* ...Ad caesarem oratio pro christiana repv...

Copts and moslems under british control : a collection of facts and a resume of authoritative opinions on the coptic question / Mikhail, Kyriakos – London: Smith, Elder, 1911 – 1mf – 9 – 0-7905-5490-9 – mf#1988-1490 – us ATLA [960]

La copulata de leyes de indias y las ordenanzas ovandinas / Pena Camara, Jose de la – Madrid: Revista de Indias, 1941 – 1 – sp Bibl Santa Ana [950]

Copway's american indian – New York NY. 1851 aug 23-sep 6 – 1r – 1 – mf#875647 – us WHS [071]

Copy : essays from an editor's drawer on religion, literature, and life / Thompson, Hugh Miller – Hartford CT: M H Mallory 1872 [mf ed 1985] – 1mf – 9 – 0-8370-5586-5 – mf#1985-3586 – us ATLA [240]

The copy correspondence, 1697-1725 : from the victoria & albert museum, forster collection, ref. 419 / Nicolson, William (Bishop) & Wake (Archbishop) – 1r – 1 – mf#96686 – uk Microform Academic [240]

Copy of a communication and other papers... from the honorable denis benjamin viger, esquire : appointed to proceed to england, and support the petitions of complaint of of the assembly of lower canada, to the imperial parliament = Copie d'une communication ainsi que d'autres documens recus...de la part de l'honorable denis benjamin viger, ecuyer, lequel a ete autorise de se rendre en angleterre... / Viger, Denis-Benjamin – [S.l: s.n, 1832?] (mf ed 2000) – 5mf – 9 – (in french and english) – mf#SEM105P3243 – cn Bibl Nat [324]

Copy of a despatch, and its enclosures : addressed to earl amherst by the earl of aberdeen, on the 2d april 1835 – [London, England: s.n, 1838] [mf ed 1991] – 1mf – 9 – mf#SEM105P1392 – cn Bibl Nat [324]

Copy of a despatch from lord goderich to lord aylmer – [London, England: s.n, 1841] (mf ed 1992) – 1mf – 9 – mf#SEM105P1379 – cn Bibl Nat [324]

Copy of a despatch from the governor-general of british north america : transmitting a return from the principal of the seminary of montreal, showing the names of those who, since the passing of the rrdinance 3 vict c 30, have commuted the tenure of their property... – [London, England: s.n, 1841 ?] (mf ed 1993) – 1mf – 9 – mf#SEM105P1157 – cn Bibl Nat [971]

Copy of a despatch from the right hon charles poulett thomson to lord john russell : dated montreal, the 13th day of may 1840, transmitting memorial from various parties respecting the estates of st sulpice – [London, England: s.n, 1840?] [mf ed 1996] – 1mf – 9 – mf#SEM105P2777 – cn Bibl Nat [324]

Copy of a letter addressed by the rt hon the earl of roseberry, chairman of the imperial federation league : to its members throughout the empire – S.l: s.n, 1886? – 1mf – 9 – mf#54551 – cn Canadiana [320]

Copy of a memorial from james stuart, esquire, his majesty's attorney general for lower canada : to the right honorable lord viscount goderich, one of his majesty's principal secretaries of state, and also, copies of certain letters relating to the same – [Quebec?: s.n, 1831?] [mf ed 1984] – 1mf – 9 – mf#SEM105P412 – cn Bibl Nat [971]

Copy of a memorial from james stuart...attorney general for the province of lower canada : to the right honorable lord viscount goderich, one of his majesty's principal secretaries of state = Copie d'un memoire de james stuart...procureur general...la province du bas-canada, adresse au tres-honorable lord vicomte goderich, un des principaux secretaires d'etat de sa majest / Stuart, James – [S.l: s.n, 1831?] [mf ed 1982] – 5mf – 9 – mf#SEM105P116 – cn Bibl Nat [971]

Copy of a memorial from james stuart...to the right honorable lord viscount goderich...and also, copies of certain letters relating to the same – [Quebec?: s.n, 1831?] [mf ed 1984] – 1mf – 9 – cn Bibl Nat [971]

Copy of a report of the minister of justice : approved by his excellency the governor in council on the 22nd day of january, 1889...on the subject of the disallowance of the act of quebec relating to district magistrates, passed in the session of 1888 / Thompson, John Sparrow David – [Ottawa?: s.n, 1888?] – 1mf – 9 – mf#91500 – cn Canadiana [340]

Copy of an act passed by the legislature of upper canada to provide for the sale of the clergy reserves : and for the distribution of the proceeds thereof; together with copy of a despatch from the governor general of canada, dated 22d january – [London, England: s.n, 1840?] (mf ed 1997) – 1mf – 9 – mf#SEM105P2798 – cn Bibl Nat [348]

Copy of an act passed by the legislature of upper canada to provide for the sale of the clergy reserves – [London, England: s.n, 1840?] (mf ed 1997) – 1mf – 9 – mf#SEM105P2799 – cn Bibl Nat [240]

Copy of an explanatory memorandum : addressed by sir francis head to lord glenelg, dated the 21st of may last / Head, Francis Bond – [London, England: s.n, 1838] (mf ed 1996) – 1mf – 9 – mf#SEM105P2741 – cn Bibl Nat [971]

Copy of communication and other papers... from the honorable denis benjamin viger, esquire : appointed to proceed to england, and support the petitions of complaint of the assembly of lower canada, to the imperial parliament = Copie d'une communication ainsi que d'autres documens recus...de la part de l'honorable denis benjamin viger, ecuyer, lequel a ete autorise de se rendre en angleterre... / Viger, Denis-Benjamin – [S.l: s.n., 1831?] (mf ed 1991) – 1mf – 9 – (in french and english) – mf#SEM105P1373 – cn Bibl Nat [971]

Copy of correspondence between the governors of the british north american provinces and the secretary of state : relative to the introduction of responsible government into those colonies – [London, England: s.n, 1848?] (mf ed 1996) – 1mf – 9 – mf#SEM105P2776 – cn Bibl Nat [324]

Copy of correspondence, etc, with messrs temperleys, carter and darke : respecting the capitation tax charged on the emigrants in 1870 – S.l: s.n, 1870? – 1mf – 9 – mf#23772 – cn Canadiana [336]

Copy of correspondence relating to the establishment of the earl of durham : as governor general of british north america and her majesty's high commissioner – [London, England: s.n, 1838] (mf ed 1991) – 1mf – 9 – mf#SEM105P1389 – cn Bibl Nat [971]

Copy of letter re history of fernandina, florida – s.l, s.l? 193-? – 1r – us UF Libraries [978]

Copy of memorial and petition from inhabitants of the red river settlement : complaining of the government of the hudson's bay company and reports and correspondence on the subject of the memorial / Hudson's Bay Company – [s.l.]: The House of Commons, 1849 [mf ed 1982] – 2mf – 9 – mf#SEM105P124 – cn Bibl Nat [380]

Copy of the charter of the corporation of saint nicolet in lower canada : with instructions of lord bathurst on the subject / Bas-Canada – [S.l: s.n, 1841?] (mf ed 1992) – 1mf – 9 – (incl text in french) – mf#SEM105P1384 – cn Bibl Nat [370]

Copy of the fourth report of the standing committee of grievances made to the assembly of lower canada : respecting the conduct of lord aylmer, while governor-general of that province / Bas-Canada. Parlement. Chambre d'Assemblee & Grande-Bretagne. Parliament. House of Commons – [London]: [s.n.], [1836] (mf ed 1989) – 1mf – 9 – mf#SEM105P1127 – cn Bibl Nat [971]

Copy of the letters patent : erecting the protestant episcopal church of montreal, in notre dame street... – Montreal: printed by W Gray, [1818?] [mf ed 1983] – 1mf – 9 – 0-665-44014-6 – mf#44014 – cn Canadiana [242]

Copy of the memorial from the board of trade at toronto to the british government regarding cheap postage : and the answer of the lords of the treasury to that memorial / Grande-Bretagne. Colonial Office – [London, England: s.n, 1846?] (mf ed 1996) – 1mf – 9 – mf#SEM105P2779 – cn Bibl Nat [380]

Copy of the minutes of the evidence taken before the select committee appointed in the year 1834 / Grande-Bretagne. Parliament. House of Commons; ed by on the affairs of lower Canada – [London, England: s.n, 1838?] (mf ed 1996) – 3mf – 9 – mf#SEM105P2771 – cn Bibl Nat [323]

Copy of the minutes of the evidence taken before the select committee appointed in the year 1834 : on the affairs of lower canada / Grande-Bretagne. Parliament. House of Commons – [London, England: s.n, 1837?] (mf ed 1982) – 3mf – 9 – mf#SEM105P117 – cn Bibl Nat [323]

Copy of the rules of the prayer-meetings which are established amon... / Hawker, Robert – London, England. 18– – 1r – us UF Libraries [240]

Copy of the speech of the governor-general to the legislative assembly of canada : and correspondence relative to certain presumed changes in the commercial policy of the empire / Cathcart, Charles Murray Cathcart, Earl [Canada (Province). Governor general) – [London, England: s.n, 1846?] (mf ed 1997) – 1mf – 9 – mf#SEM105P2802 – cn Bibl Nat [380]

Copybook / Palmer, Thomas – 1803. 3 fiches – 9 – us South Carolina Historical [025]

Copybook / Palmer, Thomas – [mf ed Spartanburg SC: Reprint Co, 1981] – 3mf – 9 – mf#51-520 – us South Carolina Historical [510]

Copybooks of george washington's correspondence with secretaries of state, 1789-1796 / U.S. Dept of State – (= ser General records of the department of state) – 1r – 1 – (with printed guide) – mf#M570 – us Nat Archives [975]

Copyright arbitration royalty panel (carp) structure and process : hearing...house of representatives, 107th congress, 2nd session, june 13, 2002 / United States. Congress. House. Committee on the Judiciary. Subcommittee on Courts, the Internet, and Intellectual Property – Washington: US GPO 2002 [mf ed 2002] – 2mf – 9 – 0-16-068802-7 – (incl bibl ref) – us GPO [346]

Copyright Association of Canada *see* [Statement issued on the canadian copyright act of 1889]

Copyright bulletin : quarterly review / Unesco – Paris. 1977-1994 (1,5,9) – ISSN: 0010-8634 – mf#11424 – us UMI ProQuest [346]

The copyright conference – 1905, 1906. 1 reel – 1 – $50.00 – us Trans-Media [346]

Copyright in congress, 1789-1904 : a bibliography and chronological record of all proceedings in congress / Solberg, Thorvald – Washington: GPO, 1905 (all publ) – 5mf – 9 – $7.50 – mf#llmc 82-312 – us LLMC [346]

Copyright, its history and its law / Bowker, Richard Rogers – New York: Houghton-Mifflin, 1912 – 3mf – 9 – $13.50 – (contains a chronological table of laws and cases, english and american) – mf#LLMC 82-309 – us LLMC [346]

Copyright law / Gorman, Robert A – 1991 – 2mf – 9 – $3.00 – mf#llmc99-032 – us LLMC [346]

Copyright law : provisions of the u.s. copyright law with summary of parallel provisions of the laws of foreign countries / U.S. Laws, Statutes, etc – Washington: GPO, 1905 (all publ) – 1mf – 9 – $1.50 – mf#llmc 82-307 – us LLMC [346]

Copyright law revision : legislative history: 89th congress 2nd session / U.S. Congress. House Committee of the Whole – Washington: GPO. rept no 2237 for HR4347. 1966 – 3mf – 9 – $4.50 – mf#llmc 82-505 – us LLMC [346]

Copyright law revision / U.S. Copyright Office – Washington: GPO. 6v. 1961-65 (all publ) – 25mf – 9 – $37.50 – mf#llmc 82-302 – us LLMC [346]

Copyright law revisions : studies / U.S. Copyright Office – Washington: GPO. 86th Congress n1-35. 1960-72 – 23mf – 9 – $34.50 – (prepared for the senate subcomm. on patents, trademarks and copyright) – mf#llmc 82-303/304 – us LLMC [346]

Copyright office annual reports / U[/nited/] S[/tates/] Copyright Office – Washington: GPO, 1912, 1928-39, 1941-71, 1973-75, 1977, 1979, 1982-84 – (= ser Copyright office decisions) – 9 – (some issues lacking; aft 1984 coverage cont by: librarian of congress annual reports) – mf#llmc 82-311 – us LLMC [346]

Copyright office decisions – US Copyright Office, Library of Congress, 1783-1985 – 365mf – 9 – $547.00 – (add vols planned) – mf#llmc 78-064 – us LLMC [346]

The copyright question : a letter to the canadian society of authors / Morang, George Nathaniel – [Toronto?: s.n, 1902?] – 1mf – 9 – 0-665-72693-7 – mf#72693 – cn Canadiana [346]

The copyright question : a letter to the toronto board of trade / Morang, George Nathaniel – Toronto: G N Morang, 1902 – 1mf – 9 – 0-665-72140-4 – (incl bibl ref) – mf#72140 – cn Canadiana [346]

Copyright Society of the USA *see* Journal of the copyright society of the u.s.a.

Coqualeetza Education Training Centre *see* Sto

Coquerel, Athanase *see*
- Conscience and faith
- The fine arts in italy in their religious aspect
- First historical transformations of christianity
- Les fordcats pour la foi
- The preacher's counsellor
- Precis de l'histoire de l'eglise reformee de paris
- Protestantism in paris
- La seule chose necessaire – what the rising from the dead should mean

Coquette corrigee / La Noue, Jean-Baptiste Sauve – Paris, France. 1808 – 1r – us UF Libraries [440]

Coquille city bulletin – Coquille OR: Eickworth & Co, -1904 [wkly] – 1 – (merged with: coquille city herald to form: semi-weekly herald [1904-05]) – us Oregon Lib [071]

Coquille city bulletin *see* Coquille City herald

Coquille city herald – Coquille City OR: J A Dean, -1904 [wkly] – 1 – (merged with: coquille city bulletin, to form: semi-weekly herald [coquille, or]) – us Oregon Lib [071]

Coquille city herald *see* Coquille city bulletin

Coquille herald – Coquille OR: D F Dean, 1905-17 [wkly] – 1 – (merged with: coquille valley sentinel [-1917] to form: coquille valley sentinel and coquille herald [1917-21]; cont: semi-weekly herald [1904-05]) – us Oregon Lib [071]
Coquille tribune – Coquille OR: B M & L J Kester, 1934- [wkly] – 1 – (cont: oregon coos district courier [1931-34]) – us Oregon Lib [071]
Coquille tribune *see* Coquille valley sentinel
Coquille valley sentinel – Coquille, Coos County, OR: L W Cates. v11 n51-v12 n33. jan 5-aug 31 1917 – 1 – (merged with: coquille herald, to form: coquille valley sentinel and the coquille herald) – us Oregon Hist [071]
Coquille valley sentinel – Coquille OR: L W Cates, -1917 [wkly] – 1 – (merged with: coquille tribune [1934-] to form: coquille valley sentinel and coquille herald [1917-21]) – us Oregon Lib [071]
Coquille valley sentinel *see* Coquille valley sentinel and coquille heraldCoquille valley sentinelCoquille herald+Coquille valley sentinel
Coquille valley sentinel and coquille herald *see* Coquille valley sentinel
Coquille valley sentinel and coquille heraldCoquille valley sentinelCoquille herald+Coquille valley sentinel – Coquille OR: H W Young, 1917-21 [wkly] – 1 – (merger of: coquille valley sentinel [-1917]; coquille herald [1905-17]; cont by: coquille valley sentinel [1921-]) – us Oregon Lib [071]
Coquille valley sentinel [coquille, or] – Coquille OR: H W Young, 1921- [wkly] – 1 – (cont: coquille valley sentinel and the coquille herald) – us Oregon Lib [071]
Coquille valley sentinelCoquille herald *see* Coquille valley sentinel and coquille heraldCoquille valley sentinelCoquille herald+Coquille valley sentinel
Cor deo devotum : iesu pacifici salomonis thronus regius s gallico p. stephani luzvic... / Luzvic, S – Douai: Ex officina Balt: Belleri, 1627 – 4mf – 9 – mf#0-91 – ne IDC [090]
Cor luac prak : pralom lo k pulis nyn kar phsan / H'ael S'umpha – Bhnam Ben: Pannagar Put Chan 2495 [1953] [mf ed 1990] – 2v in 1 on 1r with other items – 1 – (in khmer) – mf#mf-10289 seam reel 105/10 [§] – us CRL [959]
Cora velu / Myui Mran On – Ran Kun: Ba ta la ca pum nhip tuik [195-?] [mf ed 1990] – 1r with other items – 1 – (in burmese) – mf#mf-10289 seam reel 185/6 [§] – us CRL [959]
Coradin, Jean *see* Hidalgo, ou la grande aventure
Coral gables : 'the best place to live under the sun' / Wyman, Vincent D – Coral Gables, FL. 1934 – 1r – us UF Libraries [978]
Coral gables / Coral Gables (Fl) Chamber Of Commerce – Coral Gables, FL. 1927? – 1r – us UF Libraries [978]
Coral Gables (Fl) Chamber Of Commerce *see* Coral gables
Coral, Leonidas *see* Guerra de los mil das en el sur de colombia
Coral reefs : journal of the international society for reef studies – Heidelberg. 1982+ (1,5,9) – ISSN: 0722-4028 – mf#13158 – us UMI ProQuest [574]
Coral ship / Munroe, Kirk – New York, NY. 1893 – 1r – us UF Libraries [978]
Coral tribune – Key West, FL. 1954-1963 sep – 9r – (gaps) – us UF Libraries [071]
Coralli, Eugene *see* Eucharis
Corals from the gulf of california and the north pacific coast of america / Durham, John Wyatt – [New York] 1947 [mf ed 1980] – (= ser Memoir (Geological Society of America) 20) – 1r – 1 – mf#84 – us Wisconsin U Libr [590]
Coran – Buenos Aires, Argentina. 1944 – 1r – us UF Libraries [025]
Le coran : traduction selon un essai de reclassement des sourates / Blachere, R – Paris. v1-3. 1947 – 21mf – 8 – €40.00 – ne Slangenburg [260]
The coran : its composition and teaching, and the testimony it bears to the holy scriptures / Muir, William – London: SPCK; New York: E & J B Young [1878?] – (= ser Non-christian religious systems) – 1mf – 9 – 0-524-01066-8 – (rev enl ed of: the testimony borne by the coran to the jewish and christian scripture) – mf#1990-2214 – us ATLA [260]
Corani textus arabicus : ad fidem librorum manuscriptorum et impressorum et ad praecipuorum interpretum lectiones et auctoritatem – Lipsiae: Typis et sumtibus Caroli Tauchnitii, 1834 – 1mf – 9 – 0-524-04429-5 – mf#1991-0003 – us ATLA [260]
Corazon / Hernandez Cata, Alfonso – Madrid, Spain. 1923 – 1r – us UF Libraries [972]
Corazon adentro / Roda Barrios, Abelardo – Guatemala, 1959 – 1r – us UF Libraries [972]
Corazon de indio / Garcia A, J Luis – Guatemala, 1959 – 1r – us UF Libraries [972]
Corazon deshabitado / Lopez, Juse Felix – Guatemala, 1963 – 1r – us UF Libraries [972]
Corazon indio (1938-1946) / Macip, Jose – Mexico City? Mexico. 1946 – 1r – us UF Libraries [972]
Corazon y vida : organo de las iglesias evangelicas "amigos" – Chiquimula, Guatemala: Iglesias Evangelicas Amigos -1986 [v48 n521-v70 (aug 1963-86)] [mf ed 2005] – 2r – 1 – (subtitle varies; guatamala, honduras y el salvador" aug 1963-jul 1970; "centroamerica" sep 1970-86; no more publ; iss numbered consecutively; several iss lacking & some iss damaged) – mf#2005h-s002 – us ATLA [366]
Corbacher zeitung – Korbach/Arolsen DE, 1976- – ca 7r/yr – 1 – (filmed by misc inst: 1887 10 may-1945 29 mar [76r] with suppl: mein waldeck 1924-41; title varies: 1 dec 1910: waldeckische landeszeitung) – gw Misc Inst [074]
Corbeil, Sylvio *see* Chomedey de maisonneuve
Corbett, Boston *see* Papers of boston corbett
Corbett, Charles H *see* Old testament story
Corbett, Griffith Owen *see*
– An appeal to the right hon w e gladstone, mp, her majesty's prime minister
– "The red river rebellion"
Corbett, Hunter et al *see* A record of american presbyterian mission work in shantung province, china, 1861-1913
Corbett, Jim *see*
– Jungle lore
– My india
Corbett, Joseph et al *see* The reformers
Corbett, Margaret Darst *see* Help yourself to better sight
Corbett's herald – Providence, RI. 1890-1918 (1) – mf#66277 – us UMI ProQuest [071]
Corbin, David Timothy *see* The law of personal injuries in the state of illinois, and the remedies and defenses of litigants
Corbin, Francis *see* Abolition and emancipation
Corbin, William Horace *see*
– The act concerning corporations in the state of new jersey, approved april 7, 1875
– An act concerning corporations (revision of 1896)
Corbisier, Roland *see* Reforma ou revolucao?
Corblet, Jules *see*
– Histoire dogmatique, liturgique et archeologique du sacrement de baptaeme
– Histoire dogmatique, liturgique, et archeologique du sacrement de l'eucharistie
Corcho Asenjo, Leopoldo *see* Quienes son los testigos de jehova
Corchon Garcia, Justo *see* Inscripciones cacerenas ineditas
[Corcoran-] corcoran journal – CA. 1971- – 19r – 1 – $1140.00 (subs $50/y) – mf#B02137 – us Library Micro [071]
Cord and creese / De Mille, James – New York: Harper, 1869? – 3mf – 9 – 0-665-01074-5 – mf#01074 – cn Canadiana [830]
Cord, William H *see* A treatise on the legal and equitable rights of married women
Cordain, Loren *see* Influence of body fat mass on excess post-exercise oxygen consumption
Cordeiro da Matta, J D *see*
– Ensaio de diccionario kimbundu-portuguez.
– Jisabu, jiheng'ele, ifika ni jinongonongo, josoneke mu limoundu ni putu, kua mon'angola jakim ria matta
Cordell, E A *see* Course in shona
Cordell, Victor V *see* Journal of transnational management development
Cordemoy, J L de *see*
– Uveau traite de toute l'architecture...
– Uveau traite de toute l'architecture ou l'art de bastir
Corder baptist church. corder, missouri : church records – 1871-1901 – 1 – 7.11 – us Southern Baptist [242]
Corder, Susanna *see* Life of elizabeth fry
Cordero, Carmen *see*
– Agraz
– Paralelas
Cordero, Juan Luis *see*
– Hojas de arbol caidas
– Mi patria y mi dama
– Mi torre de babel
– La musa del pecado
– La musa ingenua...la musa del consuelo...la musa civica
– Regionalismo
– La romeria de la luz
– Vida y ensueno
Cordero Marina, Pedro *see* Don juan de austria
Cordero Solano, Jose Abdulio *see* Ser de la nacionalidad costarricense
Los corderos de la pascua y ninguna trujillana es guapa / Ramos Sanguino, Joaquin – Trujillo: Tip. Sobrino de B. Pena, 1919 – 1 – sp Bibl Santa Ana [946]
Cordes, Gerhard *see* Der koeker
Cordier, Adolphe *see* Histoire de l'ordre maconnique en belgique
Cordier, Antonie *see* 'N radikale en gedifferensieerde universumgerigte pentekostalisme
Cordier, Emile L *see* Les grands hommes de la france
Cordier, H *see* Cathay and the way thither
Cordier, Henri *see* A narrative of the recent events in tong-king
Cordier, M *see* Colloquiorum scholasticorum libri 4
Cordiner, James *see*
– A description of ceylon
– A voyage to india
Cordley, Richard *see* Congregationalism in kansas
Cordner, John *see*
– The american conflict
– Canada and the united states
– Christ, the son of god
– The christian idea of sacrifice
– The foundations of nationality
– Jesus
– A pastoral letter to the christian congregation
– The philosophic origin and historical progress of the doctrine of the trinity
– Righteousness exalteth a nation
– The vision of the pilgrim fathers
Cordoba, Pedro de *see* Dotrina [christ]iana p[ar]a instrucion [et] informacio[n] de los indios
Cordoba Sanchez, Jose Leon *see* Poemas
Cordobesas / Fernandez de Molina Menitez Donoso, Antonio – 1894 – 9 – sp Bibl Santa Ana [946]
Cordolino De Azevedo, Pedro *see* Marechal pego junior e a invasao do parana
Le Cordon de saint francois d'assise – [Quebec?: s.n, 1876?] [mf ed 1985] – 1mf – 9 – 0-665-10448-0 – (in french and latin) – mf#10448 – cn Canadiana [241]
Cordova, Federico *see* Enrique pineyro, historiador
Cordova Landron, Arturo *see* Salvador brau
Cordova street marker / Crowe, F Hilton – s.l, s.l? 1938 – 1r – us UF Libraries [978]
Cordovez Moure, J M *see* Reminiscencias
Cordovez Moure, Jose Maria *see* Reminiscencias de santafe y bogota
Corea, the hermit nation / Griffis, William Elliot – 6th rev enl ed. New York: Charles Scribner, 1897 [mf ed 1995] – (= ser Yale coll) – xxxi/492p (ill) – 1 – 0-524-09814-X – (with additional chapter on corea in 1897) – mf#1995-0814 – us ATLA [950]
Corea, without and within : chapters on corean history, manners and religion with hendrick hamel's narrative of captivity and travels in corea, annotated / Griffis, William Elliot – 2nd ed. Philadelphia: Presbyterian Board of Publ [1885] [mf ed 1995] – (= ser Yale coll) – 315p (ill) – 1 – 0-524-09549-3 – mf#1995-0549 – us ATLA [950]
Corefiche : anthologies listed in granger's index to poetry plus – 9 – (phase 1: contains 781 books indexed by granger's 950mf $5000 isbn: 0-8486-6000-5. phase 2: 298 add vols and other select indexes 352mf $2245 isbn: 0-8486-7043-4. phase 3: 281 new vols 330mf $2245 isbn: 0-8486-7044-2. with print index $250 isbn: 0-89609-309-3) – us Roth [810]
Corefiche : books listed in the essay and general literature index – 280mf – 9 – $1,568.00 – (phase 1: 136 bks 280mf $1570 isbn: 0-8486-4136-1. phase 2: 290 bks 334mf $2500 isbn: 0-8486-4137-x. phase 3: 260 bks 341mf $2250 isbn: 0-8486-4138-8. phase 4: 224 bks 248mf $2000 isbn: 0-8486-4139-6. phase 5: 302 bks 335mf $2500 isbn: 0-8486-4140-x. phase 6: 200 bks 334mf $2500 isbn: 0-8486-4142-6. phase 7: 300 bks 361mf $2500 isbn: 0-8486-4143-4. phase 8: 300 bks 379mf $2500 isbn: 0-8486-4144-2. phase 9: 300 bks 379mf $2500 isbn: 0-8486-4145-0. phase 10: 300 bks 350mf $2500 isbn: 0-8486-4146-9. phase 11: 300 bks 350mf $2500. with author/title index $49.95 isbn: 0-89609-286-0. with suppl $39.95 isbn: 0-89609-320-4) – us Roth [840]
Corefiche : books listed in the short story index – 345mf – 9 – $2,500.00 – (phase 1: 314v indexed 345mf $2500 isbn: 0-8486-5015-8. phase 2: 123v 127mf $1000 isbn: 0-8486-5320-3. phase 3: 125v 130mf $1000 isbn: 0-8486-5445-5. phase 4: 125v 130mf $1000 isbn: 0-8486-5552-4. phase 5: 125v 130mf $1000 isbn: 0-8486-5677-6. phase 6: ca 125v 130mf $1000 isbn: 0-8486-5789-6. with author/translator/title index $19.95 isbn: 0-89609-322-0) – us Roth [830]
Corefiche : core poetry collection – 9 – (pt 1: 90 bks 106mf $800 isbn: 0-8486-7049-3. pt 2: 101v 35mf $800 isbn: 0-8486-7048-5. with author/translator/title index $75 isbn: 0-89609-325-5) – us Roth [810]
Corefiche : great american and english essays – (2376 essays by 1239 authors on 147mf $900 isbn: 0-8486-4141-8. with author/title index $29.95 0-89609-293-3) – us Roth [420]
Corefiche : literary criticism – a basic collection – 9 – (102 bks 115mf $700 isbn: 0-8486-0014-2. with author/title/subject index $19.95 isbn: 0-89609-319-0) – us Roth [840]
Corefiche : world's best drama – 160mf – 9 – $1,000.00 – 0-8486-9000-1 – (an original compilation of 860 plays by 275 dramatists, ranging from antiquity to the 20th centurty, encompassing most nationalities (in english translation) and genres. with author/translator/title index $24.95 isbn: 0-89609-294-1) – us Roth [820]
The coregency of ramses 2 / Seele, Keith – 9 – $10.00 – us IRC [930]
Coreidae of florida / Baranowski, Richard M – Gainesville, FL. 1986 – 1r – us UF Libraries [500]
Corelator / Congress of Racial Equality – 1945 may 1-1955 feb, fall/1961, 1962 brotherhood mth, apr, jun, sept, nov-1967 apr – 2r – 1 – (with gaps; cont by: core) – mf#805245 – us WHS [322]
Corelli, Arcangelo *see*
– Concerti grossi con duoi violini e violoncello di concertino obligati e duoi altri violini, viola e basso di concerto grosso ad arbitrio, che si potranno radoppiare, parte prima-[seconda, per camera
– Corelli's celebrated twelve concertos
– Corelli's twelve solos for the violin
– Oeuvres de corelly en trio
– Six solos for a flute and a bass
– [Sonatas] the score of the four operas
– Sonate a violino e violone o cimbalo
– Sonate a violino e violone o cimbalo...opera quinta, parte prima[-seconda]
– Sonate da camera a tre
Corelli, Marie *see* Barabbas
Corelli's celebrated twelve concertos : as performed by mr.cramer at the ancient concert...adapted for the organ, harpsichord, or piano forte, by thomas billington, harpsichord & singing master, opera 9 / Corelli, Arcangelo – London: printed for Mr Billington [c1790] [mf ed 19–] – 1r – 1 – mf#pres. film 116 – us Sibley [780]
Corelli's twelve solos for the violin : with an accompaniment for the violoncello to which is added a thorough bass for the piano forte or harpsichord / Corelli, Arcangelo – London: D'Almaine & Co [between 1834 & 1867] [mf ed 19–] – 1r – 1 – mf#pres. film 63 – us Sibley [780]
Coreografia colonial : acuarelas mandadas hacer por d baltasar jaime martinez companon y bujanda, siglo 18 / Jimenez Borja, Arturo – [Lima, Peru, 1941?] – 1 – mf#*ZBD-*MGO pv9 – us Misc Inst [790]
Coret, Jacques *see* Christus, der zweite adam
Corey, Henry Bascom *see* The american agriculturist law book, a compendium of every day law, for farmers, mechanics, business men, manufacturers, etc..
Corey, Merton L *see* Florida's opportunity
Corey, Stephen Jared *see*
– Among asia's needy millions
– Among central african tribes
– Ten lessons in world conquest
Corfe, Joseph *see* Twelve glees for three and four voices
Coria. Ayuntamiento *see*
– Cultos en honor de la santisima virgen de argeme. patrona de la ciudad de coria y consagracion de su nuevo santuario
– Ferias de san juan. junio 1956
Coria (caceres) / Junta Provincial de Turismo – Vitoria: Tip. Fournier, s.a. – 1 – (fotos gudiol) – sp Bibl Santa Ana [338]
Coria compostelana y templaria / Fita, Fidel – Madrid: Tip. Fortanet, 1912. B.R.A.H. 61. pp. 346-351 – 1 – sp Bibl Santa Ana [946]
Coria, Joaquin de *see*
– Nueva gramatica tagalog
– Nueva gramatica tagalog.teorico-practica
Coria y el mantel dela sagrada cena (la ciudad, su catedral, su relicario y su gran reliquia) / Munoz de San Pedro, Miguel – Madrid: Caja Ahorros y Monte de Piedad de Caceres, 1964 – 1 – sp Bibl Santa Ana [946]
Coria y sus fiestas 1963 – Coria: Imp. Fernandez, 1963 – 1 – sp Bibl Santa Ana [390]
Coria-Caceres *see* Calendario y plan de estudios del curso academico 1972-73. seminario diocesano
Coria-Caceres. Caritas Diocesana *see* Memoria, ano 1959
Coria-caceres. diocesis. organizacion curso 1968-69 – Caceres: Edit. Extremadura, 1968 – 1 – sp Bibl Santa Ana [240]
Corinth baptist church. colleton, south carolina : church records – 1869-1919, 1922-72 – 1 – us Southern Baptist [242]
Corinth baptist church. mcquady, kentucky : church records – 1892-1969 – 1 – us Southern Baptist [242]
Corinth baptist church. stewart county. model, tennessee : church records – 1915-64 – 1 – us Southern Baptist [242]
Corinthians : introduction, authorized version, revised version, with notes, index, and map / Massie, John – New York: Henry Frowde, [1902?] – 1mf – 9 – 0-8370-4307-7 – (incl ind) – mf#1985-2307 – us ATLA [227]

Corinto a traves de la historia (1514-1933). corinto (nicaragua) / d'Arbelles, Salvador – Madrid: Razon y Fe, 1934 – 1 – sp Bibl Santa Ana [946]

Corippi africani grammatici libri qui supersunt (mgh1: 3.bd. 2.teil) / ed by Partsch, J – 1879 – (= ser Monumenta germaniae historica scriptores 1: scriptores – auctores antiquissimi) – €15.00 – ne Slangenburg [470]

O corisco : orgam contra os buchecheiros – Fortaleza, CE. 24 nov 1898 – (= ser Ps 19) – mf#P17,01,44 – bl Biblioteca [079]

O corisco : periodico critico e pitoresco – Bahia: Typ de J A de Almeida, 03 dez 1867 – (= ser Ps 19) – bl Biblioteca [079]

Cork advertiser – Cork 1799-1801, 1803-04, 1806-07, 1811-16, 1818-19, dec 1822-jun 1823 [incomplete] – 11r – 1 – ie National [072]

Cork advertiser and commercial register – Cork, Ireland. 11 dec 1810, 18 sep 1823, 16 mar 1824 – 1/4r – 1 – (aka: cork advertiser and morning intelligencer) – uk British Libr Newspaper [072]

Cork advertiser and morning intelligencer *see* Cork advertiser and commercial register

Cork advertising gazette – Cork, Ireland. 5 oct 1855-12 oct 1859 – 2r – 1 – (incorp with: cork herald) – uk British Libr Newspaper [072]

Cork and south of ireland general advertiser – Ireland. -w. 24 May 1851-2 April 1853. 1 reel – 1 – uk British Libr Newspaper [072]

Cork chronicle and munster advertiser – Cork, Ireland. 9 jul 1853-28 jan 1854 – 1/4r – 1 – uk British Libr Newspaper [072]

Cork constituntion – Cork 1822-1825, 1897-jul 1922, 1924 – 138r – 1 – ie National [072]

Cork constitution *see* Constitution

Cork country eagle and munster advertiser – Cork, aug 1927-1928 – 1r – 1 – ie National [072]

Cork county chronicle – Cork 1934-36 – 1r – 1 – ie National [072]

Cork courier – Cork, jul 1794-feb 1795, 14 mar, 21 mar, 1 apr 1795 – 1r – 1 – ie National [072]

Cork daily advertiser – Ireland. -d. 1 Oct 1836-21 Jan 1837. (1 reel) – 1 – uk British Libr Newspaper [072]

Cork daily herald and daily advertiser *see* Cork herald and southern counties advertiser

Cork evening herald – Ireland. -d. 9 Sep 1833-Mar 1841. (9 reels) – 1 – uk British Libr Newspaper [072]

Cork examiner – Cork, Ireland. jul-dec 1847, 1848-96, 1923, jul-15 sep 1926, oct-dec 1926, jan-sep 1928, apr-jun 1930, oct-dec 1930, 1986-1996 – 276 1/2r – 1 – (aka: examiner) – uk British Libr Newspaper [072]

Cork examiner – Cork 1948/1949 – 4r – 1 – ie National [072]

Cork free press – Cork jun-dec 1910 – 3r – 1 – ie National [072]

Cork gazette – Cork, jun 1790-oct 1795, 1796, 1 no 1797 – 3r – 1 – ie National [072]

Cork gazette and general advertiser – Cork, Ireland. 3 jan-30 dec 1795 – 1/2r – 1 – uk British Libr Newspaper [072]

Cork herald *see* Cork advertising gazette

Cork herald and southern counties advertiser – Cork, Ireland. 3 apr 1858-1896 – 125 1/2r – 1 – (aka: cork daily herald and daily advertiser; cork daily herald and advertising gazette; cork daily herald) – uk British Libr Newspaper [072]

Cork mercantile chronicle – Ireland. -tw. Oct 1 1823; 14 feb, 22, 27 apr 1825; 1832-11 nov 1835 4 1/2r – 1 – uk British Libr Newspaper [072]

Cork morning intelligence – Cork, Ireland. 18 nov 1815 – 1/4r – 1 – uk British Libr Newspaper [072]

The cork oak forests and the evolution of the cork industry in southern spain and portugal / Parsons, James J – 1 – sp Bibl Santa Ana [338]

Cork sentinel – Ireland. -m. 19 Jan-3 Sep 1831. (1/4r) – 1 – uk British Libr Newspaper [072]

Cork sportsman – Ireland. -w. 30 May 1908-19 Oct 1911. (1 reel) – 1 – uk British Libr Newspaper [072]

Cork standard – Ireland.10 Oct 1836-1837. -d.2 1/2 reels – 1 – uk British Libr Newspaper [072]

Cork sun – Ireland. 18 Apr 1903-30 Sep 1905.-w. 4 reels – 1 – uk British Libr Newspaper [072]

Cork weekly chronicle – Ireland. -w. 24 April-20 Nov 1909. 1 2 reel – 1 – uk British Libr Newspaper [072]

Cork weekly chronicle – Cork, Ireland. may-dec 1896, 1930 – 1 1/4r – 1 – (publ 9 may 1896-4 nov 1976 only; aka: cork weekly examiner and weekly herald) – uk British Libr Newspaper [072]

Cork weekly examiner – Cork, Ireland. may-dec 1896; 1930 – 1 1/4r – 1 – uk British Libr Newspaper [072]

Cork weekly examiner and weekly herald *see* Cork weekly chronicle

Cork weekly herald *see*
- Weekly herald

Cork weekly news – Cork, Ireland. jun 1883-1923 – 39r – 1 – uk British Libr Newspaper [072]

Cork weekly times – Ireland. -w. 4 Oct 1833-26 Sep 1834. (1/2r) – 1 – uk British Libr Newspaper [072]

Corkman – Tralee apr 1966-jun 1977 – 23r – 1 – ie National [072]

Corless, George *see* Reply to the review of a pamphlet

Corlett, William Thomas *see* American tropics

Corley, Donald *see* Notes concerning st johns bluff and the spanish a...

Corley, Karen Fortenberry *see* National democratic convention keynote speeches

Corluy, Joseph *see* Spicilegium dogmatico-biblicum, seu, commentarii in selecta sacrae scripturae loca quae ad demonstranda dogmata adhiberi solent

Cormack, John *see* Account of the abolition of female infanticide in guzerat

Cormack, Margaret *see* The hindu woman

Cormatin, Pierre M de *see* Voyage du ci-devant duc du chatelet, en portugal

Cormery. France. Benedictine Abbey *see* Le cartulaire de cormery.

Cormier, Hyacinthe-Marie *see*
- Lettre a un etudiant en ecriture-sainte

Cormiguano de Brenta, Arthur *see* Mission diplomatique de laurent de brindes...padoue, 1964

Cormon, Eugene *see* Crochets du pere martin

Corn / Rolfs, P H – Gainesville, FL. 1909 – 1r – us UF Libraries [630]

Corn diseases in florida / Eddins, A H – Gainesville, FL. 1930 – 1r – us UF Libraries [630]

Corn experiment – Lake City, FL. 1889 – 1r – us UF Libraries [630]

Corn, hay, weevil, rice, cane, texas blue grass and cotton / Depass, Jas P – Lake City, FL. 1892 – 1r – us UF Libraries [630]

Corn laws – (= ser The Goldsmiths'-Kress Library of Economic Literature) – 25r – 1 – us Primary [340]

The corn laws : goldsmiths' -kress library of economic literature – 25r – 1 – (covers in depth the corn law question and related controversies) – mf#CL999-1434 – us Primary [340]

Corn varieties and hybrids and corn improvement / Hull, Fred H – Gainesville, FL. 1941 – 1r – us UF Libraries [630]

Cornaby, William Arthur *see*
- The call of cathay
- China and its people

Cornaeus, M *see* Curriculum philosophiae peripateticae...

La corne st-luc : the "general of the indians" / Lighthall, William Douw – Montreal: [s.n.] 1908 [mf ed 1997] – 1mf – 9 – 0-665-83543-4 – mf#83543 – cn Canadiana [971]

Cornea. – Philadelphia. 1991-96 (1,5,9) – ISSN: 0277-3740 – mf#16569 – us UMI ProQuest [617]

Corneille, Pierre *see*
- Don sanche d'aragon
- Suite du menteur

Corneille qui abat des noix / Barriere, Theodore – Paris, France. 1862 – 1r – us UF Libraries [440]

Corneille, Thomas *see* Ariane

La corneja sin plumas / Ipnocausto, Paulo – 1795 – 9 – sp Bibl Santa Ana [830]

Cornejo, J *see* Discurso particular y preservativo de la gota, en que se descubre su naturaleza y se pone su propia cura

Cornelis marten y cien anja di bonaire / Nooyen, R H – Willenstad, Curacao. 1959 – 1r – us UF Libraries [972]

Cornelison, Isaac A *see* The natural history of religious feeling

Cornelison, Isaac Amada *see* The relation of religion to civil government in the united states of america

Cornelissen, Mervyn Ehrich *see* Die verteenwoordigende kleurlingraad 1969-1980

Cornelius, A E *see* Sources of stress in athletes

Cornelius, Asher Lynn *see* The law of search and seizure.

Cornelius, Carl Adolf *see*
- Die ersten jahre der kirche calvins, 1541-1546
- Geschichte des muensterischen aufruhrs
- Historische arbeiten vornehmlich zur reformationzeit
- Die muensterischen humanisten und ihr verhaeltniss zur reformation

Cornelius, Carl Alfred *see* Nagra bidrag till upsala theologiska fakultets historia

Cornelius freundt : ein beitrag zur geschichte der evangelischen kirchenmusik insbesondere der saechsischen kantoreien in der 2. haelfte des 16. jahrhunderts / Goehler, Georg – Leipzig, 1896 – 1mf – 9 – 3-89349-332-8 – gw Frankfurter [780]

Cornelius Jansen, the Elder *see*
- Augustinus seu doctrina sancti augustini de humanae naturae sanitate
- Notarum spongia quibus alexipharmacum civibus sylvae-ducensibus

Cornelius news – Cornelius OR: R W Brill, -1928 [wkly] – 1 – (merged with: banks news and: gaston news and: north plains news, to form: west washington county news) – us Oregon Lib [071]

Cornelius news *see* West washington county news

Cornelius, Peter *see*
- Gedichte

Cornelius the centurion; and, life and character of st. john the evangelist and apostle = Hauptman cornelius / Krummacher, Frederic Adolphus – Edinburgh: Thomas Clark, 1840 – (= ser The biblical cabinet) – 1mf – 9 – 0-7905-3384-7 – (in english) – mf#1987-3384 – us ATLA [240]

Cornelius times *see* News-times (forest grove, or)

Cornell and lake holcombe courier – Cornell, Lake Holcombe WI. 1971 sep 16/1972 apr 27-2002 may/aug – 38r – 1 – (with gaps; cont: cornell courier) – mf#1036250 – us WHS [071]

Cornell business / Johnson Graduate School of Management (Cornell University) – Ithaca : Johnson Graduate School of Management, Cornell University 1991- [mf ed Ithaca NY: Filmed by Challenge Industries for...1995 – 1r [v14-25 (1991/92-2001/02) – 1 – (filmed with: malott times) – mf#film 236 – us Cornell [650]

Cornell chronicle – Ithaca, NY. 1969-2000 (1) – mf#65007 – us UMI ProQuest [071]

Cornell countryman – Ithaca. 1903-1996 (1) 1970-1981 (5) 1974-1981 (9) – ISSN: 0010-8782 – mf#2175 – us UMI ProQuest [630]

Cornell courier – 1916 jun 8-1917 dec 27, 1918 jan 3-mar 15 – 2r – 1 – (cont by: chippewa valley courier) – mf#1047266 – us WHS [071]

Cornell courier – Cornell WI. 1958 nov 6-1960, 1961-63, 1964 jan-1965 may, 1965 jun-1967 jun, 1967 jul 6-1969 may 8, 1969 may 15-1971 apr 15, 1971 apr 22-sep 9 – 7r – 1 – (cont: chippewa valley courier; cont by: cornell and lake holcombe courier) – mf#1047277 – us WHS [071]

Cornell daily sun – Ithaca, NY. 1990-2000 (1) – mf#61632 – us UMI ProQuest [071]

Cornell engineer – Ithaca. 1935-1978 (1) 1970-1978 (5) 1977-1978 (9) – ISSN: 0010-8790 – mf#248 – us UMI ProQuest [620]

Cornell executive – Ithaca. 1981-1983 (1,5,9) – (cont: executive) – ISSN: 0734-192X – mf#12202,01 – us UMI ProQuest [320]

Cornell graphic – Ithaca, NY. 1923-1926 (1) – mf#69015 – us UMI ProQuest [071]

Cornell hotel and restaurant administration quarterly – Ithaca. 1960+ (1,5,9) – ISSN: 0010-8804 – mf#12801 – us UMI ProQuest [640]

Cornell international law journal – v1-34. 1968-2001 – 1,5,6 – $563.00 – (v1-25 1968-92 on reel $346. v26-34 1993-2001 on mf $217.) – ISSN: 0010-8812 – mf#102051 – us Hein [340]

Cornell, John A *see* The pioneers of beverly

Cornell, John J *see* Autobiography of john j cornell

Cornell journal of law and public policy – v1-9. 1992-2000 – 9 – $156.00 set – ISSN: 1069-0565 – mf#114481 – us Hein [342]

Cornell journal of social relations – Ithaca. 1966-1984 (1) 1971-1984 (5) 1975-1984 (9) – ISSN: 0010-8820 – mf#3135 – us UMI ProQuest [302]

Cornell law forum – Ithaca. 1979+ (1,5,9) – ISSN: 0010-8839 – mf#12145 – us UMI ProQuest [340]

Cornell law quarterly – v1-11. 1915/16-1925/26 (all publ) – 77mf – $115.00 – (add vols as copyright expires) – mf#LLMC 90-319 – us LLMC [340]

Cornell law quaterly *see* Cornell law review

Cornell law review – v1-86. 1915-2001 – 1,5,6,9 – $2033.00 – (v1-80 1915-95 on reel or mf $1782; v81-86 1995-2001 on mf $251; title varies: v1-52 1915-67 as cornell law quarterly) – ISSN: 0010-8847 – mf#102091 – us Hein [340]

Cornell plantations – Ithaca. 1972-1980 (1) 1976-1980 (5) 1977-1980 (9) – ISSN: 0010-8863 – mf#8588 – us UMI ProQuest [630]

Cornell science leaflet – Ithaca. 1952-1969 (1) – mf#836 – us UMI ProQuest [500]

Cornell University *see*
- Documentation newsletter
- Newsletter

Cornell University Libraries *see* Catalogue of the dante collection presented by willard fiske

Cornelly : briefe zweier liebenden / Cuevry, Leonie von – Berlin-Grunewald: F A Herbig, c1944 – 166p – 1 – mf#7161 – us Wisconsin U Libr [860]

Cornely, Rudolph *see*
- Commentarius in librum sapientiae
- Commentarius in s pauli apostoli epistolas. 1
- Commentarius in s pauli apostoli epistolas. 2
- Commentarius in s pauli apostoli epistolas. 3
- Historica et critica introductio in u t libros sacros. vol 1
- Historica et critica introductio in u t libros sacros. vol 2, 1
- Historica et critica introductio in u t libros sacros. vol 2, 2
- Historica et critica introductio in u t libros sacros. vol 3

Corner, C *see*
- Epistola d pavli ad galatas cvm commentario
- In epistolam d pavli ad romanos scriptam commentarivs

Corner in india / Clark, Mary Mead – Philadelphia: American Baptist Publ Society, 1907 [mf ed 1995] – (= ser Yale coll) – xvi, 168p (ill) – 1 – 0-524-09024-6 – mf#1995-0024 – us ATLA [242]

Corner, Julia *see*
- China pictorial, descriptive and historical
- The history of rome

Corner stone / Nebraska State Historical Society – 1977 jul/aug-1986 win – 1r – 1 – mf#1238621 – us WHS [978]

Corner stones of a baptist church / Hobart, Alvah Sabin – Philadelphia: American Baptist Publication Society, c1894 – 1mf – 9 – 0-524-08385-1 – mf#1993-3085 – us ATLA [242]

Corneri, Christophori *see* Cantica selecta veteris novique testamenti

The corner-stone : or, a familiar illustration of the principles of christian truth / Abbott, Jacob – Boston: W Peirce; New York: J P Haven, 1834 [mf ed 1990] – ii//[3]-360p – 1 – mf#7591 – us Wisconsin U Libr [240]

Cornerstone baptist church. lincoln county. missouri (extinct) : church records – 1872-1910. 214p – 1 – 9.63 – us Southern Baptist [242]

Cornerstone clues / Cornerstone Genealogical Society – v1 n1-v11 n4 [1975:fall-1986 nov] – 1r – 1 – mf#1609781 – us WHS [929]

Cornerstone Genealogical Society *see* Cornerstone clues

Corner-stone laying, metropolitan church : mcgill square, on wednesday, the 24th day of august, 1870 at 3:30 o'clock, pm – S.l: s.n, 1870? – 1mf – 9 – mf#38720 – cn Canadiana [240]

Corner-stones of faith, or, the origin and characteristics of the christian denominations of the united states / Small, Charles Herbert – New York: E.B. Treat, 1898 – 2mf – 9 – 0-7905-6624-9 – mf#1988-2624 – us ATLA [240]

Cornet, Charles Joseph Alexandre *see* Au tchad

Cornet, Rene Jules *see*
- Katanga
- Maniema
- Terre katangaise

Cornevin, Robert *see*
- Histoire du congo, leopoldville
- Theatre en afrique noire et a madagascar

Corney, Peter *see* Voyages in the northern pacific

The cornflower : and other poems / Blewett, Jean – Toronto: W Briggs, 1906 – 3mf – 9 – 0-665-71395-9 – mf#71395 – cn Canadiana [810]

Cornhill magazine – London. 1860-1975 (1) 1970-1972 (5) 1970-1971 (9) – ISSN: 0010-891X – mf#550 – us UMI ProQuest [073]

Cornhusk bags of the plateau indians / Cheney Cowles Memorial Museum. Eastern Washington State Historical Society – 1976 – 4 color mf – 15 – $45.00f – 0-226-68987-5 – (28p accompanying text) – us Chicago U Pr [060]

Corni, Guido *see* Somalia italiana

Cornill, Carl Heinrich *see*
- Das buch des propheten ezechiel
- Das buch jeremia
- The culture of ancient israel
- Einleitung in die kanonischen buecher des alten testaments
- History of the people of israel from the earliest times to the destruction of jerusalem by the romans
- Der prophet ezechiel
- The prophets of israel
- Zur einleitung in das alte testament

Cornill, Carl Heinrich et al *see* Das christentum

Cornille, Henri *see* Souvenirs d'orient

Cornils, Pastor *see* Sind die sittlichen forderungen jesu fuer uns verbindlich?

O cornimboque – Rio de Janeiro, RJ: Typ do Cornimboque, 09-10 out 1881 – (= ser Ps 19) – mf#P17,03,58 – bl Biblioteca [079]

[Corning-] advance – CA. 1922-28 [wkly] – 2r – 1 – $120.00 – mf#B02138 – us Library Micro [071]

Corning and blossburg advocate – Corning, NY. 1840-1843 (1) – mf#64933 – us UMI ProQuest [071]

[Corning-] corning observer – CA. 1888- – 100r – 1 – $000.00 (subs $90/y) – mf#B02141 – us Library Micro [071]

[Corning-] daily republican – CA. 1926-1956 – 11r – 1 – $660.00 – mf#B02139 – us Library Micro [071]

Corning, Iowa. Corning Baptist Church *see* Records

Corning Museum of Glass *see*
- History of glass
- Rare books from the corning museum of glass
- Trade catalogs from the corning museum of glass

[Corning-] new era – CA. 1906-20 [wkly] – 2r – 1 – $120.00 – mf#B02140 – us Library Micro [071]

[Corning-] the republican of tehama county – CA. aug 1926-aug 1927 – 1r – 1 – $60.00 – mf#B02142 – us Library Micro [071]

Corning weekly democrat – Corning, NY. 1857-1896 (1) – mf#64936 – us UMI ProQuest [071]

Cornish baptist church. cornish flat, new hampshire : church records – 1791-1802. 12p – 1 – 5.00 – us Southern Baptist [242]

Cornish echo *see* Falmouth and penryn weekly times

Cornish, George Henry *see*
- Handbook of canadian methodism
- Hand-book of canadian methodism

Cornish guardian and county chronicle [newquay ed] – Bodmin, England 7 apr 1955-24 dec 1985 – 1 – (fr jan 1986 onward this ed held on the cornish guardian) – uk British Libr Newspaper [072]

Cornish, Louisa S *see* Prince of the kings of the earth

Cornish, New Hampshire. Cornish Baptist Church *see* Records

Cornish review – Cornwall. 1949-1950 – 1 – mf#495 – us UMI ProQuest [073]

Cornish times and general advertiser – Liskeard, England. 21 Feb 1859 sic; 2 May 1857-7 May 1859. -w. 1 reel – 1 – uk British Libr Newspaper [072]

Cornish times and general advertiser – [SW England] Cornwall jan 1857-dec 1950 [mf ed 2003] – 78r – 1 – (missing: 1867, 1872, 1888, 1897, 1911) – uk Newsplan [072]

The cornishman – Penzance, England. Jul 1878-Dec 1900.-w. 20mqn reels – 1 – uk British Libr Newspaper [072]

Corn-laws defended : or, agriculture our first interest, and the main-stay of trade and commerce. – Leeds: T Harrison; London: Simpkin, Marshall, & Co [1844?] – (= ser 19th c economics) – 1mf – 9 – mf#1.1.343 – uk Chadwyck [630]

Cornoldi, Giovanni Maria *see*
- The physical system of st thomas
- Sententia sancti thomae aquinatis de immunitate b v dei parentis a peccati originalis labe

Corns, Albert Reginald *see* A bibliography of unfinished books in the english language, with annotations

Cornsilk from dekalb county, il / Genealogical Society of DeKalb County, Illinois – v1 n1-v6 n2 [1975 jun-1981 feb] – 1r – 1 – (cont by: cornsilk newsletter from dekalb co il) – mf#633300 – us WHS [929]

Cornsilk newsletter from dekalb co il / Genealogical Society of DeKalb County, Illinois – v6 n3-v7 n1 [1981 mar-1982 jan] – 1r – 1 – (cont: cornsilk from dekalb county, il; cont by: cornsilk) – mf#633424 – us WHS [929]

Cornu, Charles *see* Der vergleich bei goethe unter besonderer beruecksichtigung der vergleichenden literaturkritik

Cornu copiae : sive linguae latinae commentarii / Perotti, Niccolo – Basileae, 1521 – €71.00 – ne Slangenburg [450]

Cornu, Maurice *see* Formes surcomposees en francais

Cornucopia project newsletter – v1 n1-v4 n2 [1981 spr-1984 fall] – 1r – 1 – (cont by: regeneration) – mf#851920 – us WHS [071]

Cornuti theologiae graecae compendium – Lipsiae: In aedibus BG Teubneri, 1881 [mf ed 1990] – (= ser Bibliotheca scriptorum graecorum et romanorum teubneriana) – 1mf – 9 – 0-7905-3716-8 – mf#1989-0209 – us ATLA [250]

Cornwall, Alan G *see* Recollections of an address

The cornwall canal / Keefer, Samuel – S.l: s.n, 1889? – 1mf – 9 – mf#60707 – cn Canadiana [627]

Cornwall chronicle – Launceston, Australia. 16 jan 1836-23 dec 1837; 6 jan 1838-25 dec 1847; 1 jan-2 sep 1848; 1849; 2 jan-25 dec 1850; 1851; 1852; 1854-27 jun 1857; 16 jan 1864-1876; 16 feb-24 dec 1877; 18 jan-31 dec 1878; 1879-30 aug 1880 – 24 1/2r – 1 – uk British Libr Newspaper [072]

Cornwall chronicle – Launceston, Australia. Jan 1850-Aug 1880 (missing 1853, 1857).-w. 15mqn reels – 1 – uk British Libr Newspaper [072]

The cornwall chronicle, 1776-94 : from bristol city library – West Indies – 3r – 1 – mf#96806 – uk Microform Academic [079]

Cornwall, E *see* Present crisis and future prospects of the church of god

Cornwall parish registers : marriages / ed by Phillimore, William Phillimore Watts & Taylor, Thomas – London: iss to the subsc by Phillimore & Co 1900- [mf ed 1987] – 1r – 1 – (with: the very joyous, pleasant and refreshing history of...lord de bayard / mailles, j) – mf#6942 – us Wisconsin U Libr [929]

[Cornwallis, Caroline Frances] *see* Christian doctrine and practice in the 12th century

Cornwallis, William *see* Discourses upon seneca the tragedian

Cornwall-Jones, A *see*
- Musha unofadza
- Ngenani komasiwela amaphetheni okweluka
- Ngenani komasiwela ukudla okuhle kwempilo
- Ngenani komasiwela ukuthung izigqoko zabantwana
- Ngenani komasiwela ukuwatshwa kwempahla
- Ngenani komasiwela umama losane
- Sanganai namai chamunorwa amai nomucheche
- Sanganai namai chamunorwa kuruka
- Sanganai namai chamunorwa kusona zvipfeko zvavakuru
- Sanganai namai chamunorwa kusona zvipfeko zvevana
- Sanganai namai chamunorwa kusuka nhumbi
- Sanganai namai chamunorwa mabasa emaoko
- Sanganai namai chamunorwa mapatani okuruka
- Sanganai namai chamunorwa maresipi okubika
- Sanganai namai chamunorwa musha unofadza
- Sanganai namai chamunorwa utsanana pamusha
- Sanganai namai chamunorwa zvipfeko azinoyevedza zvevana

Corollarium cantionum sacrarum quinque...et plurium vocum – 1590 – (= ser Mssa) – 9mf – 9 – €105.00 – mfchl 141 – gw Fischer [780]

Corollarium missarum sacrarum / Regnart, Jacob – 1603 – (= ser Mssa) – 8mf – 9 – €100.00 – mfchl 387 – gw Fischer [780]

Coromandel and mercury bay gazette – Paeroa, NZ. jul 1956-jun 1973 – 9r – 1 – mf#16.10 – nz Nat Libr [079]

Corominas, Enrique Ventura *see* Puerto rico libre

Corona a la memoria de rafael heliodoro valle / Romero De Valle, Emilia – Mexico City? Mexico. 1963 – 1r – us UF Libraries [972]

Corona Baratech, Carlos E *see* Hernando cortes

[Corona-] daily independent (corona/norco) – CA. 1969- – 129r – 1 – $7740.00 (subs $340/y) – mf#R02143 – us Library Micro [071]

Corona de la inmaculada / Corredor Garcia, Antonio – Caceres: Tip. El Noticiero, SL, 1954 – 1 – sp Bibl Santa Ana [946]

Corona de la vida / Ovalle Lopez, Werner – Guatemala, 1962 – 1r – us UF Libraries [972]

[Corona del mar-] careers today – CA. 1969 – 1r – 1 – $60.00 – mf#R02144 – us Library Micro [331]

Corona di sacre canzoni : o, laude spirituali di piu divoti autori – In Firenze: da Cesare Bindi 1710 [mf ed 1990] – 1r – 1 – (incl ind) – mf#pres. film 85 – us Sibley [780]

Corona e palma militaire di artiglieria et fortificatione / Capo-Bianco, A – Venetia, 1647 – 3mf – 9 – mf#OA-255 – ne IDC [720]

Corona funebre : a la memoria de la virtuosa senora maria cristina rojas de herdocia / Bayle, Constantino – Madrid: Razon y Fe, 1926 – 1 – sp Bibl Santa Ana [920]

Corona funebre / Reyes Monroy, Jose Luis – Guatemala, 1963 – 1r – us UF Libraries [972]

Corona per la vittoria del sereniss don gio d'avstria / Gualtieri, F – Venetia, 1572 – 1mf – 9 – mf#H-8329 – ne IDC [956]

Corona poetica / Asociacion Santa Eulalia. Spain – 1875 – 9 – sp Bibl Santa Ana [810]

Corona poetica de santa eulalia, natural y patrona de la ciudad de merida – 1875 – 9 – sp Bibl Santa Ana [240]

Corona quernea [mgh schriften:6.bd] : festgabe fuer karl strecker – 1941 – (= ser Monumenta germaniae historica. schriften [mgh schriften]) – €23.00 – ne Slangenburg [931]

Coronaca baptist church. greenwood county. south carolina : church records – 1943-72. Historical statistics. 1881-1942 – 1 – us Southern Baptist [242]

Coronacion canoniga de la santisima virgen de la victoria – Trujillo: Tip. Sobrino de B. Pena, 1953 – sp Bibl Santa Ana [240]

Coronacion de la senora dona gertrudis gomez de av... / Liceo Artistico Y Literario (Havana, Cuba) – Habana, Cuba. 1860 – 1r – us UF Libraries [972]

Coronacion de la virgen de chiguinguira / Mesanza, Andres – Madrid: Razon y Fe, 1935 – 1 – sp Bibl Santa Ana [240]

La coronacion de la virgen de guadalupe / Bayle, Constantino – Madrid: Razon y Fe, 1928 – 9 – sp Bibl Santa Ana [972]

Coronacion de nuestra senora de consolacion del castillo de montanchez – Caceres: Tip. El Noticiero, s.a., (1950) ? – 1 – sp Bibl Santa Ana [240]

Coronado Aguilar, Manuel *see*
- Curso de derecho procesivo penal
- Retazos de la vida

Coronado, Carolina *see*
- Ananles del tajo lisboa
- Camoens a calderon en el centenario de este
- Introduccion a la poesia de la senorita armino
- Novelas. jarilla
- Paquita adoracion. novelas originales
- Poesias
- Poesias completas
- Rueda de la desgracia. manuscrito de un conde
- La siega

[Coronado-] coronado evening mercury – CA. May 1887-Jul 1896 (incomplete) – 4r – 1 – $240.00 – mf#C02145 – us Library Micro [071]

[Coronado-] coronado journal compass – CA. sept 1948-dec 1952 – 2r – 1 – $120.00 – mf#C02146 – us Library Micro [073]

Coronado P, K Adrian *see* Monografia del departamento de sacatepequez

Corona/norco – 1930-33; 1992- – (= ser California telephone directory coll) – 6r – 1 – $300.00 – mf#P00021 – us Library Micro [917]

Coronary artery disease among young adults under 50 in la crosse county / Gower, Elizabeth M – 1998 – 1mf – 9 – $4.00 – mf#HE 627 – us Kinesology [616]

Coronary heart disease risk factors in children ages 9 to 11 years / Brewer, Julia R & Wilson, Philip K – 1991 – 1mf – $4.00 – us Kinesology [616]

Coronation / Cameron, Mrs – London, England. 18-- – 1r – us UF Libraries [240]

Coronation book of charles 5 of france (hbs16) / Dewick, E S – 1899 – (= ser Henry bradshaw society (hbs)) – 4mf – 8 – €11.00 – ne Slangenburg [941]

The coronation book of charles 5th of france / Henry Bradshaw Society, London – London 1899 – (= ser 19th c art & architecture) – 3mf – 9 – mf#4.2.767 – uk Chadwyck [740]

The coronation book of oriental literature / ed by Shah, Ikbal Ali – London: Sampson Low, Marston & Co, [1937] – (= ser Samp: indian books) – us CRL [410]

Coronation of the queen / Legg, John Wickham – London, England. 1898 – 1r – us UF Libraries [941]

The coronation of the queen / Legg, John Wickham – London: SPCK 1898 [mf ed 1993] – (= ser Church historical society (series) 42) – 1mf – 9 – 0-524-05510-6 – mf#1990-1505 – us ATLA [941]

Coronation sermon / Roberts, George – Monmouth, England. 1838 – 1r – us UF Libraries [240]

Coronel, Ambrosio *see* Memorial ajustado de los actos acreedores...iglesia colegial y hospital de zafra

El coronel de caballeria don antonio del solar e ibanez / Solar y Taboada, Antonio – Badajoz: Florencio Ger Castro, 1916 – 1 – sp Bibl Santa Ana [350]

Coronel de Palma, Luis *see* Tres problemas y el futuro economico de espana

Coronel ordonez y cuba en 1851 / Rovira, Carlos A – Paris, France. 1867 – 1r – us UF Libraries [972]

Coroners guide and the coroners act, r. s. o. 1927, chapter 123, as amended by 1931, chapter 31. / Magone, Clifford Richard – 1st ed. Toronto: Bowman, 1936? 111p. LL-2317 – 1 – us L of C Photodup [340]

Coroner's records, 1888-1915 : custer county, colorado – [S.l. : s.n.], 1915 (mf ed 1951) – 1r – 1 – ISSN: 0 – mf#MF C967c – us Colorado Hist [978]

Coroner's society of england and wales reports – 2 bound vols. 1890-1903 (all publ) – 36mf – 9 – $54.00 – mf#LLMC 84-449 – us LLMC [340]

Coronet – Chicago. 1936-1961 – 1 – mf#892 – us UMI ProQuest [073]

Coronica moralizada del orden de san augustin en el peru : con sucesos egenplares en esta monarquia... / Calancha, Antonio de la – en Barcelona: Por Pedro Lacavalleria...ano 1639 – (= ser Books on religion...1543/44-c1800: ordenes, etc: agostinos) – 5mf – 9 – mf#crl-169 – ne IDC [241]

Corowa chronicle – Corowa, oct 1905-dec 1907 – 1r – A$68.64 vesicular A$74.14 silver – at Pascoe [079]

Corowa free press – Corowa, jan 1969-dec 1995 – at Pascoe [079]

Corpeno V, Roberto S *see* Asilo diplomatico

Corpo clip : le bulletin des bibliothecaires professionnels du quebec / Corporation des bibliothecaires professionnels du Quebec – Montreal: la Corporation. n84 mars/avril 1988- [mf ed 1989-] – 9 – (cont: bulletin argus) – mf#SEM105P1120 – cn Bibl Nat [020]

Corporate accounting – Boston. 1983-1988 (1,5,9) – (cont by: financial manager) – ISSN: 0745-5119 – mf#13452 – us UMI ProQuest [650]

Corporate advantages without incorporation. / Warren, Edward Henry – New York, Baker, Voorhis, 1929. 1012 p. LL-1642 – 1 – us L of C Photodup [346]

Corporate business taxation monthly – New York. 1999+ (1,5,9) – ISSN: 1528-5294 – mf#29967 – us UMI ProQuest [336]

Corporate cashflow – Atlanta. 1988-1995 (1,5,9) – (cont: cashflow) – ISSN: 1040-0311 – mf#15757,01 – us UMI ProQuest [332]

Corporate communications – Bradford. 2001+ (1,5,9) – ISSN: 1356-3289 – mf#31583 – us UMI ProQuest [380]

Corporate controller – Boston. 1997+ (1,5,9) – (cont: small business controller) – ISSN: 1092-1672 – mf#16636,02 – us UMI ProQuest [650]

Corporate co-optation of sport : the case of snowboarding / Crissey, Joy C – 1999 – 2mf – 9 – $8.00 – mf#PE 4067 – us Kinesology [650]

Corporate counsel's annual – 1966-85 (all publ) – 1,5,6 – $725.00 – mf#102101 – us Hein [340]

Corporate counsel's quarterly – v1-13. 1984-97 – 9 – $380.00 set – mf#111081 – us Hein [346]

Corporate design – Des Plaines. 1987-1987 (1,5,9) – (cont: corporate design and realty) – ISSN: 0894-3575 – mf#14869,02 – us UMI ProQuest [720]

Corporate design and realty – Boston. 1986-1986 (1,5,9) – (cont by: corporate design) – ISSN: 8750-8206 – mf#14869,01 – us UMI ProQuest [720]

Corporate financing – London. 1968-1973 (1) – ISSN: 0010-8960 – mf#9634 – us UMI ProQuest [332]

Corporate governance – Bradford. 2001+ (1,5,9) – ISSN: 1472-0701 – mf#31578 – us UMI ProQuest [650]

Corporate growth – Santa Barbara. 1988-1989 (1,5,9) – (cont: buyouts and acquisitions. cont by: corporate growth report) – ISSN: 0898-8390 – mf#14431,02 – us UMI ProQuest [332]

Corporate growth report – Santa Barbara. 1989-1992 (1,5,9) – (cont: corporate growth) – ISSN: 1050-320X – mf#14431,03 – us UMI ProQuest [332]

Corporate growth report weekly – Santa Barbara. 1992-1996 (1,5,9) – (cont by: weekly corporate growth report) – mf#14431,04 – us UMI ProQuest [332]

Corporate practice review – v1-4. 1928-32 (all publ) – 12mf – 9 – $54.00 – mf#LLMC 84-450 – us LLMC [346]

Corporate reorganization releases / U.S. Securities and Exchange Commission – n1-312. 17 aug 1938-12 jul 1972 (all publ) – (= ser The sec release series preceding the sec docket) – 35mf – 9 – $52.00 – mf#LLMC 84-360 – us LLMC [346]

Corporate reputation review – London, 1997+ [1,5,9] – ISSN: 1363-3589 – mf#31698 – us UMI ProQuest [650]

Corporate responsibility / Perowne, Edward Henry – Cambridge, England. 1862 – 1r – us UF Libraries [240]

Corporate security digest *see* Washington crime news services' corporate security digest

Corporate social-responsibility and environmental management – Chichester. 2002+ (1,5,9) – ISSN: 1535-3958 – mf#23178,01 – us UMI ProQuest [650]

Corporate sponsorship of women's sport / Thorp, Sarah – 1999 – 2mf – 9 – $8.00 – mf#PE 3950 – us Kinesology [650]

Corporate sponsorships in ncaa division 1 athletes with emphasis on men's basketball / Mistler, Michael D – 1997 – 1mf – 9 – $4.00 – mf#PE 3812 – us Kinesology [650]

Corporate taxation – New York, 2001+ [1,5,9] – (cont: journal of corporate taxation) – ISSN: 1534-715X – mf#10073,01 – us UMI ProQuest [336]

Corporate taxation – New York. 1990-1992 (1,5,9) – ISSN: 0898-798X – mf#18366 – us UMI ProQuest [336]

Corporate travel management – Toronto. 1998+ (1,5,9) – ISSN: 1481-594X – mf#33077 – us UMI ProQuest [650]

Corporation act books of the city of wells, 1377-1835 : from wells city council, the town hall, wells – 6r – 1 – mf#97522 – uk Microform Academic [941]

Corporation assessment lists, 1909-1915 / U.S. Internal Revenue Service – (= ser Records of the internal revenue service) – 82r – 1 – (with printed guide) – mf#M667 – us Nat Archives [336]

Corporation de quebec, aux entrepreneurs d'acqueducs : avis est par le present donne que des soumissions cachetees portant a l'endos les mots "soumission pour l'acqueduc de quebec"... / Baillairge, Charles P Florent – S.l: s.n, 1883? – 1mf – 9 – mf#54472 – cn Canadiana [350]

Corporation des Arpenteurs-Geometres de la Province de Quebec *see* Statuts et reglements du bureau de direction de la corporation...

Corporation des Bibliothecaires Professionnels du Quebec *see* Argus

Corporation des bibliothecaires professionnels du Quebec *see*
- Argus
- Argus journal
- Bulletin argus
- Bulletin de nouvelles
- Corpo clip

Corporation des pilotes pour le havre de Quebec et au-dessous *see* Reglements de la...
Corporation digest *see* Indiana law magazine
Corporation, finance and business law section journal *see* Michigan business law journal
Corporation journal – New York. 1908-1994 (1) 1973-1994 (5) 1973-1994 (9) – ISSN: 0045-8597 – mf#6951 – us UMI ProQuest [338]
The corporation journal – v1-11. 1913-35 (all publ) – 45mf – 9 – $67.50 – (lacking: v1) – mf#LLMC 84-451 – us LLMC [073]
Corporation law review – Boston. 1978-1986 (1,5,9) – ISSN: 0149-8827 – mf#11464 – us UMI ProQuest [346]
The corporation laws of 1883; being a supplement to the general corporation laws of pennsylvania / Freedley, Angelo Tillinghast – Philadelphia, 1883. 30p. LL-75 – 1 – us L of C Photodup [348]
The corporation laws of new jersey / New Jersey Corporation Guarantee and Trust Company. Camden, NJ – Camden, 1896 16p. LL-973 – 1 – (ibid. camden, nj., 1896. 119p. ll-228) – us L of C Photodup [348]
Corporation of kingston : sealed tenders will be received at the office of the clerk of the common council...for the laying down of flagging on the market square... – S.l: s.n, 1844? – 1mf – 9 – mf#38201 – cn Canadiana [690]
Corporation of liverpool. walker art gallery : the 25th autumn exhibition of pictures / Walker Art Gallery, Liverpool – [Liverpool] 1895 – (= ser 19th c art & architecture) – 2mf – 9 – mf#4.2.1697 – uk Chadwyck [700]
Corporation reporter *see* Indiana law magazine
Corporations in pennsylvania / Murphy, Walter – Philadelphia: Welsh, 1891. 2v. LL-663 – 1 – us L of C Photodup [346]
Corps Bukit Barisan *see* Berkala berita
Les corps gras industriels *see* Journal de l'exploitation des corps gras industriels
Corps intendans angkatan darat / Madjallah intendans – Djakarta, 1955-1957(6) – 13mf – 9 – (missing: 1955(1-3, 5, 9); 1956(5, 8-9, 11); 1957(3/4)) – mf#SE-1813 – ne IDC [950]
Corps kehakiman angkatan darat – Djakarta, 1962-1966 – 3mf – 9 – mf#SE-1378 – ne IDC [950]
Corpus antiphonalium officii / ed by Hesbert, R J – Roma. v1-2. 1963 – 2v on 43mf – 8 – €82.00 – ne Slangenburg [241]
Corpus apologetarum christianorum saeculi secundi / ed by Otto, J C Th – Ienae. v1-9. 1847-1872 – 62mf – 8 – €118.00 – ne Slangenburg [240]
Corpus byzantinae historiae (cbh) – Parisiis. 1(1648)-27(1711) – 734mf – 9 – €1400.00 coll – (individual vols also listed separately) – ne Slangenburg [240]
Corpus Christi College. Cambridge *see* Anglo-saxon and mediaeval manuscript collection
Corpus christi college, cambridge 1 : mss 41, 57, 191, 302, 303, 367, 383, 422 / ed by Graham, Timothy – [mf ed Binghamton NY, 2003] – (= ser ASMMF) – 1253 folios – 0-86698-308-2 – mf#mr265 – us MRTS [090]
[Corpus christi-] el paladia – TX. oct 1929-jan 1939 – 2r – 1 – $120.00 – mf#R05000 – us Library Micro [071]
Corpus christi gazette – Corpus Christi TX. 1846 feb 12 – 1r – 1 – mf#858547 – us WHS [071]
El corpus de los neofitos americanos / Bayle, Constantino – Madrid: Razon y Fe, 1944 – 1 – sp Bibl Santa Ana [972]
Corpus der altdeutschen originalurkunden bis zum jahr 1300 / ed by Wilhelm, Friedrich – Lahr/Baden: M Schauenburg 1932- [mf ed 1993] – 3r – 1 – (incl bibl ref & ind. middle high german texts & some latin texts, int in german) – mf#3369p – us Wisconsin U Libr [430]
Corpus documentorum inquisitionis haereticae / ed by Fredericq, P – Gent. v1-5. 1889-1902 – €136.00 – ne Slangenburg [240]
Corpus glossariorum latinorum / ed by Loewe, G – Lipsiae. v1-7. 1889-1901 – 84mf – 8 – mf#644 – ne IDC [450]
Corpus ignatianum : a complete collection of the ignatian epistles, geniune, interpolated, and spurious = Epistles / Ignatius, Saint, Bishop of Antioch – London: F. & J. Rivington, 1849 – 1r – 1 – mf#1984-B070 – us ATLA [240]
Corpus inscriptionum latinarum – Berolini. v1-15. 1869-1926 – 829mf – 8 – mf#121c – ne IDC [450]
Corpus inscriptionum latinarum / Huebner, Emilio – 2v 1869 – 9 – (suppl 2v 1892) – sp Bibl Santa Ana [410]
Corpus iuris canonici 1-2 / Friedberg, A – Leipzig, 1879 – 2v on 63mf – 8 – €120.00 – ne Slangenburg [240]
Corpus juris : being a complete and systematic statement of the whole body of the law as embodied in and developed by all reported decisions – New York: American Law Book Co. v1-71. 1914-35 – 41v on 585mf – 9 – $877.00 – (v42-71 filmed annually as they fall out of copyright between 2001 and 2011) – mf#llmc97-590 – us LLMC [340]
Corpus juris – St Paul: West Publishing Co. v1-72. 1989 – 9 – $2395.00 set – 0-89941-709-4 – (with annotation vols 1921-55) – mf#401690 – us Hein [340]
Das corpus juris civilis : in's deutsche uebersetzt = Corpus juris civilis, german / ed by Otto, Carl Eduard – Leipzig: C Focke. v1-7. 1831-39 – (= ser Civil law 3 coll) – 79mf – 9 – (vol 1-2, durchaus verb., von dr. sintenis besorgte, aufl. 1839. vol 2. dated 1831. incl bibl ref and indexes) – mf#LLMC 96-563 – us LLMC [348]
Corpus of prehistoric pottery and palettes / Petrie, W M – London, 1921 – 3mf – 9 – mf#NE-20371 – ne IDC [930]
Corpus poeticum de la obra de juan dieguez / Dieguez, Juan – Guatemala, 1959 – 1r – us UF Libraries [972]
Corpus reformatorum / ed by Bretschneider, C G – Halis, Saxonum, 1834-1925. v1-96 – 1401mf – 8 – mf#1176c – ne IDC [240]
Corpus scriptorum ecclesiasticorum latinorum – Wien. v1-68. 1866-1936 – 68v on 686mf – 8 – €1327.00 – (individual titles in this series not listed separately; apply to publ) – ne Slangenburg [220]
Corpus scriptorum ecclesiasticorum latinorum – Vindobonae. v1-70. 1866-1942 – 717mf – 8 – mf#790c – ne IDC [450]
Corpus scriptorum historiae byzantinae (cshb) / Niebuhrii, B G [comp] – Bonnae. v 1-50. 1828-1897 – €1109.00 coll – (individual vols also listed separately) – ne Slangenburg [240]
Corpus studiosorum bandungens : officieel orgaan van het bandoengse studenten corps en aangesloten verenigingen – Bandung, 1948/49-1954 – 12mf – 9 – (missing: 1949 v14(7-9); 1950/51 v16(1-end); 1952/53 v17(2-6); 1954 v19(3)) – mf#SE-364 – ne IDC [959]
Corpus tannaiticum : abt 3 halachische midraschim. teil 3 siphre zu deuteronomium / ed by Finkelstein, L – Breslau, 1935 – 12mf – 8 – €23.00 – ne Slangenburg [270]
Corpus theologiae christianae in quindecim locos digestum / Heidanus, A – Lugdunum Batavorum, 1687. 2v – 14mf – 9 – mf#PBA-186 – ne IDC [240]
Corpvs et syntagma confessionvm fidei – Geneuae, 1612 – 11mf – 9 – mf#PBU-695 – ne IDC [240]
Corpvs poetarvm latinorvm / Postgate, John Percival – London, England. v1-5. 1893-1905 – 1r – us UF Libraries [025]
Corpvscvlvm poesis epicae graecae lvdibvndae – Lipsiae, Germany. v1-2. 1888 – 1r – us UF Libraries [450]
Corral Acedo, Francisco et al *see* 1st congreso sindical agrario de extremadura 2. ponencia reginen de precios y mercados en la agricultura
Corral, Juan del *see* Nobleza de cataluna
Corrales Lazaro, Juan Luis *see* Memoria de las obras de restauracion y ampliacion de la iglesia parroquial de nuestra senora de los angeles de fuentes de leon. anos 1940-1943
Corrales Vicente, Manuel *see* Los problemas de la fabricacion del queso
Corraliza, Jose V *see*
- Hernando cortes
- Ideal de los conquistadores
- El puente de alcantara

Correa, Antonio Augusto Mendes *see* Cariocas e paulistas
Correa, Luiz De Miranda *see* Guia de manaus
Correa, Ramon C *see* Guia historico-geografica de los 126 municipios
Correa, Viriato *see*
- Balaiada
- Terra de santa cruz

Correal y Frente de Andrade, Narciso *see* El venerable barrantes
Correas, Gonzalo *see*
- Commentatio seu declaratio adillud geneseos
- Ortografia kastellana...el manual epikteto
- Trilingue de tres arts de las tres lenguas
- Trilingue de tres...castellana, latina y griega

Correccion del lenguaje... / Obando, Luis de – Madrid: Razon y Fe, 1940 – 1 – sp Bibl Santa Ana [946]
Correcciones al trillo inventado por don juan alvarez guerra, executadas por don juan francisco gutierrez / Alvarez Guerra, Juan – 1817 – 9 – sp Bibl Santa Ana [621]
Correct system of chanting, made easy / Fawcett, J – Bolton, England. 1841 – 1r – us UF Libraries [240]
The corrected english new testament : a revision of the "authorised" version (by nestle's resultant text) – New York: G P Putnam, 1905 – 2mf – 9 – 0-524-08206-5 – mf#1993-0001 – us ATLA [225]
Corrected report of the speech / Canning, George – London, England. 1825 – 1r – us UF Libraries [240]
Correctionnelle / Rougemont, Michel-Nicolas Balisson – Paris, France. 1840 – 1r – us UF Libraries [440]
Corrections digest – Washington. 1970+ (1,5,9) – ISSN: 0010-9045 – mf#10586 – us UMI ProQuest [360]
Corrections magazine – New York. 1974-1983 (1) 1974-1983 (5) 1974-1983 (9) – ISSN: 0095-4594 – mf#12447 – us UMI ProQuest [360]
The corrections of mark adopted by matthew and luke / Abbott, Edwin Abbott – London: Adam and Charles Black, 1901 – (= ser Diatessarica) – 1mf – 9 – 0-7905-1560-1 – (incl ind) – mf#1987-1560 – us ATLA [225]
Corrections perspective – St. Paul. 1975-1979 (1) 1975-1979 (5) 1975-1979 (9) – (cont by: perspective) – mf#10684 – us UMI ProQuest [360]
Corrections today – Lanham. 1989+ (1,5,9) – ISSN: 0190-2563 – mf#12976,03 – us UMI ProQuest [360]
Corrective and social psychiatry and journal of applied behavior – Atascadero. 1973-1973 (1) 1973-1973 (5) 1973-1973 (9) – (cont by: corrective and social psychiatry and journal of behavior technology methods and therapy) – ISSN: 0091-2611 – mf#12943,02 – us UMI ProQuest [616]
Corrective and social psychiatry and journal of behavior technology methods and therapy – Atascadero. 1974-1996 (1) 1974-1996 (5) 1974-1996 (9) – (cont: corrective and social psychiatry and journal of applied behavior) – ISSN: 0093-1551 – mf#12943,03 – us UMI ProQuest [616]
Corrective church discipline / Mell, Patrick H – Reprint from ed. of 1860 – 1 – 5.00 – us Southern Baptist [242]
Corrector – New York. 1804-1804 (1) – mf#3571 – us UMI ProQuest [320]
Corredor, Antonio *see*
- Devociones antonianas
- Dialogo con un turista. reportaje grafico-historico sobre san pedro de alcantara. madrid, 1969
- En el alcazar de la reina. antologia poetica guadalupense
- Leyendas marianas recopiladas...

Corredor Garcia, Antonio *see*
- Anecdotas misionales...
- Aqui la codosera
- Bordon de peregrino. poemas
- Corona de la inmaculada
- Eucaristicas
- Hispaniarum regine. poemas guadalupenses
- Lucia de fatima dice...
- Milagros de fatima
- Milagros eucaristicos y flores...
- Poemas marianos
- Santa clara de asis. antologia poetica
- Stabat mater dolorosa
- La virgen de las lagrimas de siracusa

Corredor Rodriguez, Ulpiano *see* Regimen legal de impuestos a las sucesiones, donac...
[Correggio] Affo, I *see* Ragionamento...sopra una stanza dipinta dal...a allegri da correggio nel monistero di s paolo in parma
Los corregidores y subdelegados del c. / Morales Grinazu, Fernando – Madrid: Razon y Fe, 1940 – 1 – sp Bibl Santa Ana [946]
Correia, Leoncio *see* A verdade historica sobre o 15 de novembro
Correio catharinense : jornal commercial, noticioso e litterario – Desterro, SC: Typ Catharinense, 15 dez 1852-22 nov 1854 – (= ser Ps 19) – mf#UFSC/BPESC – bl Biblioteca [073]
O correio caxiense – Caxias, MA: Typ Imparcial, 26 ago-set, nov-04 dez 1854 – (= ser Ps 19) – mf#P17,02,49 – bl Biblioteca [079]
Correio da assemblea provincal – Ceara, CE: Typ Patriotica, 19 jan 1839-14 out 1840 – (= ser Ps 19) – mf#P18B,03,18 – bl Biblioteca [350]
Correio da bahia – Bahia, 29-31 dez 1871; abr-jun,set 1872; jan, ago, out, dez 1873; fev, maio 1874; nov-dez 1876; jan-dez 1877; jan-18 set 1878 – (= ser Ps 19) – mf#P11,02,72 – bl Biblioteca [079]
Correio da manha : propriedade de uma associacao – Manaos, AM. 22 maio 1885 – (= ser Ps 19) – 1,5,6 – mf#P11B,06,19 – bl Biblioteca [079]
Correio da noite : jornal noticioso – Ouro Preto, MG. 01 jan-11 mar 1890 – (= ser Ps 19) – bl Biblioteca [079]
Correio da semana – Rio de Janeiro, RJ. 21 ago 1908 – (= ser Ps 19) – mf#DIPER – bl Biblioteca [079]
O correio da soledade – Pernambuco, 31 ago 1865 – (= ser Ps 19) – bl Biblioteca [079]
Correio da zambezia – Quelimane: Tip do Correio, dec 1-15 1886; jan 26-feb 10 1887 – us CRL [079]
Correio das modas : jornal critico e litterario das modas, bailes, theatros... – Rio de Janeiro, RJ: Typ de Laemmert, 05 jan-jun 1839; jul-31 dez 1840 – (= ser Ps 19) – mf#P12,05,12-15 – bl Biblioteca [073]
Correio de africa – Lisboa: "Correio de Africa", [may 22-oct 13 1921; oct 27 1921-aug 1 1923; sep 10-nov 25 1924 – us CRL [079]
Correio de blumenau – Blumenau, SC. 21 maio, out 1932; 10 jan 1933 – (= ser Ps 19) – mf#UFSC/BPESC – bl Biblioteca [079]
Correio de lages : orgam do partido republicano catharinense – Lajes, SC. 28 jun, ago 1924; fev 1925; abr 1926; jan 1927; 20 abr 1929 – (= ser Ps 19) – mf#UFSC/BPESC – bl Biblioteca [325]
O correio de lavras – Lavras, MG. 04 ago 1894 – (= ser Ps 19) – bl Biblioteca [079]
O correio de macau – Macau: Manuel Joaquim dos Santos, jan 7, feb 18-20 1883 – (filmed with: correio macaense) – us CRL [079]
Correio de mafra – Mafra, SC. 12 jan, 10 jun 1933 – (= ser Ps 19) – mf#UFSC/BPESC – bl Biblioteca [079]
Correio de manaos : orgao conservador – Manaus, AM. 07 set-dez 1869; 26 out 1881 – (= ser Ps 19) – mf#P11B,06,20 – bl Biblioteca [079]
Correio de manha – Rio de Janeiro Brazil, 19 nov 1939-1946; 9 oct 1954-3 feb 1956 – 21r – 1 – uk British Libr Newspaper [079]
Correio de minas – Juiz de Fora, MG. 01 jun 1894; maio-ago 1896; jan-out 1897; jul-dez 1898; 20 mar 1904 – (= ser Ps 19) – mf#P11B,03,67 – bl Biblioteca [079]
Correio de monte santo : orgam popular – Monte Santo, MG. 05 ago 1894 – (= ser Ps 19) – bl Biblioteca [079]
Correio de petropolis – Petropolis, RJ. 17 dez 1913-28 fev 1914 – (= ser Ps 19) – bl Biblioteca [079]
Correio de s jose : folha dedicada aos interesses sociais – Alem Paraiba, MG: [s.n.] 29 jun 1881; 22 jun 1884 – (= ser Ps 19) – mf#P11B,03,64 – bl Biblioteca [079]
Correio de valenca – Valenca, RJ: Typ do Correio de Valenca, 12 jan, out, 23 dez 1908; 01 jan 1909-03 nov 1910; 05 jan-28 dez 1911 – (= ser Ps 19) – mf#DIPER – bl Biblioteca [079]
Correio do assu : periodico politico, moral e noticioso – Assu, RN: Typ do Correio do Assu, 07-13 set 1873; set 1874; fev, jun 1875; 25 out 1876 – (= ser Ps 19) – mf#P11A,08,05 – bl Biblioteca [321]
Correio do assu *see* Correio do natal
Correio do madeira : orgao democrata – Manaus, AM: Typ Largo da Matriz, 20 set 1885; 24 jan 1890 – (= ser Ps 19) – mf#P11B,06,21 – bl Biblioteca [321]
Correio do natal : periodico politico, moral e noticioso – Natal, RN: Typ do Correio do Natal, 18-26 out, dez 1878, jun-nov 1879, jan, mar, maio, nov 1882, jan, set 1883, mar 1884, out 1885, fev-mar, maio 1886, mar-out, 14, 21 dez 1888 – (= ser Ps 19) – mf#P11A,08,07 – bl Biblioteca [321]
Correio do norte : jornal dedicado aos interesses da provincia do amazonas – Manaus, AM: Typ do Correio do Norte, 22 jun 1877 – (= ser Ps 19) – mf#P11B,06,22 – bl Biblioteca [073]
Correio do povo – Porto Alegre, Brazil. 1983 feb-may – 3r – (gaps) – us UF Libraries [079]
Correio do povo – Porto Alegre, Brazil. 1944-1984 (1) – mf#67646 – us UMI ProQuest [079]
Correio do sul : jornal independente e noticioso – Laguna, SC. 01 jan 1932; abr 1934; set 1937; jan, jul-15 set 1939 – (= ser Ps 19) – bl Biblioteca [079]
Correio iberico : organo de los interesses portugueses y espanoles en amerca – Rio de Janeiro, RJ: Typ de J F A Aranha, 05 abr-17 maio 1871 – (= ser Ps 19) – bl Biblioteca [079]
Correio luso-brasileiro – Rio de Janeiro, RJ. 04 maio 1882 – (= ser Ps 19) – mf#DIPER – bl Biblioteca [079]
O correio macaense – Macau: A de Silva Telles, nov 3 1885 – 1r – (filmed with: correio de macau) – us CRL [074]
Correio mercantil – Maceio, AL: [s.n.] 02 set 1894; fev, 28 abr 1895 – (= ser Ps 19) – mf#P18B,01,62 – bl Biblioteca [079]
Correio nacional *see* Opiniao liberal
Correio noticioso – Paraiba do Norte, PB: Typ de J Joaquim da Silva Braga, 17 ago 1872; nov-dez 1876; 16 fev 1877 – (= ser Ps 19) – mf#P11B,04,02 – bl Biblioteca [079]
O correio official : de santa catharina – Desterro, SC: Typ Catharinense, 07 set 1860-02 nov 1861 – (= ser Ps 19) – bl Biblioteca [079]
Correio official da provincia de sao pedro – Porto Alegre, RS: Typ de C Dulrevil & Comp, 07 jan, mar, maio-12 set 1835 – (= ser Ps 19) – mf#P17,02,195 – bl Biblioteca [321]
Correio operario norteamericano / AFL-CIO [American Federation of Labor and Congress of Industrial Organizations] – v15 n1-v17 n1 [1978 jan-1980 jan] – 1r – 1 – mf#641596 – us WHS [331]
Correio paulistano – Sao Paulo, Brazil. Jan-July 1908 – 1 – us CRL [079]

Correio sergipense : folha official, politica e litteraria – Aracaju, SE: Typ Provincial, 16 set, 30 dez 1840; 12 nov 1845; 20 jun 1846; 07 jul 1849; 23 ago 1854; 10 nov 1855; 20 jan 1858; 07 dez 1858; 28 jun 1862, em PR SOR 02881 [1]; fev-dez 1840; jan, jun-ago 1841; jan 1842 – dez 1843; mar-jun 1844; fev-maio, set-dez 1845; jun, set 1846; jan 1847-dez 1864; maio-jun, nov 1865; jan-27 jun 1866, em PR SOR 00144[1-8]; 19 fev 1859; 22 out 1864; 27 maio-21 jun, nov 1865, em PR SOR 03668 [1] – (= ser Ps 19) – mf#P25,02,16-21 – bl Biblioteca [321]
Correio universal : suplemento ilustrado de minas do sul – Campanha, MG. 19 fev 1892, jan-29 dez 1935 – (= ser Ps 19) – bl Biblioteca [079]
Correl smith diary *see* Smith, correl, diary, ms 3285
Correlation between muscle relaxation and sarcoplasmic reticulum ca2+-atpase during fatigue : an in-situ model / Biedermann, Michael C & Klug, Gary A – 1992 – 1mf – $4.00 – us Kinesology [612]
Correlation of abdominal accessory expiratory muscle strength and pulmonary functions in older adults / Gannon, Edward K – 1999 – 1mf – 9 – $4.00 – mf#PH 1649 – us Kinesology [612]
The correlation of federalism and liberalism in massachusetts, from 1775-1825 / Spring, Chadbourne Arnold – Chicago, 1935. Chicago: Dep of Photodup, U of Chicago Lib, 1971 (1r); Evanston: American Theol Lib Assoc, 1984 (1r) – 1 – 0-8370-0279-6 – mf#1984-B167 – us ATLA [240]
Correlation of laboratory tests to distance running performance during a cross-country track season / Mosenthal, Teese M – 1988 – 105p 2mf – 9 – $8.00 – us Kinesology [617]
The correlation of the pre-karroo succession in northern rhodesia with that of adjacent territories / Hays, J – Lusaka 1950 – us CRL [960]
The correlation of the vital and physical forces : a prize thesis...may 2, 1862 / Bucke, Richard Maurice – [Montreal?: s.n, 1862?] – 1mf – 9 – 0-665-92607-3 – (incl bibl ref) – mf#92607 – cn Canadiana [574]
Corren soendag – Linkoeping, Sweden. 2000-2001 – 2r – 1 – sw Kungliga [078]
Correns, P *see* Die dem boethius...zugeschriebene abhandlung des dominici gundisalvi de unitate
Correo argentino. extremadura – Buenos Aires, 1940. 1 numero – 5 – sp Bibl Santa Ana [073]
Correo de extremadura – Badajoz, 1891-1894 – 5 – sp Bibl Santa Ana [074]
Correo de la manana – Badajoz, 1915. 1 numero – 5 – sp Bibl Santa Ana [073]
El correo espanol – Madrid, Spain. 21 sep 1914-10 aug 1919 [daily] – 15r – 1 – (imperfect) – uk British Libr Newspaper [074]
Correo literario – Madrid. Spain. -w. 15 Mar, 1 Jun 1951-Mar 1955. (3 reels) – 1 – uk British Libr Newspaper [800]
Correo mexicano : el diario de la raza – Chicago, IL: Francisco Huerta, 4 sep 1926-4 mar 1927 – 1r – 1 – us CRL [071]
El correo placentino – Plasencia, 1901 – 5 – sp Bibl Santa Ana [073]
Correoso Y Miranda, Ricardo *see* Ricardo correoso y miranda, patriota, poeta, perio...
Correspondance / Bres, G de – 2 nov 1561; juil 1563; 10 juil 1565 – 1mf – 9 – mf#PBA-438 – ne IDC [240]
Correspondance – Paris. 1906, 1908-juin 1914, 1920-23, 1925-avr 1939 – 1 – (devenu: union pour la verite; correspondance puis bulletin) – fr ACRPP [074]
Correspondance agricole et politique – Paris. juin 1898-1921 – 1 – fr ACRPP [630]
Correspondance autographiee du bureau international de la paix – 1892-1940 – (= ser The library of world peace studies) – 85mf – 9 – $510.00 – (cont as: correspondance bi-mensuelle; mouvement pacifiste; peace movement) – us UPA [320]
Correspondance de don pedre premier – Paris, France. 1827 – 1r – us UF Libraries [972]
Correspondance de victor jacquemont avec sa famille et plusieurs de ses amis : pendant son voyage dans l'inde [1828-1832] / Jacquemont, Victor – Paris 1833 [mf ed Hildesheim 1995-98] – 2v on 6mf – 9 – €120.00 – 3-487-27426-4 – gw Olms [915]
Correspondance des reformateurs dans les pays de langue francaise / Herminjard, A L – Geneve, Bale, Lyon, Paris, 1866-1893. 9 v – 53mf – 9 – mf#ZWI-89 – ne IDC [242]
Correspondance diplomatique du comte pozzo di borgo : ambassadeur de russie en france, et du comte de nesselrode depuis la restauration...1814-18 / Pozzo di Borgo, Carlo Andrea, conte – 2nd ed. Paris: Calmann Levy 1890-97 [mf ed 1987] – 2v on 1r – 1 – (with: the art of invigorating and prolonging life / kitchiner, w) – mf#1821 – us Wisconsin U Libr [327]
Correspondance, documents, temoignages et procedes dans l'enquete de messrs lafrenaye et doherty... = Correspondence, documents, evidence and proceedings in the enquiry of messrs lafrenaye and doherty... – Montreal: impr de la Minerve, 1864 [mf ed 1983] – 2mf – 9 – (with ind) – mf#SEM105P321 – cn Bibl Nat [345]
Correspondance d'orient 1830-31 / Michaud, Joseph – Paris 1833-35 [mf ed Hildesheim 1995-98] – 7v on 23mf – 9 – €230.00 – 3-487-27694-1 – gw Olms [915]
Correspondance du cardinal de bernis, ministre d'etat : avec m paris-du-verney, conseiller d'etat, depuis 1752 jusqu'en 1769 precedee d'une notice historique – Londres [u a] 1790 [mf ed Hildesheim 1995-98] – 2v on 4mf – 9 – €120.00 – 3-487-25842-0 – gw Olms [920]
Correspondance du duc d'enghein (1801-1804) et documentes – Paris, France. t1-4. 1904 – 3r – us UF Libraries [025]
Correspondance entre le gouvernement francais et les gouverneurs et intendants du canada : relative a la tenure seigneuriale, demandee par une adresse de l'assemblee legislative, 1851 = Correspondence between the french government and the governors and intendants of canada, relative to the seigniorial tenure... – Quebec: impr de E R Frechette, 1853 [mf ed 1999] – 2mf – 9 – mf#SEM105P3203 – cn Bibl Nat [971]
[Correspondance entre pietro et alexandre verri pendant les annees 1774 et 1775] / Verri, Pietro; ed by Bissoulet, J – [Paris? 1938] – 1 – mf#*ZBD-*MGO pv16 – Located: NYPL – us Misc Inst [790]
Correspondance et memoires d'un voyageur en orient / Bore, E – Paris, 1840. 2v – 11mf – 9 – mf#AR-1649 – ne IDC [915]
Correspondance havas – Paris. 2 nov-dec 1852 – 1 – fr ACRPP [073]
Correspondance inedite de henri 4, roi de france et de navarre : avec maurice-le-savant, landgrave de hesse – Paris 1840 [mf ed Hildesheim 1995-98] – 1v on 3mf – 9 – €90.00 – 3-487-26106-5 – gw Olms [920]
Correspondance inedite de marie antoinette / Marie Antoinette – Paris [mf ed Hildesheim 1995-98] – 2v on 3mf – 9 – €90.00 – 3-487-26187-1 – gw Olms [920]
Correspondance inedite officielle et confidentielle de napoleon bonaparte : tome deuxieme – Paris 1809 [i e 1819]-1820 [mf ed Hildesheim 1995-98] – 7v on 24mf – 9 – €240.00 – 3-487-26249-5 – gw Olms [920]
La correspondance internationale – Berlin, 1921-23; Wien, 1923-26; Paris, 1926-39 – (= ser Communist international periodicals from the feltrinelli archives) – 274mf – 9 – $1240.00 – us UPA [335]
La correspondance internationale – Berlin puis Vienne, Paris. oct 1921-aout 1939 – 1 – fr ACRPP [073]
Correspondance internationale ouvriere – Paris puis Nimes. sept 1932-avr 1933 – 1 – fr ACRPP [073]
Correspondance litteraire, philosophique et critique par grimm, diderot, raynal, meister, etc / ed by Tourneux, Maurice – Paris. v1-16. 1877-82 – 1 – $216.00 – mf#0590 – us Brook [410]
Correspondance maritime de nantes – Nantes. 1782 – 1 – fr ACRPP [073]
Correspondance on church and religion : selected and arranged by d c lathbury / Gladstone, William Ewart – London, 1910 – 8 – (v1 9mf. v2 9mf) – ne Slangenburg [240]
Correspondance originale des emigres : ou les emigres peints par eux-memes(cette correspondance, deposee aux archives de la convention nationale... – Paris 1793 [mf ed Hildesheim 1995-98] – 1v on 3mf [ill] – 9 – €90.00 – 3-487-26270-3 – gw Olms [920]
Correspondance politique de paris et des departemens – no. 216-315. Paris. janv-avr 1794 (II) – 1 – fr ACRPP [320]
Correspondance politique des veritables amis du roi et de la patrie – Paris. janv-aout 1792 – 1 – (devenu: nouvelle correspondance politique, pour servir de suite aux 52 premiers numeros de la correspondance) – fr ACRPP [073]
La correspondance regionaliste – Paris. n1-4. mars-aout 1901 – 1 – fr ACRPP [073]
Correspondance relative a la reunion des hopitaux d'avila. textes en prose inedits publies...par georges demerson / Melendez Valdes, Juan – Bordeaux: Feret and Fils, 1964 – 1 – sp Bibl Santa Ana [944]
Correspondance secrete de la courpendant le regne de louis 16 : ci-devant roi des francois – Paris 1793 [mf ed Hildesheim 1995-98] – 3v on 12mf – 9 – €120.00 – 3-487-26196-0 – gw Olms [920]
Correspondance socialiste internationale : Le mensuel de l'internationalisme militant – n1-106. Paris. 1951-avr mai 1960. – 1 – (lacking: n1-4, 77, 80, 88-89) – fr ACRPP [325]
Correspondance syndicale internationale *see* L'internationale syndicale rouge
Correspondance universelle – Paris. n4817-4885. 2 janv-11 mars 1884 – 1 – fr ACRPP [073]
Correspondances parlementaires de l'echo de levis – Levis, Quebec?: s.n, 1875 – 1mf – 9 – mf#24057 – cn Canadiana [325]
Correspondant – London, UK. 8 Aug 1873 – 1 – uk British Libr Newspaper [072]
Le correspondant – Paris. mars 1829-fevr 1831 – 1 – fr ACRPP [073]
Le correspondant – Paris. oct 1855-sept 1870 – 1 – fr ACRPP [073]
Le correspondant – Paris. v1-333. 1843-oct 1933 – 1 – (religion, philosophie, politique, histoire, sciences, economie sociale, voyages, litterature, beaux-arts) – us NY Public [073]
Le correspondant *see* Revue europeenne
Le correspondant des departements – Paris: Maulde et Renou, apr 30 1850 – us CRL [944]
Le correspondant. recueil periodique *see* Le correspondant
Correspondence / Arundel, John T – 1897-1912 – 1r – 1 – mf#pmb493 – at Pacific Mss [338]
Correspondence / Atchison, Topeka and Santa Fe Railroad Company – 1873-93 – 1 – us Kansas [025]
Correspondence / Baptist State Conventions. (Southern Baptist). Virginia – 1790-1820 – 1 – 5.00 – us Southern Baptist [242]
Correspondence / Cary, Orland R – 1946-53. 600p – 1 – us Southern Baptist [242]
Correspondence / Collocott, E E V – 1921-59 – 1r – 1 – mf#pmb28 – at Pacific Mss [920]
Correspondence / Columbian College – Baptist Records in Library of Congress, 1822-36. 144p – 1 – 5.04 – us Southern Baptist [242]
Correspondence / Denison, George Anthony – London, England. 1853 – 1r – us UF Libraries [240]
Correspondence / Furman, Richard – 1755-1825. 1,232p – 1 – us Southern Baptist [242]
Correspondence / Graton, John R – 1838-1910 – 1 – us Kansas [920]
Correspondence / Green, John – 1892-1896 – 3r – 1 – mf#pmb420 – at Pacific Mss [980]
Correspondence / Hill, Kate Alexander – 1896-1936 – 5 – $50.00 – us Presbyterian [920]
Correspondence / Jackson, Sheldon – 1856-1908. Calendars – 1 – $350.00 – us Presbyterian [978]
Correspondence / Jackson, Sheldon – 1856-1908. Orig – 1 – $800.00 – us Presbyterian [240]
Correspondence / Jackson, Sheldon – 1856-1908. Transcripts – 1 – $250.00 – us Presbyterian [240]
Correspondence / Koehler, Robert – 1888-1927. 1 roll including filmed inventory – 1 – $30.00r – us Minn Hist [920]
Correspondence / Metcalfe, John R – 1919-1968 – 1r – 1 – mf#pmb82 – at Pacific Mss [242]
Correspondence / Moore, Charles – 1924-1927 – 1r – 1 – mf#pmb85 – at Pacific Mss [242]
Correspondence / Presbyterian Church in the U.S.A. Board of Foreign Missions. Furrukhabad Mission – 1839-64 – 1 – $50.00 – us Presbyterian [240]
Correspondence, 16 october 1873-15 october 1878 / Fison, Lorimer – 1r – 1 – (material and supply restricted) – mf#PMB1041 – at Pacific Mss [980]
Correspondence 1621-79 : from the forster collection, victoria & albert museum / Boyle, Roger – 1r – 1 – mf#96644 – uk Microform Academic [920]
Correspondence, 1766-1820 : from the royal botanical gardens, kew / Banks, Joseph – 3v – 4r – 1 – mf#96785 – uk Microform Academic [920]
Correspondence, 1829-92 / Manly Family – 5158p – 1 – us Southern Baptist [242]
Correspondence 1830-1865 / South India (Madras) Mission. Church Missionary Society – London, Kodak Ltd, Recordak Division, 1967 – us CRL [240]
Correspondence, 1830-1865 / Church Missionary Society. South India Mission. Madras – London: Kodak Ltd, Recordak Div, 1967 (mf ed) – 1 – us CRL [240]
[Correspondence] 1864-1888 / Swan, James Gilchrist – 1r – 1 – us UW Libraries [305]
Correspondence, 1890-1942 / South Sea Evangelical Mission (formerly Queensland Kanaka Mission) – 5r – 1 – (available for reference) – mf#pmb1150 – at Pacific Mss [242]
Correspondence 1908-1930 / Evangelical Lutheran Joint Synod of Ohio and Other States. Board of Foreign Missions – [mf ed 2004] – 1r – 1 – (comprises handwritten & typewritten correspondence for officers) – mf#a0089r – us ATLA [242]
Correspondence, 1910-1924 / Gardiner, Robert Hallowell – [S.l.: s.n.], 1910-1924 – 9r – 1 – 0-8370-0022-X – mf#1984-S075 – us ATLA [920]
Correspondence 1914-1918 / Baksh, Ahmad – New Delhi, National Archives of India, 1969 – (filmed with: national archives of india: selections from the home political files jul 1914-1916, 1918 and the army department files 1914-may 1919) – us CRL [954]
Correspondence, 1943-1947 / Bengal Relief Committee – [s.l: s.n, –] – 1 – us CRL [954]
Correspondence, 1964-67 / BEST (Baptist Education Study Task) – 1103p – 5 – us Southern Baptist [242]
Correspondence addressed to the right revd dr lewis, bishop of ontario, and others, (through the ottawa press) : on the subject of the romanizing and ritualistic practices and teaching of the church of england, and the reformed episcopal church, now firmly established in ottawa, and moncton, new brunswick – [Ottawa: s.n.], 1874 [mf ed 1984] – 1mf – 9 – 0-665-46797-4 – mf#46797 – cn Canadiana [242]
Correspondence and diary of ba campaign / James, John Hall – 1870-1875 – 1r – 1 – mf#pmb125 – at Pacific Mss [630]
The correspondence and literary manuscripts of arthur hugh clough *see* Nineteenth century literary manuscripts
Correspondence and literary manuscripts of margaret oliphant *see* Oliphant
Correspondence and other documents relating to the position of the... – London, England. 1871 – 1r – us UF Libraries [240]
Correspondence and other papers related to their service with the methodist overseas mission / Simpson, Thomas Nevison & Simpson, Nellie – New Hanover, PNG: 1936-42 – 1r – 1 – (restricted access) – mf#PMB1114 – at Pacific Mss [920]
Correspondence and other papers related to their service with the methodist overseas mission, new hanover, png / Simpson, Thomas Nevison et al – 1936-1942 – 1r – 1 – mf#pmb1114 – at Pacific Mss [242]
Correspondence and other papers relating to the family of seymour – 16th-17th c – (= ser Archives of the marquess of bath, longleat house, warminster, wiltshire) – 19r – 1 – mf#96702 – uk Microform Academic [929]
Correspondence and papers / Hyatt, Thaddeus – 1858-1901 – 1 – (papers. 1843-98. 1) – us Kansas [920]
Correspondence and papers : on the formation of png workers association and the pangu pati / Kiki, Albert Maori – 1948-76 – 1r – 1 – mf#PMB1119 – at Pacific Mss [331]
Correspondence and papers, 1825-56 : from the co-operative union library, manchester / Owen, Robert – 1r – 1 – (int by peter d a jones) – mf#95915 – uk Microform Academic [970]
Correspondence and papers of david nelson / Nelson, David – 1835-1857 – 1 – $50.00 – us Presbyterian [240]
Correspondence and papers of dr whitney, general o'ryan and other members of the economic and trade mission *see* Japan and america, c1930-1955 – the pacific war and the occupation of japan, series 2
The correspondence and papers of john gibson lockhart *see* Nineteenth century literary manuscripts
Correspondence and papers of sir ernest satow *see* China through western eyes
Correspondence and record cards of the military intelligence division relating to general, political, and military conditions in central america, 1918-1941 / U.S. War Dept. Military Intelligence Division – (= ser Records of the war department general and special staffs) – 12r – 1 – (with printed guide) – mf#M1488 – us Nat Archives [355]
Correspondence and record cards of the military intelligence division relating to general, political, economic, and military conditions in cuba and the west indies, 1918-1941 / U.S. War Dept. Military Intelligence Division – (= ser Records of the war department general and special staffs) – 10r – 1 – (with printed guide) – mf#M1507 – us Nat Archives [355]
Correspondence and record cards of the military intelligence division relating to general, political, economic, and military conditions in italy, 1918-1941 / U.S. War Dept. Military Intelligence Division – (= ser Records of the war department general and special staffs) – 26r – 1 – (with printed guide) – mf#M1446 – us Nat Archives [355]
Correspondence and record cards of the military intelligence division relating to general, political, economic, and military conditions in poland and the baltic states, 1918-1941 / U.S. War Dept. Military Intelligence Division – (= ser Records of the war department general and special staffs) – 10r – 1 – (with printed guide) – mf#M1508 – us Nat Archives [355]
Correspondence and records, 1944-55 / Kiamichi Baptist Assembly. Oklahoma – 649p – 1 – us Southern Baptist [242]

CORRESPONDENCE

The correspondence and records of smith, elder and co *see* Nineteenth century literary manuscripts

Correspondence and reports / Presbyterian Church in the U.S.A. Board of Aid for Colleges and Academies. Henry Kendall College-University of Tulsa – 1892-1938 – 1 – $100.00 – us Presbyterian [378]

Correspondence and reports / Presbyterian Church in the U.S.A. Board of Foreign Missions. Korea, Independence movement – 1919-20 – 1 – $50.00 – us Presbyterian [951]

Correspondence and reports / Presbyterian Church in the U.S.A. Board of Foreign Missions. Korea, Japanese Colonial Government, Religious education controversy – 1915-19 – 1 – $50.00 – us Presbyterian [951]

Correspondence and reports of the confederate treasury department, 1861-1865 / U.S. War Dept. Confederate Records – (= ser War department coll of confederate records) – 2r – 1 – mf#T1025 – us Nat Archives [324]

Correspondence and reports on missions to the chinese in california / Loomis, Augustus Ward – 1863-1873 – 1 – $50.00 – us Presbyterian [240]

Correspondence between a hindu raja, a reverend father and a member of parliament on the policy of lord ripon / Rajendra Narayana Bahadur, raja – Calcutta 1884 – (= ser 19th c british colonization) – 2mf – 9 – (repr fr the indian mirror. int by sham loll mitter) – mf#1.1.1087 – uk Chadwyck [954]

Correspondence between donald moodie..and the rev. john philip...relative to the production for publication of alleged "official authority" : for the statement that in the year 1774 the whole race of bushmen or hottentots who had not submitted to servitude was ordered to be seized or extirpated / Moodie, Donald – Cape Town. 1841 – 1 – us CRL [960]

Correspondence between goethe and carlyle – London, New York: Macmillan, 1887 [mf ed 1999] – xix/362p – 1 – mf#10129 – us Wisconsin U Libr [860]

Correspondence between john quincy adams, esquire, president of the united states, and several citizens of massachusettes : concerning the charge of a design to dissolve the union alleged to have existed in that state – 2nd ed. Boston: Press of The Boston Daily Advertiser, 1829 [mf ed 1984] – 1mf – 9 – 0-665-44360-9 – mf#44360 – cn Canadiana [323]

Correspondence between mr jonas and mr cobden : repr from the morning herald of jan 17th and feb 13th, 1850 / Cobden, Richard – [s.l: s.n, 1850?] [mf ed 1984] – 1mf – 9 – 0-665-45205-5 – mf#45205 – cn Canadiana [380]

Correspondence between pliny the consul and the emperor trajan, respecting the early christians *see* The rise and early progress of christianity

Correspondence between sir henry mcmahon, his majesty's high commissioner at cairo : and the sherif hussein of mecca, july 1915-march 1916 – London, 1939 – 1mf – 9 – (incl map) – mf#J-28-89 – ne IDC [956]

Correspondence between the bishop of aberdeen and the clergy and church / Suther, Thomas G – Aberdeen, Scotland. 1864 – 1r – us UF Libraries [242]

The correspondence between the committee on church unity of the general assembly of the presbyterian church in the u.s.a. and the commission on christian unity of the general convention of the protestant episcopal church in the u.s – Philadelphia: Stated Clerk, 1896 – 1mf – 9 – 0-524-06656-6 – mf#1991-2711 – us ATLA [242]

Correspondence between the company and the dominion government respecting advances to the canadian pacific railway company / Chemin de fer du grand tronc – [S.l: s.n, 1884?] [mf ed 1980] – 1mf – 9 – 0-665-00115-0 – mf#00115 – cn Canadiana [380]

Correspondence between the conference committees of the presbyterian general assemblies (north and south) / Presbyterian Church in the USA. Committee of Conference – St Louis: Democrat Lithographing and Printing Co, 1875 – 1mf – 9 – 0-524-05551-3 – mf#1990-5155 – us ATLA [242]

Correspondence between the french government and the governors and intendants of canada : relative to the seigniorial tenure, required by an address of the legislative assembly, 1851 – Quebec: printed by E R Frechette, 1853 [mf ed 2000] – 2mf – 9 – mf#SEM105P3204 – cn Bibl Nat [971]

Correspondence between the general assembly of the presbyterian church in the united states of america and the general assembly of the presbyterian church in the united states : commonly known as the northern and southern general assemblies / Presbyterian Church in the USA. General Assembly – Brooklyn: Daily Union Job Print Establishment, 1870 – 1mf – 9 – 0-524-06880-1 – mf#1990-5299 – us ATLA [242]

Correspondence between the most rev dr machale...and most rev dr murray / MacHale, John – Dublin, 1885 – (= ser 19th c ireland) – 1mf – 9 – mf#1.1.409 – uk Chadwyck [941]

Correspondence between the most rev the metropolitan i e ashton oxenden : and the rev the rector of the parish of montreal i e maurice scollard baldwin – Montreal?: J Lovell, 1874 – 1mf – 9 – mf#23917 – cn Canadiana [242]

Correspondence between the rev dr james kidd of the church of scotland – Aberdeen, Scotland. 1830 – 1r – us UF Libraries [242]

Correspondence between the right rev bishop gleig / Craig, Edward – Edinburgh, Scotland. 1820 – 1r – us UF Libraries [242]

Correspondence between the right rev c h terrot, bishop of the s... – Edinburgh, Scotland. 1842 – 1r – us UF Libraries [242]

Correspondence between the roman catholic bishop of toronto and the chief superintendent of schools : on the subject of separate common schools, in upper canada – Toronto: printed & publ by Thomas Hugh Bentley, 1853 [mf ed 1983] – 1mf – 9 – mf#SEM105P332 – cn Bibl Nat [377]

Correspondence between the roman catholic bishop of toronto and the chief superintendent of schools : on the subject of separate common schools in upper canada – Toronto: T H Bentley, 1853 – 1mf – 9 – (with app) – mf#22432 – cn Canadiana [370]

Correspondence between the secretary of state for the colonies and the governors of canada and mr w b felton : relative to lands granted to the said w b felton – [London, England: s.n, 1836] (mf ed 1992) – 1mf – 9 – mf#SEM105P1395 – cn Bibl Nat [971]

Correspondence celebrating 20 years with harper brothers / Grey, Zane – 1r – 1 – mf#B25937 – us Ohio Hist [790]

Correspondence concerning lac qui parle mission, dakota and sioux indians / Riggs, Stephen R – 1837-51 – 1 – $50.00 – us Presbyterian [240]

Correspondence de fernand cortes avec 1 empereur charles quint. sur la conquete de mexique / Cortes, Hernando – Tradiste por M le vi comte de Flavigny. Suisse, Libraires Associes, 1779 – sp Bibl Santa Ana [350]

Correspondence de napoleon 1 : publiee par ordre de l'empereur... – v1-32. 1858-70 – 1 – $360.00 – (in french) – mf#0387 – us Brook [940]

Correspondence des premiers missionaires : annales des missions de l'oceanie – 1837-55 – 1r – 1 – mf#pmb doc180 – at Pacific Mss [240]

Correspondence, documents, evidence and proceedings in the enquiry of messrs lafrenaye and doherty... : followed by the remarks of messrs delisle and schiller... / Canada (Province) – Montreal: M Longmoore & Co, 1864 [mf ed 1983] – 2mf – 9 – (with ind) – mf#SEM105P322 – cn Bibl Nat [345]

Correspondence etc : warden's letter to civil secretary, containing complaints against the district treasurer – [s.l: s.n. 1845?] [mf ed 1987] – 1mf – 9 – 0-665-55472-9 – mf#55472 – cn Canadiana [350]

Correspondence etc, between colonel mackenzie fraser, major magrath and mr maitland : on matters connected with june race meeting, 1839 – [Toronto?: s.n.] 1840 [mf ed 1983] – 1mf – 9 – 0-665-44547-4 – mf#44547 – cn Canadiana [790]

Correspondence, etc, relating to the montreal, ottawa and georgian bay canal – s.l: s.n, 1897? – 1mf – 9 – mf#00249 – cn Canadiana [380]

Correspondence files, 1892-1904 / J T Arundel and Co et al – r1-8 – 1 – (presscopy letter-books of outward letters to business associates mainly in the uk) – mf#pmb1174 – at Pacific Mss [338]

Correspondence files, 1896-1908 / Pacific Islands Co Ltd et al – r1-15 – 1 – (incl general correspondence, shipping details, telegrams, machinery details, financial details) – mf#pmb1175 – at Pacific Mss [338]

Correspondence files of corresponding secretaries, baptist sunday school board / Frost, J M & Bell, T P – 1889-1916. 71,273p – 1 – us Southern Baptist [242]

Correspondence for the introduction of cochineal insects from america... / Anderson, J – Madras, 1791 – 1mf – 9 – mf#HT-1156 – ne IDC [590]

Correspondence from lewis henry morgan and some others,1870-81 / Fison, Lorimer – 1r – (permission to quote required) – mf#PMB1043 – at Pacific Mss [980]

Correspondence from the solomon islands and vanuatu / Leishman, Helen – 1930-1948 – 1r – 1 – mf#pmb1244 – at Pacific Mss [242]

Correspondence in connection with the protest against the consecration of rev w j boone as missionary bishop of the protestant episcopal church of america in china : also letters referring to the wretched management of the mission / McKeige, Ferdinand – Shanghai: [s.n.] 1885 [mf ed 1995] – (= ser Yale coll) – vi/91p – 1 – 0-524-10073-X – mf#1995-1073 – us ATLA [242]

Correspondence inedite de la comtesse de sabran et du chevalier de boufflers, 1778-88 / Sabran, Francoise Eleonore – 2e ed. Paris: E Plon, 1875 – 1 – us CRL [960]

Correspondence, journal, notes / Roman Catholic Mission, New Hebrides – 1897-1926 – 1r – 1 – mf#pmb61 – at Pacific Mss [241]

Correspondence, minutes, circulars, clippings, etc / Congress of Democrats – Johannesburg, Microfile, [19–?] – 1 – us CRL [960]

Correspondence occasioned by the refusal of thomas wilson esq to a... – London, England. 1812 – 1r – us UF Libraries [242]

Correspondence of a.d. bache, superintendent of coast and geodetic survey, 1843-1865 / U.S. Coast and Geodetic Survey – (= ser Records Of The Coast And Geodetic Survey) – 281r – 1 – (with printed guide) – mf#M642 – us Nat Archives [550]

Correspondence of bishop c j nicolas, sm / Roman Catholic Mission, Fiji – 1919-1930 – 1r – 1 – mf#pmb438 – at Pacific Mss [241]

Correspondence of bishop julian vidal, sm : with father c j nicolas, sm, and other priests / Roman Catholic Mission, Fiji – 1893-1920 – 1r – 1 – mf#pmb445 – at Pacific Mss [241]

Correspondence of daniel o'connell, the liberator / O'Connell, Daniel; ed by Fitzpatrick, William John – London: J Murray 1888 [mf ed 1988] – 2v on 1r – 1 – (with notices of his life & times. with: list of vertebrated animals) – mf#9767 – us Wisconsin U Libr [920]

Correspondence of fraeulein guenderode and bettine von arnim – Boston: T O H P Burnham, 1861 [mf ed 1991] – xii/344p – 1 – mf#7532 – us Wisconsin U Libr [880]

Correspondence of gerrit smith with albert barnes, 1868 – New York: American News Co [dist] [1868?] [mf ed 1991] – 1mf – 9 – 0-7905-8587-1 – mf#1989-1812 – us ATLA [210]

The correspondence of hans sloane : from the british library, sloane mss. 4036-4069 – 33v – 16r – 1 – (with ind) – mf#95961 – uk Microform Academic [920]

The correspondence of isaac basire : archdeacon of northumberland and prebendary of durham in the reigns of charles 1. and charles 2 / Basier, Isaac – London: John Murray, 1831 – 1mf – 9 – 0-7905-4128-9 – (in english) – mf#1988-0128 – us ATLA [920]

The correspondence of john flamsteed : from the royal society, london – 1r – 1 – mf#96886 – uk Microform Academic [920]

Correspondence of john henry hobart / Hobart, John Henry – New York: priv print, 1911-12 [mf ed 1993] – (= ser Archives of the general convention 1-6; Anglican/episcopal coll) – 6v on 38mf – 9 – 0-524-07431-3 – (incl bibl ref) – mf#1991-3091 – us ATLA [242]

The correspondence of john ray / ed by Lankester, Edwin – 1848 – (= ser Ray society publication 14) – 1r – 1 – mf#1776 – uk Microform Academic [920]

The correspondence of joseph priestley *see* Miscellaneous papers

Correspondence of lieut-general the hon sir george cathcart – New York, NY. 1969 – 1r – us UF Libraries [960]

Correspondence of major tod, war of 1812 : history of northfield – [s.l: s.n. 1873?] [mf ed 1987] – 1mf – 9 – 0-665-58982-4 – mf#58982 – cn Canadiana [975]

Correspondence of marc-michel rey, 1747-1778 : publisher of the enlightenment – 11mf – 9 – €160.00 – (with p/g & concordance) – mf#m479 – ne MMF Publ [070]

The correspondence of marc-michel rey, 1747-1778 : publisher of the enlightenment / Bibliotheek van de Koninklijke Vereeniging ter bevordering van de belangen des Boekhandels – [mf ed 2001] – 11mf – 9 – €160.00 – (with printed guide) – mf#M479 – ne MMF Publ [070]

Correspondence of ralph w brown *see* Minutes

The correspondence of richard baxter : 1615-91 – (mf ed 1999) – 3r – 1 – £150.00 – mf#DWM – uk World [240]

Correspondence of robert h gardiner *see* Minutes

The correspondence of samuel richardson, 1748-62 : from the forster collection, victoria & albert museum, ref. 457 – 2r – 1 – mf#96654 – uk Microform Academic [920]

Correspondence of secretary of state bryan with president wilson, 1913-1915 / U.S. Dept of State – (= ser General records of the department of state) – 4r – 1 – mf#T841 – us Nat Archives [977]

Correspondence of sir robert peel, 1841-46 : from the royal archives and library at windsor castle – 6r – 1 – mf#95709 – uk Microform Academic [920]

Correspondence of the archbishop of canterbury and the bishop of ex... / Sumner, John Bird – London, England. 1850 – 1r – us UF Libraries [241]

Correspondence of the coke family : letters from sir edward coke, sir walter raleigh, lord clarendon and others – 1595-1750 – (= ser Holkham library family and political papers 746) – 1r – 1 – (suppl to coke family papers, codex 727) – mf#96787 – uk Microform Academic [920]

Correspondence of the eastern division pertaining to cherokee removal, april-december 1838 / U.S. Army. Continental Commands – (= ser Records of united states army continental commands, 1821-1920) – 2r – 1 – (with printed guide) – mf#M1475 – us Nat Archives [355]

Correspondence of the mennonite woman's missionary society – n1-10. 1919-21 [complete] – (= ser Mennonite serials coll) – 1r – 1 – mf#ATLA 1994-S038 – us ATLA [242]

Correspondence of the military intelligence division correspondence relating to "negro subversion", 1917-1941 / U.S. War Dept. Military Intelligence Division – (= ser Records of the war department general and special staffs) – 6r – 1 – (with printed guide) – mf#M1440 – us Nat Archives [355]

Correspondence of the military intelligence division relating to general, political, and military conditions in scandanavia and finland, 1918-1941 / U.S. War Dept. Military Intelligence Division – (= ser Records of the war department general and special staffs) – 12r – 1 – (with printed guide) – mf#M1497 – us Nat Archives [355]

Correspondence of the military intelligence division relating to general, political, economic, and military conditions in china, 1918-1941 – 19 rolls – 9 – $437.00 – Dist. us Scholarly Res – us L of C Photodup [355]

Correspondence of the military intelligence division relating to general, political, economic, and military conditions in china, 1918-1941 / U.S. War Dept. Military Intelligence Division – (= ser Records of the war department general and special staffs) – 19r – 1 – (with printed guide) – mf#M1444 – us Nat Archives [355]

Correspondence of the military intelligence division relating to general, political, economic, and military conditions in japan, 1918-1941 / U.S. War Dept. Military Intelligence Division – (= ser Records of the war department general and special staffs) – 31r – 1 – (with printed guide) – mf#M1216 – us Nat Archives [355]

Correspondence of the military intelligence division relating to general, political, economic, and military conditions in russia and the soviet union, 1918-1941 / U.S. War Dept. Military Intelligence Division – (= ser Records of the war department general and special staffs) – 23r – 1 – (with printed guide) – mf#M1443 – us Nat Archives [355]

Correspondence of the military intelligence division relating to general, political, economic, and military conditions in spain, 1918-1941 / U.S. War Dept. Military Intelligence Division – (= ser Records of the war department general and special staffs) – 12r – 1 – (with printed guide) – mf#M1445 – us Nat Archives [355]

Correspondence of the office of civil afairs of the district of texas, the 5th military district, and the department of texas, 1867-1870 / U.S. Office of Civil Affairs – (= ser Records of united states army continental commands, 1821-1920) – 40r – 1 – (with printed guide) – mf#M1188 – us Nat Archives [350]

Correspondence of the reverend ezra fisher : pioneer missionary of the american baptist home mission society in indiana, illinois, iowa and oregon / Fisher, Ezra; ed by Henderson, Sarah Fisher et al – [Portland, Or?: s.n., 1919?] – 6mf – 9 – 0-524-07417-8 – mf#1991-3077 – us ATLA [975]

The correspondence of the right hon john beresford : illustrative of the last thirty years of the irish parliament / ed by Beresford, William – London, 1854 [mf ed 1986] – 2v – 1 – (notes by ed) – mf#8500 – us Wisconsin U Libr [941]

Correspondence of the secretary of alaska, 1900-1913 / Alaska. Office of the Secretary – 20r – 1 – mf#T1201 – us Nat Archives [324]

Correspondence of the secretary of the navy relating to african colonization, 1819-1844 / U.S. Navy – (= ser Records of the office of the secretary of the navy) – 2r – 1 – (with printed guide) – mf#M205 – us Nat Archives [355]

Correspondence of the secretary of the treasury with collectors of customs, 1789-1833 / U.S. Treasury Dept. Office of the Secretary – (= ser General records of the department of the treasury) – 39r – 1 – (with printed guide) – mf#M178 – us Nat Archives [324]

Correspondence of the surveyors general of utah, 1874-1916 / U.S. Bureau of Land Management – (= ser Records of the bureau of land management) – 86r – 1 – (with printed guide) – mf#M1110 – us Nat Archives [324]

Correspondence of the us mint at philadelphia with the branch mint at dahlonega, georgia, 1835-1861 / U.S. Bureau of the Mint – (= ser Records of the united states mint) – 3r – 1 – mf#T646 – us Nat Archives [332]

Correspondence of the u.s. naval astronomical expedition to the southern hemisphere, 1846-1861 / U.S. Naval Observatory – (= ser Records Of The U.S. Naval Observatory) – 1r – 1 – mf#T54 – us Nat Archives [520]

Correspondence of the war department relating to indian affairs, military pensions, and fortifications, 1791-1797 / U.S. War Dept – (= ser Records of the office of the secretary of war) – 1r – 1 – (with printed guide) – mf#M1062 – us Nat Archives [355]

Correspondence of thomas carlyle and janet welsh – 1r – 1 – mf#96812 – uk Microform Academic [920]

The correspondence of thomas carlyle and ralph waldo emerson, 1834-1872 = Correspondence. Selections / Carlyle, Thomas & Emerson, Ralph Waldo; ed by Norton, Charles Eliot – Boston: James R Osgood, 1883 – 2mf – 9 – 0-524-01717-4 – mf#1990-4109 – us ATLA [420]

Correspondence of william pitt, earl of chatham – London, England. v1-4. 1838-1840 – 1r – us UF Libraries [025]

Correspondence on church and religion of william ewart gladstone / Gladstone, William Ewart – New York: Macmillan, 1910 – 3mf – 9 – 0-7905-4651-5 – mf#1988-0651 – us ATLA [240]

Correspondence on infallability : between a father jesuit and general alexander kireeff, an eastern orthodox = Zur unfehlbarkeit des papstes / Kireev, Alexsandr Alekseevich – New York: [s.n.], 1896 – 1mf – 9 – 0-8370-8096-7 – mf#1986-2096 – us ATLA [241]

Correspondence re his books the lost caravel and the lost caravel re-explored / Langdon, Robert – 1986-1998 – 1r – 1 – mf#pmb1231 – at Pacific Mss [070]

Correspondence re leper asylum, makogai : and sundry related papers / Roman Catholic Mission, Fiji – 1908-1930 – 2r – 1 – mf#pmb448 – at Pacific Mss [241]

Correspondence re mission work among the indians of fiji / Roman Catholic Mission, Fiji – 1909-1919 – 1r – 1 – mf#pmb462 – at Pacific Mss [241]

Correspondence received by the surveyors general of new mexico, 1854-1907 / U.S. Bureau of Land Management – (= ser Records of the bureau of land management) – 11r – 1 – mf#M1288 – us Nat Archives [324]

Correspondence relating to father emmanual rougier, sm / Roman Catholic Mission, Fiji – 1906-1907 – 1r – 1 – mf#pmb441 – at Pacific Mss [241]

Correspondence relating to the civil list and military expenditure in canada : and to the projected railway from halifax to quebec / Canada (Province). Gouverneur general – London: Harrison & Son, 1851 [mf ed 1982] – 2mf – 9 – mf#SEM105P123 – cn Bibl Nat [355]

Correspondence relating to the eastern boundary of the province : from lieut-colonel d r cameron, royal artillery, to the right honourable the earl of derby...secretary of state for the colonies – Victoria, BC?: R Wolfenden, 1885? – 1mf – 9 – mf#29882 – cn Canadiana [971]

Correspondence relating to the enforcement of the "passenger acts," 1852-1857 / U.S. Dept of Justice – (= ser General records of the department of justice) – 1r – 1 – mf#M2010 – us Nat Archives [340]

Correspondence relating to the filibustering expedition against the spanish government of mexico, 1811-1816 / U.S. Dept of State – (= ser General records of the department of state) – 1r – 1 – mf#T286 – us Nat Archives [324]

Correspondence relating to the inter-colonial railway : laid before the legislature by command of his excellency the lieutenant governor; in continuation of correspondence laid before the legislature in 1863 – Fredericton [NB]: G E Fenety, 1864 [mf ed 1983] – 1mf – 9 – 0-665-44297-1 – mf#44297 – cn Canadiana [380]

Correspondence relative to a meeting at quebec of delegates appointed to discuss the proposed union of the british north american provinces – London: printed by George Edward Eyre & William Spottiswoode, 1865 [mf ed 2000] – 1mf – 9 – mf#SEM105P3225 – cn Bibl Nat [971]

Correspondence relative to the accounts of the indian department in canada west : together with the reports of the different accountants / Jarvis, Samuel Peters – Montreal: printed by Rollo Campbell, 1847 [mf ed 1996] – 1mf – 9 – mf#SEM105P2763 – cn Bibl Nat [971]

Correspondence relative to the affairs of canada / Grande-Bretagne. Parliament. House of Commons – London, [England]: printed by William Clowes & Sons. 4v. 1840 [mf ed 1982] – 7mf – 9 – mf#SEM105P107 – cn Bibl Nat [323]

Correspondence relative to the affairs of canada, 1841 / Grande-Bretagne. Parliament – London: printed by William Clowes & Sons, 1841 [mf ed 1990] – 1mf – 9 – mf#SEM105P1144 – cn Bibl Nat [323]

Correspondence relative to the affairs of canada, 1846 / Grande-Bretagne. Parliament – London [England]: printed by William Clowes & Sons, 1847 [mf ed 1982] – 1mf – 9 – mf#SEM105P119 – cn Bibl Nat [323]

Correspondence relative to the prospects of christianity : and the means of promoting its reception in india – Cambridge: UP – Hilliard & Metcalf, 1824 [mf ed 1995] – (= ser Yale coll) – 138p – 1 – 0-524-10247-3 – (bound with: an appeal to liberal christians for the cause of christianity in india [boston, 1825]) – mf#1996-1247 – us ATLA [240]

Correspondence relative to the recent disturbances in the red river settlement : presented to both houses of parliament by command of her majesty, aug 1870 – London: printed by W Clowes for HMSO, 1870 [mf ed 1984] – 3mf – 9 – 0-665-30620-2 – mf#30620 – cn Canadiana [971]

Correspondence relative to the refusal of sites for churches, manse... – Edinburgh, Scotland. 1846 – 1r – us UF Libraries [242]

Correspondence, reports / Presbyterian Church in the U.S.A. Board of Foreign Missions. Korea, Conspiracy Case, Japanese Colonial government – 1912-20 – 1 – $50.00 – us Presbyterian [951]

Correspondence respecting h m s "resolute," and the arctic expedition – [London?: s.n, 1858?] [mf ed 1983] – 1mf – 9 – 0-665-44035-9 – mf#44035 – cn Canadiana [917]

Correspondence respecting the dismissal of mr p m partridge, superintendent of woods and forests : by the honble alex campbell, commissioner of crown lands – Quebec: s.n, 1867 – 1mf – 9 – mf#23477 – cn Canadiana [346]

Correspondence respecting the proposed union of the british north american provinces : (in continuation of papers presented 7th february 1865) – London: George Edward Eyre & William Spottishwoode, 1867 [mf ed 1993] – 2mf – 9 – mf#SEM105P1759 – cn Bibl Nat [323]

Correspondence respecting the spiritual condition of catholics in t... / Collingridge, Ignatius – London, England. 1861 – 1r – us UF Libraries [241]

Correspondence respecting the turco-egyptian frontier in the sinai peninsula – London, 1906 – 2mf – 9 – mf#J-28-62 – ne IDC [956]

The correspondence series and speeches series of the personal papers of john foster dulles (1888-1959) / Challener, Richard – 1994 – 67r – 1 – $5,360.00 – (includes guide) – mf#D3300 – us L of C Photodup [320]

Correspondence ("top secret") of the manhattan engineer district, 1942-1946 / U.S. War Dept. Office of the Chief of Engineers – (= ser Records of the office of the chief of engineers) – 5r – 1 – (with printed guide) – mf#M1109 – us Nat Archives [355]

Correspondence with french consulate-general, sydney / Roman Catholic Mission, Fiji – 1919, 1927-1929 – 1r – 1 – mf#pmb433 – at Pacific Mss [241]

Correspondence with french high commissioner, noumea : and with politicians in paris / Catholic Mission, Wallis Island – 1936-1966 – 2r – 1 – mf#pmb962 – at Pacific Mss [241]

Correspondence with german administration, samoa / London Missionary Society – Samoan District – 1905-1915 – 1r – 1 – mf#pmb143 – at Pacific Mss [242]

Correspondence with government / Roman Catholic Mission, Fiji – 1856-1890 1999-1900 – 1r – 1 – mf#pmb434 – at Pacific Mss [241]

Correspondence with government / Roman Catholic Mission, Fiji – 1901-13 1916-30 [pmb435], 1891-1898 [pmb436] – 2r – 1 – mf#pmb435-436 – at Pacific Mss [241]

Correspondence with government re education / Roman Catholic Mission, Fiji – 1899-1912, 1920-1936 – 1r – 1 – mf#pmb432 – at Pacific Mss [241]

Correspondence with his grace the duke of newcastle, the hudson's bay company, and the delegates from canada : (with other documents) in reference to the establishment of overland passenger and telegraphic communication between the atlantic and british columbia and the pacific – [London?: s.n, 1863?] [mf ed 1983] – 1mf – 9 – 0-665-44034-0 – mf#44034 – cn Canadiana [380]

Correspondence with his sister, eliza thurston : and related family papers and photographs / Thurston, John Bates – 1843-1937 – r1-2 – 1 – (available for reference) – mf#pmb1142 – at Pacific Mss [920]

Correspondence with hy. s.l. polak, 1917-1929?, charles roberts, 1917-1919, and william wedderburn, 1904?-1918 / Natesan, Ganapati Agraharam – New Delhi: Nehru Memorial Museum & Lib, [19–?] – 1 – (filmed with: masani, r p: correspondence with reed stanley and drafts of two articles... correspondence with roberts and wedderburn filmed later on the reel) – us CRL [950]

Correspondence with kaisar bagh 1914-1917 / Mahmud, Syed – New Delhi, Nehru Memorial Museum and Library, [19–?] – (filmed with: masani, r p: correspondence...) – us CRL [920]

Correspondence with lms agent, suva / London Missionary Society – Samoan District – 1907-1946 – 1r – 1 – mf#pmb142 – at Pacific Mss [242]

Correspondence with lms stations in the pacific islands / London Missionary Society – Samoan District – 1877-1947 – 2r – 1 – mf#pmb141 – at Pacific Mss [242]

Correspondence with methodist mission office, sydney / Methodist Church Of New Zealand, Foreign Missions Department – 1934-1952 [pmb949], 1919-1954 [pmb950] – 2r – 1 – mf#pmb949-950 – at Pacific Mss [242]

Correspondence with mission stations at savusavu and tunuloa / Roman Catholic Mission, Fiji – 1919-1930 – 1r – 1 – mf#pmb456 – at Pacific Mss [241]

Correspondence with missionaries at rotuma / Roman Catholic Mission, Fiji – 1868-1888 – 1r – 1 – mf#pmb160 – at Pacific Mss [241]

Correspondence with new zealand administration, samoa / London Missionary Society – Samoan District – 1915-1946 – 1r – 1 – mf#pmb144 – at Pacific Mss [242]

Correspondence with other religious bodies : including overseas auxiliaries of the lms / London Missionary Society – Samoan District – 1908-1946 – 1r – 1 – mf#pmb128 – at Pacific Mss [242]

[Correspondence with reed stanley and drafts of two articles, 1915-1933?] / Masani, Rustom Pestonji – New Delhi: Nehru Memorial Museum & Lib, [19–?] – 1 – us CRL [950]

Correspondence with the government, 1926-1931 and with dr clifford james on clothes, 1931 / Binet, Vincent le Cornu – 1926-1931 – (= ser James, Clifford S) – 1r – 1 – mf#PMB1108 – at Pacific Mss [360]

Correspondence with the missions of the board of foreign missions of the presbyterian church in the u.s.a. regarding the distinct missionary responsibility of the presbyterian church – New York City: Board of Foreign Missions of the Presbyterian Church in the USA, [1907?] – 1mf – 9 – 0-524-07637-5 – mf#1991-3244 – us ATLA [242]

Correspondence with the palestine arab delegation and the zionist organisation – London, 1922 – 1mf – 9 – mf#J-28-67 – ne IDC [956]

Correspondence with the united states / Great Britain Foreign Office – London, England. 1856 – 1r – us UF Libraries [327]

Correspondence with various mission stations / Roman Catholic Mission, Fiji – 1912-1930 [pmb457]; 1920-1930 [pmb461] – 2r – 1 – mf#pmb457, 461 – at Pacific Mss [241]

Correspondence with william henry jackson / Taft, Robert – undated – 1 – us Kansas [920]

Correspondence...concerning issac stevens' survey of a northern route for the pacific railroad, 1853-1861 / U.S. Office of Explorations and Surveys – (= ser Records of the office of the secretary of the interior) – 1r – 1 – mf#M126 – us Nat Archives [380]

Correspondence...relating to the administration of trust funds for the chickasaw and other tribes ("s" series), 1834-1872 / U.S. Treasury Dept. Office of the Secretary – (= ser General records of the department of the treasury) – 1r – 1 – (with printed guide) – mf#M749 – us Nat Archives [336]

Correspondencia / Figueiredo, Jackson De – Rio de Janeiro, Brazil. 1946 – 1r – us UF Libraries [972]

Correspondencia con juan alcaide sanchez / Carrasco, Castulo – Caceres: Imprenta Moderna, 1955 – 1 – sp Bibl Santa Ana [920]

La correspondencia de espana – Madrid, Spain. -d. 7 Oct 1914-26 May 1916; 8 Sept 1916-9 Aug 1919. Imperfect. 20 reels – 1 – uk British Libr Newspaper [074]

Correspondencia del doctor benito arias montano con el licenciado juan de ovando / Jimenez de la Espada, Marcos – Madrid: Fortanet, 1891. B.R.A.H. 19, pp. 476-498 – sp Bibl Santa Ana [946]

Correspondencia diplomatica cambiada entre el gobierno de los estados unidos mexicanos y los de varias potencias extranjeras / Mexico. Secretaria de Relaciones Exteriores – Mexico. 1882-92. 6v – 1 – $46.00 – us L of C Photodup [972]

Correspondencia diplomatica cruzada entre la... / Cuba Secretaria De Estado – Habana, Cuba. 1940 – 1r – us UF Libraries [972]

Correspondencia diplomatica de la delegacion cuban / Partido Revolucionario Cubano – Habana, Cuba. v1-5. 1943-1948 – 1r – us UF Libraries [972]

La correspondencia diplomatica entre los duques de parma y sus agentes o ambajadores en la corte de madrid, durante los siglos xvi, xvii y xviii / Perez Bustamante, Ciriaco – Madrid. 1934 – 1 – us CRL [940]

Correspondencia do conselheiro manuel p de souza – Rio de Janeiro, Brazil. 1962 – 1r – us UF Libraries [972]

Correspondencia entre d pedro ii e o barao do rio / Pedro 2 – Sao Paulo, Brazil. 1957 – 1r – us UF Libraries [972]

Correspondencia entre la nunciatura... / Olarra y Garmendia, Jose de et al – Madrid: Archivo Ibero Americano, 1965 – 1 – sp Bibl Santa Ana [946]

Correspondencia entre los obispos de mallorca y la – Minorca, Spain. no date – 1r – us UF Libraries [324]

Correspondencia epistolar del p. andres marcos burriel, existente en la biblioteca real de bruselas / Reymondez del Campo, Jesus – Madrid: Est. Tip. Fortanet, 1908. B.R.A.H. Tomo 52, 1908, pp. 273-286 – 1 – sp Bibl Santa Ana [946]

Correspondencia y diario militar. 1810-1814. tomo 3 / Iturbide, Agustin – Mexico, 1930; Madrid: Razon y Fe, 1932 – 1 – sp Bibl Santa Ana [355]

Correspondencies of faith and views of madame guyon : being a devout study of the unifying power and place of faith in the theology and church of the future / Cheever, Henry Theodore – London: Elliot Stock, 1887 [mf ed 1985] – 1mf – 9 – 0-8370-3214-8 – (incl bibl ref) – mf#1985-1214 – us ATLA [240]

Correspondent – 1983 jan-1987 dec – 1r – 1 – mf#1110952 – us WHS [071]

Correspondent – Dublin, Ireland. 12, 17, 29 sep, 1 oct, 5, 24 nov 1807, 31 may 1813, 1823, 13 mar-13 nov 1824, 4 jan, 26 feb, 15, 22 oct, 24 nov, 31 dec 1825 – 1 3/4r – 1 – (aka: dublin correspondent) – uk British Libr Newspaper [072]

Correspondent – Dublin 1806-22 – 32r – 1 – ie National [072]

Correspondent – New York. 1827-1829 (1) – mf#3968 – us UMI ProQuest [070]

Correspondent / New York State American Revolution Bicentennial Commission – v1-v8 n5 [1970 sum-1978 spr] – 1r – 1 – mf#462826 – us WHS [975]

Correspondent / Scioto Co. Portsmouth – jan 1894-dec 1908 [wkly] – 7r – 1 – (in german) – mf#B9960-9966 – us Ohio Hist [071]

Correspondent and advocate – Toronto, ON. 1833-37 – 2r – 1 – cn Library Assoc [071]

Correspondent von und fuer schlesien – Liegnitz (Legnica PL), 1816-18, 1824, 1833 jan-jun, 1835 – 1 – gw Misc Inst [074]

Correspondentie met de raad van indie over het politiek verslag / Politiek verslag 1852, 1 – 6mf – 8 – mf#SD-100 mf 8-13 – ne IDC [950]

Correspondenzblatt der afrikanischen gesellschaft : 1. bd, berlin 1873-1876; 2. bd. dresden 1877-1878 / Deutsche Gesellschaft zur Erforschung Aequatorialafrikas – [mf ed 1994] – 5mf – 9 – €70 diazo €84 silver – 3-89131-175-3 – gw Fischer [960]

Correspondenzblatt der general-kommission der gewerkschaften deutschlands – Berlin DE, 1891-1923 – 12r – 1 – (filmed by misc inst: 1901-19) – mf#1283 – gw Mikropress; gw Misc Inst [331]

Correspondenzblatt des kreises eupen – Eupen [B], 1854 4 jan-1868 [gaps], 1900 12 may-1907, 1920 3 jan-1921 29 aug – 1 – (aka: korrespondenzblatt des kreises eupen) – gw Misc Inst [074]

Correspondenz-blatt und kieler wochenblatt *see* Kieler correspondenzblatt fuer die herzogthuemer schleswig, holstein und lauenburg

Correspondiente de la real academia de la historia / Prado, Eladio del – Madrid: Razon y Fe, 1926 – 1 – sp Bibl Santa Ana [946]

Correspondiente en caceres de la r.a. de la historia / Acedo, Federico – Madrid: Fortanet, 1919. B.R.A.H. 74. p. 101 – 1 – sp Bibl Santa Ana [946]

Correspondiente en merida de la r.a. de la historia / Gonzalez y Gomez de Soto, Juan Jose – Madrid: Ed. Reus, 1922. B.R.A.H. 80. p. 93 – 1 – sp Bibl Santa Ana [946]
Un corresponsal extranjero y unas teorias / Bayle, Constantino – Madrid: Razon y Fe, 1947 – 1 – sp Bibl Santa Ana [946]
Corretjer, Juan Antonio *see*
– Alabanza en la torre de ciales
– Distancias
– Don diego en el carino
– Genio y figura
– Llorens
– Lucha por la independencia de puerto rico
– Yerba bruja
O corretor de petas – Rio de Janeiro, RJ: Imprensa Americana de I P da Costa, 09 nov 1841 – (= ser Ps 19) – mf#P12,05,22 n02 – bl Biblioteca [079]
Corrette, Michel *see*
– L'art de se perfectionner dans le violon
– Le maitre de clavecin pour l'accompagnement, methode theorique et pratique, qui conduit en tres peu de tems a accompagner a livre ouvert
– Methode pour apprendre aisement a jouer de la flute traversiere
– Methode theorique et pratique
Correyero, M *see* 1x2 las quinielas al alcance de todos
Corrido at buhay na pinagdaanan nang dalauang magkapatid na si juan at si maria sa reinong espana – Manila: P Sayo [191-?] [mf ed Bloomington IN: Indiana Uni Lib, Preservation Dept 1984] – (= ser Coll...in the tagalog language 1) – 1r – 1 – us Indiana Preservation [490]
Corrido at buhay na pinagdaanan nang princesa florentina sa cahariang alemania – Maynila: P Sayo 1919 [mf ed Bloomington IN: Indiana Uni Lib, Preservation Dept 1984] – (= ser Coll...in the tagalog language 1) – 1r – 1 – us Indiana Preservation [490]
Corrido at buhay na pinagdaanan nang principe orontis at nang reina talestris sa caharian nang temesita – [Maynila: P Sayo 191-?] [mf ed Bloomington IN: Indiana Uni Lib, Preservation Dept 1984] – (= ser Coll...in the tagalog language 2) – 1r – 1 – (aka: principe orontis at nang reina talestris sa caharian nang temesita) – us Indiana Preservation [490]
Corrido at pinagdaanang buhay nang principe baldovino sa kaharian nang dacia at nang princesang si sevilla sa reino nang sansuena – [Maynila: P Sayo 191-?] [mf ed Bloomington IN: Indiana Uni Lib, Preservation Dept 1984] – (= ser Coll...in the tagalog language 1) – 1r – 1 – us Indiana Preservation [490]
Corrido qng bienang delanan dona marcela ampon ning metung a mercader qng cayarian portugal – Manila: Quiapo 1903 [mf ed Bloomington IN: Indiana Uni Lib, Preservation Dept 1984] – (= ser Coll of philippine literature in the pampanga language) – 1r – 1 – (aka: dona marcela) – us Indiana Preservation [490]
Corrido qng bienang delanang don juan de berbana ila ning asauanang y dona maria blanca anac ning aring salermo qng cayarian cristalino – Manila: M Reyes [190-?] [mf ed Bloomington IN: Indiana Uni Lib, Preservation Dept 1984] – 1r – 1 – us Indiana Preservation [490]
Corrido qng bierang [sic] delanan ning conde irlos ila ning condesang asauana qng cayarian francia – Manila: Santa Cruz de A Nam 1902 [mf ed Bloomington IN: Indiana Uni Lib, Preservation Dept 1984] – (= ser Coll of philippine literature in the pampanga language) – 1r – 1 – (aka: conde irlos) – us Indiana Preservation [490]
Corrie, D *see* Memoirs of the rev daniel corrie, first bishop of madras
Corrieo portugues – Canada. jan 1963-dec 1977 – 15r – 1 – (in portuguese) – cn Commonwealth Imaging [071]
Il corriere – Salerno. Italy. apr. 20-June 20, 1944 – 1 – us NY Public [074]
Corriere canadese – Canada. jun 1954-99 – 112r – 1 – (in italian) – cn Commonwealth Imaging [071]
Corriere d' italia – Omaha, NE: A R Rizzuto & Co, 1911 [mf ed anno2 n26. 28 giugno 1913 filmed 1997] – 1r – 1 – us NE Hist [071]
Corriere d'america – New York, NY: Tiber Pub Corp, [1923-1934; 1935, jan 1-jul 2, dec 15,22,29; 1936-apr 18 1937 (sunday issues only] – 156r – us CRL [071]
Corriere del niagara – Canada. jan 1960-dec 1968 – 6r – 1 – (in italian) – cn Commonwealth Imaging [071]
Il corriere del popolo – San Francisco, CA. 1916-1962 (1) – mf#62272 – us UMI ProQuest [071]
Il corriere del popolo – San Francisco, CA: Pedretti Bros, jan 13 1919-dec 30 1926 – 1 – us CRL [071]
Corriere del quebec – Canada, jan 1960-dec 1968 – 6r – 1 – (in italian) – cn Commonwealth Imaging [071]
Corriere del sabato – London, UK. 19 Jun 1943-23 Sept 1944; 16 Dec 1944-21 Jul 1945 – 1 – uk British Libr Newspaper [072]
Il corriere della dalmazia – Zadar, Yugoslavia. Jul 1919-Feb 1920 – 1r – 1 – us L of C Photodup [949]
Corriere della sera – Milan: [s.n, 1919-26; 1934-35; jul 1938-apr 25 1945] – 1 – us CRL [074]
Corriere della sera – Milano: [Filli Crespi & C, may 10 1959-1985 – us CRL [074]
Corriere della sera – Milano, 1876- – 1 – (yrly reel count varies. ind 1901-95 available 1r per yr. newspaper also available on cd-rom. inquire about 16mm film) – us UMI ProQuest [074]
Corriere della somalia = Somalia courier – Mogadiscio. Somali Republic. -d. Jan 1947-Mar 1950. (4 reels) – 1 – uk British Libr Newspaper [079]
Il corriere di chicago – Chicago IL, 1907, 1917 – 1r – 1 – (italian newspaper) – us IHRC [071]
Corriere di roma – Rome, Italy. 8 jun-31 dec 1944 – 1 – (wanting: n10,11,15,16,54,55,154,155) – mf#m.f.876.d – uk British Libr Newspaper [074]
Corriere di sicilia – Catania. Italy. feb. 1-Apr. 21, 1944 – 1 – us NY Public [074]
Corriere di trinidad *see* Miscellaneous newspapers of las animas county, reel 2
Corriere d'informazione – Milano: [Filli Crespi & C, may 22 1945-may 5 1946; 1952-1955; 1956-feb 1962 – (issues for 1952-may 9 1959 filmed consecutively with: nuovo corriere dellasera; issues for may 10 1959-1962 filmed consecutively with: corriere della sera (milan, italy)) – us CRL [074]
Corriere diplomatico e consolare – Rome, Italy.10 jan 1923 – 15 may 1940 – 1 – mf#m.f.875 – uk British Libr Newspaper [074]
Corriere illustrato – Canada. jan 1958-dec 1984 – 23r – 1 – (in italian) – cn Commonwealth Imaging [071]
Il corriere israelitico – Trieste. v. 1-53. 1862-1914. Incomplete – 1 – us NY Public [074]
Corriere italiano di londra – London, UK. 6 Jan-30 Mar 1872 – 1 – uk British Libr Newspaper [072]
Corriere libertario – Barre VT, 1914-15* – 1r – 1 – (italian newspaper) – us IHRC [071]
Corriere siciliano – New York. N.Y. v. 1-4. Mar 1931-Jun 1934 – 1 – us NY Public [071]
Il corrierre del rhode island – Providence, RI. 1915-1924 (1) – mf#66326 – us UMI ProQuest [071]
Corrig school record – Victoria, BC: Record Pub Co, [1887?-1890?] – 9 – ISSN: 1190-7304 – mf#P04497 – cn Canadiana [370]
Corrigan, Ann E *see* A descriptive analysis of corporate health promotion activity evaluations
Corrigan, Felicia *see* Some social principles of orestes a. brownson
Corrivault, Blaise *see* Bibliographie analytique de l'histoire d'acadie
Corrodi, Heinrich *see* Beytraege zur befoerderung vernuenftigen denkens in der religion
Corrosion – Houston. 1945+ (1) 1968+ (5) 1975+ (9) – ISSN: 0010-9312 – mf#3049 – us UMI ProQuest [660]
Corrosion science – Oxford. 1961+ (1,5,9) – ISSN: 0010-938X – mf#49056 – us UMI ProQuest [660]
Corrozet, G *see* Hecatomgraphie c'est...dire les declarations de plusieurs apophtegmes...
Corruption and reform – Dordrecht. 1991-1992 (1,5,9) – ISSN: 0169-7528 – mf#16778 – us UMI ProQuest [340]
Corruptions of the church of rome / Bull, George – London, England. 1836 – 1r – us UF Libraries [240]
Corry herald – Corry, PA. -w 1899-1900. 6 rolls – 13 – $25.00r – us IMR [071]
Corry, J *see* Observations upon the windward coast of africa
Corry, John *see* The life of joseph priestly
Corsair : a gazette of literature, art, dramatic criticism, fashion and novelty – New York. 1839-1840 – 1 – mf#3969 – us UMI ProQuest [073]
Le corsaire : journal des spectacles, de la litterature, des arts, des moeurs et modes – Paris.11 juil-dec 1823, 1826-8 sept 1852, 29 aout, 4 oct-14 nov 1858, n.s. 1902 – 1 – fr ACRPP [073]
Le corsaire – n1-52, 9 aout 1879-juil 1880. nn2-12, 14. 25 dec/1 janv-26 mars/3 avr 1881 – 1 – fr ACRPP [073]
O corsario : jornal litterario e de critica theatral – Rio de Janeiro, RJ: Typ Guanabarense de L A F Menezes, 08 mar-22 abr 1851 – (= ser Ps 19) – mf#P15,01,63 – bl Biblioteca [790]
O corsario : orgao de critica imparcial e propaganda social – Rio de Janeiro, RJ. 26 nov 1903; 09 mar 1904 – (= ser Ps 19) – mf#P11,08,24 – bl Biblioteca [079]
Corsario bahiano – Rio de Janeiro, RJ. 1886 – (= ser Ps 19) – mf#P05,04,68 – bl Biblioteca [321]
Corsario junior : periodico critico e noticioso – Rio de Janeiro, RJ: Typ Progresso, 10 jun 1882-21 jan 1883 – (= ser Ps 19) – mf#P05,04,70 – bl Biblioteca [079]
O corsario vermelho : critico theatral – Rio de Janeiro, RJ: Typ de Cremiere, 31 maio 1851 – (= ser Ps 19) – mf#P17,01,92 – bl Biblioteca [790]
Corse, Carita Doggett *see*
– Evolution of the american flag
– Fort caroline
– Fort george island
– Ft caroline
– Picture of fort carolina
– Picturesque and beautiful fort george island comes
– Shrine of the water gods
Corsica / Gregorovius, Ferdinand – Stuttgart [u a] 1854 [mf ed Hildesheim 1995-98] – 2v on 4mf – 9 – €120.00 – 3-487-25943-5 – gw Olms [914]
Corskery, Thomas *see* Plymouth-brethrenism
Corson, Hiram *see* Chaucer's legende of goode women
Corssen, Peter *see*
– Monarchianische prologe zu den vier evangelien
Corswarem, P de *see* De liturgische boeken van de kollegiale kerk van o l vrouw van tongeren voor het concilie van trente
Cort adeler i venedig : ballet i tre akter og et slutningstableau / Bournonville, August – Kobenhavn: J H Schubothes, boghandel, 1870 – 1 – mf#*ZBD-*MGTZ pv1-Res – Located: NYPL – us Misc Inst [790]
Cort begrijp, inhoudende de voornaemste hooftstucken der christelijcker religie / Marnix van S Aldegonde, P van – Leyden, 1599 – 1mf – 9 – mf#PBA-250 – ne IDC [240]
Cort, Cyrus *see* Woman preaching viewed in the light of god's word and church history
Cort onderwys van de vijf colommen / Bosboom, S – Amsterdam, 1657 – 2mf – 9 – mf#OA-85 – ne IDC [720]
Cort onderwys van de vyf colomen door vinsent scamozzy... / Bosboom, S – Amsterdam, 1670 – 2mf – 9 – mf#OA-22 – ne IDC [720]
Cortambert, Richard *see*
– L'amerique et les travaux americains en 1866
– Nouvelle historie des voyages et des grandes decouvertes geographiques dans tous les temps et dans tous les pays, vol 1
Corte de cuentas / El Salvador. Courts – San Salvador. Tomo 1. No. 1-33, no. 115. Jan 1940-Dec 1954 – 1 – us NY Public [340]
Corte de d joao no rio de janeiro / Costa, Luiz Edmundo Da – Rio de Janeiro, Brazil. v1-3. 1957 – 1r – us UF Libraries [972]
Corte de portugal no brasil / Norton, Luiz – Sao Paulo, Brazil. 1938 – 1r – us UF Libraries [972]
Een corte ondersoekinghe des gheloofs over die ghene die hen totter ghemeynte begheven willen / Micronius, M – n.p, 1566 – 1mf – 9 – mf#PBA-265 – ne IDC [240]
La corte suprema federale nel sistema costituzionale degli stati uniti d'america / Catinella, Salvatore – Padova, Milani, 1934. 452p. LL-226 – 1 – us L of C Photodup [340]
Corte y cortijo / Hurtado, Antonio – 1870 – 9 – sp Bibl Santa Ana [830]
Corteccia, Francesco *see*
– Canticorum liber primus cum quinque vocibus
– Canticorum liber primus cum sex vocibus
Les corte-real et leurs voyages au nouveau-monde : d'apres des documents nouveaux ou peu connus tires des archives de lisbonne et de modene... / Harrisse, Henry – Paris: Leroux, 1883 – 4mf – 9 – (incl ind) – mf#08414 – cn Canadiana [910]
Cortes, Alfonso *see* 30 [i.e. treinta] poemas
Cortes and montezume / Collis, Maurice – London: Faber and Faber, s.a. – sp Bibl Santa Ana [910]
Cortes and the aztec conque consultent. gordon eckholm / Blacker, Irwir R – New York: American Heritage. Pub. Co., Inc, 1965 – sp Bibl Santa Ana [350]
Cortes and the conquest of mexico / Hamlyn, Raul – Londres: Golden Pleasure Books LTD, 1967 – sp Bibl Santa Ana [350]
Cortes Castro, Leon, Pres, Costa Rica *see* Presidente cortes a traves de su correspondencia
Las cortes de cadiz y el obispo de orense (episodio curioso de las cortes constituyentes / Risco, A – Madrid: Razon y Fe, 1926 – 1 – sp Bibl Santa Ana [946]
Cortes de la muerte / Carvajal, Micael & Hurtado de Mendoza, Luis – Madrid: Rivadeneyra, 1855 – 1 – sp Bibl Santa Ana [920]
Cortes, Hernando *see*
– Carta
– Cartas y documentos
– Cartas y otros documentos novisimamente descubiertos en el archivo general de indias de sevilla
– Cartas y relaciones de hernando cortes al emperador carlos v
– Correspondence de fernand cortes avec 1 empereur charles quint. sur la conquete de mexique
– Hernan cortes
– Hernan cortes, letters from mexico
– Historia de nueva espana...y notas del ilmo. d.f.a. lorenzana..
– Lettres de fernand cortes a charles v
Cortes Vazquez, Luis *see* Viaje literario al norte cacereno
Cortes y la evangelizacion de nueva espana / Bayle, Constantino – Madrid: Ediciones Jura, 1948 – sp Bibl Santa Ana [946]
Cortes y la evangelizacion de nueva espana / Bayle, Constantino – Madrid: Missionalia Hispanica, 1948 – 1 – sp Bibl Santa Ana [240]
Cortes Y Larraz, Pedro *see* Descripcion geografico-moral de la diocesis de goa
Cortes y Zedeno, Jeronimo Tomas de Aquino *see* Arte, vocabulario y confessionario en el idioma mexicano
Cortesao, Jaime *see*
– Cabral e as origens do brasil
– Introducao a historia das bandeiras
– Raposo tavares e a formacao territorial do brasil
Cortex – Varese. 1987+ (1,5,9) – ISSN: 0010-9452 – mf#16570 – us UMI ProQuest [616]
Corteza Collantes, Alfonso *see* Resumen de una discusion acerca de los hijos ilegitimos ante la sociedad y el derecho
Corteza y la savia / Gomez, Jose Jorge – Habana, Cuba. 1959 – 1r – us UF Libraries [972]
Corthell, Elmer Lawrence *see* Canals and railroads, ship canals and ship railways
The cortical and subcortical efferent and afferent connections of a proposed cingulate motor cortex and its topographical relationship to the primary and supplementary motor cortices of the rhesus monkey / Morecraft, Robert Jet al – 1989 – 4mf – 9 – $16.00 – us Kinesology [612]
Corticelli home needlework – St Johns, Quebec: Corticelli Silk, [1899-19–] – 9 – mf#P04846 – cn Canadiana [640]
Cortijo Valdes, A *see* Biografia de lexcmo. d. vicente barrantes
Cortina, Jose Antonio *see* Escrito de expresion de agravios de d jose moreno
Cortina, Jose Manuel *see*
– Nuevo mundo despues de la guerra
– Periodista, el diplomatico y la nacionalidad
Cortines, Ruiz *see* Discursos...pronunciados...1951 al 1952
Cortland gazette : weekly general newspaper – Cortland, OH. 20 Aug 1880 – 1r – 1 – us Western Res [071]
The cortland herald – Cortland, NE: M A Blizzard (wkly) [mf ed v5 n9. feb 8 1888-1900 (gaps)] – 1r – 1 – (cont: cortland journal) – us NE Hist [071]
Cortland journal – Cortland, NE: [John Bloom] (wkly) [mf ed jan 8 1886] – 1r – 1 – (cont by: cortland herald.) – us NE Hist [071]
Cortland news – Cortland, NE: Chris Baker. 44v. v1 n1. apr 6 1933-v43 n21. jul 15 1976 (wkly) [mf ed with gaps filmed -1977] – 12r – 1 – (absorbed by: arbor state. publ at cortland ne, apr 6 1933-apr 11 1963; at wymore ne, apr 18 1963-jul 15 1976) – us NE Hist [071]
Cortland news – Cortland, NE: F C Wilson. v18 n40. apr 13 1916-v9 n40. apr 24 1925 (wkly) [mf ed with gaps filmed 1957] – 2r – 1 – (cont: cortland sun (1899). issues for apr 12 1917-apr 24 1925 called v2 n1-v9 n40. suspended foll sep 26 1918; resumed mar 7 1919) – us NE Hist [071]
Cortland sun – Cortland, NE: M E Kerr & Co. 1v. v1 n1. jul 23 1897-v1 n32. feb 25 1898 [wkly] [mf ed with gaps filmed 1957] – 5r – 1 – (cont by: cortland weekly sun) – us NE Hist [071]
Cortland sun – Cortland, NE: M E Kerr & Co, sep 1899- . v18 n39. apr 6 1916 (wkly) [mf ed with gaps filmed 1957] – 5r – 1 – (cont: cortland weekly sun. cont by: cortland news (1916). sep 21 1899 issue has no enumeration) – us NE Hist [071]
Cortland weekly sun – Cortland, NE: M E Kerr & Co. v1 n33. mar 4 1898-1899// (wkly) [mf ed with gaps filmed 1957] – 1r – 1 – (cont: cortland sun. cont by: cortland sun (1899)) – us NE Hist [071]
Corton, Antonio *see* Antillas
Corts Grau, Jose *see* Perfil actual de donoso cortes
Coruja theatral – Rio de Janeiro, RJ: Typ de C Ogier & C, 29 dez 1840-22 jan 1841 – (= ser Ps 19) – mf#P12,05,24 – bl Biblioteca [790]
Corumbas, romance / Fontes, Amando – Rio de Janeiro, Brazil. 1946 – 1r – us UF Libraries [972]
Corvalan, Octavio *see* Postmodernismo
Corvalan-Groessling, Veronica *see* The physiological and perceived effects of drafting on a group of highly trained distance runners

Corvallis gazette – Corvallis OR: Odeneal & Carter, [wkly] – 1 – (began in 1862? ceased in 1899; merged with: oregon union [corvallis, or: 1897] to form: union gazette [corvallis, or]) – us Oregon Lib [071]

Corvallis gazette *see* Oregon union

Corvallis gazette [corvallis, or: 1900] – Corvallis OR: [s.n.] 1900-09 [semiwkly] – 1 – (cont: union gazette [corvallis, or]; cont by: corvallis weekly gazette [1909-09]; related to: corvallis daily gazette) – us Oregon Lib [071]

Corvallis gazette-times – Corvallis OR: Ingalls, Moore & Hurd, 1921- [daily ex sun] – 1 – (related to wkly ed: weekly gazette-times, 1921, mid-valley sunday, sep 13 1998-; cont: daily gazette-times) – us Oregon Lib [071]

Corvallis gazette-times *see* Mid-valley sunday

Corvallis times – Corvallis OR: B F Irvine, -1909 [semiwkly] [mf ed 1969] – 6r – 1 – (merged with: corvallis weekly gazette [1909] to form: gazette-times [corvallis, or]; absorbed: leader [1895-1903]) – us Oregon Lib [071]

Corvallis weekly gazette – Corvallis OR: C L Springer, 1909 – 1 – (merged with: corvallis times, to form: gazette times [corvallis or]; related to: corvallis daily gazette; cont: corvallis gazette [corvallis, or: 1900]) – us Oregon Lib [071]

Corvallis weekly gazette *see* Corvallis times

Corvey – fuerstliche bibliothek corvey : sachliteratur / ed by Barckow, Klaus et al – 10047v on 20169mf – 9 – €48,000.00 set – gw Olms [020]

Corvey – fuerstliche bibliothek corvey : sachliteratur [deutschsprachige werke] / ed by Barckow, Klaus et al – Hildesheim 1995-98 – 1083 titles on 5332mf – 9 – €19,480.00 set – gw Olms [020]

Corvey – fuerstliche bibliothek corvey : sachliteratur [englischsprachige werke] / ed by Barckow, Klaus et al – Hildesheim 1995-98 – 1031 titles on 4685mf – 9 – €17,480.00 – gw Olms [020]

Corvey – fuerstliche bibliothek corvey : sachliteratur [franzoesischsprachige werke] / ed by Barckow, Klaus et al – Hildesheim 1995-98 – 1219 titles on 10124mf – 9 – €29,800.00 set – gw Olms [020]

Corvey – fuerstliche bibliothek corvey-sachliteratur : microedition der buechersammlung mit rara und unikaten / ed by Barckow, Klaus et al – 10,047v on 20,169mf – 9 – silver €48,000.00 – gw Olms [019]

Corvey – fuerstliche bibliothek corvey-sachliteratur [deutschsprachige werke] / ed by Barckow, Klaus et al – [mf ed Hildesheim 1995-98] – 1083 titles on 5332mf – 9 – €19,480.00 – gw Olms [019]

Corvey – fuerstliche bibliothek corvey-sachliteratur [englischsprachige werke] / ed by Barckow, Klaus et al – [mf ed Hildesheim 1995-98] – 1031 titles on 4685mf – 9 – €17,480.00 – gw Olms [019]

Corvey – fuerstliche bibliothek corvey-sachliteratur [franzoesischsprachige werke] / ed by Barckow, Klaus et al – [mf ed Hildesheim 1995-98] – 1219 titles on 10,124mf – 9 – €29,800.00 – gw Olms [019]

Corvin von Skibniewski, Stephan Leo, Ritter *see* Geschichte des roemischen katechismus

Corvington, Hermann *see*
- Caonabo
- Deux caciques de xaragua
- Etude sur la condition juridique de letranger en h...
- Guayacuya

Corvinus, A *see*
- Der 128 psalm vom glueck, segen, gedeien der eheleut
- Acta handlungen
- Bericht ob man on die tauffe vnd empfahungen des leibs vnd bluts christi allein durch den glauben kuenne selig werden
- De integro sacramento corporis et sanguinis domini
- Expositio decalogi, symboli, apostolici, sacramentoru, et dominicae praecationis

Corvinus, Gottlieb S [Amaranthes] *see* Nutzbares, galantes und curioeses frauenzimmer-lexicon...von amaranthes

Corvinus, J A *see* Petri molinaei novi anatomici mala encheiresis seu censura anatomes arminianismi

Corwin, Edward A *see*
- The constitution and what it means today
- John marshall and the constitution

Corwin, Edward Henry Lewinski *see* Political history of poland

Corwin, Edward Tanjore *see* A manual of the reformed church in america

Corwin, Edward Tanjore et al *see* The history of the reformed church, dutch, the reformed church, german, and the moravian church in the united states

Corwin, George W *see* Study of barca di venezia per padova by adriano banchieri

Corwin, Rebecca *see* The verb and the sentence in chronicles, ezra, and nehemiah

Corwin, Robert N *see* Vetter gabriel

Cory, Charles B *see*
- Hunting and fishing in florida
- Southern rambles

Cory, Charles Barney *see*
- The birds of eastern north america known to occur east of the ninetieth meridian
- The birds of eastern north america known to occur east of the ninetieth meridian, pt 1
- The birds of eastern north america known to occur east of the ninetieth meridian, pt 2
- How to know the ducks, geese and swans of north america
- How to know the shore birds (limicolae) of north america (south of greenland and alaska)

Cory, Geo E *see* Diary of the rev francis owen, m a [the]

Cory, Hans *see* African figurines

Corydon : a trilogy in commemoration of matthew arnold: with lyric interludes / Carman, Bliss – Fredericton, NB: L S MacNutt, 1888 – 1mf – 9 – mf#06090 – cn Canadiana [810]

Cosack, Konrad *see* Das sachenrecht mit ausschluss des besonderen rechts der unbeweglichen sachen im entwurf eines buergerlichen gesetzbuches fuer das deutsche reich

Cosantoir – Dublin. 1979-1980 (1) 1979-1980 (5) 1979-1980 (9) – mf#8738 – us UMI ProQuest [355]

Les cosaques a paris – Paris [1848?] – us CRL [944]

Cosas de la india : hojas arrancadas del diario de un misionero / Pilar, Placido Maria del – Burriana: Imprenta de A Monreal, 1915 [mf ed 1995] – (= ser Yale coll) – 149p (ill) – 1 – 0-524-09921-9 – (in spanish) – mf#1995-0921 – us ATLA [241]

Cosas de santafe de bogota / Ortega Ricaurte, Daniel – Bogota, Colombia. 1959 – 1r – us UF Libraries [972]

Cosas del mundo / Hurtado, Antonio – 1846 – 9 – sp Bibl Santa Ana [830]

Cosas del tapete verde / Roso de Luna, Mario – Madrid: Editorial Atlantida, 1930 – 1 – sp Bibl Santa Ana [946]

Cosas que usted debe conocer / Andreu, Enrique – Habana, Cuba. 1950 – 1r – us UF Libraries [972]

Cosas y gentes de antano / Fernandez Guardia, Ricardo – San Jose, Costa Rica. 1939 – 1r – us UF Libraries [972]

Cosas y gentes de antano... / Fernandez Guardia, Ricardo – Madrid: Razon y Fe, 1941 – 1 – sp Bibl Santa Ana [240]

Cosbuc, George *see* Balade si idile

Cosecha, ensayos y articulos / Fonfrias, Ernesto Juan – San Juan, Puerto Rico. 1956 – 1r – us UF Libraries [972]

Cosgrove, John Joseph *see* Principles and practice of plumbing

Coshocton Co. Coshocton *see*
- Daily times
- Practical preacher
- Times-age
- Tribune
- Tribune and times age

Coshocton county atlas, 1872 : by lake/titus – 1r – 1 – mf#B30575 – us Ohio Hist [978]

Cosmae pragensis chronica boemorum (mgh6:2.bd) / ed by Bretholz, B – 1923 – (= ser Monumenta germaniae historica 6: scriptores rerum germanicarum, nova series (mgh6)) – €15.00 – ne Slangenburg [240]

Cosmann, M *see* Essais de paleoconchologie

Cosmas Le Pretre *see* Le traite contre les bogomiles

O cosme : folha recopiladora e politica – Rio de Janeiro, RJ. 15 dez 1849-12 jan 1850 – (= ser Ps 19) – mf#P15,01,47 n02 – bl Biblioteca [321]

Cosmetic technology – Cleveland. 1979-1980 (1,5,9) – mf#12344 – us UMI ProQuest [640]

Cosmetic world news – London. 1989-1990 (1) – ISSN: 0305-0319 – mf#10392 – us UMI ProQuest [640]

Cosmetics and perfumery – Oak Park. 1906-1975 (1) 1971-1975 (5) – ISSN: 0090-6581 – mf#2538 – us UMI ProQuest [660]

Cosmetics and toiletries – Carol Stream. 1976+ (1) 1976+ (5) 1977+ (9) – ISSN: 0361-4387 – mf#2538,01 – us UMI ProQuest [660]

Cosmic consciousness : a paper read before the american medico-psychological association in philadelphia, 18 may 1894 / Bucke, Richard Maurice – Philadelphia: Conservator, 1894 – 1mf – 9 – mf#10204 – cn Canadiana [130]

Cosmic ethics or the mathematical theory of evolution : showing the full import of the doctrine of the mean, and containing the principia of the science of proportion / Thomas, William Cave – London, 1896 – (= ser 19th c evolution & creation) – 4mf – 9 – mf#1.1.9367 – uk Chadwyck [520]

Cosmic landscape / Infinity Books, Ltd – n1-2 [1992 feb 5-23], also spr 1992 catalog – 1r – 1 – mf#2797360 – us WHS [071]

Cosmic research – New York. 1975-1976 (1) 1975-1976 (5) – ISSN: 0010-9525 – mf#10818 – us UMI ProQuest [520]

Cosmic stories – New York. v1 n1-3. mar-jul 1941 [all publ] – (= ser Science fiction periodicals, 1926-1978. series 1) – 1r – 1 – $95.00 – us UPA [830]

A cosmic view of religion / Halstead, William Riley – Cincinnati: Jennings & Graham, c1913 [mf ed 1991] – 1mf – 9 – 0-7905-7749-6 – mf#1989-0974 – us ATLA [200]

The cosmogony of the vedas / Phillips, Maurice – Madras: printed by R Hill, at the Govt Press, [1887?] [mf ed 1996] – (= ser Yale coll) – 20p – 1 – 0-524-10263-5 – mf#1996-1263 – us ATLA [280]

Die cosmographiae introductio des martin waldseemueller (ilacomilus) in faksimiledruck / ed by Wieser, Fr R v – Strassburg: J H Ed Heitz, 1907 – (incl bibl ref. latin text with introduction in german) – us Wisconsin U Libr [430]

Cosmographie de levant / Thevet, A – Lion, 1556 – 3mf – 9 – mf#H-8291 – ne IDC [956]

The cosmology of the rigveda : an essay / Wallis, Henry White – London: Williams and Norgate, 1887 – 1mf – 9 – 0-524-01386-1 – mf#1990-2398 – us ATLA [280]

Cosmometapolis / Relgis, Eugen – Bucuresti, Romania. 1935 – 1r – us UF Libraries [025]

Cosmopolis : an international monthly review – London. 1896-1898 – 1 – mf#5293 – us UMI ProQuest [073]

El cosmopolita – Kansas City, KS: Cosmopolita Pub Co, [aug 22 1914-nov 15 1919] – 2r – 1 – us CRL [071]

Cosmopolitan : great britain edition – London. 1975-1993 (1) 1975-1993 (5) 1976-1993 (9) – ISSN: 0141-0555 – mf#10472 – us UMI ProQuest [073]

Cosmopolitan – New York. 1886-1925 (1) – mf#3117 – us UMI ProQuest [740]

Cosmopolitan – New York. 1925+ (1) 1964+ (5) 1976+ (9) – ISSN: 0010-9541 – mf#5555 – us UMI ProQuest [740]

Cosmopolitan – Providence, RI. 1878-1879 (1) – mf#66278 – us UMI ProQuest [071]

Cosmopolitan herald – Girard, PA., 1910-1912 – 13 – $25.00r – us IMR [071]

The cosmopolite – Girard, PA., 1889-1905 – 13 – $25.00r – us IMR [071]

Cosmopolite herald – Girard, PA. 1910-2000 (1) – mf#65909 – us UMI ProQuest [071]

Un cosmopolite suisse : jacques-henri meister / Grubenmann, Yvonne de Athayde – Geneve: E Droz, 1954 – 1r – 1 – (incl bibl ref) – us Wisconsin U Libr [440]

Cosmopolite's statistical chart of nova scotia gold mines, 1862-1866, inclusive – Halifax, NS: J Bowes, 1867 – 1mf – 9 – 0-665-00728-0 – mf#00728 – cn Canadiana [622]

O cosmorana na bahia – Rio de Janeiro, RJ: Typ Brasiliense, 02 out-22 dez 1849 – (= ser Ps 19) – mf#P14,02,29 – bl Biblioteca [320]

The cosmos and the logos : being the lectures for 1901-2 on the l.p. stone foundation... / Minton, Henry Collin – Philadelphia:Westminster Press, 1902 – 1mf – 9 – 0-8370-4448-0 – (incl bibl ref and index) – mf#1985-2448 – us ATLA [210]

Cosnier, Henri Charles *see* L'ouest africain francais

Cospar information bulletin – Oxford. 1977-1994 (1,5,9) – ISSN: 0045-8732 – mf#49289 – us UMI ProQuest [620]

Cosquin, Emmanuel *see* Etudes folkloriques

Cossack fairy tales and folk-tales / Bain, R Nisbet – London, England. 1894 – 1r – us UF Libraries [390]

The cossacks herald – Prague XR, 1941-45 – 2r – 1 – (ukrainian periodical) – us IHRC [073]

Cossarin, Mark A *see* Joyride

Cossette, Angele *see* Bibliographie de l'oeuvre de louis-philippe audet

Cossette, Raymond *see* Melanges offerts a me. raymond cossette

Cossigny, J F C *see* Voyage...canton, capitale de la province de ce nom,...la chine

Cossitt, Jennie *see* Influence of the old masters on bach

Cossman, Alexandre E *see* Essais de paleoconchologie comparee

Cossmann, Werner *see* Worte der erinnerung an den am 22 juni 1918 aus dem leben...

Cosson, E S C *see* Exploration scientifique de l'algerie

Cost and management – Hamilton. 1926-1985 (1) 1971-1985 (5) 1976-1985 (9) – (cont by: cma: the management accounting magazine) – ISSN: 0010-9592 – mf#5716 – us UMI ProQuest [650]

Cost and returns on sixty poultry farms in florida / Young, Martin Greene – s.l, s.l? 1931 – 1r – us UF Libraries [636]

Cost benchmarking als instrument des kostenmanagements / Baur, Thorsten – (mf ed 1995) – 1mf – 9 – €30.00 – 3-8267-2241-8 – mf#DHS 2241 – gw Frankfurter [650]

Cost engineering : a publication of the american association of cost engineers / American Association of Cost Engineers – Morgantown. 1978+ (1) 1978+ (5) 1978+ (9) – ISSN: 0274-9696 – mf#8768,01 – us UMI ProQuest [620]

Cost of freedom : bklyn y i p newsletter / Youth International Party – v1 n1-2 [1974 mar-jun] – 1r – 1 – mf#365085 – us WHS [335]

Cost of handling citrus fruit from the tree to the car in florida / Hamilton, H G – Gainesville, FL. 1929 – 1r – us UF Libraries [634]

Cost of health supervision in industry : august, 1917 / Alexander, Magnus Washington [comp] – [S.l: s.n, 1917?] (mf ed 19–) – 1v – mf#Z-1768 – us Harvard [360]

Cost of intemperance – London, England. 18– – 1r – us UF Libraries [240]

Cost of living for urban africans, johannesburg 1959 / De Gruchy, Joy – Johannesburg, South Africa. 1960 – 1r – us UF Libraries [339]

Cost of living on florida farms (a survey) / Wray, Robert – s.l, s.l? 1925 – 1r – us UF Libraries [630]

Cost of living on one hundred farms columbia county, florida, year... / Graham, George Ransom – s.l, s.l? 1929 – 1r – us UF Libraries [630]

Cost of producing potatoes in the hastings area / Scarborough, Chaffie Aldred – s.l, s.l? 1927 – 1r – us UF Libraries [630]

Cost of producing strawberries in the plant city area for the season / Larson, Lawrence John – s.l, s.l? 1932 – 1r – us UF Libraries [634]

The cost of production : being specimens of the pages and type in more common use, with estimates of the cost of composition, printing, paper, binding, etc, for the production of a book / Incorporated Society of Authors – 3rd enl ed. [London]: publ for the Incorporated Society of Authors, 1891 – (= ser 19th c publishing...) – 1mf – 9 – mf#3.1.71 – uk Chadwyck [680]

The cost of production : being specimens of the pages and type in more common use, with estimates of the cost of composition, printing, paper, binding, etc., for the production of a book / Incorporated Society of Authors – [London] 1889 – (= ser 19th c publishing...) – 1mf – 9 – mf#3.1.67 – uk Chadwyck [680]

The cost of regulation to small business : hearing...house of representatives, 107th congress, 2nd session, washington dc, june 6 2002 / United States. Congress. House. Committee on Small Business. Subcommittee on Workforce, Empowerment, and Government Programs – Washington: US GPO 2002 [mf ed 2002] – 2mf – 9 – (incl bibl ref) – us GPO [338]

Costa / Salasar, Jose – Habana, Cuba. 1964 – 1r – us UF Libraries [972]

Costa Aguiar, Antonio Augusto Da *see* Vida do marquez de barbacena

Costa, Angyone *see*
- Indiologia
- Introducao a arqueologia brasileira
- Migracoes e cultura indigena

Costa, Antonio Pedro da, Bishop of Damao *see* Relatorio da nova diocese de damao

Costa, Avelino de Jesus da *see* Liber fidei... tomo 1. braga, 1965

Costa, Benjamin Franklin de *see* Father joques at the lake of the holy sacrament

[La costa-] blade-citizen – CA. 1990- – 36r – 1 – $2160.00 (subs $300y) – mf#H04044 – us Library Micro [071]

Costa, Carl *see*
- Das erbe des wucherers
- Wir demokraten

Costa, Didio Iratym Affonso Da *see*
- Marcilio dias, imperial-marinheiro
- Saldanha
- Tamandare, almirante joaquim marques lisboa

Costa Duran, Maria *see* Las alas rotas

Costa, Edgard *see* Legislacao eleitoral brasileira

Costa, Eduardo Augusto Ferreira Da *see* Districto de mocambique em 1898

Costa, Esdras Borges *see* Cerrado e retiro

Costa, Francesco *see*
- Immaculate conception

Costa Gomez, Moises Frumencio Da *see*
- Naar nieuw arbeidsrecht
- Wetgevend orgaan van curacao

Costa, Isaac da *see* The four witnesses

Costa, J *see* Mesuae medici. clarissimi opera...

Costa, Joaquim Ribeiro *see* Toponimia de minas gerais

Costa, Licurgo *see* Cidadao do mundo

Costa, Luiz Edmundo Da *see*
- Corte de d joao no rio de janeiro
- Recordacoes do rio antigo
- Rio de janeiro do meu tempo

Costa, Manuel Goncalves da *see*
- Ignacio de azcuedo...
- Inacio de azevedo...

Costa, Mario Augusto Da *see* Como fizeram os portuguezes em mocambique

COSTA

Costa, Melanie Sandra Fernandes da *see* Risk management in health care in south africa
[Costa mesa-] costa mesa daily pilot – CA. 1966; 1968 – 11r – 1 – $660.00 – mf#H03190 – us Library Micro [071]
[Costa mesa-] costa mesa news – CA. 1984- – 14r – 1 – $840.00 (subs $70/y) – mf#HR04018 – us Library Micro [071]
Costa mesa herald *see* [Orange county-] balboa times
Costa, Michael *see* Eli
Costa, Octavio R *see* Suma del tiempo
Costa, Octavio Ramon *see*
– Diez cubanos
– Santovenia, historiador y ciudadano
Costa Rica *see*
– Codigo de educacion
– Codigo penal y codigo de policia
– Documentos relativos a la guerra nacional de 1856
– La gaceta
– Gaceta diario oficial
– Legislacion dictada durante 1962 en relacion con c...
Costa Rica Archivos Nacionales *see* Documentos relativos a la independencia
Costa rica ayer y hoy, 1800-1939 / Quijano Quesada, Alberto – San Jose, Costa Rica. 1939 – 1r – us UF Libraries [972]
Costa Rica Constitucion Politica (1949) *see*
– Constituciones de costa rica
– Digesto constitucional de costa rica
Costa rica de don tomas de acosta / Estrada Molina, Ligia Maria – San Jose, Costa Rica. 1965 – 1r – us UF Libraries [972]
Costa Rica Direccion General De Estadistica Y Cen... *see*
– Censo agropecuario, 1963
– Censo agropecuario de 1950
Costa Rica. Direccion General de Estadistica y Censo *see*
– Anuario estadistico 1883-1969
– Informa
Costa Rica. Direccion General de Obras Publicas *see* Informe...
Costa rica en el siglo 19 / Fernandez Guardia, Ricardo – San Jose, Costa Rica. 1929 – 1r – us UF Libraries [972]
Costa rica en la segunda guerra mundial / Rojas Suarez, Juan Francisco – San Jose, Costa Rica. 1943 – 1r – us UF Libraries [972]
Costa rica et son avenir / Biolley, Paul – Paris, France. 1889 – 1r – us UF Libraries [972]
Costa Rica Laws, Statutes, Etc *see*
– Codigo de comercio y sus reformas
– Codigo de procedimientos civiles
– Codigo de trabajo, 26 de agosto de 1943
Costa Rica. Ministerio de Economia y Hacienda *see* Memoria anual...
Costa Rica Ministerio De Educacion Publica *see* Informe sobre el estado actual de los trabajos de...
Costa Rica. Ministerio de Educacion Publica *see* Memoria
Costa Rica. Ministerio de Fomento *see* Memoria...
Costa Rica. Ministerio de Guerra, Marina y Policia *see* Memoria presentada al congreso constitucional por el...
Costa Rica. Ministerio de Guerra y Marina *see*
– Informe....
– Memoria...
Costa Rica. Ministerio de Hacienda *see*
– Informe del ministro de hacienda al congreso de...
– Informe presentado por el secretario de estado en el despacho de hacienda al congreso nacional de costa-rica en...
Costa Rica. Ministerio de Hacienda, Guerra, Marina i Educacion Publica *see* Informe...
Costa Rica. Ministerio de Hacienda, Guerra, Marina y Caminos *see*
– Informe del secretario de estado...
– Memoria leida por el...en la sesion celebrada por el congreso nacional el dia...
Costa Rica. Ministerio de Hacienda, Guerra y Caminos *see* Informe dirigido al congreso legislativo de...
Costa Rica. Ministerio de Hacienda, Guerra y Marina *see*
– Informe de hacienda en...
– Memoria...al congreso de...
Costa Rica. Ministerio de Hacienda y Guerra *see*
– Informe...
– Informe de hacienda y guerra al congreso de costa-rica en...
– Memoria...
Costa Rica. Ministerio de Instruccion Publica *see* Informe presentado por el secretario de estado en el despacho de instruccion publica al congreso nacional de costa-rica en...
Costa Rica. Ministerio de lo Interior *see*
– Informe del secretario del interior...encargado accidentalmente de los despachos de guerra, marina y obras publicas, presenta al congreso constitucional de costa-rica en el ano de...
– Memoria que el ministro del interior encargado de la cartera de relaciones presento al excmo congreso nacional de costa-rica en sus sesiones ordinarias de...
Costa Rica Ministerio De Relaciones Exteriores *see* Documentos relativos a la controversia
Costa Rica. Ministerio de Relaciones Exteriores *see* Memoria...
Costa Rica. Ministerio de Relaciones Exteriores e Instruccion Publica *see*
– Informe del ministro de estado en el despacho de relaciones exteriores e instruccion publica de costa-rica al congreso constitucional de...
– Memoria presentada al congreso legislativo de...
Costa Rica. Ministerio de Relaciones Exteriores, Justicia, Gracia, Culto y Beneficencia *see* Memoria...
Costa Rica. Ministerio de Relaciones Exteriores y Culto *see* Memoria...presentada a la asamblea legislativa por el...
Costa Rica. Ministerio de Relaciones y de lo Interior *see* Memoria...
Costa Rica. Ministerio de Relaciones y Gobernacion *see* Memoria presentada...a la representacion nacional de...
Costa Rica. Ministerio de Seguridad Publica *see* Memoria...presentada a la asamblea legislativa por el...
Costa rica, nicaragua y panama en el siglo 16 / Peralta, Manuel M(Aria) De – Madrid, Spain. 1883 – 1r – us UF Libraries [972]
Costa Rica Oficina De Planificacion *see* Caracteristicas de la actividad agropecuaria en co...
Costa Rica Oficina Del Presupuesto Seccion De Or... *see* Manual de organizacion de la administracion public
Costa Rica Oficina Nacional Del Censo *see* Poblacion de la republica de costa rica segun el c...
Costa Rica Patronato Nacional de la Infancia *see* 10 anos de labor, 1930-1940
Costa Rica. Secretaria de Educacion Publica *see*
– Memoria
– Memoria de educacion publica correspondiente al ano...
– Memoria...correspondiente al...
Costa Rica. Secretaria de Fomento *see*
– Memoria...
– Memoria de fomento...
– Memoria de fomento presentada al congreso constitucional por el secretario de estado en el despacho de esa cartera...
– Memoria presentada al congreso constitucional por el...
Costa Rica. Secretaria de Fomento y Agricultura *see*
– Informes de las dependencias de fomento correspondientes al ano...
– Memoria de fomento y agricultura correspondiente al ano... presentada al congreso constitucional por...secretario en el despacho de esas carteras
Costa Rica Secretaria De Gobernacion *see* Guanacaste
Costa Rica. Secretaria de Gobernacion, Gracia y Justicia *see* Memoria presentada al congreso constitucional por el...
Costa Rica. Secretaria de Gobernacion, Policia y Fomento *see*
– Informes de gobernacion, policia y fomento correspondientes al ano de...
– Memoria...
– Memoria...presentada al congreso constitucional por el ex-secretario de estado en esas carteras...
– Memoria...presentada al congreso constitucional por el secretario de estado en esas carteras...
Costa Rica. Secretaria de Gobernacion y Policia *see* Memoria...presentada al congreso constitucional de...por el senor secretario de estado en el despacho de esas carteras...
Costa Rica. Secretaria de Guerra, Marina, Gobernacion, Fomento y Justicia. *see* Informe...presentado al congreso nacional de costa rica en...
Costa Rica. Secretaria de Guerra y Marina *see* Memoria...
Costa Rica. Secretaria de Hacienda, Relaciones Exteriores, Culto e Instruccion Publica *see* Informes presentados por el secretario de estado en los despachos...al congresonacional de costa-rica en...
Costa Rica. Secretaria de Hacienda y Comercio *see*
– Informe...
– Memoria...
Costa Rica. Secretaria de Instruccion Publica *see* Memoria...
Costa Rica. Secretaria de Relaciones Esteriores, Instruccion Publica, Culto y Beneficencia *see*
– Informe...
– Informe presentado al congreso constitucional de la republica de costa-rica por el...
– Memoria presentada a la convencion nacional constituyente de la republica de costa-rica por...
– Memoria presentada al congreso constitucional de la republica de costa-rica en su periodo ordinario de...
– Memoria presentada al congreso constitucional por el...
Costa Rica. Secretaria de Relaciones Exteriores *see* Informe presentado por el secretario de estado en el despacho de relaciones exteriores, al congreso nacional de costa-rica en...
Costa Rica. Secretaria de Relaciones Exteriores e Instruccion Publica *see* Informe..
Costa Rica. Secretaria de Relaciones Exteriores, Gracia, Justicia y Culto *see*
– Memoria correspondiente al ano...presentada al congreso constitucional por el...
– Memoria...presentada al congreso constitucional por el... secretario de estado en el despacho de esas carteras
– Memoria...presentada al congreso constitucional por...secretario de estado en el despacho de esas carteras
Costa Rica. Secretaria de Relaciones Exteriores, Instruccion Publica, Justicia y Gracia, Culto y Beneficencia *see* Memoria...presentada al excmo congreso nacional por el honorable senor secretario de estado...
Costa Rica. Secretaria de Relaciones Exteriores, Justicia, Instruccion Publica, Culto y Beneficencia *see* Informe presentado al excelentisimo senor presidente de la republica de costa-rica por el...
Costa Rica. Secretaria de Relaciones Exteriores, Justicia, Instruccion Publica, Culto y Beneficencia. *see* Memoria...
Costa Rica. Secretaria de Relaciones Exteriores, Justicia y Gracia, Culto y Beneficencia *see*
– Informe presentado por el...al congreso constitucional de...
– Memoria...
Costa Rica. Secretaria de Relaciones Exteriores, Justicia y Gracia, Culto y Beneficencia. *see* Memoria...presentada al congreso constitucional por...secretario de estado en el despacho de esas carteras
Costa Rica. Secretaria de Seguridad Publica *see* Memoria...presentada al congreso...
Costa rica y panama / Alfaro, Ricardo J – Panama, Panama. 1927 – 1r – us UF Libraries [972]
Costa rica y su folklore / Nunez, Evangeline – San Jose? Costa Rica. 1956? – 1r – us UF Libraries [390]
Costa rican life / Biesanz, John Berry – New York, NY. 1944 – 1r – us UF Libraries [972]
Costa rican public security forces / Worthington, Wayne Lamond – Gainesville, FL. 1966 – 1r – us UF Libraries [360]
Costa rica-panama arbitration / Matamoros, Luis – Washington, DC. 1913 – 1r – us UF Libraries [972]
Costa, Sergio Correa Da *see*
– Diplomacia do marechal
– Every inch a king
Costa y Martinez, Joaquin *see*
– La ignorancia del derecho, con un amplio estudio preliminar
– Introduccion a un tratado de politica
Costadoni, A *see* Annales camaldulenses o s b
Costard, Jean *see* L'ame d'un bon roi
Costa-rica und seine zukunft / Biolley, Paul – Berlin, Germany. 1890 – 1r – us UF Libraries [972]
Costa-rica y nueva granada / Molina, Felipe – Washington, DC. 1852 – 1r – us UF Libraries [972]
A cost/benefit analysis of declining numbers of women coaches : a social exchange theory perspective / Stevens, S C – 1989 – 2mf – 9 – $8.00 – us Kinesology [790]
Cost-effective delivery of managed nurse-based primary health care in a selected medical scheme / Seymore, Martha Magarieta – Uni of South Africa 2001 [mf ed Johannesburg 2001] – 4mf – 9 – (incl bibl ref) – mf#mfm15014 – sa Unisa [610]
Coste-Floret, Alfred *see* Les problemes fondamentaux du droit
Costenismos colombianos / Revollo, Pedro Maria – Barranquilla, Colombia. 1942 – 1r – us UF Libraries [972]
Coster, F *see*
– Enchiridion controversiarum praecipuarum...
– Libellus sodalitatis
Coster, Geraldine *see* Yoga and western psychology
Costigan, George Purcell *see* Handbook on american mining law
Costigan, John *see* Discours de m john costigan mp sur l'adresse
Costigan, S P *see* Diary
Costill, David L *see*
– Effect of sodium and water intake on plasma aldosterone during prolonged exercise in warm environment
– The relationship between physiological measurements and cross-country running performance
– Reliability in the measurement of muscle fiber composition and the histochemical staining for glycogen
La costituzione etiopica : studio sequito dalla versione della costituzione... / ed by Cerulli, Enrico – 2. ed. Roma: Instituto per l'Oriente, 1936 – 1 – us CRL [960]
Costumbres cacerenas. madrid / Ramon y Fernandez, Jose – G. Bermejo Impresor, 1950. Separata de Revista de Dialectologia y Tradiciones Populares – sp Bibl Santa Ana [306]
Costumbres y tradicionalismos de mi tierra... / Cadilla De Martinez, Maria – Puerto Rico, Puerto Rico. 1938 – 1r – us UF Libraries [390]
Le costume / Ruppert, L – Paris. v1-5. 1942-1947 – €19.00 – ne Slangenburg [790]
Le costume historique / Racinet, A – Paris, 1877-1888. 6v with 500 pls – 25mf – 9 – mf#AR-1895 – ne IDC [956]
Costume of the ancients / Hope, Thomas – London 1809 – (= ser 19th c art & architecture) – 4mf – 9 – mf#4.2.844 – uk Chadwyck [390]
The costume of the theatre / Komisarjevsky, Theodore – London: G. Bles, 1931.xii,178p. plates – 1 – us Wisconsin U Libr [740]
The costume of yorkshire / Walker, Gerry – London 1814 – (= ser 19th c art & architecture) – 5mf – 9 – mf#4.2.1474 – uk Chadwyck [740]
Costume prints in the british museum : authorities for artists / British Museum. London – [mf ed 1991] – 145mf – 9 – $1125.00 – 0-907006-29-9 – (over 7200 historical fashion prints listed in chronological order (55bc-1900); over 8500 reproductions) – uk Mindata [740]
Costumes et vues de la chine : avec des explications traduites de l'anglais / Alexander, William – Paris 1815 [mf ed Hildesheim 1995-98] – 1v on 1mf [ill] – 9 – €40.00 – 3-487-27578-3 – gw Olms [390]
Costumes of italy switzerland and france / Bridgens, Richard – [London? 1821?] – (= ser 19th c art & architecture) – 2mf – 9 – mf#4.2.1582 – uk Chadwyck [740]
Coswiger tageblatt – Coswig Dresden DE, 1906 3 jan-28 jun, 1907-1919, 1921 – 16r – 1 – (with suppl: unsere heimat 1919 1 nov-1933 [1r]) – gw Misc Inst [074]
Cot and cradle stories / Traill, Catherine Parr; ed by FitzGibbon, Mary Agnes – Toronto: W Briggs; Montreal: C W Coates; Halifax NS: S F Huestis, 1895 – 3mf – 9 – (incl publ list) – mf#34022 – cn Canadiana [830]
Cotallo, Jose Luis *see* Las realidades sociales contemporaneas
Cotallo Sanchez, Jose Luis *see* Vivir en cristiano
Cotard, Charles *see* Richard wagner: tristan et iseult
Cotarelo Y Mori, Emilio *see* Avellaneda y sus obras
Cotati clarion *see* [Rohnert park-] cotati-the community voice
[Cotati-] cotatian – CA. 1946-1947; 1951-1964 – 6r – 1 – $360.00 – mf#B03589 – us Library Micro [071]
Cote, Antonia *see* Bio-bibliographie de georgina lefaivre
Cote, Athanase *see* Bibliographie analytique des ouvrages de langue francaise sur l'histoire de la ville de quebec au 19e siecle
Cote, Berthe *see* Bio-bibliographie analytique de monsieur henri turgeon
Cote de chez swann / Proust, Marcel – Paris, France. v1-2. 1919 – 1r – us UF Libraries [960]
Cote de la bourse et de la banque : journal politique, economique et financier – Paris. juil-dec 1907, juil-dec 1913 – 1 – fr ACRPP [300]
Le cote des esclaves et le dahomey / Bouche, Pierre Bertrand – Paris, 1885 – 1 – us CRL [916]
La cote d'ivoire – Corbeil, France: E Crete, 1906 – 1 – us CRL [960]
Cote du cameroun dans l'histoire et la cartographie des origines... / Bouchaud, Joseph – Paris, France. 1952 – 1r – us UF Libraries [960]
Cote, Georges Pierre *see* Notice biographique sur le reverend j auclair
Cote green, jubilee methodist church – (= ser Cheshire monumental inscriptions) – 1mf – 9 – £2.50 – mf#12a – uk CheshireFHS [929]
La cote libre : journal economique et financier quotidien – Brussels, Belgium 8 oct-28 nov 1944, 10 jan-10 jul 1945, 27 feb, 11/12 may 1947 (imperfect) – 1/2r – 1 – uk British Libr Newspaper [330]
Cote, Marielle *see* Bio-bibliographie de mme jeanne l'archeveque-duguay
Cote, Narcisse Omer *see* Political appointments, parliaments and the judicial bench in the dominion of canada, 1867 to 1895
Cote, Paul *see* Livres sur les beaux-arts, l'architecture, la danse, le dessin, la musique, la numismatique, et la peinture
Cote, Stanislas *see* La chasse a l'heritage
Cote, Thomas *see* Trois etudes
Cote, Wolfred Nelson *see* The archaeology of baptism
Cotejo de las eglogas que ha premiado la real academia de la lengua / Forner Segarra, Juan Pablo – Salamanca: CSIC, 1951 – 1 – sp Bibl Santa Ana [946]

Cotes, Everard *see* Signs and portents in the far east

Cotes, Everard, mrs [Sara Jeanette Duncan] *see*
- The burnt offering
- Cousin cinderella
- The crow's nest
- Hilda
- His honour and a lady
- The imperialist
- On the other side of the latch
- The pool in the desert
- Social departure
- The story of sonny sahib
- Those delightful americans
- Vernon's aunt

Cotes, Everard, mrs [Sara Jeannette Duncan] *see* Daughter of to-day

Cothren, William *see* History of ancient woodbury, connecticut

Cotillion figures / Watkins, Joel H – New York, Washington: The Neale Publ Co, 1911 – 1 – mf#*ZBD-*MGO pv3 – Located: NYPL – us Misc Inst [790]

Cotman, John Sell *see*
- Architectural antiquities of normandy
- Engravings of the most remarkable of the sepulchral brasses in norfolk
- A series of etchings illustrative of the architectural antiquities of norfolk

The cotner collegian – Bethany, NE: The students of Cotner University. v1 n1. sep 1902- (wkly during school yr) [mf ed 1903-33 (gaps) filmed [1975?]-1988] – 5r – 1 – (publ in lincoln ne sep 29 1927-may 26 1933. sep 29 1908 issue called v1 n1 but constitutes v7 n1) – us NE Hist [071]

Coton 1538-1950 – (= ser Cambridgeshire parish register transcript) – 4mf – 9 – £5.00 – uk CambsFHS [929]

Coton, P *see*
- Du tres-sainct et tres-auguste sacrement, et sacrifice de la messe
- Geneve plagiaire...
- Institution catholique, o- est declaree et confirmee la verite de la foy
- Recheute de geneve plagiaire

Coton plagiaire ou la verite de dieu et la fidelite de geneve maintenue / Tronchin, T – Geneve, 1620 – 11mf – 9 – mf#PFA-180 – ne IDC [240]

Cotorra – Montijo.1895. Solo no. 1 – 9 – sp Bibl Santa Ana [070]

La cotorra – Montijo, 1895. 1 numero – 5 – sp Bibl Santa Ana [073]

Cotorrona / Bitullareaga, Mario – Guatemala, 1947 – 1r – us UF Libraries [972]

Cotswold standard *see* North cotswold standard

Cotta, C Friedrich *see* Teutsche stats-literatur

Cotta, F Chr *see* Strasburgisches politisches journal

Cotta y Marquez de Prado, Fernando de *see* Bibliografia manchega. bibliografia de las provincias de albacete, ciudad real, cuenca y toledo

Cotta y Marquez de Prado, Fernando de et al *see* Catalogo de las labras heraldicas de la ciudad de villanueva de la serena (badajoz)

Cotta y Marquez de Prado, Ventura de *see* Fuero de poblacion otorgado por...don carlos 3 a las localidades formadas en la sierra morena por la llamada "colonizacion interior"...

Cottage building : and hints for improved dwellings for the labouring classes / Allen, Charles Bruce – [6th ed] London 1867 – (= ser 19th c art & architecture) – 2mf – 9 – mf#4.2.1193 – uk Chadwyck [720]

Cottage conversations – London, England. 1846 – 1r – us UF Libraries [240]

Cottage funeral / Hawker, Robert – London, England. 1841 – 1r – us UF Libraries [240]

Cottage grove and lemati echo=leader – Cottage Grove OR: E P Thorp, 1895 [wkly] – 1 – (cont: cottage grove echo=leader; cont by: leader [cottage grove, or]) – us Oregon Lib [071]

Cottage grove echo=leader – Cottage Grove OR: E P Thorp, -1895 [wkly] – 1 – (merger of: leader [1895-1903]; drain echo; cont by: cottage grove and lemati echo=leader [1895]) – us Oregon Lib [071]

Cottage grove echo=leader *see* Drain echo

Cottage grove leader – Cottage Grove OR: Leader Pub Co, 1905-15 [wkly] – 1 – (cont: lane county leader [1903-05]; absorbed by: cottage grove sentinel [1909-]) – us Oregon Lib [071]

Cottage grove leader *see* Bohemia nugget

Cottage grove sentinel – Cottage Grove OR: L A Cates, 1909- [wkly] – 1 – (cont: western oregon [1905-09]; absorbed: cottage grove leader [1905-15]) – us Oregon Lib [071]

Cottage industries : and what they can do for ireland / Hart, Alice Marion (Rowlands) – London, 1885 – (= ser 19th c ireland) – 1mf – 9 – mf#1.1.8989 – uk Chadwyck [338]

Cottage questions for clerical visitors / Thorn, William – London, England. 18-- – 1r – us UF Libraries [240]

Cottager's friend – London, England. 1820 – 1r – us UF Libraries [240]

The cottager's friend and guide of the young – Toronto: Printed by T H Bentley, for J Donogh, [1854-18-?] – 9 – mf#P04202 – cn Canadiana [240]

Cottager's wife / Richmond, Legh – Glasgow, Scotland. 1814 – 1r – us UF Libraries [240]

Cottam Bird Seed (Firm) *see* Canaries vs chickens

Cottam, John *see* Birdland reasons

Cottbuser anzeiger *see* Anzeiger

Cotte, Paul Vincent *see* Regardons vivre une tribu malgache, les betsimisaraka.

Cotteau, E *see*
- En oceanie
- Promenade dans l'inde et...ceylan

Cottenham 1572-1950 – 14mf – 9 – £17.50 – uk CambsFHS [929]

Cotter, Joseph R *see* Antichrist dethroned

Cotter, Richard *see* Sketches of bermuda, or somers' islands

Cotterill, H B *see* Travels and researches among the lakes and mountains of eastern and central africa

Cotterill, Henry *see* Revealed religion expounded by its relations to the moral being of god

Cotterill, Henry Bernard *see* Italy from dante to tasso (1300-1600), its political history..

Cotterman, Michael L *see* Comparison of muscle force production for the smith machine and free weight modes using similar exercises

Cottiaux, J *see* L'office liegeois de la fete-dieu, sa valeur et son destin

Cottineau, L H *see* Repertoire topo-bibliogrcounselrapique des abbayes et prieures

Cottingham, Lewis Nockalls *see*
- Catalogue of the museum of mediaeval art
- The smith and founder's director containing a series of designs
- Working drawings for gothic ornaments

Cotton : its cultivatin and fertilization / Persons, A A – Lake City, FL. 1896 – 1r – us UF Libraries [630]

Cotton, Arthur Thomas *see* Reply to the report of the committee of the house of commons on indian public works

Cotton diseases in florida / Walker, M N – Gainesville, FL. 1930 – 1r – us UF Libraries [630]

Cotton end baptist monumental inscriptions – Bedfordshire Family HS 2000 – (= ser Bedfordshire parish register series) – 1mf – 9 – £1.25 – uk BedsFHS [929]

Cotton experiment with long or black-seed cotton : weeds of florida / Neal, James Clinton – Lake City, FL. 1890 – 1r – us UF Libraries [630]

Cotton factory times – Manchester ALS 16 jan 1885-2 jul 1937, 9 jul 1937-7 jul 1967 – 1 – (discontinued; fr 27 sep 1907 onward publ at ashton-under-lyne) – uk MLA; uk Newsplan [072]

Cotton, George *see* Revelation, christianity and the bible

Cotton gin and oil mill press – Mesquite. 1950+ (1) 1970+ (5) 1977+ (9) – ISSN: 0010-9800 – mf#433 – us UMI ProQuest [540]

Cotton, Henry *see*
- Editions of the bible and parts thereof in english
- Memoir of a french new testament, in which the mass and purgatory are found in the sacred text...
- Memoir of a french translation of the new testament
- Rhemes and doway
- The succession of the prelates and members of the cathedral bodies in ireland. vol. 2, the province of leinster
- The succession of the prelates and members of the cathedral bodies in ireland, vol 3
- The succession of the prelates and members of the cathedral bodies in ireland, vol 4
- The succession of the prelates and members of the cathedral bodies of ireland, vol 1

Cotton, Henry John Stedman *see* New india

Cotton is king, and pro-slavery arguments : comprising the writings of hammond, harper, christy, stringfellow, hodge, bledsoe, and cartwright on this important subject / Christy, David et al; ed by Elliott, E N – Augusta, GA: Pritchard, Abbott & Loomis, 1860 [mf ed 1984] – 10mf – 9 – 0-8370-1040-3 – (incl bibl ref; essay on slavery by ed) – mf#1984-4377 – us ATLA [976]

Cotton, John *see*
- Gods mercie mixed with his justice, or his peoples deliverance in times of danger
- The keyes of the kingdom of heaven, and power thereof, according to the word of god

Cotton, Joseph Jr *see* Constitutional decisions of john marshall

Cotton market, reports on the... 1848-63 : from liverpool public library – (= ser British records relating to america in microform) – 1r – 1 – (with int by g l rees) – mf#95788 – uk Microform Academic [975]

Cotton mather, the puritan priest / Wendell, Barrett – New York: Dodd, Mead c1891 [mf ed 1991] – (= ser Makers of america) – 1mf – 9 – 0-524-01027-7 – (incl bibl ref) – mf#1990-0304 – us ATLA [243]

Cotton states : weekly confederate newspaper – Gainesville, FL. apr 16 1864 – (= ser Confederate newspapers) – 1r – 1 – us Western Res [071]

The cotton supply question : in relation to the peculiarities and resources of india / Smith, Ronald M – London 1862 – (= ser 19th c british colonization) – 1mf – 9 – mf#1.1.2884 – uk Chadwyck [333]

Cotton varieties for florida / Carver, W A – Gainesville, FL. 1935 – 1r – us UF Libraries [630]

Cotton, Walter Aidan *see* Racial segregation in south africa

Cotton, William *see*
- A catalogue of the portraits painted by sir joshua reynolds
- Sir joshua reynolds, and his works

Cotton's weekly – n25-n172 [1909 mar 4-1911 dec 12], n204 [1912 aug 8], n213 [1912 oct 10], n317-n340 [1915 jan 7-jul 8] – 1r – 1 – (cont by: canadian forward) – mf#700497 – us WHS [071]

Cottonwood press – Anderson, CA. 1972-1976 (1) – mf#62078 – us UMI ProQuest [071]

Cottony cushion scale / Gossard, H A – Lake City, FL. 1901 – 1r – us UF Libraries [630]

Cottrell, Kent *see*
- Sunburnt africa
- Sunburnt sketches of africa south, east and west in pencil paint

Cottrell, Randall R *see* An analysis of the motivational impact of a health risk appraisal and a college health education course utilizing a lifestyle theme on selected health behaviors

Cottrell, Stuart P *see* Predictors of responsible environmental behavior among boaters on the chesapeake bay

Cottu, Charles *see* De la necessite d'une dictature

Couanier de Launay, Etienne-Louis *see* Histoire des religieuses hospitalieres de saint-joseph (france et canada)

Couard, Hermann *see* Der brief pauli an die roemer

Couard, Ludwig *see*
- Altchristliche sagen ueber das leben jesu und der apostel
- The life of christians during the first three centuries of the church
- Die religioesen und sittlichen anschauungen der alttestamentlichen apokryphen und pseudepigraphen

Couat, Auguste Henri *see* Etude sur catulle

Coube, Stephen *see*
- Au pays des castes
- The great supper of god

Couch, Lorinda C *see* Restenosis rate after percutaneous transluminal coronary angioplasty in cardiac rehabilitation program participants

Couchaud, A *see* Choix d'eglises bysantines en grece

The coucher book of the cistercian abbey of kirkstall / ed by Lancaster, W T & Baildon, W Paley – Leeds, 1904 – 8mf – 8 – €17.00 – ne Slangenburg [241]

Coucheron-Aamot, William *see*
- Die chinesen und die christliche mission
- Kineserne og den kristne mission

Couchery, Jean *see* Le moniteur secret

Coudenhove-Kalergi, Richard Nicolaus *see*
- Antisemitismo (version del aleman) y el antisemitismo
- Ethik und hyperethik

Coudreau, Henri Anatole *see*
- Explorations en guyane
- Richesses de la guyane francaise
- Viagem ao tapajos

Coues, Elliott *see* Key to north american birds

The cougar cry – Kramer, ND: Camp Ding, BF-4, Co 766, Civilian Conservation Corps, feb 12 1933? (mthly) – 1 – mf#04624 – us North Dakota [071]

Cougnard, J *see* Ce qui sauve

Couillard-Despres, Azarie *see*
- Histoire de la seigneurie de st-ours
- Louis hebert
- La premiere famille francaise au canada

Couissin, P *see* Les armes romaines

Coulanges, Philippe E de *see* Memoires de m de coulanges

Coulbeaux, J B *see* Histoire politique et religieuse d'abbysinie

Couldrey, Oswald *see* South indian hours

Coulee gazette – La Crosse WI. 1976 jan 1-1978 jun 28, 1978 jul 5-aug 30, 1978 sep 6-dec 27, 1979 jan 10-aug 22 – 4r – 1 – mf#824641 – us WHS [071]

Coulee region highlights / Parents Without Partners – v3 n6-v5 n2 [1977 apr-1978 dec] – 1r – 1 – mf#619629 – us WHS [305]

Coulet, Jules *see* Etudes sur l'ancien poeme francais du voyage de charlemagne en orient

Couleur du temps / Normand, Michelle le – Montreal:edition du Devoir, 1919 [mf ed 1998] – 2mf – 9 – 0-665-66944-5 – mf#66944 – cn Canadiana [830]

Couleurs de marguerite / Bayard, Jean-Francois-Alfred – Paris, France. 1845 – 1r – us UF Libraries [440]

Coulie, B *see*
- Thesaurus amphilochii iconiensis
- Thesaurus asterii amaseni et firmi caesariensis
- Thesaurus basilii caesariensis, 1 et 2
- Thesaurus conciliorum oecumenicorum
- Thesaurus procopii caesariensis
- Thesaurus pseudo-nonni
- Thesaurus pseudo-nonni quondam panopolitani, paraphrasis evangelii s ioannis
- Thesaurus theophanis confessoris. chronographia

Couling, Samuel *see*
- The encyclopaedia sinica
- Sailor's hope, or, sweeter by-and-by

Coulombe, Marguerite *see* Bio-bibliographie du reverend pere francis goyer

Coulombe, Marie-Anne *see* Bibliographie analytique partielle de la cote-nord

Coulon, Emile *see*
- Nouvelle grammaire francaise
- Sequel to poetical leisure hours and torontonian descriptions

Coulon, R *see* Scriptores ordinis praedicatorum

Couloubaly, Pascal Baba F *see*
- Les associations bambara et leurs chants recreatifs, tome 1
- L'enfance bambara

Coulsdon and purley times *see* Purley and coulsdon times

Coulsdon purley and district news – London UK, nov 1931-feb 1933 – 1/4r – 1 – uk British Libr Newspaper [072]

Coulson, Charles Alfred *see* Waves

Coulter David L *see* Journal of religion, disability and health

Coulton, George Gordon *see*
- Catholic truth and historical truth
- From st francis to dante

Coulvier-Gravier, Remi Armand *see* Recherches sur les meteores

Coulvier-Gravier, Remi Armand et al *see* Recherches sur les etoiles filantes

Councellor *see* Reasoner series, 1846-72

Council / ed by Aldred, Guy A – [Scotland] Glasgow: Bakunin Press oct 1931-10 may 1933 (bimthly) [mf ed 2003] – 1r – 1 – uk Newsplan [072]

Council / Tanana Chiefs Conference – 1976 mar-1980 sep – 1r – 1 – mf#626303 – us WHS [307]

The council *see* The works of guy aldred

Council 12 banner / Air Line Pilots' Association – 1975 jan-1993 aug/sep – 1r – 1 – mf#1062796 – us WHS [380]

Council 66 news / American Federation of State, County and Municipal Employees [AFSCME] – v1 n1-v5 n1 [1981 aug-1986 jan/feb] – 1r – 1 – (cont by: afscme council 66 news) – mf#1289108 – us WHS [350]

Council at rome for war with the lamb / Stuart, A Moody – London, England. 1869 – 1r – us UF Libraries [240]

Council bluffs beilage – Council Bluffs, IO: [s.n.], jan 25-mar 22 1923 – 1r – us CRL [071]

Council bluffs bugle – City of Council Bluffs, IA: J E Johnson, 1853-v7 n10 jun 23 1857 (wkly) – 1 – (cont: western bugle; absorbed: omaha arrow; cont by: weekly council bluffs bugle) – us Bell [071]

Council bluffs bugle (1853) – City of Council Bluffs, IA: J E Johnson, 1853-v7 n10 jun 23 1857 (wkly) – 1r – 1 – (cont: western bugle; absorbed: omaha arrow; cont by: weekly council bluffs bugle) – us Eastman [071]

Council bluffs bugle (1860) – Council Bluffs, IA: Babbitt & Carpenter. v10 n7 jun 6 1860-v18 n8 jul 23 1868 (wkly) – 3r – 1 – (cont: weekly council bluffs bugle; cont by: council bluffs weekly bugle; suppl accompany some iss) – us Eastman [071]

Council Bluffs Central Labor Union *see* Farmer-labor press

Council bluffs freie presse und wochentliche omaha tribune – Council Bluffs, IO, and Omaha, NE: Philip Andres 1920-1923. 1922-jan 3 1923 – 1r – us CRL [071]

Council bluffs weekly bugle – Council Bluffs, IA: C H Babbitt. v18 n9 jul 30 1868-v19 n29 dec 23 1869 (wkly) – 1 – (cont: council bluffs bugle (council bluffs, ia: 1860). cont by: council bluffs bugle (council bluffs, ia: 1870)) – us Eastman [071]

Council fire – v1 n2-v3 n4 [1977 jul 17-1980 apr] – 1r – 1 – mf#604898 – us WHS [307]

Council fire... : a monthly journal devoted to the civilization and rights of the american indian – v1-8. 1878-85 – 1 – $54.00 – mf#0168 – us Brook [322]

Council fire... devoted to the civilization and rights of the american indian – v.1-12. 1878-89. (LC lacks v.5, no.10-12) – 1 – us L of C Photodup [970]

Council fires / Coeur d'Alene Tribal Council – v4 n9-v5 n2 [1980 jul-1981 apr] – 1r – 1 – mf#604883 – us WHS [307]
Council for Advancement and Support of Education *see* Case currents
Council for American Indian Ministry *see* Caim news
Council for National Cooperation in Aquatics *see* Archives, records, reference material and conference reports
Council for national cooperation in aquatics : archives, records, reference material, conference reports 1951-1972 – 32mf – 9 – $64.00 – us Kinesology [790]
Council for national cooperation in aquatics : biennial conference reports 1974-1980 – 8mf – 9 – $16.00 – us Kinesology [790]
Council for Research in Music Education *see* Bulletin – council for research in music education
Council for sciences of indonesia / Indonesian abstracts – Djakarta, 1958-1965. v1-7(4) – 17mf – 9 – mf#SE-568 – ne IDC [959]
Council for the Defense of Freedom [US] *see* Washington inquirer
Council for the Investigation of Vatican Influence and Censorship *see* Memorandum on the penetration of the trades unions by roman catholic guilds
Council for the Spanish Speaking *see* Universal
Council for world mission archives, 1795-1941 – 23,348mf – 9 – (incl coll of 115 letters written by david livingston; with p/g) – ne IDC [240]
Council for world mission archives 1941-1950 – 1 title on 2223mf – 9 – (also incl are files of the commonwealth missionary society which merged with lms in 1966; africa files [502mf], asia files [1188mf], oceania-australia files [304mf], board files [103mf], commonwealth missionary society archives [126mf]) – ne IDC [240]
Council minutes, 1660-1800 *see* Collections from the royal society
Council minutes of the royal college of physicians *see* Archives of the royal college of physicians, 1518-1988
Council of Action for Peace and Reconstruction *see* News bulletin
The council of advice at the cape of good hope, 1825-1834 : a study in colonial government / Donaldson, Margaret E – [s.l.: s.n.], 1974 – us CRL [960]
Council of Educational Facility Planners *see* Cefp journal
Council of Energy Resource Tribes *see* Cert report
Council of Europe. Directorate of Legal Affairs *see* Survey of the laws of the member states on payment of maintenance between divorced spouses
Council of legal education calendar, 1901-1925/26 / Inns of Court. England. Council of Legal Education – 111mf – 9 – $166.00 – (the council was established by the societies of lincoln's inn, the inner temple, the middle temple and gray's inn) – mf#LLMC 95-261 – us LLMC [340]
Council of Micronesia *see* Proceedings
Council of New Jersey State College Locals *see* New jersey voice of higher education
Council of Nicaea (2nd : 787) *see* The seventh general council, the second of nicaea
Council of Pacific Teachers Organisations *see* Minutes of annual meetings, reports and women's network files
Council of Planning Librarians *see* Exchange bibliographies
Council of planning librarians : bibliographies – n1-168. 1978-1986 – 9 – $435.00 set – 0-89941-268-8 – mf#400320 – us Hein [020]
Council of planning librarians exchange bibliographies – Chicago: Council of Planning Librarians, n1-1565, 1958-1978 – 9 – $2995.00 set – 0-89941-267-X – mf#400111 – us Hein [020]
Council of Social Agencies [Milwaukee WI] et al *see* Community service news
Council of state governments publications – Lexington: Council of State Governments, 1930-2000 n1 update – 9 – $6471.00 set – 0-89941-271-8 – (annual update ca $195) – mf#400121 – us Hein [350]
Council of the Great City Schools [US] *see* Urban educator
Council of the law society annual reports – Law Society of Great Britain, 1896-1953 – 88mf – 9 – $132.00 – (lacking: 1903) – mf#LLMC 84-452 – us LLMC [340]
Council of the Thirteen Original States *see*
– Historical quarterly of the council of the thirteen original states, inc
– Newsletter of the great american achievements program
Council on Abandoned Military Posts *see* Headquarters heliogram
Council on Christian Medical Work *see* Council on christian medical work
Council on christian medical work : bulletin / Council on Christian Medical Work – n37-41. nov 1947-oct 1948* – 1r – 1 – mf#ATLA S0707E – us ATLA [240]
Council on Environmental Quality *see* Environmental quality
Council on environmental quality annual reports – 1970-93 – 130mf – 9 – $195.00 – (lacking: 1992) – mf#llmc 81-221 – us LLMC [333]
Council on foreign relations : publications – New York: Council on Foreign Relations, v1-80. 1928- – 9 – $3865.00 set – 0-89941-212-2 – (collection incl foreign affairs periodicals) – mf#400130 – us Hein [327]
Council on Technology Teacher Education (US) *see* Yearbook council on technology teacher education (us)
Council on the study of religion bulletin – Waterloo. 1978-1985 (1,5,9) – ISSN: 0002-7170 – mf#11746 – us UMI ProQuest [200]
Council paper / Trinidad Legislative Council – Port-of-Spain, Trinidad and Tobago. na. 1891 jan-1947 – 14r – us UF Libraries [324]
Council records / Ontario College of Art – 18v. 1912-jun 1979 – 6r – 1 – Can$425.00 – (records incl biographical information on artists assoc with the oldest college of art in canada. (in 1996, name changed to ontario college of art and design)) – cn McLaren [700]
Council signals : a publication from the office of the coordinator of indian affairs – Montana. v1 n1-9, v2 n1, 6-17 [1986 apr-dec, 1987 jan, jun-1988 nov/dec] – 1r – 1 – mf#1671992 – us WHS [307]
Councilor / Citizens' Council of Louisiana – 1973 sep 12-1981 aug, v2 n24-? [1964 dec 31-1973 sep 11] – 2r – 1 – mf#711779 – us WHS [350]
Councils, ancient and modern : from the apostolical council of jerusalem, to the oecumenical council of nicaea, and to the last papal council in the vatican / Rule, William Harris – London: Hodder & Stoughton, 1870 [mf ed 1990] – 1mf – 9 – 0-7905-6615-X – mf#1988-2615 – us ATLA [240]
Councils and ecclesiastical documents : relating to great britain and ireland / ed by Haddan, Arthur West & Stubbs, William – Oxford. v1-3. 1869-71 – €61.00 – (vl 1869 13mf. v2 1873 7mf. v3 1871 12mf) – ne Slangenburg [240]
Councils and ecclesiastical documents relating to great britain and ireland / ed by Haddan, Arthur West & Stubbs, William – Oxford: Clarendon Press, 1869-1878 – 5mf – 9 – 0-7905-4858-5 – (incl bibl ref) – mf#1988-0858 – us ATLA [240]
Council's voice / Greater Johnstown Regional Central Labor Council – v1 n2-3 [1979 aug-dec] – 1r – 1 – (cont: labor council newsletter) – mf#646920 – us WHS [350]
Counsel and encouragement : discourses on the conduct of life / Ballou, Hosea – Boston: Universalist Pub House, 1866 – 5mf – 9 – 0-524-07847-5 – mf#1991-3392 – us ATLA [240]
Counsel & care = Berading & besorgdheid / South African Foundation for Mental Health – Johannesburg: S A Foundation for Mental Health [1981]- [mf ed Pretoria, RSA: State Library [199-]] – v1 n1 (sep 1981)- on 1r with other items – 5 – mf#r25 – us CRL [362]
Counsel for the times / Scales, Thomas – London, England. 1841 – 1r – us UF Libraries [240]
Counsel of chalcedon / Chalcedon Presbyterian Church [Atlanta GA] – v7 n1-v11 n4 [1985 mar-1989 jun], 1989 jul-1991 dec – 2r – 1 – mf#1054919 – us WHS [242]
Counsel to new missionaries / McGilvary, Daniel et al – New York: Board of Foreign Missions of the Presbyterian Church in the USA, 1905 – 2mf – 9 – 0-524-07441-0 – mf#1991-3101 – us ATLA [240]
Counseling and human development – Denver. 1977+ (1) 1977+ (5) 1977+ (9) – (cont: focus on guidance) – ISSN: 0193-7375 – mf#10348,01 – us UMI ProQuest [370]
Counseling and values – Falls Church. 1956+ (1) 1975+ (5) 1975+ (9) – ISSN: 0160-7960 – mf#10437 – us UMI ProQuest [370]
Counseling psychologist – Thousand Oaks. 1969+ (1) 1975+ (5) 1975+ (9) – ISSN: 0011-0000 – mf#10945 – us UMI ProQuest [150]
Counsellor : the new york law school law journal – v1-5. 1891-96 (all publ) – (= ser Historical legal periodical series) – 1 – $60.00 set – mf#102121 – us Hein [340]
The counsellor : journal of the new york law school – v1-5. 1891-96 (all publ) – 15mf – 9 – $22.50 – mf#LLMC 82-918 – us LLMC [340]
Counselor education and supervision – Washington. 1961+ (1) 1971+ (5) 1971+ (9) – ISSN: 0011-0035 – mf#2492 – us UMI ProQuest [370]
Counselor motivations for choosing summer resident camp employment / Roark, Mark F – 2000 – 86p on 1mf – 9 – $5.00 – mf#RC – us Kinesology [650]
Counsels to those who are living in the world / Fenelon, Francois De Salignac De La Mothe- – London, England. 1851 – 1r – us UF Libraries [240]
Count campello : an autobiography: giving his reasons for leaving the papal church / Campello, Enrico di – London: Hodder and Stoughton, 1881 – 1mf – 9 – 0-8370-9694-4 – mf#1986-3694 – us ATLA [920]
Count of monte cristo / Dumas, Alexandre – Boston, MA. v1-3. 1899 – 1r – us UF Libraries [440]
Count your blessings : a record of bible promise and of answered prayer / Simpson, Albert B – Nyack: Christian Alliance, [1900?] [mf ed 1991] – (= ser Christian & missionary alliance coll) – 1mf – 9 – 0-524-01751-4 – mf#1990-4143 – us ATLA [220]
Counter address to the protestants of great britain and ireland / Le Mesurier, Thomas – London, England. 1813 – 1r – us UF Libraries [242]
Counter attack / New Haven Panther Defense Committee – v1 n1 [1970 may] – 2r – 1 – mf#720834 – us WHS [320]
Counter manifesto to the annexationists of montreal / Kirby, William – Niagara [i.e. Niagara-on-the-Lake, Ont]: printed & publ by J A Davidson, 1849 [mf ed 1984] – 1mf – 9 – 0-665-45329-9 – mf#45329 – cn Canadiana [380]
Counter news / Retail Clerk's Union – v19 n2-v22 n4 [1971 mar/apr-1974 fall] – 1r – 1 – mf#365084 – us WHS [380]
Counter pentagon / Central Committee for Conscientious Objectors – [1974 apr], v4 n4-v9 n6 [1977 sep 15-1982 dec/1983 jan] – 1r – 1 – mf#688794 – us WHS [355]
Counter-attack / Movement for a Democratic Military – v1-4 [1970 jun ?-dec] – 2r – 1 – mf#720837 – us WHS [355]
Counterattack : facts to combat communism and those who aid its cause / American Business Consultants, Inc – 1947 may 16-1954 dec 10, 1954 dec 17-1973 jul 9 – 2r – 1 – mf#1054922 – us WHS [650]
Counterdraft / Peace Action Council – 1968 jan-1971 mar-may – 1r – 1 – mf#1110964 – us WHS [303]
Counter/measures – Bedford. 1972-1974 (1) – ISSN: 0070-1246 – mf#8217 – us UMI ProQuest [400]
Counterpoint – 1968 feb-1969 nov – 1r – 1 – mf#1110965 – us WHS [071]
Counterpoint – 1979 jul 30-1981 jan – 1r – 1 – mf#665539 – us WHS [071]
Counterpoint / GI-Civilian Alliance for Peace – 1968 nov-1969 sep 20 – 2r – 1 – mf#720831 – us WHS [355]
Counterpoint in the music of claude debussy / Caravan, Ronald L – U of Rochester 1973 [mf ed 19–] – 3mf – 9 – mf#fiche707, 708 – us Sibley [780]
A counter-poyson : modestly written for the time, to make answere to the obiections and reproches... / Fenner, D – London: Robert Waldegrave, [1584] – 3mf – 9 – mf#PW-61 – ne IDC [240]
Counterspy – Washington. 1973-1984 (1) 1973-1984 (5) 1973-1984 (9) – (cont by: national reporter) – ISSN: 0739-4322 – mf#10225 – us UMI ProQuest [320]
Countess of huntingdon – London, England. no date – 1r – us UF Libraries [240]
The countess tekla / Barr, Robert – London: Methuen, 1899 – 6mf – 9 – (previously issued in 1898 under title: tekla) – mf#41933 – cn Canadiana [830]
The countesse of pembroke's arcadia, examined and discussed / Harman, Edward George – London: Cecil Palmer (1924). x,233p., front., facsim – 1 – us Wisconsin U Libr [440]
Counties courier – Papakura, NZ. jan-jun 1989; jan-jun 1990 – 2r – 1 – mf#15.50 – nz Nat Libr [079]
Countries, nations, and languages of the oceanic region – [London]: [Chapman & Hall] [1834] – (= ser Whsb) – 1mf – 9 – €20.00 – (fr: the foreign quarterly review n28) – mf#Hu 485a – gw Fischer [954]
The countries of the western world : the governments and people of north, south and central america, from the landing of columbus to the present time / Lossing, Benson John et al – New York: Gay Bros, 1890 [mf ed 1984] – 9mf – 9 – mf#34499 – cn Canadiana [910]
Country – Adelaide, Australia. 2 sep 1893-5 may 1894; 1895; 11 jan-26 dec 1896 – 2r – 1 – uk British Libr Newspaper [072]
The country – Adelaide, Australia. 2 Sep 1893-5 May 1894; 12 Jan 1895-26 Dec 1896.-w. 2 reels – 1 – uk British Libr Newspaper [072]
Country americana magazine – v1-v4 n6 [1975 jan/feb-1978 dec] – 1r – 1 – mf#499211 – us WHS [071]
Country architecture : a work, designed for the use of the nobility and country gentlemen / Birch, John – Edinburgh 1874 – (= ser 19th c art & architecture) – 2mf – 9 – mf#4.2.1016 – uk Chadwyck [720]
Country beautiful – Waukesha. 1966-1967 – 1 – ISSN: 0011-0116 – mf#2163 – us UMI ProQuest [073]
"Country before party" / Dominion National League – Hamilton ON: The League, 1878 – 1mf – 9 – mf#02735 – cn Canadiana [330]
Country chronicle – Denmark WI. 1982 apr 14/dec-1996/97 – 14r – 1 – (with gaps; cont: new farmer's friend and rural reporter) – mf#601100 – us WHS [630]
The country church and the rural problem / Butterfield, Kenyon Leech – Chicago, IL: University of Chicago Press, c1911 – (= ser Carew Lectures) – 1mf – 9 – 0-524-05431-2 – mf#1990-1463 – us ATLA [240]
Country clergyman's observations upon some letters which have been... – Chelmsford, England. 18– – 1r – us UF Libraries [240]
Country courier – New York. 1816-1817 (1) – mf#3731 – us UMI ProQuest [420]
Country courier – St. Stephens Church, VA. 1992-2000 (1) – mf#68917 – us UMI ProQuest [071]
Country cross roads – [v1 n1]-[v2 n5] [1978 jan-1979 may] – 1r – 1 – (cont by: Spotlight [Madison WI: 1979]) – mf#674014 – us WHS [071]
Country dance and song – Northampton. 1968-1975 (1) 1968-1975 (5) 1974-1975 (9) – ISSN: 0070-1262 – mf#7658 – us UMI ProQuest [390]
The country from cape palmas to river congo : from the royal commonwealth society library / Adams, J – 1823 – 6mf – 7 – (with app) – mf#2987 – uk Microform Academic [916]
Country gentleman – Albany, NY. 1853-1910 (1) – mf#64877 – us UMI ProQuest [071]
Country gentleman – Indianapolis. 1949-1981 (1) 1975-1981 (5) 1975-1981 (9) – ISSN: 0147-4928 – mf#127 – us UMI ProQuest [073]
Country gentleman – Philadelphia, PA. 1911-1947 (1) – mf#66018 – us UMI ProQuest [071]
Country guide : eastern edition – Winnipeg, CN. 1980-89 – 17r – 1 – cn Commonwealth Imaging [073]
Country guide : western edition – Winnipeg, CN. 1980-89 – 17r – 1 – cn Commonwealth Imaging [073]
Country guide – Winnipeg, CN. 1908-79 – 47r – 1 – cn Commonwealth Imaging [073]
Country handcrafts – Greendale. 1989-1990 (1) – ISSN: 0745-3116 – mf#15134 – us UMI ProQuest [790]
The country house / Fox, Mary – London 1843 – (= ser 19th c art & architecture) – 1mf – 9 – (with designs) – mf#4.2.1131 – uk Chadwyck [720]
Country independent – at Pascoe [079]
Country journal – Harrisburg. 1986-2001 [1,5,9] – (cont: blair and ketchum's country journal) – ISSN: 0898-6355 – mf#10275,01 – us UMI ProQuest [073]
Country journal : or the craftsman – London. 1726-1733 (1) – mf#4847 – us UMI ProQuest [740]
Country journal or craftsman – [London & SE] Hackney 7 oct 1749-dec 1752 [mf ed 2003] – 2r – 1 – uk Newsplan [740]
Country life – London. 1897-1996 (1) 1988-1996 (5) 1988-1996 (9) – ISSN: 0045-8856 – mf#823 – us UMI ProQuest [073]
Country life – Sydney, Australia. 7 dec 1951-12 jun 1953 – 3 1/2r – 1 – uk British Libr Newspaper [072]
Country Line Primitive Baptist Association *see* Minutes of country line primitive baptist association
The country literary chronicle and weekly review (london) – 1 jul 1820-30 dec 1820 – (= ser 19th c british periodicals) – r9 – 1 – us Primary [410]
The country literary chronicle and weekly review (london) – 3 jan 1824-25 dec 1824 – (= ser 19th c british periodicals) – r13 – 1 – us Primary [410]
The country literary chronicle and weekly review (london) – 4 jan 1823-27 dec 1823 – (= ser 19th c british periodicals) – r12 – 1 – us Primary [410]
The country literary chronicle and weekly review (london) – 5 jan 1822-28 dec 1822 – (= ser 19th c british periodicals) – r11 – 1 – us Primary [073]
The country literary chronicle and weekly review (london) – 6 jan 1821-17 dec 1821 – (= ser 19th c british periodicals) – r10 – 1 – us Primary [410]
Country living – 1978 oct 18-1979, 1980-81, 1982 jan-jun, 1982 jul-oct 27 – 4r – 1 – (cont by: country news [la crosse wi]) – mf#500716 – us WHS [640]
Country living – New York. 1982+ (1,5,9) – ISSN: 0732-2569 – mf#14188,01 – us UMI ProQuest [640]
Country merchant – Lincoln, NE: Country Merchant Pub. v3 n17. oct 19 1901- (wkly) [mf ed -sep 18 1909 (gaps) filmed 1998] – 4r – 1 – (cont: produce weekly reporter and prices current. multiple date and numbering errors) – us NE Hist [630]

Country music – West Port. 1986-1999 (1) 1986-1999 (5) 1986-1999 (9) – ISSN: 0090-4007 – mf#15085 – us UMI ProQuest [780]

The country of the neutrals : (as far as comprised in the county of elgin), from champlain to talbot / Coyne, James Henry – St Thomas, Ont: Times Print, 1895 – 1mf – 9 – mf#03619 – cn Canadiana [305]

Country Pastor *see* Long-lost brother

Country people – v1 n1-v7 n1 [1980 nov-1986 nov/dec] – 1r – 1 – (cont by: country [greendale wi]) – mf#1218312 – us WHS [071]

Country press – v1 n16, 18-20 [1972 aug, oct-dec] – 1r – 1 – mf#1583061 – us WHS [071]

Country reports from the eiu on microfiche / Economic Intelligence Unit – 1952-97+ – £25,500.00 set – (cont: quarterly economic reviews; provide accurate up-to-the-minute assessments of the general economic situation in approx 190 countries; purchase by country or yr possible; contents list available; coll also available on 16mm) – mf#QER – uk World [330]

Country senses – 1971 may, [nov] – 1r – 1 – mf#1054927 – us WHS [071]

Country spectator – Gainsborough. 1792-1793 (1) – mf#4231 – us UMI ProQuest [630]

Country statement for african population conference, accra, 9-18 december 1971 / Swaziland. Central Statistical Office – Mbabane: The Government 1971? [mf ed Pretoria: State Library 1973] – 1mf – 9 – 0-7978-0532-X – mf#mf.186 – sa State Libr [316]

Country today – 1983 nov 3-1984 mar, 1984 apr-jun, jul-dec, 1985 jan-mar, apr-jun, jul-sep 25 – 6r – 1 – (cont by: country today [eau claire wi: state ed, swe]) – mf#715796 – us WHS [071]

Country today – 1985 oct 2/1986 mar-1988 jan/mar – 9r – 1 – (with gaps; cont: country today [baraboo wi: southern ed]) – mf#1011800 – us WHS [071]

Country today – 1979 jan 17/aug 29-1990 jul-sep – 41r – 1 – (with small gaps) – mf#499208 – us WHS [071]

Country wide – Palmerston North, NZ. 1985-86 – 1r – 1 – mf#45.16 – nz Nat Libr [079]

Country wide – Palmerston North, NZ. oct 1978-dec 1982; mar-dec 1983; feb-dec 1984 – 1 – mf#ZP 3 – nz Nat Libr [079]

Country women – v1 n7-17 [1973 jun-1975 oct] – 1r – 1 – mf#1002615 – us WHS [305]

Countryman – Sun Prairie WI. 1877 dec 6-1880 dec 30, 1881 jan 1-1882 sep 28 – 2r – 1 – (cont by: sun prairie countryman) – mf#933503 – us WHS [071]

Countryman – Coburg OR: [s.n.] -1979 [wkly] – 2r – 1 – (cont: coburg countryman [-1971]; cont by: countryman community news [1979-198?]) – us Oregon Lib [071]

Countryman community news – Coburg OR: [s.n.] 1979- [semimthly] – 1r – 1 – (cont: countryman [1971-79]) – us Oregon Lib [071]

Countryside – Barron WI. v1 n1-v3 n15 [1981 mar 3-aug 30] – 1r – 1 – mf#610344 – us WHS [071]

Countryside – Waterloo. 1985-1985 (1,5,9) – ISSN: 0363-8723 – mf#14799,06 – us UMI ProQuest [636]

Countryside and small stock journal – Waterloo. 1985+ (1,5,9) – ISSN: 8750-7595 – mf#14799,07 – us UMI ProQuest [636]

Countryside churches = Hsiang-ts'un chiao-hui – n1-3. dec 1940-dec 1942; n8-9. dec 1946-jun 1947 [complete] – (= ser Chinese christian coll 54) – 1r – 1 – (in chinese) – mf#ATLA S0296J – us ATLA [915]

Countryside miscellaneous – Barrington, IL. 1959-1974 (1) – mf#68646 – us UMI ProQuest [071]

Countryside, past and present / Molyakov, Vasilii Fedorovich – Moscow: Foreign Languages Pub House 1939 [mf ed 2000] – 1r – 1 – mf#29077 – us Harvard [630]

Countryside reminder news – Barrington, IL. 1977-1984 (1) – mf#62508 – us UMI ProQuest [071]

Countryside series / Lucas Co. Toledo – (jul 1974-jan 1975) [semimthly] – 1r – 1 – mf#B34500 – us Ohio Hist [071]

Counts, Charlene L M *see* The effect of swim training on plasma somatomedin-c levels in 8- to 10-year-old children

County and district records : aiken (city) / South Carolina. Dept of Archives and History – 1895-1916. Treasurer's Tax Duplicates. 1 reel – 1 – $30.00r – us South C Archives [025]

County and district records : charleston county / South Carolina. Dept of Archives and History – 1735-1915. 31 reels. More details on request – 1 – $30.00r – us South C Archives [025]

County and district records : florence county / South Carolina. Dept of Archives and History – 1874-1981. 58 reels. More details on request – 1 – $30.00r – us South C Archives [025]

County and district records : marion county / South Carolina. Dept of Archives and History – 1790-1955. 123 reels. More details on request – 1 – $30.00r – us South C Archives [025]

County and district records : spartanburg county / South Carolina. Dept of Archives and History – 1785-1968. 275 reels. More details on request – 1 – $30.00r – us South C Archives [025]

County and district records : st paul's parish / South Carolina. Dept of Archives and History – Road Commissioner's Minutes, 1783-1839. 1 reel – 1 – $30.00r – us South C Archives [025]

County and district records: abbeville county / South Carolina. Dept of Archives and History – 1872-1925. 46 reels. More details on request – 1 – $30.00r – us South C Archives [025]

County and district records: aiken county / South Carolina. Dept of Archives and History – 1872-1974. 154 reels. More details on request – 1 – $30.00r – us South C Archives [025]

County and district records: anderson county / South Carolina. Dept of Archives and History – Includes Pendleton District. 1790-1972.319 reels. More details on request – 1 – $30.00r – us South C Archives [025]

County and district records: barnwell county / South Carolina. Dept of Archives and History – 1784-1954. 12 reels. More details on request – 1 – $30.00r – us South C Archives [025]

County and district records: beaufort county / South Carolina. Dept of Archives and History – 1864-1980. 60 reels. More details on request – 1 – $30.00r – us South C Archives [025]

County and district records: berkeley county / South Carolina. Dept of Archives and History – 1881-1954. 17 reels. More details on request – 1 – $30.00r – us South C Archives [025]

County and district records: camden district / South Carolina. Dept of Archives and History – 1784-1841. 6 reels. More details on request – 1 – $30.00r – us South C Archives [025]

County and district records: cherokee county / South Carolina. Dept of Archives and History – 1897-1960. 85 reels. More details on request – 1 – $30.00r – us South C Archives [025]

County and district records: chester county / South Carolina. Dept of Archives and History – 1785-1839. 9 reels. More details on request – 1 – $30.00r – us South C Archives [025]

County and district records: chesterfield county / South Carolina. Dept of Archives and History – 1847-1978. 250 reels. More details on request – 1 – $30.00r – us South C Archives [025]

County and district records: clarendon county / South Carolina. Dept of Archives and History – 1857-1980. 106 reels. More details on request – 1 – $30.00r – us South C Archives [025]

County and district records: colleton county / South Carolina. Dept of Archives and History – 1802-1974. 91 reels. More details on request – 1 – $30.00r – us South C Archives [025]

County and district records: darlington county / South Carolina. Dept of Archives and History – 1803-1980. 106 reels. More details on request – 1 – $30.00r – us South C Archives [025]

County and district records: dorchester county / South Carolina. Dept of Archives and History – 1757-1949. 61 reels. More details on request – 1 – $30.00r – us South C Archives [025]

County and district records: eau claire (town) / South Carolina. Dept of Archives and History – Board of Health Minutes.1931-54. 1 reel – 1 – $30.00r – us South C Archives [025]

County and district records: edgefield county / South Carolina. Dept of Archives and History – 1785-1976. 286 reels. More details on request – 1 – $30.00r – us South C Archives [025]

County and district records: fairfield county / South Carolina. Dept of Archives and History – 1785-1970. 165 reels. More details on request – 1 – $30.00r – us South C Archives [025]

County and district records: georgetown county / South Carolina. Dept of Archives and History – 1783-1963. 176 reels. More details on request – 1 – $30.00r – us South C Archives [025]

County and district records: greenville county / South Carolina. Dept of Archives and History – 1787-1976. 281 reels. More details on request – 1 – $30.00r – us South C Archives [025]

County and district records: hampton county / South Carolina. Dept of Archives and History – 1879-1930. 33 reels. More details on request – 1 – $30.00r – us South C Archives [025]

County and district records: horry county / South Carolina. Dept of Archives and History – 1803-1950. 90 reels. More details on request – 1 – $30.00r – us South C Archives [025]

County and district records: irmo (town) / South Carolina. Dept of Archives and History – 1899-1975. Town Council Minutes. 1 reel – 1 – $30.00r – us South C Archives [025]

County and district records: kershaw county / South Carolina. Dept of Archives and History – 1782-1980. 112 reels. More details on request – 1 – $30.00r – us South C Archives [025]

County and district records: lancaster county / South Carolina. Dept of Archives and History – 1762-1960. 148 reels. More details on request – 1 – $30.00r – us South C Archives [025]

County and district records: laurens county / South Carolina. Dept of Archives and History – 1785-1969. 286 reels. More details on request – 1 – $30.00r – us South C Archives [025]

County and district records: lee county / South Carolina. Dept of Archives and History – 1902-20.20 reels. More details on request – 1 – $30.00r – us South C Archives [025]

County and district records: lexington county / South Carolina. Dept of Archives and History – 1806-1964. 237 reels. More details on request – 1 – $30.00r – us South C Archives [025]

County and district records: lexington (town) / South Carolina. Dept of Archives and History – 1917-71.Details on request. 2 reels – 1 – $30.00r – us South C Archives [025]

County and district records: marlboro county / South Carolina. Dept of Archives and History – 1785-1973. 143 reels. More details on request – 1 – $30.00r – us South C Archives [025]

County and district records: newberry county / South Carolina. Dept of Archives and History – 1776-1953. 98 reels. More details on request – 1 – $30.00r – us South C Archives [025]

County and district records: ninety six district / South Carolina. Dept of Archives and History – Plats. 1784-1803. 3 reels – 1 – $30.00r – us South C Archives [025]

County and district records: oconee county / South Carolina. Dept of Archives and History – 1868-1959. 95 reels. More details on request – 1 – $30.00r – us South C Archives [025]

County and district records: orangeburg county / South Carolina. Dept of Archives and History – 1775-1975. 217 reels. More details on request – 1 – $30.00r – us South C Archives [025]

County and district records: pickens county / South Carolina. Dept of Archives and History – 1828-1971. 81 reels. More details on request – 1 – $30.00r – us South C Archives [025]

County and district records: pinckney district (york) / South Carolina. Dept of Archives and History – 1792-1872. 39 reels. More details on request – 1 – $30.00r – us South C Archives [025]

County and district records: richland county / South Carolina. Dept of Archives and History – 1803-1962. 190 reels. More details on request – 1 – $30.00r – us South C Archives [025]

County and district records: saluda county / South Carolina. Dept of Archives and History – 1896-1979. 113 reels. More details on request – 1 – $30.00r – us South C Archives [025]

County and district records: spartanburg (town) / South Carolina. Dept of Archives and History – Town Council Ordinances, 1832-42. 1 reel – 1 – $30.00r – us South C Archives [025]

County and district records: st stephen's parish / South Carolina. Dept of Archives and History – Road Commissioner's Minutes, 1789-99. 1 reel – 1 – $30.00r – us South C Archives [025]

County and district records: sumter county / South Carolina. Dept of Archives and History – 1800-1955. 22 reels. More details on request – 1 – $30.00r – us South C Archives [025]

County and district records: union county / South Carolina. Dept of Archives and History – 1785-1974. 243 reels. More details on request – 1 – $30.00r – us South C Archives [025]

County and district records: williamsburg county / South Carolina. Dept of Archives and History – 1806-1981. 81 reels. More details on request – 1 – $30.00r – us South C Archives [025]

County and district records: york county / South Carolina. Dept of Archives and History – 1786-1974. 409 reels. More details on request – 1 – $30.00r – us South C Archives [025]

County and regional histories and atlases – 660r coll – 1 – (includes tables and list of vital statistics, military service records, municipal and county officers, chronologies, portraits of individuals, and views of urban and rural life. based on materials fr various libraries and research centres in 8 states. california 33r; ilinois 96r; indiana 67r; michigan 88r; new york 116r; ohio 91r; pennyslvania 125r; wisconsin 44r) – us Primary [978]

County banner – Barnwell, SC. 1987-1990 (1) – mf#68330 – us UMI ProQuest [071]

County borough of west ham gazette *see* Forest gate gazette and upon chronicle

County borough of west ham guardian *see* West ham guardian

County boundaries / Florida Laws, Statutes – s.l, s.l? no date – 1r – us UF Libraries [350]

County breeze – Sarasota, FL. 1950-1953 (1) – mf#62445 – us UMI ProQuest [071]

County by county listing of employers : subject to wisconsin's unemployment compensation law as of the first quarter of... / Industrial Commission of Wisconsin – 1957-1972 v2 – 18r – 1 – mf#708369 – us WHS [344]

County chronicle and mark lane journal *see* County chronicle and weekly advertiser for essex herts kent surrey middlesex

County chronicle and weekly advertiser for essex herts kent surrey middlesex – London, UK. 1834, 1837, 1865, 1877, Lewes from jan 1879 – 3r – 1 – (aka: county chronicle surrey herald and weekly advertsier for kent etc; county chronicle and mark lane journal) – uk British Libr Newspaper [072]

County chronicle surrey herald and weekly advertiser for kent... *see* County chronicle and weekly advertiser for essex herts kent surrey middlesex

County clarion – Lapeer, MI. 1880-1923 (1) – mf#63787 – us UMI ProQuest [071]

County commissioners' legal guide, (for the state of texas). / Texas. County Commissioners – St. Louis, Barnard 1894 340 p. LL-506 – 1 – us L of C Photodup [340]

County courier – Camden, NJ. 1880-1893 (1) – mf#64801 – us UMI ProQuest [071]

County courier – Seneca Falls, NY. 1859-1944 (1) – mf#65230 – us UMI ProQuest [071]

County courts chronicle and gazette of bankruptcy – London. 1847-1920. (13-20, 1860-67) wanting) – 1 – us L of C Photodup [941]

County democrat and press – Lapeer, MI. 1879-1901 (1) – mf#63788 – us UMI ProQuest [071]

County derry liberal – Limavardy, Ireland. 13 oct 1888-4 may 1889 – 3/4r – 1 – (incorp with: brotherhood fr 11 may 1889) – uk British Libr Newspaper [072]

County donegal general advertiser *see* Ballyshannon herald

County down reporter – Bangor, Ireland. Jun 1904-Dec 1916; 1925-79 (1930 imperfect).-w. 91mqn reels – 1 – uk British Libr Newspaper [072]

County down spectator and ulster standard – jun 3 1904-1916, 1925-29, mar-nov 1930, 1931-dec 1998 – 161 1/2r – 1 – (aka: ulster standard) – uk British Libr Newspaper [072]

County down spectator, and ulster standard – [Northern Ireland] Belfast jan 1917-dec 1918, 15 jan 1921-dec 1924 [mf ed 2002-04] – 6r – 1 – uk Newsplan [072]

County echo : fishguard and north pembrokeshire advertiser – [Wales] Pembrokeshire jan 1902-dec 1950 [mf ed 2003] – 46r – 1 – (missing: 1897; cont as: the county echo: fishguard, goodwick and the north of pembrokeshire general advertiser [jan 1908-dec 1950]) – uk Newsplan [072]

County employee : official publication / Los Angeles County Employees Association – 1974 jan 1-1976 dec 1, 1975 jan 1-1976 dec 1 – 2r – 1 – (cont: los angeles county employee; cont by: voice [los angeles ca: 1977]) – mf#579301 – us WHS [350]

County express [cheadle and gatley ed] – [NW England] Cheadle 15 sep-dec 1960 – 1 – uk MLA; uk Newsplan [072]

County express for worcestershire and staffordshire – Stourbridge, England. 1920-39.-w. 32 reels – 1 – uk British Libr Newspaper [072]

County farmer / Clinton County Farmers Union [IA] – 1945 dec 4 – 1r – 1 – mf#3925644 – us WHS [630]

County gazette / Athens Co. Athens – dec 1879-jun 1909 [wkly] – 5r – 1 – mf#B10519-10523 – us Ohio Hist [071]

County gazette – Cape May Court House, NJ. 1975-1976 (1) – mf#64810 – us UMI ProQuest [071]

County gazette / Paulding Co. Paulding – 1883, may 1887-jan 1888 [wkly] – 1r – 1 – mf#B6863 – us Ohio Hist [071]

County gazette *see* [San mateo-] times gazette

County government / Whittaker, Ray C – s.l, s.l? 1936 – 1r – us UF Libraries [350]

County hall records, 1719-1890 / Glamorgan County Council – 71r – 1 – (incl quarter sessions minute books; 1719-1890, quarter sessions rolls; 1727-1800, land tax assessment returns, 1783-1831) – mf#96842 – uk Microform Academic [941]

County herald – Holywell, Wales. -w. 15 July 1887-Dec 1947. Lacking 1897, 1911. 48 reels – 1 – uk British Libr Newspaper [072]

County herald – [Wales] Flintshire jan 1910-dec 1950 [mf ed 2003] – 60r – 1 – (missing: 1896-97; cont as: flintshire county herald [jan 1946-dec 1950]) – uk Newsplan [072]

County herald – Redgranite WI. 1939 apr 12 – 1r – 1 – mf#1012053 – us WHS [071]

County herald and lantern – Cape May Court House, NJ. 1996-2000 (1) – mf#68231 – us UMI ProQuest [071]

County herald and weekly advertiser etc – London, UK. 1818-25 apr; 6 jun-27 jun 1828; 1 aug 1829-31 jul 1830; 4 sep 1830-1843; 1858-1860; 1862-1865 – 20r – 1 – uk British Libr Newspaper [072]

County herald for staffordshire & worcestershire – [West Midlands] Dudley 1919-28 feb 1925 [mf ed 2002] – 12r – 1 – uk Newsplan [072]

County histories / New York – v. 1-256 – 9 – us AMS Press [978]

County independent – Choteau, MT. 1910-1925 (1) – mf#64321 – us UMI ProQuest [071]

County independent *see* [Santa cruz-] independent

County journal – Monroe WI. 1890 feb 4-dec 16, 1892 feb 2-1893 sep 26, 1893 oct 3-1894 jan 30, 1895 feb 5-1896 oct 22, 1896 nov 3-1898 jul 19 – 4r – 1 – (cont by: monroe sun-gazette; journal-gazette) – mf#1125627 – us WHS [071]

County journal – Cable WI. 1988 nov 3/1989 jun-1994 jul-dec – 12r – 1 – (cont: south county journal [cable wi]) – mf#5498239 – us WHS [071]

County ledger-press – Balsam Lake WI. 1984 apr 26/jun-1998 jan/jun – 35r – 1 – (cont: polk county ledger and the standard-press) – mf#1211345 – us WHS [071]

County line – Ontario WI. 2000 jan-jun, 2001 jul-dec – 2r – 1 – (cont: county line connection) – mf#4726291 – us WHS [071]

County line / Republican National Committee [US] – v2 n5-v7 n3 [1983 jun-1988 oct/nov] – 1r – 1 – mf#1110967 – us WHS [325]

County line baptist church. oglethorpe county. georgia : church records – 1919-54 – 1 – us Southern Baptist [242]

County line connection – Ontario WI. 1985 aug 29/1986-1999 jan/jun – 14r – 1 – (cont by: county line [ontario wi]) – mf#1048933 – us WHS [071]

The county mail *see* Miscellaneous newspapers of mesa county

County news – Statesville NC. 1998 jan-dec, 1999 jan 7-jun 24, 1999 jul 1-dec 30, 2000 jan-jun – 5r – 1 – (cont: iredell county news) – mf#4033093 – us WHS [071]

County news / Delaware Co. Delaware – dec 1864-jul 1866 [wkly] – 1r – 1 – mf#B1486 – us Ohio Hist [071]

County news / Delaware Co. Delaware – v1 n1. dec 1864-jul 1866 [wkly] – 1r – 1 – mf#B29852 – us Ohio Hist [071]

County news / National Association of Counties – ?-1973 jul 27, 1973 aug 10-1975 dec 22, 1980-81, 1982-84, 1985 jan 14-1987 dec 14, 1988 jan-1989 sep, 1996 jan 22-1998 dec 21 – 7r – 1 – (cont: naco news and views) – mf#3646469 – us WHS [071]

County news – Rotorua, NZ. 1980-1984; jan-jun 1988 – 4r – 1 – mf#17.8 – nz Nat Libr [079]

County news – Washington, DC. 1973-1978 (1) – mf#62388 – us UMI ProQuest [071]

County observer – Warren, PA. 1966-1966 (1) – mf#66133 – us UMI ProQuest [071]

County of middlesex chronicle *see* Staines and district chronicle

County of middlesex chronicle hounslow ed *see* Middlesex chronicle etc hounslow chronicle

County of middlesex chronicle (staines ed) : staines & sunbury chronicle – [London & SE] Hounslow jan 1914-dec 1932, jan 1942-dec 1950 [mf ed 2004] – 37r – 1 – uk Newsplan [072]

County of middlesex independent – [London & SE] Hounslow 3 jan 1885-29 aug 1942 [mf 1906-8, 1910-15, 1918-50] [mf ed 2004] – 1 – (cont: brentford & ealing & county of middlesex independent [2 jul-31 dec 1884; cont by: middlesex independent [5 sep 1942-29 may 1964]) – uk Newsplan [072]

County of museum art catalogue – Los Angeles, 1932-1976 – (= ser Art exhibition catalogues on microfiche) – 93 catalogues on 123mf – 9 – £775.00 – (individual titles not listed separately) – uk Chadwyck [700]

County of ontario : short notes as to the early settlement and progress of the county... / Farewell, J E – Whitby [ON]: Gazette-Chronicle Press, 1907 [mf ed 1998] – 3mf – 9 – 0-665-81578-6 – mf#81578 – cn Canadiana [971]

County of wexford express and wexford new ross enniscorthy and gorey mail *see* Wexford and kilkenny express

County pioneer press – Manistee, MI. 1989-1992 (1) – mf#68557 – us UMI ProQuest [071]

County post – Millsboro, DE. 1972-1978 (1) – mf#62380 – us UMI ProQuest [071]

County press : for northamptonshire, bedfordshire, buckinghamshire and huntingdonshire – [East Midlands] Northampton, Northamptonshire n1 [23 jan 1808]-n179 [26 jun 1811] [mf ed 2002] – 2r – 1 – uk Newsplan [072]

County press – Lapeer, MI. 1990-1999 (1) – mf#61533 – us UMI ProQuest [071]

County record – Blountstown, FL. 1952 dec 12-1997 – 43r – (gaps) – us UF Libraries [071]

County reports to the board of agriculture : 18th and 19th centuries – 242 – 7 – mf#87115 – uk Microform Academic [630]

County republican / Paulding Co. Paulding – (sep 1888-dec 1921) [wkly] – 20r – 1 – mf#B27653-27672 – us Ohio Hist [071]

County republican – Sevierville, TN. 1909-1915 (1) – mf#66568 – us UMI ProQuest [071]

County sentinel – Asotin, WA. 1891-1942 (1) – mf#66929 – us UMI ProQuest [071]

County telephone and salford district review – [NW England] Salford, Manchester ALS 9 feb 1889-1892 – 1 – us MLA; uk Newsplan [072]

County times / Lorain Co. Lorain – dec 1978-apr 1986,jul 1986-dec 1987 [wkly] – 5r – 1 – mf#B29619-29623 – us Ohio Hist [071]

County times and express – Welshpool, Wales, Apr 4-Aug 29 1981; Sep 1981-Mar 1982; 1983-96 – 47r – 1 – uk British Libr Newspaper [072]

County tipperary independent etc – Clonmel, Ireland. 11 nov 1882-18 feb 1893; apr 1894-mar 1895; 11 may-21 dec 1895; 11 jan-18 jan 1896; 21 mar-12 dec 1896 – 14r – 1 – (from 12 sep 1891 publ in waterford) – uk British Libr Newspaper [072]

County view / Inter-league Council of the Leagues of Women Voters of Milwaukee County – v2 n7-v4 n6 [i.e. 7] [1970 dec-1973 jan], v2 n8-v4 n10 [1973 mar-jun], v4 n11 [1973 nov] – 1r – 1 – (cont: county page) – mf#626003 – us WHS [325]

County wexford independent – Wexford, mar 1906-aug 1908 – 4r – 1 – ie National [072]

County whig – Oswego, NY. 1838-1844 (1) – mf#65144 – us UMI ProQuest [071]

Countyline / Williams Co. Montpelier – oct 1977-feb 1980 [wkly] – 4r – 1 – mf#B29698-29701 – us Ohio Hist [071]

Coup de clairon : probleme politique contemporain / Laroche, Dejoie – Cap-Haitien, Haiti. 1908 – 1r – us UF Libraries [972]

Le coup de trique – Paris: Ad Blondeau, [1850]. mar-apr 1850 – us CRL [074]

Coup d'oeil – Beloeil: L'Oeil regional inc. 1re annee n1 4 avril 1984, 1re annee n1 ete 1986-3e annee n11 hiver 1988/1989 [qrtly] [mf ed 1988] – 9 – (suppl to: oeil regional, suspended between 1984-86) – mf#SEM105P892 – cn Bibl Nat [073]

Coup d'oeil politique sur l'avenir de la france / Dumouriez, Charles Francois – Hambourg 1795 [mf ed Hildesheim 1995-98] – 1v on 2mf – 9 – €60.00 – 3-487-26261-4 – gw Olms [944]

Coup d'oeil sur la politique de toussaint-louvertu... / Laurent, Gerard M – Port-Au-Prince, Haiti. 1949 – 1r – us UF Libraries [972]

Coup d'oeil sur l'anthropologie du cambodge : rapport presente da la societe d'anthropologie dans la seance du 7 septembre 1871... / Hamy, Ernest Theodore – [Paris: A Hennuyer 1871] [mf ed 1989] – 1r with other items – 1 – mf#mf-10289 seam reel 009/02 [§] – us CRL [301]

Coup d'oeil sur les arts en nouvelle-france / Morisset, Gerard – Quebec: [s.n.], 1941 [mf ed 1991] – 3mf – 9 – (with ind) – mf#SEM105P1469 – cn Bibl Nat [700]

Coup d'oeil sur les ressources productives et la richesse du canada : suivi d'un "plan d'organisation" complet et detaille, relatif a la colonisation, destine a faire suite aux "etudes sur la colonisation du bas-canada depuis dix ans" / Drapeau, Stanislas – [Quebec?: s.n.], 1865 [mf ed 1984] – 1mf – 9 – 0-665-23174-1 – mf#23174 – cn Canadiana [330]

Coup d'oeil sur les soeurs de l'esperance / Tremblay, Laurent – Montreal; Sillery: [s.n, 1954?] (mf ed 1998) – 1mf – 9 – mf#SEM105P3008 – cn Bibl Nat [241]

Coup d'oeil sur l'ile de java et les autres possessions neerlandaises dans l'archipel des indes / Hogendorp, Carel S van – Bruxelles 1830 [mf ed Hildesheim 1995-98] – 1v on 3mf [ill] – 9 – €90.00 – 3-487-27436-1 – gw Olms [959]

Coupal, Louis *see*
- Ceux qui souffrent
- Graphitologie
- Les lucioles
- Toujours mieux

Coup-d'oeil sur le christianisme / Falcioni, Zeffirino – Paris: Les principaux libraires 1879 – 4mf – 9 – mf#vrl-142 – ne IDC [366]

Coup-d'oeil sur lisbonne et madrid en 1814 : suivi d'un memoire politique concernant la constitution promulguee par les cortes a cadix, et d'une notice sur l'etat moderne des sciences mathematiques et physiques en espagne / Hautefort, Charles V d' – Paris 1820 [mf ed Hildesheim 1995-98] – 3mf – 9 – €90.00 – 3-487-29871-6 – gw Olms [946]

Coupe enchantee / La Fontaine, Jean De – Paris, France. 1911 – 1r – us UF Libraries [440]

Le coupe-papier – Paris. n1-7. nov 1936-38 – 1 – fr ACRPP [073]

Couper, David *see* Scottish prelacy and tractarianism

Couperin, Francois *see*
- L'art de toucher le clavecin...organish [!] du roi...
- Les gouts-reunis
- Pieces de clavecin...[second livre]

Coupez, A *see*
- Esquisse de la langue holoholo
- Litterature de cour au rwanda

Coupland, R *see* East africa and its invaders

Coupland, Reginald *see*
- India
- The indian problem

Coupland, William Chatterton *see* Thoughts and aspirations of the ages

Coupon clipper / Central Wisconsin Center for the Developmentally Disabled – 1976 sep-1981 may – 1r – 1 – (cont: colony coupon clipper) – mf#552335 – us WHS [360]

Coups d'oeil et coups de plume / Lusignan, Alphonse – Ottawa?: s.n, 1884 – 4mf – 9 – mf#08506 – cn Canadiana [440]

Cour d'appel : william holmes, appellant sic et francois languedoc et jean belanger, intimes: cas de l'appellant sic – S.l: s.n, 1817? – 1mf – 9 – mf#55182 – cn Canadiana [340]

La cour d'appel de dijon – an huit – mille huit cent cinquante deux / Avril, Chantal – 2mf – 9 – (10443) – fr Atelier National [340]

La cour et la ville, paris et coblentz, ou l'ancien regime et le nouveau : considerees sous l'influence des hommes illustres et des femmes celebres, depuis charles 9, henri 4 et louis 14, jusqu'a napoleon, louis 18 et charles 10 / Toulotte, Eustache – Paris 1828 [mf ed Hildesheim 1995-98] – 13mf – 9 – €130.00 – ISBN-10: 3-487-26121-9 – ISBN-13: 978-3-487-26121-8 – gw Olms [944]

La cour et la ville sous louis 14, louis 15 et louis 16 : ou revelations historiques / ed by Barriere, Jean Francois – Paris 1830 [mf ed Hildesheim 1995-98] – 1v on 3mf – 9 – €90.00 – 3-487-26088-3 – gw Olms [944]

Courant / Minnesota Federation of Women's Clubs et al – 1899 may 4-oct 1, 1899 nov 1-1904 dec 4 – 2r – 1 – mf#345244 – us WHS [305]

Courant : [official newspaper of bottineau county and city of bottineau 1981-] – Bottineau, ND: Hills & Plains Free Press. v96 n52 dec 15 1981- (wkly) – 1 – (special "100 years old: dunseith, nd" suppl: jul 6 1982; cont: bottineau courant (bottineau, nd: 1969); currently publ) – mf#08605-08611++ – us North Dakota [071]

Le courant : levis, quebec – Levis: College de Levis. v1 n1 3 oct 1977-v1 n7 mai 1978 [mf ed 1988] – 9 – (cont: d'octobre; cont by: mosaique) – mf#SEM105P1041 – cn Bibl Nat [378]

The courant – Eddyville, NE: J J Tooley, jun 1892 (wkly) [mf ed v1 n3. jul 1 and jul 15 1892 filmed [1973]] – 1r – 1 – us NE Hist [071]

Les courants statiques induits de morton et quelques-unes de leurs applications en medecine / Blois, Charles N de – [Poitiers, France?: s.n.] 1908 [mf ed 1999] – 1mf – 9 – 0-659-90139-0 – mf#9-90139 – cn Canadiana [615]

Courasche *see* Grimmelshausens courasche

Courbiere-blaetter – Goerlitz DE, 1925-26 – 1r – 1 – gw Misc Inst [074]

Courcelle, P *see* Les lettres greques en occident

Courcy, B W De [comp] *see* A genealogical history of the milesian families of ireland

Courcy, Frederic De *see*
- Ange dans le monde et le diable a la maison
- Maitresse de poste, ou, l'homme de la famille
- Olivier basselin
- Vocation

Courcy, Henri de *see*
- History of the catholic church in the united states
- Les servantes de dieu en canada

Couret, A *see*
- La palestine sous les empereurs grecs 326-636
- La prise de jerusalem par les perses en 614. trois documents nouveaux

Couret de Villeneuve, Louis P *see* Recueil amusant de voyages

Courier – Hornsby, dec 1948-dec 1968 – 10r – A$737.04 vesicular A$792.04 silver – (aka: ku-ring-gai courier; advocate courier) – at Pascoe [079]

Courier / American Federation of Government Employees – v6 n1-v21 n1 [1975 nov-1988 jan] – 1r – 1 – (cont by: local news [lathorp ca]) – mf#1314407 – us WHS [350]

Courier / Columbiana Co. Leetonia – v1 n1. jul 1958-mar 1975 [wkly] – 4r – 1 – mf#B11916-11919 – us Ohio Hist [071]

Courier – Ballarat, Australia 4 aug 1951-19 jul 1952 (imperfect) – 1 – (cont: ballarat courier [20 jun 1887, 6 nov 1913-18 may 1920, 10 nov 1920-30 jun 1922]) – uk British Libr Newspaper [079]

Courier – Thorp WI. 1883 nov 23-1885 may 29, 1885 apr 3, 1885 jun 5-1887 may 27, 1887 jun 3-1888 dec 28, 1889 jan 4-1890 jul 31, 1890 aug 7-1891 jun 11, 1891 jun 18-1894 feb 15, 1894 feb 22-1895 may 23 – 8r – 1 – (cont by: thorp courier) – mf#1159288 – us WHS [071]

Courier – Norwich CT. 1808 jun 1 – 1r – 1 – (cont: chelsea courier; cont by: norwich courier [norwich ct]) – mf#846300 – us WHS [071]

Courier – Pittsburgh PA. 1950 aug 19-dec 31 – 1r – 1 – (cont: pittsburgh courier [national ed: 1910]; pittsburgh courier [national ed: 1910]; cont by: pittsburgh courier [national ed: 1955]; pittsburgh courier [national ed: 1955]) – mf#830052 – us WHS [071]

Courier – Lincoln, NE: Courier Pub Co. 11v. v9 n14. mar 17 1894-v19 n13. apr 4 1903 (wkly) – 4r – 1 – (cont: saturday morning courier. absorbed by: nebraska state journal) – us NE Hist [071]

Courier – Paris, France 1986-88 – 1,5,9 – (cont: unesco courier) – ISSN: 0041-5278 – mf#2350,02 – us UMI ProQuest [341]

Courier – Waterloo WI. [1987 oct 29/dec]-[2005 jul/dec] [gaps] – 45r – 1 – (cont: waterloo courier [waterloo wi]) – us WHS [071]

Courier – Plant City, FL. 1988-1998 apr – 18r – (gaps) – us UF Libraries [071]

Courier / Grand Army Home for Veterans [King WI] – v3 n1 [1953 may]-v8 n9 [1958] dec, 1982 jan-1989 dec – 2r – 1 – mf#401832 – us WHS [305]

Courier : manukau edition – jan 1977-aug 1986, 1987-88 – 1 – (incl: franklin and central ed) – mf#11.35 – nz Nat Libr [079]

Courier / Montgomery Co. West Carrollt – jan-feb 1963 (short) [wkly] – 1r – 1 – mf#B5001 – us Ohio Hist [071]

Courier – Narrabri. jan 1969-dec 1978, jun 1983-dec 1987, jan 1992-dec 1996 – at Pascoe [079]

Courier : official news voice of the... / Letter Carriers' Union of Canada – 1979 dec-1990 sep – 1r – 1 – mf#1110969 – us WHS [380]

Courier : or manchester advertiser-saturday's manchester courier – [NW England] Manchester jan 1817-14 nov 1821 (incomplete) [mf ed 2003] – 1r – 1 – uk Newsplan [072]

Courier / Pickaway Co. Circleville – mar 1950-jul 1951 [wkly] – 1r – 1 – mf#B29846 – us Ohio Hist [071]

Courier / Portage Co. Kent – apr 1918-apr 1920,jan 1928-sep 1929 [wkly] – 2r – 1 – mf#B29880-29881 – us Ohio Hist [071]

Courier / Portage Co. Kent – nov 1889-nov 1904, feb-sep 1913 [wkly] – 3r – 1 – mf#B2016-2018 – us Ohio Hist [071]

Courier / Sandusky Co. Fremont – v1 n1. 1859-apr 1861; (mar 1866-jan 1907) damaged [wkly] – 17r – 1 – mf#B32966-32982 – us Ohio Hist [071]

Courier – St Albans, Christchurch, NZ. 1982 – 1 – (title changes to: weekly courier jan 1982) – mf#70.10 – nz Nat Libr [079]

Courier / Westmorland Community Association, Madison, WI – v1 n1-6 [1943 jan-dec] – 1r – 1 – (cont: dope; cont by: westmorland courier) – mf#436813 – us WHS [360]

Courier – Prairie Du Chien WI. 1870 jan 25/1873 feb 25-1955 jan-dec 28 – 37r – 1 – (with gaps; cont: prairie du chien courier; crawford county press; cont by: courier-press) – mf#1011568 – us WHS [071]

Courier *see* New brunswick courier

Der courier *see* Der kurier

The courier – Pittsburgh. New York Edition. Nov 2 1957-Apr 25 1959; Nov 7 1959-Aug 23 1969 – 1 – us NY Public [071]

The courier *see*
- Courier news
- Hendon courier

courier *see* Manukau courier

Der courier an der weser – Bremen DE, 1846-93, 1894 [gaps], 1895-1906 – 122r – 1 – (title varies: courier sep 9 1863; bremer courier mar 18 1886; filmed by other misc inst: 1849 [1r]) – gw Misc Inst [074]

Courier and advertiser – Dundee: J Leng & D C Thomson 1926- (daily ex sun) [mf ed 1937-45, 1997-] – 1 – (cont: dundee advertiser & the courier, joint issue; newsplan 2000 [1926-50] 170r) – ISSN: 0307-5869 – uk Scotland NatLib; uk Newsplan [072]

Courier and church reform gazette – London, UK. 4 Mar 1854-31 Jul 1855. -irr. 1 reel – 1 – uk British Libr Newspaper [072]

Courier and east london advertiser – Tower Hamlets, London 8 may 1874-12 dec 1879 – 1 – uk British Libr Newspaper [072]
Courier and london and middlesexcounties gazette *see* Hendon courier
Courier [barnet] : courier and london & middlesex counties gazette – Barnet, England 19 may 1887-24 dec 1890 – 1 – (cont: hendon courier [10 feb-12 may 1887]; cont by: middlesex courier [2 jan 1891-3 sep 1897]) – uk British Libr Newspaper [072]
Courier [brantford] – Brantford, Canada 6 jan-3 mar 1913 (imperfect) – 1 – (cont by: brantford daily courier [4 mar 1913-28 dec 1918]) – uk British Libr Newspaper [071]
Courier de boston – La Salle IL 1789 – 1 – mf#3520 – us UMI ProQuest [071]
Courier de l'art *see* The artist 1880-82 – l'artist et courier de l'art
Courier de l'europe echo du continent – London, UK. 1867. -w. 1 reel – 1 – uk British Libr Newspaper [072]
Courier de londres – London, UK. 2 Jan 1818-30 Dec 1825 – 1 – uk British Libr Newspaper [072]
L'courier de l'ouest – Calgary, Alberta, CN. 1905-16 – 4r – 1 – cn Commonwealth Imaging [071]
Le courier de metz – Metz (F), 1892, 1899, 1901-11 – 1 – (with gaps) – gw Misc Inst [074]
Courier des pays bas – Brussels Belgium. 6 aug 1821-1837 – 33r – 1 – (aka: courrier; courrier belge) – uk British Libr Newspaper [074]
Courier du bas-rhin – Kleve DE, 1769-85, 1791-97 – 17r – 1 – gw Misc Inst [074]
Courier eastern and city editions – Auckland, NZ. nov 1984-dec 1985 – 5r – 1 – mf#11.48 – nz Nat Libr [079]
Courier highlights – Jupiter Island, FL. v10 n1-v11 n52. 1967-1968 – 2r – us UF Libraries [071]
Courier hub – Stoughton WI. 1979 nov 8-1980 jun 26, 1980 jul 1-1981 mar 31, 1981 apr 1-jul 2 – 3r – 1 – (cont: stoughton courier [stoughton wi: 1954]; stoughton hub [stoughton wi: 1954]; cont by: stoughton courier hub [stoughton wi: 1981]) – mf#939082 – us WHS [071]
Courier, isle of man – [NW England] Ramsey, Manx NHL 1958-70 – 1 – uk MLA; uk Newsplan [072]
Courier [london] – London, England 20 apr 1804-6 jul 1842 – 1 – (cont: courier & evening gazette [1 jan 1801-19 apr 1804]) – uk British Libr Newspaper [072]
Courier [madras] – Madras, India 1 jan-29 apr 1795 – 1 – (cont: madras courier [12 may, 16 jun 1790-25 apr 1793, 11 oct-8 nov 1793]; cont by: madras courier [6 may 1795-28 dec 1813, 3 jan-21 jul, 10 oct 1815-12 aug 1817, 6 jan-29 dec 1818, 19 jan 1819]) – uk British Libr Newspaper [079]
Courier mail – Brisbane, jun 1846-jul 1997 – at Pascoe [079]
Courier mail – Brisbane, Australia 28 aug 1933-4 sep 1972, jan 1978-31 dec 2004 – 1 – (wanting: nov 1970) – uk British Libr Newspaper [079]
Courier (manukau and central edition) – apr 1984-jul 1986 – 10r – 1 – mf#11.35b – nz Nat Libr [079]
Courier, manukau and franklin edition – apr 1984-jul 1986 – 1 – mf#11.35c – nz Nat Libr [079]
Courier news – Eastbourne, NZ. 1977-81 – 5r – 1 – (title changes to: the courier on 31 mar 1981) – mf#49.5 – nz Nat Libr [079]
Courier of baker *see* Haines record
Courier of baker county, or – Huntington OR: E S McCormick, 1931 [wkly] – 1 – (merged with: haines record, to form: haines record-the courier) – us Oregon Lib [071]
Courier of baker county, or *see* Courier of huntington
Courier of huntington – Huntington OR: H Grytdahl, 1930-31 [wkly] – 1 – (merged with: courier of baker county, or) – us Oregon Lib [071]
Courier of liberty / Adams Co. West Union – v1 n1. mar 1831-apr 1832 [wkly] – 1r – 1 – mf#B6738 – us Ohio Hist [071]
Courier of the mines – Bendigo, Australia 24 oct 1855-17 oct 1857 (wkly) (imperfect) – 1 – uk British Libr Newspaper [622]
Courier oxford *see* Oxford city courier
Courier, Pablo Luis *see* Historia de una mancha de tinta (el manuscrito de longo)
Courier [reedsport, or] – Reedsport OR: E A Sykes & M I Sykes, 1963- [wkly] – 1 – (cont: port umpqua courier) – us Oregon Lib [071]
Courier republicain – Algiers. 21 feb-30 oct 1944 – 1r – 1 – uk British Libr Newspaper [072]
Courier series / Mahoning Co. Canfield – jan 1978-dec 1986 [wkly] – 7r – 1 – mf#B29191-29197 – us Ohio Hist [071]
Courier series / Muskingum Co. Zanesville – apr 1846-dec 1847 [twice wkly, daily, twice wkly] – 1r – 1 – mf#B12410 – us Ohio Hist [071]
Courier series / Pike Co. Waverly – (1890-93,96-04,9/05-1906) [wkly] – 7r – 1 – mf#B9270-9276 – us Ohio Hist [071]
Courier south oxfordshire *see* South oxfordshire courier
Courier [st john's, newfoundland] – St John's, Canada 6 jan 1864-23 nov 1870 – 1 – uk British Libr Newspaper [071]
Courier times advertiser – Papakura, NZ. apr 1972, jun 1972 – 1 – mf#11.5 – nz Nat Libr [079]
Courier times advertiser *see* South auckland courier
Courier warwick – jun 15 1990-96 – 29 1/2r – 1 – (aka: warwick courier) – uk British Libr Newspaper [072]
Courier-journal – Louisville KY 7/8 mar 1869-30 dec 1879 – 1 – (cont: louisville courier-journal [ns] 14 nov 1868-6 mar 1869) – uk British Libr Newspaper [071]
Courier-journal – Louisville KY. 1879 may 19, 1888 jul 30-1889 jul 29 – 1r – 1 – (cont: louisville weekly courier; louisville weekly journal) – mf#850775 – us WHS [071]
Courier-news – Fargo ND. 1917 apr/may 17-1922 nov 21/apr 17 – 30r – 1 – (with gaps; cont: fargo daily courier-news; cont by: fargo daily tribune and courier-news) – mf#874234 – us WHS [071]
Courier-press – Prairie Du Chien WI. 1957/59-2004 jul-sep – 99r – 1 – (with gaps; cont: courier [prairie du chien wi]; kickapoo papoose) – mf#1011574 – us WHS [071]
Courier-times – Sutherland, NE: Jack Pollock. v74 n32. may 22 1969- (wkly) [mf ed filmed 1975-] – 1 – (cont: sutherland courier) – us NE Hist [071]
The courier-tribune – Callaway, NE: R E Briga, mar 2 1905-apr 1914// (wkly) [mf ed v18 n[38 (mar 9 1905]-1914] – 4r – 1 – (formed by the union of: callaway courier and: weekly tribune (1903)) – us NE Hist [071]
Courier/w / Mahoning Co. Canfield – jan 1987-may 1990 [wkly] – 2r – 1 – mf#B31318-31319 – us Ohio Hist [071]
Courier-wedge – Durand WI. 1918 jul 1/1920 jan 29-2004 jan/apr – 92r – 1 – (with gaps; cont: entering wedge and the pepin co courier; pepin herald) – mf#1009468 – us WHS [071]
Courlander, Harold *see*
– Piece of fire
– Uncle bouqui of haiti
Le courrier : l'hebdomadaire de terrebonne et de la region – Saint-Eustache: La Victoire. v1 n1 26 mars 1969- (wkly) [mf ed 1973] – 1r – 1 – (ceased 1969?) – mf#SEM35P16 – cn Bibl Nat [071]
Le courrier – Lawrence, MA: Wood Press, [apr 15, 1932-may 6, 1938] – 1 – us CRL [071]
Le courrier – Port-au-Prince, Haiti: Imp de l'Abeille, [1902-] sep 8-dec 26, 1902 – 1 – us CRL [079]
Le courrier – Providence, RI. 1906-1907 (1) – mf#66341 – us UMI ProQuest [071]
Le courrier alsacien-lorrain – Paris (F), 1904-06 – 1r – 1 – gw Misc Inst [074]
Le courrier artistique – Beaux-arts, expositions, musique, theatre, arts industriels, ventes. 15 juin 1861-nov 1865. devenu: Les Fantaisies parisiennes. Courrier artistique. Paris. dec 1865-13 mai 1866 – 1 – fr ACRPP [073]
Courrier belge – London, UK. 8, 15 May 1920 – 1 – uk British Libr Newspaper [072]
Courrier belge *see* Courier des pays bas
Courrier belge. de belgische koerier – London, UK. Organe edite pur les refugies belge en Derby. 22 Oct 1914-31 Jul 1915 – 1 – uk British Libr Newspaper [072]
Le courrier canadien : litterature, science, arts, economie politique, etc – Montreal: G A Dumont. v1 n1 14 dec 1889- (wkly) [mf ed 1986] – 1mf – 9 – (ceased 1890?) – mf#SEM105P667 – cn Bibl Nat [073]
Courrier consulaire de la haute-volta – [Ouagadougou], may-jun 1962; aug 1962-jul 1963; sep-oct 1963 – us CRL [960]
Le courrier d'afrique – Leopoldville: [s.n.], [jun-jul 22 1960] – (issues for june-jul 22 1960 at mf-691 and neg. mf-at lab filmed as pt of the bartlett collection) – us CRL [079]
Le courrier d'afrique – Leopoldville: [s.n, dec 31 1955-jan 23 1970] – 43r – 1 – us CRL [079]
Le courrier d'avignon – Avignon. oct-dec 1733, avr-dec 1734, 1736, 1756-60, 1778-83 – 1 – mf#IC. 10736 – fr ACRPP [073]
Le courrier de calais – Calais. 29 juin 1794-19 avr 1795 – 1 – (puis du pas-de-calais. nouvelles politiques, litteraires, de commerce et de marine) – fr ACRPP [073]
Courrier de chibougamau = Chibougamau courier – Chibougamau. v1 n1 29 sep 1956-(bimthly) [mf ed 1992] – 9 – (ceased 195-?) – mf#SEM105P1627 – cn Bibl Nat [073]
Courrier de finances de londres – London, UK. 30 May-14 Nov 1889 – 1 – uk British Libr Newspaper [072]
Courrier de finances de londres – London, UK. Mar 1891-Aug 1892 – 1 – uk British Libr Newspaper [072]
Le courrier de france – 2 ed. Paris. juil-dec 1875 – 1 – fr ACRPP [944]
Le courrier de france – Paris. 4 dec 1871-2 mars 1873 – 1 – (journal quotidien, politique, economique et litteraire.) – fr ACRPP [073]
Courrier de la bourse et de la banque – Brussels Belgium, 1 oct 1944-10 jul 1945 – 1r – 1 – uk British Libr Newspaper [074]
Courrier de la colere – Paris. nov 1957-oct 1958 – 1 – (puis de la nation) – fr ACRPP [073]
Courrier de la conference de la paix / ed by Stead, William T – La Haye: Maas & van Suchtelen. 109v. n1 15 juin-n109 20 oct 1907 (daily ex mon) [mf ed 1966] – 1r – 1 – (publ under the auspices of the fondation pour l'internationalisma la haye) – mf#*ZAN-2039 – us NY Public [327]
Le courrier de la gironde – Bordeaux. juil 1870-juin 1871. BM. Bordeaux – 1 – fr ACRPP [073]
Courrier de la Louisiane – New Orleans LA. 1808 apr 29 – 1r – 1 – (cont: louisiana courier [new orleans la: 1807]; cont by: courier [new orleans la: 1859]) – mf#1388719 – us WHS [071]
Le courrier de la nouvelle orleans – New Orleans, LA. 1934-1941 (1) – mf#68882 – us UMI ProQuest [071]
Courrier de lachine : journal hebdomadaire – Lachine: Lamarche & Tourigny, [c1911]- (wkly) [mf ed 1987] – 1mf – 9 – (with text in english; ceased 191-?) – mf#SEM105P800 – cn Bibl Nat [071]
Courrier de l'air – London, UK. 6 Apr 1917-25 Jan 1918 – 1 – uk British Libr Newspaper [072]
Courrier de l'art – Paris, 1881-90 – (= ser Architectural periodicals at avery library, columbia university) – 6r – 1 – $815.00 – us UPA [700]
Le courrier de l'assemblee nationale – Paris: Impr de E Marc-Aurel, jun 4 1848 – us CRL [944]
Le courrier de lawrence – Lawrence, MA: Courrier Pub [feb 1922-apr 8, 1932] – 1 – us CRL [071]
Le courrier de lawrence – Lawrence, MA: Wood Press, [may 13, 1938-aug 1946] – 1 – us CRL [071]
Courrier de l'egypte – Cairo. no. 1-116. Aug 29 1798-June 8 1801 – 1 – us NY Public [072]
Courrier de l'egypte – Le Caire. n1-116. aout 1798-juin 1801 – 1 – fr ACRPP [073]
Courrier de l'escaut – Tournai, France. 12 oct 1944-10 juil 1945 – 1r – 1 – uk British Libr Newspaper [072]
Le courrier de l'est – Nancy. 1889-sept 1892, avr-mai 1898 – 1 – fr ACRPP [073]
Courrier de l'europe – London, UK. 6 jun 1840-31 mar 1883 – 1 – (courrier de l'europe, semaine francaise 5 jan 1884-2 feb 1889; semaine francaise et le courrier de l'europe 7 apr-29 dec 1883) – uk British Libr Newspaper [072]
Courrier de l'europe et des spectacles – Paris. 1er juin 1807-30 sept 1811 – 1 – (puis et memorial europeen reunis.) – fr ACRPP [073]
Courrier de l'europe, semaine francaise *see* Courrier de l'europe
Le courrier de l'hymen, journal des dames – Nos 1-2, 4-7, 19, 22, 32, 41-42, 45. Paris. Feb-Jul 1791 – 1 – fr ACRPP [640]
Courrier de londres – London, UK. 13 May 1911-6 Dec 1912 – 1 – uk British Libr Newspaper [072]
Courrier de londres – London, UK. 31 Oct 1885-14 May 1898 – 1 – uk British Libr Newspaper [072]
Courrier de londres – London, UK. 6 May-3 Jul 1879 – 1 – uk British Libr Newspaper [072]
Courrier de londres et de paris – London, UK. 15 Jul 1899-19 May 1900 – 1 – uk British Libr Newspaper [072]
Courrier de londres et paris *see* Londres et paris
Le courrier de l'ouest (chicago, il) – Chicago: [s.n.] [ca 1896]- (wkly) [mf ed 1988] – 9 – (ceased 1897?) – mf#SEM105P1972 – cn Bibl Nat [071]
Le courrier de lyon / Valoris, Maxime – Paris: J. Rouff, 189?. 3v – 1 – us Wisconsin U Libr [830]
Le courrier de marseille commercial et politique – Marseille. 1848, 1850-51 – 1 – fr ACRPP [073]
Courrier de paris – Paris. 29 janv-31 mars 1851 – 1 – fr ACRPP [073]
Le courrier de paris *see* Le courrier de versailles a paris et de paris a versailles
Courrier de provence – Paris. n1-350. mai 1789-sept 1791 – 1 – (suite de: lettres du comte de mirabeau a ses commettants) – fr ACRPP [073]
Le courrier de smyrne : journal politique, commercial et litteraire – Smyrne. sept 1828-aout 1830 – 1 – fr ACRPP [073]
Le courrier de st hyacinthe – St Hyacinthe, QC. 1853-1900 – 41r – 1 – cn Library Assoc [071]
Le courrier de terrebonne – Terrebonne: Chambre de commerce. v1 n1 22 oct 1949-v40 n3 14 nov 1968 (bimthly) [mf ed 1973] – 6r – 1 – mf#SEM35P17 – cn Bibl Nat [071]
Le courrier de versailles a paris et de paris a versailles – Paris. juil 1789-mai 1793 – 1 – (devenu: le courrier de paris. devenu: le courrier des departements.) – fr ACRPP [944]
Courrier des comptoirs – London, UK. 7 Nov 1883 – 1 – uk British Libr Newspaper [072]
Le courrier des departements *see* Le courrier de versailles a paris et de paris a versailles
Courrier des etats-unis – New York. Mar 1 1828-Dec 25 1937. Not collated – 1 – us NY Public [071]
Le courrier des etats-unis – Ed. hebdomadaire. New York [NY: Ch Lassale, [1855-jul 1857] – 1r – us CRL [071]
Courrier des expositions : edition illustree du courrier de londres et de l'europe – London, UK. 23 jun-18 aug 1889 – 1 – uk British Libr Newspaper [072]
Courrier des planetes : ou correspondance du cousin jacques avec le firmament folie periodique dediee a la lune. – Paris. n1-74. 1788-89 – 1 – fr ACRPP [073]
Courrier des provinces maritimes – Bathurst, NB. 1885-1903 – 7r – 1 – cn Library Assoc [971]
Courrier d'ethiopie – Addis Ababa, [23 aug 1929-28 apr 1936] – 1 – us CRL [960]
Le courrier d'ethiopie – Addis-Abeba [Ethiopia]: Courrier d'Ethiopie [aug 23 1929-apr 28 1936] (semiwkly) – 2r – 1 – us CRL [079]
Le courrier d'haiphong – Vietnam. -d. 19 Sep 1886-31 Dec 1922. (Imperfect). (71 reels) – 1 – uk British Libr Newspaper [072]
Le courrier d'indochine – Hanoi: Impr Express 1908: oct 23,27, oct 31-nov 4, nov 11,14,16-18,20-26, nov 28-dec 11, dec 14-15,17-18,22-23,29,30-31; 1909: jan 4-feb 3, feb 9-13, mar 17, apr 4, 25, apr 29-may 9, may 20-jun 10, jun 17-20, jun 27-jul 11; 1910: jul 2-oct 1, oct 6-20, oct 26-nov 17, nov 22-29; 1912: feb 24, mar 16-27, apr 4-13, 20-25, may 1-4 – 1r – 1 – mf#mf-11811 seam – us CRL [079]
Le courrier d'oran : journal politique, commercial, agricole et litteraire – n1-200. Oran. 24 fevr 1850-14 fevr 1852 – 1 – fr ACRPP [073]
Courrier du canada – Quebec City, QC. 1857-73 – 18r – 1 – cn Library Assoc [071]
Le courrier du dimanche : journal politique, litteraire et financier – Paris. 1860-29 juil 1866 – 1 – fr ACRPP [073]
Le courrier du livre – Quebec: L Brousseau, [1896-1901?] – 9 – (incl english text) – mf#P04139 – cn Canadiana [010]
Le courrier du pacifique – Lima: E-A Le Roux, jun 21 1919-jun 10 1920 – (filmed consecutively with: west coast leader 1919-20) – us CRL [079]
Courrier du soir – Port-au-Prince: [s.n., [feb-dec 1919; apr 10 1920-dec 29 1921] – 27 sheets – us CRL [079]
Le courrier du soir : dernieres nouvelles de paris jusqu'a neuf heures du soir – Paris. 21 fevr 1878-87 – 1 – fr ACRPP [073]
Le courrier du sud – Les Cayes [Haiti]: Imp Ach Bonnefil, [apr 28 1890-aug 28 1891] – 2 sheets – us CRL [079]
Le courrier du sud = The south shore courier – Longueuil: [s.n.] v1 n1 27 mars 1947- (wkly) [mf ed 1982-] – (= ser Courrier plus) – 1 – (suppl: courrier plus) – mf#SEM35P179 – cn Bibl Nat [071]
Courrier extraordinaire : ou le premier arrive – Londres, Paris. mars 1790-aout 1792 – 1 – fr ACRPP [073]
Courrier financier de londres et de paris – London, UK. 3-24 Mar 1904 – 1 – uk British Libr Newspaper [072]
Courrier francais – 1890 apr-1891 apr – 1r – 1 – mf#2697690 – us WHS [071]
Courrier francais : ou tableau periodique et raisonne des operations de l'assemblee nationale – Paris. juin-dec 1789 – 1 – fr ACRPP [325]
Courrier francais – Paris, France. 16 nov 1884-31 dec 1893; 1 jul 1894 – 4 1/2r – 1 – uk British Libr Newspaper [072]
Le courrier francais – Paris. nov 1884-14 mars 1907, nov 1909-2 juil 1910, janv-oct 1911, janv-8 mars 1913, avr-mai 1914 – 1 – (litterature, beaux-arts, theatre, medecine, finances) – fr ACRPP [073]
Le courrier francais – Paris. 1 juil 1819-14 mars 1851 – 1 – (mq: 11 mai-1 juil 1843, 1845, 1846-47) – fr ACRPP [944]
Le courrier francais – Paris. -w 16 Nov 1884-31 Dec 1893 – 5r – 1 – uk British Libr Newspaper [072]
Le courrier francais – 3sept 1864-24 juin 1868 avec un prosp. du 8 aout 1868 – 1 – (suite de: revue de l'empire. 14 dec 1862-27 aout 1864) – fr ACRPP [944]

COURRIER

Courrier francais du temoignage chretien – [Paris: s.n. n8-78. 1944?-nov 23 1945 – 1r – us CRL [074]
Courrier franco-americain – Bay City MI, Chicago IL etc. [1905-13] – 1r – 1 – (cont: courrier-canadien) – mf#871540 – us WHS [071]
Le courrier haitien – Port-au-Prince: Imp aug A Heraux, nov 15 1920-sep 8/9,18/19/20-28/29/30, nov 18-28/29, dec 29 1922 – 16 sheets – us CRL [079]
Le courrier international – no. spec., no. 1-57. Paris. 28 nov 1866-30 juin 1867 – 1 – (mq n50) – fr ACRPP [073]
Le courrier missionnaire – Lausanne, Switzerland. 1901-03 [mf ed 2001] – (= ser Christianity's encounter with world religions, 1850-1950) – 1r – 1 – (in french. iss as suppl to: liberte chretienne) – mf#2001-s216 – us ATLA [240]
Le courrier picard – Amiens. 1959-juil 1975, nov 1982-1991 – 1 – fr ACRPP [073]
Le courrier plus – Longueuil: Courrier du Sud. 26 fevr 1995- (mthly) [mf ed 1995-] – 1 – mf#SEM35P179 – cn Bibl Nat [073]
Le courrier quotidien de l'exposition de 1889 *see* La bataille
Courrier royal / Maison de France – Paris. dec 1934-37 – 1 – fr ACRPP [073]
Courrier sud : hebdo dominical au service du sud de l'ontario francophone – Toronto. v1-4 n22. 24 juin 1973-12 nov 1976// – 4r – 1 – Can$440.00 – cn McLaren [071]
Courrier sud – Nicolet: Armand Bouchard. v1 n1 23 sep 1964- (wkly) [mf ed 1977-] – 1 – (with suppl) – mf#SEM35P152 – cn Bibl Nat [071]
Cours abrege d'histoire ancienne : contenant l'histoire de tous les peuples de l'antiquite jusqu'a jesus christ, a l'usage des institutions et des autres etablissements d'instruction pulique / Drioux, abbe (Claude-Joseph) – Montreal: J B Rolland, 1871 – 4mf – 9 – 0-665-90912-8 – (with ind) – mf#90912 – cn Canadiana [900]
Cours abrege d'histoire ancienne : contenant l'histoire de tous les peuples de l'antiquite jusqu'a jesus-christ a l'usage des institutions et des autres etablissements d'instruction publique / Drioux, abbe (Claude-Joseph) – Quebec: J A Langlais, 1877 – 4mf – 9 – mf#56034 – cn Canadiana [900]
Cours complet de cosmographie, de geographie, de chronologie, et d'histoire ancienne et moderne / Mentelle, Edme – Paris 1804 [mf ed Hildesheim 1995-98] – 3v on 12mf – 9 – €120.00 – 3-487-29968-2 – gw Olms [900]
Cours complet d'histoire d'haiti a l'usage des eco... / Dorsainvil, Jean Baptiste – Paris, France. 1912 – 1r – us UF Libraries [972]
Cours d'antiquitees monumentales / Caumont, A de – Paris. 6v. 1830-1841 – 33mf – 9 – mf#OA-128 – ne IDC [720]
Cours d'architecture enseigne dans l'academie royale d'architecture / Blondel, Jacques Francois – Paris, 1675-1683. 3v – 28mf – 9 – mf#O-154 – ne IDC [720]
Cours d'architecture enseigné dans l'academie royale d'architecture / Blondel, Jacques Francois – Ed 2. Paris, Amsterdam, 1698 – 9mf – 9 – mf#OA-41 – ne IDC [720]
Cours d'arithmetique : arithmetique mentale – Quebec: N S Hardy, 1882 [mf ed 1986] – 3mf – 9 – 0-665-52380-7 – mf#52380 – cn Canadiana [510]
Cours de cambodgien / Meyer, Roland Theodore – Phnom-Penh: Impr nouvelle Albert Portail 1912- [mf ed 1989] – 1r with other items – 1 – (t1: notice sur l'etude de l'ecriture khmer; t2: premier livre de lectures cambodgiennes) – mf#mf-10289 seam reel 024/06 [§] – us CRL [480]
Cours de chymie contenant la maniere de faire les operations... / Lemery, Nicolas – 11e rev corr augm ed. Paris: Chez Jean-Baptiste Delespine, 1730 [mf ed 1976] – 1r – 5 – mf#SEM16P259 – cn Bibl Nat [540]
Cours de chymie contenant la maniere de faire les operations... / Lemery, Nicolas – 8e rev corr augm ed. Paris: Chez Estienne Machelet, 1696 [mf ed 1976] – 1r – 5 – mf#SEM16P258 – cn Bibl Nat [540]
Cours de droit administratif / Price, Hannibal – Port-Au-Prince, Haiti. 1906 – 1r – us UF Libraries [972]
Cours de droit civil francais, d'apres l'ouvrage allemand de c.-s. zachariae / Aubry, Charles – 3 ed., entierement refondue et completee. Paris: Cosse. v1-6. 1856-63 – (= ser Civil law 3 coll) – 45mf – 9 – (v2. is dated 1863 and v6 is dated 1858. the first edition was publ in 1839-46. basically a translation of k.s. zachariae von lingenthal's handbuch des franzoesischen civilrechts) – mf#LLMC 96-389 – us LLMC [346]
Cours de droit de l'indochine / Dureteste, A – Paris, Larose Editeurs, 1938. 240 p. LL-10003 – 1 – us L of C Photodup [340]
Cours de francais pour les cambodgiens – [Paris?]: Centre inter-missions CAMAF [197-?] [mf ed 1989] – 1r with other items – 1 – (ce cours utilise essentiellement la langue parlee) – mf#mf-10289 seam reel 026/04 [§] – us CRL [480]
Cours de morale / Payot, Jules – 2e ed. Paris: Armand Colin, 1904 – 1mf – 9 – 0-8370-7975-6 – (incl bibl ref) – mf#1986-1975 – us ATLA [170]
Cours de pedagogie : ou principes d'education / Langevin, Jean – 2e rev augm ed. Rimouski: Impr de la Voix de (sic) Golfe, 1869 [mf ed 1974] – 1r – 5 – mf#SEM16P157 – cn Bibl Nat [370]
Cours de peinture par principes / Piles, R de – Paris, 1708 – 6mf – 9 – mf#O-402 – ne IDC [700]
Cours d'eloquence parlee d'apres delsarte / Hamel, Thomas Etienne – Quebec: Impr de la Compagnie de "L'Evenement", 1906 [mf ed 1988] – 4mf – 9 – (pref by camille roy) – mf#SEM105P656 – cn Bibl Nat [440]
Cours d'etudes historiques / Daunou, Pierre Claude Francois – [Paris?: s.n.] 1842 [mf ed 1985] – 20v on 1mf – 9 – 0-665-48509-3 – mf#48509 – cn Canadiana [900]
Cours d'histoire d'haiti a l'usage... / Dorsainvil, J B – Port-Au-Prince, Haiti. 1898 – 1r – us UF Libraries [972]
Cours d'histoire d'haiti a l'usage de... / Dorsainvil, J B – Port-Au-Prince, Haiti. 1910 – 1r – us UF Libraries [972]
Cours d'histoire du canada / Ferland, Jean-Baptiste-Antoine – 2e ed. Quebec: N S Hardy, 1882 [mf ed 1981] – 2v on 1mf – 9 – 0-665-09254-7 – mf#09255 – cn Canadiana [971]
Cours elementaire d'astronomie / Delaunay, Charles – 4. ed. Paris: V Masson 1865 [mf ed 1998] – 1r – 1 – mf#film mas 28212 – us Harvard [520]
Cours elementaire d'instruction civique / Devot, Justin – Paris, France. 1894 – 1r – us UF Libraries [972]
Cours entier de philosophie, ou systeme general selon les principes de m descartes / Regis, P S – Amsterdam, 1691. 3v – 22mf – 9 – mf#CA-57 – ne IDC [241]
Cours oral de franc-maconnerie symbolique : en douze seances / Cauchois, H – Paris: E Dentu 1863 – 3mf – 9 – mf#vrl-25 – ne IDC [366]
Cours pratique de franc-maconnerie / Chemin-Dupontes, Jean-Baptiste – Paris: L'auteur 1841 – 3mf – 9 – (filmed: 3e cahier-5e cahier [1841]) – mf#vrl-28 – ne IDC [366]
Cours preliminaire du courtier d'assurances – [Montreal?]: l'Association des courtiers d'assurances du Quebec, [entre 1963 et 1969] (mf ed 1993) – 1mf – 9 – mf#SEM105P1771 – cn Bibl Nat [368]
Course a l'etoile / Verneuil, Louis – Paris, France. 1928 – 1r – us UF Libraries [440]
The course and chains of roman catholicism : a controversy / Wharton, H M – Baltimore: R H Woodward, 1888 – 1mf – 9 – 0-8370-8074-6 – mf#1986-2074 – us ATLA [241]
Course and syllabus materials, publications on education in papua new guinea and other rare publications relating to png / International Training Institute Library – 1941-71 – 4r – 1 – mf#pmb doc474 – at Pacific Mss [370]
A course in civil government : based on the "the government of the people of the united states" / Thorpe, Francis N – 2nd ed. Philadelphia: Eldredge & Bro, 1894 – 3mf – 9 – $4.50 – mf#LLMC 96-059 – us LLMC [323]
Course in lugbara / Barr, L I – Nairobi, Kenya. 1965 – 1r – us UF Libraries [470]
Course in shona / Cordell, E A – Bulawayo, Zimbabwe. 1963 – 1r – us UF Libraries [470]
Course in the english bible : lectures 3. and 4., creation. with questions on four lectures / Carroll, Benajah Harvey – [S.l: s.n., 1902?] (Waco: Kellner Printing Co) – 1mf – 9 – 0-7905-1030-8 – mf#1987-1030 – us ATLA [220]
Course in tswana / Cole, Desmond T – Washington? DC. 1962 – 1r – us UF Libraries [470]
A course of bible study for adolescents : dealing with decision, duty, and discipline / Garvie, Alfred E – London: Sunday School Union [1913?] [mf ed 1990] – 1mf – 9 – 0-7905-3338-3 – mf#1987-3338 – us ATLA [220]
The course of creation / Anderson, John – London, 1850 – (= ser 19th c evolution & creation) – 5mf – 9 – mf#1.1.11578 – uk Chadwyck [575]
The course of divine revelation : a brief outline of the communications of god's will to man, and of the evidences and doctrines of christianity / Muir, John – Calcutta: Baptist Mission Press, 1846 [mf ed 1995] – (= ser Yale coll) – 1 – 0-524-09983-9 – (with allusions to hindu tenets. in sanskrit, hindi and english) – mf#1995-0983 – us ATLA [210]
Course of hebrew study adapted to the use of beginners / Stuart, Moses – Andover: Flagg & Gould, 1830 [mf ed 1989] – 1mf – 9 – 0-7905-2155-5 – (in english & hebrew; sequel to: hebrew chrestomathy) – mf#1987-2155 – us ATLA [470]
Course of lectures on drawing, painting, and engraving : considered as...elegant education / Craig, William Marshall – London 1821 – (= ser 19th c art & architecture) – 5mf – 9 – mf#4.2.1335 – uk Chadwyck [700]
A course of lectures on painting : delivered at the royal academy of fine arts / Howard, Henry – London 1848 – (= ser 19th c art & architecture) – 5mf – 9 – mf#4.2.1509 – uk Chadwyck [750]
Course of lectures on sabbath schools – Glasgow, Scotland. 1841 – 1r – us UF Libraries [240]
A course of lectures to young men : on science, literature, and religion: delivered in glasgow, by ministers of various denominations – 2nd ser. Glasgow, 1842 – (= ser 19th c evolution & creation) – 4mf – 9 – mf#1.1.8154 – uk Chadwyck [100]
Course of modern analysis / Whittaker, Edmund Taylor – Cambridge, England. 1915 – 1r – us UF Libraries [500]
Course of sermons / Sharpe, William – Cambridge, England. 1817 – 1r – us UF Libraries [240]
Course of study and syllabus in american and irish history : for the elementary schools of the archdiocese of new york – [New York]: New York Catholic School Board, 1911 [mf ed 1986] – 1mf – 9 – 0-8370-8042-8 – mf#1986-2042 – us ATLA [975]
Course of study and syllabus in drawing for the elementary schools of the archdiocese of new york – [New York]: New York Catholic School Board, 1911 – 1mf – 9 – 0-8370-8043-6 – mf#1986-2043 – us ATLA [377]
Course of study and syllabus in elementary science, applied geometry, bookkeeping for the elementary schools of the archdiocese of new york – [New York]: New York Catholic School Board, 1911 – 1mf – 9 – 0-8370-8044-4 – mf#1986-2044 – us ATLA [500]
Course of study and syllabus in english for the elementary schools of the archdiocese of new york – [New York]: New York Catholic School Board, 1911 – 1mf – 9 – 0-8370-8045-2 – mf#1986-2045 – us ATLA [377]
Course of study and syllabus in geography and mathematics for the elementary schools of the archdiocese of new york – [New York]: New York Catholic School Board, 1911 – 1mf – 9 – 0-8370-8046-0 – mf#1986-2046 – us ATLA [377]
Course of study and syllabus in music for the elementary schools the archdiocese of new york – [New York]: New York Catholic School Board, 1911 – 1mf – 9 – 0-8370-8047-9 – mf#1986-2047 – us ATLA [377]
Course of study and syllabus in physical culture, physiology and hygiene, nature study for the elementary schools of the archdiocese of new york – [New York]: New York Catholic School Board, 1911 – 1mf – 9 – 0-8370-8048-7 – mf#1986-2048 – us ATLA [377]
Course of study and syllabus in religion for the elementary schools of the archdiocese of new york – [New York]: New York Catholic School Board, 1911 – 1mf – 9 – 0-8370-8049-5 – mf#1986-2049 – us ATLA [240]
Course of study for probationers in the ministry of the canadian wesleyan methodist new connexion church : adopted by the conference held in toronto, june 1858 – [London, Ont?: s.n.], 1858 – 1mf – 9 – 0-665-89723-5 – mf#89723 – cn Canadiana [242]
Course of study in vocational agriculture for individualized instruction / Cooper, R F – s.l, s.l? 1940 – 1r – us UF Libraries [331]
Course of the exchange – England.1895. -w. 1/2 reel – 1 – uk British Libr Newspaper [072]
Course of the exchange – London, UK. 11 Mar 1825-1828; 1873; 1882-83; 1886-30 Jun 1908. 13 reels – 1 – uk British Libr Newspaper [072]
Course of the history of modern philosophy = Cours de l'histoire de la philosophie moderne / Cousin, Victor – [2nd ed] New York: D Appleton, 1852 – 3mf – 9 – 0-524-00015-8 – (in english) – mf#1989-2715 – us ATLA [190]
Courses in agriculture for adult farmers of florida by districts / Howard, Ar – Gainesville, FL. 1943 – 1r – us UF Libraries [630]
Courson cousins – v2 [iss1]-v2 iss2 [1981 jan-1981 apr], 1981 jul-1984 fall – 1r – 1 – (cont: courson organization newsletter) – mf#922703 – us WHS [071]
Coursons, R Des *see* La rebellion armenienne
Court – Henley-on-Thames, England 1980 – 1,5,9 – ISSN: 0308-3764 – mf#12493 – us UMI ProQuest [347]
Court apercu des principes : de l'organisation et de la formation de... / Meston, W – Croix-Rousse, France. 1847 – 1r – us UF Libraries [240]
Court circular & court news – London, England. -w. Jan-Dec 1893. 1 reel – 1 – uk British Libr Newspaper [072]
Court decisions on teacher tenure – Washington DC: Committee on Tenure, National Education Assoc of the US, 1936-56 [annual] [mf ed 1999] – 1r – 1 – (cont: recent court decisions on teacher tenure; filmed with: recent court decisions on teacher tenure [[1932/1934]]. issue for 1954 & 1955 publ in combined form; vols for 1935-43 comp by the association's committee on tenure, 1944-1953 by the association's committee on tenure and academic freedom, 1954/1955 by the association's research division) – mf#4677 – us Wisconsin U Libr [344]
Court decisions on teacher tenure *see* Recent court decisions on teacher tenure
Court decisions relating to the national labor relations act / U.S. Courts – Washington: GPO. v1-38 + index. 1939-86 – 551mf – 9 – $826.00 – (additional vols to be filmed) – mf#LLMC 79-432 – us LLMC [344]
Court documents including orders, rules of procedure, and copies of the indictment and motions of the defense, 1946-1948 / World War 2. Defense Section – (= ser Records of allied operational and occupation headquarters, world war 2) – 3r – 1 – mf#M1699 – us Nat Archives [347]
Court exhibits in english and japanese, international prosecution section, 1945-1947 / World War 2. International Prosecution Section – (= ser Records of allied operational and occupation headquarters, world war 2) – 48r – 1 – mf#M1688 – us Nat Archives [341]
Court, H *see* An exposition of the relations of the british government with the sultaun and state of palembang
The court journal, court circular and fashionable gazette – London. no. 1-4950. 1829-1925. (Scattered issues wanting) – 1 – us L of C Photodup [340]
Court, Juergen *see*
– Catalogue de la bibliotheque historique et scientifique de feu m le docteur j court
– Catalogue de la precieuse bibliotheque de feu m le docteur j court
Court magazine : or royal chronicle of news, politics and literature for town and country – London; England 1761-65 – 1 – mf#4232 – us UMI ProQuest [340]
Court martial reports – v1-50. 1951-75 – 9 – $925.00 set – mf#402410 – us Hein [347]
Court martial reports of the judge advocate general [jag] of the air force...the judicial council, and boards of review / U[/nited/] S[/tates/] Army. Air Force. Office of the Judge Advocate General – v1-4. 1948-51 [all publ] – 41mf – 9 – $61.50 – (covers: acm 2 to acm 2891; cont by: court-martial reports) – mf#llmc82-212 – us LLMC [355]
Court minute books of the borough of neath, 1759-1853 – 1r – 1 – mf#96344 – uk Microform Academic [941]
The court of appeals and its relation to the methods by which public speculators have evaded justice – New York, Polhemus, 1876. 53 p. LL-972 – 1 – us L of C Photodup [347]
The court of arbitration: its advantages and importance to business men / Shepard, Elliott Fitch – New York: Press of the Chamber of Commerce, 1875. 16p. LL-1389 – 1 – us L of C Photodup [347]
Court of customs appeals reports *see* Us court of customs amd patent appeals reports
Court of honor – 1910 may-1915 dec, 1916 jan-1918 sep, 1918 oct-1924 oct – 3r – 1 – mf#1054934 – us WHS [071]
Court of justice of the european coal and steel community *see* Court of justice of the european communities
Court of justice of the european communities : report of cases befrore the court – 1954-2002 update n1 – 9 – $5155.00 set – (title varies: 1954-58 as court of justice of the european coal and steel community) – ISSN: 0378-7591 – mf#401960 – us Hein [347]
Court papers, journal, exhibits, and judgments of the international military tribunal for the far east, 1900-1948 / International Military Tribunal for the Far East – (= ser National archives coll of world war 2 war crimes records) – 61r – 1 – mf#T918 – us Nat Archives [347]
Court records, 1811-1834 / Federal Writers' Project (FL) – Pensacola, FL. 193-? – 1r – us UF Libraries [347]
Court reporting; a manual of legal dictation and forms for teachers, stenographers and typewriters / Robinson, Addie M – New York: Excelsior 1904. 290p. LL-1091 – 1 – us L of C Photodup [347]

Court reports, county reports 1885-1921, district reports 1892-1921, district and county reports, 1922-50 / Pennsylvania – 51 reels – 1 – $1,290.00 – us Trans-Media [347]
Court rolls – billingford, castleacre, elmham, longham, wellingham, west lexham, c1300-1800 – [irreg] – (= ser Holkham library early estate records) – 7r – 1 – mf#97269 – uk Microform Academic [343]
Court rolls – holkham, and miscellaneous documents, c1200-1500 – (= ser Holkham library early estate records) – 6r – 1 – (with: tittleshall court rolls) – mf#96504-9 – uk Microform Academic [343]
Court rolls – minster lovell, oxfordshire, 1560-1627 – (= ser Holkham library early estate records) – 1r – 1 – mf#96707 – uk Microform Academic [343]
Court rules and general orders / Supreme Court of the FSM – 1990 – 13mf – 9 – $19.50 – mf#llmc82-100h, title 23 – us LLMC [347]
Court studies : a collection of surveys and analyses of judicial systems in the united states – 28r – 1 – $3915.00 – 0-89093-650-1 – (with p/g) – us UPA [347]
Court to court procedure, with forms. / Smith, Frank Charles – Springfield, Ill.: Fiske, 1910. 664p. LL-1133 – 1 – us L of C Photodup [347]
Court traite sur l'art epistolaire / Meilleur, Jean Baptiste – 2e ed. Montreal: P Gendron, 1849 [mf ed 1983] – 2mf – 9 – 0-665-38236-7 – mf#38236 – cn Canadiana [390]
Court, William B *see*
– Moody et sankey
– The story of my connection with the chiniquy movement in montreal, of 1874-77
Court-annexed arbitration in ten district courts / Meierhoefer, Barbara S – Washington: FJC, 1990 – 2mf – 9 – $3.00 – mf#LLMC 95-380 – us LLMC [347]
Court-appointed experts / Willging, Thomas E – Washington: FJC, 1986 – 1mf – 9 – $1.50 – mf#LLMC 95-330 – us LLMC [347]
Courtauld institute of art periodicals subject index – 277mf – 9 – $2800.00 – 0-907006-84-1 – (brief synopses of articles on art-historical related subjects appearing in almost 200 journals fr the 1930s to 1983) – uk Mindata [700]
Courte reponse aux dernieres attaques contre la brochure calvin a geneve / Fleury, Francois – Geneve: Pfeffer et Puky, 1864 – 1mf – 9 – 0-524-08715-6 – mf#1993-1085 – us ATLA [242]
Courteault, Paul *see* La vie economique a bordeaux pendant la guerre
Courtemanche, Joseph Israel *see* Histoire de la famille courtemanche, 1663-1895
Courtenay, Baudouin de *see*
– Dal', v.i. tolkovii slovar' zhivogo velikorusskogo iazika
– Tolkovyi slovar' zhivogo velikorusskogo iazyka
Courtenay, C L *see* Hidden life
Courtenay, Charles Leslie et al *see* Sermons on the re-union of christendom. second series
Courtenay free press – British Columbia, CN. jan 1927-jul 1931 – 2r – 1 – cn Commonwealth Imaging [071]
Courtenay, William Ashmead *see* Collection, 1781-1906
Courtes explications a mes concitoyens / Lafontant, Nicolas Stephen – Charlotte Amalie, St Thomas. 1904 – 1r – us UF Libraries [972]
Courtet, M *see* Etude sur le senegal
Courteville, Raphael *see* Solo for a flute and a bass
Courthope, William *see* Memoir of daniel chamier, minister of the reformed church
Courtier, monk and martyr : a sketch of the life and sufferings of blessed sebastian newdigate of the london charterhouse / Camm, Bede – London: Art and Book Co, 1901 – 1mf – 9 – 0-524-03912-7 – mf#1990-1171 – us ATLA [240]
Courtilz de Sandras, Gatien de *see* Testament politique de messire jean-bapt colbert, ministre & secretaire d'etat
Courtisans : ou, la barbe de neptune / Dupin, Henri – Paris, France. 1821 – 1r – us UF Libraries [440]
Court-martial reports : holdings, and decisions of the judge advocate generals, boards of reviews, and the court of military appeals / U.S. Army. Judge Advocate General – v1-50. 1951-75 – 573mf – 9 – $859.00 – (with index vols a (covering v1-25) and b (covering v26-50). cont by: military justice reporter, west publ co. v1-1977- not offered by llmc) – mf#LLMC 84-214 – us LLMC [355]
Courtney, William Leonard *see*
– The art annual for 1892 professor hubert herkomer royal academician his life and work by w.l. courtney
– Constructive ethics
– The metaphysics of john stuart mill
Courtney, Wilshire S *see* Gold fields of st domingo
Courtois *see* L'urne electorale
Courtois, Victor Joseph *see*
– Diccionario portuguez-cafre-tetense
– Elementos de grammatica tetense
Courtot, Lieutenant-Colonel *see* Du golfe des syrtes au golfe du benin par le lac tchad
Courtroom / Reynolds, Quentin James – London: Gollancz, 1950. 396p. LL-397 – 1 – us L of C Photodup [340]
Courts and criminals / Train, Arthur – New York: Chas Scribners & Sons 1925 – 4mf – 9 – $6.00 – mf#llmc92-223 – us LLMC [345]
The courts and legal profession of iowa. / Ebersole, Ezra Christian – Chicago, Cooper, 1907. 2 v. LL-767 – 1 – us L of C Photodup [347]
The courts of new jersey; their origin, composition and jurisdiction / Clevenger, William May – Plainfield: New Jersey Law Journal, 1903. 143p. LL-323 – 1 – us L of C Photodup [347]
The courts of quebec : mr casgrain's measure for their re-organization / Casgrain, Thomas Chase – S.l: s.n, 1893? – 1mf – 9 – mf#02541 – cn Canadiana [347]
The courts, the constitution and parties : studies in constitutional history and politics / McLaughlin, Andrew Cunningham – Chicago: University of Chicago Press [c1912] [mf ed 1970] – (= ser Library of American civilization 13445) – vii/299p on 1mf – 9 – us Chicago U Pr [323]
Courveille, Xavier De *see* Reve de cinyras
Courville, Louis-Leonard Aumasson, sieur de *see* Memoires sur le canada depuis 1749 jusqu'a 1760
Cousar, Robert Moore *see* Digest of the laws and decisions relating to the appointment, salary, and compensation of the officials of the united states courts.
Cousas diplomaticas / Lobo, Helio – Rio de Janeiro, Brazil. 1918 – 1r – us UF Libraries [327]
Cousens, Henry *see*
– The antiquities of sind
– The architectural antiquities of western india
– Bijapur and its architectural remains
– Bijapur, old capital of the adil shahi kings
– The chalukyan architecture of the kanarese districts
– Mediaeval temples of the dakhan
– Somanatha and other mediaeval temples in kathiawad
Cousin cinderella : a canadian girl in london / Cotes, Everard, mrs [Sara Jeanette Duncan] – Toronto: Macmillan Co of Canada, 1908 – 5mf – 9 – 0-665-74141-3 – (incl publ list) – mf#74141 – cn Canadiana [830]
Cousin d'Avallon, Charles *see* Resume de la vie du prisonnier de sainte-helene
Cousin, Jean *see*
– The book of fortune
– Le livre de fortune
– Le livre de fortune, recueil de deux cents dessins inedits de jean cousin publie d'apres le manuscrit original de la bibliotheque de l'institut par ludovic lalanne
[Cousin, Louis] *see* La morale de confucius
Cousin, Paul *see* Book of reference of the city of quebec and village of saint sauveur
Cousin, V *see* Opera
Cousin, Victor *see*
– Course of the history of modern philosophy
– Lectures on the true, the beautiful, and the good
– Ueber franzoesische und deutsche philosophie
– The youth of madame de longueville, or new revelations of court and convent in the seventeenth century
Cousin, William *see* Issue at stake in the alternative submitted to the presbyteries
Cousineau, Laurent Etienne *see* Nos societes de bienfaisance
Cousins, George *see*
– Isles afar off
– The story of the south seas
Cousins, George [comp] *see* From island to island in the south seas
Cousins, Henry Thomas *see* Tiyo soga
Cousins, James Henry *see*
– Asit kumar haldar
– The cultural unity of asia
– The faith of the artist
– Footsteps of freedom
– The garland of life
– The hound of uladh
– The renaissance in india
– A study in synthesis
– Surya-gita
– Work and worship
Cousins, John *see* Inquiry into the reported miraculous cure of mathew breslin
Cousins, Margaret *see* The awakening of asian womanhood
Cousins, Margaret E *see*
– Indian womanhood to-day
– The music of orient and occident
Cousins, Norman *see* Talks with nehru
Cousins, William Edward *see* Malagasy customs
Coussemaker, E de *see*
– Drames liturgiques du moyen-age
– Histoire de l'harmonie au moyen age
Coustau, P *see*
– Le pegme de pierre coustau
– Petri costalii pegma
Cousteau kids – Hampton VA 2005- – 1,5,9 – mf#16163.01 – us UMI ProQuest [071]
Cousturier, Lucie *see*
– La foret du haut-niger
– Mes inconnus chez eux
Coutinho, Afranio *see*
– Critica e poetica
– Filosofia de machado de assis
– Machado de assis na literatia brasileira
Coutinho, Edilberto *see* Rondon, o civilizador da ultima fronteira
Coutinho, Galeao *see* Ultimo dos morungabas
Coutinho, Jose Joaquim De Cunha De Azeredo *see* Obras economicas, 1794-1804
Coutinho, Lourival *see* General goes depoe
Coutlee, Louis William *see* An alphabetical index of the code of civil procedure of lower canada
Coutlee's supreme court cases – 1v. 1875-1906 (all publ) – 5mf – 9 – $7.50 – (coll of notes of unreported cases) – mf#LLMC 81-010 – us LLMC [347]
Couto De Magalhaes, Agenor *see* Encantos do oeste
Couto De Magalhaes, Jose Vieira *see*
– Selvagem
– Viagem ao araguaia
Couto, Miguel *see* Allocucoes do presidente da academia de medicina
Coutts, John *see* Forms of religion
Coutu, Debra L *see* The effect of the presence of the coach on pain perception and pain tolerance of athletes
Coutumes des arabes au pays de moab / Jaussen, Antonin – Paris: Victor Lecoffre 1908 [mf ed 1989] – (= ser Etudes bibliques) – 2mf – 9 – 0-7905-1010-3 – (in french & arabic; incl bibl ref & ind) – mf#1987-1010 – us ATLA [305]
Coutumier du 11 [sic] siecle de l'ordre de saint-ruf (chanoines reguliers de saint-augustin) en usage a la cathedrale de maguelone / Carrier, Albert – Sherbrooke: Apostolat de la presse, 1950 [mf ed 2001] – (= ser Etudes et documents sur l'ordre de saint-ruf) – 9 – (text in french and latin) – cn Bibl Nat [241]
Coutumiers juridiques de l'afrique occidentale francaise – Paris: Larose, 1939 – 1 – us CRL [340]
Coutumiers liturgiques de premontre du 13th et 14th siecle / Lefevre, F – Louvain, 1953 – 3mf – 8 – €7.00 – ne Slangenburg [240]
Couture, Joseph Alphonse *see*
– Choix des vaches laitieres d'apres le systeme guenon
– Precis de medicine vetennaire
Couture, Joseph-Alphonse *see*
– Le betail canadien
– The french canadian cattle
– Traite sur l'elevage et les maladies des bestiaux
Couture, Linea *see* Activities, adaptation and aging
Couture, Marguerite *see* Bibliographie analytique de luc lacourciere
Couture, Suzanne *see* Bibliographie de lotbiniere
Couturieres : ou, le ciniquieme au-dessus de l'entr... / Desaugiers, Marc-Antoine – Paris, France. 1837 – 1r – us UF Libraries [440]
Couve, Daniel *see* Petite histoire des missions chretiennes
Le couvent – [Joliette, Quebec: F A Baillairge, 1886?-1899] – 9 – ISSN: 1190-7770 – mf#P04659 – cn Canadiana [640]
Le couvent : publication mensuelle a l'usage des jeunes filles / ed by Baillairge, Frederic-Alexandre – Joliette: [s.n.] 1re annee n1 janv 1886-14e annee n9 mai/juin 1899 (mthly) [mf ed 1984] – 1r – 5 – mf#SEM16P315 – cn Bibl Nat [376]
Les couvents pres de schag / Monneret de Villard, Ugo – Milan, 1925 – 6mf – 8 – €14.00 – ne Slangenburg [240]
Couvin, Leger *see* Discours politiques et euvres diverses
Couvreur, Seraphin *see*
– Chou king
– Geographie ancienne et moderne de la chine
– Les quatre livres
Cova Garcia, Luis *see* Fundamento juridico del nuevo ideal nacional
Covarrubias Orozco, S de *see* Emblemas morales de don sebastian de covarrubias orozco
Covarrubias y Leyra, D de *see* Opera omnia
Covarsi, Antonio *see* Narraciones de un montero y practica de caza mayor
Cove creek baptist church. sherwood, north carolina : church records – 1799-May 1963 – 1 – 46.08 – us Southern Baptist [242]
Cove ledger – Cove OR: Cove Pub Co, 1898-1900 [wkly] – 1 – us Oregon Lib [071]
Covek danas : [zbornik filozofskih ogleda] / Grlic, Danko – Beograd: Nolit 1964 [mf ed 1980] – 1r – 1 – us UW Libraries [303]
Covell, Daniel D *see* "To keep a proper perspective on the role of athletics"
[Covelo-] covelo review – CA. Dec 1904-Apr 1907 – 1r – 1 – $60.00 – mf#B02147 – us Library Micro [071]
[Covelo-] frontier gazette – CA. 1963-1964 – 1r – 1 – $60.00 – mf#B06017 – us Library Micro [071]
[Covelo-] round valley news – CA. 1972-1993 – 16r – 1 – $960.00 – mf#B06018 – us Library Micro [071]
[Covelo-] the courier – CA. 1971 – 1r – 1 – $60.00 – mf#B03130 – us Library Micro [071]
The covenant : sworn and subscribed by the synod of the reformed presbyterian church in north america, at pittsburgh, pennsylvania... and, pastoral letter – Pittsburgh: Bakewell & Marthens, 1872 – 1mf – 9 – 0-524-06660-4 – mf#1991-2715 – us ATLA [242]
Covenant, constitution, by-laws, certificate of incorporation, deed / Pickens County. South Carolina. Crescent Hill Baptist Church – c1956. Miscellaneous Papers – 1 – 5.00 – us Southern Baptist [242]
Covenant name of god / Groser, Thomas – London, England. 1860 – 1r – us UF Libraries [240]
Covenant names and privileges / Newton, Richard – New York: Robert Carter, c1882 – 1mf – 9 – 0-8370-5975-5 – mf#1985-3975 – us ATLA [210]
Covenant of god : the hope of man / Lockhart, John – Edinburgh, Scotland. 1812 – 1r – us UF Libraries [240]
The covenant of grace / Wallace, Henry – Edinburgh: Andrew Elliot, 1874 [mf ed 1985] – 1mf – 9 – 0-8370-5702-7 – mf#1985-3702 – us ATLA [240]
The covenant of peace / Vincent, Marvin Richardson – New York: Anson D F Randolph, c1887 – 1mf – 9 – 0-7905-2393-0 – mf#1987-2393 – us ATLA [240]
The covenant of salt : as based on the significance and symbolism of salt in primitive thought / Trumbull, Henry Clay – New York: Charles Scribner, 1899 – 1mf – 9 – 0-8370-5578-4 – (incl bibl ref & indexes) – mf#1985-3578 – us ATLA [390]
Covenant of the Goddess *see*
– Themis
– Thesmophoria
Covenant people of god *see* Yung yuan te yueh (ccm324)
The covenant theology of francis roberts / Lim, Won Taek – Grand Rapids MI: Calvin Theological Seminary, 2000 [mf ed 2001] – 1r – 1 – $130.00 – mf#2001-B001 – us ATLA [242]
Covenant weekly / Evangelical Mission Covenant Church of America – 1934 jan 16/1935 may 28-1956 may 4/1958 jun 20 – 17r – 1 – (with gaps; cont: forbundets veckotidning; cont by: covenant companion) – mf#927167 – us WHS [243]
Covenant with the eyes – London, England. 18– – 1r – us UF Libraries [240]
The covenanter pastor / George, Robert James – New York: Christian Nation Pub Co, 1911 – 3mf – 9 – 0-524-07420-8 – mf#1991-3080 – us ATLA [240]
The covenanter vision / George, Robert James – New York: Christian Nation Pub Co, 1917 – 4mf – 9 – 0-524-07421-6 – mf#1991-3081 – us ATLA [240]
Covenanters / Beveridge, John – Edinburgh: T & T Clark [1905?] – (= ser Bible class primers) – 1mf – 9 – 0-524-06241-2 – mf#1990-5196 – us ATLA [220]
The covenanters : a history of the church in scotland from the reformation to the revolution / Hewison, James King – rev corr ed. Glasgow: J Smith, 1913 – 3mf – 9 – 0-7905-5839-4 – (incl bibl ref) – mf#1988-1839 – us ATLA [240]
Covenanter's manual / Culbertson, Robert – Edinburgh, Scotland. 1808 – 1r – us UF Libraries [240]
The covenant...explained – Saipan: Office of the Plebiscite Commissioner, may 1975 – 1mf – 9 – $1.50 – mf#llmc82-100j, title 17 – us LLMC [342]
Covenanting struggle : what was gained by it? / Begg, James – Edinburgh, Scotland. 18– – 1r – us UF Libraries [240]
The covenants / Howell, Robert Boyte Crawford – Charleston: Southern Baptist Publication Society, 1855 – 1mf – 9 – 0-524-00370-X – mf#1989-3070 – us ATLA [240]
The covenants and the covenanters : covenants, sermons, and documents of the covenanted reformation / Kerr, James – Edinburgh: RW Hunter, [1895?] – 1mf – 9 – 0-524-00995-3 – mf#1990-0272 – us ATLA [243]
The covenants of scotland / ed by Lumsden, John – Paisley: A Gardner, 1914 – 1mf – 9 – 0-7905-8111-6 – mf#1988-6073 – us ATLA [240]
Covent garden chronicle *see* Eighteenth century journals

Covent garden journal – Dublin, Ireland.23 Jan 1752-22 Nov 1753; 10 Jun 1756. -irr. 1/2 reel – 1 – uk British Libr Newspaper [072]

Covent guardian and west end advertiser – London, UK. 1986-16 mar 1989 – 5 1/2r – 1 – uk British Libr Newspaper [072]

Coventry chronicle, bedworth & foleshill news & tribune – Bedworth, England 9 may 1930-14 may 1948 – 1 – (cont: coventry chronicle, bedworth & foleshill news & warwickshire county graphic [10 jan 1925-2 may 1930]; cont by: bedworth & foleshill news, coventry chronicle & tribune [21 may 1948-1 feb 1957]) – uk British Libr Newspaper [072]

Coventry evening telegraph – England.May 1947-1959.-d. 59 reels – 1 – uk British Libr Newspaper [072]

Coventry herald – England. -w. 1829-36. (4 reels) – 1 – uk British Libr Newspaper [072]

Coventry mercury – (Coventry Standard). England. -w. 1829-36. 3 reels – 1 – uk British Libr Newspaper [072]

Coventry papers, 17th century – v1-121 – (= ser Archives of the marquess of bath, longleat house, warminster, wiltshire) – 83r – 1 – (with app, ind & catalogue) – mf#96699 – uk Microform Academic [941]

Coventry standard – England, 1940-64 – 30r – 1 – uk British Libr Newspaper [072]

Coventry times / coventry weekly times – [West Midlands] Coventry 29 jun 1855-2 dec 1914 – 54r – 1 – (missing: 1869, 1880; cont as: coventry times & warwickshire journal [jan 1881-dec 1914]) – uk Newsplan [072]

Coventry trades council records, 1890-1992 – (= ser Labour party in britain, origins and development at local level. series 2) – 4r – 1 – (with p/g. int by richard stevens) – mf#97565 – uk Microform Academic [331]

Cover crop program for florida pecan orchards / Blackmon, G H – Gainesville, FL. 1936 – 1r – us UF Libraries [634]

Coverage / Assembly of Governmental Employees [Washington DC] – v25 n2-v29 n1 [1976 dec-1980 sep] – 1r – 1 – mf#665253 – us WHS [350]

Coverage of african american basketball athletes in sports illustrated (1954 to 1986) / Francis, Mark E – 1990 – 107p 2mf – 9 – $8.00 – us Kinesology [305]

Coverage of the spiritual dimension of health in personal health textbooks in higher education / Allen, Donna – Texas Woman's University, 1993 – 2mf – 9 – $8.00 – mf#HE 561 – us Kinesology [613]

Coverdale, Miles *see* Remains of myles coverdale, bishop of exeter

Covered employment trends in new jersey by geographical areas of the state / New Jersey. Dept. of Labor and Industry. Division of Planning and Research – 1944-78 – (= ser Harvard law school library coll) – 36mf – 9 – us Harvard Law [314]

Covert from the tempest / Abbott, W – London, England. 18– – 1r – us UF Libraries [240]

Covertaction information bulletin – 1978 jul-1986 win – 1r – 1 – (cont by: covert action quarterly) – mf#1022247 – us WHS [071]

Covertaction information bulletin – Washington. 1978-1992 (1,5,9) – (cont by: covertaction quarterly) – ISSN: 0275-309X – mf#11787 – us UMI ProQuest [320]

Covertaction quarterly – Washington. 1992+(1,5,9) – (cont: covertaction information bulletin) – ISSN: 1067-7232 – mf#11787,01 – us UMI ProQuest [320]

Covetsky patriot – Paris, France. 17 aug 1945-13 jun 1947; 5 sep 1947-16 jan 1948 – 1/4r – 1 – uk British Libr Newspaper [072]

[Covina-] covina argus citizien – Ca. 1921- – 100r – 1 – $6000.00 (subs $90/y) – mf#R02148 – us Library Micro [071]

[Covina-] covina citizen – CA. 1940-45 – 6r – 1 – $360.00 – mf#R04019 – us Library Micro [071]

[Covina-] covina sentinel – CA. 1951-1979 – 28r – 1 – $1680.00 – mf#R04020 – us Library Micro [071]

[Covina-] covina valley courier – CA. 1961-1980 – 3r – 1 – $180.00 – mf#R04021 – us Library Micro [071]

[Covina-] highlander press courier – CA. 1981-1986 – 6r – 1 – $360.00 – mf#RH03191 – us Library Micro [071]

Covington first baptist church. covington, louisiana : church records – 1904-72. 1338p – 1 – 60.21 – us Southern Baptist [242]

Covington, Perry Decatur *see*
- Perry decatur covington diary, 1864-1866

Covington weekly news – Covington, NE: Erwin Wood (wkly) [mf ed v2 n46. feb 1 1872 filmed 1958] – 1r – 1 – (absorbed by: north nebraska eagle) – us NE Hist [071]

Covjek i prostor – Zagreb, Yugoslavia. Feb 1954-Dec 1970 – 4r – 1 – (lacking: jan, feb 1957) – uk British Libr Newspaper [949]

Covodonga / Rubinos, Jose – Habana, Cuba. 1950 – 1r – us UF Libraries [972]

Cow and dairy register / Topeka. Kansas. Clerk – 1908-14. Automotive vehicle register, 1916-25 – 1 – us Kansas [978]

Cow country courier – Valentine, NE: George B Gross. v14 n3. oct 22 1942- [wkly] [mf ed -1943] – 2r – 1 – (cont: monitor) – us NE Hist [071]

The cow of the barricades and other stories / Raja Rao – London; New York: Oxford University Press, 1947 – (= ser Samp: indian books) – us CRL [830]

Cow question in india : with hints on the management of cattle – Madras: Christian Literature Society 1894 – (= ser Assorted materials dealing with india filmed at the british museum) – 1r – 1 – us CRL [280]

Cowan, Christopher *see* Christopher cowan papers, ms 1328

Cowan clan newsletter – v1 n1-2 [1977 apr-aug] – 1r – 1 – mf#400900 – us WHS [071]

Cowan, David *see* Anecdotes of a life on the ocean

Cowan, Henry *see*
- Influence of the scottish church in christendom
- John knox
- Landmarks of church history to the reformation
- Sub corona

Cowan, James *see* Samoa and its story

Cowan, Minna Galbraith *see* The education of the women of india

Cowan, R *see* News to perfection

Cowan, Robert *see*
- I will never leave thee, nor forsake thee
- John flockhart, esq
- Remember your leaders
- Six piano concertos by contemporary british composers

Coward, Noel Pierce *see* Play parade

Cowbridge, borough cemetery, monumental inscriptions – 1mf – 9 – £1.25 – uk Glamorgan FHS [929]

Cowbridge, glamorgan, parish church of holy cross : baptisms 1718-1887, burials 1735-1873, marriages 1721-1837 – 2mf – 9 – £2.50 – uk Glamorgan FHS [929]

Cowbridge, holy cross; ramoth baptist, monumental inscriptions – 2mf – 9 – £2.50 – uk Glamorgan FHS [929]

Cowdenbeath & district advertiser – [Scotland] Cowdenbeath: J Westwater & Sons 8 apr 1927-dec 1950 (wkly) [mf ed 2004] – 542v on 18r – 1 – (cont by: cowdenbeath & district advertiser and kelty news [jan 1938-dec 1950]) – uk Newsplan [072]

Cowdenbeath, lochgelly, and kelty news – [Scotland] Fife, Lochgelly: J Greenhill 2 may 1906-23 apr 1918 (wkly) [mf ed 2004] – 4r – 1 – (missing: 1911, 1915; cont by: lochgelly & kelty news [jan 1916-apr 1918]) – uk Newsplan [072]

Cowdenbeath & lochgelly times and advertiser – [Scotland] Fife, Lochgelly: J Westwater jan 1896-28 dec 1950 (wkly) [mf ed 2004] – 36r – 1 – (missing: 1895 not filmable; cont by: times (cowdenbeath and lochgelly times and advertiser) [may 1919-dec 1926]; times for lochgelly, bowhill, dundonald, cardenden, glencraig and lochore [6 apr 1927-dec 1950]) – uk Newsplan [072]

Cowdenbeath mail and west fife record – [Scotland] Fife, Cowdenbeath: W F Simpson 13 jan 1912-7 nov 1914 (wkly) [mf ed 2004] – 1r – 1 – uk Newsplan [072]

Cowdery, Jabez Franklin *see*
- Cowdery's forms and precedents.
- Cowdery's new book of forms...
- The law of insolvency
- A treatise on the law and practice in justices' courts, as determined by the statutes and decisions of the states of california, colorado, nevada, and oregon...

Cowdery's forms and precedents. : being legal forms and precedents for court proceedings and business transactions... / Cowdery, Jabez Franklin – San Francisco: Bancroft-Whitney Co, 1895 – 673p – 1 – mf#LL 327 KF – us L of C Photodup [347]

Cowdery's new book of forms... : especially adapted to the codes and statutes of alaska, arizona, california, colorado, idaho, montana, nevada, new mexico, north dakota, oregon, south dakota, utah, washington and wyoming / Cowdery, Jabez Franklin – San Francisco, Bancroft-Whitney, 1905. 1087 p. LL-872 – 1 – us L of C Photodup [348]

Cowdrey-cowdry-cowdray genealogy / Mehling, Mary B A – Frank Allaben Genealogical Co, 1911 – 1r – 1 – us Western Res [920]

Cowdroy's manchester gazette : and weekly advertiser – [NW Englnad] Manchester ALS 1796-19 jun 1824* – 1 – (title change: manchester gazette [7 aug 1824-23 may 1829]) – uk MLA; uk Newsplan [072]

Cowell, D *see* The interpreter of words and terms

Cowell, E B *see* The buddha-karita of asvaghosha

Cowell, Edward Byles *see* The divyavadana

Cowell, John *see* Cowell's law dictionary

Cowell, Peter *see* Liverpool free public libraries

Cowell rebel *see* [Santa cruz-] miscellaneous titles

Cowell's law dictionary : a law dictionary on the interpretation of legal words and terms... with an appendix containing ancient names of places / Cowell, John – London, 1708 – 4mf – 9 – $6.00 – mf#LLMC 95-410 – us LLMC [340]

Cowen, Denis Victor *see*
- Constitution-making for a democracy
- Foundations of freedom
- Swaziland

Cowen, Esek *see* A treatise on the civil jurisdiction of justices of the peace in the state of new-york

Cowey, Catherine *see* The validity of the polar smartedge owncal monitor

Cowichan leader – Duncan British Columbia, Canada. 4 oct 1951-14 may 1953 – 2r – 1 – uk British Libr Newspaper [071]

Cowie, Isaac [comp] *see* The grain, grass and gold fields of south-western canada

Cowie, J S, Mrs [comp] *see* Band of hope ritual

The cowles enterprise – Cowles, NE: Karl L Spence, nov 22 1907-v1 n48. oct 16 1908 (wkly) – 2r – 1 – (publ in bladen ne, mar 6-oct 16 1908) – us NE Hist [071]

Cowles, Henry *see*
- Acts of the apostles
- A defence of ohio congregationalism and of oberlin college
- The epistle to the hebrews
- Hebrew history from the death of moses to the close of the scripture narrative
- Isaiah
- Jeremiah and his lamentations
- Matthew and mark
- The pentateuch
- The psalms
- The shorter epistles

Cowles, Henry Trask *see* Normal water losses from species and varieties of citrus

Cowles shopper – Hammond, Roberts WI. 1972 nov 9-dec 28 – 1r – 1 – (cont by: central shopper) – mf#1278403 – us WHS [380]

Cowley, A E *see* The samaritan liturgy

Cowley, Abraham *see* The works of mr abraham cowley

Cowley, Arthur Ernest *see*
- The original hebrew of a portion of ecclesiasticus (39. 15 to 49. 11)
- The samaritan liturgy

Cowley, Cecil *see* Kwa zulu

Cowley, Clive *see* Fabled tribe

Cowley County Genealogical Society *see* Buffalo trails

Cowley progress – v1 n1-141 [1967 feb 24-1985 may 1] – 1r – 1 – (cont: cowley progress) – mf#619549 – us WHS [071]

Cowley, Rafael Angel *see* Tres primeros historiadores de le isla de cuba

Cowper *see* The olney hymns and a baptist hymn book

Cowper, Benjamin Harris *see* Apocryphal gospels

Cowper, William *see*
- The anatomy of humane bodies: with figures drawn after the life...and curiously engraven..
- Life of william cowper, bishop of galloway

Cowper's task, books 3 and 4, the garden and the winter evening : and coleridge's friend, essays 3-6, life of sir alexander ball – Toronto: Copp, Clark, 1887 – 3mf – 9 – 0-665-11874-0 – (with int etc by by h i strang and a j moore) – mf#11874 – cn Canadiana [420]

Cowperthwaite, Sarah A *see* Physiological comparison of chair aerobics and cycle ergometry in young female subjects

Cowra free press – Cowra, jan 1924-may 1931 – 24r – 9 – A$1578.72 vesicular A$1710.72 silver – at Pascoe [079]

Cowra guardian – Cowra, jan 1969-jun 1995 – 69r – 9 – at Pascoe [079]

Cowtan, Robert *see*
- A biographical sketch of sir anthony panizzi...
- Memories of the british museum

Cox, David *see* A treatise on landscape painting and effect in water colours

Cox, E K *see* Christian stewardship, morristown, tenn

[Cox, E W] *see* Early promoted

Cox, Edward W *see* Cox's magistrates cases

Cox, Edward W et al *see* Cox's criminal cases

Cox, Edward Young *see* The art of garnishing churches at christmas and other festivals

Cox, Eleanor Rogers *see* Hosting of heroes

Cox, Francis Augustus *see*
- The baptists in america
- Female scripture biography
- History of the baptist missionary society: from 1792 to 1842 – a sketch of the general baptist mission
- Posthumous testimony

Cox, Francis Augustus et al *see* Religion in america

Cox, George Henry *see* The history of the evangelical lutheran synod and ministerium of north carolina

Cox, George William *see*
- The greeks and the persians
- Hughes's historical readers, standard 3
- Hughes's historical readers, standard 4
- Hughes's historical readers, standard 5
- Hughes's historical readers, standard 6
- The life of john william colenso, d.d
- The life of saint boniface
- The mythology of the aryan nations
- School history of greece
- Tales from greek mythology

Cox, H *see* Journal of a residence in the burmhan empire, and more particularly at the court of amarapoorah

Cox, Hiram *see* Journal of a residence in the burmhan empire and more particularly at the court of amarapoorah

Cox, J *see*
- Disappointing dream
- Growing evil
- Noisy sins and quiet sins
- Poisonous fountains and the life-giving spring
- Profit of punctuality
- Race of the rain-drops

Cox, J E *see* Writings and disputations

Cox, Jacob Dolson *see*
- Atlanta
- The march to the sea, franklin and nashville
- Military reminiscences of the civil war, vol 1
- Military reminiscences of the civil war, vol 2
- Military reminiscences of the civil war, vols 1 and 2

Cox, James *see* Historical and biographical record of the cattle industry and the cattlemen of texas and adjacent territory

Cox, John *see*
- Christian experience
- Drawing cheques
- Drop and the ocean
- Early and late
- Guilty, or not guilty?
- New year's sermon
- Noisome pestilence
- Our mighty all

Cox, John Charles *see*
- Churchwardens' accounts
- English church furniture
- The english parish church
- How to write the history of a parish
- The parish registers of england
- Pulpits, lecterns and organs in english churches
- The sanctuaries and sanctuary seekers of mediaeval england

Cox, John Charles et al *see* Curious church customs and cognate subjects

Cox, John Edmund *see*
- Brotherly love
- The old constitutions belonging to the ancient and honourable society of free and accepted masons of england and ireland

Cox, Kelly M *see* The effect of an incentive-based wellness challenge program on physical fitness in industrial workers

Cox, Kimberley A *see* Effect of a psychological skills training program on competition anxiety and performance of selected national youth sports program campers

The cox library collection : county, state and local histories – 1 – $50.00r – (states listed individually) – us Library Micro [978]

Cox, Lori M *see* Comparison of television viewing, arcade game play, and resting metabolic rates in youth

Cox, Marian Roalfe [comp] *see* Cinderella

Cox, Norman W *see* Manuscripts, personal papers, pastoral records, published articles, etc

Cox, Palmer *see*
- Another brownie book
- The brownies abroad
- The brownies around the world
- The brownies at home
- The brownies, their book
- The brownies through the union
- How columbus found america
- Squibs of california

Cox, Philip *see* The rani of jhansi

Cox, Richard J *see* The calvert papers

Cox, Ross *see* The columbia river

Cox, Rowland *see*
- American trademark cases
- Cox's manual of trademark cases
- A manual of trademark cases

Cox, Samuel *see*
- The book of ecclesiastes
- The hebrew twins
- Miracles
- The pilgrim psalms
- Salvator mundi

Cox, Walter Smith *see*
- Common law practice in civil actions
- Questions for the use of students in the junior law class of columbian university

Cox, William Charles *see* The development of a comprehensive plan for ministry for first baptist church, nevada, missouri

Cox, William M *see* The social and civil status of woman

Coxe, A Cleveland *see* The impossibility of the immaculate conception as an article of faith

Coxe, Arthur Cleveland *see*
- An apology for the common english bible
- Holy writ and modern thought

Coxe, Arthur Cleveland et al *see* The history and teachings of the early church as a basis for the re-union of christendom
Coxe, R C *see*
- False pleas and deceptive pretences
- Pleasures of taste incentives to devotion

Coxe, Robertson D *see* Legal philadelphia, comments and memories
Coxe, WCoxe, William *see* Travels into poland, russia, sweden, and denmark
Coxe, William *see*
- Account of the russian discoveries between asia and america
- Account of the russian discoveries between asia and america microform
- Travels in poland, russia, sweden, and denmark
- Travels in switzerland, and in the country of the grisons
- Voyage en pologne, russie, suede, dannemarc, etc
- Voyage en suisse

Coxey's sound money – v3 n97 [1897 apr 10] – 1r – 1 – mf#945445 – us WHS [332]
Cox-George, N A *see* Some problems of financing development in sierra leone, west africa
Cox-phillips family newsletter – 1981 mar-1984 jun – 1r – 1 – mf#933546 – us WHS [929]
Cox's criminal cases : reports of cases in criminal law in all the courts of england and ireland / Great Britain. England; ed by Cox, Edward W et al – v1-31. 1843-1941. London: J Crockford/Butterworth, 1846-1948 (all publ) – 215mf (only v1-25) – 9 – $322.00 – (add vols planned as they fall out of copyright) – mf#LLMC 95-285 – us LLMC [324]
Cox's magistrates cases : reports of all the cases decided by all the superior courts relating to magistrates, municipal and parochial law / Great Britain. England; ed by Cox, Edward W – London: Law Times Office. v1-27. 1862-1919 (all publ) – 205mf – 9 – $307.00 – (includes 1v of digest of cases for the yrs 1856-69) – mf#LLMC 95-226 – us LLMC [324]
Cox's manual of trademark cases : includes "sebastian's digest of trademark cases" covering all the cases reported prior to 1879, together with leading cases decided since... / Cox, Rowland – 2nd rev enl ed. Boston: Houghton-Mifflin, 1892 (all publ) – (= ser Sebastian's Digest Of Trademark Cases) – 2mf – 9 – $9.00 – mf#LLMC 84-339 – us LLMC [346]
Coy, Owen C *see* Humboldt bay region, 1850-1895
Coychurch, glamorgan, parish church of st crallo : baptisms 1733-1900, burials 1736-1900, marriages 1723-1837 – 2mf – 9 – £2.50 – uk Glamorgan FHS [929]
Coyer, Gabriel *see* Lettre au docteur maty, secretaire de la societe royale de londres
Coyle, Grace Longwell *see* Social process in organized groups
Coyle, John Patterson *see* The spirit in literature and life
Coyle, Robert F *see* Workingmen and the church
Coyne, James H *see* Exploration of the great lakes, 1669-1670
Coyne, James Henry *see*
- The country of the neutrals
- Record of the celebration of the centenary of the talbot settlement
- Richard maurice bucke
- The talbot papers

Coyne, James Henry et al *see* Memorial to u e loyalists
Coyote / University of California, Davis – 1970 spr-1971 spr – 1r – 1 – mf#1100338 – us WHS [071]
Coyote's journal – Brunswick. 1964-1972 – 1 – ISSN: 0011-0736 – mf#8541 – us UMI ProQuest [073]
Coyoti prints / Caribou Indian Education and Training Centre – 1982 apr-1983 mar, may-jul, oct, 1984 feb-mar, may-jun, 1985 mar-1986 feb – 1r – 1 – mf#1119242 – us WHS [740]
Coyuntura economica – Bogota: Fundacion para la Educacion Superior y el Desarrollo. v18 n1-2. mar-jun 1988 – us CRL [330]
Coyuntura economica andina – Bogota: Fundacion para la Educacion Superior y el Desarrollo. n9. jun 1988 – 1r – us CRL [330]
Cozad citizen – Cozad, NE: Markwood Holmes, 1892 (wkly) [mf ed v1 n29. dec 3 1892 filmed 1973] – 1r – 1 – (merged with: lexington clipper to form: clipper-citizen) – us NE Hist [071]
Cozad local – Cozad, NE: C F & M S Kleinhans. 39v. v34 n88. mar 7 1930-72nd yr. may 19 1971 (wkly) [mf ed with gaps filmed -1975] – 29r – 1 – (suppl called: platte valley farmer-stockman. cont: semiweekly local) – us NE Hist [071]
The cozad local – Cozad, NE: Sam I Stefens. 30v. v1 n1. jul 16 1897-v30 n79. feb 19 1926 (semiwkly) [mf ed with gaps] – 10r – 1 – (cont: meridian star. cont by: semi-weekly local) – us NE Hist [071]
The cozad messenger – Cozad, NE: R O Willis (wkly) [mf ed v2 n47. may 5 1887] – 1r – 1 – us NE Hist [071]
Cozad republic – Cozad, NE: J A Holmes, apr 10 1908 (wkly) [mf ed -1909 (gaps) filmed 1957] – 1r – 1 – (absorbed: cozad herald and: cozad tribune. cont: cozad tribune (1909)) – us NE Hist [071]
Cozad tribune – Cozad, NE: J A Holmes. v1 n1. sep 14 1909- (wkly) [mf ed -sep 211909] – 1r – 1 – (cont: cozad republic) – us NE Hist [071]
Cozad tribune – Cozad, NE: F P Corrick, jul 1892-jul 1908// (wkly) [mf ed 1892,1894-1907 (gaps) filmed 1957] – 3r – 1 – (cont: news-reporter. absorbed by: cozad republic. issues for oct 16 1896-nov 29 1907 called v5 n16-v15 n48) – us NE Hist [071]
Cozza, Laurentius *see* Historia polemica de graecorum schismate ex ecclesiasticis monumentis concinnata
Cozza-Luzi, Giuseppe *see* Sacrorum bibliorum vetustissima fragmenta graeca et latina ex palimpsestis codicibus bibliothecae cryptoferratensis eruta
Cozza-Luzi, J *see* Novum patrum bibliotheca
Cpa consultant – New York. 1998+ (1) – mf#19163,01 – us UMI ProQuest [650]
Cpa journal – New York. 1930+ (1) 1971+ (5) 1976+ (9) – ISSN: 0732-8435 – mf#3470 – us UMI ProQuest [650]
Cpa letter – New York. 1989-1991 (1) – ISSN: 0094-792X – mf#15418,02 – us UMI ProQuest [650]
Cpap dunman khluan, cpap baky cas nyn cpap vidhur pandit : cpap puran samrap thnak di 3 damnoep – Bhnam Ben: Pannagar Jyn Nuan Huat 1959 [mf ed 1990] – 1r with other items – 1 – (in khmer) – mf#mf-10289 seam reel 126/6 [§] – us CRL [170]
Cpap gorab mata pita / Qu Cun – Bhnam Ben: Ron Bumb Gim Sen 2510 [1967] [mf ed 1990] – 1r with other items – 1 – (in khmer) – mf#mf-10289 seam reel 113/2 [§] – us CRL [306]
Cpap kun cua, cpap kram, cpap kertikal, cpap tri neti : cpap buran samrap thnak di 4 da.mnoep – Bhnam Ben: Pannagar Jyn Nuan Huat 1959 [mf ed 1990] – 1r with other items – 1 – (in khmer) – mf#mf-10289 seam reel 129/10 [§] – us CRL [305]
Cpap phsen phsen : camlan cen bi sastra slyk ryt – Paris: Centre de documentation et de recherche sur la civilisation khmere 1980 [mf ed 1990] – 1r with other items – 1 – (repr. originally publ: chbab divers. 6e ed. phnom-penh: institut bouddhique 1973) – mf#mf-10289 seam reel 101/07 [§] – us CRL [959]
Cpap sri : camlan cen bi sastra slyk ryt – Bhnam Ben: Pannagar Put Nan 1959 [mf ed 1990] – 1r with other items – 1 – (in khmer) – mf#mf-10289 seam reel 130/8 [§] – us CRL [305]
Cpap suati – Kamban Cam: Pannagar Nuan Hen 2501 [1958] [mf ed 1990] – 1r with other items – 1 – (in khmer) – mf#mf-10289 seam reel 127/9 [§] – us CRL [170]
Cpcu journal / Society of Chartered Property and Casualty Underwriters – Malvern. 1987+ (1,5,9) – ISSN: 0162-2706 – mf#15929,02 – us UMI ProQuest [366]
Cpi purchasing – Boston. 1983-1994 (1,5,9) – (cont by: purchasing cpi edition) – ISSN: 0746-9012 – mf#14868 – us UMI ProQuest [660]
Cpst comments – Washington. 1995-1996 (1) 1995-1996 (5) 1995-1996 (9) – (cont: scientific, engineering, technical manpower comments) – mf#6443,01 – us UMI ProQuest [620]
Cq researcher – Washington. 1991+ (1,5,9) – (cont: congressional quarterly's editorial research reports) – ISSN: 1056-2036 – mf#18751 – us UMI ProQuest [320]
Cq weekly – Washington. 1998+ (1,5,9) – (cont: congressional quarterly weekly report) – mf#858,01 – us UMI ProQuest [320]
Cr : the new centennial review – East Lansing. 2001+ [1,5,9] – (cont: cr. the centennial review) – ISSN: 1532-687X – mf#32020 – us UMI ProQuest [410]
Cr – the centennial review – East Lansing. 1957-1999 [1]; 1971-1999 [5]; 1977-1999 [9] – ISSN: 0162-0177 – mf#1765 – us UMI ProQuest [400]
Crab – Baltimore. 1971+ (1) 1989+ (5) 1989+ (9) – ISSN: 0300-7561 – mf#6723 – us UMI ProQuest [400]
Crab orchard herald – Crab Orchard, NE: S Grey Howe. -v46 n35. aug 21 1936 (wkly) [mf ed 1892,1894-1936 (gaps)] – 12r – 1 – (issues for nov 9 1894-mar 29 1895 called also eagle v7 n18-38) – us NE Hist [071]
Crabb, James *see* Decorative art society
Crabbe, W H *see* Crabbe's reports of cases for the eastern district of pennsylvania, 1836-1846
Crabbe's reports of cases for the eastern district of pennsylvania, 1836-1846 / Crabbe, W H – Philadelphia: Johnson. 1v. 1853 (all publ) – (= ser Early federal nominative reports) – 7mf – 9 – $10.50 – mf#LLMC 81-444 – us LLMC [340]
Crabites, Pierre *see* The winning of the sudan
Crabtree, Alice Ione *see* Marriage and family life among the educated africans in the urban areas of the gold coast
Crabtree, William *see* Prosperity of a gospel church considered
Crace collection of london views in the british museum / British Museum. Dept of Prints & Drawings – 8r – 1 – $1100.00 – 0-907006-39-6 – (almost 8000 engravings, drawings & watercolours of london through the ages; arr topographically in chronological sequence, 1236-1800s; with guides & 2 printed ind) – uk Mindata [700]
Cracked stem of celery caused by a boron deficiency in the soil / Purvis, E R – Gainesville, FL. 1937 – 1r – us UF Libraries [630]
The cracker *see* Political pamphlets... 19th c
Cracknell, J E *see* Go to joseph
Cradle of the deep / Treves, Frederick – New York, NY. 1913 – 1r – us UF Libraries [972]
The cradle of the semites : two papers read before the philadelphia oriental club / Brinton, Daniel Garrison – Philadelphia: [s.n.], 1890 – 1mf – 9 – 0-524-01255-5 – mf#1990-2291 – us ATLA [900]
Cradle roll home – Jul 1925-46 – 1 – us Southern Baptist [242]
Cradle tales of hinduism / Noble, Margaret E – London: Longmans, Green, 1907 – 1mf – 9 – 0-524-01204-0 – mf#1990-2280 – us ATLA [390]
Cradock : south africa collected papers 1929-1962 – [New Haven CT: Yale University Library 197-?] – 13r – 1 – us CRL [960]
Cradock and tarkastad register *see* The cradock register
Cradock news *see* The cradock register
The cradock register – Cradock SA, 5 jan 1858-29 dec 1899 – 30r – 1 – (Title varies: Cradock News, 1858-1862; Cradock and Tarkastad Register, 1863) – sa National [079]
Craft horizons with craft world – New York. 1942-1979 (1) 1969-1979 (5) 1970-1979 (9) – (cont by: american craft) – ISSN: 0164-9191 – mf#773 – us UMI ProQuest [740]
Craft/midwest – Prairie View. 1971-1974 (1) – mf#9545 – us UMI ProQuest [790]
Crafts – London. 1975+(1,5,9) – ISSN: 0306-610X – mf#10760 – us UMI ProQuest [740]
Crafts 'n things – Des Plaines. 1987+ (1,5,9) – ISSN: 0146-6607 – mf#16460 – us UMI ProQuest [640]
Crafts report – Wilmington. 1978+ (1,5,9) – ISSN: 0160-7650 – mf#11896 – us UMI ProQuest [740]
Crafts, Wilbur Fisk *see*
- The bible and the sunday school
- Childhood
- Helps to the study of the versions of the new testament
- Must the old testament go?
- National perils and hopes
- The sabbath for man
- Talks to boys and girls about jesus

Crafts, Wilbur Fisk et al *see*
- Intoxicants and opium in all lands and times
- Intoxicants & opium in all lands and times

Craftsman : an illustrated monthly magazine in the interest of better art, better work, and a better and more reasonable way of living – Eastwood. 1901-1916 (1) – mf#2879 – us UMI ProQuest [700]
Craftsman – [Scotland] NLS feb-1 dec 1883 [mf ed 2004] – 1r – 1 – uk Newsplan [072]
Craftsman *see* Printer's circular and stationers' and publishers' gazette, 1866-1888 / american model printer, 1879-1882 / craftsman, 1884-1888
Craftsman and british american masonic record – Hamilton, C W [Ont]: T & R White, 1866-1869 – 9 – (cont by: the craftsman and canadian masonic record. incl ind) – mf#P04073 – cn Canadiana [366]
Craftsman and canadian masonic record – Hamilton, Ont: T & R White, 1869-1877 – 9 – (cont: the craftsman and british american masonic record; cont by: the canadian craftsman and masonic record; incl ind) – mf#P04074 – cn Canadiana [366]
Craftsmen / Gordon, Thomas – London, England. 1839 – 1r – us UF Libraries [240]
Crag and canyon – Banff Hot Springs, Alta: I Byers, [1900-19–] – 9 – mf#P04909 – cn Canadiana [790]
Cragg, Kenneth *see* The call of the minaret
Craib, Alexander *see* America and the americans
Craig advertiser and the burt county news – Craig, NE: Chas E Brooks (wkly) [mf ed v19 n52. jan 4 1906=v2 n26-sep 21 1906] – 1r – 1 – (formed by the union of: craig advertiser and: burt county news. cont by: craig news) – us NE Hist [071]
Craig, Austin *see* The gospel of luke the apostles' creed
Craig, David Irwin *see* A history of the development of the presbyterian church in north carolina
Craig, Edward *see*
- Appointment and promise of messiah
- Christian circumspection
- Correspondence between the right rev bishop gleig

Craig, Edward Thomas *see* The irish land and labour question, illustrated in the history of ralahine and co-operative farming
Craig, J A *see* Assyrian and babylonian religious texts
Craig, J D [comp] *see* The 1st canadian division in the battles of 1918
Craig, J J *see* Old colonists' jubilee, auckland, nz 1842-1892
Craig, James Alexander *see* The history of babylonia and assyria
Craig, Kenneth R *see* Procedure for serial section
Craig, Laura Gerould *see* The centennial campfire
Craig, Lulu Alice *see* Glimpses of sunshine and shade in the far north
Craig, Maria G [comp] *see* Information relating to municipal legislation of the liquor traffic
Craig, Neville B *see* Estrada de ferro madeira-mamore
Craig news – Craig, NE: W D Smith & Sons. -v74 n36. jan 5 1961 (wkly) [mf ed 1909-14,1918-61 (gaps)] – 10r – 1 – (cont: craig advertiser and the burt county news. absorbed by: oakland independent and republican. publ in craig ne, aug 12 1909-mar 5 1959; in oakland ne, mar 12 1959-jan 5 1961) – us NE Hist [071]
Craig, Peter S *see* Organized baseball
Craig, R T *see* The mammillaria handbook
Craig, Robert *see* Regeneration
Craig times – Craig, NE: Ira Thomas, -1895// (wkly) [mf ed v4 n33. oct 11-nov 29 1895 (lacks oct 18) filmed 1958] – 1r – 1 – (absorbed by: oakland independent) – us NE Hist [071]
Craig, Virginia Judith *see* Teaching of high school english
Craig, William Marshall *see* Course of lectures on drawing, painting, and engraving
Craig, Willis Green et al *see* Twentieth century addresses
Craigavon echo – Craigavon, Ireland. 31 jan-19 dec 1990, 1991, 8 jan-16 dec 1992 – 3r – 1 – (amalg: craigavon echo [lurgan ed] and craigavon echo [portadown ed]) – uk British Libr Newspaper [072]
Craigavon echo (lurgan ed) *see* Craigavon echo
Craigavon echo (portadown ed) *see* Craigavon echo
Craigavon times *see* Portadown times
Craigdam and its ministers / Walker, George – Aberdeen, Scotland. 1886 – 1r – us UF Libraries [240]
Craige, John Houston *see*
- Black bagdad
- Cannibal cousins

The craighead family : a genealogical memoir / Craighead, James Geddes – 1876 – 1 – $50.00 – us Presbyterian [920]
Craighead, James Geddes *see*
- The craighead family
- Scotch and irish seeds in american soil
- The story of marcus whitman

Craigie, William Alexander *see* The religion of ancient scandinavia
Craigmyle gazette – Alberta, CN. jan 1921-dec 1922 – 1r – 1 – cn Commonwealth Imaging [071]
Craignons d'etre un jour l'ethiopie / Nemours, Alfred – Port-Au-Prince, Haiti. 1945 – 1r – us UF Libraries [960]
Craik, Dinah Maria *see*
- John halifax, gentleman
- Sermons out of church
- A woman's thoughts about women

Craik, George *see* New zealanders
Craik, Henry *see*
- Authority of scripture considered in relation to christian union
- The hebrew language
- Impressions of india
- Improved renderings of those passages in the english version of the new testament...

Craik, James *see* Progress
Craik, Robert, 1829-1906 *see* The nature of the morbid poisons and the diseases to which they give rise
Crain's chicago business – Chicago. 1978+ (1) 1980+ (5) 1980+ (9) – ISSN: 0149-6956 – mf#12607 – us UMI ProQuest [338]
Crain's cleveland business – Cleveland. 1980+ (1,5,9) – ISSN: 0197-2375 – mf#12608 – us UMI ProQuest [338]
Crain's detroit business – Detroit. 1985+ (1,5,9) – ISSN: 0882-1992 – mf#14395 – us UMI ProQuest [338]
Crain's new york business – New York. 1985+ (1,5,9) – ISSN: 8756-789X – mf#14119 – us UMI ProQuest [338]
Crake, Augustine David *see* History of the church under the roman empire, a.d. 30-476
Craker, Lyle E *see* Journal of herbs, spices and medicinal plants

CRALLE

Cralle, Richard K *see* A disquisition on government
Cram, Ralph Adams *see*
- Church building
- Heart of europe
- Impressions of japanese architecture and the allied arts
- The ministry of art
- The ruined abbeys of great britain
Cram, Ralph Adams et al *see* Six lectures on architecture
Cramb, John Adam *see* Reflections on the origins and destiny of imperial britain
Cramer, B *see* Latomia
Cramer, Barbara J *see*
- Self-esteem and adolescent pregnancy
- Stress and job satisfaction of nurse managers in hospital settings
Cramer, Charles *see* Etwas ueber die natur wunder in nord amerika
Cramer, D *see*
- De distingvendo decalogo
- Emblemata moralia nova
- Emblematum sacra
- In natalitiam memoriam r patris d martini lvtheri
- Neun fasten
- Societas jesu et roseae crucis vera
- Wolgegruendeter beweiss von der vollstaendigkeit und vollkommenheit der heiligen bibel
Cramer, Ernst Ludwig *see* Wir kommen wieder
Cramer, F A *see* Nordische beytraege zum wachsthum der naturkunde und der wissenschaften wie auch der nuetzlichen und schoenen kuenste ueberhaupt
Cramer, Floyd *see*
- Our neighbor nicaragua
Cramer, J *see* Bibliotheca reformatoria neerlandica
Cramer, J A *see* Catenae graecorum patrum in novum testamentum
Cramer, Joh Andreas *see* Der nordische aufseher
Cramer, Johann Andreas *see* Beytraege zur befoerderung theologischer und andrer wichtigen kenntnisse von kielischen und auswaertigen gelehrten
Cramer, Johann Andreas et al *see* Neue beytraege zum vergnuegen des verstandes und des witzes
Cramer, Johann Baptist *see*
- Grande sonate pour le pianoforte
- Quintuor pour pianoforte, violin, alto, violoncelle et basse, op 69
Cramer, John Anthony *see* Catena in acta ss apostolorum e cod nov coll
Cramer, Karl Friedrich *see*
- Magazin der musik
Cramer, Maria *see* Thomas de quincey und john wilson (christopher north)
Cramer, Rose Fulton *see* Author headings for the official publications of oklahoma
Cramer-Hamman, Buffy *see* Predicting successful adjustment to disengagement from collegiate athletics
Crammer's first litany 1544 : and merbecke's book of common prayer noted 1550 / Hunt, J E – London, 1939 – 4mf – 8 – €11.00 – ne Slangenburg [242]
Crammer's liturgical projects (hbs50) / Wickham Legg, J – 1915 – (= ser Henry bradshaw society (hbs)) – 5mf – 8 – €12.00 – ne Slangenburg [240]
Cramp, J M *see* The colonial protestant and journal of literature and science
Cramp, John Mockett *see*
- Baptist history
- A memoir of madame feller
- The reformation in europe
- A text-book of popery
Crampton papers, the american material in the... 1844-1856 : from the bodleian library, oxford – (= ser British records relating to america in microform) – 17r – 1 – (with guide. int by colin bonwick) – mf#97404 – uk Microform Academic [975]
Cranach-Sichart, Eberhard von *see* Wach auf, mein herz
Cranberry eagle – Butler, PA. 1987-2000 (1) – mf#68639 – us UMI ProQuest [071]
Cranberry grower / Wisconsin State Cranberry Growers Association – 1903 jan-1905 jul – 1r – 1 – mf#2798646 – us WHS [634]
Cranbrook, James *see*
- Discourses in memoriam of the rev james cranbrook
- On responsibility
- On the existence of evil
- On the hindrances to progress in theology
Cranch, John *see* Inducements to promote the fine arts in great britain
Cranch, William *see* Cranch's reports of cases in the district of columbia, 1804-1841
Cranch's reports of cases in the district of columbia, 1804-1841 / Cranch, William – Boston: Little-Brown. v1-6. 1852-53 (all publ) – (= ser Early federal nominative reports) – 47mf – 9 – $70.00 – mf#LLMC 81-443 – us LLMC [340]
Crandall, John J *see* Leading cases, american and english, on the law of legal tender and money.
Crane american – Crane OR: Gallagher & Carter, 1916-35 [wkly] [mf ed 1966] – 2r – 1 – (cont by: harney county american [1935-]) – us Oregon Lib [071]
Crane, Charles *see* Necessity of advancement in christian knowledge and practice
Crane, F G *see* Journal of promotion management
Crane flock – n1-10 [1979 spr-1981 sum] – 1r – 1 – mf#671680 – us WHS [071]
Crane, Frank *see* The religion of to-morrow
Crane island / Youngblood, Alice P – s.l, s.l? 1940 – 1r – us UF Libraries [978]
Crane, J Willard *see* "The origin and cause of the british-american order of good templars", reviewed
Crane, Jonathan Townley *see* Holiness
Crane, Lucy *see* Art and the formation of taste
Crane, Richard Teller *see* The futility of technical, industrial, vocational and continuation schools
[Crane, Robert I] *see* Calendar of items microfilmed at the india office, london
Crane, Stephen *see*
- Maggie, a girl of the streets
- The o'ruddy
- War dispatches of stephen crane
Crane, Walter *see*
- The bases of design
- Cartoons for the cause, 1886-1896
- The claims of decorative art
- India impressions
- Line and form
- Of the decorative illustration of books old and new
Crane, William Carey *see* Collection: diaries, 1832-85; notebooks, manuscripts and correspondence with contemporaries
Cranfield – (= ser Bedfordshire parish register series) – 2mf – 9 – £5.00 – uk BedsFHS [929]
Cranfield, independent baptist monumental inscriptions – Bedfordshire Family HS 1988 – (= ser Bedfordshire parish register series) – 1mf – 9 – £1.25 – uk BedsFHS [929]
Cranfield, st peter and st paul monumental inscriptions – Bedfordshire Family HS 1988 – (= ser Bedfordshire parish register series) – 1mf – 9 – £1.25 – uk BedsFHS [929]
Cranford / Gaskell, Elizabeth Cleghorn – New York, NY. no date – 1r – us UF Libraries [025]
Crank – Pardeeville WI. 1898 mar 30-1901 oct 9 – 1r – 1 – mf#961046 – us WHS [071]
Crankshaw, Edward *see* Forsaken idea
Crankshaw, James *see*
- An analytical synopsis of the criminal code and of the canada evidence act
- The criminal code of canada and the canada evidence act
- The criminal code of canada and the canada evidence act, 1893
- The criminal code of canada and the canada evidence act as amended to date
- A practical guide to police magistrates and justices of the peace
Cranmer and the reformation in england / Innes, Arthur Donald – New York: Scribner 1900 [mf ed 1992] – (= ser The world's epoch-makers) – 1mf – 9 – 0-524-04878-9 – mf#1990-5076 – us ATLA [242]
Cranmer, David *see* European music manuscripts, series 3
Cranmer, T, Archbishop of Canterbury *see* Writings and disputations
Cranmer, Thomas et al *see* The reformation of the ecclesiastical laws
Cranmer's liturgical projects / ed by Legg, John Wickham – London: [s.n.] 1915 (London: Harrison) [mf ed 1992] – (= ser Henry bradshaw society 50) – 1mf [ill] – 9 – 0-524-03703-5 – (text in latin, int in english) – mf#1990-4808 – us ATLA [242]
Cranswick, G H *see* A new deal for papua
Cranz, August Fr *see* Fragmente ueber verschiedene gegenstaende der neuesten zeitgeschichte
Cranz, D *see* Historie von groenland
[Cranz, D] *see* Greenland missions
Cranz, David *see* The history of greenland
Craon, Princesse de *see* Sir thomas more
Le crapouillot : magazine non-conformiste – Paris. 1948-1993 – 1 – fr ACRPP [073]
Crapsey, Algernon Sidney *see*
- The re-birth of religion
- Words of farewell
La craque : journal de mobilisation des etudiants (es) de montmorency – [Laval]: [Association generale des etudiants de Montmorency], [ca 1976]-1982 (irreg) [mf ed 1988] – 9 – (cont: quidam; cont by: sang froid) – mf#SEM105P1014 – cn Bibl Nat [378]
Crary, Cristopher G *see* Pioneer and personal reminiscences
Crary's new york practice : special proceedings / New York. (State) – v1-2. 1866 (all publ) – 16mf – 9 – $24.00 – mf#LLMC 80-022 – us LLMC [340]
Crash, boom, bang : oder hits und flops der deutschen finanzmaerkte / Nycz, Krzysztof – (mf ed 2001) – 98p 1mf – 9 – €30.00 – 3-8267-2764-9 – mf#DHS 2764 – gw Frankfurter [332]
Crasset, J S J *see* La veritable devotion envers la sainte vierge
Crassulaceen-saeurestoffwechsel bei epiphytischen und epilithischen orchideen madagaskars : untersuchungen auf oekophysiologischer und biochemischer ebene / Vinson, Bettina – (mf ed 1996) – 2mf – 9 – €40.00 – 3-8267-2302-3 – mf#DHS 2302 – gw Frankfurter [574]
Crata repoa : oder einweihungen in der alten geheimen gesellschaft der egyptischen priester / [Koeppen, K F] – [Berlin, 1782] – 1mf – 9 – mf#VR-40.3 – ne IDC [930]
Cratippi hellenicorum fragmenta oxyrhynchia / ed by Lipsius, Justus Hermann – Bonn: A Marcus & E Weber, 1916 [mf ed 1992] – (= ser Kleine texte fuer vorlesungen und uebungen 138) – 1mf – 9 – 0-524-04698-0 – (text in greek; int & notes in latin) – mf#1990-3407 – us ATLA [450]
[Cratippus of Athens] *see* Cratippi hellenicorum fragmenta oxyrhynchia
Craufurd, Alexander Henry *see*
- Christian instincts and modern doubt
- Recollections of james martineau
Craufurd, C H *see* Law of the mind and the law of the members
Craufurd, Quentin *see* Histoire de la bastille
La cravache parisienne : journal satirique illustre – Paris. sept 1881-mars 1898 – 1 – fr ACRPP [870]
Craveiro Costa, Joao *see*
- Conquista do deserto ocidental
- Visconde de sinimbu
Craven, Alfred W Hamilton *see* Report of messrs j b jervis and alfred w craven, esq's, civil engineers, new york
Craven, Dunnill and Co *see*
- Dado tiling manufactured by craven, dunnill and co
- Price list of encaustic, geometrical, and mosaic tile pavements, hearth tiles...etc
- Tile pavements, geometrical and encaustic
Craven, Elijah Richardson *see*
- Address to the presbyteries of the presbyterian church in the united states of america
- The revelation of john
Craven herald – [Yorkshire & Humberside] North Yorkshire 24 apr 1875-25 jun 1937 [mf ed 2003] – 72r – 1 – (cont as: craven herald and wensleydale standard [jan 1879-dec 1922]; craven herald [jan 1923-jun 1937]) – uk Newsplan [072]
Craven herald and pioneer – [NW England] Barnoldswick jan 1853-sep 1857, oct 1874 – 1 – uk MLA; uk Newsplan [072]
Craven, Pauline *see* Reminiscences
Crawdaddy – New York. 1975-1978 (1) 1975-1978 (5) 1977-1978 (9) – (cont by: feature) – ISSN: 0011-0833 – mf#10595 – us UMI ProQuest [780]
Crawford, A W et al *see* Centennial addresses
Crawford, Alexander *see* Believer immersion as opposed to unbeliever sprinkling
Crawford, Alexander Crawford Lindsay, Earl of *see* Scepticism
Crawford, Alexander Wellington *see* Hamlet, an ideal prince and other essays in Shakespearean interpretation
Crawford, Alexander William Crawford Lindsay. 25th earl of, and 8th earl of Balcarres *see* Sketches of the history of christian art
Crawford, Ann Caddell *see* Customs and culture of vietnam
Crawford baptist church. crawford, georgia : church records – 1871-Feb 1913 – 1 – us Southern Baptist [242]
Crawford beacon – Crawford, NE: Beacon Pub Co, 1896// (wkly) [mf ed 1895-96] – 1r – 1 – (absorbed by: crawford gazette) – us NE Hist [071]
Crawford bulletin – Crawford, NE: Con Lindeman, 1897-aug 26 1904// (wkly) [mf ed with gaps] – 2r – 1 – us NE Hist [071]
Crawford Burkitt, F *see* Fragments of the books of kings
Crawford clipper – Crawford, NE: A J Enbody, -1892// (wkly) [mf ed v4 n51. jan 8-29 1892] – 1r – 1 – (cont by: crawford tribune) – us NE Hist [071]
Crawford clipper – Crawford, NE: Bob and Jann Reichenberg. v1 n1. oct 11 1979- (wkly) [mf ed filmed 1982-] – 1 – (cont: harrison sun. has suppls: northwest nebraska post oct 1979-nov 1983 and: crawford clipper's northwest nebraska post dec 1983-jun/jul 1992. dist with each issue: harrison sun v80 [n]15 may 7 1981-) – us NE Hist [071]
Crawford clipper's northwest nebraska post – Crawford, NE: Crawford Clipper. v5 n4. dec 1983- (mthly) [mf ed1983-92 filmed 1983-92] – 2r – 1 – (cont: crawford clipper (1979), harrison sun (1969) [and also suppl to:] and: northwest nebraska post) – us NE Hist [071]
Crawford Co. Bucyrus *see*
- Crawford county news
- Daily critic
- Daily evening forum
- Daily forum
- Evening telegraph
- Evening times
- Telegraph-forum
- Weekly journal
Crawford Co. Crestline *see*
- Advocate
- Citizen
- Daily news
- News-democrat
Crawford Co. New Washington *see*
- Herald
Crawford county advance – Soldiers Grove WI. 1894 jul 4-1897 nov 12, 1897 nov 19-1898 jan 28 – 2r – 1 – (cont: kickapoo transcript; cont by: advance [soldiers grove wi]) – mf#954653 – us WHS [071]
Crawford county courier – Prairie Du Chien WI. 1852 may 19-1853 may 21 – 1r – 1 – (cont by: prairie du chien weekly courier) – mf#1011560 – us WHS [071]
Crawford County Genealogical Society *see* Panning for nuggets of old
Crawford County Genealogical Society of Southeast Kansas *see* Seeker
Crawford county genealogy – 1978-1988 fall/winter – 1r – 1 – mf#1051935 – us WHS [929]
Crawford county independent – Gays Mills, Wauzeka WI. 1941 jul 24/1942 may 14-1979 jan/sep 13 – 16r – 1 – (with gaps; cont: independent [gays mills wi]; wauzeka chief; cont by: crawford county independent and the kickapoo scout) – mf#1042724 – us WHS [071]
Crawford county independent and the kickapoo scout – Gays Mills, Soldiers Grove WI. 1979 sep 20/dec-1995 – 17r – 1 – (cont: crawford county independent; kickapoo scout) – mf#1042737 – us WHS [071]
Crawford county journal – Soldiers Grove WI. 1883 apr 4-1885 jul 27, 1885 aug 3-1888 dec 31 – 2r – 1 – mf#931607 – us WHS [071]
Crawford county news / Crawford Co. Bucyrus – nov 1883-dec 1889 [wkly] – 2r – 1 – mf#B8445-8446 – us Ohio Hist [071]
Crawford county press – Prairie Du Chien WI. 1871 sep 8-1873 may 23 – 1r – 1 – mf#961804 – us WHS [071]
Crawford county press – Prairie Du Chien WI. 1904 jan 13/1905 jul 5-1952/1954 sep 2 – 30r – 1 – (with gaps; cont by: courier [prairie du chien wi]) – mf#961804 – us WHS [071]
Crawford county standard bearer – Prairie Du Chien WI. 1864 mar 25-jun 24 – 1r – 1 – (cont by: prairie du chien union) – mf#962580 – us WHS [071]
Crawford courier – Crawford, NE: Karl L Spence, jan 1906-nov 21 1929// (wkly) [mf ed v9 n22. may 30-oct 16 1914] – 1r – 1 – (cont by: northwest nebraska news) – us NE Hist [071]
Crawford crescent – Crawford, NE: Short & Edgar (wkly) [mf ed v1 n52. jul 14 1887] – 1r – 1 – us NE Hist [071]
Crawford, Daniel *see* Thinking black
Crawford, David Lindsay *see* Evolution of italian sculpture
Crawford, E May Grimes *see* By the equator's snowy peak
Crawford exchange newsletter – v1 n1-v2 n3 [1979:[aug]-1980/1981 feb] – 1r – 1 – (cont by: crawford families exchange newsletter) – mf#697542 – us WHS [978]
Crawford families exchange – v4 n1-v8 n1 [1982 aug-1986 aug] – 1r – 1 – (cont: crawford families exchange newsletter; cont by: crawford exchange) – mf#1336286 – us WHS [978]
Crawford families exchange newsletter – v2 n4-v3 n4 [1980/may 1981-may 1982] – 1r – 1 – (cont: crawford exchange newsletter; cont by: crawford families exchange) – mf#697549 – us WHS [978]
Crawford, Francis J *see* Horae hebraicae
The crawford gazette – Crawford, NE: Frank E Wingfield (wkly) [mf ed 1893-94,1902] – 1r – 1 – (cont: alliance boomerang; absorbed: crawford beacon) – us NE Hist [071]
Crawford, George Johnson Adair *see* Local and other rhymes
Crawford, Hanford et al *see* Proceedings of the third oecumenical methodist conference
Crawford, Isabella Valancy *see* The collected poems of isabella valancy crawford
Crawford, J *see* Pocket dictionary
Crawford, J Law *see* Proposed scheme of imperial commercial union
Crawford, John *see*
- The canadian pacific railway company and its extraordinary telegraphic and telephonic privileges
- "Social science"
Crawford, John Howard *see* Calvinism taught in the thirty-nine articles

Crawford, John Thomas *see*
- High school algebra
- The new brunswick school algebra

Crawford journal – Meadville, PA. -w 1885-1917 – 13 – $25.00r – us IMR [071]

Crawford, Kenneth C *see* Forms of oaths for use in the u.s. district courts

Crawford, Mary Caroline *see*
- Goethe and his woman friends
- The romance of old new england churches

Crawford, Nelson A *see* The ethics of journalism

Crawford, Tarleton Perry *see*
- Evolution in my mission views
- The patriarchal dynasties from adam to abraham

Crawford, Thomas Jackson *see*
- The doctrine of holy scripture respecting the atonement
- The fatherhood of god
- The mysteries of christianity
- The preaching of the cross and other sermons

Crawford tribune – Crawford, NE: Wm H Ketchal, 1892-v92 n37. sep 20 1979 (wkly) [mf ed 1892,1895-1929,1931-79 (gaps) filmed -1979] – 32r – 1 – (cont: crawford clipper. merged with: chadron record to form: chadron record and crawford tribune) – us NE Hist [071]

Crawford, William *see* Receipt book, 1793-1802

Crawford, William Rex *see* Panorama da cultura norte-americana

Crawford, William Saunders *see* Synesius, the hellene

Crawford's weekly – Norton, VA. 1921-1935 (1) – mf#69175 – us UMI ProQuest [071]

Crawfordsville review – Crawfordsville IN. 1878 apr 20 – 1r – 1 – (cont by: new review) – mf#855963 – us WHS [071]

Crawfordville baptist church. taliaferro county. georgia : church records – 1831-1900 – 1 – us Southern Baptist [242]

Crawfurd, J *see*
- A descriptive dictionary of the indian islands and adjacent countries
- History of the indian archipelago
- Journal of an embassy from the governor-general of india to the court of ava, in the year 1827

Crawfurd, John *see*
- History of the indian archipelago
- Journal of an embassy from the governor-general of india to the court of ava
- Journal of an embassy from the governor-general of india to the courts of siam and cochin china
- Tagebuch der gesandtschaft an die hoefe von siam und cochin-china

Crawley, Alfred Ernest *see*
- The idea of the soul
- The mystic rose
- The tree of life

Crawley and district observer – England, 1948-53; 1973- – 86+ r – 1 – uk British Libr Newspaper [072]

Crawley, Edmund Albern *see* A treatise on baptism as appointed by our lord jesus christ

Crawley, G J Ll *see* Reasons for leaving the church of england

Crawley, William *see*
- Charge delivered to the clergy and churchwardens of the archdeaconry...
- Charge delivered to the clergy of the archdeaconry of monmouth

Crawley, William John Chetwode *see* Caementaria hibernica

Crawshaw, C J *see* First kafir course

Crawshay, George *see* The immediate cause of the indian mutiny

Crayne, Janet *see* Yugoslav telephone directories

The crayon : a journal devoted to the graphic arts and the literature related to them – v1-8. 1855-61 – 1 – us AMS Press [760]

Crazy horse news / Pine Ridge Indian Reservation [SD] – v1 n3&4-v3 [i.e. 4] n4 [1973 apr 11-1976 sep 10], n?, n5-7 [1980 dec, 1981 apr/may-jul] – 2r – 1 – mf#365087 – us WHS [307]

The crazy quilt series, no 1, vol 1 : a compendium of wit, humour and pathos / Twain, Matthew – Toronto: s.n, 1888 – 1mf – 9 – (possibly written by nicholas flood davin) – mf#41474 – cn Canadiana [880]

Crazy shepherd – v1 n1-v6 n9 [1982 may-1986 jul] – 1r – 1 – (cont by: milwaukee shepherd) – mf#1048859 – us WHS [071]

Crazzolara, J Pasquale *see* Study of the acooli language

Cre information – Geneva. 1977-1980 – 1,5,9 – (cont by: creaction) – ISSN: 0007-9049 – mf#11197 – us UMI ProQuest [370]

Crea reporter / Canadian Real Estate Association – Don Mills. 1971-1980 (1) – ISSN: 0315-3843 – mf#7205 – us UMI ProQuest [333]

Creacion / Paniagua Santizo, Benjamin – Guatemala, 1958 – 1r – us UF Libraries [972]

Creacion / Yanez, Agustin – Mexico City? Mexico. 1959 – 1r – us UF Libraries [972]

Creacion del mundo / Acevedo, Alonso de – Madrid: Rivadeneyra, 1854 – 1 – sp Bibl Santa Ana [210]

Creacion filosofica y creacion poetica / Frutos Cortes, Eugenio – Barcelona: Juan Flors, editor, 1958 – sp Bibl Santa Ana [810]

Creacion poetica (j. guillen, salinas, a. machado, d. alonso, s.j. de la cruz, m. pinillos) / Frutos Cortes, Eugenio – Madrid: Edic. Jose Purrua Turanzas S.A., 1976 – sp Bibl Santa Ana [810]

Creacion y revolucion / Marinello, Juan – Havana, Cuba. 1973 – 1r – us UF Libraries [972]

Creaction – Geneva. 1989-1991 – 1 – (cont: cre information) – ISSN: 1011-9019 – mf#11197,01 – us UMI ProQuest [370]

Creagh, Pierse *see* Catholic oath

Cream City Business Association *see* Newsletter

Cream city courier – Milwaukee WI. 1874 sep 26-1876 jul 22, 1877 jan 6-1879 sep 20 – 2r – 1 – (cont: milwaukee enterprise; cont by: milwaukee gazette [milwaukee wi: 1879]) – mf#1126383 – us WHS [071]

Cream of the law – v1-3. 1905-07 (all publ) – 94mf – 9 – $141.00 – mf#LLMC 82-919 – us LLMC [340]

Cream raising by the centrifugal and other systems : compared and explained... / Barre, Stanislas Morrier – Montreal: Senecal, 1884 – 2mf – 9 – (incl ind) – mf#03381 – cn Canadiana [630]

Creasy, Edward Shepherd *see* The imperial and colonial institutions of the britannic empire

Creath, Jacob *see* Memoir of jacob creath, jr

Creatine and acute hypohydration : effect on plasma volume, mineral, and electrolyte balance / McArthur, Patrick D – 1999 – 2mf – 9 – $8.00 – mf#PH 1655 – us Kinesology [612]

Creatine does not enhance strength development in male college students : during a 10-week weigth lifting program / Stefl, Davis P – 1999 – 66p on 1mf – 9 – $5.00 – mf#PH 1703 – us Kinesology [612]

Creating a graduate dance curriculum model for the beijing dance academy in china / Huang, Jiamin – 1998 – 2mf – 9 – $8.00 – mf#PE 3917 – us Kinesology [790]

Creating a medicare prescription drug benefit : assessing efforts to help america's low-income seniors: hearing...house of representatives, 107th congress, 2nd session, april 17 2002 / United States. Congress. House. Committee on Energy and Commerce. Subcommittee on Health – Washington: US GPO 2002 [mf ed 2002] – 2mf – 9 – 0-16-068766-7 – (incl bibl ref) – us GPO [362]

Creating the federal judicial system / Wheeler, Russell R & Harrison, Cynthia – 2nd ed. Washington: GPO, 1994 – 1mf – 9 – $1.50 – mf#LLMC 95-839 – us LLMC [340]

Creating the federal judicial system / Wheeler, Russell R & Harrison, Cynthia – Washington: GPO, 1989 – 1mf – 9 – $1.50 – mf#LLMC 95-344 – us LLMC [340]

Creation / Duncan, Robert Dick – Edinburgh, Scotland. 1854 – 1r – us UF Libraries [240]

Creation : god in time and space / Foster, Randolph Sinks – New York: Hunt & Eaton; Cincinnati: Cranston & Curts 1895 [mf ed 1990] – (= ser Studies in theology) – 1mf – 9 – 0-7905-3682-X – mf#1989-0175 – us ATLA [210]

Creation / Goodwin, Harvey – London: Cassell 1886 [mf ed 1993] – (= ser Helps to belief) – 1mf – 9 – 0-524-05672-2 – mf#1992-0522 – us ATLA [210]

Creation – London, England. 18– – 1r – us UF Libraries [240]

Creation : or, the biblical cosmogony in the light of modern science / Guyot, Arnold – New York: Charles Scribner, 1887, c1884 – 1mf – 9 – 0-8370-3435-3 – mf#1985-1435 – us ATLA [210]

Creation : an oratorio arranged as quartettos for two violins, tenor and violoncello / Haydn, Joseph – London: Preston [179-?] [mf ed 1989] – 4pt on 1r – 1 – mf#pres. film 41 – us Sibley [780]

Creation according to the book of genesis and the confession of faith : speculative natural science and theology. two lectures / Duns, John – Edinburgh, 1877 – (= ser 19th c evolution & creation) – 1mf – 9 – mf#1.1.11587 – uk Chadwyck [210]

Creation according to the book of genesis and the confession of faith... / Duns, J – Edinburgh, Scotland. 1877 – 1r – us UF Libraries [221]

Creation and man / Hall, Francis Joseph – New York: Longmans, Green 1912 [mf ed 1990] – (= ser Dogmatic theology 5) – 1mf – 9 – 0-7905-3890-3 – (incl bibl ref) – mf#1989-0383 – us ATLA [240]

The creation and the early developments of society / Chapin, James Henry – New York: Putnam, c1880 – 1mf – 9 – 0-524-06460-1 – mf#1992-0888 – us ATLA [210]

Creation and the fall : a defence and exposition of the first three chapters of genesis / Macdonald, Donald – Edinburgh: Thomas Constable; London: Hamilton, Adams, 1856 – 2mf – 9 – 0-7905-1352-8 – (incl bibl ref and indexes) – mf#1987-1352 – us ATLA [220]

The creation and the scripture : the revelation of god / Monell, Gilbert Chichester – New York: G P Putnam, 1882 – 1mf – 9 – 0-524-05735-4 – mf#1992-0578 – us ATLA [220]

Creation centred in christ / Guinness, Henry Grattan – London: Hodder & Stoughton 1896 [mf ed 1998] – 2v on 1r – 1 – (v2 has subtitle: tables of vernal equinoxes and new moons for 3555 years...; incl bibl ref & ind) – mf#film mas 28408 – us Harvard [210]

Creation myths of primitive america : in relation to the religious history and mental development of mankind / Curtin, Jeremiah – Boston: Little, Brown, 1898 – 1mf – 9 – 0-524-02013-2 – mf#1990-2788 – us ATLA [290]

The creation of man : a sermon preached in whitehall chapel...on the 4th sunday after trinity, july 9th 1865... / Stanley, Arthur Penrhyn – [Oxford] 1865 – (= ser 19th c evolution & creation) – 1mf – 9 – mf#1.1.5092 – uk Chadwyck [210]

Creation of man by the triune god / Sibly, Manoah – London, England. 1796 – 1r – us UF Libraries [210]

Creation of the bible / Adams, Myron – Boston: Houghton, Mifflin, 1892 – 1mf – 9 – 0-7905-3062-7 – mf#1987-3062 – us ATLA [220]

The creation of the world : addressed to r j murchison, esq and dedicated to the geological society / Cockburn, William – London, 1840 – (= ser 19th c evolution & creation) – 1mf – 9 – mf#1.1.735 – uk Chadwyck [577]

Creation (omphalos) : an attempt to untie the geological knot / Gosse, Philip Henry – London: J Van Voorst, 1857 – 1mf – 9 – 0-524-05036-8 – mf#1992-0289 – us ATLA [210]

Creation or evolution? : a philosophical inquiry / Curtis, George Ticknor – London, 1887 – (= ser 19th c evolution & creation) – 6mf – 9 – mf#1.1.9248 – uk Chadwyck [110]

Creation or evolution? : a philosophical inquiry / Curtis, George Ticknor – New York: D Appleton, 1887 – 2mf – 9 – 0-7905-9176-6 – mf#1989-2401 – us ATLA [210]

Creation research society quarterly – Ann Arbor. 1985+ (1,5,9) – ISSN: 0092-9166 – mf#15310 – us UMI ProQuest [100]

The creation story / Gladstone, William Ewart – Philadelphia: Henry Altemus, c1896 – 1mf – 9 – 0-8370-9470-4 – (incl bibl ref) – mf#1986-3470 – us ATLA [220]

Creation, time and eternity : a book devoted to the unfolding of the great fundamental truths as found in science, nature and revelation... / Secrist, Jacob S – Elgin, IL: Brethren Pub House, 1911 – 1mf – 9 – 0-524-03804-X – mf#1990-4876 – us ATLA [220]

Creation with development or evolution / Hewitt, James Dudley Ryde – London, 1897 – (= ser 19th c evolution & creation) – 3mf – 9 – mf#1.1.9153 – uk Chadwyck [210]

Creation's testimony to its god: the accordance of science, philosophy, and revelation : a manual of the evidences of natural and revealed religion, with especial reference to the progress of science, and advance of knowledge / Ragg, Thomas – 11th rev enl ed. London: Charles Griffin, 1867 – 1mf – 9 – 0-8370-4826-5 – (incl bibl ref, indexes) – mf#1985-2826 – us ATLA [210]

The creation-story of genesis 1 : a sumerian theogony and cosmogony / Radau, Hugo – Chicago: Open Court; London: Kegan Paul, Trench, Truebner, 1902 – 1mf – 9 – 0-8370-4820-6 – (incl bibl ref) – mf#1985-2820 – us ATLA [221]

Creative Ad-Ventures *see* Onion

The creative art of life : studies in education / Munshi, Kanaiyalal Maneklal – Bombay: Published for Bharatiya Vidya Bhavan by Padma Publications, 1946 – (= ser Samp: indian books) – us CRL [370]

The creative attitude : fusion of facts and values / Maslow, Abraham S – [New York: Ethical Culture Publications, c1966] – us CRL [975]

Creative child and adult quarterly – Cincinnati. 1976-1991 (1,5,9) – ISSN: 0098-7565 – mf#11306 – us UMI ProQuest [640]

Creative Consciousness Center *see* Oak tree

Creative corner / Newspapers, Inc [Milwaukee WI] – 1979 jul-1981 may – 1r – 1 – mf#653626 – us WHS [071]

Creative ideas for living – Birmingham. 1984-1991 (1,5,9) – (cont: decorating and craft ideas) – ISSN: 0747-4768 – mf#14830,02 – us UMI ProQuest [640]

Creative moment – Sumter. 1973-1975 (1) 1974-1975 (5) 1974-1975 (9) – (cont by: creative moment and poetry eastwest) – ISSN: 0045-897X – mf#8407 – us UMI ProQuest [420]

Creative moment : world poetry and criticism – Sumter. 1976-1980 (1) 1976-1980 (5) 1976-1980 (9) – (cont: creative moment and poetry eastwest) – ISSN: 0045-897X – mf#8407,02 – us UMI ProQuest [420]

Creative moment and poetry eastwest – Sumter. 1975-1976 (1) 1975-1976 (5) 1975-1976 (9) – (cont: creative moment. cont by: creative moment: world poetry and criticism) – mf#8407,01 – us UMI ProQuest [420]

Creative problem solving in the learning of certain topics in the technology learning area / Xaba, Irene Nomathemba – Pretoria: Vista University 2002 [mf ed 2002] – 2mf – 9 – (incl bibl ref) – mf#mfm15220 – sa Unisa [370]

Creative table-top photography / Heimann, Ernest – London, England. 1949 – 1r – us UF Libraries [770]

The creative use of the lord's supper as a teaching tool for the doctrine of atonement / Stephens, David Ellis – 1982 – 1 – 6.16 – us Southern Baptist [242]

The creative workman : an address delivered before the technical association of the pulp and paper industry... / Wolf, Robert Bunsen – New York: The Association, 1918 (mf ed 19–) – 13p – mf#ZT-TB pv186 n1 – us Harvard [338]

Creator and creation : or, the knowledge in the reason of god and his work / Hickok, Laurens Perseus – Boston: Lee and Shepard; New York: Lee, Shepard and Dillingham, 1872 – 1mf – 9 – 0-8370-4879-6 – mf#1985-2879 – us ATLA [210]

The creator, and what we may know of the method of creation : the fernley lecture of 1887 / Dallinger, W H – London: T Woolmer, 1888 – 1mf – 9 – 0-7905-1511-3 – mf#1987-1511 – us ATLA [210]

The creator, and what we may know of the method of creation : the fernley lecture of 1887 / Dallinger, William Henry – London, 1887 – (= ser 19th c evolution & creation) – 2mf – 9 – mf#1.1.11607 – uk Chadwyck [110]

Creatures of the sea : being the life stories of some sea birds, beasts, and fishes / Bullen, Frank Thomas – Toronto: McClelland & Goodchild, 1909 – 6mf – 9 – 0-665-72251-6 – (incl ind. ill by theo carreras) – mf#72251 – cn Canadiana [590]

La creche d'youville et l'adoption des enfants – [Montreal: s.n.], 1933 [mf ed 1992] – 1mf – 9 – (incl english text) – mf#SEM105P1641 – cn Bibl Nat [360]

Creciendo con la hierba / Suarez, Clementina – San Salvador, El Salvador. 1957 – 1r – us UF Libraries [972]

Crede-Hoerder, Carl *see* Vom corpsstudenten zum sozialisten

Credentials – v3 n12-14,17-18,21,23-26 [1983 jun 13-jul 11, aug 22-sep 5, oct 17, nov 15-dec 27], v4 n1,3,6 [1984 jan 10, feb 7, mar 20 – 1r – 1 – mf#1054961 – us WHS [071]

Credentials of cardiac rehabilitation personnel / Bennett, Susan B – Springfield College, 1995 – 2mf – 9 – $8.00 – mf#HE553 – us Kinesology [616]

Credentials of christianity : a course of lectures delivered at the request of the christian evidence society / Goodwin, Harvey et al – London: Hodder & Stoughton, 1876 [mf ed 1984] – 4mf – 9 – 0-8370-1099-3 – (pref by earl of harrowby) – mf#1984-4486 – us ATLA [240]

The credentials of science the warrant of faith / Cooke, Josiah Parsons – 2nd ed. New York: Appleton, 1893, c1888 – 1mf – 9 – 0-8370-2735-7 – mf#1985-0735 – us ATLA [210]

The credentials of the catholic church / Bagshawe, John B – London: R Washbourne, 1885 – 1mf – 9 – 0-8370-7360-X – mf#1986-1360 – us ATLA [230]

The credentials of the gospel : a statement of the reason of the christian hope / Beet, Joseph Agar – London: Wesleyan Methodist Bk Rm 1889 [mf ed 1985] – 1mf – 9 – 0-8370-2248-7 – mf#1985-0248 – us ATLA [226]

La credibilite des evangiles : conferences donnees aux facultes catholiques de lyon / Jacquier – Paris: Lecoffre, 1913 – 1mf – 9 – 0-7905-3266-2 – mf#1987-3266 – us ATLA [220]

The credibility of the book of the acts of the apostles / Chase, Frederic Henry – London: Macmillan, 1902 – (= ser The Hulsean Lectures) – 1mf – 9 – 0-8370-2634-2 – (incl ind) – mf#1985-0634 – us ATLA [226]

The credibility of the evangelical history illustrated : with reference to the leben jesu of dr. strauss = Glaubwuerdigkeit der evangelischen geschichte / Tholuck, August – London: John Chapman, 1844 – 1mf – 9 – 0-524-06685-X – (in english) – mf#1992-0938 – us ATLA [220]

Credit – Washington. 1986-1992 (1) 1986-1992 (5) 1986-1992 (9) – (cont: consumer credit leader) – ISSN: 0097-8345 – mf#15089 – us UMI ProQuest [332]
Credit : roman / Niemann, August – 3. aufl. Dresden: E Pierson 1908 [mf ed 1995] – 1r – 1 – (filmed with: des konigs leibwache / gustav nieritz & other titles) – mf#3699p – us Wisconsin U Libr [830]
Credit and financial management / National Association of Credit Management – New York. 1902-1986 [1]; 1971-1986 [5]; 1976-1986 [9] – (cont by: credit and financial management: c&fm) – ISSN: 0011-0973 – mf#1980 – us UMI ProQuest [332]
Credit and financial management (c&fm) – New York. 1986-1987 (1,5,9) – (cont: credit and financial management. cont by: business credit) – ISSN: 0011-0973 – mf#1980,01 – us UMI ProQuest [332]
Le credit foncier / Boucherville, Georges de – Quebec: impr pour les entrepreneurs par Hunter, Rose & Lemieux, 1863 [mf ed 1983] – 2mf – 9 – mf#SEM105P240 – cn Bibl Nat [332]
The "credit foncier" : annexed to the report of the special committee appointed by the legislative assembly, 3rd march, 1863, to enquire into the expediency of establishing it in lower canada = Le credit foncier / oucherville, Georges de – Quebec: Hunter, Rose & Co, 1863 [mf ed 1992] – 2mf – 9 – (with ind) – mf#SEM105P1523 – cn Bibl Nat [332]
Le credit foncier a athenes / Caillemer, Exupere – [Paris: Impr imperiale, 1866?] (mf ed 19–) – (= ser Etudes sur les antiquites juridiques d'athenes) – (imperfect: t-p missing) – mf#ZT-TB pv489 n10 – us NY Public [930]
Credit international – London, UK. 12 Jun 1869-16 Jul 1870 – 1 – uk British Libr Newspaper [072]
Credit national – London, UK. 28 Nov 1872-15 May 1873 – 1 – uk British Libr Newspaper [072]
Credit union journal – New York. 1998+ (1,5,9) – ISSN: 1521-5105 – mf#26651 – us UMI ProQuest [332]
Credit union magazine – Madison. 1936+ (1) 1971+ (5) 1975+ (9) – ISSN: 0011-1066 – mf#523 – us UMI ProQuest [332]
Credit union management – Madison. 1989+ (1,5,9) – ISSN: 0273-9267 – mf#15930 – us UMI ProQuest [332]
Credit Union National Association *see*
– Bridge
– Everybody's money
The credit valley railway application for right of way and crossings at the city of toronto : before the railway committee of the privy council, ottawa, thursday, june 19th, 1879 / Holland, George C & Holland, A [comp] – [Ottawa?: s.n.], 1879 [mf ed 1986] – 1mf – 9 – 0-665-57940-3 – mf#57940 – cn Canadiana [380]
Credit Valley Railway Co *see*
– La difficulte de l'esplanade
– Railway commission
Credit world – St. Louis. 1914-1999 (1) 1971-1999 (5) 1977-1999 (9) – ISSN: 0011-1074 – mf#190 – us UMI ProQuest [332]
El credito extremeno : estatutos y reglamento de el credito extremeno (caja rural) sociedad de socorros mutuos y caja de ahorros... / Villafranca de los Barros – Badajoz: tipografia el noticiero extremeno, 1906 – 1 – sp Bibl Santa Ana [332]
Credito nacional o sea hacienda publica / Alvarez Guerra, Andres – 1820 – 9 – sp Bibl Santa Ana [336]
Crediton & north devon chronicle & west of england advertiser – [SW England] Devon 16 jul 1881-dec 1950 [mf ed 2002] – 58r – 1 – (cont: crediton & north devon chronicle; cont as: crediton chronicle and north devon gazette [nov 1891-27 dec 1930]; cont as: crediton chronicle and taw vale gazette [jan 1931-30 nov 1944]; became a local ed of: tiverton gazette & cont as: crediton gazette, east devon herald and county press [7 dec 1944-25 oct 1966]; cont as: crediton gazette [1 nov 1966 to date]) – uk Newsplan [072]
Credner, Karl August *see* Geschichte des neutestamentlichen kanon
El credo y la razon / Elola, Jose – Madrid: Razon y Fe, 1929 – 1 – sp Bibl Santa Ana [240]
Creed and character / Holland, Henry Scott – New York: Scribner, 1887 – 1mf – 9 – 0-7905-7763-1 – mf#1989-0988 – us ATLA [240]
Creed and constitution of the cumberland presbyterian church with a historical introduction – Nashville, TN: Cumberland Presbyterian Pub House, c1892 [mf ed 1993] – (= ser Presbyterian coll) – 1mf – 9 – 0-524-06177-7 – mf#1991-2433 – us ATLA [242]
Creed and the creeds : their function in religion / Skrine, John Huntley – London; New York: Longmans, Green, 1911 – (= ser Bampton lectures) – 1mf – 9 – 0-7905-9662-8 – mf#1989-1387 – us ATLA [240]
The creed and the prayer / Johnston, J Wesley – New York: Eaton & Mains, 1896 – 1mf – 9 – 0-524-03953-4 – mf#1991-2007 – us ATLA [240]
The creed and the year : a manual of instruction for sunday schools / Howe, Reginald Heber – New York: EP Dutton, 1887 – 1mf – 9 – 0-524-04073-7 – mf#1991-2018 – us ATLA [240]
The creed explained : or, an exposition of catholic doctrine: according to the creeds of faith and the constitutions and definitions of the church / Devine, Arthur – 4th ed. London: R & T Washbourne; New York: Benziger Bros, 1903 [mf ed 1986] – 2mf – 9 – 0-8370-8336-2 – (incl bibl ref) – mf#1986-2336 – us ATLA [241]
A creed for christian socialists : with expositions / Stubbs, Charles William – London: William Reeves, 1897 [mf ed 1991] – (= ser Bellamy library 30) – 1mf – 9 – 0-7905-9640-7 – mf#1989-1365 – us ATLA [240]
The creed in the pulpit / Henson, Hensley – London; New York: Hodder and Stoughton, [1912?] – 1mf – 9 – 0-7905-5943-9 – mf#1988-1943 – us ATLA [240]
The creed of a layman : apologia pro fide mea / Harrison, Frederic – New York: Macmillan, 1907 [mf ed 1993] – 1mf – 9 – 0-524-08346-0 – mf#1993-2036 – us ATLA [240]
The creed of andover theological seminary / Fiske, Daniel Taggart – Newburyport: MH Sargent, 1882 – 1mf – 9 – 0-7905-7626-0 – mf#1989-0851 – us ATLA [240]
The creed of buddha / Holmes, Edmond Gore Alexander – New York: J Lane, c1908 – 1mf – 9 – 0-524-00896-5 – mf#1990-2119 – us ATLA [280]
The creed of christendom : its foundations contrasted with its superstructure / Greg, William Rathbone – 5th ed. Boston: James R Osgood 1877 [mf ed 1993] – (= ser The english and foreign philosophical library 5-6) – 2v on 2mf – 9 – 0-524-07750-9 – (first printed in 1851) – mf#1991-3318 – us ATLA [240]
The creed of half japan : historical sketches of japanese buddhism / Lloyd, Arthur – London: Smith, Elder, 1911 – 1mf – 9 – 0-524-00925-2 – mf#1990-2148 – us ATLA [290]
Creed of pope pius 4 – London, England. 18– – 1r – us UF Libraries [240]
The creed of presbyterians / Smith, Egbert Watson – 7th ed. New York: Baker and Taylor, 1902 – 1mf – 9 – 0-524-04778-2 – (incl bibl ref) – mf#1991-2164 – us ATLA [242]
The creed of the christian / Gore, Charles – London: Wells Gardner, Darton, 1905 – 1mf – 9 – 0-8370-4815-X – mf#1985-2815 – us ATLA [240]
The creed or a philosophy / Mozley, Thomas – London, New York: Longmans, Green, 1893 [mf ed 1991] – 1mf – 9 – 0-7905-8527-8 – mf#1989-1752 – us ATLA [242]
Creed rebellion alias bible rebellion / Calvin, J – Glasgow, Scotland. 1877 – 1r – us UF Libraries [220]
Creed revision in the presbyterian churches / Schaff, Philip – 2nd enl ed. New York: Scribner 1890 [mf ed 1991] – 1mf – 9 – 0-524-01090-0 – mf#1990-4055 – us ATLA [242]
Creeds : the foes of heavenly faith, the allies of worldly policy / Giles, Henry – Liverpool, England. 1839 – 1r – us UF Libraries [240]
The creeds : an historical and doctrinal exposition of the apostles', nicene, and athanasian creeds / Mortimer, Alfred Garnett – London: Longmans, Green, 1902 – 1mf – 9 – 0-524-05326-X – (incl bibl ref) – mf#1990-1444 – us ATLA [240]
Creeds and churches : studies in symbolics / Stewart, Alexander; ed by Morrison, John – London, New York: Hodder & Stoughton 1916 [mf ed 1991] – (= ser Croall lectures 1901-02) – 1mf – 9 – 0-524-00112-X – (incl bibl ref) – mf#1989-2812 – us ATLA [240]
Creeds and churches in scotland : with an appendix / Moncreiff, Henry Wellwood – Edinburgh: Edmonston & Douglas 1869 [mf ed 1990] – 1mf – 9 – 0-7905-5072-5 – mf#1988-1072 – us ATLA [230]
Creeds and consistency / Begg, James – Edinburgh, Scotland. 1877 – 1r – us UF Libraries [240]
The creeds and platforms of congregationalism / ed by Walker, Williston – New York: Charles Scribner, 1893 – 2mf – 9 – 0-8370-9836-X – (incl bibl ref and index) – mf#1986-3836 – us ATLA [242]
The creeds of christendom : with a history and critical notes / Schaff, Philip – 4th rev and enl. New York: Harper, c1877 – 6mf – 9 – 0-524-04971-8 – (incl bibl ref) – mf#1990-1374 – us ATLA [240]
The creeds of the church : in their relations to the word of god and to the conscience of the christian / Swainson, Charles Anthony – Cambridge: Macmillan, 1858 – (= ser Hulsean Lectures) – 1mf – 9 – 0-7905-6575-7 – (incl bibl ref) – mf#1988-2575 – us ATLA [240]
Creegan, Charles Cole *see*
– Great missionaries of the church
– Pioneer missionaries of the church
Creek nation advocate – v1 n1 [1980 jul 16] – 1r – 1 – mf#941636 – us WHS [071]
Creel, David B *see* Effect of exercise intensity and duration on postexercise metabolism in obese adults
Creel, James Cowherd *see* The plea to restore the apostolic church
Creem – Los Angeles. 1985-1987 (1) 1985-1987 (5) 1985-1987 (9) – ISSN: 0011-1147 – mf#15028 – us UMI ProQuest [780]
Creeny, William Frederick *see* A book of facsimiles of monumental brasses on the continent of europe
Creeping socialist : newsletter of the madison local / Democratic Socialists of America – v4 n7-v5 n7 [i.e. 8] [1983 sep-1984 nov] – 1r – 1 – (cont: big red news [madison wi]) – mf#710892 – us WHS [335]
Creer, John W *see* Die darstellung der haftung des amerikanischen staates und der beamten sowie ein rechtsvergleichender uberblick uber das deutsche recht
Crees, James Harold Edward *see* Claudian as an historical authority
Crefelder zeitung *see* Rheinischer verfassungsfreund
Crehuet, Diego Maria *see*
– La pena de muerte, como tema literario
– Resumen de una discusion acerca de los hijos ilegitimos ante la sociedad y el derecho
– La tutela...discurso y contestacion...
Creighton courier – Creighton, NE: A C Logan & Co. v1 n1. jun 13 1889-1905// (wkly) [mf ed 1889,1895-1905 (gaps) filmed 1958] – 2r – 1 – (cont by: nebraska liberal) – us NE Hist [071]
Creighton, D H *see* Last enemy destroyed
Creighton, James Edwin *see* Some problems of lotze's theory of knowledge
Creighton, John Thomas *see* Biology and life history of the palm-leaf skeletonizer
Creighton law review – Omaha. 1974+ (1) 1974+ (5) 1975+ (9) – ISSN: 0011-1155 – mf#9997 – us UMI ProQuest [340]
Creighton law review – v1-34. 1968-2001 – 9 – $748.00 set – ISSN: 0011-1155 – mf#102141 – us Hein [340]
Creighton liberal – Creighton, NE: Liberal Pub Co. v25 n19. oct 18 1912- (wkly) [mf ed -1917 (lacks nov 51914)] – 3r – 1 – (cont: nebraska liberal) – us NE Hist [071]
Creighton, Louise *see*
– The church and the nation
– Historical essays and reviews
– Life and letters of mandell creighton
– Missions
– Thoughts on education
– University and other sermons
The creighton mail – Creighton, NE: C J Stockwell. v1 n1. jun 13 1901- (wkly) [mf ed jun 13-20 1901 filmed 1958] – 1r – 1 – us NE Hist [071]
Creighton, Mandell *see*
– The abolition of the roman jurisdiction
– Cardinal wolsey
– The church and the nation
– Historical essays and reviews
– Historical lectures and addresses
– A history of the papacy
– The idea of a national church
– Life of simon de montfort
– Persecution and tolerance
– Queen elizabeth
– Thoughts on education
– University and other sermons
Creighton, Mandell et al *see* Lectures on archbishop laud
Creighton news – Creighton, NE: W L Kirk and J B Lucas. v15 n14. jun 10 1904- [wkly] [mf ed with gaps filmed 1958-] – 1 – (cont: people's news) – us NE Hist [071]
Creighton pioneer – Creighton, NE: Ed A Fry, 1882-87// (wkly) [mf ed 1885-87 (gaps)] – 1r – 1 – (cont: niobrara pioneer. absorbed creighton transcript. cont by: pioneer. pioneer (daily) publ during the 2nd annual fair of the knox county agricultural association sep 22-25 1885) – us NE Hist [071]
Creighton transcript – Creighton, NE: Benner & Peirce (wkly) [mf ed v2 n16. apr 7 1887 filmed 1958] – 1r – 1 – (absorbed by: creighton pioneer) – us NE Hist [071]
Creighton, William Steel *see* New world species of the genus solenopsis (hymenop formicidae)
Creizenach, M *see* Schulchan aruch oder encyclopaedische darstellung des mosaischen gesetzes
Creizenach, Wilhelm Michael Anton *see*
– Die buehnengeschichte des goethe'schen faust
– Versuch einer geschichte des volksschauspiels vom doctor faust
Crell, Johann *see* Ethica aristotelica, ad sacrarum literarum normam emendata
Cremation, ancient and modern / Wotherspoon, George – London, England. 1886 – 1r – us UF Libraries [306]
Cremazie, Octave *see* Oeuvres completes de octave cremazie
Cremer, Ernst *see*
– Rechtfertigung und wiedergeburt
– Die stellvertretende bedeutung der person jesu christi
Cremer, Hermann *see*
– Beyond the grave
– Biblico-theological lexicon of new testament greek
– Die paulinische rechtfertigungslehre im zusammenhange ihrer geschichtlichen voraussetzungen
– A reply to harnack on the essence of christianity
– Supplement to biblico-theological lexicon of new testament greek
– Taufe, wiedergeburt und kindertaufe in kraft des heiligen geistes
– Ueber das recht und die geltung des kirchlichen bekenntnisses – ueber arbeit und eigentum nach christlicher anschauung
– Die vergottungslehre des athanasius und johannes damascenus – die grundwahrheiten der christlichen religion nach d. r. seeberg
– Vernunft, gewissen und offenbarung
– Warum koennen wir das apostolische glaubensbekenntnis nicht aufgeben?
– Weissagung und wunder im zusammenhange der heilsgeschichte
– Zum kampf um das apostolikum
Cremers, Paul Joseph *see* Die Marneschlacht
Cremieux, A *see* C'est la faute a cremieux
Cremona – London. 1906-1911 (1) – mf#5296 – us UMI ProQuest [700]
Crenshaw Christian Center *see* Ever increasing faith messenger
Creole – Georgetown, Guyana. 1856-1905 (1) – mf#67647 – us UMI ProQuest [079]
Creole haitien / Comhaire-Sylvain, Suzanne – Weteren, Belgium. 1936 – 1r – us UF Libraries [972]
Le Creole patriote *see* La revue du patriote
Les creoles : ou, la vie aux antilles / Levilloux, J – Paris: H Souverain. 2v. 1835 (mf ed 1969) – 1r – 1 – mf#Sc Micro R-1314 – us NY Public [890]
Creonte / Veloz Maggiolo, Marcio – Santo Domingo, Dominican Republic. 1963 – 1r – us UF Libraries [972]
Il crepuscolo dei filosofi / Papini, Giovanni – 4th ed., riveduta. Firenze 1921 – 1 – us Wisconsin U Libr [190]
Crepuscule / Carrie, Pierre – Port-Au-Prince, Haiti. 1948 – 1r – us UF Libraries [972]
Crepusculo : orgam litterario e noticioso – Desterro, SC. 23 abr, 17 set 1888 – (= ser Ps 19) – bl Biblioteca [079]
O crepusculo – Laguna, SC. 08 mar, 27 dez 1903 – (= ser Ps 19) – bl Biblioteca [410]
O crepusculo : litterario, critico e noticioso – Laguna, SC. 01 abr 1902 – (= ser Ps 19) – bl Biblioteca [079]
O crepusculo : periodico instructivo e moral – Bahia: Typ do Correio Mercantil, 02 ago 1845-fev 1847 – (= ser Ps 19) – mf#P19,01,68 – bl Biblioteca [079]
Crequillon, Thomas *see* Opus sacrarum cantionum...
Crescent – Marion, SC. 1866-1870 (1) – mf#66503 – us UMI ProQuest [071]
Crescent / Meigs Co. Pomeroy – jan 1870-dec 1871 – 1r – 1 – mf#B122 – us Ohio Hist [071]
Crescent – Orrville, OH. 1872-1975 (1) – mf#65625 – us UMI ProQuest [071]
Crescent / Wayne Co. Orrville – jun 1870-dec 1871 [wkly] – 1r – 1 – mf#B6705 – us Ohio Hist [071]
The crescent and the cross : romance and realities of eastern travel / Warburton, Eliot – 5th ed. London: H. Colburn, 1846. 2v. ill – 1 – us Wisconsin U Libr [910]
Crescent city call – Crescent City, FL. 1923 feb 21-1925 jul 10 – 1r – us UF Libraries [071]
[Crescent city-] crescent city news – CA. 1906-1911 – 2r – 1 – $120.00 – mf#C03590 – us Library Micro [071]
Crescent city currents – New Orleans LA. v1 n7-v4 n19 [1985 mar 29-1988 oct 21] – 1r – 1 – (cont by: currents [new orleans la]) – mf#1546573 – us WHS [071]
[Crescent city-] del norte record – CA. 1891-1911 – 6r – 1 – $360.00 – mf#C03192 – us Library Micro [071]
[Crescent city-] the triplicate – CA. 1912- – 96r – 1 – $8760.00 (subs $160/y) – mf#BC02149 – us Library Micro [071]
[Crescent city-] western tax collector – CA. 16 Dec 1939 – 1r – 1 – $60.00 – mf#B02150 – us Library Micro [071]
The crescent in india : a study in medieval history / Sharma, sri Ram – Bombay: Karnatak Pub House, -1937 – (= ser Samp: indian books) – us CRL [954]
Crescent international – Toronto, Canada. 15 aug 1980-dec 1985 – 1r – 1 – uk British Libr Newspaper [071]
The crescent moon / Tagore, Rabindranath – London: Macmillan and Co, 1929 – (= ser Samp: indian books) – us CRL [490]

Die crescentialegende in der deutschen dichtung des mittelalters / Baasch, Karen – Stuttgart: J B Metzler, 1968 [mf ed 1993] – (= ser Germanistische abhandlungen (stuttgart, germany) 20) – viii/249p – 1 – (incl bibl ref and ind) – mf#8168 – us Wisconsin U Libr [430]
Crescimbeni, G M *see* L'istoria della basilica diaconale collegiata
Crescimbeni, Giovanni Mario *see* Le vite degli arcadi illustri-scritte da diversi autori
Cresi, Domenico *see* Discussioni e documenti di storia francescana...
Cresival / Labrador Ruiz, Enrique – Habana, Cuba. 1936 – 1r – us UF Libraries [972]
Crespel, Emmanuel *see* Voiages du r p emmanuel crespel, dans le canada et son naufrage en revenant en france
Crespo, Jose D *see* Geografia de panama
Crespo, Manuel Maria *see* Catecismo social a sea la enciclica "rerum novarum" del papa leon 13 puesta en preguntas y respuetas para su mejor inteligencia
Crespo Marquez, Jose Maria *see* Pregon de la semana santa cacerena, 1976
Crespo, Nicasio *see*
– Aritmetica, las cuatro operaciones fundamentales
– Nociones elementales de aritmetica
Crespo, Pedro *see* El paso del guadiana
Crespo y Escoriaza, Benito *see*
– Brave resena de las aguas sulfurado-sodi cas termales de montemayor o banos
– Discurso inaugural...academia cientifico
Cressolles, Loius de *see* Theatrum veterum rhetorum, oratorum, declamatorum, quos in graecia nominabant...expositum libris 5
Cresswell, Samuel Gurney *see* Dedicated
Cresswell's nottingham and newark journal, 1772-75 – 1r – 1 – uk Microform Academic [072]
Cressy, Earl Herbert *see* Christian higher education in china
Crested butte citizen *see*
– Gunnison county miscellaneous newspapers
Cresthill baptist church. bowie, maryland : church records – 16 Oct 1966-Jun 1979 – 1 – us Southern Baptist [242]
[Crestline-] crestline courier – CA. 1967-79 – 9r – 1 – $540.00 – mf#R02151 – us Library Micro [071]
[Crestline-] mountain courier – CA. 1980-90 – 11r – 1 – $660.00 – mf#R02152 – us Library Micro [071]
[Crestline-] rim of the world – CA. 1975-1982 – 3r – 1 – $330.00 – mf#R03193 – us Library Micro [071]
Creston news – Creston, NE: C H Swallow, -1892 (wkly) [mf ed dec 2 1892 filmed 1980] – 1r – 1 – (merged with: leigh advocate to form: leigh world and creston news) – us NE Hist [071]
The creston statesman – Creston, NE: C E Wagner. -v36 n39. jun 1 1932 (wkly) [mf ed 1902-32 filmed 1977] – 10r – 1 – (absorbed by: leigh world) – us NE Hist [071]
Crestview news leader – Crestview, FL. v2 n3-v6 n27. 1994-1998 jun – 5r – (gaps) – us UF Libraries [071]
Creswell chronicle – Creswell OR: G H Baxter, 1909-17 [wkly] [mf ed 1964] – 1r – 1 – us Oregon Lib [071]
Creswell chronicle [creswell, or] – Creswell OR: G G Sittser, -1971 [wkly] [mf ed 1969-76] – 2r – 1 – (cont by: chronicle [creswell, or]) – us Oregon Lib [071]
Creswell, Creswell *see* Barnewell and cresswell's reports
Creswell new era – Creswell OR: I S Wilson, -1950 [wkly] [mf ed 1964] – 1r – 1 – us Oregon Lib [071]
Cresy, Edward *see* Architecture of the middle ages in italy
Cretaceous canadian crustacea / Woodward, Henry – London: Dulau, [1900?] [mf ed 1981] – 1mf – 9 – 0-665-26130-6 – (incl bibl ref) – mf#26130 – cn Canadiana [550]
The cretan collection in oxford and the dictean cave and iron age crete : from the ashmolean museum, oxford / Boardman, John – 3mf – 9 – mf#87381 – uk Microform Academic [930]
Cretan sun / Iraklion Air Station [Crete US] – 1987 nov 20-1989 jun 30, 1987 dec 11, 1989 jan 20, Jul 7-sep 29, 1989 oct 6-1991 dec 20, 1992 jan 10-1994 mar 24 – 4r – 1 – (cont: island ally; cont by: cretan sunset) – mf#1703347 – us WHS [355]
Crete daily globe – Crete, NE: Globe Pub Co. v1 n1-121. aug 11-dec 21 1884 (daily ex sun) [mf ed with gaps filmed 1957] – 1r – 1 – (weekly ed: crete globe) – us NE Hist [071]
Crete daily standard – Crete, NE: [Standard Pub Co] v1 n1. oct 22 1883- (daily ex sun) [mf ed -nov 7 1883] – 1r – 1 – (weekly ed: saline county standard) – us NE Hist [071]
The crete democrat – Crete, NE: Secord & Overcash, apr 25 1889-v48 n25. apr 12 1922 (wkly) [mf ed with gaps)] – 11r – 1 – (cont: opposition. some irregularities in numbering) – us NE Hist [071]
The crete globe – Crete, NE: Globe Pub Co. v1 n1. jan 3 1884-1891// (wkly) [mf ed -1891 (gaps) filmed 1957] – 2r – 1 – (formed by the union of: saline county union and: saline county standard. cont: crete vidette. v1 n3 jan 17 1884-v6 n52 dec 28 1889) called also whole n662-1102, cont whole numbering of saline county union. daily ed: crete daily globe aug 11-dec 31 1884) – us NE Hist [071]
Crete herald – Crete, NE: A A Hatch. 9v. v1 n1. dec 21 1892-v9 n18. apr 25 1902 (wkly) [mf ed 1893-1902 (gaps) filmed 1957] – 3r – 1 – (merged with: crete vidette to form: crete vidette-herald) – us NE Hist [071]
Crete news – Crete, NE: Shepherd Bros. v1 n1. mar 14 1908- (wkly) [mf ed with gaps] – 1 – (absorbed: crete vidette. issues for feb 8 1940-called v71 n1-) – us NE Hist [071]
Crete sentinel – Crete, NE: W S Walker, 1875-76// (wkly) [mf ed v1 n13. feb 26-apr 15 1876 (gaps) filmed 1973] – 1r – 1 – us NE Hist [071]
Crete university park star – Chicago Heights, IL. 1984-1990 (1) – mf#68362 – us UMI ProQuest [071]
Crete vidette – Crete, NE: J H Walsh. 11v. 46th yr n26. jun 6 1918-v57 n17. jun 14 1928 (wkly) [mf ed with gaps] – 4r – 1 – (cont: crete vidette-herald. absorbed by: crete news) – us NE Hist [071]
The crete vidette – Crete, NE: Wells & Chapman, 1891-v32 n3. apr 24 1902 (wkly) [mf ed with gaps] – 5r – 1 – (cont: state vidette. absorbed: crete globe. merged with: crete herald to form: crete vidette-herald. issues for may 12 1892-apr 24 1902 called v22 n1-v32 n3) – us NE Hist [071]
Crete vidette-herald – Crete, NE: Goodwin & Wells. 15v. v32 n4. may 1 1902-46th yr n26. may 30 1918 (wkly) [mf ed with gaps] – 8r – 1 – (formed by the union of: crete vidette and crete herald. cont by: crete vidette) – us NE Hist [071]
Crete-a-pierrot / Rosemond, Jules – Port-Au-Prince, Haiti. 1913 – 1r – us UF Libraries [972]
Cretesky pokrok – Omaha, NE: Pokrok Pub Co. roc1 cis 1. 8 unora 1905- (wkly) [mf ed 1905-10 (gaps) filmed 1975?-80?] – 4r – 1 – (in czech. cont by: salinsky pokrok. local ed of: pokrok zapadu) – us NE Hist [071]
Cretineau-Joly, Jacques *see* Histoire religieuse, politique, et litteraire de la compagnie de jesus
Creuze De Lesser, Augustin Francois *see*
– Secret du menage
Creuzer, F *see* Plotini enneades
Creuzer, Georg Friedrich *see*
– Friedrich creutzer's deutsche schriften
– Friedrich creuzer und karoline von guenderode
– Symbolik und mythologie der alten voelker
Crevasse, J M *see*
– Ground covers for florida gardens
– Study of various ground covers in florida gardens
[Crevecoeur, M G J Saint-John de] *see* Letters from an american farmer
Crevecoeur, Michel G de *see*
– Lettres d'un cultivateur americain, ecrites a w s ecuyer, depuis l'annee 1770, jusqu'a 1781
– Voyage dans la haute pensylvanie et dans l'etat de new-york, par un membre adoptif de la nation oneida
Crevier, Joseph Alexandre *see* Le cholera
Creviston, Todd A *see* The effects of a stair climbing program on leg strength, flexibility and functional mobility in men and women aged 76 to 86 years
Crew lists of vessels arriving at ashland, kenosha, marinette, sheboygan, sturgeon bay, and washburn, wisconsin, 1926-1956 / U.S. Immigration and Naturalization Service – (= ser Records Of The Immigration And Naturalization Service) – 1r – 1 – mf#M2044 – us Nat Archives [975]
Crew lists of vessels arriving at boston, ma, 1917-1943 / U.S. Immigration and Naturalization Service – (= ser Records of the immigration and naturalization service, 1891-1957) – 269r – 1 – mf#T938 – us Nat Archives [975]
Crew lists of vessels arriving at gloucester, ma, 1918-1943 / U.S. Immigration and Naturalization Service – (= ser Records of the immigration and naturalization service, 1891-1957) – 13r – 1 – mf#T941 – us Nat Archives [975]
Crew lists of vessels arriving at manitowoc, wisconsin, 1925-1956 / U.S. Immigration and Naturalization Service – (= ser Records Of The Immigration And Naturalization Service) – 1r – 1 – mf#M2045 – us Nat Archives [975]
Crew lists of vessels arriving at new bedford, ma, 1917-1943 / U.S. Immigration and Naturalization Service – (= ser Records of the immigration and naturalization service, 1891-1957) – 2r – 1 – mf#T942 – us Nat Archives [975]
Crew lists of vessels arriving at new orleans, la, 1910-1945 / U.S. Immigration and Naturalization Service – (= ser Records of the immigration and naturalization service, 1891-1957) – 311r – 1 – mf#T939 – us Nat Archives [975]
Crew lists of vessels arriving at san francisco, ca, dec 28, 1905-oct 30, 1954 / U.S. Immigration and Naturalization Service – (= ser Records of the immigration and naturalization service, 1891-1957) – 174r – 1 – mf#M1416 – us Nat Archives [975]
Crew lists of vessels arriving at seattle, washington, 1903-1917 / U.S. Immigration and Naturalization Service – (= ser Records of the immigration and naturalization service, 1891-1957) – 15r – 1 – mf#M1399 – us Nat Archives [975]
Crewdson, Isaac *see*
– A beacon to the society of friends
– A defence of the beacon
Crewe and nantwich chronicle and north wales advertiser – [NW England] Crewe, Cheshire Record Off 1874, 17 apr 1875-21 jul 1883 – 1 – (title change: crewe & nantwich chronicle & west cheshire advertiser [28 jul 1883-28 jun 1890]; crewe chronicle & west cheshire advertiser [5 jul 1890-23 jul 1904]; crewe chronicle [30 jul 1904-1927, sep 1972]) – uk MLA; uk Newsplan [072]
Crewe burkeville journal – Crewe, VA. 1988-2000 (1) – mf#66692 – us UMI ProQuest [071]
Crewe, christ church – (= ser Cheshire monumental inscriptions) – 2mf – 9 – £4.00 – mf#94 – uk CheshireFHS [929]
Crewe, christ church: marriage index – (= ser Cheshire church registers) – 1mf – 9 – £2.50 – mf#102a – uk CheshireFHS [929]
Crewe green, st michael – (= ser Cheshire monumental inscriptions) – 1mf – 9 – £2.50 – mf#320 – uk CheshireFHS [929]
Crewe guardian – [NW England] Crewe Lib oct 1869-1897, 1899-1920 – 1 – uk MLA; uk Newsplan [072]
Crewe, weston cemetery – (= ser Cheshire monumental inscriptions) – 1mf – 9 – £2.50 – mf#325 – uk CheshireFHS [929]
Crewe/monks coppenhall (rg12/2850) – (= ser 1891 census surname/location indexes [cheshire]) – 9 – £2.50/mf – (f1-17 [1mf]; f28-49 [1mf]; f50-71 [1mf]; f72-95 [1mf]; f96-121 [1mf]; f122-147 [1mf]) – mf#161-166 – uk CheshireFHS [929]
Crewe/monks coppenhall (rg12/2851) – (= ser 1891 census surname/location indexes [cheshire]) – 9 – £2.50/mf – (f4-27 [1mf]; ff31-48 [1mf]; f52-68 [1mf]; f72-87 [1mf]; f91-108 [1mf]; f112-130 [1mf]; f133-150 [1mf]) – mf#167-173 – uk CheshireFHS [929]
Crews, Cynthia Mary Jopson *see* Recherches sur le judeo-espagnol dans les pays balkaniques
Crews, Debra J *see* Effect of heart rate deceleration biofeedback training on golf putting performance
Crha news report – Montreal: the Association. n106 dec 1959-n134 jun 1962 (mthly) [mf ed 1995] – 1r – 1 – (with ind) – mf#SEM35P416 – cn Bibl Nat [380]
Cri de londres – London, UK. 17 Aug 1914-Feb 1916 – 1 – uk British Libr Newspaper [072]
Le cri de l'ouvrier – Douai. nov 1884-avr 1885 – 1 – fr ACRPP [073]
Cri des flandres – Hazrbrouck, France. 1 may 1910-19 nov 1911; 31 dec 1911; 5 apr 1912-27 sep 1914 – 2r – 1 – uk British Libr Newspaper [072]
Le cri des flandres – Hazebrouck, france. 1 may 1910-19 nov, 31 dec 1911; 5 apr 1912-27 sep 1914 – 1 – (not publ between 19 nov and 31 dec 1911 or between 31 dec 1911 and 5 apr 1912) – mf#m.f.8 – uk British Libr Newspaper [074]
Le cri des jeunes – n1-114. Lille. mars 1928-34 – 1 – (mq n9, 59, 79, 95, 111) – fr ACRPP [073]
Le cri des negres – Paris: La Cooptypography, aug 1931-jan/feb 1933; july/aug 1933-apr/may 1936 – (filmed with les continents and 11 other titles) – us CRL [074]
Le cri des negres – Paris. oct 1931, juil 1933-nov 1934 – 1 – fr ACRPP [305]
Le cri du forcat – Lille. avr-aout 1884 – 1 – fr ACRPP [073]
Le cri du paysan *see* La voix du paysan
Le cri du peuple – Paris dec 1929-31 [wkly] – 1 – (hebdomadaire syndicaliste revolutionnaire publie sous le controle du comite pour l'independance du syndicalisme.) – fr ACRPP [320]
Le cri du peuple – Paris, 1937 – 1 – (in french) – us UMI ProQuest [934]
Le cri du peuple – Paris: Jules Valles, [mar 26-may 23 1871] – (issues filmed as pt of: commune de paris newspapers; newspapers on these reels are filmed chronologically, not alphabetically) – us CRL [074]
Le cri du peuple : organe republicain – Fort-de-France. dec 1933-mai 1934 – 1 – fr ACRPP [073]
Le cri du peuple – Paris. 22 fevr-23 mai 1871, 28 oct 1883-9 avr 1890, 27 janv 1891-15 dec 1896, 15 fevr-15 juil 1897, 4 juin 1898-17 sept 1899, 15 juin 1901-6 dec 1908, 15 juin, 1er nov 1910, 1er mai 1912, 25 oct 1913, 18 janv-10 juin 1914, juil-dec 1919, 30 mai, 10 juin, 20 juil 1922 – 1 – fr ACRPP [073]
Le cri du peuple – Verviers. Belgium. -w. 7 Jul 1878-21 Jun 1879. (14 ft) – 1 – uk British Libr Newspaper [949]
le cri du peuple *see* La loire republicaine
Le cri du peuple de paris – Paris. 19 oct 1940-16 aout 1944 – 1 – fr ACRPP [073]
Cri du peuple organe socialiste revolutionnaire – Verviers Belgium, 7 jul 1878-21 jun 1879 – 1/4r – 1 – uk British Libr Newspaper [074]
Le cri du peuple socialiste – Organe sous le controle de la Federation du Finistere. (Section francaise de l'Internationale ouvriere). Brest. oct 1908-16 – 1 – fr ACRPP [335]
Le cri du sol : le journal de dorgeres – Lyon. aout 1940-juil 1944 – 1 – fr ACRPP [073]
Le cri du travailleur : organe du parti ouvrier de la region nord – Lille. juil 1887-sept 1891 – 1 – fr ACRPP [073]
Le cri populaire – Bordeaux. 1929-34 – 1 – (socialisme, syndicalisme, cooperation) – fr ACRPP [325]
Los criaderos de hierro de burguillos (badajos) / Gascon y Miramon, Antonio – Madrid: Imp. Ricardo Rojas, 1904 – 1 – sp Bibl Santa Ana [946]
Criado Valcarcel, Vicente *see* Luis de morales en arroyo de la luz
Cric? crac! / Sylvain, Georges – Port-Au-Prince, Haiti. 1929 – 1r – us UF Libraries [972]
Crichton, Michael *see* Andromeda strain
Crichton, W J *see* The microscope of the new testament
Crick, F [comp] *see* Holkham accounts, 1789-1814
Cricket – La Salle. 1973+ (1) 1976+ (5) 1976+ (9) – ISSN: 0090-6034 – mf#9930 – us UMI ProQuest [790]
Cricket – London. May 1882-Dec 1913.-m. 16mqn reels – 1 – uk British Libr Newspaper [072]
Cricket and i / Constantine, Learie Nicholas – London, England. 1933 – 1r – us UF Libraries [790]
Cricket argus – Bradford, England 1 may 1909-27 aug 1910, 2 may 1914-31 aug 1926 [mf 1926] – 1 – (discontinued; wanting: 1917, 1918; publ during the cricket season only) – uk British Libr Newspaper [790]
Cricklewood & kilburn leader – Brent, England 10 sep 1992-24 jun 1993 – 1 – uk British Libr Newspaper [072]
Crickmer, William Burton *see* Story of the planting of the english church in columbia
Le cri-cri – Paris. juil 1872-juin 1873, juil-dec 1876 – 1 – fr ACRPP [073]
O cri-cri : jornal noticioso e imparcial – Teresina, PI: Typ da Epoca, 14 ago, 10 set 1883 – (= ser Ps 19) – mf#P17,02,134 – bl Biblioteca [073]
Cricri et ses mitrons / Carmouche, Pierre-Frederic-Adolphe – Bruxelles, Belgium. 1829 – 1r – us UF Libraries [440]
Cridge, Edward *see* As it was in the beginning
Crie : centro regional de informaciones ecumenicas – Mexico, DF: El Centro [n76/77-360 (julio 1981-dic 1997)] (mthly) – 1 – us CRL [241]
Crieff advertiser – [Scotland] Perthshire, Crieff: R P Band 14 mar 1907-may 1915 (wkly) [mf ed 2004] – 3r – 1 – (cont by: crieff advertiser and west perthshire news [apr 1909-may 1915]) – uk Newsplan [072]
Crieff advertiser – Scotland, UK. 7 Jan-10 Jun 1865. -w. 23 feet – 1 – uk British Libr Newspaper [072]
Crieff herald – Scotland, UK. 1 Nov 1856-1862. -w. 2 reels – 1 – uk British Libr Newspaper [072]
Crieff journal and general advertiser for the central district of the county of perth – [Scotland] Stirling: George McCulloch mar 1858-dec 1874, 1878, 1889-11 may 1900 (wkly) [mf ed 2003] – 15r – 1 – (imprint varies; title changes to: crieff journal and general advertiser for the central district of perthshire [jan 1893-may 1900]) – uk Newsplan [072]
Criegern, Hermann Ferdinand von *see* Johann amos comenius als theolog
Crier – Plymouth, MI. 1981-1988 (1) – mf#68027 – us UMI ProQuest [071]
Crigler, Sara Gossett *see* Education for girls and women in upper south carolina prior to 1890
Crijghs-architecture ende fortificatien / Gerbier, B – Delft, 1652 – 2mf – 9 – mf#OA-147 – ne IDC [720]
Crile, george w, papers, ms 2806 – 1864-1943 – 71r – 1 – (correspondence, telegrams, diaries and journals, memoranda books, articles, speeches, and financial papers of dr. george w. crile, noted surgeon, author, and co-founder of the cleveland clinic) – us Western Res [617]
Crime and delinquency – New York. 1919+ (1) 1971+ (5) 1975+ (9) – ISSN: 0011-1287 – mf#2109 – us UMI ProQuest [360]
Crime and delinquency – v1-47. 1955-2001 – 5,6,9 – $1145.00 set – (v1-30 1955-84 on reel $400; v31-47 1985-2001 on mf $745; title varies: v1-6 1955-60 as nppa journal) – ISSN: 0011-1287 – mf#102151 – us Hein [360]

Crime and delinquency abstracts / U[/nited/] S[/tates/] Dept of Health and Human Services – HEW, National Clearinghouse for Mental Health Info. v1-8. 1963-72 [all publ] – 9 – (1st 3v entitled: international bibliography on crime and delinquency) – mf#llmc 81-208 – us LLMC [364]
Crime and delinquency abstracts – Washington. 1963-1972 (1) – ISSN: 0045-902X – mf#6294 – us UMI ProQuest [360]
Crime and delinquency literature *see* Criminal justice abstracts
Crime and government at hong kong : a letter to the editor of the "times" newspaper / Anstey, Thomas Chisholm – London: Effingham Wilson, 1859 – (= ser 19th c books on china) – 2mf – 9 – mf#7.1.39 – uk Chadwyck [951]
Crime and insanity / Mercier, Charles Arthur – New York, NY. 1911 – 1r – us UF Libraries [025]
Crime and justice – Chicago. 1989-1991 (1) – ISSN: 0192-3234 – mf#17098 – us UMI ProQuest [360]
Crime and justice : a review of research – v1-28. 1979-2001 – (filming in process) – ISSN: 0192-1714 – mf#111041 – us Hein [345]
Crime and justice international – Chicago. 1997+ (1,5,9) – (cont: cj international) – mf#14964,01 – us UMI ProQuest [360]
Crime and punishment / Hanson, W Stanley – s.l, s.l? 1936 – 1r – us UF Libraries [025]
Crime and punishment. the mark system / Maconochie, Alexander – London. 1846 – 1 – us CRL [360]
Crime and social justice – San Francisco. 1974-1987 (1) 1974-1987 (5) 1974-1987 (9) – ISSN: 0094-7571 – mf#11140 – us UMI ProQuest [360]
The crime control and fine enforcement acts of 1984 : a synopsis / Partridge, Anthony – Washington: FJC, Jan 1985 – 1mf – 9 – $1.50 – mf#LLMC 95-393 – us LLMC [345]
Crime control digest – Washington. 1967+ (1) 1975+ (5) 1975+ (9) – ISSN: 0011-1295 – mf#10587 – us UMI ProQuest [360]
Crime crapuleux de l'imperialisme – [Conakry]: Impr Nationale "Partice Lumumba" 1973 – us CRL [960]
Crime, criminology and civil liberties : archives of the howard league for penal reform, the howard journal, 1921-1976 – 4r – 1 – (coll includes "archives of the howard league for penal reform: the howard journal, 1921-1976) – mf#CL999-14000 – us Primary [345]
Crime de antonio vieira / Calmon, Pedro – Sao Paulo, Brazil. 1931 – 1r – us UF Libraries [972]
Crime de injurias / Monetnegro, Manoel Januario Bezerra – Recife, Brazil. 1875 – 1r – us UF Libraries [972]
Crime in india – New Delhi: Intelligence Bureau, Ministry of Home Affairs, Govt of India [1954-1955,1958-1960,1962-1963,1967-1970] (annual) – 1r – 1 – us CRL [315]
Crime in the united states – Washington. 1998+ (1,5,9) – mf#5785,01 – us UMI ProQuest [365]
Crime in wisconsin – 1973 jan/jun-? – 1r – 1 – mf#583754 – us WHS [364]
Crime, law and social change – Dordrecht. 1991+ (1,5,9) – (cont: contemporary crises) – ISSN: 0925-4994 – mf#16037,01 – us UMI ProQuest [360]
Crime of apartheid / Davidson, Apollon Borisovich – Moscow, Russia. 1966 – 1r – us UF Libraries [960]
Crime of being white / Eeden, Guy Van – Cape Town, South Africa. 1965 – 1r – us UF Libraries [320]
Crime of cuba / Beals, Carleton – Philadelphia, PA. 1934 – 1r – us UF Libraries [972]
Crime of francisco franco – New York City: North American Cttee to Aid Spanish Democracy [1938?] [mf ed 1977] – (= ser Blodgett coll) – 1mf – 9 – mf#w819 – us Harvard [946]
The crime of guernica / Spanish Information Bureau. New York – N.Y., 1937. Fiche W1200. (Blodgett Collection of Spanish Civil War Pamphlets) – 9 – us Harvard [946]
Crime of wilson in santo domingo / Fiallo, Fabio – Habana, Cuba. 1940 – 1r – us UF Libraries [972]
Crime on the road malaga-almeria : narrative with graphic documents revealing fascist cruelty / Bethune, Norman – [s.l]: Publicaciones Iberia [1937] – (= ser Blodgett coll) – 9 – mf#w754 – us Harvard [946]
Crime prevention review – v1-6. 1973-79 (all publ) – 5,6 – $58.00 set – (available in reel only) – mf#102191 – us Hein [360]
Crime prevention review – Los Angeles. 1976-1979 (1,5,9) – ISSN: 0093-044X – mf#11239 – us UMI ProQuest [360]
Le crime rituel *see* L'assassinat maconnique / le crime rituel / la trahison juive
Crime victims digest – Fairfax. 1991-1994 (1,5,9) – ISSN: 1081-0293 – mf#15224,01 – us UMI ProQuest [365]
The crimea and transcaucasia / Telfer, J B – London, 1876. 2v – 8mf – 9 – mf#AR-1422 – ne IDC [915]
Crimen de berruecos / Perez Y Soto, Juan Bautista – Roma, Italy. v1-4. 1924 – 1r – us UF Libraries [972]
Un crimen de hernando cortes. la muerte de dona catalina xuares marcayda (estudio historico y medico legal) / Toro, Alfonso – Mexico: Editorial Patria S.A., 2nd ed 1947 – sp Bibl Santa Ana [362]
Crimen de la magdalena / Gomez, Laureano – Bogota, Colombia. 1943 – 1r – us UF Libraries [972]
El crimen del camino malaga-almeria / Bethune, Norman – n.p. 1937? Fiche W 755. (Blodgett Collection of Spanish Civil War Pamphlets) – 9 – us Harvard [946]
Crimen del capitolio / Moreno Arango, Sebastian – Bogota, Colombia. 1940 – 1r – us UF Libraries [972]
Crimenes del imperialismo norteamericano / Blanco Fombona, Horacio – Mexico: D F, Ediciones Churubusco, 1927 [mf ed 1987] – 144p – 1 – mf#8624 – us Wisconsin U Libr [327]
Crimes by national bank officers and agents under sections 5208 and 5209 revised statutes of the united states / Terrell, Henry – Chicago, Callaghan, 1906. 193 p. LL-1498 – 1 – us L of C Photodup [348]
Crimes of atheism / Poulson, Edward – London, England. 188-? – 1r – us UF Libraries [210]
Crimes of christianity pts 1-2 / Foote, G W – London, England. 1885 – 1r – us UF Libraries [240]
Les crimes secrets de napoleon buonaparte : faits historiques / Cuisin, J P – Bruxelles [u.a. 1815 [mf ed Hildesheim 1995-98] – 1v on 2mf – 9 – €60.00 – ISBN-10: 3-487-26362-9 – ISBN-13: 978-3-487-26362-5 – gw Olms [944]
Criminal abortion and the new english criminal evidence act / Bell, Clark, 1832-1918 – New York, 1898? 162-171 p. LL-1452 – 1 – us L of C Photodup [345]
Criminal and civil dockets / Ellis County. Kansas. Big Creek Township. Justice of the Peace – 1872-1957 – 1 – us Kansas [978]
Criminal case files of the us circuit court for the district of maryland, 1795-1860 / U.S. Circuit and District Courts – (= ser Records of district courts of the united states) – 4r – 1 – (with printed guide) – mf#M1010 – us Nat Archives [345]
Criminal case files of the u.s. circuit court for the eastern district of pennsylvania, 1791-1840 / U.S. Circuit and District Courts – (= ser Records of district courts of the united states) – 7r – 1 – (with printed guide) – mf#M986 – us Nat Archives [345]
Criminal case files of the u.s. circuit court for the southern district of new york, 1790-1853 / U.S. Circuit and District Courts – (= ser Records of district courts of the united states) – 6r – 1 – (with printed guide) – mf#M885 – us Nat Archives [345]
The criminal code of canada and the canada evidence act : with their amendments, including the amending acts of 1900 and 1901... / Crankshaw, James – Montreal: C Theoret, 1902 – 14mf – 9 – 0-665-77732-9 – (1st publ 1894) – mf#77732 – cn Canadiana [345]
The criminal code of canada and the canada evidence act, 1893 : with an extra appendix containing the extradition act... / Crankshaw, James – Montreal: Whiteford & Theoret 1894 – 12mf – 9 – 0-665-90881-4 – (incl ind and bibl ref) – mf#90881 – cn Canadiana [345]
The criminal code of canada and the canada evidence act as amended to date : with commentaries, annotations, forms, etc, etc and an appendix... / Crankshaw, James – Toronto: Carswell, 1910 – 17mf – 9 – 0-665-74958-9 – mf#74958 – cn Canadiana [345]
The criminal code of canada and the canada evidence act as amended to date : with commentaries, annotations, forms, etc, etc and an appendix... / Crankshaw, James – Toronto: Carswell, 1915 – 18mf – 9 – 0-665-74840-X – mf#74840 – cn Canadiana [345]
The criminal conspiracy in the japanese war crimes trials – n.p., n.d. – 5mf – 9 – $7.50 – (memorandum for the hon joseph b keenan, chief of counsel, acting on behalf of the us, of the international military tribunal for the far east) – mf#LLMC 97-005 – us LLMC [327]
Criminal defense – University of Houston. v1-11. 1973-84 (all publ) – 5,6 – $107.00 set – (available in reel only) – mf#102161 – us Hein [345]
Criminal defense – Macon. 1981-1982 (1) 1981-1982 (5) 1981-1982 (9) – ISSN: 0093-8610 – mf#11943 – us UMI ProQuest [360]
Criminal forms for the state of indiana : complete under the criminal statutes of the state down to 1903 / Horner, Francis Asbury – 2d ed. Rochester: Lawyers' Co-Operative Publishing Co., 1903. 530p. LL-968 – 1 – us L of C Photodup [345]
Criminal justice – Chicago. 1975-1985 (1) 1976-1985 (5) 1976-1985 (9) – (cont by: criminal justice) – ISSN: 0092-2498 – mf#10633 – us UMI ProQuest [345]
Criminal justice – London. 1989-1997 (1,5,9) – (cont by: hlm: the howard league magazine) – ISSN: 0264-987X – mf#13457 – us UMI ProQuest [360]
Criminal justice – Chicago. 1986+ (1,5,9) – (cont: criminal justice) – ISSN: 0887-7785 – mf#16748 – us UMI ProQuest [345]
Criminal justice *see* Journal of religion and spirituality in social work
Criminal justice (aba) – v1-16. 1986-2002 – 9 – $261.00 set – ISSN: 0087-7785 – mf#110951 – us Hein [345]
Criminal justice / aba standards / military practice : report of the ad hoc committee, comparative analysis of the aba standards for the administration of criminal justice and military practice and procedure – n.p. sep 1976? – (= ser The Moran Committee Report) – 7mf – 9 – $10.50 – mf#LLMC 97-010 – us LLMC [345]
Criminal justice abstracts – Monsey. 1989-1996 (1) – ISSN: 0146-9177 – mf#17152,01 – us UMI ProQuest [345]
Criminal justice abstracts – v1-31. 1968-99 – 5,6,9 – $1080.00 set – (v1-16 1968-84 on reel $366; v17-31 1985-99 on mf $714; title varies: v2-8 1970-76 as crime and delinquency literature; information review on crime and delinquency and selected highlights of crime & delinquency literature merged with v2 as crime and delinquency literature) – ISSN: 0146-9177 – mf#102171 – us Hein [345]
Criminal justice and behavior – Thousand Oaks. 1974+ (1,5,9) – ISSN: 0093-8548 – mf#12643 – us UMI ProQuest [360]
Criminal justice and behavior – v1-28. 1974-2001 – 5,6,9 – $965.00 set – (v1-11 1974-84 on reel $201. v12-28 1985-2001 on mf $764) – ISSN: 0083-8548 – mf#400560 – us Hein [360]
Criminal justice digest – Springfield. 1973-76 (1,5,9) – mf#10588 – us UMI ProQuest [345]
Criminal justice digest – Springfield. 1985-94 (1,5,9) – ISSN: 0889-5724 – mf#14118,01 – us UMI ProQuest [345]
Criminal justice ethics – New York. 1982+ (1,5,9) – ISSN: 0731-129X – mf#13091 – us UMI ProQuest [170]
Criminal justice journal : solutions to contemporary management problems / Washington Crime News Services – Annandale. 1982-1985 [1,5,9] – mf#14118 – us UMI ProQuest [360]
Criminal justice journal – Western State University. v1-14 1976-92 [all publ] – $215.00 set – (v1-7 1976-84 on reel $88; v8-14 [1985-92] on mf $127; cont by: san diego justice journal; also sold as v1-14 of thomas jefferson law review, item 117321) – ISSN: 0145-4226 – mf#102201 – us Hein [345]
Criminal justice journal *see* Thomas jefferson law review
Criminal justice review – Atlanta. 1976+ (1,5,9) – ISSN: 0734-0168 – mf#12764 – us UMI ProQuest [360]
The criminal law and sexual offenders; a report / British Medical Association. Joint Committee on Psychiatry and the Law – London: British Medical Association 1954. 24p. LL-2257 – 1 – us L of C Photodup [345]
Criminal law bulletin – Boston. 1965+ (1) 1971+ (5) 1977+ (9) – ISSN: 0011-1317 – mf#3485 – us UMI ProQuest [360]
The criminal law, including the federal penal code. / Grigsby, James Edward – Chicago: Smith, 1922. 1440p. LL-552 – 1 – us L of C Photodup [345]
The criminal law magazine – Jersey City: F D Linn & Co. v1-18. 1880-96 + index 1880-94 (all publ) – 65mf – 9 – $292.00 – mf#LLMC 82-917 – us LLMC [345]
Criminal law magazine and reporter – Jersey City. v1-18 1880-96 (all publ) – (= ser Historical legal periodical series) – 1 – $275.00 set – mf#408970 – us Hein [345]
Criminal law reports : being reports of cases determined in the federal and state courts of th united states, and in the courts of england, ireland, canada, etc / Green, Nicholas John – New York: Hurd and Houghton, 1874-75. 2v. LL-329 – 1 – us L of C Photodup [345]
Criminal law review – London. 1989-1996 (1) – ISSN: 0011-135X – mf#17360 – us UMI ProQuest [360]
Criminal pleading and practice / Bassett, James – 2d ed. Chicago, 1885. 563p. LL-1468 – 1 – us L of C Photodup [345]
Criminal pleading and practice / Bassett, James – Chicago, Myers, 1870. 248 p. LL-1469 – 1 – us L of C Photodup [345]
Criminal procedure of united states courts / Roe, Edward Thomas – Chicago: Callaghan, 1887. 274p. LL-1492 – 1 – us L of C Photodup [345]
Criminal statistics, england and wales : statistics relating to crime and criminal proceedings 1926-1977 – [mf ed Chadwyck-Healey] – (= ser British government publications...1801-1977) – 129mf – 9 – uk Chadwyck [364]
Criminal statistics, scotland : statistics relating to police apprehensions and criminal proceedings 1938-1977 – [mf ed Chadwyck-Healey] – (= ser British government publications...1801-1977) – 35mf – 9 – uk Chadwyck [364]
Criminales de guerra / Universidad de La Habana. Federacion Estudiantil Universitaria – La Habana: Federacion Estudiantil Universitaria 1945 [mf ed 1977] – (= ser Blodgett coll) – 1mf – 9 – mf#w820 – us Harvard [946]
Los crimines del caciquismo. la tragedia de el pobo. defensa del medico... / Albinana Sanz, J – Madrid, 1916 – 2mf – 9 – sp Cultura [360]
Criminology – Beverly Hills. 1963+ (1) 1972+ (5) 1975+ (9) – ISSN: 0011-1384 – mf#6881 – us UMI ProQuest [360]
Criminology and public policy – Columbus. 2001+ (1,5,9) – mf#31900 – us UMI ProQuest [364]
Crimmitschauer anzeiger und tageblatt *see* Crimmitzschauer anzeiger
Crimmitschauer anzeiger und wochenblatt *see* Crimmitzschauer anzeiger
Crimmitschauer nachrichten *see* Werdauer-crimmitschauer wochenblatt
Crimmitschauer stadt- und land-zeitung *see* Stadt- und land-zeitung
Crimmitzschauer anzeiger – Crimmitschau DE, 1848 1 apr-1941 1 jan [gaps] – 162r – 1 – (title varies: 1854?: crimmitschauer anzeiger und wochenblatt, 1939: crimmitschauer anzeiger und tageblatt) – gw Misc Inst [074]
Crimp : a nematode disease of strawberry / Brooks, A N – Gainesville, FL. 1931 – 1r – us UF Libraries [634]
Crimson and gold – Nevada, MO. 1926-1975 (1) – mf#64193 – us UMI ProQuest [071]
Crippen, Thomas George *see*
– A popular introduction to the history of christian doctrine
– The story of congregationalism in surrey
Cripple creek crusher *see* Miscellaneous newspapers of teller county
Cripple creek mining news *see* Miscellaneous newspapers of teller county
Cripple creek sun *see* Miscellaneous newspapers of teller county
Cripple john : of the barony poorhouse, glasgow – London, England. 18– – 1r – us UF Libraries [240]
Cripps, Wilfred Joseph *see* Old english plate, ecclesiastical, decorative, and domestic
Crisana – Oradea, Romania. 1962-Jun 1980; Apr-Oct 1981; 1984-90 – 36r – 1 – us L of C Photodup [949]
La crise de conscience du canada francais : rapport du congres des anciens de la faculte des sciences sociales de laval tenu a la maison montmorency – Quebec: [s.n.], 1957 [mf ed 1994] – 1mf – 9 – mf#SEM105P2077 – cn Bibl Nat [330]
La crise du clerge / Houtin, Albert – Paris: Librairie E Nourry, 1907 – 1mf – 9 – 0-8370-8828-3 – (incl bibl ref and index) – mf#1986-2828 – us ATLA [240]
La crise du clerge *see* The crisis among the french clergy
La crise montaniste / Labriolle, Pierre Champagne de – Paris: E. Leroux, 1913 – (= ser [Bibliotheque de la fondation thiers]) – 2mf – 9 – 0-7905-5361-9 – (incl bibl ref) – mf#1988-1361 – us ATLA [240]
Crise politica brasileira / Bonavides, Paulo – Rio de Janeiro, Brazil. 1969 – 1r – us UF Libraries [321]
La crise politique de quebec : notes et precedents / Dansereau, Arthur – Quebec: [s.n.], 1879 – 1mf – 9 – 0-665-90901-2 – mf#90901 – cn Canadiana [971]
La crise religieuse du 16e siecle (he16) – Paris, 1950 – (= ser Histoire de l'eglise (he)) – €23.00 – ne Slangenburg [240]
La crise religieuse en hollande : souvenirs et impressions / Chantepie de la Saussaye, Daniel – Leyde: De Breuk & Smits, 1860 – 1mf – 9 – 0-7905-5863-7 – (incl bibl ref) – mf#1988-1863 – us ATLA [240]
La crise revolutionnaire (1789-1846) (he20) – Paris, 1951 – (= ser Histoire de l'eglise (he)) – €27.00 – ne Slangenburg [240]
Crises actuelles, leurs causes, le remede / Mathelier, Clement – Port-Au-Prince, Haiti. 1929 – 1r – us UF Libraries [972]
Crisis – Aberdeen, Scotland. 1862 – 1r – us UF Libraries [240]
Crisis – Columbus OH. 1861, 1862 jan 2-1864 jan 20 – 2r – 1 – mf#772390 – us WHS [071]
Crisis / Committee to End the War in Vietnam – 1965 mar 4-1971 feb 9 – 1r – 1 – mf#674234 – us WHS [959]

Crisis – New York. 1910-1996 (1) 1976-1996 (5) 1977-1996 (9) – (cont by: new crisis) – ISSN: 0011-1422 – mf#10994 – us UMI ProQuest [320]

Crisis – 1910 nov-1914, 1914-v13 n6=n78 [1917 apr], 1990-91, 1992-93, 1994-1996 feb/mar – 5r – 1 – (cont by: new crisis [baltimore md]) – mf#1725558 – us WHS [305]

Crisis : devoted to the support of the democratic principles of jefferson – Richmond. 1840-1840 (1) – mf#3971 – us UMI ProQuest [320]

Crisis / Franklin Co. Columbus – jan 1864-may 1871 [wkly] – 3r – 1 – mf#B1462-1464 – us Ohio Hist [071]

Crisis : a record of the darker races / National Association for the Advancement of Colored People – New York. v1-47. 1910-40 – (= ser Black journals, series 1) – 173mf – 9 – $1655.00 – us UPA [305]

The crisis – New york. nov 1910-1963 – 1 – us NY Public [073]

The crisis – Columbus, OH: S Medary, 1861-1871 – 1 – (weekly democratic journal) – mf#34 F3.1 028 – us Western Res [071]

The crisis! : giving an account of the great (metaphorical) eclipse, occultation of certain stars, and other useful knowledge not published in any almanac, for 1877 / Victor – [Hamilton, Ont?: s.n, 1877? [mf ed 1994] – 1mf – 9 – 0-665-94665-1 – mf#94665 – cn Canadiana [055]

The crisis among the french clergy = La crise du clerge / Houtin, Albert – London: David Nutt, 1910 [mf ed 1986] – 1mf – 9 – 0-8370-8680-9 – (trans by f thorold dickson. incl bibl ref & ind) – mf#1986-2680 – us ATLA [241]

Crisis and national co-operative trades union gazette / Owen, Robert – v1-4 n2,20. 1832-34 [all publ] – (= ser Radical periodicals of great britain, 1794-1914. period 1) – 11mf – 9 – $115.00 – us UPA [331]

Crisis christology – v1-5. 1943-48 [complete] – Inquire – 1 – mf#ATLA 1993-S516 – us ATLA [240]

Crisis de la alta cultura en cuba / Manach, Jorge – Habana, Cuba. 1925 – 1r – us UF Libraries [972]

Crisis de la civilizacion. la guerra europea / Trigo, Felipe – Madrid: Renacimiento, 1915 – sp Bibl Santa Ana [946]

Crisis de la democracia en colombia y 'el tiempo' / Santos, Eduardo – Mexico City? Mexico. 1955 – 1r – us UF Libraries [321]

Crisis del lujo / Borrero Y Pierra, Ana Maria – Habana, Cuba. 19-- – 1r – us UF Libraries [972]

Crisis espiritual de unamuno y su evasion a extremadura / Sanchez Morales, Narciso – Badajoz: Imp. Diputacion Prov., 1970 – sp Bibl Santa Ana [240]

Crisis files : pt 1: berlin, 1957-1963 / U.S. State Dept. Office of the Executive Secretary – 10r – 1 – $1760.00 – 1-55655-988-7 – (with p/g) – us UPA [327]

Crisis historica de la ciudad de badajoz / Morales, Ascencio – Badajoz: Tip.y Libr. de A.Arqueros, 1908 – 1 – sp Bibl Santa Ana [360]

Crisis hupfeldiana : being an examination of hupfeld's criticism on genesis: as recently set forth in bishop colenso's fifth part / Kay, William – Oxford: John Henry and James Parker, 1865 – 1mf – 9 – 0-8370-3855-3 – (includes appendixes) – mf#1985-1855 – us ATLA [220]

The crisis in morals : an examination of rational ethics in the light of modern science / Bixby, James Thompson – Boston: Roberts, 1891 – 1mf – 9 – 0-8370-7123-2 – (incl bibl ref) – mf#1986-1123 – us ATLA [170]

The crisis in south africa : some answers (from authoritative sources) to sir a milner's reflections upon colonial loyalty. [issued by friends of south africa in the interests of truth] – [London], [1899] – (= ser 19th c books on british colonization) – 1mf – 9 – mf#1.1.3493 – uk Chadwyck [960]

Crisis in the congo / Lefever, Ernest W – Washington, DC. 1965 – 1r – us UF Libraries [960]

Crisis intervention – Buffalo. 1969-1985 (1) 1972-1985 (5) 1975-1985 (9) – ISSN: 0045-9046 – mf#6664 – us UMI ProQuest [150]

Crisis matrimoniales / Perez Lozano, Jose Maria – Madrid: Edita PPC, S.A., 1965 – 1 – sp Bibl Santa Ana [306]

Crisis of authority in education in south africa : a critical analysis / Rampedi, Makgwana Arnaus – U of the Western Cape 1992 [mf ed S.l: s.n. 1992] – 2mf – 9 – (abstract in afrikaans & english; incl bibl) – sa Misc Inst [370]

The crisis of indian civilisation in the eighteenth and early nineteenth centuries : the genesis of indo-muslim civilisation / Goetz, Hermann – Calcutta: University of Calcutta, 1938 – (= ser Samp: indian books) – us CRL [954]

The crisis of missions : or, the voice out of the cloud / Pierson, Arthur Tappan – New York: Fleming H Revell, c1886 [mf ed 1986] – 1mf – 9 – 0-8370-6598-4 – (incl ind) – mf#1986-0598 – us ATLA [240]

Crisis of the church of england / Stirling, Charles – London, England. 1888 – 1r – us UF Libraries [241]

The crisis of the deeper life / Pardington, George P – New York: Alliance Press, c1906 [mf ed 1985] – 1mf – 9 – 0-8370-5974-7 – mf#1985-3974 – us ATLA [240]

Crisis of the west indian family / Matthews, Basil – s.l, s.l? 1953 – 1r – us UF Libraries [306]

The crisis of unitarianism in boston : as concerned with the twenty-eighth congregational society, with some account of the origin and decline of that organization / Sargent, John Turner – Boston: Walker, Wise, 1859 – 1mf – 9 – 0-524-08810-1 – mf#1993-3302 – us ATLA [243]

Crisis on the high plains: a study of amarillo baptists, 1920 to 1940 / Hickman, James Thomas – 1981 – 1 – 7.60 – us Southern Baptist [242]

Crisis politica / Colombia Ministerio De Gobierno – Bogota, Colombia. 1944 – 1r – us UF Libraries [320]

Crisis series / Columbiana Co. East Liverpool – 10/1884-9/92,10/97-9/03,1904 [wkly, semiwkly] – 7r – 1 – mf#B1702-1708 – us Ohio Hist [071]

El crisol – Madrid, Spain. -d. 4 April 1931-6 Jan 1932. 3 reels – 1 – uk British Libr Newspaper [072]

Crisol de la verdad ilustrado con divinas y humanas letras, padres y doctores... : repuesto al auto del motomedicatio / Aldrete y Soto, L – Madrid, 1683 – 3mf – 9 – sp Cultura [610]

Crisol de libertad – Miami, FL. 1967 aug 08-1971 oct 10 – 1r – (1968 apr) – us UF Libraries [071]

Crisp, Joseph *see* Protestant compendium, being a record of the most important...

Crisp, May Flower *see* Letters to british shipowners, etc

Crisp, William *see* The bechuana of south africa

Crisp, William Finch *see*
- The printers', lithographers', engravers', and bookbinders' business guide, with ready-reckoned general price lists
- The printers' universal book of reference and every-hour office companion

Crisp-head lettuce in florida – Gainesville, FL. 1941 – 1r – us UF Libraries [630]

Crispin, William Frost *see* Universalism and problems of the universalist church

Crispo Acosta, Osvaldo *see* Ruben dario y jose enrique rodo

Crispolti, C *see* Perugia augusta descritta

"Criss-cross" suburban montreal street-address directory = "Criss-cross" annuaire rue-adresse des banlieues de montreal – Montreal: John Lovell & Son, 1973/1974-1977 (annual) [mf ed 1997] – 2r – 1 – (cont: lovell's montreal suburban "criss-cross" cross reference directory, 0703-4644) – mf#SEM35P455 – cn Bibl Nat [971]

Crissey, Joy C *see* Corporate co-optation of sport

Crist, Raymond E *see* Etude geographique des llanos du venezuela coccide...

Cristal – Caceres, 1935 y 1936 – 5 – sp Bibl Santa Ana [073]

Cristal de epoca / Gonzalez Y Contreras, Gilberto – Habana, Cuba. 1944 – 1r – us UF Libraries [972]

Cristal de gruta / Quintana, Caridad – Habana, Cuba. 1955 – 1r – us UF Libraries [972]

Os cristaos-novos em portugal no seculo 20 / Schwarz, Samuel – Lisboa, Portugal. 1925 – 1r – us UF Libraries [939]

El cristian – Santiago, Chile: [s.n.] [ano 80 n5-ano 95 n1 (mayo 1975-1990)] – 2r – 1 – us CRL [242]

Il cristianesimo in egitto / Naldini, M – Firenze, 1968 – 9mf – 8 – €18.00 – ne Slangenburg [241]

Cristiani, Leon *see* Luther et la question sociale

Cristianismo y sociedad – Montivideo: Iglesia y Sociedad en America Latina. v1 n1-v29 n110. 1963-1991 – 6r – us CRL [972]

Cristianos cada dia / Perez Lozano, Jose Maria – Madrid: Propaganda Popular Catolica, 1962 – 1 – sp Bibl Santa Ana [240]

Cristo de espaldas / Caballero Calderon, Eduardo – Bogota, Colombia. 1958? – 1r – us UF Libraries [972]

El cristo de la reja / Casquete, Antonio – Sevilla: imp alvarez gonzalez, 1924 – 1 – sp Bibl Santa Ana [240]

Cristo negro / Salarrue – San Salvador, El Salvador. 1955 – 1r – us UF Libraries [972]

Cristo rey en mexico – Leon, Gto, Mexico: Centro General de Propaganda del Monumento Votivo Nacional a Cristo Rey de La Paz. v1-30 n1-4. 1953-jan/feb, mar/apr 1983 – 15r – us CRL [972]

Cristobal colon / Bayle, Constantino & Poch Noguer, Jose – Madrid: Razon y Fe, 1944 – 1 – sp Bibl Santa Ana [910]

Cristobal colon / Justiniano Arribas, Juan – 1897 – 9 – sp Bibl Santa Ana [910]

Cristobal colon y el descubrimiento de america. 2 vol. barcelona, 1945 / Ballesteros Beretta, Antonio – Madrid: Razon y Fe, 1946 – 1 – sp Bibl Santa Ana [910]

Cristobal colon y la isla espanola / Inchaustegui Cabral, Joaquin Marino – Santiago, Dominican Republic. 1942 – 1r – us UF Libraries [972]

Cristobal de olid / Valle, Rafael Heliodoro – Mexico City? Mexico. 1948 – 1r – us UF Libraries [972]

Cristobal de san antonio, ofm...en notas bibliografia franciscana / Castro, Manuel – Madrid: Graf. Calleja, 1969 – 1 – sp Bibl Santa Ana [240]

Cristoforo colombo – New York. jan 6-june 1891; jan-dec 1892; july-dec 1893; july-dec 1894 – 1 – us NY Public [073]

Cristoforo colombo : romantisches gedicht / Frankl, Ludwig August – Stuttgart: Fr Brodhag 1836 [mf ed 1993] – 1r – 1 – (filmed with: auf dem heimweg / j g fischer) – mf#8576 – us Wisconsin U Libr [810]

Les criteres theologiques : ouvrage traduit de l'italien par un pretre de l'oratoire de rennes sur la seconde edition revue et amelioree par l'auteur = Criteri teologici / Bartolo, Salvatore di – Paris: Berche et Tralin, 1889 – 1mf – 9 – 0-8370-8418-0 – (incl bibl ref and index) – mf#1986-2418 – us ATLA [230]

Criteria for a south african course in intercultural business communication : the case of iscor in japan / Zywotkiewicz, Elize – Uni of South Africa 2001 [mf ed Johannesburg 2001] – 9 – (incl bibl ref) – mf#mfm14750 – sa Unisa [302]

Criteria of diverse kinds of truth as opposed to agnosticism : being a treatise on applied logic / McCosh, James – New York: Charles Scribner's 1882 [mf ed 1984] – 1r – 1 – 0-8370-0305-9 – mf#1984-b339 – us ATLA [160]

Criterio – Buenos Aires. 1950-1954 – 1 – ISSN: 0011-1473 – mf#650 – us UMI ProQuest [073]

Criterio – Buenos Aires. v1-27. nos. 1-1225. march 8, 1928-december 23, 1954 – 1 – us L of C Photodup [073]

Criterion : or, how to detect error and arrive at truth = Criterio / Balmes, Jaime Luciano – New York: P O'Shea, 1875 – 1mf – 9 – 0-7905-9129-4 – (in english) – mf#1989-2354 – us ATLA [100]

Criterion / Hawaii Foundation for American Freedoms, Inc – n1-18 [1970 jul-1973 nov] – 1r – 1 – mf#1110977 – us WHS [322]

Criterion – Karachi. 1975-1978 (1) 1976-1978 (5) 1976-1978 (9) – ISSN: 0011-1481 – mf#7822 – us UMI ProQuest [260]

The criterion : a quarterly review – London. v. 1-18. oct 1922-jan 1939 – 1 – us NY Public [073]

The criterion – Toronto: J W Treen, [1885-18--?] – 9 – mf#P05972 – cn Canadiana [071]

Le criterium a l'usage de la nouvelle exegese biblique : reponse au r. p. m.-j. lagrange / Delattre, Alphonse J – Liege: H Dessain, [1907?] – 1mf – 9 – 0-524-05801-6 – mf#1992-0628 – us ATLA [220]

Critic – Halifax, Nova Scotia. 27 apr 1888-10 jan 1896 – 7r – 1 – (aka: canadian colliery guardian etc, canadian colliery guardian and critic) – uk British Libr Newspaper [071]

Critic – Allentown, PA. 1883-1894 (1) – mf#65831 – us UMI ProQuest [071]

Critic – Chicago. 1942-1996 [1]; 1971-1996 [5]; 1976-1996 [9] – ISSN: 0011-149X – mf#2029 – us UMI ProQuest [000]

Critic – Johannesburg SA, Critic Offices, 1 mar 1890-22 may 1896 (wkly) – 6r – 1 – mf#MS00072 – sa National [079]

Critic – London. 1843-1863 – 1 – mf#4632 – us UMI ProQuest [073]

Critic – Toronto: [s.n., 1883-18-- or 19--] – 9 – (merger of: pulpit criticism; merger of: medical criticism) – mf#P05974 – cn Canadiana [170]

Critic – New York. 1881-1906 (1) – mf#3875 – us UMI ProQuest [700]

Critic – Philadelphia. 1820-1820 (1) – mf#3732 – us UMI ProQuest [700]

Critic : a weekly review of literature, fine arts, and the drama – New York. 1828-1829 – 1 – mf#3970 – us UMI ProQuest [073]

The critic – Washington, DC: [The Critic], 1868-72 – (= ser Daily critic) – 1r – 1 – (filmed with: daily critic (washington dc) jan-oct 16 1872) – us CRL [073]

The critic – Halifax, Canada. Canadian Colliery Guardian. -w. 27 April 1888-10 Jan 1896. 6 reels – 1 – uk British Libr Newspaper [072]

The critic – Halifax, NS: Critic Pub Co, [1884-1894] – 9 – mf#P04952 – cn Canadiana [420]

The critic – Manila: M de Gracia Concepcion. v1 n1. oct 18 1934- (fortnightly) [mf ed 1986-2001] – 2r – 1 – (v1 n8 & n10-12/13 are filmed fr a photocopy. v1 n12/13 (sep 1 1935) lacks p14-15) – mf#6610 – us Wisconsin U Libr [073]

The critic – Wynberg. v1-5 n5. sep 1932-1939 – 1 – us CRL [073]

Critic [ayr, scotland] : official organ of ayr labour council – [Scotland] South Ayrshire, Ayr: Ayr Labour Council apr-nov 1919 (mthly) [mf ed 2004] – 1r – 1 – uk Newsplan [350]

Critica / Campos, Humberto De – Rio de Janeiro, Brazil. 1935 – 1r – us UF Libraries [972]

Critica / Campos, Humberto De – Rio de Janeiro, Brazil. 1940 – 1r – us UF Libraries [972]

Critica / Campos, Humberto De – Rio de Janeiro, Brazil. 1940 – 1r – us UF Libraries [972]

Critica : diario ilustrado de la noche, impersonal e independiente – Buenos Aires: [s.n, [oct1 1914-dec4 1916; jun 1 1919-mar11 1932] (daily) – 130r – 1 – us CRL [079]

Critica : todos os esportes, humorismo, noticioso elegante – Itajai, SC. 29 nov 1931; 26 jun 1932 – (= ser Ps 19) – 8r – 1 Biblioteca [079]

Critica americana / Brenes Mesen, Roberto – San Jose, Costa Rica. 1936 – 1r – us UF Libraries [972]

Critica de nuestra moderna / Mejia Ricart, Gustavo Adolfo – Santiago, Dominican Republic. 1938 – 1r – us UF Libraries [410]

Critica e poetica / Coutinho, Afranio – Rio de Janeiro, Brazil. 1968 – 1r – us UF Libraries [410]

Critica en la literatura cubana / Iraizoz Y De Villar, Antonio – Habana, Cuba. 1930 – 1r – us UF Libraries [440]

Critica fascista – Rome. v1-21. 1923-43 – 1 – $270.00 – mf#0170 – us Brook [945]

Critica fascista – Rome. v1-21. June 1923-July 15 1943 – 1 – us NY Public [325]

Critica historica / Camacho Carrizosa, Guillermo – Bogota, Colombia. 1951 – 1r – us UF Libraries [972]

Critica interna / Anderson Imbert, Enrique – Madrid, Spain. 1961 – 1r – us UF Libraries [972]

Critica literaria / Babin, Maria Teresa – San Juan, Puerto Rico. 1960 – 1r – us UF Libraries [410]

Critica literaria : ensayo sobre el catolicismo... por d juan donoso – 1851 – 9 – sp Bibl Santa Ana [241]

Critica literaria / Rojas Vincenzi, Ricardo – San Jose, Costa Rica. 1929 – 1r – us UF Libraries [410]

Critica litteraria / Machado De Assis – Rio de Janeiro, Brazil. 1942 – 1r – us UF Libraries [410]

Critica marxista – v1-5. 1963-70 – 1 – us AMS Press [335]

Critica musica : d i grundrichtige untersuch- und beurtheilung vieler theils vorgefassten theils einfaeltigen meinungen argumenten und einwuerffe so in alten und neuen gedruckten und ungedruckten musicalischen schrifften zu finden / Mattheson, Johann – Hamburg: auf unkosten des autoris [etc] 1722-25 [mf ed 19--] – 2v [v1 7mf v2 9mf] – 9 – mf#fiche 794 [v1] fiche 795 [v2] – us Sibley [780]

Critica musica – Hamburg DE, 1722-23 pt1-4 – 1 – gw Misc Inst [780]

La critica musicale – v. 1-6. 1918-23 – 1 – us L of C Photodup [780]

Critica sacra : edita in lucem studio & opera j cappel / Cappel, L – Lutetiae Parisiorum, 1650 – 21mf – 8 – €40.00 – ne Slangenburg [220]

Critica sacra... / Cappel, L – Lutetiae Parisiorum, 1650 – 14mf – 9 – mf#PRS-124 – ne IDC [240]

Critica sacra sive animadversiones in loca quaedam difficiliora veteris et novi testamenti... / Dieu, L de – Amstelaedami, 1693 – 16mf – 9 – mf#PBA-181 – ne IDC [240]

Critica y antologia de la poesia puertorriquena / Congreso De Poesia Puertorriquena – San Juan, Puerto Rico. 1958 – 1r – us UF Libraries [440]

Critica y compendio especulativo-practico de la arquitectura civil... / Losada, M – Madrid, 1740 – 11mf – 9 – sp Cultura [624]

Critica y doctrina / Lozano Y Lozano, Carlos – Bogota, Colombia. 1944? – 1r – us UF Libraries [972]

A critical account of the philosophy of kant : with an historical introduction / Caird, Edward – Glasgow: J Maclehose, 1877 [mf ed 1991] – 2mf – 9 – 0-7905-9363-7 – (incl bibl ref) – mf#1989-2588 – us ATLA [190]

A critical account of the philosophy of lotze : the doctrine of thought / Jones, Henry – Glasgow: James Maclehose, 1895 [mf ed 1991] – 1mf – 9 – 0-524-00277-0 – mf#1989-2977 – us ATLA [190]

Critical analysis of adolescent reproductive health services in gauteng province / Magwentshu, Beatrice Makgoale – Uni of South Africa 2000 [mf ed Johannesburg 2000] – 5mf – 9 – (incl bibl ref) – mf#mfm15053 – sa Unisa [360]

A critical analysis of elements of educational management : with reference to performance appraisal of educators in secondary schools in port elizabeth / Ferreira, Ignatius Wilhelm – Pretoria: Vista University 2000 [mf ed 2000] – 3mf – 9 – mf#mfm15125 – sa Unisa [373]

A critical analysis of the poetry of m i mogodi / Tjatji, Ramogohlo Magdeline – Pretoria: Vista University 2001 [mf ed 2001] – 3mf – 9 – (incl bibl ref) – mf#mfm15211 – sa Unisa [470]

Critical analysis of the treaty-making powers of the union of south africa and the republic of south africa / Schaffer, Rosalie Pam – U of the Witwatersrand 1978 [mf ed S.l: s.n. 1978] – 7mf – 9 – sa Misc Inst [341]

A critical and commercial dictionary of the works of painters / Seguier, Frederick Peter – London 1870 – (= ser 19th c art & architecture) – 3mf – 9 – mf#4.1.282 – uk Chadwyck [057]

A critical and doctrinal commentary upon the epistle of st paul to the romans / Shedd, William Greenough Thayer – New York: Charles Scribner, c1879 [mf ed 1985] – 2mf – 9 – 0-8370-5246-7 – (incl bibl ref) – mf#1985-3246 – us ATLA [227]

A critical and exegetical commentary on deuteronomy / Driver, Samuel Rolles – New York: Charles Scribner, 1895 [mf ed 1985] – (= ser The international critical commentary on the holy scriptures of the old and new testaments 5) – 2mf – 9 – 0-8370-2968-6 – (incl ind) – mf#1985-0968 – us ATLA [221]

A critical and exegetical commentary on the acts of the apostles / Gloag, Paton J – Edinburgh: T & T Clark, 1870 [mf ed 1989] – 3mf – 9 – 0-7905-1817-1 – (incl bibl ref) – mf#1987-1817 – us ATLA [226]

A critical and exegetical commentary on the book of ecclesiastes / Barton, George Aaron – New York: Scribner, 1908 [mf ed 1986] – (= ser International critical commentary on the holy scriptures of the old and new testaments 17) – xiv/212p – 1 – (incl bibl) – mf#1792 – us Wisconsin U Libr [221]

A critical and exegetical commentary on the book of ecclesiastes / Barton, George Aaron – New York: Charles Scribner 1908 [mf ed 1989] – (= ser The international critical commentary on the holy scriptures of the old and new testaments 17) – 3mf – 9 – 0-7905-2940-8 – mf#1987-2940 – us ATLA [221]

A critical and exegetical commentary on the book of esther / Paton, Lewis Bayles – New York: Charles Scribner, 1908 [mf ed 1986] – (= ser The international critical commentary on the holy scriptures of the old and new testaments 13) – 1mf – 9 – 0-8370-6297-7 – (incl bibl ref & ind) – mf#1986-0297 – us ATLA [221]

A critical and exegetical commentary on the book of esther / Paton, Lewis Bayles – New York: C Scribner's Sons, 1908 [mf ed 1986] – (= ser International critical commentary [on the holy scriptures of the old and new testaments]) – xvii/339p – 1 – mf#1648 – us Wisconsin U Libr [221]

A critical and exegetical commentary on the book of exodus / Murphy, James Gracey – 1868 – 9 – $12.00 – us IRC [221]

A critical and exegetical commentary on the book of exodus / Murphy, James Gracey – Andover: Warren F Draper, 1881 [mf ed 1985] – 1mf – 9 – 0-8370-4544-4 – mf#1985-2544 – us ATLA [221]

A critical and exegetical commentary on the book of genesis / Murphy, James Gracey – Andover: Warren F Draper, 1866 [mf ed 1989] – 2mf – 9 – 0-7905-1478-8 – (pref by j p thompson) – mf#1987-1478 – us ATLA [221]

A critical and exegetical commentary on the book of leviticus / Murphy, James Gracey – Andover: Warren F Draper, 1874 [mf ed 1985] – 1mf – 9 – 0-8370-4545-2 – mf#1985-2545 – us ATLA [221]

A critical and exegetical commentary on the book of psalms : with a new translation / Murphy, James Gracey – Edinburgh: T & T Clark, 1875 [mf ed 1989] – 2mf – 9 – 0-7905-3355-3 – mf#1987-3355 – us ATLA [221]

A critical and exegetical commentary on the books of chronicles / Curtis, Edward Lewis & Madsen, Albert Alonzo – New York: Charles Scribner, 1910 [mf ed 1989] – (= ser The international critical commentary on the holy scriptures of the old and new testaments 11) – 2mf – 9 – 0-7905-1647-0 – (incl ind) – mf#1987-1647 – us ATLA [221]

A critical and exegetical commentary on the epistles of st peter and st jude / Bigg, Charles – 2nd ed. Edinburgh: T & T Clark, 1902 [mf ed 1986] – (= ser The international critical commentary on the holy scriptures of the old and new testaments 42) – 1mf – 9 – 0-8370-6018-4 – (incl ind) – mf#1986-0018 – us ATLA [227]

A critical and exegetical commentary on the epistles to the ephesians and to the colossians / Abbott, Thomas Kingsmill – New York: Charles Scribner 1897 [mf ed 1985] – (= ser The international critical commentary on the holy scriptures of the old and new testaments 36) – 1mf – 9 – 0-8370-2029-8 – (incl ind) – mf#1985-0029 – us ATLA [227]

A critical and exegetical commentary on the epistles to the philippians and to philemon / Vincent, Marvin Richardson – New York: Charles Scribner, 1897 [mf ed 1985] – (= ser The international critical commentary on the holy scriptures of the old and new testaments 37) – 1mf – 9 – 0-8370-5639-X – (incl bibl & ind) – mf#1985-3639 – us ATLA [227]

A critical and exegetical commentary on the first epistle of st paul to the corinthians / Robertson, Archibald & Plummer, Alfred – 2nd ed. Edinburgh: T & T Clark, 1914 [mf ed 1989] – (= ser International critical commentary on the holy scriptures of the old and new testaments 33) – 2mf – 9 – 0-7905-2867-3 – (incl bibl ref and ind) – mf#1987-2867 – us ATLA [227]

Critical and exegetical commentary on the gospel / Gould, Ezra Palmer – New York, NY. 1896 – 1r – us UF Libraries [226]

Critical and exegetical commentary on the gospel according... / Plummer, Alfred – New York, NY. 1907 – 1r – us UF Libraries [226]

A critical and exegetical commentary on the gospel according to s matthew / Allen, Willoughby Charles – Edinburgh: T & T Clark 1912 [mf ed 1989] – (= ser The international critical commentary on the holy scriptures of the old and new testaments 26) – 2mf – 9 – 0-7905-2820-7 – mf#1985-2820 – us ATLA [226]

A critical and exegetical commentary on the gospel according to st mark / Gould, Ezra Palmer – New York: Charles Scribner, 1907, c1896 [mf ed 1985] – (= ser The international critical commentary on the holy scriptures of the old and new testaments 27) – 1mf – 9 – 0-8370-3353-5 – (incl ind) – mf#1985-1353 – us ATLA [226]

Critical and exegetical commentary on the new testament / Meyer, Heinrich August Wilhelm – Edinburgh, Scotland. v1-20. 1877-1881 – 5r – us UF Libraries [225]

Critical and exegetical handbook to the acts of the apostles = Kritisch exegetisches handbuch ueber die apostelgeschichte / Meyer, Heinr Aug Wilh – 2nd ed. New York: Funk & Wagnalls, 1889, c1883 – 2mf – 9 – 0-7905-3037-6 – (incl bibl ref. in english) – mf#1987-3037 – us ATLA [226]

Critical and exegetical hand-book to the epistle to the ephesians = Kritisch exegetisches handbuch ueber den brief an die epheser / Meyer, Heinr Aug Wilh – New York: Funk & Wagnalls, 1884 – 1mf – 9 – 0-7905-3038-4 – (incl ind. in english) – mf#1987-3038 – us ATLA [227]

Critical and exegetical hand-book to the epistle to the galatians = Kritisch exegetisches handbuch ueber den brief an die galater / Meyer, Heinr Aug Wilh – New York: Funk & Wagnalls, 1884 – 1mf – 9 – 0-7905-3039-2 – (incl ind. in english) – mf#1987-3039 – us ATLA [227]

Critical and exegetical handbook to the epistles of st paul to the thessalonians = Kritisch exegetisches handbuch ueber die briefe an die thessalonicher / Luenemann, Gottlieb – Edinburgh: T & T Clark 1880 [mf ed 1985] – 1mf – 9 – 0-8370-4193-7 – (trans fr 3rd ed of the german by paton j gloag) – mf#1985-2193 – us ATLA [227]

Critical and exegetical handbook to the epistles of st paul to timothy and titus = Kritisch exegetisches handbuch ueber die briefe an timotheus und titus / Huther, Johann Eduard – Edinburgh: T & T Clark 1881 [mf ed 1985] – 1mf – 9 – 0-8370-3706-9 – (trans fr 4th impr enl edition of the german by david hunter; incl bibl ref) – mf#1985-1706 – us ATLA [227]

Critical and exegetical hand-book to the epistles to the corinthians = Kritisch exegetisches handbuch ueber den ersten brief an die korinther / Meyer, Heinr Aug Wilh – New York: Funk & Wagnalls, 1884 – 2mf – 9 – 0-7905-3084-8 – (in english) – mf#1987-3084 – us ATLA [227]

Critical and exegetical hand-book to the epistles to timothy and titus = Kritisch exegetisches handbuch ueber die briefe an timotheus und titus / Huther, Joh Ed – New York: Funk & Wagnalls, 1890, c1885 – 2mf – 9 – 0-7905-3079-1 – (in english) – mf#1987-3079 – us ATLA [227]

Critical and exegetical handbook to the general epistles of james, peter, john, and jude / Huther, Joh Ed – New York: Funk & Wagnalls, 1887 – 2mf – 9 – 0-7905-3265-4 – mf#1987-3265 – us ATLA [227]

Critical and exegetical handbook to the general epistles of peter and jude = Kritisch exegetisches handbuch ueber den 1. brief des petrus, den brief des judas und den 2. brief des petrus / Huther, Johann Eduard – Edinburgh: T & T Clark 1893 [mf ed 1986] – 9 – 0-8370-6191-1 – (trans fr german by d b croom & paton j gloag; incl bibl ref) – mf#1986-0191 – us ATLA [227]

Critical and exegetical hand-book to the gospel of matthew = Kritisch exegetisches handbuch ueber das evangelium des matthaeus / Meyer, Heinr Aug Wilh – New York: Funk & Wagnalls, 1884 – 2mf – 9 – 0-7905-3040-6 – (incl bibl ref. in english) – mf#1987-3040 – us ATLA [226]

Critical and exegetical hand-book to the gospels of mark and luke = Kritisch exegetisches handbuch ueber die evangelien des markus und lukas / Meyer, Heinr Aug Wilh – New York: Funk & Wagnalls, 1884 – 2mf – 9 – 0-7905-3041-4 – (incl bibl ref. in english) – mf#1987-3041 – us ATLA [226]

Critical and exegetical handbook to the revelation of john = Kritisch exegetisches handbuch ueber die offenbarung johannis / Duesterdieck, Friedrich – New York: Funk & Wagnalls, 1887, c1886 – 2mf – 9 – 0-7905-3014-7 – (incl bibl ref. in english) – mf#1987-3014 – us ATLA [220]

A critical and exigetical commentary on the book of daniel / Charles, Robert Henry – 1929 – 9 – $18.00 – us IRC [221]

A critical and grammatical commentary on st paul's epistle to the ephesians / Ellicott, Charles John – Andover: Warren F Draper, 1865 [mf ed 1985] – 1mf – 9 – 0-8370-3046-3 – mf#1985-1046 – us ATLA [227]

Critical and grammatical commentary on st paul's epistle to the ephesians / Ellicott, Charles J – Andover: Warren F Draper, 1865.1 fiche – 9 – us ATLA [240]

A critical and grammatical commentary on the pastoral epistles / Ellicott, Charles John – Andover: Warren F Draper, 1865 [mf ed 1985] – 1mf – 9 – 0-8370-3048-X – mf#1985-1048 – us ATLA [227]

A critical and historical introduction to the canonical scriptures of the old testament / Wette, Wilhelm Martin Leberecht de – 3rd ed. Boston: Rufus Leighton, 1859 [mf ed 1984] – (= ser Biblical crit – us & gb 69) – 2v on 12mf – 9 – 0-8370-1245-7 – (with app) – mf#1984-1069 – us ATLA [221]

A critical and historical review of fox's book of martyrs : shewing the inaccuracies, falsehoods, and misrepresentations in that work of deception / Andrews, William Eusebius – London: W E Andrews, 1824-26 [mf ed 1992] – 3v on 14mf – 9 – 0-524-03630-6 – mf#1990-1058 – us ATLA [242]

The critical and miscellaneous writings of theodore parker / Parker, Theodore – Boston: Horace B Fuller, c1843 – 1mf – 9 – 0-524-03019-7 – mf#1990-4541 – us ATLA [240]

A critical and philosophical enquiry into the causes of prodigies and miracles, as related by historians : with an essay towards restoring a method and purity in history... / Warburton, William, Bishop of Gloucester – London: T Corbett, 1727 – xxii/137/[7]p – 1 – (incl bibl ref) – mf#2193 – us Wisconsin U Libr [900]

Critical asian studies – Basingstoke, 2001+ [1,5,9] – (cont: bulletin of concerned asian scholars) – ISSN: 1467-2715 – mf#6049,01 – us UMI ProQuest [327]

Critical care clinics – Philadelphia. 1985+ (1,5,9) – ISSN: 0749-0704 – mf#14731 – us UMI ProQuest [610]

Critical care medicine – v1-24. 1973-96 – 24r – 1,5,6,9 – $110.00r – (also available on cd-rom: 1984-88 $395.00) – us Lippincott [610]

Critical care nurse – Bridgewater. 1986+ (1,5,9) – ISSN: 0279-5442 – mf#15177 – us UMI ProQuest [610]

Critical care nursing clinics of north america – Philadelphia. 1994+ (1,5,9) – ISSN: 0899-5885 – mf#20824 – us UMI ProQuest [610]

Critical care nursing quarterly – Frederick. 1986+ (1,5,9) – (cont: ccq: critical care quarterly) – ISSN: 0887-9303 – mf#12728,01 – us UMI ProQuest [610]

Critical challenges confronting national security : continuing encroachment threatens force readiness: hearing...house of representatives, 107th congress, 2nd session, may 16 2002 / United States. Congress. House. Committee on Government Reform – [Washington DC: US GPO] 2002 [mf ed 2002] – 5mf – 9 – us GPO [341]

A critical commentary on the book of daniel : designed especially for students of the english bible / Prince, John Dyneley – Leipzig: J C Hinrichs; New York: Lemcke & Buechner, 1899 [mf ed 1985] – 1mf – 9 – 0-8370-4800-1 – (incl bibl ref & ind) – mf#1985-2800 – us ATLA [221]

A critical commentary on the epistle to the hebrews / Sampson, Francis Smith – New York: Robert Carter, 1856 [mf ed 1989] – 2mf – 9 – 0-7905-2054-0 – (in english and greek) – mf#1987-2054 – us ATLA [227]

Critical description and analytical review of "death on the pale horse" painted by benjamin west / Carey, William Paulet – London 1817 – (= ser 19th c art & architecture) – 2mf – 9 – mf#4.1.365 – uk Chadwyck [750]

Critical description of the procession of chaucer's pilgrims to canterbury, painted by thomas stothard / Carey, William paulet – [2nd ed] London 1818 – (= ser 19th c art & architecture) – 2mf – 9 – mf#4.2.1704 – uk Chadwyck [750]

Critical digest – New York. 1949-1984 (1) 1974-1984 (5) 1976-1984 (9) – (cont by: critical digest, nyc and london theatre) – ISSN: 0045-9070 – mf#6872 – us UMI ProQuest [790]

Critical digest, nyc and london theatre – New York. 1984-1984 (1) 1984-1984 (5) 1984-1984 (9) – (cont: critical digest) – mf#6872,01 – us UMI ProQuest [790]

Critical enquiries into the various editions of the bible / Simon, Richard – 1684 – 1 – $50.00 – us Presbyterian [240]

A critical essay on oilpainting : proving that the art of painting in oil was known before the pretended discovery of john and hubert van eyck... / Raspe, R E – London, 1781 – 4mf – 9 – mf#0-1177 – ne IDC [700]

A critical essay on the gospel of st luke / Schleiermacher, Friedrich [Ernst Daniel] – London: John Taylor 1825 [mf ed 1984] – 6mf – 9 – 0-8370-0230-3 – (incl bibl ref) – mf#1984-0048 – us ATLA [226]

Critical essays : contributed to the eclectic review / Foster, John; ed by Ryland, Jonathan Edwards – London: Henry G Bohn 1856 [mf ed 1993] – (= ser Bohn's standard library) – 2v on 1mf – 9 – 0-524-08258-8 – mf#1993-3013 – us ATLA [100]

Critical essays of the seventeenth century / ed by Spingarn, Joel Elias – Oxford: Clarendon Press; Toronto: H Frowde, 1908-1909 – 13mf – 9 – 0-665-85505-2 – (v2: 0-665-85506-0. v3: 0-665-85507-9. incl app) – mf#85505-85507 – cn Canadiana [840]

Critical essays on a few subjects connected with the history and present condition of speculative philosophy / Bowen, Francis – Boston: HB Williams, 1842 – 1mf – 9 – 0-7905-7559-0 – mf#1989-0784 – us ATLA [190]

Critical evaluation of the ontario department of education concert plan : and its general effect on music education in the province of ontario / Abray, William Ewart – 1953 [mf ed 1981] – 3mf – 9 – (with app & bibl) – us Sibley [780]

A critical examination and complete catalogue of the works of art now exhibiting in westminster hall / Clarke, Henry Green – London 1847 – (= ser 19th c art & architecture) – 1mf – 9 – mf#4.1.320 – uk Chadwyck [700]

A critical examination of some of the principal arguments for and against darwinism / MacLaren, James – London, 1876 – (= ser 19th c evolution & creation) – 6mf – 9 – mf#1.1.1503 – uk Chadwyck [575]

Critical examination of some passages in gen 1 : with remarks on difficulties that attend some of the present modes of geological reasoning / Stuart, Moses – [Andover: Gould & Newman, 1836] [mf ed 1984] – (= ser Biblical crit – us & gb 90) – 1mf – 9 – 0-8370-1584-7 – (incl bibl ref) – mf#1984-1090 – us ATLA [221]

A critical examination of the cartoons, frescos, and sculpture, exhibited in westminster hall : to which is added the history and practice of fresco painting / Clarke, Henry Green – London: H G Clarke & Co, 1844 – (= ser 19th c art & architecture) – 1mf – 9 – mf#4.1.190 – uk Chadwyck [740]

A critical examination of the evidences for the doctrine of the virgin birth / Thorburn, Thomas James – London: SPCK; New York: E S Gorham, 1908 [mf ed 1985] – 1mf – 9 – 0-8370-5527-X – (incl app & ind) – mf#1985-3527 – us ATLA [240]

The critical examination of the philosophy of religion / Santinatha, Sadhu – Amalner: Institute of Philosophy, 1938 – (= ser Samp: indian books) – us CRL [280]

A critical examination of the question in regard to the time of our saviour's crucifixion : showing that he was crucified on thursday the 14th day of the jewish month nisan, a d 30 / Aldrich, Jeremiah Knight – Boston: [s.n.] 1882 (London: Rand, Avery & Co) [mf ed 1985] – 1mf – 9 – 0-8370-2059-X – mf#1985-0059 – us ATLA [220]
A critical exposition of the third chapter of paul's epistle to the romans : a monograph / Morison, James – London: Hamilton, Adams; Glasgow: T D Morison, 1866 [mf ed 1985] – 1mf – 9 – 0-8370-4491-X – mf#1985-2491 – us ATLA [227]
Critical greek and english concordance of the new testament / Hudson, Charles Frederic & Abbot, Ezra – 7th ed. Boston: HL Hastings, London: S Bagster, 1885 [mf ed 1989] – 2mf – 9 – 0-7905-0715-3 – (rev & completed by ezra abbot; add: green's greek-english lexikon to the new testament; incl ind) – mf#1987-0715 – us ATLA [225]
Critical handbook / Mitchell, E C – Andover, England. 1880 – 1r – us UF Libraries [030]
The critical handbook of the greek new testament / Mitchell, Edward Cushing – new enl ed. New York: Harper, 1896 – 1mf – 9 – 0-8370-4451-0 – (incl bibl ref) – mf#1985-2451 – us ATLA [225]
Critical, historical, and admonitory letter to the right reverend f... / Wedderburn, R – London, England. 1820? – 1r – us UF Libraries [240]
Critical, historical, and miscellaneous essays and poems / Macaulay, Thomas Babington Macaulay, Baron – Philadelphia, PA. v1-3. 191-? – 1r – us UF Libraries [420]
Critical history and defence of the old testament canon / Stuart, Moses – rev ed. Andover:Warren F Draper, 1872 [mf ed 1986] – 1mf – 9 – 0-8370-5460-5 – (incl app) – mf#1985-3460 – us ATLA [221]
A critical history of christian literature and doctrine / Donaldson, J – London. v1-3. 1864 – €35.00 – ne Slangenburg [240]
A critical history of free thought in reference to the christian religion / Farrar, Adam Storey – London: J Murray, 1862 [mf ed 1990] – (= ser Bampton lectures 1862) – 2mf – 9 – 0-7905-5033-4 – (incl bibl ref) – mf#1988-1033 – us ATLA [140]
Critical history of the christian doctrine of justification and reconciliation = Christliche lehre von der rechtfertigung und versoehnung. erster band, geschichte der lehre / Ritschl, Albrecht – Edinburgh: Edmonston & Douglas, 1872 [mf ed 1991] – 2mf – 9 – 0-7905-9465-X – (trans fr german into english by john s black; incl bibl ref) – mf#1989-2690 – us ATLA [240]
A critical history of the doctrine of a future life : with a complete bibliography on the subject / Alger, William Rounseville – Philadelphia: George W Childs, 1864 [mf ed 1993] – 3mf – 9 – 0-524-08225-1 – mf#1993-2000 – us ATLA [210]
A critical history of the doctrine of a future life in israel, in judaism and in christianity / Charles, Robert Henry – London, 1899 – 8mf – 8 – €17.00 – ne Slangenburg [939]
A critical history of the doctrine of a future life in israel, in judaism, and in christianity : or, hebrew, jewish, and christian eschatology from pre-prophetic times till the close of the new testament canon / Charles, Robert Henry – 2nd rev ed. London: A & C Black, 1913 [mf ed 1989] – (= ser Jowett lectures 1898-99) – 2mf – 9 – 0-7905-0682-3 – (incl bibl ref & ind) – mf#1987-0682 – us ATLA [270]
A critical history of the evolution of trinitarianism : and its outcome in the new christology / Paine, Levi Leonard – Boston: Houghton, Mifflin, 1900 [mf ed 1985] – 1mf – 9 – 0-8370-3952-5 – (incl bibl ref and ind) – mf#1985-1952 – us ATLA [240]
Critical inquiry – Chicago. 1974+ (1) 1976+ (5) 1976+ (9) – ISSN: 0093-1896 – mf#11000 – us UMI ProQuest [073]
A critical introduction to the new testament / Peake, Arthur Samuel – London: Duckworth, 1909 [mf ed 1989] – 1mf – 9 – 0-7905-3046-5 – mf#1987-3046 – us ATLA [225]
A critical introduction to the old testament / Gray, George Buchanan – New York: Scribner 1913 [mf ed 1989] – (= ser Studies in theology) – 1mf – 9 – 0-7905-2964-5 – mf#1987-2964 – us ATLA [221]
Critical mass – v1 n1-v3 n1 [1975 apr-1977 apr] – 1r – 1 – (cont by: critical mass journal) – mf#803822 – us WHS [302]
Critical mass : public citizen's energy journal – v8 n6-15 [1982 dec-1983 sep] – 1r – 1 – (cont: critical mass energy journal; cont by: critical mass bulletin) – mf#852887 – us WHS [360]
Critical mass bulletin / Public Citizen's Energy Project [US] – v1 n1-v2 n1 [1983 nov-1984 nov] – 1r – 1 – (cont: critical mass [washington dc: 1982]; cont by: connections [washington dc]) – mf#870915 – us WHS [333]
Critical mass energy journal / Ralph Nader's Public Citizen, Inc – v6 n5-v8 n5 [1980 aug-1982 nov] – 1r – 1 – (cont: critical mass journal; cont by: critical mass [washington dc: 1982]) – mf#829798 – us WHS [333]
Critical mass journal – v3 n2-v6 n4 [1977 may-1980 jul] – 1r – 1 – (cont: critical mass; cont by: critical mass energy journal) – mf#806482 – us WHS [333]
Critical notes on old testament history : the traditions of saul and david / Cook, Stanley Arthur – London; New York: Macmillan, 1907 – 1mf – 9 – 0-7905-2732-2 – (incl ind) – mf#1985-0732 – us ATLA [221]
Critical notes on the authorised english version of the new testament : being a companion to the author's new testament, translated from griesbach's text / Sharpe, Samuel – 2nd ed. London: John Russell Smith, 1867 – 1mf – 9 – 0-524-06581-0 – mf#1992-0924 – us ATLA [225]
Critical notes on the international sunday-school lessons from the pentateuch for 1887 (january 2-june 26) / Driver, Samuel Rolles – New York: Charles Scribner, 1887 – 1mf – 9 – 0-7905-0986-5 – mf#1987-0986 – us ATLA [221]
Critical practice / International Workers Party – v1 n1-2 [1975 spr-fall] – 1r – 1 – mf#365074 – us WHS [335]
Critical quarterly – Hull. 1985+ (1,5,9) – ISSN: 0011-1562 – mf#15388 – us UMI ProQuest [410]
Critical questions : being a course of sermons / Kirkpatrick, Alexander Francis – London: S C Brown, Langham, 1903 – 1mf – 9 – 0-7905-0134-1 – (includes bibliographies) – mf#1987-0134 – us ATLA [220]
Critical realism : a study of the nature and conditions of knowledge / Sellars, Roy Wood – Chicago: Rand McNally, c1916 – 1mf – 9 – 0-7905-8887-0 – mf#1989-2112 – us ATLA [120]
Critical remarks on dr tregelles' greek text of the revelation : and his two english versions compared with the received text and authorised translation / Tomlin, Jacob – Liverpool: Arthur Newling, 1865 [mf ed 1985] – 1mf – 9 – 0-8370-5553-9 – (incl app) – mf#1985-3553 – us ATLA [225]
Critical remarks on the discussion on the indiscriminate circulatio... / Andrews, William Eusebius – London, England. 1828? – 1r – us UF Libraries [240]
Critical remarks on the hebrew scriptures : corresponding with a new translation of the bible / Geddes, Alexander – London: printed... and sold by R Faulder & J Johnson, 1800 [mf ed 1984] – (= ser Biblical crit – us & gb 16) – 6mf – 9 – (incl bibl ref. no more publ) – mf#1984-1016 – us ATLA [221]
Critical researches in philology and geography / Bell, James & Bell, John – Glasgow, 1824 – (= ser 19th c books on linguistics) – 3mf – 9 – mf#2.1.49 – uk Chadwyck [400]
Critical review – Astoria. 1987+ – 1,5,9 – ISSN: 0891-3811 – mf#18034 – us UMI ProQuest [073]
Critical review : or, annals of literature – London. 1756-1817 (1) – mf#4233 – us UMI ProQuest [500]
Critical review : or annals of literature – London. v1-24. (3 Ser.); 1-6. (4 Ser.); 1-5. (5 Ser.). 1804-Jun 1817 – 1 – us NY Public [410]
A critical review of chu chih sin's "what thing is jesus" *see* Kuan yu chu chih-hsin yeh-su shi shen mo tung hsi te tsa p'ing (ccm11)
A critical review of teachers' perceptions of the developmental appraisal system : a case study / Slingers, Joseph William McCarthy – Port Elizabeth 2004 [mf ed Pretoria: Vista University 2004] – 2mf – 9 – mf#mfm15255 – sa Unisa [370]
Critical review of theological and philosophical literature – London. 1891-1904 (1) – mf#2880 – us UMI ProQuest [200]
A critical review of validity in alignment and dance performance studies using imagery training / Fall, Lynn A – 1998 – 2mf – 9 – $8.00 – mf#PE 3897 – us Kinesology [790]
A critical review of wesleyan perfection : in twenty-four consecutive arguments... / Franklin, Samuel – Cincinnati: Methodist Book Concern, 1875 [mf ed 1993] – (= ser Methodist coll) – 2mf – 9 – 0-524-06537-3 – mf#1991-2621 – us ATLA [242]
Critical sociology – Eugene. 1988+ (1,5,9) – (cont: insurgent sociologist) – ISSN: 0896-9205 – mf#10656,01 – us UMI ProQuest [301]
Critical studies in mass communication: csmc : a publication of the speech communication association – Annandale. 1984-1999 (1) 1984-1999 (5) 1984-1999 (9) – (cont by: critical studies in media communication) – ISSN: 0739-3180 – mf#14016 – us UMI ProQuest [380]
Critical studies in media communication – Annandale. 2000+ (1,5,9) – (cont: critical studies in mass communication: csmc) – ISSN: 1529-5036 – mf#14016,01 – us UMI ProQuest [380]
Critical studies in the phonetic observations of indian grammarians / Varma, S – 1929 – (= ser Royal asiatic society. j g forlong fund) – 1r – 1 – mf#658 – uk Microform Academic [400]
Critical study of books on the clarinet / Sidorfsky, Frank Mendenhall – U of Rochester 1973 [mf ed 19–] – 4mf – 9 – (with bibl) – mf#fiche 1049 – us Sibley [780]
A critical study of christian eschatology in the light of marxist thought / Iileka, David – Uni of South Africa 2001 [mf ed Johannesburg 2001] – 3mf – 9 – (incl bibl ref) – mf#mfm14982 – sa Unisa [240]
A critical study of current theories of moral education / Hart, Joseph Kinmont – Chicago: University of Chicago Press, 1910 [mf ed 1986] – 1mf – 9 – 0-8370-8677-9 – mf#1986-2677 – us ATLA [230]
A critical study of the first four chapters of rethinking missions *see* Ping hsuan chiao shih yeh ping i (ccm77)
A critical study of the historical method of samuel rawson gardiner : with an excursus on the historical conception of the puritan revolution from clarendon to gardiner / Usher, Roland Greene – [St Louis?: Washington University? 1915?] [mf ed 1990] – (= ser Washington university studies 3/2/1 (oct 1915)) – 1mf – 9 – 0-7905-6513-7 – (incl bibl ref) – mf#1988-2513 – us ATLA [941]
A critical study of the life and novels of bankimcandra / Dasagupta, Jayantakumara – Calcutta: Calcutta University, 1937 – (= ser Samp: indian books) – us CRL [410]
Critical survey – Manchester. 1962-1968 – 1 – ISSN: 0011-1570 – mf#5131 – us UMI ProQuest [370]
Critical survey of printed vocal arrangements of afro-american religious folk songs / Kerr, Thomas H – U of Rochester 1939 [mf ed 19–] – 1r / 3mf – 1,9 – mf#film 1106 / fiche 1162 – us Sibley [780]
Critical survey of south african poetry in english / Miller, G M – Cape Town, South Africa. 1957 – 1r – us UF Libraries [410]
Criticas – New York. 2001+ (1,5,9) – ISSN: 1535-6132 – mf#32354 – us UMI ProQuest [020]
Criticas de sinceridad y exactitud / Vallenilla Lanz, Laureano – Caracas, Venezuela. 1956 – 1r – us UF Libraries [972]
Critici sacri libri 4 / Rivetus, Andr – ed 4a. Genevae, 1642 – 11mf – 8 – €22.00 – ne Slangenburg [220]
Criticism – Detroit. 1959+ (1) 1971+ (5) 1975+ (9) – ISSN: 0011-1589 – mf#2486 – us UMI ProQuest [000]
Criticism and faith : a lecture delivered before the theological union of the london conference, jun 1910 / Knight, J F – Toronto: W Briggs, 1911 [mf ed 1996] – 1mf – 9 – 0-665-80970-0 – mf#80970 – cn Canadiana [220]
A criticism of montague-chelmsford proposals of indian constitutional reforms / Malaviya, Madan Mohan, 1861-1946 – Allahabad: Printed by C Y Chintamani, 1918 – (= ser Samp: indian books) – us CRL [323]
Criticism of mr lesueur's pamphlet, entitled defence of modern thought / Armstrong, William Dunwoodie [i.e. Vindex] – Ottawa?: Woodburn, 188-? – 1mf – 9 – mf#02445 – cn Canadiana [230]
A criticism of systems of hebrew metre : an elementary treatise / Cobb, William Henry – Oxford: Clarendon Press, 1905 [mf ed 1985] – 1mf – 9 – 0-8370-2694-6 – (incl ind) – mf#1985-0694 – us ATLA [470]
A criticism of the critical philosophy / McCosh, James – New York: C Scribner 1884 [mf ed 1991] – 1mf – 9 – 0-7905-8519-7 – mf#1989-1744 – us ATLA [140]
The criticism of the fourth gospel : eight lectures on the morse foundation, delivered in the union seminary, new york... / Sanday, William – New York: Charles Scribner's, 1905. Beltsville, Md: NCR Corp, 1978 (4mf); Evanston: American Theol Lib Assoc, 1984 (4mf) – (= ser Biblical crit – us & gb) – 9 – 0-8370-0208-7 – (incl bibl ref and ind) – mf#1984-1037 – us ATLA [226]
The criticism of the fourth gospel / Sanday, William – New York: Charles Scribner, 1905 – (= ser The Morse Lectures) – 1mf – 9 – 0-8370-5039-1 – (incl ind) – mf#1985-3039 – us ATLA [226]
Criticism of the new testament : lectures for 1902 / Sanday, William et al – 2nd ed. London: John Murray 1903 [mf ed 1985] – (= ser St margaret's lectures 1902) – 1mf – 9 – 0-8370-2777-2 – mf#1985-0777 – us ATLA [225]
Criticismo y libertad / Chacon Y Calvo, Jose Maria – Habana, Cuba. 1939 – 1r – us UF Libraries [972]
Criticisms on contemporary thought and thinkers / Hutton, Richard Holt – London; New York: Macmillan, 1894 – 2mf – 9 – 0-7905-9284-3 – mf#1989-2509 – us ATLA [190]
O critico : jornal critico, satyrico, litterario, poetico e jacoso – Rio de Janeiro, RJ: Typ Imparcial de Francisco de Paula Britto, 15 jan 1842 – (= ser Ps 19) – mf#P12,05,22 n03 – bl Biblioteca [410]
Critico literario / Lima, Alceu Amoroso – Rio de Janeiro, Brazil. 1945 – 1r – us UF Libraries [410]
Critiks : being papers upon the times – London. 1718-1718 (1) – mf#5297 – us UMI ProQuest [070]
Critique / Cleveland Teachers Union – v18 n2 [1977 dec], v19 [i.e. 20] n2 [1979 jan] v21 n1 [1980 oct] – 1r – 1 – mf#678878 – us WHS [370]
Critique / Robert Schadewald Collection on Pseudo-Science – v2 n4=7/8-v4 n1/2=13/14 [1982 spr/summer-1983/84 fall/winter] – 1r – 1 – mf#846572 – us WHS [080]
Critique – Washington. 1956+ (1) 1977+ (5) 1977+ (9) – ISSN: 0011-1619 – mf#1639 – us UMI ProQuest [400]
La critique – Paris. avr 1895-1913. mq 1907 – 1 – fr ACRPP [073]
La critique des traditions religieuses chez les grecs : des origines au temps de plutarque / Decharme, Paul – Paris: A Picard, 1904 – 5mf – 9 – 0-524-02014-0 – (incl bibl ref) – mf#1990-2789 – us ATLA [250]
Critique historique du n t (etb) / Lagrange, Marie Joseph – Paris, 1937 – 5mf – 8 – €12.00 – ne Slangenburg [225]
La critique internationale – n1-13. Paris. 15 mai 1902-sept 1903 – 1 – (devenu: anthologie. n.s., n1-12. oct 1903-sept 1904) – fr ACRPP [073]
A critique of design-arguments : a historical review and free examination of the methods of reasoning in natural theology / Hicks, Lewis Ezra – New York: Charles Scribner, 1883 [mf ed 1985] – 1mf – 9 – 0-8370-3583-X – (incl ind) – mf#1985-1583 – us ATLA [210]
A critique of kant / Fischer, Kuno – London: Swan Sonnenschein, Lowrey & Co, 1888 [mf ed 1987] – 188p – 1 – (republ fr journal of speculative philosophy. trans fr german by w s hough) – mf#1909 – us Wisconsin U Libr [190]
A critique of some philosophical aspects of the mysticism of jacob boehme / Alleman, George Mervin – Philadelphia: [s.n.], 1932 [mf ed 1989] – 128p – 1 – (incl bibl) – mf#7044 – us Wisconsin U Libr [140]
A critique of the industrialists' plan / Agarwala, Amar Narain – Benares: Nand Kishore & Bros, 1944 – (= ser Samp: indian books) – us CRL [338]
Critique of three schumann piano works / King, Helen Jane – U of Rochester 1941 [mf ed 19–] – 2mf – 9 – mf#fiche 92 – us Sibley [780]
La critique sociale : revue des idees et des livres – n1-11. Paris. mars 1931-mars 1934 – 1 – (sociologie, economie politique, histoire, philosophie, droit public, demographie puis pedagogie. Mouvement ouvrier. Lettres et arts. Paraissant six fois par an) – fr ACRPP [073]
Critique textuelle du n t (etb) / Lagrange, Marie Joseph – Paris, 1937 – 12mf – 8 – €23.00 – ne Slangenburg [221]
Critiques de l'economie politique – Paris. 1973-1980 (1) 1976-1980 (5) 1976-1980 (9) – ISSN: 0045-9097 – mf#8345 – us UMI ProQuest [330]
Critische nachrichten / ed by Daehnert, Johann Carl – Greifswald 1750-54 [mf ed 1998] – (= ser Greifswalder zeitschriften 1743-1807) – 5v on 15mf – 9 – €150.00 – 3-89131-247-4 – (cont: pommersche nachrichten von gelehrten sachen) – gw Fischer [074]
Critischer musikus / Scheibe, Johann Adolph – neue verm verb auflage, Leipzig: B C Breitkopf 1745 [mf ed 19–] – 17mf – 9 – mf#fiche 935 – us Sibley [780]
Crito : eine monatsschrift / ed by Bodmer, Joh Jakob – Zuerich 1751 – (= ser Dz. abt literatur) – 1v on 2mf – 9 – €60.00 – mf#k/n4413 – gw Olms [410]
Critolaus *see* Growls from uganda
The crittenden commercial arithmetic and business manual... / Groesbeck, John – Philadelphia: Eldredge & Bro, 1869 – 4mf – 9 – $6.00 – mf#LLMC 96-064 – us LLMC [650]
Crittenden county times – West Memphis, AR. 1931-1960 (1) – mf#62071 – us UMI ProQuest [071]
Crittenden, John Jordan *see* Papers
Crittenton, Charles Nelson *see* The brother of girls
Crivellati, Cesare *see* Discorsi mvsicali, nelli quali si contengono non solo cose pertinenti alla teorica, ma etiandio alla pratica
Crivelus, Antonius [aut or III?] *see* Livius, books 9 and 21-22

Crivitz advocate – Crivitz Wl. 1917 mar 9-1918 mar 1, 1918 mar 8-aug 30 – 2r – 1 – mf#964292 – us WHS [071]
Crkva u svijetu – Split: Splitska nadbiskupija i biskupije, dubrovacka, hvarska, kotorska i sibenska (4 iss/yr) [mf ed 1986] – 1r – 1 – mf#2100 – us Wisconsin U Libr [241]
Crkva u svijetu – v9-16. 1974-81; v20-25. 1985-90 – 4r – 1 – mf#ATLA S0846 – us ATLA [073]
Crn – Jericho, 2000+ [1,5,9] – (cont: computer reseller news) – mf#19186,02 – us UMI ProQuest [000]
Les croates sous le joug magyar : address given on april 27 1915 / Hinkovic, Henrik – Paris: Plon-Nourrit 1915 [mf ed 1984] – (= ser Conferences du foyer. conferences d'education nationale) – 1r – 1 – (with: la macedoine / a beli'c [paris: bloud & gay 1919] & other titles) – mf#1010 – us Wisconsin U Libr [943]
Croatia – Geneva, Switzerland: Chief Committee of the Croatian Peasants' Party in Chicago, USA – us CRL [949]
Croatia. (Federated Republic, 1945-) Laws, Statutes, etc *see* Zakon o sumama i zakon o lovu
Croatian Genealogical Society et al *see*
– Bulletin of the croatian...
– Quarterly bulletin
Croatica chemica acta – Zagreb. 1971-1996 (1) 1971-1996 (5) 1974-1996 (9) – ISSN: 0011-1643 – mf#6559 – us UMI ProQuest [540]
Croce, Benedetto *see*
– Aesthetic as science of expression and general linguistic
– Goethe
– Indagini su hegel, e schiarimenti filosofici
– The philosophy of giambattista vico
– Philosophy of the practical
– What is living and what is dead of the philosophy of hegel
Croce, Giovanni *see*
– Canzonette a quattro voci
– Motetti a otto voci
– Il primo libro de madrigali a cinque voci
– Il quatro libro de madrigali a cinque et sei voci
Crochets du pere martin / Cormon, Eugene – Paris, France. 1858 – 1r – us UF Libraries [440]
Crociata missionaria – 1930-56 [complete] – 4r – 1 – mf#ATLA S0546 – us ATLA [240]
Crocius, L *see* Paraeneticus de theologia cryptica...
Crocker, Hannah Mather *see* Observations on the real rights of women
Crocker, Henry *see* History of the baptists in vermont
Crocker, Uriel Haskell *see*
– The history of a title
– Notes on common forms: a book of massachusetts law
– Notes on the public statutes of massachusetts
Crocker, William Andrew *see* Studies in the prophecy of daniel
Crocker, Zebulon *see* The catastrophe of the presbyterian church, in 1837
Crockery and glass journal – New York: G Whittemore & Co. v2 n27-v46 n6 jul 8 1875-dec 30 1897 (7r); v122-147 (1938-50); v151-152 (jul 1952-jun 1953); v155 n1-5 (jul-nov 1954) – 31r – 1 – us CRL [660]
Crockery journal – New York: G Whittemore & Co. v1 n2-26. dec 12 1874-jul 1 1875 – 15r – us CRL [640]
Crockett creek baptist church. model, tennessee : church records – 1863-Aug 1964 – 1 – us Southern Baptist [242]
Crockett, George B *see* Consolidated history of the churches of the oxford baptist association, state of maine
Crockett, William Day *see*
– A harmony of samuel, kings and chronicles
– Satchel guide to spain and portugal
Crockford's clerical directory – 1888-1982 – 85r – 1 – £3750.00 – mf#CRO – uk World [240]
Crocodile – Gainesville, FL. 1966 may 4-jun 29 – 1r – us UF Libraries [071]
Crocombe, Ron *see* Land tenure coursebook
Croenert, Wilhelm *see* Memoria graeca herculanensis
Croft, Aloysius *see* Continuity of the english church
Croft, Terrell *see* Electrical machinery
Croft, W R *see* The history of the factory movement
Croft, William *see* Dr croft's service in b, and evening service in e
Croft-Cooke, Rupert *see* Songs of a sussex tramp
Crofton, Denis *see* Genesis and geology
Crofton, Francis Blake *see*
– The bewildered querists and other nonsense
– Haliburton, the man and the writer
– The major's big-talk stories
– Sombre tints
Crofton, H T *see* History of the ancient chapel of stretford
Crofton journal – Crofton, NE: Peterson & Alwine. v1 n1. jun 7 1906- (wkly) [mf ed with gaps filmed 1958-] – 1 – (absorbed: center register oct 18 1906, crofton progress dec 7 1911 and: crofton citizen oct 19 1933) – us NE Hist [071]
Crofton, Walter Cavendish *see* Brief sketch of the life of charles, baron metcalfe, of fernhill, in berkshire...
Crogman, William Henry *see* An address on the occasion of the laying of the corner-stone
Crohn, Harris Nathaniel *see* Study of the opera katja kabanova by leos janacek
Crohns, Hjalmar *see* Die summa theologica des antonin von florenz und die schaetzung des weibes im hexenhammer
Croil, James *see*
– Dundas
– Genesis of churches in the united states of america, in newfoundland and the dominion of canada
– Gleanings from the nineteenth century
– A historical and statistical report of the presbyterian church in canada
– Life of the rev alex mathieson...minister of st. andrew's church, montreal
– The noble army of martyrs
– Practical agriculture
– A souvenir
La croisade canadienne : cantate: dedie aux zouaves canadiens / Bellemare, Alphonse & Labelle, Jean-Baptiste – [Montreal?]: [s.n.], [1870?] [mf ed 1984] – 1mf – 9 – 0-665-36733-3 – mf#36733 – cn Canadiana [780]
Croisee des chemins / Dalencour, Franciois Stanislas Ranier – Port-Au-Prince, Haiti. 1923 – 1r – us UF Libraries [972]
Une croisiere autour de la mer morte / Abel, Felix-Marie – Paris: J Gabalda 1911 [mf ed 1992] – (= ser Etudes palestiniennes et orientales) – 1mf – 9 – 0-524-04085-0 – (incl bibl ref) – mf#1992-0043 – us ATLA [915]
Croissans-crescens i sredneviekovyia legendy o polovoi metamorfozie / Veselovskii, Aleksandr Nikolaevich – Sanktpeterburg: Tip Imperatorskoi akademii nauk, 1881 [mf ed 2002] – (= ser Prilozhenie k 39 tomu zapisok imp akademii nauk 4) – 1r – 1 – (filmed with: k biografii adama mitiskevicha v 1821-1829 godakh / fedor verzhbovskii [teodor wierzbowski], (1898). incl bibl ref) – mf#5239 – us Wisconsin U Libr [390]
Croissant-Rust, Anna *see* Felsenbrunner hof
Croix – Paris, France. jan 1944-20 jun 1945 – 1r – 1 – uk British Libr Newspaper [072]
La croix – 1988- – 3r per y – 5 – us UMI ProQuest [070]
La croix – Paris. 16 juin 1883-1993 – 1 – fr ACRPP [073]
La croix – Paris: La Croix, 1953-61 – us CRL [074]
La croix – Quebec: J U Begin, [1897-189- ou 19–] – 9 – mf#P04116 – cn Canadiana [241]
Croix, C la *see* Theologia moralis...
Croix contre l'asson / Peters, Carl Edward – Port-Au-Prince, Haiti. 1960 – 1r – us UF Libraries [972]
Croix, de la *see* Etat present des nations et eglises grecque, armenienne, et maronite en turquie
La croix, l'epee et la charrue : ou les trois symboles du peuple canadien / Thibault, Charles – Montreal: Cadieux & Derome, 1884? – 2mf – 9 – mf#43619 – cn Canadiana [390]
Croix, Louis Antoine Nicolle de la *see* Geographie moderne
La croix presentee aux membres de la societe de temperance / Mailloux, Alexis – [Quebec?: s.n.] 1850 [mf ed 1983] – 3mf – 9 – 0-665-37789-4 – mf#37789 – cn Canadiana [230]
Croizette, Armand *see* Masque tombe
Crokaert, Jacques *see* Mediterranee americaine
Croker, John W *see* A sketch of the state of ireland
Croker's boswell and boswell. studies in the "life of johnson" / Fitzgerald, Percy Hetherington – London: Chapman & Hall 1880 [mf ed 1987] – 1r – 1 – (filmed with: a greek and english lexicon to the new testament / parkhurst j & other titles) – mf#2089 – us Wisconsin U Libr [920]
Croly, David O *see* Index to the tracts for the times
Croly, George *see*
– Divine origin, appointment, and obligation, of marriage
– England
– England, turkey, and russia
– Reformation a direct gift of divine providence
– Sermon on the death of the duke of wellington
– The theory of baptism
Croly, Jane Cunningham *see* The history of the women's club movement in america
Crombie, Frederick *see* Manual of biblical archaeology
Crombie, John *see* Confession of the patriarchs, that they were strangers and pilgrims
Crome, Carl *see*
– Erbrecht
– Handbuch des franzoesischen civilrechts
– Immaterialgueterrechte
– Recht an sachen und an rechten
– Recht der schuldverhaeltnisse
Cromhout, Emile Henri Antoine *see* Skeireins aivaggeljons pairh johannen
Cromitos cubanos / Cruz, Carlos Manuel De La – Habana, Cuba. 1892 – 1r – us UF Libraries [972]
The cromlech on howth : a poem / Ferguson, Samuel – London [1864?] – (= ser 19th c art & architecture) – 2mf – 9 – mf#4.2.1005 – uk Chadwyck [810]
Crommelynck, Fernand *see*
– Amants puerils
– Cocu magnifique
Cromos – Bogota, Colombia. -w. 1916-82; 6 nov-18 dec 1984; 1985 (1930-32; 1975-76 imperfect) – 146 1/2r – 1 – uk British Libr Newspaper [079]
Crompton and royton chronicle – [NW England] Crompton, Oldham Lib jan 1936-jun 1958 – 1 – uk MLA; uk Newsplan [072]
Crompton, Arnold *see* Unitarianism in the middlewest
Crompton, Richard *see* Star chamber cases; showing what cases properly belong to the cognizance of that court
Crompton, Sarah *see*
– Life of robinson crusoe in short words
Crompton, Thomas *see* The herald of zion
Cromwell : eine trilogie / Raupach, Ernst Benjamin Salomo – Hamburg: Hoffmann & Campe 1841-44 [mf ed 1995] – 1r – 1 – (filmed with: herman stark / oscar von redwitz) – mf#3708p – us Wisconsin U Libr [820]
Cromwell argus – 1876-78; 1885-93; 1895-jul 1896; 1897-99; 1902-03; 1905-23; 1925-1939 – 45r – 1 – mf#83.3 – nz Nat Libr [079]
Cromwell, John Wesley *see* The negro in american history
Cromwell, Oliver *see*
– Cromwell's second speech
– Oliver cromwell's letters and speeches
Cromwell, Thomas *see* Excursions in the county of suffolk
Cromwellian settlement of ireland / Prendergast, John Patrick – Dublin, Ireland. 1875 – 1r – us UF Libraries [941]
Cromwell's army : a history of the english soldier during the civil wars, the commonwealth and the protectorate / Firth, Charles Harding – London: Methuen, 1902 – (= ser Ford Lectures) – 2mf – 9 – 0-524-03399-4 – (incl bibl ref) – mf#1990-0953 – us ATLA [941]
Cromwell's place in history : founded on six lectures / Gardiner, Samuel Rawson – London; New York: Longmans, Green, 1897 – 1mf – 9 – 0-7905-4644-2 – mf#1988-0644 – us ATLA [941]
Cromwell's second speech : speech at the opening of the first protectorate parliament, september 4, 1654 = Second speech / Cromwell, Oliver – [Boston: Directors of the Old South Work, 189-?] – 1mf – 9 – 0-524-04131-8 – mf#1990-1201 – us ATLA [323]
Cromwell's soldier's bible : being a reprint, in facsimile, of "the soldier's pocket bible",... – Boston: Roberts, 1895 – 1mf – 9 – 0-8370-9203-5 – mf#1986-3203 – us ATLA [220]
Cronaca – London, UK. 5 jun 1920-24 jun 1922 – 1 – (eco d'italia 1 jul 1922-7 jul 1928) – uk British Libr Newspaper [072]
Cronaca bizantina – Jun 1883-Dec 1884 – 1 – us CRL [945]
Cronaca sovversiva – Lynn MA, 1917 – 1r – 1 – (italian newspaper) – us IHRC [071]
Cronaca sovversiva – Lynn MA, Barre VT, jun 6 1903-19, nov 1933 – 5r – 1 – (italian newspaper) – us IHRC [071]
Cronache etiopiche / Zoli, C – Roma, 1930 – 5mf – 9 – mf#NE-20229 – ne IDC [956]
Cronache illustrate della azione italiana in a.o – Roma: Tumminelli. fasc1-12. apr 25-oct 10 1936 – 1 – us CRL [960]
Le cronache italiano nel medio evodescritte *see* Italy
Crone, Curt *see* Sonnette
Crone, G R *see* The voyages of cadamosto and other documents on western africa in the second half of the fifteenth century
Cronenwett, Emanuel *see* The calvinistic conception in lutheran theology
Cronica – Jassy, Romania. -w. 1967-70. 4 reels – 1 – uk British Libr Newspaper [949]
Cronica – Los Angeles CA. [v14 n52-v15 n8,10-19 [1885 dec 26-1886 jan 2-may 8]] scattered iss – 1r – 1 – mf#919837 – us WHS [071]
La cronica – Badajoz, 1890. 1 numero – 5 – sp Bibl Santa Ana [073]
La cronica – Badajoz, 1871-1881 – 5 – (es segunda epoca de la 48) – sp Bibl Santa Ana [073]
Cronica cubana – Miami, FL. 1967 jun 15-1968 oct 10 – 2r – us UF Libraries [071]
Cronica de badajoz – Badajoz, 1866-1870 – 5 – sp Bibl Santa Ana [073]
Cronica de cavallers catalans / Tarafa, Francisco – [Barcelona]: Asociacion de Bibliofilos de Barcelona 1952-54 [mf ed 1987] – 2v on 1r [ill] – 1 – mf#2135 – us Wisconsin U Libr [929]
Cronica de isidoro pacense / Martinez de Escobar, T – 1870 – 9 – sp Bibl Santa Ana [920]
Cronica de la guerra de cuba (1895) / Guerrero, Rafael – Barcelona, Spain. v1-4. 1895-97 – 2r – us UF Libraries [972]
Cronica de la orden de alcantara / Torres Tapia, Alfonso – 1763 – 9 – sp Bibl Santa Ana [946]
Cronica de la orden de n p s augustin en las prouincias de la nueua espana : en quatro edades desde el ano de 1533 hasta el de 1592 / Grijalva, Juan de – Mexico:...S Augustin, y impr Ioan Ruyz, ano de 1624 – (= ser Books on religion...1543/44-c1800: ordenes, etc: agostinos) – 5mf – 9 – mf#crl-168 – ne IDC [241]
Cronica de la primera asamblea misional diocesana de badajoz celebrada en zafra... / Fuertes, Damaso – Zafra: Imp. de Eulalio Morera Perez, 1924 – 1 – sp Bibl Santa Ana [240]
Cronica de la provicia de san gregorio magno : de religiosos descalzos de n s p san francisco en las islas filipinas, china, japon, etc / Santa Ines, Francisco de – Manila: Tipo-litografia de Chofre, 1892 [mf ed 1995] – (= ser Yale coll; Biblioteca historica filipina 2-3) – 2v – 1 – 0-524-09667-8 – (in spanish) – mf#1995-0667 – us ATLA [950]
Cronica de la provincia de albacete / Blanch e Illa, Narciso – 1867 – 9 – sp Bibl Santa Ana [946]
Cronica de la provincia de caceres / Perez de Guzman, Juan – 1870 – 9 – sp Bibl Santa Ana [946]
Cronica de la provincia de murcia / Bisso, Jose – 1870 – 9 – sp Bibl Santa Ana [946]
Cronica de la provincia del santisimo / Vazquez, Francisco – Guatemala, v1-4. 1937-1944 – 1r – us UF Libraries [972]
Cronica de la provincia franciscana de los apostoles san pedro y san pablo de michoacan. 2nd ed. mexico, 1946 / Espinosa, Isidro Felix – Madrid: Razon y Fe, 1947 – 1 – sp Bibl Santa Ana [240]
Cronica de la...caceres / Perez de Guzman, Juan – 1870 – 9 – sp Bibl Santa Ana [946]
Una cronica de los moctezuma, madrid...1954 / Cerralbo, Marques de – Hidalguia, 2, Abril-Junio, 1954 – 1 – sp Bibl Santa Ana [946]
Cronica de nuevo mexico / Historical Society of New Mexico – 1976 jun-1980 mar – 1r – 1 – mf#492872 – us WHS [978]
La cronica de plasencia – Plasencia, 1899 – 5 – sp Bibl Santa Ana [073]
La cronica de plasencia – Plasencia. 1899. Solo no. 11 – 9 – sp Bibl Santa Ana [079]
Cronica de san pedro de alcantara – Madrid: Archivo Ibero Americana, 1960 – 1 – sp Bibl Santa Ana [240]
Cronica de uma embaixada luso-brasileira / Pires, Vicente Ferreira – Sao Paulo, Brazil. 1957 – 1r – us UF Libraries [972]
Cronica del alba / Sender, Ramon Jose – Madrid, Spain. 1942 – 1r – us UF Libraries [940]
Cronica del congreso eucaristico nacional / ed by Bayle, Constantino – Madrid: Razon y Fe, 1926 – 1 – sp Bibl Santa Ana [240]
Cronica del muy magnifico capitan d gonzalo suare / Garcia Samudio, Nicolas – Bogota, Colombia. 1952 – 1r – us UF Libraries [972]
La cronica del peru / Cieza Leon, Pedro de – Ediciones de la Revista Ximenez de Quesada, 1971 – 1 – sp Bibl Santa Ana [972]
La cronica di giovanni villani: annotata ad uso della gioventu / Villani, Giovanni – By Prof. Celestino Durando. Torino, 1880. 8v in 4 – 1 – us Wisconsin U Libr [945]
Cronica di pisa, 1276-1389... – 15th, 18th c – (= ser Holkham library manuscript books 556,575,600) – 1r – 1 – (filmed with: bishop of rimini: memorie della citta di rimini. petruccio ubaldini: relazione a cristoforo hatton) – mf#2195 – uk Microform Academic [945]
Cronica el emperador carlos 5th. edicion de juan sanchez montes. prologo de peter rasow. madrid, 1964 / Mesaguer Fernandez, J & Giron, Pedro – Madrid: Graf. Calleja, 1967 – 1 – sp Bibl Santa Ana [920]
Cronica et cartularium monasterii de dunis / ed by Putte, F van de – Brugis, 1864-70 – €90.00 – ne Slangenburg [241]
Cronica fratris salimbene de adam ordinis minorum (mgh5:32.bd) – 1905-1913 – (= ser Monumenta germaniae historica 5: scriptores in folio (mgh5)) – €38.00 – ne Slangenburg [240]
Cronica johannis vitodurani (mgh6:3.bd) / ed by Brun, B & Baethgen, F – 1924 – (= ser Monumenta germaniae historica 6: scriptores rerum germanicarum, nova series (mgh6)) – €14.00 – ne Slangenburg [240]

Cronica militar / Lima Junior, Augusto De – Belo Horizonte, Brazil. 1960 – 1r – us UF Libraries [355]

La cronica oficial de las indias occidentales. la plata 1934 / Carbra, Romulo D – Madrid: Razon y Fe, 1935 – 1 – sp Bibl Santa Ana [970]

Cronica procesal / Bello Lozano, Humberto – Caracas, Venezuela. 1965 – 1r – us UF Libraries [972]

Cronica proceso diocesano de beatificacion de siete franciscanos... – Madrid: Arch. Ibero Americano, 1964 – 1 – sp Bibl Santa Ana [240]

Cronica y el suceso / Matas, Julio – Habana, Cuba. 1964 – 1r – us UF Libraries [972]

Cronica...badajoz / Henao y Munoz, Manuel – 1870 – 9 – sp Bibl Santa Ana [520]

Cronicas / Pita Rodriguez, Felix – Habana, Cuba. 1961 – 1r – us UF Libraries [972]

Cronicas biograficas / Sobarzo, Horacio – Hermosillo, Mexico. 1949 – 1r – us UF Libraries [920]

Cronicas de antano / Manzano, Lucas – Caracas, Venezuela. 1951 – 1r – us UF Libraries [972]

Cronicas de ayer / Gallardo Diaz, Fernando – Bayamo, Puerto Rico. 1950? – 1r – us UF Libraries [972]

Cronicas de la antigua guatemala / Mencos F, Agustin – Guatemala, 1956 – 1r – us UF Libraries [972]

Cronicas de la constituyente del 45 / Marroquin Rojas, Clemente – s.l, s.l? 1955 – 1r – us UF Libraries [323]

Cronicas de la guerra / Manobens, Enrique – Valencia, 1937. Fiche W1019. (Blodgett Collection of Spanish Civil War Pamphlets) – 9 – us Harvard [946]

Cronicas de la guerra / Zamacois, Eduardo – Recopilacion de articulos periodisticos. Valencia, 1937. Fiche W1260. (Blodgett Collection of Spanish Civil War Pamphlets) – 9 – us Harvard [946]

Cronicas de los reyes de castilla, desde don alfonso de sabio, hasta los catolicos don fernando y dona isabel / ed by Rosell, Cayetano – Madrid: M Rivadeneyra 1875-78 [mf ed 1985] – (= ser Biblioteca de autores espanoles 66,68,70) – 3v on 1r – 1 – mf#6439 – us Wisconsin U Libr [946]

Cronicas de rionegro / Tobon, Ernesto – Medellin, Colombia. 1964 – 1r – us UF Libraries [972]

Cronicas de santiago de cuba / Bacardi Y Moreau, Emilio – Santiago, Cuba. v1-10. 1923-25 – 2r – us UF Libraries [972]

Cronicas de washington / Pagan, Bolivar – San Juan, Puerto Rico. 1949 – 1r – us UF Libraries [975]

Cronicas del bocono de ayer / Baptista, Jose Maria – Caracas, Venezuela. 1962 – 1r – us UF Libraries [972]

Cronicas del centenario / Febles, Horacio A A – Ciudad Trujillo, Dominican Republic. 1944 – 1r – us UF Libraries [972]

Cronicas habaneras / Casal, Julian Del – Santa Clara, Cuba. 1963 – 1r – us UF Libraries [972]

Cronicas indigenas de guatemala / Recinos, Adrian – Guatemala, 1957 – 1r – us UF Libraries [972]

Cronicas ligeras / Nieto Caballero, Agustin – Bogota, Colombia. 1964 – 1r – us UF Libraries [972]

Cronicas misionales del excmo y revmo sr dr dn... / Builes G, Miguel Angel – Medellin, Colombia. 1947 – 1r – us UF Libraries [972]

Cronicas y 105 sentencias leves / Guasp, Ignacio – San Juan, Puerto Rico. 1956 – 1r – us UF Libraries [972]

Cronicas y apuntes / Munoz Meany, Enrique – Guatemala, 1961 – 1r – us UF Libraries [972]

Cronicas y devaneos / Fernandez Cabrera, Manuel – Habana, Cuba. 1913 – 1r – us UF Libraries [972]

Cronica...san gabriel / Trinidad, Juan de la – 1652 – 9 – sp Bibl Santa Ana [946]

Cronicl cymru – Bangor, Wales. -f. Jan 1866-June 1872. 2 1 2 reels – 1 – uk British Libr Newspaper [072]

Cronicon de la causa del muerto resucitado y guia de su vista publica con la biografia de los protagonistas – Plasencia: Imp. Jose Hontiveros, 1888 – 1 – sp Bibl Santa Ana [920]

Cronicon...muerto resucitado / Paredes Guillen, Vicente – 1888 – 9 – sp Bibl Santa Ana [830]

Cronin, H S *see* Codex purpureus petropolitanus (codex n of the gospels) (ts5/4)

Cronin, Harry Stovell *see* Codex purpureus petropolitanus

Croniquillas de mi ciudad / Mora, Luis Maria – Bogota, Colombia. 1936 – 1r – us UF Libraries [972]

Cronise, Titus Fey *see* Natural wealth of california

El cronista – Panama City, Panama. 1904-08 (incomplete) – 2r – 1 – us L of C Photodup [079]

El cronista – Serradilla, 1928-1932 – 5 – sp Bibl Santa Ana [073]

Cronista e historiadores de la conquista de mexico, fondo de cultura economica / Iglesia, Ramon – 1942 – 1 – sp Bibl Santa Ana [350]

Cronje, Suzanne *see* Witness in the dark

Cronologia critica de la guerra hispano-cubanoamer / Martinez Arango, Felipe – Habana, Cuba. 1950 – 1r – us UF Libraries [972]

Cronologia della vita di s. francesco / Terzi, Arduino – Madrid: Archivo Ibero Americano, 1964 – 1 – sp Bibl Santa Ana [946]

Cronologia herediana / Gonzalez Del Valle Y Ramirez, Francisco – Habana, Cuba. 1938 – 1r – us UF Libraries [972]

Cronopios – Madison. 1966-1970 (1) – ISSN: 0011-1821 – mf#9799 – us UMI ProQuest [400]

Cronulla observer – Cronulla, aug 1939-dec 1976 – 26r – A$1001.00 vesicular A$1144.00 silver – at Pascoe [079]

Cronulla sutherland advocate – Cronulla, jan 1927-aug 1939 – 2r – A$77.00 vesicular A$88.00 silver – at Pascoe [079]

Cronyn Memorial Church (London, Ont) *see* The old bell rings in the twenty-seventh anniversary of the memorial church, december, 1900

Crook county journal – Prineville OR: Hugh Gourlay, -1921 [wkly] – 1 – (merged with: call [prineville, or] to form: central oregonian) – us Oregon Lib [071]

Crook county journal *see* Central oregonian

Crook county news – Prineville OR: Steve H Bailey, 1937-39 [wkly] – 1 – (absorbed by: central oregonian) – us Oregon Lib [071]

Crook county news *see* Central oregonian

Crook, William *see* Ireland and the centenary of american methodism

Crookall, L *see* British guiana

Crookall, Robert *see* Ecstasy

Crooke, William *see*
- Annals and antiquities of rajasthan
- Natives of northern india
- Observations on the mussulmauns of india
- The people of india
- The popular religion and folk-lore of northern india

Crooked house / Christie, Agatha – New York, NY. 1949 – 1r – us UF Libraries [830]

Crooked run baptist church (called cedar creek, 1826-1836). fairfield county. south carolina : church records – 1834-1877, 1883-1907, 1944-72 – 1 – us Southern Baptist [242]

Crooker, Joseph Henry *see*
- Different new testament views of jesus
- Problems in american society
- Religious freedom in american education
- The unitarian church

Crooks, Adam *see* Reform government in ontario

Crooks, George Richard *see*
- Life and letters of the rev. john m'clintock
- Sermons
- The story of the christian church
- Theological encyclopedia and methodology

Crooks, John Joseph *see* History of the colony of sierra leone, western africa

Crookston herald – Crookston, NE: J A Manhalter. 7v. 1913-v7 n42. may 21 1920 (wkly) [mf ed with gaps filmed [1972]] – 3r – 1 – (absorbed: searchlight. merged with: cherry county messenger to form: herald-messenger) – us NE Hist [071]

Crookwell gazette – Crookwell – 11r – A$819.41 vesicular A$879.91 silver – at Pascoe [079]

Crookwell gazette – Crookwell – 9r – A$590.13 vesicular A$639.63 silver – at Pascoe [079]

Crookwell gazette – Crookwell, jan 1969-dec 1979; jan-dec 1992 – 21r – at Pascoe [079]

'Crooter – 1983 feb, aug, 1984 jan-mar, jun-aug, 1985 nov-dec, 1986 mar-aug, oct-dec, 1987 jan, mar, aug-oct, 1988 jan, apr, jun-jul,sep-1991 feb – 1r – 1 – mf#1054981 – us WHS [071]

Crop and weather report / Wisconsin Agriculture Reporting Service – v1 n1-23 [1977 dec 5-1978 jul 10] – 1r – 1 – (cont: wisconsin snow and frost depth report; weekly crop and weather report; cont by: crop, weather report) – mf#348612 – us WHS [630]

Crop life – Willoughby. 2001+ (1) – (cont: farm chemicals) – ISSN: 1535-3923 – mf#1414,02 – us UMI ProQuest [630]

Crop news / Christian Rural Overseas Program – 1977 aug-1982 feb – 1r – 1 – (cont by: winir crop news) – mf#654836 – us WHS [338]

Crop protection – Kidlington. 1982-1996 (1,5,9) – ISSN: 0261-2194 – mf#17220 – us UMI ProQuest [630]

Crop, weather report / Wisconsin Agriculture Reporting Service – v1 n24-v16 n22 [1978 jul 17-1993 jul 12] – 1r – 1 – (cont: crop and weather report; cont by: wisconsin crop weather) – mf#577278 – us WHS [630]

Cropley's celestial spectator and york advertiser – Fredericton, NB: H A Cropley, [1880?] – 9 – mf#P04535 – cn Canadiana [073]

Croplife – Willoughby. 1955-1970 (1) – ISSN: 0574-4814 – mf#1136 – us UMI ProQuest [630]

Croquer Cabezas, Emilio *see* Centenario de la independencia espanola. noticia genealogica y biografica y biografia del mariscal campo...

Croquis de brousse / Davesne, Andre – Paris: Editions du Sagittaire, [1946] – 1 – us CRL [960]

Croquis de chine / Serviere, Joseph de la – Paris: Gabriel Beauchesne, 1912 [mf ed 1995] – (= ser Yale coll) – 200p (ill) – 1 – 0-524-10008-X – (in french) – mf#1995-1008 – us ATLA [951]

Croquis haitiens : etudes de moeurs locales / Laforest, Antoine – Port-Au-Prince, Haiti. 1906 – 1r – us UF Libraries [972]

Croquis montrealais / Morin, Victor – [Montreal]: Pacifique canadien, 1929 [mf ed 1987] – 1mf – 9 – (ill by charles w simpson) – mf#SEM105P784 – cn Bibl Nat [917]

Cros, Leonard-Joseph-Marie *see* Saint francois de xavier de la compagnie de jesus

Cros, Louis *see*
- L'afrique francaise pour tous
- L'afrique francaise pour tous comment aller & que faire en afrique francaise?

Crosby, Alpheus *see*
- A grammar of the greek language
- The second advent

Crosby, C Russell, Jr *see* Denkmaeler deutscher tonkunst

Crosby, Fanny *see* Memories of eighty years

Crosby, Frank *see*
- Everybody's lawyer and counselor in business...
- Life of abraham lincoln

Crosby, Howard *see*
- A bible manual
- The bible view of the jewish church
- The book of nehemiah
- Expository notes on the book of joshua
- The seven churches of asia
- Social hints for young christians
- Thoughts on the decalogue

Crosby, J *see* Morphology of the substantive in ncheu

Crosby, Percy Leo *see* Dear sooky

Crosby, Thomas *see* Up and down the north pacific coast by canoe and mission ship

Croskery, Thomas *see*
- Catechism on the doctrines of the plymouth brethren

Croslegh, Charles *see* Christianity judged by its fruits

Cross – 1834 – 1 – 5.00 – us Southern Baptist [242]

Cross – Dublin. 1973-1974 (1) – ISSN: 0011-1899 – mf#8577 – us UMI ProQuest [240]

The cross : being a course of sermons preached in holy trinity church, halifax, on the sunday evenings in lent, 1879 / Ancient, William Johnson – Halifax: W Macnab, 1879 – 1mf – 9 – mf#06168 – cn Canadiana [240]

The cross : a discourse / Fuller, Richard – Philadelphia: American Baptist Publication and SS Society, 1841 – 1mf – 9 – 0-7905-9206-1 – mf#1989-2431 – us ATLA [240]

The cross – Halifax [NS]: J P Walsh, [1843-1850] – 9 – mf#P04966 – cn Canadiana [241]

The cross / Howell, Robert Boyte Crawford – Charleston, S C: Southern Baptist Publication Society, 1854 – 1mf – 9 – 0-7905-9227-4 – mf#1989-2452 – us ATLA [240]

The cross : the report of a misgiving / Ross, George Alexander Johnston – New York: FH Revell, c1912 – 1mf – 9 – 0-7905-9858-2 – mf#1989-1583 – us ATLA [240]

Cross and chrysanthemum : an episode of japanese history = Kreuz und chrysanthemum / Spillmann, Jos – London: R & T Washbourne; St Louis, Mo: B Herder, 1906 – 1mf – 9 – 0-8370-7192-5 – (in english) – mf#1986-1192 – us ATLA [950]

Cross and crown – Chicago. 1949-1977 (1) 1971-1977 (5) 1975-1977 (9) – (cont by: spirituality today) – ISSN: 0011-1910 – mf#2189 – us UMI ProQuest [240]

Cross and crown : stories of the chinese martyrs / Bryson, Mary Isabella – London: London Missionary Society [1904] [mf ed 1995] – (= ser Yale coll) – 207p (ill) – 1 – 0-524-09311-3 – mf#1995-0311 – us ATLA [240]

Cross and journal / Franklin Co. Columbus – (nov 1840-jul 42), jan 44-mar 1847 [wkly] – 1r – 1 – mf#B29856 – us Ohio Hist [240]

The cross and the calumet – 1962-70 – (= ser American indian periodicals... 1) – 2mf – 9 – $95.00 – us UPA [305]

The cross and the dragon : or, light in the broad east / Henry, Benjamin Couch – London: S W Partridge, [1885] [mf ed 1995] – (= ser Yale coll) – 507p (ill) – 1 – 0-524-09527-2 – (int note by joseph cook) – mf#1995-0527 – us ATLA [951]

The cross and the dragon : or, light in the broad east / Henry, Benjamin Couch – New York: Anson D F Randolph, c1885 – 2mf – 9 – 0-8370-6263-2 – mf#1986-0263 – us ATLA [951]

The cross and the dragon : or, the fortunes of christianity in china... / Kesson, J – 4mf – 9 – mf#HTM-95 – ne IDC [915]

Cross and the flag – Los Angeles. 1967-1977 (1) 1970-1977 (5) 1976-1977 (9) – ISSN: 0011-1929 – mf#2245 – us UMI ProQuest [320]

The cross and the kingdom : as viewed by christ himself and in the light of evolution / Walker, William Lowe – Edinburgh: T & T Clark, 1902 – 1mf – 9 – 0-8370-5685-3 – (incl ind) – mf#1985-3685 – us ATLA [240]

Cross, Arthur Lyon *see*
- The anglican episcopate and the american colonies
- A history of england and greater britain

Cross county genealogical publication – v1 n1-v4 n2 [1982 sep-1985 oct] – 1r – 1 – mf#1238899 – us WHS [929]

Cross creek baptist church. cumberland association. stewart county. tennessee : church records – 1851-1969 – 1 – us Southern Baptist [242]

Cross currents – New Rochelle. 1950+ (1) 1970+ (5) 1976+ (9) – ISSN: 0011-1953 – mf#1091 – us UMI ProQuest [100]

Cross, F L *see*
- Studia evangelica, vol 2
- Studia evangelica, vol 4
- Studia evangelica, vol 5
- Studia patristica
- Studia patristica, vol 3
- Studia patristica, vol 3 pt 2
- Studia patristica, vol 4
- Studia patristica, vol 5
- Studia patristica, vol 6
- Studia patristica, vol 7
- Studia patristica, vol 8
- Studia patristica, vol 9
- Studia patristica, vol 10
- Studia patristica, vol 11

Cross Family *see* Papers

Cross, Frank M *see* Early hebrew orthography

Cross, George *see*
- The theology of schleiermacher
- What is christianity?

The cross in japan : a study in achievement and opportunity / Hagin, Fred Eugene – New York: Fleming H Revell, c1914 – 1mf – 9 – 0-524-06417-2 – (incl bibl ref) – mf#1991-2539 – us ATLA [240]

The cross in ritual, architecture and art / Tyack, George Smith – 2nd rev and greatly enl ed. London: William Andrews, [1900?] – 1mf – 9 – 0-524-01026-9 – mf#1990-0303 – us ATLA [240]

The cross in the land of the trident / Beach, Harlan Page – New York: Fleming H Revell, c1895 – 1mf – 9 – 0-8370-6644-1 – (includes bibliographies and statistical appendixes) – mf#1986-0644 – us ATLA [954]

The cross in the old testament / Robinson, H Wheeler – SCM, 1955 – 9 – $10.00 – us IRC [221]

The cross in the old testament / Wheeler Robinson, H – London, 1955 – €11.00 – ne Slangenburg [221]

The cross in tradition, history, and art / Seymour, William Wood – New York: G P Putnam, 1898, c1897 – 2mf – 9 – 0-7905-2428-7 – (incl ind) – mf#1987-2428 – us ATLA [240]

Cross index to selected city street and enumeration districts, 1910 census / U.S. Bureau of the Census – (= ser Records of the bureau of the census – other census records) – 50mf – 9 – mf#M1283 – us Nat Archives [317]

Cross index to the central files of the adjutant general's office, 1917-1939 / U.S. War Dept. Adjutant General's Office – (= ser Records of the adjutant general's office, 1917-) – 1956r – 1 – mf#T822 – us Nat Archives [355]

Cross, John A *see*
- Introductory hints to english readers of the old testament
- Some notes on the book of psalms

Cross, Joseph *see* Journals of several expeditions made in western australia

Cross, Kimball Allyn *see* A treatise, analytical, critical and historical, on successions

The cross moves east : a study in the significance of gandhi's "satyagraha" / Hoyland, John Somervell – London: George Allen & Unwin Ltd, 1931 – (= ser Samp: indian books) – us CRL [320]

Cross, Nigel [comp] *see* Archives of the royal literary fund

Cross of baron samedi / Dohrman, Richard – Boston, MA. 1958 – 1r – us UF Libraries [972]

The cross of christ : studies in the history of religion and the inner life of the church = Kreuz christi / Zoeckler, Otto – London: Hodder and Stoughton, 1877 – 2mf – 9 – 0-7905-0778-1 – (incl ind. in english) – mf#1987-0778 – us ATLA [240]

The cross of christ / Simpson, Albert B – Brooklyn, NY: Christian Alliance Pub Co, c1910 [mf ed 1992] – (= ser Christian & missionary alliance coll) – 2mf – 9 – 0-524-02268-2 – mf#1990-4274 – us ATLA [220]
The cross of christ in bolo-land / Dean, John Marvin – Chicago: FH Revell, 1902 – 3mf – 9 – 0-524-07861-0 – mf#1991-3406 – us ATLA [240]
Cross of osiris / Jones, Eustace Hinton – London, England. 1878 – 1r – us UF Libraries [240]
Cross, Peter *see* Plyometric treatment and whole-body movement times
Cross report *see* Report of the commissioners appointed to inquire into the working of the elementary education acts, england wales, (cross report), 1886-1888
Cross river natives : being some notes on the primitive pagans of obubura hill district, southern nigeria / Partridge, Charles – London: Hutchinson, 1905 – 1 – (including a description of the circles of upright sculptured stones on the left bank of the aweyong river) – us CRL [305]
Cross roads baptist church. chesterfield county. ruby, south carolina : church records – 1877-1900, 1947-1977 – 1 – us Southern Baptist [242]
Cross roads baptist church. newberry county. south carolina : church records – 1839-59, 1859-1925, 1872-77 – 1 – us Southern Baptist [242]
Cross roads baptist church. pickens county. easley, south carolina : church records – 1838-1977 – 1 – us Southern Baptist [242]
Cross roads baptist church. rotan, texas : church records – 1909-78. 1050p – 1 – 47.25 – us Southern Baptist [242]
Cross, Robert Craigie *see* Plato's republic
Cross saber newsletter : a journal of the... / U[/nized/] S[/tates/] Horse Cavalry Association – v1 n1-2 [1977 sep 1-dec 1], v2 n2-3 [1978 may 15-sep 1], v3 n1-v11 n3 [1979 jan 1-1987 sep 1] – 1r – 1 – (cont by: crossed sabers) – mf#1819193 – us WHS [355]
Cross section / American Red Cross – 1982 jul-1988 sum – 1r – 1 – (cont: capital communique [madison wi]) – mf#1110985 – us WHS [360]
A cross sectional examination of the training habits : and lifestyle characteristics of triathletes / Stewart, Laura K – 2000 – 81p 0n 1mf – 9 – $5.00 – mf#PE 4148 – us Kinesology [790]
The cross, the plus sign in our minus lives – [S.l.]: Diocesan Missionary Committee of the Diocese of New York, 1914 – 1mf – 9 – 0-524-06304-4 – mf#1991-2477 – us ATLA [240]
The cross triumphant / Kingsley, Florence Morse – Toronto: W Briggs; Montreal: C W Coates, 1899? – 4mf – 9 – mf#29362 – cn Canadiana [240]
Cross-bench views of current church questions / Henson, Hensley – London: Edward Arnold, 1902 – 1mf – 9 – 0-7905-5944-7 – mf#1988-1944 – us ATLA [240]
A cross-cultural analysis of achievement motivation in anglo american and japanese marathon runners / Hayashi, CT – 1991 – 2mf – 9 – $8.00 – us Kinesology [150]
A cross-cultural analysis of children's attitudes toward physical activity and patterns of participation / Liu, Z – 1990 – 1mf – 9 – $4.00 – us Kinesology [150]
Cross-cultural research / Human Relations Area Files, Inc – Thousand Oaks. 1993+ (1,5,9) – (cont: behavior science research) – ISSN: 1069-3971 – mf#2084,02 – us UMI ProQuest [300]
Crosscurrents – n1-165 [1975 aug-1990 oct] – 1r – 1 – mf#1699148 – us WHS [071]
La Crosse County Historical Society *see* Historical notes
Crosse, Gordon *see* A dictionary of english church history
Crosse, T F *see* On the giving of the hebrew law
Crossed swords : a canadian-american tale of love and valor / Alloway, Mary Wilson – Toronto: W Briggs, 1912 – 5mf – 9 – 0-665-71037-2 – mf#71037 – cn Canadiana [830]
Cross-education following single-limb eccentric and concentric training on the biodex isokinetic dynamometer / Mahler, Erik B – 1994 – 1mf – $4.00 – us Kinesology [617]
Crossface – v12 n1-v14 n5 [1983 dec-1986 apr] – 1r – 1 – (cont by: wisconsin crossface) – mf#1409836 – us WHS [071]
Crossing boundaries : gender and genre dislocations in selected texts by samuel r delany / Hope, Gerhard Ewoud – Uni of South Africa 2001 [mf ed Johannesburg 2001] – 5mf – 9 – (incl bibl ref) – mf#mfm14769 – sa Unisa [420]
Crosskey, Henry William *see* The method of creation
Crosskill, Herbert *see* Nova scotia
Crosskill, John H *see* A complete narrative of the celebration of the nuptials of her most gracious majesty queen victoria with his royal highness prince albert of saxe coburg and gotha
Crosskill vs the morning herald printing and publishing company – Halifax, NS?: Herald, 1880? – 1mf – 9 – mf#67258 – cn Canadiana [346]
Crosskill, William Hay *see* Prince edward island, garden province of canada
Crossley, Hugh Thomas *see*
– How to become a child of god
– Practical talks on important themes
– Songs of salvation
Crossman, R H S (Richard Howard Stafford) *see* Plato today
Crossroad / Spear and Shield Publications – 1987 may/sep-2000 fall – 1r – 1 – mf#1066114 – us WHS [071]
Crossroad trails / Effingham County Genealogical Society – v1 n1-v4 n4 [1980 sum-1985] – 1r – 1 – mf#1573052 – us WHS [929]
Crossroads – Fort Chaffee AR. v3 n1-12 [1980 may 22-aug 7] – 1r – 1 – mf#512962 – us WHS [071]
Crossroads / Lajes Field [Azores] US – Lajes, Azores. 1st ed [1981 jun-1985 dec 20] – 1r – 1 – mf#1043870 – us WHS [071]
Crossroads / New Jersey Historical Society – v16 n3-v22 n1 [1978 dec-1984 oct] – 1r – 1 – mf#801749 – us WHS [978]
Crossroads – Pittsburgh. 1972-1977 [1]; 1972-1977 [5]; 1977-1977 [9] – ISSN: 0011-2054 – mf#7470 – us UMI ProQuest [327]
Crossroads / Verissimo, Erico – New York, NY. 1943 – 1r – us UF Libraries [972]
Crossroads chronicle – Vandalia, OH. 1969-1969 (1) – mf#65702 – us UMI ProQuest [071]
Crossroads of the buccaneers / De Leeuw, Hendrik – Philadelphia, PA. 1937 – 1r – us UF Libraries [972]
Crossroads of the caribbean sea / De Leeuw, Hendrik – New York, NY. 1935 – 1r – us UF Libraries [972]
Cross-validation of a quarter-mile walk test for college males and females / Greenhalgh, Heidi A – 2000 – 1mf – 9 – $4.00 – mf#PE 4079 – us Kinesology [612]
Crosta, Clino *see* L'assunta nell'odierna teologia cattolica
Crosthwaite, Charles *see* Sketches by different hands of the revision debates in the irish synod, 1873
Crosthwaite, John Clarke *see*
– First rejection of christ
– Modern hagiology
Croston, Amanda L *see* Team cohesion and gender-role orientation
Croston, J *see* The mirrour of maiestie
Croswell, Simon Greenleaf *see*
– Croswell's collection of patent cases in the u.s. courts
– A treatise on the law relating to electricity
Croswell's collection of patent cases in the u.s. courts / Croswell, Simon Greenleaf – Boston: Little-Brown. 1v. 1888 (all publ) – 3mf – 9 – $4.50 – mf#LLMC 84-326 – us LLMC [346]
Crotalaria for forage – Gainesville, FL. 1941 – 1r – us UF Libraries [630]
Crotch, William *see* Substance of several courses of lectures on music
Crothers, Samuel McChord *see*
– Among friends
– The gentle reader
– The making of religion
– Oliver wendell holmes
– The understanding heart
Crothers, Thomas Davison et al *see* The centenary of the methodist new connexion, 1797-1897
Crottet, Alexandre Ceasar *see* Huguenot records, 1578-1787
Crotus Rubeanus *see* Epistolae obscurorum virorum
Crouch, F Nicholls *see* Sheila! my darling colleen
Crouch, Joseph *see*
– The apartments of the house
– Puritanism and art
Crouch, Marjorie *see* Collected field reports on the phonology of vagala
Crouch, Nathaniel *see* The history of the house of orange
Crouse, Nellis Maynard *see* French struggle for the west indies
Crousle, Leon *see* Bossuet et la protestantisme
Crouter, John Wesley *see*
– Boiled-down essays
– My policy for the construction of the canada pacific railway
– The solution of the great mystery
Crouter, Scott E *see* Physiological comparison of incremental treadmill exercise and free running excercise
Crouzat, Henri *see* Azizah de niamkoko
Crow, Charles L *see* East florida seminary
Crow, John Finley *see* Abolition intelligencer and missionary magazine
Crow valley news *see*
– Miscellaneous newspapers of weld county
Crow wing county genealogy newsletter – 1979 mar-jul, 1980 apr-may – 1r – 1 – (cont by: crow wing county genealogy society newsletter) – mf#2607589 – us WHS [929]
Crow wing county genealogy society newsletter – 1980 jul-1983 feb – 1r – 1 – (cont: crow wing county genealogy newsletter) – mf#2607594 – us WHS [929]
Crowded out! : and other sketches / Harrison, Susie Frances – Ottawa? : Evening Journal Office, 1886 – 2mf – 9 – mf#06401 – cn Canadiana [880]
Crowder, Todd A *see* The effects of various exercise modalities on serum cholesterol and triglyceride concentrations
Crowe, Catherine *see* The night-side of nature; or, ghosts and ghost seers
Crowe, Eyre Evans *see* History of the reigns of louis 18 and charles 10
Crowe, F Hilton *see*
– Cordova street marker
– Edison and ford in fort myers
– Fishing hazards
– Flora in fort myers
– Florida lighthouses
– Huguenot cemetery
– Lee county
– Lee county fauna
– Mr harris' school house
– Old plaza market
– Points of interest in fort myers and vicinity
– St francis barracks
Crowe, Frederick *see* The gospel in central america, published in london in 1850
Crowe, James W *see* Smokeless tobacco use among big ten wrestlers and factors associated with use
Crowe, Joseph Archer *see*
– The early flemish painters
– A history of painting in north italy...
– A new history of painting in italy from the 2nd-16th century
– Raphael
– Titian
Crowe, Sophia Bennett *see* Diary
Crowell, Dean H *see* The effect of fatigue on postural stability and neuropsychological function
Crowell, William *see*
– The church member's hand-book
– The church member's manual of ecclesiastical principles, doctrine, and discipline
Crowest, Frederick J *see* The story of the art of music
Crowfoot, J W *see*
– The buildings at samaria
– Churches at bosra and samaria-sebaste
– Churches at jerash
– Early churches in palestine
– Early ivories from samaria
– Explorations in the tyropoeon valley
– Exvacations in the tyropoeon valley, jerusalem 1927
Crowley, Jeremiah J *see* The parochial school
Crown and realm : a review of the british empire, its builders and rulers: souvenir of the coronation of king george 5 / Burroughs Wellcome and Co – London, Montreal: Burroughs Wellcome, [1911] – 6mf – 9 – 0-665-66882-1 – mf#66882 – cn Canadiana [941]
The crown colonies of great britain : an inquiry into their social condition and methods of administration...with a chapter on "the black and the brown landholder of jamaica" / Salmon, C S – [London] [1887] – (= ser 19th c british colonization) – 2mf – 9 – mf#1.1.7002 – uk Chadwyck [350]
Crown jeremiad *see* [Santa cruz-] miscellaneous titles
Crown jewels : or, gems of literature, art and music: being choice selections from the writings and musical productions of the most celebrated authors from the earliest times / Northrop, Henry Davenport [comp] – London [Ont]: McDermid & Logan, [1888?] [mf ed 1983] – 8mf – 9 – (incl ind) – mf#27784 – cn Canadiana [800]
Crown jewels of the wire – 1985 jul, sep-1987 jul, 1987 aug-1988 dec, 1989-90 – 3r – 1 – (cont: insulators) – mf#1219254 – us WHS [071]
The crown lands of australia : being an exposition of the land regulations, and of the claims and grievances of the crown tenants... / Campbell, William – [Glasgow], 1855 – (= ser 19th c books on british colonization) – 3mf – 9 – mf#1.1.7013 – uk Chadwyck [333]
The crown of hinduism / Farquhar, John Nicol – London: Oxford University Press, 1913 – 2mf – 9 – 0-524-05843-1 – (incl bibl ref) – mf#1990-3507 – us ATLA [280]
The crown of hinduism / Farquhar, John Nicol – London; New York: Oxford University Press, 1913 – (= ser Samp: indian books) – us CRL [280]
Crown paper *see* [Santa cruz-] miscellaneous titles
Crown, R M *see* Utilization of dried grapefruit meal as a feed for growing swine
Crown servants – [mf ed Marlborough 1994] – 3 ser – 1 – (ser 1: papers of thomas wentworth, 1st earl of strafford, 1593-1641 fr sheffield archives [20r] £1850; ser 2: papers of the wynns of gwydir, 1515-1690 fr national library of wales [23r] £2150; ser 3: the lauderdale papers, c1647-82, fr british library, london [10r] £925; with d/g) – uk Matthew [324]
Crown vs. adams and 29 others / Adams, Faried – Documents introduced in evidence in the trial for treason in a special criminal court, Pretoria, 1959-1961 – 1 – us CRL [960]
Crown vs. adams and 29 others : preparatory examination on a charge of high treason, in the magistrate's court of johannesburg, 1956-1958 / Adams, Faried – 1 – us CRL [960]
Crowned in palm-land : a story of african mission life / Nassau, Robert Hamill – Philadelphia: J Lippincott & Co 1874 [mf ed 1986] – 1r – 1 – (with: the indian christians of st thomas / richards, w j) – mf#1738 – us Wisconsin U Libr [240]
The crow's nest / Cotes, Everard, mrs [Sara Jeanette Duncan] – New York: Dodd, Mead, 1901 – 3mf – 9 – 0-665-77154-1 – mf#77154 – cn Canadiana [830]
Crow's nest farm : a true tale / Addison, Julia – London: Saunders, Otley & Co, 1861 – (= ser 19th c women writers) – 4mf – 9 – mf#5.1.37 – uk Chadwyck [420]
Crowther, Bryan *see* Practical remarks on insanity: to which is added a commentary on the dissection of the brains of maniacs; with some account of diseases incident to the insane
Crowther, Jonathan *see* A true and complete portraiture of methodism
Crowther, Joseph Stretch *see* An architectural history of the cathedral church of manchester
Crowther, S *see* The gospel on the banks of the niger
Crowther, Samuel *see*
– The gospel on the banks of the niger
– Romance and rise of the american tropics
Crowther, Samuel Adjai *see* A vocabulary of the yoruba language, etc etc
Crowthorne times. – Crowthorne, England 6 jan 1983-1 mar 1990 [mf 1986-] – 1 – (variant ed of: bracknell times; cont by: crowthorne & s&hurst times [8 mar 1990-]) – uk British Libr Newspaper [072]
Crowton, christ church – (= ser Cheshire monumental inscriptions) – 1mf – 9 – £2.50 – mf#13 – uk CheshireFHS [929]
Crow-Wing *see* Pueblo indian journal, 1920-1921
Croxton 1538-1950 – (= ser Cambridgeshire parish register transcript) – 4mf – 9 – £5.00 – uk CambsFHS [929]
La croyance a la vie future et le culte des morts dans l'antiquite israelite / Lods, Adolphe – Paris: Fischbacher, 1906 [mf ed 1992] – 2v on 2mf – 9 – 0-524-04104-0 – (incl bibl ref) – mf#1992-0062 – us ATLA [270]
La croyance generale et constante de l'eglise touchant l'immaculee conception de la bienheureuse vierge marie / Gousset, cardinal (Thomas Marie Joseph) – Paris: Jacques Lecoffre, 1855 [mf ed 1986] – 2mf – 9 – 0-8370-8260-9 – (in french. incl bibl ref and ind) – mf#1986-2260 – us ATLA [241]
Croyances et legendes de l'antiquite : essais de critique appliquee a quelques points d'histoire et de mythologie / Maury, Louis-Ferdinand-Alfred – 2. ed. Paris: Didier, 1863 – (= ser Librairie Academique (Paris, France: Didier)) – 1mf – 9 – 0-524-01912-6 – (incl bibl ref) – mf#1990-2725 – us ATLA [200]
Croyances et legendes du moyen age / Maury, Louis-Ferdinand-Alfred; ed by Longnon, Auguste & Bonet-Maury, Gaston – nouv ed des Fees du Moyen Age et des Legendes pieuses. Paris: Honore Champion, 1896 – 2mf – 9 – 0-524-04528-3 – (incl bibl ref) – mf#1990-3362 – us ATLA [240]
Croyances et pratiques religieuses des barundi / Zuure, Bernard – Bruxelles, Belgium. 1929 – 1r – us UF Libraries [960]
Croydon advertiser *see* Croydon guardian and surrey county gazette
Croydon advertiser and east surrey reporter – [London & SE] Croydon feb 1869-dec 1871, jan 1874-27 dec 1991; BLNL 1873, 1897, 1910, 1986- – 1 – uk Newsplan [072]
Croydon chronicle – London. -w. 7 jul 1855-26 dec 1857; 1859-17 jul 1869; 1870-25 mar 1871; 6 apr-6 jul 1872; 10 may-27 dec 1873; 1875-1896; 1898-1908; 4 mar 1909-19 oct 1912 47 3/4r – 1 – uk British Libr Newspaper [072]
Croydon cum clopton 1599-1950 – (= ser Cambridgeshire parish register transcript) – 4mf – 9 – £5.00 – uk CambsFHS [929]
Croydon echo norwood crystal palace and penge observer – London UK, 29 sep 1887-10 oct 1889 – 1r – 1 – uk British Libr Newspaper [072]
Croydon express – Croydon UK, 1879-95; 1897-18 mar 1916 – 21 1/4r – 1 – uk British Libr Newspaper [072]

Croydon free press norwood chronicle and surrey advertiser – London UK, 28 jun-13 sep 1865; 26 feb 1866 – 1/2r – 1 – uk British Libr Newspaper [072]
Croydon guardian and surrey county gazette – London UK, 1843, 1 sep 1877-96, 1898-1911, 13 jan 1912-25 mar 1916 – 38r – 1 – (wanting 1897: amalgamated with: croydon advertiser) – uk British Libr Newspaper [072]
Croydon journal – London UK, 1877; 1889; 1892 – 2r – 1 – uk British Libr Newspaper [072]
Croydon midweek – Croydon, England. 2 may 1967-16 may 1972 – n1-253 – 1 – (aka: croydon midweek post; midweek post; croydon post; 1986 on purchased microfilm from microform; missing: 2 may 1967) – uk British Libr Newspaper [072]
Croydon midweek post *see* Croydon midweek
Croydon monthly standard – Croydon UK, sep 1891-mar 1892 – 1/2r – 1 – (aka: croydon monthly standard) – uk British Libr Newspaper [072]
Croydon monthly standard *see* Croydon monthly standard
Croydon news – London UK, missing: 29 dec 1929-oct 1933 – 1 – uk British Libr Newspaper [072]
Croydon news – London UK, missing: jan 1951-jul 1953 – 1 – uk British Libr Newspaper [072]
Croydon north free press – Streatham UK, 7 aug 1986-88, 12 jan-16 mar 1989 – 5 1/4r – 1 – (aka: south london free press, after mar 1989 see: south london news) – uk British Libr Newspaper [072]
Croydon north free press *see* South london free press (croydon north ed)
Croydon observer and local and county advertiser – London UK, 1844; 1863-95; 1898-1904 – 29r – 1 – uk British Libr Newspaper [072]
Croydon penge anerley and norwood telephone – London UK, 10 may-14 jun 1912 – 1/4r – 1 – uk British Libr Newspaper [072]
Croydon post *see* Croydon midweek
Croydon review *see* Croydon weekly review
Croydon review and railway time table – London UK, apr 1880-nov 1895 – 13r – 1 – uk British Libr Newspaper [072]
Croydon standard – Croydon, England. sep-dec 1891 [mf 1891] – 1 – (cont as: croydon monthly standard; discontinued) – uk British Libr Newspaper [072]
Croydon star – London UK, 20 nov 1889-5 nov 1892 – 1r – 1 – uk British Libr Newspaper [072]
Croydon times – London, England. 29 jun 1861-1868, 2 sep-29 sep 1869, 1 jan-19 oct 1870, 1875-1899, 10 mar 1900-1909, 1911-1918, 1920, 1921, 14 jun 1922-19 jun 1959, 14 aug 1959-28 apr 1967 [wkly] – 150r – 1 – (aka: croydon times & surrey county mail) – uk British Libr Newspaper [072]
Croydon times – Croydon UK, 26 mar-17 sep 1992 – 1r – 1 – uk British Libr Newspaper [072]
Croydon times & surrey county mail *see* Croydon times
Croydon weekly review – London UK, 1738 – 1 – (aka: croydon review; missing: 23 aug 1967-2 oct 68) – uk British Libr Newspaper [072]
Croze, Maturin Veyssiere la *see* Lexicon aegyptiaco-latinum
Crozier, Dorothy *see*
- Research papers on the western pacific, particularly tonga and fiji

Crozier, John Beattie *see* History of intellectual development on the lines of modern evolution
Crucero lirico, poesias / Riancho, Providencia – San Juan, Puerto Rico. 1939 – 1r – us UF Libraries [972]
Cruchet, Rene *see* La conquete pacifique du maroc et du tafilalet
The crucial race question, or, where and how shall the color line be drawn / Brown, William Montgomery – 1st ed. Little Rock, Ark.: Arkansas Churchman's Pub. Co., 1907 – 1mf – 9 – 0-7905-6285-5 – mf#1988-2285 – us ATLA [240]
The cruciality of the cross / Forsyth, Peter Taylor – New York: Eaton & Mains; Cincinnati: Jennings & Graham, [19–] [mf ed 1985] – 1mf – 9 – 0-8370-3848-0 – mf#1985-1848 – us ATLA [240]
The crucible : or, tests of a regenerate state. designed to bring to light suppressed hopes, expose false ones, and confirm the true / Goodhue, J A – Boston: Gould and Lincoln, 1860, c1859 – 1mf – 9 – 0-8370-4914-8 – mf#1985-2914 – us ATLA [240]
The crucible, 1962-92 : the quarterly journal of the general synod board for social responsibility – 5r 11mf – 1,9 – mf#97483 – uk Microform Academic [240]
Crucificado / Forastieri De Flores, Marines – San Juan, Puerto Rico. 1963 – 1r – us UF Libraries [972]
Cruden, Alexander *see* Complete concordance to the holy scriptures of the old and new
La crue – Montreal: Jeunesse etudiante catholique inc. 1re annee n1 15 sep 1963-(bimthly) [mf ed 1984] – 1r – 1 – (ceased 1964?) – mf#SEM35P195 – cn Bibl Nat [241]
Crueger, J *see* Psalmodia sacra
Crueger, Johann *see* D m luthers wie auch anderer gottseligen und christlichen leute geistliche lieder und psalmen
Cruel persecutions of the protestants in the kingdom of france = Plaintes des protestans cruellement opprimez dans le royaume de france / Claude, Jean – Boston: [s.n.] 1893 [mf ed 1992] – 1mf – 9 – 0-524-03814-7 – (in english) – mf#1990-1130 – us ATLA [242]
Cruel pirate captain teach once used miami as his... – s.l, s.l? 193? – 1r – us UF Libraries [978]
Cruelty and christianity / Graham, Allen D – London, England. 1873 – 1r – us UF Libraries [240]
Cruelty in convents – London, England. no date – 1r – us UF Libraries [240]
Cruewell, Gottlieb August *see* Schoenwiesen
Cruickshank, B *see* Eighteen years on the gold coast of africa
Cruikshank, B *see* Eighteen years on the gold coast
Cruikshank, Ernest Alexander *see*
- The administration of lieut-governor simcoe viewed in his official correspondence
- The administration of sir james craig
- Battle of fort george
- The battle of fort george
- The battle of lundy's lane, 25th july, 1814
- Blockade of fort george, 1813
- Camp niagara
- Campaigns of 1812-1814
- The documentary history of the campaign on the niagara frontier in 1814 (pt 1)
- The documentary history of the campaign on the niagara frontier in 1814 (pt 2)
- The documentary history of the campaign upon the niagara frontier in the year 1812 (pt 3)
- The documentary history of the campaign upon the niagara frontier in the year 1812 (pt 4)
- The documentary history of the campaign upon the niagara frontier in the year 1813, part 2 (1813), june to august, 1813 (pt 6)
- The documentary history of the campaign upon the niagara frontier in the year 1813, part 3 (1813), august to october, 1813 (pt 7)
- The documentary history of the campaign upon the niagara frontier in the year 1813, part 4 (1813), october to december, 1813 (pt 8)
- The documentary history of the campaign upon the niagara frontier in the year 1813, pt 1 (1813), january to june, 1813 (pt 5)
- The documentary history of the campaigns upon the niagara frontier in 1812-4, vol 9 december, 1813 to may, 1814 (pt 9)
- Documents relating to the invasion of the niagara peninsula by the united states army
- Drummond's winter campaign, 1813
- The employment of indians in the war of 1812
- The fight in the beechwoods
- A historical and descriptive sketch of the county of welland in the province of ontario, in the dominion of canada
- Notes on the history of the district of niagara, 1791-1793
- The origin and official history of the thirteenth battalion of infantry
- Queenston heights
- The siege of fort erie
- The siege of fort erie, august 1st-september 23rd, 1814
- Some letters of robert nichol
- The story of butler's rangers and the settlement of niagara
- Ten years of the colony of niagara, 1780-1790

Cruikshank, Ernest Alexander [comp] *see*
- A century of municipal history
- A century of municipal history, 1792-1841
- A century of municipal history, 1792-1892

A cruise : or, three months on the continent – London 1818 [mf ed Hildesheim 1995-98] – 1v on 1mf – 9 – €40.00 – 3-487-27833-2 – gw Olms [910]
Cruise news / United States Naval Academy – 1947 jun 6-aug 11 – 1r – 1 – mf#4765401 – us WHS [355]
The cruise of the alice may in the gulf of st lawrence and adjacent waters / Benjamin, Samuel Greene Wheeler – New York: D Appleton, 1885, c1884 – 2mf – 9 – (ill by m j burns) – mf#00134 – cn Canadiana [917]
The cruise of the brooklyn : a journal of the principal events of a three years' cruise...in the south atlantic station / Beehler, William Henry – Philadelphia: Lippincott, 1885 [1884] [mf ed 1986] – 341p/pl – 1 – mf#8757 – us Wisconsin U Libr [910]
The cruise of the "cachalot" : round the world after sperm whales / Bullen, Frank Thomas – London: Smith, Elder & Co, 1898 – 5mf – 9 – mf#38461 – cn Canadiana [639]
Cruise of the montauk to bermuda / Mcquade, James – New York, NY. 1885 – 1r – us UF Libraries [919]
Cruise of the 'port kingston' / Caine, William Ralph Hall – London, England. 1908 – 1r – us UF Libraries [910]
The cruise of the "tomahawk" : the story of a summer's holiday in prose and rhyme / Laffan, Bertha Jane (Grundy) – [London], Sydney: Eden, Remington & Co Publ, 1892 – (= ser 19th c women writers) – 2mf – 9 – mf#5.1.9 – uk Chadwyck [830]
Cruise of the u.s. revenue marine steamer bear / Jackson, Sheldon – 1894-1896 – 1 – $50.00 – us Presbyterian [240]
Cruise, Richard *see* Journal of a ten months' residence in new zealand
Cruise travel magazine – Evanston. 1986-1993 (1) 1986-1986 (5) 1986-1986 (9) – ISSN: 0199-5111 – mf#15057 – us UMI ProQuest [910]
Cruise, William *see* A digest of the law of real property
Cruiser : forest history / Forest History Society – v1 n5-v3 n4 [1978 oct-1980 dec] – 1r – 1 – (cont: forest history cruiser [1978]; cont by: forest history cruiser [1981]) – mf#1161027 – us WHS [634]
Cruises o'er the golden caribbean / United Fruit Company Steamship Service – New York, NY. 1927 – 1r – us UF Libraries [918]
Cruising off mozambique / Karlsson, Elis – London, England. 1969 – 1r – us UF Libraries [916]
Cruising world – Newport. 1982+ (1,5,9) – ISSN: 0098-3519 – mf#13438 – us UMI ProQuest [790]
Cruls, Gastao *see*
- Amazonia que eu vi
- Aparencia do rio de janeiro
- Contos reunidos

Crum, W E *see*
- The conflict of severus patriarch of antioch by athanasius
- The monastery of epiphanius at thebes

Crum, Walter Ewing *see*
- The canons of athanasius of alexandria
- Catalogue of the coptic manuscripts in the british museum

Crumbling idols: twelve essays on art and literature / Garland, Hamlin – 1894 – 9 – us Scholars Facs [420]
Crumbs from an old dutch closet / Van Loon, Lawrence Gwyn – Hague, Netherlands. 1938 – 1r – us UF Libraries [025]
Crump, Charles George *see* The history of the life of thomas ellwood
Crump family newsletter – v2 n3-v7 n1 [1983 may-1988 jun] – 1r – 1 – mf#1712342 – us WHS [929]
Crump, J, Fr *see* Pneuma in the gospels
Crump's historical chronicle – 1974 may 13-jun 10 – 1r – 1 – mf#1054999 – us WHS [978]
Crusade / Byrne, Donn – Boston, MA. 1928 – 1r – us UF Libraries [025]
Crusade / Wisconsin Anti-Tuberculosis Association – 1928 jan-1932 may – 1r – 1 – (cont: crusader of the wisconsin anti-tuberculosis association [1910]; cont by: crusader of the wisconsin anti-tuberculosis association [1934]) – mf#1110989 – us WHS [360]
A crusade of brotherhood : a history of the american missionary association / Beard, Augustus Field – Boston: Pilgrim Press, c1909 [mf ed 1986] – 1mf – 9 – 0-8370-6013-3 – (incl ind) – mf#1986-0013 – us ATLA [240]
The crusade of fidelis, a knight of the order of the cross : being the history of his adventures, during his pilgrimage to the celestial city / Stonehouse, William – Derby 1828 [mf ed Hildesheim 1995-98] – 1v on 3mf – 9 – €90.00 – 3-487-27686-0 – gw Olms [915]
Crusade or class war? / Gallegos Rocafull, Jose Manuel – Washington, DC. 193? Fiche W 904. (Blodgett Collection of Spanish Civil War Pamphlets) – 9 – us Harvard [946]
Crusade or class war? the spanish military revolt / Gallegos Rocafull, Jose Manuel – London, 1937. Fiche W 905. (Blodgett Collection of Spanish Civil War Pamphlets) – 9 – us Harvard [946]
The crusade that lassoed spanish hearts / Whitten, Indy – 1 – 5.00 – us Southern Baptist [242]
Crusader – 1952 sep 12, nov 21, 1953 jul 3-1957 may 29, – 1r – 1 – mf#846398 – us WHS [071]
Crusader – 1968 sep 6-1971 jun 2, – 1r – 1 – mf#846398 – us WHS [071]
Crusader – Bad Kissingen, Schweinfurt DE. 1983 jul 20-1987 nov 25 – 1r – 1 – mf#1044785 – us WHS [074]
Crusader – Beloit WI, Rockford IL. 1952 sep 12, nov 21, 1953 jul 3-1957 may 29, 1968 sep 6-1971 jun 2 – 2r – 1 – mf#846398 – us WHS [074]
Crusader – Chicago IL. 1949 march 26 – 1r – 1 – mf#874812 – us WHS [071]
Crusader – Chicago, IL. 1962-2000 (1) – mf#62570 – us UMI ProQuest [071]
Crusader – Gary, IN. 1964-2000 (1) – mf#62788 – us UMI ProQuest [071]
Crusader / Ku Klux Klan [1915-] – 1977 jul-1981 jan – 1r – 1 – (cont by: white patriot) – mf#676197 – us WHS [320]
Crusader : newsletter / Eldridge Cleaver Crusades – 1977 jun 1 – 1r – 1 – mf#2847443 – us WHS [071]
Crusader : urban news – Cleveland OH. 1995 sep 7/20 – 1r – 1 – mf#3421695 – us WHS [071]
The crusader – Negro Labor Relations League. Chicago. june 16, 1945; apr. 10, 17; nov. 1954 – 1 – us NY Public [305]
The crusader – New Orleans, LA: Crusader Co, feb 16 1889 (wkly) [mf ed 1947] – (= ser Negro Newspapers on Microfilm) – 1r – 1 – us L of C Photodup [071]
The crusader – New York. Jan.-Feb.; Nov. 1921 – 1 – us NY Public [071]
Crusader service and representation / United Food and Commercial Workers International Union – 1991 jan/feb-1992 jul/aug – 1r – 1 – mf#2687069 – us WHS [331]
The crusaders in the east : a brief history of the wars of islam with the latins in syria during the twelfth and thirteenth centuries / Stevenson, William Barron – Cambridge: University Press; New York: Putnam [distributor], 1907 – 1mf – 9 – 0-7905-5973-0 – (incl bibl ref) – mf#1988-1973 – us ATLA [956]
Crusaders of the twentieth century : or, the christian missionary and the muslim / Rice, Walter Ayscoughe – London: W A Rice, 1910 [mf ed 1986] – 2mf – 9 – 0-8370-6696-4 – (incl bibl ref & ind) – mf#1986-0696 – us ATLA [230]
Crusading in the west indies / Jordon, William F – New York, NY. 1922 – 1r – us UF Libraries [972]
Crusan newsletter – n1-16 [1986] – 1r – 1 – mf#1098751 – us WHS [071]
Crusco, Romualdo *see*
- Esquila
- Motivos versos

[Cruse, Francis] *see* Romanism, protestantism, anglicanism
Cruse, Henri Pierre *see* Die opheffing van die kleurlingbevolking
Crusell, Bernhard Henrik *see* Divertimento per l'oboe con accompagnamento di 2 violini, viola et violoncello, op 9
Crushed stone journal – Washington. 1926-1962 (1) – mf#1123 – us UMI ProQuest [690]
Crusius, M *see* Tvrcograeciae libri octo a' martino crvsio, in academia tybingensi graeco and latino professore, vtraque lingua edita
Crusoe's island : a ramble in the footsteps of alexander selkirk with sketches of adventure in california and washoe, nevada / Browne, J Ross – 1864 – 1 – $15.00 – us Library Micro [978]
Crusoe's island in the caribbean / Bowman, Heath – Indianapolis, IN. 1939 – 1r – us UF Libraries [972]
Crusoes of guiana / Boussenard, Louis – London, England. 1883 – 1r – us UF Libraries [972]
Crussemeyer, Jill A *see* Determination of control parameters in pronation curve behavior during running
Crustula juris : being a collection of leading cases on contract done into verse / Fletcher, Mary E & Russell, Bernard Wallace – Toronto: Carswell, [1915?] [mf ed 1999] – 1mf – 9 – 0-659-91519-7 – (pref by humphrey mellish & int by mr justice russell) – mf#9-91519 – cn Canadiana [810]
Cruttwell, Ch Th *see* A literary history of early christianity
Cruttwell, Charles Thomas *see*
- A literary history of early christianity
- The saxon church and the norman conquest
- Six lectures on the oxford movement

Cruttwell, Clement *see*
- A concordance of parallels
- A tour through the whole island of great britain divided into journeys

Cruttwell, Maud *see* Luca signorelli
Cruz, Carlos Manuel De La *see*
- Brega
- Cromitos cubanos
- Episodios de la revolucion cubana
- Escrito de replica en los autos
- Proceso historico del machadato

Cruz Costa, Joao *see*
- Contribuicao a historia das ideias no brasil
- Esbozo de una historia de las ideas en el brasil
- Pequena historia da republica
- Positivismo na republica

Cruz Diaz, Rigoberto *see* Postales de mi pueblo
Cruz, Eddy Dias de *see* Pequena historia de amor
Cruz, Eddy Diaz de *see* Stella me abriu a porta
La cruz en la conquista de america / Bayle, Constantino – Madrid: Razon y Fe, 1933 – 1 – sp Bibl Santa Ana [917]
Cruz, Fernando *see* Instituciones de derecho civil patrio
Cruz, Guillermo Feliu *see* Imagenes de chile vida y costumbres chilenas en los siglos 18 y 19 a traves de...

Cruz Guzman, Emilio *see* Los pastos

Cruz, L M *see* A multi-case study of beginning physical education teachers

Cruz, Maria *see* Maria cruz a traves de su poesia

Cruz Marin, Eugenio de la *see* Flores y frutos de mi corazon dedicacdos a ti

Cruz Marquez Espinosa, J *see* Flores de otono

Cruz Monclova, Lidio *see* Historia del ano de 1887

Cruz Nieves, Antonio *see* Versos

Cruz Rebosa, Maximo *see* Homenaje al maestro don..., caballero de la orden de alfonso 10th el sabio septiembre, 1958

Cruz, San Juan de la *see* Cartas (siecle 16)

Cruz Santos, Abel *see* Presupuesto colombiano

Cruz Valero, Antonio *see* Ponencia acerca de "procedimientos de elaboracion de aceites de olivas en sus distintos aspectos"

Las cruzadas del corazon de jesus / Alcaniz, Florentino – Badajoz: Tip J Sanchez, s a – 1 – sp Bibl Santa Ana [240]

El cruzado extremeno – Plasencia, 1903 – 5 – sp Bibl Santa Ana [073]

Los cruzados del corazon de jesus. avisos practicos para su fundacion y organizacion / Alcaniz, Florentino – Badajoz: Tip. Grafica Corporativa, 1938 – 1 – sp Bibl Santa Ana [240]

O cruzeiro : jornal politico, litterario e noticioso – Desterro, SC: Typ Catharinense, 01 mar-30 dez 1860 – (= ser Ps 19) – bl Biblioteca [073]

O cruzeiro : orgam independente e noticioso – Itajai, SC. 24 maio, 14 jul 1918 – (= ser Ps 19) – mf#UFSC/BPESC – bl Biblioteca [079]

O cruzeiro : orgao do partido catholico por deus e pela patria – Tubarao, SC. 05 abr 1932; jul 1932; 16 abr 1933 – (= ser Ps 19) – mf#UFSC/BPESC – bl Biblioteca [079]

O cruzeiro : orgao imparcial – Sao Joaquim da Costa da Serra, SC. 11 dez 1892 – (= ser Ps 19) – mf#UFSC/BPESC – bl Biblioteca [079]

O cruzeiro do sul : jornal d'instruccao publica, litterario e noticioso – Desterro, SC: Typ Catharinense de G A Maia, mar 1858-fev 1860 – (= ser Ps 19) – mf#UFSC/BPESC – bl Biblioteca [079]

Cruzeiro tem cinco estrelas / Martins, Fran – Fortaleza, Brazil. 1950 – 1r – us UF Libraries [972]

Cruzerio do sul : orgam hebdomadario – Lajes, SC. 18 jun-set, dez 1902; 06 jan 1904 – (= ser Ps 19) – mf#UFSC/BPESC – bl Biblioteca [079]

Crv / Committee of Returned Volunteers – v3 n6-v5 n6 [1969 sep-1971 aug] – 1r – 1 – (cont: committee of returned volunteers newsletter) – mf#684920 – us WHS [327]

Cry california – San Francisco. 1965-1982 (1) 1972-1982 (5) 1975-1982 (9) – (cont by: california tomorrow) – ISSN: 0011-2224 – mf#6916 – us UMI ProQuest [639]

Cry for freedom – v2 n1-4 [1979 jan-sum] – 1r – 1 – (cont: don't mourn, organize!; cont by: new york alliance) – mf#499199 – us WHS [071]

A cry for justice : a study in amos / McFadyen, John Edgar – New York: Charles Scribner, 1912 [mf ed 1989] – (= ser The short course series) – 1mf – 9 – 0-7905-1533-4 – (incl ind) – mf#1987-1533 – us ATLA [221]

A cry from the land of calvin and voltaire : a sequel to "the white fields of france": records of the mcall mission – London: Hodder & Stoughton, 1887 [mf ed 1990] – 1mf – 9 – 0-7905-5953-6 – (int by horatius bonar) – mf#1988-1953 – us ATLA [240]

The cry of "justice to ireland" / Goschen, George Joachim Goschen, 1st viscount – [London], [1886] – (= ser 19th c ireland) – 1mf – 9 – mf#1.1.249 – uk Chadwyck [941]

The cry of the outlander *see* How the french captured fort nelson

Cry of the perishing / Paterson, Nathaniel – Edinburgh, Scotland. 1842 – 1r – us UF Libraries [240]

Cry out – v1 n1,3-4 [1972 jan, may-jun], v1 n1,3-4 [1972 jan, may-jun] – 2r – 1 – mf#721940 – us WHS [071]

Cry, the beloved country / Paton, Alan – New York, NY. 1948 – 1r – us UF Libraries [830]

Cryogenics – Kidlington. 1982+ (1,5,9) – ISSN: 0011-2275 – mf#13327 – us UMI ProQuest [530]

The cryptic rite : its origin and introduction on this continent / Robertson, John Ross – Toronto?: Hunter, Rose, 1883 – 3mf – 9 – mf#09313 – cn Canadiana [360]

The cryptogram : a novel / De Mille, James – New York: Harper, 1871 – 3mf – 9 – mf#06208 – cn Canadiana [830]

Cryptologia – West Point. 1977+ (1,5,9) – ISSN: 0161-1194 – mf#12055 – us UMI ProQuest [400]

Crystal ball – iss1-38 [1979 mar-1982 may] – 1r – 1 – (cont by: larsen file) – mf#637805 – us WHS [071]

[Crystal bay-] the villager – NV. 1961-1962; 1963 – 1r – 1 – $60.00 – mf#UN04483 – us Library Micro [071]

Crystal engineering – Oxford. 1998+ (1,5,9) – mf#42821 – us UMI ProQuest [540]

Crystal lake countryside – Barrington, IL. 1982-1985 (1) – mf#68145 – us UMI ProQuest [071]

Crystal palace : or, the half not told / Overton, Charles – Hull, England. 1851? – 1r – us UF Libraries [941]

Crystal palace : ought it to be open on sunday? / Le Blond, Robert – London, England. 1853 – 1r – us UF Libraries [941]

Crystal palace and norwood advertiser *see* Crystal palace district advertiser and railway indicator

Crystal Palace Co *see* Evolution in history, language, and science

Crystal palace district advertiser and railway indicator – London UK, 1889, 1896, 1951 – 2r – 1 – (aka: crystal palace district times and advertiser; crystal palace and norwood advertiser) – uk British Libr Newspaper [072]

Crystal palace district times – London UK, missing: 9 dec 1882-13 jun 1885 – 1 – uk British Libr Newspaper [072]

Crystal palace district times and advertiser *see* Crystal palace district advertiser and railway indicator

Crystal palace free press – London UK, 7 aug 1986-15 dec 1988, 12 jan-16 mar 1989 – 4 1/4r – 1 – (aka: south london free press [crystal palace ed]; the south london news) – uk British Libr Newspaper [072]

Crystal river current *see* – Gunnison county miscellaneous newspapers

Crystal river empire *see* Garfield county miscellaneous newspapers

Crystal river herald – Crystal River, FL. v1 n22. 1924 oct 10 – 1r – us UF Libraries [071]

Crystal river mirror – Crystal River, FL. v11 n14. 1927 apr 12 – 1r – us UF Libraries [071]

Crystal river news – Crystal River, FL. v4 n42-v8 n17. 1905 aug 18-1916 jan 07 – 1r – (filmed only: 1905: aug 18 (spec ed); 1911 feb 24; mar 10; apr 21; sep 22; nov3; dec 8; 1912 apr 5,12; dec 6; 1914 may 22,29; jul 4,10,31; aug 28; sep 11; oct 23,30; nov 6; 1915: jan 29; feb 5,19; mar 12,26; apr 2,23,30; may 7,21; 1916:jan 7) – us UF Libraries [071]

Crystal silver lance *see* Miscellaneous newspapers of gunnison county

Crystal springs first baptist church. crystal springs, mississippi : church records; directory – 1908 – 1 – 5.00 – us Southern Baptist [242]

Crystallography reports – v1- 1956- – 1,5,6 – us AIP [530]

Csa, community support association : [newsletter] – v9 n10-32 – 1r – 1 – mf#3206532 – us WHS [366]

Csac journal = Journal de l'ascc / Civil Service Association of Canada – 1958 jun-1964 dec, 1965-1966 aug – 2r – 1 – (cont by: argus; argus-journal) – mf#1053809 – us WHS [350]

Csaplovics, Janos *see* Gemaelde von ungern

Csea news / Connecticut State Employees Association – 1980 sep-1984 dec – 1r – 1 – (cont: government news) – mf#825792 – us WHS [350]

Csera news : a bi-monthly newsletter from the center for studies of ethnicity and race in america [csera] / University of Colorado, Boulder – 1990 dec-1991 mar – 1r – 1 – (cont: center for studies of ethnicity and race in america [series]) – mf#4882446 – us WHS [305]

Csokor, Franz Theodor *see* Die gewalten

Csoma, Alexandre, de Koros *see* Analyse du kandjour

Csongrad megyei hirlap – Hodmezovasarhely, Hungary. 1962-68 – 14r – 1 – us L of C Photodup [079]

Cstg press / Civil Service Technical Guild – 1979 nov 1-1980 autumn – 1r – 1 – (cont: pstg press; cont by: guild newsletter [new york ny]) – mf#662305 – us WHS [350]

Csu magazine / Chicago State University – 1992 fall/winter – 1r – 1 – mf#4717707 – us WHS [378]

Ct bulletin / Ontario English Catholic Teachers' Association – 1977 sep-1985 feb – 1r – 1 – (cont by: initiatives [toronto on]) – mf#839052 – us WHS [377]

Ct reporter / Ontario English Catholic Teachers' Association – v8 n1-v11 n8 [1982 oct-1986 jun] – 1r – 1 – (cont: reporter [ontario english catholic teachers' association]; reporter [ontario english catholic teacher's association : 1986]) – mf#922058 – us WHS [377]

Cta journal / California Teachers' Association – San Francisco. 1905-1970 – 1,5,9 – mf#1891 – us UMI ProQuest [370]

Ctm – St. Louis. 1973-1974 (1) 1973-1973 (5) – (cont: concordia theological monthly) – ISSN: 0090-9823 – mf#1536,01 – us UMI ProQuest [240]

Ctu newsletter / Chicago Teachers Union – 1976 jul 29-1980 jul – 1r – 1 – mf#671697 – us WHS [370]

Cuadernillo de miguel picazo, manuscrito / Garcia Lorca, Federico – 9 – sp Cultura [440]

Cuaderno de actividades – 1961. 70p – 1 – 5.00 – us Southern Baptist [242]

Cuaderno de concordancias y oraciones gramaticales latinas para uso particular de los estudiantes de dicha lengua / Romero de Castilla, Pedro – Badajoz: Tip. El Progreso, 1880 – sp Bibl Santa Ana [440]

Cuaderno de lenguaje 4-1 – Plasencia: Editorial Sanchez Rodrigo, S.A., 1970 – 1 – sp Bibl Santa Ana [440]

Cuaderno de lenguaje curso 2-1. otono / Hijosa del Valle, Gregorio – Plasencia: Edit. Sanchez Rodrigo, S.A. 1971 – 1 – (tambien n1-3) – sp Bibl Santa Ana [440]

Cuaderno de lenguaje. curso no 2. pt. 2 / Hinojosa del Valle, Gregorio – Plasencia: Editorial Sanchez Rodrigo, S.A., 1971 – 1 – sp Bibl Santa Ana [440]

Cuaderno de lenguaje no 4, 2 – Plasencia, Caceres: Editorial Sanchez Rodrigo, S.A., 1970 – 1 – sp Bibl Santa Ana [440]

Cuaderno de matematicas 3 no 2 – Plasencia: Edit. Sanchez Rodrigo S.A., 1970 – 1 – (tambien n3; 3-n4; 3-n1) – sp Bibl Santa Ana [510]

Cuaderno de matematicas 4th 2 / Serradillo Calvo, Martin & Serrano, A – Plasencia: Ed. Sanchez Rodrigo, s.a. – 1 – sp Bibl Santa Ana [510]

Cuaderno de matematicas. curso 1st / Rivera Casas, Francisco Moises – (Plasencia): Editorial Sanchez Rodrigo, 1970 – sp Bibl Santa Ana [510]

Cuaderno de matematicas no 1 / Rivera Casas, Francisco Mioses – Plasencia: Edit. Sanchez Rodrigo, S.A. 1971 – 1 – sp Bibl Santa Ana [510]

Cuaderno de matematicas no 3, 1 – Plasencia: Edit. Sanchez Rodrigo, 1971 – 1 – sp Bibl Santa Ana [530]

Cuaderno de matematicas no 4. 1 / Serradilla Calvo, Martin & Serrano, A – Plasencia, Caceres: Editorial Sanchez Rodrigo, S.A., 1971 – 1 – sp Bibl Santa Ana [510]

Cuaderno de matematicas no 4 b / Serradilla Calvo, Martin & Serrano, A – Plasencia, Caceres: Editorial Sanchez Rodrigo, S.A., 1971 – 1 – sp Bibl Santa Ana [510]

Cuaderno de unidades didacticas 1-3 / Rojo Cerezo, Virgilio – Plasencia: Edit. Sanchez Rodrigo, 1971 – 1 – (tambien n1-4) – sp Bibl Santa Ana [370]

Cuaderno de unidades didacticas no 3 – Plasencia: Edit. Sanchez Rodrigo S.A., 1970 – 1 – (tambien n4) – sp Bibl Santa Ana [370]

Cuaderno de unidades didacticas no 4. naturaleza 1 / Garcia Carrasco, Francisco A & Garcia Carrasco, Florencio – Plasencia, Caceres: Editorial Sanchez Rodrigo, S.A., 1970 – 1 – sp Bibl Santa Ana [500]

Cuaderno de unidades didacticas no 4 naturaleza 2 / Garcia Carrasco, Francisco A & Garcia Carrasco, Florencio – Plasencia, Caceres: Editorial Sanchez Rodrigo, S.A., 1970 – 1 – sp Bibl Santa Ana [500]

Cuaderno de unidades didacticas no 4 naturaleza 3 / Garcia Carrasco, Francisco A & Garcia Carrasco, Florenico – Plasencia, Caceres: Editorial Sanchez Rodrigo, S.A., 1970 – 1 – sp Bibl Santa Ana [500]

Cuaderno de unidades didacticas no 4. vida social 1 / Garcia Carrasco, Francisco A & Garcia Carrasco, Florencio – Plasencia, Caceres: Editorial Sanchez Rodrigo, S.A., 1970 – 1 – sp Bibl Santa Ana [301]

Cuaderno de unidades didacticas no 4 vida social 2 / Garcia Carrasco, Francisco A & Garcia Carrasco, Florencio – Plasencia, Caceres: Editorial Sanchez Rodrigo, S.A., 1970 – 1 – sp Bibl Santa Ana [301]

Cuaderno de unidades didacticas no 4 vida social 3 / Garcia Carrasco, Francisco A & Garcia Carrasco, Florencio – Plasencia, Caceres: Editorial Sanchez Rodrigo, S.A., 1970 – 1 – sp Bibl Santa Ana [301]

Cuaderno homenaje a don miguel melendez munoz – San Juan, Puerto Rico. 1957 – 1r – us UF Libraries [972]

Cuadernos americanos – v1-189. 1942-50 – 1 – $1080.00 – mf#0171 – us Brook [972]

Cuadernos de escritura escolar : adaptados al metodo rayas – Plasencia: Ed. Sanchez Rodrigo, 1961. n3,4,5,6 – 1 – sp Bibl Santa Ana [370]

Cuadernos de escritura escolar – Plasencia: Edit. Sanchez Rodrigo, 1975 – 1 – sp Bibl Santa Ana [946]

Cuadernos de escritura escolar – Plasencia: Edit. Sanchez Rodrigo, S.A. 1976 – 1 – sp Bibl Santa Ana [370]

Cuadernos de historia del peru... / Barrenechea, Raul Porras – Madrid: Razon y Fe, 1941 – sp Bibl Santa Ana [972]

Cuadernos de historia economica de cataluna – Barcelona. 1968-1978 (1) 1974-1978 (5) 1974-1978 (9) – ISSN: 0045-9186 – mf#8166 – us UMI ProQuest [330]

Cuadernos de teologia – [Buenos Aires: s.n.] v1-11. 1970-91 – 2r – us CRL [200]

Cuadernos de testimonio para reuniones comunitarias – [Santiago, Chile: Conferencia de Superiores Mayores Religiosos de Chile 1968?- – 1r – (no 1 filmed with: testimonio (santiago, chile), dec 1968-1987) – us CRL [200]

Cuadernos franciscanos – Santiago, Chile: CEFEPAL [ano 15 n57-ano 30 n128 (marzo 1982-oct/dic 1999)] (qrtly) – 4r – 1 – us CRL [073]

Cuadernos franciscanos de renovacion – Santiago, [Chile]: CEFEPAL -1981] [n16-56 (dic 1971-dic 1981)] (qrtly) – 2r – 1 – us CRL [073]

Cuadernos marxistas / Spartacist League of the US – n1-3 [1977] – 1r – 1 – (cont by: spartacist [spanish ed]) – mf#689547 – us WHS [335]

Cuadernos monasticos – Buenos Aires, Argentina: Conferencia de Comunidades Monasticas del Cono Sur. v7 n22 jul/sep 1972; v8 n25 apr/jun 1973; v9 n29 apr/jun 1974; v10 n33/34-v31 n119 apr/sep 1975-oct/dec 1996 – us CRL [240]

Cuadernos...gramaticales...estudiantes / Romero de Castilla, Pedro – 1880 – 9 – sp Bibl Santa Ana [440]

Cuadra Downing, Orlando *see* Nueva poesia nicaraguense

Cuadra, Manolo *see* Almidon

Cuadra, Pablo Antonio *see* Tierra prometida

Cuadra Pasos, Carlos *see* Historia de medio siglo

Cuadrado Ceballos, Juan *see* La voz de dios

Cuadrado Retamosa, Joaquin *see* Cartilla agraria en verso para uso de las escuelas de primera ensenanza

Cuadrilatero / Gomez, Laureano – Bogota, Colombia. 1935 – 1r – us UF Libraries [972]

Cuadro general del comercio exterior de espana con sus posesiones de ultramar y potencias estrangeras en 1849-1855 – Madrid, 1852-1856 – 51mf – 9 – sp Cultura [380]

Cuadro historico de las indias / Madariaga, Salvador De – Buenos Aires, Argentina. 1945 – 1r – us UF Libraries [972]

Cuadros americanos / Llorente Vazquez, Manuel – Madrid, Spain. 1891 – 1r – us UF Libraries [972]

Cuadros bucolicos y otros poemas / Mejia, Francisco R – Buenos Aires, Argentina. 1948 – 1r – us UF Libraries [972]

Cuadros de la historia militar y civil de venezuela / Duarte Level, Lino – Madrid, Spain. 1917 – 1r – us UF Libraries [972]

Cuadros de viaje / Heine, Heinrich – Madrid: Calpe 1920- [mf ed 1990] – (= ser Coleccion universal 269) – 1r – 1 – (trans fr german by manuel pedroso. filmed with: friedrich hebbel und die gegenwart / wilhelm tideman) – mf#2706p – us Wisconsin U Libr [910]

Cuadros del evangelio / Nolasco, Flerida De – Santiago, Dominican Republic. 1947 – 1r – us UF Libraries [972]

Cuadros sinopticos de teologia moral / Serrano Serrano, Ildefonso – Segura de Leon: Imp. Ntra. Sra. de Gracia, 1927 – 1 – sp Bibl Santa Ana [290]

Cualidades y riquezas del nuevo reino de granada / Oviedo, Basilio Vicente De – Bogota, Colombia. 1930 – 1r – us UF Libraries [972]

Cuando cantan las pisadas / Geada, Rita – Buenos Aires, Argentina. 1967 – 1r – us UF Libraries [972]

Cuando el arbol cae / Najera Farfan, Mario Efrain – Guatemala, 1958 – 1r – us UF Libraries [972]

Cuando el cielo sonrie / Ceide, Amelia – San Jose, Costa Rica. 1946 – 1r – us UF Libraries [972]

Cuando la isla era doncella / Bermudez, Ricardo J – Panama, Panama. 1961 – 1r – us UF Libraries [972]

Cuando la luz se quiebra / Stolk, Gloria – Caracas, Venezuela. 1961 – 1r – us UF Libraries [972]

Cuando la razon se vuelve inutil / Diaz Verson, Salvador – Mexico City? Mexico. 1962 – 1r – us UF Libraries [972]

Cuando pinto zurbaran los cuadros de la cartuja de jerez de la frontera? / Bravo, Luis & Peman, Cesar – Badajoz: Imp. Diput. Provincial, 1963 – sp Bibl Santa Ana [946]

Cuando reinaron las sombras / Landaeta, Federico – Madrid, Spain. 1955 – 1r – us UF Libraries [972]

Cuando y donde se ordeno bartolome de las casas / Bayle, Constantino – Madrid: Missionalia Hispanica, 1944 – 1 – sp Bibl Santa Ana [240]

Cuantas estrellas en mi cuarto / Lopez Suria, Violeta – San Juan, Puerto Rico. 1957 – 1r – us UF Libraries [972]

Cuarenta dias en el vaupes / Builes G, Miguel Angel – Santa Rosa Osos, Colombia. 1951 – 1r – us UF Libraries [972]

Cuaresma y semana santa / Sanchez Aliseda, Casimiro – Madrid: Art. G. Euroamerica, 1957 – 1 – sp Bibl Santa Ana [946]

Cuarta asamblea general del instituto panamericano / Pan American Institute Of Geography And History – Mexico City? Mexico. 1946 – 1r – us UF Libraries [972]

Cuarta conferencia internacional americana / Dominican Republic Delegacion En La Cuarta – Sevilla, Spain. 1912 – 1r – us UF Libraries [327]

Cuartero, Baltasar *see* Indice de la coleccion salazar, tomos 20, 8-37. madrid, 1961-1966

Cuartero, baltasar y vargas zuniga, antonio. indice de la coleccion salazar. tomos 23-27. madrid, 1959-1960 / Uribe, Angel – Madrid: Graf. Calleja, 1967 – 1 – sp Bibl Santa Ana [946]

Cuarterona / Tapia Y Rivera, Alejandro – San Juan, Puerto Rico. 1944 – 1r – us UF Libraries [972]

Cuartillas / Groizard y Coronado, Carlos – 1886 – 9 – sp Bibl Santa Ana [810]

Cuarto censo nacional agropecuario, 1950 / Dominican Republic Direccion General De Estadisti – Ciudad Trujillo, Dominican Republic. 1950 – 1r – us UF Libraries [972]

Cuarto centenario del nacimiento de d. benito arias montano / Perez Goyena, A – Madrid: Razon y Fe, 1927 – 1 – sp Bibl Santa Ana [946]

Cuatro anos bajo la media luna... / Nogales, Rafael de – Madrid: Razon y Fe, 1940 – 1 – sp Bibl Santa Ana [946]

Cuatro anos en la cienaga de zapata / Consculluela Y Barreras, Juan Antonio – Habana, Cuba. 1918 – 1r – us UF Libraries [972]

Cuatro articulos y un prologo / Jimenez Rodriguez, Manuel Antonio – Ciudad Trujillo, Dominican Republic. 1957 – 1r – us UF Libraries [972]

Los cuatro caminos del toreo / Mahizflor – Badajoz: Tip. Vda. de Arqueros, 1947 – 1 – sp Bibl Santa Ana [946]

Cuatro decretos basicos para el desarrollo agrario de la provincia / Duputacion Provincial – Caceres: Imprenta Diputacion Provincial, 1973 – 1 – sp Bibl Santa Ana [630]

Cuatro ensayos sobre administracion postal : brasil / Instituto Centroamericano De Administracion Public – San Jose, Costa Rica. 1970 – 1r – us UF Libraries [380]

Cuatro figuras colombianas / Rivas, Raimundo – Bogota, Colombia. 1933 – 1r – us UF Libraries [972]

Cuatro leyendas cacerenas / Arias Corrales, Juan – Caceres: Tip. El Noticiero hacia, 1973 – 1 – sp Bibl Santa Ana [946]

Cuatro meses de barbarie, mallorca bajo el terror fascista / Perez, Manuel – Valencia, 1937. Fiche W1102. (Blodgett Collection of Spanish Civil War Pamphlets) – 9 – us Harvard [946]

Cuatro poemas de eugenio florit – Habana, Cuba. 1940 – 1r – us UF Libraries [810]

Cuatro poemas en china / Jamis, Fayad – Habana, Cuba. 1961 – 1r – us UF Libraries [951]

Cuatro poetas cubanos / Baeza, Flores – Barcelona, Spain. 1956 – 1r – us UF Libraries [972]

Cuatro suertes / Samayoa Chinchilla, Carlos – Guatemala, 1936 – 1r – us UF Libraries [972]

Cuatrocientas espinelas / Sanchez Arjona, Vicente – Sevilla: Graficas T., 1952 – 1 – sp Bibl Santa Ana [810]

Cub : news [and] facts / United Farm Equipment and Metal Workers of America – 1946 sep 17-1955 jan 26 – 1r – 1 – mf#1110996 – us WHS [331]

Cub prints / Citizens Utility Board [WI] – v1 n1-v5 n3 [1980 aug 28-1984 dec], v6 n1-2 [1985:winter-spring], v7 n1-3 [1986 spr-fall], v8 n1 [1987 sum] – 1r – 1 – (cont by: cub informer) – mf#1613091 – us WHS [350]

Cuba / Bachiller Y Morales, Antonio – Habana, Cuba. 1962 – 1r – us UF Libraries [972]

Cuba / Cancio Villa-Amil, Mariano – Madrid, Spain. 1883 – 1r – us UF Libraries [972]

Cuba / Deckert, Emil – Bielefeld, Germany. 1899 – 1r – us UF Libraries [972]

Cuba / Fairford, Ford – London, England. 1926 – 1r – us UF Libraries [972]

Cuba / Fergusson, Erna – New York, NY. 1946 – 1r – us UF Libraries [972]

Cuba : internal affairs and foreign affairs, 1945-jan 1963 / U.S. State Dept – (= ser Confidential u s state department special files) – 1 – $26,445.00 coll – (internal affairs & foreign affairs, 1945-49 29r isbn 0-89093-952-7 $5610. 1950-54 39r isbn 0-89093-953-5 $7535. 1955-59 25r isbn 0-89093-548-3 $4840. internal affairs, 1960-jan 1963 39r isbn 1-55655-797-3 $7545. foreign affairs, 1960-jan 1963 12r isbn 1-55655-796-5 $2320. with p/g) – us UPA [327]

Cuba / Miro Argenter, Jose – Habana, Cuba. 1945 – 1r – us UF Libraries [972]

Cuba / Pierra, Fidel G – New York, NY. 1896 – 1r – us UF Libraries [972]

Cuba / Quesada, Gonzalo De – Washington, DC. 1905 – 1r – us UF Libraries [972]

Cuba / Sedano Y Cruzat, Carlos De – Madrid, Spain. 1872 – 1r – us UF Libraries [972]

Cuba – Toronto, ON. 1904? – 1r – us UF Libraries [972]

Cuba – Washington, DC. 1897 – 1r – us UF Libraries [972]

Cuba – Washington, DC. 1949 – 1r – us UF Libraries [972]

Cuba / Wright, Irene Aloha – New York, NY. 1910 – 1r – us UF Libraries [972]

Cuba *see*
- Codigo civil
- Compilacion de decretos del sr presidente
- Gaceta de la habana
- Gaceta oficial
- Legislacion fiscal (impuestos generales del estado
- Legislacion hipotecaria vigente en la republica de...
- Legislacion y practica consulares
- Ley constitucional de la republica de cuba
- Ley de enjuiciamiento criminal para las islas de c...
- Ley de impuestos municipales y procedimiento de co...
- Ley electoral de cuba de septiembre 11 de 1908
- Leyes civiles de la republica de cuba
- Recopilacion de todas las disposiciones vigentes
- Translation of general regulations
- Translation of the code of commerce in force in cuba

Cuba: 196 photos / Verger, Pierre – Habana, Cuba. 1958 – 1r – us UF Libraries [972]

Cuba and her people of to-day / Forbes-Lindsey, Charles Harcourt Ainslie – Boston, MA. 1911 – 1r – us UF Libraries [972]

Cuba and its international relations / Stuart, Graham Henry – New York, NY. 1923 – 1r – us UF Libraries [327]

Cuba and porto rico / Hill, Robert Thomas – New York, NY. 1903 – 1r – us UF Libraries [972]

Cuba and the bay of pigs invasion *see* Foreign office files for cuba

Cuba and the cubans / Cabrera, Raimundo – Philadelphia, PA. 1896 – 1r – us UF Libraries [972]

Cuba and the peninsualr and occidental s s co / Peninsular And Occidental Steamship Company – Chicago, IL. 1928 – 1r – us UF Libraries [972]

Cuba and the united states / Fitzgibbon, Russell Humke – Menasha, WI. 1935 – 1r – us UF Libraries [327]

Cuba, ayer y hoy: dos novelas – Buenos Aires, Argentina. 1965 – 1r – us UF Libraries [972]

Cuba before the united states – New York, NY. 1869 – 1r – us UF Libraries [972]

Cuba before the world / Alfonso, Manuel F – Havana, Cuba. 1915 – 1r – us UF Libraries [972]

Cuba, castro, and communism / Stein, Edwin C – New York, NY. 1962 – 1r – us UF Libraries [972]

Cuba city news – Cuba City, Hazel Green WI. [1894 sep 21-1918] – 1r – 1 – mf#967030 – us WHS [071]

Cuba city news-herald – Cuba City, Hazel Green WI. [1914 jan 9-mar 6] – 1r – 1 – (cont by: cuba city news-herald and the hazel green tribune) – mf#967050 – us WHS [071]

Cuba city news-herald – Cuba City WI. [1917 feb 2-1918 may 3], 1918 feb, 1918 may 10-1919 mar 28 – 3r – 1 – (cont: cuba city news-herald and the hazel green tribune; cont by: news herald [cuba city wi]) – mf#967054 – us WHS [071]

Cuba city news-herald – Cuba City WI. 1935 jan 25-1936 nov 5, 1936 nov 5-1939 dec 28, 1940 jan 4-1942 nov 19 – 3r – 1 – (cont: news-herald [cuba city wi]; cont by: cuba city news-herald and the hazel green tribune-reporter) – mf#967112 – us WHS [071]

Cuba city news-herald and the hazel green tribune – Cuba City, Hazel Green WI. [1914 mar 13-1917 jan 26] – 1r – 1 – (cont: cuba city news herald [cuba city wi: 1900]; cont by: cuba city news herald [cuba city wi: 1917]) – mf#967051 – us WHS [071]

Cuba city news-herald and the hazel green tribune-reporter – Cuba City, Hazel Green WI. 1942 nov 26-dec 31, 1943 sep 23-1945 dec 27, 1946 jan 3-1948 dec 30, 1949-57, 1958-1959 mar 26 – 8r – 1 – (cont: cuba city news-herald [cuba city wi: 1935]; cont by: tri-county press) – mf#967118 – us WHS [071]

Cuba Comision Nacional De La Unesco *see* Capital extranjero en la america latina

Cuba Congreso Camara De Representantes *see*
- Diario de sesiones del congreso de la republica de...
- Memoria de los trabajos realizados

Cuba Congreso Senado *see*
- Diario de sesiones del congreso de la republica de...
- Memoria de los trabajos realizados durante las y s...

Cuba Constitution *see* Nueva constitucion cubana y su jurisprudencia

Cuba contra espana / Varona, Enrique Jose – New York, NY. 1895 – 1r – us UF Libraries [972]

Cuba Convencion Constituyente, 1928 *see* Diario de sesiones de la convencion constituyente

Cuba Departamento De Estado *see* Documentos internacionales referentes al reconocim...

Cuba, die perle der antillen : reisedenkwuerdigkeiten und forschungen / Sivers, Jegor von – Leipzig 1861 [mf ed Hildesheim 1995-98] – 1v on 3mf – 9 – €90.00 – 3-487-26937-6 – gw Olms [918]

Cuba. Direccion Central de Estadistica *see*
- Anuario 1972-1974
- Compendio estadistico de cuba 1965-1966, 1968, 1976

Cuba Direccion De Cultura *see* Navidades para un nino cubano

Cuba. Direccion General de Estadistica *see*
- Anuario estadistico de cuba 1952, 1956-1957
- Boletin estadistico 1964-1965, 1968, 1971

Cuba Direccion General Del Censo *see*
- Censo de 1943
- Censo de la republica de cuba
- Census of the republic of cuba 1919

Cuba Ejercito Inspeccion General *see* Indice alfabetico y defunciones

Cuba en 1858 / Alcala, Galiano, Dionisio – Madrid, Spain. 1859 – 1r – us UF Libraries [972]

Cuba en america / Santovenia Y Echaide, Emeterio Santiago – Mexico City? Mexico. 1947 – 1r – us UF Libraries [327]

Cuba en el bandera / Piedra-Bueno, Andres De – Habana, Cuba. 1950 – 1r – us UF Libraries [972]

Cuba en la exposicion pan americana de buffalo, 19... / Wood, Leon – Habana, Cuba. 1901 – 1r – us UF Libraries [972]

Cuba en la mano – Habana, Cuba. 1940 – 1r – us UF Libraries [972]

Cuba en la oea / Meeting Of Consultation Of Ministers Of Foreign Affairs – Habana, Cuba. 1960 – 1r – us UF Libraries [327]

Cuba espanola / Reverter Delmas, Emilio – 6v. 1896-99 – 1 – us L of C Photodup [972]

Cuba et haiti / Fignole, Daniel – Port-Au-Prince, Haiti. 1949 – 1r – us UF Libraries [972]

Cuba for invalids / Gibbes, Robert Wilson – New York, NY. 1860 – 1r – us UF Libraries [972]

Cuba Fuerzas Armadas Revolucionias Direccion Pol... *see* Historia militar de cuba

Cuba Gobierno Y Capitania General *see* Inmigracion de trabajadores espanoles

Cuba illustrated / Prince, John C – New York, NY. 1894 – 1r – us UF Libraries [972]

Cuba in war time / Davis, Richard Harding – New York, NY. 1897 – 1r – us UF Libraries [972]

Cuba indigena / Fort Y Roldan, Nicolas – Madrid, Spain. 1881 – 1r – us UF Libraries [972]

[Cuba-] informador guerrilero – GT. 1982-85 – 1r – 1 – $50.00 – mf#R04211 – us Library Micro [079]

Cuba international – Havana, Cuba. -m. Jan-apr 1970 – 1/4r – 1 – uk British Libr Newspaper [072]

Cuba, la america latina, los estados unidos / Scott, James Brown – Habana, Cuba. 1926 – 1r – us UF Libraries [972]

Cuba Laws, Statutes, Etc *see*
- Aranceles de aduanas para los puertos de la isla d...
- Codigo civil interpretado por el tribunal supremo...
- Codigo de comercio vigentes en la republica de cuba
- Codigos de cuba
- Legislacion muncipal de la republica de cuba
- Projet de code criminel cubain
- Reglamentos de las agencias diplomaticas
- Translation of the penal code in force in cuba and...
- Translations of the law of criminal procedure

Cuba, Laws, Statutes, Etc *see* Ley de emergencia economica

Cuba mexicana – Mexico City? Mexico. 1896 – 1r – us UF Libraries [972]

Cuba Military Governor, 1899 (John R Brooke) *see* General orders and circulars

Cuba Ministerio De Educacion *see* Informacion ante el senado

Cuba Ministerio De Hacienda *see* Aportaciones para una politica economica cubana

Cuba no debe su independencia a los estados unidos / Roig De Leuchsenring, Emilio – Habana, Cuba. 1950 – 1r – us UF Libraries [972]

Cuba of today / Verrill, A Hyatt – New York, NY. 1931 – 1r – us UF Libraries [972]

Cuba Oficina Del Censo *see*
- Censo de la republica de cuba
- Cuba: population, history and resources 1907

Cuba, old and new / Robinson, Albert Gardner – New York, NY. 1915 – 1r – us UF Libraries [972]

Cuba, pais de poca memoria / Baroni, Aldo – Mexico City, Mexico. 1944 – 1r – us UF Libraries [972]

Cuba para los cubanos / Calzadilla, Rafael S De – Habana, Cuba. 1928 – 1r – us UF Libraries [972]

Cuba past and present / Verrill, A Hyatt – New York, NY. 1914 – 1r – us UF Libraries [972]

Cuba; politica–guerra–autonomia – Madrid, Spain. 1897 – 1r – us UF Libraries [972]

Cuba: population, history and resources 1907 / Cuba Oficina Del Censo – Washington, DC. 1909 – 1r – us UF Libraries [972]

Cuba por fuera : (apuntes del natural) / Gallego Y Garcia, Tesifonte – Habana, Cuba. 1892 – 1r – us UF Libraries [972]

Cuba Provisional Governor, 1906-1909 (Charles E...) *see* Decree and proclamations

Cuba puede ser independiente / Ferrer De Couto, Jose – Nueva York, NY. 1872 – 1r – us UF Libraries [972]

Cuba, receipts and expenditures, votes, / United States Congress – Washington, DC. 1900 – 1r – us UF Libraries [972]

Cuba Secretaria De Agricultura, Comercio Y Trabaj *see* Cuba, what she has to offer

Cuba. Secretaria de Agricultura, Industria y Comercio *see* Memoria de los trabajos y servicios de este departamento correspondiente al periodo de tiempo transcurrido desde el...

Cuba Secretaria De Estado *see*
- Bandera
- Correspondencia diplomatica cruzada entre la...
- Manual del diplomatico cubano

Cuba. Secretaria de Hacienda *see* Informe dirigido al honorable senor presidente de la republica [...] por el secretario de hacienda [...] sobre los trabajos realizados por el departamento desde el [...]

Cuba Secretaria De Instruccion Publica Y Bellas A *see* Iconografia del apostol jose marti

Cuba Secretaria De Instruccion Publica Y Bellas A... *see* Iconografia del apostol jose marti

Cuba. Secretaria de Obras Publicas *see* Memoria de obras publicas correspondiente al periodo de...

Cuba. Secretario de Agricultura, Comercio y Trabajo *see* Memoria general de los trabajos realizados desde el...

Cuba Servicio Femenino Para La Defensa Civil *see* Gertrudis gomez de avellaneda

Cuba, ses ressources, son administration, sa popul / Cuba Superintendencia General Delegada De Real Ha – Paris, France. 1851 – 1r – us UF Libraries [972]

Cuba. Superindendencia de Escuelas *see* Informe del superintendente de escuelas de cuba

Cuba Superintendencia General Delegada De Real Ha *see* Cuba, ses ressources, son administration, sa popul

Cuba Treaties, Etc *see* Convenios y tratados celebrados

Cuba Treaties, Etc, 1934-1936 (Mendieta) *see* Tratado de reciprocidad concertado

Cuba; un ano de republica / Mestre Y Amabile, Vicente – Paris, France. 1903 – 1r – us UF Libraries [972]

Cuba under spanish rule / Rochas, Victor De – New York, NY. 1869? – 1r – us UF Libraries [972]

Cuba, what she has to offer / Cuba Secretaria De Agricultura, Comercio Y Trabaj – Habana, Cuba. 1915 – 1r – us UF Libraries [972]

Cuba y el problema del caribe / Dihigo Y Lopez-Trigo, Ernesto – Habana, Cuba. 1950 – 1r – us UF Libraries [972]

Cuba y espana / Gutierrez, Valeriano G – Habana, Cuba. 1909 – 1r – us UF Libraries [327]

Cuba y la conferencia de educacion y desarrollo ec / Conference On Education And Economic And Social De... – Habana, Cuba. 1962 – 1r – us UF Libraries [972]

Cuba y la opinion publica / Amer, Carlos – Madrid, Spain. 1897 – 1r – us UF Libraries [972]

Cuba y las costumbres cubanas / Ewart, Frank Carman – Boston, MA. 1919 – 1r – us UF Libraries [306]

Cuba y los estados unidos / Torriente Y Peraza, Cosme De La – Habana, Cuba. 1929 – 1r – us UF Libraries [972]

Cuba y pi y margall / Conangla Fontanilles, Jose – Habana, Cuba. 1947 – 1r – us UF Libraries [972]

Cuba y puerto rico / Dupierris, Martial – Madrid, Spain. 1866 – 1r – us UF Libraries [972]

Cuba y su futuro / Garcia Montes Y Angulo, Jose – Miami, FL. 1964 – 1r – us UF Libraries [972]

Cubagua / Nunez, Enrique Bernardo – Paris, France. 1931 – 1r – us UF Libraries [972]

Cuban church in a sugar economy / International Missionary Council – New York, NY. 1942 – 1r – us UF Libraries [972]

The cuban dancer's bible : rumba, mambo / Luis, Robert – [New York: Latin Dance Studio, c1953] – 1 – mf#*ZBD-*MGO pv17 – Located: NYPL – us Misc Inst [790]
Cuban economic standard / Casanova, Jose Manuel – Habana, Cuba. 1949 – 1r – us UF Libraries [972]
Cuban expedition / Bloomfield, J H – London, England. 1896 – 1r – us UF Libraries [972]
Cuban Information Bureau, Washington, DC *see* Ambassador guggenheim and the cuban revolt
Cuban League Of The United States *see* Present condition of affairs in cuba
Cuban missile crisis *see* Foreign office files for cuba
The cuban missile crisis, 1962 – [mf ed Chadwyck-Healey] – (= ser National security archive, washington dc: the making of us policy) – 586mf – 9 – (with 2v p/g & ind) – uk Chadwyck [327]
Cuban missle crisis newspapers – s.l, s.l? 1960 aug 7-1968 may 21 – 1r – (misc titles) – us UF Libraries [079]
Cuban patriots' cause is just / Matthews, Claude – Philadelphia, PA. 1895 – 1r – us UF Libraries [972]
Cuban question in the spanish parliament / Macias, Juan Manuel – London, England. 1872 – 1r – us UF Libraries [972]
Cuban Refugee Resettlement Operation *see* Mercurio de mccoy
Cuban Refugee Resettlement Operation [Fort McCoy WI] *see*
– Daily minor
– Daily status report
Cuban sideshow / Phillips, Ruby Hart – Havana, Cuba. 1935 – 1r – us UF Libraries [972]
Cuban sketches / Steele, James William – New York, NY. 1881 – 1r – us UF Libraries [972]
Cuban story / Matthews, Herbert Lionel – New York, NY. 1961 – 1r – us UF Libraries [972]
Cuban tapestry / Clark, Sydney – New York, NY. 1936 – 1r – us UF Libraries [972]
Cubana ejemplar: marta abreu de estevez / Perez Cabrera, Jose Manuel – Habana, Cuba. 1945 – 1r – us UF Libraries [972]
Cubania de fray candil / Entraigo, Elias Jose – Habana, Cuba. 1957 – 1r – us UF Libraries [972]
Lo cubano en la poesia / Vitier, Cintio – s.l, s.l? 1958 – 1r – us UF Libraries [972]
Cubano libre nacionalista – Miami, FL. 1972 oct-1973 jan – 1r – us UF Libraries [071]
Cubano libre: organo del consejo revolucionario de cuba – Miami, FL. 1963 jul – 1r – us UF Libraries [071]
Cubano libre: organo del gobierno revolucionario cubano en el exilio – Miami, FL. 1963 jul – 1r – us UF Libraries [071]
Cubano olvidado / Lopez Y Garcia, Gustavo – Habana, Cuba. 1893 – 1r – us UF Libraries [972]
Cubans in florida – s.l, s.l? 193-? – 1r – us UF Libraries [978]
Cuba's fight for freedom and the war with spain / Beck, Henry Houghton – Philadelphia, PA. 1898 – 1r – us UF Libraries [972]
Cuba's great struggle for freedom / Quesada, Gonzalo De – s.l, s.l? 1898 – 1r – us UF Libraries [972]
Cuba's greatest struggle for freedom / Quesada, Gonzalo De – n.p, n.p? 1898 – 1r – us UF Libraries [972]
Le cubilot – Journal international d'education, d'organisation et de lutte ouvriere. no. 37-40, 42-44. Colonie d'Aiglemont (Ardennes). nov-dec 1907 – 1 – fr ACRPP [335]
Cubism : subject collections – (= ser Art exhibition catalogues on microfiche) – 117 catalogues on 145mf – 9 – £915.00 – (individual titles not listed separately) – uk Chadwyck [700]
Cubitt, James *see* Church design for congregations
Cuc jham / Samn Hael – Bhnam Ben: Ron Bumb Camroen Ratth 2515 [1970] [mf ed 1990] – 1r with other items – 1 – (in khmer) – mf#mf-10289 seam reel 129/8 [§] – us CRL [480]
Cucaracha – 1983 apr 15/18-dec – 1r – 1 – mf#1218087 – us WHS [071]
Cucheval, Victor *see* Ciceron orateur
Cuchi Coll, Isabel *see*
– 13 [i.e. trece] novelas cortas
– Arras de cristal y clara lair
Cuchipanda sonora / Suaree, Octavio De La – Cardenas, Cuba. 1920 – 1r – us UF Libraries [972]
Cucina, Irene M *see* Specificity of feedback using alternative assessment techniques in a secondary physical education badminton class
A cuckoo in kenya : the reminiscences of a pioneer police officer in british east africa / Foran, William Robert – London: Hutchinson, [1936] – 1 – us CRL [960]
Cucuel, Ernst *see* Die eingangsbuecher des parzival und das gesamtwerk
Cucumber diseases in florida / Weber, George F – Gainesville, FL. 1929 – 1r – us UF Libraries [634]
Cucumber rot / Burger, O F – Gainesville, FL. 1914 – 1r – us UF Libraries [634]
Cucurullo, Oscar *see*
– Geografia de santo domingo
– Hoya de enriquillo
Cudahy enterprise – Cudahy WI. 1912 sep 6-1914 jun 20, 1914 jun 27-1916 mar 18, 1916 mar 25-1918 nov 9, 1919 aug 23, sep 20, 1924-30, 1944-46, 1947-1952 feb 14 – 7r – 1 – mf#966863 – us WHS [071]
Cudahy reminder-enterprise – Cudahy, St Francis WI. 1971 nov 4/1972 mar 9-1980 jan/feb 28 – 19r – 1 – (with gaps; cont: reminder-enterprise [cudahy wi: 1955]; cont by: reminder-enterprise [cudahy wi: 1980]) – mf#1002215 – us WHS [071]
Cudahy, s[/ain/]t francis advisor press – Cudahy, Saint Francis WI. 1982 apr 15/sep-1987 jan/jul – 10r – 1 – (with gaps; cont: cudahy, st francis free press) – mf#999732 – us WHS [071]
Cudahy, s[/ain/]t francis free press – Cudahy, Saint Francis, South Milwaukee WI. 1973 jan 10/1974 feb 27-1982 jan/apr 8 – 14r – 1 – (with gaps; cont by: cudahy, st francis advisor press) – mf#999733 – us WHS [071]
Cudahy times – Cudahy WI. 1893 oct 29-1894 mar 21 – 1r – 1 – mf#960671 – us WHS [071]
Cuddesdon college, 1854-1904 : a record and memorial – London: Longmans, Green, 1904 – 1mf – 9 – 0-524-03546-6 – mf#1990-4741 – us ATLA [378]
Cudell, C A *see* Udinji (chez les riverains de la buschimaie)
The cudgel – Bowmanville [Ont: s.n., 1859?-18–?] – 9 – mf#P04453 – cn Canadiana [071]
Cuellar, Enrique *see* Ingenieria de carreteras
Cuellar Grajera, Antonio *see*
– La formacion profesional del jurista
– La intervencion del abogado en la constitucion de las sociedades mercantiles
– El problema de la defensa de los derechos e intereses legitimos de las minorias de accionistas en las sociedades mercantiles
Cuellar Vargas, Enrique *see* 13 (i e trece) anos de violencia
Cuellar Vizcaino, Manuel *see* 12 [i.e. doce] muertes famosas
Cuello, Julio A *see* Poemas del instinto
Cuenca, Abel *see* Salvador
Cuenco collection – Manila, Philippines: Microfilm Corporation of the Philippines, 1983 – 9r – us CRL [959]
Cuenod, R *see* Tsonga-english dictionary
Cuenta detallada y documentada que presenta el ministro de hacienda e industria ciudadano pedro garcia, como ajente financiero en el exterior, nombrado por el supremo gobierno de bolivia – La Paz: Impr de La Libertad, 1872 – us CRL [972]
Cuenta general de rentas y gastos y cuenta de gastos del departmento de hacienda / Venezuela. Ministerio de Hacienda – Caracas. 1912 13-1940 41 – 1 – us L of C Photodup [972]
Cuenta que el alcalde...de 1850 / Badajoz – 1851 – 9 – sp Bibl Santa Ana [946]
Cuenta que presenta al congreso nacional de los estados unidos de venezuela el ministro de hacienda en ... – Caracas: Impr de la "Gaceta oficial," 1880-1881 – us CRL [972]
Cuenta que presenta al congreso nacional de los estados unidos de venezuela en... – Caracas: Impr de la vapor de "la opinion nacional," 1882-90 – us CRL [972]
Cuenta...memorias criticas y apologeticas... carlos 4 de borbon / Godoy, Manuel – 1838 – 9 – sp Bibl Santa Ana [946]
Cuentas de propios (anno 1511-1555) – Caceres – 1r – 5,6 – sp Cultura [946]
Cuentero / Cardosa, Onelio Jorge – Santa Clara, Cuba. 1958 – 1r – us UF Libraries [972]
Cuentistas cubanos y la reforma apraria / Lorenzo, Jose – Habana, Cuba. 1960 – 1r – us UF Libraries [972]
Cuento costarricense / Menton, Seymour – Mexico City? Mexico. 1964 – 1r – us UF Libraries [972]
Cuento de amor / Barrientos, Alfonso Enrique – Mexico City? Mexico. 1956 – 1r – us UF Libraries [972]
El cuento de tristan de leonis : Northup, George Tyler – Chicago: University of Chicago Press, 1928 – (= ser The modern philology monographs of the university of chicago) – ix/298p – 1 – us Wisconsin U Libr [390]
Cuento en costa rica / Portuguez De Boianos, Elizabeth – San Jose, Costa Rica. 1964 – 1r – us UF Libraries [972]
Cuento en panama / Miro, Rodrigo – Panama, Panama. 1950 – 1r – us UF Libraries [972]
Cuento puertorriqueno en el siglo xx / Puerto Rico University College Of Arts And Scien – Rio Piedras, Puerto Rico. 1963 – 1r – us UF Libraries [972]
Cuentos / Fernandez, Aristides – Habana, Cuba. 1959 – 1r – us UF Libraries [972]
Cuentos / Gagini, Carlos – San Jose, Costa Rica. 1963 – 1r – us UF Libraries [972]
Cuentos / Gonzalez Garcia, Matias – San Juan, Puerto Rico. 1960 – 1r – us UF Libraries [972]
Cuentos / Gonzalez Zeledon, Manuel – San Jose, Costa Rica. 1947 – 1r – us UF Libraries [972]
Cuentos / Hauff, Wilhelm – Madrid, Barcelona: Calpe, 1920 – (= ser Coleccion universal 159,160) – 1r – 1 – (trans by c gallardo de mesa. filmed with: der frosch / otto erich hartleben) – mf#7443 – us Wisconsin U Libr [830]
Cuentos / Henriquez Y Carvajal, Federico – Ciudad Trujillo, Dominican Republic. 1950 – 1r – us UF Libraries [972]
Cuentos / Marti, Jose – Habana, Cuba. 1961 – 1r – us UF Libraries [972]
Cuentos / Nunez Quintero, Jose Maria – Panama, Panama. 1956 – 1r – us UF Libraries [972]
Cuentos / Nunez Quintero, Jose Maria – Panama, Panama. 1959 – 1r – us UF Libraries [972]
Cuentos / Pinera, Virgilio – La Habana, Cuba. 1964 – 1r – us UF Libraries [972]
Cuentos / Samayoa Chinchilla, Carlos – San Salvador, El Salvador. 1963 – 1r – us UF Libraries [972]
Cuentos / Zeno Gandia, Manuel – New York, NY. 1958 – 1r – us UF Libraries [972]
Cuentos absurdos / Collado, Martell A – Madrid, Spain. 1931 – 1r – us UF Libraries [972]
Cuentos breves y maravillosos / Menen Desleal, Alvaro – San Salvador, El Salvador. 1963 – 1r – us UF Libraries [972]
Cuentos cimarrones / Nolasco, Socrates – Ciudad Trujillo, Dominican Republic. 1958 – 1r – us UF Libraries [972]
Cuentos color sepia / Miller Otero, Fredy – Ciudad Trujillo, Dominican Republic. 1957 – 1r – us UF Libraries [972]
Cuentos completos / Cardoso, Onelio Jorge – Habana, Cuba. 1962 – 1r – us UF Libraries [972]
Cuentos completos / Dario, Ruben – Mexico City? Mexico. 1950 – 1r – us UF Libraries [972]
Cuentos completos / Pita Rodriguez, Felix – Habana, Cuba. 1963 – 1r – us UF Libraries [972]
Cuentos contemporaneos / Ibarzabal, Frederico De – Habana, Cuba. 1937 – 1r – us UF Libraries [972]
Cuentos cubanos (antologia) / Perez, Emma – Habana, Cuba. 1945 – 1r – us UF Libraries [972]
Cuentos cubanos contemporaneos / Portuondo, Jose Antonio – Mexico City? Mexico. 1946 – 1r – us UF Libraries [972]
Cuentos de adli y luas / Perera, Hilda – Havana, Cuba. 1960 – 1r – us UF Libraries [972]
Cuentos de barro / Salarrue – Lima, Peru. 1959 – 1r – us UF Libraries [972]
Cuentos de barro / Salarrue – San Salvador, El Salvador. 1962 – 1r – us UF Libraries [972]
Cuentos de barro / Salarrue – Santiago, Chile. 1943 – 1r – us UF Libraries [972]
Cuentos de batey / Torriente Brau, Pablo De La – Havana, Cuba. 1962 – 1r – us UF Libraries [972]
Cuentos de belice / Barrientos, Alfonso Enrique – Guatemala, 1961 – 1r – us UF Libraries [972]
Cuentos de ciencia-ficcion / Cabada, Carlos – Habana, Cuba. 1964 – 1r – us UF Libraries [972]
Cuentos de francisco mendez y raul carrillo meza / Mendez, Francisco – Guatemala, 1957 – 1r – us UF Libraries [972]
Cuentos de guatemala, 1952 / Grupo Saker-Ti – Guatemala, 1953 – 1r – us UF Libraries [972]
Cuentos de hoy y de manana, cuento / Chavez Velasco, Waldo – San Salvador, El Salvador. 1963 – 1r – us UF Libraries [972]
Cuentos de la abuelita / Gallardo de Alvarez, Isabel – Madrid: Sociedad de Educacion Atenas S.A., Tomo 1 – 1 – sp Bibl Santa Ana [946]
Cuentos de la abuelita / Gallardo de Alvarez, Isabel – Madrid: Sociedad de Educacion Atenas S.A., Tomo 2. 1947 – 1 – sp Bibl Santa Ana [390]
Cuentos de la carretera central / Melendez Munoz, Miguel – Barcelona, Spain. 1963 – 1r – us UF Libraries [972]
Cuentos de la conquista / Hernandez De Alba, Gregorio – Bogota, Colombia. 1937 – 1r – us UF Libraries [972]
Cuentos de la tierra / Pinto, Julieta – San Jose, Costa Rica. 1963 – 1r – us UF Libraries [972]
Cuentos de la universidad / Belaval, Emilio S – San Juan, Puerto Rico. 1944 – 1r – us UF Libraries [972]
Cuentos de mi tia panchita / Lyra, Carmen – San Jose, Costa Rica. 1956 – 1r – us UF Libraries [972]
Cuentos de trapiche / Jimenez Canossa, Salvador – San Jose, Costa Rica. 1954 – 1r – us UF Libraries [972]
Cuentos del mar y otras paginas / Rodriguez Escudero, Nestor A – San Juan, Puerto Rico. 1959 – 1r – us UF Libraries [972]
Cuentos del viejo quilques / Maciel, Santiago – Buenos Aires, Argentina. 1928 – 1r – us UF Libraries [972]
Cuentos fragiles / Fiallo, Fabio – Madrid, Spain. 1929 – 1r – us UF Libraries [972]
Cuentos fragiles / Fiallo, Fabio – New York, NY. 1908 – 1r – us UF Libraries [972]
Cuentos ingenuos / Trigo, Felipe – Madrid: Renacimiento, 1920 – sp Bibl Santa Ana [946]
Cuentos insulares / Henriquez Ureena, Max – Buenos Aires, Argentina. 1947 – 1r – us UF Libraries [972]
Cuentos, leyendas e historietas minimas / Oliveros, Augusto Cesar – Guatemala, 1964 – 1r – us UF Libraries [972]
Cuentos mexicanos de autores contemporaneas / Mancisidor, Jose – Mexico City? Mexico. 1947? – 1r – us UF Libraries [972]
Cuentos nicaraguenses / Calero Orozco, Adolfo – Managua, Nicaragua. 1957 – 1r – us UF Libraries [972]
Cuentos panamenos de la ciudad y del campo / Valdes Alvarez, Ignacio De Jesus – Panama, Panama. 1928 – 1r – us UF Libraries [972]
Cuentos para fomentar el turismo / Belaval, Emilio S – San Juan, Puerto Rico. 1946 – 1r – us UF Libraries [972]
Cuentos para mi carmencita / Calderon Ramirez, Salvador – San Salvador, El Salvador. 1958 – 1r – us UF Libraries [972]
Cuentos pasionales / Hernandez Cata, Alfonso – Paris, France. 1910 – 1r – us UF Libraries [972]
Cuentos pinoleros / Calero Orozco, Adolfo – Managua, Nicaragua. 1945 – 1r – us UF Libraries [972]
Cuentos populares...extremadura / Hernandez de Soto, Sergio – 1886 – 9 – sp Bibl Santa Ana [830]
Cuentos pouplares cubanos / Feijoo, Samuel – Santa Clara, Cuba. v1-2. 1960 – 1r – us UF Libraries [972]
Cuentos realistas / Villalba Dieguez, Fernando – Caceres: Imprenta Mordena, 1954 – sp Bibl Santa Ana [946]
Cuentos viejos / Noguera, Maria De – San Jose, Costa Rica. 1952 – 1r – us UF Libraries [972]
Cuentos y articulos varios / Tapia Y Rivera, Alejandro – San Juan, Puerto Rico. 1938 – 1r – us UF Libraries [972]
Cuentos y chascarrillos andaluces – Madrid: F Fe 1898 [mf ed Bloomington IN: Indiana Uni Lib, Preservation Dept 1984] – 1r – 1 – us Indiana Preservation [390]
Cuentos y estampas / Melendez Munoz, Miguel – San Juan, Puerto Rico. 1958 – 1r – us UF Libraries [972]
Cuentos y leyendas costarricenses / Rodriguez Gutierrez, Rafael Armando – San Jose, Costa Rica. 1960 – 1r – us UF Libraries [972]
Cuentos y narraciones / Fernandez Juncos, Manuel – San Juan, Puerto Rico. 1907 – 1r – us UF Libraries [972]
Cuentos y narraciones / Gavidia, Francisco – San Salvador, El Salvador. 1961 – 1r – us UF Libraries [972]
Cuentos y poemas / Dario, Ruben – San Salvador, El Salvador. 1958 – 1r – us UF Libraries [972]
Cueppens, Dr Henry D *see* Paraguay ano 2000
Cuerda menor (1937-1939) / Feijoo, Samuel – Santa Clara, Cuba. 1964 – 1r – us UF Libraries [972]
El cuerdo en su casa / Vega Carpio, Lope de – Madrid: Rivadeneyra, 1857 – 1 – sp Bibl Santa Ana [946]
Cuerpo amoroso / Fabrega, Demetrio – Panama, Panama. 1962 – 1r – us UF Libraries [972]
Cuerpo de documentos del siglo 16 sobre los derechos de espana en las indias y las filipinas descubiertos y contados por... / Hanke, Lewis – Madrid: Missionalia Hispanica, 1947 – 1 – sp Bibl Santa Ana [946]
Cuerpo de documentos del siglo 17 sobre los derechos de espana en las indias y las filipinas, descubiertos y anotados por lewis hanke...mexico, 1943 / Hanke, Lewis – Madrid: Razon y Fe, 1947 – 1 – sp Bibl Santa Ana [946]
Cuerto Marquez, Luis *see* Independencia de las colonias hispano-americanas
Cuervo, Angel *see* Como se evapora un ejercito
Cuervo, Fray usto *see* Biografia de fr. luis de granada...demuestra...autor del libro de oracion
Cuervo, Justo OP *see* Fr luis de granada, verdadero y unico autor del libro de la oracion
Cuervo, Luis Augusto *see* Seleccion de discursos
Cuervo Marquez, Carlos *see*
– Estudios arqueologicos y etnograficos
– Prehistoria y viajes
Cuervo Marquez, Emilio *see* Introduccion al estudio de la filosofia de la hist...

Cuervo, Rufino Jose *see*
– Apuntaciones criticas sobre el lenguaje bogotano
– Disquisiciones sobre filologia castellana
– Obras ineditas de rufino j cuervo
Cuesta, I F de la *see* Breviarium gothicum, el de silos
Cuesta, M *see* Madrid. biblioteca nacional. catalogo de obras de iberoamerica y filipinas
Cuesta Mendoza, Antonio *see* Dominicos en el puerto rico colonial, 1521-1821
Cuestion africana en la isla de cuba / Santos Suarez, Joaquin – Madrid, Spain. 1863 – 1r – us UF Libraries [972]
Cuestion de actualidad / Castel, Joaquin – 1899 – 9 – sp Bibl Santa Ana [000]
Cuestion de belice (conferencia) / Alvarado, Rafael – Quito, Ecuador. 1949 – 1r – us UF Libraries [972]
Cuestion de cuba / Ablanedo, Juan Bautista – Sevilla, Spain. 1897 – 1r – us UF Libraries [972]
Cuestion de cuba en 1884 / Gomez, Juan Gualberto – Madrid, Spain. 1885 – 1r – us UF Libraries [972]
Cuestion de las religiones acatolicas en colombia – s.l, Colombia. 1956 – 1r – us UF Libraries [240]
La cuestion electoral / Pastor Diaz, Nicomedes – 1839 – 9 – sp Bibl Santa Ana [946]
Cuestion entre mexico y guatemala / Martinez Martin, Francisco Miguel – Mexico City? Mexico. 1882 – 1r – us UF Libraries [972]
Cuestion fronteriza dominico – haitiana / Machado, Manuel Arturo – Berlin, Germany. 1912 – 1r – us UF Libraries [972]
La cuestion religiosa en america / Bayle, Constantino – Madrid: Razon y Fe, 1922 – 1 – sp Bibl Santa Ana [240]
La cuestion romana y el marques de comillas / ed by Bayle, Constantino – Madrid: Razon y Fe, 1927 – 1 – sp Bibl Santa Ana [946]
Cuestion social / Anton, Fernando De – Sevilla, Spain. 1891 – 1r – us UF Libraries [025]
La cuestion social en diciembre de 1839 y enero de 1840 / Pastor Diaz, Nicomedes – Caceres: Imprenta de Don Lucas de Burgos, 1839 – 1 – sp Bibl Santa Ana [946]
La cuestion social en extremadura a la luz de las enciclicas rerum novarum y quadregesimo anno / Fernandez Santana, Ezequiel – Los Santos de Maimona: Imprenta Boletin Parroquial, 1935 – sp Bibl Santa Ana [946]
Cuestionario resumen de las ordenes y circulares sobre enlaces sindicales y juntas de jurados. cartillas del enlace sindical (secciones sindicales) / Delegacion Provincial de Sindicatos Vicesecretaria Provincial de Ordenacion Social – Caceres: Tip. El Noticiero, 1944 – sp Bibl Santa Ana [946]
Cuestiones actuales de doctrina y practica. buenos aires, 1927 / Macklin, J M – Madrid: Razon y Fe, 1930 – 1 – sp Bibl Santa Ana [240]
Cuestiones colombianas / Lopez Michelsen, Alfonso – Mexico City? Mexico. 1955 – 1r – us UF Libraries [972]
Cuestiones legislativas / Andreve, Guillermo – Leipzig, Germany. 1924 – 1r – us UF Libraries [323]
Cuestiones medico-legales y criminologicas / Uribe Cualla, Guillermo – Bogota, Colombia. 1951 – 1r – us UF Libraries [360]
Cuestriner zeitung oderblatt – Kuestrin (Kostrzyn PL), 1930 nov-dec – 1r – 1 – gw Misc Inst [077]
Cueto Y Mena, Juan De *see* Obras
Cueva sin quietud, cuentos / Monteforte Toledo, Mario – Guatemala, 1949 – 1r – us UF Libraries [972]
Cuevas, Juan Pablo *see* Doce gaviotas para una sola tierra
Cuevas, Manuel *see* Testamento de hernan cortes
Cuevas, Mariano *see* La virgen de guadalupe en mejico
Cuevas Zegueira, Sergio *see* Manuel de zegueira y arango y los albroes de la li...
Cuevas Zequeira, Sergio *see*
– En la contienda
– Ultima verba
Cuevry, Leonie von *see* Cornelly
Cugnet, Francois Joseph *see*
– An abstract of the several royal edicts and declarations, and provincial regulations and ordinances
– Traite de la police
Cuhna, Euclydes Da *see* Margem da historia
Cui Lhuin, U *see* U punna metta ca
Cui Mon, U *see* Van sui co bhva kri
Cui Mran, U *see* Buddha ron khrann nhan pu gam mrui
Cui Ra, Mon *see* Nve chann cha vatthu tui mya
Cui Tan, U *see* Khve chon cha ra kri e sa ma can ku thum kyam
Cui Van, U *see* Ca cuam ait chon sat pum
Cuinet, V *see* La turquie d'asie geographie administrative statistique descriptive et raisonnee de chaque province de l'asie-mineure
Cuisin, J P *see*
– Bonaparte
– Les crimes secrets de napoleon buonaparte
Cuisinieres / Brazier, Nicholas – Paris, France. 1824 – 1r – us UF Libraries [440]
Cuisset, Octave *see*
– Le pere coulange
– Popular treatise on the beet root culture and sugar fabrication in canada
– Traite populaire de la culture de la betterave et de la fabrication du sucre en canada
Le cuivre / Federation Nationale des Syndicats du Cuivre et Similaires – nos 1-105. Lyon. Oct 1894 – Apr 1903 – 1 – fr ACRPP [622]
Cuk na pal'ci – Buenos Aires AG, 1930* – 1r – 1 – (slovenian periodical) – us IHRC [073]
Culbert, David *see* Leni riefenstahl's 'triumph of the will'
Culbertson banner – Culbertson, NE: J H Corrick. -v16 n38. may 27 1921 (wkly) [mf ed 1908-21 (gaps) filmed 1957] – 3r – 1 – (absorbed by: palisade times) – us NE Hist [071]
Culbertson era – Culbertson, NE: Ira Cole (wkly) [mf ed 1895-1904 (gaps) filmed 1958] – 2r – 1 – (absorbed: hitchcock county herald) – us NE Hist [071]
The culbertson globe – Culbertson. NE: Nat L Baker, 1879 (wkly) [mf ed v1 n16. oct 4 1879 filmed 1973] – 1r – 1 – us NE Hist [071]
Culbertson, Michael Simpson *see*
– Darkness in the flowery land
– The religious condition of the chinese, and their claims on the church
Culbertson progress – Culbertson, NE: Hunsaker and Travis. 3v. v9 n19. oct 11 1928-v11 n28. dec 25 1930 (wkly) [mf ed with gaps] – 2r – 1 – (cont: progress (1920). merged with: tri-state news to form: progress of southwest nebraska) – us NE Hist [071]
Culbertson progress – Culbertson, NE: L W Mourer. 31v. v16 n42. may 21 1936-v46 n6. aug 5 1965 (wkly) [mf ed filmed -1978] – 7r – 1 – (cont: progress (1932), trenton register and: palisade times. absorbed by: trenton register. publ in culbertson ne, may 21 1936-jan 31 1952; in trenton ne, feb 7-28 1952; in culbertson ne, mar 6 1952-aug 5 1965) – us NE Hist [071]
Culbertson, Robert *see*
– Consolation to the church
– Covenanter's manual
– Pillar of rachel's grave
Culbertson sentinel – Culbertson, NE: R Knowles (wkly) [mf ed v5 n11. nov 15-dec 27 1895 (gaps) filmed 1974] – 1r – 1 – (cont: sentinel. cont by: people's sentinel) – us NE Hist [071]
Culbertson sun – Culbertson, NE: [s.n.] (wkly) [mf ed v12 n3. aug 21 1890 filmed 1999] – 1r – 1 – (sun (culbertson ne). issues for v12 n3- also called whole n1027-) – us NE Hist [071]
The culdees of the british islands as they appear in the history / Reeves, W – Dublin, 1864 – €17.00 – ne Slangenburg [941]
Cull, Edward Lefrey *see*
– Beet-root and beet-root sugar
– The whole history and mystery of beet-root and beet-root sugar
Cullen, Countee *see* The papers of 1921-1969
Cullen, Lucy Pope *see* Beyond the smoke that thunders
Culler, David D *see*
– Memories of old sandstone
– Problems of pulpit and platform
Culler's friend = L'ami du mesureur / Ladurantaye, J D – [Ottawa?: s.n.] c1920 [mf ed 1996] – 1mf – 9 – 0-665-80419-9 – (int in english/french) – mf#80419 – cn Canadiana [670]
Culley, David E *see* Hebrew-english vocabulary to the book of genesis
Culligan, C T *see* Exercise adherence
Cullinan, Bernice E *see* Research on children's and young adult literature
Cullingham, Aggie *see* Kitchen wisdom
Cullom Association. North Carolina. Warrenton Baptist Church *see* Church minutes
Cullum, George Washington *see*
– Biographical register of the officers and graduates of the us military academy at west point, ny since its establishment in 1802
– Biographical sketch of major-general richard montgomery
– Campaigns of the war of 1812-15, against great britain, sketched and criticised
Culmann, Hellmut *see* Teufelsmueller
Culmer zeitung – Kulm (Chelmno PL), 1925 13 oct-1926 24 dec, 1927 8 feb-9 sep, 1928 16 jul-19 sep – 1 – gw Misc Inst [077]
Culpepper, James Edward *see* Inductive preaching: an analysis of contemporary theory and practice
Culross, James *see*
– The resurrection and the life
– The three rylands: a hundred years of various christian service
Cult observer / American Family Foundation – 1984 jun-1986 feb – 1r – 1 – (cont: advisor [weston ma]; cont by: cultic studies journal; cultic studies review) – mf#1010722 – us WHS [071]
The cult of ali / Sell, Edward – London: Christian Literature Society for India, 1910 – (= ser The Islam Series) – 1mf – 9 – 0-524-02613-0 – mf#1990-3063 – us ATLA [260]
The cult of othin : an essay in the ancient religion of the north / Chadwick, Hector Munro – London: CJ Clay, 1899 – 1mf – 9 – 0-524-00708-X – mf#1990-2036 – us ATLA [290]
Le culte catholique : ou, exposition de la foi de l'eglise romaine... / Begin, Louis-Nazaire – Quebec: A Cote, 1875 – 3mf – 9 – mf#02334 – cn Canadiana [241]
Le culte de cybele, mere des dieux : a rome et dans l'empire romain / Graillot, Henri – Paris: Fontemoing, 1912 – 2mf – 9 – 0-524-06228-5 – (incl bibl ref) – mf#1991-0021 – us ATLA [250]
Le culte de la b. vierge marie, mere de dieu : nouvelles conferences prechees a paris, a lyon, en belgique, etc... / Combalot, Abbe – Lyon: Imprimerie catholique de perisse freres; Paris: R Ruffet. 2v. 1865 – 4mf – 9 – 0-8370-8973-5 – (incl bibl ref) – mf#1986-2973 – us ATLA [240]
Le culte de la sainte vierge en afrique : d'apres les monuments archeologiques / Delattre, Alphonse J – Paris: Societe St-Augustin; Lille: Desclee, De Brouwer [1907?] [mf ed 1986] – xii/232p (ill) on 1mf – 9 – 0-8370-8334-6 – (in french. incl bibl ref) – mf#1986-2334 – us ATLA [960]
Le culte des ancetres et le culte des morts chez les arabes / Goldziher, Ignac – Paris: E Leroux, 1885 [mf ed 1991] – 1mf – 9 – 0-524-01486-8 – (incl bibl ref) – mf#1990-2462 – us ATLA [260]
Le culte des morts dans le celeste empire et l'annam : compare au culte des ancetres dans l'antiquite occidentale / Bouinais, Albert & Paulus, A – Paris: Ernest Leroux 1893 [mf ed 1991] – (= ser Annales du musee guimet) – 1mf – 9 – 0-524-01682-8 – (incl bibl ref) – mf#1990-2584 – us ATLA [390]
Le culte des saints musulmans dans l'afrique du nord : et plus specialement au maroc / Montet, Edouard Louis – Geneve: Georg, 1909 – 1mf – 9 – 0-524-01849-9 – (incl bibl ref) – mf#1990-2684 – us ATLA [260]
Un culte dynastique : avec evocation des morts chez les sakalaves de madagascar, le "tromba" / Rusillon, Henry – Paris: A Picard, 1912 [mf ed 1992] – 1mf – 9 – 0-524-02050-7 – mf#1990-2825 – us ATLA [290]
Un culte dynastique avec evocation des morts chez les sakalaves de madagascar, le "tromba" / Rusillon, Henry – Paris: A Picard, 1912 – 1 – us CRL [960]
Le culte et les fetes d'adaonis-thammouz dans l'orient antique / Vellay, Charles – Paris: Ernest Leroux 1904 [mf ed 1991] – (= ser Annales du musee guimet 16) – 1mf [ill] – 9 – 0-524-01633-X – (incl bibl ref) – mf#1990-2572 – us ATLA [290]
Les cultes paiens dans l'empire romain / Toutain, Jules – Paris: E Leroux, 1907-17 [mf ed 1993] – (= ser Bibliotheque de l'ecole des hautes etudes) – 13mf – 9 – 0-524-07958-7 – (in french) – mf#1991-0208 – us ATLA [250]
Cultivation of citrus groves / Hume, H Harold – Lake City, FL. 1904 – 1r – us UF Libraries [634]
Cultivation of flax : practical hints on the cultivation and treatment of the flax plant, expressly for the use and benefit of the canadian farmer / Donaldson, J A – [Toronto?: s.n.], 1865 [mf ed 1985] – 1mf – 9 – 0-665-49008-9 – mf#49008 – cn Canadiana [630]
The cultivation of sugar beets / Shuttleworth, Arthur E – [Toronto?: Ontario Dept of Agriculture, 1900?] – 1mf – 9 – 0-665-93764-4 – mf#93764 – cn Canadiana [636]
Cultivator – Albany. 1834-1865 (1) – mf#3972 – us UMI ProQuest [630]
El culto a la eucaristia en la espana roja / Bayle, Constantino – Burgos: Razon y Fe, 1938 – 1 – sp Bibl Santa Ana [240]
El culto a ma-bellona en la espana romana / Garcia y Bellido, Antonio – Madrid, 1956 – 1 – sp Bibl Santa Ana [946]
Culto antiguo de san masona metropolitano de merida / Smedt, C de – Madrid: Fortanet, 1885. B.R.A.H. vi/pp. 141-142 – 1 – sp Bibl Santa Ana [240]
Culto as letras : periodico scientifico e litterario – Recife, PE: Typ da Provincia, 20 maio-ago 1873; jul 1875 – (= ser Ps 19) – mf#P17,02,143 – bl Biblioteca [079]
Il culto privato di roma antica / Marchi, Attilio de – Milano: U Hoepli, 1896-1903 – 2mf – 9 – 0-524-04862-2 – (incl bibl ref) – mf#1990-3424 – us ATLA [250]
Cultos emeritenses de serapis y de mithras / Melida, Jose Ramon – Madrid: Tip. Fortanet, 1914. B.R.A.H. LIV, pp. 439-456 – sp Bibl Santa Ana [946]
Cultos en honor de la santisima virgen de argeme. patrona de la ciudad de coria y consagracion de su nuevo santuario / Coria. Ayuntamiento – Plasencia: Imp. La Victoria, 1972 – 1 – sp Bibl Santa Ana [240]
Cultos profanos / Gomez Carrillo, Enrique – Paris, France. 1910? – 1r – us UF Libraries [972]
Cults, customs and superstitions of india : being a revised and enlarged edition of indian life, religious and social / Oman, John Campbell – London: TF Unwin, 1908 – 1mf – 9 – 0-524-02432-4 – mf#1990-3016 – us ATLA [390]
Cults, myths and religions = Cultes, mythes et religions / Reinach, Salomon – London: David Nutt, 1912 – 1mf – 9 – 0-524-01293-8 – (in english) – mf#1990-2329 – us ATLA [200]
The cults of ostia / Taylor, Lily Ross – Bryn Mawr, Pa: Bryn Mawr College, 1912 – (= ser Bryn mawr college monographs) – 1mf – 9 – (incl bibl ref) – mf#1990-2344 – us ATLA [250]
The cults of the greek states / Farnell, Lewis Richard – Oxford: Clarendon Press, 1896-1909 – 6mf – 9 – 0-524-06227-7 – (incl bibl ref) – mf#1991-0020 – us ATLA [250]
Cultura – Tegucigalpa. v.1-11. July 1939-Dec 1951 – 1 – us L of C Photodup [370]
Cultura amenazada : la luso-brasilena / Freyre, Gilberto – Buenos Aires, Argentina. 1943 – 1r – us UF Libraries [972]
Cultura brasileira / Azevedo, Fernando De – Sao Paulo, Brazil. 1944 – 1r – us UF Libraries [972]
Cultura colonial en panama (ensayos) / Miro, Rodrigo – Mexico City? Mexico. 1950 – 1r – us UF Libraries [972]
Cultura como empresa multinacional / Mattelart, Armand – Mexico City? Mexico. 1976 – 1r – us UF Libraries [025]
Cultura e fe – Porto Alegre, Brasil: Instituto de Desenvolvimento Cultural. v1-12 n51. apr/jun 1978-90 – us CRL [972]
Cultura e opulencia do brasil / Antonil, Andre Joao – Salvador, Brazil. 1950 – 1r – us UF Libraries [972]
Cultura en la espana republicana / Marinello, Juan – N.Y., 1937. Fiche W 1501. (Blodgett Collection of Spanish Civil War Pamphlets) – 9 – us Harvard [946]
La cultura en mexico – [Mexico: Editorial Siempre! n1-814. feb 21 1962-sep 30 1977 (81r); n827-1342 1978-1987 (32r) – (suppl to: siempre (mexico city, mexico). filmed with main title) – us CRL [972]
La cultura espanola medieval : datos bio-bibliograficos para su historia, tomo 1, a-g / Vera, Francisco – Madrid: Imprenta Gongora, 1933 – 1 – sp Bibl Santa Ana [946]
La cultura espanola medieval : datos bio-bibliograficos para su historia. tomo 2. h-z / Vera, Francisco – Madrid: Imprenta Gongora, 1934 – 1 – sp Bibl Santa Ana [946]
Cultura indigena de guatemala / Seminario De Integracion Social Guatemalteca – Guatemala, 1956 – 1r – us UF Libraries [972]
Cultura indigena de guatemala / Seminario De Integracion Social Guatemalteca – Guatemala, 1959 – 1r – us UF Libraries [972]
Cultura literaria / Facio, Justo A – San Jose, Costa Rica. 1930 – 1r – us UF Libraries [440]
Cultura maya : caracter y creaciones de esta gran civilizaciíon precolombina / Soto-Hall, Maximo – Buenos Aires: Atlantida, s.a, [1941] (mf ed 19–) – (= ser Coleccion Antorcha; Harvard anthropology preservation microfilm project) – mf#ZH-429 – us NY Public [972]
Cultura turcica – Ankara, 1964-1967. v1-4 – 21mf – 8 – mf#NE-114 – ne IDC [956]
Cultura y economia en colombia, ecuador, venezuela – Bogota, Colombia. 1956 – 1r – us UF Libraries [972]
La cultura y la imprenta europeas en el japon durante los siglos 16 y 17. la iniciativa espanola base de la importante gesta (1548-1616) / Vindel, Francisco – Madrid: Razon y Fe, 1944 – 1 – sp Bibl Santa Ana [306]
Cultura y las letras coloniales en santo domingo / Henriquez Urena, Pedro – Buenos Aires, Argentina. 1936 – 1r – us UF Libraries [972]
Cultural and historical geography of southwest guam / Mcbryde, Felix Webster – Washington, DC. 1947 – 1r – us UF Libraries [972]
A cultural and historical study of selected women's dance from herat, afganistan : 1970-1980 / St John, Katherine – 1993 – 3mf – $12.00 – us Kinesology [790]
Cultural anthropology – Washington. 1986+ (1,5,9) – ISSN: 0886-7356 – mf#16333 – us UMI ProQuest [301]
Cultural conflicts in northern sotho dramas / Madiga, Raofa Philemon – Uni of South Africa 2000 [mf ed Johannesburg 2000] – 3mf – 9 – (incl bibl ref) – mf#mfm14789 – sa Unisa [470]

Cultural correspondence – 1975 aug-1979 fall, 1981 sum, 1982 apr-1986 – 3r – 1 – mf#669078 – us WHS [071]

Cultural creations. / Comite Tecnico de Ayuda a los Espanoles en Mexico – Mexico, 1940. Fiche W 821. (Blodgett Collection of Spanish Civil War Pamphlets) – 9 – us Harvard [946]

Cultural critique – New York. 1989+ (1,5,9) – ISSN: 0882-4371 – mf#17994 – us UMI ProQuest [700]

Cultural diversity program / Human Relations Area Files – 1988. 46 cultural files constituting a teaching sample – 9 – apply for price – us HRAF [306]

The cultural east – v1 n1-2. jul 1946-aug 1947 [complete] – 1r – 1 – (filmed with: eastern buddhist) – mf#ATLA S0068B – us ATLA [280]

The cultural east *see* Eastern buddhist...

The cultural heritage of india : sri ramakrishna centenary memorial – Calcutta: Sri Ramakrishna Centenary Committee, [1932?]-1953 – (= ser Samp: indian books) – us CRL [954]

A cultural history of assam : early period / Barua, Birinchi Kumar – Nowgong, Assam: K K Barooah, 1951- – (= ser Samp: indian books) – us CRL [954]

A cultural history of india during the british period / Ali, Abdullah Yusuf – Bombay: D B Taraporevala Sons & Co, 1940 – (= ser Samp: indian books) – us CRL [954]

Cultural nationalism and colonialism in nineteenth-century irish horror fiction / Glisson, Silas Nease – Uni of South Africa 2000 [mf ed Johannesburg 2000] – 5mf – 9 – (incl bibl ref) – mf#mfm14770 – sa Unisa [420]

Cultural post / National Endowment for the Arts – iss17-v9 n1 [1978 may/jun-1983 jun] – 1r – 1 – (cont by: arts review [washington dc]) – mf#330895 – us WHS [700]

Cultural practices for root-knot control of annual crops of cigar-wrapper tobacco / Kincaid, Randall R – Gainesville, FL. 1943 – 1r – us UF Libraries [630]

Cultural types program / Human Relations Area Files – 1988. Modules of 3 to 6 cultural files each. Limited to two-year colleges – 9 – us HRAF [306]

The cultural unity of asia / Cousins, James Henry – Adyar, Madras: Theosophical Pub House, 1922 – (= ser Samp: indian books) – us CRL [950]

Cultural values – Oxford. 1997+ (1) – ISSN: 1362-5179 – mf#25718 – us UMI ProQuest [306]

Culture : revue trimestrielle, sciences religieuses et profanes au canada / Association des recherches sur les sciences religieuses et profanes au Canada – Quebec: l'Association. v5 [i.e. 1] n1 mars 1940-v32 n1 mars 1971 [mf ed 1976] – 5r – 5 – (cont: nos cahiers; with ind; in french & english) – mf#SEM16P260 – cn Bibl Nat [073]

Culture – s.l, s.l? 193-? – 1r – us UF Libraries [978]

Culture and cytological development of psilocybe cubensis / Can, Ngo Hu – Gainesville, FL. 1974 – 1r – us UF Libraries [500]

Culture and marketing of tea / Harler, Campbell R – London, England. 1956 – 1r – us UF Libraries [630]

Culture and practical power : an address delivered at the opening of lansdowne college, portage la prairie, november 11th, 1889 / Davin, Nicholas Flood – Regina: Leader Co, 1889 – 1mf – 9 – mf#30433 – cn Canadiana [306]

Culture and religion in some of their relations / Shairp, John Campbell – New York: Hurd and Houghton; Cambridge: Riverside Press, 1872 – 1mf – 9 – 0-8370-5240-8 – mf#1985-3240 – us ATLA [306]

Culture and the gospel : or, a plea for the sufficiency of the gospel to meet the wants of an enlightened age / McCall, Salmon – New York: A D F Randolph, 1871, c1870 [mf ed 1985] – 1mf – 9 – 0-8370-4334-4 – mf#1985-2334 – us ATLA [230]

Culture des idees / Gourmont, Remy De – Paris, France. 1910 – 1r – us UF Libraries [025]

La culture du ginseng : traite complet et illustre / Grignon, Wilfrid – Sainte-Adele, Que[bec: s.n, 1907?] – 1mf – 9 – 0-665-74365-3 – mf#74365 – cn Canadiana [631]

La culture du tabac / Lippens, Bernard – Montreal: E Senecal, 1882 – 1mf – 9 – mf#09121 – cn Canadiana [630]

Culture et preparation du tabac : a l'usage de l'amateur et du cultivateur de tabac en particulier... / Laroque, G – Levis, Quebec: Mercier, 1881 – 1mf – 9 – mf#08539 – cn Canadiana [630]

Culture, fertilizer requirements and fiber yields of ramie in the florida everglades / Neller, J R – Gainesville, FL. 1945 – 1r – us UF Libraries [630]

La culture francaise en russie (1700-1900) / Haumant, E – Paris, 1913 – 571p 11mf – 9 – mf#R-18551 – ne IDC [947]

La culture fruitiere dans la province de quebec : traite complet de la propagation des arbres et arbustes fruitiers cultives dans la province de quebec... / Leopold, pere – La Trappe, Que[bec]: Institut agricole d'Oka, [1914] [mf ed 1998] – 3mf – 9 – 0-665-65178-3 – mf#65178 – cn Canadiana [634]

Culture – loisirs – Quebec *see* Loisirs quebec

Culture, medicine and psychiatry – Dordrecht. 1984+ (1,5,9) – ISSN: 0165-005X – mf#14744 – us UMI ProQuest [306]

The culture of ancient israel / Cornill, Carl Heinrich – Chicago: Open Court, 1914 – 1mf – 9 – 0-7905-1748-5 – (incl ind) – mf#1987-1748 – us ATLA [939]

Culture of cities / Mumford, Lewis – New York, NY. 1938 – 1r – us UF Libraries [307]

The culture of justice : a mode of moral education and of social reform / Du Bois, Patterson – New York: Dodd, Mead, 1907 [mf ed 1986] – 1mf – 9 – 0-8370-8504-7 – (incl bibl ref & ind) – mf#1986-2504 – us ATLA [230]

The culture of personality / Randall, John Herman – New York: Dodge, c1912 – 2mf – 9 – 0-7905-9077-8 – mf#1989-2302 – us ATLA [100]

The culture of religion : elements of religious education / Wilm, Emil Carl – Boston: Pilgrim Press, c1912 – 1mf – 9 – 0-7905-9763-2 – (incl bibl ref) – mf#1989-1488 – us ATLA [200]

The culture of simplicity / McLeod, Malcolm James – 4th ed. New York: Fleming H Revell, c1904 – 1mf – 9 – 0-8370-6333-7 – mf#1986-0333 – us ATLA [240]

The culture of the best : and the manufacture of beet sugar / Child, David Lee – Boston: Weeks, Jordan, 1840 – 1 – us CRL [631]

The culture of the soul among western nations / Ramanathan, Ponnambalam – New York: Putnam, 1906 [mf ed 1992] – 1mf – 9 – 0-524-03309-9 – mf#1990-3194 – us ATLA [230]

Culture of tobacco / Moodie, F B – Lake City, FL. 1895 – 1r – us UF Libraries [630]

La culture physique de la femme / Parnet, Max – Paris: Nilsson, 1913 – (= ser Les femmes [coll]) – 2mf – 9 – mf#11418 – fr Bibl Nationale [790]

Culture vivante / Quebec (Province). Ministere des affaires culturelles – Quebec: le Ministere. n1 mars 1966-n29/30 sep 1973 [mf ed 1977] – 1r – 1 – mf#SEM35P147 – cn Bibl Nat [700]

Culturele Voorlichting RVD *see* Indonesie cultureel

Culturele voorlichting rvd – Batavia, 1948-1949 – 5mf – 9 – (missing: 1948(1, 3-8); 1949(10, 13, 15)) – mf#SE-785 – ne IDC [959]

The cultures of prehistoric egypt : from the ashmolean museum, oxford / Baumgartel, Elise J – 2v on 5mf – 9 – mf#87380 – uk Microform Academic [930]

Culturgeschichte der israeliten der ersten halfte... / Jost, Isaak Markus – Breslau, Germany. 1846 – 1r – us UF Libraries [939]

Culturgeschichtliche streifzuege auf dem gebiete des islams / Kremer, Alfred, Freiherr von – Leipzig: FA Brockhaus, 1873 – 1mf – 9 – 0-524-01613-5 – mf#1990-2552 – us ATLA [260]

Culver, April D *see* Comparison of behavior modification techniques used in physical education with institutionalized profoundly mentally retarded students at different ages

Culver city – 1947-48; 1972-92 – (= ser California telephone directory coll) – 27r – 1 – $1350.00 – mf#P00022 – us Library Micro [917]

[Culver city-] culver city independent – CA. Oct 1957- – 270r – 1 – $16,200.00 (subs $120/y) – mf#R02154 – us Library Micro [071]

[Culver city-] culver city news – CA. 1958- – 144r – 1 – $8640.00 (subs $200/y) – mf#H04041 – us Library Micro [071]

[Culver city-] culver city star news – CA. 1941-1985 – 394r – 1 – $23,640.00 – (various titles including: crenshaw-news, hawthorne-press tribune, inglewood-news, la brea/inglewood-the daily news, lawndale-tribune, lennox-citizen, rancho-chevoit hills-news, torrance-news, venice-vanguard, westchester-star news) – mf#H03195 – us Library Micro [071]

[Culver city-] evening Star News – CA. 1945-56 – 53r – 1 – $3180.00 – mf#C02153 – us Library Micro [071]

Culver crossings – v1-2, 3-4 [1985 apr-may, jul-aug], v5 [1986 jul] – 1r – 1 – (cont by: colver/culver crossings) – mf#1703966 – us WHS [071]

Culver sity/marina del rey – 1993- – (= ser California telephone directory coll) – 2r – 1 – $100.00 – mf#P00023 – us Library Micro [917]

Culver/colver crossings – v7 [1987 feb], v8 [1988 mar] – 1r – 1 – (cont: colver/culver crossings) – mf#2604902 – us WHS [071]

Culverwel, Nathanael *see* of the light of nature

Culwick, A T *see* Good out of africa

Culwick, Arthur Theodore *see*
- Back to the trees
- Britannia waives the rules
- Don't feed the tiger

Culy, David *see* Glory of the two crown's heads, adam and christ, unveill'd

Cum thok u than kyo : [short stories] / Rvhe Myha, Dagun – Ran Kun: Do Cin Nu Pru [Mra Sve] 1980 [mf ed 1990] – 1r with other items – 1 – (in burmese) – mf#mf-10289 seam reel 179/1 [§] – us CRL [830]

Cumaneses ilustres / Sanabria, Alberto – Caracas, Venezuela. 1965 – 1r – us UF Libraries [972]

Cumberland advocate – Cumberland WI. 1885 apr 2/1885, 1887 jul/1890-2002 jan/dec – 98r – 1 – (with gaps; cont: cumberland herald) – mf#982935 – us WHS [071]

Cumberland and reg. – Camp Hill, PA., 1805-1809 – 13 – $25.00r – us IMR [071]

Cumberland and westmoreland, ancient and modern / Sullivan, Jeremiah – London: Whittaker; Kendal: J Hudson, J Dawson, and J Robinson 1857 [mf ed Bloomington IN: Indiana Uni Lib, Preservation Dept 1984] – iii 171p on 1r – 1 – us Indiana Preservation [390]

The Cumberland and Westmoreland Antiquarian and Archaeological Society *see* Transactions

Cumberland and Westmoreland Antiquarian and Archaeological Society *see* Publications of the cumberland and westmoreland antiquarian and archaeological society

Cumberland argus – Parramatta, sep 1888-oct 1962 – 55r – A$3156.03 vesicular A$3458.53 silver – at Pascoe [079]

Cumberland, Barlow *see*
- The northern lakes of canada
- A sketch of how "the diamond anthem" was sung around the world through the colonies of the empire on the 20th june, 1897

Cumberland city first baptist church. cumberland city, tennessee : church records – 1914-64 – 1 – 8.82 – us Southern Baptist [242]

The cumberland coal fields, nova-scotia : report of j campbell, practical geologist, on the coals of the south shore of chignecto channel, or joggins coal, october 1871 – S.l: s.n, 1871? – 1mf – 9 – mf#00961 – cn Canadiana [550]

Cumberland County Association for Indian People [NC] *see* Saponi

Cumberland County Historical Society *see*
- Cumberland patriot
- News letter

Cumberland flag / Colored Cumberland Presbyterian Church – 1936 aug 31-1937 aug 31, 1967 feb 15-1968 apr 15, 1978 mar-1989 dec, 1990-93, 1994-96 – 4r – 1 – mf#2362424 – us WHS [242]

[Cumberland, G] *see* Thoughts on outline, sculpture, and the system that guided the ancient artists in composing their figures and groupes...

Cumberland, George *see*
- An essay on the utility of collecting the best works...engravers
- Outlines from the antients

Cumberland herald – Cumberland WI. [1882 jan 18-1885 mar 26] – 1r – 1 – (cont by: cumberland advocate) – mf#1005868 – us WHS [071]

Cumberland islander – British Columbia, CN. jan 1910-jul 1931 – 9r – 1 – cn Commonwealth Imaging [071]

Cumberland journal – Cumberland WI. 1912 jan-1913 jun, 1913 jul-1914 dec – 2r – 1 – mf#963564 – us WHS [071]

Cumberland law review – Birmingham. 1975+ (1,5,9) – (cont: cumberland-samford law review) – ISSN: 0360-8298 – mf#7980,01 – us UMI ProQuest [340]

Cumberland maps : from the royal archives and library at windsor castle – (= ser Military maps and papers of william augustus, duke of cumberland, 18th c) – 10r – 1 – mf#96766 – uk Microform Academic [910]

Cumberland maps / Royal Library. Windsor Castle – [mf ed 1987] – 692 col mf – 15 – $12,000.00 – 0-907716-19-9 – (over 4500 military maps & docs fr 15th-19th c; with printed chronological & geographical ind) – uk Mindata [912]

Cumberland maps: catalogue and index : from the royal archives and library at windsor castle – (= ser Military maps and papers of william augustus, duke of cumberland, 18th c) – 1r – 1 – mf#96750 – uk Microform Academic [941]

Cumberland mercury – Parramatta, jan 1875-apr 1895 – 7r – A$479.91 vesicular A$518.41 silver – at Pascoe [079]

Cumberland metropolitan – Carlisle, PA., 1823 – 13 – $25.00r – us IMR [071]

Cumberland news – Carlisle, England. Jun 1910-1913; 1973- – 85+ r – 1 – uk British Libr Newspaper [072]

Cumberland news [evening ed] – aug 12 1914-68, 1979-feb 1983, nov 1984-90, jan 4 1991-92, jan 2-dec 1993, jan 4 1994-jul, aug-nov [incl sunday news], dec 1995-jun 1997 – 293r – 1 – (aka: cumberland evening news; cumberland evening news and star; news and star) – uk British Libr Newspaper [072]

Cumberland pacquet, the... : or ware's whitehaven advertiser, 1774-83 – 4r – 1 – mf#583 – uk Microform Academic [072]

Cumberland pacquet & ware's whitehaven advertiser – [NE England] Cumbria 1774-1910 [mf ed 2004] – 90r – 1 – (missing: 1811, 1812, 1884, 1897; cont as: west cumberland post etc [1898]; cumberland pacquet etc [jan 1899-1910]) – uk Newsplan [072]

Cumberland papers : from the royal archives and library at windsor castle – (= ser Military maps and papers of william augustus, duke of cumberland, 18th c) – 102r – 1 – mf#96633 – uk Microform Academic [920]

Cumberland papers, index : from the royal archives and library at windsor castle – (= ser Military maps and papers of william augustus, duke of cumberland, 18th c) – 1r – 1 – mf#96651 – uk Microform Academic [920]

Cumberland patriot / Cumberland County Historical Society – 1978 spr-1982 sum – 1r – 1 – mf#615804 – us WHS [978]

Cumberland Presbyterian Church *see* Cumberland presbyterian church

Cumberland presbyterian church : minutes / Cumberland Presbyterian Church – 1966-90 [complete] – Inquire – 1 – ISSN: 0011-2976 – mf#ATLA S0509 – us ATLA [242]

Cumberland Presbyterian Church. General Assembly *see* Minutes, 1848-1942

Cumberland presbyterian review – St. Louis. 1846-1884 (1) – mf#4841 – us UMI ProQuest [242]

Cumberland. Presbytery (Cum. Pres. Ch.) *see* Minutes, 1810-13; cumberland. synod

Cumberland, R *see* Anecdotes of eminent painters in spain

Cumberland register – Carlisle, PA. -w 1805-09. 1 roll – 13 – $25.00r – us IMR [071]

Cumberland, Richard *see*
- Few plain reasons why we should believe in christ and adhere to his...
- Sanchoniatho's phoenician history

Cumberland road : selected engineering bridges contracts, 1825-30 – 1r – 1 – mf#B27150 – us Ohio Hist [380]

Cumberland sound as a harbor of naval rendezvous... – Washington, DC. 1886? – 1r – us UF Libraries [978]

Cumberland times – Clintwood, VA. 1988-1992 (1) – mf#68306 – us UMI ProQuest [071]

Cumberland valley journal – Mechanicsburg, PA. -w 1860-63. 1 roll – 13 – $25.00r – us IMR [071]

Cumberland-samford law review – Birmingham. 1970-1975 (1) 1972-1975 (5) 1973-1975 (9) – (cont by: cumberland law review) – ISSN: 0045-9275 – mf#7980 – us UMI ProQuest [340]

Cumbernauld news and kilsyth chronicle – Cumbernauld: F Johnston & Co Ltd 1966- (wkly) [mf ed 5 jan 1994-] – 1 – (cont: cumbernauld news; sister publ to: kilsyth chronicle; foll suppls are currently entered separately: cumbernauld & kirkintilloch today) – ISSN: 0964-3885 – uk Scotland NatLib [072]

Cumhuriyet – Iskece, GR. Siyasi Tuerkce gazetedir. Sahibi ve Mesul Mueduerue: Ibrahim Demir. n1. 2 subat 1934 – (= ser O & t journals) – 1mf – 9 – $25.00 – us MEDOC [956]

Cumhuriyet – Istanbul: Cumhuriyet, 1956- – 1 – us CRL [949]

Cumhuriyet – Istanbul, Turkey. 1924-1942 (1) – mf#67867 – us UMI ProQuest [079]

Cumhuriyet – [Turkey], 1980- – 1 – enquire for prices – (yrly reel count varies) – us UMI ProQuest [079]

Cuming county advertiser – West Point, NE: M O Gentzke. 12v. v1 n1. apr 24 1889-12th yr n33. nov 27 1900 (wkly) – 4r – 1 – (merged with: west point republican to form: west point republican and cuming county advertiser) – us Bell [071]

Cuming county democrat – West Point, NE: Thiele & Miller. v23 n30. feb 10 1899-v99 n31. dec 26 1974 [wkly] [mf ed jun 13 1902-dec 26 1974 [gaps] filmed 1975] – 30r – 1 – (cont: west point progress; absrobed: bancroft bugle; merged with: west point republican [917] to form: west point newspapers; "twin wkly newspaper" of west point republican [1917] mar 1 1973-dec 26 1974) – us NE Hist [071]

Cuming, G J *see* The durham book

Cumming, Evelyn Ficarra *see* Linear properties of the writings of ralph vaughan williams

Cumming, J *see*
- Almost protestant, and the almost romanist
- God in history
- Invocation and intercession of saints
- Pope, the man of sin
- Protestant objections

Cumming, James Elder *see* Through the eternal spirit
Cumming, John *see*
- Apocalyptic sketches
- Benedictions
- Challenge to cardinal wiseman
- Christ our passover
- Cumming's minor works. [third series]
- Foreshadows
- God in history
- The great preparation
- The great tribulation
- The hammersmith protestant discussion
- The immaculate conception
- is christianity from god
- Lord taketh away
- Notes on the cardinal's manifesto
- On doing what one does with one's might
- Present state of the church of scotland
- Redemption draweth nigh
- Redemptorists at clapham
- Revealing india's past
- Romish miracles
- Signs of the times
- Synopsis papismi
- Teach us to pray
- Tent and the altar...sketches from patriarchal life

Cumming, Robert Cushing *see*
- The annotated corporation laws of all the states, generally applicable to stock corporations
- The insurance laws of the state of new york.
- The laws of new york state relating to general, religious and non-business corporations, taxation and exemption, sunday observance, marriage and divorce

Cumming, W P *see*
- Revelationes of saint birgitta
- The revelationes of saint birgitta

Cummings, Anson Watson *see* The early schools of methodism
Cummings, Edward Estlin *see* Tulips and chimneys
Cummings, Ephraim Chamberlain *see* Nature in scripture
Cummings, John N *see* Letters
Cumming's minor works. [third series] = Selections. 1859 / Cumming, John – Philadelphia: Lindsay and Blakiston, 1859 – 1mf – 9 – 0-7905-9259-2 – mf#1989-2484 – us ATLA [240]
Cummings-Danson, G *see* The differences in educational preparation and athletic experience
Cummington 1751-1902 – Oxford, MA (mf ed 1988) – (= ser Massachusetts vital records) – 43mf – 9 – 0-87623-083-4 – (mf 1-6: town & vital records 1762-1856. mf 7: proprietors 1762-79. mf 8-12: vital records 1744-1844. mf 13-18: town & vital records 1770-1835. mf 19-20: b,m,d 1844-52. mf 21-23: index to intentions 1780-1986. mf 24-28: intentions of marriage 1780-1986. mf 29: rebellion records 1861-65. mf 30-31: index to births 1853-1986. mf 32-33: index to marriages 1853-1986. mf 34-35: index to deaths 1853-1986. mf 36-40: deaths 1853-1987. mf 41: marriages 1853-94. mf 42-43: births 1853-1902) – us Archive [978]
Cummins, Alexandrine Macomb *see* Memoir of george david cummins
Cummins, George David *see* Memoir of george david cummins
Cummins, Henry I *see* Declaration of the clergy against alteration of the book of common prayer
Cummins, James Sheldon *see* State and territorial general statutes
Cummins' territory reports / Idaho. Supreme Court – 1v. 1866-1867 (all publ) – (= ser Idaho Supreme Court Reports) – 3mf – 9 – $4.50 – (a pre-nrs title) – mf#LLMC 91-031 – us LLMC [347]
Cumnock chronicle – Cumnock: A Guthrie & Sons Ltd 1979- (wkly) [mf ed 2 jan 1992-] – 1 – (not publ: 13 mar 1980, 9 jul-13 aug 1981, 24 jul 1998; cont: cumnock chronicle & muirkirk advertiser; imprint varies) – ISSN: 1357-9967 – uk Scotland NatLib [072]
Cumnock express and mauchline and catrine advertiser – [Scotland] East Ayrshire, Cumnock: G McMillan jun 1889-feb 1909 (wkly) [mf ed 2003] – 20r – 1 – (began in 1871) – uk Newsplan [072]
Cumont, Fr *see*
- Lux perpetua
- Les religions dans le paganisme romain
- Textes et monuments figures relatifs aux mysteres de mithra

Cumont, Franz Valery Marie *see*
- Astrology and religion among the greeks and romans
- The mysteries of mithra
- The oriental religions in roman paganism
- Recherches sur le manicheisme

Cumper, George E *see* Social structure of jamaica
El cumplimiento de las ogligaciones / Lumbreras Valiente, Pedro – Valencia, 1974 – 1 – sp Bibl Santa Ana [946]
Cumtux : quarterly / Clatsop County Historical Society – 1980 win-1989 fall – 1r – 1 – mf#1055019 – us WHS [978]
Cumulated index kewensis / Kew. Royal Botanical Gardens – 484mf – 9 – $2000.00 – 1-900853-25-6 – (consolidation of over 1 million taxonomic entries in the index kewensis publ in printed form fr 1893-1975) – uk Mindata [580]
Cumulated index medicus – Bethesda. 1960+ [1]; 1970+ [5]; 1975+ [9] – ISSN: 0090-1423 – mf#1417 – us UMI ProQuest [610]
The cumulative annual index to the current digest of the post-soviet press – v28 1976-v52 2000 – 9,1 – ($49.50v for v28-47 (1976-1995) $8v for v48-52 (1996-2000)) – ISSN: 1074-0007 – us Current [077]
Cumulative bulletin, income tax rulings / U.S. Treasury Dept – Bureau of Internal Revenue. [irregular] – 21mf – 9 – $31.50 – (cont by: irs cumulative bulletin) – mf#llmc 82-707 – us LLMC [336]
The cumulative effect of multiple phonophoreis treatments on dexamethasone and cortisol concentrations in the blood / Strapp, Edward J – 1999 – 1mf – 9 – $4.00 – mf#PE 4046 – us Kinesology [617]
The cumulative effects of multiple exercise bouts of equal caloric expenditure on excess post-exercise oxygen consumption / Williams, Jeffrey P – 1989 – 64p 1mf – 9 – $4.00 – us Kinesology [612]
Cumulative index to a selected list of periodicals : annual – v1-3. 1896-98 [complete] – 1r – 1 – mf#ATLA S0926 – us ATLA [073]
Cumulative index to a selected list of periodicals : monthly – v4-8. 1899-1903* – 2r – 1 – mf#ATLA S0927 – us ATLA [073]
Cumulative index to nursing and allied health literature – Glendale. 1977+ (1) 1977+ (5) 1977+ (9) – (cont: cumulative index to nursing literature) – ISSN: 0146-5554 – mf#9956,01 – us UMI ProQuest [610]
Cumulative index to nursing literature – Glendale. 1956-1976 (1) 1974-1976 (5) 1974-1976 (9) – (cont by: cumulative index to nursing and allied health literature) – ISSN: 0011-3018 – mf#9956 – us UMI ProQuest [610]
Cuna comun / Sinan, Rogelio – Panama, Panama. 1963 – 1r – us UF Libraries [972]
Cundall, Frank *see*
- Aborigines of jamaica
- Biographical annals of jamaica
- Catalogue of the portraits in the jamaica
- Chronological outlines of jamaica history, 1492-19...
- Darien venture
- Governors of jamaica in the...
- Governors of jamaica in the first half...
- Governors of jamaica in the seventeenth century
- Historic jamaica
- History of printing in jamaica from 1717 to 1834
- Jamaica in 1905
- Jamaica place-names
- Jamaica under the spaniards
- Jamaica's part in the great war, 1914-1918
- Life of enos nuttall
- Place-names of jamaica
- Reminiscences of the colonial and indian exhibition
- Some notes on the history of secondary education...
- Studies in jamaica history

Cundall, Joseph *see* Examples of ornament...from works of art in the british museum
Cundinamarca (Colombia) *see* Doce codigos del estado soberano de cundinamarca
Cundinamarca Regimiento De Milicias De Infanteria *see* Copiador de ordenes del regimiento de milicias de...
The cuneiform inscriptions and the old testament = Keilinschriften und das alte testament / Schrader, Eberhard – London: Williams & Norgate, 1885-88 [mf ed 1988] – 2v on 2mf – 9 – 0-7905-0230-5 – (english by owen c whitehouse. incl ind & int pref) – mf#1987-0230 – us ATLA [221]
The cuneiform inscriptions from western asia...chaldaea, assyria, and babylonia / Rawlinson, H C; ed by Smith, G et al – London, 1861-1884. v1-5 – 23mf – 9 – mf#NE-381 – ne IDC [956]
Cuneiform parallels to the old testament / ed by Rogers, Robert William – New York: Eaton & Mains, 1912 – 2mf – 9 – 0-524-05692-7 – (incl bibl ref) – mf#1992-0542 – us ATLA [930]
Cuneiform supplement (autographed) to the author's ancient persian lexicon and texts : with brief historical synopsis of the language / Tolma, Herbert Cushing – New York: American Book Co, c1910 [mf ed 1989] – (= ser The vanderbilt oriental series 7) – 1v on 1mf – 9 – 0-7905-2750-2 – (text in old persian. pref in english) – mf#1987-2750 – us ATLA [490]
Cuneiform text of a recently discovered cylinder of nebuchadnezzar king of babylon / Nebuchadnezzar 2, King of Babylonia – [S.l.]: Woodstock-College, 1885 – 1mf – 9 – 0-8370-7722-2 – (texts in english and akkadian; commentary in english) – mf#1986-1722 – us ATLA [470]
Cuneiform texts from babylonian tablets in the british museum – London, 1921 v36 – 2mf – 9 – mf#NE-20059 – ne IDC [956]
Cuneo-Vidal, Romulo *see*
- El capitan don gonzalo pizarro, padre de pizarro hernando, juan y gonzalo pizarro, conquistadores del peru
- Dona ines munoz, la mujer extremena, cunada de francisco pizarro, quetrajo el trigo y el olivo al peru
- Los hijos americanos de los pizarros de la conquista
- Las leyendas geograficas del peru de los incas
- Vida del conquistador del peru don francisco pizarro y de sus hermanos hernando, juan y gonzalo pizarro y francisco martin de alcantara

Cunha, Amadeu *see* Sertoes e fronteiras do brazil
Cunha, Celso Ferreira Da *see* Lingua portuguesa de realidade brasileira
Cunha, Euclides da *see* Los sertones (3-4)
Cunha, Euclydes Da *see*
- Canudos (diario de uma expedicao)
- Contrastes e confrontos
- Marjem da historia
- Rebellion in the backlands
- Rebellion in the backlands (os sertoes)
- Sertoes
- Sertoes (campanha de canudos)

Cunha, Eucyldes Da *see* Antologia euclidiana
Cunha, Heitor Xavier Pereira Da *see* Revolta na esquadra brasileira em novembro e dezem...
Cunha, Jose Antonio Flores Da *see* Campanha de 1923
Cunha, Lygia Da Fonseca Fernandes Da *see* Rio de janeiro atraves das estampas antiguas
Cunha, Rui Vieira Da *see* Cadetes
Cuninggim, J L *see* A plan for better religious instruction
Cuningham, Granville Carlyle *see*
- Energy and labor
- On the energy of fuel in locomotive engines, pt 1
- On the energy of fuel in locomotive engines, pt 2
- A scheme for imperial federation
- Snow slides in the selkirk mountains

Cuninghame, William *see* On the chronological characters marking the year eighteen hundred a...
Cunnabell's nova scotia almanac for the year of our lord... – Halifax, NS: W Cunnabell, 1841-1849 – 9 – mf#A01153 – cn Canadiana [030]
Cunnabell's nova-scotia almanac and farmer's manual for the year of our lord... – Halifax, NS: W Cunnabell, [1850?-1860?] – 9 – ISSN: 1191-3150 – mf#A01154 – cn Canadiana [030]
Cunningham, Alexander *see*
- The ancient geography of india
- Mahabodhi
- The stupa at bharut

Cunningham, Alfred *see* History of szechuen riots (may-june, 1895)
Cunningham, Allan *see*
- The life of sir david wilkie
- The lives of the most eminent british painters
- Songs chiefly in the rural language of scotland

Cunningham, George Godfrey *see* Lives of eminent and illustrious englishmen: from alfred the great to the latest times, on an original plan.
Cunningham, Henry Stewart *see* British india and its rulers
Cunningham, J W *see*
- Cautions to continental travellers
- Conciliatory suggestions on the subject of regeneration
- Observations
- Sermon on the church establishments in general...
- Sermon preached in the parish church of harrow on the hill
- To provide a refuge for the criminal is to give a bounty on the cri...

Cunningham, James *see* Robert hall
Cunningham, John *see*
- The church history of scotland
- The growth of the church in its organization and institutions
- The quakers

Cunningham, Joseph Davey *see* Anglo-sikh relations
Cunningham, K M *see* St segment changes due to handrail support during graded exercise treadmill testing
Cunningham, Lynda F *see* Factors associated with the recall of physical activity
Cunningham, Michael Frank *see* Conversion in american unitarianism
Cunningham, Peter *see* Two years in new south wales
Cunningham, W *see*
- Christianity and politics
- S austin and his place in the history of christian thought

Cunningham, William *see*
- Animadversions upon sir william hamilton's pamphlet
- Christian civilisation
- Christianity and economic science
- Christianity and social questions
- The churches of asia
- Defence of the rights of the christian people in the appointment of...
- Discussions on church principles
- A dissertation on the epistle of s barnabas
- Dr cunningham and dr bryce on the "circa sacra" power of the civil...
- Lecture on the nature and lawfulness of union between church and st...
- Letter to john hope, esq, dean of faculty
- Letters on the church question
- The reformers and the theology of the reformation
- Reply to the statement of certain ministers and elders, published i...

Cunningham, William Bennett *see* Law of forcible entry and detainer, in the state of illinois
Cunningham, William et al *see* Essays on some theological questions of the day
Cunningham-Craig, Edward Hubert *see* Report on the oilfields of barbados
Cunninghame Graham, Robert Bontine *see* Bernal diaz del castillo
Cunnison, Ian George *see* Luapula peoples of northern rhodesia
Cunnison, Ias George *see* History on the luapula
Cunnyngham, William George Etler *see*
- The foreign missionary and his work
- New life without notes

Cuno, John B *see* Use of wood by the fruit and vegetable industries
Cuntz, Otto *see* Die chronik des hippolytos im matritensis graecus 121
Cuoc tinh trong nguc that : truyen dai / Nguyen, Thi Hoang – [Saigon] Nguyen-Dinh Vuong, 1974 [mf ed 1993] – on pt of 1r – 1 – mf#11052 r453 n5 – us Cornell [830]
Cuoq, Jean Andre *see* Lexique de la langue algonquine
Cuore e critica – Savona, Bergamo. v1 n1-v4 n24. 1887-90 – (= ser Important periodicals of italian and international socialism, 1868-1917) – 12mf – 9 – $95.00 – us UPA [335]
Cup of blessing *see* Ku pei li te tien kuo (ccm152)
Cupa journal – Washington. 1987-2000 (1) 1987-2000 (5) 1987-2000 (9) – (cont: journal of the college and university personnel association) – ISSN: 1046-9508 – mf#3027,01 – us UMI ProQuest [378]
Cupa-hr journal – Washington, 2000+ – 1,5,9 – (cont: cupa journal) – mf#3027,02 – us UMI ProQuest [378]
Cuperi, G *see* Tractatus de patriarchis constantinopolitanis (cbh32)
[Cupertino-] courier – CA. 1973-76 [wkly] – 11r – 1 – $660.00 – mf#B02155 – us Library Micro [071]
[Cupertino-] valley journal – CA. Jul 1970-79 – 24r – 1 – $1440.00 – mf#B02156 – us Library Micro [073]
Cupid en route / Barbour, Ralph Henry – Boston: R G Badger; Toronto: Bell & Cockburn, c1912 – 3mf – 9 – 0-665-66408-7 – (ill by F Foster Lincoln) – mf#66408 – cn Canadiana [830]
Cupidoos mengelwerken of minnespiegel der deugden : bestaande uit stightelyke zinnebeelden... / Hesman, Gerrit – Amsterdam: Joh van Septeren; Jan Kouwe, 1728 – 2mf – 9 – mf#O-3081 – ne IDC [090]
Cupido's lusthof ende der amoureuse boogaert... / [Breughel, G H van] – Amsterdam: Jan Evertsz Cloppenburch, [1613] – 2mf – 9 – mf#O-3210 – ne IDC [090]
Cupola / International Union, United Automobile, Aircraft, and Agricultural Implement Workers of America – v1 n4 [1951 aug 9] – 1r – 1 – mf#3629231 – us WHS [331]
Cuppiramaniya Pillai, Ji *see* Introduction and history of saiva siddhanta
Cupw perspective = Perspective spc / Canadian Union of Postal Workers – v18 n1-v22, n6 [1988 jan/feb-1994 nov/dec] – 1r – 1 – (cont: perspective cupw) – mf#1066643 – us WHS [380]
Cur templa ruunt antiiqua? – London, England. 1844 – 1r – us UF Libraries [240]
El cura de santa cruz : san sebastian, 1928 / Urquijo, Julio de La Cruz de Sangre; ed by Bayle, Constantino – Madrid: Razon y Fe, 1928 – 9 – sp Bibl Santa Ana [946]
El cura santa cruz / ed by Bayle, Constantino – Madrid: Razon y Fe, 1926 – 1 – sp Bibl Santa Ana [946]
Cura y mil veces cura. barcelona, 1928 / Merino, Eugenio; ed by Bayle, Constantino – Madrid: Razon y Fe, 1929 – 9 – sp Bibl Santa Ana [240]

Curacao : the netherlands west indies / Curacao Commissie Van Toerist – Willenstad? Curacao. 1950 – 1r – us UF Libraries [972]
Curacao – n.p, n.p? n.d. – 1r – us UF Libraries [972]
Curacao – Willemstad, Curacao. 18 mar-28 oct 1944 – 1r – 1 – uk British Libr Newspaper [072]
Curacao Commissie Van Toerist *see* Curacao
De curacaosche courant – (Willemstad). jul 1853-aug 1893; 1949-69 – 1 – us NY Public [972]
Curacaosche courant – Willemstad, Netherlands Antilles. 1812-1948 (1) – mf#68600 – us UMI ProQuest [079]
Curado Garcia, Blas *see* Abc del alcoholismo
Curaeus, J *see* Exegesis perspicva et ferme integra contraversiae de sacra coena
Los curas / Aradillas Agudo, Antonio – Barcelona: Dopesa, 1978 – 1 – sp Bibl Santa Ana [946]
Curate's appeal and farewell / Maddock, S – London, England. 1818? – 1r – us UF Libraries [240]
Curative comments / Curative Rehabilitation Center – Wauwatosa WI. v5 n1-v9 n2 [1977-1982 fall], 1985 win, 1986 fall, 1987 sum, 1988 win-spring – 1r – 1 – (cont by: compass directions of the future) – mf#1712422 – us WHS [360]
Curative Rehabilitation Center *see* Curative comments
Curative Workshop [Green Bay WI] et al *see* Independence daily
Curato da Matteo Campori *see* Epistolario...
Curator – New York. 1958+ (1) 1970+ (5) 1975+ (9) – ISSN: 0011-3069 – mf#3029 – us UMI ProQuest [060]
Curator / Thomas Gilcrease Institute of American History and Art – 1972 dec-1978 jun – 1r – 1 – (cont by: gilcrease magazine of american history and art) – mf#379361 – us WHS [060]
Curatulo, Giacomo Emilio *see* Il dissidio tra mazzini e garibaldi: la storia senza veli
The curch-idea : an essay towards unity / Huntington, William Reed – New York: E P Dutton, 1870 – 1mf – 9 – 0-7905-4819-4 – mf#1988-0819 – us ATLA [240]
Curci, Carlo Maria *see* Il vaticano regio
Cure, E Capel *see* Righteousness, temperance, and judgment to come
A cure for our sherman act troubles : popular misconceptions-legal, economic, political: an address / Williams, James Harvey – New York City: American Hardware Manufacturers Assoc, 1931 (mf ed 19–) – 30p – mf#ZT-TN pv105 n1 – us Harvard [340]
Le cure labelle / Bodard, Auguste – Montreal?: s.n, 1891? – 1mf – 9 – mf#55886 – cn Canadiana [241]
Curel, Francois De *see*
– Ame en folie
– Fille sauvage
– Fossiles
– Ivresse du sage
Curentul – Bucuresti: Imp "Curentul" SAR, apr 16 1939-1942 – 10r – 1 – us CRL [070]
Cureton, Kirk J *see*
– Peak oxygen deficit as a predictor of sprint and middle-distance track performance
– Validation of fitnessgram one-mile run/walk criterion-referenced standards in men and women 18 to 25 years of age
Cureton, Thomas Kirk *see* How to teach swimming and diving
Cureton, W *see*
– The festal letters of athanasius in an ancient syriac version
– Spicilegium syriacum
Cureton, William *see*
– Ancient syriac documents relative to the earliest establishment of christianity in edessa and the neighbouring countries
– The ancient syriac version of the epistles of saint ignatius to saint polycarp, the ephesians, and the romans
– Remains of a very antient [sic] recension of the four gospels in syriac
– Spicilegium syriacum
– The third part of the ecclesiastical history of john, bishop of ephesus
– Vindiciae ignatianae
Curiae sigilli secreti (anno 1479-1516) / Fernando 2 – Barcelona – 1r – 5,6 – sp Cultura [946]
Curieuse bibliothec : oder fortsetzung der monatlichen unterredungen (repositorium 1-3) – Frankfurt/Leipzig 1704-06 [mf ed 1993] – (= ser Aus den anfaengen des zeitschriftenwesens: fruehe deutsche zeitschriften) – 13mf – 9 – €200.00 – 3-89131-087-0 – gw Fischer [020]
Le curieux antiquaire, ou recueil geographique et historique : des choses les plus remarquables qu'on trouve dans les quatre parties de l'univers; tirees des voiages de divers hommes celebres / Berkenmeyer, Paul – Leiden 1729 [mf ed Hildesheim 1995-98] – 9mf – 9 – €180.00 – 3-487-29987-9 – gw Olms [900]
Curio, C A *see*
– Caelii avgvstini cvrionis sarrracenicae historiae libri tres...
– ...historiae libri tres, ab autore innumeris locis emendati atque expoliti
– A notable historie of the saracens
Curiosidades : noticias e variedades historicas brazileiras / Moreira de Azevedo, Manuel Duarte – Rio de Janeiro: B L Garnier 1873 [mf ed 1987] – 1r – 1 – (with: ten months in brazil / codman, j) – mf#1884 – us Wisconsin U Libr [972]
Curiosidades gramaticales : gramatica ampliada del idioma espanol y sus dialectos / Martinez Garcia, Ramon – 3rd corr enl ed. Madrid: Viuda de Hernando 1896 [mf ed 1987] – 1r – 1 – mf#2034 – us Wisconsin U Libr [440]
Curiosites theologiques / Brunet, Gustave – Paris: Adolphe Delahays, 1861 – 1mf – 9 – 0-524-01683-6 – mf#1990-2585 – us ATLA [200]
Les curiositez de paris, de versailles, de marly, de vincennes, de s cloud, et des environs : avec les antiquitez justes & precises sur chaque sujet... / LeRouge, George – Paris 1723 [mf ed Hildesheim 1995-98] – 2v on 6mf – 9 – €120.00 – 3-487-29673-X – gw Olms [914]
The curiosities and law of wills / Proffatt, John – San Francisco: Whitney, 1876. 216p. LL-1023 – 1 – us L of C Photodup [340]
The curiosities of ale and beer: an entertaining history / Cook, Charles Henry – London: Swan Sonnenschein, 1889.xii,449p. illus – 1 – us Wisconsin U Libr [390]
Curiosities of christian history prior to the reformation / Paterson, James – London: Methuen, 1892 – 1mf – 9 – 0-524-01465-5 – mf#1990-0414 – us ATLA [240]
Curiosities of law and lawyers / James, Croake – New York: Funk & Wagnalls Co, 1899 – 9mf – 9 – $13.50 – (lacking: p178) – mf#LLMC 95-151 – us LLMC [340]
Curiosities of law and lawyers / James, Croake – London: Sampson, Low, Marston, Searle & Rivington, 1891 – 9mf – 9 – $13.50 – mf#LLMC 91-063 – us LLMC [340]
Curiosities of the church : studies of curious customs, services and records / Andrews, William – 2nd ed. Hull: W Andrews, 1891 – 1mf – 9 – 0-524-03570-9 – mf#1990-1030 – us ATLA [240]
Curiosities of the law reporters / Heard, Franklin Fiske – San Francisco: Bancroft, 1876. 212p. LL-1670 – 1 – us L of C Photodup [340]
Curiosities of the law reporters / Heard, Franklin Fiske – San Francisco: Sumner Whitney & Co, 1885 – 3mf – 9 – $4.50 – mf#LLMC 95-164 – us LLMC [340]
O curioso : "cara feia nao nos mete medo" – Florianopolis, SC. 31 jan, 13 mar 1932 – (= ser Ps 19) – bl Biblioteca [079]
El curioso averiquador – Valencia de Alcantara, 1907-1909 – 5 – sp Bibl Santa Ana [073]
El curioso extremeno – Llerena, 1905-1906 – 5 – sp Bibl Santa Ana [073]
Un curioso manuscrito sobre el convento de san onofre de la lapa (badajoz) (su biblioteca y sacristia en el siglo 16) / Alvarez, Arturo – Badajoz: Dip. Prov., 1958. Sep. REE – sp Bibl Santa Ana [946]
Curioso tratado de la naturaleza y calidad...del chocolate... / Colmenero Ledesma, A – Madrid, 1631 – 1mf – 9 – sp Cultura [630]
Curiosos aspectos de la terapeutica calchaqui / Rosenberg, Tobias – Tucuman, Argentina. 1939 – 1r – us UF Libraries [025]
Curious cases : a collection of american and english decisions selected for their readability / Milburn, Benjamin Arrell – Charlottesville: Michie, 1902 – 5mf – 9 – $7.50 – mf#LLMC 95-048 – us LLMC [340]
Curious cases; a collection of american and english decisions, selected for their readability / Milburn, Benjamin Arrell – Charlottesville, VA: Michie Co., 1902. 441p. LL-567 – 1 – us L of C Photodup [340]
Curious church customs and cognate subjects / Cox, John Charles et al; ed by Andrews, William – Hull: W Andrews, 1895 – 1mf – 9 – 0-524-02337-9 – mf#1990-0593 – us ATLA [240]
Curious questions / Brann, Henry Athanasius – Newark, NJ: JJ O'Connor, 1866 – 1mf – 9 – 0-7905-3694-3 – mf#1989-0187 – us ATLA [100]
Curipeschitz, B *see*
– Ein disputation oder besprech zwayer stalbuben
– Wegraysz keyserlicher maiestat legation im 32. jar zu dem turcken geschickt wie vnd was gestalt sie hinein vnd widerumb herausz komen ist...
Curle, J *see* A roman frontier post and its people. the fort of newstead
Curley, Jeffrey J *see* The effects of plyometric training on sprinting performance of collegiate males
Curling in canada and the united states : a record of the tour of the scottish team, 1902-3, and of the game in the dominion and the republic / Kerr, John – Edinburgh: G A Morton; Toronto: Toronto News Co, 1904 [mf ed 1994] – 9mf – 9 – 0-665-73271-6 – (with app) – mf#73271 – cn Canadiana [790]
Curling, W *see*
– Funeral sermon of the rev joseph brown
– Gospel and the doctrine which is not the gospel
Curme, George Oliver *see* Grammar of the german language
Curnock, Nehemiah *see* The journal of the rev. john wesley, a.m
Curr, Edward *see* An account of the colony of van diemen's land principally designed for the use of emigrants
Curr, Edward Micklethwaite *see* Recollections of squatting in victoria then called the port phillip district
Curragh news – Curragh Camp, Ireland. 7 feb-26 sep 1891 – 1/2r – 1 – uk British Libr Newspaper [072]
Curran, Edward Lodge *see* Franco: who is he, what does he fight for?
Curran, John Joseph *see* Golden jubilee of the reverend fathers dowd and toupin
Curran, Thomas Michael *see* Present home financing methods
Currency and finance in time of war : a lecture / Edgeworth, Francis Ysidro – Oxford: Clarendon Press, London: H Milford, 1918 (mf ed 19–) – 48p – mf#Z-BTZE pv289 n5 – us Harvard [332]
Currency lad – Sydney, 1832-33 – 1r – 1 – A$27.50 vesicular A$33.00 silver – at Pascoe [079]
The currency of china : (a short enquiry) / Morrison, James K – London: Effingham Wilson; Hong Kong, Shanghai: Kelly & Walsh Ltd, 1895 – (= ser 19th c books on china) – 1mf – 9 – mf#7.1.42 – uk Chadwyck [332]
The currency of india : with a letter on bi-metallism / Douglas, William – Manchester, 1886 – (= ser 19th c books on british colonization) – 1mf – 9 – mf#1.1.4299 – uk Chadwyck [332]
Current – Canton, NY. 1985-1991 (1) – mf#68754 – us UMI ProQuest [071]
Current – Johannesburg. 1983-1985 (1,5,9) – mf#12147,01 – us UMI ProQuest [621]
Current – Washington. 1960+ [1]; 1971+ [5,9] – ISSN: 0011-3131 – mf#1416 – us UMI ProQuest [370]
The current – North Platte, NE: Wm H Mullane, jan 4 1890 (wkly) [mf ed v1 n2. jan 11-nov 8 1890 (gaps) filmed 1957] – 1r – 1 – us NE Hist [071]
Current advances in applied microbiology and biotechnology – Oxford. 1992+ (1,5,9) – (cont: current advances in microbiology) – ISSN: 0964-8712 – mf#49448,01 – us UMI ProQuest [576]
Current advances in biochemistry – Oxford. 1984-1991 (1,5,9) – (cont by: current advances in protein biochemistry) – ISSN: 0741-1618 – mf#49443 – us UMI ProQuest [574]
Current advances in cancer research – Elmsford. 1988+ (1,5,9) – ISSN: 0895-9803 – mf#49518 – us UMI ProQuest [616]
Current advances in cell and developmental biology – Oxford. 1984+ (1,5,9) – ISSN: 0741-1626 – mf#49444 – us UMI ProQuest [574]
Current advances in clinical chemistry – Oxford. 1974+ (1) 1974+ (5) 1977+ (9) – ISSN: 0885-1980 – mf#49059 – us UMI ProQuest [610]
Current advances in ecological and environmental sciences – Oxford. 1989+ (1,5,9) – (cont: current advances in ecological sciences) – ISSN: 0955-6648 – mf#49057,01 – us UMI ProQuest [333]
Current advances in ecological sciences – Elmsford. 1975-1988 (1,5,9) – (cont by: current advances in ecological and environmental sciences) – ISSN: 0306-3291 – mf#49057 – us UMI ProQuest [333]
Current advances in endocrinology – Oxford. 1984-1987 (1) 1984-1987 (5) 1984-1987 (9) – ISSN: 0741-1634 – mf#49445 – us UMI ProQuest [616]
Current advances in endocrinology and metabolism – Oxford. 1992+ (1,5,9) – (cont: current advances in physiology) – ISSN: 0964-8720 – mf#49451,01 – us UMI ProQuest [612]
Current advances in genetics – Oxford. 1976-1980 (1,5,9) – (cont by: current advances in genetics and molecular biology) – ISSN: 0360-8360 – mf#49257 – us UMI ProQuest [575]
Current advances in genetics and molecular biology – Oxford. 1984+ (1,5,9) – (cont: current advances in genetics) – ISSN: 0741-1642 – mf#49446 – us UMI ProQuest [575]
Current advances in immunology – Oxford. 1984-1991 (1,5,9) – (cont by: current advances in immunology and infectious diseases) – ISSN: 0741-1650 – mf#49447 – us UMI ProQuest [616]
Current advances in immunology and infectious diseases – Oxford. 1992+ (1,5,9) – (cont: current advances in immunology) – ISSN: 0964-8747 – mf#49447,01 – us UMI ProQuest [616]
Current advances in microbiology – Oxford. 1984-1991 (1,5,9) – (cont by: current advances in applied microbiology and biotechnology) – ISSN: 0741-1669 – mf#49448 – us UMI ProQuest [576]
Current advances in neuroscience – Oxford. 1984+ (1,5,9) – ISSN: 0741-1677 – mf#49449 – us UMI ProQuest [612]
Current advances in pharmacology and toxicology – Oxford. 1984-1991 (1,5,9) – (cont by: current advances in toxicology) – ISSN: 0741-1685 – mf#49450 – us UMI ProQuest [615]
Current advances in physiology – Oxford. 1984-1991 (1,5,9) – (cont by: current advances in endocrinology and metabolism) – ISSN: 0741-1693 – mf#49451 – us UMI ProQuest [612]
Current advances in plant science – Oxford. 1972+ (1,5,9) – ISSN: 0306-4484 – mf#49058 – us UMI ProQuest [574]
Current advances in protein biochemistry – Oxford. 1992+ (1,5,9) – (cont: current advances in biochemistry) – ISSN: 0965-0504 – mf#49443,01 – us UMI ProQuest [574]
Current advances in toxicology – Oxford. 1992+ (1,5,9) – (cont: current advances in pharmacology and toxicology) – ISSN: 0965-0512 – mf#49450,01 – us UMI ProQuest [615]
Current affairs – Johannesburg: Broadcast House, jan-may 30 1978, 1977, 1976, 1975, 1974, jan 29-dec 1973 (2r); may 31 1978-mar 2 1981 (1r) – (issues for may 31 1978-mar 2 1981 filmed with: comment (south african broadcasting corporation), mar 3 1981-jun 21 1982) – us CRL [070]
Current affairs – London. 1992-1993 (1,5,9) – mf#25564 – us UMI ProQuest [321]
Current affairs bulletin – Sydney. 1972-1995 (1) 1972-1995 (5) 1976-1995 (9) – ISSN: 0011-3182 – mf#7081 – us UMI ProQuest [321]
Current affairs translations bulletin – Jakarta. 1976-1976 (1) 1976-1976 (5) 1976-1976 (9) – (cont: indonesian current affairs translation bulletin) – mf#9950,01 – us UMI ProQuest [321]
Current anthropology – Chicago. 1960+ [1]; 1971+ [5]; 1977+ [9] – ISSN: 0011-3204 – mf#1895 – us UMI ProQuest [301]
Current applied physics : physics, chemistry and materials science – Amsterdam, 2001+ [1,5,9] – ISSN: 1567-1739 – mf#42833 – us UMI ProQuest [621]
Current awareness in biological sciences: cabs – Oxford. 1983-1984 (1,5,9) – (cont: international abstracts of biological sciences) – ISSN: 0733-4443 – mf#49263,01 – us UMI ProQuest [574]
Current background / U.S. Consulate General. Hong Kong – 13 Jun 1950-74 – (= ser Hong kong press summaries) – 1 – $32.00y $503.00 – us L of C Photodup [951]
Current bibliography of epidemiology – Washington. 1974-1976 (1) 1974-1976 (5) 1975-1976 (9) – ISSN: 0011-3247 – mf#7375 – us UMI ProQuest [614]
Current coin / Haweis, Hugh Reginald – London: H S King, 1876 [mf ed 1984] – 5mf – 9 – 0-8370-0912-X – mf#1984-4260 – us ATLA [360]
Current comment and legal miscellany – v1-3. 1889-91 (all publ) – 20mf – 9 – $30.00 – mf#LLMC 82-920 – us LLMC [340]
Current concepts of cerebrovascular disease : stroke – Dallas. 1976-1991 (1,5,9) – ISSN: 0884-4194 – mf#11108 – us UMI ProQuest [616]
Current consumer – Highwood. 1976-1982 (1,5,9) – ISSN: 0199-8196 – mf#11651 – us UMI ProQuest [380]
Current consumer and lifestudies – Northbrook. 1982-1991 (1,5,9) – (cont by: challenges) – ISSN: 0745-0265 – mf#11651,01 – us UMI ProQuest [380]
The current crisis in south asia : hearing... house of representatives, 107th congress, 2nd session, june 6 2002 / United States. Congress. House. Committee on International Relations. Subcommittee on the Middle east and south asia – Washington: US GPO 2002 [mf ed 2002] – 1mf – 9 – 0-16-068879-5 – us GPO [327]
Current decisions...with notes and digest / U.S. Supreme Court – B. R. Curtis, 1790-1854, Boston, Little, 1855-56, 22 vols., (cited as Cur. Dec.) 81-421 – 9 – $116.00 – us LLMC [348]
Current digest of the post-soviet press – v45-feb 1992- – 1,9 – $325.00v diazo $10.00v extra silver – (cont: the current digest of the soviet press. with ind. mf ed prepared approx 5mths after the end of the vol year) – ISSN: 1067-7542 – us Current [077]

Current digest of the soviet press – v1-44. 1949-1992 – 1,9 – (annual compilation of wkly translations and or abstracts of significant articles from numerous russian-language publ (publ in english). quarterly and annual ind in hard copy available separately. cont as: the current digest of the post-soviet press) – us Current [077]
Current directions in psychological science – Malden. 1995+ (1,5,9) – ISSN: 0963-7214 – mf#21156 – us UMI ProQuest [150]
Current documents *see* American foreign policy series
Current energy and ecology – Northbrook. 1978-1980 (1,5,9) – ISSN: 0194-5572 – mf#11652 – us UMI ProQuest [333]
Current events – Stamford. 1965+ [1]; 1970+ [5]; 1974+ [9] – ISSN: 0011-3492 – mf#1858 – us UMI ProQuest [070]
Current events – v15 n8-13 [1915 nov 5-dec 17]-v18 n621-630, 632-639, 641-652 [1919 jan 3-10, 24-mar 14, 28-jun 13] – 1r – 1 – mf#676149 – us WHS [321]
Current genetics – Heidelberg. 1981-1996 (1,5,9) – ISSN: 0172-8083 – mf#13159 – us UMI ProQuest [575]
Current geographical publications – Milwaukee. 1938+ [1]; 1971+ [5]; 1977+ [9] – ISSN: 0011-3514 – mf#1486 – us UMI ProQuest [900]
Current health – Highwood. 1974-1977 (1) 1974-1977 (5) 1974-1977 (9) – (cont by: current health 2) – mf#11639 – us UMI ProQuest [613]
Current health 1 – Highwood. 1977+ (1,5,9) – ISSN: 0199-820X – mf#11653 – us UMI ProQuest [613]
Current health 2 – Highland Park. 1976+ (1,5,9) – (cont: current health) – ISSN: 0163-156X – mf#11639,01 – us UMI ProQuest [613]
Current history – Philadelphia. 1941+ (1) 1968+ (5) 1970+ (9) – ISSN: 0011-3530 – mf#880 – us UMI ProQuest [321]
Current history and forum – New York. 1914-1941 [1,5,9] – mf#2028 – us UMI ProQuest [900]
Current inquiry into language and linguistics – Edmonton. 1971-1973 (1) 1971-1971 (5) (9) – mf#5845 – us UMI ProQuest [410]
Current issues in higher education – Washington. 1947-1983 (1) 1975-1983 (5) 1975-1983 (9) – mf#10461 – us UMI ProQuest [378]
Current law : a complete encyclopaedia of law – St Paul: West Publ Co. v1-16. 1903-11 (all publ) – 100mf – 9 – $492.00 – mf#LLMC 84-382 – us LLMC [340]
Current left and labour press, 1978-81 – 8mf – 9 – (with ind) – mf#87272 – uk Microform Academic [073]
Current legal thought – New York. v1-14. 1935-48 (all publ) – 31mf – 9 – $139.00 – mf#LLMC 84-453 – us LLMC [340]
Current lifestudies – Highwood. 1978-1982 (1,5,9) – ISSN: 0199-8218 – mf#11654 – us UMI ProQuest [150]
Current lines / International Brotherhood of Electrical Workers – 1974 aug, oct, 1975 mar, may-dec, 1976-1977 aug, oct-dec, 1978 jan-nov, 1979 feb-dec, 1980-1982 jan, mar, sep-1983 feb, 1984 oct-nov – 1r – 1 – mf#1269034 – us WHS [621]
Current list of medical literature – Washington. 1941-1959 (1) – mf#5764 – us UMI ProQuest [610]
Current literature – London. 1967-1970 (1) – mf#1385 – us UMI ProQuest [400]
Current literature in traffic and transportation – Evanston. 1960+ (1) 1972+ (5) 1975+ (9) – ISSN: 0011-3654 – mf#7016 – us UMI ProQuest [380]
Current media – Highland Park. 1980-1981 (1) 1980-1981 (5) – (cont by: writing) – ISSN: 0194-5475 – mf#11655 – us UMI ProQuest [302]
Current medical practice – Bombay. 1972-1972 (1) 1972-1972 (5) (9) – ISSN: 0011-3700 – mf#7079 – us UMI ProQuest [610]
Current medical research and opinion – Newbury. 1979+ (1,5,9) – ISSN: 0300-7995 – mf#12032 – us UMI ProQuest [610]
Current medicine for attorneys – v1-17. 1953-70 – 1 – us AMS Press [340]
Current microbiology – Heidelberg. 1981-1996 (1,5,9) – ISSN: 0343-8651 – mf#13160 – us UMI ProQuest [576]
Current musicology – New York. 1965+ [1]; 1971+ [5]; 1976+ [9] – ISSN: 0011-3735 – mf#2023 – us UMI ProQuest [780]
Current national statistical compendiums on microfiche – groups 1- 1970- – 9 – apply for prices – (key data from national governments around the world. may be purchased either as a complete coll or select world-region subsets. a printed bibliography is included with purchase) – us CIS [310]
Current news from the world church – v1-2 n4 [1966 dec-1968 apr] – 1r – 1 – mf#1055028 – us WHS [240]
Current opinion – New York. 1888-1925 (1) – mf#4603 – us UMI ProQuest [070]
Current podiatry – Bearsville. 1972-1982 (1) 1975-1982 (5) 1975-1982 (9) – ISSN: 0011-3824 – mf#8233 – us UMI ProQuest [617]
Current prices of grain at dublin corn exchange – Dublin, Ireland. jan-may; 27, 30 aug; 24, 27, 31 dec 1853; 1854-58; 10 jan 1860-jan 1862; 15, 22 mar 1862 – 2 1/2r – 1 – uk British Libr Newspaper [072]
Current problems in american samoa : hearing before the subcommittee on territorial and insular affairs of the house committee on interior and insular affairs / American Samoa. US Congress – 93rd Congress 2nd sess 2 Apr 1974. Washington: GPO, 1974 – 2mf – 9 – $3.00 – mf#LLMC 82-100C Title 16 – us LLMC [327]
Current problems in cancer – Chicago. 1980+ (1,5,9) – ISSN: 0147-0272 – mf#13053 – us UMI ProQuest [616]
Current problems in cardiology – Chicago. 1991+ (1,5,9) – ISSN: 0146-2806 – mf#13054 – us UMI ProQuest [616]
Current problems in obstetrics, gynecology and fertility – St. Louis. 1991-1996 (1,5,9) – ISSN: 8756-0410 – mf#13055 – us UMI ProQuest [618]
Current problems in pediatrics – Chicago. 1991+ (1,5,9) – ISSN: 0045-9380 – mf#13057 – us UMI ProQuest [618]
Current problems in pediatrics and adolescent health care – St Louis. 2001+ (1,5,9) – ISSN: 1538-5442 – mf#13057,01 – us UMI ProQuest [618]
Current problems in surgery – Chicago. 1964+ (1,5,9) – ISSN: 0011-3840 – mf#13058 – us UMI ProQuest [617]
Current psychological research and reviews – New Brunswick. 1984-1987 (1,5,9) – (cont by: current psychology: research and reviews) – ISSN: 0737-8262 – mf#15428 – us UMI ProQuest [150]
Current psychology : research and reviews – New Brunswick. 1988+ (1,5,9) – (cont: current psychological research and reviews) – ISSN: 1046-1310 – mf#15428,01 – us UMI ProQuest [150]
Current questions for thinking men / MacArthur, Robert Stuart – Philadelphia: American Baptist Publ Society, 1898 – 1mf – 9 – 0-8370-9010-5 – mf#1986-3010 – us ATLA [240]
Current religious perils : with preludes and other addresses on leading reforms and a symposium on vital and progressive orthodoxy / Cook, Joseph – Boston: Houghton, Mifflin 1888 [mf ed 1985] – (= ser Boston monday lectures) – 2mf – 9 – 0-8370-2305-X – (incl ind) – mf#1985-0305 – us ATLA [240]
Current science – Bangalore. 1932-1996 (1) 1973-1996 (5) 1973-1996 (9) – ISSN: 0011-3891 – mf#8637 – us UMI ProQuest [500]
Current science – Stamford. 1965+ [1]; 1970+ [5]; 1977+ [9] – ISSN: 0011-3905 – mf#1852 – us UMI ProQuest [500]
Current slang – Vermillion. 1966-1971 (1) – ISSN: 0011-3913 – mf#8807 – us UMI ProQuest [400]
Current state of catalog card reproduction – 22 papers. 1973. RLMS Micro-File Series, v. 1 – 9 – $13.00 – us L of C Photodup [020]
Current state of catalog card reproduction: supplement 1 – 9 papers. 1974. RLMS Micro-File Series, v. 1, suppl. 1 – 1 – $8.00 – us L of C Photodup [020]
Current studies *see* Excerpta indonesica
Current surgery – Philadelphia. 1978+ (1) 1978+ (5) 1978+ (9) – (cont: review of surgery) – ISSN: 0149-7944 – mf#149,01 – us UMI ProQuest [617]
Current surgery – v49-53. 1992-1996 – 1,5,6,9 – $80.00r – us Lippincott [617]
Current titles in immunology, transplantation and allergy – London. 1979-1980 (1) 1973-1980 (5,9) – ISSN: 0301-0007 – mf#49060 – us UMI ProQuest [616]
Current topics in radiation research quarterly – Amsterdam. 1970-1977 (1) 1970-1977 (5) (9) – ISSN: 0375-880X – mf#42249 – us UMI ProQuest [574]
Current trends in the reconstruction and rehabilitation of the anterior cruciate ligament / Bovee, Kristin K & Thorland, William G – 1993 – 1mf – $4.00 – us Kinesology [617]
Current wage developments – Washington. 1972-1991 (1) 1972-1991 (5) 1972-1991 (9) – (cont by: compensation and working conditions) – ISSN: 0192-8163 – mf#7346 – us UMI ProQuest [331]
Current washington history / Washington State Historical Society – 1943 mar 8-1949 feb – 1r – 1 – mf#2785776 – us WHS [071]
Current world leaders – Santa Barbara. 1978+ (1) 1982+ (5) 1982+ (9) – ISSN: 0192-6802 – mf#13453 – us UMI ProQuest [920]
Current world leaders : speeches and reports – Santa Barbara. 1972-1977 (1) 1972-1977 (5) 1975-1977 (9) – ISSN: 0092-1386 – mf#6448 – us UMI ProQuest [080]
Current world leaders almanac – Pasadena. 1957-1978 (1) 1972-1978 (5) 1975-1978 (9) – ISSN: 0002-6255 – mf#3248 – us UMI ProQuest [327]
Current world leaders biography and news – South Pasadena. 1971-1978 (1) 1976-1978 (5) 1976-1978 (9) – ISSN: 0002-6263 – mf#3251 – us UMI ProQuest [920]
Current world wide web use in park and recreation departments / Jackson, Kristin M – 1999 – 1mf – 9 – $4.00 – mf#RC 529 – us Kinesology [790]
Currents – Washington. 1983+ – 1,5,9 – (cont: case currents) – ISSN: 0748-478X – mf#10671,01 – us UMI ProQuest [378]
Currents : international trade law journal – v1-9. 1991-2000 + summer suppl – 9 – $110.00 set – mf#115531 – us Hein [343]
Currents / US Army Laboratory Command – may 1981-oct/nov 1985 – 1r – 1 – (cont by: focus [adelphi md]) – mf#1045642 – us WHS [355]
The currents in belle isle strait : from investigations of the tidal and current survey in the seasons of 1894 and 1906 / Dawson, William Bell – Ottawa: Dept of Marine and Fisheries, 1907 – 1mf – 9 – 0-665-71000-3 – mf#71000 – cn Canadiana [550]
The currents in the gulf of st lawrence : from investigations of the tidal and current survey in the seasons of 1894, 1895, 1896, 1906, 1908, 1911 and 1912 / Dawson, William Bell – Ottawa: Dept of the Naval Service, 1913 – 1mf – 9 – 0-665-71002-X – mf#71002 – cn Canadiana [550]
Currents in theology and mission – Chicago. 1974+ (1) 1977+ (5) 1977+ (9) – ISSN: 0098-2113 – mf#10212 – us UMI ProQuest [240]
Curricular and pedagogical vision in dance teacher preparation programs in higher education : toward a partnership in general national and arts education reform / Friedlander, Joy L – 1997 – 4mf – 9 – $16.00 – mf#PE 3753 – us Kinesology [378]
Curriculum – Driffield. 1988-1991 – 1,5,9 – ISSN: 0143-8689 – mf#12536 – us UMI ProQuest [370]
Curriculum and examinations in the secondary school : report of the committee of the secondary school examination council (norwood report), 1941 – 2mf – 9 – mf#86953 – uk Microform Academic [324]
Curriculum development for exercise behavioral change / Rehor, Peter R & Jewett, Ann E – 1991 – 2mf – 9 – $8.00 – us Kinesology [150]
Curriculum inquiry – New York. 1976+ – 1,5,9 – (cont: curriculum theory network) – ISSN: 0362-6784 – mf#11052,01 – us UMI ProQuest [370]
Curriculum journal – Nashville. 1929-1943 – 1 – mf#2266 – us UMI ProQuest [370]
Curriculum latinum ad usum juventutis : a course of latin reading for the use of schools – Montreal: Armour & Ramsay, 1850 [mf ed 1986] – 2v on 1mf – 9 – mf#57402 – cn Canadiana [450]
Curriculum philosophiae peripateticae... / Cornaeus, M – Herbipoli, 1657 – 12mf – 9 – mf#CA-13 – ne IDC [100]
Curriculum product review – Belmont. 1974-1980 – 1 – ISSN: 0273-7418 – mf#8562 – us UMI ProQuest [370]
Curriculum review – Metuchen. 1976+ – 1,5,9 – ISSN: 0147-2453 – mf#11873,03 – us UMI ProQuest [370]
Curriculum theory network – Toronto. 1968-1976 (1) 1968-1976 (5) 1968-1976 (9) – (cont by: curriculum inquiry) – ISSN: 0078-4931 – mf#11052 – us UMI ProQuest [370]
Currie, David P *see* Judicial review under the clean air act and federal water pollution control act
Currie, Donald *see*
– South africa
– Thoughts upon the present and future of south africa, and central and eastern africa
Currie, Duncan Dunbar *see* A catechism of baptism
Currie, James *see* Journal of james currie, 1776
Currie, James George *see* Speeches delivered by the hon messrs currie, seymour and simpson, members of the legislative council
Currie, John *see* Jus populi divinum
Currie, John Allister *see* The red watch
Currie, Lauchlin Bernard *see*
– Ensayos sobre planeacion
– Operacion colombia
Currie, Margaret Gill *see* Gabriel west
Currier, Charles Warren *see*
– Carmel in america
– History of religious orders
Currus triumphales adventum clarissimorum moschoviae principum paul petrovitz et mariae theodorownae conjugis...in divi marci venetiarum foro die 22 januarii an 1782 / Fossatis, G & Fossatis, D – Venezia, [1782] – 1mf – 9 – mf#O-1123 – ne IDC [700]
Curry coastal pilot – Brookings OR: R W & P W Keusink, 1978- [semiwkly] – 1 – (cont: brookings-harbor pilot [1946-78]) – us Oregon Lib [071]
Curry county reporter – Gold Beach OR: W E Hassler, 1923- [wkly] – 1 – (cont: gold beach reporter [-1923]; 1923-24 incl newspapers devoted to brookings or, wedderburn or, and southern curry county) – us Oregon Lib [071]
Curry, Daniel *see* Fragments, religious and theological
Curry, Fred S *see* Reminiscences
Curry, J L M *see*
– Letters
– The southern states of the american union
Curry, J LM *see* Virginia baptists, struggles and triumphs
Curry, James *see* History of the san francisco theological seminary of the presbyterian church in the u.s.a. and its alumni association
Curry, Samuel Silas *see* Vocal and literary interpretation of the bible
Der cursaechsische landphysikus / ed by Weiz, Friedrich August – Naumburg [v3: Leipzig] 1771-73 – (= ser Dz) – 3jge on 6mf – 9 – €120.00 – mf#k/n3559 – gw Olms [500]
The curse at farewell / Tagore, Rabindranath – London: George G Harrap & Co, 1924 – (= ser Samp: indian books) – (trans by edward thompson) – us CRL [490]
The curse of central africa : ...with which is incorporated: a campaign amongst cannibals; by e canisius / Burrows, G – London, 1903 – 4mf – 9 – mf#HT-19 – ne IDC [916]
Curse of conventionalism – London, England. 18-- – 1r – us UF Libraries [240]
The curse of rome : a frank confession of a catholic priest, and a complete expose of the immoral tyranny of the church of rome / MacGrail, Joseph F – [S.l.: s.n.], c1907 (New York City: Nyvall Press) – 1mf – 9 – 0-8370-8358-3 – mf#1986-2358 – us ATLA [241]
Cursiefen, Claus *see* Untersuchungen zur substrathaftung von gefaessendothelzellen und fluessigkeitsscherstress und atp-verarmung
Cursillo de criminologia y derecho penal / Bernaldo De Quiros, Constancio – Ciudad Trujillo, Dominican Republic. 1940 – 1r – us UF Libraries [360]
Cursillo de derecho constitucional americano compa... / Ireland, Gordon – Ciudad Trujillo, Dominican Republic. 1941 – 1r – us UF Libraries [972]
Cursillo sobre explotaciones ovinas en su aspecto de produccion de lana / Junta Provincial de Fomento Pecuario. Badajoz – Madrid: Ed. Espasa-Calpe, 1947. Publ. no 9 – sp Bibl Santa Ana [946]
Curso de derecho international publico / Sanchez I Sanchez, Carlos Augusto – Ciudad Trujillo, Dominican Republic. 1943 – 1r – us UF Libraries [972]
Curso de derecho procesivo penal / Coronado Aguilar, Manuel – Guatemala, 1943 – 1r – us UF Libraries [345]
Curso de geometria y matematicas / Ciruelo, P – SL, SA – 4mf – 9 – sp Cultura [510]
Curso de historia de espana / Arenas Lopez, Anselmo – 1892 – 9 – sp Bibl Santa Ana [946]
Curso de historia de espana, tomo 1 / Arenas Lopez, Anselmo – 1881 – 9 – sp Bibl Santa Ana [946]
Curso de historia de la america central / Villacorta Calderon, Jose Antonio – Guatemala, 1926 – 1r – us UF Libraries [972]
Curso de historia de la america central / Villacorta Calderon, Jose Antonio – Guatemala, 1940 – 1r – us UF Libraries [972]
Curso de historia general / Arenas Lopez, Anselmo – 1886 – 9 – sp Bibl Santa Ana [900]
Curso de instructores en higiene y seguridad del trabajo / Hidroelectrica Espanola – Caceres: Imp. La Minerva, 1973 – 1 – (tambien 1974) – sp Bibl Santa Ana [360]
Curso de instructores en higiene y seguridad del trabajo. 199th curso de monitores de h.e. gabriel y galan. caceres, julio 1977 / Hidroelectrica Espanola – Plasencia: Graf. Sandoval, 1977 – 1 – sp Bibl Santa Ana [360]
Curso de introduccion a la historia de cuba – Habana, Cuba. 1938 – 1r – us UF Libraries [972]
Curso de introduccion a la historia de cuba.. – [Habana]: Municipio de la Habana 1938 [mf ed 1984] – (= ser Coleccion historica cubana y americana) – 1r – 1 – (incl bibl) – mf#1173p – us Wisconsin U Libr [972]
Curso de lengua griega : morfologia. fasciculo 2. ejercicios y antologia / Morillo Trivino, Santiago – Cadiz: Establ. Ceron y Libreria Cervantes, 1942 – 1 – sp Bibl Santa Ana [946]
Curso de lengua griega / Morillo Trivino, Santiago – Cadiz: Establec. Ceron. Libreria Cervantes, 1942 – 1 – sp Bibl Santa Ana [450]
Curso de lengua griega / Morillo Trivino, Santiago – Madrid: Edit. Garcia Enciso, 1941 – 1 – sp Bibl Santa Ana [946]

Curso de lengua griega. morfologia. fasciculo 1 : preceptiva gramatical. tratado de etmologia / Morillo Trivino, Santiago – Cadiz: Establec. Ceron y Libreria Cervantes, 2nd ed 1943 – 1 – sp Bibl Santa Ana [946]
Curso de lengua griega. morfologia. fasciculo 2. ejercicio y antologia : libro del alumno / Morillo Trivino, Santiago – Cadiz: Establecimiento Ceron y Libreria Cervantes, 1942 – 1 – sp Bibl Santa Ana [946]
Curso de literatura hispanoamericana / Amuchastegui, Carlos J – Buenos Aires, Argentina. 1961 – 1r – us UF Libraries [440]
Curso de matematicas : tratado octavo de arquitectura civil / Martel, C – Barcelona, 1778. MS – 3mf – 9 – sp Cultura [510]
Curso de procedimiento penal colombiano / Rendon Gaviria, Gustavo – Bogota, Colombia. 1962 – 1r – us UF Libraries [345]
Curso de procediminetos penales / Castellanos Romero, Carlos – Guatemala, 1938 – 1r – us UF Libraries [345]
Curso de teologia pastoral. barcelona, 1933 / Lithard, Victor – Madrid: Razon y Fe, 1934 – 1 – sp Bibl Santa Ana [240]
Curso general de didactica : (metodologia y organizacion escolar) / Floriano Cumbreno, Antonio C – Oviedo: Editorial Supra, 1947 – 1 – sp Bibl Santa Ana [370]
Curso general de paleografia y diplomatica espanolas (texto) / Doriano Cumbreno, Antonio C – Oviedo: Universidad, 1946 – 1 – sp Bibl Santa Ana [440]
Curso general de paleografia y diplomatica espanoles. seleccion diplomatica / Floriano Cumbreno, Antonio C – Oviedo: Universidad, 1946 – 1 – sp Bibl Santa Ana [440]
Cursory reflections on the seasons of advent and christmas – Manchester, England. 1844 – 1r – us UF Libraries [240]
Cursory thoughts : on the present state of the fine arts / Carey, William Paulet – Liverpool [1810] – (= ser 19th c art & architecture) – 1mf – 9 – mf#4.2.205 – uk Chadwyck [700]
A cursory view of the assignats and remaining resources of french finance, september 6 1795 : drawn from the debates of the convention / Ivernois, Francis d' – London: printed for P Elmsly...1795 [mf ed 1984] – 1mf – 9 – 0-665-45126-1 – (trans fr french. incl bibl ref) – mf#45126 – cn Canadiana [332]
Cursos de lengua panayana / Lozano, Raymundo – Manila: Impr del Colegio de Santo Tomas, 1876 – 1 – us CRL [490]
Cursus completus sive bibliotheca universalis... : series graeca / Patrologiae; ed by Migne, J P – Paris, 1857-1868. v1-161+ind – 1309mf – 8 – mf#392 – ne IDC [700]
Cursus completus sive bibliotheca universalis... series graeca : patrologiae / ed by Migne, J P – Paris. v1-161. 1857-1868+ind – 1309mf – 8 – mf#392 – ne IDC [450]
Cursus completus sive bibliotheca universalis... series latina / Patrologiae; ed by Migne, J P – Paris, 1844-1864. v1-221 – 1658mf – 9 – mf#393 – ne IDC [700]
Cursus completus sive bibliotheca universalis... series latina : patrologiae / ed by Migne, J P – Paris. v1-221. 1844-1864 – 1658mf – 9 – mf#393 – ne IDC [450]
Cursus fhilosophiu tomus secundus partis secundae / San Pedro de Alcantara, Domingo de – 1 – sp Bibl Santa Ana [100]
Cursus philosophici / San Pedro de Alcantara, Domingo de – 1729 $1.50f; 1731 $1.50f; 1734 $1.50f – 9 – sp Bibl Santa Ana [190]
Cursus philosophicus / Arriaga, R de – Antverpiae, 1632 – 15mf – 9 – mf#CA-2 – ne IDC [241]
Cursus philosophicus... / Lingen, B – Coloniae Agrippinae, 1718 – 12mf – 9 – mf#CA-22 – ne IDC [100]
Cursus philosophicus ad usum studentium totius ordinis minoram / Petrus a s Catharina & Thomas a s Joseph – Venetiis. 3v. 1732 – 24mf – 9 – mf#CA-30 – ne IDC [240]
Cursus philosophicus thomisticus, secundum exactam, veram et genuinam aristotelis... / Johannes a s thoma – Lugduni, 1663 – 19mf – 9 – mf#CA-20 – ne IDC [241]
Cursus quattor mathematicorum artium liberalium / Ciruelo, P – 4mf – 9 – sp Cultura [510]
Cursus theologici in primam secundam partem d thomae / Johannes a s thoma – Lugduni, 1663. 4v – 68mf – 9 – mf#CA-21 – ne IDC [240]
Curtain call : the voice of the arts in canada – Toronto. v1-13 n2. nov 9 1929-nov/dec 1941// (mthly) – 2r – 1 – Can$198.00 – cn McLaren [790]
Curtain-up on south africa / Allighan, Garry – London, England. 1960 – 1r – us UF Libraries [960]
Curteis, George Herbert *see*
– Bishop selwyn of new zealand and of lichfield
– Dissent in its relation to the church of england
– The scientific obstacles to christian belief
Curti-Forrer, Eugen *see* Schweizerisches zivilgesetzbuch
Curtin, Gregory G *see* Journal of e-government
Curtin, Jeremiah *see*
– Creation myths of primitive america
– Myths and folk-lore of ireland
– Myths of the modocs
Curtis, Anson Bartie *see* Back to the old testament for the message of the new
Curtis, B R *see*
– Curtis' decisions
– Curtis' reports of cases in the first circuit, 1851-1856
Curtis, Benjamin Robbins *see* Jurisdiction, practice, and peculiar jurisprudence of the courts of the united states
Curtis, Charles Boyd *see* Velazquez and murillo
Curtis, Charles George *see* Broken bits of byzantium by c g curtis...
Curtis, Charles Ticknor *see*
– Equity precedents
– A treatise on the law of patents for useful inventions, as enacted and administered in the united states of america
The curtis courier – Curtis, NE: S R Razee (wkly) [mf ed 1892,1895-16 (gaps) filmed 1958] – 5r – 1 – (issues for jun 24 1892-apr 23 1909 called also whole n373-1284. whole n1284 repeated aug 21 1908-apr 23 1909. suspended with jul 23 1915; resumed with aug 27 1915) – us NE Hist [071]
Curtis' decisions : reports of decisions of the supreme court of the u.s., 1790-1854 / Curtis, B R – Boston: Little Brown. v1-21. 1855-56 (all publ) – 171mf – 9 – $256.00 – (with notes and digest) – mf#LLMC 81-421 – us LLMC [347]
Curtis, Edward Lewis *see*
– The book of judges
– A critical and exegetical commentary on the books of chronicles
The curtis enterprise – Curtis, NE: W F Seward, 1891-v74 n20. may 27 1965 (wkly) [mf ed 1892-1965 (gaps) filmed -1977] – 20r – 1 – (absorbed: faber (1913). merged with: hi-line reporter to form: hi-line enterprise) – us NE Hist [071]
Curtis, George Ticknor *see*
– The american conveyancer; containing a large variety of legal forms and instruments, adapted to popular wants and professional use throughout the united states.
– Constitutional history of the united states
– Creation or evolution?
– Equity precedents
– History of the origin, formation and adoption of the constitution of the united states
Curtis, James *see* A journal of travels in barbary, in the year 1801
Curtis, Neil *see* A multi-case study of first year athletic trainers at the high school level
Curtis, Olin Alfred *see*
– The christian faith
– Elective course of lectures in systematic theology
Curtis poster – Portland OR: Mongtomery Printing Co, 1937-38 [mthly] – 1 – (cont by: poster [1938-39]) – us Oregon Lib [071]
Curtis, R *see* Particulars of the country of labradore...
The curtis reporter – Curtis, NE: P Edgar Adams, jan 1 1914 (wkly) [mf ed -1917 (gaps)] – 1r – 1 – us NE Hist [071]
Curtis' reports of cases in the first circuit, 1851-1856 / Curtis, B R – Boston: Little-Brown. v1-2. 1854-56 (all publ) – (= ser Early federal nominative reports) – 14mf – 9 – $21.00 – mf#LLMC 81-445 – us LLMC [324]
Curtis, Thomas *see* Brief biographical sketch of dr. thomas curtis founder for limestone college
Curtis, Thomas Fenner *see*
– The human element in the inspiration of the sacred scriptures
– The progress of baptist principles in the last hundred years
Curtis, William Alexander *see* A history of creeds and confessions of faith in christendom and beyond
Curtis, William Eleroy *see*
– To-day in syria and palestine
– The yankees of the east
Curtis, William Willis *see* Applied christianity in the hokkaido
Curtiss advance – Curtiss WI. 1923 jun 20-1924 jun 18, 1924 jun 25-1930, 1931-1934 mar 14 – 3r – 1 – mf#963546 – us WHS [071]
Curtiss, Frank Homer *see* Realms of the living dead
Curtiss, George Lewis *see*
– Arminianism in history
– Manual of methodist episcopal church history
Curtiss, Harriette Augusta *see* Realms of the living dead
Curtiss, Samuel Ives *see*
– The date of our gospels in the light of the latest criticism
– De aaronitici sacerdotii atque thorae elohisticae origine
– Ingersoll and moses
– The levitical priests
– The name machabee
– A plea for a more thorough study of the semitic languages in america
– Primitive semitic religion to-day
– Ursemitische religion im volksleben des heutigen orients
Curtius, Ernst *see* Peloponnesos
Curtius, Georg *see*
– The greek verb
– Griechische schulgrammatik
– Grundzuege der griechischen etymologie
– Principles of greek etymology
Curtius, Lorens *see* Politische antisemitismus von 1907-1911
Curtius Rufus, Quintus *see*
– Epistola ad quintum fratrem...
– Livius, books 31-40/dictys...
– Res gestae alexandri magni
Curtois, Rowland Grove *see* Ministerial character considered in a sermon
O curupira : jornal litterario e instructivo – Rio de Janeiro, RJ. 03 out 1852-mar 1853 – (= ser Ps 19) – mf#P14,02,31 – bl Biblioteca [440]
Curwen, John *see* Observations on the state of ireland
Curwen, John Spencer *see*
– Studies in worship music. first series
– Studies in worship music. second series
Curwen, Maskell E *see* A manual upon the searching of records and the preparation of abstracts of title to real property
Curwen, Samuel *see*
– Journal and letters of the late samuel curwen, judge of admiralty, etc
Curzon, George Nathaniel, Marquis of *see* British government in india
Curzon, india and empire : the papers of lord curzon [1859-1925] from the oriental and india office collections at the british library, london – 6pt – 1 – (pt1: demi-official correspondence & summary of administration by topic c1898-1905 [29r] £2700; pt2: private correspondence c1898-1905, and official papers on india: internal affairs c1898-1905 [c17r] $2210; pt3: official papers on india: foreign and frontier policy c1898-1905 [c20r] $2600; pt4: post vice regal correspondence regarding india c1906-25, and papers on the kitchener controversy [c25r] $3250; pt5: correspondence and papers on foreign affairs 1906-24 [c30r] $3900; pt6: indian scrapbooks, travel diaries and files on the far east [c15r] $1950) – uk Matthew [954]
Curzon, R *see*
– Armenia
– Visits to monasteries in the levant
Curzon, Sarah Anne *see*
– The battle of queenston heights, october 13th, 1812
– Canada in memoriam, 1812-14
– Centennial poem
– The story of laura secord, 1813
Cusack, Mary *see* Woman's work in modern society
Cusack, Mary Frances *see* Life of daniel o'connell, the liberator
Cusack, Mary Francis *see*
– Life inside the church of rome
– The nun of kenmare
– What rome teaches
Cusanus, N *see* De auctoritate presidendi in concilio generali
Cushing, Caleb *see*
– A discourse on the social influence of christianity
– The right of the national life insurance co. to establish agencies in the state of new york
Cushing, Caleb, 1800-1879 *see* Arguments in behalf of the united states, with supplement and appendix
Cushing, Catherine Chisholm *see* Pollyanna
Cushing, James Stevenson *see* The genealogy of the cushing family
Cushing, Josiah Nelson *see* Christ and buddha
Cushing, Luther Stearns *see*
– An introduction to the study of roman law
– An introduction to the study of the roman law
– Manual of parliamentary practice
Cushing's controverted election reports – 1v. 1780-1852 – 3mf – 9 – $13.50 – mf#LLMC 84-155 – us LLMC [340]
Cushing's election cases / Massachusetts. Supreme Court – 1v. 1780-1852 (all publ) – (= ser Massachusetts supreme court reports) – 3mf – 9 – $13.50 – mf#LLMC 84-155 – us LLMC [347]
Cushman, Herbert Ernest *see*
– The truth in christian science
– What is christianity?
Cusick, David *see* David cusick's sketches of ancient history of the six nations
Cuspinianus, J *see*
– De caesaribus atque imperatoribus romanis opus insigne...
– De tvrcorvm origine, religione, ac immanissima eorum in christianos tyrannide...
Cust, L G A *see* The status quo in the holy places
Cust, Lionel Henry *see*
– Anthony van dyck
– History of the society of dilettanti
– The master e s and the 'ars moriendi'...
Cust, Robert Needham *see*
– Africa rediviva
– Clouds on the horizon
– Essay on the common features which appear in all forms of religious belief
– Essay on the prevailing methods of the evangelization of the non-christian world
– Essays on the languages of the bible and bible-translations
– The gospel message
– Language as illustrated by bible-translation
– Normal addresses on bible-diffusion
– Notes on missionary subjects
– The shrines, or, chief places of pilgrimage of the adherents of the church of rome
– Three lists of bible-translations actually accomplished
Custead, William W *see* Catalogue of fruit and ornamental trees, flowering shrubs, garden seeds and green-house plants, bulbous roots and flower seeds
Custer chronicles – v1 iss1-4 [1981 feb-nov], v2 iss1-2 [1982], bk2 v3 [1982 fall], bk3 v1-bk7 v3 [1983 spr-1987 fall] – 1r – 1 – mf#1567444 – us WHS [975]
Custer county beacon – Broken Bow, NE: C W Beal. -v21 n52. sep 7 1911 (wkly) – 5r – 1 – (cont by: custer county herald) – us Bell [071]
The custer county chief – Broken Bow, NE: Purcell Bros. v1 n1. apr 22 1892- (semiwkly) [mf ed with gaps] – 103r – 1 – (absorbed: custer county republican 1921, sandhill news 1956, and: seven valleys farmer 1967. vol numbering dropped with jan 7 1932 iss; resumed with v64 n14 feb 16 1956. accompanied by mthly suppls: western outlook jun 1968-feb 1971 and: magazine of the grasslands may 1981-feb 1984; by a wkly suppl: weekender jun 17 1974-dec 18 1978) – us NE Hist [071]
Custer county herald – Broken Bow, NE: Horace M Davis. v22 n1. sep 14 1911- (wkly) – 1r – 1 – (cont: custer county beacon; cont by: herald (ansley ne)) – us Bell [071]
Custer county miscellaneous newspapers – Silver Cliff, CO [mf ed 1991] – 1r – 1 – (daily miner [nov 20 1879], silver cliff miner [jul 18 1879-mar 12 1880]) – ISSN: 0 – mf#MF Z99 C967 – us Colorado Hist [071]
Custer county republican – Broken Bow, NE: Robt H Miller. v1 n1. jun 29 1882-jan 20 1921// (wkly) [mf ed -jun 4 1896 (gaps) filmed 1985] – 4r – 1 – (absorbed by: custer county chief. issue numbering dropped feb 1 1917; vol numbering dropped jul 1 1920) – us NE Hist [071]
Custer, Elizabeth B *see* Collection
Custer, George Armstrong *see* Custer in texas
Custer in texas / Custer, George Armstrong – undated, Records of the Department of Texas, 2nd Cavalry Division, and general court-martial of Lt. Col. Nicholas H. Dale, 2nd Wisconsin Cavalry, at Rolla, Missouri, during the Civil War – 1 – us Kansas [350]
Custine, Astolphe L de *see* Memoires et voyages
The custodian / Eyre, Archibald – Toronto: Langton & Hall, 1904 [mf ed 1997] – 4mf – 9 – 0-665-85264-9 – mf#85264 – cn Canadiana [830]
Custodio, M *see* Disertacion eucaristica sobre la precisa obligacion de recibir todo enfermo la sagrada eucaristia en ayuno
Custom and government in the lower congo / Macgaffey, Wyatt – Berkeley, CA. 1970 – 1r – us UF Libraries [960]
Custom and myth / Lang, Andrew – New York: Harper, 1885 – 1mf – 9 – 0-7905-7359-8 – (incl bibl ref) – mf#1989-0584 – us ATLA [390]
Custom builder – San Francisco. 1987-1998 (1) 1987-1998 (5) 1987-1998 (9) – (cont: progressive builder) – ISSN: 0895-2493 – mf#16391 – us UMI ProQuest [620]
Custom house papers, port of philadelphia, 1704-1789 – ca 5r – 1 – ca $650.00 – (with guide) – mf#S3361 – Historical Society of Pennsylvania – us Scholarly Res [380]
Customary law of the people of sudan / Makec, John Wuol – London, England. 1988 – 1r – us UF Libraries [960]
Customary of the benedictine monasteries of st augustine, canterbury and st peter, westminster, vol 1 (hbs23) / Thompson, E M – 1902 – (= ser Henry bradshaw society (hbs)) – 7mf – 8 – €15.00 – ne Slangenburg [241]
Customary of the benedictine monasteries of st augustine, canterbury and st peter, westminster, vol. 2 (hbs28) / Thompson, E M – 1904 – (= ser Henry bradshaw society (hbs)) – 6mf – 8 – €14.00 – ne Slangenburg [241]
The customary of the cathedral church of norwich (hbs82) / Tolhurst, J B L – 1948 – (= ser Henry bradshaw society (hbs)) – 5mf – 8 – €12.00 – ne Slangenburg [241]
Customer inter@ction solutions – Norwalk, 2000+ [1,5,9] – (cont: call center crm solutions) – mf#18353,04 – us UMI ProQuest [380]

Customer interface – Duluth. 2000+ (1) – mf#23085,01 – us UMI ProQuest [650]
Customs 3, 1696-1780 – London: Public Record Office – (= ser BRRAM series) – 52r – 1 – £3484 / $6968 – (with p/g; int by w e minchinton & c j french) – mf#r96564 – uk Microform Academic [337]
Customs 16 : america, 1768-1772 – London: Public Record Office – (= ser BRRAM series) – 1r – 1 – £67 / $134 – (int by rupert c jarvis) – mf#r96182 – uk Microform Academic [337]
Customs 17 : states of navigation, commerce & revenue, 1772-1808 – London: Public Record Office – (= ser BRRAM series) – 9r – 1 – £603 / $1206 – (int by r s craig) – mf#r96565 – uk Microform Academic [337]
Customs and culture of vietnam / Crawford, Ann Caddell – Rutland, VT. 1966 – 1r – us UF Libraries [959]
Customs and excise tariff : with list of warehousing ports in the dominion, sterling exchange, franc, german rixmark, and the principal foreign currencies at canadian customs values – Montreal: Morton, Phillips, 1890 – 3mf – 9 – mf#56366 – cn Canadiana [336]
Customs and excise tariff with list of warehousing ports in the dominion : sterling exchange and franc tables, compiled from official sources, 21st february, 1877 – Montreal: Dawson, 1877 – 1mf – 9 – mf#56010 – cn Canadiana [336]
Customs and manners of the women of persia : and their domestic superstitions / Kulsum Nah'nah – London 1832 [mf ed Hildesheim 1995-98] – 1v on 1mf [ill] – 9 – €40.00 – 3-487-27599-6 – gw Olms [306]
Customs bulletin : 1967-1986 – v1-26 – 363mf – 9 – $544.00 – mf#LLMC 78-234B – us LLMC [346]
Customs bulletin and decisions – Washington. 1967-1996 (1) 1972-1996 (5) 1972-1996 (9) – ISSN: 0162-6442 – mf#6295 – us UMI ProQuest [336]
Customs dublin bill of entry and shipping list – Dublin, Ireland. jan-may 1853, 30 aug 1853, jul-dec 1925 – 3/4r – 1 – (aka: dublin bill of entry and shipping list) – uk British Libr Newspaper [380]
Customs, excise and commercial laws of canada / Canada (Province) – Toronto: printed by Stewart Derbishire & George Desbarats... 1859 [mf ed 1983] – 3mf – 9 – (with ind) – mf#SEM105P201 – cn Bibl Nat [346]
Customs journals of the danish government of the virgin islands – (= ser Records Of The Government Of The Virgin Islands) – 22 r – 1 – mf#T39 – us Nat Archives [336]
Customs, limerick, bill of entry and commercial list – Limerick, Ireland 8 feb-22 feb 1850 – 1 – uk British Libr Newspaper [336]
The customs of primitive churches / Edwards, Morgan – 1774 – 1 – 5.00 – us Southern Baptist [242]
Customs passenger lists of vessels arriving at port townsend and tacoma, washington, 1894-1909 – (= ser Records of the immigration and naturalization service, 1891-1957) – 1r – 1 – mf#M1484 – us Nat Archives [975]
Customs passenger lists of vessels arriving at san francisco, ca jan 2 1903-april 1 1918 – (= ser Records of the immigration and naturalization service, 1891-1957) – 13r – 1 – mf#M1412 – us Nat Archives [975]
Cutchet, Luis *see* Republica cubana
Cutforth, Nicholas J *see* Policies and provisions for adapted physical education in the ordinary school: a cross-cultural comparison between the united states of america and england and wales
Cuthberht of lindisfarne : his life and times / Fryer, Alfred Cooper – London: SW Partridge, 1880 – 1mf – 9 – 0-524-03581-4 – mf#1990-1041 – us ATLA [240]
Cuthbert, Ross *see*
– An apology for great britain
– L'areopage
– New theory of the tides
Cuthbert, W Nelson *see*
– Answers to cuthbert's exercises in arithmetic, pts 1 and 2
– Cuthbert's exercises in arithmetic
– Exercises in arithmetic for use in the junior classes of public schools, pt 1
Cuthbert's exercises in arithmetic : for use in the senior classes of public schools, pt 2 / Cuthbert, W Nelson – Toronto: Copp, Clark, 1896 – 3mf – 9 – mf#32020 – cn Canadiana [510]
Cuthbertus Butler, D *see* Benedicti regula monasteriorum, sancti
Cutis – Chatham. 1973+ (1) 1973+ (5) 1976+ (9) – ISSN: 0011-4162 – mf#7405 – us UMI ProQuest [616]
Cutler, James E *see* Lynch-law
Cutler, Thomas William *see* A grammar of japanese ornament and design
Cutten, George Barton *see* The psychological phenomena of christianity
Cutter, Charles Ammi *see* Alfabetic order table
Cutteridge, Joseph Oliver *see* Nelson's geography of the west indies
Cutting edge / International Association of Machinists and Aerospace Workers – 1983 jul 14-1991 sep – 1r – 1 – mf#1055033 – us WHS [331]
Cutting experiments with bahia grass grown in lysimeters / Leukel, W A – Gainesville, FL. 1935 – 1r – us UF Libraries [630]
Cutting, Sewall Sylvester *see* Historical vindications
Cutting tool engineering – Northfield. 1979+ (1,5,9) – ISSN: 0011-4189 – mf#12056 – us UMI ProQuest [621]
Cuttings notebooks relating to english artists : from the british museum / Whitley, W T – 14v – 8r – 1 – mf#96638 – uk Microform Academic [700]
Cutts, Edward Lewes *see*
– Augustine of canterbury
– Charlemagne
– Christians under the crescent in asia
– Constantine the great
– An essay on church furniture and decoration
– History of early christian art
– History of the church of england
– A manual for the study of the sepulchral slabs and crosses of the middle ages
– Parish priests and their people in the middle ages in england
– Saint augustine
– Saint jerome
– Scenes and characters of the middle ages
– Turning points of english church history
Cutts, J Madison *see* An american continental commercial union or alliance
Cutw voice / Connecticut Union of Telephone Workers – v44 n9-v48 n7 [1982 sep-1986 jul], v49 n1-v51 n11 [1987 jan/feb-1989 dec] – 1r – 1 – (cont: union voice; cont by: voice [connecticut union of telephone workers]) – mf#976533 – us WHS [380]
Cuvelier, J *see*
– Actes des etats generaux des anciens pays-bas
– Cartulaire de l'abbaye du val-benoit
Cuvelier, Jean-Guillaume-Antoine *see* Tribunal invisible
Cuvelier, J-G-A (Jean-Guillaume-Antoine) *see*
– Nain jaune
– Petit poucet
Cuvelier, J-G-G (Jean-Guillaume-Antoine) *see* Jean sbogar
Cuveliere, Jean *see* Oud-koninkrijk kongo
Cuvier, F *see* Dictionnaire des sciences naturelles dans lequel on traite methodiquement des differents etres de la nature
Cuvier, G *see* Voyages dans l'amerique meridionale,...depuis 1781 jusqu'en 1801
Cuvier, G L C *see* Histoire naturelle des poissons
Cuvier, Georges Leopold Chretien Frederic Dagobert de, Baron *see* Essay on the theory of the earth
Cuvier, Georges Leopold Chretien Frederic Dagobert de, baron *see* Cuvier's animal kingdom
Cuvier's animal kingdom : arranged according to its organisation, forming the basis for a natural history of animals, and an introduction to comparative anatomy – London, 1840 – (= ser 19th c evolution & creation) – 12mf – 9 – mf#1.1.6727 – uk Chadwyck [590]
Cuvillier, Louis Andrew *see* Views in reference to the simplification of civil practice in the courts of new york state
Cuxhavener tageblatt *see* Neptunus
Cuxhavener tageblatt und zeitung fuer das amt ritzebuettel *see* Neptunus
Cuyahoga Co. Berea *see*
– Advertiser series
– Enterprise
– News series
Cuyahoga Co. Brecksville *see* News
Cuyahoga Co. Cleveland *see*
– America
– Americke delnicke listy
– L'araldo
– Catholic universe
– Catholic universe bulletin
– Catholic univrs bulletin
– Citizen
– Convoy (and successor)
– Cuyahoga county legal bulletin
– Cuyahoga county news
– Daily advertiser
– Daily cleveland herald
– Denny hlas
– Deutsch-ungarisches volkblatt
– Dirva
– East side
– Echo
– Enakopravnost
– Evening plain dealer
– Grist mill
– Herald
– Herald series
– Jednosc polek
– Jutrzenka
– Leader
– Leader – index
– Locomotive engineer
– Locomotive firemen's magazine
– Mail and news
– Monitor
– Monitor clevelandzki
– National
– Nationale gazette
– Neue hiemat
– Novy svet
– Ohio farmer
– Our voice
– Polonia w ameryce
– Reserve battery
– Siebenburgisch-amerikanische volksblatt
– Six I's news happenings
– Standard of cross
– Svet
– Svet american
– Toiler
– Trainman news
Cuyahoga Co. East Cleveland *see*
– Citizen
– East clevelander
– East end journal
– Leader
– Monitor
– News
– Press
– Signal
Cuyahoga Co. Olmsted Falls *see*
– Towne crier
Cuyahoga Community College *see*
– Black ascensions
– Mosaic
Cuyahoga county auditor's tax duplicate – 1819-69 – 56r – 1 – (contains record of real estate taxes levied in cleveland and other municipalities, and certain information on companies and corporations enumerated for assessment, icluding real and personal property values) – us Western Res [978]
Cuyahoga county legal bulletin / Cuyahoga Co. Cleveland – may-jul 1922, (jan 1928-dec 1932) [wkly] – 1r – 1 – mf#B29882 – us Ohio Hist [340]
Cuyahoga county news / Cuyahoga Co. Cleveland – jul 1894-dec 1896 [wkly] – 1r – 1 – mf#B7588 – us Ohio Hist [071]
Cuyahoga County. Ohio. Auditor *see* Tax duplicates
The cuyahoga county, ohio, marriage record indexes, 1810-1973 – 27r – 1 – (name index to the cuyahoga co. marriage records) – mf#D3492 – us Western Res [978]
The cuyahoga county, ohio, marriage records, 1810-1949 – 110r – 1 – (chronological record of filing of marriage returns with the county. one guide for both the marriage records and marriage index available: d3492-3.g $15) – mf#D3493 – us Western Res [978]
Cuyahoga County. Ohio. Probate Court *see* Marriage records
Cuyahoga current – n1-n4 [1972 sep 13-oct 25] – 1r – 1 – mf#705513 – us WHS [071]
Cuyler, Theodore Ledyard *see* Beulah-land
Cuza Male, Belkis *see* Viento en la pared
Cvam Rann, Mon *see* Ba ma ca pe ba lai, bay lai
C-v-zeitung – Berlin. may 4 1922-nov 3 1938 – 1 – us NY Public [074]
C-v-zeitung : central-verein deutscher staatsbuerger juedischen glaubens – Berlin DE, 1922 4 may-1938 3 sep – 7r – 1 – gw Misc Inst [270]
Cw – canadian welfare – Ottawa. 1924-1977 [1]; 1971-1977 [5]; 1976-1977 [9] – ISSN: 0008-5332 – mf#1994 – us UMI ProQuest [360]
Cwa news / Communications Workers of America – 1971-74, 1975-80, 1981-1986 sep, 1986 oct-1994 mar – 4r – 1 – (cont: telephone worker) – mf#1110434 – us WHS [380]
Cwa news / Communications Workers of America – Washington. 1975+ (1) 1980+ (5) 1980+ (9) – ISSN: 0007-9227 – mf#10619 – us UMI ProQuest [331]
Cwa voice / Communications Workers of America – 1983 oct-1987 sep – 1r – 1 – (cont by: informer [virginia beach va]) – mf#1053815 – us WHS [380]
Cwa weekly news letter / Communications Workers of America – v1-5 n14 [1947 jun 13-1951 apr 13] – 1r – 1 – mf#1053816 – us WHS [380]
Cwa wire tap / Communications Workers of America – 1972 nov-1974 may, 1974 nov-1984 – 1r – 1 – mf#968093 – us WHS [380]
Cwa-cio coast coordinator / Communications Workers of America – 1950 apr 3-1953 nov 16 – 1r – 1 – mf#1053814 – us WHS [380]
Cwc news / Communications Workers of Canada – v1 n2-v4 n5 [1980 dec/1981 jan-1984 sep/oct] – 1r – 1 – (cont by: cwc connections) – mf#957531 – us WHS [380]
Cwic : [newsletter] / Clearinghouse on Women's Issues in Congress [US] – 1979 nov 1, dec 15, 1980 mar, 1981 jun/jul, 1982 dec 15, 1983 jan 15 – 1r – 1 – mf#1211308 – us WHS [323]
Cwm taf, llwyn-on, st mary & bethel welsh baptist, monumental inscriptions – 1mf – 9 – £1.25 – uk Glamorgan FHS [929]
Cwmbran and pontypool news – Newport, Wales 3-4-25 feb 1993 – 1 – (cont by: cwmbran & pontypool news & weekly argus [4 mar 1993-27 aug 1998]; cwmbran news & weekly argus [3 sep 1998-20 may 1999]) – uk British Libr Newspaper [072]
Cwojdrak, Guenther *see*
– Mit eingelegter lanze
– Wegweiser zur deutschen literatur
Cy whittaker's place / Lincoln, Joseph Crosby – Toronto: McLeod & Allen, c1908 [mf ed 1995] – 5mf – 9 – 0-665-74878-7 – mf#74878 – cn Canadiana [830]
Cyac pa ta yok thai : [a novel] / On Sin, Chan phru kyvan – Ran Kun: Ca pe Biman 1963 [mf ed 1990] – 1r with other items – 1 – (in burmese; title on spine: chit hta ta yauk hte) – mf#mf-10289 seam reel 161/2 [§] – us CRL [830]
Cyanen – Wien: Pfautsch 1839-43 (annual) [mf ed 1993] – 5v on 1r [ill] – 1 – (began with 1839, ceased with 1843. filmed with: die deutsche lyrik in ihrer geschichtlichen entwicklung von herder bis zur gegenwart / emil ermatinger & other titles) – mf#3383p – us Wisconsin U Libr [430]
Cyankali [paragraph symbol] 218 / Wolf, Friedrich – Berlin: Internationaler Arbeiter-Verlag 1929 [mf ed 1991] – 1r – 1 – (publ as v1 of the series das neue drama; v2 never publ. filmed with: volk, ich breche deine kohle! / otto wohlgemuth) – mf#2964p – us Wisconsin U Libr [820]
La cybelline : a new dance for a girl / Firbank, Mr – [London]: Writt by Mr Pemberton [173-?] [mf ed 1988] – 1r – 1 – mf#pres. film 40 – us Sibley [790]
Cybernetics – New York. 1965-1977 (1) 1965-1977 (5) – ISSN: 0011-4235 – mf#10904 – us UMI ProQuest [000]
Cybernetics and systems – Washington. 1980-1996 (1,5,9) – (cont: journal of cybernetics) – ISSN: 0196-9722 – mf#11137,01 – us UMI ProQuest [000]
Cybernetics in the u.s.s.r / U.S. Foreign Broadcast Information Service – 1966-Aug 1974 – 1 – us L of C Photodup [947]
Cycle – New York. 1971-1991 (1) 1971-1991 (5) 1970-1991 (9) – ISSN: 0574-8135 – mf#5982 – us UMI ProQuest [790]
Cycle d'erosion sous les differents climats / Birot, Pierre – Rio de Janeiro, Brazil. 1960 – 1r – us UF Libraries [550]
A cycle of cathay : or, china, south and north: with personal reminiscences / Martin, William Alexander Parsons – 3rd ed. New York: Fleming H Revell, 1900 [mf ed 1992] – 2mf – 9 – 0-524-04527-5 – mf#1990-3361 – us ATLA [915]
The cycle of spring / Tagore, Rabindranath – London: Macmillan & Co, 1917 – (= ser Samp: indian books) – us CRL [490]
Cycle world – New York. 1973+ (1,5,9) – ISSN: 0011-4286 – mf#10041 – us UMI ProQuest [790]
Cyclical deluges : an explication of the chief geological phenomena of the globe, by proofs of periodical changes of the earth's axis / Walker, William Basset – London, 1871 – (= ser 19th c evolution & creation) – 2mf – 9 – mf#1.1.6336 – uk Chadwyck [550]
Cycling – Toronto: W H Miln, C B Robinson, [1890-189- or 19–] – 9 – ISSN: 1190-6219 – mf#P04177 – cn Canadiana [790]
Cycling! / Robertson, William Norrie – Stratford, Ont?: F Pratt, 1894 – 4mf – 9 – mf#27852 – cn Canadiana [790]
Cycling for old and young / Adelpha – S.l: s.n, 187-? – 1mf – 9 – mf#07020 – cn Canadiana [790]
O cyclismo – Rio de Janeiro, RJ, 20-27 out 1900 – (= ser Ps 19) – mf#DIPER – bl Biblioteca [079]
The cyclists' road guide of canada / ed by Bryers, Fred – Toronto: W. Miln – 9 – mf#33674 – cn Canadiana [790]
Cyclone – Walton, NY. 1885-1886 (1) – mf#65265 – us UMI ProQuest [071]
Cyclone and fayette republican / Fayette Co. Washington Court House – jun 1888-apr 1889 [wkly] – 1r – 1 – mf#B27617 – us Ohio Hist [071]
Cyclone and fayette republican series / Fayette Co. Washington Court House – (jun 1889-nov 1905) [wkly] – 8r – 1 – mf#B9236-9243 – us Ohio Hist [071]
Cyclopaedia : or, an universal dictionary of arts and sciences / Chambers, Ephraim – 1st ed. London 1728 [mf ed 1992] – (= ser AEL 1/2) – 2v on 22mf – 9 – €120 diazo €144 silver – 3-89131-053-6 – gw Fischer [030]
Cyclopaedia : or, universal dictionary of arts, sciences and literature / ed by Rees, A – London: Longham, Hurst, Rees [1802]-1819-1820 – v1-39 [tl-2/8755] on 1074mf – 9 – mf#495/2 – ne IDC [030]

Cyclopaedia bibliographica : a library manual of theological and general literature, and guide to books for authors, preachers, students, and literary men / Darling, James – London: James Darling, 1854-1859 – 28mf – 9 – 0-524-02772-2 – mf#1987-6466 – us ATLA [012]

Cyclopaedia of biblical literature / ed by Alexander, William Lindsay – 3rd ed. Edinburgh: A & C Black, 1869 [mf ed 1990] – 7mf – 9 – 0-8370-1743-2 – (orignally ed by john kitto; incl bibl ref) – mf#1987-6139 – us ATLA [052]

Cyclopaedia of biblical, theological, and ecclesiastical literature / McClintock, John & Strong, James – New York: Harper, 1867-1887 – 29mf – 9 – 0-8370-1746-7 – (incl supplement (2 v.) and bibliographical references) – mf#1987-6142 – us ATLA [052]

A cyclopaedia of commerce, mercantile law, finance, commercial geography, and navigation / Waterston, William – 2nd ed. London: Henry G Bohn, 1863 – (= ser 19th c economics) – 11mf – 9 – (with suppl by p l simmonds) – mf#1.1.276 – uk Chadwyck [380]

Cyclopdia of history and geography : being a dictionary of historical and geographical antonomasias, origin of sects, etc / Borthwick, John Douglas – Montreal: R & A Miller, 1859 – 3mf – 9 – mf#48001 – cn Canadiana [059]

Cyclope / Alibert, Francois Paul – Carcassonne, France. 1932 – 1r – us UF Libraries [440]

Cyclopedia of law and procedure / ed by Mack, William et al – New York, London: American Law Book Co/Butterworths. v1-40 + index. 1901-18 (all publ) – 762mf – 9 – $1143.00 – (with annotations vol in 4 sections, and 1st + 2nd "permanent annotations" vols covering 1901-13 and 1914-18) – mf#LLMC 82-509 – us LLMC [340]

A cyclopedia of missions : containing a comprehensive view of missionary operations through out the world / Newcomb, Harvey – New York: Charles Scribner, 1858, c1854 [mf ed 1986] – 2mf – 9 – 0-8370-6233-0 – (incl ind) – mf#1986-0233 – us ATLA [240]

Cyclopedia of new zealand – Christchurch, 1897-1906 – 9 – (v1: wellington province 17mf isbn 0-908797-05-2 nz$44. v2: auckland province 12mf isbn 0-908797-03-6 nz$32. v3: canterbury province 13mf isbn 0-908797-06-0 nz$34. v4: otago and southland province 13mf isbn 0-908797-07-9 nz$34. v5: nelson, marlborough and westland provinces 7mf isbn 0-908797-04-4 nz$19. v6: taranaki, hawkes' bay and wellington provinces 9mf isbn 0-908797-02-8 nz$24.) – mf#NZNB c1958-c1964 – nz BAB [980]

The cyclopedia of the colored baptists of alabama / Boothe, Charles Octavius – 1895. 280p – 1 – 9.80 – us Southern Baptist [242]

Cyclopedia of the law of private corporations / Fletcher, William Meade – Chicago: Callaghan. v1-9. 1917-20 – 172mf – 9 – $258.00 – (suppl vols for 1921, 1924, 1927, 1930) – mf#llmc97-587 – us LLMC [346]

Cyclopedic manual of the united presbyterian church of north america : comprising a brief history of her ancestral branches, ministry, congregations, institutions, courts, boards, missions, periodicals, etc... / Glasgow, William Melancthon – Pittsburgh: United Presbyterian Board of Publ, 1903 – 2mf – 9 – 0-7905-4652-3 – mf#1988-0652 – us ATLA [242]

Cyclopedic review of current history – v7 [1897] – 1r – 1 – (cont: quarterly register of current history; modern culture; cont by: current history and modern culture) – mf#1409305 – us WHS [900]

Cyclorama : front and york sts toronto, open daily, 9 am to 10 pm, admission 50@, every saturday night, 7 to 10.30 pm, admission, 25 cents...battle of sedan – S.l: s.n, 1888? – 1mf – 9 – mf#59711 – cn Canadiana [790]

Le cyclorama universel : journal d'illustrations – Montreal: [s.n.] v1 n1 21 sep 1895-v4 n14 10 juil 1897 (wkly) [mf ed 1984] – 2r – 5 – mf#SEM16P338 – cn Bibl Nat [073]

Cyfaill y werin – [Wales] LLGC 13 rhag 1861-13 mehefin 1867 (incomplete) [mf ed 2003] – 5r – 1 – (cont as: y byd cymreig-y byd cymreig: newyddiadur wythnosol a hysbysydd cyffredinol i genedl y cymry [2 hydref 1862-rhag 1863]; y byd cymreig: newyddiadur wythnosol a hysbysydd cyffredinol i genedl y cymry [ion 1864-mehefin 1867]) – uk Newsplan [072]

Y cyfnod – [Wales] Gwynedd 5,12 ion 1934 & ion-rhag 1935-rhag 1950 – 12r – 1 – (cont by: y cyfnod a'r merioneth express [ion 1948-rhag 1950]) – uk Newsplan [072]

Cymbal – Carmel, CA. 1926-1942 (1) – mf#62114 – us UMI ProQuest [071]

Y cymro – [Wales] LLGC gor 1914-rhag 1950 [mf ed 2004] – 69r – 1 – (cont by: y cymro a'r ford gron [ion 1936-meh 1939]; y cymro y brython a'r ford gron [gor 1939-meh 1946]; y cymro [gor 1946-rhag 1950]) – uk Newsplan [072]

Y cymro – [Wales] University of Wales-Bangor ion 1848-rhag 1854 [mf ed 2004] – 3r – 1 – uk Newsplan [072]

Y cymro: a national newspaper for welshmen at home and abroad = The welshman – Liverpool, England 22 may 1890-11 mar 1909 [mf 1890-1906] – 1 – uk British Libr Newspaper [072]

Cymro america – [Wales] LLGC ion-15 mehefin 1832 [mf ed 2002] – 1r – 1 – uk Newsplan [072]

Y cymro [bangor] – Bangor, Wales 1848-54 – 1 – (discontinued; in 1851 publ in london, fr 1852-60 at holywell, fr 1861-66 at denbigh) – uk British Libr Newspaper [072]

Y cymro [dolgellau] – Dolgellau, Wales jun 1914-23 sep 1931 [mf 1914-31] – 1 – uk British Libr Newspaper [072]

Y cymro (lerpwl a'r wyddgrug) = Welshman – [Wales] Flintshire 22 mai 1890-11 mar 1909 [mf ed 2002-03] – 19r – 1 – uk Newsplan [072]

Y cymro (liverpool edition) – [NW England] Liverpool 10 may-dec 1900, 4 mar 1939-dec 1945 [mf ed 2003] – 13r – 1 – (missing: jan-feb 1939, jan-apr 1941; cont as: y cymro y brython a'r ford gron [mar 1939-dec 1945]) – uk Newsplan [072]

Y cymro [oswestry] – Oswestry, England. 3 dec 1932- [mf 1986-] – 1 – (publ at wrexham 1932-sep 1947, at oswestry fr oct 1947-) – uk British Libr Newspaper [072]

Cyn kyn cin nhan lak tve bhava / Mra Ke Si – Yan Kun: Naan Lan Sa Aup Tuik 1976 [mf ed 1990] – [ill] 1r with other items – 1 – (in burmese; incl index & bibl) – mf#mf-10289 seam reel 133/10 [§] – us CRL [280]

Cynick – Philadelphia. 1811-1811 (1) – mf#3733 – us UMI ProQuest [790]

Cypern : reiseberichte ueber natur und landschaft, volk und geschichte / Loeher, Franz von – Stuttgart 1878 [mf ed Hildesheim 1995-98] – 3mf – 9 – €90.00 – 3-487-29095-2 – gw Olms [914]

Cypress baptist church. cypress, florida : church records – 1888-1965 – 1 – us Southern Baptist [242]

Cypress Creek Association of Primitive Baptists *see* Annual session

[Cypress-] cypress news – CA. 1993-1994 – 6r – 1 – $360.00 – mf#R04022 – us Library Micro [071]

Cyprian : his life, his times, his work / Benson, Edward White – London, New York: Macmillan, 1897 – 2mf – 9 – 0-7905-4086-X – (incl bibl ref) – mf#1988-0086 – us ATLA [240]

Cyprian *see* Cyprian's tracts on the african pestilence

Cyprian, Saint *see* Historia...

Cyprian, Saint, Bishop of Carthage *see* Cyprian's tracts on the african pestilence

Cyprian, Saint (Cyprianus) *see* Briefe, 2. bd (bdk60 1.reihe)

Cyprian the churchman / Faulkner, John Alfred – Cincinnati: Jennings and Graham, c1906 – (= ser Men of the Kingdom) – 1mf – 9 – 0-524-05434-7 – (incl bibl ref) – mf#1990-1466 – us ATLA [240]

Cyprian und der roemische primat : eine kirchen- und dogmengeschichtliche studie / Koch, Hugo – Leipzig: J C Hinrichs, 1910 – (= ser Tugal) – 1mf – 9 – 0-7905-1722-1 – (incl bibl ref and ind) – mf#1987-1722 – us ATLA [240]

Cyprian und der roemische primat / Koch, H – Leipzig, 1910 – (= ser Tugal 3-35/1) – 3mf – 9 – €7.00 – ne Slangenburg [240]

Cyprian von karthago und die verfassung der kirche : eine kirchengeschichtliche und kirchenrechtliche untersuchung / Ritschl, Otto – Goettingen: Vandenhoeck & Ruprecht, 1885 – 1mf – 9 – 0-7905-6946-9 – (incl bibl ref) – mf#1988-2946 – us ATLA [240]

Die cyprianische briefsammlung : geschichte ihrer entstehung und ueberlieferung / Soden, Hans, Freiherr von – Leipzig, 1904 – (= ser Tugal 2-25/2) – 4mf – 9 – €11.00 – ne Slangenburg [240]

Die cyprianische briefsammlung : geschichte ihrer entstehung und ueberlieferung / Soden, Hans Otto Arthur Maria Roderich Ulrich – Leipzig: J C Hinrichs, 1904 – (= ser Tugal) – 1mf – 9 – 0-7905-4054-1 – (incl bibl ref) – mf#1988-0054 – us ATLA [240]

Cyprianische untersuchungen (akg4) / Koch, H – Bonn, 1926 – (= ser Arbeiten zur kirchengeschichte (akg)) – 9mf – 8 – €18.00 – ne Slangenburg [240]

Cyprian's tracts on the african pestilence / Cyprian – London, England. 1832 – 1r – us UF Libraries [240]

Cyprianus (Cyprian, Saint) *see* De lapsis (fp21)

Cyprise, 10 milostnych pisni pro smyccovy kvartet, dle basni gustava pflegra moravskeho = Cypresses, 10 love songs / Dvorak, Antonin – Prague: Hudebni matice umelecke besedy 1921 [mf ed 1989] – 1r – 1 – mf#pres. film 57 – us Sibley [780]

Cyprivm bellvm, inter venetos, et selymvm tvrcarvm imperatorem... / Bizari, P – [Basle, 1573] – 4mf – 9 – mf#H-8342 – ne IDC [956]

Cyprus – Larnaca, Cyprus. 29 aug-24 oct 1878; 12 jul 1880-18 aug 1882 – 3/4r – 1 – uk British Libr Newspaper [072]

Cyprus / Rogers, John Henry – Richmond, England. 1878? – 1r – us UF Libraries [240]

Cyprus *see* Statistical blue books 1880-1946

Cyprus herald – Limassol, Cyprus. 14 oct 1881-22 jan 1887 – 1r – 1 – uk British Libr Newspaper [072]

Cyprus mail – Nicosia: The "Cyprus Mail" Co, 1956- – 1 – us CRL [949]

Cyprus times – Nicosia, Cyprus. 1880-81, 1886-95, 30 may-26 oct 1955, 20 may 1957-aug 1959, oct 1959-aug 1960 [imperfect] – 13 1/2r – 1 – (aka: times of cyprus) – uk British Libr Newspaper [072]

Cyrano – Goteborg, Sweden. 1918-20 – sw Kungliga [078]

Cyrano – Stockholm, Sweden. 1918-20 – 1r – 1 – sw Kungliga [078]

Cyrano De Bergerac *see* Histoire comique des etats et empires de la lune et du soleil

Cyrano de bergerac : a play in five acts / Rostand, Edmond – New York: R H Russell 1899 [mf ed 1987] – 1r – 1 – (trans fr french by gladys thomas & mary f guillemard. with: the materialism of the present time / janet, p) – mf#1955 – us Wisconsin U Libr [820]

Cyrano's journal / New England Communications Task Force – v1 n1 [1982 fall] – 1r – 1 – mf#853303 – us WHS [380]

The cyrenaica gazette / Cyrenaica. Territory under British Occupation – Benghazi. 1944-1954 – 1 – us NY Public [956]

Cyrenaica. Territory under British Occupation *see* The cyrenaica gazette

Cyril, Saint, Patriarch of Alexandria *see*
– Commentary on the gospel according to s john
– Five tomes against nestorius

Cyrill und methodius : die lehrer der slaven / Bonwetsch, Gottlieb Nathanael – Erlangen: A Deichert, 1885 [mf ed 1989] – 1mf – 9 – 0-7905-4380-X – (incl bibl ref) – mf#1988-0380 – us ATLA [240]

Cyrille et methode : etude historique sur la conversion des slaves au christianisme / Leger, Louis – Paris: A Franck, 1868 – 1mf – 9 – 0-7905-5475-5 – (incl bibl ref) – mf#1988-1475 – us ATLA [240]

Cyrillus von Alexandrien (Cyril of Alexandria, Saint) *see* Ausgewaehlte schriften (bdk12 2.reihe)

Cyrillus von Jerusalem (Cyril of Jerusalem, Saint) *see* Katechesen (bdk41 1.reihe)

Die cyropaedie in wielands werken / Herchner, Hans – Berlin: R Gaertner. 2v in 1. 1892 – 1 – us Wisconsin U Libr [430]

Cyrpian und das papsttum / Ernst, Johann – Mainz, 1912 – 3mf – 8 – €7.00 – ne Slangenburg [240]

Cyrus, Enoch *see* The light of the sun as revealed in the book of genesis

Cyrus und tomyris : ein heroisch-pantomimisches ballett von der erfindung und ausfuehrung des herrn joseph trafieri. aufgefuehrt in den k.k. hoftheatern 1797 / Trafieri, Giuseppe – Wien: Gedruckt bey Mathias Andreas Schmidt [1797?] – 1 – (in german and italian) – mf#*ZBD-*MGTZ pv5-Res – Located: NYPL – us Misc Inst [790]

Cysarz, Herbert *see*
– Deutsche barockdichtung
– Von schiller zu nietzsche

O cysne : orgam litterario mineiro – Ouro Preto, MG: Imp H Lambaerts & C, 25 out 1894 – (= ser Ps 19) – bl Biblioteca [622]

Cystic fibrosis = Sistifibrose / South Africa. Department of Health [Departement van Gesondheid] – [Pretoria]: Dept of Health [1976?] [mf ed Pretoria, RSA: State Library [199-]] – 18p [ill] on 1r with other items – 5 – mf#op 06630 r25 – us CRL [616]

Cystic tumors of the brain following traumatism – jackson epilepsy – operation – perfect recovery / Armstrong, George E – S.l: s.n, 1896? – 1mf – 9 – mf#40971 – cn Canadiana [617]

Cytogenetic and genome research – Basel. 2002+ (1,5,9) – ISSN: 1424-8581 – mf#2053,02 – us UMI ProQuest [575]

Cytogenetics – Basel. 1965-1972 (1) 1970-1972 (5) – (cont by: cytogenetics and cell genetics) – ISSN: 0011-4537 – mf#2053 – us UMI ProQuest [574]

Cytogenetics and cell genetics – Basel. 1973-1974 (1) – (cont: cytogenetics) – ISSN: 0301-0171 – mf#2053,01 – us UMI ProQuest [610]

Cytokine and growth factor reviews – Oxford. 1996+ (1,5,9) – (cont: progress in growth factor research) – ISSN: 1359-6101 – mf#42795 – us UMI ProQuest [574]

Cytopathology – Oxford. 1991-1994 (1,5,9) – ISSN: 0956-5507 – mf#17973 – us UMI ProQuest [574]

Cytotechnology – Dordrecht. 1989-1996 (1,5,9) – ISSN: 0920-9069 – mf#16779 – us UMI ProQuest [574]

The czar unmasked : being the secret and confidential communications between the emperor of russia and the government of england, relative to the ottoman empire – [2nd ed] London, 1854 – (= ser 19th c books on british colonization) – 1mf – 9 – mf#1.1.6226 – uk Chadwyck [941]

The czardas from coppelia, act 1 : character dance for a boy and girl / Sergeev, Nikolai Grigor'evich – [n.p., n.d.] – 1 – mf#*ZBD-*MGO pv25 – Located: NYPL – us Misc Inst [790]

Czarnikow-Rionda Company, New York *see* Weekly sugar report

Czas – Krakow, Poland. 28 Dec 1918-1 Jan 1919; 1929; Aug 1939 – 3r – 1 – us L of C Photodup [947]

Czas : organ urzedowy zjednoczenia polsko narodowego w brooklynie = Times: the official organ of the polish alliance in brooklyn – Brooklyn, NY: Czas Pub Assoc, 1906-29; 1931-aug 29 1975 – 37r – 1 – us CRL [071]

Czas – Warszawa: Drukarnia Polska, jan, mar, may-jun 1939 – 4r – 1 – us CRL [947]

Czech and slovakian biographical archive = Cesky biograficky archiv a slovensky biograficky archiv (csba) / Kramme, Ulrike [comp] – [mf ed 1993-99] – 687mf (1:24) in 12 installments – 9 – diazo €10,060.00 (silver €11,080 isbn: 978-3-598-33431-3) – ISBN-10: 3-598-33430-3 – ISBN-13: 978-3-598-33430-6 – (with printed ind) – gw Saur [947]

Czech-American Heritage Center, Inc *see* Hlas naroda

Czech-Jochberg, Erich *see* Im osten feuer

Czechoslovak republic press review / Great Britain. Embassy. Prague – Jul 1958-1976 – 1 – us L of C Photodup [943]

Czechoslovak Society of America. Cleveland, Ohio *see* Records, ms p.p.

Czechoslovakia. Federalni Statisticky Urad, Cesky Statisticky Urad, Slovensky Statisticky Urad *see* statisticka rocenka ceskoslovenske socialisticke republiky 1957-1970

Czechoslovakia. Ministerstvo zahranicnich veci *see*
– Cechoslovak
– Cechoslovak v zahranici

Czechoslovakia. Narodni shromazdeni Poslanecka snemova *see*
– Tisky k tesnopiseckym zpravam o schuzich poslanecke snemovny narodniho shromazdoni..
– Zapisy o schuzich

Czechoslovakia. Narodni shromazdeni. Senat *see*
– Tesnopisecke zpravy o schuzich
– Tisky k tesnopiseckym zpravam o schuzich

Czechoslovakia. L'Office Statistique de La Republique Tchecoslovaque *see* Annuaire statistique de la republique tchecoslovaque 1934-1938

Czechoslovakia. L'Office Statistique d'Etat *see* Manuel statistique de la republique tchecoslovaque 1920-1932

Czechoslovakia. Statni Urad Statisticky *see*
– Cenove zpravy
– Mesicni prehled zahranicniho obchodu
– Mitteilungen des statisitschen staatsamtes der cechoslovakischen republik

Czechowski, Heinz *see* Sieben rosen hat der strauch

Czechowski, Heinz [comp] *see* Bruecken des lebens

Czechowski, Michael Belina *see* Thrilling and instructive developments

Czermak, W *see* Die laute der aegyptischen sprache

Czermak, Wilhelm *see* Zur sprache der ewe-neger

Czernowitzer allgemeine zeitung – Czernowitz (Cernauti RO), 1920 26 oct-1932 6 jul – 19r – 1 – (lacking: 27 mar-10 nov 1926) – gw Misc Inst [079]

Czernowitzer deutsche tagespost – Czernowitz (Cernauti RO), 1924 1 feb-1940 28 jun – 22r – 1 – (lacking: 1924) – gw Misc Inst [077]

Czernowitzer morgenblatt – Czernowitz (Cernauti RO), 1921 4 jan-1933 31 mar – 22r – 1 – gw Misc Inst [077]

Czernowitzer sonn- und montagszeitung – Czernowitz (Cernauti RO), 1930 2 mar-17 aug – 1r – 1 – gw Misc Inst [079]

Czernowitzer tagblatt – Czernowitz (Cernauti RO), 1936-37 – 1 – gw Misc Inst [079]

Czerski, Johannes *see* Johannes czerski

Czerwinski, Frank *see* Lasergestuetzte abloesung von embolisationsspiralen

Czerwony sztandar – Vilna, U.S.S.R. -d. Jan 1957-Dec 1958. 6 reels – 1 – uk British Libr Newspaper [947]

Czestochowski, Joseph S *see* The works of arthur b davies

Czibulka, Alfons von *see*
– Das abschiedskonzert
– Das lied der standarte caraffa
– Der muenzturm

Czikann, Johann *see* Oesterreichische nationalenzyklopaedie (ael1/10)

Czinar, M *see* Monasteriologiae regni hungariae

Czoefanjan profeta koenyve : bevezetes en magyarazat / Sipos, Istvan – Budapest: Voersvary Sokszorositoipar, 1937 – 2mf – 9 – 0-524-08141-7 – mf#1993-9047 – us ATLA [220]

Czubatynski, Uwe see Kirchlicher zentralkatalog beim evangelischen zentralarchiv in berlin

Czytania dla szkol powszechnych / Zlobicka, Jadwiga – 2nd ed. Lwow 1933 [mf ed 1986] – 1r – 1 – (with: evolution and religion / osbonn, h f & other titles) – mf#1798 – us Wisconsin U Libr [460]

D : the magazine of dallas – Dallas TX 1975-77 – 1,5,9 – ISSN: 0362-451X – mf#10319.02 – us UMI ProQuest [071]

D – New York NY 1978- – 1,5,9 – ISSN: 0164-8292 – mf#10319.02 – us UMI ProQuest [071]

d see Stanfield standard with which the umatilla spokesman is combined

D a burgerzeitung : organ des d a burgerbundes – Chicago: Beobachter Publ Co, [apr 10 1920-dec 30 1921] – 1r – 1 – us CRL [071]

D a w n / Determined Action for Women Now (Organization) – v7 n9 [1984 nov/dec], v8 n1-v10 n8 [1985 mar-1987 dec], v11 n1-3, 6 [1988 feb/mar-may/jun, 1989 feb] – 1r – 1 – (cont: new dawn [ypsilanti mi: 1983]) – mf#1223867 – us WHS [305]

D and J Sadlier and Co see Catalogue of school books stationery, etc, etc

D antonio manuel roxano mudarra : colegial antiguo de opposision del real colegio de san ignacio de la compania de jesus...oppositor a la canongia lectoral vacante en esta santa iglesia cathedral... / Roxana Mudarra, Antonio Manuel – [Puebla: s.n. 1760?] – (= ser Books on religion...1543/44-c1800: iglesias, catedrales) – 1mf – 9 – mf#crl-444 – ne IDC [241]

D at m oi = New land – Seattle WA. 1978 apr-1980, 1981-82, 1983-84, 1985 jan-1987 jun – 4r – 1 – mf#1611896 – us WHS [071]

D B Lewis & Co see Federal income tax record for individuals

D b ray's text book on campbellism exposed / Hand, George R – St Louis: Christian Pub Co, 1880 – 1mf – 9 – 0-524-06262-5 – mf#1991-2453 – us ATLA [240]

D balthasar hubmaier als theologe / Sachsse, Carl – Berlin: Trowitzsch 1914 [mf ed 1991] – (= ser Neue studien zur geschichte der theologie und der kirche 20) – 1mf – 9 – 0-524-00314-9 – mf#1989-3014 – us ATLA [242]

D c 9 newsletter / Brotherhood of Painters, Decorators and Paperhangers of America – v2 n6-v6 n6 [1968 nov/dec-1972 jul] – 1r – 1 – mf#1055045 – us WHS [690]

D C Bicentennial Commission see Sentry post

D c gazette – 1969 sep 25-1972 dec 20, 1973 jan 3-1975 aug, 1975 oct-1978 apr, 1978 may-1982 dec – 4r – 1 – mf#701174 – us WHS [071]

D C Student Coalition against Apartheid and Racism see Scar news

D carl daub's system der theologischen moral = System der theologischen moral / Daub, Carl; ed by Marheineke, Philipp & Dittenberger, Theophor Wilhelm – Berlin: Duncker und Humblot, 1840-1843 – (= ser Philosophische und theologische Vorlesungen) – 3mf – 9 – 0-524-00018-2 – mf#1989-2718 – us ATLA [170]

D carl daub's vorlesungen ueber die philosophische anthropologie / ed by Marheineke, Philipp – Berlin: Duncker & Humblot, 1838 [mf ed 1994] – 6mf – 9 – 0-524-00019-0 – mf#1989-2719 – us ATLA [100]

D carl daub's vorlesungen ueber die prolegomena zur dogmatik : und ueber die kritik der beweise fuer das daseyn gottes = Vorlesungen ueber die prolegomena zur dogmatik / Daub, Carl; ed by Marheineke, Philipp & Dittenberger, Theophor Wilhelm – Berlin: Duncker und Humblot, 1839 – (= ser Philosophische und theologische Vorlesungen) – 2mf – 9 – 0-7905-7927-8 – mf#1989-1152 – us ATLA [240]

D carl daub's vorlesungen ueber die prolegomena zur theologischen moral : und ueber die principien der ethik = Vorlesungen ueber die prolegomena zur theologischen moral / Daub, Carl; ed by Marheineke, Philipp & Dittenberger, Theophor Wilhelm – Berlin: Duncker und Humblot, 1839 – (= ser Philosophische und theologische Vorlesungen) – 2mf – 9 – 0-7905-7928-6 – mf#1989-1153 – us ATLA [170]

D chipman's reports / Chipman, D – v1-2. 1789-1825 (all publ) – (= ser Vermont Supreme Court Reports) – 8mf – 9 – $12.00 – (a pre-nrs title) – mf#LLMC 90-308 – us LLMC [347]

D d d / American Servicemen's Union – 1969 oct/nov – 2r – 1 – mf#915920 – us WHS [355]

D d brevis introductio in historiam litterariam mineralogicam... / Wallerius, Johan Gottschalk – Holmiae: In Officinis Libr Reg Ac Biblop M Swederi 1779 [mf ed 1984] – 1r – 1 – (original title: lucubrationum academicarum specimen primum de systematibus mineralogicis et systemate mineralogico rite condendo; incl ind) – mf#1230p – us Wisconsin U Libr [550]

D d eisenhower home and family / Endacott, J Earl – 1 – us Kansas [920]

D d ioannis oecolampadii et hvldrichi zvinglii epistolarum libri 4, etc / Oecolampadius, J – Basileae, [R Winter], 1536 – 10mf – 9 – mf#PBU-392 – ne IDC [242]

D F de Viu see Coleccion de inscripciones y antiguedades de extremadura

D fr strauss' alter und neuer glaube und seine literarischen ergebnisse : zwei kritische abhandlungen / Rauwenhoff, Lodewijk Willem Ernst & Nippold, Friedrich – Leipzig: Richter & Harrassowitz; Leiden: SC van Doesburgh, 1873 [mf ed 1986] – 1mf – 9 – 0-8370-7325-1 – (incl bibl ref) – mf#1986-1325 – us ATLA [140]

D francisco de rojas, embajador de los reyes catolicos / Rodriguez Villa, Antonio – Madrid: Fortanet, 1896, pp. 180-202, 295-339, 364-402, 440-474 y 1896, pp. 5-36. Boletin Real Academie de la Historia 28 y 29, 1896 – 1 – sp Bibl Santa Ana [320]

D francisco fernandez de la cueva / Fernandez Duro, Cesareo – 1885 – 9 – sp Bibl Santa Ana [920]

D gustav warneck, 1834-1910 : blaetter der erinnerung / Kaehler, Martin – Berlin: Martin Warneck, 1911 – 1mf – 9 – 0-7905-2126-1 – mf#1987-2126 – us ATLA [370]

D gysberti voetii selectarum disputationum fasciculus / Voet, Gijsbert – Amstelodami: JA Wormser, 1887 [mf ed 1993] – (= ser Bibliotheca reformata 4; Presbyterian coll) – 5mf – 9 – 0-524-07468-2 – mf#1991-3128 – us ATLA [242]

D hermann hupfeld : lebens- und charakterbild eines deutschen professors / Riehm, Eduard – Halle: Julius Fricke, 1867 – 1mf – 9 – 0-7905-2688-3 – mf#1987-2688 – us ATLA [920]

D i a radio – Sao Paulo (BR), 1935 n1-1939 n7 – 2r – 1 – gw Misc Inst [790]

D i dimanche illustre see Dimanche illustre

D ioannes aepini in psalmum 16 commentariu / Aepinus, J – Francofvrti, [1544] – 1mfmf – 9 – mf#TH-1 mf 5 – ne IDC [242]

D johann philipp treibers...sonderbare invention : eine arie in einer einzigen melodey aus allen tonen und accorden... / Treiber, Johann Philipp – Jena: C Junghansen 1702 [mf ed 199-] – 1r – 1 – mf#pres. film 138 – us Sibley [780]

D johannes hinrich wicherns lebenswerk in seiner bedeutung fuer das deutsche volk / Henning, M – Hamburg 1908 (mf ed 1994) – 1mf – 9 – €38.00 – 3-89349-659-9 – mf#DHS-AR 659 – gw Frankfurter [360]

D juan melendez valdes correspondance... hopitaux d'avila...bordeaux, 1964 / Demerson, Georges – Madrid: Graf. Calleja, 1964 – 1 – sp Bibl Santa Ana [946]

D juan ruiz de alarcon y mendoza / Fernandez-Guerra Y Orbe, Luis – Madrid, Spain. 1871 – 1r – us UF Libraries [972]

D luiz i : jornal de interesses portuguezes – Rio de Janeiro, RJ: Typ Progresso, 21 nov-dez 1868; fev-28 abr 1869 – (= ser Ps 19) – mf#P19,02,63 – bl Biblioteca [440]

D m, disease-a-month : disease-a-month – Oxford, England 1954+ – 1,5,9 – ISSN: 0011-5029 – mf#1653 – us UMI ProQuest [610]

D m luthers wie auch anderer gottseligen und christlichen leute geistliche lieder und psalmen : wie sie bisher in evangelischen kirchen dieser landen gebraucht werden / Crueger, Johann – Berlin: Christoff Runge 1657 [mf ed 19–] – 1r – 1 – mf#pres. film 120 – us Sibley [780]

D M thornton : a study in missionary ideals and methods / Gairdner, William Henry Temple – London; New York: Hodder and Stoughton, [1908?] – 1mf – 9 – 0-8370-6494-5 – mf#1986-0494 – us ATLA [920]

D m z [deutsche montags-zeitung] see Deutsche montagszeitung

D magni ausonii opuscula (mgh1:5/2) / ed by Schenkl, C – 1883 – (= ser Monumenta germaniae historica 1: scriptores – auctores antiquissimi) – €19.00 – ne Slangenburg [240]

D martin luthers deutsche bibel / Reichert, O – Tuebingen: J C B Mohr 1910 [mf ed 1989] – (= ser Religionsgeschichtliche volksbuecher fuer die deutsche christliche gegenwart 4/13) – 1mf – 9 – 0-7905-1675-6 – mf#1987-1675 – us ATLA [220]

D morse jr's advertiser – Empire City OR: D Morse Jr [mf ed v1 n2 (winter ed, 1874)] – 1r – 1 – us Oregon Lib [071]

D n a u newsletter / Denver Native Americans United – v2 n1-v4 n1 [1979 jan-1981 jan] – 1r – 1 – mf#609270 – us WHS [307]

D o m a gymnasii ethici : quam praeside henrico nicolai / Nicolai, Heinrich – Dantisci: Typis viduae Georgii Rhetii, anno 1649 – (= ser Ethics in the early modern period) – [92]lea on 3mf – 9 – mf#pl-359 – ne IDC [170]

D o t c news / Dakota Ojibway Tribal Council – v1 n1-v2 n2 [1985 nov-1987 mar/apr] – 1r – 1 – (cont: d o t c news [1982]) – mf#1312963 – us WHS [071]

D o t c news / Dakota Ojibway Tribal Council – v8 n8-v9 n3 [1982 sep-1984 apr] – 1r – 1 – (cont: dakota ojibway tribal council news) – mf#685725 – us WHS [307]

D pedro 1 / Lamego, Luiz – Rio de Janeiro, Brazil. 1939? – 1r – us UF Libraries [972]

D pedro 2 e o conde de gobineau (correspondencia) / Gobineau, Arthur – Sao Paulo, Brazil. 1938 – 1r – us UF Libraries [972]

D pedro 2 nos estados unidos / Segadas Machado-Guimaraes, Argeu De – Rio de Janeiro, Brazil. 1961 – 1r – us UF Libraries [972]

D pedro na regencia / Gama, Annibal – Rio de Janeiro, Brazil. 1948 – 1r – us UF Libraries [972]

D philipp fermins ausfuehrliche historisch-physikalische beschreibung der kolonie surinam : auf veranlassung der gesellschaft naturforschender freunde in berlin aus dem franzoesischen uebersetzt, und mit anmerkungen begleitet – Berlin 1775 [mf ed Hildesheim 1995-98] – 2v on 8mf – 9 – €160.00 – 3-487-26885-X – gw Olms [959]

D philipp jacob speners erklaerung der christlichen lehre : nach der ordnung des kleinen katechismus d. martin luthers = Einfaeltige erklaerung der christlichen lehr / Spener, Philipp Jakob – 2. Aufl. Berlin: Der Verein, 1849 – 1mf – 9 – 0-8370-9420-8 – (incl ind) – mf#1986-3420 – us ATLA [242]

El d raphael ruiz calado : presbytero domeciliario, y patrimonial del obispado de la puebla de los angeles, abogado de la real audiencia, y opositor... / Calado, Raphael Ruiz – [Puebla: s.n. 1753?] – (= ser Books on religion...1543/44-c1800: iglesias, catedrales) – 1mf – 9 – mf#crl-440 – ne IDC [241]

D sofonias salvatierra y su 'comentario polemico' / Chamorro, Pedro Joaquin – Managua, Nicaragua. 1950 – 1r – us UF Libraries [972]

D thomae aquinatis. de essentia et potentiis animae in generali (ia, q 75-77). una cum guilelmi de la mare correctorii art 28 (fp14) / ed by Geyer, B – 1920 – (= ser Florilegium patristicum (fp)) – €5.00 – ne Slangenburg [241]

D thomae aquinatis quaestiones disputatae de veritate. q 11 (fp13) / ed by Dyroff, A – 1921 – (= ser Florilegium patristicum (fp)) – €3.00 – ne Slangenburg [241]

D w funk see Z-j-funk

D w griffith papers, 1897-1954 / ed by Martin, Ann – (= ser Cinema history microfilm series) – 36r – 1 – $6100.00 – 0-89093-687-0 – (with p/g & ind) – us UPA [790]

D W griffith's the birth of a nation – 24mf – 9 – (coll gives 2000 individual images from that film along with accompanying guide: birth of a nation – a formal shot-by-shot analysis by john cuniberti. guide also provides detailed descriptive and technical references) – mf#C39-29300 – us Primary [790]

Da asia de joo de barros dos feitos, que os portuguezes fizeram no descubrimento : et conquista dos mares, e terras do oriente / Barros, J de – Lisboa: na regia officina typographica 1777 [3rd decada, pt2] – 6mf – 9 – mf-773/1 – ne IDC [915]

Da c san xuan giap dan 1974 / Trung Tam Giao Duc Le Qui Don – [Saigon]: Trung Tam Giao Duc Le Qui Don [1974] [mf ed 1993] – on pt of 1r – 1 – mf#11052 r485 n11 – us Cornell [959]

Da c-san xuan at ma-o / Truong Trung Hoc Do Thi Tan Dinh. – [Saigon]: Trung-Hoc Do-Thi Tan Dinh [1975] [mf ed 1993] – on pt of 1r – 1 – mf#11052 r484 n3 – us Cornell [959]

Da draussen vor dem tore : heimatliche naturbilder / Loens, Hermann – Hannover: A Sponholtz 1923, c1911 [mf ed 1995] – 1r – 1 – (filmed with: bert brecht / willy haas) – mf#3941p – us Wisconsin U Libr [880]

Da gong bao – Tianjin: [Da gong bao guan], 1902-66 [daily] [mf ed 1976] – 163r – 1 – (ceased publ on sep 10 1966; publ in hankou, chongqing 1937-45, in beijing 1949-66; publ suspended nov 27 1925-aug 31 1926, july 26 1937-nov 30 1945) – cc Misc Inst [079]

Da gong bao – Shanghai. 26 sep 1946-mar 1947; 1 apr, 19 may 1947; 23 sep 1947-29 dec 1948 – 8r – 1 – uk British Libr Newspaper [072]

Da gong bao – Tienbin, China. 12 jan-18 sep 1953 – 1 1/2r – 1 – uk British Libr Newspaper [072]

Da governacao de angola / Monteiro, Armindo Rodrigues – Lisboa, Portugal. 1935 – 1r – us UF Libraries [960]

Da gun khan khan le e ca chui to vatthu / Khan Khan Le, Da gun – Ran kun: Ca pe mve' su 1975 [mf ed 1993] – on pt of 1r – 1 – mf#11052 r635 n2 – us Cornell [959]

Da gun Nat Rhan see Da gun nat rhan e bhava khari can

Da gun nat rhan e bhava khari can / Da gun Nat Rhan – Run Kun: Ca pe Biman 1974 [mf ed 1990] – (= ser Ca pe biman thut prann s lak cvai ca can) – 1r with other items – 1 – (in burmese) – mf#mf-10289 seam reel 163/2 [§] – us CRL [920]

Da gun Rvhe Myha see Rhve pan ku

Da gun rvhe myha e bhava nhan a tve' a krum – Ran kun: Ca pay u Ca pe Phran khyi re 1975 [mf ed 1993] – on pt of 1r – 1 – mf#11052 r638 n4 – us Cornell [480]

Da gun Tara see
- Ca lum, che cak, con krui, nhan kattipa ka lip
- Sajan san pran to

Da mensch muass a freud habn : boehmerwaelder schnaderhuepfeln gesammelt / Jungbauer, Gustav – Reichenberg: E Ullmann [194-?] [mf ed 1990] – 1r – 1 – (filmed with: ernst junger / wulf dieter muller) – mf#2749p – us Wisconsin U Libr [810]

Da monarchia para a republica (1870-1889) / Morais, Evaristo De – Rio de Janeiro, Brazil. 1936 – 1r – us UF Libraries [972]

Da ng, Thai Mai see Van tho cach mang viet-nam dau the ky xx, 1900-1925

Da qing quan shu / Shen, Hongzhao – [Beijing]: Long shu fang, Kangxi kui hai [1683] [mf ed 1966] – (= ser Tenri coll of manchu-books in manchu-characters. series 1, linguistics 31-32) – 12v on 2+ r – 1 – (in chinese & manchu; title also romanized: daiicing gurun-i yooni bithe) – ja Yushodo [480]

Da zeila alle frontiere del caffa : viaggi pubblicati a cura e spese della societ... geografica italiana / Cecchi, A – Roma, 1885-1887. 3v – 22mf – 9 – mf#NE-20198 – ne IDC [910]

Daab, Ursula see
- Die althochdeutsche benediktinerregel des cod sang 916
- Drei reichenauer denkmaeler der altalemannischen fruehzeit

Daabsminder fra herrens tjeneste i kirke og mission / Andersen, Rasmus – Cedar Falls, Iowa: Dansk, 1912 – 1mf – 9 – 0-524-05186-0 – mf#1991-2222 – us ATLA [240]

Daaku, K Y see Oral traditions of assin-twifo

Daanson, Edouard see Mythes and legendes

Dabbs, Norman H see Dawn over the bolivian hills

Dabeisein – mitgestalten : schriftsteller ueber ihr leben und schaffen / ed by Christ, Richard et al – Berlin: Verlag Tribuene, 1960 [mf ed 1993] – 223p – 1 – (incl bibl ref) – mf#8155 – us Wisconsin U Libr [920]

Dabel, Gerhard see Mit krad und karabiner

Daber el ha-'am / Halperin, Shim'on – s.l, s.l? 1929 or 30 – 1r – us UF Libraries [939]

Daber 'ivrit! / Jardeni, M – Kovnah, Lithuania. 1928 – 1r – us UF Libraries [939]

Dabin, Jean see
- Theorie generale du droit
- A treatise on the law of negotiable instruments, including bills of exchange; promissory notes; negotiable bonds and coupons.

Dabistan – Mashhad. sal-i 1, shumarah-i 1-12. 1 rabi al-sani 1341-1 rabi al-avval 1342 [21 nov 1922-12 oct 1923] – 1r – 1 – $175.00 – us MEDOC [079]

Dabistan – Baku. numrah-'i 1. 4 rabi' al-avval-20 jumada al-avval 1324 [16 apr 1906] – 1r – 1 – $53.00 – (missing: n2, 3) – us MEDOC [079]

The dabistan : or, school of manners: the religious beliefs, observances, philosophic opinions and social customs of the nations of the east / [Fani (Muhsin, Muhammad)]; ed by Troyer, Anthony – Washington: MW Dunne, c1901 [mf ed 1993] – (= ser Universal classics library) – 1mf – 9 – 0-524-08403-3 – (english trans fr persian by david shea and anthony troyer. int by a v williams jackson) – mf#1993-4013 – us ATLA [200]

Dabney, Michael see Twelve minuets and twelve dances for a violin, hautboy, and harpsichord

Dabney, Robert Lewis see
- The believer born of almighty grace
- The christian sabbath
- A defence of virginia
- Discussions
- The five points of calvinism
- The sensualistic philosophy of the nineteenth century considered
- Syllabus and notes of the course of systematic and polemic theology

Dabry de Thiersant, Philibert see
- Le catholicisme en chine au 8 siecle de notre ere
- De l'origine des indiens du nouveau monde et de leur civilisation

Dabt al-nasl : abaduhu wa-atharuhu al-dimughrafiyah wa-al-iqtisadiyah wa-al -ijtimaiyah 101=abd al-qadir salih, hasan – al-Kuwayt: [s.n.], 1981 – us CRL [956]
Dabula, Tolakele *see* Perceived occupational stress amongst female nurses working in a general hospital
The dacca gazette / Pakistan. East – Dacca. 1961-1966 – 1 – us NY Public [079]
Dacca news, 1856-58 – 2r – 1 – mf#4894 – uk Microform Academic [079]
Dach, Simon *see*
– Gedichte
Dach, Walter *see* Volksgenosse mueller 2
Dachauer nachrichten – Dachau DE, 1988- – 14r/yr – 1 – (bezirksausgabe von muenchner merkur, muenchen) – gw Misc Inst [074]
d'Achery, L *see* Acta sanctorum o s b
Dacheux, Leon *see* Jean geiler de kaysersberg, predicateur a la cathedrale de strasbourg, 1478-1510
Dacia – Bucharest, Romania. -d. 31 Jan-24 May 1920. Imperfect. 1 reel – 1 – uk British Libr Newspaper [949]
D'Acres, R *see* The art of water drawing, 1659-60
Le dactylographe canadien : methode francaise et anglaise la plus rationnelle et la plus efficace... / Nadeau, Wilfrid – 4e ed. [Quebec (Province)?]: en vente chez tous les libraires; Ste-Marie: ou chez l'auteur, [1926?] (mf ed 1993) – 1mf – 9 – (with english text) – mf#SEM105P1774 – cn Bibl Nat [650]
Dacy, George H *see* Four centuries of florida ranching
Der dada – Berlin-Charlottenburg DE, 1917 jul-1921 sep – 1 – fr ACRPP [700]
Dadachanji, Bahran Edulji *see* History of indian currency and exchange
Dadaismus und religion : hugo balls "weg zu gott!" / Steinbrenner, Manfred – Frankfurt a.M., 1983 (mf ed 1994) – 2mf – 9 – 3-89349-880-X – mf#DHS-AR 880 – gw Frankfurter [700]
Dadd, George H *see* American reformed horse book
Dade city banner – Dade City, FL. 1925-1972 – 45r – (gaps) – us UF Libraries [071]
Dade country, florida – Miami, FL. 1937? – 1r – us UF Libraries [630]
Dade county and the citrus fruit industry of the state of florida / Rosser, Lillian Evelyn – s.l, s.l, s.l? 1939 – 1r – us UF Libraries [634]
Dade county election report, 1843 – s.l, s.l? 193-? – 1r – us UF Libraries [978]
Dadie, Bernard Binlin *see* Legendes africaines
Dado tiling manufactured by craven, dunnill and co / Craven, Dunnill and Co – [Jackfield? 1880?] – (= ser 19th c art & architecture) – 1mf – 9 – mf#4.2.1394 – uk Chadwyck [730]
Dadre, Emile *see* Etude dogmatique sur la predestination dans calvin
Daechsel, August *see*
– St markus's og st lukas's evangelier
– St matthaeus's evangelium
– St paulus's breve
Daedalus – Cambridge MA 1846+ (1) 1973+ (5) 1955+ (9) – ISSN: 0011-5266 – mf#4587 – us UMI ProQuest [500]
Daedalus : or, the causes and principles of the excellence of greek sculpture / Falkener, Edward – London 1860 – (= ser 19th c art & architecture) – 5mf – 9 – mf#4.2.1578 – uk Chadwyck [720]
Daehlen, Ingvald *see* The united norwegian lutheran mission field in china
Daehne, August Ferdinand *see* Entwickelung des paulinischen lehrbegriffs
Daehnert, Joh Karl *see* Neueste critische nachrichten
Daehnert, Johann Carl *see*
– Critische nachrichten
– Neue critische nachrichten
– Pommersche nachrichten von gelehrten sachen
Daehnhardt, Oskar *see*
– Die goldene gans
– Griechische dramen in deutschen bearbeitungen
Daeleman, Jan *see* Morfologie van naamwoord en werkwoord in het kongo (ntandu)
Daellenbach, C Charles *see* Investigation of the use of video tape recorder techniques in the identification of behavioral characteristics of music teachers
Daemmerung : roman / Finckenstein, Ottfried, Graf – Jena: E Diederichs 1944, c1942 [mf ed 1989] – 1r – 1 – (filmed with: double, double, toil and trouble / lion feuchtwanger) – mf#7237 – us Wisconsin U Libr [830]
Daemmrich *see* The challenge of german literature
Daemon faust : wie goethe ihn schuf / Ammon, Hermann – Berlin: F Duemmler, 1932 [mf ed 1990] – 344p – 1 – (incl bibl ref and ind) – mf#7340 – us Wisconsin U Libr [430]
Die daemonen und ihre abwehr im alten testament / Jirku, Anton – Leipzig: A Deichert, 1912 – 1mf – 9 – 0-524-04464-3 – (incl bibl ref) – mf#1992-0133 – us ATLA [221]
Daemonenbeschwoerung bei den babyloniern und assyrern : eine skizze / Weber, Otto – Leipzig: JC Hinrichs, 1906 [mf ed 1989] – (= ser Der alte orient 7/4) – 1mf – 9 – 0-7905-2094-X – mf#1987-2094 – us ATLA [130]
Daemonentaenzer der urzeit : ein roman aus den wildnissen der zweiten eisenzeit / Achermann, Franz Heinrich – 5. aufl. Olten (Switzerland): Otto Walter, c1935 [mf ed 1995] – 240p – 1 – mf#8917 – us Wisconsin U Libr [830]
Daemonologia : a discourse on witchcraft as it was acted in the family of mr. edward fairfax... / Fairfax, Edward – Harrogate: R Ackrill, 1882 – 1mf – 9 – 0-524-01438-8 – mf#1990-2433 – us ATLA [130]
Daemonologia sacra : or, a treatise of satan's temptations / Gilpin, Richard; ed by Grosart, Alexander Balloch – Edinburgh: J Nichol 1867 [mf ed 1987] – (= ser Nichol's series of standard divines. puritan period) – 1r – 1 – (memoir by ed. filmed with: modern rationalism / mccabe, joseph) – mf#1945 – us Wisconsin U Libr [130]
Daenemark und wir / Scheel, Otto – Tuebingen: Kloeres 1915 [mf ed 1987] – (= ser Durch kampf zum frieden 7) – 1r – 1 – mf#6840 – us Wisconsin U Libr [327]
Daenische blaetter – Hamburg-Altona DE, 1795-96 – 1 – gw Misc Inst [074]
Daeubler, Theodor *see*
– Hesperien
– Hymne an italien
– Mit silberner sichel
– Der sternhelle weg
– Die treppe zum nordlicht
– Wir wollen nicht verweilen
Daf yomi review – Staten Island, NY. Kesubos, Nedorim, Nozir, Sotah, Gitin, Kiddushin, Bava Kama, Minachos – 1 – us AJPC [071]
Daffodils / Jacob, Joseph – London, England. 1910? – 1r – us UF Libraries [580]
Dafforne, James *see*
– The albert memorial
– Leslie and maclise
– The life and works of edward matthew ward
– Modern art
– Pictures by clarkson stanfield
– Pictures by daniel maclise
– Pictures by john phillip
– Pictures by sir a w callcott
Daftar buku Indonesia *see* "Gedung buku nasional"
Daftar buku-buku – Djakarta, 1955 – 1mf – 9 – mf#SE-633 – ne IDC [959]
Daftar cronologisch surat keputusan bupati, kepala daerah kabupaten tulungagung – Tulungagung, 1970. v1-5 – 5mf – 9 – mf#SE-1970 – ne IDC [950]
Daftar nama buku2, tulisan2 dan alat2 lain jang dilarang untuk dipergunakan disekolah2, kursus2 dan balai2 pendidikan berita-negara ri suppl 6 pertjetakan negara ri : indonesia – Djakarta, 1950-1956 – 13mf – 9 – (missing: 1950(1-22, 24-53, 55-75, 77-end); 1951; 1952; 1953(1-20, 22-79, 81-end); 1954(1-14, 16-79, 81-89, 91-end); 1955(1-12, 14-95, 97-end); 1956(1-59, 61-76, 78-end); 1957; 1958; 1959(1-8)) – mf#SE-220 – ne IDC [959]
Daftari, Kesho Laxman *see* The social institutions in ancient india
Daftary, Ali Akkbar Khan *see* Geschichte und system des iranischen strafrechts
I dag – Goeteborg, 1990-95 – 1 – sw Kungliga [078]
I dag – Malmoe, 1990-95 – 1 – sw Kungliga [078]
Dag – Antwerp Belgium, 16 jul 1942-jun 1944 – 3r – 1 – uk British Libr Newspaper [074]
Dagarcik – Istanbul, 1872-73. Muharriri: Ahmed Midhat. n1-10. 1872-73 – (= ser O & t journals) – 5mf – 9 – $75.00 – us MEDOC [956]
Het dagblad – Paarl SA, jan 31 1883-sep 30 1898 – 17r – 1 – (fr jan 31 1883-may 1891 as: paarl district advertentieblad. fr jun 1891-jun 1896 as: de paarl. absorbed by: de kolonist; cont by: de paarl) – mf#MS00300 – sa National [079]
Het dagblad – Batavia, 1945-1949 – 7r – 1 – (missing: 1946(jun 5, 20); 1948(nov 20, dec 17)) – mf#SEF-2 – ne IDC [950]
Dagbladet – Copenhagen, Denmark. -d. 20 jul 1859-31 Dec 1863. 9 reels – 1 – uk British Libr Newspaper [072]
Dagbladet – Goteborg, Sweden. 1955- – 1 – sw Kungliga [078]
Dagbladet *see* Nya samhallet
Dagboeger fra hans rejser i gronland 1739-1753 / Walloes, P O; ed by Bobe, L – Kobenhavn, 1927 – 3mf – 9 – mf#N-446 – ne IDC [917]
Dagboek van h a l hamelberg, 1855-1871 / Hamelberg, Hendrik Anthony Lodewijk – Kaapstad, South Africa. 1952 – 1r – us UF Libraries [960]
Die dagboek van hendrik witbooi, kaptein van die witbooi-hottentotte, 1884-1905 : bewerk na die oorspronklike dokumente in die regeringsargief, windhoek / Witbooi, Hendrik – Cape Town: The Van Riebeeck Society 1929 – (= ser [Travel descriptions from south africa, 1711-1938]) – 3mf – 9 – mf#zah-67 – ne IDC [916]
Dagboger fra 1792 / Bournonville, Antoine; ed by Clausen, Julius – Kobenhavn, Gyldendal, 1924 – 1 – mf#*ZBD-*MGO pv15 – Located: NYPL – us Misc Inst [790]
Dagbok 'fver en ostindisk resaren 1750-1752 / Osbeck, P – Stockholm, 1757 – 5mf – 9 – mf#247 – ne IDC [915]
Dagen – Stockholm, Sweden. 1888-89 [2r]; 1896-1915 [58r] – 1 – sw Kungliga [078]
Dagen – Stockholm, Sweden. 1945- – 1 – sw Kungliga [078]
Dagen – Stockholm, Sweden. 2004- – 1 – sw Kungliga [078]
Dagenham & barking advertiser – Barking, England 14 apr 1989-31 jan 1992 [mf 1977-] – 1 – (cont: dagenham & barking advertiser & barking recorder [21 aug 1981-7 apr 1989]; cont by: barking & dagenham recorder [7 feb 1992-]) – uk British Libr Newspaper [072]
Dagenham post and barking and rainham guardian – Barking, London 20 jan 1928-26 dec 1930, 1 jan 1932-29 jan 1964 [mf 1928-30, 1932-may 1935] – 1 – (cont by: barking & dagenham post [5 feb 1964-6 oct 1999]) – uk British Libr Newspaper [072]
Dagens fragor – Goeteborg, Sweden. 1879-80 – 1r – 1 – sw Kungliga [078]
Dagens industri – Stockholm, Sweden. 1983- – 1 – sw Kungliga [078]
Dagens nyheter – Stockholm, Sweden. Ed. A., 1864-1977 – 1 – (ed. a-b.1875-1900. 20 reels. ed. b. 1875-1966. 600 reels. ed. a. newsbills. 1924-63. 34 reels. ed. b. newsbills. 1931-52. 14 reels. semiweekly ed. 1913-24.19 reels) – sw Kungliga [078]
Dagens nyheter – Stockholm: Dagens Nyheters Tryckeri, 1952 – 1 – (filmed from uppl a (stockholms-uppl.)) – us CRL [079]
Dagens nyheter – Stockholm, Sweden. 1979- – 1 – sw Kungliga [078]
Dagens nyheter newsbills – Stockholm, 1923-78 – 86r – 1 – sw Kungliga [078]
Dagens politik – Stockholm, 1995-97 – 2r – 1 – sw Kungliga [078]
Dageraad – London, UK. Jun 1918-15 Feb 1919 – 1 – uk British Libr Newspaper [072]
Dagestansik filial, makhach-kala / Akademiia Nauk. SSSR – Institut Istorii, Yazyka I Literatury. Uchenyye Zapiski. Makhach-Kala. v. 1-13, 15-18. 1956-1968 – 1 – us NY Public [490]
Dagestanskaia pravda : respublikanskiaia obshchestvenno-politicheskaia gazeta – Makhachkala: Izd-vo Dagestanskogo respublikanskogo komiteta KP RSFSR [1991-95] [mf ed Minneapolis MN: East View Publ [199-] – 10r – 1 – mf#mf-11785 – us CRL [077]
Dagg, John L *see* Autobiography and other materials by him
Dagg, John Leadley *see*
– The elements of moral science
– A treatise on christian doctrine
– A treatise on church order
The dagger – Quebec: [s.n, 1863-1864] – 9 – mf#P04927 – cn Canadiana [071]
Daggett, L H *see* Historical sketches of woman's missionary societies in america and england
Dagh-register gehouden int casteel batavia van passerende daer plaetse als over geheel nederlandts-india – 's Gravenhage 1887-1931 [mf ed 2004] – (= ser Rare printed sources and reference works for the history of dutch colonialism) – 30v – 9 – €1685.00 – (mmp116/1-2 [191mf] €1815 set) – mf#mmp116/2 – ne Moran [959]
Dagligt allehanda – Stockholm, Sweden. 1767-1849 – 1 – sw Kungliga [078]
Dagobert : roi des francs / Barroux, R – Paris, 1938 – €11.00 – ne Slangenburg [944]
Dagslyset / Fremskridsforening et al – Chicago IL. 1 aarg n1-8 aarg n8 [1869 apr-1878 feb] – 1r – 1 – mf#683100 – us WHS [071]
Dagsposten – Stockholm, 1941-51 – 9 – sw Kungliga [078]
Dagsposten newsbills – Stockholm, 1941-51 – 1r – 1 – sw Kungliga [078]
The daguerreian journal / humphrey's journal – 1 nov 1850-jul 1870 – (= ser Photographic periodicals of the daguerreian era) – 20v on 6r – 1 – $735.00 – us UPA [770]
Daguerreotype : a magazine of foreign literature and science – La Salle IL 1847-49 – 1 – mf#3973 – us UMI ProQuest [071]
O daguerrotypo – Rio de Janeiro, RJ: Typ do Daguerrotypo, 18 jul 1845 – (= ser Ps 19) – mf#P17,01,142 – bl Biblioteca [321]
D'Aguilar, George C D *see* Observations on the practice and forms of district, regimental and detachment courts-martial
Dahat elohim / Bernfeld, Simon – Warsaw, Poland. 1922 – 1r – us UF Libraries [939]
Dahauron, R *see* Il giardiniero francese, ovvero trattato del tagliare gl'alberi...
Daheim – Chicago: Free Press Printing Co, feb 11-may 26 1872 – 1r – 1 – us CRL [071]
Daheim – Chicago: [s.n.] 1891-jun 2 1901 – us CRL [071]
Daheim – Fond Du Lac WI. 1917 nov25-1918 aug 4 – 1r – 1 – mf#943342 – us WHS [071]
Daheim – Leipzig DE, 1892 apr-sep – 1r – 1 – gw Misc Inst [074]
Daheim (iz3) / ed by Estermann, Alfred – Leipzig/Bielefeld 1865-1943 [mf ed 2001] – (= ser Illustrierte zeitschriften (iz) 3) – ca 85,000p – 9 – diazo €5100 silver €6500 – 3-89131-348-9 – gw Fischer [640]
Dahiel, der konvertit : roman / Voss, Richard – Stuttgart: Deutsche Verlags-Anstalt 1889 [mf ed 1991] – 1r – 1 – (filmed with: poetische werke / johann heinrich voss) – mf#2969p – us Wisconsin U Libr [830]
Dahir, James Safady *see* Five contemporary liberal preachers
Dahiya, Bhim S *see* The hero in hemingway: a study in development
Dahl, H V *see* Eesti baptismi ajalugu: i. arkamise aeg
Dahl, Theodor H *see* Den forenede kirke
Dahlberg sugar cane industries / Dahlberg Corporation Of America – Chicago, IL. 1929 – 1r – us UF Libraries [630]
Dahlia *see* Oh, nasib!
Dahlke, Paul *see*
– Buddhism and its place in the mental life of mankind
– Buddhism and science
– Buddhist essays
– Buddhist stories
Dahlmann, Friedrich Christoph *see*
– Dahlmann-waitz
– The history of the english revolution
Dahlmann, Friedrich Wilhelm *see* Philosophie des sichselbstbewussten
Dahlmann, Joseph *see*
– Buddha
– Der idealismus der indischen religionsphilosophie im zeitalter der opfermystik
– Nirvana
Dahlmann-waitz : quellenkunde der deutschen geschichte / Dahlmann, Friedrich Christoph & Waitz, Georg; ed by Herre, Paul – 8. aufl. Leipzig: K F Koehler, 1912 – 3mf – 9 – 0-7905-8024-1 – mf#1988-6005 – us ATLA [019]
Dahlschen, Edith *see* Women in zambia
Dahme-kurier – Koenigs Wusterhausen DE, 1963-65 [single iss] – 1r – 1 – gw Misc Inst [074]
Dahmen, Hans *see*
– E T A hoffmanns weltanschauung
– Lehren ueber kunst und weltanschauung im kreise um stefan george
Dahmen, Johan Arnold *see* Trois trios pour violon, alto et violoncelle
Dahn, Felix *see*
– Attila
– Die bataver
– Bissula
– Chlodovech
– Dichtungen
– Felicitas
– Felix dahn's saemtliche werke poetischen inhalts
– Fredigundis
– Gedichte
– Julian der abtruennige
– Ein kampf um rom
– Kleine nordische erzaehlungen
– Koenig roderich
– Odhins rache
– Odhin's trost
– Romane
– Schaubuehne
– Die schlimmen nonnen von poitiers
– Sind goetter?: die halfred sigskaldsaga
– Die staatskunst der frau'n
– Stilicho
– Urgeschichte der germanischen und romanischen voelker
– Vom chiemgau
Dahne, Gerhard *see* Westdeutsche prosa
Dahnke, Hans-Dietrich *see*
– Erbe und tradition in der literatur
– Geschichte der deutschen literatur 1789 bis 1806
Le dahome : souvenirs de voyage et de mission / Laffitte, J – 4e ed. Tours: A Mame, 1876 – 1 – us CRL [960]
Dahomean narrative / Herskovits, Melville Jean – Evanston, IL. 1958 – 1r – us UF Libraries [960]
Dahomey *see*
– Journal officiel
– Journal officiel de la republique du dahomey
Le dahomey – Corbeil, France: E Crete, 1906 – 1 – us CRL [960]
Le dahomey : histoire, geographie, moeurs, coutumes, commerce, industrie, expeditions francaises, 1891-1894 / Foa, Edouard – Paris: A Hennuyer, 1895 – 1 – us CRL [960]
Le dahomey, a l'assaut du pays des noirs / Grandin, Leonce – Paris: R Haton, 1895 – 1 – us CRL [960]

Dahomey and the dahomans : being the journals of two missions to the king of dahomey, and residence at his capital, in the years 1849 and 1850 / Forbes, F E – London, 1851. 2v – 10mf – 9 – mf#A-306 – ne IDC [916]
Dahomey. Porto Novo *see* Journal officiel
Dai gu zhu *see* Manju gisun-i yongkiyame toktobuha bithe
Dai nippon orhanisasi kebun pertaruhan keluarkan sajur-sajuran dan lain-lain hasil bumi : kema'moeran masjarakat kaoem tani indonesia. persatoean balatentara dan ra'jat. membantoe pembangoenan asia raja – (Djakarta?: Dai Nippon Orhanisasi PKS, Bagian Propaganda, 2602) – 1mf – 9 – mf#SE-2002 mf208 – ne IDC [959]
Dai viet tap chi – Saigon. v1,n1-7. janv-juil 1918; n.s., n1-33. 1er oct 1942-16 fevr 1944 – 1 – fr ACRPP [073]
Daiber, Albert Ludwig *see* Elf jahre freimaurer!
Daiches, Samuel *see*
– Altbabylonische rechtsurkunden aus der zeit der hammurabi-dynastie
– The jews in babylonia in the time of ezra and nehemiah
Daigle, Louise *see* Bibliographie analytique de l'oeuvre de m emile castonguay
Daiicing gurun-i yooni bithe *see* Da qing quan shu
Daiigu [Dai gu zhu] *see* Manju gisun-i yongkiyame toktobuha bithe
Die daikshaa : oder, weihe fuer das somaopfer / Lindner, Bruno – Leipzig: Poeschel & Trepte, 1878 – 1mf – 9 – 0-524-07219-1 – mf#1991-0081 – us ATLA [280]
Dailey, William Nelson Potter *see* The history of montgomery classis, r.c.a
Daill, Jean *see* An exposition of the epistle of saint paul to the philippians
Daille, Jean *see*
– Apologie des eglises reformees...
– An exposition of the epistle of saint paul to the colossians
– Sermons sur le chatechisme
– Treatise on the right use of the fathers
D'Ailly, Pierre *see*
– Quaestiones super 1, 3 et 4 librorum sententiarum
– Recommendatio sacrae scripturae. quaestio in vesperiis. – quaestio de resumpta
– Tractatus et sermones
Daily – Beaver Dam WI. v1 n1-311 [1887 may 23-1888 may 22], 1888 may 23-1889 may 22, 1889 may 23-1890 may 22, 1891 may 22 – 1r – 1 – mf#2395127 – us WHS [071]
Daily advance – Cleveland OH. v2 n129,153-154,199,276,278,282-283 [1979 jan 11, feb 8-10, apr 3, jul 2,4,9-10] – 1r – 1 – mf#1009775 – us WHS [071]
Daily advertiser – London, Canada. jan 1886-3 jan 1887 [daily] – 6r – 1 – (aka: london advertiser) – uk British Libr Newspaper [072]
Daily advertiser – Milwaukee WI. 1874 oct 12 – 1r – 1 – (cont by: milwaukee daily advertiser) – mf#1165885 – us WHS [071]
Daily advertiser – New York NY. 1788 jan 1-jul 31, 1788 jul 1-dec 31, 1790 – 3r – 1 – (cont: daily advertiser: political, historical, and commercial; cont by: people's friend and daily advertiser) – mf#851131 – us WHS [071]
Daily advertiser / Cuyahoga Co. Cleveland – sep 1837-mar 1838 [daily] – 1r – 1 – mf#B29894 – us Ohio Hist [071]
Daily advertiser – Kingston. Jamaica. -d. 1 Jan-29 Dec 1790 – 1r – 1 – uk British Libr Newspaper [079]
Daily advertiser – Philadelphia, PA. -d. 7 Feb-11 Sept 1797. 1 reel – 1 – uk British Libr Newspaper [071]
Daily advertiser – Wagga wagga, jan 1969-aug 1996 – at Pascoe [079]
Daily advertiser – Wagga Wagga, oct 1968-dec 1968 – 171r – 9 – A$9,748.51 vesicular A$10,689.01 silver – at Pascoe [079]
Daily advertiser *see* The port elizabeth advertiser
The daily advertiser – Hongkong: H P C Lassen, [oct 2-30 1871] – 1r – 1 – us CRL [079]
Daily advertiser and journal / Hamilton Co. Cincinnati – jan 1839-mar 1841 (poor quality) [daily] – 5r – 1 – mf#B1244-1248 – us Ohio Hist [071]
The daily advertiser and shipping gazette – Hongkong: H P C Lassen, jan 2 1872-apr 30 1873 – 3r – 1 – us CRL [079]
Daily advertiser (waimate) – oct 1921-dec 1941; jul 1946-apr 1972 – 1 – mf#75.13 – nz Nat Libr [079]
Daily advetiser *see* Miscellaneous newspapers of teller county
Daily advocate – Green Bay WI. 1898 sep 28 – 1r – 1 – (cont: evening advocate [green bay, wis.]; cont by: advocate [green bay wi]) – mf#947506 – us WHS [071]
Daily advocate / Darke Co. Greenville – 1969-77 (centerfold shadow) [daily] – 51r – 1 – mf#B7504-7554 – us Ohio Hist [071]
Daily advocate / Darke Co. Greenville – apr 1953-dec 1966 [daily] – 55r – 1 – mf#B6515-6569 – us Ohio Hist [071]
Daily advocate / Darke Co. Greenville – (jan 1926-jun 1930) [daily] – 10r – 1 – mf#B4120-4129 – us Ohio Hist [071]
Daily advocate / Darke Co. Greenville – jul 1930-jun 1939 (centerfold shadow) [daily] – 18r – 1 – mf#B8350-8367 – us Ohio Hist [071]
Daily advocate / Darke Co. Greenville – jul 1939-mar 1953 [daily] – 44r – 1 – mf#B6063-6107 – us Ohio Hist [071]
Daily advocate – Gainesville, FL. 1888 jun 26; jul 27; nov 13; dec 07 18 – 1r – us UF Libraries [071]
Daily advocate / Licking Co. Newark – may-aug 1884 [daily] – 1r – 1 – mf#B30908 – us Ohio Hist [071]
Daily age – Philadelphia PA. 1863 jun 1-oct 31 – 1r – 1 – (cont: age [philadelphia, pa. : 1863]; cont by: age [philadelphia pa: 1866]) – mf#889847 – us WHS [071]
Daily alaska empire – Juneau AL. 1926 dec 8/1927 mar 25-1940 oct 1/dec 31 – 56r – 1 – (with small gaps; cont: alaska daily empire; cont by: juneau alaska empire) – mf#854123 – us WHS [071]
The daily alaska empire – Juneau, Alaska: Empire Printing Co, jan 2 1952-60; jul 1963-jul 21 1964 – us CRL [071]
Daily albany democrat – Albany OR: Brown & Stewart, 1876- [daily ex mon] – 1 – (related to: state rights democrat [1865-1900]; cont: albany evening democrat) – us Oregon Lib [071]
Daily albany democrat *see* State rights democrat (albany, or)
Daily alhambra advocate *see* [Alhambra-] post-advocate
Daily alice echo – Alice TX. 1948 sep 10-16,27-29, oct 26,28-29, dec 1 – 1r – 1 – mf#860481 – us WHS [071]
Daily aljamiat – Delhi, India. 1965-Jun 1966 – 2r – 1 – us L of C Photodup [079]
Daily american unionist – Salem OR: William Morgan, [daily ex mon] – 1 – (began in 1868; related to: american unionist [1866-69]; cont by: daily oregon unionist [1869-69]) – us Oregon Lib [071]
Daily american unionist *see* American unionist
The daily arbor state – Wymore, NE: J R Dodds & Liss L Mason (daily) [mf ed v1 n150. oct 29 1895-oct 21 1896 (gaps)] – 1r – 1 – (issues for nov 12 1895- called v3 n14-) – us NE Hist [071]
Daily argosy – Georgetown. Guyana. -d. 1 Oct 1908-31 Jul 1917; 13 Nov 1919-31 May 1940; 23 Feb 1941-18 Aug 1963. (215 reels) – 1 – uk British Libr Newspaper [079]
Daily argus – Birmingham, England 9 nov 1891-31 jan 1902 [mf jul-dec 1898] – 1 – (incorp with: birmingham evening dispatch) – uk British Libr Newspaper [072]
Daily argus – Madison, WI. 1852 jan 8-apr 20 – 1r – 1 – mf#916670 – us WHS [071]
Daily argus *see*
– [Alameda-] alameda times star
– [San jose-] santa clara argus
The daily argus – Red Cloud, NE: [s.n.] (daily) [mf ed sep 17 1891] – 1r – 1 – us NE Hist [071]
Daily argus and democrat – Madison WI. 1852 jun 15/1853 jan 18-1861 oct 8/oct 22 – 19r – 1 – (with gaps; cont: evening argus and democrat; cont by: wisconsin daily argus) – mf#920627, 920608 – us WHS [071]
Daily argus and forfar, perth, and fife advertiser – [Scotland] Angus, Dundee: Park, Sinclair & Co 23 may 1859-20 apr 1861 (daily) [mf ed 2004] – 6r – 1 – (merged with: dundee courier to form: dundee courier and daily argus) – uk Newsplan [072]
Daily argus observer – Ontario OR, 1970-86 [daily ex sat] – 1 – (cont by: argus observer [1986-]; cont: ontario argus-observer [1947-1970]) – us Oregon Lib [071]
Daily arizona silver belt – Miami AZ. 1928 oct 29, 1929 feb 6,16, jul 5 – 1r – 1 – (cont by: arizona silver belt) – us WHS [071]
Daily astorian – Astoria OR: Astorian-Budget Pub Co, 1961- [daily ex sun & hols] – 1 – (related to wkly ed: weekly astorian [astoria or], 1961-1965, and: weekly budget [astoria or] 1981-1989; cont: daily astorian evening budget) – us Oregon Lib [071]
Daily astorian *see*
– Weekly astorian
– Weekly astorian (astoria, or)
The daily astorian *see* Weekly budget
Daily astorian [astoria, or] – Astoria OR: D C Ireland, -1883 [daily ex mon] – 1 – (related to: weekly astorian; cont by: daily morning astorian [1883-99]) – us Oregon Lib [071]
Daily astorian evening budget – Astoria OR: Astorian-Budget Pub Co, 1961 [daily ex sun & hols] – 1 – (related to: weekly astorian; cont: astorian budget [1960]; cont by: daily astorian [1961-]) – us Oregon Lib [071]
Daily astorian evening budget *see* Weekly astorian (astoria, or)
Daily atlas / Hamilton Co. Cincinnati – jul 1847-jun 1848 [daily] – 2r – 1 – mf#B14862-14863 – us Ohio Hist [071]
Daily atlas / Hamilton Co. Cincinnati – jul 1848-jun 1849 – 1r – 1 – mf#B37396 – us Ohio Hist [071]
Daily atlas / Hamilton Co. Cincinnati – v1 n1. nov 1843-apr 1844 [daily] – 1r – 1 – mf#B33795 – us Ohio Hist [071]
The daily beatrice republican – Beatrice, NE: Hill & Davis. v1 n1. mar 31 1890- (daily) [mf ed 1973] – 1r – 1 – us NE Hist [071]
Daily bee – Gainesville, FL. 1883 jun 21 – 1r – us UF Libraries [071]
Daily black hawk journal *see* Gilpin county miscellaneous newspapers
Daily blade / Scioto Co. Portsmouth – oct-dec 1898, jan-mar 1904 [daily] – 1r – 1 – mf#B11152 – us Ohio Hist [071]
Daily blade series / Lucas Co. Toledo – jan 1862-dec 1863, jan-jun 1878 [daily] – 6r – 1 – mf#B34418-34423 – us Ohio Hist [071]
Daily blotter – n52-65 [1976 nov 9-1977 jul 7] – 1r – 1 – mf#329579 – us WHS [071]
Daily bristol times and mirror – Bristol, England (ns) 5 jan 1865-31 dec 1883 – 1 – (fr 7 jan 1865-29 dec 1883 sat ed entitled: saturday bristol times & mirror; cont: bristol times, & felix farley's bristol journal [2 apr 1853-31 dec 1864]; cont by: bristol times & mirror [1 jan 1884-29 jan 1932]) – uk British Libr Newspaper [072]
Daily british colonist – Victoria, Canada 19 mar, 3-14 may, 28 jun, 7,9 jul, 5 aug, 2,29 dec 1864; 18 jan-23 feb, 10 mar-1 may, 30 jun, 6 jul, 25 nov, 1,12 dec 1865; 25 jan, 10,26 mar 1866 – 1 – (1864-1921 imperfect; cont by: daily british colonist & victoria chronicle [3,6 aug,15 sep-2,22 oct 1866; 8 oct 1867; 13 feb, 3,28 apr 1868]; daily british colonist [30 aug 1877, 1 jan 1885, 1 jan 1886]; daily colonist [1 jan 1896, [1] jan 1898, 2 jan 1899, 20 jan 1901, 20 mar, 30 jun 1904, 28 apr 1911-31 dec 1942]) – uk British Libr Newspaper [072]
Daily british whig – Kingston, Canada. 9 mar 1858; 19 jul 1864-5 feb 1877; 19 may 1896-30 dec 1899; 3 sep-31 dec 1901 – 16 1/2r – 1 – uk British Libr Newspaper [071]
Daily budget *see* Miscellaneous newspapers of ouray county
Daily bulletin – North Platte, NE: [City Printery] 15v. v1 n1. apr 13 1932-v15 n134. oct 31 1946 (daily ex sun & mon) – (= ser North Platte Telegraph; North Platte Daily Telegraph-Bulletin) – 6r – 1 – (merged with: north platte telegraph (1938) to form: north platte daily telegraph-bulletin) – us Bell [071]
Daily bulletin – North Platte, NE: [City Printery] 15v. v1 n1. apr 13 1932-v15 n134. oct 31 1946 (daily ex sun & mon) [mf ed 1939-46 (gaps)] – 19r – 1 – (merged with: north platte telegraph (1938) to form: north platte daily telegraph-bulletin) – us NE Hist [071]
Daily bulletin / Montgomery Co. Dayton – sep 1944-sep 1946 [daily] – 3r – 1 – mf#B5424-5426 – us Ohio Hist [976]
Daily bulletin – n1-4 [1980 oct 27-27] – 1r – 1 – mf#941615 – us WHS [071]
Daily bulletin (bend, or) *see* Bend bulletin (bend, or: 1903)
Daily bulletin [bend, or] – Bend OR: G P Putnam, 1916-17 [daily] [mf ed 1966] – 1r – 1 – (related to weekly ed: bend bulletin [bend, or: 1903]; cont by: bend bulletin [bend, or: 1917]) – us Oregon Lib [071]
Daily bumble bee – Omaha, NE: Prohibition and Non-partisan County Central Comm. v1 n1. oct 30 1890- (daily) [mf ed oct 30-nov 3 1890 filmed 1986] – 1r – 1 – (lacks: oct 31) – us NE Hist [071]
Daily cairo bulletin – Cairo IL. 1884 mar 14 – 1r – 1 – mf#1159478 – us WHS [071]
Daily calumet – Chicago IL. 1928 mar 28 – 1r – 1 – mf#1159500 – us WHS [071]
Daily calumet – Chicago IL. v47 n148 [1928 mar 28] – 1r – 1 – (cont: south chicago daily calumet; cont by: chicago daily calumet) – mf#4924432 – us WHS [071]
Daily camel's hump : official jubilee publication – Bandon OR: Abd-uhl-Atef Temple, n117, Dramatic Order Knights of Khorasson [mf ed jul 28-jul 29 1921] – 1r – 1 – us Oregon Lib [071]
Daily camera – Boulder CO. 1914 apr 22-23,27, may 12,14 – 1r – 1 – mf#796496 – us WHS [071]
Daily campaign – McMinnville OR: Campaign Pub Co, 1886 [daily] – 1 – us Oregon Lib [071]
Daily capital journal [salem, or: 1896] – Salem OR: Hofer Bros, 1896-99 [daily] – 1 – (cont: capital journal [salem, or: 1893]; cont by: daily journal [salem, or: 1899]) – us Oregon Lib [071]
Daily capital journal [salem, or: 1903] – Salem OR: Hofer Bros, 1903-19 [daily] – 1 – (cont: daily journal [salem or: 1899]; cont by: capital journal [salem or: 1919]) – us Oregon Lib [071]
Daily capitol fact / Franklin Co. Columbus – jul 1851-may 1864 [daily] – 16r – 1 – mf#B1439-1454 – us Ohio Hist [071]
The daily caymanian compass *see* Caymanian compass
Daily central city register – Central City, CO: Collier & Hall, feb 9 1869-jan 5 1870; dec 9 1870-nov 5 1871 – 2r – 1 – us CRL [071]
Daily challenge – New York. 1972 mar 6 – 1r – 1 – mf#3319649 – us WHS [071]
Daily champion – Deadwood, Black Hills, DT [S D]: Charles Collins, 1877- [jul 4 1877] (daily) – 1r – 1 – us CRL [071]
Daily chief union / Wyandot Co. Upper Sandusk – may 1977-dec 1982 [daily] – 21r – 1 – mf#B12489-12509 – us Ohio Hist [071]
Daily chief-union / Wyandot Co. Upper Sandusk – jul 1945-sep 1950, dec 1950-aug 1973 [daily] – 64r – 1 – mf#B7423-7486 – us Ohio Hist [071]
Daily chronicle : daily newspaper – Warren, OH. 17 July 1868 – 1r – 1 – us Western Res [071]
Daily chronicle / Delaware Co. Delaware – jul-dec 1881, jul 1884-85 [daily] – 2r – 1 – mf#B11268-11269 – us Ohio Hist [071]
Daily chronicle / Hamilton Co. Cincinnati – v1 n1. dec 1839-sep 1840 [daily] – 1r – 1 – mf#B34127 – us Ohio Hist [071]
Daily chronicle – Kingston, Jamaica. 29 Jun 1914-29 Dec 1917 (imperfect) – 42r – 1 – uk British Libr Newspaper [072]
Daily chronicle – London, Jul-Dec 1888; Jan 1890-May 1921 – 182r – 1 – uk British Libr Newspaper [072]
Daily chronicle – Nairobi: P P Sheth, 1956-mar 1959; may 1959-may 1962 – 3r – 1 – us CRL [079]
Daily chronicle *see*
– Demerara daily chronicle
– Miscellaneous newspapers of lake county
Daily chronicle and argus – Bath, England 15 feb 1900-7 oct 1903 [mf jul-dec 1896, 1911, sep-dec 1912, 1986-] – 1 – (cont: bath daily chronicle [7 aug 1883-14 feb 1900]; cont by: bath daily chronicle & argus [8 oct 1903-15 sep 1911]) – uk British Libr Newspaper [072]
Daily chronicle and atlas / Hamilton Co. Cincinnati – jan 1850-dec 1850 – 1r – 1 – mf#B36963 – us Ohio Hist [071]
Daily chronicle and sentinel – Augusta, GA. jan 3 1855-jun 30 1857 – 4r – 1 – $460.00 – mf#D3490P02 – us Western Res [071]
Daily chronicle and sentinel – Augusta GA. 1861 nov5-1862 apr 29, oct 29, 1863 apr 5, 1864 mar 18, 30, oct, 18-nov 15, 1865 apr 22 – 1r – 1 – (cont: augusta chronicle; cont by: daily chronicle and constitutionalist) – mf#846254 – us WHS [071]
Daily cincinnati atlas – Cincinnati. Nov. 1, 1843-May 2, 1844 – 1 – us NY Public [071]
Daily cincinnati atlas / Hamilton Co. Cincinnati – jan 1854-jun 1854 – 1r – 1 – mf#B37395 – us Ohio Hist [071]
Daily cincinnati chronicle / Hamilton Co. Cincinnati – jul 1847-may 1848 – 3r – 1 – mf#B37449-37451 – us Ohio Hist [071]
Daily cincinnati commercial – Cincinnati: J W S Browne & Co, 1846-54 – us CRL [071]
Daily cincinnati enquirer / Hamilton Co. Cincinnati – apr-sep 1841, jan 1842-nov 1843 [daily] – 5r – 1 – mf#B1249-1253 – us Ohio Hist [071]
Daily cincinnati gazette / Hamilton Co. Cincinnati – jan, apr-jul 1855 [daily] – 1r – 1 – mf#B13552 – us Ohio Hist [071]
Daily cincinnati republican / Hamilton Co. Cincinnati – 9,1840-11,1840/1,1841-8,1842 – 3r – 1 – mf#B37405-37407 – us Ohio Hist [071]
Daily cincinnati republican / Hamilton Co. Cincinnati – (jul 1833-dec 1837) [daily] – 6r – 1 – mf#B6784-6789 – us Ohio Hist [071]
Daily citizen – Beloit WI. 1890 aug 20 – 1r – 1 – mf#955522 – us WHS [071]
Daily citizen / Champaign Co. Urbana – jun-dec 1917, jun-aug 1928 [daily] – 2r – 1 – mf#B33597-33598 – us Ohio Hist [071]
Daily citizen – Beaver Dam WI. [1915 jun 14/dec 14]-[1930 sep 2/1931 mar 31] [small gaps] – 23r – 1 – (cont: beaver dam daily citizen [beaver dam wi: 1911]; cont by: beaver dam daily citizen [beaver dam wi: 1930]) – mf#1139733 – us WHS [071]
Daily citizen – Vicksburg MS. 1863 jun 18, jul 2 – 1r – 1 – (cont by: vicksburg herald) – mf#780661 – us WHS [071]
Daily citizen – [NW England] Manchester ALS 8 oct 1912-5 jun 1915 – 1 – uk MLA; uk Newsplan [072]
Daily city item – New Orleans LA. 1891 mar 14 – 1r – 1 – (cont: daily item [new orleans la]) – mf#862786 – us WHS [071]
Daily clarion *see* Clarion
Daily cleveland herald / Cuyahoga Co. Cleveland – jul 1857-dec 1858 [daily] – 3r – 1 – mf#B12968-12970 – us Ohio Hist [071]
Daily cleveland herald – Cleveland, OH, jun 23 1857-dec 31 1868 – 23r – 1 – (daily whig and republican newspaper) – us Western Res [071]
Daily cleveland herald – Cleveland, OH, apr 11 1853-jul 20 1856 – 7r – 1 – (daily whig newspaper) – us Western Res [071]

Daily cleveland herald – Cleveland, OH, may 30 1835-mar 21 1837 – 2r – 1 – (daily whig newspaper, the first publ in cleveland) – us Western Res [071]

Daily cleveland herald – Cleveland, OH, jan 2 1869-dec 31 1870 – 4r – 1 – (evening ed of this daily republican newspaper) – us Western Res [071]

Daily cleveland herald – Cleveland, OH, jul 1 1873-may 30 1874 – 2r – 1 – (evening ed of this daily republican newspaper) – us Western Res [071]

Daily coast mail – Marshfield OR: Mail Publ Co, [daily ex sun] – 1 – (began in 1902? ceased in 1906; related to weekly ed: coast mail, 1902, and: weekly coast mail, 1902-06; merged with: weekly coast mail [1902-06] and: advertiser to form: coos bay times) – us Oregon Lib [071]

Daily coast mail *see* Coast mail

Daily coast mail bulletin – Marshfield OR: [s.n.] 1898- [daily] – 1 – us Oregon Lib [071]

Daily collegian *see* [Fresno-] the collegian

Daily colonist – Victoria, BC. 1858-71 – 13r – 1 – cn Library Assoc [071]

Daily colorado herald *see* Gilpin county miscellaneous newspapers

Daily colorado miner *see* Gilpin county miscellaneous newspapers

Daily columbian / Hamilton Co. Cincinnati – 6/1854-6/1855; 7-9,1856 – 3r – 1 – mf#B36964-36966 – us Ohio Hist [071]

Daily columbian / Hamilton Co. Cincinnati – apr-jun 1856 (very short) [daily] – 1r – 1 – mf#B5898 – us Ohio Hist [071]

The daily columbian – Chicago, Illinois. May-Oct 1993 – 2r – 1 – us L of C Photodup [071]

Daily commercial / Hamilton Co. Cincinnati – jul-dec 1862 [daily] – 1r – 1 – mf#B1316 – us Ohio Hist [071]

The daily commercial – Cincinnati: Curtiss & Hastings, oct 2 1843-sep 28 1844; [feb 11 1845-may 9 1846] – us CRL [071]

The daily commercial – Cincinnati [OH]: Curtiss & Hastings [oct 2 1843-may 9 1846] (daily ex sun) – 6r – 1 – us CRL [071]

Daily commercial bulletin – St Louis MO. 1838 jun 18-jul 31, aug 1-dec 31 – 2r – 1 – (cont: daily commercial bullein and missouri literary register) – mf#852978 – us WHS [071]

Daily commercial bulletin and missouri literary register – St. Louis MO. 1835 may 18-1836 apr 29, may 2-dec 31, 1837 jan 2-jul 31, aug 1-1838 jan 31, feb 1-jun 12 – 5r – 1 – (cont: st louis commercial bulletin and missouri literary register; cont by: daily commercial bulletin [st louis mo: 1838]) – mf#852977 – us WHS [071]

Daily commercial news – Sydney, jan 1892-dec 1956 – 112r – A$7,847.53 vesicular A$8,463.53 silver – at Pascoe [079]

Daily commercial news – Sydney. jan-jun 1976, sep 1979-aug 1980, jan 1988-jun 1993 – 30r – at Pascoe [079]

Daily commercial times – Milwaukee WI. 1875 jan 2-jun 30, jul 1-dec 31, 1876 jan 3-jul 10, jul 11-1877 jan 18, jan 19-jul 28, jul 30-1878 feb 9, feb 11-may 2 – 7r – 1 – (cont: milwaukee evening times; cont by: milwaukee daily news) – mf#1165015 – us WHS [071]

Daily commonwealth – Fond Du Lac WI. 1872 feb 26-sep 20, sep 21-1873 may 1, may 3-nov 30, dec 1-1874 jul 2, jul 3-1875 feb 5, feb 6-sep 9, sep 10-nov 25 – 7r – 1 – (cont by: fond du lac daily commonwealth [fond du lac wi: 1875]) – mf#928611 – us WHS [071]

Daily commonwealth – Fond Du Lac WI. 1866 oct 6-1867 apr, may-oct, nov-1868 jun – 3r – 1 – (cont: fond du lac daily commonwealth [fond du lac wi: 1866]) – mf#941481 – us WHS [071]

Daily commonwealth – Fond Du Lac WI. 1885 sep 1/1886 mar 27-1899 mar 22/sep 20 – 25r – 1 – (cont: fond du lac daily commonwealth [fond du lac wi: 1875]; cont by: daily news [fond du lac wi]; daily commonwealth and the daily news) – mf#941516 – us WHS [071]

Daily commonwealth – Ripon WI. 1881 mar 5 – 1r – 1 – mf#967881 – us WHS [071]

Daily commonwealth – Ripon WI. 1888 apr 13-17 – 1r – 1 – mf#967886 – us WHS [071]

Daily commonwealth – Fond Du Lac WI. 1912 dec 16/dec-1926 aug 10/sep 30 – 47r – 1 – (with gaps; cont: daily commonwealth and the daily bulletin; cont by: fond du lac daily reporter; fond du lac commonwealth reporter) – mf#941528 – us WHS [071]

Daily commonwealth and the daily bulletin – 1908 sep 15/1909 mar 2-1912 sep/dec 14 – 13r – 1 – (cont: daily commonwealth and the daily news; daily bulletin [fond du lac wi]; cont by: daily commonwealth [fond du lac wi: 1908]) – mf#941523 – us WHS [071]

Daily commonwealth and the daily news – Fond Du Lac WI. 1902 feb 27/sep 9-1908 mar 23/sep 5 – 14r – 1 – (with small gaps; cont: daily commonwealth [fond du lac wi: 1885]; daily news [fond du lac wi]; cont by: daily bulletin [fond du lac wi]; daily commonwealth and the daily bulletin) – mf#941520 – us WHS [071]

Daily compass – New York NY. 1949 dec 18/1950 feb 17-1952 aug 1/nov 3 – 13r – 1 – (with small gaps) – mf#851354 – us WHS [071]

Daily constitutionalist – Augusta GA. 1864 mar 18-19,26, may 17, jul 26,29, sep 3, 1866 mar 14 – 1r – 1 – mf#787163 – us WHS [071]

Daily construction journal / Atchison, Topeka and Santa Fe Railroad Company – 1887 – 1 – us Kansas [380]

Daily courier / Muskingum Co. Zanesville – jan 1884-jan 1915 [daily] – 79r – 1 – mf#B10193-10271 – us Ohio Hist [071]

Daily courier – Oshkosh WI. 1854 may 11-nov 21, 1855 sep 3-1856 sep 20, 1856 sep 22-1857 jun 16 – 3r – 1 – mf#958771 – us WHS [071]

Daily courier – Toronto. v1 n1-29. oct 17-nov 19 1914// – 1r – 1 – Can$70.00 – (pictorial tabloid filled mainly with war news) – cn McLaren [071]

Daily courier tribune / Portage Co. Ravenna – 1951-55 [daily] – (= ser West Eve Record) – 28r – 1 – (incl: west eve record) – mf#B4282-4309 – us Ohio Hist [071]

Daily court reporter / Montgomery Co. Dayton – 1/1971-73, 5/74-12/87, 6/88-11/1992 [daily] – 27r – 1 – mf#B33599-33625 – us Ohio Hist [347]

Daily court reporter / Montgomery Co. Dayton – 1937-jun 1953, jul 1954-70 [daily] – 26r – 1 – mf#B5217-5242 – us Ohio Hist [347]

Daily crisis / Columbiana Co. East Liverpool – apr 1887-mar 1900,oct 1900-sep 1903 [daily] – 29r – 1 – mf#B1709-1737 – us Ohio Hist [071]

Daily critic – Washington DC. 1880 apr 15 – 1r – 1 – (cont: critic; cont by: evening critic) – mf#846250 – us WHS [071]

Daily critic / Crawford Co. Bucyrus – jun 1885-oct 1886 [daily] – 2r – 1 – mf#B11759-11760 – us Ohio Hist [071]

Daily critic – Washington, DC: Ringwalt, Hack & Co, 1872-81 – (= ser Critic) – 1r – 1 – (filmed with: critic (washington, dc) oct 17-dec 1872) – us CRL [071]

Daily data – Green Bay WI. 1882 jan 6-jul 26, jul 27-dec 9 – 2r – 1 – (cont by: data [green bay wi: 1882]) – mf#918676 – us WHS [071]

Daily dayton gazette / Montgomery Co. Dayton – nov 1850-51 1853-sep 1859 [daily] – 7r – 1 – mf#B5311-5316 – us Ohio Hist [071]

Daily dayton transcript / Montgomery Co. Dayton – jan-oct 1850 [daily] – 1r – 1 – mf#B34540 – us Ohio Hist [071]

Daily de witt times – De Witt, NE: W H Stout. v1 n1. nov 2 1886- (daily) – 1r – 1 – (weekly ed: de witt times) – us NE Hist

Daily democrat – Boyertown, PA., 1890-1891 – 13 – $25.00r – us IMR [071]

Daily democrat / Champaign Co. Urbana – jan 1923-may 1928 [daily] – 15r – 1 – mf#B33408-33422 – us Ohio Hist [071]

Daily democrat / Clark Co. Springfield – oct 13-oct 28 1896 [daily] – 1r – 1 – mf#B10819 – us Ohio Hist [071]

Daily democrat – La Crosse WI. 1867 dec 6/7-12, 1868 jan 1-dec 2, dec 3-1869 sep 11 – 3r – 1 – (cont by: la crosse evening democrat) – mf#928229 – us WHS [071]

Daily democrat – Madison WI. 1851 jan 10-1852 jun 11 – 1r – 1 – (cont by: weekly wisconsin argus and democrat; daily wisconsin argus and democrat) – mf#921356 – us WHS [071]

Daily democrat – Gainesville, FL. 1888 aug 30 – 1r – us UF Libraries [071]

Daily democrat – Gainesville, FL. 1896 mar 16,17,24,28 – 1r – us UF Libraries [071]

Daily democrat / Harding Co. Kenton – 1946-48 1950-jul 1953 [daily] – 16r – 1 – mf#B11992-12007 – us Ohio Hist [071]

Daily democrat / Harding Co. Kenton – jan 1920-dec 1945 [daily] – 63r – 1 – mf#B9337-9399 – us Ohio Hist [071]

Daily democrat / Montgomery Co. Dayton – nov 1874-feb 1890 [daily] – 18r – 1 – mf#B5104-5121 – us Ohio Hist [071]

Daily democrat / Summit Co. Akron – apr 1892-apr 96, apr 99-dec 1902 [daily] – 21r – 1 – mf#B10528-10548 – us Ohio Hist [071]

Daily democrat *see* [Marysville-] evening democrat

The daily democrat – Doylestown, PA. -d 1891-1893 – 13 – $25.00r – us IMR [071]

[The daily] democrat – Beatrice, NE: G P Marvin. v1 n5. oct 29 1886-6th yr n197. jun 30 1892 (daily ex sun) [mf ed -1990] – 5r – 1 – (merged with: beatrice republican, to form: beatrice daily times) – us NE Hist [071]

Daily democrat series / Lorain Co. Lorain – oct 1900-jun 1903 [daily] – 4r – 1 – mf#B33266-33269 – us Ohio Hist [071]

Daily diary of president johnson [1963-69] – (= ser Presidential documents series) – 14r – 1 – $2250.00 – 0-89093-359-6 – (with p/g) – us UPA [977]

Daily digest of the arabic press – Cairo: USIS, News Dep, [jan 5 1945-oct 1946] – 3r – 1 – us CRL [079]

Daily dispatch – East London, South Africa: East London Daily Dispatch Ltd, jan 2-mar 1925; jul 1925-nov 1966 – 1 – us CRL [960]

Daily dispatch / Hamilton Co. Cincinnati – jun 1849-apr 1850 – 1r – 1 – mf#B37404 – us Ohio Hist [071]

Daily dispatch – East London SA, 1920-81 – 233r – 1 – (wkly [1872-1876]; twice/wk [1877-1897]; daily [1898-]; jul 2 1920-oct 8 1981 publ by state library; fr sep 10 1872-dec 1897 as: east london dispatch; fr jan 1898-dec 1925 as east london daily dispatch [and frontier advertiser]) – sa National [079]

Daily dispatch and manchester morning chronicle – [NW England] Manchester ALS 5 feb-26 may 1900 – 1 – (title change: daily dispatch [2 jun 1900-19 nov 1955]; news chronicle & daily dispatch [21 nov 1955-17 oct 1960]) – uk MLA; uk Newsplan [072]

Daily drovers journal – South Omaha, NE: Drovers Journal Co (daily ex sun) [mf ed 1891-92 (gaps)] – 1r – 1 – (cont by: drovers journal) – us NE Hist [636]

Daily drovers journal and stockman – South Omaha, NE: Journal-Stockman Co. v12 n131-148. nov 21-dec 9 1898 (daily ex sun) – 1r – 1 – (formed by the union of: south omaha daily stockman and: drovers journal (1895). cont by: daily drovers journal-stockman) – us NE Hist [636]

Daily drovers journal-stockman : official paper south omaha live stock exchange / South Omaha Live Stock Exchange – South Omaha, NE: Journal-Stockman Co. v12 n149. dec 10 1898-v35 n242. mar 22 1924 (daily ex sun) – 2r – 1 – (cont: daily drovers journal and stockman; cont by: omaha daily journal-stockman) – us SD Archives [636]

Daily drovers journal-stockman : [official paper south omaha live stock exchange] / South Omaha Live Stock Exchange – South Omaha, NE: Journal-Stockman Co. v12 n149. dec 10 1898-v35 n242. mar 22 1924 (daily ex sun) [mf ed with gaps] – 22r – 1 – (cont: daily drovers journal and stockman. cont by: omaha daily journal-stockman) – us NE Hist [636]

Daily eagle – Marinette WI. 1892 may 2/oct 31-1903 jul 6/aug 5 – 27r – 1 – (cont by: marinette daily star; daily eagle-star) – mf#1107303 – us WHS [071]

Daily eagle / Fairfield Co. Lancaster – 4/1890-05, 4/06-9/06, 07-1914 [daily] – 57r – 1 – mf#B9836-9892 – us Ohio Hist [071]

Daily eagle / Fairfield Co. Lancaster – jan 20-mar 4 1936 gap filler [daily] – 1r – 1 – mf#B10614 – us Ohio Hist [071]

Daily eagle-star – Marinette WI. 1903 aug 6/oct 19-1913 may 22/aug 25 – 37r – 1 – (cont: daily eagle; marinette daily star; cont by: marinette eagle-star) – mf#1107981 – us WHS [071]

Daily echo – Accra: Independent Press Ltd, [mar 20-dec 1954] – us CRL [079]

Daily empire / Montgomery Co. Dayton – (nov 1850-jun 1867) [daily] – 11r – 1 – mf#B5138-5148 – us Ohio Hist [071]

Daily enquirer / Hamilton Co. Cincinnati – jan 1845-dec 1846 (poor quality) [daily] – 4r – 1 – mf#B1256-1259 – us Ohio Hist [071]

Daily enquirer *see* Miscellaneous newspapers of rio grande county

Daily enterprise *see* [Riverside-] press enterprise

The daily enterprise *see* The city enterprise

Daily eugene guard – Eugene OR: I L Campbell, [daily ex sun] – 1 – (related to wkly ed: eugene city guard, 1891-1899, and: eugene weekly guard, 1899-1903; cont by: eugene daily guard [1904-24]) – us Oregon Lib [071]

Daily eugene guard *see* Eugene weekly guard

Daily evening albany democrat – Albany OR: Stites & Nutting, 1888 [daily ex sun] [mf ed 1965] – 1r – 1 – (related to: state rights democrat [1865-1900]; cont by: albany daily democrat [1888-1920]) – us Oregon Lib [071]

Daily evening bulletin – San Francisco CA. 1855 oct 8/1856 mar 25-1891 jul 1/dec 31 – 77r – 1 – (cont by: bulletin [san francisco ca: 1895]) – mf#881714 – us WHS [071]

Daily evening bulletin – San Francisco, 9 oct 1855-29 nov 1871 – 33r – 1 – uk British Libr Newspaper [073]

Daily evening chief – Red Cloud, NE: A C Hosmer, 1887 (daily ex sun) [mf ed 1st yr n47. nov 11-jun 25 1887-88 (gaps) filmed 1965] – 1r – 1 – (publ as: daily chief dec 12 1887-apr 13 1888) – us NE Hist [071]

Daily evening courier – Madison IN. 1858 oct 15 – 1r – 1 – mf#856365 – us WHS [071]

Daily evening democrat – Bloomington IL. 1868 aug 15 – 1r – 1 – mf#874099 – us WHS [071]

Daily evening forum / Crawford Co. Bucyrus – v1 n1. jul 12-nov 13, 1880 [daily] – 1r – 1 – mf#B11759 – us Ohio Hist [071]

Daily evening gazette / Highland Co. Hillsboro – may-dec 1883 [daily] – 1r – 1 – mf#B8868 – us Ohio Hist [071]

Daily evening journal – Berlin WI. 1881 jan 1-may 2 – 1r – 1 – (cont by: evening journal [berlin wi: 1881]) – mf#961578 – us WHS [071]

Daily evening journal – Portland OR: Evening Journal Pub Co, 1875- [daily ex sun] – 1 – us Oregon Lib [071]

Daily evening ledger – Gainesville, FL. 1895 jun 03,06 – 1r – us UF Libraries [071]

Daily evening mercury *see* Quebec mercury

Daily evening news – Lincoln, NE: Thos H Hyde & Co. 4v. v2 n200 may 23 1883-v5 n28. oct 27 1885 (daily ex sun) [mf ed with gaps] – 3r – 1 – (cont: lincoln daily news. cont by: lincoln daily news (1885). weekly ed: lincoln weekly news) – us NE Hist [071]

Daily evening news – Portland OR: Bellinger, Curry & Co, 1873 [daily ex sun] – 1 – (related to companion publ: weekly news [1883-84]) – us Oregon Lib [071]

Daily evening news *see* Weekly news (portland, or: 1874)

Daily evening republican – Broken Bow, NE: D M Amsberry. v8 n1. may 9 1898-1911// (daily ex sun) [mf ed -jul 26 1898 (gaps)] – 1r – 1 – (cont: broken bow daily republican (1892). cont by: broken bow daily republican (1911)) – us NE Hist [071]

Daily evening standard – Portland, OR: A Noltner. v1 n1-v2 n130. jul 29 1876-jul 2 1877 – 1 – us Oregon Hist [071]

Daily evening standard [portland, or] – Portland OR: A Noltner [daily ex sun] – 1 – (began in 1876; ceased in 1877; cont by: daily standard [portland, or]; related to: weekly standard [portland, or]) – us Oregon Lib [071]

Daily evening star *see* Miscellaneous newspapers of pueblo county

Daily evening telegram – Portland OR: Evening Telegram Pub Co [daily ex sun] – 1 – (began in 1878?; cont: evening telegram [1877-78]; cont by: evening telegram [-1918]) – us Oregon Lib [071]

Daily evening times – Grand Island, NE: C P R Williams. v1 n1. oct 4 1873- (daily ex sun) – 1r – 1 – (weekly ed: grand island time) – us NE Hist [071]

Daily evening transcript – Boston, MA: Dutton & Wentworth, jul-dec 1845 – 1 – us CRL [071]

Daily evening tribune / Scioto Co. Portsmouth – jun 1853-feb 1856,apr 1857-nov 1860 [daily] – 6r – 1 – mf#B4171-4176 – us Ohio Hist [071]

Daily evening voice – Boston MA. 1864 dec 2-1865 jun 30, jul 1-dec 30, 1866 jan 1-jul 25, jul 25-dec 31, 1867 jan 1-oct 16 – 5r – 1 – us WHS [071]

Daily examiner – Grafton, 1915-34; 1959-68 – at Pascoe [079]

Daily examiner – Grafton, jan 1969-aug 1997 – at Pascoe [079]

Daily examiner *see* Ulster examiner

The daily exchange – Baltimore, Maryland. Feb 22 1958-Sept 14 1961 – 8r – 1 – us L of C Photodup [071]

The daily experiences of older adults residing in institutional environments / Voelkl, Judith E – 1989 – 209p 3mf – 9 – $12.00 – us Kinesology [305]

Daily express – Dublin, Ireland. 1855-58, 1861-1906, jan-jun 1921 – 225 1/2r – 1 – (aka: daily express and irish daily mail) – uk British Libr Newspaper [072]

Daily express – Petersburg VA. 1861 sep 17, 1863 jan 13, oct 3, nov 24, 1864 jan 29, apr 13,21, jun 10,16, aug 1, 1865 jan 9, apr 15, may 4, jun 7,9-10,19, oct 6 – 1r – 1 – (cont by: daily courier [petersburg va]) – mf#881667 – us WHS [071]

Daily express / Defiance Co. Defiance – jun 1918-apr 1920 [daily] – 4r – 1 – mf#B29556-29559 – us Ohio Hist [071]

Daily express – Dublin, feb 1851-54, 1859-60, 1923-24 – 13r – 1 – ie National [072]

Daily express – [NW England] Manchester ALS 24 apr 1900-1953, 1974-89 – 1 – uk MLA; uk Newsplan [072]

Daily express – Petersburg, VA. sep 7 1861-jun 16 1865 – (= ser Confederate newspapers) – 1r – 1 – us Western Res [071]

Daily express *see*
- Popular newspapers during world war 1
- Popular newspapers during world war 2

Daily express and irish daily mail *see* Daily express

Daily express / orange free state advertiser – Pretoria: State Library Corporate Communication, 22 jul 1882-22 jul 1900 – 18r – 1 – mf#MS00044 – sa National [079]

Daily fair journal – Freeport IL. 1860 sep 27-28 – 1r – 1 – us WHS [071]

Daily fair record – Bloomington WI. 1893 sep 7,8 – 1r – 1 – mf#1004011 – us WHS [071]

The daily flail – Fremont, NE: Flail Pub Co, 1888 (daily ex sun) [mf ed v3 n5. apr 1891, may 28 1892 filmed [1979]] – 2r – 1 – us NE Hist [071]

Daily florence courier – Florence, NE: Richard H See (daily) [mf ed jan 15 1858] – 1r – 1 – us NE Hist [071]

Daily florida citizen – Jacksonville, FL. 1897 apr-jun 8 – 1r – us UF Libraries [071]

Daily florida democrat – Gainesville, FL. 1895 nov 07; oct 24 – 1r – us UF Libraries [071]

Daily florida standard – Jacksonville, FL. 1890-1892 – 2r – (gaps) – us UF Libraries [071]

Daily fond du lac press – Fond Du Lac WI. 1866 jul 10,19,23-25, 1866 sep 2-1866 nov 3 – 2r – 1 – mf#916630 – us WHS [071]

Daily forum / Crawford Co. Bucyrus – may-jun 1915, dec 1915-feb 1916 [daily] – 1r – 1 – mf#B11761 – us Ohio Hist [071]

Daily forum – Philadelphia. Oct. 12, 1843, Oct. 10, 28, Nov. 2, 1844 – 1 – us NY Public [071]

Daily forum *see* [Sparks-] nevada forum

Daily free democrat – Milwaukee WI. 1850 sep 16/1851 apr 30-1856 jul/nov 25 – 12r – 1 – (cont by: milwaukee daily free democrat) – mf#1159609 – us WHS [071]

Daily free press – Aberdeen: A Marr 1874-1900 (daily ex sun) [mf ed 1996] – 1 – (cont: aberdeen daily free press; cont by: aberdeen free press) – uk Scotland NatLib [072]

Daily free press – Eau Claire WI. 1873 jan 1 [v1 n1] – 1r – 1 – (cont by: daily telegram [eau claire wi]) – mf#875321 – us WHS [071]

Daily free press – Eau Claire WI. 1890 jan-feb 28 – 1r – 1 – (cont: eau claire daily free press) – mf#964290 – us WHS [071]

Daily free press *see* Miscellaneous newspapers of las animas county, reel 3

The daily free press – De Witt, NE: W H Stout. v1 n1. nov 4 1879- (daily) [mf ed filmed [1974?]] – 1r – 1 – (weekly ed: free press) – us NE Hist [071]

Daily gate city – Keokuk IA. 1857 dec 31 – 1r – 1 – (cont: gate city; daily whig [keokuk ia: 1854]; cont by: constitution-democrat; daily gate city and constitution-democrat) – mf#851190 – us WHS [071]

Daily gazette – Janesville WI. [1894 jan 6/apr 4]-[1901 mar 25/may 31] – 30r – 1 – (cont: janesville daily gazette [janesville wi: 1880]; cont by: janesville daily gazette [janesville wi: 1901]) – mf#1145230 – us WHS [071]

Daily gazette – Gainesville, FL. 1891 nov 03,04,10,20; dec 22 1892; jan – 1r – us UF Libraries [071]

Daily gazette / Guantanamo Bay Naval Base [Cuba] – 1983 mar 21-1987 feb 2/5 – 1r – 1 – (with gaps; cont by: guantanamo gazette) – mf#1476946 – us WHS [355]

Daily gazette / Hamilton Co. Cincinnati – 7,1827-12,1838 (scattered) – 13r – 1 – mf#B36936-36948 – us Ohio Hist [071]

Daily gazette / Hamilton Co. Cincinnati – jan-dec 1859, jul-dec 1862 [daily] – 3r – 1 – mf#B1317-1319 – us Ohio Hist [071]

Daily gazette / Hamilton Co. Cincinnati – jun-dec 1827 [daily] – 1r – 1 – mf#B12043 – us Ohio Hist [071]

Daily gazette / Jefferson Co. Steubenville – jul 1914-feb 1916 [daily] – 5r – 1 – mf#B10991-10995 – us Ohio Hist [071]

The daily gazette – Ashland, NE: T J Pickett, Jr, sep-sep 22 1881 (daily) – 1r – 1 – us Bell [071]

Daily gazette and commercial advertiser *see* Denver county miscellaneous newspapers, reel 2

Daily gazette series / Delaware Co. Delaware – (1928-apr 1933), mar 1936-66 [daily] – 93r – 1 – mf#B25405-25497 – us Ohio Hist [071]

Daily gazette series / Fairfield Co. Lancaster – 1901-15,17-9/25,4/26-28,7/29-5/1935 [daily] – 96r – 1 – mf#B7589-7683 – us Ohio Hist [071]

Daily gazetteer – La Salle 1735-45 – 1 – mf#4234 – us UMI ProQuest [071]

Daily gazette-times – Corvallis OR: N R Moore, 1909-21 [daily ex sun] – 13r – 1 – (began in 1909; related to wkly ed: gazette-times [corvallis, or] 1909, and: weekly gazette-times, 1909-21; cont by: corvallis gazette-times [1921-]) – us Oregon Lib [071]

Daily gazette-times *see* Gazette-times (corvallis, or)

Daily gleaner – Fredericton, Canada. 7 oct 1914-29 jun 1940; 17 mar 1941-1952 – 224r – 1 – uk British Libr Newspaper [071]

Daily gleaner – Fredericton. New Brunswick. Canada. Ja 1976- – 1 – cn Commonwealth Imaging [071]

Daily gleaner – Kingston, Jamaica. 10 Sep 1879-31 Dec 1888; 19 Aug 1891-Dec 1899 – 52r – 1 – uk British Libr Newspaper [072]

The daily gleaner – Kingston, Jamaica: Gleaner Co Ltd, 1902-1956-dec 6 1992 – 1 – us CRL [079]

Daily globe / Richland Co. Shelby – jan 1950-dec 1975 [daily] – 90r – 1 – mf#B1583-1673 – us Ohio Hist [071]

Daily globe / Richland Co. Shelby – jan 1976-dec 1978 [daily] – 12r – 1 – mf#B8426-8437 – us Ohio Hist [071]

Daily globe / Richland Co. Shelby – jan 1979-dec 1985 [daily] – 28r – 1 – mf#B27687-27714 – us Ohio Hist [071]

Daily globe / Richland Co. Shelby – jan 1986-sep 1987 [daily] – 7r – 1 – mf#B34774-34780 – us Ohio Hist [071]

Daily globe / Richland Co. Shelby – nov 1919-37, jul 1938-49 [daily] – 86r – 1 – mf#B27531-27616 – us Ohio Hist [071]

Daily globe / Richland Co. Shelby – v1 n1. may 1900-apr 1918,aug 1918-oct 1919 [daily] – 33r – 1 – mf#B1550-1582 – us Ohio Hist [071]

The daily globe – Washington, DC: Francis P Blair, John Rives, James C Pickett, dec 5 1871-jun 19 1872 – 2r – 1 – us CRL [071]

Daily gospel preacher – Ashland OH: [s.n] jun 7-10. 1881 [daily] [mf ed 2003] – 1r – 1 – (related title: the gospel preacher) – mf1039b – us ATLA [242]

Daily gospel preacher *see* The gospel preacher

Daily grants pass courier – Grants Pass OR: A E Voorhies, 1931-34 [daily ex sun] – 1 – (cont: grants pass daily courier [1919-31]; cont by: grants pass courier [1934-41]) – us Oregon Lib [071]

Daily graphic – Accra: Graphic Corp, 1994- [daily (ex sun)] – 80r – 1 – us CRL [079]

Daily graphic – Accra: West African Graphic, 1956-dec 30 1982 – 1 – us CRL [079]

Daily graphic – Accra: West African Graphic, -1982. [may 24 1954-sep 25 1965] – us CRL [079]

Daily graphic – New York NY. 1884 may 15 – 1r – 1 – mf#780664 – us WHS [071]

The daily graphic – New York. Mar 4 1873-Sept 23 1889. v. 1-50 no. 5129 – 1 – us NY Public [071]

Daily graphic and daily sketch – [NW England] Manchester ALS apr 1948-3 jan 1953 – 1 – uk Newsplan; uk MLA [072]

Daily guardian – Freetown, Sierra Leone. Jan 1941-Aug 1958 – 17r – 1 – (lacking: july-dec 1946. jan-june 1954. feb-may 1958) – uk British Libr Newspaper [072]

Daily hampshire independent *see* Hampshire independent

Daily herald – Adelaide, Australia. 11 Sep 1912-31 Aug 1916 – 21r – 1 – uk British Libr Newspaper [079]

Daily herald – Carlisle, PA., 1889 – 13 – $25.00r – us IMR [071]

Daily herald – Chambersburg, PA., 1886 – 13 – $25.00r – us IMR [071]

Daily herald / Columbiana Co. Salem – v1 n1. may 1891-jan 1919 [daily] – 61 – 1 – mf#B8289-8349 – us Ohio Hist [071]

Daily herald / Delaware Co. Delaware – (4-5/1879,6/94-9/1897) very scattered [daily] – 1r – 1 – mf#B155 – us Ohio Hist [071]

Daily herald – London, 1911-64 – 268r – 1 – uk British Libr Newspaper [072]

Daily herald – London: "Limited" Print. & Pub Co, 1939-55; 1956-sep 14 1964 – 1 – us CRL [072]

Daily herald – Monongahela, PA., 1970-1972 – 13 – $25.00r – us IMR [071]

Daily herald / Vanwert Co. Delphos – jun 1894-dec 1910 [daily] – 19r – 1 – mf#B6309-6327 – us Ohio Hist [071]

Daily herald *see*
– Miscellaneous newspapers of mesa county
– Miscellaneous newspapers of san juan county

Daily herald and empire / Montgomery Co. Dayton – apr 1874-aug 1876 [daily] – 2r – 1 – mf#B5149-5150 – us Ohio Hist [071]

Daily herald and gazette – Cleveland, OH, mar 22 1837-aug 5 1839 – 4r – 1 – (daily whig newspaper) – us Western Res [071]

Daily herald & general advertiser – [Scotland] Glasgow: Ross and Watson 5 may 1858-4 jun 1858 (daily) [mf ed 2004] – 27v on 1r – 1 – uk Newsplan [072]

Daily herald [nanaimo] – Nanaimo, Canada 25 apr 1913-26 apr 1916; 11 jan 1917-2 nov 1920 – 20r – 1 – uk British Libr Newspaper [071]

Daily herald news – Punta Gorda, FL. 1971 jul-1973 – 15r – (gaps) – us UF Libraries [071]

Daily herald series / Morgan Co. McConnelsvill – (7/1906-08,7/14-1/15,2-8/1917) [daily] – 6r – 1 – mf#B11222-11227 – us Ohio Hist [071]

Daily hot blast – Anniston AL. v3 n123 [1888 sep 3] – 1r – 1 – (cont: hot blast; cont by: anniston evening star; anniston evening star and daily hot blast) – mf#912100 – us WHS [071]

Daily illustrated mirror – [NW England] Manchester ALS – 1 – (title change: daily mirror [1973]) – uk MLA; uk Newsplan [072]

Daily independent / Belmont Co. Bellaire – (mar 1914-16, 1917-jun 1919) [daily] – 5r – 1 – mf#B6274-6278 – us Ohio Hist [071]

Daily independent – Chippewa Falls WI. [1887 oct 2/1888 aug 10]-1915 oct 22/1916 mar 10 – 49r – 1 – (cont by: evening independent) – mf#923785 – us WHS [071]

Daily independent *see* [Elko-] weekly elko independent

The daily independent – Kimberley SA, 29 sept 1875-30 jun 1893 – 46r – 1 – (title varies: the independent, 1875-79; 1893) – sa National [079]

Daily independent press – [East Midlands] Cambridge, Cambridgeshire n1 [2 jan]-n138 [21 jul 1892] [mf ed 2004] – 1r – 1 – (discontinued) – uk Newsplan [072]

Daily intelligencer – Atlanta GA. 1863 jul 9, jan 20, 1864 mar 18 – 1r – 1 – mf#845811 – us WHS [071]

Daily intelligencer – Atlanta, GA. oct 7 1858-sep 18 1859; jan 4 1860-dec 31 1860; jan 1 1861-apr 1 1862; apr 2 1862-jun 30 1863; jul 1 1863-may 29 1864; may 31 1864-dec 31 1864 – (= ser Confederate newspapers) – 6r – 1 – us Western Res [071]

Daily intelligencer – Belleville, Ontario, CN. 13 oct 1913-10 nov 1915 – 12r – 1 – uk British Libr Newspaper [072]

Daily inter ocean – Chicago IL. 1879 nov 12/dec 4-1913 jan 13-31 – 93r – 1 – (with gaps; cont: inter ocean [chicago, ill.: 1872: daily]; our herald [lafayette in]; inter ocean [chicago il: 1872: daily]; cont by: inter ocean [chicago il: daily]; inter ocean [chicago il: 190-: daily]) – mf#992254 – us WHS [071]

The daily inter ocean – Chicago: Inter Ocean Pub Co, 1879-1902. jan 4 1882; apr 30, may 3-6,13,20, aug 10-31 1893 – us CRL [071]

Daily inter-ocean – Superior WI. 1885 mar 16-may 26 – 1r – 1 – mf#933838 – us WHS [071]

Daily irontonian / Lawrence Co. Ironton – v1 n1. aug-dec 1888 [daily] – 1r – 1 – mf#B32859 – us Ohio Hist [071]

Daily item – Steelton, PA. -d 1883-1885 – 13 – $25.00r – us IMR [071]

Daily jefferson county union – Fort Atkinson WI. 1946 mar-nov 13, 1946 nov 14-1947 aug 21, 1947 aug 22-1948 feb 2 – 3r – 1 – (cont: jefferson county union [lake mills wi]; cont by: fort daily news; daily jefferson county union and fort daily news) – mf#943861 – us WHS [071]

Daily jefferson county union and fort daily news – 1948 feb 3/may 26-1963 jul/dec – 28r – 1 – (cont: daily jefferson county union [fort atkinson wi: 1948]; fort daily news; cont by: daily jefferson county union [fort atkinson wi: 1969]) – mf#943620 – us WHS [071]

Daily jeffersonian / Guernsey Co. Cambridge – sep 1892-jun 1954 [daily] – 146r – 1 – mf#B1064-1209 – us Ohio Hist [071]

The daily jewish courier – Chicago. Ill. 1893; 1906-47 – 1 – us AJPC [071]

Daily jewish express – London. 1908-1922 – 1 – us NY Public [072]

Daily jewish press – Chicago. Ill. 1913 – 1 – us AJPC [071]

Daily, John Riley *see* Daily-throgmorton debate

Daily journal – Milwaukee WI. 1882 nov 16-1883 may 11 – 1r – 1 – (cont by: milwaukee daily journal) – mf#1167145 – us WHS [071]

Daily journal – Superior, NE: N C Pickard. 1v. feb 27 1888-feb 1889// (daily ex sun) [mf ed v1 n2. feb 28 1888-feb 14 1889 (gaps) filmed 1995] – 1r – 1 – (cont by: superior daily journal) – us NE Hist [071]

Daily journal – Plattsmouth, NE: Kirkham & Green, may 1898-v17 n187. jun 18 1898 (daily ex sun) [mf ed v17 n166. may 23-jun 18 1898 (gaps)] – 1r – 1 – (cont: plattsmouth daily journal (1888)) – us NE Hist [071]

Daily journal – Plattsmouth, NE: Sherman & Cutright (daily ex sun) [mf ed 1884-85 (gaps)] – 1r – 1 – (cont: plattsmouth daily journal. cont by: plattsmouth daily journal (1888)) – us NE Hist [071]

Daily journal / Delaware Co. Delaware – apr 1900-mar 1902 [daily] – 3r – 1 – mf#B9592-9594 – us Ohio Hist [071]

Daily journal – Mechanicsburg, PA. -d 1901-11. 12 rolls – 13 – $25.00r – us IMR [071]

Daily journal / Montgomery Co. Dayton – jan-mar 1905 [daily] – 1r – 1 – mf#B31149 – us Ohio Hist [071]

Daily journal / Shelby Co. Sidney – mar 1914-feb 1916 [daily] – 4r – 1 – mf#B8657-8660 – us Ohio Hist [071]

Daily journal / Washington Co. Marietta – jul 1914-oct 1919 [daily] – 13r – 1 – mf#B11612-11624 – us Ohio Hist [071]

Daily journal – Wilmington NC. 1856 sep 9-dec 31, 1857 jan 1-may 30, jun 1-sep 2, 1858 sep 24-dec 31, 1859 jan 1-may 31, jun 1-aug 31 – 6r – 1 – mf#861214 – us WHS [071]

Daily journal *see* Miscellaneous newspapers of san miguel county

Daily journal herald / Delaware Co. Delaware – 3/1902-6/1927, 10/1928-3/1929 [daily] – 67r – 1 – mf#B9595-9661 – us Ohio Hist [071]

Daily journal [salem, or: 1899] – Salem OR: Hofer Bros, 1899-1903 [daily] – 1 – (cont: daily capital journal [salem, or: 1896]; cont by: daily capital journal [salem, or: 1903]) – us Oregon Lib [071]

Daily labor bulletin / Chicago Federation of Labor and Industrial Union Council – v1 n1-8,10 [1905 may 24-jun 2,5] – 1r – 1 – mf#1111019 – us WHS [331]

Daily labor report / Bureau of National Affairs [Washington DC] – 1982 jan/feb 26-1989 nov/dec – 50r – 1 – (cont: daily report on labor-management problems) – mf#1055079 – us WHS [331]

Daily leader / Belmont Co. Bellaire – 1-6/1914, 10/1920-12/1929 (damaged) [daily] – 30r – 1 – mf#B32136-32165 – us Ohio Hist [071]

Daily leader / Belmont Co. Bellaire – (1913, 1915-20), feb 1935 [daily] – 14r – 1 – mf#B6845-6858 – us Ohio Hist [071]

Daily leader / Belmont Co. Bellaire – jan 1930-jun 1937 [daily] – 22r – 1 – mf#B32166-32187 – us Ohio Hist [071]

Daily leader / Belmont Co. Bellaire – jun 1937-may 1942 [daily] – 14r – 1 – mf#B6813-6826 – us Ohio Hist [071]

Daily leader / Columbiana Co. East Palestine – 3/1937-1/43,12/46-6/78,9/78-8/1980 [daily] – 54 – 1 – mf#B12694-12747 – us Ohio Hist [071]

Daily leader / Columbiana Co. East Palestine – may 1915-apr 1926 [daily] – 14r – 1 – mf#B11334-11367 – us Ohio Hist [071]

Daily leader – Eau Claire WI. [1881 apr 27-1887 jun 30]-[1886 jul-dec] – 11r – 1 – (cont by: eau claire daily leader) – mf#1145819 – us WHS [071]

Daily leader – Superior WI. 1890 may-aug, sep-dec, 1891 feb 15-jun 23 – 3r – 1 – (cont by: superior leader) – mf#937547 – us WHS [071]

Daily leader – Grand Rapids, Wisconsin Rapids WI. [1914 jan/apr]-[1919 jul 28/oct 15] – 13r – 1 – (cont: daily reporter [wisconsin rapids wi]; cont by: grand rapids leader) – mf#1144680 – us WHS [071]

Daily leader – Gainesville, FL. 1892 sep 14; oct 05,26,29; nov 02,18; dec 13 – 1r – us UF Libraries [071]

Daily leader / Washington Co. Marietta – jan 1896-jun 1902 [daily] – 10r – 1 – mf#B32886-32895 – us Ohio Hist [071]

The daily leader – Neligh, NE: E T & C Jebest (daily) [mf ed oct 14 1885-sep 30 1886 (gaps) filmed 1999 – 1r – 1 – (issues for oct 14 1885 lack numeric designation) – us NE Hist [071]

Daily ledger – Kenosha WI. 1851 oct 30-nov 1 – 1r – 1 – mf#876765 – us WHS [071]

Daily ledger / Montgomery Co. Dayton – jul 1867-nov 1869 [daily] – 3r – 1 – mf#B5155-5157 – us Ohio Hist [071]

Daily legal news and cleveland recorder – Cleveland. On film: v48, no. 131-v75, no.313; Ju 1935-Dec 1962. LL-04 – 1 – us L of C Photodup [340]

Daily liberal – Dubbo, jan 1968-jun 1995 – 162r – at Pascoe [079]

Daily liberal Democrat – La Crosse WI. [1872 jul 26-1873 oct 22], [1873 oct 23-1875 jul 19], 1875 jul 20-1876 apr 15 – 3r – 1 – (cont: la crosse daily liberal democrat; cont by: morning liberal democrat) – mf#928309 – us WHS [071]

Daily life – Milwaukee WI. 1861 aug 17-1865 apr 1 – 1r – 1 – (cont by: daily wisconsin) – mf#1130292 – us WHS [071]

Daily life and work in india / Wilkins, W J – London, 1888 – 4mf – 9 – mf#HTM-214 – ne IDC [915]

Daily life in the kingdom of the kongo / Balandier, Georges – New York, NY. 1968 – 1r – us UF Libraries [960]

Daily lokmat – Nagpur, India. 5 May-7 Dec 1950; 2 Jan-9 Dec 1951; Jan-12 Dec 1952; 1953 – 4r – 1 – us L of C Photodup [079]

Daily madison patriot – Madison WI. 1876 apr 19-oct 12, 1876 oct 13-1877 mar 8 – 2r – 1 – (cont by: wisconsin patriot [madison wi: 1877]) – mf#939274 – us WHS [071]

Daily mail – Brisbane, Australia. 24 oct 1903-4 nov 1905, 19 feb 1906-18 nov 1916, 2 oct 1926-26 aug 1933 – 156 3/4r – 1 – (aka: brisbane daily mail) – uk British Libr Newspaper [072]

Daily mail – Brisbane, Australia. oct 1903-nov 1916, may 1920-jun 1922, oct 1926-aug 1933 [imperfect] – 163r – 1 – (imperfect) – uk British Libr Newspaper [079]

Daily mail – London, Feb 1896-1979; 1986- – 833+ r – 1 – uk British Libr Newspaper [072]

Daily mail – Silver jubilee issue. London, England. 6 May 1935 – 3ft – 1 – uk British Libr Newspaper [072]

Daily mail – Sydney, oct 1923-jan 1924 – 1r – A$70.49 vesicular A$75.99 silver – at Pascoe [079]

Daily mail *see* Daily news and leader [london daily news], 1914-1919

Daily mail and empire – Toronto, Canada. Sept-Oct 1918 – 2r – 1 – uk British Libr Newspaper [072]

Daily mail and empire *see* Toronto daily mail

Daily mail (sat and sun) – Brisbane, Australia. 15 may 1920-jun 1922 – 6 1/2r – 1 – uk British Libr Newspaper [072]

Daily mail, the... may 1896-2000 – 1080r – 1 – mf#96990 – uk Microform Academic [072]

Daily mercury *see* [Oroville-] mercury

Daily messenger – Astoria OR: Franklin Press & Pub Co, 1931- [daily ex mon] [mf ed 1957] – 5r – 1 – us Oregon Lib [071]

Daily messenger galignanis messenger (afternoon ed) *see* Galignani's messenger

Daily messenger series / Sandusky Co. Fremont – (jul 1900-dec 1916) [daily] – 22r – 1 – mf#B33202-33223 – us Ohio Hist [071]

Daily milwaukee news – Milwaukee WI. 1870 nov 17 – 1r – 1 – (cont: daily milwaukee press and news; cont by: milwaukee daily news [milwaukee wi: 1874]) – mf#1133880 – us WHS [071]

Daily milwaukee press and news – Milwaukee WI. 1861 oct 20 – 1r – 1 – (cont: daily people's press and news; cont by: daily milwaukee news [milwaukee wi: 1861]) – mf#1173970 – us WHS [071]

Daily miner *see* Custer county miscellaneous newspapers

Daily mining record *see* Denver county miscellaneous newspapers, reel 2

Daily minor / Cuban Refugee Resettlement Operation [Fort McCoy WI] – v1 n1-7 [1980 oct 13-27] – 1r – 1 – mf#512964 – us WHS [071]

Daily mirror – London, jul 1903- – 744+ r – 1 – (aka: mirror 1985) – uk British Libr Newspaper [072]

Daily mirror / Marion Co. Marion – jan 1907-jun 1912 [daily] – 15r – 1 – mf#B8142-8156 – us Ohio Hist [071]

Daily mirror – Sydney, Australia. May-Sept 3 1945; Jan-Aug 1946 – 5r – 1 – us L of C Photodup [071]

Daily mirror *see*
- Popular newspapers during world war 1
- Popular newspapers during world war 2

Daily mississippian – Jackson, Meridian MS, Selma AL. 1862 dec 20 – 1r – 1 – (cont: semi-weekly mississippian; daily southern crisis; cont by: vicksburg daily herald [vicksburg ms: 1864]; herald and mississippian) – mf#865501 – us WHS [071]

Daily morning advocate – Racine WI. 1854 jan 30, feb 1, may 13,24, jul 4,6, dec 6, 1854 jan 31-dec – 2r – 1 – (cont: daily racine advocate; cont by: daily racine advocate [racine wi: 1855]) – mf#965203 – us WHS [071]

Daily morning astorian – Astoria OR: J F Halloran & Co, 1883-99 [daily ex mon] – 1 – (related to: weekly astorian [1874-]; cont by: morning astorian [1899-1930]) – us Oregon Lib [071]

Daily morning astorian *see* Weekly astorian

Daily morning chronicle – Washington DC. 1864 feb 8, 1865 feb 1-4, apr 15,20,27, 1868 apr 22 – 1r – 1 – (cont by: washington chronicle [1874]) – mf#852883 – us WHS [071]

Daily morning democrat – Baker City OR: Bowen & Small, [wkly] – 1 – (related to: weekly bedrock democrat; cont by: morning democrat [-1929]) – us Oregon Lib [071]

Daily morning democrat *see* Weekly bedrock democrat

Daily morning news – Savannah GA. 1852 nov 12 – 1r – 1 – mf#846402 – us WHS [071]

Daily morning news *see* Morning news

Daily morning oasis – Nogales AZ. 1920 jul 1-2,8-14,16,18-24,27,29-30, aug 1,4,7,11,14,20,24-sep 3,5-15,18-28,30-oct 1,21 – 1r – 1 – mf#853927 – us WHS [071]

Daily morning record – Gainesville, FL. 1888 aug 25,oct 02 – 1r – us UF Libraries [071]

Daily morning standard – Williamsport, PA., 1872 – 13 – $25.00r – us IMR [071]

Daily morning times – Roseburg OR: Roseburg Pub Co Inc, 1935- [daily ex sun] – 1 – (cont: douglas county times; ceased in 1936) – us Oregon Lib [071]

Daily morning union *see* [Grass valley-] the union

Daily musalman – Madras, India. 2 Jul 1944-Dec 1965 – 17r – 1 – us L of C Photodup [079]

Daily nation – Nairobi, Kenya. n9736-n12168. 1992-1999 – 119r – (gaps) – us UF Libraries [079]

Daily nation – Nairobi: East African Newspapers Ltd, jun 1962. mon-sat issues filmed only – 1 – us CRL [079]

Daily national – Milwaukee WI. 1859 sep 10-nov 6 – 1r – 1 – mf#1167235 – us WHS [071]

Daily national journal – Washington, DC. Aug 9 1824-Dec 31 1831 – 13r – 1 – us L of C Photodup [071]

Daily national journal – Washington, DC, jan-jun 1831 – 1r – 1 – us CRL [071]

Daily nawan zamana – Jullundur, India. Jul-Sept 1966 – 1r – 1 – us L of C Photodup [079]

Daily nebraska city news – Nebraska City, NE: Thomas Morton (daily ex sun) [mf ed v2 n68. dec 3 1864-jul 21 1865 (gaps)] – 1r – 1 – (cont by: nebraska city news (1867). weekly ed: nebraska city news (1858)) – us NE Hist [071]

Daily nebraska city news – Nebraska City, NE: Hubner & Mainell. 6v. nov 14 1882-aug 29 1888 (daily) [mf ed with gaps filmed 2000] – 1 – (cont: nebraska city daily news. cont by: nebraska city daily news (1888). issues lack vol and issue designation) – us NE Hist [071]

Daily nebraska city news / ed by Morton, Julius Sterling – Nebraska City, NE: Wm M Ricklin. 1v. v16 n1. dec 1 1869-70// (daily ex sun) [mf ed with gaps] – 1r – 1 – (cont: nebraska city news (1867). absorbed by: nebraska city daily times. weekly ed: nebraska city news (1858)) – us NE Hist [071]

The daily nebraska commonwealth – Lincoln, NE: Gere & Carder. v1 n1. jan 11 1869-(daily) [mf ed with gaps filmed 1958] – 1r – 1 – (weekly ed: nebraska commonwealth) – us NE Hist [071]

Daily nebraska herald – Plattsmouth, NE: J A Mac Murphy. 1v. v1 n1-131. jun 24-nov 27 1872 (daily ex sun) [mf ed with gaps filmed 1979] – 1r – 1 – us NE Hist [071]

Daily nebraska press – Nebraska City, NE: Irish, Price & Co. 27v. v10 n90. aug 3 1868-v36 aug 9 1894 (daily ex mon) [mf ed 1873-94 (gaps) filmed -1971] – 8r – 1 – (cont: tri-weekly nebraska press. cont by: nebraska daily press. numbering irregular. some issues lack vol and issue numbering) – us NE Hist [071]

Daily nebraska state journal – Lincoln, NE: [Gere & Brownlee], jul 20-[v1] n2. jul 21 1870 (daily) [mf ed filmed 1958] – 1r – 1 – (cont by: daily state journal) – us NE Hist [071]

Daily nebraska state journal – Lincoln, NE: [State Journal Co] 16v. 8th yr n240. may 3 1878-23rd yr n105. dec 4 1892 (daily) [mf ed lacks sep 9-13 1879] – 43r – 1 – (cont: daily state journal. cont by: nebraska state journal. weekly ed: nebraska state journal (1878); weekly nebraska state journal 1878-91. semiwkly ed: semi-weekly nebraska state journal 1891-92) – us NE Hist [071]

The daily new era – Lancaster, PA., 1884 – 13 – $25.00r – us IMR [071]

The daily new republic – Lincoln, NE: A Roberts. v1 n1. mar 31 1888- (daily) [mf ed -apr 4 1888 filmed [1973]] – 1r – 1 – us NE Hist [071]

Daily news – London. jan 1846-oct 1960 [daily] – 581r – 1 – (aka: news chronicle) – uk British Libr Newspaper [072]

Daily news – Ashland WI. 1887 mar 29 – 1r – 1 – mf#1221711 – us WHS [071]

Daily news / Auglize Co. Wapakoneta – 2-12/1937,3/71-4/1982,12/1982-84 [daily] – 60r – 1 – mf#B25688-25747 – us Ohio Hist [071]

Daily news / Auglize Co. Wapakoneta – (jul 1913-jun 1970) poor quality [daily] – 139r – 1 – mf#B28871-29009 – us Ohio Hist [071]

Daily news / Auglize Co. Wapakoneta – jul 1913-mar 1914 (damaged) [daily] – 2r – 1 – mf#B32912-32913 – us Ohio Hist [071]

Daily news – Bridgetown, Barbados 13 dec 1963-26 sep 1965, 2 jan-31 jul 1966 – 1 – (cont: barbados daily news [jun 1960-12 dec 1963]; cont by: barbados daily news [27 jun-14 sep 1967]) – uk British Libr Newspaper [072]

Daily news – Birmingham, England 11 aug 1987-17 may 1991 [mf 1986-] – 1 – (cont: birmingham daily news [5 nov 1985-7 aug 1987]; cont by: metro news [ns] 21 may 1991-9 sep 1993) – uk British Libr Newspaper [072]

Daily news – Green Bay WI. [1972 nov 13/dec 30]-[1976 mar 1/apr 10] – 24r – 1 – (cont by: brown county chronicle; green bay news-chronicle) – mf#1144876 – us WHS [071]

Daily news – La Crosse WI. 1880 jul 13-oct 26, 1880 oct 27-1881 may 14, 1881 may 16-1881 jul 16 – 3r – 1 – (cont by: la crosse daily news) – mf#930259 – us WHS [071]

Daily news / Crawford Co. Crestline – apr-may 1902 [daily] – 1r – 1 – mf#B10573 – us Ohio Hist [071]

Daily news / Darke Co. Greenville – v1 n1. jun 1921-mar 1923 [daily] – 3r – 1 – mf#B9774-9776 – us Ohio Hist [071]

Daily news – Fond Du Lac WI. 1899 mar 10 – 1r – 1 – mf#918296 – us WHS [071]

Daily news – Gaborone: [s.n.], sep 17 1975-81 – us CRL [071]

Daily news – Geelong, Australia. 6 May 1858-30 Jun 1859.-d 3mqn reels – 1 – uk British Libr Newspaper [072]

Daily news / Greene Co. Beavercreek – apr 1977-dec 1984 [daily] – 66r – 1 – mf#B25612-25677 – us Ohio Hist [071]

Daily news / Greene Co. Beavercreek – aug 1986-nov 1989 [daily] – 17r – 1 – mf#B31300-31316 – us Ohio Hist [071]

Daily news / Greene Co. Beavercreek – jan 1985-aug 1986 [daily] – 15r – 1 – mf#B28774-28788 – us Ohio Hist [071]

Daily news – Huntingdon, PA. -d 1922-1990; 1959-1970; 1983-1984 – 13 – $25.00r – us IMR [071]

Daily news / Jefferson Co. Steubenville – mar-sep 1873 [daily] – 1r – 1 – mf#B3990 – us Ohio Hist [071]

Daily news – Kingston, ON. 1862-73 – 7r – 1 – cn Library Assoc [071]

Daily news / Lucas Co. Toledo – jan-dec 1901 [daily] – 3r – 1 – mf#B34640-34642 – us Ohio Hist [071]

Daily news – Mauch Chunk, PA., 1914 – 13 – $25.00r – us IMR [071]

Daily news – Murwillumbah, 1946-69 – at Pascoe [079]

Daily news – Murwillumbah, jan 1969-jun 1997 – at Pascoe [079]

Daily news – Nelson, Canada. -d. 1 jan 1905-1 jan 1906; 21 dec 1907; 3 jun-31 jul 1910; aug 1910-1922 – 60 1/2r – 1 – uk British Libr Newspaper [072]

Daily news – New Plymouth, NZ. 30 mar 1962-18 aug 1962; 12 jun 1963-18 sep 1963; 1 aug 1967-15 sep 1967; 1 aug 1969-28 oct 1969; jun 1974-jul 1974; sep 1974; nov 1974-aug 1977; oct 1977-feb 1978; apr 1978-apr 2002 – 1 – mf#21.1 – nz Nat Libr [079]

Daily news – Pensacola, FL. 1900 feb 2-apr 19 – 2r – us UF Libraries [071]

Daily news / Perry Co. New Lexington – aug-dec 1935, mar 1936-40// [daily] – 8r – 1 – mf#B14125-14132 – us Ohio Hist [071]

Daily news / Perry Co. New Lexington – jan 1-jun 30 1936 [daily] – 1r – 1 – mf#B32092 – us Ohio Hist [071]

Daily news – Perth, Australia. Jul 1882-Dec 1888.-w. 13 reels – 1 – uk British Libr Newspaper [079]

Daily news – Prince Rupert, Canada. -d. 10 jun 1912-15 jul 1920; 13 oct 1920- 3 dec 1921; jan-30 jun 1922; 1952-2 feb 1953 – 33 3/4r – 1 – uk British Libr Newspaper [071]

Daily news – St. John's. Canada. -d. 1911-1913, 1917-22 dec 1921; 1922-jun 1940; 18 mar 1941-25 aug 1948; 2 sep, 24 dec 1948; 21 mar 1949-1952; jan-14 jul 1953; 26 sep 1957-25 sep 1967 – 344 1/2r – 1 – uk British Libr Newspaper [072]

Daily news – Tarpon Springs, FL. 1918 – 1r – us UF Libraries [071]

Daily news – St. John, NF. 1894-1920 – 70r – 1 – (trailer reel covers issues missing fr 1908-20 [1r]) – ISSN: 0839-4180 – cn Library Assoc [071]

Daily news / Trumbull Co. Niles – (aug 1912-aug 1916) [daily] – 8r – 1 – mf#B28758-28765 – us Ohio Hist [071]

Daily news / Trumbull Co. Niles – (sep 1908-jun 1923) damaged [daily] – 23r – 1 – mf#B32069-32091 – us Ohio Hist [071]

Daily news / Wayne Co. Wooster – jun 1919-jan 1920 [daily] – 2r – 1 – mf#B9958-9959 – us Ohio Hist [071]

Daily news *see*
- Miscellaneous newspapers of las animas county, reel 2
- Miscellaneous newspapers of mesa county
- Morning news
- North-west news
- Southland news, daily news

The daily news – Athens, PA., 1889-1890 – 13 – $25.00r – us IMR [071]

The daily news – York, NE: Duncan M Smith. v1 n287. dec 1 1894? (daily) [mf ed filmed 1998] – 1r – 1 – us NE Hist [071]

Daily news advertiser – Vancouver, Canada. 1912; jul-aug 1917 – 3r – 1 – uk British Libr Newspaper [072]

Daily news and leader [london daily news], 1914-1919 – 22r – 1 – (combined in 1930 with: news chronicle wh subsequently amalgamated with the: daily mail) – mf#96734 – uk Microform Academic [072]

Daily news and the times – Menasha, Neenah WI. 1919 jul 19/oct 24-1924 may 17/jun 11 – 20r – 1 – (cont by: daily news-times) – mf#1042888 – us WHS [071]

Daily news bulletin / Jewish Telegraphic Agency, Inc – New York: The Agency. [v4 n248-v6 n300. 1924-26] – 2r – us CRL [071]

Daily news bulletin – New York, NY. 7 Nov 1923-15 Oct 1924. Incomplete – 1 – us AJPC [071]

Daily news bulletin = Wakalat al-anba al-qatariyah – Qatar: The Agency, aug 4 1986-apr 18 1987 – 1r – 1 – us CRL [079]

Daily news chief – Winter Haven, FL. 1922 apr-1977 oct – 220r – (gaps) – us UF Libraries [071]

Daily news era – Boyertown, PA., 1908-1915 – 13 – $25.00r – us IMR [071]

Daily news huntingdon saxon – Mount Union, PA., 1982 – 13 – $25.00r – us IMR [071]

Daily news [medford, or: 1926] – Medford OR: News Pub Co, 1926-29 [daily ex mon] – 1 – (cont: jackson county news; cont by: medford daily news) – us Oregon Lib [071]

Daily news [medford, or: 1933] – Medford OR: News Pub Co, 1933 [daily ex mon] – 1 – (cont: daily news [medford, or: 1933]; cont by: medford news) – us Oregon Lib [071]

Daily news mt. carmel – Mount Carmel, PA., 1892-1912 – 13 – $25.00r – us IMR [071]

Daily news [portland, or] – Portland OR: [s.n.] 1907-12 [daily ex sun] – 1 – (cont: eastside news; cont by: portland news [portland, or]) – us Oregon Lib [071]

Daily news review – Vienna, Austria. -d. 21 June 1949-31 March 1950. 2 reels – 1 – uk British Libr Newspaper [072]

Daily news/d / Lorain Co. Lorain – (jul 1903-jun 1917) [daily] – 19r – 1 – mf#B33144-33162 – us Ohio Hist [071]

Daily news-register – McMinnville OR: News Register Pub Co, 1953-58 [daily ex sun] – 1 – (cont: news register [1953]; cont by: news-register [1958-]) – us Oregon Lib [071]

Daily news-times – Menasha, Neenah WI. 1924 dec 19/1925 apr 16-1949 jul/1949 sep 17 – 73r – 1 – (cont: daily news and the times; cont by: menasha record; twin city news-record) – mf#1042889 – us WHS [071]

Daily northern argus – Rockhampton, Australia 10-30 oct 1879, 22 jan 1880-29 dec 1888, 2 may 1894-31 dec 1896 – 1 – (numeration irreg; cont: northern argus [6 apr, 4 may 1872]) – uk British Libr Newspaper [079]

Daily northern standard – Townsville, Australia. 13 oct 1833 – 1/4r – 1 – uk British Libr Newspaper [072]

Daily northwestern – Oshkosh WI. 1868 jan 6-jul 6, 1871 aug 30-dec 30, 1872 nov 16, 1874 jan 31, dec 7, 1875 jan 2, apr 26 – 3r – 1 – (cont by: oshkosh daily northwestern [oshkosh wi: 1875]) – mf#1146257 – us WHS [071]

Daily northwestern – Oshkosh WI. [1886 jun 3, dec 10]-[1933 dec 23/dec 30] – 253r – 1 – (cont: oshkosh daily northwestern [oshkosh wi: 1875]; cont by: oshkosh northwestern [oshkosh wi: 1934: daily]) – mf#1141126 – us WHS [071]

Daily northwestern – Oshkosh WI. 1861 jan 12-may 18, may 20-aug 28 – 2r – 1 – mf#1146269 – us WHS [071]

Daily notes – Canonsburg, PA. 1879-1973 – 13 – $25.00r – us IMR [071]

Daily ohio state democrat / Franklin Co. Columbus – v1 n1. apr 1853-may 1854 [daily] – 2r – 1 – mf#B1218-1219 – us Ohio Hist [071]

Daily ohio statesman – Columbus OH. 1850 jan 2-jun 29, jul 1-dec 31, 1851 jan 2-jun 30, jul 1-dec 31, 1852 jul 1-dec 31 – 5r – 1 – (cont by: daily ohio state democrat; daily ohio statesman and democrat) – mf#951981 – us WHS [071]

Daily ohio statesman – Columbus OH. 1837 sep 18 [v1 n12]-dec 29, 1838 jan 1-apr 17 [v1 n191] – 2r – 1 – (cont: daily statesman [columbus oh]) – mf#920454 – us WHS [071]

Daily ohio statesman / Franklin Co. Columbus – jan 1861-jul 1872 [daily] – 21r – 1 – mf#B28803-28823 – us Ohio Hist [071]

Daily ohio statesman / Franklin Co. Columbus – oct 1876-mar 1877,mar 1878-nov 1878 [daily] – 2r – 1 – mf#B7084-7085 – us Ohio Hist [071]

Daily okeechobee news – Okeechobee, FL. 1992 nov 21-1999 nov – 39r – (gaps) – us UF Libraries [071]

Daily okeechobee news – Okeechobee, FL. v82 n59-106. 1992 jul-sep – 1r – us UF Libraries [071]

Daily oklahoman – Oklahoma City OK. 1917 jan 1/jan 31-1939 apr 23 – 67r – 1 – (with gaps; cont: oklahoma press-gazette [oklahoma city ok: 1894]; cont by: oklahoma city times [oklahoma city ok: 1908]; daily oklahoman oklahoma city times) – mf#634354 – us WHS [071]

Daily oregon herald – Portland OR: T Patterson & Co, 1869- [daily ex mon] – 1 – (related to: weekly oregon herald; cont: oregon herald [portland, or: daily]; ceased in 1873?) – us Oregon Lib [071]

Daily oregon herald *see* Weekly oregon herald

Daily oregon statesman – Salem OR: S A Clarke, 1869-77 [daily] – 1 – (related to: oregon statesman [oregon city, or]; american unionist, oregon weekly statesman [salem, or: 1869]; weekly oregon statesman; cont: daily oregon unionist; cont by: oregon daily statesman) – us Oregon Lib [071]

Daily oregon statesman *see* Oregon weekly statesman (salem, or: 1869)

Daily oregon statesman (salem, or: 1864) *see* Oregon statesman (oregon city, or)

Daily oregon statesman (salem, or: 1889) *see* Oregon statesman (salem, or: 1884)

Daily oregon statesman [salem, or: 1889] – Salem OR: [s.n.] [daily ex mon] – 1 – (cont: oregon statesman [salem, or: 1888]; cont by: oregon daily statesman [salem, or: 1896]) – us Oregon Lib [071]

Daily oregon statesman (salem, or: 1900) *see* Weekly oregon statesman (salem, or: 1900)

Daily oregon statesman [salem, or: 1900] – Salem OR: Statesman Pub Co, 1900-16 [daily ex mon] – 1 – (related to wkly ed: weekly oregon statesman; cont: oregon statesman [salem, or: 1898]; cont by: oregon statesman [salem, or: 1916]. occasional suppl) – us Oregon Lib [071]

Daily oregon unionist – Salem OR: S A Clarke, [daily ex mon] – 1 – (related to: weekly oregon unionist [1869]. cont: daily american unionist [1868-69]; cont by: daily oregon statesman [1869-77]) – us Oregon Lib [071]

Daily oregon unionist *see* Weekly oregon unionist

Daily orleanian – New Orleans LA. 1848 sep 12 – 1r – 1 – mf#861333 – us WHS [071]

Daily outlook – Beloit WI. 1882 jan 5 – 1r – 1 – mf#955489 – us WHS [071]

Daily palo alto times – Palo Alto CA. v49 n57 [1941 mar 7] – 1r – 1 – mf#931320 – us WHS [071]

Daily paragraph / Tuscarawas Co. Dennison – (6/1912-11/1933, 2/1935-4/1943) fire damage [daily] – 21r – 1 – mf#B30886-30906 – us Ohio Hist [071]

Daily patriot / Columbiana Co. Lisbon – sep 1907-feb 1908 [daily] – 1r – 1 – mf#B30150 – us Ohio Hist [071]

Daily patriot – Madison WI. 1854 nov 1-27 – 1r – 1 – (cont by: daily wisconsin patriot) – mf#939266 – us WHS [071]

Daily patriot. (daily wisconsin patriot.-wisconsin daily patriot) – Madison, Wisconsin. 1 Nov 1854-31 Dec 1858; 31 Mar 1859-9 Nov 1864.-d. 24 reels – 1 – uk British Libr Newspaper [071]

The daily people – Lincoln, NE: W A Howard, jan 1891 (daily) [mf ed jan22-25 1891 filmed [1979]] – 1r – 1 – us NE Hist [071]

Daily people's champion *see* Gunnison county miscellaneous newspapers

The daily people's champion *see* Gunnison county miscellaneous newspapers

Daily people's press – Milwaukee WI. 1860 sep 10, oct 7,20, nov 13,17, 1960 aug 16-dec 4 – 2r – 1 – (cont by: daily milwaukee news [milwaukee wi: 1856]; daily people's press and news) – mf#1138096 – us WHS [071]

Daily phoenix – Saskatoon, Canada. 26 jan 1912-27 aug 1914, 2 oct 1914-8 mar 1915, 20 may 1915-14 feb 1917, 23 aug 1917-21 may 1918, jun 1919-31 jul 1920, 13 jan 1921-30 jun 1922 – 65 1/2r – 1 – (aka: saskatoon phoenix) – uk British Libr Newspaper [071]

Daily phoenix – Saskatoon. Canada. Oct 1902-Sept 1928 – 1 – cn Commonwealth Imaging [071]

Daily picayune – New Orleans LA. 1893 feb 13, 14, 1909 feb 22 – 1r – 1 – (cont: picayune; cont by: times-democrat; times-picayune; times-democrat and the daily picayune; times-democrat) – mf#839978 – us WHS [071]

Daily plebian – New York. Jun. 27, 1842-May 12, 1845 – 1 – us NY Public [071]

Daily post – Bangalore, India. jan 1887-dec 1888 [wkly] – 6r – 1 – uk British Libr Newspaper [079]

Daily post – Hobart, Australia, 11 Jan 1910-22 Feb 1917.-d. 28 reels – 1 – uk British Libr Newspaper [072]

The daily post – Pittsburgh, PA., 1869 – 13 – $25.00r – us IMR [071]

The daily post – Plattsmouth, NE: Fellows & Marshall. v1 n1. mar 12 1898-dec 1900// (daily ex sun) [mf ed with gaps filmed 1979] – 2r – 1 – us NE Hist [071]

The daily post – Rotorua, NZ. 27 dec 1957-17 may 1958; 5 oct 1960-28 feb 1961; 21 mar 1962-14 sep 1962; jan 1973-dec 1973; jun 1974-dec 1976; feb 1977-mar 2002 – 1 – (previous title: rotorua post) – mf#17.1 – nz Nat Libr [079]

Daily post (a) – [NW England] Liverpool Record Off jun 1855-oct 1879 – 1 – (title change: liverpool daily post (a) [29 oct 1879-oct 1904]; liverpool daily post & mercury [14 nov 1904-5 jun 1916]; liverpool post & mercury [6 jun 1916-24 jan 1935]; liverpool daily post (b) [25 jan 1935-1980]; daily post (b) [1981]) – uk MLA; uk Newsplan [072]

Daily post reminder *see* Garfield county miscellaneous newspapers

The daily pratap – Jullundur, India. Jul 1966-1980 – 52r – 1 – us L of C Photodup [079]

Daily press – Ashland WI. 1966 may 2/jul 28-2000 dec – 317r – 1 – (cont: ashland daily press) – mf#1137351 – us WHS [071]

Daily press – La Crosse WI. 1891 jan 12-1891 sep 12, dec 30, 1893 feb 16-1893 oct 23 – 4r – 1 – (cont: la crosse daily press [la crosse wi: 1889]; cont by: la crosse daily press [la crosse wi: 1893]) – mf#934798 – us WHS [071]

Daily press – Dannevirke, NZ. dec 1904-jun 1906; oct 1906-dec 1906; apr 1907-dec 1907, jul-sep 1908 – 19r – 1 – mf#35.8 – nz Nat Libr [079]

Daily press / Hamilton Co. Cincinnati – mar-jun 1859, mar 1860-feb 1862 [daily] – 3r – 1 – mf#B5611-5613 – us Ohio Hist [071]

Daily press / Montgomery Co. Dayton – jul-dec 1902 [daily] – 2r – 1 – mf#B34412-34413 – us Ohio Hist [071]

Daily press *see* [Santa barbara-] santa barbara news press

Daily prout star – Kansas City MO. v1 n1-v4 n87 [1983 sep 2-1987 jun 12] – 1r – 1 – (cont: prout star [kansas city, mo.: 1982]; cont by: prout star [kansas city mo: 1986]) – mf#1313950 – us WHS [071]

Daily racine advocate – Racine WI. 1855 jan 1-mar 23 – 1r – 1 – (cont: daily morning advocate) – mf#3476592 – us WHS [071]

Daily ranchero – Brownsville TX. 1866 feb 2,9, nov 4 – 1r – 1 – (cont by: daily ranchero and republican) – mf#858170 – us WHS [071]

The daily rebel – aug 9 1862 – (= ser Confederate newspapers) – 1 – us Western Res [071]

Daily record – Chicago IL. 1938 aug 31-oct 22 [v1 n166-210], 1938 oct 24-1939 jul 31, 1939 aug 1-nov 13 [v2 n140-228] – 3r – 1 – (cont: midwest daily record [chicago il: home ed]; continued by: record weekly) – mf#1010763 – us WHS [071]

Daily record – Rockhampton, Australia. Jan-27 apr 1894; may 1897-8 nov 1901; 5 apr 1902-1 nov 1912; 1913-23 dec 1915; 1916-29 jul 1922 – 83 1/2r – 1 – uk British Libr Newspaper [072]

Daily record – Stroudsburg, PA. -d 1909-1920; 1943-1950 – 13 – $25.00r – us IMR [071]

The daily record – Alma, NE: Ed J Mock. 1v. oct 1-v1 n35. nov 10 1894 (daily ex sun) – 1r – 1 – us NE Hist [071]

Daily record abstract : official organ of the master builders' association – Portland OR: Master Builders' Assoc, [daily ex sun] – 1 – (began in 1927?) – us Oregon Lib [071]

The daily record and judicial news – Montreal: J Daoust and Fulton and Richard, [1889?-189- or 19–] – 9 – mf#P04745 – cn Canadiana [348]

Daily recorder / Methodist Church of Canada. General Conference – Toronto: [s.n, 1874-18- or 19–] – 9 – (cont: the wesleyan daily recorder) – mf#P04352 – cn Canadiana [242]

Daily reese river reveille – Austin NV. 1864 aug 5,11 – 1r – 1 – mf#860070 – us WHS [071]

Daily register – Augusta GA. 1864 sep 16,29, oct 1,4-5,10 – 1r – 1 – mf#846064 – us WHS [071]

Daily register – Marietta OH. 1898 oct 13 – 1r – 1 – (cont: marietta tri-weekly register; cont by: marietta daily leader; register-leader) – mf#5699707 – us WHS [071]

Daily register – Portage WI. [1982 may 3/15]-[2004 dec 16/31] – 333r – 1 – (cont: portage daily register [portage wi: 1960]) – mf#114085 – us WHS [071]

Daily register / Lawrence Co. Ironton – (1/1912-3/13,1-3/16,7/20-1925) [daily] – 20r – 1 – mf#B2281-2300 – us Ohio Hist [071]

Daily register / Lawrence Co. Ironton – 1901, may 1905, (jul 1912-dec 1920) damaged [daily] – 22r – 1 – mf#B32939-32960 – us Ohio Hist [071]

Daily register / Lawrence Co. Ironton – (1916-sep 1920), oct 1922-mar 1924 (fire damaged) [daily] – 17r – 1 – mf#B29586-29602 – us Ohio Hist [071]

Daily register / Washington Co. Marietta – jan-dec 1898 [daily] – 3r – 1 – mf#B32883-32885 – us Ohio Hist [071]

Daily register-mail – Galesburg IL. 1911 jun 17, 1940 may 16 – 1r – 1 – us WHS [071]

Daily report – 1943 jan 16-dec 3, 1944 apr 24-may 12, mar 10-29, mar 30-apr 22 – 4r – 1 – (cont: daily report [united states. foreign broadcast monitoring service]; cont by: daily report, foreign radio broadcasts. european section; daily report, foreign radio broadcasts. far eastern section; daily report, foreign radio broadcasts. latin america section) – mf#2991526 – us WHS [071]

Daily report *see* London stock market report

The daily report – Fullerton, NE: R G Adams. v1 n1. amr 13 1903- (daily ex sun) [mf ed filmed [1990]] – 1r – 1 – us NE Hist [071]

Daily report: supplement / U.S. Foreign Broadcast Information Service – 1960-69 – 1 – us L of C Photodup [324]

Daily reporter – Grand Rapids, Wisconsin Rapids WI. 1906 oct/dec-1916 dec 28/1917 may 31 – 27r – 1 – (cont by: daily leader [wisconsin rapids wi]) – mf#952193 – us WHS [071]

Daily reporter – Fond Du Lac WI. 1901 apr/jun-1921 jul 1/25 – 45r – 1 – (cont: daily reporter and the fond du lac daily journal; cont by: fond du lac daily reporter) – mf#941539 – us WHS [071]

Daily reporter / Franklin Co. Columbus – jan 1984-dec 1985 [daily] – 6r – 1 – mf#B30330-30335 – us Ohio Hist [340]

Daily reporter / Franklin Co. Columbus – jan 1986-dec 1987 – 8r – 1 – mf#B29610-29617 – us Ohio Hist [340]

Daily reporter / Franklin Co. Columbus – mar 1897-1901,(1902-33),1933-83 1 – 18r – 1 – mf#B22913-23020 – us Ohio Hist [340]

Daily reporter / Franklin Co. Columbus – v1 n1. oct 1896-feb 1897 [daily] – 1r – 1 – mf#B23117 – us Ohio Hist [340]

Daily reporter / Tuscarawas Co. Dover – (11/1905-6/1921, 7/1922-2/1954) [daily] – 117r – 1 – mf#B4500-4618 – us Ohio Hist [071]

Daily reporter – Milwaukee WI. 1903 jan 3/apr 3-1995 nov/dec – 205r – 1 – (with small gaps) – mf#821955 – us WHS [071]

Daily reporter – Fond Du Lac WI. 1883 mar 31/oct 12-1893 nov 23/1894 jun 7 – 20r – 1 – (with small gaps; cont by: fond du lac daily journal; daily reporter and the fond du lac daily journal) – mf#941534 – us WHS [071]

Daily reporter *see* Miscellaneous newspapers of las animas county, reel 2

Daily reporter and the fond du lac daily journal – Fond Du Lac WI. 1894 may 11/jun7-1901 jan/mar 30 – 15r – 1 – (cont: daily reporter [fond du lac wi: 1883]; fond du lac daily journal; cont by: daily reporter [fond du lac wi: 1901]) – mf#941536 – us WHS [071]

Daily reporter (mcminnville or) *see* Yamhill reporter

Daily reporter [mcminnville, or] – McMinnville OR: D C Ireland & Co, 1886- [daily ex sun] – (ceased in 1887; related to: yamhill reporter; absorbed by: yamhill county reporter [mcminnville, or]) – mf#2144709619 – us Oregon Lib [071]

Daily reports / U.S. Foreign Broadcast Information Service – 4Sep 1941-Mar 1974. Scattered issues wanting – 1 – 8,576.00 – us L of C Photodup [324]

Daily reports of u.s. secret service agents, 1875-1936 / U.S. Secret Service – (= ser Records Of The U.S. Secret Service) – 836r – 1 – mf#T915 – us Nat Archives [360]

Daily representative – Queenstown SA, 4 nov 1865-30 dec 1939 – 117r – 1 – sa National [079]

Daily republican – Broken Bow, NE: D M Amsberry (daily ex sun) [mf ed v4 n233. dec 21 1891) filmed [1989]] – 1r – 1 – (cont: broken bow daily republican. cont by: broken bow daily republican (1892)) – us NE Hist [071]

Daily republican / Greene Co. Xenia – feb 1914-jul 1915 [daily] – 4r – 1 – mf#B10379-10382 – us Ohio Hist [071]

Daily republican / Lawrence Co. Ironton – 1892-94, jul 1895-99 (damaged) [daily] – 18r – 1 – mf#B32860-32877 – us Ohio Hist [071]

Daily republican – Monongahela, PA. -d 1881-1970 – 13 – $25.00r – us IMR [071]

Daily republican / Portage Co. Ravenna – jun 1886-jan 1888 [daily] – 2r – 1 – mf#B4194-4195 – us Ohio Hist [071]

Daily republican *see* [Fresno-] morning republican

The daily republican – Auburn, NE: J H Dundas. v1 n1. apr 19-may 1886// (daily ex sun) [mf ed 1973] – 1r – 1 – us NE Hist [071]

The daily republican – Charleston: [s.n.], aug 19 1869-sep 8 1871 – 3r – (filmed with: charleston daily republican) – us CRL [071]

The daily republican – Pawnee City, NE: A E & Roy D Hassler. 1v. may 18-v1 n149 [ie 150] nov 10 1894 (daily ex sun) [mf ed with gaps filmed [1968]] – 1r – 1 – (weekly ed: pawnee republican 1894) – us NE Hist [071]

The daily republican – Weeping Water, NE: J K Keithley. 1st yr n1. aug 21 1894- (daily) [mf ed -aug 23 1894 filmed 1999] – 1r – 1 – us NE Hist [071]

Daily republican and leader – La Crosse WI. 1871 aug 16/dec 6-1880 apr 22-may 29 – 17r – 1 – (cont: republican and leader [la crosse wi: 1871: daily]; cont by: la crosse republican-leader) – mf#930857 – us WHS [071]

Daily republican and leader – La Crosse WI. 1898 jul 16/1898 dec 20-1903 jan 20/1903 jun 13 – 9r – 1 – (cont: republican and leader [la crosse wi: 1890]; cont by: la crosse daily press [la crosse wi: 1893]; la crosse leader and press) – mf#934685 – us WHS [071]

Daily republican and news – Milwaukee WI. 1881 jan 3-apr 19, apr 20-aug 6, aug 8-nov 23, nov 24-1882 mar 13, mar 14-may 20 – 5r – 1 – (cont: milwaukee daily news [milwaukee wi: 1874]; cont by: milwaukee daily sentinel [milwaukee wi: 1862]; daily republican-sentinel) – mf#1125327 – us WHS [071]

Daily republican (corvallis, or) *see* – Benton county republican (corvallis, or: 1906)

Daily republican [corvallis, or] – Corvallis OR: W E Smith, -1915 [daily] – 3r – 1 – (related to wkly ed: benton county republican [corvallis, or: 1906], 1912-1915; merged with: benton county republican [corvallis, or: 1906], to form: benton county courier [corvallis, or: 1915]) – us Oregon Lib [071]

Daily republican-news – Hamilton, OH. 9 oct 1915-6 sept 1919 (scattered) – 1r – 1 – (daily republican newspaper) – us Western Res [071]

Daily republican-sentinel – Milwaukee WI. 1882 may 22-jul 18, jul 19-sep 26, sep 27-dec 8, dec 9-dec 30 – 4r – 1 – (cont: milwaukee daily republican and news; milwaukee daily sentinel [milwaukee wi: 1873]; cont by: milwaukee sentinel [milwaukee wi: 1883]) – mf#1109225 – us WHS [071]

The daily review – Towanda, PA. -d 1944-79; 1972-76. 18 rolls – 13 – $25.00r – us IMR [071]

Daily review of the arabic press – Cairo: American Embassy, 1947-nov 23 1954 – 12r – us CRL [079]

Daily review of the arabic press / U.S. Embassy. United Arab Republic – 1956-61. 10 reels. Scattered issues lacking – 1 – us L of C Photodup [956]

Daily review of the baghdad press – Baghdad: Embassy of the USA, sep 1-12, sep 30-dec 19 1951; jan 30-feb 18, mar 2-3 1952; sep 14,16,18-20,22 1954 – 1r – us CRL [079]

Daily richmond enquirer – Richmond VA. 1862 aug 5-1865 mar 25 [incomplete] – 1r – 1 – (cont by: daily richmond examiner; daily enquirer and examiner) – mf#887874 – us WHS [071]

Daily richmond examiner – Richmond VA. 1861 jul 16-1866 aug 29, 1862 feb 24, 1863 oct 31, nov 12, 1864 oct 3,26 – 2r – 1 – (cont by: richmond examiner; daily enquirer and examiner; daily richmond enquirer) – mf#851609 – us WHS [071]

Daily richmond whig – Richmond: J H Pleasants, M Smith and W Ramsay, mar 3 1829-feb 10 1831 – 2r – us CRL [071]

Daily roseburg review *see* Roseburg review (roseburg, or)

Daily saratogian – Saratoga Springs NY. 1883 may 21-30 – 1r – 1 – mf#860966 – us WHS [071]

Daily scioto gazette – Chillicothe, OH. feb 18, 1850-feb 28, 1857 – 13r – 1 – (daily edition of this whig, later republican newspaper) – us Western Res [071]

Daily scotsman – Edinburgh. Scotland. -d. 1855-59. (11 reels) – 1 – uk British Libr Newspaper [072]

The daily senator – Scottsbluff, NE: Carpenter Pub Co. 2v. v9 n24. aug 31 [1936]-v10 n84. aug 14 1937 (daily ex sun) [mf ed with gaps filmed 1977] – 3r – 1 – (cont: lyman ledger. absorbed by: scottsbluff daily star-herald) – us NE Hist [071]

Daily sentinel – Milwaukee WI. 1844 dec 10-27 – 1r – 1 – (cont by: milwaukie daily sentinel) – mf#1109145 – us WHS [071]

Daily sentinel / Jackson Co. Wellston – jul-dec 1910 [daily] – 1r – 1 – mf#B25296 – us Ohio Hist [071]

Daily sentinel / Meigs Co. Pomeroy – jan-mar 1971 [daily] – 1r – 1 – mf#B31486 – us Ohio Hist [071]

Daily sentinel / Perry Co. New Lexington – jan, sep 1929-sep 1931 [daily] – 4r – 1 – mf#B11280-11283 – us Ohio Hist [071]

Daily sentinel *see* Miscellaneous newspapers of las animas county, reel 2

Daily sentinel and gazette – Milwaukee WI. 1846 feb 16-aug 6, aug 7-1847 feb 1, feb 2-aug 5, aug 6-1848 feb 10, feb 11-aug 17, aug 18-sep 30 – 6r – 1 – (cont: milwaukie daily sentinel; cont by: milwaukee sentinel and gazette [milwaukee wi: 1848]) – mf#1109181 – us WHS [071]

Daily sentinel series / Jackson Co. Wellston – (1901-jun 1940) [daily] – 33r – 1 – (about 50% missing) – mf#B8106-8138 – us Ohio Hist [071]

Daily service – Lagos, Nigeria. -d. 1 Jan-15 May 1954; 1 Jan 1955-17 Sept 1960. Imperfect. 19 reels – 1 – uk British Libr Newspaper [079]

The daily service – Lagos [Nigeria]: Service Press Ltd., [-1960] – (issues filmed as pt of: st clair drake collection of africana (apr 22 1954; aug 20, sep 5 1955; oct 18 1957; dec 8,12-13,30 1958)) – us CRL [071]

Daily sheboygander – Sheboygan WI. 1894 may 8 – 1r – 1 – mf#927018 – us WHS [071]

Daily shipping and commercial news – Shanghai, China. -d. 27 jan-13 dec 1862 – 1/2r – 1 – uk British Libr Newspaper [072]

Daily shipping news – Portland, OR: Daily Shipping News, dec 17 1956-apr 29 1994 – 1 – (aka: shipping news) – us Oregon Hist [380]

Daily shipping news – Portland OR, 1888-1995 – 1 – us Oregon Lib [380]

Daily silver state *see* [Winnemucca-] silver state

Daily sketch [irish insurrection] – London 1916, 1922 – 1r – 1 – ie National [072]

The daily southern crisis – Jackson, MS: Jones, Wisely & Co, jan 2-mar 30 1863 – 1r – 1 – us CRL [071]

Daily southern cross *see* Southern cross

Daily southern standard – New Orleans LA. 1849 jul 10 – 1r – 1 – mf#861797 – us WHS [071]

Daily southwest *see* Miscellaneous newspapers of la plata county, colorado

Daily spectator – Hamilton, Canada. oct-dec 1875, 3 jul 1876-30 nov 1877, 2 apr-28 apr 1883, feb-26 mar 1952 – 6 1/4r – 1 – (aka: hamilton spectator) – uk British Libr Newspaper [071]

Daily standard – Ontario, Canada. 25 jun 1912-24 jan 1918, 6 nov-31 dec 1918, 2 jan 1920-oct 1921 [imperfect] – 49 1/2r – 1 – (aka: kingston daily standard) – uk British Libr Newspaper [071]

Daily standard – Crete, NE: M A Daugherty, jun 25 1883 (daily) [mf ed v1 n2, jun 26 1883 filmed 1957] – 1r – 1 – us NE Hist [071]

Daily standard / Mercer Co. Celina – jan 1959-dec 1980 [daily] – 15r – 1 – mf#B12510-12614 – us Ohio Hist [071]

Daily standard / Mercer Co. Celina – jan 2 1981-dec 31 1990 [daily] – 60r – 1 – mf#B31492-31551 – us Ohio Hist [071]

Daily standard / Mercer Co. Celina – jul 1935, jul 1936-dec 1958 [daily] – 59r – 1 – mf#B8593-8651 – us Ohio Hist [071]

Daily standard – Regina, Canada. -d. 20 dec 1906-1912; 15 jan-5 aug; 5 sep 1913 – 40r – 1 – uk British Libr Newspaper [072]

Daily standard [portland, or] – Portland [Or]: Standard Pub Co [daily ex mon] – 1 – (began in apr 1877; cont: daily evening standard [portland, or]; cont by: morning standard [portland, or]; related to: weekly standard [portland, or]) – us Oregon Lib [071]

Daily star / Hancock Co. Findlay – v1 n1. aug 1882-apr 1883,oct 1883-apr 1884 [daily] – 2r – 1 – mf#B10502-10503 – us Ohio Hist [071]

Daily star – Madison WI. 1877 mar 19-nov 17 – 1r – 1 – mf#939556 – us WHS [071]

Daily star / Marion Co. Marion – jul-nov 1906, dec 1907-mar 1908 [daily] – 2r – 1 – mf#B2641-2642 – us Ohio Hist [071]

Daily star – [NW England] Manchester ALS 2 nov 1978-jun 1985 – 1 – (title change: star [jul 1985-2 may 1987) – uk MLA; uk Newsplan [072]

The daily star = Bintang harian – Brunei Town, Brunei: The Daily Star [1966- (daily) [mf ed 2005 Singapore: National Library] – 23r – 1 – (in english & malay; sunday iss publ with title: sunday star) – mf#mf-9863 seam – us CRL [079]

Daily star series / Hamilton Co. Cincinnati – jan 1875-jun 1880 [daily] – 8r – 1 – mf#B10171-10178 – us Ohio Hist [071]

Daily state democrat – Lincoln, NE: Daily State Democrat Co, jun 9 1879-v10 n19. jun 30 1888 (daily ex sun) [mf ed 1879-88 (gaps)] – 5r – 1 – (cont by: lincoln daily call. weekly ed: weekly state democrat) – us NE Hist [071]

Daily state journal – Lincoln, NE: Gere & Brownlee. v1 n3. jul 22 1870-8th yr n239. may 2 1878 (daily ex mon) [mf ed with gaps filmed 1958] – 10r – 1 – (cont: daily nebraska state journal. cont by: daily nebraska state journal. monthly suppls issued occasionally. weekly ed: nebraska state journal) – us NE Hist [071]

Daily state journal – Richmond VA. 1871 apr 1-sep 20 – 1r – 1 – (cont: daily state journal [alexandria, va.: 1868]; cont by: evening journal [richmond va: 1874]) – mf#885184 – us WHS [071]

Daily state sentinel – Indianapolis IN. 1861 jul 19 – 1r – 1 – (cont: daily indiana state sentinel [indianapolis in: 1861]; cont by: indianapolis daily herald) – mf#1147230 – us WHS [071]

Daily statesman – Columbus OH. v1 n1-11 [1837 sep 2-16] – 1r – 1 – (cont by: daily ohio statesman [columbus oh: 1837]) – mf#920048 – us WHS [071]

Daily status report / Cuban Refugee Resettlement Operation [Fort McCoy WI] – 1980 may 23-jun 19, jun 22-aug 29, aug 28-nov 1 – 3r – 1 – mf#512400 – us WHS [360]

Daily strength for daily living : twenty sermons on old testament themes / Clifford, John – London: E Marlborough, 1885 – 2mf – 9 – 0-7905-7707-0 – mf#1989-0932 – us ATLA [240]

Daily strike bulletin / International Union, United Automobile, Aircraft, and Agricultural Implement Workers of America – 1954 apr 5-1955 dec, 1956 jan 1-1959 aug 28 – 2r – 1 – (cont by: kohler strike and boycott bulletin) – us WHS [331]

Daily summary of the japanese press / U.S. Embassy. Tokyo, Japan – Feb 2, 1952- – 1 – us L of C Photodup [324]

Daily sumpter reporter – Sumpter OR: J Nat Hudson, [daily ex sun] – 1 – us Oregon Lib [071]

Daily sun – Columbus GA. 1855 aug 16 – 1r – 1 – mf#846099 – us WHS [071]

Daily sun – Gainesville, FL. 1905 sep 1 – 1r – us UF Libraries [071]

Daily sun – St John, NB. 1878-1910 – 113r – 1 – cn Library Assoc [071]

The daily sun – South Omaha, NE: South Omaha Printing Co, nov 1895-may 1901// (daily ex sun) [mf ed with gaps filmed 1975-79] – 5r – 1 – (cont by: south omaha daily times. numbering ceased with jun 27 1896. weekly ed: weekly sun (south omaha) 1900-01) – us NE Hist [071]

The daily sun – Humboldt, NE: [s.n.] n1. sep 3 1919- (daily) [mf ed -sep 6 1919 filmed 1999] – 1r – 1 – (issued in the interest of county seat removal to humboldt) – us NE Hist [071]

Daily sun and press – Jacksonville, FL. 1877 jun 16-dec 21 – 1r – (missing: 1877 dec 20) – us UF Libraries [071]

Daily sun and press – Jacksonville, FL. 1877 dec 28-1878 jun 12 – 1r – (missing: 1878 jan 1-5; apr 1) – us UF Libraries [071]

Daily sun journal – Brooksville, FL. 1988 apr-1992 jun – 11r – (gaps) – us UF Libraries [071]

Daily sun-journal – Brooksville, FL. 1981 jan-1989 dec – 26r – (gaps) – us UF Libraries [071]

Der daily telegraf – New York. N.Y. The daily telegraph. 1894 – 1 – us AJPC [071]

Daily telegram / Champaign Co. Mechanicsburg – jan 2 1936-aug 31 1938 – 3r – 1 – mf#B 41417-41419 – us Ohio Hist [071]

Daily telegram – Columbus, NE: Telegram Co. 23v. [60th yr] n12. jan 16 1939-82nd yr n126. may 29 1961 (daily ex sun) – 14r – 1 – (cont: columbus daily telegram; cont by: columbus daily telegram (1961)) – us Bell [071]

Daily telegram – Superior WI. 1995 nov 1/15-2003 dec 16/31 – 165r – 1 – (cont: evening telegram [superior wi]) – mf#3356628 – us WHS [071]

Daily telegram – Eau Claire WI. [1897 jun 21/nov 17]-[1952 jan/feb] – 298r – 1 – (cont: morning telegram [eau claire wi]; daily free press [eau claire wi]; cont by: eau claire leader; eau claire leader-telegram) – mf#1139761 – us WHS [071]

Daily telegram – [Scotland] Glasgow: J McCallum 19 sep-1 oct 1864 (daily) [mf ed 2003] – 12v on 1r – 1 – uk Newsplan [072]

Daily telegram / Summit Co. Akron – v1 n1. oct 1888-sep 1889 [daily] – 2r – 1 – mf#B11297-11298 – us Ohio Hist [071]

Daily telegram – Washington DC. 1877 aug 8 – 1r – 1 – mf#851210 – us WHS [071]

The daily telegram – Columbus, NE: Telegram Co. 23v. [60th yr] n12. jan 16 1939-82nd yr n126. may 29 1961 (daily ex sun) [mf ed -1957 (lacks mar 23 1946)] – 56r – 1 – (cont: columbus daily telegram. cont by: columbus daily telegram (1961)) – us NE Hist [071]

The daily telegram – Columbus, NE: Telegram Co. 23v. [60th yr] n12. jan 16 1939-82nd yr n126. may 29 1961 (daily ex sun) [mf ed -1957 (lacks mar 23 1946)] – 56r – 1 – (cont: columbus daily telegram. cont by: columbus daily telegram (1961)) – us NE Hist [071]

Daily telegram for cambridgeshire and isle of ely, norfolk, lincolnshire, etc – [East Midlands] Wisbech, Cambridgeshire n1 [24 apr]-n123 [15 sep 1877] [mf ed 2004] – 1r – 1 – (discontinued) – uk Newsplan [072]

Daily telegrams – Port Blair, India. Nov 1968-JUl 1994 – 43r – 1 – us L of C Photodup [079]

Daily telegraph – Sydney, Australia. mar 1941 – 1r – 1 – (aka: daily telegraph and daily news/ telegraph) – uk British Libr Newspaper [072]

Daily telegraph – Launceston, Australia. 11 aug 1833-88, 2 jun 1894, 13 jun 1900-29 feb 1928 [daily] – 125r – 1 – (aka: daily telegraph news pictorial) – uk British Libr Newspaper [079]

Daily telegraph – Kitchener, Canada. 31 may 1913-15 aug 1914, 28 sep 1914-sep 1916, 19 feb 1917-sep 1920 – 38 1/2r – 1 – (aka: kitchener daily telegraph) – uk British Libr Newspaper [071]

Daily telegraph – Canada. -d. 12 jan 1903-2 feb 1919, 3 mar-21 aug 1919, sep 1919-25 feb 1922, mar 1922-1924 – 128 1/2r – 1 – (aka: quebec telegraph) – uk British Libr Newspaper [071]

Daily telegraph / Ashtabula Co. Ashtabula – v1 n2. may 14-sep 11 1884 [daily] – 1r – 1 – mf#B20722 – us Ohio Hist [071]

Daily telegraph – Napier, NZ. jan-dec 1881, sep-dec 1882, jan 1883-jun 1887, jan 1888-dec 1889, jul 1890-dec 1900, jul-nov 1901, sep 1916, jan 1917-dec 1923, jul 1924-dec 1934;1 jul-29 aug 1966, 30 sep-dec 1966, 15 may-30 jun 1967, apr 1975, oct 1976, apr 1977-may 1999// – 1 – (ceased publ 1 may 1999; merged with hawkes bay herald tribune to form hawkes bay today) – mf#31.1 – nz Nat Libr [079]

Daily telegraph – North Platte, NE: H W Hill (daily ex sun) [mf ed v4 n31. jan 31 1899-jun 2 1900] – 1r – 1 – (cont: north platte daily telegraph. cont by: evening telegraph) – us NE Hist [071]

Daily telegraph – Lagos, Nigeria. -d. 12 July 1958-31 Dec 1959; 2 Jan 1963-31 Dec 1965. Imperfect. 11 reels – 1 – uk British Libr Newspaper [072]

Daily telegraph – London. -d. 1855-1927. (491 reels) – 1 – uk British Libr Newspaper [072]

Daily telegraph – Melbourne, Australia. Feb 1869-May 1892.-d. 56mqn reels – 1 – uk British Libr Newspaper [072]

Daily telegraph – Napier. New Zealand. -d. Jan 1902-Sep 1915. (56 reels) – 1 – uk British Libr Newspaper [072]

Daily telegraph – Savannah GA. 1840 sep 29 – 1r – 1 – mf#846044 – us WHS [071]

Daily telegraph – Sydney, jul 1879-aug 1942 – 271r – A$8943.00 vesicular A$10,433.50 silver – at Pascoe [079]

Daily telegraph – Sydney, sep 1942-feb 1954, nov 1954-dec 1964 – 223r – 1 – A$14,076.77 vesicular A$15,303.16 silver – at Pascoe [079]

Daily telegraph *see* St john daily telegraph and morning journal

The daily telegraph – Waihi, NZ. jul 1904-dec 1907; jul-dec 1908; jani-jun 1909; jul 1918-jun 1923; jul 1923-41;1943-44 – 40r – 1 – (previous title: the waihi daily telegraph. title change: the daily telegraph on 2 nov 1908; title change: the waihi telegraph on 9 jun 1923) – mf#16.44 – nz Nat Libr [079]

Daily telegraph and daily news/telegraph *see* Daily telegraph

Daily telegraph and deccan herald – Bangalore, India. jan 1878-dec 1889 – 41r – 1 – (cont: deccan herald) – uk British Libr Newspaper [079]

Daily telegraph and morning post – [NW England] Manchester ALS oct 1940-1954 – 1 – (title change: daily telegraph [1955]) – uk MLA; uk Newsplan [072]

Daily telegraph and witness. (daily telegraph) – Montreal, Canada. 12 jul-29 nov 1913 – 4 1/2r – 1 – uk British Libr Newspaper [071]

Daily telegraph colour supplement – London, UK. 25 Sept 1964-1969; 1974. -w. 26 reels – 1 – uk British Libr Newspaper [072]

Daily telegraph & deccan herald *see* Deccan herald

Daily telegraph news pictorial *see* Daily telegraph

The daily telegraph / the sunday telegraph – London. England. 1945- mthly updates – (= ser The Sunday Telegraph) – 1 – (the sunday telegraph was first publ in 1961) – us Primary [072]

Daily tidings – Ashland OR, 1970-93 [daily ex sun] – 1 – (cont: ashland daily tidings [1919-70]; cont by: ashland daily tidings [1993-]) – us Oregon Hist [071]

Daily times / Belmont Co. Martins Ferry – jan 1926-mar 1943 (not B2697-2698) [daily] – 61 – 1 – mf#B2658-2720 – us Ohio Hist [071]

Daily times – Green Bay WI. 1899 nov26-1900 jun 8 – 1r – 1 – (cont by: green bay news and social mirror; green bay daily herald) – mf#918905 – us WHS [071]

Daily times – Leavenworth KS. 1864 feb 23 – 1r – 1 – (cont by: leavenworth daily conservative; times and conservative) – us WHS [071]

Daily times – Neenah WI. 1882 aug 8-9 – 1r – 1 – (cont by: neenah daily times) – mf#1042885 – us WHS [071]

Daily times – South Omaha, NE: Mayfield Bros & Worley. n1. sep 3 1900-1900// (daily ex sun) [mf ed -nov 16 1900 (gaps) filmed 1982] – 1r – 1 – (cont by: south omaha daily times) – us NE Hist [071]

Daily times – South Omaha, NE: Festner Print Co. 1v. v16 n84-[312]. apr 4-dec 31 1902 (daily ex sun) [mf ed filmed 1975-[79]] – 3r – 1 – (cont: south omaha daily times. absorbed by: nebraska daily democrat. weekly ed: magic city hoof and horn apr-dec 1902) – us NE Hist [071]

Daily times / Coshocton Co. Coshocton – jan 1910-sep 1913 [daily] – 9r – 1 – mf#B34428-34436 – us Ohio Hist [071]

Daily times / Franklin Co. Columbus – mar 1880-jun 1888 [daily] – 20r – 1 – mf#B11762-11781 – us Ohio Hist [071]

Daily times / Highland Co. Greenfield – dec 1952-oct 1958, mar 1919-82 [daily] – 56r – 1 – mf#B12638-12693 – us Ohio Hist [071]

Daily times / Highland Co. Greenfield – jan 3 1983-dec 31 1991 [daily] – 28r – 1 – mf#B31712-31739 – us Ohio Hist [071]

Daily times – Lagos: The Nigerian Print. & Pub Co Ltd, [1949-jan 2 1956-] – 1 – us CRL [079]

Daily times / Montgomery Co. Dayton – mar 1890-dec 1897 [daily] – 14r – 1 – mf#B5122-5135 – us Ohio Hist [071]

Daily times – Stroudsburg, PA. -d 1894-1908 – 13 – $25.00r – us IMR [071]

Daily times / Trumbull Co. Niles – (8/1924-8/1927), 1/1933-11/1957 (damaged iss) [daily] – 99r – 1 – mf#B31804-31902 – us Ohio Hist [071]

Daily times / Trumbull Co. Niles – oct 1927-dec 1932 [daily] – 16r – 1 – mf#B27933-27948 – us Ohio Hist [071]

Daily times *see* [Los gatos-] times

The daily times – Lehighton, PA. -d 1921-26 – 13 – $25.00r – us IMR [071]

The daily times – Mauch Chunk, PA. -d 1913-1915. 1 roll – 13 – $25.00r – us IMR [071]

Daily times and bonanza *see* [Tonopah-] times-bonanza and goldfield news

Daily times [baltimore, md] – Baltimore [MD]: F X Lipp & Co, [daily ex sun] [mf ed 1997] – 1r – 1 – (began with apr 26 1852 iss? ceased in oct 1852; filmed with: herald of freedom and torch light) – us Oregon Lib [071]

Daily times journal – Fort William, Canada. -d. 2 mar 1914-18 mar 1919; 5 jun 1919-30 jun 1922 – 49r – 1 – uk British Libr Newspaper [071]

Daily times [portland, or] – Portland OR: R D Austin & Co, [daily ex sun] – 1 – (ceased in 1863; cont by: oregon daily times [1863-]) – us Oregon Lib [071]

Daily times recorder / Muskingum Co. Zanesville – jan-jun 1892 [daily] – 1r – 1 – mf#B32918 – us Ohio Hist [071]

Daily times series / Champaign Co. Urbana – v1 n1. mar 1893-nov 1895 [daily] – 6r – 1 – mf#B9565-9570 – us Ohio Hist [071]

Daily times series / Highland Co. Greenfield – 1/1935-52,7/58-6/59,83-85,92-12/93 [daily] – 35r – 1 – mf#B33973-34007 – us Ohio Hist [071]

Daily times-record – Valley City, ND: Greenwood & Houghtaling. v8 n80 mar 1 1915-v23 n254 apr 25 1928 (daily ex sun) – 1 – (official paper of barnes county 1916-1917, 1924-1925; official paper of barnes county and valley city 1926-1928; other ed available: weekly times-record; cont: evening times=record; cont by: valley city times-record (valley city, nd: 1928)) – mf#03526-03543 – us North Dakota [071]

Daily tribune – Plattsmouth, NE: G F S Burton, oct-dec1895// (daily ex sun): [mf ed v1 n6. oct 28-dec 20 1895 (gaps) filmed 1979] – 1r – 1 – (cont by: plattsmouth daily tribune) – us NE Hist [071]

Daily tribune – Manitowoc WI. 1858 may 31-dec 8, dec 9-1859 jun 25, jun 27-dec 19, dec 20-1860 may 31, 1860 jun-dec 3, dec 4-1861 may 31, jun-nov 5 – 7r – 1 – (cont by: tri-weekly tribune) – mf#1107955 – us WHS [071]

Daily tribune – South Omaha, NE: John M Tanner. 2v. 10th yr. jul 10 1900-11th yr. may 20 1901 (daily ex sun) [mf ed with gaps filmed 1973] – 2r – 1 – (cont: south omaha daily tribune. absorbed by: south omaha daily times) – us NE Hist [071]

Daily tribune – South Omaha, NE: A L Dennett and John M Tanner. -9th yr n121. jun 24 1899 (daily ex sun) [mf ed 1896-99 (gaps) filmed 1973-79] – 6r – 1 – (cont: south omaha daily tribune. cont by: south omaha daily tribune) – us NE Hist [071]

Daily tribune : daily newspaper – Warren, OH. 1 Dec 1885 – 1r – 1 – us Western Res [071]

Daily tribune / Darke Co. Greenville – (sep 1914-feb 1916),jun 1919-22 [daily] – 6r – 1 – mf#B11601-11606 – us Ohio Hist [071]

Daily tribune / Gallia Co. Gallipolis – 1895-1921, 1923-jun 1973 [daily] – 148r – 1 – mf#B1867-2015 – us Ohio Hist [071]

Daily tribune / Gallia Co. Gallipolis – jan 1990-sep 1993 [daily] – 15r – 1 – mf#B34837-34851 – us Ohio Hist [071]

Daily tribune / Gallia Co. Gallipolis – jul 1973-jun 1983 [daily] – 40r – 1 – mf#B3355-3394 – us Ohio Hist [071]

Daily tribune / Gallia Co. Gallipolis – jul 1983-dec 1989 [daily] – 26r – 1 – mf#B30595-30620 – us Ohio Hist [071]

Daily tribune / Gallia Co. Gallipolis – oct 1993-mar 1997 [daily] – 14r – 1 – mf#B36905-36918 – us Ohio Hist [071]

Daily tribune – Manitowoc WI. 1909 jun 26-1910 apr 7 – 1r – 1 – mf#1107107 – us WHS [071]

Daily tribune / Meigs Co. Pomeroy – jan 1931-jan 1941 [daily] – 20r – 1 – mf#B7918-7937 – us Ohio Hist [071]

Daily tribune – Nebraska City, NE: Horace G Whitmore. v1 n1. may 27 1901-v8 n126. nov 25 1907 (daily ex sun) [mf ed with gaps filmed 1977] – 7r – 1 – (merged with: nebraska daily press to form: nebraska daily press and the nebraska city daily tribune. publ as: nebraska city daily tribune jun 2-sep 6 1902. semi-wkly ed: nebraska city weekly) – us NE Hist [071]

Daily tribune *see*
– Miscellaneous newspapers of san miguel county

Daily true delta – New Orleans LA. 1862 jan 30, mar 11,18, apr 3,27, may 20-21,28,30, jun 29, jul 27, aug 2,5,19,31, sep 23, 1863 apr 2 – 1r – 1 – mf#859836 – us WHS [071]

The daily truth – Williamsport, PA., 1898 – 13 – $25.00r – us IMR [071]

Daily twin city news – Menasha, Neenah WI. 1881 sep 4 – 1r – 1 – mf#1043494 – us WHS [071]
Daily union / Columbiana Co. Wellsville – 1891-93,95-98,11/99-2/1904 (damaged) [daily] – 16r – 1 – mf#B7350-7365 – us Ohio Hist [071]
Daily union – Ottawa, ON. 1861-66 – 7r – 1 – cn Library Assoc [071]
The daily union – Crete, NE: M B C True. v1 n1-15. jun 25-sep 1 1883 (daily ex sun) – 1r – 1 – (v1 n1-8 jun 25-jul 3 1883 constitute the 1st installment; v1 n9-15 aug 22-sep 1 1883 constitute the 2nd installment. weekly ed: saline county union) – us NE Hist [071]
Daily union and democrat – La Crosse WI. 1859 nov 16-1860 jun 5 – 1r – 1 – (cont: la crosse daily union; la crosse national democrat; cont by: la crosse tri-weekly union and democrat) – mf#928175 – us WHS [071]
Daily union series / Columbiana Co. Wellsville – mar-dec 1904, 1906-13 (damaged) [daily] – 14r – 1 – mf#B6295-6308 – us Ohio Hist [071]
Daily union series / Wyandot Co. Upper Sandusk – 2/1896-2/97,3/18-2/26,5/29-1/1938 [daily] – 43r – 1 – mf#B7366-7408 – us Ohio Hist [071]
Daily universe – Provo UT 1977-78 – 1 – mf#9123 – us UMI ProQuest [378]
Daily variety – New York NY 1998+ – 1,5,9 – ISSN: 0011-5509 – mf#19166 – us UMI ProQuest [071]
Daily walk / Greater Holy Temple COGIC [Jacksonville FL] – 1983 nov – 1r – 1 – mf#4026179 – us WHS [243]
Daily war telegraph & general advertiser – [NW England] Manchester 21 oct 1854-30 nov 1855 [mf ed 2002] – 4r – 1 – (cont as: war telegraph & general daily advertiser [jan-apr 1855]; manchester daily telegraph & northern counties advertiser [may-nov 1855]) – uk Newsplan [072]
Daily watermelon *see* Miscellaneous newspapers of otero county
Daily west bend news – West Bend WI. 1970 oct 5/28-1972 sep 7/27 – 14r – 1 – (cont: west bend news [west bend wi: 1903]; cont by: west bend news [west bend wi: 1972]) – mf#1046848 – us WHS [071]
Daily wisconsin – Milwaukee WI. [1847 jul 26-27,29, aug 11, 1848 jan 25, feb 18-19], 1852 apr 15-1854 dec 29, 1863 feb 26, 1864 may 14,19, jun 10,11,13-15,25, aug 1,9, sep 12,26 – 2r – 1 – (cont: evening courier [milwaukee wi]; milwaukee daily free democrat; daily life; cont by: evening wisconsin) – mf#1127825 – us WHS [071]
Daily wisconsin capitol – Madison WI. 1865 apr 15-1865 jun 5 – 1r – 1 – (cont by: wisconsin daily capitol) – mf#940595 – us WHS [071]
Daily wisconsin patriot – 1854 nov28-1855 may 16, may 17-nov 19, nov 20-1856 jun 6, jun 7-dec 22, dec 23-1857 jun, jul 2-1858 jan 15, jan 16-aug 4, aug 5-1859 feb 19, feb 21-jul 22 – 9r – 1 – (cont: daily patriot [madison wi]; cont by: wisconsin daily patriot) – mf#939267 – us WHS [071]
Daily wisconsin union – Madison WI. 1866 apr 17-1866 sep 29, oct 1-1867 mar 23, mar 25-1867 sep 10, sep 11-1868 feb 5 – 4r – 1 – (cont: wisconsin daily democrat; wisconsin daily capitol; cont by: madison daily union) – mf#940610 – us WHS [071]
Daily witness – Montreal: John Dougall, aug 13 1860-jul 11 1913 (mf ed 1999) – 1r – 1 – (cont by: daily telegraph and daily witness) – mf#SEM35P467 – cn Bibl Nat [071]
Daily witness – Montreal, Canada. -d. 3 jan 1873; 16 jan 1878; 4 oct 1884; 1885-30 jun 1913 – 212 1/2r – 1 – uk British Libr Newspaper [071]
Daily witness – Montreal, QC. 1862-79 – 35r – 1 – ISSN: 0841-7164 – cn Library Assoc [071]
Daily word – Unity Village MO 1924-92 – 1,5,9 – ISSN: 0011-5525 – mf#7768 – us UMI ProQuest [240]
Daily worker / Communist Party of the United States of America – Chicago IL, New York NY. 1924 jan 13/jul 23-1950 jul 3/dec 29 – 83r – 1 – (with gaps; cont: worker [new york ny: 1922]) – mf#1000447 – us WHS [335]
Daily worker – London, England. -d. Jan 1930-Dec 1945. 59 1 2 reels – 1 – uk British Libr Newspaper [072]
Daily worker – Madison WI. 1875 jan 12-mar 1 – 1r – 1 – mf#939966 – us WHS [071]
Daily worker – New York, NY. -d. 1 Feb 1943-2 Dec 1944; 11 April 1945-29 Dec 1955. Imperfect. 35 reels – 1 – uk British Libr Newspaper [071]
The daily worker – 1924-40 – (= ser Newspapers of the american communist party) – 51r – 1 – $6250.00 – us UPA [331]
The daily worker – 1941-49 – (= ser Newspapers of the american communist party) – 75r – 1 – $9225.00 – us UPA [331]
The daily worker – 1950-57 – (= ser Newspapers of the american communist party) – 90r – 1 – $8045.00 – us UPA [331]
The daily worker 1930-1966 / morning star 1966-1998 – London, 1930-98+ – 257r – 1 – mf#DWS – uk World [072]
Daily world / Communist Party of the United States of America – 1976 jan 2/apr 30-1983 oct/dec – 26r – 1 – (cont: worker [new york ny: 1942]) – mf#969335 – us WHS [335]
The daily world – 1968-77 – (= ser Newspapers of the american communist party) – 71r – 1 – $3955.00 – us UPA [321]
The daily world : 1978-1986 – 1978-86 – (= ser Newspapers of the american communist party) – 34r – 1 – $4185.00 – us UPA [321]
The daily world – Lincoln, NE: H C Wheeler, jan 1879 (daily) [mf ed feb13-apr 8 1879 (gaps)] – 1r – 1 – us NE Hist [071]
Daily-throgmorton debate : held at ewing, illinois, august 13-16, 1912 / Daily, John Riley & Throgmorton, William Pinckney – Marion, IL: Jas H Felts, 1913 – 1mf – 9 – 0-524-08278-2 – mf#1993-3033 – us ATLA [240]
Daimond, A I *see* Biography of edwin james turpin, an earlier settler in fiji
Dainik asha – Cuttack, India. 18 Sept 1946-1950 – 5r – 1 – (oriya language) – us L of C Photodup [079]
Dainik jagran – Kanpur, India. Jul-Sept 1966 – 1r – 1 – us L of C Photodup [079]
Dainik maratha – Bombay, India. Jul-Sept 1966 – 1r – 1 – us L of C Photodup [072]
Dainik sambad – Agartala, India. 1974-93 – 40r – 1 – us L of C Photodup [079]
Dainik vir pratap – Jullundur, India. Jul-Sept 1966 – 1r – 1 – us L of C Photodup [079]
Dainika asama – Gauhati, India. 1968-1993 – 64r – 1 – (assamese language) – us L of C Photodup [079]
Dai-nippon : essai sur les moeurs et les institutions = Le japon / Hitomi, I – Paris: Societe du Recueil general des lois et des arrets, L. Larose, 1900 [mf ed 1995] – (= ser Yale coll) – 306p (ill) – 1 – 0-524-10079-9 – (in french) – mf#1995-1079 – us ATLA [950]
Dainippon teikoku naimu-sho tokei hokoku : statistical reports of the department of the interior of japanese empire, 1884-1942 – 1st-52nd reports, 1884-1942 – 26r – 1 – Y207,000 – (in japanese) – ja Yushodo [315]
Dainippon teikoku tokei tekiyo : statistical abstracts of japanese empire, 1887-1939 – 1st-53rd reports. 1887-1939 – 13r – 1 – Y110,000 – (in japanese) – ja Yushodo [315]
The daipavamsa : an ancient buddhist historical record / ed by Oldenberg, Hermann – London: Williams and Norgate, 1879 – 1mf – 9 – 0-524-06888-7 – mf#1991-0031 – us ATLA [280]
Da'irat ma'arif al-sinima – Cairo: al-Sayyid Hasan Jum'ah, 1934-35. pts1-2. ?1934-1 sep 1934 – (= ser Arabic journals and popular press) – 1r – 1 – $75.00 – us MEDOC [956]
Dairies / Allen, L – s.l, s.l? 1936 – 1r – us UF Libraries [630]
Dairy and food sanitation – Des Moines IA 1981-88 – 1 – (cont by: dairy, food & environmental sanitation) – ISSN: 0273-2866 – mf#12756.02 – us UMI ProQuest [630]
Dairy council digest – Rosemont IL 1993-98 – 1,5,9 – ISSN: 0011-5568 – mf#19330 – us UMI ProQuest [630]
Dairy exchange – v7 n2-4 [1985 feb-apr] – 1r – 1 – (cont: dairy farmer's exchange; cont by: dairy farmer exchange) – mf#1278588 – us WHS [630]
Dairy express / Wisconsin Dairies Cooperative – 1977 apr-1981, 1982-84 – 2r – 1 – (cont by: foreward [baraboo wi]) – mf#1042799 – us WHS [630]
Dairy farmer exchange – v7 n6-v9 n1 [1985 jul-1987 jan] – 1r – 1 – (cont: dairy exchange) – mf#1278589 – us WHS [630]
Dairy farmer's exchange – v6 n9-v7 n1 [1984 may 17-1985 jan] – 1r – 1 – (cont: farmer's exchange [verona wi: central ed]; cont by: dairy exchange) – mf#1278586 – us WHS [630]
Dairy field [1985] – Deerfield IL 1985-90 – 1,5,9 – (cont by: dairy field today) – ISSN: 0198-9995 – mf#15495.05 – us UMI ProQuest [630]
Dairy field [1991] – Deerfield IL 1991+ – 1,5,9 – (cont: dairy field today) – ISSN: 1055-0607 – mf#15495,05 – us UMI ProQuest [630]
Dairy field today – Deerfield IL 1990 – 1,5,9 – (cont: dairy field; cont by: dairy field) – ISSN: 1053-9425 – mf#15495.05 – us UMI ProQuest [630]
Dairy, food and environmental sanitation – Des Moines IA 1989+ – 1,5,9 – (cont: dairy and food sanitation) – ISSN: 1043-3546 – mf#12756.02 – us UMI ProQuest [630]
Dairy foods – Troy MI 1986+ – 1,5,9 – (cont: dairy record) – ISSN: 0888-0050 – mf#12974,01 – us UMI ProQuest [630]
Dairy foods : the national dairy news – v5 n38-v6 n21 [1985 nov15-1986 jul 18] – 1r – 1 – (cont: national dairy news; cont by: cheese market news) – mf#1042651 – us WHS [630]
Dairy herd management – Lincolnshire IL 1973-2000 – 1,5,9 – ISSN: 0011-5614 – mf#6971 – us UMI ProQuest [630]
Dairy industry news – New York NY 1950-52 – 1 – ISSN: 0011-5649 – mf#187 – us UMI ProQuest [630]
Dairy industry newsletter – Chicago IL 1980-81 – 1,5,9 – mf#8357,02 – us UMI ProQuest [630]
Dairy journal – 1896 jul, aug, oct – 1r – 1 – mf#4756282 – us WHS [630]
Dairy outlook and situation – Washington DC 1975-85 – 1,5,9 – (cont: dairy situation) – mf#9159,01 – us UMI ProQuest [630]
Dairy record – Troy MI 1981-85 – 1,5,9 – (cont by: dairy foods) – ISSN: 0011-5673 – mf#12974.01 – us UMI ProQuest [630]
Dairy topics / Antigo Milk Products Co-operative – v4-13 [1953 jan-1961 feb] – 1r – 1 – mf#1055086 – us WHS [630]
Dairy world – Millbury MA 1986-87 – 1,5,9 – ISSN: 0736-4962 – mf#15063 – us UMI ProQuest [630]
Dairying in florida / Scott, John M – Gainesville, FL. 1918 – 1r – us UF Libraries [630]
Dairyland agri-view – Marshfield WI. 1976 may 7, may 21-oct 29, nov 5-1977 apr 29, may 6-oct 28, nov 4-1978 apr 28, may 5-aug 25, sep 1-oct 13 – 6r – 1 – (cont by: agri-view [marshfield wi: northern ed]) – mf#689629 – us WHS [630]
Dairyland current matters – 1953-1955 – 1r – 1 – mf#1055087 – us WHS [071]
Dairyland Fertilizers, Inc *see* Farm news
Dairyland news – [1940 jan 24-1947 dec] incomplete holdings, 1948 jan-1954 aug – 2r – 1 – (cont: wisconsin dairymens news) – mf#1055088 – us WHS [630]
Dairyland review – Denmark WI. 1948 jul 22-1950, 1951-1952 jun 5 – 2r – 1 – (cont: denmark press [denmark wi: 1914]; cont by: dairyland review and the denmark press) – mf#1004632 – us WHS [630]
Dairyland review – v1 n9-v5 n7 [1964 aug 24-1968 jul 30] – 1r – 1 – (cont: midlands review) – mf#1111022 – us WHS [630]
Dairyland review and the denmark press – Denmark WI. 1952 jun 12-dec, 1953 jan-may 18 – 2r – 1 – (cont: dairyland review; cont by: denmark press and the dairyland review) – mf#1004636 – us WHS [630]
The dairyman – Montreal: J Cheesman, [1885?-1886?] – 9 – mf#1190-6871 – cn Canadiana [630]
Dairyman-gazette – Clintonville WI. 1923 feb 8/1924 jun 12-1939 jan 7/1940 jul 11 – 13r – 1 – (with gaps; cont: clintonville gazette) – mf#966868 – us WHS [630]
Daish, John Broughton *see*
– Memorandum. the write of certiorari in the district of columbia in relation to the interstate commerce commission
– Procedure in interstate commerce cases, with illustrative precedents and forms
Daisy beresford / Aldrich, Annie Charlotte Catharine – London: Hurst & Blackett Publ. 3v. 1882 – (= ser 19th c women writers) – 12mf – 9 – mf#5.1.126 – uk Chadwyck [420]
Daisy first baptist church. daisy, tennessee : church records – 1878-1906 – 1 – 5.00 – us Southern Baptist [242]
Daja sosial *see* Jajasan dana kesedjahteraan sosial
Dakar : outpost of two hemispheres / Lengyel, Emil – Garden City, NY. 1943 – 1r – us UF Libraries [960]
Dakota – Fargo, Grand Forks ND, Moorhead MN. 1895 oct 16-1897 mar 10 – 1r – 1 – (cont: fargo posten; vesten; cont by: fjerde juli; fjerde juli og dakota) – mf#901911 – us WHS [071]
Dakota American Indian Movement *see* Oyate wicaho
Dakota city democrat – Dakota City, NE: Daniel McLaughlin. v2 n1. mar 9 1861- (wkly) [mf ed -apr 13 1861 (lacks mar 16) filmed 1958] – 1r – 1 – (cont: dakota city herald. cont by: nebraska north) – us NE Hist [071]
Dakota city herald – Dakota City, NE: John L Dailey, jul 15 1857-1860// (wkly) – 1r – 1 – (cont by: dakota city democrat. suspended: sep 1858-mar 1859) – us L of C Photodup [071]
Dakota city herald – Dakota City, NE: John L Dailey, jul 15 1857-60// (wkly) [mf ed 1859-60 (gaps) filmed 1958] – 1r – 1 – (cont by: dakota city democrat. suspended sep 1858-mar 1859) – us NE Hist [071]
Dakota city mail – Dakota City, NE: MacDonagh & O'Sullivan. 7v. v1 n1. jul 29 1870-v7 n26. jan 12 1877 (wkly) [mf ed with gaps filmed [1973]-90] – 4r – 1 – (cont by: dakota county mail. issue for dec 27 1872 not publ) – us NE Hist [071]
Dakota conflict of 1862 : manuscript collections – 4r – 1 – $770.00 – 1-55655-855-4 – (with p/g) – us UPA [978]
Dakota conflict of 1862 manuscripts collections – 4r – 1 – $120.00 $30.00r – us Minn Hist [978]
Dakota connection : a news magazine of the dakota southern baptist fellowship – Decatur. 1972+ (1) 1947+ (5) 1974+ (9) – 1 – mf#6377 – us Southern Baptist [242]
Dakota county democrat – South Sioux City, NE: Harry A McCormick. v4 n45. feb 20 1891-jul 30 1897// (wkly) [mf ed with gaps filmed -1990] – 2r – 1 – (cont: sun and news (south sioux city, ne). cont by: south sioux city press. issues for feb 20-may 8 1891 called new ser: v1 n1-12) – us NE Hist [071]
Dakota county enterprise – Emerson, NE: S E Cobb, 1902 (wkly) [mf ed v1 n[6] apr 18 1902-jun 13 1902 (gaps)] – 1r – 1 – us NE Hist [071]
Dakota county herald – Dakota City, NE: J L McKean, aug 28 1891-v74 n27. feb 25 1965 (wkly) [mf ed 1899-1965 (gaps) filmed 1958-77] – 21r – 1 – (cont: homer herald. issues for apr 17-24 1952 not publ due to flood) – us NE Hist [071]
Dakota county mail – Dakota City, NE: John T Spencer. 1v. v7 n27. jan 19-v7 n11. sep 28 1877 (wkly) [mf ed filmed [1973]-90] – 2r – 1 – (cont: dakota city mail. cont by: mail (convington ne)) – us NE Hist [071]
Dakota county record – South Sioux City, NE: Eimers & Keefer, 1887-1919// (wkly) [mf ed 1895-1902,1908-19 (gaps) filmed 1958-[1974?]] – 6r – 1 – (numbering begins with: v29 n9 nov 14 1908. absorbed: argus) – us NE Hist [071]
Dakota county star – South Sioux City, NE: Mrs H N Wagner. 27v. v33 n5. jul 2 1942-v59 n52. apr 17 1969 (wkly) [mf ed with gaps filmed 1974?-76?] – 17r – 1 – (cont: homer star. cont by: south sioux city star) – us NE Hist [071]
Dakota duster – [Valley City, ND]: Company 2770, CCC 1 undated iss: possibly jun 1935 (wkly) – 1 – mf#11461 – us North Dakota [071]
Dakota eagle : [official paper of bottineau county 1886-1889] – Willow City, Bottineau Co, Dakota [i.e. ND]: Jacob P Hager, sep 17 1886-mar 8 1889 (wkly) – 1 – (cont by: north dakota eagle) – mf#06899 – us North Dakota [071]
Dakota freie presse – Yankton, Dakota [ie SD]: Chas F Rossteuscher, 1874-laufende n2080. 7 apr 1914; jahrg 41 n1. 14 apr 1914-jahrg 79 n49. 24 feb 1954 (wkly) [mf ed 1909-16 (gaps) filmed 1975] – 3r – 1 – (in german. absorbed: dakota rundschau. absorbed by: america-herold und lincoln freie presse. publ at aberdeen sd, 8 jul 1909-24 feb 1920; new ulm mn, 2 mar 1920-25 okt 1932; bismarck nd and winona mn, 1 nov 1932-23 dez 1953; winona mn, 30 dez 1953-24 feb 1954. issues for 14 apr 1914-19 sep 1916 also called laufende n2081-2208) – us NE Hist [071]
Dakota freie presse – New Ulm MN (USA), 1920 10 aug-1939 29 nov [gaps] – 11r – 1 – gw Misc Inst [071]
Dakota law review – North Dakota School of Law. v1-4. 1927-32 (all publ) – 6mf – 9 – $27.00 – mf#LLMC 84-454 – us LLMC [340]
Dakota mail – Covington, NE: John T Spencer. 1v. v8 n17-22. jan 25-mar 1 1878 (wkly) [mf ed filmed 1990] – 1r – 1 – (cont: mail (covington ne). absorbed by: north nebraska eagle. publ in dakota city feb 22-mar 1 1878) – us NE Hist [071]
Dakota odowan = Dakota hymns / ed by Williamson, John Poage & Riggs, Alfred Longley – New York: American Tract Soc, 1879 [mf ed 1993] – 2mf – 9 – 0-524-06675-2 – (pref by ed) – mf#1991-2730 – us ATLA [780]
Dakota Ojibway Tribal Council *see*
– D o t c news
Dakota press – Valley City, ND: C C Morgan, W L Morgan, F R Crowe. v1 n1 may 3 1946-v3 n10 jul 2 1948 (wkly) – 1 – (special ed: valley city trade news of the dakota press, jun 21, jul 19, 1946; publ as: valley city trade news and dakota press, aug 16-dec 6 1946; absorbed by: valley city times-record and the barnes county news) – mf#10527-10528 – us North Dakota [071]
The dakota reports / Dakota. Territory – v1-6. 1867-89 (all publ) – 36mf – 9 – $54.00 – (pre-nrs: v1 1867-77 6mf $9.00. cont by: north dakota and south dakota official reports series) – mf#LLMC 81-801 – us LLMC [340]
Dakota sun / Standing Rock Community College – 1979 may 24-1981 dec, 1982-83 – 2r – 1 – mf#612483 – us WHS [373]
Dakota territory – 1862-1889 – 9 – $119.00 set – mf#402570 – us Hein [975]
Dakota. Territory *see*
– The dakota reports
– Reports, pre-nrs
Dakota weekly union – Yankton SD. 1864 jun 21-aug 23 – 1r – 1 – (cont by: dakotian; union dakotaian) – mf#827554 – us WHS [071]
Dakota-rundschau *see* Eureka rundschau

Dakotian – Yankton SD. 1862 jun 3-1864 sep 24 – 1r – 1 – (cont: weekly dakotian; cont by: dakota weekly union; union dakotaian) – mf#854150 – us WHS [071]
Dal protezionismo al sindacalismo : ruinisco nel presente volumetto tre discorsi, che lessi, negli ultimi anni, in tre massime universita italiane / Ricci, Umberto – Bari: G Laterza 1926 [mf ed 1980] – 1r – 1 – (incl bibl ref & ind) – mf#148 – us Wisconsin U Libr [330]
Dal "sabotaggio" massonico dell'italia alla nota pontificia : storia d'una polemica – Roma: Francesco Ferrari 1917 – 3mf – 9 – mf#vrl-171 – ne IDC [366]
Dal', V I *see* Tolkovyi slovar' zhivogo velikorusskogo iazyka
Dal', v.i. tolkovii slovar' zhivogo velikorusskogo iazika = Vldimir dal's explanatory dictionary of the living great russian language / ed by Courtenay, Baudouin de – 4th ed. St Petersburg, Moscow: M Wolf Publ. v1-4. 1912 – 44mf – 9 – $290.00 – us UMI ProQuest [460]
Dala vilayeti – Omsk, Russia 1888-1902 [mf ed Norman Ross] – 6r – 1 – mf#nrp-1254 – us UMI ProQuest [077]
Dalabladet – Falun, Sweden. 1906-17 – 12r – 1 – sw Kungliga [078]
Dalabladet – Falun, Sweden. 1917-24 – 9r – 1 – sw Kungliga [078]
Dalabygden – Borlange, Sweden. 1979- – 1 – sw Kungliga [078]
Dalademokraten – Falun, Sweden. 1917-78 – 388r – 1 – sw Kungliga [078]
Dalademokraten – Falun, Sweden. 1979- – 1 – sw Kungliga [078]
Dalal, Manockji Nadirshaw *see* Whither minorities?
Dalal, Vaman Somnarayan *see* A history of india from the earliest times
Dalarnas nyheter – Hedemora, Borlaenge, Sweden. 1902-05 – 3r – 1 – sw Kungliga [078]
Dalarne – Falun, Sweden. 1884-85 – 1r – 1 – sw Kungliga [078]
Dalarnes allehanda – Falun, Gaevle, Sweden. 1887-94 – 4r – 1 – sw Kungliga [078]
Dalayrac, Nicolas *see*
– Adele et dorsan
– Alexis
– Le chateau de montenero
– Gulistan
– La maison a vendre
– Ouverture de la dot
– Ouverture, romance et choeur de l'opera de nina
– Picaros et drego
– La soiree orageuse
Dalbemar, Jean Joseph *see*
– Haiti
– Notes sur le retrait et...
Dalberg, Karl Theodor von *see* Gedanken von bestimmung moralischen werths
Dalbian, Denyse *see* Dom pedro, empereur du bresil, roi de portugal, 17...
Dalby, Henry *see*
– The index of current events
– The index of current events, 1889
Dale, Alfred William Winterslow *see*
– History of english congregationalism
– The synod of elvira and christian life in the fourth century
Dale, Daphne *see* The story of the bible in poetry and song
Dale, Desmond *see* Shona companion
Dale, Edgar Thorniley *see* Canadian workmen's compensation acts and cases
Dale, George Allan *see* Education in the republic of haiti
Dale, Godfrey *see* Peoples of zanzibar
Dale, James Wilkinson *see*
– Classic baptism
– An inquiry into the usage of [baptizo] and the nature of judaic baptism
Dale, R W *see*
– The epistle of james and other discourses
– Nonconformity in 1662 and in 1862
Dale recorder – Dale WI. 1899 aug 3-1901, 1902-14, 1915-1918 jul 26 – 6r – 1 – mf#961583 – us WHS [071]
Dale, Rev Canon *see* Christian soldier
Dale, Robert William *see*
– The atonement
– Christ and the controversies of christendom
– Christian doctrine
– The epistle to the ephesians
– Essays and addresses
– History of english congregationalism
– The jewish temple and the christian church
– The life and letters of john angell james
– A manual of congregational principles
– Protestantism
– The ten commandments
Dale, Thomas *see*
– Introductory lecture upon the study of theology and of the greek te...
– Mind stayed on god
– Probation for the christian ministry practically considered
– Widow of the city of nain
Dale, Thomas Nelson *see* The outskirts of physical science
Dale, William Henry *see* Study of the musico-psychological dramas of vladimir ivanovich rebikov
Dalekaia okraina / ed by Panteleev, Dm et al – Vladivostok [Primor obl]: [s n] 1918-19 [1907 22 marta-1919 [26 fevr] – (= ser Asn 1-3) – n3514 [1918]-n3759 [1919] [gaps] item 107, on reel n22 – 1 – mf#asn-1 107 – ne IDC [077]
d'Alembert, M *see*
– Encyclopedie
Dalencour, Franciois Stanislas Ranier *see*
– Croisee des chemins
– Drapeau national haitien
– Fondation de la republique d'haiti
– Francisco de miranda et alexandre petion
– Philosophie de la liberte comme introduction a la...
Dalencour, Francois Stanislas Ranier *see*
– Precis methodique d'histoire d'haiti
– Principes d'education nationale, comprenant l'ense...
– Projet de constitution conforme a la destinee glor...
Daleyville doings – Daleyville, Mt Horeb, Perry WI. 1908 apr 1-1917 mar 7 – 1r – 1 – mf#967553 – us WHS [071]
Dalgairns, John Bernard *see* The holy communion
D'alger a tombouctou : des rives de la loire aux rives du niger, avec une carte / More, Rene la – Paris, Plon-Nourrit, 1913 – us CRL [912]
Dalgety's review – Sydney, Australia. Sep 1898-Sep 1916.-m. 33mqn reels – 1 – uk British Libr Newspaper [072]
Dalgety's review (australasia) – Sydney, Australia. 13 Feb 1920-30 Jun 1933 (missing Jan 1923-Jun 1925; Apr-Dec 1927).-w. 5mqn reels – 1 – uk British Libr Newspaper [072]
Dalgetys review (monthly) – Sydney, Australia. Sep-16 dec 1898; feb-15 dec 1899; 1900-sep 1916 – 33 1/4r – 1 – uk British Libr Newspaper [072]
Dalgetys review [weekly] – Sydney, Australia. 13 feb 1920-mar 1927, jan-21 dec 1928, 1929-31, 8 jan 1932-jun 1933 [missing jan 1923-jun 1925, apr-dec 1927] – 5 5 1/2r – 1 – (aka: australasia) – uk British Libr Newspaper [079]
Dalgliesh, William *see* Importance of true religion and the care of god to preserve it
Dalhberg Corporation Of America *see* Dahlberg sugar cane industries
Dalhousie, George Ramsay, Earl *see* Observations on the petitions of grievance
Dalhousie muniments, 1748-1759 – Edinburgh: Scottish Record Office – (= ser BRRAM series) – 2r – 1 – £134 / $268 – (int by t c smout) – mf#r96297 – uk Microform Academic [975]
Dalhousie review – Halifax, Canada 1989+ – 1,5,9 – ISSN: 0011-5827 – mf#18186 – us UMI ProQuest [071]
Dali, Salvador *see* Secret life of salvador dali
Dalibard, Thomas Francois *see* Histoire des incas
Dalimilova *see* Dalimils chronik von boehmen
Dalimils chronik von boehmen / ed by Hanka, Venceslav – Stuttgart: Literarischer Verein, 1859 [mf ed 1993] – (= ser Blvs 48) – 253p – 1 – (middle high german text) – mf#8470 reel 10 – us Wisconsin U Libr [880]
Dalka – Mogadishu: Yousuf Jama Ali Duhul. v1 n2-v2 n10 aug 1 1965-apr 1 1967. Ser 2: v1 n1-2 oct 1-15 1969. Special issues: jul 1, oct 21 1981 – us CRL [073]
Dalkeith advertiser – Scotland. Jul 1869-Dec 1953.-w. 31 reels – 1 – uk British Libr Newspaper [072]
Dall, Caroline Healey *see* The caroline h dall papers, 1811-1917
Dall, Caroline Wells Healey *see* Fog bells
Dall, W H *see*
– Illustrations and descriptions of new, unfigured, or imperfectly known shells, chiefly american, in the u.s. national museum
– Reports on the results of dredging by the us coast survey steamer blake, n15
– Reports on the results of dredging by the us coast survey steamer blake, n29 1886
– Scientific results of explorations by the us fish commission steamer albatross, n7
– Synopsis of the family tellinidae and of the north american species
– Synopsis of the luciniacea and of the american species
Dalla Torre, Giovanni *see* Dialogo della giosta fatta in trivigi l'anno 1597
Dallaeus, Ion *see*
– De cultibus religiosis latinorum
– De imaginibus libri 4
Dallaire, O-E *see* Considerations sur les cercles agricoles et les societes d'agriculture avec instructions aux juges dans les concours et les expositions
Dallam's reports / Texas. Supreme Court – 1v. 1840-1844 (all publ) – 6mf – 9 – $9.00 – (a pre-nrs title) – mf#LLMC 95-213 – us LLMC [340]
Dallas / Baptist Association. Texas – 1980-87 – 1 – 78.48 – us Southern Baptist [242]
Dallas – Dallas TX 1922-88 – 1,5,9 – ISSN: 0011-5835 – mf#3028 – us UMI ProQuest [338]
Dallas, Angus *see*
– Appeal on the common school law
– The common school system
– Outlines of chemico-hygiene and medicine
– Statistics of the common schools
– Suggestions on the organization of a system of common schools adapted to the circumstances and state of society in canada
Dallas business journal – Charlotte NC 1988-2000 – 1,5,9 – ISSN: 0899-4129 – mf#16691 – us UMI ProQuest [650]
Dallas County Libertarian Party *see* Newsletter
Dallas craftsman – 1933 nov 24/1936-1985 jan 18/1994 dec 9 – 16r – 1 – (with small gaps; cont: craftsman [dallas tx]; cont by: union craftsman) – mf#3358894 – us WHS [071]
Dallas daily herald – Dallas TX. v5 n95-96 [1877 jun 24, 26] – 1r – 1 – (cont by: times [dallas tx: 1887]; daily times herald [dallas tx: 1888]) – mf#858511 – us WHS [071]
Dallas dispatch – Dallas TX. 1929 jun 4 – 1r – 1 – mf#4364695 – us WHS [071]
Dallas district crusader – Fort Worth TX. 1944 dec 22 – 1r – 1 – mf#5019876 – us WHS [071]
The dallas express – Dallas, TX: W E King. v7 n14. jan 13 1900 (wkly) [mf ed 1947] – (= ser Negro Newspapers on Microfilm) – 1r – 1 – us L of C Photodup [071]
Dallas first baptist church. dallas, texas : church records – 1877-1913 – 1 – 53.10 – us Southern Baptist [242]
Dallas Genealogical Society *see* Dallas quarterly
Dallas, Helen Alexandrina *see* Evidence for a future life
Dallas itemizer – Dallas OR: E Casey, -1879 [wkly] – 1 – (cont by: polk county itemizer [1879-1927]) – us Oregon Lib [071]
[Dallas-] kennedy assassination – TX. v1-10 L.A. Metro, A-Z; International A-Z – 8r – 1 – $480.00 – mf#R05003 – us Library Micro [071]
Dallas news – Dallas TX. v1 n1-3, 5-11 [1970 aug 12-sep 9-22, oct 14/27-1971 jan 13/26] – 1r – 1 – (cont by: iconoclast [dallas tx]) – mf#1055102 – us WHS [071]
Dallas notes – Dallas TX. v1 n26 [1968 mar 3-17], v1 n26-v4 n21 [1968 mar 3-1971 feb 6] – 2r – 1 – (cont: notes from the underground) – mf#766030 – us WHS [071]
Dallas quarterly / Dallas Genealogical Society – 1984 mar-1988 dec – 1r – 1 – (cont: quarterly [dallas genealogical society]; cont by: dallas journal) – mf#1111023 – us WHS [978]
Dallas' reports / Pennsylvania. Superior Court – v1-4. 1754-1806 (all publ) – (= ser Pre-nrs nominative reports) – 24mf – 9 – $36.00 – mf#LLMC 84-191 – us LLMC [340]
Dallas, Robert *see* Geschichte der maronen-negern auf jamaika
Dallas weekly – Dallas TX. 1991 jul 11/dec 21-1997 jul 1/dec 30 – 12r – 1 – (with gaps; cont by: weekly free press [dallas tx]) – mf#2175978 – us WHS [071]
Dallas weekly reader – 1998 may-aug, sep-dec, 1999 sep 7-dec 21 – 3r – 1 – mf#3401053 – us WHS [071]
Dallaway, James *see*
– Anecdotes of the arts in england
– Observations on english architecture
– Of statuary and sculpture among the antients
Dalles chronicle – Bellinzona. 1950-1955 (1) – 9r – mf#466 – us Oregon Lib [071]
Dalles chronicle – The Dalles, OR: Western Pub Co, aug 3 1947-aug 31 1999 – (= ser Newspapers in microform, 1948-1983) – 1 – (sunday ed publ as: the chronicle; also publ in a weekly ed called: the dalles chronicle, 1890-1952; 60th anniversary ed publ on july 7 1950) – us Oregon Hist [071]
Dalles daily chronicle – The Dalles, OR. daily -1998. 12/15/1890-3/11/1998 – 149r – mf#6425297 – us Oregon Lib [071]
Dalles daily chronicle *see* Dalles weekly chronicle
Dalles of the saint croix – St Croix Falls WI. 1881 jan 21-1882 feb 24, 1882 mar 3-1884 nov 28 – 2r – 1 – mf#928238 – us WHS [071]
Dalles optimist – The Dalles OR: Bennett & Davenport, 1906-66 [wkly] – 18r – 1 – (absorbed: dufur dispatch [1891-1941]; cont by: dalles optimist mid-columbian [1966-70]) – us Oregon Lib [071]
The dalles optimist *see* Dufur dispatch
Dalles optimist mid-columbian – The Dalles OR: R C Wellman, 1966-70 [wkly] – 2r – 1 – (cont: dalles optimist [1906-66]) – us Oregon Lib [071]
Dalles weekly chronicle : official paper of wasco county – The Dalles OR: [s.n.] 1890-1947 [wkly] – 1 – (absorbed by: dalles daily chronicle) – us Oregon Lib [071]
Dallet, Charles *see* Histoire de l'eglise de coree
Dallinger, W H *see* The creator, and what we may know of the method of creation
Dallinger, William Henry *see* The creator, and what we may know of the method of creation
Dallmann, William *see*
– The lord's prayer
– Portraits of jesus
– The ten commandments
Dallo y Zavala, Manuel Romualdo *see*
– Sermon de la gloriosa madre santa monica
– Sermon moral de la santa veronica
– Sermon panegyrico
D'Almada, Andre Alvares *see* Relacao e descripcao de guine
Dalman, Gustaf *see*
– Arbeit und sitte in palaestina
– Christianity and judaism
– Grammatik des juedisch-palaestinischen aramaeisch
– Hundert deutsche fliegerbilder aus palaestina
– Jerusalem und sein gelaende
– Jesaja 53
– Der leidende und der sterbende messias der synagoge
– Der leidende und der sterbende messias der synagoge im ersten nachchristlichen jahrtausend
– Studien zur biblischen theologie
– Traditio rabbinorum veterrima de librorum veteris testamenti ordine atque origine
– The words of jesus
Dalman, Gustav Herman *see* Juedischdeutsche volkslieder aus galizien und russland
Dalmatia. Laws, Statutes, etc *see* Sveucilistni zakoni i provedbena naredba o drzavnim ispitima
Dalmatien und montenegro : mit einem ausfluge nach der herzegowina und einer geschichtlichen uebersicht der schicksale dalmatiens und ragusa's / Wilkinson, John – Leipzig 1849 [mf ed Hildesheim 1995-98] – (= ser Fbc) – 2v on 5mf – 9 – €100.00 – 3-487-29381-1 – gw Olms [949]
Dalmatien und seine inselwelt : nebst wanderungen durch die schwarzen berge / Noe, Heinrich – Wien [u.a.] 1870 [mf ed Hildesheim 1995-98] – (= ser Fbc) – 3mf – 9 – €90.00 – 3-487-29144-4 – gw Olms [914]
La dalmazia – Sibenik, Yugoslavia. Feb-May 1919 – 1r – 1 – us L of C Photodup [949]
La dalmazia la voce dalmatica – Zadar, Yugoslavia. Jun 1919-Feb 1920 – 1r – 1 – us L of C Photodup [949]
Dalmiro y silvano / Salas, Francisco Gregorio de – 1780 – 9 – sp Bibl Santa Ana [810]
Dalnevostochnaia kooperatsiia – Khabarovsk, 1928-1929(2) – 20mf – 9 – mf#COR-581 – ne IDC [335]
Dal'ne-vostochnaia respublika : organ pravitel'stva dal'ne-vost resp / ed by Perlin, B G – Chita [Zabaik obl]: [s n] 1920-21 [1920 29 apr-] – (= ser Asn 1-3) – n17 [1920]-n327 [1921] [gaps] item 108, on reel n22,23 – 1 – mf#asn-1 108 – ne IDC [077]
Dal'nevostochnoe obozrenie = The fareastern review / ed by Novitskii, N K – Vladivostok [Primor obl]: N D Kriudener-Sal'ko 1919 [1919 27 [14] fevr-[1921 ianv] – (= ser Asn 1-3) – n1-184 [1919] [gaps] item 109, on reel n23,24 – 1 – mf#asn-1 109 – ne IDC [077]
Dal'ne-Vostochnyi Aktsionernyi Bank (Dal'Bank) *see* Otchet dal'ne-vostochnogo aktsionernogo banka za 2-oi operatsionnyi god, ianv-sent 1923 g prilozhenie
Dal'nevostochnyi komsomolets – Komsomol'skna-Amure, Khabarovsk Krai 1975-85 [mf ed Norman Ross] – 1r – 1 – mf#nrp-770 – us UMI ProQuest [077]
Dal'ne-vostochnyi telegraf : vnepartiin, polit, obshchestv i lit gaz / ed by Khitrenko, N I – Chita [Zabaik obl]: D V koop izd-vo 1921-22 [1921 3 avg-1922 1 dek – (= ser Asn 1-3) – n18 [1921]-n394 [1922] [gaps] item 110, on reel n24,25 – 1 – mf#asn-1 110 – ne IDC [077]
Dal'ne-Vostochnyj kraj sovet rk i kd *see* Izvestiia dal'ne-vostochnogo kraevogo komiteta sovetov rs i kr. deputatov i khabarovskogo soveta r i s deputatov
Dal'nii vostok : organ nats mysli: ezhedn gaz, posviashch interesam priamur kraia / ed by Panov, V A – Vladivostok [Primor obl]: E A Panova 1918-19 [1892 25 okt-1922 [?] – (= ser Asn 1-3) – n1-16 [1918]-n1-172 [1919] [gaps] item 111, on reel n25,26 – 1 – mf#asn-1 111 – ne IDC [077]
Dalpilen – Falun, Sweden. 1854-1926 – 63r – 1 – sw Kungliga [078]
Dalry & kilbirnie herald & vale of garnock news – [Scotland] North Ayrshire, Irvine: C Murchland 1 jun 1894-9 jul 1926 (wkly) [mf ed 2003] – 1715v on 27r – 1 – uk Newsplan [072]
Dalrymple, A *see* Memoir concerning the passages to and from china
Dalrymple, J *see* Memoirs of great britain and ireland (from charles 2 to the sea-battle of la hogue)
Dalrymple-Champneys, Weldon *see* Undulent fever
Dals dagblad *see* Bohuslaningen

Dalsland – Amal, 1872 – 1r – 1 – sw Kungliga [078]
Dalton – Cambridge, England 2000+ – 1,5,9 – (cont: journal of the chemical society. dalton transactions) – ISSN: 1470-479X – mf#7183.02 – us UMI ProQuest [540]
Dalton 1772-1849 – Oxford, MA (mf ed 1996) – (= ser Massachusetts vital record transcripts to 1850) – 2mf – 9 – 0-87623-235-7 – (mf 1t: marriage intentions 1782-1802; births 1772-1820; marriages 1787-1816; births 1843-46. mf 2t: births 1846-49; marriages, deaths 1843-49) – us Archive [978]
Dalton, Deborah S *see* Negotiated rulemaking sourcebook
Dalton delegate – Dalton, NE: Tom W and Bertha A Lally. v38 n11. sep 28 1951 (wkly) [mf ed v5 n9. sep 27 1918-sep 28 1951 (gaps)] – 9r – 1 – (absorbed by: bridgeport news-blade. suspended foll aug 28 1942; resumed with v32 n34 on mar 8 1946. v38 n11. sep 28 1951 misnumbered and misdated v38 n10. sep 21 1951) – us NE Hist [071]
Dalton, Hermann *see*
– Auf missionspfaden in japan
– Beitraege zur geschichte der evangelischen kirche in russland
– Die evangelische kirche in russland
– Evangelische stroemungen in der russischen kirche der gegenwart
– Ferienreise eines evangelischen predigers
– Geschichte der reformirten kirche in russland
– Indische reisebriefe
– Johannes calvin
– Johannes gossner
– John a lasco
– Lasciana
– Lasciana nebst den aeltesten evangelischen synodalprotokollen polens 1555-1561
– Nathanael
– On religious liberty in russia
– Reisebilder aus griechenland und kleinasien
– Der social aussatz
Dalton Hill, Andrea *see* The relationship between anthropometry and body composition assessed by duel-energy X-ray absorptiometry in women 75-80 years old
Dalton, Hugh *see* Fabian economic and social thought, series 2
Dalton, J N *see*
– Ordinale exon, vol 3
– Ordinale exon, vol 1-2
Dalton, Leonard Victor *see* Venezuela
Dalton, O M *see* Notes on an ethnographical collection from the west coast of north america (more especially california), hawai and tahiti
Dalton, Richard B *see* Effects of exercise and vitamin b12 supplementation on the depression scale scores of a wheelchair confined population
Dalton, Roque *see*
– Mar
– Testimonios
– Turno del ofendido
– Ventana en el rostro
Dalton, William *see* Farewel sermon
Daltons – v1 n1-v7 n2 [1974 jun-1982 jun] – 1r – 1 – mf#971054 – us WHS [071]
Daluz, Eusebio T *see* Filipino-english vocabulary, with practical examples of filipino and english grammars
Daly, Charles Patrick *see*
– The common law; its origin, sources, nature and development, and what the state of new york has done to improve upon it
– Settlement of the jews in north america
[Daly city-] daly city record – CA. 1913- [wkly] – 86r – 1 – $5160.00 (subs $120/y) – mf#B02158 – us Library Micro [071]
[Daly City-] post – CA. 1963-67 [wkly] – 8r – 1 – $480.00 – mf#B02157 – us Library Micro [071]
[Daly city-] san mateo post (north county ed.) – CA. 1963- – 65r – 1 – $3900.00 – mf#B02159 – us Library Micro [071]
[Daly city-] shopping news – CA. 1939-44 – 2r – 1 – $120.00 – mf#B02160 – us Library Micro [071]
Daly, Robert *see*
– Letters on the subject of the scotch episcopal church
– Sermon on the scripture doctrine of miracles...
Daly, Thomas G *see* The relationship between muscle fiber type and serum lactate dehydrogenase
Dalyell, John *see* Shipwrecks and disasters at sea
Daly's Fifth Avenue Theatre. New York *see*
– Bill of the play
– Collection of programs
– Scrapbooks
Dalzel, A *see* The history of dahomey
Dalzell, Georg W *see* The law of the sea
Dam chronicle – Cascade Locks OR: Cummins & Shields, 1934 [wkly] – 1 – (cont by: bonneville dam chronicle [1934-39]) – us Oregon Lib [071]
Dam construction : and failures during the last thirty years / Baillairge, Charles P Florent – [Montreal?: s.n, 1903?] – 1mf – 9 – 0-665-99724-8 – mf#99724 – cn Canadiana [627]
Dama de arcon / Rodriguez Acosta, Ofelia – Mexico City, Mexico. 1949 – 1r – us UF Libraries [972]
Dama del los lebreles / Mateizan, Roberto – Manzanillo, Cuba. 1918 – 1r – us UF Libraries [972]
Dama rakanaka ra yesu kristo rakanyorwa na mariko – Gwelo, Zimbabwe. 1964 – 1r – us UF Libraries [220]
Dama rakanaka rayesu kristo rakanyorwa namateo – Gwelo, Zimbabwe. 1964 – 1r – us UF Libraries [220]
Dama rakanaka rayesu kristo rakanyorwa naruka – Gwelo, Zimbabwe. 1964 – 1r – us UF Libraries [220]
Dama rakanaka rayesu kristo rakanyorwa nayowane – Gwelo, Zimbabwe. 1964 – 1r – (in shona) – us UF Libraries [220]
Damage assessment reports, 1945 / U.S. Strategic Bombing Survey – 1993 (mf ed) – (= ser Records of the united states strategic bombing survey) – 17r – 1 – mf#M1721 – us Nat Archives [355]
Damages in international law / Whiteman, Marjorie M – Washington: GPO. 3v. 1937-43 [all publ] – 25mf – 9 – $37.50 – mf#llmc 80-910 – us LLMC [341]
Damals in weimar : erinnerungen und briefe / Schopenhauer, Johanna; ed by Houben, Heinrich Hubert – Leipzig: Klinkhardt & Biermann 1924 [mf ed 1991] – 1r [ill] – 1 – (incl bibl ref & ind. filmed with: hans heiner roselieb's ewiger sonntag / heinrich schotte) – mf#2938p – us Wisconsin U Libr [920]
Dama-ntsoha *see* Dictionnaire etymologique de la langue malgache
Damantsoha *see* La langue malgache et les origines malgaches-razafintsamala
Damariscotta Mills, Maine. Damariscotta Mills Baptist Church *see* Records
Damarjian, Nicole M *see*
– Effect of heart rate deceleration biofeedback training on golf putting performance
– The short-term training effects of practice variability on posttraining performance of three golf skills with experienced golfers
Damascus road – Wescosville PA 1961-72 – 1,5 – ISSN: 0418-3142 – mf#6209 – us UMI ProQuest [400]
Damas-Hinard, Jean Joseph *see* Memoires d'une femme de qualite, depuis la mort de louis 18 jusqu'a la fin de 1829
Damaskenos, ho Stoudites, Metropolitan of Naupaktos and Arte *see* Sokrovishche damaskina studita v novom russkom perevodie
Damaskin semenov-rudnev, episkop nizhegorodskii [1737-1795] : ego zhizn i trudy / Gorozhanskii, I I – Kiev, 1894 – 294p 6mf – 8 – mf#R-6036 – ne IDC [243]
Damaso de la presentacion. vida del p. jose ma del montecarmelo / Bayle, Constantino – Burgos, 1931; Madrid: Razon y Fe, 1933 – 1 – sp Bibl Santa Ana [920]
Damaso velazquez / Arraz, Antonio – Caracas, Venezuela. 1944 – 1r – us UF Libraries [972]
Damaso zapata / Zapata, Ramon – Bogota, Colombia. 1961 – 1r – us UF Libraries [972]
Damasus, bischof von rom : ein beitrag zur geschichte der anfaenge des roemischen primats / Rade, Martin – Freiburg i.B.: J C B Mohr (Paul Siebeck), 1882 – 1mf – 9 – 0-8370-7902-0 – mf#1986-1902 – us ATLA [920]
Damaze de Raymond *see* Tableau historique, geographique, militaire et moral de l'empire de russie
Damberger, C F *see* Travels in the interior of africa from the cape of good hope to morocco
Dame aux giroflees / Varin – Paris, France. 1867 – 1r – us UF Libraries [025]
Dame des belles cousines / Dartois, Achille – Paris, France. 1823 – 1r – us UF Libraries [440]
Dame du second / Beauplan, Amedee De – Paris, France. 1840 – 1r – us UF Libraries [440]
Dame henriette brown (demanderesse en cour inferieure) : appelante, et les cure et marguilliers de l'oeuvre et fabrique de la paroisse de montreal (defendeurs en cour inferieure), intimes / Laflamme, Rodolphe & Doutre, Joseph – Montreal?: s.n, 1870? – 1mf – 9 – mf#04392 – cn Canadiana [347]
Die dame in schwarz / Langewiesche, Marianne – Muenchen: Deutscher Volksverlag, 1942 – 1r – 1 – us Wisconsin U Libr [830]
Dame la mano / Grutter, Virginia – n.p, n.p? 1954 – 1r – us UF Libraries [972]
Die dame mit der maske / Auernheimer, Raoul – Wien: Wiener Verlag, [1905?] [mf ed 1995] – 287p – 1 – mf#8920 – us Wisconsin U Libr [820]
Dame pinfold – London, England. 18– – 1r – us UF Libraries [240]
Damen, Arnold *see*
– La bible ne suffit pas pour enseigner les verites necessaires au salut
– Conference donnee a la cathedrale d'ottawa par le r p damen...1871
– Conferences du rev pere damen, sj
– Lecture by father damen, on the real presence
– Reponses aux objections populaires contre la religion catholique
– The rule of faith
– Verbatim report of a sermon
Damen-conversations-lexikon / ed by Herlosssohn, C – Leipzig 1834-38 [mf ed 1993] – (= ser HQ 13) – 20mf – 9 – €220 diazo €264 silver – 3-89131-125-7 – gw Fischer [305]
Dames, Gordon Ernest *see* 'N ondersoek na die stand en behoeftes van kategese in die nederduitse gereformeerde sendingkerk
Damian : oder, das grosse schermesser: roman / Stehr, Hermann – Leipzig: P List c1944 [mf ed 1991] – 1r – 1 – (filmed with: vermischte schriften, und, amerikanische gedichte / friedrich spielhagen) – mf#2936p – us Wisconsin U Libr [830]
Damiani, Petrus *see*
– Sermones
Damianus, J *see* Ianida miani senensis ad leonem x pont max de expeditione in turcas elegeia, cu argutissimis doctissimorum uirorum epigrammatibus
D'Amico, Silvio *see* Scoperta dell'america cattolica
Damiron, Rafael *see*
– De nuestro sur remoto
– Resumen a los enemigos de trujillo
Dammbruch : novellen / Blunck, Hans Friedrich – Leipzig: P Reclam, 1940 [mf ed 1989] – 77p – 1 – mf#7036 – us Wisconsin U Libr [830]
Dammertz, V *see* Das verfassungsrecht der benediktinischen moenkskongregation
Damnation de faust / Berlioz, Hector – Paris, France. 1846 – 1r – us UF Libraries [440]
Damoiseau de Monfort, Marie-Charles-Theodore, baron de *see* Tables ecliptiques des satellites de jupiter
Damp, Alice Bancroft *see* Fusion of sacred and secular elements in benjamin britten's vocal and choral literature
Dampf als verfahren zur oberflaechendekontamination in der zahnaerztlichen routinebehandlung / Sturz, Philipp – (mf ed 1998) – 1mf – 9 – €30.00 – 3-8267-2546-8 – mf#DHS 2546 – gw Frankfurter [617]
Das dampfboot *see*
– Coepenicker dampfboot
– Danziger dampfboot
Das dampfboot-berliner ostzeitung *see* Coepenicker dampfboot
Das dampfschiff – Bremen DE, 1838 1 apr-7 oct [gaps] – 1r – 1 – gw Misc Inst [074]
Dampier, Margaret Georgiana *see*
– History of the orthodox church in austria-hungary
– The organization of the orthodox eastern churches
Dampier, William *see*
– A collection of voyages in four volumes, vol 1
– A collection of voyages in four volumes, vol 2
– A collection of voyages in four volumes, vol 3
Dampier's voyages / ed by Masefield, J – London, 1906. 2v – 15mf – 9 – mf#H-6183 – ne IDC [590]
Damrongrachanuphap, Prince, son of Mongkut, King of Siam *see* Iconographie bouddique
Damskii listok – St Petersburg 1910 – 3mf – 9 – mf#nrp-1583 – us UMI ProQuest [305]
Damskii zhurnal – New York. 1959+ (1) 1965+ (5) 1970+ (9) – 128mf – 9 – mf#1191 – ne IDC [077]
Damson, R L *see* The effect of daily bean ingestion on the lipid profiles of normocholesterolemic college students
Dan ba – Hanoi: Impr TBTV. v1 n16, 20-25, 28-31, 34-35 [jul 7, aug 4-sep 9, sep 29-oct 20, nov 17-24, 1939]; v2 n42, 52-54, 56-60, 62-63, 68-70, 72-77, 82 [jan 19, apr 19-may 3, may 17-jun 14, jun 28-jul 5, aug 9-30, sep 13-oct 18, nov 22, 1940] [mf ed Hanoi, Vietnam: National Library of Vietnam 1995] – 1r – 1 – (master neg held by crl) – mf#mf-11830 seam – us CRL [079]
Dan ba moi [ho chi minh city, vietnam] – Saigon: Impr J Viet, feb 17-24, mar 16-dec 28, 1936 – 1r – 1 – mf#mf-11120 seam – us CRL [079]
Dan bao – Hanoi. n2-14. 2-16 mai 1927 – 1 – fr ACRPP [073]
Dan, D *see* Die voelkerschaften der bukowina
Dan, F *see* Rabochie deputaty v i-oi gosudarstvennoi dume
Dan leno, hys booke / Leno, Dan – London, England. 1899 – 1r – us UF Libraries [960]
Dan moi [ho chi minh city, vietnam: 1938] – [Ho Chi Minh City]: Impr speciale du "Dan-moi" [1938- [1938: sep 5-oct 29, dec 31. 1939 1 iss] – 1r – 1 – mf#mf-11123 seam – us CRL [079]
Dan, P *see* Le tresor des merveilles de la maison royale de fontainebleau...
Dan qing yu *see* Gargata manju gisun-i bithe
Dan sha / Ye Sa To Communications Society – v13 n1-v16 n7 [1986 jan 10-1989 aug] – 1r – 1 – (cont: yukon indian news; cont by: dannzha) – mf#1580239 – us WHS [071]
Dan smoot report – Dallas TX 1958-71 – 1 – ISSN: 0011-5975 – mf#2311 – us UMI ProQuest [071]
Dan song – Saigon: [s.n.] 1973- [mf ed Ithaca NY: microfilmed by Challenge Industries [for] Cornell U Libr John M Echols Coll – 1r with 4 other items – 1 – (v1 n1-31 [apr 15-may 17 1973]) – mf#12566 n3 – us Cornell [959]
Dan to beersheba : work and travel in four continents / Colquhoun, Archibald Ross – London: William Heinemann 1908 – (= ser [Travel descriptions from south africa, 1711-1938]) – 4mf – 9 – (frontispiece fr painting by herman g herkomer; facs letter of cecil rhodes) – mf#zah-80 – ne IDC [910]
Dana : an irish magazine of independent thought – Dublin. v. 1 no. 1-12. May 1904-Apr 1905 – 1 – us NY Public [941]
Dana : an irish magazine of independent thought – Killen TX 1904-05 – 1 – mf#5146 – us UMI ProQuest [071]
Dana 1801-1849 – Oxford, MA (mf ed 1996) – (= ser Massachusetts vital record transcripts to 1850) – 4mf – 9 – 0-87623-236-5 – (mf 1t: marriage intentions 1801-18. mf 2t: marriage & intentions 1817-31. mf 3t: marriages & intentions 1831-43. mf 4t: births, marriages 1843-49; deaths 1844-47) – us Archive [978]
Dana 1801-1890 – Oxford, MA (mf ed 1984) – (= ser Massachusetts vital records) – 57mf – 9 – 0-931248-72-8 – (mf 1: marriages & intentions 1801-18. mf 2-3: marriages & intentions 1818-36. mf marriages & intentions 1837-43. mf 5-6: b,m,d 1843-67. mf 7-10: b,m,d 1855-93. mf 11-21: index to births 1843-1938. mf 22-29: index to marriage intentions 1801-1938. mf 30-46: index to marriages 1801-1938. mf 47-57: index to deaths 1844-1938. mf 58: b,m,d 1891-93) – us Archive [978]
Dana, Arnold Guyot *see* Porto rico's case
Dana, Charles Anderson *see* Haifa
Dana, J D *see* Geology
Dana, James Dwight *see* The genesis of the heavens and the earth and all the host of them
Danab : warsidaha jabhadaha gobanimadoonka soomaaliyeed = Bulletin of somali liberation fronts – Mogadisho: WSLF: SALF. n16-376 jan 2-sep 27 1978. n519,577,614 mar 21, sep 27, nov 15 1979 – 1r – 1 – us CRL [321]
Danache, B *see* President dartiguenave et les americains
Danache, Berthomieux *see* Choses vues
Danbury, New Hampshire. Danbury Baptist Church *see* Records
Danbury news – Danbury, NE: Smith Bros. - v24. jun 15 1922 (wkly) [mf ed v2 n26. sep 27 1895-jun 15 1922 (gaps) filmed 1978] – 6r – 1 – (absorbed: lebanon advertiser oct 3 1918. cont by: news-advertiser. issues for jun 21 1900- called v2 n46- . issue numbering dropped with may 4 1922) – us NE Hist [071]
Danbury topics – Danbury, NE: A C Furman, 1897 (wkly) [mf ed apr 5-may 5 1898] – 1r – 1 – us NE Hist [071]
Danby, Herbert *see* The mishnah
Dance and doctrine : Shaker and mormon dancing as a manifestation of doctrinal views of the physical body / Cielslewicz, Lindsy S – 2000 – 113p on 2mf – 9 – $10.00 – mf#PE 4190 – us Kinesology [790]
Dance and the lived experience : a phenomenological account of a performer's journey / Davis, Amanda J – 2000 – 101p on 2mf – 9 – $10.00 – mf#PE 4191 – us Kinesology [790]
Dance caprice : a colonial dance for any even number of boys and girls / Pratz, Edith – Franklin, OH: Eldridge Entertainment House [1920] – 1 – mf#*ZBD-*MGO pv18 – us Misc Inst [790]
Dance chronicle – Monticello NY 1993+ – 1,5,9 – (cont: dance perspectives) – ISSN: 0147-2526 – mf#16745 – us UMI ProQuest [790]
Dance concert / Arslanian, Sharon Park – [Oakland]: Mills College, 1976 – 1r – r – us Misc Inst [790]
Dance for people with disabilities : guidelines / Imperial Society of Teachers of Dancing – London: ISTD, 1982 (mf ed 1988) – 1mf – 9 – mf#*XMD-43 – us NY Public [615]
Dance gazette – London, England 1968-91 – 1,5,9 – mf#8688 – us UMI ProQuest [790]
Dance history and computer courseware : the design, development, and production of interdisciplinary multimedia courseware for introductory-level instruction in late renaissance european court dance / Manning, Keitha D – Temple University, 1996 – 3mf – 9 – $12.00 – mf#PE 3663 – us Kinesology [790]
The dance in classical times : illustrated from the collections of the walters art gallery – Baltimore, 1945 – 1 – mf#*ZBD-*MGO pv19 – Located: NYPL – us Misc Inst [790]
Dance in india / Venkatachalam, Govindraj – Bombay: Nalanda Publications: Chief distributors, NM Tripathi, [194-] – (= ser Samp: indian books) – us CRL [790]

The dance in india / Bowers, Faubion – New York: Columbia University Press, 1953 – (= ser Samp: indian books) – us CRL [790]

Dance is...?: a lecture demonstration for children / Raabe, Julie L – 1981 – 2mf – 9 – $8.00 – us Kinesology [790]

Dance kolos with john filcich – [Oakland, CA: Slav-Art Music Co c1952] – 1 – mf#*ZBD-*MGO pv17 – Located: NYPL – us Misc Inst [790]

Dance magazine – New York NY 1927+ – 1,5,9 – ISSN: 0011-6009 – mf#1478 – us UMI ProQuest [790]

Dance news – New York NY 1972-83 – 1 – ISSN: 0011-6017 – mf#6667 – us UMI ProQuest [790]

Dance notation conversation; facts on the only successful system of recording human movement : labanotation / Cohen, Selma Jeanne – [New York, 195-] – 1 – mf#*ZBD-*MGO pv16 – Located: NYPL – us Misc Inst [790]

The dance of death : should christians indulge? / Straton, John Roach – 10th ed. New York: The Religious Literature Dept, Calvary Baptist Church [1921?] – 1 – mf#*ZBD-*MGO pv4 – Located: NYPL – us Misc Inst [790]

Dance of india / Banerji, Projesh – Allahabad: Kitabistan, 1942 – (= ser Samp: indian books) – us CRL [790]

The dance of modern society / Wilkinson, William Cleaver – New York: Oakley, Mason, 1869 – 1mf – 9 – 0-524-07841-6 – mf#1991-3388 – us ATLA [240]

The dance of shiva : fourteen indian essays / Coomaraswamy, Ananda Kentish – Bombay: Asia Pub House, 1952 – (= ser Samp: indian books) – (int pref by romain rolland) – us CRL [280]

Dance of the millions / Fluharty, Vernon Lee – Pittsburgh, PA. 1957 – 1r – us UF Libraries [972]

The dance of the reed pipes : from the nutcracker, act 2 / Ivanov, Lev – [n.p., n.d.] – 1 – mf#*ZBD-*MGO pv25 – Located: NYPL – us Misc Inst [790]

Dance on the volcano / Chauvet, Marie – New York, NY. 1959 – 1r – us UF Libraries [972]

Dance perspectives – Monticello NY 1971-73 – 1,5 – (cont by: dance chronicle) – ISSN: 0011-6033 – mf#6065 – us UMI ProQuest [790]

Dance research – 1990+ – 1,5,9 – ISSN: 0264-2875 – mf#18488 – us UMI ProQuest [790]

Dance scope – New York NY 1965-81 – 1,5,9 – ISSN: 0011-6041 – mf#6766 – us UMI ProQuest [790]

Dance spirit – New York NY 2001+ – 1,5,9 – ISSN: 1094-0588 – mf#31422 – us UMI ProQuest [790]

Dance teacher – New York NY 2000+ – 1,5,9 – ISSN: 1524-4474 – mf#31420,01 – us UMI ProQuest [790]

Dance Theatre of Harlem *see* Newsletter

Dancers in the dark : implementation of dance movement instruction on five visually-impaired female adults in the salt lake and utah county areas / Scheel, Dana Potts – 1996 – 2mf – 9 – $8.00 – mf#PE 3802 – us Kinesology [790]

The dancer's quest; essays on the aesthetic of the contemporary dance / Selden, Elizabeth S – Berkeley, CA: Univ. of CA. Press, 1935. xv,215p. illus – 1 – us Wisconsin U Libr [790]

Dances; dramatic, classic, artistic, and descriptive dances; court dances of the 16th, 17th, and 18th centuries; folk and national dances of all countries; modern dances / Tanner, Virginia – [Boston, 1912] – 1 – mf#*ZBD-*MGO pv1 – Located: NYPL – us Misc Inst [790]

Dances in lino cut / Khastagira, Sudhira – Dehra Dun: Chandbagh, 1945 – (= ser Samp: indian books) – us CRL [760]

Dances of korea / Cho, Won-Kyung – [New York: Dance Notation Bureau, 1962?] – 1 – mf#*ZBD-*MGO pv24 – Located: NYPL – us Misc Inst [790]

Dancing : steps posed by miss anna west, charles f burgess / Burgess, Charles F – New York: M King inc [191-?] – 1 – mf#*ZBD-*MGO pv4 – Located: NYPL – us Misc Inst [790]

Dancing and dancing parties / Stock, John – London, England. 18-- – 1r – us UF Libraries [240]

Dancing and its relations to education and social life / Dodworth, Allen – new enl ed. New York: Harper & Bros 1888 [mf ed 1987] – 1r [ill] – 1 – mf#10120 – us Wisconsin U Libr [790]

Dancing and the public schools / Hughes, Matthew Simpson – New York: Methodist Book Concern [c1917] – 1 – mf#*ZBD-*MGO pv1 – Located: NYPL – us Misc Inst [370]

Dancing as an amusement for christians : a sermon delivered in the brainerd presbyterian church, new york, feb 14 1847... / Smith, Asa Dodge – New York: Printed by Leavitt, Trow, 1847 – 1 – mf#*ZBD-*MGO pv19 – Located: NYPL – us Misc Inst [230]

Dancing denounced by prominent leaders of the liberian pulpit / Phillips, James E – Monrovia: Liberia, College of West Africa Press, 1907 – 1r – 1 – mf#*ZBD-*MGO pv6 – Located: NYPL – us Misc Inst [230]

The dancing english / English Folk Dance and Song Society – [London, 1949?] – 1 – mf#*ZBD-*MGO pv17 – Located: NYPL – us Misc Inst [790]

Dancing in dixie's land : theatrical dance in new orleans, 1860-1870 / Akins, Ann S & LaPointe-Crump, Janice D – 1991 – 4mf – $16.00 – us Kinesology [790]

Dancing news – Sydney, nov 1953-may 1955 – 1r – at Pascoe [079]

Dancing times, 1910-1951 – 2pts – 30r – 1 – (pt 1: 1910-sep 1930 15r. pt 2: oct 1930-51 15r) – us Primary [790]

Danckers, J *see* Architectura chivilis

Danckerts, C *see* Architectura moderna ofte bouwinge van onsen tyt

Dancoisne, Louis *see* Histoire des etablissements religieux britanniques fondes a douai avant la revolution francaise

Dandadzo rengano / Wakatama, Pius – Salisbury, Zimbabwe. 1967 – 1r – us UF Libraries [960]

Dandekar, Ramchandra Narayan *see* Vedic bibliography

Dandelion – v1 n1-v5 n18 [1977 spring-1982 spring/winter] – 1r – 1 – mf#962164 – us WHS [071]

Dandin *see*
- Dandin's dasha-kumara-charita
- The dasakumaracharita of dandin

Dandin's dasha-kumara-charita : the ten princes / Dandin – Chicago, IL: University of Chicago Press, 1927 – (= ser Samp: indian books) – (trans fr sanskrit by arthur w ryder) – us CRL [830]

Dandouau, Andre *see*
- Contes populaires des sakalava et des tsimihety de la region d'analalava
- Histoire des populations de madagascar

Dandre, Paul [Pseud] *see* Fin mot

Dandridge baptist church. jefferson county. tennessee : church records – 1786-1966 – 1 – us Southern Baptist [242]

Dandridge first baptist church. jefferson county. dandridge, tennessee : church records – 1786-1966 – 1 – us Southern Baptist [242]

Dandurand, J L *see* Le cinema canadien

Dandurand, Josephine *see*
- La carte postale
- Ce que pensent les fleurs
- Contes de noel
- Nos travers

Dandurand, Raoul *see*
- Les ecoles primaires et l'enseignement obligatoire
- Manuel de police a l'usage de la police de montreal
- Traite theorique et pratique de droit criminel

Dandurand, Therese *see* Bio-bibliographie de monsieur le chanoine victor tremblay

Dane bridge, st paul: burials 1846-1956 – (= ser Cheshire church registers) – 1mf – 9 – £2.50 – mf#106 – uk CheshireFHS [929]

Dane county advocate – Madison WI. 1902 oct 2-dec 19 – 1r – 1 – mf#921468 – us WHS [071]

Dane County Bicentennial Committee [WI] *see* Dane county crier

Dane County Childcare Union, District 65, UAW *see* Bulletin board

Dane County Conservation League [WI] *see* Newsletter

Dane county crier / Dane County Bicentennial Committee [WI] – v1 n1-2 [1975 aug 14-sep 23] – 1r – 1 – mf#365077 – us WHS [071]

Dane county democrat – Sun Prairie WI. 1895 sep 18-1896 aug 27 – 1r – 1 – mf#933710 – us WHS [071]

Dane county dialogue – 1975 jul-1978 mar – 1r – 1 – (cont: newsletter [dane county [wis.]. dept. of social services]; cont by: dialogue [madison wi]) – mf#1002552 – us WHS [071]

Dane county digest – Mcfarland WI. 1972 mar 5-apr 15 – 1r – 1 – mf#942025 – us WHS [071]

Dane county directory / Directory Service Co – 1903 may, 1904, 1907 aug, 1908 jul, 1918, 1935/36 – 4r – 1 – mf#2697695 – us WHS [917]

Dane county historical society – v1 n1-v2 n4 [1983 nov-1985 jul] – 1r – 1 – mf#1094165 – us WHS [978]

Dane county news – Black Earth WI. 1915 sep 3/1917 feb 16-1961 jan/dec 29 – 26r – 1 – (with small gaps; cont: black earth news; cont by: mazomanie sickle; dane county news and mazomanie sickle) – mf#938318 – us WHS [071]

Dane county news and mazomanie sickle – Black Earth, Mazomanie WI. 1962 jan 5-1963, 1964 jan-aug 27 – 2r – 1 – (cont: dane county news; mazomanie sickle; cont by: cross plains arrow; dane county news, mazomanie sickle, cross plains arrow) – mf#938322 – us WHS [071]

Dane county news, mazomanie sickle, cross plains arrow – Black Earth, Cross Plain, Mazomanie WI 1964 sep 3/dec-1991 jul 4/1992 jan 9 – 36r – 1 – (cont: dane county news and mazomanie sickle; cross plains arrow; cont by: news-sickle-arrow) – mf#938329 – us WHS [071]

Dane county populist – 1892 sep 10-nov 1 – 1r – 1 – (cont by: wisconsin populist) – mf#917567 – us WHS [071]

Dane county sun – Mount Horeb WI. 1892 jan 28-1893 may 25 – 1r – 1 – (cont: sun [mount horeb wi]) – mf#1093478 – us WHS [071]

Dane County WIC Nutrition Program *see* Wic news

Dane County Women's Political Caucus *see* Newsletter

Dane, Nathan *see* Dane's general abridgement of law and equity

Daneau, Lambert *see*
- Ad bellarmini disputationes responsio
- Assertio contra scriptum de adoratione carnis christi
- Christianae isagoges ad locos communes, libri 2...
- Commentarii in prophetas minores
- Compendium sacrae theologiae...
- Confirmatio doctrinae orthodoxae contra genebardi scriptum
- Controversiarum roberti bellarmi responsio
- Deux traitez
- Ethices christianae
- Isagoges pars altera...de angelis...et de ecclesia
- Isagoges pars quinta quae est de homine
- Isagoges...pars quarta de salutaribus dei donis erga ecclesiam...
- Opuscula omnia theologica
- Politices christianae libri septem...

Daneel, M L *see*
- God of the matopo hills
- Zionism and faith-healing in rhodesia

Danemarks recht auf gronland / Berlin, Knud Kugleberg – Konigsberg, Germany. 1932 – 1r – us UF Libraries [948]

Dane's general abridgement of law and equity / Dane, Nathan – Boston, MA: Cummins, Hilliard & Co. v1-8. 1823-1824 (all publ) – 64mf – 9 – $96.00 – mf#LLMC 81-404 – us LLMC [340]

Danfelt, Edwin Douglas *see*
- Clarinet choir
- Clarinet choir: a means of teaching and performing music

Danford, J *see* Diary of the siege of quebec, 1775

D'Anfreville de la salle *see* Sur la cote d'afrique

D'Angel, Arnaldo *see* San vicente paul, director de conciencia

Danger lines in the deeper life / Simpson, Albert B – South Nyack, NY: Christian Alliance Pub Co, c1898 [mf ed 1992] – (= ser Christian & missionary alliance coll) – 1mf – 9 – 0-524-02143-0 – mf#1990-4209 – us ATLA [221]

Danger of delay – London, England. no date – 1r – us UF Libraries [240]

The danger of delay : and the safety and practicability of immediate emancipation, from the evidence before the parliamentary committees on colonial slavery – London, 1833 – (= ser 19th c books on british colonization) – 1mf – 9 – mf#1.1.7440 – uk Chadwyck [331]

Danger of opposing christianity and the certainty of its final triu... / Brown, John – Edinburgh, Scotland. 1816 – 1r – us UF Libraries [240]

Danger signals : the enemies of youth, from the business man's standpoint (containing advice to the young on the evils of the day from many merchants of Boston) / Clark, Francis Edward – Boston: Lee & Shepard; New York: C T Dillingham, 1885 – 3mf – 9 – mf#27440 – cn Canadiana [170]

Dangerous nature of popery / Paterson, Nathaniel – Glasgow, Scotland. 1836 – 1r – us UF Libraries [240]

A dangerous occupation – [Toronto?: Dominion Alliance for the Suppression of the Liquor Traffic, 189-] [mf ed 1992 – 1mf – 9 – 0-665-90931-4 – (original iss in ser: campaign leaflets) – mf#90931 – cn Canadiana [360]

A dangerous plot discovered... : by a discourse... / Wotton, A – London: Nickolas Bourne, 1626 – 4mf – 9 – mf#PW-60 – ne IDC [240]

Dangerous properties of industrial materials report – Bemidji MN 1980-96 – 1,5,9 – ISSN: 0270-3777 – mf#13078 – us UMI ProQuest [360]

The dangers and safeguards of modern theology : containing 'suggestions offered to the theological student under present difficulties' (a revised edition), and other discourses / Tait, Archibald Campbell – London: John Murray, 1861 – 1mf – 9 – 0-8370-2243-6 – mf#1985-0243 – us ATLA [240]

Dangers of the apostolic age / Moorhouse, James – New York: Thomas Whittaker, 1891 – 1mf – 9 – 0-8370-4484-7 – mf#1985-2484 – us ATLA [240]

Dangers of the day / Vaughan, John Stephen – Notre Dame, IN: Ave Maria Press, 1909 – 1mf – 9 – 0-8370-7197-6 – mf#1986-1197 – us ATLA [240]

D'Angerville *see* Vie privee de louis 15 [quinze]

Danglade, Ernest *see*
- Condition and extent of the natural oyster beds an...
- Flatworm as an enemy of florida oysters

Dangond Uribe, Alberto *see* Charlas con el presbitero jeronimo

Daniaud, J M *see* The wonders of arithmetic

Danica hrvatska *see* Hrvatski list and danica hrvatska

Danicheff / Newsky, Pierre – Paris, France. 1881 – 1r – us UF Libraries [440]

Daniel : an exposition of the historical portion of the writings of the prophet daniel / Payne Smith, Robert – Cincinnati: Cranston & Curts; New York: Hunt & Eaton, [18-?] – 1mf – 9 – 0-8370-9725-8 – mf#1986-3725 – us ATLA [221]

Daniel : gespraeche von der verwirklichung / Buber, Martin – Berlin: Schocken [19--] [mf ed 1989] – 1r – 1 – (filmed with: hofische spuren im protestantischen schuldrama um 1600 / hildegard schaefer) – mf#7093 – us Wisconsin U Libr [080]

Daniel : his life and times / Deane, Henry – New York: Anson D F Randolph, [1888] – 1mf – 9 – 0-8370-2855-8 – mf#1985-0855 – us ATLA [221]

Daniel and his prophecies / Wright, Charles Henry Hamilton – London: Williams and Norgate, 1906 – 1mf – 9 – 0-7905-0473-1 – (incl bibl ref and indexes) – mf#1987-0473 – us ATLA [221]

Daniel and its critics : being a critical and grammatical commentary / Wright, Charles Henry Hamilton – London: Williams and Norgate, 1906 – 1mf – 9 – 0-8370-6550-X – (incl indes) – mf#1986-0550 – us ATLA [221]

Daniel and john : or, the apocalypse of the old and that of the new testament / Desprez, Philip Soulbien – London: C Kegan Paul, 1878 [mf ed 1993] – 1v on 2mf – 9 – 0-524-05717-6 – (incl bibl ref) – mf#1992-0560 – us ATLA [220]

Daniel axtell account book and letter entries, 1700-1711 – Boston MA: Massachusetts Historical Soc [197-?] [mf ed South Carolina Historical Soc] – 1r – 1 – (not to be repr without the permission of the mhs) – mf#45-363 – us South Carolina Historical [978]

Daniel, Bishop Of Calcutta *see*
- Prayer, the refuge of a distressed church
- Prince of peace

Daniel casper von lohenstein's trauerspiele : mit besonderer beruecksichtigung der cleopatra: beitrag zur geschichte des dramas im 17. jahrhundert / Kerckhoffs, August – Paderborn: F Schoeningh 1877 [mf ed 1990] – 1r – 1 – (incl bibl ref. filmed with: der dichter siegfried lipiner (1856-1911) / comp by hartmut von hartungen) – mf#2827p – us Wisconsin U Libr [430]

Daniel, Charles *see* One of the jesuits

Daniel, Charles T *see* William and annie

Der daniel der roemerzeit : ein kritischer versuch zur datierung einer wichtigen urkunde des spaetjudentums / Hertlein, Eduard – Leipzig: M Heinsius Nachfolger, 1908 – 1mf – 9 – 0-8370-3573-2 – mf#1985-1573 – us ATLA [221]

Daniel discoverer and documenter – n1-52 [1976 jan/feb-1988 nov] – 1r – 1 – mf#1494916 – us WHS [071]

Daniel ellis byrd papers – ca 6r – 1 – ca $780.00 – (from the coll of the amistad research center. guide also sold separately $25 s3519.g) – mf#S3519 – us Scholarly Res [305]

Daniel, Evan *see* How to teach the church catechism

Daniel, F de F *see* History of katsina

Daniel, Francois *see*
- La famille de salaberry
- Les francais dans l'amerique de nord
- Histoire des grandes familles francaises du canada
- D'iberville
- Nos gloires nationales
- Officiers de l'acadie, plaisance et ile-royale
- Precis historique
- Le vicomte c de lery

Daniel, Gabriel *see* Histoire apologetique de la conduite des jesuites de la chine

Daniel, H A *see*
- Codex liturgicus ecclesiae universa
- Thesaurus hymnologicus

Daniel, Hermann A *see* Deutschland nach seinen physischen und politischen verhaeltnissen
Daniel, Hermann Adalbert *see* Codex liturgicus ecclesiae lutheranae
Daniel in the critics' den : a reply to dean farrar's 'book of daniel' / Anderson, Robert – Edinburgh: William Blackwood, 1895 [mf ed 1984] – (= ser Biblical crit – us & gb 1) – 2mf – 9 – 0-8370-0175-7 – (incl bibl ref & ind) – mf#1984-1001 – us ATLA [221]
Daniel, J *see* Die lehre von der unfehlbarkeit des papstes aus der geschichte
Daniel, J C *see* A history of the baptists of hill county, texas
Daniel, John S *see* Principes de traitements thermiques notes et objectifs du cours 5.417
Daniel, John Warwick *see* The law and practice of attachment, under the code of virginia.
Daniel mac-kinnen's, esq. reise nach dem brittischen westindien und besonders nach den bahama-inseln : in den jahren 1802 und 1803; aus dem englischen – Weimar 1805 [mf ed Hildesheim 1995-98] – 1v on 2mf – 9 – €60.00 – 3-487-26574-5 – gw Olms [918]
Daniel o'connell upon american slavery : with other irish testimonies – New York: American Anti-slavery Society, 1860 [mf ed 1991] – (= ser Anti-slavery tracts. new series 5) – 1mf – 9 – 0-524-01235-0 – mf#1990-0374 – us ATLA [976]
Daniel, Pete *see* The peonage files of the u s department of justice, 1901-1945
Daniel pfund : [a novel] / Huggenberger, Alfred – Frauenfeld: Huber 1917, c1916 [mf ed 1995] – (= ser Schweizerische erzaehler 3) – 1r – 1 – (filmed with: dem bollme sy boes wuche & other titles) – mf#3884p – us Wisconsin U Libr [830]
Daniel quorm : and his religious notions / Pearse, Mark Guy – London: Wesleyan Conference Off, 1875 [mf ed 1984] – 3mf – 9 – 0-8370-0836-0 – mf#1984-4230 – us ATLA [240]
Daniel quorm : and his religious notions: second series / Pearse, Mark Guy – London: Charles H Kelly, 1890 [mf ed 1984] – 3mf – 9 – 0-8370-0837-9 – mf#1984-4231 – us ATLA [240]
Daniel sweetland : [novel] / Phillpotts, Eden – Toronto: McLeod & Allen, c1906 – 4mf – 9 – 0-659-90444-6 – (incl aut's list. ill by frank parker) – mf#9-90444 – cn Canadiana [830]
Daniel the beloved / Taylor, William MacKergo – New York: Harper, c1878 – 1mf – 9 – 0-8370-9311-2 – (incl bibl ref and index) – mf#1986-3311 – us ATLA [920]
Daniel the fearless / Royer, Galen Brown – Elgin, IL: Brethren Pub House, 1901 – 1mf – 9 – 0-524-03859-7 – mf#1990-4906 – us ATLA [221]
Daniel the prophet : nine lectures delivered in the divinity school of the university of oxford, with copious notes / Pusey, Edward Bouverie – 2nd ed. Oxford: sold by James Parker, 1868 [mf ed 1984] – (= ser Biblical crit – us & gb 34) – 9mf – 9 – 0-8370-0213-3 – (incl bibl ref) – mf#1984-1034 – us ATLA [221]
Daniel und die griechische gefahr / Bertholet, Alfred – Tuebingen: J C B Mohr (Paul Siebeck) 1907, c1905 [mf ed 1989] – (= ser Religionsgeschichtliche volksbuecher fuer die deutsche christliche gegenwart 2/17) – 1mf – 9 – 0-7905-0727-7 – mf#1987-0727 – us ATLA [221]
Daniel vs darwinism / Riley, William Bell – Minneapolis: HM Hall [1918?] – 1mf – 9 – 0-524-07589-1 – mf#1991-3209 – us ATLA [230]
Daniel webster / Lodge, Henry Cabot – Boston, MA. 1911 – 1r – us UF Libraries [975]
Daniel webster / Lodge, Henry Cabot – Boston & NY: Houghton, Mifflin & Co, 1899 – (= ser The american statesmen series) – 5mf – 9 – $7.50 – mf#LLMC 96-030 – us LLMC [975]
Daniel...expositvs homilijs 66. epitome temporvm / Bullinger, Heinrich – Tigvri, C[hristoph] Froschover, 1565 – 10mf – 9 – mf#PBU-225 – ne IDC [240]
Danielis prophetae explicatio brevis / Wigand, J – Ienae, 1571 – 1mf2 – 9 – mf#TH-1 mf 1490-1501 – ne IDC [242]
Daniell, Clarmont *see* Gold in the east
Daniell, Samuel (drawings) *see* Sketches representing the native tribes, animals and scenery of southern africa
Danielli, Mary Guy *see* The map of the world
Danielo, Julien *see* Histoire de toutes les villes de france
Danielou, Alain *see*
– Northern indian music
– Yoga
Danielowski, Emma *see* Richardsons erster roman
Daniels, A *see*
– Eine lateinische rechtfertigungsschrift des meister eckhart
– Quellenbeitraege und untersuchungen zur geschichte der gottesbeweise im dreizehnten jahrhundert
Daniels, Augustus Thatcher *see* Diary
Daniels bok – Upsala: Almqvist & Wiksell, 1894 – 1mf – 9 – 0-7905-2142-3 – mf#1987-2142 – us ATLA [221]
Daniel's great prophecy : the eastern question / West, Nathaniel – New York: Hope of Israel Movement, 1898 – 1mf – 9 – 0-8370-5800-7 – mf#1985-3800 – us ATLA [221]
Daniels, William H *see* The bank of india
Daniels, William Haven *see*
– The illustrated history of methodism
– Moody
Danielson, Vernon J *see* Mormonism exposed
Danielstudien / Bayer, Edmund – Muenster i W: Aschendorff, 1912 [mf ed 1993] – (= ser Alttestamentliche abhandlungen 3/5) – 1mf – 9 – 0-524-07056-3 – mf#1992-1019 – us ATLA [221]
Danik basumati – Calcutta, India. 14 Oct 1946-4 Sept 1947; Sept 1950-Aug 1952; 12 Sept 1953-57; 1960-Aug 1966; Aug 1968-Sept 1970; Sept 1971-76; May 1977-Aug 1992 – 103r – 1 – us L of C Photodup [079]
Danilova, E *see* Zakonodatelstvo o trudovykh arteliakh
Danilovaia, E N *see* Trudovye arteli
Danim, Fitnat *see* The divan project
Danish – Tehran. shumarah-i 1-30. 10 ramazan 1328-27 rajab 1329 [16 sep 1910-25 jul 1911] – 1r – 1 – $300.00 – us MEDOC [079]
Danish Baptist General Conference of America *see* Papers
Danish Evangelical Lutheran Church Association in America *see* Kirke-bladet
Danish Evangelical Lutheran Church in North America et al *see* Misions-budet
Danish islands / Parton, James – Boston, MA. 1869 – 1r – us UF Libraries [948]
Danish medical bulletin – Copenhagen, Denmark 1954+ – 1,5,9 – ISSN: 0907-8916 – mf#9773 – us UMI ProQuest [610]
Danish west indies under company rule (1671-1754) / Westergaard, Waldemar Christian – New York, NY. 1917 – 1r – us UF Libraries [972]
Danishkadah – Isfahan. sal-i 1, shumarah-i 2-5,7. 2 hut va hamal 1304-4 bahman 1305 [22 mar 1925-17 jan 1926]; sal-i 2, shumarah-i 1-5. 31 tir 1313-farvardin 1314 [22 july 1934-mar 1935]) – 1r – 9 – $125.00 – us MEDOC [079]
Danishkadah – Tehran, 1918- . sumarah 1-11/12 [21 apr 1918-20 apr 1919] – 1r – 1 – $53.00 – us MEDOC [079]
Danishnamah – Tehran. shumarah-i 2 aban 1326 [oct 1947] – 1r – 1 – $53.00 – us MEDOC [079]
Daniyel, ezra u-nehemieh : textum masoreticum accuratissime expressit e fontibus masorae codicumque varie illustravit adumbrationem chaldaismi biblici adjecit = Libri danielis, ezrae, et nehemiae – Lipsiae [Leipzig]: Bernhardi Tauchnitz, 1882 – 3mf – 9 – 0-7905-8330-5 – mf#1987-6429 – us ATLA [221]
D'anjou, Rene *see* Traite des tournois (cima32)
Dank an stalingrad : dichtungen / Becher, Johannes Robert – Moskau: Verlag fuer Fremdsprachige Literatur, 1943 [mf ed 1989] – 119p – 1 – mf#6993 – us Wisconsin U Libr [810]
Dank und dienst : reden und aufsaetze / Alverdes, Paul – Muenchen: A Langen/G Mueller, 1939 [mf ed 1988] – 292p – 1 – mf#6939 n4 – us Wisconsin U Libr [850]
Der dankbare patient / Penzoldt, Ernst – 7.-9. aufl. Berlin: S Fischer 1940 [mf ed 1991] – 1r – 1 – (filmed with: der mensch an der wege / rudolf paulsen) – mf#2859p – us Wisconsin U Libr [880]
Dankevich, Vladimir *see* Pouchenya v ograzhdenie pravoslavnykh ot shtundistskikh zabluzhdeni
Danmark ekspeditionen til gronlands nordostkyst / Friis, A – [Kobenhavn], 1909 – 13mf – 9 – mf#H-476 – ne IDC [917]
Danmark i fest og glaede – [Copenhagen]: C Erichsen 1935-36 [mf ed Bloomington IN: Indiana Uni Lib, Preservation Dept 1984] – 6v on 2r – 1 – us Indiana Preservation [390]
Dann, George James *see*
– First lessons in urdu
– An introduction to hindi prose composition
Dann, Henry E *see* The new york jewish citizen
Dann, Mary G *see* Harmonic technique of edward elgar based upon the dream of gerontius
Danna, Joseph G *see* Division i-a football recruiting violations reported by the national collegiate athletic association from 1980 through 19[9]6
Dannaud, Jean Pierre *see* Cambodge
Dannebrog news – Dannebrog, NE: J M Erickson, 1898-v49 n53. jan 1 1959 (wkly) [mf ed 1902-59 (gaps) filmed [1972]] – 16r – 1 – (absorbed by: phonograph (1911). publ in dannebrog -aug 12 1943; in st paul, aug 19 1943-) – us NE Hist [071]
Dannebrog sentinel – Dannebrog, NE: P Ebbeson. v1 n1. m[ay] 12 1888-jun 15 1889 (wkly) [mf ed 2000] – 1r – 1 – us NE Hist [071]
Dannenberg, Carl *see* Harmlose betrachtungen
Danner, Tracy *see* Running economy following an intense cycling bout in trained female duathletes and triathletes
Dannevirke : haderslev avis – Hadersleben (DK), 1838-1840/41, 1844/45, 1860-63, 1868 1 jul-1874, 1876-78, 1883-93 – 1 – (haderslev avis) – gw Misc Inst [074]
Dannevirke – Cedar Falls, IA: Holst & Christiansen, 1880-72 aarg n39. 10 okt 1951 [wkly] [mf ed 1885-95 [gaps] filmed [1966?]] – 1r – 1 – (in danish; cont: dannevirke [racine wi]; absorbed by: decorah-posten og ved arnen) – us NE Hist [071]
Dannevirke – Ceder Falls IA. 1899 jan 4/1900 may 20-1949/1951 oct 10 – 35r – 1 – (with gaps; cont by: decorah posten og ved arnen) – mf#853916 – us WHS [071]
Danou et al *see* Recueil des historiens des gaules et de la france
Dans- en bewegingsterapie as groepterapietegniek in die opvoedkundige sielkunde = Dance/movement therapy as a group therapy technique in educational psychology / Brand, Lucia – Uni of South Africa 2000 [mf ed Johannesburg 2000] – 4mf [ill] – 9 – (text in afrikaans; abstract in afrikaans and english; incl bibl ref) – mf#mfm14841 – sa Unisa [150]
Dans la haute-gambie : voyage d'exploration scientifique, 1891-1892 / Rancon, Andre – Paris: Societe d'Editions Scientifiques, 1894 – 1 – us CRL [916]
Dans la jungle du gabon / Weite, Pierre – Paris, France. 1954 – 1r – us UF Libraries [960]
Dans la melee / Paret, Timothee – Paris, France. 1932 – 1r – us UF Libraries [972]
Dans le brousse : sensations du soudan / Bonnetain, Paul (Madame) – Paris, New York: A Lemerre, 1895 – 1 – us CRL [916]
Dans les brousses africaines – Port-Au-Prince, Haiti. 1935 – 1r – us UF Libraries [972]
Dans les marches tibetaines : autour du dokerla, novembre 1906-janvier 1908 / Bacot, Jacques – Paris: Plon-Nourrit, 1909 [mf ed 1995] – (= ser Yale coll) – iii/215p (ill) – 1 – 0-524-09858-1 – (in french) – mf#1995-0858 – us ATLA [951]
Dans les montagnes rocheuses / Mandat-Grancey, Edmond, Baron de – Paris: E Plon, Nourrit & cie, 1884 – 4mf – 9 – mf#09614 – cn Canadiana [917]
Dans l'ouest africain / Segonzac, M R de – In "Revue des deux mondes". 1891. Paris. 1891 – 1 – us CRL [960]
Dans notre empire noir : preface du general de trentinian / Rondet-Saint, Maurice – Paris: Societe d'Editions Geographiques, Maritimes et Coloniales, 1929 – 1 – us CRL [960]
Dans og kvaddigining paa faeroerne / Thuren, Hjalmar – Kobenhavn: A F Host 1901 [mf ed Bloomington IN: Indiana Uni Lib, Preservation Dept 1984] – 1r – 1 – us Indiana Preservation [390]
Dansbeskrivningar – Helsingfors: [s.n.] 1938 [mf ed Bloomington IN: Indiana Uni Lib, Preservation Dept 1984] – 1r – 1 – us Indiana Preservation [390]
Dansby, George William *see* Example of how an enterprise of citrus is analyzed for use in teaching
La danse : dancing, paris-dancing, danse de nos jours reunis – Paris, [1920- (mthly) – 1r – 1 – (nov/dec 1924 lacks subtitle; absorbed: dancing, paris-dancing and danse de nos jours, jan 1921?) – mf#*ZAN-*MD20 – Located: NYPL – us Misc Inst [790]
La danse aux miroirs : essai de reconstitution d'une danse pharaonique de l'ancien empire / Hickmann, Hans – Le Caire: Impr de l'Institut Francais d'Archeologie Orientale, 1956 – 1 – (incl bibl footnotes) – mf#*ZBD-*MGO pv20 – Located: NYPL – us Misc Inst [790]
Danse des vagues / Laleau, Leon – Port-Au-Prince, Haiti. 1919 – 1r – us UF Libraries [972]
Danse sur le volcan / Chauvet, Marie – Paris, France. 1957 – 1r – us UF Libraries [972]
Dansereau, Arthur *see* La crise politique de quebec
Dansereau, Clement Arthur *see* Reponse a une adresse de l'assemblee legislative, en date du 18 mars 1885
Danses folkloriques haitiennes / Honorat, Michel Lamartiniere – Port-Au-Prince, Haiti. 1955 – 1r – us UF Libraries [390]
Danses populaires au cambodge : d'apres les travaux d'etudes et de recherches de la commission des moeurs et coutumes aves le concours des membres correspondants redige par chap pin...[et al]; illustrations, chap nou – Phnom-Penh: Editions de l'Institut bouddhique 1964 [mf ed 1990] – 1r with other items – 1 – (at head of title: commission des moeurs & coutumes du cambodge; in khmer; title also in khmer: rapam prajapriy khmaer) – mf#mf-10289 seam reel 122/1 [§] – us CRL [390]
Dansey, William *see* Letter to the archdeacon of sarum on ruri-decanal chapters
Dansk botanisk arkiv – Copenhagen, Denmark 1976-79 – 1,5,9 – mf#9146 – us UMI ProQuest [580]
Dansk luthersk kirkeblad / Norsk-Danske Evangelisk-Lutherske Kirke i Amerika – v1 n1-v6 n18 [1877 aug-1883 sep 15] – 1r – 1 – (cont by: kirkebladet [blair ne]) – mf#1321573 – us WHS [242]
Dansk luthersk kirkeblad / Norsk-danske evangelisk-lutherske kirke i Amerika & Norwegian-Danish Conference. Danish pastors – Racine, WI: udgivet af de danske prœster i konferentsen for den norsk-danske evangelisk-lutherske kirke i Amerika. 7v. v1 n1. aug 1877-7de aarg n18. 15de sep 1884 (semimthly) [mf ed 1879-84 (gaps) filmed 1975?] – 1r – 1 – (in danish. cont by: kirke-bladet. publ in argo ne, 1ste nov 1879-15de marts 1883; blair ne, 1ste apr 1883-1ste apr 1884; st paul ne, 15de apr-15de sep 1884) – us NE Hist [071]
Dansk luthersk kirkeblad : udgivet of den forenede danske evangelisk lutherske kirke i amerika / United Danish Evangelical Lutheran Church in America – Blair, NE: Danish Lutheran Publ House. 25v. 1 aarg n1. 1ste nov 1896-25 aarg n52. 29 dec 1920 [wkly] [mf ed filmed 1975?] – 10r – 1 – (in danish; formed by the union of: kirke-bladet and: missions-budet; merged with: danskeren to form: lutharsk ugeblad) – us NE Hist [071]
Dansk luthersk mission i amerika : i tiden foer 1884 / Vig, Peter Sorensen; ed by United Danish Evangelical Lutheran Church in America – Blair NE: Danish Lutheran Pub House 1917 [mf ed 1992] – 1mf – 9 – 0-524-02172-4 – mf#1990-4238 – us ATLA [242]
Dansk ornithologisk forenings tidsskrift / ed by Helms, O et al – Kjobenhavn 1906-46 – v1-40 on 174mf – 9 – mf#z-2017c/2 – ne IDC [590]
Dansk pennig magazin – Kopenhagen (DK), 1834 aug-1838 jul – 1r – 1 – gw Misc Inst [074]
Dansk pioneer – Elmwood Park IL, Omaha NE. 1895 oct 17/1896 aug 20-2000/01 – 64r – 1 – (cont: dansk tidende) – mf#856474 – us WHS [071]
Dansk skovforening *see* Tidsskrift for skovvaesen
Dansk tidende – Chicago: Dansk Tidende, mar 20 1931-mar 1952 – 11r – us CRL [071]
Dansk tidende – Chicago IL. 1921 sep 25-1922 dec 30, 1923 jan 6-1924 jul 27, 1924 jul 4-1925 dec 4, 1942 may 7-1944 feb 10, 1944 feb 17-1947 nov 20, nov 27-1949 dec 23, 1952 jan 4-mar 28 – 7r – 1 – (cont: dansk tidende og revyen; cont by: danske pioneer) – mf#1422619 – us WHS [071]
Dansk tidende og revyen – Chicago: Dansk Tidende, Inc, 1921-31. 1925-mar 13 1931 – 7r – us CRL [071]
Dansk vestindien / Larsen, Kay – Kobenhavn, Denmark. 1928 – 1r – us UF Libraries [948]
De danske baptisters historie i amerika / Lawdahl, Nels Soerensen – Morgan Park, IL: Forfatterens Forlag, 1909 [mf ed 1992] – (= ser Baptist coll) – 544p on 2mf – 9 – 0-524-06956-5 – mf#1990-5320 – us ATLA [242]
Det danske bibelselskabs arbog – 1966-92 [complete] – 1 – (cont by: nyt fra bibelselskabet) – mf#atla s0746 – us ATLA [220]
Den danske diakonissestiftelses arbog – 1962-86 [complete] – 2r – 1 – mf#ATLA S0869 – us ATLA [240]
De danske domkapitler : deres oprindelse, indretning og virksomhed, fyr reformationen / Helveg, Ludvig – Kobenhavn: C G Iversen, 1855 [mf ed 1990] – 123p on 1mf – 9 – 0-7905-7242-7 – (incl bibl ref) – mf#1988-3242 – us ATLA [241]
Danske folkemaal – [Kobenhavn]: Udvalg for folkemaal 1927-99 [mf ed Bloomington IN: Indiana Uni Lib, Preservation Dept 1983] – 1 – us Indiana Preservation [390]
Danske herold = Danish herald of canada – Kentvillw NS. v1-6. mar 15 1932-dec 28 1937// [wkly] – 2r – 1 – Can$220.00 – (in danish) – cn McLaren [071]
Den danske kirke under besaettelsen – Koebenhavn: H Hirschsprung, 1945 – 1mf – 9 – 0-524-08137-9 – mf#1993-9043 – us ATLA [240]
Danske kirkelove samt udvalg af andre bestemmelser vedrorende kirken : skolen og de fattiges forsorgelse fra reformationen indtil christian v's danske lov, 1536-1683 / ed by Rordam, Holger Frederik – Kjobenhavn: Selskabet for Danmarks kirkehistorie...1883-89 [mf ed 1988] – 3v on 1r – 1 – mf#2157 – us Wisconsin U Libr [344]
Den danske konebaads-expedition til gronlands ostkyst / Holm, G & Garde, V – Kobenhavn, 1887 – 7mf – 9 – mf#N-253 – ne IDC [919]
Danske kvinders missionsfond *see* Aarsskrift for danske kvinders missionsfond

Den danske mission i ostindien i de seneste aar : en samling af breve / Danske missionsselskab; ed by Kalkar, Christian Andreas Hermann – Kjobenhavn: Graebes Bogtrykkeri, 1870 [mf ed 1995] – (= ser Yale coll) – xxi/274p – 1 – 0-524-09027-0 – (in danish) – mf#1995-0027 – us ATLA [240]
Danske missionsselskab *see*
– Den danske mission i ostindien i de seneste aar
– Den nyere danske mission blandt tamulerne
Danske mormoner : et bidrag til belysning af mormonismens komme til danmark / Kent, Harald Jensen – [Koebenhavn]: Udvalget for Utahmissionen, 1913 – 1mf – 9 – 0-524-05254-9 – (incl bibl ref) – mf#1991-2246 – us ATLA [243]
Den danske pioneer : a danish-american weekly – Omaha, NE: Mark Hansen, aug 1 1872 issue [biwkly] – 36r – 1 – (in danish & english; absorbed: dansk tidende; publ in omaha ne 1872-jul 10 1958, elmwood park il oct 16 1958-jun 11 1984, and in hoffman estates il jun 25 1984-; suspended dec 29 1955-jan 12 1956 and jul 24-oct 2 1958) – us NE Hist [071]
Den danske pioneer – Nebraska, NE. 1942 nov-1944 oct – 2r – us UF Libraries [025]
Danske ugeblad – Minneapolis, Tyler MN. 1929 mar 14-1930 dec 25 – 1r – 1 – (cont: ugebladet [minneapolis mi: 1890]) – mf#766033 – us WHS [071]
Danske vestindien / Cavling, Henrik – Kobenhavn, Denmark. 1894 – 1r – us UF Libraries [948]
Danskeren – Neenah, WI: Jersild Pub Co, 1892-29 aarg n52. 29 dec 1920 [wkly] [mf ed 1899-1920 [gaps] filmed 1974] – 8r – 1 – (in danish; merged with: dansk luthersk kirkeblad to form: luthersk ugeblad. publ in blair ne, 20 apr 1899-29 dec 1920) – us NE Hist [071]
Danskeren – Neenah WI. 1894 oct 11-1898 dec 29 – 1r – 1 – mf#1097642 – us WHS [071]
Danskerens magazin – 1898 apr-oct – 1r – 1 – mf#1330461 – us WHS [071]
Danskunst / Hartong, C – Leiden, 1948 – €11.00 – ne Slangenburg [790]
Dantas, Mercedes *see* Forca nacionalizadora do estado novo
Dantas, Paulo *see* Cidade enferma
Dantas, Raymundo Souza *see* Solidao nos campos
Dante / Eliot, T S – Modena, Italy. 1942 – 1r – us UF Libraries [440]
Dante Alighieri *see*
– Divina commedia
Dante, Alighieri *see* Il convivio: the banquet of dante alighieri
Dante and aquinas / Wicksteed, Philip Henry – London: J M Dent; New York: E P Dutton, 1913 – (= ser Jowett Lectures) – 1mf – 9 – 0-7905-6799-7 – mf#1988-2799 – us ATLA [180]
Dante and the mystics : a study of the mystical aspect of the divina commedia and its relations with some of its mediaeval sources / Gardner, Edmund Garratt – London: JM Dent; New York: EP Dutton, 1913 – 1mf – 9 – 0-7905-7739-9 – (incl bibl ref) – mf#1989-0964 – us ATLA [440]
Dante gabriel rossetti : his work and influence / Tirebuck, William Edwards – London 1882 – (= ser 19th c art & architecture) – 1mf – 9 – mf#4.2.208 – uk Chadwyck [750]
Dante gabriel rossetti / Megroz, Rodolphe Louis – London, England. 1928 – 1r – us UF Libraries [920]
Dante gabriel rossetti / Stephens, Frederic George – London 1894 – (= ser 19th c art & architecture) – 2mf – 9 – mf#4.2.375 – uk Chadwyck [700]
Dante rossetti and the pre-raphaelite movement / Wood, Esther – London: Sampson Low, Marston & Co Ltd, 1894 – (= ser 19th c art & architecture) – 4mf – 9 – mf#4.1.108 – uk Chadwyck [750]
Dante studies – Albany NY 1972-93 – 1,5,9 – ISSN: 0070-2862 – mf#6691 – us UMI ProQuest [440]
Dantes-Castillo, Porfirio *see* Rocio
Danthony, Marie-Josephe *see* Europaeische studiengaenge in der bundesrepublik deutschland
Dantin Cereceda, Juan *see* Exploradores y conquistadores de indias
Danton, George H *see* Tieck's essay on the boydell shakespeare gallery
Danton und robespierre : tragoedie in fuenf aufzuegen / Hamerling, Robert – Hamburg: J F Richter, 1873 [mf ed 1996] – 179p – 1 – mf#9669 – us Wisconsin U Libr [820]
Dantwala, Mohanlal Lalloobhai *see* Gandhism reconsidered
Dantz, Antonie *see* Goethe und die wirtschaft
Dantzig, A van *see* Dutch documents relating to the gold coast and translations of letters and papers collected in the algemeen rijks archief (ara), state archives of the netherlands at the hague
Danvers 1652-1849 – Oxford, MA (mf ed 1996) – (= ser Massachusetts vital record transcripts to 1850) – 26mf – 9 – 0-87623-237-3 – (mf 1t-6t: births & deaths 1652-1835. mf 6t-7t: marriages 1752-73. mf 7t-10t: vital records 1720-1857. mf 10t-15t: vital records 1695-1860. mf 15t-17t: marriages 1752-1818. mf 17t: out-of-town marriages 1752-99; marriage intentions 1752-1819. mf 17t-19t: marriages 1773-1828. mf 19t-20t: marriage intentions 1819-49. mf 20t-21t: marriages 1827-49. mf 22t-24t: births 1843-49. mf 24t: marriages 1774-1849. mf 25t-26t: deaths 1830-49) – us Archive [978]
D'anvers a bruxelles via le lac kivu / Wauters, Arthur – Bruxelles, Belgium. 1929 – 1r – us UF Libraries [949]
Danvers Historical Society *see* Old anti-slavery days
Danvers, Massachusetts. Danvers Baptist Church *see* Records
Danvila y Collado, Manuel *see* El poder civil en espana
Danville banner / Prohibition Party of Illinois – v13 n20-v20 n49 [1903 jul 2-1910 dec 1] – 1r – 1 – (cont: state leader; cont by: illinois banner) – mf#939583 – us WHS [325]
Danville baptist church. vermont : church records – 1792-1842 – 1 – 6.32 – us Southern Baptist [242]
Danville intelligencer – Danville, PA. -w 1904-1907 – 13 – $25.00r – us IMR [071]
Danville journal *see* [Walnut creek-] courier journal
Danville quarterly review – La Salle IL 1861-64 – 1 – mf#5299 – us UMI ProQuest [071]
Danville textile worker / Textile Workers Union of America – v1 n1-7 [1948 jul 23-1949 jul 11] – 1r – 1 – mf#694493 – us WHS [680]
[Danville-] the valley pioneer – CA. 1961-81; 1982-87 – 60r – 1 – $3600.00 – mf#B02162 – us Library Micro [071]
[Danville-] the village pioneer – CA. 1960-1963 – 1r – 1 – $60.00 – mf#B02163 – us Library Micro [071]
Danville Trades and Labor Council *see* Labor leader
[Danville-] tri-valley news – CA. Sep 1973-Mar 1980 – 95r – 1 – $5700.00 – mf#B02161 – us Library Micro [071]
Danville weekly advertiser – Danville WI. 1850 oct 5 – 1r – 1 – (cont by: advertiser [danville in: 1853]) – mf#856271 – us WHS [071]
Danviller herold und zeitung – Danville, IL. jul 1917-jul 13 1919 – 3r – us CRL [071]
Danylchuk, Karen *see* Marketing structures, activities and outcomes amongst selected national sport organizations
Danza del sacrificio, y otros estudios / Chinchilla Aguilar, Ernesto – Guatemala, 1963 – 1r – us UF Libraries [972]
Danza del venado en guatemala / Paret-Limardo De Vela, Lise – Guatemala, 1963 – 1r – us UF Libraries [972]
Danzas clasicas : primer ano preparatorio / Flores, Elsa Mercedes – Buenos Aires: Ricordi Americana [c1961] – 1 – mf#*ZBD-*MGO pv27 – Located: NYPL – us Misc Inst [790]
Danzel, Theodor Wilhelm *see*
– Gesammelte aufsaetze
– Ueber goethe's spinozismus
Danzi, Franz *see*
– Trois quatuors pour 2 violons, alto et violoncelle, oeuvre 44
– Trois quatuors pour deux violons, alto & violoncelle, oeuvre 6
– Trois quatuors pour flute, violon, alto & violoncelle, oeuvre 56
– Trois quintetti pour deux violons, deux altos, et violoncello...oeuvre 66
Danzig. Regierungsbezirk *see* Amtsblatt der koeniglichen regierung zu danzig
Danziger allgemeine zeitung *see* Mennonitische blaetter
Danziger buergerzeitung – Danzig (Gdansk PL), 1913-1914 aug – 1 – gw Misc Inst [077]
Danziger courier – Danzig (Gdansk PL), 1897 jul-dec – 1 – gw Misc Inst [077]
Danziger dampfboot – Danzig [Gdansk PL], 1846-49 – 2r – 1 – (title varies: 1838: das dampfboot; filmed by other misc inst: 1834-1836 okt, 1837-43, 1858 jul-dez [17r], 1831 nov-1832, 1835, 1838 jan-jun) – gw Misc Inst [380]
Danziger echo – Danzig (Gdansk PL), 1934 25 may & 1935 [single iss], 1936 15 feb & 21 mar – 1r – 1 – gw Misc Inst [077]
Danziger intelligenzblatt 1739 – Danzig [Gdansk PL], 1844 [gaps], 1848 [gaps] – 2r – 1 – (title varies: 5 jan 1818: intelligenzblatt fuer den bezirk der koeniglichen regierung zu danzig, 1850: danziger intelligenzblatt; filmed by other misc inst: 1816 & 1818 jan-jun, 1821 jan-jun, 1823 jul-dez, 1825 jan-jun, 1828 apr-jun, 1828 okt-1829 mar, 1829 okt-1831, 1833 jan-jun, 1834 apr-jun, 1835 jan-mar, 1836 jan-jun, 1836 okt-1837, 1838 jul-dez, 1843, 1844 jul-dez, 1847 jul-dez, 1848 jul-dez, 1849 jul-dez) – gw Misc Inst [077]
Danziger landes-zeitung – Danzig (Gdansk PL), 1927, 1928 [gaps], 1929-34 – 16r – 1 – gw Misc Inst [077]
Danziger nachrichten – Danzig (Gdansk PL), 1752-57 – 3r – 1 – gw Misc Inst [077]
Danziger neueste nachrichten – Danzig (Gdansk PL), 1929 mai-jun – 1 – 1 – (filmed by other misc inst: 1924 [gaps], 1925-43 [69r]) – gw Misc Inst [077]
Danziger neueste nachrichten – Gdansk, Poland. Jan-Aug 1944 (scattered issues) – 1r – 1 – us L of C Photodup [943]
Danziger sonntags-zeitung – Danzig (Gdansk PL), 1930 feb-1936 28 jun, 1936 4 oct-1942 – 26r – 1 – (with suppls) – gw Misc Inst [077]
Danziger volksstimme *see* Volkswacht
Danziger volks-zeitung – Danzig (Gdansk PL), 1935 [gaps], 1936-37 – 13r – 1 – gw Misc Inst [077]
Danziger vorposten – Gdansk, Poland. Jan-Mar 1938; 1940-Jun 1943; Apr 1944-Mar 1945 – 10r – 1 – (some missing issues) – us L of C Photodup [943]
Der danziger vorposten – Danzig (Gdansk PL), 1933-44 – 33r – 1 – (with gaps) – gw Misc Inst [077]
Danziger zeitung 1796 – Danzig (Gdansk PL), 1796-1801 – 1 – gw Misc Inst [077]
Danziger zeitung 1858 – Danzig (Gdansk PL), jul-dec 1859, jul 1865-66, jan-jun 1868, jul-dec 1871, apr 1878-mar 1881, apr-jun, oct-dec 1885, apr-jun 1886, apr-dec 1892, jul-sep 1893, 1894, jul 1895-mar 1897, feb 1921-jan 31 1930 – 62r – 1 – gw Misc Inst [077]
Danzig-westpreussischer kirchenbrief / ed by Hilfskomitee fuer die evangelischen aus Danzig-Westpreussen – Luebeck. n1- 1948- [3 or 4 times/yr] [mf ed 1988-] – 1 – mf0840 – us ATLA [242]
Dao dara yuk sayam – Bangkok, Thailand. 1974-76 – 14r – 1 – us L of C Photodup [079]
Dao, Duy Anh *see* Tu die n truyen kieu
Dao Vu *see* Hoa lua
Daoust, Charles-Roger *see* Cent-vingt jours de service actif
Daoust, Daniele et al *see* Etude comparative de deux index de periodiques
Dape 'aliyah / Jewish Agency For Israel. Dept For Aliyah And Absorption – Jerusalem, Israel. 1950 – 1r – us UF Libraries [939]
Daphnis et chloe / Clairville, M – Paris, France. 1849? – 1r – us UF Libraries [440]
Dapper, O *see*
– Description de l'afrique
– Gedenkwaerdig bedryf der nederlandsche oost-indische maetschappye, op de kuste en het keizerrijk van taising of sina...
– Historische beschryving der stadt amsterdam: waer in de voornaemite geschiedeniffen..
Dapper, Olfert *see* Naukeurige beschrijvinge der afrikaensche gewesten van egypten, barbaryen, libyen, biledulgerid, negroslant, guinea, ethiopien, abyssinie
Dar, Bashir Ahmad *see* A study in iqbal's philosophy
Dar es Salaam. African Conference on Local Courts and Customary Law *see* Record of the proceedings...under the chairmanship of the minister of justice of tanganyika, sheikh amri abedi
Dar Es Salaam University College Department Of History *see* Maji maji research project
Dar mdo'i gsar 'gyur – Kangting, China. July-Oct 1956 – 1r – 1 – us L of C Photodup [079]
Dara shukoh / Kanunago, Kalika Ranjana – Calcutta: MC Sarkar & Sons, [1934]- – (= ser Samp: indian books) – (foreword by r c majumdar) – us CRL [954]
Dara shukoh / Qanungo, Kalika-Ranjan – 2nd ed. Calcutta: S C Sarkar 1952- [mf ed 1987] – 1r – 1 – (foreword by r c majumdar. with: facts & fancies about java / wit, a & other titles) – mf#1823 – us Wisconsin U Libr [954]
Darah-Maluku *see* Speciale editie ter herdenking van het vijftienjarig bestaan van de repoeblik maluku selatan
D'aranda : ou, les grandes passions / Scribe, Eugene – Paris, France. 1847 – 1r – us UF Libraries [440]
Darbar – Ajmer, India. 6 Apr 1944-53; 1957-60; 1962-63 – 8r – 1 – us L of C Photodup [079]
Darbas = Labor – New York NY. jan 4 1929-dec 6 1930 – 1r – 1 – (lithuanian newspaper) – us IHRC [331]
Darbininkas : amerikos lietuviu r k svento juozapo darbininku sajunga – South Boston: 1919-20 – 2r – us CRL [071]
Darbininku zodis : the only lithuanian weekly in canada = The worker's word – Toronto. v1-5. sep? 1932-dec 31 1936// [mf ed oct 1932 [oct 18 1934-nov 19 1936] – 2r – 1 – Can$185.00 – (in lithuanian; cont by: liaudies balsas) – cn McLaren [331]
D'arblay, Frances *see* Diary and letters of madame d'arblay
Darby, J N *see* On ministry
Darby, John Nelson *see* Notes on the book of revelations
Darby, William Arthur *see* Church vestments
Darby's monthly, geographical, historical and statistical repository – La Salle IL 1824 – 1 – mf#3734 – us UMI ProQuest [071]
Darbyshire, Alfred *see* An architect's experiences
Darcey, Barbara Berry *see* Biographical sketch of george e sebring, sr
D'Archery, L *see* Opera omnia
Darcom news – v11 n2 [1982 dec], v12 n1-2 [1983 nov-dec] – 1r – 1 – (cont by: amc news) – mf#1477266 – us WHS [071]
D'Arcy, Charles Frederick *see* Christianity and the supernatural
Dard, Jean *see* Dictionnaire francais-wolof et francais-bambara
Dardanus / Guillard, Nicolas Francois – Paris, France. 1802 – 1r – us UF Libraries [440]
Dardier, Charles *see* Michel servet d'apres ses plus recents biographes
El dardo – Plasencia, 1901-1904 – 5 – sp Bibl Santa Ana [073]
Dare to struggle / Great Lakes Movement for a Democratic Military – v1 iss1-iss3,7-v2 iss1 [1970 aug-oct 9, 1971 feb?-may – 1r – 1 – mf#721540 – us WHS [355]
Dare we be christians / Rauschenbusch, Walter – Boston: Pilgrim Press, c1914 – 1mf – 9 – 0-7905-9601-6 – mf#1989-1326 – us ATLA [240]
Daredevil – iss n1-5. jul-nov 1941 – 15 – mf#006GL – us MicroColour [740]
Daresbury, all saints (new section) – (= ser Cheshire monumental inscriptions) – 1mf – 9 – £2.50 – mf#14 – uk CheshireFHS [929]
Dares-studien / Schissel Von Fleschenberg, Otmar – Halle a.S., Germany. 1908 – 1r – us UF Libraries [960]
Daressy, G *see* Textes et dessins magiques
Dareste, R *see*
– Dix ans de la vie de francois hotman
– Francois hotman, sa vie et sa correspondance
Daretis phrygii de excidio troiae historia / ed by Meister, Ferdinand Otto – Lipsiae: in aedibus B G Teubneri 1873 [mf ed 1979] – 1r – 1 – mf#film mas 8376 – us Harvard [450]
Darey, P J *see* The dominion phrase book
Darfur al-jadidah = Darfur al-gadida – Sudan: s.n., jul 4 1992-aug 28 1993 – 1r – us CRL [079]
Dargan, Edwin Charles *see*
– Commentary on the epistle to the colossians
– The doctrines of our faith
– Ecclesiology
– A history of preaching
– Papers and sermons of e.c. dargan
– Society, kingdom, and church
Dari sekolah kemedan perang / Takato, A – Djakarta: Poesat Keboedajaan, 2604 – 67p on 2mf – 9 – mf#SE-2002 mf174-175 – ne IDC [370]
Darias y Padron, Dacio V *see* A proposito de los condes de la gomera "fue antano regular la sucesion de este titulo"
Darien company records : papers of the company of scotland trading to africa and the indies, 1696-1707. from the royal bank of scotland, edinburgh – Edinburgh: Royal Bank of Scotland – (= ser BRRAM series) – 3r – 1 – £201 / $402 – (with p/g; int by john simpson) – mf#r96801 – uk Microform Academic [380]
Darien news – Darien WI. 1859 apr 26 [v1 n38] – 1r – 1 – mf#875009 – us WHS [071]
Darien venture / Cundall, Frank – New York, NY. 1926 – 1r – us UF Libraries [972]
Daring mystery – iss n1-8. jan 1940-jan 1942 – ea set incl 4mf – 15 – mf#054MV-055MV – us MicroColour [740]
Dario, Ruben *see*
– Antologia
– Antologia chilena
– Antologia poetica
– Autobiografia
– Azul
– Baladas y canciones
– Blanco
– Canto a la argentina
– Cantos de vida y esperanza
– Caravana pasa
– Cartas de ruben dario
– Cuentos completos
– Cuentos y poemas
– Emelina
– Mundo de los suenos
– Obras completas
– Obras escogidas
– Parisiana
– Poema del otono y otros poemas
– Poemas escogidos
– Prosas profanas y otros poemas
– Raros
– Ruben dario, critico literario
– Ruben dario, ensayo biografico y breve antologia
– Selected poems
– Selections from the prose and poetry
– Sol del domingo
– Sus mejores cuentos
– Sus mejores poemas
– Viaje a nicaragua

Darius I, King of Persia *see* Die grabschrift des darius zu nakschi rustam
The darjeeling disaster : its bright side, the triumph of the six lee children / Warne, Francis Wesley – Calcutta: Methodist Publ House, 1900 [mf ed 1995] – (= ser Yale coll) – viii/216p (ill) – 1 – 0-524-09855-7 – mf#1995-0855 – us ATLA [954]
The dark ages : a series of essays intended to illustrate the state of religion and literature in the 9th, 10th, 11th, and 12th centuries / Maitland, Samuel Roffey – 2nd ed. London: F & J Rivington, 1845 [mf ed 1991] – 2mf – 9 – 0-524-01887-1 – (incl bibl ref & ind) – mf#1990-0514 – us ATLA [931]
Dark blue – La Salle IL 1871-73 – 1 – mf#5305 – us UMI ProQuest [420]
Dark county democrat / Darke Co. Greenville – aug 1855-jun 1862 (centerfold shadow) [wkly] – 1r – 1 – mf#B6293 – us Ohio Hist [071]
Dark eye in africa / Van Der Post, Laurens – New York, NY. 1955 – 1r – us UF Libraries [960]
The dark huntsman (a dream) / Heavysege, Charles – Montreal: "Witness" Steam Print House, 1864? – 1mf – 9 – mf#35998 – cn Canadiana [810]
Dark, Phillip John Crosskey *see* Bush negro art
The dark room : a novel / Narayan, R K – London: Macmillan and Co, 1938 – (= ser Samp: indian books) – us CRL [830]
The dark well / Chattopadhyaya, Harindranath – Madras: Kalakshetra, 1939 – (= ser Samp: indian books) – us CRL [490]
Darke Co. Ansonia *see*
– Ansonian
– Mirror
Darke Co. Arcanum *see*
– Early bird
– Early bird series
– Times
– Times series
Darke Co. Greenville *see*
– Advocate
– Daily advocate
– Daily news
– Daily tribune
– Dark county democrat
– Darke county boy
– Democrat
– Democratic advocate
– Democratic herald / telegraph
– Journal
– Transcript
– Tribune / news tribune series
Darke Co. Hollansburg *see*
– Independent
– News
Darke Co. New Madison *see*
– Herald
– Miscellaneous
– Times
Darke county boy / Darke Co. Greenville – v1 n1. dec 1910-apr 1912 [mthly] – 1r – 1 – mf#B6783 – us Ohio Hist [071]
Darkening days : being a narrative of famine striken bengal / Sen, Ela – Calcutta: Susil Gupta, 1944 – (= ser Samp: indian books) – (drawings from life by zainul abedin) – us CRL [954]
Darkest england gazette / Salvation Army – (The Social Gazette). Official newspaper of the social operations of the Salvation Army. London. -w. Jul. 1893-Mar. 1917. 9 reels – 1 – uk British Libr Newspaper [072]
Darkest india / Booth-Tucker, Frederick de Latour – Bombay: Bombay Gazette Steam Print Works, 1891 – 1mf – 9 – 0-524-03756-6 – mf#1990-1103 – us ATLA [360]
The darkness and the dawn in india : two missionary discourses / Sheshadri, Narayan & Wilson, John – Edinburgh: W Whyte, 1853 – 1mf – 9 – 0-524-07499-2 – mf#1991-0120 – us ATLA [240]
Darkness fleeing before light / Fleming, James – London, England. no date – 1r – us UF Libraries [240]
Darkness in the flowery land : or, religious notions and popular superstitions in north china / Culbertson, Michael Simpson – New York: Charles Scribner, 1857 – (= ser 19th c books on china) – 3mf – 9 – mf#7.1.29 – uk Chadwyck [230]
Darkness in the flowery land : or, religious notions and popular superstitions in north china / Culbertson, Michael Simpson – New York: Charles Scribner, 1857 [mf ed 1995] – (= ser Yale coll) – xii/235p (ill) – 1 – 0-524-09254-0 – mf#1995-0254 – us ATLA [280]
Darkroom photography – Beverly Hills CA 1979-90 – 1,5,9 – (cont by: camera & darkroom) – ISSN: 0163-9250 – mf#11830.01 – us UMI ProQuest [770]
Darlegung der dichterischen technik und litterarhistorischen stellung von goethes elegie "alexis und dora" / Kassewitz, Joseph – Leipzig: G Fock 1893 [mf ed 1990] – 1r – 1 – (filmed with: goethe / c h herford) – mf#7387 – us Wisconsin U Libr [430]
Darley, George Marshall *see* Pioneering in the san juan
Darley, Mary *see* Cameos of a chinese city
Darling, Charles John Darling *see* Scintillae juris
Darling, Charles William *see*
– Historical account of some of the more important versions and editions of the bible
– Versions of the bible
Darling downs gazette – Toowoomba, Australia. 16 Jun 1864; 14 Oct 1865; 4 Jan, 30 Oct 1866; 6 Jul-12 Sep, 24 Dec 1867, 11 Jan, 4-25 Apr 1868; 14 Apr 1875; 30 May 1877; 6 Feb 1888-15 Jul 1891 (imperfect).-w. 5 reels – 1 – uk British Libr Newspaper [072]
Darling, Henry *see* The closer walk
Darling, Hon. Mr. Justice *see* Scintillae juris
Darling, James *see* Cyclopaedia bibliographica
Darling, Malcolm Lyall *see*
– The punjab peasant in prosperity and debt
– Wisdom and waste in the punjab village
Darling, Mark Roger *see* Effects of cannabis, tobacco, methaqualone smoking on the oral environment
Darling, Sharon S *see* Decorative and architectural arts in chicago, 1871-1933
Darling, William Stewart *see* Papers on the unpopularity of religious truth
Darlington and stockton times – [NE England] Darlington oct 1847-dec 1950 [mf ed 2004] – 127r – 1 – (missing: 1894, 1897, 1911; cont by: darlington & stockton times, south durham & north york advertiser [jan 1874-dec 1884]) – uk Newsplan [072]
Darlington, Charles F *see* African betrayal
Darlington democrat – Darlington WI. 1894 jan 5/jun 28-1947 may 8 – 39r – 1 – (with gaps; cont: democrat and register; cont by: lafayette county news) – mf#1003455 – us WHS [071]
Darlington first baptist church. darlington, south carolina : church records – Deacons' Meetings. 1856-96, 1923-52 – 1 – us Southern Baptist [242]
Darlington journal – Darlington WI. 1887 sep 28/1889 jun 19-1899 jun 21/1900 jan 3 – 8r – 1 – (with small gaps) – mf#961907 – us WHS [071]
Darlington republican – Darlington WI. 1879 jul 4/1880 nov 12-1898 dec 2/1900 jan 5 – 10r – 1 – (cont: republican [darlington wi: 1869]; cont by: republican-journal [darlington wi: 1900]) – mf#1011104 – us WHS [071]
Darlington & stockton times – Darlington, Stockton, England. -w. 1847-54. 3 reels – 1 – uk British Libr Newspaper [072]
Darlite – Dar-es-Salaam, Tanzania 1966-70 – 1 – ISSN: 0418-3797 – mf#2429 – us UMI ProQuest [079]
Darlow, Thomas Herbert *see* Historical catalogue of the printed editions of holy scripture in the library of the british and foreign bible society
Darmabakti *see* Dewan mahasiswa institut agama islam negeri
Darmesteter, James *see* Selected essays of james darmesteter
Darmstaedter echo – Darmstadt DE, 1945 21 nov-29 dec, 1946 27 feb-1954 31 mar, 1955-67 – 72r – 1 – (filmed by misc inst: 1946 1 jan-27 feb, 1949 18 jul-13 aug [gaps], 1954 1 apr-31 dec; 1968- ca 11r/yr]) – gw Mikrofilm; gw Misc Inst [074]
Darmstaedter journal – Darmstadt DE, 1843-1845 sep, 1848 12 mar-1852 29 jun – 6r – 1 – gw Misc Inst [074]
Darmstaedter neue presse *see* Darmstaedtisches frag- und anzeigungs-blaettgen
Darmstaedter tagblatt *see* Darmstaedtisches frag- und anzeigungs-blaettgen
Darmstaedter wochenblatt *see* Darmstaedtisches frag- und anzeigungs-blaettgen
Darmstaedter zeitung *see* Hessen-darmstaedtische privilegirte landes-zeitung
Darmstaedtisches frag- und anzeigungs-blaettgen – Darmstadt DE, 1848-49 – 1r – 1 – (several title changes then fr 26 sep 1835: darmstaedter frag- und anzeigeblatt, 1 jan 1874: darmstaedter tagblatt, 18 dec 1949: darmstaedter wochenblatt, 30 may 1950: darmstaedter neue presse, 1 jul 1950: darmstaedter tagblatt; filmed by other misc inst: 1964 mai-jun, 1978 1 sep-1986 30 sep [ca 8r/yr]) – gw Misc Inst [074]
Darnel, Bw *see* Success and failure in tropical land settlement
Darnell, Ermina Jett *see* Forks of elkhorn church
Darnhall, weaver methodist church – (= ser Cheshire monumental inscriptions) – 1mf – 9 – £2.50 – mf#78 – uk CheshireFHS [929]
Darondel, Louis *see* Legendes et traditions dans l'histoire de saint-do...
Darques, F Martin *see* Miscellaneous manuscripts, 1968-1983
Darquitain, Victor *see* Notice sur le guyane francaise
Darracott, Charles R 3. *see* Ratings of perceived exertion as a determinant of physical activity in 9 to 11 year-old children
Darracott, Risdon *see* Scripture marks of salvation
Darracott, Shirley H *see* Individual differences in variability and pattern of performance
Darragh, Patricia M *see* The relationship of fat patterning to peripheral resistance following supine submaximal exercise
Darras, Joseph Epiphane *see* A general history of the catholic church
Darras, Maxime *see* Le nouveau code civil du tonkin.
Darrell, Frederick *see* Should i succeed in south africa?
Darrell st claire, senate service 1933-1939, 1949-1977 : assistant secretary to the senate – (= ser Us senate historical office oral history coll) – 3mf – 9 – $15.00 – us Scholarly Res [323]
Darrington, Melissa *see* Use of a stage-based nutrition intervention at an industrial workplace
Darroch, John *see* Chinese self-taught
Darrouzes, J *see* Litterature et histoire des textes byzantins (1950-1965)
Darrow, Clarence S *see*
– An eye for an eye
– Plea in defence of loeb and leopold
– Resist not evil
Darrow, Heather *see* The effects of decadron phonophoresis on serum levels of dexamethasone sodium phosphate
Darsey, Barbara *see* De soto city
Darsey, Barbara Berry *see*
– Avon park
– Avon Park, Florida
– Florida squatters
– Hollywood guide
– Jason and lily iby
– Lake placid
– Legend of the orange grove in highlands hammock
– Life history of albert denman and family
– Sebring
– Topics and observations relative to life history
– Virginia suffolk
Die darstellbarkeit des oesophago-gastralen ueberganges des erwachsenen durch die transkutane sonographie – normalbefunde und pathologie - : eine prospektive, endoskopisch kontrollierte studie / Lehnhardt, Martina Ulrike – (mf ed 2000) – 1mf – 9 – €30.00 – 3-8267-2710-X – mf#DHS 2710 – gw Frankfurter [616]
Darstellung der arabischen verskunst : mit sechs anhaengen / Freytag, Georg Wilhelm [comp] – Bonn: Cnobloch 1830 – (= ser Whsb) – 6mf – 9 – €70.00 – (ind by georg wilhelm freytag) – mf#Hu 214 – gw Fischer [460]
Die darstellung der haftung des amerikanischen staates und der beamten sowie ein rechtsvergleichender uberblick uber das deutsche recht / Creer, John W – n.p. 1969. 166 p. LL-4207 – 1 – us L of C Photodup [340]
Darstellung der moralphilosophischen anschauungen des philosophen hermann samuel reimarus / Richardt, Hermann – Leipzig: Bereiter & Meissner, [1906?] – 1mf – 9 – 0-7905-9458-7 – mf#1989-2683 – us ATLA [170]
Darstellung des einflusses von erfahrung im umgang mit einem hightech-verfahren in der medizin : eine analyse der ergebnisse nach carbonlasergestuetzter konisation der cervix uteri / Johst, Petra – (mf ed 1998) – 1mf – 9 – €30.00 – 3-8267-2571-9 – mf#DHS 2571 – gw Frankfurter [618]
Die darstellung des sichtbaren in der dichterischen prosa um 1900 / Iskra, Wolfgang – Muenster: Aschendorff, c1967 [mf ed 1993] – (= ser Muenstersche beitraege zur deutschen literaturwissenschaft v2) – 1 – mf#8272 – us Wisconsin U Libr [430]
Die darstellung des wahnsinns im englischen drama bis zum ende des 18. jahrhunderts / Berghaeuser, Wilhelm – Mainz, 1914 (mf ed 1994) – 2mf – 9 – €31.00 – 3-8267-3099-2 – mf#DHS-AR 3099 – gw Frankfurter [420]
Darstellung und beurteilung der theologie ritschls / Haug, Ludwig – 3. aufl. Stuttgart: D Gundert, 1895 [mf ed 1990] – 1mf – 9 – 0-7905-3948-9 – (incl bibl ref) – mf#1989-0441 – us ATLA [242]
Darstellung und eigenschaften der metagermanate des mangans, eisens und kobalts / Forwerg, Walter – Frankfurt a.M. 1962 (mf ed 1993) – 1mf – 9 – €24.00 – 3-89349-655-6 – mf#DHS-AR 655 – gw Frankfurter [540]
Darstellung und geschichte des geschmacks der vorzueglichsten voelker in beziehung auf die innere auszierung der zimmer und auf die baukunst / Racknitz, J F – Leipzig, 1796 – 7mf – 9 – mf#OA-108 – ne IDC [720]
Darstellung und kritik der ansicht wellhausens von geschichte und religion des alten testaments / Finsler, Rudolf – Zuerich: Friedrich Schulthess, 1887 – 1mf – 9 – 0-7905-0991-1 – (incl bibl ref) – mf#1987-0991 – us ATLA [221]
Darstellung und kritik der lehre des descartes von der bildung des universums / Kratzmoeller, Wilhelm – Rostock, 1903 (mf ed 1994) – 1mf – 9 – €24.00 – 3-8267-3105-0 – mf#DHS-AR 3105 – gw Frankfurter [110]
Darstellung und kritik der schleiermacherschen dogmatik / Weissenborn, Georg – Leipzig: TO Weigel, 1849 – (= ser Vorlesungen ueber Schleiermachers Dialektik und Dogmatik) – 1mf – 9 – 0-524-00204-5 – mf#1989-2904 – us ATLA [240]
Darstellung und kritik der von herder gegebenen ergaenzung und fortbildung der ansichten lessings in seinem laokoon / Hoffmann, G F – Augsburg: Reichel 1901 [mf ed 1990] – 1r – 1 – (incl bibl ref. filmed with: die freunde machen den philosophen...) – mf#2823p – us Wisconsin U Libr [430]
Darstellung und kritik des hegelschen systems aus dem standpunkt der christlichen philosophie / Staudenmaier, A, Fr – Mainz, 1844 – €25.00 – ne Slangenburg [140]
Die darstellung von krieg und frieden in der deutschen barockdichtung / Weithase, Irmgard – Weimar: H Boehlaus Nachfolger, 1953 – 128p/[3]pl (ill) – 1 – (incl bibl ref) – us Wisconsin U Libr [430]
Darstellungen aus dem preussischen rhein- und mosellande / Storck, Adam – Essen [u.a.] 1818 [mf ed Hildesheim: 1995-98] – (= ser Fbc) – 2v on 4mf – 9 – €120.00 – 3-487-29545-8 – gw Olms [914]
Darstellungen aus dem steyermaerk'schen oberlande / Weidmann, Franz – Wien 1834 [mf ed Hildesheim: 1995-98] – (= ser Fbc) – xviii/228p on 2mf [ill] – 9 – €60.00 – 3-487-29442-7 – gw Olms [914]
Darstellungen aus der schweitz / Eichholz, Johann – Leipzig 1808 [mf ed Hildesheim 1995-98] – (= ser Fbc) – 227p on 2mf [ill] – 9 – €60.00 – 3-487-29365-X – gw Olms [914]
Darstellungen aus italien / Meyer, Friedrich – Berlin 1792 [mf ed Hildesheim 1995-98] – 3mf [ill] – 9 – €90.00 – 3-487-29320-X – gw Olms [914]
Darstellungen aus russland's kaiserstadt und ihrer umgegend bis gross-nowgorod : im sommer 1828 / Meyer, Friedrich – Hamburg 1829 [mf ed Hildesheim 1995-98] – (= ser Fbc) – 3mf [ill] – 9 – €90.00 – 3-487-29008-1 – gw Olms [914]
Die darstellungsweise lessings in seinen prosaischen schriften : schulnachrichten von dem direktor der anstalt heinrich vockeradt / Mummenhoff, Wilhelm – Recklinghausen: J Bauer, 1903 – 1r – 1 – us Wisconsin U Libr [430]
Dart bulletin – Red Bank NJ 1963-71 – 1,5 – ISSN: 0011-6742 – mf#2290 – us UMI ProQuest [790]
Dart & midland figaro – Birmingham, England 30 sep,18 nov 1881, 10 nov 1882, 12 jan 1883, 4 nov 1884-20 mar 1896 [mf 1884,1885,1887,1891] – 1 – (wanting: jul-dec 1895; cont: dart [11 nov 1876]; cont by: birmingham pictorial & dart [27 mar 1896-1 sep 1911]) – uk British Libr Newspaper [072]
Dartein, F de *see* Etude sur l'architecture lombarde et sur les origines de l'architecture roma-byzantine
Dartford advance – Green Lake WI. 1902 dec 15, 1903 jan 19, may 4, 1907 feb 25-mar 4 – 1r – 1 – mf#916486 – us WHS [071]
Dartford and district free press – [London & SE] Kent 24 oct 1925-6 nov 1937 [mf ed 2003] – 9r – 1 – (cont as: dartford and district free press and entertainment guide [19 oct 1929-6 nov 1937]) – uk Newsplan [072]
Dartigue, Maurice *see*
– Conditions rurales en haiti
– Geographie locale
– Probleme de la communaute
Dartmoor prison. / Rhodes, Albert John – London. 1933 – 1 – us CRL [941]
Dartmouth 1650-1849 – Oxford, MA (mf ed 1996) – (= ser Massachusetts vital record transcripts to 1850) – 28mf – 9 – 0-87623-238-1 – (mf 1t-6t: vital records 1650-1791. mf 1t: marriage intentions 1727, 1739-44. mf 7t-8t: births 1710-80. mf 7t-9t: marriages 1722-88. mf 7t-15t: marriage intentions 1742-1821. mf 15t-20t: marriages 1749-1844. mf 17t: marriage intentions 1787-88. mf 20t: out-of-town marriages 1698-1799. mf 20t-23t: marriage intentions 1821-50. mf 23t-26t: births & deaths 1722-1847. mf 26t-27t: births 1843-49. mf 27t-28t: marriages 1843-49. mf 28t: deaths 1848-49) – us Archive [978]
Dartmouth 1650-1900 – Oxford, MA (mf ed 1992) – (= ser Massachusetts vital records) – 111mf – 9 – 0-87623-147-4 – (mf 1-13: vital records 1650-1847. mf 14-19: vitals 1667-1788. mf 20-23: vitals 1722-1864. mf 24-30: town, vitals 1674-1779. mf 31-48: town records 1777-1844. mf 49-58: town records 1789-1867. mf 59-60: taxes 1882, 1883, 1885. mf 61-74: marrs & ints 1699-1850. mf 75-76: town & marrs 1829-39. mf 77-78: ints index 1848-77+. mf 79-81: intentions 1848-77. mf 82: baptist records 1831-78. mf 83-85: birth index 1843-1922. mf 86-88: marr index 1844-91. mf 89-91: death index 1850-1923. mf 92-94: births 1843-64. mf 95: marriages 1844-55, 1869. mf 95-96: deaths 1848-51. mf 97-98: births 1843-64. mf 98: marriages

1844-55. mf 98-99: deaths 1848-61. mf 100-102: deaths 1862-91. mf 103-105: marriages 1855-91. mf 106-107: births 1865-95. mf 108-109: deaths 1892-1903. mf 110-111: marriages 1892-1906) – us Archive [978]
Dartmouth bi-monthly – Killen TX 1905-08 – 1 – mf#4815 – us UMI ProQuest [378]
Dartmouth papers, the... : the american papers of william legge, the second earl of dartmouth. from the staffordshire record office – (= ser British records relating to america in microform) – 16r – 1 – (with guide. int by colin bonwick) – mf#97518 – uk Microform Academic [975]
Dartmouth review – 1986 may 7-1989 mar 15 – 1r – 1 – mf#1614480 – us WHS [071]
Dartmouth western guardian – [SW England] Devon jan 1902-dec 1907, jan 1928-dec 1929, jan 1933-dec 1934, 1937, 1939-41, 1943, jan 1945-dec 1949 [mf ed 2003] – 20r – 1 – uk Newsplan [072]
Dartois, Achille *see* Dame des belles cousines
Daru, Pierre Antoine *see* Histoire de bretagne
Dar'uel-elhan mecmuasi – Istanbul: Evkaf-i Islamiyye Matbaasi, 1924-26. Nesreden: Dar'uel-Elhan Heyet-i Tedrisiyyesi; Sahib-i Imtiyaz: Fazil Hakki. sene 1-2. n1-7. 1 subat 1340 [1924]-1 subat 1926 – (= ser O & t journals) – 5mf – 9 – $85.00 – us MEDOC [956]
Dar'ues-s'afaka – n1-12. 1325-26 [all publ] – (= ser O & t journals) – 10mf – 9 – $165.00 – us MEDOC [956]
Dar-ul-islam : a record of a journey through ten of the asiatic provinces of turkey / Sykes, Mark – London: Bickers, 1904 – 6mf – 9 – 0-524-07384-8 – mf#1991-0104 – us ATLA [915]
Darussalam, Banda Atjah, Lembaga Penjelidikan Ekonomi dan Sosial, Fakultas Ekonomi Universitas Sjiah Kuala *see* Bulletin fakta-fakta ekonomi daerah istimewa atjah
Daruty de Grandpre, Jean Emile, marquis *see* Recherches sur le rite ecossais ancien accepte
Darveau, C *see* Specimen de photo-gravure de l'imprimerie de c darveau, quebec
Darveau, Louis-Michel *see* Nos hommes de lettres
Darwen news – [NW England] Darwen Lib 11 mar 1876-1948* – 1 – uk MLA; uk Newsplan [072]
Darwen post – [NW England] Darwen Lib oct 1885-1888 – 1 – (title change: darwen gazette (a) [1900-7 nov 1903]; darwen & county gazette [14 nov 1903-1 oct 1910]; darwen gazette (b) [8 oct 1910-15, 1918-20]) – uk MLA; uk Newsplan [072]
Darwen weekly advertiser and courier of coming events – [NW England] Darwen Lib 21 apr 1893-17 mar 1916 – 1 – (title change: darwen advertiser & courier of coming events [24 mar 1916-4 feb 1971]; darwen advertiser & times [1 feb 1971-1980]) – uk MLA; uk Newsplan [072]
Darwin, and after darwin : an exposition of the darwinian theory and a discussion of post-darwinian questions / Romanes, George John – London, 1892 – (= ser 19th c evolution & creation) – 13mf – 9 – mf#1.1.6489 – uk Chadwyck [575]
Darwin and darwinism pure and mixed : a criticism, with some suggestions / Alexander, P Y – London, 1899 – (= ser 19th c evolution & creation) – 4mf – 9 – mf#1.1.1535 – uk Chadwyck [575]
Darwin and hegel with other philosophical studies / Ritchie, David George – London, 1893 – (= ser 19th c evolution & creation) – 4mf – 9 – mf#1.1.955 – uk Chadwyck [140]
Darwin and the humanities / Baldwin, James Mark – Baltimore: Review Pub., 1909 – (= ser Psychological Review Publications) – 1mf – 9 – 0-7905-3755-9 – (incl bibl ref) – mf#1989-0248 – us ATLA [100]
Darwin, C *see*
– The zoology of the voyage of hms beagle... during the years 1832-1836
Darwin, Charles *see*
– Descent of man
– The descent of man and selection in relation to sex
– The diary and correspondence of charles darwin written during the voyage of the beagle, 1831-36
– Journal of the linnean society
– The life and letters of charles darwin
– Naturwissenschaftliche reisen nach den inseln des gruenen vorgebirges, suedamerika, dem feuerlaende, den falkland-inseln...
– Notebooks compiled during the voyage of "the beagle", 1831-1836
– Sochineniia charl'za darvina polnye perevody
Darwin, Charles Robert *see*
– The descent of man
– The different forms of flowers on plants of the same species
– The effects of cross and self fertilisation in the vegetable kingdom
– The expression of the emotions in man and animals
– Insectivorous plants
– Journal of researches into the natural history and geology of the countries visited during the voyage of hms beagle round the world
– On the origin of species by means of natural selection
– On the various contrivances by which british and foreign orchids are fertilised by insects
– The variation of animals and plants under domestication
Darwin christ church marriages *see* Anglican church registers index 1902-1953
Darwin, Erasmus *see* Commonplace book
Darwin, Francis *see* The life and letters of charles darwin
Darwin, huxley and the natural sciences : manuscripts and rare printed works – 3pt-coll – 183r in 5 units – 1 – $19,215.00 coll – (units 1 and 2:the huxley papers from imperial college library, london 35r and 34r c39-28851 and c39-28852 respectively. units 3: the darwin papers from cambridge university library mss vols 1-119 34r c39-28853. units 4,5: the darwin papers from cambridge university library and down house, kent mss vols 120-186, and 187-226 33r and 47r c39-28854 and c39-28855 respectively. printed guide available) – mf#C39-28850 – us Primary [500]
Darwin inquest book 1875-1905 – [mf ed 1990] – 1mf – 9 – A$5.50 – 0-949124-64-8 – at Northern [980]
The darwinian theory and the law of the migration of organisms / Wagner, Moritz – London, 1873 – (= ser 19th c evolution & creation) – 1m – 9 – (trans fr german by james l lair) – mf#1.1.1516 – uk Chadwyck [574]
The darwinian theory of the origin of the species / Pascoe, Francis Polkinghorne – London, 1890 – (= ser 19th c evolution & creation) – 2mf – 9 – mf#1.1.9066 – uk Chadwyck [575]
Darwiniana : essays / Huxley, Thomas Henry – London: Macmillan, 1893 – 2mf – 9 – 0-7905-3920-9 – mf#1989-0413 – us ATLA [575]
Darwiniana : essays and reviews pertaining to darwinism / Gray, Asa – New York: D Appleton, 1889, c1876 – 1mf – 9 – 0-7905-7393-8 – mf#1989-0618 – us ATLA [575]
Darwinischen theorien und ihre stellung zur philosophie, religion und moral = The theories of darwin and their relation to philosophy, religion, and morality / Schmid, Rudolf – Chicago:Jansen, McClurg, 1883, c1882 – 1mf – 9 – 0-8370-5107-X – (in english. incl bibl ref) – mf#1985-3107 – us ATLA [210]
Die darwinischen theorien und ihre stellung zur philosophie, religion und moral / Schmid, Rudolf – Stuttgart:Paul Moser, 1876 – 1mf – 9 – 0-8370-5105-3 – (incl bibl ref) – mf#1985-3105 – us ATLA [210]
Darwinism : an exposition of the theory of natural selection with some of its applications / Wallace, Alfred Russel – London, 1889 – (= ser 19th c evolution & creation) – 6mf (ill) – 9 – mf#1.1.4254 – uk Chadwyck [575]
Darwinism a fallacy / Pocock, William Willmer – London, 1891 – (= ser 19th c evolution & creation) – 2mf – 9 – mf#1.1.1587 – uk Chadwyck [575]
Darwinism and design : or, creation by evolution / Saint Clair, George – London, 1873 – (= ser 19th c evolution & creation) – 3mf – 9 – mf#1.1.1528 – uk Chadwyck [575]
Darwinism and other essays / Fiske, John – London, 1879 – (= ser 19th c evolution & creation) – 4mf – 9 – mf#1.1.11303 – uk Chadwyck [575]
Darwinism and politics / Ritchie, David George – London, 1889 – (= ser 19th c evolution & creation) – 2mf – 9 – mf#1.1.1083 – uk Chadwyck [140]
Darwinism and race progress / Haycraft, John Berry – London, 1895 – (= ser 19th c evolution & creation) – 3mf – 9 – mf#1.1.8301 – uk Chadwyck [572]
Darwinism in morals : and other essays / Cobbe, Frances Power – London: Williams and Norgate, 1872 – 1mf – 9 – 0-7905-3821-0 – mf#1989-0314 – us ATLA [240]
Darwinism refuted : an essay on mr darwin's theory of "the descent of man" / Laing, Sidney Herbert – London, 1871 – (= ser 19th c evolution & creation) – 1mf – 9 – mf#1.1.1500 – uk Chadwyck [575]
Darwinism refuted out of darwin's book / Walduck, Henry – London [1885] – (= ser 19th c evolution & creation) – 1mf – 9 – mf#1.1.9073 – uk Chadwyck [575]
Darwinism tested by language / Bateman, Frederic – [London] 1877 – (= ser 19th c evolution & creation) – 3mf – 9 – (pref by edward meyrick goulburn) – mf#1.1.1531 – uk Chadwyck [575]
Darwinism tested by the science of language / Schleicher, August – London, 1869 – (= ser 19th c evolution & creation) – 1mf – 9 – (trans fr german of august schleicher. pref & additional notes by alex v w bikker) – mf#2.1.230 – uk Chadwyck [575]
Das, Abinas Chandra *see* Rig-vedic india
Das, Adhar Chandra *see*
– Negative fact, negation, and truth
– Sri aurobindo and the future of mankind
Das, Bhagavan *see*
– Bhagavad-gita
– Krshna
– The philosophy of non-co-operation and of spiritual-political swaraj
– The science of peace
– The science of social organisation
– The science of the sacred word
Das, Bhagavan [comp] *see* The essential unity of all religions
Das, Bishnu Charan *see* Life of vijaykrishna
Das, Devandra Nath *see* Sketches of hindoo life
Das die beicht ainem christen menschen nitt burdlich oder schwer sey / Oecolampadius, J – Basel: Andreas Cratander, 1519 – 2mf – 9 – mf#PBU-348 – ne IDC [240]
Das die evangelischen kilchen weder kaetzerische noch abtruennige...syend gruntliche erwysung / Bullinger, Heinrich – Zuerych: Andreas Geszner d.j. vnd Ruodolff Wyssenbach, 1552 – 2mf – 9 – mf#PBU-173 – ne IDC [240]
Das, Frieda Hauswirth *see* Purdah
Das Gupta, Debendra Chandra *see* Jaina system of education
Das Gupta, Harendra Mohan *see* Studies in western influence on nineteenth century bengali poetry, 1857-1887
Das Gupta, J N *see*
– Bengal in the sixteenth century ad
– India in the seventeenth century
Das Gupta, Tamonash Chandra *see* Aspects of bengali society from old bengali literature
Das kleine journal 1883 – Berlin DE, 1883 may-1884 jun, 1892-1897 mar, 1897 jul-1902, 1903 apr-1905 sep, 1906-14, 1915 jul-1919, 1925-1935 mar – 66r – 1 – (title varies: 19 nov 1918: berliner mittagszeitung, 1920: das kleine journal) – gw Misc Inst [074]
Das, Matilal *see* Bankim chandra
Das, Nabagopal *see*
– Banking and industrial finance in india
– Industrial enterprise in india
Das, Rajani Kanta *see*
– The industrial efficiency of india
– Principles and problems of indian labour legislation
Das, Rashvihari *see*
– A handbook to kant's critique of pure reason
– The philosophy of whitehead
Das, Santosh Kumar *see* The educational system of the ancient hindus
Das, Sarat Chandra *see* A tibetan-english dictionary with sanskrit synonyms
Das Socialistische Bund *see* Turn-zeitung
Das, Sudhendu Kumar *see* Sakti or divine power
Das, Sudhir Ranjan *see* Folk religion of bengal
Das, Tarakchandra *see* The purums
Das, Taraknath *see* Foreign policy in the far east
Das volk 1945 – Berlin DE 1946 15 mar-21 apr – 1 – (filmed by mikropress: 1945 7 jul-1946 21 apr [1r] order#1012; filmed by misc inst: 1945 7 jul-1946 21 apr [1r]) – gw Mikrofilm; gw Mikropress; gw Misc Inst [074]
Dasa, Gobinda *see*
– Hindu ethics
– Hinduism and india
Dasa, Harihara *see* Life and letters of toru dutt
Dasagupta, Jayantakumara *see* A critical study of the life and novels of bankimcandra
The dasakumaracharita of dandin : with a commentary / ed by Ka'le, M R – Bombay: Gopal Narayan & Co, 1925 – (= ser Samp: indian books) – (with various readings, a literal english trans, explanatory and critical notes, and an exhaustive int by ed) – us CRL [280]
Dasaparamikatha : traduit de pali et annote par preas krou vimalapanna oum-sou / Narada Bhikkhu of Ceylon. – Phnom-Penh: Impr du gouvernement 1928 [mf ed 1990] – 1r with other items – 1 – (title & text in khmer; added t.p. in french) – mf#mf-10289 seam reel 114/5 [§] – us CRL [280]
Dasar pendidikan dan pengadjaran / Yamin, M – Djakarta, 1955 – 4mf – 9 – mf#SE-815 – ne IDC [959]
Dase cuenta del estado de lo sucedido en los pleytos de la religion de nuestra padre s. geronimo desde 22 de octubre de 1641 hasta 16 de abril de 1642 / Davila, Andres – S.l., s.i., s.a. 1642 – 1 – sp Bibl Santa Ana [946]
Das dasein gottes / Kroening, G – Kleinere Ausg. Milwaukee: G Kroening, 1907 – 1mf – 9 – 0-8370-4002-7 – (incl bibl ref) – mf#1985-2002 – us ATLA [210]
"Dasein heisst eine rolle spielen" : studien zur deutschen literaturgeschichte / Burger, Heinz Otto – Muenchen: C Hanser Verlag, c1963 [mf ed 1993] – (= ser Literatur als kunst) – 303p – 1 – (incl bibl ref) – mf#8135 – us Wisconsin U Libr [430]
Daseinsangst als ursprung menschlichen fehlverhaltens : eine theologisch-ethische untersuchung zu fritz riemanns tiefenpsychologischer studie ueber die antinomien des lebens und das wesen der menschlichen angst / Schumacher, Stefan – (mf ed 1992) – 2mf – 9 – €49.00 – 3-89349-562-2 – mf#DHS 562 – gw Frankfurter [230]
Dasent, George Webbe *see*
– Collection of popular tales from the norse and north german
Daser, Ludwig *see* Passionis domini nostri jesu christi historia...
Dasgupta, Amar Prasad *see* Studies in the history of the british in india
Dasgupta, Amiya Kumar *see* The conception of surplus in theoretical economics
Dasgupta, Hemendra Nath *see* The indian national congress
Dasgupta, Jyotiprova *see* Girls' education in india
Dasgupta, S N *see* A history of sanskrit literature
Dasgupta, Surendranath *see*
– Hindu mysticism
– A history of indian philosophy
– Indian idealism
– Philosophical essays
– Rabindranath, the poet and the philosopher
– The study of patanjali
– Yoga as philosophy and religion
Dashar, M *see* The revolutionary movement in spain
Dass, Ishuree *see* Domestic manners and customs of the hindoos of northern india
Dassanah khnum : cambokh kar karabar jati liddhi prajadhipateyy nai prades khmaer / Bryn Dumm – Bhnam Ben: [s.n.] 2517 [1974] [mf ed 1990] – 1r with other items – 1 – (in khmer; incl bibl ref) – mf#mf-10289 seam reel 130/9 [§] – us CRL [959]
Dassanavatti mahasamaggi kram dan ranasirs – [Bhnam Ben]: Ganakammadhikar Majjhim Ranasirs Samaggi Sangroh Jati Kambuja 1980- [mf ed Ithaca NY: Photo Services Cornell University Lib [1983] – lekh 1-3 [1980] lekh 4 [1981] – 1 – us Cornell [490]
Dassanavijja / Li Jani et al – Bhnam Ben: Khemarayanakamm 1974 [mf ed 1990] – 1r with other items – 1 – (in khmer) – mf#mf-10289 SEA M reel 113/1 [§] – us CRL [180]
The dassenaike family of hapitigam korale / Dassenaike, Louis Arthur – Colombo: Colombo Apothecaries, 1923 – 1 – us CRL [954]
Dassenaike, Louis Arthur *see* The dassenaike family of hapitigam korale
Dassler, Charles Frederick William *see* Dassler's book of forms, and conveyancers' manual
Dassler's book of forms, and conveyancers' manual / Dassler, Charles Frederick William – Topeka, Kan., Crane, 1894. 446 p. LL-768 – 1 – us L of C Photodup [340]
Dastan hazrat amir hamzah razi allah anah ki – Bamba'i: Khoja, Khanmahamui Rahm, 1269 A H [1852?]. v2-4 – 1r – us CRL [956]
Daswen padshah ka granth *see* The religion of the sikhs
Dat Co Don *see* Tinh hoi lam sao vui
Dat trong lang : tieu thuyet / Dinh Quang Nha – [s.l]: Van Nghe Giai Phong 1974 [mf ed 1992] – on pt of 1r – 1 – mf#11052 r387 n1 – us Cornell [959]
Data – Green Bay WI. 1882 dec 21 – 1r – 1 – (cont: daily data; cont by: green bay data) – mf#918679 – us WHS [071]
Data – Washington DC 1957-70 – 1 – mf#3137 – us UMI ProQuest [355]
Data and knowledge engineering – Oxford, England 1988+ – 1,5,9 – ISSN: 0169-023X – mf#42526 – us UMI ProQuest [000]
Data base for advances in information systems – Atlanta GA 2000+ – (= ser Advances in information systems; Data base) – 1,5,9 – (cont: data base) – mf#28527,01 – us UMI ProQuest [000]
Data book of the nolachucky baptist association / Reeves, Thomas H – Morristown, Tennessee. 1914. 242p – 1 – 8.47 – us Southern Baptist [242]
Data communications – Manhasset NY 1974-99 – 1,5,9 – ISSN: 0363-6399 – mf#10139 – us UMI ProQuest [000]
Data from his papers on baptist international youth conference, 1931-1947 / Leavell, Frank H – 1 – $5.00 – us Southern Baptist [242]
Data management [1963] – Chicago IL 1963-74 – 1,5 – (cont by: dm, data management) – ISSN: 0022-0329 – mf#2680.04 – us UMI ProQuest [000]
Data management [1984] – Chicago IL 1984-87 – 1,5,9 – (cont: dm, data management) – ISSN: 0148-5431 – mf#2680.04 – us UMI ProQuest [000]
The data of ethics / Spencer, Herbert – New York: Hurst, [1879?] – 1mf – 9 – 0-8370-6411-2 – mf#1986-0411 – us ATLA [170]

The data of jurisprudence / Miller, William G – Edinburgh/London: William Green & Sons, 1903 – 6mf – 9 – $9.00 – mf#LLMC 95-174 – us LLMC [340]

The data of modern ethics examined / Ming, John Joseph – New York: Benziger, 1894 – 1mf – 9 – 0-8370-6284-5 – (incl bibl ref) – mf#1986-0284 – us ATLA [170]

Data processing – Oxford, England 1963-86 – 1,5,9 – (cont by: information & software technology) – ISSN: 0011-684X – mf#1323.01 – us UMI ProQuest [000]

Data processing & communications security – Madison WI 1984-91 – 1,5,9 – (cont: assets protection; cont by: computing & communications protection) – ISSN: 0749-1484 – mf#11931.03 – us UMI ProQuest [000]

Data processing digest – Los Angeles CA 1955-96 – 1,5,9 – ISSN: 0011-6858 – mf#1500 – us UMI ProQuest [000]

Data processing magazine – Philadelphia PA 1958-72 – 1,5 – ISSN: 0276-2684 – mf#1183 – us UMI ProQuest [000]

Data processor – Yorktown Heights NY 1970-81 – 1,5,9 – ISSN: 0011-6890 – mf#8223 – us UMI ProQuest [000]

Data sheets to microfilmed captured german records / National Archives Trust Fund Board. Washington, D.C – (= ser National archives coll of foreign records seized, 1941-) – 34r – 1 – mf#T176 – us Nat Archives [943]

Data systems – Midland Park NJ 1960-71 – 1,5,9 – ISSN: 0011-6939 – mf#2035 – us UMI ProQuest [000]

Database – Medford NJ 1978-99 – 1,5,9 – (cont by: e content) – ISSN: 0162-4105 – mf#14367.01 – us UMI ProQuest [000]

Database programming & design – Manhasset NY 1987-98 – 1,5,9 – ISSN: 0895-4518 – mf#16328 – us UMI ProQuest [000]

Database technology – Oxford, England 1990-92 – 1,5,9 – ISSN: 0951-9327 – mf#49614 – us UMI ProQuest [000]

Datamation – New York NY 1955-98 – 1,5,9 – ISSN: 0011-6963 – mf#1520 – us UMI ProQuest [000]

Date et destinaire de l'"histoire auguste" / Stern, Henri – Paris, France. 1953 – 1r – us UF Libraries [025]

Date historique, 4 octobre 1916 : son honneur le juge e lafontaine presentant les delegues des ligues antialcooliques de la province de quebec a sir lomer gouin et a ses collegues / Lafontaine, Eugene – [Quebec (Province)?: s.n, 1916?] [mf ed 1996] – 1mf – 9 – 0-665-79038-4 – mf#79038 – cn Canadiana [344]

The date of our gospels in the light of the latest criticism / Curtiss, Samuel Ives – Chicago: F H Revell, 1881 [mf ed 1985] – 1mf – 9 – 0-8370-2794-2 – mf#1985-0794 – us ATLA [226]

The date of st paul's epistle to the galatians / Round, Douglass – Cambridge: University Press, 1906 – 1mf – 9 – 0-7905-0270-4 – (incl bibl ref) – mf#1987-0270 – us ATLA [227]

The date of the exodus in the light of external evidence / Jack, J W – T. & T. Clark, 1925 – 9 – $12.00 – us IRC [270]

Date, S R *see* Bhaganagar struggle

The dated events of the old testament : being a presentation of old testament chronology / Beecher, Willis Judson – Philadelphia: Sunday School Times Company, c1907 – 1mf – 9 – 0-8370-2239-8 – (includes appendix and index. includes chronological tables comparing near east events with israelite history) – mf#1985-0239 – us ATLA [221]

Dateline / AFL-CIO [American Federation of Labor-Congress of Industrial Organizations] Community File [San Francisco CA] – San Francisco CA. 1980 aug-1987 oct – 1r – 1 – mf#1055166 – us WHS [331]

Dateline world jewry – (New York, NY), May 1988-December 1993 – 25ft – (missing: june 1990) – us AJPC [939]

Daten deutscher dichtung : chronologischer abriss der deutschen literaturgeschichte von den anfaengen bis zur gegenwart / Frenzel, Herbert Alfred & Frenzel, Elisabeth – 2nd rev enl ed. Koeln: Kiepenheuer & Witsch [1959, c1953] [mf ed 1992] – 1r – 1 – (incl ind. filmed with: introductory studies in german literature / richard hochdoerfer) – mf#3157p – us Wisconsin U Libr [430]

Dates and distances : showing what may be done in a tour of sixteen months through various parts of europe, as performed in the years 1829 and 1830 – London 1831 [mf ed Hildesheim 1995-98] – 1v on 3mf – 9 – €90.00 – 3-487-27832-4 – gw Olms [914]

Dates and events / Spanish Committee in Defense of Democracy – Washington, DC, 1936? Fiche W1198. (Blodgett Collection of Spanish Civil War Pamphlets) – 9 – us Harvard [946]

The dates of genesis : a comparison of the biblical chronology with that of other ancient nations: with an appendix on chronological astronomy / Jones, Frederick Augustus – London: Kingsgate Press, 1909 – 1mf – 9 – 0-8370-3795-6 – mf#1985-1795 – us ATLA [221]

The dathavansa : or, the history of the tooth-relic of gotama buddha = dathavamsa / Dhammakitti, Polonnaruve – London: Truebner, 1874 – 1mf – 9 – 0-524-07298-1 – (in english) – mf#1991-0084 – us ATLA [280]

Dathenus, P *see*
- Ad bartholomaei latomi rhetoris calumnias...
- Bestendige antwort etlicher fragstueck
- Brevis ac perspicua
- Compendiosa et diserta ad annotationes papistae cuiusdem anonymi...
- Historie v.d. spaensche inquisitie...
- Kurtze und warhafftige erzehlung...
- Libellus supplex imperatoriae maiestati...

Dati, C R *see*
- Esequie della maesta christianiss
- Vite de pittori antichi...

Die datierung der psalmen salomos : ein beitrag zur juedischen geschichte / Frankenberg, Wilhelm – Giessen: J Ricker, 1896 – (= ser Beihefte zur zeitschrift fuer die alttestamentliche wissenschaft) – 1mf – 9 – 0-8370-3181-8 – mf#1985-1181 – us ATLA [240]

Datillo, John P *see* Attitudes of therapeutic recreation professionals toward persons with aids and the relationship of their attitude to their knowledge of aids

Dato y su vida : notas recopiladas / Peris, Ramon – Madrid: Razon y Fe, 1926 – 1 – sp Bibl Santa Ana [324]

Datos biograficos del general e ingeniero miguel y... / Zirion, Grace H De – Guatemala, 1961 – 1r – us UF Libraries [972]

Datos curiosos sobre la demarcacion politica de gu... / Reyes M, Jose Luis – Guatemala, 1951 – 1r – us UF Libraries [972]

Datos estadisticos / Argentine Republic. Direccion de Economica Rural y Estadistica – 2v. 1898-1900 – 1 – us L of C Photodup [318]

Datos etnograficos de venezuela / Alvarado, Lisandro – Caracas, Venezuela. 1945 – 1r – us UF Libraries [972]

Datos historico-culturales sobre las tribus de la... / Reichel-Dolmatoff, Gerardo – Bogota, Colombia. 1951 – 1r – us UF Libraries [972]

Datos historicos sobre la frontera dominico-haitia / Mclean, James J – Santo Domingo, Dominican Republic. 1921 – 1r – us UF Libraries [972]

Datos ineditos para la biografia del capitan armando de montenegro, companero de pizarro en la conquista del peru / Rafal, Marques de – Madrid: Tip. de Archivos, 1932. B.R.A.H. 100, pp.801-813 – sp Bibl Santa Ana [350]

Datos informativos / Subsecretaria de turismo – Badajoz: Imp. Campini, 1964 – sp Bibl Santa Ana [338]

Datos para el estudio de las antiguedades de merida / Alvarez Saenz de Buruaga, Jose – Badajoz: Imprenta Provincial, 1950 – 1 – sp Bibl Santa Ana [946]

Datta, Bhupendranatha *see* Dialectics of hindu ritualism

Datta, D C *see*
- Exegi monumentum and lyrics
- Vidyapati, renderings in english verse

Datta, Dhirendra Mohan *see*
- The chief currents of contemporary philosophy
- The philosophy of mahatma gandhi
- The six ways of knowing

Datta, Dhirendramohan *see* An introduction to indian philosophy

Datta, Hirendranath *see*
- Indian culture, its strands and trends
- Philosophy of the gods

Datta, Kalikinkar *see*
- Alivardi and his times
- Education and social amelioration of women in pre-mutiny india
- Studies in the history of the bengal subah, 1740-70

Datta, Narendranath *see*
- The life of swami vivekananda
- The life of vivekananda and the universal gospel
- The master as i saw him
- The science and philosophy of religion

Datta, Roby *see* Echoes from east and west

Datta, Surendra Kumar *see* The desire of india

Dattelner morgenpost – Datteln DE, 1964 2 jan-22 nov [lokalteil], 1965 31 may-31 dec [lokalteil] – 2r – 1 – (regional ed of recklinghaeuser zeitung) – gw Mikrofilm [074]

Dattilo, John P *see* The relationship between shared family recreation time and expressed parent-adolescent conflict

Datum – v2 n1-v7 n2 [1976 summer-1982 spring] – 1r – 1 – mf#637068 – us WHS [071]

Datus / Finn, George – 2mf – 9 – NZ$8.00 – 0-908797-13-3 – (a chronology of nz from the time of the moa. account of important events in history from c925ad-1910) – nz BAB [980]

Datus, Augustinus *see* In orationes quasdam ciceronis...

Dau cuh qun! jat neh la hoey : [pralom lok knun gruasar] / Thu Dhan – Kraceh: Pannagar Ghun Thai San 2508 [1965] [mf ed 1990] – 1r with other items – 1 – (in khmer; at head of title: ryan) – mf#mf-10289 seam reel 109/4 [§] – us CRL [959]

Dau, William Herman Theodore *see*
- Four hundred years
- Luther examined and reexamined

Daub, Carl *see*
- D carl daub's system der theologischen moral
- D carl daub's vorlesungen ueber die philosophische anthropologie
- D carl daub's vorlesungen ueber die prolegomena zur dogmatik
- D carl daub's vorlesungen ueber die prolegomena zur theologischen moral

Daub, Karl et al *see* Studien

Daube, Anna *see* Der aufstieg der muttersprache im deutschen denken des 15. und 16. jahrhunderts

Daube, D *see* Studies in biblical law

Daube, Johann Friedrich *see*
- Anleitung zur erfindung der melodie und ihrer fortsetzung
- Der musikalische dillettant

Daube, Julie *see* La femme pauvre au 19e siecle

Daubentonia seed poisoning of poultry / Shealy, A L – Gainesville, FL. 1928 – 1r – us UF Libraries [636]

Daubeny, Charles Giles Bridle *see* Remarks on the final causes of the sexuality of plants

D'aubigne's "history of the great reformation in germany and switzerland" reviewed : or, the reformation in germany examined in its instruments, causes, and manner... / Spalding, Martin John – Baltimore: J Murphy; Pittsburg: G Quigley, 1844 – 1mf – 9 – 0-524-01131-1 – mf#1990-0345 – us ATLA [242]

D'aubigne's miscellany = Essays. selections / Merle d'Aubigne, Jean Henri – New York: John S Taylor, 1845 – 1mf – 9 – 0-7905-9515-X – (in english) – mf#1989-1220 – us ATLA [240]

Daubigny, Eugene *see* Choiseul et la france d'outre-mer apres le traite de paris

Daubney, William Heaford *see*
- The three additions to daniel
- The use of the apocrypha in the christian church

Daudet, Alphonse *see*
- La belle-nivernaise
- Oeuvres completes
- Sidonie
- Tartarin de tarascon
- Tartarin sur les alpes

Daudin, F M *see* Histoire naturelle, generale et particuliere des reptiles

Dauebler, Theodor *see* Die akte theodor daeubler

Die dauer der lehrtaetigkeit jesu : nach dem evangelium des hl. johannes / Pfaettisch, Ioannes Maria – Freiburg i.B, St Louis MO: Herder, 1911 – 1mf – 9 – 0-7905-1371-4 – (incl bibl ref) – mf#1987-1371 – us ATLA [220]

Die dauer der oeffentlichen wirksamkeit jesu / Fendt, Leonhard – Muenchen: J.J. Lentner, 1906 – (= ser Veroeffentlichungen Aus Dem Kirchenhistorischen Seminar Muenchen) – 1mf – 9 – 0-8370-3114-1 – (incl bibl ref) – mf#1985-1114 – us ATLA [240]

Die dauer der oeffentlichen wirksamkeit jesu : eine patristisch-exegetische studie / Homanner, Wilhelm – Freiburg im Breisgau; St Louis, MO: Herder, 1908 – (= ser Biblische studien) – 1mf – 9 – 0-7905-2413-9 – (incl bibl ref and indexes) – mf#1987-2413 – us ATLA [240]

Die dauer der oeffentlichen wirksamkeit jesu / Zellinger, Johann B – Muenster i.W.: Aschendorff, 1907 – 1mf – 9 – 0-8370-9356-2 – (incl bibl ref and index) – mf#1986-3356 – us ATLA [240]

Daugavas vanagi : [meneskrats latvijas brivibai unlabakai nakotnei] – 1958-61, 1962-64 – 2r – 1 – (cont by: daugavas vanagu maeneesraksts) – mf#681861 – us WHS [071]

Daugavas vanagu maeneesraksts – 1965-1967 oct, nov-1970 aug, sep-1973 jun, jul-1976 feb, mar-1978, 1979-81, 1982 mar/apr-nov/dec, 1983-87, 1988-1989 oct – 10r – 1 – (cont: daugavas vanagi) – mf#681876 – us WHS [071]

Daugherty Family Association *see* Newsletter

Daugherty family newsletter – v3 n1-4 [1986 jan-dec] – 1r – 1 – (cont: newsletter [daugherty family association]) – mf#1336856 – us WHS [929]

The daughter at school / Todd, John – Northampton: Hopkins, Bridgman, 1854, c1853 – 1mf – 9 – 0-8370-7673-0 – mf#1986-1673 – us ATLA [376]

The daughter of affliction : a memoir of the protracted sufferings and religious experience of miss mary rankin / Rankin, Mary – 2nd ed. Dayton, Ohio: Printed for the author at the United Brethren Print Establishment, 1871 – 1mf – 9 – 0-524-06497-0 – mf#1991-2597 – us ATLA [240]

Daughter of to-day : a novel / Cotes, Everard, mrs [Sara Jeannette Duncan] – London: Chatto & Windus, 1895 – 4mf – 9 – (incl publ list) – mf#32300 – cn Canadiana [830]

Daughters of america, or, women of the century / Hanaford, Phebe Ann – Augusta, Me.: True and Co., 1883 – (= ser Women & the church in america) – 2mf – 9 – 0-8370-1571-5 – mf#1984-2200 – us ATLA [920]

Daughters of china : or, sketches of domestic life in the celestial empire / Bridgman, Eliza Jane Gillett – New York: Robert Carter & Bros, 1853 [mf ed 1995] – (= ser Yale coll) – x/234p (ill) – 1 – 0-524-09346-6 – mf#1995-0346 – us ATLA [951]

Daughters of india / Campbell, Mary J – Monmouth IL: Republican-Atlas Printing, 1908 [mf ed 1995] – (= ser Yale coll) – 121p (ill) – 1 – 0-524-09371-7 – mf#1995-0371 – us ATLA [305]

The daughters of india : their social condition, religion, literature, obligations, and prospects / Robinson, Edward Jewitt – Glasgow: T Murray, 1860 – 1mf – 9 – 0-524-01575-9 – mf#1990-2529 – us ATLA [390]

Daughters of Sarah – v2 n3-v9 [1976 may-1983 dec] – 1r – 1 – mf#969830 – us WHS [939]

Daughters of sarah – Chicago IL 1985-96 – 1,5,9 – ISSN: 0739-1749 – mf#15242 – us UMI ProQuest [366]

Daughters of the American colonists *see* Yearbook

Daughters of the American Revolution *see*
- Lineage books
- Wisconsin news letter

Daughters of the American Revolution. Lakewood. Ohio Chapter *see* Cemetery inscriptions fairview park, ohio

Daughters of the american revolution magazine – Washington DC 1892+ – 1,5,9 – ISSN: 0011-7013 – mf#6765.01 – us UMI ProQuest [366]

Daughters of vienna / Adolph, Karl – London, New York: The International Editor, c1922 [mf ed 1995] – 230p – 1 – (adpated by jo sternberg) – mf#8918 – us Wisconsin U Libr [820]

Daulah islamyah – Djakarta, 1957. v1(1-11) – 9mf – 9 – mf#SE-356 – ne IDC [950]

Dauley, Patricia A *see* The effects of acute exercise of varying intensities on subjects with type 1 diabetes mellitus

Daulte, Henri *see* L'eglise d'apres l'institution chretienne de jean calvin

Daumann, Rudolf Heinrich *see*
- Duenn wie eine eierschale
- Die insel der 1000 wunder

Daumas, F *see* Relation d'un voyage d'exploration au nord-est de la colonie du cap du bonne-esperance en 1836

Daumont, Alexandre *see* Voyage en suede

Daunais, Jean *see* Les 12 coups de mes nuits

Dauner kreisblatt – Daun DE, 1866 1 apr-1870 – 2r – 1 – gw Misc Inst [074]

Daunou, Pierre Claude Francois *see* Cours d'etudes historiques

Daunt, Achilles *see* The person and offices of the holy spirit

Dauphine libre – Veurey, Rhone-Alpes, France 1944 [mf ed Norman Ross] – 1 – mf#nrp-560 – us UMI ProQuest [074]

Dauphin-Meunier, Achille *see*
- Le cambodge de sihanouk
- Histoire du cambodge

Daurignac, J M S *see* Histoire du bienheureux pierre claver de la compagnie de jesus

Dausch, Petrus *see*
- Christus in der modernen sozialen bewegung
- Die inspiration des neuen testamentes
- Jesus und paulus
- Das johannesevangelium
- Der kanon des neuen testamentes
- Kirche und papsttum, eine stiftung jesu
- Das leben jesu
- Lebensbejahung und aszese jesu
- Die synoptische frage
- Die wunder jesu
- Die zweiquellentheorie und die glaubwuerdigkeit der drei aelteren evangelien

Dauster, Frank N *see* Breve historia de la poesia mexicana

Dautant, Caius *see*
- Gouverenment haitien
- Ministre faussaire and assassin

Dauth, Gaspard *see* Le diocese de montreal a la fin du dix-neuvieme siecle

Dauthendey, Max *see*
- Die acht gesichter am biwasee
- Das maerchenbriefbuch der heiligen naechte im javanerlande
- Mich ruft dein bild
- Das rauschen der grossen muschel

Dautriche, Gregoire *see* Le proces dautriche

Dauvert-Romilly *see* Petits cadets

Dauxion-Lavaysse, Jean *see*
- Reise nach den inseln trinidad, tabago und margaretha
- Voyage aux iles de trinidad, de tabago, de la marguerite, et dans diverses parties de venezuela, dans l'amerique meridionale

Dauxion-Lavaysse, Jean F *see* A statistical, commercial, and political description of venezuela, trinidad, margarita, and tobago
Dauzat, Albert *see* Noms de personnes, origine et evolution
Dauzats, Andre *see*
– Elements de langue peule du nord-cameroun
– Lexique francais-peul et peul-francais
Davalos, Pedro Maria *see* Colombia en el sur
Davanne, J B *see* Garcon, l'addition?
Davar latzofeh – New York, NY. Nov 1945-Jun 1946 – 1 – us AJPC [071]
Da'vat al-islam – Bombay, India: Anjuman-i Da'vat al-Islam. sal-i 1, shumarah-'i 1-24. ramazan 1324-sha'ban 1325 [oct 1906-sep 1907] – 1r – 1 – $53.00 – us MEDOC [079]
Davenant, John *see* A treatise of justification
Davenham, st wilfred – (= ser Cheshire monumental inscriptions) – 2mf – 9 – £4.00 – mf#237 – uk CheshireFHS [929]
Davenport, Mrs *see* Journal of a fourteen days' ride through the bush from quebec to lake st john
Davenport, Cyril *see* The book
Davenport, Ernest Harold *see* The false decretals
Davenport, Frances G *see* A guide to the manuscript materials for the history of the united states to 1783
Davenport, Frederick Morgan *see*
– Primitive traits in religious revivals
Davenport, John *see* An apology for mohammed and the koran
Davenport, John M *see* Papal infallibility
Davenport lancet – Davenport, NE: F E Matson. 3v. jul 6 1888-v3 n27. jan 2 1891 (wkly) [mf ed with gaps filmed 1974?] – 1r – 1 – (cont by: people's journal) – us NE Hist [071]
Davenport, Montague *see* Under the gridiron
Davenport news – Davenport, NE: Madeline Sorensen. 3v. v1 n1. aug 31 1951-v3 n4. jul 24 1953) (wkly) [mf ed 1974?] – 1r – 1 – (absorbed by: hebron journal-register) – us NE Hist [071]
Davenport news – Davenport, NE: Madeline Sorensen. 3v. v1 n1. aug 31 1951-v3 n4. jul 24 1953 (wkly) [mf ed filmed [1974?]] – 1r – 1 – (absorbed by: hebron journal-register) – us NE Hist [071]
Davenport news – Davenport, NE: Ray A Wild. v1 n1. jun 18 1879-v1 n12. sep 3 1897 (wkly) [mf ed lacks aug 20-27 filmed 1958] – 1r – 1 – (cont by: oak news) – us NE Hist [071]
Davenport, Samuel *see* Some new industries for south australia...
Davenport, T R H *see* Afrikaner bond
Davenport, William *see* Papers of william davenport and co, 1745-1797
Daventriae illustratae sive historiae urbis daventriensis, libri sex / Revius, J – Lugduni Batavorum, 1651 – €32.00 – ne Slangenburg [240]
Davesne, Andre *see* Croquis de brousse
Davey, Anthony S *see* Morphology of the substantive in subiya
Davey mirror – Lincoln, NE: Inter-State Newspaper Co, 1891-apr 1934 (wkly) [mf ed 1892,1894-1921 (gaps) filmed [1972?]] – 8r – 1 – (absorbed by: lancaster county weekly. some irregularities in numbering) – us NE Hist [071]
Davey, Richard Patrick Boyle *see* Furs and fur garments
Davey, William G *see* A validation study of the q-plex 1 cardio-ulmonary exercise system
Davey, William Harrison *see* Articuli ecclesiae anglicanae
Davey, William Harrison et al *see* Historical books, joshua to esther
Davezies, Robert *see* Angolais
David : shepherd, psalmist, king / Meyer, Frederick Brotherton – New York: FH Revell, c1895 – 1mf – 9 – 0-524-05621-8 – mf#1992-0476 – us ATLA [220]
David : tragedie biblique de forme classique / Abellard, Alexandre Charles – Port-Au-Prince, Haiti. 1950 – 1r – us UF Libraries [972]
David : virtutis exercitatissimae probatum deo spectaculum... / Arias Montanus, B – [Frankfort]: Ex Officina M. Zachariae Pathenii, 1597 – 2mf – 9 – mf#0-1817 – ne IDC [090]
David, Anan ben *see* Mitsvot
David and solomon : investigating the archaeological evidence / Thompson, Lynn – Uni of South Africa 2001 [mf ed Johannesburg 2001] – 3mf [ill] – 9 – (incl bibl ref) – mf#mfm15050 – sa Unisa [220]
David, Arthur Evan *see* Australia
The david bailie warden papers / ed by Marks, Bayly Ellen – 1972 – 8r – 1 – $1040.00 – (guide sold separately $10.00) – mf#S1623 – us Scholarly Res [362]
David baptist church. kings mountain, north carolna : church records – 1938-63 – 1 – 5.00 – us Southern Baptist [242]
David brainerd : the apostle to the north american indians / Page, Jesse – London: S. W. Partridge, [18–] Chicago: Dep of Photodup, U of Chicago Lib, 1973 (1r); Evanston: American Theol Lib Assoc, 1984 (1r) – 1 – 0-8370-0356-3 – mf#1984-B366 – us ATLA [240]
David, C *see* Is a russian invasion of india feasible?
David, [C M] *see* Funafuti, or three months on a coral island
David chytraeus : nach gleichzeitigen quellen / Pressel, Theodor – Elberfeld: RL Friderichs, 1862 – (= ser Leben und ausgewaehlte Schriften der Vaeter und Begruender der lutherischen Kirche) – 1mf – 9 – 0-524-00585-0 – (incl bibl ref) – mf#1990-0085 – us ATLA [240]
David city news – David City, NE: A H Betzer, 1891-1901// (wkly) [mf ed 1895,1898-1901 (gaps) filmed [1972?]] – 2r – 1 – us NE Hist [071]
David city republican – David City, NE: E Heath, C Patterson, 1877 (wkly) [mf ed v8 n33. sep 11 1884 filmed [1983]] – 1r – 1 – us NE Hist [071]
David city tribune – David City, NE: A H Betzer, jan 1884-1893// (wkly) [mf ed v1 n3. jan 31 1884-88, 1892 (gaps) filmed [1974?]] – 1r – 1 – (absorbed: ulysses monitor. cont by: people's banner and david city tribune) – us NE Hist [071]
David, Claude *see*
– Von richard wagner zu bertolt brecht
– Zwischen romantik und symbolismus, 1820-1855
The david cobb papers, 1708-1833 – [mf ed 1980] – 3r – 1 – (with p/g. coll consists primarily of papers relating to cobb's private interests in land and fisheries development in maine) – us MA Hist [978]
David cusick's sketches of ancient history of the six nations : comprising first- a tale of the great island, [now north america,] the two infants born, and the creation of the universe, second- a real account of the early settlers of north america, and their dissensions... – [Lockport, NY?: s.n.], 1848 [mf ed 1982] – 1mf – 9 – mf#33363 – cn Canadiana [390]
David d wallace papers, 1822-1967 – 35 linear ft also on microfilm – 1 – (consist of research notes, correspondence, financial records, mss & publ of wallace's articles & other writings, family papers, & other items) – mf#1261.00 – us South Carolina Historical [920]
David Dunlap Observatory *see* Publications of the david dunlap observatory, university of toronto
David, Felicien *see* Lalla-roukh
David, Francois *see* Methode nouvelle
David friedrich strauss / Eck, Samuel – Stuttgart: JG Cotta, 1899 – 1mf – 9 – 0-524-00633-4 – (incl bibl ref) – mf#1990-0133 – us ATLA [240]
David friedrich strauss : ein lebens- und literaturbild / Hettinger, Franz – Freiburg im Breisgau; St Louis, MO: Herder, 1875 – 1mf – 9 – 0-524-00866-3 – mf#1990-0251 – us ATLA [240]
David friedrich strauss als denker und erzieher / Kohut, Adolf – Leipzig: A Kroener, 1908 – 1mf – 9 – 0-524-04466-X – (incl bibl ref) – mf#1992-0135 – us ATLA [240]
David friedrich strauss in his life and writings = David friedrich strauss in seinem leben und seinen schriften / Zeller, Eduard – London: Smith, Elder, 1874 – 1mf – 9 – 0-524-04320-5 – (in english) – mf#1990-1246 – us ATLA [240]
David g swaim letters, 1861-1874 / Swaim, David G – [mf ed 2000] – 2r – 1 – us Western Res [976]
D h lawrence review – San Marcos TX 1968-92 – 1,5,9 – ISSN: 0011-4936 – mf#7050 – us UMI ProQuest [420]
David hill : missionary and saint / Barber, William Theodore Aquila – London: Charles H Kelly, 1898 [mf ed 1995] – (= ser Yale coll) – 331p (ill) – 1 – 0-524-10158-2 – mf#1995-1158 – us ATLA [920]
David, his life and times / Deane, William John – New York: Fleming H Revell, [18–?] [mf ed 1986] – 1mf – 9 – 0-8370-9856-4 – (incl bibl ref) – mf#1986-3856 – us ATLA [220]
David, hoy / Aradillas Agudo, Antonio – Madrid: Ediciones Studium, 1963 – 1 – sp Bibl Santa Ana [946]
David hudson family papers *see* Family papers, ms 3893
David hume / Calderwood, Henry – Edinburgh: Oliphant Anderson & Ferrier, [1898?] – (= ser Famous Scots Series) – 1mf – 9 – 0-7905-3709-5 – mf#1989-0202 – us ATLA [140]
David hume / Macnabb, D G C – London, England. 1951 – 1r – us UF Libraries [140]
David hunter miller papers – [mf ed ProQuest] – 6r – 1 – us UMI ProQuest [327]
David, J *see*
– Christeliicken waerseggher
– Duodecim specula deum aliquando videre desideranti concinnate
– Occasio arrepta, neglecta, huius commoda
– Paradisus sponsi et sponsae
– Veridicus christianus
David, Jakob Julius *see*
– Anzengruber
– Vom schaffen
David joris : bibliografie / Linde, Antonius van der – 's Gravenhage: M Nijhoff, 1867 [mf ed 1990] – 1mf – 9 – 0-7905-8041-1 – (in dutch. incl bibl ref) – mf#1988-6022 – us ATLA [242]
David, king of israel : his life and its lessons / Taylor, William Mackergo – New York: Harper, c1874 [mf ed 1986] – 1mf – 9 – 0-8370-9989-7 – (incl bibl ref & ind) – mf#1986-3989 – us ATLA [920]
David, Laurent Olivier *see*
– Biographie avec portrait de m l'abbe mercier
– Le clerge canadien
– Le drapeau de carillon
– Les patriotes de 1837-38
David, Laurent-Olivier *see*
– Biographies et portraits
– Esquisse biographique de sir george-etienne cartier
– Le heros de chateauguay
– Histoire du canada depuis la confederation, 1867-1887
– L'honorable p-j-o chauveau
– Laurier
– Laurier et son temps
– Melanges historiques et litteraires
– Mgr ignace bourget et mgr alexandre tache
– Monseigneur alexandre-antonin tache
– Monseigneur bourget
– Monseigneur joseph-octave plessis
– Monsieur isaac s desaulniers
– Sir ls-h lafontaine
– Souvenirs et biographies, 1870-1910
David leib magdeburger / Kohn, S – Tel-Aviv, Israel. 1930-1939? – 1r – us UF Libraries [939]
David livingstone / Horne, Charles Silvester – New York: Macmillan, 1913 – 1mf – 9 – 0-524-08428-9 – mf#1993-1038 – us ATLA [910]
David livingstone and the rovuma – Edinburgh, Scotland. 1965 – 1r – us UF Libraries [960]
David lloyd george / Edwards, John Hugh – New York, NY. v1-2. 1929 – 1r – us UF Libraries [941]
David, M *see*
– Die adoption im altbabylonischen recht
– Assyrische rechtsurkunden
David morton, a biography / Hoss, Elijah Embree – 2nd ed. Louisville, KY: Board of Church Extension of the ME Church, South, c1916 [mf ed 1993] – 1mf – 9 – 0-524-08387-8 – mf#1993-3087 – us ATLA [242]
David outlaw papers / Outlaw, David – 1847-66. University of North Carolina Library. Guide – 1 – $18.00 – us CIS [920]
David, P *see* Les monasteres du diocese de grenoble a l'epoque merovingienne
David, Placide *see*
– Heritage colonial en haiti
– Sur les rives du passe
David reubeni und salomo molcho : ein beitrag zur geschichte der messianischen bewegung im judentum in der ersten haelfte des 16. jahrhunderts / Voos, Julius – [S.l.]: J Voos, [1933?]. (Berlin [Germany]: Michel) (mf ed 19–) – 1 – (incl bibl ref) – mf#*ZP-*PBM pv218 n7 – us NY Public [270]
The david robison collection of materials related to the war in the southern sudan – Chicago: U of Chicago, Photodup Dep, 1972 (mf ed) – us CRL [355]
David, roi, psalmiste, prophete : avec une introduction sur la nouvelle critique / Meignan, Guillaume Rene – Paris: Victor Lecoffre, 1889 [mf ed 1993] – (= ser Propheties messianiques 3; Ancien testament dans ses rapports avec le nouveau et la critique moderne 3) – 2mf – 9 – 0-524-05683-8 – (incl bibl ref) – mf#1992-0533 – us ATLA [221]
David, Siegfried *see* August wilhelm ifflands schauspielkunst bis zum abschluss der mannheimer zeit (1796)
David swing : a memorial volume: ten sermons – Chicago: F T Neely, 1894 [mf ed 1991] – 1mf – 9 – 0-7905-9643-1 – mf#1989-1368 – us ATLA [242]
David swing, poet-preacher / Newton, Joseph Fort – Chicago: Unity Pub Co, 1909 – 1mf – 9 – 0-524-08490-4 – (incl bibl ref) – mf#1993-3135 – us ATLA [240]
David, Tennant *see* Notary's manual containing instructions for notaries
David, the king of israel : a portrait drawn from bible history and the book of psalms = David, der koenig von israel / Krummacher, Friedrich Wilhelm – New York: Harper, 1874 [mf ed 1986] – 2mf – 9 – 0-8370-9709-6 – (trans by matthew george easton) – mf#1986-3709 – us ATLA [920]
David, Uta *see* Herzkraft und koerpermasse
David, V D *see* Een geheiligd leven
David worthington simon / Powicke, Frederick James – London; New York: Hodder and Stoughton, [1912?] – 1mf – 9 – 0-7905-5670-7 – (incl bibl ref) – mf#1988-1670 – us ATLA [920]
David zeisberger and his brown brethren / Rice, William Henry – Bethlehem, Pa: Moravian Publication Concern, c1897 – 1mf – 9 – 0-8370-6343-4 – mf#1986-0343 – us ATLA [242]
David zeisberger, der indianer apostel – Watertown, WI: D Blumenfeld, [19–?] – (= ser Tannenreiser) – 1mf – 9 – 0-524-07208-6 – mf#1990-5366 – us ATLA [240]
David zeisberger's history of the northern american indians / ed by Hulbert, Archer Butler & Schwarze, William Nathaniel – [Columbus?]: Ohio State Archaeological & Historical Society [1910?] [mf ed 1990] – 1mf – 9 – 0-7905-7098-X – (trans fr german; incl bibl ref) – mf#1988-3098 – us ATLA [975]
Davidenko, Vasilii *see* Sovremennoe sostoyanie rakolo sektantstva v kharkovskoi eparkii
David-frederic strauss, la vie et l'oeuvre / Levy, Albert – Paris: Felix Alcan, 1910 – (= ser Coll historique des grands philosophes) – 1mf – 9 – 0-7905-8826-9 – (incl bibl ref) – mf#1989-2051 – us ATLA [100]
Davidis regis...psalmi / Arias Montano, Benito – 1574 – 9 – sp Bibl Santa Ana [240]
Die davidische abkunft der mutter jesu : 1. teil, die ausserbiblischen nachrichten / Fischer, Joseph – Wien: A Opitz, 1910 – 1mf – 9 – 0-7905-1385-4 – (incl bibl ref) – mf#1987-1385 – us ATLA [241]
Den david-jorischen gheest in leven ende leere : breeder ende wijdt-loopigher ontdect... / Emmius, U – 's Graven Hage, 1603 – 5mf – 9 – mf#PBA-173 – ne IDC [240]
David-Neel, Alexandra *see*
– Buddhism
– Le modernisme bouddhiste et le bouddhisme du bouddha
– Socialisme chinois
Davids, Arthur Lumley *see* Kitabuel-ilm uen-nafi fi tahsil-isarf ve nahv-i tuerki
David's, Bishop Of St... *see* Protestant's catechism on the origin of popery
David's blessed man : or, a short exposition on the first psalm: directing a man to true happiness / Smith, Samuel – Edinburgh: James Nichol, 1868 [mf ed 1985] – (= ser Nichol's series of commentaries) – 1mf – 9 – 0-8370-5304-8 – (incl biogr info) – mf#1985-3304 – us ATLA [242]
Davids, Caroline *see* Psalms of the early buddhists
Davids, Caroline Augusta Foley Rhys *see*
– The birth of indian psychology and its development in buddhism
– Buddhism
– Buddhist psychology
– Gotama the man
– Indian religion and survival
– Kindred sayings on buddhism
– Manual of a mystic
– Outlines of buddhism
– Poems of cloister and jungle
– Psalms of the early buddhists
– Wayfarer's words
– What was the original gospel in 'buddhism'?
David's harp in song and story / Clokey, Joseph Waddell – Pittsburgh: United Presbyterian Board of Publication, 1896 – 1mf – 9 – 0-524-01718-2 – mf#1990-4110 – us ATLA [221]
David's harp new tunes *see* Hymn and tune collection from library of edmond d. keith
Davids psalmer / ed by British and Foreign Bible Society – Kristiania [Oslo]: Groendahl, 1885 – 2mf – 9 – 0-524-06787-2 – mf#1992-0950 – us ATLA [221]
Davids, Thomas William Rhys *see*
– Buddhism
– Buddhist india
– Early buddhism
– Lectures on the origin and growth of religion
– On yuan chwang's travels in india, 629-645 a.d
Davidson, A B *see*
– The book of the prophet ezekiel
– The called of god
– Old testament prophecy
Davidson, Alex Dyce *see* Lectures, expository and practical
Davidson, Alexander *see*
– The canada spelling book
Davidson, Andrew Bruce *see*
– Biblical and literary essays
– The book of the prophet ezekiel
– The book of the prophet ezekiel: in the revised version
– The book of the prophet isaiah
– The books of nahum, habakkuk and zephaniah
– Called of god
– A commentary, grammatical and exegetical, on the book of job
– The epistle to the hebrews

- Hebrew syntax
- An introductory hebrew grammar
- Outlines of hebrew accentuation, prose and poetical
- The theology of the old testament

Davidson, Anne Jane *see* The autobiography and diary of samuel davidson
Davidson, Apollon Borisovich *see* Crime of apartheid
Davidson, Basil *see*
- African awakening
- The african awakening
- Black mother
- Growth of african civilization
- Old africa rediscovered

Davidson, Charles Peers *see* A compilation of the statutes passed since confederation relating to banks and banking
Davidson, Dana *see* Journal of religion in disability and rehabilitation
Davidson, Donna Morris *see* "The morrises as pioneers"
Davidson, Edward *see* The railways of india
Davidson, Frederic Joseph Arthur *see* Le village
Davidson, H S *see* De lagardes ausgabe der arabischen uebersetzung der genesis
Davidson, Hannah Frances *see* South and south central africa
Davidson, Israel *see* Saadia's polemic against hiwi al-balkhi
Davidson, James *see*
- Assurance of salvation practically considered
- Bible and the school board

Davidson, James Wightman *see* Northern rhodesian legislative council
Davidson, James Wood *see* Florida of today
Davidson, Judson D *see* Enoch
Davidson, Judson France *see*
- Biographical sketch of the famous and brilliant canadian evangelist
- A lapful of lyrics and merry muse-whangs
- Muse whangs

Davidson, P *see* Catechumen
Davidson, Randall Thomas *see*
- Captains and comrades in the faith
- The character and call of the church of england
- The christian opportunity
- Kikuyu
- The lambeth conferences of 1867, 1878, and 1888

Davidson, Randall Thomas et al *see* The scotch church crisis
Davidson, Robert *see*
- Historical sketch of the synod of philadelphia
- History of the presbyterian church in the state of kentucky
- Lectures on grammar, rhetoric

Davidson, Robert John *see* Life in west china
Davidson, Samuel *see*
- The canon of the bible
- The doctrine of last things
- Facts, statements, and explanations
- An introduction to the new testament
- An introduction to the old testament
- An introduction to the study of the new testament
- The new testament
- On a fresh revision of the english old testament
- Sacred hermeneutics developed and applied
- The text of the old testament considered
- A treatise on biblical criticism
- Vorlesungen ueber die apokalypse

Davidson, Thomas *see*
- Giordano bruno and the relation of his philosophy to free thought; a lecture...new york liberal club, oct. 30, 1885
- Glory of god displayed in the building up of zion

Davidsonia – Vancouver, Canada 1970-80 – 1,5,9 – ISSN: 0045-9739 – mf#7726 – us UMI ProQuest [580]
Davidts, Hermann *see* Die erstlingsnovellen heinrich von kleist
Davie, John *see*
- Letters from buenos ayres and chili
- Letters from paraguay

Davie, John G *see* Introduction to the solution of the problems of the pyramid
Davie, Oliver *see* An egg check list of north american birds
Davie, W Galsworthy *see* Architectural studies in france
Davies, Alfred Thomas *see* A theological and behavioral analysis of the practice of ministerial authority and leadership strategy
Davies, Arthur *see* Proposals for uniting the british colonies with their mother-country, by making them "integral portions of the empire"
Davies, Charles F *see* The church chant book
Davies, Charles Maurice *see*
- Heterodox london
- Mystic london, or, phases of occult life in the metropolis

Davies, Courtney C *see* A study of the relationship between selfreported stress-related physical symptomology and spiritual wellness
Davies, David Charles *see* The atonement and intercession of christ
Davies, E *see* Memoir of the rev samuel dyer
Davies, Edward *see* Celtic researches, on the origin, traditions & language of the ancient britons
Davies, Edward Owen *see*
- The miracles of jesus
- Prolegomena to systematic theology
- Theological encyclopaedia

Davies, Evan *see* Lectures on christian theology
Davies, George Jennings *see* Successful preachers
Davies, Gerald Stanley *see* St paul in greece
Davies, H W *see* Bernhard von breydenbach and his journey to the holy land, 1843-1844
Davies, Hannah *see* Among hills and valleys in western china
Davies, Henry *see* Hours in the picture gallery of thirlestane house, cheltenham
Davies, Henry D *see* A new proposal for the gradual creation of a farmer proprietary in ireland
Davies, Henry William *see* An analytical and practical grammar of the english language
Davies, Horton *see* Great south african christians
Davies, J *see* Christian preaching as exemplified in the conduct of st paul
Davies, J G *see* The biu book
Davies, J Llewelyn *see* St paul and modern thought
Davies, John *see*
- Davies' patent cases
- Encouragement to the faithful ministers of christ
- Grammar of the tahitian dialect of the polynesian language
- Ministerial commission

Davies, John Llewelyn *see*
- The christian calling
- Epistles of st paul to the ephesians, colossians, and philemon
- The epistles of st paul to the ephesians, the colossians, and philemon
- The gospel and modern life
- Morality according to the sacrament of the lord's supper
- Order and growth
- The purpose of god
- St paul and modern thought
- Sermons on the manifestation of the son of god
- Social questions from the point of view of christian theology
- Spiritual apprehension
- Theology and morality
- Warnings against superstition
- The work of christ

Davies, Katherine Currie *see* Charles tomlinson griffes and his music
Davies, M J *see* A bibliography of nineteenth-century legal literature
Davies, Michael J *see* Gender differences in running economy
Davies, N de G *see*
- Rock tombs of deir el gebrawi
- Rock tombs of el amarna
- Rock tombs of sheikh said

Davies, Nigel *see* Aztecs
Davies' patent cases : 1624-1816 (british and american) / Davies, John – London, 1816 (all publ) – 5mf – 9 – $7.50 – mf#LLMC 84-324 – us LLMC [346]
Davies, Richard *see*
- Minsiterial gift and its faithful exercise
- Public worship

Davies, Richard Newton *see* Doctrine of the trinity
Davies, Robert *see*
- Extracts from the municipal records of the city of york
- A memoir of the york press

Davies, Roy *see* "Marching rule"
Davies, Samuel *see* Selected works, manuscript and published
Davies, Sydney John *see* Heat pumps and thermal compressors
Davies, Thomas Alfred *see*
- Am i jew or gentile?
- Answer to hugh miller and theoretic geologists
- Genesis disclosed

Davies, Thomas Lewis Owen *see* Bible english
Davies, Thomas Witton *see*
- Ezra, nehemiah and esther
- Heinrich ewald
- The psalms

Davies, William Walter *see* The codes of hammurabi and moses
Davies-Crawford, Gillian Peta *see* Dilemma of the progressive black parent regarding school choice
Davila, Andres *see*
- Consulta...sobre asunto p. caceres y respuesta de andres barbosa
- Dase cuenta del estado de lo sucedido en los pleytos de la religion de nuestra padre s. geronimo desde 22 de octubre de 1641 hasta 16 de abril de 1642
- Informe ajustado al nuncio por parte del p. general (caceres) en el pleyto con los padres diputados

Davila, Enrico C *see* Dell'istoria delle guerre civili di francia
Davila Garibi, Jose Ignacio *see* La sociedad de zacatecas en los albores del regimen colonial
Davila, Jose Antonio *see*
- Vendimia

Davila Olivo, Guillermo *see* Resonancia en el deseo
Davila, Vicente *see*
- Don sancho briceno, su monumento en trujillo. el arbol de los bricenos. caracas, 1929
- Encomiendas, tomo 1
- Investigaciones historicas
- Investigaciones historicas. tomo 2. caracas, 1927
- Labores culturales

Davila y Figueroa, Marino *see* Catalogo de las obras
Davila y Heredia, A *see* Parecer de d. andres davila...que no hay medicina universal...
Daviler, A C *see* Ausfuehrliche anleitung zu der gantzen civil-bau-kunst...
Davin, Nicholas Flood *see*
- Album verses
- The british empire
- British versus american civilization
- The crazy quilt series, no 1, vol 1
- Culture and practical power
- The demands of the north-west!
- Dual language and federal government
- The earl of beaconsfield
- Eos
- The fair grit
- For the leader company, limited, et al
- Great speeches
- Home rule, a speech
- Homes for millions
- In memory of the queen
- The jesuits' estates act
- Mr davin on "fanning in church"
- The new tariff
- On the address
- Queen's jubilee, boston
- Report on industrial schools for indians and half-breeds
- The session of 1891
- Sonnet to e w
- Speech of n f davin, mp, on the address
- Speech of nicholas flood davin, mp on inquiry into election frauds
- Speech of nicholas flood davin, mp on the review of the financial situation
- Speeches of n f davin, mp on duty on agricultural implements
- "The springs of national progress"
- Strathcona horse

Davis : ou, le bonheur d'etre fou / Fournier, Narcisse – Paris, France. 1842 – 1r – us UF Libraries [440]
Davis, A G *see* Investigations on the action of certain soil constituents
Davis, Abraham N *see* Protecting the lambs
Davis, Alexander *see* Native problem in south africa
Davis, Amanda J *see* Dance and the lived experience
Davis, Andrew Jackson *see* The principles of nature, her divine revelations, and a voice to mankind.
Davis, Beale *see* Goat without horns
[Davis-] california aggie – CA: University Campus Davis, 1922- – 42r – 1 – $2520.00 (subs $90y) – mf#B02163 – us Library Micro [378]
Davis, Charles *see* A description of the works of art...of alfred de rothschild
Davis, Charles Henry *see* A practical defence of the evangelical clergy
Davis, Charles Henry Stanley *see* Greek and roman stoicism and some of its disciples
Davis, Cynthia M *see* The relationship between knowledge and expertise in breaststroke swimming
Davis, Daniel *see* A practical treatise upon the authority and duty of justices of the peace in criminal prosecutions.
Davis, Edward Maurice *see* 'N ondersoek na die betekenis van die konsep "gemeenskap" en die implikasies daarvan vir demokratiese opvoeding in suid-afrika
Davis, Emerson *see*
- Church extension
- The half century

Davis family association – 1979 mar-may – 1r – 1 – mf#668967 – us WHS [929]
Davis family newsletter / ed by James, Catherine – 1980/81 winter-1981/82 winter – 1r – 1 – mf#655244 – us WHS [929]
Davis family quarterly – 1980 sep-1981 spring – 1r – 1 – mf#669627 – us WHS [929]
Davis, Frederick Hadland *see*
- The persian mystics, jalalud-din rumi
- The persian mystics, jami

Davis, Frederick W *see* Predictors of overall job satisfaction among public school physical educators
Davis, George B *see* The elements of international law
Davis, George Edward *see* A handbook of chemical engineering
Davis, George Thompson Brown *see*
- Korea for christ
- Torrey and alexander
- Twice around the world with alexander

Davis, George Thompson Brown et al *see* Dwight l. moody
Davis, George W *see* Treatise on the culture of the orange
Davis, Gladys Mary Norman *see* The asiatic dionysos
Davis, Harold Lenoir *see* Honey in the horn
Davis, Harold Palmer *see* Black democracy
Davis, Howard (Mrs) *see* Early marriages in geauga county [ohio]
Davis, Isaac *see* An historical discourse on the 50th anniversary of the first baptist church in worcester, mass
Davis, J *see* History of the welsh baptists, from the year 1763 to the year 1770
Davis, J D *see* Maker of new japan
Davis, J F *see*
- La chine, ou description generale des moeurs et des coutumes, du gouvernment, des lois, des religions, des sciences, de la literature, des productions naturelles, des arts, des manufactures et du commerce de l'empire chinois
- The chinese

Davis, Jack E *see* Lynching of jesse james payne
Davis, Jackson *see* Africa advancing
Davis, Jefferson *see* Papers
Davis, Jerome Dean *see*
- Hand-book of christian evidences
- Sketch of the life of rev joseph hardy neesima, ll.d, president of doshisha

Davis, John *see*
- The chinese
- Papers of john davis

Davis, John A *see* Choh lin
Davis, John Chandler Bancroft *see* The committees of the continental congress, chosen to hear and determine appeals from courts of admiralty
Davis, John D *see* Genesis and semitic tradition
Davis, John Francis *see* Chinese moral maxims
Davis, John Francis, 1st Bart. *see* Chinese moral maxims
Davis, John Philip *see* The royal academy, and the national gallery
Davis, Kingsley *see* The population of india and pakistan
Davis' land court cases / Massachusetts. Supreme Court – 1v. 1898-1908 – (= ser Massachusetts supreme court reports) – 4mf – 9 – $6.00 – mf#LLMC 84-158 – us LLMC [347]
Davis, Lynn M, Jr *see* Decade of progress
Davis, Mary A *see* History of the free baptist woman's missionary society
Davis, Mary B *see* Stockbridge indian papers in the huntington free library
Davis, Mary Irene *see*
- General letter
- Historical
- History of daytona beach
- History of flagler county
- Picnic grounds
- Raids
- Recreation and amusement
- Supplementary data
- Supplementary to manufacturing and industry

Davis, Minnie K *see* The lollard
Davis, Morrison Meade *see*
- First principles
- How the disciples began and grew
- The restoration movement of the nineteenth century

Davis, Nicholas Darnell *see* Cavaliers and roundheads of barbados
Davis, Noah *see*
- Concise account of the late rev. noah davis
- Report of hon. noah davis, referee, to the surrogate's court, city and county of new york, on the american surety company, of new york

Davis, Noah Knowles *see* Elements of ethics
Davis, Owen *see* Robin hood or the merry outlaws of sherwood forest
Davis, Owen William *see*
- Art and work
- The rudiments of decorative painting

Davis, Ozora S *see* John robinson
Davis, Ozora Stearns *see*
- John robinson
- The pilgrim faith
- Using the bible in public address
- Vocabulary of new testament words

Davis, Reginald Charles *see* Canadian folk songs for american schools
Davis, Reuben *see* Speech of hon reuben davis
Davis, Richard Harding *see*
- Cuba in war time
- Three gringos in venezuela and central america

Davis, Robert *see* The canadian farmer's travels in the united states of america
Davis, Samuel T *see* Caribou shooting in newfoundland
Davis, Shaun A *see* Biomonitoring as a means to determine the pollution level in stellenbosch
Davis, Stephen Brooks *see* The law of radio communication

Davis, T N et al *see* Excavations
Davis, Tamar *see* A general history of the sabbatarian churches
[Davis-] the davis enterprise – CA. 1898- – 229r – 1 – $13,740.00 (subs $400/y) – mf#B02164 – us Library Micro [071]
[Davis-] the weekly agricole – CA. 1915-22 – (= ser Wine & agriculture coll) – 3r – 1 – $330.00 – mf#B02165 – us Library Micro [071]
Davis, Thomas Frederick *see*
- Climatology of jacksonville
- Macgregor's invasion of florida
- Narrative history of the orange in the floridian peninsula
- Newspaper clippings about curacoa, dutch west indies...

Davis, Valentine David *see*
- Minister of god
- Proceedings and papers

Davis, Warren Jefferson *see* Putting laws over wings
Davis, William Jafferd *see* A dictionary of the kaffir language
Davis, William Morris *see* Lesser antilles
Davis, William Stearns *see*
- The friar of wittenberg
- The influence of wealth in imperial rome

Davis, William Watson *see* Civil war and reconstruction in florida
Davis, Winfield J *see* History of political conventions in california
[Davis-] woodland daily democrat : davis edition – CA. 1972-1984 – 143r – 1 – $8580.00 – mf#B02166 – us Library Micro [071]
Davison, Annie *see* Georgia baptist association
Davison, Carolyn J *see* The boys and girls clubs of nova scotia
Davison, David *see* On the consolations of the gospel
Davison, John *see*
- Discourses on prophecy
- Reply to an article in the last number, viz lxiv, of the edinburg

Davison, Newman and Co, Ltd *see* Papers of davison, newman and co ltd 1753-1897
Davison, Patricia Joan *see* Material culture of the lobedu
Davison, Thomas *see* Mechanics' institutes and the best means of improving them
Davison, Thomas Raffles *see* International exhibition glasgow 1888
Davison, William Theophilus *see*
- The chief corner-stone
- The christian conscience
- The christian interpretation of life and other essays
- The indwelling spirit
- The praises of israel
- The psalms 1-72
- Strength for the way
- The wisdom-literature of the old testament

Davison, William Theoplilus et al *see* The chief corner-stone
Davitt, Michael *see*
- Impressions of the canadian north-west
- Within the pale

Davosskii vestnik / ed by Avrashov, G – Davos, 1908 – 1mf – 9 – mf#R-18039 – ne IDC [077]
Davul – n1-24. 1325 [1908/09] – (= ser O & t journals) – 8mf – 9 – $130.00 – (in french and ottoman. satirical articles and cartoons publ under the direction of hasan vasif in istanbul) – us MEDOC [870]
Davy, Christopher *see* Architectural precedents
Davy, Humphry *see* Consolations in travel
Davy jones : or, harlequin and mother carey's chickens. performed at the theatre royal, drury lane, dec 27th 1830 / Barrymore, William – London: W Kenneth at his Dramatic Repository, 1830 – 1 – mf#*ZBD-*MGTZ pv2-Res – Located: NYPL – us Misc Inst [790]
Davy, M M *see* Les sermons universitaires parisiens de 1230-1231 (ephm15)
Davydov, Gavrilo *see* Reise der russisch-kaiserlichen flott-officiere chwostow und dawydow
Daw, Jessica L *see* Goal involvements, goal orientation, and perceptions of parent- and coach-initiated motivational climates among youth sport participants
Dawani, Jalal-al-Din *see* Practical philosophy of the mohammedan people akhlaq jalali
Dawbarn, Robert *see* History of a forgotten sect of baptised believers heretofore known as johnsonians
Dawe, Charles G *see* Essays and speeches
Dawes county journal – Chadron, NE: E E Egan. -v13 n16. feb 12 1897 (wkly) [mf ed 1885, 1887-97 (gaps) – 3r – 1 – (absorbed: chadron advocate. cont by: journal) – us NE Hist [071]
Dawes county journal – Chadron, NE: Journal Pub Co. v14 n20. mar 11 1898-v17 n2, nov 2 1900 (wkly) [mf ed with gaps filmed -1989] – 2r – 1 – (cont: chadron journal. cont by: chadron journal (1900)) – us NE Hist [071]
Dawley, William Wallace *see* Truths that abide
Dawlish gazette general advertiser, visitors and local directory for dawlish and district – [SW England] Devon 24 apr-dec 1897; jan 1898-dec 1950 [mf ed 2003] – 42r – 1 – uk Newsplan [072]
Dawlish times, etc – [SW England] Devon oct 1868-6 jun 1903 – 33r – 1 – (missing: 1877, 1897, 1912) – uk Newsplan [072]
Dawn – 1889 may 15-1896 mar – 1r – 1 – mf#1111054 – us WHS [071]
Dawn – A Baptist Magazine. Apr, Jul, Oct 1916, Jan 1917 – 1 – 5.00 – us Southern Baptist [242]
Dawn – Sydney, jan 1952-jul 1975 – (= ser New dawn) – 4r – A$258.68 vesicular A$280.68 silver – (aka: new dawn) – at Pascoe [079]
Dawn – Blantyre: GrafoPrint Works, jul 27, aug 19/23, aug 31/sep 6-oct 24, dec 1 1995; apr 20-29 1996 – 1r – 1 – us CRL [072]
Dawn – Delhi, India. 26 Oct 1941-3 Sept 1947 – 13r – 1 – us L of C Photodup [079]
Dawn / Determined Action for Women Now [Organization] – 1982 apr-1983 feb – 1r – 1 – (cont: new dawn [ypsilanti, mich.]; cont by: new dawn [ypsilanti, mich.: 1983]) – mf#1223813 – us WHS [305]
Dawn – Karachi: Pakistan Herald Press by A A Khan, 1951-66 – 1 – us CRL [079]
Dawn : a semi-monthly magazine containing original and selected essays, anecdotes, etc, in prose and poetry – La Salle IL 1822 – 1 – mf#3735 – us UMI ProQuest [420]
Dawn – Sydney, 1888-1905 – 3r – 1 – A$115.50 vesicular A$132.00 silver – at Pascoe [079]
The dawn : al-fajr – Jerusalem, 1980-1993. v1-14+ind 1980-1985 – 13r – 1 – (missing: 1981(83, 84); 1982/1983(139); 1983(161); 1983/1984(191); 1984/1985(243); 1986(343-346); 1990 v11) – mf#J-91-117 – ne IDC [956]
Dawn baptist church. dawn, missouri : church records – 1865-1968 – 1 – us Southern Baptist [242]
Dawn ginsbergh's revenge / Perelman, Sidney Joseph – New York: H Liveright [c1929] [mf ed 1984] – 1r [ill] – 1 – (with: four years of irish history, 1845-1849 / duffy, c g & other titles) – mf#1285 – us Wisconsin U Libr [870]
Dawn in india : british purpose and indian aspiration / Younghusband, Francis Edward – London: John Murray, 1930 – (= ser Samp: indian books) – us CRL [954]
Dawn in the dark continent : or, africa and its missions / Stewart, James – Edinburgh: Oliphant Anderson & Ferrier 1903 [mf ed 1986] – (= ser Duff missionary lectures 1902) – 1mf – 9 – 0-8370-6618-2 – (incl bibl ref & ind) – mf#1986-0618 – us ATLA [242]
Dawn in toda land : a narrative of missionary effort on the nilgiri hills, south india / Ling, C F – London: Morgan & Scott, 1910 [mf ed 1995] – (= ser Yale coll) – 90p (ill) – 1 – 0-524-09123-4 – (foreword by amy wilson-carmichael) – mf#1995-0123 – us ATLA [240]
Dawn magazine – Baltimore MD. 1987 aug – 1r – 1 – mf#4025046 – us WHS [071]
Dawn of a new day in venezuela / Williams, William – Lurgan, Northern Ireland. 1948 – 1r – us UF Libraries [972]
The dawn of a new religious era : and other essays / Carus, Paul – rev enl ed. Chicago: Open Court Pub Co, 1916 [mf ed 1991] – 1mf – 9 – 0-524-00706-3 – (1st printed 1899) – mf#1990-2034 – us ATLA [210]
Dawn of african history / Oliver, Roland Anthony – London, England. 1968 – 1r – us UF Libraries [960]
The dawn of christianity / Martin, Alfred Wilhelm – New York: D Appleton, 1914 – 1mf – 9 – 0-524-03288-2 – (incl bibl ref) – mf#1990-0899 – us ATLA [240]
The dawn of christianity : or studies of the apostolic church / Vedder, Henry Clay – Philadelphia: American Baptist Pub Soc, c1894 [mf ed 1985] – 1mf – 9 – 0-8370-5626-8 – (incl app and bibl) – mf#1985-3626 – us ATLA [226]
The dawn of civilization; egypt and chaldaea / Maspero, Gaston – Ed. by A.H. Sayce, trans. by M.L. McClure. London: Society for Promoting Christian Knowledge; New York, Toronto: Macmillan Co., 1922. xiv,800p. illus.maps – 1 – us Wisconsin U Libr [900]
The dawn of history / Myres, John Linton – New York: H Holt, c1911 – (= ser Home university library of modern knowledge) – 1mf – 9 – 0-524-05623-4 – mf#1992-0478 – us ATLA [930]
The dawn of indian freedom / Winslow, Jack Copley & Elwin, Verrier – London: George Allen & Unwin, 1932 – (= ser Samp: indian books) – us CRL [280]
The dawn of magic / Pauwels, Louis – London: A. Gibbs & Phillips c1963 – 1 – us Wisconsin U Libr [150]
The dawn of manhood : twelve sermons / Clifford, John – London: Christian Commonwealth: Hodder & Stoughton, 1886 – 1mf – 9 – 0-7905-7708-9 – mf#1989-0933 – us ATLA [240]
The dawn of modern england : being a history of the reformation in england, 1509-1525 / Lumsden, Carlos Barren – London, New York: Longmans, Green, 1910 – 1mf – 9 – 0-7905-4891-7 – (incl bibl ref) – mf#1988-0891 – us ATLA [941]
The dawn of the catholic revival in england, 1781-1803 / Ward, Bernard – London; New York: Longmans, Green, 1909 – 2mf – 9 – 0-7905-7263-X – mf#1988-3263 – us ATLA [241]
The dawn of the modern mission / Stevenson, William Fleming – New York: A.C. Armstrong, 1888 – (= ser Duff Missionary Lectures) – 1mf – 9 – 0-8370-6950-5 – (incl ind of persons and places) – mf#1986-0950 – us ATLA [240]
Dawn of tomorrow : the national negro weekly devoted to the darker races – London, ON, jul 1923-easter 1972 – 3r – 1 – Can$285.00 – (official organ of the canadian league for the advancement of colored people, 1925-29) – cn McLaren [305]
Dawn on the hills of t'ang : or, missions in china / Beach, Harlan Page – New York: Student Volunteer Movement for Foreign Missions, 1898 – 1mf – 9 – 0-8370-6561-5 – (incl indes) – mf#1986-0561 – us ATLA [240]
The dawn over asia / Richard, Paul – Madras: Ganesh & Co, 1920 – (= ser Samp: indian books) – (trans fr french by aurobindo ghose) – us CRL [950]
Dawn over the bolivian hills / Dabbs, Norman H – 286p – 1 – us Southern Baptist [242]
The dawn, socialism – Ilkeston, England. -m. Jan 1902-Feb 1905. 13 ft – 1 – uk British Libr Newspaper [072]
Dawning lights : an inquiry concerning the secular results of the new reformation / Cobbe, Frances Power – London: Edward T Whitfield, 1868 – 1mf – 9 – 0-524-00252-5 – mf#1989-2952 – us ATLA [240]
The dawning of music in kentucky, or the pleasures of harmony in the solitude of nature / Heinrich, Anthony Philipp – Opera prima. Philadelphia: Bacon & Hart, 1820. A miscellaneous collection of piano, vocal and instrumental music. MUSIC 136, Item 2 – 1 – us L of C Photodup [780]
Dawnings of light in the east : with biblical, historical, and statistical notices of persons and places visited during a mission to the jews, in persia, coordistan, and mesopotamia / Stern, Henry Aaron – London: CH Purday, 1854 – 1mf – 9 – 0-7905-7259-1 – mf#1988-3259 – us ATLA [915]
Dawson, Aeneas McDonell *see*
- The catholics of scotland
- Lament for the right reverend james gillis... bishop of edinburgh, etc
- Lines for october
- Lines for the day of prince arthur patrick's arrival at ottawa
- The north-west territories and british columbia
- Pius 9 and his time
- [Poems]
- Queen victoria's jubilee
- St vincent of paul
- Sermon at the requiem of the hon thomas d'arcy mcgee
- Sermon delivered in the cathedral of ottawa at the funeral of the late h j friel, esq mayor of ottawa
- The temporal sovereignty of the pope with relation to the state of italy

Dawson, Aeneas McDonell [comp] *see* Our strength and their strength
Dawson, Alfred *see* The life of henry dawson
Dawson, C R *see* Comparison of soy bean silage and alfalfa hay for milk production
Dawson, Charles F *see*
- Equine glanders and its eradication
- Salt sick
- Texas cattle fever and salt-sick

Dawson, Christopher *see* Los origenes de europa. traduccion de francisco elias de tejada. madrid, 1945
Dawson, Coningsby *see* Living bayonets
Dawson county enterprise – Lexington, NE: Devinny & Sage, mar 22 1895-1897// (wkly) [mf ed with gaps filmed -1989] – 3r – 1 – (absorbed by: clipper-citizen) – us NE Hist [071]
Dawson county gazette – Plum Creek, NE: A A Signor. 2v. v5 n48. feb 8 1889-v6 n49. feb 14 1890 (wkly) [mf ed filmed 1979] – 1r – 1 – (cont: plum creek gazette. cont by: lexington gazette) – us NE Hist [071]
Dawson county herald – Lexington, NE: Clyde H Taylor. 56v. v37 n29. jan 6 [1938]-v92 n32. nov 30 1991 (wkly) [mf ed with gaps -1991] – 28r – 1 – (cont: overton herald. lexington clipper and dawson county pioneer, lexington clipper (1983). merged with: lexington clipper (1983) to form: clipper-herald. subsc also received: lexington clipper and dawson county pioneer feb 27 1964-dec 30 1982 and: lexington clipper jan 6 1983-nov 30 1991) – us NE Hist [071]
Dawson county pioneer – Lexington, NE: Fred Jas Pearson, nov 20 1873-n33. aug 13 1937 (wkly) [mf ed 1874,1877-1937 (gaps) filmed 1974-86] – 22r – 1 – (merged with: lexington clipper (1922) to form: lexington clipper and dawson county pioneer. daily ed: daily pioneer (may 1898)) – us NE Hist [071]
Dawson creek star – Dawson Creek, British Columbia, CN. nov 1949-nov 1962 – 9r – 1 – cn Commonwealth Imaging [071]
Dawson, E C *see*
- In and out of chanda
- The last journals of bishop hannington

Dawson, Edwin Collas *see* In and out of chanda
Dawson, Francis Warrington *see* Letters, 1881-1889
Dawson, G M *see* Rapport sur un voyage d'exploration fait dans la region du yukon, t n-o et dans...colombie-anglaise...en 1887
Dawson, George M *see*
- Glaciation of high points in the southern interior of british columbia
- Notes on the locust invasion of 1874
- Notes on the ore-deposit of the treadwell mine, alaska / On the microscopical character of the ore of the treadwell mine, alaska
- On foraminifera from the gulf and river st lawrence
- Preliminary note on the geology of the bow and belly river, n w territory
- Report on the geology and resources of the region in the vicinity of the forty-ninth parallel
- The superficial geology of british columbia and adjacent regions

Dawson, Grace D *see* Gerontology and geriatrics education
The dawson herald – Dawson, NE: Vern Gibbens. v1 n1. may 19 1921-v26 n45. nov 6 1947 (wkly) [mf ed with gaps filmed [1974?]] – 5r – 1 – (cont: verdon delphic. absorbed by: humboldt standard (1899). some irregularities in numbering) – us NE Hist [071]
Dawson, J M *see* Collection, five notebooks.
Dawson, John *see* Prepare to meet thy god!
Dawson, John E *see* The life and services of john e. dawson
Dawson, John William *see*
- Address [to the british association for the advancement of science]
- Appendix to memoranda of june, 1892
- Archaia
- Association of protestant teachers of the province of quebec
- Canadian pleistocene
- The canadian student
- The constitution of mcgill university
- Continental and island life, a review of wallace
- Day of rest in relation to the world that now is and that which is...
- The duties of educated young men in british america
- Eden lost and won
- Educated women
- Egypt and syria
- Facts and fancies in modern science
- Fossil men and their modern representatives
- Fossil sponges
- The future of mcgill university
- Handbook of zoology
- The historical deluge
- An ideal college for women
- James mcgill and the origin of his university
- Lectures, notes on geology, and outline of the geology of canada
- Loyalty
- Mcgill university
- Memoranda and statements
- Modern ideas of evolution
- Modern ideas of evolution as related to revelation and science
- Nature and the bible
- Note on carboniferous entomostraca from nova scotia, in the peter redpath museum
- On portions of the skeleton of a whale from gravel on the line of the canada pacific railway, near smith's falls, ontario
- On the genus lepidophloios
- On the silurian and devonian rocks of nova scotia
- The origin of the world
- The origin of the world according to revelation and science
- Peter redpath
- A plea for the extension of university education in canada
- The recent history of mcgill university
- The relation of mcgill university to legal education
- Report of j w dawson...principal of mcgill university, montreal
- Report on the higher education of women
- Review of the evidence for the animal nature of eozoon canadense, pt 1
- Review of the evidence for the animal nature of eozoon canadense, pt 2
- Revision of the bivalve mollusks of the coal-formation of nova scotia
- The story of the earth and man

– The testimony of the holy scriptures respecting wine and strong drink
– Thirty-eight years of mcgill
Dawson, Miles M *see* Elements of life insurance
Dawson news – Dawson, YK. 1899-1954 – 76r – 1 – cn Library Assoc [071]
Dawson news boy – Dawson, NE: E W Buser, jun 28 1888-1909// (wkly) [mf ed with gaps] – 4r – 1 – (issues for sep 16 1892-apr 30 1909 also called whole n222-1086) – us NE Hist [071]
Dawson, Oswald *see* An indictment of darwin
Dawson outlook – Dawson, NE: Ike W Watson, jul 1909-10// (wkly) [mf ed v1 n28. jan 28 1910 filmed [1993]] – 1r – 1 – (absorbed by: falls city tribune) – us NE Hist [071]
Dawson, P K *see* Effects of training on resting blood pressure in men at risk for coronary heart disease
The dawson reporter – Dawson, NE: J R Harrah, 1913-may 1920// (wkly) [mf ed 1914-20 (gaps)] – 1r – 1 – (suspended with mar 21 1919; resumed with may 2 1919) – us NE Hist [071]
Dawson, Robert *see* Present state of australia
Dawson, Samuel Edward *see*
– Champlain
– The discovery of america by john cabot in 1497
– Episcopal elections
– Hand-book for the city of montreal and its environs
– Handbook for the dominion of canada
– A study
– The voyages of the cabots in 1497 and 1498
Dawson, Simon James *see*
– Rapport sur l'exploration de la contree situee entre le lac superieur
– Rapport sur l'exploration de la contree situee entre le lac superieur et la colonie de la riviere rouge
Dawson springs first baptist church. dawson springs, kentucky : church records – November 1909-May 1989. 1,666p – 1 – $74.97 – us Southern Baptist [242]
Dawson, Walter Robert *see* Equipping the laity as worship leaders in the ministry of big bethel baptist church
Dawson, William Bell *see*
– The currents in belle isle strait
– The currents in the gulf of st lawrence
– Masonry arches for railway purposes
– Mean sea level at quebec and new york
– A new method for the design of retaining walls
– The paroy reservoir
– Tables of hourly direction and velocity of the currents and time of slack water in the bay of fundy and its approaches as far as cape sable
– Tide levels and datum planes in eastern canada
– Tide levels and datum planes on the pacific coast of canada
– The tides and tidal streams with illustrative examples from canadian waters
– Tides at the head of the bay of fundy
Dawson, William Francis *see* Christmas
Dawson, William James *see* The life of christ
Dawud, Yusuf *see* Grammaire de la langue arameenne
Daxer, Georg *see* Der subjektivismus in franks "system der christlichen gewissheit"
Day – London, UK. 1809. -d. 2 1/2 reels – 1 – uk British Libr Newspaper [072]
Day – London, UK. 19 Mar-4 May 1867. 1/2 reel – 1 – uk British Libr Newspaper [072]
Day : a morning journal of literature, politics, arts and fashion – [Scotland] Glasgow: J Wylie 2 jan-30 jun 1832 (daily, wkly) [mf ed 2003] – 112v on 1r – 1 – uk Newsplan [072]
The day – New York, NY. 1914-57 – (= ser The Day Jewish Journal) – 1 – us AJPC [071]
The day – New York. N.Y. Der tog. 1952-57 – 1 – us AJPC [071]
The day after death : or, our future life, according to science / Figuier, Louis – London, 1872 – (= ser 19th c evolution & creation) – 4mf – 9 – (trans fr french of louis figuier. ill by 10 astronomical plates) – mf#1.1.6897 – uk Chadwyck [500]
Day, Alfred *see* Treatise on harmony
Day book / Indian Home Guards. First Regiment – 1862-63 – 1 – us Kansas [978]
Day book / Miami Trading Post – 1847-49 – 1 – us Kansas [978]
The day book – (The Evening Day Book). New York. Jan 19-Dec 1852; Jan-Nov 1856; Jan-June, Oct-Dec 1857; Apr 1858-June 1860; and Jan-Aug 1861. Not collated – 1 – us NY Public [071]
Day book of the register's office of the treasury, 1789-1791 / U.S. Register of the Treasury – (= ser Records of the accounting officers of the department of the treasury) – 1r – 1 – mf#T964 – us Nat Archives [336]
Day by day with jesus : a book for holy week / Barton, William Eleazar – Oak Park, IL: Puritan Press, 1913 – 1mf – 9 – 0-524-04784-7 – mf#1992-0204 – us ATLA [240]
Day care and early education – Dordrecht, Netherlands 1973-96 – 1,5,9 – (cont by: early childhood education journal) – ISSN: 0092-4199 – mf#11177.01 – us UMI ProQuest [640]
Day, Charles William *see* Five years' residence in the west indies
Day, Clarence B *see* Chinese peasant cults
Day, Clinton D *see* Baptizing
Day dawn in africa : or, progress of the protestant episcopal mission at cape palmas, west africa / Scott, Anna Miller – New York: Protestant Episcopal SPEK, 1858 [mf ed 1986] – 1mf – 9 – 0-8370-6369-8 – mf#1986-0369 – us ATLA [242]
Day, Edward *see* Social life of the hebrews
Day, Edwin *see*
– Doctrine of election
– A letter to chief justice draper, the very rev dean grasett
Day, Francis A *see* Diary, ms p.p.
Day, Henry Cyril *see* Catholic democracy
Day in and day out in korea : being some account of the mission work that has been carried on in korea since 1892 by the presbyterian church in the united states / Nisbet, Anabel – Richmond: Presbyterian Committee of Publ [1919] [mf ed 1995] – (= ser Yale coll) – 199p (ill) – 1 – 0-524-09666-X – mf#1995-0666 – us ATLA [242]
A day in capernaum = Ein tag in kapernaum / Delitzsch, Franz – NY: Funk & Wagnalls, 1892, c1887 [mf ed 1985] – 1mf – 9 – 0-8370-2867-1 – (english trans fr 3rd german ed by george h schodde) – mf#1985-0867 – us ATLA [240]
Day in court : or the subtle arts of great advocates / Wellamn, Francis L – New York: The Macmillan Co, 1910 – 3mf – 9 – $4.50 – mf#LLMC 92-181 – us LLMC [347]
Day in court; or, the subtle arts of great advocates / Wellman, Francis Lewis – New York, Macmillan, 1914. 257 p. LL-1568 – 1 – us L of C Photodup [340]
A day in the temple / Maas, Anthony John – St Louis, MO: Herder, 1892 [mf ed 1986] – 1mf – 9 – 0-8370-6997-1 – mf#1986-0997 – us ATLA [270]
Day, John *see*
– International nationalism
– The valley people
Day, John Percival *see* Report on the economic conditions of the canadian plumbing and heating industry
Day journal / Heritage Researchers Enterprises – v1 n1-v3 n4 [1983 jun-1986 mar, jun] – 1r – 1 – mf#1289662 – us WHS [071]
Day, Kathleen L *see* Reliability of a fitness plan scoring rubric
Day, Lal Behari *see*
– Folk-tales of bengal
– Recollections of alexander duff...
Day, Lewis Foreman *see*
– The anatomy of pattern
– The application of ornament
– Art in needlework
– Instances of accessory art
– The planning of ornament
– Some principles of every-day art
– Windows a book about stained and painted glass
Day, Mary A *see* Letters
Day, Maurice Fitzgerald *see* The person and offices of the holy spirit
Day of adversity / Woodd, Basil – London, England. no date – 1r – us UF Libraries [240]
Day of judgement / Pusey, E B – Oxford, England. 1839 – 1r – us UF Libraries [240]
Day of national humiliation and prayer on the occasion of war with... – London, England. 18– – 1r – us UF Libraries [240]
The day of our visitation / Littleboy, William – [London]: Pub for the Woodbrooke Extension Committee by Headley Bros, 1917 – (= ser Swarthmore Lecture) – 1mf – 9 – 0-524-06818-6 – mf#1991-2805 – us ATLA [240]
Day of rest in relation to the world that now is and that which is... / Dawson, John William – London, England. 18– – 1r – us UF Libraries [240]
Day of small things / Ridley, William Henry – London, England. 1846 – 1r – us UF Libraries [240]
Day of supplication, war in south africa : a sermon preached in st peter's, brockville... february 11th, 1900 / Bedford-Jones, T – [Brockville, Ont?: s.n, 1900?] – 1mf – 9 – 0-665-94266-4 – mf#94266 – cn Canadiana [240]
Day of tears / While, Henry Gostling – London, England. 1817 – 1r – us UF Libraries [240]
Day release: report of the ministry of education committee, 1964 – 1mf – 9 – mf#87033 – uk Microform Academic [324]
Day researcher – v1 n1-3 [1983 apr-oct], v2 n1-v7 n4 [1984 jan-1989 oct] – 1r – 1 – mf#1055182 – us WHS [071]
Day, Samuel Phillips *see*
– English america, vol 1
– English america, vol 2
– English america, vols 1-2
Day star / Cincinnati, OH – Cincinnati, OH. 1843-47 – 1 – (publ as: western midnight cry 1843-44) – us Western Res [071]
Day star – Cincinnati, OH. Feb 1845-Jul 1847 – 1 – us Western Res [071]
A day to remember – 1958 oct-1961 apr/jun – 1 – mf#615752 – us WHS [071]
Day to remember – 1958 oct-1983 sep – 1r – 1 – (cont: aday to remember) – mf#615719 – us WHS [071]
Day two – Napier, NZ. 31 dec 1985-dec 1987 – 2r – 1 – mf#31.4 – nz Nat Libr [079]
Day, Warren *see* Letters
The daya-crana-sangraha, an original treatise on the hindoo law of inheritance / Tarkalankara, Krsna – Calcutta, Pereira, 1818. 138 p. LL-29 – 1 – us L of C Photodup [340]
Dayal, Leela Row *see*
– Manipuri dances
– Nritta manjari
Dayan, Shmuel *see*
– 'Im Hatsi Yovel
– Nahalal
Dayananda Sarasvati *see*
– The five great duties of the aryans
– Maharshi swami dayanand saraswati's exposition of vedic religion
Dayananda Sarasvati, Swami *see* Introduction to the commentary on the vedas
The dayanandi interpretation of the word "deva" in the rig veda / Griswold, Hervey De Witt – Lodiana: Lodiana Mission Press, 1897 – 1mf – 9 – 0-524-01490-6 – mf#1990-2466 – us ATLA [280]
Daybook / Western Bakery. Lawrence, Kansas – 1861 – 1 – us Kansas [380]
Daybook, ms 2067 / Parker, James M – 1840-41 – 1r – 1 – (entries made by parker while employed at president andrew jackson's plantation, the hermitage. lists include hands employed, and material possessions of the plantation) – us Western Res [920]
Daybook of the department of state for miscellaneous and contingent expenses, feb 1 1798-nov 3 1820 / U.S. Dept of State – (= ser General records of the department of state) – 1r – 1 – mf#T903 – us Nat Archives [324]
Daybreak : a poem / Beck, George Fairley – Toronto: s.n, 1873 – 1mf – 9 – mf#03536 – cn Canadiana [810]
Daybreak *see* Church missionary society archive, section 2
Daybreak dispatch / Lucas Co. Toledo – v1 n1. aug 1977-feb 1978 [daily] – 2r – 1 – mf#B34110-34111 – us Ohio Hist [071]
Daybreak in korea : a tale of transformation in the far east / Baird, Annie Laurie Adams – New York: Fleming H Revell, c1909 [mf ed 1986] – 1mf – 9 – 0-8370-6642-5 – mf#1986-0642 – us ATLA [950]
Daybreak in livingstonia : the story of the livingstonia mission, british central africa / Jack, James William – Edinburg: Oliphant, Anderson & Ferrier, 1901 – 1mf – 9 – 0-7905-6297-9 – mf#1988-2297 – us ATLA [240]
Daybreak in the dark continent / Naylor, Wilson Samuel – New York: Laymen's Missionary Movt c1908 [mf ed 1986] – (= ser Forward mission study courses) – 1mf – 9 – 0-8370-6286-1 – (incl bibl ref & ind) – mf#1986-0286 – us ATLA [960]
Daybreak star / United Indians of All Tribes Foundation [US] – v7 n1-8 [1981 oct-1982 may] – 1r – 1 – mf#648867 – us WHS [307]
Daybreak Star Press *see* Quarter moon
Day-breaking of the gospel with the indians / Wilson, John – Boston: Directors of the Old South Work, [1903?] – 1mf – 9 – 0-524-04154-7 – mf#1990-1224 – us ATLA [240]
Day-dawn in dark places / Mackenzie, John – New York, NY. 1969 – 1r – us UF Libraries [960]
Daydreams by a butterfly : in nine parts / Allen, Joseph Antisell – Kingston, Ont?: J M Creighton, 1854 – 2mf – 9 – mf#43112 – cn Canadiana [810]
Daye, Pierre *see* Congo et angola
Dayes, Edward *see*
– A picturesque tour through the principal parts of yorkshire and derbyshire
– The works of the late edward dayes
Day-federation review / Montgomery Co. Dayton – (feb 1945-dec 1969) scattered [irreg] – 1r – 1 – mf#B10191 – us Ohio Hist [331]
Dayfis, Uri et al *see* Al-siyasah al-maiyah li-israil
Daykin herald – Daykin, NE: Albert W Koepff. v1 n1. oct 14 1927-v22 n40. dec 31 1954 (wkly) [mf ed 1928-54 (gaps)] – 5r – 1 – (absorbed: tobias times. some irregularities in numbering) – us NE Hist [071]
The daykin herald – Daykin, NE: [M G King] 1900 (wkly) [mf ed -1902 (gaps) filmed 1979] – 1r – 1 – us NE Hist [071]
Dayman, A J *see* Houses of god
Dayringer, Richard *see* American journal of pastoral counseling
Day's doings – Omro WI. 1886 jun 19-1888 aug 8 – 1r – 1 – (cont by: omro weekly journal) – mf#959727 – us WHS [071]
"A days march nearer home", / Miller, J Graham – v5, 6, 12 – 2r – 1 – (docts on the presbyterian teachers training institute, tangoa, vanuatu, 1947-52; rev miller's lecture notes fr the tti; and documents fr the presbyterian bible college, tangoa, 1971-73. available for reference) – mf#PMB1140 – at Pacific Mss [242]
The days of advance : british civilization as shown in some institutions, free libraries and water supply systems, interesting facts gleaned by alderman hallam while across the sea / Hallam, John – [Toronto?: s.n, 188-?] – 1mf – 9 – 0-665-89789-8 – mf#89789 – cn Canadiana [350]
Days of blessing in inland china : being an account of meetings held in the province of shan-si / Taylor, James Hudson – 2nd ed. London: Morgan & Scott, 1887 – 1mf – 9 – 0-8370-6533-X – mf#1986-0533 – us ATLA [240]
The days of bruce : a story from scottish history / Aguilar, Grace – London: Groombridge & Sons, 1852 – (= ser 19th c women writers) – 7mf – 9 – mf#5.1.25 – uk Chadwyck [830]
Days of crisis in rhodesia / Horrell, Muriel – Johannesburg, South Africa. 1965 – 1r – us UF Libraries [960]
Days of god's right hand : our mission tour in australasia and ceylon / Cook, T – London, 1896 – 4mf – 9 – mf#HTM-43 – ne IDC [915]
Days of grace in india : a record of visits to indian missions / Newman, Henry Stanley – London: SW Partridge, [1882?] [mf ed 1993] – (= ser Society of friends (quakers) coll) – 1mf – 9 – 0-524-07029-6 – mf#1991-2882 – us ATLA [240]
Days of heaven upon earth : a year book of scripture texts and living truths / Simpson, Albert B – New York: Christian Alliance, c1897 [mf ed 1992] – (= ser Christian & missionary alliance coll) – 1mf – 9 – 0-524-02144-9 – mf#1990-4210 – us ATLA [240]
Day's reports / Connecticut. Supreme Court – v1-5. 1830-33 (all publ) – (= ser Connecticut Supreme Court Reports) – 30mf – 9 – $45.00 – (a pre-nrs title) – mf#LLMC 90-004 – us LLMC [347]
Dayspring / Episcopal Church – 1997 sep/oct – 1r – 1 – (cont: episcopalian. mountain dayspring) – mf#4888687 – us WHS [242]
Dayspring – Boston MA: American Board of Commissioners for Foreign Missions, 1842-49 [mf ed 2001] – (= ser Christianity's encounter with world religions, 1850-1950) – 1r – 1 – (small periodical paper auxiliary to "missionary herald") – mf#2001-s143 – us ATLA [242]
Dayspring : [unitarian monthly for children] – Boston MA: Unitarian Sunday-School Society, 1872-83 [mf ed 2001] – (= ser Christianity's encounter with world religions, 1850-1950) – 1r – 1 – mf#2001-s145 – us ATLA [243]
Dayspring digest – 1992 sep – 1r – 1 – mf#4878436 – us WHS [071]
Dayspring / the mission dayspring – Boston MA: American Board and Women's Board of Missions, 1882-1913 [mthly] [mf ed 2001] – (= ser Christianity's encounter with world religions, 1850-1950) – 3r – 1 – mf#2001-s019-020 – us ATLA [242]
Dayton – Dayton OH 1989 – 1,5,9 – (cont: dayton usa) – ISSN: 0199-9214 – mf#7584,01 – us UMI ProQuest [380]
Dayton, A C *see*
– Gospel holiness
– How children may be brought to christ
– Perseverance of all saints
Dayton afl-cio news – Dayton OH. v16 n15-v16 n17 [1959 oct 23/nov 15-dec 16] – 1r – 1 – mf#686523 – us WHS [331]
Dayton cio news / Montgomery County Industrial Union Council – Dayton OH. 1944 mar 17-1946, 1947-52, 1953-1959 sep 15 – 3r – 1 – (cont: dayton union news; cont by: dayton afl-cio labor news) – mf#686517 – us WHS [331]
Dayton, Danielle M *see* Case study analysis of teacher change with the sport education model
Dayton defender – Dayton OH. v5 n6-64 [1991 mar 28/apr 11-may 9/23] – 1r – 1 – (cont: new dayton defender) – mf#2626932 – us WHS [071]
Dayton directory and gridiron revivdus / Montgomery Co. Dayton – jun-oct 1839 [wkly] – 1r – 1 – mf#B5001 – us Ohio Hist [071]
Dayton herald – Dayton, OR: Herald Printing and Pub Co. v6 n51-v22 n34. aug 4 1893-oct 26 1906 – 1 – us Oregon Hist [071]

Dayton herald – Dayton OR: Herald Printing & Pub Co [wkly] – 1 – us Oregon Lib [071]

Dayton j.c.c. news – Dayton, Ohio – 1 – (vol. 27, no. 11 (1 july 1970); v. 28, no. 1 (1 sept. 1970); v. 28, no 10 (1 june 1971); v. 28, no. 12 (2 aug. 1971); comes on a reel with other titles) – us AJPC [978]

Dayton jewish chronicle – Dayton. Ohio. 1964-68 – 1 – us AJPC [071]

The dayton jewish life – Dayton, OH. 1918 – 1 – us AJPC [071]

Dayton journal and advertiser / Montgomery Co. Dayton – sep 1827-8/1829, (sep 29-oct 1830) [wkly] – 1r – 1 – mf#B3227 – us Ohio Hist [071]

[Dayton-] lyon county times – NV. jan 1889; may 1891; oct 1897-apr 1902 [wkly] – 2r – 1 – $120.00 – mf#U04484 – us Library Micro [071]

[Dayton-] news reporter – NV. mar-dec 1886 [wkly] – 1r – 1 – $60.00 – mf#U04485 – us Library Micro [071]

Dayton, OH *see* Selections (1796-1932)

Dayton. Presbytery (Pres. Church in the USA) *see* Records, 1838-1917

Dayton tribune – Dayton OR: F T Mellinger [wkly ex 1st 2 wks in jul] – 1 – (1963-69 incl newspaper pub during school terms by dayton high school students. publ suspended aug 24 1923; resumed aug 28 1924) – us Oregon Lib [071]

Dayton union news / Congress of Industrial Organizations [CIO] [US] – Dayton OH. v3 n1-v4 n5 [1943 jan 6-1944 mar 3] – 1r – 1 – (cont by: dayton cio news) – mf#686511 – us WHS [331]

Dayton usa – Dayton OH 1972-79 – 1,5,9 – (cont by: dayton) – ISSN: 0011-7137 – mf#7584.01 – us UMI ProQuest [380]

Dayton workers' voice / Revolutionary Union – v1-4 n1[1972 may/jun-1975 jun/jul] – 1r – 1 – mf#1111059 – us WHS [331]

Daytona : points of interest – s.l, s.l? 193-? – 1r – us UF Libraries [978]

Daytona beach : jack jones rewrite – s.l, s.l? 193-? – 1r – us UF Libraries [978]

Daytona beach and environs / Jones, Jack – s.l, s.l? 193-? – 1r – us UF Libraries [978]

Daytona beach art school / Goebel, Rubye K – s.l, s.l? 1936 – 1r – us UF Libraries [978]

Daytona beach evening news – Daytona, FL. 1925 nov-1950 mar – 70r – us UF Libraries [071]

Daytona beach, florida – s.l, s.l? 193-? – 1r – us UF Libraries [978]

Daytona beach news – Daytona Beach, FL. 1925 jul-1937 feb – 7r – us UF Libraries [071]

Daytona beach observer – Daytona Beach, FL. 1934-1973 – 19r – (gaps) – us UF Libraries [071]

Daytona daily news – Daytona, FL. 1905-1924 – 25r – (gaps) – us UF Libraries [071]

Daytona local guide – s.l, s.l? 1937 – 1r – us UF Libraries [978]

Daytona times – Daytona Beach FL. [1992 sep 24/30-dec 31/jan 6]-[1999 jan 7/13-jun 24/30] – 14r – 1 – mf#2589084 – us WHS [071]

Daytshland / Heine, Heinrich – Moskve, Russia. 1936 – 1r – us UF Libraries [939]

Daz buoch von dem uebeln wibe / ed by Ebbinghaus, Ernst A – 2nd rev ed. Tuebingen: Niemeyer, 1968 [mf ed 1993] – (= ser Altdeutsche textbibliothek n46) – x/33p – 1 – (incl bibl ref) – mf#8193 reel 4 – us Wisconsin U Libr [890]

Daz buoch von guoter spise *see* Ein buch von guter speise

Daz deutsche allgemeine zeitung im bilde – Berlin DE, 1924 6 jul-1927 26 jun [gaps] – 1r – 1 – gw Mikrofilm [074]

Dazey commercial (1911) : [official paper of barnes county," 1917-1918] – Dazey, Barnes Co, ND: Leo Ratcliff, 1911; -v8 n20 aug 9 1918 (wkly) [mf ed with gaps] – 1 – (combined with: rogers citizen (1913) for christmas iss, dec 12 1913; regular vol numbering cont for each; merged with: rogers citizen (1913) to form: commercial citizen) – mf#11029-11030 – us North Dakota [071]

Dazey commercial (1922) – Dazey, Barnes Co, ND: Victor Phillips. v1 n1 feb 2 1922-oct 18 1923?// (wkly) – 1 – (missing: 1922 feb 23, sep 28, oct 19; 1923 jan 18, sep 26) – mf#11031 – us North Dakota [071]

The dazey herald – Dazey, Barnes Co, ND: A W Klein, 1906?-1909?// (wkly) [mf ed with gaps] – 1 – mf#07580 – us North Dakota [071]

Dazumal : vier novellen / Heimburg, W – Stuttgart: Union Deutsche Verlagsgesellschaft [1912] [mf ed 1995] – 1r – 1 – (filmed with: fragmente / gottfried benn & other titles) – mf#3784p – us Wisconsin U Libr [830]

Db – Commack NY 1972-93 – 1,5,9 – ISSN: 0011-7145 – mf#6581 – us UMI ProQuest [620]

Dbms – Manhasset NY 1988 – 1,5,9 – ISSN: 1041-5173 – mf#16925 – us UMI ProQuest [000]

Dc bar journal *see* Journal of the bar association of the district of columbia

Dci – Carol Stream IL 1998 – 1,5,9 – (cont: drug & cosmetic industry; cont by: global cosmetic industry) – ISSN: 1096-4819 – mf#2551.02 – us UMI ProQuest [640]

Dcl journal of international law – Michigan State University. v1-9. 1992-2000 – 9 – $140.00 set – (title varies: v1-8 n2 1992-99 as journal of international law and practice) – mf#114621 – us Hein [341]

Dclp news / Libertarian Party of Texas – v4 n8-v11 n6 [1980 dec-1983 apr] – 1r – 1 – (cont: newsletter; cont by: calendar...dallas chapter) – mf#1013712 – us WHS [325]

Ddp arizona newsletter / Doctors for Disaster Preparedness – v2 n2-v6 n6 [1986 jan-1990 sep] – 1r – 1 – (cont by: civil defense perspectives) – mf#1818211 – us WHS [610]

Ddr-landtagsprotokolle – Thueringen, 21 Nov 1946-1948 – 4r – 1 – (sachsen jun 1946-49. mecklenburg nov 1946-49. brandenburg nov 1946-49. sachsen-anhalt nov 1946-48) – gw Mikropress [074]

Ddt treatment for control of mole-crickets in seedbeds / Kelsheimer, E G – Gainesville, FL. 1947 – 1r – us UF Libraries [630]

De 6 a 6 / Dominguez Navarro, Ofelia – Mexico City? Mexico. 1937 – 1r – us UF Libraries [972]

De 27ste october 1553 / Hofstede de Groot, C P – Rotterdam: DJP Storm Lotz, 1876 – 1mf – 9 – 0-524-03231-9 – mf#1990-0859 – us ATLA [240]

De 1867 a 1871 – S.l: s.n, 1872? – 1mf – 9 – mf#03780 – cn Canadiana [971]

De aaronitici sacerdotii atque thorae elohisticae origine : dissertatio historico-critica / Curtiss, Samuel Ives – Lipziae [Leipzig]: J C Hinrichs, 1878 – 1mf – 9 – 0-8370-2795-0 – mf#1985-0795 – us ATLA [270]

De abassionorum rebus... / [Godinho, N] – Lugduni, 1615 – 5mf – 9 – mf#SEP-36 – ne IDC [956]

De abusu philosophiae cartesianae / Desmarets, S – Groningue, Everts, 1670 – 2mf – 9 – mf#PFA-144 – ne IDC [240]

De accentibvs, et orthographia, lingvae hebraicae / Reuchlin, Johann – [Hagenoae: in aedibus Thomae Anshelmi Badensis anno 1518 Mense Februario] – 4mf – 9 – mf#fiche 236, 653 – us Sibley [470]

De adentro / Calderon Ramirez, Salvador – San Salvador, El Salvador. 1956 – 1r – us UF Libraries [972]

De adiaphoristicis corrvptelis, in magno libro actorum interimisticorum, sub conficto titulo professorum vuitebergensium aedito, repetitis, admonitiones / Wigand, J – np, 1559 – 1mf – 9 – mf#TH-1 mf 1502 – ne IDC [242]

De adoranda vnitione dvarvm natvrarvm christi inseparabili in vnam personam confessio pia et orthodoxa / Musculus, A – np, 1591 – 1mf – 9 – mf#TH-1 mf 1215 – ne IDC [242]

De aedificiis dn iustiniani libri 6 (cbh3,2) / Procopii Caesariensis; ed by Maltret, Cl – Parisiis, 1663 – (= ser Corpus byzantinae historiae (cbh)) – €14.00 – ne Slangenburg [243]

De aeterna dei praedestinatione, qua in salutem alios ex hominibus elegit, alios suo exitio reliquit : item de providentia qua res humanas gubernat, consensus pastorum genevensis ecclesiae, a jo calvino expositus / Calvin, J – Genevae: Ex officina Joannis Crispini, 1552 – 3mf – 9 – mf#CL-8 – ne IDC [242]

De aeterna praedestinatione filiorvm dei ad salvtem propositiones theologicae / Hunnius, A – Francoforti ad Moenvm, 1594 – 1mf – 9 – mf#TH-1 mf 749 – ne IDC [242]

De aeternitate considerationes coram ser.mis utriusque bavariae principibus maximiliano et elizabetha explicatae : eisdem inscriptae et dedicatae / Drexelius, H – Monachii: Apud Raphaelem Sadelerum, 1621 – 6mf – 9 – mf#O-1552 – ne IDC [090]

De aeternitate poenarum deque igne inferno : commentarii / Passaglia, Carlo – Ratisbonae: Manz, 1854 – 1mf – 9 – 0-524-00551-X – mf#1993-2076 – us ATLA [240]

De aeterno dei filio...adversus... antitrinitarios / Simler, J – Tigvri, Christoph Froschover, 1568 – 8mf – 9 – mf#PBU-329 – ne IDC [240]

De afflictione tam captivorvm quam etiam sub turcae tributo viuentium christianorum... / Bartholomaeus, G – Antverpiae, 1544 – 1mf – 9 – mf#H-8270 – ne IDC [956]

De albuminas / Obregon Y Garcia, J G – Ciudad Trujillo, Dominican Republic. 1949 – 1r – us UF Libraries [972]

De alcaei et sapphonis copia vocabulorum / Gerstenhauer, Arthurius – Halis Saxonum: Max Niemeyer, 1894 – 1mf – 9 – 0-8370-9286-8 – (incl bibl ref and index) – mf#1986-3286 – us ATLA [450]

De algunas glorias de la raza y gente de santander / Reyes Rojas, Luis – Bucaramanga, Colombia. 1939 – 1r – us UF Libraries [972]

De Alpartil, Martin *see* Martin de alpartils chronica actitatorum temporibus domini benedicti 13. band 1, einleitung, text der chronik, anhang ungedruckter aktenstuecke

De alquilan cuartos amueblados / Aguilar Derpich, Juan – Habana, Cuba. 1962 – 1r – us UF Libraries [972]

De Alwis, James *see* Buddhist nirvana

De amicitia... / Cicero, Marcus Tullius – 15th c – (= ser Holkham library manuscript books 386,383,375) – 1r – 1 – (filmed with: s hieronymus: vita b monachi, vita b pauli heremitae. cicero: de officiis, paradoxa, de amicitia, de senectute. cicero: de inventione rhetorica) – mf#96543 – uk Microform Academic [450]

De amissa dicendi rationi... / Sturm, J – Argentorati, 1538 – 2mf – 9 – mf#PPE-137 – ne IDC [240]

De anabaptismi exordio, erroribvs, historijs abominandis, confutationibus adiectis libri duo... / Gast, J – Basileae, [1544] – 6mf – 9 – mf#ZWI-33 – ne IDC [242]

De anabaptismo grassante adhvcin mvltis germaniae, poloniae, prvssiae, belgicae et italiis / Wigand, J – Lipsiae, 1582 – 7mf – 9 – mf#TH-1 mf 1508-1514 – ne IDC [242]

De angelis angelicoqve hominvm praesidio atqve cvstodia meditatio... / Stucki, J W – Tigvri, Ioannes Vvolph, 1595 – 4mf – 9 – mf#PBU-640 – ne IDC [240]

De anthropologie van zwingli / Oorthuys, G – Leiden, 1905 – 3mf – 9 – mf#ZWI-68 – ne IDC [242]

De antinomia veteri et nova, collatio et commonefactio / Wigand, J – Ienae, 1571 – 3mf – 9 – mf#TH-1 mf 1503-1505 – ne IDC [242]

De antiquis ecclesiae ritibus / Martene, Edmond – Antverpiae. v1-4. 1763-64 – 4v on 71mf – 8 – €136.00 – ne Slangenburg [240]

De antiquis monachorum ritibus / Martene, Edmond – Antverpiae, 1764 – 20mf – 8 – €38.00 – ne Slangenburg [240]

De antiquitate benedictinorum in regno angliae / Reynerus, Cl – Duaci, 1626 – €50.00 – ne Slangenburg [241]

De Anza College *see* California history center foundation newsletter

De apellis gnosi monarchica : commentatio historica / Harnack, Adolf von – Lipsiae: E Bidder, 1874 – 1mf – 9 – 0-7905-6475-0 – (incl bibl ref) – mf#1988-2475 – us ATLA [240]

De aqui para alla / Aguero, Luis – Habana, Cuba. 1962 – 1r – us UF Libraries [972]

De aqui y de alla / Garcia Godoy, Federico – Santo Domingo, Dominican Republic. 1916 – 1r – us UF Libraries [972]

De aramaismis libri ezechielis / Selle, Fridericus – Halis Saxonum: Formis Kaemmererianis, 1890 – 1mf – 9 – 0-8370-6373-6 – (incl bibl ref) – mf#1986-0373 – us ATLA [221]

De arbore consanguinitatis liber / Nicasia de Voerda – Coloniae apud Quentell, 1504 – €5.00 – ne Slangenburg [240]

De arcanis catholici veritatis libri 12 / Galatinus, P – Barii, 1603 – 14mf – 8 – €27.00 – ne Slangenburg [241]

De architectura / Vitruvio, M – SL, SA – 5mf – 9 – sp Cultura [720]

De architectura, alcala de henares, 1582 / Vitruvio, M – Alcala de Henares, 1582 – 7mf – 9 – sp Cultura [720]

De architectvra libri decem... / Vitruvius Pollio, M – Argentorati, 1543 – 4mf – 9 – mf#OA-23 – ne IDC [720]

De ark gods : het oud-israelitische heiligdom / Sevensma, Tietse Pieter – Amsterdam: J Clausen, 1908 – 1mf – 9 – 0-8370-5233-5 – (in dutch) – mf#1985-3233 – us ATLA [221]

De arte cabalistica libri tres / Reuchlinus, Ioan – Basilae, 1603 – €12.00 – ne Slangenburg [270]

De arte canendi : ac vero signorvm in cantibvs, vsv, libri duo / Heyden, Sebald – Norimbergae: apud Ioh Petreium 1540 [mf ed 19–] – 1r / 3mf – 1,9 – (1st ed publ [nuremberg 1537] under title: musicae, id est, artis canendi libri duo) – mf#film 732 / fiche 557 – us Sibley [780]

De arte curativa libri quator quibus sanandi morbos brenis traditur satis... / Lopez de Corella, A – Estella, 1555 – 8mf – 9 – sp Cultura [610]

De arte rhythmica hebraeorum *see* Commentarius in proverbia

De articulari morbo commentarius / Laguna, A de – Roma, 1551 – 1mf – 9 – sp Cultura [610]

De attaque et de la defense des places... / Vauban, S – La Haye, 1737. 2v – 7mf – 9 – mf#OA-199 – ne IDC [720]

De auctoritate presidendi in concilio generali / Cusanus, N; ed by Kallen, G – 1935 – 3mf – 8 – €7.00 – ne Slangenburg [200]

De auctorum graecorum versionibus et commentariis syriacis, arabicis, armeniacis, persicisque... / Wenrich, J G – Lipsiae, 1842 – 4mf – 9 – mf#AR-1580 – ne IDC [956]

De auxiliis divinae gratiae / Alvares, Didac – Col. Agrippinae, 1621 – 9mf – 8 – €18.00 – ne Slangenburg [240]

De band tussen ambon en nederland / Graaf, H J de – 's-Gravenhage, 1969 – 1mf – 8 – mf#SE-1600 – ne IDC [959]

De Bary, Richard *see* The spiritual return of christ within the church

De bazuin – Kampen, 1853-1964 – 567mf – 9 – (missing: 1962(52)) – mf#H-2030 – ne IDC [240]

De Beers Consolidated Mines *see* Annual report

De bekeeringsgeschiedenis van een japanner / Jonker, Gerrit Jan Abraham – [Rotterdam: J M Bredee, 1905] [mf ed 1995] – (= ser Yale coll; Lichtstralen op den akker der wereld [11. jaarg 1905] 5) – 1 – 0-524-09759-3 – (in dutch) – mf#1995-0759 – us ATLA [950]

De belgen in canada / Verbist, Pascal Joseph – Turnhout Belgium: A van Genechten, 1872 – 1mf – 9 – (also available in french) – mf#39462 – cn Canadiana [304]

...De bello a christianis contra barbaros gesto pro christi sepvlchro et ivdaea recvperandis / Accolti, B – Venetiis, 1532 – 2mf – 9 – mf#H-8240 – ne IDC [956]

De bello melitensi : et eius euentu francis imposito, ad carolus caesarem v nicolai villagagnonis commentarius / Durand de Villegagnon, N – Parisiis, 1553 – 1mf – 9 – mf#H-8286 – ne IDC [956]

De bello melitensi historia / Viperanus, J A – Pervsiae, 1567 – 1mf – 9 – mf#H-8307 – ne IDC [956]

De bello pannonico : per illvstrissimvm principem dominum ac dominum fredericum comitem... / Soiterus, M – Vindelicorum, 1538 – 2mf – 9 – mf#H-8255 – ne IDC [956]

De bello rhodio, libri tres, clementi 7 pont max dedicati... / Fontanus, J – Haganoae, 1527 – 2mf – 9 – mf#H-8236 – ne IDC [956]

...De bello turcis inferendo, oratio / Sadoleto, J – Basileae, 1538 – 3mf – 9 – mf#H-8254 – ne IDC [956]

...De bello turcis inferendo, oratio grauissima... / Callimachus, P – Haganoae, 1533 – 2mf – 9 – mf#H-8241 – ne IDC [956]

De benedictionibus patriarcharum iacob et moysi : formae tplila 76 / Radbertus, Pascasius – 1993 – (= ser ILL – ser a; Cccm 96) – 4mf+39p – 9 – €30.00 – 2-503-63962-3 – be Brepols [400]

De benguella as terras do iacca... / Capello, Hermenegildo – Lisboa: Imprensa nacional 1881 [mf ed 1990] – 2v on 1r [ill] – 1 – (filmed with: retrospects and prospects of indian policy / bell, e & other titles) – mf#7493 – us Wisconsin U Libr [910]

De beque bugle *see* Miscellaneous newspapers of mesa county

De beque news *see* Miscellaneous newspapers of mesa county

De beschrijving der handschriften van jan van ruusbroec's werken / Vreese, W de – Gent, 1900-1902 – 13mf – 8 – €25.00 – ne Slangenburg [240]

De bibliorum sacrorum vulgatae editionis graecitate / Saalfeld, Guenther Alexander – Quedlinburgi [Germany (East)]: Christianus Fridericus, 1891 – 1mf – 9 – 0-8370-5011-1 – mf#1985-3011 – us ATLA [450]

De bijbel, de koran en de veda's : tafereel van britsch-indie en van den opstand des inlandschen legers aldaar / Parve, Daniel Couperus Steyn – Haarlem: J J Weeveringh, 1858-59 [mf ed 1995] – (= ser Yale coll) – 2v (ill) – 1 – 0-524-09724-0 – (in dutch) – mf#1995-0724 – us ATLA [954]

De Blij, Harm J *see* Africa south

De Blois, Austen Kennedy *see* John mason peck and one hundred years of home missions, 1817-1917

De boeren courant voor de noordelike districten *see* Der boeren bode

De bogota al atlantico / Perez Triana, Santiago – Bogota, Colombia. 1945 – 1r – us UF Libraries [972]

De bonis ecclesiae temporalibus / Vromant, G – Bruxelles, 1953 – 6mf – 8 – €14.00 – ne Slangenburg [240]

De bonis et malis germaniae, admonitio / Wigand, J – [Francoforti], 1566 – 2mf – 9 – mf#TH-1 mf 1506-1507 – ne IDC [242]

De bonorvm opervm : et novitatis vitae libertate explicatio / Musculus, A – np, 1562 – 3mf – 9 – mf#TH-1 mf 1217-1219 – ne IDC [242]

De bonorvm opervm et novae obedientiae necessitate testimonia / Praetorius, A – Francofordiae ad Oderam, 1562 – 2mf – 9 – mf#TH-1 mf 1277-1278 – ne IDC [242]

De bosporo thracio libri 3 / Gilles, P – Lvgdvni, 1561 – 3mf – 9 – mf#H-8297 – ne IDC [956]

De bow's review : devoted to the restoration of the southern states and the development of wealth and resources of the country – La Salle IL 1846-80 – 1 – mf#3974 – us UMI ProQuest [630]
De brailes bible leaves *see* Selected illuminations from manuscripts in the fitzwilliam museum, cambridge
De brevitate vitae *see* Divinae institutiones...
De Broglio, Chris *see* South africa
De Bunsen, Ernest *see* The chronology of the bible, connected with contemporaneous events in the history of babylonians, assyrians, and egyptians
De caesaribus atque imperatoribus romanis opus insigne... / Cuspinianus, J – Argentorati, 1540 – 14mf – 9 – mf#H-8262 – ne IDC [956]
De cain a pilatos / Saldarriaga Betancur, Jose Manuel – Medellin, Colombia. 1955 – 1r – us UF Libraries [972]
Una de cal y otra de arena / Ortega, Gregorio – Habana, Cuba. 1957 – 1r – us UF Libraries [972]
De california a alaska. madrid, 1945 / Ibarra y Berge, Javier de – Madrid: Razon y Fe, 1947 – 1 – sp Bibl Santa Ana [975]
De canonicae scripturae et catholicae ecclesiae autoritate... / Cochlaeus, J – Ingolstadt, 1543 – 1mf – 9 – mf#PBU-700 – ne IDC [241]
De capienda ex inimicis utilitate *see* Tract of plutarch
De capta constantinopoli, anno 1453 : oratio recitata cum decerneretur gradus doctoris reuerendo d paulo ab eizen... / Meier, G – Vitebergae, 1556 – 1mf – 9 – mf#H-8290 – ne IDC [956]
De caritate annonae ac fame conciones tres...lvdovico lavatero conscriptae ac nunc demum...in latinum conuersae... / Lavater, L – Tigvri, in officina Frosch[oviana], 1587 – 2mf – 9 – mf#PBU-604 – ne IDC [240]
De Carli, Gileno *see* Drama do acucar
De Cass, Anthonius de Petrianis *see* Exposite in terentium...
De casu apostoli, seu, fidei privilegio / Vermeersch, Arthur – Brugis: Beyaert, 1911 – 1mf – 9 – 0-524-01999-1 – mf#1990-4167 – us ATLA [240]
De catechizandis rudibus *see* Treatise of saint aurelius augustine, bishop of hippo
De catholicis seu patriarchis chaldaeorum et nestorianorum : commentarius historico-chronologicus / Assemani, J A – Romae, 1775 – €32.00 – ne Slangenburg [240]
De catholijke en protestantsche zendelingen in indie – Utrecht: J R van Rossum, 1852 [mf ed 1995] – (= ser Yale coll) – 142p – 1 – 0-524-09973-1 – (in dutch) – mf#1995-0973 – us ATLA [240]
De causa dei... / Heidanus, A – Leyden, 1645 – 11mf – 9 – mf#PBA-185 – ne IDC [240]
De causalitate sacramentorum iuxta scholam franciscanam (fp26) / ed by Lampen, W – 1931 – (= ser Florilegium patristicum (fp)) – €5.00 – ne Slangenburg [241]
De causis magnitudinis imperii tvrcici : et virtvtis ac felicitatis tvrcarum in bellis perpetuae / Foglietta, U – Rostochi, 1594 – 1mf – 9 – mf#H-8381 – ne IDC [956]
De celsi, adversari christianorum, philosophandi genere / Philippi, Friedrich Adolph – Berolini [Berlin]: Apud Gustavum Eichler, 1836 – 1mf – 9 – 0-524-04316-7 – mf#1990-1242 – us ATLA [180]
De certaldo isignis opus de claris mulieribus / Boccaccio, Giovanni – Bernae, [1539] – 5mf – 9 – mf#O-1013 – ne IDC [700]
De Cesare, Raffaele *see* The last days of papal rome, 1850-1870
De cheribonsche opstand van 1806 / Broek, J A van den – 's-Gravenhage, 1891 – 1mf – 8 – mf#SE-1593 – ne IDC [959]
De christiana expeditione apud sinas suspecta ab societatis iesu / Ricci, M – Augustae Vind.: apud Christoph Mangium, 1615 – 8mf – 9 – mf#HT-910 – ne IDC [910]
De christlicke ordinancien der nederlantscher gemeynten christi...te london... / Micronius, M – Emden, 1560 – 2mf – 9 – mf#PBA-267 – ne IDC [240]
De chronologia librorum regum : dissertatio critico-historica / Hellmann, Othmar – Romae: Ex typographia pontificia in Instituto Pii 9, 1914 – 2mf – 9 – 0-524-06738-4 – mf#1992-0941 – us ATLA [220]
...De ciuili and bellica fortitudine liber, ex myseriis poetae vergilii neune primum depromptus... / Balbi, G – Roma, [1526] – 2mf – 9 – mf#H-8232 – ne IDC [450]
De clachte pavli : over sijn natuerlijcke verdorventheyt... / Teelinck, W – Dordrecht, 1620 – 2mf – 9 – mf#H-2500 – ne IDC [240]
De claris mulieribus / Boccaccio, Giovanni; ed by Drescher, Karl – Stuttgart: Litterarischer Verein, 1895 (Tuebingen: H Laupp, Jr) [mf ed 1993] – (= ser Blvs 205) – lxxvi/341p – 1 – (early modern german trans of latin text by stainhoewel. incl bibl ref and ind) – mf#8470 reel 42 – us Wisconsin U Libr [920]
De clausulis minucianis et de ciceronianis quae quidem / Ausserer, Alois – Ad Aenipontem, Austria. 1906 – 1r – us UF Libraries [960]
De clemente presbytero alexandrino : homine, scriptore, philosopho, theologo liber / Reinkens, Joseph Hubert – Vratislaviae: G. Ph. Aderholz, [1851?] – 1mf – 9 – 0-7905-8087-X – mf#1988-8023 – us ATLA [242]
De clementis romani epistola ad corinthios priore disquisitio / Lipsius, Richard Adelbert – Lipsiae [Liepzig]: FA Brockhaus, 1855 – 1mf – 9 – 0-524-03902-X – (incl bibl ref) – mf#1990-1161 – us ATLA [240]
De clerico medico curiosa disceptatio, sive interpretatio ad testum... / Tristan Valentin, G – Valencia, 1606 – 4mf – 9 – sp Cultura [610]
De codice sancti evangelii, libri 3 / Catalanus, Josepho – Roma, 1738 – 5mf – 8 – €12.00 – ne Slangenburg [220]
De codicibus mss. graecis pii 2 : in bibliotheca alexandrino-vaticana schedas / Duchesne, Louis – Lutetiae Parisiorum: E. Thorin, 1880 – (= ser Bibliotheque des ecoles francaises d'athenes et de rome) – 1mf – 9 – 0-7905-8103-5 – mf#1988-6065 – us ATLA [012]
De coena domini confessio / Westphal, J aus Hamburg – Vrsellis, 1558 – 2mf – 9 – mf#TH-1 mf 1480-1481 – ne IDC [242]
De coena domini sermo / Bullinger, Heinrich – [Tiguri, Christoph Froschauer], 1558 – 1mf – 9 – mf#PBU-203 – ne IDC [240]
De communibus omnium rerum naturalium principiis et affectionibus / Pererius, B – Romae, 1576 – 10mf – 9 – mf#CA-28 – ne IDC [240]
De como o mulato porciuncula descarregou seu defunto / Amado, Jorge – Lisboa: Distribuicao da Editorial Organizacoes, 1962 – us CRL [074]
De compendiosa doctrina / Marcellus, Nonius – (= ser Holkham library manuscript books 407,408) – 2r – 1 – mf#97364, 97365 – uk Microform Academic [450]
De compositione hominis / Wycliffe, John; ed by Beer, Rudolf – London: Published for the Wiclif Society by Truebner, 1884 – 1mf – 9 – 0-524-00221-5 – mf#1989-2921 – us ATLA [240]
De conciliis / Bullinger, Heinrich – Tigvri, Christoph Froschouer, 1561 – 4mf – 9 – mf#PBU-214 – ne IDC [240]
De conciliis...in primitiva ecclesia / Bullinger, Heinrich – Tiguri, 1561 – 4mf – 9 – mf#PBU-107 – ne IDC [240]
De concordia ecclesiae occidentalis et orientalis / Arcudius, P – Parisiis, 1626 – 29mf – 8 – €56.00 – ne Slangenburg [240]
De confessione avgvstana : ...et de concordia m lvtheri cum m bucero et helvetiis ecclesiis / [Hardesheim, C] – n.p, 1579 – 2mf – 9 – mf#PBU-595 – ne IDC [240]
De confutatione latina : quae apologiae concionatorum evangelicorum in comitiis hauniensibus anno 1530 traditae opposita est / Engelstoft, Christian Thorning – Hauniae [Copenhagen]: Typis Schultzianis, 1847 [mf ed 1992] – 1mf – 9 – 0-524-02853-2 – (incl bibl ref) – mf#1990-0710 – us ATLA [242]
De congregationibus clericorum in communi viventium / Miraeus, A – Colonia Agrippina, 1632 – 4mf – 8 – €12.00 – ne Slangenburg [240]
De consolatione philosophiae = [Consolation of philosophy] / Boethius, Anicius Manlius Severinus – 14th c – (= ser Holkham library manuscript books 402) – 1r – 1 – mf#96533 – uk Microform Academic [180]
De constantino imperatore, pontificio maximo : dissertationem / Aube, Benjamin – Lutetiae: F Didot: A Durand, 1861 – 1mf – 9 – 0-7905-4015-0 – (incl bibl ref) – mf#1988-0015 – us ATLA [240]
De constructione libri 2 : from the monastery of san salvatore, venice / Priscianus [Priscian: Priscianus Caesariensis] – 14th c – (= ser Holkham library manuscript books 405) – 1r – 1 – mf#96811 – uk Microform Academic [450]
De constructione octo partium orationis liber emmanuelis alvari – Hispali Seville: Alonsi a Barrera, 1590 [mf ed 1983] – 54lea – 1 – mf#1041 – us Wisconsin U Libr [450]
De controversiis in coena / Beza, Theodor de – Geneve, Le Preux, 1594 – 2mf – 9 – mf#PFA-120 – ne IDC [240]
De conversione indorum et gentilium libri duo / Hoornbeek, J – Amstelodami, 1669 – 4mf – 9 – mf#PBA-201 – ne IDC [240]
De convincendis et convertendis judaeis et gustilibus / Hoornbeek, J – Lugduni Batavorum, 1655 – 7mf – 9 – mf#PBA-198 – ne IDC [240]
De cosas extremenas y de algo mas / Sancho y Gonzalez, Javier – Badajoz: Vicente Rodriguez, 1912 – 1 – sp Bibl Santa Ana [946]
De Coverly, Roger *see* Novanglus, and massachusettensis
De criticae sacrae argumento e linguae legibus repetito : ratione ducta maxime geneseos capp. 1-11 eius historiam, naturam, vim / Koenig, Eduard – Lipsiae [Leipzig]: JC Hinrichs, 1879 – 1mf – 9 – 0-8370-3964-9 – (incl bibl ref) – mf#1985-1964 – us ATLA [220]
De criticis vet. gr. et latinis : from the british library copy (836i 3(1)) of the 1587 paris edition / Estienne, Henri – 1r – 1 – mf#97087 – uk Microform Academic [450]
De cultibus religiosis latinorum / Dallaeus, lon – Genevae, 1671 – 23mf – 8 – €44.00 – ne Slangenburg [240]
De cultu adorationis libri tres / Vazquez, Gabriel – Compluti, 1594 – 11mf – 8 – €21.00 – ne Slangenburg [241]
De cymbalis vetervm libri tres : in quibus quaecunque ad eorum nomina, differentiam, originem, historiam, ministros, ritus pertinent, elucidantur... / Lampe, Friedrich Adolf – Trajecti ad Rhenum: ex [officina] bibliopolae Guilelmi a Poolsum 1703 [mf ed 19–] – 3mf – 9 – mf#fiche 773 – us Sibley [780]
De d joao vi a independencia / Moura Romeiro, Joao Marcondes De – Sao Paulo, Brazil. 1962 – 1r – us UF Libraries [972]
De d n jesu christi divinitate : adversus hujus aetatis incredulos, rationalistas & mythicos / Perrone, Giovanni – Taurini [Turin]: Ex typis stereotypis Hyacinthi Marietti, 1870 [mf ed 1991] – 3v on 12mf – 9 – 0-524-00305-X – mf#1989-3005 – us ATLA [240]
De deo : disputationes metaphysicae: quas excipit dissertatio de mente sancti anselmi in proslogio / Piccirelli, Josephus M – Lutetiae Parisiorum: Victorem Lecoffre, 1885 – 2mf – 9 – 0-8370-7252-2 – (incl bibl ref) – mf#1986-1252 – us ATLA [210]
De deo creante : praelectiones scholastico-dogmaticae / Mazzella, Camillo – Woodstock, Md: Ex Officina Typographica collegii, 1877 – 3mf – 9 – 0-7905-9031-X – mf#1989-2256 – us ATLA [210]
De deo creatore, de angelis, de homine et de gratia divina / Mancini, Jerome Marie – Romae: SC de Propaganda Fide, 1903 [mf ed 1991] – (= ser Theologia dogmatica 2) – 2mf – 9 – 0-7905-8697-5 – mf#1989-1922 – us ATLA [210]
De deo uno et trino / Mancini, Jerome Marie – Romae: SC de Propaganda Fide, 1903 – (= ser Theologia Dogmatica) – 2mf – 9 – 0-7905-8698-3 – mf#1989-1923 – us ATLA [210]
De deorum romanorum cognominibus : quaestiones selectae / Carter, Jesse Benedict – Lipsiae: In aedibus BG Teubneri, 1898 – 1mf – 9 – 0-524-00874-4 – (incl bibl ref) – mf#1990-2097 – us ATLA [250]
D+E [design and environment] – New York NY 1970-76 – 1,5,9 – (cont by: urban design) – ISSN: 0011-930X – mf#10257.01 – us UMI ProQuest [710]
De desterrado a presidente / Bedoya Cardona, Ernesto – Medellin, Colombia. 1950 – 1r – us UF Libraries [972]
De dienst der baalim in israel : naar aanleiding van het geschrift van dr. r. dozy "de israelieten te mekka" / Oort, Henricus – Leiden: P Engels, 1864 – 1mf – 9 – 0-8370-4620-3 – (incl bibl ref) – mf#1985-2620 – us ATLA [220]
De digamia episcoporum : ein beitrag zur lutherforschung / Kawerau, Gustav – Kiel: E Homann, 1889 – 1mf – 9 – 0-7905-6192-1 – (incl bibl ref) – mf#1988-2192 – us ATLA [242]
De dioscuris / Greebe, Cornelius Aleidus Arnoldus Ioannes – Lugduni-Batavorum [Leiden]: EJ Brill, 1905 [ie 1895] – 1mf – 9 – 0-524-01361-6 – (incl bibl ref) – mf#1990-2373 – us ATLA [250]
De disciplinae arcani, quae dicitur, in ecclesia christiana origine : commentatio, quam pro munere professoris publici ordinarii in facultate theologica academiae ruperto-carolae rite suscipiendo / Rothe, Richard – Heidelbergae: JCB Mohr, 1841 – 1mf – 9 – 0-7905-7608-2 – (incl bibl ref) – mf#1989-0833 – us ATLA [240]
De distingvendo decalogo / Cramer, D – Wittebergae, 1598 – 2mf – 9 – mf#TH-1 mf 370-371 – ne IDC [242]
De diversis ministrorum evangelii gradibus... / Saravia, H – Londini, 1590 – 2mf – 9 – mf#PBA-303 – ne IDC [240]
De divina praedestinatione : formae tplila 4 / Eriugena, Ioannes Scotus – 1982 – (= ser ILL – ser a; Cccm 50) – 3mf+35p – 9 – €20.00 – 2-503-60502-8 – be Brepols [400]
De divino...jesus...orationem / Martinez Siliceo, Juan – 1551 – 9 – sp Bibl Santa Ana [240]
De divisione philosophiae / Gundissalinus, Dominicus; ed by Baur, Ludwig – Muenster: Aschendorff, 1903 – (= ser Beitraege zur geschichte der philosophie des mittelalters) – 1mf – 9 – 0-7905-8655-X – (incl bibl ref) – mf#1989-1880 – us ATLA [100]
De doctrina christiana / Augustinus, St – 1982 – (= ser ILL – ser a; Cccm 32) – 4mf+52p – 9 – €20.00 – 2-503-60322-X – be Brepols [400]
De [domestic engineering] – Chicago IL 1900-92 – 1,5,9 – (cont by: plumbing, heating, piping) – ISSN: 0147-6998 – mf#1099.01 – us UMI ProQuest [690]
De dominio divino libri tres – de pauperie salvatoris / Wycliffe, John & FitzRalph, Richard; ed by Poole, Reginald Lane – London: Published for the Wyclif Society by Truebner, 1890 – 2mf – 9 – 0-524-00810-8 – mf#1990-0242 – us ATLA [210]
De Dominis, Marco Antonio *see* My motives for renouncing the protestant religion
De donde son los cantantes / Sarduy, Severo – Mexico City? Mexico. 1967 – 1r – us UF Libraries [972]
De dubio solvendo in re morali / Waffelaert, Gustave Joseph – Lovanii: Valinthout, [1880?] – 1mf – 9 – 0-8370-6535-6 – (incl bibl ref) – mf#1986-0535 – us ATLA [230]
De duodecum abusivis saeculi *see* Pseudo-cyprianus de 7 abusivis saeculi
De dvabvs natvris in christo de hypostatica earvm vnione / Chemnitz d A, M – Lipsiae, 1578 – 7mf – 9 – mf#TH-1 mf 204-210 – ne IDC [242]
De dvabvs natvris in christo; de hypostatica earvm vnione, de commvincatione idiomatvm, et de aliis qaestionibvs independentibvs; libellvs ex scriptvra sententijs & ex pvrioris antiqvitatis testimonijs ... cvm praefatione nicolai selnecceri / Chemnitz, Martin – Lipsiae, 1580 – 1r – 1 – 0-8370-1477-8 – mf#1984-B018 – us ATLA [220]
De dvabvs natvris in christo, earvmqve vnione hypostatica tractatvs / Hesshusen, T – Magdebvrgi, 1590 – 4mf – 9 – mf#TH-1 mf 613-616 – ne IDC [242]
De ebionitarum origine et doctrina : ab essenis repetenda / Baur, Ferdinand Christian – Tubingae: Hopferi de L'Orme, 1831 [mf ed 1989] – 1mf – 9 – 0-7905-3004-X – (incl bibl ref) – mf#1987-3004 – us ATLA [240]
De ecclesia christi : commentariorum libri quinque / Passaglia, Carlo – Ratisbonae [Regensburg]: Sumptus fecit G Iosephus Manz, 1853-1856 – 10mf – 9 – 0-524-00300-9 – mf#1989-3000 – us ATLA [240]
De ecclesia christi : praelectiones novae in seminario sancti sulpitii habitae: cum multis annotationibus in ulteriora cujusque studia et praedictionis usus profuturis / Brugere, Lud-Fred – ed nova. Parisiis: A Roger et F Chernoviz, 1878 – 2mf – 9 – 0-8370-8408-3 – (in latin and french. incl bibl ref and index) – mf#1986-2408 – us ATLA [240]
De ecclesia christi / Straub, Anton – Oeniponte [Innsbruck]: Typis et sumptibus Feliciani Rauch, 1912 – 4mf – 9 – 0-524-00151-0 – mf#1989-2851 – us ATLA [240]
De ecclesia christi ut infallibili revelationis divinae magistra / Ottiger, Ignaz – Friburgi Brisgoviae; S Ludovici Americae: Herder, 1911 – (= ser Theologia Fundamentalis) – 1mf – 9 – 0-7905-8541-3 – (incl bibl ref) – mf#1989-1766 – us ATLA [240]
De ecclesia libri sex / Lubbertus, S – Franekerae, 1607 – 5mf – 9 – mf#PBA-237 – ne IDC [240]
De ecclesiae occidentalis et orientalis perpetua consensione / Allatius, L – 1648 – 18mf – 8 – €44.00 – ne Slangenburg [240]
De ecclesiasticis officiis : fragmentos (siecle 9-10) / Isidore de Seville, Saint – Barcelona – 1r – 5,6 – sp Cultura [240]
De ecclesiasticis officiis / Hispalensis, Isidorus – 1989 – (= ser ILL – ser b; Ccsl 113) – 2mf+27p – 9 – €30.00 – 2-503-71132-4 – be Brepols [400]
De ecclesiasticis officiis / Hispalensis, Isidorus – 1989 – (= ser ILL – ser a; Ccsl 113) – 3mf+41p – 9 – €30.00 – 2-503-61132-X – be Brepols [400]
De economische politiek van het nieuwe indonesie / Fruin, T A – Batavia-C: M Vervoort, [194-?] [mf ed 1999] – 1r – 1 – (filmed with: wir lernen deutsch / leo kober) – mf#4690 – us Wisconsin U Libr [339]
De eekboom *see* Uns eekboom
De el remedio de el amor impuro / Ovidio Nason, P – 1732 – 9 – sp Bibl Santa Ana [450]
De elohistae pentateuchici sermone : commentatio historico-critica / Ryssel, Victor – Lipsiae [Leipzig]: L Fernau, 1878 – 1mf – 9 – 0-8370-5009-X – (in latin. incl ind of hebrew words) – mf#1985-3009 – us ATLA [221]
De el...!siglo pasado! / Sanchez Arjona, Vicente – Sevilla: Graficas Tirvia, 1954 – 1 – sp Bibl Santa Ana [810]
De emblemata van hadrianus junius : herdruk der plantijnsche... / Junius, H – Antwerpen: Museum Plantin, 1902 – 1mf – 9 – mf#O-3234 – ne IDC [090]
De ente praedicamentali: from the unique vienna ms.; quaestiones 13 logicae et philosophicae: from the unique prague ms / Wycliffe, John; ed by Beer, Rudolf – London: Published for the Wyclif Society by Truebner, 1891 – 1mf – 9 – 0-524-00811-6 – mf#1990-0243 – us ATLA [180]

De enuntiationibus relativis semiticis. pars prior, praemisso ibn jaoi si in zamach sarii, de pronominibus relativis locum commentario, de enuntiationibus relativis arabicis agens : dissertatio linguistica = Sharh al-mufassal / Ibn Yaoish, Abu al-Baqa Yaoish ibn Ali – Bonnae ad Rhenum: Tobiae Habichtii, 1868 – 1mf – 9 – 0-8370-7324-3 – (text in arabic; notes in latin. no more publ. incl bibl ref) – mf#1986-1324 – us ATLA [470]

De enuntiatis finalibus apud graecorum rerum scriptores posterioris eatatis / Diel, Heinrich – Muenchen: J Fuller, 1895 – 1mf – 9 – 0-8370-9226-4 – mf#1986-3226 – us ATLA [450]

De epistolae quae barnabae tribuitur authentia / Henke, Ernst Ludwig Theodor – Ienae [Jena]: Prostat in Libraria Croekeriana, 1827 – 1mf – 9 – 0-7905-8306-2 – (incl bibl ref) – mf#1987-6411 – us ATLA [240]

De eruditione puerorum regalium *see* Liber gratiae. liber laudum virginis mariae. de sto johanne. de eruditione puerorum regalium

De eruditione solida : superficiaria et falsa, libri tres / Poiret, P – Amstelodami, 1692 – 9mf – 9 – mf#PPE-210 – ne IDC [240]

De eruditione triplici, solida, superficiaria et falsa libri tres : praemittitur vera methodus inveniendi verum / Poiret, P – Amsterdam, 1707 – 8mf – 9 – mf#PPE-209 – ne IDC [240]

De escultura e imageneria / Perez Comendador, Enrique – Madrid: (Blas), 1957 – sp Bibl Santa Ana [946]

De espana al japon / Oteyza, Luis de – Madrid: Editorial Pueyo, S.L., 1927 – sp Bibl Santa Ana [946]

De essentia originalis ivstitiae et inivstitiae seu imaginis dei et contrariae / Flacius Illyricus d A, M – Basileae, 1568 – 4mf – 9 – mf#TH-1 mf 413-416 – ne IDC [242]

De este lado del mar / Cabral, Manuel Del – Ciudad Trujillo, Dominican Republic. 1949 – 1r – us UF Libraries [972]

De estherae libro et ad eum quae pertinet vaticiniis et psalmis : libri tres / Nickes, Johannes Anselm – Romae: Typis S C de Propaganda Fide, 1856-1858 – 2mf – 9 – 0-524-06849-6 – mf#1992-0991 – us ATLA [220]

De eucharistia tractatus maior : accedit tractatus de eucharistia et poenitentia sive de confessione / Wycliffe, John; ed by Loserth, Johann – London: Published for the Wyclif Society by Truebner, 1892 – 1mf – 9 – 0-524-00222-3 – mf#1989-2922 – us ATLA [240]

De eucharistiae sive coenae dominicae sacramento libri tres / Albertinus, Edm – Daventriae, 1655 – 27mf – 8 – €52.00 – ne Slangenburg [241]

"De evenaar" chattulistiwa – Amsterdam, 1947-1960 – 25mf – 9 – (missing: 1948, v1; 1949, v2(11-12); 1951, v4(11-12)-1951, v5(1, 5, 8-12)-1952, v6(1-2, 4, 8-12); 1954, v7(aug-sep); 1957, v10(4)-1960(jan-nov)) – mf#SE-748 – ne IDC [959]

De facultatibus naturalibus disputationes medicae et phylosophicae / Fernandez, F – Granada, 1619 – 7mf – 9 – sp Cultura [610]

De Faehrkrog : en dramatisch gliknis in dree akten / Bossdorf, Hermann – Hamborg: R Hermes, 1920 [mf ed 1989] – (= ser Nedderduetsch boekerei 63) – 70p – 1 – mf#7053 – us Wisconsin U Libr [820]

De fatis monarchiae romanae somnium vaticanum esdrae prophetae, quod theodorus bibliander interpretatus est... / Bibliander, T – Basileae, [Johannes Oporinus, 1553] – 2mf – 9 – mf#PBU-584 – ne IDC [240]

De febrium differentiis... / Mercado, P – Granada, 1581 – 6mf – 9 – sp Cultura [610]

De fide : synopsis praelectionum, quas in c.r. universitate oenipontana / Stentrup, Ferdinandus Aloisius – Oeniponte [Innsbruck]: Feliciani Rauch, 1890 – 1mf – 9 – 0-8370-7022-8 – mf#1986-1022 – us ATLA [240]

De fide et symbolo : documenta quaedam nec non aliquorum ss patrum tractatus / ed by Heurtley, Charles Abel – 5th ed. Oxonii [Oxford]: Apud Parker, 1909 [mf ed 1992] – 1mf – 9 – 0-524-04016-8 – (text in greek & latin. notes in english) – mf#1990-1188 – us ATLA [240]

De fide, eujusque ortu, et natura, plana ac dilucida explicatio : adjecta sunt alia quedam ejusdum authoris, de codem argumento, qua sequens pagina indicabit / Baro, P – Londini: Apud Richardum Dayum, 1580 – 3mf – 9 – mf#PW-3 – ne IDC [240]

De fide, spe et caritate : formae tplila 61 / Radbertus, Pascasius – 1990 – (= ser ILL – ser a; Cccm 97) – 4mf+44p – 9 – €60.00 – 2-503-63972-0 – be Brepols [400]

De fide, spe et charitate et de incarnatione / Mancini, Jerome Marie – Romae: SC de Propaganda Fide, 1904 – (= ser Theologia Dogmatica) – 2mf – 9 – 0-7905-8699-1 – mf#1989-1924 – us ATLA [240]

De fidei notione ethica paulina / Schnedermann, Georg – Lipsiae: J C Hinrichs, 1880 – 1mf – 9 – 0-7905-0379-4 – (in latin and greek. incl bibl ref) – mf#1987-0379 – us ATLA [220]

De fine seculi and iudicio... / Bullinger, Heinrich – Basileae, [Ioannes Oporinus], 1557 – 2mf – 9 – mf#PBU-193 – ne IDC [240]

De finibus bonorum et malorum / Cicero, Marcus Tullius – New York, NY. 1931 – 1r – us UF Libraries [025]

De flavii iosephi elocutione : observationes criticae / Schmidt, Wilhelm – Lipsiae [Leipzig]: B.G. Teubner, 1893 – 1mf – 9 – 0-8370-5144-4 – (incl bibl ref, summary index and indexes of words, subjects, and authors cited) – mf#1985-3144 – us ATLA [450]

De fonseka family of kalutara (ceylon) / Abeyesooriya, Samson – Colombo: Independent Press, [n.d.] – 1 – us CRL [954]

De forest times – Deforest WI. [1896 apr 10/1898 jul 15]-1945/1948 oct 29 – 32r – 1 – (cont by: morrisonville tribune; de forest times-tribune) – mf#938755 – us WHS [071]

De forest times-tribune – Deforest WI. 1948 nov 5/dec-1994 jul/dec – 31r – 1 – (cont: de forest times; morrisonville tribune) – mf#918807 – us WHS [071]

De formandis consionibus sacris... / Hyperius, A – Marpurgi, 1553 – 3mf – 9 – mf#PBA-209 – ne IDC [240]

De formulae concordiae rabisbonensis origine atque indole / Brieger, Theodor – Halis Saxonum: Formis Hendeliis, 1870 – 1mf – 9 – 0-524-08333-9 – mf#1993-2023 – us ATLA [240]

De fornicatione cavenda admonitio : sive, adhortatio ad pudicitiam et castitatem / Beverland, Adriaan – Ed. nova & ab autore correcta. Juxta exemplar Londinense [Amstelodami?, s.n.] 1698 – 1r – 1 – 0-8370-0053-X – mf#1984-6012 – us ATLA [240]

De fractione panis evcharistici theses / Pelargus, C – Hanoviae, 1607 – 1mf – 9 – mf#TH-1 mf 1631 – ne IDC [242]

De fundamenten omgestooten : predikatie. gehouden op 12 januari 1913, in de alpine av. chr. geref. kerk te grand rapids, mich / Lonkhuijzen, Jan van – Grand Rapids, MI: L Kregel, 1913 – 1mf – 9 – 0-524-06638-8 – mf#1991-2693 – us ATLA [240]

De gaspe et garneau / Casgrain, Henri-Raymond – Montreal: Librairie Beauchemin, 1924 [mf ed 1986] – 2mf – 9 – mf#SEM105P749 – cn Bibl Nat [920]

De gaudio resurrectionis sermo... / Oecolampadius, J – [Augustae Vindelicorum, Sigismundus Grimm & Marcus Vuyrsung, 1521] – 1mf – 9 – mf#PBU-349 – ne IDC [240]

De gentes del otro mundo / Roso de Luna, Mario – Madrid: Libreria Viuda de Pueyo, 1917 – 1 – sp Bibl Santa Ana [240]

De genvina verborum domini, hoc est corpus meum...expositione liber / Oecolampadius, J – [Strassburg, Johann Knobloch, 1525] – 2mf – 9 – mf#PBU-365 – ne IDC [240]

De geographia vniuersali / Muhammad ibn Muhammad – Rome, 1592 – 4mf – 9 – mf#H-8434 – ne IDC [956]

De gestis mirabilibus regis edwardii tertii *see* Continuatio chronicarum [rs93]

De gestis pontificum anglorum libri quinque [rs52] / William of Malbesbury; ed by Hamilton, N E S A – 1870 – (= ser The rolls series [rs]) – €23.00 – ne Slangenburg [241]

De gestis regum anglorum libri quinque [rs90] : historiae novellae libri tres / William of Malbesbury; ed by Stubbs, W – (= ser The rolls series [rs]) – (v1 1887 €17. v2 1889 €18) – ne Slangenburg [931]

De gestis siculorum sub frederico 2 rege / Specialis, Nicolaus – 15th c – (= ser Holkham library manuscript books 495) – 1r – 1 – mf#96132 – uk Microform Academic [090]

De gibraltar a lisboa : viaje historico / Espronceda, Jose de – 1852 – 9 – sp Bibl Santa Ana [914]

De gl'eroici furori : al molto illustre et eccelente cavaliero signor filippo sidneo / Bruno, Giordano – Parigi: Appresso Antonio Baio, 1585 – 3mf – 9 – mf#O-84 – ne IDC [090]

De gonzalo ximenez de quesada a don pablo morillo / Restrepo Tirado, Ernesto – Paris, France. 1928 – 1r – us UF Libraries [972]

De graecitate patrum apostolicorum librorumque apocryphorum novi testamenti quaestiones grammaticae / Reinhold, Henricus – [s.l: s.n, 18–?] [mf ed 1986] – 1mf – 9 – 0-8370-9651-0 – (incl bibl ref) – mf#1986-3651 – us ATLA [450]

De gratia dei iustificante nos propter christum : per solam fidem absqz operibus bonis, fide interim exuberante in opera bona libri 3... / Bullinger, Heinrich – Tiguri: Froschoviana, 1554 – 1r – 1 – 0-8370-0004-1 – mf#1984-B464 – us ATLA [240]

De gratia dei ivstificante / Bullinger, Heinrich – Tigvri, Officina Froschoviana, 1554 – 3mf – 9 – mf#PBU-182 – ne IDC [240]

De Gruchy, Joy *see* Cost of living for urban africans, johannesburg 1959

De Gubernatis, Angelo *see*
– Materiaux pour servir a l'histoire des etudes orientales en italie
– Zoological mythology

De Guingand, Francis Wilfred *see* South africa and the world in 1968

De gulden throen / Otten van Passau (Otto of Passau) – Utrecht, 1480 – €32.00 – ne Slangenburg [240]

De habacuci prophetae : vita atque aetate / Delitzsch, Franz – ed auctior et emandatior. Lipsiae: R Beyer, 1842 – 1mf – 9 – 0-7905-3428-2 – (incl bibl ref) – mf#1987-3428 – us ATLA [221]

De habitu et colore aethiopum qui vulgo nigritae... / Pechlin, Johann Nicolas – Kiloni: Impensis J Reumanni, 1677 – (filmed with: armistead, w. a tribute for the negro) – us CRL [960]

De haeresis anglicanae intrusione et progressu / Rinuccini, Giovani Battista, archbishop of Fermo and Nuncio – 17th c – (= ser Holkham library manuscript books 200) – 1r – 1 – (aka: nuncio's memoirs) – mf#96786 – uk Microform Academic [241]

De harmonia musicorum instrumentorum opus / Gaffurio, Franchino – 1518 – (= ser Mssa) – 3mf – 9 – €50.00 – mfchl 55 – gw Fischer [780]

De Hart, William Chetwood *see* Observations on military law, and the constitution and practice of courts martial.

De Hartog, Jan *see* Maitre apres dieu

De hebdomadis, qvae apvd danielem sunt, opusculum / Bullinger, Heinrich – Tigvri, Christopherus Froschouer, 1530 – 1mf – 9 – mf#PBU-108 – ne IDC [240]

De hebraeorum leviratu / Benary, Ferdinand – Berolini: Impensis F Duemmleri, 1835 – 1mf – 9 – 0-7905-3303-0 – mf#1987-3303 – us ATLA [220]

De heraut – Amsterdam, 1877-1945 – 22r – 1 – mf#SF-3 – ne IDC [074]

De hexateuch *see* Historico-critical inquiry into the origin and composition of the hexateuch (pentateuch and book of joshua)

De heyr-baene des cruys : waer-langhs alle soorten van menschen worden ghewesen, ende sekerlijck gheleert... / Haeften, B van – Brugghe: Lucas vanden Kerchove, 1667 – 5mf – 9 – mf#O-3076 – ne IDC [090]

De hierarchia anglicana : dissertatio apologetica / Denny, Edward & Lacey, Thomas Alexander – Londini: Veneunt apud CJ Clay, 1895 – 1mf – 9 – 0-524-04039-7 – mf#1990-4947 – us ATLA [240]

De historia de badajoz / Lozano Rubio, Tirso – Badajoz: Ed.Arqueros, Tomo 2(Mateos). 1930 – 1 – sp Bibl Santa Ana [946]

De historia. relaciones de badajoz con la corte y con don manuel godoy durante el valimiento de este / Guerra Guerra, Arcadio – Badajoz: Imp. Dip. Provincial, 1958 – sp Bibl Santa Ana [946]

De historia textus actorum apostolorum / Coppieters, Honoratus – Lovanii [Louvain]: J van Linthout, 1902 – 1mf – 9 – 0-524-06514-4 – (incl bibl ref) – mf#1992-0898 – us ATLA [225]

De historiae byzantinae scriptoribus emittendis protrepticon (cbh1,1) : apparatus historiae byzantinae delineatio prima / Labbe, Ph – Parisiis, 1648 – (= ser Corpus byzantinae historiae (cbh)) – €7.00 – ne Slangenburg [243]

De historias americanas / Bayle, Constantino – Madrid: Razon y Fe, 1925 – 1 – sp Bibl Santa Ana [972]

De homine integro corrvpto renato glorificato / Wigand, J – Francoforti, 1562 – 3mf – 9 – mf#TH-1 mf 1557-1559 – ne IDC [242]

De huidige stand van het nationalisme *see* Voordracht gehouden op de indonesische conferentie te hardenbroek, den 17en april 1929

De hydrophobiae natura, causis atque medela... / Bravo de Piedrahita, J – Salamanca, 1571 – 4mf – 9 – sp Cultura [610]

De iesu christo servatore... / Lubbertus, S – Radaeus, 1611 – 7mf – 9 – mf#PBA-236 – ne IDC [240]

De illustratione urbis florentiae / Verino, U – Parigi, 1790 – 4mf – 9 – mf#O-1056 – ne IDC [700]

De imaginibus libri 4 / Dallaeus, Ion – Lugd. Batavorum, 1642 – 6mf – 8 – €14.00 – ne Slangenburg [240]

De immaculata deiparae conceptione hymnologia graecorum : ex editis et manuscriptis codicibus cryptoferratensibus: latina et italica interpretatione – Romae: Typis S Congreg de Propaganda Fide, 1862 – 1mf – 9 – 0-8370-9032-6 – mf#1986-3032 – us ATLA [780]

De immaculato b. v. mariae conceptu an dogmatico decreto definiri possit : disquisitio theologica / Perrone, Giovanni – Editio undecima, Taurinensis prima, aucta atque emendata Taurini: Speirani et Tortone, 1854 – 1mf – 9 – 0-8370-8288-9 – mf#1986-2288 – us ATLA [240]

De impedimentis magnorum auxiliorum in morborum curatione, libri 3 / Ponce de Santa Cruz, A – Madrid, 1629 – 4mf – 9 – sp Cultura [610]

De imperatorum constantinopolitanorum seu de inferioris aevi : vel imperii numismatibus dissertatio / Cange, C du – Parisiis, 1766 – €12.00 – ne Slangenburg [243]

De imperio et rebus gestis iustiniani (cbh4) / Agathiae Scholastici; ed by Vulcanii, B – Parisiis, 1660 – (= ser Corpus byzantinae historiae (cbh)) – €19.00 – ne Slangenburg [243]

De incantamentis nonnullis sumerico-assyriis / Jensen, Petrus – Monachii: F Straub, [1884?] – 1mf – 9 – 0-8370-7640-4 – (text in latin and sumerian; preface and notes in latin) – mf#1986-1640 – us ATLA [230]

De incarnatione filii dei item de officio et maiestate christi tractus : martini chemnicii de incarnatione filii dei item de officio et maiestate christi tractatus / Chemnitz, Martin – Berolini:Gust. Schlawitz, 1865 – 1mf – 9 – 0-8370-3222-9 – (incl ind) – mf#1985-1222 – us ATLA [240]

De incarnatione verbi dei : together with three essays subsidiary to the same / Hawkesworth, Alan S – Albany, NY: Riggs, 1897 – 1mf – 9 – 0-8370-4528-2 – (incl bibl ref) – mf#1985-2528 – us ATLA [240]

De incarnatione veri et aeterni filii dei... / Gwalther, R – Zuerich, Froschouer, 1572 – 5mf – 9 – mf#PBU-299 – ne IDC [240]

De incubatione : capita quattuor / Deubner, Ludwig – Lipsiae [Leipzig]: In aedibus BG Teubneri, 1900 – 1mf – 9 – 0-524-02803-6 – mf#1990-3133 – us ATLA [200]

De indiae utriusque re naturali et medica / Piso, G – Amstelaedami: Apud L et D Elzevirios, 1658 – 14mf – 9 – mf#I-1063 – ne IDC [590]

...De initere svo constantinopolitano, epistola / Dousa, G – Lugduno Batavae, 1599 – 2mf – 9 – mf#H-8405 – ne IDC [956]

De inspiratione sacrae scripturae / Pesch, Christian – Friburgi Brisgoviae [Freiburg i B]: Herder, 1906 – 2mf – 9 – 0-524-05932-2 – (incl bibl ref) – mf#1992-0689 – us ATLA [220]

De inspiratione scripturae sacrae quid statuerint patres apostolici et apologetae secundi saeculi : commentatio dogmatico-historica / Delitzsch, Johannes – Lipsiae [Leipzig]:Prostat apud A Lorentz Bibliopolam, 1872 – 1mf – 9 – 0-8370-2883-3 – (incl bibl ref and index) – mf#1985-0883 – us ATLA [220]

De institutionis theologicae via ac ratione : oratio academica solemnis, quam pro anni scholastici cursu fauste finiendo auspiciis rev. episcopi paderbornensis in auditorio almae scholae theodorianae maximo / Oswald, H – Paderbornae: Apud Ferdinandum Schoeningh, 1850 – 1mf – 9 – 0-8370-8461-X – mf#1986-2461 – us ATLA [240]

De integro sacramento corporis et sanguinis domini / Corvinus, A – [Hannoverae, 1544] – 2mf – 9 – mf#TH-1 mf 357-358 – ne IDC [242]

De interpretatione scripturarum sacrarum / Patrizi, Francesco Saverio – Romae: Typis Ioannis Baptistae Marini et Soc, 1844 – 2mf – 9 – 0-524-05994-2 – (incl bibl ref) – mf#1992-0731 – us ATLA [220]

De ira liber / Philodemus – Lipsiae, Germany. 1864 – 1r – us UF Libraries [960]

De isidori pelusiotae vita : scriptis et doctrina commentatio, historica theologica / Niemeyerus, H A – Hallae, 1825 – 5mf – 8 – €12.00 – ne Slangenburg [240]

De itinere terrae sanctae liber *see* Ludolphi, rectoris ecclesiae parochialis in suchem, de itinere terrae sanctae liber

De itinere terrae sanctae...nach alten handschriften berichtigt / Deycks, Ferdinand – Stuttgart, 1851 – 2mf – 9 – mf#H-3082 – ne IDC [915]

De iure sacrorum / Becerra y Valcarzel, Diego – 1673. 2 tomas – 9 – sp Bibl Santa Ana [240]

De iustificatione doctrina universa / Vega, Andreas (Andres) de – Coloniae, 1572 – 39mf – 8 – €75.00 – ne Slangenburg [240]

De jaerfschopp : erzaehlung in muensterlaender mundart / Wibbelt, Augustin – Essen: Fredebeul & Koenen, [1911?] [mf ed 1992] – 322p – 1 – mf#7821 – us Wisconsin U Libr [390]

De Jager, E J *see* Select bibliography of the anthropology of the cape nguni tribes

De johanneiska smabrefvens ursprung : undersoekt med saerskild haensyn till presbyterhypotesen: akademisk afhandling / Hjelt, Arthur – Helsingfors [Helsinki]: Frenckellska Tryckeri-Aktiebolaget, 1901 [mf ed 1985] – 1mf – 9 – 0-8370-3600-3 – (in swedish) – mf#1985-1600 – us ATLA [225]

De jure et justitia / Gariepy, Charles-Napoleon – Quebeci [Quebec]: Action sociale, 1913 – 5mf – 9 – 0-665-86321-7 – (in latin) – mf#86321 – cn Canadiana [170]
De jure et officiis bellicis et disciplina militari, libri 3 / Ayala, Balthazar – Washington: Carnegie Endowment. 2v. 1912 – 9mf – 9 – $13.50 – mf#LLMC 88-108 – us LLMC [355]
De justini martyris scriptis et doctrina / Otto, Johannes Carl Theodor – Jenae: F Maukium, 1841 – 1mf – 9 – 0-7905-5854-8 – (incl bibl ref) – mf#1988-1854 – us ATLA [240]
De K, Emma *see* Holly grange
De kartuize sint-anna-ter-woestijne 1350-1792 / Ydewalle, St d' – Brugge, 1945 – €39.00 – ne Slangenburg [241]
De Kiewiet, C W (Cornelius William) *see* Imperial factor in south africa
De Kiewiet, Cornelius W *see* Imperial factor in south africa
De Kiewiet, Cornelius William *see* History of south africa
De Kock, Victor *see* Those in bondage
De kometen van de jaren 1556, 1264, en 975 : en hare vermeende identitert / Hoek, Martin – 's Gravenhage: De Gebroeders van Cleef 1857 [mf ed 1998] – 1r [pl/ill] – 1 – (incl bibl ref) – mf#film mas 28292 – us Harvard [520]
De konst des wijsheid / Gracian – Den Haag, 1696 – €12.00 – ne Slangenburg [100]
De koulikoro a tombouctou a bord du "mage", 1889-1890 / Jaime, G – Paris: E Dentu, [1892] – 1 – us CRL [960]
De Koven, James *see* Sermons
De kroon aller koningen / Bukhari von Johor – Batavia: Lands drukkerij 1827 – (= ser Whsb) – 5mf – 9 – €60.00 – mf#Hu 317 – gw Fischer [490]
De Kruif, Paul *see* Our medicine men
De l h 'a 3 h / Dreyfus, Abraham – Paris, France. 1891 – 1r – us UF Libraries [440]
De la antiguedad, y universalidad del bascuenze en espana : de sus perfecciones, y ventajas sobre otras muchas lenguas, demonstracion previa al arte, que se dara a luz desta lengua / Larramendi, Manuel de – Salamanca: Garcia 1728 – (= ser Whsb) – 2mf – 9 – €30.00 – mf#Hu 043 – gw Fischer [440]
De la Beche, Henry Thomas *see* The geological observer
De la capacite presidentielle sous le regime parli / Thoby, Armand – Port-Au-Prince, Haiti. 1888 – 1r – us UF Libraries [972]
De la caracterologia individual a la colectiva / Frutos Cortes, Eugenio – Badajoz: Imp. Dip. Provincial, 1960. Sep. Revista de Estudios Extremenos – sp Bibl Santa Ana [946]
De la carpa a la gloria / Giavi – Remedios, Cuba. 1959 – 1r – us UF Libraries [972]
De la charge des gouverneurs des places / Ville, A – Amsterdam, 1674 – 6mf – 9 – mf#OA-202 – ne IDC [720]
De la chine, ou description generale de cet empire : redigee d'apres les memoires de la mission de pe-kin; ouvrage qui contient la description topographique des quinze provinces de la chine... / Grosier, Jean B – Paris 1818-20 [mf ed Hildesheim 1995-98] – 7v on 22mf – 9 – €220.00 – 3-487-27572-4 – gw Olms [951]
De la Condamine *see* Neue reisen nach guiana, peru und durch das suedliche amerika
De la condition internationale de l'egypte depuis la declaration anglaise de 1922 / Liu, M C – Lyon, 1925 – 2mf – 9 – mf#ILM-2128 – ne IDC [956]
De la condition juridique des etrangers au cambodge / Hoeffel, Ernest – [Strasbourg: Charles Hiller 1932] [mf ed 1989] – 1r with other items – 1 – (with bibl) – mf#mf-10289 seam reel 007/04 [§] – us CRL [959]
De la condition legale des societies etrangeres dans l'empire ottoman / Polyvios, P J – Paris, 1913 – 3mf – 9 – mf#ILM-3515 – ne IDC [956]
De la condition legale du culte israelite en france et en algerie / Baugey, Georges – Paris, France. 1899 – 1r – us UF Libraries [939]
De la condition sociale des femmes au temps present / Rameau, Marcel – Asnieres: La Revue des Independants, 1927 – (= ser Les femmes [coll]) – 2mf – 9 – mf#7332 – fr Bibl Nationale [360]
De la connaissance de dieu *see* Guide to the knowledge of god
De la contre-revolution en france : ou de la restauration de l'ancienne noblesse et des anciennes superiorites sociales dans la france nouvelle / Ganilh, Charles – Paris [u a] 1823 [mf ed Hildesheim 1995-98] – 1v on 2mf – 9 – €60.00 – 3-487-26330-0 – gw Olms [944]
De la croyance a l'immaculee conception de la sainte vierge *see* The impossibility of the immaculate conception as an article of faith
De la croyance en dieu / Piat, Clodius – 3e ed. Paris: Felix Alcan, [1908?] – 1mf – 9 – 0-8370-5459-1 – (incl bibl ref) – mf#1985-3459 – us ATLA [210]
De la Cruz Napoli, Jose *see* Changes in blood resistivity over a sub-maximal exercise bout
De la democratie en amerique / Tocqueville, Alexis de – 13e rev corr augm ed. Paris: Pagnerre. 2v. 1850 [mf ed 1985] – 2v on 1mf – 9 – (with app) – mf#49751 – cn Canadiana [320]
De la democratie en france (janvier 1849) / Guizot, Francois Pierre Guillaume – Paris 1849 [mf ed Hildesheim 1995-98] – 1v on 1mf – 9 – €40.00 – 3-487-26010-7 – gw Olms [944]
De la democratie representative / Dorsainvil, Jean Baptiste – Port-Au-Prince, Haiti. 1900 – 1r – us UF Libraries [972]
De la dictadura al comunismo / Saldarriaga Betancur, Juan Manuel – Medellin, Colombia. 1962 – 1r – us UF Libraries [972]
De la distribution des maisons de plaisance et de la decoration desedifices en general / Blondel, Jacques Francois – Paris. 2v. 1737-1738 – 16mf – 9 – mf#O-153 – ne IDC [720]
De la division du travail social : etude sur l'organisation des societes superieures / Durkheim, Emile – Paris: F Alcan, 1893 – 2mf – 9 – 0-7905-6993-0 – mf#1988-2993 – us ATLA [301]
De la filosofia morale libri dieci : sopra li dieci libri de l'ethica d'aristotile / Figliucci, Felice – in Roma: Impresso Vincenzo Valgrisi [1551] – (= ser Ethics in the early modern period) – 6mf – 9 – mf#pl-345 – ne IDC [170]
De la fin du 2e siecle a la paix constantinienne (he2) – Paris, 1935 – (= ser Histoire de l'eglise (he)) – €25.00 – ne Slangenburg [243]
De la flore pharaonique / Schweinfurth, G – n.p, 1883 – 1mf – 9 – mf#13062 – ne IDC [956]
De la france et des etats-unis : ou de l'importance de la revolution de l'amerique pour le bonheur de la france des rapports de ce royaume et des etats-unis, des avantages reciproques qu'ils peuvent retirer de leurs liaisons de commerce... / Claviere, Etienne – Paris 1791 [mf ed Hildesheim 1995-98] – 1v on 3mf – 9 – €90.00 – 3-487-27066-8 – gw Olms [327]
De la grece moderne et de de ses rapports avec l'antiquite / Quinet, Edgar – Paris [u.a.] 1830 [mf ed Hildesheim 1995-98] – (= ser Fbc) – 3mf – 9 – €90.00 – 3-487-29076-6 – gw Olms [949]
...De la gverra di giustiniano imperatore contra i persiani... / Procopius of Caeserea – Vinegia, 1547 – 5mf – 9 – mf#H-8278 – ne IDC [956]
De la lamentacion de la virgen maria / Bravo Riesco, Agustin – Sp. Calatrava, 1954 – 1 – sp Bibl Santa Ana [240]
De la latinite des sermons de saint augustin / Regnier, Louis-Adolphe – Paris: Hachette, 1886 [mf ed 1990] – 1mf – 9 – 0-7905-7073-4 – (incl bibl ref) – mf#1988-3073 – us ATLA [240]
De la legislacion romana en las relaciones con la de los pueblos europeos / Chaves y Manso, Rafael – Madrid, 1851 – 1 – sp Bibl Santa Ana [350]
De la legislation fonciere ottomane / Padel, Wilhelm – Paris: Pedone, 1904. 350p. LL-12029 – 1 – us L of C Photodup [340]
De la litterature allemande / Frederick 2, King of Prussia; ed by Geiger, Ludwig – Heilbronn: Henninger, 1883 [mf ed 1993] – (= ser Deutsche litteraturdenkmale des 18. und 19. jahrhunderts 16) – xxx/37p – 1 – (french text. int in german. int by ludwig geiger) – mf#8676 reel 2 – us Wisconsin U Libr [430]
...De la longa and astra guerra de gothi... / Procopius of Caeserea – Venetia, 1544 – 6mf – 9 – mf#H-8273 – ne IDC [956]
De la metrica en ruben dario / Sanchez, Juan Francisco – Ciudad Trujillo, Dominican Republic. 1955 – 1r – us UF Libraries [972]
De la monarchie pontificale a propos du livre de mgr l'eveque de sura / Gueranger, Prosper – Paris: V Palme; Au Mans: Leguicheux-Gallienne, 1870 [mf ed 1990] – 1mf – 9 – 0-7905-6291-X – mf#1988-2291 – us ATLA [240]
De la monarchie pontificale a propos du livre de mgr l'eveque de sura *see* Die hoechste lehrgewalt des papstes
De la monarchie selon la charte / Chateaubriand, Francois R de – Bruxelles 1816 [mf ed Hildesheim 1995-98] – 1v on 2mf – 9 – €60.00 – 3-487-26324-6 – gw Olms [323]
De la mort de theodose a l'election de gregoire le grand (he4) – Paris, 1937 – (= ser Histoire de l'eglise (he)) – €31.00 – ne Slangenburg [240]
de la Motte Fouque *see* Frauentaschenbuch
De la musique religieuse : les congres de malines [1863 et 1864] et de paris [1860] et la legislation de l'eglise sur cette matiere / Vroye, Theodore Joseph de – Paris: Lethielleux [etc] 1866 [mf ed 19–] – 6mf – 9 – mf#fiche 52 – us Sibley [780]
De la nation *see* Courrier de la colere
De la nationalite dans l'empire ottoman specialement en egypte / Arminjon, P – Paris, 1901 – 1mf – 9 – mf#ILM-3435 – ne IDC [956]
De la nationalite suivant la legislation serbe / Peri'c, Zivojin M – Paris: Marchal & Billard 1900 [mf ed 1984] – 1r – 1 – (with: les croates sous le joug magyar / h hinkovitch [paris: plon-nourrit 1915]) – mf#1010 – us Wisconsin U Libr [323]
De la natural historia de las indias. sumario de historia natural de las indias. con un estudio preliminar y notas por enrique alvarez lopez / Fernandez de Oviedo Valdes, Gonzalo – Madrid: Editorial S., 1942 – 1 – sp Bibl Santa Ana [970]
De la nature de nos idees et de l'ontologisme en general / Ubaghs, Gerard Casimir – Tirlemont: P-J Merckx, 1854 – 1mf – 9 – 0-524-00407-2 – mf#1989-3107 – us ATLA [110]
De la necessite de l'assolement dans la culture du tabac / Chevalier, Omer – Ottawa: Ministere de l'agriculture, 1909 – 1mf – 9 – 0-665-65833-8 – mf#65833 – cn Canadiana [630]
De la necessite d'une dictature / Cottu, Charles – Paris 1830 [mf ed Hildesheim 1995-98] – 1v on 1mf – 9 – €40.00 – 3-487-26051-4 – gw Olms [320]
De la neutralidad vigilante a la mediacion con gua... / El Salvador Secretaria De Informacion – San Salvador, El Salvador. 1954 – 1r – us UF Libraries [972]
De la paix constantinienne a la mort de theodose (he3) – Paris, 1936 – (= ser Histoire de l'eglise (he)) – €27.00 – ne Slangenburg [240]
De la passion du jeu, de l'infidelite des joueurs, et de leurs ruses : ouvrage anecdotique / Aureville, J A d' – Paris 1824 [mf ed Hildesheim 1995-98] – 1v on 2mf – 9 – €60.00 – 3-487-25875-7 – gw Olms [364]
De la pirotechnia libri 10... / Biringuccio, V – Veneto, 1540 – 5mf – 9 – mf#O-1031 – ne IDC [700]
De la police de paris, de ses abus, et des reformes dont elle est susceptible : avec documens anecdotiques et politiques, pour servir a l'histoire judiciaire de la restauration / Claveau, Antoine G – Paris 1831 [mf ed Hildesheim 1995-98] – 1v on 4mf – 9 – €120.00 – 3-487-25896-X – gw Olms [360]
De la predestination eternelle de dieu : par laquelle les uns sont eleuz a salut, les autres laissez en leur condemnation / [Calvin, J] – [Geneve: De l'imprimerie Jehan Crespin], 1552 – 3mf – 9 – mf#CL-57 – ne IDC [240]
De la primaute en l'eglise / Blondel, D – Geneve, 1641 – 23mf – 9 – mf#CA-115 – ne IDC [241]
De la prostitution dans la ville de paris?? : consideree sous le rapport de l'hygiene publique, de la morale et de l'administration / Parent-Duchatelet, Alexandre Jean Baptiste – Paris, Londres: J B Bailliere, 1836 – (= ser Les femmes [coll]) – 2v on 13mf – 9 – mf#6077-78 – fr Bibl Nationale [360]
De la prueba en derecho / Rocha, Antonio – Bogota, Colombia. 1951 – 1r – us UF Libraries [972]
De la recherche de la paternite naturelle / Fanfant, J E – Port-Au-Prince, Haiti. 1934? – 1r – us UF Libraries [972]
De la rehabilitation de la race noire par la repub... / Price, Hannibal – Port-Au-Prince, Haiti. 1900 – 1r – us UF Libraries [972]
De la religion : consideree dans sa source, ses formes et ses developpements / Constant, Benjamin – Paris: Pichon et Didier, 1830-1831 – 6mf – 9 – 0-524-03829-5 – (incl bibl ref) – mf#1990-3267 – us ATLA [200]
De la republica a la dictadura / Lleras Restrepo, Carlos – Bogota, Colombia. 1955 – 1r – us UF Libraries [972]
De la repvblique des turcs : and la ou l'occasion s'offrera, des meurs and loys de tous muhamedistes... / Postel, G – Poitiers, [1565]. 3 pts – 4mf – 9 – mf#H-8303 – ne IDC [956]
De la revolucion al orden nuevo / Azula Barrera, Rafael – Bogota, Colombia. 1956 – 1r – us UF Libraries [972]
De la revolucion y de las cubanas / Rodriguez Garcia, Jose Antonio – Habana, Cuba. 1930 – 1r – us UF Libraries [972]
De la rochefaucauld [sic] liancourt : reisen in den jahren 1795, 1796 und 1797 durch alle an der see belegenen staaten der nordamerikanischen republik... – Hamburg: Bei Benjamin Gottlob Hoffmann. 1799 [mf ed 1983] – (= ser Neuere geschichte der see- und land-reisen 9) – 3v on 1mf – 9 – 0-665-37088-1 – (trans of voyage dans les etats-unis d'amerique, fait en 1795, 1796, et 1797) – mf#37088 – cn Canadiana [917]
De la saincte cene de nostre seigneur jesus / Farel, Guillaume – [Geneve], Crespin, 1553 – 3mf – 9 – mf#PFA-160 – ne IDC [240]
De la senegambie francaise / Carrere, Frederic & Holle, Paul – Paris: F Didot, 1855 – 1 – us CRL [960]
De la sevle foy en christ ivstifiante / Bullinger, Heinrich – Lyon, Iean Savgrain, 1565 – 1mf – 9 – mf#PBU-145 – ne IDC [240]
De la sombra / Read, Horacio – Ciudad Trujillo, Dominican Republic. 1959 – 1r – us UF Libraries [972]
De la tierra y el cielo. poesias liricas y religiosas. caceres, 1925 / Lopez Cruz, Lorenzo – Madrid: Razon y Fe, 1930 – 1 – sp Bibl Santa Ana [810]
De la treshevrevse victoire des chrestiens a l'encontre de l'armee du grand turc – Anuers, 1571 – 1mf – 9 – mf#H-8176 – ne IDC [956]
De la verite de la religion chrestienne / Plessis-Mornay, P du – Anvers, 1581 – 10mf – 9 – mf#PRS-165 – ne IDC [240]
De la vertu et usage du ministere de la parolle de dieu et des sacremens / Viret, P – [Geneve, Gerard], 1548 – 8mf – 9 – mf#PFA-197 – ne IDC [240]
De la vida, sentires y saberes de d. jose moreno nieto / Rodriguez y Rivero, Jose – Sevilla: Imp. J. Peralto, 1929, 1 lam – sp Bibl Santa Ana [920]
De la vie intime des dogmes et de leur puissance d'evolution = The vitality of christian dogmas and their power of evolution / Sabatier, Auguste – London: Adam & Charles Black, 1898 – 1mf – 9 – 0-8370-5018-9 – (in english) – mf#1985-3018 – us ATLA [240]
De laatste dagen van het pauselijk leger / Gerlache, Eugene de – Mechlin, Belgium? : H Dessain, 1870 – 1mf – 9 – mf#58765 – cn Canadiana [940]
De laatste jaren in medan / Romme, RFEM – Amsterdam, 1962 – 1mf – 9 – mf#SE-1442 – ne IDC [950]
De l'abolition des droits feodaux et seigneuriaux du canada : et sur le meilleur mode a employer pour accorder une juste indemnite aux seigneurs / Dumesnil, Clement – Montreal: Impr J Starke & cie, 1849 [mf ed 1983] – 1mf – 9 – mf#SEM105P247 – cn Bibl Nat [323]
De l'abolition du regime feodal en canada : et de l'indemnite due aux seigneurs pour la suppression des droits et devoirs feodaux...et ouverte a quebec le quatre septembre 1855 – Quebec: typographie d'Augustin Cote, 1855 [mf ed 1983] – 2mf – 9 – mf#SEM105P275 – cn Bibl Nat [323]
De l'admissibilite de la preuve par temoins en droit civil / Dorion, Charles-Edouard – Montreal: Whiteford & Theoret, 1894 – 2mf – 9 – 0-665-91672-8 – mf#91672 – cn Canadiana [347]
De lagardes ausgabe der arabischen sbersetzung des pentateuchs / Hughes, J C – Leipzig, J C Hinrich, 1920 – 1mf – 9 – mf#NE-20120 – ne IDC [956]
De lagardes ausgabe der arabischen uebersetzung der genesis / Davidson, H S – Leipzig, 1919 – 1mf – 9 – mf#NE-20115 – ne IDC [956]
De lagarde's ausgabe der arabischen uebersetzung des pentateuchs (cod leiden arab 377) / Hughes, John Caleb – Leipzig: August Pries, 1914 – 1mf – 9 – 0-524-08203-0 – mf#1992-1172 – us ATLA [221]
De l'amitie commentaire de saint thomas sur les livres 8 et 9 de l'ethique a nicomaque d'aristote – [Sherbrooke]: Faculte des arts, Seminaire de Sherbrooke, [196-] [mf ed 1999] – (= ser In decem libros ethicorum aristotelis ad nicomachum expositio) – 2mf – 9 – (trans of: in decem libros ethicorum aristotelis ad nicomachum expositio) – mf#SEM105P3109 – cn Bibl Nat [170]
De lanae in antiquorum ritibus usu / Pley, Jakob – Giessen: A Toepelmann, 1911 – (= ser Religionsgeschichtliche Versuche und Vorarbeiten) – 1mf – 9 – 0-524-00959-7 – (incl bibl ref) – mf#1990-2182 – us ATLA [930]
De Lancey, Edward Floyd *see*
– The capture of mount washington
– Marshall s bidwell, a memoir
– New york and admiral sir peter warren at the capture of louisbourg, 1745
– Origin and history of manors in the province of new york and in the county of westchester
De Land, Helen Parce *see* Story of de land and lake helen, florida
De lapsis (fp21) / Cyprianus (Cyprian, Saint); ed by Martin, J – Bonn, 1932 – (= ser Florilegium patristicum (fp)) – 2mf – 8 – €5.00 – ne Slangenburg [240]
De Lara, D Laurent *see* Elementary instruction in the art of illuminating and missal painting on vellum
De l'art de regner, au roy / Moyne, P le – Paris: Sebastien Cramoisy, & Sebastien Mabre Cramoisy, 1665 – 9mf – 9 – mf#O-1351 – ne IDC [090]
De l'art des devises / Moyne, P le – Paris: Antoine Dezallier, 1688 – 7mf – 9 – mf#O-86 – ne IDC [090]

De l'art des devises / Moyne, P le – Paris: Sebastien Cramoisy, & Sebastien Mabre Cramoisy, 1666 – 11mf – 9 – mf#O-668 – ne IDC [090]
De l'art en allemagne / Fortoul, H – Paris, 1842. 2v – 14mf – 9 – mf#OA-131 – ne IDC [720]
De las andanzas de unamuno por tierras extremenas / Garcia Blanco, Manuel – Madrid, 1956 – 1 – sp Bibl Santa Ana [946]
De las cinco ordenes de architectura / Vignola, J – Madrid, 1593 – 1mf – 9 – sp Cultura [720]
De las deudas amortizables...cupones / Bravo Murillo, Juan – 1864 – 9 – sp Bibl Santa Ana [336]
De las ordenes militares de calatrava, alcantara y montesa / Guillamas, Manuel de – 1852 – 9 – sp Bibl Santa Ana [355]
De las siete palabras...de manuel benitez sanchez cortes / Perez Embrid, Florentino – Madrid: Arbor, 1949 – 1 – sp Bibl Santa Ana [240]
De l'atlantique au fleuve congo / Sautter, Gilles – Paris, France. v1-2. 1966 – 1r – us UF Libraries [960]
De l'atlantique au niger par le foutah-djallon : carnet du voyage / Olivier, Aime, Vicomte de Sanderval – Paris: P Ducrocq, 1882 – 1 – us CRL [960]
De laudando in maria deo / Oecolampadius, J – [Basileae, Andreas Cratander, 1519] – 1mf – 9 – mf#PBU-351 – ne IDC [240]
De l'autorite imperiale en matiere religieuse a byzance / Gasquet, A – Paris: Ernest Thorin, 1879 – 1mf – 9 – 0-8370-8181-5 – (incl bibl ref) – mf#1986-2181 – us ATLA [240]
De l'avenir des peuples catholiques / Laveleye, Emile de – Paris: G Fischbacher; Geneve: Stapelmohr, [1898?] – 1mf – 9 – 0-8370-9004-0 – mf#1986-3004 – us ATLA [241]
De laviolencia a la paz / Colombia Ejercito 8 Brigada – Manizales, Colombia. 1965 – 1r – us UF Libraries [972]
De l'ecole d'alexandrie : rapport a l'academie des sciences morales et politiques, precede d'un essai sur la methode des alexandrins et le mysticisme, et suivi d'une traduction de morceaux choisis de plotin / Barthelemy Saint-Hilaire, Jules – Paris: Ladrange, 1845 [mf ed 1992] – 1mf – 9 – 0-524-04186-5 – mf#1990-1225 – us ATLA [180]
De l'economie du salut : etude sur le dogme dans ses rapports avec la morale / Weber, Alfred – Strasbourg: Treuttel et Wurtz, 1864 – 1mf – 9 – 0-8370-5720-5 – mf#1985-3720 – us ATLA [210]
De l'edit de nantes execute selon... / Meynier, B – Paris, 1670 – 2mf – 9 – mf#CA-140 – ne IDC [240]
De l'education / Dupanloup, Felix – 9e ed. Paris: Charles Douniol. 3v. 1872 [mf ed 1986] – 5mf – 9 – 0-8370-9057-1 – (in french. incl bibl ref) – mf#1986-3057 – us ATLA [377]
De l'education des femmes / Choderlos de Laclos, P A F – Paris: L Vanier, 1903 – (= ser Les femmes [coll]) – 2mf – 9 – mf#5039 – fr Bibl Nationale [376]
De Leeuw, Hendrik *see*
– Crossroads of the buccaneers
– Crossroads of the caribbean sea
– Onze west
De l'egalite des deux sexes : discours physique et morale ou l'on voit l'importance de se defaire des prejugez / La Barre, Francois Poullain de – Paris: Jean du Pais, 1676 – (= ser Les femmes [coll]) – 3mf – 9 – mf#11335 – fr Bibl Nationale [306]
De legatione evangelica ad indos capessenda admonitio / Heurnius, I – Lugduni Batavorum, 1618 – 4mf – 9 – mf#PBA-194 – ne IDC [242]
De legibus et consuetudines angliae libri quinque in varios tractatus distincti [rs70] / Henry de Bracton; ed by Twiss, Travers – (= ser The rolls series [rs]) – (v1 1878 €27. v2 1879 v3 1880 v4 1881 €25 ea. v5 1882 €23. v6 1883 €23) – ne Slangenburg [240]
De legibus o tratado de leyes... / Leon, Luis de – Madrid: Arch. Ibero Americano, 1964 – 1 – sp Bibl Santa Ana [946]
De legitima uindicatione christianismi ueri et sempiterni...libri antisophistici tres scripti... / Bibliander, T – Basileae, [Ioannes Oporinus, 1553] – 3mf – 9 – mf#PBU-585 – ne IDC [240]
De l'embaumement avant et apres jesus christ / Reutter de Rosemont, Louis – Paris: Vigot Freres [1912?] [mf ed 1986] – 1r [ill] – 1 – mf#1763 – us Wisconsin U Libr [390]
De l'empire du bresil : considere sous ses rapports politiques et commerciaux / LaBeaumelle, Victor L de – Paris 1823 [mf ed Hildesheim 1995-98] – 1v on 2mf – 9 – €60.00 – 3-487-26881-7 – gw Olms [972]
De l'enfance a la jeunesse / Marcelin, Emile – Port-Au-Prince, Haiti. 1934 – 1r – us UF Libraries [972]
De l'enseignement en haiti / Lubin, Maurice Alcibiade – Port-Au-Prince, Haiti. 1947 – 1r – us UF Libraries [972]
De leon first baptist church. de leon, texas : church records – 1877-1942 – 1 – 59.04 – us Southern Baptist [242]
De l'etablissement en canada de la fabrication du sucre de betterave : considerations pratiques sur les nombreux avantages qui en seraient le resultat au point de vue de l'agriculture / Bran, Telesphore – Montreal: E Senecal, 1876 – 1mf – 9 – mf#24134 – cn Canadiana [635]
De l'etat de la france : present et a venir / Calonne, Charles Alexandre de – Londres 1790 [mf ed Hildesheim 1995-98] – 1v on 3mf – 9 – €90.00 – 3-487-26183-9 – gw Olms [944]
De l'etat des partis dans les chambres : et des alliances possibles entre eux / Saint-Chamans, Auguste L de – Paris 1828 [mf ed Hildesheim 1995-98] – 1v on 2mf – 9 – €60.00 – 3-487-26136-7 – gw Olms [323]
De l'etat present et de l'avenir de l'islam : six conferences / Montet, Edouard Louis – Paris: P Geuthner, 1911 [mf ed 1991] – 1mf – 9 – 0-524-01850-2 – (in french) – mf#1990-2685 – us ATLA [260]
De leyte de cavalleros...cavallos / Maestre de San Juan, Lucas – 1735 – 9 – sp Bibl Santa Ana [340]
De l'habitation du saint-esprit dans les aames justes d'apres la doctrine de saint thomas d'aquin / Froget, Barthelemy – Paris: P Lethielleux, [1898?] – 1mf – 9 – 0-7905-7822-0 – (incl bibl ref) – mf#1989-1047 – us ATLA [241]
De l'histoire de la vulgate en france : lecon d'ouverture / Berger, Samuel – Paris: Fischbacher, 1887 – 1mf – 9 – 0-8370-1784-X – (incl bibl ref) – mf#1987-6172 – us ATLA [220]
De l'hygiene en haiti / Perpignand-Lafontant, – Paris, France. 1896 – 1r – us UF Libraries [972]
De l'hypotheque legale de la femme mariee / Dufoussat, Henry – Paris: Rousseau, 1898 – (= ser Les femmes [coll]) – 4mf – 9 – mf#11017 – fr Bibl Nationale [305]
De libero arbitrio hominis, integro corrvpto in rebus externis mortvo in rebus spiritualibus renato doctrina solide ac methodice ex verbo dei tradita et explicata / Wigand, J – Vrsellis, 1572 – 5mf – 9 – mf#TH-1 mf 1529-1533 – ne IDC [242]
De libero arbitrio...elenchus / Oecolampadius, J – Basileae, Thomas Volffius, 1524 – 4mf – 9 – mf#PBU-360 – ne IDC [240]
De libris revolutionum...nicolai copernici... narratio prima...,una cum encomio borussiae scripta / Rheticus, G I – Basileae, Robert Winter, 1541 – 2mf – 9 – mf#PBU-496 – ne IDC [240]
De liggeren en andere historische archieven der antwerpsche sint lucas gilde / ed by Rombouts, P & Lerius, T van – Antwerpen, 1872-1876. 2v – 41mf – 9 – mf#O-124 – ne IDC [700]
De l'immaculee conception de la sainte vierge : examen critique des articles du journal des debats / Sisson, A – Paris: Libr de litterature religieuse de Ch Douniol, 1854 [mf ed 1986] – 1mf – 9 – 0-8370-8310-9 – (in french) – mf#1986-2310 – us ATLA [241]
De l'independance des rites maconniques : ou, refutation des pretentions du g o de france: sur le rit ecossais ancien et accepte – Paris: F Setier 5827 [1827] – 1mf – 9 – mf#vrl-1 – ne IDC [366]
De l'influence d'aristote et de ses interpretes sur la decouverte du nouveau monde / Jourdain, Charles – Paris: P Dupont, 1861 [mf ed 1984] – 1mf – 9 – 0-665-45219-5 – (incl bibl ref) – mf#45219 – cn Canadiana [180]
De l'influence eucharistique sur l'apostolat des premiers missionnaires au canada : conference lue au congres eucharistique de montreal, septembre 1910 / Emard, Joseph-Medard – [Montreal?: s.n, 1910?] [mf ed 1994] – 1mf – 9 – 0-665-73216-3 – mf#73216 – cn Canadiana [240]
De lingua et literis veterum aegyptiorum : cum permultis tabulis lithographicis literas aegyptiorum...accedunt grammatica atque glossarium aegyptiacum. pars prima cum imagine vitaque spohnii / Spohn, Friedrich August Wilhelm; ed by Seyffarth, Gustav – Lipsiae: Weidemann 1825 – (= ser Whsb) – 2mf – 9 – €30.00 – mf#Hu 372 – gw Fischer [470]
De Lisle, Edwin *see*
– Life and letters of ambrose phillipps de lisle
De Lisser, Herbert George *see*
– In jamaica and cuba
– Twentieth century jamaica
De literarum ludis recte aperiendis / Sturm, J – Argentorati, 1538 – 1mf – 9 – mf#PPE-138 – ne IDC [240]
De liturgia gallicana libri 3 / Mabillon, Jean – Parisiis, 1685 – 9mf – 8 – €18.00 – ne Slangenburg [242]
De lo mas hondo / Galvez Y Del Monte, Wenceslao – Habana, Cuba. 1925 – 1r – us UF Libraries [972]
De lo pasado : estampas breves / Pineda, Ramon – Badajoz: Imp. Ind. Graficas, 1958 – sp Bibl Santa Ana [946]
De l'organisation – Conakry: Imp nationale Patrice Lumumba, [1970] – us CRL [079]
De l'organisation judiciaire en haiti / Justin, Joseph – Havre, France. 1910 – 1r – us UF Libraries [972]
De l'orgue et de son architecture / Cavaille-Coll, Aristide – 2nd rev augm ed, Paris: Ducher & Sie 1872 [mf ed 19–] – 1r – 1 – mf#film 1299 – us Sibley [780]
De l'origine des cultes arcadiens : essai de methode en mythologie grecque / Berard, Victor – Paris: Thorin, 1894 – (= ser Bibliotheque des ecoles francaises d'athenes et de rome) – 1mf – 9 – 0-524-01166-4 – (incl bibl ref) – mf#1990-2242 – us ATLA [250]
De l'origine des indiens du nouveau monde et de leur civilisation / Dabry de Thiersant, Philibert – Paris: E Leroux, 1883 – 4mf – 9 – mf#02254 – cn Canadiana [305]
De l'origine du pentateuque *see* Introduction a la critique generale de l'ancien testament. de l'origine du pentateuque
De l'origine et de l'etablissement des mouvements astronomiques / Lagrange, Charles – Bruxelles: F Hayez 1878-79 [mf ed 1998] – 2v on 1r [ill] – 1 – (extracts fr v42 of: memoires couronnes et memoires des savants etrangers; incl bibl ref) – mf#film mas 28292 – us Harvard [520]
De l'Orme, Ph *see*
– Architecture...
– Le premier tome de l'architecture
– Le premier tome de l'architectvre de philibert de l'orme conseillier et avmous-nier ordinaire du roy...
– Uvelles inventions pour bien bastir et a petits fraiz...
De los caminos o vias militares fabricadas por los romanos en espana / Morales, Ambrosio – Madrid: Oficina de D. Benito Cano, 1792 – 1 – sp Bibl Santa Ana [946]
De los hechos de la conquista durante la fundacion / Briceno Perozo, Ramon – Merida, Mexico. 1955 – 1r – us UF Libraries [972]
De los maleficios y los demonios / Montoto, Jose Maria – 1884 – 1 – sp Bibl Santa Ana [946]
De los menores descalcos de la mas estrecha observancia regular de n s p s francisco en esta nueva-espana *see* Constituciones de la provincia de san diego de mexico / de los menores descalcos de la mas estrecha observancia regular de n s p s francisco en esta nueva-espana
De los nombres atribuidos a trujillo / Acedo, Federico – Caceres: Tip. Enc. y Lib. de N.M. Jimenez, 1900 – 1 – sp Bibl Santa Ana [946]
De l'oubanghi a fachoda : marchand et la mission congo-nil. ouvrage orne de gravures / Poirier, Jules – Paris: J Lefort, [1899] – 1 – us CRL [960]
De l'universalite du deluge / Schoebel, Charles – Paris: Benjamin Duprat, 1858 – 1mf – 9 – 0-8370-4634-3 – (incl bibl ref) – mf#1985-2634 – us ATLA [220]
De macario magnete et scriptis ejus / Duchesne, Louis – Parisiis: Fr. Klincksieck, 1877 – 1mf – 9 – 0-7905-4414-8 – (incl bibl ref) – mf#1988-0414 – us ATLA [240]
De madrid a lisboa / Diaz Perez, Nicolas – 1887 – 9 – sp Bibl Santa Ana [914]
De magisterio vivo et traditione / Bainvel, Jean Vincent – Paris: Gabriel Beauchesne, 1905 – 1mf – 9 – 0-7905-9124-3 – (incl bibl ref) – mf#1989-2349 – us ATLA [240]
De manichaeismo apud latinos quinto sextoque saeculo : atque de latinis apocryphis libris / Dufourcq, Albert – Paris: A. Fontemoing, 1900 – 1mf – 9 – 0-7905-6226-X – (incl bibl ref) – mf#1988-2226 – us ATLA [240]
De manichaeismo renovato / Wigand, J – Lipsiae, 1587 – 7mf – 9 – mf#TH-1 mf 1515-1521 – ne IDC [242]
De massua a saati : narrazione della spedizione italiana del 1888 in abissinia – Milan, Treves, 1888 – 1 – (con un' appendice contenente il testo completo del libro verde presentado al parlamento, la relazione ufficiale sul combattimento di saganeiti e tutte le note crispi e goblet sull' incidente di massua) – us CRL [960]
...De Medendis Humani Corporis. Enchiridion Vulgo Veni Mecum... / Bayro, P – Coimbra, 1689 – 14mf – 9 – sp Cultura [610]
De mente concilii viennensis in definiendo dogmate unionis animae humanae cum corpore : deque unitate formae substantialis in homine iuxta doctrinam s thomae praemissa theoria scholastica de corporum compositione / Zigliara, Thoma Maria – Romae: SC de Propaganda Fide, 1878 – 1mf – 9 – 0-524-00236-3 – mf#1989-2936 – us ATLA [241]
De metallicis libri tres andrea caesalpino aretino / Cesalpino, Andrea – Noribergae: Recusi, curante Conrado Agricola 1602 [mf ed 1980] – 1r – 1 – (with: dispvtationvm de medicina nova philippi paracelsi para prima / erastus, thomas) – mf#726 – us Wisconsin U Libr [660]
De mi album de firma. semblanzas de ilutres varones / Sanchez Arjona, Vicente – Sevilla: Imprenta Zambrano, Tomos 1-3. 1956 – 1 – sp Bibl Santa Ana [810]
De mi barrio y otros cuentos / Carrera, Carlos – Guatemala, 1965 – 1r – us UF Libraries [972]
De mi lira popular / Rodriguez V, Severino A – Ciudad Trujillo, Dominican Republic. 1952 – 1r – us UF Libraries [972]
De mi vida; memorial politicas / Reyes, Rodolfo – Madrid. 1929-30. 2 vols – 1 – us CRL [320]
De mi vieja extremadura (paginas montanchegas) / Galan y Galan, FG – Sevilla: Tip. Divina Pastora, 1935 – 1 – sp Bibl Santa Ana [946]
De mi yo, poemario / Labarthe, Pedro Juan – Mexico City? Mexico. 1956 – 1r – us UF Libraries [972]
De michael et androico palaeologis libri 13 (cshb24,25) / Georgii Pachymeris; ed by Bekkerus, Imm – Bonnae. v1-2. 1835 – (= ser Corpus scriptorum historiae byzantinae (cshb)) – €56.00 – ne Slangenburg [243]
De michaelis serveti doctrina : commentationum dogmatico-historicam / Puenjer, Georg Christian Bernhard – Jenae: Sumptibus Hermanni Dufft, 1876 – 1mf – 9 – 0-7905-6944-2 – mf#1988-2944 – us ATLA [240]
De middellandsche afrikaander – Cradock [South Africa]: White and Boughton, [jan 10-dec 21 1928] (wkly) – 2r – 1 – us CRL [079]
De militia eqvestri antiqva et va ad regem philippum 4 libri qvinqve / Hugo, H – Antverpiae, 1630 – 4mf – 9 – mf#OA-151 – ne IDC [720]
De militia romana libri quique... / Lipsius – Antwerpen, 1598 – 4mf – 9 – mf#OA-152 – ne IDC [720]
De Mille, James *see*
– Among the brigands
– The babes in the wood
– Le baron americain
– Behind the veil
– The boys of grand pre school
– A castle in spain
– A comedy of terrors
– Cord and creese
– The cryptogram
– The dodge club
– The early english church
– The elements of rhetoric
– Fire in the woods
– Helena's household
– James de mille's works, vol 1
– The living link
– Lost in the fog
– Old garth
– An open question
– Picked up adrift
– The seven hills
– A strange manuscript found in a copper cylinder
– The treasure of the seas
– A week at forestdale being a summer idyll
– The winged lion
De minnende siele gheschoncken aen alle lief-hebbers godts voor een nieuw-iaer / [Smidt, Ae de] – t'Antwerpen: Michiel Cnobbaert, 1665 – 1mf – 9 – mf#O-3167 – ne IDC [090]
De miraculis libri duo : formae tplila 44 / Cluniacensis, Petrus – 1988 – (= ser ILL – ser a; Cccm 83) – 5mf+58p – 9 – €40.00 – 2-503-63832-5 – be Brepols [400]
De mis cantares ineditos / Sanchez Arjona, Vicente – Sevilla: Imprenta Carlos Acuna, Tomo 1. 1948 – 1 – sp Bibl Santa Ana [810]
De mis ocho mil conetos / Sanchez Arjona, Vicente – Sevilla: Imprenta Alvarez, 1955 – 1 – sp Bibl Santa Ana [810]
De mis poesias ineditas / Sanchez Arjona, Vicente – Sevilla: Imp. Carlos Acuna, Tomo 1. 1948 – 1 – sp Bibl Santa Ana [810]
De mis poesias ineditas / Sanchez-Arjona, Vicente – Sevilla: Imp. Carlos Acuna, Tomo 2. 1949 – 1 – sp Bibl Santa Ana [810]
De mis soledades vengo / Sanchez-Arjona, Vicente – Sevilla: Imprenta Alvarez, 1959 – 1 – sp Bibl Santa Ana [810]
De mis veinte mil cantares / Sanchez-Arjona, Vicente – Sevilla: Editorial Franciscana San Antonio, 1950 – 1 – sp Bibl Santa Ana [810]
De missione legatorvm iaponensium ad romanam curiam... / Saude, E de – Macaensi, 1590 – 5mf – 9 – mf#H-8425 – ne IDC [914]
De modernismo : tractatus et notae canonicae cum actis s. sedis: a 17 aprilis 1907 ad 25 septembris 1910 / Vermeersch, Arthur – Brugis: Sumptibus C. Beyaert, 1910 – 1mf – 9 – 0-8370-8072-X – (incl bibl ref) – mf#1986-2072 – us ATLA [240]

De modernistarum doctrinis : tractatus philosophico-theologicus ad cleri scholarumque penitiorem institutionem / Carbone, Caesar – Romae: Desclee et Socii Editores, 1909 – 2mf – 9 – 0-8370-8969-7 – (incl bibl ref) – mf#1986-2969 – us ATLA [240]

De monachico statu iuxta diciplinam byzantinam / Meester, Placidus de – Romae, 1942 – 18mf – 8 – €35.00 – ne Slangenburg [243]

De Montfaucon, Bern *see* Collectio nova patrum et scriptorum graecorum

De montreal a victoria par le transcontinental canadien : conference faite par honore beaugrand, ancien maire de montreal, devant la chambre de commerce du district de montreal / Beaugrand, Honore – Montreal?: s.n, 1887? – 1mf – 9 – mf#30009 – cn Canadiana [917]

De morali principis institutione : formae tplila 88 / Belvacensis, Vincentius – 1995 – (= ser ILL – ser a; Cccm 137) – 4mf+52p – 9 – €40.00 – 2-503-64372-8 – be Brepols [400]

De morbo postulato sive centicvlari / Lopez de Corella, A – Zaragoza, 1574 – 4mf – 9 – sp Cultura [610]

De mozaeische oorsprong van de wetten in de boeken exodus, leviticus en numeri : lezingen over de moderne schrift-critiek des ouden testaments / Hoedemaker, Philippus Jacobus – Leiden: D A Daamen, 1895 – 1mf – 9 – 0-8370-3611-9 – (incl bibl ref and indexes) – mf#1985-1611 – us ATLA [221]

De multro, traditione et occisione gloriosi karoli comitis flandriarum : formae tplila 83 / Brugensis, Galbertus notarius – 1995 – (= ser ILL – ser a; Cccm 131) – 5mf+54p – 9 – €40.00 – 2-503-64312-4 – be Brepols [400]

De mysteriis liber / Iamblichus Chalcidensis; ed by Gale, Thomas – Oxonii, 1678 – €31.00 – ne Slangenburg [100]

De mysterijs salutiferae passionis et mortis iesv messiae : expositiones historicae libri tres... / Bibliander, T – Basileae: Ioannes Oporinus, [1555] – 2mf – 9 – mf#PBU-586 – ne IDC [240]

De natura novi orbis libri duo et de promulgatione evangeli apud... / Acosta, J – Salamanca, 1589 – 11mf – 9 – sp Cultura [240]

De naturis rerum / Neckham, A; ed by Wright, Thomas – London, 1863 – 14mf – 7 – mf#690 – uk Microform Academic [120]

De navidad : historia de un billete premiado / Iglesia, Alvaro de la – Habana: Imp "La Universal", 1900 (mf ed 19–) – (= ser Cuba literaria; Coleccion de novelas populares; Harvard College Library preservation microfilm program) – 64p – mf#Z-NPW pv18 n3 – us NY Public [830]

De necessaria secessione nostra ab ecclesia romana... / Turrettini, F – Genevae, 1589 – 3mf – 9 – mf#PFA-1 – ne IDC [240]

De necessaria secessione nostra ab ecclesia romana... / Turrettini, F – Lugduni Batavorum, 1696 – 6mf – 9 – mf#PFA-184 – ne IDC [240]

[De nomi corde del monochordo] / Mei, Girolamo – Italy late 16th c [mf ed 19–] – 1r – 1 – mf#film 1083 – us Sibley [780]

De nominum analogia : de conceptu entis / Cajetan, Tommaso de Vio Gaetani – Roma, 1934 – 2mf – 8 – €5.00 – ne Slangenburg [241]

De non habendo pauperum delectu / Oecolampadius, J – Basileae, [Cratander], 1523 – 1mf – 9 – mf#PBU-354 – ne IDC [240]

De nonnulli porphyrii...scholae / Sanchez de las Brozas, Francisco – 1597 – 9 – sp Bibl Santa Ana [450]

De nonullis porpliyry, alriorumq; in dialectica er roribus schola dialecttica / Sanchez de las Brozas, Francisco – Salamanca: Miguel Serrona de Vargas, 1588 – 1 – sp Bibl Santa Ana [946]

De noodzakelijkheid van de instelling eener indische volksvertegenwoordiging met wetgevende macht / Dwidjo Sewojo, M N G – 1mf – 8 – mf#SE-1282 – ne IDC [959]

De norte a sur / Caban Soler, Jose – Washington, DC. 1963 – 1r – us UF Libraries [972]

De nos institutions communales / Narcisse, Franck D – Port-Au-Prince, Haiti. 1919 – 1r – us UF Libraries [972]

De nuestro antano historico / Diaz Vasconcelos, Luis Antonio – Guatemala, 1948 – 1r – us UF Libraries [972]

De nuestro sur remoto / Damiron, Rafael – Ciudad Trujillo, Dominican Republic. 1947 – 1r – us UF Libraries [972]

De nuevo acerca del condado de la gomera / Regulo Perez, Juan – Madrid, s.i., 1956 – 1 – sp Bibl Santa Ana [946]

De nuptiis philologiae et mercurii / Notker, Labeo; ed by Sehrt, Edward Henry & Starck, Taylor – Halle/S: M Niemeyer, 1935 [mf ed 1993] – (= ser Altdeutsche textbibliothek 37) – viii/220p – 1 – (in old high german and latin. incl bibl ref) – mf#8193 reel 3 – us Wisconsin U Libr [430]

De obitu reverendi viri..., philippi melanthonis / Praetorius, Christoph – 1560 – (= ser Mssa) – 1mf – 9 – €20.00 – mfchl 374 – gw Fischer [780]

De oeconomia foederum dei cum hominibus... / Witsius, H – Leovardiae, 1685 – 8mf – 9 – mf#PBA-403 – ne IDC [240]

De officialibus palatii constantinopolitani (cshb37) : et de officiis magnae ecclesiae liber / Codini Curopalatae; ed by Bekkeri, Imm – Bonnae, 1839 – (= ser Corpus scriptorum historiae byzantinae (cshb)) – €17.00 – ne Slangenburg [243]

De officiis ecclesiasticis de jean d'arvranches / Delmare, R Le – Paris, 1923 – 5mf – 8 – €12.00 – ne Slangenburg [241]

De officiis magnae ecclesiae et aulae constantinopo-litanae (cbh18) / Georgii Codini; ed by Gretser, J & Goar, J – Parisiis, 1648 – (= ser Corpus byzantinae historiae (cbh)) – €38.00 – ne Slangenburg [243]

De officiis, paradoxa, de amicitia, de senectute *see* De amicitia...

De omni rerum fossilium genere, gemmis, lapidibus, metallis,... / Gessner, C – Tiguri: I Gesnerus, 1565. 8v – 14mf – 9 – mf#Z-2271 – ne IDC [590]

De omnibus illiberalibus sive mechanicis artibus... / Schopperus, H – Francofurti ad Moenum, 1574 – 3mf – 9 – mf#O-1009 – ne IDC [700]

De omnibvs sanctae scriptvrae libris... / Bullinger, Heinrich – [Tigvri, 1539] – 1mf – 9 – mf#PBU-675 – ne IDC [240]

De onderwerping van djambi, 1901-1907 : beknopte geschiedenis, naar officieele gegevens samengesteld / Velds, G J – Batavia, 1909 – (= sp Indisch Militair Tijdschrift. Extra Bijlage, n24) – 3mf – 8 – mf#SE-1598 – ne IDC [959]

De on-ghemaskerde liefde des hemels : tot wederliefde door verscheyden beweeghredenen, aenspraecken ende betrachtinghen / Castro, Joannes a – Antwerpen: Wed. van I Willemsens, 1686 – 6mf – 9 – mf#O-192 – ne IDC [090]

De Onis, Harriet *see*
– Golden land
– Spanish stories and tales

De optima legendorum ecclesiae patrum methodo / Bonaventura Argonensis – Augustae Taurinorum, 1742 – 8mf – 8 – €17.00 – ne Slangenburg [240]

De optimo imperio sive josuae / Arias Montano, Benito – 1583 – 9 – sp Bibl Santa Ana [240]

De oraculis sibyllinis a iudaeis compositis, pars 1 / Badt, Benno Guilelmus – Vratislaviae [Wroclaw]: Henricus Lindner, [1869?] – 1mf – 9 – 0-8370-2147-2 – (in latin) – mf#1985-0147 – us ATLA [240]

De orbe novo / Angleria, Pedro Martir – 1587 – 9 – sp Bibl Santa Ana [970]

De orbe novo / Angleria, Pedro Martir – v. 1-2. 1892 – 9 – sp Bibl Santa Ana [970]

De orbis terrae concordia libri quatuor... / Postel, G – [Basle, 1544] – 8mf – 9 – mf#H-8272 – ne IDC [956]

De l'organisation de la justice repressive aux principales epoques historiques / Becot, Joseph – Paris, Durand, 1860. 308 p. LL-4065 – 1 – us L of C Photodup [340]

De origine, continuatione, usu, autoritate... ministerii verbi dei / Viret, P – [Geneve], R Estienne, 1554 – 9mf – 9 – mf#PFA-205 – ne IDC [240]

De origine erroris et de conciliis / Bullinger, Heinrich – Heidelberg, [Johannes Maier], 1574 – 9mf – 9 – mf#PBU-106 – ne IDC [240]

De origine erroris, in divorum ac simvlachrorvm cvltv / Bullinger, Heinrich – [Basileae, Thomas Wolffius], 1529 – 2mf – 9 – mf#PBU-104 – ne IDC [240]

De origine erroris, in negocio evcharistiae... / Bullinger, Heinrich – [Basileae, Thomas Wolffius], 1528 – 1mf – 9 – mf#PBU-103 – ne IDC [240]

De origine erroris libri duo / Bullinger, Heinrich – Tiguri, 1568 – €13.00 – ne Slangenburg [240]

De origine erroris libri dvo... / Bullinger, Heinrich – Tigvri, 1539 – 6mf – 9 – mf#PBU-872 – ne IDC [240]

De origine erroris libri dvo... / Bullinger, Heinrich – Tigvri, 1568 – 7mf – 9 – mf#PBU-673 – ne IDC [240]

De origine erroris libri octo / Heidanus, A – Amstelodami, 1678 – 7mf – 9 – mf#PBA-190 – ne IDC [240]

De origine et avtoritate uerbi dei, et quae pontificum, patrum et conciliorum admonitio hoc tempore, quo de concilio congregando agitur, ualde necessaria / Major, G – Wittembergae, 1550 – 2mf – 9 – mf#TH-1 mf 926-927 – ne IDC [242]

De origine et gestis francorum compendium : impr lugdunis impensis joh trechsl alemanni et diligenti accuratione jodoci badii ascensii / Robertus Gaguinus – 1491 – €15.00 – ne Slangenburg [241]

De originibus : sev, de varia et potissimum orbi latino ad hanc diem incognita, aut incosyderata historia, quu totius orientis, tum maxime tartarorum, persarum, turcarum... / Postel, G – Basileae, [1553] – 2mf – 9 – mf#H-8414 – ne IDC [956]

De originibus et fatis ecclesiae christianae in india orientali : disquisitio historica, ad finem seculi decimi quinti perducta, quam pro summis in philosophia honoribus rite obtinendis / Hohlenberg, Matthias Haquinus – Havniae: Typis Hartv Frid Popp, 1822 [mf ed 1995] – (= ser Yale coll) – 165p – 1 – 0-524-09900-6 – (in latin) – mf#1995-0900 – us ATLA [240]

...De originibus seu de hebraicae linguae et gentis antiquitate... / Postel, G – np, 1538 – 1mf – 9 – mf#H-8253 – ne IDC [470]

De ortho seu matutino byzantino / Mateos, J – 3mf – 8 – €7.00 – ne Slangenburg [243]

De ortu et progressu ordinis...de monte carmelo / Trithemius, Ioan & Miraeus A – Colonia Agrippina, 1643 – 8mf – 8 – €30.00 – ne Slangenburg [240]

De ortu et tempore antichristi. opera hagiographica : formae tplila 151 / Adso Dervensis – [mf ed 2003] – (= ser ILL – ser a; Cccm 45) – 5mf – 9 – €30.00 – 2-503-64982-3 – be Brepols [450]

De ortu, vita et obitu d conradi pellicani... – Zuerich, Froschauer, 1582 – 1mf – 9 – mf#PBU-566 – ne IDC [240]

De osiandrismo : dogmata et argvmenta, stvdiose ac fideliter collecta / Wigand, J – [lenae], 1586 – 5mf – 9 – mf#TH-1 mf 1522-1526 – ne IDC [242]

De oude tijd – Haarlem: A C Kruseman 1869-74 [mf ed Bloomington IN: Indiana Uni Lib, Preservation Dept 1984] – 2r – 1 – us Indiana Preservation [949]

De papa romano libri decem... / Lubbertus, S – Franekerae, 1594 – 11mf – 9 – mf#PBA-235 – ne IDC [240]

De partu virginis. de assumptione sanctae mariae virginis : formae tplila 30 / Radbertus, Paschasius – 1985 – (= ser ILL – ser a; Cccm 56c) – 3mf+32p – 9 – €20.00 – 2-503-30568-7 – be Brepols [400]

De paso por la vida / Rodriguez-Embil, Luis – Paris, France. 1913 – 1r – us UF Libraries [972]

De passauer anonymus (mgh schriften:22.bd) : ein sammelwerk ueber ketzer, juden, antichrist aus der mitte des 13. jahrhunderts / Patschovsky, A – 1968 – (= ser Monumenta germaniae historica. schriften (mgh schriften)) – €12.00 – ne Slangenburg [931]

De pastillis viperinis theriacae... / Monsalvo, O – SL, SA – 2mf – 9 – sp Cultura [616]

...De patriarche armenien de constantinople / Brosset, M – 1mf – 8 – mf#R-5819 mf23 – ne IDC [240]

De peccato in spiritum sanctum qua cum eschatologia christiana contineatur ratione, disputatio / Oettingen, Alexander von – Dorpati Livonorum [Dorpat]: Typis Henrici Laakmanni, 1856 – 1mf – 9 – 0-524-07060-1 – mf#1992-1023 – us ATLA [240]

De peccato originis repetitio doctrinae sanae ex verbo dei, corpore thuringicae ecclesiae / Wigand, J – lenae, 1572 – 2mf – 9 – mf#TH-1 mf 1527-1528 – ne IDC [242]

De pentateuch : naar zijne wording onderzocht / Bolland, Gerardus Johannes Petrus Josephus – Batavia: Albrecht & Rusche, 1892 – 1mf – 9 – 0-524-06330-3 – mf#1992-0868 – us ATLA [221]

De pentateucho samaritano ejusque cum versionibus antiquis nexu / Kohn, Samuel – Lipsiae: G Kreysing, 1865 – 1mf – 9 – 0-7905-3028-7 – (incl bibl ref) – mf#1987-3028 – us ATLA [221]

Une de perdue, deux de trouvees / Boucherville, Georges Boucher de – Montreal: E Senecal, 1874 [mf ed 1980] – 2v on 1mf – 9 – 0-665-06599-X – mf#06599 – cn Canadiana [830]

De pere facts – De Pere WI. 1879 feb 6-1881 dec 8 – 1r – 1 – mf#962672 – us WHS [071]

De pere journal – De Pere WI. 1966 jun 16/1967 aug 31-2002 oct/dec – 37r – 1 – (cont: de pere journal-democrat) – mf#982644 – us WHS [071]

De pere journal-democrat – De Pere WI. 1919 mar 20/1920 jan 22-1966 jan 6/jun 9 – 35r – 1 – (cont: brown county democrat [de pere wi]; brown county journal-news; cont by: de pere journal) – mf#1006353 – us WHS [071]

De pere news – De Pere, Nicollet WI. 1885 jan 24/1887 jan 1-1917 nov 1/1918 apr 11 – 24r – 1 – (cont by: brown county journal) – mf#1224249 – us WHS [071]

De pere news – De Pere WI. 1871 apr 8-1873 mar 29, 1873 jan 4-1876 apr 22, 1876 apr 29-1878 sep 28 – 4r – 1 – (cont by: de pere news and brown county herald) – mf#874874 – us WHS [071]

De pere news and brown county herald – De Pere WI. 1878 oct 5-1879 sep 13, sep 20-1883 jan 27, feb 3-nov 17 – 3r – 1 – (cont: de pere news; brown county herald [green bay wi]) – mf#1224246 – us WHS [071]

De pere reporter – Allouez, De Pere WI. 1981 jan 6-oct 27 – 1r – 1 – mf#962669 – us WHS [071]

De perhimpoenan indonesia en de indonesische nationalistische beweging / Salim, A – De Socialist, 19 oct 1929 n55 – 1mf – 8 – mf#SE-1300 – ne IDC [959]

De persecvtionibvs ecclesiae christianae / Bullinger, Heinrich – Tigvri, Christoph Froschover, 1573 – 3mf – 9 – mf#PBU-249 – ne IDC [240]

De pestilentia concio lvdovici lavateri in qua ostenditur vnde sit & quare immitatur... / Lavater, L – Tigvri, Ex officina Frosch[oviana], 1586 – 1mf – 9 – mf#PBU-602 – ne IDC [240]

De philippi melanchthonis ortv, totivs vitae cvrricvlo et morte / Camerarius, J – Lipsiae, [1566] – 5mf – 9 – mf#TH-1 mf 194-198 – ne IDC [242]

De philosophia morali : praelectiones quas in collegio georgiopolitano soc. jesu anno 1889-90 / Russo, Nicolaus – Ed altera. Neo-Eboraci [New York]: Benziger, 1891 – 1mf – 9 – 0-8370-6361-2 – (incl ind) – mf#1986-0361 – us ATLA [170]

De phrynicho sophista / Kaibel, George – Gottingae: Vandenhoeck et Ruprecht, [1899?] – 1mf – 9 – 0-8370-9798-3 – (discussion in latin; citations in greek) – mf#1986-3798 – us ATLA [450]

De pictura praestantissima, et numquam satis laudata arte libri tres... / Alberti, L B – Basileae, 1540 – 2mf – 9 – mf#O-104 – ne IDC [700]

De pictura veterum libri tres... / Junius, F – Roterodami, 1694 – 14mf – 9 – mf#O-317 – ne IDC [700]

De piscibus et aquatilibus omnibus libelli 3. novi / Gessner, C – Tiguri: Apud A Gesnerum, 1556 – 3mf – 9 – mf#Z-2269 – ne IDC [590]

De piscinis... / Dubravius, Janus; ed by Gessner, C – Tiguri, 1559 – 3mf – 9 – mf#Z-2270 – ne IDC [590]

De placitis hippocratis et platonis libri novem : vol 1: prolegomena critica, textum graecum, adnotationem criticam versionemque latinam continens / Galenus; ed by Mueller, Iwanus – 1874 [mf ed 1979] – 1r – 1 – (no more publ) – mf#mflm1837 – us Harvard [180]

De planctu ecclesiae / Alvarus Pelagius (Alvaro Pelayo) – Reutlingen, 1474 – €65.00 – (ulmae per joh zeiner de reutlingen) – ne Slangenburg [241]

De poenitentia / Medina, Ioannes (Juan de Medina) – Brixiae, 1606 – 9mf – 8 – €18.00 – ne Slangenburg [240]

De praedestinatione saluandorum : et huic opposita reprobatione damnandorum, disputatio / Hunnius, A – Marpurgi, 1588 – 1mf – 9 – mf#TH-1 mf 798 – ne IDC [242]

De praedestinatione theses in scholis theologorum discvtiendae / Pappus, J – Argentorati, 1589 – 1mf – 9 – mf#TH-1 mf 1232 – ne IDC [242]

De praesentia corporis et sangvinis christi iesv domini nostri, in administratione eucharistiae vel coenae dominicae : dictatus in academia ihenensi, anno domini 1553 / Strigel, V – [Francofortí], 1576 – 1mf – 9 – mf#TH-1 mf 1449 – ne IDC [242]

De praestantia musicae veteris libri tres totidem dialogis comprehensi : in quibus vetus ac recens musica, cum singulis earum partibus... / Doni, Giovanni Battista – Florentiae: typis Amatoris Massae Foroliuien 1647 [mf ed 19–] – 3mf – 9 – mf#fiche 408, 718 – us Sibley [780]

De principe / Pontanus, Jovianus [Pontano, Giovanni] – (= ser Holkham library manuscript books 492) – 1r – 1 – mf#97597 – uk Microform Academic [450]

De principiis astronomiae libri dvo... / Simler, J – Tigvri, [Christoph] Froschover, ivnior, 1559 – 2mf – 9 – mf#PBU-625 – ne IDC [240]

De principiis christianorum dogmatum libri 7... / Lubbertus, S – Franekerae, 1591 – 9mf – 9 – mf#PBA-234 – ne IDC [240]

De' principj dell' armonia musicale contenuta nel diatonico genere / Tartini, Giuseppe – Padova: Stamperia del Seminario 1767 [mf ed 19–] – 3mf – 9 – mf#fiche 675, 850 – us Sibley [780]

De priore et posteriore forma kantianae critices rationis purae / Ueberweg, Friedrich – Berolini: Mittler, 1862 – 1mf – 9 – 0-7905-8742-4 – (incl bibl ref) – mf#1989-1967 – us ATLA [100]

De probatis sanctorum historiis / Surius, L – Coloniae. v1-6. 1570-75 – 6v on 266mf – 8 – €507.00 – ne Slangenburg [240]

De problemas filosoficos... / Huerta, J – Madrid, 1628 – 6mf – 9 – sp Cultura [100]

De prohibitione et censura librorum : dissertatio canonico-moralis / Vermeersch, Arthur – Quarta editio auctior, accuratior, et novo ordine disposita. Romae: Typis Societatis Sanctis Joannis [a] Desclee, Lefebvre, 1906 – 1mf – 9 – 0-524-00797-7 – mf#1990-0229 – us ATLA [240]

De pronominibus, pars generalis = On pronouns / Apollonius, Dyscolus; ed by Maas, Paul – Bonn: A Marcus & E Weber, 1911 [mf ed 1992] – (= ser Kleine texte fuer vorlesungen und uebungen 82) – 1mf – 9 – 0-524-04691-3 – (in greek; notes in latin) – mf#1990-3400 – us ATLA [450]

De prophetae officio... / Bullinger, Heinrich – [Tygvri, Christof. Froschouer, 1532] – 1mf – 9 – mf#PBU-111 – ne IDC [240]

De proverbiorum quae dicuntur aguri et lemuelis origine atque indole henricus ferdinandus muehlau / Muehlau, Ferdinand – Lipsiae [Leipzig]: Metzger & Wittig, [c1869] – 1mf – 9 – 0-8370-4521-5 – mf#1985-2521 – us ATLA [220]

De provincia remensi 1 (gc9) – Parisiis, 1751 – (= ser Gallia christiana (gc)) – €63.00 – ne Slangenburg [240]

De provincia remensi 2 (gc10) – Parisiis, 1751 – (= ser Gallia christiana (gc)) – €61.00 – ne Slangenburg [240]

De publijke intrede van william de 3...gedaen in 's gravenhage op den 5 februarij 1691... / [Bidloo, G] – 's Gravenhage: Meyndert Uytwerf, 1691 – 1mf – 9 – mf#O-1154 – ne IDC [090]

De puertas adentro / Sanchez Arjona, Vicente – Sevilla: Graficas T., Tomo 2. 1954 – 1 – sp Bibl Santa Ana [810]

De puertas adentro / Sanchez Arjona, Vicente – Sevilla: Graficas Tirvia, Tomo 4. 1954 – 1 – sp Bibl Santa Ana [810]

De puertas adentro / Sanchez Arjona, Vicente – Sevilla: Graficas Tirvia, Tomo 5. 1954 – 1 – sp Bibl Santa Ana [810]

De puertas adentro / Sanchez Arjona, Vicente – Sevilla: Imprenta Crufer, Tomo 7-8. 1958 – 1 – sp Bibl Santa Ana [810]

De puertas adentro / Sanchez Arjona, Vicente – Sevilla: Vda. de J. Mejias. Impresor, Tomo 6. 1956 – 1 – sp Bibl Santa Ana [810]

De puertas adentro. poesias intimas / Sanchez Arjona, Vicente – Sevilla: Imprenta San Antonio, Tomo 1. 1954 – 1 – sp Bibl Santa Ana [810]

De puertas adentro (todo en broma) / Sanchez Arjona, Vicente – Sevilla: Graficas Tirvia, Tomo 3. 1954 – 1 – sp Bibl Santa Ana [810]

De Puy, William Harrison *see* The methodist centennial year-book for 1884

De quattordecim regionibus urbis romae earumdemque aedificiis / Pancirolus, G – Lugduni, 1608 – €12.00 – ne Slangenburg [241]

De quattuor evangeliorum codicibus origenianis / Hautsch, Ernestus – Gottingae: Officina Academica Dieterichiana, 1907 – 1mf – 9 – 0-7905-0498-7 – (incl bibl ref) – mf#1987-0498 – us ATLA [220]

...De que el aforismo primero de hipocrates... sirve a la milicia como a la medicina... / Gomez, M – Antverpiae, 1643 – 3mf – 9 – sp Cultura [610]

De quebec a mexico, souvenirs de voyage, de garnison, de combat et de bivouac / Faucher de Saint-Maurice – Montreal: Duvernay & Dansereau, 1874 [mf ed 1980] – 2v on 1mf – 9 – 0-665-03741-4 – mf#03741 – cn Canadiana [917]

De quebec aux antilles : notes de voyage / Montminy, Theophile – Quebec: J A Langlais, 1888 – 3mf – 9 – mf#11201 – cn Canadiana [918]

De Quincey, Thomas *see* Collected writings of thomas de quincey

De ratione communi omnium linguarum literaru commentarius...explicatio doctrinae recte... vivendi / Bibliander, T – Tigvri, Christoph Frosch[ouer], 1548 – 3mf – 9 – mf#PBU-481 – ne IDC [240]

De ratione temporvm, christianis rebus et cognoscendis et explicandis accomodata, liber unus : demonstrationum chronologicarum liber alius, eodem autore... / Bibliander, T – Basileae, [Ioannes Oporinus, 1551] – 4mf – 9 – mf#PBU-580 – ne IDC [240]

De rationibus festorum mobilium utriusque ecclesiae occidentalis atque orientalis : commentarius usui clericorum accomadatus / Nilles, Nicolaus – Viennae: Mayer, 1868 – 1mf – 9 – 0-8370-7009-0 – (incl bibl ref and ind) – mf#1986-1009 – us ATLA [240]

De rationibus festorum sacratissimi cordis jesu et purissimi cordis mariae : libri 4 / Nilles, Nicolaus – ed quinta, novis accessionibus adornata. Oeniponte [Innsbruck]: Libraria Academica Wagneriana, 1885 – 4mf – 9 – 0-524-06699-X – (incl bibl ref) – mf#1990-5270 – us ATLA [240]

De rationis auctoritate tum in se, tum secundum sanctum anselmum considerata / Vacherot, Etienne – Cadomi [Caen]: Pagny, 1836 – 1mf – 9 – 0-524-00356-4 – mf#1989-3056 – us ATLA [240]

De re aedificatoria... / Alberti, L B – Strasbourg, 1541 – 4mf – 9 – mf#OA-4 – ne IDC [720]

De re aedificatoria dece : opus integru et absolutu: diligenterqz recognitum... / Alberti, L B – Parrhisijs, [1512] – 4mf – 9 – mf#OA-31 – ne IDC [720]

De re bibliographica : la autobibliografia de juan antonio munoz gallardo / Fernandez Serrano, Francisco – Badajoz: Dip. Provincial, 1975. Sep. REE – 1 – sp Bibl Santa Ana [920]

De re diplomatica libri 6 / Mabillon, Jean – 2nd rev ed. Luteciae Parisiorum, 1709 – 37mf – 8 – €71.00 – ne Slangenburg [240]

De re metrica hebraeorum / Gietmann, Gerhard – Friburgi Brisgoviae: Herder, 1880 – 1mf – 9 – 0-8370-3276-8 – (incl ind) – mf#1985-1276 – us ATLA [470]

De re militari et bello tractatus / Belli, Piero – Washington: Carnegie Endowment for International Peace, Division of International Law. 2v. 1936 – 9mf – 9 – $13.50 – mf#LLMC 90-460 – us LLMC [355]

De re militari libri quatuor / Vegetius, F – Paris, 1535 – 5mf – 9 – mf#OA-262 – ne IDC [720]

De rebus constantinopolitanis libri 4 (cbh30,1) / Genesius, J – Venetiis, 1733 – (= ser Corpus byzantinae historiae (cbh)) – €11.00 – ne Slangenburg [243]

De rebus fidei hoc tempore... / Valentia, Greg de – Lugduni, 1591 – 42mf – 8 – €80.00 – ne Slangenburg [240]

De rebus hispaniae memorabili / Marineo Siculo, Lucio – 1533? – 9 – sp Bibl Santa Ana [946]

De rebus oceanicis & orbe novo decada tres / Angleria, Pedro Martir – 1533 – 9 – sp Bibl Santa Ana [970]

De rebus oceanicis & orbe novo decada tres / Angleria, Pedro Martir – 1553 – 9 – sp Bibl Santa Ana [970]

De rebus turcaru... / Richier, C – Parisiis, 1540 – 2mf – 9 – mf#H-8265 – ne IDC [956]

De rebvs : emmanvelis, lvsitaniae regis invictissimi, virtvte et avspicio... / Osorio da Fonesca, J – Coloniae, 1586 – 10mf – 9 – mf#H-8422 – ne IDC [956]

De rebvs natvralibvs libris 30 : qvibus qvaestiones.... / Zabarella, Jacopo – 4th ed. Coloniae: Sumptibus L Zetzneri 1602 [mf ed 1980] – 1r – 1 – (with: dispvtationvm de medicina nova philippi paracelsi pars prima / erastus, thomas) – mf#726 – us Wisconsin U Libr [180]

De rebvs tvrcicis commentarii dvo accvratissimi / Camerarius, J – Francofvrti, 1598 – 3mf – 9 – mf#H-8396 – ne IDC [956]

...De rebvs tvrcicis liber / Balbi, G – Romae, 1526 – 1mf – 9 – mf#H-8233 – ne IDC [956]

De reclusis/speculum humanae salvationis : york minister library ms. 16.k5 / Hilton, W – 1r – 1 – (filmed with: speculum humanae salvationis) – mf#4287 – uk Microform Academic [240]

De reformationis ecclesiae anglicanae annalibus...edendis consilium / Simler, J J – N p, 1780 – 1mf – 9 – mf#PBU-425 – ne IDC [240]

De rege et regendi ratione / Lopez Bravo, Mateo – 1616 – 9 – sp Bibl Santa Ana [321]

De regentenpositie / Soeria Nata Atmadja, R A A A – Bandoeng, 1940 – 4mf – 8 – mf#SE-1285 – ne IDC [959]

De regno, civitate : et domo dei ac domini nostri iesu christi... / Lambert, F – Worms, 1538 – 2mf – 9 – mf#PPE-123 – ne IDC [240]

De reizende chinees : op bevel en kosten van zynen keizer... – Amsterdam, 1727. 2v – 10mf – 9 – mf#HT-1180 – ne IDC [915]

De religione et ecclesia praelectiones scholastico-dogmaticae / Mazzella, Camillo – Editio 5. Romae: Officina Typographica Forzani et Socii, 1896 – 3mf – 9 – 0-8370-8201-3 – mf#1986-2201 – us ATLA [240]

De religionum indagationis comparativae vi ac dignitate theologica / Spiess, Edmund – Jenae: Typis F Frommann, 1871 – 1mf – 9 – 0-524-02808-7 – (incl bibl ref) – mf#1990-3138 – us ATLA [200]

De republica ecclesiastica, libri 10 / Dominis, M A de – Londini. v1-3. 1617-1622 – €197.00 – ne Slangenburg [240]

De repvblica helvetiorvm libri duo...auctore iosia simlero... / Simler, J – Tigvri, Christoph Froschouer, 1576 – 5mf – 9 – mf#PBU-500 – ne IDC [240]

De rerum creatione ex nihilo / Hoonacker, Albin van – Lovanii: VanLinthout, 1886 – 1mf – 9 – 0-8370-3815-4 – (incl bibl ref) – mf#1985-1815 – us ATLA [210]

De rerum usu et abusu / Furmer, Bernardo – Antverpiae: Ex officina Christophori Plantini, 1575 – 1mf – 9 – mf#O-615 – ne IDC [090]

De rerum varietate libri 17... : post alias omnes editiones, nunc recogniti, castigati, infinitisque mendis repurgati... / Cardano, Girolamo – Lugduni: Apud Stephanum Michaelem 1580 [mf ed 19–] – 1r [ill] – 1 – us OmniSys [900]

De restitutione et contractibus / Medina, Ioannes (Juan de Medina) – Brixiae, 1606 – 8mf – 8 – €17.00 – ne Slangenburg [240]

De revelatione supernaturali / Ottiger, Ignaz – Friburgi Brisgoviae; S Ludovici Americae: Herder, 1897 – (= ser Theologia Fundamentalis) – 3mf – 9 – 0-7905-8542-1 – (incl bibl ref) – mf#1989-1767 – us ATLA [240]

De Ricci, James Herman *see*
– The fisheries dispute and annexation of canada

De ritibus et institutis ecclesiae tigvrinae opusculum / Lavater, L – Tiguri, [Zuerich, Christoph Froschauer, 1566-1567] – 1mf – 9 – mf#PBU-305 – ne IDC [240]

De ritibus et institutis eddlesiae tigurianae opusculum / Lavater, Ludwig – [S.l.: s.n., 1559?] – 1r – 1 – 0-8370-1565-0 – mf#1984-B480 – us ATLA [100]

De roberval : a drama; also the emigration of the fairies; and the triumph of constancy, a romaunt / Hunter-Duvar, John – Saint John, NB: J & A McMillan, 1888 – 3mf – 9 – mf#07306 – cn Canadiana [410]

De romanensibus helvetiae et teriolis gentibus / Walter, Friedrich Ludwig – Berlin 1832 – (= ser Whsb) – 1mf – 9 – €20.00 – mf#Hu 146 – gw Fischer [410]

De rome a l'evangile : quelques pionniers du dernier siecle / Marsault, F – Paris: Fischbacher, 1908 – 1mf – 9 – 0-8370-8766-X – mf#1986-2766 – us ATLA [240]

De Ruyter, John A *see* The star of the twentieth century

De rvssorvm moscovitarvm et tartarorvm religione, sacrificiis... – n.p, 1582 – 4mf – 9 – mf#H-8202 – ne IDC [956]

De, S C *see* Kalidasa and vikramaditya

De sacra poesi hebraeorum / Lowth, Robert – Goettingae: Pockwizii & Berm., 1758 – 1r – 1 – 0-8370-0896-4 – mf#1984-B508 – us ATLA [810]

De sacra praedicationes in o.f.m... / Belluco, Bartolome – Madrid: Archivo Ibero Americano, 1960 – 1 – sp Bibl Santa Ana [240]

De sacra traditione contra novam haeresim evolutionismi / Billot, Louis – Romae: A S Iosepho, 1904 – 1mf – 9 – 0-8370-2343-2 – (incl ind) – mf#1985-0343 – us ATLA [210]

De sacramentarijsmo, dogmata et argvmenta ex qvatvor patriarchis sacramentariorum, carlstadio, zvvinglio, oecolampadio, calvino / Wigand, J – Lipsiae, 1584 – 13mf – 9 – mf#TH-1 mf 1534-1546 – ne IDC [242]

De sacramentis in communi et in speciali / Mancini, Jerome Marie – Romae: SC de Propaganda Fide, 1905 – (= ser Theologia Dogmatica) – 1mf – 9 – 0-7905-8700-9 – mf#1989-1925 – us ATLA [240]

De sacramento eucharistiae contra oecolampadium opusculum... / Clichtove, J – Parisiis, 1526 – 4mf – 9 – mf#CA-82 – ne IDC [241]

De sacramento extremae unctionis : tractatus dogmaticus / Kern, Josephus – Ratisbonae [Regensburg]; Neo Eboraci [New York]: Friderici Pustet, 1907 – 1mf – 9 – 0-8370-9162-4 – (incl ind) – mf#1986-3162 – us ATLA [240]

De sacris aedificiis a constanti mag constructis... / Ciampini, J – Romae, 1693 – 9mf – 9 – mf#O-1141 – ne IDC [720]

De sacris rhodiorum commentatio / Dittenberger, Wilhelm – Halis: Formis Gebauero-Schwetetschkeianis 1886-87 [mf ed 1987] – 2pt on 1r – 1 – (iss in "index scholarum in universitate litteraria fridericiana halensi cum vitebergensi consociata" 1886. with: handbook for canoeing councillors / deming, e) – mf#1860 – us Wisconsin U Libr [450]

De sacro conjugio / Lambert, F – Nuremberg, 1525 – 3mf – 9 – mf#PPE-120 – ne IDC [240]

De sacro foedere in selimvm : libri quattor / Foglietta, U – Genvae, 1587 – 4mf – 9 – mf#H-8368 – ne IDC [956]

De sacrosancta coena domini nostri iesv christi / Bullinger, Heinrich – Tigvri, Christ[oph] Froschouer, 1553 – 1mf – 9 – mf#PBU-179 – ne IDC [240]

De saint domingue a haiti : essai sur la culture... / Price Mars, Jean – Paris, France. 1959 – 1r – us UF Libraries [972]

De saint-louis...tripoli par le lac tchad / Monteil, P L – Paris, 1896 – 13mf – 9 – mf#A-180 – ne IDC [916]

De Salviac, M *see* Un peuple antique au pays de menelik

De sancto cypriano et de primaeva carthaginiensi ecclesia / Blampignon, Emile – Paris: F Didot, 1861 – 1mf – 9 – 0-7905-8637-1 – (in latin & greek; incl bibl ref) – mf#1989-1862 – us ATLA [240]

De sanctorum invocatione et intercessione... adversus henricum bullingerum helvetium / Cochlaeus, J – Ingolstadt, 1544 – 1mf – 9 – mf#PBU-702 – ne IDC [240]

De Santi, Angelo *see* Les litanies de la sainte vierge

De santo tomas a krause? / Fernandez Valbuena, Ramiro – 1882 – 9 – sp Bibl Santa Ana [240]

De satisfactione christi disputationes... / Turrettini, F – Genevae, 1666 – 4mf – 9 – mf#PFA-18 – ne IDC [240]

De satisfactione christi disputationes... / Turrettini, F – Genevae, de Tournes, 1691 – 5mf – 9 – mf#PFA-185 – ne IDC [240]

De scandalis, quibus hodie plerique absterrentur, nonnulli etiam alienantur a pura evangelii doctrina : joannis calvini libellus apprime utilis. ad laurentium normandium / Calvin, J – Genevae: Apud Joannem Crispinum, 1550 – 2mf – 9 – mf#CL-30 – ne IDC [242]

De scholasticorum sententia philosophiam esse theologiae ancillam commentatio / Clemens, Franz Jakob – Monasterii Guestphalorum [Muenster in Westphalen]: Academica Aschendorffiana, [1856?] – 1mf – 9 – 0-524-00013-1 – (incl bibl ref) – mf#1989-2713 – us ATLA [100]

De schvvenckfeldismo dogmata et argvmenta : cum succinctis solutionibus / Wigand, J – Lipsiae, 1586 – 5mf – 9 – mf#TH-1 mf 1547-1551 – ne IDC [242]

De Schweinitz, Edmund *see*
– The history of the church known as the unitas fratrum or the unity of the brethren
– The life and times of david zeisberger
– Some of the fathers of the american moravian church

De scriptoribus scholasticis s 14 ex ordine carmelitarum / Xiberta, B M – Louvain, 1931 – €25.00 – ne Slangenburg [241]

De scriptura sacra / Bainvel, Jean Vincent – Paris: Gabriel Beauchesne, 1910 – 1mf – 9 – 0-524-05786-9 – (incl bibl ref) – mf#1992-0613 – us ATLA [220]

De scriptvrae sanctae avthoritate...deque episcoporum...institutione / Bullinger, Heinrich – Tigvri, Officina Froschoviana, 1538 – 4mf – 9 – mf#PBU-136 – ne IDC [240]

De scriptvrae sanctae praestantia, dignitate... / Bullinger, Heinrich – Tigvri, Christ[oph] Frosch[auer], 1571 – 2mf – 9 – mf#PBU-239 – ne IDC [240]

De secessione ab ecclesia romana deque ratione pacis inter evangelicos... / Amyraut, M – Salmurii, 1647 – 3mf – 9 – mf#PRS-103 – ne IDC [242]

De' secreti del r d alessio piemontese / Ruscelli, Girolamo – Venetia: Appresso Angelo Bodio 1674 [mf ed 1980] – 1r – 1 – mf#60 – us Wisconsin U Libr [615]

De servetianismo, sev de antitrinitariis / Wigand, J – Regiomonti, 1575 – 2mf – 9 – mf#TH-1 mf 1552-1553 – ne IDC [242]

De sevilla a batalha. excursion...de sevilla a merida y badajoz.. / Cascales Munoz, Jose – 1892 – 9 – sp Bibl Santa Ana [914]

De sevilla a guadalupe. breves apuntes tomados a vuela pluma / Gestoso y Perez, Jose – Sevilla: Oficina del Correo de Andalucia, 1913 – 1 – sp Bibl Santa Ana [946]

De sidste ti aar i japan / Winther, J M T – Koebenhavn: O Lohre, 1915 – 1mf – 9 – 0-524-05895-4 – mf#1991-2345 – us ATLA [240]

De signis...morbi suffocantis... / Villarreal, J – Alcala de Henares, 1611 – 5mf – 9 – sp Cultura [610]

De Silva Barahona e Costa, Henrique Cesar *see* Apontamentos para a historia de guerra de zambezia, 1871-1875

De sinarum magnaeque tartariae rebus commentario alphabetica... / Mueller, A – [Francofurti ad Oderam: apud Johannem Volcker], n.d. – 1mf – 9 – mf#HT-628 – ne IDC [915]

De sizilische grosshof unter kaiser friedrich 2 (mgh schriften:4.bd) : eine verwaltungs-geschichtliche studie / Heupel, W E – 1940 – (= ser Monumenta germaniae historica. schriften (mgh schriften)) – €12.00 – ne Slangenburg [931]

De sociaal-psychologische aspecten van het zuid-molukse vraagstuk / Bouwan, J C – Eindhoven, [1955] – 1mf – 8 – mf#SE-1293 – ne IDC [959]

De Solla, Jacob Mendes *see*
– Jewish student's companion
– Khol divre ha-torah

De somno scipionis *see* De vita pomponi secunda

De soto chronicle – De Soto Wl. 1886 jun 12-1889 aug 16 – 1r – 1 – mf#962705 – us WHS [071]

De soto city / Darsey, Barbara – s.l, s.l? 1936 – 1r – us UF Libraries [978]

De Soysa, A H T *see* Ancient kaurawa flags

The de soysa charitaya : or, the life of charles henry de soysa, esq, j p / Bastian, C Don – Colombo: "Sinhalese Daily News" Press, 1904 – 1 – us CRL [954]

De Spain, Kent S *see* Solo movement improvisation

De spectris, lemuribus et...fragoribus... / Lavater, L – Genevae, Anchora Crispiniana, 1570 – 4mf – 9 – mf#PBU-312 – ne IDC [240]
De ss sacramentis secundum ritum aethiopicum / Abba Tecle Mariam Semhary Selam – Romae, 1931 – 3mf – 8 – €7.00 – ne Slangenburg [243]
De stabilitate et progressu dogmatis / Lepicier, Alexius Maria – Romae: Typographia Editrix Romana, 1908 – 1mf – 9 – 0-8370-8196-3 – (incl ind) – mf#1986-2196 – us ATLA [240]
De stad palembang in 1935 : 275 jaar geleden als een phoenix uit haar asch herrezen / Wellan, J W J – Groningen, 1935 – 1mf – 8 – mf#SE-1280 – ne IDC [959]
De stancarismo, dogmata et argvmenta cvm solvtionibvs, qvibvs / Wigand, J – Lipsiae, 1585 – 2mf – 9 – mf#TH-1 mf 1554-1555 – ne IDC [242]
De standaard – Amsterdam, 1872-1944. sunday suppl 1872-1876 – 213r – 1 – mf#SF-1 – ne IDC [074]
De statu religionis christianae / Miraeus, A – Colonia Agrippina, 1619 – 5mf – 8 – €12.00 – ne Slangenburg [240]
De statu religionis et reipublicae carolo v caesare commentarii / Sleidan, J – new ed. Francofurti ad Moenum, 1785-86. 3v – 21mf – 9 – mf#PPE-124 – ne IDC [240]
De stem der ambonnezen : rede / Lokollo, P W; ed by Bureau Zuid-Molukken – 's-Gravenhage. n3. 1950 – 1mf – 8 – mf#SE-1298 – ne IDC [959]
De sterrekunde en de mensheid / Minnaert, Marcel Gilles Jozef – Den Haag: N.V. Servire 1946 [mf ed 2006] – (= ser Servire's encyclopaedie a2; Afdeling-sterrekunde 5) – 1r – 1 – (incl bibl ref & ind) – mf#film mas 37452 – us Harvard [520]
De st-lin a san-francisco : ou journal de voyage, 1894 / Legault, Philomene – Joliette, Quebec?: s.n, 1897 – 4mf – 9 – mf#08639 – cn Canadiana [917]
De sto johanne *see* Liber gratiae. liber laudum virginis mariae. de sto johanne. de eruditione puerorum regalium
De stolatae virginitatis jure lucubratio academica / Beverland, Adriaan – Lugduni in Batavis: Typis J. Lindani, 1680 – 1r – 1 – 0-8370-0054-8 – mf#1984-6013 – us ATLA [240]
De substantia orbis tractatus / Averrois Cordubensis – Venetiis, 1560 – €3.00 – ne Slangenburg [110]
De suikerhandel van java / Tio, Poo Tjiang – Amsterdam: J H de Bussy, 1923 (mf ed 19–) – 139p – (incl bibl footnotes) – mf#ZT-TB pv343 n9 – us Harvard [380]
De sumatra post 1898-1923 – Medan, 1924 – 1mf – 8 – mf#SE-1448 – ne IDC [079]
De summa trinitate et fide catholica... / Bibliander, T – Basileae, [Ioannes Oporinus, 1555] – 2mf – 9 – mf#PBU-587 – ne IDC [240]
De suprema romani pontificis in ecclesiam potestate disputatio quadripartita / Val, Andr du – Parisiis, 1614 – 14mf – 8 – €27.00 – ne Slangenburg [241]
De, Sushil Kumar *see*
- Early history of the vaisnava faith and movement in bengal
- History of bengali literature in the nineteenth century, 1800-1825

De svbstantia hominis de viribvs hominis de depravationibvs hominis contra manichaeos / Wigand, J – Ratisponae, 1575 – 1mf – 9 – mf#TH-1 mf 1556 – ne IDC [242]
De symbolica aegyptiorum sapientia in quo symbola, parabolae, historiae selectae... / Caussin, N – Coloniae Agrippinae: Sumptibus Ioannis lost, 1631 – 9mf – 9 – mf#O-54 – ne IDC [090]
De symbolis heroicis / Pietrasancta, Silvestro – Antverpiae: Ex officina Plantiniana Balthasaris Moreti, 1634 – 11mf – 9 – mf#O-717 – ne IDC [090]
De symbolo foederis confessio / Lambert, F – n.p, 1530 – 1mf – 9 – mf#PPE-122 – ne IDC [240]
De synode in het oude bisdom doornik gesitueerd in de europese ontwikkeling / Lambecht, D – Gent, 1976 – €65.00 – ne Slangenburg [240]
...De tartaris diarivm / Brussius, G – Francofvrti, 1598 – 1mf – 9 – mf#H-8395 – ne IDC [956]
De templo : et de iis quae ad templum pertinent, libri quinque / Ribera, Fr – Antverpiae, 1593 – 30mf – 8 – €58.00 – ne Slangenburg [221]
De Terra, Helmut *see* Man and mammoth in mexico
De testamento sev foedere...expositio / Bullinger, Heinrich – Tigvri, Christoph Frosch[auer], 1534 – 2mf – 9 – mf#PBU-121 – ne IDC [240]
De theologia generatim : commentarius in sacram theologiam hodegos / Schrader, Clemens – Pictaviis [Poitiers]: H Oudin, 1874 – 3mf – 9 – 0-524-00337-8 – (incl bibl ref) – mf#1989-3037 – us ATLA [240]
De theologia gentili et physiologia christiana; sive de origine ac progressu idololatriae... / Vossius, G – Amsterdami, 1641. 4v – 22mf – 9 – mf#CA-154 – ne IDC [240]
De theologo seu de ratione studii theologici / Hyperius, A – Basil, 1559 – 9mf – 9 – mf#PBA-211 – ne IDC [240]
De Thierry, C *see* Imperialism
De thoma bradwardino commentatio / Lechler, Gotthard Victor – Lipsiae: Apud A Edelmannum, 1862 [mf ed 1990] – 1mf – 9 – 0-7905-5368-6 – (incl bibl ref) – mf#1988-1368 – us ATLA [240]
De timotheo 1 nestorianum patriarcha (728-823) et christianorum orientalium condicione sub chaliphis abbasidis : accedunt 99 eiusdem timothei definitiones canonicae e textu syriaco inedito nunc primum latine redditae / Labourt, Hieronymus – Parisiis: Victor Lecoffre, 1904 – 1mf – 9 – 0-8370-8032-0 – (incl bibl ref) – mf#1986-2032 – us ATLA [450]
De todos : organo de difusion de la federacion argentina de iglesias evangelicas – Buenos Aires: La Federacion, n1-17. aug 1987-1992 – 1r – 1 – us CRL [240]
De tonto que soy / Fresquet, Fresquito – Habana, Cuba. 1964 – 1r – us UF Libraries [972]
De topographia constantinopoleos, et de illivs antiqvitatibvs libri qvatvor / Gilles, P – Lvgdvni, 1561 – 3mf – 9 – mf#H-8298 – ne IDC [956]
De Totanes, Sebastian *see* Arte de la lengua tagala; y, manual tagalog
De translatione imperii romani ad germanos / Flacius Illyricus d A, M – Basileae, 1566 – 8mf – 9 – mf#TH-1 mf 417-424 – ne IDC [242]
De tribord a babord : trois croisieres dans le golfe saint laurent: nord et sud / Faucher de Saint-Maurice – Montreal: Duvernay et Dansereau, 1877 – 5mf – 9 – mf#03063 – cn Canadiana [917]
De triginta sex decanis : formae tplila 78 / Trismegistus, Hermes – 1994 – (= ser ILL – ser a; Cccm 144) – 4mf+35p – 9 – €30.00 – 2-503-64442-2 – be Brepols [400]
De trinitate. praefatio. libri 1-7 (ccsl 62) / Pictaviensis, Hilarius; ed by Smulders, P – 1979 – (= ser Corpus christianorum series latina (ccsl)) – 12mf+400p – 9 – €90.00 – 2-503-00621-3 – be Brepols [940]
De trognas frihet ifran lagen / Rosenius, Carl Olof – 2. uppl. Stockholm: Evangeliska Fosterlands-Stiftelsens foerlag, [1874?] – 1mf – 9 – 0-524-05710-9 – mf#1991-2324 – us ATLA [340]
...De turcarum origine, moribus, and rebus gestis commentarius / Cervarius Tubero, L – Florentiae, 1590 – 2mf – 9 – mf#H-8424 – ne IDC [956]
De turcarum ritv et caeremoniis... / Bartholomaeus, G – Antverpiae, 1544 – 1mf – 9 – mf#H-8271 – ne IDC [956]
De tvrcarvm moribvs epitome / Bartholomaeus, G – Lvgdvni, 1567 – 3mf – 9 – mf#H-8306 – ne IDC [956]
De tvrcorvm origine, religione, ac immanissima eorum in christianos tyrannide... / Cuspinianus, J – Antverpiae, 1541 – 3mf – 9 – mf#H-8267 – ne IDC [956]
De una especie de garrotillo o esquilencia mortal / Figueroa, F – Lima, 1616 – 1mf – 9 – sp Cultura [616]
De unione ecclesiarum ac totius christianae societatis congressu (vulgo the world congress) pro quaestionibus ad fidem ordinemque ecclesiae spectantibus rite explorandis et perpendendis – [S.l.: s.n., 1917?] – 1mf – 9 – 0-524-08657-5 – (incl bibl ref) – mf#1993-2117 – us ATLA [240]
De usu astrolabi compedium, schematibus commodissimis illustratum / Poblacion, Juan Martinez – Parisiis: Ioannis Barbaei, 1546 [mf ed 1988] – 2mf – 9 – mf#SEM105P878 – cn Bibl Nat [623]
De usura et foenore / Thysius, A – Trajecti ad Rhenum, 1658 – 1mf – 9 – mf#PBA-354 – ne IDC [240]
De utraque potestate papali et regali / Johannes Parisiensis – Parisiis, 1506 – €11.00 – ne Slangenburg [241]
De va dat a ja ta sat : [a novel] / Can Nve, U – Ran Kun: Van Mo U Ca Pe 1979 [mf ed 1990] – 1r with other items – 1 – (in burmese) – mf#mf-10289 seam reel 152/3 [§] – us CRL [830]
De vanitate mundi *see* Soliloquium de arrha animae. de vanitate mundi (kit123)
De varia commesuracion para la esculptura y architectura / Arfe y Villafane, J – Sevilla, 1585 – 4mf – 9 – sp Cultura [720]
De varia republica sive commentaria in librum judicum / Arias Montano, Benito – 1592 – 9 – sp Bibl Santa Ana [320]
De varios colores / Sanchez Arjona, Vicente – Sevilla: Imp. Zambrano, 1955 – 1 – sp Bibl Santa Ana [810]
De veneratione sanctorum, opusculum duos libros coplectens / Clichtove, J – Parisiis, 1523 – 3mf – 9 – mf#CA-86 – ne IDC [241]
De vera hominis christiani iustificatione : vera item et iusta bonorum operum ratione / Bullinger, Heinrich – Tiguri, 1543 – €5.00 – ne Slangenburg [240]
De vera iesu christi...praesentia / Simler, J – Tigvri, officina Froschoviana, 1574 – 3mf – 9 – mf#PBU-331 – ne IDC [240]
De vera religione : praelectiones novae in seminario sancti sulpitii habitae: cum multis annotationibus in ulteriora cujusque studia et praedicationis usus profuturis / Brugere, Lud-Fred – ed nova. Parisiis: A Roger et F Chernoviz, 1878 – 1mf – 9 – 0-8370-8409-1 – (in latin and french. incl bibl ref and index) – mf#1986-2409 – us ATLA [240]
De vera religione et apologetica / Bainvel, Jean Vincent – Paris: G Beauchesne, 1914 [mf ed 1991] – 1mf – 9 – 0-7905-9125-1 – (incl bibl ref) – mf#1989-2350 – us ATLA [241]
De verbo incarnato : commentarius in tertiam partem s. thomae / Billot, Louis – ed 3, novis additionibus adornata. Romae: Ex typographia Polyglotta, 1900 – 2mf – 9 – 0-524-06009-6 – (incl bibl ref) – mf#1991-2369 – us ATLA [220]
De verbo incarnato : pro manuscripto / Groot, J F de – Romae: Ex Pontificia Universitate Gregoriana, [19–?] – 1mf – 9 – 0-524-06183-1 – mf#1991-2439 – us ATLA [240]
De verborgenheid des evangelies / Pauptit, G J – Den Haag, 1935 – €5.00 – ne Slangenburg [240]
De verborum 'religio' atque 'religiosus' usu apud romanos : quaestiones selectae / Kobbert, Maximilian – Regimonti [Koenigsberg]: Hartungiana, 1910 – 1mf – 9 – 0-524-02022-1 – (incl bibl ref) – mf#1990-2797 – us ATLA [450]
De Vere, Aubrey Thomas *see* Ireland and proportional representation
De veri precetti della pittura... / Armenini, G B – Ravenna, 1587 – 4mf – 9 – mf#O-129 – ne IDC [700]
De veritate historica libri judith : a liisque ss. scripturarum locis specimen criticum exegeticum / Palmieri, Domenico – Galopiae: M Alberts, 1886 – 1mf – 9 – 0-524-05929-2 – mf#1992-0686 – us ATLA [221]
De vero verbi dei sacramentorum, et ecclesiae ministerio / Viret, P – [Geneve], R. Estienne, 1553 – 5mf – 9 – mf#PFA-204 – ne IDC [240]
De veteris latinae ecclesiastici capitibus 1-43 : una cum notis ex eiusdem libri translationibus aethiopica, armeniaca, copticis, latina altera, syro-hexaplari depromptis / Herkenne, Henr – Leipzig: J C Hinrichs, 1899 – 1mf – 9 – 0-8370-3566-X – mf#1985-1566 – us ATLA [221]
De veteris testamenti locis a paulo apostolo allegatis / Kautzsch, E – Lipsiae [Leipzig]: Metzger & Wittig, 1869 – 1mf – 9 – 0-7905-1122-3 – (incl bibl ref) – mf#1987-1122 – us ATLA [220]
De vi et usu praepositionum epi, meta, para, peri, pros, hupo apud aristophanem / Iltz, Johannes – Halis Saxonum: Formis Kaemmererianis, 1890 – 1mf – 9 – 0-8370-9159-4 – (incl bibl ref) – mf#1986-3159 – us ATLA [450]
De vi percussionis liber / Borelli, Giovanni Alfonso – Bononiae [Bologna]: Ex typographia Iacobi Montij 1667 [mf ed 1987?] – 1r [ill] – 1 – (incl bibl ref) – mf#2258p – us Wisconsin U Libr [530]
De Villiers, Hertha *see* Skull of the south african negro
De viris illustribus urbis romae, a romulo ad augustum : ad usum sextae scholae / Lhomond, Charles F – Quebeci: apud Joannem Neilson, 1809 [mf ed 1971] – 1r – 5 – mf#SEM16P56 – cn Bibl Nat [450]
De visibili monarchia ecclesiae libri octo / Sanderus, Nic – Lovanii, 1571 – 37mf – 8 – €71.00 – ne Slangenburg [240]
De visitatione ss. liminum et dioeceseon ac de relatione s. sedi exhibenda : commentarium in decretum "a remotissima ecclesiae aetate" iussu p2 10. pont., o.m., a s. congregatione consistoriali die 31 decembris 1909 / ed by Cappello, Felice M – Rome: F Pustet, 1912-1913 – 4mf – 9 – 0-524-04997-1 – (incl bibl ref) – mf#1990-5085 – us ATLA [240]
De vita et honestate clericorum / Fabrotus, C A – Parisiis, 1651 – €11.00 – ne Slangenburg [241]
De vita et moribus sacerdotum opusculum... / [Clichtove, J] – [Parisiis], 1519 – 2mf – 9 – mf#CA-79 – ne IDC [241]
De vita et scriptis aphraatis, sapientis persae / Forget, J – Lovanii, 1882 – €15.00 – ne Slangenburg [240]
De vita et scriptis aphraatis, sapientis persae / Forget, Jacques – Lovanii: Vanlinthout, 1882 [mf ed 1991] – (= ser Universitas catholica in oppido lovaniensi, s facultas theologica (series) 514) – 1mf – 9 – 0-7905-9924-4 – mf#1989-1649 – us ATLA [240]
De vita et scriptis sancti jacobi, batnarum sarugi in mesopotamia episcopi : cum ejusdem syriacis carminibus duobus integris ac aliorum aliquot fragmentis, necnon georgii ejus discipuli oratione panegyrica / Abbeloos, Jean Baptiste – Lovanii: Vanlinthout, 1867 – 1mf – 9 – 0-524-03450-8 – mf#1990-0993 – us ATLA [240]
De vita moribvs ac rebvs praecipve adversvs tvrcas... / Barletius, M – Argentorati, 1537 – 7mf – 9 – mf#H-8247 – ne IDC [956]
De vita pomponi secunda / Pliny The Younger [Gaius Plinius Caecilius Secundus] – (= ser Holkham library manuscript books 395,399,433) – 1r – 1 – (filmed with: leodrisius crivellus: poems, with extracts from virgil, ovid and lactantius; macrobius: de somno scipionis) – mf#97104 – uk Microform Academic [450]
De vita recessuoli... / Bluma, Daciano – Madrid: Arch. Ibero Americano, 1963 – 1 – sp Bibl Santa Ana [240]
De Vliegende tering *see* Extra cambodja-nummer
De vocabvlo fidei et aliis qvibvsdam vocabvlis, explicatio uera et utilis, sumta ex fontibus ebraicis / Flacius Illyricus d A, M – Vitebergae, 1549 – 3mf – 9 – mf#TH-1 mf 474-476 – ne IDC [242]
De Voecht *see* Narratio de inchoatione domus clericorum in zwollis
De volksraad als kristallisator van het "indische bewustzijn" / Hart, G H C – 's-Gravenhage, 1938 – 1mf – 8 – mf#SE-1284 – ne IDC [959]
De volksvriend *see* De zuid-afrikaan
De Waal, Anton *see* Papst pius 10
De Waal, Esther Aletta Susanna *see* Die onderwysstrewes van enkele politieke groeperinge in die republiek van suid-afrika
De wandschilderingen van de slangenburg : een poging tot interpretatie / Hoppenbrouwers, H – Bijdragen en Mede- delingen der Vereniging "Gelre", deel 52, 1952 – €3.00 – ne Slangenburg [740]
De Wesselow, Charles Hare Simpkinson *see* Thomas harrison, regicide and major-general
De Wet, Christiaan Rudolf *see* Three years' war
De Wette, Wilhelm Martin Leberecht *see*
- Commentar ueber die psalmen
- Human life
- Eine idee ueber das studium der theologie
- Kritischer versuch ueber die glaubwuerdigkeit der buecher der chronik
- Kurze erklaerung der briefe des petrus, judas, und jakobus
- Lehrbuch der hebraeisch-juedischen archaeologie
- Lehrbuch der historisch-kritischen einleitung in die kanonischen buecher des neuen testaments
- Lehrbuch der historisch-kritischen einleitung in die kanonischen und apokryphischen buecher des alten testaments
- Opuscula theologica
- Synopsis evangeliorum matthaei, marci et lucae cum parallelis ioannis pericopis
- Theodore

De wijsbegeerte der wetsidee / Dooyeweerd, H – 3bks – (bk1: de wetsidee als grondlegger der wijsbegeerte, amsterdam 1935 €32. bk2: de functionele zin-structuur der tijdelijke werkelijkheid en het probleem der kennis, amsterdam 1935 €32. bk3: de individualiteitsstructuren der tijdelijke werkelijkheid, amsterdam 1936 €40) – ne Slangenburg [100]
De witt advertiser – De Witt, NE: John Wehn. v1 n1. may 30 1874- // (semimthly) [mf ed sep15 1874] – 1r – 1 – (cont by: opposition) – us NE Hist [071]
De witt free press – De Witt, NE: Wm H Stout, 1877-v3 no1. may 3 1879 (wkly) [mf ed v1 n42. feb 14 1878-may 3 1879 (gaps) filmed [1974?]] – 1r – 1 – (cont by: free press) – us NE Hist [071]
De Witt, John *see*
- In memoriam, william miller paxton, d.d., ll.d., 1824-1904
- James ormsbee murray
- Praise-songs of israel
- The psalms
- What is inspiration?

De Witt, John et al *see* Ought the confession of faith to be revised?
De witt republican – De Witt, NE: J L Witters. 1v. v1 n1-39. jan 8-sep 30 1904 (wkly) [mf ed with gaps] – 1r – 1 – (merged with: dewitt record to form: dewitt eagle) – us NE Hist [071]
De witt rip-saw – De Witt, NE: John L Morrison. 1v. v1 n1-12. jul 18-nov 8 1888 (wkly) [mf ed lacks nov 1 filmed [1974?]] – 1r – 1 – (cont by: democratic guide) – us NE Hist [071]
De witt times – De Witt, NE: Suiter & Stout, aug 18 1881-v22 n46. may 7 1903 (wkly) [mf ed with gaps] – 5r – 1 – (merged with: de witt news to form: de witt times-news. numbering ceased with v19 n14 sep 13 1859; resumed with v21 n41 apr 3 1902. daily ed: daily de witt times) – us NE Hist [071]

Deacon lite / Deaconess Hospital [Milwaukee WI] – v25 n6-v28 n6 [1977 jul-1980 oct] – 1r – 1 – (cont by: center scope [milwaukee wi]) – mf#642436 – us WHS [360]

Deaconess advocate – v1-29 n2, 1-10, 1886-feb 1914 [complete] – 3r – 1 – (title varies) – mf#atla s0292 – us ATLA [240]

The deaconess and her vocation / Thoburn, James Mills – New York: Hunt & Eaton; Cincinnati: Cranston & Curtis, 1893 – 1mf – 9 – 0-7905-6739-3 – mf#1988-2739 – us ATLA [240]

The deaconess and her work / Mergner, Julie – Philadelphia, PA: United Lutheran Publication House, c1911 – 1mf – 9 – 0-524-07578-6 – (incl bibl ref) – mf#1991-3198 – us ATLA [240]

Deaconess Hospital [Milwaukee WI] *see* Deacon lite

Deaconesses, biblical, early church, european, american : with the story of the chicago training school, for city, home and foreign missions, and the chicago deaconess home / Meyer, Lucy Rider – 2nd rev enl ed. Chicago, IL: Message Publ Co, c1889 [mf ed 1984] – (= ser Women & the church in america 29) – 2mf – 9 – 0-8370-0205-2 – mf#1984-2029 – us ATLA [305]

Deaconesses in europe : and their lessons for america / Robinson, Jane Marie Bancroft – New York: Hunt & Eaton, 1890 [mf ed 1984] – (= ser Women & the church in america 205) – 1mf – 9 – 0-8370-1434-4 – mf#1984-2205 – us ATLA [240]

Deaconesses in the church of england : a short essay on the order as existing in the primitive church and on their present position and work / Howson, John Saul – London: Griffith and Farran, 1880 – 1mf – 9 – 0-524-06815-1 – mf#1991-2802 – us ATLA [241]

The deaconship / Howell, Robert Boyte Crawford – Philadelphia: American Baptist Publication Society, 1851, c1846 – 1mf – 9 – 0-7905-3915-2 – mf#1989-0408 – us ATLA [240]

The dead cities of the zuyder zee: a voyage to the picturesque side of holland / Havard, Henry – Annie Wood, trans. Illus. by Havard and van Beest. London: R. Bentley & son, 1875. xii, 363p. 9 pl – 1 – us Wisconsin U Libr [949]

Dead in christ / Wilkins, George – London, England. 1858 – 1r – us UF Libraries [240]

Dead in trespasses and sins / Simpson, J Y – London, England. 18– – 1r – us UF Libraries [240]

The dead king / Kipling, Rudyard – Toronto: Musson [1910?] [mf ed 1995] – 1mf – 9 – 0-665-77249-1 – mf#77249 – cn Canadiana [810]

Dead march and monody / Carr, Benjamin – Performed in the Lutheran Church, Philadelphia, on Thursday, 26 December 1799, being part of the music selected for funeral honours to our late illustrious cheif General George Washington. Baltimore: J. Carr 1799-1800. MUSIC 3082, Item 2 – 1 – us L of C Photodup [780]

Dead mountain echo – Oakridge OR: Dead Mountain Graphics, [wkly] – 1 – (1974-76 incl newspaper pub during school terms by oakridge high school students) – us Oregon Lib [071]

The dead pulpit / Haweis, Hugh Reginald – London: Bliss, Sands, 1896 [mf ed 1984] – 4mf – 9 – 0-8370-0911-1 – mf#1984-4261 – us ATLA [242]

The dead queen : sermon preached in st peter's, brockville, on sunday morning, jan 27th, the sunday after the death of her most gracious majesty queen victoria, on january 22nd / Bedford-Jones, T – [Brockville, Ont?: s.n, 1901?] – 1mf – 9 – 0-665-65931-8 – mf#65931 – cn Canadiana [242]

The dead sea scrolls : a personal account / Trever, John C – rev ed. Grand Rapids: W B Eerdmans, 1979, c1977 – 1mf – 9 – 0-8370-1772-6 – (incl bibl ref) – mf#1984-4497 – us ATLA [930]

Dead serious : voice of workers for safe energy and black hills alliance / Black Hills Alliance et al – v1 n1-2 [1979 summer-autumn] – 1r – 1 – mf#667578 – us WHS [366]

Deady, M P *see* Deady's reports of cases in the ninth circuit, 1859-1869

Deady's reports of cases in the ninth circuit, 1859-1869 / Deady, M P – San Francisco: Bancroft. 1v. 1872 (all publ) – (= ser Early federal nominative reports) – 8mf – 9 – $12.00 – mf#LLMC 81-447 – us LLMC [340]

Deaf american – Silver Springs MD 1888-1989 – 1,5,9 – (cont by: deaf american monograph) – ISSN: 0011-720X – mf#3209.01 – us UMI ProQuest [616]

Deaf american monograph – Silver Springs MD 1990-99 – 1,5,9 – (cont: deaf american) – ISSN: 1065-7191 – mf#3209,01 – us UMI ProQuest [616]

The deaf churchman – Syracuse NY: Conference of Church Workers among the Deaf, 1946- [bimthly ex jul & aug,. mthly [mf ed 2005] – v23-32 (1946-56) on 1r – 1 – (ceased in 1977? place of publ varies; cont: silent missionary; cont by: deaf episcopalian; some pgs damaged) – mf#2005c-s080 – us ATLA [242]

Deakin, Ralph *see* Southward ho!

Dealerscope – Philadelphia PA 2000+ – 1,5,9 – (cont: dealerscope consumer electronics marketplace) – ISSN: 1534-4711 – mf#16081,03 – us UMI ProQuest [621]

Dealerscope consumer electronics marketplace – Philadelphia PA 1996-99 – 1,5,9 – (cont: dealerscope merchandising; cont by: dealerscope) – ISSN: 1087-1055 – mf#16081.03 – us UMI ProQuest [621]

Dealerscope merchandising – Philadelphia PA 1986-94 – 1,5,9 – (cont by: dealerscope consumer electronics marketplace) – ISSN: 0888-4501 – mf#16081.03 – us UMI ProQuest [621]

Dealings with the firm of dombey and son / Dickens, Charles – London, England. 1910 – 1r – us UF Libraries [420]

Dealings with the inquisition : or, papal rome, her priests, and her jesuits. with important disclosures / Achilli, Giacinto – New York: Harper, 1851 – 1mf – 9 – 0-524-03330-7 – mf#1990-0911 – us ATLA [241]

Dealtry, T *see* Minister's parting advice and valediction

Dealtry, W *see* Gospel message

Dealtry, William *see*
- Character and happiness of them that die in the lord
- Charge delivered in the autumn of 1834, at the visitation in hampsh...
- Chruch and its endowments
- Excellence of the liturgy
- Obligations of the national church
- On contending for the faith once delivered to the saints
- On the importance of caution in the use of certain familiar words

Dean – (= ser Bedfordshire parish register series) – 2mf – 9 – £5.00 – uk BedsFHS [929]

Dean, A *see* A tour through the upper provinces of hindostan

Dean, Arnold Walker *see* Search for hafnium in florida minerals

Dean church / Lathbury, Daniel Conner – Oxford: A.R. Mowbray, 1905 – (= ser Leaders Of The Church) – 1mf – 9 – 0-7905-5246-9 – (incl bibl ref) – mf#1988-1246 – us ATLA [240]

Dean, George Alfred *see* A treatise on the land tenure of ireland

Dean, Harry *see* Pedro gorino

Dean, James *see* Religion an essential element of education

Dean, John *see* Every man's duty to be a teetotaller proved

Dean, John Marvin *see* The cross of christ in bolo-land

Dean, John Taylor *see*
- The book of revelation
- Visions and revelations

Dean, Maurice B *see* A digest of corporation cases

Dean Medical Center *see* Monthly bulletin for our patients and friends

Dean, Nina Oliver *see*
- Golden harvest
- Orlando day nursery

Dean Of Moray *see* Priests' manual

Dean, Peter *see* The life and teachings of theodore parker

Dean swamp. barnwell county. south carolina : church records – 1880-1921, 1932-44 – 1 – us Southern Baptist [242]

Dean, William *see* The china mission

Deane, Henry *see* Daniel

Deane, John C *see* Narrative of the atlantic telegraph expedition, 1865

Deane, W J *see* On the celebration of the holy eucharist

Deane, William John *see*
- Abraham, his life and times
- The book of wisdom
- David, his life and times
- Joshua, his life and times
- Pseudepigrapha
- Samuel and saul
- Samuel and saul; their lives and times

Deaner, Heather R *see* An exploratory factor analysis of collegiate athletes' perceptions of psychological adjustment to sport disengagement

The deanery magazine – Sussex, NB: C Medley, [1889] [mf ed v6 n2 feb 1889-v6 n6 jun 1889; v6 n8 aug 1889; v6 n11 nov 1889-v6 n12 dec 1889] – 9 – mf#P04524 – cn Canadiana [242]

The dean's english : a criticism on the dean of canterbury's essays on the queen's english / Moon, George Washington – 6th ed. London: Hatchard, 1868 – 1mf – 9 – 0-8370-9971-4 – mf#1986-3971 – us ATLA [420]

The dean's handbook to gloucester cathedral / Spence-Jones, Henry Donald Maurice – London: JM Dent, 1913 – 1mf – 9 – 0-7905-8153-1 – mf#1988-6100 – us ATLA [720]

Deansgate catalogue of printed books : john rylands university library – [mf ed Chadwyck-Healey] – 1081mf – 9 – uk Chadwyck [020]

Dear fatherland / Bilse, Fritz Oswald – London, New York: J Lane, 1905 [mf ed 1989] – 257p – 1 – mf#7023 – us Wisconsin U Libr [890]

Dear mr schroder – mf#ZB 34 – nz Nat Libr [079]

Dear sooky / Crosby, Percy Leo – New York, NY. 1929 – 1r – us UF Libraries [025]

Dearborn independent – Aurora IN. 1870 mar 3, oct 27, 1971 apr 13, may 25, jun 8, sep 21, 1872 jan 4-11, feb 1-may 2,16-jul 4,18-aug 29, sep 12-26, oct 10-31, nov 21-28, dec 12-19, 1900 jul 5,19 – 1r – 1 – (cont: aurora commercial) – mf#880344 – us WHS [071]

Dearborn Observatory *see* Annals of the dearborn observatory of northwestern university

Dearborn society newsletter – v1 n1-4 [1983 jan-oct] – 1r – 1 – mf#687537 – us WHS [978]

Dearing, F *see* Wesleyan and tractarian worship

Dearmer, Percy *see*
- Body and soul
- Everyman's history of the english church
- Everyman's history of the prayer book
- False gods
- Is "ritual" right?
- The parson's handbook
- Religious pamphlets
- Reunion and rome
- The story of the prayer-book in the old and new world

Death / Weaver, Richard – London, England. 18– – 1r – us UF Libraries [025]

Death abolished, and life and immortality brought to light / Hawker, Robert – London, England. 1824 – 1r – us UF Libraries [240]

Death- and after? / Besant, Annie Wood – London: Theosophical Pub Society, 1901 – (= ser Samp: indian books) – us CRL [230]

Death and burial lore in the english and scottish popular ballads / Wimberley, Lowry Charles – Lincoln, Neb: [s n] 1927 [mf ed Bloomington IN: Indiana Uni Lib, Preservation Dept 1984] – 138p on 1r – 1 – us Indiana Preservation [390]

Death and the life beyond : in the light of modern religious thought / Spurr, Frederic Chambers – New York: Hodder & Stoughton [1913?] [mf ed 1993] – 1mf – 9 – 0-524-06501-2 – mf#1991-2601 – us ATLA [240]

Death blow to corrupt doctrines : a plain statement of facts published by the gentry and people / Masteer, Calvin W – Shanghai, 1870 – (= ser 19th c books on china) – 1mf – 9 – (a chinese pamphlet against christianity) – mf#7.1.26 – uk Chadwyck [306]

Death by wrongful act...2nd ed / Tiffany, Francis Buchanan – Kansas City, Mo., Vernon, 1913. 646 p. LL-1317 – 1 – us L of C Photodup [340]

Death education – Abingdon, Oxfordshire 1977-84 – 1,5,9 – (cont by: death studies) – ISSN: 0145-7624 – mf#11134.01 – us UMI ProQuest [150]

Death education and death anxiety in student nurse aides / Kienow, Nancy L & Heit, Philip – 1992 – 2mf – 9 – $8.00 – us Kinesology [613]

Death, gain to the believer / Gloag, Paton James – Edinburgh, Scotland. 1875 – 1r – us UF Libraries [240]

Death lists and lot registers / Shawnee Center Cemetery, Shawnee County, KS – 1875-1990 – 1 – us Kansas [920]

Death of a christian soldier at the battle of barossa / Innes, William – Stirling, Scotland. 1855? – 1r – us UF Libraries [240]

The death of a nation : or, the ever persecuted nestorians or assyrian christians / Yohannan, Abraham – New York: Putnam, 1916 [mf ed 1993] – 1mf – 9 – 0-524-06235-8 – mf#1991-0028 – us ATLA [240]

Death of an infidel – London, England. 18– – 1r – us UF Libraries [240]

The death of christ : its place and interpretation in the new testament / Denney, James – 5th ed. New York: A C Armstrong, 1907 – 1mf – 9 – 0-8370-2889-2 – (incl ind) – mf#1985-0889 – us ATLA [220]

The death of death : or, a study of god's holiness in connection with the existence of evil / Patton, John Mercer – rev ed. London: Truebner, 1881 [mf ed 1992] – (= ser Unitarian/universalist coll) – 1mf – 9 – 0-524-06437-7 – mf#1991-2559 – us ATLA [210]

Death of eminently good men a source of great lamentation to the ch... / Thomson, Adam – Edinburgh, Scotland. 1820 – 1r – us UF Libraries [240]

Death of god's saints / Fraser, John – Brechin, Scotland. 1883? – 1r – us UF Libraries [240]

Death of lord rochester – London, England. no date – 1r – us UF Libraries [240]

Death of nadab and abihu / Hogg, Robert – Whitehaven, England. 1821 – 1r – us UF Libraries [240]

The death of the good man...rev john brown / Blythe, James – 1804 – 1 – $50.00 – us Presbyterian [240]

The death of the verbal theory and the unveiling of christ : or, the bible a "sufficient witness" to the "self-effacing christ" / MacInnes, George – Sydney: George Robertson, 1894 [mf ed 1985] – 1mf – 9 – 0-8370-4235-6 – mf#1985-2235 – us ATLA [220]

Death penalty reporter – v1. 1980-81 (all publ) – 5,6 – $45.00 set – (available in reel only) – mf#102341 – us Hein [345]

The death problem in the life and works of gerhart hauptmann / Klemm, Frederick Alvin – Philadelphia: [s.n.], 1939 – 1r – 1 – (incl bibl ref) – us Wisconsin U Libr [430]

Death real and apparent : in relation to the sacraments = Muerte real y la muerte aparente / Ferreres, Juan Bautista – St Louis, Mo: B Herder, 1906 – 1mf – 9 – 0-7905-7574-4 – (incl bibl ref. in english) – mf#1989-0799 – us ATLA [240]

Death scenes and other poems / Allom, Elizabeth Anne – Hackney, London: Caleb Turner; & Simpkin and Marshall, 1844 – (= ser 19th c women writers) – 1mf – 9 – mf#5.1.143 – uk Chadwyck [810]

Death ship times – [[1973] jun-1974 jun] – 1r – 1 – mf#365078 – us WHS [071]

The death song of an indian chief, from "ouabi" / Gram, Hans – Printed as supplement to the March, 1791, number of "The Massachusetts Magazine." According to Sonneck, the first orchestral score published in the U.S. MUSIC 1117 – 1 – us L of C Photodup [780]

Death squads, guerrilla wars, covert operations, and genocide : guatemala and the united states, 1954-1999 – [mf ed Chadwyck-Healey] – (= ser National security archive, washington dc: the making of us policy) – 2000+ docs on 388mf – 9 – (with p/g & ind) – uk Chadwyck [327]

Death struggles of slavery : being a narrative of facts and incidents, which occurred in a british colony, during the two years immediately preceding negro emancipation / Bleby, Henry – London, 1853 – (= ser 19th c books on british colonization) – 4mf – 9 – mf#1.1.1377 – uk Chadwyck [972]

Death studies – Abingdon, Oxfordshire 1985+ – 1,5,9 – (cont: death education) – ISSN: 0748-1187 – mf#11134,01 – us UMI ProQuest [150]

Death swallowed up in victory / Symington, Andrew – Edinburgh, Scotland. 1844 – 1r – us UF Libraries [240]

Death the last enemy / Brydie, Andrew – Edinburgh, Scotland. 1866 – 1r – us UF Libraries [240]

The death treasure of the khmers / Colam, Lance – London: Stanley Paul & Co [1939] [mf ed 1989] – 1r with other items – 1 – mf#mf-10289 seam reel 002/16 [§] – us CRL [915]

Death-bed repentance / Woollacott, Christopher – London, England. no date – 1r – us UF Libraries [240]

Death-bed scenes : or, dying with and without religion, designed to illustrate the power and truth of christianity / ed by Clark, Davis Wasgatt – New York: Lane & Scott, 1852 [mf ed 1984] – 7mf – 9 – 0-8370-1026-8 – mf#1984-4389 – us ATLA [240]

Death-bed scenes / Stock, John – London, England. no date – 1r – us UF Libraries [240]

Death-bed testimony / Medhurst, T W – London, England. 18– – 1r – us UF Libraries [240]

Debacle – Brussels Belgium, 7 jan-22 jan, 23 jul-6 aug 1893 – 1/4r – 1 – uk British Libr Newspaper [074]

La debacle – Brussels. Belgium. -m. 7 Jan-6 Aug 1893. (14 ft) – 1 – uk British Libr Newspaper [949]

Debacle du congo belge / Monstelle, Arnaud De – Bruxelles, Belgium. 1965 – 1r – us UF Libraries [960]

Debacle generale du comite de la rue de poitiers avec la biographie de ses candidats avortes – Paris [1848?] – us CRL [940]

DeBacy, Diane L *see* Development of a child injury data base for use in biomechanics research

De-bah-ji-mon : telling news / Leech Lake Band of Chippewa Indians – Cass Lake, Leech Lake Indian Reservation MN. 1979 apr-1981 sep, 1981 dec-1986 dec – 2r – 1 – mf#635032 – us WHS [307]

Debar, C C *see* Vingt quatre variations

Debat sur l'adresse, novembre 1896 : discours de m f-x lemieux – S.l: s.n, 1896? – 1mf – 9 – mf#02620 – cn Canadiana [323]

Debat tusschen ds. e.l. meinders, herder en leeraar bij de ware hollandsche gereformeerde kerk te south holland, ill., en ds. g.e. boer, docent bij de theol. school van de holl. christ. geref. kerk te grand rapids, mich : opzicht hebbende op het arminianismus en op de leer der rechtvaardiging / Meinders, E L & Boer, Geert Egberts – Holland, MI: De Grondwet Boekdrukkerij, 1894 – 1mf – 9 – 0-524-06645-0 – mf#1991-2700 – us ATLA [240]

The debatable land between this world and the next : with illustrative narrations / Owen, Robert Dale – New York: G.W. Carleton; London: Trubner, 1872, c1871 – 2mf – 9 – 0-7905-3395-2 – (incl bibl ref) – mf#1987-3395 – us ATLA [130]

Debate – Miami, FL. 1976 jul 09-1978 feb 27 – 1r – us UF Libraries [071]

El debate – Madrid. Spain. -d. 1-31 Jan 1915, 1 Jan 1917-19 Jul 1936. (76 reels) – 1 – uk British Libr Newspaper [074]

O debate : orgao popular – Manaus, AM. 27 abr-maio 1901; jan, jun 1902; abr-12 jul 1903 – (= ser Ps 19) – mf#P11B,06,23 – bl Biblioteca [321]

Debate between rev a campbell and rev n l rice : on the action, subject, design and administrator of christian baptism...lexington, ky, from the 15th nov to the 2nd dec 1843... – Lexington, KY: AT Skillman, 1844 [mf ed 1992] – (= ser Christian church (disciples of christ) coll) – 3mf – 9 – 0-524-03136-3 – mf#1990-4585 – us ATLA [242]

The debate between the church and science : or, the ancient hebraic idea of the six days of creation / Lewis, Tayler – Andover: Warren F Draper, 1860 – 1mf – 9 – 0-7905-0174-0 – mf#1987-0174 – us ATLA [210]

El debate constitucional. discursos en la asamblea 1931-1933. lima 1933 / Belaunde, Victor Andres – Madrid: Razon y Fe, 1934 – 1 – sp Bibl Santa Ana [972]

Debate in the senate on the public expenditure of the dominion, march 1878 : speeches of the hon messrs macpherson, mclelan and campbell / Macpherson, (David Lewis – Ottawa: s.n, 1878 [mf ed 1984] – 1mf – 9 – 0-665-02432-0 – mf#02432 – cn Canadiana [336]

Debate on baptism and kindred subjects : between elder james m. mathes, of the church of christ, and reverend t.s. brooks, of the m.e. church. held in the town-hall in bedford, ind... / Mathes, James Madison – Cincinnati: HS Bosworth, 1868 – 1mf – 9 – 0-524-07632-4 – mf#1991-3239 – us ATLA [242]

A debate on baptism and the witness of the holy spirit : held in fairview, ia., nov 1847 / Terrell, Williamson & Pritchard, Henry Russell – Milton [i.e. Indiana]: Franklin & Smith, 1848 [mf ed 1992] – (= ser Christian church (disciples of christ) coll) – 3mf – 9 – 0-524-02507-X – mf#1990-4366 – us ATLA [242]

A debate on christian baptism : between the rev w l maccalla, a presbyterian teacher, and alexander campbell, held at washington, ky... 15th...21st oct 1823... / Campbell, Alexander & McCalla, William Latta – Buffaloe: Campbell & Sala, 1824 [mf ed 1993] – (= ser Christian church (disciples of christ) coll) – 4mf – 9 – 0-524-06749-X – mf#1990-5275 – us ATLA [242]

Debate on first day adventism : between albert t fitts and g c minor...oct 16th-18th, 1900 – Knoxville, TN: SB Newman, 1901 [mf ed 1993] – (= ser Christian church (disciples of christ) coll) – 1mf – 9 – 0-524-07748-7 – mf#1991-3316 – us ATLA [242]

Debate on the action of baptism : the design of baptism, the subjects of baptism, the work of the holy spirit, the discipline of the m.e. church, and human creeds. held in vienna, johnson county, ill... / Braden, Clark & Hughey, George Washington – Cincinnati: Published by Franklin & Rice for C Braden, c1870 – 2mf – 9 – 0-524-02107-4 – mf#1990-4173 – us ATLA [242]

A debate on the beginning of messiah's reign, the abrogation of the mosaic law, and first proclamation of the gospel : baptism, action, subject, and design / Brooks, John A & Fitch, J W – Cincinnati: RW Carroll, 1870 [mf ed 1992] – (= ser Christian church (disciples of christ) coll) – 1mf – 9 – 0-524-02110-4 – mf#1990-4176 – us ATLA [242]

Debate on the fisheries bill of the hon alex campbell, commissioner of crown lands : in the legislative council on the 9th and 10th march, 1865 – Quebec?: "Daily News", 1865 – 1mf – 9 – mf#34363 – cn Canadiana [639]

A debate on the roman catholic religion : ...cincinnati, from the 13th to the 21st of jan 1837 / Campbell, Alexander & Purcell, John Baptist – Cincinnati: UP James, 1855 [mf ed 1993] – (= ser Christian church (disciples of christ) coll) – 4mf – 9 – 0-524-07857-2 – mf#1991-3402 – us ATLA [241]

A debate on total depravity, election, the polity or church government of the regular baptist church, free moral agency... / Thompson, Gregg M & Burgess, Otis Asa – Indianapolis: Levi Pennington, 1868 [mf ed 1993] – (= ser Methodist coll) – 1mf – 9 – 0-524-07046-6 – mf#1991-2899 – us ATLA [240]

Debate on trine immersion, the lord's supper, and feet-washing : between elder james quinter, of ohio (german baptist), and elder n a m'connell, of iowa (disciple)...14th to the 18th of oct 1867 – Cincinnati: H S Bosworth c1868 [mf ed 1991] – 1mf – 9 – 0-524-00855-8 – mf#1990-4015 – us ATLA [242]

Debates – 1846-74 – 5r – 1 – (debates of parliament as recorded in newspapers) – cn Library Assoc [971]

Debates / Great Britain. Parliament – 36v. 1066-1803. (Hansard) – 1 – us AMS Press [941]

Debates / Nigeria. Eastern Region. House of Assembly – 11 Jul 1952-30 Nov 1965 – 1 – us L of C Photodup [960]

Debates / Nigeria. House of Representatives – Jan 1952-May 1965 – 1 – us L of C Photodup [960]

Debates / Nigeria. Legislative Council – Feb 1924-Aug 1951 – 1 – us L of C Photodup [960]

Debates / Nigeria. Western Provinces. House of Assembly – 1947-50 – 1 – us L of C Photodup [960]

Debates / Nigeria. Western Region. House of Assembly – 1st-7th sessions. 1952-59 – 1 – us CRL [960]

Debates / Nigeria. Western Region. House of Chiefs – 2nd, 4th-6th sessions. 1953, 1956-58 – 1 – us CRL [960]

Debates / Rhodesia. (Northern). Legislative Council – May 1924-Dec 1945 – 1 – us L of C Photodup [324]

Debates / Rhodesia. Southern. Legislative Council (later Legislative Assembly) – 1899-1964 65. Lacking only fourth session of the third council for 1907 – 1 – us L of C Photodup [960]

Debates and proceedings of legislative council / Prince Edward Island – 1867-93 – (= ser Legislative Council) – 2r – 1 – cn Library Assoc [971]

The debates in the federal convention of 1789 : which framed the constitution of the united states / Madison, James; ed by Hunt, Galliard & Scott, James B – Buffalo, NY: Prometheus Books. 2v. 1987 – 8mf – 9 – $12.00 – mf#LLMC 90-360 – us LLMC [323]

The debates in the several state conventions on the adoption of the federal constitution, as recommended by the general convention at philadelphia, in 1787 / Elliot, Jonathan – 2d ed. Philadelphia, Lippincott, 1896. 5 v. LL-166 – 1 – us L of C Photodup [323]

Debates of both houses of parliaments – Cape Town: Printed & publ under contract with the Union Parliament by Cape Times, sep 9/14 1914 – us CRL [324]

Debates of the federal constitution : in the several state conventions – 2nd ed. Philadelphia: Lippincott. v1-5. 1836 – 33mf – 9 – $49.50 – mf#LLMC 84-243 – us LLMC [323]

Debates of the house of commons in the year 1774 on the bill for making more effectual provision for the government of province of quebec / Cavendish, Henry – London: Ridgway Piccadilly, 1839 [mf ed 1982] – 4mf – 9 – mf#SEM105P86 – cn Bibl Nat [323]

Debates: official report / Nigeria. North-Central States Legislature. House of Assembly – Feb 1952-Aug 1959 – 6 – $40.00 – us L of C Photodup [960]

Debats / European Coal and Steel Community. Common Assembly – Luxemburg. Sept 10 13 1952-Feb 28 1958. Incomplete – 1 – us NY Public [324]

Debats / France. Assemblee consultative provisoire – Algers puis Paris. 4 nov 1943-20 oct 1945 – 1 – fr ACRPP [323]

Debats / France. Assemblee de l'Union francaise – 10 dec 1947-29 mai 1958 – 1 – fr ACRPP [323]

Debats / France. Assemblee puis Nationale Constituante – 7nov 1945-27 nov 1946 – 1 – fr ACRPP [323]

Les debats – Montreal, QC: P LeMoyne de Martigny, 1899-1904 – 2r – 1 – cn Library Assoc [971]

Les debats – Port-au-Prince: [s.n.], [1951-]. [1 annee, n1-6 annee, n203 1er mai 1951-23 dec 1956] – us CRL [079]

Debats dans l'assemblee legislative sur la tenure seigneuriale / Canada (Province). Parlement. Assemblee legislative – Quebec: E R Frechette, 1853 [mf ed 1983] – 1mf – 9 – mf#SEM105P263 – cn Bibl Nat [323]

Debats dans l'assemblee legislative sur la tenure seigneuriale / Canada (Province). Parlement. Assemblee legislative – Quebec: E R Frechette...1853 [mf ed 1983] – 1mf – 9 – mf#SEM105P263 – cn Bibl Nat [323]

Debats de la convention nationale : ou analyse complete des seancesavec les noms de tous les membres, petitionnaires ou personnages qui ont figure dans cette assemblee / Thiesse, Leon – Paris 1828 [mf ed Hildesheim 1995-98] – 5v on 17mf – 9 – €170.00 – 3-487-26284-3 – gw Olms [323]

Debats de la legislature de la province de quebec / Desjardins, Alphonse – Quebec: [s.n.] (Quebec: impr de L J Demers & frere) 1881-1890 [mf ed 1994] – 10r – 1 – (cont: debats de la legislature provinciale de la province de quebec; cont by: debats de l'assemblee legislative de la province de quebec) – mf#SEM35P401 – cn Bibl Nat [323]

Debats de la legislature provinciale de la province de quebec / Desjardins, Alphonse – Quebec: impr du Canadien. dec 1877/mars 1878-1880 [mf ed 1994] – 1r – 1 – (cont by: debats de la legislature de la province de quebec) – mf#SEM35P400 – cn Bibl Nat [323]

Debats de l'assemblee legislative de la province de quebec / Desjardins, Louis Georges – Quebec: impr de L J Demers, 1895 [mf ed 1994] – 1r – 1 – (cont: debats de la legislature de la province de quebec) – mf#SEM35P402 – cn Bibl Nat [323]

Debats de l'assemblee nationale : series compte rendu et questions / France. Assemblee Nationale – 1974- – ca 200mf per yr – 9 – €153.21y – (1881-1910 eur365.88. 1911-1940 eur457.35. 1944-1973 eur524.49. 1974-1983 eur868.96. 1881-1983 eur591.63) – fr Journal Officiel [944]

Les debats du senat : series compte rendu et questions / France. Senat – 1987- – 9 – €116.60y – (backfile: 1881-1910 €30.49. 1911-40 €30.49. 1947-76 €53.36 1977 €71.65; complete coll: 1881-1910 €335.39 1911-1940 €426.86 1947-1976 €1494 1977-1986 €640.29 1881-1986 €2515.41) – fr Journal Officiel [944]

Debats parlementaires / France. Assemblee Nationale – 29 nov 1946-85 – 1 – fr ACRPP [323]

Debats parlementaires / France. Chambre des Deputes – 1881-10 JUIL 1940 – 1 – fr ACRPP [323]

Debats parlementaires / France. Conseil de la Republique – 26 dec 1946-58 – 1 – fr ACRPP [323]

Debats parlementaires / France. Senat – 1881-10 juil 1940 – 1 – fr ACRPP [323]

Debats parlementaires / France. Senat – 1959-86 – 1 – fr ACRPP [323]

Debats parlementaires / France. Senat – Compte rendu in extenso des comites secrets des 14 mars et 16 avr 1940. no. special du 2 aout 1948 – 1 – fr ACRPP [323]

Debats parlementaires sur la question de la confederation des provinces de l'amerique britannique du nord : 3e session, 8e parlement provincial du canada = Parliamentary debates on the subject of the confederation of the british north american provinces / Canada (Province). Parlement – Quebec: Hunter, Rose & Lemieux, 1865 [mf ed 1984] – 11mf – 9 – mf#SEM105P391 – cn Bibl Nat [323]

Die debatte – Koeln DE, 1956 n5, 1956 n14-1971 jan 1971, 1978 n1 – 1 – gw Misc Inst [320]

Debatterna sasom de framsta under diskussionen af lutherska och methodist episkopal kyrkans laera : wid moetet hallit i milwaukee, wisconsin, februari 1866 – Chicago: Poe & Hitchcock 1866 [mf ed 1993] – 1mf – 9 – 0-524-06607-8 – (in swedish) – mf#1991-2662 – us ATLA [242]

Deben derogarse los escritos de replica, duplica y extracto de litis del codigo de procedimientos civiles del estado de la baja california / Guerrero Meza, Hector Emilio – Guadalajar Mexico Universidad Autonoma de Guadalajara, 1969. 52 3 p. LL-8026 – 1 – us L of C Photodup [340]

Debenham, Frank *see*
- Nyasaland
- Study of an african swamp

El deber del estado con relacion a la riqueza intelectual de un pueblo / Perez Bueno, Fernando – Madrid: est tip de jaime rates, 1917 – sp Bibl Santa Ana [946]

El deber y la pasion / Solar y Taboada, Antonio – Badajoz: tip del nuevo diario s.a. – 1 – sp Bibl Santa Ana [946]

Deberard, Philip E *see* Promoting florida

Deberes de centroamerica con guatemala ante el cas... / Chica, Luis Alonso – San Salvador, El Salvador. 1964 – 1r – us UF Libraries [972]

Deberes y facultades de los alcaldes de barrio / Alonso, Longinos – Santiago, Chile. 1939 – 1r – us UF Libraries [972]

Debidour, Antonin *see*
- L'eglise catholique et l'etat sous la troisieme republique
- Histoire des rapports de l'eglise et de l'etat en france de 1789 a 1870

Debien, Gabriel *see*
- Colons de saint-domingue et la revolution
- Peuplement des antille francaises au 17e siecle

Debit / Industrial Insurance Agents Union et al – v1 n2, 4 [1937 dec, 1938 may], v4 n6-v5 n1 [1938 jun-1939 jan] – 2r – 1 – (cont by: ledger [new york ny]; uopwa news) – mf#3565611 – us WHS [331]

Deblois, Isidore Gregoire *see* Theoretical and practical system of book-keeping by single and double entry

Debogrii-Mokrievich, V *see* Svobodnaia rossiia

Die deborah – Cincinnati. Ohio. 1855-1900 – 1 – us AJPC [071]

Deborah and barak : an oratorio [for solo voices, chorus, and orchestra] / Greene, Maurice – [1732?] [mf ed 19–] – 1r – 1 – mf#pres. film 16 – us Sibley [780]

Debouge, Xavier *see* L'immaculee conception

Debout *see* Vie de saint camille de lellis...paris, 1932

Debout, Jacques *see* D'apres les paraboles histoires vraies...

DeBow, Samuel P *see* Scrapbook, 1915-1936

Debow's review – 1 – (formerly: the commercial review of the south and the west, etc v1-34 1846-64) – us AMS Press [073]

Debrecen – Debrecen, Hungary. 1 Feb-30 Apr 1950 – 1r – 1 – us L of C Photodup [079]

Debret, Jean Baptiste *see* Viagem pitoresca e historica ao brasil

Debrett's baronetage and knightage : to which is added much information respecting the immediate family connections of baronets... / ed by Mair, Robert H – library ed. London: Dean & son 1880 [mf ed 1984] – 1r – 1 – (with: autobiography of dean merivale / merivale, c) – mf#8664 – us Wisconsin U Libr [929]

Debs, Joseph *see* Protestation en faveur de la perpetuelle orthodoxie des maronites

Deb's magazine – Chicago. v1 n2-v2 n20. sept 1921-apr 1923. (incomplete) – 1 – us NY Public [073]

Debs magazine – v1-2. 1921-23 [all publ] – (= ser Radical periodicals in the united states, 1881-1960. series 1) – 5mf – 9 – $95.00 – us UPA [073]

Debt payer – Richmond VA. 1881 jul 22 – 1r – 1 – mf#883900 – us WHS [332]

Debtor's journal – La Salle IL 1820-21 – 1 – mf#3736 – us UMI ProQuest [332]

Debu, K I *see* Kollektivnoe ispolzovanie selskokhoziaistvennykh mashin i orudii

Debunking the so-called spanish mission near new smyrna beach, volusia county, florida / Coe, Charles H – Daytona Beach, FL. 1941 – 1r – us UF Libraries [978]

Deburaux, Edouard *see* Du tchad au dahomey en ballon.

Debussy, Claude *see*
- Marche des anciens comtes de ross; marche pour piano a quatre mains
- La mer
- Minstrels. transcription pour piano et hartmann

Les debuts de la litterature allemande du 8e au 12e siecles / Fuchs, Albert – Paris: Les Belles Lettres, 1952 [mf ed 1993] – (= ser Publications de la faculte des lettres de l'universite de strasbourg 118) – 172p – 1 – (incl bibl ref and ind) – mf#8156 – us Wisconsin U Libr [430]

Les debuts de l'oeuvre africaine de leopold 2 1875-1879 / Roeykens, Auguste – Bruxelles, 1955 – 1 – us CRL [960]

Les debuts du lyrisme en allemagne : des origines a 1350 / Moret, Andre – Lille: Bibliotheque universitaire, 1951 [mf ed 1993] – (= ser Travaux et memoires de l'université de lille 27) – 356p – 1 – (incl bibl ref and ind) – mf#8173 – us Wisconsin U Libr [430]

Una decada de progreso en badajoz / Duarte Insua, Lino – Badajoz: Dip. Provincial, 1945 – 1 – sp Bibl Santa Ana [946]

Decada republicana – Rio de Janeiro, Brazil. vi-8. 1899-1901 – 1r – us UF Libraries [972]

Decadas de una cultura / Felice Cardot, Carlos – Caracas, Venezuela. 1951 – 1r – us UF Libraries [972]

La decade egyptienne – Journal litteraire et d'economie politique. Le Caire. 1799-1800 (I-III) – 1 – fr ACRPP [960]

A decade of christian endeavor, 1881-1891 / Pratt, Dwight Mallory – New York: Fleming H Revell, c1891 [mf ed 1991] – 1mf – 9 – 0-524-01128-1 – mf#1990-0342 – us ATLA [240]

A decade of civic development / Zueblin, Charles – Chicago: University of Chicago Press, 1905 [mf ed 1970] – (= ser Library of american civilization 10368) – vii/188p on 1mf – 9 – us Chicago U Pr [710]

A decade of foreign missions, 1880-1890 / Tupper, Henry Allen – Richmond, VA: Foreign Mission Board of the Southern Baptist Convention, 1891 [mf ed 1990] – 3mf – 9 – 0-7905-8160-4 – mf#1988-6107 – us ATLA [242]

Decade of progress / Davis, Lynn M, Jr – A history of the State Convention of Baptists in Ohio. 1954-64 – 1 – 5.00 – us Southern Baptist [242]

La decade philosophique, litteraire et politique *see* La revue philosophique, litteraire et politique

Decadence and other essays on the culture of ideas / Gourmont, Remy De – New York, NY. 1921 – 1r – us UF Libraries [025]

La decadence artistique et litteraire – no. 1-3. Paris. oct 1886. mq no. 4. Supplement de: Le Scapin voir a ce titre – 1 – fr ACRPP [800]

Decadencia del contrato / Buen Lozano, Nestor De – Mexico City? Mexico. 1965 – 1r – us UF Libraries [972]

Le decadent litteraire et artistique – Dir. A. Baju. no. 1-35. avr-dec 1886. devenu: Le Decadent. Revue litteraire bimensuelle. 2e s., no. 1-32. dec 1887-avr 1889. devenu: La France litteraire. Philosophie, critique, sociologie. n.s., no. 33-35. Paris. mai 1889 – 1 – fr ACRPP [800]

The decades : the fifth decade / Bullinger, Heinrich; ed by Harding, T – Cambridge, 1852 – 7mf – 9 – mf#PBU-682 – ne IDC [240]

The decades : the first and second decades / Bullinger, Heinrich; ed by Harding, T – Cambridge, 1849 – 5mf – 9 – mf#PBU-679 – ne IDC [240]

The decades : the fourth decade / Bullinger, Heinrich; ed by Harding, T – Cambridge, 1851 – 5mf – 9 – mf#PBU-681 – ne IDC [240]

The decades : the third decade / Bullinger, Heinrich; ed by Harding, T – Cambridge, 1850 – 5mf – 9 – mf#PBU-680 – ne IDC [240]

The decalogue and criticism : or, the place of the decalogue in the development of the hebrew religion / Robinson, George Livingstone – Chicago: R R Donnelley, 1899 – 1mf – 9 – 0-7905-2868-1 – (incl bibliographic references) – mf#1987-2868 – us ATLA [220]

Decalogue journal – Chicago IL 1950-92 – 1,5,9 – ISSN: 0011-7250 – mf#7481 – us UMI ProQuest [340]

Decameron / Boccaccio, Giovanni; ed by Keller, Adelbert von – Stuttgart: Literarischer Verein, 1860 [mf ed 1993] – (= ser Blvs 51) – 704p – 1 – (middle high german text. trans attr to heinrich leubing) – mf#8470 reel 11 – us Wisconsin U Libr [830]

Decamerone / Boccaccio, Giovanni – 14th c – (= ser Holkham library manuscript books 531) – 1r – 1 – (1 col reel [ill only] c524. illuminated by taddeo crivell. ann by w o hassall) – mf#2193 – uk Microform Academic [830]

Decatur advertiser – Decatur, NE: Hemphill Bros. 5v. v1 n1. jul 7 1955-v5 n40. mar 31 1960 [wkly] [mf ed lacks dec 5 1957] – 2r – 1 – (absorbed by: burt county plaindealer; publ in tekamah ne, jul 17 1958-60) – us NE Hist [071]

Decatur Baptist College. Decatur, Texas *see* Catalogs and college records

The decatur herald – Decatur, NE: F A Shepherd. 41v. v1 n1. aug 21 1902-v41 n14. dec 10 1942 (wkly) [mf ed with gaps] – 5r – 1 – us NE Hist [071]

Decatur news – Decatur, NE: Albert O Higgins. 1v. v1 n1. apr 3 1947-v1 n29. oct 16 1947 (wkly) – 1r – 1 – (cont by: lyons mirror-sun) – us NE Hist [071]

The decatur news – Decatur, NE: A P De Milt, 1894 (wkly) [mf ed 1895-96 (gaps)] – 1r – 1 – us NE Hist [071]

The decay of the church of rome / McCabe, Joseph – New York: E P Dutton, 1909 – 1mf – 9 – 0-8370-7889-X – (incl bibl ref and index) – mf#1986-1889 – us ATLA [240]

Deccan chronicle – Secunderabad, India. 1962-Nov 1994 – 199r – 1 – us L of C Photodup [079]

Deccan herald – Poona, India. 1876-89 [daily] – 45r – 1 – (aka: daily telegraph & deccan herald) – uk British Libr Newspaper [079]

Deccan herald – Bangalore, India. Mar 1949-Jul 1995 – 247r – 1 – us L of C Photodup [079]

Deccan times – Madras, India. 16 Apr 1944-8 Jul 1956 – 4r – 1 – us L of C Photodup [079]

The deccan times – Madras: Deccan Print and Pub Co, 1949 – us CRL [079]

The deccan times – Madras: Deccan Print & Pub Co, 1949 – us CRL [079]

Decedencia cubana / Ortiz, Fernando – Habana, Cuba. 1924 – 1r – us UF Libraries [972]

Deceitfulness of sin / Wilberforce, Samuel – Oxford, England. 1853 – 1r – us UF Libraries [240]

DeCelles, Alfred Duclos *see*
- A la conquete de la liberte
- L'abbe bourassa
- Cartier et son temps
- A la conquete de la liberte en france et au canada
- Les constitutions du canada
- Discours de sir wilfrid laurier de 1889 a 1911
- Discours de sir wilfrid laurier de 1911 a 1919
- Les etats-unis
- The habitant
- Les hommes du jour
- Lafontaine et son temps
- Laurier et son temps
- Papineau, 1786-1871
- The "patriotes" of '37
- Scenes de moeurs electorales
- Visite de son honneur le lieutenant-gouverneur l'hon t robitaille au seminaire de ste-therese

DeCelles, Alfred Duclos [comp] *see* Discours de sir wilfrid laurier

December – Highland Park IL 1958-98 – 1,5,9 – ISSN: 0070-3141 – mf#6616 – us UMI ProQuest [400]

Decent fellow doesn't work / Green, Lawrence George – Cape Town, South Africa. 1963 – 1r – us UF Libraries [960]

Decentralize! / Non-Violent Radical Decentralist Strategy – n1-12 [1986 jul/sep-1989 apr/jun] – 1r – 1 – mf#1542243 – us WHS [320]

Dechado y reformacion de todas las medicinas compuestas, renales... / Jubera, A – Valladolid, 1578 – 12mf – 9 – sp Cultura [610]

Dechambre, A *see* Dictionnaire encyclopedique des sciences medicales (ael3/15)

Dechamps, Victor Auguste *see*
- Appeal and a defiance
- La cause catholique
- Entretiens sur la demonstration catholique de la revelation chretienne
- First letter to the rev. father gratry
- La franc-maconnerie
- Lettres theologiques sur la demonstration de la foi
- Les masques bibliques
- Pie 9 et les erreurs contemporaines
- Die unfehlbarkeit des papstes und das allgemeine concil

Decharme, Paul *see* La critique des traditions religieuses chez les grecs

Decheniana – Title varies. v1-62, 1844-1905; index 1844-83 – 13r – 1 – $270.00; outside North America add $1.25r – us L of C Photodup [574]

Dechent, Hermann *see* Goethes schoene seele susanna katharina v. klettenberg

Dechet, Arlette *see* Strategie pour l'enseignement de l'anglais de l'informatique

Dechiaratione di vn salmo fatto sopra la felissima vittoria de l'armata christiana... – n p, [1571] – 1mf – 9 – mf#H-8182 – ne IDC [956]

Dechy, M von *see* Kaukasus, reisen und forschungen im kaukasischen hochgebirge

Deciding cases without argument : a description of procedures in courts of appeals / Cecil, Joe S & Steinstra, Donna – Washington: FJC, 1985 – 1mf – 9 – $1.50 – mf#LLMC 95-323 – us LLMC [347]

Deciding cases without argument : an examination of four courts of appeals / Cecil, Joe S & Steinstra, Donna – Washington: FJC, 1987 – 3mf – 9 – $4.50 – mf#LLMC 95-334 – us LLMC [347]

The deciding voice of the monuments in biblical criticism / Kyle, Melvin Grove – Oberlin, OH: Bibliotheca Sacra, 1912 – 1mf – 9 – 0-7905-1423-0 – (incl ind) – mf#1987-1423 – us ATLA [220]

Decima culta en cuba / Feijoo, Samuel – Havana, Cuba. 1963 – 1r – us UF Libraries [972]

Decima popular / Feijoo, Samuel – Havana, Cuba. 1961 – 1r – us UF Libraries [972]

A decimal currency – weights and measures : 3rd and 4th reports of the standing committee on public accounts – [Quebec?: s.n.] 1855 [mf ed 1983] – 1mf – 9 – 0-665-44213-0 – mf#44213 – cn Canadiana [332]

Decimals and decimalisation : a study and sketch / Harvey, Arthur – Toronto: Hunter, Rose, 1901 – 1mf – 9 – 0-665-99719-1 – mf#99719 – cn Canadiana [332]

Decimas / Alix, Juan Antonio – Ciudad Trujillo, Dominican Republic. 1953 – 1r – us UF Libraries [972]

Decimas en honor...de la virgen / Barco Perez, Paulina – 188? – 9 – sp Bibl Santa Ana [946]

Decimas por el jubilo martiano / Ballagas, Emilio – Habana, Cuba. 1953 – 1r – us UF Libraries [972]

Decimas sobre la era de trujillo / Perez, Manuel Ramon – Ciudad Trujillo, Dominican Republic. 1955 – 1r – us UF Libraries [972]

Decimos – Caceres, 1933-1934 – 5 – sp Bibl Santa Ana [073]

Decir del propio ser / Morales, Jorge Luis – New York, NY. 1954 – 1r – us UF Libraries [972]

Decision / Billy Graham Evangelistic Association – 1960 nov-1968 dec, 1969 jan-1975jun, jul-1978 jul, aug-1981 dec – 4r – 1 – mf#1010193 – us WHS [243]

Decision – Charlotte NC 1960+ – 1,5,9 – ISSN: 0011-7307 – mf#3274 – us UMI ProQuest [240]

Decision – New York NY (USA), 1942 n1/2 – 1r – 1 – gw Misc Inst [071]

Decision for christ and its results – London, England. 18-- – 1r – us UF Libraries [240]

Decision makers of the pasadena area – Los Angeles Co, CA. 1969-89 – 1r – 1 – $50.00 – mf#B06096 – us Library Micro [978]

Decision of a general congress convened to agree on terms of commun... / Evans, Christmas – London, England. 18-- – 1r – us UF Libraries [240]

Decision on federal rules of civil procedure. bulletins / U.S. Dept of Justice – Washington. v1-167. 1938-43 – 1 – $324.00 – mf#0617 – us Brook [347]

Decision sciences – Oxford, England 1970+ – 1,5,9 – ISSN: 0011-7315 – mf#13439 – us UMI ProQuest [650]

Decision support systems – Oxford, England 1985+ – 1,5,9 – ISSN: 0167-9236 – mf#42544 – us UMI ProQuest [650]

Decisiones del tribunal de contribuciones de puerto rico : (tax court) 13 august 1943 to 12 december 1947 – San Juan: Govt Press. v1-5. 1946-51 – 56mf – 9 – $84.00 – mf#LLMC 92-411 – us LLMC [343]

Decisiones s. rotae romanae. opera omnia / Gutierrez, Juan – 1730 – 9 – sp Bibl Santa Ana [240]

Decision-making processes in four west javanese villages diss / Hofsteede, W M F – Nijmegen, 1971 – 3mf – 9 – mf#SE-20017 – ne IDC [959]

Decisions / California. Public Utilities Commission – v1-. Jan. 1, 1911-75. 818 fiches. (Harvard Law School Library Collection.) – 9 – $ – us Harvard Law [336]

Decisions / Pennsylvania. Public Utility Commission – Jul 26, 1913-78. 495 fiches. (Harvard Law School Library Collection.) – 9 – $ – us Harvard Law [336]

Decisions / U.S. Employee's Compensation Appeals Board – v1-33 plus 9 suppl and index vols. 1946-82. 81-219 – 9 – us LLMC [324]

Decisions / U.S. Indian Claims Commission – v1-43. 1948-78.80-510 – 9 – us LLMC [324]

Decisions / U.S. Patent Office – v.1-100, 1869-1968. 80-700 – 9 – us LLMC [324]

Decisions / Wisconsin. Employment Relations Board – No1-8879, 1939-67. 130 fiches. (Harvard Law School Library Collection.) – 9 – $ – us Harvard Law [330]

Decisions constitucionales de los tribunales federales de estados unidos desde 1789. / Bump, Orlando Franklin – 2ed. Buenos Aires, Klingelfuss, 1887. 2 v. in 1. LL-157 – 1 – us L of C Photodup [340]

Decisions, findings, orders, and stipulations *see* Us federal trade commission. decisions, findings, orders, and stipulations

Decisions of the agricultural labor relations board of the state of california (alrb) – v1-20. 1975-94 – 9 – $750.00 set ($55.00y update service) – (incl in coll: case digest which is a summary of cases arranged by topical classification number with cross reference to alrb citation number; an alphabetical case table and other cross references. coll also incl: subsequent history table listing decisions affected by a new decision by the board or by a state court ruling on appeal) – mf#B50572 – us Library Micro [344]

Decisions of the appellat sports, and facilities of the state of washington / Maynard, D N – 1992 – 2mf – 9 – $8.00 – us Kinesology [790]

The decisions of the court of session of scotland – Edinburgh, 1801-23 – 24v on 9r – 1 – $1250.00 – 0-89093-029-5 – us UPA [347]

Decisions of the department of interior in cases related to public lands *see* Decisions of the department of the interior

Decisions of the department of the interior / U[/nited/] S[/tates/] Dept of the Interior – Washington: GPO. v1-101. 1881-1994 – 749mf – 9 – $1123.00 – (v1-52 entitled: decisions of the dept of interior in cases related to public lands; updates planned) – mf#llmc 78-020 – us LLMC [340]

Decisions of the department of the interior *see*
- Department of the interior annual reports
- Index/digests of decisions of the department of the interior

Decisions of the department of the Interior in cases related to public lands *see* Digest of decisions of the department of the interior in cases related to public lands

Decisions of the general master workman / Knights of Labor – 1885, 1887, 1890 – 1r – 1 – mf#3185224 – us WHS [331]

Decisions of the interior department in public land cases and land laws passed, 1838-1870 / Lester, William W – Philadelphia: Small. 2v. 1860; 1870 – (= ser Land laws of the united states, 1776-1938) – 13mf – 9 – $19.50 – mf#LLMC 82-101-2 – us LLMC [343]

Decisions of the speakers of the legislative assembly and house of commons of canada : from 1841 to june 1872 / Lapierre, Augustin – Ottawa: Times Printing & Publ Co, 1872 [mf ed 1984] – 3mf – 9 – mf#SEM105P439 – cn Bibl Nat [323]

Decisions of the treasury department, 1857-1865 – Washington: GPO. 1v. 1876 [all publ] – 4mf – 9 – $6.00 – (publ as suppl to: synopsis of sundry decisions of the treasury department, 1868-1898) – mf#llmc 84-365a – us LLMC [340]

Decisions of the treasury department on appeals, 1865-1867 – Washington: GPO. 1v [all publ] – 2mf – 9 – $3.00 – (publ as suppl to: synopsis of sundry decisions of the treasury department, 1868-1898) – mf#llmc 84-365b – us LLMC [340]

Decisions on the law of patents for inventions rendered by english courts (v.1-3) and by the u.s. supreme court (v.4-20) / U.S. Supreme Court – 1662-1890. Washington: GPO, 1887-92, C.R. Brodix. 81-420 – 9 – $96.00 – us LLMC [346]

Decisions...on appeal from courts of requests. pt. 1 / Ceylon. Supreme Court – Kandy: Industrial School, 1871? 39p. L.C. copy imperfect: cover title wanting. LL-8 – 1 – us L of C Photodup [347]

The decisive hour of christian missions / Mott, John Raleigh – New York: Laymen's Missionary Movement, 1910 – 1mf – 9 – 0-8370-6687-5 – (incl ind) – mf#1986-0687 – us ATLA [240]

Deck, J Northcote *see* Papers on the south sea evangelical mission in the solomon islands

Deck, James B *see* Second letter on receiving and rejecting brethren, and on the principles of the church of god

Deck, Louis *see* Syphilis et reglementation de la prostitution

Decken, Carl C von der *see* Baron carl claus von der decken's reisen in ost-afrika

Decker, A *see* Die passion des herrn nach den vier evangelien synoptisch dargestellt fuer die gebildeten in der gemeinde

Decker, Charles L *see* Colonel charles l decker's collection of records relating to military justice and the revision of military law, 1948-1956

Decker, Marlene *see* Gestaltungselemente im bildwerk von otto mueller

Decker, P *see* Fuerstlicher baumeister

Deckert, Emil *see* Cuba

...Declamatio, de bello turcis inferendo / Nannius, P – Lovanii, 1536 – 1mf – 9 – mf#H-8244 – ne IDC [956]

Declamationes in omnes solemnitates... / Orozco, Alfonso de – Salamanca: Simonis Portonarys, 1573 – 1 – sp Bibl Santa Ana [946]

Declamationes quadragesimales. / Alfonso de Orozco, Beato – 1576 – 9 – sp Bibl Santa Ana [890]

Declamationes vintiginque in evangelia. / Orozco, Alfonso de – 1571 – 9 – sp Bibl Santa Ana [240]

Declamationes...dominicis / Orozco, Alfonso de – 1571 – 9 – (1573 ed) – sp Bibl Santa Ana [240]

Declaracion breve...y sumaria del valor del oro... / Gallo, A – Madrid, 1613 – 3mf – 9 – sp Cultura [330]

Declaracion copiosa de las quatro partes mas essenciales, y necessarias de la doctrina christiana = Dichiarazione piu copiosa della dottrina cristiana / Bellarmino, Roberto Francesco Romolo, Saint – en Lima: Por lorge Lopez de Herrera...ano de 1649 – (= ser Books on religion...1543/44-c1800: doctrina cristiana, obras de devocion) – 4mf – 9 – (in spanish & quechua) – mf#crl-46 – ne IDC [241]

Declaracion funebre...clerigos / Lozano, Antonio – 1742 – 9 – sp Bibl Santa Ana [240]

Declaracion magistral sobre las emblemas de andres alciato con todas las historias... / Lopez, D – Najera: Ian de Mongaston, 1615 – 11mf – 9 – mf#O-1482 – ne IDC [090]

Declaracion magistral sobre los emblemas de andres alciato. / Lopez, Diego – 1655 – 9 – sp Bibl Santa Ana [946]

Declaracion magistral sobre los emblemas de andres alciato. / Lopez, Diego – 1670 – 9 – sp Bibl Santa Ana [946]

Declaracion...emblemas de andres alciato / Lopez, Diego – 1655 – 9 – (1670) – sp Bibl Santa Ana [740]

Declaraciones del pater noster y ave maria / Martinez Siliceo, Juan – Toledo, 1551 – 1 – sp Bibl Santa Ana [240]

Declaraciones...noster / Martinez Siliceo, Juan – 1551 – 9 – sp Bibl Santa Ana [240]

Declaracion...satiras de iuvenal / Lopez, Diego – 1642 – 9 – sp Bibl Santa Ana [450]

Declaracion...sobre las satiras de.. / luvenal – 1642 – 9 – sp Bibl Santa Ana [410]

DEDHAM

Declaration adressee au nom du roi a tous les anciens francois de l'amerique septentrionale / Estaing, Charles Henri – A Bord du Languedoc: De l'impr de F P Demauce...1778? – 1mf – 9 – mf#20565 – cn Canadiana [975]
The declaration against catholic doctrines which accompanies the coronation oath of the british sovereign / Fallon, Michael Francis – Ottawa: St Joseph's Branch of the Catholic Truth Society, 1899 – 1mf – 9 – 0-665-90170-4 – mf#90170 – cn Canadiana [240]
The declaration against catholic doctrines which accompanies the coronation oath of the british sovereign / Fallon, Michael Francis – Ottawa: St Joseph's Branch of the Catholic Truth Society of Ottawa, 1899 – 1mf – 9 – 0-665-92150-0 – mf#92150 – cn Canadiana [241]
Declaration and message of marshal lon nol, president of the khmer republic before the parliament [january 28, 1973] / Lon Nol – [Phnom Penh: Ministry of Information, Khmer Republic 1973] [mf ed 1989] – 1r with other items – 1 – (in khmer, english & french) – mf#mf-10289 seam reel 020/14 [§] – us CRL [323]
Declaration de j. de l... : contenant les raisons qui l'ont oblige a quitter la communion de l'eglise romaine... / Labadie, Jean de – Geneve, 1666 – 6mf – 9 – mf#PPE-147 – ne IDC [240]
Declaration de l'universite laval faite par mgr le recteur devant le comite des bills prives – S.l: s.n, 1890? – 1mf – 9 – mf#62212 – cn Canadiana [378]
Declaration des intellectuels patriotes / Front uni national du Kampuchea – [n.p.] Diffuse par le Gouvernement royal d'union nationale du Cambodge [GRUNC] 1972 [mf ed 1989] – 1r with other items – 1 – (at head of title: royaume du cambodge) – mf#mf-10289 seam reel 016/26 [§] – us CRL [959]
La declaration du droit... / Chevalley, L – Le Caire, Barbey, [1912] – 2mf – 9 – mf#ILM-451 – ne IDC [956]
Declaration du parlement cambodgien : statement made by the cambodian parliament – Phnom-Penh: Ministere de l'information [1970?] [mf ed 1989] – 1r with other items – 1 – (in french & english) – mf#mf-10289 seam reel 016/11 [§] – us CRL [323]
Declaration et observations presentees : par j b ch bedard, ptre du seminaire de montreal a mr rioux [i.e. roux], superieur de cette maison...au sujet du gouvernement ecclesiastique du district de montreal, juin 1824. – [s.l: A Ouimet, 1872?] [mf ed 1984] – (= ser La comedie infernale: pieces justificatives) – 1mf – 9 – 0-665-10313-1 – mf#10313 – cn Canadiana [241]
Declaration issued in the preface to the catalogue / British Institution for Promoting the Fine Arts in the United Kingdom, London – London [1815-16?] – (= ser 19th c art & architecture) – 1mf – 9 – mf#4.2.367 – uk Chadwyck [700]
Declaration made by cheng heng, vice president of the high political council at the press conference held on september 5 1973 / Cheng-Heng – [Phnom-Penh 1973?] [mf ed 1989] – 1r with other items – 1 – (in khmer, english & french) – mf#mf-10289 seam reel 020/15 [§] – us CRL [323]
Declaration made by marshal lon nol, president of the khmer republic [at the august 29 press conference] / Lon Nol – [Phnom-Penh 1973?] [mf ed 1989] – 1r with other items – 1 – (in khmer, english & french) – mf#mf-10289 seam reel 028/06 [§] – us CRL [323]
A declaration of faith of the english people remaining at amsterdam in holland / Helwys, Thomas – York Minster Library, 1611 – 1r – 1 – mf#95708 – uk Microform Academic [941]
The declaration of faith of the society of friends in america – New York City: Henry H Mosher Fund of New York Yearly Meeting [1912?] – 1mf – 9 – 0-524-06608-6 – mf#1991-2663 – us ATLA [240]
The declaration of independence : illustrated story of its adoption, with biographies and portraits of the signers / Casey, Robert E – Fredericksburg, VA: printed for the Citizen's Guild of Washington's Boyhood Home, n.d. – 2mf – 9 – $3.00 – mf#LLMC 92-193 – us LLMC [340]
Declaration of intent / Palau Political Status Commission – Koror: Political Status Commission, mar 1977 – (= ser Republic Of Belau (Palau) – Status Negotiations With The U.S.) – 1mf – 9 – $1.50 – mf#LLMC 82-100G, Title 14 – us LLMC [323]
Declaration of the clergy against alteration of the book of common prayer / Scott, William & Cummins, Henry I – London: Bell & Daldy, 1860 – 1mf – 9 – 0-524-03189-4 – mf#1990-4638 – us ATLA [240]
A declaration on biblical criticism by 1725 clergy of the anglican communion / ed by Handley, Hubert – London: Adam & Charles Black, 1906 [mf ed 1989] – 1mf – 9 – 0-7905-1326-9 – mf#1987-1326 – us ATLA [242]
The declaration on kneeling and the new irish rubric / Maturin, Basil William – Dublin, 1874 – (= ser 19th c ireland) – 1mf – 9 – mf#1.1.2628 – uk Chadwyck [240]
Declaration pour maintenir la vraye foy que tiennent tous chrestiens de la trinite des personnes en un seul dieu... / Calvin, J – Geneva: Chez Jean Crespin, 1554 – 4mf – 9 – mf#CL-32 – ne IDC [240]
Declaration sommaire dv faict de cevx de la ville de vallencienne / Bres, G de – [Vianen, A. van Hasselt pour Chr. Plantin], 1566 – 1mf – 9 – mf#H-2500 – ne IDC [240]
Declaration volontaire de m charles hindenlang : general de brigade dans l'armee des rebelles – [Montreal?: s.n, 1838] [mf ed 1983] – 1mf – 9 – 0-665-44942-9 – mf#44942 – cn Canadiana [971]
Declarations by mr. irujo – n.p. 1937. Fiche W 824. (Blodgett Collection of Spanish Civil War Pamphlets) – 9 – us Harvard [946]
Declareuil, Jean *see* Les systemes de transportation et de main-d'oeuvre penale aux colonies dans le droit francais
Declassified documents reference system – 1975- (ongoing) – 6553mf – 9 – (information on post-world war 2 us domestic and international relations. backfiles 1975-97 c39-28882) – mf#C39-28880 – us Primary [324]
The decline and fall of keewatin : or, the free trade redskins: a satire – Toronto: Grip, 1876 – 1mf – 9 – (ill by j w bengough) – mf#24108 – cn Canadiana [870]
The decline and fall of the kingdom of judah / Cheyne, T K – London: Adam and Charles Black, 1908 – 1mf – 9 – 0-7905-1580-6 – (incl bibl ref and indexes) – mf#1987-1580 – us ATLA [270]
Decline of england / Stirling, Charles – London, England. 1897 – 1r – us UF Libraries [240]
The decline of popery and its doctrinal diversities : two discourses delivered november 24th, and december 1st, 1850, in reply to the lecture of archbishop hughes on "the decline of protestantism and its cause" / Hatfield, Edwin Francis – New-York: Mark H Newman, 1851 – 1mf – 9 – 0-8370-8185-8 – mf#1986-2185 – us ATLA [242]
The decline of the saljuqid empire / Sanaullah, Mawlawi Fadil – Calcutta: University of Calcutta, 1938 – (= ser Samp: indian books) – (int by sir edward denison ross) – us CRL [954]
Decolonization : special issue on american samoa / United Nations Dept of Political Affairs, Trusteeship and Colonization – Oct 1978 – 1mf – 9 – $1.50 – mf#LLMC 82-100C Title 40 – us LLMC [341]
Decolonization : a special issue on the trust territory of the pacific islands / United Nations Dept of Political Affairs, Trusteeship and Decolonization – no 16. apr 1880 – 1mf – 9 – $1.50 – mf#LLMC 82-100F Title 101 – us LLMC [324]
Deconcentration and modernization of economic power *see* The occupation of japan
Deconstructing dominant realities : and the co-creation of hope at an hiv/aids baby sanctuary / Roesch, Ilse – Pretoria: Vista University 2002 [mf ed 2002] – 3mf – 9 – (incl bibl ref) – mf#mfm15180 – sa Unisa [362]
Deconstructing identity in a landscape of ideology, culture, belief and power / Myburg, Johannes Lodewikus – Uni of South Africa 2000 [mf ed Johannesburg 2000] – 4mf [ill] – 9 – (incl bibl ref) – mf#mfm14986 – sa Unisa [240]
Decorah postenog ved arnen – Decorah IA. 1903 nov 10/1904 feb 26-1942 – 60r – 1 – (cont: dannevirke [ceder falls ia]; decorah posten; minneapolis tidende; skandinaven [chicago il]) – mf#769708 – us WHS [071]
Decorah republican – Decorah IA. 1885 jun 4 – 2r – 1 – (cont by: decorah public opinion; decorah public opinion and decorah republican) – mf#851255 – us WHS [071]
Decorah-posten – Decorah IA. 1895 oct 15/1896 oct 16-1903 aug 14-nov 6 – 1r – 1 – (cont by: decorah posten og ved arnenn) – mf#881149 – us WHS [071]
Decorah-posten og ved arnen – Decorah, IA. 1943 apr-1944 apr – 1r – us UF Libraries [025]
La decoration *see* Art and decoration
Decoration and furniture of town houses / Edis, Robert William – London 1881 – (= ser 19th c art & architecture) – 4mf – 9 – mf#4.2.113 – uk Chadwyck [740]
The decoration of houses / Wharton, Edith Newbold & Codman, Ogden – London 1898 – (= ser 19th c art & architecture) – 4mf – 9 – mf#4.2.338 – uk Chadwyck [640]
The decoration of metals chasing, repousse and saw-piercing / Harrison, John – London 1894 – (= ser 19th c art & architecture) – 2mf – 9 – mf#4.2.1183 – uk Chadwyck [730]
The decorations of the garden-pavilion in the grounds of buckingham palace / Gruner, Wilhelm Heinrich Ludwig – London: John Murray; Longman & Co.; P & D Colnaghi etc, 1846 – (= ser 19th c art & architecture) – 2mf – 9 – mf#4.1.34 – uk Chadwyck [720]
Decorative and architectural arts in chicago, 1871-1933 : an illustrated guide to the ceramics and glass exhibition / Darling, Sharon S – 1982 – 4 color mf – 15 – $95.00f – 0-226-68884-4 – us Chicago U Pr [740]
Decorative art / Caisse Nationale des Monuments Historiques et des Sites. Paris – 1900-25 – (= ser Fine and decorative arts in france) – 153mf – 9 – $1080.00 – 0-907006-75-2 – (major collections fr the louvre & the musee cluny, & oriental coll fr musee guimet are incl; over 2000 tapestries; over 9000 reproductions) – uk Mindata [740]
Decorative art, as applied to the ornamentation of churches / Harrison, William Randle – London 1871 – (= ser 19th c art & architecture) – 1mf – 9 – mf#4.2.339 – uk Chadwyck [740]
Decorative art in the victoria and albert museum / Victoria and Albert Museum. London – 9 – $5180.00 complete coll – 0-907006-30-2 – (pictorial record of principal objects in the museum, in 5 sects; also listed separately) – uk Mindata [740]
Decorative art society : design. a paper / Crabb, James – [London] 1844 – (= ser 19th c art & architecture) – 1mf – 9 – mf#4.2.118 – uk Chadwyck [740]
The decorative arts...of the middle ages / Shaw, Henry – London 1851 – (= ser 19th c art & architecture) – 3mf – 9 – mf#4.2.584 – uk Chadwyck [740]
Decorative furniture english, italian, german, flemish, etc / Arundel Society, London – London 1871 – (= ser 19th c art & architecture) – 2mf – 9 – mf#4.1.398 – uk Chadwyck [740]
The decorative painters' and glaziers' guide / Whittock, Nathaniel – London 1827 – (= ser 19th c art & architecture) – 5mf – 9 – mf#4.1.313 – uk Chadwyck [740]
The decorator and furnisher – v. 1-32. Oct 1882-Aug 1898 – 1 – us L of C Photodup [740]
Decorator and painter for australia and new zealand – Sydney, Australia 1 jun 1924-30 dec 1946 – 1 – (fr 1 jun 1924- publ also in melbourne; cont: australasian decorator & painter [1 oct 1908-1 may 1924]) – uk British Libr Newspaper [640]
DeCosta, B F *see* Memoirs of the protestant episcopal church in the united states of america
DeCosta, Benjamin Franklin *see* The moabite stone
Decoud, Diogenes *see* Atlantida
Decourt, Fernand *see* La famille kerdalec au soudan
Decouverte de l'amerique par les normands au 10e siecle / Gravier, Gabriel – Rouen: E Cagniard, 1874 [mf ed 1971] – 1r – 1 – mf#SEM35P66 – cn Bibl Nat [917]
Decouverte du congo / Stanley, Henry Morton – Paris, France. no date – 1r – us UF Libraries [960]
La decouverte du mississipi : avec notice sur les explorateurs de soto, jolliet, marquette et de la salle: suivies du recit des voyages et decouvertes du r p jacques marquette, de la compagnie de jesus / Bois, Louis-Edouard – Quebec?: A Cote, 1873 [mf ed 1985] – 2mf – 9 – 0-665-05247-2 – mf#05247 – cn Canadiana [917]
La decouverte du nouveau monde par les irlandais et les premieres traces du christianisme en amerique avant l'an 1000 / Beauvois, Eugene – [Nancy, France?: s.n] 1875 [mf ed 1985] – 1mf – 9 – 0-665-05101-8 – mf#05101 – cn Canadiana [240]
Decouvertes des portugais en amerique au temps de christophe colomb / Gaffarel, Paul & Gariod, Charles – S.l: C Chadenat, 1892 – 1mf – 9 – (incl bibl ref) – mf#58599 – cn Canadiana [910]
Decouvertes et etablissements des francais dans l'ouest et dans le sud de l'amerique septentrionale, 1614-1754 / ed by Margry, Pierre – 1879-88 – 1 – us AMS Press [978]
Decouvreurs et pionniers : histoire du canada: cahier d'exercices sur le manuel de 4e et de 5e annee / Brisebois, Raymond – Montreal: Lidec inc, [1961?] (mf ed 1992) – 2mf – 9 – (with ind) – mf#SEM105P1693 – cn Bibl Nat [917]
Les decouvreurs francais du 14e au 16e siecle : cotes de guinee, du bresil, et de l'amerique du nord / Gaffarel, Paul – Paris: Challamel, 1888 – 4mf – 9 – mf#06234 – cn Canadiana [910]
The decree and commission of the almighty appointing jeremiah and his representatives the ministers of religion to intoxicate the nations / Miller, James – [Toronto?: s.n, 186-?] [mf ed 1984] – 1mf – 9 – 0-665-45719-7 – mf#45719 – cn Canadiana [240]
Decree and proclamations / Cuba Provisional Governor, 1906-1909 (Charles E...) – Havana, Cuba. v1-7. 1907-09 – 7r – us UF Libraries [323]
The decree of redemption is in effect a covenant : david dickson and the covenant of redemption / Williams, Carol Ann – 2005 [mf ed 2006] – 1r – 1 – 0-524-10550-2 – (incl bibl ref) – mf#d00011 – us ATLA [240]
The decree on daily communion : a historical sketch and commentary = derecho sacramental / Ferreres, Juan Bautista – Edinburgh: Sands, 1909 – 1mf – 9 – 0-7905-7575-2 – (incl bibl ref. in english) – mf#1989-0800 – us ATLA [240]
Decreta concilii plenarii baltimorensis tertii : a.d. 1884 / Catholic Church. Plenary Council of Baltimore – Baltimorae: Joannis Murphy, 1886 – 1mf – 9 – 0-8370-9850-5 – (incl bibl ref) – mf#1986-3850 – us ATLA [240]
Decretales cum glossis (siecle 14) / Gregorio 10 – Barcelona – 1r – 5,6 – sp Cultura [240]
Decretales et constituciones papales (siecle 14) – Barcelona – 1r – 5,6 – sp Cultura [240]
Decret-loi reglementant / Haiti Laws, Statutes, Etc – Port-Au-Prince, Haiti. 1945 – 1r – us UF Libraries [323]
Decreto 203 / Guatemala Laws, Statutes, Etc – Guatemala, 1963 – 1r – us UF Libraries [323]
Decreto de 25 de abril de 1938 y reglamento del mis mo mes, reorganizando el servicio del subsidio al combatiente / Comision Provincial de subsidio el Combatiente – Instrucciones sobre Inspeccion. Caceres: Imp. Moderna, 1938 – 1 – sp Bibl Santa Ana [060]
El decreto de 25 de junio de 1856,o sea, ecsamen sobre la legalidad y conveniencia de la llamada, ley de desamortizacion de bienes raices de las corporaciones civiles y eclesiasticas : coleccion de articulos publicados por el lic sabino flores en "la nacionalidad", periodico oficial del gobierno de estado de guanajuato – Mexico: impr de i cumplido, 1856 – us CRL [972]
Decreto y resolucion de la direccion general de expansion comercial sobre organizacion del registro general de exportadores y de los registros especiales / Camara Oficial de Comercio e Industria de Badajoz – Badajoz: Tip. A. Mangas Cuenda, 1966 – sp Bibl Santa Ana [380]
Decretos de caracter extradorinario / Colombia Laws, Etc – Bogota, Colombia. 1942 – 1r – us UF Libraries [323]
Decretos del congreso nacional, 1946-1947 / Honduras – Tegucigalpa, Mexico. 1947 – 1r – us UF Libraries [323]
Decretos del libertador / Colombia – Caracas, Venezuela. v1-3. 1961 – 1r – us UF Libraries [323]
Decretos-leyes del actual gobierno (emitidos hasta...) / Guatemala Laws, Statutes, Etc – Guatemala, 1963 – 1r – us UF Libraries [323]
Das decretum gelasianum / Dobschuetz, Ernst von – Leipzig, 1912 – (= ser Tugal 3-38/4) – 6mf – 9 – €14.00 – ne Slangenburg [240]
Decretum Gelasianum *see* Das decretum gelasianum de libris recipiendis et non recipiendis
Das decretum gelasianum de libris recipiendis et non recipiendis : in kritischem text / Decretum Gelasianum – Leipzig: J C Hinrichs, 1912 – (= ser Tugal) – 1mf – 9 – 0-7905-1751-5 – (incl bibl ref and ind) – mf#1987-1751 – us ATLA [220]
Decroix, F W *see* Historical, industrial, and commercial data of mia...
Decroix, Jacques Joseph Marie *see* L'ami des arts
Decroux, Paul *see* La femme dans l'islam moderne
Decuscope – Shrewsbury MA 1962-83 – 1,5,9 – ISSN: 0011-7447 – mf#5743 – us UMI ProQuest [000]
Dede / Willemetz, Albert – Paris, France. 1921 – 1r – us UF Libraries [025]
Dede, Galib *see* Huesn ve ask
Dederich, Hermann *see* Ludwig uhland als dichter und patriot
Dedham 1635-1905 – Oxford, MA (mf ed 2001) – (= ser Massachusetts vital records) – 195mf – 9 – 0-87623-412-0 – (mf1-9, 193-195: town records 1636-59. mf10-31: land grants 1636-1806. mf32-44:town/land grants 1636-1813. mf43: deaths & marriages 1792-1858. mf45-85: town records 1672-1818. mf86-90: town records index 1773-1875. mf91-109: town accounts 1757-89. mf110-116: history,topography 1636-1836. mf117-123: vitals & index 1635-1777. mf124-128,192: vital

records 1727-1847. mf129-137: vitals & index 1727-1852. mf138-141: vital records 1844-53. mf142: vital records index 1844-49. mf143-146: birth index 1849-92. mf147-152: births 1853-70. mf153-156, 146: births 1867-77. mf157-162: births 1871-1905. mf163-165: death index 1849-93. mf166-171: deaths 1853-94. mf172-177: deaths 1895-1916. mf178-179: intentions index 1850-79. mf180-182: marriage index 1849-92. mf183-188: marriages 1854-91. mf189-191: marriages 1892-1908. mf192: vital records 1757-88. mf193-195: town records 1636-53) – us Archive [978]

The dedham historical register – v. 1-14. 1890-1903 – 1 – us L of C Photodup [978]

Dedicated : by special permission, to her most gracious majesty the queen, a series of eight sketches in colour (together with a chart of the route) / Cresswell, Samuel Gurney – London: Day and Son, 1854 – 3mf – 9 – mf#16725 – cn Canadiana [917]

Dedicated by special permission to the hon the minister of agriculture : and prefaced with a highly commendatory introduction by prof h mccandless, principal of the ontario school of agriculture, guelph... / Whitcombe, Charles Edward – Toronto: J Adam [1874?] [mf ed 1987] – 1mf – 9 – 0-665-27490-4 – mf#27490 – cn Canadiana [630]

Dedication of the library building, gammon theological seminary, may 26th 1889, 2:30 p m, atlanta, georgia – [Georgia: s.n. 1889] [mf ed 2006] – 1r [complete] – 1 – (reel also incl: catalogue of the gammon theological seminary, & two addresses) – mf#2006-s007 – us ATLA [080]

Dedication of the new synagogue of the congregation mikve israel : at broad and york streets on sep 14, 1909, elul 29, 5669 / Rosenbach, Abraham Simon Wolf – Philadelphia: [s.n.], 1909 – 1mf – 9 – 0-8370-7051-1 – (includes the form of service in english and hebrew) – mf#1986-1051 – us ATLA [939]

Dedications and patron saints of english churches : ecclesiastical symbolism / Bond, Francis – London, New York: OUP 1914 [mf ed 1989] – 1mf – 9 – 0-7905-4488-1 – (incl bibl ref) – mf#1988-0488 – us ATLA [240]

Dedications and patron saints of english churches: ecclesiastical symbolism : saints and their emblems / Bond, Francis – London; New York: Oxford University Press, 1914 – 1mf – us ATLA [240]

Dedications and patron saints of english churches, ecclesiastical symbolism, saints and their emblems / Bond, Francis – London, etc, 1914 – 7mf – 8 – mf#H-1243 – ne IDC [700]

Dedicatorias de mis libros / Sanchez Arjona, Vicente – Sevilla: Graficas Tirvia, Tomo 1. 1956 – 1 – sp Bibl Santa Ana [810]

Dedicatorias de mis libros / Sanchez Arjona, Vicente – Sevilla: Graficas Tirvia, Tomo 2. 1956 – 1 – sp Bibl Santa Ana [810]

Dedicatorias en mis libros / Sanchez Arjona, Vicente – Sevilla: Imprenta Alvarez, Tomo 4-10. 1957 – 1 – sp Bibl Santa Ana [810]

Dedicatory exercises at the unveiling of bronze tablets in memory... / John P Altgeld Memorial Association Of Chicago – Chicago, IL. 1910 – 1r – us UF Libraries [025]

Dedo ajeno : cuentos inutiles / Ozores, Renato – Panama, Panama. 1954 – 1r – us UF Libraries [972]

Dedos de la mano / Laguerre, Enrique A – Mexico City? Mexico. 1951 – 1r – us UF Libraries [972]

Dedreux, R *see* Der suezkanal im internationalen rechte

Dee, John *see*
- A letter, containing a most briefe discourse apologeticall
- Renaissance man: the reconstructed libraries of european scholars, 1450-1700, series 1

Dee, Simon Pieter *see* Het geloofsbegrip van calvijn

Deed of settlement... / Farmers' Joint Stock Banking Co – [Toronto: s.n.] 1835 [mf ed 1987] – 1mf – 9 – 0-665-57331-6 – mf#57331 – cn Canadiana [332]

Deeds, ms 3193 / Ashland County & Wayne County. Ohio – 1833-93 Ashland County 1853-93; Wayne County 1833-41 Excerpts – 1r – 1 – us Western Res [920]

Deeds of the borough of neath, 1566-1826 – 1r – 1 – mf#505 – uk Microform Academic [941]

Deegan, Mabel Alice *see* Contributions to the technique of violin playing made by nicolo paganini

Deems, Charles Force *see*
- Annals of southern methodism for 1856
- The gospel of common sense
- The gospel of spiritual insight

Deenbhandhu – Poona, India. 14 Mar 1947-19 Dec 1950; 7 Jan 1951-4 Dec 1953; Feb 1950-7 Jul 1978 – 24r – 1 – us L of C Photodup [079]

Deep creek review *see* Miscellaneous newspapers of san miguel county

Deep furrows / Ben-Shalom, Avraham – New York, NY. 1937 – 1r – us UF Libraries [939]

Deep south patriot / Southern Conference Educational Fund – 1966 jun – 1r – 1 – mf#1111066 – us WHS [370]

Deep waters / Gimenez, Joseph Patrick – Charlotte Amalie, St Thomas. 1939 – 1r – us UF Libraries [972]

Deep West Peace Press (US) *see* Heartland

Deepening of the spiritual life / Forbes, Alex Penrose – Leeds, England. 1872 – 1r – us UF Libraries [240]

Deepika – Kottayam, India. Apr 1944-Apr 1948; Apr 1949-1953; Mar-Sept 1966 – 12r – 1 – us L of C Photodup [079]

Deep-sea research – Oxford, England 1953+ – 1,5,9 – ISSN: 0198-0149 – mf#49061 – us UMI ProQuest [550]

Deere and Co *see* Open door

Deerfield 1675-1898 – Oxford, MA (mf ed 1987) – (= ser Massachusetts vital records) – 18mf – 9 – 0-87623-039-7 – (mf 1: index to births & deaths 1675-1844 (a). mf 2: births & deaths 1675-1844 (a-c). mf 3: births & deaths 1675-1844 (c-h). mf 4: births & deaths 1675-1844 (j-s). mf 5: births & deaths 1675-1844 (s-w). mf 5-6: marriages 1689-1844. mf 7: births 1844-60. mf 8: births 1861-63; marriages/deaths 1844-54. mf 9: deaths 1855-59. mf 10-12: births 1864-97. mf 13-15: marriages 1853-97. mf 15-18: deaths 1860-98) – us Archive [978]

Deerfield 1676-1849 – Oxford, MA (mf ed 1996) – (= ser Massachusetts vital record transcripts to 1850) – 12mf – 9 – 0-87623-239-X – (mf 1t-7t: births & deaths by family 1676-1865. mf 7t-8t: marriages 1689-1833. mf 8t-11t: intentions of marriage 1752-1850. mf 10t: out-of-town marriages 1703-99. mf 10t-11t: marriages 1832-1843. mf 12t: b,m,d 1844-49) – us Archive [978]

Deerfield enterprise – Deerfield WI. 1892 may 21-1895 mar 15, mar 23-1896 aug 28, sep 4-1897 dec 31, 1898 jan 7-dec 22 – 4r – 1 – (cont by: Leader [Deerfield wi]; Enterprise-leader) – mf#936888 – us WHS [071]

Deerfield independent – Deerfield WI. 1944 jan 14-1947 dec 26, 1948-65, 1966 jan 6-1968 feb 29, mar 7-1969 aug 21, aug 28-1971 feb 25, mar 4-jul 1 – 10r – 1 – (cont by: independent [deerfield wi]) – mf#938656 – us WHS [071]

Deerfield news – Deerfield WI. 1917 sep 14 [v13 n48] – 1r – 1 – mf#938649 – us WHS [071]

Deerfield tobacco herald – Deerfield WI. 1885 sep 18-1888 apr 6 – 1r – 1 – mf#936875 – us WHS [071]

Deerslayer / Cooper, James Fenimore – New York, NY. no date – 1r – us UF Libraries [025]

Deerslayer / Cooper, James Fenimore – Philadelphia, PA. no date – 1r – us UF Libraries [830]

Deeside advertiser – [NW England] Hoylake, Birkenhead Lib – 1 – (title change: hoylake news & advertiser [apr 1974-1975]; hoylake & west kirby news [1976-83]) – uk MLA; uk Newsplan [072]

Deeside and buckley leader. (mold, deeside & buckley leader) – Mold, Wales. 8 Dec 1922-Dec 1933.-w. 14 reels – 1 – uk British Libr Newspaper [072]

Deeside piper and herald – Forfar: Angus County Press Ltd 1989- [mf ed 1 jul 1994-] – 1 – (cont: deeside piper) – uk Scotland NatLib [072]

La deesse anat (mr-s vol 4) / Virolleaud, Ch – Paris, 1938 – (= ser Mission des ras-shamra (mr-s)) – 5mf – 8 – €12.00 – ne Slangenburg [270]

[Deeth-] the commonwealth – NV. 1910-14 [wkly] – 2r – 1 – $120.00 – mf#U04486 – us Library Micro [071]

Deetjen, Werner *see*
- Die goechhausen
- Das haus am frauenplan seit goethes tod

Deetjen, Werner et al *see* Funde und forschungen

Deewa roka mana dsihwe, 1904-1928 = God's hand in my life, 1904-1928 / Lauberts, Peter – Liepaja: P. Lauberts, 1928. Publ. No. 6349 c. One of trhee items on reel – 1 – us Southern Baptist [242]

La defaite des anglais a tanger en 1664 / Rouard de Card, E – Paris, 1912 – 1mf – 9 – mf#ILM-1692 – ne IDC [956]

Defauconpret, Auguste *see*
- Une annee a londres
- Londres en...
- Memoires et anecdotes sur la cour de napoleon bonaparte

Defeat of the spanish armada / Vine, F T – London, England. 1880 – 1r – us UF Libraries [946]

Defects, civil and military, of the indian government / Napier, Charles James; ed by Napier, W F P – London 1853 – (= ser 19th c british colonization) – 5mf – 9 – mf#1.1.2293 – uk Chadwyck [350]

Defects of our system of government : delivered by mr edward miall before the literary and historical society of ottawa, on 3rd february, 1877 / Miall, Edward – Ottawa?: C W Mitchell, 1892 – 1mf – 9 – mf#10101 – cn Canadiana [320]

Defektive kinder in der yidisher literatur / Rubin Rivkai, Israel – Vilne, Lithuania. 1928 – 1r – us UF Libraries [470]

Defence – London, England. 1821? – 1r – us UF Libraries [240]

Defence and confirmation of the faith : six lectures...1885 / Taylor, William Mackergo – New York: Funk & Wagnalls 1885 [mf ed 1985] – 1mf – 9 – 0-8370-2860-4 – mf#1985-0860 – us ATLA [240]

A defence and exposition of truth : a book for this time / Foote, LeRoy – Ottawa: s.n, 1879 – 2mf – 9 – mf#28297 – cn Canadiana [240]

Defence and letter of resignation / Ross, Alexander Johnstone – Brighton: Henry S King, 1852 – 1mf – 9 – 0-524-06102-5 – mf#1991-2415 – us ATLA [240]

Defence de la reformation : (contre le livre intitule prejugez legitimes contre les calvinistes) / Claude, J – Leeuwarde, 1745 – 11mf – 9 – mf#PRS-131 – ne IDC [242]

A defence of christianity : against the work of george b english...entitled, the grounds of christianity examined, by comparing the new testament with the old / Everett, Edward – Boston: Publ by Cummings & Hilliard, no 1, Cornhill; Cambridge: Hilliard & Metcalf, 1814 [mf ed 1984] – (= ser Biblical crit – us & gb 15) – 6mf – 9 – 0-8370-0668-6 – (incl bibl ref) – mf#1984-1015 – us ATLA [240]

Defence of civil establishments of religion / Mackray, William – Aberdeen, Scotland. 1833 – 1r – us UF Libraries [240]

Defence of civil establishments of religion / Ritchie, Ebenezer – Edinburgh, Scotland. 1835 – 1r – us UF Libraries [240]

A defence of columbia college from the attack of samuel b ruggles / Ogden, Morris – New York: JP Wright, 1854 [mf ed 1992] – 1mf – 9 – 0-524-03649-7 – mf#1990-1077 – us ATLA [378]

Defence of fort m'henry-star-spangled banner / Key, Francis Scott – Broadside, 1814. First printed edition of the words with the tune indicated (Anacreon in Heaven). MUSIC 1121 – 1 – us L of C Photodup [780]

A defence of gospel baptism : with a brief historical sketch of the origin of infant baptism and sprinkling / French, James – Holyoke: ABF Hildreth, 1854 [mf ed 1993] – 1mf – 9 – 0-524-08465-3 – mf#1993-3110 – us ATLA [240]

A defence of liberal christianity / Norton, Andrews – [Cambridge, MA: William Hilliard, 1812] [mf ed 1984] – (= ser Biblical crit – us & gb 77) – 1mf – 9 – 0-8370-1593-6 – mf#1984-1077 – us ATLA [240]

A defence of luther and the reformation : against the charges of john bellinger and others... / Bachman, John – Charleston: William Y Paxton, 1853 [mf ed 1991] – 2mf – 9 – 0-524-00740-3 – mf#1990-0172 – us ATLA [242]

A defence of ohio congregationalism and of oberlin college : in reply to kennedy's plan of union / Cowles, Henry – [s.l: s.n, 1857?] [mf ed 1992] – 1mf – 9 – 0-524-02955-5 – mf#1990-4507 – us ATLA [242]

A defence of "our fathers" and of the original organization of the methodist episcopal church against the rev alexander m'caine and others : with historical and critical notices of early american methodism / Emory, John – 5th ed. New York: T Mason & G Lane...1838 [mf ed 1990] – 1mf – 9 – 0-7905-5086-5 – mf#1988-1086 – us ATLA [242]

A defence of philosophic doubt : being an essa on the foundations of belief / Balfour, Arthur James, 1st Earl of – London: Macmillan, 1879 [mf ed 1985] – 1mf – 9 – 0-8370-2164-2 – (incl app) – mf#1985-0164 – us ATLA [110]

The defence of professor briggs before the presbytery of new york, december 13, 14, 15, 19, and 22, 1892 / Briggs, Charles Augustus – New York: Scribner, 1893 – 1mf – 9 – 0-524-02447-2 – mf#1990-4306 – us ATLA [242]

Defence of the associate synod against the charge of sedition / Peddie, James – Edinburgh, Scotland. 1800 – 1r – us UF Libraries [240]

Defence of the athanasian creed / Chevalier, Thomas Wm – London, England. 1830 – 1r – us UF Libraries [240]

A defence of the beacon : or, a supplement to the reply to the statement of the yearly meeting's committee... / Crewdson, Isaac – London: Hamilton, 1836 [mf ed 1993] – (= ser Society of friends (quakers) coll) – 1mf – 9 – 0-524-07561-1 – mf#1991-3181 – us ATLA [220]

Defence of the catholic faith : concerning the satisfaction of christ against faustus socinus = Defensio fidei catholicae de satisfactione christi / Grotius, Hugo – Andover: Warren F Draper, 1889 [mf ed 1985] – 1mf – 9 – 0-8370-3847-2 – (english trans with notes & int by frank hugh foster; incl ind) – mf#1985-1847 – us ATLA [241]

Defence of the church missionary society against the objections / Wilson, Daniel – London, England. 1818 – 1r – us UF Libraries [240]

A defence of the churches and ministery of englande : written in two treatises, against the reasons and obiections of maister francis iohnson, and others of the separation commonly called brownists / Jacob, H – Middelburgm: Richard Schilders, 1599 – 2mf – 9 – mf#PW-69 – ne IDC [240]

A defence of the deity and atonement of jesus christ : in reply to ram-mohun roy of calcutta / Marshman, Joshua – London: Kingsbury, Parbury & Allen 1822 [mf ed 1993] – 3mf – 9 – 0-524-07900-5 – mf#1991-3445 – us ATLA [240]

A defence of "the eclipse of faith" : the "reply"... / Rogers, Henry & Newman, Francis William – Boston: Crosby, Nichols 1854 [mf ed 1990] – 1mf – 9 – 0-7905-3470-3 – mf#1987-3470 – us ATLA [240]

A defence of the elkhorn association in sixteen letters : addressed to elder henry toler, entitled union-no union / Fishback, James – 1822 – 1 – $6.65 – us Southern Baptist [242]

Defence of the hebrew grammar of gesenius against prof. stuart's translation / Conant, Thomas J – New-York: D Appleton; Philadelphia: Geo S Appleton, 1847 – 1mf – 9 – 0-7905-2101-6 – mf#1987-2101 – us ATLA [470]

Defence of the illustration of the hypothesis proposed in the disse... / Marsh, Herbert – Cambridge, England. 1804 – 1r – us UF Libraries [240]

Defence of the jesuits / Ward, William Perceval – Dublin, Ireland. 1848 – 1r – us UF Libraries [241]

A defence of the landlords of ireland : with remarks on the relation between landlord and tenant / Simpson, William Wooler – London, 1844 – (= ser 19th c ireland) – 1mf – 9 – mf#1.1.3978 – uk Chadwyck [333]

Defence of the latest form of infidelity examined : a second letter to andrews norton occasioned by his defence of a discourse on the latest form of infidelity / Ripley, George – Boston: James Munroe, 1840 – 1mf – 9 – 0-524-07460-7 – mf#1991-3120 – us ATLA [240]

Defence of the latest form of infidelity examined : a third letter to andrews norton, occasioned by his defence of a discourse on the latest form of infidelity / Ripley, George – Boston: James Munroe, 1840 – 2mf – 9 – 0-524-07461-5 – mf#1991-3121 – us ATLA [240]

A defence of the ministers reasons : for refusall of subscription to the booke of common prayer, and of conformitie... / Hieron, S – n.p., 1607 – 6mf – 9 – mf#PW-48 – ne IDC [240]

A defence of the missionary organizations of the baptist denomination : being a review of a pamphlet entitled "thoughts on the missionary organizations of the baptist denomination," by francis wayland – Boston: John M Hewes, 1859 [mf ed 1993] – 1mf – 9 – 0-524-08283-9 – mf#1993-3038 – us ATLA [242]

Defence of the patronage act of 1874 / Bannatyne, Alexander M – Aberdeen, Scotland. 1875 – 1r – us UF Libraries [240]

Defence of the rev. charles voysey, vicar of healaugh : on the hearing of the charges of heresy preferred against him in the chancery court of york, on the 1st december, 1869 / Voysey, Charles – London: Truebner, 1869 – 1mf – 9 – 0-524-02528-2 – mf#1990-0628 – us ATLA [240]

Defence of the rev. rowland williams, d.d., in the arches' court of canterbury / Stephen, James Fitzjames, Sir – London: Smith, Elder, 1862 – 1mf – 9 – 0-7905-8912-5 – (incl bibl ref) – mf#1989-2137 – us ATLA [241]

Defence of the right reverend the lord bishop of bangor / Hughes, Rice – London, England. 1796 – 1r – us UF Libraries [241]

Defence of the rights of the christian people in the appointment of... / Cunningham, William – Edinburgh, Scotland. 1840 – 1r – us UF Libraries [240]

Defence of the roman church against father gratry = Defense de l'eglise romaine / Gueranger, Prosper – London: R Washbourne, 1870 [mf ed 1986] – 1mf – 9 – 0-8370-8577-2 – (english trans by romuald w woods; int by r b vaughan) – mf#1986-2577 – us ATLA [241]

Defence of the roman church against father gratry / Gueranger, Prosper – London, England. 1870 – 1r – us UF Libraries [241]

A defence of the truth : as set forth in the "history and mystery of methodist episcopacy", being a reply to john emory's "defence of our fathers" / M'Caine, Alexander – Baltimore: Matchett 1829 – 1r – 1 – $35.00 – (p124-125 missing) – mf#um-15 – us Commission [242]
Defence of the universality and perpeptuity of the sabbath / Oliver, Alexander – Edinburgh, Scotland. 1852 – 1r – us UF Libraries [240]
A defence of the wesleyan methodist missions in the west indies... / Watson, R – London, 1817 – 2mf – 9 – mf#HTM-201 – ne IDC [918]
A defence of trine immersion : being a review of elder e adamson's treatise on (against) trine immersion / Quinter, James – Columbiana OH: Office of the Gospel visitor 1862 [mf ed 1993] – 1mf – 9 – 0-524-06071-1 – mf#1990-5185 – us ATLA [242]
A defence of virginia : and through her, of the south, in recent and pending contests against the sectional party / Dabney, Robert Lewis – New York: EJ Hale, 1867 [mf ed 1990] – 1mf – 9 – 0-7905-3777-X – mf#1989-0270 – us ATLA [976]
The defences of norumbega and a review of the reconnaissances of col t w higginson, professor henry w haynes, dr justin winsor, dr francis parkman, and rev edmund f slafter : a letter to judge daly / Horsford, Eben Norton – Boston, New York: Houghton, Mifflin, 1891 – 2mf – 9 – mf#12222 – cn Canadiana [910]
Defences to crime. the adjudged cases in the american and english reports wherein the different defences to crimes are contained – San Francisco, Whitney, 1874-92. 6 v. LL-667 – 1 – us L of C Photodup [345]
The defender – Scranton, PA. -w 1904-1905 – 13 – $25.00r – us IMR [071]
The defender – Philadelphia, Bryn Mawr [PA]: H C C Astwood. v2 n27. jan 27 1900 [mf ed 1947] – (= ser Negro Newspapers on Microfilm) – 1r – 1 – (suspended: nov 18 1905) – us L of C Photodup [071]
Defender la independencia de la patria / Spain. Presidencia del Consejo de Ministros – Barcelona, 19?? Fiche W825. (Blodgett Collection of Spanish Civil War Pamphlets) – 9 – us Harvard [946]
Defenders – Washington DC 1976+ – 1,5,9 – (cont: defenders of wildlife) – ISSN: 0162-6337 – mf#6904,02 – us UMI ProQuest [639]
Defenders of new zealand / Gudgeon, T W – 5mf – 9 – NZ$20.00 – 0-908797-69-9 – (publ in 1887 as "defenders of new zealand and maori history of the war". contains biographies, lists recipients of the war medal and members of colonial forces killed 1860-70) – mf#NZNB 2371 – nz BAB [355]
Defenders of wildlife – Washington DC 1964-73 – 1,5 – (cont by: defenders) – ISSN: 0011-7528 – mf#6904.02 – us UMI ProQuest [639]
Defenders of wildlife news – Washington DC 1964-73 – 1,5 – ISSN: 0011-7528 – mf#6904.02 – us UMI ProQuest [639]
Defensa ciudadana / Juez, Antonio – Badajoz: Tip.Vda.de A.Arqueros, 1936 – 1 – sp Bibl Santa Ana [350]
Defensa continental / Haya De La Torre, Victor Raul – Buenos Aires, Argentina. 1942 – 1r – us UF Libraries [972]
Defensa de badajoz – Badajoz.1887-89. No. sueltos – 9 – sp Bibl Santa Ana [074]
Defensa de cuba / Sanguily, Manuel – Habana, Cuba. 1948 – 1r – us UF Libraries [972]
Defensa de d. luis calderon...presos de confesion / Calzado Pedrilla, Felipe – 1836 – 9 – sp Bibl Santa Ana [946]
Defensa de don fernando perez – 1790. Por un amigo de Don Fernando – 9 – sp Bibl Santa Ana [920]
Defensa de don fernando perez...paracuellos / Forner Segarra, Juan Pablo – 1790 – 9 – sp Bibl Santa Ana [946]
Defensa de la astrologia y conjeturas / Aldrete y Soto, L – Madrid, 1861 – 1mf – 9 – sp Cultura [130]
Defensa de la china y verdadera respuesta a las falsas razones ove para su reprobacion trae el doctor don jose colmenero : china y verdadera respuesta a las falsas razones... / Fernandez, T – Madrid, 1689 – 3mf – 9 – sp Cultura [610]
Defensa de la hispanidad / Maeztu, Ramiro de – Madrid: Razon y Fe, 1934 – 1 – sp Bibl Santa Ana [946]
Defensa de la naturaleza / Carrion Marquez, Jesus – Caceres: Imp. D. Rodriguez, 1972 – 1 – sp Bibl Santa Ana [946]
Defensa de la poesia (siecle 17) / Sidney, Philip – Madrid – 1r – 5,6 – sp Cultura [090]
Defensa de los derechos del hombre y del ciudadano / Torriente Y Peraza, Cosme De La – Habana, Cuba. 1930 – 1r – us UF Libraries [972]
Defensa de madrid. relato historico / Lopez Fernandez, Antonio et al – Mejico: Editorial A.P. Marquez, 1945 – 1 – sp Bibl Santa Ana [355]
Defensa de un fuero historico / Tafur Garces, Leonardo – Cali, Colombia. 1939 – 1r – us UF Libraries [972]
Defensa del tratado de limites entre yucatan y bel... – Guatemala, 1958 – 1r – us UF Libraries [972]
Defensa Institucional Cubana *see* Tres anos
Defensa nacional y la escuela / Guerra, Ramiro – Habana, Cuba. 1923 – 1r – us UF Libraries [972]
Defensa oral...manuel blanco / Valcarcel, Joaquin, Ma – 1850 – 9 – sp Bibl Santa Ana [946]
Defensa y respuesta...de la medicina racional y philosophica...contra... / Delgado de Vera, J – Madrid, 1687 – 5mf – 9 – sp Cultura [610]
Defensa y verdadero manifiesto de la via curativa que d. manuel pellaz y espinosa... / Pellaz y Espinosa, Manuel – SL, SA – 1mf – 9 – sp Cultura [610]
Defensa...diocesis de plasencia...vicario capitular / Ros Biosca, Godofredo – 1872 – 9 – sp Bibl Santa Ana [240]
Defensa...fernando perez...paracuellos / Sanchez, Tomas Antonio – 1790 – 9 – sp Bibl Santa Ana [440]
Defensa...orden de alcantara / Valencia y Bravo, Alonso – 1818 – 9 – sp Bibl Santa Ana [946]
Defensas omitidas en el memorial ajustado hecho por el relator en la causa de don juan de ovando... / Consejo de Hacienda – 1 – sp Bibl Santa Ana [946]
Defense – Washington DC 1980-97 – 1,5,9 – (cont: command policy) – ISSN: 0737-1217 – mf#12637 – us UMI ProQuest [355]
Defense : m stewart et les finances haitiennes / Firmin, Antenor – Paris, France. 1892 – 1r – us UF Libraries [332]
La defense – Alger. 1934-aout 1939 – 1 – fr ACRPP [073]
La defense – Paris. 1933-juil 1934 – 1 – (puis organe du secours populaire de france.) – fr ACRPP [073]
Defense and disarmament news / Institute for Defense and Disarmament Studies [US] – 1985 mar/apr-1988 feb – 1r – 1 – (cont by: defense and disarmament alternatives) – mf#1155371 – us WHS [355]
Defense counsel journal – Chicago IL 1987+ – 1,5,9 – (cont: insurance counsel journal) – ISSN: 0895-0016 – mf#2143,01 – us UMI ProQuest [347]
Defense counsel journal – v1-67. 1934-2000 – 5,6,9 – $1229.00 set – (v1-51 1934-85 on reel $635; v52-67 1985-2000 on mf $594; title varies: v1-53 1934-85 as insurance counsel journal) – ISSN: 0895-0016 – mf#103491 – us Hein [340]
Defense d'afficher / Passeur, Steve – Paris, France. 1931 – 1r – us UF Libraries [440]
Defense de calvin contre l'outrage fait...sa memoire / Drelincourt, C – Geneve, 1667 – 5mf – 9 – mf#PRS-139 – ne IDC [242]
Defense de la fidelite des traductions de la bible faites...geneve opposee au pere coton / Turrettini, B – Geneve: de Tournes, 1618 – 8mf – 9 – mf#PFA-181 – ne IDC [240]
Defense de la france – Paris, France. -m. 15 dec 1942-jun 1944 (imperfect) – 1/4r – 1 – uk British Libr Newspaper [072]
Defense de l'occident – Paris-Cedex 05, France 1952-80 – 1,5,9 – ISSN: 0011-7552 – mf#8397 – us UMI ProQuest [327]
Defense Depot Ogden (UT) *see* Hub
Defense des nouveaux chrestiens et des missionaires de la chine... / Le Tellier, M – Paris, 1687 – 7mf – 9 – mf#HTM-227 – ne IDC [915]
Defense des resumes historiques / Bodin, Felix – Paris 1825 [mf ed Hildesheim 1995-98] – 1v on 1mf – 9 – €40.00 – 3-487-25986-9 – gw Olms [900]
Defense documents rejected as evidence before the international military tribunal for the far east, 1946-1947 / World War 2. Defense Section – (= ser Records of allied operational and occupation headquarters, world war 2) – 16r – 1 – mf#M1693 – us Nat Archives [355]
Defense du capitaine charles gariepy : contre les accusations du lieutenant colonel bourdages, commandant la division de milice a saint denis – Montreal: Impr par C B Pasteur...1819 [mf ed 1983] – 1mf – 9 – 0-665-44539-3 – mf#44539 – cn Canadiana [343]
Defense du culte exterieur de l'eglise / Brueys, D-A – Paris, 1686 – 6mf – 9 – mf#CA-118 – ne IDC [241]
Defense electronics – Shawnee Mission KS 1979-94 – 1,5,9 – ISSN: 0278-3479 – mf#10742.03 – us UMI ProQuest [621]
Defense history program studies prepared during the korean war period / U.S. Office of Price Stabilization – (= ser Records of the office of price stabilization) – 3r – 1 – mf#T460 – us Nat Archives [324]
Defense indicators – Washington DC 1974-78 – 1,5,9 – ISSN: 0418-5013 – mf#7347 – us UMI ProQuest [071]
Defense industry bulletin – Washington DC 1965-72 – 1 – ISSN: 0418-5021 – mf#7423 – us UMI ProQuest [355]
Defense Language Institute (US) *see* Globe
Defense Language Institute [US] *see* Fortnightly fogbank
Defense law journal – Charlottesville VA 1967+ – 1,5,9 – ISSN: 0011-7587 – mf#2465 – us UMI ProQuest [347]
Defense management journal – Washington DC 1972-87 – 1,5,9 – ISSN: 0011-7595 – mf#7348 – us UMI ProQuest [355]
Defense manual – Chicago IL 1978-79 – 1,5,9 – ISSN: 0191-877X – mf#10689,01 – us UMI ProQuest [350]
Defense monitor / Center for Defense Information [Washington DC] – 1972 may-1982 – 1r – 1 – mf#780433 – us WHS [355]
The defense monitor – 1972-82 – (= ser The library of world peace studies) – 10mf – 9 – $105.00 – us UPA [355]
Defense nationale – London, UK. 21 Nov 1870 – 1 – uk British Libr Newspaper [072]
La defense nationale dans le nord en 1870-1871 : recueil methodique de documents / Levi, Camille – Paris: Charles-Lavauzelle [1904]-21 [mf ed 1980] – 4v on 1r [ill] – 1 – (subtitle varies) – mf#film mas 9396 – us Harvard [940]
Defense news – Springfield VA 1986+ – 1,5,9 – ISSN: 0884-139X – mf#15321 – us UMI ProQuest [355]
Defense news bulletin / Industrial Workers of the World – n1-12 [1917 nov 3-1918 feb 2], n6,19,24-33,48-51 [1918 mar 2,23, apr 27-jun 29, oct 19-nov 9] – 1r – 1 – (cont: solidarity [new castle pa]; cont by: new solidarity [chicago il]) – mf#1009823 – us WHS [355]
A defense of christian perfection : or, a criticism of dr james mudge's growth in holiness toward perfection / Steele, Daniel – New York: Hunt & Eaton, 1896 [mf ed 1992] – (= ser Methodist coll) – 1mf – 9 – 0-524-04441-4 – mf#1991-2106 – us ATLA [242]
A defense of judaism versus proselytizing christianity / Wise, Isaac Mayer – Cincinnati: American Institute, 1889 [mf ed 1986] – 1mf – 9 – 0-8370-6467-8 – mf#1986-0467 – us ATLA [230]
The defense of the aunswere to the admonition : against the replie of t c / Whitgift, J – London: Henry Binneman, 1574 – 15mf – 9 – mf#PW-31 – ne IDC [240]
A defense of the bible against the charges of modern infidelity : consisting of the speeches of elder jonas hartzel, made during a debate conducted by him and mr joseph barker, in july 1853 / Hartzel, Jonas – Cincinnati: Columbian Printing Co, 1854 [mf ed 1993] – (= ser Christian church (disciples of christ) coll) – 4mf – 9 – 0-524-07007-5 – mf#1991-2860 – us ATLA [230]
Defense of the jessey records and kiffin manuscript with a review of dr. john t. christian's work entitled: "baptist history vindicated" / Lofton, George A – 1899 – 1 – 5.11 – us Southern Baptist [242]
Defense Personnel Support Center [US] *see*
– Provider
La defense sociale – Organe de la Federation socialiste de la Vienne (S.F.I.O.). Chatellerault. 1916-17 – 1 – fr ACRPP [335]
La defense sociale de saone-et-loire – Chalon-sur-Saone. nov 1904-05 [biwkly] – 1 – fr ACRPP [073]
Defense transportation journal – Alexandria VA 1973+ – 1,5,9 – ISSN: 0011-7625 – mf#8396 – us UMI ProQuest [355]
Le defenseur de la patrie : faisant suite a l'ami du peuple – n1-54. Paris. juin-aout 1799 – 1 – fr ACRPP [073]
Defensio abbatiae imperialis s maximini trevirensis / Zyllesius, N – Treviris, 1638 – €60.00 – ne Slangenburg [241]
Defensio confessionis ministrorvm iesv christi, ecclesiae antuerpiensis, quae augustanae confessioni adsentitur, contra ivdoci tiletani uaria sophismata / [Flacius Illyricus d A, M] – Basileae, 1567 – 4mf – 9 – mf#TH-1 mf 430-433 – ne IDC [242]
Defensio doctrinae...de sacrosancto eucharistiae sacramento / Vermigli, P M – [Tigvri, Froschouer, 1559] – 10mf – 9 – mf#PBU-281 – ne IDC [240]
Defensio fidei catholicae de satisfactione christi *see* Defence of the catholic faith
Defensio orthodoxae fidei de sacra trinitate, contra prodigiosos errores michaelis serveti hispani : ubi ostenditur haereticos iure gladii coercendos esse, et nominatim de homine hoc tam impio juste et merito sumptum genevae fuisse supplicium / Calvin, J – [Geneva]: Robert Estienne, 1554 – 3mf – 9 – mf#CL-31 – ne IDC [240]
Defensio sanae doctrinae de originali ivstitia ac iniustitia, aut peccato / Flacius Illyricus d A, M – Basileae, 1570 – 2mf – 9 – mf#TH-1 mf 465-466 – ne IDC [242]
Defensio sanae et orthodoxae doctrinae de sacramentis, eorumque natura, vi, fine, usu, et fructu : quam pastores et ministri tigurinae ecclesiae et genevensis antehac brevi consensionis mutuae formula complexi sunt... / Calvin, J – [Geneva]: Robert Estienne, 1555 – 1mf – 9 – mf#CL-33 – ne IDC [240]
Defensio sanae et orthodoxae doctrinae de servitute et liberatione humani arbitrii, adversus calumnias alberti pighii campensis / Calvin, J – Geneva: Jean Girard, 1543 – 3mf – 9 – mf#CL-21 – ne IDC [240]
Defensio verae semperque in ecclesia receptae doctrinae de christi dom. incarnatione : adversus mennonem simonis / Lasco, J – Bonnae, 1545. – 2mf – 9 – mf#PBA-223 – ne IDC [240]
Defensio veritatis hebraicae sacrarum scripturarum / Levita, Iohannes Isaac – Coloniae, 1559 – 4mf – 8 – €11.00 – ne Slangenburg [221]
Defensio...ad...smythaei duos libellos de caelibatu sacerdotum & votis monasticis / Vermigli, P M – Basilea, Petrus Perna, 1559 – 6mf – 9 – mf#PBU-282 – ne IDC [240]
Defensiones theologiae divi thomae aquinatis / Capreolus, Johannes (Capreolus, Jean); ed by Paban, C & Pegues, T – Turonibus. v1-7. 1900-1908 – 7v on 94mf – 8 – €179.00 – ne Slangenburg [241]
El defensor de la verdad – Valencia de Alcantara, 1911 – 5 – sp Bibl Santa Ana [073]
El defensor de su agravio : noticia satirica sobre la peste en el puerto de santa maria / Diez de Leiva, F – S.l, s.a. – 1mf – 9 – sp Cultura [616]
O defensor do commercio – Rio de Janeiro, RJ: Typ Carioca de J I da Silva & Comp, 05-15 jun 1850 – (= ser Ps 19) – mf#P15,01,44 – bl Biblioteca [320]
Defensor pacis (mgh leges 4:7.bd) / Marsilius von Padua – 1932 – (= ser Monumenta germaniae historica leges 4. fontes iuris germanici antiqui in usum scholarum separatim editi (mgh leges 4)) – €25.00 – ne Slangenburg [320]
A defesa – S Tome: H J Assumpcao, sep 25 oct 25-dec 25 1915; jan 8-mar 10 1916 – us CRL [079]
Defesa da economia nacional / Gasparian, Fernando – Rio de Janeiro, Brazil. 1966 – 1r – us UF Libraries [330]
Deffke, Paul *see* Der laeroverks-lehrer
Deffontaines, Pierre *see*
– El brasil
– Brasil
Defiance / National Socialist Liberation Front [US] – v1 iss9-10, and undated sample iss – 1r – 1 – (cont: defiance [buffalo ny: 1977]) – mf#1207823 – us WHS [325]
Defiance Co. Defiance *see*
– Daily express
– Herold series
Defiance Co. Hicksville *see* News-tribune/w
Deficiency symptoms in growing pigs fed on a peanut ration / Kirs, W Gordon – Gainesville, FL. 1942 – 1r – us UF Libraries [636]
DeFilippo, G J *see* Effect of training frequency on cervical rotation strength
Define your terms / Dowden, John – Edinburgh, Scotland. 1900 – 1r – us UF Libraries [240]
Definiciones de la orden y cavalleria de alcantara – 1663 – 9 – sp Bibl Santa Ana [946]
Definiciones y establecimientos de la orden de alcantara – Madrid: Luis Sanchez, 1609 – 1 – sp Bibl Santa Ana [240]
Defining and limiting the jurisdiction of courts sitting in equity: hearing. / U.S. Congress. Senate. Committee on the Judiciary – Washington, Govt. Print. Off., 1930. 36 p. LL-1358 – 1 – us L of C Photodup [347]
Defining national purpose in lesotho / Weisfelder, Richard F – Athens, OH. 1969 – 1r – us UF Libraries [960]
Definite reform in english land law / Hopkinson, Alfred. – London, 1880 – (= ser 19th c ireland) – 1mf – 9 – mf#1.1.1803 – uk Chadwyck [340]
Definition du folklore : suivi de notes sur folklore et psychotechnique et sur l'agriculture temporaire, la prehistoire et le folklore / Varagnac, Andre – Paris : Societe d'editions geographiques, maritimes et coloniales 1938 [mf ed Bloomington IN: Indiana University Libraries, Preservation Dept 1984] – viii 66p on 1r [ill] – 1 – us Indiana Preservation [390]
Definitiones theologicae secundum ordinem locorum communium traditae / Alsted, J H – Hanoviae, 1631 – 2mf – 9 – mf#PBA-105 – ne IDC [240]
Definitions geometriques appliquees au dessin lineaire / Saint-Theotiste, soeur – Montreal: Congregation de Notre-Dame, 1878 – 1mf – 9 – mf#56293 – cn Canadiana [740]

The definitions of faith and canons of discipline of the six oecumenical councils, with the remaining canons of the code of the universal church / Hammond, William Andrew – Oxford: JH Parker, 1843 – 1mf – 9 – 0-7905-9217-7 – mf#1989-2442 – us ATLA [240]
Definitionum medicarum libri 24 literis graecis distincti (ael3/1) / Gorraeus, Ioannis – Paris 1564 [mf ed 1993] – (= ser Archiv der europaeischen lexikographie: fach-enzyklopaedien) – 27mf – 9 – €70.00 – 3-89131-145-1 – (int by michael stolberg) – gw Fischer [610]
Definitive treaty of peace and friendship between his britannic majesty and the united states of america : signed at paris, the 3rd of sep 1783 – [s.l: s.n, 1783?] [mf ed 1984] – 1mf – 9 – 0-665-44215-7 – mf#44215 – cn Canadiana [341]
Deflorationes patrum : sive excerptiones ex patrum doctrina per wernerum abbatem s blasii in nigra silva – Baslilee, 1494 – €27.00 – ne Slangenburg [240]
Defluorinated superphosphate for livestock – Gainesville, FL. 1944 – 1r – us UF Libraries [636]
Defoe, Daniel *see*
– The adventures of robert drury, during fifteen years captivity on the island of madagascar
– Adventures of robinson crusoe
– Adventures of robinson crusoe of york, mariner
– Life and adventures of robinson crusoe
– Life and adventures of robinson crusoe of york
– Life and adventures of robinson crusoe, of york
– Life and adventures of robinson crusoe of york
– Life and adventures of robinson crusoe, of york
– Life and adventures of robinson crusoe of york
– Life and most surprising adventures of robinson crusoe
– Life and strange adventures of robinson crusoe
– Life and strange surprising adventures of robinson...
– Life and strange surprizing adventures of robinson...
– Life and surprising adventures of robinson crusoe,
– Life and surprising adventures of robinson crusoe
– Life and surprising adventures of robinson crusoe,
– Life and surprising adventures of robinson crusoe
– Life and surprising adventures of robinson crusoe,
– Life and surprising adventures of robinson crusoe
– Robinson crusoe
– Robinson crusoe, and a journal of the plague year
– Surprising adventures of robinson crusoe
Defoe's review : review of the state of the british nation – v1-9. 1704-13 – 1 – us AMS Press [941]
Defoy, Henri *see* Le citoyen
Defoy, Louisa *see* Bibliographie analytique de monseigneur yvon theriault
Defremery *see* Fragments de geographes et historiens arabes et persans inedits, relatifs aux anciens peuples du caucase et de la russie meridionale
Defries, Amelia Dorothy *see* Fortunate islands
Defuniak herald – Defuniak Springs, FL. 1932 nov 24-1992 dec – 41r – (gaps) – us UF Libraries [071]
Defuniak herald – Defuniak Springs, FL. v106 n1-v111 n18. 1993-1998 apr – 11r – (gaps) – us UF Libraries [071]
Defuniak herald/breeze – Defuniak Springs, FL. 1957-1992 jun – 7r – (gaps) – us UF Libraries [071]
Deganwy sentinel – conway sentinel – [Wales] LLGC 19 feb 1915-nov 1916 [mf ed 2004] – 1r – 1 – uk Newsplan [072]
DeGarmo, James M *see* The hicksite quakers and their doctrines
Degel ha-torah *see* Bet ya'akov
Degel yehudah / Lazarow, Judah Loeb – New York, NY. 1914 – 1r – us UF Libraries [939]
Degen, Bruce N *see* Oliver shaw
Degeneration : a chapter in darwinism / Lankester, Edwin Ray – London, 1880 – (= ser 19th c evolution & creation) – 1mf – 9 – mf#1.1.9043 – uk Chadwyck [575]
Degenhart, Friedrich *see* Studien ueber zacharias werners stil
Degering, Edward Franklin *see* Outline of organic nitrogen compounds
Deggendorfer zeitung – Deggendorf DE, 1978 1 sep- – ca 9r/yr – 1 – (bezirksausgabe von passauer neue presse, passau) – gw Misc Inst [074]
Degnan, Frank *see* Scuba diving for divers with special needs
Die degradationshypothese und die alttestamentliche geschichte / Giesebrecht, Friedrich – Leipzig: A Deichert, 1905 – 1mf – 9 – 0-8370-3272-5 – mf#1985-1272 – us ATLA [221]
Degrandpre, L M J *see* Voyage...la cote occidentale d'afrique, fait dans les annees 1786 et 1787...

The degrees of the spiritual life : a method of directing souls according to their progress in virtue = degres de la vie spirituelle / Saudreau, Auguste – London: R & T Washbourne, 1907 – 2mf – 9 – 0-524-08242-1 – (in english) – mf#1993-2017 – us ATLA [240]
Deguchi, Madoka *see* Influence of caffeine on substrate utilization
Deguileville, Guillaume de *see* Die pilgerfahrt des traeumenden moenchs
Deguilleville, Guillaume de *see*
– Le pelerinage de vie humaine
– The pilgrimage of the life of man
Deguise, Charles *see* Le cap au diable
DeGuise, Charles et al *see* Chroniques litteraires publiees dans "l'union liberale" de quebec
Deh numi pietosi = Gli giochi d'agrigento / Gederici, V – London: T Skillern, 1793 – 1 – us Sibley [780]
Dehaisnes, [C C A] *see* Documents et extraits divers
Deharbe, Joseph *see* Catechism of christian doctrine
Dehart, Mehgan M *see* Relationship between the talk test and ventilatory threshold
Dehasse, Jean *see* Role politique des associations de ressortissants a leopoldville
Dehaven memorial baptist church. oldham county. lagrange, kentucky : church records – March 1867-May 1977 – 1 – us Southern Baptist [242]
Y deheuwr – [Wales] Swansea 7 hydref-rhag 1886 [mf ed 2003] – 1r – 1 – uk Newsplan [072]
Dehio, G G *see* Die kirchliche baukunst des abendlandes, historisch und systematisch dargestellt
Dehmel, Richard *see*
– Ausgewaehlte briefe aus den jahren 1883 bis 1902
– Ausgewaehlte briefe aus den jahren 1902 bis 1920
– Bekenntnisse
– Die goetterfamilie
– Schoene wilde welt
Dehn, Mura *see* Moved by the spirit
Dehn, Paul *see* Unfallstatistisches zur unfall versicherung
Dehn, Siegfried Wilhelm *see* Caecilia
Dehner, Walter *see* Hessisches nachbarrecht
Dehnert, Max *see*
– Die dominante
– Karlmann
Deho, Ettore *see* La condanna del modernismo
Dehon, Leon Gustave *see* Le plan de la franc-maconnerie en italie et en france d'apres de nombreux temoignages
Dehors – Orleans, France. 12 mar-dec 1925, aug 1926-oct 1939 – 4r – 1 – (aka: en dehors) – uk British Libr Newspaper [074]
Dehoux, Jean Baptiste *see* Rapport au gouvernement
Dehoux, Lorrain *see* L'accord americano-haitien du 7 aout 1933
Dehri *see* The divan project
Dei aekerjagd tau vorigeslewen am baerensee : eine humoristisch-plattduetsche vertellung / Deumeland, Heinrich – Braunschweig: H Sievers, 1875 [mf ed 1989] – 83p – 1 – mf#7174 – us Wisconsin U Libr [870]
Dei apologie des aristides ((tugal / Hennecke, E – Leipzig, 1893 – (= ser Tugal 1-4/3) – 2mf – 9 – €5.00 – ne Slangenburg [230]
Dei concilii ecumenici : in generale ed in specie: del concilio ecumenico vaticano / Coppola, Raffaele – Roma: Fratelli Pallotta Tipografi, 1869 – 1mf – 9 – 0-8370-8976-X – (incl bibl ref) – mf#1986-2976 – us ATLA [240]
Dei gesta per francos : formae tplila 97 / Guitbertus Abbas Novigenti – [mf ed 2002] – (= ser ILL – ser a; Cccm 127a) – 8mf+viii/166p – 9 – €74.00 – 2-503-64274-8 – be Brepols [400]
Deianiia pervykh dvukh vserossiiskikh sezdov russkikh liudei – 1906 – 42p 1mf – 9 – mf#RPP-161 – ne IDC [325]
Deianiia petra velikogo, mudrogo preobrazitelia rossii, sobrannye iz dostovernykh istochnikov i raspolozhennye po godam / Golikov, I I – 1837-1843. v1-15 – 162mf – 8 – mf#R-6033 – ne IDC [947]
Deianiia znamenitykh polkovodtsev i ministrov sluzhivshikh v tsarstvovanie gosudaria imperatora petra velikogo / Bantysh-Kamenskii, D N – 1821. v1-2 – 10mf – 8 – mf#R-5980 – ne IDC [947]
Deiatelnost moskopromsoiuza v 1924-25 godu – 1926 – 69p 1mf – 9 – mf#COR-421 – ne IDC [335]
Deiatel'nost' moskovskogo narodnogo banka limited za 1925 g / Moskovskii Narodnyi Bank Limited – London, 1926 – 1mf – 9 – mf#REF-84 – ne IDC [332]

O deiatel'nosti krest'ianskogo pozemel'nogo banka po samarskoi gubernii 109=dokl vn l'vova samarskomu gubernskomu chrezvychainomu dvorianskomu sobraniiu ot 4-go sentiabria 1909 g / L'vov, VN – N p, n d – 1mf – 9 – mf#REF-256 – ne IDC [332]
Deibel, Franz *see* Goethe im gespraech
Deibler, Lisa K *see* A three-year plan for the university of north carolina at chapel hill department of athletics
Der deichgraf : erzaehlung / Buchheld, Kurt – Prag: Noebe 1944 [mf ed 1989] – 1r – 1 – (filmed with: hofische spuren im protestantischen schuldrama um 1600 / hildegard schaefer) – mf#7093 – us Wisconsin U Libr [880]
Les deicides : examen de la vie de jesus et des developpements de l'eglise chretienne dans leurs rapports avec le judaeisme / Cohen, Joseph – new ed. Paris: Michel Levy, 1864 – 1mf – 9 – 0-8370-9690-1 – (incl bibl ref) – mf#1986-3690 – us ATLA [240]
The deicides : analysis of the life of jesus, and of the several phases of the christian church in their relation to judaism / Cohen, Joseph – London: Simpkin, Marshall, 1872 [mf ed 1985] – 1mf – 9 – 0-8370-2703-9 – (trans by anna maria goldsmid) – mf#1985-0703 – us ATLA [230]
Deicke, Guenther *see* Deutsches gedichtbuch
Deile, Gotthold *see*
– Goethe als freimaurer
Deimann, Wilhelm *see*
– Hermann loens
Deimel, Anton *see* Veteris testamenti chronologia monumentis babylonico-assyriis
Deinard, Ephraim *see*
– 'Atidot Yisra'el
– Masa' be-eropa
– Milhamah la-'adonai ba-'amalek
Deindoerfer, Johannes *see*
– Geschichte der evangel.-luth. synode von iowa und anderen staaten
– Kurzgefasste geschichte der evangel.-luth. synode von iowa und andern staaten
Deine heimat, kamerad! / Dietrich, Stephan – Hartenstein-Sachsen, Leipzig: E Matthes, [1944?] [mf ed 1989] – 71p – 1 – mf#7177 – us Wisconsin U Libr [800]
Deinert, Katja *see* Sonographisch gesteuerte eswl (extrakorporale stosswellenlithotripsie) von pankreasgangsteinen
Deinhardstein, Johann Ludwig *see*
– Gedichte
– Hans sachs
Deinzer, Johannes *see* Liturgy for christian congregations of the lutheran faith
Deisinger, Barbara *see* Reinigung der mitochondrialen atp-synthase aus rinderherzen funktionelle rekonstitution und rekoppelung synthetisierender f1-partikel an den membranintetralen f0-teil neue medizinische bibliothek
Der deismus in der religions- und offenbarungskritik des hermann samuel reimarus / Engert, Joseph – Wien: Verlag der Oesterreichischen Leo-Gesellschaft, 1916 – (= ser Theologische Studien der Oesterr. Leo-Gesellschaft) – 1mf – 9 – 0-7905-7816-6 – (incl bibl ref) – mf#1989-1041 – us ATLA [210]
Deissmann, Gustav Adolf *see*
– Bibelstudien
– Bible studies
– Deutscher schwertsegen
– The epistle of psenosiris
– Die hellenisierung des semitischen monotheismus
– Johann kepler und die bibel
– Der krieg und die religion
– Der lehrstuhl fuer religionsgeschichte
– A light from the ancient east
– Neue bibelstudien
– Die neutestamentliche formel st "in christo jesu"
– New light on the new testament
– St paul
– Die septuaginta-papyri
– Die urgeschichte des christentums im lichte der sprachforschung
Deissmann, Gustav Adolf et al *see* Beitraege zur weiterentwicklung der christlichen religion
The deist, or, moral philosopher : being an impartial inquiry after moral and theological truths – London: R. Carlile, 1819-1826 – 1r – 1 – 0-8370-0064-5 – mf#1984-B407 – us ATLA [210]
Deister- und weserzeitung – Hameln DE, 1949 21 oct-1968 [gaps] – 71r – 1 – (title varies: 22 sep 1997: dewezet; filmed by misc inst: 1969- [ca 8r/yr]) – gw Mikrofilm; gw Misc Inst [074]
Deistviia nizhegorodskoi gubernskoi uchenoi arkhivnoi komissii – Nizhnii Novgorod, 1888-1916. v1-18 – 106mf – 9 – (missing: 1888-1894(1); 1899-1908(4-8); 1912, v13(1-2); 1913, v15(1-2); 1913, v16(1); 1914, v17(1)) – mf#1703 – ne IDC [077]

Deistvuiushchee kooperativnoe zakonodatelstvo : sistematicheskii, khronologicheskii i predmetnyi ukazatel zakonov o kooperatsii / Berdichevskii, N G – 1926, 1927 – 208p 7mf – 9 – mf#COR-712 – ne IDC [335]
Deistvuiushchee zakonodatelstvo o potrebitelskoi kooperatsii / Berdichevskii, N G – Rostov n/D, 1925 – 60p 1mf – 9 – mf#COR-287 – ne IDC [335]
The deity of christ : an address delivered at northfield, with three supplementary notes / Speer, Robert Elliott – NY: Fleming H Revell, c1909 – 1mf – 9 – 0-8370-5394-3 – mf#1985-3394 – us ATLA [240]
The deity of jesus, and other sermons / Kellems, Jesse Randolph – St Louis, MO: Christian Board of Publ, c1919 – 1mf – 9 – 0-524-07627-8 – mf#1991-3234 – us ATLA [240]
"Deixai vir a mim os pequeninos" : seminario ecumenico sobre os problemas dos menores carentes e marginalizados e a participacao das igrejas na sua solucao, sao paulo, 20 a 26 de junho de 1980 – Rio de Janeiro: Centro Ecumenico de Documentacao e Informacao, 1982 – us CRL [972]
Deixis : cibles et ordre des operations dans la structuration de l'enonce en anglais contemporain / Augustin, Catherine – 1mf – 9 – (10007) – fr Atelier National [420]
Dejanija trex svjatyx bliznecov muchenikov spevsipa, elasina i melasina / Marr, N – (= ser zapiski vostochnogo otdel) – 2mf – 8 – (zapiski vostochnogo otdel. imp russ arkh obshchestva. v17 1906 p285-344) – mf#1267 mfB145-146 – ne IDC [243]
[Dejean, M] *see* Anecdotes americaines
Dejeuner d'employes / Gabriel, M – Paris, France. 1823 – 1r – us UF Libraries [440]
Dekaden-blatt fuer den landmann – Strassburg (Strasbourg F), o.J, v1 n1-15, v2 n1-15 – 1 – fr ACRPP [074]
Das dekaden-blatt zum unterricht des landvolks im oberrheinischen departement – Colmar / Elsass (F), 1794 n1, 2, 11-19 – 1 – fr ACRPP [350]
Dekadenz in der neueren deutschen prosadichtung = Decadence in modern german fiction / Eickhorst, William – [Jackson], MS: W Eickhorst; Delmenhorst [Germany]: Kommissions-Verlag, S Rieck c1953 [mf ed 1993] – 1r – 1 – (incl bibl ref & ind. filmed with: das deutsche geschichtsdrama / friedrich sengle) – mf#3382p – us Wisconsin U Libr [430]
Dekadenz und heroismus : zeitroman und voelkisch-nationalsozialistische literaturkritik / Geissler, Rolf – Stuttgart: Deutsche Verlags-Anstalt, c1964 [mf ed 1992] – (= ser Schriftenreihe der vierteljahrshefte für zeitgeschichte n9) – 168p – 1 – (incl bibl ref) – mf#8271 – us Wisconsin U Libr [430]
Dekalb literary arts journal – Clarkston GA 1966-89 – 1,5,9 – ISSN: 0011-7714 – mf#7603 – us UMI ProQuest [400]
Dekalb-sycamore labor news – 1949 jan 28-1953 dec, 1956 jan 20-dec [v4 n3-v14 n1] – 2r – 1 – mf#1055214 – us WHS [331]
Dekalog *see* Die zehn gebote (mxt3)
Der dekalog als katechetisches lehrstuck / Achelis, Ernst Christian – Giessen: Alfred Toppelmann, 1905 – (= ser Vortrage Des Hessischen Theologischen Ferienkurses, Heft 1) – 9 – 0-8370-2035-2 – mf#1985-0035 – us ATLA [220]
DeKay, James *see* Sketches of turkey in 1831 and 1832
Deken, Constant de *see* Dwars door azie
Dekker, C et al *see* Album paleographicum 17 provinciarum
Dekker, R M *see* Ego documents from the netherlands, 16th century-1814
Dekle, George Wallace *see* Florida armored scale insects
Dekonstruksie van die teologiese diskoers liefde / Pretorius, Hendrik Erasmus Sterrenberg – Uni of South Africa 2000 [mf ed Pretoria: UNISA 2000] – 3mf – 9 – (incl bibl; text in afrikaans) – mf#mfm14883 – sa Unisa [240]
Dekorative kunst – Muenchen DE, 1898-99 – 1r – 1 – gw Misc Inst [740]
Dekret o potrebitelskoi kooperatsii, 20 maia 1924 g : prak post komment / Berdichevskii, N G – 1925 – 56p 1mf – 9 – mf#COR-288 – ne IDC [335]
Dekrety i postanovleniia po finansam / Moskovskii Sovet Rabochikh Deputatov – M, 1918 – 1mf – 9 – mf#REF-1 – ne IDC [332]
Dekrety o gosudarstvennom strakhovanii / Glavnoe Pravlenie Gosudarstvennogo Strakhovaniia (Gosstrakh) – M, 1922 – 1mf – 9 – mf#REF-115 – ne IDC [332]
Del 4 pleno del consejo economico sindical provincial / Organizacion Sindical – Badajoz: Graficas Jimenez, 1964 – sp Bibl Santa Ana [330]
Del 13 le trece de junio al 10 le diez de / Canal Ramirez, GonzalO – Bogota, Colombia. 1958 – 1r – us UF Libraries [972]

Del amor, del dolor, y del vicio / Gomez Carrillo, Enrique – Paris, France. 1901 – 1r – us UF Libraries [972]

Del amor i del dolor / Henriquez Y Carvajal, Federico – Barcelona, Spain. 193- – 1r – us UF Libraries [972]

Del antiguo cucuta / Febres Cordero, Luis – Bogota, Colombia. 1950 – 1r – us UF Libraries [972]

Del arbol de las hesperides / Roso de Luna, Mario – Madrid: Editorial Pueyo, 1923 – 1 – sp Bibl Santa Ana [240]

Del avila al pichincha / Yanes M, Julio – Caracas, Venezuela. 1940 – 1r – us UF Libraries [972]

Del calor hogareno / Cardona, Jenaro – San Jose, Costa Rica. 1929 – 1r – us UF Libraries [972]

Del cercado ajeno / Diez-Canedo, Enrique – Madrid: M. Perez. Villavicencio, Editor, 1907 – 1 – (versiones poeticas) – sp Bibl Santa Ana [810]

Del congreso de panama a la conferencia de caracas / Lopez Maldonado, Ulpiano – Quito, Ecuador. 1954 – 1r – us UF Libraries [972]

Del congreso de panama a la conferencia de caracas / Yepes, Jesus Maria – Caracas, Venezuela. v1-2. 1955 – 1r – us UF Libraries [972]

Del conocimiento de dios / Noguera, Rodrigo – Bogota, Colombia. 1953 – 1r – us UF Libraries [972]

Del contorno hacia el dintorno / Rosa-Nieves, Cesareo – San Juan, Puerto Rico. 1961 – 1r – us UF Libraries [972]

Del dialetto napoletano / Galiani, Ferdinando – 2. corr accr ed. Napoli: Porcelli 1789 – (= ser Whsb) – 3mf – 9 – €40.00 – (collezione di tutti i poemi in lingua napoletana; 28) – mf#Hu 135 – gw Fischer [410]

Del epistolario de heredia / Gonzalez Del Valle Y Ramirez, Francisco – Habana, Cuba. 1937 – 1r – us UF Libraries [972]

Del estado de salud y enfermedad y sus consecuencias juridico-sociales; memoria de prueba para optar al grado de licenciado en la facultad de ciencias juridicas y sociales de la universidad de chile / Perez Grille, Ramon – Santiago de Chile: Carrera, 1936. 101 6p. LL-4081 – 1 – us L of C Photodup [340]

Del' historia della chinas... / Gonzalez de Mendoza, J – Roma: Appresso Bartolomeo Grassi, 1586 – 5mf – 9 – mf#HT-517 – ne IDC [915]

Del jardin de la leyenda / Soto Hall, Maximo – Buenos Aires, Argentina. 1929 – 1r – us UF Libraries [972]

Del keresztje or southern cross – Sydney, Australia. 28 feb 1951-15 dec 1956, 15 jan 1957-15 dec 1959, 1960-oct 1967 – 4r – 1 – (aka: fueggetlen magyarorszag) – uk British Libr Newspaper [072]

Del lenguaje dominicano / Jimenez, Ramon Emilio – Ciudad Trujillo, Dominican Republic. 1941 – 1r – us UF Libraries [972]

[Del mar-] surfcomber – CA. 1958-1983; 1984 – 50r – 1 – $3000.00 (subs $75/y) – mf#H03198 – us Library Micro [071]

Del modo di fortificar le citta / Zanchi, G B – Venetia, 1556 – 1mf – 9 – mf#OA-209 – ne IDC [720]

[Del norte county-] del norte, lassen, modoc, plumas, shasta, sierra, siskyou, tehama and trinity counties – CA. 1885 – 1r – 1 – $50.00 – mf#015 – us Library Micro [978]

Del norte enquirer *see* Miscellaneous newspapers of rio grande county

Del norte prospector *see* Miscellaneous newspapers of rio grande county

Del pais de la quimera. historias y paisajes / Manzano de Garias, Antonio – Madrid: Patr.Social de Buenas Lecturas S.A. Bib.de Cultura Popular. Tomo 40 – 1 – sp Bibl Santa Ana [946]

Del panico al ataque / Galich, Manuel – Guatemala, 1949 – 1r – us UF Libraries [972]

Del pasado y del presente / Sanchez Arjona, Vicente – Sevilla: Graficas Sevillanas, 1954 – 1 – sp Bibl Santa Ana [810]

Del periodo marxista / Sanchez Arjona, Vicente – Sevilla: Imprenta Alvarez, 1958 – 1 – sp Bibl Santa Ana [810]

Del periodo marxista / Sanchez Arjona, Vicente – Sevilla: Imprenta de la Divina Pastora, 1938 – 1 – sp Bibl Santa Ana [810]

Del romancero dominicano / Rodriguez Demorizi, Emilio – Santiago, Chile. 1943 – 1r – us UF Libraries [972]

Del rosal del arte / Estevez, Andres Maria – Habana, Cuba. 1957 – 1r – us UF Libraries [972]

Del Rosario, Marissa E *see* Bagong aklat sa pilipino

Del siglo 17 extremeno. contienda entre torre de miguel sesmero y almendral / Rodriguez Amaya, Esteban – Badajoz: Dip. Provincial, 1948. Sep. REE – 1 – sp Bibl Santa Ana [946]

Del tiempo en que fuisteis angeles / Sanchez Arjona, Vicente – Sevilla: Imprenta Carlos Acuna, 1953 – 1 – sp Bibl Santa Ana [810]

Del tiempo y su figura / Franco Oppenheimer, Felix – Puerto Rico. 1956 – 1r – us UF Libraries [972]

Del unico modo de atraer a todos los pueblos a la verdadera religion. advertencia...mexico, 1942 / Casas, Bartolome de las – Madrid: Razon y Fe, 1947 – 1 – sp Bibl Santa Ana [200]

Del unico modo de atraer todos los pueblos a la verdadera religion / Casas, Bartolome de las – Mejico, 1942; Madrid: Missionalia Hispanica, 1947 – 1 – sp Bibl Santa Ana [240]

Del viaggio di terra santa : da venetia,... tripoli, di soria per mare, et di l...per terra... gierusaleme... / Alcarotti, G F – Novara, 1596 – 4mf – 9 – mf#H-8384 – ne IDC [910]

Del viento y de las nubes / Jimenez Canossa, Salvador – San Jose, Costa Rica. 1953 – 1r – us UF Libraries [972]

Del yunque a los andes / Guevara Castaneira, Josefina – San Juan, Puerto Rico. 1959 – 1r – us UF Libraries [972]

Dela repvblique des turcs... / Postel, G – Poitiers, 1560. 3 pts – 2mf – 9 – mf#H-8407 – ne IDC [956]

Delachaux, Theodore *see* Pays et peuples d'angola

Delacour, M (Alfred) *see*
- Monsieur va au cercle
- Phoque

Delacroix, Henri *see*
- Essai sur le mysticisme speculatif en allemagne au quatorzieme siecle
- Etudes d'histoire et de psychologie du mysticisme

Delafield gazette – Pewaukee WI. 1946 jul 4-1948 dec 30, 1949-51, 1952-53 – 3r – 1 – (cont by: pewaukee post; hartland news; lake country reporter) – mf#945136 – us WHS [071]

Delafons, John *see* Treatise on naval courts-martial

Delafosse, Maurice *see*
- Chroniques de fouta senegalais de sire-abbas-soh
- Enquete coloniale dans l'afrique francaise occidentale et equatoriale sur l'organisation de la famille indigene, les financailles, le mariage
- Les frontieres de la cote d'ivoire, de la cote d'or et du soudan
- La langue mandingue et ses dialectes (malinke, bambara, dioula)
- Traditions historiques et legendaires du soudan occidental
- Vocabulaires comparatifs de plus de 60 langues ou dialectes parles a la cote d'ivoire et dans les regions limitrophes

[Delamar-] daily lode – NV. mar-jun 1898 – 1r – 1 – $60.00 – mf#U04487 – us Library Micro [071]

[Delamar-] delamar roaster – MI – 1r – 1 – $110.00 – mf#U04489 – us Library Micro [071]

[Delamar-] lode – NV. 1892-96; 1898-1900; 1901-06 [wkly; biwkly] – 5r – 1 – $300.00 – mf#U04488 – us Library Micro [071]

Delamotte, Freeman Gage *see* Mediaeval alphabets and initials for illuminators

Delamotte, Philip Henry *see*
- The art of sketching from nature
- Choice examples of art workmanship

Delamotte, William Alfred *see* Views of the colleges chapels and gardens, of oxford

DeLand, Charles Edmund *see* The mis-trials of jesus

Deland, Charles Edmund *see* Annotated statutes and rules of trial practice and appellate procedure in south dakota and north dakota

Deland daily news – Deland, FL. 1915 jan 4-1923 – 5r – (gaps) – us UF Libraries [071]

Deland, florida – s.l, s.l? 19- – 1r – us UF Libraries [978]

Deland, Margaret Wade Campbell *see* Florida days

Deland news – Deland, FL. 1910-1920 – 3r – (missing: 1910 jan 7; nov 18; dec 23, 30; 1916 mar 22; 1917 jan 3; aug 29; nov 28) – us UF Libraries [071]

Deland sun – Deland, FL. 1923 jan 5-mar 30; oct 5-dec 28 – 1r – us UF Libraries [071]

Deland sun news – Deland, FL. 1946 jan-1992 jun – 167r – (gaps) – us UF Libraries [071]

Deland weekly news – Deland, FL. 1904; 1906 – 1r – (missing: 190? sep 21; nov 2) – us UF Libraries [071]

Deland weekly news – Deland, FL. 1903 feb 13-may 29; oct 23-dec 18 – 1r – (missing: 1903 nov 27) – us UF Libraries [071]

Delandine, Antoine *see* Tableau des prisons de lyon

Delanne, Gabriel *see* Evidence for a future life

Delano baptist church. delano, tennessee : church records – Sept 1923-Sept 1979 – 1 reel – 1 – $37.17 – (formerly prendergast baptist church 826p) – us Southern Baptist [242]

Delano, Charles G *see* Outline of the law of landlord and tenant in massachusetts

[Delano-] el malcriado – CA. 1971 – (= ser Chicano studies library serial) – 1r – 1 – $60.00 – mf#R02167 – us Library Micro [071]

Delano, Isaac O *see* The singing minister of nigeria

Delany, Martin R *see* Principia of ethnology

Delany, Martin Robison *see* Principia of ethnology

Delany, Selden Peabody *see*
- Difficulties of faith
- The ideal of christian worship

Delapierre, Andre *see* Faux billet

Delaplanche, abbe *see*
- Le pelerin de terre sainte
- La voie douloureuse

Delaporte, Louis *see*
- Ankoru tosako
- Voyage au cambodge; l'architecture khmer

Delaporte, P-Henry *see* Vie de mahomet d'apres le coran et les historiens arabes

Delapree, Louis *see*
- Le martyre de madrid, temoignages inedits
- Das martyrium von madrid; ein unveroeffentlichtes zeugnis

Delarc, Odon *see* L'eglise de paris pendant la revolution francaise, 1789-1801

Delashmit, S J *see* The effects of game stress situations on the heart rates of selected high school football coaches

Le delassement de montceau-les-mines et du canton – n1-4. Paris. oct-nov 1889 – 1 – fr ACRPP [073]

Les delassements militaires : the favorite divertissment composed by mon gallet as danced at the king's theatre; adapted for the piano forte by joseph mazzinghi / Mazzinghi, Joseph – London: Goulding [1797?] [mf ed 1991] – 1r – 1 – mf#pres. film 108 – us Sibley [780]

Delattre, Alphonse J *see*
- Autour de la question biblique
- Un catholicisme americain
- Le criterium a l'usage de la nouvelle exegese biblique
- Le culte de la sainte vierge en afrique
- Les inscriptions historiques de ninive et de babylone

Delaunay, Charles *see* Cours elementaire d'astronomie

Delauney, Honore *see* Origine de la tapisserie de bayeux

Delavan enterprise – Delavan WI. 1900 sep 13/1901 nov 14-1959 jan 1-1959 apr 30 – 36r – 1 – (cont: enterprise [delavan wi]; delavan republican; cont by: delavan enterprise and the delavan republican) – mf#1139463 – us WHS [071]

Delavan enterprise – Delavan WI. 1878 aug 15-1881 dec 30, 1882 jan 6-1884 dec 31, 1885 jan 7-1887 dec 28, 1888 jan 4-1890 dec 31, 1891 jan 7-1893 oct 19 – 5r – 1 – (cont: sharon inquirer; delavan tribune; cont by: enterprise [delavan wi]) – mf#1139458 – us WHS [071]

Delavan enterprise and the delavan republican – Delavan WI. 1959 may 7/sep 30-2001 oct/dec – 152r – 1 – (cont: Delavan enterprise [Delavan wi: 1900]; Delavan republican) – mf#1139464 – us WHS [071]

Delavan messenger – Delavan WI. 1857 feb 25-apr 1, may 6, jun 3 – 1r – 1 – (cont: wisconsin messenger) – mf#962665 – us WHS [071]

Delavan patriot – Delavan WI. 1862 mar 13 – 1r – 1 – mf#962671 – us WHS [071]

Delavan republican – Delavan WI. 1868 apr 23/1870-1953/1959 apr 30 – 43r – 1 – (cont by: delavan enterprise [delavan wi: 1900]) – mf#986166 – us WHS [071]

Delavec – Chicago IL, 1926-27* – 1r – 1 – (slovenian newspaper) – us IHRC [071]

Delavec – Detroit MI, 1928* – 1r – 1 – (slovenian newspaper) – us IHRC [071]

Delavigne, Casimir *see*
- Comediens
- Ecole des vieillards
- Paria
- Vepres siciliennes

Delavignette, Robert Louis *see*
- Les paysans noirs

Delaville Le Roulx, J *see* Les hospitaliers en terre sainte et a chypre (1100-1810)

Delaville le Roulx, Joseph Marie Antoine *see* Cartulaire de l'ordre des hospitaliers de saint-jean de jerusalem (1100-1310)

Delavska enotnost – Ljubljana, Yugoslavia. -w. 1958-70. 16 reels – 1 – uk British Libr Newspaper [949]

Delavska politika – Maribor, Yugoslavia. Sept 1939 – 1r – 1 – us L of C Photodup [949]

Delavska slovenija – Milwaukee WI, 1922, 1925-26* – 1r – 1 – (slovenian newspaper) – us IHRC [071]

Delawarde, Jean Baptiste *see* Prehistoire martiniquaise

Delaware : session laws of american states and territories – 1776-2000 – 9 – $1345.00 set – mf#402580 – us Hein [348]

Delaware *see*
- Delaware chancery reports
- Opinions
- Reports, post-nrs
- Reports, pre-nrs

Delaware advertiser *see* American watchman and delaware advertiser

Delaware alternative press – v1 n2-v8 n2 [1979 nov-1986 winter] – 1r – 1 – (cont: delaware free press [newark de]) – mf#1277612 – us WHS [071]

Delaware and Shawnee Indian Tribes *see* Registers, rolls, and publications

Delaware attorney general reports and opinions – 1963-2001 – 6,9 – $268.00set – (1963-77 on reel $105. 1978-2001 on mf $163) – mf#408170 – us Hein [340]

Delaware chancery reports / Delaware – v1-13. 1814-1912 – (= ser Delaware Supreme Court Reports) – 84mf – 9 – $126.00 – (add vols planned) – mf#LLMC 84-128 – us LLMC [347]

Delaware Co. Delaware *see*
- College transcript
- County news
- Daily chronicle
- Daily gazette series
- Daily herald
- Daily journal
- Daily journal herald
- Democrat-herald
- Early newspapers
- Gazette
- Gazette series
- Herald
- Journal-herald series
- Loco foco / democratic standard
- Ohio wesleyan transcript
- Practical student
- Transcript
- Weekly delaware herald
- Weekly journal
- Western collegian

Delaware code annotated – St Paul: West Pub Co, 1953-apr 2002 update – 9 – $1809.00 set – mf#401151 – us Hein [348]

Delaware county american – Media, PA. -w 1890-1912 – 13 – $25.00r – us IMR [071]

Delaware county atlas, 1866 – 1r – 1 – mf#B7070 – us Ohio Hist [978]

Delaware. Courts *see* Rules of the superior court, court of chancery, orphans' court, court of general sessions, and supreme court of the state of delaware

Delaware free press – v1 n1 [1979 oct] – 1r – 1 – (cont by: delaware alternative press) – mf#1277654 – us WHS [071]

Delaware gazette – Wilmington DE. 1814 apr 22 – 1r – 1 – (cont by: delaware gazette and peninsula advertiser) – mf#854553 – us WHS [071]

Delaware gazette and peninsula advertiser – Wilmington DE. 1818 nov 25 – 1r – 1 – (cont: delaware gazette [wilmington, del.: 1814]; cont by: delaware gazette [wilmington, del.: 1820]) – mf#846096 – us WHS [071]

Delaware Heritage Commission *see* Fully, freely and entirely

Delaware indians dictionary – 1r – 1 – (vocabulary and numerical terms) – mf#B26377 – us Ohio Hist [490]

Delaware journal – Ilmington DE. 1828 jun 24, oct 3 – 1r – 1 – (cont by: delaware state journal, advertiser and star) – mf#854551 – us WHS [071]

Delaware journal of corporate law – Widener University: v1-25. 1976-2000 – 9 – $564.00 set – (v1-16 1976-91 on reel or mf $308. v17-25 1992-2000 on mf $256) – ISSN: 0364-9490 – mf#102391 – us Hein [340]

Delaware journal of corporate law – Wilmington DE 1978+ – 1,5,9 – ISSN: 0364-9490 – mf#11928 – us UMI ProQuest [346]

Delaware lawyer – v1-19. 1982-2001 – 9 – $223.00 set – ISSN: 0735-6595 – mf#110191 – us Hein [340]

Delaware loan office records relating to the loan of 1790 / U.S. Treasury Dept. Bureau of the Public Debt – (= ser Records of the bureau of the public debt) – 1r – 1 – mf#T784 – us Nat Archives [336]

Delaware medical journal – Wilmington DE 1929+ – 1,5,9 – ISSN: 0011-7781 – mf#399 – us UMI ProQuest [610]

Delaware republican – Wilmington, DE: Allerdice, Jeandell & Miles, jan 2 1860-jun 1865 – 2r – 1 – us CRL [071]

Delaware. Supreme Court *see* Delaware supreme court reports

Delaware supreme court reports / Delaware. Supreme Court – v1-31. 1832-1922 – 231mf – 9 – $346.00 – (pre-nrs: v1-11 1832-85 80mf $120.00. updates planned) – mf#LLMC 84-127 – us LLMC [347]

Delaware valley advance – Langhorne, PA. -w 1972 – 13 – $25.00r – us IMR [071]

Delaware valley prout news / Proutist Universal – v3 n6-10 [1982 jun-dec] – 1r – 1 – (cont: prout news) – mf#656891 – us WHS [071]

Delayed onset muscle soreness and damage in relation to electromyographic activity during concentric and eccentric contraction / Harper, Elizabeth – 1988 – 173p 2mf – 9 – $8.00 – us Kinesology [612]
Delbare, Francois *see* Nouveaux eclaircissemens sur la conspiration du 20 mars
Delbet, J *see* Exploration archeologique de la galatie et de la bithynie
Delbos, V *see* Etudes de la philosophie de malebranche
Delbrueck, Ferdinand *see* Der verewigte schleiermacher
Delbrueck, Hans *see* Numbers in history
Delburne times – Alberta, CN. 1912-67 – 12r – 1 – cn Commonwealth Imaging [071]
Delco antenna / International Union, United Automobile, Aerospace and Agricultural Implement Workers of America – v23 n2 [1973 sep], v26 n5-v33 n8 [1974 dec-1984 sep] – 1r – 1 – (cont by: local 292 antenna) – mf#689142 – us WHS [331]
Delco electronics broadcaster / General Motors Corporation – v28 n10 [oct 1974], v29 n2-v40 n6 [1975 feb-1986 jul 1] – 1r – 1 – mf#1222784 – us WHS [621]
Delco sparks / International Union, United Automobile, Aerospace and Agricultural Implement Workers of America – 1963 oct 24-1966 oct 27, nov 10-1969 dec 18, 1970 jan 1-1978 may 25, 1978 jun-1985 jun, 1985 jul-1989 dec – 5r – 1 – (cont by: sparks [anderson in]) – mf#365080 – us WHS [331]
Delcoitre, Patrick Antoine *see* Vanua scope
Deleage, Paul *see* Haiti en 1886
Delegacion de Deportes *see* 2nd gala del deporte comarcal
Delegacion de la Juventud *see* Aventura 77
Delegacion de Mutualidades Laborales *see* Memoria, 1974
Delegacion Local de la Juventud *see* Semana de la juventud. madronera, 20 al 27 mayo 1972
Delegacion Nacional de la Juventud *see* Compromiso y responsabilidad. campamento nacional "francisco franco". cavaleda (soria) 74
Delegacion Nacional de Prensa, Propaganda y Radio *see* Espana y francisco franco
Delegacion Nacional de Sindicatos de FET y de las JONS *see* Reglamento del grupo sindical de colonizacion no 2049 de gabriel y galan
Delegacion Prov. de la Org. Sindical *see* Convenio colectivo sindical provincial para las industrias del aceite y sus derivados de badajoz
Delegacion Provincial Consejo Superior de Deportes *see* 16th gala del deporte provincial
Delegacion Provincial de Educacion Fisica y Deportes *see* 13th gala del deporte provincial. 1977. homenaje a la excma. diputacion provincial
Delegacion Provincial de Educacion y Ciencia *see* Memoria 1974
Delegacion Provincial de la Juventudes *see*
– 4th juegos nacionales de la educacion general basica
– 10th certamen de experiencias teatrales para la juventud
– Campamentos 1971. caceres
– Campana de campamentos 1974
– Fiestas de la juventud. caceres 26 abril 30 mayo 1973
– Normas y orientaciones para la ensenanza primaria
Delegacion Provincial de Ministerio de Informacion y Turismo *see* Carro de la alegria en las localidades de hervas, jarandilla...
Delegacion Provincial de Mutualidades Laborales (de Badajoz) *see* Memoria de actividades
Delegacion Provincial de Sindicatos *see*
– 1 consejo sindical comarcal. delegacion comarcal de azuaga
– 2 congreso sindical agrario de extremadura
– Ciclo
– Convenio colectivo sindical de trabajo agricola, aprobado y suscrito por los representantes...de villamiel
– Convenio colectivo sindical de trabajo agricola de alcantara (caceres) y sus agregados de estorninos y piedras albas
– Convenio colectivo sindical de trabajo agricola... de casas de millan caceres
– Convenio colectivo sindical de trabajo agricola... de guadalupe
– Convenio colectivo sindical de trabajo...de la empresa textil lanera sobrino de benito matas en hervas
– Estatutos de la cooperativa de viviendas de proteccion oficial "san carlos barromero" del sindicato provincial de banca, bolsa y ahorro de caceres
– Obra sindical de colonizacion. granja escuela sindical agraria. nuestra senora de botoa
– Primer consejo comarcal sindical de olivenza. conclusiones definitivas
– Primer consejo sindical comarcal de don benito
– Primer consejo sindical comarcal. delegacion comarcal de fregenal de la sierra
– Proyecto de reglamento de las casas sindicales comarcales dependientes de la delegacion provincial sindical de caceres
Delegacion Provincial de Sindicatos de FET y de las JONS *see* Primer consejo sindical comarcal. delgacion comarcal de castuera
Delegacion Provincial de Sindicatos Vicesecretaria Provincial de Ordenacion Social *see* Cuestionario resumen de las ordenes y circulares sobre enlaces sindicales y juntas de jurados. cartillas del enlace sindical (secciones sindicales)
Delegacion Provincial del Frente de Juventudas, Seccion de Centros de Trabajo *see* 2 concurso provincial de formacion profesional, mayo 1948
Delegacion Provincial del Frente de Juventudes *see*
– Dia de la cancion
– Formacion de las falanges juveniles de franco
– Normas que ha de tener en cuenta el que manda
– Plan de trabajo de educacion preliminar para los meses de febrero y marzo
Delegacon Provincial de Sindicatos *see* Convenio colectivo sindical de trabajo agricola, aprobado y suscrito por los representantes...de coria
Delegate / Iowa State AFL-CIO – 1961 jun – 1r – 1 – mf#3925714 – us WHS [331]
Delegate proposals : n1-162 – Saipan, n.p, 1975 – (= ser Micronesian constitutional convention, 1975) – 6mf – 9 – $9.00 – (various pagination) – mf#LLMC 82-100F, Title 90 – us LLMC [323]
Delegation directory – Saipan, n.p, 1975 – (= ser Micronesian constitutional convention, 1975) – 2mf – 9 – $3.00 – (incl biographies; unpag) – mf#LLMC 82-100F, Title 89 – us LLMC [323]
Delegation en france de la societe saint-jean-baptiste (societe nationale des canadiens-francais) : et de la societe historique de montreal sous la presidence conjointe de mm leon trepanier et victor morin / Thomas Cook et fils – Montreal: Thos Cook & fils, [1926?] (mf ed 1987) – 1mf – 9 – (pref by leon trepanier) – mf#SEM105P813 – cn Bibl Nat [914]
La Delegation Syro-Palestinienne. aupres de la Societe des Nations *see* La nation arabe
Delegue / Sermet, Julien – Paris, France. 1891 – 1r – us UF Libraries [440]
Le delegue du luxembourg – [Paris]: Typ F Malteste et cie. n1. may 1849 – us CRL [074]
Delehaye, Hippolyte *see*
– The legends of the saints
– Les origines du culte des martyrs
Deleseluse, A *see* Chartes inedites de l'abbaye d'orval
Deletang, Luis F *see* Contribucion al estudio de nuestra toponimia
Delevskii, I L *see* Izdanie gruppy sotsialistov-revoliutsionerov
Delevsky, J *see* Sotsyaler ideal un zayne visnshaftlekhe yesoydes
Deleyte de cavalleros y placer de los cavallos / Maestre de San Juan, Lucas – Madrid: Franciso Martin Abad, 1736 – 1 – sp Bibl Santa Ana [946]
Delff, Heinrich Karl Hugo *see*
– Die geschichte des rabbi jesus von nazareth
– Die hauptprobleme der philosophie und religion
Delgado, Antonio *see* Memoria historica...historia
Delgado de Vera, J *see* Defensa y respuesta...de la medicina racional y philosophica...contra...
Delgado del Pino, Francisco *see* Discurso
Delgado, Emilio *see* Tiempos del amor breve
Delgado Fernandez, Manuel *see*
– Mensajes de sol y luna
– Romancero del coronel villalba
Delgado Fernandez, Rufino *see* Breviario sentimental
Delgado Gomez, Enrique *see* La iglesia y la patria. la accion catolica
Delgado, Jose *see* Relato oficial de la mentisima expedicion carlista dirigida por el general andaluz, don miguel gomez...
Delgado, Jose Matias *see* Historia geral das guerras angolanas, 1680
Delgado, Luis Humberto *see* Cartas de america
Delgado, Luiz *see* Rui barbosa
Delgado Moreno, Mateo *see* Instruccion pastoral establecidas..
Delgado, P J *see* En plena polemica...
Delgado Solis, Sebastian *see*
– El buho del ribero
– Yo he visitado fragosa, el gasco y martilandran
Del-gen-data bank – v1 n1-4 [1986?] – 1r – 1 – mf#1288480 – us WHS [000]
Delhaise *see* Les warego (congo belge)
Delhaise, Arnould M L *see* Amedra
Delhaye, P *see* Siger de brabant
Delhi, agra, and rajpootana : illustrated by eighty photographs / Impey, Eugene Clutterbuck – London 1865 – (= ser 19th c art & architecture) – 4mf – 9 – mf#4.1.329 – uk Chadwyck [770]
[Delhi-] delhi express – CA. 1967- – 14r – 1 – $840.00 (subs $50/y) – mf#B02168 – us Library Micro [071]
[Delhi-] delhi record – CA. 1926-1932 – 3r – 1 – $180.00 – mf#B06019 – us Library Micro [071]
Delhi diary : prayer speeches from 10-9-47 to 30-1-48 / Gandhi, Mahatma – Ahmedabad: Navajivan Pub House, [1948] – (= ser Samp: indian books) – us CRL [850]
Delhi gazette – Agra, India. jan 1837-dec 1838, jan 1842-may 1857, dec 1859, jan 1859-jan 1877 [wkly] – 118r – 1 – (cont by: delhi gazette and north west englishman; cont: delhi gazette and north west englishman) – uk British Libr Newspaper [079]
Delhi gazette / Delhi. India. (State) – 1958-1966. Incomplete – 1 – us NY Public [954]
Delhi gazette and north-west englishman – Delhi, India. jan 1839-dec 1841 – 3r – 1 – (cont: delhi gazette; cont by: delhi gazette) – uk British Libr Newspaper [079]
Delhi. India. (State) *see* Delhi gazette
Delhi mission news – London, 1895-1918 – 31mf – 8 – (missing: 1886(2-4)) – mf#I-526 – ne IDC [954]
Deli courant – [Medan]: J Deen, mar 28 1885-mar 16 1940 [daily ex sun, semiwkly] – 142r – 1 – (lacks: nov 28-dec 31 1917; began in 1885; publ varies; publ in various chronological eds) – mf#mf-4997 seam – us CRL [079]
Deli hirlap – Budapest, Hungary. -d. 1 Feb 1918-29 March 1919. Imperfect. 2 reels – 1 – uk British Libr Newspaper [072]
Deli hirlap – Timisoara. Rumania. -d. 1 Jan-20 Aug 1944. (Imperfect). (1 reel) – 1 – uk British Libr Newspaper [949]
Deli in woord en beeld – Medan, 1925-1933 – 73mf – 9 – (missing: 1926(9, 19-21, 46-49), 1927(2, 17, 22, 32, 35), 1928(7, 20), 1929(8, 10-11, 29, 32-33), 1930(1-52), 1931(50), 1932(14, 51), 1933(11, 16)) – mf#SE-20117 – ne IDC [954]
Deli niu – Bangkok, Thailand. 1976 – 5r – 1 – us L of C Photodup [079]
Deli Spoorweg Maatschappij Bah Telefoondienst *see* Penundjuk telepon lokal dan interlokal untuk medan, belawan, bindjai, dolok merangir, galang, gunung melaju, kisaran, kwala, labuan ruku, lubuk pakam, perlanaan, prapat, rampah, siantar
Delia : the blue-bird of mulberry bend / Whittemore, Emma Mott – New York: Fleming H Revell, c1914 [mf ed 1993] – (= ser Christian & missionary alliance coll) – 1mf – 9 – 0-524-07368-6 – mf#1990-5405 – us ATLA [920]
Delibes, Leo *see* Lakme
Delicado, Juan M *see* Cartas sobre quintos
Delices du brabant et de ses campagnes : ou description des villes, bourgs & principales terres seigneuriales de ce duche accompagnee des evenemens les plus remarquables jusqu'au tems present / Cantillon, Philippe de – Amsterdam 1757 [mf ed Hildesheim: 1995-98] – (= ser Fbc) – 4v on 9mf – 9 – €180.00 – 3-487-29651-9 – gw Olms [914]
Deliciae emblematicae : oder, anmuthige sinnbilds-ergoetzligkeiten... / Dexelio, G – Dresden, 1701 – 9mf – 9 – mf#O-1242 – ne IDC [090]
Deliciae juveniles / Friderici, Daniel – 1654 – (= ser Mssa) – 2mf – 9 – €35.00 – mfchl 226 – gw Fischer [780]
Deliciarum juvenilium ander theil / Friderici, Daniel – 1654 – (= ser Mssa) – 2mf – 9 – €35.00 – mfchl 227 – gw Fischer [780]
Delicie musicali / Priuli, Giovanni – 1625 – (= ser Mssa) – 5mf – 9 – €70.00 – mfchl 380 – gw Fischer [780]
Deliciosa jujuy / Gonzalez Arrili, Bernardo – Jujuy, Argentina. 1926 – 1r – us UF Libraries [972]
Delienne, Castera *see* Souvenirs d'epopee
Deligne, Gaston Fernando *see*
– Galaripsos
– Paginas olvidadas
DeLigney, Francis *see* Catholic gems
Delile, A R *see* Fragments d'une flore de l'arabie petree
Delile, A R et al *see* Description de l'egypte
Delile, A [R] et al *see* Description de l'egypte
Delimitacao de manica / Sociedade De Geografia De Lisboa – Lisboa, Portugal. 1893 – 1r – us UF Libraries [960]
Delineations, historical and topographical, of the isle of thanet and the cinque ports / Brayley, Edward W – London 1817-18 [mf ed Hildesheim 1995-98] – 2v on 5mf – 9 – €100.00 – 3-487-27916-9 – gw Olms [914]
Delineations of american scenery and character / Audubon, John James – New York, NY. 1926 – 1r – us UF Libraries [975]
Delineations of the north western division of the county of somerset : and of its antediluvian bone caverns; with a geographical sketch of the district / Rutter, John – Shaftesbury 1829 [mf ed Hildesheim 1995-98] – 1v on 3mf [ill] – 9 – €90.00 – 3-487-27918-5 – gw Olms [914]
The delineator – Toronto: Delineator Pub Co, [1849?-19–] – 9 – mf#P04228 – cn Canadiana [640]
Delitiae historicae et poeticae das ist : historische und poetische kurzweil / Sandrub, Lazarus; ed by Milchsack, Gustav – Halle a.S: M Niemeyer 1878 [mf ed 1993] – (= ser Neudrucke deutscher literaturwerke des 16. und 17. jahrhunderts 10-11) – 11r – 1 – mf#3387p – us Wisconsin U Libr [410]
Delitos contra la vida y la integridad personal / Gutierrez Anzola, Jorge Enrique – Bogota, Colombia. 1952 – 1r – us UF Libraries [972]
Delitos politicos – Bogota, Colombia. 1948 – 1r – us UF Libraries [972]
Delitsch, Otto *see* Westindien und die sudpolar-lander
Delitzsch, Franz *see*
– Die bayerische abendmahlsgemeinschaftsfrage
– Behold the man!
– Beitraege zur assyriologie und vergleichenden semitischen sprachwissenschaft
– Biblical commentary on the prophecies of isaiah
– Biblical commentary on the proverbs of solomon
– Die biblisch-prophetische theologie
– Commentar zum briefe an die hebraeer
– Commentary on the book of psalms
– Commentary on the epistle to the hebrews
– Commentary on the song of songs and ecclesiastes
– A day in capernaum
– De habacuci prophetae
– Fuer und wider kahnis
– Die genesis
– Das hohelied
– Iris
– Jewish artisan life in the time of our lord
– Joshua, judges, ruth
– Judentum und christentum
– Juedisch-arabische poesien aus vormuhammedischer zeit
– Juedische theologie
– Messianic prophecies
– Messianische weissagungen in geschichtlicher folge
– Neue untersuchungen ueber entstehung und anlage der kanonischen evangelien
– Neueste traumgesichte des antisemitischen propheten
– A new commentary on genesis
– New commentary on genesis
– Old testament history of redemption
– The ologische briefe der professoren delitzsch und v. hofmann
– Paulus des apostels brief an die roemer
– Sifre ha-berit ha-hadashah
– Studien zur entstehungsgeschichte der polyglottenbibel des cardinals ximenes
– A system of biblical psychology
– Ein tag in kapernaum
– Vier buecher von der kirche
– Zur geschichte der juedischen poesie
Delitzsch, Frederic *see* The hebrew language viewed in the light of assyrian research
Delitzsch, Friedrich *see*
– Assyrian grammar
– Assyrische grammatik
– Assyrische lesestuecke
– Assyrische thiernamen
– Assyrisches handwoerterbuch
– Babel and bible
– Die babylonische chronik
– Das babylonische weltschoepfungsepos
– Die entstehung des aeltesten schriftsystems
– Judentum und entwicklungslehre
– Das land ohne heimkehr
– Prolegomena eines neuen hebraeisch-aramaeischen woerterbuchs zum alten testament
– Die sprache der kossaeer
– Wo lag das paradies?
Delitzsch, Johannes *see*
– De inspiratione scripturae sacrae quid statuerint patres apostolici et apologetae secundi saeculi
– Das grunddogma des romanismus
Delius, Eduard *see* Wanderungen eines jungen norddeutschen durch portugal, spanien und nord-amerika
Delius, Heinrich Friedrich *see* Fraenkische sammlungen von anmerkungen aus der naturlehre arzneygelahrtheit oekonomie und den damit verwandten wissenschaften
Delius, W *see* Geschichte der irischen kirche von anfang bis zum 12. jahrhundert
Deliverance : the freeing of the spirit in the ancient world / Taylor, Henry Osborn – New York: Macmillan, 1915 [mf ed 1990] – 1mf – 9 – 0-7905-6696-6 – mf#1988-2696 – us ATLA [180]
The deliverance / Cattopadhyaya, Saratcandra – Bombay: Nalanda Publications, 1944 – (= ser Samp: indian books) – (trans fr original bengali by dilip kumar roy; rev by sri aurobindo; pref by rabindranath tagore) – us CRL [280]
Deliverance Evangelistic Church (Philadelphia PA) *see* Bible alive
Delivery and development of christian doctrine : the fifth series of the cunningham lectures / Rainy, Robert – Edinburgh: T & T Clark, 1874 – 1mf – 9 – 0-8370-4829-X – (incl ind) – mf#1985-2829 – us ATLA [240]
Deliz, Monserrate *see* Renadio del cantar folklorico de puerto rico

Delke, James Almerius *see* History of the north carolina chowan baptist association, 1806-1881
Dell, William *see*
- Doctrine of baptisms
- Testimony from the word against divinity degrees in the university
Della architettura... : con centosessanta figure dissegnate dal medesimi, secondo i precetti di vitruvio... / Rusconi, G A – Venetia, 1590 – 5mf – 9 – mf#O-1029 – ne IDC [720]
Della architettura militare, libri 3 / Marchi, F de – Brescia, 1599 – 13mf – 9 – mf#OA-206 – ne IDC [720]
Della cometa apparsa in luglio del 1819 : osservazioni e risultati / Cacciatore, Niccolo – [s.n.]: Dalla Reale Stamperia 1819 [mf ed 1998] – 1r [pl/ill] – 1 – mf#film mas 28407 – us Harvard [520]
Della divisione del tempo nella musica : nel ballo e nella poesia / Sacchi, Giovenale – Milano [per Giuseppe Mazzucchelli nella stamperia Malatesta] 1770 [mf ed 19–] – 4mf – 9 – mf#fiche 663, 843 – us Sibley [780]
Della espugnatione et difesa delle fortezze libri due / Busca, G – Turi, 1585 – 3mf – 9 – mf#OA-254 – ne IDC [720]
Della fallibilit a dei pontefici nel dominio temporale / Muratori, Lodovico Antonio – Modena: Andrea Rossi, 1872 – 1mf – 9 – 0-8370-7815-6 – mf#1986-1815 – us ATLA [240]
Della guerra...fatta per difesa de religione... / Campana, C – Vicenza, 1602. 3v – 14mf – 9 – mf#OA-203 – ne IDC [720]
Della gverra di rhodi libri 3 aggiunta la discrittione dell 'isola di malta concessa a caualieri, dopo che rhodi fu preso / Fontanus, J – Vinegia, 1545 – 2mf – 9 – mf#H-8274 – ne IDC [956]
Della historia...delle cose dell' imperio di constantinopoli libri 7 / Nicetas, A C – Venetia, [1562] – 3mf – 9 – mf#H-8300 – ne IDC [956]
Della luce intellettuale e dell'ontologismo : secondo la dottrina de' santi agostino, bonaventura e tommaso di aquino / Zigliara, Thoma Maria – Roma: F Chiapperini, 1874 – 3mf – 9 – 0-524-08854-3 – mf#1993-2139 – us ATLA [120]
Della natura e perfezione della antica musica de greci : e della utilita che ci potremmo noi promettere dalla nostra applicandola secondo il loro esempio alla educazione de giovani / Sacchi, Giovenale – Milano [per Antonio Mogni nella stamperi?a Malatesta] 1778 [mf ed 19–] – 3mf – 9 – mf#fiche 664, 842 – us Sibley [780]
Della pittura veneziana e delle opere pubbliche de' veneziani maestri / Zanetti, A M – Venezia, 1771 – 8mf – 9 – mf#O-1067 – ne IDC [700]
Della piu che novissima iconologia di cesare ripa perugino cavalier di s.s. mauritio et lazaro / Ripa, Cesare – Padova: Per Donato Pasquardi. 3v. 1630 – 11mf – 9 – mf#O-1267 – ne IDC [090]
Della realt...et perfettione delle imprese di hercole tasso... – Bergamo: C Ventura, 1614 – 6mf – 9 – mf#O-1271 – ne IDC [090]
Della realt...et perfettione delle imprese di hercole tasso... – Bergamo: Per Comine Ventura, 1612 – 6mf – 9 – mf#O-855 – ne IDC [090]
Della storia civile e politica del papato : dal primo secolo dell'aera cristiana fino all'imperatore teodosio / Nobili-Vitelleschi, F – Bologna: Nicola Zanichelli, 1900 – 2mf – 9 – 0-8370-7896-2 – mf#1986-1896 – us ATLA [240]
Della vita di antonio rosmini-serbati : memorie / Paoli, Francesco – Torino: Accademia di Rovereto: G B Paravia, 1880 – 2mf – 9 – 0-8370-6928-9 – (incl bibl ref) – mf#1986-0928 – us ATLA [920]
Della vita e delle opere di terenzio mamiani : con l'aggiunta dell'idillio i patriarchi e dell'inno a s terenzio / Bianchi, Nerino – Pesaro: Federici 1896 [mf ed 1991] – 1r – 1 – mf#c1738 – us Harvard [920]
DellaCella, Paolo *see*
- Narrative of an expedition from tripoli in barbary, to the western frontier of egypt, in 1817, by the bey of tripoli
- Reise von tripolis an die graenzen von aegypten
DellaValle, Pietro *see* Voyages de pietro della valle, gentilhomme romain, dans la turquie, l'egypte, la palestine, la perse, les indes orientales, & autres lieux
Delle acvtezze che altrimenti spiriti, vivezze, e concetti, volgarmente si appellano : trattato / Pellegrini, Matteo – Genoua: C Ferroni 1639 [mf ed 1983] – 1r – 1 – mf#6345 – us Wisconsin U Libr [390]
Delle allusioni, imprese, et emblemi del sig. principio fabricii... / Fabricii, P – Roma: Appresso Bartolomeo Grassi, 1588 – 5mf – 9 – mf#O-1854 – ne IDC [090]
Delle benemerenze di s tommaso d'aquiverso le arti belle / Marchese, V – Genova, 1874 – 2mf – 9 – mf#O-992 – ne IDC [700]
Delle cinque piaghe della santa chiesa = Of the five wounds of the holy church / Rosmini, Antonio; ed by Liddon, H P – London: Rivingtons, 1883 – 1mf – 9 – 0-8370-8466-0 – (in english) – mf#1986-2466 – us ATLA [240]
Delle cose de tvrchi : libri tre – Vinegia, 1541 – 1mf – 9 – mf#H-8147 – ne IDC [956]
Delle cose gentilesche e profane trasportate ad uso et adornamento delle chiese / Marangoni, G – Roma, 1744 – 10mf – 9 – mf#O-359 – ne IDC [700]
Delle demostrationi degli errori della setta macomettana libri cinque / Pientini da Corsignano, A – Firenze, 1588 – 5mf – 9 – mf#H-8372 – ne IDC [956]
Delle fortificationi...libri cinque / Lorini, B – Venetia, 1597 – 4mf – 9 – mf#OA-277 – ne IDC [720]
Delle imprese sacre con utili e dilettevoli discorsi accompagnate, libro prima / Aresi, P – Verona: Appresso Angelo Tamo, [1615] – 4mf – 9 – mf#O-856 – ne IDC [090]
Delle imprese, trattato di giulio cesare capaccio : in tre libri diviso / Capaccio, G C – Napoli: Apresso Gio. Giacomo Carlino, & Antonio Pace, 1592 – 9mf – 9 – mf#O-578 – ne IDC [090]
Delle navigationi et viaggi / Ramusio, G B – Venetia: Stamperia de Givnti, 1554-1559. 3v – 42mf – 9 – mf#HT-680 – ne IDC [910]
Delle navigationi et viaggi / Ramusio, Giovanni Battista – Venetia: Appresso i Giunti, 1606 [mf ed 1988] – 27mf – 9 – mf#SEM105P879 – cn Bibl Nat [910]
Delle navigationi et viaggi raccolte da m gio battista ramusio : volume terzo nel quale si contiene le navigationi al mondo nuovo, a gli antichi incogniti, fatta da don christoforo colombo genovese... – In Venetia [Italy]: Appresso i Giunti, 1606 – 11mf – 9 – 0-665-90330-8 – mf#90330 – cn Canadiana [910]
Delle quinte successive nel contrappunto e delle regole degli accompagnamenti / Sacchi, Giovenale – Milano: per Cesare Orena, stamperia Malatesta 1780 [mf ed 19–] – 3mf – 9 – mf#fiche 665, 844 – us Sibley [780]
Delle ville di plinio il giovane, opera di d pietro marquez messica...vitruvio / Marquez, P J – Roma, 1796 – 3mf – 9 – mf#GDI-16 – ne IDC [700]
Dellen, Idzerd van *see* Kerkelijk handboek ten dienste der chr ger kerk in noord-amerika
Dellenbaugh, Frederick Samuel *see* The north-americans of yesterday
Dell'immacolato concepimento di maria e della sua dogmatica definizione : dialogo polemico famigliare / Finazzi, Giovanni – Milano: Boniardi-Pogliani di E Besozzi, 1855 – 1mf – 9 – 0-8370-8019-3 – (incl bibl ref) – mf#1986-2019 – us ATLA [240]
Dell'imprese di scipion bargagli gentil'huomo sanese : alla prima parte, la seconda, e la terza nuovamente aggiunte / Bargagli, S – Venetia: Appresso Francesco de' Franceschi Senese, 1594 3pts – 7mf – 9 – mf#O-547 – ne IDC [090]
Delling, G *see*
- Bibliographie zur juedisch-hellenistischen und intertestamentarischen literatur 1900-1965
- Paulus' stellung zu frau und ehe
Dell'istoria della compagnia di ges- l'asia / Bartoli, D – Torino, 1825. 2v – 6mf1 – 8 – mf#1507 – ne IDC [956]
Dell'istoria delle guerre civili di francia / Davila, Enrico C – Milano 1807 [mf ed Hildesheim 1995-98] – 6v on 18mf – 9 – €180.00 – 3-487-26120-0 – gw Olms [944]
Dellon, Gabriel *see* Dellon's account of the inquisition at goa
Dellon's account of the inquisition at goa : a new translation, from the french, with an appendix, containing an account of the escape of archibald bower, (one of the inquisitors,) from the inquisition at macerata in italy = Relation de l'inquisition de goa / Dellon, Gabriel – 2nd corr ed. London: Printed for Baldwin, Cradock, and Joy, 1815 [mf ed 1995] – (= ser Yale coll) – viii/187p – 1 – 0-524-09747-X – mf#1995-0747 – us ATLA [241]
Dell'orgine e delle regole della musica colla storia del suo progresso, decadenza, e rinnovazione / Eximeno y Pujades, Antonio – Roma: Stamperia di M A Barbiellini 1774 [mf ed 19–] – 9mf – 9 – mf#fiche 455 – us Sibley [780]
Del-magyarorszag – Szeged, Hungary. 1962-79 – 36r – 1 – us L of C Photodup [079]
Delmare, R Le *see* De officiis ecclesiasticis de jean d'arvranches
Delmas, Louis *see* The huguenots of la rochelle
Delmenhorster kreisblatt – Delmenhorst DE, 1977- – ca 9r/yr – 1 – gw Misc Inst [074]
Delmenhorster kurier – Delmenhorst DE, 1971 16 dec- – 1 – (regional ed of bremer weser-kurier) – gw Misc Inst [074]
Delmenhoster kurier – Delmenhorst DE, 1971 16 dec- – 1 – (bezirksausgabe von bremer weser-kurier) – gw Misc Inst [074]
Delmond, Stany *see* Jeunesse aux antilles
Delmonico Hotel. Dodge City, Kansas *see* Register of guests
Delmonte Y Aponte, Domingo *see*
- Escritos de domingo del monte
- Humanismo y humanitarismo
Delnicka besidka – V Praze: Josef Schuster [mf ed Bloomington IN: Indiana Uni Lib, Preservation Dept 1990] – 3r – 1 – us Indiana Preservation [073]
Delnicke listy – Omaha, NE: Typograficka Unie. -roc5 cis28. 12 cerven 1898 (wkly) [mfe ed roc2 cis26. 12 srp 1895-1898 (gaps) filmed 1978] – 2r – 1 – (in czech. cont by: osveta) – us NE Hist [071]
Delnicke listy – Vienna, Austria may 1890-dec 1934 [mf ed Norman Ross] – 68r – 1 – mf#nrp-1936 – us UMI ProQuest [077]
Delo – Beograd: A M Stanojevic [1894-1915 [mf ed Bloomington IN: Indiana Uni Lib, Preservation Dept 1989] – 37r – 1 – us Indiana Preservation [073]
Delo – Ljubljana, 1976-1992ff – 90r – 1 – gw Mikropress [949]
Delo – Ljubljana, Yugoslavia. May 1959-1976 – 92r – 1 – us L of C Photodup [949]
Delo borisa savinkova : sbornik materialov i dokumentov – 1924 – 161p 2mf – 9 – mf#RPP-227 – ne IDC [325]
Delo [irkutsk: 1918] : ezhedn obshchestv -polit i lit gaz / ed by Anisimov, V A – Irkutsk: O-va potrebit[elei] sluzh[ashchikh] i rab[ochikh] Zabaik zh d i o-va potrebit[elei] "Truzhenik-Kooperator" 1918 [1918 6 avg-24 noiab – (= ser Asn 1-3) – n2-87 [1918] (gaps) item 112, on reel n26 – 1 – (cont by: nashe delo) – mf#asn-1 112 – ne IDC [077]
Delo iv kaliaeva – 1906 – 60p 1mf – 9 – mf#RPP-226 – ne IDC [325]
Delo naroda – (Delo narodnoe Delo narodov Delo Dela naroda). Leningrad. USSR. -d. 2 May 1917-30 Jan 1918. (Imperfect). (1 reel) – 1 – uk British Libr Newspaper [947]
Delo [odessa: 1919] : organ odes kom partii sotsialistovrevoliutsionerov: obshchestv -polit i lit gaz / ed by Egorova, N F – Odessa: [s n] 1919 [1919 23 marta-] – (= ser Asn 1-3) – n11 [1919] item 113, on reel n26 – 1 – mf#asn-1 113 – ne IDC [077]
Delo [semipalatinsk: 1918] : obshchestv -polit gaz : organ semipalat obl komissariata – Semipalatinsk: [s n] 1918 [1917 4 [17] iiunia-] – (= ser Asn 1-3) – n1-17 [1918] item 114, on reel n26 – 1 – (cont: izvestiia semipalatinskogo oblastnogo ispolnitel'nogo komiteta) – mf#asn-1 114 – ne IDC [077]
Delo sibiri : organ akmol gub kom partii sotsialistov-revoliutsionerov / ed by Rasner, S M – Omsk [Akmol obl]: [s n] 1918 [1917 1 okt-1918 27 sent – (= ser Asn 1-3) – n1-105 [1918] item 115, on reel n26,27 – 1 – (lacks: n12; cont by: put' sibiri) – mf#asn-1 115 – ne IDC [335]
Delo truda : organ akmol gub kom partii sotsialistov-revoliutsionerov / ed by Novgorodtseva, A O – Omsk [Akmol obl]: [s n] 1918 [1918 10 okt-] – (= ser Asn 1-3) – n1-6 [1918] item 116, on reel n27 – 1 – (cont: ponedel'nik) – mf#asn-1 116 – ne IDC [325]
Delog-post – Berlin DE, 1916 oct-1917 – 1 – gw Mikrofilm [074]
Deloproizvodstvo i korrespondentsiia selskokhoziaistvennykh kooperativnykh organizatsii / Khudiakov, P – 1930 – 125p 2mf – 9 – mf#COR-523 – ne IDC [335]
Delorme, D *see*
- Misere au sein des richesses
- Paisibles
Delorme, Demesvar *see*
- 1842 [mil huit cent quarante-deux] au cap tremblement de terre
- Memoires sur la question des frontieres
- Theoriciens au pouvoir
Delorme, Ferdinand Marie *see*
- Dialogus de gestis sanctorum fratrum minorum
- Meditatio pauperis in solitudine
O delormista : orgao consagrado ao theatro fluminense e ao grupo delormista – Rio de Janeiro, RJ. 31 mar 1889 – (= ser Ps 19) – mf#P17,01,126 – bl Biblioteca [790]
Delormois *see* Instruction generale pour la teinture des laines et manufactures de laines de toutes couleurs, & pour la culture des drogues ou ingrediens qu'on y employe
Delort, Joseph *see*
- Essai critique sur l'histoire de charles 7, d'agnes sorelle et de jeanne d'arc
- Mes voyages aux environs de paris
Delos – Austin TX 1968-71 – 1 – ISSN: 0011-7951 – mf#7259 – us UMI ProQuest [400]
Delovoi mir – Moscow, 1991-98 – 19r – 1 – (ceased publ) – us East View [330]
Delovoi spravochnik "moskva" / Rudometov, N M – 1922 – 182p 2mf – 9 – mf#COR-293 – ne IDC [335]
Delozier, Eric P *see* Health care on the internet
Delpech, Jacques *see* Le christianisme en koree
Delphick oracle : set forth through correspondence held with the most learned scholars in the most famous universities of europe – La Salle 1719-20 – 1 – mf#4235 – us UMI ProQuest [420]
Delprat, Guillaume Henri Marie *see*
- Verhandeling over de broederschap van geert groote
- Verhandeling over de broedreschap van g. groote en over den invloed der fraterhuizen
Delrieu, Etienne Joseph Bernard *see* Artaxerce
Delta / Delta Sigma Theta Sorority – 1976 fall/winter-1982 spring – 1r – 1 – mf#670055 – us WHS [378]
Delta – Madison WI 1968-76 – 1,5,9 – ISSN: 0011-801X – mf#6931 – us UMI ProQuest [510]
Delta bulletin / Delta Sigma Theta Sorority – 1959 nov, 1960 jan, jun – 1r – 1 – mf#4851421 – us WHS [378]
Delta Canal Company *see* Rich farmlands in florida
Delta daily news / Delta Sigma Theta Sorority – 1994 jul 18-19 – 1r – 1 – mf#4877105 – us WHS [378]
Delta devils gazette / Mississippi Valley State University – v12 n8 [1994 may] – 1r – 1 – mf#3149762 – us WHS [378]
Delta first baptist church. delta, missouri : church records – 1900-Oct 1921 – 1 – 5.76 – us Southern Baptist [242]
Delta kappa gamma bulletin – Austin TX 1934+ – 1,5,9 – ISSN: 0011-8044 – mf#12915 – us UMI ProQuest [378]
Delta newsletter / Delta Sigma Theta Sorority – 1976 summer-1977 summer, v64 n6-v65 n4 [1977 convention-1978 projects], 1983 feb, 1983 jul, dec-1992 spring/summer, 1974 sep/oct-1994 summer – 3r – 1 – (with gaps; cont by: delta update) – mf#3683315 – us WHS [378]
Delta pi epsilon journal – Little Rock AR 1957+ – 1,5,9 – ISSN: 0011-8052 – mf#5739 – us UMI ProQuest [378]
Delta Sigma Theta Sorority *see*
- Delta
- Delta bulletin
- Delta daily news
- Delta newsletter
- Delta update
- Newsletter
Delta update / Delta Sigma Theta Sorority – 1992 mar-sep, 1993 jan-may – 1r – 1 – (cont: delta newsletter) – mf#3683430 – us WHS [378]
Delteil, H L *see* Le peintre-graveur illustre
Deltion plerophorion – Athens, Greece. 3 Jul-4 Oct 1940 (imperfect) – 1r – 1 – (in french, german, italian and english) – uk British Libr Newspaper [949]
Delubom, Nosiphiwo Ethel *see* Learning problems of learners
Deluge – Glasgow, Scotland. no date – 1r – us UF Libraries [240]
Deluge / Javeri, Shanti – Bombay: S Javeri: Sole selling agents, Padma Publ, 1944 – (= ser Samp: indian books) – us CRL [954]
The deluge, history or myth / Townsend, Luther Tracy – New York: American Tract Society, c1907 – 1mf – 9 – 0-524-03997-6 – mf#1992-0040 – us ATLA [210]
Deluxe and illuminated manuscripts : containing technical and literary texts / ed by Doane, A N & Grade, Tiffany J – [mf ed Tempe AZ, 2001] – (= ser ASMMF) – 1226 folios – 8 – $120.00v / £76.00v [institution] ($96v / £60v if part of subsc) – 0-86698-267-1 – mf#mr225 – us MRTS [090]
Delvaille, C *see* Guia higienica y medica del maestro
Delvau, Alfred *see* Jacques bonhomme
Delvert, Jean *see*
- Geographie du cambodge
- Le paysan cambodgien
Delvin estate papers relating to montserrat, leeward islands, 1812-38 : from the archives of sir everard radcliffe at rudding park, harrogate – 1r – 1 – mf#96770 – uk Microform Academic [025]
Delvolve, Jean *see* Rationalisme et tradition
Dem bollme si boes wuche : lustspiel in drei akten / Huggenberger, Alfred – Frauenfeld: Huber 1914 [mf ed 1990] – 1r – 1 – (filmed with: ricarda huch / gertrud baumer) – mf#2734p – us Wisconsin U Libr [820]
Dem hebraeisch-phoenizischen sprachzweige angehoerige lehnwoerter in hieroglyphischen und hieratischen texten / Bondi, J H – Leipzig, 1886 – 2mf – 9 – mf#NE-476 – ne IDC [956]
Dem, Marc *see* Insolite colombie
Dem. rep. and farmers museum – Carlisle, PA., 1824 – 13 – $25.00r – us IMR [071]
"Dem traum folgen" : innere raeume und eigene fremdheit in den bildern des anderen bei jean-jacques rousseau, joseph conrad und karl may / Trinkaus, Stephan – (mf ed 1995) – 3mf – 9 – €49.00 – 3-8267-2170-5 – mf#DHS 2170 – gw Frankfurter [410]

Demachi, Giuseppe *see* Sei quartetti

Demachy, Jacques Francois *see*
- Instituts de chymie, ou principes elementaires de cette science
- Recueil de dissertations physico-chymiques, presentees a differentes academies

DeMaere, Jodi Michelle *see* Effects of deep water and treadmill running on oxygen uptake and energy expenditure in seasonally trained cross country runners

Demain : journal socialiste et de defense syndicaliste – n1-91. Alger. avr 1919-fevr 1921 – 1 – fr ACRPP [325]

Demain – Lyons, France. 27 oct 1905-26 jul 1907; 31 jan 1943-9 jul 1944 – 1 1/2r – 1 – uk British Libr Newspaper [074]

Demain : pages et documents. Organe du groupe communiste francais de Moscou – n1-31. Geneve. 1916-sept 1919 – 1 – fr ACRPP [325]

Deman, Daniela *see* An exploratory study of grasping in preterm, low birthweight infants

Demand for an educated ministry 1865 / Thompson, H A – 1r – 1 – $35.00 – mf#um-312 – us Commission [242]

Demand for freedom – n2 [1970 nov 16], n2 reprint [1979 nov 16], n5 [?] – 1r – 1 – mf#721942 – us WHS [071]

Demande au public en reparation d'honneur contre la demoiselle petit – [Paris, 1741] – 1 – mf#*ZBD-*MGO pv30 – Located: NYPL – us Misc Inst [340]

The demands of darwinism on credulity... : dedicated by permission to her most gracious majesty the queen / Morris, Francis Orpen – London, [1890 – (= ser 19th c evolution & creation) – 1mf – 9 – mf#1.1.11613 – uk Chadwyck [575]

The demands of the north-west! : a speech delivered in the house of commons, ottawa, on wednesday, february 27th, 1889, by mr n f davin, mp – S.l: s.n, 1889? (S.l: A Senecal) – 1mf – 9 – mf#30128 – cn Canadiana [971]

Los demanes en la conquista de america, buenos aires, 1943 / Bayle, Constantino & Arcimegas, German – Madrid: Razon y Fe, 1944 – 1 – sp Bibl Santa Ana [972]

Demantin / Holle, Berthold von; ed by Bartsch, Karl – Stuttgart: Litterarischer Verein 1875 (Tuebingen: H Laupp) [mf ed 1993] – (= ser Blvs 123) – 58r – 1 – mf#3420p – us Wisconsin U Libr [430]

Demantin / Holle, Berthold von; ed by Bartsch, Karl – Stuttgart: Litterarischer Verein, 1875 (Tuebingen: H Laupp) [mf ed 1993] – (= ser Blvs 123) – 400p – 1 – mf#8470 reel 26 – us Wisconsin U Libr [890]

Demantius, Christoph *see*
- Isagoge artis musicae ad incipientium captum maxime accommodata
- Tirades, sioniae introtium missarum ex prosarum...

Demar, Carmen *see*
- Derrumbe
- Vuelo intimo y mar del sargazo

Demar, Claire *see* Appel d'une femme du peuple, sur l'affranchissement de la femme

Demarcacion politica de la republica de guatemala / Guatemala Direccion General De Estadistica – Guatemala, 1893 – 1r – us UF Libraries [972]

Demare, Sophie *see* La femme dans le droit penal du proche-orient ancien

Demarest, David D *see* The reformed church in america

Demarest, Gerherdus Langdon *see* Songs of joy

Demarest, John Terhune *see*
- A commentary on the catholic epistles
- A commentary on the second epistle of the apostle peter
- A translation and exposition of the first epistle of the apostle peter

DeMatteo, Edward D *see* History and development of the double bass

Demaus, R *see* William tindale, a biography

Demaus, Robert *see* Hugh latimer

Dembetembe, N C *see* Verbal constructions in korekore dialect

Dembitzer, Phinehas Elijah *see* Giv'at pinhas

Dembo, Isaak Aleksandrovich *see*
- Schachten
- Schaechten im vergleich mit anderen schlacht-methoden

De'medici, Andrea *see* Sonetti e poesie italiane di vari autori

Dementev, AG et al *see* Russkaia periodicheskaia pechat (1702-1894)

Dementev, B A *see* Trudovye arteli

Dementev, P A *see* Ezhemesiachnoe politicheskoe izdanie, posviashchennoe tekushchim russkim delam

Dementia and geriatric cognitive disorders – Basel, Switzerland 2001+ – 1 – ISSN: 1420-8008 – mf#20599,01 – us UMI ProQuest [618]

Dementin – Vaestervik, Sweden. 1923-23 – 1r – 1 – sw Kungliga [078]

Demenz vom alzheimer typ : diagnostik und schweregradbestimmung mit psychometrischen testverfahren und elektroenzephalographie / Ihl, Ralf – (mf ed 1998) – 2mf – 9 – €40.00 – 3-8267-2581-6 – mf#DHS 2581 – gw Frankfurter [616]

Demerara after 15 years of freedom / Brunnell, John – London, England. 1853 – 1r – us UF Libraries [025]

Demerara daily chronicle – Georgetown, Guyana. nov 1881-jun 1922, 1930-51, 1958-apr 1966 – 292 1/2r – 1 – uk British Libr Newspaper [079]

The demerara martyr : memoirs of the rev john smith, missionary to demerara / Wallbridge, Edwin Angel – London: C Gilpin, 1848 – 1mf – 9 – 0-7905-6850-0 – (with pref by william garland barrett. also available in film b00640 [mf ed 2002]) – mf#1988-2850 – us ATLA [240]

Demers, Benj *see* Quelques notes historiques

Demers, Benjamin *see*
- Une branche de la famille amyot-larpiniere
- Un des premiers colons d'etchemin, p q, jean dumet ou demers
- La famille demers d'etchemin, p q
- Monographie
- Notes sur la paroisse de st francois de la beauce
- Quelques notes historiques

Demers, Denise S *see* A prediction equation for estimating body fat percentage using noninvasive measures

Demers, Jerome *see* Institutiones philosophicae ad usum studiosae juventutis...

Demers, Louis-Philippe *see* Jean du met, 1662...a jacques demers, 1965

Demers, Philippe *see* Des privileges [sic] sur les biens meubles

Demerson, Georges *see* D juan melendez valdes correspondance...hopitaux d'avila...bordeaux, 1964

Demeter – v5 n9-v8 n4 [1983 jan-1985 summer] – 1r – 1 – mf#1131357 – us WHS [071]

Demetrio ramos perez. el tratado de limites de 1750 y la expedicions de iturriaga al orinoco... / Mateos, Francisco – Madrid: Missionalia Hispanica, 1950 – 1 – sp Bibl Santa Ana [240]

Demetrius : roman / Worms, Carl – 1-3. aufl. Stuttgart: J G Cotta 1919 [mf ed 1993] – 1r – 1 – (filmed with: briefe nach dem westwall / hans woerner & other titles) – mf#7968 – us Wisconsin U Libr [830]

Demetrius : trauerspiel in fuenf acten / Schiller, Friedrich von – Dresden: E Pierson 1897 [mf ed 1991] – 1r – 1 – (filmed with: schillers demetrius / martin greif) – mf#2872p – us Wisconsin U Libr [820]

Demetz, Peter *see* Goethes 'die aufgeregten'

Demian, Johann *see* Geographie und statistik des grossherzogthums baden nach den neuesten bestimmungen bis zum 1 maerz 1820

Demian, Johann A *see* Briefe aus paris

Dem'ianenko, I I *see* Slovo naroda

Demidov, A N *see* Voyage dans la russie meridionale et la crimee par la hongrie, la valachie et la moldavie

Demidov, I *see* Novyi put' [eniseisk: 1918]

Demimuid, Maurice *see* Vie du bienheureux francois-regis clet

Deming, Eleanor *see* Handbook for canoeing councillors

Un demi-siecle d'apostolat en chine *see* Le reverend pere joseph gonnet de la compagnie de jesus

La democracia : diario del partido filipino federal – Manila: [s.n.], jan 16 1901 – (filmed with other miscellaneous titles from duke university) – us CRL [079]

La democracia – Manila: [s.n.], aug 11 1899 – us CRL [079]

Democracia colectivista. lecciones de sociologia sobre una nueva politica a la antigua espanola...por... / Cascales Munoz, Jose – Madrid: Sociedad Espanola de Libreria, s.a. – 1 – sp Bibl Santa Ana [320]

Democracia e nacao / Lacerda, Jorge – Rio de Janeiro, Brazil. 1960 – 1r – us UF Libraries [972]

Democracia e parlamentarismo / Azevedo, Fay De – Porto Alegre, Brazil. 1934 – 1r – us UF Libraries [321]

Democracia no brasil / Castilho, Augusto Ferreira De – Sao Paulo, Brazil. 1929 – 1r – us UF Libraries [972]

Democracia representativa / Assis Brazil, Joaquim Francisco De – Rio de Janeiro, Brazil. 1893 – 1r – us UF Libraries [972]

Democracia y el poder militar / Sarria, Eustorgio – Bogota, Colombia. 1959 – 1r – us UF Libraries [972]

Democracia y redentorismo / Lopez Pineda, Julian – Managua, Nicaragua. 1942 – 1r – us UF Libraries [972]

Democracia y socialismo, seguido de otros breves e... / Fortin Magana, Romeo – San Salvador, El Salvador. 1953 – 1r – us UF Libraries [972]

Democracias y tiranias en el caribe / Krehm, William – Habana, Cuba. 1960 – 1r – us UF Libraries [972]

Democracy – Bangkok, Thailand. Jan 1946-47 – 1r – 1 – uk British Libr Newspaper [072]

Democracy / Flack, A G – New York: Cochrane, 1910 – 1mf – 9 – 0-8370-7942-X – mf#1986-1942 – us ATLA [320]

Democracy – New York NY 1981-83 – 1,5,9 – ISSN: 0272-6750 – mf#12850 – us UMI ProQuest [320]

Democracy *see* Ethical world series, 1898-1916

Democracy and anti-monopoly : an address... thomas jefferson club of brooklyn, april 16th, 1883 / Thurber, Francis Beatty – [New York: [s.n], 1883] (mf ed 19–) – 18p – mf#Z-134 n24 – us Harvard [332]

Democracy and christian doctrines : an essay in reinterpretation / Carnegie, William Hartley – London: Macmillan, 1914 – 1mf – 9 – 0-7905-9164-2 – mf#1989-2389 – us ATLA [240]

Democracy and diplomacy : a plea for popular control of foreign policy / Ponsonby, Arthur Ponsonby – London: Methuen [1915] [mf ed 1986] – 1r – 1 – (filmed with: philosophical works / locke, j) – mf#1601 – us Wisconsin U Libr [327]

Democracy in america / Tocqueville, Alexis De – New York, NY. 1945 – 1r – us UF Libraries [977]

Democracy in france / Thomson, David – London, England. 1958 – 1r – us UF Libraries [944]

Democracy in india / Appadorai, Angadipuram – London; New York: Oxford University Press, 1943 – (= ser Samp: indian books) – us CRL [954]

Democracy in spain / Dingle, Reginald James – London. I937. Fiche W 840. (Blodgett Collection of Spanish Civil War Pamphlets) – 9 – us Harvard [946]

Democracy in the dominions / Brady, Alexander – Toronto, ON. 1947 – 1r – us UF Libraries [320]

Democracy in the dominions / Brady, Alexander – Toronto, ON. 1952 – 1r – us UF Libraries [320]

Democracy in the dominions / Brady, Alexander – Toronto, ON. 1956 – 1r – us UF Libraries [320]

Democracy not suited to india / Allahabad, Oudh – Pioneer Press, 1888 – (filmed with: a historical sketch of fyzabad tehsil/p carnegy. lucknow, 1870) – us CRL [321]

The democracy of christianity : or, equality in the dealings of god with men / White, Lorenzo – NY: Hunt & Eaton, 1892 – 1mf – 9 – 0-8370-5759-0 – mf#1985-3759 – us ATLA [240]

Democracy or revolution in spain? / Matteo, Johan – London, 1937. Fiche W 1033. (Blodgett Collection of Spanish Civil War Pamphlets) – 9 – us Harvard [946]

"Democrat" – 1882 oct 13-1883 oct 11; 1883 oct 18-1885 apr 9; 1885 apr 16-1886 sep 30; 1886 oct 7-1888 may 4 – 1 – mf#1003416 – us WHS [071]

Democrat – 1948 apr-1952 nov, 1964 jan 13-1969 jan – 1r – 1 – mf#1111080 – us WHS [071]

Democrat – feb 5 1919-dec 24 1920, 1921, 1923, 1986-87, jan 7-jun 16 1988, sep 8 1988, sep 15 1988-jan 5 1989, jan 12-jun 29 1989, jul 6 1989-jan 4 1990, jan 11 1990-dec 1990, jan 10-jan 27 1991, jul 4 1991-92, jan 14-jun 24 1993, jul 1993-jan 6 1994, jan 13-dec 1994, jan 12-jun 29 1995, jul 6-dec 21 1995, 1996 – 25 1/2r – 1 – (aka: dungannon democrat [n ireland]) – uk British Libr Newspaper [072]

Democrat / Ashland Co. Loudonville – (june 1895-oct 1900) [wkly] – 1r – 1 – mf#B10373 – us Ohio Hist [071]

Democrat / Belmont Co. Bellaire – 1928-29, (1931-apr 1933) [wkly] – 2r – 1 – mf#B12016-12017 – us Ohio Hist [071]

Democrat – Grand Island, NE: Hall & Jaques, jul 1884-v17 n21. nov 29 1901 (wkly) [mf ed 1891-92,1895-1901 (gaps) filmed 1978] – 2r – 1 – (cont by: grand island democrat) – us NE Hist [071]

Democrat – York, NE: L D Woodruff (wkly) [mf ed v2 n22. may 29 1884-85 (gaps)] – 2r – 1 – (cont by: york democrat) – us NE Hist [071]

Democrat – Humphrey, NE: H P Walker, 1893-v9 n32. sep 27 1895 (wkly) [mf ed v7 n20. jul 7 1893-sep 27 1895 (gaps) filmed 1974-75] – 2r – 1 – (cont: humphrey democrat. cont by: humphrey democrat (1895)) – us NE Hist [071]

Democrat – Sturgeon Bay WI. 1892 dec 29-1895 mar 7 – 1r – 1 – (cont: republican [sturgeon bay wi]; cont by: door county democrat) – mf#934421 – us WHS [071]

Democrat / Darke Co. Greenville – 1867-84, 1888-mar 1908 (center shadow) [wkly] – 9r – 1 – mf#B7579-7587 – us Ohio Hist [071]

Democrat / Darke Co. Greenville – 3/1908-10/21,4/23-25,1-5/1927 [wkly] – 8r – 1 – mf#B9198-9205 – us Ohio Hist [071]

Democrat / Darke Co. Greenville – v1 n1. 8/1864-11/66,5/83-4/89,1905-07 [wkly] – 4r – 1 – mf#B12081-12084 – us Ohio Hist [071]

Democrat – Dungannon, Ireland. 1986-97 – 25 1/2r – 1 – uk British Libr Newspaper [072]

Democrat / Franklin Co. Columbus – v1 n1. dec 1878-mar 1880 [daily] – 4r – 1 – mf#B28770-28773 – us Ohio Hist [071]

Democrat / Lawrence Co. Ironton – v1 n1. nov 1874-dec 1877 [wkly] – 1r – 1 – mf#B33858 – us Ohio Hist [071]

Democrat / Meigs Co. Pomeroy – sep 1927-dec 1929,jan 1932-aug 1941 (damaged) [wkly] – 6r – 1 – mf#B32763-32768 – us Ohio Hist [071]

Democrat / Mercer Co. Celina – 1898-1902, 1904-18 [wkly] – 9r – 1 – mf#B9093-9101 – us Ohio Hist [071]

Democrat / Miami Co. Troy – (5/1927-3/28,9/29-11/34,7/35-8/35) [wkly] – 3r – 1 – mf#B9277-9279 – us Ohio Hist [071]

Democrat / Morgan Co. McConnelsvill – aug 1871-dec 1874 [wkly] – 2r – 1 – mf#B5514-5515 – us Ohio Hist [071]

Democrat / New Democratic Party of British Columbia – 1967 aug 31-1978 may, 1978 jun-1984 oct – 2r – 1 – (cont: ccf news for british columbia and the yukon) – mf#515471 – us WHS [325]

Democrat – Pensacola, FL. 1846 jan-jul – 1r – us UF Libraries [071]

Democrat / Pike Co. Waverly – 1907-10, feb 1911-may 1913 [wkly] – 3r – 1 – mf#B10898-10900 – us Ohio Hist [071]

Democrat / Pike Co. Waverly – jan-aug 1867 [wkly] – 1r – 1 – mf#B5581 – us Ohio Hist [071]

Democrat / Preble Co. Eaton – apr 1935-dec 1936 [wkly] – 1r – 1 – mf#B28692 – us Ohio Hist [071]

Democrat / Preble Co. Eaton – aug 1914-apr 1915 (damaged material) [wkly] – 1r – 1 – mf#B29857 – us Ohio Hist [071]

Democrat / Preble Co. Eaton – dec 1908-dec 1926 [wkly, semiwkly] – 11r – 1 – mf#B32411-32421 – us Ohio Hist [071]

Democrat / Preble Co. Eaton – jun 1854-dec 1856 [wkly] – 1r – 1 – mf#B32396 – us Ohio Hist [071]

Democrat / Preble Co. Eaton – may 1875-oct 1899, (1/1900-2/1902) [wkly] – 11r – 1 – mf#B32397-32407 – us Ohio Hist [071]

Democrat : [republican series] / Portage Co. Ravenna – sep 1858-jul 1924, 1925-feb 1928 [wkly] – 27r – 1 – mf#B4196-4222 – us Ohio Hist [071]

Democrat / Sandusky Co. Clyde – v1 n1. apr 1899-apr 1907 [wkly] – 4r – 1 – mf#B32961-32964 – us Ohio Hist [071]

Democrat – Townsville, dec 1895-aug 1897 – 1r – A$30.23 vesicular A$35.73 silver – at Pascoe [079]

Democrat / Vanwert Co. VanWert – mar 1898-jun 1900,nov 1900-jul 1904 [wkly] – 3r – 1 – mf#B8773-8775 – us Ohio Hist [071]

Democrat / Vinton Co. McArthur – sep 1853-aug 1862, mar 1863-65 [wkly] – 3r – 1 – mf#B199-201 – us Ohio Hist [071]

The democrat – Basseterre, St. Kitts. -w. 10 Jan 1959-22 Oct 1966; 28 Jan 1967-14 Dec 1968; 2 Jan 1971-27 Oct 1973. Imperfect. 6 reels – 1 – uk British Libr Newspaper [072]

The democrat – Columbus, NE: A B Coffroth. -v6 n16. jul 17 1885 (wkly) [mf ed 1883-85 (gaps)] – 2r – 1 – (cont by: columbus democrat) – us NE Hist [071]

The democrat – Kuala Lumpur. Malaysia. -w. 9 Mar-26 May 1946. (17 ft) – 1 – uk British Libr Newspaper [072]

The democrat – Lilongwe: [s.n.], 1994. jul 20/27-nov 21; dec 6-21; 1995: jan-mar 10, 23-apr 6, 21-28; may 11-Jul 21; aug 5, 18-nov 24; dec 21; 1996: jan 13-apr 9, 19-27; may 10-jun 15; jul-aug 2 – 2r – us CRL [079]

The democrat *see* Miscellaneous newspapers of lake county

Democrat and argus – Easton, PA., 1833-1841 – 13 – $25.00r – us IMR [071]

Democrat and labour advocate – Birmingham, England. -w. 3 Nov-8 Dec 1855. 3 ft – 1 – uk British Libr Newspaper [072]

Democrat and register – Darlington WI. 1888 may 11-jun 1, jun 8-1889 nov 29, 1889 dec 6-1891 jun 5, jun 12-1892 dec 30, 1893 jan 6-dec 29 – 5r – 1 – (cont: democrat [darlington wi]; cont by: darlington democrat) – mf#1003450 – us WHS [071]

Democrat and register-start democrat – Mifflintown, PA. -w 1889-1906 – 13 – $25.00r – us IMR [071]

Democrat and watchman / Pickaway Co. Circleville – jan 1875-may 1916 [wkly] – 19r – 1 – mf#B9788-9806 – us Ohio Hist [071]

Democrat and watchman / Pickaway Co. Circleville – may 1916-dec 1926 [wkly] – 6r – 1 – mf#B6953-6958 – us Ohio Hist [071]

Democrat enterprise – Sparta WI. 1885 jan 24-nov 14 – 1r – 1 – (cont: enterprise [sparta wi]; cont by: sparta democrat [sparta wi: 1885]) – mf#1044263 – us WHS [071]
Democrat herald – Butler, PA., 1889-1898 – 13 – $25.00r – us IMR [071]
Democrat herald [baker, or] – Baker OR: L C Bollinger, 1963-90 [daily ex sun & hols] – 1 – (cont: baker democrat-herald [1929-63]; cont by: baker city herald [baker city, or]) – us Oregon Lib [071]
Democrat kongolais – Leopoldville: G Masiala, aug-sep 1961 – us CRL [079]
The democrat. (labour world.-sunday world) – London. Nov 1884-May 1891.-w,m. 2mqn reels – 1 – uk British Libr Newspaper [072]
Democrat news / Greene Co. Xenia – 1880-82, 1884-94 [wkly] – 6r – 1 – mf#B10364-10369 – us Ohio Hist [071]
Democrat press – Reading, PA., 1835-1840 – 13 – $25.00r – us IMR [071]
Democrat sentinel / Harrison Co. Cadiz – dec 1911-dec 1917 [wkly] – 3r – 1 – mf#B7022-7024 – us Ohio Hist [071]
Democrat series / Champaign Co. Urbana – mar 1879-mar 1883 [wkly] – 2r – 1 – mf#B9523-9524 – us Ohio Hist [071]
Democrat series / Meigs Co. Pomeroy – 1889-jan 1918, sep 1918-jun 1919 [wkly] – 14r – 1 – mf#B6279-6292 – us Ohio Hist [071]
Democrat series / Portage Co. Ravenna – 1875-aug 1893, aug 1894-aug 1908 [wkly] – 14r – 1 – mf#B11000-11013 – us Ohio Hist [071]
Democrat state journal – Harrisburg, PA., 1835-1836 – 13 – $25.00r – us IMR [071]
Democrat union – Harrisburg, PA., 1853-1855 – 13 – $25.00r – us IMR [071]
Democrat union / Perry Co. Somerset – 5-10/1858, 12/58-6/60, 9/60-4/1866 [wkly] – 2r – 1 – mf#B5516-5517 – us Ohio Hist [071]
El democrata – Mexico, 1916-19 – 8r – 1 – us CRL [079]
El democrata – Mexico. -d. 12 May 1916-10 April 1917; 2-31 March, 21 Oct-31 Dec 1918. 3 reels – 1 – uk British Libr Newspaper [079]
O democrata : orgao propagandista deste "restaurant" – Rio de Janeiro, RJ: [s.n.] maio, out 1889 – (= ser Ps 19) – mf#P17,01,150 – bl Biblioteca [079]
O democrata : publicacao para o povo – Rio de Janeiro, RJ. 23 jan-10 fev 1892 – (= ser Ps 19) – mf#DIPER – bl Biblioteca [079]
El democrata fronterizo – Laredo, TX: J Cardenas, dec 8 1917-jun 6 1919 – 1 – us CRL [071]
Democrate *see* Le drapeau blanc
Le democrate – Cap-Haitien: Impr National – us CRL [079]
Le democrate – Le Paysan libre. suite de: Le Paysan libre. Le democrate de Tarn-et-Garonne. Montauban. juin 1966-67 – 1 – fr ACRPP [073]
Le democrate egalitaire – [Paris]: Impr d'Ed Bautruche, apr 1848 – 1r – us CRL [074]
Democrate et paysan *see* L'unite paysanne
Democrate kongolais – Leopoldville: G Masiala, 1961 [aug 15-sep 5 1961] (mthly) – (= ser Herbert j [i.e. f] weiss coll on the belgian congo) – 1r – 1 – us CRL [079]
Le democrate savoyard *see* Le reveil des gauches
Le democrate vendeen – La Roche-sur-Yon. 14 dec 1850-20 dec 1851 – 1 – (journal politique, agricole, commercial et d'annonces. mq n1-3, 8, 18, 60, 101, 106) – fr ACRPP [073]
Democrat-herald / Delaware Co. Delaware – 11/1885-90,1/92-10/97,98-1/1900 [wkly, semiwkly, wkly] – 8r – 1 – mf#B9009-9016 – us Ohio Hist [071]
Democratic advocate / Darke Co. Greenville – 1884-99, 1902-07 [center shadows] [wkly] – 9r – 1 – mf#B7570-7578 – us Ohio Hist [071]
Democratic advocate / Darke Co. Greenville – 1908-6/1915,2/1920-9/1921,1926-29 [wkly] – 5r – 1 – mf#B10565-10569 – us Ohio Hist [071]
Democratic advocate – Westminster, Maryland. 1866-1928 – 1 – us MD Archives [071]
Democratic age/the age – York, PA., 1889 – 13 – $25.00r – us IMR [071]
Democratic banner / Knox Co. Mount Vernon – nov 1847-apr 1853 [wkly] – 2r – 1 – mf#B197-198 – us Ohio Hist [071]
Democratic banner – Williamsport, PA., 1874-1876 – 13 – $25.00r – us IMR [071]
Democratic banner series / Knox Co. Mount Vernon – jan 1895-dec 1922 [wkly, semiwkly] – 22r – 1 – mf#B9982-10003 – us Ohio Hist [071]
The democratic blade – [Valentine, NE]: Robt O Fink. [v1 n1] sep 18 1885- (wkly) [mf ed -1889 (gaps) filmed [1974-79]] – 2r – 1 – us NE Hist [071]
Democratic call / Franklin Co. Columbus – sep 1894-aug 1896 – 1r – 1 – mf#B12018 – us Ohio Hist [071]
Democratic citizen / Warren Co. Lebanon – oct 1854-may 1859 [wkly] – 2r – 1 – mf#B2450-2451 – us Ohio Hist [071]
Democratic citizen – Lebanon, OH. aug 25, 1859-aug 8, 1862 – 1r – 1 – (weekly democratic newspaper) – us Western Res [071]
Democratic companion / Carroll Co. Carrollton – feb 1854-oct 1855 [wkly] – 1r – 1 – mf#B3982 – us Ohio Hist [071]
Democratic crisis – Corvallis OR: T B Odeneal, 1859 [wkly] – 1r – 1 – (cont: occidental messenger [1857-59]; cont by: oregon union [1859-18-?]) – us Oregon Lib [071]
Democratic digest / Woman's National Democratic Club – v5-30. 1930-195-? – 1 – $162.00 – mf#0177 – us Brook [322]
Democratic enquirer – Bedford, PA. -w 1827-1833 – 13 – $25.00r – us IMR [071]
Democratic enquirer / Vinton Co. McArthur – feb 1867-jan 1870, mar 1870-jan 1873 [wkly] – 2r – 1 – mf#B147-148 – us Ohio Hist [071]
Democratic forum / Democratic Party [WI] – 1969 apr-1970 jul/aug – 1r – 1 – (cont: badger bulletin; badger bulletin; insight [madison wi]; cont by: wisconsin democrat [madison wi: 1973]) – mf#3564471 – us WHS [325]
Democratic free press / Ashtabula Co. Ashtabula – v1 n1. feb 1834-jan 1835 [wkly] – 1r – 1 – mf#B11213 – us Ohio Hist [071]
Democratic front for the liberation of palestine publications – Chicago – (filmed by the university of chicago library photoduplication laboratory for the middle eastern microfilm project at the center for research libraries, 1994) – us CRL [071]
Democratic guide – De Witt, NE: J S Culbertson. 2v. v1 n1. nov 15 1888-v2 n14. feb 13 1890 (wkly) [mf ed lacks dec 5 1889 filmed [1966?]] – 1r – 1 – (occasional articles in german. cont: de witt rip-saw. absorbed by: saline county democrat. publ in wilber ne, apr 25 1889-90) – us NE Hist [071]
Democratic herald – Eugene City OR: A Blakely, [wkly] [mf ed 1974] – 1r – 1 – (absorbed by: oregon weekly union [1859-62]) – us Oregon Lib [071]
Democratic herald / Montgomery Co. Dayton – v1 n1. (6-12/1834), 1/35-8/37, 1-2/1840 [wkly] – 1r – 1 – mf#B5524 – us Ohio Hist [071]
Democratic herald series / Montgomery Co. Dayton – dec 1836-mar 1839; (1839-41) scattered [wkly] – 1r – 1 – mf#B33692 – us Ohio Hist [071]
Democratic herald series / Perry Co. New Lexington – 1875-apr 1879, 1881-aug 1888 [wkly] – 5r – 1 – mf#B11607-11611 – us Ohio Hist [071]
Democratic herald series / Perry Co. New Lexington – dec 1867-nov 1869, nov 1870-dec 1874 [wkly] – 3r – 1 – mf#B5511-5513 – us Ohio Hist [071]
Democratic herald / telegraph / Darke Co. Greenville – jun 1850-may 1853 (center shadows) [wkly] – (= ser Telegraph) – 1r – 1 – (title changes) – mf#B6294 – us Ohio Hist [071]
Democratic impulse in jewish history / Silver, Abba Hillel – New York, NY. 1928 – 1r – us UF Libraries [939]
Democratic / jackson herald / Jackson Co. Jackson – jan 1867-dec 1869, jan 1873-dec 1874 [wkly] – (= ser Jackson Herald) – 2r – 1 – (title changes) – mf#B118-119 – us Ohio Hist [071]
Democratic left / Democratic Socialist Organizing Committee [US] – 1979 feb-1982 jun, sep/oct-1991 nov/dec – 2r – 1 – (cont: newsletter of the democratic left) – mf#620555 – us WHS [335]
Democratic messenger / Sandusky Co. Fremont – 10/1877-79, 82-97, 99-12/1900 [wkly, semiwkly] – 9r – 1 – mf#B33484-33492 – us Ohio Hist [071]
Democratic messenger / Sandusky Co. Fremont – may 1871-jan 1873 [wkly] – 1r – 1 – mf#B1801 – us Ohio Hist [071]
The democratic messenger – Waynesburg, PA. -d 1972-76. 16 rolls – 13 – $25.00r – us IMR [071]
Democratic mirror / Marion Co. Marion – (jan 1843-aug 1844) [irreg] – 1r – 1 – mf#B29848 – us Ohio Hist [071]
Democratic monthly magazine and western review – v1. may 1844 – 1r – 1 – mf#B26307 – us Ohio Hist [073]
Democratic National Convention, Houston, Texas *see* Official report of the proceedings
Democratic news : weekly democratic newspaper – Warren, OH. 17 Feb 1938-22 May 1941 – 1r – 1 – us Western Res [071]
Democratic northwest – Napoleon, OH. may 8, 1869-jan 11, 1894 – 11r – 1 – (weekly democratic newspaper) – us Western Res [071]
Democratic northwest and henry co news – Napoleon, OH. jan 18, 1894-jan 26, 1905 – 6r – 1 – (weekly democratic newspaper) – us Western Res [071]
Democratic Party *see*
– Wisconsin democrat
Democratic party and philippine independence / Storey, Moorfield – Boston: Press of G H Ellis Co 1913 [mf ed 1994] – on pt of 1r – 1 – mf#11052 r1657 n9 – us Cornell [959]
Democratic Party. National Committee *see* Campaign text books
Democratic Party. National Convention *see* Official report of the proceedings
Democratic party. national convention proceedings – 1832-1988 – 1 – $390.00 – (with ind) – mf#0179 – us Brook [325]
Democratic party of dane county / [calendar] – 1980 aug, 1981 jan-mar, may-jun, oct-1983 aug, oct-1984 oct, dec-1985 dec, 1986 feb, 1988 aug-oct, dec-1989 jan – 1r – 1 – (cont: dane county democrat [1969]; cont by: common ground [madison wi]) – mf#1498116 – us WHS [325]
Democratic Party [US] *see*
– Fact
– National democrat
Democratic Party [WI] *see*
– Democratic forum
– New wisconsin democrat
Democratic pharos – Logansport IN. 1861 jan 9, 1864 jun 15 – 1r – 1 – (cont by: logansport pharos) – mf#856342 – us WHS [071]
Democratic Policy Committee *see*
– Economic data review
– Special report
Democratic press – Fond Du Lac WI. 1858 jun 5, jul 9, dec 15, 1859 apr 27-1860 jul 4 – 2r – 1 – (cont: fond du lac union; fond du lac journal [fond du lac wi: 1857]) – mf#916666 – us WHS [071]
Democratic press / Portage Co. Ravenna – v1 n1. sep 1868-dec 1874 [wkly] – 3r – 1 – mf#B6339-6341 – us Ohio Hist [071]
Democratic press / Preble Co. Eaton – sep 1860-apr 1865 [wkly] – 1r – 1 – mf#B32603 – us Ohio Hist [071]
Democratic press – York, PA. 1 Jul 1839-31 Dec 1862 – 5r – 1 – us L of C Photodup [071]
The democratic process / Prasad, Beni – London: Oxford University Press, 1935 – (= ser Samp: indian books) – us CRL [325]
Democratic progress / Institute for Democratic Analysis – n1-19 [1974 nov4-1976 jan] – 1r – 1 – mf#365082 – us WHS [320]
Democratic record : official organ of the democracy of geauga county – Chardon, OH: C.L. King. v3 n6. feb 2 1889-aug 30 1890-91 – 1r – 1 – (weekly democratic newspaper. other titles: geauga record 1889-aug 16 1890; geauga record (chardon, ohio: 1888); geauga county record) – mf#34 G2.1 015 – us Western Res [071]
Democratic register – Eugene City OR: A Noltner, -1862 [wkly] – 1 – (cont by: eugene city review) – us Oregon Lib [071]
Democratic register – Lawrenceburg IN. 1872 sep 19 – 1r – 1 – (cont by: lawrenceburg register) – mf#856335 – us WHS [071]
Democratic register – Darlington WI. 1885 nov12-1888 may 4 – 1r – 1 – mf#961912 – us WHS [071]
Democratic republican and agriculture register – Carlisle, PA. -w 1829-30. 1 roll – 13 – $25.00r – us IMR [071]
Democratic review – Jacksonville OR: Linn, Crutcher & Jackson, [wkly] [mf ed may 25-jun 15 1872] – 1r – 1 – (incl on reel: jacksonville, or. miscellaneous newspapers, 1863-1963) – us Oregon Lib [071]
Democratic secretary – Sheboygan WI. 1853 oct 7 – 1r – 1 – mf#927092 – us WHS [071]
Democratic sentinel – Kittanning, PA. -w 1889-1910. 7 rolls – 13 – $25.00r – us IMR [071]
Democratic socialism : being the manifesto of the action group of nigeria for an independent nigeria – [Lagos]: Action Group Bureau of Information, 1960 – us CRL [960]
Democratic Socialist Organizing Committee [US] *see*
– Democratic left
– Newsletter of the democratic left
Democratic socialist report and review / Socialist Party, USA [1973-] – v1 n12 [1985 jul/aug], v2 n1-v5 n23 – 1r – 1 – (cont: may day! [miami beach, fla.]; cont by: social issues essays) – mf#1548382 – us WHS [325]
Democratic Socialists of America *see*
– Creeping socialist
– Socialist standard
– Women organizing
Democratic standard – Portland OR: Alonzo Leland, [wkly] – 1 – (began with jul 19 1854. ceased with jun 6 1859. publishers: leland, northrop & co, 1855-oct 1857; a leland & co, nov-dec 1857; j O'meara, 1858-59. suspended jan 4-feb 1859) – us Oregon Lib [071]
Democratic standard – Portland, Or.T [i.e. OR]: Alonzo Leland. v1 n12-v5 n38. sep 27 1854-may 11 1859 – 1 – (began with july 19 1854 issue. cf. turnbull, g.s. history of oregon newspapers. ceased with june 6 1859 issue. cf. scott, h.w. history of portland, or, 1890) – us Oregon Hist [071]
Democratic standard / Brown Co. Georgetown – v1 n1. aug 1840-feb 1845 [wkly] – 1r – 1 – mf#B12474 – us Ohio Hist [071]
Democratic standard – Janesville WI. 1851 oct 11-1854 oct 11, 1854 apr 19-1856 apr 16 – 2r – 1 – (cont by: weekly democratic standard) – mf#926643 – us WHS [071]
Democratic standard – Holidaysburg, PA. -w 1889-1912 – 13 – $25.00r – us IMR [071]
Democratic State Central Committee of Nebraska *see* Nebraska democrat
Democratic state register – Juneau, Watertown WI. 1850 mar-1853 apr 9 – 1r – 1 – (cont by: wisconsin weekly register) – mf#944246 – us WHS [071]
Democratic test – New Bloomfield, PA., 1860 – 13 – $25.00r – us IMR [071]
Democratic times / Auglize Co. Wapakoneta – aug 1888-jul 1891 (poor quality) [wkly] – 2r – 1 – mf#B25349-25350 – us Ohio Hist [071]
Democratic times – Jacksonville OR: J N T Miller & Co, 1871- [wkly] – 1 – (ceased in 1907; cont: democratic news [jacksonville, or]; merged with: southern oregonian [medford, or: 1902]; southern oregonian and jacksonville times) – us Oregon Lib [071]
Democratic times *see* Southern oregonian (medford, or: 1902)
Democratic transcript / Star Co. Canton – v1 n1. apr 1853-sep 1854 [wkly] – 1r – 1 – mf#B8492 – us Ohio Hist [071]
Democratic union / Adams Co. West Union – feb 1860-oct 1864, jan-oct 1865 [wkly] – 1r – 1 – mf#B6611 – us Ohio Hist [071]
Democratic union / Perry Co. New Lexington – jun 1866-nov 1867 [wkly] – 1r – 1 – mf#B5511 – us Ohio Hist [071]
Democratic vindicator – Tionesta, PA. -w 1904-1912 – 13 – $25.00r – us IMR [071]
Democratic vistas and other papers / Whitman, Walt – London: Routledge; New York: Dutton 1906? [mf ed 1987] – (= ser New universal library) – 1r – 1 – (with: the power and beauty of superb womanhood / macfadden, b) – mf#10645 – us Wisconsin U Libr [840]
Democratic voice – Salem OR: Democrat Pub Co Inc [wkly] – 1 – (began in 1950) – us Oregon Lib [071]
Democratic watchman – Bellefonte, PA. -w 1861-1932 – 13 – $25.00r – us IMR [071]
Democratic west side telephone – McMinnville OR: H L Heath, 1887-89 [wkly] – 1 – (cont: west side semi-weekly telephone; merged with: oregon register [lafayette, or] to form: mcminnville telephone=register) – us Oregon Lib [071]
Democratic west side telephone *see* Oregon register (lafayette, or)
Democratic whig standard / Harrison Co. Cadiz – feb 1844-feb 1845 [wkly] – 1r – 1 – mf#B6993 – us Ohio Hist [071]
Democratic Workers Party *see*
– Our socialism
– Plain speaking
La democratie – Paris. 17 aout 1910-2 aout 1914 – 1 – fr ACRPP [320]
La democratie – Point-a-Pitre, Guadeloupe. 1900-1906 (1) – mf#67938 – us UMI ProQuest [079]
La democratie au cambodge / Baruch, Jacques – Bruxelles: Editions Thanh-Long 1967 [mf ed 1989] – (= ser Etudes orientales 2) – 1r with other items – 1 – mf#mf-10289 seam reel 001/07 [§] – us CRL [320]
La democratie chretienne – Lille. mai 1894-1908 – 1 – (puis revue sociale d'etudes et d'action) – fr ACRPP [325]
La democratie jurassienne – Salins. 31 dec 1848-30 juil 1850 – 1 – (journal politique et litteraire puis journal de la revolution sociale) – fr ACRPP [073]
Democratie nouvelle – Paris, France. 27 oct 1918-12 aug 1919 – 2 1/2r – 1 – uk British Libr Newspaper [072]
Democratie nouvelle : revue mensuelle de politique mondiale / ed by Duclos, J – Paris. 1947-59 – 1 – fr ACRPP [320]
La Democratie Socialiste des Arrondissements de Saint-Claude (Jura) et de Nantua (Ain) *see* Le montagnard
La Democratie Socialiste du 20e Arrondissement *see* Paris-demain
Democraties – [Dakar?]: Impr Tandian-Yoff, [feb 1992-apr 1993] – 1r – 1 – us CRL [320]
Democrat-sentinel / Hocking Co. Logan – apr 1906-dec 1909 [wkly] – 2r – 1 – mf#B8549-8550 – us Ohio Hist [071]
Democrat-sentinel / Hocking Co. Logan – (jan 1910-dec 1931) [wkly] – 8r – 1 – mf#B11251-11258 – us Ohio Hist [071]
Democrat-tribune – Mineral Point WI. 1958 apr 8/1960-1999 – 32r – 1 – (cont: iowa county democrat and the mineral point tribune) – mf#1131448 – us WHS [071]
Democriti germanici [...] – Lauban (Luban PL), 1732-33 – 1 – gw Misc Inst [077]
O democrito : jornal para ser lido – Rio de Janeiro, RJ. 05 fev-01 mar 1881 – (= ser Ps 19) – mf#P05,04,196 – bl Biblioteca [073]
Democritus [pseud] *see* Faction defeated

Democritus ridens : or, comus and momus, a new jest and earnest pratling concerning the times – La Salle IL 1681 – 1 – mf#4236 – us UMI ProQuest [320]
Demographic structure of seventeen villages in the peri-urban area of blantyre-limbe, nyasaland / Bettison, David G – Lusaka: Rhodes-Livingstone Institute 1958 – (= ser Rhodes-livingstone institute, livingstone. communications 11) – (master microform held by: gmc) – sa Misc Inst [304]
Demographics, employment status, and employment satisfaction of graduates from nata approved undergraduate athletic training programs in the state of pennsylvania / Marks, Melissa A – 1994 – 2mf – $8.00 – us Kinesology [378]
Demography – Silver Spring MD 1971+ – 1,5,9 – ISSN: 0070-3370 – mf#6061 – us UMI ProQuest [304]
Demoiselle a marier / Scribe, Eugene – Paris, France. 1826 – 1r – us UF Libraries [440]
Demokracija – Trieste. Italy. -f. 1 Sep 1958-17 May 1963. (1 reel) – 1 – uk British Libr Newspaper [949]
Demokrasi – Surabaja, 1950-1951. nos 1-42/43 – 11mf – 9 – (missing: 1950(18, 21-22, 25-27); 1951(31, 34-36, 39-41)) – mf#SE-1398 – ne IDC [950]
Der demokrat : mitteilungsblatt der demokratischen volkspartei – Stuttgart DE, 1946 14 jun-1951 – 1 – gw Misc Inst [325]
Der demokrat – Rostock, Schwerin DE, 1947 [gaps] – 1 – mf#6522 – gw Mikropress [074]
Der demokrat – Temeschburg (Timisoara RO), 1925-27 – 1 – gw Misc Inst [077]
Demokraticke hlasy – Zvolen, Czechoslovakia. Jul-Nov 1945 – 1r – 1 – us L of C Photodup [077]
Demokratija – Belgrade, Yugoslavia. Sept-Nov 1945 – 1r – 1 – us L of C Photodup [949]
Demokratische correspondenz – Stuttgart DE, 1868-69 – 1r – 1 – gw Misc Inst [320]
Das demokratische deutschland – Berlin DE, 1920, 1923 – 1 – gw Misc Inst [074]
Die demokratische gemeinde – Hannover, Bonn DE, 1949 oct-1950 – 1r – 1 – mf#3771 – gw Mikropress [074]
Demokratische post : organo de los alemanes democratas de mexico y centro-america – Mexiko-Stadt (MEX), 1943 15 aug-1952 feb/mar – 1r – 1 – gw Misc Inst [079]
Die demokratische republik – Heidelberg DE, 1849 1 may-6 jun – 1r – 1 – gw Misc Inst [074]
Demokratische rundschau – Berlin DE, 1919 4 may-23 nov – 1r – 1 – gw Misc Inst [074]
Demokratische zeitung *see*
– Die zukunft 1867
– Volksmund
Demokratisches volksblatt – Salzburg, Austria. 12 dec 1945-12 feb 1948 – 4r – 1 – uk British Libr Newspaper [072]
Demokratisches wochenblatt – Leipzig, Stuttgart, Hamburg DE, 1868-1878 21 oct [gaps] – 7r – 1 – (title varies: 2 oct 1869: der volksstaat; 1 oct 1876: vorwaerts; filmed by misc inst: 1868-76) – mf#165 – gw Mikropress; gw Misc Inst [074]
Demokratisch-zionistischen fraktion : programm und organisations-statut – Berlin, 1902 – 1mf – 9 – mf#J-28-29 – ne IDC [956]
Demokratizatsiya – Washington DC 2001+ – 1,5,9 – ISSN: 1074-6846 – mf#21826 – us UMI ProQuest [077]
Demokratsiia – Sofia, Bulgaria. 12 feb 1990-30 dec 1996 [mf 1992-6] – 1 – (in cyrillic) – mf#[1992-6:] mf.677.a – uk British Libr Newspaper [077]
Demolins, Edmond *see*
– Anglo-saxon superiority
– Quoi tient la superiorite des anglo-saxons
The demon alcohol, the great man-slayer : a sermon preached december 9th, 1888, in the first baptist church, yarmouth, ns / Adams, Henry – Yarmouth, NS?: s.n, 1888? – 1mf – 9 – mf#09102 – cn Canadiana [230]
Demon possession and allied themes : being an inductive study of phenomena of our own times / Nevius, John Livingston – 2nd ed with corr and supplement. Chicago: Fleming H Revell, 1896, c1894 – 2mf – 9 – 0-524-00845-0 – (incl bibl ref) – mf#1990-2091 – us ATLA [130]
Demonic possession in the new testament : its relations historical, medical, and theological / Alexander, William Menzies – Edinburgh: T & T Clark, 1902 – 1mf – 9 – 0-8370-2070-0 – (incl ind) – mf#1985-0070 – us ATLA [225]
The demonism of the ages : spirit obsessions so common in spiritism, oriental and occidental occultism / Peebles, James Martin – 3rd ed. Battle Creek, MI: Peebles Medical Institute, c1904 [mf ed 1992] – 1mf – 9 – 0-524-03603-9 – mf#1990-3247 – us ATLA [130]
Demonology and devil-lore / Conway, Moncure Daniel – London: Chatto and Windus, 1879 – 3mf – 9 – 0-7905-1811-2 – (incl bibl ref and index) – mf#1987-1811 – us ATLA [210]
Demons of the dust / Wheeler, William Morton – New York, NY. 1930 – 1r – us UF Libraries [500]
Demonstracion y vindicacion de las injusticias...por acusacion del delator don joaquin rodriguez leal / Ceresoles, Mauricio – 1841 – 9 – sp Bibl Santa Ana [360]
Demonstratio evangelica : dat is, de evangelische waerheyd van de gereformeerde godsdienst / Leydecker, M – Utrecht, 1684 – 7mf – 9 – mf#PBA-230 – ne IDC [242]
Demonstratio fallaciarum johannis calvini, in doctrina de coena domini qvibvs vsvs est in libro institutionis christianae, et ex quo suum calvinismum in omnem egurgitavit christianum orbem / Huber, S – Witebergae, 1593 – 1mf – 9 – mf#TH-1 mf 720 – ne IDC [242]
Demonstratio imposturarum et fraudum, quibus egidius hunius... / Pezelius, C – Bremae, 1591 – 3mf – 9 – mf#PBA-288 – ne IDC [240]
Demonstration de l'immaculee conception de la bienheureuse vierge marie, mere de dieu / Parisis, Pierre-Louis – Paris: Jacques Lecoffre, 1849 [mf ed 1986] – 1mf – 9 – 0-8370-8135-1 – (text of encyclical in french & latin. comm in french. incl bibl ref) – mf#1986-2135 – us ATLA [241]
Demonstration du principe de l'harmonie : servant de base a tout l'art musical theorique & pratique / Rameau, Jean Philippe – Paris: Durand [etc] 1750 [mf ed 19–] – 2mf – 9 – mf#fiche 506 – us Sibley [780]
A demonstration of the trueth of that discipline which christ hath prescribed... / Udall, J – n.p., [1588] – 1mf – 9 – (missing: title pg) – mf#PW-54 – ne IDC [240]
Demonstration of the truth of the christian religion / Keith, Alexander – New York: Harper, 1855 – 1mf – 9 – 0-7905-7864-6 – (incl bibl ref) – mf#1989-1089 – us ATLA [240]
Demonstration philosophique du catholicisme : a tous ceux qui tiennent a la verite, a leurs semblables, a leur patrie / Polge, Abbe – Paris: Jacques Lecoffre, 1851 – 1mf – 9 – 0-8370-8371-0 – (incl bibl ref) – mf#1986-2371 – us ATLA [241]
Demonstrationes evidentissimae doctrinae de essentia imaginis dei et diaboli / Flacius Illyricus d A, M – Basileae, [1507] – 5mf – 9 – mf#TH-1 mf 425-429 – ne IDC [242]
Demonstrations catholiques, ou l'art de reunir les pretendus reformez / Regourd, A – Paris, 1630 – 11mf – 9 – mf#CA-144 – ne IDC [241]
Demonstrator : a periodical of fact, thought, and comment – v1-5 n14. 1903-08 [all publ] – (= ser Radical periodicals in the united states, 1881-1960. series 2) – 1r – 1 – $115.00 – us UPA [320]
Demont, Louise *see* Journal of the visit of her majesty the queen, to tunis, greece, and palestine
Demoret family newsletter – 1986 jul-1989 jun – 1r – 1 – mf#1712997 – us WHS [929]
Demorgny, G *see* La question persane et la guerre
DeMoss, Lucy King *see* With hammer and hoe in mission lands
Demostenes, Manuel *see* Estudos sobre a nova capital do brasil
Demosthenis orationes publicae – London, England. 1885? – 1r – us UF Libraries [930]
Demosthenis quae supersunt opera – Lipsiae, Germany. v1-3. 1822-22 – 1r – us UF Libraries [930]
Demostracion...engano...fray marcos de alcala... fundacion descalzas / Velasco, Matias – 1737 – 9 – sp Bibl Santa Ana [240]
Demostractiones palmarias.. / Bachiller Reganadientes – 1787 – 9 – sp Bibl Santa Ana [946]
Das demotische totenbuch der pariser nationalbibliothek : papyrus des pamonthes / Lexa, F – Leipzig, 1910 – 2mf – 9 – mf#NE-20408 – ne IDC [956]
Demotz de la Salle, abbe *see* Methode de musique selon un nouveau systeme tres-court, tres-facile & tres-sur
DeMounteney, Thomas *see* Selections from the various authors who have written concerning brazil
Demoustier, Charles Albert *see* Conciliateur
Dempf, A *see* Das unendliche in der mittelalterlichen metaphysik und in der kantischen dialektik
Dempsey, Richard *see* Magistrate's hand-book
Dempwolff, Otto *see* Papers regarding pacific linguistics
Dems-studie zum elektrochemischen reaktionsverhalten ungesaettigter kohlenwasserstoffe an platin-einkristalloberflaechen sowie an polykristallinen platin- und palladium-elektroden / Mueller, Ulrich – (mf ed 1996) – 2mf – 9 – €40.00 – 3-8267-2390-2 – mf#DHS 2390 – gw Frankfurter [540]
The demurrer: or, proofs of error in the decision of the supreme court of the state of new york, requiring faith in particular religious doctrines as a legal qualification of witnesses. / Herttell, Thomas – New York: Conrad, 1828. 158p. LL-624 – 1 – us L of C Photodup [347]
Demus, Otto *see* The medieval mosaics of san marco, venice
Demystification of the learning of mathematics : analysis of narratives from feminist perspective / Mathamela, Maureen Matshingwana – Pretoria: Vista University 2003 [mf ed 2003] – 3mf – 9 – (incl bibl) – mf#mfm15161 – sa Unisa [510]
Den' – Vodessa. 1869-71 – 1 – (in russian.) – us NY Public [073]
Den = Day – New York: Orient Publ Co, dec 30 1922-feb 5 1926 – 1 – us CRL [071]
Den – Sofia, Bulgaria. Aug 1945-Apr 1946 – 1r – 1 – us L of C Photodup [949]
Denatured africa / Streeter, Daniel Willard – New York, NY. 1926 – 1r – us UF Libraries [960]
Denault, Joseph Marie Amedee *see* Mgr p n bruchesi
Denbigh journal, and general advertising medium for the vale of clwyd and north wales, etc – [Wales] LLGC 18 jul 1853-1 mar 1854 [mf ed 2004] – 1r – 1 – uk Newsplan [072]
Denbigh, ruthin and vale of clwyd free press – Denbigh, Wales. Denbighshire Free Press. -w. 11 Nov 1882-Dec 1920. Lacking 1896, 1897. 33 reels – 1 – uk British Libr Newspaper [072]
Denbigh, ruthin & vale of clwyd free press – [Wales] Denbighshire 11 nov 1882-dec 1900 [mf ed 2004] – 67r – 1 – (cont as: denbighshire free press etc [jan 1889-dec 1893, dec 1900]) – uk Newsplan [072]
Denderah : description generale du grand temple de cette ville / Mariette, A – Paris, 1870-1874. v1-4+suppl – 21mf – 9 – mf#NE-365 – ne IDC [956]
Dendereh / Flinders Petrie, W M – London, 1900 – (= ser Mees 17) – 10mf – 8 – €19.00 – ne Slangenburg [930]
Dendereh, 1898 / Petrie, W M – London, 1900 – 5mf – 9 – mf#NE-20347 – ne IDC [956]
Dendy, A *see* Report on the sponges collected by prof. herdman at ceylon in 1902
Dendy D R *see* The use of lights in christian worship
Dene, E de *see* De warachtighe fabulen der dieren. psalm 8 a.7
Dene nation newsletter – 1980 jun-1983 aug, 1984 may 4-1985 apr 30 – 1r – 1 – mf#698113 – us WHS [071]
Denecke, Rolf *see* Gestalten deutscher dichtung
Denezhnaia politika sovetskoi vlasti (1917-1927) / Iurovskii, L N – M, 1928 – 5mf – 9 – mf#REF-66 – ne IDC [332]
Denezhnaia reforma / Kamenev, L B – Rostov n/D, M, 1924 – 1mf – 9 – mf#REF-53 – ne IDC [332]
Denezhnaia reforma : materialy dlia agitatorov i propagandistov – L, 1924 – 1mf – 9 – mf#REF-51 – ne IDC [332]
Denezhnaia reforma / Shleifer, I O – M, 1924 – 1mf – 9 – mf#REF-61 – ne IDC [332]
Denezhnaia reforma / Sokol'nikov, G Ia – M, 1925 – 2mf – 9 – mf#REF-57 – ne IDC [332]
Denezhnaia reforma : svod mnenii i otzyvov – Spb, 1896 – 7mf – 9 – mf#REF-234 – ne IDC [332]
Denezhnaia reforma (bor'ba za ustoichivye den'gi) / Rozentul, S – Khar'kov, 1924 – 1mf – 9 – mf#REF-59 – ne IDC [332]
Denezhnaia reforma i puti ee zakrepleniia / Sokol'nikov, G Ia – M, [1924] – 1mf – 9 – mf#REF-56 – ne IDC [332]
Denezhnaia reforma, snizhenie tsen, zarabotnaia plata : dekrety i postanovleniia / ed by Finansovaia Gazeta – M, 1924 – 1mf – 9 – mf#REF-50 – ne IDC [332]
Denezhnaia reforma v rossii, 1895-1898 gg / Vlasenko, VE – Kiev, 1949 – 4mf – 9 – mf#REF-235 – ne IDC [332]
Denezhnaia reforma v sovetskoi rossii / Bernatskii, M – Praga, 1925 – 1mf – 9 – mf#REF-60 – ne IDC [332]
Denezhnoe obrashchenie i emissionaia operatsiia v rossii 1917-1918 gg : gosudarstvennyi kreditnyi bilet-banknota / Zak, Aleksandr N – Pg, 1918 – 2mf – 9 – mf#REF-183 – ne IDC [332]
Denezhnoe obrashchenie i kredit sssr / ed by Atlas, Z V & Bregel', E Ia – M, 1947 – 5mf – 9 – mf#REF-34 – ne IDC [332]
Denezhnoe obrashchenie i kreditnaia sistema soiuza ssr za 20 let : sbornik vazhneishikh zakonodatel'nykh materialov za 1917-1937 gg / ed by D'iachenko, VP & Rovinskii, N N – M, 1939 – 5mf – 9 – mf#REF-4 – ne IDC [332]
Denezhnoe obrashchenie rossii 1914-1924 / Katsenelenbaum, Z S – M, L, 1924 – 3mf – 9 – mf#REF-16 – ne IDC [332]
Denezhnoe obrashchenie v rossii / Kashkarov, M – Spb, 1898. 2v – 10mf – 9 – mf#REF-233 – ne IDC [332]
O denezhnoi reforme, proektirovannoi ministerstvom finansov / Bortkevich, I – Spb, 1896 – 1mf – 9 – mf#REF-228 – ne IDC [332]
Denezhnye kursy i tovarnye tseny : prakticheskoe posobie dlia gosudarstvennykh uchrezhdenii, birzh, trestov, promyshlennykh i torgovykh predpriiatii, khoziaistvennykh i finansovykh rabotnikov, bukhgalterov i proch / Derevenko, N N & Iakushkin, N V – M, 1923 – 2mf – 9 – mf#REF-52 – ne IDC [332]
Dengel, DR *see* Effects of dehydration on ratings of perceived exertion at the lactate and ventilatory thresholds
Den'gi i banki / Zhukovskii, Iu G – Spb, 1906 – 4mf – 9 – mf#REF-177 – ne IDC [332]
Den'gi i denezhnye obiazatel'stva / Lunts, L A – M, 1927 – 2mf – 9 – mf#R-9437 – ne IDC [332]
Dengo, Gabriel *see* Estudio geologico de la region de guanacaste, cost...
Denham, D *see*
– Narrative of travels and discoveries in northern and central africa in the years 1822, 1823 and 1824
– Narrative of travels and discoveries in northern and central africa in the years 1822, 1823, and 1824...
– Travels and discoveries in africa
Denham, Dixon *see*
– Beschreibung der reisen und entdeckungen im noerdlichen und mittlern africa in den jahren 1822 bis 1824
– Narrative of travels and discoveries in northern and central africa, in the years 1822, 1823, and 1824
Denials and beliefs of unitarians / Wright, John – London: P Green, 1901 – 1mf – 9 – 0-524-07646-4 – mf#1991-3253 – us ATLA [243]
Denifle, Heinrich *see*
– Auctarium chartularii universitatis parisiensis
– Chartularium universitatis parisiensis
Denifle, Heinrich Seuse *see*
– Humanity
– Luther in rationalistischer und christlicher beleuchtung
– Luther und luthertum in der ersten entwickelung
– Luther und luthertum in der ersten entwicklung
– Taulers bekehrung
– Die universitaeten des mittelalters bis 1400
– Les universites francaises au moyen-age
Denike, D P *see* Novoe kazanskoe slovo
Deniker, Joseph *see* Races of man
Deniliquin chronicle – Deniliquin. jul 1864-dec 1879, 1894, jan 1896-jun 1902, may 1904-dec 1907 – 6r – A$359.08 vesicular A$392.08 silver – at Pascoe [079]
Deniliquin independent – Deniliquin – 11r – A$539.22 vesicular A$599.72 silver – at Pascoe [079]
Denio, Francis Brigham *see*
– The supreme leader
– The supreme need
Denis, A-M *see* Concordance latine des pseudepigraphes d'ancien testament
Denis, Charles *see* Les vrais perils
Denis diderot / Gillot, Hubert – Paris, France. 1937 – 1r – us UF Libraries [440]
Denis, Ferdinand *see*
– Brazil
– Les californies, l'oregon, et l'amerique russe
Denis, Jean *see* La guyane
Denis, Jean F *see* Buenos ayres et le paraguay
Denis, Lorimer *see*
– Avenir du pays el l'action nefaste de m foisset
– Porbleme des classes a travers l'histoire d'haiti
– Problema de clases en la historia de haiti
Denis, Serge *see* Nos antilles
Denise de montmidi : roman / Ompteda, Georg, Freiherr von – Berlin, Wien: Ullstein [1910?] [mf ed 1991] – 1r – 1 – (filmed with: das passions-schauspiel in oberammergau) – mf#2857p – us Wisconsin U Libr [830]
Denison, Edward *see*
– Church
– Difficulties in the church
Denison, Frank Napier *see* Victoria, "the city of sunshine"
Denison, Frederic *see* The evangelist
Denison, George Anthony *see*
– Appeal to the clergy and laity of the church of england to combine
– Confession, absolution, and holy communion
– Correspondence
– Episcopate with two voices
– Fifty years at east brent
– Mr. gladstone
– Notes of my life, 1805-1878
– Real presence
– "Ritualism" and the real presence
– Supplement to "notes of my life," 1879, and "mr. gladstone," 1886
Denison, George Taylor *see*
– Canada, is she prepared for war?
– A chronicle of st john's cemetery on the humber
– The fenian raid on fort erie
– The german peace offer
– A history of cavalry from the earliest times
– A history of the cavalry from the earliest times
– History of the fenian raid on fort erie

– Letter
– Manual of outpost duties
– The national defences
– The petition of george taylor denison, jr
– Recollections of a police magistrate
– Reminiscences of the red river rebellion of 1869
– Soldiering in canada
– Speech delivered by the president, lt-col george t denison
– The struggle for imperial unity

Denison, Henry Mandeville *see* Lectures to business men

Denison, Henry Phipps *see*
– Prayer-book ideals
– The true religion

Denison, Jacob *see* She'erit ya'akov

Denison, John Hopkins *see* Beside the bowery

Denison, John Ledyard *see* An illustrated history of the new world

Denison Library (TX) *see* Grayson gateway

Denison, Louisa Evelyn *see* Fifty years at east brent

Denison, Samuel Dexter *see* A history of the foreign missionary work of the protestant episcopal church

Denison university. journal of the scientific laboratories – Granville OH 1885-1980 – 1,5,9 – ISSN: 0096-3755 – mf#6596 – us UMI ProQuest [500]

Denison, William Thomas *see* An attempt to approximate to the antiquity of man by induction from well established facts

Denisonian / Licking Co. Granville – feb 1897-oct 1899,sep 1901-jun 1907 [wkly] – 1r – 1 – mf#B12967 – us Ohio Hist [370]

Denisov, L I *see* Pravoslavnye monastyri rossiiskoi imperii

Denizens of the deep / Bullen, Frank Thomas – New York, Toronto: F H Revell, c1904 – 6mf – 9 – 0-659-90617-1 – (ill by charles livingston [and] theodore carreras) – mf#9-90617 – cn Canadiana [574]

Denjoy, Paul *see* Etude pratique de la legislation civile annamite

Denkblaetter aus jerusalem / Tobler, Titus – St Gallen 1853 [mf ed Hildesheim 1995-98] – 1v on 5mf [ill] – 9 – €100.00 – 3-487-27675-5 – gw Olms [880]

Denker, Fred Herman *see*
– Study of the transition from the cantus firmus mass to the parody mass

Denker, Sven *see* Tumorinfiltrierende leukozyten (til) und das sekretorische immunglobulin a (siga) bei pharynx- und larynxkarzinomen

Denkler, Horst *see* Der deutsche michel

Denkmacher deutscher tonkunst – 1folge. Herausgegeben von der Musikgeschichtlichen Kommission. Leipzig. 65 v. 1892-1931. (Called "Erste Folge" from bd. 4, 1900) – 1 – us L of C Photodup [780]

Denkmaehler der deutschen baukunst / Moller, G – Darmstadt, 1815-1821. 3v – 14mf – 9 – mf#OA-138 – ne IDC [720]

Denkmaeler aus aegypten und aethiopien / ed by Lepsius, C R – Berlin, 1849-1913. v1-13 – 172mf – 8 – mf#H-109 – ne IDC [956]

Denkmaeler aus lykaonien, pamphylien und isaurien / Swoboda, H et al – Prag, 1935 – 5mf – 8 – mf#H-650 – ne IDC [956]

Denkmaeler der aelteren deutschen literatur fuer den litteraturgeschichtlichen unterricht an hoeheren lehranstalten / ed by Boetticher, Gotthold & Kinzel, Karl – Halle/S: Verlag der Buchhandlung des Waisenhauses, 1890-1913 [mf ed 1993] – 14v – 1 – (each vol also has sep title. incl bibl ref) – mf#8185 – us Wisconsin U Libr [430]

Denkmaeler der entstehungsgeschichte des byzantinischen ritus / Baumstark, Anton – 1927 – 1mf – 8 – €3.00 – ne Slangenburg [243]

Denkmaeler der provenzalischen litteratur / ed by Bartsch, Karl – Stuttgart: Litterarischer Verein, 1856 [mf ed 1993] – (= ser Blvs 39) – xxv/356p – 1 – (provencal text. int in german) – mf#8470 reel 9 – us Wisconsin U Libr [440]

Denkmaeler der tonkunst in oesterreich = Monuments of music in austria / ed by Adler, Guido et al – Vienna: Oesterreichischer Bundesverlag. 83v. 1894-1938. Repr: Graz ed 1959- – 11 – $680.00 set – us Univ Music [780]

Denkmaeler deutscher tonkunst : first series / ed by Moser, Hans Joachim & Crosby, C Russell, Jr – Graz: Akademische Druckund Verlagsantalt. 65v + 2 suppls. repr 1957-61 – 11 – $590.00 set – (revisions involve changes of musical text which are explained in detail) – us Univ Music [780]

Denkmal der dritten jubelfeier der concordienformel im jahre des heils 1877 : enthaltend beschreibungen dieser feier, auf dieselbe bezuegliche predigten, auszuege aus solchen, predigtdispositionen und lieder – St Louis, Mo: MC Barthel, 1877 – 5mf – 9 – 0-524-07873-4 – mf#1991-3418 – us ATLA [240]

Ein denkmal memphitischer theologie / Erman, A – Berlin, 1911 – 1mf – 9 – mf#NE-20389 – ne IDC [290]

Denkschrift ueber das verhaeltniss des staates zu den saetzen der paepstlichen constitution vom 18. juli 1870: gewidmet den regierungen deutschlands und oesterreichs / Schulte, Johann Friedrich von – Prag: Friedrich Tempsky, 1871 – 1mf – 9 – 0-8370-8382-6 – mf#1986-2382 – us ATLA [241]

Denkschrift zum entwurf eines buergerlichen gesetzbuchs : nebst drei anlagen / Germany. Bundesrat – Berlin: J Guttentag, 1896 – (= ser Civil law 3 coll) – 5mf – 9 – mf#LLMC 96-512 – us LLMC [346]

Denkschrift zum entwurf eines buergerlichen gesetzbuchs nebst drei anlagen : dem reichstage vorgelegt in der vierten session der neunten legislaturperiode / Germany. Reichsjustizamt und Reichsjustizministerium – 2., unveraend Aufl. Berlin: C Heymann, 1896 – (= ser Civil law 3 coll) – 7mf – 9 – mf#LLMC 96-507 – us LLMC [346]

Denkschrift zum entwurf eines buergerlichen gesetzbuchs nebst drei anlagen, ergaenzt durch hinweise auf die beschluesse des reichstages sowie auf die paragraphen des buergerlichen gesetzbuchs und seiner nebengesetze / Jaentsch, H – Berlin: J Guttentag, 1899 – (= ser Civil law 3 coll) – 5mf – 9 – mf#LLMC 96-515 – us LLMC [346]

Denkschrift zur einweihung des knappschafts-verwaltungsgebaudes am 18. juni 1910 – 1 – gw Mikropress [330]

Denkschriften der kaiserlichen adakemie / KAISERL AKADEMIE DER WISSENSCHAFTEN IN WIEN – Wien, Austria. 1899 – 1r – us UF Libraries [500]

Denkschriften der kaiserlichen akademie der wissenschaften : mathematisch-naturwissenschaftliche klasse – Wien 1850-1946 – v1-107 on 2091mf – 9 – (ind 1913; v75(2) not publ) – mf 1070c/2 – ne IDC [500]

Denkschriften der kaiserlichen akademie der wissenschaften : philosophisch-historische klasse – Wien, 1896. v44 – 15mf – 8 – mf#H-632 – ne IDC [956]

Denkwuerdigkeiten / Menzel, Wolfgang; ed by Menzel, Konrad – Bielefeld, Leipzig, 1877 (mf ed 1992) – 4mf – 9 – €37.50 – 3-89349-102-3 – mf#DHS-AR 57 – gw Frankfurter [430]

Denkwuerdigkeiten, aufgezeichnet zur befoerderung des edlen und schoenen / ed by Moritz, Karl Phillipp et al – Berlin 1786-88 – (= ser Dz) – 2v on 5mf – 9 – €100.00 – mf#k/n5129 – gw Olms [074]

Denkwuerdigkeiten aus dem leben der fuerstin amalia von gallitzin / Katerkamp, Theodor – Muenster, 1828 (mf ed 1993) – 2mf – 9 – €24.00 – 3-89349-206-2 – mf#DHS-AR 95 – gw Frankfurter [920]

Denkwuerdigkeiten aus der geschichte des christenthums und des christlichen lebens *see* Memorials of christian life in the early and middle ages

Denkwuerdigkeiten aus der philosophischen welt / ed by Caesar, Carl Adolph – Leipzig 1785-88 – (= ser Dz. abt philosophie) – 6v on 15mf – 9 – €150.00 – mf#k/n553 – gw Olms [100]

Denkwuerdigkeiten der franzoesischen revolution [...] – Kopenhagen (DK), 1794-95, 1797, 1801, 1803 – 3r – 1 – gw Misc Inst [933]

Denkwuerdigkeiten der natur und kunst, religion und geschichte, schiffahrt und handlung in den koeniglich preussischen niederrheinisch-westf provinzen : ein lesebuch fuer alle staende / Engels, Johann – Elberfeld 1818 [mf ed Hildesheim: 1995-98] – (= ser Fbc) – 248p on 2mf [ill] – 9 – €60.00 – 3-487-29541-5 – gw Olms [880]

Denkwuerdigkeiten der scharfrichterfamilie sanson / ed by Sanson, Henry – Muenchen: Rosl & Cie, 1924 – us CRL [920]

Denkwuerdigkeiten des eigenen lebens / Varnhagen von Ense, Karl August; ed by Leutner, Karl – 3. erw aufl. Berlin: Verlag der Nation [1954] [mf ed 1995] – 1r – 1 – (incl ind.filmed with: a morte de camoes / luis tieck) – mf#3754p – us Wisconsin U Libr [880]

Denkwuerdigkeiten einer reise nach dem russischen amerika, nach mikronesien und durch kamschatka / Kittlitz, F H von – Gotha, 1858. 2v – 16mf – 9 – mf#N-281 – ne IDC [915]

Denkwuerdigkeiten einer reise nach dem russischen amerika, nach mikronesien und durch kamtschatka / Kittlitz, Friedrich H von – Gotha 1858 [mf ed Hildesheim 1995-98] – 2v on 6mf – 9 – €120.00 – 3-487-27077-3 – gw Olms [910]

Denkwuerdigkeiten ueber die mongolei / Iakinf [Bichurin, Ia] – Berlin: G Reimer, 1832 – 5mf – 9 – mf#HT-635 – ne IDC [915]

Denkwuerdigkeiten und erinnerungen aus dem orient / Prokesch-Osten, Anton – Stuttgart 1836/37 [mf ed Hildesheim 1995-98] – 3v on 13mf – 9 – €130.00 – 3-487-29107-X – gw Olms [880]

Denkwuerdigkeiten und reisen des verstorbenen herzoglich braunschweigischen obristen von nordenfels, commandanten der stadt wolfenbuettel, ritter des guelphen-ordens... – Braunschweig [u a] 1830 [mf ed Hildesheim 1995-98] – 1v on 3mf – 9 – €90.00 – 3-487-27810-3 – gw Olms [880]

Denkwuerdigkeiten und vermischte schriften / Varnhagen von Ense, Karl August – 2. aufl. Leipzig: F Brockhaus 1843-59 [mf ed 1991] – 9v on 3r – 1 – (v8-9 ed by ludmilla assing) – mf#2961p – us Wisconsin U Libr [800]

Denkwuerdigkeiten von peru und chili – Muenster [1817] [mf ed Hildesheim 1995-98] – 1v on 2mf – 9 – €60.00 – 3-487-26835-3 – gw Olms [918]

Denkwurdigkeiten der gluckel von hameln – Berlin, Germany. 1920 – 1r – us UF Libraries [390]

Denmark – London, England 1973-74 – 1 – ISSN: 0011-8400 – mf#8911 – us UMI ProQuest [320]

Denmark *see* Statstidende

Denmark delineated : or, sketched of the present state of that country / Feldborg, Andreas – Edinburgh 1824 [mf ed Hildesheim 1995-98] – (= ser Fbc) – 3mf [ill] – 9 – €90.00 – 3-487-28949-0 – gw Olms [948]

Denmark, Maine. First Free Will Baptist Church *see* Records

Denmark press – Denmark WI. [1920 apr 1-1946 dec], 1948 jan-jul 15 – 2r – 1 – (cont by: dairyland review) – mf#1004627 – us WHS [071]

Denmark press – 1953 nov 5/1955-1999 jul/dec – 50r – 1 – (with gaps; cont: denmark press and the dairyland review) – mf#1004643 – us WHS [071]

Denmark press and the dairyland review – Denmark WI. 1953 jun 4-oct 29 – 1r – 1 – (cont: dairyland review and the denmark press; cont by: denmark press [denmark wi: 1953]) – mf#1004639 – us WHS [071]

Denmark. Statistiske Bureau *see*
– Bureau de statistique
– Sammendrag af statistiske oplysninger angaaende kongeriget danmark 1869-1893
– Statistisk arbog 1896-1965

Dennery, Germaine *see* Chants du souvenir

Dennett, Mary Ware *see* The sex side of life

Dennett, Richard Edward *see*
– At the back of the black man's mind
– Nigerian studies
– Notes on the folklore of the fjort (french congo)

Denney, James *see*
– The atonement and the modern mind
– The church and the kingdom
– The death of christ
– The epistles to the thessalonians
– Factors of faith in immortality
– Gospel questions and answers
– The second epistle to the corinthians
– The way everlasting

Dennica : uradny organ evanjelickej slovenskej zenskej jednoty v am = Morning star – Pittsburgh, PA: Jednota, sep 15 1924-1930, 1932-33, 1935-45 – 3r – 1 – us CRL [071]

Dennice novoveku : organ bratrstva cesko-slovansych podporujicich spolku – Cleveland, OH: V Snajdr & Korizek, oct 10 1877-sep 1 1910; The Svet Print & Pub Co, sep 8 1910-1917 – 14r – 1 – (weekly czech language, free thinker newspaper. suppls accompany some issues. has literary suppl 1898-1901: zert a pravda. absorbed by: svet) – us Western Res [071]

Dennice novoveku / Snajdr, Vaclav – Cleveland, OH: Snajdr & Korizek. roc1 cis1. rij 10 1877-(wkly) – 14r – 1 – (absorbed: pokrok (chicago, il); absorbed by: svet (cleveland, oh)) – us Ohio Hist [071]

Dennie, Joseph *see* The lay preacher

Dennig-Zettler, Regina *see* Translatio sancti marci

Dennill, Ingrid *see* Stress as a source of injury among a group of professional ballet dancers

Denning, Margaret Beahm *see* Mosaics from india

Dennis 1710-1890 – Oxford, MA (mf ed 1986) – (= ser Massachusetts vital records) – 40mf – 9 – 0-931248-98-1 – (mf 1-4: births & deaths 1710-1855 pt 1. mf 5-8: births & deaths 1710-1855 pt 2. mf 9-16: births & deaths early 1700's-1890. mf 17-22: index to births & deaths 1710-1855. mf 23-24: marriage records 1814-55. mf 25-27: b,m,d 1855-62. mf 28-29: index to b,m,d 1855-62. mf 30-31: births 1862-90 bk 2. mf 32-33: marriage record 1863-90 bk 2. mf 34-35: deaths 1863-90 bk 2. mf 36-40: index to b,m,d 1863-1966) – us Archive [978]

Dennis, D *see* The chronology of effects of caffeine during prolonged cycle ergometry

Dennis, F G *see* Second international workshop on temperate zone fruits in the tropics and subtropics

Dennis, George *see* The cities and cemeteries of etruria

Dennis, James S *see*
– The modern call of missions
– Social evils of the non-christian world

Dennis, James Shepard *see*
– Centennial survey of foreign missions
– Christian missions and social progress
– Foreign missions after a century
– The message of christianity to other religions
– The new horoscope of missions
– A sketch of the syria mission

Dennis, Jonas *see* Character of the king

Dennis, Robert *see* Industrial ireland

Dennis Wire and Iron Co *see* The dennis wire and iron co metal workers and designers

The dennis wire and iron co metal workers and designers : manufacturers of architectural and ornamental iron and wire work, art metal work, wire and iron specialities: illustrated catalogue n7 / Dennis Wire and Iron Co – [London, Ont?: s.n.], 1900 [mf ed 1986] – 1mf – 9 – 0-665-60270-7 – (incl ind) – mf#60270 – cn Canadiana [680]

Dennison, George Taylor *see* Canada and her relations to the empire

Denniston, J M *see*
– Exodus
– The perishing soul according to scripture

Dennistoun, James *see* Memoirs of sir robert strange

Denny, Edward *see*
– Companion to a chart
– De hierarchia anglicana
– The english church and the ministry of the reformed churches
– Retributive justice

Denny hlas / Cuyahoga Co. Cleveland – jun 1918-dec 1923 [wkly] – 7r – 1 – (in slovak) – mf#B7063-7069 – us Ohio Hist [071]

Denny hlas = Daily voice – Cleveland: John Pankuch, -1925. dec 1917-jun 1918; 1924 – 5r – 1 – us CRL [071]

Denny, J K H *see* Toward the sunrising

Denny, Karen L *see* A biomechanical analysis of the effects of hand weights on the arm-swing while walking and running

Denny, S R *see* Up and down the great north road

Denny, William *see* The worth of wages

Denolf, Prosper *see* Aan de rand de dibese

A denominational offering from the literature of universalism : in twelve parts / Hodgdon, Norris C – Boston: Universalist Pub House, 1871 [mf ed 1992] – (= ser Unitarian/universalist coll) – 1mf – 9 – 0-524-03159-2 – mf#1990-4608 – us ATLA [243]

Denon, Dominique *see* Travels in upper and lower egypt during the campaigns of general bonaparte

Denon, V *see* Voyage dans la basse et haut egypte pendant les campagnes du general bonaparte

Denoncourt, Louise *see* Bio-bibliographie de yves leclerc, 1953-1961

Denoon, Donald *see* Collected papers on aspects of png history

Denosa – Saskatchewan. 1981 jun-1983 fall – 1r – 1 – mf#230829 – us WHS [071]

Dent, George Robinson *see*
– Compact zulu dictionary
– Scholar's zulu dictionary

Dent, John Charles *see*
– Canadian notabilities
– Canadian notabilities, vol 1
– Canadian notabilities, vol 2
– Canadian notabilities, vols 1-2
– The canadian portrait gallery
– The last forty years
– Prospectus
– Sir James Douglas
– The story of the upper canadian rebellion
– Toronto

Dent, William *see* Various views of the higher christian life

Dental abstracts – Oxford, England 1956+ – 1,5,9 – ISSN: 0011-8486 – mf#7635 – us UMI ProQuest [617]

Dental assistant [1931] – Chicago IL 1931-91 – 1,5,9 – (cont by: dental assistant journal) – ISSN: 0011-8508 – mf#8091.02 – us UMI ProQuest [617]

Dental assistant [1994] – Chicago IL 1994+ – 1,5,9 – (cont: dental assistant journal) – ISSN: 1088-3886 – mf#8091,02 – us UMI ProQuest [617]

Dental assistant journal – Chicago IL 1992-93 – 1,5,9 – (cont: dental assistant [1931]; cont by: dental assistant [1994]) – ISSN: 1072-754X – mf#8091.02 – us UMI ProQuest [617]

Dental clinics of north america – Philadelphia PA 1957+ – 1,5,9 – ISSN: 0011-8532 – mf#2698 – us UMI ProQuest [617]

Dental digest – Tulsa OK 1895-1971 – 1,5 – ISSN: 0011-8567 – mf#2320 – us UMI ProQuest [617]

Dental economics – Tulsa OK 1911+ – 1,5,9 – ISSN: 0011-8583 – mf#2342 – us UMI ProQuest [617]

Dental Guidance Council for Cerebral Palsy *see* Bulletin of the dental guidance council for cerebral palsy

Dental guidance council on the handicapped. journal – New York NY 1976 – 1 – (cont: bulletin of the dental guidance council for cerebral palsy) – ISSN: 0147-3972 – mf#5061,01 – us UMI ProQuest [617]

Dental health – London, England 1962+ – 1,5,9 – ISSN: 0011-8605 – mf#2377 – us UMI ProQuest [617]

Dental health attitudes and knowledge levels of rural and suburban texas / Russell, Linda M & Cissell, William B – 1991 – 1mf – $4.00 – us Kinesology [613]

Dental hygiene – Chicago IL 1973-87 – 1,5,9 – (cont: journal of the american dental hygienists' association; cont by: journal of dental hygiene) – ISSN: 0091-3979 – mf#2274.02 – us UMI ProQuest [617]

Dental hygienists' knowledge, attitudes and infection control practices in relation to aids and aids patients / Snyder, Gail A & Levy, Marvin R – 1992 – 2mf – 9 – $8.00 – us Kinesology [613]

Dental journal of australia – St Leonards, Australia 1950-54 – 1 – mf#578 – us UMI ProQuest [617]

Dental management – Cleveland OH 1961-83 – 1,5,9 – ISSN: 0011-8680 – mf#1566 – us UMI ProQuest [617]

Dental news / Milwaukee County Dental Society – 1936 jan 1-apr 1 – 1r – 1 – (cont by: dental news of the milwaukee county dental society) – mf#832417 – us WHS [617]

Dental practice – Epsom, England 1975+ – 1,5,9 – ISSN: 0011-8710 – mf#10603 – us UMI ProQuest [617]

Dental practitioner and dental record – Oxford, England 1950-72 – 1,5 – ISSN: 0011-8729 – mf#2344 – us UMI ProQuest [617]

Dental student – Dallas TX 1975-85 – 1,5,9 – ISSN: 0011-877X – mf#10222 – us UMI ProQuest [617]

Dental student news – Washington DC 1971-74 – 1 – ISSN: 0045-995X – mf#7846 – us UMI ProQuest [617]

Dental students views and actions – Washington DC 1971 – 1 – mf#8429 – us UMI ProQuest [617]

Denti, Giuseppe *see* Quattro parole in confidenza al popolo riguardo all'enciclica del papa leone 13 contro la framassoneria

Dentler, Eberhard *see* Die auferstehung jesu christi nach den berichten des neuen testamentes

Dento-maxillo-facial radiology – Ann Arbor MI 1989-96 – 1,5,9 – ISSN: 0250-832X – mf#17221 – us UMI ProQuest [616]

Denton baptist church. greenup association : church records – Denton, KY. oct 1965-73; feb 1974-may 1987 (scattered) – 1 – $5.00 – us Southern Baptist [242]

Denton, haughton and district weekly news – [NW England] Denton, Stalybridge Lib jun 1873-20 mar 1874 – 1 – (title change: denton & haughton weekly news, & audenshaw, hooley hill, & dukinfield advertiser [27 mar 1874-15 oct 1875]; denton examiner, audenshaw, hooley hill & dukinfield advertiser [23 oct 1875-23 feb 1878]; denton & haughton examiner [2 mar 1878-6 aug 1892]) – uk MLA; uk Newsplan [072]

Denton, John *see* Account of the county of cumberland

Denton, Julia C *see* The effects of short-term exercise on lipid and lipoprotein metabolism in obese males with abnormal glucose

Denton post : incorporating the dentonian – [NW England] Denton, Stalybridge Lib may 1965-dec 1969 – 1 – (title change: denton & hyde post [jan-aug 1970]) – uk MLA; uk Newsplan [072]

Le denturo / Association des denturologistes du Quebec – Montreal: l'Association. v1 n1 juil 1970-v10 n3 nov 1979 (irreg) [mf ed 1985] – 1r – 1 – (cont: bulletin des denturologistes du quebec; cont by: denturo plus) – mf#SEM35P219 – cn Bibl Nat [617]

Le denturo (1989) / Association des denturologistes du Quebec – Montreal: l'Association. v20 n1 juin 1989- (qrtly) [mf ed 1990-91] – 1r – 5 – (cont: denturo plus) – mf#SEM16P180 – cn Bibl Nat [617]

Le denturo (1989) / Association des denturologistes du Quebec – Montreal: l'Association. v20 n1 juin 1989- (qrtly) [mf ed 1992-] – 9 – (cont: denturo plus) – mf#SEM105P1707 – cn Bibl Nat [617]

Denturo plus / Syndicat professionnel des denturologistes du Quebec – Montreal: le Syndicat. v11 n1 mars 1980-v14 n1 mars 1983; v15 n1 1er trim 1984-v19 n4 (qrtly) [mf ed 1985-] – 5 – (cont: denturo (0384-8000); cont by: denturo (1989)) – mf#SEM16P351 – cn Bibl Nat [617]

Denuncia / Hermandad de Trabajadores de Servicios Sociales [PR] – v1 n4, 6/7-8/9 [1979jun, sep/oct-nov/dic]-v10 n1,3/6-7 [1988 jan/feb, jul-sep] – 1r – 1 – (with gaps; incl: boletin especial 1983 nov, 1985 may-jun) – mf#1055268 – us WHS [360]

Denver and rio grande railway : official timetables – Denver, CO: A B Eads Co, 1887 [mf ed 1969] – 1r – 1 – (also incl: the rocky mountain official railway guide [jul 1892, jul 1896, may 1897, jul-oct 1897]) – mf#MF R598m – us Colorado Hist [380]

Denver argonaut *see* Denver city and county miscellaneous newspapers

Denver Art Museum *see* Leaflet

Denver call *see* Denver county miscellaneous newspapers, reel 3

Denver city and county miscellaneous newspapers – Denver, CO [mf ed 1964] – 1r – 1 – (contains miscellaneous dates for the foll newspapers: colorado exchange journal, denver argonaut, denver mercury, denver newsletter) – mf#MF Z99 cde2 – us Colorado Hist [071]

Denver city and county miscellaneous newspapers – Denver, CO. 1859-1944 (mf ed 1964) – 1r – 1 – mf#MF Z99 cde1 – us Colorado Hist [071]

Denver city and county miscellaneous newspapers – Denver CO, 1865 – (incl miscellaneous dates for the foll papers: la stella, rocky mountain tourist, weekly denver gazette) – mf#MF Z99 cde3 – us Colorado Hist [071]

Denver Civic and Commercial Association *see* Commercial

Denver county miscellaneous newspapers – Denver, CO (mf ed 1991) – 1r – 1 – mf#MF Z99 D437 – us Colorado Hist [071]

Denver county miscellaneous newspapers, reel 2 : the denver daily gazette – Denver, CO: F J Stanton, 1865 [mf ed 1991] – 1r – 1 – (denver daily gazette [scattered iss 1865-67, jan 1-apr 16 1869], daily gazette and commercial advertiser [apr 20-may 5 1869]; missing issue of daily mining record [mar 9 1907] and colorado worker [nov 10 1913] can be found as retakes at the beginning of this roll; added entries: the denver daily gazette, weekly denver gazette, daily gazette and commercial advertiser) – mf#MF Z99 D437 Reel 2 – us Colorado Hist [071]

Denver county miscellaneous newspapers, reel 3 – Denver, CO [mf ed 1994] – 1r – 1 – (denver call [jan 4 1957-jan 18 1957, oct 16 1959, clarion [oct 3 1951], denver saturday night [nov 1936], denver star [dec 1-dec 8 1961, apr 13 1962, aug 31 1962], denver catholic register [jul 16 1931, apr 27 1933], el gallo [apr 1973-may 1980], el reportero [scattered issues nov 1991-apr 28 1993], monitor [sep 7, 21 1951], montelibre monthly [may 1988-jun 1993]) – mf#MF Z99 D437 Reel 3 – us Colorado Hist [071]

Denver county miscellaneous newspapers, reel 4 – Denver, CO [mf ed 1994] – 1r – 1 – (national defense news [sep 1938], office park news [aug 5 1992-jan 1993], play fair [aug 10 1919], register [apr 19 1931, may 3 1931, jul 12 1931], rocky mountain christian advocate [sep 20 1888], service record [scattered issues sep 6 1948-nov 28 1953], washington park press [aug 1 1935], west side hustler [apr 16 1943, oct 8 1943], western news [scattered iss jan 13 1972-mar 9 1978], western farm life [sep 1 1926-oct 1 1926]) – mf#MF Z99 D437 Reel 4 – us Colorado Hist [071]

[Denver-] el gallo – CO. 1968-88 – 11r – 1 – $1210.00 – mf#R04001 – us Library Micro [071]

Denver Firefighters Union *see* Local 858 alarm

Denver journal of international law and policy – Denver CO 1980+ – 1,5,9 – ISSN: 0196-2035 – mf#12808 – us UMI ProQuest [341]

Denver journal of international law and policy – v1-28. 1971-2000 + 10 yr Ind – 5,6,9 – $528.00 set – (v1-13 1971-84 on reel $163. v14-28 1985-2000 on mf $365) – ISSN: 0196-2035 – mf#102401 – us Hein [341]

Denver labor bulletin / Colorado State Federation of Labor – 1914 dec 12-1917 may 26, jun 2-1918 jun 29, jul 6-1919 jun 28, jul 5-1920 jun 26, jul 3-1921 jul 31, aug 6-1922 dec 30, 1923 jan 6-1925 jan 31, feb 6-1927 aug 27, sep 3-1937 may – 9r – 1 – (cont: united labor bulletin) – mf#467583 – us WHS [331]

Denver law center journal *see*
- Denver university law review
- Dicta

Denver law journal *see* Dicta

The denver law journal – v1-2. 1883-84 – 9 – $8.00 – mf#LLMC 82-921 – us LLMC [340]

Denver legal news – v1-2. 1887-89 (all publ) – 1 – $50.00 set – mf#408980 – us Hein [340]

Denver Museum of Natural History *see* Proceedings

Denver Native Americans United *see* D n a u newsletter

Denver post – Denver CO. 1914 apr 24, may 3,10 – 1r – 1 – (cont: denver evening post) – mf#350182 – us WHS [071]

Denver quarterly – Denver CO 1966+ – 1,5,9 – ISSN: 0011-8869 – mf#2554 – us UMI ProQuest [000]

Denver republican – Denver CO. 1896 oct 2, 1906 apr 20 – 1r – 1 – (cont: denver tribune republican) – mf#854371 – us WHS [071]

Denver saturday night *see* Denver county miscellaneous newspapers, reel 3

The denver saturday night – Denver, CO: Ray McGovern, 1934-37 – 1r – 1 – mf#MF D437d – us Colorado Hist [073]

The denver star – Denver. Colo. aug. 7, 1926; Feb. 14, 1942 – 1 – us NY Public [071]

Denver university law review – v1-76. 1923-99 – 5,6,9 – $1152.00 set – (v1-62 1923-85 on reel $770; v63-76 1986-99 on mf $382; title varies: v1-39, 1923-62 as dicta. v40-42 1963-65 as denver law center journal; v43-61 1966-83 as denver law journal) – ISSN: 0883-9409 – mf#102421 – us Hein [340]

Denver vecko-blad – Denver CO. 1888 nov 15 – 1r – 1 – (cont by: svenska korrespondenten) – mf#944417 – us WHS [071]

Denver weekly news – Denver CO. v17 n869 [1987 nov 26]-v22 n2040-2046, 2048-2049, 2116-2118 [1993 feb 4-mar 18, apr 1-8, jul 28-aug 18] – 1r – 1 – (with gaps) – mf#2293274 – us WHS [071]

Deny, J *see* Grammaire de la langue turque

Denys, George Williams *see* Chaldean account of genesis

Denzinger, Heinrich *see* Enchiridion symbolorum et definitionum

Deonier, Marshall T *see*
- Identification of the leading citrus rootstocks by microscopical an
- Identification of the leading citrus rootstocks by microscopical an...

Deontologia administrativa / Walker, Harvey – San Jose, Costa Rica. 1961 – 1r – us UF Libraries [972]

Deparcieux, M *see* Essay on the probabilities of the duration of human life

Departament okladnykh sborov, 1863-1913 – Spb, 1913 – 7mf – 9 – mf#REF-207 – ne IDC [332]

Departamento de Intercambio y Propaganda Exterior *see* Falange exterior

Departed glory : the deserted cities of india / Slater, Arthur R – London: Epworth Press, 1937 – (= ser Samp: indian books) – us CRL [720]

Departed gods : the gods of our fathers / Fradenburgh, Jason Nelson – Cincinnati: Cranston & Stowe, 1891 – 1mf – 9 – 0-524-04336-1 – (incl bibl ref) – mf#1990-3320 – us ATLA [200]

Departed worth and greatness lamented / Symington, William – Paisley, Scotland. 1853 – 1r – us UF Libraries [240]

Departemen agama / Masjarakat Islam – Djakarta, 1970-1972. v1-2(1-13) – 9mf – 9 – mf#SE-1795 – ne IDC [959]

Departemen agama agenda kementerian agama bagian publikasi dan redaksi, djawatan penerangan agama / Indonesia – Djakarta, 1951-1952 – 12mf – 9 – mf#SE-788 – ne IDC [959]

Departemen agama konperensi dinas / Indonesia – Djakarta, 1950-1961 – 67mf – 9 – (missing: 1954/1955(3-4); 1961(3)) – mf#SE-789 – ne IDC [959]

Departemen agama laporan kementerian agama, bagian penerbitan / Indonesia – Djakarta, 1954. v1-2 – 15mf – 9 – (missing: 1954 v1) – mf#SE-790 – ne IDC [959]

Departemen agrirprop cc pki : kehidupan partai – Djakarta, 1955-1960 – 11mf – 9 – (missing: 1955(jun-dec); 1956(jan-sep); 1957(4, 9-12); 1959(1-8, 11, 12); 1960(5)) – mf#SE-379 – ne IDC [959]

Departemen anggaran negara laporan tahunan (ekonomi-keuangan) / Indonesia – Djakarta, 1965 – 3mf – 9 – mf#SE-1547 – ne IDC [959]

Departemen angkatan darat daftar singkatan-singkatan istilah resmi dalam angkatan darat / Indonesia – Djakarta, 1960-1961 – 2mf – 9 – mf#SE-1548 – ne IDC [959]

Departemen angkatan darat suad : berkala irian barat – Djakarta, 1962 – 4mf – 9 – mf#SE-1356 – ne IDC [959]

Departemen angkatan udara lembaran keamanan penerbangan assisten direktorat keamanan terbang / Indonesia – Djakarta, 1963 – 1mf – 9 – (missing: 1963 v1(1-3)) – mf#SE-1549 – ne IDC [959]

Departemen Dalam Negeri *see* Pemerintahan

Departemen dalam negeri mimbar departemen dalam negeri bagian hubungan dan penerangan masjarakat / Indonesia – Djakarta, 1969-1971. v1-3(17/18) – 11mf – 9 – (missing: 1969, v1(1); 1971, v2(11/12); 1971, v3(16)) – mf#SE-155=0 – ne IDC [959]

Departemen dalam negeri sektor chusus irian-barat himpunan peraturan2 pemerintah tentang masalah pengurusan daerah propinsi irian barat / Indonesia – Djakarta, 1966-1970 – 19mf – 9 – mf#SE-155=1 – ne IDC [959]

Departemen dalam negeri sektor chusus irian-barat laporan pembangunan irian barat / Indonesia – Djakarta, 1969 – 3mf – 9 – mf#SE-155=2 – ne IDC [959]

Departemen kehakiman / Mimbar kehakiman – Djakarta, 1967-1972 – 60mf – 9 – (missing: 1970 v4(59)) – mf#SE-1771 – ne IDC [959]

Departemen Kesehatan *see*
- Berita tuberculosea indonesiensis
- Madjalah kesehatan

Departemen kesehatan pedoman dan berita / Indonesia – Djakarta, 1952-1970 – 109mf – 9 – (missing: 1952, v1(1, 4); 1953; 1954, v3(1-4); 1956, v5(2); 1958, v7(4); 1959, v8(3-4); 1960, v9(3-4); 1961, v10(3-4); 1962, v11(1-4); 1963, v12(2-4); 1964; 1965(jan-mar); 1970(1)) – mf#SE-851 – ne IDC [360]

Departemen Luar Negeri *see* Pewarta kemlu

Departemen luar negeri department of foreign affairs / Indonesia – Djakarta, 1961-1964. v1-4 – 23mf – 9 – mf#SE-887 – ne IDC [327]

Departemen luar negeri department of foreign affairs / Indonesia – Djakarta, 1970 – 2mf – 9 – (missing: 1970(1-3, 5-7)) – mf#SE-1688 – ne IDC [327]

Departemen luar negeri direktorat asia timur laut dan pasifik malaysia masalah dan perkembangan selandjatnja / Indonesia – Djakarta 1963-1965 – 149mf – 9 – (missing: 1963) – mf#SE-989 – ne IDC [959]

Departemen luar negeri direktorat research biro research umum research kronologi dan dokumentasi / Indonesia – Djakarta, 1968-1969 – 15mf – 9 – (missing: 1968-1969(1-9a)) – mf#SE-155-9 – ne IDC [959]

Departemen luar negeri direktorat research pewarta dan kronologi bulanan seksi perentjanaan dan penerbitan / Indonesia – Djakarta, 1968 – 27mf – 9 – (missing: 1968 v1(7, 9)) – mf#SE-155-4 – ne IDC [959]

Departemen Luar Negeri Direktorat Research Rentjana kerdja *see* Indonesia

Departemen Luar Negeri Direktorat Research Research brief *see* Indonesia

Departemen Luar Negeri Direktorat Research Research diplomatik *see* Indonesia

Departemen Luar Negeri Direktorat Research Research dokumentasi *see* Indonesia

Departemen Luar Negeri Direktorat Research Research landasan *see* Indonesia

Departemen Luar Negeri Direktorat Research Research publikasi *see* Indonesia

Departemen Luar Negeri Direktorat Research Research reconnaissance *see* Indonesia

Departemen Pekerdjaan Umum dan Tenaga *see* Indonesia membangun

Departemen pekerdjaan umum dan tenaga berita dep-pu-t / Indonesia – Kebajoran Baru, Djakarta, [1968]-1971 – 16mf – 9 – (missing: [1968], v1; 1969, v2; 1970, v3(1-3); 1971, v4(3)) – mf#SE-1563 – ne IDC [959]

Departemen pekerdjaan umum dan tenaga progress report; tahun kerdja 1967 / Indonesia – [Djakarta, 1968] – 3mf – 9 – mf#SE-8479 – ne IDC [959]

Departemen pendidikan dan kebudajaan perpustakaan sedjarah politik dan sosial press index / Indonesia – Djakarta, 1969-1970 – 41mf – 9 – mf#SE-1564 – ne IDC [959]

Departemen pendidikan pengadjaran dan kebudajaan buku alamat sekolah landjutan dan kursus-kursus negeri subsidi dan bantuan / Indonesia – Djakarta, 1957-1958 – 6mf – 9 – mf#SE-470 – ne IDC [959]

Departemen pendidikan pengadjaran dan kebudajaan daftar adanja sekolah landjutan negeri, subsidi dan bantuan dimasing-masing kabupaten dan kota / Indonesia – Djakarta, 1958 – 3mf – 9 – mf#SE-471 – ne IDC [959]

Departemen Penerangan *see*
- Dunia internasional
- Pantjaran ampera

Departemen penerangan – Djakarta, [1955]-1965 – 4mf – 9 – (missing: [1955]-1962, v1-7; 1963, v8(1-93, 97-end); 1964, v9(5-end); 1965, v10(1-35)) – mf#SE-532 – ne IDC [959]

Departemen penerangan daftar harian dan madjalah seluruh indonesia / Indonesia – Djakarta, 1951-1955 – 9mf – 9 – (missing: 1952-1953 v2) – mf#SE-537 – ne IDC [959]

Departemen penerangan department of information facts & figures / Indonesia – Djakarta, [1966]-1972 – 15mf – 9 – (missing: [1966], v1; 1967, v2; 1968, v3(1-6, 10-end); 1969, v4(1-10, 15-16, 19-end); 1970, v5(1, 4-end); 1971, v6(1-4, 9)) – mf#SE-1568 – ne IDC [959]

Departemen penerangan department of information information bulletin / Indonesia – Djakarta, 1965-1971 – 30mf – 9 – (missing: 1965, v1; 1966, v2(1-3, 5-11, 13-17, 19-23, 26-27, 32-34); 1969, v4(61, 70-71); 1970, v5(78, 84, 87, 90-91); 1971, v6(96)) – mf#SE-1570 – ne IDC [959]

Departemen penerangan detik peristiwa dalam negeri / Indonesia – Djakarta, 1968 – 12mf – 9 – mf#SE-1567 – ne IDC [959]

Departemen penerangan direktorat publisiteit & penerangang daerah, bagian dokumentasi madjelis permus jawaratan rakjat sementara / Indonesia – Djakarta, [1961] (Kronik dokumentasi NS no 11) – 4mf – 9 – mf#SE-4610 – ne IDC [959]

Departemen penerangan dpp "masjumi" : berita masjumi – Djakarta, [1951]-1954 v1-4 – 2mf – 9 – (missing: [1951]-1953, v1-3; 1954, v4(4-6)) – mf#SE-349 – ne IDC [959]
Departemen penerangan hmi tjabang – Jogjakarta, 1968 – 1mf – 9 – mf#SE-1414 – ne IDC [950]
Departemen penerangan ichtisar harian tanah air selama 24 djam / Indonesia – Djakarta, 1958-1961 – 3mf – 9 – (missing: 1958-1961(1-114, 119-140)) – mf#SE-539 – ne IDC [959]
Departemen penerangan ministry of informatin foreign observers on the question of west irian / Indonesia – Djakarta, 1952 – 1mf – 9 – mf#SE-1569 – ne IDC [959]
Departemen penerangan pengurus besar hmi / Media mahasiswa – Djakarta, Dec, 1970-1971 – 4mf – 9 – mf#SE-1785 – ne IDC [959]
Departemen penerangan penlugri miscellany / Indonesia – Djakarta, 1967 – 1mf – 9 – mf#SE-1572 – ne IDC [959]
Departemen penerangan press and broadcast releases / Indonesia. Department of Information – Djakarta, 1967 – 7mf – 9 – mf#SE-4995 – ne IDC [350]
Departemen penerangan ri / Mimbar kabinet pembangunan – Djakarta, 1967-1972. v1-3(1-34) – 42mf – 9 – mf#SE-1770 – ne IDC [959]
Departemen penerangan senat mahasiswa fak hukum ui – Djakarta, 1966-1969 – 3mf – 9 – (missing: 1966, v1(4-end); 1967, v2; 1968, v3; 1969, v4(1)) – mf#SE-1726 – ne IDC [959]
Departemen penerangan seri amanat / Indonesia – Djakarta, 1968-1972 – 25mf – 9 – (missing: 1968(12); 1969(19); 1970(30); 1971(49)) – mf#SE-1573 – ne IDC [959]
Departemen penerangan siaran departemen penerangan melalui siaran rri pusat rtd / Indonesia – Djakarta, 1968-1972 – 12mf – 9 – (missing: 1971(1-42); 1972(1-25, 30)) – mf#SE-1574 – ne IDC [959]
Departemen penerangan siaran pemerintah / Indonesia – Djakarta, 1956 – 10mf – 9 – mf#SE-1575 – ne IDC [959]
Departemen penerangan special issue / Indonesia – Djakarta, 1958-1965 – 50mf – 9 – (missing: 1958(1-6); 1962/63(94, 96-98, 100-104, 109-224); 1965(226-651)) – mf#SE-373 – ne IDC [959]
Departemen penerangan special issue : reference edition / Indonesia – Djakarta, 1963-1964 – 8mf – 9 – (missing: 1963, v1(6328-6330); 1963-1964, v1-2(6340-6406, 6408-6412, 6414)) – mf#SE-540 – ne IDC [959]
Departemen penerangan tanja djawab / Indonesia – Djakarta, 1957-1962 – 22mf – 9 – (missing: 1957 v2-3) – mf#SE-541 – ne IDC [959]
Departemen penerangan the fourth asian games / Indonesia – Djakarta, 1962(7-8) – 3mf – 9 – mf#SE-538 – ne IDC [959]
Departemen penerangan upe / Indonesia – Djakarta, 1967-1968 – 4mf – 9 – (missing: [1967], (1-151, 153-231, 233-end); 1968(1-34)) – mf#SE-1576 – ne IDC [959]
Departemen penerangan uraian departemen penerangan ri melalui siaran rri pusat udp / Indonesia – Djakarta, 1968-1972 – 53mf – 9 – (missing: 1968(5-6, 8); 1969(11); 1970(54, 77-78, 117); 1971(5-20, 23, 25, 27, 29, 31-32, 40, 46-47, 52-53, 56, 60-66, 70-74, 84); 1972(22, 24, 40, 42)) – mf#SE-1577 – ne IDC [959]
Departemen perburuhan laporan / Indonesia – Djakarta, 1959 – 3mf – 9 – mf#SE-1579 – ne IDC [959]
Departemen perdagangan dalam negeri beserta urusan perdagangan luar negeri dari kompartimen luar negeri/heln dan perdagangan luar negeri / Warta perdagangan – Djakarta, [1948]-1968. v1-21(26) – 53mf – 9 – (missing: [1948]-1964, v1-17; 1966, v18(20); 1966, v19(9-10, 14-15, 17-18, 20, 22); 1967, v20(1-end); 1968, v21(9-16, 22)) – mf#SE-699 – ne IDC [959]
Departemen perdagangan data2 perdagangan laporan semester / Indonesia – Djakarta, 1969-1970 – 3mf – 9 – mf#SE-1580 – ne IDC [959]
Departemen perdagangan himpunan peraturan2 dibidang perdagangan jajasan penjuluhan dan penerangan perdagangan / Indonesia – Djakarta, 1968-1971 – 31mf – 9 – mf#SE-1581 – ne IDC [959]
Departemen perdagangan perwakilan sumatera utara laporan tahunan / Indonesia – Medan, 1966-1967 – 8mf – 9 – mf#SE-1583 – ne IDC [959]
Departemen perdagangan progress report / Indonesia – Djakarta, 1968 – 2mf – 9 – mf#SE-1582 – ne IDC [959]
Departemen perhubungan bulletin perhubungan untuk dinas bagian hubungan masjarakat / Indonesia – Djakarta, 1970-1971 – 29mf – 9 – (missing: 1970(1-13); 1970(22); 1971(1)) – mf#SE-1584 – ne IDC [959]
Departemen perhubungan laporan bidang organisasi dan personil departemen perhubungan biro organisasi & personil, sekretariat djenderal departemen perhubungan / Indonesia – Np, 1969 – 2mf – 9 – mf#SE-1585 – ne IDC [959]
Departemen Perhubungan Laut *see* Suluh nautika
Departemen perindustrian : berita industri – Djakarta, 1968-1972 v1-5(4/5) – 70mf – 9 – (missing: 1968 v1(1) is missing) – mf#SE-1350 – ne IDC [959]
Departemen perindustrian rakjat buku laporan tahunan / Indonesia – Djakarta, 1959-1960 – 9mf – 9 – mf#SE-288 – ne IDC [959]
Departemen perindustrian rakjat laporan team departeman perindustrian rakjat kedaerah tahun 1960 kantor penjuluhan perindustrian, departemen perindustrian rakjat / Indonesia – [Djakarta, 1960] – 2mf – 9 – mf#SE-10923 – ne IDC [959]
Departemen perindustrian rakjat laporan team departemen perindustrian rakjat kedaerah tahun 1961 kantor penjuluhan perindustrian, departemen perindustrian rakjat / Indonesia – [Djakarta, 1961] – 2mf – 9 – mf#SE-4122 – ne IDC [959]
Departemen perindustrian tekstil dan keradjinan rakjat warta deptekra; bulletin bulanan bagian humas deptekra / Indonesia – Djakarta, 1967-1968 – 10mf – 9 – (missing: 1967 v1(1-2)) – mf#SE-1625 – ne IDC [959]
Departemen pertahanan keamanan pusat perlawanan dan keamanan rakjat laporan kegiatan tahun kerdja / Indonesia – Djakarta, 1968-1969 – 16mf – 9 – (missing: 1968) – mf#SE-1626 – ne IDC [959]
Departemen Pertanian *see* Sari warta pertanian
Departemen Pertanian dan Agraria *see* Penjuluh landreforn
Departemen Pertanian, Lembaga Perpustakaan Biologi dan Pertanian "Bibliotheca Bogoriensis" *see* Indeks biologi dan pertanian di indonesia
Departemen tenaga kerdja buku hasil raker departemen tenaga kerdja sekretariat raker / Indonesia – Djakarta, 1970 – 7mf – 9 – mf#SE-1627 – ne IDC [959]
Departemen tenaga kerdja laporan / Indonesia – Djakarta, 1967 – 3mf – 9 – mf#SE-1628 – ne IDC [959]
Departemen transmigrasi dan koperasi feasibility studies pembangunan koperasi / Indonesia – Djakarta, 1970-1971 – 5mf – 9 – mf#SE-1629 – ne IDC [959]
Departemen Transmigrasi Koperasi, Bagian Hubungan Masjarakat *see* Transkop
Departemen transmigrasi, koperasi dan pembangunan masjarakat desa / Desa membangun – Djakarta, 1961-1962 – 4mf – 9 – (missing: 1961 v1(1)) – mf#SE-824 – ne IDC [959]
Departemen transmigrasi, koperasi dan pembangunan masjarakat desa / Indonesia – Djakarta, 1961 – 2mf – 9 – mf#SE-826 – ne IDC [959]
Departemen transmigrasi, koperasi dan pembangunan masjarakat desa biro pembangunan masjarakat desa marilah membangun masjarakat / Indonesia – Djakarta, 1961 – 1mf – 9 – mf#SE-3152 – ne IDC [959]
Departemen transmigrasi, koperasi dan pembangunan masjarakat desa biro pembukaan tanah setelah enam bulan bekerdja / Indonesia – Djakarta, 1961 – 1mf – 9 – mf#SE-3401 – ne IDC [959]
Departemen transmigrasi, koperasi dan pembangunan masjarakat desa djadikan koperasi sebagai alat untuk mentjapai masjarakat sosialis indonesia atas dasar usdek; himpunan pidato pada peringatan koperasi di istana negara 12 djuli 1960 / Indonesia – Djakarta, 1960 – 1mf – 9 – mf#SE-2864 – ne IDC [959]
Departemen transmigrasi, koperasi dan pembangunan masjarakat desa koperasi dalam alam sosialisme indonesia; pidato / Indonesia – Djakarta, 1960 – 1mf – 9 – mf#SE-3341 – ne IDC [959]
Departemen transmigrasi, koperasi dan pembangunan masjarakat desa koperasi indonesia berdasarkan pantja sila dan usdek / Indonesia – Djakarta, 1961 – 1mf – 9 – mf#SE-2865 – ne IDC [959]
Departemen transmigrasi, koperasi dan pembangunan masjarakat desa pembangunan masjarakat desa dalam hubungan internasional / Indonesia – Djakarta, 1961 – 1mf – 9 – mf#SE-10020 – ne IDC [959]
Departemen transmigrasi, koperasi dan pembangunan masjarakat desa pola kerdja sama departemen transkopemada dengan departemen2 lain; himpunan keputusan2 bersama / Indonesia – [Djakarta, 1961] – 1mf – 9 – mf#SE-4123 – ne IDC [959]
Departemen transmigrasi, koperasi dan pembangunan masjarakat desa pola pelaksanaan tugas departemen transkopemada th dinas 1962 / Indonesia – [Djakarta, 1961] – 1mf – 9 – mf#SE-4124 – ne IDC [959]
Le departement – (city unknown) 1944 – 1 – (in french) – us UMI ProQuest [934]
Departement des finances and du commerce d'haiti, 18... / Haiti Departement Des Finances – Paris, France. 1895 – 1r – us UF Libraries [336]
Departement d'information du M P L A *see* Vitoria ou morte
Departement van economische zaken grafieken behorende bij de economische toestand van indonesie / Indonesia – Batavia – 5mf – 9 – mf#SE-279 – ne IDC [959]
Departementsblatt des werra-departements *see* Marburger anzeigen auf das jahr...1789
Department convention proceedings / Veterans of Foreign Wars of the United States – 1958-60 – 1r – 1 – (cont: department encampment proceedings...; cont by: convention proceedings...) – mf#2836713 – us WHS [305]
Department encampment proceedings / Veterans of Foreign Wars of the United States – 1952-57 – 1r – 1 – (cont: proceedings of the... annual encampment of the ladies' auxiliaries of the...; cont by: department convention proceedings...) – mf#2836691 – us WHS [305]
Department of education : st croix county, hammond, wi / Saint Croix County [WI] – 1935/36-1938/39, 1941/42, 1947/48, 1950/51-1951/52, 1957/58-1958/59 – 1r – 1 – (cont: annual school directory for saint croix county, wisconsin) – mf#5192594 – us WHS [370]
Department of energy's freedomcar : hurdles, benchmarks for progress, and role in energy policy: hearing...house of representatives, 107th congress, 2nd session, june 6 2002 / United States. Congress. House. Committee on Energy and Commerce. Subcommittee on Oversight and Investigations – Washington: US GPO 2002 [mf ed 2002] – 8mf – 9 – 0-16-068827-2 – (incl bibl ref) – us GPO [629]
Department of health, education and welfare (1963-1969) : official history and documents – 2pt – (= ser Presidential documents series) – 1 – (pt1: history 6r isbn 0-89093-390-1 $935. pt2: documents 11r isbn 0-89093-390-1 $1725. with p/g) – us UPA [350]
Department of health, education and welfare annual reports / U[/nited/] S[/tates/] Dept of Health and Human Services – 1953-70 [all publ] – 71mf – 9 – $106.00 – (cont: social security administration, federal security agency annual reports; agencies within hew publ reports subsequent to 1970) – mf#llmc 81-207 – us LLMC [360]
Department of housing and urban development annual reports / U.S. Dept of Housing and Urban Development – 1st-18th. 1965-82 – 25mf – 9 – $37.50 – (updates planned) – mf#llmc 81-210 – us LLMC [360]
Department of information, Republic of Indonesia *see* Pantjasila
Department of information republic of indonesia / Indonesia – Djakarta, 1960-1961. v1-2 – 3mf – 9 – mf#SE-1689 – ne IDC [959]
Department of justice investigative files – 2pt – (= ser Research colls in american radicalism) – 1 – $6060.00 coll – (pt1: the industrial workers of the world ed by melvyn dubofsky [15r] isbn 1-55655-055-3 $2345; pt2: the communist party ed by mark naison [28r] isbn 1-55655-056-1 $4360; with p/g) – us UPA [322]
Department of labor annual reports / U.S. Dept of Labor – 1913-80 – 401mf – 9 – $742.00 – mf#LLMC 81-219 – us LLMC [331]
Department of state bulletin – Washington DC 1939-89 – 1,5,9 – (cont by: us department of state dispatch) – ISSN: 0041-7610 – mf#891 – us UMI ProQuest [323]
Department of state bulletin : the official weekly record of the united states foreign – Washington, DC: US GPO, US Dept of State. v1-89. 1939-89 [all publ] – 9 – $4750.00 set – (cont as: dispatch; v1-77 $90v v78-89 $120v) – mf#107590 – us Hein [341]
Department of state bulletin / U[/nited/] S[/tates/] Dept of State – v1-89. 1939-89 [all publ] – 1029mf – 9 – $1543.00 – (incl ind n1-2141; cont by: department of state dispatch) – mf#llmc 79-446 – us LLMC [327]
Department of state dispatch – v1-10. 1990-99 – 156mf – 9 – $82.00 – (cont: department of state bulletin; updates planned) – mf#llmc 79-446B – us LLMC [327]
Department of state office of the legal adviser : opinions and reports, 1866-1950 – card ind on 97mf. 1993 – 39r – 5 – $2995.00 set – mf#402280 – us Hein [327]
Department of the interior and related agencies appropriations for 2003 : hearings...house of representatives, 107th congress, 2nd session / subcommittee on the department of the interior and related agencies [pt 1] / United States. Congress. House. Committee on Appropriations. Subcommittee on Dept. of the Interior and Related Agencies – Washington: US GPO 2002 [mf ed 2002] – 9 – 0-16-066940-5 – (pt3: isbn: 0-16-066908-1 pt4: isbn: 0-16-066938-3) – us GPO [350]
Department of the interior annual reports – 1921-63 – 9 – mf#llmc 94-297 – us LLMC [340]
Department reports of the state of new york : containing decisions, opinions and rulings of the state officers, departments, boards and commissions and messages of the governor / New York State. Executive Branch – Albany: Lyon Co/State. v1-74. 1914-1954 – 636mf – 9 – $954.00 – (indexes for v1-4 (1931)) – mf#LLMC 94-108 – us LLMC [340]
Department Store Employees Union, Local 21 *see* Local 21 guide
Department Store Employees Union, Local 1100 *see* Local 1100 report
Departmental cooperation in state government / Ellingwood, Albert R – New York: The Macmillan Co, 1918 – us CRL [350]
Departure / Campbell, Wilfred – S.l: s.n, 1899? – 1mf – 9 – mf#06080 – cn Canadiana [810]
Departure – London, England. no date – 1r – us UF Libraries [240]
Depass, Jas P *see*
– Annual report of the director to the board of trustees of experimental station for the year
– Corn, hay, weevil, rice, cane, texas blue grass and cotton
– Peach growing in florida
– Tobacco
– Tobacco and its cultivation
Depass, Jas. P *see* Agricultural experiments
Depatie, Caroline *see* Employment equity in canadian newspaper sports journalism
Depatie, Francine *see* Analyse statistique des donnees
Depaul business law journal – v1-13. 1989-2001 – 9 – $214.00 set – ISSN: 1049-6122 – mf#112191 – us Hein [346]
Depaul journal of health care law – v1-3. 1996-2000 – 9 – $128.00 set – mf#117561 – us Hein [344]
Depaul law review – v1-50. 1951-2001 – 5,6,9 – $930.00 set – (v1-34 1951-84 on reel $517. v35-50 1985-2001 on mf $413) – ISSN: 0011-7188 – mf#102431 – us Hein [340]
Depaul lca journal of art and entertainment law – v1-11. 1991-2001 – 9 – $220.00 set – mf#113301 – us Hein [346]
DePauw, Karen P *see*
– The effects of glasnost and perestroika on the soviet sport system
– Femininity and masculinity
Depeche – London, UK. 23 Sept 1914-28 Jan 1915 – 1 – uk British Libr Newspaper [072]
La depeche africaine – Paris: Maurice Satineau, feb 1928-jul 1931; feb 1932; jan 1938 – 1r – (filmed with les continents and 11 other titles) – us CRL [074]
La depeche africaine – no. 8-9, 11-14. Paris. oct 1928-avr 1932 – 1 – fr ACRPP [960]
Depeche algerienne – Algeria. 8 nov 1939-12 dec 1942; 31 jan 1943-28 jul 1945 – 4r – 1 – uk British Libr Newspaper [072]
La depeche algerienne – Alger. 1908-21 – 1 – fr ACRPP [073]
La depeche coloniale – Paris. aout 1896-juin 1940 – 1 – (puis et maritime. suite de: tablettes coloniales. devenu: france – outre-mer.) – fr ACRPP [073]
Depeche coloniale illustree – 1903-1905 – 1 – us CRL [073]
Depeche d'algerie – Algeria. 12 jun 1961-17 sep 1963 – 14r – 1 – uk British Libr Newspaper [072]
Depeche de constantine – Algeria. 24 feb 1943-7 may 1945 – 3r – 1 – uk British Libr Newspaper [072]
La depeche de constantine – Constantine. 15 nov 1908-fevr 1914, 1915, 28-29 nov 1920, juil 1922-1939, 1954-62 – 1 – fr ACRPP [949]
Depeche de londres – London, UK. 5 Jul 1903 – 1 – uk British Libr Newspaper [072]
Depeche de paris – Paris, France. 28 feb-10 dec 1945 – 1r – 1 – uk British Libr Newspaper [072]
La depeche de tahiti – aug-dec 1964 – 2r – 1 – mf#pmb doc261-262 – at Pacific Mss [079]
La depeche de tahiti – jan 1965-dec 1965 – 4r – 1 – mf#pmb doc263-266 – at Pacific Mss [079]
La depeche de tahiti – jan-dec 1967 – 4r – 1 – mf#pmb doc271-274 – at Pacific Mss [079]
La depeche de tahiti – jan-dec 1967 – 4r – 1 – mf#pmb doc267-270 – at Pacific Mss [079]
La depeche de tahiti – jan-dec 1968 – 4r – 1 – mf#pmb doc275-78 – at Pacific Mss [079]
La depeche de tahiti – jan-dec 1969 – 1r – 1 – mf#pmb doc279-282 – at Pacific Mss [079]
La depeche de tahiti – jan-dec 1970 – 4r – 1 – mf#pmb doc283-86 – at Pacific Mss [079]
La depeche de tahiti – jan-dec 1971 – 4r – 1 – mf#pmb doc287-90 – at Pacific Mss [079]

La depeche de tahiti – jan-dec 1972 – 6r – 1 – mf#pmb doc291-96 – at Pacific Mss [079]

La depeche de tahiti – Papeete. aout 1964-1993 – 1 – fr ACRPP [073]

La depeche de toulouse – Toulouse. 1885-1940 – 1 – fr ACRPP [073]

La Depeche democratique *see* Le memorial

La depeche democratique – Saint-Etienne. 4 sept 1944-45 – 1 – fr ACRPP [073]

La depeche d'indochine – Saigon: Impr de la Depeche 1940 jan 2-mar 30 – 1r – 1 – mf#mf-11770 seam – us CRL [079]

La depeche du midi – Toulouse. avr 1969-1987 – 1 – fr ACRPP [074]

La depeche du midi – Toulouse: [s.n.], 1953-mar 1969 – 130r – 1 – us CRL [074]

La depeche marocaine – Tangier, Morocco. 5 aug 1916-23 dec 1918 [daily] – 4r – 1 – (imperfect) – uk British Libr Newspaper [079]

Depeches du secretaire de sa majeste pour les colonies et autres documents relatifs a l'union federale des colonies britanniques de l'amerique du nord... / Canada (Province). Parlement. Assemblee legislative – Toronto: Impr par Stewart Derbishire & George Desbarats, 1859 [mf ed 1983] – 1mf – 9 – mf#SEM105P165 – cn Bibl Nat [323]

Depeches du secretaire de sa majeste pour les colonies, et autres documents relatifs au siege du gouvernement... : 22 victoriae, appendice (no 2) a. 1859 / Canada (Province). Parlement. Assemblee legislative – Toronto: Impr par Stewart Derbishire & George Desbarats, 1859 [mf ed 1983] – 1mf – 9 – mf#SEM105P164 – cn Bibl Nat [323]

Depechos legales...tierra santa / Garcia, Manuel – 1814 – 9 – sp Bibl Santa Ana [340]

Les dependances de l'abbaye de saint-germain-des-pres (afm3) : tom 1: seine et seine-et-marne / Anger, D – 1906 – (= ser Archives de la france monastique (afm)) – €17.00 – ne Slangenburg [241]

Les dependances de l'abbaye de saint-germain-des-pres (afm6) : tom 2: seine-et-oise / Anger, D – Paris, 1907 – (= ser Archives de la france monastique (afm)) – €15.00 – ne Slangenburg [241]

Les dependances de l'abbaye de saint-germain-des-pres (afm8) : tom 3 / Anger, D – 1909 – (= ser Archives de la france monastique (afm)) – €19.00 – ne Slangenburg [241]

Dependence on god / Macleod, John – Edinburgh, Scotland. 1810 – 1r – us UF Libraries [240]

Dependence, or, the insecurity of the anglican position / Rivington, Luke – London: K. Paul, Trench, 1889 – 1mf – 9 – 0-7905-6671-0 – mf#1988-2671 – us ATLA [241]

El dependiente extremeno – Merida, 1921. 1 numero – 5 – sp Bibl Santa Ana [073]

Depere advertiser – De Pere WI. 1851 feb 5-1852 mar 3 – 1r – 1 – mf#962662 – us WHS [071]

Depere standaard – De Pere WI. 1878 jan 18-1883 mar 8, 1880 jan 22-1881 jan 6, 1884 feb 28 – 3r – 1 – (cont by: onze standaard) – mf#874832 – us WHS [071]

Deperthes, Jean Louis Hubert Simon *see* Histoire des naufrages

Die depeschen des nuntius aleander vom wormser reichstage 1521 / Aleandro, Girolamo – Halle: Verein fuer Reformationsgeschichte, 1886 [mf ed 1990] – (= ser [Schriften des vereins fuer reformationsgeschichte] 17) – 1mf – 9 – 0-7905-4600-0 – (english trans by paul kalkoff. incl bibl ref) – mf#1988-0600 – us ATLA [241]

Depestre, Edourard *see* Faillite d'une democratie

Depestre, Rene *see*
- Etincelles
- Gerbe de sang

Depew, Chauncy M *see* My memories of eighty years

DePeyster, John Watts *see* Waterloo: the campaign and battle

DePiano, Frank *see* Journal of child and adolescent substance abuse

Depince, M Ch *see* Compte-rendu des travaux

Depoimento perante a comissao de inquerito sobre a... / Prestes, Luis Carlos – Rio de Janeiro, Brazil. 1948 – 1r – us UF Libraries [972]

Depoin, J *see*
- Histoire et cartulaire de l'abbaye demalbuisson
- Recueil de chartes et documents de saint-martin-des-champs

Depont, O *see* Les confreries religieuses musulmanes

Depont, Octave *see* Les confreries religieuses musulmanes

Deporte a cayenne / Jusselain, Armand – Paris, France. 1865 – 1r – us UF Libraries [972]

Depositaire / Duport, Paul – Paris, France. 1839 – 1r – us UF Libraries [440]

Depossessions / Union Nationaliste, Port-Au-Prince – Port-Au-Prince, Haiti. 1930 – 1r – us UF Libraries [972]

Depot dispatch / Elkhart Lake Area Chamber of Commerce – 1980 may 22-1987 jun – 1r – 1 – mf#1269031 – us WHS [338]

Depot echo – v1 n12-43 [1944 dec-1945 aug 7] – 1r – 1 – mf#2891678 – us WHS [071]

Deppe, Oskar *see* Schillers "xenien" und "tabulae votivae" im musenalmanach fuer 1797

Deppen, R I *see* Lukisan dwikora

Deppert, Fritz *see* Schuld und ueberwindung der schuld in den dramen ernst barlachs

Depping, Georges *see*
- L'angleterre ou description historique et topographique du royaume uni de la grande-bretagne
- Erinnerungen aus dem leben eines deutschen in paris
- Geographie de la jeunesse, ou nouveau manuel de geographie
- Merveilles et beautes de la nature en france

Depravados / Canaan Fernandez, Euridice – Santo Domingo, Dominican Republic. 1964 – 1r – us UF Libraries [972]

The depressed classes : their economic and social condition / Singh, Mohinder – Bombay: Hind Kitabs, 1947 – (= ser Samp: indian books) – (int by radhakamal mukherjee) – us CRL [305]

Depression and anxiety – Hoboken NJ 2001+ – 1,5,9 – ISSN: 1091-4269 – mf#25578 – us UMI ProQuest [150]

Depression and new deal, 1933-1939 – (= ser Morgenthau diaries, 1933-1945; Research colls in american politics) – 63r – 1 – $10,955.00 – 1-55655-563-6 – (with p/g) – us UPA [977]

Depressive verstimmungen, verminderte verhaltenskompetenzen und drogennahes verhalten bei jugendlichen : eine empirische untersuchung an 10 berliner schulen / Vockrodt-Scholz, Viola – (mf ed 1996) – 9mf – 9 – €71.00 – 3-8267-2317-1 – mf#DHS 2317 – gw Frankfurter [150]

Deprey, Teresa M *see* Nurse practitioners' views on menopause

Dept of Geography Gadjah Mada University *see* Madjalah geografi indonesia

Dept of Justice, Office of Information and Privacy *see* Freedom of information caselist

Dept of Justice, Office of Justice Programs *see* Task force on felon identification in firearm sales report to the a.g.

Dept of Public Information of the RMS *see* The forgotten war

Dept Penerangan IPNU wil Djk Raya *see* Chazanah

Deptford and peckham mercury – London, England. -w. 15 april 1981-24 dec 1991; 1992-25 may 1995 40r – 1 – uk British Libr Newspaper [072]

Deptford "better" times – [London & SE] Lewisham jun-oct 1934 (incomplete) [mf ed 2003] – 1r – 1 – uk Newsplan [072]

Deputado no exilio / Coelho, Jose Saldanha – Rio de Janeiro, Brazil. 1965 – 1r – us UF Libraries [972]

Der abend 1892 – Berlin DE, 1892 24 mar-dec – 2r – 1 – gw Misc Inst [074]

Der abend 1946 – Berlin DE, 10 oct 1946-30 sep 1950 – 5r – 1 – (filmed by bnl: 1946 14 oct-1952 22 mar [17r]; filmed by misc inst: 1976-1981 23 jan [21r]) – gw Mikrofilm; uk British Libr Newspaper; gw Misc Inst [074]

Der arbeiter 1927 – New York NY (USA), 1927 15 sep-1937 13 feb – 1 – gw Misc Inst [331]

Der beobachter 1789 – Stuttgart DE, 1789 jan-jun, 1790 jul, dec – 1r – 1 – gw Misc Inst [074]

Der Cicerone *see* Halbmonatsschrift fuer die interessen des kunstforschers und sammlers

Der morgen 1901 – Berlin DE, 1901 1 feb-15 jun – 2r – 1 – gw Misc Inst [074]

Der morgen 1945 – Berlin DE, 1945 3 aug-1952, 1991 2 jan-11 jun – 13r – 1 – (filmed by misc inst: 1945 3 aug-1948 [7r]; 1945 3 aug-1990 [87r]) – gw Mikropress; gw Misc Inst [074]

Der morgen 1945 / republik-ausgabe – Berlin DE, 1965 7 jul-1979 29 jun – 28r – 1 – gw Misc Inst [074]

Der Neue Welt-Bott *see* Mit allerhand nachrichten deren missionarien soc iesu

Der sozialdemokrat 1894 – Berlin DE, 1894 3 feb-1895 – 2r – 1 – mf#205 – gw Mikropress [325]

Der sozialdemokrat 1946 – Berlin DE, 1946 1 oct-1947 30 sep, 1948 2 jan-30 sep, 1949 1 jan-30 sep, 1981-1988 10 sep – 6r – 1 – (filmed by misc inst: may-aug 1948 [1r], 1950-29 apr 1951 [3r], filmed by bnl: 1946 3 jun-1949; title varies: 6 mar 1948: sozialdemokrat, 10 mar 1950: bs, 11 mar 1950: bs. berliner sozialdemokrat, 24 mar 1950: berliner stadtblatt, 1 apr 1950: bs. das berliner stadtblatt, 3 jun 1950: berliner sozialdemokrat, 3 may 1951: stadtblatt, 20 oct 1951: berliner stimme, [3 jun 1946-19 may 1951 & 20 oct 1951-10 sep 1988?]) – gw Mikrofilm; gw Misc Inst; uk British Libr Newspaper [074]

Der sturm 1930 – Kassel DE, 1930 1 feb-1944 – 43r – 1 – (title varies: 1931: hessische volkswacht; 1 sep 1933: kurhessische landeszeitung; with suppl) – gw Misc Inst [074]

Der sturm 1932 – Kassel DE, 1932 2 apr-31 dec – 1r – 1 – gw Misc Inst [074]

Der tag 1948 – Berlin DE, 1948 23 mar-1957 – 28r – 1 – (filmed by other misc inst: 1951 1 nov-1953 10 may, 1953 1 aug-1963 31 mar. with suppl: bilder tag 1950 2 apr-1952 24 feb filmed by mfa) – uk British Libr Newspaper; gw Misc Inst [074]

Deraismes, Maria *see* Eve dans l'humanite

Deramey, J-P *see* Precis du mouvement catholique-liberal dans le jura bernois, 1873-74

Derap *see* Ikatan buruh pantjasila dpwx

Derashot / Nissenbaum, Isaac – Warszawa, Poland. 1922 – 1r – us UF Libraries [939]

Derbentskij rajonnyj sovet rabochikh i voennykh deputatov *see* Izvestiia soveta rabochikh i voennykh deputatov derbentskogo rajona

Derby chronicle – [East Midlands] Derby, Derbyshire n1 [7 may]-n23 [8 oct 1881] [mf ed 2002] – 1r – 1 – uk Newsplan [072]

Derby comet – [East Midlands] Derby, Derbyshire n1 [22 jul 1893]-n76 [29 dec 1894] [mf ed 2002] – 6r – 1 – uk Newsplan [072]

Derby, Edward George Geoffrey Smith Stanley, 14th earl of *see* The canada corn bill

Derby, Edward Henry Smith Stanley, 15th Earl of *see* The irish question

Derby, Elias Hasket *see*
- The catholic
- A preliminary report on the treaty of reciprocity with great britain

Derby evening gazette – [East Midlands] Derby, Burton on Trent, Derbyshire n1 [28 jul 1879]-30 dec 1884 [mf ed 2002] – 11r – 1 – (in jun 1881 it split into two eds: derby & burton evening gazette [jan-jun 1881] and: derby & burton gazette [jul 1881-dec 1884]) – uk Newsplan [072]

Derby exchange gazette – [East Midlands] Derby, Derbyshire n2 [14 jan 1860-1871], 1873-96, 1899-n2294 [3 mar 1899] [mf ed 2002] – 37r – 1 – (missing: 1872, 1897, 1898; cont by: derby gazette and county journal & advertiser [jan 1862-dec 1865]; derby and derbyshire gazette [jan 1866-dec 1892]; derby & south derbyshire gazette [jan 1893-dec 1894]; derby & derbyshire gazette [jan 1895-96 & jan-3 mar 1899]) – uk Newsplan [072]

Derby express – Derby, Derbyshire, 19 jun 1986-to date – 1 – (fr 12 jan 1989 split into 5 ed: derby, belper, heanor ilkeston and ripley express) – uk Newsplan [072]

Derby express [daily] – [East Midlands] Derby, Derbyshire n1 [22 oct 1884-1897], 1899-n14197 [29 jan 1932] (daily) [mf ed 2004] – 270r – 1 – (missing: 1897-1898, 1911; cont by: derby daily express [apr 1909-jan 1932]) – uk Newsplan [072]

Derby free press and district advertiser – [East Midlands] Derby, Derbyshire n29 [17 feb 1904]-n31 [18 mar 1904] [mf ed 2002] – 1r – 1 – uk Newsplan [072]

Derby, George Horatio *see* Phoenixiana; or, sketches and burlesques

Derby herald – [East Midlands] Derby, Derbyshire n1 [29 jun-3 aug 1906] [mf ed 2002] – 1r – 1 – uk Newsplan [072]

Derby, John Sayward *see* A legal monograph upon provisional remedies under the code

Derby leader – [East Midlands] Derby, Derbyshire n1 [8 mar 1932]-n12 [24 may 1932] [mf ed 2002] – 1r – 1 – (fr 10 may 1932: derby echo) – uk Newsplan [072]

Derby mission, november, 1873 / Aitken, William Hay Macdowall Hunter – London, England. 1873? – 1r – us UF Libraries [240]

Derby morning post – [East Midlands] Derby, Derbyshire n1 [16 nov 1885]-n654 [5 jul 1887] [mf ed 2002] – 7r – 1 – uk Newsplan [072]

Derby telegraph and monthly advertiser – [East Midlands] Derby, Derbyshire, 11 aug 1855-n763 [9 jul 1869] (mthly, wkly) [mf ed 2002] – 7r – 1 – (missing: 1862) – uk Newsplan [072]

Derby weekly observer – [East Midlands] Derby, Derbyshire n1-4 [3 mar-24 mar 1903] [mf ed 2002] – 1r – 1 – uk Newsplan [072]

Derbyshire and leicestershire examiner – Derby, Burton-on-Trent, Wolverhampton, 13 sep 1873-17 aug 1877 (wkly) – 3r – 1 – uk Newsplan [072]

Derbyshire chronicle – [East Midlands] Derby, Derbyshire jun 1855-22 feb 1873 [mf ed 2004] – 14r – 1 – uk Newsplan [072]

Derbyshire courier – Chesterfield, Derbyshire, 12 dec 1829-18 feb 1922 (wkly) – 1 – (cont: chesterfield gazette) – uk British Libr Newspaper [072]

Derbyshire patriot : or, repository of politics, news, literature, etc – [East Midlands] Chesterfield, Derbyshire n1-3 [4 may-18 may 1833] [mf ed 2002] – 1r – 1 – uk Newsplan [072]

Derbyshire post – [East Midlands], Bakewell, Derby, Derbyshire, 26 apr-20 sep & 31 jan-26 dec 1885, jan 1886-2 jul 1887 [mf ed 2004] – 2r – 1 – (not publ fr 20 sep 1884-31 jan 1885) – uk Newsplan [072]

Derbyshire telephone and people's advocate – [East Midlands] Belper, Derbyshire n1 [16 sep 1898-n152 [9 aug 1901] [mf ed 2004] – 3r – 1 – (incorp with: belper news) – uk Newsplan [072]

Derbyshire worker and labour leader for ripley, alfreton, belper – [East Midlands] Ripley 27 jun 1919-27 aug 1920 [mf ed 2004] – 1 – (cont by: derbyshire worker for ripley [3 sep 1920-29 sep 1922]) – uk Newsplan; uk British Libr Newspaper [331]

Derchak, P A *see* Expiratory flow limitation and ventilatory responsiveness interact to determine exercise ventilation

Derecho agrario colombiano / Aguilera Camacho, Alberto – Bogota, Colombia. 1962 – 1r – us UF Libraries [972]

Derecho colonial venezolano / Venezuela Laws, Statutes, Etc (Indexes) – Caracas, Venezuela. 1952 – 1r – us UF Libraries [323]

Derecho constitucional / Zamora Y Lopez, Juan Clemente – Habana, Cuba. 1925 – 1r – us UF Libraries [323]

Derecho constitucional / Zamora Y Lopez, Juan Clemente – Habana, Cuba. 1925 – 1r – us UF Libraries [323]

Derecho constitucional colombiano / Perez, Francisco De Paula – Bogota, Colombia. 1962 – 1r – us UF Libraries [323]

Derecho constitucional colombiano / Perez, Francisco De Paula – Bogota, Colombia. v1-2. 1952 – 1r – us UF Libraries [323]

Derecho constitucional comparado / Garcia-Pelayo, Manuel – Madrid, Spain. 1950 – 1r – us UF Libraries [323]

Derecho constitutional colombiano / Perez, Francisco De Paula – Bogota, Colombia. 1962 – 1r – us UF Libraries [323]

Derecho consular guatemalteco / Moreno, Laudelino – Guatemala, 1946 – 1r – us UF Libraries [972]

El derecho de beligerancia : la voz del estado espanol contra la mediacion; no cabe mediacion en la guerra espanola; jugando con fuego / Kindelan, Alfredo – [Spain?: s.n. 1938] – (= ser Blodgett coll) – 9 – mf#w828 – us Harvard [946]

Derecho de familia y la legislacion guatemalteca / Cabrera Munoz, Rosalinda – Guatemala, 1964 – 1r – us UF Libraries [360]

Derecho del trabajo : revista critica mensual de jurisprudencia – Buenos Aires. Ano. 2, no. 1-4, no. 3; Ano. 7-26. Jan 1942-Mar 1944; 1947-1966 & Repertorio 1956 65 – 1 – us NY Public [340]

Derecho fiscal / Menocal Y Barreras, Juan Manuel – Habana, Cuba. 1953 – 1r – us UF Libraries [332]

Derecho hipotecario / Aguirre, Agustin – Habana, Cuba. 1939 – 1r – us UF Libraries [610]

Derecho hispanico y "common law" en puerto rico / Mouchet, Carlos – Buenos Aires: Perrot, 1953. 134p. LL-8017 – 1 – us L of C Photodup [346]

Derecho hispanico y 'common law' en puerto rico / Mouchet, Carlos – Buenos Aires, Argentina. 1953 – 1r – us UF Libraries [346]

Derecho internacional privado / Colombia. Congreso Senado – Medellin, Colombia. 1938 – 1r – us UF Libraries [346]

Derecho internacional privado / Munoz Meany, Enrique – Guatemala, 1953 – 1r – us UF Libraries [346]

Derecho minero de mexico y vocabulario con definic... / Becerra Gonzalez, Maria – Mexico City? Mexico. 1963 – 1r – us UF Libraries [550]

Derecho panal salvadoreno / Castro Ramirez, Manuel – San Salvador, El Salvador. 1947 – 1r – us UF Libraries [972]

Derecho penal / Zecena, Oscar – Quezaltenango, 1933 – 1r – us UF Libraries [360]

Derecho penal islamico, escuela malekita / Arevalo, Rafael – Tanger, F. Erola, 1939. 188 p. LL-12021 – 1 – us L of C Photodup [340]

Derecho procesal fiscal / Briseno Sierra, Humberto – Mexico City? Mexico. 1964 – 1r – us UF Libraries [332]

Derecho publico interno de colombia / Samper, Jose Maria – Bogota, Colombia. v1-2. 1951 – 1r – us UF Libraries [350]

Derechos y garantias del procesado / Afanador, Gonzalo – Bogota, Colombia. 1964 – 1r – us UF Libraries [972]

Derechos y prestaciones del trabajador oficial / Martinez Munoz, Enrique – Bogota, Colombia. 1965 – 1r – us UF Libraries [972]

Deregulator – v1 n1-14 [1986 aug-1987 dec] – 1r – 1 – mf#1289559 – us WHS [071]

Dereham & fakenham times – [East Midlands] Norfolk 7 feb 1880-dec 1950 [mf ed 2003] – 90r – 1 – (missing: 1881, 1898, 1911; cont as: dereham & fakenham times & journal [jan 1923-dec 1950]) – uk Newsplan [072]

Derekh avraham / Sochen, Abraham – New York, NY. 1930 – 1r – us UF Libraries [939]
Derekh emunah, ha-nikra sefer ha-hakirah / Schneersohn, Menahem Mendel – Poltava, Ukraine. 1912 – 1r – us UF Libraries [939]
Derekh tsedakah – Warsaw, Poland. 1895 – 1r – us UF Libraries [939]
Derenbourg, Hartwig *see*
– Quelques observations
– La science des religions et l'islamisme
Derenbourg, Joseph *see*
– Essai sur l'histoire et la geographie de la palestine
– Manuel du lecteur, d'un auteur inconnu
Derevenko, N N *see*
– Denezhnye kursy i tovarnye tseny
– Vsemirnyi ekonomicheskii, finansovyi i politicheskii spravochnik 1923 g
Derevenskaia bednota – Moscow, Russia 1917-18 [mf ed Norman Ross] – 2r – 1 – mf#nrp-91 – us UMI ProQuest [077]
Derevenskaia kommuna : izdanie sankt-peterburgskogo komiteta rkp – St Petersburg, Russia 1918-21 [mf ed Norman Ross] – 12r – 1 – mf#nrp-1584 – us UMI ProQuest [077]
Derevenskaia kooperatsiia – Ivano-Voznesensk, 1925-1926(1) – 8mf – 9 – mf#COR-582 – ne IDC [335]
Derevenskaia kooperatsiia / Kulyzhnyi, A E – [1909] – 274p 3mf – 9 – mf#COR-55 – ne IDC [335]
Derevenskaia pravda – St Petersburg, Russia 1917-18 [mf ed Norman Ross] – 1r – 1 – mf#nrp-92 – us UMI ProQuest [077]
Derevenskaya bednota – Leningrad. USSR. -d. 9, 17 Nov-14 Dec 1917. (1 reel) – 1 – uk British Libr Newspaper [947]
Derevenskie vpechatleniia (iz zapisok zemskogo statistika) / Belokonskii, I P – Spb, 1906 – 6mf – 8 – mf#RZ-143 – ne IDC [314]
Derevitskii, V A *see*
– Kommunal'nye banki
– Sbornik zakonopolozhenii po kreditnym uchrezhdeniiam
Dergah – Istanbul. 1-4. cilt n1-42. 15 nisan 1337-5 kanunisani 1339 [apr 1921-jan 1924] – (= ser O & t journals) – 14mf – 9 – $230.00 – us MEDOC [956]
Derham, Walter *see* A visit to cape colony and natal in 1879
Deriabina, A A *see* Osnovnye teoreticheskie problemy planovogo tsenoobrazovaniia
DeRicci, Caterina, Saint *see* Le lettere di santa caterina de'ricci
Derigo, G A *see* Biography of william sherman wiley
Dering, E *see* A sparing restraint, of many lavishe untruthes...
Dering, Richard *see* Manuscripts
Derishat tsiyon / Kalischer, Zevi Hirsch – Jerusalem, Israel. 1919 – 1r – us UF Libraries [939]
Derivation techniques in the chanson-mass "au travail" of johannes ockeghem / Sorenson, Katherine Murray – U of Rochester 1973 [mf ed 19–] – 4mf – 9 – (with bibl) – mf#fiche 938 – us Sibley [780]
Derleth, Kurt *see* Die religioese entwicklung gottfried kellers im hinblick auf den natur-kultur-begriff seines gereiften weltbildes
Dermatologic clinics – Philadelphia PA 1983+ – 1,5,9 – ISSN: 0733-8635 – mf#13380 – us UMI ProQuest [616]
Dermatologica – Basel, Switzerland 1966-74 – 1,5,9 – ISSN: 0011-9075 – mf#2054.01 – us UMI ProQuest [616]
Dermatology digest – Northfield IL 1976-77 – 1,5,9 – ISSN: 0011-9105 – mf#10284.02 – us UMI ProQuest [616]
Dermoncourt, Paul *see* La vendee et madame
Dernburg, Heinrich *see*
– Die allgemeinen lehren des buergerlichen rechts des deutschen reichs und preussens
– Deutsches erbrecht
– Deutsches familienrecht
– Pandekten
– Das sachenrecht des deutschen reichs und preussens
– Die schuldverhaeltnisse nach dem rechte des deutschen reichs und preussens
Derner lokal-anzeiger – Dortmund DE, 1937 2 jan-1937 30 jun, 1938 1 jul-1941 31 may – 8r – 1 – gw Misc Inst [074]
La dernier anneau de la queue de robespierre – Paris: Imp de Lacour, 1848 – us CRL [944]
Le dernier chant des serins de laval / Beausoleil, Joseph Maxime – Montreal: s.n, 1890 – 1mf – 9 – mf#03529 – cn Canadiana [378]
La dernier conseil du pere duchesne aux electeurs – Montmartre: Imp Pilloy freres et Ce, 1849 – us CRL [944]
Dernier mot sur la presidence : or, candidature compares de mm bonaparte et cavaignac – Paris, [1848?] – us CRL [944]
Un dernier mot sur les biens des jesuites – Montreal: [s.n.], 1888 [mf ed 1985] – 1mf – 9 – mf#SEM105P454 – cn Bibl Nat [241]
Le dernier mot sur les femmes / Larcher, Louis Julien – Paris: Achille Faure, 1864 – (= ser Les femmes [coll]) – 2mf – 9 – mf#8497 – fr Bibl Nationale [360]
Le dernier rapport d'un europeen sur ghat et les touareg de l'air : journal de voyage d'erwin de bary, 1876-1877 / Bary, Erwin de – Paris: Librarie Fischbacher, 1898 – 1 – us CRL [914]
Dernier tableau de paris : ou recit historique de la revolution du 10 aout 1792 des causes qui l'ont produite, des evenemens qui l'ont precedee, et des crimes qui l'ont suivie / Peltier, Jean Gabriel – Londres 1793-94 [mf ed Hildesheim 1995-98] – 4v on 6mf – 9 – €120.00 – 3-487-26290-8 – gw Olms [944]
Dernier voyage du capitaine cook autour du monde : ou se trouvent les circonstances de sa mort / Zimmermann, Heinrich, de Wiesloch – Berne [Suisse]: Chez la Nouvelle societe typographique, 1783 [mf ed 1984] – 3mf – 9 – 0-665-44925-9 – (trans fr german) – mf#44925 – cn Canadiana [910]
Derniere correspondance entre s e le cardinal barnabo et l'hon m dessaulles – Montreal?: s.n, 1871 – 1mf – 9 – mf#23737 – cn Canadiana [360]
La derniere guerre des betes : fables pour servir a l'histoire du 18. siecle [sic] / Falques, Marianne-Agnes Pillement – Londres: Chez C G Seyffert... 1758 [mf ed 1986] – 2v on 1mf – 9 – 0-665-51285-6 – mf#51285 – cn Canadiana [830]
Derniere heure – Brussels Belgium, 12 sep 1944-10 sep 1945; 27 oct 1946 – 1r – 1 – uk British Libr Newspaper [074]
La derniere journee de sappho / Faure, Gabriel – Paris: Mercure de France, 1901 – (= ser Les femmes [coll]) – 3mf – 9 – mf#8834 – fr Bibl Nationale [830]
Dernieres annees du regne et de la vie de louis 16 / Hue, Francois – Paris 1816 [mf ed Hildesheim 1995-98] – 1v on 3mf [ill] – 9 – €90.00 – 3-487-26192-8 – gw Olms [944]
Dernieres conversations avec anatole france / Segur, Nicolas – Paris, France. 1927 – 1r – us UF Libraries [025]
Dernieres decouvertes dans l'amerique septentrionale de m de la salle / Tonti, Henri de – Paris au Palais: chez Jean Guignard...1697 [mf ed 1979] – 4mf – 9 – mf#SEM105P17 – cn Bibl Nat [917]
Dernieres nouvelles – Algeria. 8 dec 1942; 31 jan 1943-25 oct 1944 – 2r – 1 – uk British Libr Newspaper [072]
Les dernieres nouvelles d' alsace *see* Strassburger neueste nachrichten
Les dernieres nouvelles d'alsace – Strasbourg, France. Toutes eds.21 dec 1944-1987 – 1 – (bilingue. toutes eds. 1959-81. 1. strasbourg-ville. 1969-81. 1) – fr ACRPP [074]
Les dernieres nouvelles du lundi – Strasbourg, France. Toutes eds.1961-84 – 1 – fr ACRPP [074]
Les dernieres persecutions du troisieme siecle (gallus, valerien, aurelien) / Allard, Paul – 3e ed., rev. et augm. Paris: V. Lecoffre, 1907 – 2mf – 9 – 0-7905-5501-8 – (incl bibl ref) – mf#1988-1501 – us ATLA [240]
Les dernieres potieres de sainte-anne, martinique / Roo Lemos, Noelle de – [Montreal]: Centre de recherches caraibes, Universite de Montreal, [1979?] (mf ed 1995) – 1mf – 9 – mf#SEM105P2397 – cn Bibl Nat [660]
Derniers adieux de graziella : suivis de quelques autres poesies detachees / Baillairge, Maurice – Quebec?: C Darveau, 1879 – 1mf – 9 – mf#27028 – cn Canadiana [810]
Les derniers jours : cahier politique et litteraire – Paris. n1-7. fevr-juil 1927 – 1 – fr ACRPP [073]
Les derniers vestiges du christianisme preche du 10e au 14e siecle dans le markland et la grande irlande : les porte-croix de la gaspesie et de l'acadie (domination canadienne) / Beauvois, Eugene – Paris?: Impr Moquet, 1877 – 1mf – 9 – mf#24203 – cn Canadiana [240]
Deroche, F Louis *see* Abreje istoua, daiti, 1492-1945
DeRoos, John *see* Personal narrative of travels in the united states and canada in 1826
Derose, Rodolphe *see* Caractere, culture, vodou
Derouet, Camille *see*
– La federation nationale des canadiens-francais
– Les francais du canada
– Les metis canadiens-francais
Derouin, Barbara *see* Administrative structure of athletic departments and the impact of title ix
Deroux, Jean *see* Livre et complainte sur l'execution de cleophas lachance le meurtrier
Derpich Aguilar, Juan *see* Canto de bronce
Derrick / Fairfield Co. Bremen – (may 1911-73) scattered dates, brittle condition [wkly] – 17r – 1 – mf#B2937-2953 – us Ohio Hist [071]
Derrick / Fairfield Co. Bremen – sep 1974-jul 1976,nov 1976-jan 1977 [wkly] – 1r – 1 – mf#B29201 – us Ohio Hist [071]
Derrick / Sandusky Co. Gibsonburg – 4,1890-8,1948 (scattered) – 10r – 1 – mf#B36949-36958 – us Ohio Hist [071]
Derriere le sourire khmer / Meyer, Charles – [Paris] Plon [1971] [mf ed 1989] – 1r with other items – 1 – (incl bibl ref) – mf#mf-10289 seam reel 018/08 [§] – us CRL [320]
Derrota / Villaveces, Jorge – Bogota, Colombia. 1963 – 1r – us UF Libraries [972]
Derrota de una batalla / Marroquin Rojas, Clemente – s.l, s.l? 1957? – 1r – us UF Libraries [972]
Derrotados / Andreu Iglesias, Cesar – Mexico City? Mexico. 1956 – 1r – us UF Libraries [972]
Derrotados del llanto / Moncada Luna, Jose Antonio – Panama, 1961 – 1r – us UF Libraries [972]
Derrotero...costas de espana...atlantico / Tofino de San Miguel, Vicente – 1849 – 9 – sp Bibl Santa Ana [914]
Derrumbe / Andreu Iglesias, Cesar – Mexico City? Mexico. 1960 – 1r – us UF Libraries [972]
Derrumbe / Demar, Carmen – San Juan, Puerto Rico. 1948 – 1r – us UF Libraries [972]
Derrumbe / Soler Puig, Jose – Santiago, Cuba. 1964 – 1r – us UF Libraries [972]
Derry journal *see*
– Londonderry journal
Derry people – Tyrone jun-21st dec 1921, 1929 – 1r – 1 – ie National [072]
Derry people and donegal news – Londonderry, Ireland. 16 feb-28 dec 1929, 11 jan 1986-1989, 13 jan 1990-1991, 11 jan-dec 1992 – 20 1/2r – 1 – (aka: derry people and tirconaill news) – uk British Libr Newspaper [072]
Derry people and donegal news – [Northern Ireland] Belfast 20 sep 1902-dec 1950 [mf ed 2002] – 44r – 1 – (missing: 1915, 1921, 1929; cont by: derry people and tirconaill news [jan 1926-dec 1950]) – uk Newsplan [072]
Derry people and tirconaill news *see* Derry people and donegal news
Derry standard – [Northern Ireland] Belfast jan 1930-dec 1950 [mf ed 2002] – 58r – 1 – uk Newsplan [072]
Derry standard *see*
– Londonderry standard
Derry weekly news and tyrone herald – Londonderry, Ireland. 1950; 1952 – 1 1/2r – 1 – uk British Libr Newspaper [072]
Derry weekly news and tyrone herald – [Northern Ireland] Belfast aug 1906-dec 1950 [mf ed 2002] – 36r – 1 – uk Newsplan [072]
Dershay, F K *see* Ucheno-literaturnyi zhurnal
Dertigduizend ballingen willen terug naar hun tropisch vaderland / Weltje, H C – [1966] – 1mf – 9 – mf#SE-1441 – ne IDC [950]
Dertli – Bolu, 1919-19? Sahib-i Imtiyaz: Ilyaszade Suekrue; Mueduer-i Mes'ul: Yaglioglu Ahmed Resad. n127. 1 agustos 1338 [1922], 147. 26 kanuniewel 1338 [1922] – (= ser O & t journals) – 1mf – 9 – $25.00 – us MEDOC [956]
Dertli *see* The divan project
Derush ve-hidush 'al ha-torah / Frankel, Menahem Mordecai – Yerushalayim, Israel. 1929 – 1r – us UF Libraries [939]
Derushe maharshim / Margolin, J Joseph – New York, NY. 1921 – 1r – us UF Libraries [939]
The dervishes, or, oriental spiritualism / Brown, John Porter – London: Truebner, 1868 – 1mf – 9 – 0-524-00699-7 – mf#1990-2027 – us ATLA [290]
Derwacter, Frederick Milton *see* Preparing the way for paul
Derwein, Herbert *see* Hoffmann von fallersleben und johanna kapp
Dery, Marie-Claire *see*
– Bibliographie de mme marthe lemaire-duguay
– Biobibliographie de mme marthe lemaire-duguay
Derzhava – Astrakhan, Russia. n4(ian)-n7(okt '98) – 1 – mf#mf-12248 (reel 2) – us CRL [077]
Des 20. psalm auslegung : inn reim gefast zu beten und zu singen vor die loeblichsten gottfuerchtigen herrn den churfuersten zu sachsen und landgrauen zu hessen... / Jonas, Justus – Wittemberg: Rhaw 1546 – (= ser Hqab. literatur des 16. jahrh.) – 1mf – 9 – €20.00 – mf#1546d – gw Fischer [780]
Des abbe i i barthelemy reise durch italien : nach dem originalbriefen des grafen von caylus abgedruckt; nebst einen anhang von noch ungedruckten schriften – Paris [u a] 1802 [mf ed Hildesheim 1995-98] – 3mf [ill] – 9 – €90.00 – 3-487-29316-1 – gw Olms [914]
Des adelard von bath traktat de eodem et diverso = De eodem et diverso / Adelard of Bath; ed by Willner, Hans – Muenster: Aschendorff, 1903 – (= ser Beitraege zur geschichte der philosophie des mittelalters) – 1mf – 9 – 0-7905-3988-8 – (incl bibl ref) – mf#1989-0481 – us ATLA [180]
Des adelard von bath traktat de eodem et diverso / Wilner, H – Muenster, 1903 – (= ser Bgphma 4/1) – 2mf – 8 – €5.00 – ne Slangenburg [140]
Des alfred von sareshel (alfredus anglicus schrift de motu cordis / Baeumker, Cl – 1923 – (= ser Bgphma 23/1-2) – €7.00 – ne Slangenburg [180]
Des alpes au niger : souvenir d'un marsouin (1868-1891) / Descastes, Francois – Paris: F Juven, 1898 – 1 – us CRL [910]
Des andern theyls, viler kurtzweyliger, frischer teutscher liedlein – 1553 – (= ser Mssa) – 6mf – 9 – €80.00 – mfchl 98 – gw Fischer [780]
Des associations religeuses chez les grecs : thiases, eranes, orgeons, avec le texte des inscriptions relatives a ces associations / Foucart, Paul Francois – Paris: Klincksieck, 1873 – 1mf – 9 – 0-524-01701-8 – mf#1990-2603 – us ATLA [250]
Des aufrichtigen hermogensis apocalypsis spagyrica et philosophica : oder, wahrhaffter und untrueglicher weg zu der hoechsten medicin / Hermogenes – Leipzig: In Johann Samuel Heinsii Buchladen 1739 [mf ed 1988] – 1r – 1 – mf#8544 – us Wisconsin U Libr [540]
Des augsburger patriciers philipp hainhofer beziehungen zum herzog philipp 2 von pommern-stettin / ed by Doering, O – Wien. v6. 1894 – 5mf – 9 – mf#O-517 – ne IDC [700]
Des augustinerpropstes ioannes busch : chronicon windeshemense und liber de reformatione monasteriorum / ed by Grube, K – Halle, 1886 – €29.00 – ne Slangenburg [241]
Des averroes abhandlung : ueber die moeglichkeit der conjunktion, oder, ueber den materiellen intellekt = Maamar efsharut ha-devekut / ed by Hannes, Ludwig – Halle a S: CA Kaemmerer, 1892 [mf ed 1992] – 1mf – 9 – 0-524-03058-8 – (incl bibl ref; in german & hebrew) – mf#1990-3161 – us ATLA [110]
Des ballets anciens et modernes selon les regles du theatre / Menestrier, Claude-Francois – Paris: R Guignard 1682 [mf ed 19–] – 6mf – 9 – mf#fiche 522 – us Sibley [790]
Des bamberger fuerstbischofs johann gottfried von aschhausen gesandtschafts-reise : nach italien und rom 1612 und 1613 / ed by Haeutle, Christian – Stuttgart: Litterarischer Verein in Stuttgart, 1881 (Tuebingen: H Laupp) [mf ed 1993] – (= ser Blvs 156) – 204p – 1 – (incl bibl ref) – mf#8470 reel 32 – us Wisconsin U Libr [910]
Des bamberger fuerstbischofs johann gottfried von aschhausen gesandtschafts-reise nach italien und rom 1612 und 1613 / ed by Haeutle, Christian – Stuttgart: Litterarischer Verein, 1881 (Tuebingen: H Laupp) – (incl bibl ref) – us Wisconsin U Libr [914]
Des beruehmten meister hans blumen von lor am main nuezlichs seulenbuch / Blum, Hans – Zuerich, 1668 – 3mf – 9 – mf#OA-60 – ne IDC [720]
Le des betsileo (madagascar) / Dubois, Henri M – Paris, France. 1938 – 1r – us UF Libraries [960]
Des bildhauergesellen franz ferdinand ertinger' reisebeschreibung durch oesterreich und deutschland / Tietze-Conrad, E – Wien, 1907. v14 – 2mf – 9 – mf#O-517 – ne IDC [914]
Des blasphemes et imprecations : extraits divers des meilleurs auteurs / Laporte, Stanislas – [Mile-End, Quebec?: s.n.] 1887 [mf ed 1984] – 2mf – 9 – 0-665-46410-X – (incl bibl ref) – mf#46410 – cn Canadiana [230]
Des boehmischen freiherrn, loew von rozmital und blatna : denkwuerdigkeiten und reisen durch deutschland, england, frankreich, spanien, portugal und italien / Sasek z Birkova, Vaclav – Bruenn 1824 [mf ed Hildesheim 1995-98] – 2v on 4mf – 9 – €120.00 – 3-487-27715-8 – gw Olms [914]
Des burschen heimkehr : oder, der tolle hund / Niebergall, Ernst Elias; ed by Fuchs, Georg – Darmstadt: A Bergstraesser, 1894 – 1r – 1 – us Wisconsin U Libr [820]
Des colonies : particulierement da la guyane francaise, en 1821 / Saint-Amant, Pierre C de – Paris 1822 [mf ed Hildesheim 1995-98] – 1v on 2mf – 9 – €60.00 – 3-487-26893-0 – gw Olms [972]
Des commencements de l'eglise du canada / Verreau, Hospice Anthelme Baptiste – Montreal?: Dawson, 1885 – 1mf – 9 – mf#35797 – cn Canadiana [241]
Des commencements de montreal / Verreau, Hospice Anthelme Baptiste – S.l: s.n, 1887? – 1mf – 9 – mf#28657 – cn Canadiana [971]
Des conflits de lois en matiere de mariage au maroc / Franchassin, L – Toulouse, 1936 – 2mf – 9 – mf#ILM-3011 – ne IDC [956]
Des courtisans aux partisans : essai sur la crise cambodgienne / Pomonti, Jean Claude – [Paris] Gallimard [1971] [mf ed 1989] – (= ser Coll idees 230. idees actuelles) – 1r with other items – 1 – (with bibl) – mf#mf-10289 seam reel 019/07 [§] – us CRL [959]
Des critischen musicus an der spree / Marpurg, Friedrich Wilhelm – Berlin: A Haude & J C Spener 1750 [mf ed 19–] – 10mf – 9 – (iss called also: 1. bd) – mf#fiche 426 – us Sibley [780]

Des deutschen spiessers wunderhorn : gesammelte novellen / Meyrink, Gustav – Muenchen: A Langen, 1913 – 1r – 1 – us Wisconsin U Libr [830]

Des deux cotes de la barricade – Paris [1848?] – us CRL [074]

Des diakons pontius leben des hl cyprianus / cyprians traktate, 1. bd (bdk34 1.reihe) – (= ser Bibliothek der kirchenvaeter. 1. reihe (bdk 1.reihe)) – €15.00 – ne Slangenburg [240]

Des doctrines religieuses des juifs pendant les deux si ecles anterieurs a l'ere chretienne / Nicolas, Michel – Paris: Michel Levy, 1860 – 1mf – 9 – 0-7905-2192-X – (incl bibl ref) – mf#1987-2192 – us ATLA [270]

Des dodes danz : nach den luebecker drucken von 1489 und 1496 = Dance of death – Stuttgart: Litterarischer Verein, 1876 (Tübingen: H Laupp) [mf ed 1993] – (= ser Blvs 127) – 145p – 1 – mf#8470 reel 27 – us Wisconsin U Libr [830]

Des dodes danz : nach den luebecker drucken von 1489-1496 / ed by Baethcke, Hermann – Stuttgart: Litterarischer Verein, 1876 (Tuebingen: H Laupp) – (incl bibl ref and ind) – us Wisconsin U Libr [430]

Des dominicus gundissalinus schrift "von der unsterblichkeit der seele" / Buelow, G – Muenster, 1897 – (= ser Bgphma 2/3) – 3mf – 9 – €7.00 – ne Slangenburg [110]

Des dritten teyls viler schoener teutscher liedlein – 1552 – (= ser Mssa) – 5mf – 9 – €70.00 – mfchl 99 – gw Fischer [780]

Des fuersten von ruegen wizlaw's des vierten sprueche und lieder in niederdeutscher sprache : nebst einigen kleinern niederdeutschen gedichten / ed by Ettmueller, Ludwig – Quedlinburg, Leipzig: G Basse, 1852 [mf ed 1993] – (= ser Bibliothek der gesammten deutschen national-literatur von der aeltesten bis auf die neuere zeit sect1/33) – 99p – 1 – mf#8438 reel 7 – us Wisconsin U Libr [810]

Des gedultigen joben glauben vnnd bekanntnuss... / Lavater, L – Zuerich, Christoffel Froschower, 1577 – 1mf – 9 – mf#PBU-319 – ne IDC [240]

Des gorilles, des nains et meme...des hommes : histoire de la grande foret, de la brousse et de la cote africaines / Gouzy, Rene – Lausanne, 1919 – 1 – us CRL [590]

Des grafen wolrad von waldeck tagebuch waehrend des reichstages zu augsburg 1548 / ed by Tross, C L P – Stuttgart: Litterarischer Verein, 1861 [mf ed 1993] – (= ser Blvs 59) – 271p – 1 – (text in latin, notes & ind in german) – mf#8470 reel 12 – us Wisconsin U Libr [880]

Des gregorius abulfarag, gen bar-hebraeus, scholien zum buche daniel = Horreum mysteriorum. selections / Bar Hebraeus; ed by Freimann, Jacob – Bruenn: B Epstein, 1892 – (= ser Beitraege zur geschichte der bibelexegese) – 1mf – 9 – 0-524-02761-7 – mf#1987-6455 – us ATLA [221]

Des h eustathius beurtheiling des origenes / Jahn, A – Leipzig, 1886 – (= ser Tugal 1-2/4) – 2mf – 9 – €5.00 – ne Slangenburg [240]

Des h eustathius, erzbischofs von antiochien, beurtheilung des origenes betreffend die auffassung der wahrsagerin, 1 koen [sam] 28 : und die bezuegliche homilie des origenes = On the witch of endor against origen – Leipzig: J C Hinrichs, 1886 [mf ed 1989] – (= ser Tugal 2/4) – 1mf – 9 – 0-7905-2895-9 – (text in greek; int in german) – mf#1987-2895 – us ATLA [221]

Des h hippolytus von rom commentar zum buche daniel / Bardenhewer, Otto – Freiburg i. Br, 1877 – 2mf – 8 – €5.00 – ne Slangenburg [221]

Des heil. gregor von nyssa lehre vom menschen / Hilt, Franz – Koeln: JP Bachem, 1890 – 1mf – 9 – 0-7905-9964-3 – (incl bibl ref) – mf#1989-1689 – us ATLA [240]

Des heiligen augustinus schriften als liturgiegeschichtliche quelle / Roetzer, W – Muenchen, 1930 – €12.00 – ne Slangenburg [241]

Des heiligen augustinus speculative lehre von gott dem dreieinigen : ein wissenschaftlicher nachweis der objectiven begruendetheit dieses christlichen glaubensgegenstandes... / Gangauf, Theodor – Augsburg: B Schmid, 1865 – 2mf – 9 – 0-7905-7297-4 – mf#1989-0522 – us ATLA [240]

Des heiligen epiphanius von salamis / Epiphanius, Saint, Bishop of Constantia in Cyprus – 1919. Trans. by Joseph Hoermann – 1 – us Wisconsin U Libr [240]

Des heiligen hippolytus von rom commentar zum buche daniel : ein literaergeschichtlicher versuch / Bardenhewer, Otto – Freiburg im Breisgau, St Louis, MO: Herder, 1877 – 1mf – 9 – 0-7905-4321-4 – (incl bibl ref) – mf#1988-0321 – us ATLA [221]

Des heiligen irenaeus schrift zum erweise der apostolischen verkuendigung : eis epideixin tou apostolikou kerygmatos: in armenischer version entdeckt / Irenaeus; ed by Ter-Mekerttschian, Karapet & Ter-Minassiantz, Erwand – Leipzig: J C Hinrichs, 1907 – (= ser Tugal) – 1mf – 9 – 0-7905-4039-8 – (incl ind) – mf#1988-0039 – us ATLA [240]

Des histoires orientales et principalement des turkes ou turchikes et schitiques et tartaresques et aultres qui en sont descendues... / Postel, G – Paris, 1575 – 6mf – 9 – mf#HU-131 – ne IDC [956]

Des hocherleuchteten lehrers, herrn johann arndts, weiland general-superintendenten des fuerstenthums lueneburg, sechs buecher vom wahren christenthum : welche handeln von heilsamer busse, herzlicher reue und leid ueber die suende, und wahren glauben, auch heiligem leben und wandel der rechten wahren christen = Sechs buecher vom wahren christenthum / Arndt, Johann – Reutlingen: BG Kurtz, 1835 – 3mf – 9 – 0-524-08666-4 – mf#1993-3191 – us ATLA [240]

Des hochgelerte erasmi von roterdam vn doctor luthers maynung vom nachtmal... / [Jud, L] – [Zuerich, Christoph Froschauer, 1526] – 1mf – 9 – mf#PBU-534 – ne IDC [242]

Des johann neudoerfer nachrichten von kuenstlern und werkleuten in nuernberg / Lochner, G W K – Wien, 1875. v10 – 4mf – 9 – mf#O-517 – ne IDC [700]

Des kaisers soldaten : schauspiel in drei aufzuegen / Essig, Hermann – Stuttgart: J G Cotta, 1915 (mf ed 1990) – 1r – 1 – (filmed with: bozena) – us Wisconsin U Libr [820]

Des knaben plunderhorn : [essays] / Bergengruen, Werner – Berlin: O Schlegel, 1934 [mf ed 1989] – 175p – 1 – mf#7009 – us Wisconsin U Libr [840]

Des knaben wunderhorn : alte deutsche lieder / Arnim, Ludwig Achim, Freiherr von & Brentano, Clemens – Heidelberg: Mohr und Zimmer, 1808-19 [mf ed 1993] – 3v (ill) – 1 – mf#8357 – us Wisconsin U Libr [780]

Des koenigs leibwache : eine jugend-erzaehlung / Nieritz, Gustav – 4. Aufl. Guetersloh: C Bertelsmann, [1909?] – 1r – 1 – us Wisconsin U Libr [830]

Des kronprinzen regiment : roman / Samarow, Gregor – 2. Aufl. Stuttgart: Deutsche Verlags-Anstalt, [18–] – 1r – 1 – us Wisconsin U Libr [830]

Des langst gewunschten und versprochenen chymisch-philosophischen probier-steins, erste classe, in welcher der wahren und achten adeptorum und anderer wurdig erfundenen schrifften... / Fictuld, Hermann – Dresden: Hilscher, 1784 – 1 – us Wisconsin U Libr [540]

Des lydens jesu cristi gantze...historia / Jud, L – Zuerich, Christoffel Froschouer, 1539 – 3mf – 9 – mf#PBU-277 – ne IDC [240]

Des Marchais, Etienne Renaud *see* Journal de navigation du voyage de la coste de guinee.

Des marqves des enfans de dieu : et des consolations en levrs afflictions / Taffin, J – Ed 3. Amsterdam, 1588 – 2mf – 9 – mf#PBA-423 – ne IDC [240]

Des meeres und der liebe wellen : trauerspiel in fuenf aufzuegen / Grillparzer, Franz – Leipzig: Hesse & Becker, [19–?] – 1r – 1 – us Wisconsin U Libr [430]

Des meeres und der liebe wellen : trauerspiel in fuenf aufzuegen / Grillparzer, Franz; ed by Schuetze, Martin – 2nd rev ed. New York: H Holt, 1926 (1930 printing), c1912 [mf ed 1993] – lxxxvi/156p – 1 – (incl bibl ref. german text. int and notes in english) – mf#8701 – us Wisconsin U Libr [820]

Des menschen begin : midden en einde / Luyken, Jan – Amsteldam: Wed P Arentz, en K vander Sys, 1712 – 3mf – 9 – mf#O-349 – ne IDC [090]

Des minnesangs fruehling / Vogt, Friedrich – Leipzig: S Hirzel, 1930 – (incl bibl ref. text in middle hihg german; notes in german) – us Wisconsin U Libr [430]

Des minnesangs fruehling / Vogt, Friedrich; ed by Lachmann, Karl & Haupt, Moritz – Leipzig: S Hirzel, 1882 – (incl bibl ref and index) – us Wisconsin U Libr [430]

Des moines dispatch – Des Moines IA. 1938 feb 13 – 1r – 1 – mf#3925698 – us WHS [071]

Des moyens de gouvernement et d'opposition dans l'etat actuel de la france / Guizot, Francois Pierre Guillaume – Paris 1821 [mf ed Hildesheim 1995-98] – 1v on 3mf – 9 – €90.00 – 3-487-26031-X – gw Olms [325]

Des nordischen mercurii extraordinaire relation *see* Nordischer mercurius

Des pater alexander von rhodes aus der gesellschaft jesu missionsreisen in china, tonkin, cochinchina und anderen asiatischen reichen / Rhodes, Alexandre de – Freiburg im Breisgau: Herder, 1858 [mf ed 1995] – (= ser Yale coll) – xi/345p – 1 – 0-524-09591-4 – (in german) – mf#1995-0591 – us ATLA [240]

Des patriarchen gennadios von konstantinopel confession : nebst einem excurs ueber arethas' zeitalter / Otto, Johannes Carl Theodor – Wien: Wilhelm Braumueller, 1864 – 1mf – 9 – 0-8370-7494-0 – (in greek and german. incl bibl ref and index) – mf#1986-1494 – us ATLA [240]

Des peuples du caucase et des pays au nord de la mer noire et de la mer caspienne, dans le dixieme siecle : ou voyage d'abou-el-cassim / Ohsson, Constantin d' – Paris 1828 [mf ed Hildesheim 1995-98] – 1v on 2mf – 9 – €60.00 – 3-487-27668-2 – gw Olms; ne IDC [914]

Des phrases non des faits – Madrid, 193? Fiche W 829. (Blodgett Collection of Spanish Civil War Pamphlets) – 9 – us Harvard [946]

Des poemes latins attribues a saint bernard / Haureau, Barthelemy – Paris: C Klincksieck 1890 [mf ed 1990] – 1mf – 9 – 0-7905-6293-6 – (incl bibl ref) – mf#1988-2293 – us ATLA [810]

Un des premiers colons d'etchemin, p q, jean dumet ou demers / Demers, Benjamin – [Quebec?: s.n.], 1914 – 1mf – 9 – 0-665-73926-5 – mf#73926 – cn Canadiana [929]

Des Pres, Francois Marcel-Turenne *see* Children of yayoute

Des principes de l'architecture, de la sculpture, de la peinture, et des autres arts qui en dependent... / Felibien, A Sieur des Avaux et de Javercy – Paris, 1676 – 9mf – 9 – mf#OA-43 – ne IDC [720]

Des privileges [sic] sur les biens meubles : these pour le doctorat presentee et soutenue le 12 janvier 1889 / Demers, Philippe – Montreal: A Periard, 1889 [mf ed 1984] – 2mf – 9 – 0-665-07897-8 – (incl ind) – mf#07897 – cn Canadiana [346]

Des progres de la revolution et de la guerre contre l'eglise / Lamennais, Felicite Robert de – Paris [u a] 1829 [mf ed Hildesheim 1995-98] – 1v on 3mf – 9 – €90.00 – 3-487-26137-5 – gw Olms [944]

Des puits et des aqueducs : investigations a faire / Beaudry, Joseph Alphonse Ubalde – [Quebec (Province)?: s.n, 1908?] – 1mf – 9 – 0-665-98274-7 – mf#98274 – cn Canadiana [627]

Des rabbi israel ben elieser, genannt baal-shem-tow / Ba'al Shem Tov – Berlin, Germany. 1935 – 1r – us UF Libraries [270]

Des reiches kommen / Kroeger, Timm – Hamburg: Janssen, 1909 – 1r – 1 – us Wisconsin U Libr [830]

Des religions comparees au point de vue sociologique / La Grasserie, Raoul de – Paris: V Giard & E Briere 1899 – (= ser Bibliotheque sociologique internationale) – 1mf – 9 – 0-524-01614-3 – mf#1990-2553 – us ATLA [230]

Des representations en musique anciennes et modernes / Menestrier, Claude-Francois – Paris: Chez R Guignard 1681 [mf ed 19–] – 4mf – 9 – mf#fiche 614, 798 – us Sibley [780]

Des Rochers, Guy *see* Bio-bibliographie analytique de marcel trudel

Des roemischen kaisers lieb- lob- und gluecks-werther davidischer achilles, der baeyrische mars / Bartsch, J J – Wienn: Bey Johann Christoph Cosmerovij, [1685] – 1mf – 9 – mf#O-85 – ne IDC [090]

Des samuel al-magrebi abhandlungen ueber die pflichten der priester und richter bei den karaeern / ed by Cohn, Julius – Berlin: H Itzkowski, 1907 [mf ed 1985] – 1mf – 9 – 0-8370-5035-9 – (german trans with ann by ed) – mf#1985-3035 – us ATLA [270]

Des scandales qui empeschent aujourd'huy beaucoup de gens de venir a la pure doctrine de l'evangile, et en desbauchent d'autres : traicte compose de nouvellement par jehan calvin / Calvin, J – Geneve: De l'imprimerie de Jehan Crespin, 1550 – 2mf – 9 – mf#CL-78 – ne IDC [242]

Des schwaebischen ritters georg von ehingen reisen nach der ritterschaft / ed by Pfeiffer, Franz – Stuttgart: Literarischer Verein, 1842 [mf ed 1993] – (= ser Blvs 1/2) – vii/28p – 1 – (original ed (augsburg, 1600) has title: itinerarium, das ist: historische beschreibung, weylund herrn georgen von ehingen raisenn nach der ritterschafft...incl bibl ref) – mf#8470 reel 1 – us Wisconsin U Libr [910]

Des sepultures nationales et particulierement de celles des rois de france / Roquefort, Jean Baptiste de – Paris 1824 [mf ed Hildesheim 1995-98] – 1v on 3mf – 9 – €90.00 – 3-487-25906-0 – gw Olms [944]

Des societes secretes en allemagne, et en d'autres contrees : de la secte des illumines, du tribunal secret, de l'assassinat de kotzebue, etc / Lombard de Langres, Vincent – Paris: Gide, fils 1819 – 3mf – 9 – mf#vrl-207 – ne IDC [366]

Des teufels netz : satirisch-didaktisches gedicht aus der ersten haelfte des 15. jahrhunderts / ed by Barack, K A – Stuttgart: Litterarischer Verein, 1863 [mf ed1993] – (= ser Blvs 70) – 467p – 1 – (incl bibl ref and ind) – mf#8470 reel 14 – us Wisconsin U Libr [430]

Des teufels netz *see* Der edelstein / des teufels netz / sibyllenweissagung [cima7]

Des theodor abu kurra traktat ueber den schoepfer und die wahre religion / Graf, G – 1913 – (= ser Bgphma 14/1) – €5.00 – ne Slangenburg [240]

Des vaters fluch : historisches drama in drei acten, aus der zeit des letzten hohenstaufen / Bieleck, Rudolph – Wien: Im Selbstverlage des Verfassers, 1874 [mf ed 1993] – 105p – 1 – mf#8518 – us Wisconsin U Libr [820]

Des vedas / Barthelemy Saint-Hilaire, Jules – Paris: B Duprat: A Durand, 1854 – 1mf – 9 – 0-524-01412-4 – mf#1990-2407 – us ATLA [280]

Des verwegenen chirurgus weltberuehmbt wunder-doktor johann andreas eisenbart : tugenden und laster getreulich mitgeteilt und vorgestellt / Winckler, Josef – Berlin: Deutsche Buch-Gemeinschaft, 1933 – 1 – us Wisconsin U Libr [920]

Des verwegenen chirurgus weltberuehmbt wunder-doktor johann andreas eisenbart : zahnbrechers, baenkelsaengers, okulisten,... / Winckler, Josef – Berlin: Deutsche Buch-Gemeinschaft, 1933 – 1 – us Wisconsin U Libr [617]

Des Voeux, George William *see*
– Experiences of a demarara magistrate
– My colonial service in british guiana, st lucia,...

Des voies d'execution des jugements a rome / Lemoine, Charles Amedee – Nancy, Crepin-Leblond. 1881. 344p. LL-4042 – 1 – us L of C Photodup [340]

Des vortrefflichen religionsverbesserer ulrich zwingli... : anmerkungen ueber des evangelisten matthaeus lebensgeschichte jesu bis zum anfang der letzten leiden / Kuester, K D – Halle, 1783 – 9mf – 9 – mf#ZWI-41 – ne IDC [242]

Des wagners e ch doebel wanderungen im morgenlande / Doebel, Ernst C – [Berterode bei Eisenach] 1845 [mf ed Hildesheim 1995-98] – 2v on 4mf – 9 – €120.00 – 3-487-26698-9 – gw Olms [915]

Des wereldts proef-steen ofte de ydelheydt door de waerheyd beschuldight ende overtuyght van valscheydt / Burgundia, Antonium – t'Antwerpen: Gedruckt bij de weduwe ende erfgenamen van Ian Cnobbaert, 1643 – 5mf – 9 – mf#O-3043 – ne IDC [090]

Desa membangun *see* Departemen transmigrasi, koperasi dan pembangunan masjarakat desa

Desabhimani varika – Cochin, India. Jul-Sept 1966; 1981-May 1995 – 30r – 1 – us L of C Photodup [079]

Desabrati – Calcutta, India. Jul 1968-Apr 1970 – 3r – 1 – us L of C Photodup [079]

Desafio a historia (mitos e homens na historia do... / Pinto, Wilson – Rio de Janeiro, Brazil. 1969 – 1r – us UF Libraries [972]

Desafio a pecuaria brasileira / Medeiros Neto, Jose Bernardo De – Porto Alegre, Brazil. 1970 – 1r – us UF Libraries [972]

Desafios y retos de varios caballeros del siglo 15 – Madrid, Biblioteca Nacional [19–] – us CRL [074]

A desafronte – [Sao Tome]: Tip do Anunciador, feb 9 1924-sep 3 1925 – us CRL [079]

Desai, Akshayakumar Ramanlal *see* Introduction to rural sociology in india

Desai, Kanu *see*
– Mahatma gandhi
– Water colours

Desai, Mahadeo *see* My early life, 1869-1914

Desai, Mahadev Haribhai *see*
– Gandhiji in indian villages
– Maulana abul kalam azad, the president of indian national congress
– A righteous struggle
– The story of bardoli
– With gandhiji in ceylon

Desai, Mohanalala Dalicanda *see* Shrimad yashovijayaji

Desai, Shantaram Anant *see* A study of the indian philosophy

Desai, Valaji Govindaji [comp] *see* A gandhi anthology

Desai, Yunus *see* From 'coolie location' to group area

Desalination – Oxford, England 1966+ – 1,5,9 – ISSN: 0011-9164 – mf#42250 – us UMI ProQuest [540]

Desa-madju *see* Djawatan penerangan ri kabupaten modjokerto

La desamortizacion de las propiedades... valencia de alcantara / Garcia Perez, Juan – 9 – sp Bibl Santa Ana [336]

Desani, Govindas Vishnoodas *see* Hali

Desapande, Panduranga Ganesa [comp] *see* Gandhi sahitya suci

DeSapio, Vincent *see* Abrasives

Desa-rapporten cheribon, mogjokerto, semarang, surabaja, probolinggo, kraksaan / Ranneft, J W Meyer – 6mf – 8 – mf#SD-109 mf 1-6 – ne IDC [959]

Desarrollo del programa oficial / Caceres Tinoco De Giron, Ela – Tegucigalpa, Mexico. 1942 – 1r – us UF Libraries [972]

Desarrollo del sistema de transportes en colombia / Ardila B, Jose Joffre – Santander, Spain. 1949 – 1r – us UF Libraries [380]
Desarrollo industrial de el salvador / Hoselitz, Berthold Frank – Nueva York, NY. 1954 – 1r – us UF Libraries [338]
Desarrollo literario de el salvador / Toruno, Juan Felipe – San Salvador, El Salvador. 1958 – 1r – us UF Libraries [972]
Desarrollo socio-economico de la duodecima region de chileh / ed by Ramirez Ceballos, Valeria – Santiago, Chile: Centro de Estudios del Desarrollo, 1985 (mf ed 1987) – 1mf – 9 – (incl bibl ref) – mf#*XME-12973 – us NY Public [300]
Desarrollo y resoluciones : informe final / Seminario Nacional Sobre Problemas De La Educacion – Guatemala, 1961 – 1r – us UF Libraries [370]
Desassure [i e desaussure] on fraser 1885-1897 *see* Historical notes
Desaubrys, John Phillip *see* Sonatas, for two violins and a violoncello, opera 1
Desaugiers, Marc Antoine *see* Air des deux jumeaux de bergame...
Desaugiers, Marc-Antoine *see*
– Chacun son tour
– Couturieres
– Deux boxeurs
– Deux voisines
– Epoux avant le mariage
– Heure de folie
– Hotel garni
– Juif
– Monsieur croque-mitaine
– Retour des lys
Desaulniers, Francois Lesieur *see*
– Charles lesieur et la fondation d'yamachiche
– Le fondateur des religieuses de l'assomption
– La genealogie des familles gouin et allard
– La genealogie des familles richer de la fleche et hamelin
– Notes historiques sur la paroisse de saint-guillaume d'upton
– Recherches genealogiques
– Les vieilles familles d'yamachiche
– Les vieilles familles d'yamachiche, vol 1
– Les vieilles familles d'yamachiche, vol 2
– Les vieilles familles d'yamachiche, vol 3
– Les vieilles familles d'yamachiche, vol 4
– Les vieilles familles d'yamachiche, vols 1-4
DeSaussure, Wilmot Gibbes *see*
– Historical notes
– Papers
Desaussure's cases in equity / South Carolina. Supreme Court – v1-4. 1784-1816 (all publ) – (= ser Pre-nrs nominative equity reports) – 29mf – 9 – $44.00 – mf#LLMC 94-026 – us LLMC [342]
Desautels, Adrien *see* Bibliographie des oeuvres litteraires publiees au canada de robert rumilly
Desbarats, George Edouard *see* L'esclavage dans l'antiquite et son abolition par le christianisme
Desboulmiers, Jean-Augustin-Julien *see*
– Histoire anecdotique et raisonnee du theatre italien
– Histoire du theatre de l'opera comique...
Descamps, J B *see* La vie des peintres flamands, allemands et hollandais...
Descanso / Pineiro, Abelardo – Habana, Cuba. 1962 – 1r – us UF Libraries [972]
Descarries, Alfred *see* Le pardon revolutionnaire en 1 acte represente au theatre national francais
Descartes / Mahaffy, John Pentland, Sir – Edinburgh: William Blackwood, 1880 – (= ser Philosophical Classics for English Readers) – 1mf – 9 – 0-7905-8840-4 – mf#1989-2065 – us ATLA [100]
Descartes, par ch adam, e brehier, l brunschvicg – Paris, France. 1937 – 1r – us UF Libraries [190]
Descartes, Rene *see*
– The meditations and selections from the principles of rene descartes
– Method, meditations and philosophy of descartes
– Rene descartes
Descartes, spinoza and the new philosophy / Iverach, James – Edinburgh: T & T Clark 1904 [mf ed 1991] – (= ser The world's epoch-makers) – 1mf – 9 – 0-7905-9230-4 – mf#1989-2455 – us ATLA [190]
Descastes, Francois *see* Des alpes au niger
Descaves, Lucien *see* Coeur ebloui
Descaves, Pierre *see* Cite des voix
The descendants of john eliot from 1598-1905 / Emerson, Wilimena H – 1905 – 1r – 1 – $50.00 – mf#B50010 – us Library Micro [920]
Descendants of philip henry / Lawrence, Sarah – 1844 – 1 – $50.00 – us Presbyterian [920]
Descendants of William Dawes Who Rode Association *see* Newsletter
Descendents of the late queen victoria / McNaughton, A – 1r – 7 – mf#96027 – uk Microform Academic [929]
Descendez! / Letraz, Jean De – Paris, France. 1946 – 1r – us UF Libraries [440]
Descenso, y humillacion de dios, para el ascenso, y exaltacion del hombre : baxa dios a la tierra para que el hombre suba a el cielo, por medio de la passion del verbo divino hecho hombre: poema heroico / Ximenes y Frias, Joseph Antonio – en Mexico: impr Phelipe de Zuniga y Ontiveros...1769 – (= ser Books on religion...1543/44-c1800: doctrina cristiana, obras de devocion) – 5mf – 9 – mf#crl-52 – ne IDC [810]
Descent into hell / Williams, Charles – New York, NY. 1949 – 1r – us UF Libraries [025]
Descent of man : and selection in relation to sex / Darwin, Charles – New York, NY. 1874 – 1r – us UF Libraries [575]
The descent of man : and selection in relation to sex / Darwin, Charles Robert – London, 1871 – (= ser 19th c evolution & creation) – 2v on 10mf – 9 – (with ill) – mf#1.1.4270 – uk Chadwyck [575]
The descent of man and selection in relation to sex / Darwin, Charles – London: J Murray, 1871 – 3mf – 9 – 0-524-00017-4 – (incl bibl ref) – mf#1989-2717 – us ATLA [575]
Descent of the danube, from ratisbon to vienna, during the autumn of 1827 : with anecdotes and recollections, historical and legendary, of the towns, castles, monasteries, etc, upon the banks of the river, and their inhabitants and proprietors, ancient and modern / Planche, James – London 1828 [mf ed Hildesheim: 1995-98] – (= ser Fbc) – xv/320p on 2mf [ill] – 9 – €60.00 – 3-487-29406-0 – gw Olms [914]
La descente du christ aux enfers : d'apres les apaotres et d'apres l'eglise / Bruston, Charles – Paris: Fischbacher, 1897 – 1mf – 9 – 0-7905-0559-2 – (incl bibl ref) – mf#1987-0559 – us ATLA [240]
Deschamps, Clement E *see*
– Liste des municipalites dans la province de quebec
– Municipalites et paroisses dans la province de quebec
Deschamps, Clement E [comp] *see* List of municipalities in the province of quebec
Deschamps, Edouard *see* Africa
Deschamps, Enrique *see*
– Republica dominicana
– La republica dominicana; directorio y guia general
Deschamps, Hubert Jules *see*
– Antaisaka
– Eveil politique africain
Deschamps, J B *see* Voyages pittoresques de la flandre et du brabant, avec des reflexions relativement aux arts et quelques gravures
Deschamps, Nicolas *see* Les societes secretes et la societe
Deschenes, Jean-Claude *see* Bibliographie de l'oeuvre du reverend pere hector-l bertrand
Deschènes, R [comp] *see* Charte et reglements de la cite de st-hyacinthe
Deschler, Lewis et al *see* Deschler's precedents of the house of representatives
Deschler's precedents of the house of representatives / Deschler, Lewis et al – Washington: GPO. v1-15. 1965- – 17mf (1:42) 43mf (1:24) – 9 – $235.00 – (active series. add vols to be filmed) – mf#llmc 84-108 – us LLMC [323]
Deschmann, Ida Maria *see* Der buesser
Deschutes echo – Bend OR: A C Palmer, [wkly] [mf ed 1971] – 1r – 1 – (began in 1902) – us Oregon Lib [071]
Descobrimento da ilha de sam thome – [S.l: s.n., 19–?] – (filmed with pinto, m r relacas) – us CRL [079]
Descobrimento do brasil / Marcondes De Souza, Thomas Oscar – Sao Paulo, Brazil. 1946 – 1r – us UF Libraries [972]
Descobrimentos do rio das amazonas / Carvajal, Gaspar De – Sao Paulo, Brazil. 1941 – 1r – us UF Libraries [972]
Desconhecido niassa / Santos, Nuno Beja Valdez Thomaz Dos – Lisboa, Portugal. 1964 – 1r – us UF Libraries [960]
Un desconocido cedulario del siglo 16. mexico, 1944 / Carreno, Alberto Maria – Madrid: Razon y Fe, 1947 – 1 – sp Bibl Santa Ana [972]
Descourtilz, M E *see* Voyage d'un naturaliste en haiti
Descourtilz, Michel *see* Voyages d'un naturaliste, et ses observations
O descrido : periodico critico e litterario – Aracaju, SE: Typ do Democrata, 15 out, dez 1881; jan-03 out 1882 – (= ser Ps 19) – mf#P11A,03,44 – bl Biblioteca [320]
Descripcao de serra-leoa e seus contornos : escripta em doze cartas, a qual se ajuntao os trabalhos da commissao-mixta portugueza e ingleza, estabelecida naquella colonia – Lisbon: J Baptista Morando, 1822 – 1 – us CRL [960]
Descripcao dos rios parnahyba e gurupy / Dodt, Gustavo Luiz Guilherme – Sao Paulo, Brazil. 1939 – 1r – us UF Libraries [972]
Descripcao historica, topographica e ethnographica do districto de s. joao baptista d'ajudia e do reino de dahome na costa da mina / Bettencourt Vasconcellos Corte Real do Canto, Vital de – Lisboa: Typographia Universal de T Q Antunes, 1869 – 1 – us CRL [916]
Descripcion de algunos moluscos del mioceno del va... / Ramirez, Ricardo – Ciudad Trujillo, Dominican Republic. 1949 – 1r – us UF Libraries [972]
Descripcion de la canada...montemolin – 1856 – 9 – sp Bibl Santa Ana [971]
Descripcion de la ciudad y obispado de plasencia por... / Toro, Luis de – Plasencia: La Victoria, 1961 – 1 – sp Bibl Santa Ana [240]
Descripcion de la parte espanola de santo domingo / Moreau De Saint-Mery, M L E – Ciudad Trujillo, Dominican Republic. 1944 – 1r – us UF Libraries [972]
Descripcion de las fiestas que hicieron los diputados de la ciudad de tehuacan : en celebridad de la dedicacion del templo de nuestra senora del carmen, rasgo epico / Soria, Francisco de – en Mexico: impr D Joseph de Jauregui, ano de 1783 – (= ser Books on religion...1543/44-c1800: iglesias, catedrales) – 1mf – 9 – mf#crl-356 – ne IDC [241]
Descripcion de las honras que se hicieron a la catholica magd. de d. phelippe... / Rodriguez de Monforte, P – Madrid: Por Francisco Nieto, 1666 – 4mf – 9 – mf#O-2020 – ne IDC [090]
Descripcion de las indias occidentales... / Herrera, Antonio de – Madrid: Imprenta Real, 1601 – 1 – sp Bibl Santa Ana [959]
Descripcion de las provincias del rio de la plata / Borrero, Fernando – Buenos Aires, Argentina. 1911 – 1r – us UF Libraries [972]
Descripcion del reyno de santa fe de bogota / Silvestre, Francisco – Bogota, Colombia. 1950 – 1r – us UF Libraries [972]
Descripcion geografica de los reinos de la nueva galicia... / Motay Escobar, Alonso – Madrid: Razon y Fe, 1940 – sp Bibl Santa Ana [918]
Descripcion geografico-moral de la diocesis de goa / Cortes Y Larraz, Pedro – Guatemala, v1-2. 1958 – 1r – us UF Libraries [960]
Descripcion historical y moral del yermo de san migvel, de las cvevas en el reyno de la nueva-espana : y invencion de la milagrosa imagen de christo nuestro senor crucificado, que se venera en ellas / Florencia, Francisco de – Impresso en Cadiz: En la imprenta de la Compania de lesus, por Christoval de Requena, [1683?] – (= ser Books on religion...1543/44-c1800: vidas y cultos de santos) – 4mf – 9 – mf#crl-114 – ne IDC [241]
Descripcion proclama...badajoz...al trono...rey d. fernando 6 / Gallardo Bonilla, Leandro – 1747 – 9 – sp Bibl Santa Ana [946]
Descripcion que hace la ilustre provincial del sr s raphael del peru, y reyno de chile : del sagrado hospitalario orden de n p s juan de dios, a su meritissimo... / Brothers Hospitallers of St John of God – [Lima: s.n. 1759?] – (= ser Books on religion...1543/44-c1800: orden de san juan de dios) – 1mf – 9 – mf#crl-245 – ne IDC [241]
Descripcion y diseno del trillo / Alvarez Guerra, Andres – 1815 – 9 – sp Bibl Santa Ana [621]
Descripcion...casa de aguallo / Ramos, Antonio – 1781 – 9 – sp Bibl Santa Ana [946]
Descripcion...corridas de toros...caceres / Sevillano, Casimiro – 1846 – 9 – sp Bibl Santa Ana [790]
Descripcion...de las enfermedades mas comunes del exercito con un nuevo metodo de curar el mal venereo / Van Swieten – Madrid, 1761 – 5mf – 9 – sp Cultura [616]
Descripcion...del estandarte real – 1700 – 9 – sp Bibl Santa Ana [920]
Descripcion...del santuario...de guadalupe en extremadura. – 1878 – 9 – sp Bibl Santa Ana [946]
Descripciones de indias occidentales / Herrera, Antonio de – 1730 – 9 – sp Bibl Santa Ana [972]
Descripsion del reyno de santa fe de bogota / Silvestre, Francisco – Panama, 1927 – 1r – us UF Libraries [972]
Descriptio ac delineatio geographica detectionis freti, sive, transitus ad occasum, sufra terras americanas, in chinam atq / Hudson, Henry; ed by Gerritsz, Hessell – Amsterodami: ex officina Hesselij Gerardi, 1612 [mf ed 1988] – 1mf – 9 – (trans of original dutch by ed) – mf#SEM105P881 – cn Bibl Nat [910]
Descriptio de situ helvetiae et vicinis gentibus...de quatuor helvetiorum pagis...cum commentariis osualdi myconii...ad maximilianum augustum...panegyricon / Glareanus, H – [Basileae, Io[annes] Frobenius, 1519] – 1mf – 9 – mf#PBU-489 – ne IDC [240]
Descriptio itineris per helvetiam galliam et germaniae partem / Schmidel, C C – Zagreb. 1980-1980 (1,5,9) – 2mf – 9 – mf#12491 – ne IDC [914]
Descriptio magnae ecclesiae seu sanctae sophiae (cbh11,2) / Pauli Silentiarii; ed by Gange, C du – Parisiis, 1670 – (= ser Corpus byzantinae historiae (cbh)) – €14.00 – ne Slangenburg [243]
Descriptio publicae gratulationis, spectaculorum et ludorum, in adventu sereniss principis ernesti archiducis austriae... / Bochius, J – Antverpiae: Ex officina Plantiniana, 1595 – 5mf – 9 – mf#0-1116 – ne IDC [700]
Descriptio templi sanctae sophiae [cshb32] / Pauli Silentiarii; ed by Bekkerus, Imm – Bonnae, 1837 – (= ser Corpus scriptorum historiae byzantinae [cshb]) – €21.00 – (incl: georgii pisidae: expeditio persica, bellum avaricum, heraclias, and: sancti nicephori patriarchae cp: breviarium rerum post mauricium gestarum) – ne Slangenburg [243]
Description and history of the great eastern – [Quebec?: s.n.] 1861 [mf ed 1983] – 1mf – 9 – 0-665-44208-4 – mf#44208 – cn Canadiana [623]
A description and list of the lighthouses of the world, 1863 / Findlay, Alexander George – 3rd ed. London: publ for R H Laurie, 1863 [mf ed 1984] – 1mf – 9 – 0-665-44540-7 – (incl bibl ref) – mf#44540 – cn Canadiana [623]
Description de la cathedrale de strasbourg – Strasbourg 1817 [mf ed Hildesheim 1995-98] – 1mf [ill] – 9 – €40.00 – 3-487-29709-4 – gw Olms [720]
Description de la chine – Paris, 1749-1761. 21-23v – 16mf – 9 – mf#HT-678 – ne IDC [915]
Description de la chine... / Martini, M – Paris: Andre Cramoisy, 1672 – 5mf – 9 – mf#HT-613 – ne IDC [915]
Description de la defaicte des tvrcz estans entrez dans l'isle de malte – Paris, 1565 – 1mf – 9 – mf#H-8165 – ne IDC [956]
Description de la nigritie / Pruneau de Pommegorge, Antoine E – Amsterdam [u a] 1789 [mf ed Hildesheim 1995-98] – 1v on 2mf [ill] – 9 – €60.00 – 3-487-27273-3 – gw Olms; ne IDC [910]
Description de la ville de peking : pour servir...l'intelligence du plan de cette ville, grave par les soins de m de l'isle / L'Isle, J n d e – Paris, 1765 – 2mf – 9 – mf#HT-638 – ne IDC [915]
Description de la ville de strasbourg : contenant des notices topographiques et historiques sur l'etat ancien et actuel de cette ville... / Farges-Mericourt, P J – Strasbourg [u a] 1831 [mf ed Hildesheim 1995-98] – 3mf [ill] – 9 – €90.00 – 3-487-29706-X – gw Olms [914]
Description de l'afrique : contenant les noms, la situation et les confins de toutes ses parties, leurs rivieres,... / Dapper, O – Amsterdam: Wolfgang, Waesberge, Boom & van Someren, 1686 – 1 – us CRL [960]
Description de l'afrique et de l'espagne / Idrisi – Leyde: EJ Brill, 1866 – 2mf – 9 – 0-524-08000-3 – (incl bibl ref) – mf#1991-0222 – us ATLA [910]
Description de l'aile de patmos et de l'aile de samos / Guerin, V – Paris: Auguste Durand, 1856 – 1mf – 9 – 0-7905-1523-7 – (incl bibl ref) – mf#1987-1523 – us ATLA [919]
Description de l'art de fabriquer les canons / Monge, Gaspard – Paris: Impr. du Comite de Salut public, (1794) – 2p/leaf/viii/231p. 70 folding plates, 4 folding tables – 1 – us Wisconsin U Libr [623]
Description de l'egypte / France. Commission des Monuments d'Egypte – 21v. 1809-28 – 1 – us AMS Press [956]
Description de l'egypte : ou recueil des observations et des recherches qui ont ete faites en egypte pendant l'expedition de l'armee francaise / ed by Panckoucke, Charles L – Paris 1821-26 [mf ed Hildesheim 1995-98] – 36v on 95mf – 9 – €950.00 – 3-487-27384-5 – gw Olms [916]
Description de l'egypte : ou recueil des observations et des recherches qui ont ete faites en egypte pendant l'expedition de l'armee francaise...histoire naturelle / Delile, A [R] et al – Paris, 1812. v2 – 30mf – 9 – mf#78 – ne IDC [916]
Description de l'egypte : ou recueil des observations et des recherches...histoire naturelle. v1: [zoology] / Delile, A R et al – Peiping. 1949-1958 (1) – 45mf – 9 – mf#2615 – ne IDC [916]
Description de l'hotel de ville d'amsterdam : avec l'explication de tous les emblemes, figures, tableaux, statues... – Amsterdam 1751 [mf ed 1995-98] – 2mf [ill] – 9 – €60.00 – 3-487-29659-4 – gw Olms [914]
Description de l'isle formosa en asie : du gouvernement, des loix, des moeurs & de la religion des habitans... / Psalmanaazaar, G – Amsterdam: d'Estienne Roger, 1705 – 6mf – 9 – mf#HT-664 – ne IDC [915]

Description de tovte l'isle de cypre : et des roys, princes, et seigneurs, tant payens que chrestiens... / Lusignano, S di – Paris: G Chaudiere, 1580 – 7mf – 9 – mf#H-8356 – ne IDC [914]

Description de valence : ou tableau de cette province, de ses productions, de ses habitans, de leurs moeurs, de leurs usages, etc; pour faire suite au voyage en espagne, du meme auteur / Fischer, Christian – Paris 1804 [mf ed Hildesheim 1995-98] – 3mf – 9 – €90.00 – 3-487-29872-4 – gw Olms [914]

Description des bains de titus, o- collection des peintures trouvees dans les ruines des thermes de cet empereur / Ponce, M – Paris, 1786 – 7mf – 9 – mf#O-415 – ne IDC [700]

Description des monumens de rhodes / Rottiers, Bernard E – Bruxelles 0 [mf ed Hildesheim 1995-98] – 3v on 5mf – 9 – €100.00 – 3-487-27646-1 – gw Olms [720]

Description des pyrenees : considerees principalement sous les rapports de la geologie, de l'economie politique, rurale et forestiere, de l'industrie et du commerce; ouvrage où l'on traite de la nature... / Dralet – Paris 1813 [mf ed Hildesheim 1995-98] – 2v on 4mf – 9 – €120.00 – 3-487-29754-X – gw Olms [914]

Description du bosphore / Ingigian, L – Paris: J B Sajou, 1813 – 2mf – 9 – mf#AR-1732 – ne IDC [915]

Description du cabinet de mr paul de praun... nuremberg / Murr, C T de – Neuremberg, 1797 – 7mf – 9 – mf#O-994 – ne IDC [700]

Description du cap de bonne-esperance : ou l'on trouve tout ce qui concerne l'histoire-naturelle du pays, la religion, les moeurs & les usages des hottentots, et l'etablissement des hollandois = [Caput bonae spei hodiernum] / Kolb, Peter – a Amsterdam: Chez Jean Catuffe 1741 – (= ser [Travel descriptions from south africa, 1711-1938]) – 3v on 5mf – 9 – mf#zah-34 – ne IDC [916]

Description du cap de bonne-esperance. / Kolb, Peter – Amsterdam. 1741 – 1 – us CRL [960]

Description du jubil : de sept cens ans de s. macaire... – Gand: J Meyer, [1767] – 2mf – 9 – mf#O-3242 – ne IDC [090]

Description du parnasse francois execute en bronze : a la gloire de la france et de louis le grand, et a la memoire perpetuelle des illustres poetes et des fameux musiciens francois... / Titon du Tillet, Evrard – A Paris: [s.n.] 1760 – 2pt in 1v – 9 – (incl: le parnasse francois [1732], suppls [1743-55] and description du parnasse [1760]) – mf#fiche 956 – us Sibley [780]

Description du pegu et de l'isle de ceylan : renfermant des details exacts et neufs sur le climat, les productions, le commerce, le gouvernement, les moeurs et les usages de ces contrees / Hunter, William – Paris 1793 [mf ed Hildesheim 1995-98] – 1v on 3mf – 9 – €90.00 – 3-487-27450-7 – gw Olms [915]

Description du tibet, d'apres la relation des lamas tangoutes, etablis parmi les mongols... / Reuilly, J de – Paris: Bossange, Masson et Besson, 1808 – 2mf – 9 – mf#HT-646 – ne IDC [915]

Description du tubet... / Iakinf [Bichurin, Ia] – Paris: Imprimerie Royale, 1831 – 4mf – 9 – mf#HT-642 – ne IDC [910]

Description et details des arts du meunier, du vermicelier et du boulanger : avec une histoire abregee de la boulangerie, et un dictionnaire de ces arts / Malouin, Paul Jacques – [Paris]: [Saillant et Nyon], 1767 [mf ed 1974] – 1r – 5 – mf#SEM35P116 – cn Bibl Nat [660]

Description et histoire naturelle du groenland / Egede, H – Copenhague, Geneve, 1763 – 6mf – 9 – mf#H-473 – ne IDC [917]

Description et histoire naturelle du groenland / Egede, Hans – Copenhague, Geneve: Chez les freres C & A Philibert, 1763 [mf ed 1983] – 3mf – 9 – 0-665-44494-X – (french trans by jean-baptiste des roches de parthenay) – mf#44494 – cn Canadiana [919]

Description et plan d'un nouveau calorifer a air chaud : sur le systeme tubulaire pour chauffer les edifices prives et publics / Baillairge, Charles P Florent – Quebec?: Bureau et Marcotte, 1853 – 1mf – 9 – mf#34204 – cn Canadiana [621]

Description generale de la chine,... / Grosier, J B G A – Paris, 1787. 2v – 15mf – 9 – mf#HT-521 – ne IDC [590]

Description generale de l'afrique... / Avity, Pierre d' – 1 – us CRL [916]

Description generale, historique, geographique et physique de la colonie de surinam : contenant ce qu'il y a de plus curieux et de plus remarquable, touchant sa situation, ses rivieres, ses forteresses... / Fermin, Philippe – Amsterdam 1769 [mf ed Hildesheim 1995-98] – 2v on 5mf – 9 – €100.00 – 3-487-26886-8 – gw Olms [918]

Description geographique de l'empire de la chine : paris, 1696. v2 / Martini, M – 5mf – 9 – mf#HT-682 – ne IDC [915]

Description geographique, historique, chronologique, politique et physique de l'empire de la chine et de la tartarie chinoise... / Du Halde, J B – Paris: P G Le Mercier, 1735. 4v – 75mf – 9 – mf#HT-509 – ne IDC [915]

Description geographique, historique et archeologique de la palestine / Guerin, v – Paris, 1868-1896. 3v – 43mf – 9 – mf#H-2835 – ne IDC [956]

Description geographique, historique, militaire et routiere de l'espagne : contenant des details sur tous les lieux remarquables, et les particularites les plus interessantes de l'histoire de cette monarchie... / DuRozoir, Charles – Paris 1823 [mf ed Hildesheim 1995-98] – 3mf – 9 – €90.00 – 3-487-29841-4 – gw Olms [946]

Description historique de la ville de paris et de ses environs / Piganiol de LaForce, Jean – Paris 1765 [mf ed Hildesheim 1995-98] – 10v on 36mf – 9 – €360.00 – 3-487-29770-1 – gw Olms [914]

Description historique de paris : et de ses plus beaux monumens pour servir d'introduction a l'histoire de paris and de la france / Beguillet, Edme – Paris 1779 [mf ed Hildesheim: 1995-98] – (= ser Fbc) – 4mf [ill] – 9 – €120.00 – 3-487-29669-1 – gw Olms [914]

Description historique des prisons de paris : pendant et depuis la revolution, avec des anecdotes curieuses et peu connues, et des notices sur les personnages celebres qui y ont ete renfermes / Bourg, Edme Theodore – Paris 1828 [mf ed Hildesheim 1995-98] – 1v on 4mf – 9 – €120.00 – 3-487-25922-2 – gw Olms [365]

Description nouvelle des merveilles de ce monde / Parmentier, J – Paris, 1531 – 2mf – 9 – mf#SEP-4 – ne IDC [910]

Description of a series of thin sections of typical rocks / Adams, Frank Dawson – Montreal: s.n, 1896 – 1mf – 9 – mf#06635 – cn Canadiana [550]

A description of a singular aboriginal race : inhabiting the summit of the neilgherry hills, or blue mountains of coimbatoor, in the southern peninsula of india / Harkness, Henry – London 1832 [mf ed Hildesheim 1995-98] – 1v on 2mf – 9 – €60.00 – 3-487-27519-8 – gw Olms [305]

Description of a suite of sculptured decorative furniture : illustrative of irish history and antiquities, manufactured of irish bog yew / Arthur J Jones, Son and Co – Dublin: Hodges & Smith, 1853 – (= ser 19th c art & architecture) – 1mf – 9 – mf#4.1.203 – uk Chadwyck [740]

Description of a view of canton, the river tigress, and the surrounding country... / Burford, R – London: T Brettell, 1838 – 1mf – 9 – mf#HT-650 – ne IDC [915]

Description of a view of macao in china... / Burford, R – London: Geo Nicols, 1840 – 1mf – 9 – mf#HT-651 – ne IDC [915]

Description of a view of the city of cairo / Burford, Robert – London 1847 – (= ser 19th c art & architecture) – 1mf – 9 – mf#4.2.822 – uk Chadwyck [710]

Description of a view of the city of mexico / Burford, Robert – London 1853 – (= ser 19th c art & architecture) – 1mf – 9 – mf#4.2.823 – uk Chadwyck [710]

Description of a view of the city of quebec : now exhibiting at the panorama, leicester square / Burford, Robert – London?: J & C Adlard, 1830 – 1mf – 9 – mf#21314 – cn Canadiana [917]

Description of a view of the continent of boothia : discovered by captain ross in his late expedition to the polar regions / Burford, Robert – London?: s.n, 1884 – 1mf – 9 – mf#17097 – cn Canadiana [919]

Description of a view of the falls of niagara : now exhibiting the panorama, leicester square / Burford, Robert – London?: T Brettell, 1834 – 1mf – 9 – mf#46524 – cn Canadiana [917]

Description of a view of the island and bay of hong kong... / Burford, R – London: J Mitchell and Co, 1844 – 1mf – 9 – mf#HT-652 – ne IDC [915]

Description of british guiana / Schomburgk, Robert H – London, England. 1840 – 1r – us UF Libraries [972]

Description of census enumeration districts, 1900 / U.S. Bureau of the Census – (= ser Records of the bureau of the census – other census records) – 10r – 1 – mf#T1210 – us Nat Archives [317]

A description of ceylon : containing an account of the country, inhabitants, and natural productions; with narratives of a tour round the island in 1800, the campaign in candy in 1803, and a journey to ramisseram in 1804 / Cordiner, James – London 1807 [mf ed Hildesheim 1995-98] – 2v on 9mf – 9 – €180.00 – 3-487-27258-X – gw Olms [915]

A description of china : containing the geography, with the civil and natural history – London, 1745-1747. v4 – 15mf – 9 – mf#A-271 – ne IDC [915]

A description of fonthill abbey, wiltshire / Storer, James Sargant – London 1812 – (= ser 19th c art & architecture) – 1mf – 9 – mf#4.2.1304 – uk Chadwyck [720]

Description of irish priests / O'Connor, Arthur – London, 1852 – (= ser 19th c ireland) – 1mf – 9 – mf#1.1.9706 – uk Chadwyck [241]

A description of leg muscle pain and the effect of acetylsalicyclic acid on the perception of pain and effort during and after cycle ergometry / Cook, Dane B – 1995 – 1mf – 9 – $4.00 – mf#PE 3748 – us Kinesology [616]

Description of messrs marshall's grand peristrephic panorama of the polar regions : which displays the north coast of spitzbergen, baffin's bay, arctic highlands, etc... – Leith, Scotland: printed by William Heriot...1821 – 1mf – 9 – (incl bibl) – mf#45628 – cn Canadiana [919]

A description of some important theatres : and other remains in crete / Belli, Onorio – London 1854 – (= ser 19th c art & architecture) – 1mf – 9 – mf#4.1.438 – uk Chadwyck [720]

Description of summer and winter views of the polar regions : as seen during the expedition of capt james clark ross, kt., frs in 1848-9 / Burford, Robert – London?: s.n, 1850 – 1mf – 9 – mf#17098 – cn Canadiana [919]

Description of summer tours via south eastern railway quebec and gulf tours : and saguenay line of steamers, embracing routes to lake memphremagog, white and franconia mountains... – [Montreal?: s.n, 1877?] – 1mf – 9 – 0-665-92171-3 – mf#92171 – cn Canadiana [380]

The description of swedland, gotland, and finland / North, George – 1561 – 9 – us Scholars Facs [914]

A description of tartary : subject to china – London, 1745-1747. v4 – 1mf – 9 – mf#A-271 – ne IDC [915]

Description of the "annular" or "ring oven" : now erecting at glen brick works, tanneries west, near montreal / Leeming, John – Montreal?: s.n, 1869? [mf ed 1986] – 1mf – 9 – 0-665-42723-9 – mf#42723 – cn Canadiana [660]

A description of the antiquities and other curiosities of rome : from personal observation during a visit to italy in the years 1818-19; with illustrations from ancient and modern writers / Burton, Edward – London 1828 [mf ed Hildesheim 1995-98] – (= ser Fbc) – 2v on 5mf – 9 – €100.00 – 3-487-29243-2 – gw Olms [914]

Description of the approach to the florida keys / Miner, Frances H – s.l, s.l? 1936 – 1r – us UF Libraries [978]

A description of the belgo-canadian fruit lands companies' irrigation works, near kelowna, bc / Stoess, Charles A – [Montreal?: s.n, 1912?] – 1mf – 9 – 0-665-99725-6 – mf#99725 – cn Canadiana [627]

A description of the causal attributions made to perceived teaching behavior across three elementary physical education contexts / Mros, M – 1990 – 2mf – 9 – $8.00 – us Kinesology [150]

Description of the chapel of the annunziata dell'arena...padua / Callcott, Maria (Dundas) Graham, lady – London 1835 – (= ser 19th c art & architecture) – 1mf – 9 – mf#4.2.315 – uk Chadwyck [720]

Description of the character, manners, and customs of the people of india : and of their institutions, religious and civil / Dubois, Jean A – London 1817 [mf ed Hildesheim 1995-98] – 1v on 7mf – 9 – €140.00 – 3-487-27255-5 – (trans fr french) – gw Olms [306]

Description of the city of canton : with an appendix, containing an account of the population of the chinese empire, chinese weights and measures, and the imports and exports of canton / Bridgman, E C – Canton, 1834 – 2mf – 9 – mf#HT-649 – ne IDC [915]

Description of the climate, soil, and products of suwanee county... – Savannah, GA. 1871 – 1r – us UF Libraries [630]

Description of the coast of guinea : from the royal commonwealth society library / Bosman, Willem – 1907 – 11mf – 7 – mf#2988 – uk Microform Academic [916]

A description of the coasts of north and south guinea / Barbot, J – London, 1746 – 29mf – 9 – mf#A-128 – ne IDC [916]

A description of the collection of ancient terracottas in the british museum / Combe, Taylor – London 1810 – (= ser 19th c art & architecture) – 3mf – 9 – mf#4.2.1533 – uk Chadwyck [730]

Description of the country from lake superior to cook's river : extract of a letter from...of quebec, to a friend in london – [s.l: s.n, 1790?] [mf ed 1984]] – 1mf – 9 – 0-665-41457-9 – mf#41457 – cn Canadiana [917]

A description of the east : and some other countries / Pococke, R – London, 1743-1745. 2v – 55mf – 9 – mf#H-314 – ne IDC [910]

Description of the egyptian court : erected in the crystal palace / Jones, Owen et al – London 1854 – (= ser 19th c art & architecture) – 1mf – 9 – mf#4.1.218 – uk Chadwyck [720]

A description of the empire of china and chinese-tartary : together with the kingdoms of korea and tibet: containing the geography and history (natural as well as civil) of those countries / Du Halde, J B – London, 1738-1741. 2v – 36mf – 9 – (with maps and index) – mf#CH-1206 – ne IDC [590]

A description of the feroe islands : containing an account of their situation, climate, and productions; together with the manners, and customs, of the inhabitants, their trade etc / Landt, Jorgen – London 1810 [mf ed Hildesheim 1995-98] – (= ser Fbc) – 3mf – 9 – €90.00 – 3-487-28933-4 – (trans fr danish) – gw Olms [914]

A description of the great bible, 1539 : and the six editions of cranmer's bible, 1540 and 1541 / Fry, Francis – London: Willis & Sotheran; Bristol: Lasbury, 1865 [mf ed 1989] – 2mf – 9 – 0-7905-1045-6 – mf#1987-1045 – us ATLA [220]

Description of the international bridge : constructed over the niagara river, near fort erie, canada, and buffalo, us of america / Gzowski, Casimir Stanislaus – Toronto: Copp, Clark, 1873 – 2mf – 1 – mf#05136 – cn Canadiana [624]

A description of the island of jamaica : with the other isles and territories in america to which the english are related, viz barbadoes, st christophers, nievis, or mevis, antego... – London: printed for Dorman Newman...1678 [mf ed 1982] – 2mf – 9 – mf#35035 – cn Canadiana [918]

A description of the island of st helena : containing observations on its singular structure and formation; and an account of its climate, natural history, and inhabitants / Duncan, Francis – London 1805 [mf ed Hildesheim 1995-98] – 1v on 2mf – 9 – €60.00 – 3-487-27284-9 – gw Olms [914]

A description of the island of st michael : comprising an account of its geological structure; with remarks on the other azores or western islands; originally communicated to the linnaean society of new-england / Webster, John W – Boston 1821 [mf ed Hildesheim 1995-98] – 2mf – 9 – €60.00 – 3-487-29811-2 – gw Olms [914]

A description of the royal colosseum – 22nd ed. London 1848 – (= ser 19th c art & architecture) – 2mf – 9 – mf#4.2.526 – uk Chadwyck [720]

A description of the shetland islands : comprising an account of their geology, scenery, antiquities, and superstitions / Hibbert-Ware, Samuel – Edinburgh 1822 [mf ed Hildesheim 1995-98] – 1v on 7mf – 9 – €140.00 – 3-487-27549-X – gw Olms [914]

Description of the township of rankin, sault ste marie ontario – [s.l: s.n. 1890?] [mf ed 1987] – 1mf – 9 – 0-665-35003-1 – mf#35003 – cn Canadiana [917]

Description of the township of rankin, sault ste marie ontario – [s.l: s.n. 1890?] [mf ed 1987] – 1mf – 9 – 0-665-38545-5 – mf#38545 – cn Canadiana [917]

A description of the western islands of scotland / MacCulloch, John – London 1819 [mf ed Hildesheim 1995-98] – 3v on 8mf – 9 – €160.00 – 3-487-27886-3 – gw Olms [914]

Description of the wilton house diptych, containing a contemporary portrait of king richard the second / Scharf, George – (London): Printed for the Arundel Society, 1882.vi,(7),99p. ill – 1 – us Wisconsin U Libr [760]

A description of the work of the manchester art museum / Horsfall, Thomas Coglan – Manchester 1891-95 – (= ser 19th c art & architecture) – 2mf – 9 – mf#4.2.1618 – uk Chadwyck [060]

A description of the works of art...of alfred de rothschild / Davis, Charles – London 1884 – (= ser 19th c art & architecture) – 8mf – 9 – mf#4.2.251 – uk Chadwyck [700]

A description of tibet, or tibbet – London, 1745-1747. v4 – 3mf – 9 – mf#A-271 – ne IDC [915]

A description of trends emerging during years three and four : of the wisconsin comprehensive elementary health education pilot project / Knutson-Kaske, Jill A – University of Wisconsin-La Crosse, 1995 – 1mf – 9 – $4.00 – mf#HE556 – us Kinesology [370]

Description of two pictures : 1. the banishment of aristeides, and 2. the burning of rome by nero / Haydon, Benjamin Robert – London [1846] – (= ser 19th c art & architecture) – 1mf – 9 – mf#4.2.1636 – uk Chadwyck [750]

Description of yacht basin – s.l, s.l? 193-? – 1r – us UF Libraries [550]

Description routiere et geographique de l'empire francais : divise en quatre regions / Vaysse de Villiers, Regis – Paris 1813 [mf ed Hildesheim 1995-98] – 6v on 12mf – 9 – €120.00 – 3-487-29738-8 – gw Olms [910]

Description technique des manuscrits grecs relatifs au nouveau testament : conserves dans les bibliotheques de paris / Martin, Jean Pierre Paulin – Paris: F & C Leclerc, 1884 [mf ed 1994] – xix/205p on 3mf – 9 – 0-524-08781-4 – (in french) – mf#1993-0056 – us ATLA [225]

Description topographique de la province du bas-canada : avec des remarques sur le haut-canada... / Bouchette, Joseph – Londres: W Faden, 1815 [mf ed 1971] – 1r – 1 – mf#SEM35P58 – cn Bibl Nat [917]

Description tripartita medico-astronomica que toca... sobre la constitucion epidemica... de espana, con especialidad en la villa de orgaz en 1735 y 1737 / Aranda y Marzo, J – Madrid, 1737 – 4mf – 9 – sp Cultura [610]

A description...of the great picture by paul delaroche / Jameson, Anna Brownell [Murphy] – London [1853?] – (= ser 19th c art & architecture) – 1mf – 9 – mf#4.2.442 – uk Chadwyck [750]

Descriptions des arts et metiers / Academie Royale des Sciences – Paris, 1761-1788. 74v – 452mf – 9 – mf#O-2126 – ne IDC [700]

Descriptions of census enumeration districts, 1830-1890 and 1910-1950 / U.S. Bureau of the Census – (= ser Records of the bureau of the census – other census records) – 146r – 1 – mf#T1224 – us Nat Archives [317]

Descriptions of the plates of fresco decorations / Gruner, Ludwig – new ed. London 1854 – (= ser 19th c art & architecture) – 2mf – 9 – mf#4.2.100 – uk Chadwyck [720]

Descriptions of the province of fars in persia / Ibn-al-Balkhi – 1912 – (= ser Royal asiatic society monograph) – 1r – 1 – mf#722 – uk Microform Academic [915]

A descriptive analysis of aerobic instructor behaviors and related student responses / Fuller, Catherine J – 1998 – 2mf – 9 – $8.00 – mf#PE 3921 – us Kinesology [790]

A descriptive analysis of corporate health promotion activity evaluations / Corrigan, Ann E & Kaplan, Leah E – 1992 – 2mf – $8.00 – us Kinesology [613]

A descriptive analysis of selected personality traits of student teachers in physical education / Lu, Chunlei – 2000 – 63p on 1mf – 9 – $5.00 – mf#PE 4185 – us Kinesology [150]

A descriptive analysis of sports commissions in the united states / Pennell, BL – 1990 – 1mf – 9 – $4.00 – us Kinesology [790]

Descriptive and classified catalogue of hindi christian literature : published up to 1916-17 – 2nd ed. Allahabad: North Indian Christian Tract & Book Society, 1917 [mf ed 1995] – (= ser Yale coll) – v/44p – 1 – 0-524-09315-6 – mf#1995-0315 – us ATLA [240]

Descriptive and historical catalogue...japanese and chinese / Anderson, William – London 1886 – (= ser 19th c art & architecture) – 7mf – 9 – mf#4.2.571 – uk Chadwyck [700]

Descriptive and historical view of burr's moving mirror of the lakes, the niagara, st lawrence, and saguenay rivers : embracing the entire range of border scenery of the united states and canadian shores... – [Boston?: s.n.] 1850 [mf ed 1983] – 1mf – 9 – 0-665-43092-2 – mf#43092 – cn Canadiana [917]

Descriptive and predictive discriminant analysis of the golf ability of college males / Joyner, A Barry & Baumgartner, Ted A – 1992 – 1mf – 9 – $4.00 – us Kinesology [790]

Descriptive catalogue of a collection of [prince albert's] byzantine, early italian, german, and flemish pictures / Waagen, Gustav Friedrich – London 1854 – (= ser 19th c art & architecture) – 1mf – 9 – mf#4.2.1717 – uk Chadwyck [750]

Descriptive catalogue of a collection of the economic minerals of canada, and of its crystalline rocks : sent to the london international exhibition for 1862 / Canada. Exploration Geologique – Montreal: John Lovell, 1862 [mf ed 1983] – 1mf – 9 – mf#SEM105P228 – cn Bibl Nat [550]

Descriptive catalogue of a loan exhibition of canadian historical portraits and other objects relating to canadian archaeology / Numismatic and Antiquarian Society of Montreal – Montreal?: The Gazette, 1887 – 1mf – 9 – (incl ind) – mf#11421 – cn Canadiana [750]

A descriptive catalogue of agricultural, garden and flower seeds : for sale by john d roberts, king street west, cobourg, cw – [Cobourg, Ont?: s.n, 186-?] [mf ed 1984] – 1mf – 9 – 0-665-45718-9 – mf#45718 – cn Canadiana [635]

Descriptive catalogue of art works in japanese lacquer : forming the third division of the japanese collection in the possession of james l bowes... / Audsley, George Ashdown – [London]: printed...at the Chiswick Press, 1875 – (= ser 19th c art & architecture) – 2mf – 9 – mf#4.1.163 – uk Chadwyck [740]

Descriptive catalogue of articles of church decoration / French, Gilbert James – 31st ed. Manchester 1864 – (= ser 19th c art & architecture) – 1mf – 9 – mf#4.2.1795 – uk Chadwyck [740]

Descriptive catalogue of materials relating to the history of great britain and ireland to the end of the reign of henry 7 (rs26) / Hardy, T D – 3v – (= ser The rolls series (rs)) – (v1/1 + v1/2 1862 €37. v2 1865 €23. v3 1871 €23) – ne Slangenburg [941]

Descriptive catalogue of sanskrit mss in the library of the asiatic society of bengal : part first, grammar / ed by Mitra, Rajendralala, Raja – Calcutta: Printed by C B Lewis, at the Baptist mission press, 1877 [mf ed 1995] – (= ser Yale coll) – vii/171p/lvii – 1 – 0-524-096570 – mf#1995-0657 – us ATLA [490]

A descriptive catalogue of the bronzes of european origin in the south kensington museum / Fortnum, Charles Drury Edward – London 1876 – (= ser 19th c art & architecture) – 6mf – 9 – mf#4.2.1555 – uk Chadwyck [730]

A descriptive catalogue of the collection of pictures belonging to the earl of northbrook / Weale, William Henry James & Richter, Jean Paul – London 1889 – (= ser 19th c art & architecture) – 4mf – 9 – mf#4.2.1034 – uk Chadwyck [700]

Descriptive catalogue of the great historical picture painted by mr george hayter : representing the trial of her late majesty queen caroline of england, with a faithful interior view of the house of lords... – [London?: s.n.] 1823 [mf ed 1983] – 1mf – 9 – 0-665-44209-2 – mf#44209 – cn Canadiana [347]

A descriptive catalogue of the historical pictures...of the first reformed house of commons / Hayter, George – [London] 1843 – (= ser 19th c art & architecture) – 1mf – 9 – mf#4.2.1712 – uk Chadwyck [720]

A descriptive catalogue of the maiolica...in the south kensington museum / Fortnum, Charles Drury Edward – London 1873 – (= ser 19th c art & architecture) – 13mf – 9 – mf#4.2.1724 – uk Chadwyck [730]

A descriptive catalogue of the manuscripts... fitzwilliam museum / James, Montague Rhodes – Cambridge 1895 – (= ser 19th c art & architecture) – 7mf – 9 – mf#4.2.859 – uk Chadwyck [060]

Descriptive catalogue of the north american hepaticae, north of mexico / Underwood, Lucien Marcus – [s.l: s.n, 1883?] [mf ed 1984] – 2mf – 9 – 0-665-32248-8 – mf#32248 – cn Canadiana [580]

A descriptive catalogue of the western mediaeval manuscripts in edinburgh university library / Borland, C R – Edinburgh, 1916 – 11mf – 8 – €21.00 – ne Slangenburg [090]

A descriptive catalogue of the works of rembrandt, and his scholars, bol, lievens and van vliet... / [Rembrandt] Daulby, D – Liverpool, 1796 – 5mf – 9 – mf#O-1199 – ne IDC [700]

Descriptive commentaries from the medical histories of posts / U.S. Army – 5r – 1 – (with printed guide) – mf#M903 – us Nat Archives [355]

A descriptive dictionary of the indian islands and adjacent countries / Crawfurd, J – London, 1856 – 5mf – 9 – mf#SE-20156 – ne IDC [915]

Descriptive essays contributed to the quarterly review, vol 1 / Head, Francis Bond – London: J Murray, 1857 – v1 on 5mf – 9 – mf#47579 – cn Canadiana [300]

Descriptive essays contributed to the quarterly review, vol 2 / Head, Francis Bond – London: J Murray, 1857 – v2 on 4mf – 9 – mf#47580 – cn Canadiana [300]

Descriptive essays contributed to the quarterly review, vols 1 and 2 / Head, Francis Bond – London: J Murray, 1857 – 2v on 1mf – 9 – mf#47578 – cn Canadiana [300]

A descriptive handbook for the national pictures in the westminster palace / Gullick, Thomas John – London 1865 – (= ser 19th c art & architecture) – 2mf – 9 – mf#4.2.1460 – uk Chadwyck [700]

Descriptive handbook of the cape colony : its condition and resources / Noble, John – Cape Town 1875 – (= ser 19th c british colonization) – 4mf – 9 – mf#1.1.3756 – uk Chadwyck [916]

Descriptive handbook of the cape colony: its condition and resources / Noble, John – Cape Town: J.C. Juta, 1875. 315p. illus – 1 – us Wisconsin U Libr [960]

Descriptive list catalogue of the disston lands... / Florida Land And Improvement Company – Philadelphia, PA. 1885 – 1r – us UF Libraries [978]

Descriptive list of pictures at government house, calcutta – Calcutta 1897 – (= ser 19th c art & architecture) – 1mf – 9 – mf#4.2.1504 – uk Chadwyck [700]

Descriptive list of the hebrew and samaritan mss. in the british museum / ed by Margoliouth, George – London: [s.n.], 1893 – 1mf – 9 – 0-524-05798-2 – mf#1992-0625 – us ATLA [090]

Descriptive lists of documents relating to the history of lousiana / Wright, Irene Aloha – Los Angeles, CA. v1-3. 1939 – 1r – us UF Libraries [978]

Descriptive newsletters from the solomon islands / Metcalfe, John R – 16 sep 1920-jan 1950 – 1r – mf#pmb68 – at Pacific Mss [980]

Descriptive notes...accompanying the stereographs of madura / Tracy, W – [London?] 1858 – (= ser 19th c art & architecture) – 1mf – 9 – mf#4.2.722 – uk Chadwyck [770]

Descriptive sketches of nova scotia in prose and verse / Frame, Elizabeth – Halifax, NS: A & W MacKinlay, 1864 – 3mf – 9 – mf#37908 – cn Canadiana [917]

A descriptive study examining twenty years of football related injury research at the high school and college levels / Peters, Arlene L – University of North Carolina at Chapel Hill, 1995 – 2mf – 9 – $8.00 – mf#PE3617 – us Kinesology [617]

A descriptive study of collegiate arena managers / Barajas, Gonzalo – 2001 – 82p on 1mf – 9 – $5.00 – mf#PE 4201 – us Kinesology [650]

Descriptive study of intramural activity offerings and entry rates in college/ university intramural programs with a student population between 10,001-30,000 / Dierks, Tamara J – 1998 – 1mf – 9 – $4.00 – mf#PE 3881 – us Kinesology [790]

A descriptive study of north american exercise programs for persons infected with the human immunodeficiency virus / Petersen, Carolyn – 1998 – 2mf – 9 – $8.00 – mf#HE 617 – us Kinesology [616]

A descriptive study of sexual health attitudes and practices among adolescent and young adult male county health department clients / Robbins, R D – 1991 – 2mf – 9 – $8.00 – us Kinesology [613]

Descriptive study of state statute regulation of athlete agents / Martyak, Christina M – 2000 – 1mf – 9 – $4.00 – mf#PE 4066 – us Kinesology [790]

A descriptive study of the current status of physical leisure activities in community bases residential facilities in wisconsin / Tarrell, Julie M – 1996 – 2mf – 9 – $8.00 – mf#RC 510 – us Kinesology [307]

A descriptive study of the current status of physical leisure activities in wisconsin nursing homes / McKenna, Lisa – 1997 – 1mf – 9 – $4.00 – mf#RC 506 – us Kinesology [790]

Descrittione *see* Extracts from regole brievi della volgare grammatica

Descrittione degli apparati fatti in bologna per la venuta di n s papa clemente 8... / Benacci, V – Bologna, 1598 – 1mf – 9 – mf#O-1036 – ne IDC [700]

Descrittione dell' apparato fatto nella festa di s. giovanni dal fedelissimo popolo napolitano / Giuliani, G B – Napoli: Per Domenico Maccarano, 1631 – 2mf – 9 – mf#O-1872 – ne IDC [090]

Descrittione della pompa funerale fatta nelle essequie del...cosimo de medici gran duca di toscana... – Fiorenza: Appresso i Giunti, 1574 – 1mf – 9 – mf#O-1836 – ne IDC [090]

Descrittione di tutta italia / Alberti, Fra Leandro – Venetia, 1581 – 14mf – 9 – mf#O-103 – ne IDC [720]

Descrizion delle feste fatte in firenze per la canonizzazione di s. to andrea corsini / [Buommattei, B] – Firenze: Nella stamperia di Zanobi Pignoni, 1632 – 1mf – 9 – mf#O-1531 – ne IDC [090]

Descrizione degl'istromenti armonici d'ogni genere, del padre bonanni / Buonanni, Filippo – 2. riv corr ed, Roma: A spese di V Monaldini 1776 [mf ed 19–] – 8mf – 9 – (in italian & french) – mf#fiche 367 – us Sibley [780]

Descrizione dell' apparato funebre per le esequie celebrate dalla nazione spagnuola nella sua chiesa di s giacomo in roma alla memoria di carlo iii... 101=azara, g n de – Roma, 1789 – 2mf – 9 – mf#O-1117 – ne IDC [700]

Descrizione dell' arco trionfale ed altre decorazioni architettoniche inalzate in roma nella piazza del popolo : per solennizare nel di 3 luglio 1800. il primo glorioso ingresso nella dominante della santita' di nostro signore papa pio vii – Roma, 1800 – 1mf – 9 – mf#O-1092 – ne IDC [700]

Descrizione dell' arco trionfale eretto nella pubblica piazza di vicenza la tte 12 vembre 1758 per...antonio mari priuli... – Vicenza, 1758 – 1mf – 9 – mf#O-1121 – ne IDC [720]

Descrizione della regia villa, fontane, e fabbriche di pratolino / Sgrilli, B S – Firenze, 1742 – 1mf – 9 – mf#GDI-25 – ne IDC [700]

Descrizione delle feste celebrate in venezia per la venuta di s m i r napoleone il massimo... / Morelli, I – Venezia, 1808 – 2mf – 9 – mf#O-1126 – ne IDC [700]

Descrizione delle feste fatte nelle reali nozze de' serenissimi principi di toscana d. cosimo de' medici, e maria maddalena archiduchessa d'austria – Firenze: Apresso i Giunti, 1608 – 2mf – 9 – mf#O-1986 – ne IDC [090]

Descrizione delle pompe e delle feste fatte nella venuta all citta di firenze del sereniss. don vincenzio gonzaga principe di mantova... – Firenze: Nella stamperia di Bartolomeo Sermartelli, 1584 – 1mf – 9 – mf#O-1987 – ne IDC [090]

Descrizione dell'imperiale giardidi boboli... / Cambiagi, G – Firenze, 1757 – 2mf – 9 – mf#O-961 – ne IDC [700]

Descrizione ragionata della galleria doria preceduta da un breve saggio di pittura... / Tonci, S – Roma, 1794 – 3mf – 9 – mf#O-1051 – ne IDC [700]

Descrizione storica delle pitture del regio-ducale palazzo del te fuori della porta di mantova detta pusterla – Mantova, 1783 – 1mf – 9 – mf#O-1174 – ne IDC [700]

Descrizzione delle immagini dipinte da rafaelle d'urbinelle camere del palazzo apostolico vaticano / [Raphael] Bellori, G P – Roma, 1695 – 3mf – 9 – mf#O-143 – ne IDC [700]

Descubridores jesuitas del amazonas / Bayle, Constantino – Madrid: Instituto Gonzalo Fernandez de Oviedo, 1940 – 1 – sp Bibl Santa Ana [241]

Descubridores jesuitas del amazonas, breve descripcion / Bayle, Constantino – Madrid: Revista de Indias, 1940 – 1 – sp Bibl Santa Ana [972]

Descubrimiento de america y sus vinculaciones / Pena Batlle, Manuel Arturo – Madrid, Spain. 1931 – 1r – us UF Libraries [972]

Descubrimiento de los restos de frey nicolas de ovando, primer governador de las indias / Munoz de San Pedro, Miguel – Sevilla, 1948. Separata del T.V. del Anuario de Est. Americanos – 1 – sp Bibl Santa Ana [946]

Descubrimiento de puerto rico / Gonzalez Ginorio, Jose – San Juan, Puerto Rico. suppl. 1936 – 1r – us UF Libraries [972]

Descubrimiento del amazonas / Acuna, Cristobal de – Buenos Aires, Argentina. 1942 – 1r – us UF Libraries [972]

Descubrimiento del rio amazonas / Carvajal, Gaspar de – 1895 – 9 – sp Bibl Santa Ana [918]

Descubrimiento y colonizacion de la nueva granada / Acosta, Joaquin – Bogota, Colombia. 1942 – 1r – us UF Libraries [972]

Descubrimiento y conquista de chile / Carrasco, Adolfo – 1892 – 9 – sp Bibl Santa Ana [972]

Descubrimiento y conquista de chile. barcelona, 1946 / Estave Barba, Francisco – Madrid: Razon y Fe, 1948 – 1 – sp Bibl Santa Ana [972]

Descubrimiento y conquista del peru / Reyna y Reyna, Tomas de la – 1892 – 9 – sp Bibl Santa Ana [972]

Descubrimiento...minas / Sabido y Martinez, Antonio – 1875 – 9 – sp Bibl Santa Ana [972]

Descubrimientos geometricos / Molina Cano, Juan A – 1598 – 9 – sp Bibl Santa Ana [510]

El descuento de las clases privadas / Diaz Perez, Nicolas – 1879 – 9 – sp Bibl Santa Ana [300]

Desde colon a fidel / Blanes, Nilo – Habana, Cuba. 1960 – 1r – us UF Libraries [972]

Desde el exilio / Gomez, Laureano – s.l, s.l? 1955? – 1r – us UF Libraries [972]

Desde la barra / Hernandez Poveda, Ruben – San Jose, Costa Rica. 1953 – 1r – us UF Libraries [972]

Desde la formacion profesional a las facultades o escuelas universitarias e industrias y servicios / Caceres. Delegacion Provincial del Ministerio, de Educacion – Aldea Moret: Imp. Linea 21, 1981 – 1 – sp Bibl Santa Ana [378]

DESDE

Desde mi belvedere / Varona, Enrique Jose – Habana, Cuba. 1938 – 1r – us UF Libraries [972]
Desde mi cigarral / Santovenia, Emeterio Santiago – Habana, Cuba. 1951 – 1r – us UF Libraries [972]
Desde mi tonel / Barrantes Molina, Luis – Madrid: Razon y Fe, 1934 – 1 – sp Bibl Santa Ana [972]
Desdevises du Dezert, Georges *see* L'eglise et l'etat en france
Desdevises du Dezert, Theophile *see* L'amerique avant les europeens
The desecrated bones and other stories / Habib, Mohammad – [London?]: Oxford University Press, 1925 – (= ser Samp: indian books) – us CRL [830]
Desempeno al metodo racional en la curacion de las calenturas tercianas / Flores, S L de – Sevilla, 1698 – 3mf – 9 – sp Cultura [616]
O desengano das papeletas – Rio de Janeiro, RJ: Typ Liberal de F J S Ramalho, 05 maio 1849 – (= ser Ps 19) – mf#P15,01,65 n06 – bl Biblioteca [321]
Desengano del abuso de la sangria... / Romero, L – Tarragona, 1623 – 6mf – 9 – sp Cultura [610]
Desenvolvimento da civilizacao material no brasil / Arinos De Melo Franco, Afonso – Rio de Janeiro, Brazil. 1944 – 1r – us UF Libraries [972]
Desenvolvimento e cultura / Mello, Mario Vieira De – Sao Paulo, Brazil. 1963 – 1r – us UF Libraries [972]
Desenvolvimento economico do sao francisco / Serebrenick, Salomao – Rio de Janeiro, Brazil. 1961 – 1r – us UF Libraries [330]
Desenvolviemiento educacional de costa rica con la... / Karsen, Sonja – San Jose, Costa Rica. 1954 – 1r – us UF Libraries [370]
Desenvovimento economico e social dos municipios / Cavalcanti, Araujo – Rio de Janeiro, Brazil. 1959 – 1r – us UF Libraries [350]
Deseo / Leiva, Raul – Mexico City? Mexico. 1947 – 1r – us UF Libraries [972]
Deseret news / Church of Jesus Christ of Latter-Day Saints – Fillmore, Salt Lake City UT. 1855 mar 14-1857 dec 30, 1858 jan 6-1861 feb 27, 1861 mar 6-1864 jun 29, 1864 jul-1868 feb 5 – 4r – 1 – (cont by: deseret weekly) – us WHS [243]
Deseret sampler – Dugway, Tooele UT. 1980 jan 11-1982 mar 12 – 1r – 1 – (cont: test run; cont by: sampler [dugway, utah]) – mf#611798 – us WHS [071]
Desert – Palm Desert CA 1937-81 – 1,5,9 – ISSN: 0011-9237 – mf#988 – us UMI ProQuest [574]
Desert airman – 1981 may 1-1982, 1983-84, 1985, 1986 jan 10-dec 18, 1987, 1988 – 6r – 1 – mf#660878 – us WHS [071]
Desert edge and desert / Retail Clerks Union, Local 1167 [Colton CA] – 1973 nov-1977 dec, 1978 jan-1991 apr – 2r – 1 – (cont by: desert edge) – mf#674016 – us WHS [331]
[Desert hot springs-] desert sentinel – CA. 1946- – 40r – 1 – $2400.00 (subs $90/y) – mf#R02169 – us Library Micro [071]
The desert of sinai : notes of a spring-journey from cairo to beersheba / Bonar, Horatius – New York: R Carter, 1857 – 1mf – 9 – 0-524-08173-5 – mf#1992-1159 – us ATLA [916]
The desert of the exodus : journeys on foot in the wilderness of the forty years' wanderings / Palmer, Edward Henry – Cambridge: Deighton, Bell, 1871. Chicago: Dep of Photodup, U of Chicago Lib, 1972 (1r); Evanston: American Theol Lib Assoc, 1984 (1r) – 1 – 0-8370-0337-7 – (incl bibl ref and ind) – mf#1984-B312 – us ATLA [910]
Desert plant life – Pasadena CA 1950-52 – 1 – ISSN: 0891-4907 – mf#116 – us UMI ProQuest [574]
Desert post / Dugway Proving Ground [UT] – 1991 nov27-1993 dec 16 – 1r – 1 – (cont: desert sun; cont by: dugway dispatch) – mf#4722754 – us WHS [071]
Desert star – Salt Lake City, Tooele Ordnance Depot, Tooele UT. 1981 jun-1993 dec – 1r – 1 – mf#1073901 – us WHS [071]
Desert sun / Dugway Proving Ground [UT] – 1987 mar 19-1988 aug 25 – 1r – 1 – (cont: sampler [dugway ut]; cont by: desert post [dugway ut]) – mf#1546693 – us WHS [071]
Desert wings – 1981 may 1-1982 jun, jul-1983 nov 4, 1984 feb 17-1985 apr, may-1986 may 2, may 9-1987 may 29 – 5r – 1 – mf#627624 – us WHS [071]
Le deserteur : drame en trois actes...represente par les comediens italiens ordinaires du roi le 6 mars 1769... / Monsigny, Pierre-Alexandre – Paris: C Herisant [1769] [mf ed 19–] – 1r – 1 – mf#pres. film 101 – us Sibley [780]
Desertores / Jurado, Ramon H – Lima, Peru. 1959? – 1r – us UF Libraries [972]
Desertores / Jurado, Ramon H – Panama, 1955 – 1r – us UF Libraries [972]
Deserve to be great / Gale, William Daniel – Bulawayo, Zimbabwe. 1960 – 1r – us UF Libraries [960]
O deseseis de dezembro – Ceara: Typ Constitucional, 05 out-nov 1839; fev-mar, 02 set 1840 – (= ser Ps 19) – mf#P18B,03,21 – bl Biblioteca [079]
O desespero – Bahia, maio 1878 – (= ser Ps 19) – bl Biblioteca [079]
Le desespoir de jocrisse : ou les folies d'une journee piece comique en un acte / Doin, Ernest – Montreal: Librairie Beauchemin ltee, [entre 1908 et 1924] (mf ed 1985) – 1mf – 9 – mf#SEM105P460 – cn Bibl Nat [830]
Le desespoir de jocrisse ou les folies d'une journee : piece comique en un acte / Doin, Ernest – Montreal?: Beauchemin, 188-? – 1mf – 9 – mf#11787 – cn Canadiana [820]
Le desespoir d'une jeune mere : experience de la vie reelle / Morel de la Durantaye, A – Montreal: Impr Jacques-Cartier, 1896 – 1mf – 9 – mf#04704 – cn Canadiana [971]
Deset, Enoch *see* Banque agricole and fonciere d'haiti
O desfalque – Santos, SP: Typ do Diario de Santos, 03 jun 1877 – (= ser Ps 19) – mf#P18,01,111 – bl Biblioteca [079]
Desfile de gobernadores de puerto rico / Todd, Roberto Henry – San Juan, Puerto Rico. 1943 – 1r – us UF Libraries [972]
Desforges, Pierre-Jean-Baptiste *see* Sourd
Desgodins, C H *see* Le thibet d'apres la correspondance des missionnaires
Deshasheh / Flinders Petrie, W M – London, 1898 – (= ser Mees 15) – 6mf – 8 – €14.00 – ne Slangenburg [930]
Deshasheh, 1897 / Petrie, W M – London, 1898 – 3mf – 9 – mf#NE-20346 – ne IDC [956]
Deshayes, Charles *see* Histoire de l'abbaye royale de jumieges
Deshdoot – Allahabad, India. 3 Oct 1948-18 Jun 1950 – 1r – 1 – us L of C Photodup [079]
Deshler chronicle – Deshler, NE: N Ernest Bottom. 7v. v1 n1. jun 23 1899-v7 n36. apr 27 1906 (wkly) [mf ed with gaps] – 1r – 1 – (cont by: chronicle) – us NE Hist [071]
Deshler citizen – Deshler, NE: Chas G Low, 1893-v4 n33. mar 19 1897 [wkly] [mf ed nov 1 1895-mar 191897 with gaps] – 1r – 1 – (absorbed: people's champion [hebron, ne]) – us NE Hist [071]
The deshler rustler – Deshler, NE: Jas Pontius, 1906 (wkly) [mf ed with gaps] – 1 – (cont: chronicle. vol numbering irregular: v17 repeated) – us NE Hist [071]
Deshoulieres, Antoinette du L de la G *see* Oeuvres de madame et de mademoiselle deshoulieres
Deshpande, R R *see* Kalidas's meghadutam
Deshusses, J *see* Le sacramentaire gregorien
Desiat let na kooperativnoi rabote : vologod obshchestvo selskogo khoziaistva (1908-1918) / Shevtsov, A – Vologda, 1918 – 55p 1mf – 9 – mf#COR-143 – ne IDC [335]
Desiat' let raboty moskovskogo gorodskogo banka / Moskovskii Gorodskoi Bank; ed by Kumbler, G – M, 1933 – 1mf – 9 – mf#REF-96 – ne IDC [332]
Desiat let sovetskoi kooperatsii / Tikhomirov, V A – 1927 – 52p 1mf – 9 – mf#COR-190 – ne IDC [335]
Desiat let sovetskoi promyslovoi kooperatsii, 1917-1927 / Frommett, B R – 1927 – 72p 1mf – 9 – mf#COR-454 – ne IDC [335]
Desiat let trudovoi gruppy / Khiriakov, A – 1916 – 14p 1mf – 9 – mf#RPP-207 – ne IDC [325]
Desideri, I *see* Lettre...au pere ildebrand grassi
Desideri, Ippolito *see* Il tibet
Il desiderio, overo, de' concerti di varii strumenti musicali : venetia, 1594 / Bottrigari, Ercole – Berlin: M Breslauer 1924 [mf ed 19–] – (= ser Veroeffentlichungen der musik-bibliothek paul hirsch) – 2mf – 9 – mf#fiche 1111 – us Sibley [780]
Desiderius erasmus : concerning the aim and method of education / Woodward, William Harrison – Cambridge: University Press, 1904 – 1mf – 9 – 0-8370-7599-8 – mf#1986-1599 – us ATLA [370]
Desiderius erasmus of rotterdam / Emerton, Ephraim – New York: G P Putnam, c1899 – (= ser Heroes of the reformation) – 240 – 9 – 0-7905-4511-X – (incl bibl ref) – mf#1988-0511 – us ATLA [100]
Desiertos y campinas / Izaguirre, Carlos – Tegucigalpa, Mexico. 1950 – 1r – us UF Libraries [972]
Design – Washington DC 1899-1977 – 1,5,9 – (cont by: design for arts in education) – ISSN: 0011-9253 – mf#824.02 – us UMI ProQuest [700]
Design – London, England 1949-94 – 1,5,9 – ISSN: 0011-9245 – mf#1283 – us UMI ProQuest [740]
Design / Victoria and Albert Museum. London – (= ser Fine art and design in the victoria and albert museum) – 172mf – 9 – $1300.00 – 0-907006-45-0 – (over 10,000 reproductions; with printed ind) – uk Mindata [700]
Design abstracts international – Oxford, England 1977-81 – 1,5,9 – ISSN: 0145-2118 – mf#49258 – us UMI ProQuest [020]
Design adn development of a solar powered heliotropic fluid / Wiggins, David Bruce – s.l, s.l? 1975 – 1r – us UF Libraries [530]
Design and construction of induction coils / Collins, A Frederick – New York, NY. 1908 – 1r – us UF Libraries [530]
Design and darwinism / Carmichael, James – Toronto: Hunter, Rose, 1880 – 1mf – 9 – mf#05849 – cn Canadiana [210]
The design and implementation of a program of ministry to non-participating resident members of berea baptist church / Jenkins, Floyd Thomas, jr – 1981 – 1 – 5.00 – us Southern Baptist [242]
The design and importance of christian baptism / Hall, Alexander Wilford – New York: Thomas Holman, 1848 – 1mf – 9 – 0-524-03155-X – mf#1990-4604 – us ATLA [242]
Design centre selection – [mf ed 1984] – 159mf – 9 – $875.00 – 1-900853-16-7 – (7000 british-made products with names, specifications, dimensions, prices, manufacturers & designers; incl design council awards since 1973; with printed ind) – uk Mindata [740]
Design cost & data – Tampa FL 1991-95 – 1,5,9 – (cont: design cost & data for management of building design; cont by: design cost data) – ISSN: 1054-3163 – mf#9904.05 – us UMI ProQuest [720]
Design cost data – Tampa FL 1997+ – 1,5,9 – (cont: design cost & data) – ISSN: 1093-846X – mf#9904,05 – us UMI ProQuest [720]
Design cost & data for management of building design – Tampa FL 1983-89 – 1,5,9 – (cont: design cost & data for building design management; cont by: design cost & data) – ISSN: 0739-3946 – mf#9904.05 – us UMI ProQuest [720]
Design cost & data for the construction industry – Pasadena CA 1979-80 – 1,5,9 – (cont: architectural design cost & data; cont by: design cost & data for building design management) – ISSN: 0192-0227 – mf#9904.05 – us UMI ProQuest [720]
Design criteria for lateral dikes in estuaries / Berger, Rutherford C – Vicksburg MS: US Army Corps of Engineers...1993 [mf ed 1994] – 1mf – 9 – us GPO [627]
Design for arts in education – Washington DC 1978-92 – 1,5,9 – (cont: design; cont by: arts education policy review) – ISSN: 0732-0973 – mf#824.02 – us UMI ProQuest [700]
"Design" in nature : replies to the christian guardian and christan advocate / Pringle, Allen – Toronto: s.n, 1881 – 1mf – 9 – mf#12172 – cn Canadiana [210]
Design issues – Cambridge MA 1988+ – 1,5,9 – ISSN: 0747-9360 – mf#17786 – us UMI ProQuest [740]
Design management – Shawnee Mission KS 1989-92 – 1,5,9 – (cont by: design technologies) – ISSN: 1042-8534 – mf#14112.04 – us UMI ProQuest [600]
Design news – New York NY 1961+ – 1,5,9 – ISSN: 0011-9407 – mf#1407 – us UMI ProQuest [600]
The design of baptism, viewed in its doctrinal relations : the leading passages in which it is taught exegetically treated and explained / Kirtley, James Addison – Cincinnati: Geo E Stevens, 1873 – 1mf – 9 – 0-524-07692-8 – mf#1991-3277 – us ATLA [242]
The design of buildings / Woodley, William F – London 1894 – (= ser 19th c art & architecture) – 2mf – 9 – mf#4.2.1017 – uk Chadwyck [720]
Design of god in blessing us / Styles, John – London, England. 1812 – 1r – us UF Libraries [240]
Design of solvents for extractive distillation / Dyk, Braam van – Stellenbosch: U of Stellenbosch 1998 [mf ed 1998] – 8mf – 9 – mf#mf.1288 – sa Stellenbosch [660]
Design of supersonic nozzles for viscous flow of gas mixtures with variabble properties / Merwe, Deon van der – Stellenbosch: U of Stellenbosch 1991 [mf ed 1998] – 6mf – 9 – mf#mf.1291 – sa Stellenbosch [620]
Design of the death of christ explained / Ward, William – London, England. 1820 – 1r – us UF Libraries [240]
Design of the times : official publication / Friends of Janet – v1 n1-v2 n2 [1995 oct 23-1997 mar 20] – 1r – 1 – mf#3958943 – us WHS [071]
Design quarterly – Cambridge MA 1946-96 – 1,5,9 – ISSN: 0011-9415 – mf#536 – us UMI ProQuest [740]
Design studies – Oxford, England 1979+ (1,5,9) – 1,5,9 – ISSN: 0142-694X – mf#17222 – us UMI ProQuest [720]
Design technologies – Shawnee Mission KS 1992-93 – 1,5,9 – (cont: design management) – ISSN: 1066-7504 – mf#14112,04 – us UMI ProQuest [600]
Design today – High Point NC 1985-86 – 1,5,9 – mf#15114 – us UMI ProQuest [740]
Design von locus- und gewebespezifischen retroviralen vektoren fuer eine in vivo gentherapie / Saller, Robert Michael – (mf ed 1995) – 2mf – 9 – €40.00 – 3-8267-2151-9 – mf#DHS 2151 – gw Frankfurter [574]
Designing and painting scenery for the theatre / Melvill, Harald – London, England. 1948 – 1r – us UF Libraries [790]
Designing procedures for the provision of new facilities for central baptist association / Overton, Carl McKinely – 1982 – 1 – 5.00 – us Southern Baptist [242]
Designs and sketches for furniture / Smith, Bernard E – London [1875] – (= ser 19th c art & architecture) – 1mf – 9 – mf#4.2.67 – uk Chadwyck [740]
Designs for coloured ornamental windows / Chance Bros & Co – [Paris] 1853 – (= ser 19th c art & architecture) – 1mf – 9 – mf#4.2.128 – uk Chadwyck [740]
Designs for cottage and villa architecture / Brooks, Samuel H – London [1839?] – (= ser 19th c art & architecture) – 6mf – 9 – mf#4.1.453 – uk Chadwyck [720]
Designs for cottages, cottage farms : and other rural buildings; including entrance gates and lodges / Gandy, Joseph – London: printed for John Harding, 1805 – (= ser 19th c art & architecture) – 1mf – 9 – mf#4.1.79 – uk Chadwyck [720]
Designs for country churches / Truefitt, George – London 1850 – (= ser 19th c art & architecture) – 1mf – 9 – mf#4.2.1769 – uk Chadwyck [720]
Designs for gothic ornaments and furniture / Gibbs, John – London 1854 – (= ser 19th c art & architecture) – 2mf – 9 – mf#4.2.857 – uk Chadwyck [740]
Designs for iron and brass work in the style of the 15th and 16th centuries / Pugin, Augustus Welby Northmore – London: Ackermann & Co, 1836 – (= ser 19th c art & architecture) – 1mf – 9 – mf#4.1.51 – uk Chadwyck [730]
Designs for mosaic and tessellated pavements / Jones, Owen – London 1842 – (= ser 19th c art & architecture) – 1mf – 9 – mf#4.2.763 – uk Chadwyck [740]
Designs for ornamental plate / Tatham, Charles Heathcote – London 1806 – (= ser 19th c art & architecture) – 3mf – 9 – mf#4.2.197 – uk Chadwyck [740]
Designs for parsonage houses, alms houses, etc / Hunt, Thomas Frederick – London 1827 – (= ser 19th c art & architecture) – 1mf – 9 – mf#4.2.1127 – uk Chadwyck [720]
Designs for public and private buildings / Soane, John – [London] 1828 – (= ser 19th c art & architecture) – 6mf – 9 – mf#4.2.1358 – uk Chadwyck [720]
Designs for public improvements in london and westminster / Soane, John – London 1827 – (= ser 19th c art & architecture) – 4mf – 9 – mf#4.2.1357 – uk Chadwyck [720]
Designs for rural churches / Hamilton, George E – London 1836 – (= ser 19th c art & architecture) – 1mf – 9 – mf#4.2.1129 – uk Chadwyck [720]
Designs for schools and school houses : parochial and national / Kendall, Henry Edward Jr – London 1853 – (= ser 19th c art & architecture) – 1mf – 9 – mf#4.2.1606 – uk Chadwyck [720]
Designs for the decoration of rooms / Cooper, George – London [1807] – (= ser 19th c art & architecture) – 1mf – 9 – mf#4.2.56 – uk Chadwyck [740]
Designs for the proposed new houses of parliament / Thompson, Peter – London 1836 – (= ser 19th c art & architecture) – 1mf – 9 – mf#4.2.1752 – uk Chadwyck [720]
Designs for villas / Jackson, J G – London 1829 – (= ser 19th c art & architecture) – 1mf – 9 – mf#4.2.1114 – uk Chadwyck [720]
Designs for villas and other rural buildings / Aikin, Edmund – London 1808 – (= ser 19th c art & architecture) – 1mf – 9 – mf#4.1.245 – uk Chadwyck [720]
Designs for villas in the italian style of architecture / Wetten, Robert – [London] 1830 – (= ser 19th c art & architecture) – 1mf – 9 – mf#4.2.1113 – uk Chadwyck [720]
Designs for works in stained glass – London [1864] – (= ser 19th c art & architecture) – 2mf – 9 – mf#4.2.529 – uk Chadwyck [740]
Designs from orissan temples / Ghosh, D P – Calcutta: Thacker's Press and Directories, 1950 – (= ser Samp: indian books) – (int by kim christen; text by dp ghosh, nirmal kumar bose, and yd sharma; line drawings by gopal ghose and phoni bhusan; an album of photographs publ by a goswami; produced by mv gough-govia) – us CRL [720]

Designs of cabinet and upholstery furniture in the most modern style / Whitaker, Henry – London [1825-27] – (= ser 19th c art & architecture) – 1mf – 9 – mf#4.2.31 – uk Chadwyck [740]
Designs of christian baptism / Wilkes, Lanceford Bramblet – Louisville, KY: Guide Printing & Pub Co, 1895 – 1mf – 9 – 0-524-02173-2 – mf#1990-4239 – us ATLA [242]
Designs of furniture illustrative of cabinet furniture / Shoolbred and Co, James – London 1876 – (= ser 19th c art & architecture) – 2mf – 9 – mf#4.2.833 – uk Chadwyck [740]
Designs of stoves, ranges, virandas, railings, belconets / Skidmore, G & Skidmore, M – London [1811?] – (= ser 19th c art & architecture) – 3mf – 9 – mf#4.2.698 – uk Chadwyck [730]
The designs of william burges – [London? 1886] – (= ser 19th c art & architecture) – 1mf – 9 – mf#4.1.305 – uk Chadwyck [720]
Designs. sketches / Head, Edith – 1934-65 – 1 – $162.00 – mf#0258 – us Brook [740]
The desinamamala of hemacandra / ed by Banerjee, Muralydhar – Calcutta: University of Calcutta, 1939- – (= ser Samp: indian books) – (with int, ind to text and commentary and english trans of text and extracts fr comm of hemachandra with a complete gloss of desi words from all sources with ref, derivation, and meanings) – us CRL [490]
Desinor, Yvan M *see* Bombes sur le guatemala
Le desir de voir l'hostie et les origines de la devotion au saint-sacrement / Dumoutet, E – Paris, 1926 – 2mf – 8 – €5.00 – ne Slangenburg [241]
The desire of india / Datta, Surendra Kumar – London: Young People's Missionary Movt [1908] – 1mf – 9 – 0-8370-6103-2 – (incl ind) – mf#1986-0103 – us ATLA [240]
The desire of the eyes : and other stories / Allen, Grant – New York: R F Fenno, c1895 – 4mf – 9 – mf#05021 – cn Canadiana [830]
Desjardins, Alphonse *see*
– Debats de la legislature de la province de quebec
– Debats de la legislature provinciale de la province de quebec
Desjardins, Jeannette *see* Bibliographie analytique de l'oeuvre de du docteur jean-baptiste jobin
Desjardins, Joseph *see* Guide parlementaire historique de la province de quebec, 1792 a 1902
Desjardins, Louis Georges *see*
– Debats de l'assemblee legislative de la province de quebec
– Grande assemblee a levis dimanche prochain, le 15 octobre
Desjardins, Louis-Georges *see* Considerations sur l'annexion
Desjardins, Maurice *see* Momo s'en va-t'en guerre
Desjardins, Paul *see* Catholicisme et critique
Desjardins, soeur *see* Esquisse bio-bibliographique de monsieur le notaire leonidas bachand
Desjardin's speaker's decisions / Canada. Quebec – 1v. 1867-1901; L'Assemblee Legislative, 1902 (all publ) – 13mf – 9 – $19.50 – mf#LLMC 81-072 – us LLMC [324]
Desk book for chief judges of u.s. district courts / Wheeler, Russell R – Washington: FJC, 1984 – 3mf – 9 – $4.50 – (including revision pages through nov. 1985) – mf#LLMC 95-809 – us LLMC [347]
Desk diaries / Lansing, Robert – 1915-22 – 1 – $59.00 – us L of C Photodup [920]
Desktop computing – Peterborough NH 1981-83 – 1,5,9 – ISSN: 0731-3616 – mf#12907 – us UMI ProQuest [000]
Deslandes, Germain *see* Climatologie, pedologie et ecologie forestieres au canada, 1937-1956
Deslandes, Paulin *see* Enfant du faubourg
Deslandres, Henri Alexandre *see* Etude spectrale des cometes et de leur queue
Deslauriers, Francoise *see* Bio-bibliographie analytique de m aime plamondon
Desloge, T *see* Etudes sur la signification des choses liturgiques
Desmarais, Cyprien *see*
– Ephemerides historiques et politiques du regne de louis 18 depuis la restauration
– Voyage pittoresque dans l'interieur de la chambre des deputes
Desmarais, Louis Elie *see* Album des peres du concile oecumenique du vatican commence le 8 decembre 1869
Desmarais, Odilon *see* Discours de m desmarais, depute de st-hyacinthe
Desmarest, Henri *see* La femme future
Desmarest, Pierre Marie *see* Temoignages historiques
Desmarets, S *see*
– Collegium theologicum
– Concordia discors et antichristus revelatus
– De abusu philosophiae cartesianae
– Sylloge disputationum aliquot selectiorum...
– Sylloge disputationum aliquot selectiorum
Desmazures, Adam Charles Gustave *see*
– Eglise de st francois d'assise
– Entretien sur les arts industriels
– Histoire du chevalier d'iberville, 1663-1706
– M faillon, pretrede st sulpice
– M flavien martineau, pretre de st sulpice
– Mr e picard pretre de saint-sulpice
Desmond, Humphrey Joseph *see*
– The apa movement
– Mooted questions of history
Desmoulins, Camille *see* Le vieux cordelier de camille-desmoulins
Desnitsa / Russkii Fond – Moscow, Russia. n0 (mar 1996)-n1(iiul 1996), n1(1997), n3(1997)-n7(1997), suppl n1[3](1997)-n2[4](1997), n1(1998) – 1 – mf#mf-12248 (reel 2) – us CRL [077]
Desnoes, Edmundo *see*
– Cataclismo
– Memorias del subdesarrollo
– No hay problema
Desnoes, Gustave *see* Dictionnaire de mekeo
Desnoyer, Charles *see*
– Casimir,
– Julien et justine
Desnoyers, Paul-Henri *see* Le patronage
Desobediencia, estudio de este delito / Tejera Y Garcia, Diego Vicente – Habana, Cuba. 1933 – 1r – us UF Libraries [972]
Le desoeuvre : ou l'espion du boulevard du temple / Mayeur de Saint-Paul, Francois – Londres [i.e. Paris 1781 [mf ed Hildesheim 1995-98] – 1v on 1mf – 9 – €40.00 – ISBN-10: 3-487-25880-3 – ISBN-13: 978-3-487-25880-5 – gw Olms [944]
Desolate marches / Nesbitt, Ludovico Mariano – London, England. 1935 – 1r – us UF Libraries [972]
Desordem / Santa Rosa, Virginio – Rio de Janeiro, Brazil. 1932 – 1r – us UF Libraries [972]
Les desordres de l'amour : british library, ref. 12511 / Villedieu, Marie C H des Jardins – Paris: Claude Barbin, 1676 – 1r – 1 – mf#96652 – uk Microform Academic [810]
Desoto : first tourist to lee county / Borchardt, Bernard F – s.l, s.l? 1936 – 1r – us UF Libraries [978]
Desoto county – s.l, s.l? 193-? – 1r – us UF Libraries [978]
Desoto county news – Arcadia, FL. 1905 aug 11-1924 oct – 5r – (gaps) – us UF Libraries [071]
Desoto pilot – DeSoto, NE: John E Parrish, apr? 1857-sep? 1858// (wkly) – 1r – 1 – (not publ jun 20, aug 8-22, oct 10-24 and nov 24 1857. chiefly advertisements) – us L of C Photodup [071]
DeSoto republican – Desoto WI. 1870 dec 15-1872 jan 18 – 1r – 1 – mf#1003412 – us WHS [071]
Despair and hope / Donna, Rose Bernard – Washington, DC. 1948 – 1r – us UF Libraries [025]
Despard, G *see* Memory
Despatch *see* The uitenhage times
Despatch from sir john thompson on canadian copyright, may, 1894 : with notes and observations on each paragraph – [S.l: s.n, 1894?] [mf ed 1981] – 1mf – 9 – mf#24883 – cn Canadiana [346]
A despatch from the right honorable lord glenelg, his majesty's secretary of state for the colonies : to his excellency sir francis bond head, lieutenant governor of upper canada: containing his majesty's answer to the separate addresses and representations... – [Kingston, Ont: s.n.], 1836 [mf ed 1985] – 1mf – 9 – 0-665-39227-3 – mf#39227 – cn Canadiana [971]
Despatches and other documents on the subject of the intercolonial railway – Quebec: Hunter, Rose & Lemieux, 1863 [mf ed 1987] – 1mf – 9 – mf#SEM105P647 – cn Bibl Nat [380]
Despatches from the ssem / South Sea Evangelical Mission. Melbourne and Sydney – n1-128. mar 1932-jul 1956 – 1r – 1 – (incl early iss publ as: prayer notes. available for ref) – mf#pmb doc440 – at Pacific Mss [240]
Despatches from u.s. consular representatives in puerto rico, 1821-1899 / U.S. Dept of State – (= ser General records of the department of state) – 31r – 1 – mf#M76 – us Nat Archives [327]
Despatches from US consuls in [...] : (t publications : t1 through t492) / U.S. Dept of State – (= ser General records of the department of state) – 1 – (paris, france 1790-1906 32r t1. havana, cuba 1783-1906 133r t20. winnipeg, canada 1869-1906 10r t24. lauthala, fiji islands 1844-1890 7r t25. apia, samoa 1843-1906 27r t27. kingston jamaica, british west indies 1796-1906 40r t31. san jose, costa rica 1852-1906 7r t35. cairo, egypt 1864-1906 24r t41. alexandria, egypt 1835-73 7r t45. boma, congo 1888-95 1r t47. bay of islands and auckland, new zealand 1839-1906 13r t49. santiago de cuba, cuba 1799-1906 17r t55. santo domingo, dominican republic 1837-1906 19r t56. stettin, germany 1830-1906 10r t59. tamatave, madagascar 1853-1906 11r t60. tangier, morocco 1797-1906 27r t61. maracaibo, venezuela 1824-1906 20r t62. genoa, italy 1799-1906 13r t64. butaritari, gilbert islands 1888-92 1r t89. ponape, caroline islands 1890-92 1r t90. noumea, new caledonia island 1887-1905 2r t91. newcastle, australia 1887-1906 6r t92. lahaina, hawaii 1850-71 3r t101. melbourne, australia 1852-1906 16r t102. saigon, vietnam 1889-1906 1r t103. petropavlovsk, russia 1875-8 1r t104. levuka and suva, fiji islands 1891-1906 1r t108. iloilo, philippine islands 1878-86 1r t109. brunei, borneo 1862-68 1r t110. amoor river, russia 1856-74 2r t111. vancouver, canada 1890-1906 5r t114. talcahuano, chile 1836-95 5r t115. bogota, colombia 1851-1906 4r t116. barcelona, spain 1797-1906 15r t121. christiania, norway 1869-1906 5r t122. hobart, australia 1842-1906 4r t127. st john's, newfoundland, canada 1852-1906 9r t129. victoria, canada 1862-1906 16r t130. hilo, hawaii 1853-72 4r t133. rio grande do sul, brazil 1829-97 7r t145. trinidad, west indies federation, 1824-1906 11r t148. galveston, texas 1832-46 2r t151. san juan del sur, nicaragua 1847-81 4r t152. texas 1825-44 1r t153. tetuan, morocco 1877-88 1r t156. berlin, germany 1865-1906 27r t163. bordeaux, france 1783-1906 13r t164. bradford, uk 1865-1906 8r t165. brussels, belgium 1863-1906 4r t166. london, uk 1790-1906 64r t168. lyon, france 1829-1906 14r t169. milan, italy 1874-1906 4r t170. lourenco marques, mozambique, portuguese africa 1854-1906 6r t171. rio de janeiro, brazil 1811-1906 33r t172. st. george, british west indies 1878-1906 1r t173. turin, italy 1877-1906 2r t174. lisbon, portugal 1791-1906 11r t180. antwerp, belgium 1802-1906 14r t181. bilbao, spain 1791-1875 1r t183. bremen, germany 1794-1906 21r t184. bristol, uk 1792-1906 16r t185. cadiz, spain 1791-1904 20r t186. cagliari, italy, 1802-25 1r t187. aleppo, syria 1835-40 1r t188. canea, crete, greece 1832-74 2r t190. cape town, cape colony 1800-1906 22r t191. cartagena, colombia 1822-1906 14r t192. colon, panama 1852-1906 19r t193. constantinople, turkey 1820-1906 24r t194. copenhagen, denmark 1792-1906 11r t195. cork, ireland 1800-1906 12r t196. curacao, netherlands west indies 1793-1906 13r t197. dublin, ireland 1790-1906 11r t199. dundee, uk 1834-1906 8r t200. elsinore, denmark 1792-1874 6r t201. falmouth, british west indies 1790-1905 12r t202. fayal, azores 1795-1897 11r t203. florence, italy 1824-1906 10r t204. funchal, madeira, portugal 1793-1906 9r t205. gibraltar, spain 1791-1906 17r t206. glasgow, uk 1801-1906 12r t207. guadeloupe, french west indies 1802-1906 8r t208. guayaquil, ecuador 1826-1906 13r t209. hamburg, germany 1790-1906 35r t211. havre, france 1789-1906 21r t212. hesse-cassel, germany 1835-69 3r t213. leghorn, italy 1793-1906 10r t214. leipzig, germany 1826-1906 12r t215. londonderry, uk 1835-76 3r t216. malaga, spain, 1793-1906 17r t217. malta 1801-1906 13r t218. manchester, uk 1847-1906 7r t219. marseilles, france 1790-1906 20r t220. montreal, canada 1850-1906 22r t222. nantes, france 1790-1906 8r t223. naples, italy 1796-1906 12r t224. new orleans, louisiana 1r 1798-1807 t225. paramaribo, brazil 1799-1897 8r t226. plymouth, uk 1793-1906 7r t228. puerto cabello, venezuela 1823-1906 12r t229. stockholm, sweden 1810-1906 9r t230. rome, italy 1801-1906 20r t231. rotterdam, netherlands 1802-1906 13r t232. st. croix, virgin islands 1791-1876 8r t233. st christophe) – us Nat Archives [327]
Despatches from us consuls in [...] : (m publications : m9 through m199) / U.S. Dept of State – (= ser General records of the department of state) – 1 – (cap haitien, haiti 1797-1906 17r m9. buenos aires, argentina 1811-1906 27r m70. montevideo, uruguay 1821-1906 15r m71. st. bartholomew, french west indies 1799-1899 3r m72. st petersburg, russia 1803-1906 18r m81. la guaira, venezuela 1810-1906 23r m84. amoy, china 1844-1906 15r m100. canton, china 1790-1906 20r m101. chefoo, china 1863-1906 9r m102. chinkiang, china 1864-1902 7r m103. chungking, china 1896-1906 1r m104. foochow, china 1849-1906 10r m105. hangchow, china, 1904-06 1r m106. hankow, china 1861-1906 8r m107. hong kong, 1844-1906 21r m108. macao, china 1849-69 2r m109. nanking, china 1902-06 1r m110. ningpo, china 1853-96 7r m111. shanghai, china 1847-1906 53r m112. swatow, china 1860-81 4r m113. tientsin, china 1868-1906 8r m114. newchwang, manchuria, china 1865-1906 7r m115. tamsui, formosa 1898-1906 1r m117. nagasaki, japan 1860-1906 7r m131. kanagawa, japan 1861-97 22r m135. yokohama, japan 1897-1906 5r m136. monterey, upper california 1834-48 1r m138. panama city, panama 1823-1906 7r m139. buenaventura, colombia 1867-85 1r m140. liverpool, united kingdom 1790-1906 55r m141. acapulco, mexico 1823-1906 8r m143. honolulu, hawaii 1820-1903 22r m144. valparaiso, chile 1812-1906 14r m146. venice, italy 1830-1906 7r m153. lima, peru 1823-54 6r m154. callao, peru 1854-1906 17r m155. mazatlan, mexico, 1826-1906 7r m159. frankfort on the main, germany, 1829-1906 30r m161. monterrey, mexico 1849-1906 7r m165. seoul, korea 1886-1906 2r m167. bombay, india 1838-1906 8r m168. monrovia, liberia, 1852-1906 7r m169. grand bassa, liberia 1868-1882 1r m171. sydney, new south wales, australia, 1836-1906 18r m173. veracruz, mexico 1822-1906 18r m183. ciudad juarez (paso del norte), mexico, 1850-1906 6r m184. santa fe, new mexico, 1830-46 1r m199) – us Nat Archives [327]
Despatches from US consuls in [...] (contd) : (m publications : m280 through m486) / U.S. Dept of State – (= ser General records of the department of state) – 1 – (nuevo laredo, mexico 1871-1906 4r m280. matamoras, mexico 1826-1906 12r m281. la paz, mexico 1855-1906 5r m282. nogales, mexico 1889-1906 4r m283. guaymas, mexico 1832-96 5r m284. aguascalientes, mexico 1901-06 1r m285. campeche, mexico 1820-80 1r m286. merida, 1843-97, and progreso, mexico 1897-1906 4r m287. camargo, mexico 1870-80 1r m288. chihuahua, mexico 1830-1906 3r m289. durango, mexico 1886-1906 1r m290. ensenada, mexico 1888-1906 1r m291. guerrero, mexico 1871-88 1r m292. hermosillo, mexico 1905-06 1r m293. jalapa enriquez, mexico 1905-06 1r m294. manzanillo, mexico 1855-1906 2r m295. mexico city, mexico 1822-1906 15r m296. mier, mexico 1870-78 1r m297. minatitlan, mexico 1853-81 2r m298. piedras negras, mexico 5r 1868-1906 m299. saltillo, mexico 1876-1906 1r m300. san blas, mexico 1837-92 1r m301. san luis potosi, mexico 1869-86 1r m302. tabasco, mexico 1832-74 2r m303. tampico, mexico 1824-1906 8r m304. tehuantepec, mexico 1850-67 1r m305. tuxpan, mexico 1879-1906 2r m306. zacatecas, mexico 1860-84 1r m307. ciudad del carmen, mexico 1830-72 1r m308. oaxaca, mexico 1869-78 1r m328. san dimas, mexico 1871-73 1r m442. amsterdam, the netherlands 1790-1906 7r m446. antung, manchuria, china 1904-06 1r m447. bangkok, siam 1856-1906 6r m448. batavia, java, netherlands east indies, 1818-1906 6r m449. calcutta, india 1792-1906 7r m450. colombo, ceylon, 1850-1906 4r m451. hakodate, japan 1856-78 1r m452. jerusalem, palestine, 1856-1906 5r m453. mahe, seychelles islands, indian ocean 1868-88 1r m454. manila, philippine islands 1817-99 6r m455. moscow, russia 1857-1906 2r m456. mukden, manchuria, china 1904-06 1r m457. novorossisk, russia, 1883-84 1r m458. odessa, russia, 1831-1906 7r m459. osaka and hiogo (kobe), japan 1868-1906 6r m460. padang, sumatra, netherlands east indies 1853-98 1r m461. port louis, mauritius, mascarene islands, indian ocean 1794-1805, 1817-1906 8r m462. st denis, reunion island, mascarene islands, indian ocean 1880-92 1r m463. singapore, straits settlements 1833-1906 16r m464. tahiti, society islands, french oceania 1836-1906 5r m465. tripoli, libya 1796-1885 7r m466. warsaw, poland, 1871-1906 3r m467. zanzibar, british africa 1836-1906 5r m468. in archangel, russia 1833-61 1r m481. batum, russia 1890-1906 1r m482. helsingfors, finland 1851-1906 1r m483. reval, estonia, 1859-70 1r m484. riga, latvia 1811-72 and 1890-1906 1r m485. vladivostok, russia 1898-1906 1r m486) – us Nat Archives [327]
Despatches from US consuls in [...] (contd) : (t publications : t502 through t711) / U.S. Dept of State – (= ser General records of the department of state) – 1 – (aarau, switzerland 1898-1902 1r t502. aden 1880-1906 3r t503. alexandretta, turkey 1896-1906 1r t504. antofagasta, chile 1893-1906 1r t505. baghdad, iraq 1888-1906 2r t509. bamberg, germany 1892-1906 1r t510. baracoa, cuba 1827-46, 1878-99 3r t511. barranquilla, colombia 1883-1906 6r t512. belgrade, serbia 1883-1906 1r t513. belleville, canada 1878-1906 2r t514. bern, switzerland 1882-1906 3r t528. brockville, canada 1885-1906 1r t530. budapest, hungary 1876-1906 4r t531. breslau, poland 1878-1906 3r t532. burslem, uk 1905-06 1r t533. piraeus, greece 1864-74 3r t534. campbellton, canada 1897-1906 1r t535. plauen, germany 1887-1906 3r t536. cannes, france 1891 1r t537. cape gracias a dios, nicaragua 1903-06 1r t538. gloucester, uk 1879-86 1r t539. carlsbad, czechoslovakia 1902-06 1r t540. carlsruhe, germany 1854-74 3r t541. otranto, italy 1861-67 1r t542. castellammare di stabia, italy 1878-1906 3r t543. catania, italy 1883-1906 4r t544. ceiba, honduras 1902-06 1r t545. chatham, canada 1879-1906 3r t546. chaudiere junction, canada 1898-1905 1r t547. cienfuegos, cuba 1876-1906 8r t548. clifton, canada 1864-1906 7r t549. coaticook, canada 1864-1906 4r t550. coburg, germany 1898-1906 2r t551. cognac, france 1883-98 2r t552. collingwood, canada 1879-1906 3r t553. san andres, colombia 1870-78 1r t554. cologne, germany 1876-1906 5r t555.

colonia, uruguay 1870-1906 1r t556. cordoba, argentina 1870-1906 1r t557. cornwall, canada 1901-06 2r t558. crefeld, germany 1878-1906 5r t559. dawson city, canada 1898-1906 4r t560. dunfermline, uk 1877-1906 3r t561. dusseldorf, germany 1881-1906 3r t562. erfurt, germany 1892-94 1r t563. st. ubes, portugal 1835-42 1r t564. eibenstock, germany 1902-06 1r t565. elberfeld, lubeck, and rostock, germany 1804-49, 1883-89 2r t566. porto principe and xibara, cuba 1828-43 1r t567. erzerum, turkey 1895-1904 2r t568. freiburg, germany 1892-1906 2r t569. galway, ireland 1834-63 1r t570. garrucha, spain 1877-97 1r t571. glauchau, germany 1891-1906 2r t572. goree dakar, french africa 1883-1906 2r t573. grenoble, france 1893-1906 1r t574. grenville, canada 1904-06 1r t575. guelph, canada 1883-1906 2r t578. harput, turkey 1895-1906 1r t579. mannheim, germany 1874-1906 7r t582. cardenas, cuba 1843-45, 1879-98 5r t583. san juan de los remedios, cuba 1844-46, 1879-98 2r t584. ludwigshafen am rhein, germany 1858-74 1r t585. nuevitas, cuba 1842-47, 1892-98 1r t588. amapala, honduras 1873-86 1r t589. amherstburg, canada 1882-1906 2r t590. horgen, switzerland 1882-98 3r t591. annaberg, germany 1882-1906 3r t592. huddersfield, uk 1890-1906 1r t593. hull, uk 1879-1906 4r t594. iquique, chile, 1877-1906 5r t595. jerez de la frontera, spain 1903-06 1r t596. kehl, germany 1882-1906 5r t597. koeningsberg (formerly east prussia, germany, now kaliningrad, ussr), 1879-81 1r t598. caracas, venezuela, 1868-85 1r t599. paita, peru 1833-74 3r t600. limoges, france 1887-1906 4r t601. edinburgh, uk 1893-1906 1r t602. lindsay, canada 1891-92 1r t603. london, canada 1885-1906 2r t604. lucerne, switzerland 1902-06 1r t605. luxembourg city, luxembourg 1893-96 1r t607. manzanillo, cuba 1844-46 1r t613. madrid, spain 1882-84, 1891-1906 2r t632. magdeburg, germany 1890-1906 2r t633. managua, nicaragua 1884-1906 5r t634. mayence (mainz), germany 1871-1906 7r t635. moncton, canada 1885-1905 2r t636. morrisburg, canada 1882-1901 1r t637. muscat, oman 1880-1906 2r t638. carlisle, uk 1867-69 1r t639. nottingham, uk 1877-1906 3r t641. orillia, canada 1893-1906 1r t642. ottawa, canada 1877-1906 12r t643. palmerston, canada 1892-1900 1r t647. patras, greece 1874-1906 4r t648. peterborough, canada 1905-06 1r t649. port antonio, british west indies 1895-1906 1r t650. port hope, canada 1882-1906 2r t651. port limon, costa rica 1902-06 1r t656. port rowan, canada 1882-1906 1r t657. port said, egypt 1870-76 1r t658. port stanley, canada, and st thomas, canada 1878-1906 4r t659. pretoria, transvaal 1898-1906 3r t660. puerto cortes, honduras 1902-06 2r t661. puerto plata, dominican republic 1875-1906 3r t662. prague, czechoslovakia 1869-1906 4r t663. liberec, czechoslovakia 1886-1906 3r t664. rimouski, canada 1897-1906 1r t666. roubaix, france 1890-1906 2r t667. rouen, france 1790, 1878-1906 4r t668. chios, greece 1862-71 1r t669. samana, dominican republic 1873-1905 2r t670. st etienne, france 1877-1906 3r t672. st gall, switzerland 1878-1906 6r t673. st hyacinthe, canada 1882-1906 2r t674. st michael island, azores 1897-1906 1r t675. st stephen, canada 1882-1906 2r t676. sagua la grande, cuba 1878-1900 6r t678. sault ste marie, canada 1891-1906 1r t679. sherbrooke, canada 1879-1906 3r t680. sivas, turkey 1886-1906 2r t681. sofia, bulgaria 1901-04 1r t682. solingen, germany 1898-1905 2r t683. sorel, canada 1882-98 1r t684. stanbridge station, canada 1878-1906 2r t685. stavanger, norway 1905-06 1r t686. stratford, canada 1887-1906 2r t687. swansea, uk 1892-1906 1r t688. furth, germany 1890-98 1r t689. teneriffe, canary islands 1795-1906 10r t690. three rivers, canada 1881-1906 3r t691. trinidad, cuba 1824-76 9r t699. trebizond, turkey 1904-06 1r t700. utila, honduras 1894-1906 2r t701. wallaceburg, canada 1888-1905 1r t702. waubaushene, canada 1890-93 1r t704. weimar, germany 1893-1906 1r t705. windsor, nova scotia, canada 1872-1906 3r t706. woodstock, canada 1882-1906 1r t707. yarmouth, canada 1886-1906 3r t708. zittau, germany 1897-1906 1r t709. brusa (brousa), turkey 1837-40 1r t711) – us Nat Archives [327]

Despatches from us ministers to [...] : (m publications : m10 through m223) / U.S. Dept of State – (= ser General records of the department of state) – 1 – (chile 1823-1906 52r m10. great britain 1791-1906 200r m30. spain 1792-1906 134r m31. france 1789-1906 128r m34. russia 1808-1906 66r m35. denmark 1811-1906 28r m41. the netherlands 1794-1906 46r m42. portugal 1790-1906 41r m43. german states and germany 1799-1801, 1835-1906 107r m44. sweden and norway 1813-1906 28r m45. turkey 1818-1906 77r m46. argentina 1817-1906 40r m69. venezuela 1835-1906 60r m79. haiti 1862-1906 47r m82. italian states 1832-1906 44r m90. china 1843-1906 131r m92. dominican republic 1883-1906 15r m93. mexico 1823-1906 179r m97. brazil 1809-1906 74r m121. paraguay and uruguay 1858-1906 19r m128. japan 1855-1906 82r m133. korea 1883-1905 22r m134. liberia 1863-1906 14r m170. siam 1882-1906 9r m172. belgium 1832-1906 37r m193. central america 1824-1906 93r m219. persia 1883-1906 11r m223) – us Nat Archives [327]

Despatches from us ministers to [...] : (t publications : t30 through t728) / U.S. Dept of State – (= ser General records of the department of state) – 1 – (hawaii 1843-1900 34r t30. colombia 1820-1906 64r t33. ecuador 1848-1906 19r t50. bolivia 1848-1906 22r t51. peru 1826-1906 66r t52. switzerland 1853-1906 35r t98. austria 1838-1906 51r t157. cuba 1902-1906 18r t158. greece 1868-1906 18r t159. montenegro mar 12 1905-june 14 1906 1r t525. serbia july 5 1900-july 31 1906 1r t630. morocco 1905-06 1r t725. panama 1903-06 5r t726. rumania 1880-1906 5r t727. texas 1836-45 2r t728) – us Nat Archives [327]

Despatches received by the department of state from the u.s. commission to central and south america, 14 july 1884-26 dec 1885 / U.S. Dept of State – (= ser General records of the department of state) – 1r – 1 – mf#T908 – us Nat Archives [327]

Despencer : newsletter of the... / Spencer Family Association – 1978 sep-1986 oct – 1r – 1 – (cont: bulletin) – mf#1573059 – us WHS [929]

Desperate insecurity of the sinner and the certainty of his ultimat... / Scott, John – Hull, England. 1849 – 1r – us UF Libraries [240]

Despertador : jornal critico e noticioso – Fortaleza, CE: Typ de Odorico Cotas, 24 set 1871; 30 jul-20 set 1891 – (= ser Ps 19) – mf#P17,01.45 – bl Biblioteca [321]

O despertador – Desterro, SC: Typ de Jose Joaquim Lopes, 03 jan 1863-26 ago 1885 – (= ser Ps 19) – mf#P11A,04,07 – bl Biblioteca [079]

O despertador : diario commercial, politico, scientifico e litterario – Rio de Janeiro, RJ: Typ da Associacao do Despertador, 27 mar 1838-dez 1840; abr-18 out 1841 – (= ser Ps 19) – mf#P25,03,13-22 – bl Biblioteca [073]

O despertador : jornal politico, litterario e noticiador – Paraiba do Norte, PB: Typ Liberal Parahybana de F T Brito, 21 set 1861; abr, set 1866; fev 1869; nov 1874; nov-dez 1876; jun, jul 1877; 22 ago 1888 – (= ser Ps 19) – mf#P11B,04,03 – bl Biblioteca [073]

Despertador brasiliense – Rio de Janeiro, RJ: Typ Nacional, dez 1821 – (= ser Ps 19) – mf#P17,01,129 – bl Biblioteca [079]

O despertador constitucional – Maranhao: Typ de Torres, 14 ago 1828 – (= ser Ps 19) – mf#P17,02,50 – bl Biblioteca [972]

O despertador constitucional extraordinario – Rio de Janeiro, RJ: Typ de Silva Porto e C, 01-25 fev, maio 1825; jan 1826; maio 1827; 01 maio 1828 – (= ser Ps 19) – mf#P15,01,79 – bl Biblioteca [323]

O despertador municipal – Rio de Janeiro, RJ: Typ Torres, 11 jan-04 nov 1850 – (= ser Ps 19) – mf#P01B,05,05 – bl Biblioteca [350]

Despertador republicano : que por las letras del a b c compendia los dos compendios del primero, y segundo tomo del despertador de noticias theologicas morales co[n] varias adiciones necessarias... / Ledesma, Clemente de – en Mexico: Por Dona Maria de Benavides Viuda de Juan de Ribera, ano de 1700 – (= ser Books on religion...1543/44-c1800: doctrina cristiana, obras de devocion) – 6mf – 9 – mf#crl-48 – ne IDC [241]

Despertar de un pueblo / Geigel Polanco, Vicente – San Juan, Puerto Rico. 1942 – 1r – us UF Libraries [972]

El despido de obreros y las demas causas de determinacion del contrato de trabajo / Menayo Garcia, Alfonso – Badajoz: imp campini, 1958 – 1 – sp Bibl Santa Ana [331]

El despido del trabajador (comentarios del decreto de 26 de octubre de 1956) : manual de la empresa. 1 / Elias Perez, Alberto – 2nd ed. Madrid: editorial pizarro, 1957 – sp Bibl Santa Ana [331]

Despins, Simonne *see* Essai bibliographique

Despite everything – 1963 feb-1969 jun – 1r – 1 – mf#154177 – us WHS [071]

Despite everything – Berkeley CA 1963-69 – 1 – ISSN: 0011-9482 – mf#2518 – us UMI ProQuest [071]

Despradel I Batista, Guido *see* Duarte (bosquejo historico)

Desprez, Claude Aime *see* Retournons a paris

Desprez, Philip Soulbien *see* Daniel and john

Desptaches from u.s. ministers in [...] : (t publications) / U.S. Dept of State – (= ser General records of the department of state) – 1 – us Nat Archives [327]

Despues de la zeta / Herrera, Mariano – Havana, Cuba. 1964 – 1r – us UF Libraries [972]

Despues de mi / Pocaterra, Jose Rafael – Caracas, Venezuela. 1965 – 1r – us UF Libraries [972]

Despues del brocal / Gallardo, Luis F – Cienfuegos, Cuba. 1956 – 1r – us UF Libraries [972]

Desputationes physiologico-theologicae : de humanae generationis oeconomia: de embryologia sacra: da abortu medicali et de embryotomia: de colenda castitate / Eschbach, Alphons – Parisiis: Vict Palme, 1884 – 2mf – 9 – 0-8370-7040-6 – (incl bibl ref) – mf#1986-1040 – us ATLA [240]

Desquiron, Antoine Toussaint *see* Haitiade

Desrochers-Leduc, Lucienne *see* Bibliographie analytique de la reliure au canada francais

Desroches, Charles *see* Matieres a reflexion pour les revolutionnaires

Desroches, Joseph Israel *see*
- Catechism of private and public hygiene
- Catechisme d'hygiene privee
- Catechisme d'hygiene privee et publique
- Cholera
- L'homme et l'hygiene
- Mort apparente et mort reelle
- Preceptes de l'hygiene scolaire
- Quelques reflexions sur le bureau de sante et sur l'assainissement de montreal
- Traite elementaire d'hygiene privee

Dessalines a parle / Bellegarde, Dantes – Port-Au-Prince, Haiti. 1948 – 1r – us UF Libraries [972]

Dessauer zeitung – Dessau DE, 1962 28 sep-1970 – 1r – 1 – gw Misc Inst [074]

Dessaulles, L A *see*
- A messieurs les electeurs de la division de rougemont
- A sa grandeur monseigneur charles larocque
- Discours sur l'institut canadien
- Galilee, ses travaux scientifiques et sa condamnation
- La grande guerre ecclesiastique
- La guerre americaine, son origine et ses vraies causes
- Papineau et nelson
- Reponse honnete a une circulaire assez peu chretienne
- Six lectures sur l'annexion du canada aux etats-nis

Dessert to the true american – La Salle IL 1798-99 – 1 – mf#3521 – us UMI ProQuest [740]

Le dessin a l'ecole primaire : rapport presente a l'honorable secretaire de la province / Lefevre, Charles Albert – Quebec: C F Langlois, 1892 – 1mf – 9 – mf#08662 – cn Canadiana [370]

Dessoir, Max *see*
- Karl philipp moritz als aesthetiker
- Outlines of the history of psychology

Destin de la jeune litterature / Blanchet, Jules – Port-Au-Prince, Haiti. 1939 – 1r – us UF Libraries [972]

Destin des caraibes / Morisseau-Leroy, Felix – Port-Au-Prince, Haiti. 1941 – 1r – us UF Libraries [972]

Destinata literaria et fragmenta lusatica [...] – Luebben DE, 1738 pt1-4 1738, ca 1742 pt5-12, 1747 pt1-3 – 1 – gw Misc Inst [074]

Destination image of taiwan as an international tourism destination / Wang, Li-Shaun L – 1998 – 1mf – 9 – $4.00 – mf#RC 522 – us Kinesology [338]

The destination of works of art and the use to which they are applied : considered with regard to their influence on the genius and taste of artists, and the sentiment of amateurs = Considerations morales sur la destination des ouvrages de l'art (1815) / Quatremere de Quincy, Antoine Chrysostome – London: John Murray, 1821 – (= ser 19th c art & architecture) – 2mf – 9 – mf#4.1.29 – uk Chadwyck [700]

Les destinees du congo belge / Vermeersch, Arthur – Bruxelles: A Dewit, 1906 – 1 – us CRL [960]

Destino de un continente / Ugarte, Manuel – Madrid, Spain. 1923 – 1r – us UF Libraries [972]

Destino historico de un pueblo / Parra Caro, Julio Daniel – Tunja, Colombia. 1964 – 1r – us UF Libraries [972]

Destinon, Justus Von *see* Die chronologie des josephus

Destiny / Anglo-Saxon Federation of America – v7 n1-v9 n12 [1936 jan-1938 dec] – 1r – 1 – (cont: messenger of the covenant; cont by: destiny editorial newsletter) – mf#154178 – us WHS [071]

Destiny editorial letter – v43 n9-v48 n12 [1972 sep-1977 dec] – 1r – 1 – (cont by: destiny editorial letter service. special alert; destiny editorial letter service. news in brief) – mf#505634 – us WHS [071]

Destiny editorial letter – Merrimac MA 1939-73 – 1,5 – mf#2294 – us UMI ProQuest [240]

Destiny editorial letter service – n12-32 [1977 dec-1980 sep] – 1r – 1 – (cont: destiny editorial letter) – mf#665279 – us WHS [071]

The destiny of man viewed in the light of his origin / Fiske, John – 14th ed Boston: Houghton, Mifflin, 1889, c1884 – 1mf – 9 – 0-7905-3672-2 – mf#1989-0165 – us ATLA [210]

The destiny of mankind : or, what do the scriptures teach respecting the final condition of the human family? / Tillotson, Obadiah H – Boston: J M Usher, 1851 [mf ed 1992] – 1mf – 9 – 0-524-04280-2 – mf#1991-2064 – us ATLA [242]

Destiny of the british empire as revealed in the scriptures – London, 1865 – (= ser 19th c books on british colonization) – 1mf – 9 – mf#1.1.2769 – uk Chadwyck [230]

The destiny of the human race : a scriptural inquiry / Dunn, Henry – new rev ed. London: Simpkin, Marshall, [1872?] – 1mf – 9 – 0-7905-8645-2 – mf#1989-1870 – us ATLA [220]

Destiny of the veda in india / Renou, Louis – Delhi, India. 1965 – 1r – us UF Libraries [280]

The destitute alien in great britain : a series of papers dealing with the subject of foreign pauper immigration / White, Arnold Henry – London 1892 – (= ser 19th c british colonization) – 3mf – 9 – mf#1.1.4187 – uk Chadwyck [304]

Destitution and suggested remedies : with preface / King, T G et al – London: PS King, 1911 – (= ser Catholic Studies in Social Reform) – 1mf – 9 – 0-524-03014-6 – mf#1990-4536 – us ATLA [240]

Destouches, Nericault *see*
- Dissipateur
- Fausse agnes
- Philosophe marie

Destouches, Philippe Nericault *see* Oeuvres dramatiques

Destree, Jules *see* Le socialisme en belgique

Destree, Olivier Georges *see* The renaissance of sculpture in belgium

Destroyer / Philadelphia Resistance – v1 n2-4 [1970 sep 18-1971 nov/dec], 1971 aug – 1r – 1 – mf#721527 – us WHS [978]

Destruction of babylon : its nature and effects / Sibly, Manah – London, England. 1796 – 1r – us UF Libraries [240]

Destruction of st pierre and st vincent / Morris, Charles – Philadelphia, PA. 1902 – 1r – us UF Libraries [972]

Destruction of the last enemy considered and a tribute / Bosworth, Newton – London, England. 1831 – 1r – us UF Libraries [240]

Destructive and poor man's conservative – London, UK. 1833-34 – 1r – 1 – uk British Libr Newspaper [072]

Desty, Robert *see*
- A compendium of american criminal law
- A manual of practice in the courts of the united states
- The removal of causes from state to federal courts

Desultory exposition of an anti-british system of incendiary publication / Carey, William Paulet – London 1819 – (= ser 19th c art & architecture) – 4mf – 9 – mf#4.2.1081 – uk Chadwyck [700]

Desvelado silencio (1956-1958) / Geada, Rita – Habana, Cuba. 1959 – 1r – us UF Libraries [972]

Desvios de la naturaleza o tratado del origen de los monstruos a queva anadido su compendio de curaciones chyrurgicas... / Rivilla y Bonet y Pueyo, J – Lima, 1695 – 5mf – 9 – sp Cultura [610]

Details – [mthly] – 2r/yr – 1 – $200.00/yr – us Fairchild Micro [305]

Details historiques des tremblemens de terre arrives en italie depuis le 5 fevrier jusqu'en mai 1783 / Hamilton, William – Paris 1783 [mf ed Hildesheim 1995-98] – (= ser Fbc) – 1mf – 9 – €40.00 – 3-487-29323-4 – gw Olms [945]

Details of antient timber houses of the 15th and 16th centuries : selected from those existing at rouen, caen, beauvais... / Pugin, Augustus Welby Northmore – [London]: Ackermann & Co, 1837 – (= ser 19th c art & architecture) – 1mf – 9 – mf#4.1.1 – uk Chadwyck [720]

Details of elizabethan architecture / Shaw, Henry – London 1839 – (= ser 19th c art & architecture) – 2mf – 9 – mf#4.2.1124 – uk Chadwyck [720]

Details particuliers sur la journee du 10 aout 1792 : suivis de deux notices historiques, l'une sur s a s mgr le duc d'enghien, l'autre sur s a s mgr le prince de conti / Durand, Camille H – Paris 1822 [mf ed Hildesheim 1995-98] – 1v on 2mf – 9 – €60.00 – 3-487-26289-4 – gw Olms [944]

Detatom : roman / Sieg, Paul Eugen – Berlin: Scherl, 1944 – 1r – 1 – us Wisconsin U Libr [830]

Detchevery, Joseph *see* Resume de l'histoire du roussillon, (pyrenees-orientales)

Detector – Dublin, Ireland. 30 jan-20 may 1800 – 1/4r – 1 – uk British Libr Newspaper [072]

The detector – Dublin. Ireland. -w. 30 Jan-20 May 1800. (3 ft) – 1 – uk British Libr Newspaper [072]

Detering, Soeren *see* Technikunterstuetzung fuer virtuelle unternehmen

Die determinanten des wirtschaftlichen erfolgs in der entwicklung der angewandten kunst am beispiel vorindustrieller moebel- und glasentwuerfe bis zum 2. weltkrieg / Schnaufer-Arens, Iris – (mf ed 1997) – 4mf – 9 – €56.00 – 3-8267-2457-7 – mf#DHS 2457 – gw Frankfurter [338]

Determinants of critical power and anaerobic work capacity in young and elderly men / Overend, T J – 1992 – 2mf – 9 – $8.00 – us Kinesology [612]

Determinants of intrinsic motivation among female and male adolescent students in physical education / Ferrer Caja, Emilio – 1997 – 3mf – 9 – $12.00 – mf#PSY 2035 – us Kinesology [150]

Determinatio compendiosa de iurisdictione imperii auctore anonymo ut videtur tholomeo lucensi o p (mgh leges 4:1.bd) – 1909 – (= ser Monumenta germaniae historica leges 4. fontes iuris germanici antiqui in usum scholarum separatim editi (mgh leges 4)) – €5.00 – ne Slangenburg [342]

Determination des ascensions droites des etoiles : de culmination lunaire et de longitudes / Loewy, Maurice – [Paris: s.n. 1887?] [mf ed 1998] – 1r – 1 – mf#film mas 28401 – us Harvard [520]

Determination of control parameters in pronation curve behavior during running / Crussemeyer, Jill A – 1998 – 2mf – 9 – $8.00 – mf#PE 3882 – us Kinesology [612]

A determination of member satisfaction at three different types of fitness center / Streff, L L – 1991 – 2mf – 9 – $8.00 – us Kinesology [790]

Determination of occupational stress and coping strategies of mediators utilizing the delphi technique / Arabyazdi, Behjat – 1989 – 139p 2mf – 9 – $8.00 – us Kinesology [150]

Determination of the precision, accuracy and resolution of a video-based motion analysis system / Wisner, David M & Widule, Carol J – 1992 – 1mf – 9 – $4.00 – us Kinesology [612]

Determination of the winter survival of the cotton boll weevil by field counts / Grossman, Edgar F – Gainesville, FL. 1931 – 1r – us UF Libraries [630]

Determined Action for Women Now (Organization) *see* D a w n

Determined Action for Women Now [Organization] *see*
– Dawn
– New d a w n
– New dawn

Determining matrimonial property rights on divorce : an appraisal of the legal regimes in botswana / Quansah, Emmanuel Kwabena – 12mf – 9 – (incl bibl ref) – mf#mfm15072 – sa Unisa [346]

Determining ministry priorities for macedonia baptist church during a transitional period / Duvall, Terry Glenn – 1982 – 1 – 6.56 – us Southern Baptist [242]

Determining the essential elements of golf swings used by elite golfers / Fujimoto-Kanatani, Koichiro – Oregon State University, 1995 – 8mf – 9 – $32.00 – mf#PE 3643 – us Kinesology [612]

Determining the presence of lifeguards during competitive swimming events at mid-american conference universities / Beumer, Rebecca L – 2001 – 62p on 1mf – 9 – $5.00 – mf#PE 4202 – us Kinesology [790]

Determining the validity and reliability of the nicholas manual muscle tester as a measure of isometric strength in women with arthritis / Sierra, Nelson – Oregon State University, 1995 – 1mf – 9 – mf#PH 1508 – us Kinesology [612]

Determining types and typal profiles of adult amateur theatre participants through q-technique / Haner, Janet A S – 1989 – 129p 2mf – 9 – $8.00 – us Kinesology [790]

Detert, Richard A *see* The relationship between personality type prefernces and levels of coping resources among cardiac rehabilitation participants at the university of wisconsin-a crosse

Dethmar, Friedrich *see* Vertraute briefe auf einer reise von hannover ueber braunschweig durch die harzgegenden

Detlefsen, Hans *see* Die namengebung in den dramen der vorgaenger shakespeares

Detlev, Christian Ulrich, Freiherr von Eggers *see* Deutsches magazin

Detlev liliencron / Remer, Paul – Berlin: Schuster and Loeffler, [1904] – 1r – 1 – us Wisconsin U Libr [920]

Detlev von liliencron : vortrag gehalten in der literarhistorischen gesellschaft im januar 1910 / Litzmann, Berthold – Bonn: F Cohen, [1910?] – 1r – 1 – us Wisconsin U Libr [430]

Detmer, Heinrich *see*
– Hermanni a kerssenbroch anabaptistici furoris
– Zwei schriften des muensterschen wiedertaeufers bernhard rothmann

Detmers, Arthur C, Jr *see* Peter b porter papers in the buffalo and erie county historical society

Detmolder buergerblatt – Detmold DE, jun 28 1850-nov 6 1851 – 1 – gw Misc Inst [074]

Detracteurs de la race noire et de la republique d... – Paris, France. 1882 – 1r – us UF Libraries [972]

Detroicki dziennik ludowy – Detroit; Chicago: The Polish Peoples Pub Co, 1919-feb 1923 – 13r – us CRL [071]

Detroit and Milwaukee Railroad Co *see* Letter to the bondholders of the detroit and milwaukee railroad company

Detroit azuwer – v1 n1-20 [i.e. 21] [1919 feb 8-jul 2] – 1r – 1 – mf#681383 – us WHS [071]

Detroit Building Trades Council *see*
– Building tradesman
– Detroit michigan building tradesman

[Detroit-] cadillac manual – MI. pts 1 + 2. 1964 – 2r – 1 – $120.00 – mf#R04146 – us Library Micro [071]

Detroit college of law at michigan state *see* Law review of michigan state university

Detroit college of law at michigan state university entertainment and sports law journal *see* Entertainment and sports lawyer

Detroit college of law, entertainment and sports law forum *see* Entertainment and sports lawyer

Detroit college of law review *see* Law review of michigan state university

Le detroit de belle-isle / Fortin, Pierre – [S.l: s.n, 1877?] – 1mf – 9 – 0-665-06956-1 – mf#06956 – cn Canadiana [550]

Detroit Federation of Labor et al *see* Detroit labor news

Detroit Federation of Teachers *see* Detroit teacher

Detroit fire fighters – 1975 dec-1981 jun, 1981 jul-1984 sep, oct-1985 mar, 1986 may-1989 jun – 4r – 1 – mf#819028 – us WHS [360]

Detroit gay liberator – v1,no.5-v1,no.10. Sept-Oct 1970-Mar 1971. Detroit: Gay Liberation Front of Detroit, 1970-71. Continued by: Gay Liberator – 1 – us Wisconsin U Libr [305]

Detroit independent – Detroit MI. 1923 jan 13 – 1r – 1 – mf#819028 – us WHS [071]

Detroit Indian Center et al *see* Native sun

The detroit informer – Detroit, MI: Detroit Informer Co. v3 n7. jan 13 1900 [mf ed 1947] – (= ser Negro Newspapers on Microfilm) – 1r – 1 – us L of C Photodup [071]

Detroit institute of arts. bulletin of the detroit institute of arts – Detroit MI 1919+ – 1,5,9 – ISSN: 0011-9636 – mf#1804 – us UMI ProQuest [700]

The detroit jewish chronicle – Detroit. Mich. 1916-51 – 1 – us AJPC [071]

Detroit jewish herald – Detroit. Mich. 1927-28 – 1 – us AJPC [071]

Detroit labor news / Detroit Federation of Labor et al – 1917 mar 30/1920 dec 31-1985/88 – 24r – 1 – (cont by: metro detroit labor news) – mf#3412329 – us WHS [331]

Detroit labor news – Detroit MI 1975-97 – 1,5,9 – ISSN: 1072-1525 – mf#8724.01 – us UMI ProQuest [331]

Detroit law journal – Detroit. On film: 1906-17. Incomplete. LL-015 – 1 – us L of C Photodup [340]

Detroit law review – Detroit College of Law. v1-9. 1931-48 – 8mf – 9 – $36.00 – mf#LLMC 84-456 – us LLMC [340]

Detroit lawyer – v1-59. 1931-92 + v1-4 n2 1994-97 – 9 – $605.00 set – (temporarily suspended with v59 n2 1992, restarted with v1 1994-95) – ISSN: 0011-9652 – mf#102461 – us Hein [340]

Detroit legal news – v1-23. 1895-1916. Devoted to Michigan Court Opinions, Supreme and Lower Courts – 272mf – 9 – $4008.00 – (lacking: v1-2) – mf#LLMC 84-457 – us LLMC [340]

Detroit marine historian : journal / Marine Historical Society of Detroit – 1947 sep-1950 aug [repr], 1950 sep-1967 aug, ind v1-10 [1947 sep-1957 aug] – 1r – 1 – mf#1494410 – us WHS [978]

Detroit MI *see*
– Convoy dispatch
– Uaw-cio administrative letter

Detroit (MI). Board of Street Railway Commissioners *see* Report to hon john c lodge, mayor

Detroit (MI). Rapid Transit Commission *see* Proposed financial plan for a rapid transit system for the city of detroit

Detroit michigan building tradesman / Detroit Building Trades Council – 1955 jul 22-1957 jun 28, jul 2-1959 feb 13 – 2r – 1 – (cont by: building tradesman [detroit, mich.]) – mf#3320994 – us WHS [690]

Detroit monthly – Detroit MI 1986-96 – 1,5,9 – (cont: monthly detroit) – ISSN: 0888-0867 – mf#12097,01 – us UMI ProQuest [071]

[Detroit-] news and letters – MI. 1976-1983 – 3r – 1 – $180.00 – mf#R04390 – us Library Micro [071]

Detroit society for genealogical research magazine – Detroit MI 1937+ – 1,5,9 – ISSN: 0011-9687 – mf#7153 – us UMI ProQuest [929]

Detroit sunday journal – Detroit MI. 1998 jan-jun, jul 5/11-dec 27/jan 2, 1999 jan-jul – 3r – 1 – mf#3542499 – us WHS [071]

Detroit teacher / Detroit Federation of Teachers – v33 n7 [n314] [1974 feb 25], v35 n1-? [1975 sep 16-1983 dec 20] – 1r – 1 – mf#708638 – us WHS [370]

Detroiter – Detroit MI 1990+ – 1,5,9 – ISSN: 0011-9709 – mf#15190 – us UMI ProQuest [380]

Detroiter abend-post – Detroit MI (USA), 1922 7 nov-1926 30 sep, 1927 1 apr-1932 30 sep, 1933-1934 30 sep, 1935-1939 12 oct, 1972 – 34r – 1 – (with gaps) – gw Misc Inst [071]

Detroiti magyar ujsag = Detroit hungarian news – Detroit: Metropolitan Hungarian News, Inc, sep 1971-75 – 5r – us CRL [071]

Detroiti ujsag = Detroit hungarian news – Detroit, MI: Julius Fodor, nov 17 1933-aug 1971 – 1 – us CRL [071]

Detskaia entsiklopediia = The children's encyclopedia / ed by Vagner, Y et al – Moscow: I Sytin Publ. v1-10. 1914 – 31mf – 9 – $200.00 – us UMI ProQuest [030]

Detskoe chtenie – Spb., 1869-1874(6) – 109mf – 9 – (missing: 1869, v7-8; 1871, v12) – mf#R-2271 – ne IDC [077]

Detskoe chtenie dlia serdtsa i razuma – M., 1785-1789. v1-18 – 60mf – 9 – mf#R-2287 – ne IDC [077]

Detskoe chtenie dlia serdtsa i razuma – M., 1819. v1-18 – 50mf – 9 – mf#R-1581 – ne IDC [077]

Dettaer zeitung – Detta (Deta RO), 1921 2 oct-1939 11 nov – 4r – 1 – gw Misc Inst [077]

Dette, Guido *see* Kursbildung am deutschen aktienmarkt unter besonderer beruecksichtigung verhaltensorientierter ueberlegungen

Dette skal siges, saa vaere det da sagt / Kierkegaard, Soeren – Kobenhavn: CA Reitzels Bo og [sic] Arvinger, 1855 [mf ed 1989] – 1mf – 9 – 0-7905-7414-4 – mf#1989-0639 – us ATLA [240]

Dettman, Eduard Johann Karl *see* Brasiliens aufschwung in deutscher beleuchtung

Detzer, Johann Andreas *see* Evangelisches concordienbuch

Deuber, Walter *see* Realismus der arbeiterliteratur

Deubner blaetter : arbeitsmaterialien des zirkels schreibender arbeiter bkw "erich weinert", deuben, kreis hohenmoelsen – Halle (Saale): Mitteldeutscher Verlag, 1961 – 163p/2pl (ill) – 1 – us Wisconsin U Libr [430]

Deubner, Ludwig *see* De incubatione

Deuda sagrada / Sanchez Arjona, Vicente – Fregenal de la Sierra (Badajoz): Imprenta de Angel Verde, 1924 – 1 – sp Bibl Santa Ana [810]

Deuda sagrada / Sanchez Arjona, Vicente – Sevilla: Imprenta Alvarez, 1960 – 1 – sp Bibl Santa Ana [810]

Der deudsch psalter : mit den summarien / Luther, D M – Wittemberg: Lufft 1541 – (= ser Hqab. literatur des 16. jahrh.) – 6mf – 9 – €70.00 – mf#1541a – gw Fischer [780]

Deuel county herald – Big Springs, NE: Marie Berges. 16v. v5 n31. apr 11 1935-v20 n1. jul 29 1949 (wkly) [mf ed with gaps] – 5r – 1 – (cont: big spring news. cont by: big springs news (1935) and: panhandle press) – us NE Hist [071]

Deuel county news – Lewellen, NE: L M Warner. v1 n1. jul 3 1909-jan 1910// (wkly) [mf ed jul 3 1909-jan 1 1910 (gaps) filmed 1970] – 1r – 1 – (cont by: garden county news) – us NE Hist [071]

Deuetsche predigten des 12. und 13. jahrhunderts / ed by Roth, Karl – Quedlinburg; Leipzig: G Basse, 1839 – us Wisconsin U Libr [430]

Deugden-spoor : in de on-deughden des werelts aff-gebeeldt / Baart, Peter A – Leeuwarden: Hans Willems Coopman, 1645 – 5mf – 9 – mf#O-3206 – ne IDC [090]

Deugden-spoor : in de on-deughden des werelts aff-gebeeldt / Baart, Peter A – Leeuwarden: Steffen Geerts, 1645 – 4mf – 9 – mf#O-3207 – ne IDC [090]

Deugden-spoor : in de on-deughden des werelts aff-gebeeldt... / Baart, Peter A – Leewaarden: Hans Willems Coopman, 1645 – 3mf – 9 – mf#O-131 – ne IDC [090]

Deulig scala – Berlin DE, 1921 n8 – 1 – gw Mikrofilm [074]

Deumeland, Heinrich *see* Dei aekerjagd tau vorigeslewen am baerensee

Deusinger, Ingrid M *see* Untersuchungen zum problem der normvorstellung

Deussen, Paul *see*
– Allgemeine geschichte der philosophie
– Erinnerungen an friedrich nietzsche
– The philosophy of the upanishads
– The system of the vedaanta

Deustsh-sudwet afrikaansche zeitung – Pretoria: State Library Corporate Communication, 12 oct 1898-15 oct 1914 – 9r – 1 – mf#MS00140 – sa National [079]

Die deuterocanonischen stuecke des buches esther : eine biblisch-kritische abhandlung / Langen, Joseph – Freiburg i.B: Herder, 1862 – 1mf – 9 – 0-7905-0315-8 – (incl bibl ref) – mf#1987-0315 – us ATLA [221]

Deuterographs : duplicate passages in the old testament: their bearing on the text and compilation of the hebrew scriptures / Girdlestone, Robert Baker – Oxford: Clarendon Press, 1894 – 1mf – 9 – 0-7905-1402-8 – (incl ind) – mf#1987-1402 – us ATLA [221]

Deuterojesaia : hebraeisch und deutsch mit anmerkungen / Klostermann, August – Muenchen: C H Beck, 1893 – 1mf – 9 – 0-8370-3927-4 – (incl ind of hebrew words) – mf#1985-1927 – us ATLA [221]

The deuteronomical writers and the priestly documents / Addis, William Edward – New York: G P Putnam, 1898 – (= ser Documents Of The Hexateuch) – 2mf – 9 – 0-7905-1569-5 – (incl ind to 2-vol set) – mf#1987-1569 – us ATLA [221]

Das deuteronomium / Schultz, Fr. W – Berlin: Gustav Schlawitz, 1859 – 2mf – 9 – 0-7905-2064-8 – mf#1987-2064 – us ATLA [221]

Das deuteronomium : eine schutzschrift wider modern-kritisches unwesen / Zahn, Adolf – Guetersloh: C Bertelsmann, [1890?] – 1mf – 9 – 0-7905-3239-5 – mf#1987-3239 – us ATLA [221]

Das deuteronomium : sein inhalt und seine literarische form / Staerk, Willy – Leipzig: J C Hinrichs, 1894 – 1mf – 9 – 0-8370-5357-9 – (incl bibl ref) – mf#1985-3357 – us ATLA [221]

Das deuteronomium *see* Deuteronomy

Das deuteronomium und der deuteronomiker : untersuchungen zur alttestamentlichen rechts- und literaturgeschichte / Kleinert, Paul – Bielefeld: Velhagen & Klasing, 1872 – 1mf – 9 – 0-8370-3920-7 – (incl indes) – mf#1985-1920 – us ATLA [221]

Deuteronomy : or, the fifth book of moses = Das deuteronomium / Schroeder, Friedrich Wilhelm Julius – New York: Charles Scribner, 1900, c1879 [mf ed 1985] – (= ser A commentary on the holy scriptures. old testament 3/21) – 1mf – 9 – 0-8370-4529-0 – (trans & enl by abraham gosman) – mf#1985-2529 – us ATLA [221]

Deuteronomy *see* Leviticus and numbers

Deuteronomy and joshua : introductions, revised version with notes, map and index / ed by Robinson, Henry Wheeler – New York: Oxford University Press, American Branch, [1907?] – (= ser The New-Century Bible) – 1mf – 9 – 0-524-05901-2 – (incl bibl ref) – mf#1992-0658 – us ATLA [221]

Deuther-Conrad, Winnie *see* Studien an strukturell unterschiedlichen ps ii partikeln aus spinacea oleracea l. zur wirkung von endogenem calcium auf ausgewaehlte prozesse im photosystem ii

Deutliche und moeglichst vollstaendige uebersicht ueber das theologische system dr. friedrich schleiermachers : und ueber die beurtheilungen, welche dasselbe theils nach seinen eigenen grundsaetzen, theils aus den standpunkten des supranaturalism, des rationalism, der fries'schen und der hegel'schen philosophie erhalten hat / Gess, Friedrich Wilhelm – 2. stark verm und verb Aufl. Reutlingen: Ensslin und Laiblin, 1837 – 1mf – 9 – 0-524-00366-1 – mf#1989-3066 – us ATLA [240]

Deutsch als fremdsprache – Leipzig, Germany 1978-99 – 1,5,9 – ISSN: 0323-3766 – mf#11272 – us UMI ProQuest [430]

Deutsch amerikaner / Butler Co. Hamilton – jul 10 1914- jun 30 1916 – 1r – 1 – (in german) – mf#B35448 – us Ohio Hist [071]

Deutsch amerikaner / Hamilton Co. – aug 1914-jun 1916 – 1r – 1 – (in german) – mf#B35448 – us Ohio Hist [071]

Deutsch, Emanuel *see*
– Literary remains of the late emanuel deutsch
– The talmud

Deutsch, Gotthard *see* Philosophy of jewish history

Deutsch kirchenampt : gedruckt zu erffurdt durch merten von dolgen zu den dreyen guelden kronen bey sanct joergen – Erffurdt, Merten von Dolgen [1539?] [mf ed 19–] – 1r – 1 – (also contains: so man jtzundt (gott zu lob)inn den kirchen zu singen pfleget) – mf#film 2570 – us Sibley [780]

Deutsch, Lev Grigorevich *see* Sixteen years in siberia

Deutsch, Otto Erich *see* Ferdinand kuernbergers briefe an eine freundin, 1859-1879

Die deutsch revolution / Blum, Hans – Leipzig, Germany. 1897 – 1r – us UF Libraries [943]

Deutsch, S M *see* Lehrbuch der kirchengeschichte

Deutsch, Samuel Martin *see*
– Drei actenstuecke zur geschichte des donatismus
– Lehrbuch der kirchengeschichte
– Peter abaelard

Deutsch schweizerische courier – New Glarus WI. 1897 sep 6-1901 nov 5, 1905 jan 24-1908, 1909-1912 sep 24 – 3r – 1 – (cont: new glarus bote; cont by: deutsch schweizerischer courier) – mf#1097594 – us WHS [071]

Deutsch schweizerischer courier – New Glarus WI 1912 oct-dec, 1913-1917 apr 3 – 2r – 1 – (cont: deutsch schweizerische courier) – mf#1097590 – us WHS [071]

Deutsch, Solomon *see*
– a key to the pentateuch explanatory of the text and the grammatical forms
– A new practical hebrew grammar

Deutsch-amerika / New Yorker Staats-Zeitung – 1915 feb 27/dec 4-1928 nov 3/dec 29 – 15r – 1 – (with small gaps) – mf#1055302 – us WHS [071]

Deutsch-amerikaner / German-American National Congress – 1977 aug-1981 dec, 1982-86 – 2r – 1 – (cont by: german-american journal) – mf#966183 – us WHS [071]

Deutsch-amerikaner – Neillsville WI. 1916 feb 17-1917, 1918-1920 oct 7 – 2r – 1 – mf#1097640 – us WHS [071]

Deutsch-amerikanisch illustrierte zeitung / Hamilton Co. Cincinnati – oct 16-dec 4 1886 – 1r – 1 – mf#B37533 – us Ohio Hist [071]

Deutsch-amerikanische baecker zeitung – New York: H Weismann. v11, n1-4. may 1895] – us CRL [071]

Deutsch-amerikanische baecker zeitung – New York NY (USA), 1886-91 [gaps] – 1r – 1 – gw Misc Inst [640]

Deutsch-amerikanische balladen und gedichte / Doernenburg, Emil – [Philadelphia: The author, c1933] – 1r – 1 – us Wisconsin U Libr [810]

Deutsch-amerikanische buchdrucker-zeitung : offizielles organ / Deutsch-Amerikanischen Typographie – New york. jahrg. 2, n22; 4, n1-67, n13. 15 may 1875; 1 jul 1876-jul 1940 – 1 – us NY Public [680]

Deutsch-amerikanische buergerzeitung – Chicago IL (USA), feb 8 1924-dec 26 1940 [gaps] – 1r – 1 – gw Misc Inst [071]

Deutsch-amerikanische burger-zeitung : offizielles organ des deutsch-amerikanischen burgerbundes – Forest Park, IL: H Kaul, feb 8 1924-nov 1 1927; feb 16, mar 13, jun 3-oct 1 1929: jan-aug 30 1930; nov 1930-aug 1 1931 – 1r – 1 – us CRL [071]

Deutsch-Amerikanische Typographia *see* Buchdrucker-zeitung

Deutsch-Amerikanischen Typographie *see* Deutsch-amerikanische buchdrucker-zeitung

Deutsch-amerikanischer farmer und der hausfreund – 1899 dec 13, v11 n26-v12 n26 [1900 jan 10-1901 jan 16] – 1r – 1 – mf#1055307 – us WHS [630]

Deutsch-amerikanisches conversations-lexicon (ael1/11) : mit spezieller ruecksicht auf das beduerfnis der in amerika lebenden deutschen / Schem, Alexander J – New York 1869-74 [mf ed 1993] – (= ser Archiv der europaeischen lexikographie, abt 1: enzyklopaedien) – 11v on 55mf – 9 – €500.00 – 3-89131-075-7 – gw Fischer [040]

Deutsch-asiatische warte – Tsingtau (Tsingtao VR) 1898 21 nov-1902 19 dec – 1 – gw Misc Inst [079]

Deutschbein, Max *see* Grammatik der englischen sprache auf wissenschaftlicher grundlage

Deutsch-belgische rundschau – Bruessel (B), 1931 1 jul-1934/35 15 dec – 1 – gw Misc Inst [380]

Deutsch-chinesische nachrichten – Tientsin [VR], 1930 oct-1941 2 may – 25r – 1 – (title varies: 2 oct 1939: deutsche zeitung in nordchina) – gw Misc Inst [079]

Die deutsch-daenische dichterin friederike brun : ein beitrag zur empfindsam-klassizistischen stilperiode / Olbrich, Rosa – [S.l.: s.n.], 1932 (Breslau: M C Wolf) – 1r – 1 – us Wisconsin U Libr [920]

Der deutsche : tageszeitung der christlichen gewerkschaftsbewegung – Berlin DE, 1921 apr-1935 31 jan [gaps] – 29r – 1 – (filmed by misc inst: 1925 oct-dec [1r]) – gw Mikrofilm; gw Misc Inst [331]

Der deutsche aar – Weimar, Ilmenau DE, 1935 1 jul-1940 30 jun, 1940 1 oct-1941 31 mar, 1941 1 oct-1944 – 29r – 1 – (title varies: 21 mar 1925: der nationalsozialist; 1 apr 1933: thueringische staatszeitung; 18 jan 1936: thueringer gauzeitung; 2 jan 1937: thueringer gauzeitung-der nationalsozialist; publ in ilmenau, fr 28 mar 1925 in weimar) – gw Mikrofilm; gw Misc Inst [074]

Der deutsche abrogans / ed by Baesecke, Georg – Halle/S: M Niemeyer Verlag, 1931 [mf ed 1993] – (= ser Altdeutsche textbibliothek 30) – xix/77p – 1 – (latin-old high german dictionary) – mf#8193 reel 3 – us Wisconsin U Libr [054]

Deutsche acta eruditorum : oder geschichte der gelehrten, welche den gegenwaertigen zustand der litteratur in europa begreiffen / ed by Rabener, Justus Gotthard et al – Leipzig 1712-39 [mf ed 1993] – (= ser Aus den anfaengen des zeitschriftenwesens: fruehe deutsche zeitschriften) – v1-20=t1-240 on 76mf – 9 – €1080.00 – 3-89131-095-1 – gw Fischer [410]

Das deutsche ahnenbuch / Finckh, Ludwig – Goerlitz: C A Starke 1934 [mf ed 1989] – 1r [ill] – 1 – (filmed with: double, double, toil and trouble / lion feuchtwanger) – mf#7237 – us Wisconsin U Libr [890]

Deutsche Akademie der Wissenschaften zu Berlin *see* Veroeffentlichungen der sternwarte in sonneberg

Deutsche Akademie der Wissenschaften zu Berlin. Institut fuer deutsche Sprache und Literatur *see* Studienausgaben zur neueren deutschen literatur

Deutsche akademiker-zeitung *see* Deutsche hochschulstimmen aus der ostmark

Deutsche aksum-expedition, 1906 – Berlin, 1913. 4v – 20mf – 9 – mf#NE-20310 – ne IDC [919]

Deutsche allgemeine handwerks-zeitung *see* Nordwestdeutsche handwerks-zeitung

Deutsche allgemeine zeitung – Berlin: Deutscher Verlag, mar 13-apr 19 1945 – us CRL [074]

Deutsche allgemeine zeitung – Berlin, Germany. 1914-34 – 79r – 1 – us L of C Photodup [074]

Deutsche allgemeine zeitung – Berlin: Norddeutsche Buchdruckerei und Verlagsanstalt, [1922-] Reichsausg. nov 10 1939-mar 9 1945 – 1 – us CRL [074]

Deutsche allgemeine zeitung – Berlin, Germany 12 nov 1918-19 apr 1945 – 1 – (cont: internationale [10 nov 1918]) – uk British Libr Newspaper [074]

Deutsche allgemeine zeitung – Leipzig, Germany 1 apr 1843-29 jun 1879 (daily) – 1 – (cont: leipziger allgemeine zeitung [1 oct 1837-31 mar 1843]) – uk British Libr Newspaper [074]

Deutsche allgemeine zeitung – Stuttgart DE, 1831 27 jun-1832 28 sep – 3r – 1 – (until 19 nov 1831: stuttgarter allgemeine zeitung) – gw Misc Inst [074]

Deutsche allgemeine zeitung *see* Der fortschritt

Deutsche allgemeine zeitung der russlanddeutschen *see* Freundschaft

Deutsche allgemeine zeitung / reichsausgabe – Berlin DE, 1922 sep-1923 apr, 1925 1 oct-1945 28 feb – 1 – gw Misc Inst [074]

Deutsche Angestelltenschaft *see*
– Rechenschaftsbericht
– Schriften des dhv

Deutsche Angestelltenschaft. Gau Bayern *see* Standes-rundschau und jahresbericht

Deutsche Angestelltenschaft. Gau Brandenburg-Pommern *see* Jahresbericht

Deutsche arbeit – Neusatz (Novi Sad YU), 1943 1 jan-24 dec – 1r – 1 – gw Misc Inst [331]

Deutsche arbeit 1936 – Berlin DE, 1936-37, 1939-40, 1942 n11, 12 – 1 – gw Misc Inst [331]

Deutsche arbeit in amerika : erinnerungen / Francke, Kuno – Leipzig: F Meiner, 1930 – 1r – 1 – us Wisconsin U Libr [880]

Deutsche arbeit in rio grande do sul / Porto, Aurelio – Sao Leopoldo, Brazil. 1934 – 1r – us UF Libraries [331]

Deutsche arbeiten der universitaet koeln / ed by Bertram, Ernst & Leyen, Friedrich von der – Jena: E Diederich, 1930-1941 [mf ed 1993] – 17v – 1 – (individual titles also listed separately) – mf#8215 – us Wisconsin U Libr [430]

Der deutsche arbeiter – Chicago IL (USA), 1869 28 aug-1870 1 aug – 1r – 1 – gw Misc Inst [071]

Der deutsche arbeiter in politik und wirtschaft – Berlin. no. 1-4. 1925 – 1 – us NY Public [331]

Deutsche arbeiterhalle – Hannover DE, 1851 jan-jun – 1 – gw Misc Inst [331]

Deutsche arbeiterhalle – Mannheim DE, 1867-68 [gaps] – 1 – gw Misc Inst [331]

Die deutsche arbeiterin – v1-12. 1909-20,16 [mf ed 2003] – (= ser Hq 55) – 67mf – 9 – €340.00 – 3-89131-441-8 – (filmed with: deutsche arbeiterin [v2-14 1921-33]; wort und werk [v15-20 1934-1939,6]) – gw Fischer [331]

Deutsche arbeiterpresse – Wien (A), Muenchen DE, 1928 28 jul-1933 2 dec – 1r – 1 – gw Misc Inst [331]

Deutsche arbeiter-zeitung – Berlin DE, 1848 8 apr-24 jun – 1r – 1 – gw Misc Inst [331]

Deutsche arbeiter-zeitung : organ der deutschen arbeiter und farmer in kanada – Winnipeg. v1 n5-v8 n222. oct 1930-jul 14 1937//? – 2r – 1 – Can$145.00 – cn McLaren [071]

Deutsche arbeiter-zeitung – Duchcov, Czechoslovakia. Aug 1922-Apr 1924 – 1r – 1 – (scattered issues) – us L of C Photodup [077]

Deutsche arbeiter-zeitung – Usti Nad Labem, Czechoslovakia. 1904-05 – 1r – 1 – us L of C Photodup [077]

Deutsche arbeitgeber-zeitung : zentralblatt der deutschen arbeitgeber-verbaende – Berlin, Oct 1905-06; 1909; 1911-Sep 1922 – 8r – 1 – gw Mikropress [074]

Die deutsche arbeitgeber-zeitung : zentralblatt der deutschen arbeitgeber-verbaende – Berlin DE, 1902 5 oct-1903 27 dec, 1910 2 jan-28 aug – 1r – 1 – (filmed by mikropress: 1905 oct-1906, 1909, 1911-1922 24 sep [8r] order#6651) – gw Mikrofilm; gw Mikropress [331]

Deutsche architekturbuecher zur zivilbaukunst aus dem 16. und 17. jahrhundert = German architectural books on civil engineering of the 16th and 17th century / ed by Schuette, Ulrich – [mf ed 2000] – (= ser Nachschlagewerke und quellen zur kunst 3) – 121mf (1:24) – 9 – silver €1896.00 – ISBN-10: 3-598-34548-8 – ISBN-13: 978-3-598-34548-7 – gw Saur [720]

Deutsche architekturbuecher zur zivilbaukunst des 18. jahrhunderts = German architectural books on civil engineering of the 18th century / ed by Schuette, Ulrich – [mf ed 2003-04] – (= ser Nachschlagewerke und quellen zur kunst 5) – 2pt on 389mf (1:24) – 9 – €4200.00 – ISBN-10: 3-598-34557-7 – ISBN-13: 978-3-598-34557-9 – (pt also sold separately) – gw Saur [624]

Deutsche architekturbuecher zur zivilbaukunst des 18. jahrhunderts : teil 1: 1700 bis 1749 = German architectural books on civil engineering of the 18th century, pt 1: 1700 to 1749 – [mf 2003] – (= ser Nachschlagewerke und quellen zur kunst 5) – 124mf (1:24) – 9 – silver €1850.00 – ISBN-10: 3-598-34558-5 – ISBN-13: 978-3-598-34558-6 – gw Saur [624]

Deutsche architekturbuecher zur zivilbaukunst des 18. jahrhunderts : teil 2: 1750 bis 1800 = German architectural books on civil engineering of the 18th century, pt 2: 1750 to 1800 – [mf ed 2004] – (= ser Nachschlagewerke und quellen zur kunst 5) – 265mf (1:24) in 3 installments – 9 – silver €2490.00 – ISBN-10: 3-598-34561-5 – ISBN-13: 978-3-598-34561-6 – gw Saur [624]

Deutsche aussenpolitik (erschienen bis 1983) – 1956-1983 – 1,015mf – 1 – gw Mikropress [943]

Deutsche auswanderer-zeitung – Bremen DE, 1852-67 [gaps], 1869-75 – 6r – 1 – (suppl: beiblatt) – gw Misc Inst [074]

Deutsche balkan-zeitung – Sofia (BG) 1917 2 mar & 2 nov-31 dec, 1918 19 jun-23 sep [mf ed 2004] – 1r – 1 – gw Mikrofilm [077]

Die deutsche ballade : eine auslese aus der gesamten deutschen balladen-, romanzen-, und legenden-dichtung, unter besonderer beruecksichtigung des volksliedes / ed by Benzmann, Hans – 2. aufl. Leipzig: Hesse & Becker, 1925 [mf ed 1993] – 2v in 1/pl (ill) – 1 – (incl bibl ref and ind) – mf#8354 – us Wisconsin U Libr [780]

Deutsche balladen / Miegel, Agnes – Jena: E Diederichs, 1939, c1935 – 1r – 1 – us Wisconsin U Libr [780]

Deutsche balladen : von buerger bis brecht / Berger, Karl Heinz & Pueschel, Walter – Berlin: Verlag Neues Leben, 1956 [mf ed 1993] – 470p (ill) – 1 – (incl ind. wood engravings by ursula wendorff-weidt) – mf#8354 – us Wisconsin U Libr [810]

Deutsche barockdichtung : renaissance, barock, rokoko / Cysarz, Herbert – Leipzig: H Haessel, 1924 – 311p – 1 – (incl bibl ref and ind) – us Wisconsin U Libr [430]

Der deutsche bauerkrieg / Engels, Friedrich – Leipzig, 1875 – 1 – gw Mikropress [335]

Deutsche bauernstimme – Apatin (YU), 1939 5 jan-26 oct – 1r – 1 – gw Misc Inst [630]

Deutsche bauernstimme – Sombor (YU), 1938 21 apr-25 sep – 1r – 1 – gw Misc Inst [630]

Deutsche bauern-zeitung – Koeln DE, 1949 13 mar-1961 21 dec – 7r – 1 – (filmed by other misc inst: 1959 2 jul-1988) – gw Misc Inst [630]

Deutsche bauzeitung – Berlin, 1867-1942. v1-76+suppl+ind – 1357mf – 9 – mf#OA-300 – ne IDC [720]

Die deutsche beichte vom 9. jahrhundert bis zu reformation / Zimmermann, Charlotte – Weida, 1934 (mf ed 1993) – 1mf – 9 – €24.00 – 3-89349-270-4 – mf#DHS-AR 127 – gw Frankfurter [240]

Deutsche beobachter / Tuscarawas Co. New Philadelp – v1 n1. 5/1869-83,85-05,07-12/1910 [wkly] – 16r – 1 – (in german) – mf#B33581-33596 – us Ohio Hist [071]

Der deutsche berg im osten : ein volksdeutscher roman / Bremen, Carl von – Stuttgart: A Spemann, c1938 [mf ed 1989] – 217p – 1 – mf#7082 – us Wisconsin U Libr [830]

Deutsche berg- und huetten-arbeiter-zeitung – Bochum, 1889-1933 – 13r – 1 – (ab 1903: deutsche bergarbeiter-zeitung; ab 1905: bergarbeiter-zeitung; ab 1931: die bergbau-industrie; ab 1933: der deutsche bergknappe) – gw Mikropress [622]

Deutsche berg- und huettenarbeiter-zeitung *see* Glueckauf!

Deutsche bergarbeiter-zeitung *see* Glueckauf!

Der deutsche bergknappe *see* Glueckauf!

Deutsche bergwerks-zeitung – Essen, Duesseldorf DE, 1 apr-30 jun 1923, 27 jul 1923, 20 nov 1923-29 jun 1924, 1 sep 1927-31 aug 1944 – 1 – (publ in duesseldorf fr 1 sep 1927. filmed by mikropress: 1924-25, 1929 jul [Jubilaumsausg], 1931-32 [6r]; filmed by misc inst: 1901 mar-1927 aug, 1932 jan-mar & jul-sep, 1933 jan-mar [29r]) – gw Mikrofilm; gw Misc Inst [622]

Deutsche bergwerkszeitung – Bochum DE, 1924-25, jul 1929 [ann ed], 1931-32 – 6r – 1 – mf#7216 – gw Mikropress [622]

Deutsche beskidenzeitung – Friedek-Friedberg (Frydek-Mistek CZ), 1934 14 apr-1937 feb – 1r – 1 – gw Misc Inst [077]

Die deutsche bibel in ihrer geschichtlichen entwickelung / Risch, Adolf – Berlin: Edwin Runge, 1907 – 1mf – 9 – 0-7905-0511-8 – mf#1987-0511 – us ATLA [220]

Die deutsche bibel vor luther : sein verhaeltniss zu derselben und seine verdienste um die deutsche bibeluebersetzung / Krafft, Wilhelm – Bonn: Carl Georgi, 1883 – 1mf – 9 – 0-8370-3990-8 – mf#1985-1990 – us ATLA [220]

Die deutsche bibeluebersetzung des mittelalterlichen waldenser in dem codex teplensis und der ersten gedruckten deutschen bibel nachgewiesen : mit beitraegen zur kenntnis der romantischen bibeluebersetzung und dogmengeschichte der waldenser / Haupt, Herman – Wuerzburg: Stahel, 1885 – 1mf – 9 – 0-8370-3528-7 – (incl ind & appendixes) – mf#1985-1528 – us ATLA [220]

Deutsche bibliothek der schoenen wissenschaften / ed by Klotz, Christian Adolf – Halle 1767-71 – (= ser Dz) – 6v[zu je 4st] on 30mf – 9 – €300.00 – mf#k/n261 – gw Olms [500]

Deutsche bildnisse : dichter- und gelehrtenportraets / Scherer, Wilhelm – Berlin: Deutsche Bibliothek, [1874?] [mf ed 1993] – 228p – 1 – (incl bibl ref) – mf#8155 – us Wisconsin U Libr [430]

Deutsche blaetter – Leipzig, Altenburg DE, 1813 14 oct-1814 7 mar – 2r – 1 – (began in altenburg. with suppl) – gw Misc Inst [074]

Deutsche blaetter : fuer ein europaeisches deutschland – gegen ein deutsches europa – Santiago de Chile (RCH), 1943-46 – 2r – 1 – gw Misc Inst [079]

Deutsche blaetter *see* Die gartenlaube

Deutsche blaetter in polen – Posen (Poznan PL), 1928, 1930-31 – 1 – gw Misc Inst [077]

Das deutsche blatt – Berlin DE, 1891-97, 1898 apr-1903 sep, 1904-1907 sep, 1908-1910 mar, 1910 jul-1911 sep, 1912-14, 1915 apr-sep, 1916-18, 1919 may-aug, 1920-1921 apr, 1921 sep-1924, 1924 sep-1927 jun, 1927 oct-dec, 1928 jul-1932 jun, 1932 oct-1933 sep, 1934-1936 jun, 1936 oct-1937 jun, 1937 oct-1938, 1939 apr-1943 feb [gaps] – 149r – 1 – (title varies: 1 jul 1908: berliner allgemeine zeitung; incl suppl: das deutsche jugendblatt 1908) – gw Misc Inst [074]

Deutsche boettcher-zeitung – Bremen DE, 1913-16 – 2r – 1 – gw Misc Inst [680]

Der deutsche bote – Klattau (Klatovy CZ), 1942 may-dec – 1r – 1 – gw Misc Inst [077]

Deutsche briefe aus paris / Helfferich, Adolph – Pforzheim 1848 [mf ed Hildesheim 1995-98] – 7v on 7mf – 9 – €140.00 – 3-487-26020-4 – gw Olms [860]

Deutsche bruesseler zeitung – Bruessel (B), 1847 3 jan-1848 27 feb – 1r – 1 – mf#3722 – gw Mikropress [074]

Das deutsche buch : ersatzpublikation fuer die zeitschrift aktion – Porto Alegre (BR), 1937 may-jul – 1r – 1 – gw Misc Inst [430]

Deutsche buehnenspiele : ausgabe in einem bande / Holz, Arno – Dresden: C Reissner, [1922?] – 1r – 1 – us Wisconsin U Libr [430]

Deutsche chansons / Bierbaum, Otto Julius et al – Leipzig: Insel-Verlag, 1917 [mf ed 1993] – 249p – 1 – mf#8368 – us Wisconsin U Libr [780]

Deutsche Chemische Gesellschaft *see* Berichte der deutschen chemischen gesellschaft

Deutsche chronik – Weyauwega WI. 1898-1901, 1902-05 – 2r – 1 – (cont by: appleton volksfreund) – mf#959837 – us WHS [071]

Deutsche chronik / ed by Schubart, Christian Friedrich Daniel – Augsburg 1774-78 [mf ed Hildesheim 1992-98] – (= ser Dz. historisch-politische abt) – 20mf – 9 – €200.00 – mf#k/n1071 – gw Olms [943]

Deutsche chroniken : (scriptores qui vernacula lingua usu sunt) / Monumenta Germaniae Historica. Scriptores – v3-6 – 141mf – 8 – mf#371 – ne IDC [700]

Deutsche classiker des mittelalters : mit wort- und sacherklaerungen / ed by Pfeiffer, Franz – Leipzig: Brockhaus, 1866-72 [mf ed 1993] – 12v – 1 – (incl bibl ref and ind) – mf#8189 – us Wisconsin U Libr [430]

Deutsche constitutionelle zeitung – Muenchen DE, 1848 jul-1849 7 oct – 3r – 1 – gw Misc Inst [323]

Der deutsche correspondent – Baltimore MD [USA], 1923 3 jul-1924, 1926, 1928-1937 30 sep [gaps] [7r], 1972-75 – 1 – (title varies: mai 1918: baltimore correspondent; v30 nov 1935-6 jun 1941: taeglicher baltimore correspondent) – gw Misc Inst [074]

Deutsche demokratische republik im aufbau – Berlin DE, 1952-1956 n6, 1957-59 – 1mf=2df – 1 – (title varies: 1957: deutsche demokratische republik : ddr) – gw Mikrofilm [323]

Deutsche demokratische zeitung *see* Volksmund

Deutsche dichter, 1700-1900 : eine geistesgeschichte in lebensbildern / Ermatinger, Emil – 2nd rev ed. Frankfurt/ Main: Anthenaeum, 1961 [mf ed 1993] – 855p (ill) – 1 – (incl bibl ref) – mf#8177 – us Wisconsin U Libr [430]

Deutsche dichter, denker und wissensfuersten im 18. und 19. jahrhundert : in lebensbildern fuer jungend und volk / ed by Spamer, Franz Otto – 2nd enl ed. Leipzig: O Spamer, 1877 [mf ed 1993] – (= ser Galerie der meister in wissenschaft und kunst. meister der wissenschaft und dichtkunst) – 3pts, x/360p (ill) – 1 – mf#8231 – us Wisconsin U Libr [430]

Deutsche dichter der gegenwart : ihr leben und werk / ed by Wiese, Benno von – Berlin: E Schmidt, 1973 [mf ed 1993] – 686p – 1 – (incl bibl ref and ind) – mf#8262 – us Wisconsin U Libr [430]

Deutsche dichter des 18. und 19. jahrhunderts und ihre politik : ein vaterlaendischer vortrag / Roethe, Gustav – Berlin: Weidmannsche Buchhandlung, 1919 – 30p – 1 – us Wisconsin U Libr [850]

Deutsche dichter und schriftsteller unserer zeit : einzeldarstellungen zur schoenen literatur in deutscher sprache / Lennartz, Franz – 10. erw aufl. Stuttgart: A Kroener, 1969 [mf ed 1993] – (= ser Kroeners taschenausgabe 151) – vi/783p – 1 – mf#8154 – us Wisconsin U Libr [430]

Deutsche dichtung : eine darstellung ihrer geschichte / Vogelpohl, Wilhelm; ed by Hafner, Gotthilf – Stuttgart: E Klett, [1957] – 1r – 1 – (incl ind) – us Wisconsin U Libr [430]

Deutsche dichtung / ed by George, Stefan & Wolfskehl, Karl – 3. Aufl. Berlin: G Bondi, 1923-1932 – 1r – 1 – us Wisconsin U Libr [810]

Deutsche dichtung : kurzgefasste literaturgeschichte / Mueller, Friedrich von & Valentin, G – Paderborn: F Schoeningh, 1957 – 1r – 1 – (incl indes) – us Wisconsin U Libr [430]

Die deutsche dichtung : grundriss der deutschen literaturgeschichte / Heinemann, Karl – Leipzig: A Kroener, 1927 – 1 – (incl ind) – us Wisconsin U Libr [430]

Die deutsche dichtung : grundriss der deutschen literaturgeschichte / Heinemann, Karl – Leipzig: Kroener, [1930] – 1 – (incl ind) – us Wisconsin U Libr [430]

Die deutsche dichtung : vom ausgang des barocks bis zum beginn des klassizismus, 1700-1785 / Schneider, Ferdinand Josef – Stuttgart: J B Metzler, 1924 [mf ed 1993] – (= ser Epochen der deutschen literatur v3) – x/492p – (incl bibl ref and ind) – mf#8231 – us Wisconsin U Libr [430]

Die deutsche dichtung 1936-1937 / ed by Elsner, Richard – Berlin: West-Ost-Verlag, 1936-1937 – 1r – 1 – us Wisconsin U Libr [790]

Die deutsche dichtung der aufklaerungszeit / Schneider, Ferdinand Josef – 2nd rev ed. Stuttgart: J B Metzler, 1948 [mf ed 1993] – (= ser Epochen der deutschen literatur v3/1) – 368p – 1 – (incl bibl ref and ind) – mf#8237 – us Wisconsin U Libr [430]

Die deutsche dichtung der geniezeit / Schneider, Ferdinand Josef – Stuttgart: J B Metzler, 1952 [mf ed 1993] – (= ser Epochen der deutschen literatur v3/2) – viii/367p – 1 – (incl bibl ref and ind) – mf#8210 – us Wisconsin U Libr [430]

Die deutsche dichtung des 19. jahrhunderts in ihren bedeutenderen erscheinungen : populaere vorlesungen / Schroeer, Karl Julius – Leipzig: F C W Vogel, 1875 [mf ed 1993] – vi/496p – 1 – (incl bibl ref and ind) – mf#8243 – us Wisconsin U Libr [430]

Die deutsche dichtung im mittelalter, 800 bis 1500 / Golther, Wolfgang – Stuttgart: J B Metzler, 1912 – 1r – 1 – (incl bibl ref and index) – us Wisconsin U Libr [430]

Die deutsche dichtung in der schule : geschichte und probleme 1750-1860 / Boehnke, Frieda – Frankfurt a.M., 1967 – 3mf – 9 – 3-89349-697-1 – gw Frankfurter [430]

Deutsche dichtung in ihren geschichtlichen grundzuegen / Lienhard, Friedrich – Leipzig: Quelle, 1917 – 141p – 1 – us Wisconsin U Libr [430]

Die deutsche dichtung in ihren sozialen, zeit- und geistesgeschichtlichen bedingungen : eine skizze / Kleinberg, Alfred – Berlin: J H W Dietz, 1927 – 1r – 1 – (incl bibl ref and indexes) – us Wisconsin U Libr [430]

Die deutsche dichtung seit goethes tod / Walzel, Oskar Franz – 2. aufl. Berlin: Askanischer Verlag, 1920 [mf ed 1993] – xiv/527p – 1 – (incl ind) – mf#8243 – us Wisconsin U Libr [430]

Deutsche dichtung von der aeltesten bis auf die neueste zeit / Menzel, Wolfgang; ed by Garber, Klaus – Stuttgart 1858-59 – 3v on 16mf – 9 – diazo €98.00 – gw Olms [430]

Deutsche dichtungen / Frischlin, Nicodeums; ed by Strauss, David Friedrich – Stuttgart: Litterarischer Verein, 1857 [mf ed 1993] – (= ser Blvs 41) – 201p – 1 – (incl bibl ref) – mf#8470 reel 9 – us Wisconsin U Libr [810]

Deutsche dichtungen des mittelalters / ed by Bartsch, Karl – Leipzig: F A Brockhaus. 7v in 6. 1872-88 – (incl bibl ref and ind. middle high german and middle low german texts with introductions in german) – us Wisconsin U Libr [810]

Deutsche dienstbotenzeitung – Berlin DE, 1909-10 – 1r – 1 – gw Misc Inst [640]

Das deutsche drama, 1880-1933 / ed by Steinhauer, H – New York: W W Norton, c1938 [mf ed 1993] – (= ser Gateway books) – 2v – 1 – (incl bibl ref) – mf#8187 – us Wisconsin U Libr [430]

Das deutsche drama, 1880-1933 / ed by Steinhauer, Harry – New York: W W Norton, c1938 [mf ed 1993] – (= ser Gateway books) – 2v – 1 – (incl bibl ref) – mf#8187 – us Wisconsin U Libr [430]

Das deutsche drama in den litterarischen bewegungen der gegenwart : vorlesungen, gehalten an der universitaet bonn / Litzmann, Berthold – 4. aufl. Hamburg: L Voss 1897 [mf ed 1993] – 1r – 1 – (filmed with: christliches erbe und lyrisches gestaltung / hans giesecke) – mf#8297 – us Wisconsin U Libr [430]

Das deutsche drama vom barock bis zur gegenwart : interpretationen / ed by Wiese, Benno von – Duesseldorf: A Bagel, 1958 [mf ed 1993] – 2v – 1 – (incl bibl ref) – mf#8191 – us Wisconsin U Libr [430]

Deutsche dramaturgie : von barock bis zur klassik / ed by Wiese, Benno von – Tuebingen: M Niemeyer, 1956 [mf ed 1993] – (= ser Deutsche texte 4) – vii/144p – 1 – (incl bibl ref) – mf#8282 – us Wisconsin U Libr [790]

Deutsche dramaturgie von gryphius bis brecht / Dietrich, Margret & Stefanek, Paul – Muenchen: List Verlag, 1965 [mf ed 1993] – (= ser List taschenbuecher 287) – 170p – 1 – (incl bibl ref) – mf#8297 – us Wisconsin U Libr [430]

Deutsche drucke des barock 1600-1720 : mit den ergaenzungen polnische drucke und polonica 1501-1700 sowie ungarische drucke und hungarica 1480-1720. katalog der herzog august bibliothek wolfenbuettel / ed by Bircher, Martin et al – [mf ed 1996] – 157mf (1:24) + 4 ind vols – 9 – silver €2600.00 – ISBN-10: 3-598-32187-2 – ISBN-13: 978-3-598-32187-0 – (also sold individually: hungarica €158; polonica €240) – gw Saur [430]

Deutsche einheit – Bonn, Muenchen DE, 1956 6 oct-1988 may – 1 – gw Misc Inst [074]

Deutsche einsamkeiten : der roman unseres volkes / Erbt, Wilhelm – Berlin: Verlag der Taeglichen Rundschau, 1921 – 1r – 1 – us Wisconsin U Libr [830]

Der deutsche eisenbahner : organ der gewerkschaft der eisenbahner deutschlands – [Frankfurt am Main: s.n.] 1. jahrg [15 aug 1948]-47. jahrg [dec 1994] [semimthly] [mf ed 1987-] – 1 – (subtitle varies; cont by: gded inform) – mf#1837 – us Wisconsin U Libr [380]

Deutsche eisenzeitung und taeglicher anzeiger – Duesseldorf DE, 1896 jan-feb, 1904-1911 30 jun – 23r – 1 – (also: taeglicher anzeiger, 1895?: duesseldorfer neueste nachrichten; with suppl: duesseldorfer illustrierte zeitung 1891 5 apr-1892 25 dec [1r], duesseldorfer radschlaeger 1889-90 [1r] publ in berlin, familienblatt [fr 13 jun 1907: unterhaltungs-beilage] 1904-10 [gaps] [6r], rhein und duessel 1904-1911 jun [3r]) – gw Misc Inst [074]

Deutsche encyclopaedie : oder allgemeines realwoerterbuch aller kuenste und wissenschaften / ed by Koester, H M G & Roos, J F – Frankfurt 1778-1807 [mf ed 1992] – (= ser AEL 1/1) – 24v on 212mf – 9 – €730 diazo €876 silver – 3-89131-052-8 – gw Fischer [700]

Deutsche entomologische zeitschrift – Berlin, 1906-40 – 9 – $720.00 – mf#0181 – us Brook [580]

Deutsche epigramme / ed by Hofmannsthal, Hugo von – Muenchen: Verlag der Bremer Presse, 1923 – 1r – 1 – us Wisconsin U Libr [430]

Deutsche erzaehler / ed by Hofmannsthal, Hugo von – Leipzig: Insel Verlag. 3v. 1921 – 1r – 1 – us Wisconsin U Libr [430]

Deutsche erzaehler des achtzehnten jahrhunderts / ed by Fuerst, Rudolf – Leipzig: G J Goeschen, 1897 [mf ed 1993] – (= ser Deutsche litteraturdenkmale des 18. und 19. jahrhunderts 66-69. n f n16-19) – xxix/178p – 1 – mf#8676 reel 5 – us Wisconsin U Libr [830]

Deutsche erzaehlungen / Varnhagen von Ense, Karl August – 2. Aufl. Stuttgart: J G Cotta, 1879 – 1r – 1 – us Wisconsin U Libr [830]

Der deutsche erzieher – Wien (A), 1939-42 [gaps] – 1 – gw Misc Inst [074]

Deutsche evangelische kirchenzeitung – 1(1887)-16(1902) – 240mf – 9 – €458.00 – (lacking: 5(1891)-7(1893)) – ne Slangenburg [242]

Die deutsche expedition der loango-kueste : nebst aelteren nachrichten ueber die zu erforschenden laendern / Bastian, A – Jena, 1874-1875. 2v – 9mf – 9 – mf#H-6174 – ne IDC [916]

Deutsche expressionistische dichtung : im lichte der philosophie der gegenwart / Stuyver, Wilhelmina – Amsterdam: H J Paris, 1939 [mf ed 1992] – 222p – 1 – (incl bibl ref) – us Wisconsin U Libr [430]

Deutsche fackel – Berlin DE, 1922 jan-nov – 1r – 1 – gw Misc Inst [074]

Deutsche feste und volksbraeuche / Fehrle, Eugen – Leipzig: B G Teubner 1920 [mf ed Bloomington IN: Indiana Uni Lib, Preservation Dept 1984] – 1r – 1 – us Indiana Preservation [390]

Deutsche feuerwehrzeitung – Stuttgart DE, 1860 12 oct-1923 1 sep – 11r – 1 – gw Misc Inst [360]

Der deutsche film – Berlin DE, 1919 2 jul-23 dec – 1 – gw Mikrofilm [790]

Der deutsche film in wort und bild – Muenchen, Berlin DE, 1919 n7 & 11, 1920 7 oct, 25 nov, 9 & 17 dec, 1921 28 jan-30 dec [gaps], 1922 13 jan-24 mar – 1 – (missing: 1922 n5) – gw Mikrofilm [790]

Deutsche filmgewerkschaft – Berlin DE, 1920 15 apr-15 dec, 1921 1 feb-15 dec, 1922 1 jan-15 nov, 1923 1 jan-1 aug – 2r – 1 – gw Mikrofilm [790]

Deutsche filmzeitung – Muenchen DE, 1922 10 nov-1941 – 10r – 1 – (title varies: 1922-27: sueddeutsche filmzeitung) – gw Mikrofilm [790]

Deutsche finanzpolitik / Baumgarten, Dietrich – 1924-28 – 1 – gw Mikropress [336]

Deutsche finanzwirtschaft – Berlin. v1-23, jun 1947-jun 1969 [mnthly], jun 1947-sept 1949 [semimnthly]oct 1949-69 – 1 – us Wisconsin U Libr [332]

Der deutsche finckh : gedichte / Finckh, Ludwig; ed by Seibold, Karl – Muenchen: Deutscher Volksverlag 1941 [mf ed 1989] – 1r – 1 – (aft by ed. filmed with: das goldene erbe) – mf#7241 – us Wisconsin U Libr [810]

Deutsche forschungen / ed by Panzer, Friedrich Wilhelm & Petersen, Julius – Frankfurt am Main: M Diesterweg. 33v. 1921-40 – 1 – (v1 was publ in 1925. cf. union list of serials. each vol also has a distinctive title) – us Wisconsin U Libr [943]

Die deutsche frau in der sozialen kriegsfuersorge / ed by Baeumer, Gertrud – Gotha: Perthes, 1916 [mf ed 1987] – vi/60p (ill) – 1 – mf#6839 – us Wisconsin U Libr [305]

Die deutsche frau und der nationalsozialismus *see* Nsdap (national socialist german workers party) nazi publications

Deutsche frauen-zeitung : central-organ der... / Verein zur Verbesserung der Lage der Frauen – 1852 oct 15 – 1r – 1 – mf#1166615 – us WHS [305]

Deutsche frauenzeitung – Duesseldorf DE, 1933-43, 1944 [gaps] – 5r – 1 – (title varies: 1933 n4: voelkische frauenzeitung) – gw Misc Inst [640]

Deutsche freiheit : einzige unabhaengige tageszeitung deutschlands – Saarbruecken DE, 1933 21 jun-1935 17 jan – 3r – 1 – (with suppl) – gw Misc Inst [074]

Deutsche freiheit = La liberte allemande – Paris (F), 1937 3 dec-1939 apr – 1 – gw Misc Inst [074]

Deutsche freiheit – Muenchen DE, 1956 1 oct-1 dec, 1957 15 jan-1958 8 jun, 1959 1 feb-1960 3 nov – 1r – 1 – gw Misc Inst [074]

Deutsche freiheitsbriefe – Paris (F), 1937, 1938 [gaps] – 1 – gw Misc Inst [860]

Deutsche front – Saarbrucken, Germany. Nov 1934-Jan 1935 – 1r – 1 – us L of C Photodup [074]

Deutsche front – Metz [F], 1940 1 aug-1944 1 dec – 1 – (title varies: 1 dec 1940: nsz-westmark) – gw Misc Inst [943]

Deutsche front – Metz [F], 1940 1 aug-1944 1 dec – 1 – (title varies: 1 dec 1940: nsz-westmark) – gw Misc Inst [074]

Der deutsche frontsoldat : mythos udn gestalt / Kalkschmidt, Till – Berlin: Junker & Duennhaupt 1938 [mf ed 1992] – (= ser Neue deutsche forschungen. abteilung neuere deutsche literaturgeschichte 15) – 2r – 1 – (incl bibl ref) – mf#3185p – us Wisconsin U Libr [430]

Deutsche fuehrerbriefe – Berlin DE, 1928 2 feb-19 aug, 27 oct, 1929 4 jan-31 may, 4 jun-1 oct, 1929 8 oct-29 oct, 1929 5 nov-1933 – 2r – 1 – gw Misc Inst [860]

Das deutsche fuehrerlexikon 1934/1935 – Berlin: Otto Stollberg c1934 [mf ed 1985] – 1r – 1 – (certain biogr sketches have been expunged, leaving blank spaces in the text) – mf#6566 – us Wisconsin U Libr [943]

Deutsche fuersten als dichter und schriftsteller : mit einer auswahl ihrer dichtungen: von den hohenstaufen biz zur gegenwart: ein beitrag zur deutschen litteraturgeschichte / Seidl, Franz Xaver – Regensburg: Alfred Coppenrath, 1883 – cii/194p – 1 – (incl bibl ref) – us Wisconsin U Libr [430]

Deutsche fuerstenlieder : von einem rheinpreussen – Bern: Jenni, 1844 [mf ed 1989] – 63p – 1 – mf#7174 – us Wisconsin U Libr [810]

Deutsche funk illustrierte – Berlin DE, 1932 11 mar-1935, 1938-1941 31 may – 7r – 1 – (title varies: 22 oct 1933: deutsche radio illustrierte) – gw Mikrofilm [380]

Das deutsche funkprogramm – Berlin DE, 1933 n4-1936 n180 – 5r – 1 – (1934 n56 publ as fachfunk) – gw Misc Inst [380]

Deutsche gaertnerpost – Berlin DE, 1955 14 jan-1990 – 19r – 1 – (title varies: 1975: gaertnerpost) – gw Misc Inst [635]

Deutsche gassenlieder / Hoffmann von Fallersleben, August Heinrich – Zuerich: Literarisches Comptoir, 1843 – 1r – 1 – us Wisconsin U Libr [780]

Das deutsche gebet : [poems] / Boehme, Herbert – Muenchen: Zentralverlag der NSDAP, F Eher, [1936?] [mf ed 1989] – 24p – 1 – mf#7043 – us Wisconsin U Libr [810]

Der deutsche gedanke bei jakob grimm : in grimms eignen worten / Matthias, Theodor – Leipzig: R Voigtlaender 1915 [mf ed 1990] – 1r – 1 – mf#2692p – us Wisconsin U Libr [306]

Das deutsche gedicht : ein jahrtausend deutscher lyrik / Scholz, Wilhelm von [comp] – Berlin: T Knaur c1941 [mf ed 1993] – 1r – 1 – (filmed with: dichtergruesse / elise polko [comp]) – mf#3348p – us Wisconsin U Libr [810]

Deutsche gedichte des 12. jahrhunderts und der naechstverwandten zeit / ed by Massmann, H F – Quedlinburg; Leipzig: G Basse. 2v in 1. 1837 – us Wisconsin U Libr [810]

Deutsche gedichte in handschriften – Leipzig: Insel-Verlag, [1935] – 1 – us Wisconsin U Libr [810]

Deutsche gegenreformation und deutsches barock : die deutsche literatur im zeitraum des 17. jahrhunderts / Hankamer, Paul – 2. aufl. Stuttgart: J B Metzler, 1947, c1946 [mf ed 1993] – (= ser Epochen der deutschen literatur. geschichtliche darstellungen 2/2) – viii/542p – 1 – (incl bibl ref and ind) – mf#8174 – us Wisconsin U Libr [430]

Deutsche gegenwart : ein informationsbrief – New York NY (USA), 1947-48 – 1r – 1 – gw Misc Inst [071]

Die deutsche gegenwartsdichtung im kampf um die deutsche lebensform / Kindermann, Heinz – Wien: Wiener Verlagsgesellschaft, 1942 [mf ed 1993] – (= ser Kleinbuchreihe Suedost n41) – 54p – 1 – mf#8245 – us Wisconsin U Libr [430]

Deutsche geister : aufsaetze / Braun, Felix – Wien: Rikola Verlag 1925 [mf ed 1993] – 1r – 1 – (filmed with: uber den tag hinaus / harry bergholz [comp]; with int by c f w behl) – mf#3204p – us Wisconsin U Libr [430]

Deutsche geistesheroen : in ihrer wirksamkeit auf dem gebiete der freimaurerei / Fischer, Robert – Leipzig: B Zechel 1881 – 2mf – 9 – mf#vrl-56 – ne IDC [366]

Deutsche gesangskunst – Leipzig, Berlin DE, 1900-02 – 1r – 1 – gw Misc Inst [074]

Deutsche geschichte / Lamprecht, Karl – Berlin: R. Gaertner, 1902-04. 2v in 3 – 1 – us Wisconsin U Libr [943]

Das deutsche geschichtsdrama : geschichte eines literarischen mythos / Sengle, Friedrich – Stuttgart: J B Metzler 1952 [mf ed 1993] – 1r – 1 – (incl bibl ref. filmed with: volksspiel und feier & other titles) – mf#3382p – us Wisconsin U Libr [430]

Deutsche Gesellschaft, Leipzig *see* Beytraege zur critischen historie der deutschen sprache, poesie und beredsamkeit

Deutsche Gesellschaft zur Erforschung Aequatorialafrikas *see*
– Correspondenzblatt der afrikanischen gesellschaft
– Mittheilungen der afrikanischen gesellschaft in deutschland

Deutsche gewerbezeitung – Leipzig DE, 1845, 1846 [gaps] – 1r – 1 – gw Misc Inst [380]

DEUTSCHE

Deutsche gewerkschaft – [Wuppertal-] Elberfeld DE, 1923-25, 1927-30 – 1r – 1 – (with gaps) – gw Mikrofilm [331]
Deutsche gewerkschafts-zeitung *see* Die neue front
Deutsche grammatik / Grimm, Jacob – Gottingen, Germany. v1-4. 1826-1840 – 4r – us UF Libraries [430]
Deutsche grenzstimmen – Tachau (Tachov CZ), 1938 8 jul-30 sep – 1r – 1 – gw Misc Inst [074]
Deutsche grenzwacht – Landskron (Lanskroun CZ), 1922-1924 sep – 1r – 1 – gw Misc Inst [077]
Deutsche grenzwacht – Lanskroun, Czechoslovakia. Dec 1940-Mar 1941 – 1r – 1 – us L of C Photodup [077]
Deutsche handels-archiv – Berlin. Jan 1856-June 1939 – 1 – us L of C Photodup [380]
Deutsche Handels-und Plantagen Gesellschaft *see* Registers of melanesian indentured labourers in samoa
Deutsche hausfrau – 1905 sep-1906 may/jun – 1r – 1 – (cont by: modernes journal; deutsche hausfrau und modernes journal) – mf#3164559 – us WHS [640]
Deutsche hausfrau – 1908 feb-oct, nov-1910 dec, 1916 jan-1917 dec, 1918 jan-sep – 4r – 1 – (cont: deutsche hausfrau und modernes journal; cont by: hausfrau) – mf#3164589 – us WHS [305]
Deutsche hausfrau und modernes journal – 1906 jul-1908 jan – 1r – 1 – (cont: modernes journal; deutsche hausfrau [1904]; cont by: deutsche hausfrau, modernes journal) – mf#569146 – us WHS [640]
Deutsche hausfrauen-zeitung – Berlin DE, 1878 6 jan-29 dec, 1880-1888 10 jun, 1888 18 nov-1907 30 jun [gaps] – 20r – 1 – (title varies: 17 sep 1905: frauen-reich) – gw Mikrofilm [640]
Das deutsche heldenbuch / ed by Keller, Adelbert von – Stuttgart: Litterarischer Verein, 1867 [mf ed 1993] – (= ser Blvs 87) – 788p – 1 – (incl bibl ref and ind) – mf#8470 reel 18 – us Wisconsin U Libr [390]
Das deutsche heldenbuch : nach dem muthmasslich aeltesten drucke neu herausgegeben / ed by Keller, Adelbert von – Stuttgart: Litterarischer Verein, 1867 [mf ed 1993] – (= ser Blvs 87) – 58r – 1 – (incl bibl ref & ind. filmed by loc: ed by emil henrici [berlin, stuttgart: w spemann [1887]] [mf ed 1971] 1r) – mf#3420p – us Wisconsin U Libr; us L of C Photodup [810]
Die deutsche heldensage des mittelalters / Gunther, Ernst A W – Hannover: C Brandes 1870 [mf ed Bloomington IN: Indiana Uni Lib, Preservation Dept 1984] – 1r – 1 – us Indiana Preservation [390]
Deutsche heldensage im breisgau / Panzer, Friedrich Wilhelm – Heidelberg: C Winter's Universitaetsbuchhandlung, 19904 – 90p – 1 – us Wisconsin U Libr [390]
Deutsche heldensagen des mittelalters – Leipzig: F Brandstetter 1877 [mf ed Bloomington IN: Indiana Uni Lib, Preservation Dept 1984] – 2v on 1r – 1 – us Indiana Preservation [390]
Deutsche hobelspaene : stossseufzer und stammbuchblaetter / Vierordt, Heinrich – Heidelberg: C Winter, 1909 – 1r – 1 – us Wisconsin U Libr [943]
Deutsche hochschulstimmen aus der ostmark – Wien [A], 1931, 1936 12 dec – 1r – 1 – (filmed by other misc inst: 1914-41; title varies: 1916: deutsche hochschulzeitung, 1925 n27: deutsche akademiker-zeitung) – gw Misc Inst [378]
Der deutsche holzarbeiter *see* Holzarbeiter-zeitung
Deutsche humoristen aus alter und neuer zeit / ed by Riffert, Julius – Altenburg: O Bonde. 3v in 1. [188-?] – us Wisconsin U Libr [430]
Deutsche hutmacher-zeitung – Berlin DE, 1922-28 – 11r – 1 – uk British Libr Newspaper [680]
Deutsche illustrierte – Berlin DE, Wien (A), 1939 4 jul-1940 18 may – 1 – gw Misc Inst [074]
Deutsche illustrierte zeitung – Berlin DE, 1910 n34-1913 6 sep, 1915 n33 – 2r – 1 – gw Misc Inst [074]
Der deutsche im auslande – Hamburg DE, 1934 n6, 1936 n18, 1937-1939 n9 – 1 – gw Misc Inst [900]
Der deutsche in argentinien – Buenos Aires (RA), 1938-39 [gaps] – 1 – gw Misc Inst [305]
Der deutsche in canada – Hamilton, Ont: Marrhausen'schen Buchhandlung, [1872-18– or 19–] – 9 – ISSN: 1190-724X – mf#P04154 – cn Canadiana [073]
Der deutsche in der landschaft / Borchardt, Rudolf – 1.-4. aufl. Berlin: Suhrkamp Verlag, 1953 [mf ed 1993] – 491p – 1 – (incl bibl ref) – mf#8373 – us Wisconsin U Libr [430]
Der deutsche in heimath und fremde – Kassel DE, 1841 2 jan-9 jun – 1r – 1 – gw Misc Inst [074]
Deutsche in nahost 1946-1965 : sozialgeschichte nach akten und interviews, bd 1 / Schwanitz, Wolfgang G – (mf ed 1998) – 7mf – 9 – €65.00 – 3-8267-2553-0 – mf#DHS 2553 – gw Frankfurter [943]
Deutsche in nahost 1946-1965 : sozialgeschichte nach akten und interviews, bd 2 / Schwanitz, Wolfgang G – (mf ed 1998) – 5mf – 9 – €59.00 – 3-8267-2554-9 – mf#DHS 2554 – gw Frankfurter [943]
Deutsche in ohio / Star Co. Canton – dec 1861-dec 1868 [wkly] – 2r – 1 – (in german) – mf#B6447-6448 – us Ohio Hist [071]
Der deutsche in polen – Kattowitz [Katowice PL], 1934 4 feb-1939 27 aug – 5r – 1 – (in 1934 also entitled: schlesische warte mit anzeiger fuer den kreis pless; filmed by other misc inst: 1934 4 feb-1939 27 aug [3r]) – gw Misc Inst [077]
Der deutsche industrie-arbeiter – Duesseldorf DE, oct 19 1923-mar 14 1925 – 1r – 1 – gw Misc Inst [331]
Deutsche industrie-zeitung – Chemnitz DE, 1870-86 – 9r – 1 – gw Misc Inst [338]
Deutsche industrie-zeitung : organ des centralverbandes deutscher industrieller – Berlin DE, 1901 18 jul-1905, 1908-1914 6 aug – 7r – 1 – mf#6644 – gw Mikropress [338]
Deutsche informationen – Paris. mars 1936-mars 1937 – 1 – (journal periodique paraissant trois fois par semaine.) – fr ACRPP [073]
Deutsche inlandsberichte – London (GB), 1939 5 feb-25 aug, 1939 17 oct-1941 14 dec – 2r – 1 – (filmed by other misc inst: 1939 n55-1941 n65) – gw Misc Inst [943]
Deutsche innerlichkeit / Ulrich, Christoffel – Muenchen: R Piper, c1940 – 1r – 1 – (incl bibl ref) – us Wisconsin U Libr [430]
Deutsche inselzeitung – Jersey (GB), 1941 1 jul-1931 dec – 1 – gw Misc Inst [072]
Deutsche installateur- und klempner-zeitung – Duesseldorf DE, 1912-16, 1918-19, 1931 [single iss], 1932-mar 15 1934 – 6r – 1 – gw Misc Inst [621]
Deutsche instrumentenbau-zeitung – Berlin. 1933-41. Lacking: v.34; 42, n.6-8. 2 reels – 1 – 47.00 – us L of C Photodup [780]
Deutsche interlinearversionen der psalmen : aus einer windberger handschrift zu muenchen (12. jahrhundert) und einer handschrift zu trier (13. jahrhundert) / ed by Graff, Eberhard Gottlieb – Quedlinburg, Leipzig: G Basse, 1839 [mf ed 1993] – (= ser Bibliothek der gesammten deutschen national-literatur von den aeltesten bis auf die neuere zeit sect1/10) – vi/670p – 1 – (middle high german and latin) – mf#8438 reel 3 – us Wisconsin U Libr [221]
Deutsche israelitische zeitung – Regensburg DE, 1913-15, 1918, 1924, 1930, 1933 n5, 1937-38 – 1 – gw Misc Inst [939]
Die deutsche jakobinische literatur und publizistik, 1789-1800 / Voegt, Hedwig – Berlin: Ruetten & Loening, 1955, c1954 [mf ed 1993] – 244p (ill) – 1 – (incl bibl ref and ind) – mf#8168 – us Wisconsin U Libr [430]
Deutsche jugendbewegung und jugendarbeit in polen 1919-1939 / Nasarski, Peter Emil – Wuerzburg: Holzner-Verlag, 1957 – 10r – 1 – (incl bibl ref and indexes) – us Wisconsin U Libr [320]
Deutsche jugendkraft *see* Freiburger tagespost
Deutsche justiz / a – Berlin DE, 1933-45 [gaps] – 1 – gw Misc Inst [340]
Deutsche kaempfe / Frenzel, Karl – Hannover: C Ruempler, 1873 (mf ed 1990) – 1r – 1 – (filmed with: ein glaubensbekenntnis) – us Wisconsin U Libr [943]
Deutsche katastrophe / Meinecke, Friedrich – Wiesbaden, Germany. 1949 (c1946) – 1r – us UF Libraries [943]
Deutsche kino-rundschau – Berlin DE, 1915 – 1r – 1 – gw Mikrofilm [790]
Deutsche kino-rundschau – Muenchen DE, 1914 n1, 20-32 – 1 – gw Mikrofilm [790]
Das deutsche kirchenlied der schweiz im reformationszeitalter / Odinga, T – Frauenfeld, 1889 – 2mf – 9 – mf#ZWI-95 – ne IDC [242]
Die deutsche klassik und die franzoesische revolution / Mehring, Franz; ed by Raddatz, Fritz J – Darmstadt, Neuwied: Luchterhand, 1974 [mf ed 1993] – (= ser Sammlung luchterhand 170; Mehring werkauswahl 1) – 329p – 1 – (incl bibl ref and ind) – mf#8140 – us Wisconsin U Libr [430]
Deutsche klassik und romantik : oder, vollendung und unendlichkeit: ein vergleich / Strich, Fritz – 3rd rev enl ed. Muenchen: Meyer & Jessen, c1928 [mf ed 1993] – 428p – 1 – (incl ind) – mf#8231 – us Wisconsin U Libr [430]
Deutsche kolonialzeitung – Muenchen, Berlin, Frankfurt/M DE, 1885-89, 1899, 1901, 1903, 1905, 1911 – 7r – 1 – gw Misc Inst [074]
Deutsche kolonialzeitung – v. 1-39, no. 5. Jan 1884-15 Sep 1922 – 1 – us L of C Photodup [943]
Deutsche kolonisation in ostafrika. : aus briefen und tagebuechern des am 24. september 1888 zu kilwa umgefommenen beamten der deutsch-ostafrikanischen gesellschaft heinrich hessel / Hessel, Karl – Bonn: E Weber, 1889 – 1 – us CRL [960]
Der deutsche kolumbus-brief : in faksimile-druck / ed by Haebler, Konrad – Strassburg: J H E Heitz 1900 [mf ed 1993] – (= ser Drucke und holzschnitte des 15. und 16. jahrhunderts in getreuer nachbildung 6) – 1r – 1 – (filmed with: kleines deutsches sagenbuch / will-erich peuckert [ed]) – mf#3367p – us Wisconsin U Libr [430]
Deutsche kommentare – Heidelberg, Stuttgart, Berlin DE, 1949 3 oct-1956 27 oct – 8r – 1 – uk British Libr Newspaper [074]
Deutsche kommentare *see* Die buecher-kommentare
Die deutsche komoedie unter der einwirkung des aristophanes : ein beitrag zur vergleichenden literaturgeschichte / Hille, Curt – Leipzig: Quelle & Meyer, 1907 [mf ed 1992] – (= ser Breslauer beitraege zur literaturgeschichte. neue folge 2) – vi/180p – 1 – mf#8014 reel 2 – us Wisconsin U Libr [410]
Deutsche korrespondenz – London, UK. 15 nov 1901-31 mar 1913 – 1 – (aka: allgemeine correspondenz, 1 apr 1913-5 aug 1914) – uk British Libr Newspaper [074]
Der deutsche krieg 1870-71 : ein heldengedicht aus dem nachlass des seligen philipp ulrich schartenmayer [pseud] / Vischer, Friedrich Theodor – 4. verm aufl. Noerdlingen: C H Beck 1874 [mf ed 1995] – 1r [ill] – 1 – (filmed with: a morte de camoes / luis tieck) – mf#3754p – us Wisconsin U Libr [810]
Die deutsche kriegfuehrung in belgien und die mahnungen benedict 15 *see* Pan-germanism versus christendom
Deutsche kriegsopfer-zeitung – Bonn DE, 1951 nov-1988 – 1 – gw Misc Inst [934]
Deutsche kriegszeitung – Berlin DE, 1914 16 aug-1918 – 2r – 1 – (missing: 19 may 1918. filmed by misc inst: 1915, 1917-1919 27 jul) – gw Mikrofilm; gw Misc Inst [933]
Die deutsche kriminalerzaehlung von schiller bis zur gegenwart / ed by Greiner-Mai, Herbert & Kruse, Hans-Joachim – Berlin: Neue Berlin. 3v. 1967-69 – (incl bibl ref) – us Wisconsin U Libr [430]
Deutsche kritik – 1-30. Oct 1924-Dec 1925 – 1 – us L of C Photodup [780]
Deutsche kulturgeschichte im abriss / Jordan, E L – New York, NY. 1937 – 1r – us UF Libraries [943]
Deutsche kulturwacht – Berlin-Schoeneberg DE, 1932-33 – 1r – 1 – gw Mikrofilm [074]
Deutsche kulturzeitschriften des 19. jahrhunderts : mikrofiche-volltextausgabe zu alfred estermanns inhaltsanalytischer bibliographie deutscher kulturzeitschriften des 19. jahrhunderts (ibdk) – [mf ed 2002-04] – 1742mf (1:24) in 6 installments – 9 – diazo €9000.00 (silver €11,260 isbn: 978-3-598-35111-2) – ISBN-10: 3-598-35110-0 – ISBN-13: 978-3-598-35110-5 – gw Saur [074]
Deutsche kunst und dekoration – Darmstadt. v1-70. 1897-1932 – 9 – $828.00 – mf#0182 – us Brook [700]
Deutsche kunst-theater-musik-film-woche – Muenchen DE, 1919 mar, apr? – 1 – gw Mikrofilm [790]
Das deutsche land in seinen charakteristischen zuegen und seinen beziehungen zu geschichte und leben der menschen : zur belebung vaterlaendischen wissens und vaterlaendischer gesinnung / Kutzen, Joseph A – Breslau 1867 [mf ed Hildesheim: 1995-98] – (= ser Fbc) – 2v on 6mf – 9 – €120.00 – 3-487-29606-3 – gw Olms [943]
Der deutsche landarbeiter – Eisleben DE, 1913-34 n31 [many iss missing] – 10r – 1 – gw Misc Inst [630]
Deutsche landheimat – Leitmeritz (Litomerice CZ), 1928 23 may-1931 mar [gaps] – 3r – 1 – gw Misc Inst [077]
Die deutsche landnahme / Voigt, Bernhard – Potsdam: L Voggenreiter, 1936 [mf ed 1992] – (= ser Suedafrikanischer lederstrumpf 2) – 373p – 1 – mf#7758 – us Wisconsin U Libr [943]
Deutsche landpost – Boehmisch-Leipa (Ceska Lipa CZ), 1923 jan-mar – 1 – gw Misc Inst [077]
Deutsche landpost – Prag (CZ), 1923 4 apr-1938 31 mar – 29r – 1 – gw Misc Inst [077]
Deutsche la-plata-zeitung – Buenos Aires (RA), 1895-1916 (single iss), 1919 6 jul-1939 nov [gaps], 1941 3 apr-2 sep [gaps] – 107r – 1 – gw Misc Inst [079]
Deutsche latern *see* Frankfurter latern (sz2)
Deutsche lehrerinnenzeitung *see* Die lehrerin
Deutsche lehrerzeitung – Apr 1954-. semiweekly, -w – 1 – us Wisconsin U Libr [370]
Deutsche lehrerzeitung 1914 – Berlin DE, 1914 [mpf], 1923 – 1r – 1 – gw Misc Inst [370]
Deutsche lehrerzeitung 1954 – Berlin DE, 1954 3 apr-1990 nov – 25r – 1 – (filmed with suppl) – gw Misc Inst [370]
Deutsche leipaer zeitung – Boehmisch-Leipa (Ceska Lipa CZ), 1938 24 aug-nov – 1 – gw Misc Inst [077]
Die deutsche library – 1881-1892. 244 issues – 1 – $293.00 – us L of C Photodup [430]
Der deutsche lichtbildtheater-besitzer – Berlin DE, 1910 6 jan-1911 29 jun – 2r – 1 – (title varies: 1911: deutscher lichtspieltheater-besitzer) – gw Mikrofilm [790]
Deutsche lichtspiel-zeitung – Muenchen, Berlin DE, 1919 7 jun-26 jun, 1920-21 – 2r – 1 – gw Mikrofilm [790]
Deutsche liebe : aus den papieren eines fremdlings / Mueller, Friedrich Max – 11. Aufl. Leipzig: F A Brockhaus, 1898 – 1r – 1 – us Wisconsin U Libr [430]
Das deutsche lied der neuzeit : sein geist und wesen / Honegger, Johann Jakob – Leipzig: W Friedrich [1891] [mf ed 1993] – 1r – 1 – (filmed with: das deutsche volkslied / otto boeckel) – mf#8295 – us Wisconsin U Libr [430]
Deutsche lieder / Matthaey, Heinrich – Winterthur: Hegner, 1847 – 1r – 1 – us Wisconsin U Libr [810]
Deutsche lieder aus italien / Jacoby, Leopold – Muenchen: M Poessl, 1892 – 1r – 1 – us Wisconsin U Libr [780]
Deutsche liederdichter des 12. bis 14. jahrhunderts : eine auswahl / Bartsch, Karl – Berlin: B Behr, 1914 [mf ed 1993] – xciv/414p – 1 – (incl bibl ref and ind. middle high german text. int in german) – mf#8397 – us Wisconsin U Libr [430]
Deutsche literatur : das 19. und 20. jahrhundert : epochen, gestalten, gestaltungen / Urbanek, Walter – 2. verb. aufl. Bamberg: C C Buchners, 1971 [mf ed 1993] – 596p/[32pl] (ill) – 1 – (incl bibl ref and ind) – mf#8241 – us Wisconsin U Libr [430]
Deutsche literatur : eine geschichtliche darstellung ihrer hauptgestalten / Clauss, Walter – 3., durchgesehene Aufl. Zuerich: Schulthess & Co., 1945 – 1r – 1 – (incl bibl ref and index) – us Wisconsin U Libr [430]
Die deutsche literatur : geschichte und hauptwerke in den grundzuegen / Schulze, Erich; ed by Henning, Hans – Wittenberg: A Ziemsen Verlag, [1923] – 1r – 1 – (includes bilbiographical references and index) – us Wisconsin U Libr [430]
Die deutsche literatur / Menzel, Wolfgang – Stuttgart: Gebrueder Franckh. 2v. 1828 – 1 – us Wisconsin U Libr [430]
Deutsche literatur 1770-1870 / Grisebach, Eduard – Wien, Austria. 1876 – 1r – us UF Libraries [430]
Die deutsche literatur des barock : eine einfuehrung / Szyrocki, Marian – Stuttgart: Reclam, 1979 [mf ed 1993] – (= ser Rowohlts deutsche enzyklopaedie. sachgebiet literaturwissenschaft) – 268p – 1 – (incl bibl ref and ind) – mf#7848 – us Wisconsin U Libr [430]
Die deutsche literatur des barock : eine einfuehrung / Szyrocki, Marian – Reinbek bei Hamburg: Rowohlt Taschenbuch Verlag, 1968 – 1r – 1 – (incl bibl ref and index) – us Wisconsin U Libr [430]
Die deutsche literatur des lateinischen mittelalters in ihrer geschichtlichen entwicklung / Langosch, Karl – Berlin: De Gruyter, 1964 [mf ed 1993] – vi/284p – 1 – (incl bibl ref and ind) – mf#8162 – us Wisconsin U Libr [430]
Deutsche literatur des mittelalters : ein abriss / Wapnewski, Peter – Goettingen: Vandenhoeck & Ruprecht, c1960 [mf ed 1993] – (= ser Kleine vandenhoeck-reihe 96/97) – 127p – 1 – (incl bibl ref and ind) – mf#8162 – us Wisconsin U Libr [430]
Deutsche literatur des spaetmittelalters : ergebnisse, probleme und perspektiven der forschung – Greifswald: Ernst-Moritz-Arndt-Universitaet Greifswald, 1986 [mf ed 1993] – (= ser Wissenschaftliche beitraege der ernst-moritz-arndt-universitaet greifswald. deutsche literatur des mittelalters 3) – 448p (ill) – 1 – (incl bibl ref) – mf#8161 – us Wisconsin U Libr [430]
Die deutsche literatur im 19. jahrhundert, 1832-1914 / Alker, Ernst – 2nd rev enl ed. Stuttgart: A Kroener, 1962, c1961 [mf ed 1993] – (= ser Kroeners taschenausgabe v339) – 943p – 1 – (formerly under title: geschichte der deutschen literatur von goethes tod bis zur gegenwart. incl ind) – mf#8211 – us Wisconsin U Libr [430]
Deutsche literatur im 20. jahrhundert : strukturen und gestalten, zwanzig darstellungen / ed by Friedmann, Hermann & Mann, Otto – 3rd rev enl ed. Heidelberg: W Rothe, 1959 [mf ed 1993] – 482p – 1 – (incl bibl ref and ind) – mf#8254 – us Wisconsin U Libr [430]

Deutsche literatur im 20. jahrhundert : strukturen und gestalten / ed by Friedmann, Hermann & Mann, Otto – 4th rev enl ed. Heidelberg: W Rothe, 1961 [mf ed 1993] – 2v – 1 – (incl bibl ref and ind) – mf#8255 – us Wisconsin U Libr [430]

Deutsche literatur im 20. jahrhundert : strukturen und gestalten / ed by Mann, Otto & Rothe, Wolfgang – 5th rev enl ed. Bern, Muenchen: Francke Verlag, 1967 [mf ed 1993] – 2v – 1 – (incl bibl ref and ind) – mf#8255 – us Wisconsin U Libr [430]

Deutsche literatur im dritten reich : versuch einer darstellung in polemisch-didaktischer absicht / Schonauer, Franz – Olten (Switzerland), Freiburg im Brisgau (Germany): Walter-Verlag, c1961 [mf ed 1993] – 196p – 1 – (incl bibl ref) – mf#8244 – us Wisconsin U Libr [430]

Deutsche literatur im spaeten mittelalter, 1250-1450 / Wentzlaff-Eggebert, Friedrich Wilhelm & Wentzlaff-Eggebert, Erika – Reinbek bei Hamburg: Rowohlt, 1971- [mf ed 1993] – (= ser Rowohlts deutsche enzyklopaedie) – 1 – (incl bibl and ind) – mf#8171 – us Wisconsin U Libr [430]

Deutsche literatur im zwanzigsten jahrhundert : gestalten und strukturen, dreiundzwanzig darstellungen / ed by Friedmann, Hermann & Mann, Otto – Heidelberg: W Rothe, 1954 [mf ed 1993] – 450p – 1 – mf#8254 – us Wisconsin U Libr [430]

Deutsche literatur im zwanzigsten jahrhundert : gestalten und strukturen, zwanzig darstellungen / ed by Friedmann, Hermann & Mann, Otto – 2nd rev ed. Heidelberg: W Rothe, 1956 [mf ed 1993] – 442p – 1 – mf#8254 – us Wisconsin U Libr [430]

Deutsche literatur in unserer zeit / Kayser, W et al – 2nd rev ed. Goettingen: Vandenhoeck & Rupprecht, c1959 [mf ed 1993] – (= ser Kleine vandenhoeck-reihe 73/74) – 28p – 1 – (incl bibl ref and ind) – mf#8244 – us Wisconsin U Libr [430]

Deutsche literatur seit thomas mann / Mayer, Hans – Reinbeck bei Hamburg: Rowohlt Verlag, 1968, c1967 [mf ed 1993] – (= ser Rororo taschenbuch ausgabe 1063) – 125p – 1 – (incl ind) – mf#8257 – us Wisconsin U Libr [430]

Die deutsche literatur und die revolution von 1848 / Mehring, Franz; ed by Raddatz, Fritz J – Darmstadt: Luchterhand, 1975, c1974 [mf ed 1993] – (= ser Sammlung luchterhand 177; Mehring werkauswahl 2) – 237p – 1 – (incl ind) – mf#8140 – us Wisconsin U Libr [430]

Deutsche literaturgeschichte / Biese, Alfred – 25. aufl. Muenchen: C H Beck, c1930 [mf ed 1993] – 3v on 2r – 1 – (incl ind) – mf#8117 – us Wisconsin U Libr [430]

Deutsche literaturgeschichte – [Leipzig?: s.n., 187-] – 1r – 1 – (incl ind) – us Wisconsin U Libr [430]

Deutsche literaturgeschichte / Krell, Leo & Fiedler Leonhard – Bamberg: C C Buchners Verlag, 1960 – 1 – (incl bibl ref and index) – us Wisconsin U Libr [430]

Deutsche literaturgeschichte / Storck, Karl; ed by Rockenbach, M – Stuttgart: J B Metzler, 1926 – 1r – 1 – us Wisconsin U Libr [430]

Deutsche literaturgeschichte des neunzehnten jahrhunderts : dargestellt nach generationen / Kummer, Friedrich – Dresden: C Reissner, 1909 – 1 – (incl ind) – us Wisconsin U Libr [430]

Deutsche literaturgeschichte des neunzehnten jahrhunderts / Kummer, Friedrich – Dresden: C Reissner, 1909, c1905 – 1 – (incl ind) – us Wisconsin U Libr [430]

Deutsche literaturgeschichte in einer stunde : von den aeltesten zeiten bis zur gegenwart / Henschke, Alfred (pseud. Klabund) – Leipzig: Duerr & Weber, 1923 – 1r – 1 – us Wisconsin U Libr [430]

Deutsche literaturgeschichte in tabellen / Schmitt, Fritz – Bonn: Athenaeum-Verlag. 3v. 1949-52 – 1 – (incl bibl ref and indexes) – us Wisconsin U Libr [430]

Deutsche literaturkritik im zwanzigsten jahrhundert : kaiserreich, erster weltkrieg und erste nachkriegszeit [1889-1933] / Mayer, Hans – Stuttgart: Govert, 1965 [mf ed 1993] – (= ser Neue bibliothek der weltliteratur) – 858p – 1 – (incl bibl ref & ind; cont chronologically aut's: meisterwerke deutscher literaturkritik; cont by: deutsche literaturkritik der gegenwart) – mf#8085 – us Wisconsin U Libr [430]

Deutsche litteraturdenkmale des 18. und 19. jahrhunderts : [nr] 1 (1881)-nr 151 (1924) – Heilbronn: Henninger 1881-1924 [mf ed 1993] – 87v on 9r [ill] – 1 – mf#8676 – us Wisconsin U Libr [430]

Deutsche lodzer zeitung – Lodz (PL), 1915 8 feb-1918 9 nov – 8r – 1 – (filmed by other misc inst: 1939 24 sep-29 dec [1r]) – gw Misc Inst [077]

Deutsche londoner zeitung – London (GB), 1845 4 apr-1851 14 feb – 3r – 1 – uk British Libr Newspaper [072]

Deutsche lyrik : gedichte seit 1945 / ed by Bingel, Horst – 2. aufl. Muenchen: Deutscher Taschenbuch Verlag, 1965, c1961 – 1 – (incl bibl ref) – us Wisconsin U Libr [810]

Deutsche lyrik : gedichte seit 1945 / ed by Bingel, Horst – Stuttgart: Deutsche Verlags-Anstalt, c1961 – 1 – (incl bibl ref) – us Wisconsin U Libr [810]

Deutsche lyrik des siebzehnten jahrhunderts : in auswahl / ed by Merker, Paul – Bonn: A Marcus und E Weber, 1913 – (= ser Kleine texte fuer vorlesungen und uebungen) – 1mf – 9 – 0-524-04621-2 – (incl bibl ref) – mf#1990-1281 – us ATLA [430]

Deutsche lyrik des siebzehnten jahrhunderts in auswahl (kit124) / ed by Merker, P – Bonn, 1913 – €3.00 – ne Slangenburg [810]

Deutsche lyrik in gegenstuecken von hoelty bis werfel : 60 gedichte, zum zwecke vergleichender betrachtung / Oppert, Kurt – Wiesbaden: Kesselring, 1949 – 1r – 1 – (incl bibl ref) – us Wisconsin U Libr [430]

Die deutsche lyrik in ihrer geschichtlichen entwicklung : von herder bis zur gegenwart / Ermatinger, Emil – Leipzig: B G Teubner, 1921 – (= ser Aus deutscher dichtung v19) – 2v – 1 – (incl bibl ref and ind) – mf#8409 – us Wisconsin U Libr [430]

Deutsche lyrik nach 1945 / Wolf, Gerhard; ed by Volk und Wissen Volkeigener Verlag. Kollektiv fuer literaturgeschichte – Berlin: Volk und Wissen Volkeigener Verlag, 1964 – 175p – 1 – us Wisconsin U Libr [430]

Deutsche lyrik seit 1850 / ed by Spiero, Heinrich – Wien: Manz, 1912 – 1 – us Wisconsin U Libr [810]

Deutsche lyrik von heute und morgen / ed by Tille, Alexander – Leipzig: C G Naumann, 1896 – 1 – us Wisconsin U Libr [810]

Deutsche maedchenbildung – 1925-35 [mf ed 2001] – (= ser Hq 49) – 11v on 76mf – 9 – €420.00 – 3-89131-380-2 – gw Fischer [376]

Deutsche maerchen und sagen / Bechstein, Ludwig [comp] – 7. aufl. Berlin: Aufbau-Verlag, 1969 [mf ed 1993] – 474p – 1 – mf#8300 – us Wisconsin U Libr [390]

Das deutsche maerchendrama / Kober, Margarete – Frankfurt/Main: M Diesterweg 1925 [mf ed 1993] – (= ser Deutsche forschungen 11) – 1r – 1 – (incl bibl ref) – mf#8023 reel 2 – us Wisconsin U Libr [430]

Der deutsche maler – Duesseldorf DE, 1908-22, 1924-1933 23 jun – 3r – 1 – gw Misc Inst [750]

Deutsche malererzaehlungen : die art des sehens bei heinse, tieck, hoffmann, stifter und keller / Harnisch, Kaethe – Berlin: Junker und Duennhaupt, 1938 – 1 – (incl bibl ref) – us Wisconsin U Libr [430]

Deutsche medizinische wochenschrift – Stuttgart, Germany 1975+ – 1,5,9 – ISSN: 0012-0472 – mf#10157 – us UMI ProQuest [610]

Der deutsche meistergesang : poetische technik, musikalische form und sprachgestaltung der meistersinger / Nagel, Bert – Heidelberg: F H Kerle, c1952 [mf ed 1993] – 225p – 1 – (incl bibl ref) – mf#8174 – us Wisconsin U Libr [430]

Der deutsche merkur : literarische und politische zeitschrift / ed by Wieland, Christoph Martin – Frankfurt, Leipzig, Weimar DE, 1773-1805 – 446mf – 9 – (filmed by misc inst: 1773-76 [5r]. title varies: aug 1773: der teutsche merkur) – gw Mikropress; gw Misc Inst [074]

Der deutsche merkur / ed by Wieland, Christoph Martin – Weimar 1773-89 [mf ed 1993] – 201mf – 9 – €2780 diazo €3336 silver – 3-89131-100-1 – (filmed with: der neue teutsche merkur [1790-1810]; incl: anzeiger 1783-87; intelligenzblatt 1800-08; suppl: monatsberichte 1805-08) – gw Fischer [074]

Der deutsche metallarbeiter : organ des christlich-sozialen metallarbeiterverbandes – Duisburg DE, 1903-04, 1906-1933 8 jul – 9r – 1 – mf#3239 – gw Mikropress [660]

Deutsche metallarbeiterzeitung – Nürnberg, Stuttgart, Berlin DE, 1883 15 sep-1933 n23 [gaps] – 14r – 1 – (filmed by misc inst: 1916-24, 1926, 1891 [1]; title varies: 1903: metallarbeiter-zeitung) – mf#9289 – gw Mikropress; gw Misc Inst [670]

Der deutsche michael : historischer roman / Brachvogel, Albert Emil – Milwaukee, WI: G Brumdes, [1900?] [mf ed 1989] – 329p – 1 – mf#7061 – us Wisconsin U Libr [830]

Der deutsche michel : revolutionskomoedien der achtundvierziger / ed by Denkler, Horst – Stuttgart: Reclam c1971 [mf ed 1993] – 1r – 1 – (incl bibl ref. filmed with: sieben rosen hat der strauch / heinz czechowski [ed]) – mf#3360p – us Wisconsin U Libr [820]

Der deutsche militarismus / Blume, Wilhelm von – Tuebingen: Kloeres, 1915 [mf ed 1987] – (= ser Durch kampf zum frieden 2) – 26p – 1 – mf#6840 – us Wisconsin U Libr [355]

Die deutsche mission in suedindien : erzaehlungen und schilderungen von einer mission-studienreise durch ostindien / Richter, Julius – Guetersloh: C Bertelsmann, 1902 [mf ed 1995] – (= ser Yale coll) – vii/275p – 1 – 0-524-09023-8 – (in german) – mf#1995-0023 – us ATLA [915]

Deutsche mitteilungen – Gross-Wardein (Oradea RO), 1931 10 jan-20 jun – 1r – 1 – gw Misc Inst [077]

Deutsche mitteilungen – Paris (F), 1938 10 feb-1940 7 may – 3r – 1 – gw Misc Inst [074]

Der deutsche modernismus / Engert, Thaddaeus – Wuerzburg: Memminger, 1910 – 1mf – 9 – 0-8370-8095-9 – mf#1986-2095 – us ATLA [240]

Deutsche monatshefte in norwegen – Oslo (N), 1942 n6, 8 – 1 – gw Misc Inst [074]

Deutsche monatsschrift / ed by Fischer, Gottlob Nathanael – Leipzig 1795-99 – (= ser Dz. abt literatur) – 3v on 72mf – 9 – €670.00 – mf#k/n4605 – gw Olms [430]

Deutsche monatsschrift / ed by Gentz, Friedrich von et al – Berlin 1790-94 – (= ser Dz. abt literatur) – 3v on 67mf – 9 – €670.00 – mf#k/n4580 – gw Olms [430]

Deutsche montagszeitung – Berlin DE, 1910 3 oct-1916 – 2r – 1 – (title varies: 17 oct 1910: deutsche montags-zeitung, 23 nov 1910: d m z, deutsche montags-zeitung; with suppl: illustrierte deutsche montags-zeitung 1911 13 mar-3 apr) – mf#6243 – gw Mikropress [074]

Deutsche Morgenlaendische Gesellschaft *see*
– Zeitschrift fuer indologie und iranistik
– Zeitschrift fuer semitistik und verwandte gebiete

Die Deutsche Morgenlaendische Gesellschaft *see* Wissenschaftlicher jahresbericht ueber die morgenlaendischen studien

Deutsche morgenlaendische gesellschaft. zeitschrift – v1-111. 1847-1961. – 9 – $1752.00 – (with ind) – mf#0184 – us Brook [400]

Deutsche musikbibliographie – Leipzig. 1873-1975. 19 reels – 1 – us L of C Photodup [780]

Deutsche musiker-zeitung – Berlin. v. 1-27, 63-64. Apr 1870-1896; Jan 2 1932-May 6 1933 – 1 – us NY Public [780]

Die deutsche mystik im prediger-orden (von 1250-1350) : nach ihren grundlehren, liedern und lebensbildern / Greith, Carl – Freiburg im Breisgau: Herder, 1861 [mf ed 1991] – 2mf – 9 – 0-7905-9946-5 – mf#1989-1671 – us ATLA [241]

Deutsche mythologie / Grimm, J – 3. ausgabe. Goettingen. v1-2. 1854 – €44.00 – ne Slangenburg [390]

Deutsche mythologie / Grimm, Jacob – 4. Ausg. Berlin: F Duemmler, 1875-1878 – 1mf – 9 – 0-524-04512-7 – (incl bibl ref) – mf#1990-3346 – us ATLA [430]

Deutsche mythologie / Kauffmann, Friedrich – 2. aufl. Stuttgart: G J Goeschen 1900 [mf ed 1991] – (= ser Sammlung goeschen) – 1mf – 9 – 0-524-01773-5 – (incl bibl ref) – mf#1990-2621 – us ATLA [290]

Deutsche nachrichten – Hannover DE, 1960-73 – 1 – (1974: deutsche wochen-zeitung, hannover) – gw Misc Inst [074]

Deutsche nachrichten : antifaschistische monats- bzw halbmonatsschrift – Kopenhagen DE, 1943 aug- 1949 15 nov [gaps] – 1r – 1 – (title varies: jul 1945: wochenzeitung fuer deutsche fluechtlinge aus den ostgebieten in daenemark) – gw Misc Inst [074]

Deutsche nachrichten – Bielitz-Biala (Bielsko-BiaLa PL), 1939 1jul-15/16 aug – 1 – gw Misc Inst [077]

Deutsche nachrichten – Budapest (H), 1942 3 jan-25 dec – 1r – 1 – gw Misc Inst [077]

Deutsche nachrichten – Posen (Poznan PL), 1936, 1939 jul-aug – 1 – (filmed by other misc inst: 1934 nov-1936, 193 mar-jun, 1937 sep-1939 jun [7r]) – gw Misc Inst [077]

Deutsche nachrichten – Pressburg (Bratislava, SK), 1923 20 oct-1925 14 nov – 1r – 1 – gw Misc Inst [077]

Deutsche nachrichten – Zagreb (Agram HR), 1939 4 mar-16 dec, 23 dec – 1r – 1 – gw Misc Inst [077]

Deutsche nachrichten aus kultur, wirtschaft und politik – London (GB), 1946 oct-1947 mar/apr – 1r – 1 – gw Misc Inst [072]

Deutsche nachrichten fuer litauen – Kauen (Kaunas, Kowno LT), Memel (Klaipeda LT), 1931-37, 1939-40 – 3r – 1 – (fr 1989? in klaipeda (memel). filmed by other misc inst: 1931 11 jan-1937, 1939 7 jan-1941 1 mar, 2000 apr-2001 nov [until 1941 3r]) – gw Misc Inst [077]

Deutsche nachrichten in griechenland – Athen (GR), 1942-1944 30 jun – 1 – gw Misc Inst [074]

Deutsche namenkunde : unsere familiennamen nach ihrer entstehung und bedeutung / Gottschald, Max – Muenchen: J.F. Lehmanns verlag, 1932. 423p – 1 – us Wisconsin U Libr [920]

Deutsche national- und soldatenzeitung – Muenchen, 1951-92 – 30r – 1 – (1993-dm80.00y) – gw Mikropress [074]

Deutsche National-Literatur *see* Historisch-kritische ausgabe

Die deutsche nationallitteratur des neunzehnten jahrhunderts : litterarhistorisch und kritisch dargestellt / Gottschall, Rudolf von – 3. verm. und verb. Aufl. Breslau: E Trewendt, 1872 – 1r – 1 – us Wisconsin U Libr [430]

Deutsche national-zeitung – Muenchen, Passau DE, 1951 6 jun- – ca 1r/yr – 1 – (began as: deutsche national- und soldatenzeitung, aka: deutsche national-zeitung; filmed by bnl: 1968 12 jan-dec, 1970-1971 jun, 1972 jan-jun, 1973-76 [7r], filmed by misc inst: 1954-72, 1955-1965 19 mar, 1967 7 apr-1969 21 jan) – mf#2603 – gw Mikropress; uk British Libr Newspaper; gw Misc Inst [355]

Der deutsche oekonomist – Berlin DE, 1893, 1914 – 1 – gw Misc Inst [330]

Deutsche opposition – Hamburg DE, 1951 n12-29, 1952 n1-17, 19-28 – 1 – gw Misc Inst [943]

Deutsche ostfront – Gleiwitz (Gliwice PL), 1935 sep-oct, 1935 dec-1936 feb – 1 – gw Misc Inst [077]

Deutsche passion 1933 : hoerwerk in sechs saetzen / Euringer, Richard – Oldenburg i/O: G Stalling 1933 [mf ed 1989] – (= ser Stalling-buecherei "schriften an die nation" 24) – 1r – 1 – (filmed with: die arbeitslosen) – mf#7226 – us Wisconsin U Libr [780]

Deutsche pilgerreisen nach dem heiligen lande / Roehricht, R & Meisner, H – Berlin, 1880 – 8mf – 9 – mf#H-3127 – ne IDC [956]

Der deutsche pionier : erinnerungen aus dem pionierleben der deutschen in amerika – Cincinnati: Deutschen Pionier-Verein, 1869- . v6 mar 1874-feb 1875 – 1r – 1 – us CRL [305]

Deutsche pionier – Wausau WI. 1884 oct 18-1887 oct 1, 1889 apr 13-1891 dec 31, 1892-98, 1899-1900, 1901-13, 1914 jan 1-1917 jan 6 – 12r – 1 – mf#1094633 – us WHS [071]

Deutsche polar-zeitung – Tromsoe (N), 1943-1944 27 may [gaps] – 2r – 1 – gw Misc Inst [074]

Die deutsche polenliteratur 1918-193? : stoff- und motivgeschichte / Chodera, Jan – Poznan: Uniwersytet im A Mickiewicz w Poznaniu, 1966 [mf ed 1993] – (= ser Prace wydzialu filologicznego. seria filologia germa'nska / uniwersytet im adama mickiewicza w poznaniu 3) – 357p – 1 – (incl bibl ref) – mf#8360 – us Wisconsin U Libr [430]

Deutsche post – Lodz (PL), 1915 19 jul-1918 10 nov – 1r – 1 – gw Misc Inst [077]

Deutsche post – Troppau (Opava CZ), 1923-33 – 33r – 1 – gw Misc Inst [077]

Deutsche predigten des 12. und 13. jahrhundertes / ed by Roth, Karl – Quedlinburg, Leipzig: G Basse, 1839 [mf ed 1993] – (= ser Bibliothek der gesammten deutschen national-literatur von den aeltesten bis auf die neuere zeit sect1/11-1) – xl/84p – 1 – (incl bibl ref) – mf#8438 reel 4 – us Wisconsin U Libr [430]

Deutsche predigten des 13. und 14. jahrhundertes / ed by Leyser, Herm – Quedlinburg, Leipzig: G Basse, 1838 [mf ed 1993] – (= ser Bibliothek der gesammten deutschen national-literatur von den aeltesten bis auf die neuere zeit sect1/11-2) – xxxiii/170p – 1 – (incl bibl ref) – mf#8438 reel 4 – us Wisconsin U Libr [430]

Deutsche presse – Belgrad (YU), 1938 9 oct-11 dec – 1r – 1 – gw Misc Inst [077]

Deutsche presse – Berlin DE, 1913 18 oct-1944 – 5mf=8df – 9 – gw Mikrofilm [074]

Deutsche presse – London, UK. 5 Jun-24 Jul 1841 – 1 – uk British Libr Newspaper [072]

Deutsche presse – Prag (CZ), 1925 8 may-1934 16 mar – 19r – 1 – gw Misc Inst [077]

Deutsche presse – Toronto, Ontario (CDN), 1982 21 apr-1993 25 aug, 2001- – 1 – gw Misc Inst [071]

Deutsche presse *see* Der zeitungs-verlag

Der deutsche primas : eine untersuchung zur deutschen kirchengeschichte in der ersten haelfte des neunzehnten jahrhunderts / Becher, Hubert – Kolmar [c1945] (mf ed 1992) – 2mf – 9 – €24.00 – 3-89349-079-5 – mf#DHS-AR 52 – gw Frankfurter [240]

Die deutsche prosa von mosheim bis auf unsere tage : eine mustersammlung / ed by Schwab, Gustav – Stuttgart: C Bertelsmann. 3v. 1860 – 1r – 1 – us Wisconsin U Libr [430]

Deutsche prosadichtungen der gegenwart : interpretation fuer lehrende und lernende / ed by Zimmermann, Werner – Duesseldorf: Paedagogischer Verlag Schwann, 1960-1962 [mf ed 1993] – 3v on 1r – 1 – (incl bibl ref) – mf#8263 – us Wisconsin U Libr [430]

Deutsche prosadichtungen unseres jahrhunderts : interpretationen fuer lehrende und lernende / Zimmermann, Werner – 4. aufl. Duesseldorf: Paedagogischer Verlag Schwann, 1974-76 [mf ed 1993] – 2v – 1 – (incl bibl ref) – mf#8187 – us Wisconsin U Libr [430]

Der deutsche protestantenverein : und seine bedeutung in der gegenwart / Schenkel, Daniel – Wiesbaden: C.W. Kreidel, 1868 – 1mf – us ATLA [242]
Der deutsche protestantenverein : und seine bedeutung in der gegenwart / Schenkel, Daniel – Wiesbaden: C.W. Kreidel, 1868 – 1mf – 9 – 0-7905-6678-8 – mf#1988-2678 – us ATLA [242]
Der "deutsche ptolemaeus" : aus dem ende des 15. jahrhunderts (um 1490): in faksimiledruck / ed by Fischer, Joseph – Strassburg: J H E Heitz (Heitz & Muendel) 1910 [mf ed 1993] – (= ser Drucke und holzschnitte des 15. und 16. jahrhunderts in getreuer nachbildung 13) – 1r – 1 – (int by ed; incl bibl ref. filmed with: kleines deutsches sagenbuch / will-erich peuckert [ed]) – mf#3367p – us Wisconsin U Libr [520]
Deutsche radio illustrierte *see* Deutsche funk illustrierte
Deutsche rechtsalterthuemer / Grimm, Jacob – 4., verm Ausg. Leipzig: T Weicher. v1-2. 1899 – (= ser Civil law 3 coll) – 16mf – 9 – (incl bibl ref) – mf#LLMC 96-537 – us LLMC [340]
Deutsche rechts-zeitschrift – Tuebingen DE, 1946-50 [gaps] – 1 – gw Misc Inst [342]
Deutsche redekunst im 17. und 18. jahrhundert / Stoetzer, Ursula – Halle (Saale): M Niemeyer Verlag, 1962 – 290p – 1 – (incl bibl ref) – us Wisconsin U Libr [430]
Der deutsche redner fuer recht und freiheit – Bochum DE, 1849 apr-1850 19 jul, 1850 28 aug-14 sep – 1 – (title varies: aug 1850: oeffentlicher anzeiger fuer den kreis bochum) – gw Misc Inst [074]
Deutsche reform *see* Deutsche wacht 1905
Deutsche reform 1880 – Dresden, Berlin DE, 1880-1887 22 mar, 1887 5 apr-1911 29 jun – 1 – (title varies: 5 apr 1887: deutsche wacht, 1 oct 1905: deutsche reform) – gw Misc Inst [074]
Die deutsche reform : politische zeitung fuer das constitutionelle deutschland – Berlin DE, 1849 2 jan-1851 10 mar – 4r – 1 – gw Misc Inst [323]
Die deutsche reformation / Kahnis, Karl Friedrich August – Leipzig: Doerffling und Franke, 1872 – 1mf – 9 – 0-7905-4765-1 – (incl bibl ref) – mf#1988-0765 – us ATLA [242]
Deutsche reichs-bremse *see* Die laterne (sz5)
Deutsche reichspost : zentralorgan der konservativen sueddeutschlands – Stuttgart DE, 1901-11 – 1 – gw Misc Inst [325]
Der deutsche reichstag waehrend des oesterreichischen erbfolgekrieges (1740-1748) / Meisenburg, Friedrich – Bonn 1931 (mf ed 1995) – 2mf – 9 – €31.00 – 3-8267-3165-4 – mf#DHS-AR 3165 – gw Frankfurter [943]
Deutsche reichstags-zeitung – Frankfurt/M DE, 1848 1 oct-1849 2 apr – 1r – 1 – gw Misc Inst [074]
Deutsche reichs-zeitung – Braunschweig DE, 1848 20 mar-1866 2 aug – 36r – 1 – (title varies: until 30 jun 1848: zeitung fuer das deutsche volk) – gw Misc Inst [074]
Deutsche reichszeitung – Bonn DE, 1871 17 dec-1941 21 mar – 134r – 1 – (title varies: 1934 1 oct: mittelrheinische landeszeitung; filmed by mikropress: 1872-77) – gw Misc Inst [074]
Deutsche reichszeitung *see* Godesberger volkszeitung
Deutsche republik – Frankfurt. v1-7 n39. nov 1926-june 25 1933 – 1 – us NY Public [073]
Deutsche republik – Frankfurt/M DE, 1926 nov-1933 25 jun – 5r – 1 – gw Misc Inst [074]
Deutsche revolution – Duesseldorf DE, 1934 [single iss], 1935-1937 jul – 1r – 1 – (title varies: 1935: deutsche volksschoepfung) – gw Misc Inst [943]
Die deutsche revolution : organ der schwarzen front – Prag (CZ), 1934 13 may-1937 15 nov – 1r – 1 – (previously publ in berlin) – gw Misc Inst [934]
Deutsche revue : eine monatsschrift – Stuttgart and Leipzig. v1-47. 1877-1922 – 1 – $512.00 – (wanting: scattered iss; cont by: adolf bastian: spiritisten und theosophen, breslau 1885) – us L of C Photodup [290]
Die deutsche revue / Gutzkow, Karl & Wienbarg, Ludolf; ed by Dresch, J – Berlin: B Behr, 1904 [mf ed 1993] – (= ser Deutsche litteraturdenkmale des 18. und 19. jahrhunderts 132, 3. folge n12) – xliii/39p – 1 – mf#8676 reel 7 – us Wisconsin U Libr [430]
Deutsche revue der gegenwart *see* Unsere zeit, leipzig 1857-1891
Deutsche rio-zeitung – Rio de Janeiro (BR), 1940 1 aug-1941 11 jan, 1941 27 feb-29 jun – 2r – 1 – (filmed by other misc inst: 1923 10 sep-1940 10 feb, 1940 1 aug-1941 11 jan (gaps), 1941 27 feb-29 jun (gaps) [39r]. filmed with suppl) – gw Misc Inst [079]
Der deutsche roman im 20. jahrhundert / Welzig, Werner – 2. erw aufl. Stuttgart: A Kroener, c1970 [mf ed 1993] – (= ser Kroeners taschenausgabe 367) – vii/428p – 1 – (incl bibl ref and ind) – mf#8282 – us Wisconsin U Libr [430]
Der deutsche roman im 20. jahrhundert / Welzig, Werner – 2. erw auf. Stuttgart: A Kroener c1967 [mf ed 1992] – 1r – 1 – (incl bibl ref & ind. filmed with: abriss der deutschen dichtung / hans roehl) – mf#3232p – us Wisconsin U Libr [430]
Deutsche romantik / Sallwuerk, Edmund von – 2. aufl. Frankfurt/M: Verlag von M Diesterweg, 1920 [mf ed 1993] – (= ser Diesterwegs deutsche schulausgaben 16) – 192p – 1 – (incl bibl ref and ind) – mf#8184 – us Wisconsin U Libr [430]
Die deutsche romantik / Jantzen, Hermann – new rev ed. Bielefeld [Germany]: Velhagen & Klasing, 1933 – xviii/268p/1pl (ill) – 1 – us Wisconsin U Libr [430]
Der deutsche rundfunk – Berlin DE, 1923 14 oct-1941 25 may – 23r – 1 – gw Mikrofilm [380]
Deutsche rundschau – Bydgoszcz, Poland sep-1 oct 1939, 4 apr-31 dec 1942, 2 may-18 sep 1944 (imperfect) [mf 1942, 1944] – 1 – uk British Libr Newspaper [077]
Deutsche rundschau – Bydgoszcz, Poland. Sept-Oct 1939 – 1r – 1 – us L of C Photodup [943]
Deutsche rundschau / ed by Rodenberg, Julius – Berlin/Leipzig 1874-1942 [mf ed 1993] – 271v on 951mf – 9 – €5220.00 – 3-89131-157-5 – gw Fischer [074]
Deutsche rundschau – Udora, Ontario (CDN), 2000 feb-2002 – 1r – 1 – gw Misc Inst [071]
Deutsche rustungspolitik vom beginn der genfer abrustungskonferenz bis zur wiedereinfuehrung der allgemeinen wehrpflicht / Rautenberg, Hans Juergen – 1932-1935 – 1 – gw Mikropress [943]
Deutsche saar – Saarbruecken DE, 1955 jul-1964 jun – 1 – gw Misc Inst [074]
Deutsche saar-zeitung – Bad Kreuznach DE, 1951 22 dec-1955 2 dec – 1r – 1 – gw Misc Inst [074]
Deutsche sagen / ed by Grimm, Brueder – 2. aufl. Berlin: Nicolaische Verlagsbuchhandlung, G Parthey, 1865-66 [mf ed 1993] – 2v (ill) – 1 – mf#8360 – us Wisconsin U Libr [390]
Deutsche sagen / Grimm, Wilhelm & Grimm, Jacob; ed by Ritter, Gustav A – Berlin: Verlagsdruckerei 'Merkur', 1904 [mf ed 1993] – 675p/60pl (ill) – 1 – (ill by ed bruening and h tischler) – mf#8375 – us Wisconsin U Libr [390]
Deutsche sagen / Richter, Albert – Leipzig: F Brandstetter 1878 [mf ed Bloomington IN: Indiana Uni Lib, Preservation Dept 1984] – 1r – 1 – us Indiana Preservation [390]
Deutsche sagen / Richter, Albert – Leipzig, Germany. 1878 – 1r – us UF Libraries [390]
Deutsche salonlieder / Hoffmann von Fallersleben, August Heinrich – Zuerich: Verlag des literarischen Comptoirs, 1844 – 1r – 1 – us Wisconsin U Libr [780]
Deutsche satiriker des 16. jahrhunderts / Geiger, Ludwig – Berlin: C Habel, 1878 – 1r – 1 – us Wisconsin U Libr [430]
Die deutsche schallform der letzten bluetezeit und ihrer auslaufer in dichtung und prosa / Bluemel, Rudolf – Halle, 1923 (mf ed 1994) – 2mf – 9 – €31.00 – 3-8267-3020-8 – mf#DHS-AR 3020 – gw Frankfurter [430]
Deutsche schauspieler und schauspielkunst / Kirchbach, Wolfgang – Kiel: Lipsius und Tischer, 1892 – 1r – 1 – us Wisconsin U Libr [790]
Deutsche schriften / Ellis, Manfred Maria – 2. Aufl. Berlin: Sanssouci-Verlag des Deutschen Verlags-Instituts. 3v. 1924 (mf ed 1990) – 1r – 1 – (filmed with: gesammelte werke) – us Wisconsin U Libr [800]
Deutsche schriften / Seuse, Heinrich; ed by Bihlmeyer, K – Stuttgart, 1907 – 17mf – 8 – €32.00 – ne Slangenburg [240]
Deutsche schriftsteller als richter ihrer zeit / Roch, Herbert – Berlin: Horizont, 1947 [mf ed 1993] – 142p – 1 – mf#8155 – us Wisconsin U Libr [430]
Deutsche schriftsteller in der entscheidung : wege zur arbeiterklasse 1918-1933 / Albrecht, Friedrich – 1. aufl. Berlin: Aufbau-Verlag, 1970 [mf ed 1993] – (= ser Beitraege zur geschichte der deutschen sozialistischen literatur im 20. jahrhundert v2) – 698p – 1 – (incl bibl ref) – mf#8265 – us Wisconsin U Libr [430]
Das deutsche schrifttum bis zum ausgang des mittelalters, vol 1 : von der germanischen welt zum christlichdeutschen mittelalter / Wolff, Ludwig – Goettingen: Vandenhoeck & Ruprecht 1951- [mf ed 1992] – 1r [ill] – 1 – (incl bibl ref & ind. filmed with: dichtung und dichter der zeit / albert soergel) – mf#3010p – us Wisconsin U Libr [430]
Deutsche schuetzen- und wehr-zeitung – Bremen DE, 1866-84, apr 1890-1915, jan 14 1920-nov 1922, 1924-28, 1929, 1930-32, 1935 – 21r – 1 – (with gaps) – gw Misc Inst [355]
Der deutsche schulfreund : ein nuetzliches hand- und lesebuch fuer lehrer in buergerund landschulen / ed by Zerrener, H G – Erfurt 1791-1801 – (= ser Dz) – 24v on 32mf – 9 – €320.00 – mf#k/n727 – gw Olms [370]
Der deutsche schulfreund – Erfurt DE, 1791-1817, 1820-21, 1823 – 11r – 1 – (title varies: 1801: der neue schulfreund; 1812: der neueste deutsche schulfreund) – gw Misc Inst [370]
Deutsche schulzeitung – Berlin DE, 1876-80, 1885-86, 1891-92, 1905 – 1r – 1 – gw Misc Inst [370]
Deutsche schwaenke : in einem band / ed by Albrecht, Guenter – Weimar: Volksverlag, 1959 – 1r – 1 – (includes bio-bibliographical) – us Wisconsin U Libr [430]
Die deutsche schweizerbegeisterung in den jahren 1750-1815 / Ziehen, Eduard – Frankfurt am Main: M Diesterweg, 1922 – 1r – 1 – (incl bibl ref) – us Wisconsin U Libr [943]
Der deutsche sender – Berlin DE, 1930 3 oct-1936 7 jun – 5mf=9df – 9 – gw Mikrofilm [380]
Deutsche shanghai-zeitung – Schanghai (VR), 1932 4-9 dec, 1933 31 dec-1935 – 6r – 1 – (with suppl: shanghai-illustrierte 1935 31 dec) – gw Misc Inst [079]
Deutsche sippennamen : ableitendes woerterbuch der deutschen familiennamen / Brechenmacher, Josef Karlmann – Goerlitz: C A Starke 1936 [mf ed 1986] – (= ser Sippen bucherei 5-9) – 5v on 1r – 1 – mf#8499 – us Wisconsin U Libr [430]
Deutsche soldaten-zeitung *see* Deutsche national-zeitung
Die deutsche soldaten-zeitung / soldat im volk – Passau DE, 1954-88 – 1 – gw Misc Inst [355]
Deutsche sonderrenaissance in deutscher prosa : strukturanalyse deutscher prosa im sechzehnten jahrhundert / Gumbel, Hermann – Frankfurt am Main: M Diesterweg, 1930 [mf ed 1993] – (= ser Deutsche forschungen 23) – xii/268p – 1 – (incl bibl ref) – mf#8023 reel 4 – us Wisconsin U Libr [430]
Deutsche sonntagspost – Winona MN. 1939 apr 30-1941 dec 28, 1942 jan 4-1943 apr 11 – 2r – 1 – (cont by: sonntagspost) – mf#852487 – us WHS [071]
Deutsche sprachlehre fur auslander / Griesbach, Heinz – Muenchen, Germany. 1961 – 1r – us UF Libraries [430]
Deutsche Staatsbibliothek Berlin. Musiksammlung *see* Alphabetischer katalog der staatsbibliothek zu berlin preussischer kulturbesitz
Deutsche staatsbuerger zeitung *see* Zeitung des jungdeutschen ordens
Deutsche stimmen – Bratislava, Czechoslovakia. Jan-Jun 1943 – 1r – 1 – us L of C Photodup [077]
Deutsche stimmen – Berlin DE, 1919-28 – 4r – 1 – (filmed by misc inst: 1904 23 mar [probe-nr], 1904 2 apr-1906 22 dec [1r]) – mf#2817 – gw Mikropress; gw Misc Inst [074]
Deutsche stimmen – Pressburg (Bratislava SK), 1936 jan-1943 dec – 4r – 1 – gw Misc Inst [077]
Deutsche stimmen – Pressburg (Bratislava SK), 1936-43 – 4r – 1 – gw Misc Inst [077]
Deutsche stimmen 1956 : neue prosa und lyrik aus ost und west / ed by Bruns, Marianne et al – Stuttgart: Kreuz-Verlag; Halle (Saale): Mitteldeutscher Verlag, c1956 – 367p – 1 – us Wisconsin U Libr [800]
Deutsche studenten-zeitung – Muenchen DE, 1933-1935 n17 – 1 – gw Misc Inst [378]
Deutsche studentenzeitung – Duesseldorf DE, 1951-feb 18 1957, may 8 1957-feb 23 1959 – 2r – 1 – gw Misc Inst [378]
Deutsche studien : 3. dramen und dramatiker / Scherer, Wilhelm – Wien: In Commission bei Karl Gerold's Sohn, 1878 [mf ed 1993] – 60p – 1 – mf#8034 – us Wisconsin U Libr [430]
Der deutsche sturmtrupp – Muenchen DE, 1933 1 jan-15 dec, 1934 1 jan-1 oct – 1r – 1 – gw Misc Inst [355]
Deutsche suedpolar-expedition 1901-1903... : vol 2: geographie und geologie / ed by Drygalski, E von – London. 1852-1854 (1) – 26mf – 9 – mf#2831 – ne IDC [919]
Deutsche tabakarbeiter-zeitung – Duesseldorf DE, 1906-1933 15 jul – 4r – 1 – gw Mikropress [660]
Das deutsche tabu : sunfuzius und der wind des 21. jahrhunderts und der neue preussenzyklus und die elitedaemmerung / Schuermeyer, Manfred – 1994 – 2mf – 9 – 3-89349-876-1 – gw Frankfurter [330]
Deutsche tagespost – Hermannstadt (Sibiu RO), 1919 6 dec-1925 – 8r – 1 – gw Misc Inst [077]
Deutsche tagespost *see* Augsburger tagespost
Deutsche tageszeitung – Berlin DE, 1914 1 sep-1934 30 apr [gaps] – 103r – 1 – (filmed by misc inst: 1894 1 sep-1898 aug, 1898 oct-1914 (gaps) [135r]) – mf#8547 – gw Mikropress; gw Misc Inst [074]
Deutsche tageszeitung – Hermannstadt (Sibiu RO), 1934 2 oct-1939 14 jul – 6r – 1 – gw Misc Inst [077]
Deutsche tageszeitung – Karlsbad (Karlovy Vary CZ), 1938 4 oct-dec – 1r – 1 – gw Misc Inst [077]
Deutsche tageszeitung fuer sued-brasilien – Curityba [BR], 1925 20 jun-1930 30 jan [gaps] – 4r – 1 – (title varies: 15 oct 1927: deutsche zeitung) – gw Misc Inst [079]
Deutsche tageszeitung in polen – Posen (Poznan PL) 1934 1 dec-1939 29 aug – 8r – 1 – gw Misc Inst [077]
Deutsche taschen-encyklopaedie (ael1/21) : oder handbibliothek des wissenswuerdigsten in hinsicht auf natur und kunst, staat und kirche, wissenschaft und sitte – Leipzig/ Altenburg 1816-20 [mf ed 1994] – (= ser Archiv der europaeischen lexikographie, abt 1: enzyklopaedien) – 4pt on 16mf – 9 – €180.00 – 3-89131-171-0 – gw Fischer [030]
Der deutsche telescope *see* Der froehliche botschafter
Deutsche texte des mittelalters – Berlin. v1-41. 1904-38 – 9 – $432.00 – mf#0185 – us Brook [430]
Deutsche texte des mittelalters / ed by Zentralinstitut fuer Sprachwissenschaft – Berlin: Akademie-Verlag, 1904- (irregular) [mf ed 1994] – 21r – 1 – (began 1904. v1-41 publ) – mf#8623 – us Wisconsin U Libr [430]
Deutsche tierfabeln vom 12. bis zum 16. jahrhundert / ed by Schaeffer, Richard – Berlin: Ruetten & Loening, 1955 – 1r – 1 – (incl bibl ref) – us Wisconsin U Libr [430]
Deutsche toepfer-zeitung – Leipzig DE, 1892-1898 oct, 1899 jul-1902 feb [gaps] – 8r – 1 – uk British Libr Newspaper [730]
Die deutsche tragoedie : von lessing bis hebbel / Wiese, Benno von – 4.aufl. Hamburg: Hoffmann und Campe, 1958, c1948 [mf ed 1993] – xvii/712p – 1 – (incl bibl ref and ind) – mf#8190 – us Wisconsin U Libr [430]
Die deutsche tragoedie von lessing bis hebbel *see* Sebastian brants narrenschiff
Deutsche treue – Hemme DE, 1914-16 – 1 – gw Misc Inst [074]
Die deutsche treue : vortrag, gehalten in der aula der annen-realschule zu dresden am 23. april, dem geburtstage sr. majestaet, des koenigs von sachsen / Dolch, Oskar – Dresden: R v Zahn, 1879 – 1r – 1 – (incl bibl ref) – us Wisconsin U Libr [943]
Die deutsche treue in sage und poesie : vortrag, gehalten am geburtstage seiner koeniglichen hoheit des grossherzogs von mecklenburg-schwerin friedrich franz am 28. februar 1867 / Bartsch, Karl – Leipzig: F C W Vogel, 1867 [mf ed 1993] – 28p – 1 – mf#8139 – us Wisconsin U Libr [430]
Deutsche tribuene – Muenchen, Zweibruecken, Homburg DE, 1831 1 jul-1832 21 mar [gaps] – 1r – 1 – (fr dec 1831 publ in zweibruecken, since jan 1832 in homburg) – gw Misc Inst [074]
Deutsche turnzeitung : blaetter fuer die interessen des gesamten turnwesens – Leipzig DE, 1856 jul-1943 – 32r – 1 – mf#7304 – gw Mikropress [790]
Deutsche turnzeitung fuer frauen : organ fuer die frauen- turn- spiel- und sportvereinigungen – Krefeld DE, 1899 1 jan-1902 28 dec – 1r – 1 – mf#12509 – gw Mikropress [790]
Der deutsche typus der tragoedie : dramaturgisches fundament / Bacmeister, Ernst – Berlin: Theaterverlag Albert Langen/ Georg Mueller, [1941?] [mf ed 1993] – 122p – 1 – (incl bibl ref) – mf#8409 – us Wisconsin U Libr [790]
Deutsche ukraine zeitung – Lemberg (Lwow UA), 1924 4 mar-30 jun – 1r – 1 – gw Misc Inst [077]
Deutsche ukraine-zeitung – Luzk (UA), 1943 1 jul-1944 7 jan [gaps] – 1 – (filmed by misc inst: 1942 23 jan-1943 29 jun, 1943 1 oct-31 dec) – uk British Libr Newspaper; gw Misc Inst [077]
Deutsche und englische romantik : eine gegenueberstellung / Mason, Eudo Colecestra – Goettingen: Vandenhoeck &. Ruprecht, 1959 [mf ed 1993] – (= ser Kleine vandenhoeck-reihe 85/85a) – 102p – 1 – (incl bibl ref) – mf#8245 – us Wisconsin U Libr [410]
Deutsche und franzoesische dichtung des mittelalters / Spanke, Hans – Stuttgart: W Kohlhammer, 1943 [mf ed 1993] – (= ser Frankreich: sein weltbild und europa) – viii/117p – 1 – (incl bibl ref and ind) – mf#8162 – us Wisconsin U Libr [430]
Das deutsche vaterland – Muenchen DE, 1884-88 [gaps] – 1r – 1 – gw Misc Inst [943]
Deutsche verfassungen – Berlin, Germany. 1951 – 1r – us UF Libraries [323]

Deutsche vergangenheit in deutscher dichtung : deutsche renaissance: rede bei uebernahme des rektorats der schlesischen friedrich wilhelms-universitaet zu breslau, am 30. september 1918 / Koch, Max – Stuttgart: Metzler, 1919 [mf ed 1993] – (= ser Breslauer beitraege zur literaturgeschichte. neue folge 50) – 72p – 1 – (incl bibl ref) – mf#8014 reel 5 – us Wisconsin U Libr [430]
Deutsche verkehrszeitung – Hamburg DE, 1947 10 jul-1988 – 1 – gw Misc Inst [625]
Der deutsche volksaberglaube in seinem verhaeltnis zum christentum und im unterschiede von der zauberei / Freybe, Albert – Gotha: F A Perthes, 1910 – 1mf – 9 – 0-524-00878-7 – mf#1990-2101 – us ATLA [130]
Deutsche volksbibliothek – Nuernberg DE, 1881-83 – 2r – 1 – gw Misc Inst [020]
Deutsche volks-blatter – 1860 oct-1862 apr, 1862 may-1864 feb, mar-1866 puebla n5, 1866 puebla n6 p289-1867 sep – 4r – 1 – mf#766051 – us WHS [071]
Deutsche volksbuecher : aus einer zuercher handschrift des fuenfzehnten jahrhunderts / ed by Bachmann, Albert & Singer, Samuel – Stuttgart: Litterarischer Verein, 1889 (Tuebingen: H Laupp) [mf ed 1993] – cxx/509p – 1 – (incl bibl ref and ind) – mf#8470 reel 38 – us Wisconsin U Libr [800]
Der deutsche volkserzieher – Berlin DE, 1935-39 [gaps] – 1 – gw Misc Inst [370]
Deutsche volksgemeinschaft – Kattowitz (Katowice PL), 1934 19 jan-21 dec, 1935 11 jan-5 apr, 1936 10 apr-19 dec, 1937 9 jan-23 dec, 1938 27 jan-31 dec, 1939 14 jan-5 aug [gaps] – 2r – 1 – gw Misc Inst [077]
Deutsche volksgesundheit – Nuernberg DE, 1934-35 [gaps] – 1 – gw Misc Inst [360]
Die deutsche volksgruppe in daenemark und das national-sozialistische deutschland 1933-1939 / Lenzine, Hilke – 1 – gw Mikropress [943]
Deutsche volkshalle – Konstanz DE, 1839 1 sep-1841 30 mar – 1 – gw Misc Inst [074]
Deutsche volkshalle *see* Rheinische volkshalle
Deutsche volkskunde aus dem oestlichen boehmen / Langer, Eduard – Braunau i B: Selbstverlag 1901- [mf ed Bloomington IN: Indiana Uni Lib, Preservation Dept 1984] – 2r – 1 – us Indiana Preservation [390]
Das deutsche volkslied : hilfsbuechlein fuer den deutschen unterricht / Boeckel, Otto – Leipzig: Quelle & Meyer, 1917 [mf ed 1993] – (= ser Deutschkundliche buecherei) – 103p – 1 – mf#8295 – us Wisconsin U Libr [780]
Deutsche volkspost – Temeschburg (Timisoara RO), 1934 26 jul-1936 5 nov – 1r – 1 – gw Misc Inst [077]
Die deutsche volkssage im verhaeltnis zu den mythen aller zeiten und voelker : mit ueber tausend eingeschalteten original-sagen / Henne am Rhyn, Otto – 2 rev ed. Wien: A Hartleben, 1879 – xvi/720p – 1 – (incl bibl ref) – us Wisconsin U Libr [390]
Deutsche volksschauspiele, in steiermark gesammelt – Halle: M Niemeyer 1891 [mf ed Bloomington IN: Indiana Uni Lib, Preservation Dept 1984] – 2v in 1 on 1r – 1 – us Indiana Preservation [390]
Deutsche volksschoepfung *see* Deutsche revolution
Deutsche volksstimme – Duchcov, Czechoslovakia. Feb 1912-Jul 1914 (scattered issues) – 1r – 1 – us L of C Photodup [077]
Deutsche volksstimme aus hanau – Hanau DE, 1848 19 mar-3 aug [gaps] – 1r – 1 – (astraea) – gw Misc Inst [074]
Deutsche volkstaenze aus der dobrudscha : veroeffentlicht mit unterstuetzung durch die kommission fuer volkskunde der heimatvertriebenen im verband der vereine fuer volkskunde / ed by Au, Hans von der – Regensburg: G Bosse, 1955 – (= ser Quellen und Forschungen zur musikalischen Folklore, Bd 3) – 1 – mf#*ZBD-*MGO pv29 – Located: NYPL – us Misc Inst [790]
Deutsche volkswacht – Friedberg, Hessen DE, 1894 3 oct-1907 11 dec – 7r – 1 – (with suppl: der hessische bauer, and: spinnstube) – gw Mikrofilm [074]
Deutsche volkswehr – Friedek-Friedberg (Frydek-Mistek CZ), 1930 24 may-1933 – 3r – 1 – gw Misc Inst [077]
Der deutsche volkswirt – Berlin DE, 1926 1 oct-1943 17 apr – 33r – 1 – gw Mikropress [630]
Die deutsche volkswirtschaft – Berlin DE, 1932-44 [gaps] – 1 – gw Misc Inst [630]
Deutsche volks-zeitung – Prague, Czechoslovakia. Dec 1936-1937 – 1r – 1 – us L of C Photodup [077]
Deutsche volks-zeitung – Saarbruecken DE, 1934 15 feb-31 dec – 1r – 1 – gw Misc Inst [074]
Deutsche volkszeitung : ausgabe suedhannover – Goettingen DE, 1947 25 mar-1949 8 aug – 2r – 1 – gw Mikrofilm [074]
Deutsche volkszeitung – Berlin DE, 1945 17 jun-1946 19 apr – 1 – mf#1013 – gw Mikropress [074]
Deutsche volkszeitung – Czernowitz (Cernauti RO), 1921 28 may-1922 11 mar – 1r – 1 – gw Misc Inst [077]
Deutsche volkszeitung – Karlsbad (Karlovy Vary CZ), 1938 apr-dec – 1r – 1 – gw Misc Inst [077]
Deutsche volkszeitung – Prigrevica (YU), 1939 6 aug-1940 29 dec – 1r – 1 – gw Misc Inst [077]
Deutsche volkszeitung – Saaz (Zatec CZ), 1936 apr-1938 mar – 1r – 1 – gw Misc Inst [077]
Deutsche volkszeitung – Bruex [Most CZ], 1929 apr-1938 – 1 – (title varies: 1938: tagblatt bruexer volkszeitung) – gw Misc Inst [077]
Deutsche volkszeitung – Duesseldorf DE, 1958 4 jan-1962 21 sep, 1963 28 jun-1967 3 nov, 1969 21 mar-19 dec – 1 – (title varies: end of 1983: deutsche volkszeitung / die tat, 1987: volkszeitung, 9 nov 1990 merged with: der sonntag, berlin to: freitag, until 1955 publ in fulda, later koeln; filmed by other misc inst: 1953 12 may-1982; filmed by mikropress: 1975-1990 26 oct) – gw Misc Inst; gw Mikropress [074]
Deutsche volkszeitung : wochenzeitung fuer "freiheit und recht, frieden dem deutschen volk" – Prag (CZ)/Paris/Basel/Kopenhagen, 1936-39 – 1 – (filmed by misc inst: 1936 22 mar-1939 27 aug [1r]) – fr ACRPP; gw Misc Inst [322]
Deutsche volkszeitung fuer das kuhlaendchen – Neu-Titschein [Novy Jicin CZ], 1922 nov-1938 dec – 10r – 1 – (title varies: jan 1930: deutsche zeitung fuer das kuhlaendchen; jan 1934: neu-titscheiner zeitung) – gw Misc Inst [074]
Deutsche volkszeitung in polen *see* Volkszeitung
Deutsche volkszeitung/freitag – Duesseldorf DE, 1975- – 1 – mf#6638 – gw Mikropress [074]
Deutsche volkzeitung – Berlin, Germany. Jun 1945-Apr 1946 – 1r – 1 – us L of C Photodup [074]
Das deutsche vortragsbuch : eine auswahl sprechbarer dichtungen vom mittelalter bis zur gegenwart... / ed by Gerathewohl, Fritz – Muenchen: G D W Callwey 1929 [mf ed 1993] – 1r – 1 – (incl bibl ref & ind. filmed with: romantik aus schriften, briefen, tagebuechern / richard benz [ed]) – mf#3368p – us Wisconsin U Libr [810]
Deutsche wacht – Hohenstadt (Zabreh CZ), 1929 aug-1934 18 may – 3r – 1 – gw Misc Inst [077]
Deutsche wacht : wochenschrift der deutschen vereinigung – Bonn DE, 1908 26 jan-1922 10 dec – 3r – 1 – (with suppl: westdeutsche waehlerzeitung 1921 n2-6) – mf#5221 – gw Mikropress [943]
Deutsche wacht *see* Deutsche reform 1880
Deutsche wacht 1905 – Dresden, Berlin DE, 1905 oct-1906 30 jun, 1911 1 jul-1923 30 sep – 54r – 1 – (filmed with: deutsche reform) – gw Misc Inst [074]
Deutsche warschauer zeitung – Warschau (PL), 1917 2 jan-30 jun, 1917 24 jul, 16 sep, 1918 2 jan-10 nov – 3r – 1 – (filmed by misc inst: 1915 10 aug-1916 [3r]) – gw Mikrofilm; gw Misc Inst [077]
Deutsche warte – Chicago IL. 1881 jan 3/1882 jun 12-1914 feb 18/oct 28 – 32r – 1 – (with small gaps) – mf#1219600 – us WHS [071]
Deutsche warte – Berlin DE, 1890 1 oct-1894 jun, 1894 oct-1901, 1902 apr-dec, 1903 apr-1904, 1905 apr-dec, 1906 apr-sep, 1907 jan-mar, 1907 jul-1909 sep, 1910 jan-sep, 1911 jan-mar, 1911 oct-1912, 1913 apr-1914, 1915 apr-1916 mar, 1916 jul-1918 apr, 1918 sep-1920, 1921 may-1922 – 107r – 1 – (with suppls) – gw Misc Inst [074]
Der deutsche weg – Lodz (PL), 1935-1937 28 may – 1r – 1 – (filmed by misc inst: 1937-39 (gaps) [1r]) – gw Misc Inst [077]
Der deutsche weg : katholische wochenzeitung – Oldenzaal (NL), 1934 12 aug-1940 5 may – 3r – 1 – gw Misc Inst [241]
Der deutsche weg – Moenchengladbach DE, 1928 1 jul-1930 26 jun, 1930 2 oct-1931 11 jun – 2r – 1 – gw Mikrofilm [074]
Der deutsche wegweiser – Lodz (PL), 1938 27 feb-18 dec, 1939 jan-27 aug – 1r – 1 – gw Misc Inst [077]
Deutsche wehr – Eger (Cheb CZ), 1930 mar-1931 jan, 1932 jan-10 apr – 1r – 1 – gw Misc Inst [077]
Deutsche wehrzeitung – Coburg DE, 1864/65 [single iss] – 1r – 1 – gw Misc Inst [074]
Deutsche wehrzeitung – Berlin, Potsdam DE, 1848 7 jul-1854 29 jun – 4r – 1 – (title varies: yr3: preussische wehr-zeitung, publ in potsdam with yr5) – gw Mikrofilm [355]
Deutsche welle *see* Z-j-funk
Die deutsche werbung *see* Mitteilungen des vereins deutscher reklamefachleute
Deutsche westboehmische stimmen – Plan (Plana CZ), 1928 16 may-1933 may – 4r – 1 – gw Misc Inst [077]
Deutsche wissenschaft, erziehung und volksbildung – Berlin DE, 1935-1944 n16 – 1 – gw Misc Inst [370]
Deutsche wissenschaft, erziehung, und volksbildung : amtsblatt des reichs- und preussischen ministeriums fuer wissenschaft, erziehung und volksbildung und der unterrichtsverwaltungen der anderen laender / Germany. Reichsministerium fuer Wissenschaft, Erziehung und Volksbildung – Berlin: Wiedmaensche Buchhaendlung [1935-45] jahrg1-10 [semimthly] [mf ed 1979] – 11v on 6r – 1 – (cont: zentralblatt fuer die gesamte unterrichts-verwaltung in preussen; ceased in 1945) – mf#film mas c629 – us Harvard [370]
Das deutsche wissenschaftliche schrifttum *see* Literarisches centralblatt fuer deutschland
Deutsche wissenschaftliche zeitschrift im wartheland – Posen: Im Verlag der Historischen Gesellschaft fuer Posen 1940- [heft 1-5/6] [mf ed 198-] – 1r [ill] – 1 – (cont: deutsche wissenschaftliche zeitschrift fuer polen; heft 1 iss by: historische gesellschaft fuer posen; heft 2- by: historische gesellschaft im wartheland) – mf#film mas c9225 – us Harvard [943]
Deutsche woche – Troppau (Opava CZ), 1938 – 1r – 1 – gw Misc Inst [077]
Die deutsche woche – Muenchen DE, 1958-1961 28 jun – 3r – 1 – gw Misc Inst [074]
Deutsche wochenschau – Berlin DE, 1934-35 – 2r – 1 – gw Misc Inst [074]
Deutsche wochen-zeitung *see*
– Deutsche nachrichten
– Das neue reich
Deutsche worte : politische zeitschrift fuer das deutsche volk in oesterreich – Vienna, may 1881-dec 1883 [mf ed Norman Ross] – 1r – 1 – mf#nrp-1952 – us UMI ProQuest [074]
Deutsche zeitschrift fuer geschichtswissenschaft – Leipzig DE, v2 1898 – 1 – gw Misc Inst [900]
Deutsche zeitschrift fuer nervenheilkunde – Dordrecht, Netherlands 1947-70 – 9 – (cont by: zeitschrift fuer neurologie) – ISSN: 0367-004X – mf#13121.02 – us UMI ProQuest [616]
Deutsche zeitschrift fuer philosophie – 1953-1989 – 1,209mf – 1 – gw Mikropress [100]
Deutsche zeitschrift fuer wohlfahrtspflege *see* Die fuersorge
Deutsche zeitschriften des 18. und 19. jahrhunderts. abteilung literatur – [mf ed 1992.98] – 1273mf – 9 – €10,750.00 – (vols also listed individually) – gw Olms [400]
Deutsche zeitschriften des 18. und 19. jahrhunderts. abteilung naturwissenschaft – [mf ed 1992-98] – 840mf – 9 – €7876.00 – (vols also listed individually) – gw Olms [500]
Deutsche zeitschriften des 18. und 19. jahrhunderts. abteilung philosophie – [mf ed 1992-98] – 254mf – 9 – €1320.80 – (vols also listed individually) – gw Olms [100]
Deutsche zeitschriften des 18. und 19. jahrhunderts. abteilung theologie – [mf ed 1992-98] – 1208mf – 9 – €9113.00 – (vols also listed individually) – gw Olms [240]
Deutsche zeitschriften des 18. und 19. jahrhunderts [ergaenzungslieferung] = German periodicals of the 18th and 19th century [supplementary instalment] / ed by Kulturstiftung der Laender – [mf ed 1992-98] – 67 titles on 3166mf – 9 – diazo €8800.00 silver €10,800.00 – gw Olms [074]
Deutsche zeitschriften des 18. und 19. jahrhunderts [hauptlieferung] = German periodicals of the 18th and 19th century [main delivery] / ed by Kulturstiftung der Laender – [mf ed 1992-98] – 520 titles on 20,018mf [1:24] – 9 – diazo €27,800.00 set €40/mf; silver €32,800.00 – gw Olms [074]
Deutsche zeitschriften des 18. und 19. jahrhunderts. historisch-geographische abteilung – [mf ed 1992.98] – 2057mf – 9 – €9800.00 – (vols also listed individually) – gw Olms [900]
Deutsche zeitschriften des 18. und 19. jahrhunderts. historisch-politische abteilung – [mf ed 1992-98] – 2055mf – 9 – €9800.00 – (vols also listed indivdually) – gw Olms [320]
Deutsche zeitung – Celje, Yugoslavia. Feb 1929-1931; Jul-Dec 1932; Jul-Dec 1934 – 4r – 1 – us L of C Photodup [077]
Deutsche zeitung – Vienna, Austria dec 1871-nov 1907 [mf ed Norman Ross] – 143r – 1 – (changing political orientations: founded as the organ of the german nazi party; later the organ of the democratic leftists) – mf#nrp-1953 – us UMI ProQuest [074]
Deutsche zeitung – Sombor (YU), 1931 1 jan-28 jun – 1r – 1 – (filmed by loc: jul-nov 1931 [1r]) – gw Misc Inst; us L of C Photodup [077]
Deutsche zeitung – Budapest (H), 1940 20 oct-1943 – 7r – 1 – (filmed by misc inst: 1940 20 oct-1943 30 apr, 1943 1 sep-1944 30 apr) – gw Misc Inst [077]
Deutsche zeitung – Heidelberg, Frankfurt/M, Mannheim, Leipzig DE, 1847 1 jul-1850 – 12r – 1 – (fr 1 oct 1848 publ in frankfurt/m, later in mannheim, fr 1849 in leipzig. filmed by misc inst: 1847 1 jul-1848 30 jun, 1848 sep-1850 [6r]) – mf#2103 – gw Mikropress; gw Misc Inst [074]
Deutsche zeitung – Gablonz-Neisse (Jablonec nad Nisou CZ), 1933 21 oct-1934 13 oct – 1r – 1 – gw Misc Inst [077]
Deutsche zeitung – Guatemala (GCA), 1932 17 sep-1940 5 may [gaps] – 3r – 1 – gw Misc Inst [079]
Deutsche zeitung – Klausenburg (Cluj RO), 1931 9 oct-1941 29 mar, 1942 13 jun-1943 18 dec – 3r – 1 – gw Misc Inst [077]
Deutsche zeitung – Medford WI. 1886 jul 17 – 1r – 1 – mf#1097871 – us WHS [071]
Deutsche zeitung – Neusatz (Novi Sad YU), 1932, 1934 1 apr-1935 29 sep, 1936 1 jul-30 sep, 1939 1 jul-30 sep, 1940 1 oct-31 dec – 4r – 1 – gw Misc Inst [077]
Deutsche zeitung – Novi Sad, Yugoslavia. Dec 1931-Sept 1940 – 9r – 1 – us L of C Photodup [079]
Deutsche zeitung – Olmuetz (Olomouc CZ), 1920 3 aug-1933 31 dec – 33r – 1 – gw Misc Inst [077]
Deutsche zeitung – Porto Alegre (BR), 1861 10 aug-1865, 1867 9 jan-1868, 1871-1917 27 oct – 27r – 1 – gw Misc Inst [079]
Deutsche zeitung – Pressburg (Bratislava SK), 1922 may-1932 – 1r – 1 – gw Misc Inst [077]
Deutsche zeitung – Riga (LV), 1933 1 mar-15 may – 1r – 1 – gw Misc Inst [077]
Deutsche zeitung – Sao Paulo (BR), 1920 26 jan-1939 20 oct, 1974-76, 1982 24 apr- – 1 – gw Misc Inst [079]
Deutsche zeitung – Stuttgart, Germany. 1972-79 – 15r – 1 – us L of C Photodup [074]
Deutsche zeitung – Troppau (Opava CZ), 1921 nov-1922 nov – 3r – 1 – gw Misc Inst [077]
Deutsche zeitung – Wien (A), 1873 apr-jun, 1874 jan-mar, 1892 sep-dec, 1894 may-jun – 5r – 1 – gw Misc Inst [074]
Deutsche zeitung – Wien: Druck von Kreisel & Groger, 1895-98; 1905-06 – 26r – 1 – us CRL [074]
Deutsche zeitung *see*
– Cillier zeitung
– Deutsche tageszeitung fuer sued-brasilien
– Dirschauer zeitung
– Dorpater zeitung
Deutsche zeitung 1896 – Berlin DE, 1917 2 apr-1922 30 sep, 1928 12 may-1 sep, 1934 2 jan-30 dec – 26r – 1 – (tw. reichsausg, tw. ausg a]; filmed by other misc inst: 1898-1934 [gaps] [184r tw. ausg b]. with suppls, among them: rundschau 1896 apr-1897) – gw Misc Inst [074]
Die deutsche zeitung *see* Der zeitungs-verlag
Deutsche zeitung bessarabiens – Tarutino (RO), 1919 6 nov-1940 3 feb – 7r – 1 – gw Misc Inst [077]
Deutsche zeitung / christ und welt – Bonn, 1970-1980 – (= ser Christ Und Welt) – 1 – gw Mikropress [240]
Deutsche zeitung, christ und welt *see* Christ und welt
Deutsche zeitung fuer canada – Winnipeg, Manitoba (CDN), 1935 12 jun-1939 31 may – 1 – gw Misc Inst [071]
Deutsche zeitung fuer chile – Santiago de Chile [RCH], 1914 5 aug-1918, 1919 1 jul-31 dec [gaps] – 7r – 1 – (publ in valparaiso, fr 7 sep 1914 in santiago; filmed by other misc inst: 1914 5 aug-1918, 1919 1 jul-31 dec, 1920 1 jun-1930, 1932 14 jan-9 apr, 1932 24 oct-1938 31 jan, 1938 22 nov-1940 9 jan [22r], incl suppl: der sonntag 1931 4 jan-12 apr) – gw Misc Inst [079]
Deutsche zeitung fuer china – Schanghai (VR), 1917 2 jan-21 may – 1r – 1 – gw Misc Inst [079]
Deutsche zeitung fuer den leitmeritzer kreis – Leitmeritz [Litomerice CZ], 1920 7 apr-jul – 1r – 1 – (title varies: 4 may 1920: leitmeritzer tagblatt) – gw Misc Inst [077]
Deutsche zeitung fuer die jugend und ihre freunde – Gotha DE, 1796/97, 1801-03, 1816 – 1r – 1 – (filmed by other misc inst: 1801-03, 1805-07, 1809-11 [3r]; 1784-1786 nov, 1787-95) – gw Misc Inst [305]
Deutsche zeitung fuer die krim und taurien – Simferopol (UA), 1918 1 sep-27 oct – 1r – 1 – gw Misc Inst [077]
Deutsche zeitung im ostland – Riga (LV), 1942 7 jan-1943 20 jan, 1944 19 apr-25 jul [gaps] – 12r – 1 – (filmed by misc inst: 1944 1 jan-31 mar & 11 jul-26 jul; 1941 aug-dec, 1942 jul-sep, 1943 apr-jun, 1944 1 jan-9 oct) – uk British Libr Newspaper; gw Misc Inst [077]
Deutsche zeitung in den niederlanden – Amsterdam NL. 24 oct 1940-27 mar 1945* – 14r – 1 – uk British Libr Newspaper [074]
Deutsche zeitung in den niederlanden – Amsterdam NL, 5 jun 1940-1943 – 8mf – 9 – gw Mikrofilm [074]

Deutsche zeitung in frankreich *see*
- Wochen kurier
- Wochen-kurier

Deutsche zeitung in grossbritannien – London (GB), 1938 12 feb-1939 2 sep – 1r – 1 – uk British Libr Newspaper [072]

Deutsche zeitung in kroatien – Zagreb (Agram HR), 1942 1 jan-1945 22 mar [gaps] – 12r – 1 – (filmed by misc inst: 1942 1 nov-1944 19 aug [3r]) – uk British Libr Newspaper; gw Misc Inst [077]

Deutsche zeitung in nordchina *see* Deutsch-chinesische nachrichten

Deutsche zeitung in norwegen – Oslo (N), 1942 [gaps] – 1 – gw Misc Inst [074]

Deutsche zeitung und wirtschaftszeitung *see*
- Wirtschafts-zeitung

Deutsche zeitung von mexiko – Mexiko-Stadt (MEX), 1915-18, 1921-24, 1925 2 jul-31 dec, 1940-1941 28 jun – 7r – 1 – (with gaps) – gw Misc Inst [079]

Deutsche zeitung von spanien – Barcelona (E), 1916 27 jan-1934 25 dec, 1936 10 jan-25 aug, 1942 10 jan-25 dec – 1 – (title varies: 1917: deutsche zeitung fuer spanien) – gw Misc Inst [074]

Deutsche zeitung/tageszeitung *see* Deutsche zentral-zeitung [dzz]

Deutsche zentral-zeitung [dzz] : organ der deutschen werktaetigen in der udssr / ed by Zentralorgan "Prawda" des ZK der KPDSU – Moskau [RUS], 1935-1939 12 jul – 4r – 1 – (title varies: 26 aug 1938-28 feb 1939: deutsche zeitung/tageszeitung) – gw Misc Inst [335]

Deutsche zukunft – Berlin DE, 1933 15 oct-1940 2 jun – 8r – 1 – gw Mikrofilm [074]

Deutsche zukunft *see* Die bruecke

Die deutsche zukunft – Duesseldorf DE, 1952 feb-1957 – 3r – 1 – gw Misc Inst [074]

Der deutsche zuschauer : oder archiv der denkwuerdigsten eraeugnisse, welche auf die glueckseligkeit oder das elend des menschlichen geschlechts einige beziehungen haben / ed by Winkopp, Peter Adolf – Zuerich 1785-88, 1790 [mf ed Hildesheim 1992-98] – (= ser Dz) – 9v[=25iss]+ind on 26mf – 9 – €260.00 – mf#k/n5711 – gw Olms [074]

Der deutsche zuschauer – Zuerich (CH), 1785-88, 1789 [gaps] – 4r – 1 – gw Misc Inst [074]

Die deutschen abschwoerungs-, glaubens-, beicht- und betformeln vom 8. bis zum 12. jahrhundert / ed by Massmann, Hans Ferdinand – Quedlinburg; Leipzig: G Basse, 1839 – us Wisconsin U Libr [430]

Deutschen Akademie der Kuenste zu Berlin. Sektion Dichtkunst und Sprachpflege. Abt Geschichte der sozialistischen Literatur *see* Aktionen, bekenntnisse, perspektiven

Die deutschen alpen : ein handbuch fuer reisende durch tyrol, oesterreich, steyermark, illyrien, oberbayern und die anstossenden gebiete / Schaubach, Adolph – Jena 1845-47 [mf ed Hildesheim: 1995-98] – (= ser Fbc) – 5v on 11mf – 9 – €110.00 – 3-487-29433-8 – gw Olms [914]

Die deutschen bischoefe und der aberglaube : eine denkschrift / Reusch, Franz Heinrich – Bonn: P Neusser, 1879 [mf ed 1986] – 1mf – 9 – 0-8370-8705-8 – (incl bibl ref) – mf#1986-2705 – us ATLA [241]

Die deutschen colonien in der naehe des saginaw-flusses : ein leitfaden fuer deutsche auswanderer nach dem staate michigan in nord-amerika / Koch, Friedrich C – Braunschweig [u a] 1851 [mf ed Hildesheim 1995-98] – 1v on 1mf – 9 – €40.00 – 3-487-27001-3 – gw Olms [304]

Die deutschen dominikaner im kampfe gegen luther (1518-1563) / Paulus, Nikolaus – Freiburg im Breisgau; St. Louis, Mo.: Herder, 1903 – (= ser Erlaeuterungen Und Ergaenzungen Zu Janssens Geschichte Des Deutschen Volkes) – 1mf – 9 – 0-7905-6422-X – (incl bibl ref) – mf#1988-2422 – us ATLA [943]

Die deutschen historienbibeln des mittelalters / ed by Merzdorf, J F L Theodor – Stuttgart: Litterarischer Verein, 1870 (Tuebingen: L F Fues) [mf ed 1993] – (= ser Blvs 100-101) – 2v – 1 – (incl bibl ref and ind) – mf#8470 reel 21 – us Wisconsin U Libr [221]

Die deutschen historienbibeln des mittelalters / ed by Merzdorf, J F L Theodor – Stuttgart: Litterarischer Verein, 1870 (Tuebingen: L F Fues) [mf ed 1993] – (= ser Blvs 100-101) – 2v – 1 – mf#8470 reel 21 – us Wisconsin U Libr [430]

Die deutschen im brasilischen urwald / Zoeller, Hugo – Berlin, Stuttgart: W. Spemann, 1883. 2v. illus. map – 1 – us Wisconsin U Libr [972]

Die deutschen im heiligen lande / R"hricht, R – Innsbruck, 1894 – 2mf – 9 – mf#HT-286 – ne IDC [915]

Die deutschen im staate new york waehrend des 18. jahrhunderts / Kapp, Friedrich – New York: E Steiger, 1884 [mf ed 1992] – (= ser Geschichtsblaetter 1; Lutheran coll) – 1mf – 9 – 0-524-03165-7 – mf#1990-4614 – us ATLA [978]

Deutschen juden als soldaten im kriege 1914-1918 / Segall, Jacob – Berlin, Germany. 1922 – 1r – us UF Libraries [939]

Die deutschen kriegskreditbanken / Lang, Friedrich – Fuerth i.B: J Kellermann, 1927 (mf ed 19–) – 119p – mf#Z-BTZE pv811 n3 – us Harvard [332]

Die deutschen paepste : nach handschriftlichen und gedruckten quellen / Hoefler, Karl Adolf Constantin, Ritter von – Regensburg: GJ Manz, 1839 – 2mf – 9 – 0-7905-6478-5 – (incl bibl footnotes) – mf#1988-2478 – us ATLA [240]

Die deutschen personennamen / Baehnisch, Alfred – Leipzig: B G Teubner, 1914 [mf ed 1988] – (= ser Aus natur und geisteswelt. sammlung wissenschaftlich-gemeinverstaendlicher darstellungen 296) – viii/126p – 1 – mf#2186 – us Wisconsin U Libr [929]

Die deutschen saeculardichtungen an der wende des 18. und 19. jahrhunderts / ed by Sauer, August – Berlin: B Behr (E Bock), 1901 [1993] – (= ser Deutsche litteraturdenkmale des 18. und 19. jahrhunderts 91-104, n f n41-54) – xlccii/654p – 1 – (incl bibl ref and ind) – mf#8676 reel 6 – us Wisconsin U Libr [810]

Die "deutschen sagen" der brueder grimm : ein beitrag zu ihrer entstehungsgeschichte, unter besonderer beruecksichtigung des westfaelischen anteils / Erfurth, Fritz – Muenster, 1937 (mf ed 1993) – 1mf – 9 – €24.00 – 3-89349-354-9 – mf#DHS-AR 354 – gw Frankfurter [430]

Die deutschen sagen der brueder grimm : ein beitrag zu ihrer entstehungsgeschichte, unter besonderer beruecksichtigung des westfaelischen anteils / Erfurth, Fritz – Muenster 1937 (mf ed 1993) – 1mf – 9 – €24.00 – mf#DHS-AR 354 – gw Frankfurter [390]

Die deutschen schriftstellerinnen des 19. jahrhunderts / Schindel, Carl Wilhelm Otto August von – Leipzig 1823-25 [mf ed Hildesheim 1983] – (= ser Die schriftsteller- und gelehrtenlexika des 17., 18., und 19. jahrhunderts) – 3pts in 1v on 14mf – 9 – diazo €64.00 silver €78.00 – gw Olms [430]

Deutschen Schriftsteller-Verband *see* Menschen und werke

Deutschen Schriftstellerverband *see* Neue deutsche literatur

Die deutschen schutzgebiete in afrika / Eschner, Max; ed by Bukacz, Franz – Leipzig, 1910 (mf ed 1994) – 1mf – 9 – €24.00 – 3-8267-3094-1 – mf#DHS-AR 3094 – gw Frankfurter [960]

Die deutschen stroeme in ihren verkehrs- und handels-verhaeltnissen : mit statistischen uebersichten; in vier abtheilungen / Meidinger, Heinrich – Leipzig 1854 [mf ed Hildesheim 1995-98] – 4pt on 3mf – 9 – €90.00 – 3-487-29537-7 – gw Olms [380]

Deutschen Verwaltung fuer Volksbildung in der sowjetischen Besatzungszone *see* Liste der auszusondernden literatur

Die deutschen volksbuecher : tristan und isolde, die schoene melusina, genovefa, lother und maller, die vier heymonskinder / Ernst, Paul – Berlin: Deutsche Buch-Gemeinschaft, GmbH, [192-?] – 395p (ill) – 1 – us Wisconsin U Libr [390]

Die deutschen volksbuecher *see* Elsaessische stammeskunde

Die deutschen volksnamen der pflanzen / Pritzel, G A – Hannover: P. Cohen, 1882. viii,701p. illus – 1 – us Wisconsin U Libr [580]

Der deutschen volkszahl und sprachgebiet in den europaeischen staaten : eine statistische untersuchung / Boeckh, Richard – Berlin 1869 [mf ed Hildesheim 1995-98] – 3mf – 9 – €90.00 – 3-487-29621-7 – gw Olms [430]

Deutschendorff, Jean-Jacob *see* Michel servet

Deutsch-englische hefte – Berlin DE, 1936-39 [gaps] – 1 – gw Misc Inst [074]

Deutschenspiegel und augsburger sachsenspiegel (mgh leges 3:3.bd) – 1930 – (= ser Monumenta germaniae historica leges 3. fontes iuris germanici antiqui, nova series (mgh leges 3)) – €14.00 – ne Slangenburg [943]

Deutscher adel : eine erzaehlung / Raabe, Wilhelm Karl – Braunschweig: G Westermann, 1880 – 1r – 1 – us Wisconsin U Libr [830]

Deutscher anzeiger – Muenchen DE, 1973 17 jan-24 dec, 1974 25 jan-1985 20 dec – 6r – 1 – (filmed by misc inst: 1958 6 sep-4 oct, 1962 15 may-15 aug, 1984 17 feb-1990 21 dec [4r]) – gw Mikrofilm; gw Misc Inst [074]

Deutscher Arbeitgeberbund fuer das Baugewerbe *see* Geschaeftsbericht

Deutscher Arbeitsnachweis-Kongress *see* Proceedings

Deutscher aufstieg – Magdeburg DE, 1931-33 – 1 – gw Misc Inst [943]

Deutscher Bauarbeiterverband. Hamburg *see*
- Jahrbuch
- Protokoll ueber die verhandlungen..

Deutscher Baugewerksbund *see*
- Jahrbuch
- Niederschriften ueber die verhandlungen..

Der deutscher beobachter – Rio de Janeiro, RJ. 16 abr-16 jul 1853 – (= ser Ps 19) – mf#P19A,04,02 – bl Biblioteca [079]

Deutscher beobachter – West Bend WI. 1885 dec 31, 1886 mar 18 – 1r – 1 – mf#1097650 – us WHS [071]

Deutscher beobachter – Windhoek, South West Africa, 4 jan 1939-43 – 8r – 1 – mf#MS00253 – sa National [079]

Deutscher beobachter oder privilegirte hanseatische zeitung – Bremen, Hamburg DE, apr-may 1813, jan-jul 1814, 1815, 1817-18 – 1 – (publ in hamburg since 1814. filmed by other misc inst: 1817-19 [3r]) – gw Misc Inst [074]

Deutscher bote – Klausenburg (Cluj RO), 1924 5 jan-1930 – 2r – 1 – gw Misc Inst [077]

Deutscher bote – Libau (Liepaja LV), 1938 1 apr-1939 22 jun – 2r – 1 – gw Misc Inst [077]

Deutscher bote – Mitau (Jelgava LV), 1924 27 mar-1925 – 1r – 1 – gw Misc Inst [077]

Deutscher bote – Riga (LV), 1926 1 apr-1936 18 sep – 3r – 1 – gw Misc Inst [077]

Deutscher bote – Braunau [Broumov CZ], 1922 nov-1938 – 14r – 1 – (title varies: mar 1936: ostboehmens deutscher bote) – gw Misc Inst [077]

Deutscher buehnen-almanach – Berlin: Commissions-Verlag von L Lassar 1854-60 [jahrg18-24] (annual) [mf ed 198-] – 11r – 1 – (cont: almanach fuer freunde der schauspielkunst aus das jahr...; cont by: a heinrich's deutscher buehnen-almanach) – mf#film mas c487 – us Harvard [790]

Deutscher buerger- und bauernfreund – Berlin DE, 1890 14 jun-1891 26 sep, 1892 – 1r – 1 – gw Misc Inst [074]

Deutscher buhnenspielplan mit unterstutzung des deutschen buhnenvereins – Leipzig. 1896-1941. Lacking: v.19; 25; 28 and scattered ns. 16, 17, 21, 22, 29. 10 reels – 1 – us L of C Photodup [780]

Ein deutscher cisianus fuer das jahr 1444 gedruckt von gutenberg / Wyss, Arthur Franz Wilhelm – Strassburg: J H E Heitz, 1900 – 1r – 1 – (incl bibl ref) – us Wisconsin U Libr [430]

Deutscher Eisenbahner-Verband *see* Protokoll der generalversammlung

Deutscher Forstverein *see* Jahresbericht

Deutscher gaststaetten-paechter *see* Deutscher wirtschaftspaechter

Deutscher geist im 18. jahrhundert : essays zur geistes- und religionsgeschichte / Schoeffler, Herbert; ed by Selle, Goetz von – Goettingen: Vandenhoeck & Ruprecht, c1956 [mf ed 1993] – 317p – 1 – (incl bibl ref and ind) – mf#8232 – us Wisconsin U Libr [840]

Deutscher grenzbote fuer polnisch-schlesien – Teschen (Cieszyn PL), 1922-33, 1935 – 2r – 1 – gw Misc Inst [077]

Deutscher Handels- und Industrieangestellten-Verband *see*
- Geschaeftsbericht
- Rechenschaftsbericht

Deutscher Handwerks- und Gewerbekammertag. Koenigsberg *see* Stenographischer bericht ueber die verhandlungen..

Deutscher Holzarbeiter-Verband *see*
- Protokoll...

Deutscher Holzarbeiter-Verband. Berlin *see*
- Jahrbuch
- Protokoll des kongresses..
- Tarifvertrage des deutschen..
- Verhandlungen der konferenz..
- Verhandlungsbericht ueber die reichskonferenz der bursten- und pinselmacher

Deutscher industrie- und handelstag / reichswirtschaftskammer [bestand r 11] bd 12 / ed by Facius, Friedrich & Trumpp, Thomas – 1976 – 616p – €22.00 – 978-3-89192-042-8 – gw Bundesarchiv [338]

Deutscher kinderfreund – Hamburg DE, 1879 oct-1880 sep, 1891-1892 sep, 1905 oct-1906 sep, 1907 oct-1908 sep – 1 – gw Misc Inst [370]

Ein deutscher kolonialheld : der fall peters in psychologischer beleuchtung / Giesebrecht, Franz – 2. Aufl. Zuerich: C Schmidt, 1897 – us CRL [920]

Deutscher kurier – Berlin DE, 1916 3 sep-1919 jun [gaps] – 6r – 1 – uk British Libr Newspaper [074]

Deutscher kurierdienst – Prag (CZ), 1935 4 sep-1936 25 jun – 1r – 1 – gw Misc Inst [077]

Deutscher kurzwellensender / iberoamerika-programm – Berlin DE, 1939 may-sep – 1r – 1 – gw Misc Inst [380]

Deutscher kurzwellensender / mittel- und suedamerika-programm – Berlin DE, 1938 apr-1939 apr – 1r – 1 – gw Misc Inst [380]

Deutscher kurzwellensender / suedamerika-programm – Berlin DE, 1938 apr-1939 sep – 1r – 1 – gw Misc Inst [380]

Deutscher Landarbeiter-Verband *see* Bericht

Deutscher landbote – Karlsbad (Karlovy Vary CZ), 1929 apr-1938 mar – 7r – 1 – gw Misc Inst [077]

Deutscher landruf – Bruenn (Brno CZ), oct 1923[gaps]-mar 1938 – 6r – 1 – gw Misc Inst [077]

Deutscher landruf – Eger (Cheb CZ), 1925 jul-1932 – 3r – 1 – gw Misc Inst [077]

Deutscher Lederarbeiter-Verband *see*
- Jahrbuch
- Jahresbericht ueber die taetigkeit des zentralvorstundes
- Protokoll...

Deutscher Lehrerverein *see* Jahrbuch

Deutscher lichtspieltheater-besitzer *see* Der deutsche lichtbildtheater-besitzer

Deutscher literaturkalender *see* Allgemeiner deutscher literaturkalender

Deutscher Maschinensetzerkongress *see* Protokoll

Deutscher merkur : organ fuer die katholische reformbewegung – Muenchen-Bonn, 1(1870)-46(1915) – 366mf – 9 – €835.00 – (formely: rheinischer merkur. lacking: 7(1876)) – ne Slangenburg [241]

Deutscher metallarbeiter verband verwaltungstelle berlin : jahresberichte – Berlin, Stuttgart DE, 1920-23 – 1r – 1 – mf#11261 – gw Mikropress [331]

Deutscher Metallarbeiter-Verband *see* Der ordentliche verbandstag des...

Deutscher Metallarbeiterverband *see*
- Jahr- und handbuch
- Protokoll der konferenz des reichsbeirats der petriebsrate und konzernvertreter der metallindustrie. berlin. v10. 1931. (serial

Deutscher metallarbeiter-verband : dmv jahr- und handbuch fuer verbandsmitglieder – Berlin, Stuttgart DE, 1914-23 – 2r – 1 – mf#11265 – gw Mikropress [331]

Deutscher Metallarbeiterverband. Verwaltungsstelle. Berlin *see* Jahresbericht fuer das geschaftsjahr

Deutscher morgen – Sao Paulo (BR), 1934-1939 25 aug [gaps] – 4r – 1 – (filmed by other misc inst: 1937 [1r] filmed with suppl) – gw Misc Inst [079]

Deutscher nachrichtendienst (spd) – Prag (CZ), 1935 n5a-1937 n7 [gaps] – 1 – gw Misc Inst [325]

Deutscher Nahrungs- und Genussmitterlarbeiter-verband *see*
- Jahrbuch
- Protokoll ueber die verhandlungen des verbandstages

Deutscher reichsanzeiger und koeniglich preussischer staats-anzeiger – Berlin, 1819-1945 – 553r – 1 – (teil 1 1819-71 102r. teil 2 1871-96 145r. teil 3 1897-1918 167r. teil 4 93r. teil 5 1933-45 46r) – gw Mikropress [943]

Deutscher reichs-anzeiger und preussischer staats-anzeiger – Berlin. 1875-1943 (incomplete) – 1 – us NY Public [943]

Deutscher reichs-anzeiger und preussischer staats-anzeiger / Germany – Berlin. 1875-1943 – 1 – us NY Public [943]

Deutscher reichsanzeiger und preussischer staatsanzeiger *see* Allgemeine preussische staats-zeitung

Deutscher Romisch-Katholischer Central-Verein von Nord Amerika *see* Central-blatt

Deutscher Sattler-, Tapezierer- und Portefeuiller-Verband *see* Jahrbuch

Deutscher schwertsegen : kraefte der heimat fuers reisige heer / Deissmann, Gustav Adolf – 2. Aufl. Stuttgart: Deutsche Verlags-Anstalt, 1915 – 1mf – 9 – 0-7905-7218-4 – mf#1988-3218 – us ATLA [943]

Deutscher Seefischereiverein *see* Abhandlungen

Deutscher sprachschatz geordnet nach begriffen zur leichten auffindung und auswahl des passenden ausdrucks : ein stilistisches huelfsbuch fuer jeden deutsch schreibenden / Sanders, Daniel – Hamburg: Hoffmann & Campe 1873-77 [mf ed 1979] – 2v on 1r – 1 – mf#film mas 8709 – us Harvard [430]

Deutscher Tabakarbeiter-Verband *see*
- Jahresbericht
- Protokoll ueber die verhandlungen der generalversammlung

Deutscher Textilarbeiter-Verband *see*
- Jahrbuch
- Protokoll ueber die konferenz...im volkshaus zu leipzig am sonntag, 28 juni. berlin. 1925. 77p. (serial publications of german trade unions in the memorial library, university of wisconsin-madison.)
- Protokoll vom kongress...
- Verhandlungsbericht der konferenz...

Deutscher Transportarbeiter-Kongress. First *see* Protokoll der verhandlungen

Deutscher Transportarbeiter-Verband *see* Jahrbuch

Deutscher Verkehrsbund *see*
- Jahrbuch
- Jahresbericht ueber das geschaftsjahr

Deutscher volksbote – Budapest (H), 1936-1943 – 1r – 1 – (dec iss always missing) – gw Misc Inst [077]

Deutscher volksbote – Dortmund DE, 1919 14 feb-1923 – 1 – gw Misc Inst [074]

Deutscher volksbote – Lodz (PL), 1935 28 apr-25 dec – 1r – 1 – gw Misc Inst [077]

Deutscher volksfreund – Freiwaldau [Jesenuik CZ], 1929 20 apr-1938 – 6r – 1 – (cont: volksbote [publ in freiwaldau, jaegerndorf]) – gw Misc Inst [077]

Deutscher volksfreund : ein illustriertes deutsches familienblatt – 1871 jan-dec, 1876-77, 1880, 1894 aug 18-1899 nov 25, 1899 dec 23-1901 feb 23 – 5r – 1 – mf#1055453 – us WHS [071]

Deutscher volksfreund : der volksfreund fuer stadt und land – Breslau (WrocLaw PL), 1881-84 – 4r – 1 – gw Misc Inst [077]

Deutscher volksfreund – Vrsac, Yugoslavia. Apr-Dec 1921; 1924; 1928-39 – 5r – 1 – us L of C Photodup [949]

Deutscher volksfreund – Werschetz (Vrsac YU), 1922-1927 29 dec, 1940 4 jan-22 dec – 2r – 1 – gw Misc Inst [077]

Deutscher volksfreund *see*
– Dresdner correspondent fuer literatur und tagesneuigkeiten
– Katholische volkszeitung

Deutscher volksfuehrer – Altoona PA (USA), 12 jan 1923-5 sep 1940* – 5r – 1 – Dist. gw Mikrofilm – gw Misc Inst [071]

Deutscher volksrat – Teschen (Cieszyn PL), 1919 jan, nov – 1r – 1 – gw Misc Inst [077]

Deutscher Webertag. 2nd. Berlin *see* Stenographischer bericht ueber..

Deutscher weckruf und beobachter – New York NY (USA), 1935 5 jul-30 dec, 1936 2 jul-1937 jun, 1938 30 jun-1939 22 jun – 1 – gw Misc Inst [071]

Deutscher Werkmeister-Verband *see*
– Geschaeftsbericht
– Stenographischer bericht
– Stenographischer bericht...
– Stenographischer bericht..

Deutscher wetterdienst / Germany. Federal Republic – Monatlicher Witterungsbericht. Bad Kissingen, etc.. v. 1-17. 1953-1969, and Indexes – 1 – us NY Public [943]

Deutscher Wirtschaftsbund fuer das Baugewerbe *see* Geschaeftsbericht

Deutscher wirtschaftspaechter – Dortmund, Berlin DE, 1930 1 nov-1932 15 sep – 1r – 1 – (title varies: oct 1932: deutscher gaststaetten-paechter) – gw Mikrofilm [640]

Deutscher zeitungskatalog – Leipzig DE, 1841, 1845, 1848, 1850, 1853 – 1 – (began as [?]: leipziger zeitungskatalog) – gw Misc Inst [074]

Deutscher zuschauer – Mannheim DE, Basel [CH]...1846 21 nov-1848 15 oct, 1849 13 jun, 1851 9 jul-31 dec – 1r – 1 – (filmed with: falscher zuschauer 1848 8 jul-28 jul) – gw Mikrofilm [074]

Deutsches agrarblatt – Prag (CZ), 1920 jul-1923 dec – 3r – 1 – gw Misc Inst [630]

Deutsches allgemeines sonntagsblatt – Hamburg, 1975-1994 – 30r – 1 (DM150.00y 1995ff) – gw Mikropress [943]

Deutsches allgemeines sonntagsblatt *see* Sonntagsblatt

Deutsches alpenbuch : die deutschen hochlande in wort und bild / Noe, Heinrich – Glogau 1875/88 [mf ed Hildesheim: 1995-98] – (= ser Fbc) – 2v on 8mf – 9 – €160.00 – 3-487-29463-X – gw Olms [914]

Deutsches anonymen-lexikon, 1501-1910 / Holzmann, M & Bohatta, Hans – 7v. 1902-28 – 1,9 – us AMS Press [054]

Deutsches Archaeologisches Institut. Rome *see* Index der antiken kunst und architektur

Deutsches Auslandswissenschaftliches Institut *see* Jahrbuch der weltpolitik

Deutsches bekenntnis : [poems and short prose pieces] / Eggers, Kurt – Berlin: Widukind 1934 [mf ed 1989] – 1r – 1 – (filmed with: hallo welt! / kasimir edschmid) – mf#7204 – us Wisconsin U Libr [800]

Deutsches biographisches archiv 1960-1999 (dba3) = German biographical archive 1960-1999 / Herrero Mediavilla, Victor [comp] – [mf ed 1999-2001] – 1075mf (1:24) in 12 installments – 9 – diazo €10,060.00 (silver €11,080 isbn: 978-3-598-34151-9) – ISBN-10: 3-598-34150-4 – ISBN-13: 978-3-598-34150-2 – (with printed ind) – gw Saur [943]

Deutsches biographisches archiv (dba1) = German biographical archive / Fabian, Bernhard [comp] – [mf ed 1982-85] – 1447mf (1:24) – 9 – diazo €10,060.00 (silver €11,080 isbn: 978-3-598-30421-7) – ISBN-10: 3-598-30410-2 – ISBN-13: 978-3-598-30410-1 – (with printed ind) – gw Saur [943]

Deutsches biographisches archiv (dba2) : neue folge bis zur mitte des 20. jahrhunderts = German biographical archive (dba2): a sequel up to the mid-twentieth century / Gorzny, Willi [comp] – [mf ed 1989-93] – 1457mf (1:24) – 9 – diazo €10,060.00 (silver €11,080 isbn: 978-3-598-32820-6) – ISBN-10: 3-598-32834-6 – ISBN-13: 978-3-598-32834-3 – (with printed ind) – gw Saur [920]

Deutsches buecherverzeichnis : eine zusammenstellung der im deutschen buchandel erschienenen buecher, zeitschriften und landkarten – 37v. 1916- – 1,9 – us AMS Press [010]

Deutsches buergerblatt fuer stadt und land – Luebeck DE, 1848 7 oct-30 dec – 1r – 1 – gw Misc Inst [074]

Deutsches christentum – Schwerin DE, 1937-39 – 1r – 1 – gw Misc Inst [240]

Deutsches dichten und denken vom mittelalter zur neuzeit : deutsche literaturgeschichte von 1270 bis 1700 / Mueller, Guenther – Berlin; Leipzig: W de Gruyter, 1934 [mf ed 1993] – (= ser Sammlung goeschen) – 159p – 1 – (incl bibl ref and ind) – mf#8156 – us Wisconsin U Libr [430]

Deutsches dichten und denken von der germanischen bis zur staufischen zeit : deutsche literaturgeschichte vom 5. bis 13. jahrhundert / Naumann, Hans Heinz – 2. verb aufl. Berlin: W de Gruyter, 1952 [mf ed 1993] – (= ser Sammlung goeschen 1121) – 166p – 1 – (incl ind) – mf#8166 – us Wisconsin U Libr [430]

Deutsches erbrecht / Dernburg, Heinrich – 2. Aufl. Halle (Saale): Waisenhaus, 1905 – (= ser Civil law 3 coll; Archiv fuer theorie und praxis des allegemeinen deutschen handels- und wechselrecht) – 7mf – 9 – (incl bibl ref and index) – mf#LLMC 96-583 – us LLMC [348]

Deutsches familienblatt – Berlin DE, 1880-87, 1890, 1893-1894 n13 – 1 – (filmed by other misc inst: 1905-12 [3r]; title varies: 1911 n14: sonnenstrahlen, also filmed as suppl of fulda-werra-zeitung and of: schaumburger wochenblatt, later as: koelner gerichts-zeitung, duesseldorfer gerichts-zeitung, casseler stadt-anzeiger) – gw Misc Inst [640]

Deutsches familienrecht / Dernburg, Heinrich – 3. Aufl. Halle (Saale): Waisenhaus, 1907 – (= ser Civil law 3 coll; Archiv fuer theorie und praxis des allegemeinen deutschen handels- und wechselrecht) – 6mf – 9 – (incl bibl ref and index) – mf#LLMC 96-582 – us LLMC [348]

Deutsches frauenblatt – 1926-33 [mf ed 2003] – (= ser Hq 59) – 8v on 14mf – 9 – €110.00 – 3-89131-449-3 – gw Fischer [305]

Deutsches gedichtbuch / Deicke, Guenther & Berger, Uwe – Berlin: Aufbau-Verlag, 1959 – 1r – 1 – (incl bibl ref and index) – us Wisconsin U Libr [810]

Deutsches gemeinnuetziges magazin / ed by Eggers, Chr U D von – Leipzig 1788-90 – (= ser Dz) – 4v on 17mf – 9 – €170.00 – mf#k/n409 – gw Olms [074]

Deutsches gesangbuch : eine auswahl geistlicher lieder aus allen zeiten der christlichen kirche fuer oeffentlichen und haeuslichen gebrauch – Taschenausg. Philadelphia: Lindsay und Blakiston, c1860 – 2mf – 9 – 0-524-06026-6 – mf#1991-2386 – us ATLA [240]

Deutsches handwerk : kampfblatt des deutschen handwerks, gewerbes und einzelhandels – Muenchen DE, 1932 2 apr-1933 – 1r – 1 – mf#5866 – gw Mikropress [380]

Deutsches journal – Seward, NE: Clemens Schwabe, -jahrg 19 n26. jun 28 1918 (wkly) [mf ed 1913-18 filmed [1974]] – 3r – 1 – (cont: nebraska deutsche farmer-zeitung. cont by: seward journal) – us NE Hist [071]

Deutsches kinderbuch in wort und bild / Wesendonk, Mathilde – Stuttgart: G J Goeschen 1869 – 3mf – 9 – (with wood engravings by louis ruff; drawings by ernst schweinfurth) – mf#mw-2 – ne IDC [430]

Ein deutsches kriegsschiff in der suedsee / Werner, Bartholomaeus von – Leipzig 1889 [mf ed Hildesheim 1995-98] – 1v on 4mf [ill] – 9 – €120.00 – 3-487-26753-5 – gw Olms [355]

Deutsches land und deutsches volk / Gutsmuths, Johann C – Gotha 1824-24 [mf ed Hildesheim: 1995-98] – (= ser Fbc) – 2v on 7mf – 9 – €140.00 – 3-487-29609-8 – gw Olms [943]

Deutsches landblatt – Berlin, 1991 – 4r – 1 – gw Mikropress [943]

Deutsches landblatt *see* Bauern-echo

Deutsches lesebuch / ed by Wackernagel, Wilhelm – Basel: Druck und Verlag der Schweighauserischen Buchhandlung, 1840 – 3v in 4, 2r – 1 – (incl ind) – us Wisconsin U Libr [800]

Deutsches Literaturarchiv. Marbach am Neckar *see* Das ausland (1828-1893)

Deutsches magazin / ed by Detlev, Christian Ulrich, Freiherr von Eggers – Hamburg [1792ff: Altona] 1791-1800 – (= ser Dz) – 20v on 91mf – 9 – €910.00 – mf#k/n438 – gw Olms [430]

Deutsches magazin – Hamburg DE, 1791-1800 – 4r – 1 – gw Misc Inst [073]

Deutsches magazin – v6-v7 n13 [1902 may-1904 jul] – 1r – 1 – mf#1055469 – us WHS [071]

Deutsches magazin zur unterhaltung und belehrung – Berlin DE, 1863 – 1r – 1 – gw Misc Inst [073]

Deutsches monatsblatt – Bonn, Siegen DE, 1979- – 1 – (cont as: union 1990; first publ in siegen; filmed by misc inst: 1964 1 nov-1975 [4r], 1954-87) – mf#7140 – gw Mikropress; gw Misc Inst [074]

Deutsches monatsblatt : organ der cdu – Bonn DE, 1964 1 nov-1975 – 4r – 1 – (cont by: union; filmed by other misc inst: 1954-87) – gw Misc Inst [325]

Deutsches monatsblatt : westfaelische ausgabe – Dortmund DE, 1954 feb-1959 oct, 1960-1962 apr [gaps] – 1 – (aka: westfaelisches monatsblatt) – gw Misc Inst [074]

Deutsches montagsblatt – Berlin DE, 1877 2 jul-1888 – 6r – 1 – gw Misc Inst [074]

Deutsches museum / ed by Boie, Heinrich Christian – Leipzig 1776-88 [mf ed Hildesheim 1979] – (= ser Allgemeinwissenschaftliche und literarische zeitschriften des 17. und 18. jahrhunderts) – 212mf – 9 – diazo €1092.00 – (cont as: neues deutsches museum [leipzig 1789-91]) – gw Olms [060]

Deutsches museum / ed by Boie, Heinrich Christian – Leipzig 1776-88 – (= ser Dz) – 26v on 182mf – 9 – €1092.00 – mf#k/n4495 – gw Olms [060]

Deutsches museum / ed by Schlegel, Friedrich – Wien 1812-13 – (= ser Dz. abt literatur) – 4v on 16mf – 9 – €160.00 – mf#k/n4727 – gw Olms [060]

Deutsches museum-bildarchiv = German museum picture archive / ed by Wilhelm Deutsches Museum, Muenchen – [mf ed 1987-96] – 480mf (1:24) – 9 – silver €6348.00 – ISBN-10: 3-598-30403-X – ISBN-13: 978-3-598-30403-3 – (with ind) – gw Saur [060]

Deutsches privatrecht / Bluntschli, Johann Caspar – Muenchen: Literarisch-artistische Anstalt. v1-2. 1853-54 – (= ser Civil law 3 coll) – 12mf – 9 – (incl bibl ref) – mf#LLMC 96-629 – us LLMC [346]

Deutsches privatrecht / Gierke, Otto Friedrich von – Leipzig: Duncker & Humblot. v1-3. 1895, 1905, 1917 – (= ser Civil law 3 coll; Systematisches handbuch der deutschen rechtswissenschaft) – 33mf – 9 – mf#LLMC 96-608 – us LLMC [346]

Deutsches pseudonymen-lexikon / Holzmann, M & Bohatta, Hans – 1v. 1906 – 1,9 – us AMS Press [054]

Deutsches recht – Berlin DE, 1931-33 [gaps], 1934-44 – 1 – gw Misc Inst [342]

Deutsches sagenbuch / Bechstein, Ludwig – Leipzig: G Wigand, 1853 [mf ed 1989] – xxiv/813p – 1 – (wood drawings by a ehrhardt) – mf#7001 – us Wisconsin U Libr [390]

Deutsches schicksal : tagebuchblaetter eines ausgewanderten / Francke, Kuno – Dresden: E Pierson, 1923 – 1r – 1 – us Wisconsin U Libr [880]

Deutsches schrifttum – Weimar DE, 1909-17 – 1 – gw Misc Inst [430]

Deutsches sportecho – Berlin DE, 1954 31 may-1990 – 59r – 1 – gw Misc Inst [790]

Deutsches tageblatt – Berlin DE, 1881 apr-jun, oct-dec, 1883 oct-1884 mar, 1888 apr-jun, 1889 apr-jun – 6r – 1 – gw Misc Inst [074]

Deutsches tageblatt – Temeschburg (Timisoara RO), 1900 16 dec-1903 30 aug – 6r – 1 – gw Misc Inst [077]

Deutsches tageblatt *see* Barmer zeitung 1833

Deutsches tageblatt fuer westboehmen – Eger (Cheb CZ), 1933 oct-dec – 1 – gw Misc Inst [077]

Deutsches Uebersee-Institut Hamburg *see* Laenderkatalog afrika der uebersee-dokumentation hamburg, 1971-1984

Deutsches und amerikanisches / Knortz, Karl – Glarus: Vogel, 1894 – 1r – 1 – us Wisconsin U Libr [400]

Deutsches volksblatt : bayerische antisemitische zeitung – Muenchen DE, 1892-1914 – 1 – gw Misc Inst [320]

Deutsches volksblatt – Berlin DE, 1919 3 may-1920 30 jun, 1922-1923 10 oct, 1928 7 jan-8 sep – 1r – 1 – gw Misc Inst [074]

Deutsches volksblatt – Czernowitz (Cernauti RO), 1924-25 – 1 – gw Misc Inst [077]

Deutsches volksblatt – Neusatz [Novi Sad YU], 1935 jan-1937 aug – 17r – 1 – (filmed by misc inst: 1923 1 jul-1941 30 sep [with gaps], 1937-1938 30 jun, 1938 1 oct-1940 31 mar, 1940 1 jul-1941 31 mar, 1941 1 oct-1942 30 aug, 1943-1944 31 may; title varies: 4 feb 1942: tageszeitung der deutschen suedungarns) – uk British Libr Newspaper; gw Misc Inst [077]

Deutsches volksblatt – Stuttgart DE, 1849 – 1 – (filmed by other misc inst: 1863, 1866 1 jul-30 dec, 1911-13, 1915 1 jul-1922 31 may, 1923-30, 1931 1 jul-1935 31 oct. various ed: landesausgabe 1953 1 jul-1965 31 jul [16r]; stadtausgabe 1954-1965 31 jul [16r]) – gw Misc Inst [074]

Deutsches volksblatt : german nationalist and anti-semitic – Wiener Neustadt, Austria jan 1889-sep 1922 [mf ed Norman Ross] – 174r – 1 – (aka: wiener neustadter bezirksblatt) – mf#nrp-2153 – us UMI ProQuest [074]

Deutsches volksblatt – Porto Alegre (BR), 1921 31 jan-30 jun, 1927 26 sep-1939 19 aug – 29r – 1 – (incl weekend ed [1920 4 aug-1927 5 jan] & suppl) – gw Misc Inst [079]

Deutsches volksblatt – Komotau (Chomutov CZ) 1922 7 sep-1938 31 dec – 35r – 1 – gw Misc Inst [077]

Deutsches volksblatt – Novi Sad, Yugoslavia. Feb 1920-Oct 1944 (incomplete) – 36r – 1 – us L of C Photodup [949]

Deutsches volksblatt : organ of the progressive party – Vienna, Austria jan 1883-dec 1902 [mf ed Norman Ross] – 10r – 1 – mf#nrp-1954 – us UMI ProQuest [074]

Deutsches volksblatt – Tarutino (RO), 1935 16 feb-1940 – 4r – 1 – gw Misc Inst [077]

Deutsches volksblatt – Wien (A), 1916 jun-1918, 1919 mar [gaps] – 11r – 1 – uk British Libr Newspaper [074]

Deutsches volksblatt – Zagreb (Agram HR), 1940 12 may-1941 17 mar [gaps] – 4r – 1 – uk British Libr Newspaper [077]

Deutsches volksblatt *see* Der unabhaengige

Deutsches volksblatt fuer galizien – Lemberg (Lwow UA), 1914/15-1918 n44 – 1r – 1 – gw Misc Inst [077]

Deutsches volks-echo – Paris [F], 1938 8 may-25 dec [gaps] – 1 – (with n26: das volks-echo, paris/zuerich) – gw Misc Inst [074]

Deutsches volksecho / ed by Heym, Stefan – New York NY [USA], 1937 20 feb-1939 16 sep – 1r – 1 – (title varies: 1-8 jan: deutsches volksecho und deutsch-kanadische volks-zeitung) – gw Misc Inst [071]

Deutsches volksecho und deutsch-kanadische volks-zeitung *see* Deutsches volksecho

Deutsches volkstum in glauben und aberglauben / Pfister, Friedrich – Berlin: de Gruyter 1936 [mf ed Bloomington IN: Indiana Uni Lib, Preservation Dept 1984] – 1r – 1 – us Indiana Preservation [390]

Deutsches volkstum in maerchen und sage, schwank und raetsel / Peuckert, Will Erich – Berlin: de Gruyter 1938 [mf ed Bloomington IN: Indiana Uni Lib, preservation Dept 1984] – 1r – 1 – us Indiana Preservation [390]

Deutsches volkstum in sitte und brauch / Geiger, Paul – Berlin: de Gruyter 1936 [mf ed Bloomington IN: Indiana Uni Lib, Preservation Dept 1984] – 1r – 1 – us Indiana Preservation [390]

Deutsches volkstum in volkskunst und volksstracht / Lehmann, Otto – Berlin: de Gruyter 1938 [mf ed Bloomington IN: Indiana Uni Lib, Preservation Dept 1984] – 1r – 1 – us Indiana Preservation [390]

Deutsches volkstum in volksschauspiel und volkstanz / Moser, Hans – Berlin: de Gruyter 1938 [mf ed Bloomington IN: Indiana Uni Lib, Preservation Dept 1984] – 1r – 1 – us Indiana Preservation [390]

Ein deutsches vorspiel / Neuber, Friederika Karoline Weissenborn; ed by Richter, Arthur – Leipzig: G J Goeschen, 1897 [mf ed 1993] – (= ser Deutsche litteraturdenkmale des 18. und 19. jahrhunderts 63, n f n13) – xvi/28p – 1 – (original ed, leipzig 1734. incl bibl ref) – mf#8676 reel 5 – us Wisconsin U Libr [820]

Deutsches wochenblatt – Berlin DE, 1919 3 may-1920 30 jun, 1922-1923 10 oct, 1928 7 jan-8 sep – 1r – 1 – gw Misc Inst [074]

Deutsches wochenblatt : fuer die provinz parana – Curitiba, PR: Typ D Derhandt & Co, 07 jan 1883; 13 jun 1885 – (= ser Ps 19) – mf#P16,02,27 – bl Biblioteca [079]

Deutsches wochenblatt : organ der deutschen volkspartei – Mannheim DE, 1864 22 dec [specimen copy]-1867 22 sep – 1r – 1 – mf#6506 – gw Mikropress [074]

Deutsches wollen – Berlin DE, 1935 1 feb-1936 9 apr – 1r – 1 – gw Misc Inst [074]

Deutsch-Evangelischen Frauenbund *see* Frauenkalender

Deutsche-volks-zeitung – Prague. Mar 22 1936-Aug 27 1939. Incomplete – 1 – us NY Public [077]

Deutsche-vom sturme verweht – Article in Deutsche Wochenzeitung Christ und Welt. Stuttgart, 21 Jun 1956. 8p – 1 – 5.00 – us Southern Baptist [242]

Deutsch-franzoesische monatshefte – Karlsruhe DE, 1934-41, 1943 – 1 – (with gaps) – gw Misc Inst [074]

Deutsch-gabler zeitung – Deutsch Gabel (Jablonne Podjestedi CZ), 1938 20 aug-31 dec – 1r – 1 – gw Misc Inst [077]

Deutsch-herero-worterbuch / Irle, J – Hamburg, Germany. 1917 – 1r – us UF Libraries [040]

Deutsch-indische geistesbeziehungen / Alsdorf, Ludwig – Heidelberg, Berlin, Magdeburg: K Vowinckel, 1942 [mf ed 1993] – (= ser Indien 7) – vii/111p (ill) – 1 – (incl bibl ref) – mf#8141 – us Wisconsin U Libr [410]

Deutschinoff, Gerd *see* Evaluation von scoresystemen in der intensivmedizin und deren zusammenhang mit dem langzeitueberleben

Deutsch-israelitische gemeindebund nach ablauf / Jacobsohn, B – Leipzig, Germany. 1879 – 1r – us UF Libraries [939]

Deutsch-kanadische volkszeitung : das blatt fuer die deutsch-kanadische familie...unabhaengig, progressiv, actuell – Toronto. v8 n11 oct 2 1937; v8 n16-22 nov 6-dec 25 1937// (wkly) – (= ser Deutschen volksecho) – 1r – 1 – Can$65.00 – (absorbed by: deutsches volksecho, new york) – cn McLaren [071]

Deutschland : oder briefe eines in deutschland reisenden deutschen / Weber, Carl J – Stuttgart 1826-28 [mf ed Hildesheim 1995-98] – 4v on 18mf – 9 – €180.00 – 3-487-29624-1 – gw Olms [860]

Deutschland / ed by Reichardt, Johann Friedrich – Berlin 1796 – (= ser Dz. abt literatur) – 4v on 12mf – 9 – €120.00 – mf#k/n4608 – gw Olms [430]
Deutschland : sein volk und seine sitten in geographisch-ethnographischen charakterbildern / Biffart, Max – Stuttgart 1860 [mf ed Hildesheim: 1995-98] – (= ser Fbc) – iv/576p on 4mf [ill] – 9 – €120.00 – 3-487-29626-8 – gw Olms [943]
Deutschland : ein vollstaendiges handbuch fuer die kunde des vaterlandes / Winderlich, Carl – Leipzig 1852 [mf ed Hildesheim: 1995-98] – (= ser Fbc) – xii/656p on 5mf – 9 – €100.00 – 3-487-29627-6 – gw Olms [943]
Deutschland *see* Das konstitutionelle deutschland
Deutschland am vorabend seines falles oder seiner groesse / Gutzkow, Karl – Frankfurt a M: Literarische Anstalt, 1848 [mf ed 2001] – 235p – 1 – mf#10526 – us Wisconsin U Libr [943]
Deutschland, armenien und die tuerkei 1895-1925 / Meissner, Axel – [mf ed 1998-2004] – 3pt – 9 – silver €2600.00 – ISBN-10: 3-598-34406-6 – ISBN-13: 978-3-598-34406-0 – (incl guide) – gw Saur [327]
Deutschland, deutschland ueber alles! : ein lebensbild des dichters hoffmann von fallersleben / Gerstenberg, Heinrich – Muenchen: C H Beck, 1916 [mf ed 1991] – vi/100p/4pl (ill) – 1 – mf#7480 – us Wisconsin U Libr [920]
Deutschland im jahre 2000 / Erman, G – Kiel: Lipsius und Tischer, 1891 – 1r – 1 – us Wisconsin U Libr [943]
Deutschland muss leben! / Lersch, Heinrich – Jena: E Diederichs, 1943, c1935 – 1r – 1 – us Wisconsin U Libr [810]
Deutschland nach seinen physischen und politischen verhaeltnissen / Daniel, Hermann A – Leipzig 1867-68 [mf ed Hildesheim 1995-98] – 2v on 10mf – 9 – €100.00 – 3-487-29625-X – gw Olms [943]
Deutschland und die deutschen / Beurmann, Eduard – Altona 1840 [mf ed Hildesheim: 1995-98] – (= ser Fbc) – 4v on 12mf – 9 – €120.00 – 3-487-29629-2 – gw Olms [943]
Deutschland und die nationalsozialisten in den vereinigten staaten von amerika / Graessner, Gernot Heinrich Willi – 1933-1939 – 1 – gw Mikropress [977]
Deutschland und die paepstliche weltherrschaft / Hauck, Albert – Leipzig: Alexander Edelmann, 1910 – 1mf – 9 – 0-8370-7824-5 – (incl bibl ref) – mf#1986-1824 – us ATLA [240]
Deutschland und polen, 1772-1945 / ed by Fechner, Helmuth – Wuerzburg: Holzner, Verlag, 1964 – 1r – 1 – (incl bibl ref) – us Wisconsin U Libr [327]
Deutschland und russland / Haller, Johannes – Tuebingen: Kloeres, 1915. 32p – 1 – us Wisconsin U Libr [943]
Deutschland und seine bewohner : ein handbuch der vaterlandskunde fuer alle staende / Hoffmann, Karl F – Stuttgart 1834-36 [mf ed Hildesheim: 1995-98] – 4v on 15mf – 9 – €150.00 – 3-487-29631-4 – gw Olms [943]
Deutschland unterm hakenkreuz : dichtungen gesammelt zu feiern in schule und jugendbund / ed by Hennesthal, Rudolf – 3. aufl. Frankfurt/M: M Diesterweg, 1936 [mf ed 1993] – 111p – 1 – mf#8360 – us Wisconsin U Libr [810]
Deutschland-bericht der sopade – Prag: [s.n.] 1934-40 [mf ed 1979] – 3v on 5r [ill] – 1 – (cont by: deutschland-berichte der sozialdemokratischen partei deutschlands [sopade], began with iss for apr/mai 1934, ceased in 1936) – mf#film mas c683 – us Harvard [335]
Deutschlandberichte der sopade – Prag [CZ]/Paris [F], 1934 apr/may-1940 apr – 8r – 1 – (1937: deutschland-berichte der sozialdemokratischen partei deutschlands [sopade]) – gw Misc Inst [325]
Deutschlandberichte der sopade *see* Germany
deutschland-berichte der sozialdemokratischen partei deutschlands [sopade] *see* Deutschlandberichte der sopade
Deutschland-brief – Berchtesgaden DE, 1951-54 [gaps] – 1 – gw Misc Inst [943]
Deutschlander, Leo *see* Westostliche dichterklange
Deutschland-information des zentralkomitees der kpd – Paris (F), 1938 n10, 1939 n1-6 – 1 – gw Misc Inst [325]
Deutschlands achtzehntes jahrhundert – Kempten DE, 1781-82 – 1r – 1 – gw Misc Inst [943]
Deutschlands erlauchten souverainen bei dem sturz der dynastie karls 10. koenigs von frankreich – [s. l.] 1830 [mf ed Hildesheim 1995-98] – 1mf – 9 – €40.00 – 3-487-29174-6 – gw Olms [940]
Deutschlands erneuerung – Muenchen DE, 1920-22 – 3r – 1 – gw Misc Inst [074]
Deutschlands geschichtsquellen im mittelalter / Wattenbach, W – Berlin, 1858 – €18.00 – ne Slangenburg [931]
Deutschlands politische parteien und das ministerium bismark / Parisius, L – Berlin, 1878 – 1r – 1 – mf#96374 – uk Microform Academic [943]
Deutschlands stimme – Berlin DE, 1948 25 jan-18 may [gaps] – 2r – 1 – (filmed by other misc inst: 1947 24 dec-1950, 1952 [2r]) – gw Misc Inst [074]
Deutschlands traum, kampf und sieg : geharnischte sonette nebst einem anhang vaterlaendischer gesaenge / Minckwitz, Johannes – Leipzig: M G Priber, 1870 – 1 – us Wisconsin U Libr [780]
Deutschlandsender *see* Nachrichtendienst des deutschlandsenders
Deutsch-mandschurische nachrichten : einzige deutsche tageszeitung in china und japan – Harbin (Pingkiang VR), 1919 8 dec-1930 9 jul – 1r – 1 – gw Misc Inst [079]
Deutschmann, Hayim Abraham *see* Shemu'ot tovot
Deutsch-oesterreichische theaterzeitung – Berlin DE, 1891-93 – 1r – 1 – gw Misc Inst [790]
Deutsch-ostafrika : geographie und geschichte der colonie / Forster, Brix – Leipzig: F A Brockhaus, 1890 – 1 – us CRL [960]
Deutsch-ostafrika : geschichte der gesellschaft fuer deutsche kolonisation und der deutsch-ostafrikanischen gesellschaft nach den amtlichen quellen / Wagner, J – Berlin: Verlag der Engelhardt'schen Landkartenhandlung, 1886 – 1 – us CRL [960]
Deutsch-ostafrika unverloren! : erzaehlung aus den deutschen kolonialkaempfen im weltkrieg / Viera, Josef – 4. Aufl. Stuttgart: Loewes, 1936 – 1 – us Wisconsin U Libr [960]
Deutsch-ostafrikanische zeitung – Daressalam (Dar-Es-Salaam EAT), feb 2 1899-may 16 1916 – 1 – (publ in morogoro) – gw Misc Inst [079]
Die deutschsprachige oscar-wilde-rezeption (1893-1906) : bibliographie / Haensel-Hohenhausen, Markus – Egelsbach, 1990 (mf ed 1993) – 1mf – 9 – €24.00 – 3-89349-247-X – mf#DHS-AR 104 – gw Frankfurter [430]
Deutschsprachige schriften zu japan 1477 bis 1945 *see* German books on japan 1477 to 1945, pt 1
Deutschsprachige zeitungen aus palaestina und israel = German newspapers from palestine and israel / ed by Reichenstein, Friedrich – [mf ed 2003-04] – 1395mf (1:24) 2pt in 6 installments – 9 – diazo €5900.00 (silver €7900 isbn: 978-3-598-35145-7) – ISBN-10: 3-598-35144-5 – ISBN-13: 978-3-598-35144-0 – (incl guide; also sold individually: jedioth chadashot 1935-67; yedioth hayom 1935-64) – gw Saur [321]
Deutschsprachige zeitungen aus palaestina und israel, abt 1 : palaestina 1935-1948 = German newspapers from palestine and israel, sect 1 / ed by Reichenstein, Friedrich – [mf ed 2003] – 359mf – 9 – diazo €2070.00 (silver €2750 isbn: 978-3-598-35147-1) – ISBN-10: 3-598-35146-1 – ISBN-13: 978-3-598-35146-4 – gw Saur [079]
Deutschsprachige zeitungen aus palaestina und israel, abt 2 : israel 1948-1973 = German newspapers from palestine and israel, sect 2 / ed by Reichenstein, Friedrich – [mf ed 2004] – 976mf (1:24) in 4 installments – 9 – diazo €3830.00 (silver €5150.00 isbn: 978-3-598-35153-4) – ISBN-10: 3-598-35152-6 – ISBN-13: 978-3-598-35152-5 – gw Saur [079]
Die deutschsprachigen freimaurer-zeitschriften des 18. und 19. jahrhunderts / Haensel-Hohenhausen, Markus – Egelsbach, Koeln, New York, 1993 – 3 installments – 9 – 3-89349-282-8 – (1. lieferung: s1-1251 13mf €350.00 isbn 3-89349-279-8 dhs-ar 138. 2. lieferung: s1252-2067 9mf €275.00 isbn 3-89349-280-1 dhs-ar 139. 3. lieferung: s2068-3032 10mf €275.00 isbn 3-89349-281-x dhs-ar 140) – gw Frankfurter [366]
Deutsch-sudwestafrika : drei jahre im lande nendrik witboois: schilderungen von land und leuten / Bulow, Franz Josef von – 2. Aufl. Berlin: E S Mittler, 1897 – 1 – us CRL [916]
Deutsch-sudwestafrika seit der besitzergreifung, die zuge und kriege gegen die eingeborenen / Bulow, Heinrich von – Berlin: W Susserott, 1904 – 1 – us CRL [960]
Deutsch-suedwestafrikanische zeitung – Windhuk (Windhoek NAM), 1902-1908 4 jul, 1909-11, 1913 4 jan-1914 29 jul – 1 – gw Misc Inst [079]
Das deutschtum im ausland : vierteljahrshefte des vereins fuer das deutschtum im ausland (allg deutscher schulverein) / Verein fuer das Deutschtum im Ausland [Germany] – Berlin: Der Verein. iss1-41/42 [1909-19] [mf ed 1979] – 42v on 1r – 1 – mf#film mas c619 – us Harvard [305]
Deutschtum im ausland / Weck, Hermann – Muenchen, Germany. 1916 – 1r – us UF Libraries [025]
Deutsch-ukrainische zeitung – Berlin DE, 1920 5 oct-1921 31 oct – 1r – 1 – gw Misc Inst [074]
Deutsch-ungarischer bote = German-american herald – Cincinnati: Deutsch-Ungarischer Bote Co, jan-may 23, 1918 – 1r – us CRL [071]
Deutsch-ungarischer volksfreund – Temeschburg (Timisoara RO), 1903 13 dec-1916 14 apr [gaps] – 1 – gw Misc Inst [077]
Deutsch-ungarisches volkblatt / Cuyahoga Co. Cleveland – apr 1915-jun 1917/ [wkly] – 1r – 1 – (in hungarian-german) – mf#B5903 – us Ohio Hist [071]
Deutschvoelkische gedichte / Bartels, Adolf – Zeitz: Sis-Verlag, c1918 [mf ed 1989] – 174p – 1 – mf#6980 – us Wisconsin U Libr [810]
Deutschvoelkisches jahrbuch – Weimar DE, 1920-22 – 1 – gw Misc Inst [943]
Deutschvolk *see* Hessischer vorkaempfer
Deutsch-zigeunerisches woerterbuch / Bischoff, Wilhelm Ferdinand – Ilmenau: Voigt 1827 – (= ser Whsb) – 2mf – 9 – €30.00 – mf#Hu 151 – gw Fischer [040]
Deutungen und bekenntnisse : ausgewaehlte texte zur deutschen literatur / ed by Schubert & Hoefer, Karl-Heinz – Leipzig: Verlag Enzyklopaedie, c1986 – 1r – 1 – (incl bibl ref and index) – us Wisconsin U Libr [430]
Les deux abbes de fenelon – Levis: P-G Roy, 1898 – 1mf – 9 – mf#25339 – cn Canadiana [920]
Deux amours / Brun, Amedee – Port-Au-Prince, Haiti. 1895 – 1r – us UF Libraries [972]
Deux amours d'adrien / Domingue, Jules – Corbeil, France. 1902 – 1r – us UF Libraries [972]
Deux annees a constantinople et en moree (1825-1826) : ou esquisses historiques sur mahmoud, les janissaires, les nouvelles troupes, ibrahim-pacha, solyman-bey, etc / Deval, Charles – London 1828 [mf ed Hildesheim 1995-98] – (= ser Fbc) – 2mf [ill] – 9 – €60.00 – 3-487-29111-8 – gw Olms [956]
Deux ans au se-tchouan (chine centrale) / Vigneron, Lucien – Paris: Bray et Retaux, 1881 [mf ed 1995] – (= ser Yale coll) – x/299p (ill) – 1 – 0-524-09018-1 – (in french) – mf#1995-0018 – us ATLA [951]
Deux ans de regne 1830-1832 / Pepin, Alphonse – Paris 1833 [mf ed Hildesheim 1995-98] – 1v on 3mf – 9 – €90.00 – 3-487-26057-3 – gw Olms [944]
Deux ans de sejour en abyssinie : ou vie morale, politique et religieuse des abyssiniens... / Dimotheos, P S – Jerusalem. 2v. 1871 – 4mf – 9 – mf#NE-20200 – ne IDC [916]
Deux ans et demi de ministere / Dubois, F-E – Paris, France. 1867 – 1r – us UF Libraries [972]
Deux boxeurs : ou, les anglais de falaise et de nan / Desaugiers, Marc-Antoine – Paris, France. 1814 – 1r – us UF Libraries [440]
Deux caciques de xaragua / Corvington, Hermann – Port-Au-Prince, Haiti. 194- – 1r – us UF Libraries [972]
Deux campagnes au soudan francais, 1886-1888 / Gallieni, Joseph-Simon – Paris: Hatchette, 1891 – 1 – us CRL [960]
Deux concepts d'independance a saint-domingue / Jean-Baptiste, St Victor – Port-Au-Prince, Haiti. 1944 – 1r – us UF Libraries [972]
Deux contes creoles / Dufrenois, M – Paris, France. 1936 – 1r – us UF Libraries [972]
Deux couronnes / Bayard, Jean-Francois-Alfred – s.l, s.l? 1842? – 1r – us UF Libraries [440]
Deux couronnes / Moreau, Eugene – Paris, France. 1840 – 1r – us UF Libraries [440]
Deux discours / Alvarez del Vayo, Julio – Paris, 1938. Fiche W 714. (Blodgett Collection of Spanish Civil War Pamphlets) – 9 – us Harvard [946]
Deux discours maconniques / Merzbach, Henry – Bruxelles: [s.n.] 1867 – 1mf – 9 – mf#vrl-80 – ne IDC [366]
Deux edmon / Barre, M – Paris, France. 1811 – 1r – us UF Libraries [440]
Deux edmon / Barre, M – Paris, France. 1813 – 1r – us UF Libraries [440]
Deux et deux font quatre = Two and two make four / Coler, Bird Sim – Montreal: impr au Devoir, [1919?] (mf ed 1992) – 3mf – 9 – (trans fr english by j-a fauteux; pref by louis ad paquet) – mf#SEM105P1555 – cn Bibl Nat [230]
Deux font la paire / Bayard, Jean-Francois-Alfred – Paris, France. 1832 – 1r – us UF Libraries [440]
Deux freres / Kotzebue, August Von – Paris, France. 1802? – 1r – us UF Libraries [440]
Deux gendres / Etienne, Charles Guillaume – Paris, France. 1810 – 1r – us UF Libraries [440]
Les deux jumelles, ou, la meprise : the favorite grand ballet by mr barree, as performed at the king's theatre, the music composed & arranged for the pianoforte, with an accompaniment for a violin / Bossi, Cesare – London: Goulding, Phipps & d'Almaine [1799] [mf ed 1988] – 1r – 1 – mf#pres. film 44 – us Sibley [790]
Deux modes satiriques : ou le choix fait par butler dans hudibras et swift dans a tale of a tub (le conte du tonneau) / Aniq Filali, Rabea – 2mf – 9 – (10556) – fr Atelier National [420]
Les deux mousquetaires : ou, la robe de chambre: opera comique en un acte / Berton, Henri – Paris: Vor Dufaut & Dubois [1824?] [mf ed 1990] – 1r – 1 – mf#pres. film 75 – us Sibley [780]
Deux nocturnes pour harpe & hautbois, oeuvre 50 no 1 / Bochsa, Robert Nicolas Charles – Paris: Dufaut & Dubois [182-] [mf ed 1992] – 1r – 1 – mf#pres. film 112 – us Sibley [780]
Deux papas tres bien : ou, la grammaire de chicard / Labiche, Eugene – Paris, France. 1844 – 1r – us UF Libraries [440]
Deux peres : ou, la lecon de botanique / Dupaty, Emmanuel – Paris, France. 1806 – 1r – us UF Libraries [440]
Deux peres : ou, la lecon de botanique / Dupaty, Emmanuel – Paris, France. 1809 – 1r – us UF Libraries [440]
Deux philibert / Picard, Louis-Benoit – Paris, France. 1816? – 1r – us UF Libraries [440]
Deux poemes : i e poemes couronnes par l'universite laval / Lemay, Pamphile – Quebec?: s.n, 1870 – 3mf – 9 – mf#08651 – cn Canadiana [440]
Deux points d'histoire / Cazes, Paul de – Quebec?: s.n, 1884 – 1mf – 9 – mf#08698 – cn Canadiana [917]
Deux pretres en colere : pour la liberation des chretiens / Lambert, Charles & Bouchard, Romeo – Montreal: editions du Jour, [1968] (mf ed 1974) – 1r – 5 – mf#SEM16P155 – cn Bibl Nat [241]
Deux procedes – n.p. 193? Fiche W830. (Blodgett Collection of Spanish Civil War Pamphlets) – 9 – us Harvard [946]
Deux quators, pour flute, violon, alto et violoncelle / Rolla, Alessandro – Paris: Sieber pere [17–?] [mf ed 1989] – 4pt on 1r – 1 – mf#pres. film 52 – us Sibley [780]
Deux quatuors pour deux violons, alto et violoncelle / Kreutzer, Rodolphe – Leipsic: Breitkopf & Haertel [179-] [mf ed 1988] – 4pt on 1r – 1 – mf#pres. film 39 – us Sibley [780]
Deux quintetti pour deux violons, deux alto & basse, oeuvre 5 / Pleyel, Ignaz – Paris: Imbault [179-?] [mf ed 1989] – 5pt on 1r – 1 – mf#pres. film 42 – us Sibley [780]
Deux quintetti pour deux violons, deux alto & basse, oeuvre 5 / Pleyel, Ignaz – Paris: Imbault [179-?] [mf ed 1989] – 5pt on 1r – 1 – (with bibl) – mf#pres. film 42 – us Sibley [780]
Deux recits de chasse : banmana ndoronkelen et bani-nyenema / Sidibe, Fode Moussa Balla – 1984 – us CRL [960]
Deux sentinelles / Doche, Joseph Denis – Paris, France. 1803 – 1r – us UF Libraries [440]
Deux sermons : la grace de dieu, le pardon des offenses. suivis de seize plans de sermons / Rabaut, Paul; ed by Frossard, Charles Louis – Paris: Grassart, 1886 – 1mf – 9 – 0-7905-7194-3 – mf#1988-3194 – us ATLA [240]
Deux soeurs : ou, le mentor / Fournier, Narcisse – Paris, France. 1843 – 1r – us UF Libraries [440]
Deux sonates pour le piano forte : avec accompagnement d'un violon ou flute et violoncelle, oeuvre 45 / Gyrowetz, Adalbert – Augsbourg: Gombart [180-] [mf ed 1988] – 3pt on 1r – 1 – mf#pres. film 47 – us Sibley [780]
Deux sonates pour le piano-forte : avec accompagnement de violon et basse, oeuvre dernier et posthume / Dussek, Johann Ladislaus – Leipsic: Breitkopf & Haertel [182-?] [mf ed 19–] – 3pt on 1r – 1 – mf#pres. film 91 – us Sibley [780]
Deux sonates pour le pianoforte avec accompagnement de violon et basse, oeuvre 34 / Dussek, Johann Ladislaus – Leipsic: Breitkopf & Hartel [182-?] [mf ed 1989] – 3pt on 1r – 1 – mf#pres. film 45 – us Sibley [780]
Deux sonates pour piano-forte : avec accompagnement de violon et violoncelle, oeuvre 61 / Steibelt, Daniel – Offenbach: J Andre [1805] [mf ed 1991] – 1r – 1 – mf#pres. film 110 – us Sibley [780]
Les deux testaments : esquisse de moeurs canadiennes / Duval-Thibault, Anna-Marie – Fall River, MA: impr de l'Indépendant, 1888 [mf ed 1979] – 2mf – 9 – mf#SEM105P18 – cn Bibl Nat [830]
Les deux theologies nouvelles dans le sein du protestantisme francais : etude historico-dogmatique / Astie, Jean-Frederic – Paris: Meyrueis, 1862 [mf ed 1990] – 1mf – 9 – 0-7905-6521-8 – mf#1988-2521 – us ATLA [242]
Deux traitez : l'un, de la messe et de ses parties. l'autre, de la transsubstantiation du pain et vin de la messe / Daneau, Lambert – La Rochelle, [Portau], 1589 – 5mf – 9 – mf#PFA-134 – ne IDC [240]

Deux vieux papillons / Laya, Leon – Paris, France. 1850 – 1r – us UF Libraries [440]
Deux voisines : ou, les pretes rendus / Desaugiers, Marc-Antoine – Paris, France. 1815 – 1r – us UF Libraries [440]
Deux voyages sur le saint maurice / Caron, Napoleon – Trois Rivieres, Quebec: P V Ayotte, 1889 – 4mf – 9 – mf#00467 – cn Canadiana [917]
Deuxieme centenaire de la fondation de l'institut des freres des ecoles chretiennes : sermon prononce dans l'eglise st jean baptiste de quebec le 20 octobre 1880 / Bruchesi, Louis Joseph Paul Napoleon – Quebec: C Darveau, 1880 – 1mf – 9 – mf#03735 – cn Canadiana [240]
Deuxieme conference pleniere des ordinaires des missions / Conference Pleniere Des Ordinaires Des Missions Du Congo Belge – Leopoldville, Congo. 1936 – 1r – us UF Libraries [960]
Deuxieme congres de la federation nationale saint-jean-baptiste : (section des dames de l'association saint-jean-baptiste) tenu a montreal les 23, 25, 26 juin / Federation nationale Saint-Jean-Baptiste. Congres (2e: 1909: Montreal, Quebec) – [Montreal?: s.n, 1909?] [mf ed 1995] – 2mf – 9 – 0-665-73657-6 – mf#73657 – cn Canadiana [305]
Deuxieme fantaisie, pour piano et cor ou violon / Duvernoy, Frederic – Paris: J J de Momigney [182-?] [mf ed 1989] – 1r – 1 – mf#pres. film 45 – us Sibley [780]
Deuxieme quintuor pour la flute, violon, 2 altos et violoncelle, op 60 / Brandl, Johann Evangelist – Bonn: Simrock [c1812] [mf ed 1988] – 5pt on 1r – 1 – mf#pres. film 44 – us Sibley [780]
Deuxieme rapport de la commission chargee de reviser et de modifier le code de procedure civile du bas-canada = Second report of the commission charged with the revision and amendment of the code of civil procedure of lower canada / Casgrain, Thomas Chase et al – Quebec: impr par Leger Brousseau, 1894 [mf ed 1992] – 3mf – 9 – mf#SEM105P1514 – cn Bibl Nat [350]
Deuxieme supplement au catalogue de la bibliotheque de l'apostolat des bons livres / Apostolat des bons livres. Bibliotheque – Quebec: L'Action sociale ltee, 1932 [mf ed 1996] – 1mf – 9 – (with ind) – mf#SEM105P2504 – cn Bibl Nat [020]
Deuxieme these de doctorat / Chrisphonte, Prosper – Port-Au-Prince, Haiti. 1950 – 1r – us UF Libraries [972]
Deva *see*
– Ma cim pan
– Pan cum
The devachanic plane : or, the heaven world. its characteristics and inhabitants / Leadbeater, Charles Webster – 2nd rev enl ed. London: Theosophical Pub Society, 1902 – (= ser Theosophical Manuals (London, England)) – 1mf – 9 – 0-524-02311-5 – mf#1990-2934 – us ATLA [290]
Devadhar, C R *see* Ratnavali
Deval, Charles *see* Deux annees a constantinople et en moree (1825-1826)
The devalaya, its aims and objects : with a short sketch of the life and work of its founder / Tattvabhushan, Sitanath – 2nd ed. Calcutta: Elysium Press, 1912 – 1mf – 9 – 0-524-02053-1 – mf#1990-2828 – us ATLA [280]
Devalle, Giovanni *see*
– Ascensioni sul monte kenya
– Sulle sponde del lago rodolfo
Devant la nation / Villard, Suirad – Port-Au-Prince, Haiti. 1926 – 1r – us UF Libraries [972]
Devas, Charles Stanton *see* L'eglise et le progres du monde
Devas, Raymund P *see*
– The dominican revival in the 19th century
– Island of grenada
Devassamento do piaui / Lima Sobrinho, Barbosa – Sao Paulo, Brazil. 1946 – 1r – us UF Libraries [972]
Devasthali, G V *see* Bhatta narayana's venisamharam
Devekusu – Trabzon. Sahib ve Sermuharriri: Cemal Riza. n5. 17 eyluel 1341 [1925] – (= ser O & t journals) – 1mf – 9 – $25.00 – us MEDOC [956]
Developing a marketing information systems (mkis) model for south african service organizations / Venter, Petrus – Uni of South Africa 2000 [mf ed Johannesburg 2001] – 7mf – 9 – (incl bibl ref; abstract in english & afrikaans) – mf#mfm15032 – sa Unisa [650]
Developing a ministry of lay evangelism at the first baptist church of roanoke, virginia / Peverall, Albert Arthur – 1982 – 1 – 7.36 – us Southern Baptist [242]
Developing a program of family ministry by the deacons of first baptist church, hodgenville, kentucky / McDonald, Isaac Burkhalter – 1981 – 1 – 5.60 – us Southern Baptist [242]
Developing an activity conference at the middle school level / Dickerson, Thomas A – 1997 – 1mf – 9 – $4.00 – mf#PE 3789 – us Kinesology [370]
Developing and implementing an outcomes-based computer literacy programme : in distance learning mode for south african students / Oosthuizen, Marita – Pretoria: Vista University 2003 [mf ed 2003] – 5mf [ill] – 9 – (incl bibl) – mf#mfm15189 – sa Unisa [370]
The developing food security crisis in southern africa : hearing...house of representatives, 107th congress, 2nd session, june 13 2002 / United States. Congress. House. Committee on International Relations – Washington: US GPO 2002 [mf ed 2002] – 3mf – 9 – 0-16-068979-1 – us GPO [327]
Developing the space frontier – 1983 – (= ser Advances in the astronautical sciences 52) – 9 – $45.00 – us Univelt [380]
Development : thoughts on bishop gore's "roman catholic claims" / Rickaby, Joseph – London: Catholic Truth Society, 1905 – 1mf – 9 – 0-8370-7096-1 – mf#1986-1096 – us ATLA [241]
Development : what it can do and what it cannot do / McCosh, James – New York: Scribner 1883 [mf ed 1991] – 1mf – 9 – 0-7905-9804-3 – mf#1989-1529 – us ATLA [210]
The development and application of the signature as an identification method : in the south african law / Robinson, Melanie-Jane – Pretoria: Vista University 2002 [mf ed 2002] – 4mf – 9 – (incl bibl ref) – mf#mfm15181 – sa Unisa [346]
Development and change – Oxford, England 1996+ – 1,5,9 – ISSN: 0012-155X – mf#20358 – us UMI ProQuest [303]
The development and decline of the all-american girls baseball league, 1943-1954 / Fidler, Merrie A – 1976 – 4mf – 9 – $16.00 – mf#PE 4037 – us Kinesology [790]
Development and education in the cook islands : a study of community and education in an emergent pacific islands territory / Coppell, William G – 1823-1967 – 1r – mf#pmb65 – at Pacific Mss [350]
Development and evaluation of a leisure education module for use in a college resource center / Fierle, Karen M – 1982 – 2mf – 9 – $8.00 – us Kinesology [790]
Development and evaluation of an interpretive nineteenth century american children's games program : knowledge and satisfaction attained through program participation / Bakke, P Q – 1991 – 3mf – 9 – $12.00 – us Kinesology [790]
Development [and] evaluation of computer-assisted instruction in smoking education for adolescents / Howerton, Mollie W – 1999 – 3mf – 9 – $12.00 – mf#HE 662 – us Kinesology [373]
Development and evolution : including psychophysical evolution, evolution by orthoplasy, and the theory of genetic modes / Baldwin, James Mark – New York: Macmillan Co, 1902 – 1 – us CRL [575]
The development and implementation of a computer-assisted instruction series to be utilized as an aid to curriculum methodology in physical education / Lease, Barbara J – 1981 – 2mf – 9 – $8.00 – us Kinesology [790]
The development and implementation of marketing strategies for small, medium and micro enterprises (smmes) / Mphirime, Keneilwe Florence – Pretoria: Vista University 2000 [mf ed 2000] – 3mf – 9 – mf#mfm15126 – sa Unisa [650]
The development and implementation of the total person program at the georgia tech athletic association / McGlade, Bernadette V – 1997 – 1mf – 9 – $4.00 – mf#PSY 2002 – us Kinesology [150]
Development and psychopathology – Cambridge, England 1989+ – 1,5,9 – ISSN: 0954-5794 – mf#17116 – us UMI ProQuest [616]
Development and purpose : an essay towards a philosophy of evolution / Hobhouse, Leonard Trelawney – London: Macmillan, 1913 – 1mf – 9 – 0-7905-3955-1 – mf#1989-0448 – us ATLA [100]
The development and testing of the american heart association slim for life weight-loss program / Scholes, Melissa A – 1998 – 2mf – 9 – $8.00 – mf#HE 646 – us Kinesology [613]
Development and validation of a maximal testing protocol for the nordictrack cross-country ski simulator / Haug, Rhea C – University of Wisconsin-La Crosse, 1995 – 1mf – 9 – mf#PH 1495 – us Kinesology [612]
Development and validation of a questionnaire for assessing habitual physical activity of sixth-grade students / Koehler, Karen M – 1988 – 131p on 2mf – 9 – $8.00 – us Kinesology [370]
Development and validation of a questionnaire to measure body image / Rowe, David A – 1996 – 3mf – 9 – $12.00 – mf#PSY 2003 – us Kinesology [150]
Development and validation of an instrument to measure student attitude toward physical education : a mixed method approach / Subramaniam, Prithwi R – 1998 – 3mf – 9 – $12.00 – mf#PSY 2011 – us Kinesology [790]
The development and validation of the coaching staff cohesion scale / Martin, Kathleen A – 1999 – 2mf – 9 – $8.00 – mf#PE 3957 – us Kinesology [790]
Development and validation of the wellness knowledge, attitude, and behavior instrument / Dinger, Mary K & Watts, Parris – 1993 – 2mf – $8.00 – us Kinesology [613]
Development and validity of the teachers' attitude, comfort and training scale (tacts) on sexuality education / d'Entremont, Laura S – 1999 – 1mf – 9 – $4.00 – mf#HE 633 – us Kinesology [613]
Development [cambridge] – Cambridge, England 1987-96 – 1,5,9 – (cont: journal of embryology & experimental morphology) – ISSN: 0950-1991 – mf#13598,01 – us UMI ProQuest [612]
Development digest – Washington DC 1962-83 – 1,5,9 – ISSN: 0012-1576 – mf#6296 – us UMI ProQuest [320]
The development from kant to hegel : with chapters on the philosophy of religion / Seth Pringle-Pattison, Andrew – London: Williams and Norgate, 1882 – 1mf – 9 – 0-7905-9884-1 – mf#1989-1609 – us ATLA [100]
Development genes and evolution – Dordrecht, Netherlands 2002+ – 1,5,9 – (cont: roux's archives of developmental biology) – ISSN: 0949-944X – mf#13232,07 – us UMI ProQuest [574]
Development [hants] – Hants, England 1981+ – 1,5,9 – (cont: development [rome]) – ISSN: 1011-6370 – mf#13063 – us UMI ProQuest [338]
Development in africa / Green, L P – Johannesburg, South Africa. 1962 – 1r – us UF Libraries [960]
A development model for middle-managers in the sebideng district council / Moshebi, Oupa Mochongoane – Pretoria: Vista University 2002 [mf ed 2002] – 6mf – 9 – (incl bibl ref) – mf#mfm15247 – sa Unisa [650]
Development needs in botswana and lesotho / Jacqz, Jane W – New York, NY. 1967 – 1r – us UF Libraries [337]
Development of a child injury data base for use in biomechanics research / Kelleher-Walsh, Barbara J & DeBacy, Diane L – 1992 – 1mf – 9 – $4.00 – us Kinesology [612]
Development of a collegiate licensing administrative paradigm / Irwin, R L – 1990 – 3mf – 9 – $12.00 – us Kinesology [378]
The development of a comprehensive plan for ministry for first baptist church, nevada, missouri / Cox, William Charles – 1981 – 1 – 6.00 – us Southern Baptist [242]
The development of a conceptual model and definition of quality practice from the perspectives of expert coaches / Sverduk, Kevin L – 1998 – 1mf – 9 – $4.00 – mf#PE 3885 – us Kinesology [790]
Development of a conceptual model for organizational transformation with specific reference to absa group / Dyk, Laetitia Arene van – Stellenbosch: U of Stellenbosch 1998 [mf ed 1998] – 5mf – 9 – mf#mf.1286 – sa Stellenbosch [330]
The development of a criterion-referenced health knowledge instrument for grades four, five, and six / Massey, MS – 1991 – 3mf – 9 – $12.00 – us Kinesology [613]
The development of a decision precess map : application to the snow ski market / Aukers, Steven M – 1999 – 3mf – 9 – $12.00 – mf#PE 4092 – us Kinesology [650]
The development of a design for a total evaluation system for professional baseball umpires / Janssen, Philip F – 1996 – 4mf – 9 – $16.00 – mf#PE 4006 – us Kinesology [790]
The development of a folk dance unit as a resource for the state of utah elementary sixth grade social studies core / Chamberlain, Tamara M – 1997 – 2mf – 9 – $8.00 – mf#PE 3875 – us Kinesology [790]
Development of a high school sports medicine/athletics training course / Hostetter, Karen – 189p on 2mf – 9 – $10.00 – mf#PE 4204 – us Kinesology [373]
The development of a manual for recreation leaders in southern baptist associations / Smith, Frank Hart – 1982 – 1 – 6.64 – us Southern Baptist [242]
The development of a measurement system for ball skills based on the ecological task analysis model / Yun, Joonkoo – 1998 – 3mf – 9 – $12.00 – mf#PE 3934 – us Kinesology [790]
The development of a ministry at the immanuel baptist church, toronto, canada, to integrate multicultural peoples : with special reference to west indian immigrants / Baxter, Samuel John – 1982 – 1 – 5.60 – us Southern Baptist [242]
Development of a predictive equation for maximal oxygen consumption on the steptreadmill / Carroll, K K – 1991 – 1mf – 9 – $4.00 – us Kinesology [612]
The development of a program of student financial assistance for east coast bible college / Bell, Kenneth Ray – 1982 – 1 – 5.44 – us Southern Baptist [242]
The development of a program of supplemental pastoral care for whitsitt chapel baptist church / Nail, Marvin Powell – 1982 – 1 – 5.00 – us Southern Baptist [242]
Development of a recipient based guide for coping with the process of liver transplantation / Solberg, Jim C – 1997 – 1mf – 9 – $4.00 – mf#HE 610 – us Kinesology [617]
Development of a three-octave bandwidth printed log-periodic dipole array / Wyk, M J van – Stellenbosch: U of Stellenbosch 1998 [mf ed 1999] – 2mf – 9 – mf#mf.1301 – sa Stellenbosch [621]
The development of a video-based motion analysis system / Bothner, Krisanne E & Widule, Carole J – 1992 – 1mf – 9 – $4.00 – us Kinesology [612]
Development of agricultural cooperatives, cambodia / Knobel, Fred H – [Phnom-Penh? 1961] [mf ed 1989] – 1r with other items – 1 – mf#mf-10289 seam reel 024/02 [§] – us CRL [334]
Development of agricultural education in the state of florida from 1918-1928 / Brown, J Colvin – s.l, s.l? 1931 – 1r – us UF Libraries [630]
The development of american commerce / Frederick, John Hutchinson – New York, London: D. Appleton and Company, c1932. xx,390p. Illus. Maps, tables, diagrams. With: Lehrbuch der Mechanischen Naturlehre by E.G. Fisher. 1 reel. 1292 – 1 – us Wisconsin U Libr [380]
The development of american hymnody, 1620-1900 / Burnett, Madeline L – 1946 – 1 – 5.04 – us Southern Baptist [242]
Development of an anthropometric regression equation to predict body density in african american women / Irwin, Melinda L – 1994 – 2mf – $8.00 – us Kinesology [612]
The development of an empirically grounded set of salient ski resort attributes / Aukers, Steven M – 1997 – 2mf – 9 – $8.00 – mf#RC 518 – us Kinesology [790]
The development of an instrument to assess the attitudes toward cultural diversity and cultural pluralism among preservice physical education majors / Stanley, Linda S et al – 1992 – 2mf – 9 – $8.00 – us Kinesology [150]
The development of an integrated model of risk / Briers, Steven – Uni of South Africa 2000 [mf ed Johannesburg 2000] – 7mf – 9 – (incl bibl ref) – mf#mfm15092 – sa Unisa [650]
Development of an inventory to assess multicultural education attitudes, competencies and knowledge of physical education professionals / Woods, Lydia A & Jewett, Ann E – 1992 – 3mf – 9 – $12.00 – us Kinesology [150]
The development of balance control mechanisms in infants and young children / Roncesvalles, Maria N – 1997 – 2mf – 9 – $8.00 – mf#PSY 1984 – us Kinesology [612]
The development of bantu education in the north-western cape, 1840-1947 : a historical survey / Lekhela, Ernest Plaelo – Pretoria, 1958 – us CRL [370]
The development of baptist principles in rhode island / Barrows, Comfort Edwin – Philadelphia: American Baptist Publication Society, [1875?] – 1mf – 9 – 0-524-00963-5 – (incl bibl ref) – mf#1990-4021 – us ATLA [242]
The development of capitalistic enterprise in india / Buchanan, Daniel Houston – New York: Macmillan Co, 1934 – (= ser Samp: indian books) – us CRL [338]
The development of china / Latourette, Kenneth Scott – Boston: Houghton Mifflin, 1917 – 1mf – 9 – 0-524-07628-6 – (incl bibl ref) – mf#1991-3235 – us ATLA [951]
The development of christianity = Die entwicklung des christentums / Pfleiderer, Otto – aut ed. New York: B W Huebsch, 1910 [mf ed 1990] – 1mf – 9 – 0-7905-7540-X – (trans by daniel a huebsch) – mf#1989-0765 – us ATLA [240]
The development of christianity in taiwan *see* – Chi-tu-chiao tsai tai-wan te fa chan [ccm301]
Development of christ's humanity / Bell, Henry – Aberdeen, Scotland. 1880 – 1r – us UF Libraries [240]

DEVELOPMENT

Development of commercial education in the public secondary... / Moorman, John H – s.l, s.l? 1934 – 1r – us UF Libraries [373]
Development of community recreation programs / U.S. Congress. House. Committee on Education and Labor – 1946 – 2mf – 9 – $12.00 – us Kinesology [790]
The development of competency guidelines for riding instructors and equestrian coaches / Harris, Johanna L – University of North Carolina at Chapel Hill, 1995 – 2mf – 9 – $8.00 – mf#PE3596 – us Kinesology [790]
The development of doctrine from the early middle ages to the reformation / Banks, John Shaw – London: Charles H Kelly, 1901 [mf ed 1990] – (= ser Books for bible students) – 1mf – 9 – 0-7905-3535-1 – (incl bibl ref) – mf#1989-0028 – us ATLA [240]
The development of doctrine in the early church / Banks, John Shaw – London: CH Kelly, 1900 – (= ser Books for bible students) – 1mf – 9 – (incl bibl ref) – mf#1989-0029 – us ATLA [240]
The development of doctrine in the epistles / Henderson, Charles Richmond – Philadelphia: American Baptist Publ Soc, 1896 – 1mf – 9 – 0-8370-3558-9 – (incl ind) – mf#1985-1558 – us ATLA [227]
Development of education in indonesia / Indonesia. Kementerian pendidikan, pengadjaran dan kebudajaan – Djakarta, 1955-1957 – 2mf – 9 – mf#SE-476 – ne IDC [370]
Development of education in the bechuanaland protectorate (1824-1944) : an historical survey / Seboni, Michael Ontefetse Martinus – 1946 – us CRL [370]
The development of english theology in the nineteenth century, 1800-1860 / Storr, Vernon Faithfull – London; New York: Longmans, Green, 1913 – 2mf – 9 – 0-7905-6022-4 – (incl bibl ref) – mf#1988-2022 – us ATLA [240]
The development of european polity / Sidgwick, Henry; ed by Sidgwick, Eleanor Mildred – London; New York: Macmillan, 1903 – 2mf – 9 – 0-7905-8585-5 – mf#1989-1810 – us ATLA [327]
The development of higher education in south africa, 1873-1927 / Metrowich, F C – Cape Town: M Miller, 1929 – 1 – us CRL [960]
The development of hindu iconography / Banerjea, Jitendra Nath – Calcutta: University of Calcutta, 1941 – (= ser Samp: indian books) – us CRL [280]
Development of hispanic america / Wilgus, Curtis – New York, NY. 1941 – 1r – us UF Libraries [972]
The development of interscholastic sports at seventh-day adventist academies and colleges / Sather, Brian A – 1996 – 2mf – 9 – $8.00 – mf#PE 3803 – us Kinesology [790]
The development of iron and steel technology in china / Needham, J – 1964 – (= ser Newcomen society extra publication 2) – 4mf – 7 – mf#86580 – uk Microform Academic [951]
The development of japan / Latourette, Kenneth Scott – New York: Macmillan, 1918 – 1mf – 9 – 0-524-07629-4 – mf#1991-3236 – us ATLA [950]
The development of job-related education and training in soweto, 1940-1990 / Kelm, Erwin – Uni of South Africa 2001 [mf ed Pretoria: UNISA 2000] – 3mf – 9 – (incl bibl ref) – mf#mfm15119 – sa Unisa [370]
The development of liberalism amongst the icelanders in north america / Petursson, Philip Markus – [Chicago, 1932.] Chicago: Dep of Photodup, U of Chicago Lib, 1971 (1r); Evanston: American Theol Lib Assoc, 1984 (1r) – 1 – 0-8370-0333-4 – mf#1984-B158 – us ATLA [240]
The development of long-range goals for the first baptist church of fairfield, ohio / Copeland, Edward Brent – 1982 – 1 – 5.04 – us Southern Baptist [242]
Development of major community musical activities / Duncan, Richard Edward – U of Rochester 1953 [mf ed 19–] – 2v on 10mf – 9 – mf#fiche181 – us Sibley [780]
The development of metaphysics in persia : a contribution to the history of muslim philosophy / Iqbal, Muhammad, Sir – London: Luzac, 1908 – 1mf – 9 – 0-7905-9970-8 – (incl bibl ref) – mf#1989-1695 – us ATLA [110]
The development of modern philosophy : with other lectures and essays / Adamson, Robert; ed by Sorley, William Ritchie – Edinburgh: W Blackwood, 1903 – 2mf – 9 – 0-7905-3510-6 – (incl bibl ref) – mf#1989-0003 – us ATLA [100]
The development of muscular power in swimmers / Schlagel, David A – 1997 – 1mf – 9 – $4.00 – mf#PE 3859 – us Kinesology [612]
Development of music in canada / Howell, Gordon P – U of Rochester 1959 [mf ed 19–] – 2v on 1r – 1 – mf#film 1486 – us Sibley [780]
Development of muslim theology, jurisprudence and constitutional theory / Macdonald, Duncan Black – New York: C Scribner, 1903 – (= ser The Semitic Series) – 1mf – 9 – 0-524-01789-1 – (incl bibl ref) – mf#1990-2637 – us ATLA [260]
Development of native agriculture and land tenure in southern... / Alvord, Emery Delmont – s.l, s.l? between 1955 and 1959 – 1r – us UF Libraries [333]
The development of native education in the bechuanaland protectorate (1840-1946) : an historical survey – BC Thema, 1947 – us CRL [370]
The development of naturalism in german poetry : from the hainbund to liliencron / Bohm, Erwin Herbert – [Columbus, Ohio?]: s.n, 1917 [mf ed 1993] – 61p – 1 – (incl bibl ref) – mf#8243 – us Wisconsin U Libr [430]
Development of optimized production planning and control systems for the vwsa engine manufacturing plant / Niekerk, Frank S A van – Stellenbosch: U of Stellenbosch 1998 [mf ed 1998] – 5mf – 9 – mf#mf.1284 – sa Stellenbosch [620]
Development of ornamental art in the international exhibition / Dresser, Christopher – London 1862 – (= ser 19th c art & architecture) – 3mf – 9 – mf#4.2.119 – uk Chadwyck [740]
The development of palestine exploration : being the ely lectures for 1903 / Bliss, Frederick Jones – New York: Charles Scribner, 1906 – 1mf – 9 – 0-7905-1081-2 – (incl bibl ref and index) – mf#1987-1081 – us ATLA [930]
The development of personal liberty in great britain, france and their colonies : an historical sketch / Hamilton, James Cleland – [Toronto?: s.n, 1905?] – 1mf – 9 – 0-665-72999-5 – (incl bibl ref) – mf#72999 – cn Canadiana [322]
The development of philosophy in japan / Kishinami, Tsunezo – Princeton: Princeton University Press, 1915 – 1mf – 9 – 0-524-01189-3 – mf#1990-2265 – us ATLA [100]
The development of religion : a study in anthropology and social psychology / King, Irving – New York: Macmillan, 1910 – 1mf – 9 – 0-524-01188-5 – (incl bibl ref) – mf#1990-2264 – us ATLA [150]
Development of religion and thought in ancient egypt : lectures / Breasted, James Henry – New York: Charles Scribner, 1912 [mf ed 1989] – (= ser The morse lectures) – 1mf – 9 – 0-7905-1744-2 – (incl bibl ref & ind) – mf#1987-1744 – us ATLA [290]
Development of religion in japan / Knox, George William – New York: GP Putnam, 1907 – (= ser American lectures on the history of religions) – 1mf – 9 – 0-524-00916-3 – mf#1990-2139 – us ATLA [290]
The development of religious liberty in connecticut / Greene, Maria Louise – Boston: Houghton, Mifflin, 1905 – 2mf – 9 – 0-7905-5399-6 – (incl bibl ref) – mf#1988-1399 – us ATLA [240]
The development of revelation : an attempt to elucidate the nature and meaning of old testament inspiration / Palmer, Ebenezer Reeves – London: Clement Sadler Palmer, 1892 – 1mf – 9 – 0-7905-9050-6 – mf#1989-2275 – us ATLA [221]
The development of spiritual health instructional strategies using a systems approach model / Larsen, Michelle H – Brigham Young University, 1994 – 1mf – 9 – mf#HE 568 – us Kinesology [613]
Development of strains of cigar wrapper tobacco resistant to blackshank (phytophthora nicotianae) / Tisdale, W B – Gainesville, FL. 1931 – 1r – us UF Libraries [630]
Development of the athlete satisfaction questionnaire / Riemer, Harold A – Ohio State University, 1995 – 6mf – 9 – $24.00 – mf#PSY1858 – us Kinesology [150]
Development of the attitudes toward the disabled in physical education scale / Gantz, J L – 1991 – 1mf – 9 – $4.00 – us Kinesology [150]
Development of the bechuanaland economy / Great Britain Ministry Of Overseas Development Economic Survey – Gaberone, Botswana. 1966? – 1r – us UF Libraries [330]
Development of the british army, 1899-1914 / Dunlop, John Kinninmont – London, England. 1938 – 1r – us UF Libraries [355]
Development of the canaanite dialects / Harris, Z S – 1939 – 9 – $10.00 – us IRC [470]
Development of the christian life / Clifford, William – London, England. no date – 1r – us UF Libraries [240]
The development of the cumberland presbyterian church : an address. delivered at the centennial celebration of the formation of the synod of kentucky of the presbyterian church, at lexington, ky... / McKamy, John A – Nashville, TN: Cumberland Presbyterian Pub House, 1903 – 1mf – 9 – 0-524-07207-8 – mf#1990-5365 – us ATLA [242]
Development of the doctrine of infant salvation / Warfield, Benjamin Breckinridge – New York, NY. 1891 – 1r – us UF Libraries [240]
The development of the doctrine of infant salvation / Warfield, Benjamin Breckinridge – New York: Christian Literature, 1891 – 1mf – 9 – 0-8370-5704-3 – (incl bibl ref) – mf#1985-3704 – us ATLA [240]
Development of the functional use of the supertonic seventh chord : as evidenced in representative keyboard suites of the seventeeth century / Zeyen, Mary Mark – U of Rochester 1956 [mf ed 19–] – 5mf – 9 – (with app & bibl) – mf#fiche 185 – us Sibley [780]
The development of the hymn among spanish speaking evangelicals / McConnell, Harry C – 1952 – 1 – $13.44 – us Southern Baptist [780]
Development of the leeward islands under the resto / Higham, Charles Strachan Sanders – Cambridge, England. 1921 – 1r – us UF Libraries [338]
The development of the logical method in ancient china / Hu, Shih – 2nd ed. Introd. by Hyman Kublin. New York: Paragon Book Reprint Corp., 1963 – 187p – 1 – us Wisconsin U Libr [180]
Development of the logos-doctrine in greek and hebrew thought / Walton, Frank Edward – Bristol: John Wright; London: Simpkin, Marshall, Hamilton, Kent, 1911 [mf ed 1989] – 1mf – 9 – 0-7905-3299-9 – mf#1987-3299 – us ATLA [180]
The development of the motive of protestant missions to china, 1807-1928 / Workman, George Bell – [New Haven], 1928. Chicago: Dep of Photodup, U of Chicago Lib, 1969 (1r); Evanston: American Theol Lib Assoc, 1984 (1r) – 1 – 0-8370-0400-4 – mf#1984-B119 – us ATLA [242]
Development of the psychological skills inventory for chinese athletes / Yang, Xiaochun – 1997 – 1mf – 9 – $4.00 – mf#PSY 2000 – us Kinesology [150]
Development of the root-knot nematode on beans as affected by soil temperature / Townsend, G R – Gainesville, FL. 1937 – 1r – us UF Libraries [630]
The development of the software for the wellness development process : la crosse wellness project / Xiong, Donald C – 1998 – 1mf – 9 – $4.00 – mf#HE 623 – us Kinesology [613]
The development of the sunday-school, 1780-1905 : the official report of the 11th international sunday-school convention, toronto, canada, june 23-27, 1905 / ed by International Sunday-School Convention of the United States and British American Provinces – Boston, Mass.: Executive Committee of the International Sunday-School Association, 1905 – 2mf – us ATLA [240]
The development of the sunday-school, 1780-1905 : the official report of the eleventh international sunday-school convention, toronto, canada, june 23-27, 1905 – Boston, Mass: Executive Committee of the International Sunday-School Association, 1905 – 2mf – 9 – 0-7905-4694-9 – (incl bibl ref) – mf#1988-0694 – us ATLA [240]
The development of the teaching of law in the university of edinburgh / Coldstream, John Phillips – Edinburgh, Morrison & Gibb, 1884. 20 p. LL-2250 – 1 – us L of C Photodup [340]
The development of the transportation pattern in ghana / Gould, Peter R – Evanston: Dept of Geography, Northwestern University, 1960 – us CRL [380]
The development of the wealth of india : ...with notes on the different administrative and judicial systems required for the asiatic races and the british inhabitants / Hare, Thomas – Cambridge, 1861 – (= ser 19th c books on british colonization) – 1mf – 9 – mf#1.1.5359 – uk Chadwyck [330]
The development of the young people's movement / Erb, Frank Otis – Chicago, Ill: University of Chicago Press, c1917 – 1mf – 9 – 0-524-07677-4 – (incl bibl ref) – mf#1991-3262 – us ATLA [240]
The development of theology as illustrated in english poetry from 1780 to 1830 / Brooke, Stopford Augustus – London: Philip Green, 1893 [mf ed 1991] – (= ser The essex hall lecture 1893) – 1mf – 9 – 0-7905-9150-2 – mf#1989-2375 – us ATLA [420]
The development of theology in germany since kant and its progress in great britain since 1825 = Entwicklung der protestantischen theologie in deutschland seit kant und in grossbritannien seit 1825 / Pfleiderer, Otto – 2nd ed. London: Swan Sonnenschein; New York: Macmillan, 1893 [mf ed 1986] – 2mf – 9 – 0-8370-8701-5 – (trans by j frederick smith. incl bibl ref, ind & app) – mf#1986-2701 – us ATLA [240]
The development of trinitarian doctrine in the nicene and athanasian creeds : a study in theological definition / Bishop, William Samuel – New York: Longmans, Green, 1910 – 1mf – 9 – 0-7905-3639-0 – mf#1989-0132 – us ATLA [240]
The development of unitarian thought in america from arminianism to transcendentalism / Cook, Alden Stoddard – Meadville, Pa., 1924. Chicago: Dep of Photodup, U of Chicago Lib, 1971 (1r); Evanston: American Theol Lib Assoc, 1984 (1r) – 1 – 0-8370-0377-6 – mf#1984-B149 – us ATLA [243]
The development of weight-adjusted estimates of caloric expenditures for the nordic track / Bowes, Michelle L – 1989 – 95p 1mf – 9 – $4.00 – us Kinesology [612]
The development of weightbearing skills : a longitudinal study of infants four to seven months of age / Roncesvalles, Maria N C – 1993 – 2mf – $8.00 – us Kinesology [150]
Development planning in surinam in historical pers... / Adhin, Jan Handsdew – Leiden, Netherlands. 1961 – 1r – us UF Libraries [338]
Development progress in indonesia / Japenpa – Djakarta, 1969 – 2mf – 9 – mf#SE-1401 – ne IDC [959]
Development [rome] = Desarrollo – Rome, Italy 1978-80 – 1,5,9 – (cont: revista del desarrollo internacional; cont by: development) – ISSN: 0020-6555 – mf#1617,01 – us UMI ProQuest [337]
Developmental and comparative immunology – Oxford, England 1977+ – 1,5,9 – ISSN: 0145-305X – mf#49259 – us UMI ProQuest [616]
Developmental and situational factors affecting little league participation / McAndrews, M R – 1991 – 1mf – 9 – $4.00 – us Kinesology [150]
Developmental brain research – Oxford, England 1992+ – 1,5,9 – ISSN: 0165-3806 – mf#42428 – us UMI ProQuest [616]
Developmental consequences of unrestricted trade / Vollrath, Thomas L – Washington DC: US Dept of Agriculture, Economic Research Service [mf ed 1985] – (= ser Foreign agricultural economic report 213) – 9 – (with bibl) – us GPO [337]
Developmental disabilities abstracts – Washington DC 1977-78 – 1,5,9 – ISSN: 0191-1600 – mf#3194,02 – us UMI ProQuest [616]
Developmental medicine and child neurology – London, England 1958+ – 1,5,9 – ISSN: 0012-1622 – mf#2676 – us UMI ProQuest [618]
Developmental neuropsychology – Mahwah NJ 1990+ – 1,5,9 – ISSN: 8756-5641 – mf#17603 – us UMI ProQuest [150]
Developmental psychobiology – Hoboken NY 1968+ – 1,5,9 – ISSN: 0012-1630 – mf#11053 – us UMI ProQuest [574]
Developmental psychology – Washington DC 1969+ – 1,5,9 – ISSN: 0012-1649 – mf#6028 – us UMI ProQuest [150]
Developmental relationships between throwing and striking: a pre-longitudinal test of motor stage theory / Langendorfer, Stephen – 1982 – 3mf – 9 – $12.00 – us Kinesology [790]
The developments of roman catholicism / Bain, John A – Edinburgh: Oliphant Anderson & Ferrier, [1908] – 1mf – 9 – 0-8370-8243-9 – (incl bibl ref) – mf#1986-2243 – us ATLA [240]
Le developpement de la pensee religieuse de luther jusqu'en 1517 : d'apres des documents inedits / Jundt, Andre – Paris: Fischbacher, 1906 – 1mf – 9 – 0-7905-6191-3 – (incl bibl ref) – mf#1988-2191 – us ATLA [242]
Developpement de l'enseignement populaire a cuba / Gervais, Villius – Port-Au-Prince, Haiti. 1927 – 1r – us UF Libraries [972]
Le developpement intellectuel de l'enfant de dieu : these / Guiton, J-Ph – Cahors: A Coueslant, 1909 [mf ed 1989] – 70p on 1mf – 9 – 0-7905-0427-8 – (in french) – mf#1987-0427 – us ATLA [200]
Le developpement rural – Conakry: PDG, 1970 – us CRL [600]
Developpement-quebec / Quebec (Province). Office de Planification et de Developpement du Quebec – Quebec: l'OPDQ. v1 n1 dec 1973-v8 n3 mars 1983 (mthly) [mf ed 1978-89] – 1r – 5 – (suspended: mai 1981-mars 1982; ceased: 1981?) – mf#SEM16P302 – cn Bibl Nat [330]
Devenir – Paris. n1-5. fevr-juil 1944 – 1 – fr ACRPP [073]
Devenir du metissage racial en haiti / Trouillot, Henock – Port-Au-Prince, Haiti. 1948 – 1r – us UF Libraries [972]
Le devenir social : revue internationale d'economie, d'histoire et de philosophie – Paris. avr 1895-98 – 1 – fr ACRPP [073]
Deventer : de stad van geert groote / Lugard, G J – Amsterdam, 1949 – €5.00 – ne Slangenburg [917]
Deventer, Mary Lynn van *see* Pilot study of the enhancement of musical performance through inner game strategies

Dever, Lem A *see* Masks off!
Dever, Mary *see* Woman in the pulpit
Devereux / Lytton, Edward Bulwer Lytton, Baron – Boston, MA. 189- – 1r – us UF Libraries [025]
Devereux papers, 14th-17th centuries – (= ser Archives of the marquess of bath, longleat house, warminster, wiltshire) – 10v on 9r – 1 – (incl ind) – mf#96701 – uk Microform Academic [920]
Devi, Tandra *see* Poems
Deviant behavior – Abingdon, Oxfordshire 1979+ – 1,5,9 – ISSN: 0163-9625 – mf#11990 – us UMI ProQuest [616]
Devienne, Francois *see* Neuvieme concerto, de flute principale, deux violons, alto, basse, cors et hartbois
The devil : his origin, greatness and decadence = Histoire du diable / Reville, Albert – London: Williams and Norgate, 1871 – 1mf – 9 – 0-7905-7607-4 – (in english) – mf#1989-0832 – us ATLA [210]
The devil and some of his doings. dayton, ohio : printed for the author at printing establishment of the united brethren in christ / Raber, W B – 1855 – 1r – 1 – $35.00 – mf^{s}-75 – us Commission [242]
The devil in britain and america / Ashton, John – [London?]: Ward and Downey, 1896 – 1mf – 9 – 0-524-02529-0 – (incl bibl ref) – mf#1990-3024 – us ATLA [130]
The devil in the church : his secret works exposed and his snares laid to destroy our public schools – 3rd ed. Beaver Springs, PA: American Pub Co, c1902 – 2mf – 9 – 0-8370-8709-0 – mf#1986-2709 – us ATLA [240]
Deville, Edouard *see*
- Abacus of the altitude and azimuth of the pole star
- Photographic surveying
Deville, F *see* Lettres sur le bengale, ecrites des bords du gange
Deville, V J de *see* Complement des memoires et revelations d'un page de la cour imperiale
Devilliers, John Abraham Jacob *see* Transvaal
Devils / Wall, James Charles – London: Methuen, 1904 – 1mf – 9 – 0-524-02877-X – mf#1990-3150 – us ATLA [210]
The devil's 13 – [s.l: s.n. 1888?] [mf ed 1987] – 1mf – 9 – 0-665-41031-X – mf#41031 – cn Canadiana [780]
The devils and evil spirits of babylonia : being babylonian and assyrian incantations against the demons, ghouls, vampires...which attack mankind – London: Luzac, 1903-1904 – (= ser Luzac's Semitic Text and Translation Series) – 2mf – 9 – 0-524-02325-5 – mf#1990-2948 – us ATLA [130]
Devil's garden – s.l, s.l? 193-? – 1r – us UF Libraries [978]
Devils Lake Sioux Tribe *see* E'yanpaha reservation news
The devil's parables : and other essays / Hannon, John – London: R & T Washbourne; New York: Benzinger, 1910 [mf ed 1986] – 1mf – 9 – 0-8370-6816-9 – (incl bibl ref) – mf#1986-0816 – us ATLA [241]
Devil's pi / Superior High School [WI] – v2 n5 [1919 apr 23], v4 n1-v4 1/2 n8,11-14 [1920 sep 24-1921 mar 18, apr 8-may 20] – 1r – 1 – mf#910063 – us WHS [373]
The devil's shadow / Thiess, Frank – New York: A A Knopf, 1928 – 1 – us Wisconsin U Libr [430]
Devine, Arthur *see*
- The creed explained
- The law of christian marriage according to the teaching and discipline of the catholic church
Devine, Pius *see* Eutropia
Devine, Thomas *see* [Atlas consisting of 43 maps of the counties of lower canada and 42 maps of upper canada]
Devinelli, Carlos *see* Politica brasileira (sintese e critica)
Devis Echandia, Julian *see* Ciudad vencida
La devise du roy justifiee... / Menestrier, C F – Paris: Estienne Michalet, 1679 – 3mf – 9 – mf#O-1353 – ne IDC [090]
Les devises : ou emblemes heroiques et morales, inventees... / Simeoni, G – Lyon: Guillaume Roville, 1559 – 1mf – 9 – mf#O-1914 – ne IDC [090]
Devises et emblemes anciennes et modernes : tirees de plus celebres auteurs / [Offelen, H] – Amsterdam, 1691 – 2mf – 9 – mf#O-703 – ne IDC [090]
Devises et emblemes anciennes et modernes tirees des plus celebres auteurs : oder: emblematische gemueths-vergnuegung bey betrachtung siben hundert und funffzehen der curieusesten ergaetzlichsten sinn-bildern / [Offelen, H] – Augspurg: Verlegts Lorentz Kroniger und Gottlieb Goebels seel. Erben, 1697 – 2mf – 9 – mf#O-1241 – ne IDC [090]
Devises et emblemes d'amour anciens et modernes moralisez... / Pallavicini – Amsterdam: Daniel de la Feuille, 1696 – 1mf – 9 – (in latin, italian, french, spanish, dutch, english and german) – mf#O-705 – ne IDC [090]
Devises et emblesmes d'amour moralisez / Flamen, A – Paris: Olivier de Varennes, 1658 – 2mf – 9 – mf#O-256 – ne IDC [090]
Devises heroiques / Paradin, Claude – Lion: Ian de Tournes et Guil. Gazeau, 1557 – 3mf – 9 – mf#O-1930 – ne IDC [090]
Devises heroiques / Paradin, Claude – Lyon: Ian de Tournes et Guil. Gazeau, 1551 – 2mf – 9 – mf#O-1929 – ne IDC [090]
Les devises heroiques / Paradin, Claude et al – Anvers: Veuve de Jean Stelsius, 1563 – 4mf – 9 – mf#O-1913 – ne IDC [090]
Devises heroiques, et emblemes... / Paradin, Claude – Paris: Iean Millot, [1614] – 4mf – 9 – mf#O-1931 – ne IDC [090]
Devises heroiques et morales... / Moyne, P le – Paris: Augustin Courbe, 1649 – 2mf – 9 – mf#O-667 – ne IDC [090]
Devisscher, Charles Antoine *see* New french scholastic conversations
Devitt, Edward J *see* Your honor
Devivier, Walter *see*
- Christian apologetics
- The inquisition
Devizes and wiltshire gazette – Devizes, England. 1897, 1913 – 2r – 1 – (aka: wiltshire gazette 1909-56) – uk British Libr Newspaper [072]
Devlet-i aliye osmaniyenin 1318 senesi muvazene umumiye hulasasi – Dersaadet [Istanbul]: Matbaa-yi Osmaniye, 1318 [1902] – 1mf – 9 – $25.00 – us MEDOC [350]
Devlet-i aliyenin doksan senesi muvazene defteridir – [Istanbul]: Matbaa-yi Amire, 1291 [1875] – 1mf – 9 – $25.00 – us MEDOC [350]
Devlet-i 'osmaniyye' nin buetcesi – 9 – (1325m [1909] 3mf $95; 1326 [1910] 3mf $95; 1327m [1911] 3mf $55; 1328m [1912] 2mf $40; 1330m [1913] 3mf $55; 1331m [1914] 3mf $55; 1332m [1915] 4mf $60; 1334m [1917] 4mf $60) – us MEDOC [350]
Devletler hususi hukuku / Birsen, K – Istanbul, 1936. 2v – 4mf – 9 – (missing: v2) – mf#ILM-3386 – ne IDC [956]
La devocion al papa : carta pastoral / Perez Munoz, Adolfo – Badajoz: Uceda Hnos, 1914 – 1 – sp Bibl Santa Ana [240]
Devociones antonianas / Corredor, Antonio – Caceres: Ediciones Cruzada mariana. Tip. La Minerva, 1958 – sp Bibl Santa Ana [240]
Devociones antonianas / Corredor, Antonio – Plasencia: Graf. Sandoval, 9th ed 1979 – 1 – sp Bibl Santa Ana [240]
Las devociones de mi pueblo, las santas reliquias, el santuario...alburquerque / Duarte Insua, Lino – Badajoz: Dip Prov, 1947 – 1 – sp Bibl Santa Ana [240]
Devoir – Montreal, Canada. 27 jun 1914-jul 1917; 28 sep 1917-aug 1918; 3 sep 1918-22 feb 1921; 15 apr-dec 1921 – 39r – 1 – uk British Libr Newspaper [071]
Le devoir – Port-au-Prince: Impr de l'Abeille, [1902-]. [1re annee, n1-2e annee, n2. 10 avril 1902-24 juin 1903] – 3r – 9 – us CRL [079]
Le devoir politique des catholiques / Barbier, Emmanuel – Aisne: Association Saint-Remy, 1910 – 2mf – 9 – 0-524-06237-4 – mf#1990-5192 – us ATLA [241]
Les devoirs des femmes dans la famille / Chassay, Frederic – Paris 1852 [mf ed Hildesheim 1995-98] – 1v on 2mf – 9 – €60.00 – ISBN-10: 3-487-25861-7 – ISBN-13: 978-3-487-25861-4 – gw Olms [305]
Devolution in mission administration : as exemplified by the legislative history of five american missionary societies in india / Fleming, Daniel Johnson – New York: Fleming H Revell [c1916] [mf ed 1995] – (= ser Yale coll) – 310p – 1 – 0-524-09955-3 – mf#1995-0955 – us ATLA [954]
Devon and cornwall record society – v1-25. 1906-54 – (= ser Publications of the english record societies, 1835-1972) – 162mf – 9 – (ns: v26-29 1955-70 [39mf]) – uk Chadwyck [941]
Devon booksellers and printers in the 17th and 18th centuries / Dredge, John Ingle – Plymouth: W H Luke, Printer, 1885-87 – (= ser 19th c publishing...) – 1mf – 9 – (incl 2 suppl papers) – mf#3.1.73 – uk Chadwyck [070]
Devon & somerset chronicle *see* South molton gazette / devon & somerset chronicle / west of england advertiser
The devonian fossils of canada west / Billings, Elkanah – Toronto?: Lovell & Gibson, 1860? – 2mf – 9 – mf#62285 – cn Canadiana [560]
Devonport and areas news advertiser – 1981 – 1r – 1 – mf#11.43 – nz Nat Libr [079]
Devonport city news – Devonport – 3r – at Pascoe [079]
Devonport independent & plymouth & stonehouse gazette – [SW England] Plymouth 9 feb 1833-17 jul 1869; 10 jan 1874-aug 1891 [mf ed 2004] – 46r – 1 – (cont by: royal devonport telegraph & plymouth chronicle [jan 1828-dec 1836]; devonport telegraph & plymouth chronicle etc [jan 1835-dec 1851]; devonport & plymouth telegraph etc [jan 1852-jun 1863]) – uk Newsplan [072]
Devonshire freeholder – Plymouth, England. 18 Sept 1824-31 May 1828. -w. 1 reel – 1 – uk British Libr Newspaper [072]
Devonshire, Spencer Compton Cavendish, 8th Duke of *see* The government of ireland bill
Devot, Justin *see*
- Acta et verba
- Centenaire de l'independance nationale d'haiti
- Cours elementaire d'instruction civique
- Travail intellectuel et la memoire sociale
Devota corona...a la virgen maria – 1855 – 9 – sp Bibl Santa Ana [240]
Devoted to natural history, primarily that of the prairie states / The American Midland Naturalist – Indianapolis. 1973-1980 (1) 1975-1980 (5) 1975-1980 (9) – 71mf – 9 – mf#8556 – ne IDC [590]
Devotion a saint-joseph / Liguori, Alfonso Maria de', Saint – Sainte-Anne-de-Beaupre, Quebec: [s.n.] 1915 [mf ed 1995] – 1mf – 9 – 0-665-74903-1 – mf#74536 – cn Canadiana [241]
Devotion au precieux sang : ses motifs, sa pratique / Raymond, Joseph-Sabin, 1810-1887 – Montreal: E Senecal, 1870 [mf ed 1984] – 1mf – 9 – 0-665-46416-9 – mf#46416 – cn Canadiana [240]
Devotion to the blessed virgin : being the substance of all the sermons for mary's feasts troughout the year / Bossuet, Jacques Benigne – London; New York: Longmans, Green, 1903 – 1mf – 9 – 0-8370-6886-X – mf#1986-0886 – us ATLA [240]
Devotional literature / Stowell, W H – London, England. 1854 – 1r – us UF Libraries [240]
Devotional readings from luther's works for every day of the year / Luther, Martin – Rock Island, IL: Augustana Book Concern, c1915 – 2mf – 9 – 0-524-03405-2 – mf#1990-0959 – us ATLA [242]
The devotional use of the holy scriptures / Gibson, John Monro – London: National Council of Evangelical Free Churches, 1904 – (= ser Little Books on the Devout Life) – 1mf – 9 – 0-524-04398-1 – mf#1992-0091 – us ATLA [220]
Devotions of bishop andrews / Andrewes, Lancelot – London, England. 1832 – 1r – us UF Libraries [240]
The devotions of saint anselm, archbishop of canterbury / Anselm, Saint, Archbishop of Canterbury; ed by Webb, Clement Charles Julian – London: Methuen 1903 [mf ed 1991] – (= ser The library of devotion) – 1mf – 9 – 0-7905-9355-6 – (trans fr latin) – mf#1989-2580 – us ATLA [241]
Il devotissimo viaggio die giervsalemme / Zuallart, J – Roma, 1595 – 4mf – 9 – mf#H-8383 – ne IDC [915]
Devoto quincenario a nuestra madre y senora de la asuncion en su milagrosa imagen : que se venera en la santa iglesia matriz de esta villa de oruro / Mexia y Tordoya, Francisco – [Mexico City]: En la real imprenta de los ninos expolitos, ano de 1786 – (= ser Books on religion...1543/44-c1800: milagros y culto de la virgen) – 2mf – 9 – mf#crl-424 – ne IDC [241]
Devout and moral reflections on the pious life and happy death of t... / Hammond, John – Canterbury, England. 1800 – 1r – us UF Libraries [240]
Devout loyalty / Osborn, George – Worcester, England. 1809? – 1r – us UF Libraries [240]
Devout observation of national calamities enforced / Fletcher, Joseph – Blackburn, England. 1808 – 1r – us UF Libraries [240]
Devrient, Otto *see* Luther
Devries, Gerben M *see* Chasco
DeVries, Steven N *see* Approval of agressive acts in wrestling
Devteronomivm in mosis librvm 5... / Wolf, J – Tigvri, in officina Froschoviana, 1585 – 6mf – 9 – mf#PBU-660 – ne IDC [240]
Devx livres, l'vn de la puissance & sapience de dieu, l'autre de la volonte de dieu / Hermes Trismegistus – Le tout Traduit de Grec en Francois par Gabriel du Preau. Paris, 1557 – 1 – us Wisconsin U Libr [240]
Dew drops – Toronto: W Briggs, [1897-19–] – 9 – mf#P04335 – cn Canadiana [240]
Dew of hermon / Hamilton, James – London, England. 1845 – 1r – us UF Libraries [240]
Dew, Stephen H *see* Journal of library and information services in distance learning
Dewan Dakwah Islamiyah Indonesia *see* Islamic news letter
Dewan Geredja2 Keristen Tionghoa di Indonesia *see* Hui k'an
Dewan geredja-geredja di indonesia – Djakarta, 1952-1972 – 31mf – 9 – (missing: 1952-1954(1-3, 6-8, 11-12); 1955(2-12)-1956(1-4, 9); 1957(11-12)-1965(1, 3-12)-1967(1-12); 1969(4, 9); 1971(1-12); 1972(5, 8)) – mf#SE-345 – ne IDC [959]
Dewan geredja-geredja di indonesia – Djakarta, 1960-1968. v1-9(1) – 15mf – 9 – (missing: 1962, v2(1, 4-end); 1962, v3; 1963/1964, v4(1-3); 1965, v5(4-end); 1967, v6(1-3)) – mf#SE-1893 – ne IDC [959]
Dewan kesenian djakarta / Budaja djaja. Madjalah kebudajaan umum – Djakarta, 1968-1972. v1-5 – 41mf – 9 – (missing: 1971 v4(40)) – mf#SE-1365 – ne IDC [959]
Dewan ko-operasi indonesia : almanak ko-operasi – Djakarta, 1957/1958 – 7mf – 9 – mf#SE-1307 – ne IDC [959]
Dewan mahasiswa institut agama islam negeri / Darmabakti – Jogjakarta, 1961. v1(1-6) – 4mf – 9 – mf#SE-766 – ne IDC [959]
Dewan Mahasiswa ITB *see* Gelora teknologi
Dewan Mahasiswa Universitas Gadjah Mada *see* Madjalah gama; gema intrauniversiter
Dewan mahasiswa, universitas indonesia – Djakarta, 1966-1969(2) – 11mf – 9 – (missing: 1966(1)) – mf#SE-438 – ne IDC [959]
Dewan mahasiswa universitet / Mahasiswa – Djakarta, 1954-1956 – 3mf – 9 – (missing: 1955 v1-2(8-11)) – mf#SE-728 – ne IDC [959]
Dewan Nasional Permuda Rakjat *see* Generasi baru
Dewan nasional putusan-putusan sidang dewan nasional : indonesia – Djakarta, 1958 – 9mf – 9 – mf#SE-1630 – ne IDC [959]
Dewan pemerintah daerah laporan kepada dewan perwakilan rakjat daerah – Surabaja, 1958 – 3mf – 9 – mf#SE-243 – ne IDC [950]
Dewan perniagaan dan perusahaan / Warta niaga dan perusahaan – Djakarta, 1958-1960. v1-2(1-51) – 11mf – 9 – (missing: 1959 v1(1-4), v2(44/45)) – mf#SE-1992 – ne IDC [959]
Dewan perpustakaan nasional berita berkala : indonesia – Djakarta, 1955. v1-2(1) – 1mf – 9 – (missing: v1) – mf#SE-1631 – ne IDC [959]
Dewan perwakilan rakjat daerah gotong rojong – Djombang, 1969-1970 – 29mf – 9 – mf#SE-1461 – ne IDC [950]
Dewan perwakilan rakjat daerah gotong rojong bendel dprd-gr kabupaten klaten sekretariat dprd-gr : klaten, indonesia (kabupaten) – Klaten, 1968(bendel A); 1969/1970(bendel B-F); 1970(bendel G-J) – 11mf – 9 – mf#SE-1748 – ne IDC [959]
Dewan perwakilan rakjat daerah gotong rojong bendel surat-surat keputusan dprd-gr propinsi djawa timur sekretariat dprd-gr – Surabaja, 1968 – 6mf – 9 – mf#SE-1467 – ne IDC [950]
Dewan perwakilan rakjat daerah gotong rojong bundel surat-surat keputusan dprd-gr kabupaten pekalongan – Pekalongan, 1970 – 2mf – 9 – mf#SE-1879 – ne IDC [950]
Dewan perwakilan rakjat daerah gotong rojong himpunan keputusan-keputusan, resolusi-resolusi pernjataan-pernjataan, peraturan daerah hasil sidang untuk landasan pedoman kerdja tahun dinas ngawi, sekretariat dprd-gr – Ngawi, 1968 – 7mf – 9 – mf#SE-1846 – ne IDC [950]
Dewan perwakilan rakjat daerah gotong rojong himpunan produk-produk bagian sekretariat kantor pemerintah daerah – Ponorogo, 1968 – 4mf – 9 – mf#SE-1907 – ne IDC [950]
Dewan perwakilan rakjat daerah gotong rojong himpunan resolusi : pernjataan dprd-gr kabupaten pekalongan – Pekalongan, 1968 – 8mf – 9 – mf#SE-1880 – ne IDC [950]
Dewan perwakilan rakjat daerah gotong rojong himpunan surat keputusan dprd-gr kabupaten pasuruan, djawa timur – Pasuruan, 1970 – 2mf – 9 – mf#SE-1873 – ne IDC [950]
Dewan perwakilan rakjat daerah gotong rojong himpunan surat-surat keputusan dprd-gr kabupaten semarang – Semarang, 1969 – 1mf – 9 – mf#SE-1925 – ne IDC [950]
Dewan perwakilan rakjat daerah gotong rojong kumpulan pernjataan dprd-gr kabupaten magetan / Magetan, Indonesia (Kabupaten) – Magetan, 1969 – 1mf – 9 – mf#SE-1805 – ne IDC [959]
Dewan perwakilan rakjat daerah gotong rojong laporan dprd-gr kabupaten pemalang – Pemalang, 1969 – 2mf – 9 – mf#SE-1885 – ne IDC [950]
Dewan perwakilan rakjat daerah gotong rojong laporan hasil sidang seksi a – Batang, 1970 – 2mf – 9 – mf#SE-1342 – ne IDC [950]
Dewan perwakilan rakjat daerah gotong rojong peraturan tata-tertib dewan perwakilan rakjat daerah gotong rojong / Magetan, Indonesia (Kabupaten) – Magetan, 1969 – 3mf – 9 – mf#SE-1787 – ne IDC [959]

Dewan perwakilan rakjat daerah gotong rojong resolusi dewan perwakilan rakjat daerah gotong rojong kabupaten sumenep – Sumenep, 1968 – 1mf – 9 – mf#SE-1946 – ne IDC [950]
Dewan perwakilan rakjat daerah gotong rojong risalah lengkap sidang sekretariat dprd-gr, kotamadya surakarta – Surakarta, 1968-1969 – 55mf – 9 – mf#SE-1954 – ne IDC [950]
Dewan perwakilan rakjat daerah gotong rojong risalah resmi sidang dprd-gr – Sukohardjo, 1968 – 9mf – 9 – mf#SE-1944 – ne IDC [950]
Dewan perwakilan rakjat daerah gotong rojong risalah resmi sidang paripurna dprd-gr / Magetan, Indonesia (Kabupaten) 1968-1969 – 65mf – 9 – mf#SE-1788 – ne IDC [959]
Dewan perwakilan rakjat daerah gotong rojong risalah resmi sidang paripurna dprd-gr kabupaten ngawi ngawi, sekretariat dprd-gr – Ngawi, 1968-1969 – 12mf – 9 – mf#SE-1847 – ne IDC [950]
Dewan perwakilan rakjat daerah gotong rojong risalah resmi sidang paripurna dprd-gr kabupaten sidoardjo – Sidoardjo, 1969 – 13mf – 9 – mf#SE-1928 – ne IDC [950]
Dewan perwakilan rakjat daerah gotong rojong risalah resmi sidang paripurna dprd-gr kabupaten sumenep – Sumenep, 1968-1969 – 17mf – 9 – (missing: 1968-1969(aug 13-may 11); 1969(may 12, 13)) – mf#SE-1947 – ne IDC [950]
Dewan perwakilan rakjat daerah gotong rojong risalah resmi sidang paripurna dprd-gr kabupaten trenggalik – Trenggalek, 1968(apr 24)-1970(may 9) – 77mf – 9 – (missing: 1969(jul 2a)) – mf#SE-1965 – ne IDC [950]
Dewan perwakilan rakjat daerah gotong rojong risalah resmi sidang pleno chusus dprd-gr kabupaten klaten sekretariat dprd-gr : klaten, indonesia (kabupaten) – Klaten, 1968(jan 25/feb 1, 7; jun 5, A-B, 26, A-B, sep 2, A-B, 3, A-B, 5, 9); 1969(feb 4, mar 4, A-B, 5, A-B, 6, A-B, 10, A-C, 13, jul 19, A-B, oct 7, 28); 1970(sep 24, A-B, oct 10, 28) – 67mf – 9 – mf#SE-1749 – ne IDC [959]
Dewan perwakilan rakjat daerah gotong rojong risalah resmi sidang pleno dprd-gr kabupaten klaten sekretariat dprd-gr : klaten, indonesia (kabupaten) – Klaten, 1968(may 1, A-B, jun 3, A-B, oct 22, 23, 24, nov 13, 14, 19, 20, dec 18, A-B); 1970(sep 15, A-B) – 23mf – 9 – mf#SE-1750 – ne IDC [959]
Dewan perwakilan rakjat daerah gotong rojong risalah resmi sidang pleno dprd-gr kabupaten pemalang – Pemalang, 1969 – 2mf – 9 – mf#SE-1886 – ne IDC [950]
Dewan perwakilan rakjat daerah gotong rojong risalah resmi sidang pleno istimewa dprd-gr kabupaten ponorogo sekretariat dprd-gr kabupaten ponorogo djawa-timur – Np, 1968 – 2mf – 9 – mf#SE-1908 – ne IDC [950]
Dewan perwakilan rakjat daerah gotong rojong risalah resmi sidang pleno paripurna dprd-gr kabupaten pasuruan, djawa timur – Pasuruan, 1969 – 48mf – 9 – mf#SE-1874 – ne IDC [950]
Dewan perwakilan rakjat daerah gotong rojong surat keputusan dprd-gr kabupaten pemalang – Pemalang, 1970 – 1mf – 9 – mf#SE-1887 – ne IDC [950]
Dewan perwakilan rakjat daerah laporan hasil karya dewan perwakilan rakjat daerah propinsi kalimantan selatan – Bandjarmasin, 1969 – 4mf – 9 – mf#SE-1931 – ne IDC [950]
Dewan perwakilan rakjat daerah-gotong rojong buku chronologisch kumpulan aktivitas dprd-gr, kabupaten kediri sekretariat dprd-gr : kediri, indonesia (kabupaten) – Kediri, 1969-1970 – 18mf – 9 – mf#SE-1735 – ne IDC [959]
Dewan perwakilan rakjat daerah-gotong rojong himpunan sidang-sidang seksi "b" dprd-gr kabupaten karanganjar dan sidang panitia penjelesaian status tanah : karanganjar, indonesia (kabupaten) – Karanganjar, 1968/1969 – 1mf – 9 – mf#SE-1729 – ne IDC [959]
Dewan perwakilan rakjat daerah-gotong rojong laporan sidang komisi "a" dprd-gr kabupaten kediri sekretariat dprd-gr : kediri, indonesia (kabupaten) – Kediri, 1970 – 1mf – 9 – mf#SE-1736 – ne IDC [959]
Dewan perwakilan rakjat daerah-gotong rojong risalah resmi sidang paripurna : kendal, indonesia (kabupaten) – Kendal, 1969 – 6mf – 9 – (missing: 1969 (mar-jul)) – mf#SE-1742 – ne IDC [959]
Dewan perwakilan rakjat daerah-gotong rojong risalah resmi sidang pleno : karanganjar, indonesia (kabupaten) – Karanganjar, 1968-1969 – 24mf – 9 – mf#SE-1731 – ne IDC [959]
Dewan perwakilan rakjat gotong rojong indeks risalah resmi : indonesia – Djakarta, 1960 – 1mf – 9 – mf#SE-1632 – ne IDC [950]
Dewan perwakilan rakjat gotong rojong risalah stenografis badan kesedjahteraan peg (bkp) : indonesia – Djakarta, 1967-1972 – 39mf – 9 – (missing: 1967-1971; 1972(27)) – mf#SE-1634 – ne IDC [959]
Dewan perwakilan rakjat pertanjaan anggota : indonesia (federation, 1949-1950) – Dakarta, 1950 – 7mf – 9 – mf#SE-216 – ne IDC [959]
Dewan perwakilan rakjat risalah : indonesia (federation, 1949-1950) – Djakarta, 1950 – 11mf – 9 – mf#SE-217 – ne IDC [959]
Dewan perwakilan rakjat risalah perundingan pertjetakan negara : indonesia – Djakarta, 1950-1960 – 559mf – 9 – (missing: 1954(2); 1954(25 aug); 1955(1-10, 15, 19, 21-23, 25, 28, 40-43, 47); 1959(1-7)) – mf#SE-227 – ne IDC [959]
Dewan perwakilan rakjat risalah sementara pertjetakan negara : indonesia – Djakarta, 1954(100), 1957-1959(68) – 454mf – 9 – mf#SE-228 – ne IDC [959]
Dewan pewakilan rakjat daerah-gotong rojong himpunan sidang-sidang seksi "e" dprd-gr kabupaten karanganjar : karanganjar, indonesia (kabupaten) – Karanganjar, 1968/1969 – 3mf – 9 – mf#SE-1730 – ne IDC [959]
Dewan pimpinan pusat djam'ijatul muslimin indonesia / Al-Falah – Djakarta, 1964 – 2mf – 9 – mf#SE-1477 – ne IDC [959]
Dewan pusat organisasi islam afrika-asia / Suara Muslimin – Djakarta, 1966(1-3) – 3mf – 9 – mf#SE-1936 – ne IDC [950]
Dewan research ekonomi, sosial dan budaja laporan : indonesia – Djakarta, 1962/1963 – 3mf – 9 – mf#SE-686 – ne IDC [959]
Dewan Research Fakultas Kedokteran Universitas Airlangga *see* Madjalah kedokteran surabaja
Dewar, Daniel *see*
– Glories of christ's kingdom
– The holy spirit
Dewar, Douglas *see*
– Bombay ducks
– In the days of the company
Dewart, Edward Hartley *see*
– Additional poems
– The bible under higher criticism
– Brief outlines of christian doctrine
– The children of the church
– Essays for the times
– German protestantism and the right of private judgement in the interpretation of holy scripture
– High church pretensions disproved
– Living epistles
– Lord tennyson's pessimism
– Misleading lights
– Priestly pretensions disproved
– University federation
Dewart, Edward Hartley [comp] *see* Selections from canadian poets
Dewart, William *see* [Eleven letters on free trade vs protection which appeared in the canadian illustrated news]
Dewasagajam, Nj *see* Aus meinem leben
The deweese booster – Deweese, NE: A D Scott. v1 n1. apr 16 1915-18// (wkly) [mf ed -sep 28 1918 (gaps)] – 1r – 1 – us NE Hist [071]
Dewey, Davis R *see* Financial history of the united states
Dewey, Frederic Perkins *see* Some canadian iron ores
Dewey John *see* Ethics
Dewey, John *see*
– The child and the curriculum
– The educational situation
– Ethical principles underlying education
– Ethics
– German philosophy and politics
– How we think
– The influence of darwin on philosophy, and other essays in contemporary thought
– Leibniz's new essays concerning the human understanding
– Moral principles in education
– The school and society
– Schools of to-morrow
– Studies in logical theory
Dewey, Julia M *see* Lessons on morals
Dewey, Lyster Hoxie *see* Legislation against weeds
Dewey, Orville *see*
– American unitarian association anniversary
– Discourses and discussions in explanation and defence of unitarianism
– Discourses on human life
– Discourses on various subjects
– Moral views of commerce, society and politics
– The old world and the new
– The problem of human destiny
– The two great commandments
– The unitarian's answer
Dewey, Thomas Henry *see* A treatise on contracts for future delivery and commercial wagers, including "options," "futures," and "short sales."
Dewezet *see* Deister- und weserzeitung
Dewi telaga warna / Ang, Siauw Tan – Soerabaia: Tan's Drukkerij, 1936 [mf ed 1998] – (= ser Penghidoepan 136) – 1r – 1 – (filmed with: pembalesan dendam hati / phoa gin hian) – mf#10003 – us Wisconsin U Libr [830]
Dewick, E S *see*
– Coronation book of charles 5 of france
– Facsimiles of horae de b m v 11th century
– The leofric collectar, vol 1
– The martiloge in englysshe
Dewing, Rolland *see* The fbi files on the american indian movement and wounded knee
Dewitt eagle – De Witt, NE: Eagle Pub Co. v10 n37. oct 7 1904-jul 1922// (wkly) [mf ed -sep 19 1907 (gaps)] – 2r – 1 – (formed by the union of: dewitt record and: de witt republican. absorbed by: dewitt times-news) – us NE Hist [071]
DeWitt, Jean [Jeannette] *see* Amsterdam Toonkunst-Bibliotheek BT 61/580-212-H-12
Dewitt news – DeWitt, NE: H D Rogers. 1v. v1 n1 aug 22 1902-v1 n38. may 8 1903 (wkly) [mf ed filmed [1974?]] – 1r – 1 – (split from: saline county independent. merged with: de witt times to form: dewitt times-news) – us NE Hist [071]
Dewitt record – De Witt, NE: Walter I Stout, dec 18 1903-sep 30 1904// (wkly) [mf ed v9 n49. jan 1-sep 23 1904 (gaps)] – 1r – 1 – (cont: saline county independent. merged with: de witt republican to form: dewitt eagle) – us NE Hist [071]
Dewitt times-news – De Witt, NE: H D Rogers. v22 n47. may 15 1903)- (wkly) [mf ed with gaps] – 1 – (formed by the union of: de witt times and: dewitt news. absorbed: dewitt eagle. some irregularities in numbering) – us NE Hist [071]
Dewitt's base-ball guide – New York. 1869-1885 – 1 – us NY Public [790]
Dewitz, August Karl Ludwig Von *see* I dansk verstindien
DeWolf, Charles Wesley *see* Diary
Dewolf, Lotan Harold *see* Theological dictionary
Dewolfe, Harry F *see* Diary
Dewora, Viktor Joseph *see* Briefe und gespraeche veranlasst durch die entfuehrung und gefangenschaftsreise des heiligen vaters pius des siebenten
Dewsbury, batley & district social-democrat : the local organ of the international socialist movement – Dewsbury, England feb 1907-jan 1909 – 1 – (cont: dewsbury social-democrat [jan 1907]) – uk British Libr Newspaper [335]
Dewsbury & batley herald, & heckmondwike, mirfield & ossett advertiser – Dewsbury, England 31 mar-dec 1855, 2 jan, 2 jul 1856 [mf 1855] – 1 – uk British Libr Newspaper [072]
Dewsbury reporter – England. -w. 1869-84. (15 reels) – 1 – uk British Libr Newspaper [072]
Dewsland & kemes guardian – [Wales] Pembrokeshire 16 jul 1869-dec 1905 [mf ed 2004] – 89r – 1 – (missing: 1883; cont by: pembroke county guardian & cardigan reporter [jan 1884-dec 1926]; pembroke county and west wales guardian [jan 1927-dec 1950]) – uk Newsplan [072]
Dexelio, G *see* Deliciae emblematicae
Dexippi et al *see* Historiarum quae supersunt (cshb14)
Dexter, Henry Martyn *see*
– As to roger williams and his 'banishment' from the massachusetts plantation
– Congregationalism
– The congregationalism of the last three hundred years, as seen in its literature
– The england and holland of the pilgrims
– A glance at the ecclesiastical councils of new england
– A handbook of congregationalism
– The moral influence of manufacturing towns
– The true story of john smyth, the se-baptist
– The verdict of reason upon the question of the future punishment of those who die impenitent
Dexter-Fogarty, Tracey *see* The effectiveness of the hinged golf club as a training aid to develop consistency in novice golfers
Dexter-smith's – Boston. 1-14, 1872-78. Incomplete – 1 – us L of C Photodup [780]
Dextwer, Morton *see* The england and holland of the pilgrims
Dey, Mukul *see*
– Fifteen drypoints
– My pilgrimages to ajanta and bagh
– Portraits of mahatma gandhi
– Twenty portraits
Dey, Shumbhoo Chunder *see* Hooghly
Deycks, Ferdinand *see*
– De itinere terrae sanctae...nach alten handschriften berichtigt
– Friedrich heinrich jacobi im verhaeltnis zu seinen zeitgenossen, besonders zu goethe
– Goethes faust
– Ludolphi, rectoris ecclesiae parochialis in suchem, de itinere terrae sanctae liber
Deyrieux, L *see*
– Apprenti gabriel
– Pot au lait
Dez anos no brasil / Seidler, Karl Friedrich Gustav – Sao Paulo, Brazil. 1941 – 1r – us UF Libraries [972]
[Dez, J] *see* Ad virum nobilem de cultu confucii philosophi et progenitorum apud sinas
[Dezallier d'Argenville, A J] *see* Abrege de la vie des plus fameux peintres...
Dezallier d'Argenville, Antoine *see*
– Voyage pittoresque de paris
– Voyage pittoresque des environs de paris
Dezeimeris, Jean E et al *see* Dictionnaire historique de la medecine ancienne et moderne (ael3/17)
Dezell, Robert *see*
– Fire and frost
– I sez, sez i
Dezen, A A *see* Sistema bankovskogo kreditovaniia (operatsii sovremennykh bankov sssr)
Dezenove de abril : periodico dos estudantes de medicina e pharmacia – Rio de Janeiro, RJ. 16 jun-18 ago 1882 – (= ser Ps 19) – mf#DIPER – bl Biblioteca [610]
O dezenove de dezembro – Curitiba, PR. abr 1854-mar 1856 – (= ser Ps 19) – bl Biblioteca [079]
O dezenove de dezembro – Curitiba, PR: Typ Paranaense de C M Lopes, 01 abr 1854-15 fev 1890 – (= ser Ps 19) – mf#P18,04,11 – bl Biblioteca [321]
Dezenove de outubro *see* 19 de outubro
Dezentje, J A *see* Overzicht bevolkingsgroepen in de desa's, door j a dezentje
Dezesseis de fevereiro : orgam popular – Local: Fortaleza, CE: Typ D'O Bemtevi, 16-23 fev 1893 – (= ser Ps 19) – mf#P18B,03,20 – bl Biblioteca [079]
Dezesseis de julho : orgao conservador – Rio de Janeiro, RJ: Typ Dezesseis de Julho, 04 jul-out, dez 1869; jan-02 jul 1870 – (= ser Ps 19) – mf#P25,01,06-07 – bl Biblioteca [321]
Dezessete districto *see* 17 districto
Dhail kitab al-fariq / Bachajizade, 'Abd al-Rahman Bey – Cairo: Mawsu at Bishar Press, 1322 – 1 – 0-8370-1767-X – mf#1984-6006 – us ATLA [470]
Dhalla, M N *see* Nyaishes or zoroastrian litanies
Dhalla, Maneckji Nusservanji *see*
– Our perfecting world
– Zoroastrian theology
– Zoroastrian theology: from the earliest times to the present day
The dhamma of gotama the buddha and the gospel of jesus the christ : a critical inquiry into the alleged relations of buddhism with primitive christianity / Aiken, Charles Francis – Boston: Marlier, 1900 – (= ser Universitas Catholica Americae, Washingtonii, S. Facultas Theologica (Series)) – 1mf – 9 – 0-524-00814-0 – mf#1990-2060 – us ATLA [230]
Dhamma vilasa : [a novel] / Lan Yun Sac Lvan – Ran Kun: Sac Lvan ca pe Phran khyi re 1977 [mf ed 1990] – 1r with other items – 1 – (in burmese) – mf#mf-10289 seam reel 170/1 [§] – us CRL [830]
Dhammagan bhat ca nhan puran kyam / Kyi sai Le thap Cha ra to – Ran Kun: Hamsavati Pitakat pum nhip tuik 1956 – [pl] 1r with other items – 1 – (in burmese) – mf#mf-10289 seam reel 167/3 [§] – us CRL [390]
Dhammakitti *see* A manual of buddhist historical traditions
Dhammakitti, Polonnaruve *see* The dathavansa
Dhammapada : being footprints in the way of life / Cooke, J P – Boston: CF Libbie, Jr, [1889?] – 1mf – 9 – 0-524-08901-9 – mf#1993-4036 – us ATLA [280]
The dhammapada : with introductory essays, pali text, english translation, and notes – London; New York: Oxford University Press, 1950 – (= ser Samp: indian books) – us CRL [280]
The dhammapada (stbe10) – 1881 – (= ser Sacred book of the east (sbte)) – 7mf – 8 – €15.00 – (trans fr pali by f max mueller and: the sutta-nipata trans fr pali by v fausboell; being canonical books of the buddhists) – ne Slangenburg [280]
Dhammapradip / Suvannajoto, Bhikkhu – Bhnam Ben: Buddhasasana Pandity 2500 [1957] [mf ed New Haven CT: SE Asia Coll, Yale Uni Library 1992] – 1r with other items – 1 – (in khmer & pali [in khmer script]; added t.p. in french: dhammapradipa: morceaux choisis tires de l'anguttaranikaya et du khuddakanikaya / traduit du pali en cambodgien par venerable chap-pin [suvannajoto]) – mf#mf-10393/1 – us CRL [280]
Dhammarama, P S *see* Initiation pratique au buddhisme
Dhan Van, Brah gru sangh sumedh *see* Sundarakatha
Dhar, Lakshmi *see* Padumavati
Dhar, Mohini Mohan *see*
– Krishna the charioteer
– Krishna the cowherd
The dharma : or, the religion of enlightenment, an exposition of buddhism / Carus, Paul – 6th rev enl ed. Chicago, London: Open Court Publ, 1918 [mf ed 1995] – (= ser Yale coll) – 134p (ill) – 1 – 0-524-09248-6 – mf#1995-0248 – us ATLA [280]

The dharma sa'stra : or, the hindu law codes: english translation / ed by Dutt, Manmatha Nath – Calcutta: Manmatha Nath Dutt, 1908- – (= ser Samp: indian books) – us CRL [280]
Dharma-budhi see Badan penerangan persatuan mahasiswa krishnadwipajana
Dharmapala, Anagarika see
– The arya dharma of sakya muni, gautama, buddha
– Buddhism in its relationship with hinduism
– The life and teachings of buddha
Dharmatrata see Udaanavarga
Dhawan, Gopi Nath see The political philosophy of mahatma gandhi
Dhesa nesan weekly – Penang, Malaysia. In Tamil. -w. 9 April 1933-28 Feb 1940. 5 reels – 1 – uk British Libr Newspaper [072]
Dhingra, Baldoon see A national theatre for india
Dhlomo, Rolfes Robert Reginald see African tragedy
Dhole, Heeralal see A manual of adwaita philosophy
d'Hoop, F-G see Recueil des chartes du prieure de saint-bertin a poperinghe
Dhorme, Edouard see
– Les livres de samuel
– La religion assyro-babylonienne
Dhorme, Edouard et al see Conferences de saint-etienne
Dhorme, P see Le livre job (etb)
Dhormoys, Paul see
– Sous les tropiques
– Visite chez soulouque
Dhu Vam see Gvam pum
Dhurup, Manilall see An evaluation of customer service
Di antonio il verso siciliano della citta di piazza. il primo libro de' madrigali a cinqve voci. nouamente date in luce / Il Verso, Antonio – In Palermo [pref 1590] [mf ed 19–] – 5pt on 1r – mf#film 1134 – us Sibley [780]
Di bruklin bronzvil post = The e.n.y. and brownsville post – Brooklyn [NY]: E N Y, Brooklyn and Brownsville Post Publ Co, dec 1917-1919 – 1r – us CRL [071]
Di Bruno, Joseph Faa see Catholic belief
Di cait – Vilnius, Lithuania 1924-39 [mf ed Norman Ross] – 22r – 1 – mf#nrp-2038 – us UMI ProQuest [077]
Di cipriano di rore li madrigali a cinque voci. libro secondo – 1562 / 1552 – (= ser Mssa) – 2mf/3mf – 9 – €35.00/€50.00 – mfchl 125 / 125a – gw Fischer [780]
Di cipriano et annibale madrigali a quatro voci – 1566 – (= ser Mssa) – 2mf – 9 – €35.00 – mfchl 130 – gw Fischer [780]
Di cipriano rore il terzo libro de madrigali a cinque voci – 1552 – (= ser Mssa) – 4mf – 9 – €60.00 – mfchl 100 – gw Fischer [780]
Di cipriano rore il terzo libro di madregali a cinque voci – 1560 – (= ser Mssa) – 4mf – 9 – €60.00 – mfchl 126 – gw Fischer [780]
Di dzshoyrzi shtime = The jersey voice – Bayonne, NJ: The Jersey Voice Pub Co, 1928-31 (gaps) [mf ed 197-?] – 1 – (in yiddish and english) – mf#*ZAN-*P932 – us NY Public [071]
Di gio. g. gastoldi il primo libro della musica a due voci – 1598 – (= ser Mssa) – 1mf – 9 – €20.00 – mfchl 149 – gw Fischer [780]
Di havatselet see Yidisher heftlings-kongres in bergn-belzn
Di marco polo e degli altri viaggiatori veneziani... / Zurla, P – Venezia: Gio Giacomo Fuchs, 1818-1819. 2v – 11mf – 9 – mf#HT-695 – ne IDC [910]
Di s. zaccaria papa e degli anni del suo pontificato : commentarii storico-critici / Bartolini, Domenico – Ratisbona: Federico Pustet, 1879 – 2mf – 9 – 0-8370-8161-0 – (incl bibl ref) – mf#1986-2161 – us ATLA [240]
Di Tella, Torcuato S see Estructuras sindicales
Di tim la-ng quen / Hoang Ha – [Saigon] Ve Tinh 1974 [mf ed 1992] – on pt of 1r – 1 – mf#11052 r130 n10 – us Cornell [959]
Di Venuti, Biagio see
– Banking growth in puerto rico
– Money and banking in puerto rico
Di verdelot tutti li madrigali del primo et del secondo libro a quatro voci – 1565 – (= ser Mssa) – 3mf – 9 – €50.00 – mfchl 129 – gw Fischer [780]
Di vi ac notione vocabuli elpis in novo testamento / Zoeckler, Otto – Gissae: Wilhelm Keller, 1856 – 1mf – 9 – 0-7905-0419-7 – (in latin and greek. incl bibl ref) – mf#1987-0419 – us ATLA [225]
Di vicenzo ruffo il secondo libro di madrigali a cinque voci – 1553 – (= ser Mssa) – 2mf – 9 – €35.00 – mfchl 113 – gw Fischer [780]
Di vicenzo ruffo il terzo libro di madrigali a cinque voci / Tonk. Schl. 406-410 – 1555 – (= ser Mssa) – 2mf – 9 – €35.00 – mfchl 114 – gw Fischer [780]
Di yidishe prese in der gevezener ruslandisher imperie / Kirzhnitz, A – M., 1930 – 2mf – 9 – mf#J-290-18 – ne IDC [077]
Di yudishe bibiothek – Warzawa, 1891, 1892, 1895. 3 v – 12mf – 9 – (missing: 1891, v1(p 73-104)) – mf#J-92-5 – ne IDC [077]
Dia – Hialeah, FL. 1970 oct 10-1971 nov 15 – 1r – us UF Libraries [071]
El dia – Tegucigalpa, Honduras: J L Pineda, jan, apr, jun 1954 – us CRL [079]
El dia – Tegucigalpa, Honduras: J L Pineda, [jan, apr, jun 1954] – 1 – us CRL [079]
O dia – Lisboa : O Dia. ano1 n1 (11 de dez 1975)- [mf ed 1984] – (= ser Portuguese newspapers of the 1970's, a coll) – 8r – 1 – mf#2012 – us Wisconsin U Libr [074]
O dia : orgao do partido republicano catharinense – Florianopolis, SC. 01 jan-jul, set-nov 1901; abr-nov 1902; jan-dez 1903; jan-jun, out-dez 1904; jan 1905-dez 1914; jul 1915-28 set 1918 – (= ser Ps 19) – mf#P11A,04,01 – bl Biblioteca [321]
Dia como hoy / Santovenia Y Echaide, Emeterio Santiago – Habana, Cuba. 1946 – 1r – us UF Libraries [972]
El dia de colon y de la paz : 12 de octubre de 1492-12 de octubre de 1918 / Gonzales, Jose Maria – 2nd ed. Oviedo, 1933 [Madrid: Razon y Fe, 1934] – 1 – sp Bibl Santa Ana [946]
El dia de colon y la paz / Gonzales, Jose Maria – Madrid: Razon y Fe, 1930 – 1 – sp Bibl Santa Ana [946]
Dia de la cancion / Delegacion Provincial del Frente de Juventudes – Badajoz: Tip. Vda. A. Arqueros, 1942 – sp Bibl Santa Ana [780]
Dia del seminario / Plasencia. Seminario Diocesano – Plasencia: La Victoria, 1960 – sp Bibl Santa Ana [240]
El dia deseado, relacion de la solemnidad : con que estreno la iglesia del santo cristo de los milagros, patron jurado por esta cuidad contra los temblores de que es amenazada... / Colmenares Fernandez de Cordoba, Felipe Urbano, marques de Zelada de la Fuente – en Lima: ...de la Calle de San Jacinto, ano de 1771 – (= ser Books on religion...1543/44-c1800: iglesias, catedrales) – 2mf – 9 – mf#crl-354 – ne IDC [241]
El dia familiar / Sanchez Arjona, Vicente – Sevilla: Imprenta Alvarez, 1956 – 1 – sp Bibl Santa Ana [810]
Dia festivo proprio para el culto, y rezo del senor san joachin el veinte de marzo de cada ano : concedido con motu proprio por especial privilegio a la magestad del senor don luis i rey catholico de las espanas... / Elizalde Ita y Parra, Jose Mariano Gregorio de – en Mexico: En la Imprenta de don Francisco Xavier Sanchez, en la Calle de S Francisco, ano de 1744 – (= ser Books on religion...1543/44-c1800: vidas y cultos de santos) – 1mf – 9 – mf#crl-121 – ne IDC [241]
Dia festivo...guadalupe...procesion / Zafra, Manuel de – 1755 – 9 – sp Bibl Santa Ana [240]
Dia forum / Drug Information Association – Horsham PA 1999+ – 1,5,9 – mf#22765,02 – us UMI ProQuest [615]
El dia historico : caracas, 1929 / Machado, Jose E – Madrid: Razon y Fe, 1930 – 1 – sp Bibl Santa Ana [946]
O diabete – Rio Grande do Sul: Typ Lith do Diabete, 04 jul 1875-jan 1876; jan-fev 1878; maio-jun, set 1879; 28 nov 1880 – (= ser Ps 19) – bl Biblioteca [079]
Diabetes – Alexandria VA 1952+ – 1,5,9 – ISSN: 0012-1797 – mf#6183 – us UMI ProQuest [616]
Diabetes care – Alexandria VA 1978+ – 1,5,9 – ISSN: 0149-5992 – mf#12858 – us UMI ProQuest [616]
Diabetes forecast – Alexandria VA 1995+ – 1,5,9 – ISSN: 0095-8301 – mf#19373,01 – us UMI ProQuest [616]
Diabetes literature index – Washington DC 1972-79 – 1,5,9 – ISSN: 0012-1819 – mf#7349 – us UMI ProQuest [616]
Diabetes mellitus im alter / Naurath, Hans Joachim – (mf ed 1995) – 4mf – 9 – €56.00 – 3-8267-2102-0 – mf#DHS 2102 – gw Frankfurter [616]
Diabetes: metabolism research and reviews – Hoboken NJ 1999+ – 1,5,9 – (cont: diabetes/ metabolism reviews) – ISSN: 1520-7552 – mf#14805,01 – us UMI ProQuest [616]
Diabetes: metabolism reviews – Hoboken NJ 1985-98 – 1,5,9 – (cont by: diabetes: metabolism research & reviews) – ISSN: 0742-4221 – mf#14805.01 – us UMI ProQuest [616]
Diabetes research and clinical practice – Oxford, England 1989+ – 1,5,9 – ISSN: 0168-8227 – mf#42517 – us UMI ProQuest [616]
Diabetes spectrum – Alexandria VA 1992+ – 1 – ISSN: 1040-9165 – mf#16506 – us UMI ProQuest [616]
Diabetic medicine : a journal of the british diabetic association / British Diabetic Association – Oxford, England 1984+ – 1,5,9 – ISSN: 0742-3071 – mf#14806 – us UMI ProQuest [616]
Diabetologia – Dordrecht, Netherlands 1965+ – 1,5,9 – ISSN: 0012-186X – mf#13161 – us UMI ProQuest [616]
Diable a quatre / Jaime, E – Paris, France. 1845 – 1r – us UF Libraries [440]
Le diable amoureux : ballet pantomime en trois actes et huit tableaux / Saint-Georges, Henri – Paris: Henriot, 1840 – 1 – mf#*ZBD-*MGTZ pv2-Res – Located: NYPL – us Misc Inst [790]
Le diable est aux vaches : et, vie de jeunesse de johnny cassepinette / La Glebe, Jean de – Quebec: [s.n.], 1923 [mf ed 1990] – 2mf – 9 – mf#SEM105P1309 – cn Bibl Nat [420]
Le diable est aux vaches / Glebe, Jean de la – [Quebec?: s.n.] 1911 [mf ed 1996] – 1mf – 9 – 0-665-78267-5 – mf#78267 – cn Canadiana [830]
Le diable et les elections – Paris, [1848?] – us CRL [325]
Le diable rose – [Paris]: Imp de Bureau. [n1-3. jun 15-29 1848] – us CRL [074]
Diablo cojuelo – Miami, FL. 1986 jul-1988 apr – 1r – us UF Libraries [071]
El diablo mundo / Espronceda, Jose de – 1841 – 9 – (1849 ed. 1852 ed. 1875 ed) – sp Bibl Santa Ana [830]
El diablo mundo / Espronceda, Jose de – Buenos Aires: Emece, editores, s.a. 1942 – 1 – sp Bibl Santa Ana [946]
Diablo woman – v2 n1-5 [1984 nov/dec-1985 fall] – 1r – 1 – mf#1477194 – us WHS [071]
O diabo – Lisboa: Via Norte, SARL. ano1 n1. 10 fev 1976- (wkly) [mf ed 1984] – (= ser Portuguese newspapers of the 1970's, a coll) – 1r – 1 – (Replaced (?) by: o sol between mar 9 1976-apr 27 1976) – mf#1101 – us Wisconsin U Libr [074]
O diabo see O sol
Diabo a quatro : illustracao infernal – Rio de Janeiro, RJ: Typ Lambaerts & Comp, 12 out-30 nov 1881 – (= ser Ps 19) – mf#P03A,03,34 – bl Biblioteca [870]
Diabolology : the person and kingdom of satan / Jewett, Edwin Hurtt – New York: Thomas Whittaker 1889 [mf ed 1994] – (= ser The bishop paddock lectures 1889) – 1mf – 9 – 0-8370-9876-9 – mf#1986-3876 – us ATLA [130]
Diabouniotis, C see Hippolyts schrift ueber die segnungen jakobs
O diabrete : critico,litterario e noticioso – Bahia, 23 ago, set, 15 out 1885 – (= ser Ps 19) – mf#P18B,02,54 – bl Biblioteca [079]
Diachenko, G see Polnyi tserkovno-slavianskii slovar
Diachenko, V P see
– Istoriia finansov sssr (1917-1950 gg)
– Sovetskie finansy v pervoi faze razvitiia sotsialisticheskogo gosudarstva
D'iachenko, VP see
– Denezhnoe obrashchenie i kreditnaia sistema soiuza ssr za 20 let
– Finansovo-kreditnyi slovar'
Diacin, Michael see Perceptions of male intercollegiate athletes on performance-enhancing substances in sport
Diaconi casinensis in sanctam regulam commentarium / Warnefridi, Pauli – Monte Casino, 1880 – 15mf – 8 – €53.00 – ne Slangenburg [240]
Diacono, Paulo see Liber de vita et miracles
Diaconus, Paulus [Paul The Deacon] see Historia...
Diacritics – Baltimore MD 1971+ – 1,5,9 – ISSN: 0300-7162 – mf#8377 – us UMI ProQuest [400]
Diaetisches wochenblatt fuer alle staende – Rostock DE, 1781-83 – 1r – 1 – gw Misc Inst [613]
Diaghilev's oversight : and the aftermath / Gregory, John – London: Fed of Russian Classical Ballet [1954] – 1 – mf#*ZBD-*MGO pv28 – Located: NYPL – us Misc Inst [790]
Diagne, Leon see Le systeme de parente matrilineaire serere
Diagnosa-kimia dan tafsir-kliniknja / Asikin widjaja kusumah, D Raden & Ramali, Ahmad – Djakarta: Gunseikanbu Kokumin Tosyokyoku, 2605. v1 – 235p 3mf – 9 – mf#SE-2002 mf17-19 – ne IDC [616]
Diagnose des technischen zustandes und des innenraumzustandes von gasleitungen unter der bedingung unvollstaendiger information / Heymer, Juergen – (mf ed 1993) – 2mf – 9 – €49.00 – 3-89349-674-2 – mf#DHS 674 – gw Frankfurter [621]
The diagnosis and treatment of postural defects / Phelps, Winthrop M & Kiphuth, R – 1932 – 4mf – 9 – $12.00 – us Kinesology [610]
Diagnosis of brain power : speech of lieut-colonel hon james baker...dominion national educational association, held at toronto, april 18th, 1895 – Toronto?: s.n, 1895? – 1mf – 9 – mf#17077 – cn Canadiana [150]
Diagnosis of the brazilian crisis / Furtado, Celso – Berkeley, CA. 1965 – 1r – us UF Libraries [972]
Diagnosis of the mental hygiene problems of college women by means of personality ratings / Rice, Mary Berenice – Washington, 1937 (mf ed 1994) – 2mf – 9 – €31.00 – 3-8267-3093-3 – mf#DHS-AR 3093 – gw Frankfurter [150]
Diagnostic histopathology – Hoboken NJ 1981-83 – 1,5,9 – (cont: investigative & cell pathology) – ISSN: 0272-7749 – mf#11772,01 – us UMI ProQuest [574]
Diagnostic microbiology and infectious disease – Oxford, England 1983+ – 1,5,9 – ISSN: 0732-8893 – mf#42412 – us UMI ProQuest [576]
Diagnostic reagents : a code of practice for their sale and labelling = Diagnostiese reagense: 'n verkoops en etiketterings-gebruikskode – Pretoria: Dept of National Health & Population Development 1987 [mf ed Pretoria, RSA: State Library [199-]] – 8p on 1r with other items – 5 – (in english & afrikaans) – mf#op 08561 r24 – us CRL [350]
Diagnostic study of technical incorrectness in the writings of... / Eason, Joshua Lawrence – Nashville, TN. 1929 – 1r – us UF Libraries [025]
Diagnostico do sistema estatistico fazendario / Brazil. Direcao Geral da Fazenda Nacional Assesso... – Rio de Janeiro. 1968 – 1r – us UF Libraries [310]
Diagnostics & clinical testing – New York NY 1990 – 1,5,9 – (cont: laboratory management) – ISSN: 1044-4092 – mf#1673,01 – us UMI ProQuest [619]
Diagnostischer und prognostischer wert neurologischer zusatzdiagnostik bei schwerem alkoholdelir / Baltzer, Florian – (mf ed 2000) – 1mf – 9 – €30.00 – 3-8267-2703-7 – mf#DHS 2703 – gw Frankfurter [616]
Diagrams for packing citrus fruits / Hume, H Harold – Lake City, FL. 1902 – 1r – us UF Libraries [634]
Diagrams for the black board : of ornamental and other forms / Smith, Walter – [Leeds? 1862?] – (= ser 19th c art & architecture) – 1mf – 9 – mf#4.2.502 – uk Chadwyck [740]
Diakonia : boletin del centro ignaciano de centro america – Panama: El Centro, [n1-88 (abr 1977-oct/dic 1998)] (qrtly) – 4r – 1 – us CRL [230]
Diakonia : a quarterly devoted to advancing orthodox-catholic dialogue – 1(1966)-8(1973) – 72mf – 9 – €137.00 – ne Slangenburg [243]
Diakonia – Scranton NY 1966+ – 1,5,9 – ISSN: 0012-1959 – mf#6847 – us UMI ProQuest [240]
Diakonie, festfreude und zelos : in verbindung mit der altchristlichen agapenfeier / Reicke, B – Uppsala, 1951 – 9mf – 8 – €18.00 – ne Slangenburg [243]
Das diakonissenhaus fuer die provinz sachsen zu halle a. saale, 1857-1907 : eine denkschrift zu der am 3. juli stattfindenden jubelfeier der anstalt / Jordan – Halle a S: Diakonissenhause [distributor], 1907? (CN Ploetz) – 1mf – 9 – 0-524-03232-7 – mf#1990-0860 – us ATLA [240]
Diakonov, L P see Sovetskie zakony o tserkvi
D'lakonov, M M see Rukopisi shakh-name v leningradskikh sobraniiakh
Diakov, F B see Pravovye osnovy promyslovoinogo kontrolia
Dial – Boscobel WI. 1878 may 24-1879, 1880-83, 1884-87, 1888 jan-mar 29 – 4r – 1 – (cont: boscobel dial [boscobel wi: 1872]; cont by: boscobel dial [boscobel wi: 1888]) – mf#1008755 – us WHS [071]
Dial : monthly paper for east-enders – [London & SE] Tower Hamlets may-jun 1897 [mf ed 2003] – 1r – 1 – uk Newsplan [072]
The dial : a monthly magazine for literature, philosophy and religion – n1-12. 1860 – 1 – us AMS Press [073]
Dial [1880] – La Salle IL 1880-1929 – 1 – mf#5307 – us UMI ProQuest [380]
Dial [1980] – New York NY 1980-81 – 1,5,9 – mf#12403 – us UMI ProQuest [380]
Dial: a magazine for literature, philosophy, and religion – La Salle IL 1840-44 – 1 – mf#4367 – us UMI ProQuest [071]
Dial: a monthly magazine for literature, philosophy, and religion – La Salle IL 1860 – 1 – mf#4132 – us UMI ProQuest [071]
Dial baptist church. honey grove, texas : church records – 1893-1954. Formerly Pleasant Hill Baptist Church. Name changed Jul 1904. 342p – 1 – us Southern Baptist [242]
Dial poetico da viaxe dun galego pol-os estados un... / Rubinos, Jose – Habana, Cuba. 1958 – 1r – us UF Libraries [972]
Les dialectes belgo-romans – Bruxelles. v1-3. 1937-1939 – 12mf – 9 – mf#H-10028 – ne IDC [440]
Dialectes indo-europeans / Meillet, Antoine – Paris, France. 1950 – 1r – us UF Libraries [500]

Les dialectes neo-arameens de salamas : textes sur l'etat actuel de la perse et contes populaires – Paris: F Vieweg, 1883 – 1mf – 9 – 0-8370-7627-7 – (text in aramaic and french; introduction in french. incl bibl ref) – mf#1986-1627 – us ATLA [470]

Dialectica del desarrollo / Furtado, Celso – Mexico City? Mexico. 1969 – 1r – us UF Libraries [972]

Dialectical anthropology – Dordrecht Netherlands 1986+ – 1,5,9 – ISSN: 0304-4092 – mf#16038 – us UMI ProQuest [301]

Dialectics – v1-9. 1937-39 [all publ] – (= ser Radical periodicals in the united states, 1881-1960. series 1) – 4mf – 9 – $85.00 – us UPA [430]

Dialectics of hindu ritualism / Datta, Bhupendranatha – Calcutta: Gupta Press, 1956- – (= ser Samp: indian books) – us CRL [280]

Dialectique des barundi / Makarakiza, Andre – Bruxelles, Belgium. 1959 – 1r – us UF Libraries [470]

Der dialekt in den dorfgeschichten berthold auerbachs und melchior meyrs / Glueck, Hermann – Tuebingen: H Laupp 1914 [mf ed 1988] – 1r – 1 – (incl bibl) – mf#6968 – us Wisconsin U Libr [430]

Dialektologicheski sektor / Akademiia Nauk. SSSR Institut Russkovo Yazyka – Moscow 1947-49. Byulleten'. Vyp. 1-6 – 1 – us NY Public [460]

Dial-enterprise – 1895 oct 9-1897 apr 21, apr 28-1898 nov 23, nov 30-1900 jun 27, jul 4-1902 jan 2, jan 29-1903 sep 2, sep 9-1905 mar 8, mar 15-1906 sep 19, sep 26-1908 apr 29, may 6-aug 5 – 9r – 1 – (cont: boscobel dial [boscobel wi: 1888]; enterprise [boscobel, wis]; cont by: boscobel dial-enterprise) – mf#1008759 – us WHS [071]

Dialog / Allis-Chalmers Corporation – 1st [1973-1982 may] – 1r – 1 – (cont by: reporter) – mf#638049 – us WHS [071]

Dialog – Oxford, England 1962+ – 1,5,9 – ISSN: 0012-2033 – mf#6011 – us UMI ProQuest [240]

Der dialog bei den christlichen schriftstellern der ersten vier jahrhunderte / Hoffmann, M – Berlin, 1966 – (= ser Tugal 5-96) – 4mf – 9 – €11.00 – ne Slangenburg [240]

Dialog der adamantius (gcsej1) / ed by Sande Bakhuyzen, W H van de – 1901 – (= ser Griechische christlichen schriftsteller der ersten jahr- hunderte (gcsej)) – €15.00 – ne Slangenburg [240]

Dialog mit dem juden tryphon [bdk33 1.reihe] / Justinus der Maertyrer – (= ser Bibliothek der kirchenvaeter. 1. reihe [bdk 1.reihe]) – €14.00 – (filmed with: pseudo-justinus: mahnrede an die heiden) – ne Slangenburg [240]

Dialoge in poetischer und prosaischer form / Brunnquell, Paul [comp] – Milwaukee, WI: Brunnquell & Rohde, 1885 [mf ed 1993] – 252p – 1 – (consists largely of extracts fr the dramas of goethe, schiller, lessing and others. incl bibl ref and ind. ann by comp) – mf#8361 – us Wisconsin U Libr [430]

Dialoghi a sette et otto voci / Vecchi, Orazio – 1608 – (= ser Mssa) – 3mf – 9 – €50.00 – mfchl 432 – gw Fischer [780]

Dialoghi di m. lodovico domenichi : cioe, d'amore, delle vera nobilita... / Domenichi, Lodovico – Vinegia: Appresso Gabriel Giolito de'Ferrari, 1562 – 5mf – 9 – mf#0-1548 – ne IDC [090]

Dialoghi di massimo troiano : ne' quali si narrano le cose piu notabili... / Troiano, M – Venetia: Appresso Bolognino Zaltieri, 1569 – 5mf – 9 – mf#0-1964 – ne IDC [090]

Dialoghi musicali... : a sette, otto, nove, dieci, undeci & dodeci voci – 1592 – (= ser Mssa) – 7mf – 9 – €90.00 – mfchl 147 – gw Fischer [780]

Dialoghi piacevoli del sig. stefano guazzo : gentil' huomo di casale di monferrato / Guazzo, Stefano – Venetia: Presso Gio. Antonio Bertano, 1586 – 5mf – 9 – mf#0-1595 – ne IDC [090]

Dialogi *see* Rhetorica ad herennium...

Dialogi di messer della inventione poetica : from the british library / Lionardi, Allesandro – 1554 – 1r – 1 – mf#2237 – uk Microform Academic [410]

Dialogical preaching in the local church : a model for a shared approach to preaching / Durkee, Robert Peter – Princeton, New Jersey, 1976. Chicago: Dep of Photodup, U of Chicago Lib, 1976 (1r); Evanston: American Theol Lib Assoc, 1984 (1r) – 1 – 0-8370-1287-2 – mf#1984-T013 – us ATLA [240]

Dialogische strukturen und lernerbezogenheit beim freien schreiben in der fremdsprache deutsch : beitraege zu einer didaktik des schreibens fuer deutschlernende, bezogen auf texte polnischer linguistikstudenten / Thiel, Klaus – (mf ed 2000) – 5mf – 9 – €59.00 – 3-8267-2747-9 – mf#DHS 2747 – gw Frankfurter [430]

El dialogo con los hijos / Perez Lozano, Jose Maria – Madrid: PPC, 1966 – 1 – sp Bibl Santa Ana [946]

Dialogo con un turista. reportaje grafico-historico sobre san pedro de alcantara. madrid, 1969 / Corredor, Antonio – Madrid: Graf. Calleja, 1969 – 1 – sp Bibl Santa Ana [240]

Dialogo de' giuochi che nelle vegghie sanesi si usano di fare / [Bargagli, G] – Venetia: Appresso Alessandro Gardane, 1581 – 4mf – 9 – mf#0-1805 – ne IDC [090]

Dialogo de la infancia y adolescencia / Montero y Santaren, Eulogio – 1892 – 9 – sp Bibl Santa Ana [370]

Dialogo de las empresas militares : y amorosas, compuesta en lengua italiana... – Lyons: Guillielmo Roville, 1562 – 3mf – 9 – mf#0-1835 – ne IDC [090]

Dialogo dell' imprese militari et amorose : di monsignor giovio vescovo di nocera / Giovio, Paolo & Domenichi, Ludovico – Vinegia: Appresso Gabriel Giolito de Ferrari, 1556 – 2mf – 9 – mf#0-267 – ne IDC [090]

Dialogo dell' imprese militari et amorose di monsignor giovio vescovo di nocera : et del s. gabriel symeoni fiorentino – Lyone: Appresso Guglielmo Rovillio, 1574 – 4mf – 9 – mf#0-271 – ne IDC [090]

Dialogo dell' imprese militari et amorose di monsignor paolo giovio vescovo di nucera... / Giovio, Paolo – Rome: Appresso Antonio Barre, 1555 – 2mf – 9 – mf#0-1868 – ne IDC [090]

Dialogo della giosta fatta in trivigi l'anno 1597 / Dalla Torre, Giovanni – Trivigi: Appresso evangelista Dehuchino, 1598 – 2mf – 9 – mf#0-1545 – ne IDC [090]

Dialogo dell'imprese del sig. torquato tasso / Tasso, Torquato – Napoli: Stigliola, [1594] – 1mf – 9 – mf#0-1952 – ne IDC [090]

Dialogo dell'imprese militari et amorose : di monsignor giovio vescovo di nocera / Giovio, Paolo – Vinegia: Appresso Gabriel Giolito de' Ferrari, 1557 – 2mf – 9 – mf#0-1586 – ne IDC [090]

Dialogo dell'imprese militari et amorose di monsignor giovio vescovo di nocera / Giovio, Paolo & Domenichi, Ludovico – Lione: Appresso Guglielmo Roviglio, 1559 – 3mf – 9 – mf#0-269 – ne IDC [090]

Dialogo di caracosa, e caronte, il quale gli nega il passo della sua barca – Venetia, [1572] – 1mf – 9 – mf#H-8416 – ne IDC [956]

Dialogo di vincentio galilei nobile fiorentino della musica antica, et della moderna / Galilei, Vincenzo – Fiorenza: G Marescotti 1581 [mf ed 19–] – 12mf – 9 – mf#fiche 469 – us Sibley [780]

El dialogo sexual / Aradillas Agudo, Antonio – Madrid: a q ediciones, s.a, 1976 – 1 – sp Bibl Santa Ana [306]

Dialogo sopra la miracolosa vittoria ottenvta dall' armata della santissima lega christiana, contra la turchesca / Meduna, B – Venetia, 1572 – 1mf – 9 – mf#H-8332 – ne IDC [956]

Dialogo teologico – 1973. No. 1 & No. 2. 202p – 1 – 7.07 – us Southern Baptist [242]

Dialogo teologico – Buenos Aires, Argentina: Asociacion Bautista de Instituciones Teologicas Hispanoamericanas: distribuido por la Casa Bautista de Publicaciones. n1-34. 1973-89 – 2r – 1 – us CRL [079]

Dialogorum libri 4 / Wiclef, John [John Wycliffe] – s.l, 1525 – €23.00 – ne Slangenburg [240]

Dialogos con america / Selva, Mauricio De La – Mexico City? Mexico. 1964 – 1r – us UF Libraries [972]

Dialogos con el coronel monzon / Monzon, Elfego H – Guatemala, 1958 – 1r – us UF Libraries [972]

Dialogos das grandezas do brasil / Brandao, Ambrosio Fernandes – Recife, Brazil. 1966 – 1r – us UF Libraries [972]

Dialogos das grandezas do brasil pela primeira / Brandao, Ambrosio Fernandes – Rio de Janeiro, Brazil. 1930 – 1r – us UF Libraries [972]

Dialogos de la conquista...dios / Juan de los Angeles – 1595 – 9 – sp Bibl Santa Ana [240]

Dialogos de la pintura... / Carducho, V – Madrid, 1633 – 9mf – 9 – mf#0-190 – ne IDC [700]

Dialogos de los muertos de luciano / Franco y Lozano, Francisco – 1882 – 9 – sp Bibl Santa Ana [946]

Dialogos morales de luciano / Lucian; ed by Herrera Maldonado, Francisco – 1796 – 9 – sp Bibl Santa Ana [450]

Dialogos para el futuro / Estrella Estralla, Jose Emilio – Madrid: Graf. Menor, 1966 – sp Bibl Santa Ana [946]

Dialogue – v2 n3-v5 n9 [1983 feb-1986 may/jun], n62-66 [1986 jul-nov] – 1r – 1 – (cont by: community dialogue) – mf#1231135 – us WHS [071]

Dialogue – Dane County WI. 1978 apr-1982 apr – 1r – 1 – mf#1002554 – us WHS [071]

Dialogue : an international review – Vienna, Austria 1968 – 1 – ISSN: 0012-2149 – mf#5176 – us UMI ProQuest [071]

Dialogue / Liberal Party of Canada – v1 n1-v2 n1 [1975 summer-1976 winter] – 1r – 1 – (cont by: dialogue) – mf#1543957 – us WHS [325]

Dialogue / Liberal Party of Canada – v2 n2-v4 n3 [1976 apr-1978 oct] – 1r – 1 – (cont: dialogue [ottawa on]; cont by: dialogue newsletter) – mf#1544087 – us WHS [325]

Dialogue – Venice CA 1989-2004 – 1,5,9 – ISSN: 0279-568X – mf#15419 – us UMI ProQuest [700]

Dialogue: a journal of mormon thought – Salt Lake City 1966+ – 1,5,9 – ISSN: 0012-2157 – mf#11016 – us UMI ProQuest [243]

Dialogue avec la femme endormie / Thoby-Marcelin, Philippe – Port-Au-Prince, Haiti. 1941 – 1r – us UF Libraries [972]

Dialogue between a churchman and a methodist / Gray, Robert – London, England. 1802 – 1r – us UF Libraries [242]

A dialogue between a gentleman of london... and an honest alderman of the country party : where in the grievances under which the nation at present groans are fairly and impartially laid open considered: earnestly address'd to the electors of great-britain / Fielding, Henry – London: printed for M Cooper, 1747 [mf ed 1974] – 1r – 5 – mf#SEM16P231 – cn Bibl Nat [941]

Dialogue between a minister of the church and his parishioner / Sikes, Thomas – London, England. 1802 – 1r – us UF Libraries [240]

Dialogue by way of catechism, religious, moral, and philosophical – Ramsgate, England. 1872 – 1r – us UF Libraries [240]

Dialogue entre deux mondes / Catalogne, Gerard de – Paris. 1931 – 1 – fr ACRPP [900]

A dialogue, luther and the reformation : the doctrine and government / Ritz, S – Mansfield OH: Western Branch Book Concern of the Wesleyan Methodist Connection of America 1854 [mf ed 1992] – 1mf – 9 – 0-524-04775-8 – mf#1991-2161 – us ATLA [242]

A dialogue on the distinct characters of the picturesque and the beautiful : in answer to the objections of mr. knight / Price, Uvedale – Hereford, 1801 – (= ser 19th c art & architecture) – 3mf – 9 – mf#4.1.110 – uk Chadwyck [700]

Dialogue sur la musique des anciens / Chateauneuf, Francois de Castagneres, Abbe de – Paris: N Pissot 1725 [mf ed 19–] – 1r – 1 – (pref by jacques morabin) – mf#film 252 – us Sibley [780]

Dialogues and detached sentences in the chinese language : with a free and verbal translation in english / ed by Bannerman, James – Macao: printed at the Honorable East India Company's Press, 1816 – (= ser 19th c books on linguistics) – 3mf – 9 – (trans by robert morrison) – mf#2.1.22 – uk Chadwyck [480]

Dialogues between a clergyman and a layman on family worship / Maurice, Frederick Denison – Cambridge: Macmillan, 1862 [mf ed 1990] – 1mf – 9 – 0-7905-7528-0 – mf#1989-0753 – us ATLA [230]

Dialogues between a protestant and a roman catholic / Hobson, Samuel – London, England. 1840 – 1r – us UF Libraries [240]

Dialogues between two methodists, algernon newways and samuel oldpaths : in which attendance at class meetings as a condition of church membership, is shown to be both wesleyan and scriptural and the relation of children to the visible church / Borland, John – Toronto: printed for the aut, 1856 [mf ed 1987] – 1mf – 9 – 0-665-67293-4 – mf#67293 – cn Canadiana [242]

Dialogues concerning two new sciences = Discorsi e dimostrazioni matematiche / Galilei, Galileo – New York: Macmillan, 1914 – 1mf – 9 – 0-7905-9934-1 – (in english) – mf#1989-1659 – us ATLA [530]

Dialogues familiers sur les principales objections des missionnaires de ce temps / Drelincourt, C – N.p., 1648 – 4mf – 9 – mf#CA-125 – ne IDC [240]

Dialogues intended to facilitate the acquiring of the bengalee language / Carey, William – 3rd ed. Serampore: Mission Press, 1818 [mf ed 1995] – (= ser Yale coll) – viii/113p – 1 – 0-524-09426-8 – (in bengali & english on opposite pp) – mf#1995-0426 – us ATLA [490]

The dialogues of athanasius and zacchaeus and of timothy and aquila / ed by Conybeare, Frederick Cornwallis – Oxford: Clarendon Press, 1898 [mf ed 1992] – (= ser Anecdota oxoniensia) – 1mf – 9 – 0-524-03894-5 – (in greek) – mf#1990-1153 – us ATLA [240]

Dialogues of plato – New York, NY. 1950 – 1r – us UF Libraries [180]

The dialogues of saint gregory, surnamed the great : pope of rome & the first of that name = dialogi / Gregory 1, Pope; ed by Gardner, Edmund Garratt – London: PL Warner, 1911 – 1mf – 9 – 0-7905-7398-9 – (in english) – mf#1989-0623 – us ATLA [240]

Dialogues of the buddha / Digha-Nikaya – London, England. v1-3. 1899-1921 – 1r – us UF Libraries [280]

Dialogues of the buddha : translated from the pali of the digha nikaya / Rhys Davids, C A F & Rhys Davids, T W [trans] – London: Hamphrey Milford, 1921 – (= ser Samp: indian books) – us CRL [280]

Dialogues on the church question : dialogue 3 / Morren, Nathaniel – Greenock? Scotland. 1843? – 1r – us UF Libraries [240]

Dialogues on theology : or, familiar conversations between two aged friends. a partialist and a universalist, principally relating to the doctrine of endless punishment / Prime, Daniel Noyes – Newburyport: William H Huse, 1875 – 1mf – 9 – 0-524-04273-X – mf#1991-2057 – us ATLA [240]

Dialogues par un ministre suisse / Vernes, Jacob – s.l.: s.n; 1763 – 1 – us Wisconsin U Libr [949]

Dialogues sur la musique...adresses a son amie / Villers, Clemence de – Paris: Chez Vente 1774 [mf ed 19–] – 1mf – 9 – mf#fiche 1125 – us Sibley [780]

Dialogus de gestis sanctorum fratrum minorum / Thomas de Papia; ed by Delorme, Ferdinand Marie – Ad Claras Aquas (Quaracchi) prope Florentiam: ex typographia Collegii s. Bonaventurae, 1923. Chicago: Dep of Photodup, U of Chicago Lib, 1975 (1r); Evanston: American Theol Lib Assoc, 1984 (1r) – (= ser Bibliotheca franciscana ascetica medii aevi) – 1 – 0-8370-0653-8 – (incl bibl ref & ind) – mf#1984-B419 – us ATLA [240]

Dialogus de imperio et pontifica potestate / Occam, Guilelmus de (Ockham, William of) – Lyon, 1494 – 18mf – 8 – €35.00 – ne Slangenburg [241]

Dialogus de vtraqve in christo natvra... / Vermigli, P M – Tigvri, Christoph Froschouer, 1561 – 3mf – 9 – mf#PBU-649 – ne IDC [240]

Dialogus malegranatum / Gallus A Koenigsaal – Argentorati, 1474 – 24mf – 8 – €46.00 – ne Slangenburg [241]

Dialogus metricus / Franck, Melchior – 1608 – (= ser Mssa) – 1mf – 9 – €20.00 – mfchl 224 – gw Fischer [780]

Dialogus miraculorum / Caesarii Heisterbacensis [Caesarius of Heisterbach]; ed by Strange, J – Coloniae. v1-2. 1851 – €23.00 – ne Slangenburg [241]

Dialogus miraculorum / Caesarii Heisterbacensis (Caesarius of Heisterbach); ed by Strange, J – Keulen. v1-2. 1851 – 2v on 12mf – 8 – €23.00 – ne Slangenburg [241]

Ein dialogus oder gespraech etlicher personen vom interim / [Alber, E] – [Augsburg, 1557] – 2mf – 9 – mf#TH-1 mf 6-7 – ne IDC [242]

Dialogus, sive, speculum ecclesie militantis / Wycliffe, John – London: Published for the Wyclif Society by Truebner, 1886 – 1mf – 9 – 0-524-00809-4 – mf#1990-0241 – us ATLA [240]

A dialogve philosophicall : wherein natvres secret closet is opened, and the cavse of all motion in natvre shewed ovt of matter and forme... / Tymme, Thomas – London: Printed for C Knight, 1612 [mf ed 1995] – 72p on 1r – 1 – mf#8878 – us Wisconsin U Libr [500]

Dialogvs : ein gespraech von den beyden naturen christi... / Vermigli, P M – [Zuerich: Christoph Froschauer], 1563 – 7mf – 9 – mf#PBU-652 – ne IDC [240]

Dialogvs : oder ein gespreche eines esels vnd bergknechts dem synodo auium zu lieb geschrieben / Magdeburg, J – [Luebeck], 1557 – 1mf – 9 – mf#TH-1 mf 907 – ne IDC [242]

Der diamant : eine komoedie in fuenf acten / Hebbel, Friedrich – Hamburg: Hoffmann und Campe, 1847 [mf ed 1995] – 178p – 1 – mf#8764 – us Wisconsin U Libr [820]

Diamante, Juan Bautista *see* Valor no tiene edad y sanson de extremadura. el

Diamantina : ballo in due parti e sei scene / Fusco, Federico – Napoli: Stab Tip del Cosmopolita, 1861 – 1 – mf#*ZBD-*MGTZ pv1-Res – Located: NYPL – us Misc Inst [790]

Diamantveldsche getuige en griekwaland west advertentieblad *see* Diamond fields witness and griqualand west advertiser

Diamond, A I *see* Biography of edwin james turpin

Diamond and related materials – Oxford, England 1991+ – 1,5,9 – ISSN: 0925-9635 – mf#42654 – us UMI ProQuest [550]

The diamond field – Kimberley SA, 13 oct 1870-20 jul 1877 – 2r – 1 – mf#MS00245 – sa National [079]

DIARIO

Diamond fields advertiser – Kimberley SA, 23 mar 1878-1968 – 1 – (cont by: diamond fields advertiser (regional : 1995). title shortened to: diamond fields advertiser in feb 1879) – mf#MS00107 – sa National [079]
Diamond fields advertiser – Kimberley, South Africa. -d. 1906-40. 189 reels – 1 – uk British Libr Newspaper [072]
The diamond fields express – Kimberley SA, 1886-88 (daily) [mf ed Cape Town: SA library 1982] – 4r – 1 – (first publ as: the diamond fields express and griqualand mercantile gazette) – mf#MS00425 – sa National [079]
The diamond fields express and griqualand mercantile gazette *see* The diamond fields express
Diamond fields herald – Kimberly: Josiah Angove, [187-?]-1885 [daily, wkly, 3x/wk] [mf ed 1 jan-31 dec 1885] – 1r – 1 – (title varies: v8 n977 [30 sep 1885]-n1000 [27 oct 1885]: herald) – mf#mp.1092 – sa National [079]
The diamond fields mail : and mining and commercial advertiser – Kimberley SA, n1 9 jul-n48 sep 1 1888 (daily) [mf ed Cape Town: SA library 1985] – 1r – 1 – mf#MS00377 – sa National [079]
The diamond fields mining news – Kimberley SA, 21 nov 1896-7 apr 1897 – 1r – 1 – mf#MS00247 – sa National [079]
Diamond fields news *see* Diamond news
Diamond fields times – Kimberley SA: George Garcia Wolf 1 oct 1884-13 aug 1885 [daily] – 3r – 1 – mf#mp.1094 – sa National [079]
Diamond fields witness *see* The diamond fields express
Diamond fields witness and griqualand west advertiser = Diamantveldsche getuige en griekwaland west advertentieblad – Kimberley: A A de Groot 1886 [mf ed Kimberley, Cape Province: SA Lib 1982] – 9 – sa National [079]
Diamond jubilee of her majesty queen victoria, sunday, 20th june, 1897 : sons of england service to be held in continuous succession through the british colonies around the world – S.l: s.n, 1897? – 1mf – 9 – mf#16978 – cn Canadiana [941]
Diamond lens and other stories / O'brien, Fitz-James – New York, NY. 1932 – 1r – us UF Libraries [830]
Diamond news – Kimberley SA, 16 oct 1870-1884 – 13r – 1 – (title varies: diamond fields news, diamond news and vaal advertiser) – mf#MS00261 – sa National [079]
Diamond news and vaal advertiser *see* Diamond news
Diamond river / Garavini Di Turno, Sadio – New York, NY. 1963 – 1r – us UF Libraries [972]
The diamond shoe buckles / Albert, Mary – [Westminster]: The Roxburghe Press Ltd, [1897] – (= ser 19th c women writers) – 2mf – 9 – mf#5.1.91 – uk Chadwyck [830]
The diamond sutra = chin-kang-ching : or, prajna-paramita – London: K Paul, Trench, Truebner, 1912 – 1mf – 9 – 0-524-07799-1 – (incl ind) – mf#1991-0176 – us ATLA [280]
Diamonds and dust : india through french eyes / Pellenc, Jean, Baron – London: John Murray, 1936 – (= ser Samp: indian books) – us CRL [954]
Diamonds in the desert : Professional baseball in arizona and the desert southwest, 1915 to 1958 / Rosebrook, Jeb S – 1999 – 277p on 3mf – 9 – $15.00 – mf#PE 4186 – us Kinesology [790]
Dian – Djakarta, 1953-1969. v1-17 – 59mf – 9 – (missing: 1967 v15(3)) – mf#SE-712 – ne IDC [959]
Diana : and ladies' spectator – La Salle IL 1822 – 1 – mf#3737 – us UMI ProQuest [640]
Diana : oder, gesellschaftsschrift zur erweiterung und berichtigung der natur- forst- und jagdkunde / ed by Bechstein, Johann Matthaeus – Waltershausen: In der offentlichen Lehranstalt der Forst- und Jagdkunde und Schnepfenthal in Commission bey J F Mueller 1797-1816 [v1-4(1797-1816)] [mf ed 2005] – 4v on 1r [ill] – 1 – (v4 iss also as: diana, oder, neue gesellschaftsschrift zur erweiterung und berichtigung der natur- forst- und jagdkunde; publ: gotha: c w ettinger 1801-1805; marburg und cassel: in der kriegerschen buchhandlung 1816) – mf#film mas 37255 – us Harvard [634]
Diana, Augusto *see* Youth at play
Diana, Pierre *see* Un cas de conscience
Diane la belle aventuriere / Saurel, Pierre [pseud] – Montreal: ed Police journal. n1 11 avril 1956-n303 24 janv 1962; ns: n1 fev 1962- [wkly] [mf ed 1982] – 4r – 5 – (ceased: ns: n12 janv 1963?) – mf#SEM16P320 – cn Bibl Nat [073]
Diapason – Des Plaines IL 1909+ – 1,5,9 – ISSN: 0012-2378 – mf#10720 – us UMI ProQuest [780]
The diapason : an international monthly devoted to the organ and the interests of organists – n2-720. 1910-69 – 1 – us AMS Press [780]
Diapason negro / Rosa-Nieves, Cesareo – San Juan, Puerto Rico. 1960 – 1r – us UF Libraries [972]
Diaries / Anderson, Chandler P – 1914-28, 1934 – 1 – 72.00 – us L of C Photodup [920]
Diaries / Arundel, John T – 1870-1919 – 13r – 1 – mf#pmb480-492 – at Pacific Mss [338]
Diaries / Banks, Charles W – Rarotong, 1897-1898 – 1r – 1 – mf#PMB1067 – at Pacific Mss [920]
Diaries / Banks, Charles W – Rarotonga, 1892, 1899, 1900, 1904 – 1r – 1 – mf#PMB1068 – at Pacific Mss [920]
Diaries / Bartholomew, Elam – 1871-1934 – 1 – us Kansas [920]
Diaries / Bocker Family – 1895-1952 – 1 – us Kansas [978]
Diaries / Bonekemper, Johannes – 1 – us Southern Baptist [242]
Diaries / Brewer, Mandane Williamson – 1854-60 – 1 – us Kansas [920]
Diaries / Bridwell, Arthur – 1903-44 – 1 – us Kansas [920]
Diaries / Byers, George D – 1907-24.Incomplete – 1 – $50.00 – us Presbyterian [920]
Diaries / Dimond, Susan B – 1870-73 – 1 – us Kansas [920]
Diaries / Kelsey, Dandridge E – 1853-1903, Relate to Indiana; Shawnee county, KS; and Salida, Colorado – 1 – us Kansas [920]
Diaries / Lee, Ada – 1934-1966 – 2r – 1 – mf#PMB1098 – at Pacific Mss [920]
Diaries / Nassau, Robert Hamill – 1880-1919 – 1 – $100.00 – us Presbyterian [920]
Diaries / Norton Family – 1876-1895, Accounts of life in early Pawnee County, KS. Included are many references to the fort and city of Larned. Most of the entries pertain to agricultural work. Copied and annotated by Helen Norton Starr – 1 – us Kansas [920]
Diaries / Parkerson, Harriet – 1891-1900, Diaries of the adopted daughter of Isaac Goodnow – 1 – us Kansas [920]
Diaries / Riddle, L H – 1887-1891 – 1 – us Kansas [978]
Diaries / Spotts, David L – 1868-69. Letters received, 1924-28 – 1 – us Kansas [978]
Diaries / Thompson, James – 1873-74 – 1 – us Kansas [978]
Diaries / Thornton, Anna Maria (Brodeau) – 1793-1863 – 1 – us L of C Photodup [920]
Diaries / Trego, Joseph Harrington – 1844-59 – 1 – us Kansas [978]
Diaries / Wheeler, John H – 1854-72. Includes material relating to Nicaragua – 1 – 85.00 – us L of C Photodup [920]
Diaries / Wiberg, Anders – v1-15. 1844-87. 5849p – 1 – us Southern Baptist [242]
Diaries, 1877-8 nov 1912 / Rennolds, Edwin Hansford – 2893p – 1 – us Southern Baptist [242]
Diaries, 1851-1902 / Buskirk, Philip Clayton van – 35v + 18r – 1 – (diaries describing van buskirk's experiences as a sailor in the u s navy, incl those as member of the perry expedition wh opened japan to foreign trade, service as a confederate soldier in the civil war, & settler in snohomish county, wa [1896-1902]) – us UW Libraries [920]
Diaries, 1877-1879, 1884-1900, 1902-1924 / Huegel, Friedrich, Freiherr von – [S.l.: s.n., 18–] – 11r – 1 – 0-8370-1766-1 – mf#1984-S127 – us ATLA [943]
Diaries, 1914-1919 / Russell, Charles Edward – Washington, DC: Library of Congress Photodup Service, 1979 – us CRL [320]
Diaries (abolitionist) / Hise, Daniel H – 1r – 1 – mf#B32293 – us Ohio Hist [976]
Diaries and account books / Oakley, Edward Ellsworth – 1859-61 – 1 – us Kansas [978]
Diaries and accounts / Stephenson Family – 1881-1948, Diaries and account books of several members of this Osborne County, KS, farm family – 1 – us Kansas [920]
The diaries and letters of arthur j munby and hannah culwick *see* Working women in victorian britain, 1850-1910
The diaries and memoirs, 1811-70 : from liverpool central library / Brown, George Alexander – (= ser British records relating to america in microform) – 1r – 1 – (with guide. int by john rowe) – mf#96796 – uk Microform Academic [920]
Diaries and papers of bertha and herman benke / Benke, Bertha & Benke, Herman C – 1886-93 – 1 – us Kansas [978]
Diaries and papers of elizabeth inchbald : from the folger shakespeare library and the london library – 4r – 1 – £385.00 – uk Matthew [880]
Diaries and papers of julia and charles lovejoy / Lovejoy, Julia L & Lovejoy, Charles H – 1828-64 – 1 – us Kansas [978]
Diaries and pearling logs / Hamilton, William – 1882-1905 – 1r – 1 – mf#pmb15 – at Pacific Mss [880]
Diaries and records, 1838-1967 / St Mary's College. St Mary's, Kansas – 1 – us Kansas [378]
Diaries and related records describing life in india *see* India during the raj: eyewitness accounts
Diaries, correspondence and miscellaneous papers / Gray, William – 1882-1937 – 3r – 1 – mf#PMB1046 – at Pacific Mss [920]
Diaries, correspondence and related papers from the solomon islands / Luxton, Clarence T J – 1945-47 – 1r – 1 – mf#PMB1099 – at Pacific Mss [920]
Diaries, notebooks and literary manuscripts of mary elizabeth braddon *see* Sensation fiction
Diaries of anna margaretta larpent *see* Woman's view of drama, 1790-1830
Diaries of bishop william nicolson, 1701-1714 : from carlisle public library – 2r – 1 – mf#96131 – uk Microform Academic [240]
The diaries of dwight d eisenhower, 1953-1961 – 28r – 1 – $4840.00 – 0-89093-889-X – (int by louis galambos & daun van ee. with p/g) – us UPA [920]
The diaries of edward pease, the father of english railways / Pease, Edward; ed by Pease, Alfred Edward – London: Headley Bros, 1907 – 1mf – 9 – 0-524-07107-1 – mf#1991-2930 – us ATLA [625]
The diaries of elizabeth fry : 1797-1845 – 7r – 1 – £350.00 – mf#FRY – uk World [305]
Diaries of g e morrison *see* China through western eyes
Diaries of james v forrestal, 1944-1949 : secretary of the navy, 1944-1947, and first secretary of defence, 1947-1949 – 4r – 1 – £375.00 – (complete & unexpurgated diaries fr the seeley g mudd manuscript library, princeton university; with d/g) – uk Matthew [880]
Diaries of john neville keynes *see* Economists' papers
Diaries of miss sarah hale : missionary for many years to mexico / Hale, Sarah – 612p – 1 – us Southern Baptist [242]
Diaries of sir frederic madden : 1801-73 / The Bodleian Library – 17r – 1 – £800.00 – mf#MDP – uk World [920]
The diaries of sir horace plunkett, 1881-1932 : from the plunkett foundation for co-operative studies, oxford – (= ser British records relating to america in microform) – 8r – 1 – (int by bernard crick) – mf#2662 – uk Microform Academic [920]
The diaries of streynsham master, 1675-1680 : and other contemporary papers relating thereto / ed by Temple, Richard Carnac – London: Published for the Govt of India, 1911– (= ser Samp: indian books) – us CRL [920]
Diario – La Paz Bolivia, 8 nov 1939; 16 nov 1944-oct 1945 – 5r – 1 – uk British Libr Newspaper [079]
Diario / Portugal. Cortes. Camara dos deputados – 1885-1911 12. 1889, v. 1; 1895 wanting – 1 – 605.00 – us L of C Photodup [946]
El diario – La Paz: [s.n.], 1956- – 1 – us CRL [074]
El diario – La Paz: [s.n.], [1956-] – us CRL [070]
El diario – New York, NY. 1962-1967 (1) – mf#65069 – us UMI ProQuest [071]
Diario abierto / Feijoo, Samuel – Santa Clara, Cuba. 1959 – 1r – us UF Libraries [972]
Diario constitucional *see* O constitucional
O diario da assemblea legislativa provincial de minas gerais – Ouro Preto, MG: Typ Social, 08 abr-18 out 1850 – (= ser Ps 19) – mf#P17,02,58 – bl Biblioteca [972]
Diario da justica / Minas Gerais. Brazil – 1945-1946 – 1 – us NY Public [972]
Diario da justica / Pernambuco. Brazil (State). Courts – Diario Oficial Suplemento. Racife. 1962-1963 – 1 – us NY Public [972]
Diario da justica and apenso jurisprudencia / Brazil. Courts – 1958-1962 – 1 – us NY Public [324]
Diario da manha – Lisbon. Portugal. -d. 2 Aug 1941-25 May 1945, 1 Jan-31 Aug 1946. (Imperfect). (19 reels) – 1 – uk British Libr Newspaper [072]
Diario da republica : orgao oficial da republica de angola – Luanda. [1s nov 11 1975-1977; mar 17 1981-aug 13 1984; 1s, 2s, 3s: 1978-1980] – 9r – us CRL [079]
Diario das petas – Bahia. 01 abr 1878 – (= ser Ps 19) – bl Biblioteca [079]
Diario de assembleia / Parana. Brazil. Assembleia – 1957-1958 – 1 – us NY Public [972]
Diario de badajoz – Badajoz, 1883-1886 y 1888-1892 – 5 – sp Bibl Santa Ana [079]
Diario de badajoz – Badajoz.1830-1900 – 9 – sp Bibl Santa Ana [074]
Diario de badajoz – Badajoz, 1882 – 5 – (numeros sueltos. estan relacionados otros anos en el numero 9, ya enviado) – sp Bibl Santa Ana [073]
Diario de bucaramanga / Peru De Lacroix, Luis – Caracas, Venezuela. 1949 – 1r – us UF Libraries [972]
Diario de caceres – Caceres, 1903-1935. (incomplete) – 5 – sp Bibl Santa Ana [073]
Diario de caceres – Caceres, 1910-1921 – 5 – sp Bibl Santa Ana [074]
Diario de campana del comandante luis rodolfo miranda – Habana, Cuba. 1954 – 1r – us UF Libraries [972]
Diario de campana del mayor general maximo gomez – Ceiba del Agua, Cuba. 1941 – 1r – us UF Libraries [972]
Diario de centro america – Guatemala: [s.n.], dec 31 1955-feb 1972 – 65r – 1 – us CRL [079]
Diario de centro america. guatemala / Guatemala – On film: Ag. 1972-. LL-02115 – 1 – us L of C Photodup [340]
Diario de costa rica – San Jose: J V Coto, 1956-sep 29 1964 – 53r – 1 – us CRL [079]
El diario de hoy – San salvador, El Salvador: n v altamirano, r membreno, 1936-56- – 1 – us CRL [079]
Diario de la gente – 1973 sep 14, 1980 apr/may-1982 dec, 1980 apr/may-1982 dec – 2r – 1 – mf#615817 – us WHS [071]
Diario de la marina – Havana, Cuba. 13 feb 1940; 28 mar-16 sep 1945 – 7r – 1 – uk British Libr Newspaper [072]
Diario de la palabra encadenada / Vallejo, Alejandro – Bogota, Colombia. 1949 – 1r – us UF Libraries [972]
Diario de madrid – Anos 1808-1809 (16-IV/31-I) – 12mf – 9 – sp Cultura [946]
Diario de manaos : propriedade de uma associacao – Manaus, AM. 10 dez 1890-22 mar 1894 – (= ser Ps 19) – mf#P11B,06,26 – bl Biblioteca [079]
Diario de manila – Manila: Ramirez y Giraudier, Suppl to nov 7 1897; nov 12 18, dec 30 1897; jan 5,11-12,17,28, mar 11,26, apr 12, may 4,16,26 1898 – us CRL [079]
Diario de mi prision en san carlos / Paredes, Antonio – Caracas, Venezuela. 1963 – 1r – us UF Libraries [972]
Diario de minas – Ouro Preto, MG: Typ J F de Paula Castro, 01 jun 1866-mar 1868; fev 1873-14 mar 1878 – (= ser Ps 19) – bl Biblioteca [079]
Diario de noticias – Bahia: [s.n.] 02 set,nov 1876; set 1877; out-nov 1881; maio 1882; abr 1883; ago 1884; fev 1885; jan 1886; nov 1888; jun 1893; fev-mar 1895; nov 1897; maio 1900; mar 1903; 06 ago 1909 – (= ser Ps 19) – mf#P11,02,14 – bl Biblioteca [079]
Diario de noticias – Belem, PA: Typ do Diario de Noticias, 01 jul 1881-out 1887; fev-dez 1888; jul 1889-jun 1895; jan 1896-17 maio 1898 – (= ser Ps 19) – mf#P11,05,04 – bl Biblioteca [079]
Diario de noticias : noticioso, litterario e commercial – Rio de Janeiro, RJ. 21 mar-jun 1868; jul-27 set 1872 – (= ser Ps 19) – mf#DIPER – bl Biblioteca [079]
Diario de noticias – Rio de Janeiro Brazil, jun-dec 1954; 19 jan-dec 1955; feb-jun 1956; 25 jun-28 jul 1957 – 13r – 1 – uk British Libr Newspaper [079]
Diario de noticias – Rio de Janeiro, RJ. 17 set-19 out 1881 – (= ser Ps 19) – mf#DIPER – bl Biblioteca [079]
O diario de noticias – Manaus, AM. 14 mar 1900 – (= ser Ps 19) – mf#P11,01,45 – bl Biblioteca [079]
El diario de nueva york = New york's spanish daily – Brooklyn: El Diario Publ Co, 1948-1963. feb 25 1950-mar 11 1952 – 15r – us CRL [071]
Diario de panama/panama journal – Panama City, Panama. 14 Apr 1905-9 Sept 1914. 21 Oct 1921-33 (incomplete) – (= ser Panama Journal) – 43r – 1 – us L of C Photodup [079]
Diario de petas – Bahia: [s.n.] 07 mar 1886 – (= ser Ps 19) – mf#P18B,02,55 – bl Biblioteca [079]
Diario de s luiz – [Sao Luiz], MA. 16 out 1920-jun 1925; abr-jun 1945; jan-dez 1946; abr 1947-30 set 1949 – (= ser Ps 19) – mf#P11,04,15 – bl Biblioteca [079]
Diario de sesiones / Guatemala Asamblea Constituvente (1945) – Guatemala, 1951 – 1r – us UF Libraries [324]
Diario de sesiones de la convencion constituyente / Cuba Convencion Constituyente, 1928 – Habana, Cuba. 1928 – 1r – us UF Libraries [324]
Diario de sesiones de las cortes constituyentes / Spain. Cortes, 1836-1837 – v. 1-10. 1870-77 – 1 – us L of C Photodup [946]
Diario de sesiones de las cortes constituyentes / Spain. Cortes Constituyentes, 1869-1871 – v. 1-15, Nos. 1-332. 1870-71 – 1 – us L of C Photodup [946]
Diario de sesiones del congreso de la republica de... / Cuba Congreso Camara De Representantes – Habana, Cuba. v1-95. 1902-1957 – 21r – (gaps) – us UF Libraries [324]
Diario de sesiones del congreso de la republica de... / Cuba Congreso Senado – Habana, Cuba. v1 n1-v62 n20. 1902-1929/30 – 8r – us UF Libraries [324]

Diario de sessiones de la comision de los quince e... / Guatemala Comision De Los Quince – Guatemala, 1953 – 1r – us UF Libraries [324]
Diario de tipacoque / Caballero Calderon, Eduardo – Bogota, Colombia. 1950 – 1r – us UF Libraries [972]
Diario de um confinado / Ribeiro, Mauro – Sao Paulo, Brazil. 1968 – 1r – us UF Libraries [972]
Diario de uma campanha / Quadros, Janio – Sao Paulo, Brazil. 1961 – 1r – us UF Libraries [972]
Diario de un ingeniero / Sanchez Arjona, Vicente – Sevilla: Imprenta Carlos Acuna, 1954 – 1 – sp Bibl Santa Ana [810]
Diario de un padre de familia / Perez Lozano, Jose Maria – Madrid: Propaganda Popular Catolica, 1959 – 1 – sp Bibl Santa Ana [920]
Diario de viaje indios y negros de la provincia de... / Palacios De La Vega, Joseph – Bogota, Colombia. 1955 – 1r – us UF Libraries [972]
Diario del caribe – Barranquilla, Colombia. 1990-May 1991 – 17r – 1 – us L of C Photodup [079]
Diario des las sesiones del congreso de los diputados / Spain. Congreso de los Diputados – v. 1-523. 1837 38-1936 – 1 – 5,678.00 – us L of C Photodup [946]
Diario do aracaju – Aracaju, SE: Typ de Sergipe, 11-12 mar, abr, jul, out 1885 – (= ser Ps 19) – mf#P11A,03,05 – bl Biblioteca [079]
Diario do commercio : critico, litterario, commercial e noticioso – Rio de Janeiro, RJ: Typ Fluminense de Domingos Luiz dos Santos, 01-30 jul 1866 – (= ser Ps 19) – mf#DIPER – bl Biblioteca [380]
Diario do gram-para – Belem, PA. 01 dez 1885; fev-24 mar 1886 – (= ser Ps 19) – mf#DIPER – bl Biblioteca [079]
Diario do poder legislativo / Pernambuco. Brazil (State). Assemblea Legislativo – 1959-1963 – 1 – us NY Public [972]
Diario do rio de janeiro – Rio de Janeiro, RJ: Real Typographica, 01 jan 1821-dez 1844; jan-jun, ago-dez 1845; jan 1846-31 out 1878 – (= ser Ps 19) – mf#P04A,01,01-19P05,01,01-18P06,01,01-15P30,04,01-13 – bl Biblioteca [079]
Diario e notas autobiograficas / Reboucas, Andre Pinto – Rio de Janeiro, Brazil. 1938 – 1r – us UF Libraries [972]
Diario fluminense : critico, litterario, recreativo e noticioso – Rio de Janeiro, RJ: Typ Esperanca, 01 jul-21 out 1884 – (= ser Ps 19) – mf#P18A,02,44 – bl Biblioteca [079]
Diario illustrado – Rio de Janeiro, RJ. 16 abr-11 set 1887 – (= ser Ps 19) – mf#P18A,01,37 – bl Biblioteca [079]
Diario illustrado – Rio de Janeiro, RJ. 18 out-09 nov 1910 – (= ser Ps 19) – mf#DIPER – bl Biblioteca [079]
Diario ilustrado – Santiago, Chile. 30 oct 1941; 14 feb-aug 1945; mar 1955-jul 1957 – 31r – 1 – uk British Libr Newspaper [072]
Diario intimo do engenheiro vauthier, 1840-1846 / Vauthier, Louis Leger – Rio de Janeiro, Brazil. 1940 – 1r – us UF Libraries [972]
Diario latino – San Salvador Bahamas, 8 jan 1940-20 oct 1944 – 1r – 1 – uk British Libr Newspaper [079]
Diario latino – San Salvador, El Salvadore. 1956 – 6r – 1 – us L of C Photodup [079]
Diario municipal / Rio de Janeiro. Federal District. Camara Legislativa – 1957-Apr. 20, 1960 – 1 – us NY Public [324]
Diario municipal and supplements / Rio de Janeiro. Federal District. Prefeitura – Nov. 14, 1957-Apr. 1960 – 1 – us NY Public [324]
Diario nacional – Sao Paulo, SP. 14 jul 1927-03 out 1932 – (= ser Ps 19) – mf#P11A,06,162 – bl Biblioteca [079]
El diario nacional – Panama City, Panama. 16 Jun 1920-11 Aug 1921 – 2r – 1 – us L of C Photodup [079]
Diario official do imperio do brasil – Rio de Janeiro, RJ. 01 out 1862 – (= ser Ps 19) – mf#P18A,01,35P18A,1,35A – bl Biblioteca [323]
Diario oficial / Brazil – 1823-36, 1892-1969 – 1 – us L of C Photodup [324]
Diario oficial / Brazil – Rio de Janeiro. 1900-1949 – 1 – us NY Public [324]
Diario oficial / Brazil – Section 1, Part 1. 1971- – 1 – (section 1, part 2. 1971-. 1. section 4. 1973-. 1) – us L of C Photodup [972]
Diario oficial / Chile – 1877-1969 – 1 – 7751.00 – us L of C Photodup [324]
Diario oficial / Chile – Santiago. 1952-1960 – 1 – us NY Public [324]
Diario oficial / Colombia – 1821-1969 – 1 – $6,615.00 – us L of C Photodup [340]
Diario oficial / Colombia – Bogota. On film: 1970-. LL-02086 – 1 – us L of C Photodup [340]
Diario oficial / Mexico – 1821-1945 – 1 – us NY Public [972]
Diario oficial / Mexico – 1954-59, 1970- – 1 – $501.00 – us L of C Photodup [340]
Diario oficial / Parana. Brazil – Curitiba. 1946-Feb. 1969- – 1 – us NY Public [324]
Diario oficial / Pernambuco. Brazil (State) – Feb. 1947-1963 – 1 – us NY Public [972]
Diario oficial / El Salvador – San Salvador. 1944-Nov 1948, 1958-1967 – 1 – us NY Public [972]
Diario oficial / El Salvador – San Salvador. 1970- – 1 – us L of C Photodup [340]
Diario oficial / El Salvador – 1847-1969 – 1 – $7,070.00 – (supplement. 1847-1969) – us L of C Photodup [340]
Diario oficial / Yucatan. Mexico (State) – Merida. 1946-1960 – 1 – us NY Public [972]
Diario oficial. 1890- / Uruguay – 1 – us L of C Photodup [972]
Diario oficial de la republica de chile / Chile – Santiago. On film: 1970-. LL-02085 – 1 – us L of C Photodup [340]
Diario oficial de la union europeas *see* The official journal of the european union
Diario oficial del gobierno del estado de yucatan / Yucatan. Mexico (State) – Merida. On film: 1821-1969. LL-02040 – 1 – us L of C Photodup [340]
Diario oficial. seccion avisos / Uruguay – 1906-69, 1970- – 1 – $10,902.00 – us L of C Photodup [972]
Diario oficiel / Montevideo. Uruguay – Seccion Avisos. Oct 1959-68 – 1 – us NY Public [972]
Diario politico y militar / Restrepo, Jose Manuel – Bogota, Colombia. v1-4. 1954 – 1r – us UF Libraries [972]
Diario popular : folha consagrada aos interesses da provincia – Aracaju, SE: Typ do Echo Liberal, 06 fev, 13, 27 mar 1879 – (= ser Ps 19) – mf#FUNDAJ – bl Biblioteca [079]
Diario popular – Para, 06 fev 1891 – (= ser Ps 19) – bl Biblioteca [079]
Diario portugues – Rio de Janeiro, RJ: Typ Esperanca, 11 nov 1884-04 set 1885 – (= ser Ps 19) – mf#DIPER – bl Biblioteca [079]
Diario y derrotero de lo caminado / Rivera Y Villalon, Pedro De – Mexico City? Mexico. 1946 – 1r – us UF Libraries [972]
Diario y notas autobiograficas / Reboucas, Andre – Rio de Janeiro. 1938 – 1 – us CRL [920]
Diarium spirituale. roman om en roest / Gyllensten, Lars Johan Wictor – Stockholm: Bonnier, 1968. 182p – 1 – us Wisconsin U Libr [830]
Diary / Alexander, Thomas P – 1883-1913 – 1 – us Kansas [920]
Diary / Baptist, Edward – May 1790-May 1860. 41p – 1 – 5.00 – us Southern Baptist [242]
Diary / Barnett, William T – 1899-1900 – 1 – us Kansas [920]
Diary / Billard, Louis Phillip – 1917-18 – 1 – us Kansas [920]
Diary / Bolton, C S – 1861-62 – 1 – us Kansas [920]
Diary / Bragg, Thomas – 1861-62. Guide – 1 – $18.00 – us CIS [920]
Diary / Brownfield, J M – Ione, Arkansas, 1921. 108p – 1 – 5.00 – us Southern Baptist [920]
Diary / Carter, Elizabeth (Simerwell) – 1852-61 – 1 – us Kansas [920]
Diary / Chaudoin, William – 1858-59. 76p – 1 – 5.00 – us Southern Baptist [920]
Diary : a church newsletter serving the old order society / Old Order Amish Church of America – 1978-82, 1983-85, 1986-88 – 3r – 1 – (cont: diary of the old order amish church of america) – mf#654907 – us WHS [243]
Diary / Cole, Isaac – 1759-1870 – 1 – us Southern Baptist [242]
Diary / Comer, John & Barrows, E C & Willmarth, J W – 1892 – 1 – us Southern Baptist [242]
Diary / Converse, John Melvin – Jan 1863-Dec 1864 – 1 – us Kansas [920]
Diary / Cool, Amanda & Cool, J – 1879-85 – 1 – us Kansas [920]
Diary / Copas, J V – Apr-Sept 1912 – 1 – us Kansas [978]
Diary / Costigan, S P – 1874-77 – 1 – us Kansas [978]
Diary / Crowe, Sophia Bennett – 1874 – 1 – us Kansas [920]
Diary / Daniels, Augustus Thatcher – 1865-1921 – 1 – us Kansas [920]
Diary / DeWolf, Charles Wesley – undated, Civil War diary of a Kansas soldier in Arkansas – 1 – us Kansas [976]
Diary / Dewolfe, Harry F – Dec 1862-May 1863 – 1 – us Kansas [920]
Diary / Dimond, W W – 1865 – 1 – us Kansas [920]
Diary / Evans, Eliza (Pruitt) – 1884-92 – 1 – us Kansas [920]
Diary / Fouquet, Leon Charles – 1880-82 – 1 – us Kansas [920]
Diary / Gailland, Maurice, S J – 1848-77 – 1 – us Kansas [920]
Diary / Grey, Zane – 1r – 1 – mf#B32188 – us Ohio Hist [080]
Diary / Guthrie, James H – 1861-65 – 1 – us Kansas [920]
Diary / Hall, Cyrus – 1861-64 – 1 – us Kansas [920]
Diary / Hamlin, Charles – Washington, DC: Library of Congress Photodup Service. v2-16. 1913-1929 – us CRL [320]
Diary / Hammond, OT – 1834-37. 284p – 1 – 9.94 – us Southern Baptist [920]
Diary / Hand, Julia – 1872-75 – 1 – us Kansas [920]
Diary / Harts, William Henry – 1r – 1 – mf#B33081 – us Ohio Hist [976]
Diary / Hinson, William Godber – 1864-65 [mf ed Spartanburg SC: Reprint Co, 1981?] – 2mf – 9 – mf#51-082 – us South Carolina Historical [976]
Diary / Holman, Charles – 1863-64, 1866-67 – 1 – us Kansas [920]
Diary / Holmes, Henry S – 1895-1903 [mf ed Spartanburg SC: Reprint Co, 1981] – 4mf – 9 – mf#51-083 – us South Carolina Historical [977]
Diary / Honyman, Robert – 1 – us L of C Photodup [920]
Diary : in tahitian, mangarevan and english, kept on flint island, eastern pacific / Moouga, H I N – 14 apr 1889-31 jan 1891 – 1r – 1 – mf#pmb14 – at Pacific Mss [880]
Diary / Kautenberger, Peter G – 1865 – 1r – 1 – (diary of this private soldier in the american civil war, commenting on his experiences whith the 46th illinois vounteer infantry regiment, company c, which participated in the seige of mobile, al, and later, garrison duty) – us Western Res [976]
Diary / Kelly, Edward – Pioneer Baptist Minister, Corn Island, Nicaragua. 1903-1913 – 1 – 5.00 – us Southern Baptist [242]
Diary / McCoy, Joseph Geating – Jun 1880-Jan 1881 – 1 – us Kansas [978]
Diary / McKechnie, Archie – 1882-83 – 1 – us Kansas [978]
Diary / Peelle, Will J – 1879-84, 1902 – 1 – us Kansas [978]
Diary / Pomeroy, Henry H – 1r – 1 – $50.00 – mf#C40020 – us Library Micro [920]
Diary / Raymond, Henry Hubert – 1870-72 – 1 – us Kansas [978]
Diary / Raymond, J M – 1862-65 – 1 – us Kansas [978]
Diary / Richards, James C – 1867 – 1 – us Kansas [978]
Diary / Riley, B F – 1894, 1898, 1904. 150p – 1 – 5.25 – us Southern Baptist [242]
Diary / Ruffin, Edmund – 1856-65 – 1 – us L of C Photodup [636]
Diary / Scott, Cyrus McNeely – 1867-1915 – 1 – us Kansas [978]
Diary / Smith, Margaret A – 1878-81 – 1 – $50.00 – us Presbyterian [240]
Diary / Smith, William M – 1864 – 1 – us Kansas [978]
Diary / Snyder, Edwin – 1 Dec 1864-30 Jun 1865 – 1 – us Kansas [978]
Diary / Snyder, S J H – 1848 – 1 – us Kansas [978]
Diary / Spooner, E A – 1849-50 – 1 – us Kansas [978]
Diary / Stafford, Alfred – 1864 – 1 – us Kansas [978]
Diary / Stewart, James R – 1855-60 – 1 – us Kansas [978]
Diary / Tichenor, Isaac Taylor – 1850. 486p – 1 – us Southern Baptist [242]
Diary / Trego, Joseph Harrington – 1861-63 – 1 – us Kansas [978]
Diary / Watson, Sidney O – 1899-1901 – 1 – us Kansas [978]
Diary / Welch, James E – 1832-59 – 1 – $50.00 – us Presbyterian [920]
Diary / Witts, Maurice M – 1 jan-31 dec 1905 – 1r – 1 – mf#pmb1 – at Pacific Mss [880]
Diary / Woods, Walter Hastings – 1858-59 – 1 – us Kansas [978]
The diary, 1771-94 : from whitehaven public library and museum / Bragg, John – (= ser British records relating to america in microform) – 1r – 1 – (with int by j spence) – mf#659 – uk Microform Academic [920]
Diary, 1832-1917 / Littleberry, J Haley – 170p – 1 – 5.95 – us Southern Baptist [242]
Diary, 1845-46 / Smith, Thomas – 1 – 6.09 – us Southern Baptist [242]
Diary, 1861-1865, 58th ovi / Stuber, Johann – 1r – 1 – (in german) – mf#B31281 – us Ohio Hist [355]
Diary, 1911-20 / Lockett, BL – 500p – 1 – us Southern Baptist [242]
The diary and correspondence of charles darwin written during the voyage of the beagle, 1831-36 : from the library of the royal college of surgeons, down house, kent – 1r – 1 – mf#96687 – uk Microform Academic [574]
Diary and journal kept at ioway mission in kansas / Irvin, Samuel M – 1841-49 – 1 – us Kansas [920]
The diary and journal of his grace, the archbishop of york toby mathew, 1583-1622 : york minster library, add. ms. 18 – 1r – 1 – mf#2224 – uk Microform Academic [241]
Diary and letters / Landon, R B – 1881-1916 – 1 – us Kansas [978]
Diary and letters of madame d'arblay / Burney, Fanny – London, England. v1-7. 1854 – 2r – us UF Libraries [420]
Diary and letters of madame d'arblay ed by ger niece [charlotte barrett] / Burney, Fanny; ed by Barrett, Charlotte – London: H Colburn, 1854 [mf ed 1987] – 7v – 1 – mf#8203 – us Wisconsin U Libr [880]
Diary and personal records / Younker, Bowman H – 1867-68 – 1 – us Kansas [978]
Diary and transcript / Odgers, Len – 1942-1943 – 1r – 1 – mf#PMB1061 – at Pacific Mss [920]
Diary and travelogue / Dill, Jacob Smiser – 1879-81. 420p – 1 – us Southern Baptist [910]
Diary and visitation record of the rt. rev. francis patrick kenrick : administrator and bishop of philadelphia, 1830-1851, later, archbishop of baltimore / Kenrick, Francis Patrick – [S.l.: s.n.], 1916 (Lancaster, Pa: Wickersham Print Co) – 1mf – 9 – 0-524-04214-4 – mf#1990-5005 – us ATLA [241]
Diary during a trip from papar to kimanis via tambunan, lobo and limbawan, 1882 : from the colonial office library, london / Donop, L B von – 1r – 1 – mf#6748 – uk Microform Academic [920]
Diary during an excursion across north borneo from maruda bay to sandakan, 1881 / Witti, Francis X – 1r – 1 – mf#6746 – uk Microform Academic [915]
A diary in america : with remarks on its institutions / Marryat, Frederick – Paris 1839-40 [mf ed Hildesheim 1995-98] – 2v on 6mf – 9 – €120.00 – 3-487-27221-0 – gw Olms [880]
Diary in france : mainly on topics concerning education and the church / Wordsworth, Christopher – London: F & J Rivington, 1845 – 1mf – 9 – 0-7905-8215-5 – mf#1988-8098 – us ATLA [914]
Diary, in tahitian, mangarevan and english : kept on flint island, eastern pacific / Moouga, H I N – apr 1889-jan 1891 – 1r – 1 – mf#pmb14 – at Pacific Mss [331]
Diary, journal and letters (wesleyan mission in fiji) / Jaggar, Thomas James – 1838-1846 – 1r – 1 – mf#pmb1185 – at Pacific Mss [242]
Diary, ms 3079 / Frary, George S – 1864 – 1r – 1 – (frary served in the 171th ohio national guard, and for a while, was posted at johnson's island prison, near sandusky, ohio, during the american civil war) – us Western Res [976]
Diary, ms p.p. / Day, Francis A – Apr-Dec 1864 – 1r – 1 – us Western Res [976]
Diary, ms p.p. / Mills, Henry A – 1862-65 – 1r – 1 – us Western Res [976]
Diary of a buccaneer : ms from the admiralty library, london / Sharp, Bartholomew – 1680 – 1r – 1 – mf#3443 – uk Microform Academic [920]
"Diary of a buffalo hunter, 1872-73" / Raymond, Henry Hubert – 1 – us Kansas [978]
Diary of a journey overland through the maritime provinces of china : from manchao on the south coast of hainan, to canton, in the years 1819 and 1820 / [Supercargo, R J] – London: Richard Phillips and Co, 1822 – 2mf – 9 – mf#HT-661 – ne IDC [915]
Diary of a journey to the cape of good hope and the interior / Mist, Augusta Uitenhage De – Cape Town, South Africa. 1954 – 1r – us UF Libraries [916]
The diary of a modernist / Palmer, William Scott – London: E Arnold, 1910 – 1mf – 9 – 0-7905-9562-1 – mf#1989-1287 – us ATLA [240]
Diary of a spring holiday in cuba / Levis, Richard J – Philadelphia, PA. 1872 – 1r – us UF Libraries [972]
Diary of a tour through southern india, egypt, and palestine : in the years 1821 and 1822 / Mackworth, Digby – London 1823 [mf ed Hildesheim 1995-98] – 1v on 3mf [ill] – 9 – €90.00 – 3-487-26706-3 – gw Olms [910]
The diary of an exiled nun – 2nd ed. St Louis, MO: B Herder, 1911, c1910 – 1mf – 9 – 0-8370-7293-X – mf#1986-1293 – us ATLA [920]
The diary of an invalid : being the journal of a tour in pursuit of health in portugal, italy, switzerland and france in the years 1817, 1818 and 1819 / Matthews, Henry – London 1820 [mf ed Hildesheim 1995-98] – 1v on 4mf – 9 – €120.00 – 3-487-27736-0 – gw Olms [880]
Diary of an italian commander killed at guadalajara – n.p. 1937? Fiche W 831. (Blodgett Collection of Spanish Civil War Pamphlets) – 9 – us Harvard [946]

Diary of beatrice webb 1873-1943 – [mf ed Chadwyck-Healey] – 237mf – 9 – (with ind) – uk Chadwyck [880]

Diary of courtenay hughes fenn (1866-1953) for the period 1866-1927 / Fenn, Courtenay Hughes – 1927 – 1r – 1 – mf#1984-B390 – us ATLA [240]

Diary of escape from salamaua, territory of new guinea / Melrose, Robert – 22 jan-19 feb 1942 – 1r – 1 – (available for ref) – mf#pmb1181 – at Pacific Mss [920]

The diary of george folliot, 1765-66 : from wigan public library – (= ser British records relating to america in microform) – 1r – 1 – (int by w e minchinton) – mf#96886 – uk Microform Academic [920]

Diary of george martin, 1779-1800 : from st. mary's, whitehaven – (= ser British records relating to america in microform) – 1r – 1 – (int by w e minchinton) – mf#4636 – uk Microform Academic [920]

Diary of hans frank / U.S. World War Two Crimes Records – (= ser National archives coll of world war 2 war crimes records) – 12r – 5 – mf#T992 – us Nat Archives [943]

The diary of henry edward price, 1842-48 : from islington public library – (= ser British records relating to america in microform) – 1r – 1 – (int by bernard crick) – mf#2677 – uk Microform Academic [920]

Diary of henry francis fynn – Pietermaritzburg, South Africa. 1969 – 1r – us UF Libraries [960]

The diary of henry teonge, chaplain on board his majesty's ships assistance, bristol, and royal oak, anno 1675 to 1679 : now first published from the original ms with biographical and historical notes – London 1825 [mf ed Hildesheim 1995-98] – 1v on 3mf – 9 – €90.00 – 3-487-27314-4 – gw Olms [880]

Diary of his journey to the low countries, 1755 / Smeaton, John – 1938 – (= ser Newcomen society extra publication 4) – 3mf – 7 – (int by arthur titley) – mf#86576 – uk Microform Academic [914]

The diary of james losh, 1796-1815 : from carlisle museum & art gallery – 3r – 1 – mf#96099 – uk Microform Academic [920]

Diary of john comer / Barrows, E C & Willmarth, J W – 1892. 434p – 1 – us Southern Baptist [242]

The diary of joseph farington, 1788-1821 : from the royal archives and library at windsor castle – 1 – (original ms 7r 96542. typed transcript with ind 7r 334) – uk Microform Academic [920]

Diary of lady mildmay : from northampton central library – 1r – 1 – mf#97377 – uk Microform Academic [920]

Diary of my trip to america and havana / Mark, John – Manchester, England. 1885 – 1r – us UF Libraries [918]

Diary of operations division, war department general staff, 1942-1946 / U.S. Army – 4r – 1 – $520.00 – mf#S1677 – Center of Military History – us Scholarly Res [355]

Diary of otto braun *see* They call it patriotism

The diary of otto braun : with selections from his letters and poems / ed by Vogelstein, Julie – London: W Heinemann, 1924 [mf ed 1989] – xxxii/362p – 1 – (int by havelock ellis) – mf#7065 – us Wisconsin U Libr [880]

Diary of richard reynolds, 1763 : from ketley furnaces, wellington – 1r – 1 – mf#362 – uk Microform Academic [920]

Diary of ten years eventful life of an early settler in western australia : and also a descriptive vocabulary of the language of the aborigines / Moore, George Fletcher – London 1884 – (= ser 19th c british colonization) – 6mf – 9 – mf#1.1.2805 – uk Chadwyck [980]

A diary of the home rule parliament 1892-1895 / Lucy, Henry William – [London], 1896 – (= ser 19th c ireland) – 6mf – 9 – mf#1.1.1608 – uk Chadwyck [941]

Diary of the old order amish church of america / Old Order Amish Church of America – 1969-73, 1974-77 – 2r – 1 – (cont by: diary [gordonville pa]) – mf#600000 – us WHS [243]

Diary of the rev francis owen, m a [the] : missionary with dingaan in 1837-38. together with extracts from the writings of the interpreters in zulu, messrs hulley and kirkman / Owen, Francis; ed by Cory, Geo E – Cape Town: The Van Riebeeck society 1926 – (= ser [Travel descriptions from south africa, 1711-1938]) – 3mf – 9 – mf#zah-66 – ne IDC [880]

The diary of the rev. samuel dodd / Dodd, Samuel – New York: [s.n., 1894] – 1r – 1 – 0-8370-0997-9 – mf#1984-B516 – us ATLA [240]

Diary of the siege of quebec, 1775 : from the british library, add ms 46840 / Danford, J – (= ser British records relating to america in microform) – 1r – 1 – (with int by ivor burton) – mf#96660 – uk Microform Academic [920]

Diary of thomas robbins, d.d., 1796-1854 : printed for his nephew / Robbins, Thomas; ed by Tarbox, Increase Niles – Boston: Beacon Press, 1886-1887 – 5mf – 9 – 0-7905-8088-8 – mf#1988-8024 – us ATLA [240]

Diary; papers / Grinell, DeWitt Clinton – 1867-77 – 1 – us Kansas [920]

Diary relating to the new hebrides / Witts, Maurice M – 1 jan-15 aug 1911 – 1r – 1 – mf#pmb8 – at Pacific Mss [880]

Dias, Antonio Goncalves *see*
- Primeiros cantos
- Segundos cantos e sextilhas de frei antao

Dias carneiro (o conservador) / Pizarro Jacobina, Alberto – Sao Paulo, Brazil. 1938 – 1r – us UF Libraries [972]

Dias como llamas / Benitez, Adigio – Habana, Cuba. 1962 – 1r – us UF Libraries [972]

Dias De Carvalho, Henrique Augusto *see* Methodo pratico para fallar a lingua da lunda

Dias de futuro / Martinez Matos, Jose – Habana, Cuba. 1964 – 1r – us UF Libraries [972]

Dias de nuestra angustia / Navarro, Noel – Habana, Cuba. 1962 – 1r – us UF Libraries [972]

Dias, Demosthenes de Oliveira *see* Formacao territorial do brasil; origem e evolucao

Dias Filho, Manoel A Santos *see* Canna e o assucar nas antilhas

Dias, Giuliana Zorrer *see* An investigation into nurses' anxiety when dealing with hiv/aids patients

Dias, Jorge *see* Portuguese contribution to cultural anthropology

Dias, Manuel Da Costa *see* Colonizacao dos planaltos de angola

Dias, Manuel Nunes *see* Fomento e mercantilismo

Dias, Margot *see* Maganjas da costa

Dias repartidos / Fernandez Gomez, Otto – Habana, Cuba. 1964 – 1r – us UF Libraries [972]

Dias Rollemberg, Luiz *see* Aspectos e perspectivas da economia nacional

Dias sin sol / Barrantes Moreno, Vicente – 1875 – 9 – sp Bibl Santa Ana [946]

Dias, Walter Patrick *see* Shakespeare: his tragic world

Dias y la politica / Barrios, Gonzalo – Caracas, Venezuela. 1963 – 1r – us UF Libraries [972]

Diaspora – North York, Canada 1991+ – 1,5,9 – ISSN: 1044-2057 – mf#17036 – us UMI ProQuest [327]

Diatelevi, Michael P *see* An examination of the relationships between coaching behaviors, sport confidence, and motivational orientation

The diatessaron of tatian : a harmony of the four holy gospels compiled in the third quarter of the second century / ed by Hemphill, Samuel – London: Hodder & Stoughton; Dublin: William McGee, 1888 – 1mf – 9 – 0-7905-0163-5 – (incl bibl ref) – mf#1987-0163 – us ATLA [225]

The diatessaron of tatian : a preliminary study / Harris, James Rendel – London: CJ Clay, 1890 – 1mf – 9 – 0-8370-3489-2 – (incl an appendix on codex wod) – mf#1985-1489 – us ATLA [220]

The diatessaron of tatian and the synoptic problem : being an investigation of the diatessaron for the light which it throws upon the solution of the problem of the origin of the synoptic gospels / Hobson, Alphonzo Augustus – Chicago: University of Chicago Press, 1904 – 1mf – 9 – 0-8370-3602-X – mf#1985-1602 – us ATLA [220]

Diavola : or, nobody's daughter / Braddon, Mary Elizabeth – New York: Dick & Fitzgerald, [188-?] – us CRL [920]

Diaz Alfaro, Abelardo Milton *see*
- Terrazo

Diaz, Antolin *see* A la sombra de fouche

Diaz Benzo, Antonio *see* Pequeneces de la guerra de cuba

Diaz Bustamante y Quijano, Alfonso *see* La universidad de extremadura

Diaz Castro, Tania *see* Apuntes para el tiempo

Diaz Chavez, Luis *see* Pescador sin fortuna

Diaz Daza, A *see* Libro de los provechos y danos...por bebida del agua...

Diaz de Arce, Juan *see*
- Libro de la vida del proximo evangelico, el vener padre bernardino alvarez
- Libro primero del proximo evangelico exemplificado en la vida del venerable bernardino alvares, espanol

Diaz de Entresotos, Baldomero *see* Seis meses de anarquia en extremadura

Diaz de Isla, R *see* Tractado contra el mal serpentino que vulgarmente en espana es llamado bubas

Diaz de la Carrera, Diego *see* Origen y principio...alcantara

Diaz de la Cruz, Felipe *see* El asunto de plasencia

Diaz de Montalvo, A *see* El fuero real de espana hecho por alfonso 9

Diaz De Olano, Carmen R *see* Felix matos bernier

Diaz de Vargas, Francisco *see*
- Discurso. guerra de portugal
- Discurso y sumario de la guerra en portugal

Diaz de Villar y Martinez, Juan M *see* Tratado elemental de higiene comparada del hombre y los animales domesticos

Diaz Del Castillo, Bernal *see*
- Historia verdadera de la conquista de la neuva esp...

Diaz del Castillo, Bernal *see*
- La conquista de mejico
- Discovery and conquest of mexico, 1517-1521
- Histoire de la nouvelle espagne
- Historia de la conquista de nueva espana
- Historia de la...nueva espana
- Historia verdadera de la conquista de la nueva espana
- Historia verdadera de la conquista de la nueva espana...
- Historia verdadera de la conquista de la nueva espana
- Historia verdadera de la conquista de nueva espana
- The true history of the conquest of new spain
- Verdadera historia de los sucesos de la conquista de la nueva espana
- Verdadera y notable relacion del descubrimiento y conquista de guatemala
- Verdadera y notable relacion del descubrimiento y conquista de la nueva espana y guatemala
- Verdadera...nueva espana
- Veridique...nouvelle histoire

Diaz Jordan, Jenaro *see* Discursos y conferencias

Diaz, Jose *see*
- Espana y la guerra imperialista
- Por la unidad, hacia la victoria
- Tres anos de lucha

Diaz Lozano, Argentina *see* Mayapan

Diaz Maciaz, Jose *see*
- Fabianelo
- Los hijod del mar
- La huelga
- Virtud y ciencia

Diaz Martinez, Manuel *see*
- Amor como ella
- Caminos
- Soledad y otros temas

Diaz miron a ruben dario / Meza Fuentes, Roberto – Santiago, Chile. 1940 – 1r – us UF Libraries [972]

Diaz Montero, Anibal *see*
- Biblioteca encantada
- Cerro y llanura
- Mujer y una sota
- Pedruquito y sus amigos

Diaz Montilla, Rafael *see* La alimentacion racional del ganado

Diaz Montilla, Rafael et al *see* 1st congreso sindical agrario de extremadura. ponencia 5 mutualidades agricolas

Diaz Moreno, Juan *see* Vigencia del frente e juventudes en el presente y en el futuro de espana

Diaz Nadal, Roberto *see* Contrastes, cuentos, aguafuertes, cronicas

Diaz Palacios, Santiago *see* Nuestro sistema bancario y su funcionamiento

Diaz Perez, Nicolas *see*
- Banos de banos
- Las bibliotecas de espana...publicas
- Catalogo de los objetos...exposicion
- De madrid a lisboa
- El descuento de las clases privadas
- Diccionario historico...extremenos ilustres
- Ecos perdidos
- La emigracion en baleares y canarias
- Extremadura
- Historia de talavera la real...
- Historia de talavera la real
- Influencia de extremadura en la literatura espanola
- Jose mazzini. ensayo italia
- Noticia historica de...badajoz
- El plutarco extremeno
- El poder temporal de los papas en el s 19
- Recuerdos de extremadura

Diaz Rozzotto, Jaime *see*
- Caracter de la revolucion guatemalteca
- Seis cantos a la estatua de la libertad...

Diaz Seijas, Pedro *see* Historia y antologia de la literatura venezolana

Diaz Soler, Luis M *see*
- Histoire de la esclavitud negra en puerto rico
- Historia de la esclavitud negra en puerto rico (14

Diaz Tanco, Vasco *see*
- Jardin del alma cristiana
- Palinodia

Diaz Valcarcel, Emilio *see* Asedio, y otros cuentos

Diaz Valdeparez, J *see* Generalisimo trujillo molina

Diaz Vasconcelos, Luis Antonio *see* De nuestro antano historico

Diaz Verson, Salvador *see* Cuando la razon se vuelve inutil

Diaz, Victor Miguel *see*
- Barrios ante la posteridad
- Bellas artes en guatemala
- Romantica ciudad colonial

Diaz, Vigil *see*
- Musica de ayer
- Oregano

Diaz Villar Martinez, Juan Manuel *see*
- Manual de fisiologia experimental
- Profilaxis de la fiebre carbuncosa
- Profilaxis de las enfermedades infecciosas del ganado de cerda
- Tratado elemental de higiene
- Tratado elemental de higiene comparada del hombre y de los animales domesticos

Diaz y Perez, Nicolas *see*
- Historia de talavera la real
- Historia del pueblo de alange

Diaz Y Sotelo, Manuel *see* Tierra retonada

Diaz-Ambrona, Adolfo *see* Discurso del..., en defensa de los dictamenes de los siguientes proyectos de ley: 1º de montes vecinales en mano comun. 2º de ordenacion rural. 3º regimen de tierras del instituto nacional de colonizacion

Diaz-Plaja, Guillermo *see*
- Marti desde espana
- Ruben dario

Dibble, R A *see* John h newman

Dibdin, Charles *see*
- The musical tour of mr. dibdin, in which previous to his embarkation for indian he finished his career as a public character..
- Padlock

Dibdin, Lewis Tonna *see* The endowments and establishment of the church of england

Dibdin, R W *see* Ruin of all israel

Dibelius, Martin *see*
- A fresh approach to the new testament
- Die geisterwelt im glauben des paulus
- Die lade jahves
- Die urchristliche ueberlieferung von johannes dem taeufer

Dibelius, Otto *see* Staatsgrenzen und kirchengrenzen

Dibhre hay-yamim = Words of the day – London, UK. jan-apr 1896 – 1 – (aka: dibre hayomim, may 1896-nov/dec 1909) – uk British Libr Newspaper [072]

Dibre hayomim *see* Dibhre hay-yamim

Diccion rio historico de los m s illustres profesores de las bellas artes en espana / Cean Bermudez, J A – Madrid, 1800. 6v – 24mf – 9 – mf#O-979 – ne IDC [700]

Diccionario biografico cubano / Calcagno, Francisco – s.l, s.l? 1829 – 1r – us UF Libraries [920]

Diccionario biografico de el salvador / Perez Marchant, Braulio – Nueva San Salvador, El Salvador. 1937 – 1r – us UF Libraries [920]

Diccionario cakchiquel-espanol / Saenz De Santa Maria, Carmelo – Guatemala, 1940 – 1r – us UF Libraries [025]

Diccionario critico-burlesco del que se titula – Madrid: Imprenta de Repulles, 1812 – 1 – sp Bibl Santa Ana [050]

Diccionario de anglicismos / Alfaro, Ricardo J – Panama, 1950 – 1r – us UF Libraries [420]

Diccionario de gobierno y legisalcion de indias...tomo 4, vol 1 / Ayala, Manuel Jose de – Madrid: Razon y Fe, 1929 – 1 – sp Bibl Santa Ana [340]

Diccionario de jurisprudencia *see* Gaceta del foro

Diccionario de la constitucion – Habana, Cuba. 1941 – 1r – us UF Libraries [323]

Diccionario de la musica illustrado – Barcelona. 1927-29. 2v – 1 – us L of C Photodup [780]

Diccionario de literatura puertorriquena / Rivera De Alvarez, Josefina – Rio Piedras, Puerto Rico. 1955 – 1r – us UF Libraries [054]

Diccionario de temas regionalistas en la poesia pu... / Arana Soto, Salvador – San Juan, Puerto Rico. 1961 – 1r – us UF Libraries [054]

Diccionario de terminos comunes tagalo-castellano / Serrano, Rosalio – Manila: Imprenta de Ramirez y Giraudier, 1858 – us CRL [040]

Diccionario general de americanismo. 3 vol. mexico, 1943 / Santamaria, Francisco – Madrid: Razon y Fe, 1946 – 1 – sp Bibl Santa Ana [972]

Diccionario general de bibliografia espanola / Hidalgo, Dionisio – 7v. 1862-81 – 1,9 – us AMS Press [010]

Diccionario geografico de la isla de cuba / Marquez, Jose De Jesus – Habana, Cuba. 1926 – 1r – us UF Libraries [918]

Diccionario geografico de la republica de el salvador / El Salvador Direccion General De Estadistica – San Salvador, El Salvador. 1945 – 1r – us UF Libraries [918]

Diccionario geografico de la republica de el salvador / El Salvador Direccion General De Estadistica – San Salvador, El Salvador. 1959 – 1r – us UF Libraries [918]

Diccionario geografico, estadistico e historico de... / Arocha, Jose Ignacio – Caracas, Venezuela. 1949 – 1r – us UF Libraries [059]

Diccionario geografico popular de cantares / Vergara Y Martin, Gabriel Maria – Madrid, Spain. 1923 – 1r – us UF Libraries [059]

Diccionario geografico-historico del departamento / Londono, Julio – Bogota, Colombia. 1955 – 1r – us UF Libraries [059]

Diccionario geographico da provincia de s paulo / Mendes De Almeida, Joao – Sao Paulo, Brazil. 1902 – 1r – us UF Libraries [059]

Diccionario hispano-kanaka / Arinez, Agustin Maria De – Tambobong, Philippines. 1892 – 1r – us UF Libraries [040]

Diccionario hispano-tagalog / Laktaw, Pedro Serrano – Manila: Estab tip "La Opinion" a cargo de G Bautista. v1. 1889-1914 – 2r – us CRL [040]

Diccionario historico del departamento de la paz / Aranzaes, Nicanor – La Paz, Bolivia. 1915 – 1r – us UF Libraries [972]

Diccionario historico-biografico del peru. 2nd ed. tomos 9, 10 y 11. lima 1934-1935 / Mendiburu, Manuel de – Madrid: Razon y Fe, 1935 – 1 – sp Bibl Santa Ana [972]

Diccionario historico-biografico del peru...2nd ed. publicada por evaristo sancristoval / Mendiburu, Manuel de – Madrid: Razon y Fe, v6-8. 1935 – 1 – sp Bibl Santa Ana [972]

Diccionario historico...extremenos ilustres / Diaz Perez, Nicolas – 2v. 1884 – 9 – sp Bibl Santa Ana [946]

Diccionario historico-geografico de las poblacione / Bonilla, Marcelina – Tegucigalpa, Mexico. 1945 – 1r – us UF Libraries [059]

Diccionario ibanag-espanol / Bugarin, Jose – Manila: Impr de los Amigos del Pais, 1854 – 1 – us CRL [490]

Diccionario manual de terminos comunes espanol-tagalo / Serrano, Rosalio – 2nd ed. Manila: J. Martinez, 1913. 400p – 1 – us Wisconsin U Libr [040]

Diccionario muy copioso de la lengua espanola y francesa (ael2/15) = Dictionnaire tres ample de la langue francoise et espagnole / Pallet, Jean – 1st ed. Paris 1604 [mf ed 1995] – (= ser Archiv der europaeischen lexikographie: woerterbuecher) – 8mf – 9 – €90.00 – 3-89131-198-2 – (int by brigitte lepinette) – gw Fischer [040]

Diccionario politico / Nunez, Rafael – Bogota, Colombia. 1952 – 1r – us UF Libraries [320]

Diccionario portuguez, e brasiliano : obra necessaria aos ministros do altar; primeira parte – Lisboa: Officina Patriarcal 1795 – (= ser Whsb) – 1mf – 9 – €20.00 – mf#Hu 397 – gw Fischer [440]

Diccionario portuguez-cafre-tetense / Courtois, Victor Joseph – Coimbra, Portugal. 1899 – 1r – us UF Libraries [040]

Diccionario provincial casi-razonado de vozes cuba / Pichardo Y Tapia, Esteban – Habana, Cuba. 1862 – 1r – us UF Libraries [972]

Diccionario tiruray-espanol / Bennasar, P Guillermo – Manila: Tipo-litog de Chofre, [1892-1893]. v1. 1892 – 1r – us CRL [040]

Diccionario trilingue del castellano, bascuence, y latin / Larramendi, Manuel de – San Sebastian: Riesgo y Montero 1745 – (= ser Whsb) – 3mf – 9 – €105.00 – mf#Hu 046 – gw Fischer [410]

Diccionario valenciano-castellano / Ros, Carlos – Valencia: Monfort 1764 – (= ser Whsb) – 5mf – 9 – €60.00 – mf#Hu 037 – gw Fischer [440]

Dicey, Albert Venn *see*
- A leap in the dark or our new constitution
- A treatise on the rules for the selection of the parties to an action

Dichosos en el mal / Sondereguer, Pedro – Buenos Aires, Argentina. 1924 – 1r – us UF Libraries [972]

Dicht- en zedekundige zinnebeelden en bespiegelingen / Broeckhoff, J P – Amsterdam: P J Entrop, 1770 – 6mf – 9 – mf#O-7 – ne IDC [090]

Dicht- sing- und spiel-kunst : so wohl der alten als ins besonder der hebreer / Til, Salomon van – Leipzig: Matthias Groot 1706 [mf ed 19–] – 14mf – 9 – mf#fiche 954 – us Sibley [780]

Dichtende frauen der gegenwart : mit 9 portraits / Klaiber, Theodor – Stuttgart: Strecker & Schroeder, 1907 [mf ed 1993] – 246p (ill) – 1 – (incl ind) – mf#8156 – us Wisconsin U Libr [430]

Die dichter der deutschen *see* Grimmelshausen

Die dichter des alten bundes / Ewald, Heinrich – 2. Aufl. Goettingen: Vandenhoeck & Ruprecht, 1854-1867. Chicago: Dep of Photodup, U of Chicago Lib, 1971 (1r); Evanston: American Theol Lib Assoc, 1984 (1r) – 1 – 0-8370-0102-1 – mf#1984-B267 – us ATLA [470]

Dichter des deutschen barock : weltliche und geistliche lieder des 17.jahrhunderts / Hauswedell, Ernst L [comp] – 2 veraenderte aufl. Hamburg: E Hauswedell, 1946 – 1 – (incl bibl ref) – us Wisconsin U Libr [810]

Der dichter des oberon / Seuffert, Bernhard – Prag: Verlag des deutschen Vereins zur Verbreitung gemeinnuetziger Kenntnisse in Prag 1900 [mf ed 1991] – (= ser Sammlung gemeinnuetziger vortraege 264) – 1r – 1 – (filmed with: die wahre geschichte vom wiederhergestellten kreuz / franz werfel) – mf#2958p – us Wisconsin U Libr [430]

Dichter im dienst : der sozialistische realismus in der deutschen literatur / Balluseck, Lothar von – 2nd rev enl ed. Wiesbaden: Limes Verlag, c1963 [mf ed 1993] – 286p/[32pl] (ill) – 1 – (incl bibl ref and ind) – mf#8265 – us Wisconsin U Libr [430]

Der dichter j m r lenz in livland : eine monographie nebst einer bibliographischen parallele zu m bernay's jungem goethe von 1766-1768... / Falck, P T – Winterthur: J Westfehling 1878 [mf ed 1990] – 1r – 1 – (incl bibl ref. filmed with: vier madchenleben / emma laddey) – mf#2822p – us Wisconsin U Libr [430]

Der dichter siegfried lipiner (1856-1911) / Hartungen, Hartmut von – [Munich: s.n. 1932?] [mf ed 1990] – 1r – 1 – (incl bibl ref. filmed with: friede auf erden! / richard lipinski) – mf#2827p – us Wisconsin U Libr [430]

Dichter un welten / Lieberman, Herman – Berlin, Germany. 1923 – 1r – us UF Libraries [430]

Der dichter vor der geschichte : hoelderlin, novalis / Schneider, Reinhold – 2. aufl. Heidelberg: F H Kerle 1946 [mf ed 1990] – 1r – 1 – (filmed with: friedrich holderlin / friedrich franz von unruh) – mf#2732p – us Wisconsin U Libr [430]

Dichtergruesse : neuere deutsche lyrik / Polko, Elise – Leipzig: C F Amelang, 1869 – 1r – 1 – us Wisconsin U Libr [430]

Dichterische arbeiten / Winkler, Eugen Gottlob – Leipzig-Markkleeberg: Karl Rauch Verlag, 1937 – 1 – us Wisconsin U Libr [430]

Die dichterische entwicklung j.f.w. zacharias / Kirchgeorg, Otto Hermann – Greifswald: J Abel, 1904 – 1 – (incl bibl ref) – us Wisconsin U Libr [430]

Der dichterische essay : die prosaform der englischen romantik / Egner, Fritz – Marburg, 1931 (mf ed 1994) – 1mf – 9 – €24.00 – 3-8267-3039-9 – mf#DHS-AR 3039 – gw Frankfurter [420]

Der dichterische plan des parzivalromans / Schroeder, Walter Johannes – Halle: M Niemeyer, 1953 [mf ed 1993] – viii/76p – 1 – (incl bibl ref) – mf#8448 – us Wisconsin U Libr [430]

Die dichterische selbstdarstellung im roman des jungen deutschland / Greatwood, Edward Albert – Berlin, 1935 (mf ed 1995) – 2mf – 9 – €31.00 – 3-8267-3203-0 – mf#DHS-AR 3203 – gw Frankfurter [430]

Dichterjuristen / Wohlhaupter, Eugen; ed by Seifert, H G – Tuebingen: J C B Mohr, 1953-57 [mf ed 1993] – 3v – 1 – (incl bibl ref) – mf#8144 – us Wisconsin U Libr [430]

Dichtertum und fuehrerschaft : gedenkrede gehalten an der goethefeier des lehrervereins bern-stadt / Schaeffner, Georg – Bern: A Francke, 1932 – 1r – 1 – us Wisconsin U Libr [430]

Dichtkundige bespiegelingen op 57 gepaste in koper gebragte zinnebeelden / Houbraken, A – Amsterdam: L Groenewoud, 1782 – 4mf – 9 – mf#O-3089 – ne IDC [090]

Dichtlievende verlustigingen / Bosch, Bernardus de – Amsteldam: Gerrit Tielenburg, 1742-88 – 4mf – 9 – mf#O-3037 – ne IDC [090]

Dichtung / Becher, Johannes Robert – Berlin: Aufbau-Verlag, 1951, c1949 [mf ed 1989] – 2v – 1 – (incl ind) – mf#6993 – us Wisconsin U Libr [810]

Dichtung der afrikaner / Meinhof, Carl – Berlin, Germany. 1911 – 1r – us UF Libraries [470]

Dichtung des rokoko : nach motiven geordnet / ed by Anger, Alfred – Tuebingen:M Niemeyer, 1958 – 1 – us Wisconsin U Libr [800]

Dichtung des rokoko, nach motiven geordnet *see* Romantic movement in german literature

Die dichtung richard dehmels als ausdruck der zeitseele / Kunze, Kurt – Leipzig: R Voigtlaender, 1914 [mf ed 1989] – (= ser Beitraege zur kultur- und universalgeschichte 26) – xv/120p – 1 – (incl bibl) – mf#7174 – us Wisconsin U Libr [430]

Dichtung, sprache, gesellschaft : akten des 4. internationalen germanisten-kongresses / ed by Internationaler Germanisten-Kongress – Frankfurt/M: Athenaeum Verlag, c1971 – 1 – (german and english. incl bibl ref) – us Wisconsin U Libr [430]

Die dichtung stefan georges / Morwitz, Ernst – Berlin: G Bondi, 1934 – 1r – 1 – us Wisconsin U Libr [430]

Dichtung und arete : untersuchunen zur bedeutung der musischen erziehung bei plato / Harth, Helene – Frankfurt a.M., 1965 – 3mf – 9 – 3-89349-679-3 – gw Frankfurter [180]

Dichtung und bildende kunst im zeitalter des deutschen barock / Mueller, Richard – Frauenfeld; Leipzig: Huber, 1937, c1936 [mf ed 1993] – (= ser Wege zur dichtung) – 133p – 1 – (incl bibl ref) – mf#8175 – us Wisconsin U Libr [430]

Dichtung und dichter der kirche / Schroeder, Rudolf Alexander – Berlin-Steglitz: Eckart Verlag, 1936 – 195p – 1 – us Wisconsin U Libr [240]

Dichtung und dichter der zeit : eine schilderung der deutschen literatur der letzten jahrzehnte / Soergel, Albert – 20. Auflage, 67. bis 71. Tausend. Leipzig: R Voigtlaender, 1928 – 1 – us Wisconsin U Libr [430]

Dichtung und wahrheit / Goethe, Johann Wolfgang von – Offenburg/Mainz: Lehrmittel Verlag. 2v. 1947 (mf ed 1990) – 1r – 1 – (filmed with: campagne in frankreich) – us Wisconsin U Libr [810]

Dichtung und welt im mittelalter / Kuhn, Hugo – Stuttgart: J B Metzler, 1959 [mf ed 1993] – vi/304p – 1 – (incl bibl ref and ind) – mf#8162 – us Wisconsin U Libr [430]

Die dichtung von sturm und drang im zusammenhange der geistesgeschichte : ein gemeinverstaendlicher vortragszyklus / Korff, Hermann August – Leipzig: Quelle & Meyer, 1928 [mf ed 1993] – 98p – 1 – mf#8210 – us Wisconsin U Libr [430]

Dichtungen / Allmers, Hermann – 3. stark verm aufl. Oldenburg: Schulzesche Hof-Buchhandlung und Hof-Buchdruckerei (A Schwartz), [1892?] [mf ed 1996] – 239p – 1 – mf#9580 – us Wisconsin U Libr [810]

Dichtungen / Binding, Rudolf Georg – Bielefeld: Velhagen & Klasing, 1941 [mf ed 1989] – (= ser Velhagen und klasings deutsche lesebogen 158) – 48p – 1 – (incl bibl) – mf#7024 – us Wisconsin U Libr [810]

Dichtungen / Dahn, Felix – Leipzig: Breitkopf & Haertel, 1898 – 1 – us Wisconsin U Libr [810]

Dichtungen / Dahn, Felix – Leipzig: Breitkopf & Haertel, 1898 – 4r – 1 – us Wisconsin U Libr [810]

Dichtungen / Guenderode, Karoline von; ed by Pigenot, Ludwig von – Muenchen: Hugo Bruckmann, 1922 [mf ed 1993] – 287p – 1 – mf#8701 – us Wisconsin U Libr [810]

Dichtungen der droste : eine auswahl / Castelle, Friedrich [comp] – 2. aufl. Moenchengladbach: Volksvereins-Verlag 1923 [mf ed 1993] – 1r – 1 – (filmed with: gedichte / deinhardstein) – mf#8539 – us Wisconsin U Libr [810]

Dichtungen der droste / ed by Castelle, Friedrich – 2. Aufl. Moenchengladbach: Volksvereins-Verlag, 1923 – 1 – us Wisconsin U Libr [810]

Die dichtungen der frau ava / Ava; ed by Maurer, Friedrich – Tuebingen: Niemeyer, 1966 [mf ed 1993] – (= ser Altdeutsche textbibliothek n66) – xiv/68p – 1 – (incl bibl ref) – mf#8193 reel 6 – us Wisconsin U Libr [810]

Dichtungen des 16. jahrhunderts / ed by Weller, Emil Ottokar – Stuttgart: Litterarischer Verein, 1874 (Tuebingen: L F Fues) – us Wisconsin U Libr [810]

Dichtungen des deutschen ordens – Berlin: Weidmann, 1907- [mf ed 1993] – (= ser Deutsche texte des mittelalters 8, 9, 19, 21) – 1 – (incl bibl ref and ind. middle high german poetry) – mf#8623 reel 3-4 – us Wisconsin U Libr [810]

Dichtungen des sechzehnten jahrhunderts / ed by Weller, Emil – Stuttgart: Litterarischer Verein, 1874 (Tuebingen: L F Fues) [mf ed 1993] – (= ser Blvs 119) – 126p – 1 – mf#8470 reel 25 – us Wisconsin U Libr [810]

Dicionario brasileiro de datas historicas / Teixeira De Oliveira, Jose – Rio de Janeiro, Brazil. 1950 – 1r – us UF Libraries [972]

Dicionario de bandeirantes e sertanistas do brasil / Carvalho Franco, Francisco De Assis – Sao Paulo, Brazil. 1954 – 1r – us UF Libraries [972]

Dicionario de cooperativismo / Pinho, Diva Benevides – Sao Paulo, Brazil. 1961 – 1r – us UF Libraries [334]

Dicionario geologico-geomorfologico / Guerra, Antonio Teixeira – Rio de Janeiro, Brazil. 1969 – 1r – us UF Libraries [550]

Dicionario portugues-chisena e chisena-portugues / Alves, Albano – Lisboa? Portugal. 1957 – 1r – us UF Libraries [040]

Dicionario ronga-portugues / Nogueira, Rodrigo De Sa – Lisboa, Portugal. 1960 – 1r – us UF Libraries [040]

Dick and jane's adventures on sable island / Ashley, Barnas Freeman – Chicago: Laird & Lee, 1896? – 4mf – 9 – mf#06156 – cn Canadiana [830]

Dick, Francis *see* Cheerful giver

Dick, Helene *see* Terminologiefelder als kriterium der fachlichkeit in deutschsprachigen fachtexten der geschichte und medizin

Dick, James *see*
- Authority of christ over the individual, the church, and the nation
- Instrumental music in christian worship

Dick, John *see* Qualifications and call of missionaries

Dick, Karl *see* Der schriftstellerische plural bei paulus

Dick, Robert Paine *see* Hebrew poetry

Dick, Thomas *see* Christian's hope

Dick, Vinceslas Eugene *see* Le roi des etudiants

Dick, Vinceslas-Eugene *see* L'enfant mysterieux

Dick, Vinceslas-Eugene [i.e. Caron, Napoleon] *see* Legendes et revenants

Dickason, Clifford *see* Ground-water mining in the united states

Dickason, Elizabeth L *see* Use of the health belief model in determining mammography screening practice in older women

Dickens and his illustrators / Kitton, Frederic George – London 1899 – (= ser 19th c art & architecture) – 6mf – 9 – mf#4.2.831 – uk Chadwyck [740]

Dickens, Charles *see*
- American notes for general circulation
- The annotated proofs
- Dealings with the firm of dombey and son
- The letters of charles dickens
- Original manuscripts and papers
- Original manuscripts of charles dickens
- Our mutual friend

The dickensian – 1905-74 – 216mf – 9 – (official organ of the worldwide dickens fellowship) – mf#C35-22500 – us Primary [420]

Dickens's dictionary of london, 1888 – London, England. 1888 – 1r – us UF Libraries [914]

Dickerhoff, Hans *see* Die entstehung der jobsiade

Dicker's mining record : (mining record of victoria, australia, and public companies gazette) – Melbourne, Australia. 23 Nov 1861-14 Jun 1870. -w – 3r – 1 – uk British Libr Newspaper [622]

Dickers mining record – Melbourne, Australia. 23 nov 1861-1867; 14 jan-14 jun 1870 – 3r – 1 – (aka: mining record of victoria australia and public companies gazette) – uk British Libr Newspaper [622]

Dickerson, Philip *see* Burning bush not consumed

Dickerson, Thomas A *see* Developing an activity conference at the middle school level

Dickey, Herbert Spencer *see* Misadventures of a tropical medico

Dickey, R D *see*
- Copper deficiency of tung in florida
- Flowering, fruiting, yield and growth habits of tung trees
- Grape growing in florida
- Manganese sulfate as a corrective for a chlorosis of certain ornamental plants
- Paperwhite narcissus
- Preliminary report on iron deficiency of tung in florida
- Preliminary report on little-leaf of the peach in florida

Dickey, Samuel *see* The position of greek in the theological education of today

Dickie, George *see* Typical forms and special ends in creation

Dickie, Geraldine W *see* Systematic giving

Dickie, John *see* The philosophy of witchcraft

Dickins, Frederick Victor *see* Fugaku hiyaku-kei

Dickinson Bros *see* Dickinsons' comprehensive pictures of the great exhibition of 1851

Dickinson, Charles *see* Observations on ecclesiastical legislature and church reform

Dickinson County. Kansas. Board of Commissioners *see* Journals

Dickinson, Edward *see* Music in the history of the western church

Dickinson, Elmira Jane *see* Historical sketch of the christian woman's board of missions

Dickinson, Emily *see*
- The complete poems..
- Poems

Dickinson, F H *see* Missale ad usum insignis et praeclarae ecclesiae sarum

Dickinson, Goldsworthy Lowes *see*
- Is immortality desirable?
- Letters from a chinese official
- Religion
- Religion and immortality

Dickinson, H W *see* The water supply of greater london, 1870-1952

Dickinson international law annual *see* Dickinson journal of international law

Dickinson, Jonathan *see* A sermon...september 19, 1722

Dickinson journal of international law – v1-18. 1982-2000 – 5,6,9 – $423.00 set – (v1-3 1982-85 on reel $58. v4-18 1985-2000 on mf $365. title varies: v1-2 n1 1982-83 as dickinson international law annual) – ISSN: 0887-283X – mf#109181 – us Hein [341]

Dickinson law review – Dickinson Law School. v1-24. 1897-1919/20 – (= ser The forum) – 75mf – $112.00 – (v1-12 1897-1912 are titled "the forum") – mf#LLMC 84-446 – us LLMC [340]

Dickinson law review – Dickenson Law School. v1-30. 1897-1925/26 (all publ) – (= ser The forum) – 96mf – 9 – $144.00 – (v1-12 known as "the forum") – mf#LLMC 84-466 – us LLMC [340]

Dickinson law review – v1-104. 1897-2000 – 5,6,9 – $1367.00 set – (v1-89 1897-85 on reel $913. v90-104 1985-2000 on mf $454. title varies: v1-12 1897-1909 as the forum) – ISSN: 0012-2459 – mf#102471 – us Hein [340]

Dickinson, Marguerite *see* Holmes county, ohio, cemetery records

Dickinson, Mary F *see* Seminoles of south florida

Dickinson, Richard William *see* The resurrection of jesus christ historically and logically viewed
Dickinson, Robert Latou *see*
– An american text-book of obstetrics for practioners and students
– Single woman
Dickinsons' comprehensive pictures of the great exhibition of 1851 / Dickinson Bros – London 1854 – (= ser 19th c art & architecture) – 13mf – 9 – mf#4.2.1226 – uk Chadwyck [700]
Dickison and his men / Dickison, Mary Elizabeth – Louisville, KY. 1890 – 1r – us UF Libraries [978]
Dickison, John Jackson *see* Military history of florida
Dickison, Mary Elizabeth *see* Dickison and his men
Dickmann, Fritz *see* Die notwendigkeit des religionsunterrichts in der staatsschule
Dick's monthly advertiser : for linlithgowshire, and the eastern district of stirlingshire – [Scotland] West Lothian, Linlithgow: G H Dick 5 sep, 7 nov 1843 (mthly) [mf ed 2004] – 1r – 1 – uk Newsplan [072]
Dicksee, J R *see* Perspective theoretical and practical
Dickson, A F *see* Plantation sermons, or, plain and familiar discourses for the instruction of the unlearned
Dickson, Alexander *see* All about jesus
Dickson, Andrew F *see* Plantation sermons..
Dickson, Antonia *see* History of the kinetograph, kinetoscope and kinetophonograph
Dickson, David *see*
– Elder's thoughts on union
– End of our being
Dickson, George *see* A history of upper canada college
Dickson, Hugh *see* Reasons of hugh dickson...
Dickson, James *see* Camping in the muskoka region
Dickson, James A R *see*
– Working for jesus
– Working for the children in the home and in the sunday school
Dickson, John *see* History of the presbyterian church of new zealand
Dickson, Richard W *see* John howard and the prison world of europe
Dickson, Robert *see*
– Introduction of the art of printing into scotland
– Who was scotland's first printer?
Dickson, W E *see* Letter to the lord bishop of salisbury
Dickson, William *see* Apt to teach
Dickson, William Kennedy Laurie *see* History of the kinetograph, kinetoscope and kinetophonograph
Dickson, William Purdie *see* St paul's use of the terms flesh and spirit
Dicp – Cincinnati OH 1989-91 – 1,5,9 – (cont: drug intelligence & clinical pharmacy; cont by: annals of pharmacotherapy) – ISSN: 1042-9611 – mf#6492.02 – us UMI ProQuest [615]
Dicta – Denver Bar Association. v1-39. 1923-62 – 172mf – 9 – $258.00 – (title changes: v40 1963: the denver law center journal; v43 1966: the denver law journal; v62 1985: denver university law review; for copyright reasons only the dicta portion of the title can be offered by llmc) – mf#LLMC 84-458 – us LLMC [340]
Dicta *see* Denver university law review
Dicta beati aegidii assisiensis / Giles of Assisi – ed 2. Ad Claras Aquas (Quaracchi) prope Florentiam: Ex typographia Collegii s Bonaventurae, 1939. Chicago: Dep of Photodup, U of Chicago Lib, 1975 (1r); Evanston: American Theol Lib Assoc, 1984 (1r) – (= ser Bibliotheca franciscana ascetica medii aevi) – 1 – 0-8370-0651-1 – (incl bibl ref) – mf#1984-B423 – us ATLA [240]
Dictador y yo / Samayoa Chinchilla, Carlos – Guatemala, 1950 – 1r – us UF Libraries [972]
La dictadura de o'higgins : memoria presentada a la universidad de chile en la sesión solemne celebrada el 11 de diciembre de 1853 / Amunategui, Miguel Luis – 1st ed, Santiago, 1853. Santiago de Chile: Impr Barcelona, 1914 (mf ed 2000) – 1r – 1 – mf#*Z-8520 – us NY Public [972]
El dictamen – Veracruz: [s.n. 1920-47] – 99r – 1 – us CRL [079]
Dictamen de las academia...de barcelona...sobre la frecuencia de las muertes...repentinas y apoplejias – Barcelona, 1748 – 2mf – 9 – sp Cultura [616]
Dictamen fiscal...oficiales generales – 1816 – 9 – sp Bibl Santa Ana [336]
Dictamen sobre la construccion de una linea ferrea...ha tomado la expresada corporacion – 1867 – 9 – sp Bibl Santa Ana [380]
Dictamen sobre la utilidad...de la excavacion del pozo-airon...para evitar terremotos / Vaca de Guzman, O – Granada, 1779 – 2mf – 9 – sp Cultura [946]
Dictamen torremocha – 1894 – 9 – sp Bibl Santa Ana [946]
Dictatum christianum / Arias Montano, Benito – 1575 – 9 – sp Bibl Santa Ana [240]
La dictature de la franc-maconnerie sur la france : documents / Michel, A G – Paris: Editions Spes 1924 – 1mf – 9 – mf#vrl-156 – ne IDC [366]
Dictees graduees et analyses : 1er manuel: deux cents dictees pour cours preparatoire, 1ere et 2eme annees / Theodule, frere – Montreal: les freres du Sacre-Coeur, [1932?] (mf ed 1992) – 1mf – 9 – mf#SEM105P1618 – cn Bibl Nat [440]
Dictees graduees et analyses : 3e manuel: deux cent cinquante dictees pour cours primaire, 5e et 6e annees / Theodule, frere – Montreal: les freres du Sacre-Coeur, [1932?] (mf ed 1992) – 3mf – 9 – mf#SEM105P1612 – cn Bibl Nat [440]
Dictionaire de musique : contenant une explication des termes grecs, latins, italiens, & francois les plus usitez dans la musique... / Brossard, Sebastien de – Paris: Christophe Ballard 1703 [mf ed 19–] – 8mf – 9 – mf#fiche 700 – us Sibley [780]
Dictionaire des sciences medicales (ael3/16) : biographie medicale / Jourdan, A J L – Paris 1820-25 [mf ed 1994] – (= ser Archiv der europaeischen lexikographie: fach-enzyklopaedien) – 7v on 27mf – 9 – €220.00 – 3-89131-184-2 – gw Fischer [056]
Dictionaire francois allemand et allemand francois (ael2/5) / Hulsius, Levinus – 1st ed. Nuernberg 1596 – (= ser Archiv der europaeischen lexikographie: woerterbuecher) – 21mf – 9 – €110.00 – 3-89131-065-X – (filmed with: dictionarium teutsch frantzoesisch und frantzoesisch teutsch [3rd ed 1607]; int by laurent bray) – gw Fischer [040]
Dictionaire harmonique : ou guide sur pour la vraie modulaison = Dictionarium harmonicum, of zeekere wegwyzer tot de waare modulatie / Geminiani, Francesco – Amsterdam: Aux depens de l'auteur 1756 [mf ed 19–] – 3mf – 9 – mf#fiche 473 – us Sibley [780]
Dictionaire historique et critique (ael1/45) / Bayle, Pierre – Rotterdam 1697, Rotterdam 1720, Amsterdam 1734, Amsterdam 1750-1756, english ed: London 1734-41 [mf ed 1998] – (= ser Archiv der europaeischen lexikographie, abt 1: enzyklopaedien) – 287mf – 9 – €1950.00 set – 3-89131-343-8 – (vols available individually; (ael1/45.1): 1st ed, rotterdam 1697 [2v on 30mf] isbn: 3-89131-330-6 €280; (ael1/45.2): 3rd ed, rotterdam 1720 [4v on 57mf] isbn: 3-89131-331-4 €340; (ael1/45.3): 5th ed, amsterdam 1734 [5v on 55mf] isbn: 3-89131-332-2 €500; (ael1/45.4): nouveau dictionnaire historique et critique ed by jaques george de chaufepie, amsterdam 1750-1756 [4v on 71mf] isbn: 3-89131-333-0 €370; english ed: (ael 1/45.5): a general dictionary, historical and critical...london 1734-1741 [10v on 74mf] isbn: 3-89131-334-9 €720) – gw Fischer [059]
Dictionaire [sic] de l'ancien droit du canada : ou compilation des edits, declarations royaux, et arrets du conseil d'etat des roix [sic] de france concernant le canada / McCarthy, Justin – Quebec: J Neilson, 1809 [mf ed 1974] – 1r – 5 – mf#SEM16P181 – cn Bibl Nat [340]
Dictionaire theologique, historique, poetique, cosmographique et chronologique (ael1/46) / Juigne-Broissiniere, D de – Paris 1650, Lyon 1682 [mf ed 1998] – (= ser Archiv der europaeischen lexikographie, abt 1: enzyklopaedien) – 30mf – 9 – €240.00 set – 3-89131-344-6 – (vols available individually: (ael1/46.1): 4th ed, paris 1650 [15mf] isbn: 3-89131-335-7; (ael1/46.2): nouveau dictionaire theologique, historique, poetique, cosmographique et chronologique lyon 1682 [15mf] isbn: 3-89131-336-5) – gw Fischer [052]
Dictionaire universel (ael2/17) : contenant generalement tous les mots francois tant vieux que modernes, et les termes de toutes les sciences et des arts / Furetiere, Antoine – Den Haag/Rotterdam 1690-1727 [mf ed 1997] – (= ser Archiv der europaeischen lexikographie: woerterbuecher) – 243mf – 9 – €1180.00 – 3-89131-226-1 – (gesamtedition der ausgaben 1690 [23mf] €140; 1691 [22mf] €140; 1694 [16mf] €100; 1701 [34mf] €200; 1702 [24mf] €150; 1708 [36mf] €210; den haag 1725 [42mf] €240; 1727 [46mf] €250; int by dorothea behnke; vols available individually) – gw Fischer [050]
Dictionare de la provence et du comte-venaissin : dedie a monseigneur le marechal prince de beauvau. par une societe de gens de lettres / Achard, Claude Francois – Marseille: Mossy 1785 – (= ser Whsb) – 2pt on 5mf – 9 – €115.00 – (t1: francois-provencal; t2: provencal-francois) – mf#Hu 066 – gw Fischer [440]
A dictionarie in spanish and english (ael2/6) : [...] hereunto is annexed an ample english dictionaire alphabetically... / Percyvall, Richard & Minsheu, John – London 1599 [mf ed 1993] – (= ser Archiv der europaeischen lexikographie: woerterbuecher) – 8mf – 9 – €100.00 – 3-89131-066-8 – (int by gabriele stein) – gw Fischer [040]
Dictionaries and vocabularies / Roman Catholic Mission, New Hebrides – n.d. – 1r – 1 – mf#pmb60 – at Pacific Mss [241]
Dictionariolum puerorum tribus linguis latina, gallica et germanica conscriptum... / Stephanus, R & Fries, J – Tiguri, Froschouer, 1548 – 7mf – 9 – mf#PBU-487 – ne IDC [240]
Dictionarium frantzoesicsh-teutsch (ael2/1) / Kramer, Matthias – Nuernberg 1712 [mf ed 1992] – (= ser Archiv der europaeischen lexikographie: woerterbuecher) – 5v on 30mf – 9 – €270.00 – 3-89131-057-9 – (filmed with: teutsch-frantzoesisches woerterregister ueber kramers dictionnaire [1715]; int by laurent bray) – gw Fischer [040]
Dictionarium latinogermanicum... / Fries, J – Tiguri, Christoph Froschouer, 1556 – 15mf – 9 – mf#PBU-509 – ne IDC [240]
Dictionarium medicum... (ael3/21) : vel expositiones ocum medicinalium ad verum excerptae es hippocrate, areteo etc / Estienne, Henri – Paris 1564 [mf ed 1996] – (= ser Archiv der europaeischen lexikographie: fach-enzyklopaedien) – 7mf – 9 – €80.00 – 3-89131-217-2 – (int by michael stolberg) – gw Fischer [610]
Dictionarium novum latino-armenium : ex praecipuis armeniae linguae scriptoribus concinnatum in quo, praeter adjunctos singularum vocum sensus multiplices, multa etiam theologica, physica, moralia... / Villotte, Jacques – Romae: Typ Sac Congreg de propag fide 1714 – (= ser Whsb) – 9mf – 9 – €85.00 – mf#Hu 226 – gw Fischer [410]
Dictionarium syriaco-latinum / Brun, J – Beryti Phoeniciorum: Typographia PP Soc Jesu, 1911 – 2mf – 9 – 0-8370-9765-7 – mf#1986-3765 – us ATLA [040]
Dictionarium teutsch-italiaenisch und italiaenisch-teutsch (ael2/7) / Hulsius, Levinus – Frankfurt 1605 [mf ed 1992] – (= ser Archiv der europaeischen lexikographie: woerterbuecher) – 6mf – 9 – €50.00 – 3-89131-067-6 – gw Fischer [040]
Dictionarivm latino epiroticvm : vna cum nonnullis vsitatioribus loquendi formulis / Bianchi, Francesco – Romae: Typ Sac Congr de propag fide 1635 – (= ser Whsb) – 3mf – 9 – €40.00 – mf#Hu 174 – gw Fischer [450]
Dictionarivm latino lvsitanicvm... / Calepinus, A – Amacvsa, 1595 – 10mf – 9 – mf#H-8426 – ne IDC [956]
Dictionarivm trilingve, in qvo scilicet latinis vocabvlis...respondent graeca and hebraica... / Muenster, S – Basileae, 1562 – 5mf – 9 – mf#H-8409 – ne IDC [956]
Dictionarum sive thesauri linguae japonicae / Collado, Fr. Diego – 1632 – 9 – sp Bibl Santa Ana [480]
A dictionary carnataca and english / Reeve, William – Madras: Printed at Govt Gazette Press, 1832 – 1 – us CRL [040]
Dictionary english, german and french (ael2/4) / Ludwig, Christian – Leipzig 1706 [mf ed 1992] – (= ser Archiv der europaeischen lexikographie: woerterbuecher) – 23mf – 9 – €120.00 – 3-89131-064-1 – (filmed with: teutsch-englisches lexicon [1716]; int by franz josef hausmann) – gw Fischer [040]
Dictionary, grammar and phrase-book of fanagalo (kitchen kafir) / Bold, J D – s.l, South Africa. 1958 – 1r – us UF Libraries [470]
A dictionary, hindustani and english / Shakespear, John – London: printed...by Cox & Baylis, 1817 – (= ser 19th c books on linguistics) – 9mf – 9 – mf#2.1.31 – uk Chadwyck [040]
A dictionary in bengalee and english / Tarachanda Chakravarti – Calcutta: printed at the Baptist Mission Press, 1827 – (= ser 19th c books on linguistics) – 3mf – 9 – mf#2.1.10 – uk Chadwyck [040]
A dictionary of all religions and religious denominations : jewish, heathen, mahometan and christian, ancient and modern: with an appendix... / Adams, Hannah – 4th corr enl ed. New York: J Eastburn, 1817 [mf ed 1993] – 1mf – 9 – 0-524-08036-4 – (1st publ boston 1784 under title: an alphabetical compendium of the various sects which have appeared in the world) – mf#1991-0252 – us ATLA [052]
Dictionary of american hymnology : first-line index / Hymn Society of America; ed by Ellinwood, Leonard & Lockwood, Elizabeth – New York, 1984 – 179r – 5 – $5,559.00 silver; $3,950.00 diazo – (produced by the hymn society of america. more than one million first-line citations covering 192,000 hymns publ in n and s america from 1640-1978, in all languages using roman alphabet. first lines of hymns, refrains, titles, orig. first lines of translated hymns, authors, translators, etc. explanatory printed user's guide includes denomination codes, location of hymnals, list of contributors and series of brief essays on hymns with confused authorship) – us Univ Music [780]
Dictionary of anonymous and pseudonymous english literature / Halkett, Samuel & Laing, John – 7v. 1926-34 – 1,9 – us AMS Press [420]
Dictionary of architecture – 1852-92 [mf ed Chadwyck-Healey] – (= ser Rare 19th century american art journals) – 30mf – 9 – uk Chadwyck [720]
The dictionary of architecture / Architectural Publication Society – London [1851-92] – (= ser 19th c art & architecture) – 58mf – 9 – mf#4.1.175 – uk Chadwyck [720]
A dictionary of artists who have exhibited works in the principal london exhibitions from 1760 to 1893 / Graves, Algernon – new ed. London 1895 – (= ser 19th c art & architecture) – 4mf – 9 – mf#4.1.345 – uk Chadwyck [057]
A dictionary of assyrian chemistry and geology / Thompson, R C – Oxford, 1936 – 4mf – 9 – mf#NE-439 – ne IDC [540]
A dictionary of christ and the gospels / ed by Hastings, James & Selbie, John Alexander – New York: Scribner, 1906-08 [mf ed 1992] – 2v on 18mf – 9 – 0-524-02778-1 – mf#1987-6472 – us ATLA [052]
A dictionary of christian antiquities : being a continuation of the "dictionary of the bible" / ed by Smith, William, Sir & Cheetham, Samuel – London: J Murray, 1875-80 [mf ed 1990] – 20mf – 9 – 0-7905-8229-5 – (incl bibl ref) – mf#1988-6129 – us ATLA [052]
A dictionary of christian biography and literature : to the end of the 6th century a d, with an account of the principal sects and heresies / ed by Wace, Henry & Piercy, William Coleman – London: John Murray, 1911 [mf ed 1991] – 3mf – 9 – 0-524-00613-X – (incl bibl ref) – mf#1990-0113 – us ATLA [052]
Dictionary of doctrinal and historical theology / ed by Blunt, John Henry – 2nd ed. London: Rivingtons, 1872 – 8mf – 9 – 0-7905-4844-5 – mf#1988-0844 – us ATLA [052]
The dictionary of education and instruction : a reference book and manual on the theory and practice of teaching / ed by Kiddle, Henry – 3rd ed. New York: E. Steiger & Co., 1882 – 332p – 1 – us Wisconsin U Libr [370]
A dictionary of english and welsh surnames / Bardsley, C W – London, 1901 – 1r – 1 – mf#96411 – uk Microform Academic [420]
A dictionary of english church history / ed by Ollard, Sidney Leslie & Crosse, Gordon – London: A R Mowbray; Milwaukee: The Young Churchman, 1912 [mf ed 1991] – xvi/672p (ill) – 1 – mf#7635 – us Wisconsin U Libr [242]
Dictionary of foreign phrases and classical quotations / Jones, Henry Percy – Edinburgh, Scotland. 1939 – 1r – us UF Libraries [054]
Dictionary of grammar / Hennesy, James A – New York, NY. 1917 – 1r – us UF Libraries [054]
Dictionary of greek and roman biography and mythology / Smith, William George – London, England. v1-3. 1880 – 3r – us UF Libraries [930]
Dictionary of greek and roman geography / ed by Smith, William – London: John Murray, 1878 – 24mf – 9 – 0-524-03884-8 – (incl bibl ref) – mf#1987-6497 – us ATLA [059]
A dictionary of hindu architecture : treating of sanskrit architectural terms with illustrative quotations from silpasastras, general literature and archaeological records / Acharya, Prasanna Kumar – London: Oxford University Press, 1927 – (= ser Samp: indian books) – us CRL [720]
A dictionary of hymnology : setting forth the origin and history of christian hymns of all ages and nations / ed by Julian, John – rev ed. London: J Murray, 1915 [mf ed 1993] – 17mf – 9 – 0-524-08327-4 – (incl bibl ref. with new suppl) – mf#1993-1022 – us ATLA [052]
Dictionary of indian biography / Buckland, C E – London: Swan Snnenchein & Co, 1906 – (= ser Samp: indian books) – us CRL [920]
A dictionary of london / Harben, H – 1918 – 1r – 1 – mf#457 – uk Microform Academic [941]
Dictionary of luvale / Horton, A E – El Monte, CA. 1953 – 1r – us UF Libraries [960]
A dictionary of malayalam phrases and idioms = Malayala sailinghantu / Pillai, T Ramalingam – Trivandrum: R S Pillai. v1. 1930 – 1 – us CRL [490]
A dictionary of miniaturists, illuminators, calligraphers, and copyists / Bradley, John William [comp] – London 1887-89 – (= ser 19th c art & architecture) – 13mf – 9 – mf#4.1.392 – uk Chadwyck [740]
A dictionary of miracles : imitative, realistic, and dogmatic / Brewer, Ebenezer Cobham – Philadelphia: JB Lippincott, 1884 [mf ed 1990] – 1mf – 9 – 0-7905-5381-3 – (incl bibl ref) – mf#1988-1381 – us ATLA [052]
The dictionary of needlework / Caulfeild, Sophia Frances Anne & Saward, Blanche C – London 1882 – (= ser 19th c art & architecture) – 6mf – 9 – mf#4.1.401 – uk Chadwyck [740]

A dictionary of non-classical mythology / Edwardes, Marian & Spence, Lewis – London: JM Dent [1912?] [mf ed 1993] – (= ser Everyman's library) – 1mf – 9 – 0-524-07606-5 – mf#1991-0132 – us ATLA [390]

Dictionary of philosophical terms : chiefly from the japanese / Richard, Timothy & MacGillivray, Donald – Shanghai: Christian Literature Society for China, 1913 [mf ed 1995] – (= ser Yale coll) – 71p – 1 – 0-524-10022-5 – mf#1995-1022 – us ATLA [051]

A dictionary of practical medicine (ael3/14) : comprising general pathology, the nature and treatment of diseases, morbid structures, and the disorders especially incidental to climates, to the sex, and to the different epochs of life / Copland, James – London 1858 [mf ed 1995] – (= ser Archiv der europaeischen lexikographie: fach-enzyklopaedien) – 3v on 36mf – 9 – €410.00 – 3-89131-182-6 – (int by michael stolberg) – gw Fischer [616]

Dictionary of sects, heresies, ecclesiastical parties, and schools of religious thought / ed by Blunt, John Henry – Philadelphia: JB Lippincott, 1874 – 2mf – 9 – 0-524-01645-3 – (incl bibl ref) – mf#1990-0466 – us ATLA [052]

A dictionary of terms in art / Fairholt, Frederick William – London [1854] – (= ser 19th c art & architecture) – 6mf – 9 – mf#4.1.460 – uk Chadwyck [057]

A dictionary of terms in art...with 500 engravings on wood / Fairholt, F W – [London, 1854] – 6mf – 9 – mf#O-248 – ne IDC [720]

Dictionary of the amharic language / Isenberg, C W – London, 1841. 2pts – 5mf – 9 – mf#NE-20251 – ne IDC [470]

Dictionary of the amharic language / Isenberg, Karl Wilhelm – London: printed for the Church Missionary Society, 1841 – (= ser 19th c books on linguistics) – 5mf – 9 – (in 2pt: amharic/english, and english/amharic) – mf#2.1.26 – uk Chadwyck [470]

A dictionary of the bengalee language, vol 1 : bengalee and english / Carey, William & Marshman, John Clark – Serampore. 2v. 1827-28 – (= ser 19th c books on linguistics) – 11mf – 9 – (v1 abridged, v2 [dated 1828] comp by john c marshman) – mf#2.1.18 – uk Chadwyck [490]

A dictionary of the bengalee language, vol 1 : in which the words are traced to their origin and their various meanings given / Carey, William – Serampore: printed at the Mission-Press, 1815 – (= ser 19th c books on linguistics) – 10mf – 9 – (no more publ) – mf#2.1.27 – uk Chadwyck [490]

A dictionary of the bhotanta, or boutan language / Schroeter, Friedrich Christian Gotthelf; ed by Marshman, John Clark & Carey, William – Serampore, 1826 – (= ser 19th c books on linguistics) – 8mf – 9 – mf#2.1.58 – uk Chadwyck [480]

A dictionary of the bible : dealing with its language, literature, and contents, including the biblical theology / ed by Hastings, James & Selbie, John Alexander – New York: Scribner, 1902-04 [mf ed 1990] – 45mf – 9 – 0-8370-1906-0 – mf#1987-6293 – us ATLA [220]

A dictionary of the burman language : with explanations in english / Judson, Adoniram; ed by Wade, Jonathan – Calcutta: printed at the Baptist Mission Press, 1826 – (= ser 19th c books on linguistics) – 5mf – 9 – mf#2.1.42 – uk Chadwyck [480]

Dictionary of the burman language, with explanations in english : compiled from the manuscripts of a judson and of the other missionaries in burmah / Judson, Adoniram; ed by Wade, Jonathan – Calcutta: Baptist Mission Pr 1826 – (= ser Whsb) – 5mf – 9 – €60.00 – mf#Hu 287 – gw Fischer [480]

Dictionary of the chichewa language / Scott, David Clement Ruffelle – London, England. 1970 – 1r – us UF Libraries [470]

A dictionary of the church : containing an exposition of terms, phrases and subjects connected with the external order, sacraments, worship and usages of the protestant episcopal church / Staunton, William – 2nd rev corr enl ed. Philadelphia: Herman Hooker, 1840 [mf ed 1992] – (= ser Anglican/episcopal coll) – 2mf – 9 – 0-524-03024-3 – (incl bibl ref) – mf#1990-4546 – us ATLA [242]

Dictionary of the english and benga languages – 1879 – 1 – $50.00 – us Presbyterian [490]

A dictionary of the english and singhalese, and singhalese and english languages / Clough, Benjamin – Colombo: printed...at the Wesleyan Mission Press. 2v. 1821-30 – (= ser 19th c books on linguistics) – 22mf – 9 – mf#2.1.39 – uk Chadwyck [040]

A dictionary of the english language containing all english words and phrases now in use : with their meanings, synonyms, and tamil equivalents / Pillay, A Mootootamby – Jaffna: Navalar Press, 1907 – 1 – us CRL [490]

Dictionary of the galla language / Tutschek, C – Munich, 1844 – 5mf – 9 – mf#NE-20283 – ne IDC [470]

A dictionary of the gathic language of the zend avesta : being vol 3 of a study of the five zarathushtrian gathas / Mills, Lawrence Heyworth – Leipsic: FA Brockhaus, 1913 [mf ed 1993] – 2mf – 9 – 0-524-05768-0 – mf#1991-0011 – us ATLA [490]

Dictionary of the hausa language. : with appendices of hausa literature / Schoen, James Frederick – London: Church Missionary House, 1876 – 1 – us CRL [470]

A dictionary of the holy bible : for general use in the study of the scriptures, with engravings, maps, and tables – New York: American Tract Society, c1859 [mf ed 1991] – 2mf – 9 – 0-8370-1973-7 – mf#1987-6360 – us ATLA [220]

A dictionary of the kaffir language : including the xosa and zulu dialects. part 1. kaffir-english / Davis, William Jafferd – London: Wesleyan Mission House, 1872. Chicago: Dep of Photodup, U of Chicago Lib, 1973 (1r); Evanston: American Theol Lib Assoc, 1984 (1r) – 1 – 0-8370-0007-6 – mf#1984-B372 – us ATLA [040]

A dictionary of the language of bugotu, santa isobel island, solomon islands / Ivens, W G – 1940 – (= ser Royal asiatic society. j g forlong fund) – 1r – 1 – mf#667 – uk Microform Academic [490]

A dictionary of the malagasy language / Freeman, Joseph John – Antananarivo: printed... by R Kitching, 1835 – (= ser 19th c books on linguistics) – 8mf – 9 – (in 2pts. pt1: english & malagasy. pt2 has the titlepage: ny dikisionary malagasy, mizara boa: english sy malagasy, ary malagasy sy english) – mf#2.1.12 – uk Chadwyck [470]

A dictionary of the malagasy language in two parts : pt 1: english and malagasy / Freeman, J J – Tananarive: R Kitching: Press of the London Missionary Society, 1835 – 1 – us CRL [490]

Dictionary of the maratha language : in two parts / Kennedy, Vans – Bombay: Courier Press 1827 – (= ser Whsb) – 4mf – 9 – €50.00 – (pt1: containing maratha & english; pt2: containing english & maratha) – mf#Hu 265 – gw Fischer [490]

A dictionary of the maratha language, in two parts : 1. part containing maratha and english, 2. part containing english and maratha / Kennedy, Vans, 1784-1846 – Bombay: printed at the Courier Press, 1824 – (= ser 19th c books on linguistics) – 3mf – 9 – mf#2.1.33 – uk Chadwyck [490]

A dictionary of the persian and arabic languages / Barretto, Joseph – Calcutta: printed by S Greenaway. 2v. 1804-06 – (= ser 19th c books on linguistics) – 18mf – 9 – mf#2.1.4 – uk Chadwyck [470]

Dictionary of the suahili language / Krapf, Johann Ludwig – London, England. 1882 – 1r – us UF Libraries [470]

A dictionary of the targumim / Jastrow, Morris – 1903 – 9 – $57.00 – us IRC [270]

A dictionary of the targumim, the talmud babli and yerushalmi, and the midrashic literature : with an index of scriptural quotations / Jastrow, Marcus – London: Luzac; New York: G P Putnam, 1903 [mf ed 1986] – 2v on 5mf – 9 – 0-8370-7394-4 – (incl ind) – mf#1986-1394 – us ATLA [470]

A dictionary of the teloogoo language : commonly termed the gentoo, peculiar to the hindoos of the north eastern provinces of the indian peninsular / Campbell, Alexander Duncan – Madras: printed at the College Press, 1821 – (= ser 19th c books on linguistics) – 7mf – 9 – mf#2.1.32 – uk Chadwyck [490]

Dictionary of the teloogoo language, commonly termed the gentoo : peculiar to the hindoos of the north eastern provinces of the indian peninsula / Campbell, Alexander Duncan – Madras: College Press 1821 – (= ser Whsb) – 7mf – 9 – €75.00 – mf#Hu 279 – gw Fischer [490]

A dictionary of the tiv language / Abraham, Roy Clive – London: Crown Agents for the Colonies, 1940 – 1 – us CRL [490]

Dictionary of the welsh language explained in english : with numerous illustrations, from the literary remains and from the living speech of the cymry / Pughe, William Owen – London: Williams 1803 – (= ser Whsb) – 9mf – 9 – €85.00 – (missing: v2; v1: welsh & english dictionary...to which is prefixed a welsh grammar) – mf#Hu 102 – gw Fischer [490]

Dictionary of the yao language / Sanderson, George Meredith – Zomba, Malawi. 1954 – 1r – us UF Libraries [470]

Dictionary phrase-book and grammar of fanagalo / Bold, J D – Johannesburg, South Africa. 1964 – 1r – us UF Libraries [470]

Dictionary, sanscrit and english : translated, amended and enlarged, from an original compilation prepared by learned natives for the college of fort william / Wilson, Horace Hayman – Calcutta: Hindoostancee press 1819 – (= ser Whsb) – 12mf – 9 – €100.00 – mf#Hu 240 – gw Fischer [040]

Dictionary, universal military, 1779 – 1r – 1 – mf#B26332 – us Ohio Hist [355]

Dictionnaire amarigna-francais : suivi d'un vocabulaire francais-amarigna / Baeteman, J – Dire-Dadona, 1929 – 10mf – 9 – mf#NE-20317 – ne IDC [040]

Dictionnaire biographique de musiciens et vocabulaire de termes musicaux / Soeurs de Sainte-Anne – Lachine: Mont Sainte-Anne, 1922 [mf ed 1975] – 1r – 5 – mf#SEM16P207 – cn Bibl Nat [780]

Dictionnaire cambodgien / Phnom-Penh. Institut Bouddhique – 5th ed 1968 – 2v on 38mf – 9 – $40.00 – us IASWR [490]

Dictionnaire celto-breton : ou breton-francais / Gonidec, Jean-Francois-Marie-Maurice-Agatha le – Angouleme: Tremau 1821 – (= ser Whsb) – 5mf – 9 – €60.00 – mf#Hu 111 – gw Fischer [040]

Dictionnaire chinois, francais et latin : punoie d'apres l'ordre de sa majeste l'empereur et roi napoleon le grand / Guignes, Chretien-Louis-Joseph de – Paris: Impr imperiale 1813 – (= ser Whsb) – 3mf – 9 – €105.00 – mf#Hu 291 – gw Fischer [040]

Dictionnaire classique d'histoire naturelle / ed by Bory de Saint-Vincent, J B G M – Paris 1822-31 – 17v [tl-2/669] on 324mf – 9 – mf#8330/1 – ne IDC [500]

Dictionnaire classique et universel de geographie moderne : contenant la description succincte des pays et principaux lieux du globe... / Langlois, Hyacinthe E – Paris 1826-30 [mf ed Hildesheim 1995-98] – 4v on 21mf – 9 – €210.00 – 3-487-29980-1 – gw Olms [059]

Dictionnaire critique et raisonne des etiquettes de la cour, des usages du monde, des amusemens, des modes, des moeurs, etc, des francois : depuis la mort de louis 13 jusqu'a nos jours... / Genlis, Stephanie F de – Paris 1818 [mf ed Hildesheim 1995-98] – 2v on 6mf – 9 – €120.00 – 3-487-25891-9 – gw Olms [944]

Dictionnaire d'archeologie chretienne et de liturgie – Paris, 1907-1939. v1-14 – 368mf – 8 – mf#0-186c – ne IDC [956]

Le dictionnaire d'aujourd'hui toujours a jour : langue, histoire, biographie, geographie, sciences, arts, etc – [Ed canadienne]. Tours: Maison Mame, [entre 1939 et 1947] (mf ed 2000) – 1r – 1 – mf#SEM35P485 – cn Bibl Nat [054]

Dictionnaire de la bible : contenant tous les noms de personnes, de lieux, de plantes, d'animaux mentionnes dans les saintes ecritures... / Vigouroux, Fulcran et al – Paris: Letouzey et Ane, 1895-1912 – 14mf – 9 – 0-8370-1753-X – mf#1987-6149 – us ATLA [052]

Dictionnaire de la bible : supplement / ed by Pirot et al – Paris. v1-6. 1928-1960 – €172.00 – (v6: mysteres-passion) – ne Slangenburg [220]

Dictionnaire de la bible / ed by Virgouroux et al – Paris. v1-5. 1895-1912 – €235.00 – ne Slangenburg [220]

Dictionnaire de la geographie physique et politique de la france : faisant connaitre avec beaucoup de detail les villes, bourgs et villages... / Girault de Saint-Fargeau, Eusebe – Paris 1826 [mf ed Hildesheim 1995-98] – 5mf – 9 – €100.00 – 3-487-29788-4 – gw Olms [019]

Dictionnaire de la langue bretonne : ou l'on voit son antiquite, son affinite avec les anciennes langues, l'explication de plusieurs passages de l'ecriture sainte, et des auteurs profanesm avec l'etymologie de plusieurs mots des autres langues / Pelletier, Louis le – Paris: Delaguette 1752 – (= ser Whsb) – 6mf – 9 – €70.00 – mf#Hu 107 – gw Fischer [440]

Dictionnaire de la langue francaise : lexique historique et geographique, apercu de grammaire / Azed – nouv ed. Montreal: Librairie Beauchemin, 1961 [mf ed 2000] – 1r – 1 – mf#SEM35P354 – cn Bibl Nat [054]

Dictionnaire de la noblesse de la france / Chenaye-Desbois, F A de la – 3rd ed. v1-19. 1863-77 – 1 – $360.00 – mf#0314 – us Brook [929]

Dictionnaire de l'academie francaise – Paris, France. v1-2. 1878 – 1r – us UF Libraries [440]

Dictionnaire de l'ancienne langue francaise / Godefroy, Fredric – v1-10. 1891-1902 – 1 – $216.00 – mf#0242 – us Brook [440]

Dictionnaire de mekeo / Desnoes, Gustave – 1933 – 2r – 1 – mf#pmb17 – at Pacific Mss [059]

Dictionnaire de musique / Rousseau, Jean-Jacques – Paris: Chez la veuve Duchesne 1768 [mf ed 19–] – 11mf – 9 – mf#fiche 658 – us Sibley [780]

Dictionnaire de noms hieroglyphiques en ordre genealogique et alphabetique / Lieblein, J D C – Christiania, Leipzig, 1871, 1892. 2v – 13mf – 9 – mf#NE-331 – ne IDC [470]

Dictionnaire de nos fautes contre la langue francaise / Rinfret, Raoul – 4e mille. Montreal: C O Beauchemin et fils, [1897?] (mf ed 1991) – 4mf – 9 – (in french and english) – mf#SEM105P1346 – cn Bibl Nat [440]

Dictionnaire de paleographie : de cryptographie, de dactylologie, d'hieroglyphie, de stenographie et de telegraphie / Mas Latrie, Louis de – Paris: J-P Migne, 1854 – 2mf – 9 – 0-8370-7759-1 – (incl bibl ref) – mf#1986-1759 – us ATLA [400]

Dictionnaire des artistes, dont nous avons des estampes... / Heinecken, K H V – Leipsig, 1778-1790. 4v – 38mf – 9 – mf#O-995 – ne IDC [700]

Dictionnaire des devises ecclesiastiques / Tausin, H – Paris: Emile Lechevalier, 1907 – 4mf – 9 – mf#O-1953 – ne IDC [090]

Dictionnaire des noms anciens et modernes des villes et arrondissements...dans l'empire chinois... / Biot, E C – Paris, 1842 – 4mf – 9 – mf#HT-611 – ne IDC [915]

Dictionnaire des noms geographiques contenus dans les textes hieroglyphiques / Gauthier, H – Caire, 1925. v1-7 – 23mf – 9 – mf#NE-461 – ne IDC [930]

Dictionnaire des ouvrages anonymes / Barbier, Antoine A – 4v. 1872-79 – 1,9 – us AMS Press [010]

Dictionnaire des ouvrages anonymes et pseudonymes publies par des religieux de la compagnie de jesus : depuis sa fondation jusqu' a nos jours / Sommervogel, Carlos – Paris: Librairie de la Societe bibliographique, 1884 – 2mf – 9 – 0-524-03191-6 – mf#1990-4640 – us ATLA [240]

Dictionnaire des parlementaires francais, comprenant tous les membres des assemblees francaises et tous les ministres francais depuis le 1. mai 1789 jusqu'au 1. mai 1889... / Robert, Adolphe et al – Paris: Bourloton, 1891 – 1 – us CRL [944]

Dictionnaire des sciences medicales (ael3/5) : par une societe de medecins et de chirurgiens: mm adelon, alard, alibert [et al] – Paris 1812-22 [mf ed 1993] – (= ser Archiv der europaeischen lexikographie: fach-enzyklopaedien) – 60v on 299mf – 9 – €1490.00 – 3-89131-143-5 – gw Fischer [610]

Dictionnaire des sciences naturelles dans lequel on traite methodiquement des differents etres de la nature / ed by Cuvier, F – ed 2. Paris 1816-30 – 60v [tl-2/1293] on 621mf – 9 – mf#5493/2 – ne IDC [590]

Dictionnaire des verbes irreguliers et defectifs de la langue francaise / Baillairge, Frederic-Alexandre – Joliette, Quebec?: s.n, 1887 – 1mf – 9 – mf#00079 – cn Canadiana [440]

Dictionnaire d'histoire et de geographie au japon *see* Historical and geographical dictionary of japan

Dictionnaire d'homonymes, rimes, etc / Baillairge, Charles P Florent – s.l: s.n, 1888? – 1mf – 9 – mf#02662 – cn Canadiana [440]

Dictionnaire d'hygiene publique et de salubrite (ael3/24) / Tardieu, Ambroise – Paris 1852-54 [mf ed 1995] – (= ser Archiv der europaeischen lexikographie: fach-enzyklopaedien) – 3v on 21mf – 9 – €180.00 – 3-89131-220-2 – (int by michael stolberg) – gw Fischer [614]

Dictionnaire diplomatique : ou etymologies des termes des bas siecles / Montignot, Henri – Nancy: De l'impr de C S Lamort, 1787 [mf ed 1977] – 1r – 5 – mf#SEM16P299 – cn Bibl Nat [327]

Dictionnaire djaghatai-turc / Veliaminof-Zernof, V de – Spb, 1869 – 8mf – 8 – mf#U-363 – ne IDC [470]

Dictionnaire encyclopedique des sciences medicales (ael3/15) / ed by Dechambre, A & Raige-Delorme, J – Paris 1864-89 [mf ed 1994] – (= ser Archiv der europaeischen lexikographie: fach-enzyklopaedien) – 100v in 5sects on 450mf – 9 – €3890.00 – 3-89131-183-4 – (int by michael stolberg) – gw Fischer [610]

Dictionnaire etymologique de la langue malgache / Dama-ntsoha – [Tananarive]: Ny antsiva, 1953 – 1 – us CRL [490]

Dictionnaire francais-armenien-turc / Awgerian, Y [Aucher, P] – Venice, 1840 – 8mf – 9 – mf#AR-1700 – ne IDC [040]

Dictionnaire francais-grec... / Alexandre, Charles – Paris, France. 1888 – 1r – us UF Libraries [040]

Dictionnaire francais-montagnais : avec un vocabulaire montagnais-anglais, une courte liste de noms geographiques, et une grammaire montagnaise / Lemoine, Georges – Boston: W B Cabot & P Cabot, 1901 [mf ed 1996] – 4mf – 9 – 0-665-79804-0 – mf#79804 – cn Canadiana [440]

Dictionnaire francais-volof / Guy-Grand, V J – 3e rev aug ed. Saint Joseph de Negasobil: Impr de la Mission, 1890 – 1 – us CRL [040]

Dictionnaire francais-wolof et francais-bambara : suivi du dictionnaire wolof-francais / Dard, Jean – Paris: Impr royale 1825 – (= ser Whsb) – 1mf – 9 – €20.00 – mf#Hu 360 – gw Fischer [040]

Dictionnaire francais-wolof et wolof-francais – Nouv ed. Dakar: Impr de la Mission, 1855 – 1 – us CRL [490]

Dictionnaire francois contenant les mots et les choses (ael2/16) / Richelet, Cesar-Pierre – [mf ed 1997] – (= ser Archiv der europaeischen lexikographie: woerterbuecher) – 110mf – 9 – €660.00 – 3-89131-225-3 – (gesamtedition der ausgaben genf 1680/1679: dictionnaire francois...genf: jean herman widerhold 2pts in 1 1680/1679 [12mf] €90; genf 1693: dictionaire francois...genf: ritter, miege 2v 1693 [12mf] €100; amsterdam 1709: nouveau dictionnaire francois...amsterdam: jean elzevir 2v 1709 [15mf] €120; rouen 1719: nouveau dictionnaire francois..rouen: vaultier, machuel, le boucher, benard 2v 1719 [17mf] €120; lyon 1728: dictionnaire de la langue francoise, ancienne et moderne, lyon: duplain, bruyset, estienne 3v 1728 [29mf] €200; basel 1735: dictionnaire de la langue francoise, ancienne et moderne, basel: jean brandmuller 3v 1735 [23mf] €170; int by laurent bray; vols available separately) – gw Fischer [440]

Dictionnaire francois-celtique ou francois-breton : necessaire a tous ceux qui veulent apprendre a traduire le francois en celtique... / Rostrenen, Gregoire de – Rennes: Vatar 1732 – (= ser Whsb) – 11mf – 9 – €95.00 – mf#Hu 106 – gw Fischer [040]

Dictionnaire galibi, presente sous deux formes : 1: commencant par le mot francois; 2: par le mot galibi; precede d'un essai de grammaire / LaSalle de L'etang, Simon Philibert de – Paris: Bauche 1763 – (= ser Whsb) – 2mf – 9 – €30.00 – mf#Hu 405 – gw Fischer [440]

Dictionnaire genealogique des familles canadiennes depuis la fondation de la colonie jusqu'a nos jours / Tanguay, Cyprien – Province de Quebec (Montreal): E Senecal. 7v. 1871-1890 [mf ed 1982] – 2r – 1 – mf#SEM35P175 – cn Bibl Nat [929]

Dictionnaire genealogique des familles de charlesbourg : depuis la fondation de la paroisse jusqu'a nos jours / Gosselin, David – Quebec: [s.n.], 1906 – 7mf – 9 – 0-665-72086-6 – mf#72086 – cn Canadiana [929]

Dictionnaire geographique de l'ancienne egypte : contenant par ordre alphabetique la nomenclature comparee des noms propres geographiques qui se rencontrent sur les monuments et dans les papyrus... / Brugsch, Heinrich Karl – Leipzig: JC Hinrichs, 1879 – 3mf – 9 – 0-524-08040-2 – mf#1991-0256 – us ATLA [040]

Dictionnaire geographique de l'empire ottoman / Mostras, C – Spb, 1873 – 3mf – 9 – mf#AR-1616 – ne IDC [915]

Dictionnaire geographique et descriptif de l'italie : servant d'itineraire et de guide aux etrangers qui voyagent dans ce pays... / Barzilay, Jacques – Paris 1823 [mf ed Hildesheim 1995-98] – (= ser Fbc) – 4mf – 9 – €120.00 – 3-487-29290-4 – gw Olms [059]

Dictionnaire geographique universel : contenant la description de tous les lieux du globe interessans sous le rapport de la geographie physique et politique, de l'histoire, de la statistique, du commerce, de l'industrie, etc – Paris 1823-33 [mf ed Hildesheim 1995-98] – 20v on 59mf – 9 – €590.00 – 3-487-29995-X – gw Olms [059]

Dictionnaire geographique-portatif : ou description des royaumes, provinces, villes, patriarchats, eveches, duches, comtes, marquisats...et autres lieux considerables des quatre partie du monde / Eachard, Laurence – Paris 1759 [mf ed Hildesheim 1995-98] – 5mf – 9 – €100.00 – 3-487-29986-0 – gw Olms [059]

Dictionnaire historico-artistique du portugal... / Raczynski, A – Paris, 1847 – 3mf – 9 – mf#O-1059 – ne IDC [700]

Dictionnaire historique de la medecine ancienne et moderne (ael3/17) / ed by Dezeimeris, Jean E et al – Paris 1828-39 [mf ed 2006] – (= ser Archiv der europaeischen lexikographie: fach-enzyklopaedien) – 7pt in 4v on 32mf – 9 – €280.00 – 3-89131-185-0 – gw Fischer [610]

Dictionnaire historique de la medecine ancienne et moderne (ael3/19) / Eloy, F J Nicholas – 1778 [mf ed 1996] – (= ser Archiv der europaeischen lexikographie: fach-enzyklopaedien) – 4v on 29mf – 9 – €170.00 – 3-89131-191-5 – gw Fischer [610]

Dictionnaire historique des canadiens et des metis francais de l'ouest / Morice, Adrien Gabriel – 2e augm ed. Quebec: Garneau, 1912 [mf ed 1974] – 1r – 5 – mf#SEM16P267 – cn Bibl Nat [053]

Dictionnaire historique et geographique des paroisses, missions et municipalites de la province de quebec / Magnan, Hormidas – Arthabaska: Impr d'Arthabaska Inc, 1925 [mf ed 1985] – 8mf – 9 – mf#SEM105P386 – cn Bibl Nat [059]

Dictionnaire historique et geographique du canada – Montreal: Beauchemin & Valois, 1885 – 2mf – 9 – 0-665-02661-7 – mf#02661 – cn Canadiana [971]

Dictionnaire historique, politique et geographique de la suisse / Tscharner, Vincenz B von – Geneve 1788 [mf ed Hildesheim 1995-98] – 3v on 6mf – 9 – €120.00 – 3-487-29382-X – gw Olms [019]

Dictionnaire iconologique : ou introduction a la connoissance des peintures, sculptures, estampes, medailles... / Lacombe de Prezel, H – Paris: Hardouin, 1779. – 8mf – 9 – mf#O-74 – ne IDC [090]

Dictionnaire iconologique : ou introduction a la connoissance des peintures, sculptures, estampes, medailles... / [Lacombe de Prezel, H] – Paris: Th de Hansy, 1756 – 4mf – 9 – mf#O-75 – ne IDC [090]

Dictionnaire iconologique / P[rezel, L] de – Gotha, 1758 – 4mf – 9 – mf#O-1619 – ne IDC [700]

Dictionnaire iconologique / P[rezel, L] de – Paris, 1756 – 3mf – 9 – mf#O-1258 – ne IDC [700]

Dictionnaire iconologique / Prezel, L de – Paris, 1779 – 6mf – 9 – mf#O-1260 – ne IDC [700]

Dictionnaire infernal; repertoire universel des etres, des personnages, des livres, des faits et des choses qui tiennent aux esprits. / Collin de Plancy, J A S – 6eme ed. augm. de 800 articles nouv., illus. Paris: H. Plon, 1863. 723p – 1 – us Wisconsin U Libr [150]

Dictionnaire kikingo-francais, avec une etude phonetique decrivant les dialectes les plus importants de la langue dite kikongo / Laman, Karl Eduard – [Bruxelles: G van Campenhout, 1936] – 1 – us CRL [490]

Dictionnaire languedocien-francois : contenant un recueil des principales fautes que commettent, ans la diction & dans la prononciation, les habitans des provinces meridionales / Boissier de Sauvages, Pierre A – nouv ed. Nismes: Gaude 1785 – (= ser Whsb) – 10mf – 9 – €90.00 – mf#Hu 073 – gw Fischer [440]

Dictionnaire ngbandi / Lekens, Benjamin – Tervuren, Belgium. 1952 – 1r – us UF Libraries [960]

Dictionnaire portatif et abrege des loix et regles du parlement provincial du bas-canada : depuis son etablissement...jusques et compris l'an de notre seigneur, 1805 / Perrault, Joseph Francois – Quebec: John Neilson, 1806 [mf ed 1971] – 1r – 5 – mf#SEM16P76 – cn Bibl Nat [323]

Dictionnaire raisonne de l'architecture francaise du 11e au 16e siecle / Viollet-Le-Duc, E E – Paris, 1854-1870. 10v – 69mf – 9 – mf#O-454 – ne IDC [720]

Dictionnaire tamoul-francais = Tamil porancu akarati / Mousset, Louis Marie – 3rd. ed. Pondichery. 1938-1942. 2 vols – 1 – us CRL [490]

Dictionnaire turk-oriental / Pavet de Courteille, M – Paris, 1870 – 10mf – 8 – mf#U-362 – ne IDC [470]

Dictionnaire universel : contenant generalement tous les mots francois, tant vieux que modernes, et les termes des sciences et des arts / Furetiere, Antoine – La Haye: Chez Pierre Husson etc, 1727 [mf ed 1983] – 2r – 1 – mf#SEM35P187 – cn Bibl Nat [440]

Dictionnaire universel abrege de geographie ancienne comparee : offrant la description des contrees, villes, fleuves, montagnes, monumens, et generalement de tous les lieux celebres de l'antiquite... / Dufau, Pierre A – Paris 1820 [mf ed Hildesheim 1995-98] – 2v on 6mf – 9 – €120.00 – 3-487-29906-2 – gw Olms [059]

Dictionnaire universel de geographie moderne : ou description physique, politique et historique de toutes les contrees et de tous les lieux remarquables de la terre / Perrot, Aristide M – Paris 1834 [mf ed Hildesheim 1995-98] – 2v on 8mf – 9 – €160.00 – 3-487-29981-X – gw Olms [059]

Dictionnaire universel de la france ancienne et moderne, et de la nouvelle france : traitant de tout ce qui y a rapport... / Saugrain – Paris: Chez Saugrain...3v. 1726 [mf ed 1984] – 3v on 1mf – 9 – mf#47574 – cn Canadiana [030]

Dictionnaire universel de la france ancienne et moderne, et de la nouvelle-france : traitant de tout ce qui y a rapport: soit geographie, etymologie, topographie, histoire... – Paris: Chez Saugrain etc, 1726 [mf ed 1982] – 2r – 1 – mf#SEM35P184 – cn Bibl Nat [944]

Dictionnaire universel de medecine (ael3/3.2) / James, Robert – Paris 1746-48 [mf ed 1993] – (= ser Archiv der europaeischen lexikographie: fach-enzyklopaedien) – 6v on 57mf – 9 – €270.00 – 3-89131-153-2 – (trans by denis diderot) – gw Fischer [610]

Dictionnaire universel des geographies physique, historique et politique, du monde ancien, du moyen age et des temps modernes, comparees : indispensable aux voyageurs, negocians...utile a toutes les classes de la societe, et necessaire pour l'intelligence des auteurs anciens... / Masselin, J G – Paris 1827 [mf ed Hildesheim 1995-98] – 2v on 10mf – 9 – €100.00 – 3-487-29966-6 – gw Olms [059]

Dictionnaire universel des synonymes de la langue francaise / Guizot, Francois – Paris, France. 1864 – 1r – us UF Libraries [440]

Dictionnaire universel francois et latin (ael1/48) : "dictionnaire de trevoux" – nouv ed. Paris 1732 [mf ed 1998] – (= ser Archiv der europaeischen lexikographie, abt 1: enzyklopaedien) – 5v on 100mf – 9 – €510.00 – 3-89131-339-X – gw Fischer [040]

Dictionnaire volof-francais / Kobes, Alois – Nouv. ed. Dakar. 1923 – 1 – us CRL [490]

Dictionnaire volof-francais / Kobes, Mgr – Nouv ed. Dakar, Senegal: Mission Catholique, 1923 – us CRL [040]

Les dictons du peuple et les paroles de jesus-christ *see* Die redensarten des volkes und was der herr jesus darauf antwortet

Dictorum fere omnium : quae de sacramentali verborum coenae interpret... / Pezelius, C – Bremae, 1592 – 3mf – 9 – mf#PBA-289 – ne IDC [240]

Dictys *see* Livius, books 31-40/dictys...

Dicurso...mal de urina sea el que padece diego enriquez leon... / Sanchez de Oropesa, F – Sevilla, 1594 – 4mf – 9 – sp Cultura [610]

Did christ claim to be son of god? / Hiller, H Croft – Manchester: H Croft Hiller, 1907 [mf ed 1985] – 1mf – 9 – 0-8370-3588-0 – mf#1985-1588 – us ATLA [230]

Did jesus christ teach socialism? / Veritas – Manchester, England. 18– – 1r – us UF Libraries [240]

Did jesus live 100 b.c.? : an enquiry into the talmud jesus stories, the toldoth jeschu, and some curious statements of epiphanius / Mead, G R S – London: Theosophical Publ Soc, 1903 – 2mf – 9 – 0-7905-2179-2 – (incl bibl ref) – mf#1987-2179 – us ATLA [240]

Did jesus really live? : a reply to the christ myth / Rossington, Herbert J – London: Philip Green, 1911 – 1mf – 9 – 0-524-06682-5 – mf#1992-0935 – us ATLA [240]

Did moses write the pentateuch after all? / Spencer, Frank Ernest – London: Elliot Stock, 1892 [mf ed 1989] – 1mf – 9 – 0-7905-2081-8 – mf#1987-2081 – us ATLA [221]

Did the anglican church reform herself in the sixteenth century? – London, England. 184-? – 1r – us UF Libraries [241]

Did the first church of salem originally have a confession of faith distinct from their covenant? / Felt, Joseph Barlow – Boston: Edward L. Balch, 1856 – 1mf – 9 – 0-7905-6059-3 – mf#1988-2059 – us ATLA [240]

Did the florida legislature of 1891 elect a senato... / Fleming, Francis P – Tallahassee, FL. 1891 – 1r – us UF Libraries [978]

Did they dip? : or, an examination into the act of baptism as practiced by the english and american baptists before the year 1641 / Christian, John Tyler – 2nd ed. Louisville, KY: Baptist Book Concern, c1896 [mf ed 1993] – (= ser Baptist coll) – 3mf – 9 – 0-524-07400-3 – mf#1991-3060 – us ATLA [242]

Die didache : mit kritischem apparat / ed by Lietzmann, Hans – 2. aufl. Bonn: A Marcus & E Weber 1907 [mf ed 1992] – (= ser Kleine texte fuer theologische vorlesungen und uebungen 6) – 1mf – 9 – 0-524-04674-3 – mf#1990-1301 – us ATLA [240]

Die didache des judentums und der urchristenheit / Seeberg, Alfred – Leipzig: A Deichert, 1908 – 1mf – 9 – 0-524-04853-3 – mf#1990-1345 – us ATLA [240]

Didactic books and prophetical writings / Gigot, Francis Ernest – New York: Benziger Bros, 1906 – 2mf – 9 – 0-524-05979-9 – mf#1992-0716 – us ATLA [220]

Didactica magna *see* The great didactic of john amos comenius

Die didaktische dimension der elektronischen medien fuer den fremdsprachenunterricht / Rudelt, Ulrike – 2000 – (= ser Leipziger arbeiten zur fachsprachenforschung) – 1mf – 3-8267-2680-4 – mf#DHS 2680 – gw Frankfurter [370]

Didascaliae apostolorum canonum ecclesiasticorum traditionis apostolicae : versiones latinae / Tidner, E – Berlin, 1963 – (= ser Tugal 5-75) – 4mf – 9 – €11.00 – ne Slangenburg [240]

Didascalion (smrl10) : de studio legendi / Hugh of Saint-Victor – Washington DC, 1939 – €11.00 – (a critical text by c h buttimer) – ne Slangenburg [240]

Didcot herald – Didcot, England 4 apr 1974- [mf 1986-] – 1 – (variant ed of: abingdon herald; cont: didcot advertiser [6 jan 1961-28 mar 1974]) – uk British Libr Newspaper [072]

Didcot post – England. Jul 1933-Sep 1935.-w. 1men reels – 1 – uk British Libr Newspaper [072]

Diddell, Mary W *see* Fort george island

Dide, Auguste *see*
– Heretiques et revolutionnaires
– J-J rousseau

Diderot, Denis *see* Encyclopedie

Diderot, M *see*
– Encyclopedie

Diderot's early philosophical works / ed by Jourdain, Margaret – Chicago: Open Court 1916 [mf ed 1991] – (= ser The open court series of classics of science and philosophy 4) – 1mf – 9 – 0-7905-7811-5 – (incl bibl ref; trans by ed) – mf#1989-1036 – us ATLA [100]

Didier, Jean *see* Rencontre de jean anouilh

Dido in der deutschen dichtung / Semrau, Eberhard – Berlin: W de Gruyter & Co, 1930 – 2r – 1 – (incl bibl ref) – us Wisconsin U Libr [430]

Didon / Pompignan, Jean-Jacques Lefranc – Paris, France. 1801 – 1r – us UF Libraries [440]

Didon, H *see* Jesus christ

Didon, Henri *see*
– Belief in the divinity of jesus christ
– Science without god

Didron, Adolphe Napoleon *see*
– Christian iconography

Didsbury and withington observer – [NW England] Manchester 21 may 1914 [mf ed 2003] – 1r – 1 – uk Newsplan [072]

Didsbury pioneer – Alberta, CN. jan 1903-dec 1903 – 1r – 1 – cn Commonwealth Imaging [071]

Didyme l'aveugle / Bardy, Gustave – Paris: G. Beauchense, 1910 – (= ser Etudes De Theologie Historique) – 1mf – 9 – 0-7905-5751-7 – (incl bibl ref) – mf#1988-1751 – us ATLA [240]

Didymus : der blinde von alexandria / Leipoldt, Johannes – Leipzig, 1905 – (= ser Tugal 2-29/3) – 3mf – 9 – €7.00 – ne Slangenburg [240]

Didymus der blinde von alexandria / Leipoldt, Johannes – Leipzig: J C Hinrichs, 1905 – 1mf – 9 – 0-7905-4041-X – (incl bibl ref) – mf#1988-0041 – us ATLA [240]

Die arbeit 1919 : organ der zionistischen volkssozialistischen partei hapoel-hazair – Berlin DE, 1919 15 jan-1924 sep, 1928 jul – 1r – 1 – gw Misc Inst [939]

Die arbeit 1924 : zeitschrift fuer gewerkschaftspolitik und wirtschaftskunde – Berlin DE, 1924 1 jul-1932 – 3r – 1 – mf#2249 – gw Mikropress [331]

Die casting engineer – Rosemont IL 1957-96 – 1,5,9 – ISSN: 0012-253X – mf#5193 – us UMI ProQuest [621]

Die fackel 1909 – Berlin DE, 1909, 1913, 1914 jan-jul, 1916 feb-nov, 1917 jan-nov – 1 – gw Misc Inst [074]

Die fackel 1931 – Berlin DE, 1931 4 sep-1933 8 feb – 1 – (title varies: 1932: das kampfsignal) – gw Misc Inst [074]

Die kirche 1945 : evangelische wochenzeitung – Berlin DE, 1945 9 dec-1948 5 dec [gaps] – 8r – 1 – uk British Libr Newspaper [240]

Die kirche 1956 – Berlin DE, 1956-1986 28 sep – 5r – 1 – gw Misc Inst [240]

Die neckarquelle 1880 – Villingen-Schwenningen DE, 1980-82 – 21r – 1 – (bezirksausgabe von suedwest-presse, ulm; title varies: 15 jul 1940: schwenninger tagblatt; 1 mar 1943: ns-volkszeitung; 7 feb 1947: schwaebisches tagblatt; 22 oct 1949: die neckarquelle; filmed by other misc inst: 1997- [ca 9r/yr]) – gw Misc Inst [074]

Die presse 1848 – Wien (A), 1848 3 jul-30 nov, 1856-59, 1864 1 feb-1868 sep, 1875-89 [single iss] – 15mf=30df – 9 – (filmed by mikropress: 1860-62 [8r] order#4937; filmed by misc inst: 1849 2 jan-30 jun; with suppl: an der schoenen blauen donau 1886-90 [gaps, mpf], 1892 [mpf]) – gw Mikrofilm; gw Mikropress; gw Misc Inst [074]

Die presse 1946 – Wien (A), 1953 1 sep-1969 28 feb, 1969 13 sep-1972 2 jun, 1972 1 jul-1984 18 jul, 1984 1 aug-1998 17 jun [gaps] – 225r – 1 – (filmed by misc inst: 1974 31 dec-1978 [18r], 1998 18 jun-2003 30 sep [77r]. incl suppl: das schaufenster 1987 30 apr-1990 18 jan [gaps]) – gw Mikrofilm; gw Misc Inst [074]

Die republik 1918 – Berlin DE, 1919 jan-23 jun – 1r – 1 – (filmed by other misc inst: 1918 3 dec-1919 25 apr, 1919 4 jun-24 jun [1r]) – gw Misc Inst [320]

Die tageszeitung (taz) 1978 – Berlin, Frankfurt/Main DE, 1987 2 jan-24 nov; 2002 – 11r [teils ausg frankfurt & ausg west] – 1 – (filmed by misc inst: 1979 30 apr-1980 23 oct, 1998 31 aug-1999 14 jan, 2001 30 mar-2002 2 jan; 1978 22 sep- [ca 3r/yr later 9r/yr]) – gw Mikrofilm; gw Misc Inst [074]

Die tageszeitung (taz) 1990 – Berlin DE, 1990 6 feb-1991 23 dec – 14r – 1 – (ausg berlin-ost/ddr/berliner ausg) – gw Misc Inst [074]

DIE WACHT

Die wacht 1948 – Duesseldorf DE, 1948 1 mar-1955 – 2r – 1 – gw Misc Inst [074]
Die welt 1924 – Berlin DE, 1924 n35-1926 n5 – 1 – gw Misc Inst [074]
Die welt 1951 : ausgabe berlin – Berlin DE, 1996- – 11r/yr – 1 – (filmed by misc inst: 1951 2 apr-15 may & 2 jul-15 aug, 1955 nov-dez [3r]) – gw Mikrofilm; gw Misc Inst [074]
Die zeit 1896 – Berlin DE, 1896 20 sep-1897 30 sep – 2r – 1 – (filmed by misc inst: 29 apr-30 sep) – gw Mikrofilm; gw Misc Inst [074]
Die zeit 1902 – Berlin DE, 1902 2 oct-1903 24 sep – 1r – 1 – (1923 2 okt-1925 30 jun [4r]) – mf#2984 – gw Mikropress [074]
Die zeit 1915 – Wien (A), 1915 jul-1919 6 aug [gaps] – 24r – 1 – uk British Libr Newspaper [074]
Die zeit 1921 – Berlin DE, 1922 apr-jun, 1925 mar – 2r – 1 – gw Misc Inst [074]
Die zukunft 1867 – Berlin DE, 1869 2 apr-30 jun, 1 oct-31 dec, 1870 1 apr-30 jun – 2r – 1 – (filmed by other misc inst: 1871 1 oct-1872 [1r]. title varies: 1 oct 1871: demokratische zeitung) – gw Misc Inst [320]
Die zukunft 1877 : socialistische revue – Berlin DE, 1877 oct-1878 nov – 1 – mf#2252 – gw Mikropress [074]
Die zukunft 1892 – Berlin DE, 1898 – 1 – gw Misc Inst [074]
Die zukunft 1949 : sozialistische monatsschrift fuer politik, wissenschaft und kultur – Wien (A), 1949-51 – 1r – 1 – mf#6832 – gw Mikropress [321]
Dieback of grapevines in south africa / Ferreira, Jan Hendrik Strauss – U of the Western Cape 1987 [mf ed S.l: s.n. 1987] – 2mf [ill] – 9 – (incl bibl) – sa Misc Inst [634]
Dieback or exanthema of citrus trees / Floyd, B F – Gainesville, FL. 1917 – 1r – us UF Libraries [634]
Diebold, Bernhard *see*
– Anarchie im drama
– Das reich ohne mitte
Diebow, Paul *see* Die paedagogik schleiermachers im lichte seiner und unserer zeit
Diecasting and metal moulding – Hatfield, England 1974-76 – 1,5,9 – ISSN: 0012-2548 – mf#10751 – us UMI ProQuest [621]
Dieciocho de julio – n.p. 1938. Fiche W 836. (Blodgett Collection of Spanish Civil War Pamphlets) – 9 – us Harvard [946]
Diecisiete anos / Fernandez, David – Habana, Cuba. 1959 – 1r – us UF Libraries [972]
Diecisiete meses de alcaldia / Martinez Montero, Emilio – 1898 – 9 – sp Bibl Santa Ana [830]
Dieck, Herman *see* The most complete and authentic history of the life and public services of general u s grant, "the napoleon of america"
Dieckhoff, A W *see* Das wort gottes
Dieckhoff, August Wilhelm *see*
– Der ablassstreit
– Justin, augustin, bernhard und luther
– Luthers lehre in ihrer ersten gestalt
– Die waldenser im mittelalter
– Zur lehre von der bekehrung und von der praedestination
Dieckmann, August *see* Die christliche lehre von der gnade
Dieckmann, R *see*
– Israelitische chronologie
– Judaea und die nachbarschaft im jahrhundert vor und nach der geburt christi
– Zeitordnung und zeitbestimmungen in den evangelien
Died : at the residence of thomas william birchall...on wed, the 9th mar 1859, isabella elizabeth gamble... – [s.l: s.n. 1859?] [mf ed 1983] – 1mf – 9 – 0-665-40581-2 – mf#40581 – cn Canadiana [090]
Died in manvers, on friday, feb 24, 1888, elizabeth fallis, wife of james fallis : aged 49 years, 10 mos, 13 days – S.l: s.n, 1888? – 1mf – 9 – mf#01433 – cn Canadiana [390]
Died in newmarket on the 19th of may, 1890, robert prest : aged 62 years and 19 days... – [Newmarket ON?: s.n. 1890?] [mf ed 1983] – 1mf – 9 – 0-665-38699-0 – mf#38699 – cn Canadiana [090]
Diederich, Benno *see* Prinzessin ursula
Diederich, Franz *see*
– Die haemmer droehnen
– Jungfreudig volk
Diederich von dem werder : ein beitrag zur deutschen litteraturgeschichte des siebzehnten jahrhunderts / Witkowski, Georg – Leipzig: Veit, 1887 – 1r – 1 – (incl bibl ref) – us Wisconsin U Libr [430]
Diederichs, Ernst *see* Meister eckharts reden der unterscheidung
Diederichs, Eugen *see* Die tat (mme6)
Diederichs, Joern *see* Espanol – castellano
Diefenbach, Lorenz *see* Ueber die jetzigen romanischen schriftsprachen, die spanische, portugiesische, rheoromanische (in der schweiz), franzoesische, italiaenische und dakoromanische (in mehren laendern des oestlichen europa's)
Dieffenbach, E *see* Naturwissenschaftliche reisen nach den inseln des gruenen vorgebirges, suedamerika, dem feuerlaende, den falkland-inseln...
Dieffenbach, Ernst *see* New zealand
Diego antonio feijo / Sousa, Octavio Trrquinioi De – Rio de Janeiro, Brazil. 1942 – 1r – us UF Libraries [972]
Diego de Madrid *see*
– Vida...de...san pedro de alcantara...sacada a la luz juan de la calzada...tomo 4...
– Vida...san pedro de alcantara
Diego, Eliseo *see* Nombrar las cosas
Diego garcia de paredes. hercules y sanson de espana / Munoz de San Pedro, Miguel – Madrid: Espasa-Calpe, S.A., 1946 – 1 – sp Bibl Santa Ana [946]
Diego garcia de paredes, hercules y santon de espana. madrid, 1946 / Munoz de San Pedro, Miguel – Madrid: Razon y Fe, 1948 – 1 – sp Bibl Santa Ana [946]
Diego, Jose De *see*
– Cantos de pitirre
– Jovillos (coplas de estudiante)
Diego Padro, Jose Isaac De *see*
– Minotauro se devora a si mismo
– Ultima lampara de los dioses
Diego, Sandalio *see*
– En el 4th centenario del doctor arias montano. la version metricolatina del salterio hebraico
– La version metrica del salterio de benito arias montano
Diego velazquez and his times / Justi, Karl – London 1889 – (= ser 19th c art & architecture) – 6mf – 9 – mf#4.1.266 – uk Chadwyck [750]
Diego vicente tejera / Perez Cabrera, Jose Manuel – Habana, Cuba. 1948 – 1r – us UF Libraries [972]
Diegues Junior, Manuel *see* Bangue nas alagoas
Dieguez, Juan *see* Corpus poeticum de la obra de juan dieguez
Diehl, Adolf *see*
– Dionysius dreytweins esslingische chronik
Diehl, C *see*
– Justinien et la civilisation byzantine au 6e siecle
– Le monde oriental de 395 a 1081
Diehl, Charles *see* Byzantine portraits
Diehl, Ernst *see*
– Altlateinische inschriften
– Euripides medea
– Inscriptiones latinae christianae veteres
– Pompeianische wandinschriften und verwandtes
– Res gestae divi augusti
– Svpplementum sophocleum
– Die vitae vergilianae und ihre antiken quellen
– Vulgaerlateinische inschriften
Diehl, Ernst [comp] *see*
– Lateinische christliche inschriften
– Pompeianische wandinschriften und verwandtes
– Supplementum lyricum
Diehl, George *see* History of the lutheran church of frederick, md
Diehl, Otto *see* Stefan george und das deutschtum
Diehl, Robert W *see* Diehl-vardon golf manual
Diehl, Wilhelm *see* Landgraf philipp von hessen / m butzers bedeutung fuer das kirchliche leben in hessen
Diehl-vardon golf manual / Diehl, Robert W – St Paul, MN. 1927 – 1r – us UF Libraries [790]
Diekamp, Franz *see* Die origenistischen streitigkeiten
Diekmann, Rudolf *see* Zacharias werners dramen
Diel, Heinrich *see* De enuntiatis finalibus apud graecorum rerum scriptores posterioris eatatis
Diela naroda – Petrograd: P K P s-r, 1918. n1-6. feb 12-feb 19 1918 – (library's copy imperfect: n3 has several articles cut out. filmed with: dielo naroda, dielo narodov, dielo, dielo narodnoe, diela narodnyia) – us CRL [077]
Diela narodnyia – Petrograd: Petrogradskii komitet Partii sotsialistov-revoliutsionerov. n1. feb 21 1918 – (library's copy imperfect: issue has several articles cut out. filmed with: dielo naroda, dielo narodnoe, dielo, dielo narodov, dielo, diela naroda) – us CRL [077]
Dielli – Boston, MA. Feb 15 1908-1977; Jan 16 1984-Dec 1991 – 31r – 1 – (albanian language) – us L of C Photodup [071]
Dielo – Petrograd: P K P s-r, 1918. n1. jan 23 1918 – (filmed with: dielo naroda, dielo narodnoe, dielo narodov, diela naroda, diela narodnyia) – us CRL [077]
Dielo – v. 1-21, no. 1 3. Oct 1866-Apr Mar 1888. wanting v. 21, no. 1 3, Jan Mar 1888 – 1 – us L of C Photodup [460]
Dielo naroda – Petrograd: redaktor-izdatel S P Postnikov, 1917-1918 (Tip "Viestnika Vremennago pravitelstva"). [n25-216. apr 15 1917-nov 22 1917]; [n230-251. dec 12 1917-jan 26 1918] – us CRL [077]
Dielo naroda – Petrograd: T-vo s-r v litsie D F Rakova, [1918-]. [n8-50. mar 30 1918-jun 22 1918] – us CRL [077]
Dielo narodnoe – Petrograd: TSK Partii sotsialistov-revoliutsionerov, 1918. [n1-2. jan 14-16 1918] – (filmed with: dielo naroda, dielo narodov, dielo, diela naroda, diela narodnyia) – us CRL [077]
Dielo narodov petrograd: tsk p s-r, 1918. [n1-3. jan 17-jan 20 1918] – (filmed with: dielo naroda, dielo narodnoe, dielo, diela naroda, diela narodnyia) – us CRL [077]
Diels, Hermann *see*
– Poetarum philosophorum fragmenta
– Sibyllinische blaetter
Diemaking, diecutting and converting – Philadelphia PA 1973 – 1 – ISSN: 0012-2556 – mf#8578 – us UMI ProQuest [621]
Diemel, Raymond Anthony van *see* Industriele behuising aan die diamantvelde van griekwaland-wes met spesifieke verwysing na die oop kampongs as werkershuisvesting, 1868-1880
Dienemann, Max *see* Judentum und christentum
Dieng, M Samba *see* L'epopee d'el hadj omar
Der dienst der frau in den ersten jahrhunderten der christlichen kirche / Zscharnack, Leopold – Goettingen: Vandenhoeck & Ruprecht, 1902 – 1mf – 9 – 0-7905-6279-0 – (incl bibl ref) – mf#1988-2279 – us ATLA [240]
Der dienst der frau in der christlichen kirche : geschichtlicher ueberblick mit einer sammlung von urkunden / Goltz, Eduard, Freiherr von der – 2. verm. Aufl. Potsdam: Stiftungsverlag, 1914 – 2mf – 9 – 0-7905-5885-8 – (incl bibl ref) – mf#1988-1885 – us ATLA [240]
Der dienst der frauen in der kirche / Wichern, Johann Hinrich – 3. Aufl. Hamburg: Agentur des Rauhen Hauses, 1880 – 1mf – 9 – 0-524-00805-1 – mf#1990-0237 – us ATLA [240]
Dienst van den mijnbouw bulletin of the bureau of mines and the geological survey in indonesia – Bandoeng, 1947. v1(1) – 1mf – 9 – mf#SE-1463 – ne IDC [959]
Dienstagischer / freytagischer nordischer mercurius *see* Nordischer mercurius
Dienstags=blatt – Oshkosh WI. 1905 nov 14-1908, 1909-11, 1912-14, 1915-1917 nov 13 – 4r – 1 – mf#1097581 – us WHS [071]
Das dienstboten-buch oder beispiele des guten – Augsburg DE, 1832, 1835, 1838 – 1r – 1 – gw Misc Inst [640]
Dienststellen zur vorbereitung des westdeutschen verteidigungsbeitrages 1950-1955 (bestand bw 9) bd 40 / ed by Krueger, Dieter – 1992 – 2pt cxxii/762p – €20.50 – ISBN-13: 978-3-89192-033-6 – gw Bundesarchiv [943]
Dien-xa tap chi – Saigon. n1-63. 4 fevr 1928-8 juin 1929 – 1 – fr ACRPP [073]
Diep, Brigitte et al *see* With naree in cambodia
Diep vu chong hong quan nhat : tieu thuyet gian diep, phong tac / Khach Giang Ho – ed: In lan 1. Saigon: Ngua Hong 1974 [mf ed 1992] – on pt of 1r – 1 – mf#11052 r379 n9 – us Cornell [959]
Diep vu duong day hoa luc : tieu-thuyet gian -diep, phong-tac / Khach Giang Ho – Saigon: Di-Tuong 1974 [mf ed 1992] – on pt of 1r – 1 – mf#11052 r379 n8 – us Cornell [959]
Diepenbrock, Irmgard *see* Quantitative elastizitaetsmessung der haut und ihre auswertung
Diepenhorst, Pieter Arie *see* Calvijn en de economie
Dieppe en 1826 [dix-huit cent vingt-six] : ou lettres du vicomte de...a milord... / Feret, Pierre J – Dieppe 1826 [mf ed Hildesheim 1995-98] – 2mf – 9 – €60.00 – 3-487-29714-0 – gw Olms [914]
Diere stories : soos deur hotnots vertel / Wielligh, Gideon Retief von – Pretoria: Uitgewer J L van Schaik 1917 – (= ser [Travel descriptions from south africa, 1711-1938]) – 1mf – 9 – mf#zah-19 – ne IDC [916]
Dieren, B *see* Die meistersinger von nuernberg als drama betrachtet
Dierks, Tamara J *see* Descriptive study of intramural activity offerings and entry rates in college/university intramural programs with a student population between 10,001-30,000
Dierlamm's triple wall concrete building block machine – Stratford, ON: P Dierlamm, [1904?] [mf ed 1991] – 1mf – 9 – 0-665-99527-X – mf#99527 – cn Canadiana [624]
Dies, Auguste *see* Autour de platon
Dies blatt gehoert der hausfrau – Berlin, Wiesbaden, Hamburg DE, 1891 oct-1894 22 sep, 1902 2 oct-1909 28 mar, 1910 2 oct-1911 24 sep, 1916 1 oct-1919 28 sep, 1923 oct-1936 n26, 1938/39-1941 sep, 1941 nov, 1942 mar-jun, 1943 jun, 1944 feb-may, 1949 oct-1950 n18; 1951 n1-25, 1952 n1-1953 n25, 1954 n1-26, 1955 n6-1956 n26/1, 1957 n2-23, 1958-1975 19 dec, 1976-1982 24 feb, 1982 24 mar-1993 15 dec – 198r – 1 – (filmed by misc inst: 1901, 1902, 1903 29 dec-1998 30 sep [29r]. title varies: oct 1920: das blatt der hausfrau, oct 1922: ullsteins blatt der hausfrau, oct 1927: das blatt der hausfrau. since n1 1934/35: ausgabe a ohne schnittmusterbogen. 1952 n10: brigitte: das blatt der hausfrau. apt 1954: brigitte. with suppl: modenheft oct 1906-24 mar 1907. publ in hamburg fr apr 1957; printed in itzehoe fr jun 1958) – gw Mikrofilm; gw Misc Inst [640]
Dies buch gehoert dem koenig / Arnim, Bettina von – Berlin: im Propylaeen-Verlag, 1921, c1920 [mf ed 1993] – (= ser Saemtliche werke v6) – 504p/pl – 1 – mf#8196 reel 2 – us Wisconsin U Libr [890]
Dies irae : erinnerungen eines franzoesischen offiziers an sedan / Bleibtreu, Karl – 5. aufl. Stuttgart: Carl Krabbe, [1904?] [mf ed 1989] – 108/2pl (ill) – 1 – mf#7031 – us Wisconsin U Libr [830]
The dies irae : on this hymn and its english versions / Warren, Charles Frere Stopford – London: Skeffington, 1897 – 1mf – 9 – 0-524-00799-3 – mf#1990-0231 – us ATLA [240]
Diese deutschen : drei schicksale / Bostrand, Torgerd – Karlsbad: A Kraft, 1940 [mf ed 1989] – 300p – 1 – mf#7053 – us Wisconsin U Libr [890]
Diese woche – Duesseldorf DE, 1951 17 feb-1953 19 sep, 1958 1 apr-1966 – 1 – (title varies: 1954: welt am sonnabend; 1961: neue welt am sonnabend. filmed by other misc inst: 1949 aug-1957 (gaps) [8r]) – gw Misc Inst [074]
Diese woche *see* Der spiegel
Diesel and gas turbine progress – Waukesha WI 1935-80 – 1,5,9 – (cont by: diesel progress north american) – ISSN: 0012-2602 – mf#17.03 – us UMI ProQuest [621]
Diesel equipment superintendent (des) – Laguna Beach CA 1950-97 – 1,5,9 – (cont by: truck fleet management) – ISSN: 0884-6324 – mf#427.01 – us UMI ProQuest [621]
Diesel progress engines & drives – Waukesha WI 1989-96 – 1,5,9 – (cont: diesel progress north american; cont by: diesel progress [north american ed]) – ISSN: 1040-8878 – mf#17.03 – us UMI ProQuest [621]
Diesel progress north american – Waukesha WI 1981-87 – 1,5,9 – (cont: diesel & gas turbine progress; cont by: diesel progress engines & drives) – ISSN: 0744-0073 – mf#17.03 – us UMI ProQuest [621]
Diesel progress [north american ed] – Waukesha WI 1997+ – 1,5,9 – (cont: diesel progress engines & drives) – ISSN: 1091-370X – mf#17,03 – us UMI ProQuest [621]
Dieselecho – Schoenebeck DE, 1961 12 aug-1977 nov [gaps], 1978-1984 12 dec – 3r – 1 – (dieselmotorenwerk) – gw Misc Inst [621]
Diesing, C M *see* Systema helminthum
Diesseits und jenseits der alpen : bilder von der adria, aus oberitalien und der schweiz / Rodenberg, Julius – Berlin 1865 [mf ed Hildesheim 1995-98] – 2mf – 9 – €60.00 – 3-487-29258-0 – gw Olms [914]
Diestel, Ludwig *see* Der segen jakob's in genes 49
Diestensis, Petri Dorlandi *see* Chronicon cartusiense
Diet and diet reform / Gandhi, Mahatma – Ahmedabad: Navajivan Pub House, 1949 – (= ser Samp: indian books) – us CRL [613]
Diet composition and body fat : a multivariate study of 203 adult males / Nelson, Lisa H – 1994 – 2mf – $8.00 – us Kinesology [612]
Diet compositional changes during mountaineering at high altitude in cold weather / Schneider, Allison K – University of North Carolina at Chapel Hill, 1995 – 1mf – 9 – $4.00 – mf#PH1474 – us Kinesology [612]
Diet in health and disease / Friedenwald, Julius & Ruhraeh, John – Philadelphia: W B Saunders and Co, 1905, c1904 [mf ed 1986] – 1r – 1 – (incl ind) – mf#*Z-4321 – us NY Public [613]
Dietary adequacy and changes in the nutritional status of appalachian trail through-hikers / Lutz, Karen L – 1982 – 1mf – 9 – $4.00 – us Kinesology [612]
Dietary fat and carbohydrate in relation to body fatness in lean and obese men and women / Niederpruem, Michael G & Miller, Wayne C – 1992 – 1mf – 9 – $4.00 – us Kinesology [613]
Dietary intake and energy expenditure of female collegiate swimmers during taper / Ousley, Laura J – 1999 – 1mf – 9 – $4.00 – mf#PH 1648 – us Kinesology [612]
Dietel, Margarita Leonor *see* Treatment of modality in the works of vaughan williams
Dieter und die frauen : ein roman von musik, freundschaft und liebe / Reichelt, Johannes – Dresden: Wodni & Lindecke, 1941 – 1r – 1 – us Wisconsin U Libr [830]
Dieterich, Albrecht *see*
– Abraxas
– Eine mithrasliturgie
– Mutter erde
– Vortraege und aufsaetze
Dieterich, Karl *see* Untersuchungen zur geschichte der griechischen sprache

Dieterici, Friedrich *see*
- Die anthropologie der araber im zehnten jahrhundert n chr
- Die lehre von der weltseele bei den arabern im 10. jahrhundert
- Die logik und psychologie der araber im zehnten jahrhundert n chr
- Die naturanschauung und naturphilosophie der araber im 10. jahrhundert
- Die philosophie der araber im 10. jahrhundert n. chr
- Die propaedeutik der araber im zehnten jahrhundert
- Reisebilder aus dem morgenlande

Dieterici, Wilhelm *see* Der wahre inwendige und auswendige christ

The dietetics of temperance / Watkins, Thomas C – [Hamilton, Ont?: s.n, 188-?] [mf ed 1994] – 1mf – 9 – 0-665-94626-0 – (in dble clms. original iss in ser: prohibition series) – mf#94626 – cn Canadiana [615]

Dietmann, K G *see* Lausitzisches magazin

Dietrich eckart : ein vermaechtnis / Eckart, Dietrich; ed by Rosenberg, Alfred – Muenchen: F Eher Nachf, 1928 – 1 – us Wisconsin U Libr [430]

Dietrich, Ernst *see* Die bruchstuecke der skeireins

Dietrich, H A *see* Neue medizinische bibliothek

Dietrich, Margret *see* Deutsche dramaturgie von gryphius bis brecht

Dietrich, Martin *see* Der meierhof

Dietrich schernbergs spiel von frau jutten (1480) : nach der einzigen ueberlieferung im druck des hieronimus tilesius (eisleben 1565) / Schernberg, Dietrich; ed by Schroeder, Edward – Bonn: A Marcus & E Weber 1911 [mf ed 1992] – (= ser Kleine texte fuer theologische und philologische vorlesungen und uebungen 67) – 1mf – 9 – 0-524-04687-5 – mf#1990-1314 – us ATLA [430]

Dietrich sebrandt : roman aus der zeit der schleswig-holsteinischen erhebung / Bartels, Adolf – Kiel: Lipsius & Tischer, 1899 [mf ed 1989] – 2v in 1 – 1 – mf#6979 – us Wisconsin U Libr [830]

Dietrich, Sixt *see*
- Epicedion thomae sporeri...
- Novum opus musicum

Dietrich, Stephan *see*
- Deine heimat, kamerad!
- Die melodie der heimat

Dietrich, Veit *see*
- Agend buechlein fuer die pfarrherren auff dem land
- Agend-buechlein fuer die pfarrherrn auff dem land
- Summaria und gesaeng auff alle sontag und fuernemste fest durchs jar
- Wie man das volck zur buss und ernstlichem gebet wider den tuercken auff der cantzel vermanen sol

Dietrich von bern in der neueren literatur / Altaner, Bruno – Breslau: F Hirt, 1912 [mf ed 1992] – (= ser Breslauer beitraege zur literaturgeschichte. neue folge 20) – 114p – 1 – (incl bibl ref) – mf#8014 reel 3 – us Wisconsin U Libr [430]

Dietrich, W *see* Zuege aus der missionsarbeit in china

Dietrichs erste ausfahrt / ed by Stark, Franz – Stuttgart: Literarischer Verein, 1860 [mf ed 1993] – (= ser Blvs 52) – xx/356p – 1 – mf#8470 reel 11 – us Wisconsin U Libr [890]

Dietrick, Ellen Battelle *see* Women in the early christian ministry

Dietsche warande en belfort – 10(1909) – 9 – €18.00 – ne Slangenburg [073]

Dietter, Johannes *see* Molekulardynamische simulation der freien fluessigkeitsschicht von einem quasi-stockmaier fluid, einer loesung von caesium fluorid in wasser, benzylalkohol und formamid

Dietterlin, W *see* Architectura von ausstheilung / symmetria und proportion der fuenff seulen...

Diettrich, Gustav *see*
- Ein apparatus criticus zur pesitto zum propheten jesaia
- Isaodaadh's stellung in der auslegungsgeschichte des alten testamentes
- Die oden salomons

Dietz, Gerhard *see* Interaktive frueherziehung bei entwicklungsverzoegerten und entwicklungsgefaehrdeten kindern

Dietz, Theodor *see* Aufbau, aufgaben und ergebnisse der internationalen arbeitsorganisation

Dietze, Walter *see*
- Aesthetische feldzuege
- Reden, vortraege, essays
- Schriften zur deutschen literatur

Dietzel, Uwe *see* Die komplexitaet der bedeutungsexplikation in literarischen dialogen

Dietzsch, August *see* Adam und christus

"A dieu" : november 12th, 1898 / Aberdeen and Temair, Ishbel Gordon, Marchioness of – S.l: s.n, 1898? – 1mf – 9 – mf#51882 – cn Canadiana [810]

Dieu : l'experience en metaphysique / Moisant, Xavier – Paris: Marcel Riviere, 1907 – (= ser Bibliotheque de philosophie experimentale) – 1mf – 9 – 0-8370-2924-4 – (incl bibl ref) – mf#1985-0924 – us ATLA [210]

Le dieu au coeur qui rayonne. paris, 1928 / Anizan, Felix – Madrid: Razon y Fe, 1930 – 1 – sp Bibl Santa Ana [944]

Dieu, L de *see* Critica sacra sive animadversiones in loca quaedam difficiliora veteris et novi testamenti...

Dieu le veut / Arlincourt, Charles V d' – Paris 1848 [mf ed Hildesheim 1995-98] – 1v on 1mf – 9 – €40.00 – 3-487-26019-0 – gw Olms [340]

Dieu vous benisse! / Ancelot, Francois – Paris, France. 1839 – 1r – us UF Libraries [440]

Dieulafoy, Michel *see* Portrait de michel cervantes

Diez anos de planificacion en puerto rico / Pico, Rafael – San Juan, Puerto Rico. 1954 – 1r – us UF Libraries [972]

Diez anos de politica liberal, 1892-1902 / Rodriguez Pineres, Eduardo – Bogota. 1945 – 1 – us CRL [972]

Diez blanco, alejandro evolucion del pensamiento filosofico. tomo 1 : desde tales de mileto a martin heidegger, 1942 / Marquez, Gabino – Madrid: Razon y Fe, 1943 – 1 – sp Bibl Santa Ana [100]

Diez Canedo, Enrique *see*
- Algunos versos
- El arte en la gran bretana e irlanda
- Epigramas americanos
- Imagenes (versiones poeticas) rosas del tiempo antiguo. mies de logrono
- Juan ramon jimenez en su obra
- Oracion de los debiles al comenzar el ano por...
- Unidad y diversidad de las letras hispanicas
- La visita del sol

Diez Coronel, Diego Manuel *see* Por d. joaquin topete, afronte...por si y por sus hijos...

Diez cubanos / Costa, Octavio Ramon – Habana, Cuba. 1945 – 1r – us UF Libraries [972]

Diez cuentos para un libro / Brenes, Maria – New York, NY. 1963 – 1r – us UF Libraries [972]

Diez Daza, A *see*
- Avisos y documentos para la preservacin y cura de la peste
- Libri tres de ratione cognoscendi causas et signa tam in prospera quam adversa valitudine, urinarium

Diez De Andino, Juan *see* Ronda de trompetas

Diez de Leiva, F *see*
- Antiaxiomas morales, medicos...
- El defensor de su agravio

Diez de mis cuentos / Febus, Sixto – San Juan, Puerto Rico. 1963 – 1r – us UF Libraries [972]

Diez, Genadius *see* Spain's struggle against anarchism and communism

Diez, Jorge A *see* Itenarios del tropico

Los diez libros de architectura / Alberti, L B – SL, 1582 – 7mf – 9 – sp Cultura [720]

Diez Lopez, Juan *see*
- Discurso pronunciado por el obrero en el circulo catolico de villafranca de los barros el dia 8 de diciembre de 1906
- Sueno mistico explicacion

Diez, M *see* Libro de albeiteria

Diez Olivares, J Ma *see* Compendio de gramatica castellana

Diez poetas cubanos, 1937-1947 / Vitier, Cintio – Habana, Cuba. 1948 – 1r – us UF Libraries [972]

Diez privilegios para mujeres prenadas...con un diccionario medico / Alonso y de los Ruizes de Fontecha, J – Alcala de Henares, 1606 – 14mf – 9 – sp Cultura [610]

Diez-Canedo, Enrique *see* Del cercado ajeno

Difa' az zindaniyan-i siyasi-yi iran – Jibhah-'i Milli-i Iran. shumarah-'i 1-2. isfand 1344-farvardin 1345 [feb/mar 1966-mar 1966] – 1r – 1 – $53.00 – us MEDOC [079]

La difesa – Chicago IL, 1918 – 1r – 1 – (italian newspaper) – us IHRC [071]

La difesa – Lawrence MA, aug 28 1920-sep 1922 – 1r – 1 – (italian newspaper) – us IHRC [071]

Diffenderffer, Frank Ried *see* German immigration into pennsylvania through the port of philadelphia

The difference between participation in intercollegiate athletics and academic performance based on time use / Kartschoke, Christopher – University of Wisconsin-La Crosse, 1995 – 1mf – 9 – mf#PSY 1891 – us Kinesology [150]

The difference in coach role model behaviors for male and female athletes / Killmer, Karen J – 1993 – 2mf – $8.00 – us Kinesology [150]

Differences – Durham NC 1992+ – 1,5,9 – ISSN: 1040-7391 – mf#19197 – us UMI ProQuest [320]

Differences between old and new school presbyterians / Cheeseman, Lewis – Rochester: E Darrow, 1848 [mf ed 1989] – 1mf – 9 – 0-7905-4386-9 – mf#1988-0386 – us ATLA [242]

The differences between physical activity levels and percent body fat using two methods of predicting percent body fat in male senior athletes / Hilbig, Jennifer Johnson – 1996 – 1mf – 9 – $4.00 – mf#PE 3830 – us Kinesology [612]

Differences in clinical evaluation models for first time pass rate of undergraduate athletic trainers on the nata certification examination / Sauka, Mark J – 1999 – 104p on 2mf – 9 – $10.00 – mf#PE 4187 – us Kinesology [378]

Differences in cohesion among starters and non-starters of recreational basketball teams / Kimball, Grayson T – 1998 – 2mf – 9 – $8.00 – mf#PSY 2042 – us Kinesology [150]

Differences in competitive anxiety and perceived competence for high and low level competitive youth swimmers / McNamara, Catherine M – Springfield College, 1995 – 2mf – 9 – $8.00 – mf#PSY1853 – us Kinesology [150]

The differences in educational preparation and athletic experience : between division 1 and division 2 athletic directors of national collegiate athletic association member institutions / Cummings-Danson, G – 1991 – 1mf – 9 – $4.00 – us Kinesology [790]

Differences in intrinsic risk factors for injured and non-injured athletes / Allen, Kristen L – 1997 – 2mf – 9 – $8.00 – mf#PE 3743 – us Kinesology [612]

Differences in movement speed between six year old children and adults on three motor tasks / McMillan, Monique C & Zelaznik, Howard – 1992 – 1mf – 9 – $4.00 – us Kinesology [150]

Differences in peak blood lactate values in long course versus short course swimming / Lowensteyn, I – 1991 – 1mf – 9 – $4.00 – us Kinesology [612]

Differences in physical activity attitudes and fitness knowledge between health fitness standard, sex, and grade group / Bocket, Thomas J – 1994 – 2mf – $8.00 – us Kinesology [306]

Differences in physiological and mechanical properties of stair climbing on three different apparatus / Aiello, Kimberly A – Springfield College, 1995 – 2mf – 9 – $8.00 – mf#PE1450 – us Kinesology [612]

The differences in stress and peripheral vision between injured and uninjured collegiate athletes / Weuve, Celestine M – 1998 – 2mf – 9 – $8.00 – mf#PSY 2019 – us Kinesology [790]

Differences in the academic achievement of athletes and non-athletes from intact two-parent, divorced, single-parent, and divorced/ remarried two parent families / Patterson, Aaron C – 1998 – 2mf – 9 – $8.00 – mf#PSY 2044 – us Kinesology [150]

Differences of co-dependency and self-esteem in college age male and female athletes and nonathletes / Mau, Robert E – Springfield College, 1995 – 2mf – 9 – $8.00 – mf#PSY1852 – us Kinesology [150]

Differend entre la republique d'haiti / Justin, Joseph – Port-Au-Prince, Haiti. 1912 – 1r – us UF Libraries [972]

Different conceptions of priesthood and sacrifice : a report of a conference held at oxford, dec 13 and 14, 1899 / ed by Sanday, William – London; New York: Longmans, Green, 1900 – 1mf – 9 – 0-8370-9816-5 – mf#1986-3816 – us ATLA [240]

Different drummer / Arkansas Radical Media Co-op – v1 n8 [1970 apr] – 1r – 1 – mf#1583111 – us WHS [071]

The different forms of flowers on plants of the same species / Darwin, Charles Robert – London, 1877 – (= ser 19th c evolution & creation) – 4mf – 9 – (with ill) – mf#1.1.4278 – uk Chadwyck [580]

Different new testament views of jesus / Crooker, Joseph Henry – Boston: American Unitarian Association, 1891 – 1mf – 9 – 0-524-04431-7 – (incl bibl ref) – mf#1991-2096 – us ATLA [240]

Differential effectiveness of relaxation procedures in attenuating components of anxiety in shooters / Doyle, Lauren A – 1981 – 2mf – 9 – $8.00 – us Kinesology [616]

Differential effects of strength training and endurance training on parameters realted to resistance to gravitational forces / Kim, H D – 1991 – 2mf – 9 – $8.00 – us Kinesology [613]

Differential equations – Dordrecht, Netherlands 1965+ – 1,5,9 – ISSN: 0012-2661 – mf#10905 – us UMI ProQuest [510]

Differential geometry and its applications – Oxford, England 1995+ – 1,5,9 – ISSN: 0926-2245 – mf#42646 – us UMI ProQuest [510]

Die differentialtherapie der infektioesen spondylitis : aus der orthopaedischen klinik volmarstein / Reinke, Barbara – 2000 – 2mf – 9 – 3-8267-2684-7 – mf#DHS 2684 – gw Frankfurter [617]

Differentiation – Oxford, England 1973+ – 1,5,9 – ISSN: 0301-4681 – mf#13162 – us UMI ProQuest [574]

Differentiation of ethnic culture regions using laban movement analysis : a study of bulgarian dance / Kerr, Kathleen A & Lockhart, Aileene – 1991 – 3mf – $12.00 – us Kinesology [790]

The differentiation of the religious consciousness / King, Irving – New York: Macmillan, 1905 – (= ser Psychological Review) – 1mf – 9 – 0-524-01285-7 – (incl bibl ref) – mf#1990-2321 – us ATLA [150]

Differenzierte betrachtungen des einsatzes von lyrik und karikaturen bei der bildung von umweltrelevanten einstellungen im unterricht der klassenstufen 7 bis 10 / Schilling, Kerstin – (mf ed 1992) – 3mf – 9 – €49.00 – 3-89349-628-9 – mf#DHS 628 – gw Frankfurter [373]

Differing levels of aggression and extraversion across the five categories of united states cycling federation (uscf) riders / Riley, Devin B – 1998 – 1mf – 9 – $4.00 – mf#PSY 2007 – us Kinesology [790]

Diffesa et offesa delle piazze / Floriani, P P – Ed 2. Venetia, 1654 – 6mf – 9 – mf#OA-204 – ne IDC [720]

Le difficile des chansons : second livre – 1544 – (= ser Mssa) – 1mf – 9 – €20.00 – mfchl 87 – gw Fischer [780]

La difficulte de l'esplanade : response...aux allegations des companies de fer du grand tronc et du nothern dans l'affaire de la contestation du droit d'entree du port de toronto / Credit Valley Railway Co – Ottawa?: A Bureau, 1880 – 1mf – 9 – mf#05276 – cn Canadiana [380]

Difficulte scolaire de manitoba par questions et reponses a la portee de tous / Lacasse, Zacharie – [s.l.]: [s.n.], [s.d.] (mf ed 1985) – 1mf – 9 – mf#SEM105P458 – cn Bibl Nat [370]

Les difficultes de croire / Brunetiere, Ferdinand – Paris: Librairie Academique, 1904 [mf ed 1985] – 1mf – 9 – 0-8370-2494-3 – (incl bibl ref) – mf#1985-0494 – us ATLA [210]

Difficulties about christianity : no reason for disbelieving it / Talbot, E S – London, England. 1886 – 1r – us UF Libraries [240]

Difficulties and alleged errors and contradictions in the bible / Torrey, Reuben Archer – Chicago: Bible Institute Colportage Association, c1907 – 1mf – 9 – 0-524-06056-8 – mf#1992-0769 – us ATLA [220]

Difficulties and discouragements which attend the study of the scriptures... / Hare, Francis – London, England. 1840 – 1r – us UF Libraries [220]

Difficulties and perversions are no arguments against the universal – London, England. 18– – 1r – us UF Libraries [240]

Difficulties in the church / Denison, Edward – London, England. 1853 – 1r – us UF Libraries [240]

The difficulties of arminian methodism : a series of letters addressed to bishop simpson of pittsburgh / Annan, William – 4th rev enl ed. Philadelphia: William S & Alfred Martien, 1860 – 1mf – 9 – 0-524-00501-X – mf#1990-0001 – us ATLA [240]

The difficulties of belief : in connexion with the creation and the fall, redemption and judgement / Birks, Thomas Rawson – 2nd enl ed. London: Macmillan, 1876 [mf ed 1989] – 1mf – 9 – 0-7905-0730-7 – mf#1987-0730 – us ATLA [210]

Difficulties of darwinism : read before the british association at norwich and exeter in 1868 and 1869 / Morris, Francis Orpen – London, 1869 – (= ser 19th c evolution & creation) – 1mf – 9 – (pref & correspondence with professor huxley) – mf#1.1.1522 – uk Chadwyck [575]

Difficulties of faith / Delany, Selden Peabody; ed by Grafton, Charles Chapman – Milwaukee: Young Churchman, 1906 – 1mf – 9 – 0-8370-2863-9 – mf#1985-0863 – us ATLA [240]

Difficulties of protestantism / Fletcher, John – London, England. 1829 – 1r – us UF Libraries [242]

Difficulties of the christian ministry and the means of surmounting – Birmingham, England. 1802 – 1r – us UF Libraries [240]

The difficulties of the new hypothesis / Streibert, Jacob – New York: Funk & Wagnalls, 1888 – 1mf – 9 – 0-8370-5783-3 – (incl bibl ref) – mf#1985-3783 – us ATLA [220]

Difficultueux / Rousseau-Saint-Phal – Paris, France. 1803 – 1r – us UF Libraries [440]

Diffie, Bailey Wallys *see* Porto rico

Diffusion of christianity dependent on the exertions of christians / Grey, Henry – Edinburgh, Scotland. 1818 – 1r – us UF Libraries [240]

Diffusion of divine truth / Bogue, David – London, England. 1800 – 1r – us UF Libraries [240]

Dificultades vencidas...para la limpieza y aseo de las calles de esta corte / Arce, Joan C – Madrid, 1735 – 3mf – 9 – sp Cultura [614]

Dig : the archaeological newsletter / Indian Shop [Independence KY] – 1975 sep-1980 may – 1r – 1 – mf#506311 – us WHS [930]

Dig in – 1980 oct-1982 may – 1r – 1 – mf#657283 – us WHS [071]

Digby courier, 1874-1948 and digby record, 1908-09 – Digby, NS – 28r – 1 – cn Library Assoc [079]

Digby, Kenelm Henry *see* Compitum

Digby record *see* Digby courier, 1874-1948 and digby record, 1908-09

Digby, William *see*
- The british invasion from the north
- The famine campaign in southern india (madras and bombay presidencies and province of mysore) 1876-1878
- India for the indians – and for england

Digdaja / Tan, Boen Soan – Soerabaia: Tan's Drukkerij, 1935 [mf ed 1998] – (= ser Penghidoepan 128) – 1r – 1 – (coll as pt of the colloquial malay collection. filmed with: multi-millionair / ong khing han) – mf#10002 – us Wisconsin U Libr [830]

Digest / AFL-CIO [American Federation of Labor-Congress of Industrial Organizations] – 1979 dec.-1992 oct – 1r – 1 – (cont: iud bulletin; cont by: iud action) – mf#1113496 – us WHS [331]

Digest and index of decisions *see* Us national labor relations board. digest and index of decisions

Digest of acts and deliverances / Presbyterian Church in the U.S.A. General Assembly – 1820-1976. 53v. and supplements – 1 – $550.00 – us Presbyterian [240]

A digest of all the decisions of all the courts relating to national banks, reported from 1864 to april 1, 1898 / Smith, Hal Horace – Chicago: Flood, 1899. 326p. LL-1118 – 1 – us L of C Photodup [346]

Digest of american cases relating to patents for inventions and copyrights from 1789 to 1862 : including numerous manuscript cases, decisions on appeals from the commissioners of patents and the opinions of the attorneys general of the united states under the patent and copyright laws... / Law, Stephen Dodd – New York: publ by aut & by Baker, Voorhis, 1868 [mf ed 1982] – 8mf – 9 – mf#25866 – cn Canadiana [346]

A digest of and index to the reports of cases decided in the supreme court of the gold coast colony. 1844-1931 / Griffith, William Brandford – Accra, the government printer, 1935. 194col. LL-12041 – 1 – us L of C Photodup [347]

A digest of cases : determined by the supreme court of canada from the organization of the court, in 1875, to the 1st day of may, 1893... / Cassels, Robert – Toronto: Carswell, 1893 – 11mf – 9 – (incl ind) – mf#00759 – cn Canadiana [348]

A digest of cases decided by the supreme court of canada from the organization of the court, in 1875, to the 1st day of may 1886 : comprising both reported and unreported cases, and many points of practice determined by the court and by the judges in chambers / Cassels, Robert – Toronto; Edinburgh: Carswell & Co, 1886 – 7mf – 9 – (incl ind) – mf#00758 – cn Canadiana [348]

The digest of cases determined in the court of queen's bench from michaelmas term, tenth george 4, to hilary term, third victoria / Cameron, John Hillyard – Toronto: H Rowsell, 1840 – 2mf – 9 – 0-665-91502-0 – (incl ind) – mf#91502 – cn Canadiana [348]

Digest of christian doctrine / Seiss, Joseph Augustus – Baltimore: [s.n.], 1857 – 1mf – 9 – 0-524-05268-9 – mf#1991-2260 – us ATLA [240]

A digest of corporation cases / Dean, Maurice B – New York, Banks, 1906. 1087 p. LL-1142 – 1 – us L of C Photodup [348]

A digest of criminal law of canada (crimes and punishments : founded by permission on sir james fitzjames stephen's digest of the criminal law / Burbidge, George Wheelock – Toronto: Carswell, 1890 – 7mf – 9 – mf#00331 – cn Canadiana [345]

A digest of decisions in criminal cases, contained in the reports of the federal courts. / Waterman, Thomas Whitney – New York, Baker, Voorhis, 1877. 816 p. LL-1643 – 1 – us L of C Photodup [345]

Digest of decisions of law and practice in the patent office and the united states and state courts in patents, trade-marks, copyrights, and labels / Hart, Amos Winfield – Chicago: Callaghan, 1898. 385p. LL-408 – 1 – us L of C Photodup [346]

Digest of decisions of law and practice in the patent office and the united states courts in patents, trademarks, copyrights and labels, 1912-1919 / Pollard, Willard Lacy – Washington, D.C.: Byrne, 1920. 119p. LL-1616 – 1 – us L of C Photodup [346]

Digest of decisions of the courts and the interstate commerce commission...under the act to regulate commerce / Pierce, Edward B – 1887-1908. Chicago, 1908 – 4mf – 9 – $18.00 – mf#LLMC 84-392 – us LLMC [347]

Digest of decisions of the department of the interior in cases related to public lands – Washington: GPO. 2pts. 1913 [all publ] – (= ser Native american coll) – 10mf – 9 – $15.00 – (covers v1-40 of decisions of the department of the interior in cases related to public lands. also supplied as pt of native american collection) – mf#llmc 88-007 – us LLMC [340]

Digest of decisions relating to the national banks / U.S. Comptroller of the Currency – Washington: GPO. v1-5. 1864-1936 – 28mf – 9 – $42.00 – mf#LLMC 84-394 – us LLMC [346]

Digest of decisions under the customs revenue laws / U.S. Treasury Dept. Division of Customs – Washington: GPO. 2v. 1918 (all publ) – (= ser Treasury Decisions, Internal Revenue, 1898-1942) – 6mf – 9 – $27.00 – (covers treasury dept board of general appraisers and us court decisions) – mf#LLMC 84-366 – us LLMC [336]

A digest of federal decisions and statutes, from the earliest period 1789 to the year 1880 / Rapalje, Stewart – Jersey City, Linn, 1880. 793 p. LL-1472 – 1 – us L of C Photodup [348]

Digest of general public bills and selected resolutions / U.S. Library of Congress – 74th congress 2nd session-94th congress 2nd session. 1936-76 – 9 – $894.00 – (95th congress 1977-80 $535 [0642]) – mf#0641 – us Brook [348]

Digest of income tax rulings : digest a, april 1919-december 1930 / U.S. Treasury Dept – Washington: GPO, 1932 – (= ser Treasury decisions, internal revenue, 1898-1942) – 5mf – 9 – $7.50 – mf#LLMC 82-708 – us LLMC [336]

A digest of international law / Moore, John Bassett – 8v. 1906 – 1 – us AMS Press [341]

A digest of international law / Moore, John Bassett – Washington, 1906. 8v – 3r – 1 – $100.00 – us Trans-Media [341]

Digest of international law / Hackworth, Green Haywood – 1904-44 – 3r – 1 – $100.00 – us Trans-Media [341]

Digest of international law / Hackworth, Green Haywood – 8v. 1940-1944 – 1 – us AMS Press [341]

Digest of international law / Whiteman, Marjorie M – 14 v. plus gen. ind. vol. 1963-73 – 1 – us AMS Press [341]

Digest of maryland statutes and decisions on criminal law / Gans, Edgar Hilary – Baltimore: Murphy, 1884. 170p. LL-723 – 1 – us L of C Photodup [345]

A digest of methodist law : or, helps in the administration of the discipline of the methodist episcopal church / Merrill, Stephen Mason – Cincinnati: Curts & Jennings; New York: Eaton & Mains [1896?], c1885 [mf ed 1991] – (= ser Methodist coll) – 1mf – 9 – 0-524-01518-X – (rev since general conference of 1896) – mf#1990-4098 – us ATLA [242]

A digest of mississippi railway decisions from vol. 1 to and including vol. 71, mississippi reports / Harris, James Bowmar – Galveston: Clarke & Courts, 1894. 191p. LL-953 – 1 – us L of C Photodup [343]

Digest of neurology and psychiatry – Hartford CT 1975+ – 1,5,9 – ISSN: 0012-2769 – mf#10317 – us UMI ProQuest [616]

Digest of new hampshire school law : adapted to the general laws and amendments thereto – Concord: J B Sanborn, 1881 – 2mf – 9 – $3.00 – mf#LLMC 96-046 – us LLMC [348]

A digest of parochial returns made to the select committee into the education of the poor, 1819 : together with tables showing the state of education in england, scotland and wales, 1870. command n151, 177 and 224 – 18mf – 9 – mf#87080 – uk Microform Academic [324]

Digest of patent and trade-mark cases decided by the court of appeals of the district of columbia. / Torbert, William Sydenham – Washington, Byrne, 1909. 291 p. LL-1345 – 1 – us L of C Photodup [346]

A digest of patent cases : decided in the federal and state courts from 1789-1888 / Simonds, William Edgar – New York: Strouse & Co, 1888 – 3mf – 9 – $4.50 – mf#LLMC 84-336 – us LLMC [346]

Digest of "precedents or decisions" by select committees appointed to try the merits of upper canada contested elections : from 1824 to 1849 / Patrick, Alfred – Montreal: Printed by Lovell and Gibson, 1849 [mf ed 1982] – 2mf – 9 – mf#SEM105P135 – cn Bibl Nat [325]

Digest of public general bills and resolutions / U.S. Congress – 75th-101st congress. 1937-90 [all publ] – 1044mf – 9 – $1566.00 – (lacking: 1976 pt1-2) – mf#llmc 80-036 – us LLMC [348]

Digest of public land laws of the united states – Washington: GPO, 1968 – (= ser Land laws of the united states, 1776-1938) – 12mf – 9 – $18.00 – (prepared for the public land law review commission by shepard's citations) – mf#LLMC 85-301 – us LLMC [343]

Digest of reported cases touching the criminal law of canada. / Foran, Thomas Patrick – Toronto: Carswell, 1889. 252p. LL-2377 – 1 – us L of C Photodup [345]

Digest of reports of cases decided in the court of chancery, in the court of error & appeal, on appeal from the court of chancery, and in chancery chambers. / Cooper, Charles William – Toronto: Blackburn's, 1868-73. 2v in 1. LL-2304 – 1 – us L of C Photodup [347]

A digest of scripture : consisting of extracts from the old and new testaments, on the plan of brown's "selection of scripture passages" – Maulmain [Burma]: American Baptist Mission Press 1840 – 1r with other items – 1 – (text in peguan [mon]) – mf#mf-10289 seam reel 198/7 [§] – us CRL [220]

Digest of state laws regulating fraternal beneficiary societies / Landis, Abb – Washington, D.C., Rogers, 1921 239 p. LL-1486 – 1 – (suppl 1922? 28p washington, dc ll-1486. suppl with rev to 1924 washington, dc 1924 86p ll-1486) – us L of C Photodup [348]

Digest of statistics 1959-1974 / St Vincent. Statistical Office – (= ser Latin american & caribbean...1821-1982) – 13mf – 9 – (1959-61 entitled: quarterly digest of statistics. 1959, 1961, 1964, 1966, 1974 not available) – uk Chadwyck [318]

Digest of statutes and legal decisions relating to official stenographers / National Shorthand Reporters' Association – New Haven, Conn.: Mac, 1906. 242p. LL-1259 – 1 – us L of C Photodup [348]

A digest of statutes, equity rules, and decisions, upon the jurisdiction, pleadings, and practice of the circuit courts of the united states / Thatcher, Erastus – Boston, Little, Brown, 1883. 976 p. LL-1406 – 1 – us L of C Photodup [348]

A digest of statutes, rules, and decisions relative to the jurisdiction and practice of the supreme court of the united states 1790-1879 / Thatcher, Erastus – Boston, Little, Brown, 1882. 520 p. LL-1180 – 1 – (2d ed. boston, little, brown, 1883. 602 p. ll-1454. 1) – us L of C Photodup [348]

Digest of studies and lectures in theology – Auburn: Wm J Moses, 1866 – 1mf – 9 – 0-8370-2449-8 – mf#1985-0449 – us ATLA [240]

A digest of the acts and proceedings of the general assembly of the presbyterian church in the united states : revised down to and including acts of the general assembly of 1910 / Alexander, William Addison & Nicolassen, George Frederick – Richmond, VA: Presbyterian Committee of Publ, 1911 [mf ed 1992] – (= ser Presbyterian coll) – 2mf – 9 – 0-524-02440-5 – (1st publ 1888) – mf#1990-4299 – us ATLA [242]

Digest of the american and english annotated cases – New York: Thompson Co. 1v. 1912 (all publ) – 20mf – 9 – $30.00 – mf#LLMC 84-695D – us LLMC [340]

A digest of the criminal rulings of burma, [1956-68] *see* Mran ma nuin nam rajavat tara civan thum pon khyup, [1956-68]

Digest of the decisions of the comptroller of the treasury – 1894-1920 (all publ) – (= ser Comptroller of the treasury dicisions, 1894-1921) – 26mf – 9 – $39.00 – mf#LLMC 90-373 – us LLMC [348]

A digest of the decisions of the department of the interior and the general land office in cases relating to the public lands : from july, 1881 to december, 1887 / Matthews, William Baynham – Washington, D.C.: Wilson, 1888. 579p. LL-189 – 1 – us L of C Photodup [343]

Digest of the decisions...in cases related to public lands / Heselman, George J – pts 1-2, covering v1-40 of the Dec. of the Dept. of the Int.... Washington: GPO, 1913 – 10mf – 9 – $15.00 – mf#LLMC 88-007 – us LLMC [324]

Digest of the different castes of the southern division of southern india : with descriptions of their habits, customs etc / Venkata Ramasvami, Kavali [comp] – [s.l.]: printed at the Telegraph and Courier Press, 1847 [mf ed 1995] – (= ser Yale coll) – 25p (ill) – 1 – 0-524-09929-4 – mf#1995-0929 – us ATLA [305]

Digest of the doctrinal standards of the methodist church / Shaw, William Isaac – Toronto: W Briggs; Montreal: C W Coates, 1895 – 2mf – 9 – mf#13540 – cn Canadiana [242]

Digest of the doctrine of s thomas on the incarnation = Summa theologica. pars 3. selections – London: JT Hayes [1868?] [mf ed 1991] – 2mf – 9 – 0-7905-8602-9 – (english trans by william humphrey) – mf#1989-1827 – us ATLA [240]

Digest of the evidence : before the committees of the houses of lords and commons, in the year 1837 – London, 1838 – (= ser 19th c ireland) – 3mf – 9 – mf#1.1.1134 – uk Chadwyck [370]

Digest of the international law of the united states / Wharton, Francis A – Washington. 1886. 3v – 1 – $50.00 – us Trans-Media [341]

A digest of the law of evidence in criminal cases / Roscoe, Henry – Philadelphia: Johnson, 1852. 993 (i.e.929)p. LL-1168 – 1 – us L of C Photodup [345]

A digest of the law of evidence...from the 15th ed. (1899)...with both general american notes and notes especially adapted to the state of ohio / Stephen, James Fitzjames – Hartford, Conn., Dissell, 1902. 579 p. LL-886 – 1 – (from the 5th ed. (1899) with both general american notes and notes especially adapted to the states of illinois, indiana, and michigan. hartford, conn., bissell, 1903. 719 p. ll-961) – us L of C Photodup [348]

Digest of the law of impeachment / Todd, Hiram C – n.p. 1913. 3 p., 2-41 numb. LL-1367 – 1 – us L of C Photodup [348]

A digest of the law of landlord and tenant : in the provinces subject to the lieutenant-govenor of bengal c d field / Bengal (India) – Calcutta: Printed at the Bengal Secretariat Press, 1879 – xxxv/253p – 1 – (incl bibl ref and ind) – mf#1232 – us Wisconsin U Libr [346]

Digest of the law of mines and minerals and of all controversies incident to the subject-matter of mining / Morrison, Robert Stewart – San Francisco: Bancroft, 1878. 448p. LL-578 – 1 – us L of C Photodup [622]

A digest of the law of partnership. / Montagu, Basil – 1st Amer. ed. from the 2nd and last London ed. of 1822. New York: Lamson and Collins & Hannay, 1824. 2v. LL-511 – 1 – us L of C Photodup [348]

A digest of the law of real property / Cruise, William – Boston, 1856-57 – 7v on 1r – 1 – $150.00 – 0-89093-032-5 – us UPA [346]

A digest of the law relating to district councils / Chambers, George Frederick – 9th ed. London: Stevens, 1895. 283p. LL-108 – 1 – us L of C Photodup [348]

A digest of the law relating to public health and local government / Chambers, George Frederick – 8th ed. London: Stevens, 1881 i.e. 1884?. 556p. LL-1451 – 1 – us L of C Photodup [344]

Digest of the laws and decisions relating to the appointment, salary, and compensation of the officials of the united states courts. / Cousar, Robert Moore – Washington, Govt. Print. Off., 1895. 300 p. LL-1171 – 1 – us L of C Photodup [347]

A digest of the laws of england / Comyns, John – 5th ed. London, 1822 – 8v on 2r – 1 – $285.00 – 0-89093-025-2 – us UPA [348]

Digest of the laws of the state of florida / Florida – Tallahassee, FL. 1881 – 1r – us UF Libraries [348]

Digest of the laws of virginia of a civil nature / Matthews, James Muscoe – Richmond: Randolph, 1856-57. 2v. LL-634 – 1 – us L of C Photodup [348]

Digest of the laws of virginia, of a criminal nature / Matthews, James Muscoe – Richmond: West & Johnston, 1861. 320p. LL-815 – 1 – (2nd ed. richmond: randolph and english, 1871. 376p. supplement. richmond, 1878. 57p. ll-683. 3rd ed. richmond: randolph and english, 1890. 421p. ll-887) – us L of C Photodup [345]

A digest of the mechanics' lien law of illinois / Scott, Adolphus G – Chicago: Stansbury, 1896. 23, 3p. LL-97 – 1 – us L of C Photodup [348]

Digest of the mentally retarded – Hallandale Beach FL 1963-69 – 1 – (cont by: journal for special educators of the mentally retarded) – mf#2543.02 – us UMI ProQuest [370]

A digest of the military and naval laws of the confederate states : from the commencement of the provisional congress to the end of the first congress under permanent constitution, analytically arranged / Lester, W W & Bromwell, William J – Columbia: Evans & Cogswell, 1864 – 4mf – 9 – $6.00 – mf#LLMC 96-091 – us LLMC [355]

A digest of the nova scotia common law, equity, vice-admiralty and election reports : with notes of many unreported cases and of cases appealed to the privy council and supreme court of canada from nova scotia / Congdon, Frederick Tennyson – Toronto: Carswell, 1890 [mf ed 1980] – 10mf – 9 – 0-665-00743-4 – mf#00743 – cn Canadiana [348]

Digest of the presbyterian church of korea (chosen) / Clark, Charles Allen [comp] – Seoul: Korean Religious Book & Tract Society, 1918 [mf ed 1995] – (= ser Yale coll) – viii/263p – 1 – 0-524-09511-6 – mf#1995-0511 – us ATLA [242]

Digest of the principal acts and deliverances / United Presbyterian Church of North America – 1878-1952 – 1 – $50.00 – us Presbyterian [240]
A digest of the reported decisions of the courts of the united states of america, and of great britain and her colonies, relating to the rights and liabilities of gas companies / Greenough, Charles Pelham – Boston: Little, Brown, 1883. 307p. LL-170 – 1 – us L of C Photodup [347]
A digest of theology : being a brief statement of christian doctrine according to the consensus of the great theologians of the one, holy, catholic and apostolic church / Percival, Henry Robert – Philadelphia: JJ McVey, 1893 [mf ed 1991] – 1mf – 9 – 0-7905-9837-X – (incl bibl ref) – mf#1989-1562 – us ATLA [242]
Digest of united kingdom energy statistics 1948/49-1977 – [mf ed Chadwyck-Healey] – (= ser British government publications...1801-1977) – 73mf – 9 – uk Chadwyck [333]
Digest to american negligence cases and reports – Chicago: Callaghan. 1v. 1912 (all publ) – (= ser American negligence cases and reports) – 18mf – 9 – $27.00 – (covers both american negligence cases and american negligence reports) – mf#LLMC 84-699F – us LLMC [348]
Digest to american negligence cases and reports – New York: Remick & Schilling. 1v. 1902 (all publ) – (= ser American negligence cases and reports) – 7mf – 9 – $10.50 – mf#LLMC 84-699A – us LLMC [348]
Digest to american negligence reports – New York: Remick & Schilling. 1v. 1909 (all publ) – (= ser American negligence reports) – 18mf – 9 – $27.00 – (covers v1-20 of american negligence reports) – mf#LLMC 84-699B – us LLMC [348]
Digest western law reporter, vols 1 to 24, territories law reports vols 1 to 7 : and the official reports for the provinces of alberta, british columbia, manitoba and saskatchewan... / Rolph, Thomas Taylor & Lear, Walter Edwin [comps] – Toronto: Carswell, 1915 – 16mf – 9 – 0-659-91627-4 – (in dble clms) – mf#9-91627 – cn Canadiana [348]
A digested index to the reported cases in lower canada : contained in the reports of pyke, stuart, revue de legislation, law reporter... / Ramsay, Thomas Kennedy – Quebec?: G E Desbarats, 1865 [mf ed 1983] – 5mf – 9 – 0-665-38400-9 – (incl ind) – mf#38400 – cn Canadiana [348]
Digester / Nekoosa Edwards Paper Co – v1 n2-v19 [i.e. 12] n3 [1971 jul-1982 jul] – 1r – 1 – mf#648852 – us WHS [670]
Digester's reader : monthly newsletter / Intra-Community Cooperative [Madison WI] – v1 n1-v2 n3 [1976 feb-1977 jun], v3 n1,2 [1978 may, fall] – 1r – 1 – mf#499192 – us WHS [334]
Digestible nutrient content of napier grass silage, crotalaria intermedia silage and natal grass hay / Neal, W M – Gainesville, FL. 1935 – 1r – us UF Libraries [630]
Digestion – Basel, Switzerland 1968+ – 1,5,9 – (cont: gastroenterologia) – ISSN: 0012-2823 – mf#2695 – us UMI ProQuest [611]
Digestive diseases and sciences – Dordrecht, Netherlands 1979+ – 1,5,9 – (cont: american journal of digestive diseases) – ISSN: 0163-2116 – mf#53,01 – us UMI ProQuest [612]
Digesto constitucional americano / Carranza, Arturo Bartolome – Buenos Aires, Argentina. v1-2. 1910 – 1r – us UF Libraries [323]
Digesto constitucional centroamericano / Organizacion De Estados Centroamericanos – San Salvador, El Salvador. 1962 – 1r – us UF Libraries [323]
Digesto constitucional de costa rica / Costa Rica Constitucion Politica (1949) – San Jose, Costa Rica. 1946 – 1r – us UF Libraries [972]
Digests and decisions: law and practice in the patent office – in 3 bks: Bk. 1, by D.H. Rice & L.C. Rice, 1869-80, publ. in 1880; Bk. 2, by E.S. Beach, 1880-90, publ. in 1890; Bk. 3, by L.H. Rice, 1890-1900, publ. in 1900; George B. Reed, Boston.84-331 – 9 – $12.00 – us LLMC [346]
Digests of international law series – (= ser Original monograph series; Digests of united states practice in international law) – 503mf – 9 – $754.00 – (original monograph series: cadwalader 1v 1877. wahrton 3v 1887. moore 8v 1906. hackworth 9v 1940-44. whiteman 15v 1963-73. these occasional studies followed by annuals: digests of us practice in international law: rovine v1-2 1973-1974. mcdowell v3-4 1975-1976. boyd v5 1977. cum ind for 1973-80 and 1981-88) – mf#llmc 79-448 – us LLMC [341]
The digger movement in the days of the commonwealth : as revealed in the writings of gerrard winstanley, the digger, mystic and rationalist, communist and social reformer / Berens, Lewis Henry – London: Simpkin, Marshall, Hamilton, Kent, 1906 – 1mf – 9 – 0-524-03538-5 – mf#1990-4733 – us ATLA [100]
Digger's digest / Sutter-Yuba Genealogical Society – 1980 spring-1987 – 1r – 1 – (cont: sutter-yuba digger's digest [yuba city ca: 1980]) – mf#1686807 – us WHS [929]
Diggers digest / Sutter-Yuba Genealogical Society – v4 n1-4 [1977 jan /mar-oct/dec] – 1r – 1 – (cont: sutter-yuba genealogical society's diggers digest; cont by: sutter-yuba digger's digest [yuba city ca: 1978]) – mf#517942 – us WHS [929]
Digger's friend = De delvers' vriend – Bloemhof SA, 9 jan-21 aug 1914 (wkly) [mf ed Cape Town: S A Library 1983] – 1r – 1 – (text in english or dutch) – sa National [079]
Digger's news – Bloemhof SA, 16 jan-8 may 1914 – 1r – 1 – sa National [079]
Diggers' news and witwatersrand advertiser – Johannesburg. South Africa. -w. 2 Feb 1888-28 Dec 1889. (1 reel) – 1 – uk British Libr Newspaper [072]
Digging a little deeper / Barker, William – London, England. 1862 – 1r – us UF Libraries [240]
Diggs, Paul *see*
– George and bessie derrick
– William and corneal jackson
Digha-Nikaya *see* Dialogues of the buddha
Dighanikaya : das buch der langen texte des buddhistischen kanons – Goettingen: Vandenhoeck & Ruprecht, 1913 – (= ser Quellen der Religionsgeschichte) – 5mf – 9 – 0-524-07374-0 – mf#1991-0094 – us ATLA [280]
Dighe, Vishvanath Govind *see* Peshwa bajirao 1 and maratha expansion
Dighton 1695-1890 – Oxford, MA (mf ed 1984) – (= ser Massachusetts vital records) – 54mf – 9 – 0-931248-64-7 – (mf 1-5: vital records 1687-1792. mf 6-7: vitals index 1687-1792. mf 8-15: vital records 1695-1859. mf 16-17: births 1844-54. mf 17: marriages 1844-54. mf 17-18: deaths 1844-55. mf 19-21: births 1855-91. mf 21-23: marriages 1855-90. mf 23-25: deaths 1855-89. mf 26-28: vitals index 1695-1859. mf 29-30: births index 1855-1900. mf 30-32: marriages 1841-1900. mf 32-33: deaths 1855-1900. mf 34-38: births 1700-1899. mf 39-40: brides 1700-1899. mf 41-43: grooms 1700-1899. mf 44-51: deaths 1700-1980. mf 52-54: deaths by maiden names) – us Archive [978]
Dighton, T *see*
– Certain reasons of a private christian against conformitie to kneeling in the very act of receiving the lord's supper...
– The second part of a plain discourse of an unlettered christian...
A digit of the moon : a hindoo love story – London: James Paroker & Co, 1899 – (= ser Samp: indian books) – (trans fr original by f w bain) – us CRL [830]
Digital design – Westborough MA 1982-86 – 1,5,9 – (cont by: esd: the electronic system design magazine) – ISSN: 0147-9245 – mf#13088.01 – us UMI ProQuest [000]
Digital equipment computer users society. proceedings – Shrewsbury MA 1968-86 – ISSN: 0095-2095 – mf#5742.01 – us UMI ProQuest [000]
Digital news – Boston MA 1989-92 – 1,5,9 – ISSN: 0891-9860 – mf#17094 – us UMI ProQuest [000]
Digital news & review – New York NY 1992-96 – 1,5,9 – ISSN: 1065-7452 – mf#20302 – us UMI ProQuest [000]
Digital review – New York NY 1985-92 – 1,5,9 – ISSN: 0739-4314 – mf#13553 – us UMI ProQuest [000]
Digital systems journal – Aliso Viejo CA 1992-94 – 1,5,9 – (cont: vax professional) – ISSN: 1067-7224 – mf#15074.02 – us UMI ProQuest [000]
DigitalTV – Manhasset NY 2002+ – 1,5,9 – ISSN: 1551-1928 – mf#19115.05 – us UMI ProQuest [380]
Dignan, John *see* Slave captain
Dignite nouvelle – Bukavu: Ramazani-Ngongo, Joseph M, [1961-]. mar 19-apr 2 1961 – (issues for mar 19-apr 2 1961 filmed as pt of: herbert c weiss collection on the belgian congo) – us CRL [079]
The dignity of labor : lecture to the young / Christie, Robert – S.l: s.n, 1879 – 1mf – 9 – mf#00637 – cn Canadiana [331]
Dihigo, Juan Miguel *see*
– Elogio del dr enrique jose varona y pera
– Elogio del dr jose a rodriguez garcia
– Elogio del dr mario garcia kohly
– Epigrafia en cuba
– Influencia de la universidad de la habana
– Universidad de la habana
Dihigo Y Lopez-Trigo, Ernesto *see* Cuba y el problema del caribe
Dihkhuda, Ali Akbar *see* Lughatnamah
Dijkstra, Harmen *see* Duisternis en licht
Dijon *see* Bulletin municipal
Dik Gam *see*
– Brah nan indradevi
– Kar siksa vivattan nai qaksar khmaer
– Qakkharanukram vevacanasabd khmaer
– Ram kerti
Dike, Samuel Warren *see* Sociology in the higher education of women
Dik-Keam *see* Rapid study of cambodian for foreign beginners
Dikovics, John *see* Twenty-five years of presbyterian...hungarians; our magyar presbyterians
Dikshit, Kashi Nath *see* Six sculptures from mahoba
Dikshit, Kashinath Narayan *see* Prehistoric civilization of the indus valley
Dikshit, Moreshwar Gangadhar *see* Etched beads in india
Dikshitar, V R Ramachandra *see*
– The lalita cult
– Pre-historic south india
– Studies in tamil literature and history
– War in ancient india
Diksiyonaryo ng wikang pilipino – Rosendo Ignacio. Quezon City: Samar Pub Co, 1958 – 1 – us CRL [079]
Dilber kontes *see* Afiyet
Dilectissima nobis – 3Jun 1933. English. Encyclical on Spain. New York, 1937. Fiche W 786. (Blodgett Collection of Spanish Civil War Pamphlets) – 9 – us Harvard [946]
The dilemma of humanitarian modernism / Calhoun, Robert Lowry – [s.n.: s.l., 1937] Chicago: Dep. of Photodup, U of Chicago Lib, 1971 (1r); Evanston: American Theol Lib Assoc, 1984 (1r) – 1 – 0-8370-0440-3 – mf#1984-B212 – us ATLA [240]
Dilemma of the progressive black parent regarding school choice / Davies-Crawford, Gillian Peta – U of the Western Cape 1989 [mf ed S.l: s.n. 1989] – 3mf – 9 – (incl bibl) – sa Misc Inst [370]
Dilemmas of a churchman / Lushington, Charles – London, England. 1838 – 1r – us UF Libraries [240]
Dilettanten – Sundsvall, Sweden. 1853-58 – 1 – sw Kungliga [078]
Dilettanten des lebens : roman / Viebig, Clara – Berlin: Ullstein, [19–?] – 1r – 1 – us Wisconsin U Libr [830]
Dilg, William *see* Gedichte
Dilger, Wilhelm *see*
– Der indische seelenwanderungsglaube
– Krischna oder christus?
– Salvation in hinduism and christianity
Diligence for both worlds – London, England. 18– – 1r – us UF Libraries [240]
Dilke, Charles Wentworth *see*
– The british empire
– Exhibition of the works of industry of all nations, 1851
– Papers of a critic
Dilke, Emilia Frances (Strong) *see*
– Art in the modern state
– French architects and sculptors of the 18th century
– French painters of the 18th century
Dilke, Emilia Francis (Strong), lady *see* The renaissance of art in france
Dill, E M *see* Rome and her baptism
Dill, Jacob Smiser *see* Diary and travelogue
Dill, Liesbet *see* Wir von der saar
Dill, Samuel *see*
– Roman society from nero to marcus aurelius
– Roman society in the last century of the western empire
Dillaliya – Lagos: Shu-Aibu Paiko, nov 7 1969 – (filmed with zaruma and other hausa newspapers) – us CRL [079]
Dillard capsules : a publication of the dillard university office of university relations – 1989 jul 31, sep 15-dec 15, 1990 jan 15, mar 15, may 15, nov 15, 1992 jan 15, mar 15-sep 4, nov 12-1997 may – 1r – 1 – mf#2684199 – us WHS [378]
Diller record – Diller, NE: Frank T Pearce. v15 n38 [ie 40] dec 13 1901-v66 n9. jan 7 1954 (wkly) [mf ed with gaps] – 14r – 1 – (cont: jefferson county record. absorbed by: fairbury journal. vol numbering irregular: v56 repeated. suspended with oct 5 1950; resumed in 1951) – us Misc Inst [071]
Diller record – Diller, NE: Frank T Pearce. v15 n38 [ie 40] dec 13 1901-v66 n9. jan 7 1954 (wkly) [mf ed with gaps] – 14r – 1 – (cont: jefferson county record (endicott ne). absorbed by: fairbury journal. vol numbering irregular: v56 repeated) – us NE Hist [071]
Dillinger, Stefan *see* Systemisches denken
Dillinger zeitung – Dillingen (Saar) DE, 1916 1 jul-31 dec – 1 – gw Misc Inst [074]
Dillmann, A *see*
– Catalogus codicum manuscriptorum bibliothecae bodleianae oxoniensis
– Chrestomathia aethiopica edita et glossario explanata
Dillmann, August *see*
– Das buch der jubilaeen
– Das buch henoch
– Die buecher exodus und leviticus
– Chrestomathia aethiopica edita et glossario explanata
– Die genesis
– Genesis
– Handbuch der alttestamentlichen theologie
– Hiob fuer die dritte auflage nach l. hirzel und j. olshausen
– Lexicon linguae aethiopicae
– Der prophet jesaia
– Ueber die griechische uebersetzung des qoheleth
Dillon, Arthur *see* Historical notes on the services of the irish officers in the french army
Dillon, Emile Joseph *see* A scrap of paper
Dillon first baptist church. dillon county. south carolina : church records – 1891-1902, 1905-16, 1924-41, 1947-89 – 1 – $63.59 – us Southern Baptist [242]
Dillon, Franz J *see* Grosses illustriertes frauen-lexikon
Dillon, George F *see* The war of antichrist with the church and christian civilization
Dillon, Henry Augustus Dillon-Lee, 13th viscount *see* Short view of the catholic question
Dillon, J T *see* Travels through spain
Dillon, John Forest *see* Dillon's reports of cases in the eighth circuit, 1870-1880
Dillon, John Forrest *see*
– Commentaries on the law of municipal corporations
– John marshall
– The laws and jurisprudence of england and america
– Property; its rights and duties in our legal and social systems
– Removal of causes from state courts to federal courts.
Dillon, John M *see* John marshall
Dillon, Peter *see*
– Narrative and successful result of a voyage in the south seas
– Voyage aux iles de la mer du sud
Dillon, Richard *see* Popular premises examined
Dillon's reports of cases in the eighth circuit, 1870-1880 / Dillon, John Forest – Davenport, IA: Griggs. v1-5. 1871-80 (all publ) – (= ser Early federal nominative reports) – 35mf – 9 – $52.00 – mf#LLMC 81-448 – us LLMC [340]
Dill-zeitung – Dillenburg DE, 1976- – ca 7r/yr – 1 – gw Misc Inst [074]
Dilnot, George *see* The story of scotland yard
DiLorenzo, Peter A *see* Status of physical education basic instruction programs at selected two-year colleges in the united states
Dilthey, Wilhelm *see*
– Das erlebnis und die dichtung
– Leben schleiermachers. 1. bd
– Le monde de l'esprit
Dilthey, Wilhelm et al *see* Systematische philosophie
Dilucida explicatio sanae doctrinae de vera participatione carnis et sanguinis christi in sacra coena, ad discutiendas heshusii nebulas... / Calvin, J – Genevae: Excudebat Conradus Badius, 1561 – 2mf – 9 – mf#CL-37 – ne IDC [240]
Dilucidatio in generali capitulo ordino d. hyeronymi a.. / Caceres, Diego de – 1641 – 9 – sp Bibl Santa Ana [240]
Diluvio / Lopez Suria, Violeta – San Juan, Puerto Rico. 1958 – 1r – us UF Libraries [972]
Dilworth, John *see*
– Pictorial description of the tabernacle in the wilderness
– Pictorial model of the tabernacle
Diman, Jeremiah Lewis *see*
– Memoirs of the rev. j. lewis diman, d.d
– The theistic argument
Dimanche – New Orleans LA. 1861 feb 10, 1862 jun 8-jul 6 – 1r – 1 – mf#861341 – us WHS [071]
Dimanche illustre – Marseilles, France. 1941-2 jul 1943 – 1 1/4r – 1 – (aka: d i dimanche illustre) – uk British Libr Newspaper [074]
Dimanche-matin – Montreal, Canada. 20 mar 1966-1969 – 15r – 1 – (incl "perspectives") – uk British Libr Newspaper [071]
Dimanche-turf – Paris. oct 1948-avr 1958 – 1 – fr ACRPP [073]
Dimanshtein, S *see* Kto takie mensheviki
Dimbaza, Mongezeleli Christopher *see* Perceptions about women principals of secondary schools in the port elizabeth west district
Dimbleby, Jabez Bunting *see* The appointed time
Dimboola chronicle – Dimboola, feb 1921-jan 1929 – 2r – A$135.07 vesicular A$146.08 silver – at Pascoe [079]
The dime base-ball player – New York. 1860-1862, 1864-1881 – 1 – us NY Public [790]
La dime de penitance : altfranzoesisches gedicht, verfasst im jahre 1288 / Journi, Jean de; ed by Breymann, Hermann – Stuttgart: Litterarischer Verein, 1874 (Tuebingen: L F Fues) – (incl bibl ref. old french text with synopsis in german) – us Wisconsin U Libr [440]
La dime de penitance : altfranzoesisches gedicht, verfasst im jahre 1288 / Journi, Jean de; ed by Breymann, Hermann – Stuttgart: Litterarischer Verein, 1874 (Tuebingen: L F Fues) [mf ed 1993] – (= ser Blvs 120) – 146p – 1 – (old french text. synopsis in german) – mf#8470 reel 25 – us Wisconsin U Libr [810]

Dimension – Philadelphia PA 1969-78 – 1,5,9 – ISSN: 0012-2890 – mf#8459 – us UMI ProQuest [240]
Dimension: contemporary german arts and letters – Austin TX 1968-94 – 1,5,9 – ISSN: 0012-2882 – mf#5042 – us UMI ProQuest [700]
Dimensional methods and their applications / Focken, Charles Melbourne – London, England. 1953 – 1r – us UF Libraries [500]
Dimensionen der alterung im sozialstaat : sozialpolitische und sozialgerontologische untersuchung des politikfeldes alternspolitik / Baur, Tobias – 1995 – 2mf – 9 – mf#DHS 2178 – gw Frankfurter [350]
Dimensionen der alterung im sozialstaat : sozialpolitische und sozialgerontologische untersuchung des politikfeldes alternspolitik / Baur, Tobias – (mf ed 1995) – 2mf – 9 – €40.00 – 3-8267-2178-0 – gw Frankfurter [350]
Dimensions – 1994 jan 15-feb 1, apr 1-may 1, jun 1, 1995 may 15-jun 30 – 1r – 1 – (cont by: dimensions news) – mf#2910644 – us WHS [071]
Dimensions : journal of pastoral concern – v1-10. 1969-1978 [complete] – 2r – 1 – ISSN: 0012-2890 – mf#ATLA S0899 – us ATLA [240]
Dimensions : the magazine of the national bureau of standards, us department of commerce / United States National Bureau of Standards – Washington DC 1974-81 – 1,5,9 – (cont: technical news bulletin) – ISSN: 0093-0458 – mf#5783,01 – us UMI ProQuest [600]
Dimensions / Ontario Metis and Non-Status Indian Association – v2 n1-v8 n2 [1974 jan-1980 jul] – 1r – 1 – mf#625709 – us WHS [307]
Dimensions : a publication of the wisconsin area united methodist church / United Methodist Church [US] – v1-13 [1969-81] – 1r – 1 – (cont by: united methodist record [dimensions ed]) – mf#614650 – us WHS [242]
Dimensions in american judaism – New York NY 1967-71 – 1 – ISSN: 0002-9653 – mf#3491 – us UMI ProQuest [270]
Dimensions in health service – Ottawa, Canada 1977-91 – 1,5,9 – (cont: canadian hospital) – ISSN: 0317-7645 – mf#562,01 – us UMI ProQuest [360]
Dimensions of critical care nursing: dccn – Baltimore MD 1982+ – 1,5,9 – ISSN: 0730-4625 – mf#14095 – us UMI ProQuest [610]
Dimensionstreue von abformsilikonen : eine in-vitro-untersuchung von 144 additions- und kondensationssilikonabformungen / Scheiner, Paola – (mf ed 1997) – 1mf – 9 – €30.00 – 3-8267-2495-X – mf#DHS 2495 – gw Frankfurter [617]
Dimier, Louis *see* Matres de la contre-revolution au dix-neuvieme siecle
Dimineata – Bucharest, Romania. -d. 30 Jan-22 May 1920. Imperfect. 1 reel – 1 – uk British Libr Newspaper [949]
Das diminutiv in der deutschen originalliteratur des 12. und 13. jahrhunderts / Hastenpflug, Fritz – Marburg, 1914 (mf ed 1994) – 2mf – 9 – €31.00 – 3-8267-3107-7 – mf#DHS-AR 3107 – gw Frankfurter [430]
Dimitrii (Tuptalo), Rostovskii *see* Zhitiia sviatykh...
Dimitroff, George Zakharieff et al *see* Telescopes and accessories
Dimitroff's letters from prison / Dimitrov, Georgi – New York, NY. 1935 – 1r – us UF Libraries [860]
Dimitrov, Georgi *see*
- Dimitroff's letters from prison
- Las lecciones de almeria
- Two years of heroic struggle of the spanish people

Dimitrovgradska pravda – Dimitrovgrad, Bulgaria. Dec 1954-1966 – 2r – 1 – (lacking: 1964-65) – us L of C Photodup [949]
Dimitrovsko zname – Pernik, Bulgaria. 1964-1970 – 4r – 1 – (lacking: 1965-66) – us L of C Photodup [949]
Dimitrovsko zname – Dimitrovo, Bulgaria. 1955-66 – 5r – 1 – (missing: 1964) – us L of C Photodup [949]
Dimmalaetting – Thorshavn, Faroe Islands. 1967-70 – 4r – 1 – uk British Libr Newspaper [072]
Dimmitt first baptist church. dimmitt, texas : church records – 1891-1963 – 1 – 55.17 – us Southern Baptist [242]
Dimock, A W *see* Florida enchantments
Dimock, Anthony Weston *see* Florida enchantments
Dimock, J F *see*
- Giraldi cambrensis opera, vols 5-7
- Magna vita s hugonis, episcopi lincolniensis

Dimock, James F *see* Holy communion
Dimock, Marshall Edward *see* Congressional investigating committees
Dimock, Nathaniel *see*
- The doctrine of the death of christ
- The doctrine of the sacraments in relation to the doctrines of grace
- The history of the book of common prayer in its bearing on present eucharistic controversies

Dimon, Denise *see* Latin american business review
Dimond, Susan B *see* Diaries
Dimond, W W *see* Diary
Dimostrationi harmoniche... : nelle quali realmente si trattano le cose della musica: & si risoluono molti dubij d' importanza / Zarlino, Gioseffo – Venetia: per Francesco de i Franceschi Senese 1571 [mf ed 19–] – 12mf – 9 – mf#fiche 908 – us Sibley [780]
Dimotheos, P S *see* Deux ans de sejour en abyssinie
Dimt, Peter *see* Die doktorsfamilie
The dina i main-i khyrat : or the religious decisions of the spirit of wisdom / ed by Peshotan, D D – Bombay, 1895 – 2mf – 9 – mf#NE-20156 – ne IDC [290]
Dinamani – Madras, India. 17 Aug 1950-1959; Jul-Sept 1966 – 24r – 1 – us L of C Photodup [079]
Dinamika (banjarmasin, kalimantan selatan, indonesia) – Banjarmasin: Yayasan Sapta Karya (3 times/wk) [mf ed Ithaca NY: [John M Echols Collection] Cornell University 2004] – 1r with other items [lacks: 1985 may 11,16] – 1 – (occasionally iss irregularly; iss for jan 3 1984-oct 22 1985 filmed with: dinamika (surabaya, indonesia), nov 16-dec 16 1965) – mf#mf-13948 seam – us CRL [079]
Dinamika izmeneniia polozheniia dagestanskoi zhenshchiny i semia / Gadzhieva, S – Moskva: In-t konkretnykh sotsialnykh issledovanii AN SSSR, 1972 – (filmed with various others) – us CRL [947]
Dinamika izmeneniia polozheniia zhenshchiny i semia / Sovetskaia sotsiologicheskaia assotsiatsiia & Institut konkretnykh sotsialnykh issledovanii AN SSSR & Orgkomitet XII Mezhdunarodnogo seminara po issledovaniiu semi – Moskva: the Institute, 1972 – (filmed with: dinamika izmeneniia polozheniia dagestanskoi zhenshchiny i semia/s gadzheiva) – us CRL [947]
Dinamika izmeneniia polozheniia zhenshchiny i semia / Sovetskaia sotsiologicheskaia assotsiatsiia & Institut konkretnykh sotsialnykh issledovanii AN SSSR & XII Mezhdunarodnyi seminar po issledovaniiu semi – Moskva: the Institute, 1972 – (filmed with: dinamika izmeneniia polozheniia dagestanskoi zhenshchiny i semia/s gadzhieva) – us CRL [947]
Dinamika izmeneniia polozheniia zhenshchiny i semia / Sovetskaia sotsiologicheskaia assotsiatsiia & Institut konkretnykh sotsialnykh issledovanii SSSR & Orgkomitet XII Mezhdunarodnogo seminara po issledovaniiu semi – Moskva: the Institute, 1972 – (filmed with: dinamika izmeneniia polozheniia dagestanskoi zhenshchiny i semia/s gadzhieva) – us CRL [947]
Dinamika narodnogo khoziaistva ukrainy : 1921/22-1924/25 gg – Khar'kov, 1926. 145, 25p – 2mf – 9 – mf#RHS-20 – ne IDC [314]
Dinas agama daerah bali : almanak hindu-bali – Np, 1962 – 2mf – 9 – mf#SE-327 – ne IDC [959]
Dinas Humas Pusat *see* Bulletin pertamina
Dinas pekerdjaan umum progress report sekretariat pemerintah daerah, kabupaten madiun : madiun, indonesia (kabupaten) – Madiun, 1969 – 2mf – 9 – mf#SE-1837 – ne IDC [959]
Dinas Penerbitan Balai Pustaka *see* Pustaka dan budaja
Dinas perikanan laut laporan tahunan – Djakarta, 1969-1970 – 6mf – 9 – mf#SE-1406 – ne IDC [959]
Dinas perikanan laut laporan tahunan – Surabaja, 1969-1971 – 20mf – 9 – mf#SE-1468 – ne IDC [959]
Dinas perindustrian daerah laporan kantor penjuluhan perindustrian / Indonesia – Djakarta, 1960 – 8mf – 9 – mf#SE-289 – ne IDC [959]
Dinas Pertanian dan Perikanan Laporan tahunan *see* Jogjakarta, indonesia (city)
Dinas pertanian rakjat daerah swatantra tingkat 1 / Tani – Bukittinggi, [1954]-1959(8) – 2mf – 9 – (missing: [1954]-1959 v1-6(1-3, 5-6)) – mf#SE-845 – ne IDC [950]
Dinas purbakala : warna warta kepurbakalaan / Amerta – Djakarta, 1952-1955 – 5mf – 9 – mf#SE-650 – ne IDC [959]
Dinas purbakala berita = Bulletin of the archaeological service of the republic of indonesia – Djakarta, 1955-1958 – 4mf – 9 – mf#SE-654 – ne IDC [959]
Dinasari – Madras, India. 1947-51 – 9r – 1 – us L of C Photodup [079]
Dincklage-Campe, Friedrich, Freiherr von *see* Anker geschlippt
Dindorfii, Guil *see* Georgius syncellus et nicephorus cp (cshb12,13)
Dindorfius, L *see*
- Chronicon paschale ad exemplar vaticanum
- Chronographia

Dine baa-hane – v1 n3, v2 n5-v4 n4 [1969 oct 21, 1971 jan-1973 jul] – 1r – 1 – mf#366012 – us WHS [071]
Dine baa-hani – 1969-72 – (= ser American indian periodicals... 1) – 15mf – 9 – $115.00 – us UPA [305]
Dine be'iina' / Navajo Community College – 1987 spring-1988 winter – 1r – 1 – mf#1701113 – us WHS [373]
Dine Bi Olta Association *see* Navajo education
Diner de madelon : ou, le bourgeois du marais – Paris, France. 1819 – 1r – us UF Libraries [440]
Diner de pierrot / Millanvoye, Bertrand – Paris, France. 1881 – 1r – us UF Libraries [440]
Diner en musique : fantaisie gastronomico-musicale en deux actes...14 fevrier 1931 / Morin, Victor – Montreal: Therien freres, [1931] (mf ed 1987) – 1mf – 9 – mf#SEM105P786 – cn Bibl Nat [790]
Diner en musique : fantaisie gastronomico-musicale en deux actes...28 novembre 1935 / Morin, Victor – [Montreal?]: [s.n.], [1935] (mf ed 1987) – 1mf – 9 – mf#SEM105P790 – cn Bibl Nat [790]
Diner en musique : fantaisie gastronomico-musicale en deux actes...3 fevrier 1930 / Morin, Victor – Montreal: Impr Therien freres, 1930 [mf ed 1987] – 1mf – 9 – mf#SEM105P787 – cn Bibl Nat [790]
Le diner interrompu ou nouvelle farce de jocrisse : piece comique en un acte / Doin, Ernest – Montreal: Beauchemin, 1879? – 1mf – 9 – mf#33378 – cn Canadiana [820]
Diner offert a l'honorable adelard turgeon : par ses amis de levis a l'occasion de son depart pour quebec au club de la garnison, jeudi, le 26 septembre 1901 – Levis: [s.n.], 1901 [mf ed 1985] – 1mf – 9 – mf#SEM105P462 – cn Bibl Nat [920]
El dinero de san pedro : carta pastoral que... adolfo perez munoz dirige al clero y fieles de su diocesis / Perez Munoz, Adolfo – Badajoz: tip uceda hnos, 1915 – 1 – sp Bibl Santa Ana [240]
Diner-operette en deux actes : fantaisie gastronomico-musicale offerte par la societe saint-jean-baptiste de montreal pour couronner le troisieme congres de la langue francaise au canada / Morin, Victor – [Montreal?]: [s.n.], [1952?] (mf ed 1987) – 1mf – 9 – mf#SEM105P788 – cn Bibl Nat [790]
Diner-operette en deux actes : fantaisie gastronomico-musicale offerte par la societe saint-jean-baptiste de montreal, sous les auspices de la commission du 3e centenaire de montreal / Morin, Victor – 7e ed. [Montreal]: [s.n.], [1942] (mf ed 1987) – 1mf – 9 – mf#SEM105P789 – cn Bibl Nat [790]
Diner-operette en deux actes : a gastronomico-musical fantasy offered to the fellows of the royal society of canada on the occasion of their annuel meeting in montreal 1939 / Morin, Victor – [Montreal?]: [s.n.], [1939] (mf ed 1987) – 1mf – 9 – mf#SEM105P791 – cn Bibl Nat [790]
Les diners : ou conversations politiques entre quatre deputes des differens cotes de la chambre de 1820 dedies a messieurs les electeurs de la cinquieme serie – Paris 1821 [mf ed Hildesheim 1995-98] – (= ser Fbc) – 1mf – 9 – €40.00 – 3-487-29088-X – gw Olms [325]
Diners a trente-deux sous / Cogniard, Theodore – Paris, France. 1843 – 1r – us UF Libraries [440]
Dineshon, Jacob *see* Hersheleh
Dineson, Jacob *see*
- Alter
- Krizis

Y dinesydd cymreig = Welsh citizen – [Wales] Gwynedd 8 mai 1912-10 gor 1929 [mf ed 2004] – 18r – 1 – uk Newsplan [072]
Dinge der zeit : buchausgabe der fuenf hefte / Flake, Otto – Muenchen: Roland-Verlag, 1921 (mf ed 1990) – 1r – 1 – (filmed with: a passage in the night) – us Wisconsin U Libr [840]
Dingelstedt, Franz *see* Wanderbuch
Dingelstedt, Franz, Freiherr von *see*
- Aus der briefmappe eines burgtheaterdirektors
- Eine faust-trilogie
- Die neuen argonauten
- Saemmtliche werke

Die dingelstocks : der weg einer sippe: roman / Wurtz, Johann – Belgrad: Verlags- und Vertriebs- A G "Suedost", 1943 – 1r – 1 – us Wisconsin U Libr [830]
Dinger, Hugo *see*
- Die meistersinger von nuernberg
- Versuch einer darstellung der weltanschauung richard wagners

Dinger, Mary K *see* Development and validation of the wellness knowledge, attitude, and behavior instrument
Dingfelder, S *see* Vierzig jahre israelitischer lehrer-verein fur bayern, 1880-1920
Dingle, Edwin John *see*
- Across china on foot
- China's revolution, 1911-1912

Dingle, Reginald James *see*
- Democracy in spain
- Russia's work in spain
- Second thoughts on democracy in spain

[Dinglers] polytechnisches journal – 1820-1931 [mf ed 1994] – 346v on 1167mf (text) 346mf (ill) – 9 – €7670.00 – 3-89131-169-9 – gw Fischer [600]
Dingman, Benjamin S *see* Ten years in south america
Dingolfinger anzeiger – Dingolfing DE, 1083 1 jun- – ca 9r/yr – 1 – gw Misc Inst [074]
Dingwall, fordyce and connections – Fergus, Ont?: s.n, 1884? – 2mf – 9 – mf#07122 – cn Canadiana [920]
Dinh Quang Nha *see* Dat trong lang
Dinh, Tien Luyen *see*
- Anh chi yeu dau
- Chu nhat uyen uong
- Thoi nho cua nang

The din-i-ilahi : or, the religion of akbar / Roy Choudhury, Makhan Lal – Calcutta: University of Calcutta, 1941 – (= ser Samp: indian books) – us CRL [280]
The din-i-ilahi: or, the religion of akbar / Roychoudhury, Makhanlal, Sastri – 2nd ed. Calcutta: Das Gupta, 1952. xxxiii,222p. ill – 1 – us Wisconsin U Libr [260]
Dining room employee – v14 n9-v25 n4 [1967 sep-1977 apr] – 1r – 1 – (cont by: restaurant employee) – mf#645137 – us WHS [331]
Diniz, Silvio Gabriel *see* O goncalvismo em pitangui
Dinkard / ed by Behramjee, P & Sunjana, P D B – Bombay, 1874-1928. 19 v – 15mf – 9 – mf#NE-20149 – ne IDC [956]
Dinnaga *see* Kundamala of dinnaga
Dinner, bill of fare : walker house, toronto, david walker, proprietor – [Toronto?]: R Smith [1888?] [mf ed 1987] – 1mf – 9 – 0-665-35860-1 – mf#35860 – cn Canadiana [640]
Dinschel, Kimberly M *see* The influence of agility on the mile run and pacer tests of aerobic endurance in fourth- and fifth-grade school children
Dinsmore, Charles Allen *see*
- Aids to the study of dante
- Atonement in literature and life
- The teachings of dante

Dinter, Artur *see*
- Die schmuggler
- Die suende wider das blut

Dinter, K *see* Botanische reisen in deutsch-suedwest-afrika
[Dinuba-] dinuba sentinel – CA. 1909-17; 1919-36; 1938-39; Jan-Dec 1948; 1952-79; 1981- – 85r – 1 – $5100.00 (subs $50/y) – mf#BC02170 – us Library Micro [071]
[Dinuba-] dinuba tribune : alta district special edition – CA. 15 Mar 1906 – 1r – 1 – $60.00 – mf#B02171 – us Library Micro [071]
[Dinuba-] news of orange cove – CA. 1983-1984 – 1r – 1 – $60.00 – mf#B06022 – us Library Micro [071]
[Dinuba-] orange cove and mountain news – CA. 1985 – 1r – 1 – $60.00 – mf#B06023 – us Library Micro [071]
[Dinuba-] the dinuba advocate – CA. 1907-1915 – 5r – 1 – $300.00 – mf#B06020 – us Library Micro [071]
[Dinuba-] the graphic – CA. 1895-1896 – 1r – 1 – $60.00 – mf#B06021 – us Library Micro [071]
Dinuzulu / Binns, C T – London, England. 1968 – 1r – us UF Libraries [960]
Diobounioits, C *see* Der scholien-kommentar das origenes zur apokalypse johannis
Diobouniotis, C *see* Hippolyts danielcommentar
The diocesan and parish magazine, victoria, b c – Victoria [BC: s.n., 1887?-189- or 1?] – 9 – ISSN: 1190-6383 – mf#P04660 – cn Canadiana [242]
Diocesan archives / Rarotonga and Niue. Catholic Church Diocese – 1891-1993 – 53r – 1 – (available for reference) – mf#PMB1064 – at Pacific Mss [240]
Diocesan Church Society of New Brunswick *see*
- Fiftieth report...1885
- Fifty-fifth report...1890
- Fifty-first report...1886
- Fifty-fourth report...1889
- Fifty-second report...1887
- Fifty-third report...1888
- Fortieth report...1875
- Forty first report...1876
- Forty-eighth report...1883
- Forty-fifth report...1880
- Forty-fourth report...1879
- Forty-ninth report...1884
- Forty-second report...1877
- Forty-seventh report...1882
- Forty-sixth report...1881
- Forty-third report...1878
- Seventeenth report of the proceedings...during the year 1852

– Sixteenth report of the proceedings...during the year 1851
– Thirty first report...1866
– Thirty-fifth report...1870
– Thirty-fourth report...1869
– Thirty-ninth report...1874
– Thirty-second report...1867
– Thirty-seventh report...1872
– Thirty-sixth report...1871
– Twentieth report of the proceedings...during the year 1855
– Twenty fifth report...1860
– Twenty seventh report...1862
– Twenty sixth report...1861

Diocesan Human Relations Services *see* Wabanaki alliance

Diocesan synods and convocation / Garbett, James – London, England. 1852 – 1r – us UF Libraries [240]

Le diocese de montreal a la fin du dix-neuvieme siecle : avec portraits du clerge, helio-gravures et notices historiques de toutes les eglises et presbyteres institutions d'education et de charite... / Dauth, Gaspard – Montreal: Eusebe Senecal & cie, 1900 [mf ed 1987] – 1r – 5 – (pref by raphael bellemare) – mf#SEM16P366 – cn Bibl Nat [241]

Diocese of florida annual council / Episcopal Church Diocese Of Florida – s.l, s.l? 1897-1910 – 1r – us UF Libraries [978]

Diocese of mackenzie river / Bompas, William Carpenter – London: SPCK, 1888 – (= ser Colonial church histories) – 2mf – 9 – mf#00169 – cn Canadiana [242]

The diocese of quebec : its natural features, equipment, financial system, character of work... / Balfour, Andrew Jackson – [Quebec?: s.n, 1910?] – 1mf – 9 – 0-665-86240-7 – mf#86240 – cn Canadiana [242]

The diocese of st. paul : the golden jubilee, 1851-1901 – St Paul: Pioneer Press, [1901?] – 1mf – 9 – 0-524-03836-8 – mf#1990-4883 – us ATLA [240]

La diocesis de badajoz. estadistica de 1970 / Badajoz. Diocesis – Badajoz: Imp. Espanola, 1970 – 1 – sp Bibl Santa Ana [240]

Diodor van tarsus : vier pseudojustinische schriften als eigentum diodors / Harnack, Adolf von – Leipzig, 1901 – (= ser Tugal 2-21/4) – 4mf – 9 – €11.00 – ne Slangenburg [240]

Diodor von tarsus : vier pseudojustinische schriften als eigentum diodors / Harnack, Adolf von – Leipzig: J C Hinrichs, 1901 [mf ed 1989] – (= ser Tugal 21/4) – 1mf – 9 – 0-7905-1763-9 – (in german & greek) – mf#1987-1763 – us ATLA [240]

Diodori bibliotheca historica : ex recensione et cum annotationibus ludovici dindorfii / Diodorus, Siculus – Lipsiae: B G Teubneri 1866-68 [mf ed 1973] – (= ser Bibliotheca scriptorum graecorum et romanorum teubneriana) – 5v in 7 on 1r – 1 – mf#film mas 4030 – us Harvard [900]

Diodorus, Siculus *see* Diodori bibliotheca historica

Die dioecesansynode / Phillips, George – Freiburg i B: Herder, 1849 – 1mf – 9 – 0-7905-6943-4 – (incl bibl ref) – mf#1988-2943 – us ATLA [240]

Dioeceses brixinensis, frisingensis, ratisbonensis (mgh antiquitates 2:3.bd) – 1905 – (= ser Monumenta germaniae historica antiquitates. 2. necrologia germaniae (mgh antiquitates 2)) – €27.00 – ne Slangenburg [241]

Dioecesis augustensis, constantiensis, curiensis (mgh antiquitates 2:1.bd) – 1888 – (= ser Monumenta germaniae historica antiquitates. 2. necrologia germaniae (mgh antiquitates 2)) – €40.00 – ne Slangenburg [241]

Dioecesis pataviensis (mgh antiquitates 2:4.bd) : pars 1.1: dioecesis pataviensis regio bavarica. 2: dioecesis pataviensis regio austriaca nunc lentensis – 1920 – (= ser Monumenta germaniae historica antiquitates. 2. necrologia germaniae (mgh antiquitates 2)) – €40.00 – ne Slangenburg [241]

Dioecesis pataviensis (mgh antiquitates 2:5.bd) : pars 2: austria inferior 1913 – 1913 – (= ser Monumenta germaniae historica antiquitates. 2. necrologia germaniae (mgh antiquitates 2)) – €38.00 – ne Slangenburg [241]

Dioecesis salisburgensis (mgh antiquitates 2:2.bd) – 1904 – (= ser Monumenta germaniae historica antiquitates. 2. necrologia germaniae (mgh antiquitates 2)) – €40.00 – ne Slangenburg [241]

Diogene sans-culotte – Paris: E Bautruche, jun 18/22-22/25 1848 – us CRL [074]

Diogenes – Madison, Wisc. v1 n1-2. oct nov 1940-dec 1940 jan 1941 – 1 – us NY Public [073]

Diogenes – Montreal: G Burden. v1-3. nov 13 1868-feb 4 1870// (wkly) – 1r – 1 – Can$75.00 – cn McLaren [400]

Diogenes [english ed] – London, England 1953+ – 1,5,9 – ISSN: 0392-1921 – mf#3066 – us UMI ProQuest [000]

Diogenes Laertius *see* Historia...

Diogenes, Laertius *see* The lives and opinions of eminent philosophers

Diogenes review – v5 n6,11-13,15-19 [1983 apr 15, aug 15-sep 15, oct 15-dec 15], v6 n1,3-4,7-12,14 [1984 jan 1, feb 1-15, jun 1-aug 15, sep 15] – 1r – 1 – mf#1546694 – us WHS [071]

Diogenis laertii de clarorum philosophorum – Paris, France. 1878 – 1r – us UF Libraries [180]

Diogo, Alfredo *see* Angola perante uma conspiracao internacional

Dion, Albert, abbe *see*
– Album-souvenir du 3e centenaire du quebec, 1608-1908

Dion, J O *see* Souvenir du reverend pierre marie migneault sic

Dion, Marie Berthe *see*
– Ideas sociales y politicas de arevalo

Dione, Salif *see* L'education traditionelle a travers les chants et les poemes

Dionigi da Fano, B *see* Viaggio di m cesare de i fredrici nell' india orientale et, oltra l'india...

Dionisos / Pagan, Juan Bautista – San Juan, Puerto Rico. 1957 – 1r – us UF Libraries [972]

Dionne, Narcisse Eutrope *see*
– L'abbe gabriel richard
– Les ecclesiastiques et les royalistes francais refugies au canada a l'epoque de la revolution, 1791-1802
– Fete nationale des canadiens-francais celebree a windsor, ontario, le 25 juin 1883
– Galerie historique
– Inventaire chronologique
– Inventaire chronologique des livres, brochures, journaux et revues publies en diverses langues dans et hors la province de quebec
– Jacques cartier
– Le parler populaire des canadiens francais

Dionne, Narcisse-Eutrope *see*
– Etats-unis, manitoba, et nord-ouest
– Jean-francois de la roque
– La "petite hermine" de jacques cartier et diverses monographies historiques
– Sainte-anne-de-la-pocatiere, 1672-1900
– Serviteurs et servantes de dieu en canada
– The siege of quebec and the battle of the plains of abraham
– Travaux historiques publies depuis trente ans

Dionys-bacsi : drei novellen / Schaukal, Richard von – Braunschweig: G Westermann, 1922 – 1r – 1 – us Wisconsin U Libr [830]

Dionysiac as a determining factor in nietzsche's philosophy of superman / Abrahams, Glenis Gail – Belville: U of Western Cape 1984 [mf ed 1984] – 5mf – 9 – sa Misc Inst [190]

Dionysios, proklos, plotinus / Mueller, H F – 1918 – (= ser Bgphma 20/3-4) – €5.00 – ne Slangenburg [140]

Dionysius Areopagita *see* Ueber die beiden hierarchien (bdk2 1.reihe)

Dionysius Areopagita (Dionysius the Areopagite) *see* Ausgewaehlte schriften (bdk2 2.reihe)

Dionysius de Leewis *see*
– Speculum aureum animae peccatricis
– Speculum conversionis peccatorum

Dionysius dreytweins esslingische chronik : 1548-1564 / ed by Diehl, Adolf – Stuttgart: Litterarischer Verein, 1901 (Tuebingen: H Laupp, Jr) [mf ed 1993] – (= ser Blvs 221) – xxiii/326p – 1 – (incl bibl ref and ind) – mf#8470 reel 46 – us Wisconsin U Libr [943]

Dionysius dreytweins esslingische chronik / Dreytwein, Dionysius; ed by Diehl, Adolf – Stuttgart: Litterarischer Verein, 1901 (Tuebingen: H Laupp, Jr) – (incl bibl ref and ind) – us Wisconsin U Libr [920]

Dionysius of Alexandria, Saint *see* The letters and other remains of dionysius of alexandria

Dionysius of Halicarnassus *see* Hierarchies

Dionysius The Carthusian *see* Opera omnia

Dionysius the ps-areopagite : the ecclesiastical hierarchy / Campbell, Th L – Washington, DC. n83 1955 – 2mf – 8 – €5.00 – ne Slangenburg [241]

Dionysos and immortality : the greek faith in immortality as affected by the rise of individualism / Wheeler, Benjamin Ide – Boston: Houghton, Mifflin, 1899 – (= ser Ingersoll Lecture) – 1mf – 9 – 0-524-02330-1 – mf#1990-2953 – us ATLA [250]

Dionysos and immortality; the greek faith in immortality as affected by the rise of individualism / Wheeler, Benjamin Ide – Boston, New York: Houghton, Mifflin, 1899. (Ingersoll Lectures on Immortality) – 1 – us Wisconsin U Libr [250]

Diop, Birago *see* Les contes d'amadou-koumba

Diop, Cheikn Anta *see* Nations negres et culture

Diorama de londres : ou tableau des moeurs britanniques en mil huit cent vingt-deux / Salle, Eusebe F de – Paris 1823 [mf ed Hildesheim 1995-98] – (= ser Fbc) – 3mf – 9 – €90.00 – 3-487-27948-7 – gw Olms [941]

Dios esta aqui / Jesus, Gabriel de – Madrid, 1933 – 1 – sp Bibl Santa Ana [240]

Dios immortal... / Stanihursto, Guillermo – 1826. Francisco P. Berguizas, trans – 9 – sp Bibl Santa Ana [240]

Dios inmortal...pasion de cristo / Stanihursto, Guillermo – 1826 – 9 – sp Bibl Santa Ana [240]

Dios, patria y libertad / Morillo, Gabriel A – Moca, Dominican Republic. 1926? – 1r – us UF Libraries [972]

Dios sobre todo / Spinola de Gironza, Araceli – Ediciones Ritmo, S, L. Madrid, 1940 – 1 – sp Bibl Santa Ana [946]

Dios y espana : o sea ensayo de lo que debe a la r catolica / Amado, Manuel – v. 1-2. 1831 – 9 – sp Bibl Santa Ana [241]

Dioscoro patriarcha alexandrino / Caceres, Diego de – S.l., s.i., s.a. Hacia 1641 – 1 – sp Bibl Santa Ana [240]

The dioscuri in the christian legends / Harris, James Rendel – London: C J Clay, 1903 – 1mf – 9 – 0-7905-1887-2 – (incl bibl ref) – mf#1987-1887 – us ATLA [240]

Diospolis parva : the cemeteries of abadiyeh and hu, 1898-1899 / Petrie, W M – London, 1901 – 3mf – 9 – mf#NE-20350 – ne IDC [956]

Diospolis parva : the cemetries of abadiyeh and hu / Flinders Petrie, W M – London, 1901 – (= ser Mees 20) – 7mf – 8 – €16.00 – ne Slangenburg [930]

Diosy, Arthur *see* The new far east

Dioum, Abdoulaye *see* Les exploits de masire isse dieye

Dipanagara Pangerannja *see* Babad diponagoro

Dipanda – Brazzaville, nov 1963-oct 10 1967; special issue 1967 – 1 – us CRL [079]

The dipavamsa and mahavamsa and their historical development in ceylon = dipavamsa und mahavamsa / Geiger, Wilhelm – Colombo: HC Cottle, 1908 – 1mf – 9 – 0-524-07139-X – (incl bibl ref. in english) – mf#1991-0069 – us ATLA [280]

Diplomacia do marechal / Costa, Sergio Correa Da – Rio de Janeiro, Brazil. 1945 – 1r – us UF Libraries [972]

Diplomacia en nuestra historia / Marquez Sterling, Manuel – Habana, Cuba. 1909 – 1r – us UF Libraries [972]

Diplomacy and the borderlands / Brooks, Philip Coolidge – Berkeley, CA. 1939 – 1r – us UF Libraries [978]

Diplomata do imperio / Souza, Jose Antonio Soares De – Sao Paulo, Brazil. 1952 – 1r – us UF Libraries [972]

Diplomata in folio, vol 1 – (= ser Monumenta Germaniae Historica) – 7mf – 8 – mf#380 – ne IDC [430]

Diplomata na corte da inglaterra / Mendonca, Renato – Rio de Janeiro, Brazil. 1968 – 1r – us UF Libraries [972]

Diplomata na corte de inglaterra / Mendon Ca, Renato – Sao Paulo, Brazil. 1942 – 1r – us UF Libraries [327]

Diplomata regum francorum e stirpe merovingica (mgh diplomata 1:1.bd) – 1872 – (= ser Monumenta germaniae historica diplomata. 1. diplomata in folio (mgh diplomata 1)) – €23.00 – ne Slangenburg [931]

Diplomate / Scribe, Eugene – Paris, France. 1828 – 1r – us UF Libraries [440]

Diplomates et diplomatie / Firmin, Antenor – Cap-Haitien, Haiti. 1899 – 1r – us UF Libraries [327]

Diplomatic and consular instructions...1791-1801 / U.S. Dept of State – (= ser General records of the department of state) – 5r – 1 – (with printed guide) – mf#M28 – us Nat Archives [327]

Diplomatic correspondence of british ministers to the russian court at st petersburg 1704-1776 : a detailed, comprehensive record of correspondence between the two countries – [mf ed Chadwyck-Healey] – 3342 titles on 100mf – 9 – (with p/ind. in french & english) – uk Chadwyck [327]

The diplomatic correspondence of the united states of america, (1783-1789) : from the signing of the definitive treaty of peace, 10th september, 1783, to the adoption of the constitution, march 4, 1789... – Washington: printed by Francis Preston Blair 1832-34 – v1-7 on 42mf – 9 – $63.00 – mf#llmc97-243 – us LLMC [327]

Diplomatic despatches...to haiti, 1862-1906 / U.S. Dept of State – (= ser General records of the department of state, 1910-1929 decimal file) – 47r – 1 – mf#M82 – us Nat Archives [324]

Diplomatic despatches...to liberia, 1863-1906 / U.S. Dept of State – (= ser General records of the department of state, 1910-1929 decimal file) – 14r – 1 – mf#M170 – us Nat Archives [324]

Diplomatic despatches...to the dominican republic, 1883-1906 / U.S. Dept of State – (= ser General records of the department of state, 1910-1929 decimal file) – 15r – 1 – mf#M93 – us Nat Archives [324]

Diplomatic history – Oxford, England 1995+ – 1,5,9 – ISSN: 0145-2096 – mf#18420 – us UMI ProQuest [327]

Diplomatic history of the american revolution / U.S. Dept of State; ed by Wharton, Francis A – 9 – (suppl to wharton's digest of the international law of the u.s.) – mf#LLMC 82-979 – us LLMC [976]

A diplomatic history of the congo free state / Muller, George F – Washington, DC, 1948 – us CRL [960]

Diplomatic history of the panama canal : correspondence relating to the negotiation and application of certain treaties on the subject of the construction of an interoceanic canal and accompanying papers – Senate doc no 474. 63rd Congress 2nd sess. Washington: GPO, 1914 – 7mf – 9 – $10.50 – mf#LLMC 82-100D Title 19 – us LLMC [324]

Diplomatic instructions...1801-1906 / U.S. Dept of State – (= ser General records of the department of state, 1910-1929 decimal file) – 175r – 1 – (with printed guide) – mf#M77 – us Nat Archives [327]

Diplomatic papers of john moors cabot, 1929-1978 – 1 – (pt1: latin america 5r $770 isbn 0-89093-477-0. pt2: europe 6r $920 isbn 0-89093-478-9. pt3: general political & diplomatic materials 6r $920 isbn 0-89093-479-7. pt4: diaries 5r $770 isbn 0-89093-480-0. with p/g) – us UPA [327]

Diplomatic petrel / Hohler, Thomas Beaumont – 1st ed. London: J Murray, [1942] – 1 – us CRL [960]

Diplomatic reminiscences of lord augustus loftus *see*
– Native life in south africa
– Der sentenzenkommentar peters von candia, des pisaner papstes alexanders 5

The diplomatic reminiscences of lord augustus loftus...1837-62 / Loftus, Augustus William Frederick Spencer, Lord – London: Cassell & Co., Ltd., 1892. 2v – 1 – us Wisconsin U Libr [941]

Diplomatic review – v1-25 n2,1+suppl n1-61. 1855-81 [all publ] – (= ser Radical periodicals of great britain, 1794-1914. period 1) – 6r – 1 – $470.00 – us UPA [327]

Diplomatico mexicano en paris / Mangino, Fernando – Mexico City? Mexico. 1948 – 1r – us UF Libraries [327]

Diplomatie de la france et de l'espagne depuis l'avenement de la maison de bourbon / Capefigue, Jean Baptiste Honore Raymond – Paris 1846 [mf ed Hildesheim 1995-98] – 1v on 3mf – 9 – €90.00 – 3-487-26084-0 – gw Olms [327]

Diplomatische und curieuse nachlese der historie von ober-sachsen und angraentzenden laendern – Dresden, Leipzig DE, 1730/31-1733, theil 1-12 – 3r – 1 – gw Misc Inst [943]

Ein diplomatischer briefwechsel : aus dem zweiten jahrtausend vor christo / Klostermann, August – 2. Aufl. Leipzig: A Deichert (Georg Boehme), 1902 – 1mf – 9 – 0-8370-7226-3 – (incl bibl ref) – mf#1986-1226 – us ATLA [470]

Diplomats, scientists, and politicians / Jacobson, Harold Karan – Ann Arbor, MI. 1966 – 1r – us UF Libraries [025]

Diplomatum belgicorum nova collectio / Miraeus, A; ed by Foppens, J F – Bruxellis, 1734-1748 – €128.00 – ne Slangenburg [240]

Diplome de cabaleur pour les elections, delivre par le comite anti-national – Paris, 1849 – us CRL [325]

Dipomacia en venezuela / Pulido Santana, Maria Trinidad – Caracas, Venezuela. 1963 – 1r – us UF Libraries [327]

Dippel, Horst *see*
– Constitutions of the world 1850 to the present, pt 1
– Constitutions of the world 1850 to the present, pt 2

Diptera scandinaviae disposita et descripta / ed by Zetterstedt, J W – Lundae 1842-60 – 14v on 93mf – 9 – mf#z-1740/2 – ne IDC [590]

DiPuma, Joseph J *see* Evaluation of collegiate coaches from the perspective of the student-athlete

Diputacion Provincial *see*
– Exposicion de reproducciones en color de la unesco 90 anos de pintura universal
– Memoria, ano de 1978
– Oficina provincial de inversiones informe anual. diciembre de 1978
– Ordenanza fiscal, num. 3 para la exaccion de derechos y tasas por prestaciones de servicios...sanitarios...1968
– Presupuesto ordinario de gastos e ingresos. ejercicio 1966
– Presupuesto ordinario de gastos e ingresos. ejercicio de 1963
– Presupuesto ordinario de gastos e ingresos. ejercicio de 1967
– Presupuesto ordinario de gastos e ingresos. ejercicio de 1968
– Presupuesto ordinario de ingresos y gastos
– Presupuesto ordinario de ingresos y gastos. ejercicio de 1960
– Presupuesto ordinario de ingresos y gastos. ejercicio de 1971

- Presupuesto ordinario de ingresos y gastos para el ejercicio de 1973...
- Reglamento de la caja de credito provincial
- Reglamento de los servicios benefico-sanitarios
- Reglamento del centro de estudios extremenos
- Reglamento para la concesion de becas de estudios que establece esta corporacion
- Ssmm los reyes de espana en la diputacion de caceres

Diputados pintados por sus hechos... – Madrid, 1870 – 29mf – 9 – sp Cultura [946]

Diputados por cuba en las cortes de espana / Entralgo, Elias Jose – Habana, Cuba. 1945 – 1r – us UF Libraries [972]

Diputationes in universam aristotelis logicam / Peinado, I – Alcala de Henares, 1721 – 1mf – 9 – sp Cultura [160]

Dirac, Pam (Paul Adrien Maurice) *see* Principles of quantum mechanics

Dirasat askariyah – [Lebanon?: Harakat al-Tahrir al-Watani al-Filastini, "Fath", 1970. [n1-2. jun-jul 1970] – 1r – us CRL [079]

Dirceu – Rio de Janeiro, RJ: Typ do Dirceu, 05 jul 1885 – (= ser Ps 19) – mf#P17,01,156 – bl Biblioteca [440]

Dircks, Henry *see* A biographical memoir of samuel hartlib

Direccion general de archivos y bibliotecas : exposicion... / Barrado Manzano, Arcangel – Madrid: Archivo Ibero-Americano, 1959 – 1 – sp Bibl Santa Ana [020]

Direccion General de Estadistica *see*
- Nomenclator de las ciudades, villas, lugares, aldeas y demas entidades de poblacion de espana...con referencia al 31-12-1940. provincia de badajoz
- Nomenclatura de las ciudades, villas, lugares, aldeas y demas entidades de poblacion de espana...con referencia al 31-12-1940. provincia de caceres

Direccion General de Estadistica, Ministerio de Trabajo *see* Censo de las poblaciones de espana segun la inscripcion de 31 de diciembre de 1940

Direccion General de Ganaderia. Junta Provincial de Fomento Pecuario de Badajoz *see* Cartilla divulgadora. sobre explotacion ovina en su faceta de lana

Direccion General de Turismo *see* Breve resena de badajoz

Direccion Liberal Nacional (Colombia) *see* Quince meses de politica liberal

Direct action / Industrial Workers of the World – special suppl [1978], v1 n1-5 [1978 feb-dec 8] – 1r – 1 – mf#626035 – us WHS [331]

Direct action for a non-violent world / New England Committee for Non-violent Action – n1-31 [1960 jun 2-1973 apr 9] – 1r – 1 – mf#1055557 – us WHS [327]

The direct and fundamental proofs of the christian religion : an essay in comparative apologetics...lectures for 1903 / Knox, George William – New York: Scribner 1903 [mf ed 1985] – (= ser Nathaniel william taylor lectures 1903) – 1mf – 9 – 0-8370-3949-5 – mf#1985-1949 – us ATLA [240]

Direct answers to plain questions : handbook for american churchmen / Scadding, Charles – Milwaukee, WI: Young Churchmen, c1901 [mf ed 1993] – (= ser Anglican/episcopal coll) – 1mf – 9 – 0-524-07110-1 – mf#1991-2933 – us ATLA [242]

Direct from cuba – Havana, Cuba 1979-81 – 1,5,9 – ISSN: 0046-0338 – mf#8579 – us UMI ProQuest [320]

Direct legislation record / National Direct Legislation League [US] – v1 n1-v8 n3 [1894 may-1901 sep] – 1r – 1 – (cont by: direct legislation record and the proportional representation review) – mf#1218464 – us WHS [323]

Direct legislation record and the proportional representation review / National Direct Legislation League [US] – v8 n4-v10 n4 [1901 dec-1903 dec] – 1r – 1 – (cont: direct legislation record; proportional representation review; cont by: equity series) – mf#1219282 – us WHS [071]

Direct marketing – Garden City NY 1938+ – 1,5,9 – ISSN: 0012-3188 – mf#299 – us UMI ProQuest [650]

Direct route through the north-west territories of canada to the pacific ocean : the chartered hudson's bay and pacific railway route (with a map) / Harris, Josiah [comp] – [London?: s.n.], 1897 [mf ed 1982] – 2mf – 9 – mf#30294 – cn Canadiana [380]

The direct system of ladies cutting / Holding, Thomas Hiram – London [1897] – (= ser 19th c art & architecture) – 3mf – 9 – mf#4.2.395 – uk Chadwyck [740]

Directeur de l'Ecole d'agriculture de Ste-Anne *see* Les ecoles d'agriculture de la province de quebec vengees

Direction – Los Angeles, CA.Winter 1973/74-Summer 1985. Many issues missing. In English – 1 – us AJPC [071]

Direction de l'education publique, G.M.Z.F.O. *see* Campagne in frankreich

Direction de l'enseignement primaire. rapport sur l'organisation et la situation de l'enseignement primaire public en france / France. Ministere de l'Instruction Publique et des Beaux-Arts – Paris. 1900 – 1 – fr ACRPP [324]

Direction of trade – Washington DC 1972-80 – 1,5,9 – (cont by: direction of trade statistics) – ISSN: 0012-3226 – mf#6528.02 – us UMI ProQuest [380]

Direction of trade statistics – Washington DC 1981-94 – 1,5,9 – (cont: direction of trade; cont by: direction of trade statistics quarterly) – ISSN: 0252-306X – mf#6528.02 – us UMI ProQuest [380]

Direction of trade statistics quarterly – Washington DC 1995+ – 1,5,9 – (cont: direction of trade statistics) – mf#6528,02 – us UMI ProQuest [380]

Direction one / Sir George Williams University – 1970 mar – 1r – 1 – mf#1583170 – us WHS [378]

Direction pour la culture du tabac / Schmouth, J E – Quebec?: A Cote, 1865 – 1mf – 9 – mf#47502 – cn Canadiana [630]

Direction pour la culture en vert du ble-d'inde et son ensilage / Beaubien, Louis – S.l: s.n, 1889? – 1mf – 9 – mf#53682 – cn Canadiana [630]

Directions / Madison Opportunity Center, Inc – v1 n1-3 [1983 summer-fall], v2 n1,3 [1984 spring, fall], v3 n1,2-3 [1985 spring, fall-winter] v4 n1-3 [1986 summer-winter], v5 n1 [1987 spring] – 1r – 1 – mf#1111215 – us WHS [331]

Directions 76 / Colorado Centennial-Bicentennial Commission – [v1 n4?]-v3 n11 [1974 [apr.?]-1976 dec] – 1r – 1 – mf#366017 – us WHS [975]

Directions and forms for the execution and acknowledgement of deeds to be used or recorded in other states / Butts, Isaac Ridler – Boston, Butts, 1857. 108 p. LL-275 – 1 – us L of C Photodup [340]

Directions de navigation pour l'ile de terreneuve et la cote du labrador et pour le golfe et le fleuve st-laurent / Bayfield, Henry Wolsey – Quebec: Impr Elzear Vincent, 1864 [mf ed 1983] – 3mf – 9 – (trans fr english by thomas t nesbitt) – mf#SEM105P309 – cn Bibl Nat [918]

Directions diverses donnees en 1878 par la rev mere caron : superieure generale des soeurs de charite de la providence, pour aider ses soeurs a former de bonnes cuisinieres / Caron, Mother – Montreal: s.n, 1883 – 3mf – 9 – mf#26821 – cn Canadiana [640]

Directions diverses donnees en 1878 par la reverende mere caron : alors superieure generale des soeurs de charite de la providence pour aider ses soeurs a former de bonnes cuisinieres / Caron, mere – 3e rev augm ed. Montreal: [s.n.], 1889 [mf ed 1998] – 9 – cn Bibl Nat [640]

Directions diverses donnees en 1878 par la reverende mere caron : alors superieure generale des soeurs de charite de la providence pour aider ses soeurs a former de bonnes cuisinieres / Caron, mere – 8e rev augm ed. Montreal: [s.n.], 1913 [mf ed 2001] – 9 – cn Bibl Nat [640]

Directions for prayer / Ken, Thomas – London, England. 1791 – 1r – us UF Libraries [240]

Directions for the performance of public vaccination = Voorskrifte vir die uitvoering van openbare inenting / South Africa. Department of Health [Departement van Gesondheid] [Departement van Gesondheid] – Pretoria: Govt Printer 1937 [mf ed Pretoria, RSA: State Library [199-]] – 6p [ill] on 1r with other items – 5 – mf#op 12809 r23 – us CRL [360]

Directions for the prevention and treatment of malaria and blackwater fever = Voorskrifte vir die voorkoming en behandeling van malaria en swartwaterkoors / South Africa. Department of Public Health – Pretoria: Govt Printer 1939 [mf ed Pretoria, RSA: State Library [199-]] – (= ser Malaria pamphlet 2) – 4p [ill] on 1r with other items – 5 – mf#op 12806 r23 – us CRL [614]

Directions for the worthy receiving of the lord's supper / Vaughan, J – Ashton-under-Lyne, England. 1837 – 1r – us UF Libraries [240]

Directions for young students in divinity / Owen, Henry – London, England. 1790 – 1r – us UF Libraries [240]

Directions of change in south african politics / ed by Randall, Peter – Johannesburg, Study Project on Christianity in Apartheid Society, 1971 – us CRL [321]

Directions to members of the class of anatomy, faculty of medicine, mcgill university / McGill University. Faculty of Medicine – [Montreal?: s.n.] 1890 [mf ed 1985] – 1mf – 9 – 0-665-01838-X – mf#01838 – cn Canadiana [611]

Directoire de joliette, st jacques, st lin, st jerome, terrebonne, st eustache, l'assomption, ste therese etc... – Montreal: Compagnie d'impression et de publication Lovell, 1877-(annual) [mf ed 1987] – 5mf – 9 – (ceased 187-?) – mf#SEM105P822 – cn Bibl Nat [030]

Directoire de joliette, st jacques, st lin, st jerome, terrebonne, st eustache, l'assomption, ste therese, etc : corrige jusqu'au 1er fevrier, 1877 / Watkins, John A [comp] – Montreal?: s.n, 1877 – 5mf – 9 – (with ind) – mf#33487 – cn Canadiana [030]

O director : folha politica, commercial, litteraria e juridica – Para: Typ da Sociedade Propagadora dos Conhecimentos Uteis, 28 jan 1857 – (= ser Ps 19) – bl Biblioteca [079]

Director [la salle] : a weekly literary journal – La Salle IL 1807 – 1 – mf#5308 – us UMI ProQuest [420]

Director [london] – London, England 1972+ – 1,5,9 – ISSN: 0012-3242 – mf#6955 – us UMI ProQuest [650]

Directorio – Miami, FL. 1987 jun 01-1988 jun 17 – 1r – (missing: 1987 jun 29, aug 03-17, sep 18) – us UF Libraries [071]

Directorio cathechistico...roipalda / Ortiz Cantero, Jose – 1705 – 9 – (v2 1727. v2 1766) – sp Bibl Santa Ana [240]

Directorio cathequistico del christiano ilustrado en la fe con la glosa universal de la doctrina christiana... / Ortiz Cantero, Jose – Madrid: Antonio Perez de Soto, 1766.-v1 – 1 – sp Bibl Santa Ana [241]

Directorio comercial pro-barranquilla / Rasch Isla, Enrique – Barcelona, Spain. 1928 – 1r – us UF Libraries [380]

Directorio de asociaciones sindicales de la repubI... / Mexico Departamento Del Trabajo – Mexico City? Mexico. 1935 – 1r – us UF Libraries [366]

Directorio de importadores y exportadores de venez... / Venezuela Ministerio De Relaciones Exteriores – Caracas, Venezuela. 1956 – 1r – us UF Libraries [380]

Directorio espiritual en la lengua espanola, y quichua general del inga / Prado, Pablo de – en Lima: Por lorge Lopez de Herrera, ano de 1641 – (= ser Books on religion...1543/44-c1800: doctrina cristiana, obras de devocion) – 239lea on 6mf – 9 – mf#crl-45 – ne IDC [241]

Directorio oficial de senado y de la camara de representates / Philippines. Legislature – 1917. Manila: Bureau of Printing – 1 – us Wisconsin U Libr [959]

Directorio parraquial / Ortiz Cantero, Jose – 1727 – 9 – (ed 1 1796) – sp Bibl Santa Ana [240]

Directorio parroquial, practica de concursos y de curas... / Ortiz Cantero, Jose – Madrid: Antonio Perez de Soto, 1769 – 1 – sp Bibl Santa Ana [240]

Directorium ad divinum officium...melchiore granados et ortiz – Don Benito: D. Amalio Gallardo Valades, 1871 – 1 – sp Bibl Santa Ana [946]

The directorium anglicanum : being a manual of directions for the right celebration of the holy communion, for the saying of matins and evensong, and for the performance of other rites and ceremonies of the church according to the ancient uses of the church of england / ed by Lee, Frederick George – 4th carefully rev ed, with numerous emendations. London: John Hogg, 1879 – 2mf – 9 – 0-524-02392-1 – (incl bibl ref) – mf#1990-4294 – us ATLA [241]

Directorium annuale divinum officium...pacensis – 1828 – 9 – sp Bibl Santa Ana [946]

Directorium annuale divinum officium...pacensis – 1829 – 9 – sp Bibl Santa Ana [946]

Directorium annuale divinum officium...pacensis – 1838 – 9 – sp Bibl Santa Ana [946]

Directorium asceticum : in quo de viri spiritualis eruditione tutissima sanctorum patrum documenta – Friburgi Brisgoviae [Freiburg i B]; S Ludovici Americana [St Louis]: Herder, 1893 – 1mf – 9 – 0-8370-7100-3 – (incl the de vita spirituali of saint vincentius ferrerius) – mf#1986-1100 – us ATLA [240]

Directors and boards – New York NY 1979-98 – 1,5,9 – ISSN: 0364-9156 – mf#12265 – us UMI ProQuest [650]

Directors' minutes of meetings / St Francis' Boys Home. Kansas – 1945-69 – 1 – us Kansas [360]

Directors of athletics' attitudes toward women / Hayward, Sharman L – 1996 – 2mf – 9 – $8.00 – mf#PE 3758 – us Kinesology [150]

Directory / Brotherhood of Locomotive Firemen and Enginemen – n57 [1915 mar 1], n69 [1918 feb 1], 1966 apr-oct – 1r – 1 – (cont by: directory of international headquarters and locals, united transportation union) – mf#3371867 – us WHS [380]

Directory / International Longshoremen's Association – 1900-02 – 1r – 1 – (Continued by: directory of locals, international longshoremen, marine, and transportworkers' association) – mf#3164440 – us WHS [360]

Directory / Phi Delta Phi – 8th ed. Galesburg, IL: Mail, 1909. 320p. LL-454 – 1 – us L of C Photodup [340]

Directory and guide of florida railways for shippe... – s.l, s.l? 1920? – 1r – us UF Libraries [380]

Directory of administrative hearing facilities / Administrative Conference of the US (ACUS) – 1st ed Feb 1981. Acus: np. nd. (all publ) – 3mf – 9 – $4.50 – mf#LLMC 94-343A – us LLMC [340]

Directory of administrative hearing facilities / Administrative Conference of the US (ACUS) – 2nd ed Nov 1984. Washington: GPO, 1984 (all publ) – 3mf – 9 – $4.50 – mf#LLMC 94-343B – us LLMC [340]

Directory of affiliated societies / American Historical Association – 1977-1986/87 – 1r – 1 – mf#568412 – us WHS [366]

Directory of chambly basin, chambly canton, st jean baptiste, st cesaire, rougemont, st hilaire, beloeil, st bruno, marieville, laprairie, and st edouard for... – Montreal: Lovell Printing and Publ Co, 1877/78 [mf ed 1986] – 4mf – 9 – (incl text in french) – mf#SEM105P666 – cn Bibl Nat [030]

Directory of computer software / U.S. National Technical Information Service – Annual listing of thousands of mainframe and microcomputer software. More than 1700 programs. Full indexes by subject, hardware, language, and sponsoring agency. PB88-190962 – 9 – us NTIS [000]

Directory of computerized datafiles / U.S. National Technical Information Service – Annual listing of thousands of mainframe and microcomputer datafiles. 27 subject headings. PB89-191761 – 9 – us NTIS [000]

Directory of huntingdon, beauharnois, st jean chrysostome, chateauguay, st thimothee, valleyfield, durham, howick, hemmingford, ste martine, lacolle, st remi and napierville – Montreal: Lovell, [1877?] – 9 – (incl french text; ceased 187-?) – ISSN: 1190-6065 – mf#A00202 – cn Canadiana [917]

Directory of huntingdon, beauharnois, st jean chrysostome, chateauguay, st thimothee, valleyfield, durham, howick, hemmingford, ste martine, lacolle, st remi and napierville – Montreal: Lovell Printing & Publ Co, [ca 187-?]- (annual) [mf ed 1987] – 5mf – 9 – (incl text in french; ceased 187-?) – mf#SEM105P821 – cn Bibl Nat [030]

Directory of japanese technical resources / U.S. National Technical Information Service – U.S. sources of Japanese high-technology information. PB89-158869 – 9 – us NTIS [000]

Directory of labor organizations / Massachusetts. Dept. of Labor and Industries. Division of Statistics – Aug. 1902-78. 50 fiches. (Harvard Law School Library Collection.) – 9 – us Harvard Law [324]

Directory of locals / International Longshoremen, Marine and Transportworkers' Association – 1903-05 – 1r – 1 – (cont: directory, international longshoremen's association) – mf#3164450 – us WHS [380]

Directory of member schools / Wisconsin Interscholastic Athletic Association – n5th-17th [1957/58-1969/70], n18th-31st [1970/71-1983/84] – 2r – 1 – (cont: member school directory) – mf#682903 – us WHS [790]

The directory of mines (corrected and published quarterly) : a guide for the use of investors and others interested in the mines of british columbia / ed by Begg, Alexander – Victoria, BC: Mining Record, [1897?] [mf ed 1981] – 2mf – 9 – (contains a synopsis of the mining laws of british columbia by archer martin) – mf#14922 – cn Canadiana [622]

Directory of opportunities for negro youth in florida / United States National Youth Administration (Fl) – Jacksonville, FL. 1936 – 1r – us UF Libraries [305]

A directory of oral history interviews related to the federal courts / Anthony Champagne et al [comp] – 1992 – 1mf – 9 – $1.50 – mf#llmc99-014 – us LLMC [347]

Directory of protestant indian christians / Modak, S – Ahmednagar (India): [Printed at the Bombay Education Society's Steam Press], 1900 [mf ed 1995] – (= ser Yale coll) – 2v – 1 – 0-524-09324-5 – (int by s sattianadhan) – mf#1995-0324 – us ATLA [242]

Directory of school officers and teachers, waushara county, wisonsin – Waushara County WI. 1905/06 – 1r – 1 – (cont by: school directory for waushara county; school district clerks of waushara county elected for the year beginning...) – mf#5194549 – us WHS [370]

Directory of special libraries in indonesia / Indonesian National Scientific Documentation Center – Djakarta, 1966-1969 – 8mf – 9 – mf#SE-1402 – ne IDC [959]

Directory of st johns, west farnham, granby, west shefford, waterloo, roxton falls, etc, etc... – Montreal: Lovell Printing & Publ Co, 1876 [mf ed 1986] – 4mf – 9 – mf#SEM105P635 – cn Bibl Nat [917]
Directory of st scholastique, lachute, hull, etc – Montreal: Printed by Lovell Printing & Publ Co, 1878- (annual) [mf ed 1983] – 5mf – 9 – mf#SEM105P173 – cn Bibl Nat [030]
Directory of the bar of new jersey / Pierson, Leslie Cook – 4th ed. Trenton: MacCrellish & Quigley, 1888. 42p. LL-459 – 1 – us L of C Photodup [340]
Directory of the brethren in christ : commonly called river brethren – 1880-86 [complete] – (= ser Mennonite serials coll) – 1r – 1 – mf#ATLA 1993-S026 – us ATLA [242]
Directory of the church of the brethren in christ – 1899 [complete] – (= ser Mennonite serials coll) – 1r – 1 – mf#ATLA 1993-S028 – us ATLA [242]
Directory of the city of nevada and grass valley / Thompson, Hugh H – Nevada Co, CA. 1861 – 1r – 1 – $50.00 – mf#B40247 – us Library Micro [978]
Directory of the county of hastings : containing a full and complete list of householders of each town, township, and village in the county... – Belleville, Ont?: M Bowell, 1869 – 6mf – 9 – mf#28008 – cn Canadiana [917]
The directory of the devout life : meditations on the sermon on the mount / Meyer, Frederick Brotherton – New York: Fleming H Revell, 1904 – 1mf – 9 – us ATLA [240]
Directory of the members and officials of the brethren in christ church – 1903 [complete] – (= ser Mennonite serials coll) – 1r – 1 – mf#ATLA 1993-S029 – us ATLA [242]
Directory of the members of the bar in practice in new jersey / Honeyman, Abraham Van Doren – Somerville, N.J.: Honeyman, 1888. 64p. LL-1101 – 1 – us L of C Photodup [340]
Directory of the...session of the north carolina annual conference of the african methodist episcopal zion church / African Methodist Episcopal Zion Church. North Carolina Conference – [s.l: African Methodist Episcopal Zion Church, North Carolina Conference] [annual] [mf ed 2004] – 59th (1923) [complete] on 1r – 1 – (reel incl other titles: minutes of the...session of the philadelphia and baltimore annual conference of the african, methodist, episcopal, zion church [2004-s049] and: minutes of the...annual session of the little river primitive baptist association colored [2004-s050]) – mf#2004-s048 – us ATLA [242]
Directory of wisconsin dairy plants – special bulletin n49 [1955], 65 [1957], 75 [1959], 77 [1961], 81 [1963], 500 [1965] – 1r – 1 – (cont: directory of wisconsin dairy manufacturing plants in operation; cont by: wisconsin dairy plant directory) – mf#532954 – us WHS [630]
Directory, school district officers and teachers, langlade co – Langlade County WI. 1957/58-1958/59, 1961/62 – 1r – 1 – (cont: teachers of langlade county; school districts, langlade county board members) – mf#5193101 – us WHS [370]
Directory, school district officers and teachers, shawano co – Shawano County WI. 1951/52 – 1r – 1 – (cont: public school teachers...shawano county, wisconsin; shawano county school district officers for the year ending...; cont by: directory, shawano county school district officers and teachers) – mf#5195128 – us WHS [370]
Directory Service Co *see* Dane county directory
Directory, shawan county school district officers and teachers – Shawano County WI. 1957/58-1958/59, 1961/62 – 1r – 1 – (cont: directory, school district officers & teachers, shawano county) – mf#5195133 – us WHS [370]
A directory treatise / Martin, Charles Hynes – Jacksonville, Texas: Small, 1886. 42p. LL-496 – 1 – us L of C Photodup [340]
Directory, waushara county schools – Waushara County WI. 1956/57, 1958/59 – 1r – 1 – (cont: waushara county school directory) – mf#5194575 – us WHS [370]
Directrizes de ruy barbosa / Barbosa, Ruy – Sao Paulo, Brazil. 1938 – 1r – us UF Libraries [972]
Directrizes do direito mercantil brasileiro / Ferreira, Waldemar Martins – Lisboa, Portugal. 1933 – 1r – us UF Libraries [380]
Direito do brasil / Brazil – Sao Paulo, Brazil. 1949 – 1r – us UF Libraries [972]
Direito do povo : orgao democratico, critico litterario, noticioso e commercial – Rio de Janeiro, RJ: Typ do Povo, 30 ago, 25 dez 1884 – (= ser Ps 19) – mf#P19A,04,05 – bl Biblioteca [079]
Direktorat badan pimpinan umum perusahaan perkebunan dwikora laporan kerdja perusahaan / Indonesia – Djakarta, 1965-1966 – 5mf – 9 – mf#SE-1636 – ne IDC [959]
Direktorat badan pimpinan umum perusahaan perkebunan dwikora laporan tahunan / Indonesia – Djakarta, 1965-1968 – 3mf – 9 – mf#SE-1637 – ne IDC [959]
Direktorat bahasa dan kesusasteraan, ditdjen kebudajaan, departemen p dan k : bahasa dan kesusasteraan – Djakarta, 1967-1972 v5(1) – 23mf – 9 – (missing: 1967, v1(6); 1970, v3(2-4); 1971, v4(3-4)) – mf#SE-1325 – ne IDC [959]
Direktorat djenderal bea dan tjukai himpunan peraturan/instruksi direktorat chusus / harga / laboratorium / Indonesia 1969-1971 – 87mf – 9 – mf#SE-1638 – ne IDC [959]
Direktorat djenderal kehutanan data kehutanan / Indonesia – Djakarta, 1968 – 2mf – 9 – mf#SE-1639 – ne IDC [959]
Direktorat djenderal kehutanan laporan tahun / Indonesia – Djakarta, 1969 – 3mf – 9 – mf#SE-1640 – ne IDC [959]
Direktorat djenderal kehutanan publication / Indonesia – Bogor, 1967 – 5mf – 9 – mf#SE-1641 – ne IDC [959]
Direktorat Djenderal Koperasi *see* Bulletin koperasi
Direktorat djenderal koperasi peraturan2 tentang bimas / Indonesia – Djakarta, 1970/1971 – 1mf – 9 – mf#SE-1646 – ne IDC [959]
Direktorat djenderal koperasi recording rapat kerdja departemen transmigrasi dan koperasi / Indonesia – Djakarta, 1970 – 3mf – 9 – mf#SE-1647 – ne IDC [959]
Direktorat Djenderal Minjak dan Gas Bumi *see* Bulletin bulanan industri minjak dan gas bumi indonesia
Direktorat djenderal padjak laporan triwulan / Indonesia – Djakarta, 1969-1972 – 47mf – 9 – mf#SE-1648 – ne IDC [959]
Direktorat djenderal padjak musjawarah kerdja : buku laporan / Indonesia – Djakarta, 1970 – 11mf – 9 – mf#SE-1649 – ne IDC [331]
Direktorat djenderal pembangunan masjarakat desa lembaran pmd / Indonesia – Djakarta, 1966-1967 – 5mf – 9 – (missing: 1966, v1(1-2, 13-14, 23-27)) – mf#SE-1650 – ne IDC [959]
Direktorat djenderal pembangunan masjarakat desa madjalah pembangunan masjarakat desa / Indonesia – Djakarta, 1966/1967 – 2mf – 9 – mf#SE-1651 – ne IDC [959]
Direktorat djenderal pengolahan kekajaan laut laporan tahunan / Indonesia – Djakarta, 1967 – 6mf – 9 – mf#SE-1652 – ne IDC [959]
Direktorat djenderal perguruan tinggi dan ilmu pengetahuan research journal / Indonesia – Djakarta, 1963-1969 – 2mf – 9 – (missing: 1963, v1(3); 1968/69, v2(1-3)) – mf#SE-473 – ne IDC [959]
Direktorat djenderal perindustrian kimia laporan kerdja / Indonesia – Djakarta, 1969-1970 – 13mf – 9 – mf#SE-1653 – ne IDC [959]
Direktorat djenderal perindustrian kimia laporan pelaksanaan repelita / Indonesia – Djakarta, 1969-1970 – 13mf – 9 – mf#SE-1654 – ne IDC [959]
Direktorat djenderal perindustrian ringan laporan tahunan / Indonesia – Djakarta, 1969 – 5mf – 9 – mf#SE-1655 – ne IDC [959]
Direktorat kehutanan rasionalisasi lembaga penelitian ekonomi kehutanan / Indonesia – Bogor, 1964 – 4mf – 9 – mf#SE-1657 – ne IDC [959]
Direktorat kemahasiswaan direktorat djenderal perguruan tinggi / Mahasiswa dan masjarakat – Djakarta, 1968/1970-1972. v1-2(1-6) – 5mf – 9 – mf#SE-1789 – ne IDC [950]
Direktorat landuse buku tahunan / Indonesia – Djakarta, 1969-1971 – 1mf – 9 – (missing: 1969-1970) – mf#SE-1658 – ne IDC [959]
Direktorat landuse publikasi / Indonesia – Djakarta, 1969-1971 – 18mf – 9 – (missing: 1969(1-2, 12-14); 1971(23)) – mf#SE-1659 – ne IDC [959]
Direktorat Museum *see* Varia museografia
Direktorat pembinaan lembaga sosial desa kegiatan lsd diseluruh indonesia / Indonesia – Djakarta 1968-1971 – 37mf – 9 – (missing: 1968, v1(1); 1968, v1-1969, v3(3-84)) – mf#SE-1660 – ne IDC [959]
Direktorat pembinaan perusahaan2 negara industri kimia laporan tahunan direktorat djenderal perindustrian / Indonesia – Djakarta, 1967 – 7mf – 9 – mf#SE-1661 – ne IDC [959]
Direktorat pembinaan wilajah direktorat djendral padjak : berita padjak – Djakarta, 1967-1970. v1-3 – 56 – 9 – (missing: 1967, v1(1); 1968, v1(36); 1968, v2(56, 64)) – mf#SE-1353 – ne IDC [950]
Direktorat pendidikan guru dan tenaga tehnis statistik pendidikan guru / Indonesia – Djakarta, 1969 – 3mf – 9 – mf#SE-1662 – ne IDC [315]
Direktorat penelitian & pengabtian masjarakat, direktorat djenderal perguruan tinggi / Madjalah perguruan tinggi – Djakarta, 1962-1968 – 8mf – 9 – (missing: 1964(17); 1968) – mf#SE-472 – ne IDC [950]
Direktorat perumahan rakjet laporan kerdja / Indonesia – Djakarta, 1969-1970 – 3mf – 9 – mf#SE-1663 – ne IDC [331]
Direktorat publisitet & penerangan daerah, deppen / Mimbar penerangan – Djakarta, 1950-1972. v1-23(25) – 294mf – 9 – (missing: 1950, v1(5, 8); 1965, v15(4-end)) – mf#SE-564 – ne IDC [959]
Diretrizes – Rio de Janeiro. n. 1-207. 1938-44. and Suplemento literario diretrizes. 1939-41. (Wanting n. 1-7, 33-34, 131, 144-207) – 1 – 85.00 – us L of C Photodup [972]
Diretrizes do estado novo / Galvao, Francisco – Rio de Janeiro, Brazil. 1942 – 1r – us UF Libraries [972]
Diretrizes e bases da educacao nacional / Brazil – Rio de Janeiro, Brazil. 1968 – 1r – us UF Libraries [370]
Direttorio monastico di canto fermo : per vso particolare della congregatione oliuetana in preuenire l'offitio diurno al choro / Banchieri, Adriano – Bologna: Per gli heredi di G Rossi 1615-16 [mf ed 19–] – 4mf – 9 – mf#fiche339 – us Sibley [780]
Dirigido por hermanas carmelitas de la caridad legalmente reconocido para ensenanza media / Colegio de Santa Cecilia de Caceres – Caceres: Tip. Vda. de Floriano, s.a. – 1 – sp Bibl Santa Ana [946]
Il diritto – Rome, Italy. 1 jan 1872-31 dec 1895 – 1 – (discontinued) – mf#m.f.849 – uk British Libr Newspaper [074]
Il diritto – New York NY, 1918-19* – 1r – 1 – (italian newspaper) – us IHRC [071]
Il diritto commerciale e la parte generale delle obbligazioni. – Pisa. On film: v1-58; 1883-1939. Missing: v40 & 48; 1921 & 1929. LL-0284 – 1 – us L of C Photodup [346]
Il diritto di fraterna nella giurisprudenza da accursio alla codificazione / Fumagalli, Camillo – Torino etc.: Fratelli Bocca, 1912. 178p. LL-4099 – 1 – us L of C Photodup [340]
Dirschauer zeitung – Dirschau (Tczew PL), 1922 28 nov-30 dec, 1939 8 sep-1940 17 jan – 1 – (title varies: 8 sep-3 dec 1939: deutsche zeitung) – gw Misc Inst [077]
Dirsos...!tengo madre! cosas de un recluta / Galan, Leocadio – Caceres: Tip. El Noticiero, 1972 – 1 – sp Bibl Santa Ana [946]
Diruta, Girolamo *see*
– Seconda parte del transilvano dialogo
– Il transilvano dialogo sopra il vera modo di sonar organi, & istromenti da penna
Dirva / Cuyahoga Co. Cleveland – feb 1943-dec 1951,(1/1958-12/1969) [wkly, semiwkly, twice wkly, wkly] – 16r – 1 – (in lithuanian) – mf#B30368-30383 – us Ohio Hist [071]
Dirva = Field – Cleveland: Ohio Lithuanian Pub Co, 1958-65 – 14r – us CRL [071]
Dirva – Cleveland, OH. aug 25 1971-dec 20 1990 – (= ser Ethnic newspapers) – 19r – 1 – (weekly lithuanian language newspaper) – us Western Res [071]
Disability and rehabilitation – Abingdon, Oxfordshire 1992+ – 1,5,9 – ISSN: 0963-8288 – mf#18681,02 – us UMI ProQuest [617]
Disability appeals in social security programs / Liebman, Lance – Washington: FJC, 1985 – 1mf – 9 – $1.50 – mf#LLMC 95-375 – us LLMC [344]
Disability law in the united states : a legislative history of the americans with disability act of 1990, public law 101-336 / Reams, Bernard N & Schultz, Jon S – 1995 – 6v – 9 – $210.00 set – mf#307401 – us Hein [340]
Disabled american veteran magazine – 1969 jan-1973 feb, may-1977 feb, mar-1980 dec – 3r – 1 – (cont by: wisconsin dav news) – mf#366015 – us WHS [305]
Disabled American Veterans *see* Wisconsin dav news
Disabled american veterans semi-monthly – Cincinnati OH. 1932 apr 22 – 1r – 1 – mf#4360734 – us WHS [305]
Disaggregated farm income by type of farm, 1959-1982 / Somwaru, Agapi – Washington DC: US Dept of Agriculture, Economic Research Service...1986 – (= ser Agricultural economic report 558) – 9 – (with bibl) – us GPO [630]
Disappearing bushmen of lake chrissie / Potgieter, E F – Pretoria, South Africa. 1955 – 1r – us UF Libraries [307]
Disappointing dream / Cox, J – London, England. 18– – 1r – us UF Libraries [240]
Disarmament and disbandment of the german armed forces / U.S. Army. Office of the Chief Historian, European Command – 1947 – 1 – $19.00 – us L of C Photodup [943]
Disarmament campaigns – Hague, Netherlands. ill. 11 issues yearly. 1980?- – 1 – us Wisconsin U Libr [327]
Disarmament news & international views – New York NY 1976-77 – 1,5,9 – (cont: disarmament news & views) – ISSN: 0363-3721 – mf#6870,01 – us UMI ProQuest [327]
Disarmament news & views – New York NY 1970-72 – 1,5,9 – (cont by: disarmament news & international views) – ISSN: 0275-794X – mf#6870.01 – us UMI ProQuest [327]
Disarmament Research *see* Sanity
Disarming notes / Riverside Church [New York NY] – v1 n1-7,8 [1978 sep 15-dec 15, 1979 jan 15]-v9 n1-3,4 [1987 feb/mar-jul/aug, dec] – 1r – 1 – (with gaps) – mf#1055592 – us WHS [071]
Disaster of darien / Hart, Francis Russell – Boston, MA. 1929 – 1r – us UF Libraries [972]
Disaster prevention and management – Bradford, England 2001+ – 1,5,9 – ISSN: 0965-3562 – mf#31584 – us UMI ProQuest [360]
Disasters – Oxford, England 1989+ – 1,5,9 – ISSN: 0361-3666 – mf#17417 – us UMI ProQuest [550]
Disc and music echo/music mirror – London. 1958-75. 20 reels – 1 – us L of C Photodup [780]
Le discernement d'une veritable eglise suivant l'ecriture sainte / Labadie, Jean de – Amsterdam, 1668 – 2mf – 9 – mf#PPE-228 – ne IDC [240]
Disciple – St Louis MO 1974+ – 1,5,9 – ISSN: 0092-8372 – mf#8929 – us UMI ProQuest [240]
Disciple of christ – Cincinnati, 1884-85 [mf ed 2001] – (= ser Christianity's encounter with world religions, 1850-1950) – 1r – 1 – mf#2001-s118 – us ATLA [240]
Disciple of christ and canadian evangelist – Hamilton, Ont: G Munro, [1895-1896] – 9 – (cont: the canadian evangelist. cont by: the canadian evangelist and disciple of christ) – mf#P04636 – cn Canadiana [242]
Disciples and baptists : their resemblances and differences in belief and practice / Adkins, Frank – Philadelphia: American Baptist Publication Society, c1896 – 1mf – 9 – 0-524-08247-2 – mf#1993-3002 – us ATLA [242]
Disciples indeed / Goe, F F – London, England. 1877? – 1r – us UF Libraries [240]
Disciples of christ / Gates, Errett – New York: Baker & Taylor, 1905 – 1mf – 9 – 0-7905-5143-8 – (incl bibl ref) – mf#1988-1143 – us ATLA [240]
Disciples of christ / Lowndes, Arthur – Cincinnati, Ohio: American Christian Missionary Society, 1911 – (= ser Christian Unity Foundation (Series)) – 1mf – 9 – 0-524-07256-6 – mf#1991-2997 – us ATLA [240]
The disciples of sri ramakrishna – Almora: Advaita Ashrama, [1943] – (= ser Samp: indian books) – us CRL [280]
Discipleship / Morgan, George Campbell – New York: Fleming H Revell, c1897 – 1mf – 9 – 0-8370-7313-8 – mf#1986-1313 – us ATLA [240]
Discipleship training – Nashville TN 1962-94 – 1,5,9 – (cont: church training; cont by: growing disciples) – ISSN: 1047-9449 – mf#2456,01 – us UMI ProQuest [240]
Disciplina ordinis cartusiensis tribus libris distributa / Le Masson, dom – Monstrolii, 1894 – €40.00 – ne Slangenburg [241]
Disciplina unidad triumfo / Largo Caballero, Francisco – n.p., 1937. Fiche W991. (Blodgett Collection of Spanish Civil War Pamphlets) – 9 – us Harvard [946]
La discipline – Port-au-Prince: Impr de l'Abeille, jan 27-apr 3 1909; apr 28 1909-mar 12, apr 9 1910 – 3r – 9 – us CRL [079]
La discipline des eglises reformees de france / [Huisseau, I d'] – n.p, 1656 – 2mf – 9 – mf#PRS-146 – ne IDC [242]
The discipline of iowa yearly meeting of the society of friends / Iowa Yearly Meeting of the Society of Friends – Oskaloosa, Iowa: Herald Print Co, 1883 – 1mf – 9 – 0-524-07135-7 – mf#1990-5342 – us ATLA [240]
La discipline penitentielle en gaule des origines a la fin du 7th siecle / Vogel, C – Paris, 1952 – 4mf – 8 – €11.00 – ne Slangenburg [210]
Discipline-based dance education : a translation and interpretation of discipline-based art education for the discipline of dance / Hong-Joe, Christina M & Hanstein, Penelope – 1991 – 1mf – $4.00 – us Kinesology [790]
Disciplines of the united brethren in christ / ed by Drury, Augustus Waldo – Dayton, Ohio: United Brethren Pub House, 1895 – 1mf – 9 – 0-7905-8164-7 – mf#1988-6111 – us ATLA [242]
Discipulo a quien marti amaba / Santovenia Y Echaide, Emeterio Santiago – Habana, Cuba. 1948 – 1r – us UF Libraries [972]
A disclosure of the principles, designs, and machinations of the popish revolutionary faction of ireland / Ryan, John – London, 1838 – (= ser 19th c ireland) – 2mf – 9 – mf#1.1.1557 – uk Chadwyck [941]

Discontent and authority 1820-1840 : pro class h0 64, boxes 1-19, rewards, pardons and secret service – (= ser 19th century series) – 17r – 1 – (the yrs between 1820-40 witnessed an upheaval in britain's penal system. coll gives insight into the reactions of the governing elite towards urban discontent. also includes material relating to offer of pardons and rewards leading to capture of criminals) – mf#CL999-17100 – us Primary [941]

Discontent and danger in india / Connell, Arthur Knatchbull – London, 1880 – (= ser 19th c books on british colonization) – 2mf – 9 – mf#1.1.8049 – uk Chadwyck [954]

The discontinuity between education policy and implementation in secondary school education in zambia : 1964-1998 / Sakyi, Kwesi Atta – Uni of South Africa 2000 [mf ed Johannesburg 2000] – 6mf – 9 – (incl bibl ref) – mf#mfm15018 – sa Unisa [350]

The discophile : the magazine for record information – London. n1-61. aug 1948-dec 1958 (irreg) [all publ] – (= ser Jazz periodicals, 1914-1977) – 1r – 1 – $200.00 – us UPA [780]

Las discordancias entre alberdi y sus adversarios, sobre la cultura publica / Victoria, Maximio S – Tucuman, 1925 – 1 – us CRL [972]

Discorsi della musica / Chiavelloni, Vincenzo – Roma: per Ignatio de Lazeri 1668 [mf ed 19–] – 10mf – 9 – mf#fiche 165 – us Sibley [780]

Discorsi delle fortificationi... / Theti, C – Venetia, 1588 – 3mf – 9 – mf#OA-208 – ne IDC [720]

Discorsi mvsicali, nelli quali si contengono non solo cose pertinenti alla teorica, ma etiandio alla pratica / Crivellati, Cesare – Viterbo: Appresso Agostino Discep 1624 [mf ed 19–] – 3mf – 9 – mf#fiche 164 – us Sibley [780]

Discorsi sopra l'antichita di roma / Scamozzi, V – Venetia, 1582 – 6mf – 9 – mf#O-1039 – ne IDC [720]

Discorso intorne alla scoltura et pittura... / Lamo, A – Cremona, 1584 – 2mf – 9 – mf#O-1024 – ne IDC [700]

Discount merchandiser – New York NY 1973-99 – 1,5,9 – ISSN: 0012-3579 – mf#9336.01 – us UMI ProQuest [380]

Discount store news – New York NY 1974-99 – 1,5,9 – ISSN: 0012-3587 – mf#8260.02 – us UMI ProQuest [380]

Discours a l'etranger et au canada / Laurier, Wilfrid – Montreal: Beauchemin, c1909 – 7mf – 9 – 0-665-74853-1 – (int by laurent-olivier david) – mf#74853 – cn Canadiana [971]

Discours de arthur lachance, c r : depute de quebec-centre, a l'auditorium, le 24 fevrier 1911 / Lachance, Arthur – [Quebec: s.n, 1911] [mf ed 1996] – 1mf – 9 – 0-665-77582-2 – mf#77582 – cn Canadiana [355]

Discours de c s cherrier, ecr, cr : prononce dans l'eglise paroissiale de montreal, le 26 fevrier 1860, dans la grande demonstration des catholiques en faveur de pie 9 / Cherrier, Come Seraphin – Montreal?: s.n, 1860? – 1mf – 9 – mf#48740 – cn Canadiana [241]

Discours de f x lemieux...prononces a l'assemblee legislative de quebec : sujets, chemin de fer de la baie des chaleurs, et destitution des employes publics dans le comte de bonaventure – Quebec?: s.n, 1895? – 1mf – 9 – mf#04577 – cn Canadiana [323]

Discours de la grace... / Mestrezat, J – Charenton, 1638 – 1mf – 9 – mf#PRS-160 – ne IDC [240]

Discours de la souverainete des rois / Amyraut, M – Paris, 1650 – 2mf – 9 – mf#PRS-104 – ne IDC [240]

Discours de l'hon j a chapleau a l'occasion de la motion censurant le ministere pour avoir permis l'execution de louis riel : (compte-rendu officiel): seance du 24 mars / Chapleau, Joseph Adolphe – Montreal: Impr generale...1886 [mf ed 1980] – 1mf – 9 – mf#SEM105P36 – cn Bibl Nat [971]

Discours de l'hon jos cauchon sur la question de la confederation : prononce a la seance de l'assemblee legislative du 2 mars 1865 – [Quebec]: [s.n.], [1865?] (mf ed 1974) – 1r – 5 – mf#SEM16P121 – cn Bibl Nat [971]

Discours de l'hon l p pelletier sur la question des asiles d'alienes : prononce a l'assemblee legislative le 28 fevrier 1889 / Pelletier, Louis Philippe – Quebec: La Justice, 1889 – 1mf – 9 – mf#11747 – cn Canadiana [346]

Discours de l'hon louis joseph papineau a l'occasion du 23eme anniversaire de la fondation de l'institut canadien : 17 decembre 1867 – Montreal: [Le Pays?], 1868 – 1mf – 9 – 0-665-11604-7 – mf#11604 – cn Canadiana [320]

Discours de l'hon m bellerose : prononce les 23 et 24 janvier 1884: la langue francais – S.l: s.n, 1884? – 1mf – 9 – mf#01061 – cn Canadiana [440]

Discours de l'hon w s fielding, m p : sur la situation financiere, ottawa, mercredi, 3 aout 1904 – [Ottawa?: s.n, 1904?] – 1mf – 9 – 0-665-65531-2 – mf#65531 – cn Canadiana [336]

Discours de l'honorable a w atwater...et de felix carbray...sur le budget – Quebec: L. Demers, 1899 – 1mf – 9 – mf#04481 – cn Canadiana [336]

Discours de l'honorable e j flynn, depute de gaspe : sur les resolutions de la conference interprovinciale, prononce devant l'assemblee legislative a sa seance du lundi le 12 mai 1888 – Quebec: s.n, 1888 – 1mf – 9 – mf#05575 – cn Canadiana [971]

Discours de l'honorable e-james flynn, prononce a l'assemblee legislative aux seances des 8, 13 et 16 mai 1884 : en reponse aux reproches de l'opposition d'avoir vote en 1879, contre le gouvernement joly – Quebec: impr generale, 1897 – 1mf – 9 – mf#03116 – cn Canadiana [320]

Discours de l'honorable joseph shehyn en reponse a la critique de l'honorable ex-tresorier, sur l'expose budgetaire : refutation complete de toutes les pretentions de l'opposition...14 et 15 fevrier 1899 – Quebec?: "Le Soleil", 1899 – 1mf – 9 – mf#50741 – cn Canadiana [336]

Discours de l'honorable m chapleau : subventions aux chemins de fer: reclamations de la province de quebec (compte rendu officiel), chambre des communes, 12 avril 1884 – [Ottawa?: Maclean, Roger], 1884 – 1mf – 9 – 0-665-02164-X – mf#02164 – cn Canadiana [380]

Discours de l'honorable m chapleau en proposant la vente du chemin de fer quebec, nontreal, ottawa et occidental a l'assemblee legislative, seances des 27 et 28 mars 1882 – Quebec: A Cote, 1882 – 1mf – 9 – mf#00568 – cn Canadiana [380]

Discours de l'honorable m chapleau sur les resolutions du chemin de fer canadien du pacifique / Chapleau, Joseph-Adolphe – Ottawa?: Maclean, Roger, 1885 – 1mf – 9 – mf#30095 – cn Canadiana [380]

Discours de l'honorable m l o david sur le bill d'autonomie : au cours des debats provoques par le bill d'autonomie du nord-ouest, m le senateur l o david a prononce au senat l'eloquent discours suivant – [Canada?: s.n, 19–?] – 1mf – 9 – 0-665-72344-X – mf#72344 – cn Canadiana [971]

Discours de l'honorable w s fielding... : revue de la situation financiere, ottawa, canada, 21 octobre, 1903 / Fielding, William Stevens – [Canada?: s.n, 1903?] [mf ed 1997] – 1mf – 9 – 0-665-85282-7 – mf#85282 – cn Canadiana [336]

Discours de m beausoleil, mp sur la reciprocite avec les etats-unis : ottawa, 23 mars 1888 / Beausoleil, Cleophas – Ottawa?: s.n, 1888? [mf ed 1982] – 1mf – 9 – 0-665-17986-3 – mf#17986 – cn Canadiana [337]

Discours de m desmarais, depute de st-hyacinthe : sur l'adresse en reponse au discours du trone, assemblee legislative, seance du 7 mars 1890 – [Quebec?: s.n, 1890?] – 1mf – 9 – 0-665-93455-6 – mf#93455 – cn Canadiana [323]

Discours de m j b e dorion... : sur le projet de confederation des provinces anglaises – L'Avenir, Quebec?: "Defricheur", 1865 – 1mf – 9 – mf#34931 – cn Canadiana [971]

Discours de m john costigan mp sur l'adresse : ottawa, 23 avril 1895 – S.l: s.n, 1895? – 1mf – 9 – mf#58041 – cn Canadiana [350]

Discours de m l g desjardins, depute du district electoral de montmorency : fait a l'assemblee legislative a la seance du mardi le 21 avril 1885, sur les finances de la province de quebec – Quebec: s.n, 1885 [mf ed 1985] – 1mf – 9 – 0-665-02273-5 – mf#02273 – cn Canadiana [336]

Discours de m p w ellis : president de l'association des manufacturiers canadiens, a la convention annuelle, tenue a montreal, mardi et mercredi 5-6 novembre 1901 / Ellis, Philip William – Montreal: la Cie d'Imprimerie moderne...[s.d.] (mf ed 1985) – 1mf – 9 – mf#SEM105P469 – cn Bibl Nat [670]

Discours de m t c casgrain...sur le budget : ottawa, jeudi, 19 avril 1900 – S.l: s.n, 1900? – 1mf – 9 – mf#58019 – cn Canadiana [336]

Discours de m t chase casgrain...depute du comte de quebec : sur la conference interprovinciale – Quebec: L J Demers, 1888 [mf ed 1984] – 1mf – 9 – 0-665-02052-X – mf#02052 – cn Canadiana [350]

Discours de samdech norodom sihanouk a la trentieme session de l'assemblee generale de l'o.n.u., 6 octobre 1975 / Norodom Sihanouk, Prince – Paris: Mission du Gouvernement royal d'union nationale du Cambodge [mf ed 1989] – 1r with other items – 1 – mf#mf-10289 seam reel 021/05 [§] – us CRL [959]

Discours de samdech preah norodom sihanouk upayuvareach, chef de l'etat du cambodge, a l'occasion de l'ouverture de la conference pleneriere des peuples indochinois [phnom-penh, le 25 fevrier 1965] / Norodom Sihanouk, Prince – [Phnom Penh: Impr du Minister de l'information 1965] [mf ed 1989] – 1r with other items – 1 – mf#mf-10289 seam reel 015/32 [§] – us CRL [959]

Discours de samdech sahachivin a l'ouverture du 15e congres national = Speech by h r h prince norodom sihanouk head of state at the opening of the 15th national congres / Norodom Sihanouk, Prince – [Phnom-Penh? Impr du Ministere de l'information 19–?] [mf ed 1989] – 1r with other items – 1 – (in french & english) – mf#mf-10289 seam reel 015/04 [§] – us CRL [323]

Discours de sir adolphe caron sur l'execution de louis riel – Ottawa?: s.n, 1886? – 1mf – 9 – (also available in english) – mf#58017 – cn Canadiana [971]

Discours de sir wilfrid laurier / DeCelles, Alfred Duclos [comp] – Montreal: Librairie Beauchemin ltee 1920 [mf ed 1985] – 2v on 6mf – 9 – mf#SEM105P540 – cn Bibl Nat [971]

Discours de sir wilfrid laurier de 1889 a 1911 / ed by DeCelles, Alfred Duclos – Montreal: Beauchemin, 1920 – 3mf – 9 – 0-665-72640-6 – mf#72640 – cn Canadiana [323]

Discours de sir wilfrid laurier de 1911 a 1919 / ed by DeCelles, Alfred Duclos – Montreal: Beauchemin, 1920 – 3mf – 9 – 0-665-72639-2 – mf#72639 – cn Canadiana [323]

Discours de son altesse royale le prince norodom sihanouk au 18eme congres national / Norodom Sihanouk, Prince – [Phnom-Penh: Impr du Ministere de l'information] 1959 [mf ed 1989] – 1r with other items – 1 – mf#mf-10289 seam reel 015/08 [§] – us CRL [959]

Discours de son excellence monsieur l'ambassadeur huot sambath, president de la delegation du cambodge : a la 23eme session ordinaire de l'assemblee general des nations unies, lundi, 21 octobre 1968. / Cambodia. Permanent Mission to the United Nations – [n.p. 1968] [mf ed 1989] – 1r with other items – 1 – mf#mf-10289 seam reel 015/22 [§] – us CRL [323]

Discours de...dans les cortes / Donoso Cortes, Juan Francisco – 1851 – 9 – sp Bibl Santa Ana [946]

Discours d'ouverture du c n r de labe : parti democratique de guinee – [Conakry]: Imp Nationale "P Lumumba", 1968 – us CRL [320]

Discours d'overture du president ahmed sekou toure : 12e session de la conference des ministres africains du travail – 2nd aug ed. [Conakry]: Impr nationale "Patrice-Lumumba", 18 mars 1974 – us CRL [320]

Discours du prince norodom sihanouk, chef de l'etat, a l'ouverture de la conference diplomatique des quakers / Norodom Sihanouk, Prince – [Phnom-Penh] Impr de l'information [196-] [mf ed 1989] – 1r with other items – 1 – mf#mf-10289 seam reel 015/03 [§] – us CRL [959]

Discours du songe de poliphile... / [Colonna, F] – Paris, 1546 – 9mf – 9 – mf#O-1037 – ne IDC [700]

Le discours du voyage de constantinoble, enuoye dudict lieu...une damoyselle francoyse – Lyon: Pierre de Tours, 1542 – 1mf – 9 – mf#H-8149 – ne IDC [915]

Discours et plaidoyers politiques / Gambetta, Leon M – Paris. v1-11. 1880-85 – 1 – $108.00 – mf#0228 – us Brook [944]

Discours et votes de jacques bureau contre la conscription en 1917 – [Trois-Rivieres ?]: [s.n.], [1921 ?] (mf ed 1991) – 1mf – 9 – mf#SEM105P1440 – cn Bibl Nat [355]

Discours laiques / Secretan, Charles – Paris: Sandoz et Fischbacher, 1877 – 1mf – 9 – 0-7905-8730-0 – mf#1989-1955 – us ATLA [190]

Discours, messages et proclamations de l'empereur : depuis son retour en france jusqu'au 1er janvier 1855 / Napoleon – Paris 1855 [mf ed Hildesheim 1995-98] – 1v on 2mf – 9 – €60.00 – 3-487-26026-3 – gw Olms [850]

Discours politiques et euvres diverses / Couvin, Leger – Port-Au-Prince, Haiti. 1928 – 1r – us UF Libraries [320]

Discours – programme / Mayard, Constantin – Port-Au-Prince, Haiti. 1930 – 1r – us UF Libraries [972]

Discours prononce a la cathedrale de quebec le 10 avril 1869 : cinquantieme anniversaire de la pretrise de pie 9 / Paquet, Benjamin – Quebec?: P G Delisle, 1869 – 1mf – 9 – mf#52177 – cn Canadiana [241]

Discours prononce a la salle windsor, montreal, le 16 fevrier 1892 : sur les affaires financieres de la province et critique de l'administration mercier / Hall, John Smythe – S.l: s.n, 1892? – 1mf – 9 – mf#05338 – cn Canadiana [336]

Discours prononce a la seance du 3 juin 1892 de l'assemblee legislative de la province de quebec / Beaubien, Louis – [S.l: s.n.], 1892 – 1mf – 9 – mf#03515 – cn Canadiana [630]

Discours prononce a une veture au monastere du precieux sang / Raymond, Joseph-Sabin – [s.l: s.n, 1873?] [mf ed 1984] – 1mf – 9 – 0-665-46383-9 – mf#46383 – cn Canadiana [240]

Discours prononce a vienne, france, le 12 aoaut 1909 / Guild, Curtis – [S.l.: s.n., 1909?] – 1mf – 9 – 0-524-02592-4 – mf#1990-0644 – us ATLA [240]

Discours prononce au petit seminaire de montreal, le 2 fevrier 1890 : fete de la purification de la sainte vierge / Bourassa, Gustave – Montreal?: s.n, 1890 – 1mf – 9 – mf#03807 – cn Canadiana [241]

Discours prononce le mercredi, 18 juillet 1855 : a la ceremonie de la pose de la pierre angulaire du monument dedie, par souscription nationale, a la memoire des braves tombes sur la plaine d'abraham, le 28 avril 1760 / Chauveau, Pierre-Joseph-Olivier – S.l: s.n, 1855? – 1mf – 9 – mf#46604 – cn Canadiana [971]

Discours prononce par l'abbe jul guihot pretre de st sulpice a l'occasion du cinquantenaire des oblats a montreal le 8 decembre 1891 – [Montreal?: s.n.], 1892 [mf ed 1986] – 1mf – 9 – 0-665-60950-7 – mf#60950 – cn Canadiana [241]

Discours prononce par le president / Vincent, Stenio – s.l, s.l? 1933? – 1r – us UF Libraries [972]

Discours prononce par l'hon rod lemieux a l'assemblee de ste-foy, le 30 septembre 1906 / Lemieux, Rodolphe – [Quebec?: s.n, 1906?] [mf ed 1995] – 1mf – 9 – 0-665-74389-0 – mf#74389 – cn Canadiana [320]

Discours prononce par l'honorable a t galt...en presentant le budget / Canada (Province). Departement des finances – Ottawa: G E Desbarats, 1866 [mf ed 1983] – 1mf – 9 – mf#SEM105P298 – cn Bibl Nat [350]

Discours prononce par l'honorable m edward blake, m p : dans la chambre des communes du canada...17 mars 1884 – Ottawa?: s.n, 1884? – 1mf – 9 – mf#05448 – cn Canadiana [323]

Discours prononce par l'honorable m flynn sur la deuxieme lecture du bill pour diviser les districts electoraux de montreal-est, montreal centre et montreal-ouest, quebec-est, drummond et arthabaska, chicoutimi et saguenay : il demande un me – [Quebec?: s.n, 1890?] – 1mf – 9 – 0-665-93462-9 – mf#93462 – cn Canadiana [325]

Discours prononce par l'honorable m honore mercier, premier ministre de la province : le 6 novembre 1889 au club national, montreal – S.l: s.n, 1890? – 1mf – 9 – mf#09878 – cn Canadiana [080]

Discours prononce par l'honorable thomas chapais contre l'abolition du conseil legislatif, le 22 mars 1900 – Quebec?: s.n, 1900? – 1mf – 9 – mf#03994 – cn Canadiana [323]

Discours prononce par m c larocque cure de saint jean dorchester : a l'occasion de la benediction de la premiere pierre d l'eglise des rr rr jesuites...22 mai 1864 – Montreal: E Senecal, 1864 [mf ed 1984] – 1mf – 9 – 0-665-45223-3 – mf#45223 – cn Canadiana [241]

Discours prononce par monsieur f-x lemieux... : depute de levis a l'assemblee legislative de quebec le 30 avril 1886, sur la question riel – Quebec?: La Justice, 1886 – 1mf – 9 – mf#51509 – cn Canadiana [971]

Discours prononce par monsieur le resident superieur p i richomme a l'ouverture de la session ordinaire du conseil francaise des interets economiques et financiers du cambodge, 28 octobre 1935 / Richomme – [Pnom-Penh: Impr A Portail 1935] [mf ed 1989] – 1r with other items – 1 – mf#mf-10289 seam reel 026/23 [§] – us CRL [336]

Discours prononce par mr alexandre dumas au club constitutionel, tenu a quebec le 30 mai 1792 : imprime pour l'instruction des electeurs de la province du bas-canada, aux fraix de cette societe, composee de deux a trois cens citoyens – Quebec?: Printed by Samuel Neilson sic, 1792? – 1mf – 9 – mf#57548 – cn Canadiana [325]

Discours prononces par l'hon depute de gaspe : a la seance du 28 octobre sur le vote de non confiance, et celui prononce sur la colonisation a la seance du 20 aout de l'assemblee legislative – Quebec?: s.n, 1879? – 1mf – 9 – mf#45799 – cn Canadiana [971]

Discours religieux / Pressense, Edmond de – Paris: Ch. Meyrueis, 1859 – 1mf – 9 – 0-7905-5672-3 – mf#1988-1672 – us ATLA [240]

Discours sur la castrametation et discipline militaire des romains... / Du Choul, G – Lyon, 1555 – 3mf – 9 – mf#H-8289 – ne IDC [956]

Discours sur la confederation prononces / Cherrier, Come Seraphin et al – Montreal?: s.n, 1865 – 1mf – 9 – mf#23224 – cn Canadiana [971]

Discours sur la constitution de 1889 / Cauvin, Leger – Port-Au-Prince, Haiti. 1902 – 1r – us UF Libraries [323]

Discours sur la loi de l'instruction publique : prononce par l'honorable m chapais devant le conseil legislatif, les 2 et 3 mars 1899 / Chapais, Thomas – Quebec?: L-J Demers, 1899 – 1mf – 9 – mf#02866 – cn Canadiana [350]

Discours sur la question riel : prononce le 22 mars 1886 a la chambre des communes / Thompson, John Sparrow David – [S.l.]: [s.n.], [s.d.] (mf ed 1971) – 1r – 1 – mf#SEM35P81 – cn Bibl Nat [971]

Discours sur la question riel, prononce le 17 mars 1886, a la chambre des communes / Caron, Adolphe – S.l: s.n, 1886? – 1mf – 9 – mf#30087 – cn Canadiana [345]

Discours sur le budget prononce a la chambre des communes du canada : par l'honorable mr cartwright, ministre des finances, le 16e jour de fevrier, 1875 – Ottawa?: Grison, O'Donoghue, 1875 – 1mf – 9 – mf#58018 – cn Canadiana [336]

Discours sur le budget prononce a l'assemblee legislative de quebec vendredi, le 20 mai 1892 / Hall, John Smythe – [Quebec?: "Morning Chronicle"], 1892 – 1mf – 9 – 0-665-54316-6 – mf#54316 – cn Canadiana [336]

Discours sur le budget prononce le 15 mai 1882 / Wurtele, Jonathan Saxton Campbell – Quebec?: A Cote, 1882 – 1mf – 9 – mf#26152 – cn Canadiana [336]

Discours sur le sacerdoce : aux noces d'argent de m l'abbe alphonse graton a saint-jean-baptiste de pawtucket, 2 mai 1915 – Montreal: Arbour & Dupont, 1915 – 1mf – 9 – 0-659-91907-9 – mf#9-91907 – cn Canadiana [241]

Discours sur les arcs triomphaux dresses en la ville d'Aix *see* ...L'heureuse ariv

Discours sur les arcs triomphaux dresses en la ville d'aix, a l'heureuse arriv, e de monseigneur le duc de bourgogne, et de monseigneur le duc de berry / [Gallaup de Chastueil] – Aix: Jean Adibert, 1701 – 2mf – 9 – mf#0-75 – ne IDC [090]

Discours sur l'harmonie / Gresset, J-B – Paris: J-N Le Clerc 1737 [mf ed 19–] – 2mf – 9 – mf#fiche 487 – us Sibley [780]

Discours sur l'institut canadien / Dessaulles, L A – Montreal?: s.n, 1863 – 1mf – 9 – mf#23091 – cn Canadiana [360]

Discours sur l'ouvrier : prononce par le reverend m colin...devant l'institut des artisans canadiens le 2 avril 1869 – Montreal?: Le Nouveau Monde, 1869 – 1mf – 9 – mf#23591 – cn Canadiana [331]

Discours sur l'ouvrier prononce par le reverend m colin...devant l'institut des artisans canadiens le 2 avril 1869 / Colin, Frederic Louis de Gonzague – Montreal: Typographie Le Nouveau monde...1869 [mf ed 1980] – 1mf – 9 – mf#SEM105P39 – cn Bibl Nat [331]

Discours sur l'universalite de la langue francaise / Rivarol, Antoine – Paris: Librairie Hatier, [1929] (mf ed 1989) – 1mf – 9 – mf#SEM105P1156 – cn Bibl Nat [440]

Discours veritable des propos tenus par monsieur le prince de conde, auec les seigneurs deputez par le roy: contenant les causes gui ont contraint ledict seigneur prince & autres de sa copagnie a prendre les arms / Conde, Louis 1st Prince of Bourbon – s.l.: s.n., 1567 – 1 – us Wisconsin U Libr [944]

A discourse : commemorative of the history of the church of christ in yale college, during the first century of its existence...nov 22 1857 / Fisher, George Park – New Haven: Thomas H Pease, 1858 [mf ed 1984] – 9 – 0-8370-0192-7 – mf#1984-0071 – us ATLA [240]

A discourse : containing a loving invitation both honourable and profitable to all such as shall be adventurers, either in person or purse for the advancement of his majesties most hopefull plantation in new-foundland, lately undertaken / Whitbourne, Richard – [London]: impr at London by Felix Kyngston...1622 [mf ed 1987] – 1mf – 9 – 0-665-67690-5 – mf#67690 – cn Canadiana [917]

A discourse : delivered...july 17 1855 / Eliot, William Greenleaf – Boston: Crosby, Nichols, 1855 [mf ed 1990] – 1mf – 9 – 0-7905-3666-8 – mf#1989-0159 – us ATLA [240]

A discourse : investigating the doctrine of washing the saints' feet / Brookes, Iveson L – 1830 – 1 – us Southern Baptist [242]

A discourse : on the incarnation of the word of god / Athanasius, Saint, Patriarch of Alexandria – with an english translation and copious analysis by James Ridgway. 1880 – 1 – us ATLA [240]

A discourse : opening the nature of that episcopacie, which is exercised in england / Brooke, R – London: RC, 1641 – 2mf – 9 – mf#PW-39 – ne IDC [240]

Discourse / Alison, Archibald – Edinburgh, Scotland. 1810 – 1r – us UF Libraries [240]

Discourse : delivered at the general conference, salt lake city, on sunday afternoon, april 9th, 1882 / Taylor, John – [S.l: s.n., 1882?] – 1mf – 9 – 0-524-03298-X – mf#1990-0909 – us ATLA [240]

Discourse – Detroit MI 1991+ – 1,5,9 – ISSN: 1522-5321 – mf#19652 – us UMI ProQuest [380]

A discourse about the state of true happiness delivered in certaine sermons in oxford, and at st pauls crosse / Bolton, R – Ed 7. London: Iohn Legatt, 1638 – 3mf – 9 – mf#PW-36 – ne IDC [240]

Discourse before the society for "propagating the gospel among the... / Lathrop, John – Boston, MA. 1804 – 1r – us UF Libraries [240]

Discourse commemorative of rev. rufus anderson, d.d., ll.d : late corresponding secretary of the american board of commissioners for foreign missions: together with addresses at the funeral / Thompson, Augustus Charles & Clark, Nathaniel George – Boston: American Board of Commissioners for Foreign Missions, 1880 – 1mf – 9 – 0-7905-6324-X – mf#1988-2324 – us ATLA [240]

Discourse concerning sins of infirmity and wilful sins / Kidder, Richard – London, England. 1804 – 1r – us UF Libraries [240]

Discourse delivered at great st mary's church, cambridge / Graham, John – Cambridge, England. 1837 – 1r – us UF Libraries [240]

A discourse delivered at the funeral of rev levi w leonard : late pastor of the first congregational church, dublin, n h, jan 5 1865 / Learned, John Calvin – Exeter, NH: Thomas J Whittem, 1865 [mf ed 1993] – 1mf – 9 – 0-524-08478-5 – mf#1993-3123 – us ATLA [242]

Discourse delivered at the opening session of the second provincial... / Ullathorne, William Bernard – London, England. 1855 – 1r – us UF Libraries [240]

A discourse delivered before the general assembly of the presbyterian church in the united states of america : on the opening of their session in 1820 / Rice, John Holt – Philadelphia: Tand W Bradford, 1820 [mf ed 1992] – 1mf – 9 – 0-524-05557-2 – mf#1990-5161 – us ATLA [242]

Discourse delivered before the synod of ross / Cameron, D – Inverness, Scotland. 1867 – 1r – us UF Libraries [240]

Discourse delivered in st george's chapel, montreal, on christmas day, 1861 : (the 6th and 7th companies of the prince of wales regiment being present) / Leach, William Turnbull – [Montreal?: s.n.] 1862 [mf ed 1984] – 1mf – 9 – 0-665-45326-4 – mf#45326 – cn Canadiana [240]

Discourse delivered in the catholic chapel / Husenbeth, Frederick Charles – Norwich, England. 1827? – 1r – us UF Libraries [241]

A discourse delivered on board the transport ship java, off quebec : on sabbath, the 22nd october, 1843, to the first battalion, 71st highland light infantry (en route to the west indies) / Mathieson, Alexander – Montreal?: J Starke, 1843 [mf ed 1983] – 1mf – 9 – 0-665-38228-6 – mf#38228 – cn Canadiana [240]

A discourse in commemoration of the 46th anniversary of the mite society : and the 250th anniversary of the first baptist church in america / Jackson, Henry – Providence: John K Stickney, 1854 [mf ed 1993] – (= ser Baptist coll) – 1mf – 9 – 0-524-07201-9 – (incl bibl ref) – mf#1990-5359 – us ATLA [242]

A discourse in memory of thomas harvey skinner / Prentiss, George Lewis – New York: ADF Randolph, [1871?] [mf ed 1992] – (= ser Presbyterian coll) – 1mf – 9 – 0-524-04423-6 – mf#1992-2028 – us ATLA [242]

A discourse in two parts, preached in st andrew's church, toronto : on the occasion of the commemorative services connected with the tricentennary of the scottish reformation / Barclay, John – Toronto: s.n, 1861 – 1mf – 9 – mf#59487 – cn Canadiana [242]

Discourse occasioned by the death of convers francis, d.d. delivered before the first congregational society, watertown... / Weiss, John – Cambridge: [s.n.], 1863 – 1mf – 9 – 0-524-04305-1 – mf#1992-2025 – us ATLA [242]

Discourse occasioned by the death of elizabeth prowse / Owen, John – London, England. 1810 – 1r – us UF Libraries [240]

Discourse occasioned by the death of william sharp, esq / Owen, John – London, England. 1811 – 1r – us UF Libraries [240]

A discourse of matters pertaining to religion / Parker, Theodore – 5th ed. Boston: Horace B Fuller, 1870 [mf ed 1984] – (= ser Biblical crit – us & gb 33) – 6mf – 9 – 0-8370-0200-1 – (incl bibl ref) – mf#1984-1033 – us ATLA [243]

A discourse of s athanasius on the incarnation of the word of god / Athanasius, Saint, Patriarch of Alexandria – Oxford: James Parker & Co, 1880. Chicago: Dep of Photodup, U of Chicago Lib, 1974 (1r); Evanston: American Theol Lib Assoc, 1984 (1r) – 1 – 0-8370-0018-1 – mf#1984-B392 – us ATLA [220]

A discourse of the common weal of this realm of england / Lamond, Elizabeth – Cambridge: at the University Press, 1929 – 3mf – 9 – $4.50 – (first printed in 1581 and commonly attributed to "w.s.", this manuscript describes legal and social conditions in elizabethan england) – mf#LLMC 92-150 – us LLMC [346]

A discourse of the kingdom of china : taken out of ricius [ricci] and trigautius, conteyninğ the countrey, government, religion, rites, sects, characters, studies, arts, acts... / Purchas, S – London, 1625-1626. v3 – 2mf – 9 – mf#HT-679 – ne IDC [915]

Discourse of the removal of the gospel / Charnock, Stephen – London, England. 1827 – 1r – us UF Libraries [220]

Discourse on concluding a pastorate of thirty years : june 30, 1889 / Hart, Burdett – New Haven, CT: Fiske Print Co, 1890 [mf ed 1993] – 1mf – 9 – 0-524-08376-2 – mf#1993-3076 – us ATLA [242]

A discourse on creeds and ecclesiastical machinery : delivered at peterboro, feb 21 1858 / Smith, Gerrit – Boston: John P Jewett, 1858 [mf ed 1992] – (= ser Tracts for thinking men and women 2) – 1mf – 9 – 0-524-02993-8 – mf#1990-0780 – us ATLA [240]

Discourse on divine influences and conversion / Carpenter, Lant – Bristol, England. 1822 – 1r – us UF Libraries [240]

Discourse on justification by faith / Bickersteth, Edward Henry – London, England. 1828 – 1r – us UF Libraries [210]

Discourse on metaphysics / Leibniz, Gottfried Wilhelm – La Salle, IL. 1937 – 1r – us UF Libraries [110]

Discourse on national establishments of christianity / Willis, Michael – Glasgow, Scotland. 1833 – 1r – us UF Libraries [240]

A discourse on occasion of the death of the rev wilbur fisk : president of the wesleyan university, delivered...new york...29th of mar 1839 / Bangs, Nathan – New York: Publ by T Mason & G Lane, for the Methodist Episcopal Church, 1839 [mf ed 1984] – 1mf – 9 – 0-8370-0916-2 – mf#1984-4256 – us ATLA [242]

A discourse on systematic benevolence : ...oct 15 1852; and, an address to the laity of the evangelical lutheran church / Anspach, Frederick Rinehart – Hagerstown MD: M'Kee & Robertson 1853 [mf ed 1992] – 1mf – 9 – 0-524-04367-1 – mf#1991-2071 – us ATLA [242]

Discourse on the doctrine of the trinity / Parker, Gavin – Aberdeen, Scotland. 1900 – 1r – us UF Libraries [240]

Discourse on the evidences of the american indians being the descendants of the lost tribes of israel : delivered before the mercantile library association, clinton hall / Noah, Mordecai Manuel – New York: J Van Norden, 1837 – 1mf – 9 – mf#45453 – cn Canadiana [305]

A discourse on the history, character, and design of christian baptism / Henderson, David Patterson – Louisville, KY: Morton & Griswold, 1857 [mf ed 1992] – 1mf – 9 – 0-524-03318-8 – mf#1990-4678 – us ATLA [240]

Discourse on the history, character, and prospects of the west / Drake, Daniel – 1834 – 9 – 5.00 – us Scholars Facs [978]

A discourse on the latest form of infidelity / Norton, Andrews – Cambridge: John Owen, 1839 [mf ed 1991] – 1mf – 9 – 0-524-00381-5 – mf#1989-3081 – us ATLA [230]

A discourse on the life and character of daniel webster / Boardman, Henry Augustus – Philadelphia: JM Wilson, 1852 [mf ed 1992] – 1mf – 9 – 0-524-04670-0 – mf#1990-1297 – us ATLA [920]

A discourse on the life and services of professor moses stuart / Adams, William – New York: JF Trow, 1852 [mf ed 1989] – 1mf – 9 – 0-7905-4422-9 – mf#1988-0422 – us ATLA [920]

Discourse on the modern mental philosophy viewed in its aspects on... / Barrett, Alfred – London, England. 1850 – 1r – us UF Libraries [240]

Discourse on the presence of god / Newman, Francis William – London, England. 1875 – 1r – us UF Libraries [210]

A discourse on the revival : delivered in the universalist church, portsmouth, n h...apr 18 1858 / Patterson, Adoniram Judson – Portsmouth: FW Miller, 1858 [mf ed 1992] – (= ser Unitarian/universalist coll) – 1mf – 9 – 0-524-04739-1 – mf#1991-2144 – us ATLA [243]

A discourse on the social influence of christianity / Cushing, Caleb – Andover: printed by Gould & Newman, 1839 [mf ed 1990] – 1mf – 9 – 0-7905-4672-8 – mf#1988-0672 – us ATLA [240]

A discourse on the studies of the university of cambridge / Sedgwick, Adam – London, 1850 – (= ser 19th c evolution & creation) – 8mf – 9 – mf#1.1.10890 – uk Chadwyck [500]

A discourse on the study of the law of nature and nations / Mackintosh, James – Edinburgh: T Clark, 1838 [mf ed 1984] – 1mf – 9 – 0-665-46137-2 – (incl bibl ref) – mf#46137 – cn Canadiana [170]

A discourse on the transient and permanent in christianity : preached at the ordination of mr charles c shackford, in the hawes place church in boston, may 19 1841 / Parker, Theodore – 3rd ed. Boston: B H Greene & E P Peabody, 1841 [mf ed 1984] – (= ser Biblical crit – us & gb 80) – 1mf – 9 – 0-8370-1244-9 – mf#1984-1080 – us ATLA [243]

A discourse on theological education : delivered...july 1843... / Howe, George – New York: Leavitt, Trow, 1844 [mf ed 1990] – 1mf – 9 – 0-7905-6925-6 – mf#1988-2925 – us ATLA [240]

Discourse, preached in salisbury cathedral, on king charles's marty / Bowles, William Lisle – Salisbury, England. 1836 – 1r – us UF Libraries [240]

A discourse preached in st andrew's church, toronto on the 24th of may, 1863 : being the anniversary of the birth-day of her most gracious majesty queen victoria / Barclay, John – Toronto?: s.n, 1863 – 1mf – 9 – mf#61943 – cn Canadiana [240]

Discourse preached in the episcopal chapel / Alison, Archibald – Edinburgh, Scotland. 1814 – 1r – us UF Libraries [240]

Discourse, preached in the episcopal chapel / Alison, Archibald – Edinburgh, Scotland. 1816 – 1r – us UF Libraries [240]

Discourse preached in the new north church, edinburgh / Bruce, John – Edinburgh, Scotland. 1834 – 1r – us UF Libraries [240]

A discourse preached in warren at the completion of the first century of the warren association, september 11, 1867 / Caldwell, Samuel Lunt – Providence: Hammond, Angell, 1867 [mf ed 1993] – (= ser Congregational coll) – 1mf – 9 – 0-524-07190-X – mf#1990-5348 – us ATLA [242]

A discourse...at plymouth, dec 20 1828 : on the...anniversary of the landing of the pilgrim fathers... / Green, Samuel – Boston: Pierce & Williams, 1829 (mf ed 19–) – 1 – mf#*ZH-IAG pv171 n3 – us NY Public [975]

Discourses : doctrinal and practical / Kirk, Edward Norris – Boston: American Tract Society [c1860] [mf ed 1984] – 3mf – 9 – 0-8370-1009-8 – mf#1984-4365 – us ATLA [242]

Discourses / Furness, William Henry – Philadelphia: G Collins, 1855 [mf ed 1986] – 1mf – 9 – 0-8370-9944-7 – mf#1986-3944 – us ATLA [243]

Discourses / Jack, Robert – Manchester, England. 1834 – 1r – us UF Libraries [240]

Discourses / Jackson, Abner – New York: T Whittaker, 1875 [mf ed 1984] – 3mf – 9 – 0-8370-0789-5 – mf#1984-4160 – us ATLA [242]

Discourses / Perry, Charles John; ed by Armstrong, Richard Acland – Liverpool: Henry Young, 1884 – 1mf – 9 – 0-524-00306-8 – mf#1989-3006 – us ATLA [240]

Discourses and discussions in explanation and defence of unitarianism / Dewey, Orville – Boston: pub by Joseph Dowe, 1840 [mf ed 1984] – 4mf – 9 – 0-8370-0852-2 – mf#1984-4216 – us ATLA [243]

Discourses and essays / Shedd, William Greenough Thayer – [2nd ed]. Andover: Warren F Draper, c1862 – 1mf – 9 – 0-8370-5346-3 – (incl bibl ref) – mf#1985-3346 – us ATLA [240]

The discourses and sayings of confucius – Shanghai, Hong Kong: Kelly & Walsh, 1898 [mf ed 1995] – (= ser Yale coll) – x/182p – 1 – 0-524-09327-X – (new special trans by ku hung-ming, ill with quotations from goethe and other writers) – mf#1995-0327 – us ATLA [180]

Discourses bearing upon the sonship and brotherhood of believers : and other kindred subjects / Candlish, Robert Smith – Edinburgh: Adam and Charles Black, 1872 – 1mf – 9 – 0-7905-1506-7 – mf#1987-1506 – us ATLA [240]

Discourses biological and geological : essays / Huxley, Thomas Henry – London: Macmillan, 1894 – 1mf – 9 – 0-7905-7306-7 – mf#1989-0531 – us ATLA [574]

Discourses in america / Arnold, Matthew – London: Macmillan, 1885 – 1mf – 9 – 0-7905-3751-6 – mf#1989-0244 – us ATLA [320]

Discourses in memoriam of the rev james cranbrook / Lake, John W – London, England. 1869? – 1r – us UF Libraries [240]

Discourses of epictetus – London, England. 1887 – 1r – us UF Libraries [180]

Discourses of our blessed saviour – London, England. 1820 – 1r – us UF Libraries [240]

Discourses on christian nurture / Bushnell, Horace – Boston: Massachusetts Sabbath School Society, 1847 – 1mf – 9 – 0-524-00009-3 – mf#1989-2709 – us ATLA [240]

Discourses on human life / Dewey, Orville – New York: David Felt, 1841 [mf ed 1984] – 4mf – 9 – 0-8370-0851-4 – mf#1984-4217 – us ATLA [243]

Discourses on iranian literature / Madan, Dhanjishah Meherjibhai – Bombay: Parsi Pub Co, 1909 – 1mf – 9 – 0-524-01791-3 – mf#1990-2639 – us ATLA [470]

Discourses on moral and religious subjects = Sermons. selections / Rosmini, Antonio – London: J Duffy, 1882 – 1mf – 9 – 0-7905-9857-4 – (in english) – mf#1989-1582 – us ATLA [240]

Discourses on occasion of the dedication of hope-street new church... / Madge, Thomas – London, England. 1849 – 1r – us UF Libraries [240]

Discourses on philippians / Noble, Frederick Alphonso – Chicago: Fleming H Revell, c1896 – 1mf – 9 – 0-8370-4594-0 – mf#1985-2594 – us ATLA [227]

Discourses on prophecy : in which are considered its structure, use and inspiration: being the substance of twelve sermons preached in the chapel of lincoln's inn, in the lecture founded by the right reverend william warburton, bishop of gloucester / Davison, John – new ed. Oxford: James Parker, 1875 – 1mf – 9 – 0-8370-9458-5 – (incl bibl ref) – mf#1986-3458 – us ATLA [220]

Discourses on radhasoami faith / Sastri, Brahmasankara – Benares: BP Dey, 1909 – 1mf – 9 – 0-524-01811-1 – mf#1990-2659 – us ATLA [280]

Discourses on some of the most difficult texts of scripture / Cochrane, James – Edinburgh: Paton and Ritchie, 1851 – 1mf – 9 – 0-7905-0928-8 – mf#1987-0928 – us ATLA [220]

Discourses on some theological doctrines as related to the religious character / Park, Edwards Amasa – Andover: Warren F Draper, 1885 – 1mf – 9 – 0-8370-3974-6 – (incl ind) – mf#1985-1974 – us ATLA [240]

Discourses on the beatitudes / Chapin, Edwin Hubbell – Boston: A Tompkins, 1853 – 1mf – 9 – 0-524-06393-1 – mf#1991-2515 – us ATLA [240]

Discourses on the bhagavat gita : to help students in studying its philosophy / Subba Row, Tiruvalum – Bombay: Joint-Stock Printing Press, 1888 – 1mf – 9 – 0-524-01384-5 – mf#1990-2396 – us ATLA [180]

Discourses on the christian spirit and life / Bartol, Cyrus Augustus – 2nd rev ed. Boston: Wm Crosby and HP Nichols, 1850 – 1mf – 9 – 0-524-06705-8 – mf#1991-2735 – us ATLA [240]

Discourses on the kingdom and reign of christ / Pope, William Burt – 2nd ed. Manchester: Palmer & Howe; London: Simpkin, Marshall, 1869 – 1mf – 9 – 0-7905-8556-1 – mf#1989-1781 – us ATLA [240]

Discourses on the lord's prayer / Chapin, Edwin Hubbell – Boston: A Tompkins, 1850 – 1mf – 9 – 0-524-06459-8 – mf#1992-0887 – us ATLA [240]

Discourses on the nature and extent of the atonement of christ / Wardlaw, Ralph – [2nd ed] Glasgow: James Maclehose 1844 [mf ed 1993] – 1mf – 9 – 0-524-07770-3 – mf#1991-3338 – us ATLA [240]

Discourses on the resurrection / Weaver, Jonathan – Dayton, OH: United Brethren Pub House, 1871 [mf ed 1989] – 1mf – 9 – 0-7905-2447-3 – mf#1987-2447 – us ATLA [242]

Discourses on the unity of god, and other subjects / Eliot, William Greenleaf – St Louis: Printed at the Republican Office, 1852 – 1mf – 9 – 0-7905-3725-7 – mf#1989-0218 – us ATLA [240]

Discourses on truth : delivered in the chapel of the south carolina college / Thornwell, James Henley – New York: R Carter, 1855 – 1mf – 9 – 0-524-05093-7 – mf#1991-2217 – us ATLA [240]

Discourses on various occasions = Selections. 1809 / Hyacinthe, pere – New York: G.P. Putnam; London: S. Low, Son & Marston, 1869 – 1mf – 9 – 0-8370-8748-1 – (in english) – mf#1986-2748 – us ATLA [240]

Discourses on various subjects / Dewey, Orville – 3rd ed. New York: David Felt, 1838 [mf ed 1994] – 4v on 5mf – 9 – 0-8370-0850-6 – mf#1984-4218 – us ATLA [240]

Discourses on various subjects relative to the being and attributes of god, and his works in creation, providence, and grace / Clarke, Adam – London: William Tegg, 1868 – 5mf – 9 – 0-524-08847-0 – mf#1993-2132 – us ATLA [210]

Discourses preached in st andrew's church, toronto / Barclay, John – S.l: s.n, 186-? – 2mf – 9 – mf#64049 – cn Canadiana [240]

Discourses upon seneca the tragedian / Cornwallis, William – 1601 – 9 – us Scholars Facs [450]

Discourses upon tradition and episcopacy / Benson, Christopher – London, England. 1839 – 1r – us UF Libraries [240]

Discoursos leidos en la recepcion / Garrigo, Roque E – Habana, Cuba. 1935 – 1r – us UF Libraries [972]

Discoursos leidos en la recepcion / Gay-Calbo, Enrique – Habana, Cuba. 1942 – 1r – us UF Libraries [972]

Discover puerto rico / Vandercook, John W – New York, NY. 1939 – 1r – us UF Libraries [972]

The discoverie of witchcraft...being a reprint of the first edition published in 1584 / Scot, Reginald – Ed. with explanatory notes, glossary and introd. by Brinsley Nicholson. London: E. Stock, 1886. xlvii,xxxviii,589p. illus – 1 – us Wisconsin U Libr [150]

Discoveries and adventures in central america / Gann, Thomas William Francis – London, England. 1928 – 1r – us UF Libraries [918]

Discoveries in anatolia / Von der Osten, Hans Henning – 1933 – 9 – $10.00 – us IRC [930]

Discoveries in asia minor : including a description of the ruins of several ancient cities, and especially antioch of pisidia / Arundell, Francis V – London 1834 [mf ed Hildesheim 1995-98] – 2v on 6mf – 9 – €120.00 – 3-487-27654-2 – gw Olms [930]

Discoveries in the ruins of nineveh and babylon : with travels in armenia, kurdistan, and the desert... / Layard, A H – London: J Murray, 1853 – 9mf – 9 – mf#HT-77 – ne IDC [915]

The discoveries of the norsemen in america : with special relation to their early cartographical representation / Fischer, J – Oakland. 1955-1977 (1) 1977-1977 (5) 1977-1977 (9) – 4mf – 9 – mf#2223 – ne IDC [910]

Discovery – Arlington Heights IL 1961-95 – 1,5,9 – ISSN: 0012-3641 – mf#2760 – us UMI ProQuest [910]

Discovery – Brooks Air Force Base [TX]. 1981 may 8-1986 apr 25 – 1r – 1 – mf#1048618 – us WHS [355]

Discovery – London, England 1920-66 – 1 – mf#1267 – us UMI ProQuest [500]

The discovery and conquest of florida / ed by Rye, W B – (= ser Hakluyt society. extra series 9) – 4mf – 7 – (trans fr portuguese by richard hakluyt. int & notes by ed) – mf#311 – uk Microform Academic [917]

Discovery and conquest of mexico, 1517-1521 / Diaz del Castillo, Bernal – New York, NY. 1956 – 1r – us UF Libraries [972]

The discovery and conquest of the new world : containing the life and voyages of christopher columbus / Irving, Washington – Toronto: Rose, 1892 – 10mf – 9 – (together with: a separate account of the conquest of mexico and peru by w w robertson. a perfect history of the united states from the works of bancroft, fiske, blaine, brant, sherman, johnston and others by benjamin rush. int by the hon murat halstead) – mf#02719 – cn Canadiana [970]

Discovery and disclosure practice; problems, and proposals for change : a case-based national survey of counsel in closed federal civil cases / Willging, Thomas E et al – 1997 – 1mf – 9 – $1.50 – mf#llmc99-018 – us LLMC [347]

Discovery and exploration of the mississippi valley : with the original narratives of marquette, allouez, membre, hennepin, and anastase douay / Shea, John Dawson Gilmary – New York: Redfield, 1852 – 1mf – 9 – 0-7905-6623-0 – mf#1988-2623 – us ATLA [917]

The discovery and exploration of the pelly (yukon) river / Campbell, Robert – S.l: s.n, 18– – 1mf – 9 – mf#02020 – cn Canadiana [917]

The discovery and exploration of the youcon (pelly) river / Campbell, Robert – Winnipeg?: s.n, 1885 – 1mf – 9 – mf#00932 – cn Canadiana [917]

The discovery, conquest, and organization of spanish america and oceania – Madrid, 1842-95 [mf ed Microcard Editions] – 239mf (24:1) – 9 – $1595.00 – us UPA [910]

The discovery of a north-west passage by h m s investigator, capt r m'clure : during the years 1850, 1851, 1852, 1853, 1854 / ed by Osborn, Sherard – Edinburgh, London: W Blackwood, 1865 – 5mf – 9 – mf#52737 – cn Canadiana [919]

The discovery of a world in the moone / Wilkins, John – 1638 – 9 – us Scholars Facs [520]

The discovery of america and islands adjacent, 1582 / Hakluyt, Richard [comp] – (= ser Hakluyt society. extra series 7) – 5mf – 7 – (publ by comp) – mf#302/A – uk Microform Academic [917]

The discovery of america by john cabot in 1497 : being extracts from the proceedings of the royal society of canada relative to a cabot celebration in 1897; and, the voyages of the cabots, a paper from the transactions of the society in 1896, with appendices on kindred subjects / Dawson, Samuel Edward – Ottawa?: s.n, 1896 – 1mf – 9 – mf#02612 – cn Canadiana [971]

The discovery of america by the northmen : in the tenth century / Beamish, North Ludlow – London, 1841 – (= ser 19th c ireland) – 3mf – 9 – mf#1.1.7405 – uk Chadwyck [941]

The discovery of america, vol 1 : with some account of ancient america and the spanish conquest / Fiske, John – Boston; New York: Houghton Mifflin. 2v. c1892 [mf ed 1980] – 7mf – 9 – (incl bibl ref) – mf#05664 – cn Canadiana [970]

The discovery of america, vol 2 : with some account of ancient america and the spanish conquest / Fiske, John – Boston; New York: Houghton Mifflin. 2v. c1892 [mf ed 1980] – 8mf – 9 – (incl bibl ref) – mf#05665 – cn Canadiana [970]

The discovery of america, vols 1-2 : with some account of ancient america and the spanish conquest / Fiske, John – Boston; New York: Houghton Mifflin. 2v. c1892 – 1mf – 9 – 0-665-05663-X – mf#05663 – cn Canadiana [970]

The discovery of china's earliest story of jesus christ *see* Yeh-su chi-tu tsai chung-kuo ku chi chung chih fa hsien (ccm214)

The discovery of lake superior : a study from the jesuits journals / Harvey, Arthur – S.l: s.n, 1885? – 1mf – 9 – mf#05251 – cn Canadiana [917]

The discovery of nebraska : and, a visit to nebraska in 1662 / Savage, James Woodruff – Washington: Govt Print Off, 1893 (mf ed 19–) – (= ser U.S. Congress. 53 Congress 2 Session Senate Mis Doc No 14) – 1 – mf#*ZH-IAG pv139 n10 – us NY Public [975]

A discovery of the barmudas / Jourdain, Silvester – 1610 – 9 – 5.00 – us Scholars Facs [972]

Discovery of the bermudas (1619) / Jourdain, Silvester – New York, NY. 1940 – 1r – us UF Libraries [972]

The discovery of the book of the law under king josiah : an egyptian interpretation of the biblical account / Naville, Edouard – London: SPCK, 1911 – 1mf – 9 – 0-7905-1367-6 – (incl bibl ref) – mf#1987-1367 – us ATLA [221]

The discovery of the empire of guiana, 1595 / Raleigh, Walter – (= ser Hakluyt society. extra series 3) – 5mf – 9 – mf#282 – uk Microform Academic [918]

The discovery of the north-west passage / McClure, R; ed by Osborn, S – London, 1856 – 9mf – 9 – mf#N-302 – ne IDC [919]

The discovery of the true and natural era of mankind : and the means of carrying it into effect / Edwards, George – [London]: printed for J Johnson, 1807 – (= ser 19th c economics) – 2mf – 9 – mf#1.1.103 – uk Chadwyck [330]

Discovery of tripoli : or polishing powder near st john / Allison, L C – S.l: s.n, 1881? – 1mf – 9 – mf#05936 – cn Canadiana [550]

Discovery problems in civil cases / Ebersole, Joseph L & Burke, Barlow – Washington: FJC, Apr 1980 – 2mf – 9 – $3.00 – mf#LLMC 95-822 – us LLMC [347]

Discovrs, de la bataille novvellement perdve par le tvrc, contre le roy de perse, 1586 – Paris, 1586 – 1mf – 9 – mf#H-8205 – ne IDC [956]

Discovrs de la grande et pvissante armee de soltan solyman grand empereut des turcz... – Paris, 1565 – 1mf – 9 – mf#H-8166 – ne IDC [956]

Discovrs svr la chrestienne and genereuse entreprise de hault and puissant prince monseigneur charles de lorraine...contre le grand turc, en l'an 1572 – Paris, 1572 – 1mf – 9 – mf#H-8191 – ne IDC [956]

Discovrs veritable des visions advenves av premier et second iour d'aoust 1589 a la personne de l'empereur des turcs sultan amurat, en la ville de constantinople – Paris, [1589] – 1mf – 9 – mf#H-8208 – ne IDC [956]

Discrete and computational geometry – Dordrecht, Netherlands 1986+ – 1,5,9 – ISSN: 0179-5376 – mf#16986 – us UMI ProQuest [510]

Discrete applied mathematics – Oxford, England 1979+ – 1,5,9 – ISSN: 0166-218X – mf#42163 – us UMI ProQuest [510]

Discrete event dynamic systems – Dordrecht, Netherlands 1991+ – 1,5,9 – ISSN: 0924-6703 – mf#18599 – us UMI ProQuest [621]

Discrete mathematics – Oxford, England 1971+ – 1,5,9 – ISSN: 0012-365X – mf#42180 – us UMI ProQuest [510]

The discriminate use of amalgam for filling teeth / Beers, William George – Montreal?: J Lovell, 1871 – 1mf – 9 – mf#01511 – cn Canadiana [617]

Discrimination in the criminal justice system, 1910-1955 – 2ser – (= ser Papers of the naacp 8) – 1 – (ser a: legal dept & central office records, 1910-39 17r isbn 1-55655-079-0 $3280. ser b: legal dept & central office records, 1940-55 32r isbn 1-55655-080-4 $6180. with p/g) – us UPA [322]

Discrimination in the u s armed forces, 1918-1955 – 3ser – (= ser Papers of the naacp 9) – 1 – (ser a: general office files on armed forces' affairs, 1918-55 18r isbn 1-55655-116-9 $3520. ser b: armed forces' legal files, 1940-50 30r isbn 1-55655-117-7 $5985. ser c: the veterans affairs committee, 1940-50 12r isbn 1-55655-118-5 $2330. with p/g) – us UPA [322]

Discurso / Chaves y Manso, Rafael – Universidad literaria de.... 1851 – 9 – sp Bibl Santa Ana [440]

Discurso / Delgado del Pino, Francisco – 1835 – 9 – sp Bibl Santa Ana [946]

Discurso.. / Olabarrieta, Francisco de – 1834 – 9 – sp Bibl Santa Ana [890]

Discurso a los escritores venezolanos / Marinello, Juan – Habana, Cuba. 1948 – 1r – us UF Libraries [972]

Discurso al ofrecer el banquete dedicado / Orbe, Diogenes Del – Santiago, Dominican Republic. 1946 – 1r – us UF Libraries [972]

Discurso antisofistico extractado del hombre / Forner Segarra, Juan Pablo – 1787 – 9 – sp Bibl Santa Ana [840]

Discurso de clausura pronunciado por el excmo. sr. obispo de badajoz el...1907 / Soto y Mancera, Felix – Badajoz: Tip. Lit. y Enc. de Uceda Hnos – 1 – sp Bibl Santa Ana [240]

Discurso de cosas aromaticas...de las indias... para uso de medicinas / Fragoso, J – Madrid, 1572 – 8mf – 9 – sp Cultura [615]

Discurso de la razon / Pizarro de Aragon, Juan – Madrid: Francisco Martinez, 1629 – 1 – sp Bibl Santa Ana [100]

Discurso de recepcion...en la academia de la historia : tema: el conquistador espanol, los fundadores de nuestra senora de la paz, de trujillo. caceres, 1930 / Briceno-Iregorry, Mario – Madrid: Razon y Fe, 1930 – 1 – sp Bibl Santa Ana [240]

Discurso de santos / Fuentes, Milton – Bogota, Colombia. 1938 – 1r – us UF Libraries [972]

Discurso de s.s....alas conferencias de san vicente de paul (en la audiencia de 27 de abril de 1952) – Caceres: Tip. Extremadura, s.a. – 1 – sp Bibl Santa Ana [240]

Discurso de...audiencia de caceres / Penalver, Nicolas – 1853 – 9 – sp Bibl Santa Ana [340]

Discurso del cometa del ano 1680 / Aldrete y Soto, L – Madrid, S.A. – 1mf – 9 – sp Cultura [100]

Discurso del cometa inocente... / Gamez, A – Napoles, 1681 – 2mf – 9 – sp Cultura [100]

Un discurso del dr. goebbels, ministro de propaganda de alemania / Goebbels, Joseph – n.p. 193? Fiche W922. (Blodgett Collection of Spanish Civil War Pamphlets) – 9 – us Harvard [946]

Discurso del..., en defensa de los dictamenes de los siguientes proyectos de ley: 1º de montes vecinales en mano comun. 2º de ordenacion rural. 3º regimen de tierras del instituto nacional de colonizacion / Diaz-Ambrona, Adolfo – Madrid: Suc. Rivadeneyra, 1959 – sp Bibl Santa Ana [946]

Discurso del generalisimo : no cabe transaccion idealogica / Spain. Ministerio del Interior – n.p, 1938 – (= ser Blodgett coll) – 9 – mf#fiche w843 – us Harvard [946]

Discurso del generalisimo rafael l trujillo molin – Ciudad Trujillo, Dominican Republic. 1938 – 1r – us UF Libraries [972]

Discurso del licenciado julio ortega frier – Ciudad Trujillo, Dominican Republic. 1942 – 1r – us UF Libraries [972]

Discurso del papa a 20000 obreros el 13 de junio, dia de pentecostes. plasencia / Pio 12. – Ediciones J.D. de A.C., Imprenta La Victoria, s.a. – sp Bibl Santa Ana [240]

Discurso del presidente del consejo de ministros don juan negrin / Negrin, Juan – [Barcelona: Ediciones Espanolas 1938?] [mf ed 1977] – (= ser Blodgett coll) – 1mf – 9 – mf#w1071 – us Harvard [350]

Discurso en el "gran price" de barcelona, el dia 17 de enero de 1937 / Tomas, Pascual – Barcelona, 1937? Fiche W1230. (Blodgett Collection of Spanish Civil War Pamphlets) – 9 – us Harvard [946]

DISCUSION

Discurso filosofico, medico e historial...en defensa de la medicina dogmatica y su sangria... / Gamez, A – Madrid, 1683 – 2mf – 9 – sp Cultura [610]
Discurso. guerra de portugal / Diaz de Vargas, Francisco – 1644 – 9 – sp Bibl Santa Ana [946]
Discurso inaugural / Remon de Moncada y Calderon, Jesus – 1856 – 9 – sp Bibl Santa Ana [440]
Discurso inaugural pronunciado por d. jesus reunion de moncada y calderon / Ramon de Moncada y Calderon, Jesus – Badajoz: Geronimo Orduna, 1856 – 1 – sp Bibl Santa Ana [946]
Discurso inaugural...academia cientifico / Crespo y Escoriaza, Benito – 1857 – 9 – sp Bibl Santa Ana [500]
Discurso leido en la apertura del curso academico 1943-44 / Hernandez Pacheco, Francisco – Madrid: Estades-Artes Graficas, 1943 – sp Bibl Santa Ana [370]
Discurso leido en la solemne inauguracion del curso academico 1912 a 1913 / Rivas Mateos, Marcelo – Madrid: Imp. Colonial, 1912 – 1 – sp Bibl Santa Ana [370]
Discurso leido por...artes y oficios de fregenal en la solemne inauguracion de la misma / Real, Enrique – Fregenal: Imp. Indalecio Blanco, s.a. – sp Bibl Santa Ana [946]
Discurso medicinal...en el que se declara la horden...para preservarse de la peste / Franco, M – Cordoba, 1601 – 1mf – 9 – sp Cultura [614]
Discurso medico sobre el verdadero metodo de curar las viruelas... / Adami, A – Sevilla, S.A. – 1mf – 9 – sp Cultura [616]
Discurso parlamentario / Donoso Cortes, Juan Francisco – Madrid: Imprenta Espn., 1915 – 1 – sp Bibl Santa Ana [350]
Discurso particular y preservativo de la gota, en que se descubre su naturaleza y se pone su propia cura / Cornejo, J – Madris, S.A. – 2mf – 9 – sp Cultura [600]
Discurso politico sobre la importancia de los hospicios, casas de expositos con y hospitales / Murcia, P J de – Madrid, 1798 – 3mf – 9 – sp Cultura [360]
Discurso proferido pelo general norton de matos – Loanda, Angola. 1923 – 1r – us UF Libraries [960]
Discurso pronunciado 8 noviembre de 1888 / Canovas del Castillo, Antonio – Madrid, 1888 – 1 – us CRL [946]
Discurso pronunciado en 'chickering hall' / Yero Buduen, Eduardo – New York, NY. 1896 – 1r – us UF Libraries [972]
Discurso pronunciado en la colaboracion de grados de licenciados en jurisprudencia... / Gomez Jara, Francisco – Sevilla: Imprenta del Regalo, 1850 – 1 – sp Bibl Santa Ana [340]
Discurso pronunciado en la sociedad patriotica constitucional de badajoz el dia 9 de julio de 1820 / Rocha, Manuel de la – Badajoz: Imp. de Capitania General, 1820 – 1 – sp Bibl Santa Ana [350]
Discurso pronunciado por el doctor carlos prio soc – Havana, Cuba. 1948 – 1r – us UF Libraries [972]
Discurso pronunciado por el dr horacio diaz pardo – Habana, Cuba. 1919 – 1r – us UF Libraries [972]
Discurso pronunciado por el excelentisimo senor presidente de la arepulica : generalisimo dr. rafael leonidas trujillo molina en el altar de la patria el dia 27 de febrero de 1944 en ocasion del primer centenario de la independencia nacional / Trujillo Molina, Rafael Leonidas – Ciudad Trujillo: R D [Impreso en "La Nacion"] 1944 (rd 2000) – 1r – 1 – mf#*Z-9039 – us NY Public [972]
Discurso pronunciado por el excmo. sr. gobernador civil-presidente...en la reunion...25 de febrero de 1950 / Ruiz de la Serna, Manuel – Badajoz: Graficas Iberia, 1950 – 1 – sp Bibl Santa Ana [321]
Discurso pronunciado por el obrero en el circulo catolico de villafranca de los barros el dia 8 de diciembre de 1906 / Diez Lopez, Juan – Villafranca de los Barros: Imp. Ventura Rodriguez, 1906 – 1 – sp Bibl Santa Ana [240]
Discurso pronunciado por luis companys el dia 27 de diciembre de 1936 en el palacio de bellas artes de barcelona : con motivo del 3 aniversario del fallicimiento de francisco macia / Companys, Lluis – Barcelona, 1937? Fiche W1503. (Blodgett Collection of Spanish Civil War Pamphlets) – 9 – us Harvard [946]
Discurso pronunciado por...el 30 de noviembre de 1881 en el ateneo cientifico y literario de madrid con motivo de la apertura de sus catedras / Moreno Nieto, Jose – Madrid: Imp. Central a cargo de Victor Saiz, 1881 – 1 – sp Bibl Santa Ana [946]
Discurso pronunciado por...el dia 8 de noviembre de 1877 en el ateneo cientifico literario de madrid, con motivo de la apertura de sus catedras / Moreno Nieto, Jose – Madrid: Empresa del Boletin Oficial del Ateneo, 1877 – 1 – sp Bibl Santa Ana [946]
Discurso pronunciado por...el dia 10 de noviembre de 1880 en el ateneo cientifico y literario de madrid con motivo de la apertura de sus catedras / Moreno Nieto, Jose – Madrid: Imp. Central a cargo de V. Saiz, 1880 – 1 – sp Bibl Santa Ana [946]
Discurso pronunciado por...el dia 31 de octubre de 1878 en el ateneo cientifico y literario de madrid con motivo de la apertura de sus catedras / Moreno Nieto, Jose – Madrid: Empresa del Bol. del Ateneo, 1878 – 1 – sp Bibl Santa Ana [946]
Discurso pronunciado...sobre el proyecto de reforma agraria / Teixeira, Antonio – 1931 – 1 – sp Bibl Santa Ana [630]
Discurso que...pronuncio el 23 de abril. / Alvarado, Manuel – 1821 – 9 – sp Bibl Santa Ana [946]
Discurso sobre algunas proposiciones del doctor luys de mercado / Lopez Coronel, P – SL, 1611 – 1mf – 9 – sp Cultura [610]
Discurso sobre el charlatanismo medico y quirurgico... / Arguello Castrillo, A – Valladolid, 1796 – 3mf – 9 – sp Cultura [610]
Discurso sobre que los ninos expositos consigan en las inclusas el fin de estos establecimientos / Trespalacios y Mier, J – Madrid, 1798 – 1mf – 9 – sp Cultura [360]
Discurso y sumario de la guerra en portugal / Diaz de Vargas, Francisco – Zaragoza: Pedro Vargas, 1644 – 1 – sp Bibl Santa Ana [946]
Discurso...6 noviembre de 1822...apertura de la universidad...badajoz / Rocha, Manuel de la – Imp. de la comandancia general, 1822 – 1 – sp Bibl Santa Ana [946]
Discurso...academia de la historia / Barrantes Moreno, Vicente – 1872 – 9 – sp Bibl Santa Ana [946]
Discurso...academia de la historia / Barrantes Moreno, Vicente – 1874 – 9 – sp Bibl Santa Ana [946]
Discurso...apertura de curso sobre la importancia del catolicismo / Guillen y Flores, Agustin – 1857 – 9 – sp Bibl Santa Ana [241]
Discurso...apertura de curso...caceres / Ollero, Meliton – 1849 – 9 – sp Bibl Santa Ana [440]
Discurso...audiencia...caceres. .bi.-1838 / Miguel Sanchez & Paula, Francisco de – 9 – sp Bibl Santa Ana [850]
Discurso...catedras / Moreno Nieto, Jose – 1876 – 9 – sp Bibl Santa Ana [440]
Discurso...ciencias morales / Fernandez de Soria, Rafael – 1862 – 9 – sp Bibl Santa Ana [170]
Discurso...curso / Muela, Jose de la – 1848 – 9 – sp Bibl Santa Ana [440]
Discurso...de las sesiones del ano 1853 en la real academia de medicina... / Nieto y Serrano, M – Madrid, 1853 – 1mf – 9 – sp Cultura [610]
Discurso...del sarampion y viruelas... / Samillan, L – Montilla, 1626 – 1mf – 9 – sp Cultura [616]
Discurso...distribucion de premios / Redondo y Poblacion, Pedro – 1892 – 9 – sp Bibl Santa Ana [946]
Discurso..."el tratamiento antiparasito de la tuberculosis y la linfa de koch" / Miguel y Guerra, Regino de – Badajoz: Tip. Lit. y Enc. La Industria, 1891 – 1 – sp Bibl Santa Ana [616]
Discurso...nobleza...espana / Moreno de Vargas, Bernabe – 1622 ed – 9 – (1659 ed. 1795 ed) – sp Bibl Santa Ana [946]
Discurso...pedro calderon / Fuertes Acevedo, Maximo – 1881 – 9 – sp Bibl Santa Ana [440]
Discurso-politico...colero norbo / Gomez, Florencio – 1834 – 9 – sp Bibl Santa Ana [946]
Discursos / Aramburo y Machado, Mariano – San Jose de Costa Rica: J Garcia Monge, 1922 [mf ed 1998] – (= ser Repertorio americano. biblioteca) – 1r – 1 – (int by jose maria chacon y calvo) – mf#28128 – us Harvard [440]
Discursos / Elviro Meseguer, Francisco – Toledo: Dip. Provincial, 1961 – sp Bibl Santa Ana [946]
Discursos / Fernandez-Ouesta, Raimundo – Gijon. 1939? Fiche W 880. (Blodgett Collection of Spanish Civil War Pamphlets) – 9 – us Harvard [946]
Discursos / Gavidia, Francisco – San Salvador, El Salvador. 1941 – 1r – us UF Libraries [972]
Discursos / Tenorio Cordero de Santoyo, Miguel – 1841, 1843 – 9 – sp Bibl Santa Ana [946]
Discursos / Valencia, Guillermo – Bogota, Colombia. 193-? – 1r – us UF Libraries [972]
Discursos / Varona, Enrique Jose – Habana, Cuba. 1918 – 1r – us UF Libraries [972]
Discursos.. / Moreno Nieto, Jose – 1877 – 9 – sp Bibl Santa Ana [440]
Discursos.. / Moreno Nieto, Jose – 1879 – 9 – sp Bibl Santa Ana [440]
Discursos.. / Moreno Nieto, Jose – 1880 – 9 – sp Bibl Santa Ana [440]
Discursos.. / Moreno Nieto, Jose – 1881 – 9 – sp Bibl Santa Ana [440]
Discursos a los asturianos de america / Ibarguren, Carlos – Buenos Aires: [s.n.] 1937 – (= ser Blodgett coll) – 9 – mf#w952 – us Harvard [946]
Discursos a los vascos de america / Ibarguren, Carlos – Buenos Aires, 1937. Fiche W952. (Blodgett Collection of Spanish Civil War Pamphlets) – 9 – us Harvard [946]
Discursos academicos del... / Moreno Nieto, Jose – 1882 – 9 – sp Bibl Santa Ana [440]
Discursos apologeticos, en que se defiende la ingenuidad del arte de la pintura... / Butron, I de – Madrid, 1626 – 4mf – 9 – mf#0-1184 – ne IDC [700]
Discursos de bobadilla / Bobadilla Y Briones, Tomas – Ciudad Trujillo, Dominican Republic. 1938 – 1r – us UF Libraries [972]
Discursos de la nobleza de espana / Moreno de Vargas, Bernabe – Corregidos y anadidos por el mismo autor. Madrid, Imprenta de Don Antonio Espinosa, 1795 – 1 – sp Bibl Santa Ana [946]
Discursos de la nobleza de espana / Moreno de Vargas, Bernabe – Madrid: Jose Fernandez Buendia, 1659 – 1 – sp Bibl Santa Ana [946]
Discursos del presidente de guatemala / Castillo Armas, Carlos – Guatemala, 1957 – 1r – us UF Libraries [972]
Discursos de...sobre cuestiones de caracter politico...legislatura de 1864-65 / Claros, Jose Ma de – 1865 – 9 – sp Bibl Santa Ana [320]
Discursos do j m da silva paranhos – Rio de Janeiro, Brazil. 1872 – 1r – us UF Libraries [972]
Discursos en defensa de la religion catholica / Farfan de los Godos, Antonio – 1623 – 9 – sp Bibl Santa Ana [240]
Discursos en la presidencia / Arevalo, Juan Jose – Guatemala, 1947 – 1r – us UF Libraries [972]
Discursos evangelicos / Gomez, Antonio – 1688 – 9 – (1698 ed) – sp Bibl Santa Ana [242]
Discursos filosoficos sobre el hombre / Forner Segarra, Juan Pablo – 1787 – 9 – sp Bibl Santa Ana [190]
Discursos forenses / Malendez Valdes, Juan – 1821 – 9 – sp Bibl Santa Ana [340]
Discursos historicos y literarios / Rodriguez Demorizi, Emilio – Ciudad Trujillo, Dominican Republic. 1947 – 1r – us UF Libraries [972]
Discursos leidos ante la real academia de la historia en la recepcion...el 14 de enero de 1872 con un biografia de este / Barrantes Moreno, Vicente – Madrid: Imp. Julian Pena, 3rd ed. 1873 – 1 – sp Bibl Santa Ana [946]
Discursos leidos ante la real academia espanola en la recepcion publica de.... : noticias sobre don luis zapata, contestacion de don francisco rodriguez marin / Menendez Pidal, Juan – Madrid: Tip. Rev. Arch. B. y Museos, 1915 – 1 – sp Bibl Santa Ana [370]
Discursos leidos ante la real academia sevillana de buenas letras el 3 de enero de 1897 / Perez de Guzman y Boza, Manuel & Rodriguez Marin, Francisco – Sevilla: Imp. E. Rasco, 1897 – 1 – sp Bibl Santa Ana [370]
Discursos leidos en el acto de reparto de premios...organizado por el excmo. ayuntamiento / Villafranca de los Barros – Villafranca de los Barros: Ventura Rodriguez, impresor, 1928 – 1 – sp Bibl Santa Ana [946]
Discursos leidos en la recepcion publica / Castellanos Garcia, Gerardo – Habana, Cuba. 1936 – 1r – us UF Libraries [972]
Discursos leidos en la recepcion publica / Cespedes Y Quesada, Carlos Manuel De – Habana, Cuba. 1933 – 1r – us UF Libraries [972]
Discursos leidos en la recepcion publica / Chacon Y Calvo, Jose Maria – Habana, Cuba. 1945 – 1r – us UF Libraries [972]
Discursos leidos en la recepcion publica / Manach, Jorge – Habana, Cuba. 1943 – 1r – us UF Libraries [972]
Discursos leidos en la recepcion publica / Remos Y Rubio, Juan Nepomuceno Jose – Habana, Cuba. 1949 – 1r – us UF Libraries [972]
Discursos leidos en la recepcion publica / Roig De Leuchsenring, Emilio – Habana, Cuba. 1938 – 1r – us UF Libraries [972]
Discursos leidos en la recepcion publica / Torriente Y Peraza, Cosme De La – Habana, Cuba. 1944 – 1r – us UF Libraries [972]
Discursos leidos en la recepcion publica / Trelles Y Govin, Carlos Manuel – Habana, Cuba. 1926 – 1r – us UF Libraries [972]
Discursos leidos en la recepcion publica del docto... / Valverde Y Maruri, Antonio L – Habana, Cuba. 1923 – 1r – us UF Libraries [972]
Discursos leidos en la recepcion publica del ldo... / Montoro, Rafael – Habana, Cuba. 1926 – 1r – us UF Libraries [972]
Discursos leidos...ciencias morales y politicas / Concha Castaneda, Juan de la – 1888 – 9 – sp Bibl Santa Ana [170]
Discursos leidos...san benito de villanueva / Fernandez Valbuena, Ramiro – 1886 – 9 – sp Bibl Santa Ana [240]
Discursos parlamentares / Silva, Jose Bonifacio De Andradae – Rio de Janeiro, Brazil. 1880 – 1r – us UF Libraries [324]
Discursos parlamentares, 1879-1889 / Nabuco, Joaquim – Sao Paulo, Brazil. 1949 – 1r – us UF Libraries [324]
Discursos parlamentarios : congreso nacional de 189... / Uribe Uribe, Rafael – Bogota, Colombia. 1897 – 1r – us UF Libraries [972]
Discursos parlamentarios de emilio castelar en la asamblea – Madrid, Spain. 188-? – 1r – us UF Libraries [323]
Discursos patrios...badajoz / Dosma Delgado, Rodrigo – 1661 – 9 – (1870 ed) – sp Bibl Santa Ana [946]
Discursos physico-medico, politico-moral que tratan ser toda calentura hectica contagiosa / Cerdan, F – Valencia, 1752 – 4mf – 9 – sp Cultura [610]
Discursos politicos academicos y forenses / Labra Y Cadrana, Rafael Maria De – Madrid, Spain. 1886 – 1r – us UF Libraries [972]
Discursos politicos y morales en cartas apologeticas contra los que defienden el vso de la comedias modernas que se representan en espana, en comparacion del teatro antiguo... / Navarro Castellanos, Gonzalo – Madrid: Impr. Real, 1684- – 1 – us Wisconsin U Libr [440]
Discursos pronunciados / Arias Madrid, Arnulfo – Panama, 1939 – 1r – us UF Libraries [972]
Discursos pronunciados en la sesion / Sociedad Cubana De Derecho Internacional – Habana, Cuba. 1944 – 1r – us UF Libraries [972]
Discursos pronunciados en las recepciones / Alvarado Quiros, Alejandro – San Jose, Costa Rica. 1935 – 1r – us UF Libraries [972]
Discursos y conferencias / Diaz Jordan, Jenaro – Neiva, Colombia. 1958 – 1r – us UF Libraries [972]
Discursos y conferencias / Martin, Ernesto – San Jose, Costa Rica. 1930 – 1r – us UF Libraries [972]
Discursos y conferencias : rebano servil / Buttari Guanaurd, J – Habana, Cuba. 1953 – 1r – us UF Libraries [972]
Discursos y conferencias / Sanguily, Manuel – Habana, Cuba. v1-2. 1918-19 – 1r – us UF Libraries [972]
Discursos y conferencias / Zayas Y Alfonso, Alfredo – Habana, Cuba. v1-2. 1942 – 1r – us UF Libraries [972]
Discursos y conferencias enjuiciando la... / Bonilla Atiles, Jose Antonio – Ciudad Trujillo, Dominican Republic. 1946 – 1r – us UF Libraries [972]
Discursos y sermones / Mosquera, Manuel Jose – Bogota, Colombia. 1954 – 1r – us UF Libraries [972]
Discursos...academia espanola / Barrantes Moreno, Vicente – 1876 – 9 – sp Bibl Santa Ana [946]
Discursos...academia sevillana de buenas letras / Perez de Guzman, Juan – 1897 – 9 – sp Bibl Santa Ana [440]
Discursos...academia...manuel de lo palacio / Barrantes Moreno, Vicente – 9 – sp Bibl Santa Ana [946]
Discursos...congreso de los diputados... / Bravo Murillo, Juan – 1858 – 9 – sp Bibl Santa Ana [946]
Discursos...de la real academia de medicina y cirugia de madrid en el ano. 1861 / Nieto y Serrano, M – Madrid, 1861 – 1mf – 9 – sp Cultura [610]
Discurso...segunda ensenanza de caceres / Sergio Sanchez, Luis – 1846 – 9 – sp Bibl Santa Ana [920]
Discursos...nobleza / Moreno de Vargas, Bernabe – 1636 – 9 – sp Bibl Santa Ana [946]
Discurso...sobre la epidemia de pamplona / Ortiz, M – Pamplona, 1789 – 4mf – 9 – sp Cultura [614]
Discurso...sociedad...badajoz / Rocha, Manuel de la – 1820 – 9 – sp Bibl Santa Ana [440]
Discursos...pronunciados...1951 al 1952 / Cortines, Ruiz – 1 – us CRL [946]
Discurso...tratamiento...tuberculosis / Miguel y Guerra, Regino de – 1891 – 9 – sp Bibl Santa Ana [610]
Discurso...universidad.. / Rocha, Manuel de la – 1822 – 9 – sp Bibl Santa Ana [440]
Discursus...mysteria fidei / Ovando, Juan de – 1593 – 9 – sp Bibl Santa Ana [240]
Discus newsletter / Distilled Spirits Council of the US – n337-412 [1974 sep-1985 dec] – 1r – 1 – (cont: newsletter) – mf#610205 – us WHS [660]
Discusion – Miami, FL. 1970 feb 17-1982 feb 01 – 1r – us UF Libraries [071]
La discusion – Havana, Cuba. 1924-1925 (1) – mf#67682 – us UMI ProQuest [079]

Discussing christianity with chan and tsen *see* Yu ch'en tu-hsiu shen hsuan-lu pien tao (ccm12)

Discussion bulletin / New American Movement [Organization] – 1974 may, 1975 mar-1977 spring, 1979 spring-1981 spring – 2r – 1 – mf#585659 – us WHS [320]

Discussion bulletin / Socialist Workers Party – v2 n1-3 [1949 jan 1-feb 15], v3 n1-2 [1950 jan 20-mar 5], 1951 jan – 1r – 1 – mf#669355 – us WHS [335]

Discussion bulletin / Student Peace Union [US] – v1 n1-v4 n3 [1961-1963 fall] – 1r – 1 – mf#626194 – us WHS [327]

Discussion, design, and specifications for a reinforced concrete bridge abutment / Fyshe, Thomas Maxwell – [S.l: s.n, 1907?] [mf ed 1991] – 1mf – 9 – 0-665-99512-1 – mf#99512 – cn Canadiana [624]

Discussion of the compositional material in the ludus tonalis of paul hindemith / Wahler, Harry Joe – U of Rochester 1948 [mf ed 19–] – 1r – 1 – (with bibl) – mf#film 671 – us Sibley [780]

A discussion of the constitutionality of the act of congress of march 2, 1867, authorizing the seizure of books and papers for alleged frauds upon the revenue. / Eaton, Sherburne Blake – New York, Chamber of Commerce, 1874. 56 p. LL-56 – 1 – us L of C Photodup [342]

A discussion of the doctrine of universal salvation : question, "do the scriptures teach the final salvation of all men?" / Sawyer, Thomas Jefferson & Wescott, Isaac – New York: H Lyon, 1854 [mf ed 1992] – (= ser Unitarian/universalist coll) – 1mf on 233p – 9 – 0-524-04276-4 – mf#1991-2060 – us ATLA [220]

A discussion of the doctrines of endless misery and universal salvation : in an epistolary correspondence / Campbell, Alexander & Skinner, Dolphus – Utica: CCP Grosh, 1840 [mf ed 1993] – 5mf – 9 – 0-524-08739-3 – mf#1993-3244 – us ATLA [240]

A discussion of the general epistle of st james / Parry, Reginald Saint John – London: C J Clay, 1903 [mf ed 1988] – 1mf on 100p – 9 – 0-7905-0109-0 – (in english and greek. incl bibl ref) – mf#1987-0109 – us ATLA [227]

A discussion of the mode and subjects of christian baptism : in a series of letters first published in the maysville "post-boy"... / Grundy, Robert Caldwell & Young, John – Maysville KY: Post-Boy Office, 1851 [mf ed 1993] – (= ser Christian church (disciples of christ) coll) – 1mf on 166p – 9 – 0-524-07130-6 – mf#1990-5337 – us ATLA [240]

A discussion of the question, is the roman catholic religion, in any or in all its principles or doctrines, inimical to civil or religious liberty? : and of the question, is the presbyterian religion, in any or in all its principles or doctrines, inimical to civil or religious liberty? / Hughes, John & Breckinridge, John – Philadelphia: Carey, Lea & Blanchard 1836 [mf ed 1993] – 2mf – 9 – 0-524-05951-9 – mf#1991-2351 – us ATLA [230]

Discussion of the scripturalness of future endless punishment / Adams, Nehemiah & Cobb, Sylvanus – Boston: Sylvanus Cobb, 1859, c1858 – 2mf – 9 – 0-7905-8756-4 – mf#1989-1981 – us ATLA [220]

The discussion on reunion : a review / Baird, Samuel John – 2nd enl ed. Richmond, Va: Whittet & Shepperson, 1888 – 1mf – 9 – 0-524-08667-2 – mf#1993-3192 – us ATLA [240]

Discussion on the existence of god and the authenticity of the bible / Bacheler, Origen & Owen, Robert Dale – New-York: The authors, 1832 – 7mf – 9 – 0-524-08727-X – mf#1993-3232 – us ATLA [220]

Discussion on the necessity of revising king james' version of the holy scriptures : and on the character, principles and revisions of the american bible union / Buckbee, Charles A & Buel, Frederick – San Francisco: Towne & Bacon, 1867 [mf ed 1991] – 1mf – 9 – 0-8370-1964-8 – mf#1987-6351 – us ATLA [220]

Discussion on the trinity, church constitutions and disciplines, and human depravity : between n. summerbell and j.m. flood, held in centreville, ohio, from august 2, to august 9, 1854 / Summerbell, Nicholas & Flood, J M – 4th ed. Cincinnati: Applegate, 1855 – 5mf – 9 – 0-524-07918-8 – mf#1991-3463 – us ATLA [240]

A discussion on universal salvation and future punishment / Manford, Erasmus & Sweeney, John Steele – Chicago: Rand, McNally, 1870 [mf ed 1993] – (= ser Christian church (disciples of christ) coll) – 1mf on 411p – 9 – 0-524-07322-8 – mf#1991-3037 – us ATLA [240]

Discussion, shall christians go to war? / Munnell, Thomas & Sweeney, John Steele – Cincinnati: Bosworth, Chase & Hall, 1872 – 1mf – 9 – 0-524-07026-1 – mf#1991-2879 – us ATLA [240]

Discussion sommaire sur les anciennes limites de l'acadie : et sur les stipulations du traite d'utrecht qui y sont relatives discussione... / [Pidansat de Mairobert, Mathieu-Francois] – Basle [Suisse]: Chez Samuel Thourneisan, 1755 [mf ed 1983] – 1mf – 9 – 0-665-37790-8 – (text in french and italian in dble clms) – mf#37790 – cn Canadiana [971]

Discussion sur les sept conciles oecumeniques : etudies au point de vue traditionnel et liberal / Michaud, Eugene – Berne: Jent et Reinert, 1878 – 1mf – 9 – 0-524-03768-X – (incl bibl ref) – mf#1990-1115 – us ATLA [240]

Discussioni e documenti di storia francescana... / Cresi, Domenico – Madrid: Arch. Ibero Americano, 1959 – 1 – sp Bibl Santa Ana [240]

Discussions / Dabney, Robert Lewis; ed by Vaughan, Clement Read – Richmond, Va: Presbyterian Committee of Publication, 1890-1897 – 26mf – 9 – 0-524-07408-9 – mf#1991-3068 – us ATLA [240]

Discussions and arguments on various subjects / Newman, John Henry – 2nd ed. London: Pickering, 1873 – 1mf – 9 – 0-7905-6604-4 – mf#1988-2604 – us ATLA [240]

Discussions in church polity : from the contributions to the "princeton review" / Hodge, Charles – New York: Scribner, c1878 – 2mf – 9 – 0-7905-5604-9 – (incl bibl ref) – mf#1988-1604 – us ATLA [240]

Discussions in history and theology / Fisher, George Park – New York: Scribner, 1880 – 2mf – 9 – 0-7905-4033-9 – (incl bibl ref and ind) – mf#1988-0033 – us ATLA [240]

Discussions in neuroscience – Oxford, England 1989-93 – 1,5,9 – ISSN: 0254-8852 – mf#42606 – us UMI ProQuest [612]

Discussions in theology : doctrinal and practical / Garland, Landon Cabell et al – Nashville, Tenn: Pub House of the ME Church, South, 1890 – 1mf – 9 – 0-524-06781-3 – mf#1991-2788 – us ATLA [240]

Discussions in theology / Skinner, Thomas Harvey – New York: ADF Randolph, 1868 – 1mf – 9 – 0-7905-9661-X – mf#1989-1386 – us ATLA [240]

Discussions of philosophical questions / Girardeau, John Lafayette; ed by Blackburn, George Andrew – Richmond, VA: Presbyterian Committee of Publ, c1900 – 2mf – 9 – 0-524-05009-0 – mf#1991-2179 – us ATLA [100]

Discussions of theological questions / Girardeau, John Lafayette; ed by Blackburn, George Andrew – Richmond, Va: Presbyterian Committee of Publication, c1905 – 2mf – 9 – 0-7905-9936-8 – mf#1989-1661 – us ATLA [240]

Discussions on church principles : popish, erastian, and presbyterian / Cunningham, William – Edinburgh: T & T Clark, 1863 – 2mf – 9 – 0-7905-4029-0 – mf#1988-0029 – us ATLA [242]

Discussions on damnation / Huizinga, Arnold van Couthen Piccardt – New York: Randolph R Beam, c1910 – 1mf – 9 – 0-524-00374-2 – mf#1989-3074 – us ATLA [240]

Discussions on the apocalypse / Milligan, William – London; New York: Macmillan, 1893 – 1mf – 9 – 0-8370-9886-6 – (incl ind) – mf#1986-3886 – us ATLA [221]

Discussions on the gospels : in two parts. part 1, on the language employed by our lord and his disciples, part 2, on the original language of st. matthew's gospel, and on the origin and authenticity of the gospels / Roberts, Alexander – 2nd rev enl ed. Cambridge: Macmillan, 1864 – 2mf – 9 – 0-7905-0218-6 – (incl bibl ref and index) – mf#1987-0218 – us ATLA [226]

The disease and the remedy : or, parochial and national emigration, versus parochial and national pauperism / Philo-humanitas [pseud] – London 1849 – (= ser 19th c british colonization) – 1mf – 9 – mf#1.1.529 – uk Chadwyck [304]

Disease markers – Hoboken NJ 1983+ – 1,5,9 – ISSN: 0278-0240 – mf#12918 – us UMI ProQuest [616]

Diseases and insect pests of the pecan / Matz, J – Gainesville, FL. 1918 – 1r – us UF Libraries [634]

Diseases of beans in southern florida / Townsend, G R – Gainesville, FL. 1939 – 1r – us UF Libraries [630]

Diseases of beans in southern florida / Townsend, G R – Gainesville, FL. 1947 – 1r – us UF Libraries [630]

Diseases of citrus fruits / Rolfs, P H – Gainesville, FL. 1911 – 1r – us UF Libraries [634]

Diseases of citrus in florida / Rhoads, Arthur S – Gainesville, FL. 1931 – 1r – us UF Libraries [634]

Diseases of citrus in florida / Rhoads, Arthur Stevens – Gainesville, FL. 1931 – 1r – us UF Libraries [634]

Diseases of cucumbers / Weber, George F – Gainesville, FL. 1925 – 1r – us UF Libraries [634]

Diseases of glasshouse plants / Bewley, William Fleming – London, England. 1923 – 1r – us UF Libraries [576]

Diseases of grapes in florida / Rhoads, Arthur – Gainesville, FL. 1926 – 1r – us UF Libraries [634]

Diseases of lettuce, romaine, escarole, and endive / Weber, George Frederick – Gainesville, FL. 1928 – 1r – us UF Libraries [634]

Diseases of peppers in florida / Weber, George F – Gainesville, FL. 1932 – 1r – us UF Libraries [634]

Diseases of sweet potatoes in florida / Weber, George F – Gainesville, FL. 1930 – 1r – us UF Libraries [630]

The diseases of the bible / Bennett, James Risdon – 2nd rev ed, [London]: Religious Tract Soc 1891 [mf ed 1985] – (= ser By-paths of bible knowledge 9) – 1mf – 9 – 0-8370-2266-5 – (incl ind) – mf#1985-0266 – us ATLA [220]

Diseases of the colon and rectum – Dordrecht, Netherlands 1958+ – 1,5,9 – ISSN: 0012-3706 – mf#10391 – us UMI ProQuest [616]

Diseases of the colon and rectum – v34-39. 1991-96 – 6r – $95.00r – us Lippincott [616]

Diseases of the nervous system – Memphis TN 1940-77 – 1,5,9 – (cont by: journal of clinical psychiatry) – ISSN: 0012-3714 – mf#1697.01 – us UMI ProQuest [616]

Diseases of the teeth : their diagnoses and treatment / Marshall, John Albert – 1926 – 1 – us CRL [617]

Diseases of the tomato / Rolfs, P H – Lake City, FL. 1898 – 1r – us UF Libraries [634]

Diseases of watermelons in florida / Walker, M N – Gainesville, FL. 1931 – 1r – us UF Libraries [634]

Disegpartito in piu ragionamenti, ne quali si tratta della scoltura et pittura... / Doni, [A F] – Vinetia, 1549 – 2mf – 9 – mf#O-1034 – ne IDC [700]

Disembodied state / Henry, T Shuldham – London, England. 1885 – 1r – us UF Libraries [240]

Disertacion chirurgica relativa al gobierno politico en la que se proponen los danos de la castracion vulgar segun se practica para curar ninos quebrados / Arguello Castrillo, A – Madrid, 1775 – 1mf – 9 – sp Cultura [617]

Disertacion eucaristica sobre la precisa obligacion de recibir todo enfermo la sagrada eucaristia en ayuno / Custodio, M – Sevilla, 1779 – 1mf – 9 – sp Cultura [240]

Disertacion fisico-legal de los sitios...para sepulturas / Fernandez, F – Madrid, 1783 – 3mf – 9 – sp Cultura [614]

Disertacion fisico-medica..., para preservar...de viruela / Gil, F – Madrid, 1784 – 3mf – 9 – sp Cultura [616]

Disertacion physico-medica de las virtudes medicinales, uso y abuso de las aguas termales de la villa de archena / Cerdan, F – Orihuela, 1760 – 8mf – 9 – sp Cultura [610]

Disestablishment : or, is the church of scotland worth preserving? / Munro, J D – Greenock, Scotland. 1882 – 1r – us UF Libraries [242]

Disestablishment in france = A propos de la separation des eglises et de l'etat / Sabatier, Paul – London: TF Unwin, 1906 [mf ed 1990] – 1mf – 9 – 0-7905-6676-1 – (english trans fr french with pref by robert dell) – mf#1988-2676 – us ATLA [241]

Disestablishment in wales / Rendel, Stuart – Oswestry, England. 1885 – 1r – us UF Libraries [240]

Dishman, Rodney K *see*
- The effect of acute exercise on state anxiety and acoustic startle eyeblink response in physically active and inactive men
- Ratings of perceived exertion as a determinant of physical activity in 9 to 11 year-old children
- The relationship of aerobic fitness, type a behavior pattern, and hostility to baroreflex responses and cardiovascular reactivity to nonexertional stressors

Disillusioned india / Mukerji, Dhan Gopal – New York: EP Dutton & Co, 1930 – (= ser Samp: indian books) – us CRL [954]

The disintegration of canada / Bender, Prosper – Boston: s.n, 18– – 1mf – 9 – mf#03570 – cn Canadiana [971]

The disintegration of canada / Bender, Prosper – Boston: s.n, 18– – 1mf – 9 – mf#43298 – cn Canadiana [971]

The disintegration of islam / Zwemer, Samuel Marinus – New York: Fleming H Revell, c1916 – (= ser Students' Lectures on Missions) – 1mf – 9 – 0-524-01328-4 – (incl bibl ref) – mf#1990-2364 – us ATLA [260]

Diskorina – Jogjakarta, 1961/1962. v1(1-8) – 1mf – 9 – (missing: 1961/1962 v1(1-6)) – mf#SE-878 – ne IDC [950]

Diskriminierung am arbeitsmarkt : ausgewaehlte erklaerungsansaetze mit besonderer beruecksichtigung der frauendiskriminierung / Djumena, Sascha – (mf ed 1993) – 1mf – 9 – €30.00 – 3-89349-774-9 – mf#DHS 774 – gw Frankfurter [331]

Diskussie van de waarnemingen van satellieten 1, 2 en 3 van jupiter, gedaan te johannesburg door r t a innes in de jaren 1908-1925 / Brouwer, Dirk – Leiden: E Ijdo 1927 [mf ed 2006] – 1r – 1 – (incl bibl ref) – mf#film mas 37490 – us Harvard [520]

Diskussionnyi listok – Paris, France 1910-11 [mf ed Norman Ross] – 1 – mf#nrp-1322 – us UMI ProQuest [077]

Dislexias : conceptos fundamentales clinica / Gutierrez Gomez, Diego & Pelaez Lorenzo, Luis – Badajoz: Imprenta Provincial, 1969. Ponencia a la V Reunion anual de la Sociedad espanola de neuropsiquiatria infantil – sp Bibl Santa Ana [616]

Dismantler – v1 n1-6/7 [1978 mar-1979 feb], v2 n1-v8 n3 [1980 jan-1986 oct], v8 n4/v9 n1-v9 n3/4 [1987 spring-summer], v10 n2-3 [1988 fall-1988/89 winter] – 1r – 1 – mf#1055597 – us WHS [071]

Dismembered hungary / Buday, Laszlo – London, England. 1923 – 1r – us UF Libraries [943]

Dismission, rest, and future glory of the good and faithful servant / Symington, Andrew – Paisley, Scotland. 1832 – 1r – us UF Libraries [240]

Disney, John *see*
- A catalogue of some marbles, bronzes, pictures, and gems, at the hyde, near ingatestone, essex
- Reciprocal duty of a christian minister and a christian congregatio

Disowned / Lytton, Edward Bulwer Lytton, Baron – Boston, MA. 189- – 1r – us UF Libraries [025]

Disparatario / Mendez, Jose Maria – San Salvador, El Salvador. 1957 – 1r – us UF Libraries [972]

Dispareri in materia d'architettura et perspettiva... / Bassi, M – Brescia, 1572 – 1mf – 9 – mf#O-1003 – ne IDC [720]

Un disparu / Dumont, Georges-A – Montreal: G A et W Dumont, 1894? – 1mf – 9 – (incl biogr of leandre-wilfrid tessier) – mf#02794 – cn Canadiana [440]

Dispatch – 1981 apr 30-1982, 1983-1984 mar 29 – 2r – 1 – (cont by: maxwell-gunter dispatch) – mf#652676 – us WHS [071]

Dispatch – Washington, DC: US GPO. v1-10. 1990-99 – 9 – $547.00 set – (cont: us department of state bulletin) – mf#202001 – us Hein [327]

Dispatch / Highland Co. Hillsboro – mar 1898-jan 1922 [wkly, semiwkly] – 17r – 1 – mf#B8935-8951 – us Ohio Hist [071]

Dispatch – Raleigh, NC. 1991 dec 21, 1993 mar 28, apr 1/10 – 1r – 1 – mf#2697797 – us WHS [071]

Dispatch – Raleigh, NC. 1992 jun 20, sep 19 – 1r – 1 – mf#2712709 – us WHS [071]

Dispatch *see* Era dispatch / quiver / news / dispatch

Dispatch (1986 edition) / Franklin Co. Columbus – jul 1871-dec 1871,jul 1872-dec 1947 [daily] – 710r – 1 – mf#B27979-28688 – us Ohio Hist [071]

Dispatch series / Champaign Co. Saint Paris – (may 1881-jul 1954) scattered [wkly, semiwkly, daily, wkly] – 14r – 1 – mf#B11675-11688 – us Ohio Hist [071]

Dispatcher / International Longshoremen's and Warehousemen's Union – 1944 mar 10-1946, 1947-71, 1972 jan 17-1976 dec 17, 1977-94 – 14r – 1 – (cont: ilwu dispatcher) – mf#1111228 – us WHS [331]

Displaced persons / U.S. Army. Office of the Chief Historian, European Command – 1947 – 1 – us L of C Photodup [940]

Display world – Cincinnati OH 1922-73 – 1,5 – (cont by: visual merchandising) – ISSN: 0012-3803 – mf#244.02 – us UMI ProQuest [650]

Displays – Oxford, England 1979+ – 1,5,9 – ISSN: 0141-9382 – mf#17223 – us UMI ProQuest [000]

Disposiciones complementarias de las leyes de indias / Spain. Laws, Statutes, etc – 1930 – 1 – us Wisconsin U Libr [946]

Dispositio et perioche historiae evangelicae / Bullinger, Heinrich – Tigvri, Christoph Froschouer, 1553 – 2mf – 9 – mf#PBU-180 – ne IDC [240]

Die disputacion vor den xij orten einer loblichen eidtgnoschafft...thomas murner... – Lutzern, [1527] – 4mf – 9 – mf#ZWI-24 – ne IDC [240]

Disputaciones metafisicas (1638) / Briceno, Alfonso – Caracas, Venezuela. 1955 – 1r – us UF Libraries [972]

Disputatio de natura febris / Rodriguez Guerrero, D – Sevilla, 1606 – 3mf – 9 – sp Cultura [616]

Disputatio de originali peccato et libero arbitrio : inter matthiam flacium illyricum, & victorinum strigelium, publice vinariae per integram hebdomadam, praesentibus illustriss / Flacius Illyricus, Matthias & Strigel, Victorinus – [s.l.: s.n.], 1563. Chicago: Dep of Photodup, U of Chicago Lib, 1978 (1r); Evanston: American Theol Lib Assoc, 1984 (1r) – 1 – 0-8370-0708-9 – mf#1984-T105 – us ATLA [240]

Disputatio de pericope num 22:2-24, historiam bileami continente / Oort, H – Lugduni-Batavorum [London, England]: P Engels, 1860 – 2mf – 9 – 0-7905-3045-7 – (incl bibl ref) – mf#1987-3045 – us ATLA [220]

Disputatio inauguralis de generis humani varietate / Prichard, James Cowles – Edinburgi, Excudebant Abernethy & Walker, 1808 – 1 – us Wisconsin U Libr [572]

Disputatio musica prima[-tertia]... / Lippius, Johann – Wittebergae: typis I Gormani 1609-10 [mf ed 19–] – 3v in 1 on 3mf – 9 – (subtitle varies) – mf#fiche 152-154 – us Sibley [780]

Disputatio pro religione mohammedanorum adversus christianos : textum arabicum et codice leidensi cum varr lect = Takhjil man harrafa al-injil. selections / Jafari, Salih ibn al-Husayn; ed by Ham, Frederik Jacob van den – Lugduni Batavorum [Leiden]: EJ Brill, 1890 [mf ed 1992] – 1mf – 9 – 0-524-02873-7 – mf#1990-3146 – us ATLA [230]

Disputatio theologica de jesu, e virgine maria nato / Oosterzee, Johannes Jacobus van – Trajectum ad Rhenum: Schultze & Voermans, 1840 – 1mf – 9 – 0-7905-3089-9 – (incl bibl ref) – mf#1987-3089 – us ATLA [220]

Ein disputation oder besprech zwayer stalbuben : so mit kueniglicher maye botschafft bey dem tuerckischen keyser zu constantinopel gewesen... / Curipeschitz, B – [Constantinopolis, 1531] – 1mf – 9 – mf#H-8415 – ne IDC [956]

Disputationes de controversiis christianae fidei, adversus huius temporis haereticos... / Bellarmin, R – Ingolstadii, 1599. 8v – 66mf – 9 – mf#CA-87 – ne IDC [241]

Disputationes de universa philosophia / Hurtado de Mendoza, P – Lugduni, 1617 – 15mf – 9 – mf#CA-19 – ne IDC [100]

Disputationes exegeticae in confessionem helveticam / Gernler, L – Basel, 1661-1674 – 6mf – 9 – mf#PBU-705 – ne IDC [240]

Disputationes in...aristotelis de anima / Alfonso, Francisco – 1640 – 9 – sp Bibl Santa Ana [120]

Disputationes medicae de anginorum ... / Alonso y de los Ruizes de Fontecha, J – Alcala de Henares, 1611 – 8mf – 9 – sp Cultura [616]

Disputationes meta physicae... / Pasqualigus, Z – Romae, 1634-1636. 2v – 29mf – 9 – mf#CA-27 – ne IDC [240]

Disputationes theologicae : in priman partem divi thomae. tomus secundus / Godoy, Pedro de – Burgo de Osma: Imp. Episcopal, 1670 – 9 – sp Bibl Santa Ana [240]

Disputationes theologicae in tertiam partem divi thomae / Godoy, Pedro de – Burgo de Osma: Imprenta Episcopal, 1966 – 1 – sp Bibl Santa Ana [290]

Disputationes theologicae intertiam partem divi thonae... / Godoy, Pedro de – Burgo de Osma: Imprenta Episcopal, 1668 – 1 – sp Bibl Santa Ana [290]

Disputationes theologicae intertian pastem divi thonae... / Godoy, Pedro de – Burgo de Osma: Imprenta Episcopal, 1667 – 1 – sp Bibl Santa Ana [290]

Disputationes theologicae ordinariae repetitiae / Heidanus, A – Lugduni Batavorum, 1654-59 – 7mf – 9 – mf#PBA-188 – ne IDC [240]

Disputationes tridentinae / Lainez, Diego – Oeniponte [Innsbruck]: Felicianus Rauch; Neo-Eboraci [New York]: Fr Pustet. 2v. 1886 – 4mf – 9 – 0-8370-9003-2 – (incl bibl ref and index) – mf#1986-3003 – us ATLA [240]

Disputationes...thomae eidem angelico / Godoy, Manuel – 1669, 1671, 1672. 3v – 9 – sp Bibl Santa Ana [946]

Disputations chrestiennes, touchant l'estat des trepasses... / Viret, P – Geneve, Girard, 1552 – 7mf – 9 – mf#PFA-191 – ne IDC [240]

Disputationum de sancto matrimonii sacramentum / Sanchez, T – Antverpiae, 1626. 3v – 25mf – 9 – mf#CA-68 – ne IDC [240]

Disputationum theologicarum... / Trigland, J – Lugduni Batavorum, 1642-47 – 14mf – 9 – mf#PBA-351 – ne IDC [240]

Disputationum theologicarum miscellanearum / Spanheim, F – Geneve, Chouet, 1652. 2 pts – 9mf – 9 – mf#PFA-177 – ne IDC [240]

Disputationum theologicarum...tomus primus : de deo uno / Smising, T – Antverpiae, 1627 – 17mf – 9 – mf#CA-38 – ne IDC [240]

Disputa...y averiguaciones de la enfermedad pestilente... / Valdes, F – Sevilla, 1599 – 1mf – 9 – sp Cultura [616]

Dispute resolution journal – New York NY 1993+ – 1,5,9 – (cont: arbitration journal) – ISSN: 1074-8105 – mf#2498,01 – us UMI ProQuest [347]

A dispute upon the question of kneeling in the acte of receiving the sacrementall bread and wine, proving it to be unlawfull : or a third parte of the defence of the ministers reasons, for refusall of the subscription and conformitie requyred / Hieron, S – n.p., 1608 – 2mf – 9 – mf#PW-49 – ne IDC [240]

The disputed waters of the jordan / Smith, C G – Oxford: Institute of British Geographers, 1966 – us CRL [956]

Dispvtatio d ioachimi morlini, de commvnicatione idiomatvm / Moerlin, J – np, 1571 – 1mf – 9 – mf#TH-1 mf 1176 – ne IDC [242]

Dispvtatio de originali peccato et libero arbitrio : inter matthiam flacivm illyricvm et victorinum strigelium publice vinariae / Strigel, V – np, 1563 – 5mf – 9 – mf#TH-1 mf 1451-1455 – ne IDC [242]

Dispvtatio secvnda contra calvinistas, et praesertim synopsin kimedoncij / Huber, S – Wittebergae, [1593] – 1mf – 9 – mf#TH-1 mf 721 – ne IDC [242]

Dispvtatio tertia contra calvinistas qvod faciant devm avtorem peccati / Huber, S – Witebergae, 1593 – 1mf – 9 – mf#TH-1 mf 722 – ne IDC [242]

Eine dispvtation von mitteldingen vnd von den itzigen verenderungen in kirchen die christlich vnd wol geordent sind / Gallus, N – [Magdeburg, 1550] – 1mf – 9 – mf#TH-1 mf 478 – ne IDC [242]

Dispvtationvm de medicina nova philippi paracelsi pars prima: in qva, qvae de remediis.. / Erastus, Thomas – Basileae: Apud Petrvm Pernam, 1572. With: solis e pvteo emergentis. by J. Rhenanus; de rebvs natvralibvs libri XXX. by G. Zabarella; de metalicis libri tres. by A. Cesalpini – 1 – us Wisconsin U Libr [610]

Disqualification of federal judges by peremptory challenge / Chaset, Alan J – Washington: FJC, Feb 1981 – 1mf – 9 – $1.50 – mf#LLMC 95-307 – us LLMC [340]

Le disque vert – n.s., I, no. 1-6. Bruxelles. dec 1952 (no. hors serie sur Proust), avr 1953-54, 1955 (no. special sur Jung) – 1 – fr ACRPP [800]

Le disque vert – sous ce titre sont regroupees les revues suivantes: Signaux de France et de Belgique. Revue mensuelle de litterature. no. 1-11 12. Anvers, Paris. 1er mai 1921-mars juin 1922. Le Disque vert. Revue mensuelle de litterature. I, no. 1-6. Bruxelles, Paris. mai-oct 1922. Ecrits du Nord. Revue mensuelle de litterature. I, 2e s., no. 1-3. Bruxelles et Paris. nov 1922-janv 1923. Le Disque vert. Revue mensuelle de litterature. I, 2e s., no. 4 5 6. 1923; II, 3e s., no. 1-4 5, no. special Charlie Chaplain et no. special Freud et la psychanalyse, oct 1923-24; III, 4e s., no. 1-3, no. special Le Cas Lautreamont, 1925. Nord. Cahiers litteraires trimestriels. no. 1-4. Bruxelles. avr 1929-nov 1930. Au Disque vert. no. 1. 1934. Ecrits du Nord. Revue mensuelle de litterature, d'art et de critique. no. 1-2. juin-juil 1935. Le Disque vert. Revue mensuelle de litterature. I, n.s., no. 1. 15 juil 1941. et collection privee – 1 – fr ACRPP [800]

Disquisiciones americanas 2. don martin cortes y don diego colon, caballeros de santiago / Fita, Fidel – Madrid: Fortanet, 1892. B.R.A.H. 21, pp. 374-377 – sp Bibl Santa Ana [970]

Disquisiciones sobre filologia castellana / Cuervo, Rufino Jose – Bogota, Colombia. 1950 – 1r – us UF Libraries [972]

Disquisiciones sociologicas, y otros ensyos / Brau, Salvador – Rio Piedras, Puerto Rico. 1956 – 1r – us UF Libraries [301]

Disquisicion...institucion...montes de piedad / Texeyro de Valcarcel, Francisco J – 1746 – 9 – sp Bibl Santa Ana [240]

Disquisitio exegetico-theologica de formulae paulinae pistis iesou christou significatione / Berlage, Hendrik Petrus – Lugduni-Batavorum: P Engels, 1856 – 1mf – 9 – 0-7905-0362-X – (in latin and greek. incl bibl ref) – mf#1987-0362 – us ATLA [220]

Disquisitio geographica & historica, de chataja... / Mueller, A – Berolini: Rungianis, 1671 – 2mf – 9 – mf#HT-598 – ne IDC [910]

Disquisitio historico-theologica exhibens j calvini et de ecclesia sententiarum inter se compositionem / Kuyper, A – Hagae Comitum, 1862 – €11.00 – ne Slangenburg [242]

A disquisition on government / Calhoun, John C; ed by Cralle, Richard K – New York: Peter Smith, 1853 repr 1943 – 2mf – 9 – $3.00 – mf#LLMC 95-073 – us LLMC [323]

Disquisitionis grammaticae de alphabeti gothicic ulphilani / Zacher, Julius – Lipsiae, Germany. 1854 – 1r – us UF Libraries [025]

Disquisitions upon the painted greek vases / Christie, James – London 1825 – (= ser 19th c art & architecture) – 3mf – 9 – mf#4.2.1624 – uk Chadwyck [730]

Disruption / Free Church Of Scotland General Assembly (1862) – Edinburgh, Scotland. 1862 – 1r – us UF Libraries [240]

The disruption of canada / Ewart, John Skirving – [Ottawa?: s.n, 1917?] [mf ed 1994] – 1mf – 9 – 0-665-73198-1 – (incl bibl ref) – mf#73198 – cn Canadiana [933]

The disruption of the methodist episcopal church, 1844-1846 : comprising a thirty years' history of the relations of the two methodisms / Myers, Edward Howell – Nashville, TN: AH Redford, 1875 – 1mf – 9 – 0-524-03174-6 – mf#1990-4623 – us ATLA [242]

Disruption question stated / Brown, Charles John – Edinburgh, Scotland. 1863 – 1r – us UF Libraries [240]

Disruptions and secessions in methodism : their causes, consequences, and lessons / Swallow, Thomas – London: Ralph Fenwick, 1880 – 1mf – 9 – 0-524-06294-3 – mf#1990-5223 – us ATLA [242]

Dissel, Karl *see* Philipp von zesen und die deutschgesinnte genossenschaft

Disselhoff, Julius et al *see* Vortraege fuer das gebildete publikum. vierte sammlung

Dissemination of unitarian principles recommended and enforced in a... / Lyons, James – London, England. 1808 – 1r – us UF Libraries [243]

Disseminator – Harrisburg OR: S S Train, [wkly] [mf ed 1969] – 1r – 1 – (merged with: albany herald (1879-) to form: weekly herald=disseminator (-1904)) – us Oregon Lib [071]

Disseminator (harrisburg, or) *see* Albany herald (albany, or)

Dissent – New York NY 1954+ – 1,5,9 – ISSN: 0012-3846 – mf#2008 – us UMI ProQuest [320]

Dissent – Salisbury, Rhodesia: T Ranger and J Reed. [n1-24. mar 26 1959-mar 2 1961] – us CRL [079]

Dissent in england : two lectures / Henson, Hensley – London: Rivingtons, 1900 – 1mf – 9 – 0-524-03096-0 – mf#1990-0821 – us ATLA [240]

Dissent in its relation to the church of england : eight lectures / Curteis, George Herbert – new ed. London; New York: Macmillan, 1906 – (= ser Bampton lectures) – 2mf – 9 – 0-524-01341-1 – mf#1990-0387 – us ATLA [241]

Dissenters' mutual friendly colonizing society : upon a plan embodying the new testament principles, to the exclusion of those of an anti-christian tendency / Papps, W R – London 1848 – (= ser 19th c british colonization) – 1mf – 9 – mf#1.1.524 – uk Chadwyck [330]

Dissenting academies in england : their rise and progress and their place among the educational systems of the country / Parker, Irene – Cambridge: University Press; New York: G.P. Putnam [distributor], 1914 – (= ser Contributions to the history of education) – 1mf – 9 – 0-7905-5785-1 – (incl bibl ref) – mf#1988-1785 – us ATLA [370]

Dissenting ritualism – London, England. 1873? – 1r – us UF Libraries [240]

Disseritur de praecipuis ad primas causas christianismi formaliter spectati penetrandi subsidiis / Bertholdt, Leonhard – Erlangae: ex officina Hilpertiana, 1818 – 1mf – 9 – 0-7905-3362-6 – (incl bibl ref) – mf#1987-3362 – us ATLA [240]

Dissertacion canonica sobre los justos motivos que representa el reyno de guatemala : para que el consejo se sirva de erigir en metropoli eclesiastica la s[an]ta iglesia cathedral de la ciudad de santiago, su cabeza / Rodriguez de Rivas y Velasco, Diego – [Mexico City: s.n. 1718] – (= ser Books on religion...1543/44-c1800: iglesias, catedrales) – 1mf – 9 – mf#crl-349 – ne IDC [241]

Dissertatio : de ingenii muliebris... / Schuurman, A M van – Lugduni Batavorum, 1641 – 2mf – 9 – mf#PBA-306 – ne IDC [240]

Dissertatio de consociatione evangelica reformatorum et augustanae confessionis sive de colloquio cassellano...1661 / Hoornbeek, J – Amstelodami, 1663 – 1mf – 9 – mf#PBA-197 – ne IDC [242]

Dissertatio de generatione et metamorphosibus insectorum surinamensium... / Merian, M S – Amstelaedami, 1719 – 8mf – 9 – mf#Z-2238 – ne IDC [590]

Dissertatio de gubernatione ecclesiae... / Bucer, G(?) – Middelburgi, 1618 – 7mf – 9 – mf#PBA-136 – ne IDC [240]

Dissertatio de syrorum fide et disciplina in re eucharistica : accedunt veteris ecclesiae syriacae monumenta duo... / Lamy, Thomas Joseph – Lovanii: Vanlinthout, 1859 – 1mf – 9 – 0-8370-8122-X – (incl bibl ref and ind) – mf#1986-2122 – us ATLA [240]

Dissertatio entomologica novas insectorum species, sistens, cujus partem primam / Thunberg, C P – Upsaliae, 1781-1791. 6pts – 2mf – 9 – mf#Z-2276 – ne IDC [590]

Dissertatio historico-critica de dogmatices christianae fontibus eorumque usu publico omnium examini offert albertus goswinus boon / Boon, Albertus Goswinus – Groningae:K de Waard, [c1860] – 1mf – 9 – 0-8370-2415-3 – mf#1985-0415 – us ATLA [240]

Dissertatio historico-critica de pseudoprophetismo hebraeorum / Matthes, Jan Carel – Lugduni-Batavorum [Leiden]: Fratres van der Hoek, [1859?] – 1mf – 9 – 0-524-06521-7 – mf#1992-0905 – us ATLA [221]

Dissertatio theologica de civili et ecclesiastica potestate / Trigland, J – Amstelodami, 1642 – 5mf – 9 – mf#PBA-349 – ne IDC [242]

Dissertation abregee sur le nom antique et hieroglyphique de la judee, ou traditions conservees en chine, sur l'ancien pays de tsin, pays qui fut celui des cereales et de la croix / Paravey, C H de – Paris: Treuttel et Wurtz, 1836 – 1mf – 9 – mf#HT-741 – ne IDC [930]

Dissertation de syrorum fide et disciplina in re eucharistica / Lamy, Thomas Josephus – Lovanii, 1859 – 5mf – 8 – €12.00 – ne Slangenburg [240]

A dissertation of the rule of faith / Spring, Gardiner – New York: Leavitt, Trow, 1844 [mf ed 1988] – 1mf – 9 – 0-7905-0388-3 – (incl bibl ref) – mf#1987-0388 – us ATLA [241]

A dissertation on native depravity / Spring, Gardiner – New York: Jonathan Leavitt, 1833 [mf ed 1993] – 1mf – 9 – 0-524-08588-9 – mf#1993-3173 – us ATLA [210]

A dissertation on oriental gardening... / Chambers, W – Ed 2. London, 1772 – 3mf – 9 – mf#O-1061 – ne IDC [700]

A dissertation on the coincidence between the priesthoods of jesus christ and melchisedec : in three parts, in which the passages of scripture relating to that subject... / Gray, James – Hagerstown, MD: William Stewart; Philadelphia: James M Campbell, 1845, c1844 [mf ed 1986] – 1mf – 9 – 0-8370-9950-1 – mf#1986-3950 – us ATLA [240]

A dissertation on the course and probable termination of the niger / Donkin, Rufane S – London 1829 [mf ed Hildesheim 1995-98] – 1v on 2mf – 9 – €60.00 – 3-487-27275-X – gw Olms [916]

A dissertation on the epistle of s barnabas : including a discussion of its date and authorship / Cunningham, William – London: Macmillan, 1877. Chicago: Department of Photodup, U of Chicago Lib, 1967 (1r); Evanston: American Theol Lib Assoc, 1984 (1r) – 1 – 0-8370-0448-9 – (incl ind) – mf#1984-B084 – us ATLA [227]

A dissertation on the eternal sonship of christ / Kidd, James – new ed. London: Hamilton, Adams, 1872 [mf ed 1985] – 1mf – 9 – 0-8370-4365-4 – (int by int, biogr and theological by robert s candish) – mf#1985-2365 – us ATLA [240]

Dissertation on the fable of papal antichrists / Gradwell, Robert – London, England. 1816 – 1r – us UF Libraries [240]

A dissertation on the gospel commentary of s ephraem the syrian : with a scriptural index to his works / Hill, James Hamlyn – Edinburgh: T & T Clark, 1896 [mf ed 1989] – 1mf – 9 – 0-7905-1102-9 – mf#1987-1102 – us ATLA [225]

A dissertation on the hebrew roots, intended to point out their extensive influence on all known languages / Pirie, Alexander – Edinburgh: printed for James Morrison, 1807 – (= ser 19th c books on linguistics) – 3mf – 9 – mf#2.1.11 – uk Chadwyck [470]

A dissertation on the history and prophecies of balaam *see* Dissertations on the genuineness of daniel and the integrity of zechariah

A dissertation on the law of nature : together with some observations on the roman civil law in particular; to which is added, by way of an appendix, a curious catalog of books, very useful to the students of these several laws, together with the canon law – London: J Roberts, 1723 – 2mf – 9 – $3.00 – mf#LLMC 95-200 – us LLMC [340]

A dissertation on the means of regeneration / Spring, Gardiner – New York: John P Haven, 1827 [mf ed 1988] – 1mf – 9 – 0-7905-0387-5 – (incl bibl ref) – mf#1987-0387 – us ATLA [230]

Dissertation on the origin and connection of the gospels : with a synopsis of the parallel passages in the original and authorised version, and critical notes – Edinburgh: William Blackwood, 1853 [mf ed 1990] – 1mf – 9 – 0-8370-1794-7 – mf#1987-6182 – us ATLA [226]

Dissertation on the progress of ethical philosophy : chiefly during the seventeenth and eighteenth centuries / Mackintosh, James, Sir – Edinburgh: Adam and Charles Black, 1836 – 1mf – 9 – 0-7905-8839-0 – (incl bibl ref) – mf#1989-2064 – us ATLA [170]

A dissertation on the puerperal fever : delivered at a public examination for the degree of bachelor in medicine... / Laterriere, Pierre de Sales – Boston: Printed by Samuel Hall...1789 – 1mf – 9 – mf#36164 – cn Canadiana [616]

A dissertation on the soil and agriculture of the british settlement of penang, or prince of wales island, in the straits of malacca : including province wellesley on the malayan peninsula / Low, James – [Singapore], 1836 – (= ser 19th c british colonization) – 4mf – 9 – mf#1.1.3178 – uk Chadwyck [630]

Dissertation on the theology of the chinese : with a view to the elucidation of the most appropriate term for expressing the deity, in the chinese language / Medhurst, Walter Henry – Shanghae: Mission Press, 1847 [mf ed 1995] – (= ser Yale coll) – 284p – 1 – 0-524-10096-9 – mf#1995-1096 – us ATLA [240]

Dissertation or discourse concerning a judge of controversies in ma... / Sherlock, William – London, England. 1851 – 1r – us UF Libraries [240]

Dissertation sur la nature et propagation de feu / Chatelet, Gabrielle-Emilie Le Tonnelier de Breteuil, Marquise du – Paris: Prault 1744 [mf ed 1980] – 1r – 1 – (incl bibl ref) – mf#96p – us Wisconsin U Libr [530]

Dissertation sur le canon de bronze que l'on voit dans le musee de m. chasseur a quebec / Berthelot, Amable – Quebec: Neilson & Cowan, 1830 [mf ed 1984] – 1mf – 9 – 0-665-21298-4 – mf#21298 – cn Canadiana [971]

Dissertation sur les differentes methodes d'accompagnement pour le clavecin ou pour l'orgue : avec le plan d'une nouvelle methode, etablie sur une mechanique des doigts, que fournit la succession fondamentale de l'harmonie... / Rameau, Jean Philippe – Paris: Bailleux [1732] [mf ed 19–] – 2mf – 9 – mf#fiche 507 – us Sibley [780]

Dissertation sur une deuxieme mission aux iles salomon / Verguet, Leopold, abbe – 1893 – 1r – 1 – mf#pmb996 – at Pacific Mss [241]

A dissertation upon the constitutional freedom of the press in the united states of america – Boston: printed...for Joseph Nancrede...1801 [mf ed 1984] – 1mf – 9 – 0-665-45125-3 – mf#45125 – cn Canadiana [342]

Dissertationen-katalog der universitaet tuebingen, 1500-1981 = Catalogue of dissertations of the university of tuebingen – [mf ed 1983] – 370mf (1:42) – 9 – silver €1790.00 – ISBN-10: 3-598-30449-8 – ISBN-13: 978-3-598-30449-1 – gw Saur [020]

Dissertationes de laudibus et effectibus podagrae quas sub auspiciis... – n.p, [1715] – 3mf – 9 – mf#O-1828 – ne IDC [090]

Dissertationes duae... / Wittichius, C – Amstelodami, 1653 – 4mf – 9 – mf#PBA-405 – ne IDC [240]

Dissertationes ethicae / Sinapius, Daniel – Lugduni Batavorum: Ex officina Francisci Hackii, anno 1643 – (= ser Ethics in the early modern period) – 2mf – 9 – mf#pl-470 – ne IDC [170]

Dissertationes philologae vindobonenses / Wien. Universitaet – Lipsiae etc: G Freytag 1887- [mf ed 1978] – 12v on 2r – 1 – mf#film mas c283 – us Harvard [450]

Dissertationes philologicae argentoratenses selectae – Argentorati. v1-14. 1879-1910 – 102mf – 8 – mf#H-359 – ne IDC [400]

Dissertations on subjects relating to the "orthodox" or "eastern-catholic" communion / Palmer, William – London: J. Masters, 1853 – 1mf – 9 – 0-7905-5784-3 – mf#1988-1784 – us ATLA [241]

Dissertations on the apostolic age : reprinted from editions of st. paul's epistles / Lightfoot, Joseph Barber – London; New York: Macmillan, 1892 – 2mf – 9 – 0-8370-4123-6 – (incl bibl ref and ind) – mf#1985-2123 – us ATLA [220]

Dissertations on the english language / Webster, Noah – 1789 – 9 – us Scholars Facs [420]

Dissertations on the genuineness of daniel and the integrity of zechariah : and a dissertation on the history and prophecies of balaam = Authentie des daniel und die integritaet des sacharjah / Hengstenberg, Ernst Wilhelm – Edinburgh: T & T Clark; New York: Wiley & Putnam, 1847 – 2mf – 9 – 0-8370-9545-X – (incl bibl ref and ind. in english) – mf#1986-3545 – us ATLA [221]

Dissertations on the genuineness of daniel and the integrity of zechariah / Hengstenberg, Ernst Wilhelm – Edinburgh: T & T Clark, 1847. Beltsville, Md: NCR Corp,1978 (7mf); Evanston: American Theol Lib Assoc, 1984 (7mf) – 9 – 0-8370-1057-8 – (english by benjamin plummer pratten. filmed with: a dissertation on the history and prophecies of balaam, english trans by jonathan edwards ryland. incl bibl ref) – mf#1984-4415 – us ATLA [221]

Dissertations on the genuineness of the pentateuch = Die authentie des pentateuches / Hengstenberg, Ernst Wilhelm – Edinburgh: publ...by John D Lowe, 1847 [mf ed 1988] – 2v on 4mf – 9 – 0-7905-0040-X – (incl bibl ref. in english) – mf#1987-0040 – us ATLA [221]

Dissertations read to the edinburgh royal medical society, 1750-1970 : from edinburgh royal medical society, edinburgh university library – 215v – 115r – 1 – (with ind) – mf#96935 – uk Microform Academic [500]

The dissident press of revolutionary iran : a unique collection / Behn, Wolfgang & Hoefig, Willi – 1981-82 – 19r – 1 – (with handbk and explicit descriptions) – gw Mikropress [956]

Il dissidio tra mazzini e garibaldi: la storia senza veli / Curatulo, Giacomo Emilio – Documenti inediti. Milano: A. Mondadori, 1928. 431p. illus., ports – 1 – us Wisconsin U Libr [945]

Dissipateur : ou, l'honnete friponne / Destouches, Nericault – Paris, France. 1803 – 1r – us UF Libraries [440]

Dissipateur : ou l'honnete friponne / Destouches, Nericault – Paris, France. 1808 – 1r – us UF Libraries [440]

Dissipativnye svoistva metallov i metallicheskikh splavov : o raschetakh i prognozirovanii svoistv zhidkostei i gazov termodinmiacheskie i perenosnye svoistva zhidkostei v metastabilnom sostoianii / Novikov, I I & Filippov, L P et al; ed by Filippova, L P – Moskva: In-t vysokikh temperatur AN SSSR, 1980 – us CRL [077]

Dissolving views of scenes described in the new testament – London, England. 18– – 1r – us UF Libraries [225]

Dissuasive from schism / Terrot, Charles Hughes – Edinburgh, Scotland. 1843 – 1r – us UF Libraries [240]

Dist local 340 reporter / Amalgamated Meat Cutters and Butcher Workmen of North America – v1 n1-v5 n1 [1978 sep/oct-1983 jan /feb] – 1r – 1 – mf#718268 – us WHS [660]

Distaff – v2 n8-v3 n6 [1974 dec-1975 jul] – 1r – 1 – mf#366018 – us WHS [071]

Distance education – Abingdon, Oxfordshire 1983+ – 1,5,9 – ISSN: 0158-7919 – mf#14138 – us UMI ProQuest [374]

Distancias / Corretjer, Juan Antonio – Guaynabo, Puerto Rico. 1957 – 1r – us UF Libraries [972]

Distant drummer – 1969 nov 27/dec 4-1970 sep 24 – 1r – 1 – (cont by: thursday's drummer) – mf#541763 – us WHS [071]

Distant drummer *see* Drummer

Distant drums – v7 n1-v9 n2 [1985 mar-1987 may] – 1r – 1 – mf#1553981 – us WHS [071]

Distant interactions and their effects on children's physical activity levels during fitness instruction / Patterson, Debra L – 2000 – 1mf – 9 – $4.00 – mf#PE 4088 – us Kinesology [370]

Distilled Spirits Council of the United States *see* Newsletter: the national organization of the distilled spirits industry

Distilled Spirits Council of the US *see* Discus newsletter

Distiller's magazine and spirit trade news. (distillers' and brewers' magazine.-distillers', brewers' and spirit merchants' magazine) – Glasgow, Scotland. Apr 1897-Apr 1905. -m – 4r – 1 – uk British Libr Newspaper [660]

Distillery, rectifying and wine workers international journal – 1942-47 – (= ser Labor union periodicals, pt 3: food and agricultural industries) – 1r – 1 – $210.00 – 1-55655-616-0 – us UPA [660]

Distillery, Rectifying, Wine and Allied Workers' International Union of America *see* Drwaw journal

Distinction and the criticism of beliefs / Sidgwick, Alfred – London: Longmans, Green, 1892 – 1mf – 9 – 0-7905-7372-5 – (incl bibl ref) – mf#1989-0597 – us ATLA [240]

Distinctiones et regulae theologicae ac philosophicae / Maccovius, J – Amstelodami, 1656 – 3mf – 9 – mf#PBA-242 – ne IDC [240]

Distinctiones per universum theologiam sumtae ex canone sacrarum literarum / Alsted, J H – Francofurti, 1626 – 2mf – 9 – mf#PBA-107 – ne IDC [240]

Distinctiones philosophicae / Reeb, G – Duaci, 1637 – 5mf – 9 – mf#CA-33 – ne IDC [100]

The distinctive doctrines and usages of the general bodies of the evangelical lutheran church in the united states / Loy, Matthias et al – 4th ed., rev. and enl. Philadelphia, PA: Lutheran Publication Society, c1914 – 1mf – 9 – 0-7905-6654-0 – mf#1988-2654 – us ATLA [242]

Distinctive doctrines of lutheranism / Voigt, A G – Philadelphia, Pa: United Lutheran Publication House, c1910 – (= ser Lutheran Monographs) – 1mf – 9 – 0-7905-9732-2 – mf#1989-1457 – us ATLA [242]

The distinctive doctrines of the different christian confessions in the light of the word of god : also, a presentation of the significance and harmony of evangelical doctrine and a summary of the principal unsound religious tendencies in christianity = die unterscheidungslehren der verschiedenen christlichen bekenntnisse in lichte goettlichen worts [sic] / Graul, Karl; ed by Seeberg, Reinhold – Columbus, Ohio: Lutheran Book Concern, [1897?] – 1mf – 9 – 0-7905-3939-X – (in english) – mf#1989-0432 – us ATLA [240]

Distinctive errors of romanism / Bennett, William J E – London, England. 1842 – 1r – us UF Libraries [240]

The distinctive features of the christian school = Typeerende van de gereformeerde school / Kooy, Tijmen van der – Grand Rapids, MI: Wm B Eerdmans, 1925 – 1mf – 9 – 0-524-06634-5 – mf#1991-2689 – us ATLA [240]

The distinctive messages of the old religions / Matheson, George – New York: Dodd, Mead, 1894 – 1mf – 9 – 0-524-00935-X – mf#1990-2158 – us ATLA [200]

Distinctive plea of the disciples of christ / Bradley, Ernest J – Bowling Green, Mo: Times, [1902?] – 1mf – 9 – 0-524-01207-5 – mf#1990-4065 – us ATLA [240]

The distinctive principle of the baptists / Skevington, Samuel John – [Chicago?]: Printed for private distribution, 1914 – 1mf – 9 – 0-524-07761-4 – mf#1991-3329 – us ATLA [242]

Distinctive principles and present position and duty of the free ch... / Kennedy, J – Edinburgh, Scotland. 1875 – 1r – us UF Libraries [240]

Distinctive principles of the free church / Ker, William T – Edinburgh, Scotland. 1852 – 1r – us UF Libraries [240]

The distinctive principles of the presbyterian church in the united states : commonly called the southern presbyterian church, as set forth in the formal declarations, and illustrated by extracts from proceedings of the general assembly from 1861-70 – Richmond: Presbyterian Committee of Publication, [1871?] – 1mf – 9 – 0-524-00309-2 – mf#1989-3009 – us ATLA [242]

Distinctive tenets of the church of england / Gresley, William – London, England. 1847 – 1r – us UF Libraries [241]

Distinctive tokens of christian communion / Sumner, Charles Richard – London, England. 1829 – 1r – us UF Libraries [240]

Distinguished 100: the book of eminent alumni of the university of the philippines / Gwekoh, Sol H – Manila: Apo Book Co., c1939. 147p. ill – 1 – us Wisconsin U Libr [920]

Distinguished converts to rome in america / Scannell-O'Neill, Denis James – St Louis, MO: B Herder, 1907 – 1mf – 9 – 0-524-03803-1 – mf#1990-4875 – us ATLA [240]

Distinguishing sentiments of the particular baptists – London, England. 18– – 1r – us UF Libraries [242]

Distinguo : maengel und uebelstaende im heutigen katholizismus nach professor dr. schell in wuerzburg und dessen vorschlaege zu ihrer heilung / Braun, Carl – 4. Aufl. Mainz: Franz Kirchheim, 1897 – 1mf – 9 – 0-8370-7045-7 – mf#1986-1045 – us ATLA [241]

Distler, Johann Georg *see* Six quatuors pour deux violons, alto et basse, oeuvre 6, liv 2

Distribucion de las obras ordinarias y extraordinarias del dia : para hacerlas perfectamente conforme al estado de las senoras religiosas – en Mexico: ...Miguel de Ribera Calderon, ano de 1712 – (= ser Books on religion...1543/44-c1800: ordenes, etc: congregacion de la purisima) – 2mf – 9 – mf#crl-182 – ne IDC [241]

Distributable union catalog / Harvard College. Library – Author, title and subject entries for computerized cataloging, produced by Harvard libraries since Jul 1977. Includes entries for works on order and works received but not yet cataloged. 360 fiches – 9 – 75.00 – us Harvard [010]

Distributable union catalog / Harvard University. Library – [Cambridge MA]: The Library 1981-93 – 9 – (final cumulation & suppls; ceased with final suppl in 1993) – us Harvard [020]

Distributed computing – Dordrecht, Netherlands 1986+ – 1,5,9 – ISSN: 0178-2770 – mf#16987 – us UMI ProQuest [000]

Distribution – New York NY 1993-97 – 1,5,9 – (cont: chilton's distribution) – ISSN: 1066-8489 – mf#944,04 – us UMI ProQuest [380]

Distribution and concentration of copper in the newborn calf / Rusoff, Louis L – Gainesville, FL. 1941 – 1r – us UF Libraries [636]

The distribution of aerolites in space / Harvey, Arthur – Ottawa: J Durie & Son; Toronto: Copp-Clark Co, 1896 – 1mf – 9 – mf#01185 – cn Canadiana [520]

The distribution of african population, native and immigrant, in buganda / Fortt, J M – 1953 – (filmed with: language teaching in kikuyu schools/l j beecher, and various others) – us CRL [960]

The distribution of ancient volcanic rocks along the eastern border of north america / Williams, George Huntington – Chicago: University Press, [1894?] [mf ed 1983] – 1mf – 9 – 0-665-44115-0 – (repr fr: journal of geology; incl bibl ref) – mf#44115 – cn Canadiana [550]

The distribution of canadian forest trees in its relation to climate and other causes : a paper read before the british association for the advancement of science, montreal, sep 2nd, 1884 / Drummond, Andrew Thomas – Montreal: Dawson, 1885 [mf ed 1980] – 1mf – 9 – 0-665-02767-2 – (repr fr: canadian economics) – mf#02767 – cn Canadiana [634]

The distribution of estates of deceased persons in massachusetts / Newhall, Guy – 2d ed. Boston: Jackson, 1915. 12p. LL-393 – 1 – us L of C Photodup [340]

Distribution of macro and micro elements in some soils of peninsular florida – Gainesville, FL. 1939 – 1r – us UF Libraries [630]

The distribution of power to regulate interstate carriers between the nation and the states / Reynolds, George Greenwood – New York: Columbia, 1928. 434p. LL-1298 – 1 – us L of C Photodup [340]

The distribution of stress in certain tension members / Batho, Cyril – [S.l: s.n, 1907?] [mf ed 1991] – 1mf – 9 – 0-665-99506-7 – mf#99506 – cn Canadiana [624]

The distribution of stress in riveted connections / Young, Clarence Richard – [S.l: s.n, 1906?] [mf ed 1991] – 1mf – 9 – 0-665-99511-3 – mf#99511 – cn Canadiana [624]

Distribution of "trace elements" in the newborn calf as influenced by the nutrition of the dam / Rusoff, Louis L – Gainesville, FL. 1941 – 1r – us UF Libraries [636]

Distributive worker / Distributive Workers of America et al – 1969 jul-1974 dec, 1975-1986 oct, 1987 jan-1992 aug – 3r – 1 – mf#366019 – us WHS [331]

Distributive Workers of America et al *see* Distributive worker

District 50 news / United Mine Workers of America – 1941 oct 20-1948 apr 15 – 1r – 1 – (cont: cio news [district 50 edition]; cont by: ucw news; news) – mf#3558007 – us WHS [622]

District advertiser – [NW England] Cheadle – 1 – (title change: cheadle, gatley and district advertiser [20 jul 1972]; district advertiser [1976-78, jan-15 mar 1979]; cheadle, gatley & cheadle hulme district advertiser [22 mar 1979-1981, jan-mar 1984]; cheadle, gatley, cheadle hulme & heald green district advertiser [5 apr 1984-27 sep 1985]) – uk MLA; uk Newsplan [072]

The district association a review and a plea / Ryland, Charles H – 18p – 1 – 5.00 – us Southern Baptist [242]

District census statistics / Northwestern Provinces and Oudh. India – 1891 – (= ser Census of india) – 1 – (budaun, bahraich, bara banki, bijnor, cawnpore, gorakhpur, jhansi, lucknow, moradabad, rae bareli, saharanpur and shahjahanpur districts) – us CRL [315]

District census statistics / Northwestern Provinces and Oudh. India – 1911 – (= ser Census of india) – 1 – (budaun, bahraich, dehra dun, etah, farrukhabad, hamirpur, jalaun, lucknow, manipuri, partabgarh and sitapur districts) – us CRL [315]

The district councils elections proclamation, 1959 – [S.l: s.n., 1959?] – us CRL [325]

The district court executive pilot program : a report / Eldridge, William B – Washington: FJC, 1984 – 1mf – 9 – $1.50 – mf#LLMC 95-316 – us LLMC [347]

District court implementation of amended civil rule 16 : a report / Weeks, Nancy – Washington: FJC, Apr 1984 – 1mf – 9 – $1.50 – mf#LLMC 95-317 – us LLMC [347]

District digest / Civilian Conservation Corps [US] – v1 n2 [1937 jun] – 1r – 1 – mf#1497303 – us WHS [333]

District fifty news / International Union of District 50, Allied and Technical Workers of the United States and Canada – 1954-jun 25-dec, 1955-69, 1970 jan 10-1972 mar – 8r – 1 – mf#1055607 – us WHS [331]

District lawyer – v1-10. 1976-86 (all publ) – 9 – $132.00 set – (cont by: washington lawyer) – mf#108621 – us Hein [340]

District news – v7 n4 [1983 jul/aug], v8 n2 [1984 mar/apr], v10 n6 [1986 summer] – 1r – 1 – mf#1131955 – us WHS [071]

District newsletter / Waunakee Community Schools – v1 n1-v5 n2 [1978 dec-1983 may] – 1r – 1 – (cont by: waunakee community school district's t i m e) – mf#615793 – us WHS [370]

District nursing – London, England 1958-74 – 1,5 – (cont by: queen's nursing journal) – ISSN: 0012-4044 – mf#1378.01 – us UMI ProQuest [610]
District of Columbia *see* Reports, pre-nrs
District of columbia – (= ser General education board: the early southern program) – 3r – 1 – $390.00 – us Scholarly Res [370]
District of columbia : session laws of american states and territories – 1975-86 – 9 – $160.00 set – mf#402560 – us Hein [348]
District of columbia appeals cases / U.S. Courts. District of Columbia – v1-48. 1893-1919 (all offered) – 353mf – 9 – $529.00 – (a pre-nrs title) – mf#LLMC 81-481 – us LLMC [340]
District of columbia building permits, 1877-1949, and index, 1877-1958 / District of Columbia. Govt – (= ser Records Of The Government Of The District Of Columbia) – 5 – (1877-1903 283r + index. july 1 1915-sep 7 1949 681r. these records are still being filmed) – mf#M1116 – us Nat Archives [350]
District of columbia code – 1981-mar 2001 update – 9 – $2088.00 set – mf#401910 – us Hein [348]
District of Columbia. Govt *see*
– District of columbia building permits, 1877-1949, and index, 1877-1958
– Records of the city of georgetown, dc, 1800-1879
District of columbia law review *see* University of the district of columbia law review
District of columbia laws and resolution – 1975-1986 – 9 – $160.00 set – mf#400630 – us Hein [348]
District of Columbia. Laws, Statutes, etc *see* The alcoholic beverage laws of the district of columbia, rev. to january 1, 1948
District of Pecos. Headquarters *see* Headquarters records of the district of the pecos, 1878-1881
A district office in northern india : with some suggestions on administration / Whish, Charles William – Calcutta 1892 – (= ser 19th c british colonization) – 4mf – 9 – mf#1.1.8885 – uk Chadwyck [350]
District or neighborhood architecture and housing / Goebel, Rubye K – s.l, s.l? 1936 – 1r – us UF Libraries [720]
District post – [London & SE] Hounslow feb 1911-dec 1913 [mf ed 2003] – 3r – 1 – (cont as: chiswick district post [jan 1912-dec 1913]) – uk Newsplan [072]
District post – Postville IA. 1881 aug 10 – 1r – 1 – mf#851244 – us WHS [071]
District record : official publication of district union 271, amc and bwna, afl-cio / Amalgamated Meat Cutters and Butcher Workmen of North America et al – 1975 july, nov, 1976 jan-1988 nov/dec – 1r – 1 – mf#1876035 – us WHS [660]
District review : news of the sparta ccc district / Civilian Conservation Corps [US] – v1 n1-v3 n30 [1937 jan-1940 jan] – 1r – 1 – (cont by: spartan [sparta wi]) – mf#1497304 – us WHS [333]
District school journal of education of the state of new york – La Salle IL 1840-52 – 1 – mf#3727 – us UMI ProQuest [370]
District silver advocate – Vale OR: Advocate Pub Co [wkly] – 1 – us Oregon Lib [071]
District steam supply : heating buildings by steam, from a central source / Bartlett, James Herbert – Montreal?: J Lovell, 1884? – 1mf – 9 – mf#18645 – cn Canadiana [621]
District times – London, UK. 3 may 1901-7 aug 1914 – 20 1/2r – 1 – uk British Libr Newspaper [072]
District union 427 voice / Amalgamated Meat Cutters and Butcher Workmen of North America et al – 1970 may-1983 sep – 1r – 1 – (cont by: voice [akron, ohio]; 880 news and views; voice of local 880) – mf#611749 – us WHS [660]
O districto de lourenco marques e a africa do sul / Eduardo de Noronha. Lisboa: Imprensa Nacional, 1895 – us CRL [079]
O districto de lourenco marques, no presento e no futuro / Casrilho Barreto e Noronha, Augusto Vidal de – Lisboa: Soc de Geographia, 1880 – 1 – (filmed with: apontamentos para historia de guerra de zambezia 1871-75 et al) – us CRL [946]
Districto de mocambique em 1898 / Costa, Eduardo Augusto Ferreira Da – Lisboa, Portugal. 1902 – 1r – us UF Libraries [960]
Distrikts-rabbiner nathan bamberger – Wurzburg, Germany. 1919 – 1r – us UF Libraries [939]
El distrito – Jerte, 1914. Numeros sueltos – 5 – sp Bibl Santa Ana [073]
Distrito de la audiencia de santo domingo / Malagon Barcelo, Javier – Ciudad Trujillo, Dominican Republic. 1942 – 1r – us UF Libraries [972]
Distrito federal e seus recursos naturais / Abreu, Sylvio Froes – Rio de Janeiro, Brazil. 1957 – 1r – us UF Libraries [972]
Disturbances in may, 1921 – London, 1921 – 2mf – 9 – mf#J-28-63 – ne IDC [956]
Disturbed ireland : being the letters written during the winter of 1880-81 / Becker, Bernard Henry – London: Macmillan, 1881 [mf ed 1988] – ix/338p – 1 – (republ from the london daily news) – mf#8815 – us Wisconsin U Libr [941]
Disturnell, John *see* The northern traveller
Disunion : two discourses at music hall, on january 20th, and february 17th, 1861 / Phillips, Wendell – Boston: Robert F Wallcut, 1861 – 1mf – 9 – 0-524-01125-7 – mf#1990-0339 – us ATLA [976]
Disuse of the athanasian creed advisable in the present state of th... / Hull, William Winstanley – Oxford, England. 1831 – 1r – us UF Libraries [240]
Disweek – Belize City, Belize. June 24 1983-Feb 15 1985 – 1r – 1 – us L of C Photodup [079]
The ditches and watercourses acts of ontario : with notes and references to decided cases / Cameron, Malcolm Graeme – Toronto: Carswell, 1886 [mf ed 1995] – 1mf – 9 – 0-665-94811-5 – mf#94811 – cn Canadiana [343]
Ditchfield, P H *see*
– Church in the netherlands
– The parish clerk
– Symbolism of the saints
Ditchfield, Peter Hampson *see*
– The old-time parson
– The village church
Ditfurth, Franz Wilh Freiherr v *see* Fuenfzig unterdrueckte balladen und liebeslieder des 16. jahrhunderts
Ditha dhamma vipassana nnana dassana / Lay ti Cha ra to Bhu ra kri – Ran Kun: U Khan Lat 1979 [mf ed 1990] – [pl] 1r with other items – 1 – mf#mf-10289 seam reel 133/11 [§] – us CRL [280]
Die dithmarscher : historischer roman in vier buechern / Bartels, Adolf – Hamburg: Hanseatische Verlagsanstalt, c1928 [mf ed 1989] – 526p (ill) – 1 – mf#6980 – us Wisconsin U Libr [830]
Dithmarscher anzeigenblatt – Heide, Holst DE, 1948 31 jul-1949 28 sep – 1r – 1 – gw Misc Inst [074]
Dithmarscher blaetter *see* Dithmarsische zeitung
Dithmarscher bote – Wesselburen DE, 1865, 1872, 1877-1940, 1951-56 – 1 – (further title: wesselburener marschbote) – gw Misc Inst [074]
Dithmarscher landeszeitung – Heide, Holst DE, 1949 1 oct-1968 – 71r – 1 – (filmed by misc inst: 1969- [ca 7r/yr]) – gw Mikrofilm; gw Misc Inst [074]
Dithmarscher landeszeitung *see* Meldorfer wochen-blatt
Dithmarsische zeitung – Heide, Holst DE, 1832 28 apr-1873 27 dec – 1 – (title varies: 6 jan 1849: dithmarscher blaetter) – gw Misc Inst [074]
Ditiatin, I *see* Ustroistvo i upravlenie gorodov rossii
Ditirafalo tsa merafe ya batswana ba lefatshe la tshireletso / Schapera, Isaac – Lovedale, South Africa. 1954 – 1r – us UF Libraries [960]
Ditirafalo tsa merafe ye batswana be lefatshe la tshireletso / Schapera, Isaac – Lovedale, South Africa. 1940 – 1r – us UF Libraries [960]
Ditmar, K von *see* Reisen und aufenthalt in kamtschatka in den jahren 1851-1855
Dito, Oreste *see* Massoneria, carboneria ed altre societa segrete nella storia del risorgimento italiano
Ditscheid, Aegidius *see* Matthias eberhard, bischof von trier, im kulturkampf
Ditson, Oliver *see* Catalogs of vocal music
Dittebrandt, Hazel *see* African elephant
Dittenberger, Theophor Wilhelm *see*
– D carl daub's system der theologischen moral
– D carl daub's vorlesungen ueber die prolegomena zur dogmatik
– D carl daub's vorlesungen ueber die prolegomena zur theologischen moral
Dittenberger, W *see* Orientis graeci inscriptiones selectae
Dittenberger, Wilhelm *see*
– De sacris rhodiorum commentatio
– Observationis de sacris amphiarai thebanis et oropiis
Dittmar, Manuela *see*
– Bibliografia sobre los mixtecas
– Zur genetik der wahrnehmungsgeschwindigkeit
Dittmar, Thomas *see* Entwicklung einer antisense-strategie und wanderungsdynamik c-erbb-2/egfr ueberexprimierender brust-adenokarzinomzelllinien in einer 3d-kollagen matrix
Der dittmarser und eiderstedter bote – Friedrichstadt DE, 1799 27 jun-1846, 1850-52, 1855-56, 1858-69, 1884-1941 30 may – 1 – (aka: eiderstedter wochenblatt; friedrichstaedter intelligenzblatt; friedrichstaedter wochenblatt) – gw Misc Inst [074]
Dittmer, Alma *see* Vocal polyphony of william byrd
Dittmer, Ernst *see* Ein kleiner deutscher
Dittmer, Hans *see* Spiel mit wolken und winden
Dittmer, Wilhelm *see* Te tohunga
Dittrich, Franz *see* Gasparo contarini, 1483-1542
Ditzler, Jacob *see*
– Baptism
– The graves-ditzler or great carrollton debate
– The louisville debate
Diu crone / Tuerlin, Heinrich von dem; ed by Scholl, Gottlob Heinrich Friedrich – Stuttgart: Litterarischer Verein, 1852 [mf ed 1993] – (= ser Blvs 27) – li/511p – 1 – (middle high german text. int in german) – mf#8470 reel 6 – us Wisconsin U Libr [830]
Diurnale monasticum secundum rubricam romanam [et] s[e]c[un]d[u]m ritum [et] consuetudine[m] monasterij beate marie virgi[ni]s al[ia]s scotoru[m] vie[n]ne ordinis s[an]c[t]i benedicti – Venetiis: Alantsee (Giunta) 1515 – (= ser Hqab. literatur des 16. jahrh.) – 8mf – 9 – €80.00 – mf#1515c – gw Fischer [780]
Diurnale romanum ex decreto sacrosancti concilii tridentini restitutum... – Parisiis: Societ. Typogr 1597 – (= ser Hqab. literatur des 16. jahrh.) – 9mf – 9 – €85.00 – mf#1597b – gw Fischer [780]
Diutiska : an historical and critical survey of the literature of germany, from the earliest period to the death of goethe / Solling, Gustav – London: Truebner and Co, 1863 – 1r – 1 – us Wisconsin U Libr [430]
Diuturnity : or, the comparative age of the world: showing that the human race is in the infancy of its being, and demonstrating a reasonable and rational world, and its immense future duration / Abbey, Richard – Cincinnati: Applegate, 1866 – 1mf – 9 – mf#ATLA 1985-0003 – us ATLA [230]
Diuturnity, or, the comparative age of the world : showing that the human race is in the infancy of its being, and demonstrating a reasonable and rational world, and its immense future duration / Abbey, Richard – Cincinnati: Applegate, 1866 – 1mf – 9 – 0-8370-2003-4 – mf#1985-0003 – us ATLA [210]
Divagaciones filologicas / Sanin Cano, Baldomero – Santiago, Chile. 1952 – 1r – us UF Libraries [440]
Divagar... / Sanchez Arjona, Vicente – Sevilla: Imprenta Carlos Acuna, Tomo 1. 1948 – 1 – sp Bibl Santa Ana [810]
Divagar... / Sanchez Arjona, Vicente – Sevilla: Imprenta Carlos Acuna, Tomo 3. 1949 – 1 – sp Bibl Santa Ana [810]
Divagar...(de mis archivos) / Sanchez Arjona, Vicente – Sevilla: Carlos Acuna Imprenta, Tomo 2. 1948 – 1 – sp Bibl Santa Ana [810]
Divan : [revue litteraire] – Paris, 1909-1946 – 235mf – 8 – mf#H-403c – ne IDC [410]
Le divan – Paris. 1909-31 – 1 – fr ACRPP [073]
The divan project : divan – 9 – (abduelkadir gulami 1291, aczi 1290, aribozli nu'man mahir beg 1288: istanbul 2mf ea $40 per 2mf. ahmed muesellem istanbul 1326 3mf $55. a'ma yusuf garibi efendi istanbul n.d. 2mf $40. 'arif bulak 1258 4mf $60. 'asik oemer 1341, 'avni 1303: istanbul 1mf ea $25 per mf. 'ayni istanbul 1258 8mf $130. 'azbi istanbul 1286 2mf $40. baki istanbul 1276 4mf $60. belig istanbul 1258 2mf $40. celaleddin istanbul 1317 6mf $90. dertli istanbul 1329? 1mf $25. ebue-l'kemal istanbul 1324 2mf $40. emrah istanbul 1332 1mf $25. enis dede edirne 1307 2mf $40. es'ad istanbul 1337 1mf $25. esrar dede efendi istanbul 1257 3mf $55. esrefoglu 'abdullah al-rumi 1869, fatin 1288 istanbul 2mf ea $40 per 2mf. finat (1,2) 1264,1286 istanbul 1mf ea $25 per mf. finat (3) istanbul 1291 3mf $55. fitnat danim (4) n.p.,n.d. 2mf $40. fuzuli (1) bulak 1256 3mf $55. fuzuli (2) istanbul 1268 3mf $55. fuzuli (3) tabriz 1266 2mf $40. fuzuli (4) istanbul 1308 5mf $75. fuzuli (5) istanbul 1328 6mf $90. halat efendi istanbul 1258 1mf $25. halid istanbul 1260 2mf $40. halim giry sultan istanbul 1257 1mf $25. hasmet bulak 1257 3mf $55. hazik efendi istanbul 1318 2mf $40. hersekkli 'arif hikmet beg, hikmet 'arif (seyhuelislam) 1283: istanbul 4mf ea $60 per 4mf. hilmi (1,2) 1274,1293: istanbul 1mf ea $25 per mf. hizir agazade sa'id bey (1) istanbul 1257 1mf $25. hizir agazade seyyid bey (2) n.p. 1289 1mf $25. hudayi aziz mahmud efendi n.p.,n.d. 2mf $40. huzni n.p. 1312 2mf $40. ihsan (harnamizade) istanbul 1347 2mf $40. isma'il hakki istanbul 1288 3mf $55. 'ismet istanbul 1291 1mf $25. 'izzet bulak 1255 7mf $110. 'izzet beg istanbul 1258 2mf $40. kayguli sultan n.p.,n.d. 2mf $40. kaygulu efendi n.p. 1272 1mf $25. kaygulu 1273, kazi burhaneddin 1338: istanbul 1mf ea $25. kazim pasa (1) n.p.,n.d. 1mf $25. kazim pasa (2) istanbul 1328 2mf $40. kemalpasazade istanbul 1313 3mf $55. kethudazade 'arif istanbul 1271 1mf $25. kuddusi (1,2,3) 1322,1325,1326: istanbul 3mf ea $55 per 3mf. leskofceli galib beg istanbul 1135 2mf $40. leyla hanim (1,2) 1260,1267: bulak 2mf ea $40 per 2mf. leyla hanim (3) 1299, mehmed 'ali hilmi dede baba 1327, mehmed emin 'iffet 1257: istanbul 2mf ea $40 per 2mf. mehmed sebateddin istanbul 1310 1mf $25. muhibbi istanbul 1308 4mf $60. munif istanbul 1266 2mf $40. murad emri bursa 1329 2mf $40. mustafa nuzuli al-kulavi 1331, muestak efendi 1264: istanbul 2mf ea $40 per 2mf. nabi (1) bulak 1257 10mf $165. nabi (2) istanbul 9mf $150. na'ili bulak 1253 2mf $40. naim istanbul 1257 8mf $130. necmi istanbul 1287 2mf $40. nedim istanbul 1338/40 6mf $90. nef'i istanbul 1269 4mf $60. nesimi (1,2) 1260, 1286: istanbul 3mf $55. nes'et efendi bulak 1252 2mf $40. nevres istanbul 1290 5mf $75. niyazi (1,2,3) 1254,1291,1325: istanbul 2mf ea $40 per 2mf. niyazi izmir 1291 1mf $25. 'oerfi ve 'avnuellah al-kazimi 1327, pertev pasa 1256: istanbul 2mf ea $40 per 2mf. ragib (1) n.p. 1253 3mf $55. ragib pasa (2) bulak 1252 2mf $40. refi 'i kalayi istanbul 1284 2mf $40. sabri bursa 1292 2mf $40. sabri-i sakir istanbul 1296 2mf $40. sami bulak 1253 4mf $60. selami istanbul 1946 3mf $55. senmih mevlevi 1275, sermed 1254 2mf ea $40 per 2mf. seyyid mehmed nesib istanbul 1261 1mf $25. seyyid nigari istanbul 1301 6mf $90. seyyid seyfullah istanbul 1288 3mf $55. sezayi bulak 1257 3mf $55. sueleyman sadi istanbul 1325 1mf $25. sultan bayezid sani istanbul 2mf $40. sultan hueseyin baykara istanbul 1946 5mf $75. sultan veled istanbul 1946 12mf $195. sururi bulak 1255 6mf $90. suzi istanbul 1290 3mf $55. sem'i (1,2,3) 1291,1302,1342 1343?: istanbul 1mf ea $25 per mf. sem'i (4) n.p.,n.d. 1mf $25. seref hanum (1) bulak 1284 4mf $60. seref danim (2) istanbul 1292 5mf $75. seyh galib bulak 1252 5mf $75. seyhi (1) istanbul 1291 1mf $25. seyhi (2) istanbul 1942 5mf $75. sinasi (1,2) 1287,1310: istanbul 2mf ea $40 per 2mf. turabi istanbul 1294 2mf $40. uftade istanbul 1328 1mf $25. 'ulvi istanbul 1290 2mf $40. vasif-i enderuni (1) bulak 1. ...) – Library of Congress, UCLA, and Princeton University – us MEDOC [810]
The divan project : divance – 9 – (dehri istanbul 1330 2mf $40. es'ad pasa istanbul 1268 1mf $25. fazil istanbul 1329 2mf $40. hanyevi sefik efendi 1293, hasimi 1329, 'izzet 1257: istanbul 1mf ea $25 per mf. kazim n.p.,n.d. 1mf $25. nazim istanbul 1308 1mf $25. ragib pasa n.p. 1276 1mf $25. sueleyman fehim istanbul 1262 1mf $25. sinaver istanbul 1330 2mf $40. tevfik cairo 1283 1mf $25. vak'anuevis ahmed lutfi istanbul 2mf $40) – us MEDOC [810]
The divan project : other titles – 9 – (asar-i ziver pasa bursa 6mf $90. baki'nin es'ar-i muentehabesi istanbul 1317 2mf $40. divaneliklerim ('abduelhak hamid) istanbul 1303 1mf $25. divan-i guelzar (sa'di) istanbul 1284 3mf $55. envar uel-'astkin (ahmed bican) bulak 1300 6mf $90. es'ar (mehmed memduh) istanbul 1332 4mf $60. es'ar-i nedim istanbul 1920 1mf $25. esref ues- su'ara (esref) istanbul 1278 4mf $60. habname- ' veysi istanbul 1293 1mf $25. kasa'id tevfik istanbul 1304 1mf $25. kulliyat-i es'ar ruhi' bagdadi istanbul 1287 5mf $75. kulliyat-i (divan-i) fuzuli istanbul 1291 6mf $90. kulliyat-i fuzuli istanbul 1296 5mf $75. kulliyat-i hudayi 1338/40, kulliyat-i sa'ir esref 1928: istanbul 3mf per yr $55 per 3mf. kulliyat-i ziya pasa 4mf $60. mecmu'a-i es'ar-i re'fet istanbul 1289 1mf $25. mecmu'a-i 'irfan pasa istanbul 1287 2mf $40. munse'at-i akif istanbul 1259 4mf $60. munse'at-i akif bulak 1262 3mf $55. munse'at-i 'izzet beg istanbul 1263 1mf $25. munse'at-i nu'man mahir beg istanbul 1261 2mf $40. munse'at-i rif'at efendi bulak 1254 4mf $60. muntahabat-i divan-i fevzi n.p.,n.d. 1mf $25. muntahabat-i es'ar-i sinasi istanbul 1289 1mf $25. nevhat uel-'ussak (m. ibn-i receb) istanbul 1261 2mf $40. siyer-i veysi istanbul 1286 5mf $75. tuhfe-i asim bulak 1254 1mf $25. tuerki-i sultan veled istanbul 1341 2mf $40. ufak mecmu'a-'i si'ir (tevfik) istanbul 1329 1mf $25. vahdetname (ahmed efendi) 1302 6mf 90. zade-i sa'ir (mehmed celal): istanbul 1311 2mf $40) – us MEDOC [810]
Divaneliklerim *see* The divan project
Divan-i oerfi ve avnullah kazimi / Kazimi, Avnullah [Mehmet Selim] – Dersaadet [Istanbul]: Nuemune-i Tabaat Matbaasi, 1327 [1909] – (= ser Ottoman histories and historical sources) – 3mf – 9 – $55.00 – us MEDOC [956]
Divatia, Narsinhrao Bholanath *see* Gujarati bhasha ane sahitya
Divekara, Mahadevasastri *see* Hindusaskrtipradipa
Diver / Mayo, Edward L – Minneapolis, MN. 1947 – 1r – us UF Libraries [960]
Diver, Maud *see* Kabul to kandahar
Divergence of calvinism from pauline doctrine / Newman, Francis William – Ramsgate, England. 1871 – 1r – us UF Libraries [242]
Divers : ou, les enseignements de la vie / Baillairge, Charles P Florent – Quebec: C Darveau, 1898 – 8mf – 9 – mf#00049 – cn Canadiana [440]
Divers documens adresses a l'honorable louis joseph papineau : orateur de la chambre d'assemblee...nomme pour se rendre en angleterre, et y appuyer les petitions de la chambre a sa majeste et aux deux chambres du parlement imperial = Divers documents

addressed to the honorable louis joseph papineau...to support the petitions of the house to his majesty and to the two houses of the imperial parliament / Viger, Denis-Benjamin – [S.l: s.n, 1834?] (mf ed 1991) – 4mf – 9 – (with bibl) – mf#SEM105P1375 – cn Bibl Nat [324]

Divers documens adresses a l'honorable louis joseph papineau, orateur de la chambre d'assemblee : par l'honorable denis b viger, nomme pour se rendre en angleterre...mis devant la chambre, et dont l'impression a ete ordonnee mercredi, 8 janvier, 1834 – [S.l: s.n, 1834?] [mf ed 1985] – 1mf – 9 – 0-665-18733-5 – mf#18733 – cn Canadiana [320]

Divers documens et communications adresses a l'honorable louis joseph papineau : orateur de la chambre d'assemblee, par l'honorable denis b viger et augustin norbert morin, ecuyer, nommes pour se rendre en angleterre... / Viger, Denis-Benjamin – [S.l: s.n, 1835?] (mf ed 1991) – 2mf – 9 – mf#SEM105P1377 – cn Bibl Nat [324]

Divers documents addressed to the honorable louis joseph papineau : ...by the honorable denis b viger, appointed to proceed to england, and there to support the petitions of the house of his majesty and to the two houses of the imperial parliament – [S.l: s.n, 1834 ?] (mf ed 1991) – 4mf – 9 – mf#SEM105P1374 – cn Bibl Nat [324]

Divers documents addressed to the honorable louis joseph papineau, speaker of the house of assembly : by the honorable denis b viger, appointed to proceed to england...wednesday, 8th january, 1834 – S.l: s.n, 1834? – 1mf – 9 – mf#21446 – cn Canadiana [971]

Divers voyages et missions du p alexandre de rhodes en la chine : et autres royaumes de l'orient, avec son retour en europe par la perse & l'armenie / Rhodes, A de – Paris: Sebastian Cramoisy, 1703 – 5mf – 9 – mf#HT-662 – ne IDC [915]

Diverse imprese accomodate a diverse moralit... / Alciato, Andrea – Lyons: M Bonhomme, 1549 – 2mf – 9 – mf#O-1477 – ne IDC [090]

Diverses actes de la herencia del duch de calabria (siecle 16) – Valencia – 2r – 5,6 – sp Cultura [945]

Diversiones pascuales en oriente / Olivares Figueroa, Rafael – Caracas, Venezuela. 1949 – 1r – us UF Libraries [972]

Diversity of christian holiness / Woodford, James Russell – London, England. 1854 – 1r – us UF Libraries [240]

Diversorum (anno 1479-1516) / Fernando 2 – Barcelona – 1r – 5,6 – sp Cultura [946]

Diversorum...locos de regia.. / Weijers, Henrico E – 1839 – 9 – sp Bibl Santa Ana [240]

Divertimento per l'oboe con accompagnamento di 2 violini, viola et violoncello, op 9 / Crusell, Bernhard Henrik – Lipsia: C F Peters [183-?] [mf ed 19–] – 5pt on 1mf – 9 – mf#fiche 415 – us Sibley [780]

Diverting post – Dublin, Ireland. -w. 18 25 Oct-15 22 Nov 1725. Lacking 1 8 Nov. 1/4 reel – 1 – uk British Libr Newspaper [072]

Diverting post – La Salle IL 1704-06 – 1 – mf#4237 – us UMI ProQuest [071]

[Divide city-] times – NV. 1919 (scats) [wkly] – 1r – 1 – $60.00 – mf#W04490 – us Library Micro [071]

Divina commedia / Dante Alighieri – late 14th c – (= ser Holkham library manuscript books 514) – 1r – 1 – (1 col reel [ill only] c516-8) – mf#2192 – uk Microform Academic [810]

Divina commedia / Dante Alighieri – 14th, 15th c – (= ser Holkham library manuscript books 513, 515, 516, 517) – 1r – 1 – mf#95912 – uk Microform Academic [810]

Divina commedia / Dante Alighieri – 1474 – (= ser Holkham library manuscript books 518) – 1r – 1 – (written by marabettino di tuccio manetti) – mf#95931 – uk Microform Academic [810]

Divina proportione... / Paciolo, L – Venezia, 1509 – 8mf – 9 – mf#O-1006 – ne IDC [700]

Divina proportione (die lehre vom goldenen schnitt) / [Pacioli, L] Winterberg, C – Wien, 1889. v2 – 5mf – 9 – mf#O-517 – ne IDC [700]

Divinae institutiones... / Lactantius Placidus – 14th, 15th c – (= ser Holkham library manuscript books 120, 389, 391) – 1r – 1 – (filmed with: orationes 30 by cicero; epistolae, declamationes, epistolae mutuae by seneca; de brevitate vitae by s paulo) – mf#96541 – uk Microform Academic [450]

La divination a la cote des esclaves et a madagascar : le vodou fa, le sikidy / Trautmann, Rene – Paris: E Larose, 1940 – 1 – us CRL [306]

Divine attributes : including also the divine trinity / Swedenborg, Emanuel – Philadelphia, PA. 1866 – 1r – us UF Libraries [210]

The divine authority of the bible / Wright, George Frederick – Boston: Congregational Sunday-School & Pub Soc c1884 [mf ed 1985] – 1mf – 9 – 0-8370-5924-0 – (incl bibl ref & ind) – mf#1985-3924 – us ATLA [220]

Divine authority of the holy scripture asserted : from its adaptation to the real state of human nature / Miller, John – Oxford: University Press 1817 [mf ed 1984] – (= ser Bampton lectures 1817) – 3mf – 9 – 0-8370-0203-6 – mf#1984-1027 – us ATLA [220]

The divine authority of the pentateuch vindicated / McIntyre, Daniel – London: Bell & Daldy 1863 [mf ed 1989] – 1mf – 9 – 0-7905-2932-7 – (incl bibl ref) – mf#1987-2932 – us ATLA [221]

The divine authority of the scriptures of the old testament / McIntyre, David Martin – Stirling [Scotland?]: Drummond's Tract Depot [1902?] [mf ed 1984] – (= ser Biblical crit – us & gb 24) – 2mf – 9 – 0-8370-0239-7 – (incl bibl ref) – mf#1984-1024 – us ATLA [221]

Divine brotherhood : jubilee gleanings, 1842-92 / Hall, Newman – Edinburgh: T & T Clark, 1892 [mf ed 1984] – 4mf – 9 – 0-8370-0901-4 – mf#1984-4270 – us ATLA [240]

The divine character vindicated : a review of some of the principal features of rev dr e beecher's recent work entitled, "the conflict of ages / Ballou, Moses – New York: Redfield, 1854 [mf ed 1984] – 5mf – 9 – 0-8370-1000-4 – mf#1984-4356 – us ATLA [243]

The divine classic of nan-hua : being the works of chuang tsze, taoist philosopher / Balfour, Frederic Henry – Shanghai, Hongkong: Kelly & Walsh, 1881 [mf ed 1995] – (= ser Yale coll) – xxxviii/425p – 1 – 0-524-09588-4 – (with an excursus, and copious ann in english and chinese) – mf#1995-0588 – us ATLA [180]

Divine commission and perpetuity of the christian priesthood / Watson, John James – London, England. 1839 – 1r – us UF Libraries [240]

Divine companion *see* Hymn and tune collection from library of edmond d. keith

Divine compassion / Beecher, Henry Ward – London, England. 1886? – 1r – us UF Libraries [240]

The divine covenants, their nature and design : or, the covenants considered as successive stages in the development of the divine purposes of mercy / Kelly, John – London: Jackson, Walford, & Hodder, 1861 – (= ser Congregational Lecture) – 1mf – 9 – 0-524-04907-6 – mf#1992-0250 – us ATLA [220]

The divine demonstration : a text-book of christian evidence / Everest, Harvey William – St Louis, MO: Christian Pub Co, c1884 [mf ed 1993] – (= ser Christian church (disciples of christ) coll) – 1mf – 9 – 0-524-05949-7 – mf#1991-2349 – us ATLA [240]

The divine discipline of israel : an address and three lectures on the growth of ideas in the old testament / Gray, George Buchanan – London: Adam and Charles Black, 1900 – 1mf – 9 – 0-8370-3367-5 – mf#1985-1367 – us ATLA [220]

Divine dwellers in the desert / Mallik, Gurdial – Bombay: Nalanda Publications, c1949 – (= ser Samp: indian books) – us CRL [280]

Divine emblems : embellished with etchings on copper, after the fashion of master francis quarles / Abricht, J – [London]: Thomas Ward and Co, 1838 – 1mf – 9 – mf#O-534 – ne IDC [090]

Divine emblems in genesis and exodus / Simpson, Albert B – Nyack: Christian Alliance Pub Co, [1901?] [mf ed 1992] – (= ser Christian & missionary alliance coll) – 1mf – 9 – 0-524-02145-7 – mf#1990-4211 – us ATLA [221]

Divine emblems in the book of exodus / Simpson, Albert B – New York: Word, Work & World Pub Co, 1888 [mf ed 1992] – (= ser Christian & missionary alliance coll) – 1mf – 9 – 0-524-04239-X – mf#1990-5030 – us ATLA [221]

The divine enterprise of missions : a series of lectures / Pierson, Arthur Tappan – New York: Baker & Taylor, c1891 – 1mf – 9 – 0-8370-6599-2 – (incl ind) – mf#1986-0599 – us ATLA [240]

The divine force in the life of the world / McKenzie, Alexander – Boston: Houghton, Mifflin, 1899, c1898 – (= ser Lowell institute lectures) – 1mf – 9 – 0-8370-2896-5 – (incl ind) – mf#1985-0896 – us ATLA [240]

Divine forgiveness – London, England. 18– – 1r – us UF Libraries [240]

The divine foundation of the lord's day : an address / Caven, William – Toronto?: Ontario Lord's Day Alliance, 1897? – 1mf – 9 – mf#55574 – cn Canadiana [240]

The divine glory manifested in the conduct and discourses of our lord / Ogilvie, Charles A – Oxford: Printed by S. Collingwood for the author: J.H. Parker [distributor], 1836 – (= ser Bampton lectures) – 1mf – 9 – 0-7905-1548-2 – mf#1987-1548 – us ATLA [240]

The divine glory of christ / Brown, Charles J – London, New York: T Nelson, 1868 – 1mf – 9 – 0-8370-2998-8 – mf#1985-0998 – us ATLA [240]

The divine government / Smith, Southwood – 5th ed. Philadelphia: J B Lippincott, 1866 – 1mf – 9 – 0-8370-5169-X – mf#1985-3169 – us ATLA [240]

Divine grace illustrated in the conversion of tom frost – London, England. 18– – 1r – us UF Libraries [240]

Divine guidance : memorial of allen w. dodge / Hamilton, Gail – New York: D Appleton, 1881 – 1mf – 9 – 0-524-08372-X – mf#1993-3072 – us ATLA [240]

Divine healing : or, the atonement for sin and sickness / Carter, Russell Kelso – new enl ed. New York: JB Allen 1888 [mf ed 1992] – (= ser Christian & missionary alliance coll) – 1mf – 9 – 0-524-04658-1 – (incl bibl ref) – mf#1990-5054 – us ATLA [230]

Divine healing : a series of addresses and a personal testimony / Murray, Andrew – 7th ed. New York, NY: Christian Alliance, c1900 [mf ed 1992] – 1mf – 9 – 0-524-02131-7 – mf#1990-4197 – us ATLA [230]

Divine healing in mission work / Hussey, A H – Nyack, NY: Christian Alliance Pub Co, [1902?] [mf ed 1992] – (= ser Christian & missionary alliance coll) – 1mf – 9 – 0-524-02335-2 – mf#1990-4287 – us ATLA [230]

Divine healing in the light of scripture / Oerter, John H – Brooklyn, NY: Christian Alliance, c1900 [mf ed 1992] – (= ser Christian & missionary alliance coll) – 1mf – 9 – 0-524-02134-1 – mf#1990-4200 – us ATLA [230]

Divine healing of soul and body : also how god heals the sick, and the conditions upon which they are restored, giving wonderful testimonies of his miraculous power in these last days / Byrum, Enoch Edwin – Moundsville, W. Va.: Gospel Trumpet, 1892. Chicago: Dep of Photodup, U of Chicago Lib, 1975 (1r); Evanston: American Theol Lib Assoc, 1984 (1r) – 1 – 0-8370-0502-7 – mf#1984-B460 – us ATLA [130]

[Divine healing pamphlets] / Simpson, Albert B – New York, NY: Christian Alliance Pub Co, [1885?-1913?] [mf ed 1992] – (= ser Christian & missionary alliance coll) – 1v on 1mf – 9 – 0-524-04240-3 – mf#1990-5031 – us ATLA [130]

Divine heritage of man / Abhedananda, Swami – New York: Vedaanta Society, c1903 – 1mf – 9 – 0-524-00673-3 – mf#1990-2001 – us ATLA [280]

The divine human in the scriptures / Lewis, Tayler – New York: Robert Carter, 1860 – 1mf – 9 – 0-8370-4100-7 – mf#1985-2100 – us ATLA [220]

The divine indwelling / Brown, Edmund Woodward – New York:Fleming H. Revell, c1895 – 1mf – 9 – 0-8370-3020-X – mf#1985-1020 – us ATLA [210]

Divine inspiration : or, the supernatural influence exerted in the communication of divine truth, and its special bearing on the composition of the sacred scriptures / Henderson, Ebenezer – New and uniform ed. London: Jackson and Walford, 1852 – 1mf – 9 – 0-7905-1331-5 – (incl bibl ref and index) – mf#1987-1331 – us ATLA [240]

Divine inspiration vs the documentary theory of the higher criticism / Stuart, T McK – Cincinnati: Jennings and Graham; New York: Eaton and Mains, c1904 – 1mf – 9 – 0-7905-8596-0 – mf#1989-1821 – us ATLA [220]

Divine, J A F *see* Stained glass craft

The divine law as to wines : established by the testimony of sages, physicians, and legislators against the use of fermented and intoxicating wines / Samson, George Whitefield – Philadelphia: J B Lippincott, 1885, c1884 – 2mf – 9 – 0-7905-3474-6 – mf#1987-3474 – us ATLA [230]

The divine library : suggestions how to read the bible / Smyth, John Paterson – New York: James Pott; London: Samuel Bagster, 1897, c1896 – 1mf – 9 – 0-8370-5313-7 – mf#1985-3313 – us ATLA [220]

The divine library of the old testament : its origin, preservation, inspiration and permanent value: five lectures / Kirkpatrick, Alexander Francis – London, NY: Macmillan, 1891 – 1mf – 9 – 0-8370-3905-3 – mf#1985-1905 – us ATLA [221]

The divine life in man / Brown, James Baldwin – 2nd ed. London: Ward, [1860?] – 1mf – 9 – 0-7905-9248-7 – mf#1989-2473 – us ATLA [240]

Divine Light Mission *see* Divine times

Divine love / Apostolate of Christian Action – v1 n1-v29 n97 [1957 jul/sep-1986 sep] – 1r – 1 – mf#396796 – us WHS [240]

The divine love : a series of doctrinal, practical, and experimental discourses / Eadie, John – 2nd ed. Edinburgh: W. Oliphant, 1865 – 1mf – 9 – 0-7905-1595-4 – mf#1987-1595 – us ATLA [240]

The divine mysteries : the divine treatment of sin and the divine mystery of peace / Brown, James Baldwin – New York: Carlton & Lanahan, 1869 – 1mf – 9 – 0-7905-9160-X – mf#1989-2385 – us ATLA [240]

The divine mystery : a reading of the history of christianity down to the time of christ / Upward, Allen – Boston: Houghton Mifflin, 1915 – 1mf – 9 – 0-524-02327-1 – mf#1990-2950 – us ATLA [240]

The divine name in ancient china / Inglis, James William – Shanghai: American Presbyterian Mission Press, 1910 [mf ed 1995] – (= ser Yale coll) – 21p – 1 – 0-524-10145-0 – mf#1995-1145 – us ATLA [210]

Divine Of The Church Of England *see* Christian's way to heaven

The divine office : considered from a devotional point of view = Saint office considere au point de vue de la piete / Bacuez, Nicolas – London: Burns and Oates; New York: Catholic Publ Society, [1885?] – 2mf – 9 – 0-8370-7523-8 – (in english and latin. incl bibl ref) – mf#1986-1523 – us ATLA [240]

Divine or civil obedience? / Brown, Brian et al – Johannesburg, Ravan [n.d.] – us CRL [321]

The divine order of human society / Thompson, Robert Ellis – Philadelphia: JD Wattles, 1891 – (= ser Stone Lectures) – 1mf – 9 – 0-7905-9716-0 – mf#1989-1441 – us ATLA [240]

Divine origin, appointment, and obligation, of marriage / Croly, George – London, England. 1836 – 1r – us UF Libraries [240]

The divine origin of christianity : indicated by its historical effects / Storrs, Richard Salter – New York: Anson DF Randolph, c1884 [mf ed 1988] – (= ser Ely lectures 1884) – 2mf – 9 – 0-7905-0053-1 – (incl bibl ref & ind) – mf#1987-0053 – us ATLA [240]

Divine passibility / Brake, Peter H Vande – Grand Rapids MI: Calvin Theological Seminary, 2000 [mf ed 2001] – 1r – 1 – $130.00 – mf#2001-B003 – us ATLA [210]

La divine pastorale : dix linos d'ant. de vinck – Bruxelles: Editions du Marais, 1952-55 – 1 – us CRL [490]

Divine patience exhausted through the making void the divine law / Melvill, Henry – London, England. 1835 – 1r – us UF Libraries [240]

The divine pedigree of man : or, the testimony of evolution and psychology to the fatherhood of god / Hudson, Thomson Jay – Chicago: A C McClurg, 1899 – 1mf – 9 – 0-8370-3685-2 – (incl bibl ref) – mf#1985-1685 – us ATLA [210]

Divine penology : the philosophy of retribution and the doctrine of future punishment considered in the light of reason, science, revelation, and redemption / Hartman, Levi Balmer – New York: Fleming H Revell, c1898 – 1mf – 9 – 0-524-08377-0 – mf#1993-3077 – us ATLA [210]

The divine personality : being a consideration of the arguments to prove that the author of nature a being endued with liberty and choice / Pearson, John Batteridge – Cambridge: Deighton, Bell; London: Bell and Daldy, 1865 – 1mf – 9 – 0-8370-4688-2 – (incl bibl ref) – mf#1985-2688 – us ATLA [210]

Divine providence / Weaver, Jonathan – Dayton, Ohio: United Brethren Pub. House, 1873 – 1mf – 9 – 0-7905-2448-1 – mf#1987-2448 – us ATLA [210]

The divine reason of the cross : a study of the atonement as the rationale of our universe / Mabie, Henry Clay – New York: Fleming H Revell c1911 [mf ed 1991] – 1mf – 9 – 0-7905-7911-1 – mf#1989-1136 – us ATLA [240]

The divine revelation : an essay in defence of the faith; to which is prefixed a brief memoir of the author = Die gottliche offenbarung / Auberlen, Carl August – Edinburgh: T & T Clark 1867 [mf ed 1986] – (= ser Clark's foreign theological library. 4th series 16) – 2mf – 9 – 0-8370-8722-8 – (trans fr german by a b paton; incl ind) – mf#1986-2722 – us ATLA [240]

Divine revelation explained and vindicated : a course of lectures for the times / Fairbairn, Patrick et al – Glasgow: David Bryce, 1866 – 1mf – 9 – 0-8370-2925-2 – mf#1985-0925 – us ATLA [240]

The divine right of kings / Figgis, John Neville – 2nd ed. Cambridge: University Press, 1914 – 1mf – 9 – 0-7905-4962-X – (incl bibl ref) – mf#1988-0962 – us ATLA [941]

The divine right of missions : or, christianity the world-religion and the right of the church to propagate it / Mabie, Henry Clay – Philadelphia: Griffith & Rowland Press, 1908 [mf ed 1986] – 1mf – 9 – 0-8370-6211-X – mf#1986-0211 – us ATLA [230]

The divine rule of faith and practice : or, a defence of the catholic doctrine that holy scripture has been, since the times of the apostles... / Goode, William – 2nd rev enl ed. London: John Henry Jackson, 1853 [mf ed 1989] – 4mf – 9 – 0-7905-1943-7 – (incl ind and bibl) – mf#1987-1943 – us ATLA [242]

The divine satisfaction : a review of what should, and should not, be thought about the atonement / Whiton, James Morris – 3rd ed. London: James Clarke 1889 [mf ed 1985] – 1mf – 9 – 0-8370-2970-8 – mf#1985-0970 – us ATLA [240]

The divine society : or, the church's care of large populations. six lectures on pastoral theology / Jacob, Edgar – 2nd ed. London: SPCK; New York: E S Gorham, 1903 – 1mf – 9 – 0-7905-5236-1 – mf#1988-1236 – us ATLA [240]

The divine songs of zarathushtra / Taraporewala, Irach Jehangir Sorabji – Bombay: DB Taraporevala Sons, 1951 – (= ser Samp: indian books) – us CRL [290]

Divine teacher / Humphrey, William – London, England. 1873 – 1r – us UF Libraries [240]

Divine times / Divine Light Mission – 1974 jun 1-1978 may 1, 1978 jun-1980 feb – 2r – 1 – mf#500519 – us WHS [243]

Divine transcendence : and its reflection in religious authority / Illingworth, John Richardson – London: Macmillan, 1911 – 1mf – 9 – 0-7905-7342-3 – mf#1989-0567 – us ATLA [210]

The divine unity of scripture / Saphir, Adolph – London: Hodder and Stoughton, 1896 – 1mf – 9 – 0-8370-5048-0 – mf#1985-3048 – us ATLA [220]

The divine urge to missionary service / Goddard, Dwight – Ann Arbor, Mich: [s.n.], 1917 – 1mf – 9 – 0-524-04376-0 – mf#1991-2080 – us ATLA [240]

Divine warning to the church, at this time / Bickersteth, Edward Henry – London, England. 1842 – 1r – us UF Libraries [240]

The divine wisdom of the draavida saints / Govindacharya, Alkondavilli – Madras: Printed at the CN Press, 1902 – 1mf – 9 – 0-524-03118-5 – (incl bibl ref) – mf#1990-3171 – us ATLA [280]

Divine word messenger – Bay St Louis MS: Divine Word Missionaries 1961- [mthly ex jul, aug & dec, bimthly, qrtly, biannual] [mf ed 2004] – v39-54 n4 (dec 1961/jan 1962-winter 1977/spring 1978) on 1r – 1 – (ceased in 1982; lacks: v54 (1977) n1-3; damaged: v41 (1964) n5 p157; v46 (1969) p15; publ by the divine word missionaries at st augustine's seminary in bay st louis, ms, 1962-sep/oct 1965; cont: st augustine's catholic messenger; cont by: in a word) – ISSN: 0012-4214 – mf#2004-s026 – us ATLA [241]

Divine word messenger / Society of the Divine Word – v52 n2-v58 n1 [1975 spring-1982 spring/summer] – 1r – 1 – (cont by: in a word) – mf#688767 – us WHS [243]

Divine worship in england in the thirteenth and fourteenth centuries : contrasted with and adapted to that in the nineteenth / Chambers, John David – new rev enl ed. London: B M Pickering 1877 [mf ed 1988] – 1r – 1 – mf#7812 – us Wisconsin U Libr [240]

Divine worship in england in the thirteenth and fourteenth centuries contrasted with and adapted to that in the nineteenth / Chambers, John David – London: BM Pickering, 1877 – 2mf – 9 – 0-524-03697-7 – mf#1990-4802 – us ATLA [240]

Divinely appointed mode of supporting the christian ministry / Paterson, R – Edinburgh, Scotland. 1835 – 1r – us UF Libraries [240]

The diviner immanence / McConnell, Francis John – New York: Eaton & Mains, c1906 – 1mf – 9 – 0-7905-9799-3 – mf#1989-1524 – us ATLA [210]

La divinisation du chretien d'apres les peres grecs / Gross, J – Paris, 1938 – €19.00 – ne Slangenburg [240]

Divinitas scriptuarum : adversus hodiernas novitates asserta et vindicata / Schiffini, Sancto – Augustae Taurinorum [Turin]: E typographia S Josephi, 1905 – 1mf – 9 – 0-524-00328-9 – mf#1989-3028 – us ATLA [220]

Divinity and atonement of jesus christ scripturally expounded / Beard, John Reilly – London, England. 1858 – 1r – us UF Libraries [240]

Divinity and man : an interpretation of spiritual law in its relation to mundane phenomena and to the ruling incentives and moral duties of man, together with an allegory dealing with cosmic evolution and certain social and religious problems / Roberts, William Kemuel – Rev ed. New York: GP Putnam, 1903 – 1mf – 9 – 0-7905-8725-4 – mf#1989-1950 – us ATLA [130]

The divinity of christ / Ames, Edward Scribner – Chicago: Bethany Press, c1911 – 1mf – 9 – 0-7905-3629-3 – mf#1989-0122 – us ATLA [240]

The divinity of christ in the gospel of john / Robertson, A T – New York: Fleming H Revell, c1916 – 1mf – 9 – 0-524-06857-7 – mf#1992-0999 – us ATLA [226]

The divinity of jesus christ : a new demonstration taken from the latest attacks of incredulity = divinite de jesus-christ / Nicolas, Auguste – London: Thomas Richardson, 1865 – 1mf – 9 – 0-7905-9424-2 – (in english) – mf#1989-2649 – us ATLA [240]

The divinity of jesus christ : an exposition of the origin and reasonableness of the belief of the christian church – Boston: Houghton, Mifflin; Cambridge: Riverside Press, 1894, c1893 [mf ed 1985] – 1mf – 9 – 0-8370-5388-9 – mf#1985-3388 – us ATLA [240]

The divinity of jesus christ : the truth maintained by searching the scriptures to be received as given by inspired men of god, as revealed to us in 2 pet. 1:19, 20... / Moomaw, Benjamin F – Elgin, IL: Brethren Pub House, 1899 – 1mf – 9 – 0-524-03800-7 – mf#1990-4872 – us ATLA [240]

The divinity of our lord / Alexander, William – London, New York: Cassell [1886?] [mf ed 1989] – (= ser Helps to belief) – 1mf – 9 – 0-7905-0960-1 – mf#1987-0960 – us ATLA [240]

The divinity of our lord / Funkhouser, George Absalom – Dayton, Ohio: United Brethren Pub House, 1902 – (= ser Doctrinal Series (Dayton, Ohio)) – 1mf – 9 – 0-7905-2659-X – mf#1987-2659 – us ATLA [240]

The divinity of our lord and saviour jesus christ : eight lectures. preached before the university of oxford... / Liddon, Henry Parry – 8th ed. London: Rivingtons, 1878 – (= ser Bampton lectures) – 2mf – 9 – 0-524-08638-9 – mf#1993-2098 – us ATLA [240]

El divino amor / Willemenot, Luis G – Jerez de los Caballeros: talleres tipograficos horizonte, 1957 – sp Bibl Santa Ana [946]

Divino...garcia golfin de carvajal. manifiesto / Olivas y Frances, Francisco – 1744 – 9 – sp Bibl Santa Ana [240]

Les divins herauts de la penitence du monde : ou avis de saison par forme de remonstrance et d'exhortations, addrese... / Labadie, Jean de – Amsterdam, 1667 – 3mf – 9 – mf#PPE-164 – ne IDC [240]

Divisao territorial do brasil / Conselho Nacional De Estatistica – Rio de Janeiro, Brazil. 1951 – 1r – us UF Libraries [972]

Division bipartita de la provincia franciscana de san miguel de extremadura / Barrado Manzano, Arcangel – Separata del Archivo Ibero-Americano. 19, Julio-Septiembre 1959 – 1 – sp Bibl Santa Ana [946]

Division courts and small credits / Armour, Edward Douglas – Toronto: Rose-Belford, 1879 – 1mf – 9 – mf#10226 – cn Canadiana [347]

Division i-a football recruiting violations reported by the national collegiate athletic association from 1980 through 19[9]6 / Danna, Joseph G – 1998 – 2mf – 9 – $8.00 – mf#PE 3994 – us Kinesology [790]

Division in the protestant house / Hoge, Dean R – Princeton: Princeton Theo. Sem., 1976 – 1r – 1 – 0-8370-0593-0 – mf#1984-T006 – us ATLA [242]

Division lists / Great Britain. Parliament. House of Commons – 1880-92 – 1 – us CRL [324]

Division lists, 1836-1909 : the results of every vote taken in the house of commons – [mf ed Chadwyck-Healey, 1981] – (= ser House of commons parliamentary papers, 1801-1900) – 517mf – 9 – (incl ind for 1836-75) – uk Chadwyck [324]

Division politico-administrativa de colombia / Colombia Departamento Administrativo Nacional De... – Bogota, Colombia. 1954 – 1r – us UF Libraries [972]

Divisions in the society of friends / Speakman, Thomas Henry – Philadelphia: Lippincott, 1869 – 1mf – 9 – 0-524-01396-9 – mf#1990-4092 – us ATLA [240]

Division-violist *see* Chelys, minuritionum artificio exornata

Div'n 10 voice / National Federation of Telephone Workers – v5 n9-v11 n3 [1945 oct-1951 mar] – 1r – 1 – mf#1055612 – us WHS [380]

Divo hieronymo sacrarum divae annae sacrarum – 1512 – (= ser Mssa) – 1mf – 9 – €20.00 – mfchl 77 – gw Fischer [780]

Le divorce et la separation de corps / Fremont, Joseph – Quebec: A Cote, 1886 – 3mf – 9 – mf#03255 – cn Canadiana [346]

Divorce law in ohio / May, Geoffrey – Baltimore: Johns Hopkins Press, 1932. 76p. LL-1391 – 1 – us L of C Photodup [348]

The divorce of catherine of aragon : the story as told by the imperial ambassadors resident at the court of henry 8 / Froude, James Anthony – New York: Scribner, 1891 – 2mf – 9 – 0-7905-5329-5 – mf#1988-1329 – us ATLA [941]

The divorce question : the divorce laws of every state in the united states and possessions / Collins, William A – Chicago, Penn, 1926. 104p. LL-854 – 1 – us L of C Photodup [346]

Divorciarse en espana mercado negro y corrupcion / Aradillas Agudo, Antonio – Madrid: Angel Herrero Fernandez, 1977 – 1 – sp Bibl Santa Ana [306]

Divorcias? lagarto / Merchan Vargas, Regina – Toro: Imp. Siris, 1936 – 1 – sp Bibl Santa Ana [946]

Divorcio en el salvador / Lindo, Hugo – Salvador, El Salvador. 1959 – 1r – us UF Libraries [306]

Divorcio en espana / Aradillas Agudo, Antonio – Barcelona: Luis de Caralt, Editor, S.A., 1977 – 1 – sp Bibl Santa Ana [306]

El divorcio, estudio de legislacion comparada : espana y sudamerica / Guzman Gundian, Lucila – Santiago de Chile, 1936 – 222p – 1 – mf#LL-8011 – us L of C Photodup [306]

Divre eli'ezer / Weissblum, Lazar – New York, NY. 1911 – 1r – us UF Libraries [939]

Divre hefets / Schulman, Kalman – Vilna, Lithuania. 1891 – 1r – us UF Libraries [939]

Divre kohelet – Varsha, Poland. 1904 – 1r – us UF Libraries [939]

Divre sefer / Karlin, A – Tel-Aviv, Israel. 1952 – 1r – us UF Libraries [939]

Divre shalom ve-emet / Lipschitz, Jacob Lipmann – Varsha, Poland. 1884 – 1r – us UF Libraries [939]

Divre yeme ha-tsiyonut / Hazan, Lew – Jerusalem, Israel. 1952 – 1r – us UF Libraries [939]

Divre yeme 'olam / Schulman, Kalman – Vilna, Lithuania. v1-9. 1880-1884 – 2r – us UF Libraries [939]

Divre yitshak / Weisz, Issak – Mukachevo, Czechoslovakia. 1906 – 1r – us UF Libraries [939]

Divre yosef / Klain, Yosef – Lvov, Ukraine. 1893 – 1r – us UF Libraries [939]

Divtti – Margao, India. 1968-Oct 1971 – 5r – 1 – (konkani language) – us L of C Photodup [079]

Divulgacion historica francisco pizarro / Tena Fernandez, Juan – Trujillo: Tip. Sobrino de B. Pena, 1925 – 1 – sp Bibl Santa Ana [920]

Divulgacion martiana / Universidad De La Habana – Habana, Cuba. 1953 – 1r – us UF Libraries [972]

Divus thomas – Freiburg Schw., 1(1914)-24(1946) – 205mf – 9 – €391.00 – ne Slangenburg [073]

The divyavadana : a collection of early buddhist legends, now first edited from the nepalese sanskrit mss in cambridge and paris / ed by Neil, Robert Alexander & Cowell, Edward Byles – Cambridge: University Press, 1886 [mf ed 1995] – (= ser Yale coll) – x/712p – 1 – 0-524-09688-0 – mf#1995-0688 – us ATLA [280]

Diwakar, Ranganath Ramachandra *see*
- Glimpses of gandhiji
- Satyagraha
- Satyagraha in action
- The upanisads in story and dialogue

The diwan of zeb-un-nissa : the first fifty ghazals / Zeb-un-Nissa, Princess – London: John Murray, 1913 – (= ser Samp: indian books) – (rendered fr persian by magan lal and jessie duncan westbrook; with int and notes) – us CRL [490]

Dix annees d'apostolat au pundjab (indes anglaises) : mission confiee aux freres-mineurs capucins de belgique rapport de dix annees d'apostolat au pundjab / Pelckmans, Gottfried – Bruges: C Ryckbost-Monthaye, 1900 [mf ed 1995] – (= ser Yale coll) – 140p (ill) – 1 – 0-524-10041-1 – (in french) – mf#1995-1041 – us ATLA [241]

Dix annees de lutte pour la liberte, 1915-1925 / Sylvain, Georges – Port-Au-Prince, Haiti. v1-2. 195-? – 1r – us UF Libraries [972]

Dix ans a la cote d'ivoire / Clozel, Francois Joseph – Paris: Augustin Challamel, 1906 – 1 – us CRL [916]

Dix ans a la cour du roi louis philippe et souvenirs du tems de l'empire et de la restauration / Appert, Benjamin Nicolas Marie – Berlin 1846 [mf ed Hildesheim 1995-98] – 3v on 7mf – 9 – €140.00 – 3-487-26009-3 – gw Olms [944]

Dix ans au canada de 1840 a 1850 : histoire de l'etablissement du gouvernement responsable / Gerin-Lajoie, Antoine – Quebec: Demers, 1888 – 7mf – 9 – mf#03331 – cn Canadiana [971]

Dix ans de journalisme : melanges / Dunn, Oscar – Montreal: Duvernay & Dansereau, 1876 – 4mf – 9 – mf#02799 – cn Canadiana [305]

Dix ans de la vie de francois hotman (1563-1573) / Dareste, R – 1mf876. v25 (p 529-544) – (= ser Bulletin, societe de l'histoire du protestantisme) – 1mf – 9 – mf#PBU-435 – ne IDC [240]

Dix, Arthur *see* Geografia politica. barcelona, 1929

Le dix decembre – 15 avr 1849-18 juin 1850 – 1 – (cont by: le pouvoir [paris 19 juin 1850-15 janv 1851]) – fr ACRPP [073]

Dix, G *see* The theology of conformation in the relation to baptism

Dix, John Adams *see* Speech of hon john a dix of new york on the oregon question

Dix, Morgan *see*
- The authority of the church
- Blessing and ban from the cross of christ
- Christian education the remedy for the growing ungodliness of the times
- The gospel and philosophy
- Lectures on the calling of a christian woman
- Lectures on the first prayer book of king edward 6
- Lectures on the pantheistic idea of an impersonal-substance-deity
- Lectures on the two estates
- The sacramental system considered as the extension of the incarnation
- Sermons, doctrinal and practical
- The seven deadly sins
- Three guardians of supernatural religion

Dixiana baptist church. lexington county. south carolina : church records – 1969-72 – 1 – 5.00 – us Southern Baptist [242]

Dixie – Jacksonville, FL. 1910 dec 3-1916 – 4r – (gaps) – us UF Libraries [071]

Dixie baptist church : church minutes – Seiper, LA. 1919-84 – 1 – $76.86 – mf#6882 – us Southern Baptist [242]

Dixie county advocate – Cross City, FL. 1946 oct 10-1997 – 46r – (gaps) – us UF Libraries [071]

Dixie digest – 1971-73, 1974 jan-nov, 1975, 1976-79, 1980 jan-feb, 1983-90 – 13r – 1 – mf#703133 – us WHS [071]

Dixie, Florence C *see* Bei den patagoniern

Dixie philatelist / Southern Philatelic Association – v11 n1,3-v19 n4 [1977 jan, jul-1985 winter] – 1r – 1 – mf#1125125 – us WHS [760]

Dixon, Amzi Clarence *see*
- Evangelism old and new
- The god man
- Lights and shadows of american life
- The person and ministry of the holy spirit

Dixon, Benjamin Homer *see*
- The bible and the prayer book
- The lord's supper / the east in prayer

Dixon, Charles *see* Evolution without natural selection

Dixon county advocate – Ponca, NE: G L Weed & J J McCarthy (wkly) [mf ed 1917-28 (gaps)] – 4r – 1 – us NE Hist [071]

Dixon county leader – Ponca, NE: M B Cox, E H Wills. 20v. 1894-v20 n37. mar 13 1913 (wkly) [mf ed 1895-1913 (gaps)] – 2r – 1 – (merged with: northern nebraska journal to form: nebraska journal-leader. issues for -dec 3 1895 publ every tue at ponca and every fri at wakefield. issue for jun 4 1896 called v2 n50 but constitutes v2 n47) – us NE Hist [071]

[Dixon-] dixon tribune – CA. 1883- – 68r – 1 – $40800.00 (subs $120/y) – mf#BC02172 – us Library Micro [071]

Dixon, Frederick Augustus *see*
- The maire of st brieux
- Masque entitled "canadas sic welcome"
- The mayor
- Pipandor

Dixon, Frederick Eldon [comp] *see* The volunteer's active service manual

Dixon, G *see* A voyage round the world

Dixon, George *see*
- Further remarks on the voyages of john meares, esq
- Remarks on the voyages of john meares, esq

Dixon, H N *see* Studies in the bryology of new zealand

The dixon index – Dixon, NE: N B Ecker, 1891 (wkly) [mf ed v2 n9. jun 16 1892 filmed [1973]] – 1r – 1 – us NE Hist [071]

Dixon, James *see* Narrative of a voyage to new south wales, and van dieman's land, in the ship skelton, during the year 1820

Dixon, James Henry *see* Ballads and songs of the peasantry of england

Dixon journal – Dixon, NE: P J Burke, 1917-v25 n39. may 29 1942 (wkly) [mf ed 1920-42 (gaps)] – 5r – 1 – (absorbed: concord journal. publ in dixon ne, apr 7 1921-42) – us NE Hist [071]

Dixon, L *see* Halifax to the saskatchewan

Dixon line : the dixon gayer newsletter – v7 n2-v12 n5 [1969 dec-1975 jul/aug] – 1r – 1 – (cont: Reason [Fullerton ca]; webster quimmley reader) – mf#1532594 – us WHS [071]

Dixon, M C *see* Hungering and thirsting after righteousness

Dixon, Myles C *see*
- Letter addressed to the members of the methodist society
- Letter to a wesleyan-methodist local preacher on the subject of bap...

Dixon, Richard Watson *see*
- Close of the tenth century of the christian era
- The life of james dixon, d.d

Dixon, Robert *see* Realizing faith

Dixon, S F *see* Substituted liabilities

Dixon, Thomas *see* Observations on the management of the north dublin union

Dixon, William Hepworth *see* John howard and the prison-world of europe

Dixon-Gottschild, Brenda see Aesthetic standards in old time dancing in southwest virginia

Diyarbakr – (= ser Vilayet salnames) – 9 – (1286 [1869], 1300 [1883] 2mf $75; 1302 [1885] 6mf $90; 1308 [1891] 2mf $75; 1312 [1894] 4mf $60; 1316 [1898] def'a 15 3mf $75; 1317 [1899], 1319 [1901] def'a 18, 1321 [1903], 1323 [1905] 4mf $60) – us MEDOC [956]

Diyarbekir – Diyarbekir: Vilayet Matbaasi, 1869-? n319. 22 eyluel 1927 – (= ser O & t journals) – 1mf – 9 – $25.00 – us MEDOC [956]

Diyojen – Istanbul. 1-3 sene n1-183. 12 tesrinisani 1286-12 kanunievvel 1288 [26 nov 1869-12 jan 1872] – (= ser O & t journals) – 13mf – 9 – $210.00 – us MEDOC [956]

Dizionario dei pittori dal rinnovamento dello belle arti fial 1800 / Ticozzi, S – Milano, 1818. 2v – 10mf – 9 – mf#O-1081 – ne IDC [700]

...Dizionario della lingua italiana e nubiana / Carradori, Arcangelo – Uppsala, 1931 – 1 – us CRL [040]

Dizzionario italiano-tedesco (ael2/10) : neues teutsch-italienisches woerterbuch 1732 / Leys, Franz Jacob – 1732 [mf ed 1993] – (= ser Archiv der europaeischen lexikographie: woerterbuecher) – 63mf – 9 – €750.00 – 3-89131-084-6 – (handschrift aus der universitaetsbibliothek erlangen-nuernberg; int by laurent bray) – gw Fischer [040]

Djabatan Penerangan see Indonesia merdeka

Djagoeng : sing ndjawakake wasir reksawardaja, tjap2an kapindo / Sanif, S – Djakarta: Gunseikanbu Kokumin Tosyokyoku (Balai Poestaka), 2604 (B P n1513) – 25p 1mf – 9 – mf#SE-2002 mf151 – ne IDC [959]

Djakarta. Balai Poestaka see Copernicus atau rahasia-rahasia langit

Djakarta. Barisan Propaganda see Pemimpin bahasa nippon

Djakarta dispatches – Washington, 1959-1961 – 23mf – 9 – (missing: 1959 (may 4th, may 26th-oct, dec); 1960 (jan 4th, jan 18th, nov 4th, dec 14th, dec 28th); 1961, v2(2, 11, 18, 20)) – mf#SE-526 – ne IDC [959]

The djakarta times – Djakarta: Djakarta Times Foundation, sep 1966-sep 1972 – 15r – 1 – us CRL [079]

[Djalan jang haroes dilaloei oleh pegawai negeri] / Kataoka, S Kanridoo – (Djakarta?): Gunseikanbu Soomubu Zinzika (2604?) – 92p 2mf – 9 – (disoesoen oleh s kataoka, rikugun siseikan) – mf#SE-2002 mf88-89 – ne IDC [959]

Djalan rajat see Jajasan "indonesia baru"

Djamaloedin Bin Moh Rasad, B see Boeah pikiran

Djapen kabupatan surabaja – sidoardjo-modjokerto-djombang, djapen kotapradja surabaja-modjokerto / Madjalah bahtera – Malang, 1953-1956 – 7mf – 9 – (missing: 1953, v1(1-3, 6-end); 1954(1-4, 6-12)) – mf#SE-669 – ne IDC [950]

Djapen kotapradja – Pontianak, 1951-1963 – 4mf – 9 – (missing: 1952 v2(4-8)) – mf#SE-949 – ne IDC [950]

Djasa jang ta'diloepakan, tjetakan 2 – Djakarta: Gunseikanbu Kokumin Tosyokyoku (2603) – 40p 1mf – 9 – mf#SE-2002 mf31 – ne IDC [959]

Djawa : tijdschrift van het java instituut – Weltevreden etc 1921-41 – v1-21 on 252mf – 9 – (ind 1921-40) – mf#5e-86/1 – ne IDC [590]

Djawa see Tijdschrift van het java instituut

Djawa baroe – Djakarta: Djawa Shimbun Sha, jan 1 2603-aug 1 2605 – 65mf – 9 – mf#SE-2002 mf209-273 – ne IDC [079]

Djawa Sinbun Kai see Iboe dan anak

Djawa tengah dalam angka : central java (province) – Semarang, 1969-1971 – 36mf – 9 – (missing: 1970) – mf#SE-1377 – ne IDC [959]

Djawaban bupati, kepala daerah kabupaten ponorogo pada sidang pleno dprd-gr kabupaten ponorogo sekretariat daerah pemerintah – [Ponorogo], 1969 – 9mf – 9 – mf#SE-1906 – ne IDC [950]

Djawa-koena – Djakarta: Dioesahakan oleh Djawa Gunseikanbu, 2604 (159 no daftar 1458) – 2mf – 9 – mf#SE-2002 mf32-33 – ne IDC [959]

Djawatan Bimbingan dan Perbaikan Sosial, Bagian Penjuluhan see Penjuluh sosial

Djawatan bimbingan perawatan sosial, kementerian sosial republik indonesia / Sosiawan – Jogjakarta, [1950]-1956. v1-7 – 3mf – 9 – (missing: [1950], v1-[1953], v4(1-9, 12); [1954], v5; [1955], v6(1-3, 6-12) 1956, v7(1-5)) – mf#SE-864 – ne IDC [959]

Djawatan kebudajaan laporan perwakilan djawatan kebudajaan nusa tenggara singaradja / Indonesia – Singaradja, 1954 – 2mf – 9 – mf#SE-1667 – ne IDC [959]

Djawatan kebudajaan pusat dep pdr : budaya; madjalah bulanan kebudajaan – Jogjakarta, [1951]-1964 – 80mf – 9 – (missing: [1951]-1952(1-29); 1954(2); 1961(3)) – mf#SE-652 – ne IDC [959]

Djawatan koperasi pusat lampiran statistik pada buku tahunan / Indonesia – Djakarta, 1960-1961 – 3mf – 9 – mf#SE-687 – ne IDC [315]

Djawatan koperasi pusat laporan tahunan / Indonesia – Djakarta, 1960-1961 – 8mf – 9 – mf#SE-688 – ne IDC [959]

Djawatan pendidikan kedjuruan almanak / Indonesia – Djakarta, 1960 – 4mf – 9 – mf#SE-474 – ne IDC [959]

Djawatan pendidikan kedjuruan, kementerian pendidikan, pengadjaran dan kebudajaan / Warta Kedjuruan – Djakarta, 1957-1959. v1-2(4) – 8mf – 9 – mf#SE-5473 – ne IDC [959]

Djawatan pendidikan masjarakat, bahagian pemuda / IndonesiaIndonesia. Kementerian pendidikan, pengadjaran dan kebudajaan – Djakarta, 1951-1957 – 35mf – 9 – (missing: 1951 v1(1-5, 8-9); 1954 v4(11); 1955 v5(5-7); 1956 v6(2); 1957 v7(4)) – mf#SE-761 – ne IDC [959]

Djawatan pendidikan masjarakat report of the mass education department / Indonesia – Djakarta, 1953-1954 – 3mf – 9 – (missing: 1954) – mf#SE-475 – ne IDC [370]

Djawatan penerangan agama / Pedoman guru agama honorair – Djakarta, 1954-1955 – 3mf – 9 – (missing: 1954/1955(1, 3-4)) – mf#SE-395 – ne IDC [959]

Djawatan penerangan agama, bag penjuluh masjarakat agama dan kebudajaan : gema kebudajaan agama – Djakarta, 1954-1955 – 4mf – 9 – (missing: 1954(1-2)) – mf#SE-717 – ne IDC [959]

Djawatan Penerangan Agama, Departemen Agama see Penjuluh agama

Djawatan penerangan agama departemen agama : penuntun dengan lampiran / Indonesia – Djakarta. v1-17(2). 1947-1963 – 80mf – 9 – (missing: 1947(1, 3, 8-12); 1948, v2; 1949, v3; 1950, v4(1-12); 1951, v5(2-3, 10-12); 1952, v6(3-4, 7-12); 1953, v7(12); 1958, v12(9-11); 1959, v13(7); 1961, v15(11-12); 1962, v16(1-12)) – mf#SE-401 – ne IDC [959]

Djawatan Penerangan Angkatan Laut Republik Indonesia see Putera samudera

Djawatan penerangan central sumatra / Sumatera Tengah – Bykittingi, 1950-1957(1-165) – 16mf – 9 – (missing: 1950-1953, v1-4(1-99, 107-109, 111-114, 116-122); 1954(126-134); 1954-1956(139-153); 1956-1957(156-158); 1957(161-162)) – mf#SE-969 – ne IDC [950]

Djawatan Penerangan Daerah istimewa Jogjakarta see Jogjakarta, indonesia (city)

Djawatan penerangan kabupaten djatinegara / Madjalah pewartaan – Djakarta, 1950-1952 – 7mf – 9 – (missing: 1950-1951, v1-2(1-end); 1952, v3(3-4)) – mf#SE-558 – ne IDC [959]

Djawatan penerangan kabupaten kotapradja kediri : berita penerangan – Kediri, [1951]-1954 – 6mf – 9 – (missing: [1951], v1; 1952, v2(1-11, 13-end); 1953, v3(1, 4, 6-end); 1954, v4(1-4, 7-8)) – mf#SE-1354 – ne IDC [950]

Djawatan penerangan kabupaten lamongan / Suara Lamongan – Lamongan, [1950]-1954 – 2mf – 9 – (missing: [1950]-1952 v1-2(1-15, 17)) – mf#SE-967 – ne IDC [950]

Djawatan penerangan, kabupaten pasuruan – Pasuruan, 1950-1954 – 1mf – 9 – (missing: [1950]-1954 v1-5(1-11)) – mf#SE-950 – ne IDC [950]

Djawatan penerangan kebupaten malang / Madju – Malang, 1951-1954 – 1mf – 9 – (missing: 1951, v1; 1952, v2;1953, v3; 1954, v4(1-11)) – mf#SE-920 – ne IDC [950]

Djawatan penerangan kota-besar bandung – Bandung, 1949-1951 – 1mf – 9 – (missing: 1949/1950, v1; 1950/1951, v2(1-10)) – mf#SE-948 – ne IDC [959]

Djawatan penerangan kotapradja : kumandang kotapradja probolinggo – Probolinggo, 1951-1954 – 2mf – 9 – (missing: 1951; 1953; 1954(1-22, 24)) – mf#SE-557 – ne IDC [950]

Djawatan penerangan kotapradja djakarta raya / Madjalah kotapradja – Djakarta, 1950-1959 – 52mf – 9 – (missing: 1950, v1(1, 4, 11-37, 39-end); 1951, v2(3-5, 8, 9, 13-15, 17, 18, 22, 26); 1952, v3(6, 22, 23); 1954, v4(21, 22); 1955, v5(5, 9, 11-13); 1956, v4(4-10, 14-end); 1956, v7(1, 6-end); 1957, v8(1, 2, 5-8, 20); 1958, v9(2)) – mf#SE-912 – ne IDC [959]

Djawatan penerangan, propinsi djawa barat / Madjalah penerangan daerah – Bandung, 1950-1953 – 13mf – 9 – (missing: 1950/1951, v1(1-16, 19, 21, 22, 24-33); 1952, v2(46-56)) – mf#SE-916 – ne IDC [959]

Djawatan penerangan propinsi sumatera barat / Madju terus – Padang, 1961-1964(2) – 2mf – 9 – (missing: 1961(1-2, 4-end)-1964(1)) – mf#SE-1800 – ne IDC [950]

Djawatan penerangan propinsi sumatera barat / Sari warta – Padang, 1962-1964(6/8) – 5mf – 9 – (missing: 1962(1, 3-4, 7-9, 11-[12]); 1963; 1964(1-5)) – mf#SE-1920 – ne IDC [950]

Djawatan penerangan rakjat ichtisar minggoean – Malang, 1948-1949 – 18mf – 9 – (missing: 1948, v1(1-21, 34-38); 1949, v1(41); 1949, v2(2, 5)) – mf#SE-1469 – ne IDC [950]

Djawatan penerangan ri kabupaten modjokerto / Desa-madju – Modjokerto, nd – 1mf – 9 – mf#SE-1400 – ne IDC [950]

Djawatan penerangan ri propinsi maluku-bahagian pewartaan – Ambon, 1952-1954. v1-3(31) – 3mf – 9 – mf#SE-1855 – ne IDC [950]

Djawatan Penerangan Siaran Kotamadya Jogjakarta see Jogjakarta, indonesia (city)

Djawatan penerangan, siaran penerangan-daerah – Palembang, 1969-1971(15) – 13mf – 9 – (missing: 1969(1-18, 25); 1970(7, 14); 1970-1971 v3(17-26? nov-dec); 1971(3, 12)) – mf#SE-1932 – ne IDC [950]

Djawatan penerangan suara penerangan – Medan, [1950]-1951 – 1mf – 9 – (missing: [1950], v1; 1951, v2(1-4)) – mf#SE-932 – ne IDC [950]

Djawatan penerangan, sumatera tengah : dunia seminggu – Bukit Tinggi, 1950-1952 – 40mf – 9 – (missing: 1950, v1(2, 4, 9-17, 19, 21-52); 1951, v2(3-5, 9-11); 1952, v3(53-54, 56-61, 64-65, 71-74, 76-81, 87-96, 100)) – mf#SE-529 – ne IDC [950]

Djawatan penerangan warta penerangan – Palembang, 1969-1971. v1-3(81) – 14mf – 9 – (missing: 1969, v1(1-95); 1970, v2(34, 43, 44, 46, 53, 56-58); 1971, v3(60-61, 63, 69)) – mf#SE-1933 – ne IDC [950]

Djawatan perikanan darat/laut : berita perikanan – Djakarta, 1949-1962. v1-14 – 31mf – 9 – (several issues missing) – mf#SE-1355 – ne IDC [959]

Djawatan pertanian / Madjalah pertanian – Djakarta, 1950-1972. v1-20(1) – 110mf – 9 – (missing: 1950, v1(1-2, 4-12); 1951, v2(1); 1963, v14; 1964, v15(9-12); 1965, v16(1-6); 1965, v16(10-end)-1971, v19) – mf#SE-831 – ne IDC [959]

Djawatan ppk daerah istimewa jogjakarta / Sana budaja – Jogjakarta, 1955-1964 – 11mf – 9 – mf#SE-786 – ne IDC [959]

Djawatan transmigrasi pusat / Madjalah transmigrasi – Djakarta, 1951-1957 – 5mf – 9 – (missing: 1951, v1; 1952, v2(1-11); 1953, v3(1-6, 9-12); 1954, v4; 1955-1956, v5(1-3, 5-12); 1957, v6(1-6)) – mf#SE-832 – ne IDC [959]

Djazair see Al-djazair

Djem – Istanbul. n1-33. 15 kanunievvel 1927-2 agustos 1928 [15 dec 1927-2 aug 1928] – (= ser O & t journals) – 9mf – 9 – $150.00 – (cont by: cem) – us MEDOC [956]

Djembatan : madjalah resmi palang merah indonesia – Djakarta, [1950]-1956 – 18mf – 9 – (missing: [1950]-1952, v1-3(1-3); 1952, v3(8-9); 1953, v4(2-12); 1954, v5(1-2, 5, 10-12); 1955, v6(4-12); 1956, v7(1-6, 8-10)) – mf#SE-1413 – ne IDC [959]

Djeng soepiah / Sonja – Soerabaia: Tan's Drukkery, 1934 [mf ed 1998] – (= ser Penghidoepan 120) – 1r – 1 – (coll as pt of the colloquial malay collection. filmed with: poetri satrija dewi, atawa, resia madjapait / h s t) – mf#4440 – us Wisconsin U Libr [830]

Djevalikian, Raffi see The relationship between asymmetrical leg power and change of running direction

Dji hoe eng hiong = Erhu yingxiong – Soerabaia: Tan's Drukkery, [1933] [mf ed 1998] – (= ser Penghidoepan 104) – 1r – 1 – (trans of a chinese novel "erhu yingxiong" (the two tiger heroes) into indonesian; coll as pt of the colloquial malay collection; filmed with: nona olanda sebagi istri tionghoa / [njoo cheong seng]) – mf#10000 – us Wisconsin U Libr [830]

Djibouti and the horn of africa / Thompson, Virginia McLean – Stanford CA: Stanford UP 1968 [mf ed 1996] – 4mf – 9 – us Indiana Preservation [916]

Le djin [cin] – Istanbul, 1918-? Sahib-i Imtiyaz ve Mueduer-i Mes'ul: Ahmed Cemaleddin. n1. 16 mai 1918 – (= ser O & t journals) – 1mf – 9 – $25.00 – us MEDOC [956]

Djiwa : madjalah psikiatri indonesian psychiatric quarterly / Jajasan Kesehatan Djiwa "Dharmawangsa" – Kebajoran Baru, 1968-1972 – 30mf – 9 – mf#SE-1415 – ne IDC [610]

Djiwa 45 see Angkatan 45

Djiwa Islam see Poetjoek pimpinan gpii

Djoco sumantri & co, surabaja buku penundjuk telepon interlokal djawa-timur – Surabaja, 1957 – 10mf – 9 – mf#SE-614 – ne IDC [950]

Djojoadhiningrat, A see Der pressezustand in indonesien

Djojobojo – Kediri, 1945-1948 – 12mf – 9 – (missing: 1945, v1(1, 3, 4, 8-14); 1947, v2(5, 6, 9-end); 1948, v3(1-13, 16-19)) – mf#SE-879 – ne IDC [950]

Djumena, Sascha see Diskriminierung am arbeitsmarkt

Djursholms tidning – Stockholm, Sweden. 1895-1957 – 24r – 1 – sw Kungliga [078]

Djurusan antropologi, fakultas sastra, universitas indonesia : berita antropologi – Djakarta, 1969-1972 – 9mf – 9 – mf#SE-1348 – ne IDC [959]

Djurusan bahasa dan sastra sunda fakultas keguruan : bende rantjage – Bandung, 1968-1969 – 2mf – 9 – (missing: 1968(1)) – mf#SE-1347 – ne IDC [950]

Dlad : official newsletter of the department of local affairs and development – 1970 jun-1975 feb – 1r – 1 – (cont by: dlad newsletter: official publication) – mf#543974 – us WHS [350]

Dlad newsletter : official publication/ wisconsin department o local affairs and development – 1975 apr-1980 jul – 1r – 1 – (cont: dlad, official newsletter of the department of local affairs & development) – mf#359849 – us WHS [071]

Dlamini, Charles Robinson Mandlenkosi see Juridical analysis and critical evaluation of ilibolo in a changing zulu society

Dlia chego nuzhna selskokhoziaistvennaia i promyslovaia kooperatsiia i kak ee organizovat v derevne / Kisliakov, E N – Tula, 1922 – 75p 1mf – 9 – mf#COR-423 – ne IDC [335]

Dlia chego nuzhny sovety krestianskikh deputatov : partiia sotsialistov-revoliutsionerov / Bykhovskii, N I – 1917 – 13p 1mf – 9 – mf#RPP-219 – ne IDC [325]

Dlia roditelei i vospitatelei : pedagogicheskii listok – Spb., 1871-1885 – 105mf – 9 – (missing: 1881(2)) – mf#R-4156 – ne IDC [077]

Dlova, E S M see Umvuzo wesono

Dm itb & mpmitb – Bandung, 1968-1972 – 4mf – 9 – (missing: 1968, v1; 1969, v2(5, 7-end)-1971) – mf#SE-1375 – ne IDC [959]

Dmitri iwanowitsch : drama / Wilhelmi, Adolph – Leipzig: I M Gebhardt's Verlag, 1869 – 1r – 1 – us Wisconsin U Libr [820]

Dmitriev, M M see Pridneprovskii krai

Dmitriev, N A see Sbornik tsirkuliarov ministerstva finansov kazennym palatam, kaznacheistvam i podatnym inspektoram za 1865-1894 gg

Dmitrieva, L et al see Opisanie tiurkskikh rukopisei instituta narodov azii

Dmitriev-Mamonov, VA see
- Teoriia i praktika kommercheskogo banka
- Ukazatel' deistvuiushchikh v imperii aktsionernykh predpriiatii

Dmitriev-Sadovnikov, G M see Zemlia i volia [tobol'sk: 1918]

Dmitrievskii, A see Opisanie liturgicheskikh rukopisei, khraniashchikhsia v bibliotekakh pravoslavnogo vostoka

Dn nord sollentuna/vasby – Stockholm, Sweden. 1988 – 1 – sw Kungliga [078]

Dn nord solna/sundbyberg – Stockholm, Sweden. 1988 – 1 – sw Kungliga [078]

Dn nordost taby – Stockholm, Sweden. 1988 – 1 – sw Kungliga [078]

Dn nordvast jarfalla – Stockholm, Sweden. 1988 – 1 – sw Kungliga [078]

Dn nordvast vallingby – Stockholm, Sweden. 1988 – 1 – sw Kungliga [078]

Dn sodermalm – Stockholm, Sweden. jan-oct 1988 – 1 – sw Kungliga [078]

Dn stockholmssporten – Stockholm, Sweden. 1988 – 1 – sw Kungliga [078]

Dn syd farsta – Stockholm, Sweden. 1988 – 1 – sw Kungliga [078]

Dn syd haninge – Stockholm, Sweden. 1988 – 1 – sw Kungliga [078]

Dn sydost – Stockholm, Sweden. 1988 – 1 – sw Kungliga [078]

Dn sydvast – Stockholm, Sweden. 1988 – 1 – sw Kungliga [078]

Dna newsletter – 1968-75 – (= ser American indian periodicals... 1) – 6mf – 9 – $95.00 – us UPA [305]

Dna repair – Oxford, England 2002+ – 1,5,9 – ISSN: 1568-7864 – mf#42885 – us UMI ProQuest [574]

Dneprovskii, S P see
- Kak sostavit plan deiatelnosti selskogo obshchestva potrebitelei
- Primernaia reviziia i plan deiatelnosti selskogo kooperativa

Dnes – Sofia, Bulgaria. 10 apr 1940-15 feb, 9 aug-2 sep 1941; 26 dec 1942-28 mar 1944 – 1 – (in cyrillic. imperfect) – mf#mf.686.c – uk British Libr Newspaper [077]

Dnes – Sofia, Bulgaria.May 1941-Jun 1944 (scattered issues) – 4r – 1 – us L of C Photodup [949]

Dnesek – V Praze: [sn], 1946-48 [mf ed Bloomington IN: Indiana Uni Lib, Preservation Dept 1990] – 2r – 1 – us Indiana Preservation [073]

Dnevnik – Sofia, Bulgaria. 2 jul 1917-12 jul 1919; 4 dec 1939-13 jan, 10 apr 1940; 29 sep 1941; 6,7,10 nov 1942; 1 sep-25 dec 1943 – 1 – (in cyrillic. imperfect) – mf#mf.680 – uk British Libr Newspaper [077]

Dnevnik – Novi Sad, Yugoslavia. Jan 1953-1970 – 61r – 1 – us L of C Photodup [949]

Dnevnik – Sofia, Bulgaria. Oct 1942-1943 – 2r – 1 – us L of C Photodup [949]
Dnevnik artista *see* Teatralnyi, muzykalnyi i khudozhestvennyi zhurnal
Dnevnik gosudarstvennogo sekretaria a a polovtsova – M, 1966. 2v – 21mf – 9 – mf#REF-487 – ne IDC [332]
Dnevnik novostei, otnosiashchikhsia do prosveshcheniia i obshchezhitiia : dvukhnedelnaia gazeta – Spb., 1829-1831 – 38mf – 9 – mf#R-1526 – ne IDC [077]
Dnevnik pavlodarskogo otdeleniia osvedstepi – Pavlodar, Kazakhstan 1919 [mf ed Norman Ross] – 1r – 1 – mf#nrp-1366 – us UMI ProQuest [077]
Dnevnik pavlodarskogo otdeleniia osvedstepi – Pavlodar [Semipalat obl]: [s n] 1919 [1919 [sent]-] – (= ser Asn 1-3) – n3-30 [1919] [gaps] item 123, on reel n27 – 1 – mf#asn-1 123 – ne IDC [077]
Dnevnitzi / Bulgaria. Obiknoveno Narodno Subraniye – Sofia. Feb 10 1879-Mar 28 1943. Incomplete – 1 – us NY Public [324]
Dn.i. – Paris, France. 9 sep-28 oct 1928 – 1/4r – 1 – uk British Libr Newspaper [072]
Dni – Berlin puis Paris. oct 1922-juin 1933 – 1 – (in russian) – fr ACRPP [073]
Dnipro : official organ. / Ukrainian Orthodox Church of America – Trenton, NJ. 17 may 1924-march 1942 [wkly] – 4r – 1 – (lacking dec 1926-dec 1927) – uk British Libr Newspaper [243]
Dnog : divisao naval em operacoes de guerra / Maia, Prado – Rio de Janeiro, Brazil. 1961 – 1r – us UF Libraries [972]
Dnr – New York NY 1988+ – 1,5,9 – ISSN: 1041-1119 – mf#17275,02 – us UMI ProQuest [670]
Do american adults know how to exercise for a health benefit? / Krzewinski-Malone, Jeanette A – 1998 – 1mf – 9 – $4.00 – mf#HE 638 – us Kinesology [613]
Do' Cvan, Ma *see* Khyan a myui sa myae nay khyai chan kyan re samuin
Do dominio da uniao e dos estados / Octavio, Rodrigo – Sao Paulo, Brazil. 1924 – 1r – us UF Libraries [972]
Do it / Rubin, Jerry – New York, NY. 1970 – 1r – us UF Libraries [025]
Do it loud / Black Brigade – v1 n1 [1970 feb] – 1r – 1 – mf#721627 – us WHS [071]
Do it now : madison n o w chapter newsletter / National Organization for Women – 1973 apr-jun – 1r – 1 – (cont by: equality now [madison wi]) – mf#998824 – us WHS [305]
Do it now / National Organization for Women – 1971 mar-1977 nov – 1r – 1 – mf#639338 – us WHS [305]
Do missions pay? / Goodrich, Chauncey – Oberlin, Ohio: News Print Co, 1903 [mf ed 1993] – (= ser Green leaf series 1) – 1mf – 9 – 0-524-08366-5 – mf#1993-3066 – us ATLA [240]
The "do not file" file / ed by Theoharis, Athan – (= ser Federal bureau of investigation confidential files) – 2r – 1 – $350.00 – 1-55655-134-7 – (with p/g) – us UPA [322]
Do not go down, o sun! : poems / Vijayatunga, Jinadasa – Bombay: Hind Kitabs, 1946 – (= ser Samp: indian books) – us CRL [810]
Do not say / Horsburgh, J Heywood – London, England. 189- – 1r – us UF Libraries [240]
Do obha son e kabya lvat ka kvak mya / Obha Son, Do – Ran Kun: Prann thonn cu Chuirhaylac Sammata Mran ma nuin nam to, Yan kye mhu Van kri Thana, Anupanna U ci Thana 1979 [mf ed 1990] – [ill] 1r with other items – 1 – (in burmese) – mf#mf-10289 seam reel 160/2 [§] – us CRL [390]
Do rancho ao palacio / Motta, Othoniel – Sao Paulo, Brazil. 1941 – 1r – us UF Libraries [972]
Do sentimento nacionalista na poesia brasileira / Almeida, Guilherme de – Sao Paulo, Brazil. 1926 – 1r – us UF Libraries [972]
Do, Thi Khoi Nguyen *see*
– Tan cung noi nho
– Tinh lanh
– Tinh yeu oi xin chao mi
"Do this in remembrance of me," should it be, "offer this"? / Abbott, Thomas Kingsmill – London: Longmans, Green; Dublin: Hodges, Figgis 1898 [mf ed 1989] – 1mf – 9 – 0-7905-3061-9 – mf#1987-3061 – us ATLA [220]
Do we believe? : a record of a great correspondence in "the daily telegraph," oct, nov, dec 1904 – London: Hodder & Stoughton 1905 [mf ed 1991] – 1mf – 9 – 0-7905-7721-6 – mf#1989-0946 – us ATLA [210]
Do we need christ for communion with god? = Brauchen wir christum, um gemeinschaft mit gott zu erlangen? / Lemme, Ludwig – New York: Eaton & Mains; Cincinnati: Jennings & Graham c1908 [mf ed 1989] – (= ser Foreign religious series. 2nd series) – 1mf – 9 – 0-7905-1221-1 – (in english) – mf#1987-1221 – us ATLA [240]
Do you consider sterilisation? = Oorweeg u sterilisasie? / South Africa. Department of Health [Departement van Gesondheid] – [Pretoria: Dept of Health 1977?] [mf ed Pretoria, RSA: State Library [199-]] – 4p [ill] on 1r with other items – 5 – mf#op 06687 r24 – us CRL [360]
Do you go to the prayer-meeting? – Kelso, Scotland. 18– – 1r – us UF Libraries [240]
Doan, Frank Carleton *see* Religion and the modern mind
Doan, Thach Bien *see* Vi du ta yeu nhau
Doan truong tan thanh tan khao / Tran, Cuu Chan – Saigon: Trung Tam San Xuat Hoc Lieu 1974- [mf ed 1992] – on pt of 1r – 1 – mf#11052 r226 n4 – us Cornell [959]
Doane, A N *see*
– Anglo-saxon bibles and "the book of cerne"
– Anglo-saxon gospels
– Books of prayer and healing
– Deluxe and illuminated manuscripts
– Manuscripts of durham, ripon, and york
Doane, George Washington *see* A working church
Doane, Thomas William *see*
– Bible myths
– Bible myths and their parallels in other religions
Doane, William Croswell *see*
– Evidence, experience, influence
– The manifestations of the risen jesus
Doane, William Croswell et al *see* The church in the british isles
Dob Baer *see* Torat ha-magid mi-mezritsh ve-sihotav
Dobayashi, Yoichi *see* The hermeneutical problem of yahweh war in the book of joshua 1-12
Dobbek, Wilhelm *see*
– Die akte ludwig feuerbach
– Herder
– J G herders humanitaetsidee als ausdruck seines weltbildes und seiner persoenlichkeit
Dobbin, Orlando Thomas *see*
– Christophaneia
– Tentamen anti-straussianum
Dobbins files – 1981 jun-1987 sep – 1r – 1 – mf#1055635 – us WHS [071]
Dobbins, Frank Stockton *see* Error's chains
Dobbriner, Paul *see* "Eritis sicut deus"
Dobbs, Archibald Edward *see*
– Home rule
– Philosophy and popular morals in ancient greece
– Representative reform for ireland
Dobbs, R S *see* Reminiscences of life in mysore, south africa and burmah
Dobell, Bertram *see* Catalogue of a collection of privately printed books
Dobell, P *see* Sept annees en chine, nouvelles observations sur cet empire l'archipel indo-chinois, les philippines et les iles sandwich
Dobell, Peter *see* Travels in kamtchatka and siberia
Dobiasch, Josef *see*
– Jugend vor 1914
– Volk auf dem amboss
Dobiash-Rozhdestvenskaia, Olga Antonovna *see* La vie paroissiale en france au 13e siecle
Dobie, John Shedden *see* South african journal 1862-6
Dobie, Robert *see*
– Dobie vs the temporalities board in the superior court, montreal
– In the superior court, montreal
Dobie vs the temporalities board in the superior court, montreal : judgement by the honorable mr justice jette, 29th december, 1879 / Dobie, Robert – S.l: s.n, 1880? – 1mf – 9 – mf#32227 – cn Canadiana [242]
Dobkin, Eliahu *see* 'Aliyah Veha-Hatsalah Bi-Shenot Ha-Sho'ah
Doble acento / Florit, Eugenio – Habana, Cuba. 1937 – 1r – us UF Libraries [972]
Doble, G H *see*
– Ordinale exon, vol 4
– Pontificale lanaletense
Dobles, Fabian *see*
– Aguas turbias
– Burbuja en el limbo
– Historias de tata mundo
– Lenos vivientes
– Maiju
– Rescoldera
Dobles, Gonzalo *see* Raiz profunda
Dobles, Julieta *see* Peso vivo
Dobles Segreda, Luis *see* Provincia de heredia
Dobner, Felix Jakob *see* Prager gelehrte nachrichten
Dobri vesti = Good news / ed by Angelov, Vasil G – Bulgaria. v1-2, no 10, Sept. 1946-June 1948 – 1r – 1 – $17.28 – us Southern Baptist [242]
Dobrianskii, F N *see* Opanisie rukopisei vilenskoi publichnoi biblioteki, tserkovno-slavianskikh i russkikh
Dobrizhoffer, Martin *see* An account of the abiponesan equestrian people of paraguay
Dobroe slovo : nezavisimyi, vnepartiin organ nar tserkovno-obshchestv mysli / ed by Red'kin, I – Kurgan [Tobol gub]: Izd kom 1918 [1918 19 avg [1 sent]-] – (= ser Asn 1-3) – n1-11 [1918] item 124, on reel n27 – 1 – (lacks: n6]) – mf#asn-1 124 – ne IDC [077]
Dobrof, Rose *see* Journal of gerontological social work
Dobrogea noua – Constanta, Romania. 1962-81 – 28r – 1 – us L of C Photodup [949]
Dobrokhotov, N S *see* Polozhenie ob uchrezhdeniiakh melkogo kredita
Dobroklonskii, Aleksandr Pavlovich *see* Prep. feodor, ispovednik i igumen studiiskii
Dobroklonskii, S *see* Ukazatel traktatov i snoshenii rossii s 1462 po 1826
Dobroliubov, Nikolai Aleksandrovich *see* Temnoe tsarstvo
Dobrolovskii, I A *see* Vestnik man'chzurii
Dobrovol'skii, N T *see* Narodnaia gazeta [rostov n / d: 1919]
Dobrovol'skii, V *see* Statisticheskii spravochnik po vladimirskoi gubernii za 1923-1927 gg
Dobrovol'skii, V I [comp] *see* Statisticheskii ezhegodnik vladimirskoi gubernii, 1918-1922 gody
Dobrovsky, Josef *see*
– Glagolitica
– Institutiones linguae slavicae dialecti veteris
Dobrowolski, Augustinus *see* Paradisus eucharisticus
Dobrowsky, Jos *see* Boehmische literatur
Dobrowsky, Joseph *see*
– Boehmische und maehrische literatur
– Litterarisches magazin von boehmen und maehren fuer die jahre 1781-83
Dobruca sadasi – Koestence (Konstanza): Dobruca Muesuelman Te'emin-i Maarif Cemiyeti, 1910-19? Sahib-i Imtiyaz ve Mueduer-i Mes'ul: Sueleyman Abduelhamid. n19. 11 eyluel 1326 [1910] – (= ser O & t journals) – 1mf – 9 – $25.00 – us MEDOC [956]
Dobrudzhanska tribuna – Tolbukhin, Bulgaria. 1955-1969 – 10r – 1 – us L of C Photodup [949]
Dobrynin, Mikhail Kuz'mich *see* Protiv mekhanistov i eklektikov
Dobschtz, Ernst von *see* The eschatology of the gospels
Dobschuetz, Ernst von *see*
– Die akten der edessenischen bekenner gurjas, samonas und abibos
– The apostolic age
– Das apostolische zeitalter
– Christian life in the primitive church
– Christusbilder
– Das decretum gelasianum
– Das kerygma petri
– Ostern und pfingsten
– Probleme des apostolischen zeitalters
– Studien zur textkritik der vulgata
– Die thessalonicher-briefe
Dobschuetz, Ernst von et al *see* Geschichtliche studien
Dobsevage, Abraham Baer *see* Lo dubim ve-lo ya'ar
Dobson, Austin *see*
– Thomas bewick and his pupils
– William hogarth
Dobson Collet, Sophia *see* Keshub chunder sen's england visit
Dobson, George H *see*
– Modern transportation and atlantic express tracks
– Ocean routes and modern transportation
– A pamphlet compiled and issued under the auspices of the boards of trade of pictou and cape breton
Dobson, J P *see* Resurrection of the body
Dobson, John *see* Chronological annals of the war from its beginning to the present time
Dobson, Margaret Jane *see* Memoir of john dobson
Doc savage comics – New York NY 1940-43 – 1 – mf#6132 – us UMI ProQuest [740]
Doc savage magazine – New York NY 1933-44 – 1 – mf#6131 – us UMI ProQuest [071]
Docca, Emilio Fernandes De Souza *see* Convencao preliminar de paz de 1828
Doce codigos del estado soberano de cundinamarca / Cundinamarca (Colombia) – Paris, France. v1-3. 1877-1879 – 1r – us UF Libraries [972]
Doce gaviotas para una sola tierra / Cuevas, Juan Pablo – Ciudad Trujillo, Dominican Republic. 1959 – 1r – us UF Libraries [972]
Doce nos... / Sanchis Alventosa, Joaquin – Barcelona, 1965: Madrid: Graf. Calleja, 1966 – 1 – sp Bibl Santa Ana [946]
Doce poemas / Grupo Saker-Ti – Guatemala, 1950 – 1r – us UF Libraries [810]
Doche, Joseph Denis *see* Deux sentinelles
Dock and harbour authority – London, England 1927+ – 1,5,9 – ISSN: 0012-4419 – mf#1388 – us UMI ProQuest [380]
Le docker noir / Sembene, Ousmane – Paris, Editions Debresse [1956] – us CRL [074]
Docket / Coldwater. Kansas. Police Court – 1884-1915 – 1 – us Kansas [360]
Docket – v1 n1-v11 n1 [1968 aug-1978 fall] – 1r – 1 – (cont: reflector [topeka ks]) – mf#632526 – us WHS [071]
Docket / Dodge City. Kansas. Police Court – 1885-1906 – 1 – us Kansas [978]
The docket – Toronto: Docket Publ Co, [1889?-18– or 19–] – 9 – mf#P04253 – cn Canadiana [347]
The docket – v1-2 no 5. 1897-98 – 1mf – 9 – $4.50 – mf#LLMC 82-922 – us LLMC [340]
Dockets / Hays. Kansas. Police Court – 1912-41 – 1 – us Kansas [978]
Dockets of the supreme court of the united states, 1791-1950 / U.S. Supreme Court – (= ser Records of the supreme court of the united states) – 27r – 1 – (with printed guide) – mf#M216 – us Nat Archives [347]
Docking, Virginia *see* Letters received from prominent individuals
Docklands express – London, UK. 11 apr 1987-23 dec 1989; 1990-92 – 13r – 1 – uk British Libr Newspaper [072]
Dockter, Cindy R *see* The physiological responses to walking and stepping while wearing a weighted vest
Le docteur labrie : un bon patriote d'autrefois / Gosselin, Auguste – [Quebec?: Laflamme & Proulx], 1907 – 4mf – 9 – 0-665-74323-8 – mf#74323 – cn Canadiana [610]
Docteur robin / Premaray, Jules De – Paris, France. 1842 – 1r – us UF Libraries [440]
Doctor – Sutton, England 1971+ – 1,5,9 – ISSN: 0046-0451 – mf#8440 – us UMI ProQuest [610]
El doctor alem y el radicalismo / Castellanos, Joaquin – Buenos Aires, n.d – 1 – us CRL [335]
Doctor balthasar hubmaier und die anfaenge der wiedertaufe in maehren : aus gleichzeitigen quellen und mit benuetzung des wissenschaftlichen nachlasses des hofrathes dr. josef ritter v beck / Loserth, Johann – Bruenn: Verlag der Hist-Statist Section 1893 [mf ed 1990] – 1mf [ill] – 9 – 0-7905-5003-2 – (incl bibl ref) – mf#1988-1003 – us ATLA [242]
D[/octo/]r clifford james on wearing clothes in the tropics and hygiene *see* Correspondence with the government, 1926-1931 and with dr clifford james on clothes, 1931
El doctor d antonio bustamante bustillo pable fernandez : clerigo diacono domiciliaro de el arzibispado de mexico... / Bustamente Bustillo, Antonio – [Mexico: s.n. 1766?] – (= ser Books on religion...1543/44-c1800: iglesias, catedrales) – 1mf – 9 – mf#crl-447 – ne IDC [241]
Doctor, family physician & medical answers – London, UK. 1906. -irr. 8 feet – 1 – uk British Libr Newspaper [072]
Doctor fray jose joaquin escobar de los libertodor / Ramos Hidalgo, Nicolas – Cali, Colombia. 1934 – 1r – us UF Libraries [972]
Dr friedrich muenter's, professor der theologie an der universitaet zu kopenhagen, handbuch der aeltesten christlichen dogmen-geschichte = Haandbog i den aeldste christelige kirkes dogmehistorie / Muenter, Friedrich; ed by Ewers, Johann Philipp Gustav – Goettingen: Vandenhoeck-Ruprecht 1802-06 [mf ed 1992] – 2v in 3 on 3mf – 9 – 0-524-03655-1 – (incl bibl ref; german trans fr dutch) – mf#1990-1083 – us ATLA [240]
Doctor gion : a novel / Carossa, Hans – New York: R O Ballou [1933?] [mf ed 1989] – 1r – 1 – (trans fr german by agnes neill scott; filmed with: georg buchners drama dantons tod / hans landsberg) – mf#7143 – us Wisconsin U Libr [830]
Doctor johannes faust : puppenspiel in vier aufzuegen / Simrock, Karl Joseph – Frankfurt a/M: H L Broenner, 1846 [mf ed 1990] – 1r – 1 – (filmed with: fausto) – us Wisconsin U Libr [790]
Doctor kerkhoven / Wassermann, Jakob – New York: H Liveright, c1932 – 1r – 1 – us Wisconsin U Libr [830]
Doctor lee / Broomhall, Marshall – London: Morgan & Scott; Philadelphia: China Inland Mission [1908] [mf ed 1995] – (= ser Yale coll) – 61p (ill) – 1 – 0-524-10128-0 – (pref by walter b sloan) – mf#1995-1128 – us ATLA [920]
Doctor, M *see* Die philosophie des josef (ibn zaddik
Das doctor martinus kein adiaphorist gewesen ist / Amsdorff, N von – Magdeburg, 1550 – 1mf – 9 – mf#TH-1 mf 13 – ne IDC [242]
Doctor nye of north ostable / Lincoln, Joseph Crosby – New York, NY. 1923 – 1r – us UF Libraries [025]
A doctor of philosophy / Brady, Cyrus Townsend – Toronto: Langton & Hall, 1903 – 4mf – 9 – 0-665-71633-8 – mf#71633 – cn Canadiana [880]
Doctor of tanganyika / White, Paul Hamilton Hume – London, England. 1952 – 1r – us UF Libraries [960]

Doctor of tanganyika / White, Paul Hamilton Hume – Sydney: G.M. Dash, publisher's note 1943. 244p. ill – 1 – us Wisconsin U Libr [610]

Das doctor pomer vnd doctor maior mit iren adiaphoristen ergernis vnd zurtrennung angericht vnnd den kirchen christi vnueberwintlichen schaden gethan haben / Amsdorff, N von – [Magdeburg], 1551 – 1mf – 9 – mf#TH-1 mf 14 – ne IDC [242]

Dr robinson's voice in the wilderness – v1-3 n2,2. 1917-20 [all publ] – (= ser Radical periodicals in the united states, 1881-1960. series 1) – 5mf – 9 – $95.00 – us UPA [303]

Doctor Solemnis ("Exalted Teacher") *see* Summa theologica

Doctor syntax, his three tours in search of the picturesque, of consolation, of a wife / Combe, William – London: F. Warne & Co., (n.d.). 376p – 1 – us Wisconsin U Libr [830]

Doctor tucker, priest-musician : a sketch which concerns the doings and thinkings of the rev john ireland tucker / Knauff, Christopher Wilkinson – New York: A D F Randolph, 1897 – 1mf – 9 – 0-7905-4990-5 – mf#1988-0990 – us ATLA [780]

Doctor vanderkemp / Martin, Arthur Davis – Westminster: The Livingstone Press [1931] – (= ser [Travel descriptions from south africa, 1711-1938]) – 3mf – 9 – mf#zah-62 – ne IDC [916]

Doctor y general prospero pinzon / Penuela, Cayo Leonidas – Bogota, Colombia. 1941 – 1r – us UF Libraries [972]

Doctoris seraphici s bonaventurae prolegomena ad sacrum theologiam (fp30) / ed by Soiron, Th – 1932 – (= ser Florilegium patristicum (fp)) – €3.00 – ne Slangenburg [241]

Doctoris subtilis et mariani joannis duris scoti (ofm). opera omnia... / Barrado Manzano, Arcangel – Madrid: Arch. Ibero Americano, 1965 – 1 – sp Bibl Santa Ana [780]

Doctoris...joannis dius scoti...opera omnia / Barrado Manzano, Arcangel – Madrid: Graf. Calleja, 1966 – 1 – sp Bibl Santa Ana [780]

El doctor...medico de sevilla / Saavedra, J – Malaga, SA: S. 17 – 1mf – 9 – sp Cultura [610]

The doctor's daughter / MacGeorge, David – [Galt, Ont?: Jaffray Bros], 1905 – 2mf – 9 – 0-665-76361-1 – mf#76361 – cn Canadiana [830]

Doctors for Disaster Preparedness *see*
– Ddp arizona newsletter
– Triage!

Doctors for disaster preparedness newsletter – 1983 may – 1r – 1 – (cont by: triage!) – mf#1159928 – us WHS [610]

Doctrina capreoli de influxu dei in actus voluntatis humanae : secundum principia thomismi et molinismi collata / Ude, Ioanne – Graecii: "Styria", 1905 [mf ed 1986] – 1mf – 9 – 0-8370-7196-8 – (incl ind) – mf#1986-1196 – us ATLA [241]

Doctrina christiana en le[n]gua espanola y mexicana : hecha por los religiosos de la orden de s[an]cto domingo – corr & enl ed. Mexico: E[n] casa d[e] Jua[n] Pablos...Ano. d[e] 1550 anos – (= ser Books on religion... 1543/44-c1800: doctrina cristiana, obras de devocion) – 66leav on 4mf – 9 – mf#crl-35 – ne IDC [241]

Doctrina christiana en lengua espanola y mexicana / Domingo de la Anunciacion, Father – Mexico: En casa d[e] Pedro Ocbarte por ma[n]dado d[e]l illustrissimo y revere[n]dissimo senor Alo[n]se de Mo[n]tusar...1565 anos – (= ser Books on religion...1543/44-c1800: doctrina cristiana, obras de devocion) – 84lea on 2mf – 9 – mf#crl-36 – ne IDC [241]

Doctrina christiana muy cumplida : donde se contiene la exposicion de todo lo necessario para doctrinar alos yndios, y administralles los sanctos sacramentos / Juan de la Anunciacion, fray – en Mexico: En casa de Pedro Balli 1575 – (= ser Books on religion...1543/44-c1800: doctrina cristiana, obras de devocion) – 4mf – 9 – mf#crl-38 – ne IDC [241]

Doctrina christiana, o cartilla / Catholic Church. Province of Peru. Concilio Provincial – Impr e la ciudad de los reynes: Por Antonio Ricardo... ano de 1584 – (= ser Books on religion... 1543/44-c1800: evangelizacion) – 3mf – 9 – mf#crl-336 – ne IDC [241]

Doctrina christiana, y cathecismo en lengua mexicana / Molina, Alonso de – en Mexico: Por la Viuda de Francisco de Rivera Calderon... ano de 1744 – (= ser Books on religion... 1543/44-c1800: catecismos) – 1mf – 9 – mf#crl-12 – ne IDC [241]

Doctrina christiana, y platicas doctrinales / Anaya, Jose Lucas – [Mexico]: Impressas en la Imprenta del Real...ano de 1765 – (= ser Books on religion...1543/44-c1800: doctrina cristiana, obras de devocion) – 2mf – 9 – (trans by manuel aguirre) – mf#crl-50 – ne IDC [241]

Doctrina christianae religionis per aphorismos summatim descripta / Vitringa, C – Ed 6. Arnhemiae, 1761-89. 9v – 60mf – 9 – mf#PBA-422 – ne IDC [240]

Doctrina cristiana : mas cierta y v[er]dadera pa gere sin erudicio y letras: en q[ue] se co[n]tiene el catecismo o informacio[n] para indios co[n] todo lo principal y necessario q[ue] el [christ]iano debe saber y obrar / Zumarraga, Juan de – en Mexico:...fray Juan Zummaraga..., e[n] fin di ano d' mil quinie[n]ta y quaranta y seis – (= ser Books on religion...1543/44-c1800: doctrina cristiana, obras de devocion) – 100 dble lea on 3mf – 9 – mf#crl-33 – ne IDC [241]

Doctrina cristiana / Pedro De Cordoba – Ciudad Trujillo, Dominican Republic. 1945 – 1r – us UF Libraries [972]

Doctrina cristiana cantada y amenizada, dedicada a cortijos, escuelas rurales, barriadas, etc / Lopez de Sosoaga y Borinaga, Benigno – Badajoz: Imp. Comercial, 1968 – 1 – sp Bibl Santa Ana [240]

Doctrina de la iglesia sobre el derecho de ensenar / Marquez, Gabino – Madrid: Ed. Studium de Cultura, 1951 – 1 – sp Bibl Santa Ana [240]

Doctrina de praedestinatione / Tossanus, D – Hanoviae, 1609 – 2mf – 9 – mf#H-2500 – ne IDC [240]

Doctrina del estoico...epicteto...enchiridion / Sanchez de las Brozas, Francisco – 1612 – 9 – sp Bibl Santa Ana [180]

Doctrina duodecim apostolorum. barnabae epistula (fp1) / Klauser, Th – 1940 – (= ser Florilegium patristicum (fp)) – €5.00 – ne Slangenburg [227]

Doctrina iesu christi de lege mosaica ex oratione montana / Baumgarten, Michael – Berolini: Ludewicum Oehmigke, 1838 – 1mf – 9 – 0-7905-0859-1 – mf#1987-0859 – us ATLA [240]

La doctrina que...infalibidad de la razon.. / Romero de Castillay Perosso, Francisco – 1879 – 9 – sp Bibl Santa Ana [140]

Doctrina romanensium de invocatione sanctorum : being a brief enquiry into the principles that underlie the practice of the invocation of saints / Stewart, Hugh Fraser – London: SPCK; New York: ES Gorham, 1907 – 1mf – 9 – 0-7905-8915-X – (incl bibl ref) – mf#1989-2140 – us ATLA [240]

Doctrina...antonio gomez...diego gomez cornejo / Perez Villamil, Juan – 1776 – 9 – sp Bibl Santa Ana [890]

Doctrinae christianae par theoretica *see* An elementary course of biblical theology

Doctrinal de un heroe y hombre de estado / Mola, Emilio – [Bilbao: Editora nacional 1937] [mf ed 1977] – (= ser Blodgett coll) – 1mf – 9 – mf#w1054 – us Harvard [946]

The doctrinal differences which have agitated and divided the presbyterian church : or, old and new theology / Wood, James – enl ed. Philadelphia: Presbyterian Board of Pub, 1853 [mf ed 1992] – (= ser Presbyterian coll) – 1mf – 9 – 0-524-04285-3 – mf#1991-2069 – us ATLA [242]

Doctrinal errors and practical scandals of the english prayer book / Forbes, George Henry – Burntisland, England. 1863 – 1r – us UF Libraries [240]

A doctrinal instruction on the indulgences and masses for the dead : decreed by his holiness pope leo 13, for sunday, 30th september 1888 / Cleary, James Vincent – Montreal, Toronto: J A Sadlier, 1888? – 1mf – 9 – mf#04169 – cn Canadiana [241]

The doctrinal theology of the evangelical lutheran church = dogmatik der evangelisch-lutherischen kirche / Schmid, Heinrich – 5th ed. Philadelphia: United Lutheran Publication House, c1899 – 2mf – 9 – 0-7905-8877-3 – (in english) – mf#1989-2102 – us ATLA [242]

Doctrinal thoughts / Seely, Amos W – New York: Frank McElroy, 1861 [mf ed 1985] – 1mf – 9 – 0-8370-5361-7 – mf#1985-3361 – us ATLA [240]

Doctrinas / Colombia Superintendencia De Sociedades Anonimas – Bogota, Colombia. 1958 – 1r – us UF Libraries [972]

Doctrinas de la procuraduria general de la nacion / Escallon, Rafael – Bogota, Colombia. 1945 – 1r – us UF Libraries [972]

Las doctrinas politicas de eugenio maria de hostos / Elias de Tejada, Francisco – Madrid: Ediciones de Cultura Hispanica, 1949 – sp Bibl Santa Ana [320]

Las doctrinas politicas en portugal (edad media) / Elias de Tejada Spinola, Francisco – Madrid: Escelicer, S L, 1943 – sp Bibl Santa Ana [320]

Doctrinas sociales, superadas por... / Almendrallucas, Bernardo – Caceres: Tip. El Noticiero, s.a. (1950) – sp Bibl Santa Ana [946]

Doctrine and covenants and the future / Doxey, Roy Watkins – Salt Lake City, UT. 1954 – 1r – us UF Libraries [025]

Doctrine and deed : ...in 17 sermons preached in the broadway tabernacle... / Jefferson, Charles Edward – New York: Thomas Y Crowell, c1901 [mf ed 1990] – 1mf – 9 – 0-7905-7949-9 – mf#1989-1174 – us ATLA [242]

Doctrine and development : university sermons / Rashdall, Hastings – London: Methuen, 1898 – 1mf – 9 – 0-7905-8722-X – mf#1989-1947 – us ATLA [240]

Doctrine and doctrinal disruption : being an examination of the intellectual position of the church of england / Mallock, William Hurrell – London: Adam and Charles Black, 1900 – 1mf – 9 – 0-7905-8513-8 – mf#1989-1738 – us ATLA [241]

The doctrine and history of christian baptism / Rooke, Thomas George – London: Alexander & Shepheard, 1894 – 1mf – 9 – 0-524-07590-5 – mf#1991-3210 – us ATLA [242]

Doctrine and life / ed by Brokaw, George Lewis – Des Moines, IA: Christian Index, 1898 [mf ed 1993] – 2mf – 9 – 0-524-08261-8 – mf#1993-3016 – us ATLA [242]

Doctrine and life – Dublin, Ireland 1972+ – 1,5,9 – ISSN: 0012-446X – mf#7128 – us UMI ProQuest [240]

Doctrine and life : a study of some of the principal truths of the christian religion in their relation to christian experience / Stevens, George Barker – New York: Silver, Burdett, 1895 – 1mf – 9 – 0-8370-2898-1 – (include index) – mf#1985-0898 – us ATLA [240]

The doctrine and literature of the kabalah / Waite, Arthur Edward – London: Theosophical Pub Society, 1902 – 2mf – 9 – 0-524-07801-7 – mf#1991-0178 – us ATLA [210]

The doctrine and validity of the ministry and sacraments of the national church of scotland / Macleod, Donald – Edinburgh: W Blackwood 1903 [mf ed 1991] – (= ser The baird lecture 1903) – 1mf – 9 – 0-7905-7965-0 – mf#1989-1190 – us ATLA [242]

La doctrine ascetique de saint basile de cesaree / Humbertclaude, P – Paris, 1932 – 6mf – 8 – €14.00 – ne Slangenburg [240]

Doctrine chretienne en forme de lectures de piete : ou l'on expose les preuves de la religion, les dogmes de la foi, les regles de la morale, ce qui concerne les sacrements et la priere / Lhomond – Nouv ed, rev. Tours: Alfred Mame, 1868 – 1mf – 9 – 0-8370-6752-9 – mf#1986-0752 – us ATLA [240]

La doctrine curieuse des beaux esprits de ce temps... / Garasse, F – Paris, 1624 – 12mf – 9 – mf#CA-112 – ne IDC [240]

La doctrine de la creation dans l'ecole de chartres / Parent, J M – Paris, 1938 – 4mf – 8 – €11.00 – ne Slangenburg [210]

La doctrine de la matiere chez avicebron / Brunner, F – 1956 – 1mf – 8 – €3.00 – ne Slangenburg [100]

La doctrine de la redemption dans schleiermacher / Bonifas, Francois – Paris: Ch Meyrueis: Grassart, 1865 – 1mf – 9 – 0-7905-9139-1 – mf#1989-2364 – us ATLA [240]

Doctrine de la sainte cene *see* Collected works

La doctrine de la sainte cene : essai dogmatique / Lobstein, Paul – Lausanne: Impr Georges Bridel, 1889. Chicago: Dep of Photodup, U of Chicago Lib, 1975 (1r); Evanston: American Theol Lib Assoc, 1984 (1r) – 1 – 0-8370-0554-X – (incl bibl ref) – mf#1984-6057 – us ATLA [220]

La doctrine de l'expiation et son evolution historique *see* The doctrine of the atonement and ist historical evolution

Doctrine des fonctions mediatrices du sauveur *see* Collected works

La doctrine des fonctions mediatrices du sauveur / Lobstein, Paul – Paris: Librairie Fischbacher, 1891. Chicago: Dep of Photodup, U of Chicago Lib, 1975 (1r); Evanston: American Theol Lib Assoc, 1984 (1r) – (= ser HIS etudes christologiques) – 1 – 0-8370-0561-2 – (incl bibl ref) – mf#1984-6058 – us ATLA [240]

La doctrine des moeurs, tiree de la philosophie des stoiques : representee en cent tableaux... / Roy, M le, Sieur de Gomberville – Paris: A Soubron, 1681 – 5mf – 9 – mf#O-1590 – ne IDC [090]

Doctrine drago et la deuxieme conference de la paix / Leger, Abel-Nicolas – Port-Au-Prince, Haiti. 1915 – 1r – us UF Libraries [025]

La doctrine du sacrifice dans les braahmanas / Levi, Sylvain – Paris: Ernest Leroux, 1898 – (= ser Bibliotheque de l'ecole des hautes etudes) – 1mf – 9 – 0-524-01570-8 – (incl bibl ref) – mf#1990-2524 – us ATLA [280]

La doctrine du salut (doctrina salutis) : d'apres les commentaires de jean calvin sur le nouveau testament / Goumaz, Louis – Lausanne: Librairie Payot; Paris: Librairie Fischbacher, 1917 – 1r – 1 – 0-8370-1764-5 – mf#1984-T025 – us ATLA [242]

The doctrine of a future life : from a scriptural, philosophical, and scientific point of view / Strong, James – New York: Hunt & Eaton; Cincinnati: Cranston & Stowe, 1891 – 1mf – 9 – 0-7905-0294-1 – mf#1987-0294 – us ATLA [210]

The doctrine of a future life as contained in the old testament scriptures : a discourse... camborne, july 28th 1874...the 5th [fernley] lecture... / Geden, John Dury – 2nd ed. London: Wesleyan Conference Off 1877 [mf ed 1989] – 1mf – 9 – 0-7905-2710-3 – mf#1987-2710 – us ATLA [221]

The doctrine of a future state : in nine sermons / Humphry, William Gilson – London: John W Parker, 1850 – 1mf – 9 – (incl bibl ref) – mf#1987-0432 – us ATLA [240]

Doctrine of a particular providence / Wardlaw, Ralph – Glasgow, Scotland. 1819 – 1r – us UF Libraries [240]

Doctrine of an impersonal god in its effects on morality and religion / Martin, W Todd – Belfast, Northern Ireland. 1875 – 1r – us UF Libraries [240]

The doctrine of annihilation in the light of the gospel of love / Brown, James Baldwin – 2nd ed. London: Henry S King, 1875 – 1mf – 9 – 0-7905-1505-9 – mf#1987-1505 – us ATLA [220]

The doctrine of atonement / Mozley, John Kenneth – London: Duckworth 1915 [mf ed 1991] – (= ser Studies in theology) – 1mf – 9 – 0-7905-8859-5 – mf#1989-2084 – us ATLA [240]

Doctrine of baptisms / Dell, William – Manchester, England. 1844 – 1r – us UF Libraries [242]

The doctrine of baptisms : or, the washing of regeneration restored / Thurman, William Carr – Philadelphia: J Goodyear, 1867 – 1mf – 9 – 0-524-03746-9 – mf#1990-4851 – us ATLA [242]

The doctrine of cy pres as applied to charities / McGrath, Robert Hunter, jr – Philadelphia, Johnson, 1887. 74 p. LL-310 – 1 – us L of C Photodup [340]

The doctrine of descent and darwinism... / Schmidt, Eduard Oskar – [London] 1875 – (= ser 19th c evolution & creation) – 4mf – 9 – mf#1.1.9595 – uk Chadwyck [575]

The doctrine of development and conscience : considered in relation to the evidences of christianity and of the catholic system / Palmer, William – London: F & J Rivington, 1846 – 1mf – 9 – 0-7905-7539-6 – (incl bibl ref) – mf#1989-0764 – us ATLA [240]

Doctrine of divine immutability as god's constancy / Aben, Tersur Akuma – Grand Rapids MI: Calvin Theological Seminary, 2000 [mf ed 2001] – 1r – 1 – $130.00 – mf#D00000 – us ATLA [210]

The doctrine of divine love : or, outlines of the moral theology of the evangelical church = Die lehre von der heiligen liebe / Sartorius, Ernest – Edinburgh: T & T Clark 1884 [mf ed 1985] – (= ser Clark's foreign theological library. new series 18) – 1mf – 9 – 0-8370-5381-1 – (incl bibl ref; trans fr german by sophia taylor) – mf#1985-3381 – us ATLA [230]

Doctrine of election / Day, Edwin – [Toronto?: s.n.] 1873 [mf ed 1985] – (= ser n2) – 1mf – 9 – 0-665-28170-6 – (original issued in ser: the principles of the reformation) – mf#28170 – cn Canadiana [242]

The doctrine of election : and its connection with the general tenor of christianity / Erskine, Thomas – 2nd ed. Edinburgh: David Douglas, 1878 – 1mf – 9 – 0-8370-3723-9 – mf#1985-1723 – us ATLA [240]

The doctrine of election : neither derogatory to god, nor discouraging to man / Boardman, Henry Augustus – Philadelphia: Presbyterian Bd of Publ, 1860 – 1mf – 9 – 0-8370-2713-6 – mf#1985-0713 – us ATLA [240]

Doctrine of election considered with reference to the ministerial o... / Brereton, John – London, England. 1844 – 1r – us UF Libraries [240]

The doctrine of endless punishment / Shedd, William Greenough Thayer – New York: Charles Scribner, 1886, c1885 – 1mf – 9 – 0-8370-9821-1 – (incl bibl ref) – mf#1986-3821 – us ATLA [240]

The doctrine of equity / Adams, John Coleman – 5th American ed. Phila., Johnson, 1868. 811 p. LL-729 – 1 – us L of C Photodup [342]

The doctrine of eternal punishment refuted upon natural principles. the reign of a thousand years : or, kingdom of heaven on earth. to which is added, a lecture on the architectural structure of the universe / Brown, Harvey – Portsmouth: Printed for H Brown by the Republican Printing Co, 1868 – 1mf – 9 – 0-524-06388-5 – mf#1991-2510 – us ATLA [240]

Doctrine of forgiveness / Thom, John Hamilton – Liverpool, England. 1839 – 1r – us UF Libraries [240]

The doctrine of god / Hall, Francis Joseph – 2nd ed, rev throughout. Milwaukee, Wis: Young Churchman, 1905 – (= ser Theological Outlines) – 1mf – 9 – 0-7905-9215-0 – mf#1989-2440 – us ATLA [240]

The doctrine of god in the jewish apocryphal and apocalyptic literature / Wicks, Henry J – London: Hunter & Longhurst, 1915 – 1mf – 9 – 0-524-04816-9 – (incl bibl ref) – mf#1992-0236 – us ATLA [220]

The doctrine of grace in our apostolic fathers / Torrance, Thomas F – 1947 – 9 – $10.00 – us IRC [240]

The doctrine of grace in the apostolic fathers / Torrance, Thomas F – 1959 – 9 – $10.00 – us IRC [241]

The doctrine of hell / Walworth, Clarence Augustus & Burr, William Henry – New York: Catholic Publication Society, 1873 – 1mf – 9 – 0-8370-5826-0 – mf#1985-3826 – us ATLA [240]

The doctrine of holy baptism : with remarks on the rev. w. goode's effects of infant baptism / Wilberforce, Robert Isaac – 3rd ed. London: John Murray, 1850 – 1mf – 9 – 0-524-00118-9 – mf#1989-2818 – us ATLA [242]

The doctrine of holy scripture respecting the atonement / Crawford, Thomas Jackson – 2nd ed. Edinburgh: William Blackwood; New York: Scribner, Welford & Armstrong 1875 [mf ed 1989] – 2mf – 9 – 0-7905-0877-X – (incl ind) – mf#1987-0877 – us ATLA [220]

The doctrine of immortality : its essence, relativity, and present-day aspects / Thompson, John Day – London: Edwin Dalton, 1908 – (= ser Hartley Lecture) – 1mf – 9 – 0-7905-8604-5 – mf#1989-1829 – us ATLA [240]

Doctrine of immortality in its bearing on education / Harris, Joseph Hemington – Ramsgate, England. 1871 – 1r – us UF Libraries [240]

The doctrine of immortality in the odes of solomon / Harris, James Rendel – London: Hodder and Stoughton, [1912?] – (= ser Little books on religion) – 1mf – 9 – 0-7905-1888-0 – mf#1987-1888 – us ATLA [221]

The doctrine of inspiration : being an inquiry concerning the infallibility, inspiration, and authority of holy writ / Macnaught, John – 2nd rev corr ed. London: Longman, Brown, Green, and Longmans, 1857 – 1mf – 9 – 0-8370-9966-8 – (incl bibl ref) – mf#1986-3966 – us ATLA [220]

The doctrine of inspiration : an outline historical study / Hopkins, Theodore Weld – Rochester, NY: Printed for the author, 1881 – 1mf – 9 – 0-8370-3654-2 – (incl bibl ref and index) – mf#1985-1654 – us ATLA [220]

The doctrine of intention : with special reference to the validity of ordinations in the english church – London: Harrison, [1894?] – 1mf – 9 – 0-524-06610-8 – mf#1991-2665 – us ATLA [240]

The doctrine of intervention / Hodges, Henry G – Princeton: The Banner Press, 1915 – us CRL [321]

Doctrine of jehovah addressed to the parsis / Wilson, John – Bombay, India. 1839 – 1r – us UF Libraries [240]

Doctrine of jesus inseparable from the history of jesus / Wicksteed, Charles – London, England. 1848? – 1r – us UF Libraries [240]

The doctrine of justification / Loy, Matthias – 2nd ed, rev and enl. Columbus, O[hio]: Lutheran Book Concern, 1882 – 1mf – 9 – 0-7905-7908-1 – mf#1989-1133 – us ATLA [240]

The doctrine of justification : an outline of its history in the church, and of its exposition from scripture / Buchanan, James – Edinburgh: T & T Clark, 1867 – (= ser Cunningham Lectures) – 2mf – 9 – 0-7905-9250-9 – mf#1989-2475 – us ATLA [240]

Doctrine of justification briefly stated / Bird, John – London, England. 18– – 1r – us UF Libraries [240]

The doctrine of karman in jain philosophy / Glasenapp, Helmuth von; ed by Kapadia, Hiralal R – Bombay: Bai Vijibai Jivanlal Panalal Charity Fund, 1942 – (= ser Samp: indian books) – (trans fr original german by g barry gifford) – us CRL [280]

The doctrine of last things : contained in the new testament compared with the notions of the jews and the statements of church creeds / Davidson, Samuel – London: Kegan Paul, Trench, 1882 – 1mf – 9 – 0-8370-2837-X – mf#1985-0837 – us ATLA [220]

The doctrine of man : outline notes based on luthardt / Weidner, Revere Franklin – Chicago: Wartburg, c1912 – 1mf – 9 – 0-7905-9749-7 – (incl bibl ref) – mf#1989-1474 – us ATLA [240]

The doctrine of man and of the god-man / Hall, Francis Joseph – 2nd ed, rev throughout. Milwaukee, Wis: Young Churchman, 1915 – (= ser Theological Outlines) – 1mf – 9 – 0-7905-9216-9 – mf#1989-2441 – us ATLA [240]

The doctrine of maya in the philosophy of the vedanta / Shastri, Prabhu Dutt – London: Luzac, 1911 – 1mf – 9 – 0-524-01986-X – mf#1990-2777 – us ATLA [280]

The doctrine of merits in old rabbinical literature / Marmorstein, Arthur – London, 1920 – 4mf – 8 – €11.00 – ne Slangenburg [270]

The doctrine of modernism and its refutation / Godrycz, John A – Philadelphia: John Joseph McVey, 1908 – 1mf – 9 – 0-8370-8427-X – mf#1986-2427 – us ATLA [240]

The doctrine of original sin, or, the native state and character of man unfolded / Payne, George – 2nd ed. London: Jackson and Walford, 1854 – (= ser Congregational Lecture) – 1mf – 9 – 0-7905-3095-3 – mf#1987-3095 – us ATLA [240]

The doctrine of papal infallibility stated and vindicated : with an appendix on civil allegiance, and certain historical difficulties / Walsh, John – London, Ont?: Free Press, 1875 – 1mf – 9 – mf#37421 – cn Canadiana [241]

The doctrine of passive resistance / Ghose, Aurobindo – Calcutta: Arya Pub House, 1948 – (= ser Samp: indian books) – us CRL [180]

The doctrine of plenary inspiration : and the errors of m scherer of geneva / Gasparin, Agenor, comte de – Edinburgh: Johnstone & Hunter, 1852 [mf ed 1993] – 1mf – 9 – 0-524-06201-3 – (english trans fr french by john montgomery) – mf#1992-0839 – us ATLA [220]

The doctrine of prayer / Prideaux, John – New ed. Oxford: John Henry Parker, 1841 – 1mf – 9 – 0-524-00310-6 – mf#1989-3010 – us ATLA [240]

The doctrine of probation examined : with reference to current discussions / Emerson, George Homer – Boston: Universalist Publ House, 1883 [mf ed 1989] – 1mf – 9 – 0-7905-1656-X – mf#1987-1656 – us ATLA [240]

The doctrine of proximate cause and last clear change / Peck, Melville – Richmond, Va.: Peck 1914. 181p. LL-1153 – 1 – us L of C Photodup [340]

Doctrine of reception / Blakeney, Richard Paul – London, England. 1881 – 1r – us UF Libraries [240]

The doctrine of reprobation / Ross, Frederick Augustus – Nashville, TN: Cumberland Presbyterian Publ House, 1881 – 1mf – 9 – 0-8370-5274-2 – mf#1985-3274 – us ATLA [240]

The doctrine of retribution : philosophically considered in eight lectures / Jackson, William – 3rd ed. London: Hodder and Stoughton, 1885 – (= ser Bampton lectures) – 1mf – 9 – 0-7905-7782-8 – mf#1989-1007 – us ATLA [240]

The doctrine of sacred scripture : a critical, historical and dogmatic inquiry into the origin and nature of the old and new testaments / Ladd, George Trumbull – New York: Charles Scribner. 2v. 1883 – 4mf – 9 – 0-8370-5846-5 – (incl bibl ref, indexes) – mf#1985-3846 – us ATLA [220]

Doctrine of sacrifice, deduced from the scriptures : a series of sermons / Maurice, Frederick Denison – London: Macmillan, 1879 – 1r – 1 – mf#1984-B026 – us ATLA [220]

The doctrine of saint john : an essay in biblical theology / Lowrie, Walter – New York: Longmans, Green, 1899 – 1mf – 9 – 0-8370-4188-0 – mf#1985-2188 – us ATLA [225]

The doctrine of scripture concerning the holy ghost, in its relations to ministerial education / Williams, William R – (A discourse) – 1 – 5.00 – us Southern Baptist [242]

The doctrine of scripture in the theology of john calvin and francis turretin / Allison, Leon McDill – 1958 – 1r – 1 – 0-8370-1688-6 – mf#1984-6102 – us ATLA [242]

Doctrine of substitution / Webb-Peploe, Hanmer William – London, England. 1878/ – 1r – us UF Libraries [240]

The doctrine of the ages / Cameron, Robert – New York: Fleming H Revell, c1896 [mf ed 1985] – 1mf – 9 – 0-8370-2574-5 – mf#1985-0574 – us ATLA [230]

The doctrine of the apocalypse : and its relation to the doctrine of the gospel and epistles of john = Der lehrbegriff der apocalypse und seine verhaeltnisse zum lehrbegriff des evangelium und der episteln des johannes / Gebhardt, Hermann – Edinburgh: T & T Clark 1878 [mf ed 1985] – (= ser Clark's foreign theological library. new series 58) – 2mf – 9 – 0-8370-3236-9 – (trans fr german by john jefferson) – mf#1985-1236 – us ATLA [225]

Doctrine of the atonement : explained and advocated / Robinson, Edward – London, England. 1881? – 1r – us UF Libraries [240]

The doctrine of the atonement : deduced from scripture, and vindicated from misrepresentations and objections: six discourses / Macdonnell, John Cotter – London: Rivingtons; Dublin: Hodges, Smith 1858 [mf ed 1989] – (= ser The donnellan lectures 1857) – 1mf – 9 – 0-7905-1189-4 – (incl bibl ref) – mf#1987-1189 – us ATLA [240]

The doctrine of the atonement : a historical essay = Le dogme de la redemption / Riviere, Jean – London: Kegan Paul, Trench, Truebner 1909 [mf ed 1988] – (= ser The international catholic library 12,15) – 2v on 2mf – 9 – 0-7905-0217-8 – (trans by luigi cappadelta; incl ind) – mf#1987-0217 – us ATLA [241]

The doctrine of the atonement and ist historical evolution : and, religion and modern culture = La doctrine de l'expiation et son evolution historique [and] religion et la culture moderne / Sabatier, Auguste – New York: G P Putnam; London: Williams & Norgate 1904 [mf ed 1985] – 1mf – 9 – 0-8370-5268-8 – (trans fr french by victor leuliette; incl bibl ref) – mf#1985-3268 – us ATLA [240]

The doctrine of the atonement as taught by christ himself : or, the sayings of jesus on the atonement / Smeaton, George – Edinburgh: T & T Clark 1868 [mf ed 1989] – 2mf – 9 – 0-7905-2318-3 – (incl ind) – mf#1987-2318 – us ATLA [240]

The doctrine of the atonement as taught by the apostles : or, the sayings of the apostles exegetically expounded / Smeaton, George – Edinburgh: T & T Clark 1870 [mf ed 1989] – 2mf – 9 – 0-7905-3357-X – (incl bibl ref, ind & app) – mf#1987-3357 – us ATLA [225]

Doctrine of the atonement cleared from popular errors / Mac Donnell, John Cotter – Dublin, Ireland. 1856 – 1r – us UF Libraries [240]

Doctrine of the atonement to be taught without reserve / Townsend, George – London, England. 1838 – 1r – us UF Libraries [240]

The doctrine of the brethren defended : or, the faith and practice of the brethren proven by the gospel to be true / Miller, Robert Henry – Indianapolis: Print & Pub House Print 1876 [mf ed 1992] – 1mf – 9 – 0-524-03557-1 – mf#1990-4752 – us ATLA [242]

The doctrine of the buddha : the religion of reason / Grimm, George – Leipzig: Offizin W Drugulin, 1926 – (= ser Samp: indian books) – us CRL [280]

The doctrine of the cherubim : being an inquiry, critical, exegetical, and practical, into the symbolical character and design of the cherubic figures of holy scripture / Smith, George – London: Longman, Brown, Green & Longmans, 1850 [mf ed 1991] – 1mf – 9 – 0-524-00134-0 – mf#1989-2834 – us ATLA [220]

The doctrine of the church / Hall, Arthur Crawshay Alliston – Sewanee, TN: University Press at the University of the South, [1909?] – 1mf – 9 – 0-7905-1057-X – (includes bibliographies) – mf#1987-1057 – us ATLA [240]

The doctrine of the church : a historical monograph, with a full bibliography of the subject / McElhinney, John J – Philadelphia: Claxton, Remsen & Haffelfinger, 1871 – 2mf – 9 – 0-7905-2178-4 – (incl ind) – mf#1987-2178 – us ATLA [240]

The doctrine of the church : outline notes based on luthardt and krauth / Weidner, Revere Franklin – Chicago: FH Revell, c1903 – 1mf – 9 – 0-7905-8966-4 – mf#1989-2191 – us ATLA [240]

The doctrine of the church and of last things / Hall, Francis Joseph – 2nd ed, rev throughout. Milwaukee, Wis: Young Churchman, 1915 – (= ser Theological Outlines) – 1mf – 9 – 0-7905-9214-2 – mf#1989-2439 – us ATLA [240]

The doctrine of the church in scottish theology / MacPherson, John; ed by McCrie, Charles Greig – Edinburgh: Macniven & Wallace, 1903 – (= ser Chalmers Lectures) – 1mf – 9 – 0-524-00061-1 – mf#1989-2761 – us ATLA [240]

The doctrine of the church of england : as stated in ecclesiastical documents set forth by authority of church and state in the reformation period between 1536 and 1662 – London: Rivingtons, 1868 – 1mf – 9 – 0-8370-8729-5 – (incl ind) – mf#1986-2729 – us ATLA [241]

Doctrine of the church of england as contrasted with the church of... / Ellis, Brabazon – London, England. 1840 – 1r – us UF Libraries [241]

The doctrine of the communion of saints in the ancient church : a study in the history of dogma = Lehre von der gemeinschaft der heiligen im christlichen alterthum / Kirsch, Johann Peter – Edinburgh: Sands, [1910?] – 1mf – 9 – 0-7905-8673-8 – (incl bibl ref. in english) – mf#1989-1898 – us ATLA [240]

Doctrine of the cross of christ stated and improved – Edinburgh, Scotland. 18– – 1r – us UF Libraries [240]

The doctrine of the death of christ : in relation to the sin of man, the condemnation of the law, and the dominion of satan / Dimock, Nathaniel – 2nd ed, rev. London: E Stock, 1903 – 1mf – 9 – 0-7905-3825-3 – mf#1989-0318 – us ATLA [240]

The doctrine of the death of christ : in relation to the sin of man, the condemnation of the law, and the dominion of satan / Dimock, Nathaniel – London: E. Stock, 1903 – 1mf – us ATLA [240]

The doctrine of the divine fatherhood in relation to the atonement / Brown, James Baldwin – London: Ward [1860] [mf ed 1985] – 1mf – 9 – 0-8370-2869-8 – (incl bibl ref) – mf#1985-0869 – us ATLA [225]

Doctrine of the eternal sonship of christ considered / Martin, Robert – Oxford, England. 1821 – 1r – us UF Libraries [240]

The doctrine of the higher christian life : compared with the teaching of the holy scriptures / Hovey, Alvah – Boston: Henry A Young, c1876 – 1mf – 9 – 0-7905-7767-4 – mf#1989-0992 – us ATLA [240]

The doctrine of the holy spirit : the ninth series of the cunningham lectures / Smeaton, George – 2nd ed. Edinburgh: T & T Clark, 1889 – 1mf – 9 – 0-7905-0336-0 – (incl bibl ref and indexes) – mf#1987-0336 – us ATLA [210]

The doctrine of the holy spirit : or, philosophy of the divine operation in the redemption of man / Walker, James Barr – Chicago: Church and Goodman; New York: Sheldon, 1869 – (= ser Philosophy of the Plan of Salvation) – 1mf – 9 – 0-7905-8619-3 – mf#1989-1844 – us ATLA [240]

Doctrine of the key / Lee, Samuel – London, England. 1846 – 1r – us UF Libraries [240]

The doctrine of the key, or, sacerdotal binding and loosing : as taught in holy scripture, the fathers of the primitive church, and in the united church of great britain and ireland / Lee, Samuel – London: Seeley, Burnside, and Seeley, 1846 – 1mf – 9 – 0-7905-3082-1 – mf#1987-3082 – us ATLA [240]

The doctrine of the last things : jewish and christian / Oesterley, W O E – London: John Murray, 1908 – 1mf – 9 – 0-7905-1611-X – (incl ind) – mf#1987-1611 – us ATLA [220]

The doctrine of the lord's supper : its importance and necessity / Walther, Carl Ferdinand Wilhelm – Philadelphia: Lutheran Bookstore, 1872 – 1mf – 9 – 0-7905-8962-1 – mf#1989-2187 – us ATLA [240]

Doctrine of the millennium / Morison, John – London, England. 1829 – 1r – us UF Libraries [240]

The doctrine of the ministry : outline notes based on luthardt and krauth / Weidner, Revere Franklin – Chicago: Wartburg, c1907 – 1mf – 9 – 0-7905-8967-2 – (incl bibl ref) – mf#1989-2192 – us ATLA [240]

The doctrine of the ministry as taught by the dogmaticians of the lutheran church / Jacobs, Henry Eyster – Philadelphia: Lutheran Book Store, 1874 – 1mf – 9 – 0-7905-9002-6 – mf#1989-2227 – us ATLA [242]

The doctrine of the pastorate : or, the divine institution, religious responsibilities, and scriptural claims of the christian ministry. considered with special reference to wesleyan methodism / Smith, George – London: Printed for the author, 1851 – 1mf – 9 – 0-7905-6500-5 – mf#1988-2500 – us ATLA [242]

Doctrine of the person of christ : an historical sketch / Glover, Octavius – Cambridge: Deighton, Bell, 1867 [mf ed 1984] – 2mf – 9 – 0-8370-0945-6 – mf#1984-4300 – us ATLA [240]

The doctrine of the person of jesus christ / Mackintosh, Hugh Ross – [2nd ed] New York: Scribner, 1916 – (= ser The International Theological Library) – 2mf – 9 – 0-7905-9325-4 – (incl bibl ref) – mf#1989-2550 – us ATLA [240]

Doctrine of the priesthood – v1 n3 [1981 mar], v4 n1-2 [1987 jan-feb], v5 n5 [1988 jun] – 1r – 1 – mf#3362678 – us WHS [241]

The doctrine of the prophets / Kirkpatrick, Alexander Francis – 3rd ed. London, New York: Macmillan, 1901 – 2mf – 9 – 0-8370-9395-3 – (incl bibl ref and index) – mf#1986-3395 – us ATLA [221]

The doctrine of the resurrection of the body : as taught in holy scripture / Goulburn, Edward Meyrick – Oxford: Printed and published by J. Vincent, 1850. Chicago: Dep of Photodup, U of Chicago Lib, 1978 (1r); Evanston: American Theol Lib Assoc, 1984 (1r) – (= ser Bampton lectures) – 1 – 0-8370-1216-3 – (incl bibl ref) – mf#1984-T053 – us ATLA [240]

The doctrine of the sacraments in relation to the doctrines of grace : as contained in the scriptures, taught in our formularies, and upheld by our reformers / Dimock, Nathaniel – new ed. London; New York: Longmans, Green, 1908 [mf ed 1990] – 1mf – 9 – 0-7905-3826-1 – (1st printed 1871) – mf#1989-0319 – us ATLA [242]

The doctrine of the saints infirmities : delivered in severall sermons by john preston doctor in divinity, mr. of emanuel college in cambridge / Preston, John – London: Nich and Iohn Okes, 1637 – 3mf – 9 – mf#PW-21 – ne IDC [240]

The doctrine of the saint's perseverance, vindicated and established, 1820 / Tyler, Bennet – 1 – 5.00 – us Southern Baptist [242]

Doctrine of the second advent / Hooper, John – London, England. 1829 – 1r – us UF Libraries [240]

Doctrine of the trinity : the biblical evidence / Davies, Richard Newton – Cincinnati: Cranston & Stowe; New York: Hunt & Eaton, 1891 – 1mf – 9 – 0-7905-8779-3 – mf#1989-2004 – us ATLA [220]

Doctrine of the trinity / Harris, George – London, England. 1853 – 1r – us UF Libraries [240]

The doctrine of the trinity : apologetically considered / Illingworth, John Richardson – London:Macmillan, 1907 – 1mf – 9 – 0-8370-4502-9 – mf#1985-2502 – us ATLA [210]

Doctrine of the trinity founded neither on scripture, nor on reason / Drummond, William Hamilton – Belfast? Northern Ireland. 1827 – 1r – us UF Libraries [240]

The doctrine of the will : determined by an appeal to consciousness / Tappan, Henry Philip – New York: Wiley & Putnam, 1840 [mf ed 1989] – 1mf – 9 – 0-7905-2497-X – mf#1987-2497 – us ATLA [120]

The doctrine of the will, applied to moral agency and responsibility / Tappan, Henry Philip – New-York: Wiley and Putnam, 1841 – 1mf – 9 – 0-7905-0162-7 – (incl bibl ref) – mf#1987-0162 – us ATLA [170]

Doctrine of the word of god respecting union among christians / Noel, Baptist Wriothesley – London, England. 1844 – 1r – us UF Libraries [240]

Doctrine of tradition as maintained by the church of england / Pearson, George – Cambridge, England. 1837 – 1r – us UF Libraries [241]

Doctrine of tradition as maintained by the church of england / Pearson, George – Cambridge, England. 1838 – 1r – us UF Libraries [241]

Doctrine philosophique et religieuse de michel servet / Saisset, Emile – Paris: Au bureau de la Revue des deux mondes, 1848 – 1mf – 9 – 0-524-02601-7 – mf#1990-0653 – us ATLA [240]

Doctrine scolastique du droit de guerre / Vanderpol, Alfred – Paris, France. 1919 – 1r – us UF Libraries [025]

Doctrine spirituelle de saint augustin / Martin, Jules – Paris: Lethielleaux c1901 [mf ed 1992] – 1mf – 9 – 0-524-03905-4 – (incl bibl ref) – mf#1990-1164 – us ATLA [240]

Doctrine which drops as the rain, and the speech that distils as th... / Philpot, J C – Stamford, England. 1857? – 1r – us UF Libraries [240]

The doctrines and difficulties of the christian faith contemplated from the standing ground afforded by the catholic doctrine of the being of our lord jesus christ : being the hulsean lectures for the year 1855 / Goodwin, Harvey – Cambridge: Deighton, Bell; London: Bell and Daldy, 1856 – 1mf – 9 – 0-8370-9783-5 – (incl bibl ref) – mf#1986-3783 – us ATLA [240]

The doctrines and discipline of the canadian wesleyan methodist new connexion church – [Montreal?: s.n.], 1841 [mf ed 1987] – 1mf – 9 – 0-665-34332-9 – mf#34332 – cn Canadiana [242]

The doctrines and discipline of the canadian wesleyan methodist new connexion church : revised and approved by the annual conference held at toronto, 1853 – [Toronto?: s.n.], 1854 [mf ed 1982] – 2mf – 9 – mf#32388 – cn Canadiana [242]

The doctrines and discipline of the methodist episcopal church, 1884 : with an appendix / ed by Harris, William Logan – New York: Phillips & Hunt, c1884 – 1mf – 9 – 0-524-05375-8 – mf#1991-2281 – us ATLA [242]

The doctrines and discipline of the united evangelical church : formulated by the general conference of 1894, held in naperville, ill – Harrisburg, PA: Board of Publication of the United Evangelical Church, c1895 – 1mf – 9 – 0-524-06296-X – mf#1990-5225 – us ATLA [242]

The doctrines and dogmas of mormonism / Bays, Davis H – St Louis: Christian Pub Co, c1897 – 2mf – 9 – 0-524-06382-6 – mf#1991-2504 – us ATLA [243]

Doctrines and genius of the cumberland presbyterian church / Miller, Alfred Brashear – Nashville, Tenn: Cumberland Presbyterian Pub House, 1892 – 1mf – 9 – 0-524-04266-7 – mf#1991-2050 – us ATLA [242]

The doctrines and practices of the church of rome truly represented : in answer to a book entitled "a papist misrepresented and represented" / Stillingfleet, Edward – New ed, rev. Edinburgh: Johnstone and Hunter, 1851 – 1mf – 9 – 0-8370-8150-5 – (incl bibl ref) – mf#1986-2150 – us ATLA [240]

Doctrines and practices of the jesuits / Groves, Henry Charles – London, England. 1889 – 1r – us UF Libraries [241]

Doctrines, discipline, and mode of worship of the methodists, serio... / Vipond, W – Canterbury, England. 1807 – 1r – us UF Libraries [242]

Les doctrines modernistes : lettre encyclique de notre saint-pere le pape pie 10 a tous les evaeques de l'univers catholique = Pascendi dominici gregis – Rome: Typographie Vaticane, 1907 – 1mf – 9 – 0-8370-8009-6 – (in french) – mf#1986-2009 – us ATLA [240]

Doctrines of free and sovereign grace / Gowring, John William – Northwich, England. 1834? – 1r – us UF Libraries [240]

The doctrines of friends : or, principles of the christian religion, as held by the society of friends, commonly called quakers / Bates, Elisha – 11th ed. Providence: Knowles, Anthony, 1866 – 1mf – 9 – 0-8370-8884-4 – (incl ind) – mf#1986-2884 – us ATLA [220]

The doctrines of grace : and kindred themes / Bishop, George Sayles – New York: Gospel Pub House, 1910 [mf ed 1991] – 2mf – 9 – 0-7905-9359-9 – mf#1989-2584 – us ATLA [242]

The doctrines of grace / Maclaren, Ian – London: Hodder & Stoughton, 1900 [mf ed 1985] – 1mf – 9 – 0-8370-5707-8 – mf#1985-3707 – us ATLA [240]

The doctrines of our faith : a convenient handbook for use in normal classes, sacred literature courses and individual study / Dargan, Edwin Charles – Nashville, TN: Sunday School Board, Southern Baptist Convention, c1905 [mf ed 1985] – 1mf – 9 – 0-8370-3371-3 – mf#1985-1371 – us ATLA [242]

Doctrines of personal election / Clyde, Thomas – Dundee, Scotland. 1824 – 1r – us UF Libraries [240]

The doctrines of the bible developed in the facts of the bible : with an appendix containing a catechism on each section for the use of families, scripture classes and schools / Lewis, George – Edinburgh: Thomas Constable, 1854 – 1mf – 9 – 0-524-06519-5 – mf#1992-0903 – us ATLA [220]

The doctrines of the methodist episcopal church in america : as contained in the disciplines of said church from 1788 to 1808... / ed by Tigert, John James – Cincinnati: Jennings & Pye 1902 [mf ed 1991] – (= ser Little books on doctrine) – 1mf – 9 – 0-7905-9348-3 – (int by ed) – mf#1989-2573 – us ATLA [242]

Les doctrines romaines sur le liberalisme : envisagees dans leurs rapports avec le dogme chretien... / Ramiere, Henri – Paris: Lecoffre, 1870 – 1mf – 9 – 0-8370-7094-5 – mf#1986-1094 – us ATLA [240]

Documens pour servir a l'histoire de la captivite de napoleon bonaparte a sainte-helene : ou recueil de faits curieux sur la vie qu'il y menait, sur sa maladie, et sur sa mort – Paris 1821 [mf ed Hildesheim 1995-98] – 1v on 3mf [ill] – 9 – €90.00 – 3-487-26373-4 – gw Olms [944]

Documens relatifs au king's college : soumis a l'assemblee legislative par l'honorable m le procureur-general draper, d'apres l'ordre de son excellence le gouverneur-general, le 7 mai, 1846 = Documents respecting king's college, laid before the legislative assembly by the honorable mr attorney general draper, by command of his excellency the governor general, on the 7th may, 1846 – Montreal: impr par Lovell et Gibson, [1846?] (mf ed 1996) – 1mf – 9 – (trans by m myrand) – mf#SEM105P2760 – cn Bibl Nat [378]

Document / Committee to Frame a World Constitution – Chicago: The Committee. Description based on Doc. no.2, 16 Sept 1975 – 1 – us Wisconsin U Libr [320]

Document / European Coal and Steel Community. Common Assembly – Luxemburg. Jan 1953-1957 58. Incomplete – 1 – us NY Public [324]

Document catalog : catalog of the public documents of the congress and of all departments / U.S. Government Printing Office – Washington: GPO. 53rd-76th Congresses. v1-25. 1893-1940 (all publ) – 530mf – 9 – $795.00 – (superior to monthly catalog and document index for time frame duplicated) – mf#LLMC 81-405 – us LLMC [324]

Document from the honorable denis benjamin viger, dated 25th october, 1832 : communicated...saturday, 22d december, 1832: substance of a conversation with lord goderich, relative to the indictments laid before the grand jury at the criminal court of august and september... = Document de l'honorable denis benj viger, en date du 25 octobre 1832 – [S.l: s.n, 1832?] (mf ed 1991) – 1mf – 9 – (in english and french) – mf#SEM105P1376 – cn Bibl Nat [971]

Document image automation – Westport CT 1991-92 – 1,5,9 – (cont: optical information systems) – ISSN: 1054-9692 – mf#13007.05 – us UMI ProQuest [020]

Document series...1933-1936 / U.S. National Recovery Administration – (= ser Records of the national recovery administration) – 186r – 5 – (with printed guide) – mf#M213 – us Nat Archives [324]

Document world – Silver Spring MD 1996+ – 1,5,9 – (cont: imc journal) – ISSN: 1025-9228 – mf#23933 – us UMI ProQuest [000]

Documenta ad illustrandum concilium vaticanum anni 1870 / Vatican Council 1st – Noerdlingen: C H Beck, 1871 – 2mf – 9 – 0-8370-8257-9 – mf#1986-2257 – us ATLA [240]

Documenta ad pontificiam commissionem de re biblica spectantia / ed by Fonck, Leopold – Romae: Sumptibus Pontificii Instituti Biblici, 1915 – 1mf – 9 – 0-524-07282-5 – mf#1992-1059 – us ATLA [220]

Documenta historica : officii nocturni et matutini / Mateos, J – 2mf – 8 – €5.00 – ne Slangenburg [240]

Documenta mag. joannis hus : vitam, doctrinam, causam in constantiensi concilio actam et controversias de religione in bohemia, annis 1403-1418 motas illustrantia / ed by Palacky, Frantisek – Pragae: Sumptibus Friderici Tempsky, 1869 – 2mf – 9 – 0-524-00776-4 – mf#1990-0208 – us ATLA [240]

Documenta ophthalmologica – Dordrecht, Netherlands 1991+ – 1,5,9 – ISSN: 0012-4486 – mf#16780 – us UMI ProQuest [617]

Documentacao, 27 a 31 de julho de 1970 / Encontro De Brasilia, 1970 – Brasilia, Brazil. 1970 – 1r – us UF Libraries [972]

Documentacion celam / Consejo Episcopal Latinoamericano, Secretariado General – Bogota, Colombia: El Consejo. [v1-v8 n37/38. 1976-83] – 4r – us CRL [210]

Documentacion de la capellania y enterramiento del presidente don juan de ovando / Martinez Quesada, Juan – Badajoz: Imp. Dip. Provincial, 1958. Sep. REE – sp Bibl Santa Ana [320]

Documentacion historica de diego garcia de paredes / Munoz San Pedro, Miguel – Badajoz: Imp. de la Diputacion Prov., 1949. Sep. Rev. Est. Extremenos – 1 – sp Bibl Santa Ana [946]

Documentacion social catolica latinoamericana docla / Instituto Latinoamericano de Doctrina y Estudios Sociales – Santiago, Chile: El Instituto, [1974-]. [v1 n1-v20 n106. oct 1972-92] – 3r – 1 – us CRL [210]

Documentario do nordeste / Castro, Josue De – Rio de Janeiro, Brazil. 1937 – 1r – us UF Libraries [972]

Documentary annals of the reformed church of england : being a collection of injunctions, declarations, orders, articles of inquiry, etc from the year 1546 to the year 1716 / Cardwell, Edward – Oxford: University Press, 1844 [mf ed 1990] – 2v on 3mf – 9 – 0-7905-4666-3 – mf#1988-0666 – us ATLA [242]

Documentary history of the american committee on revision : prepared by order of the committee for the use of the members – New York: [s.n.] 1885 [mf ed 1986] – 1mf – 9 – 0-8370-9201-9 – mf#1986-3201 – us ATLA [220]

Documentary history of the basotho [lesotho] : ethnographic archives of the missionary david frederic ellenberger / Ellenberger, David Frederic – ca 10th-13th c-1854 – 60mf – 9 – €465.00 – (mainly in french, but also in sesotho & english; with p/g) – ne IDC [960]

The documentary history of the campaign on the niagara frontier in 1814 (pt 1) / ed by Cruikshank, Ernest Alexander – Niagara Falls, Ont?: The Society, 1896? – 3mf – 9 – mf#05281 – cn Canadiana [971]

The documentary history of the campaign on the niagara frontier in 1814 (pt 2) / ed by Cruikshank, Ernest Alexander – Niagara Falls, Ont?: The Society, 1897 – 4mf – 9 – mf#05282 – cn Canadiana [971]

The documentary history of the campaign upon the niagara frontier in the year 1812 (pt 3) / ed by Cruikshank, Ernest Alexander – Niagara Falls, Ont?: The Society, 1899 – 4mf – 9 – (incl ind) – mf#05283 – cn Canadiana [971]

The documentary history of the campaign upon the niagara frontier in the year 1812 (pt 4) / ed by Cruikshank, Ernest Alexander – Niagara Falls, Ont?: The Society, 1900 – 4mf – 9 – mf#05284 – cn Canadiana [971]

The documentary history of the campaign upon the niagara frontier in the year 1813, part 2 (1813), june to august, 1813 (pt 6) / ed by Cruikshank, Ernest Alexander – Niagara Falls, Ont?: The Society, 1903? – 4mf – 9 – mf#05286 – cn Canadiana [971]

The documentary history of the campaign upon the niagara frontier in the year 1813, part 3 (1813), august to october, 1813 (pt 7) / ed by Cruikshank, Ernest Alexander – Niagara Falls, Ont?: The Society, 1905 – 4mf – 9 – mf#05287 – cn Canadiana [971]

The documentary history of the campaign upon the niagara frontier in the year 1813, part 4 (1813), october to december, 1813 (pt 8) : with additional documents, june to october, 1813 / ed by Cruikshank, Ernest Alexander – Niagara Falls, Ont?: The Society, 1907 – 4mf – 9 – mf#05288 – cn Canadiana [971]

The documentary history of the campaign upon the niagara frontier in the year 1813, pt 1 (1813), january to june, 1813 (pt 5) / ed by Cruikshank, Ernest Alexander – Niagara Falls, Ont?: The Society, 1902 – 4mf – 9 – mf#05285 – cn Canadiana [971]

The documentary history of the campaigns upon the niagara frontier in 1812-4, vol 9 december, 1813 to may, 1814 (pt 9) / ed by Cruikshank, Ernest Alexander – Niagara Falls, Ont?: The Society, 1908 – 5mf – 9 – (incl ind) – mf#05289 – cn Canadiana [971]

Documentary history of the florida canal / Ship Canal Authority Of The State Of Florida – Washington, DC. 1936 – 1r – us UF Libraries [978]

Documentary history of the general council of the evangelical lutheran church in north america / Ochsenford, Solomon Erb – Philadelphia: General Council Publication House, 1912 – 2mf – 9 – 0-7905-8269-4 – mf#1988-6147 – us ATLA [242]

Documentary history of the protestant episcopal church in the united states of america. south carolina / ed by Hawks, Francis Lister & Perry, William Stevens – New-York: J Pott, 1862 – 1mf – 9 – 0-7905-7239-7 – mf#1988-3239 – us ATLA [242]

Documentatie over kolonische onderwerpen: indonesie (oost-indie), 1819-1933; en west indie, 1815-1929 = Documentation register to colonial subjects: east indies and west indies / Netherlands. General State Archives – 157mf – 9 – (indonesia (oost indie), 1819-1933.-142mf.dfl1540.00; west indie (west indies), 1815-1929.-15mf.dfl155.00) – ne MMF Publ [324]

Documentation des oblats de marie immaculee – Paris: OMI, 1943-1944, fasc 11 – (issues for 1943-1944, fasc 11 filmed with: petites annales des missionaires oblats de marie immaculee (1926), 1940-jan/feb 1942; and: du pole et tropiques; and: petites annales des missionaires oblats de marie immaculee (1926) (cont), dec 1944-jan 1953; and: petites annales, pole et tropiques, mar 1953-jul 1957) – us CRL [210]

Documentation et bibliotheques / Association pour l'avancement des sciences et des techniques de la documentation – Montreal: ASTED. v19 n1 mars 1973- [mf ed 1992-] – mf#SEM105P1708 – cn Bibl Nat [020]

Documentation et bibliotheques / Association pour l'avancement des sciences et des techniques de la documentation – Montreal: ASTED. v19 n1 mars 1973- (qrtly) [mf ed 1974-92] – 4r – 5 – mf#SEM16P221 – cn Bibl Nat [020]

Documentation historique pour nos etudiants / Laurent, Gerard Mentor – Port-Au-Prince, Haiti. 1959 – 1r – us UF Libraries [972]

Documentation newsletter / Cornell University – 1975 may-1981 spring – 1r – 1 – (cont: report of the curator and archivist, cornell university. collection of regional history and university archives; newsletter of the cornell program in oral history) – mf#625720 – us WHS [378]

Documentation of emergency period in india (jun 1975 till march 77) : limaye papers – [Bombay: microfilmed for South Asia Microform Project at Center for Research Libraries by Popular Prakashan, 1981] – 1 – us CRL [954]

Documentation register to colonial subjects : east and west indies, 1818-1933 – 157mf – 9 – €1025.00 set – (east indies [142mf] €930; west indies [15mf] €95) – mf#m110 – ne MMF Publ [900]

Documentation scientifique – n1-75. paris. fevr 1932-juil aout 1939 [bimnthly] – 1 – (revue bimestrielle des laboratoires et des industries chimiques) – fr ACRPP [540]

Documente der national-juedischen christglaeubigen bewegung in suedrussland – Erlangen: A Deichert, 1884 – (= ser Schriften des Institutum Judaicum in Leipzig) – 1mf – 9 – 0-7905-3824-5 – mf#1989-0317 – us ATLA [939]

Documenti di storia italiana – Firenze. v1-15 – 9 – $234.00 – mf#0186 – us Brook [945]

Documento – Mexico, DF: CRIE [n4-n146/147, n1 (sep 1982-dic 1997)] (irreg) – 1r – 1 – us CRL [230]

Documento inedito...manuel gomez / Venegas, Francisco Javier – 1886 – 9 – (ed 1 1888) – sp Bibl Santa Ana [946]

Documento y la reconstruccion historica / Chacon Y Calvo, Jose Maria – Habana, Cuba. 1929 – 1r – us UF Libraries [972]

Documentos / Bolivar, Simon – Habana, Cuba. 1996 – 1r – us UF Libraries [972]

Documentos autografos e ineditos / Venegas, Francisco Javier – Sevilla: Imp. E. Rasco, 1888 – 1 – sp Bibl Santa Ana [946]

Documentos autografos e ineditos / Venegas, Francisco Javier – Sevilla: Sociedad del Archivo Hispalense, 1886 – 1 – sp Bibl Santa Ana [946]

Documentos de 1584 a 1595, relativos a don luis zapata de chaves, existentes en el archivo municipal de llerena / Carrasco, Antonio – Badajoz: Dip. Provincial, 1969 – sp Bibl Santa Ana [946]

Documentos de carlos 5 (anno 1529) – Caceres – 1r – 5,6 – sp Cultura [946]

Documentos de la compania de jesus en el archivo historico nacional / Guglieri, A – Madrid; 1967 – 10mf – 9 – sp Cultura [240]

Documentos de la curia romana (anno 1238-1517) – Zamora – 1r – 5,6 – sp Cultura [240]

Documentos de la union centroamericana / Herrarte, Alberto – Guatemala City, 1957 – 1r – us UF Libraries [972]

Documentos de las fundaciones religiosas y beneficas de la villa de almonte, por lorenzo cruz de fuentes / T'Serclaes, Duque de – Madrid: Fortanet, 1913. B.R.A.H. 63, pp. 162-164 – sp Bibl Santa Ana [240]

Documentos del cabildo (anno 1045-1600) – Calahorra – 1r – 5,6 – sp Cultura [946]

Documentos del cabildo (anno 1276-1312) – Calahorra – 1r – 5,6 – sp Cultura [946]

Documentos del cabildo (siecle 13-15) – Calahorra – 1r – 5,6 – sp Cultura [946]

Documentos del cabildo (siecle 13-16) – Calahorra – 1r – 5,6 – sp Cultura [946]

Documentos del cabildo (siecle 1312-1436) – Calahorra – 1r – 5,6 – sp Cultura [946]

Documentos episcopales (anno 1199-1260) – Zamora – 1r – 5,6 – sp Cultura [240]

Documentos gallegos de los siglos siii al svi / Martinez Salazar, Andres – Coruna, Spain. 1911 – 1r – us UF Libraries [025]

Documentos historicos / El Salvador Asamblea Nacional Constituyente, 1950 – San Salvador, El Salvador. 1950-51 – 1r – us UF Libraries [972]

Documentos historicos coleccionados po... seccion geografia... / Grenon, P; ed by Bayle, Constantino – Madrid: Razon y Fe, 1928 – 9 – sp Bibl Santa Ana [900]

Documentos historicos referentes a extremadura / Archivo Extremeno – Badajoz: Tip.y Lib.de Antonio Arqueros, Tomo 1. 1908 – 1 – sp Bibl Santa Ana [946]

Documentos historicos sobre las persecuciones sufridas por la masoneria en espana / Gallardo y Victor, Manuel – Cadiz: J Benitez Estudillo 1888 – 1mf – 9 – mf#vrl-105 – ne IDC [366]

Documentos historicos. testamento de don bartolome martinez, obispo de panama y arzobispo de santa fe / Rodriguez Amaya, Esteban – Badajoz: Dip. Provincial, 1948. Sep. REE – 1 – sp Bibl Santa Ana [920]

Documentos ineditos o muy raros para la historia de mexico – Mexico. 1905-11. 36v – 1 – 529.00 – us L of C Photodup [972]

Documentos internacionales referentes al reconocim... / Cuba Departamento De Estado – Habana, Cuba. 1904 – 1r – us UF Libraries [972]

Documentos. lo que han visto en madrid los parlamentarios ingleses / Spain. Ministerio de Estado – Valencia, 1936? Fiche W1188. (Blodgett Collection of Spanish Civil War Pamphlets) – 9 – us Harvard [946]

Documentos militares / Uribe Uribe, Rafael – San Cristobal, Venezuela. 1901 – 1r – us UF Libraries [355]

Documentos para la bibliografia de d manuel jose quintana / Perez de Guzman, Juan – Madrid: Fortanet, 1910. B.R.A.H. 57, 1910. pp. 376-381 – sp Bibl Santa Ana [010]

Documentos para la historia argentina : tomo 20, iglesia... / Bayle, Constantino – Buenos Aires, 1929; Madrid: Razon y Fe, 1931 – 1 – sp Bibl Santa Ana [240]

Documentos para la historia argentina. tomo 20, iglesia. cartas antiguas de la provincia del paraguay, avila y tucuman de la compania de jesus (1609-1614). buenos aires, 1927 / Bayle, Constantino – Madrid: Razon y Fe, 1929 – 1 – sp Bibl Santa Ana [946]

Documentos para la historia de la guerra de sucesion en extremadura / Munoz de San Pedro, Miguel – Badajoz: Diputacion Prov. de Badajoz, 1948. Sep. Revista Estudios Extremanos – 1 – sp Bibl Santa Ana [946]

Documentos para la historia de la vida publica del libertador de colombia, peru, y bolivia / Blanco, Jose Felix – Caracas. 1875-78. 14v – 1 – $161.00 – us L of C Photodup [972]

Documentos particulares y pontificos (siecle 13-18) – Albarracin – 1r – 5,6 – sp Cultura [240]

Documentos pontificios (anno 1210-1616) – Zamora – 1r – 5,6 – sp Cultura [240]

Documentos pontificios (anno 1468-1572) – Zamora – 1r – 5,6 – sp Cultura [240]

Documentos pro-constitucion, 1899-1926 – San Salvador, El Salvador. 1926 – 1r – us UF Libraries [972]

Documentos reales (anno 1260) – Avila – 1r – 5,6 – sp Cultura [946]

Documentos reales (anno 1520-1558) : no 1-185 – Avila – 1r – 5,6 – sp Cultura [946]

Documentos reales (anno 1558-1584) : no 185-215 – Avila – 1r – 5,6 – sp Cultura [946]

Documentos reales (anno 1636-1653) : no 1-7 – Avila – 1r – 5,6 – sp Cultura [946]

Documentos reales, eclesiasticos t particulares en pergamino (siecle 11-17) – Valencia – 1r – 5,6 – sp Cultura [240]

Documentos reales. ejecutorias...(siecle 15-16) : no 1-30 – Avila – 1r – 5,6 – sp Cultura [946]

Documentos reales (siecle 15-16) : no 1-264 – Avila – 1r – 5,6 – sp Cultura [946]

Documentos reales (siecle 16-17) – Avila – 1r – 5,6 – sp Cultura [946]

Documentos referentes a la creacion de bolivia... / Lecuna, Vicente – Madrid: Razon y Fe, 1926 – 1 – sp Bibl Santa Ana [972]

Documentos referentes a la familia topete (anno 1292-1613) – Caceres – 1r – 5,6 – sp Cultura [920]

Documentos relacionados con la recuncia del presid... / Colombia President Lopez – Bogota, Colombia. 1945 – 1r – us UF Libraries [972]

Documentos relativos a la controversia / Costa Rica Ministerio De Relaciones Exteriores – San Jose, Costa Rica. 1909 – 1r – us UF Libraries [972]

Documentos relativos a la guerra nacional de 1856 / Costa Rica – San Jose, Costa Rica. 1914 – 1r – us UF Libraries [972]

Documentos relativos a la independencia / Costa Rica Archivos Nacionales – San Jose, Costa Rica. v1-3. 1899-1902 – 1r – us UF Libraries [972]

Documentos sobre el 20 de julio de 1918 / Ortega Ricaurte, Enrique – Bogota, Colombia. 1960 – 1r – us UF Libraries [972]

Documentos sobre la expulsion de los moriscos de denia (anno 1596-1621) – Almeria – 1r – 5,6 – sp Cultura [946]

Documentos sobre la puebla de arganzon (siecle 13-15) – Calahorra – 1r – 5,6 – sp Cultura [946]

Documentos varios (anno 1138-1593) – Avila – 1r – 5,6 – sp Cultura [946]

Documentos varios (anno 1260-1518). inventario de escrituras (anno 1585) – Agreda – 1r – 5,6 – sp Cultura [946]

Documentos varios (anno 834-1566) : no 1-314 – Calahorra – 1r – 5,6 – sp Cultura [946]

Documentos varios (siecle 12) – Caceres – 1r – 5,6 – sp Cultura [946]

Documentos varios (siecle 13-16) – Caceres – 1r – 5,6 – sp Cultura [946]

Documentos varios (siecle 15-16) – Caceres – 1r – 5,6 – sp Cultura [946]

Documentos y datos historicos y estadisticos / El Salvador Biblioteca National – San Salvador, El Salvador. 1926 – 1r – us UF Libraries [972]

Documentos y estudios historicos – Santiago, Dominican Republic. v1-10. 1944 – 5r – us UF Libraries [972]

Documentos y monumentos epigraficos del museo provincial de badajoz / Carrasco Lianes, Virgilio – Badajoz: Imp. Dip. Provincial, 1976. Sep. Revista de Estudios Extremenos – sp Bibl Santa Ana [060]

Documents – Paris. v1, n1-7; v2 n1-8, 3e s., n 1; 4e s., n1. 1929-30, 1933-34 – 1 – (doctrines, archeologie, beaux-arts, ethnographie) – fr ACRPP [073]

Documents / Federation des Ouvriers des Metaux et Similaires de France. 4e-6e Congres National – 1919-23 – 1 – fr ACRPP [331]

Documents / France. Assemblee consultative provisoire – 7nov 1944-5 oct 1946 – 1 – fr ACRPP [323]

Documents / France. Assemblee de l'Union francaise – 10 dec 1947-29 mai 1958 – 1 – fr ACRPP [323]

Documents 1-758 / Maritime Law Association of the United States – 1899-2001 – 378mf – 9 – $567.00 – (lacking: doc nos 129b,146,147. updated regularly) – mf#LLMC 85-100 – us LLMC [348]

Documents administratifs / France – 1905-86 – 1 – fr ACRPP [323]

Les documents administratifs – 1981- – €42.69y – (backfile: 1906-41 €144.83. 1945-80 €144.83. 1906-80 €274.41) – fr Journal Officiel [350]

Documents algeriens – Serie culturelle. [Alger]: Service d'information du cabinet du gouverneur general. [n1-81]. mar 25 1946-sep 25 1957 – 1 – (serie economique: n1-125 oct 15 1945-sep 30 1957. serie militaire: n1-11 sep 8 1946-mar 20 1953. serie monographie: n1-21 jun 20 1948-may 10 1953. serie politique: [n1-30] sep 1 1945-aug 1 1957. serie sociale: [n1-49] jan 5 1946-aug 15 1956) – us CRL [080]

Documents and communications addressed to the honorable louis joseph papineau, speaker of the house of assembly : by the honorable denis b viger and augustin norbet morin, esquire, named to proceed to england, and support the petitions of this house to his majesty and both houses of the imperial parliament – S.l: s.n, 1835? – 1mf – 9 – mf#47669 – cn Canadiana [320]

Documents and correspondence relating to mission stations at rotuma / Roman Catholic Mission, Fiji – 1868-1930 – 1r – 1 – mf#pmb428 – at Pacific Mss [241]

Documents and correspondence relating to palestine, august 1939 to march 1940 – London, 1940 – 1mf – 9 – mf#J-28-141 – ne IDC [956]

Documents and correspondence relating to the quest / Great Britain – London, England. v1-2. 1896 – 1r – us UF Libraries [972]

The documents and facts in the irvine-talbot case : with notes on the presentment of bishop talbot / Irvine, Ingram N W & Price, William S – Philadelphia: John R McFetridge, 1902 – 1mf – 9 – 0-524-05480-0 – mf#1990-5127 – us ATLA [240]

Documents and studies on the japanese free balloons of world war 2 / U.S. Army – 5items. 1945-? – 1 – $26.00 – us L of C Photodup [950]

Documents arabes relatifs a l'histoire du soudan : tarikh es-soudan... / Abderrahman ben Abdallah ben Imran ben Amir Es-sa'di – Paris, 1898-1900. 2v – 24mf – 8 – mf#A-265 – ne IDC [956]

Documents armeniens : recueil des historiens des croisades – Paris. 2v. 1869-1906 – 97mf – 9 – mf#H-508 – ne IDC [947]

Documents assembled by the international prosecution section for use as exhibits before the international military tribunal for the far east, 1945-1947 / World War 2. International Prosecution Section – (= ser Records of allied operational and occupation headquarters, world war 2) – 34r – 1 – mf#M1680 – us Nat Archives [355]

Documents assyriens relatifs aux presages. tome premier / Boissier, Alfred – Paris: Emile Bouillon, 1894 – 3mf – 9 – 0-8370-9126-8 – mf#1986-3126 – us ATLA [470]

Documents concerning jews in the berlin document center / Germany. Berlin Documents Center – (= ser National archives coll of foreign records seized, 1941-) – 14r – 5,1 – mf#T457 – us Nat Archives [324]

Documents concerning the negotiation of a trade agreement with japan *see* Japan and the west

Documents connected with the foundation of the anglican bishopric i / Neale, J M – London, England. 1853 – 1r – us UF Libraries [241]

Les documents de l'assemblee nationale : series ordinaire – debats compte rendu / France. Assemblee Nationale – 1988- – 9 – €157.40y – (backfile: 1881-1910 €30.49 1911-1940 €30.49 1944-1973 €53.36 depuis 1974 €99.09; complete coll: 1881-1910 €365.88 1911-1940 €€457.35 1944-1973 €1524. 49 1974-1983 €868.96 1881-1983 €2591.63) – fr Journal Officiel [944]

Les documents de l'assemblee nationale : series ordinaire – debats questions – 1997- – €143.80y – (backfile: depuis 1997 11e legislature €91.47; complete coll: 1881-1940 3e-16e legislature €807.98 1943-58 1ere-3e legislature €243.92 1958-73 1ere-4e legislature €350.63 1973-78 5e legislature €457.35 1978-81 6e legislature €304.90 1981-86 7e legislature €548.82 1986-88 8e legislature €243.92 1988-93 9e legislature €548.82 1993-97 10e legislature €838.47 1881-1993 €3475.84) – fr Journal Officiel [944]

Documents de paleographie hebraique et arabe / ed by Merx, Adalbert – Leyde: EJ Brill, 1894 – 1mf – 9 – 0-524-05582-3 – (incl bibl ref) – mf#1992-0442 – us ATLA [470]

Documents diplomatiques / Haiti (Republic). Departement Des Affaires Etrang... – Port-Au-Prince, Haiti. 1921 – 1r – us UF Libraries [972]

Documents du congres / Federation nationale des Travailleurs du Sous-Sol – 1920, 1925-26, 1929, 1931-32, 1934, 1938, 1946, 1950 – 1 – fr ACRPP [331]

Les documents du progres – Paris, Berlin, Londres, Budapest, Madrid. dec 1907-janv 1914 [mnthly] – 1 – (revue internationale. paraissant tous les mois a paris, berlin, londres, budapest, madrid.) – fr ACRPP [073]

Les documents du senat / France. Senat – 1971- – 9 – €97.40y – (backfile: 1971 11e legislature €60.98; complete coll: 1881-1940 €335.39 1946-1970 €716.51 1881-1970 €960.43) – fr Journal Officiel [944]

Documents et extraits divers : concernant l'histoire de l'art dans la flandre, l'artois et le hainaut avant le 15e siecle / Dehaisnes, [C C A] – Lille, 1886. 2v – 30mf – 9 – mf#O-221 – ne IDC [700]

Documents francais – Clermont Ferrand, France. 1942-may 1944 – 1r – 1 – uk British Libr Newspaper [072]

Documents from the 3rd round of the covenant section 902 consultations : between the special representative of the president of the u s and the special representative of the commonwealth of the northern mariana islands, including the official joint press release – Washington: Macweelin et al, apr 1 1987 – 4mf – 9 – $6.00 – mf#llmc82-100j, title 26 – us LLMC [324]

Documents from the archive of the states of holland, c. 1445-1572 – 74r – 1 – ne MMF Publ [949]

Documents from the secret archives of the general secretariat of the netherlands indies government and the cabinet of the governor general – [mf ed 2003] – (= ser Dutch political conflict with the republic of indonesia, 1945-1949) – 1344mf – 9 – €9995.00 – (printed publ's guide & concordance based on m g h a de graaff & a m tempelaars "inventaris van het archief van de algemene secretarie van de nederlands-indische regering en de daarbij gedeponeerde archieven, 1942-50") – mf#mmp107 – National archives of the netherlands, the hague – ne Moran [959]

Documents illustrative of english church history – London; New York: Macmillan, 1896 – 2mf – 9 – 0-7905-5533-6 – mf#1988-1533 – us ATLA [240]

Documents illustrative of english history in the 13th and 14th centuries / ed by Cole, H – London, 1814 – €38.00 – ne Slangenburg [941]

Documents illustrative of the continental reformation / ed by Kidd, Beresford James – Oxford: Clarendon Press, 1911 – 2mf – 9 – 0-7905-5607-3 – mf#1988-1607 – us ATLA [242]

Documents illustrative of the formation of the union of the american states : selected, arranged and indexed / Tansil, Charles C – Washington: GPO, 1927 – 12mf – 9 – $18.00 – mf#LLMC 90-361 – us LLMC [323]

Documents illustrative of the oppressions and cruelties of irish revenue officers / Chichester, Edward – London, 1818 – (= ser 19th c ireland) – 1mf – 9 – mf#1.1.173 – uk Chadwyck [324]

Documents in reference to the general adoption of the twenty-four hour notation of the rail-ways of america – Ottawa: Citizen, 1887 – 1mf – 9 – mf#00792 – cn Canadiana [380]

Documents in relation to the differences which subsisted between the late commodore o h perry and captain j d elliott – Washington: [s.n.] 182; Boston: [s.n.] 1834 [mf ed 1987] – 1mf – 9 – 0-665-63569-9 – mf#63569 – cn Canadiana [355]

Documents inedits pour l'histoire / Morpeau, Louis – Port-Au-Prince, Haiti. 1920 – 1r – us UF Libraries [972]

Documents inedits pour servir a l'histoire ecclesiastique de la belgique / ed by Berliere, Ursmer – Maredsous: Abbaye de Saint-Benoait, 1894 – 1mf – 9 – 0-7905-6583-8 – mf#1988-2583 – us ATLA [240]

Documents inedits sur le colonel de longueuil – Montreal?: Desaulniers & Leblanc, 1891 – 1mf – 9 – (ann and publ by monongahela de beaujeu) – mf#56369 – cn Canadiana [971]

Documents internationaux de l'esprit nouveau – Paris. n1. printemps 1927 – 1 – (collection privee) – fr ACRPP [073]

Documents juridiques de l'assyrie et de la chalde / Menant, Joachim – Paris: Maisonneuve, 1877 [mf ed 1989] – 1mf – 9 – 0-7905-2253-5 – (in french, latin & akkadian. incl ind) – mf#1987-2253 – us ATLA [470]

Documents maconniques / Favre, Francois [comp] – Paris: Libr maconnique de A Teissier [1866?] – 7mf – 9 – mf#vrl-143 – ne IDC [366]

Documents obtenus des archives du departement de la marine et des colonies a paris : par l'entremise de m. faribault, lors de son voyage en europe en 1851 – [s.l: s.n, 1851?] [mf ed 1985] – 1mf – 9 – 0-665-01719-7 – mf#01719 – cn Canadiana [333]

Documents of the canadian constitution, 1759-1915 / ed by Kennedy, William Paul McClure – Toronto: OUP, 1918 [mf ed 1997] – 1mf – 9 – 0-665-85154-5 – mf#85154 – cn Canadiana [342]

Documents of the interdivisional country and area committee, 1943-1946 / U.S. Dept of State – (= ser General records of the department of state) – 6r – 1 – mf#T1221 – us Nat Archives [324]
Documents of the national security council / U.S. National Security Council – 1 – (1947-77 5r isbn 0-89093-311-1 $970. suppl: 1st 3r $570 isbn 0-89093-310-3. 2nd 3r $570 isbn 0-89093-536-X. 3rd 3r $570 isbn 0-89093-569-6. 4th 7r $1340 isbn 0-89093-192-5. 5th 4r $770 isbn 1-55655-161-4. 6th 10r $1935 isbn 1-55655-473-7. 7th 7r $1340 isbn 1-55655-592-X. 8th 15r isbn 1-55655-823-6 $2905. minutes of meetings of the nsc, with special advisory reports 3r isbn 0-89093-462-2 $570. suppl: 1st 5r isbn 0-89093-939-X $970. 2nd 3r isbn 1-55655-162-2 $570. 3rd 7r isbn 1-55655-600-4 $1340. ind to documents of the nsc 721p isbn 0-89093-994-2 $735. with p/g) – us UPA [327]
Documents of the post war programs committee, 1944 / U.S. Dept of State – (= ser General records of the department of state) – 4r – 1 – mf#T1222 – us Nat Archives [324]
Documents of Truth, Inc *see* Dot
Documents officiels relatifs a l'avenement du gene / Chaumette, Gustave – Port-Au-Prince, Haiti. 1909 – 1r – us UF Libraries [972]
Documents on british west indian history / Williams, Eric Eustace – Port-of-Spain, Trinidad and Tobago. 1952 – 1r – us UF Libraries [972]
[Documents on colonial and commonwealth history] : from the royal commonwealth society, london – mid 1850's-mid 1980's – 12 titles on 767mf – 9 – (with p/g) – ne IDC [320]
Documents on disarmament / U.S. Arms Control and Disarmament Agency – 1945-73 – 1 – $378.00 – mf#0592 – us Brook [327]
Documents on disarmament, 1945-1959 / U.S. Dept of State – 2v. 1960-79 [all publ?] – 281mf – 9 – $421.00 – (annuals: 1960-86) – mf#llmc 80-911 – us LLMC [327]
Documents on disarmament, 1945-1982 – 11r – 1 – $1915.00 – 0-89093-195-X – (with p/g) – us UPA [327]
[Documents on education development] – 2703mf – 9 – (sect: africa [253 titles on 944mf]; middle east/ north africa [45 titles on 196mf]; pacific [39 titles on 134mf]; latin america [111 titles on 336mf]; north america [1 title on 1mf]; europe [16 titles on 110mf]; general [26 titles on 57mf]; south east asia [133 titles on 406mf]; south asia [114 titles on 341mf]; east asia [58 titles on 169mf]; eastern europe [5 titles on 9mf]; with printed catalogue) – ne IDC [370]
Documents on modern africa / Wallbank, Thomas Walter – Princeton, NJ. 1964 – 1r – us UF Libraries [960]
Documents on the constitutional history of puerto / Puerto Rico. Office Of The Commonwealth Of Puerto – Washington, DC. 1964 – 1r – us UF Libraries [972]
Documents on the constitutional history of puerto rico – Washington: Office of Puerto Rico, n.d. – 2mf – 9 – $3.00 – (contains: spanish constitution of 1876, the self-government constitution of 1897, the treaty of paris of 1898, the military government of 1898-1900, among others) – mf#LLMC 92-404 – us LLMC [324]
Documents on the present situtation in cambodia – Paris: Information Bureau of the Provisional Revolutionary Govt of the Republic of South Vietnam [mf ed 1989] – 1r with other items – 1 – mf#mf-10289 seam reel 022/05 [§] – us CRL [959]
Documents on university apartheid and on christian national education policies : printed and mimeographed reports, etc [1931?-63] – [S.l: s.n., 19–?] – us CRL [378]
Documents para la historia economica de mexico – Mexico: Secretaria de la Economia Nacional, 1933-36. 11v – 1 – us Wisconsin U Libr [972]
Documents parlementaires / France. Assemblee Nationale – 28 nov 1946-23 mars 1971 – 1 – fr ACRPP [323]
Documents parlementaires / France. Chambre des Deputes – 1881-1938 – 1 – fr ACRPP [323]
Documents parlementaires / France. Conseil de la Republique – 26 decembre 1946-sept 1957 – 1 – fr ACRPP [323]
Documents parlementaires / France. Senat – 1881-1938 – 1 – fr ACRPP [323]
Documents pertaining to the rule of nasir al-din shah qajar – 1mf – 9 – $25.00 – (bureaucratic correspondence fr the period of muhammad shah and nasir al-din shah regarding various administrative and provincial problems) – us MEDOC [956]
Documents presented as evidence by the defense before the international military tribunal for the far east, 1945-1947 / World War 2. Defense Section – (= ser Records of allied operational and occupation headquarters, world war 2) – 19r – 1 – mf#M1692 – us Nat Archives [355]
Documents relatifs a la repression de la traite des esclaves – Bruxelles: Hayez, 1901-13 – 1 – $144.00 – mf#0187 – us Brook [305]
Documents relatifs a la suspension des relations diplomatiques entre le cambodge et la thailande / Cambodia. Ministere de l'information – Phnom-Penh: Ministere de l'information [1958?] [mf ed 1989] – 1r with other items – 1 – mf#mf-10289 seam reel 002/08 [§] – us CRL [327]
Documents relatifs a l'echange des proprietes des tanneries, pres montreal – [Quebec?: s.n, 1875?] – 6mf – 9 – 0-665-91912-3 – mf#91912 – cn Canadiana [333]
Documents relatifs a l'erection et a l'organisation de l'universite laval – [Quebec: s.n.] 1862 [mf ed 1993] – 1mf – 9 – 0-665-91750-3 – mf#91750 – cn Canadiana [378]
Documents relatifs au developpement du commerce entre les etats-unis et le canada y compris la colonie de terreneuve sic 1891 – Ottawa: S E Dawson, 1891 [mf ed 1986] – 1mf – 9 – 0-665-56032-X – mf#56032 – cn Canadiana [337]
Documents relatifs aux eglises de l'orient et a leurs rapports avec rome / Avril, Adolphe d' – 3rd rev enl ed. Paris: Challamel Aine, 1885 [mf ed 1986] – 1mf – 9 – 0-8370-7602-1 – (incl bibl ref) – mf#1986-1602 – us ATLA [243]
Documents relating to appointment and career of reginald arthur roberts, as judicial officer (vice consul, senior resident) of the niger coast protectorate, later onitsha province, 1895-1928 – Oxford: Oxford University Press, [19–?] – us CRL [920]
Documents relating to murders in telefomin, papua new guinea, 6 nov 1953 / Healey, Lionel Rhys – 1953-88 – 1r – 1 – mf#pmb1265 – at Pacific Mss [364]
Documents relating to the colonial history of the state of new jersey / New Jersey Historical Society – v1,7 [1880, 1883], v22 – 2r – 1 – (cont by: documents relating to the revolutionary history of the state of new jersey; documents relating to the colonial and revolutionary history of the state of new jersey) – mf#3921267 – us WHS [978]
Documents relating to the commercial policy of spa / Whitaker, Arthur Preston – Deland, FL. 1931 – 1r – us UF Libraries [978]
Documents relating to the construction of the parliamentary and departemental buildings at ottawa / Canada (Province). Departement des travaux publics – Quebec: [s.n.], 1862 [mf ed 1983] – 5mf – 9 – mf#SEM105P310 – cn Bibl Nat [350]
Documents relating to the invasion of the niagara peninsula by the united states army : commanded by general jacob brown, in july and august, 1814 / ed by Cruikshank, Ernest Alexander – Niagara-on-the-Lake, Ont: Niagara Historical Society, [1920?] – 2mf – 9 – 0-665-73074-8 – (incl bibl ref) – mf#73074 – cn Canadiana [355]
Documents relating to the military and naval service of blacks awarded the congressional medal of honor from the civil war to the spanish-american war – 4r – 1 – (with printed guide) – mf#M929 – us Nat Archives [355]
Documents relating to the negotiation of ratified and unratified treaties with various indian tribes, 1801-1869 / U.S. Bureau of Indian Affairs – (= ser Records Relating To Indian Treaties) – 10r – 1 – mf#T494 – us Nat Archives [324]
Documents relating to the northern mariana islands / U.S. Dept of the Interior – n.d. – 1mf – 9 – $1.50 – (docs assembled by staff of the hawaiian pacific coll univ of hawaii library) – mf#llmc82-100j, title 18 – us LLMC [324]
Documents relating to the proposed new chinese translation of the holy scriptures : memorial addressed to the british and foreign bible society... – [London: s.n. 1836] [mf ed 1995] – (= ser Yale coll) – 1 – 0-524-09642-2 – mf#1995-0642 – us ATLA [220]
Documents relating to the settlement of the church of england by the act of uniformity of 1662 : with an historical introduction / ed by Gould, George – London: W Kent, 1862 – 2mf – 9 – 0-7905-4735-X – (incl bibl ref) – mf#1988-0735 – us ATLA [941]
Documents relating to universities' mission to central africa, 1861-1929 : from the archives of the united society for the propagation of the gospel – 39r – 1 – mf#4791 – uk Microform Academic [240]
Documents relative to an outrage alleged to have b... – Washington, DC. 1853 – 1r – us UF Libraries [972]
Documents relative to central american affairs / United States. Dept Of State – Washington, DC. 1856 – 1r – us UF Libraries [972]
Documents relative to the colonial history of the state of new york / New York. (colony) – Procured in Holland, England and France. v. 1-15. 1853-87 – 1 – us AMS Press [978]
Documents relative to the erection and endowment of additional bishoprics in the colonies : with a short historical preface / Hawkins, Ernest – London, 1844 – (= ser 19th c british colonization) – 2mf – 9 – mf#1.1.6676 – uk Chadwyck [230]
Documents sur la mission des freres-precheurs a sa... / Le Ruzic, Ignace Marie – Lorient, France. 1912 – 1r – us UF Libraries [972]
Documents sur l'agression vietcong et nord-vietnamienne contre le cambodge [1970] / Cambodia. Ministere de l'information – Phnom Penh [1970] [mf ed 1989] – 1r with other items – 1 – mf#mf-10289 seam reel 016/12 [§] – us CRL [959]
Documents sur l'histoire, la geographie et le commerce de l'afrique orientale / Guillain, M – Paris: A Bertrand, [1856] – 1 – us CRL [916]
Documents towards a history of reformation in cornwall / Snell, Lawrence S [comp] – 2v on 1r – 1 – £57 / $114 – (v1: the chantry certificates for cornwall... v2: the edwardian inventories of church goods for cornwall...) – mf#R10002 – uk Microform Academic [941]
Docvmenti armonici di d angelo berardi... : nelli quali con varij discorsi... / Berardi, Angelo – Bologna: G Monti 1687 [mf ed 19–] – 4mf – 9 – mf#fiche 356 – us Sibley [780]
Docvmenti d'amore di m. francesco barberino / Barberino, F, da – [Roma: Nella stampería di Vitale Mascardi, 1640] – 6mf – 9 – mf#0-1334 – ne IDC [090]
Dod, Albert Baldwin *see* Essays, theological and miscellaneous, second series
Dodatak k sanktpeterburgskim sravniteljnim rjecnicima sviju jezika i narjecija : s osobitim ogledima bugarskog jezika / Karadzic, Vuk Stefanovic – U Becu 1822 – (= ser Whsb) – 1mf – 9 – €20.00 – mf#Hu 008 – gw Fischer [460]
Dodd, C H *see*
- The apostolic preaching and its development
- The epistle of paul to the romans
- The johannine epistles
Dodd, Charles E *see* An autumn near the rhine
Dodd, Joseph *see* A history of canon law in conjunction with other branches of jurisprudence
Dodd, Lee Wilson *see* Golden complex
Dodd, Samuel *see* The diary of the rev. samuel dodd
Dodd, Walter F *see* State government
Dodd, William *see* A full and circumstantial account of the trial of the rev. doctor dodd, at the sessions house in the old bailey, on saturday the 22nd of february, 1777.
Doddington 1600-1950 – 16mf – 9 – £21.00 – uk CambsFHS [929]
Doddridge, Philip *see*
- Evidences of christianity briefly stated
- Practical discourses on regeneration
Dodds, James *see*
- Lays of the covenanters
- Thomas chalmers
Dodecacorde contenant douze psaumes de david / Lejeune, Claude – 1598 – (= ser Mssa) – 2mf – 9 – €35.00 – mfchl 470 – gw Fischer [780]
Das dodekapropheton / Marti, Karl – Tuebingen: J C B Mohr (Paul Siebeck), 1904 – 2mf – 9 – 0-8370-9804-1 – (includes bibliographies and index) – mf#1986-3804 – us ATLA [221]
Dodekapropheton aethiopum : oder, die zwoelf kleinen propheten der aethiopischen bibeluebersetzung / ed by Bachmann, Johannes – Halle a S: M Niemeyer, 1892 – 1mf – 9 – 0-8370-1885-4 – mf#1987-6272 – us ATLA [221]
Doderer, Otto *see*
- Brentanos im rheingau
- Gruenewald und der edelmann
Dodge advertiser – Dodge, NE: C A Manville, 1889-95// (wkly) [mf ed v4 n45. nov 24 1892-dec 1 1892 filmed [1979]] – 1r – 1 – (absorbed by: dodge criterion) – us NE Hist [071]
Dodge, Arthur Pillsbury *see* Whence? why? whither?
Dodge City. Kansas. City Council *see* Records
Dodge City. Kansas. Police Court *see* Docket
The dodge club : or italy in 1859 / De Mille, James – New York: Harper, 1875 – 2mf – 9 – mf#64756 – cn Canadiana [830]
Dodge county banner – Mayville WI. 1899 oct 27-dec 8, 1902 jul 29-1904 feb 19, 1907 feb 1-1908 jun 30, jul 3-1909 dec 28, 1910 jan 4-1911 jun 30, jul 7-1912 dec 31, 1913 jan 6-1914 jun 30, jul 7-1915 dec 31, 1916 jan 7-1918 dec 26, 1919 jan 2-1919 dec 4 – 9r – 1 – mf#1108427 – us WHS [071]
Dodge county citizen – Beaver Dam WI. 1856 apr 10-1859 apr 28, 1859 may 5-1861 may 30 – 2r – 1 – mf#957709 – us WHS [071]
Dodge county citizen – Beaver Dam WI. 1862 oct 8/1863 jun 25-1923 jan 3/1924 apr 16 – 24r – 1 – (with small gaps) – mf#928614 – us WHS [071]
Dodge county democrat – Juneau WI. 1876 jan 5-1879 feb 5 – 1r – 1 – (cont by: telephone [mayville wi]) – mf#927067 – us WHS [071]
Dodge county gazette – 1852 jun 16-1853 sep 23 – 1r – 1 – (cont by: burr oak) – mf#927162 – us WHS [071]
Dodge county independent-news – Hustisford, Juneau, Reeseville WI. 1962 sep 6/1964-1996 jul-dec – 45r – 1 – (cont: independent [juneau wi]; hustisford news; reeseville review) – mf#1107388 – us WHS [071]
Dodge county pionier – Mayville WI. 1876 mar 10/1877 aug 24, 1878 mar 1-1944 may 24/1945 dec 26 – 55r – 1 – (with small gaps) – mf#1126300 – us WHS [071]
Dodge criterion – Dodge, NE: Birge E Burns, 1888 (wkly) [mf ed 1895-] – 1 – (absorbed: dodge advertiser 1895 and: snyder banner 1954) – us NE Hist [071]
Dodge, David *see* How green was my father
Dodge, David Low *see* War inconsistent with the religion of jesus christ
Dodge, Ebenezer *see* The evidences of christianity
Dodge, Joseph Smith *see* The purpose of god
Dodge, Martin Herbert *see* The government of the city of frankfort-on-the-main
Dodge, Mary Mapes *see* The irvington stories
Dodge, Norman *see* The month at goodspeed's, 1929-1969
Dodge, Walter Phelps *see* Real sir richard burton
Dodge worker / Workers [Communist] Party of America – 1926 aug-dec, 1927 apr-aug – 1r – 1 – (cont: dodge bros. workers news) – mf#1055666 – us WHS [335]
Dodgeville chronicle – Dodgeville WI. 1862 sep 18/1863 sep 10-1995 jan-jun – 115r – 1 – (with small gaps; cont: iowa county advocate) – mf#1133810 – us WHS [071]
Dodgeville star – Dodgeville WI. 1883 nov 30-1887 mar 5 – 1r – 1 – (cont by: eye and star) – mf#962733 – us WHS [071]
Dodgeville sun – Dodgeville WI. 1881 oct 13-1905 jan 7/1906 mar 3 – 11r – 1 – (with small gaps; cont by: dodgeville sun-republic) – mf#875203 – us WHS [071]
Dodgeville sun-republic – Dodgeville WI. 1906 mar 7/dec 28-1929 apr 18/1931 may 7 – 16r – 1 – (with small gaps; cont: dodgeville sun; weekly republic) – mf#998653 – us WHS [071]
Dodgson, Charles *see* Controversy of faith
Dodleston, st mary – (= ser Cheshire monumental inscriptions) – 2mf – 9 – £4.00 – mf#89 – uk CheshireFHS [929]
Dodleston, st mary: baptisms 1570-1970 – (= ser Cheshire church registers) – 4mf – 9 – £5.00 – mf#207 – uk CheshireFHS [929]
Dodleston, st mary: burials 1570-1970 – (= ser Cheshire church registers) – 3mf – 9 – £4.50 – mf#208 – uk CheshireFHS [929]
Dodleston, st mary: marriages 1570-1970 – (= ser Cheshire church registers) – 2mf – 9 – £4.00 – mf#98 – uk CheshireFHS [929]
Dods, Marcus *see*
- Anglicanus scotched
- Bearings of popery on the priesthood of christ
- The bible, its origin and nature
- Christ and man
- Early letters of marcus dods, d.d
- The epistle of our lord to the seven churches of asia
- Erasmus and other essays
- The first epistle to the corinthians
- Footsteps in the path of life
- How to become like christ
- An introduction to the new testament
- Later letters of marcus dods, d.d
- Mohammed, buddha, and christ
- The parables of our lord
- The post-exilian prophets
- Presbyterianism, older than christianity
- Revelation and inspiration
- The visions of a prophet
- Why be a christian?
Dods, Marcus et al *see* The literal interpretation of the sermon on the mount
Dod's parliamentary companion – London, England 1906-80 – 1,5,9 – ISSN: 0070-7007 – mf#1739 – us UMI ProQuest [323]
Dods, Selby Ord *see* Chief points of difference betwixt the established and the free chu...
Dodson, George Rowland *see*
- Bergson and the modern spirit
- Sympathy of religions
Dodsworth, Henrique De Toledo *see* Cem anos de ensino secundario no brasil (1826-1926
Dodsworth, W *see* Allegiance to the church
Dodsworth, William *see*
- Anglicanism considered in its results
- Few comments on dr pusey's letter to the bishop of london
- Further comments on dr pusey's renewed explanation
- Gorham case briefly considered
- Hosue divided against itself
- Romanism successfully opposed only on catholic principles
- Sermon occasioned by the appeal of the lord bishop of london for th
Dodt, Gustavo Luiz Guilherme *see* Descripcao dos rios parnahyba e gurupy
Dodwell, H H *see* The cambridge shorter history of india

Dodwell, Henry *see*
- India
- The nabobs of madras
- Treatise concerning the lawfulness of instrumental musick in holy offices

Dodwell, William *see* Athanasian creed vindicated and explained

Dodworth, Allen *see*
- Dancing and its relations to education and social life
- Dodworth's brass band school: containing instructions in the first principles of music;... together with a number of pieces of music, arranged for a full brass band

Dodworth's brass band school: containing instructions in the first principles of music;...together with a number of pieces of music, arranged for a full brass band / Dodworth, Allen – New York: H. B. Dodworth & Co. 1853. In score, playable by 6-12 instruments.MUSIC 1978 – 1 – us L of C Photodup [780]

Doe, Walter P *see* Revivals

Doe, Walter P [comp] *see* Important religious truths

Doea lobang pelor / tjoe bo kim so / Bong, Kok No & Kwo, Lay Yen – Batavia: Goedang Tjerita, 1948-1949 [mf ed 1998] – (= ser Goedang tjerita 8-13) – 1r – 1 – (coll as pt of the colloquial malay collection; "tjoe bo kim so" is an indonesian trans of the chinese novel entitled zi mu jin suo by zheng zhengyin; filmed with: lajangan biroe / im, yang tjoe) – mf#10005 – us Wisconsin U Libr [830]

Doebel, Ernst C *see* Des wagners e ch doebel wanderungen im morgenlande

Doebelner allgemeine – Doebeln DE, 1993- – 1 – (regional ed of leipziger volkszeitung, leipzig) – gw Misc Inst [074]

Doebelner anzeiger *see* Leisniger wochenblatt

Doebler, Marion *see* Zum fortpflanzungsmodus des amazonenkaerpflings (poecilia formosa girard 1859)

Doeblin, Alfred *see*
- Der blaue tiger
- Das land ohne tod
- Minotaurus
- Der oberst und der dichter
- Der schwarze vorhang
- Wallenstein

Doederlein, Julius *see* Gottes dasein bewiesen am wissen und sein

Doedes, J I *see* Manual of hermeneutics for the writings of the new testament

Doedes, Jacobus Izaak *see*
- Encyclopedie der christelijke theologie
- De leer van god

Doeg the edomite : or, the informer. a lecture on the fifty-second psalm. delivered in the first presbyterian church, philadelphia... / Barnes, Albert – Philadelphia: Henry B Ashmead, 1861 – 1mf – 9 – 0-524-05900-4 – mf#1992-0657 – us ATLA [220]

Doege, Volker *see* Aufklaerung von elektrodenprozessen der positiven masse einer li/licoo$_2$ sekundaerbatterie mittels elektrochemischer impedanzspektroskopie

Doegen, M *see* Architectura militaris moderna

Doehrn, Gisela *see* An einen geliebten soldaten

Doelger, Frans J *see* Antike und christentum

Doelker-Rehder, Margarete *see* Der schwesternsohn

Doell, Heinrich *see* Goethe und schopenhauer

Doell, M *see*
- Die benuetzung der antike in wielands moralischen briefen
- Die einfluesse der antike in wielands hermann
- Wieland und die antike

Doeller, Johannes *see*
- Abraham und seine zeit
- Geographische und ethnographische studien zum 3. und 4. buche der koenige
- Die messiaserwartung im alten testament
- Rhythmus, metrik und strophik in der biblisch-hebraeischen poesie
- Das weib im alten testament

Doellinger, Ignaz von *see*
- Beitraege zur sektengeschichte des mittelalters
- Geschichte der moralstreitigkeiten in der roemisch-katholische kirche

Doellinger, Johann Josef Ignaz von *see* Kirche und kirchen, papstthum und kirchenstaat

Doellinger, Johann Joseph Ignaz von *see*
- Addresses on historical and literary subjects
- Briefe und erklaerung von j. von doellinger ueber die vaticanischen decrete, 1869-87
- Briefe und erklaerungen von j. von doellinger ueber die vaticanischen decrete, 1869-87
- The church and the churches
- Conversations of dr. doellinger
- Dokumente vornehmlich zur geschichte der valdesier und katharer
- Dr j a moehlers...gesammelte schriften und aufsaetze
- Dr j j i von doellinger's fables respecting the popes in the middle ages
- Erwaegungen fuer die bischoefe des concilium's ueber die frage der paepstlichen unfehlbarkeit
- The first age of christianity and the church
- The gentile and the jew in the courts of the temple of christ
- Geschichte der gnostisch-manichaeischen sekten im frueheren mittelalter
- Geschichte der moralstreitigkeiten in der roemisch-katholischen kirche
- Geschichte und kirche
- Hippolytus and callistus, or, the church of rome in the first half of the third century
- History of the church
- Lectures on the reunion of the churches
- Die lehre von der eucharistie in den drei ersten jahrhunderten
- Letters from rome on the council
- Die papst-fabeln des mittelalters
- The pope and the council
- Prophecies and the prophetic spirit in the christian era
- Die reformation
- Roemische briefe vom concil
- Studies in european history
- Ueber die wiedervereinigung der christlichen kirchen
- Ungedruckte berichte und tagebuecher zur geschichte des concils von trient

Doerfler, Anton *see*
- Das neue heiligtum
- Der ruf aus dem garten
- Die schoene wuerzburgerin

Doerfler, Peter *see*
- Die alte heimat
- Das gesicht im nebel
- Der weltkrieg im schwaebischen himmelreich

Doerfliches leben : sieben erzaehlungen / Blunck, Hans Friedrich – Leipzig: Gesellschaft der Freunde der Deutschen Buecherei, 1934 [mf ed 1989] – 62p – 1 – mf#7036 – us Wisconsin U Libr [880]

Doering, Bruno *see* Heitere hamsterkiste

Doering, Georg *see* Frauentaschenbuch

Doering, Heinrich *see* Bilder aus der deutschen jesuitenmission puna

Doering, O *see*
- Der augsburger patriciers philipp hainhofer reisen nach innsbruck und dresden
- Des augsburger patriciers philipp hainhofer beziehungen zum herzog philipp 2 von pommern-stettin

Doernenburg, Emil *see*
- Deutsch-amerikanische balladen und gedichte
- Sturm und stille

Doerner, Klaus *see* Nuremberg medical trial 1946-1947

Doerner, Klaus et al *see* Der "nuernberger aerzteprozess" 1946-1947

Doerptische beitraege fuer freunde der philosophie, literatur und kunst – Dorpat, Leipzig, 1813-1814 – 17mf – 8 – mf#R-1751 – ne IDC [410]

Doerrer, Anton *see* Bozner buergerspiele, alpendeutsche prang- und kranzfeste

Doerries, H *see*
- Die 50 homilien des makarios
- Symeon von mesopotanien
- Die vita antonii als geschichtsquelle. nachrichten d akademie d wissenschaften in goettingen

Doerrlamm, Brigitte et al *see* Klassiker heute

Doersam, Edgar *see* Technische hochschule darmstadt

Does annexation follow? : commercial union and british connection: an open letter from erastus wiman to mr j redpath dougall, editor of the montreal witness – [New York: s.n, 1887?] [mf ed 1981] – 1mf – 9 – mf#25966 – cn Canadiana [971]

Does body temperature mediate anxiolytic effects af acute exercise / Youngstedt, S D – 1991 – 2mf – 9 – $8.00 – us Kinesology [150]

Does god answer prayer? / Edgar, Robert McCheyne – London: Hodder & Stoughton, 1883 [mf ed 1990] – (= ser The theological library) – 1mf – 9 – 0-7905-7512-4 – (incl bibl ref) – mf#1989-0737 – us ATLA [210]

Does god care for me? *see* Shang-ti hui kuan huai wo mo? (ccm191)

Does intercostal stretch affect respiratory muscle activities / Puckree, T – 1992 – 1mf – 9 – $4.00 – us Kinesology [612]

Does morality depend on longevity? / Neale, Edward Vansittart – Ramsgate, England. 1871 – 1r – us UF Libraries [230]

Does the established church acknowledge christ as its head? / M'cosh, James – Brechin, Scotland. 1846 – 1r – us UF Libraries [240]

Does vitamin e supplementation attenuate exercise-induced skeletal muscle injury / Warren, J A – 1991 – 1mf – 9 – $4.00 – us Kinesology [612]

Does woman represent god? : an inquiry into the true character of the movement for the emancipation of woman / Easton, Peter Zaccheus – New York: Revell [c1895] [mf ed 1984] – (= ser Women & the church in america) – 1mf – 9 – 0-8370-1230-9 – mf#1984-2069 – us ATLA [305]

Doetsch, Carlos *see*
- Benito arias montano. extractos de su vida
- Iconografia de benito arias montano
- La pena (retiro predilecto de montano)

The dog crusoe : a tale of the western prairies / Ballantyne, Robert Michael – Boston: Crosby and Nichols, 1863 – 5mf – 9 – 0-665-90786-9 – mf#90786 – cn Canadiana [830]

The dog crusoe and his master : a story of adventure in the western prairies / Ballantyne, Robert Michael – London, Edinburgh: T Nelson, 1893 – 4mf – 9 – mf#17917 – cn Canadiana [830]

Dogale granted to niccolo bernardo *see* Historia...

Dogan, Selami *see* A biomechanical analysis of canine gait before and after unilateral cemented total hip replacement

Doggett, L L *see* Life of robert r. mcburney

Doggett, Laurence Locke *see* History of the young men's christian association. volume 1, the founding of the association, 1844-1855

Dogliotti, Rosa-Luisa Amalia *see* Le theme du mariage mixte et/ou polygame comme foyer d'observation socioculturelle et interculturelle dans quatre romans francophones

Dogma and history / Krueger, Gustav – London: Philip Green, 1908 – (= ser Essex Hall Lecture) – 1mf – 9 – 0-7905-7056-4 – mf#1988-3056 – us ATLA [240]

Il dogma dell' immacolata : ragionamenti / Alimonda – 3a ed accresciuta. Genova: Giovent u, 1880 – 2mf – 9 – 0-8370-8241-2 – (incl bibl ref) – mf#1986-2241 – us ATLA [240]

Das dogma der alten kirche / Baur, Ferdinand Christian – Leipzig: Fues, 1865-1866 – 3mf – 9 – 0-7905-4133-5 – mf#1988-0133 – us ATLA [240]

Das dogma der neueren zeit / Baur, Ferdinand Christian – Leipzig: Fues, 1867 – 2mf – 9 – 0-7905-4135-1 – mf#1988-0135 – us ATLA [240]

Das dogma des mittelalters / Baur, Ferdinand Christian – Leipzig: Fues, 1866 [mf ed 1989] – (= ser Vorlesungen ueber die christliche dogmengeschichte 2) – 2mf – 9 – 0-7905-4134-3 – mf#1988-0134 – us ATLA [240]

Dogma, fact and experience / Rawlinson, Alfred Edward John – London: Macmillan, 1915 – 1mf – 9 – 0-7905-9848-5 – mf#1989-1573 – us ATLA [150]

Dogma in religion and creeds in the church / Kinross, John – Edinburgh: James Thin, 1897 – 1mf – 9 – 0-8370-3900-2 – (incl ind) – mf#1985-1900 – us ATLA [240]

Dogma moralium philosophorum, compendiose & studiose collectum / Clichtove, Josse – Argentorat: Ex Aeibus Schurerianis, Mense Iulio anno 1512 – (= ser Ethics in the early modern period) – [2]lea on 1mf – 9 – mf#pl-343 – ne IDC [170]

Dogma und schulmeinung : denkschrift in sachen der sogenannten "erhebung" von lehransichten zu "neuen glaubenswahrheiten" / Liano, Heinrich St A von – Muenchen: J J Lentner, 1869 – 1mf – 9 – 0-8370-8919-0 – (incl bibl ref) – mf#1986-2919 – us ATLA [240]

Das dogma vom heiligen abendmahl und seine geschichte / Ebrard, Johannes Heinrich August – Frankfurt a M: Heinrich Zimmer, 1845-1846 – 4mf – 9 – 0-524-00364-5 – mf#1989-3064 – us ATLA [240]

Das dogma vom neuen testament / Krueger, Gustav – Giessen: Curt von Muenchow, 1896 – 1mf – 9 – 0-8370-4417-0 – (incl bibl ref) – mf#1985-2417 – us ATLA [225]

Das dogma von christi person und werk / Gess, Wolfgang Friedrich – Basel: C Detloff, 1887 – 2mf – 9 – 0-7905-0839-7 – mf#1987-0839 – us ATLA [240]

Das dogma von der dreieinigkeit und gottmenschheit in seiner geschichtlichen entwicklung / Krueger, Gustav – Tuebingen: J C B Mohr (Paul Siebeck) 1905 [mf ed 1985] – (= ser Lebensfragen 8) – 1mf – 9 – 0-8370-4416-2 – (incl ind) – mf#1985-2416 – us ATLA [240]

Das dogma von der sichtbaren und unsichtbaren kirche : ein historisch-kritischer versuch / Muenchmeyer, August Friedrich Otto – Goettingen: Vandenhoeck und Ruprecht, 1854 – 1mf – 9 – 0-7905-5726-6 – (incl bibl ref) – mf#1988-1726 – us ATLA [240]

The dogmatic faith : an inquiry into the relation subsisting between revelation and dogma: in eight lectures / Garbett, Edward – New ed. London: Rivingtons, 1879 – 1mf – 9 – 0-8370-3229-6 – mf#1985-1229 – us ATLA [240]

The dogmatic principle in relation to christian belief : a discourse...liverpool, aug 1st 1881... 11th lecture... / Macdonald, Frederic William – London: Wesleyan-Methodist Book-room 1881 [mf ed 1991] – 1mf – 9 – 0-7905-9026-3 – mf#1989-2251 – us ATLA [240]

Dogmatic theology / Shedd, William Greenough Thayer – 3rd ed. New York: Scribner, 1891-1894 – 5mf – 9 – 0-7905-9635-0 – (incl bibl ref) – mf#1989-1360 – us ATLA [240]

Dogmaticheskow znachenie sed'mago vselenskago sobora: nauchno istoricheskow izslenovanie vazhnosti i neopkhodimosti.. / Ostroumov, G – St. Petersburg, 1884 – 1 – 9.96 – us Southern Baptist [242]

Dogmatik : akademische vorlesungen / Vilmar, August Friedrich Christian; ed by Piderit, Karl Wilhelm – Guetersloh: C Bertelsmann, 1874 – 2mf – 9 – 0-524-00192-8 – mf#1989-2892 – us ATLA [240]

Dogmatik : darstellung der christlichen glaubenslehre auf reformirt-kirchlicher grundlage / Boehl, Eduard – Amsterdam: Von Scheffer, 1887 – 2mf – 9 – 0-524-05368-5 – (incl bibl ref) – mf#1991-2274 – us ATLA [240]

Dogmatik / Kaftan, Julius – 1. und 2. Aufl. Freiburg i B: JCB Mohr, 1897 – (= ser Grundriss der theologischen wissenschaften) – 2mf – 9 – 0-7905-9984-8 – (incl bibl ref) – mf#1989-1709 – us ATLA [240]

Die dogmatik der evangelisch-lutherischen kirche : mit beruecksichtigung des dogmengeschichtlichen zunaechst den bekenntnistreuen geistlichen und den theologie-studierenden / Rohnert, Wilhelm – Braunschweig: H Wollermann, 1902 – 2mf – 9 – 0-524-00088-3 – (incl bibl ref and ind) – mf#1989-2788 – us ATLA [242]

Die dogmatik der evangelisch-reformirten kirche / Heppe, Heinrich – Elberfeld: RL Friderichs, 1861 – (= ser Schriften zur reformirten theologie) – 2mf – 9 – 0-524-00040-9 – (incl bibl ref) – mf#1989-2740 – us ATLA [242]

Die dogmatik des neunzehnten jahrhunderts : in ihrem inneren flusse und im zusammenhang mit der allgemeinen theologischen, philosophischen und literarischen entwicklung desselben / Muecke, Albert – Gotha: Friedrich Andreas Perthes, 1867 – 2mf – 9 – 0-8370-8845-3 – (incl indes) – mf#1986-2845 – us ATLA [240]

Dogmatique chretienne / Bovon, Jules – Lausanne: Georges Bridel, 1895-96 [mf ed 1989] – 3mf – 9 – 0-7905-2351-5 – (in french. incl bibl ref and ind) – mf#1987-2351 – us ATLA [240]

Dogmatische studien / Frank, Franz Hermann Reinhold – Erlangen:Andr. Deichert (Georg Boehme), 1892 – 1mf – 9 – 0-8370-3177-X – mf#1985-1177 – us ATLA [240]

Dogme de la naissance miraculeuse du christ *see* Collected works

Le dogme de la naissance miraculeuse du christ / Lobstein, Paul – Paris: Librairie Fischbacher, 1890. Chicago: Dep of Photodup, U of Chicago Lib, 1975 (1r); Evanston: American Theol Lib Assoc, 1984 (1r) – (= ser HIS etudes christologiques) – 1 – 0-8370-0560-4 – (incl bibl ref) – mf#1984-6059 – us ATLA [240]

Le dogme de la redemption *see* The doctrine of the atonement

Dogme et critique / Le Roy, Edouard – 4e ed. Paris: Bloud, 1907 – 1mf – 9 – 0-8370-9486-0 – (incl bibl ref and index) – mf#1986-3486 – us ATLA [200]

Le dogme grec / Bois, Henri – Paris: Librairie Fischbacher, 1893 – 1mf – 9 – 0-8370-7446-0 – (incl bibl ref) – mf#1986-1446 – us ATLA [250]

Die dogmen des christenthums *see* Revealed religion

Dogmengeschichte = Histoire des dogmes / Schwane, Joseph – [2nd augm et corr ed]. Paris: Gabriel Beauchesne. 6v. 1903-04 – 12mf – 9 – 0-8370-9112-8 – (french. incl bibl ref and index) – mf#1986-3112 – us ATLA [210]

Die dogmengeschichte der alten kirche : periode der patristik / Thomasius, Gottfried; ed by Bonwetsch, Gottlieb Nathanael – 2. aufl. Erlangen: A Deichert, 1886 [mf ed 1990] – (= ser Die christliche dogmengeschichte als entwicklungs-geschichte des kirchlichen lehrbegriffs 1) – 2mf – 9 – 0-7905-9645-8 – (incl bibl ref) – mf#1989-1370 – us ATLA [240]

Die dogmengeschichte der alten kirche *see* History of doctrines in the ancient church

Die dogmengeschichte der mittelalters und neuzeit *see* History of doctrines in the middle and modern ages

Die dogmengeschichte des mittelalters von christologische standpunkte : oder, die mittelalterliche christologie vom 8. bis 16. jahrhundert / Bach, Joseph – Wien: W Braumueller, 1873-75 [mf ed 1990] – 2v on 3mf – 9 – 0-7905-3753-2 – (text in german, notes in latin. incl bibl ref) – mf#1989-0246 – us ATLA [240]

Dogmengeschichtliche tabellen / Werner, Johannes – 2. stark verm Aufl. Gotha: FA Perthes, 1898 – 1mf – 9 – 0-7905-8235-X – mf#1988-6135 – us ATLA [240]

Dogmengeschichtliches lesebuch / ed by Rinn, Heinrich – Tuebingen: JCB Mohr, 1910 – 2mf – 9 – 0-7905-9612-1 – (incl bibl ref) – mf#1989-1337 – us ATLA [240]

Dogmhistoria : den kristna laerobildningens utvecklingsgang fran den efterapostoliska tiden till vara dagar / Aulen, Gustaf – Stockholm: PA Norstedt, 1917 – 1mf – 9 – 0-524-08227-8 – (incl bibl ref) – mf#1993-2002 – us ATLA [240]
Dognon, Paul *see* Les institutions politiques et administratives du pays de languedoc du 13e siecle aux guerres de religion.
Dogs of the world / Schneider-Leyer, Erich – London, England. 1970 – 1r – us UF Libraries [590]
Dohartua co : business directory of indonesia – Djakarta, 1955 – 15mf – 9 – mf#SE-615 – ne IDC [959]
Doherty, Kathryn B *see* Jordan waters conflict
Dohm, Chr Konr Wilh von *see* Materialien fuer die statistick und neuere staatengeschichte
Dohm, Chr Wilh von *see* Encyklopaedisches journal
Dohms, Evelyn *see* Die bestimmung des fachlichkeitsgrades von texten der industriesoziologie des englischen und deutschen
Dohn, Walter *see* Das jahr 1848 im deutschen drama und epos
Dohrman, Richard *see* Cross of baron samedi
Dohrmann, Hanns Arved *see* Die brandstifter von karabanowka
Doi song moi = New life / HEW Refugee Task Force [US] – v1 n1-v4 n1 [1975 aug 16/31-1978 oct/nov] – 1r – 1 – mf#601263 – us WHS [360]
Doiarenko, A G *see*
– Izbrannye raboty i stati
– Puti k podniatiiu urozhainosti ozimykh khlebov
Doin, Alexandre *see* Napoleon et l'europe
Doin, Ernest *see*
– Le conscrit ou le retour de crimee
– Le conscrit, ou, le retour de crimee
– Le desespoir de jocrisse
– Le desespoir de jocrisse ou les folies d'une journee
– Le diner interrompu ou nouvelle farce de jocrisse
– Joachim murat, roi des deux siciles
– Joachim murat roi des deux-siciles
– La mort du duc de reichtadt sic, fils de l'empereur napoleon 1er
– Le pacha trompe, ou, les deux ours
Doinel, Jules *see* Lucifer demasque
Doings in grain at milwaukee / Milwaukee Chamber of Commerce – v1 n1-12 [1912 feb-1913 jan] – 1r – 1 – mf#1055669 – us WHS [380]
Dois anos no brasil / Biard, Francois Auguste – Sao Paulo, Brazil. 1945 – 1r – us UF Libraries [972]
Dois arautos da democracia / Carneiro, Levi – Rio de Janeiro, Brazil. 1954 – 1r – us UF Libraries [972]
Dois discursos / Vargas, Getulio – Rio de Janeiro, Brazil. 1940 – 1r – us UF Libraries [972]
Os dois mundos : illustracao para portugal e brasil – Rio de Janeiro, RJ: Typ Ch Unsinger, 31 ago 1877-31 jul 1878 – (= ser Ps 19) – mf#P25,03,06 – bl Biblioteca [079]
Doistoriia, preistoriia, istoriia i myshlenie : k voprosu o metode i kadrakh po obshchestvennym naukam / Marr, Nikolai Iakovlevich – Leningrad: Izd-vo GAIMK, 1933 [mf ed 2002] – (= ser Izvestiia gosudarstvennoi akademii istorii material'noi kul'tury 74) – 1r – 1 – (filmed with: rech' stalina: rech' po radio predsedatelia... [1941] & other titles) – mf#5226 – us Wisconsin U Libr [900]
Doke, Clement Martin *see* Text book of lamba grammar
Doke, Clement Martyn *see*
– Contributions to the history of bantu linguistics
– English-lamba vocabulary
– Graded zulu exercises
– Grammar of the lamba language
– Lamba folk-lore
– Lambas of northern rhodesia
– Outline grammar of bantu
– Southern bantu languages
– Textbook of southern sotho grammar
– Textbook of zulu grammar
Doke, Clement-Martyn *see* Zulu-english dictionary
Dokhody i raskhody zemstv 34-kh gubernii po smetam – Spb, 1908-1915. 5v – 76mf – 8 – mf#RZ-165 – ne IDC [314]
Dokimion historias tes hellenikes glosses / Maurophrydes, Demetrios I – Smyrne: Ek tou Typographeiou tes Amaltheias, 1871 – 2mf – 9 – 0-524-08180-8 – mf#1992-1166 – us ATLA [450]
Dokimion symbolikes ex epopseos orthodoxou / Androutsos, Chrestos – En Athenais: Dionysios G Eustratios, 1901 – 1mf – 9 – 0-8370-7522-X – (incl bibl ref) – mf#1986-1522 – us ATLA [240]
Doklad komissii dlia sostavleniia proekta ustava vladimirskogo dvorianskogo banka – Vladimir, 1873 – 1mf – 9 – mf#REF-339 – ne IDC [332]
Doklad [orlovskoi gubernskoi zemskoi] komissii po strakhovaniiu stroenii ot ognia 18 ocherednomu gubernskomu zemskomu sobraniiu 1893 goda – Orel, 1893 – 1mf – 9 – mf#REF-428 – ne IDC [332]
Doklad po evreiskomu voprosu tsentral'nago komiteta partii k-d : prikazy vlastei; raznye dokumenty; istoriia odnogo pogroma / Konstitutsionno-demokraticheskaia partiia – [Bern?]: Izd Zagranichnago Komiteta Bunda, 1916 [mf ed 2004] – (= ser Razgrom evreev v rossii 2) – 1r – 1 – (filmed with: bog i den'gi / vl krymov (v1-2 1926)) – us Wisconsin U Libr [939]
Doklad po konstitutsionnym voprosam na 7 s"ezd sovetov soiuza ssr. 5 fevralia 1935 g / Enukidze, Avel'Sofronovich – Moskva Partizdat TsK VKP (b), 1935. 28. 4p LL-4013 – 1 – us L of C Photodup [340]
Doklad pravleniia iaroslavsko-kostromskogo zemel'nogo banka obshchemu sobraniiu gg aktsionerov marta 11 dnia 1879 goda – Iaroslavl', 1879 – 1mf – 9 – mf#REF-500 – ne IDC [332]
Doklad pravleniia pervogo obshchestva vzaimnogo kredita v leningrade godichnomu sobraniiu upolnomochennykh 21 noiabria 1926 g o deiatel'nosti obshchestva za vremia s 1 oktiabria 1925 g po 1 oktiabria 1926 g / Pervoe Obshchestvo Vzaimnogo Kredita v Leningrade – L, [1926] – 1mf – 9 – mf#REF-139 – ne IDC [332]
Doklad pravleniia po deiatel'nosti obshchestva za vremia s 1 oktiabria 1924 goda po 1 oktiabria 1925 goda / Pervoe Obshchestvo Vzaimnogo Kredita v Leningrade – L, [1925] – 1mf – 9 – mf#REF-138 – ne IDC [332]
Doklad pravleniia vtorogo rossiiskogo strakhovogo obshchestva uchrezhd : v 1935 godu, obshchemu sobraniiu gg. aktsionerov 24-go marta 1916 goda – Spb, 1916 – 1mf – 9 – mf#REF-407 – ne IDC [332]
Doklad revizionnoi komissii vsekobanka 3-mu ocherednomu obshchemu sobraniiu paishchikov za 1924-25 operats god / Vserossiiskii Kooperativnyi Bank – M, 1926 – 1mf – 9 – mf#REF-78 – ne IDC [332]
Doklad revizionnoi kommissii 40-mu sobraniiu upolnomochennykh tsentrosoiuza : reviziionnaia otsenka deiatelnosti tsentrosoiuza za vremia s 1 oktiabria 1924 g po 1 ianvaria 1936 g – 1926 – 164p 3mf – 9 – mf#COR-360 – ne IDC [335]
Doklad tsentrosoiuza sovetu narodnykh komissarov sssr – 1929 – 132p 2mf – 9 – mf#COR-325 – ne IDC [335]
Doklad viatskoi gubernskoi zemskoi upravy 2-mu ocherednomu gubernskomu zemskomu sobraniiu o zemskom banke – Viatka, 1869 – 1mf – 9 – mf#REF-349 – ne IDC [332]
Doklady. biochemistry / Akademiya nauk SSSR. Doklady – Dordrecht, Netherlands 1964+ – 1,5,9 – ISSN: 0012-4958 – mf#10879 – us UMI ProQuest [574]
Doklady. biological sciences / Akademiya nauk SSSR. Doklady – Dordrecht, Netherlands 1965+ – 1 – ISSN: 0012-4966 – mf#10819 – us UMI ProQuest [574]
Doklady. biological sciences sections – Ann Arbor MI 1962-64 – 1,5,9 – ISSN: 0886-7534 – mf#17720 – us UMI ProQuest [574]
Doklady. biophysics / Akademiya nauk SSSR. Doklady – Dordrecht, Netherlands 1964+ – 1,5,9 – ISSN: 0012-4974 – mf#10821 – us UMI ProQuest [530]
Doklady. botanical sciences / Akademiya nauk SSSR. Doklady – Dordrecht, Netherlands 1964+ – 1,5,9 – ISSN: 0012-4982 – mf#10820 – us UMI ProQuest [580]
Doklady. chemical technology / Akademiya nauk SSSR. Doklady – Dordrecht, Netherlands 1957+ – 1,5,9 – ISSN: 0012-4990 – mf#10823 – us UMI ProQuest [540]
Doklady. chemistry / Akademiya nauk SSSR. Doklady – Dordrecht, Netherlands 1956+ – 1,5,9 – ISSN: 0012-5008 – mf#10822 – us UMI ProQuest [540]
Doklady chlena-proizvoditelia del komiteta s"ezdov po voprosam, podlezhashchim obsuzhdeniiu iv-go s"ezda predstavitelei uchrezhdenii russkogo zemel'nogo kredita – Spb, 1879 – 3mf – 9 – mf#REF-323 – ne IDC [332]
Doklady i otchety : russkoe bibliologicheskoe obshchestvo – Spb., 1908, 1913. n1-2 – 6mf – 9 – mf#R-4332 – ne IDC [077]
Doklady i prigovory sostoiavshieia v pravitelstvuiushchem senate v tsarstvovanie petra velikogo – 1880-1901. 6 vols – 182mf – 8 – mf#R-3431 – ne IDC [947]
Doklady i soobshcheniia – Ikutsk: IAkutskoe knizhnoe izd-vo, 1973-1975. [v3,5. 1973] – us CRL [077]
Doklady. mathematics / Akademiya nauk SSSR. Doklady – Providence RI 1992-96 – 1,5,9 – (cont: soviet mathematics – doklady) – ISSN: 1064-5624 – mf#13418,02 – us UMI ProQuest [510]
Doklady. physical chemistry / Akademiya nauk SSSR. Doklady – Dordrecht, Netherlands 1957+ – 1,5,9 – ISSN: 0012-5016 – mf#10824 – us UMI ProQuest [540]
Doktor doellinger und die petition der bischoefe an's concil – Trier: Fr. Lintz, 1870 – 1mf – 9 – 0-8370-8215-3 – mf#1986-2215 – us ATLA [240]
Doktor johannes faust : puppenspiel in vier aufzuegen / Simrock, Karl Joseph; ed by Petsch, Robert – Leipzig: P Reclam, [1923] (mf ed 1990) – 1r – 1 – (filmed with: fausto) – us Wisconsin U Libr [790]
Doktor pomeranus, johannes bugenhagen : ein lebensbild aus der zeit der reformation / Hering, Hermann – Halle: Verein fuer Reformationsgeschichte, 1888. (Schriften des Vereins fuer Reformationsgeschichte; 6. Jahrg., 1. Stueck, Nr. 22) – 1mf – us ATLA [240]
Doktor pomeranus, johannes bugenhagen : ein lebensbild aus der zeit der reformation / Hering, Hermann – Halle: Verein fuer Reformationsgeschichte 1888 [mf ed 1990] – (= ser Schriften des vereins fuer reformationsgeschichte 6/1/22) – 1mf – 9 – 0-7905-4687-6 – (incl bibl ref) – mf#1988-0687 – us ATLA [242]
Die doktorsfamilie : novelle / Dimt, Peter – Feldpostausg. [Wien]: Wiener Verlag, 1944 [mf ed 1989] – (= ser Kleinbuchreihe suedost 81) – 60p – 1 – mf#7177 – us Wisconsin U Libr [830]
Dokumentarische information-panorama ddr – Berlin: Panorama DDR – 1 – (continued by: dokumentation (panorama ddr)) – us Wisconsin U Libr [073]
Dokumentation – Evangelischer Pressedienst. Zentralredaktion [wkly] – 1 – us Wisconsin U Libr [242]
Dokumentation der zeit (erschienen bis 1972) – 1949-1972 – 750mf – 1 – gw Mikropress [900]
Dokumentation zur juedischen kultur in deutschland 1840-1940, abt 1 : die zeitungsausschnittsammlung steininger. teil 1: bildende kuenstler; teil 2: darstellende kuenstler / ed by Archiv Bibliographia Judaica – [mf ed 1995-96] – 101mf – 9 – silver €2625.00 – ISBN-10: 3-598-33317-X – ISBN-13: 978-3-598-33317-0 – gw Saur [700]
Dokumentation zur juedischen kultur in deutschland 1840-1940, abt 2 : musiker / ed by Archiv Bibliographia Judaica e.V. – [mf ed 1996-97] – 102mf (1:24) – 9 – silver €2625.00 – ISBN-10: 3-598-33322-6 – ISBN-13: 978-3-598-33322-4 – gw Saur [780]
Dokumentation zur juedischen kultur in deutschland 1840-1940, abt 3 : schriftsteller / ed by Archiv Bibliographia Judaica e.V. – [mf ed 1998-99] – 124mf (1:24) – 9 – silver €2625.00 – ISBN-10: 3-598-33326-9 – ISBN-13: 978-3-598-33326-2 – gw Saur [939]
Dokumentation zur juedischen kultur in deutschland 1840-1940, abt 3. neue folge : schriftsteller / ed by Archiv Bibliographia Judaica e.V. – [mf ed 2004-05] – 102mf (1:24) – 9 – silver €2625.00 – ISBN-10: 3-598-34576-3 – ISBN-13: 978-3-598-34576-0 – gw Saur [939]
Dokumentation zur juedischen kultur in deutschland 1840-1940, abt 4 : publizisten und geisteswissenschaftler / ed by Archiv Bibliographia Judaica e.V. – [mf ed 1999-2000] – 108mf (1:24) – 9 – silver €2625.00 – ISBN-10: 3-598-33480-X – ISBN-13: 978-3-598-33480-1 – gw Saur [939]
Dokumentation zur juedischen kultur in deutschland 1840-1940, abt 5 : teil 1: religionswissenschaftler, philologen und lehrer; teil 2: rabbiner, zionisten, hebraisten, jargonschriftsteller / ed by Archiv Bibliographia Judaica e.V. – [mf ed 2001-02] – 124mf in 3 installments – 9 – silver €2625.00 – ISBN-10: 3-598-33484-2 – ISBN-13: 978-3-598-33484-9 – gw Saur [939]
Dokumentation zur juedischen kultur in deutschland 1840-1940, abt 6 : persoenlichkeiten des oeffentlichen lebens / ed by Archiv Bibliographia Judaica e.V. – [mf ed 2003-04] – 102mf (1:24) in 3 installments – 9 – silver €2625.00 – ISBN-10: 3-598-34570-4 – ISBN-13: 978-3-598-34570-8 – gw Saur [939]
Dokumentation zur juedischen kultur in deutschland 1840-1940, abt 7 : zionisten, wissenschaftler, juristen = Documentation on jewish culture in germany 1840-1940, vol 7 / ed by Archiv Bibliographia Judaica e.V. – [mf ed 2006] – 101mf (1:24) in 3 installments – 9 – €2625.00 – ISBN-10: 3-598-34580-1 – ISBN-13: 978-3-598-34580-7 – gw Saur [939]
Dokumentation zur juedischen kultur in deutschland 1840-1940, abt 8 : geschichte des judentums = Documentation on jewish culture in germany 1840-1940, vol 8 – [mf ed 2007/08] – 120mf (1:24) in 3 installments – 9 – €2625.00 – ISBN-10: 3-598-34584-4 – ISBN-13: 978-3-598-34584-5 – gw Saur [939]
Dokumentationen zum parlamentarismus : eine microformdokumentation zur parlamentsgeschichte – 15,288mf – 9 – (individual titles also listed separately) – gw Olms [323]
Dokumente der frauen / ed by Lang, Marie – Wien/Leipzig 1899-1902 [mf ed 1999] – (= ser Hq 33) – 7v on 35mf – 9 – €190.00 – 3-89131-291-1 – gw Fischer [305]
Dokumente der frauen *see* Frauen-rundschau
Dokumente des revolutionaeren kambodscha : reden / Norodom Sihanouk, Prince et al – 1.aufl. Heidelberg [Germany]: Hrsg vom Indochina-Komitee; Mannheim: Sendler [mf ed 1989] – 1r with other items – 1 – (nachdruck aus bulletins der koniglichen botschaft kambodschas in der ddr [unautorisierte ubersetzung]) – mf#mf-10289 seam reel 021/06 [§] – us CRL [959]
Dokumente vornehmlich zur geschichte der valdesier und katharer / ed by Doellinger, Johann Joseph Ignaz von – Muenchen: C.H. Beck, 1890 – 1mf – 9 – 0-7905-5267-1 – (incl bibl ref) – mf#1988-1267 – us ATLA [940]
Dokumente zu luthers entwicklung (bis 1519) / ed by Scheel, Otto – Tuebingen: J.C.B. Mohr, 1911 – (= ser Sammlung ausgewaehlter kirchen- und dogmengeschichtlicher Quellenschriften) – 1mf – 9 – 0-7905-8090-X – (incl bibl ref) – mf#1988-8026 – us ATLA [242]
Dokumente zum ablassstreit von 1517 / ed by Koehler, Walther – Tuebingen: Mohr, 1902 – (= ser Sammlung ausgewaehlter kirchen- und dogmengeschichtlicher Quellenschriften) – 1mf – 9 – 0-7905-5242-6 – (incl bibl ref) – mf#1988-1242 – us ATLA [240]
Dokumente zum aufbau des bayerischen staates / Bavaria (Germany). Bayerische Staatskanzlei – Muenchen, Germany. 1948 – 1r – us UF Libraries [324]
Dokumente zum weltkrieg 1914 / Bernstein, Eduard – Berlin: Buchhandlung Vorwaerts, 1914 [mf ed 1987] – 1 – mf#2037 – us Wisconsin U Libr [933]
Dokumenty zbrodni i meczenstwa / Borwicz, Michal – Krakow, Poland. 1945 – 1r – us UF Libraries [939]
Dokumenty zbrodni i meczenstwa, kolegium redakcyjne / ed by Borwicz, Michal Maksymilian – Krakow: Centralny Komitet Zydow Polskich, 1945. xv,214p – 1 – us Wisconsin U Libr [940]
Dolan, Edward F *see* Animal rights
Dolan, Gilbert et al *see* Ecclesia, the church of christ
Dolatowski, Elrun et al *see*
– Mikrofiche-edition der protokolle des politbueros des zentralkomitees der sozialistischen einheitspartei deutschlands bd 71
Dolbeare, Harwood B *see* Forewarnings of bank failure
Dolbey, Robert Valentine *see* Sketches of the east africa campaign
Dolby, Anastasia *see*
– Church embroidery ancient and modern practically illustrated
– Church vestments
Dolce, L *see* Imprese di diversi principi, duchi...
Dolce, Lodovico *see* Le prime imprese del conte orlando di m. lodovico dolce
[Dolce, Lodovico] *see*
– Aretino
– El nascimiento y primeras empressas del conde orlando
Dolch, A *see* Die verbreitung oberlaendischer mystikerwerke im niederlaendischen
Dolch, Oskar *see* Die deutsche treue
Dolci affetti madrigali a cinque voci – 1582 – (= ser Mssa) – 2mf – 9 – €35.00 – mfchl 133 – gw Fischer [780]
Dold, Alban *see*
– Das aelteste liturgiebuch der lateinischen kirche
– Getilgte paulus-und psalmtexte
– Ein hymnus abecedarius auf christus
– Die im cod vat reg lat 9 vorgeheftete liste paul lesungen fuer die messfeier
– Das irische palimpsest-sakramentar in clm 14429 [tab53-54]
– Konstanzer altlat propheten- und evang bruchstuecke mit glossen
– Lateinische fragmente der sapientialbuecher
– Lehrreiche basler brevierfragmente des 10. jahrhunderts
– Neue st galler vorhieronymianische prophtenfragmente
– Der palimpsestpsalter im cod sangallensus 91
– Das palimpsestsakramentar im cod aug 112
– Palimpsest-studien 2
– Palimpsest-studien
– Das prager sakramentar
– Prophetentexte
– Das sakramentar im schabcodex m 12 sup der bibliotheca ambrosiana
– Die vom missale romanum abweichenden lesetexte fuer die meszfeiern
– Vom sakramentar comes und capitulare zum missale
– Ein vorhadrianisches gregorianisches palimpsest-sakramentar
– Die zuericher und peterlinger messbuchfragmente
– Zwei bobbienser palimpseste
Dold, Gilbert William Frederick *see* Union of south africa

Die doldenbluetler als heilpflanzen im mittelalter : ein medizinhistorischer vergleich / Wemmer, Dagmar – (mf ed 1994) – 1mf – 9 – €30.00 – 3-8267-2054-7 – ISSN: DHS 2054 – gw Frankfurter [615]
Dole, Artemus Wood *see* Autobiography
Dole, Charles Fletcher *see*
– The coming religion
– The hope of immortality
– Jesus and the men about him
– The right and wrong of the monroe doctrine
– The theology of civilization
– What we know about jesus
Dole, William Peters *see* Carmen acadium
Dolgorukov, P *see* Listok, izdavaemyi kniazem petrom dolgorukovym
Dolgozok lapja – Tatabanya, Hungary. 1962-68; 1970-Mar 1980 – 30r – 1 – us L of C Photodup [079]
Dolin, Anton *see* Capricioso
Dolitzky, Menahem Mendel *see* Sheve sofer
Doll, Michael *see*
– Experimentelle studien und phantomuntersuchungen zur computertomographie des thorax
– Hautekzeme bei studenten der zahnheilkunde
Dollar express – Terre Haute IN. 1878 feb 21, oct 31 – 1r – 1 – (cont: terre haute express [wkly: 1867]; cont by: terre haute express [wkly: 1879]) – mf#856290 – us WHS [071]
Dollar, John E *see* Contextual interference in the motor domain
Dollar magazine [1833] : a literary, political, advertising, and miscellaneous newspaper – La Salle IL 1833 – 1 – mf#3975 – us UMI ProQuest [071]
Dollar magazine [1841] : a monthly gazette of current literature, music and art – La Salle IL 1841-42 – 1 – mf#3976 – us UMI ProQuest [071]
Dollar magazine [1848] – La Salle IL 1848-51 – 1 – mf#4444 – us UMI ProQuest [071]
Dollar newspaper – Philadelphia PA. 1846 aug 12-1848 dec 27, 1849 jan 3-1850 dec 25 – 2r – 1 – (cont by: home weekly and household newspaper) – mf#780673 – us WHS [071]
The dollar newspaper – Philadelphia, PA: A H Simmons, 1860; [1861-63] – 2r – 1 – us CRL [071]
Dollar times – Baraboo, Spring Green WI. 1878 sep 17-1880 may 4 – 1r – 1 – (cont: inter-county times) – mf#931523 – us WHS [071]
Dollar weekly / Richland Co. Bellville – mar 1872-sep 1873 [wkly] – 1r – 1 – mf#B2914 – us Ohio Hist [071]
Dollar weekly times / Hamilton Co. Cincinnati – (aug 1852-oct 1864) spotty [wkly] – 1r – 1 – mf#B3317 – us Ohio Hist [071]
Dollar weekly times / Hamilton Co. Cincinnati – jul 13 1854-nov 22 1855 – 1r – 1 – mf#B37468 – us Ohio Hist [071]
Dollard, James Bernard *see* Irish mist and sunshine
Dollars & sense – Boston MA 1986+ – 1,5,9 – ISSN: 0012-5245 – mf#15065 – us UMI ProQuest [332]
Dollar's worth / Gosse, Philip Henry – London, England. 18– – 1r – us UF Libraries [240]
Dolley, Georges *see* Heure avant
Dollhausen, Karin *see* Organisation und kultur
Dollier de Casson, Francois *see*
– Exploration of the great lakes, 1669-1670
– Histoire du montreal
– Histoire du montreal, 1640-1672
Dolling, Robert R *see* Ten years in a portsmouth slum
Dollinger, Hermann *see* Die dramatische handlung in klopstocks "der tod adams" und gerstenbergs "ugolino"
Dollman, Francis Thomas *see*
– An analysis of ancient domestic architecture
– Examples of antient pulpits existing in england
Dollmayr, Viktor *see*
– Die altdeutsche genesis
– Die geschichte des pfarrers von kalenberg
Doll's house / Isben, Henrik – London, England. 1910 – 1r – us UF Libraries [820]
Dolly Svobodny *see* Research on children's and young adult literature
Dolman, Alfred *see* In the footsteps of livingstone
Dolmatovskii, A M *see*
– Novye zakony o kooperatsii
– Zakony o kooperatsii
The dolmens of the pulney hills / Anglade, A & Newton, L V – Calcutta: Govt of India, Central Publication Branch, 1928 – (= ser Samp: indian books) – us CRL [730]
Dolmetsch, H *see* The historic styles of ornament
Dolomieu, Deodat G de *see* Voyage aux iles de lipari, fait en 1781
Dolomiten : tagblatt der suedtiroler – Bozen (I), 1954 1 jul-1989 apr – 1 – (filmed by other misc inst: 1970-) – gw Misc Inst [074]
Dolor bohemio / Ruiz de Silva, Enrique – Badajoz: Tipografia Correo de la Manana, 1916 – 1 – sp Bibl Santa Ana [946]
Dolor en la lirica cubana / Salazar Y Roig, Salvador – Habana, Cuba. 1925 – 1r – us UF Libraries [972]
El dolor supremo / Oton, Alfonso y Zaldivar, Ignacio – Caceres: imp santos floriano, s.a. – sp Bibl Santa Ana [946]
Dolores y Santa Marta, Felix *see* Oracion funebre
Doloroso novenario...jesus, de la soledad de badajoz – 1805 – 9 – sp Bibl Santa Ana [240]
Dolphin / Naval Submarine Base, New London – New London CT. 1983 oct 28-1984 mar 30, apr 6-aug 24, sep 7-1985 feb 8 – 3r – 1 – mf#918082 – us WHS [355]
Dolphin digest – v10 n1-10 [1982 may-1983 feb], v11 n1-6,8-11 [1983 apr-sep, nov-1984 feb], v12 n1-7,9-v15 n7,9-13 [1984 mar-sep, nov-1987 mar, sep-1988 mar] – 1r – 1 – (cont by: spectre [duke field fl]) – mf#1055686 – us WHS [355]
Dolphin log – Hampton VA 1987+ – 1,5,9 – ISSN: 8756-6362 – mf#16163.01 – us UMI ProQuest [370]
Dolz Y Arango, Maria Luisa *see* Liberacion de la mujer cubana por la educacion
Der dom – Paderborn DE, 1961 26 nov-1963 9 jun, 1963 16 sep-1966 – 1 – gw Misc Inst [074]
Dom a skola – Ruzomberok, Slovakia, 1886-97 – 3r – 1 – (slovak periodical) – us IHRC [073]
Dom austriaci caesares mariae annae magni caes.is ferd.di 3 filiae maximi regum phil.4 sponsae potentiss : sereniss.ae posteritatis exhibiti ab hortensio pallavicie soc iesu / Pallavicino, O – Mediolani: Typographia Ludovici Montii, 1649 – 3mf – 9 – mf#O-390 – ne IDC [090]
Dom in svet – Chicago IL, 1929* – 1r – 1 – (slovenian periodical) – us IHRC [073]
Dom joao 6 no brasil, 1808-1821 / Oliveira Lima, Manuel De – Rio de Janeiro, Brazil. v1-3. 1945 – 1r – us UF Libraries [972]
Dom pedro 1 e a marquesa de santos / Rangel, Alberto – Sao Paulo, Brazil. 1969 – 1r – us UF Libraries [972]
Dom pedro, empereur du bresil, roi de portugal, 17... / Dalbian, Denyse – Paris, France. 1959 – 1r – us UF Libraries [972]
Der dom von koeln und das muenster von strassburg / Goerres, Joseph von – Regensburg, 1842 (mf ed 1992) – 1mf – 9 – €24.00 – 3-89349-071-X – mf#DHS-AR 44 – gw Frankfurter [720]
Domaci noveny – Clarkson, NE: Ant. Odvarka, 1904-roc20 cis12. 25.brez.1924 (wkly mar 10 1911-24) [mf ed 1909-24 (gaps) filmed 1975?] – 6r – 1 – (absorbed by: narodni pokrok) – us NE Hist [071]
The domain of belief / Coke, Henry John – London: Macmillan, 1910 – 1mf – 9 – 0-7905-3658-7 – mf#1989-0151 – us ATLA [240]
Le domaine colonial de la france, ses resources et ses besoins : guide pratique de l'algerie, des colonies, des pays de protectorat et territoires a mandat... – Paris: F Alcan, 1922 – 1 – us CRL [960]
Domanig, Karl *see*
– Die fremden
Domanski, B *see* Die psychologie des nemesius
Domashniaia beseda dlia narodnogo chteniia – Spb., 1858-1877 – 443mf – 9 – (missing: 1863, v10-31; 1864, v27-52; 1867, v29; 1872, v19; 1874-1877) – mf#R-1583 – ne IDC [077]
Domashniaia biblioteka – Spb., 1869-1871. v1-12 – 88mf – 9 – mf#R-1584 – ne IDC [077]
Domasi community development scheme, 1949-54 / Thomson, T D – Zomba, Govt Print [1955] – us CRL [079]
Domaszewski, A *see* Die provincia arabia
Dombacnost – 1883, 1885, 1897 oct 6-1898 jun 29-1929/30 [small gaps] – 21r – 1 – us WHS [071]
Dombart, B *see* Zur textgeschichte der civitas dei augustins
Dombart, Bernhard *see* Zur textgeschichte der civitas dei augustins seit dem entstehen der ersten drucke
Dombay, Franz Lorenz von *see* Grammatica linguae mauro-arabicae
Dome : an illustrated magazine and review of literature, music, architecture and the graphic arts – La Salle IL 1897-1900 – 1 – mf#4187 – us UMI ProQuest [071]
Dome / Saskatchewan Government Employees' Association – 1975 jan-1986 mar – 1r – 1 – mf#603325 – us WHS [331]
Domencas africana no brasil / Freitas, Octavio – Sao Paulo, Brazil. 1935 – 1r – us UF Libraries [972]
Domenech De Calvo, Carmen *see* Alma
Domenech, Juan C *see* Veinticinco anos de ateneo, 1912-1937
Domenichi, Lodovico *see* Dialoghi di m. lodovico domenichi
Domenichi, Ludovico *see*
– Dialogo de las empresas militares
– Dialogo dell' imprese militari et amorose
– Dialogo dell' imprese militari et amorose di monsignor giovio vescovo di nocera
– Dialogo dell'imprese militari et amorose di monsignor giovio vescovo di nocera
Domenico scarlatti [1683-1757] : his harmonic equipment / Lagen, Peggy – U of Rochester 1931 [mf ed 19–] – 2mf – 9 – mf#fiche 73 – us Sibley [780]
Domenzain, Moises *see* El japon su evolucion, cultura, religiones
Domesday book and beyond / Maitland, Frederic William – Cambridge. 1907 – 1 – us CRL [941]
Domestic air news – serial n1-54 [1926 dec 18-1929 jun 15] – 1r – 1 – (cont by: air commerce bulletin) – mf#450050 – us WHS [380]
Domestic animal endocrinology – Oxford, England 1992+ – 1,5,9 – ISSN: 0739-7240 – mf#18970 – us UMI ProQuest [636]
Domestic architecture : being a second series of designs for cottages...and other residences / Goodwin, Francis – London 1834 – (= ser 19th c art & architecture) – 2mf – 9 – mf#4.2.1673 – uk Chadwyck [720]
Domestic architecture / Brown, Richard – London [1842] – (= ser 19th c art & architecture) – 10mf – 9 – mf#4.2.731 – uk Chadwyck [720]
The domestic architecture of the reign of queen elizabeth and james the first / Clarke, Thomas Hutchings – London 1833 – (= ser 19th c art & architecture) – 1mf – 9 – mf#4.2.1121 – uk Chadwyck [720]
Domestic duties of the family / Bailey, Rufus W – 1 – $50.00 – us Presbyterian [240]
Domestic experimentation / Lyons, Isabel J – s.l, s.l? 1937 – 1r – us UF Libraries [978]
Domestic letters of the department of state, 1784-1906 / U.S. Dept of State – (= ser General records of the department of state, 1910-1929 decimal file) – 171r – 1 – mf#M40 – us Nat Archives [324]
Domestic life in palestine / Rogers, Mary Eliza – Cincinnati: Poe & Hitchcock, 1867 – 1mf – 9 – 0-8370-6357-4 – mf#1986-0357 – us ATLA [956]
Domestic manners and customs of the hindoos of northern india : or, more strictly speaking, of the north west provinces of india / Dass, Ishuree – 2nd ed. Benares: E J Lazarus; London: Truebner, 1866 [mf ed 1995] – (= ser Yale coll) – xi/280p – 1 – 0-524-09268-0 – mf#1995-0268 – us ATLA [280]
Domestic manners and social condition of the white, coloured, and negro population of the west indies / Carmichael, A C – London 1833 [mf ed Hildesheim 1995-98] – 2v on 4mf – 9 – €120.00 – 3-487-26961-9 – gw Olms [306]
Domestic manners of the americans / Trollope, Frances – London 1832 [mf ed Hildesheim 1995-98] – 2v on 8mf – 9 – €160.00 – 3-487-27115-X – gw Olms [306]
The domestic slave trade of the southern states / Collins, Winfield Hazlitt – New York, 1904 – 1 – us CRL [305]
Domestic style – Toronto: New York Domestic Fashion, [1877?-189- or 19–] – 9 – mf#P05157 – cn Canadiana [740]
Domestic utensils *see* The index of american design (tiam)
Dominacion inglesa en la habana / Roig De Leuchsenring, Emilio – Habana, Cuba. 1929 – 1r – us UF Libraries [972]
Dominacion y guerras de espana en los paises bajos / Barrado Font, Francisco – Madrid: Est. Tip. El Trabajo, 1902. Rev. Tecnica de Infanteria y Caballeria – sp Bibl Santa Ana [946]
Dominance : the influence of circumstance on science in tanzania / Vitta, Paul B – Dar es Salaam: Dar es Salaam University Press, c1981 (mf ed 1995) – (= ser Inaugural lecture series (Chuo Kikuu cha Dar es Salaam) No 28) – 1mf – 9 – mf#Sc Micro F-14111 – us NY Public [960]
Dominance and defiance / Cohen, Ronald – Washington, DC. 1971 – 1r – us UF Libraries [025]
Die dominante : und andere erzaehlungen / Dehnert, Max – 2. aufl. Leipzig: H H Kreisel, 1944 [mf ed 1989] – 168p – 1 – mf#7174 – us Wisconsin U Libr [830]
Domingo cresi... / Barrado Manzano, Arcangel – Madrid: Archivo Ibero-Americano, 1959 – 1 – sp Bibl Santa Ana [240]
Domingo de la Anunciacion, Father *see* Doctrina christiana en lengua espanola y mexicana
Domingo de San Pedro de Alcantara *see* Celestial lirio
Domingue, Jules *see* Deux amours d'adrien
Dominguez Berrueta, Juan *see* Santa teresa de jesus. madrid 1934
Dominguez, Blanca *see* Arco iris
Dominguez, Franklin *see*
– Amigo desconocido nos aguarda
– Espera
– Se busca un hombre honesto
– Ultimo instante
Dominguez, Manuel *see* Instituto de segunda ensenanza de merida. memoria del curso 1934-35
Dominguez Navarro, Ofelia *see* De 6 a 6
Dominguez Peaz, Fidel *see*
– Duques de la torre
– Rasgos bibliograficos de santa teresa de jesus
Dominguez Perez, Francisco *see* Paginas libres
Dominguez, Rafael *see*
– Galeria universitaria. tomo 1. caracas, 1934
– Jose maria vargas
Dominguez Rodlan, Maria Luisa *see* Entre amor y musica
Dominguez Villagra, David *see* Odontologia sanitaria por...
Dominguez y Argaiz, Francisco Eugenio *see* Platicas de los principales mysterios de nuestra s[an]ta fee
Dominguez Y Roldan, Guillermo *see* Jesus castellanos
Dominiak, Kathleen M *see* The role of dance making for the older adult
Dominic, Saint *see* Acta capituli generalis ulyssiponae
Dominica chronicle – Roseau, Dominica. 5jan 1910-1939; 19 feb-12 nov 1941; 7 oct 1942-1952; 7 jan-7 mar 1953; 6 jul 1957-4 feb 1977; 11 mar 1967-28 dec 1968; 4 jan 1969-13 jun 1970 – 79r – 1 – uk British Libr Newspaper [079]
Dominica dial – Roseau, Dominica. 6 jan 1883-25 dec 1886; 1887-5 jul 1890 – 3r – 1 – uk British Libr Newspaper [079]
Dominica guardian – Roseau, Dominica. -w. 1 jul 1893-25 dec 1895; 1896-1913; 4 jun 1914-27 dec 1917; 3 sep 1918-15 dec 1921. 1914-19 very imperfect – 9r – 1 – uk British Libr Newspaper [072]
Dominica herald – Roseau, Dominica. -w. 10 jan 1959-31 dec 1960; 14 jan 1961-29 dec 1962; 1963-1964; 9 jan 1965-17 dec 1966; 7 jan 1967-21 dec 1968; 4 jan 1969-18 dec 1971; 8 jan 1972; 10 feb-14 apr 1973 – 6r – 1 – uk British Libr Newspaper [072]
Dominica tribune – Roseau. Dominica. -w. 1 jan-20 dec 1930; 1931-1932; 4 jul 1933-1937; 8 jan-30 jul 1938; 1939; 6 jan-21 dec 1940; 4 jan 1941-26 dec 1942; 1943-1944; 9 jun 1945-27 dec 1947; 10 jan 1948-5 may 1951. (Wanting jan-jun 1933; aug-dec 1938; dec 1944-may 1945; jul-sep 1946) – 12r – 1 – uk British Libr Newspaper [079]
Dominican – Roseau, Dominica. 30 oct 1839; 26 apr 1954; 16 jan 1856; jan 1864-dec 1866; 2 jan 1867-23 sep 1868; 11 nov 1868-29 dec 1869; 8 jan 1870-23 dec 1875; 6 jan 1877-1882; 4 jan-22 nov 1883; 3 jan 1884-10 dec 1885; 7 jan 1886-27 dec 1888; 8 oct 1891; 21 jun-31 dec 1894; 3 jan-26 dec 1895; 2 jan 1896-25 oct 1906 – 12r – 1 – uk British Libr Newspaper [079]
The dominican order and convocation : a study of the growth of representation in the church during the thirteenth century / Barker, Ernest, Sir – Oxford: Clarendon Press, 1913 – 1mf – 9 – 0-7905-7317-2 – (incl bibl ref) – mf#1989-0542 – us ATLA [240]
Dominican province of saint joseph : justice studies – 1977-80 [complete] – 1r – 1 – mf#ATLA S0458 – us ATLA [241]
Dominican province of saint joseph : news digest – v1-23. nov 1958-82 [complete] – 1r – 1 – ISSN: 0159-7345 – mf#ATLA S0457 – us ATLA [241]
Dominican reality / Balaguer, Joaquin – Mexico City? Mexico. 1949 – 1r – us UF Libraries [972]
Dominican Republic *see*
– Codigo de procedimiento civil y legislacion comple...
– Codigo de procedimiento criminal de la republica d...
– Constitucion politica y reformas constitutionales
– Gaceta oficial
– Memorandum de los ministros plenipotenciarios
Dominican republic – Archivo General de la Nacion; Archivo de la Catedral – 66r – 1 – (coll incl: archivo general de la nacion: books from the ministry of foreign relations, books from the interior and police departments, books from archives of bayaguana; cathedral archives) – Pan-American Institute of Geography and History (IPGH) – us UMI ProQuest [972]
Dominican republic : the land columbus loved most – New York, NY. 1939? – 1r – us UF Libraries [972]
Dominican republic / United States. Commission Of Inquiry To Santo Dom – Washington, DC. 1871 – 1r – us UF Libraries [972]
Dominican Republic. Administracion General de Correos *see*
– Memoria...
– Memoria que al ciudadano ministro de correos y telegrafos presenta el ciudadano admor

Dominican Republic Archivo General De La Nacion *see*
- Samana, pasado y porvenir
- San cristobal de antano

Dominican Republic Comision Para El Estudio Del I *see*
- Capacidad de la republica dominicana
- Capacity of the dominican republic to absorb refug...

Dominican Republic Constitution *see*
- Constitucion de la republica dominicana

Dominican Republic Delegacion En La Cuarta *see* Cuarta conferencia internacional americana

Dominican Republic Direccion General De Estadisti *see*
- Cuarto censo nacional agropecuario, 1950
- Estudio estadistico de algunos aspectos
- Poblacion de la republica dominicana
- Primer censo de profesionales de la republica
- Republica dominicana

Dominican Republic. Direccion General de Estadistica y Censos *see*
- 21 anos de estadisticas dominicanas 1936-1956
- Anuario estadistico 1936-1954

Dominican Republic. Direccion General De La Cedula *see* Evolucion e importancia de la cedula

Dominican Republic Direccion Nacional De Turismo *see* Tourists' guide of ciudad trujillo, capital of the...

Dominican Republic Embajada Spain *see* Mas antigua universidad de america

Dominican Republic Laws, Statutes, Etc *see*
- Ley de organizacion universitaria
- Leyes para el comercio en vigor, recopiladas de la...
- Proyecto de codigo civil de la republica dominican
- Proyecto de codigo de comercio de la republica
- Proyecto de codigo de procedimiento civil de...

Dominican Republic. Laws, Statutes, etc *see* Codigo civil de la republica dominicana.

Dominican Republic Military Governor, 1919- *see* Santo domingo

Dominican Republic. Ministerio de Correos y Telegrafos *see* Memoria que al ciudadano presidente de la republica presenta el ciudadano ministro de correos y telegrafos, correspondiente al ano...

Dominican Republic. Ministerio de Hacienda y Comercio *see* Memoria que presenta al ciudadano presidente constitucional de la republica...

Dominican Republic. Ministerio de Relaciones Exteriores *see* Memoria que al ciudadano presidente de la republica presenta el secretario de estado en el despacho de relaciones exteriores

Dominican republic of today / White, John W – Trujillo, Peru. 1945 – 1r – us UF Libraries [972]

Dominican Republic. Oficina Nacional de Estadistica *see* Republica dominicana en cifras 1964-1969

Dominican Republic Presidencia Secretaria De Est... *see* Obra politico-economica y financiera

Dominican Republic Secretaria De Educacion Public *see* Trujillo, restaurador de la independencia

Dominican Republic Secretaria De Educacion Y Bell... *see* Homenaje de los musicos al excelentisimo

Dominican Republic. Secretaria de Estado de Finanzas *see* Memoria correspondiente al ejercicio del...

Dominican Republic. Secretaria de Estado de Hacienda *see* Memoria correspondiente al ano...que al ciudadano presidente de la republica presenta el senor...

Dominican Republic. Secretaria de Estado de Hacienda y Comercio *see*
- Exposicion del ministro de hacienda y comercio, mensaje del presidente de la republica al congreso nacional e informes de la comision de hacienda y comercio
- Informe...presenta el director general de estadistica referente al movimiento del ano...
- Memoria...

Dominican Republic. Secretaria de Estado del Tesoro *see* Memoria...que a su excelencia el honorable presidente de la republica...

Dominican Republic. Secretaria de Estado del Tesoro y Credito Publico *see*
- Informe...que al senor secretario de estado del tesoro y credito publico presenta el contralor y auditor general de la republica
- Memoria...

Dominican Republic Secretaria De Finanzas *see* Boletin especial

Dominican Republic Secretaria De Relaciones Exter... *see* Cancelacion de una mision diplomatica

Dominican Republic. Secretaria de Relaciones Exteriores *see*
- Memoria...
- Memorias correspondientes a los ejercicios de...

Dominican Republic Settlement Association, Inc *see* Informe del presidente honario

Dominican Republic Treaties, Etc, 1924-1930 *see* Tratado fronterizo dominico-haitiano

The dominican revival in the 19th century : being some account of the restoration of the order of preachers throughout the world under fr jandel the 73rd master-general / Devas, Raymund P – London, New York: Longmans, Green, 1913 [mf ed 1989] – 1mf – 9 – 0-7905-4342-7 – mf#1988-0342 – us ATLA [241]

Dominicana – Washington DC 1927-68 – 1 – mf#2674 – us UMI ProQuest [240]

Dominicanidad de pedro henriquez urena / Rodriguez Demorizi, Emilio – Ciudad Trujillo, Dominican Republic. 1947 – 1r – us UF Libraries [972]

Dominicanismo y educacion / Salazar, Joaquin E – Ciudad Trujillo, Dominican Republic. 1945 – 1r – us UF Libraries [972]

Dominicanismos / Patin Maceo, Manuel Antonio – Ciudad Trujillo, Dominican Republic. 1947 – 1r – us UF Libraries [972]

Dominicanizacion de la frontera en la era gloriosa / Estrada, Enrique – Ciudad Trujillo, Dominican Republic. 1945 – 1r – us UF Libraries [972]

Dominicanizacion fronteriza / Machado Baez, Manuel A – Ciudad Trujillo, Dominican Republic. 1955 – 1r – us UF Libraries [972]

Dominicans in early florida / O'Daniel – New York, 1930; Madrid, 1931 – 1 – sp Bibl Santa Ana [240]

Dominicans in early florida / Townsend, Anselm M – s.l, s.l? 1936 – 1r – us UF Libraries [978]

Dominicans. Province of the Holy Name *see* Forma electionis prioris in ordine praedicatorum

Dominicans. Provincia de San Hipolito Martir de Oaxaca [Mexico] *see* Por la provincia de s hipolyto martyr del sagrado orden de predicadores de oaxaca

Dominicans. Provincia de Santiago de Mexico *see* Acta capituli provincialis celebrati in hoc imperiali s p n dominici mexicano conventu

Dominicis, Saverio F de *see* Galilei e kant, o, l'esperienza e la critica nella filosofia moderna

Dominick, Mary F *see* Human rights, european politics, and the helsinki accord

Dominicker settlement – s.l, s.l? 193-? – 1r – us UF Libraries [978]

Dominicker settlement – s.l, s.l? 193-? – 1r – us UF Libraries [978]

Dominicos en el puerto rico colonial, 1521-1821 / Cuesta Mendoza, Antonio – Mexico City? Mexico. 1946 – 1r – us UF Libraries [972]

Dominicus de Flandria *see*
- In 12 libros metaphysicae aristotelis
- In d thomae aq commentaria super libros posteriorum analyticorum aristotelis

Dominicus, F C *see* Het huiselik en maatschappelik leven van se zuid-afrikaner in de eerste helft der 18de eeuw

Dominicus, Foort Cornelius *see* Het huiselik en maa schappelik leven van de zuid-afrikaner in de eerste helft der 18de eeuw

Dominik, Hans *see*
- Das erbe der uraniden
- Himmelskraft
- John workman, der zeitungsboy
- Kautschuk
- Koenig laurins mantel
- Land aus feuer und wasser
- Lebensstrahlen
- Die spur des dschingis-khan
- Das staehlerne geheimnis
- Treibstoff sr
- Vom schraubstock zum schreibtisch

Dominik, Heinrich *see* Die attacke

Dominio colonial hollandez no brasil / Watjen, Hermann Julius Eduard – Sao Paulo, Brazil. 1938 – 1r – us UF Libraries [972]

Dominio entre alas / Puigdollers, Carmen – New York, NY. 1955 – 1r – us UF Libraries [972]

Dominio insular de honduras : estudio historico-geo... / Castaneda S, Gustavo A – San Pedro Sula, Honduras. 1939 – 1r – us UF Libraries [972]

Dominion – Wellington, NZ. 5 oct 1907-feb 1908; may 1908-oct 1908; jan 1909-oct 1912; jan 1913-aug 1917; nov 1917-dec 1959; jan 1979-feb 1991 – 1 – (mar-apr 1908, nov-dec 1908, nov-dec 1912 unavailable) – mf#41.14 – nz Nat Libr [079]

The dominion – Wellington, 1991- – 12r per y – 1 – us UMI ProQuest [072]

Dominion Alliance for the Total Suppression of the Liquor Traffic *see*
- Drink and crime in canada
- Minutes of the annual meeting of the council of the dominion alliance, 1897

The dominion almanac and daily remembrancer for the year... – Ottawa: J Hopes, 187-?-18– – 9 – mf#A00155 – cn Canadiana [030]

Dominion Astrophysical Observatory *see* Publications of the dominion astrophysical observatory

The dominion at the west : a brief description of the province of british columbia, its climate and resources: the government prize essay, 1872 / Anderson, Alexander Caulfield – Victoria, BC?: s.n, 1872 – 2mf – 9 – mf#14036 – cn Canadiana [333]

Dominion bazaar – Yorkville [Toronto]: Dominion Bazaar Co, [1877?-1881?] – 9 – ISSN: 1190-660X – mf#P04552 – cn Canadiana [760]

Dominion church of england temperance journal – Toronto: A C Winton, [1886-18– or 19–] – 9 – mf#P04990 – cn Canadiana [170]

Dominion Commercial Travellers' Association *see* Constitution and by-laws...

Dominion dental journal – Toronto: Dominion Dental Journal Pub. Co, [1889-1934] – 9 – ISSN: 1189-640X – mf#P04219 – cn Canadiana [617]

Dominion Exhibition (3rd : 1881 : Halifax, NS) *see* Prize list and general regulations of the third annual agricultural, industrial, and mechanical exhibition...

The dominion gazetteer *see* The international railway and steam navigation guide

Dominion illustrated : a canadian pictorial weekly – Montreal: G E Desbarats [etc]. v1-7. jul 7 1888-dec 26 1891// – 3r – 1 – Can$335.00 – (cont as: dominion illustrated monthly) – cn McLaren [971]

Dominion illustrated monthly – Montreal: Sabiston Lithographic & Pub Co. v1-2 n6. feb 1892-aug/sep 1893// – 2r – 1 – Can$165.00 – (cont: the dominion illustrated, 1888-91) – cn McLaren [971]

Dominion illustrated [special numbers] – Montreal: Sabiston Lithographic & Pub Co 1891-92 – 1r – 1 – Can$110.00 – cn McLaren [971]

Dominion Iron and Steel Co *see* Trust deeds, acts of incorporation and statutes

Dominion mechanical and milling news – Toronto: Beaver Pub. Co, [1884?-1889?] – 9 – mf#P04279 – cn Canadiana [621]

The dominion monthly journal of music and general miscellany – Toronto: Sargant & Eldridge, [1876?-18– or 19–] – 9 – mf#P06077 – cn Canadiana [780]

The dominion musical journal – Toronto: Timms, [1891?-189 or 19–] – 9 – mf#P05976 – cn Canadiana [780]

Dominion National League *see* "Country before party"

Dominion Observatory [Canada] *see*
- Publications of the dominion observatory, ottawa
- Report of the chief astronomer for the year ending...

Dominion oddfellow – Napanee, Ont: Templeton and Beeman, [1881?-189- or 19–] – 9 – mf#P04192 – cn Canadiana [360]

The dominion of canada : a study of annexation / Aitken, William Benford – New York: Van Siclen, [1890?] [mf ed 1980] – 2mf – 9 – 0-665-02282-4 – (incl bibl ref) – mf#02282 – cn Canadiana [971]

The dominion of canada : with particulars as to its extent, climate, agricultural resources, fisheries, mines, manufacturing and other industries / Patterson, William John – [Montreal?: s.n.], 1883 [mf ed 1981] – 1mf – 9 – mf#11805 – cn Canadiana [917]

The dominion of canada and the canadian pacific railway / Wilson, William – [Victoria, BC?: s.n.], 1874 [mf ed 1981] – 1mf – 9 – mf#23932 – cn Canadiana [380]

The dominion of canada and the canadian pacific railway / Wilson, William – [Victoria, BC?: s.n.], 1874 [mf ed 1982] – 1mf – 9 – mf#30554 – cn Canadiana [380]

The dominion of canada, its interests, prospects and policy : an address to his fellow citizens / Gates, Hartley Baxter – Montreal?: J Lovell, 1872 – 1mf – 9 – mf#23741 – cn Canadiana [380]

The dominion of canada, its interests, prospects and policy : an address to his fellow citizens / Gates, Hartley Baxter – Montreal?: s.n, 1872 (Montreal: J Lovell) – 1mf – 9 – mf#23741 – cn Canadiana [380]

Dominion of canada, pacific railway and north-west territories – [S.l: s.n, 1886?] [mf ed 1982] – 1mf – 9 – mf#30147 – cn Canadiana [380]

The dominion of canada with newfoundland and an excursion to alaska : handbook for travellers / Karl Baedeker (Firm) – Leipzig: K Baedeker, 1900 – 5mf – 9 – (incl ind) – mf#32549 – cn Canadiana [917]

The dominion of christ : the claims of foreign missions in the light of modern religious thought and a century of experience / Pierce, William – London: H R Allenson, 1895 – 1mf – 9 – 0-8370-6597-6 – (incl indes) – mf#1986-0597 – us ATLA [240]

The dominion philatelist : published monthly in the interests of stamp collecting – Bellevile, Ont: H F Ketcheson, [1889?-189-?] – 9 – ISSN: 1190-6456 – mf#P04546 – cn Canadiana [760]

The dominion phrase book : or, the student's companion for practically acquiring the french and english languages / Darey, P J – Montreal: Dawson, 1871 [mf ed 1984] – 2mf – 9 – 0-665-05414-9 – (in english and french) – mf#05414 – cn Canadiana [410]

Dominion presbyterian – Montreal: Mount Royal Pub. Co, 1898-[1910] – 9 – mf#P04182 – cn Canadiana [242]

The dominion review : a monthly record of events and opinions in politics, religion and science – Toronto: C M Ellis, [1896-19–] – 9 – mf#P04180 – cn Canadiana [073]

The dominion review – Montreal: W Drysdale, [1882-188-?] – 9 – mf#P04259 – cn Canadiana [971]

Dominion school of telegraphy and railroading : prospectus]: train for good positions as telegraphers, station agents, freight and ticket clerks through day, evening or home study courses, toronto, canada – [Toronto?: The School?, c1917] – 1mf – 9 – 0-665-97976-2 – mf#97976 – cn Canadiana [380]

The dominion statist : a record of canada's progress since confederation – Ottawa: Citizen Print and Pub, [1890-189-or-19–] – 9 – mf#P05064 – cn Canadiana [971]

Dominion sunday times – jan 1967-apr 1967; jan 1975-apr 1981; jan-feb 1987; feb 1987-oct 1992; nov 1992-6 mar 1994// – 1 – (previously known as: the sunday times (wellington). ceased publ apr 1981. recommenced publ jan 1987. title change to dominion sunday times feb 1987-oct 1992. title change to sunday times on oct 25 1992, nov 1992-6 ma r1994. ceased publ 6 mar 1994. amalg with sunday star (auckland) on 6 mar 1994 to form sunday star times. not publ apr 1981-jan 1987) – mf#41.3 – nz Nat Libr [079]

Dominion watchman – Hamilton, Ont: G D Griffin, [1876?-187-?] – 9 – (cont by: the dominion watchman and national reformer) – mf#P05071 – cn Canadiana [320]

Dominion watchman and national reformer : a quarterly magazine which explains the causes of national depression... – Hamilton [Ont]: G D Griffin, [187-or 18–18–] – 9 – (cont: the dominion watchman) – mf#P04436 – cn Canadiana [320]

Dominique albert azuni's, senators und richters im handels- und seewesen-tribunal zu rissa : mitglieds mehrerer akademien der wissenschaften, reisen durch sardinien in geographischer, politischer und naturhistorischer hinsicht – Hamburg [u a] 1803 [mf ed Hildesheim 1995-98] – 2v on 5mf – 9 – €100.00 – 3-487-29183-5 – gw Olms [380]

Dominique de flandre : sa metaphysique / Mahieu, L – Paris, 1942 – 10mf – 8 – €19.00 – ne Slangenburg [110]

Dominique, Joseph Biancolelli *see* Nouveau theatre italien

Dominique, L C *see* Un gouverneur general de l'algerie, l'amiral de gueydon

Dominis, M A de *see* De republica ecclesiastica, libri 10

O domino : orgam critico, litterario, noticioso e recreativo – Santa Maria Madalena, RJ. 20 fev-10 abr 1910 – (= ser Ps 19) – mf#DIPER – bl Biblioteca [079]

Dominquez Alba, Bernardo *see* Chiquilinga

Dominquez Sosa, Julio Alberto *see* Ensayo historico sobre las tribus nonualcas y su c...

Dominus domi : or, the chateau saint-louis / Harper, John Murdoch – Quebec?: s.n, 1898? – 1mf – 9 – mf#05368 – cn Canadiana [810]

Dommer selv! : til selvprovelse: samtiden anbefalet: anden raekke / Kierkegaard, Soren – Kobenhavn: CA Reitzels, 1876 – 1mf – us ATLA [190]

Domont, Jean Marie *see* Prise de conscience de l'individu en milieu rural kongo

Domovina – New York NY, jan 7 1916-oct 26 1917 – 2r – 1 – (croatian newspaper) – us IHRC [071]

Domville-Fife, Charles William *see* Guatemala and the states of central america

Don : vech gaz – Novocherkassk [Obl voiska Don]: Osved otd Don Pravitel'stva 1918 [1918 n19 [22 [9] iiunia]] – (= ser Asn 1-3) – n19 [1919] item 125, on reel n27 – 1 – mf#asn-1 125 – ne IDC [077]

Don agustin arambul / Soler Y Gabarda, Geronimo – Habana, Cuba. 1876 – 1r – us UF Libraries [972]

Don Alvaro *see* Programa oficial de cultos y festejos en honor de la santa cruz

Don alvaro / Rivas, Angel De Saavedra – New York, NY. 1928 – 1r – us UF Libraries [960]

Don alvaro de sande cronista del desastre delos gelves / Munoz, Pedro de san – Badajoz: Diputacion Provincial de Badajoz, 1955 – 1 – sp Bibl Santa Ana [946]

Don alvaro de sande y la orden de malta / Serrablo Aguareles, Eugenio – Madrid: Revista de Archivos, Bibliotecas y Museos, 1955 – 1 – sp Bibl Santa Ana [240]

Don andres de arriola and the occupation of pensac... / Leonard, Irving Albert – s.l, s.l? 1932 – 1r – us UF Libraries [978]

DONEGAL

Don andres manjon y la libertad de ensenanza / Marquez, Gabino – Madrid: Razon y Fe, 1940 – 1 – sp Bibl Santa Ana [946]
Don andres manjon y la libertad de ensenanza (conclusion) / Marquez, Gabino – Madrid: Razon y Fe, 1940 – sp Bibl Santa Ana [946]
Don bartolome jose gallardo y la critica literaria de su tiempo / Sainz y Rodriguez, Pedro – New-York, Paris. 1921. Extracto de Revue Hispanique, tome 49 – sp Bibl Santa Ana [440]
Don bell reports : a weekly commentary, methods of survival if the red terror should strike – 1964 nov 20-1966 mar 11 – 1r – 1 – mf#2862639 – us WHS [360]
Don Benito *see*
– Feria y fiestas. 1967
– Ferias y fiestas, 1945
– Folleto primero don benito...
– Ordenanzas municipales para el regimen y gobierno de don benito. ano de 1928
Don benito maria de moxo y de francoli, arzobispo de charcas / Vargas Ugarte, Ruben – Buenos Aires, 1931; Madrid: Razon y Fe, 1931 – 1 – sp Bibl Santa Ana [240]
Don Benito. Spain. Ayuntamiento *see* Ordenanzas municipales
Don bernardo marquez de la vega / Chaves, Manuel – 1896 – 9 – sp Bibl Santa Ana [920]
Don bosco / Bayle, Constantino & Joergen, Juan – Madrid: Razon y Fe, 1944 – 1 – sp Bibl Santa Ana [946]
Don bosco : poema / Piedra-Bueno, Andres De – Buenos Aires, Argentina. 1941 – 1r – us UF Libraries [810]
Don boscocon dios / Ceria, E – Madrid: Razon y Fe, 1933 – 1 – sp Bibl Santa Ana [920]
Don carlos / Maurenbrecher, Wilhelm – Berlin: C Habel, 1876 – 1 – (incl bibl ref. includes handwritten contents at end) – us Wisconsin U Libr [430]
Don carlos / Schiller, Friedrich von; ed by Ibel, Rudolf – Frankfurt am Main: M Diesterweg, [between 1957 and 1960] – 1 – (incl bibl ref) – us Wisconsin U Libr [430]
Don carlos *see* Ueber die betonungsweise in der deutschen lyrik
Don carlos und hamlet / Thomas, Anneliese – Bonn a. Rh.: L Roehrscheid, 1933 – 1r – 1 – (incl bibl ref) – us Wisconsin U Libr [400]
Don constituyentes del ano 1824. biografias de don miguel ramos arizpe y d. lorenzo zavala, mexico. museo nacional de arqueologia, 1925 / Toro, Alfonso; ed by Bayle, Constantino – Madrid: Razon y Fe, 1928 – 9 – sp Bibl Santa Ana [920]
Don cristobal / Bauza, Guillermo – Barcelona, Spain. 1963 – 1r – us UF Libraries [972]
Don diego camacho y avila... / Rubio Merino, Pedro – Madrid: Archivo Ibero Americano, 1960 – 1 – sp Bibl Santa Ana [920]
Don diego en el carino / Corretjer, Juan Antonio – San Juan, Puerto Rico. 1956 – 1r – us UF Libraries [972]
Don diego hurtado de mendoza no fue el autor de "la guerra de granada". apuntes para un libro / Torre Yfranco Romero, Lucas de – Madrid: Fortanet, 1914. B.R.A.H. 64 y 65, pp. 461-501, 557-596 y 1915, pp. 28-47, 273-302 y 369-415 – sp Bibl Santa Ana [440]
Don diego portales / Soto Hall, Maximo – Guatemala, 1950 – 1r – us UF Libraries [972]
Don diego portales / Soto Hall, Maximo – Santiago, Chile. 1935 – 1r – us UF Libraries [972]
Don diego quijada... / Schols, France V & Adams, Elenor B – Madrid: Razon y Fe, 1940 – 1 – sp Bibl Santa Ana [946]
Don dorrigo gazette – Dorrigo. 1912-16, 1918-96 – 9 – at Pascoe [079]
Don emilio blanchet / Marban Escobar, Edilberto – Habana, Cuba. 1950 – 1r – us UF Libraries [972]
Don eugenio escobar prieto. apuntes de su vida / Solar y Taboada, Antonio – Badajoz: Imp. La Minerva Extremena, 1916 – 1 – sp Bibl Santa Ana [920]
Don fernando / Giraldo Londono, Pedronel – Medellin, Colombia. 1963 – 1r – us UF Libraries [972]
Don francisco de navarra, obispo de badajoz (1545-1556). sus interveneiones en trento sobre la obligacion episcopal de residir / Camacho Macias, Aquilino – Badajoz: Dip. Provincial, 1968 – sp Bibl Santa Ana [240]
Don francisco de paula romero y palomeque / Risco, Alberto – Jerez de la Frontera: Tip. de Salido Hermanos, 1916 – 1 – sp Bibl Santa Ana [920]
Don francisco de toledo, supremo organizador del peru. 1515-1585. madrid, 1935 / Levillier, Roberto – Madrid: Razon y Fe, 1935 – 1 – sp Bibl Santa Ana [972]
Don francisco moreno / Hargis, Modeste – s.l, s.l? 1939 – 1r – us UF Libraries [978]
Don gabino de gainza y otros estudios / Cid Fernandez, Enrique Del – Guatemala, 1959 – 1r – us UF Libraries [972]
Don gabriel jose de zuloaga en la governacion de v... / Pikaza, Otto – Sevilla, Spain. 1963 – 1r – us UF Libraries [972]
Don gerardo patrullo y otros desmayos / Gabaldon Marquez,Joaquin – Caracas, Venezuela. 1952 – 1r – us UF Libraries [972]
Don hernando cortes, marques del valle de oajace / Solana y Gutierrez, Mateo – Madrid: Razon y Fe, 1940 – sp Bibl Santa Ana [240]
Don hernando cortes...marques del valle de ocijaca / Solana y Gutierrez, Mateo – Mexico: Ediciones Botas, 1938 – sp Bibl Santa Ana [920]
Don jose maria plata y su epoca / Tamayo, Joaquin – Bogota, Colombia. 1933 – 1r – us UF Libraries [972]
Don juan / Aucouturier, Michel – Paris, France. 1946 – 1r – us UF Libraries [440]
Don juan de austria : principe de la cristiandad / Cordero Marina, Pedro – Plasencia: Imp. La Victoria, 1978 – 1 – sp Bibl Santa Ana [943]
Don juan de carvajal. un espanol al servicio de la santa sede. madrid, 1947 / Cereada, F & Gomez Canedo, Lino – Madrid: Razon y Fe, 1948 – 1 – sp Bibl Santa Ana [920]
Don juan de palafox y mendoza / Garcia, Genaro – Mexico City? Mexico. 1918 – 1r – us UF Libraries [972]
Don juan decadente / Melida, Jose Ramon – 1894 – 9 – sp Bibl Santa Ana [830]
Don juan nunez garcia / Mencos F, Agustin – Guatemala, 1956 – 1r – us UF Libraries [972]
Don juan prim y su labor diploamtica en mexico / Estrada, Genaro – Mexico City? Mexico. 1928 – 1r – us UF Libraries [972]
Das don juan-problem in der neueren dichtung / Heckel, Hans – Stuttgart: Metzler, 1915 [mf ed 1993] – (= ser Breslauer beitraege zur literaturgeschichte. neue folge 47) – 171p – 1 – (incl bibl ref) – mf#8014 reel 5 – us Wisconsin U Libr [410]
Don karlos in der geschichte und in der poesie / Pappritz, Richard – Naumburg a.S.: H Sieling, 1913 – 1r – 1 – (incl bibl ref) – us Wisconsin U Libr [430]
Don karlos in der geschichte und in der poesie *see*
– Aufgaben aus "die jungfrau von orleans"
– Das lied von der glocke
Don lope diaz de azmendariz marquez de carreyta de su magestad su mayrdomo : y virrey lugar thenieme, gouernador y captan generalissimo desta nueua espana, y predidente de la audiencia y chaucilleria real... / Contreras Gallardo, Pedro de – [Mexico City: s.n. 1638] – (= ser Books on religion...1543/44-c1800: manuales de rito) – 4mf – 9 – mf#crl-69 – ne IDC [241]
Don lorenzo suarez de figueroa y de mendoza : notas sobre su descendencia / Solar y Taboada, Antonio – Badajoz: Ediciones Arqueros, 1929 – 1 – sp Bibl Santa Ana [920]
Don manuel ruiz zorrilla...noticias sobre... / Perez, Miguel – 1883 – 9 – sp Bibl Santa Ana [920]
Don mirocletes / Gonzalez-Doria, Fernando De – Paris, France. 1932 – 1r – us UF Libraries [972]
Don narciso diaz de escovar. apuntes de su vida / Solar y Taboada, Antonio – Badajoz: Imp. La Constancia s.a. – 1 – sp Bibl Santa Ana [920]
Don pedro de alvarado, conquistador del reino de guatemala. madrid, 1927 / Altolaguirre, Angel de; ed by Bayle, Constantino – Madrid: Razon y Fe, 1928 – 9 – sp Bibl Santa Ana [972]
Don pedro de alvarado. obra postuma revisada por antonio fernandez del castillo / Fernandez del Castillo, Francisco – Guatemala: Editorial Cultura, 1945 – sp Bibl Santa Ana [920]
Don pedro de valdivia conquistador de chile / Arciniega, Rosa – Santiago de Chile: Editorial Nascimento, 1925 – 1 – sp Bibl Santa Ana [972]
Don pedro de valdivia...badajoz, 1928 / Rujula, Solar y Manzano; ed by Bayle, Constantino – Madrid: Razon y Fe, 1928 – 9 – sp Bibl Santa Ana [910]
Don pepe / Estenger, Rafael – Habana, Cuba. 1940 – 1r – us UF Libraries [972]
Don placido : dialogo del p don giovenale sacchi...dove cercasi: se lo studio della musica al religioso convenga o disconvenga / Sacchi, Giovenale – Pisa: Presso L Raffaelli 1786 [mf ed 19–] – 2mf – 9 – mf#fiche 894 – us Sibley [780]
Le don quichotte montrealais sur sa rossinante ou m dessaulles et la grande guerre ecclesiastique / Pelletier, Alexis – Montreal: Societe des ecrivains catholiques, 1873 – 2mf – 9 – mf#23881 – cn Canadiana [241]
Don quijote en america / Febres Cordero, Julio – Caracas, 1930; Madrid: Razon y Fe, 1931 – 1 – sp Bibl Santa Ana [440]
Don ramiro en america / Gandia, Enrique De – Buenos Aires, Argentina. 1934 – 1r – us UF Libraries [972]
Don rodrigo de bastidas / Navarro, Nicolas – Caracas, 1931; Madrid: Razon y Fe, 1931 – 1 – sp Bibl Santa Ana [920]
Don rodrigo de torres, primer marques de matallana / Solar y Taboada, Antonio – Badajoz: Arqueros – 1 – sp Bibl Santa Ana [920]
Don sanche d'aragon / Corneille, Pierre – Paris, France. 1844 – 1r – us UF Libraries [440]
Don sancho briceno, su monumento en trujillo. el arbol de los bricenos. caracas, 1929 / Davila, Vicente – Madrid: Razon y Fe, 1930 – 1 – sp Bibl Santa Ana [946]
Don sebastien de portugal / Foucher, Paul – Paris, France. 1838 – 1r – us UF Libraries [440]
Don tika sac : [poems] / Mhuin, Sa khan Kuiy to – Mantale: Lu thu Sa tan Ca tuik 1322 [1960] [mf ed 1990] – 1r with other items – 1 – (in burmese) – mf#mf-10289 seam reel 192/1 [§] – us CRL [810]
Don tomas cipriano de mosquera / Tamayo, Joaquin – Bogota, Colombia. 1944 – 1r – us UF Libraries [972]
Don vasco de quinoga... / Aquayo Spencer, Rafael – Madrid: Razon y Fe, 1940 – sp Bibl Santa Ana [920]
Dona carolina coronado / Castelar, Emilio – 1869 – 9 – sp Bibl Santa Ana [920]
Dona ines munoz, la mujer extremena, cunada de francisco pizarro, quetrajo el trigo y el olivo al peru / Cuneo-Vidal, Romulo – Madrid: Tip. de la Rev. de Bib. Archivos y Museos, 1928 – 1 – sp Bibl Santa Ana [946]
Dona leonor de alvarado y otros estudios / Recinos, Adrian – Guatemala, 1958 – 1r – us UF Libraries [972]
Dona marcela *see* Corrido qng bienang delanan dona marcela ampon ning metung a mercader qng cayarian portugal
Dona mencia de los nidos / Cidoncha, Marques de & Solar y Taboada, Antonio – Badajoz: Arqueros, 1943 – 1 – sp Bibl Santa Ana [920]
Dona velorio / Amado Blanco, Luis – Santa Clara, Cuba. 1960 – 1r – us UF Libraries [972]
Donahue, Judy M *see* The value of cardiorespiratory field tests
Donald, Elijah Winchester *see* The expansion of religion
Donald j detwiler, senate service 1917-18 : senate page – (= ser Us senate historical office oral history coll) – 1mf – 9 – $5.00 – us Scholarly Res [323]
Donald, J M *see* Peddie settlers' outpost
Donald mcleod's gloomy memories in the highlands of scotland : versus mrs harriet beecher stowe's sunny memories in (england), a foreign land, or, a faithful picture of the extirpation of the celtic race from the highlands of scotland – Toronto: printed for the aut by Thompson, 1857 [mf ed 1984] – 3mf – 9 – 0-665-46278-6 – mf#46278 – cn Canadiana [333]
Donald record – Donald OR: H E Hodges, 1916-18 [wkly] – 1 – us Oregon Lib [071]
Donald, William John *see* The canadian iron and steel industry
Donald's baptist church. abbeville county. south carolina : church records – 1876-1912 – 1 reel – 1 – $6.35 – (141p) – us Southern Baptist [242]
Donaldson, Augustus Blair *see*
– The bishopric of truro
– Five great oxford leaders
Donaldson, G *see* The making of the scottish prayer book of 1637
Donaldson, Gualt *see* Synopseos philosophiae moralis
Donaldson, J *see* A critical history of christian literature and doctrine
Donaldson, J A *see* Cultivation of flax
Donaldson, James *see*
– The westminister confession of faith and the thirty-nine articles of the church of england
– Woman
– Woman: her position and influence in ancient greece and rome, and among the early christians
Donaldson, Joh William *see* Christian orthodoxy reconciled with the conclusions of modern biblical learning
Donaldson, John William *see*
– A comparative grammar of the hebrew language
– Farewell sermon
– Jashar
Donaldson, Margaret E *see* The council of advice at the cape of good hope, 1825-1834
Donaldson, Stuart Alexander *see* Church life and thought in north africa a.d. 200
Donaldson, Thomas *see* The public domain, its history...and disposition to 1880
Donaldson, Thomas Leverton *see*
– Architectura numismatica
– Pompeii
– Preliminary discourse...on architecture
– A review of the professional life of sir john soane
Donan, Peter *see* Memoir of jacob creath, jr
Donanma [donanma mecmuasi] – Istanbul: Matbaa-i Hayriye ve Suerekasi, Matbaa-i Ahmet Ihsan, 1910-13, 1914-19. Sahibi: Donanma-i Osmani Muavenet-i Milliye Cemiyeti Merkez-i Umumisi. n1, 16/64, 31/79, 83-84, 86, 102-103, 80/129-143/192. mart 1326-1 nisan 1335 [1910-19] – (= ser O & t journals) – 30mf – 9 – $505.00 – (publ mthly 1910-13 as: donanma mecmuasi) – us MEDOC [956]
Donat, Walter FW K *see* Die landschaft bei tieck und ihre historischen voraussetzungen
Donatelle, Rebecca J *see*
– Comparative analysis of factors influencing participation in an employee health promotion program, including characterizations of participants and nonparticipants
– Effectiveness of selected components in behavioral weight-loss interventions
Donatello, seine zeit und schule / [Donatello] Semper, H – Wien, 1875. v9 – 5mf – 9 – mf#0-517 – ne IDC [700]
[Donatello] Semper, H *see* Donatello, seine zeit und schule
Donath, Andreas *see* Der vorhang zu und alle fragen offen
Donato, Baldassare *see*
– Canto di baldassare donato il primo libro di madrigali a cinque e a sei voci con tre dialoghi a sette novamente per antonio gardano
– Il primo libro de motetti a cinque, a sei, et otto voci
– Il secondo libro de madrigali a quatro voci
Donato, Messias Pereira *see* Movimento sindical operario no regime capitalista
Donatus Ortigraphus *see* Ars grammatica
Donatus und augustinus : oder, der erste entscheidende kampf zwischen separatismus und kirche. ein kirchenhistorischer versuch / Ribbeck, Ferdinand – Elberfeld: Baedeker, 1858 – 2mf – 9 – 0-524-04145-8 – (incl bibl ref) – mf#1990-1215 – us ATLA [240]
Die donau – Apatin (YU), 1940 6 jan-31 oct – 1r – 1 – gw Misc Inst [077]
Die donau – Apatin, Yugoslavia. 1938-39; 1941-43 – 2r – 1 – us L of C Photodup [949]
Donau-bodensee-zeitung *see*
– Hohenzollerische volkszeitung
– Leutkircher wochenblatt
– Riedlinger zeitung
– Schwaebische zeitung [main edition]
– Wuerttembergisches seeblatt
Donau-bulgarien und der balkan. historisch-geographisch-etnographische reisestudien aus den jahren 1860-1879 / Kanitz, F P – Leipzig, 1882. 3v – 29mf – 9 – mf#R-3879 – ne IDC [914]
Donau-kurier – Budapest (H), 1933 15 nov-1939/40 [gaps] – 1r – 1 – gw Misc Inst [077]
Donau-kurier – Ingolstadt DE, 1945 11 dec-1968 – 70r – 1 – (filmed by misc inst: 1969- [ca 9r/yr]) – gw Mikrofilm; gw Misc Inst [074]
Donau-zeitung – Augsburg DE, 1987- – 8r/yr – 1 – gw Misc Inst [074]
Donau-zeitung – Passau DE, 1848 jan-jun, 1849 – 1r – 1 – gw Misc Inst [074]
Donau-zeitung *see* Schwaebische landeszeitung
Donauzeitung – Belgrad (YU), 1941 12 aug-1944 25 jul [gaps] – 13r – 1 – (filmed by misc inst: 1941 15 jul-1944 29 jun; 1943 [2r]) – uk British Libr Newspaper; gw Misc Inst [074]
Doncaster labour party records, 1920-51 – (= ser Labour party in britain, origins and development at local level. series 1) – 4r – 1 – (int by d e martin) – mf#97298 – uk Microform Academic [325]
Doncaster, Phebe *see* John stephenson rowntree, his life and work
Doncel, Fernando *see* Felipe 5 en moraleja, ano de 1704
Doncel y Ordaz, Jose *see*
– Escritos anejos. serios y humuristicos
– Fabulas morales satiricas y...
Donde acaban los caminos / Monteforte Toledo, Mario – Guatemala, 1953 – 1r – us UF Libraries [972]
Donde canta el tocoloro / Yanes, Leoncio – Santa Clara, Cuba. 1963 – 1r – us UF Libraries [972]
Donde llegan los pasos / Lars, Claudia – San Salvador, El Salvador. 1953 – 1r – us UF Libraries [972]
Donde renace la esperanza / Ramos, Lilia – San Jose, Costa Rica. 1963 – 1r – us UF Libraries [972]
Dondoli B, Cesar *see* Estudio geoagronomico de la region oriental de la...
Donegal democrat – Ballyshannon, Ireland. 1921; 1925; 1986-92 – 23r – 1 – uk British Libr Newspaper [072]
Donegal independent – Ballyshannon, Ireland. apr 1884; jun 1894; aug 1894-1896 – 6r – 1 – uk British Libr Newspaper [072]
Donegal peoples press – Donegal, Ireland. 1986-92 – 14r – 1 – (aka: peoples press donegal derry and tyrone news) – uk British Libr Newspaper [072]
Donegal. Presbytery (Pres. Ch. in the U.S.A. Old School) *see* Minutes, 1843-1870

Donegal. Presbytery (Pres. Church in the USA) *see* Minutes
Donegal vindicator – Ballyshannon, jun 1906-1911, jun 1921-1956 – 28r – 1 – ie National [072]
Donegal vindicator – [Northern Ireland] Belfast jan 1897-dec 1920, jan 1931-dec 1950 [mf ed 2002] – 34r – 1 – (missing: 1914-15) – uk Newsplan [072]
Donegal vindicator etc – Ballyshannon, Ireland. Feb 1889-1896; jun 1914-1915; 1930; 7 jan 1950 – 6 1/2r – 1 – uk British Libr Newspaper [072]
Donehoo, James DeQuincey *see* The apocryphal and legendary life of christ
Donehoo, James Ramsey *see* The new testament view of the old testament
Donelaitis, Kristijonas *see* Das jahr in vier gesaengen
Donelson, Andrew J *see* Papers
Donelson view baptist church. nashville, tennessee : church records – Feb 1922-Sept 1956. Formerly Seventh Baptist Church, name changed 1962 – 1 – us Southern Baptist [242]
Donetskii kolokol – Lugansk, Ukraine 1906-07 [mf ed Norman Ross] – 1 – mf#nrp-871 – us UMI ProQuest [077]
Donetskii proletarii [kharkiv] : organ oblastnogo komiteta donetskogo i krivorozhskogo basseјna i khar'kovskogo kom vkp(b) – Khar'kov, Ukraine 1917-19 [mf ed Norman Ross] – 1r – 1 – mf#nrp-691 – us UMI ProQuest [077]
Donetskii proletarii [lugansk] : organ luganskogo komiteta rsdrp(b) – Lugansk, Ukraine 1917 [mf ed Norman Ross] – 2r – 1 – mf#nrp-872 – us UMI ProQuest [077]
Donetskoe slovo : ezhedn obshchestv -polit gaz / ed by Kuleshov, P V – Lugansk [Ekaterinosl gub]: [s n] 1919 [1919-] – (= ser Asn 1-3) – n18,51,52 [1919] item 126, on reel n27 – 1 – mf#asn-1 126 – ne IDC [077]
Dong du'o'ng tap chi – n.s., no. 186-231. Hanoi, Saigon. 11 aout 1er sept 1918-15 juin 1919 – 1 – fr ACRPP [073]
Dong duong tap chi [hanoi, vietnam: 1913] – Hanoi: Impr d'Extreme-orient, [1913- [n1-52 [may 15, 1913-may 11, 1914]; n109, 114-124, 126, 128-132, 134 [mar 4?, mar 25-jun 3, jun 17, jul 1-29, aug 12, 1917] [mf ed Hanoi, Vietnam: National Library of Vietnam 1995] – 2r – 1 – (in vietnamese & french; at head of title mar 25 1917-: bibliotheque de vulgarisation; also filmed by: paris: association pour la conservation et la reproduction photographique de la presse 1975 [3r] center has: mf-11953 seam n186/189-231 [aug 11/sep 1 1918-jun 1919]; master neg held by crl) – mf#mf-11845 seam – us CRL [079]
Dong duong tap chi [hanoi, vietnam: 1937] = La revue indochinoise – Hanoi: Impr d'Extreme-orient, [1937- [v1 n1-16 [may 15-aug 28, 1937] [mf ed Hanoi, Vietnam: National Library of Vietnam 1995] – 1r – 1 – (master neg held by crl) – mf#mf-11859 seam – us CRL [079]
Dong mingduo jing zhu *see* Nikan hergen-i ubaliyambuha manju gisun-i buleku bithe
Dong phuong [hanoi, vietnam] – [Hanoi]: Impr du Dong-Phuong [1929- [1929 nov-dec 30, 1933 jan-nov 17, 1934 nov 4-dec 24] – 1r – 1 – mf#mf-11121 seam – us CRL [079]
Dong song pha ng la ng : tieu thuyet / To, Nhua n Vy – [Ha Noi]: Thanh Nien 1974- [mf ed 1992] – on pt of 1r – 1 – mf#11052 r400 n2 – us Cornell [959]
Dong tay tieu thuyet bao – Hanoi: Impr Thuy-ky, mar 15-apr 5/8 1940 [mf ed Hanoi, Vietnam: National Library of Vietnam 1996] – 1r – 1 – mf#mf-11142 seam – us CRL [079]
Dong ton tuong te pho = Tong zun xiang ji pu = bulletin semestriel – Hanoi: Tieng-Dan 1928-35 [n1-14/15 [1927/1928-35] – 1r – 1 – (iss by: association de secours et d'assistance mutuels des membres de la famille royale d'annam) – mf#mf-12543 seam – us CRL [079]
Dongeng-dongeng sasakala, kenging ngempelkeun moh / Ambri, M [comp] – Ambri. Djakarta, Gunseikanbu Kokumin Toshokyoku (Balai Poestaka) 2604. 2v. (B P 1517) v1 – 32p 1mf – 9 – mf#SE-2002 mf5 – ne IDC [959]
Dongerkery, Sunderrao Ramrao *see* Universities and national life
Dongola first baptist church. dongola, illinois : church records – 1892-Jan 1981. 2304p – 1 – us Southern Baptist [242]
Dong-phap – Hanoi: Impr Bao dong-phap [1932-[jun 6/7-30 1932] – reel 7 – 1 – mf#mf-11119 seam – us CRL [079]
Dong-phap tho'i-bao – Saigon. mai 1923-28 – 1 – fr ACRPP [073]
Doni, A F *see*
– I marmi del doni
– I marmi del doni...tre libri di lettere de doni
– I mondi del doni
– Nuova opinione sopra le imprese amorose e militari
– Pitture del doni academico pellegrino
– Tre libri di lettere del doni
Doni, [A F] *see* Disegpartito in piu ragionamenti, ne quali si tratta della scoltura et pittura...
Doni, Giovanni Battista *see* De praestantia musicae veteris libri tres totidem dialogis comprehensi
Doniphan eagle – Doniphan, NE: I M Augustine, nov 1892 (wkly) [mf ed v1 n5. dec 9 1892 filmed [1973]] – 1r – 1 – us NE Hist [071]
The doniphan enterprise – Doniphan, NE: J W Mahaffey, 1914-21// (wkly) [mf ed with gaps filmed [1972?]] – 2r – 1 – us NE Hist [071]
Doniphan herald – Doniphan, NE: Roy H Minder. v1 n1. aug 10 1923-37// (wkly) [mf ed with gaps filmed [1972?]] – 4r – 1 – (cont by: anselmo news) – us NE Hist [071]
The doniphan herald – Doniphan, NE: [Richard and Jean Mohanna] v1 n1. aug 12 1971- (wkly) [mf ed filmed 1978-] – 1 – us NE Hist [071]
Doniphan index – Doniphan, NE: Seth P Mobley, jul 1896-97// (wkly) [mf ed with gaps filmed [1965?]] – 1r – 1 – (absorbed by: grand island daily independent (regular ed)) – us NE Hist [071]
Donkin, Rufane S *see* A dissertation on the course and probable termination of the niger
La donna del lago : melodramma in due atti di andrea leone tottola / Rossini, Gioacchino – Milano: Ricordi [185-?] [mf ed 19–] – 10mf – 9 – (arr for piano by luigi truzzi) – mf#fiche 237 – us Sibley [780]
Donna, Rose Bernard *see* Despair and hope
La donna vestita di sole : a collection of sacred madrigals by orfeo vecchi / Mather-Pike, Dolores Anne – U of Rochester 1961 [mf ed 19–] – 2v on 1r – 1 – (with bibl) – mf#film 2579 – us Sibley [780]
La donna vestita di sole, coronata di stelle, calcante la lvna : in 21. madrigali cantata / Vecchi, Orfeo – In Milano: Per l'herede di Simon Tini & Gio Francesco Besozzi [1602] [mf ed 196-?] – 5pt on 1r – 1 – mf#film 413 / pres. film 186 – us Sibley [780]
Donnay, Maurice *see* Impromptu du paquetage
Donnell, Courtney Graham *see* Twentieth-century european paintings
Donnelly, Eleanor C *see* Girlhood's hand-book of woman
Donnelly, Eleanor Cecilia *see* Lot leslie's folks and their queer adventures among the french and indians, ad 1755-1763
Donnelly, Ignatius *see*
– The bryan campaign for the american people's money
– The cipher in the plays, and on the tombstone
– Papers
– Ragnarok: the age of fire and gravel
Donner, Anders Severin *see* Beobachtungen von cometen angestellt auf der sternwarte zu helsingfors im winter und fruehjahr 1885-1886
Donner, Joakim Otto Evert *see* Der einfluss wilhelm meisters auf den roman der romantiker
Donner, O [comp] *see* Lieder der lappen
Donnet, Gaston *see* En sahara a travers le pays des maures nomades
Donny, Albert *see* Manuel du voyageur et du r'esident au congo
Donohoe, William Arlington *see* History of british honduras
Donohue, Mary *see* Occupational therapy in mental health
Donop, L B von *see* Diary during a trip from papar to kimanis via tambunan, lobo and limbawan, 1882
Donoso Cortes *see* Seleccion de antonio tovar
Donoso cortes / Blanco Garcia, Francisco – Madrid: Saenz de Jubera, 1909 – 1 – sp Bibl Santa Ana [920]
Donoso cortes / Donoso Cortes, Juan Francisco – Ediciones Fe, 1940 – 1 – sp Bibl Santa Ana [946]
Donoso cortes / Galindo Herrero, Santiago – Madrid: Publicaciones espanolas, 1953 – sp Bibl Santa Ana [920]
Donoso cortes : su posicion en la historia de la filosofia del estado europeo / Schmitt, Carl – Madrid, 1930; Madrid: Razon y Fe, 1931 – 1 – sp Bibl Santa Ana [190]
Donoso cortes : su sentido trascendente de la vida / Armas, Gabriel – Madrid: Edit. E.T. Coleccion Alamo, 1953 – sp Bibl Santa Ana [200]
Donoso cortes, el profeta de la hispanidad / Gutierrez Lasanta, Francisco – Imprenta Torroba, 1953-1954 – 1 – sp Bibl Santa Ana [946]
Donoso cortes en la problematica de la espiritualidad (esbozo de biografia mistica) / Armas, Gabriel – Las Palmas de Gran Canaria: Tip. Minerva, 1950 – sp Bibl Santa Ana [240]
Donoso cortes en la ultima etapa de su vida / Galindo Herrero, Santiago – Madrid: Arbor, 1953 – 1 – sp Bibl Santa Ana [320]
Donoso cortes, en su tiempo y en el nuestro / Silio, Francisco Javier – Madrid: Arbor, 1950 – 1 – sp Bibl Santa Ana [320]
Donoso Cortes, Juan Francisco *see*
– Acivilicao catolica e os erros modernos
– Africa en el pensamiento de donoso cortes
– Coleccion...la diplomacia
– Coleccion...proyecto de ley
– Consideraciones sobre la diplomacia
– Discours de...dans les cortes
– Discurso parlamentario
– Donoso cortes
– Donoso cortes y la cuestion social
– Ensayo sobre el catolicismo, el liberalismo y el socialismo
– Ensayo sobre el catolicismo liberalismo
– Los errores de nuestro tiempo
– Essay on catholicism, liberalism and socialism
– Lecciones de derecho politico
– La ley electoral
– Memoria sobre...monarquia
– Obras
– Obras escogidas
– El pensamiento politico hispanoamericano...
– Proyecto de ley sobre estados excepcionales
– Rappel
– Saggio sul cattolicesimo, il liberalismo e il socialismo
– La venida de cristina
Donoso cortes, juan. obras completas recopiladas por...2 vol. madrid, 1946 / Errandonea, Ignacio – Madrid: Razon y Fe, 1947 – 1 – sp Bibl Santa Ana [946]
Donoso Cortes y Donoso Cortes, E et al *see* 1st congreso sindical agrario de extremadura, ponencia i estructura y fines del sindicalismo agrario
Donoso cortes y la cuestion social / Donoso Cortes, Juan Francisco – Barcelona: Casa Subirana, 1934 – 1 – sp Bibl Santa Ana [301]
Donoso cortes y la dictadura / Sevilla Andres, Diego – Madrid: Arbor, 1953 – 1 – sp Bibl Santa Ana [320]
Donoso, Ricardo *see* Un letrado del siglo 18, el dcotor jose perfecto de salas. buenos aires, 1963
Donovan, Carolyn M *see* Health attitudes and their relation to compliance and measured cholesterol levels
Donovan, D *see* Tone patterns for various conjugations
Donovan, Joseph Wesley *see*
– Skill in trials: containing a variety of civil and criminal cases won by the art of advocates.
– Tact in court
Donovan, Karen S *see* The relationship between heart rate and rate of perceived exertion among phase 2 cardiac rehabilitation patients with various modes of exercise
Donovan, Maura E *see* The economic benefits of a sporting event to a community
Donside piper and herald – Forfar: Angus County Press (wkly) [mf ed 1 jul 1994-] – 1 – (cont: piper (donside and insch ed)) – uk Scotland NatLib [072]
Donskie vedomosti / ed by Kriukov, F D – Novocherkassk [Obl voiska Don]: [s n] 1919 [1919 ianv]-] – (= ser Asn 1-3) – n11-279 [1919] [gaps] item 127, on reel n27 – 1 – mf#asn-1 127 – ne IDC [077]
Donskoi krai / ed by Cherevkov, S P – Novocherkassk [Obl voiska Don]: [s n] 1918 [1918 apr]-] – (= ser Asn 1-3) – n34 [1918] item 128, on reel n27 – 1 – mf#asn-1 128 – ne IDC [077]
Donskoi prodovolstvennik i kooperator – Rostov n/D, 1920(1) – 1mf – 9 – mf#COR-583 – ne IDC [335]
Donskoi statisticheskii ezhegodnik za 1922-1927 gg – Rostov n.d. 1922-1927 – 34mf – 9 – mf#RHS-21 – ne IDC [314]
Don't feed the tiger / Culwick, Arthur Theodore – Cape Town, South Africa. 1968 – 1r – us UF Libraries [960]
Don't mourn, organize! – v1-3 n1 [1977 feb-1979 jan] – 1r – 1 – (cont by: cry for freedom) – mf#403983 – us WHS [071]
Dontvenytar / Hungary. Courts – v1-101; 1870-1906 – 1r – 1 – us L of C Photodup [340]
Donzellini, G *see* Epistolae principvm, rervmpvblicarvm, ac sapientvm virorvm
Dookola swiata – Warsaw, Poland. -w. June 1954-Dec 1958. 5 reels – 1 – uk British Libr Newspaper [947]
Dooley bulletin / Bodak, Shirley L – 1977 mar 1-dec – 1r – 1 – mf#637877 – us WHS [071]
Den doolhof van de dwalende gheesten waer in den aenvang... – Amstelredam: Jan Evertsz Cloppenburgh, [c1620] – 1mf – 9 – mf#O-3021 – ne IDC [090]
Doolittle, J *see* Social life of the chinese
Doolittle, James R *see* Papers
Doolittle, Justus *see* Social life of the chinese
Doolittle, Quenten *see* Chamber sonatas of georg muffat
Doolittle, Thomas *see* On eyeing of eternity
Doom eternal : the bible and church doctrine of everlasting punishment / Reimensnyder, Junius Benjamin – Chicago: Funk & Wagnalls, 1887 – 1mf – 9 – 0-524-08501-3 – mf#1993-3146 – us ATLA [240]
The doom of dogma and the dawn of truth / Frank, Henry – New York:G.P. Putnam, 1901 – 1mf – 9 – 0-8370-3179-6 – (incl ind) – mf#1985-1179 – us ATLA [140]
Dooman, Isaac *see* Missionary's life in the land of the gods
Doomed religions : a series of essays on great religions of the world / ed by Reid, John Morrison – New York: Phillips & Hunt; Cincinnati: Walden & Stowe, 1884 – 2mf – 9 – 0-7905-7021-1 – (incl bibl ref) – mf#1988-3021 – us ATLA [200]
Doopsgezinde bijdragen – Leiden, 1861-1919 – 135mf – 9 – mf#H-3063 – ne IDC [240]
Door county advocate – Sturgeon Bay WI. 1862 mar 22-1866 dec 27, 1867-85, 1886-1889 jul 20, jul 27-1892 oct 22, oct 29-1895 dec 28, 1896 jan 4-1897 jan 30 – 11r – 1 – (cont by: advocate [sturgeon bay wi]) – mf#1131073 – us WHS [071]
Door county advocate – Sturgeon Bay WI. 1918 aug 2/1920 jul 30-2002 jun – 225r – 1 – (with small gaps; cont: sturgeon bay advocate; door county democrat) – mf#1127677 – us WHS [071]
Door county almanak – 1982 – 1r – 1 – mf#977853 – us WHS [071]
Door county democrat – Sturgeon Bay WI. 1893-99, 1900-15, 1916-1918 jul – 10r – 1 – (cont: democrat [sturgeon bay wi]; cont by: sturgeon bay advocate; door county advocate) – mf#934607 – us WHS [071]
Door county farmer and fruit grower – Sturgeon Bay WI. 1927 feb 9 – 1r – 1 – mf#927395 – us WHS [634]
Door county news – Sturgeon Bay WI. 1914 jul-1915, 1916-36, 1937-1939 aug – 9r – 1 – mf#934712 – us WHS [071]
Door County State Bank [Sturgeon Bay, WI] et al *see* Free press
Door de duisternis tot het licht / Henzel, J – [Rotterdam]: J M Bredee, 1915 [mf ed 1995] – (= ser Yale coll; Lichtstralen op den akker der wereld [21. jaarg 1915] 6) – 32p (ill) – 1 – 0-524-10063-2 – (in dutch) – mf#1995-1063 – us ATLA [240]
Door of hope – New York, NY. 1898 – 1r – us UF Libraries [939]
Door west-indie / Wijnaendts Francken, Cornelis Johannes – Haarlem, Netherlands. 1915 – 1r – us UF Libraries [972]
Doorgraving der landengte van suez... / Bake, R W J C – Haarlem, 1857 – 2mf – 9 – mf#ILM-1817 – ne IDC [956]
Doorknob collector / Antique Doorknob Collectors of America – 1982 jun-1986 may – 1r – 1 – mf#1130947 – us WHS [740]
Doorlugte voorbeelden der ouden... / [Leuve, R van] – Amsteldam: Henrik Bosch, 1725 – 10mf – 9 – mf#O-341 – ne IDC [090]
Doorninck, J van *see* Geslachtkundige aanteekeningen
Doornink, Adam van *see* Bijdrage tot de tekstkritiek van richteren 1-16
Doors to latin america – North Miami Beach FL 1954-81 – 1,5,9 – ISSN: 0012-5490 – mf#7093 – us UMI ProQuest [070]
Doorway in antigua / Idell, Albert Edward – New York, NY. 1949 – 1r – us UF Libraries [972]
Dooyeweerd, H *see*
– De wijsbegeerte der wetsidee
– Transcendental problems of philosophical thought
– De wijsbegeerte der wetsidee
Dop och barndop : samtal mellan natanael och timoteus / Waldenstroem, Paul – 2. uppl. Stockholm: Pietistens Expedition, 1898 – 1mf – 9 – 0-524-05023-6 – (incl bibl ref) – mf#1991-2193 – us ATLA [240]
Dop och foersamling enligt den heliga skrift / Pendleton, James Madison – Stockholm: David Lund, 1884 – 1mf – 9 – 0-524-07705-3 – mf#1991-3290 – us ATLA [240]
Dope / Westmorland Community Association, Madison WI – v1-2 n7 [1941 jun 16-1942 oct] – 1r – 1 – (cont by: courier) – mf#436816 – us WHS [360]
Dopico Y Gonzalez, Blanca *see* Salvador salazar, una vida abundante
Dopolnitelnye materialy dlia bibliografii, ili opisanie russkikh i inostrannykh knig, graviur i portretov, nakhodiashchikhsia v biblioteke liubitelia n n / Berezin-Shiriaev, I – 1876 – 7mf – 8 – mf#R-4656 – ne IDC [947]
Die doppelbearbeitungen der "raeuber", des "fiesco" und des "don carlos" von schiller : eine litterarhistorische studie / Tischler, Hermann – Leipzig: E Herrmann, 1888 – 1 – (incl bibl ref) – us Wisconsin U Libr [430]
Doppelberichte im pentateuch : ein beitrag zur einleitung in das alte testament / Schulz, Alfons – Freiburg i B, St Louis MO: Herder, 1908 – (= ser Biblische studien) – 1mf – 9 – 0-7905-2427-9 – (incl bibl ref) – mf#1987-2427 – us ATLA [221]
Die doppelehe des landgrafen philipp von hessen / Rockwell, William Walker – Marburg: NG Elwert, 1904 – 1mf – 9 – 0-524-02709-9 – (incl bibl ref) – mf#1990-0690 – us ATLA [943]
Doppelselbstmord : bauernposse mit gesang in drei akten / Anzengruber, Ludwig – 2. aufl. Stuttgart: J G Cotta, 1910 [mf ed 1996] – 110p – 1 – mf#9580 – us Wisconsin U Libr [820]

DORTMUNDER

Dopper, Cornelis *see*
- Blaeserstueck
- Prelude, scherzino, impromptu
Doppler, Josef *see*
- Pater wenzel hocke
Doprava a spoje – Prague, Czechoslovakia. -w. Jan 1959-dec 1968 – 1 1/2r – 1 – uk British Libr Newspaper [072]
Dor, Georges *see* La memoire innocente
Dor ha-haskalah he-rusiyah / Margulis, Menasheh – Vilna, Lithuania. 1910 – 1r – us UF Libraries [939]
Dor, Marlene *see* An analysis of referrals received by a psychiatric unit in a general hospital
Dora holdenrieth : roman / Bertololy, Paul – Leipzig: P List, c1939 [mf ed 1989] – 477p – 1 – mf#7013 – us Wisconsin U Libr [830]
[El dorado county-] amador, el dorado, placer and sacramento counties – CA. 1884-1885 – 1r – 1 – $50.00 – mf#D005 – us Library Micro [978]
[El dorado county-] placerville city directories – CA. 1862; 1947 – 2r – 1 – $100.00 – mf#D016 – us Library Micro [917]
El dorado daily news – El Dorado AR. 1925 aug 28 – 1r – 1 – (continued by: el dorado times (el dorado, ar: daily); el dorado news-times) – mf#4364596 – us WHS [071]
Los dorados ingleses / Bayle, Constantino – Madrid: Razon y Fe, 1930 – 1 – sp Bibl Santa Ana [240]
Doran, Carol *see* Influence of raynor taylor and benjamin carr on church music in philadelphia at the beginning of the nineteenth century
Dorat des Monts, Roger *see* La cause immorale, etude de jurisprudence
d'Orbigny, A *see* Histoire naturelle generale et particuliere des c,phalopodes ac,tabuliferes vivants et fossiles
Dorchester 1631-1869 – Oxford, MA (mf ed 1986) – (= ser Massachusetts vital records) – 72mf – 9 – 0-931248-85-X – (mf 1-4: b,m,d 1631-83. mf 5-6: b,m,d 1684-1744/45. mf 7-11: b,m,d 1745-1825. mf 12: b,m,d 1826-44. mf 13-22: b,m,d 1631-1844. mf 23-31: first church dorchester 1729-1845. mf 32-39: b,m,d 1843-49. mf 40-43: index to b,m,d 1631-1849. mf 44-45: publishments of marriages 1799-1849. mf 46-55: births 1850-69; index to births. mf 56-64: intentions, marriages, indexes 1850-69. mf 65-72: deaths 1850-69; index to deaths) – us Archive [978]
Dorchester and sherbourne journal and western advertiser – Dorchester, Sherbourne, England. -w. 21 Jan, 4 Feb, 11 March, 8, 22, 29 April, 13 May, 3, 10, 24 June, 23 Sept 1791. 9 ft – 1 – uk British Libr Newspaper [072]
The dorchester booster – Dorchester, NE: Geo Stiegelmar. 6v. v1 n1. aug 8 1941-v6 n20. dec 20 1946 (wkly) [mf ed lacks oct 18 1941 filmed 1972?] – 1r – 1 – us NE Hist [071]
Dorchester clarion – Dorchester WI. 1937 nov30-1939 jul 27, aug 3-1940 dec 26, 1941 jan 2-1943 sep 30, oct 7-1946, 1947-64, 1965 jan 7-1968 jul 18, aug 1-1973 jan 4 – 10r – 1 – mf#965367 – us WHS [071]
Dorchester county genealogical magazine – 1982 mar-1989 mar – 1r – 1 – (cont: dorchester genealogical magazine) – mf#1685358 – us WHS [929]
Dorchester County Historical Society [MD] *see* Dorchester genealogical magazine
Dorchester, Daniel *see*
- Christianity in the united states
- Christianity vindicated by its enemies
- The problem of religious progress
- Romanism versus the public school system
- The why of methodism
Dorchester genealogical magazine / Dorchester County Historical Society [MD] – v1 n1-5 [1981 may-1982 jan] – 1r – 1 – (cont by: dorchester county genealogical magazine) – mf#1055702 – us WHS [929]
Dorchester herald – Dorchester WI. 1914 jan 2-30, feb 6-1915 aug 27, sep 3-1918 aug 30, sep 6-1919 jun 6 – 4r – 1 – (cont: clark county herald) – mf#965415 – us WHS [071]
Dorchester leader – Dorchester, NE: M E Wintermute, 1933-v8 n41. jul 25 1941 (wkly) [mf ed 1934-41 (gaps) filmed [1972?]] – 2r – 1 – (absorbed by: dorchester star) – us NE Hist [071]
Dorchester star – Dorchester, NE: H C Bittenbender, 1881-1968// (wkly) [mf ed with gaps filmed -1978] – 17r – 1 – (absorbed: dorchester leader. some irregularities in numbering) – us NE Hist [071]
Dordevic, Andra *see* Porodicno pravo nasleda u danasnjim romanskim i germanskim drzavama
Dordrechti habitae anno 1618 et 1619... / Acta Synodi Nationalis... – Dordrechti, 1619-20. 2v – 14mf – 9 – mf#PBA-100 – ne IDC [240]
Dore, Henri *see* Superstitious practices
Dore, James *see*
- Harmony of divine operations
- Holy spirit, the spirit of truth
- Letter sent by mr james dore to the church at maze pond
- Sermon occasioned by the death of mr john flight
Dore, John Read *see* Old bibles
Doren, J B J van *see* Thomas matulesia
Doren, Richard E *see* Piano music of erik satie
Doren, William Howard van *see* A suggestive commentary on st luke
Dorer-Egloff, Edward *see* J m r lenz und seine schriften
Doreste, Arturo *see* Ultimos instantes de marti
Doret, Frederic *see* Comment je conjcois une constitution d'haiti
Das dorf am meer : roman / Swars, Ewald – Karlsbad: A Kraft 1944 [mf ed 1991] – 1r – 1 – (filmed with: hinter der maske / armin gimmertahl) – mf#2909p – us Wisconsin U Libr [830]
Dorf im kaukasus : roman / Strobl, Karl Hans – Budweis: Verlagsanstalt Moldavia, 1944 – 1r – 1 – us Wisconsin U Libr [830]
Das dorf in der taiga : [a novel] / Velter, Joseph Matheus – Leipzig: W Goldmann 1944, c1936 [mf ed 1991] – 1r – 1 – (filmed with: der reiter auf dem fahlen pferd / emanuel stickelberger) – mf#2953p – us Wisconsin U Libr [830]
Dorf- und schlossgeschichten; neue dorf- und schlossgeschichten; zwei komtessen / Ebner-Eschenbach, Marie von – Leipzig: H Fikentscher, H Schmidt & H Guenther, [1928] – 2r – 1 – us Wisconsin U Libr [430]
Dorfbarbier – Berlin DE, 1884 & 1888, 1890, 1893-96, 1898-1900, 1903, 1905-08, 1911, 1913-16, 1920-25, 1927-30 – 1 – gw Misc Inst [074]
Der dorfbote – Budweis (Ceske Budejovice CZ), 1933-1936 9 feb, 1937, 1939-1942 17 oct – 1 – gw Misc Inst [077]
Der dorfbote – Wiener Neustadt, Austria jul-oct 1902 [mf ed Norman Ross] – 1r – 1 – (political weekly for the landbevoelkerung) – mf#nrp-2151 – us UMI ProQuest [074]
Dorf-chronik 1848 – Moers DE, 1848, 1849 [gaps], 1852-83, 1885-1945 1 mar – 115r – 1 – (title varies: 1852?: dorf-chronik und grafschafter; 1854: dorf-chronik; 1888: dorf-chronik und grafschafter; 1914: der grafschafter. incl suppls: der grafschafter 1854-83 [fr 1885: dorf-chronik bound with der grafschafter; illustrierter familienfreund 1913-1916 26 mar [1r]; land und leute der grafschaft moers 1928-1937 jul [1r]) – gw Misc Inst [914]
Dorf-chronik und grafschafter *see* Dorf-chronik 1848
Dorfgaenge : gesammelte bauerngeschichten / Anzengruber, Ludwig – Wien: L Rosner, 1879 [mf ed 1993] – 2v in 1 – 1 – mf#8459 – us Wisconsin U Libr [880]
Dorfgenossen : neue erzaehlungen / Huggenberger, Alfred – Leipzig: L Staackmann 1922, c1913 [mf ed 1995] – 1r – 1 – (filmed with: daniel pfund) – mf#3884p – us Wisconsin U Libr [390]
Dorfpredigten / Frenssen, Gustav – Gesamtausg. Goettingen: Vandenhoeck & Ruprecht 1902-03 [mf ed 1989] – 3v in 1 on 1r – 1 – (filmed with: die drei getreuen & other titles) – mf#7264 – us Wisconsin U Libr [240]
Dorfschwalben aus oesterreich : frischer flug / Silberstein, August – Breslau: S Schotlaender, 1881 – 1r – 1 – us Wisconsin U Libr [390]
Doria, Joao De Seixes *see* Eu
Doriano Cumbreno, Antonio C *see* Curso general de paleografia y diplomatica espanolas (texto)
Dorington, John Edward *see* Endowments of the church and their origin
Dorion, Charles-Edouard *see* De l'admissibilite de la preuve par temoins en droit civil
Dorion, Eugene P *see* Historique des fonds de retraite en europe et en canada
Dorion, Jacques Edmond *see* Lecture publique
Dorion, Jean Baptiste Eric *see*
- Discours de m j b e dorion...
- Institut-canadien en 1852
- Tenure seigneuriale
Dorion, Louis Charles Wilfrid *see* Vengeance fatale
Dorion's decisions on appeal / Canada. Quebec – v1-4. 1881-84 (all publ) – 20mf – 9 – $30.00 – mf#LLMC 81-073 – us LLMC [340]
Doris, Charles, de Bourges [M Santini pseud] *see* An appeal to the british nation on the treatment experienced by napoleon buonaparte in the island of st helena
Dorman, Alain *see* Alcool, alcoolisme, milieu de travail
Dorman, Rushton M *see* The origin of primitive superstitions
Dormant and extinct peerages / Burke, J B – 1846 – 1r – 1 – mf#424 – uk Microform Academic [920]
Dorn, Gerhard *see* Schluessel der chimistischen philosophie
Dorn, Kaethe *see* Auf glaubenspfaden
Dorn, Kaethe et al *see*
- Die neue heimat
- Vom himmel hoch, da komm' ich her!
Dornas Filho, Joao *see* Ouro das gerais e a civilizacao da capitania
Dornas, Joao *see*
- Apontamentos para a historia da republica
- Capitulos da sociologia brasileira
- Silva jardim
Die dornburger schloesser : zum 28. august 1923 / Wahl, Hans – Weimar: Verlag der Goethe-Gesellschaft, 1923 [mf ed 1993] – (= ser Schriften der goethe-gesellschaft 36) – 40p (ill) – 1 – mf#8657 reel 9 – us Wisconsin U Libr [720]
Dornemann, Timothy M *see* The effect of a weight training program on the bone density of women aged 40-50 years
Der dornenweg : roman / Wilbrandt, Adolf von – 4. aufl. Stuttgart: J G Cotta 1901 [mf ed 1991] – 1r – 1 – (filmed with: jedermann / ernst wiechert) – mf#3031p – us Wisconsin U Libr [830]
Dorner, August *see*
- Augustinus
- Grundriss der dogmengeschichte
- Grundriss der religionsphilosophie
- Kirche und reich gottes
- Pessimismus, nietzsche und naturalismus
- System of christian ethics
- Ueber die principien der kantischen ethik
- Zur geschichte des sittlichen denkens und lebens
Dorner, Emil *see* Badisches landesprivatrecht
Dorner, Isaak August *see*
- Briefwechsel zwischen h. l. martensen und j. a. dorner, 1839-188l
- Die christliche lehre
- Dorner on the future state
- Geschichte der protestantischen theologie
- Grundriss der encyclopaedie der theologie
- History of protestant theology
- History of the development of the doctrine of the person of christ
- The liturgical conflict in the reformed church of north america
- Sendschreiben ueber reform der evangelischen landeskirchen
- A system of christian doctrine
- System of christian ethics
- Ueber jesu suendlose vollkommenheit
Dorner on the future state : being a translation of the section of his system of christian doctrine comprising the doctrine of the last things / Dorner, Isaak August – New York: Scribner, 1883 – 1mf – us ATLA [240]
Dorner on the future state : being a translation of the section of his system of christian doctrine comprising the doctrine of the last things = System der christlichen glaubenslehre. selections / Dorner, Isaak August – New York: Scribner, 1883 – 1mf – 9 – 0-7905-3830-X – (in english) – mf#1989-0323 – us ATLA [240]
Dorneth, J v *see* Martin luther
Dorneval, E *see* Abrege des principaux episodes de la revolution du 15 janvier 1908
Dornrosen erstlingsbluethen deutscher lyrik in amerika / Steiger, Ernst – New York, 1871. vi,159p – 1 – us Wisconsin U Libr [430]
Dornstetter, Paul *see* Abraham
Doroga na zapad – (city unknown) 1943-44 [mf ed Norman Ross] – 1 – mf#nrp-93 – us UMI ProQuest [934]
Dorogu frontu – (city unknown) 1944-45 [mf ed Norman Ross] – 1 – mf#nrp-94 – us UMI ProQuest [934]
Dorokhov, Pavel Nikolaevich *see* Kolchakovshchina
Doron basilikon sive corona imperii romani ferdinando 4... / Marx, J R – Francofurti: Typis Antonii Hummii, impensis Christiani Hermsdorffi, 1653 – 1mf – 9 – mf#O-38 – ne IDC [090]
Doronovich, M F *see* Otkliki kavkaza
Dorothea angermann : schauspiel / Hauptmann, Gerhart – Berlin: S Fischer, c1926 – 1r – 1 – us Wisconsin U Libr [820]
Dorothy south: a love story of virginia just before the war / Eggleston, George Cary – Illus. by C.D. Williams.Boston: Lothrop Pub. Co., (1902). 453p – 1 – us Wisconsin U Libr [830]
Dorothye g scott, senate service 1945-1977 : administrative assistant to the democratic secretary and the secretary of the senate – (= ser Us senate historical office oral history coll) – 4mf – 9 – $20.00 – us Scholarly Res [323]
Dorp, Frederik van et al *see* Stichtelycke gedichten
Dorpater nachrichten – Dorpat (Tartu EW), 1921-24 – 1 – gw Misc Inst [077]
Dorpater tagesblatt – Dorpat (Tartu EW), 1862 21nov, 10 dec, 15 dec, 1863-1864 20 jul – 3r – 1 – gw Mikrofilm [077]
Dorpater zeitung – Dorpat (Tartu EW), 1925 jul-1929, 1930 jul-1939 8 sep – 1 – (title varies: oct 1934: deutsche zeitung) – gw Misc Inst [077]
Dorpater zeitung – Dorpat (Tartu EW), 1925 jul-1929, 1930 jul-1939 8 sep – 1 – (title varies: oct 1934: deutsche zeitung) – gw Misc Inst [077]
Dorptische beitraege fuer freunde der philosophie, literatur und kunst – Dorpat, Leipzig, 1813-1814 – 17mf – 8 – mf#R-1751 – ne IDC [700]
Dorr, Harbottle *see* Harbottle dorr collection of annotated massachusetts newspapers, 1765-1776
Dorr, Nicolas *see* Teatro
Dorrance, John C [pseud: James Hanley] *see* Yankee consul and caniibal king
[Dorris-] booster – CA. 1908-11 [wkly] – 1r – 1 – $60.00 – mf#B02173 – us Library Micro [071]
[Dorris-] butte valley star – CA. 1924; 1927; 1930-39; 1983- – 9r – 1 – $540.00 (subs $50/y) – mf#B02174 – us Library Micro [071]
Dorris, Charles Ellis *see* A program of outreach through recreation at the huffman baptist church
[Dorris-] the dorris times – CA. 1915-19 – 2r – 1 – $120.00 – mf#B02175 – us Library Micro [071]
[Dorris-] weekly advocate – CA. 1911-12 – 1r – 1 – $60.00 – mf#B02176 – us Library Micro [071]
Dorsainvil, J B *see*
- Cours d'histoire d'haiti a l'usage...
- Cours d'histoire d'haiti a l'usage de...
- Essai sur l'histoire et l'etablissement
- Petite histoire d'haiti
Dorsainvil, J C *see*
- Essais de vulgarisation scientifique et questions
- Organisons nos partis politiques
- Quelques vues politiques et morales
Dorsainvil, Jc *see* Lectures historiques
Dorsainvil, Jean Baptiste *see*
- Cours complet d'histoire d'haiti a l'usage des eco...
- De la democratie representative
- Elements de droit constitutionnel
Dorschel, Gotthold *see* Maria theresias staats- und lebensanschauung
Dorschner-Lanz, Friedrich *see* Freie klaenge
Dorset and Somerset, England *see* Dorset and somerset papers 18th and 19th century
Dorset and somerset papers 18th and 19th century / Dorset and Somerset, England – Various dates from 4 Apr 1748-3 Mar 1797 (14 feet) and 25 Apr 1806-18 Jul 1822 (34 feet) – 1 – uk British Libr Newspaper [072]
Dorset county chronicle – Dorchester, England. -w. 1830-62. 17 1 2 reels – 1 – uk British Libr Newspaper [072]
Dorset daily echo and weymouth dispatch – Weymouth, England. 1955-83. -d. 187 reels – 1 – uk British Libr Newspaper [072]
D'Orsey, Alexander James Donald [comp] *see* Portuguese discoveries, dependencies and missions in asia and africa
Dorsey dreams – v1 n1-v4 n4 [1982 oct/dec-1984 jul/sep] – 1r – 1 – mf#1507345 – us WHS [071]
Dorsey, George Amos *see*
- The oraibi soyal ceremony
- Traditions of the arikara
- Traditions of the caddo
Dorsinville, Luc *see*
- Abrege d'histoire d'haiti
- Chambre des deputes
Dorsinville, Roger *see* Lettre a mon ami serge corvington
"Dort drueben in westfalen" : hoelderlins reise nach bad driburg mit wilhelm heinse und diotima / Hock, Erich – Muenster: Regensberg, 1949 [mf ed 1991] – (= ser Der schatzkamp 9) – 82p/[5pl] (ill) – 1 – (incl bibl ref) – mf#7486 – us Wisconsin U Libr [920]
Dortch, John Douglas *see* An analysis of the hermeneutic of the southern baptist convention sermon in selected periods of biblical controversy
Dortmunder anzeiger *see* Dortmunder wochenblatt
Dortmunder general-anzeiger 1879 – Dortmund DE, 1879, 1882 4 jan-30 dec, 1884 6 jan-1886 4 sep, 1887-89 – 1 – (title varies: 1 jan 1852: dortmunder volkszeitung; 13 aug 1885: dortmunder general-anzeiger; 1 jul 1886: dortmunder volkszeitung; after 4 spe 1886: neue westfaelische zeitung. fr 1 jul 1886 publ in (dortmund-) barop) – gw Misc Inst [074]
Dortmunder kreisblatt – Dortmund DE, 1858 – 1 – gw Misc Inst [074]
Dortmunder kreisblatt *see* Schwerter wochenblatt
Dortmunder nachrichten [main edition] – Dortmund DE, 1888-97, 1898 10 mar-1915 30 nov, 1916-1924 jun, 1926 1 feb-31 aug & 1 oct-31 dec, 1927 1 feb-1930 1 dec, 1931-1932 31 oct, 1932 1 dec-1945 13 apr – 185r – 1 – (title varies: 2 oct 1889: general-anzeiger; 3 may 1933: general-anzeiger / rote erde, auch: rote erde / general-anzeiger; 30 jan 1934: westfaelische landeszeitung / rote erde. regional ed: e=essen 1933 feb 3 & 3 may, 1934 2 jan-apr [gaps], 1934 jun-1936 30 mar (gaps) [nur lokalteil]; g=herne, gelsenkirchen, gladbeck, recklinghausen 1933 1 dec-31 dec, 1935 1 aug-31 aug [gaps], 1936-41 (gaps) [nur lokalteil]; s=hamm (westf), soest, lippstadt 1933 1 dec-1934 31 may, 1934 1 jul-31 jul, 1934 1 oct-1938, 1940-42 [10r]) – gw Misc Inst [074]
Dortmunder nord-west-zeitung *see* Amtszeitung
Dortmunder nordwest-zeitung *see* Mengeder zeitung

DORTMUNDER

Dortmunder tageblatt *see* Kleiner local-anzeiger fuer die kreise dortmund und hoerde
Dortmunder tageblatt 1883 – Dortmund DE, 1883 1 feb-1885 – 1 – gw Misc Inst [074]
Dortmunder volksblatt – Dortmund DE, 1932 16 dec-1935 27 jun, 1937 1 jul-1938 – 1 – (covers: (dortmund-) luetgendortmund. regional ed of volksblatt, (dortmund-) hoerde) – gw Misc Inst [074]
Dortmunder volkszeitung *see* Dortmunder general-anzeiger 1879
Dortmunder wochenblatt : amtliches kreisblatt – Dortmund DE, 1869 2 jan-21 dec, 1871 10 jan-1873, 1875 5 oct, 1876-77 – 1 – gw Misc Inst [074]
Dortmunder wochenblatt – Dortmund DE, 1828 4 oct-1839, 1841-46, 1847 26 may-1877 30 may, 1878-85, 1886 1 jul-1905, 1906 20 mar-1926 31 aug [gaps], 1926 1 oct-1934 31 oct [gaps], 1934 1 dec-31 dec, 1935 1 mar-30 jun – 236r – 1 – (title varies: 2 jan 1841: wochenblatt fuer die stadt und den kreis dortmund; 26 may 1847: dortmunder anzeiger; 12 jul 1848: anzeiger; 12 mar 1853: dortmunder anzeiger; 4 apr 1855: dortmunder amtliches kreisblatt; 1860: dortmunder anzeiger; 1 jul 1874: dortmunder zeitung; publ banned between 4 oct 1828-30 apr 1939, due to french occupation. with suppl: mussestunden 1905 1 apr-1914, 1915 1 apr-25 nov) – gw Misc Inst [074]
Dortmundische vermischte nachrichten – Dortmund DE, 1869 14 jan-1771 – 1 – gw Misc Inst [074]
Dortmund-mengeder lokalanzeiger *see* Mengeder zeitung
Dorval, Paul *see* Notions elementaires de morphologie et de physiologie des insectes
Dorvigny, M *see* Nitouche et guignolet
Dorville, A *see* Voyage...la chine
Dorvo, Hyacinthe *see* Vernon de kergalek
Dorweiler, Joachim *see* Karl von raumer und sein beitrag zur volksbildung im 19. jahrhundert
Dos anos de guerra *see* 18 de julio
Dos anos de labor municipal / Havana Alcalde, 1947 – Castellanos Y Rivero – Habana, Cuba. 1949 – 1r – us UF Libraries [972]
Dos anos en america / Zamacois, Eduardo – Barcelona, Spain. 1913 – 1r – us UF Libraries [972]
Los dos artistas / Lopez de Ayala, Adelardo – 1882 – 9 – sp Bibl Santa Ana [820]
Dos aventuras en el lejano oriente / Sinan, Rogelio – Panama, 1947 – 1r – us UF Libraries [972]
Dos barcos / Montenegro, Carlos – Habana, Cuba. 1934 – 1r – us UF Libraries [972]
Dos brujitos mayas / Rodriguez Beteta, Virgilio – San Salvador, El Salvador. 1958 – 1r – us UF Libraries [972]
Las dos ciudades / Pla y Deniel, Enrique – Salamanca, 1936 – 1 – us CRL [240]
Dos discursos al servicio de la causa popular / Marinello, Juan – Paris, 1937. Fiche W 1025. (Blodgett Collection of Spanish Civil War Pamphlets) – 9 – us Harvard [946]
Dos documentos para la historia de la seguridad so... – Guatemala, 1950 – 1r – us UF Libraries [972]
Dos ensayos arqueologicos; estudios de investigaci... / Morales Patino, Oswaldo – Habana, Cuba. 1939 – 1r – us UF Libraries [972]
Dos entrevistas sensacionales con ramon grau san m... – Habana, Cuba. 1942 – 1r – us UF Libraries [972]
Dos estudios sobre el derecho puertorriqueno de la... / Velazquez, Guaroa – Rio Piedras, Puerto Rico. 1954 – 1r – us UF Libraries [972]
Dos fraie vort – Bulnes, Chile. jan 1962-jan 1968 – 1/4r – 1 – uk British Libr Newspaper [072]
Dos goteras, y otros cuentos / Valenzuela Oliva, Wilfredo – Guatemala, 1962 – 1r – us UF Libraries [972]
Dos grandes aventureros espanoles del siglo 16 / Dusmet de Arizcun, Xavier – Madrid: Tipografia "Hesperia" 1927 [mf ed 1989] – 1r with other items – 1 – (with bibl) – mf#mf-10289 seam reel 021/04 [§] – us CRL [920]
Los dos guzmanes / Lopez de Ayala, Adelardo – 1851 – 9 – sp Bibl Santa Ana [820]
Dos hemisferios – London, UK. 2 Apr, 1 May 1883 – 1 – uk British Libr Newspaper [072]
Dos historias, la una de la sancta casa de nuestra senora de guadalupe, y su principio y fundacion y cosas notables dellas : y la otra del principio y fundacion de la casa del senor sanctiago de galizia, patron de espania y de las cosas notables desta sancta casa – Valencia: Art. Graf. Soler, 1965 – 1 – sp Bibl Santa Ana [946]
Dos horas de literatura colombiana / Arango Ferrer, Javier – Medellin, Colombia. 1963 – 1r – us UF Libraries [972]
Dos informaziones: una dirijida al emperador carlos 5 / Sleidanus, Johannes – Madrid, Impr. de Alegria, 1857. Various pagings – 1 – us Wisconsin U Libr [946]
Dos medicos extremenos de ayer / Enriquez Anselmo, Juan – Badajoz: Dip. Provincial, 1970 – 1 – sp Bibl Santa Ana [610]
Dos misioneros franciscanos hermanos en el colegio de chillan (chile) / Barrado Manzano, Arcangel – Madrid: Archivo Ibero-Americano, 1959 – 1 – sp Bibl Santa Ana [240]
Dos momentos dramaticos de una sola cuestion / Lopez, Francisco Marcos – Guatemala, 1959 – 1r – us UF Libraries [972]
Dos musas cubanas: / Martinez Bello, Antonio M – Habana, Cuba. 1954 – 1r – us UF Libraries [972]
Dos naje wort – Warsaw, Poland 1935-37 [mf ed Norman Ross] – 4r – 1 – (in yiddish) – mf#nrp-2096 – us UMI ProQuest [939]
Dos obras...fiel cristiano / Pedro de Vitoria, Fray – 1626 – 9 – sp Bibl Santa Ana [240]
Dos palabras sobre el presente numero-homenaje. benito arias montano / Fernandez de Castro, Eduardo Felipe – Malaga, 1928 – 1 – sp Bibl Santa Ana [946]
[Dos palos-] dos palos star – CA. 1911-22; 1923- – 50r – 1 – $3000.00 (subs $50/y) – mf#B02177 – us Library Micro [071]
Las dos partidas de bautismo de don luis de salazar y castro / Siete Iglesias, Antonio de – Madrid: Hidalguia, 1958 – 1 – (aparte rev hidalguia, marzo-abril 1958 n27) – sp Bibl Santa Ana [946]
Dos Passos, Benjamin Franklin *see* The law of collateral and direct inheritance, legacy and succession taxes, embracing all american and many english decisions.
Dos Passos, John *see* Brazil on the move
Dos Passos, John Randolph *see*
– The inter-state commerce act; an analysis of its provisions
– A treatise on the law of stock-brokers and stock exchanges.
Dos pesos de agua, cuentos / Bosch, Juan – Habana, Cuba. 1941 – 1r – us UF Libraries [972]
Dos pueblos amigos – Guatemala? 1955? – 1r – us UF Libraries [972]
Las dos republicas: el 11 de febrero y el 14 de abril / Castrovido, Roberto – Barcelona, 1938? Fiche W 782. (Blodgett Collection of Spanish Civil War Pamphlets) – 9 – us Harvard [946]
Dos robinsones – 1860. 2v – 9 – sp Bibl Santa Ana [830]
Las dos romas : impresiones de un peregrino en el ano santo / Ruano, Jesus Maria – Madrid: Razon y Fe, 1926 – 1 – sp Bibl Santa Ana [910]
Dos telegraf – Warsaw, 1906 (jan 16-jul 16). v1 – 1r – 1 – mf#J-92-18 – ne IDC [077]
Dos testigos abrumadores contra los nacionales / Bayle, Constantino – Burgos: Razon y Fe, 1938 – 1 – sp Bibl Santa Ana [946]
Dos tsvantsigste yahrhundert – Warsaw, Poland. 1900 – 1r – us UF Libraries [939]
Dos veces retono / Ramirez De Arellano De Nolla, Olga – San Juan, Puerto Rico. 1965 – 1r – us UF Libraries [972]
Dos viajes / Agostini, Victor – Habana, Cuba. 1965 – 1r – us UF Libraries [972]
Dos vidas no ejemplares / Miramon, Alberto – Bogota, Colombia. 1962 – 1r – us UF Libraries [972]
Dos vokhenblat – Copenhagen, Denmark 1915-18 [mf ed Norman Ross] – 1r – 1 – (in yiddish. with: lebens-fragen [warsaw] 1915-20) – mf#nrp-452 – us UMI ProQuest [939]
Dos wort – Bialystock, Poland 1934-35 [mf ed Norman Ross] – 1r – 1 – (in yiddish; with: sztegn [stanislawow, ukraine] 1934-38; with: pruzaner sztime [pruzany, poland] 1935) – mf#nrp-334 – us UMI ProQuest [939]
Dosch, Margaret *see* The effects of acquaintance rape prevention programming on male athletes' sexual and dating attitudes
Doscientas obras...!en una! / Sanchez Arjona, Vicente – Sevilla: Imprenta Zambrano, 1957 – 1 – sp Bibl Santa Ana [810]
Dose, Johannes *see* Der muttersohn
Dosis-wirkungsbeziehung von unretadiertem isosorbiddinitrat in kleinen dosen bei patienten mit koronarer herzkrankheit / Clement, Richard Gray – Mainz: Gardez 1993 (mf ed 1996) – (= ser Neue medizinische Bibliothek 1) – 1mf – 9 – €24.00 – 3-8267-9654-3 – mf#DHS 9654 – gw Frankfurter [615]
Dosker, Henry Elias *see* Outline studies in church history
Dosker, Nina Ellis *see* Pioneer preacher in the treasure state
Dosma Delgado, Rodrigo *see* Discursos patrios... badajoz
Doss, Lal Mohun *see* The law of riparian rights, alluvion and fishery
Doss, Leland *see* Speech delivered by leland doss
Dossier del ordre de la penitence... / Meerssernan, G G – Madrid: Archivo Ibero Americano, 1963 – 1 – sp Bibl Santa Ana [240]
Dossier des agressions des forces americano-sud-vietnamiennes a chantrea, province de svayrieng [cambodge] le 19 mars 1964, soumis au conseil de securite de l'o n u / Cambodia. Ministry of Foreign Affairs – [Phnom Penh 1964] [mf ed 1989] – 1r with other items – 1 – mf#mf-10289 seam reel 015/06 [§] – us CRL [327]
Dossiers de l'agitateur – Paris. v1-3 n7. nov 1933-july 1935 – 1 – us NY Public [073]
Les dossiers de l'agitateur – Paris. juin 1933-juin 1935 – 1 – (supplement aux cahiers du bolchevisme) – fr ACRPP [320]
Doster, Frank *see* Papers
Dostert, Klaus *see*
– Prinzipien der informationsuebertragung ueber elektrische energieversorgungsnetze
– Reports on industrial information technology
Dostizheniia sovetskoi vlasti za 40 let v tsifrakh : statisticheskii sbornik – M, 1957. 370p – 5mf – 9 – mf#RHS-22 – ne IDC [314]
Dostoevskii, Fedor M *see* Zhurnal literaturnyi i politicheskii
Dostoievsky / Woodhouse, Christopher Montague – London, England. 1951 – 1r – us UF Libraries [460]
Dot / Documents of Truth, Inc – v2 n5,8,10 [1968 mar, jun/jul, aug/sep], v3 n17 [1969 jun] – 1r – 1 – mf#1583185 – us WHS [071]
Die dotationsansprueche und der nothstand der evangelischen kirche im koenigreich preussen / Gerlach, Hermann – Leipzig: E Bidder, 1874 – 1mf – 9 – 0-524-08325-8 – mf#1993-1020 – us ATLA [240]
Dothel, Niccolo *see*
– Sei duetti notturni, per due flauti, 3e oeuvre
– Sonates pour une flute traversiere et un violincelle, 2e oeuvre
Dotrina breue muy p[ro]uechosa delas cosas q[ue] p[er]tenecen ala fe catholica y a n[uest]ra cristiandad en estilo llano p[ar]a comu[n] intelige[n]cia / Zumarraga, Juan de – Imp[re]ssa e[n] la misma ciudad d[e] Mexico: Por su ma[n]dado y a su costa, ano d[e] 1544 – (= ser Books on religion...1543/44-c1800: doctrina cristiana, obras de devocion) – 2mf – 9 – mf#crl-30 – ne IDC [241]
Dotrina [christ]iana p[ar]a instrucion [et] informacio[n] de los indios : por manera de hystoria / Cordoba, Pedro de – en Mexico: Por mandado del muy R S do[n] fray Jua[n] Cumarraga p[ri]mer obispo desta ciudad...ano de 1544 – (= ser Books on religion...1543/44-c1800: doctrina cristiana, obras de devocion) – 1mf – 9 – mf#crl-32 – ne IDC [241]
Dots and dashes : official publication of the... / Morse Telegraph Club – 1973 mar-1985 mar – 1r – 1 – mf#941882 – us WHS [380]
Dotson, Floyd *see* Indian military of zambia, rhodesia, and malawi
Dotted crotchet : an insignificant, trenchant, coruscant, petulant, poignant, fulminant & itinerant mendicant – [Scotland] Dunfermline: [s.n.] 18 mar 1892 (irreg) [mf ed 2004] – 1r – 1 – uk Newsplan [072]
Dottings on the roadside, in panama, nicaragua, an... / Pim, Bedford Clapperton Trevelyan – London, England. 1869 – 1r – us UF Libraries [972]
Le dottrine moderniste : lettera enciclica della santit a di nostro signore papa pio10 a tutti i vescovi dell'orbe cattolico = Pascendi dominici gregis – Roma: Tipografia Vaticana, 1907 – 1mf – 9 – 0-8370-8010-X – (in italian) – mf#1986-2010 – us ATLA [240]
Doty, James Duane *see* James duane doty papers, ms 1090
Dou, J P *see*
– Practijck des landmetens...
– Practijck des landmetens
– De ses eerste boucken euclidis...
– Van het gebruick der geometrische instrumenten
Douai Abbey *see* Nobility of england – with their descent, the...
Douai, Adolf *see* Abc des wissens fuer die denkenden
Douais, Celestin *see* Essai sur l'organisation des etudes dans l'ordre des freres precheurs
Double bay-rose bay courier – Double Bay – 3r – A$135.21 vesicular A$151.72 silver – at Pascoe [079]
Double blind clinical efficacy study of dexamethasone-lidocaine pulsed phonophoresis on perceived pain associated with symptomatic tendinitis / Penderghest, Caroline E – Temple University, 1995 – 1mf – 9 – $4.00 – mf#PE3616 – us Kinesology [617]
Double branches baptist church. lincolnton, georgia : church records – 1826-Jun 1952. Lacking: Oct 1923-Jul 1938 – 1 – us Southern Baptist [242]
The double cure : or, echoes from national camp-meetings / Pike, J M et al – Boston: Christian Witness Co, 1894 – 2mf – 9 – 0-524-06757-0 – mf#1990-5283 – us ATLA [240]
A double dilemma in darwinism... : read before the british association at glasgow, 1870 / Morris, Francis Orpen – London [1877] – (= ser 19th c evolution & creation) – 1mf – 9 – (with additions) – mf#1.1.11609 – uk Chadwyck [575]
The double doctrine of the church of rome / Zedtwitz, Mary Elizabeth Caldwell, Baroness von – New York: Fleming H Revell, c1906 – 1mf – 9 – 0-8370-9039-3 – mf#1986-3039 – us ATLA [230]
Double, double, toil and trouble = Zauberer / Feuchtwanger, Lion – New York: The Viking press, 1943 (mf ed 1990) – 1r – 1 – (filmed with: spukflieger) – us Wisconsin U Libr [430]
Double eagle report / Golden Mean Society – n1-6 [1984 mar-aug] – 1r – 1 – (cont by: journal from the golden mean society; journal-report from the golden mean society) – mf#956960 – us WHS [071]
Double fortresse sacramentale de foy : from the collection of chief justice coke, holkham hall, norfolk / Roscius, L – London, 1590 – 1r – 1 – mf#96446 – uk Microform Academic [025]
The double game as played by the big interests : religious prejudices a favorite pastime... / Smith, Alexander – [Ottawa?: s.n.], c1919 – 1mf – 9 – 0-665-86423-X – mf#86423 – cn Canadiana [320]
The double search : studies in atonement and prayer / Jones, Rufus Matthew – Philadelphia: J C Winston 1906 [mf ed 1991] – 1mf – 9 – 0-7905-9980-5 – mf#1989-1705 – us ATLA [240]
Double shoals baptist church. shelby, norht carolina : church records – 1954-62 – 1 – 5.00 – us Southern Baptist [242]
Double springs baptist church. shelby, north carolina : church records – 1854-1963 – 1 – 82.40 – us Southern Baptist [242]
Double talk / Madison Area Mothers of Multiples – Madison WI. v7 n8,10-11 [1983 aug, nov-dec]-v14 n1-3 [1990 jan-mar] – 1r – 1 – (with gaps) – mf#1062874 – us WHS [071]
The double text of jeremiah (massoretic and alexandrian) compared : together with an appendix on the old latin evidence / Streane, A W – Cambridge: Deighton Bell; London: George Bell, 1896 [mf ed 1986] – 1mf – 9 – 0-8370-7268-9 – (incl bibl ref & ind) – mf#1986-1268 – us ATLA [221]
The double witness of the church / Kip, William Ingraham – New York: D Appleton, 1843 – 1mf – 9 – 0-524-02477-4 – mf#1990-4336 – us ATLA [240]
Double-cropping wheat and soybeans in the southeast : input use and patterns of adoption / Marra, Michele C & Carlson, Gerald A – Washington DC: US Dept of Agriculture, Economic Research Service...1986 – (= ser Agricultural economic report 552) – 9 – (incl bibl) – us GPO [635]
Doubleday, Charles William *see* Reminiscences of the 'filibuster' war in nicaragua
The double-dealer – New Orleans. v. 1-8. 1921-May 1926 – 1 – us NY Public [410]
Doublier, Roger *see*
– La propriete fonciere en aof
– La propriete fonciere en a.o.f. regime en droit prive
Doubt and faith : being donnellan lectures / Hardy, Edward John – London: T Fisher Unwin, 1899 [mf ed 1985] – 1mf – 9 – 0-8370-5109-6 – (incl bibl ref) – mf#1985-3109 – us ATLA [230]
A doubter's doubts about science and religion / Anderson, Robert – 2nd ed. London: Kegan Paul, Trench, Truebner, 1894 [mf ed 1985] – 1mf – 9 – 0-8370-2096-4 – mf#1985-0096 – us ATLA [210]
Doubts dispelled / Woollacott, Christopher – London, England. 18– – 1r – us UF Libraries [240]
Doucet, Camille *see* Consideration
Doucet, Louis-Joseph *see*
– A la memoire de charles gill
– Les intermedes
Doucet, Stanislas Joseph *see* Dual language in canada
Doudart de Lagree, Ernest Marc Louis de Gonzague *see* Explorations et missions de doudart de lagree
Dou-el-fakar – Alger. 1913-14 – 1 – fr ACRPP [073]
Douen, O *see* Clement marot et le psautier huguenot
Douen, Orentin *see* Les premiers pasteurs de desert (1685-1700)
Dougall, James *see* The canadian fruit-culturist
Dougall, Lily *see*
– Absente reo
– The christ that is to be
– Madonna of a day
– The practice of christianity
– Pro christo et ecclesia

Dougherty collection of military newspapers : newspapers from the us armed services – [mf ed State Historical Society of Wisconsin] – c2500 titles on 58r – 1 – (with p/g. as one of the largest coll of its kind ever amassed, the walter s and esther dougherty coll of military newspapers of the us brings together more than 4000 iss of military papers) – us WHS [355]
Dougherty, Peter *see* Letters
Doughty, Arthur G *see* The siege of quebec and the battle of the plains of abraham
Doughty, Arthur George *see* The fortress of quebec 1608-1903
Doughty, Charles Montagu *see* Travels in arabia deserta
Doughty, Oswald *see* The castle of otranto
Doughty, William Lamplough *see* John wesley, preacher
Douglas 1718-1890 – Oxford, MA (mf ed 1986) – (= ser Massachusetts vital records) – 21mf – 9 – 0-87623-004-4 – (mf 1-4: b,m,d 1718-1843. mf 5-7: b,m,d 1844-60. mf 8-10: births 1860-90. mf 11-13: index to births 1844-1939. mf 14-15: marriages 1860-90. mf 16-17: index to marriages 1844-1938. mf 18-19: deaths 1860-90. mf 20-21: index to deaths 1844-1939) – us Archive [978]
Douglas 1724-1849 – Oxford, MA (mf ed 1996) – (= ser Massachusetts vital record transcripts to 1850) – 10mf – 9 – 0-87623-240-3 – (1t-5t: births & deaths 1724-1825. mf 1t: out-of-town marriages 1759-75. mf 1t-3t: marriage intentions 1751-90; marriages 1748-97. mf 5t-9t: marriages & intentions 1789-1849. mf 9t-10t: births 1841-49. mf 10t: marriages, deaths 1844-49) – us Archive [978]
Douglas, Andrew Halliday *see* Five sermons
Douglas, C *see* Memorials of rev carstairs douglas...,missionary of the presbyterian church of england at amoy, china, 1877
Douglas, C R *see* Summary judgement practice in three district courts
Douglas, Charles *see* The ethics of john stuart mill
Douglas, Charles H *see* The government of the people in the state of connecticut
Douglas, Christopher *see* Review of recent publications, regarding the proposed reform of the bankruptcy laws of scotland
[Douglas city-] douglas city gazette – CA. May 1861-Apr 1862 – 1r – 1 – $60.00 – mf#B02178 – us Library Micro [071]
Douglas county [city directory] : listing] – 1971 p1-52, p53-end – 2r – 1 – mf#3193678 – us WHS [917]
Douglas county daily morning times – Roseburg, OR. 5/31/1935-6/13/1936 – 1r – mf#833985132 – us Oregon Lib [071]
The douglas county gazette – Waterloo, NE: Frank B Cox. Vol. 43, no. 1 (Jan. 5, 1934)-1983// (wkly) [mf ed -jul 25 1983 (lacks oct 4 1957) filmed 1972-83] – 26r – 1 – (publ in omaha mar 9 1978-jul 25 1983. formed by the union of: waterloo gazette, millard courier, elkhorn exchange and: bennington herald. merged with: elkhorn valley post to form: douglas county post-gazette. other ed: metro jun 28 1973-sep 30 1977) – us NE Hist [071]
Douglas county, georgia, genealogy – v1 n1-v4 n6 [1978 sep-1981/82 win/spr] – 1r – 1 – mf#669883 – us WHS [929]
Douglas county post-gazette – Blair, NE: Post-Gazette Pub Corp, 1983 (wkly) [mf ed v54 n38. sep 20 1983- filmed 1984-] – 1 – (formed by the union of: elkhorn valley post and: douglas county gazette) – us NE Hist [071]
Douglas county times – Roseburg OR: Times Pub Co, 1934-35 [semiwkly] – 1 – (cont by: daily morning times (roseburg, or)) – us Oregon Lib [071]
The douglas enterprise – Douglas, NE: Wren & Walker. 46v. v1 n1. mar 15 1889-v46 n35. sep 27 1934 (wkly) – 1 – (cont as a suppl in: peru enterprise) – us NE Hist [071]
Douglas, George Cunningham Monteath *see*
- The book of jeremiah
- The book of joshua
- The book of judges
- Isaiah one and his book one
- Samuel and his age
- Why i still believe that moses wrote deuteronomy

Douglas, James *see*
- Biographical sketch of thomas sterry hunt
- Canadian independence, annexation and british imperial federation
- Facts and reflections bearing on annexation, independence and imperial federation
- The gold fields of canada
- The structure of prophecy

Douglas jerrold's weekly newspapers – London, UK. Jul 1846-1848. -w.3 reels – 1 – uk British Libr Newspaper [072]
Douglas, Kenneth C *see* Manuscripts relating to papua new guinea, [1943]
Douglas, Lawrence *see* Two christmas gifts
Douglas, Marjorie Stoneman *see* Through blood to gold
Douglas, Robert K *see* Confucianism and taouism
Douglas, Robert Kennaway *see* Society in china
Douglas, Robert Langton *see* Fra angelico
Douglas, Stephen Arnold *see* An american continental commercial union or alliance
Douglas, William *see* The currency of india
Douglass, Benjamin *see* Four short lectures on the book of revelation
Douglass, Frederick *see*
- Life of frederick douglass
- My bondage and my freedom
- Papers

Douglass, Harlan Paul *see*
- Christian reconstruction in the south
- Congregational missionary work in porto rico
- The new home missions

Douglass, Jacqueline A *see* An examination of two theoretical distributions using three methods of scoring criterion-referenced measures of motor performance
Douglass' monthly – Killen TX 1859-63 – 1 – mf#3345 – us UMI ProQuest [976]
Douglass' monthly – Rochester NY. v1-5. 1858-63 [all publ] – (= ser Black journals, series 1) – 9mf – 9 – $105.00 – us UPA [073]
Douglass, R *see* Infant baptism
Douglass, Truman Orville *see* The pilgrims of iowa
Douglass, William *see*
- A summary, historical and political, of the first planting, progressive improvements, and present state of the british settlements in north-america
- A summary, historical and political, of the first planting, progressive improvements, and present state of the british settlements in north-america, vol 1
- A summary, historical and political, of the first planting, progressive improvements, and present state of the british settlements in north-america, vol 2

Douglasville first baptist church. douglasville, georgia : church records – Nov 1944-88. 4,258p – 1 – us Southern Baptist [242]
Douhet-Rathail *see* L'accusateur revolutionnaire
Douin, Georges *see*
- Histoire de regne du khedive ismail. l'empire africain. tome 3, le et 2e parties
- Histoire du regne du khedive ismail

Doull, Alexander *see* Extensive and systematic colonisation in connection with the construction of the intercolonial railway through the canada dominion
Doull, Alexander John *see* Ordination vows
Doulton and co at chicago exhibition, 1893 / Doulton and Co Ltd – London [1893] – (= ser 19th c art & architecture) – 1mf – 9 – mf#4.2.862 – uk Chadwyck [730]
Doulton and Co Ltd *see*
- Architectural designs
- Doulton and co at chicago exhibition, 1893

Doumergue, Emile *see*
- Calomnies anti-protestantes
- Calvijn in het strijdperk
- Calvijn's jeugd, jongelingsjaren, omzwervingen, bekeering, en eerste optreden als reformator
- Calvin, le predicateur de geneve
- L'emplacement du baucher de michel servet
- Essai sur l'histoire du culte reforme principalement au 16e et au 19e siecle
- Geneva, past and present
- La hongrie calviniste
- Iconographie calvinienne
- La piete reformee d'apres calvin
- La veille de la loi de l'an 10, 1763-1802

Doumergue, Emile et al *see* Calvin and the reformation
Dourado, Mecenas *see* Mecenas
Dourille, Joseph *see* Resume de l'histoire de napoleon et des armees qui ont ete sous son commandement
Dousa, G *see* ...De initere svo constantinopolitano, epistola
Dousdebes, Pedro Julio *see* Trayectoria militar de santander
Dousman index – Dousman WI. 1949 jan 28-1951, 1952-61, 1962-1965 mar 25, apr 1-1968 jun 13, jun 20-1970 may 28 – 6r – 1 – (cont: weekly index [dousman wi]; cont by: index [dousman wi]) – mf#999839 – us WHS [071]
Douthit family tree – v1 n1-v6 n1 [1981 sum-1986 jun] – 1r – 1 – mf#1289671 – us WHS [929]
Douthwaite, William R *see* Gray's inn
Doutre, Gonzalve *see*
- Le principe des nationalites
- Proces ruel-boulet

Doutre, Joseph *see*
- Constitution of canada the british north america act, 1867; its interpretation.
- Dame henriette brown (demanderesse en cour inferieure)
- Les fiances de 1812

Doutressoulle, Georges *see* Elevage in afrique occidentale francaise
Doutrina do padre feijo e suas relacoes com a sede / Talassi, Luis – Sao Paulo, Brazil. 1954 – 1r – us UF Libraries [972]
Doutte, Edmond *see*
- L'islam algerien en l'an 1900
- Magie & religion dans l'afrique du nord

Douville, J B *see*
- 30 mois de ma vie, quinze mois avant et quinze mois apres mon voyage au congo
- Voyage au congo et dans l'interieur de l'afrique equinoxiale

Douville, Jean-Baptiste *see* Voyage au congo et dans l'interieur de l'afrique equinoxiale
Douwen, Wiebe Jans van *see* Socinianen en doopsgezinden
Douwes, Jan *see* Het leven en werken van dr. william carey, evangeliebode onder de heidenen in bengalen
Doux au bec / La Roussie, Roger De – Paris, France. 1913 – 1r – us UF Libraries [440]
Douze bacchanales pour le forte-piano : avec accompagnement de tambourin ad-libitum / Steibelt, Daniel – Paris: Chez Mlles Erard [1801?-] [mf ed 19–] – 1mf – 9 – mf#fiche 27 – us Sibley [780]
Les douze jeunes filles : ou, l'histoire de neang kangrey / Pavie, Auguste – [Phnom Penh] Institut bouddhique 1969 [mf ed 1989] – (= ser Series de culture et civilisation khmeres 1) – 1r with other items – 1 – mf#mf-10289 seam reel 006/06 [§] – us CRL [959]
Douze lecons avec leur doigte pour le piano-forte : tirees de la methode de piano / Steibelt, Daniel – Offenbach sur le Mein: Jean Andre [1809] [mf ed 1992] – 1r – 1 – mf#pres. film 113 – us Sibley [780]
Douze nouveaux quatuors, 4e livraison / Pleyel, Ignaz – Paris: Le Duc [18–?] [mf ed 1989] – 4pt on 1r – 1 – mf#pres. film 50 – us Sibley [780]
Les douze petits prophetes / Hoonacker, Albin van – Paris: Victor Lecoffre 1908 [mf ed 1908] – (= ser Etudes bibliques) – 2mf – 9 – 0-7905-0578-9 – (incl bibl & ind) – mf#1987-0578 – us ATLA [221]
Douze romances : avec accompagnement de pianoforte / Baumbach, Friedrich August – [1790?] [mf ed 1988] – 1v on 1r – 1 – mf#pres. film 44 – us Sibley [780]
Dov Baer Ben Samuel *see* Shivhe ha-besht
Dov shefer – Kaunas, Lithuania. 1939 – 1r – us UF Libraries [939]
The dove and the leopard : more uraon poetry / Archer, William George – Bombay: Orient Longmans, 1948 – (= ser Samp: indian books) – us CRL [490]
Dove of christ described – London, England. 1824 – 1r – us UF Libraries [240]
The dove; or, passages of cosmography; a poem...reprinted from the original ed. of 1613. / Zouch, Richard – Memoir, notes, coll. and arr. by Richard Walker.Oxford: H. Slatter; London: T. Rodd, 1839.xiii,82p. incl. geneal. tab – 1 – us Wisconsin U Libr [920]
Dover 1749-1910 – Oxford, MA (mf ed 2003) – (= ser Massachusetts vital records) – 24v on 118mf – 9 – 0-87623-429-5 – (mf 1-4: precinct records 1749-84. mf 4-24: district records 1784-1836. mf 24-33: town meetings 1836-65. mf 34-45: town meetings 1866-97. mf 46-57: treasurers records 1749-1854. mf 58-67: tax invoices 1811-28. mf 68-79: tax valuations 1829-47. mf 80-84: payments 1841-91. mf 85-86: church records 1812-1904. mf 87-88: rebellion records 1861-65. mf 89-92: voters 1877-1920. mf 93-95: birth index 1753-2002. mf 95-97: death index 1764-2002. mf 98-100: marriage index 1785-2002. mf 101-107: vital records 1753-1844. mf 103,107: out-town marrs 1785-99. mf 103-104,108-110: intents 1785-1852. mf 111-112: births 1843-1900. mf 112, 118: marriages 1844-1915. mf 113-114: deaths 1844-1901. mf 115-117: marr intentions 1852-1935) – us Archive [978]
Dover 1753-1849 – Oxford, MA (mf ed 2003) – (= ser Massachusetts vital record transcripts to 1850) – 1v on 5mf – 9 – 0-87623-241-1 – (mf 1t-2t: births by family 1753-1844. mf 2t: marriages 1785-1844. mf 2t-3t: deaths 1774-1844. mf 3t-4t: intentions 1786-1849. mf 5t: births 1843-49. mf 5t: marriages 1844-49. mf 5t: deaths 1844-49) – us Archive [978]
Dover Baptist Association. Virginia *see* Church discipline
Dover baptist church. lagrange, missouri : church records – 1838-1965 – 1 – 46.80 – us Southern Baptist [242]
Dover baptist church. shelby, north carolina : church records – 1934-63 – 1 – us Southern Baptist [242]
Dover baptist church. tennessee : church records – 1924-58 – 1 – us Southern Baptist [242]
Dover, New Hampshire. Dover Free Will Baptist Church *see* Records
Dover. Ohio. Baptist Church *see* Church records, ms 668
The dover selection of spiritual songs / Broadus, Andrew – Philadelphia, 1828 – 1 – us Southern Baptist [242]
Dover, Thomas Birkett *see*
- The hidden word
- The ministry of mercy

Doveri dell'uomo / Mazzini, Giuseppe – Torino, Italy. 194- – 1r – us UF Libraries [025]
Dovetail / Iowa Peace Network – v5 n1-v11 n2 [1981 sum-1987 spr] – 1r – 1 – (cont by: dovetail/cipar peaces) – mf#1321042 – us WHS [327]
Dovnar-Zapol'skii, M V *see* Tainoe obshchestvo dekabristov
Dow air strip – Bangor, ME. dec 20 1951-jun 13 1952 – 1 – us CRL [071]
Dow field observer – Bangor, ME. jul 4 1942-1945; jan 9,30 1946 – 1 – us CRL [071]
Dow thunderjet – Bangor, ME. [jan 30-dec 17 1948]; jan 12-sep 26 1949 – 1 – us CRL [071]
Dow, William *see*
- Church's hope
- Elements of Unity
- Former and the latter rain

The dowager lady tremaine / Alliott, James Bingham (Mrs) – London: Elliot Stock, 1895 – (= ser 19th c women writers) – 2mf – 9 – mf#5.1.97 – uk Chadwyck [830]
Dowbiggin, Herbert *see*
- Confidential report on the northern rhodesia police
- Report on the northern rhodesia police

Dowdakin, James Daniel *see* Analysis of the dumbarton oaks concerto for chamber orchestra by igor stravinsky
Dowden, Edward *see* Puritan and anglican
Dowden, John *see*
- The annotated scottish communion office
- The bishops of scotland
- The celtic church in scotland
- Church year and kalendar
- Define your terms
- Further studies in the prayer book
- Helps from history to the true sense of the minatory clauses of the...
- I–the church of rome and recent projects for re-union
- I-the church of rome and recent projects for re-union
- Lord's faithful servant
- The medieval church in scotland
- Outlines of the history of the theological literature of the church of england
- Quaestiunculae liturgicae
- Relation of christian ethics to philosophical ethics
- Relations of the church of england and the episcopal church
- The workmanship of the prayer book in its literary and liturgical aspects

Dowding, William Charles *see*
- Blind apostle
- The life and correspondence of george calixtus

Dowdy, Homer E *see* Out of the jaws of the lion
Dowell, Linus J *see* The effect of the lead leg plant on factors affecting distance on kickoffs in football
Dower, John *see* New british gold fields
Dowkontt, George D *see* Murdered millions
Dowlais boys elementary schools admission register – 1905-11 – 1mf – 9 – £1.25 – uk Glamorgan FHS [370]
Dowlais cottage leases – 1818-77 – 1mf – 9 – £1.25 – uk Glamorgan FHS [333]
Dowlais gellifaelog schools admission register – 1878-1907 – 1mf – 9 – £1.25 – uk Glamorgan FHS [370]
Dowlais, glamorgan, parish church of st john the baptist : baptisms 1839-1925, burials 1839-1920 – 3mf – 9 – £3.75 – uk Glamorgan FHS [929]
Dowlais, glamorgan, r c parish church of st illtyd : baptisms, 1836-1900; burials, 1856-1890; marriages, 1836-1900 – [Glamorgan]: GFHS [mf ed c2003] – 3mf – 9 – £3.75 – uk Glamorgan FHS [929]
Dowlais iron co sick workers & other lists – 1845-59 – 1mf – 9 – £1.25 – uk Glamorgan FHS [331]
Dowlais poor rate book – 1894 – 2mf – 9 – £2.50 – uk Glamorgan FHS [350]
Dowlais st illtyd – girls admission register – 1904-24 – 1mf – 9 – £1.25 – uk Glamorgan FHS [370]
Dowlais times and merthyr, aberdare, pontypridd echo – [Wales] LLGC 22 may 1891-6 oct 1899 [mf ed 2004] – 8r – 1 – (missing: 1897; cont by: merthyr times & dowlias times & aberdare echo [8 jan 1892-oct 1899]) – uk Newsplan [072]
Dowling, Joannis Goulter *see* Notitia scriptorum ss. patrum aliorumque veteris ecclesiae monumentorum
Dowling, John *see*
- The judson memorial
- The judson offering
- Missionary inquiry

Dowling, John Goulter *see*
- Introduction to the critical study of ecclesiastical history
- Notitia scriptorum ss. patrum aliorumque veteris ecclesiae monumentorum

Dowling, Theodore Edward *see*
– The abyssinian church
– The armenian church
– The egyptian church
– Gaza, a city of many battles
– Hellenism in england
– The patriarchate of jerusalem
– Six branches of the missionary work of the church set forth as subjects for meditation during the week of intercession for missions, 1878
– Sketches of caesarea
– Sketches of georgian church history

Dowling, Theodore Edward [comp] *see* Subjects for meditation during the week of intercession for missions 1877

Dowling, William Worth *see*
– The christian psalter
– Lesson commentary on the international bible studies of 1901
– The lesson helper

Down beat – Elmhurst IL 1937+ – 1,5,9 – ISSN: 0012-5768 – mf#2032 – us UMI ProQuest [780]

Down home music newsletter – 1989 jan/mar – 1r – 1 – mf#4875479 – us WHS [780]

Down in water street : a story of sixteen years life and work in water street mission / Hadley, Samuel Hopkins – memorial ed. New York: Fleming H Revell, c1906 [mf ed 1986] – 1mf – 9 – 0-8370-6188-1 – mf#1986-0188 – us ATLA [240]

Down independent – Downpatrick, Ireland. 5 oct 1878-mar 1882 – 2 1/2r – 1 – uk British Libr Newspaper [072]

Down recorder *see* Downpatrick recorder

Down second avenue / Mphahlele, Ezekiel – London, England. 1965 – 1r – us UF Libraries [420]

Down that pan american highway / Stephens, Roger – New York, NY. 1948 – 1r – us UF Libraries [972]

Down the fairway : the golf life and play / Jones, Bobby – New York, NY. 1931 – 1r – us UF Libraries [790]

Down the village street scenes in a west country hamlet by christopher hare / Andrews, Marian – [Edinburgh], London: William Blackwood & Sons, 1895 – (= ser 19th c women writers) – 4mf – 9 – mf#5.1.99 – uk Chadwyck [420]

Down the water and weekly chit chat – [Scotland] [Glasgow: s.n.] 4 jun 1887 (wkly) [mf ed 2003] – 1r – 1 – uk Newsplan [072]

Down the yukon and up the mackenzie : 3,200 miles by foot and paddle / Ogilvie, William – Toronto: Toronto Pub Co, 1893 [mf ed 1981] – 1mf – 9 – 0-665-15789-4 – (fr: the canadian magazine of politics, science, art and literature) – mf#15789 – cn Canadiana [917]

Down with it! / Sowter, G Arthur – London, England. 18– – 1r – us UF Libraries [240]

Downame, J *see* The christian warfare against the devill world and flesh

Downeast ancestry – 1977 jun-1982 dec, 1983 feb-1987 apr – 2r – 1 – mf#687377 – us WHS [929]

Downeaster – Bangor, ME. [feb 1960-apr 13 1962] – 1 – us CRL [071]

Downer, Arthur Cleveland *see* The mission and ministration of the holy spirit

Downes, George *see*
– Letters from continental countries
– Letters from mecklenburg and holstein

Downes, Leonard Stephen *see* Introduction to modern brazilian poetry

Downey – 1931; 1933; 1935-52; 1955-74 – (= ser California telephone directory coll) – 61r – 1 – $3050.00 – mf#P00024 – us Library Micro [917]

Downey/norwalk – 1984 – (= ser California telephone directory coll) – 2r – 1 – $100.00 – mf#P00025 – us Library Micro [917]

Downham in the isle (aka little downham) 1558-1950 – (= ser Cambridgeshire parish register transcript) – 13mf – 9 – £16.25 – uk CambsFHS [929]

Downie, David *see*
– The lone star
– The lone star, the history of the telugu mission of the american baptist missionary union

Downie, William *see*
– Explorations in jarvis inlet and desolation sound, british columbia
– Hunting for gold

[Downieville-] mountain messenger – CA. 1865-66; 1872-1977 – 39r – 1 – $2340.00 – mf#BC02179 – us Library Micro [071]

[Downieville-] sierra advocate – CA. 1866-1867 – 1r – 1 – $60.00 – mf#C03200 – us Library Micro [071]

[Downieville-] sierra age – CA. 1871 – 1r – 1 – $60.00 – mf#C03591 – us Library Micro [071]

Downing, C T *see* The fan-qui in china

Downing day – Downing WI. 1903 aug 22 [v1 n2] – 1r – 1 – mf#1139735 – us WHS [071]

Downing, Elijah Hedding *see* Remains of rev joshua wells downing

Downing herald – Downing WI. 1910 jul 7, 1918 may 25 – 1r – 1 – mf#965389 – us WHS [071]

Downing, Hugh Urquhart *see* Cases in georgia reports that have been overruled, doubted, criticised, or modified

Downing, William *see* Observations on the constitution, customs, and usage of middle temple

Downingtown bulletin / Downingtown Industrial and Agricultural School of Downingtown, Pennsylvania – 1938 jul, 1939/40-1941/42, 1943/44, 1946/47, 1950/51-1951/52, 1954/55 [mutilated] – 1r – 1 – mf#5266581 – us WHS [630]

Downingtown Industrial and Agricultural School of Downingtown, Pennsylvania *see* Downingtown bulletin

Downpatrick recorder – Downpatrick, Ireland. 31 dec 1836; may 1856; jul 1856-1929; 1931-dec 1998 – 129 1/2r – 1 – (aka: down recorder) – uk British Libr Newspaper [072]

Downpatrick recorder – Dec 31 1836-Dec 28 1839; 1840-77; Jan 5 1878-Dec 24 1880; 1881-1929; 1931-33; Jan 6 1934-Dec 19 1936; 1937-47; Jan 3 1948-Dec 24 1949; 1950; Jan 6 1951-Dec 24 1953; Jan 9 1954-Dec 24 1955; 1956-Sep 1990, Oct 3-Dec 19 1990; 1991-Dec 22 1992; Jan-Sep 1993; Oct 6-Dec 22 1993; Jan-Dec 21 1994; Jan-Dec 20 1995; 1996 – 121r – 1 – uk British Libr Newspaper [072]

Downriver reporter – Lincoln Park MI. 1979 jun 7-oct 30 – 1r – 1 – mf#851707 – us WHS [071]

Down's syndrome (mongolism) = Down-sindroom (mongolisme) / South Africa. Department of Health [Departement van Gesondheid] – Pretoria: Dept of Health [1975?] [mf ed Pretoria, RSA: State Library [199-]] – 10p [ill] on 1r with other items – 5 – (in english & afrikaans) – mf#op 06430 r23 – us CRL [616]

Downshire chronicle – Downpatrick. Ireland. -w. 2 Feb 1839-13 Jun 1840. (1/2 reel) – 1 – uk British Libr Newspaper [072]

Downshire protestant – Downpatrick, Ireland. 6 jul 1855-27 sep 1862 – 2 1/2r – 1 – uk British Libr Newspaper [072]

Downside : the history of st. gregory's school from its commencement at douay to the present time / Birt, Henry Norbert – London: K. Paul, Trench, Truebner, 1902 – 1mf – 9 – 0-7905-5569-7 – mf#1988-1569 – us ATLA [941]

The downside review – 1(1880)-91(1973) – 623mf – 9 – €1188.00 – ne Slangenburg [241]

Downto business / Bank of Middleton [WI] – v10 n8-v11 n12 [1983 aug-1984 dec] – 1r – 1 – mf#957565 – us WHS [332]

Downtown edition – v2 n1-51 [1990 may 21-1991 may 13] – 1r – 1 – (cont by: milwaukee's downtown edition) – mf#1856820 – us WHS [071]

Downtown shopping guide *see* [Fresno-] guide

Downy mildew (blue mold) of tobacco / Kincaid, Randall R – Gainesville, FL. 1939 – 1r – us UF Libraries [630]

Dowson, John *see*
– A classical dictionary of hindu mythology and religion, geography, history, and literature

Dox, Thurston J *see* Hendrik f andriessen

Doxey, G V *see* High commission territories and the republic of south africa

Doxey, Roy Watkins *see* Doctrine and covenants and the future

Doyle, Andrew *see* A very interesting selection of important mathematical problems, with solutions

Doyle, Arthur Conan *see*
– Guerre dans l'afrique australe
– My friend the murderer

Doyle, Charles W *see* Briefe aus aegypten

Doyle, James *see* Letters on the state of ireland

Doyle, James William Edmund *see* The official baronage of england, showing the succession, dignities, and offices of every peer from 1066 to 1885

Doyle, John Andrew *see* The english in america

Doyle, Lauren A *see* Differential effectiveness of relaxation procedures in attenuating components of anxiety in shooters

Doyle, Mike N *see* The effect of phase 2 cardic rehabilitation on self-efficacy and quality of life

Doyle, Sherman Hoadley *see* Presbyterian home missions

Doylestown democrat – Doylestown, PA. -w 1889-1912 – 13 – $25.00r – us IMR [071]

Doyon limited newsletter – v1 n4-v7 n10 [1973 apr-1979 dec] – 1r – 1 – (cont: tanana chiefs conference newsletter; cont by: doyon newsletter) – mf#610638 – us WHS [071]

Doyon newsletter – v8 n1-v9 n1 [1980 jan/feb-1981 jan], v10 n1-10, [1982 feb-nov] – 1r – 1 – (cont: doyon limited newsletter; cont by: doyon) – mf#610636 – us WHS [071]

Doze estudos / Rosenfeld, Anatol – Sao Paulo: Conselho Estadual de Cultura, Commissao de Literatura, [1959] – 1r – 1 – (incl bibl ref and index) – us Wisconsin U Libr [440]

Dozier, H C *see* Historical data

Dozor – (city unknown) 1942-44 [mf ed Norman Ross] – 1 – mf#nrp-95 – us UMI ProQuest [934]

Dozy, R *see*
– Commentaire historique sur le poeme d'ibn-abdoen
– The history of the almohades

Dozy, R P A *see* Notices sur quelques manuscrits arabes

Dozy, Reinhart Pieter Anne *see*
– De israelieten te mekka van davids tijd tot in de vijfde eeuw onzer tijdrekening
– Spanish islam

Dpp pni / Siaran PNI – Djakarta, 1957 – 1mf – 9 – mf#SE-408 – ne IDC [959]

Dpp pni dep pen prop / Suara Marhaenis – Djakarta, [1950]-1958 – 22mf – 9 – (missing: [1950]-1954, v1-4; 1957, v7(9, 11)) – mf#SE-417 – ne IDC [959]

DPP Sarbupri *see* Warta sarbupri

DPPPNI *see* Suara marhaen

Dr. a. neander's katholicismus und protestantismus = Katholicismus und protestantismus / Neander, August; ed by Messner, Hermann – Berlin: Wiegandt & Grieben, 1863 – 1mf – 9 – 0-524-03292-0 – mf#1990-0903 – us ATLA [240]

Dr. abbott and christian evolution : 1., the irresistible conflict between two world-theories / Savage, Minot Judson – Boston: George H Ellis, 1892 – 1mf – 9 – 0-8370-5055-3 – mf#1985-3055 – us ATLA [240]

Dr allan aubrey boesak : 'n bree agtergrondskets en sy aandeel in die stigting van die united democratic front / Pedro, Enrico Graham – U of the Western Cape 1989 [mf ed S.l: s.n. 1989] – 2mf – 9 – (incl bibl) – sa Misc Inst [325]

Dr. ambrosius moibanus : ein beitrag zur geschichte der kirche und schule schlesiens im reformationszeitalter / Konrad, Paul – Halle: Verein fuer Reformationsgeschichte, 1891 – (= ser [Schriften des Vereins fuer Reformationsgeschichte]) – 1mf – 9 – 0-7905-4702-3 – (incl bibl ref) – mf#1988-0702 – us ATLA [943]

Dr anna s kugler papers / Kugler, Anna Sarah – 1868-1983[mf ed 2004] – (= ser Archival coll) – 8r – 1 – (organized in foll subseries: diaries 1877, 1883-1930; notes 1890-1928 [1907-09]; correspondence 1882-1931 [1922-31]; subject files 1883-1983 [1883-1930]; photographs 1868-1930. with finding aid) – mf#xa0085r – us ATLA [242]

Dr anton friedrich bueschings neue erdbeschreibung / Buesching, Anton – Hamburg 1807 [mf ed Hildesheim 1995-98] – 2mf – 9 – €60.00 – 3-487-28942-3 – gw Olms [550]

Dr at pierson on the evangelisation of the world / Pierson, Arthur T – London, England. 1890? – 1r – us UF Libraries [240]

Dr bell's system of instruction / Bell, Andrew – London, England. 1821 – 1r – us UF Libraries [240]

Dr bloch's oesterreichische wochenschrift – Vienna, Austria 1902 [mf ed Norman Ross] – 12r – 1 – mf#nrp-1964 – us UMI ProQuest [939]

Dr buchner's report : repertorium fur die pharmacie / Bley – Nurnberg. 1834, 1837 – 22mf – 7 – mf#731/2 – uk Microform Academic [900]

Dr bushnell's orthodoxy : or, an inquiry whether the factors of the atonement are recognized in his vicarious sacrifice / Taylor, Oliver S – New Haven: E Hayes 1867 [mf ed 1993] – 1mf – 9 – 0-524-06109-2 – mf#1991-2422 – us ATLA [240]

Dr carpenter at sion college / M, M – London, England. 1874 – 1r – us UF Libraries [240]

Dr chalmers – Edinburgh, Scotland. 1847 – 1r – us UF Libraries [240]

Dr chase's family physician, farrier, bee-keeper, and second receipt book : being an entirely new and complete treatise, pointing out, in plain and familiar language, the cause, symptoms, and treatment of the leading diseases of persons, horses and cattle, upon common sense principles... – Toledo, OH: Chase, 1875 [mf ed 1985] – 7mf – 9 – 0-665-01767-7 – (incl ind) – mf#01767 – cn Canadiana [640]

Dr chase's new receipt book : or, information for everybody: the life-long observations of the author... – Toronto: Rose, 1889 [mf ed 1985] – 5mf – 9 – 0-665-01769-3 – (incl ind) – mf#01769 – cn Canadiana [640]

Dr chase's recipes, or, information for everybody : an invaluable collection of about eight hundred practical recipes for merchants, grocers, saloon-keepers... / Chase, Alvin Wood – 23rd enl ed. London, CW [Ont]: J Moffat, 1865 [mf ed 1985] – 5mf (ill) – 9 – 0-665-01773-1 – (with remarks and full explanations) – mf#01773 – cn Canadiana [030]

Dr chase's third, last and complete receipt book and household physician : or, practical knowledge for the people: from the life-long observations of the author... – Detroit, Windsor, Ont: F B Dickerson, 1889 [mf ed 1985] – 10mf – 9 – 0-665-01772-3 – (incl ind) – mf#01772 – cn Canadiana [640]

Dr clifford on "inspiration" examined and criticised / Varley, Henry – London, England. 18– – 1r – us UF Libraries [240]

Dr croft's service in b, and evening service in e / Croft, William – [1748] [mf ed 1988] – 1r – 1 – mf#pres. film 12 – us Sibley [780]

Dr cumming's concludin lecture : and a note by the editor – London, England. 18– – 1r – us UF Libraries [240]

Dr cunningham and dr bryce on the "circa sacra" power of the civil... / Cunningham, William – Edinburgh, Scotland. 1843? – 1r – us UF Libraries [240]

El dr d anotonio joachin de urizar y bernal : colegial huesped en el mayor de santa maria de todod santos, cathedratico jubilado de prima de leyes... / Urizal y Bernal, Amtonio Joachin de – [Puebla: s.n. 1764?] – (= ser Books on religion...1543/44-c1800: iglesias, catedrales) – 1mf – 9 – mf#crl-446 – ne IDC [241]

El dr d antonio manuel roxana mudarra : domiciliario de este obispado de la puebla de los angeles... / Roxana Mudarra, Antonio Manuel – [Puebla: s.n. 1767?] – (= ser Books on religion...1543/44-c1800: iglesias, catedrales) – 1mf – 9 – mf#crl-457 – ne IDC [241]

El dr d gregorio pelayo de la granda y junco, colegial antiguo del eximio theo-jurista de s pablo : vura por su magestad vicario foranco, y juez ecclesiastico... / Pelayo de la Granda, Greogorio – [Puebla: s.n. 1758?] – (= ser Books on religion...1543/44-c1800: iglesias, catedrales) – 1mf – 9 – mf#crl-441 – ne IDC [241]

El dr d joseph isidro montana, tenorio de la vanda : cura beneficiado por s mag de la villa de carrion en el valle de atlixco... / Montana, Joseph Isidro – [Puebla: s.n. 1767?] – (= ser Books on religion...1543/44-c1800: iglesias, catedrales) – 1mf – 9 – mf#crl-454 – ne IDC [241]

El dr d joseph maria lasso de la vega : colegial real de oposicion en el m illustre, y real de san ignacio, y opositor a la canongia magistral vacante, representa...sus corjosephrvicios – [Puebla: s.n. 1767?] – (= ser Books on religion...1543/44-c1800: iglesias, catedrales) – 1mf – 9 – mf#crl-452 – ne IDC [241]

El dr d joseph martinez de la canal : colegial antiguo del eximio de san pablo, examinador synodal de este obispado... / Martinez de la Canal, Joseph – [Puebla: s.n. 1767?] – (= ser Books on religion...1543/44-c1800: iglesias, catedrales) – 1mf – 9 – mf#crl-453 – ne IDC [241]

El dr d joseph perez calama : familiar del ilmo sr obispo de esta ciudad de la puebla de los angeles... / Perez Calama, Joseph – [Puebla: s.n. 1767?] – (= ser Books on religion...1543/44-c1800: iglesias, catedrales) – 1mf – 9 – mf#crl-456 – ne IDC [241]

El dr d m primo de ribera : colegial huesped del insigne, vicio, y mayor de santa maria de todos santos... / Primo de Ribera, Miguel – [Puebla: s.n. 1766?] – (= ser Books on religion...1543/44-c1800: iglesias, catedrales) – 1mf – 9 – mf#crl-450 – ne IDC [241]

El dr d manuel antonio manzanedo, colegial antiguo del eximio thelogu de s pablo : cura por su magestad, vicario... / Manzanedo, Manuel Antonio – [Puebla: s.n. 1758?] – (= ser Books on religion...1543/44-c1800: iglesias, catedrales) – 1mf – 9 – mf#crl-442 – ne IDC [241]

El dr d manuel ignacio de gorospe y padilla : prebendado de esta santa iglesia cathedral... / Gorospe, Manuel Ignacio de – [Puebla: s.n. 1766?] – (= ser Books on religion...1543/44-c1800: iglesias, catedrales) – 1mf – 9 – mf#crl-448 – ne IDC [241]

El dr d manuel joseph de olmedo : araziel, martin, y zintra, colegial antiguo del eximio theologo de s pablo... / Olmedo, Manuel Joseph de – [Puebla: s.n. 1767?] – (= ser Books on religion...1543/44-c1800: iglesias, catedrales) – 1mf – 9 – mf#crl-455 – ne IDC [241]

El dr d sebastian sanchez pareja : cura beneficiado por su magestad, vicario, y juez ecclesiastico de la doctrina de s juan de los llanos... / Sanchez Pareja, Sebastian – [Puebla: s.n. 1758?] – (= ser Books on religion...1543/44-c1800: iglesias, catedrales) – 1mf – 9 – mf#crl-443 – ne IDC [241]

El dr d thadeo gavino de la puerta sanchez de tagle : quo lo es en sagrados canones por la real universidad de mexico... / Puerta, Thadeo Gavino de la – [Puebla: s.n. 1766?] – (= ser Books on religion...1543/44-c1800: iglesias, catedrales) – 1mf – 9 – mf#crl-451 – ne IDC [241]

Dr david livingstone / Tanguy, F – Cape Town, South Africa. 1957 – 1r – us UF Libraries [960]

Dr davidson's removal from the professorship of biblical literatur / Nicholas, Thomas – London, England. 1860 – 1r – us UF Libraries [220]

Dr e zachariae's, correspondirenden mitgliedes des archaeologischen instituts zu rom, reise in den orient in den jahren 1837 und 1838 : ueber wien, venedig, florenz, rom, neapel, malta, sicilien und griechenland nach saloniki, dem berge athos, konstantinopel und trapezunt – Heidelberg 1840 [mf ed Hildesheim 1995-98] – 1v on 3mf – 9 – €90.00 – 3-487-26699-7 – gw Olms [910]

Dr. e.l. th. henke's nachgelassene vorlesungen ueber liturgik und homiletik = Liturgik und homiletik / Henke, Ernst Ludwig Theodor; ed by Zschimmer, Wilhelm – Halle a/S: Lippert, 1876 – 2mf – 9 – 0-7905-4862-3 – (incl bibl ref) – mf#1988-0862 – us ATLA [240]

Dr. e.l. th. henke's neuere kirchengeschichte : nachgelassene vorlesungen = Neuere kirchengeschichte / Henke, Ernst Ludwig Theodor; ed by Gass, Wilhelm – Halle a/S.: Lippert, 1874-1880 – 3mf – 9 – 0-7905-8057-8 – (incl bibl ref) – mf#1988-6038 – us ATLA [240]

Dr f c kolbe : priest, patriot and educationist / Boner, Kathleen – Pretoria: Unisa 1979 [mf ed 1980] – 7mf – 9 – (incl bibl & summary) – sa Unisa [370]

Dr farrar's "life of christ" – London, England. 1874 – 1r – us UF Libraries [240]

Dr. foote's health monthly – New York: Murray Hill Pub Co, [1876?-18– or 19–] – 9 – mf#P04178 – cn Canadiana [613]

Dr friedrich schleiermacher's philosophische und vermischte schriften : zweiter band – Berlin: G Reimer 1838 [mf ed 1990] – 2mf – 9 – 0-7905-7311-3 – mf#1989-0536 – us ATLA [140]

Dr girardeau's anti-evolution : the logic of his reply / Martin, James L – Columbia, SC: Presbyterian Pub House, 1889 [mf ed 1985] – 1mf – 9 – 0-8370-4299-2 – mf#1985-2299 – us ATLA [210]

Dr grant and the mountain nestorians / Laurie, T – Boston, 1853 – 5mf – 9 – mf#HT-168 – ne IDC [910]

Dr. grant and the mountain nestorians / Laurie, Thomas – Boston: Gould and Lincoln, 1853 – 1mf – 9 – 0-7905-5103-9 – mf#1988-1103 – us ATLA [240]

Dr. grenfell's parish : the deep sea fishermen / Duncan, Norman – 6th ed. New York: Fleming H Revell, c1905 – 1mf – 9 – 0-524-08319-3 – mf#1993-1014 – us ATLA [910]

Dr griffith john, fra hankow / Morthensen, Eilert – Kobenhavn: Kirkelig forening for den indre mission i Danmark, 1908 [mf ed 1995] – (= ser Yale coll; Missions-biblioteket 4,5) – 56p – 1 – 0-524-09589-2 – (in danish) – mf#1995-0589 – us ATLA [920]

Dr hampden's theology other than the catholic faith / Mayow, Mayow Wynell – London, England. 1847 – 1r – us UF Libraries [241]

Dr haydn's 6 original canzonettas : for the voice with accompaniment for the pianoforte / Haydn, Joseph – 1st ed, London: printed for aut [1794] [mf ed 1991] – 1r – 1 – (italian & english words) – mf#pres. film 108 – us Sibley [780]

Dr hevesi simon / Wertheimer, Adolf – Budapest, Hungary. 1935 – 1r – us UF Libraries [939]

Dr hook's test of controversy examined / Ellis, Brabazon – London, England. 1840 – 1r – us UF Libraries [240]

D'r huesfrind *see* Der hausfreund

Dr. isaac watts, the bard of the sanctuary : his birthplace and personality, his literary and philosophical contributions, his life and times, hymnology and bible / Wills, Joshua Edwin – [S.l.: s.n., 1913?] – 1mf – 9 – 0-524-05124-0 – mf#1992-2077 – us ATLA [240]

Dr iuliu barasch : iunie 1815 – 30 april 1863 / Schwarzfeld, Moses – Bucuresti, Romania. 1919 – 1r – us UF Libraries [939]

Dr j a guldenstadts beschreibung der kaukasischen laender : aus seinen papieren gaenzlich umgearbeitet, verbessert herausgegeben... / Klaproth, J [H von] – Berlin, 1834 – 3mf – 9 – mf#AR-1601 – ne IDC [914]

Dr j a moehlers...gesammelte schriften und aufsaetze / ed by Doellinger, Johann Joseph Ignaz von – Regensburg: G J Mans, 1839-40 [mf ed 1990] – 2v on 2mf – 9 – 0-7905-4957-3 – (incl bibl ref) – mf#1988-0957 – us ATLA [240]

Dr j a moehlers...patrologie, oder, christliche literaergeschichte / ed by Reithmayr, Franz Xaver – Regensburg: G J Manz, 1839-40 [mf ed 1990] – 2v on 3mf – 9 – 0-7905-4956-5 – (no more publ. incl bibl ref) – mf#1988-0956 – us ATLA [240]

Dr j j i von doellinger's fables respecting the popes in the middle ages – New York: Dodd & Mead, 1872 [mf ed 1984] – 6mg – 9 – 0-8370-0956-1 – (trans by alfred plummer. together with: dr doellinger's essay on the prophetic spirit and the prophecies of the christian era trans with int & notes by henry b smith. incl bibl ref) – mf#1984-4319 – us ATLA [240]

Dr j schusters handbuch zur biblischen geschichte : erster band: das alte testament = Handbuch zur biblischen geschichte. erster band, das alte testament / Schuster, Ignaz – 4. verm verb aufl. Freiburg im Breisgau; St Louis, MO: Herder, 1886, c1877 – 3mf – 9 – 0-8370-1369-0 – (incl bibl ref) – mf#1987-6057 – us ATLA [221]

Dr jan bouws 1902-1978 : 'n bibliografie / Boehmer, Elizabeth Wilhelmina – Stellenbosch: U van Stellenbosch 1978 – 1mf – 9 – (incl ind & bibl ref) – mf#mf.348 – sa Stellenbosch [016]

Dr johann albrecht bengels auslegung des neuen testaments : oder, kleiner gnomon = Gnomon novi testamenti / Bengel, Johann Albrecht; ed by Werner, C F – Basel: Ferd Riehm 1867 [mf ed 1992] – 2mf – 9 – 0-524-05205-0 – (in german) – mf#1992-0338 – us ATLA [225]

Dr. johann eck, professor der theologie an der universitaet ingolstadt : eine monographie / Wiedemann, Theodor – Regensburg: F Pustet, 1865 – 2mf – 9 – 0-524-04604-2 – (incl bibl ref) – mf#1992-2038 – us ATLA [240]

Dr. johannes bugenhagens briefwechsel : im auftrage der gesellschaft fuer pommersche geschichte und alterthumskunde = Correspondence / Bugenhagen, Johann; ed by Vogt, Otto – Stettin [Szczecin]: Leon Saunier, 1888 – 2mf – 9 – 0-524-03633-0 – (incl bibl ref) – mf#1990-1061 – us ATLA [240]

Dr Johannes-Lepsius-Archiv an der Martin-Luther Universitaet Halle-Wittenberg *see* Deutschland, armenien und die tuerkei 1895-1925

Dr john abercrombie / Wilson, George – London, England. 18– – 1r – us UF Libraries [240]

Dr john walker and the sufferings of the clergy / Tatham, Geoffrey Bulmer – Cambridge: University Press, 1911 [mf ed 1990] – (= ser Cambridge historical essays 20) – 1mf – 9 – 0-7905-7151-X – (incl bibl ref) – mf#1988-3151 – us ATLA [242]

Dr karl burney's nachricht von georg haendel's lebensumstaenden : und der ihm zu london im mai und jun 1784 angestellten gedaechtnissfeyer / Burney, Charles – Berlin, Stettin: F Nicolai 1785 [mf ed 19–] – 1mf – 9 – (trans fr english by johann joachim eschenburg) – mf#fiche 702 – us Sibley [780]

Dr kaufmann david emlkezete – Budapest, Hungary. 1899 – 1r – us UF Libraries [939]

Dr kenealy's lecture on temperance / Kenealy, Dr – London, England. 18– – 1r – us UF Libraries [240]

Dr l g blanc's handbuch des wissenswuerdigsten aus der natur und geschichte der erde und ihrer bewohner : zum gebrauch beim unterricht in schulen und familien, vorzueglich fuer hauslehrer auf dem lande, sowie zum selbstunterricht – Halle 1849 [mf ed Hildesheim 1995-98] – 3v on 14mf – 9 – €140.00 – 3-487-29972-0 – gw Olms [550]

Dr l wieger's moral tenets and customs in china / Wieger, Leon – Ho-Kien-fu, Catholic Mission Press, 1913 [mf ed 1995] – (= ser Yale coll) – [2]/iii/604p (ill)/34pl – 1 – 0-524-09423-3 – (chinese text in latin type in parallel clmns with english text, trans and ann by l davrout) – mf#1995-0423 – us ATLA [230]

Dr leopold plaschkes / Sahawi-Goldhammer, Arjeh – Tel-Aviv, Israel. 1943 – 1r – us UF Libraries [939]

Dr. liddon / Russell, George William Erskine – London: AR Mowbray, 1905 – 1mf – 9 – 0-7905-6619-2 – mf#1988-2619 – us ATLA [240]

Dr. liddon's tour in egypt and palestine in 1886 : being letters descriptive of the tour / King, Annie Liddon – London: Longmans, Green, 1891 – 1mf – 9 – 0-524-05378-2 – mf#1991-2284 – us ATLA [915]

Dr. martin luther : lebensbild des reformators den glaubensgenossen in amerika / Graebner, Augustus Lawrence – Milwaukee, Wis: Geo Brumder, 1895 – 2mf – 9 – 0-524-00754-3 – mf#1990-0186 – us ATLA [242]

Dr. martin luther als erzieher der jugend : seine grundsaetze ueber die kunderzucht und seine erziehungsweise im eignen hause aus seinen schriften / Lindemann, Johann Christoph Wilhelm – St Louis, Mo: A Wiebusch, 1866 – 1mf – 9 – 0-524-04554-2 – mf#1991-2118 – us ATLA [377]

Dr. martin luthers krankheiten und deren einfluss auf seinen koerperlichen und geistigen zustand / Ebstein – Stuttgart, 1908 (mf ed 1993) – 1mf – 9 – €24.00 – 3-89349-204-6 – mf#DHS-AR 93 – gw Frankfurter [242]

Dr. martin luthers paedagogische schriften und aeusserungen = Paedagogische schriften und aeusserungen / Luther, Martin – Langensalza: H Beyer, 1888 – (= ser Bibliothek paedagogischer klassiker) – 1mf – 9 – 0-524-03646-2 – mf#1990-1074 – us ATLA [377]

Dr. martin luther's small catechism : a history of its origin, its distribution and its use / Reu, Johann Michael – Chicago, Wartburg pub house, 1929 – 1r – 1 – 0-8370-1481-6 – mf#1984-B022 – us ATLA [242]

Dr max neuda – Wien, Austria. 1911? – 1r – us UF Libraries [939]

Dr mccave (a roman priest in kidderminster) on the reformation / Collette, Charles Hastings – London, England. 1877 – 1r – us UF Libraries [242]

Dr. middleton's letter from rome, showing an exact conformity between popery and paganism : or, the religion of the present romans derived from that of their heathen ancestors with the author's defence against a roman catholic opponent / Middleton, Conyers – New-York: American and Foreign Christian Union, 1854, c1847 – 1mf – 9 – 0-8370-8040-1 – (incl bibl ref) – mf#1986-2040 – us ATLA [241]

Dr. nevin's theology : based on manuscript class-room lectures / Nevin, John Williamson; ed by Erb, William Harvey – Reading, Pa: IM Beaver, 1913 – 2mf – 9 – 0-524-04080-X – mf#1991-2025 – us ATLA [240]

Dr paley's works / Whately, Richard – London, England. 1859 – 1r – us UF Libraries [240]

Dr phillipe : first settler of safety harbor / Cannella, Felix – s.l, s.l? 1936 – 1r – us UF Libraries [978]

Dr r maurice bucke on the functions of the great sympathetic nervous system – [Sarnia, Ont?: s.n, 1875?] – 1mf – 9 – 0-665-94264-8 – mf#94264 – cn Canadiana [611]

Dr. ralph wardlaw thompson / Mathews, Basil Joseph – London: Religious Tract Society, 1917 – 1mf – 9 – 0-524-03238-6 – mf#1990-0866 – us ATLA [240]

Dr. richard rothe's geschichte der predigt : von den anfaengen bis auf schleiermacher = Geschichte der predigt / Rothe, Richard; ed by Truempelmann, August – Bremen: M Heinsius, 1881 – 2mf – 9 – 0-7905-9471-4 – (incl bibl ref and ind) – mf#1989-2696 – us ATLA [240]

Dr rigsby's (ie rigby's) papers on florida / Rigby, T C – Cincinnati, OH. 1876 – 1r – us UF Libraries [630]

Dr robert morrison : den forste evangeliske missionaer i kina. en kort levnetsskildring / Morthensen, Eilert – Kobenhavn: Kirkelig forening for den indre mission i Denmark, 1907 [mf ed 1995] – (= ser Yale coll; Missions-biblioteket 1) – 1 – 0-524-09779-8 – (in danish) – mf#1995-0779 – us ATLA [920]

Dr ryerson's letters in reply to the attacks of foreign ecclesiastics against the schools and municipalities of upper canada : including the letters of bishop charbonnel, mr bruyere, and bishop pinsoneault – Toronto: Lovell & Gibson, 1857 – 2mf – 9 – mf#35202 – cn Canadiana [241]

Dr ryerson's letters in reply to the attacks of the hon george brown... : "editor-in-chief" and proprietor of the 'globe' – Toronto: Lovell & Gibson, 1859 – 2mf – 9 – (ed with notes and app) – mf#35203 – cn Canadiana [230]

Dr ryerson's reply to the recent pamphlet of mr langton and dr wilson on the university question : in five letters to the hon m cameron... – Toronto?: s.n, 1861 (Toronto: "Guardian") – 1mf – 9 – mf#22899 – cn Canadiana [378]

Dr s kierkegaard mod dr h martensen : et indlaeg / Kofoed-Hansen, Hans Peter – Kobenhavn: C G Iversen, 1856 – 1mf – 9 – 0-524-00445-5 – mf#1989-3145 – us ATLA [190]

Dr s radhakrishnan / ed by Singh, Jagannath – Allahabad: [J Singh], 1953 – (= ser Samp: indian books) – us CRL [920]

Dr schlesinger samuel emlekezete – s.l, s.l? 1937? – 1r – us UF Libraries [939]

Dr thomson's two last letters to the editor of the perthshire cour... – Edinburgh, Scotland. 1829 – 1r – us UF Libraries [240]

Dr tillotson's letter to mr nicholas hunt, of canterbury – London, England. 1799 – 1r – us UF Libraries [241]

Dr titus toblers zwei buecher topographie von jerusalem und seinen umgebungen – Berlin 1853-54 [mf ed Hildesheim 1995-98] – 2v on 11mf – 9 – €110.00 – 3-487-27676-3 – gw Olms [915]

Dr valentin thalhofers, weil paepstl hausprealaten und dompropstes in eichstaett : erklaerung der psalmen und der im roemischen brevier vorkommenden biblischen cantica = Erklaerung der psalmen / Thalhofer, Valentin; ed by Schmalzl, Peter – 7. verb aufl. Regensburg: G J Manz, 1904 – 3mf – 9 – 0-7905-2548-8 – mf#1987-2548 – us ATLA [220]

Dr. william smith's dictionary of the bible : comprising its antiquities, biography, geography, and natural history = Dictionary of the bible / Smith, William; ed by Hackett, Horatio Balch & Abbot, Ezra – Boston: Houghton, Mifflin, 1881 – 35mf – 9 – 0-524-03885-6 – (incl bibl ref) – mf#1987-6498 – us ATLA [052]

Dr. Williams's Library *see* Henry crabb robinson diaries, travel journals and reminiscences 1790-1867

Dr wiseman's popish literary blunders exposed first series / Collette, Charles Hastings – London, England. 1858 – 1r – us UF Libraries [240]

Drach, George, Kuder, Calvin F *see* The telugu mission of the general council of the evangelical lutheran church in north america

Drach, Margaret *see* Principes guidant l'elaboration d'une methode d'enseignement du francais oral pour des etudiants anglophones

Der drache – Leipzig DE, 1919-1924/25 n27 – 7r – 1 – gw Misc Inst [074]

Draconti carmina / Arevalo, Faustino – 1791 – 9 – sp Bibl Santa Ana [780]

Dracopoli, J L *see* Sir andries stockenstrom, 1792-1864

Dracut 1687-1849 – Oxford, MA (mf ed 1996) – (= ser Massachusetts vital record transcripts to 1850) – 14mf – 9 – 0-87623-242-X – (mf 1t: marriage intentions 1710-22. mf 1t-4t: births & deaths 1687-1822. mf 2t: marriages 1730-38. mf 2t-7t: marriage intentions 1736-1822. mf 4t-8t: marriages 1765-1822. mf 8t: births & deaths 1776-1825. mf 9t: births 1801-40. mf 9t-10t: deaths 1796-1840. mf 10t: marriages 1822-40. mf 10t-11t: marriage intentions 1822-40. mf 11t-13t: births 1843-49. mf 13t: marriages 1844-49. mf 14t: deaths 1844-49) – us Archive [978]

Dracut 1697-1900 – Oxford, MA (mf ed 1995) – (= ser Massachusetts vital records) – 92mf – 9 – 0-87623-379-5 – (mf 1: proprietors 1710-34. mf 2-7: town & vitals 1697-1751. mf 8-9: vitals index 1710-1840. mf 10-57: town records 1710-1822. mf 13-15: intentions 1715-51. mf 13-15: births, deaths 1715-50. mf 20-21: births, deaths 1722-1820. mf 23-23,25: intents, marriages 1749-90. mf 31-32: births, deaths 1756-1821. mf 33: marriages & intents 1785-94. mf 41-46: births, deaths 1757-1822. mf 42,45-48: marriages, intents 1786-1822. mf 51-53: births, deaths 1757-1822. mf 53-56: marriages, intents 1800-22. mf 58-59: births, deaths 1776-1840. mf 63-64: marriages, intents 1822-32. mf 65-72: town records 1835-50. mf 71-72: marriages, intents 1831-40. mf 73-75: military 1807-19, 1862. mf 75-76: paupers 1855-1900. mf 77-80: vitals index 1848-1905. mf 81-84: vitals 1840-61. mf 85-86: deaths 1861-96. mf 87-88: marriages 1853-1900. mf 89: deaths 1896-1905. mf 90-92: births 1861-1900) – us Archive [978]

Drady, drawdy, droddy, drody, drude and variants [o'grady, a variant of draddy] : the journal of research and reporting for the genealogical association / Genealogical Association for Uncommon Surnames – 1985 jun 1-dec, 1986 jul 4-dec, 1987 dec – 1r – 1 – (cont: family association newsletter, droddy, drody, drawdy & variants) – mf#1832798 – us WHS [929]

Draeger, Otto *see* The odor mundt und seine beziehungen zum jungen deutschland

Draeseke, J *see* Apollinarios von laodicea

Draeseke, Johannes *see*
- Apollinarios von laodicea
- Der brief an diognetos
- Gesammelte patristische untersuchungen
- Johannes scotus erigena und dessen gewaehrsmaenner in seinem werke de divisione naturae libri 5

Der draeumling : [a novel] / Raabe, Wilhelm Karl – 2. aufl. Berlin: Otto Janke 1893 [mf ed 1995] – 1r – 1 – (filmed with: deutscher adel) – mf#3707p – us Wisconsin U Libr [830]

Draft analysis of the us joint resolution for approval of the compact of free association : addressed by the chief secretary and adviser to the president and cabinet to the president, the cabinet, the nitijela, and the department secretaries / Marshall Islands – 9 jan 1986 – 1mf – 9 – $1.50 – mf#LLMC 82-100I Title 4 – us LLMC [324]

The draft bill to constitute the commonwealth of australia : as adopted by the convention of 1891 / Australian Federal Convention; ed by Barton, G B – Sydney, 1891 – (= ser 19th c books on british colonization) – 1mf – 9 – mf#1.1.4932 – uk Chadwyck [323]

Draft civil and commercial code for the kingdom of siam. book on obligations / Thailand. Laws, Statutes, etc – Bangkok, Bangkok Daily Mail, 1914. 5 p. l., 284, lxxiii p. LL-10016 – 1 – us L of C Photodup [348]

Draft compact of free association : presented by the joint committee on future status to the cong of micronesia. 4th cong, 2nd spec sess, aug 1972 / Joint Committee on Future Status [TTPI (U.S.)] – n.p, n.d. – (= ser Micronesia: prelude to the constitutional convention) – 1mf – 9 – $1.50 – (various pagination) – mf#LLMC 82-100F, Title 45 – us LLMC [323]
Draft constitution of ralik ratak, 1977 / Committee on Convention Procedure and Jurisdiction, Marshall Islands Constitutional Convention – Majuro: the Committee sep 22 1977 – 3mf – 9 – $4.50 – mf#llmc82-100i, title 7 – us LLMC [342]
Draft constitutions. alternatives 1 and 2 / Constitutional Convention, Koror – Koror: Constitutional Convention, may 30, 1975 – (= ser Republic Of Belau (Palau) – Constitution And Laws) – 2mf – 9 – $3.00 – (various pagination) – mf#LLMC 82-100G, Title 11 – us LLMC [323]
Draft counselor's newsletter / Central Committee for Conscientious Objectors – 1968 jun 25-1973 mar 18, 1975 nov 17 – 1r – 1 – (cont by: draft counselor's news; draft counselor's newsletter [san francisco ca]) – mf#405104 – us WHS [355]
Draft environmental impact statement for the compact of free association / Micronesia. (U.S.) – Washington: Office for Micronesian Status Negotiations, 1984 – (= ser Micronesia: evolution to separate political entities) – 2mf – 9 – $3.00 – (covers fsm, rmi and palau) – mf#LLMC 82-100F Title 16 – us LLMC [324]
The draft guam commonwealth act : prepared for the house committee on international and insular affairs / Zafren, Daniel H – Washington: Library of Congress, Congressional Research Service, 27 May 1986 – 1mf – 9 – $1.50 – mf#LLMC 82-100B Title 29 – us LLMC [324]
Draft of an act designed to simplify and improve transfers of land and titles in massachusetts and to enlarge the jurisdiction of the land court / Rackemann, Charles Sedgwick – Boston, Mudge, 1908. 35 p. LL-1378 – 1 – us L of C Photodup [347]
Draft of the revised canons of the diocese of ontario : adopted by the synod of the diocese on the 20th june 1889 / Church of England. Diocese of Ontario – [Ottawa?: s.n.], 1889 [mf ed 1983] – 1mf – 9 – mf#01079 – cn Canadiana [242]
Draft resistance clearing house memorandum – n1 [1967 may 8] – 2r – 1 – mf#721502 – us WHS [355]
Draft resistance-seattle newsletter – v1-v2 n4 [1968 feb-1969 nov] – 1r – 1 – mf#1055723 – us WHS [355]
Drafting federal grant statutes / Administrative Conference of the US (ACUS) – Acus study. no90-1. n.p. 1990? (all publ) – 4mf – 9 – $6.00 – mf#LLMC 94-351 – us LLMC [348]
Drafting of indian wills covering trust and restricted property / U.S. Dept of the Interior. Office of Hearings and Appeals – 2nd ed. Washington, OHA, 1988 – 2mf – 9 – $3.00 – mf#llmc 88-001 – us LLMC [346]
Drag on up-hill – London, England. 18– – 1r – us UF Libraries [240]
Drag racing – Los Angeles CA 1984-87 – 1,5,9 – ISSN: 0894-5187 – mf#15157 – us UMI ProQuest [790]
[Drage] *see* An account of a voyage for the discovery of a north-west passage by hudson's streights...
Dragendorff, Georg *see* Die heilpflanzen der verschiedenen voelker und zeiten
Dragmaticon philosophiae (cccm 152) / Conchis, Guillelmus de – 2001 – (= ser Corpus christianorum continuatio mediaevalis (cccm)) – 5mf+64p – 9 – €34.00 – 2-503-64522-4 – be Brepols [400]
Die dragomanatsassistenz vor den tuerkischen gerichten / Ziemke, K – Berlin, 1912 – 1mf – 9 – mf#ILM-2404 – ne IDC [340]
Dragon, Antonio *see* A la rencontre du christ
The dragon, image, and demon : or, the three religions of china, confucianism, buddhism, and taoism: giving an account of the mythology, idolatry and demonolatry of the chinese / DuBose, Hampden C – New York: A C Armstrong 1887 [mf ed 1993] – 2mf [ill] – 9 – 0-524-05842-3 – mf#1990-3506 – us ATLA [230]
Dragon seed – Baltimore MD. v1 n7-9 [1972 oct 13/nov 14, dec, jan/feb] – 1r – 1 – mf#1583194 – us WHS [071]
Dragons and dragon slayers / Hackwood, Frederick William – Illus. by Gordon Brown. London: Religious Tract Society, n.d. 175p – 1 – us Wisconsin U Libr [240]
Dragon's seed / Elegant, Robert S – New York, NY. 1959 – 1r – us UF Libraries [025]
Dragontea / Blanco, Tomas – San Juan, Puerto Rico. 1956 – 1r – us UF Libraries [972]
Dragt / ed by Norlund, Poul – Stockholm: Bonnier 1941 [mf ed Bloomington IN: Indiana Uni Lib, Preservation Dept 1984] – 171p on 1r – 1 – us Indiana Preservation [390]
Draheim, Christopher C *see* Cardiovascular disease risk in adults with mental retardation and down syndrome
Drahn, Hermann *see* Das werk stefan georges
Drahomaniv, Mykhailo Petrovych *see* Rozvidky mykhaila drahomanova pro ukrains ku narodniu slovesnist i pis menstvo
Drain echo – Drain OR: Kuykendall Bros, 1885-[wkly] – 1 – (merged with: leader (1895-1903) to form: cottage grove echo=leader (18-?-1895)) – us Oregon Lib [071]
Drain enterprise – Drain OR: W A Priaulx [wkly] – 1 – (began in 1922) – us Oregon Lib [071]
Drain enterprise – Drain, Douglas County, OR: W A Priaulx. v1 n14-v54 n33. aug 3 1922-dec 30 1976 – 1 – us Oregon Hist [071]
Drain nonpareil – Dain, Duglas [i.e. Drain, Douglas] OR: O L Williams, -1914 [wkly] – 1 – (began in 1901; cont by: north douglas herald (1914-)) – us Oregon Lib [071]
The drain of silver to the east : and the currency of india / Lees, William Nassau – London, 1864 [i.e. 1863] – (= ser 19th c british colonization) – 3mf – 9 – mf#1.1.5940 – uk Chadwyck [332]
Drain watchman – Drain OR: Watchman Pub Co [wkly] – 1 – (ceased in 1901) – us Oregon Lib [071]
Drakard's paper – London, England. 10 jan-26 dec 1813 – n1-51 – 1 – (cont as: the champion; the investigator) – uk British Libr Newspaper [072]
Drake, Allison Emery *see* The authorship of the west saxon gospels
Drake, Benjamin *see* Sketch of a journey through the western states of north america
Drake, Benjamin C *see* Effects of an intercollegiate sport season on selected personality traits and mental preparation skills
Drake, Brent M *see* Is winning the only thing
Drake, Charles Daniel *see*
– Address, delivered may 8, 1878, at the annual commencement of the cincinnati law school
– A treatise on the law of suits by attachment in the united states
Drake, Charles Frederick *see* Unexplored syria
Drake, Daniel *see* Discourse on the history, character, and prospects of the west
Drake, Durant *see* Problems of conduct
Drake, Francis Samuel *see* The indian tribes of the united states
Drake law review – v1-49. 1951-2001 – 1,5,6 – $893.00 set – (v1-42 1951-93 on reel $693. v43-49 1994-2001 on mf $200) – ISSN: 0012-5938 – mf#102481 – us Hein [340]
Drake, N M *see* The history of english glass painting
Drake, S B *see* Among the dark-haired race in the flowery land
Drake, Samuel Adams *see*
– A book of new england legends and folk lore
– The border wars of new england
– Burgoyne's invasion of 1777
Drake, Samuel B *see* Among the dark-haired race in the flowery land
Drake, Samuel Gardner *see* Annals of witchcraft in new england
Dralet *see* Description des pyrenees
Drama – London, England 1919-89 – 1,5,9 – ISSN: 0012-5946 – mf#1263 – us UMI ProQuest [790]
Drama : martin county / Lyons, Isabel J – s.l, s.l? 1936 – 1r – us UF Libraries [978]
The drama – London. v. 1-2 no. 1-17, 24, 27, 30. Sept. 13-Dec. 27 1883; Jan 3, Feb 21, Mar 13, Apr 3 1884 – 1 – us NY Public [790]
Drama critique – Racine WI 1958-68 – 1 – ISSN: 0419-7119 – mf#5853 – us UMI ProQuest [790]
El drama de la vida / Henao y Munoz, Manuel – 1878 – 9 – sp Bibl Santa Ana [830]
Drama do acucar / De Carli, Gileno – Rio de Janeiro, Brazil. 1941 – 1r – us UF Libraries [972]
Das drama heinrich von kleists / Meyer-Benfey, Heinrich – Goettingen: O Hapke 1911-13 [mf ed 1995] – 2v on 1r – 1 – (incl bibl ref & ind. filmed with: florentine nights / heinrich heine) – mf#3682p – us Wisconsin U Libr [430]
Drama in sanskrit literature / Jagirdar, R V – Bombay: Popular Book Depot, 1947 – (= ser Samp: indian books) – us CRL [490]
Drama no 666 : daytona beach / Goebel, Rubye K – s.l, s.l? 1936 – 1r – us UF Libraries [978]
The drama of isaiah / Whitman, Eleanor Wood – Boston: Pilgrim Press, c1917 – 1mf – 9 – 0-524-04600-X – mf#1992-0188 – us ATLA [221]
The drama of spain from the proclamation of the republic to the civil war, 1931-36 / Ramos-Oliveira, Antonio – London, 1936? Fiche W1127. (Blodgett Collection of Spanish Civil War Pamphlets) – 9 – us Harvard [946]
The drama of the apocalypse in relation to the literary and political circumstances of its time / Palmer, Frederic – New York: Macmillan, 1903 – 1mf – 9 – 0-8370-4659-9 – (includes appendix) – mf#1985-2659 – us ATLA [221]
Drama on the world stage *see* Prompt books of the english and american stage
Drama, or, theatrical pocket magazine – La Salle IL 1821-26 – 1 – mf#5309 – us UMI ProQuest [790]
Drama review – Cambridge MA 1955-85 – 1,5,9 – (cont by: tdr: the drama review) – ISSN: 0012-5962 – mf#7339.01 – us UMI ProQuest [790]
Das drama richard wagner's : eine anregung / Chamberlain, Houston Stewart – 3. aufl. Leipzig: Breitkopf & Haertel 1908 [mf ed 1991] – 1r – 1 – (filmed with: lohengrin) – mf#2973p – us Wisconsin U Libr [790]
Drama survey – Killen TX 1961-69 – 1 – ISSN: 0419-7127 – mf#1637 – us UMI ProQuest [790]
Drama & theatre – Fredonia NY 1961-75 – 1,5 – ISSN: 0012-5954 – mf#6491 – us UMI ProQuest [790]
Het drama van indie / Notosoetarso – 's-Gravenhage, 1945 – 1mf – 8 – mf#SE-1287 – ne IDC [959]
Das drama zacharias werners : entwicklung und literaturgeschichtliche stellung / Stuckert, Franz – Frankfurt/Main: M Diesterweg 1926 [mf ed 1993] – (= ser Deutsche forschungen 15) – 1r – 1 – (incl bibl ref) – mf#8023 reel 3 – us Wisconsin U Libr [430]
Dramagraphische woche – Vienna, Austria aug 1912-dec 1913 [mf ed Norman Ross] – 1r – 1 – mf#nrp-1965 – us UMI ProQuest [074]
Dramagraphische woche – Wien (A), 1912 30 aug-1913 6 jun – 1r – 1 – gw Mikrofilm [790]
The dramas and dramatic dances of non-european races in special reference to the origin of greek tragedy.. / Ridgeway, William – With an appendix on the origin of Greek comedy. Cambridge: The University Press, 1915. xv,448p. illus – 1 – us Wisconsin U Libr [450]
The dramas of shri harsha / Harsavardhana, King of Thanesar and Kanauj – Allahabad: Ketabistan, 1948 – (= ser Samp: indian books) – (trans into english by bela bose) – us CRL [820]
Dramatic and musical law. / Strong, Albert Ambrose – London, "The Era," 1898. 155 p. LL-129 – 1 – us L of C Photodup [340]
The dramatic art of shakespeare : with especial reference to "a midsummer night's dream": being an inaugural lecture delivered at the mcgill university, montreal / Moyse, Charles Ebenezer – Montreal?: s.n, 1879 – 1mf – 9 – mf#11170 – cn Canadiana [420]
Dramatic censor : or, weekly theatrical report – La Salle IL 1800-01 – 1 – mf#5310 – us UMI ProQuest [790]
Dramatic magazine : embellished with numerous engravings of the principal performers – La Salle IL 1829-31 – 1 – mf#4238 – us UMI ProQuest [790]
Dramatic mirror : and literary companion devoted to the stage and fine arts – La Salle IL 1841-42 – 1 – mf#4368 – us UMI ProQuest [790]
Dramatic notes – La Salle IL 1879-92 – 1 – mf#5311 – us UMI ProQuest [790]
Dramatic opinions and essays / Shaw, Bernard – New York, NY. v1-2. 1907 – 1r – us UF Libraries [080]
The dramatic reader : comprising a selection of pieces for practice in elocution, with introductory hints to readers – Montreal: Dawson, 1869 [mf ed 1985] – 5mf – 9 – 0-665-10176-7 – (incl ind) – mf#10176 – cn Canadiana [850]
Dramatic selections / Schell, Stanley – New York, NY. 1915 – 1r – us UF Libraries [790]
Dramatic times – London. -w. Feb-Sep 1895, Apr-Jul 1919. (1 reel) – 1 – uk British Libr Newspaper [790]
Dramatic works of wycherley, congreve, vanbrugh, and farquhar – London, England. 1840 – 1r – us UF Libraries [420]
Dramatica vida de ruben dario / Torres, Edelberto – Mexico City? Mexico. 1956 – 1r – us UF Libraries [440]
Dramatics – Cincinnati OH 1929+ – 1,5,9 – ISSN: 0012-5989 – mf#2161 – us UMI ProQuest [370]
Dramatische dichtungen : ernst von schwaben; ludwig der baier / Uhland, Ludwig – Heidelberg: C F Winter, 1846 – 1r – 1 – us Wisconsin U Libr [810]
Dramatische eindruecke : aus dem nachlasse / Auerbach, Berthold – Stuttgart: J G Cotta, 1893 [mf ed 1988] – (= ser Bibliothek der deutschen literatur 1401-1402) – xiv/326p – 1 – (incl ind) – mf#6968 – us Wisconsin U Libr [430]
Dramatische elemente in hebbels jugendballaden, 1829-1839 : eine studie zur entwicklungsgeschichte friedrich hebbels / Jahn, Walter – Halle: H John, 1915 – 1r – 1 – (incl bibl ref) – us Wisconsin U Libr [430]
Die dramatische handlung in gerhart hauptmanns webern / Rabl, Hans – Halle (Saale): M Niemeyer, 1928 – 1r – 1 – (incl bibl ref) – us Wisconsin U Libr [430]
Die dramatische handlung in klopstocks "der tod adams" und gerstenbergs "ugolino" / Dollinger, Hermann – Halle: Niemeyer, 1930 – 1r – 1 – (incl bibl ref) – us Wisconsin U Libr [790]
Die dramatische handlung in sophokles' "koenig oidipus" und kleists "der zerbrochene krug" / Gordon, Wolff von – Halle: M Niemeyer, 1926 – 1r – 1 – us Wisconsin U Libr [430]
Dramatische handlung und aufbau in hebbels herodes und mariamne / Weichenmayr, Franz – Halle (Saale): M Niemeyer, 1929 – 1r – 1 – (incl bibl ref) – us Wisconsin U Libr [430]
Dramatische werke / Bleibtreu, Karl – Leipzig: W Friedrich, [19–?] [mf ed 1989] – 3v – 1 – mf#7031 – us Wisconsin U Libr [820]
Dramatische werke / Freytag, Gustav – 3. Aufl. Leipzig: S Hirzel. 2v. 1874 – 1r – 1 – us Wisconsin U Libr [820]
Dramatische werke / Freytag, Gustav – 4. Aufl. Leipzig: S Hirzel. 2v in 1. 1881 (mf ed 1990) – 1r – 1 – (filmed with: raetsel um herta) – us Wisconsin U Libr [820]
Die dramatischen versuche des jungen grillparzer : auf ihre entstehung geprueft und in zusammenhang gebracht mit der inneren entwickelung des dichters / Keidel, Heinrich – Muenster i.W.: Theissing, 1911 – 1r – 1 – (incl bibl ref) – us Wisconsin U Libr [430]
Die dramatischen werke des luzerners zacharias bletz : nach der einzigen handschrift zum erstenmal gedruckt / ed by Steiner, E – Frauenfeld: Huber, 1916 [mf ed 1989] – (= ser Schweizerische lustspiele des 16. jahrhunderts 41-42; Die schweiz im deutschen geistesleben) – 1 – mf#7031 – us Wisconsin U Libr [820]
Die dramatischen werke des peter probst / ed by Kreisler, Emil – Halle: M Niemeyer, 1907 – 11r – 1 – (incl bibl ref) – us Wisconsin U Libr [820]
Dramatisches / Gugler, Julius – Milwaukee: Selbstverlag des Verfassers, c1892 – 1r – 1 – us Wisconsin U Libr [830]
Dramaturgische monate / ed by Schink, Johann Friedrich – Schwerin 1790 – (= ser Dz) – 8mf – 9 – €160.00 – mf#k/n4229 – gw Olms [790]
Drame, Anja *see* Entwicklung der terminologie in der afrikanische sprache der xhosa (Suedafrika)
Le drame de dankori : mission voulet-chanoine, mission joalland-meynier / Joalland, Jules – Paris: Editions Argo, [1930] – 1 – us CRL [960]
Drame du 6 decembre 1897 / Vieux, Isnardin – Port-Au-Prince, Haiti. 1963 – 1r – us UF Libraries [972]
Le drame d'une guerre imposee a un peuple pacifique : dans ses 1000 jours de resistance nationale – Phnom Penh: Republique Khmere [mf ed 1989] – 1r with other items – 1 – mf#mf-10289 seam reel 018/06 [§] – us CRL [959]
Le drame maconnique : la conjuration juive contre le monde chretien / Copin-Albancelli, Paul – 12e ed. Paris: Renaissance francaise 1909 – 6mf – 9 – mf#vrl-131 – ne IDC [366]
Le drame maconnique : le pouvoir occulte contre la france / Copin-Albancelli, Paul – 6e ed. Paris: Renaissance francaise 1908 – 5mf – 9 – mf#vrl-130 – ne IDC [366]
Le drame musical / Schure, Edouard – Paris: Sandoz et Fischbacher 1875 – 2v on 10mf – 9 – mf#wa-102 – ne IDC [790]
Dramen / Pinski, David – Varshe, Poland. 1909 – 1r – us UF Libraries [939]
Dramen von ackermann und voith / ed by Holstein, Hugo – Stuttgart: Litterarischer Verein, 1884 (Tuebingen: H Laupp) – us Wisconsin U Libr [820]
Dramen von ackermann und voith / ed by Holstein, Hugo – Stuttgart: Litterarischer Verein, 1884 (Tuebingen: H Laupp) [mf ed 1993] – (= ser Blvs 170) – 340p – 1 – mf#8470 reel 35 – us Wisconsin U Libr [430]
Dramenformen des barock : die funktion von rollen, reyen und buehne bei joh chr hallmann (1640-1704) / Beheim-Schwarzbach, Eberhard – Jena: [s.n.], 1931 [mf ed 1990] – 33p – 1 – (incl bibl ref) – mf#7429 – us Wisconsin U Libr [430]
Drames de la vie reelle : roman canadien / Barthe, Georges Isidore – Sorel: J A Chenevert, [1896?] (mf ed 1984) – 1mf – 9 – mf#SEM105P378 – cn Bibl Nat [830]
Drames liturgiques du moyen-age / Coussemaker, E de – Rennes, 1860 – €23.00 – ne Slangenburg [931]
Drane, Augusta Theodosia *see*
– The history of st dominic
– Letters of archbishop ullathorne
– The morality of tractarianism
– The spirit of the dominican order
– The three chancellors
Drapeau – Paris, France. 29 dec 1881-25 dec 1886 – 2 1/2r – 1 – uk British Libr Newspaper [072]

Le drapeau – Paris, France. -w. 29 Dec 1881-25 Dec 1886. 3 reels – 1 – uk British Libr Newspaper [072]
Le drapeau – Port-au-Prince: [s.n.], aug 14-nov 6, dec 6-13, 1897 – 2r – 9 – us CRL [079]
Le drapeau : revue hebdomadaire illustree – 29 dec 1881-88, 1898-99, 11 mai-9 dec 1901, 1917-20 nov 1927, 31 oct 1930, 20 sept 1931, 1932, mars-nov 1933, mai-nov 1935 – 1 – (Tir, gymnastique, secours aux blesses, sauvetage, escrime, equitation, histoire militaire, etc. paris. le sous-titre et la periodicite varient.) – fr ACRPP [073]
Le drapeau blanc – Paris. 16 juin 1819-1er fevr 1827, 1er juin 1829-28 juil 1830 – 1 – (a paru sous le titre: democrate. 1er juin-15 juil 1829) – fr ACRPP [073]
Le drapeau de carillon : drame historique en trois actes et deux tableaux / David, Laurent Olivier – Montreal: C O Beauchemin, 1902 – 2mf – 9 – 0-665-72548-5 – mf#72548 – cn Canadiana [820]
Le drapeau de l'union : bulletin interieur de l'union de la jeunesse democratique algerienne – n1-2. Alger. avr 1953-janv fevr 1954 – 1 – fr ACRPP [320]
Le drapeau fantome : episode historique / Frechette, Louis – Montreal?: Typ de La Patrie, 1884 – 1mf – 9 – mf#07128 – cn Canadiana [890]
Drapeau, Julien *see* Essai de bibliographie sur le regime municipal dans la province de quebec
Le drapeau national – Port-au-Prince: [s.n.], dec 21-dec 24?, 1877; jan 11-mar 8, apr 5-aug 16, 1878; jan 11, feb 5-21, 1879 – 2r – 9 – us CRL [079]
Drapeau national haitien / Dalencour, Franciois Stanislas Ranier – Port-Au-Prince, Haiti. 1939 – 1r – us UF Libraries [972]
Drapeau rouge – Brussels Belgium, 1 aug 1936; 6 sep 1944; 29 jun 1945; jul 1945-9 aug 1946; jan-1 oct 1947; 2 jun 1948; 20 may, 25 nov 1963 – 5r – 1 – uk British Libr Newspaper [074]
Drapeau rouge – n1. n.s., n16 33. Paris. 11 dec 1936-1er sept 1937 – 1 – (journal communiste puis marxiste revolutionnaire) – fr ACRPP [325]
Le drapeau rouge – Brussels. Belgium. -w. 6 Sep 1944-9 Aug 1946, 1 Jan 1947-2 Jan 1948. (6 reels) – 1 – uk British Libr Newspaper [949]
Drapeau, Stanislas *see*
– Biographie de sir n f belleau
– Canada
– La colonisation du canada envisagee au point de vue national
– Coup d'oeil sur les ressources productives et la richesse du canada
– Etudes sur les developpements de la colonisation du bas-canada depuis dix ans, (1851 a 1861)
– Histoire des institutions de charite de bienfaisance et d'education du canada
Drapelul rosu – Timisoara, Romania. 1962-1985 – 33r – 1 – us L of C Photodup [949]
Draper, Andrew Sloan *see* Rescue of cuba, and episode in the growth of free...
Draper, B H *see* Poor blind jane
Draper Collection *see* Manuscripts
Draper, John William *see*
– History of the conflict between religion and science
– History of the intellectual development of europe
Draper manuscript, 24 nov. 1842-27 oct. 1869 / Peck, J M & Piggott, Isaac N & Reynolds, J – 364p – 1 – us Southern Baptist [242]
The draper manuscripts : one of the most famous collections of historical records of the revolution and the westward expansion – [mf ed Chadwyck-Healey] – c500v on 123r – 1 – (with p/g by josephine l harper. pts available and listed separately) – uk Chadwyck [975]
Draper of australasia – Sydney, Australia. 27 Jul 1903-10 Feb 1965.-w. 67 1 2 reels – 1 – uk British Libr Newspaper [072]
Draper, Warwick Herbert *see* Alfred the great
Draper, William George *see* The rules of the courts of queen's bench and common pleas, the municipal council rules, the county courts' equity extension and the new division court rules.
Draper's upper canada king's bench reports / Ontario. Canada – 1v. 1829-31 (all publ) – 6mf – 9 – $9.00 – mf#LLMC 81-043 – us LLMC [340]
Dra-po / Holly, Arthur – Port-Au-Prince, Haiti. 1928 – 1r – us UF Libraries [972]
Die drau – Osijek, Yugoslavia. Oct 1923-Jun 1928; Jun 1932-Nov 1933 (scattered issues) – 5r – 1 – us L of C Photodup [949]
Draugas = Lithuanian daily friend – Chicago, IL: Draugas Pub Co Inc, [jul-dec 1909; 21 nov 1917-17 jul 1929; jan-apr 15 1931; 1936; 1940; jul 1945-1955] – 1 – us CRL [071]
Draugas / Wilkes-Barre: Draugas, 1909-1916 – 1r – us CRL [070]
A draught of the blue – London: Medici Society, 1914 – (= ser Samp: indian books) – (trans fr original mss by f w bain) – us CRL [830]
Dravert, P *see* Narodnaia armiia
Dravida and kerala in the art of travancore / Kramrisch, Stella – Ascona [Switzerland]: Artibus Asiae, 1953 – (= ser Samp: indian books) – us CRL [700]
The dravidian element in indian culture / Slater, Gilbert – Foreword by H.J. Fleure.London: E. Benn, 1924 – 1 – us Wisconsin U Libr [900]
The dravidian element in indian culture / Slater, Gilbert – London: Ernest Benn Ltd, 1924 – (= ser Samp: indian books) – (foreword by h j fleure) – us CRL [930]
Dravidian gods in modern hinduism : a study of the local village deities of southern india / Elmore, Wilber Theodore – Hamilton, N.Y.: Published by the author, 1915. Chicago: Dep of Photodup, U of Chicago Lib, 1967 (1r); Evanston: American Theol Lib Assoc, 1984 (1r) – 1 – 0-8370-0363-6 – (incl ind) – mf#1984-B065 – us ATLA [280]
Drawbar / United Transportation Union – 1975 aug-1984 mar – 1r – 1 – mf#1289935 – us WHS [380]
Drawbridge, C L *see* The training of the twig (religious education of children)
Drawert, Ernst Arno *see* Moerikes maler nolten
The drawing book of the government school of design / Great Britain. School of Design – London 1842 – (= ser 19th c art & architecture) – 5mf – 9 – mf#4.2.438 – uk Chadwyck [740]
Drawing cheques / Cox, John – London, England. 18– – 1r – us UF Libraries [240]
Drawing in public, high and normal schools : a letter from emil vossnack...to the council of public instruction for the province of nova scotia, and the school commissioners of the city of halifax / Vossnack, Emil – Halifax, NS: J Burgoyne, 1879 – 1mf – 9 – mf#35015 – cn Canadiana [370]
The drawing of geometric patterns in saracenic art / Hankin, E H – Calcutta: Govt of India, Central Publication Branch, 1925 – (= ser Samp: indian books) – us CRL [700]
Drawing the net : for prayer meeting workers: familiar letters / Clark, Francis Edward – Boston: United Society of Christian Endeavor, 1890 – 1mf – 9 – mf#27441 – cn Canadiana [240]
Drawings : subject collections – (= ser Art exhibition catalogues on microfiche) – 241 catalogues on 284mf – 9 – £1,490.00 – (individual titles not listed separately) – uk Chadwyck [700]
Drawings and plans for holkham, c1729 / Kent, William – (= ser Holkham library, the house, park & art colls) – 1r – 1 – mf#2075 – uk Microform Academic [720]
Drawings and specifications / Mills, Robert – c1804 [mf ed Spartanburg SC: Reprint Co, 1981] – 1mf – 9 – (proposal for a protestant episcopal church building for john's island, sc. also, a descriptive article by samuel lapham from the architectural record (mar 1923)) – mf#51-095 – us South Carolina Historical [720]
Drawings and watercolours in the fitzwilliam museum, cambridge / Turner, John Mallord William – 1col r – 14 – mf#C97297 – uk Microform Academic [740]
The drawings collection : microfilms from the general catalogues, 1780-1840 / Royal Institute of British Architects – 1834- – 14 – (phase e: 1590-1780 a-z complete £1370 12r. phase f: 1780-1840 a-d £2100 21r. phase g: 1780-1840 e-p £1150 11r. phase h: 1780-1840 r-z £600 6r. phase i: 1840-1914 a-b £560 5r. phase j: 1840-1914 bentley to burges £1370 13r. phase k: 1840-1914 b-f £1640 14r. phase m: 1840-1914 g-lethaby £2380 21r. phase n: 1840-1914 l-m £1720 12r. phase o: 1840-1914 m-p £760 6r. phase p: 1840-1914 p-s £1370 11r. phase q: 1840-1914 salvin-simpson £1060 8r. phase r: 1840-1914 s-w £1900 15r. phase s: 1840-1914 webster-young £930 7r. phases a-s complete £25,000 (updated annually). phase t: 1840-1914 supplement-drawings not available at time of original filming £530 4r. phase u: 1914-40 allen-bilson £995 8r. phase v: 1914-40 £2400 19r) – uk World [720]
The drawings collection : microfilms from the specialist catalogues / Royal Institute of British Architects – 1834- – 1,14 – (phase a: colen campbell, jacques gentilhatre, inigo jones and john webb, alfred stevens, antonio visentini, c f a voysey £1370 14r (9r b/w 5r col). phase b: the pugin family, the wyatt family and j b papworth £1420 18r (16r b/w 2r col). phase c: the drawings of sir edwin lutyens and the scott family drawings £3000 50r – available separately: lutyens drawings £850 12r scott drawings £2600 38r. phase d: the palladio drawings, the adam drawings and the smythson collection £440 4r (1r b/w 3r col). phase l: the charles holden coll £1470 13r. phases a-l complete £7500 coll) – uk World [720]
Drawings in provincial and other museums / Caisse Nationale des Monuments Historiques et des Sites. Paris – (= ser Fine and decorative arts in france) – 126mf – 9 – $885.00 – 0-907006-70-1 – (over 7500 reproductions) – uk Mindata [740]
Drawings in the louvre and national museums : together with public and private modern art collections in paris / Caisse Nationale des Monuments Historiques et des Sites. Paris – (= ser Fine and decorative arts in france) – 128mf – 9 – $900.00 – 0-907006-65-5 – (over 7500 reproductions) – uk Mindata [740]
Drawings of johan tobias sergel / Stockholm. Nationalmuseum; ed by Bjurstrom, Per – 1979 – x/68p on 1 color 3 b/w mf – 15 – $55.00f – 0-226-69420-8 – us Chicago U Pr [740]
Drawings of raphael in the ashmolean museum / Ashmolean Museum – [mf ed 1986] – 4 col mf – 15 – $180.00 – 0-907716-12-1 – (largest & most repr coll of raphael's drawings; with printed ind) – uk Mindata [740]
Drawings of robert and james adam in sir john soane's museum – [mf ed Chadwyck-Healey] – 10r b/w 2r col – 1,14 – (with catalogue comp by walter l spiers [1r]) – uk Chadwyck [740]
The drawings of robert and james adam in sir john soane's museum / London. Sir John Soane's Museum – 2r col + 10r b/w – 1 – £1,380.00 – (includes catalogue 1r £63) – uk Chadwyck [740]
Drawings, paintings and sculptures / Reddy, P T – Bombay: New Book Co, [19–] – (= ser Samp: indian books) – us CRL [700]
Drawn in color : african contrasts / Jabavu, Noni – New York, NY. 1962 – 1r – us UF Libraries [960]
Drawn in colour : african contrasts / Jabavu, Noni – London, England. 1960 – 1r – us UF Libraries [960]
Draysig yor yidishe literatur in rumenye / Kara, I – Yas, Romania. 1947 – 1r – us UF Libraries [470]
Drayton, John *see*
– The carolinian florist
– Drayton's view of south carolina
– John drayton's gouverneurs und oberkommandanten von sued-carolina, beschreibung von sued-carolina
Drayton valley western review – Drayton Valley, Alberta, CN. 1980-89 – 25r – 1 – cn Commonwealth Imaging [071]
Drayton's view of south carolina – 1802 [mf ed 1981] – 7mf – 9 – (publ as: a view of south carolina, charleston: printed by w p young, 1802) – mf#51-046 – us South Carolina Historical [917]
Drc (dutch reformed church) africa news – v1-10. 1976-85 [complete] – 1r – 1 – mf#ATLA S0786 – us ATLA [242]
The dread voyage : poems / Campbell, Wilfred – Toronto: W Briggs; Montreal: C W Coates; Halifax: S F Huestis, 1893 – 3mf – 9 – 0-665-00407-9 – mf#00407 – cn Canadiana [810]
Dreadful shipwreck : by the bomb-ketch, "observer," john carey... – S.l: s.n, 18–? – 1mf – 9 – mf#43702 – cn Canadiana [380]
A dream is life : dramatic fantasy in four acts = Traum, ein leben / Grillparzer, Franz – Yarmouth Port, MA: Register Press, 1946 [mf ed 1993] – 128p – 1 – (trans by henry h stevens) – mf#8711 – us Wisconsin U Libr [820]
The dream of a church mouse – St John, NB?: s.n, 1874 – 1mf – 9 – mf#33731 – cn Canadiana [880]
The dream of columbus : a poem / Wright, Robert Walter – Toronto: W Briggs; Montreal: C W Coates, 1894? – 1mf – 9 – mf#09292 – cn Canadiana [810]
The dream of dante : an interpretation of the inferno / Henderson, Henry F – Cincinnati: Jennings and Graham; Edinburgh and London: Oliphant, Anderson and Ferrier, [1903]. Beltsville, Md: NCR Corp, I977 (2mf); Evanston: American Theol Lib Assoc, 1984 (2mf) – 9 – 0-8370-0165-X – mf#1984-0042 – us ATLA [440]
A dream of the past, present, and future / McLean, Thomas Alexander – Calgary?: s.n, 189-? – 1mf – 9 – mf#30403 – cn Canadiana [880]
Dream psychology / Nicoll, Maurice – 2nd ed. London: H. Frowde; Hodder & Stoughton, 1920. xv,194p – 1 – us Wisconsin U Libr [150]
Dream world cruise destinations – London, England 2001+ – 1,5,9 – mf#32368 – us UMI ProQuest [338]
Dream World Dragon Press *see* No limits!
Dreams / Bergson, Henri – New York: BW Huebsch, 1914 – 1mf – 9 – 0-7905-3580-7 – mf#1989-0073 – us ATLA [150]
Dreams / Walters, W – London, England. 1858 – 1r – us UF Libraries [240]
Dreams and ghosts / Zerffi, George Gustavus – London, England. 1875 – 1r – us UF Libraries [240]
Dreams and myths : a study in race psychology / Abraham, Karl – New York: Journal of Nervous & Mental Disease Pub Co, 1913 (mf ed 19–) – (= ser Nervous and mental disease monograph series; Harvard medicine preservation microfilm project) – 74p – (transl fr german into english. incl bibl ref) – mf#Z-1489 – us NY Public [150]
Dreams of a spirit-seer : illustrated by dreams of metaphysics = Traeume eines geistersehers / Kant, Immanuel; ed by Sewall, Frank – London: S Sonnenschein; New York: Macmillan, 1900 – (= ser The Philosophy at Home Series) – 1mf – 9 – 0-7905-9009-3 – (incl bibl ref. in english) – mf#1989-2234 – us ATLA [110]
Dreamworks – Ann Arbor MI 1980-88 – 1,5,9 – ISSN: 0192-2890 – mf#12182 – us UMI ProQuest [150]
Drechsler, Christoph Moritz Bernhard Julius *see* Grundlegung zur wissenschaftlichen konstruktion des gesammten woerter- und formenschatzes
Drechsler, Moritz *see* Die unwissenschaftlichkeit im gebiete der alttestamentlichen kritik
Dredge, James *see* The paris international exhibition of 1878
Dredge, John Ingle *see*
– Devon booksellers and printers in the 17th and 18th centuries
– A few sheaves of devon bibliography gleaned
Dredgeman / International Union of Operating Engineers – 1970 apr-1975 apr – 1r – 1 – mf#647861 – us WHS [627]
Dreesbach, E *see* Der orient in der altfranzoesischen kreuzzugeliteratur
Dreesen, Willrath *see* Romantische elemente bei theodor storm
Dreger, Gottfried von *see* Neue skizzen einer sommer-reise durch italien, unter-oesterreich, steyermark, salzburg, tyrol...
Drei : drama in drei aufzuegen / Dreyer, Max – Stuttgart, Leipzig: Deutsche Verlagsanstalt, 1905 [mf ed 1989] – 80p – 1 – mf#7185 – us Wisconsin U Libr [820]
Drei abhandlungen zur geschichte der alten philosophie und ihres verhaeltnisses zum christenthum / Baur, Ferdinand Christian; ed by Zeller, Eduard – Leipzig: Fues, 1876 [mf ed 1989] – 2mf – 9 – 0-7905-4068-1 – (incl bibl ref) – mf#1988-0068 – us ATLA [180]
Drei actenstuecke zur geschichte des donatismus / ed by Deutsch, Samuel Martin – Berlin: W Weber, 1875 – 1mf – 9 – 0-7905-6989-2 – (incl bibl ref) – mf#1988-2989 – us ATLA [240]
Die drei aeltesten bearbeitungen von goethe's iphigenie / ed by Duentzer, Heinrich – Stuttgart: J G Cotta, 1854 [mf ed 1993] – viii/372p – 1 – (incl bibl ref) – mf#8606 – us Wisconsin U Libr [820]
Die drei aeltesten martyrologien / ed by Lietzmann, Hans – Bonn: A Marcus & E Weber 1903 [mf ed 1992] – (= ser Kleine texte fuer theologische vorlesungen und uebungen 2) – 1mf – 9 – 0-524-04679-4 – mf#1990-1306 – us ATLA [240]
Die drei aergsten erznarren in der ganzen welt / Weise, Christian – Halle: M Niemeyer, 1878 – 11r – 1 – (incl bibl ref) – us Wisconsin U Libr [430]
Drei akademische reden / Kaftan, Julius – Tuebingen: JCB Mohr, 1908 [mf ed 1990] – 1mf – 9 – 0-7905-7654-6 – mf#1989-0879 – us ATLA [170]
Die drei begegnungen des baumeisters wilhelm : roman / Ehrler, Hans Heinrich – Muenchen: A Langen, G Mueller 1935, c1934 [mf ed 1989] – 1r – 1 – (filmed with: menschen und affen / albert ehrenstein) – mf#7207 – us Wisconsin U Libr [830]
Die drei briefe des apostels johannes *see* The epistles general of john
Drei buecher deutscher prosa in sprach- und stylproben : von ulphilas bis auf die gegenwart, 360-1837 / ed by Kuenzel, Heinrich – Frankfurt/M: J D Sauerlaender, 1838 [mf ed 1993] – 3v in 2 – 1 – mf#8362 – us Wisconsin U Libr [430]
Drei deutsche pyramus-thisbe-spiele : (1581-1607) / ed by Schaer, Alfred – Stuttgart: Litterarischer Verein, 1911 (Tuebingen: H Laupp, Jr) [mf ed 1993] – xix/237p – 1 – (incl bibl ref and ind. int by ed) – mf#8470 reel 52 – us Wisconsin U Libr [790]
Drei deutsche pyramus-thisbe-spiele (1581-1607) / ed by Schaer, Alfred – Stuttgart: Litterarischer Verein, 1911 (Tuebingen: H Laupp, Jr) – (incl bibl ref and ind) – us Wisconsin U Libr [430]
Die drei ersten evangelien und die apostelgeschichte / Ewald, Heinrich – 2. vollst Ausg. Goettingen: Dieterich. 2v. 1871-72 – 4mf – 9 – 0-7905-0124-4 – (incl bibl ref) – mf#1987-0124 – us ATLA [225]
Die drei esel der doktorin loehnefink : [a novel] / Beste, Konrad – Wilhelmshaven: Hera Verlag, [19–] [mf ed 1995] – 244p – 1 – mf#8978 – us Wisconsin U Libr [830]
Die drei fassungen von wielands agathon / Freise, Otto – Goettingen: W Fr Kaestner, 1910 – 1 – (incl bibl ref) – us Wisconsin U Libr [430]

Die drei fassungen von wielands agathon / Freise, Otto – Goettingen: W Fr Kaestner, 1910 – 1r – 1 – (incl bibl ref) – us Wisconsin U Libr [430]
Drei federn / Raabe, Wilhelm Karl – 2. Aufl. Berlin: Otto Janke, 1895 – 1r – 1 – us Wisconsin U Libr [830]
Drei frauen : novellen / Musil, Robert – Zuerich: Pegasus Verlag, 1944 – 1r – 1 – us Wisconsin U Libr [830]
Drei georgisch erhaltene schriften von hippolytus : der segen jakobs, der segen moses, die erzaehlung von david und goliath / Bonwetsch, G N – Leipzig, 1904 – (= ser Tugal 2-26/1a) – 2mf – 9 – €5.00 – ne Slangenburg [240]
Die drei gerechten kammacher / Keller, Gottfried – Stuttgart: J G Cotta'sche Buchhandlung Nachfolger, [1903] – 1 – us Wisconsin U Libr [830]
Die drei getreuen : roman / Frenssen, Gustav – Berlin: G Grote 1910 [mf ed 1990] – (= ser Grote'sche sammlung von werken zeitgenoessischer schriftsteller 62) – 1r – 1 – (filmed with: dorfpredigten) – mf#7264 – us Wisconsin U Libr [830]
Drei goethe-reden / Koch, Franz – Weimar: H Boehlau, 1932 [mf ed 1990] – 86p – 1 – (incl bibl of aut's works) – mf#7390 – us Wisconsin U Libr [850]
Drei jahrzehnte deutscher pioniermissionsarbeit in sued-china 1852-1882 / Schmidt, Sauberzweig – Berlin: Berliner evangelischen Missionsgesellschaft, 1908 [mf ed 1995] – (= ser Yale coll; Seines literarischen nachlasses 1) – 129p – 1 – 0-524-09745-3 – (in german) – mf#1995-0745 – us ATLA [951]
Die drei lachenden geschichten / Grimm, Hans – Muenchen: A Langen/G Mueller, 1939 – 1r – 1 – us Wisconsin U Libr [830]
Drei maerchen fuer alt und jung / Ebers, Georg – Stuttgart: Deutsche Verlags-Anstalt, [1893-97?] [mf ed 1993] – (= ser Georg ebers gesammelte werke 21) – 226p – 1 – mf#8554 reel 4 – us Wisconsin U Libr [390]
Drei monate in der libyschen wueste / Rohlfs, G – Washington. 1969+ (1,5,9) – 6mf – 9 – (missing: map) – mf#11685 – ne IDC [915]
Drei monate in spanien / Ringseis, Bettina – Freiburg im Breisgau [u a] 1875 [mf ed Hildesheim 1995-98] – 2mf – 9 – €60.00 – 3-487-29854-6 – gw Olms [914]
Die drei motive und gruende des glaubens / Fechner, Gustav Theodor – Leipzig: Breitkopf und Haertel, 1863 – 1mf – 9 – 0-7905-3740-0 – mf#1989-0233 – us ATLA [240]
Die drei naechte : liebeslieder / Marie Madeleine – Leipzig: F Moeser Nachf, [1901?] [mf ed 1996] – 1r – 1 – (filmed with: yoshiwara / von hermione v preuschen) – mf#9239 – us Wisconsin U Libr [780]
Drei novellen / Ebner-Eschenbach, Marie von – Berlin: Gebrueder Paetel, 1892 – 1r – 1 – us Wisconsin U Libr [430]
Drei psychologische fragen zur spanischen thronkandidatur leopolds von hohenzollern, mit geheimdepeschen bismarcks / Hesselbarth, Hermann – Leipzig: B.G. Teubner, 1913. 130p – 1 – us Wisconsin U Libr [920]
Drei reden jakob grimms : friedrich schiller; ueber das alter; wilhelm grimm / Grimm, Jacob; ed by Mendheim, Max – Leipzig: P Reclam, [19–?] – 1r – 1 – us Wisconsin U Libr [850]
Drei reichenauer denkmaeler der altalemannischen fruehzeit / ed by Daab, Ursula – Tuebingen: M Niemeyer, 1963 [mf ed 1993] – (= ser Altdeutsche textbibliothek n57) – xi/268p – 1 – (parallel latin and old high german text. int in german. incl bibl ref) – mf#8193 reel 5 – us Wisconsin U Libr [430]
Drei schauspiele vom sterbenden menschen : das muenchner spiel von 1510; macropedius, hecastus, 1539; [und] naogeorgus, mercator, 1540 / ed by Bolte, Johannes – Leipzig: K W Hiersemann, 1927 – 1 – (first play and editorial matter in german; the other plays in latin. incl bibl ref) – us Wisconsin U Libr [820]
Drei schauspiele vom sterbenden menschen : das muenchner spiel von 1510, macropedius, hecastus, 1539 [und] naogeorgus, mercator, 1540 / ed by Bolte, Johannes – Leipzig: K W Hiersemann, 1927 – (incl bibl ref) – us Wisconsin U Libr [450]
Drei sommer in norwegen : reiseerinnerungen und kulturstudien / Passarge, Louis – Leipzig 1881 [mf ed Hildesheim 1995-98] – 3mf – 9 – €90.00 – 3-487-28913-X – gw Olms [914]
Die drei urspruenglichen, noch ungeschriebenen evangelien : zur synoptischen frage / Holsten, Carl – Karlsruhe, 1883 – 2mf – 8 – €5.00 – ne Slangenburg [220]
Drei versuchungsgeschichten : zarathustra, buddha, christus / Pietilae, Antti J – Helsinki: Suomalaisen Tiedeakatemian Kustantama, 1910 – (= ser Suomalaisen Tiedeakatemian Toimituksia) – 1mf – 9 – 0-524-00957-0 – (incl bibl ref) – mf#1990-2180 – us ATLA [230]
Drei wenig beachtete cyprianische schriften und die "acta pauli" / Harnack, Adolf von – Leipzig, 1899 – (= ser Tugal 2-19/3b) – 1mf – 9 – €3.00 – ne Slangenburg [240]
Drei-groschen-blatt – Düsseldorf 1949 4 mar-1950 [mf ed 2004] – 2r – 1 – gw Mikrofilm [074]
Der dreigroschenroman / Brecht, Bertolt – Muenchen: K Desch, 1949 [mf ed 1989] – 474p – 1 – mf#7065 – us Wisconsin U Libr [820]
Dreiguds un noschens / Fuchs, Meik – Milwaukee, Wis.: M H Wiltzius, 1898 – 1r – 1 – us Wisconsin U Libr [830]
Dreiheit und dreifache wiederholung im deutschen volksmaerchen : ein beitrag zur technik des maerchens ueberhaupt / Lehmann, Alfred – Leipzig, 1914 (mf ed 1994) – 1mf – 9 – €24.00 – 3-8267-3106-9 – mf#DHS-AR 3106 – gw Frankfurter [390]
Dreilaendereck *see* Am dreilaendereck
Dreiling, R *see* Der konzeptualismus in der universalienlehre des franziskaner- erzbischofs petrus aureoli (pierre d'auriole)
Dreisel, Hermann O *see* Gesammelte schriften
Dreisig yor in argentine / Alperschn, Marcos – Buenos Aires, Argentina. v1-3. 1923 – 1r – us UF Libraries [939]
Dreissig jahre : hilfsverein der deutschen juden / Hilfsverein Der Deutschen Juden (Germany) – Berlin, Germany. 1931 – 1r – us UF Libraries [939]
Dreissig jahre missionsarbeit in wuesten und wildnissen / Flierl, Johann – Neuendettelsau: Missionshauses, 1910 [mf ed 1995] – (= ser Yale coll) – 134p (ill) – 1 – 0-524-10142-6 – (in german) – mf#1995-1142 – us ATLA [240]
Dreissig jahre protestantischer mission in japan *see* History of protestant missions in japan
Dreissig neue ausserlesene padovane und galliard / Johann Ghro – 1604 – (= ser Mssa) – 2mf – 9 – €35.00 – mf#466 – gw Fischer [780]
Dreissig neue erzaehler des neuen deutschland : junge deutsche prosa / ed by Herzfelde, Wieland – Berlin: Malik-Verlag, 1932 – us Wisconsin U Libr [430]
Dreistaedte-bote : amtliches blatt fuer die staedte viersen, duelken, suechteln und umgebung – Viersen DE, 1949 9 jul-26 oct – 1r – 1 – gw Misc Inst [350]
Dreistaedte-zeitung – Viersen DE, 1958 2 jan-31 jul – 1 – (filmed by other misc inst: 1950-57 [18r]. title varies: 1 apr 1955: grenzland-kurier) – gw Misc Inst [074]
Dreistaendige sinnbilder zu fruchtbringendem nuetze : und belieben der ergetzlichkeit, ausgefertiget durch den geheimen / [Knesebeck, F J von dem] – Braunschweig: Bei Conrad Buno, 1643 – 2mf – 9 – mf#0-16 – ne IDC [090]
Dreiviertel stund vor tag : roman aus dem nieder-saechsischen volksleben / Voigt-Diederichs, Helene – Jena: E Diederichs, 1906 [mf ed 1989] – 311p – 1 – mf#7176 – us Wisconsin U Libr [830]
Dreizehn jahre in indien / Woerrlein, Johann – Hermannsburg: Missionshausdruckerei, 1885 [mf ed 1995] – (= ser Yale coll) – vi/248p – 1 – 0-524-10187-6 – (in german) – mf#1995-1187 – us ATLA [920]
Dreizehnlinden / Weber, Friedrich Wilhelm – 120. Aufl. Paderborn: F Schoningh, 1904 – 1 – us Wisconsin U Libr [890]
Drelincourt, C *see*
- Les consolations de l'ame fidele, contre les frayeurs de la mort
- Defense de calvin contre l'outrage fait...sa memoire
- Dialogues familiers sur les principales objections des missionnaires de ce temps
- Traitte des iustes causes de la separation des protestans d'avec l'eglise romaine

[Drelincourt, C] *see* Avertissement sur les disputes et le procede des missionnaires
Dremsa, Catherine J *see* Handrail support versus free arm wing treadmill fitness test
Drennan, Meredith L *see* Incentive motivation of female basketball players across three age levels
Dresch, J *see* Die deutsche revue
Drescher, Birgit *see* Anatomische, histologische, histomorphologische und ethologische untersuchungen zur tiergerechtheit am beispiel des kaninchens
Drescher, Karl *see*
- De claris mulieribus
- Das gemerkbuechlein des hans sachs, 1555-1561
- Joachim rauchels satyrische gedichte
- Johann hartliebs uebersetzung des dialogus miraculorum von caesarius von heisterbach
- Nuernberger meistersinger-protokolle von 1575-1689

Drescher, Martin *see* Gedichte
Drescher, Max *see* Kambodscha
The dresden gallery : fifty of the finest examples of the old masters of this famous gallery / Gemaelde-Galerie, Dresden – London 1875 – (= ser 19th c art & architecture) – 4mf – 9 – mf#4.2.1062 – uk Chadwyck [700]
Die dresdener romantik und heinrich von kleist / Luetteken, Anton – [S.l.: s.n.], 1917 – 1r – 1 – (incl bibl ref) – us Wisconsin U Libr [430]
Dresdener stadt-rundschau – Dresden DE, 1963 2 may-1971 9 sep – 1r – 1 – gw Misc Inst [074]
Dresdensia – Dresden DE, 1892 27 nov-1934 10 mar – 25r – 1 – (title varies: 1893: dresdener rundschau) – gw Misc Inst [074]
Dresdner abendzeitung *see* Dresdner volkszeitung 1854
Dresdner anzeigen *see* Woechentliche dressdnische frag- und anzeigen
Dresdner anzeiger *see* Woechentliche dressdnische frag- und anzeigen
Dresdner anzeiger und tageblatt *see* Woechentliche dressdnische frag- und anzeigen
Dresdner correspondent fuer literatur und tagesneuigkeiten – Dresden DE, 1845-1850 31 aug – 82r – 1 – (title varies: 4 apr 1848: deutscher volksfreund; 28 sep 1848: dresdner zeitung fuer saechsische und allgemeine deutsche zustaende; 1 apr 1849: dresdner zeitung. filmed with: dresdner zeitung 1869 & 1878) – gw Misc Inst [410]
Dresdner haide-zeitung – Klotzsche DE, 1894 22 sep-1935, 1937-40 – 45r – 1 – (title varies: sep 1902: dresdner heide-zeitung) – gw Misc Inst [074]
Dresdner heide-zeitung *see* Dresdner haide-zeitung
Dresdner journal *see* Dresdner tageblatt
Dresdner journal und anzeiger *see* Dresdner tageblatt
Dresdner morgenpost – Dresden DE, 1993- – 2r/yr – 1 – (regional ed: chemnitz, chemnitzer morgenpost 1993- [2r/yr]) – gw Misc Inst [074]
Dresdner nachrichten – Dresden DE, 1872-77 – 5r – 1 – (with suppl: belletristische sonntagsbeilage, fr 27 aug 1914: unterhaltungsbeilage 1856 5 oct-1865 22 jan, 1868-1918 20 nov; humoristisches fr 1 jul 1899: humoristische beilage 1884-1914 1 aug [3r]) – gw Misc Inst [074]
Dresdner neue presse – Dresden, Freital DE, 1925 8 nov-1932 25 dec – 13r – 1 – gw Misc Inst [074]
Dresdner neueste nachrichten – Dresden DE, 1991 [gaps], 1992- – 3r/yr – 1 – (filmed by other misc inst: 1990 2 sep-31 dec [1r]) – gw Misc Inst [074]
Dresdner neueste nachrichten *see* Neueste nachrichten
Dresdner salonblatt – Dresden DE, 1906-13, 1915-1922 n19 – 19r – 1 – (title varies: 1907 n37: salonblatt) – gw Misc Inst [074]
Dresdner stadtblatt *see* Dresdner stadtblatt 1895
Dresdner stadtblatt 1895 – Dresden DE, 1885 18 jan-1893 – 18r – 1 – (title varies: 1 jan 1889: neues dresdner tageblatt; 17 feb 1891: dresdner tageblatt und lokalanzeiger; 29 mar 1891: dresdner stadtblatt; 17 sep 1891: dresdner tageblatt und boersenzeitung) – gw Misc Inst [074]
Dresdner stadtrundschau – Dresden DE, 1963 2 may-1971 9 sep – 1r – 1 – gw Misc Inst [074]
Dresdner tageblatt – Dresden DE, 1903 oct-dec – 1r – 1 – (title varies: 1 apr 1848: dresdner journal; 1 oct 1848: dresdner journal und anzeiger; 1 oct 1851: dresdner journal; 7 sep 1914: saechsische staatszeitung. filmed by other misc inst: 1846 1 jul-1932 31 mar [gaps]. with suppl) – gw Misc Inst [074]
Dresdner tageblatt und boersenzeitung *see* Dresdner stadtblatt 1895
Dresdner tageblatt und deutsche reform *see* Dresdner tageblatt und elbthalbote
Dresdner tageblatt und elbthalbote – Dresden DE, 1883 31 dec-1888 – 13r – 1 – (title varies: 23 mar 1887: dresdner tageblatt und deutsche reform; 1 oct 1887: saechsische landeszeitung) – gw Misc Inst [074]
Dresdner tageblatt und lokalanzeiger *see* Dresdner stadtblatt 1895
Der dresdner volksbote *see* Dresdner volkszeitung 1854
Dresdner volkszeitung *see* Dresdner volkszeitung 1854
Dresdner volkszeitung 1854 – Dresden DE, 1890-1900, 1904-05, 1907-1933 2 mar – 127r – 1 – (title varies: 1859?: saxonia; 1871: der dresdner volksbote; 1 apr 1877: dresdner volkszeitung; dec 1878: dresdner abendzeitung; 1883: saechsisches wochenblatt; 25 dec 1889: saechsische arbeiter-zeitung; 1 may 1908: dresdner volkszeitung. filmed by other misc inst: 1890-1915 apr, 1915 sep-1933 2 mar [152r]) – gw Misc Inst [074]
Dresdner zeitung *see*
- Dresdner correspondent fuer literatur und tagesneuigkeiten
- Woechentliche dressdnische frag- und anzeigen

Dresdner zeitung 1869 – Dresden DE, 1869 3 oct-1871 26 mar – 1 – gw Misc Inst [074]
Dresdner zeitung fuer saechsische und allgemeine deutsche zustaende *see* Dresdner correspondent fuer literatur und tagesneuigkeiten
Dresdner zeitung nebst boersen- und handelsblatt – Dresden DE, 1878 30 jun-1907 18 aug – 1 – gw Misc Inst [332]
Dresdnisches magazin : oder ausarbeitungen und nachrichten zum behufe der naturlehre, der arzneykunst, der sitten und der schoenen wissenschaften – Dresden 1760-65 – (= ser Dz) – 2v on 7mf – 9 – €140.00 – mf#k/n225 – gw Olms [615]
Dreske, O *see* Zwingli und das naturrecht
Dress and address / Stockdale, John Joseph – [2nd ed]. London 1819 – (= ser 19th c art & architecture) – 3mf – 9 – mf#4.2.1014 – uk Chadwyck [740]
Dress and address...dedicated to the merveilleux of either sex / Stockdale, John Joseph – [2nd ed]. London 1819 – (= ser 19th c art & architecture) – 3mf – 9 – mf#4.2.1399 – uk Chadwyck [740]
The dress reform problem : a chapter for women / E Ward & Co – London: Hamilton, Adams & Co; Bradford: John Dale & Co, 1886 – (= ser 19th c art & architecture) – 2mf – 9 – mf#4.1.61 – uk Chadwyck [740]
Dressdnische frag- und anzeigen *see* Woechentliche dressdnische frag- und anzeigen
Dressdnische [sp: dresdner] gelehrte anzeigen – Dresden 1749-1802 – (= ser Dz) – 410mf – 9 – €2460.00 – mf#k/n162 – gw Olms [074]
Dressdnische woechentliche frag- und anzeigen *see* Woechentliche dressdnische frag- und anzeigen
Dresser, Christopher *see*
- The art of decorative design
- Development of ornamental art in the international exhibition
- Japan
- Modern ornamentation
- Principles of decorative design
- Studies in design

Dresser, Frank Farnum *see* The employers' liability acts and the assumption of risks in new york, massachusetts, indiana, alabama, colorado, and england
Dresser, Horatio Willis *see*
- Handbook of the new thought
- Health and the inner life
- A physician to the soul

Dreuillette, Gabriel *see* Narre du voyage faict pour la mission des abnaquiois
Dreves, G M *see* Ein jahrtausend lateinischer hymnendichtung
Dreves, Guido Maria *see*
- Analecta hymnica medii aevi
- Cantiones bohemicae
- Conradus gemnicensis
- Hymnarius moissiacensis

Drevnee pskovsko-novgorodskoe pismennoe nasledie : obozrenie pergamennykh rukopisei tipografskoi i patriarshei bibliotek v sviaza s voprosom o vremeni obrazovaniia etikh knigokhranilishch / Pokrovskii, A A – 1916 – 5mf – 9 – mf#R-11175 – ne IDC [243]
Drevnerusskiia zhitiia sviatykh kak istoricheskii istochnik / Kliuchevskii, V O – 1871 – 6mf – 9 – mf#R-18303 – ne IDC [243]
Drevniaia i novaia rossiia : sistematicheski ukazatel' statei – St. Petersburg, 1875-1881, 1893 – 1 – us NY Public [073]
Drevniaia rossiiskaia vivliofika soderzhashchaia v sebie sobranie drevostei rossiiskiia kasaiushchikhsia – Moscow. v1-20.. 1788-91 – 9 – $234.00 – mf#0189 – us Brook [947]
Drevnie grobnitsy vo vladimirskom kafedralnom uspenskom sobore i uspenskom kniagininom devicheskom monastyre i pogrebennye v nikh kniazia, kniagini i sviatiteli – Vladimir, 1903 – 2mf – 9 – mf#R-18404 – ne IDC [243]
Drevnie i nyneshnie bolgare v politicheskom, narodopisnom, istoricheskom i religioznom ikh otnosheniI k rossianam / Venelin, I I – 1829-1841. 2v – 10mf – 8 – mf#R-117 – ne IDC [243]
Drevnii slavianskii perevod apostola i ego sudby do 15 veka : opyt izsledovaniia iazyka i teksta / Voskresenskii, G A – 1879 – 4mf – 9 – mf#R-10232 – ne IDC [243]
Drevnii slavianskii perevod psaltyri : izsledovanie ego teksta i iazyka po rukopisiam 11-14 v / Sreznevskii, V I – 1877 – 5mf – 8 – mf#R-10063 – ne IDC [243]
Drevniia siriiskiia obiteli i proslavivshie ikh sviatye podvizhniki / Sladkopevtsev, Petr – 1902. 2v – 3mf – 9 – (missing: 1902 v2) – mf#R-18241 – ne IDC [243]
Drevon, I F *see* Voyage en suede
Drew, G S *see* The son of man
Drew gateway – Madison NJ 1972-92 – 1,5,9 – ISSN: 0012-6152 – mf#7677 – us UMI ProQuest [240]
Drew, George Smith *see* Nazareth
Drew, Richard D *see* Ecological characterization of the caloosahatchee river

Drewry, William Sidney *see* Southampton insurrection

Drews, Arthur *see*
- Der ideengehalt von richard wagners dramatischen dichtungen
- Plotin und der untergang der antiken weltanschauung
- The witnesses to the historicity of jesus

Drews, Paul *see*
- Beitraege zu luthers liturgischen reformen
- Das kirchliche leben der evangelisch-lutherischen landeskirche des koenigreichs sachsen
- Petrus canisius
- Untersuchungen ueber die sogen. clementinische liturgie im 8. buch der apostolischen konstitutionen
- Wilibald pirkheimers stellung zur reformation
- Zur entstehungsgeschichte des kanons in der roemischen messe

Drexel library quarterly – Philadelphia PA 1965-85 – 1,5,9 – ISSN: 0012-6160 – mf#1992 – us UMI ProQuest [020]

Drexelius, H *see*
- Aeternitatis prodromus mortis nuntius...
- Antigrapheus sive conscientia hominis coram s.s.mo maximiliano
- De aeternitate considerationes coram ser.mis utriusque bavariae principibus maximiliano et elizabetha explicatae
- Gymnasium patientiae
- Heliotropium seu conformatio humanae voluntatis cum divine
- Helipotropium seu conformatio humanae voluntatis eum divine
- Nicetas seu triumphata incontinentia
- Nicetas seu triumphata incontinentia...editio tertia
- Orbis phaeton
- Recta intentio omnium humanarum actionum amussis
- De sonne-bloeme
- Zodiacus christianus seu duodecim signa praedestinationis...
- Zodiacus christianus seu signa 12. divinae praedestinationis
- Zodiacus christianus seu signa 12 divinae praedestinationis

[Drexelius, H] *see* Zodiacus christianus seu signa 12. divinae praedestinationis una cum 12. symbolis quibus signa illa adumbrantur

Drey christliche vnd in gottes wort vnd der alten lehrer schrifften wolgegruendte predigten / Gedik, S – Magdeburg, 1591 – 2mf – 9 – mf#TH-1 mf 519-520 – ne IDC [242]

Drey schrifften : i protestation samuel hubers doctor vnd professors der h schrifft zu wittemberg wider johan wilhelm stuck d johan jacob gryneum johan jetzlern welche fuer abraham maeusslein vnd peter hybener in jrer legation zu bern haben falsche kundtschafft geredt / Huber, S – Wittemberg, 1593 – 1mf – 9 – mf#TH-1 mf 730 – ne IDC [242]

Drey vnd dreissig predigen von den fuernembsten spaltungen in der christlichen religion / Andreae d A, J – Tuebingen, 1568 – 10mf – 9 – mf#TH-1 mf 45-54 – ne IDC [242]

Dreydorff, Johann Georg *see*
- Pascal
- Zum neubau auf altem grunde

Dreyer, Aloys *see*
- Franz von kobell

Dreyer, Ernst Adolf *see* Sicht des werkes

Dreyer, J L E *see* Tycho brahe's scientific achievements

Dreyer, Johannes Machiel *see* Onderwysersopleiding vir uitkomste-gebaseerde onderwys in suid-afrika

Dreyer, Karl *see* Die religioese gedankenwelt des salomo ibn gabirol

Dreyer, Max *see*
- Drei
- Erdkraft
- Das gymnasium von st juergen
- Liebestraeume
- Der probekandidat
- Unter blonden bestien
- Winterschlaf

Dreyer, Otto *see* Undogmatisches christentum

Dreyfus, Abraham *see*
- Amis
- De l h 'a 3 h

The dreyfus affair in the making of modern france : from the holdings of the houghton library collection, harvard university – 66r – 1 – $7,990.00 – (coll covers the dreyfus affair fr 1894-1908 in over 1000 vols. coll's predominant language is french, and it inlcudes works in english, german, italian, dutch, spanish and swedish as well. complete listing available in print) – mf#C39-28710 – us Primary [944]

Dreyfus, Alfred *see* La revision du proces dreyfus

Dreyfus, Hippolyte *see* The universal religion, bahaism

Dreyfus, Pierre *see* Le cambodge economique

Dreytwein, Dionysius *see* Dionysius dreytweins esslingische chronik

Dreyzehender theil americae, das ist, fortsetzung der historien von der newen welt : oder nidergaengischen indien, waran es auff diese zeit noch anhero ermangelt... – Franckfurt: gedruckt bey Caspar Rötel, in Verlegung Matthei Merian, 1628 [mf ed 1994] – (= ser America 13) – 2mf – 9 – 0-665-94749-6 – mf#94749 – cn Canadiana [970]

Drezen, A K *see* Burzhuaiia i pomeshchiki v 1917 godu

Dri Sun Mun *see* Khyal ratuv rin ramn

Drie en zestig jaren prediker : gedenkschriften / Hulst, Lammert J – Grand Rapids, Mich: Eerdmans Sevensma, 1913 – 1mf – 9 – 0-524-07569-7 – mf#1991-3189 – us ATLA [240]

Drie evangeliedienaren uit den tijd der hervorming / Sepp, Christiaan – Leiden: EJ Brill, 1879 – 1mf – 9 – 0-524-04150-4 – (incl bibl ref) – mf#1990-1220 – us ATLA [242]

Drie jaarige reize naar china, te lande gedaan door den moskovischen afgezant : van moskou af, over groot ustiga, siriania, permia, siberien, daour, groot tartaryen tot in china / [Ides, E I] – Amsterdam: Pieter de Coup, 1710 – 4mf – 9 – mf#HT-251 – ne IDC [915]

Drie leerredenen / Raalte, Albertus C van – Kalamazoo, MI: C Kriekard, [1863?] – 1mf – 9 – 0-524-06101-7 – mf#1991-2414 – us ATLA [220]

De drie punten in alle deelen gereformeerd / Berkhof, Louis – Grand Rapids, MI: Wm B Eerdmans, 1925 [mf ed 1993] – 64p on 1mf – 9 – 0-524-06080-0 – mf#1991-2393 – us ATLA [240]

Driele, F van *see* De twee reisgezellen

Driesch, Hans *see*
- The history and theory of vitalism
- Leib und seele
- The problem of individuality
- The science and philosophy of the organism

Driesener zeitung – Driesen (Drezdenko PL), 1933-34, 1935 1 apr-30 sep, 1936 1 jul-1937 30 jun, 1937 1 oct-1940, 1941 apr-14 sep – 1 – gw Misc Inst [077]

Driessen, A *see* Dziko la nyasaland ndi anthu ace

Driessen, Helmut *see* Ermittlung von grundlagen zur ultrafiltration

Drift / Driftless Bioregional Network [WI] – v1 n1-18 [1985 win-1989 sum] – 1r – 1 – mf#1055732 – us WHS [574]

Drift / Willson, Beckles – London: Gay and Bird, 1895 – 1mf – 9 – 0-665-93952-3 – mf#93952 – cn Canadiana [890]

Drift *see* Pacific monthly

The drift toward religion / Palmer, Albert Wentworth – Boston: Pilgrim Press, c1914 – 1mf – 9 – 0-7905-9555-9 – mf#1989-1280 – us ATLA [240]

Drifted in / Carleton, Will – New York: Moffat, Yard & Co 1908 [mf ed 1987] – 1r – 1 – mf#1959 – us Wisconsin U Libr [830]

Drifting away : a few remarks on professor drummond's search for "natural law in the spiritual world" / Hill, Philip Carteret – London: Bemrose, 1885? – 1mf – 9 – mf#08448 – cn Canadiana [210]

Drifting away / Hill, Philip Carteret – London, England. 18– – 1r – us UF Libraries [240]

Driftless Bioregional Network [WI] *see* Drift

Drijfhout, A E *see* 24 emblemata dat zijn zinnebeelden

Drill and rifle instruction for the corps of rifle volunteers / Grande-Bretagne. War Office – Quebec: printed by Stewart Derbishire & George Desbarats, 1862 [mf ed 1993] – 2mf – 9 – mf#SEM105P1791 – cn Bibl Nat [355]

Drin Van *see* Khnum man dos broh qvi?

Drink and crime in canada / Dominion Alliance for the Total Suppression of the Liquor Traffic – [Toronto?: s.n, 189-] [mf ed 1992 – 1mf – 9 – 0-665-90995-0 – (original iss in ser: campaign leaflets) – mf#90995 – cn Canadiana [362]

Drink, drugs and gambling / Gandhi, Mahatma; ed by Kumarappa, Bharatan – Ahmedabad: Navajivan Pub House, 1962 – (= ser Samp: indian books) – us CRL [360]

Drinker, Henry Sandwith *see* Legal ethics

Drinking in college / Straus, Robert – New Haven, CT. 1953 – 1r – us UF Libraries [025]

Drinking water needs and infrastructure : hearing...house of representatives, 107th congress, 2nd session, april 11 2002 / United States. Congress. House. Committee on Energy and Commerce. Subcommittee on Environment and Hazardous Materials – Washington: US GPO 2002 [mf ed 2002] – 2mf – 9 – 0-16-068797-7 – (incl bibl ref) – us GPO [350]

Drinkwater-Bethune, C R *see* Sir richard hawkins

Drioux, abbe (Claude-Joseph) *see*
- Cours abrege d'histoire ancienne
- Precis elementaire d'histoire ecclesiastique

Dripps, Joseph Frederick *see* Historical sketch of the missions in siam and among the laos

Drischath zion, oder zions herstellung / Kalischer, H – Berlin, 1905 – 2mf – 9 – mf#J-72-506 – ne IDC [956]

Drischath zion, oder zions herstellung / Kalischer, Z H – Thorn, 1865 – 1mf – 9 – mf#J-28-58 – ne IDC [956]

Driscoll, Lori *see* Journal of access services

Driskell, David C *see* Peter clarke

Drita – Tirane Albania, 1976-1996 – 21r – 1 – gw Mikropress [077]

Drita e vertete = The true light / Albanian Orthodox Diocese of America – 1978 oct-1982 feb – 1r – 1 – mf#652531 – us WHS [243]

Das dritte buch esdras : und sein verhaeltnis zu den buechern esra-nehemia / Bayer, Edmund – Freiburg im Breisgau, St Louis MO: Herder 1911 [mf ed 1989] – (= ser Biblische studien 16/1) – 2mf – 9 – 0-7905-2521-6 – (in german, greek & hebrew) – mf#1987-2521 – us ATLA [221]

Die dritte gattung der achaemenischen keilinschriften / Stern, Moriz Abraham – Goettingen: Dieterich, 1850 – 1mf – 9 – 0-8370-7670-6 – (incl bibl ref and index. text in german and akkadian; introduction and commentary in german) – mf#1986-1670 – us ATLA [470]

Das dritte geschlecht – Berlin DE, 1928-1929 n5 1929 – 1r – 1 – gw Misc Inst [306]

Der dritte humanismus im werke stefan georges und thomas manns / Maier, Hans Albert – [s.l: s.n.] 1946 [mf ed 1989] – 1r – 1 – (incl bibl ref. filmed with: stefan george und thomas mann & other titles) – mf#7296 – us Wisconsin U Libr [430]

Das dritte reich : documentarische aufstellung des aufbaus der nation ohne ortsbezeichnung – 1933-1938 – 2r – 1 – gw Mikropress [943]

Der dritte theil schoener, newer, teutscher lieder, mit 5 stimmen / Lasso, Orlando di – 1576 – (= ser Mssa) – 2mf – 9 – €60.00 – mfchl 297 – gw Fischer [780]

Dritte wanderung nach palaestina im jahre 1857 / Tobler, T – Gotha, 1859 – 6mf – 9 – mf#H-6151 – ne IDC [915]

Dritter theil der clavier uebung bestehend in verschiedenen vorspielen ueber die catechismus- und andere gesaenge, vor die orgel... : denen liebhabern, und besonders denen kennern von dergleichen arbeit, zur gemueths ergezung verfertiget von johann sebastian bach / Bach, Johann Sebastian – Leipzig: In Verlegung des Authoris [1739] [mf ed 19–] – 1r – 1 – mf#film 1343 – us Sibley [780]

Driu liet von der maget (cima62) : farbmikrofiche-edition der handschrift berlin, ehem preussische staatsbibliothek, ms germ oct 109 (z.zt. krakow, biblioteka jagiellonska, depositum) – (mf ed 2001) – (= ser Codices illuminati medii aevi (cima)) – 4 color mf – 15 – €290.00 – 3-89219-062-3 – (description & comm by elisabeth radaj) – gw Lengenfelder [090]

Drive – Basingstock, England 1967-73 – 1 – ISSN: 0046-0710 – mf#8902 – us UMI ProQuest [380]

Drive news / Four Wheel Drive Auto Co – v1 n2-v11 n2 [1943 nov 8-1954 feb] – 1r – 1 – mf#1055736 – us WHS [629]

Driver – Washington DC 1972-86 – 1,5,9 – ISSN: 0002-2373 – mf#7909 – us UMI ProQuest [355]

Driver, G R *see*
- The assyrian laws
- Babylonian laws
- Canaanite myths and legends
- Problems of the hebrew verbal system
- Semitic writing

Driver, I D *see* Biblical lectures

Driver, Samuel Rolles *see*
- Additions and corrections to the book of genesis
- The book of daniel
- The book of exodus
- The book of genesis
- The book of job in the revised version
- The book of leviticus
- The book of the prophet jeremiah
- The books of joel and amos
- Christianity and other religions
- A critical and exegetical commentary on deuteronomy
- Critical notes on the international sunday-school lessons from the pentateuch for 1887 (january 2-june 26)
- Hebrew tenses
- The higher criticism
- The ideals of the prophets
- An introduction to the literature of the old testament
- Isaiah
- Isaiah, his life and times
- The minor prophets
- Modern research as illustrating the bible
- Notes on the hebrew text of samuel
- Sermons on subjects connected with the old testament
- Studies in the psalms
- A treatise on the use of the tenses in hebrew

Driver, Samuel Rolles et al *see*
- Authority and archaeology, sacred and profane
- The international critical commentary on the old and new testaments
- Studia biblica

Driver, William *see* Logbook and memoir

Driving digest magazine – n20-50 [1982 jun-1988 jun – 1r – 1 – mf#1055737 – us WHS [629]

Drobisch, Max Wilhelm *see* Neue darstellung der logik nach ihren einfachsten verhaeltnissen

Drobisch, Moritz Wilhelm *see* Grundlehren der religionsphilosophie

Drobnitzky-Eickhoff, Barbara *see*
- Heilpaedagogische moeglichkeiten der musik in der sonderpaedagogik
- Struktur und aufbau ausgewaehlter stuecke darstellender musik (tierdarstellungen) in verbindung mit instrumentenkunde

Drochon, Jean-Emmanuel B *see* Un chevalier apotre

Droescher, Georg *see* Gustav freytag in seinen lustspielen

Droese, Miss *see* Indian gems for the master's grown

Drogheda advertiser – Drogheda 1929 – 1r – 1 – ie National [072]

Drogheda advertiser – Drogheda, Ireland 7 oct 1908-14 dec 1929 [mf 1915] – 1 – (wanting: 12,20,27 may; 1,8,15,22 jul; 12 aug-30 dec 1922; 13 oct 1923-19 apr 1924) – uk British Libr Newspaper [072]

Drogheda argus and leinster journal – Drogheda, Louth 1882, 1936, 1943-47 – 1r – 1 – (cont as: drogheda argus, and advertiser [21 dec 1929-24 jun 1944]; cont as: argus [jul 1944-ca aug 1954-7 nov 1959] a monaghan ed was publ) – ie National [072]

Drogheda argus and leinster journal – Drogheda, Ireland 19 sep 1835-14 dec 1929 [mf 1851-96] – 1 – (cont by: drogheda argus, & advertiser [21 dec 1929-7 nov 1936]) – uk British Libr Newspaper [072]

Drogheda conservative, and general advertiser for the counties of meath, louth, dublin, monaghan and cavan – Drogheda, Ireland 27 feb 1864-3 oct 1908 [mf 1849-96] – 1 – (cont: conservative, & general advertiser for drogheda, meath, louth, monaghan & cavan [17 sep 1853-20 feb 1864]) – uk British Libr Newspaper [072]

Drogheda conservative journal : or meath louth monaghan and cavan advertiser – Drogheda, Ireland. 24 jun 1837-30 dec 1848 – 4r – 1 – uk British Libr Newspaper [072]

Drogheda independent – Louth 1924-48 – 21r – 1 – ie National [072]

Drogheda independent etc – Drogheda, Ireland.1890-1923; 3 jun-16 may 1925; 1950; 1986-1992 – 52 1/4r – 1 – uk British Libr Newspaper [072]

Drogheda journal – Louth 1793-1820 [odd] – 0.5 – 1 – ie National [072]

Drogheda journal : or meath and louth advertiser – Drogheda, Ireland. 1823-23 may 1840; 31 jul-4 mar 1843 – 17 1/2r – 1 – (aka: drogheda journal or: meath louth cavan and monaghan general advertiser) – uk British Libr Newspaper [072]

Drogheda journal *see* Drogheda journal

Drogheda news letter – Drogheda, Ireland. 29 may 1813 – 1/4r – 1 – uk British Libr Newspaper [072]

Drogheda news-letter – Drogheda 1801-07, 1810, 1813 [partial] – 3r – 1 – ie National [072]

Drogheda reporter and general advertiser etc – Drogheda, Ireland. 6 jul 1861-1 apr 1865 – 3r – 1 – uk British Libr Newspaper [072]

Drogheda sentinel etc – Drogheda, Ireland. -w. 19 apr-5 jul 1834 – 1/4r – 1 – uk British Libr Newspaper [072]

Drogin, E B *see* Evaluating the boater experience

Drogueria jorge garces b y su radio de accion en... – Cali, Colombia. 1929 – 1r – us UF Libraries [972]

Die drohende sichel / Nord, F R – 7.-9. Aufl. Leipzig: P List, 1935 – 1r – 1 – us Wisconsin U Libr [830]

Die drohende spaete metabolische azidose der frueh- und neugeborenen / Kalhoff, Hermann – (mf ed 1999) – 2mf – 9 – €40.00 – 3-8267-2632-4 – mf#DHS 2632 – gw Frankfurter [618]

Drohojowska, Antoinette *see* Les femmes pieuses de la france

Les droicts, avtoritez et prerogatives qve pretendent au royavme de hierusalem... / Lusignano, S di – Paris, 1586 – 1mf – 9 – mf#H-8367 – ne IDC [956]

Le droit : bulletin des tribunaux – Paris. dec 1836-juin 1837, 15 oct-31 dec 1845, 1848, 1850-51, 1853-54, janv-juin 1856, juil 1857-juin 1860, 1865-juin 1867, juil 1868-juin 1869, 1870-73, juil 1880-81, 1888, 1891, 1894-95 – 1 – fr ACRPP [944]

Le droit – Port-au-Prince: [s.n.], [1892-]. [1re annee, n1-2e annee, n26 25. fevr 1892-19 aout 1893] – 6r – 9 – us CRL [079]

Droit administratif : ou manuel des paroisses et fabriques / Langevin, Hector – Quebec: Desbarats et Derbishire, 1863 [mf ed 1984] – 3mf – 9 – 0-665-45226-8 – (incl ind) – mf#45226 – cn Canadiana [240]
Le droit civil canadien avec revue de la jurisprudence de nos tribunaux / Mignault, Pierre Basile – Montreal: Wilson & Lafleur, 1909 – 6mf – 9 – mf#10087 – cn Canadiana [347]
Le droit de l'art et des lettres / Savatier, Rene – Paris: Librairie generale de droit et de jurisprudence, 1953. 224p. LL-4116 – 1 – us L of C Photodup [340]
Le droit de preemption en matiere civile et commerciale. / Mauzac, Louis – Montpellier, 1935. 151p. LL-4047 – 1 – us L of C Photodup [346]
Droit des assurances : recueil de textes legislatifs, reglementaires et jurisprudentiels / Lluelles, Didier – [Montreal]: Librairie de l'Universite de Montreal, [1978] (mf ed 1999) – 1r – 1 – (incl english text) – mf#SEM35P463 – cn Bibl Nat [340]
Le droit des femmes – Revue politique, litteraire et d'economie sociale. Red. en chef Leon Richer. Paris. 1869-91. A paru sous le titre de: L' Avenir des femmes de 1871 a 1879 – 1 – fr ACRPP [322]
Le "droit d'oblat" (afm49) / Marchal, J – 1955 – (= ser Archives de la france monastique (afm)) – €15.00 – ne Slangenburg [241]
Droit du peuple – Grenoble, France. 11 jun 1916-29 nov 1917 – 2r – 1 – uk British Libr Newspaper [072]
Le droit du peuple – Journal des interets sociaux. Red. en chef Jean-Jacques Danduran. no. spec. Paris. fevr 1850 – 1 – fr ACRPP [073]
Droit du travail en haiti / Latortue, Francois – Port-Au-Prince, Haiti. 1961 – 1r – us UF Libraries [331]
Droit et liberte : contre le racisme et l'antisemitisme, et pour la paix – Paris. 20 fevr 1946-juil 1974 – 1 – fr ACRPP [325]
Droit et pratique du commerce international = International trade law and practice – Paris, France 1987-88 – 1,5,9 – ISSN: 0335-5047 – mf#16578 – us UMI ProQuest [341]
Droit et science dans la pensee de hans kelsen (contribution a la theorie pure du droit) / Acka, Sohuily Felix – 2mf – 9 – (10212) – fr Atelier National [320]
Le droit immobilier marocain... / Menard, A – Rabat, 1934 – 4mf – 9 – mf#ILM-3041 – ne IDC [956]
Le droit maritime francais – Paris. 1949-67 – 1 – fr ACRPP [341]
Le droit musulman explique / Savvas, Pasha – Paris: Marchal et Billard, 1896. 161p. LL-12030 – 1 – us L of C Photodup [340]
Le droit paroissial : etude historique et legale de la paroisse catholique... / Mignault, Pierre Basile – [Montreal: C O Beauchemin, 1893] – 1mf – 9 – 0-665-92514-X – mf#92514 – cn Canadiana [342]
Le droit paroissial de la province de quebec : precede d'un formulaire par wilfrid camirand / Pouliot, Jean-Francois – [Fraserville, Quebec?: s.n, 1919?] – 8mf – 9 – 0-665-97455-8 – mf#97455 – cn Canadiana [342]
Droit public ou gouvernement des colonies francoises : d'apres les loix faites pour ces pays / Petit, Emilien – Paris: Chez Delalain...2v. 1771 [mf ed 1984] – 2v on 1mf – 9 – mf#47041 – cn Canadiana [320]
Droit social : textes et documents annotes concernant les rapports professionnels et l'organisation de la production – Paris. 1938-40 – 1 – fr ACRPP [073]
Le droit social – n1. Marseille. mai 1885 – 1 – fr ACRPP [073]
Le droit social *see* L'etendard revolutionnaire
Les droits de la femme devant la loi francaise / Neulat, L – Paris: Librairie mondiale, 1907 – (= ser Les femmes [coll]) – 2mf – 9 – mf#9689 – fr Bibl Nationale [340]
Les droits de la langue francaise meconnus : humilante position de la province de quebec – S.l: s.n, 1880? – 1mf – 9 – mf#02293 – cn Canadiana [323]
Les droits de l'homme : journal politique et quotidien – Paris. 1898-17 mars 1900, 27 oct-8 dec 1901, 8 juin-5 nov 1910 – 1 – fr ACRPP [322]
Les droits de l'homme : liberte, egalite, fraternite – Paris. 11 fevr 1876-15 fevr 1877, 25 mars-3 juin 1878 – 1 – fr ACRPP [073]
Les droits de l'homme – Paris: Blondeau, jan 1849 – us CRL [074]
Les droits du saint-siege : alexandre 6 et cesar borgia / La Rochelle, E – Paris: E Dentu, 1861 – 1mf – 9 – 0-8370-7882-2 – mf#1986-1882 – us ATLA [940]
Drolet, Bernadette *see* Bibliographie analytique de la federation des guides catholiques de la province de quebec
Dromore leader – 1978-1984; 1986- dec 1998 – 39 1/2r – 1 – (aka: leader /dromore edt); leader (dromore and lisburn ed)) – uk British Libr Newspaper [072]
Dromore leader – [Northern Ireland] Belfast 21 oct 1916-dec 1950 [mf ed 2002] – 17r – 1 – (cont by: leader [jan 1931-dec 1950]) – uk Newsplan [072]
Dromore star – Dromore, Ireland. 23 aug 1991-1992 – 4r – 1 – uk British Libr Newspaper [072]
Dromore weekly mail – Dromore, Ireland. 7 jan 1905 – 1/4r – 1 – uk British Libr Newspaper [072]
Dromore weekly times – Ireland. 13 May 1905-1929; 1931-4 Oct 1952 (missing 1930).-w. 31 1/2 reels – 1 – uk British Libr Newspaper [072]
Drona parva – Calcutta: Bharata Press, 1888 – 2mf – 9 – 0-524-08012-7 – mf#1991-0234 – us ATLA [280]
Droop, Fritz *see*
– Emil goetts vermaechtnis
– Otto julius bierbaum
Drop and the ocean / Cox, John – London, England. 18– – 1r – us UF Libraries [240]
Dropper, Eaves *see* In re corney v father evangelicus
Dross, Friedrich *see* Der gestohlene mond
Drossbach, Maximilian *see* Die harmonie der ergebnisse der naturforschung mit den forderungen des menschlichen gemuethes, oder, die persoenliche unsterblichkeit als folge der atomistischen verfassung der natur
Drost, Willi *see* Goethe als zeichner
Die droste : der lebensroman der annette von droste-huelshoff / Karwath, Juliane – Stuttgart: Deutsche Verlags-Anstalt, 1929 – 1r – 1 – us Wisconsin U Libr [430]
Droste, Georg *see*
– For de fierstunnen
– Ottjen alldag un sien lehrtied
– Ottjen alldag un sien moorhex
– Sunnenschien un wulken
Droste-Huelshoff, Annette von *see*
– Annette von droste-huelshoff
– Briefe
– Die briefe der annette von droste-huelshoff
– Die briefe der dichterin annette v droste-huelshoff
– Briefe der freiin annette von droste-huelshoff
– Der freiin annette elisabeth von droste-huelshoff gesammelte werke
– Gedichte
– Das geistliche jahr / geistliche lieder
– Geistliches jahr
– Die judenbuche
– Letzte gaben
– Lyrische gedichte
– Saemtliche werke
– Ungedrucktes
Droste-Huelshoff, Elisabeth, Freiin von *see* Der freiin annette elisabeth von droste-huelshoff gesammelte werke
Drott line / Drott Manufacturing – 1977 apr-1978 hols – 1r – 1 – (cont by: drottline) – mf#665562 – us WHS [670]
Drott Manufacturing *see* Drott line
Drottline / J I Case Co – 1979 spr-1980 sum – 1r – 1 – (cont: drott line; cont by: web [schofield wi]) – mf#665810 – us WHS [670]
Drouet, Etienne-Francois *see* Le grand dictionnaire historique (ael1/44.8)
Drouin de Bercy *see*
– L'europe et l'amerique comparees
Drouville, Gaspard *see* Voyage en perse, fait en 1812 et 1813
Drover's journal – Lincolnshire IL 1989-93 – 1 – ISSN: 0012-6454 – mf#17149,05 – us UMI ProQuest [636]
Drovers journal – South Omaha, NE: Drovers Journal Co (daily ex sun) [mf ed 1893 (gaps) filmed 1976] – 1r – 1 – (cont: daily drovers journal. cont by: south omaha drovers journal) – us NE Hist [636]
Drovers journal – South Omaha, NE: Drovers Journal Co. v8 n21 [ie 214] aug 28 1895-v11 n284. nov 19 1898 (daily ex sun) [mf ed with gaps] – 5r – 1 – (cont: south omaha drovers journal. cont in pt by: magic city hoof and horn. merged with: south omaha daily stockman to form: daily drovers journal and stockman) – us NE Hist [636]
Drown, Edward Staples *see* The apostles' creed to-day
Drowned / Grosart, Alexander Balloch – Kinross, Scotland. 1864 – 1r – us UF Libraries [240]
Droylsden express – [NW England] Tameside 23 apr 1870-1 apr 1871 [mf ed 2003] – 1r – 1 – uk Newsplan [072]
Droylsden literary and advertising journal – [NW England] Droylsden, Stalybridge Lib may 1854-jan 1855 – 1 – uk MLA; uk Newsplan [072]
Droysen, Gustav *see* Geschichte der gegenreformation
Droysen, H *see* Eutropi breviarium ab urbe condita (mgh1:2.bd)
Droysen, Johann Gustav *see* Geschichte des hellenismus
Drozd, Leslie *see* Journal of child custody
Druck, David *see* Meforshim fun der torah
Druck und papier *see* Der korrespondent fuer deutschlands buchdrucker und schriftgiesser
Die drucke der bachsoehne der hoboken-sammlung musikalischer erst- und fruehdrucke / Oesterreichische Nationalbibliothek Wien. Musiksammlung – (= ser Die europeaische musik) – 107mf – 9 – diazo €648.00 silver €748.00 – gw Olms [780]
Drucke und holzschnitte des 15. und 16. jahrhunderts in getreuer nachbildung – Strassburg: J H E Heitz. 15v. 1899-1922 – 1r – 1 – us Wisconsin U Libr [730]
Druckhaus echo – Halle S DE, 1975-1989 – 1r – 1 – (druckhaus "freiheit") – gw Misc Inst [074]
Druehe-Wienholt, Christiane-Maria *see* Der veraenderte wiconsins kartensortiertest und seine relevanz fuer neuropsychologische diagnostik und therapie
Druffel, August von *see*
– Kaiser karl 5. und die roemische curie, 1544-46
– Von der sendung der legaten nach trient (maerz 1545) bis zum beginn des schmalkaldischen krieges (juni 1546)
Drug abuse and alcoholism newsletter – San Diego CA 1980-97 – 1,5,9 – ISSN: 0160-0028 – mf#12667 – us UMI ProQuest [362]
Drug abuse and the use and abuse of medicine / South Africa. Department of Health [Departement van Gesondheid] – Pretoria: [Dept of Health 1978?] [mf ed Pretoria, RSA: State Library [199-]] – 12p [ill] on 1r with other items – 5 – (Dwelmmiddelmisbruik en die gebruik en misbruik van medisyne) – mf#op 06718 r23 – us CRL [362]
Drug and alcohol dependence – Oxford, England 1975+ – 1,5,9 – ISSN: 0376-8716 – mf#42181 – us UMI ProQuest [362]
Drug and alcohol use by freshman at siuslaw high school and their opinions regarding potentially effective drug and alcohol education programs / Byrd, Marcia J – Oregon State University, 1995 – 1mf – 9 – mf#HE 564 – us Kinesology [613]
Drug and therapeutics bulletin – London, England 1967-73 – 1,5 – ISSN: 0012-6543 – mf#2325 – us UMI ProQuest [615]
Drug & cosmetic industry – Carol Stream IL 1926-96 – 1,5,9 – (cont by: dci) – ISSN: 0012-6527 – mf#2551.02 – us UMI ProQuest [660]
Drug enforcement / U.S. Bureau of Narcotics and Dangerous Drugs – v1-12. 1973-85 – 27mf – 9 – $40.50 – (lacking: v10 n1. cont: bndd bulletin. updates available) – mf#LLMC 82-200 – us LLMC [360]
Drug enforcement – Washington DC 1973-85 – 1,5,9 – ISSN: 0098-3470 – mf#11241 – us UMI ProQuest [362]
Drug Information Association *see* Dia forum
Drug intelligence & clinical pharmacy – Cincinnati OH 1967-88 – 1,5,9 – (cont by: dicp) – ISSN: 0012-6578 – mf#6492.02 – us UMI ProQuest [615]
Drug law reporter – National College, Houston. v1. 1981-1982 (all publ) – 5,6 – $49.00 – (available in reel only) – mf#105031 – us Hein [344]
Drug metabolism and disposition – v1-24. 1973-96 – 22r – 1,5,6,9 – $110.00r – us Lippincott [150]
Drug naroda : organ slavgor zemstva i koop soiuzov / ed by Frizen, I I – Slavgorod [Alt gub]: Koop zemstvo 1919 [1919-] – (= ser Asn 1-3) – n31-79 [1919] [gaps] item 129, on reel n27 – 1 – (cont: vestnik slavgorodskogo zemstva) – mf#asn-1 129 – ne IDC [077]
Drug store news – New York NY 1979+ – 1,5,9 – ISSN: 0191-7587 – mf#12564 – us UMI ProQuest [071]
Drug therapy – Parsippany NJ 1971-94 – 1,5,9 – ISSN: 0001-7094 – mf#6560 – us UMI ProQuest [615]
Drug topics – Montvale NJ 1972+ – 1,5,9 – ISSN: 0012-6616 – mf#7505 – us UMI ProQuest [615]
Drug usage by athletes related to performance / Master, Ronald R – 1978 – 1mf – 9 – $4.00 – us Kinesology [790]
Drugs : special studies, 1972-1986 – 14r – 1 – $2705.00 – 0-89093-990-X – (with p/g) – us UPA [360]
Drugs and society / ed by Segal, Bernard – v1- 1986- – 1, 9 ($175.00 in US $245.00 outside hardcopy subsc) – us Haworth [360]
Drugs in current use and new drugs – New York NY 1969-73 – (cont by: drugs in current use and new drugs) – ISSN: 0070-7392 – mf#8195.01 – us UMI ProQuest [615]
Drugs under experimental and clinical research – Geneve, Switzerland 1979+ – 1,5,9 – ISSN: 0378-6501 – mf#12199 – us UMI ProQuest [615]
Druid – [Wales] LLGC 23 may 1907-1 jun 1939 [mf ed 2002] – 19r – 1 – (cont as: welsh-american [jan 1914-dec 1919]; druid [jan 1920-jun 1939]) – uk Newsplan [072]
Druker, I *see* Sholem-aleykhem
Drum / Ohio State University – 1977 win qrt – 1r – 1 – mf#5297699 – us WHS [378]
Drum *see* The african drum
Drum corps america – Racine WI 1973-76 – 1,5,9 – mf#8305 – us UMI ProQuest [780]
Drum nou – Brasov, Romania. 1962-1989 – 38r – 1 – us L of C Photodup [949]
Drumbeat – v3 n1 [i.e. 2] [[1980] apr 1?] – 1r – 1 – (cont: screaming eagle) – mf#615837 – us WHS [071]
Drumbeat : a mau mau kraal publication – v1 n4 [1995 jun] – 1r – 1 – mf#3400176 – us WHS [071]
Drumbeats / Institute of American Indian Arts – v6 n5-v6 n8 v7 n2, block 2 [1973 jan 26-may 25, 1974 mar] – 1r – 1 – mf#705518 – us WHS [740]
Drumgoole, Edward *see* Edward drumgoole papers
Drumheller mail – Alberta, CN. may 1918-dec 1996 – 51r – 1 – cn Commonwealth Imaging [071]
Drumheller review – Alberta, CN. jan 1914-dec 1940 – 4r – 1 – cn Commonwealth Imaging [071]
Drumheller sun – Alberta, CN. jan 1978-dec 1979 – 2r – 1 – cn Commonwealth Imaging [071]
Drummer : a contemporary newsweekly – n1-60. 1967-69 – 1 – (formerly: distant drummer, thursday's drummer) – us AMS Press [073]
Drummer – Jackson MS. v1-v2 n4 [1971 apr 30, jul 15, aug 1-15, oct 23-29-1972 feb 23 – 1r – 1 – mf#875560 – us WHS [071]
Drummer – 1971 aug 19/-1979 jan 2/aug 1 – 10r – 1 – (with gaps; cont: thursday's drummer) – mf#507516 – us WHS [071]
Drummer boy / Tappantown Society – 1976 jul 4-1979 may – 1r – 1 – mf#641470 – us WHS [071]
Drummond, A L *see* Edward irving and his circle, including some consideration of the "tongues" movement in the light of modern psychology
Drummond, Andrew Thomas *see*
– Canadian timber trees
– The distribution of canadian forest trees in its relation to climate and other causes
– Railway accidents
Drummond, D T K *see*
– Reply to resolutions of the clergy of the scottish episcopal church
– Sermon for the times, preached in trinity chapel
Drummond, David Thomas Kerr *see*
– Historical sketch of episcopacy in scotland
– The parabolic teaching of christ
Drummond, Dtk *see* Scottish communion office, examined and proved to be repugnant to scripture
Drummond, Henry *see*
– Addresses
– A college of colleges
– Dwight l moody
– The greatest thing in the world
– The ideal life
– Letter on the payment of the roman catholic clergy
– Love, the supreme gift
– The lowell lectures on the ascent of man
– Natural law in the spiritual world
– The new evangelism
– The new evangelism and other addresses
– Principles of ecclesiastical buildings and ornaments
– Programme of christianity
– Stones rolled away
– Tropical africa
Drummond, James *see*
– The epistles of paul the apostle to the thessalonians, corinthians, galatians, romans and philippians
– An inquiry into the character and authorship of the fourth gospel
– The jewish messiah
– Johannine thoughts
– Philo judaeus
– Via, veritas, vita
Drummond, Lewis Henry *see* The french element in the canadian northwest
Drummond, Robert *see*
– Illustrations of the grammatical parts of the guzerattee, mahratta and english languages
– Sabbath and the christian
Drummond, Robert Blackley *see* Free will in relation to statistics
Drummond, Robert J *see* The relation of the apostolic teaching to the teaching of christ
Drummond, William Hamilton *see* Doctrine of the trinity founded neither on scripture, nor on reason
Drummond, William Henry *see*
– Johnnie courteau
– Montreal in halftone
– Phil-o-rum's canoe and madeleine vercheres
– Pioneers of medicine in the province of quebec
– Politics in Ontario
– The voyageur and other poems
Drummond's winter campaign, 1813 / Cruikshank, Ernest Alexander – S.l: Lundy's Lane Historical Society, 1900? – 1mf – 9 – mf#06406 – cn Canadiana [355]
Drumont, Edouard Adolphe *see* France juive

Drums – 1971 jan-1975 mar 1 – 1r – 1 – mf#384135 – us WHS [071]

Drums – 1972-75 – (= ser American indian periodicals... 1) – $95.00 – us UPA [305]

Drums – Bethel AL. 1983 dec 22-1984 sep 13 – 1r – 1 – (cont by: tundra drums) – mf#920020 – us WHS [071]

Drums in bahia / Eskelund, Karl – London, England. 1960 – 1r – us UF Libraries [972]

Drums of affliction / Turner, Victor Witter – Oxford, England. 1968 – 1r – us UF Libraries [420]

Drum-taps (1865) and sequel to drum-taps (1865-66) / Whitman, Walt – 9 – us Scholars Facs [810]

Drumul socialismului – Deva, Romania. 1962-Jul 1979 – 28r – 1 – us L of C Photodup [949]

A drunkard's experience at home and abroad : written by himself, henry adams / Adams, Henry – [Saint John, NB?: E J Armstrong, 188-?] – 1mf – 9 – 0-665-94493-4 – mf#94493 – cn Canadiana [230]

A drunkard's experience at home and abroad : written by himself, henry adams – [Saint John, NB?: s.n, 188-?] – 1mf – 9 – 0-665-94493-4 – mf#94493 – cn Canadiana [920]

Drunkenness is madness / Russom, J – Middlewich, England. 18– – 1r – us UF Libraries [240]

Drupa presse – Duesseldorf DE, 1951 1 apr-10 jun, 1953 15 jun-1954 30 may – 1r – 1 – gw Misc Inst [074]

Drury, Allen *see* Very strange society

Drury, Augustus Waldo *see*
- Disciplines of the united brethren in christ
- The life of rev philip william otterbein
- Minutes of the annual and general conferences of the church of the united brethren in christ, 1800-1818
- Outlines of doctrinal theology

Drury, B Paxson *see* A fruitful life

Drury, Belle Paxson *see* A fruitful life

Drury, Clifford Merrill *see*
- Christian missions and foreign relations in china
- Four hundred years of world presbyterianism

Drury, James Westbrook *see* Changes made by the 1951 legislature in kansas library laws

Drury lane under sheridan : manuscript plays and correspondence, 1776-1812 – 16r – 1 – (from the british library, london. coll offers 130 plays submitted to r b sheridan during his yrs at the theatre royal, drury lane) – mf#C35-12700 – us Primary [790]

The druses of the lebanon : their manners, customs and history, with a translation of their religious code / Chasseaud, George Washington – London: Richard Bentley, 1855 – 1mf – 9 – 0-524-01477-9 – mf#1990-2453 – us ATLA [290]

Drut : roman / Bahr, Hermann – Wien: H Bauer, c1946 [mf ed 1989] – 830 – 1 – mf#530/[1]p – us Wisconsin U Libr [830]

Drut : roman / Bahr, Hermann – Wien: H Bauer, c1946 [mf ed 1989] – 530/[1]p – 1 – mf#6973 – us Wisconsin U Libr [830]

Druyanow, Alter *see* Pinsker u-zemano

The druzes and the maronites under the turkish rule from 1840 to 1860 / Churchill, Charles Henry – London: B Quaritch, 1862 – 1mf – 9 – 0-524-03525-3 – mf#1990-3230 – us ATLA [956]

Druzhba – Sofia, bulgaria. 12 feb-14 dec 1947 – 1 – (in cyrillic) – mf#mf.685.l – uk British Libr Newspaper [077]

Druzhba – Pekin, China. -d. 27 feb-29 sep 1957 (imperfect) – 1r – 1 – uk British Libr Newspaper [079]

Druzhba – Sofia, Bulgaria. Oct 1946-1947 – 1r – 1 – us L of C Photodup [949]

Druzhinin, V G *see* Pisaniia russkikh staroobriadtsev, perechen spiskov, sostavlennykh po pechatnym opisaniiam rukopisnykh sobranii

Drwaw journal / Distillery, Rectifying, Wine and Allied Workers' International Union of America – v5 n6-v8 n1 [1965 apr-1968 jun] – 1r – 1 – mf#1055051 – us WHS [660]

Dry creek baptist church. edgefield county. south carolina : church records – 1825-58, 1880-1924 – 1 – us Southern Baptist [242]

Dry drayton 1564-1950 – 6mf – 9 – £7.50 – uk CambsFHS [929]

Dry, Wakeling *see* Wagner's die meistersinger

Dryander, Ernst von *see*
- Commentary on the first epistle of st john
- Das vaterunser

Y drych – Utica, NY. 1908, 1911, 1915-21 [incomplete] – 1 – (in welsh) – us ABHS [071]

Drych *see* Y drych

Y drych americanaidd = Mirror – [Wales] LLGC 1851, ion 1856-rhag 1857, ion 1867-rhag 1950 [mf ed 2003] – 71r – 1 – (cont as; y drych a'r gwyliedydd [ion 1856-rhag 1857]; y drych [ion 1867-rhag 1877]; y drych a baner america [ion 1878-rhag 1890]; y drych y wasg a baner america [ion 1891-rhag 1894]; y drych a'r columbia [ion 1895-rhag 1923]; y drych [ion 1924-rhag 1950]) – uk Newsplan [072]

Y drych a'r columbia – Utica, NY, Chicago, IL: Thomas J Griffiths, jan 1-dec 30 1926; jan 13 1927-dec 27 1928; jan 3-dec 26 1929 – 1 – us CRL [071]

Drych a'r columbia – Chicago IL, De Pere WI etc. 1957 nov 15-1959 jun 15, 1960 jan 15-mar 15, 1961 jan-1964 dec, 1964 nov-1973 oct, 1973 nov-1981 dec, 1982-97 – 8r – 1 – (cont by: ninnau) – mf#519407 – us WHS [071]

Drycleaning world – Midland Park NJ 1965-73 – 1,5 – ISSN: 0012-6829 – mf#1674 – us UMI ProQuest [660]

Dryden, D A *see* They rise

Dryden, John *see* Amphitryon

Drygalski, E von *see* Deutsche suedpolar-expedition 1901-1903...

Drying up of the euphrates : and the kings of the east / Jukes, Andrew John – London, England. 1845 – 1r – us UF Libraries [240]

Drysdale, Alexander Hutton *see*
- Early bible songs
- History of the presbyterians in england

Drysdale and co's canadian farmer's almanac for the year of our lord... – Montreal: W Drysdale, 188-?-18– or 19– – 9 – (ceased 18–?) – mf#A00118 – cn Canadiana [630]

Ds. willem hendrik frieling : levensschets / Beets, Henry – [S.l.: s.n., 1903?] – 1mf – 9 – 0-524-06077-0 – mf#1991-2390 – us ATLA [240]

Die dschinn, teufel und engel im koran / Eichler, P A – Leipzig, 1928 – 3mf – 8 – €7.00 – ne Slangenburg [260]

Dsh abstracts – Rockville MD 1960-85 – 1,5,9 – ISSN: 0011-5150 – mf#5892 – us UMI ProQuest [020]

Dsn retailing today – New York NY 2000+ – 1,5,9 – (cont: discount store news) – mf#8260.02 – us UMI ProQuest [650]

Dsuq'wub'siatsub = The suquamish news – Suquamish WA. [1979 jun/jul-1996 may] scattered iss – 3r – 1 – mf#941841 – us WHS [071]

Dttp : documents to the people – Chicago IL 1983+ – 1,5,9 – ISSN: 0270-5095 – mf#14137,01 – us UMI ProQuest [020]

Du Bartas, Guillaume de Salluste *see* Bartas: his devine weekes and workes

Du Bois, Cora Alice *see* People of alor

Du Bois, Patterson *see* The culture of justice

Du Bois, W E B *see* Souls of black folk

Du Bois, William Edward Burghardt *see*
- The philadelphia negro
- The suppression of the african slave-trade to the united states of america, 1638-1870

Du Bois-Reymond, Emil Heinrich *see*
- Goethe und kein ende
- Ueber das nationgefuehl; friedrich 2 und jean-jacques rousseau

Du Bose, Horace Mellard *see* Life of joshua soule

Du Boys, Albert *see* Catharine of aragon and the sources of the english reformation

Du brahmanisme et de ses rapports avec le judaisme et le christianisme / Laouenan, Francois – Pondichery: Impr de la mission catholique, 1884-1885 – 3mf – 9 – 0-524-08301-0 – mf#1993-4006 – us ATLA [230]

Du Calvet, Pierre *see* The case of peter du calvet, esq of montreal in the province of quebeck

Du Cange, Charles d. Fresne *see* Glossarium mediae et infimae latinatis conditum a carclo du fresne

Du caractere religieux de la royaute pharaonique / Moret, Alexandre – Paris: E Leroux 1902 [mf ed 1991] – (= ser Annales du musee guimet 15) – 1mf – 9 – 0-524-01798-0 – (incl bibl ref) – mf#1990-2646 – us ATLA [930]

Du Chaillu, Paul Belloni *see* The land of the midnight sun: summer and winter journeys through sweden, norway, lapland and northern finland. ..bi.new york: harper and brothers, 1881. 2v. illus. map. with: the germans by i.a.r. wylie. 1 reel. 1260

Du Choul, G *see* Discours sur la castrametation et discipline militaire des romains...

Du concile general et de la paix religieuse : premiere partie, la constitution de l'eglise et la periodicite des conciles generaux / Maret, Henri-Louis-Charles – Paris: Henri Plon. 2v. 1869 – 4mf – 9 – 0-8370-9080-6 – (no more publ) – mf#1986-3080 – us ATLA [240]

Du conflit irano-irakien / Toure, Ahmed Sekou – Conakry, RPRG: Impr nationale "Patrice Lumumba" [1981] – (= ser Revolution democratique africaine, No 166) – 1 – mf#Sc Micro F-11309 – Located: NYPL – us Misc Inst [956]

Du culte des dieux fetiches : ou, parallele de l'ancienne religion de l'egypte avec la religion actuelle de nigritie / Brosses, Charles de – [Paris?: s.n.], 1760 – 1mf – 9 – 0-524-08039-9 – mf#1991-0255 – us ATLA [210]

Du droit de cite a rome / Lesterpt de Beauvais, Henri – Poitiers, Oudin, 1882. 175 p. LL-4045 – 1 – us L of C Photodup [340]

Du droit internationale : discours prononce a l'universite laval, a montreal, le 22 juin 1886 / Chapleau, Joseph-Adolphe – Ottawa: "Canada", 1888 – 1mf – 9 – 0-665-02863-6 – mf#02863 – cn Canadiana [341]

Du dynamisme : considere en lui-meme et dans ses rapports avec la sainte eucharistie / Ubaghs, Gerard Casimir – Louvain: Vanlinthout, 1852 – 1mf – 9 – 0-524-00408-0 – mf#1989-3108 – us ATLA [110]

Du findest hier jeden montag all das, was dich interessiert – Karlsruhe DE, 1947 28 apr-1948 15 nov – 1r – 1 – gw Misc Inst [074]

Du golfe des syrtes au golfe du benin par le lac tchad : journal de marche de la mission tunis-tchad / Courtot, Lieutenant-Colonel – Tunis: A Guenard, 1926 – 1 – us CRL [960]

Du gouvernement arabe et de l'institution qui doit l'exercer / Richard, Charles Louis Florentin – Alger: Bastide, 1848 – 1 – us CRL [960]

Du Halde, J B *see*
- Description geographique, historique, chronologique, politique et physique de l'empire de la chine et de la tartarie chinoise...
- A description of the empire of china and chinese-tartary
- The general history of china
- Geographical and historical observations upon the map of thibet

Du iuge des controverses / Du Moulin, P – Sedan, 1630 – 8mf – 9 – mf#CA-129 – ne IDC [240]

Du Jon (Junius), F *see* Eirenicum de pace ecclesiae catholicae

Du Maurier, George Louis Palmella Busson *see*
- English society at home
- Social pictorial satire

Du mein vaterland : worte und gedichte / Arndt, Ernst Moritz – Leipzig: Reclam, 1944 [mf ed 1988] – (= ser Reclams reihenbaendchen 41) – 19p – 1 – mf#6954 – us Wisconsin U Libr [810]

"Du, mein volk" : bekenntnis und zwiesprache / Kremer, Hannes – [4.aufl] Muenchen: F Eher, 1943 [mf ed 1992] – 109p – 1 – mf#7524 – us Wisconsin U Libr [840]

Du mode de filiation des racines semitiques et de l'inversion / Cazet, Cl – Paris: Maisonneuve, 1882 – 1mf – 9 – 0-8370-8493-8 – mf#1986-2493 – us ATLA [470]

Du Mont, Henry *see* Airs a quatre parties

Du Moulin, Gabriel *see*
- Les conquestes et les trophees des normand-francois aux royaumes de naples et de sicile
- Histoire generale de normandie

Du Moulin, P *see*
- Accomplissement des propheties...
- Du iuge des controverses

Du mouvement de la population catholique dans l'amerique anglaise la revue francaise / Rameau, Edme – [Paris?: s.n.] 1890 [mf ed 1983] – 1mf – 9 – 0-665-41336-X – mf#41336 – cn Canadiana [241]

Du niger au golfe de guinee par le pays de kong et le mossi, 1887-1889 / Binger, Louis Gustave – Paris, 1892. 2v – 22mf – 9 – mf#A-278 – ne IDC [916]

Du niger au golfe du guinee par le pays de kong et le mossi / Binger, Louis Gustave – Paris, 1892 – 1 – us CRL [916]

Du Perron, J D *see*
- Replique a la response du serenissime roy de la grande bretagne
- Traitte du sainct sacrement de l'eucharistie...

Du Plessis, Izak David *see*
- Die bydrae van die kaapse maleier tot die afrikaanse volkslied
- Cape malays
- Kaapse moppies
- Malay quarter and its people
- Maleise liederskat
- Die maleise samelewing aan die kaap
- Tales from the malay quarter

Du Plessis, Johannes Christiaan *see*
- Economic fluctuations in south africa, 1910-1949
- Evangelisation of pagan africa
- History of christian missions in south africa

Du pole et tropiques – Paris: OMI, 1942 – 2r – 1 – (vols for 1942 filmed with: petites annales des missionaires oblats de marie immaculee (1926), 1940-jan/feb 1942; and: documentation des oblats de marie immaculee, 1943-1944, n11; and: petites annales des missionaires oblats de marie immaculee (1926) (cont), dec 1944-jan 1953; and: petites annales, pole et tropiques, mar 1953-jul 1957) – us CRL [240]

Du pont agricultural news letter – Wilmington DE 1934-67 – 1 – mf#56 – us UMI ProQuest [630]

Du Pont, Samuel Francis *see* The samuel francis du pont papers

Du Preez, Andries Bernardus *see* Inside the south african crucible

Du Prel, Carl *see*
- Immanuel kants vorlesungen ueber psychologie
- The philosophy of mysticism

Du premier concile du latran a l'avenement d'innocent 3 (1123-1198) (he9) – Paris, 1946 – (= ser Histoire de l'eglise (HE)) – €29.00 – ne Slangenburg [241]

Du premier esprit de l'ordre de cisteaux / Paris, Julian – Paris, 1664 – 23mf – 8 – €44.00 – ne Slangenburg [241]

Du pretendu polytheisme des hebreux : essai critique sur la religion du peuple d'israel suivi, d'un examen de l'authenticite des ecrits prophetiques / Vernes, Maurice – Paris: Ernest Leroux, 1891 – 1mf – 9 – 0-7905-2159-8 – mf#1987-2159 – us ATLA [939]

Du protestantisme au catholicisme : john-henry newman, 1801-1845 / Gout, Raoul – Geneve: J.-H. Jeheber, 1906 – 1mf – us ATLA [242]

Du protestantisme au catholicisme : john-henry newman, 1801-1845 / Gout, Raoul – Geneve : J.-H. Jeheber, 1906 – 1mf – 9 – 0-7905-6594-3 – (incl bibl ref) – mf#1988-2594 – us ATLA [240]

Du Puigaudeau, Odette *see* La piste: maroc-senegal

Du recrutement des gens de mer; etude historique et critique / Zoete, Robert – Bordeaux, Impr. de l'Universite, 1919. 160 p. LL-4007 – 1 – us L of C Photodup [340]

Du renouveau catholique et des dispositions que les protestants doivent avoir devant lui / Sabatier, Paul – Saint-Blaise: Foyer Solidariste, 1908 – 1mf – 9 – 0-524-00313-0 – mf#1989-3013 – us ATLA [241]

Du reve a la realite / Jaloux, Edmond – Paris: Editions R-A Correa, 1932 – 1r – 1 – us Wisconsin U Libr [430]

Du rhin au nil : tyrol – hongrie – provinces danubiennes – syrie – palestine – ..gypte / Marmier, X – Paris, [1846]. 2v – 11mf – 9 – mf#HT-282 – ne IDC [910]

Du Rieu, Willem Nikolaas *see* De portretten en het testament van josephus justus scaliger

Du sol a l'arbre : transformations de la matiere / Anguenot, Joelle; ed by Dubois, Jean-Louis – [mf ed 2002] – 30mf – 9 – €13.50 – mf#250b0174 – fr CRDP [574]

Du sucre de betteraves et de sa production economique dans la province de quebec / Barnard, Edouard-Andre – Quebec?: s.n, 1877? [mf ed 1984] – 1mf – 9 – 0-665-02437-1 – mf#02437 – cn Canadiana [338]

Du tchad au dahomey en ballon. : voyage aerien au long cours / Deburaux, Edouard – nouvelle ed. Paris: Hachette, 1903 – 1 – us CRL [916]

Du the : ou nouveau traite sur sa culture, sa recolte, sa preparation et ses usages / Marquis, F – Paris 1820 [mf ed Hildesheim 1995-98] – 1v on 1mf [ill] – 9 – €40.00 – 3-487-27576-7 – gw Olms [630]

Du theatre italien et de son influence sur le gout musical francois / Ortique, Joseph Louis d' – Paris: Au Depot central des meilleures productions de la presse 1840 [mf ed 19–] – 5mf – 9 – (publ in 1839 under title: de l'ecole italienne et de l'administration de l'academie royale de musique a l'occasion de l'opera de m berlioz [benvenuto cellini]; "fragments de la satyre de benedetto marcello, intitulee il teatro alla moda" [p[321]-347]) – mf#fiche 1028 – us Sibley [790]

Du Toit, Jacoba Johanna *see* Riglyne vir die psigoterapeutiese hantering van die vigslyer en sy gesin

Du tres-sainct et tres-auguste sacrement, et sacrifice de la messe / Coton, P – Avignon, 1600 – 7mf – 9 – mf#CA-124 – ne IDC [240]

Du Tu Le *see* Voi nhau mot ngay nao

Du und deutschland / Schwarz, Hans – Breslau: W G Korn, c1933 – 1r – 1 – us Wisconsin U Libr [810]

Du verbe incarne (agnus dei) / Boulgakof, S – Paris, 1943 – 7mf – 8 – €15.00 – ne Slangenburg [240]

Du vray usage de la croix... / Farel, Guillaume – [Geneve], Rivery, 1560 – 4mf – 9 – mf#PFA-161 – ne IDC [240]

Du vray usage de la croix de jesus christ / Farel, Guillaume – Geneve: Jules-Guillaume Fick, 1865 – 1mf – 9 – 0-524-05144-5 – mf#1990-1400 – us ATLA [240]

Du wunderliches kind – bettine und goethe : aus dem briefwechsel zwischen goethe und bettine von arnim / Kantorowicz, Alfred [comp] – Schwerin [Germany]: Petermänken-Verlag, 1955 [mf ed 1993] – 210p – 1 – (int by comp) – mf#8464 – us Wisconsin U Libr [860]

Duae orationes de ssae theologiae...praestantia et certitudine / Polyander, J – Lugduni Batavorum, 1614 – 1mf – 9 – mf#PBA-302 – ne IDC [240]

A dual concordance to leibniz's philosophische schriften, teil 2 : konkordanz des vollstaendigen vokabulars vom typ key-word-in-context / Finster, Reinhard et al – [mf ed Hildesheim 1988] – 65mf – 9 – €368.00 – 3-487-09150-X – gw Olms [140]

Dual language and federal government : a speech delivered in the house of commons, on february 12th, 1890 / Davin, Nicholas Flood – S.l: s.n, 1890 – 1mf – 9 – mf#03648 – cn Canadiana [306]

Dual language in canada : its advantages and disadvantages: a lecture delivered before the professors and students of the university of new brunswick, fredericton, march 18, 1896 / Doucet, Stanislas Joseph – [Saint John, NB?: s.n.], 1896 [mf ed 1980] – 1mf – 9 – 0-665-03937-9 – mf#03937 – cn Canadiana [306]
Dual-city tribune – Clintonville, New London WI. 1888 aug 18-dec, 1889-1891 mar 13 – 2r – 1 – (cont: clintonville tribune [clintonville wi: 1885]; cont by: clintonville tribune [clintonville, wis: 1891]) – mf#1009205 – us WHS [071]
Der dualismus ludwig tiecks als dramatiker und dramaturg / Kaiser, Oscar – Leipzig: Sturm & Koppe 1885 [mf ed 1991] – 1r – 1 – (incl bibl ref. filmed with: der junge tieck und seine marchenkomoedien / kathe brodnitz) – mf#2913p – us Wisconsin U Libr [430]
The dualistic conception of nature / Murray, John Clark – Montreal: [s.n.], 1896 – (= ser Mcgill university. papers from the department of philosophy 1) – 1mf – 9 – 0-665-89780-4 – (incl bibl ref) – mf#89780 – cn Canadiana [110]
Die dualistiese arbeidsmarkteorie / Uys, Marthina Dorathea – Uni of South Africa 2000 [mf ed Johannesburg 2000] – 4mf – 9 – (text in afrikaans) – mf#mfm14923 – sa Unisa [331]
Duane, William *see* A hand book for riflemen
An duang mag-amigo na si d alejandre asin d luis sa reinong aragon asin moscobia – [s l: s n 191-?] [mf ed Bloomington IN: Indiana Uni Lib, Preservation Dept 1984] – (= ser Coll...in the tagalog language 2) – 3v on 1r – 1 – us Indiana Preservation [490]
Duarte / Henriquez Y Carvajal, Federico – Ciudad Trujillo, Dominican Republic. 1945 – 1r – us UF Libraries [972]
Duarte (bosquejo historico) / Despradel I Batista, Guido – La Vega, Dominican Republic. 1937 – 1r – us UF Libraries [972]
Duarte, Candido *see* Organizacao municipal no governo getulio vargas
[Duarte-] duarte dispatch – CA. 1950-1956 – 5r – 1 – $300.00 – mf#H04006 – us Library Micro [071]
[Duarte-] duartean – CA. 1963-1973 – 11r – 1 – $660.00 – mf#H04007 – us Library Micro [071]
[Duarte-] duarte-bradbury journal – CA. 1966-1967 – 2r – 1 – $120.00 – mf#H04005 – us Library Micro [073]
Duarte, Fausto *see* Aua
Duarte Filho, Joao *see* Sertao e o centro
Duarte Insua, Lino *see*
– Las alcabolas de alburquerque o los celebres baldios
– Antiguedades extremenas
– Una decada de progreso en badajoz
– Las devociones de mi pueblo, las santas reliquias, el santuario...alburquerque
– Historia de alburquerque
– La propiedad en alburquerque
– Valencia del rey
Duarte Level, Line *see* Historia patria
Duarte Level, Lino *see* Cuadros de la historia militar y civil de venezuela
Duarte, Manoel *see* Provincia e nacao
Duarte, Nestor *see* Ordem privada e a organizacao politica nacional
Duarte, Paulo *see* Que e que ha?
Duarte, Teofilo *see* O rei de timor
Duas inconfidencias / Oliveira, Almir De – Juiz de Fora, Brazil. 1970 – 1r – us UF Libraries [972]
Dub catcher : the soul voice of jamaican music – 1992 jul, 1993 jul/aug, win, dec/1994 jan, jul/aug – 1r – 1 – mf#2901391 – us WHS [780]
Dubbio di don antonio eximeno sopra il saggio fondamentale pratico di contrappunto del reverendissimo padre maestro giambattista martini / Eximeno y Pujades, Antonio – Roma: Michelangelo Barbiellini 1775 [mf ed 19–] – 3mf – 9 – mf#fiche 454 – us Sibley [780]
Dubbo dispatch – Dubbo, jan 1969-dec 1970 – 2r – at Pascoe [079]
Dubbo dispatch – Jan-dec 1942 – 1r – 9 – A$38.28 vesicular A$43.78 silver – at Pascoe [079]
Dubbo liberal – Dubbo. jan 3 1928-dec 30 1948; jan 2 1951-dec 31 1954; jan 4 1960-dec 30 1960 – 40r – 9 – A$2319.50 vesicular A$2539.50 silver – at Pascoe [079]
Dubbs, Joseph Henry *see*
– Historic manual of the reformed church in the united states
– Leaders of the reformation
– The reformed church in pennsylvania
Dube, B J *see* Inkinga yomendo
Dube, Charles *see* Constitution haitienne de 1889 et sa revision
Dube, J L *see* Jeqe, the bodyservant of king tshaka
Dube, Joseph-Edmond *see* La situation hospitaliere a montreal
Dube, Marcel *see* Le temps des lilas
Dube, Shumirai *see* An investigation of the current shona orthography
Dube, Shyama Charan *see*
– Field songs of chhattisgarh
– Indian village
Dubeau, Jean *see* Bio-bibliographie analytique de reine malouin
Duberstein, Murray W *see* Outline of the law of wills, based on new york cases and statutes
Lo dubim ve-lo ya'ar / Dobsevage, Abraham Baer – Berdichev, Ukraine. 1890 – 1r – us UF Libraries [939]
Dubin, M H *see* The jewish community news
Dublin Abbey of St Thomas the Martyr *see* Register of the abbey of st thomas the martyr, dublin (rs94)
Dublin advertising gazette – Dublin, aug 1858-sep 1859 – 0.5r – 1 – ie National [072]
Dublin advertising gazette – Dublin, Ireland. 21 apr 1858-5 aug 1871; 14 oct 1871-31 mar 1877 – 10r – 1 – uk British Libr Newspaper [072]
Dublin advertising gazette *see* Commercial journal and family herald
Dublin and london magazine – La Salle IL 1825-28 – 1 – mf#4239 – us UMI ProQuest [072]
Dublin argus : or trades' gazette – Dublin, Ireland 17 jan-29 aug 1846 – 1 – uk British Libr Newspaper [072]
Dublin bill of entry and shipping list *see* Customs dublin bill of entry and shipping list
Dublin bill of entry & shipping list – Dublin, Ireland 1 apr 1909-30 jun 1926 [mf jul-dec 1925] – 1 – (wanting: 1 apr-8 may,1-7,9-31 aug 1916; cont: customs, dublin, bill of entry & shipping list [1 jan-30 mar,30 apr-30 may,30 aug 1853, 1 jan 1892-31 mar 1909]) – uk British Libr Newspaper [380]
Dublin builder – Ireland. Irish Builder. -m, -w. Jan 1859-Dec 1900 – 18r – 1 – uk British Libr Newspaper [072]
Dublin chronicle – Dublin 1815-17 [incomplete] – 1r – 1 – ie National [072]
Dublin chronicle – Dublin, Ireland. may 1787-dec 1893 (imperfect) – 1 – uk British Libr Newspaper [072]
Dublin chronicle – Ireland. -w. 1 Jan 1816-20 Apr 1817. (1 reel) – 1 – uk British Libr Newspaper [072]
Dublin chronicle *see* South dublin chronicle / dublin chronicle
Dublin chronicle [hunter's] – Dublin 1770-71 – 1r – 1 – ie National [072]
Dublin chronicle [sleater/byrne] – Dublin, may 1787-1788, 1789-dec 1793 – 8r – 1 – ie National [072]
Dublin correspondent *see* Correspondent
Dublin correspondent / evening packet – Dublin 1823-27, 1829 – 6r – 1 – ie National [072]
Dublin courant – Ireland.24 Apr 1744-30 Dec 1746. -d. 1 reel – 1 – uk British Libr Newspaper [072]
Dublin courant – Dublin, Ireland. 24 apr 1744-24 mar 1750 – 2 1/2r – 1 – (publ 24 apr 1744-24 mar 1750 only) – uk British Libr Newspaper [072]
Dublin courier – Dublin, 27 oct, 17 nov, 8 dec 1934 – 0.25r – 1 – ie National [072]
Dublin courier – Dublin, Ireland. 4 jan 1760-1764; 30 dec 1765-1766 – 3r – 1 – uk British Libr Newspaper [072]
Dublin daily advertiser – Ireland. -d. 14 Oct-2 Dec 1736. 1/4 reel – 1 – uk British Libr Newspaper [072]
Dublin evening express – Dublin, jan-apr 1824 – 0.5r – 1 – ie National [072]
Dublin evening herald – Dublin, Ireland. 3 nov 1846-26 mar 1853 – 6 1/2r – 1 – uk British Libr Newspaper [072]
Dublin evening herald – Ireland. -d. 30 Jan 1821-29 Jan 1822. (1 reel) – 1 – uk British Libr Newspaper [072]
Dublin evening journal – Dublin, feb-jul 1778 – 0.5r – 1 – ie National [072]
Dublin evening mail – Dublin, Ireland. 1824; 1826-28; 1831; 1833; 1838; 1840-1907; sep-dec 1911; jan-apr 1914; jan-apr 1916; jul-dec 1920; jan-16 apr 1926; sep-dec 1926; 26 may-30 aug 1930; 1952 – 149r – 1 – (aka: evening mail) – uk British Libr Newspaper [072]
Dublin evening mail – Dublin 1823, 1829, 1830, 1832, 1834, 1836, 1837, 1839, 1881-84, 1886-93 – 32r – 1 – ie National [072]
Dublin evening mail – Ireland. -d. 1897-1907. 33 reels – 1 – uk British Libr Newspaper [072]
Dublin evening packet – Dublin, Ireland. 27 nov 1770; 25-27 jun 1771 – 1/4r – 1 – uk British Libr Newspaper [072]
Dublin evening post – Dublin, Ireland. 5 jul 1737-11 jul 1741 – 2r – 1 – uk British Libr Newspaper [072]
Dublin evening post – Dublin. Ireland. -d. 1780; 1829; 1860-65; 1870. (10 reels) – 1 – uk British Libr Newspaper [072]
Dublin evening post – Dublin, jul 1778-1805, 1811-14, 1816, 1824, 1859 – 36.5r – 1 – ie National [072]
Dublin evening post – Dublin, Ireland. Aug 1778-1781; 1783-85; 1787; 1789-90; 1792; 1794-97; 1804-10; 5 jan-16 feb, 22 apr, 29 may, 24 jun 1813; 1814; 1815; 2 jan-28 apr, 4 jun-31 oct 1816; 1817-58; 1860-21 aug 1875 (1859 missing) – 94r – 1 – (publ aug 1778-aug 1875 only) – uk British Libr Newspaper [072]
Dublin evening post *see* Independent irishman
Dublin evening press – Dublin, Ireland. 13 jul 1811; 22 mar 1824 – 1/2r – 1 – uk British Libr Newspaper [072]
Dublin evening standard – Dublin, Ireland. 10 jan-23 may 1870 – 1r – 1 – (incorp with: dublin evening mail) – uk British Libr Newspaper [072]
Dublin. Exhibition of Art and Art-industry, 1853 *see*
– The art-journal
– Record of the great industrial exhibition 1853
Dublin figaro *see* Irish life
Dublin hospital gazette – Dublin, Ireland. 15 feb-15 dec 1845; 15 jan-15 apr 1846; 15 jul-15 dec 1854; 15 jul-15 dec 1855; 1 jan-1 mar, 1 oct-22 dec 1856; 1 jan-15 dec 1857; 1858-61; 1 feb-1 mar 1862 – 4 1/4r – 1 – (wanting mar-sep 1856) – uk British Libr Newspaper [072]
Dublin intelligence – Ireland. -sw. 10, 31 Aug, 11 Sept 1708; 4 Jan 1709-30 Dec 1712. Imperfect. 1 reel – 1 – uk British Libr Newspaper [072]
Dublin intelligence (supplements) – Dublin, Ireland. 21 mar 1722-21 may 1724 – 1/4r – 1 – uk British Libr Newspaper [072]
Dublin intelligencer – Ireland. 17 June 1756. 1 ft – 1 – uk British Libr Newspaper [072]
Dublin. International Exhibition of Arts and Manufactures, 1865 *see* Official catalogue
Dublin journal – Dublin, Ireland. 1748-49; 1754-68; 17 oct 1782-12 dec 1799; 4 aug 1803; 27 oct 1804-1 feb 1816; 1820-4 oct 1824 – 14 1/2r – 1 – uk British Libr Newspaper [072]
Dublin journal *see* Parnellite
Dublin journal [t t faulkner] – Dublin 1783-86 – 4r – 1 – ie National [072]
Dublin latern – Dublin, Ireland. 17 aug 1895-26 dec 1896; 1901 – 2r – 1 – (aka: rathmines news and dublin latern. wanting: 11 aug, 22 sep-27 oct 1906) – uk British Libr Newspaper [072]
Dublin literary journal and select family visitor – Dublin, Ireland. jan-jun 1844; mar 1845-feb 1846 – 1/2r – 1 – uk British Libr Newspaper [072]
Dublin local advertiser – Ireland.Sept 1858-Mar 1861. -w. 1 reel – 1 – uk British Libr Newspaper [072]
Dublin medical press – Dublin, Ireland. 1846-48; 1850-68 – 29r – 1 – (aka: medical press; medical press and circular) – uk British Libr Newspaper [072]
Dublin mercantile advertiser – Ireland.1851-Jan 1865.-w. 5 reels – 1 – uk British Libr Newspaper [072]
Dublin mercantile advertiser and weekly price current – Dublin, Ireland. 1823-42; 1844-2 jan 1865 – 15r – 1 – uk British Libr Newspaper [072]
Dublin mercury – (Hoey's Dublin Mercury). Ireland. -sw. 18 Mar 1766-1 Apr 1773. (5 reels) – 1 – uk British Libr Newspaper [072]
Dublin mercury – Ireland. -sw. 23 Jan-21 Sep 1742. Imperfect. (1/4 reel) – 1 – uk British Libr Newspaper [072]
Dublin mercury – Dublin, Ireland. 18 mar 1766-1 apr 1773 – 4r – 1 – (publ 18 mar 1766-1 apr 1773. aka: hoeys dublin mercury) – uk British Libr Newspaper [072]
Dublin monitor – Ireland. -w. 6 Nov 1838-11 Jul 1845. (6 1/2 reels) – 1 – uk British Libr Newspaper [072]
Dublin morning post – Dublin 1784-85 – 1r – 1 – ie National [072]
Dublin morning post and daily advertiser – 16 feb 1824; 11 may 1825; 1830-5 may 1832 – 6r – 1 – uk British Libr Newspaper [072]
Dublin morning press – Ireland. -d. 9 Feb-2 Apr 1842. (1/4 reel) – 1 – uk British Libr Newspaper [072]
Dublin news – Dublin, Ireland. 22 mar-1 sep 1858 – 1/2r – 1 – uk British Libr Newspaper [072]
Dublin news letter – Dublin, Ireland. 31 jan 1743-7 apr 1744 (imperfect) – 1/4r – 1 – uk British Libr Newspaper [072]
Dublin newsletter or dublin gazette – Ireland.31 Jan 1743-7 Apr 1744. -irr.1 reel – 1 – uk British Libr Newspaper [072]
Dublin observer *see* Sunday observer
Dublin penny journal – Dublin, jun 1832-jun 1836 – 2r – 1 – ie National [072]
Dublin post boy – Dublin, Ireland. 25, 31 jul; 8, 15, 22, 29 aug 1734 – 1/4r – 1 – uk British Libr Newspaper [072]
Dublin postal guide – Dublin, Ireland. 1894-96 – 1/4r – 1 – (aka: post office guide for dublin and district) – uk British Libr Newspaper [072]
Dublin post-man – Ireland. -sw. 14 Dec 1724, 4, 11, 15 Mar, 12 Apr, 13 May, 14 Jun, 29 Jul, 27 Aug, 2, 23 Sep, 10 Nov, 1, 2, 16, 20 Dec 1725. (1/4 reel) – 1 – uk British Libr Newspaper [072]
Dublin record – Dublin, Ireland. 13 feb 1835-oct 1846 – 12r – 1 – (aka: statesman; statesman and dublin christian record) – uk British Libr Newspaper [072]
Dublin record *see* Statesman
Dublin review – London, England 1836-1969 – 1 – mf#493 – us UMI ProQuest [072]
Dublin saturday magazine – La Salle IL 1865-67 – 1 – mf#4188 – us UMI ProQuest [071]
Dublin saturday post – Dublin, Ireland. 3 jan-1 may 1920 – 1r – 1 – uk British Libr Newspaper [072]
Dublin saturday post – Dublin, jun 1910-may 1920 – 9r – 1 – ie National [072]
Dublin shipping and mercantile gazette – Dublin, Ireland. 13 jul 1869-22 aug 1871; 12 oct 1871-24 feb 1872 – 1 1/4r – 1 – (incorp with: commercial journal) – uk British Libr Newspaper [380]
Dublin sporting news – Ireland. 5 feb 1889-17 oct 1891; 17 mar-27 sep 1892; 18 apr-29 dec 1893. -w 4r – 1 – uk British Libr Newspaper [072]
Dublin. St Mary's Abbey *see* Chartularies of st mary's abbey, dublin (rs80)
Dublin standard – Ireland. -sw. 4 Oct 1836-4 Aug 1837. (1 reel) – 1 – uk British Libr Newspaper [072]
[Dublin-] the news – CA. 1981-82 – 38r – 1 – $2280.00 – (formerly: tri-valley news) – mf#B02181 – us Library Micro [071]
Dublin times – Ireland.5 Feb-8 May 1845; 22 Mar-1 Sept 1858. -w – 1 1/2r – 1 – uk British Libr Newspaper [072]
Dublin times – Dublin, Ireland. 16 mar 1831-8 oct 1833 – 5 1/2r – 1 – (publ 16 mar 1831-8 oct 1833. aka: dublin times and the dublin morning post) – uk British Libr Newspaper [072]
Dublin times – Dublin, Ireland. feb-8 may 1845 – 1r – 1 – (publ feb-8 may 1845) – uk British Libr Newspaper [072]
Dublin times and the dublin morning post *see* Dublin times
Dublin trades council papers, 1893-1951 – (= ser Labour party in britain, origins and development at local level. series 1) – 5r – 1 – (int by seamus cody) – mf#97303 – uk Microform Academic [331]
[Dublin-] tri-valley herald – CA. 1974- – 273r – 1 – $16,380.00 – mf#B02180 – us Library Micro [071]
[Dublin-] tri-valley news – CA. Dec 1974-Mar 1981 – 88r – 1 – $5280.00 – (cont: the news) – mf#B02182 – us Library Micro [071]
[Dublin-] valley times – CA. Aug 1971-Feb 1978 – 79r – 1 – $4740.00 – (cont by: times, livermore) – mf#B02183 – us Library Micro [071]
Dublin verses by members of trinity college / ed by Hinkson, Henry Albert – London, England. 1895 – 1r – us UF Libraries [810]
Dublin weekly herald – Ireland. -w. 10 nov 1838-2 apr 1842; 14 may-28 may 1842 1 1/2r – 1 – uk British Libr Newspaper [072]
Dublin weekly messenger – Dublin 1808-12 – 2r – 1 – ie National [072]
Dublin weekly news *see* Weekly news
Dublin weekly post – Dublin, sep 1959-nov 1960 – 1r – 1 – ie National [072]
Dublin weekly programme of events – Dublin, Ireland. oct 1893-2 apr 1896 – 4 1/2r – 1 – (aka: legal and commercial advertiser and weekly programme of events) – uk British Libr Newspaper [072]
Dublin weekly register – Dublin 1818-44 [vol incomplete] – 1r – 1 – ie National [072]
Dublin weekly register – Dublin, Ireland. Oct 1818-23; 1827-35; 1837-14 sep 1850 – 27 1/2r – 1 – uk British Libr Newspaper [072]
Dubnova-Erlikh, Sofiia *see*
– Garber-bund un bershter-bund
– Lebn un shafn fun shimen dubnov
Dubnow, Simon *see* Fun "zshargon" tsu yidish
Dubois *see* 1re [-12me] feuille [d'allemandes]
Dubois, Augustus Jay *see* Science and the spiritual
Dubois de Montpereux, F *see* Voyage autour du caucase, chez les tcherkesses et les abkhases, en colchide, en georgie, en armenie et en crimee...
Dubois, Emile *see* Chez nos freres les acadiens
Dubois, Emile [comp] *see*
– Cantiques et prieres
– La priere chantee
Dubois, F-E *see* Deux ans et demi de ministere
Dubois, Felix *see*
– Notre beau niger
– Timbuctoo the mysterious
Dubois, Henri M *see* Le des betsileo (madagascar)
Dubois item – DuBois, NE: O M Backus, 1891-times ser. v10 n22. aug 14 1896=item ser. v6 n17 (wkly) [mf ed 1892-96 (gaps) filmed [1973]] – 2r – 1 – (cont: dubois times. cont by: dubois times) – us NE Hist [071]

DUENTZER

Dubois, J B (Jean Baptiste) *see* Marton et frontin, ou, assaut de valets
Dubois, Jean A *see* Description of the character, manners, and customs of the people of india
Dubois, Jean Antoine *see*
– Hindu manners, customs, and ceremonies
– Hindu manners, customs and ceremonies
– Letters on the state of christianity in india
Dubois, Jean-Louis *see* Du sol a l'arbre
DuBois, Louis *see* Resume de l'histoire de normandie
The dubois paper – DuBois, NE: H J McCoy. 2v. v1 n1. mar 13 1941-v2 n13. jun 4 1942 (wkly) [mf ed filmed [1974?] – 1r – 1 – (absorbed by: pawnee chief) – us NE Hist [071]
Dubois, Philip *see* Administrative structure in large district courts
Dubois press – DuBois, NE: A J Kirkpatrick, 1904 (wkly) [mf ed 1912-38 (gaps)] – 9r – 1 – (cont by: du bois weekly press) – us NE Hist [071]
Dubois times – DuBois, NE: F N Merwin. v10 n24. aug 28 1896-1900// (wkly) [mf ed with gaps] – 1r – 1 – (cont: dubois item) – us NE Hist [071]
Dubois weekly press – DuBois, NE: L D Stanek (wkly) [mf ed 1939 (gaps)] – 1r – 1 – (cont: dubois press) – us NE Hist [071]
Dubois-Fontanelle, Joseph *see* Naufrage et aventures de m pierre viaud, natif de bordeaux, capitaine de navire
[Dubois-Goibaud, P] *see* Conformite de la conduite de l'eglise de france...avec celle de l'eglise d'affrique...
Duboius, L L *see* Lettres aux missionaires
DuBose, Hampden C *see*
– The dragon, image, and demon
– Memoirs of rev. john leighthon wilson
– Preaching in sinim
Dubose heyward / Durham, Frank – 1954 [mf ed Spartanburg SC: Reprint Co, 1981] – 4mf – 9 – mf#51-047 – us South Carolina Historical [420]
DuBose, William Porcher *see* High priesthood and sacrifice
Dubose, William Porcher *see*
– The ecumenical councils
– The gospel in the gospels
– The reason of life
– Turning points in my life
Dubowy, Ernst *see* Klemens von rom ueber die reise pauli nach spanien
Dubravius, Janus *see* De piscinis...
Dubreuil, Guy *see* La famille martiniquaise
Dubreuil, J P *see* Histoire des francs-macons
Dubreuil, Joseph Fereol *see* Index to the criminal & penal statutes of canada, as affecting the province of quebec.
Dubroca *see* Vida de j j dessalines
Dubroca, Louis *see* L'itineraire des francais dans la louisiane
Dubrovin, A I *see* Kuda vremenshchiki vedut soiuz russkogo naroda
Dubrovina, V F *see* Sinaiskii paterik
Dubrovskii, V A *see* Utrenniaia zvezda
Dubruel, Marc *see* En plein confict
Dubuque County Greenback Club *see* Statesman
Dubuque herald – Dubuque IA. 1885 jul 4 – 1r – 1 – (cont by: dubuque daily herald) – mf#846347 – us WHS [071]
Dubuque leader – 1919/20-1989 jul 3/1991 dec 27 – 25r – 1 – (with gaps; cont: labor leader [dubuque ia]) – mf#802009 – us WHS [071]
Dubuque semi-weekly times – Dubuque IA. 1865 apr 21 – 1r – 1 – mf#851138 – us WHS [071]
Le duc de reichstadt / Montbel, Guillaume I de – Paris 1832 [mf ed Hildesheim 1995-98] – 1v on 3mf – 9 – €90.00 – ISBN-10: 3-487-26347-5 – ISBN-13: 978-3-487-26347-2 – gw Olms [940]
Duc job / Laya, Leon – Paris, France. 1862 – 1r – us UF Libraries [440]
Duc Sidym *see* Bralyn puspa
Ducae, Michaelis Ducae Nepotis *see* Historia byzantina (cshb21)
Ducae Michaelis Nepotis *see* Historia byzantina (cbh15)
DuCange, Charles DuFresne *see* Histoire de l'empire de constantinople sous les empereurs francais jusqu'a la conquete des turcs
Ducange, Victor *see*
– Lisbeth
– Il y a seize ans
Ducaruik a prac bhe kan lvat re ta ra to / Sundara, A rhan, Prvan tan cha – Ran Kun: Si ri ratana Ca pe tuik, Si ri Ca pe Phran khyi re 1984 [mf ed 1990] – [ill] 1r with other items – 1 – (in burmese) – mf#mf-10289 seam reel 203/1 [§] – us CRL [280]
Ducas-hippolyte / Marcelin, Frederic – Havre, France. 1878 – 1r – us UF Libraries [972]
Ducasse, Angel Braulio *see* Estrindencias poesias
Ducasse, Raymond *see* Mahomet dans son temps
Duch casu – Chicago IL. 1892 oct 9/1893 dec 31-1936 jan 5/1939 jul 23 – 16r – 1 – (with small gaps) – mf#772874 – us WHS [071]
Duchatelard, Auguste *see*
– Eustache
– Vieux de la vieille
Duchatellier, Armand *see*
– Excursion dans l'amerique du sud
– La mort de louis 16
Duchaussois : rose du canada... / Bayle, Constantino – Paris, 1932; Madrid: Razon y Fe, 1933 – 1 – sp Bibl Santa Ana [999]
Duchaussois, Pierre *see* Femmes heroiques!
Duchemin Descepeaux, Jacques *see* Lettres sur l'origine de la chouannerie et sur les chouans du bas-maine
Duchene, John D *see* L'elevage du cheval en canada
Duchenet, Edouard *see* Histoires somalies
Duchesne, Andre *see* Historiae francorum scriptores coaetanei
Duchesne, J *see* L'expedition de madagascar
Duchesne, Louis *see*
– The beginnings of the temporal sovereignty of the popes, a d 754-1073
– The churches separated from rome
– De codicibus mss. graecis pii 2
– De macario magnete et scriptis ejus
– Early history of the christian church from its foundation to the end of the fifth century
– Early history of the christian church from its foundation to the end of the third century
– Eglises separees
– Fastes episcopaux de l'ancienne gaule, tome 1
– Fastes episcopaux de l'ancienne gaule, tome 2
– Fastes episcopaux de l'ancienne gaule, tome 3
– Liber pontificalis
– Le liber pontificalis, vol 3
– Le liber pontificalis, vol 1-2
Duchesne, Louis et al *see* Etude sur le liber pontificalis – recherches sur les manuscrits archeologiques de jacques grimaldi: archiviste de la basilique de la vaticane au seizieme siecle – etude sur le mystere de sainte agnes
Duchesne-Fournet, J *see* Mission en ethiopie (1901-1903)
The duchess : a story / Hungerford, Margaret Wolfe – 2nd ed. London: Hurst & Blackett, 1889 [i.e. 1888] – (= ser 19th c women writers) – 4mf – 9 – mf#5.1.105 – uk Chadwyck [830]
The duchess of powysland : a novel / Allen, Grant – London: Chatto & Windus. 3v. 1892 – 1mf – 9 – mf#05023 – cn Canadiana [830]
The duchess of powysland, vol 1 : a novel / Allen, Grant – London: Chatto & Windus, 1892 – 3mf – 9 – (pt of cihm set) – mf#05024 – cn Canadiana [830]
The duchess of powysland, vol 2 : a novel / Allen, Grant – London: Chatto & Windus, 1892 – 4mf – 9 – (pt of cihm set) – mf#05025 – cn Canadiana [830]
The duchess of powysland, vol 3 : a novel / Allen, Grant – London: Chatto & Windus, 1892 – 4mf – 9 – (pt of cihm set) – mf#05026 – cn Canadiana [830]
Ducis, Jean-Francois *see* Hamlet
Duck book 2 / Robert White, Inc – 1982 feb-jun – 1r – 1 – (cont: robert white's duck book; cont by: duck book digest) – mf#1495177 – us WHS [071]
Duck book digest / Robert White, Inc – 1982 nov, 1983 jan-mar – 1r – 1 – (cont: duck book 2; cont by: duck book) – mf#1494909 – us WHS [071]
Duck club news – v1 n2-8 [1982 mar-sep] – 1r – 1 – (cont by: dcn educational actionpaper) – mf#1145790 – us WHS [071]
Duck club news digest – v2 n7-v6 [final] [1983 jul-1987 jan] – 1r – 1 – (cont: dcn educational actionpaper) – mf#1140842 – us WHS [071]
Duck lake : stories of the canadian backwoods / Young, Egerton Ryerson – Toronto: Musson, [19-?] – 3mf – 9 – 0-659-92180-4 – mf#9-92180 – cn Canadiana [830]
Duck power / GI's Against Fascism – v1 n3-v2:iss8 [1969 sep 24-1980 jul 10], v1 n3-v2 iss8 [1969 sep 24-1980 jul 10] – 2r – 1 – mf#721519 – us WHS [355]
Duck, Simeon *see* Budget speech delivered in the provincial legislature
Duckesz, Eduard *see* Sefer ivah le-moshav
Duckett's dispatch – (Duckett's Paper). London. -w. 4 Jan-3 May 1818. (18 ft) – 1 – uk British Libr Newspaper [072]
Duckworth, Henry Thomas Forbes *see*
– The church of cyprus
– Greek manuals of church doctrine
– Some pages of levantine history
Duclos, Charles *see* Voyage en italie
Duclos, Charles Pinot *see* Memoires secrets sur les regnes de louis 14 et de louis 15
Duclos, J *see* Democratie nouvelle
DuContant de La Molette, Philippe *see* Traite sur la poesie et la musique des hebreaux
Ducoudray, Alexandre *see* Nouveaux essais historiques sur paris
Ducoudray, J H *see* Bajo la egida del generalisimo
Ducreux, Louis *see* Part du feu
Duda, Joan L *see* Psychological antecedents of the frequency and intensity of flow in golfers
Duda, Mark Damian *see* Factors related to hunting and fishing participation in the united states
Dudas acerca de las ceremonias sanctas de la missa : resueltas por los clerigos de la congregacion de nuestra senora, fundada con autoridad apostolica en el collegio de la compania de iesus de mexico / Mexico [City]. Congregacion de nuestra senora – en Mexico: Por Henrico Martinez, ano de 1602 – (= ser Books on religion...1543/44-c1800: jesuitas) – 3mf – 9 – mf#crl-213 – ne IDC [241]
Dudas acerca de las ceremonias sanctas de la missa: resueltas por los clerigos de la congregacion de nuestra senora, fundada con autoridad apostolica en el colegio de la compania de iesus de mexico / Mexico [City]. Congregacion de Nuestra Senora – en Mexico: Por Henrico Martinez, ano de 1602 – (= ser Books on religion...1543/44-c1800: teologia "culta"; derecho canonico) – 3mf – 9 – mf#crl-266 – ne IDC [241]
Dudden, Frederick Homes *see*
– The future life
– Gregory the great
Duddon, st peter – (= ser Cheshire monumental inscriptions) – 1mf – 9 – £2.50 – mf#15 – uk CheshireFHS [929]
Dudenhausen, Wolfgang *see* Buergschaft und garantievertrag unter besonderer beruecksichtigung der bankpraxis
Dudenhofen, Petra *see* Rechtsextremismus in der bundesrepublik deutschland
Dudin, K *see* Organizatsiia snabzhencheskoi raboty v sisteme selskokhoziaistvennoi kooperatsii i voprosy proizvodstvennogo kooperirovaniia
Dudley 1725-1849 – Oxford, MA (mf ed 1996) – (= ser Massachusetts vital record transcripts to 1850) – 13mf – 9 – 0-87623-243-8 – (mf 1t-5t: vital records 1725-1812. mf 5t-9t: births & deaths 1728-1849. mf 9t: out-of-town marriages 1732-99. mf 9t-10t: births 1844-49. mf 10t: marriages, deaths 1844-49. mf 10t-12t: intentions 1796-1849. mf 12t-13t: marriages 1796-1843. mf 13t: colored intents & marriages 1821-27) – us Archive [978]
Dudley 1725-1891 – Oxford, MA (mf ed 1986) – (= ser Massachusetts vital records) – 41mf – 9 – 0-87623-010-9 – (mf 1-4: births & deaths 1725-1849. mf 5-8: vital & land records 1725-1887. mf 9-12: marriage intentions 1788-1882. mf 13-14: index to intentions 1788- 1882. mf 15-16: b,m,d 1844-54. mf 17-21: births 1854-91. mf 22-23: marriages 1854-91. mf 24-27: deaths 1854-91. mf 28-32: index to births 1844-91. mf 33-37: index to marriages 1844-91. mf 38-40: index to deaths 1844-91. mf 41: soldiers & officers 1861-65) – us Archive [978]
Dudley and district news – England. -w. 3 Jan 1880-10 Jan 1885. (5 reels) – 1 – uk British Libr Newspaper [072]
Dudley and east worcestershire gazette – England. -w. 3 Jul-25 Sep 1869. (9 ft) – 1 – uk British Libr Newspaper [072]
Dudley and midland counties express – England. -w. 19 Sep 1857-21 Aug 1858. (1 reel) – 1 – uk British Libr Newspaper [072]
Dudley chronicle – England. -w. 1 Jan 1910-4 Apr 1935. (25 reels) – 1 – uk British Libr Newspaper [072]
Dudley chronicle – England. -w. 14 Feb-11 Jul 1885. (28 ft) – 1 – uk British Libr Newspaper [072]
Dudley guardian – England. -w. 24 Jul 1865-24 Jul 1875. (4 reels) – 1 – uk British Libr Newspaper [072]
Dudley herald – England. -w. 1912-20. (Wanting 1914). (8 reels) – 1 – uk British Libr Newspaper [072]
Dudley herald and wednesbury borough news – [West Midlands] Dudley jul 1869-1950 [mf ed 2003] – 95r – 1 – (missing: 1866-68, 1872-74, 1896, 1880, 1898, 1900, 1910-11, 1913-14; cont as: dudley herald [jan 1901-dec 1950]) – uk Newsplan [072]
Dudley mercury – England. -w. 5 Feb 1887-8 Feb 1890. (2 reels) – 1 – uk British Libr Newspaper [072]
Dudley news – England. -w. 20 Aug-1 Oct 1857. (5 ft) – 1 – uk British Libr Newspaper [072]
Dudley papers, 16th century – v1-5 – (= ser Archives of the marquess of bath, longleat house, warminster, wiltshire) – 3r – 1 – mf#96700 – uk Microform Academic [920]
Dudley, Richard M et al *see* Baptist, why and why not
Dudley, Robert, Earl of Leicester *see* Dudley papers, 16th century
Dudley times – 1987 jun-aug/sep, 1988 spr-1989 spr, 1992 win, 1993 win, sum – 1r – 1 – mf#4712895 – us WHS [071]
Dudley, W M *see* Christ the author and end of civil government
Dudley weekly times – (Dudley Times). England. -w. 20 Dec 1856-18 Dec 1858. (1 reel) – 1 – uk British Libr Newspaper [072]
Dudley's cases in equity / South Carolina. Supreme Court – 1v. 1837-1838 (all publ) – (= ser Pre-nrs nominative equity reports) – 3mf – 9 – $4.50 – mf#LLMC 94-031 – us LLMC [342]
Dudley's law reports / South Carolina. Supreme Court – 1v. 1837-1838 (all publ) – (= ser Pre-nrs nominative law reports) – 9mf – 9 – $13.50 – mf#LLMC 94-019 – us LLMC [340]
Dudley's reports / Georgia. Supreme Court – 1v. 1821-1833 (all publ) – (= ser Georgia supreme court reports) – 4mf – 9 – $6.00 – (a pre-nrs title) – mf#LLMC 91-022 – us LLMC [347]
Dudok collection of architectural plans and drawings of the city of hilversum / Hilversum. City and Regional Archives Gooi- en Vechtstreek, The Netherlands – [mf ed 2001] – 10,000 plans on 20r – 1 – €5900.00 – (with p/g & concordance) – mf#m491 – ne MMF Publ [720]
Due lezioni, nella prima delle quali si dichiara un sonetto di m michelangolo buonarotti : nella seconda si disputa quala sia piu nobile arte la scultura, o la pittura... / Varchi, [B] – Fiorenza, 1549 – 2mf – 9 – mf#O-1025 – ne IDC [700]
Due novelline toscane / ed by Pitre, Giuseppe – Palermo: Giornale di Sicilia 1890 [mf ed Bloomington IN: Indiana Uni Lib, Preservation Dept 1984] – 1r – 1 – (contents: la novella di oime; le fate) – us Indiana Preservation [830]
Due observance of the lord's day / Shepherd, Richard Herne – London, England. 1819 – 1r – us UF Libraries [240]
Le due regole della prospettiva pratica / Vignola, J – Roma, 1583 – 3mf – 9 – sp Cultura [720]
Le due regole della prospettiva pratica... / Vignola, J B da – Roma, 1583 – 5mf – 9 – mf#O-1027 – ne IDC [700]
Due trattati uintoralle otto principali arti dell' oreficeria : l'altro in materia dell' arte della scultura... / Cellini, B – Fiorenza, 1568 – 2mf – 9 – mf#O-197 – ne IDC [700]
Dueerkob, Monika *see* Trauern
Duehring, Eugen Karl *see*
– Der ersatz der religion durch vollkommeneres und die abstreifung alles asiatismus
– Die judenfrage
– Logik und wissenschaftstheorie
– Die ueberschaetzung lessing's und seiner befassung mit literatur
Duehring, Hans *see* Das gymnasium marienwerder
Duel / Lavedan, Henri – Paris, France. 1906 – 1r – us UF Libraries [440]
Duel' : gazeta bor'by obshchestvennykh ideii – Moscow, Russia. n1(fev 1996)-n13(iiul 1996), n15(sen 1996)-n22(dek 1996), n1[23](iian 1997)-n 44[91](dek 1998) – 1 – mf#mf-12248 (reel 3-4) – us CRL [077]
Un duel a poudre : comedie en trois actes / Fontaine, Raphael Ernest – Montreal: s.n, 1881 – 1mf – 9 – mf#33672 – cn Canadiana [820]
Duelberg, Franz *see* Korallenkettlin
Dueling (a sermon) / Kendrick, J R – 1853 – 1 – 5.00 – us Southern Baptist [242]
Das duell wegen ems : gedanken ueber den frieden / Gutzkow, Karl – Berlin: Puttkammer & Muehlbrecht, 1870 [mf ed 2001 – 15p – 1 – mf#10526 – us Wisconsin U Libr [940]
Duelmener zeitung – Duelmen (DE), 1957-88 – 127r – 1 – (filmed by misc inst: 1988-) – gw Mikrofilm; gw Misc Inst [074]
Duelo de mi vecino / Meza Y Suarez Inclan, Ramon – Habana, Cuba. 1961 – 1r – us UF Libraries [972]
Duelos en cuba / Cervantes, Augustin – Habana, Cuba. 1894 – 1r – us UF Libraries [972]
Duemichen, J *see*
– Altaegyptische tempelinschriften in den jahren 1863-1865 an ort und stelle gesammelt
– Der grabpalast des patuamenap in der thebanischen nekropolis
Der duemmste sibiriak : erzaehlung / Brehm, Bruno – Leipzig: P Reclam 1939 [mf ed 1989] – 1r – 1 – (aft by herbert guenther; filmed with other titles) – mf#7066 – us Wisconsin U Libr [830]
Duenn wie eine eierschale : roman / Daumann, Rudolf Heinrich – Berlin: Schuetzen-Verlag, 1937 [mf ed 1989] – 363p – 1 – mf#7170 – us Wisconsin U Libr [830]
Duentzer, Heinrich *see*
– Abhandlungen zu goethes leben und werken
– Aus goethe's freundeskreise
– Aus herders nachlass
– Briefwechsel zwischen goethe und staatsrath schultz
– Charlotte von stein, goethe's freundin
– Charlotte von stein und corona schroeter
– Die drei aeltesten bearbeitungen von goethe's iphigenie
– Erlaeuterungen zu goethes werken
– Frauenbilder aus goethe's jugendzeit
– Freundesbilder aus goethes leben
– Friederike von sesenheim im lichte der wahrheit
– Goethe, karl august und ottokar lorenz

– Goethe und karl august
– Goethes faust
– Goethes goetz von berlichingen
– Goethes leben
– Goethe's lyrische gedichte
– Goethes stammbaeume
– Goethes tagebuecher der sechs ersten weimarischen jahre
– Goethes tasso
– Herders cid
– Herders legenden
– Herders reise nach italien
– Life of goethe
– Schillers lyrische gedichte
– Uhlands dramen und dramenentwuerfe
– Von und an herder
– Wielands oberon

Duenyaya ikinci gelis yahut istanbul'da neler olmus / Cevdet, Mehmed – Istanbul: Sark Matbaasi, 1921 – (= ser Ottoman literature, writers and the arts) – 2mf – 9 – $40.00 – us MEDOC [470]

Duer, John *see* A lecture on the law of representations in marine insurance.

Duerbeck, Ernst *see* Kursachsen und die durchfuehrung des prager friedens 1635

Dueren, Wilhelm *see* Ueber goethe und spengler

Duerener zeitung *see* Duerener zeitung 1875

Duerener zeitung 1875 – Dueren DE, 1957 2 nov-1959 30 jun [nur lokalseiten] – 1 – (title varies: 3 jul 1889: general-anzeiger; 1896: duerener zeitung; 1946 as regional ed of aachener volkszeitung, aachen. filmed by other misc inst: 1886, 1889 2 jan-29 jun, 1891 1 apr-1916, 1917 2 jul-1919, 1978 1 sep- [ca 7r/yr]) – gw Misc Inst [074]

Duerer, A *see*
– Etliche underricht...
– Institutionum geometricarum libri quatuor...
– Underweysung der messung mit dem zirkel und richtscheyt in linien ebnen und gantzen corporen zu samen gezoge und zu nutz alle kunst-liebhabenden mit zu gehoerigen figuren in truck gebracht
– Vier buecher von menschlicher proportion

Duerer als fuehrer / Langbehn, J & Nissen, M – Muenchen, 1928 – €7.00 – ne Slangenburg [750]

[Duerer] Thausing, M *see* Duerers briefe, tagebuecher und reime...

Duerers briefe, tagebuecher und reime... / [Duerer] Thausing, M – Wien, 1872. v3 – 4mf – 9 – mf#O-517 – ne IDC [700]

Duerfen wir noch christen bleiben? : kritische betrachtungen zur theologie der gegenwart / Heinrici, Carl Friedrich Georg – Leipzig: Duerr, 1901 – 1mf – 9 – 0-8370-2080-8 – mf#1985-0080 – us ATLA [240]

Duerksen, Rosella R *see* Hymody of the 16th century anabaptists

Duerler, Josef *see* Die bedeutung des bergbaus bei goethe und in der deutschen romantik

Duerrenmatt, Nelly *see* Das nibelungenlied im kreis der hoefischen dichtung

Duers past and present – v1 n1-7, v2 n1-6, v3 n7-v6 n12, v5 n20-v6 n24 [1983 nov-1984 nov, 1985 jan-nov, 1986 jan-nov, 1987 jan-nov] – 1r – 1 – mf#1713459 – us WHS [071]

Duesseldorf express – Duesseldorf DE, 1998- – 6r/yr – 1 – gw Misc Inst [074]

Duesseldorfer abendblatt *see* Duesseldorf-gerresheimer zeitung

Duesseldorfer allgemeine beamten-zeitung – Duesseldorf DE, 1926 3 sep-1927 16 jul – 1r – 1 – gw Misc Inst [350]

Duesseldorfer amtsblatt – Duesseldorf DE, 1946-80 – 7r – 1 – gw Misc Inst [074]

Duesseldorfer anzeigen-blatt – Duesseldorf DE, 1946-47 – 1r – 1 – gw Misc Inst [074]

Duesseldorfer anzeiger – Duesseldorf DE, 1857-mar 1893 – 45r – 1 – gw Misc Inst [074]

Duesseldorfer arbeiterzeitung – Duesseldorf DE, 1890 oct-dec, 1892-1896 31 jul, 1898-1899 26 jan, 1899 sep-1902 jul, 1903-1933 27 feb – 79r – 1 – (title varies: spd, regional ed of freie presse, elberfeld; 2 jan 1892: niederrheinische volkstribuene, regional ed of freie presse, elberfeld; 1 apr 1901: duesseldorfer volkszeitung; 6 aug 1902: volkszeitung, with suppl: der kinderfreund 1925 [gaps], 1926-27 [1r], also suppl to other social democratic daily newspapers) – gw Mikropress [331]

Duesseldorfer bau-zeitung *see*
– Duesseldorfer handelszeitung fuer kapital, baugewerbe und grundstuecksmarkt

Duesseldorfer beobachter – Duesseldorf DE, 1921 oct-1923 aug [gaps], 1924-1929 sep – 2r – 1 – (title varies: 10 feb 1923: der beobachter) – gw Misc Inst [074]

Duesseldorfer blaetter [...] – Duesseldorf DE, 1932 dec-1941 may – 4r – 1 – (title varies: mai 1933: westdeutsche woche) – gw Misc Inst [074]

Duesseldorfer buerger-zeitung – Duesseldorf DE, 1892 24 mar-1901 31 may – 13r – 1 – (title varies: 16 sep 1892: buerger-zeitung. suppl xanthippus as individual title) – gw Misc Inst [074]

Duesseldorfer chronik – Duesseldorf DE, 1888-1889 mar – 1r – 1 – gw Misc Inst [074]

Duesseldorfer freie presse – Duesseldorf DE, feb 1918-oct 1922 – 16r – 1 – gw Misc Inst [074]

Duesseldorfer general-anzeiger – Germany. Duesseldorfer Nachrichten. -d. 1 Sept 1916-6 Aug 1919. 16 reels – 1 – uk British Libr Newspaper [072]

Duesseldorfer general-anzeiger *see* General-anzeiger fuer duesseldorf und umgegend

Duesseldorfer gerichts-zeitung 1905 – Duesseldorf DE, 1905-1915 6 mar – 4r – 1 – (title varies: jul 1909: rheinisch-westfaelische gerichts-zeitung) – gw Misc Inst [347]

Duesseldorfer gerichts-zeitung 1924 – Duesseldorf DE, 1924 28 sep-1926 31 jan – 1r – 1 – (title varies: 1 mar 1925: westdeutsche gerichts-zeitung) – gw Misc Inst [347]

Duesseldorfer handelszeitung fuer haus- und grundbesitz, bauwesen und staedtische angelegenheiten *see* Duesseldorfer handelszeitung fuer kapital, baugewerbe und grundstuecksmarkt

Duesseldorfer handelszeitung fuer kapital, baugewerbe und grundstuecksmarkt – Duesseldorf DE, 1905-1943 mar, 1945 apr-1949 – 19r – 1 – (title varies: 23 feb 1907: duesseldorfer handelszeitung fuer haus- und grundbesitz, bauwesen und staedtische angelegenheiten; 6 apr 1907: duesseldorfer bau-zeitung; 5 apr 1913: haus- und grundbesitzer-zeitung; jul 1913: duesseldorfer haus- und grundbesitzer-zeitung; jan 1915: duesseldorfer bau-zeitung; oct 1915: duesseldorfer haus- und grundbesitzer-zeitung; apr 1946: rundschreiben. haus- und grundbesitzerverein duesseldorf e v; apr 1947: haus und grund; 1964: duesseldorfer hausbesitzer-zeitung) – gw Misc Inst [333]

Duesseldorfer haus- und grundbesitzer-zeitung *see* Duesseldorfer handelszeitung fuer kapital, baugewerbe und grundstuecksmarkt

Duesseldorfer hausbesitzer-zeitung *see* Duesseldorfer handelszeitung fuer kapital, baugewerbe und grundstuecksmarkt

Duesseldorfer illustrirte zeitung *see* Westdeutsche illustrirte zeitung

Duesseldorfer journal *see* Duesseldorfer kreisblatt und taeglicher anzeiger

Duesseldorfer journal und kreisblatt *see* Duesseldorfer kreisblatt und taeglicher anzeiger

Duesseldorfer kreisblatt und taeglicher anzeiger – Duesseldorf, Koeln DE, 1848-49 – 2r – 1 – (title varies: 1848 n120: duesseldorfer journal und kreisblatt; 1 jan 1856: duesseldorfer journal; 12 sep 1860: niederrheinische volks-zeitung; 1 jan 1863: rheinische zeitung. since 1863 n238 publ in koeln. filmed by mikropress: 1860-68 [13r]) – gw Misc Inst; gw Mikropress [074]

Duesseldorfer leben – Duesseldorf DE, 1921 n1-9 – 1r – 1 – gw Misc Inst [074]

Duesseldorfer lokal-zeitung – Duesseldorf DE, oct 13 1906-sep 11 1937 – 32r – 1 – (with suppl: areal-anzeiger 1910 1 jun-1916 [1r]) – gw Misc Inst [074]

Duesseldorfer merkur – Duesseldorf DE, oct 30 1880-mar 1881, jul-dec 1882 – 2r – 1 – (incl suppl: der erzaehler 1882 1 jul-30 dec) – gw Misc Inst [074]

Duesseldorfer monatshefte – Duesseldorf DE, 1849 – 1r – 1 – gw Misc Inst [074]

Duesseldorfer morgenpost – Duesseldorf DE, jun 2 1920-may 31 1921 – 1r – 1 – (title varies: 3 jan 1921: westdeutsche zeitung) – gw Misc Inst [074]

Duesseldorfer mostert – Duesseldorf DE, sep 27 1902-nov 7 1903 – 1r – 1 – gw Misc Inst [074]

Duesseldorfer nachrichten *see* General-anzeiger fuer duesseldorf und umgegend

Duesseldorfer neueste nachrichten *see* Deutsche eisenzeitung und taeglicher anzeiger

Duesseldorfer post *see* Der gewerksvereinsbote

Duesseldorfer rundschau – Duesseldorf DE, 1922 7 jan-4 mar – 1r – 1 – gw Misc Inst [074]

Duesseldorfer sonntagsblatt – Duesseldorf DE, 1867 6 oct, 1868-69, 1870 [single iss], 1871-1941 31 may – 218r – 1 – (title varies: 2 jul 1871: duesseldorfer volksblatt; 15 jun 1904: duesseldorfer tageblatt; with suppl: der feuerreiter 1928 23 jun-1941 28 jun [9r] publ in koeln; sonntagsblatt 1882 11 jun-31 dec [1r]) – gw Misc Inst [074]

Duesseldorfer stadtanzeiger – Duesseldorf DE, may 8 1926-may 10 1933 – 43r – 1 – (with suppl: rheinische illustrierte 1927-30 [3r]; illustrierte sonntagspost oct 29 1930-oct 28 1933 [3r]) – gw Misc Inst [074]

Duesseldorfer tageblatt *see* Duesseldorfer sonntagsblatt

Duesseldorfer volksblatt *see* Duesseldorfer sonntagsblatt

Duesseldorfer volkszeitung *see*
– Buergermeisterblatt
– Duesseldorfer arbeiterzeitung

Duesseldorfer wirtezeitung – Duesseldorf DE, 1925-1931 23 nov – 2r – 1 – gw Misc Inst [640]

Duesseldorf-gerresheimer zeitung – Duesseldorf DE, jan-jun 1911 [gaps], 1912-14 – 6r – 1 – (title varies: apr 1913: duesseldorfer abendblatt. with suppl: sonntagsblatt, berlin (since 1914 n8: illustriertes unterhaltungsblatt), 1911-14 [1r]; unterhaltungsblatt: der erzaehler jan-jun 1911, 1912-14 (gaps) [4r]) – gw Misc Inst [074]

Duesselthaler jugendblaetter – Duesseldorf-(Duesselthal) DE, 1869-75 – 1r – 1 – gw Misc Inst [074]

Duesterdieck, Friedrich *see* Critical and exegetical handbook to the revelation of john

Duet for two performers on one grand pianoforte [op 14 [koech verz 521]] / Mozart, Wolfgang Amadeus – London: printed for Monzani & Cimador [180-?] [mf ed 1989] – 1r – 1 – mf#pres. film 50 – us Sibley [780]

Duetsch, Gerald *see* Tarifvertraege im koenigreich bayern

Dufau, Pierre A *see* Dictionnaire universel abrege de geographie ancienne comparee

Dufau, Pierre Armand *see* Histoire generale de france

Dufay, Jean *see* Publications de l'observatoire de lyon

Dufays, Felix *see* Jours troubles, pages d'epopee africaine

Dufey, Pierre *see* Nouveau dictionnaire historique des environs de paris

Dufey, Pierre Joseph Spiridion *see*
– La bastille
– Histoire des communes de france et legislation municipale
– Resume de l'histoire de bourgogne

Duff, A *see* India, and india missions

Duff, Alexander *see*
– Bombay in april, 1840
– Cause of christ and the cause of satan
– Church of scotland's india mission
– Explanatory statement addressed to the friends of the India mission
– Farewell address
– Foreign missions
– India und india missions
– The indian rebellion
– Jesuits
– Letter from alexander duff
– Liberality as a means of sanctification
– Mutual duties and responsibilities of pastor and people
– Proposed modes of extending the foreign mission operations of the f...

Duff, Archibald *see*
– Abraham and the patriarchal age
– Hints on old testament theology
– History of old testament criticism

Duff, David *see*
– The early church

Duff, David, jr *see* The early church

Duff, E C *see* Gazetteer of the kontagora province

Duff, E S *see* Redeemed by the blood

Duff, Edward Macomb *see* Psychic research and gospel miracles

Duff green papers / Green, Duff – 1810-1902. University of North Carolina Library. Guide – 1 – $450.00 – us CIS [920]

Duff, Hector Livingston *see* Nyasaland under the foreign office

Duff, James Grant *see* A history of the mahrattas

Duff, Robert C *see* The attitude of the texas banker to texas railroads

Duff, William *see* An essay on original genius and its various modes of exertion in philosophy and the fine arts, particularly in poetry

Duffels, Arnold *see* Het leven van den gelukzaligen martelaar carolus spinola

Dufferin and Ava, Frederick Temple Blackwood, Marquis of *see*
– A yacht voyage

Dufferin and Ava, Frederick Temple Hamilton-Temple-Blackwood, Marquis *see* Mr mill's plan for the pacification of ireland examined

Dufferin and Ava, Hariot Georgina (Hamilton) Hamilton-Temple-Blackwood, marchioness of *see*
– My canadian journal 1872-8
– Our viceregal life in india

Dufferin and Ava, Hariot Georgina Hamilton-Temple-Blackwood, marchioness of *see* My canadian journal, 1872-'78

Duffey, Frank M *see* Early cuadro de costumbres in colombia

Duffey, Thelma *see* Journal of creativity in mental health

Duffield, George *see* The bible rule of temperance

Duffield, Mary Elizabeth (Rosenberg) *see* The art of flower painting

Duffield, Samuel Willoughby *see*
– English hymns
– The latin hymn-writers and their hymns

Duffy, Charles Gavan *see*
– Four years of irish history, 1845-1849
– Thomas davis: the memoirs of an irish patriot, 1840-1846

Duffy's hibernian magazine : a monthly journal of legends, tales, and stories, irish antiquities, biography, science and art – La Salle IL 1860-64 – 1 – mf#4716 – us UMI ProQuest [390]

Dufort, Giovanni Battista *see* Trattato del ballo nobile di giambatista dufort...

Dufougere, William *see* Madinina

Dufour, Helene *see* Marie-claire blais

Dufourcq, Albert *see*
– De manichaeismo apud latinos quinto sextoque saeculo
– Saint irenee
– Saint irenee (2e siecle)

Dufoussat, Henry *see* De l'hypotheque legale de la femme mariee

Dufrenois, M *see* Deux contes creoles

Dufresne, Ferdinand *see* Quatuor brillant pour deux violons, alto et basse, op 20, 1er livre

Dufresne, Guy *see* Kebec

Dufresne, Lise *see* L'institution des sourds-muets de montreal (1848-1948)

Dufresne, Roger *see* Bibliographie des ecrits de freud

Dufton, H *see* Narrative of a journey through abyssinia in 1862-1863

Dufur dispatch – Dufur, Wasco County, OR: W H Brooks. v16 n8-v47 n24. jun 22 1910-aug 29 1941 – 1 – (began in 1891. 1925-1938 incl newspaper publ during school terms by dufur high school. suspended from late 1892-may 1 1896) – us Oregon Hist [071]

Dufur dispatch – Dufur OR: W H Brooks, -1941 [wkly] – 8r – 1 – (began in 1891. absorbed by: dalles optimist (1906-66). related to: school daze (1925-38). suspended fr late 1892-may 1 1896) – us Oregon Lib [071]

Dufur dispatch (dufur, or) – Dufur OR: Creston Creek College Press, 1970 [wkly] – us Oregon Lib [071]

Dufur dispatch (dufur, or: 1982) – Dufur OR: Fort Dufur Pub Co [various dates] – 1 – us Oregon Lib [071]

Duga – Belgrade, Yugoslavia. -w. Jan-Dec 1960. 2 reels – 1 – uk British Libr Newspaper [949]

Dugai-trouin : prisonier a plymouth / Barre, M – Paris, France. 1804 – 1r – us UF Libraries [440]

Dugal, Armand-J *see* Voyage en zigzag a travers la publicite et le commerce deux amis inseparables

Dugas, Georges *see*
– The canadian west
– Etablissement des soeurs de charite a la riviere rouge
– Histoire de l'ouest canadien de 1822 a 1869
– Legendes du nord-ouest
– Manitoba et ses avantages pour l'agriculture
– Monseigneur provencher et les missions de la riviere-rouge
– L'ouest canadien
– La premiere canadienne du nord-ouest
– Quelques erreurs historiques a corriger
– Un voyageur des pays d'en haut

Dugast de Bois-Saint-Just, Jean *see* Paris, versailles et les provinces, au dix-huitieme siecle

Dugat, Gustave *see* Histoire des philosophes et des theologiens musulmans (de 632 a 1258 j.-c.)

Dugdale, James *see* The new british traveller, or, modern panorama of england and wales

Dugdale, William *see* Origines juridiciales

Duggan, James *see*
– The life of christ
– Steps towards reunion

Dugger, Gordon Leslie *see* Lithium bromide-methyl alcohol

Dugmore, Henry Hare *see* Reminiscences of an albany settler

Dugway Proving Ground [UT] *see*
– Desert post
– Desert sun

Duhamel du Monceau, Henri Louis *see*
– Art du cirier
– Art du couvreur
– L'art du tuilier et du briquetier
– Elemens de l'architecture navale
– Traite des arbres fruitiers

Duhautcours / Picard, Louis-Benoit – Paris, France. 1801 – 1r – us UF Libraries [440]

Duhem, P *see* Le systeme du monde

Duhigg, Barthelomew Thomas *see* A letter to the right hon. lord manners, &c. &c. &c. on the expediency of an immediate and separate record commission, to investigate, illustrate, and arrange the records of ireland.

Duhm, Bernhard *see*
– Das buch hiob
– Das buch jesaia
– Die entstehung des alten testaments
– Das geheimnis in der religion
– Die psalmen
– The twelve prophets

Duhr, Bernhard *see*
– Geschichte der jesuiten in den laendern deutscher zunge im 16. jahrhundert
– Geschichte der jesuiten in den laendern deutscher zunge in der ersten haelfte des 17. jahrhunderts

DUMPLIN

– Jesuiten-fabeln
– Ratio studiorum et institutiones scholasticae societatis jesu
– Die stellung der jesuiten in den deutschen hexenprozessen
– Die studienordnung der gesellschaft jesu
Duhr, J *see* Apercus sur l'espagne chretienne du 4th siecle
Duhring, Julia *see* Philosophers and fools
Dui im ni khyan – Ran Kun: Bhasa pran Ca pe A san 1955 [mf ed 1990] – [ill] 1r with other items – 1 – mf#mf-10289 seam reel 192/3 [§] – us CRL [480]
Dui yin ji zi – [Beijing]: Fan yi zong xue, Guangxu geng yin [1890] [mf ed 1966] – (= ser Tenri coll of manchu-books in manchu-characters. series 1, linguistics 48; Mango bunkenshu. 1, gogaku hen) – 2v on 1r – 1 – (in manchu and chinese. with app) – ja Yushodo [480]
Duin hacin-i hergen kamciha buleku bithe = Dorben zuil-un usug qabsurugsan toli bicig – [China: s.n, 17–] [mf ed 1966] – (= ser Tenri coll of manchu-books in manchu-characters. series 1, linguistics 27-28; Mango bunkenshu. 1, gogaku hen) – 8v on 2r – 1 – (in manchu, mongolian, tibetan and chinese) – ja Yushodo [480]
Duindui language, new hebrides : vocabulary, primer and hymn book – n.d. – 1r – 1 – mf#pmb46 – at Pacific Mss [490]
Duisberg, Adolf von *see* Primer of kanuri grammar
Duisburger general-anzeige *see* Duisburger tageblatt 1881
Duisburger tageblatt 1881 – Duisburg DE, 1958-1961 18 sep [gaps], 1962 1 sep-1964 1 oct [gaps], 1965 15 mar-1966 30 nov [restfilm schon waz] – 1 – (title varies: 1 sep 1893: general-anzeiger; 27 apr ?1914: duisburger general-anzeige; takeover by waz, essen. filmed by other misc inst: 1951 18 jun-1957 [23r]) – gw Misc Inst [074]
Duisternis en licht : die zending in oost-afrika en op madagaskar / Dijkstra, Harmen – Leiden: D. Donner, [1880] – 1r – 1 – 0-8370-0408-X – mf#1984-B209 – us ATLA [240]
Duitsche lier : draayende veel van de nieuwste, deftige, en dertelende toonen / Luyken, Jan – 's Gravenhage: HH van Drecht, 1783 – 2mf – 9 – mf#0-3238 – ne IDC [090]
De duitsche orde : of, beknopte geschiedenis, indeeling en statuten der broeders van het duitsche huis van st marie van jerusalem / Ablaing van Giessenburg, Willem Jan, baron d' – 's Gravenhage: Martinus Nijhoff 1857 – 4mf – 9 – mf#vrl-34 – ne IDC [366]
Dujardin, Edouard *see*
– Les predecesseurs de daniel
– The source of the christian tradition
Dujarric, Gaston *see*
– L'etat mahdiste du soudan
– Vie de mahomet d'apres la tradition
– La vie du sultan rabah
– La vie du sultan rabah; les francais au tchad
Dujon, F *see* Opera theologica
Duk Rasi *see* Vasana prades jati
Duka, Tivadar *see* Life and works of alexander csoma de koros
Duke and no duke : a farce as it is acted by their majesties servants...with several songs set to music, with thorow basses for the theorbo, or bass viol / Tate, Nahum – London: Henry Bonwicke...1685 [mf ed 19–] – 2mf – 9 – mf#fiche 676, 951 – us Sibley [780]
Duke bar association journal – v1-10. 1933-42 – 18mf – 9 – $27.00 – (Duke law school, student bar association proceedings and notes and comments on current decisions) – mf#LLMC 84-459 – us LLMC [340]
Duke bar journal *see* Duke law journal
Duke environmental law and policy forum – v1-11. 1991-2001 – 9 – $173.00 set – mf#114161 – us Hein [344]
Duke law journal – v1-7. 1951-58; 1959-v50. 1959-2001 + Ind – 5,6,9 – $1347.00 set – (v1-7, 1959-84 1951-84 in reel or mf $671. 1985-v50 1985-2001 in mf $676. v1-22 1951-73 cum ind inquire for price. title varies: v1-6 1951-57 as duke bar journal. v unnumbered 1959-60.) – ISSN: 0012-7086 – mf#102521 – us Hein [340]
Duke mathematical journal – Durham NC 1935+ – 1,5,9 – ISSN: 0012-7094 – mf#7149 – us UMI ProQuest [510]
The duke of wellington : a funeral sermon preached on sunday the 21st of nov... / Smithurst, John – [Elora ON?: s.n.] 1852 [mf ed 1987] – 1mf – 9 – 0-665-63593-1 – mf#63593 – cn Canadiana [240]
Duke University *see* Harambee
Duker, A C *see* Gisbertus voetius
Duker, Arnoldus Cornelius *see* Gisbertus voetius
Dukes, Clement *see* Model woman
Dukes, Edwin Joshua *see* Alltagsleben in china
Duke's funeral – London, England. 1852? – 1r – us UF Libraries [240]
Dukes, Hugh *see* Textural and color characteristics of some important red and yellow...
Dukes, Leopold *see* Philosophisches aus dem zehnten jahrhundert
Dukh khristianina : dukhovno-literaturnyi zhurnal – v2 n1-7, 9-12 1861; n1-3 1865 – (= ser Corpus of russian orthodox periodicals) – 2r – 1 – (lacking: v2 n1 1861) – mf#ATLA S0193C – us ATLA [243]
Dukhovenstvo i obshchestvo v sovremennom religioznom dvizhenii / Tikhonirov, L A – 1893 – 36p 1mf – 8 – mf#R-106=38 – ne IDC [243]
Dukhovnaia beseda, ezhenedeleno izdavaemaia pri sankt-peterburgskoi dukhovnoi seminarii – Spb., 1858-1869 – 273mf – 9 – (missing: 1863(20); 1865; 1866(1); 1867(2); 1869(2)) – mf#R-1586 – ne IDC [077]
Dukhovnaia besieda – n1,16-43,46-55. 1866 (complete) – (= ser Corpus of russian orthodox periodicals) – 1r – 1 – mf#ATLA S0193D – us ATLA [243]
Dukhovnaia politsiia v rossii / Reisner, M A – 1907 – 107p 2mf – 9 – mf#R-10090 – ne IDC [243]
Dukhovnaia tsenzura v rossii : 1799-1855 gg / Kotovich, A – 1909 – 608p 12mf – 8 – mf#R-9790 – ne IDC [243]
Dukhovnoe nasledie / Vserossiiskoe obshchestvenno-politicheskoe dvizhenie "Dukhovnoe nasledie" – Moscow, Russia. iiun'(1996)-iiul'(1996), [spetsvyp] (avg 1996), n4(avg 1996)-n1[8](ian 1997), n3[10](apr 1997)-n5[12](apr 1997), n7[14](iiul 1997), n0[8](mai 1998)-n1[9](iiiun 1998), Sankt-Peterburgskii vyp (okt 1998), spetsial'nyi moskovskii vypusk [no number] – 1 – mf#mf-12248 (reel 4) – us CRL [077]
Dukhovno-nravstvennyi zhurnal : organ russkikh baptistov – Rostov-na-Donu, Baku, Odessa, 1907-1912, 1914 – 130mf – 9 – (missing: 1907(2); 1910(49); 1911(22, 49-52); 1914(9-10)) – mf#R-1529 – ne IDC [077]
Dukhovnye shkoly v rossii do reformy 1808 goda / Znamenskii, P – Kazan, 1881 – 806p 15mf – 8 – mf#R-7988 – ne IDC [243]
Dukhovnyi vestnik – Kharkov, 1862-1867 – 193mf – 9 – mf#R-3399 – ne IDC [077]
Dukhovnyi zhurnal sovremennoi zhizni, nauki i literatury – Ann Arbor. 1948+ (1) 1948+ (5) 1948+ (9) – 1805mf – 9 – (missing: 1890, v3(9); 1899, v1(4), v2(6), v3(10-12); 1900, v1(2-4), v2-3; 1901-1908; 1916, v1(2-3), v3(11-12); 1917) – mf#1912 – ne IDC [077]
Dukinfield reporter – Dukinfield, England 21 jul 1967-28 feb 1986 [mf jan-feb 1986] – 1 – (cont of dukinfield ed of the ashton-under-lyne reporter; amalg with: stalybridge reporter & subsequently publ as the stalybridge & dukinfield reporter) – uk British Libr Newspaper [072]
Dukun as referrer of family planning acceptors : a study in east java / Pardoko, R H & Soemodinoto, Soekanto – Surabaya: National Institute of Public Health, [1972-1975] – us CRL [300]
Dulaney, N M *see* The effects of a flexibility training program on flexibility test scores in elementary school children
Dulaure, Jacques *see*
– Histoire de la revolution francaise
– Histoire physique, civile et morale de paris
– Histoire physique, civile et morale des environs de paris
– Singularites historiques
DuLaurens, Henri J *see* Imirce
Dulce domum : george moberly (d.c.l., headmaster of winchester college, 1835-1866, bishop of salisbury, 1869-1885), his family and friends / Moberly, Charlotte Anne Elizabeth – London: J Murray, 1911 – 1mf – 9 – 0-7905-4955-7 – mf#1988-0955 – us ATLA [920]
Dulcissimae quaedam cantiones numero 32... / Knoefel (Knefel), Johannes – 1571 – (= ser Mssa) – 4mf – 9 – €60.00 – mfchl 272 – gw Fischer [780]
Duling, Anton *see* Cithera melica, vel opus musicum plane novum...
Dulk, Albert Friedrich Benno *see* Gedichte
Dulken, G van *see*
– Het gereinigt herte door 't geloof
Dull brass – 1969 apr 14-jul, 1970 may – 1r – 1 – mf#721511 – us WHS [071]
Duller, Eduard *see* Franz von sickingen
Dulles, John Foster *see* Papers of john foster dulles and of christian a herter, 1953-1961
Dulles, Joseph Heatly *see* Princeton theological seminary biographical catalogue, 1909
Dullingham 1558-1950 – 8mf – 9 – £10.00 – uk CambsFHS [929]
Duluth Central Labor Body [Duluth MN] *see* Labor world
Duluth directory / R L Polk and Co – 1882/83 – 1r – 1 – (cont by: r l polk & co's duluth directory) – mf#802140 – us WHS [917]
Duluth press – Duluth MN. 1893 mar 4-jul 22, dec 2-23 – 1r – 1 – (cont: people's press [duluth mi: 1892]) – mf#766070 – us WHS [071]
Duluth volksfreund – Duluth, MN: Josef Grahamer, jul 1894-mar 17 1898 – 2r – us CRL [071]
Dulwich guardian – London, UK. 28 feb 1991-1992; 1993 – 5 1/2r – 1 – uk British Libr Newspaper [072]
Dulwich labour party records, 1924-86 – (= ser Labour party in britain, origins and development at local level. series 2) – 6r – 1 – (with p/g. int by nick tiratsoo) – mf#97568 – uk Microform Academic [325]
Dulwich picture gallery – [mf ed 1987] – 14 col mf – 15 – $740.00 – 0-907716-22-9 – (complete coll of this fine but little known gallery; 750 paintings illustrated, with 250 details) – uk Mindata [750]
Duma – Sofia, Bulgaria. 4 apr 1990-30 dec 1995 – 1 – (in cyrillic) – mf#mf.685.f – uk British Libr Newspaper [077]
Duma *see* Rabotnichesko delo
Dumaine, Jacques *see* Quai d'orsay
Dumais, A *see* Index alphabetique des noms de 3400 familles de douze enfants vivants
Dumaniant, Antoine-Jean *see*
– Guerre ouverte
– Laure et fernando
DuManoir, Guillaume *see* Le mariage de la musique avec la dance
Dumanoir, Philippe *see*
– Escadron volant de la reine
– Exposition des produits de la republique
Dumaresq's daughter : a novel / Allen, Grant – London: Chatto & Windus, 1893 – 4mf – 9 – mf#17939 – cn Canadiana [830]
Dumas, A [pere] *see* Le caucase, nouvelles impressions de voyage
Dumas, Alexandre *see*
– Ami des femmes
– El conde de montecristo
– Count of monte cristo
– Discours prononce par mr alexandre dumas au club constitutionel, tenu a quebec le 30 mai 1792
– Fils naturel
– Honneur est satisfait
– Idees de mme aubray
– Lady of the camellias
– Laird of dumbiky
– Pere prodigue
– Question d'argent
– Teresa
– Das weib des claudius
Dumas, Alexandre (pere) *see*
– Impressions de voyage
– Paris et les parisiens au 19 [dix-neuvieme] siecle
– Quinze jours au sinai
Dumas, Antoine Joseph *see* L'art de la musique enseigne et pratique par la nouvelle methode du bureau typografique etablie sur une seule cle
Dumas' art annual : an illustrated record of the exhibitions of the world 1882 / Dumas, Francois Guillaume – London 1882 – (= ser 19th c art & architecture) – 4mf – 9 – mf#4.2.1391 – uk Chadwyck [700]
Dumas, Francois Guillaume *see*
– Dumas' art annual
– Modern artists
Dumas, Gabriel-Marie *see* Bibliographie analytique de l'oeuvre du reverend pere alexis de barbezieux, capucin
Dumas, Norbert *see* Cadastre abrege du fief vieuxpont...
Dumas, Petrus *see* Viridarium humilitatis
Dumas, Rollande *see* Bibliographie analytique de la psychologie infantile, 1948 a 1952
Dumazedier, Joffre *see* Television and rural adult education
Dumbar, Gerhard *see* Het kerkelyk en wereltlyk deventer, deel 1
Dumbarton, Alfred *see* Light in the dark jungles
Dumbarton and vale of leven reporter – Dumbarton: C M Jeffrey 1987- (wkly) [mf ed 1 jan 1997-] – 1 – (cont: reporter for dumbarton, vale of leven, old kilpatrick, bowling and milton) – ISSN: 1356-8647 – uk Scotland NatLib [072]
Dumbarton argus : or, lennox magazine – [Scotland] West Dunbartonshire, Dumbarton : Dumbarton Printing Office 1832-nov 1834 (irreg) [mf ed 2004] – 35v on 1r – 1 – uk Newsplan [072]
Dumbarton herald, and county advertiser, for dumbartonshire, argyllshire, and buteshire – [Scotland] Dumbarton: S Bennett n1 [sep 25 1851]-n4565 [dec 27 1933] (wkly) [mf ed 2002] – 66r – 1 – (merged with: lennox herald and weekly advertiser to form: lennox herald for western dumbartonshire) – uk Newsplan [072]
Dumbarton Oaks Collection *see* Pre-columbian art
Dumeril, A et al *see* Mission scientifique au mexique et dans l'amerique centrale
Dumeril, A H A *see* Histoire naturelle des poissons
Dumeril, Edmond *see* Le lied allemand et ses traductions poetiques en france
Dumersan, Theophile Marion *see*
– Fete d'un bourgeois de paris
– Macedoine, ou, les etrennes et le carnaval
Dumesnil, Alexis *see* Le regne de louis 11
Dumesnil, Clement *see* De l'abolition des droits feodaux et seigneuriaux du canada
Dumfries and galloway courier and herald – [Scotland] Dumfries: D Mitchell Miller jan 1895-15 nov 1939 (twice(wk) [mf ed 2004] – 90r – 1 – (formed by union of: dumfries & galloway courier [1809-84] and: dumfriesshire & galloway herald and register [184?-1884]) – uk Newsplan [072]
Dumfries and galloway news review – [Scotland] Dumfries: Allardyce, Balfour & Co jan 1947-mar 1949 (mthly) [mf ed 2004] – 4r – 1 – (cont by: dumfries & galloway review [apr 1947-mar 1949]) – uk Newsplan [072]
Dumfries and galloway review, and south-western counties advertiser – [Scotland] Dumfries: D Miller mar 1876-sep 1877 (wkly) [mf ed 2004] – 1r – 1 – uk Newsplan [072]
Dumfries and galloway review [dumfries, scotland : 1901] – [Scotland] Dumfries: D M'Millan 23 feb-2 mar 1901 (wkly) [mf ed 2004] – 1r – 1 – uk Newsplan [072]
Dumfries and galloway standard (dumfries, scotland : 1993) – Dumfries: Dumfries & Galloway standard 1993- (semiwkly) [mf ed 1999-] – 1 – (formed by union of: dumfries & galloway standard (wed ed) and: dumfries & galloway standard (fri ed)) – uk Scotland NatLib [072]
Dumfries courier (annan, scotland : 1977) – Annan: Dumfriesshire Newspapers Group 1977- (wkly) [mf ed 1999-] – 1 – uk Scotland NatLib [072]
Dumfries & galloway standard news bulletin – [Scotland] [Dumfries: s.n.] 5-13 may 1926 (daily) [mf ed 2004] – 1r – 1 – (produced during the 1926 general strike, some iss typescript) – uk Newsplan [331]
Dumfries mercury : containing and account of the most remarkable occurrences both foreign and domestick... – [Scotland] Dumfries: printed by Robert Rae...1-8 may 1721 [mf ed 2004] – 1r – 1 – uk Newsplan [072]
Dumfriesshire bulletin – [Scotland] Dumfries: D C Howie sep 1935 [mf ed 2004] – 1v on 1r – 1 – (election bulletin issue on behalf of john downie, the labour candidate) – uk Newsplan [325]
Duminy-dagboeke, duminy diaries / Franken, Johan Lambertus Machiel – Kaapstad, South Africa. 1938 – 1r – us UF Libraries [960]
Dumke, Charles L *see* Protective mechanism of estradiol on eccentrically induced muscle damage
Der dumme gaertner : oder die beyden anton, ein comisches singspiel in zwey aufzugen fuers clavier gesetzt / Neefe, Christian Gottlob – Bonn: N Simrock [1795?] [mf ed 1989] – 1r – 1 – mf#pres. film 50 – us Sibley [780]
Dummhans : roman / Frenssen, Gustav – Berlin: G Grote 1930 [mf ed 1989] – (= ser Grote'sche sammlung von werken zeitgenoessischer schriftsteller 181) – 1r – 1 – (filmed with: dorfpredigten) – mf#7264 – us Wisconsin U Libr [830]
Dummitt orange grove / Kerce, Red – s.l, s.l? 193-? – 1r – us UF Libraries [634]
Dumolard, Henri Francois *see*
– Mari instituteur, ou, les nouveaux epoux
– Memoires et correspondance litteraires, dramatiques et anecdotiques...
– Philinte de destouches
Dumon, Frederic *see* Bresil
Dumonchau, Charles *see* Trio pour le forte piano avec accompagnement de violon et violoncelle obligee, 2e oeuvre
Dumont, Auguste *see*
– La franc-maconnerie: sa politique et son oeuvre...
– Les propos de lucius
Dumont d'Urville, Jules *see*
– Voyage de decouvertes autour du monde et a la recherche de la perouse
– Voyage de la corvette l'astrolabe
Dumont, Emile *see* Les conditions de l'enseignement religieux dans les eglises nationales de la suisse romande
Dumont, Georges-A *see* Un disparu
Dumont, Jean *see* Voyages de mr du mont, en france, en italie, en allemagne, a malthe, et en turquie
Dumont, Pierre *see* Histoire de l'esclavage en afrique (pendant trente-quatre ans) de p j dumont, natif de paris, maintenant a l'hospice royal des incurables
Dumoulin, Stephane *see* Le tonkin: exploration du mekong
Dumouriez, Charles Francois *see*
– Coup d'oeil politique sur l'avenir de la france
– La vie et les memoires du general dumouriez
Dumoutet, E *see* Le desir de voir l'hostie et les origines de la devotion au saint-sacrement
Dumpfe trommel und berauschtes gong / nachdichtungen chinesischer kriegslyrik / Henschke, Alfred (pseud. Klabund) – Leipzig: Insel-Verlag, 1915 – 1r – 1 – us Wisconsin U Libr [480]
Dumplin creek baptist church. jefferson county. tennessee : church records – 1797-1938 – 1 – us Southern Baptist [242]

Dumskie vystupleniia a s viazigina – Kharkov, 1913 – 72p 1mf – 9 – mf#RPP-197 – ne IDC [325]
Dumy : gaz sotsialist / ed by Skal'nenkov, N G – Biisk [Alt gub]: [s n] 1918-19 [1918 27 [14] noiab-1919 19 ianv – (= ser Asn 1-3) – n1-28 [1918] n6-11 [1919] [gaps] item 130, on reel n27 – 1 – (cont: dumy altaia; cont by: altaiskii krai) – mf#asn-1 130 – ne IDC [335]
Dumy altaia : gaz sotsialist / ed by Skal'nenkov, N G – Biisk [Alt gub]: [s n] 1918 [1918 22 [9] iiunia-24 [11] noiab – (= ser Asn 1-3) – n1-102 [1918] [gaps] item 131, on reel n27,28 – 1 – (cont: svobodnyi altai; cont by: dumy) – mf#asn-1 131 – ne IDC [335]
Dumy kooperatora – Penza, 1920(1) – 1mf – 9 – mf#COR-584 – ne IDC [335]
Dun and bradstreet, inc. d & b reports – Short Hills NJ 1987-94 – 1,5,9 – ISSN: 0746-6110 – mf#15935,03 – us UMI ProQuest [338]
Dun echt circular / Dun Echt Observatory – Aberdeen: The Observatory 1879-90 (irreg) [mf ed 2001] – n1-179(1879-90) on 1r – 1 – (cont by: royal observatory, edinburgh. circular) – mf#film mas c5062 – us Harvard [520]
Dun Echt Observatory *see* Dun echt circular
Dun echt observatory publications – Dun Echt, Aberdeen: The Observatory 1876-85 (irreg) [mf ed 1999] – v1-3(1876-85) on 1r – 1 – mf#film mas c4316 – us Harvard [520]
Dun, Finlay *see* Landlords and tenants in ireland
Dun, John *see* British banking statistics
Duna zeitung – Riga, Latvia 1888-1908 [mf ed Norman Ross] – 38r – 1 – (in german) – mf#nrp-1460 – us UMI ProQuest [077]
Dunaets – zaschitnik rodiny – (city unknown) 1944-45 [mf ed Norman Ross] – 1 – mf#nrp-96 – us UMI ProQuest [934]
Dunaev, Boris Ivanovich *see* Skazaniia pro khrabrago vitezia pro bovu korolevicha
Dunantuli naplo – Pecs, Hungary. 1962-Jun 1991 – 60r – 1 – (cont as: uj dunantuli naplo as of 3 apr 1990) – us L of C Photodup [079]
Dunavska pravda – Ruse, Bulgaria. 1951-Jul 1990 – 55r – 1 – us L of C Photodup [949]
Dunavski otechestven front – Ruse, Bulgaria. Feb-Aug 1945 – 1r – 1 – us L of C Photodup [949]
Dunbar, George *see* A history of india
Dunbar, Helen Flanders *see* Symbolism in medieval thought and its consummation
Dunbar, Hugh *see* The christian record
Dunbar, Paul Laurence *see* Complete poems
Dunbar review – Dunbar, NE: [C F Collins] 1899-1938// (wkly) [mf ed 1900-36 (gaps)] – 3r – 1 – (some irregularities in numbering) – us NE Hist [071]
Dunbar, W *see* Farewell sermon
Dunbar, William *see* Travels in the interior parts of america
Dunbarton, New Hampshire. Dunbarton Baptist Church *see* Records
Duncalf, Frederic *see* Parallel source problems in medieval history
Duncan, Annie N *see* The city of springs
Duncan, Archibald *see* The mariner's chronicle
Duncan, B M *see* Letter to mr h chamberlin
Duncan, Daniel Wendell *see* The effect of a study of the biblical concept of church on establishing long range planning goals
Duncan, David *see*
– The law of moses
– The life and letters of herbert spencer
Duncan, Francis *see*
– Beschreibung der insel st helena
– Canada in 1871
– A description of the island of st helena
– Our garrisons in the west
Duncan, George *see*
– Baptism and the baptists
– The epistle of paul to the galatians
– Paedobaptism
Duncan, George M *see* The protection of the foreshore at dallas road, victoria, b c
Duncan, Graham Alexander *see* "Coercive agency"
Duncan, Henry *see* Sacred philosophy of the seasons
Duncan, Irma *see* Agenda and diaries
Duncan, Isadora *see* My life
Duncan, J G *see* Exvacations on the hill of ophel, jerusalem, 1923-1925...
Duncan, J T *see* The internal parasites of the horse (entozoa)
Duncan, John *see* Travels through part of the united states and canada in 1818 and 1819
Duncan I clinch papers – s.l. 1819-1864 – 1r – us UF Libraries [300]
Duncan, Meg *see* W c groves
Duncan, Moir B *see* The missionary mail
Duncan, Norman *see* Dr. grenfell's parish
Duncan, P *see* A narrative of the wesleyan mission to jamaica
Duncan, Patrick *see* South africa's rule of violence
Duncan, Peter *see* A narrative of the wesleyan mission to jamaica
Duncan, Richard Edward *see* Development of major community musical activities
Duncan, Robert Dick *see* Creation
Duncan, Robert Samuel *see* A history of the baptists in missouri
Duncan, Sara Jeanette *see*
– The crow's nest
– Hilda
– On the other side of the latch
– Social departure
Duncan, Sara Jeannette *see*
– An american girl in london
– The burnt offering
– Cousin cinderella
– Daughter of to-day
– His honour and a lady
– The imperialist
– The pool in the desert
– The story of sonny sahib
– Those delightful americans
– Vernon's aunt
– A voyage of consolation
Duncan, Susan C *see* The role of cognitive appraisal and friendship provisions in children's experience of affect in physical activity
Duncan, W T *see* Fort george island
Duncan, William Cecil *see*
– A brief history of the baptists and their distinctive principles and practices
– The tears of jesus of nazareth
Duncan, William Wallace *see* A new hebrew grammar
Duncan-Jones, Arthur Stuart *see* Chichester customary
Duncannon record – Duncannon, PA. 1825-1942; 1942-1975 – 13 – $25.00r – us IMR [071]
Duncker, Albert *see* Emanuel geibel's briefe an karl freiherrn von der malsburg und mitglieder seiner familie
Duncker, Dora *see* Ernst von wildenbruch
Duncker, Maximilian Wolfgang *see* Geschichte des alterthums
Duncombe, Edward *see*
– Guide to church-reform
– Letter to the hierarchy of the church of england
Duncumb, Thomas *see* The british emigrant's advocate
Dundalk and newry express and louth observer – Dundalk, Ireland. 30 jun 1860-1 jan 1870 – 4 1/2r – 1 – (aka: dundalk express; dundalk express louth meath monaghan and armagh observer) – uk British Libr Newspaper [072]
Dundalk democrat – Dundalk. Ireland. -w. 20 Oct 1849-Dec 1860 – 10r – 1 – uk British Libr Newspaper [072]
Dundalk democrat and peoples journal – Dundalk, Ireland. 20 oct 1849-24 dec 1840; 1881-96; 1921; 1922; 30 jun-4 dec 1926; 1986-90; 12 jan-dec 1991 – 65r – 1 – uk British Libr Newspaper [072]
Dundalk examiner and louth advertiser – Newry, Ireland. 1881-86; 1890-96; 1897-1901; 1902-15; 1916-29; jan-jun 1930 – 38r – 1 – (cont: newry examiner. aka: examiner) – uk British Libr Newspaper [072]
Dundalk express *see* Dundalk and newry express and louth observer
Dundalk express louth meath monaghan and armagh observer *see* Dundalk and newry express and louth observer
Dundalk herald – Ireland.1869-78; 1880-86; 1888-90; 1892-95. -w. 15 reels – 1 – uk British Libr Newspaper [072]
Dundalk herald etc – Dundalk, Ireland. Oct 1868-96; 1919 – 18 1/2r – 1 – uk British Libr Newspaper [072]
Dundalk patriot and ulster and leinster reporter – Dundalk, Ireland. -w. 11 dec 1847-19 aug 1848 – 1/4r – 1 – uk British Libr Newspaper [072]
Dundas : or, a sketch of canadian history: and more particularly of the county of dundas, one of the earliest settled counties in upper canada / Croil, James – Montreal: B Dawson, 1861 – 4mf – 9 – mf#48477 – cn Canadiana [971]
Dundas, Charles *see* Problem territories of southern africa
Dundee advertiser and courier – Dundee, Scotland 10-25 may 1926 [mf 1926-50, 1986-] – 1 – (formed by amalg of: dundee advertiser and: courier & argus; cont by: dundee courier and advertiser [26 may-2 jun 1926]) – uk British Libr Newspaper [072]
Dundee and perth saturday post and general advertiser for the midland counties of scotland – [Scotland] Dundee: A Fraser 19 may 1855-dec 1856 (wkly) [mf ed 2004] – 238v on 1r – 1 – uk Newsplan [072]
Dundee and perth weekly express – [Scotland] Dundee: J Irvine 2 jan-21 aug 1858 (wkly) [mf ed 2003] – 1r – 1 – (cont by: weekly express for the counties of forfar, perth and fife [24 apr-21 aug 1858]) – uk Newsplan [072]
Dundee and west omaha sun – Omaha, NE: David Blacker, nov 6 1958-v70 n39. sep 14 1967 (wkly) [mf ed 1959.67 (gaps) filmed -1970] – 23r – 1 – (cont: dundee and west omaha news. cont by: west omaha and dundee sun) – us NE Hist [071]
Dundee, Charles Roger *see* Collation of the sacred scriptures
Dundee christian monitor – [Scotland] Dundee: A Ewan sep-dec 1874 (mthly) [mf ed 2004] – 4v on 1r – 1 – (cont: christian monitor (dundee, scotland)) – uk Newsplan [240]
Dundee commercial gazette and shipping register – [Scotland] Dundee: J P Mathew & Co 10 oct 1860-5 oct 1861 (twice/wk) [mf ed 2003] – 104v on 1r – 1 – uk Newsplan [380]
Dundee commercial list and tay shipping register – [Scotland] Dundee: printed...by D Annan 1 feb 1840 (wkly) [mf ed 2004] – 1r – 1 – uk Newsplan [380]
Dundee courier – Dundee, Scotland 29 oct 1839, 2 jan 1844-19 apr 1861 – 1 – (cont: dundee courier & daily argus [22 apr 1861-14 apr 1862]; cont by: dundee courier & argus, & northern warder [28 feb-11 mar 1873]; dundee courier & argus [12 mar 1873-15 nov 1899 (mf 1882-83, 1885, 1887)]; courier & argus [16 nov 1899-4 may 1926]) – uk British Libr Newspaper [072]
Dundee distributor – [Scotland] London: F E Longley jan-dec 1876 (mthly) [mf ed 2004] – 12v on 1r – 1 – uk Newsplan [072]
Dundee free press [dundee, scotland : 1900] – [Scotland] Dundee: K Burke 13 nov 1900-21 aug 1903 (wkly) [mf ed 2003] – 2r – 1 – uk Newsplan [072]
Dundee free press [dundee, scotland : 1926] : a non-party paper for the community – [Scotland] Dundee: printed...by Harley & Cox 11 jun 1926-3 mar 1933 (wkly) [mf ed 2003] – 13r – 1 – uk Newsplan [072]
Dundee magazine : or a history of the present times – [Scotland] [Dundee: T Colvill 11 aug 1775-jun 1777 (wkly) [mf ed 2004] – 1v on 3r – 1 – (cont by: dundee weekly magazine; or, a history of the present times) – uk Newsplan [941]
Dundee magazine and caledonian review – [Scotland] Dundee: A M Sandeman jul 1822-apr 1823 (mthly) [mf ed 2004] – 1r – 1 – (cont by: caledonian magazine and review) – uk Newsplan [072]
Dundee magazine, and journal of the times – [Scotland] Dundee: printed by T Colvill & Son jan 1799-may 1802 (semiannual, annual) [mf ed 2004] – 4r – 1 – (ceased in 1802? apparently a revival of the dundee weekly magazine of the 1770s; cont: dundee weekly magazine; or, a history of the present times) – uk Newsplan [072]
Dundee reformer and lochee observer – [Scotland] Dundee: W Blair 14 mar 1885 (mthly) [mf ed 2004] – 1r – 1 – uk Newsplan [072]
Dundee repository : of political and miscellaneous information – [Scotland] Dundee: printed by T Colbill feb 1793-feb 1794 (semiannual) [mf ed 2004] – 2v on 1r – 1 – uk Newsplan [072]
Dundee sun – Omaha, NE: Stanford Lipsey, mar 10 1077-v83 n135. aug 31 1983 (wkly) [mf ed 1979-83 (gaps) filmed 1983] – 2r – 1 – (cont: dundee edition of the sun. merged with: south omaha sun, benson sun, north omaha sun, northwest sun, and west omaha sun to form: omaha sun (1983)) – us NE Hist [071]
Dundee warder – Scotland, UK. 9 Feb 1841-45 – 2r – 1 – uk British Libr Newspaper [072]
Dundee weekly news – [Scotland] Dundee: R Parkn jun 1855-dec 1950 – 173r – 1 – (missing: 1864; cont by: weekly news [sep 1856-dec 1862]; weekly news and telegraph [jan 1863-dec 1866]; weekly news [jan 1867-dec 1885]; dundee weekly news (town edition) [jan 1886-dec 1888]; dundee weekly news (city edition) [jan 1889-jun 1893]; dundee weekly news (city edition) [jul 1893-jun 1904]; weekly news (city edition) [jul 1904-dec 1950]) – uk Newsplan [072]
Dundee weekly news (edinburgh and the south edition) – [Scotland] Dundee: W & D C Thomson 6 jan 1894-jun 1941 (wkly) [mf ed 2003] – 1 – (cont by: dundee weekly news for edinburgh and the south [jan 1895-dec 1900]; dundee weekly news and edinburgh weekly news [jan 1901-jun 1903]; weekly news for edinburgh and the south [jul 1903-jun 1941]) – uk Newsplan [072]
Dundee weekly news (fifeshire and kinross-shire edition) – [Scotland] Dundee: W & D C Thomson jan 1894-jun 1941 (wkly) [mf ed 2003] – 106r – 1 – (cont by: dundee weekly news for fife and kinross [jan 1895-jun 1903]; weekly news for fife and kinross [jul 1903-jun 1941]) – uk Newsplan [072]
Dundee weekly news for forfarshire – [Scotland] Dundee: W & D C Thomson 20 jan 1894-21 feb 1903 (wkly) [mf ed 2004] – 12r – 1 – (cont: dundee weekly news (forfarshire ed); cont by: weekly news for forfarshire) – uk Newsplan [072]
Dundee weekly news for glasgow and the west – [Scotland] Dundee: W & D C Thomson 27 jan 1894-dec 1950] [mf ed 2003] – 125r – 1 – (cont as: dundee weekly news and the glasgow weekly news [jan 1897-dec 1898]; glasgow weekly news and the dundee weekly news for glasgow and the west [1899]; glasgow weekly news and the dundee weekly news [jan 1900-dec 1901]; glasgow weekly news [jan 1902-dec 1903]; glasgow weekly news city edition [jan-jun 1904]; glasgow weekly news counties edition [jul 1904-jun 1910]; glasgow weekly news [jul 1910-dec 1950]) – uk Newsplan [072]
Dundee weekly news for perthshire – [Scotland] Dundee 20 jan 1894-26 mar 1910 [mf ed 2003] – 24r – 1 – (cont as: weekly news for perthshire [jul 1903-mar 1910]) – uk Newsplan [072]
Dundee weekly news for stirling district – [Scotland] Dundee: W & D C Thomson jan 1894-jun 1941 (wkly) [mf ed 2003] – 105r – 1 – (missing: 1920; cont as: weekly news for stirling district [jul 1903-jun 1941]) – uk Newsplan [072]
Dundee weekly news (forfar ed) – [Scotland] Dundee: W & D C Thomson 13 jan 1894-dec 1911 [mf ed 2004] – 29r – 1 – (cont as: dundee weekly news for forfarshire [jan 1895-jun 1903]; weekly news for forfarshire [jul 1903-dec 1911]) – uk Newsplan [072]
Dundy county pioneer – Benkelman, NE: Frank Israel & Son. v1 n1. apr 30 1885 (wkly) [mf ed -aug 12 1892 (gaps)] – 1r – 1 – us NE Hist [071]
The dundy democrat – Benkelman, NE: Howard & Andrews (wkly) [mf ed v3 n2. may 17-jun 7 1889 (gaps) filmed 1973] – 1r – 1 – us NE Hist [071]
Dunedin, florida / Phillips, Roland – s.l, s.l? 1936 – 1r – us UF Libraries [978]
Dunedin star midweek – jan 1982-dec 1987; jul-dec 1989 – 14r – 1 – mf#81.6 – nz Nat Libr [079]
Dunedin star weekender – jul 1980-dec 1987 – 15r – 1 – mf#81.7 – nz Nat Libr [079]
Dunedin times – Dunedin, FL. 1988 mar 21-1999 apr – 3r – (gaps) – us UF Libraries [071]
Dunets, Kh *see* Af literarishe temes
Dunfermline citizen and west of fife mail – [Scotland] Fife, Dunfermline: L Macbean 9 jan 1895-2 jan 1910 (wkly) [mf ed 2004] – 9r – 1 – (cont by: dunfermline citizen and fife mail [9 jan 1907-22 jan 1908]; fife mail (kirkcaldy, scotland) [28 jan 1908-12 jan 1910]) – uk Newsplan [072]
Dunfermline co-operative citizen – [Scotland] Fife, Dunfermline: National Co-operative Publ Soc jul 1932-jun 1933 (mthly) [mf ed 2004] – 12v on 1r – 1 – uk Newsplan [072]
Dunfermline express – [Scotland] Fife, Dunfermline: W Clark & Sons 3 oct 1900-11 nov 1925 (wkly) [mf ed 2004] – 14r – 1 – (absorbed by: rosyth & forth mail fr apr 23 1918-oct 30 1919) – uk Newsplan [072]
Dunfermline herald and post – Edinburgh: Scotsman Communications -2001 (wkly) [mf ed 6 jan 1994-] – 1 – (cont by: herald & post. fife) – uk Scotland NatLib [072]
Dunfermline monthly advertiser – [Scotland] Fife, Dunfermline: J Miller & Son, Printers 15 jan 1858-15 may 1863 (mthly) [mf ed 2004] – 1r – 1 – (missing: may 1858, jan 1861, apr 1862, oct 1862; cont: j miller & son's monthly advertiser for the western district of fife) – uk Newsplan [072]
Dunfermline press, and west of fife advertiser – Dunfermline: A Romanes 1897- (wkly) [mf ed 1 jul 1994-] – 1 – (cont: dunfermline saturday press, & west of fife advertiser; suppl: dp lifestyle) – uk Scotland NatLib [072]
Dunfermline saturday press, & west of fife advertiser – [Scotland] Fife, Dunfermline: A Romanes jan 1893-dec 1950 (wkly) [mf ed 2004] – 69r – 1 – (cont: saturday press (dunfermline, scotland); cont by: dunfermline press, & west of fife advertiser) – uk Newsplan [072]
Dung [Saigon] *see*
– Loi co mu suong
– Mot chut yeu tren vanh moi uot
Dungan, D R *see* Hermeneutics
Dungan, David Roberts *see* Lectures on the modern phases of skepticism
Dungannon democrat and nationalist weekly – Dungannon, Ireland. 12 feb 1913-1918; feb 1919-1923 – 7r – 1 – uk British Libr Newspaper [072]
Dungannon democrat [n ireland] *see* Democrat
Dungannon news and county tyrone advertiser – Dungannon, Ireland. 6 jul 1893-13 may 1915 – 11 1/2r – 1 – uk British Libr Newspaper [072]
Dungannon news and tyrone courier *see* Tyrone courier
Dungannon observer – Dungannon, Ireland – 4r – 1 – uk British Libr Newspaper [072]
Dungannon observer – [Northern Ireland] Belfast jun 1935-50 [mf ed 2002] – 15r – 1 – uk Newsplan [072]

Dungarvan leader – Dungarvan 1943-57 – 7r – 1 – ie National [072]
Dungarvan leader *see* Dungarvan leader and southern democrat
Dungarvan leader and southern democrat – Dungarvan, Ireland. 19 apr 1958-21 dec 1991; 1992-97 – 39 1/4r – 1 – (aka: dungarvan leader) – uk British Libr Newspaper [072]
Dungarvan observer and munster industrial advocate – Dungarvan, Ireland. 10 feb 1912-1921; mar 1925-10 oct 1927; 25 apr 1936-2 jan 1993 – 60 3/4r – 1 – uk British Libr Newspaper [072]
Dunghen, Henry *see* Opera omnia
Dungog chronicle – Dungog, jan 1969-dec 1993 – 12r – at Pascoe [079]
Dungog chronicle – Jun 12 1888-dec 20 1968 – 34r – at Pascoe [079]
Dungravan observer – Waterford 1927-36 – 8.5r – 1 – ie National [072]
Dunham, Lowell *see* Romulo gallegos, vida y obra
Dunia internasional / Departemen Penerangan – Djakarta, 1950/51-1959 – 101mf – 9 – (missing: 1950, v1(1-7, 9, 11-12); 1951, v2(1-2); 1959, v10(10-12)) – mf#SE-528 – ne IDC [959]
Dunia madrasah – Djakarta, 1954-1956 – 5mf – 9 – (missing: 1954/55, v1(2-7, 9, 10); 1955, v2(13-16)) – mf#SE-361 – ne IDC [959]
Dunia wanita – Medan, 1949-1966 – 47mf – 9 – (missing: 1949, v1(2-end); 1950, v2(1-26, 28-end); 1951, v3; 1952, v4(1-3, 5-16, 18-end); 1956, v8(1-4, 6, 9-end); 1957, v9(1-3, 11, 16, 17, 19-end); 1960-1961, v12(1-19, 21-end); 1962, v13(1-2, 5-end); 1964, v15(1-2)) – mf#SE-880 – ne IDC [950]
Dunigan's american catholic almanac and list of the clergy, for the year of our lord... – New York: Edward Dunigan and Brother, 1858-59 – (= ser List Of The Clergy, For The Year Of Our Lord) – 1r – 1 – $40.00r – us Notre Dame [240]
'Dunkelheit' und freiheit : ursachen und wirkungen des spekulativen im politischen denken hegels / Reiter, Raimond – (mf ed 2000) – 3mf – 9 – €49.00 – 3-8267-2704-5 – mf#DHS 2704 – gw Frankfurter [110]
The dunkers : a sociological interpretation / Gillin, John Lewis – New York: [s.n.], 1906 – 1mf – 9 – 0-524-03260-2 – (incl bibl ref) – mf#1990-4663 – us ATLA [240]
Dunkin, Christopher *see* Speech delivered in the legislative assembly during the debate on the subject of the confederation of the british north american provinces
Dunkle, William Frederick *see* Memorial methodist episcopal church, south...
Ein dunkler punkt / Fabri, Friedrich – Gotha: FA Perthes, 1880 – 1mf – 9 – 0-524-03519-9 – mf#1990-1024 – us ATLA [943]
Ein dunkles loos : volkserzaehlung / Bechstein, Ludwig – Nuernberg: F Korn, 1850 [mf ed 1993] – 3v – 1 – mf#8534 – us Wisconsin U Libr [880]
Dunklin County Genealogical Society *see* Semo neark record
Dunkmann, Karl *see*
- Der historische jesus, der mythologische christus und jesus der christ
- Das religioese apriori und die geschichte
- Das sakramentsproblem in der gegenwaertigen dogmatik
- Die theologische prinzipienlehre schleiermachers nach der kurzen darstellung und ihre begruendung durch die ethik

Dunlap, Erik M *see* An assessment of the nature and prevalence of sport psychology service provision in professional sports
Dunlap, James Eugene *see* Office of the grand chamberlain in the later roman and byzantine
Dunlap, Samuel Fales *see*
- The ghebers of hebron
- Sod, the mysteries of adoni
- Sod, the son of the man
- Vestiges of the spirit-history of man

Dunlap, Susan *see* Letters
Dunlap's maryland gazette : or, the baltimore general – Baltimore MD. 1775 may 2-1779 jan 5 – 1r – 1 – (cont by: maryland gazette, and baltimore general advertiser) – mf#908464 – us WHS [071]
Dunlavy, John *see* The manifesto
Dunlevie, Horace G [comp] *see* Our volunteers in the north-west
Dunlevy genealogical history, 1901 – 1r – 1 – (also spelled dunlavy) – mf#B27438 – us Ohio Hist [978]
Dunlop, Alexander *see* Answer to the dean of faculty's "letter to the lord chancellor"
Dunlop, John *see*
- Compulsory drinking usages
- Memories of gospel triumphs among the jews during the victorian era

Dunlop, John Kinninmont *see* Development of the british army, 1899-1914
Dunlop, Robert *see* Life of henry grattan
Dunlop, William *see* The uses of creeds and confessions of faith
Dunn, Arthur William *see* An analysis of the social structure of a western town
Dunn county lumberman – Menomonie WI. 1862 apr 19-1865 apr 15, nov 25-1866 mar 31 – 1r – 1 – (cont by: dunn county news) – mf#1127213 – us WHS [634]
Dunn county news – Menomonie WI. 1866 apr 7/1867 aug 10-2003 nov/dec – 217r – 1 – (cont: dunn county lumberman) – mf#1127214 – us WHS [071]
Dunn county pictorial messenger – Menomonie WI. 1938 may 4-dec 8 – 1r – 1 – mf#1097268 – us WHS [071]
Dunn County School of Agriculture and Domestic Economy *see* Bulletin of the dunn...
Dunn county schools – v1 n3-5 [1919 oct 13-nov 10], v1 n7-16 [1919 dec 8-1920 may 3 – 1r – 1 – mf#5195292 – us WHS [370]
Dunn, Henry *see*
- The destiny of the human race
- Guatimala, or, the republic of central america, in 1827-8
- The kingdom of god, or, what is the gospel?
- Liber librorum
- Reply to the misrepresentations of the rev francis close and other...
- The study of the bible

Dunn, James B *see* The pope's last veto in american politics
Dunn, Lewis Romaine *see*
- The angels of god
- A manual of holiness and review of dr james b mudge
- The mission of the spirit
- Sermons on the higher life

Dunn, Martin *see* Martin dunn's descriptive circular of florida, groves, residences a...
Dunn, Nathan *see* "Ten thousand chinese things"
Dunn, Oscar *see*
- L'amerique avant christophe colomb
- Catalogue d'une bibliotheque canadienne
- Dix ans de journalisme
- Glossaire franco-canadien et vocabulaire de locutions vicieuses usitees au canada
- Lecture pour tous
- L'union des catholiques
- L'union des partis politiques dans la province de quebec

Dunn, Ransom *see* Lectures on systematic theology
Dunn, Thomas William Shea [comp] *see* Almanach judiciaire de la province de Quebec
Dunn, W *see* Paper read before the ruri-decanal chapter of leeds
Dunn, William Edward *see* Spanish and french rivalry in the gulf region of t...
Dunne, Edmund Michael *see* Memoirs of zi pre'
Dunne, Finley Peter *see*
- Mr dooley
- Mr dooley in peace and in war
- Mr dooley says

Dunning, Albert Elijah *see*
- Children's sunday
- The sunday-school library

Dunning, Albert Elijah et al *see* Congregationalists in america
Dunning, Nelson A *see* The philosophy of price
Dunning, William A *see* A history of political theories from luther to montesquieu
Dunnington *see* Music part books
Dunolly and betbetshire express – Australia, 1 Jun 1875-12 Jun 1917 (imperfect) – 41r – 1 – uk British Libr Newspaper [072]
Dunoon advertiser and district courier – [Scotland] Argyll & Bute, Dunoon: R Craig 15 jan 1903-30 dec 1926 (wkly) [mf ed 2003] – 12r – 1 – uk Newsplan [072]
Dunoon herald and cowal advertiser – [Scotland] Argyll & Bute, Dunoon: D Lawson 18 mar 1876-dec 1890 (wkly) [mf ed 2003] – 4479v on 35r – 1 – uk Newsplan [072]
Dunoon observer & argyllshire standard – [Scotland] Argyll & Bute, Dunoon: J, E & R Inglis 5 jan 1895-dec 1950 (wkly) [mf ed 2003] – 37r – 1 – (formed by the union of: dunoon observer and cowal watchman and: argyllshire standard, and advertiser for the coast; absorbed: dunoon and district bulletin; suppl: dunoon and district bulletin, cowal's pictorial weekly – accompany some iss) – uk Newsplan [072]
Dunoon observer & argyllshire standard – Dunoon: J E & R Inglis 1895- (wkly) [mf ed 7 jan 1995-] – 1 – (formed by union of: dunoon observer & cowal watchman, and: argyllshire standard, and advertiser for the coast; absorbed: dunoon & district bulletin; suppl: dunoon & district bulletin, cowal's pictorial weekly – accompany some iss; newsplan 2000 [mf ed 2003] 37r) – uk Scotland NatLib; uk Newsplan [072]
Dunoon observer & cowal watchman – [Scotland] Argyll & Bute, Sandbank: J, E & R Inglis 20 oct 1886-26 dec 1894 (wkly) [mf ed 2003] – 7r – 1 – (cont: cowal watchman, and advertiser for the counties of argyll, ayr, bute, dumbarton, and renfrew; merged with: argyllshire standard, and advertiser for the coast to form: dunoon observer and argyllshire standard) – uk Newsplan [072]
Dunoon telegraph – [Scotland] Argyll & Bute, Dunoon: Harvey & Co 22 mar 1889-21 mar 1890 (wkly) [mf ed 2003] – 1r – 1 – (began in 1887) – uk Newsplan [072]
Dunord, Charles *see* Aux urnes, citoyennes!
Dunoyer, Anne Marguerite *see* Memoires de mme dunoyer ecrits par elle-meme
Dunraven, William Thomas Wyndham-Quin *see* Notes on irish architecture
Dunraven, Windham Thomas Wyndham-Quin, 4th Earl of *see* The irish question examined in a letter to the "new york herald"
Dun's business month – Boston MA 1982-86 – 1,5,9 – (cont: dun's review; cont by: business month) – ISSN: 0279-3040 – mf#202.02 – us UMI ProQuest [650]
Duns, J *see* Creation according to the book of genesis and the confession of faith...
Duns, John *see*
- Christianity and science
- Creation according to the book of genesis and the confession of faith

Dun's review – Boston MA 1893-1980 – 1,5,9 – (cont by: dun's business month) – ISSN: 0012-7175 – mf#202.02 – us UMI ProQuest [650]
Dun's review – v1-29. 1893-1921 – 1 – us L of C Photodup [380]
Duns Scotus, John *see* Opera omnia
Dun's statistical review – Ann Arbor MI 1950-57 – 1 – mf#201 – us UMI ProQuest [332]
Dunscomb, J W [comp] *see* The provincial laws of the customs
Dunscombe, Aubrey Elsworth *see* Root system of the tung oil tree
Dunshee, Henry Webb *see* History of the school of the reformed protestant dutch church
[Dunsmuir-] dunsmuir dispatch – CA. 1910-11 – 1r – 1 – $60.00 – mf#B02184 – us Library Micro [071]
[Dunsmuir-] dunsmuir news – CA. 1890-1917; 1919- – 68r – 1 – $4080.00 (subs $90/y) – mf#B02187 – us Library Micro [071]
[Dunsmuir-] herald – CA. 1897-98 – 1r – 1 – $60.00 – mf#B02185 – us Library Micro [071]
[Dunsmuir-] mott north star – CA. 1887-90 [wkly] – 1r – 1 – $60.00 – mf#B02186 – us Library Micro [071]
[Dunsmuir-] plain dealer – CA. Mar-Dec 1912 [wkly] – 1r – 1 – $60.00 – mf#B02188 – us Library Micro [071]
[Dunsmuir-] tribune – CA. 1926-27 [wkly] – 1r – 1 – $60.00 – mf#B02189 – us Library Micro [071]
Dunstable – (= ser Bedfordshire parish register series) – 2mf – 9 – £5.00 – uk BedsFHS [929]
Dunstable 1679-1849 – Oxford, MA (mf ed 1996) – (= ser Massachusetts vital record transcripts to 1850) – 11mf – 9 – 0-87623-244-6 – (mf 1t: births & deaths 1679-1746. mf 1t-2t: marriages 1680-1839. mf 2t-4t: births 1730-1847. mf 4t: deaths 1742-1821. mf 4t-5t: marriages 1779-84, 1843-44. mf 5t-9t: births & deaths 1724-1848. mf 6t: out-of-town marriages 1682-1799. mf 9t: marriages 1757-73. mf 9t-10t: marriages & intentions 1790-1849. mf 10t: births & deaths 1757-1804; births 1830-49. mf 11t: marriages & deaths 1844-49) – us Archive [978]
Dunstable 1679-1900 – Oxford, MA (mf ed 1995) – (= ser Massachusetts vital records) – 99mf – 9 – 0-87623-380-9 – (mf 1-4: vital records 1679-1847. mf 5-6: vital records 1724-1802. mf 6-15: town records 1743-1790. mf 10,20: marriages 1757-1773. mf 14-15: vitals 1724-1801. mf 16-24: town records 1743-90. mf 25-32: town records 1790-1822. mf 35-36: town records 1792-1823. mf 33: out-of-town marriages 1682-1799. mf 33: intentions 1888-90. mf 33-36: vitals 1724-1847. mf 37-42: town records 1823-53. mf 43-48: town records 1824-88. mf 49-55: taxes 1800-18. mf 56-62: taxes 1819-38. mf 63-70: taxes 1839-58. mf 71-79: accounts 1785-1880. mf 80-86: accounts 1800-78. mf 87-88: voters 1884-1915. mf 89-93: church records 1834-98. mf 93: vital records 1834-84. mf 94: marriages 1790-1843. mf 94-95: intentions 1826-87. mf 96: births 1830-92. mf 97-98: marriages, deaths 1844-92. mf 99: vital records 1893-1900) – us Archive [978]
Dunstable chronicle – Dunstable, England. 5 Jan 1856-28 Jul 1860 – 1 1/2r – 1 – uk British Libr Newspaper [072]
Dunstable, priory st peter monumental inscriptions – Arthur Weight Matthews 1912 – (= ser Bedfordshire parish register series) – 1mf – 9 – £1.25 – uk BedsFHS [929]
Dunstan times – 1890-1939 – 59r – 1 – mf#83.7 – nz Nat Libr [079]
Dunstane, William *see* [Butte county-] history of wyandotte, butte county, california
Dunton – (= ser Bedfordshire parish register series) – 2mf – 9 – £5.00 – uk BedsFHS [929]
Dunton, st mary monumental inscriptions monumental inscriptions – Arthur Weight Matthews 1914 – (= ser Bedfordshire parish register series) – 1mf – 9 – £1.25 – uk BedsFHS [929]
Dunya – Stockholm: Kumitah-'i Markazi-i Hizb-i Tudah-'i Iran, 1974-79. dawrah-'i 3, sal-i 1, shumarah-'i 2-sal-i 5, shumarah-'i 12 murdad 1353-isfand 1357 [jul 1974-mar 1979]; wh incl sal-i 1, shumarah-'i 2,3,5-8; sal-i 2, shumarah-'i 1,2-5,8-10,12; sal-i 3, shumarah-'i 1-2,4-5,8-10,12; sal-i 4, shumarah-'i 1-3,5-8; sal-i 5, shumarah-'i 5,8-12 – 2r – 1 – $106.00 – (missing: sal-i 1, shumarah-i 4,9-12; sal-i 2, shumarah-i 6-7,11; sal-i 3, shumarah-i 3,6-7,11; sal-i 4, shumarah-i 4,9-12; sal-i 5, shumarah-i 1-4,6-7) – us MEDOC [079]
Dunya – [Tehran]: Hizb-i Tudah-'i Iran, 1941-1946/47. dawrah-'i 2, sal-i 2, shumarah-i 1-4; sal-i 3, shumarah-i 1-4; sal-i 4, shumarah-i 3-4; sal-i 6, shumarah-i 1-4; sal-i 7, shumarah-i 1-4. bahar 1340-zimistan 1345 [spr 1961-win 1966/67] – 2r – 1 – $106.00 – us MEDOC [079]
Dunya-yi sukhan – Tehran. dawrah-'i jadid, sal-i 1, shumarah-i 1-26. bahman 1364-urdibihisht 1368 – 1r – 1 – $53.00 – (missing: n14,21) – us MEDOC [079]
Duo : pour le violon et viola / Kaczkowski, Joachim – [18–?] [mf ed 19–] – 2pt on 1r – 1 – mf#pres. film 33 – us Sibley [780]
Duo concertant pour cor et viola / Makoweczky – Leipsic: Breitkopf & Haertel [179-?] [mf ed 1989] – 2pt on 1r – 1 – mf#pres. film 59, 39 – us Sibley [780]
Duo dans le roi et le fermier... / Monsigny, Pierre-Alexandre – Paris: Les Freres Gaveaux [c1805] [mf ed 1991] – 1r – 1 – (french words) – mf#pres. film 108 – us Sibley [780]
Duo de felicie : musique de catrufo. partie de piano [violon] / Catrufo, Giuseppe – [182-?] [mf ed 19–] – 2pt on 1r – 1 – (for violin & piano; originally for soprano, baritone & orchestra) – mf#pres. film 33 – us Sibley [780]
Duo del matrimonio per raggiro... : avec accompagnement de piano ou harpe / Cimarosa, Domenico – Paris: Carli [1812?] [mf ed 1991] – 1r – 1 – (italian & french words) – mf#pres. film 108 – us Sibley [780]
Duo pour deux pianos op 8 bis : d'apres les duos pour piano & orgue [op 8] / Saint-Saens, Camille – Paris: Vve E Girod [c1898?] [mf ed 1989] – 1r – 1 – mf#pres. film 57 – us Sibley [780]
...Duo sermones apologetici de dignitate eucharistiae / Oecolampadius, J – Tiguri, Froschoverus, [1550] – 1mf – 9 – mf#PBU-398 – ne IDC [240]
Duo tractatus, quorum alter vocatur florigerus / Augustinus (Augustine, Saint, Bishop of Hippo) [comp] – Coloniae, c1480 – €7.00 – ne Slangenburg [241]
Duoc nha nam – Saigon. 26 sept 1928-juil 1937 – 1 – (n'a probablement pas paru entre le 29 oct 1929, n153 et le 16 avr 1930, n1) – fr ACRPP [073]
Duoc-tue – Hanoi. dec. 1935-aout 1945 – 1 – fr ACRPP [073]
Duodecim prophetarum minorum libros : in lingua aegyptiaca vulgo coptica seu memphitica / ed by Tattam, Henry – Oxonii: E typographeo academico, 1836 – 1mf – 9 – 0-8370-1978-8 – mf#1987-6365 – us ATLA [220]
Duodecim specula deum aliquando videre desideranti concinnate / David, J – Antverpiae: Ex officina Plantiniana, apud Ioannem Moretum, 1610 – 3mf – 9 – mf#O-220 – ne IDC [090]
Duoi anh sang duong loi van nghe cua dang : tieu lua n, phe binh / Phan Nhan – Ha Noi: Van Hoc 1974 [mf ed 1992] – on pt of 1r – 1 – mf#11052 r74 n6 – us Cornell [959]
Duophile : ou, le plaisir de se voir deux / Ruppierre – Paris, France. 1805 – 1r – us UF Libraries [440]
Dupac de Bellegarde, M G *see* Histoire abregee de l'eglise metropolitaine d'utrecht
DuPage County [IL] Genealogical Society *see* Review of the dupage county [il] genealogical society
Dupanloup, Felix *see*
- Les alarmes de l'episcopat justifiees par les faits
- The child
- Convention du 15 septembre et l'encyclique du 8 decembre
- De l'education
- Etude sur la franc-maconnerie
- Instruction pastorale de monseigneur l'eveque d'orleans
- Ueber das naechste allgemeine concil

Dupasquier, S *see* Summa philosophiae scholasticae et scotisticae...
Dupaty, Emmanuel *see*
- Deux peres
- Prison militaire
- Triomphe du mois de mars

Duperrey, L I *see* Voyage autour du monde...
Duperrey, M L I *see* Voyage autour du monde sur la corvette "la coquille", 1822-25
Dupes et demagogues : a souvenir / Albyn [i.e. Andrew Shiels] – S.l: s.n, 1879 – 1mf – 9 – mf#05826 – cn Canadiana [810]

Dupeuty, Charles *see*
- Bonaventure
- Campagne a deux
- Humoriste

Dupeuty, M (Charles) *see*
- Anacreon
- Perruquier de l'empereur

Dupierris, Martial *see* Cuba y puerto rico

Dupin, Andre Marie Jean Jacques *see* Consultation de m dupin

Dupin, Andre-Marie-Jean-Jacques *see*
- Jesus devant caiphe et pilate
- Principia juris civilis tum romani tumgallici seu

Dupin, Charles *see*
- The commercial power of great britain
- View of the historical and actual state of the military force of great britain

Dupin, Henri *see*
- Courtisans
- Farinelli

Dupin, M (Henri) *see*
- Mort et le bucheron
- Roger-bontems, ou, la fete des fous

Dupla defesa, resposta ao pamphleto / Humphrey, Henry M – 1897 – 1 – $50.00 – us Presbyterian [240]

Duplatre, Louis *see* Essai sur la condition de la femme au siam

Duplessis, Claude *see* Oeuvres de mr duplessis

Duplessis donne a sa province une saine legislation agricole – [Quebec (Province): Union nationale, 1948?] (mf ed 1992) – 1mf – 9 – mf#SEM105P1659 – cn Bibl Nat [340]

Duplessis, Francois-Xavier *see* Lettres du p f x duplessis de la compagnie de jesus

Duplessis, G *see*
- Les emblemes d'alciat
- Le livre des peintres et graveurs

Duplessis, Georges *see* The wonders of engraving

Duplessis, T *see* Bibliographie necrologique des religieuses hospitalieres de saint-joseph

Duplessis-mornay considere comme theologien et principalement comme apologiste / Schaeffer, Adolphe – Strasbourg: Berger-Levrault, 1849 [mf ed 1991] – 1mf – 9 – 0-524-00322-X – (incl bibl ref) – mf#1989-3022 – us ATLA [230]

Dupont, Edouard *see* Histoire de la rochelle

Dupont, Henry Bonaventure *see* Principes de violon par demandes et par reponce par le quel toutes personne

Dupont, Jerry *see* The law library to the year 2000

Dupont, Louis E *see* On hastening the natural coloration of citrus fruits

Duport, J H *see* Outlines of a grammar of the susu language.

Duport, Jean Louis *see* Trois nocturnes en duo pour piano et violoncelle [ou violon]...ler livre

Duport, Paul *see*
- Depositaire
- Ecrin

Duport, Pierre Landrin *see* Us country dances with figures

Duppa, Richard *see*
- Miscellaneous observations and opinions on the continent
- Travels in italy, sicily, and the lipari islands

Dupre, Adrien *see* Voyage en perse, fait dans les annees 1807, 1808 et 1809, en traversant la natolie et la mesopotamie

Dupre, Alphonse *see* Relation d'un voyage en italie

Dupre de Saint-Maur, Emile *see*
- L'hermite en russie
- Petersbourg, moscou et les provinces

Duproix, Paul *see* Kant et fichte et le probleme de l'education

Dupuis, J *see*
- Journal of a residence in ashantee
- Journal of a residence in ashanti

Dupuis, Joseph *see* Journal of a residence in ashantee

Dupuis, Monique *see* Bibliographie analytique de la chanson de folklore

Duputacion Provincial *see* Cuatro decretos basicos para el desarrollo agrario de la provincia

Dupuy, Paul *see*
- L'enseignement manuel de l'enfant dans l'ecole primaire
- Enseignement pratique et technique
- Les illustrations canadiennes
- Madame de la peltrie
- Sanctuaire de sainte anne de beaupre
- Villemarie

Dupuy, Pierre *see* Le nouvel anacharsis dans la nouvelle grece, ou l'hermite d'epidaure

Dupuy, Starke *see* Hymns and spiritual songs

Duque Botero, Guillermo *see* Apuntes para la historia del clero de caldas

Duque de t'serclaes toma posesion academico numero real de la historia. noticias / Fita, Fidel & Rodriguez Villa, Antonio – Madrid: Fortanet, 1909. B.R.A.H. 54, 1909, p. 438 – sp Bibl Santa Ana [946]

Duque Fuentes, Martin *see*
- Programa de latin 1er.curso
- Programa de latin 2nd curso
- Programa de latin 3rd curso

Duque Gomez, Luis *see*
- Colombia
- Historia de pereira

Duques de endor / Arevalo Martinez, Rafael – Guatemala, 1940 – 1r – us UF Libraries [972]

Duques de la torre / Dominguez Peaz, Fidel – 1883 – 9 – sp Bibl Santa Ana [946]

Duquesne law review – v1-39. 1963-2001 – 5,6,9 – $777.00 set – (v1-23 1963-85 on reel $403. v24-39 1985-2001 on mf $374) – ISSN: 0093-3058 – mf#102531 – us Hein [340]

Duquesne science counselor for better science training – Pittsburgh PA 1935-67 – 1,5,9 – mf#1501 – us UMI ProQuest [500]

Duquet, Joseph-Norbert *see* Le miroir des caracteres

Duquoin first baptist church. duquoin, illinois : church records – 30 May 1857-1966 – 1 – us Southern Baptist [242]

Dur und moll – v1-10. 1922-32 – (= ser Muenchener musik) – 1 – $40.00 – us L of C Photodup [780]

Durability in art / Wilkins, William Noy – London 1875 – (= ser 19th c art & architecture) – 1mf – 9 – mf#4.2.1564 – uk Chadwyck [700]

Durach, Moritz *see* Christian fuerchtegott gellert

Duran, Andre *see* Le mysticisme de calvin d'apres l'institution chretienne

Duran, Augusto *see* Voceros del pueblo en el parlamento

Duran Castillo, Benito *see* Ecos de silencio; poesia y cuentos

Duran de Montijo, Juan *see*
- Adviento y sermones varios
- Santoral seraphico
- Sermones de cuaresma...siete sabios de grecia
- Sermones panegiricos de santos

Duran, Diego *see* Aztecs

Duran i Jorda, Frederic *see* The service of blood transfusion at the front

Duran, Miguel Angel *see*
- Ausencia y presencia de jose matial delgado
- Historia de la universidad de el salvador

Duran Munoz Garcia y, Alonso Buron Francisco *see* Ramon y cajal, tomo 1

Duran Ramas, Maria de los Angeles *see* Arias montano y su tratado "de optimo imperio"

Duranczyk, Denise M *see* Mechanical energy analysis of walking in elderly men

Durand, A *see* The making of a frontier

Durand, Auguste *see*
- Trois airs varies pour le violon avec accompagnement de basse, oeuvre 5e
- Trois duos concertants pour deux violons

Durand, Camille H *see* Details particuliers sur la journee du 10 aout 1792

Durand, Charles *see* Reminiscences of charles durand of toronto, barrister

Durand de troarn et les origines de l'heresie berengarienne / Heurtevent, Raoul – Paris: Gabriel Beauchesne, 1912 [mf ed 1992] – (= ser Etudes de theologie historique 5) – 1mf – 9 – 0-524-04015-X – (incl bibl ref) – mf#1990-1187 – us ATLA [240]

Durand de Villegagnon, N *see* De bello melitensi

Durand, Elliott *see* Week in cuba

Durand, Guillaume *see* Durandus on the sacred vestments

Durand, Henry Mortimer *see*
- Central india in 1857
- Charm of persia

Durand, J B L *see* Voyage au senegal

Durand, J N L *see* Precis des lecons d'architecture donnees a l'ecole polytechnique...

Durand, Jean *see* Voyage au senegal

Durand, Jean B *see* A voyage to senegal

Durand, Jennings F *see* Comparative graduation rates and grade point averages among regular admit, non-competitive admit, and admissions exception student-athletes of the university of north carolina at chapel hill

Durand, Joseph-Pierre *see* Apercus de taxinomie generale

Durand, Laurent *see* Cantiques de marseilles accommodes a des airs vulgaires

Durand, Louis *see* L'infallibilite papale prise en manifeste et flagrant delit de mensonge

Durand, Sophie *see*
- Memoires sur napoleon
- Mes souvenirs sur napoleon

Durand times – Durand WI. [1865 dec 12-1869 aug 27], 1870 mar 19-1871 apr 18 – 2r – 1 – (cont by: durand weekly times) – mf#964281 – us WHS [071]

Durand, U *see*
- Thesaurus novus anecdotorum
- Veterum scriptorum et monumentorum historicorum, dogmaticorum, moralium amplissima collectio

Durand, Valentin *see* Le jansenisme au 18e siecle et joachim colbert evaeque de montpellier (1696-1738)

Durand weekly times – Durand WI. 1871 may 2-1873 jun 20, 1873 jun 27-1877 oct 5, 1877 oct 12-1878 dec 13 – 3r – 1 – (cont: durand times) – mf#927375 – us WHS [071]

Durand, William *see* The symbolism of churches and church ornaments

Durandus de S Porciano (Durandus of Saint-Pourcain) *see* In sententias theologicas petri lombardi commentarium libri quattuor

Durandus, Gulielmus *see* Rationale divinorum officiorum

Durandus on the sacred vestments : an english rendering of the third book of the rationale divinorum officiorum of durandus, bishop of mende, a.d. 1287 = Rationale divinorum officiorum. book 3 / Durand, Guillaume – London: Thomas Baker, [1899?] – 1mf – 9 – 0-8370-6898-3 – (in english. incl bibl ref and index) – mf#1986-0898 – us ATLA [240]

Durango, 1937 – n.p., 1937. Fiche W 848. (Blodgett Collection of Spanish Civil War Pamphlets) – 9 – us Harvard [946]

Durango klansman *see* Miscellaneous newspapers of la plata county, colorado

Durango, martyrstaden : ett tyskt bombardemang dess orsaker och verkningar – Stockholm, 1937. Fiche W 849. (Blodgett Collection of Spanish Civil War Pamphlets) – 9 – us Harvard [946]

Durango. Mexico (State) *see*
- Periodico oficial
- Periodico oficial del gobierno del estado de durango

Durango telegraph *see* Miscellaneous newspapers of la plata county, colorado

Durango, ville martyre; ce que furent les bombardements de la ville de durango par les avions allemands / Comite Franco-Espagnole – Paris, 1937? Fiche W 809-810. (Blodgett Collection of Spanish Civil War Pamphlets) – 9 – us Harvard [946]

Durant first baptist church. durant, oklahoma : church records; bulletins – 1942-47 – 1 – $52.74 – us Southern Baptist [242]

Durant le premier...sovietique (1917-1920). paris / Vidal, J M Moscou – Madrid: Razon y Fe, 1934 – 1 – sp Bibl Santa Ana [946]

Durant, Thomas *see* Sermon occasioned by the death of the rev james weston

Durant, Will *see* The case for india

Durant, William *see* The church and its polity

Durante el semestre pasado fallecieron... tambien d. pedro maria plano, en merida... / Fita, Fidel & Rodriguez Villa, Antonio – Madrid: Est.Tip. Fortanet, 1901. B.A.R.H. 38, 1901, p. 75 – sp Bibl Santa Ana [946]

Durantis, Gulielmus, Bishop of Mende *see* The symbolism of churches and church ornaments

Duratin, Armand *see* Heloise paranquet

Duration and nature of future punishment / Constable, Henry – London, England. 1868 – 1r – us UF Libraries [240]

The duration and nature of future punishment / Constable, Henry – New Haven, Conn: Chas C Chatfield, 1871 – 1mf – 9 – 0-524-07842-4 – (incl bibl ref) – mf#1992-1108 – us ATLA [240]

Duration of future punishments / Barker, William – London, England. 1865 – 1r – us UF Libraries [240]

Duray, Nicholas A *see* Age of introduction and current frequency of participation in league and casual bowlers

Durban : fifty years' municipal history / Henderson, W P M – Durban: Robinson and Co, 1904 – us CRL [978]

Durban / Kuper, Leo – London, England. 1958 – 1r – us UF Libraries [960]

Durban advocate / general advertiser – Durban, 1852 – 1r – 1 – sa National [079]

Durban free press / hotel advertiser – Durban, SA. oct 1905 – 1r – 1 – sa National [079]

Durban housing survey / University Of Natal Dept Of Economics. Research Section – Pietermaritzburg, South Africa. 1952 – 1r – us UF Libraries [360]

Durban mercantile shipping gazette *see* Natal times / durban mercantile shipping gazette

Durban observer – Pretoria: State Library Corporate Communication, [22 aug?] 1851-[1852?] – (= ser State library south africa newspaper microfilm project) – 1r – 1 – mf#MS00282 – sa National [079]

Durban star – Durban, SA. 1897 – 1 – sa National [079]

Durch armenien : eine wanderung und der zug xenophons bis zum schwarzen meere / Hoffmeister, E von – Leipzig, Berlin, 1911 – 3mf – 9 – mf#AR-1430 – ne IDC [915]

Durch central-asien : die kirgisensteppe, russisch-turkestan – bochara – chiwa, das turkmenenland und persien; reiseschilderungen / Moser, Heinrich – Leipzig 1888 [mf ed Hildesheim 1995-98] – 1v on 5mf [ill] – 9 – €100.00 – 3-487-27619-4 – gw Olms [915]

Durch central-brasilien : expedition zur erforschung des schingú im jahre 1884 / Steinen, Karl von den – Leipzig 1886 [mf ed Hildesheim 1995-98] – 1v on 5mf [ill] – 9 – €100.00 – 3-487-26878-7 – gw Olms [918]

Durch chinas suedprovinz : bericht ueber die visitation des missionsinspektors sauberzweig schmidt in suedchina 1904-06 / Schmidt, Sauberzweig; ed by Schlunk, Martin – Berlin: Berliner evangelischen Missionsgesellschaft, 1908 [mf ed 1995] – (= ser Yale coll; Seines literarischen nachlasses 2) – 170p (ill) – 1 – 0-524-10160-4 – (in german) – mf#1995-1160 – us ATLA [915]

Durch das drama hauptmanns / Bab, Julius – Berlin: Oesterheld, [1922?] [mf ed 1990] – 23p – 1 – mf#7445 – us Wisconsin U Libr [430]

Durch den kaukasus zur wolga / Nansen, F – Leipzig, 1930 – 3mf – 9 – mf#AR-2006 – ne IDC [914]

Durch deutsch-kiautschou : aus den aufzeichnungen des missionsinspektors sauberzweig schmidt ueber seine visitation in nordchina im jahre 1905 / Schmidt, Sauberzweig – Berlin: Berliner evangelischen Missionsgesellschaft, 1909 [mf ed 1995] – (= ser Yale coll; Seines literarischen nachlasses 3) – 100p (ill) – 1 – 0-524-09785-2 – (in german) – mf#1995-0785 – us ATLA [915]

Durch gosen zum sinai : aus dem wanderbuche und der bibliothek / Ebers, G – Leipzig: Wilhelm Engelmann, 1881 – 7mf – 9 – mf#HT-279 – ne IDC [916]

Durch kampf zum frieden – Tuebingen. Heft 1-16. 1914-17.-irr. Each vol. has distinctive title – 1 – us Wisconsin U Libr [940]

Durch massai-land : forschungsreise in ostafrika zu den schneebergen und wilden staemmen zwischen dem kilima-ndjaro und victoria-njansa in den jahren 1883 und 1884 / Thomson, Joseph – Leipzig 1885 [mf ed Hildesheim 1995-98] – 1v on 4mf [ill] – 9 – €120.00 – 3-487-27320-9 – gw Olms [916]

Durch nacht zum licht *see* Through night to light

Durch nord-afrika und spanien : reisestudien / Wernick, Fritz – Leipzig 1881 [mf ed Hildesheim 1995-98] – 1v on 3mf – 9 – €90.00 – 3-487-27327-6 – gw Olms [910]

Durch sturm und not : roman / Baudissin, Ida, Graefin – Reutlingen: Ensslin & Laiblin, [19–?] [mf ed 1995] – (= ser Ensslins mark-baende 32) – 320p – 1 – mf#8972 – us Wisconsin U Libr [830]

Durch syrien und kleinasien : reiseschilderungen und studien / Oberhummer, R & Zimmerer, H – Berlin, 1899 – 6mf – 9 – mf#AR-2000 – ne IDC [915]

Der durchbruch – Sangerhausen DE, 1952 5 jul-1958 25 jul [gaps] – 1r – 1 – (veb kupfererz) – gw Misc Inst [660]

Der durchbruch *see* Das neue bewusstsein

Durchbruch : kampfblatt fuer deutschen glauben, rasse, volkstum – Stuttgart DE, 1934-38 – 4r – 1 – gw Misc Inst [943]

Durchbruch anno achtzehn : ein fronterlebnis / Wittek, Erhard – 36. Aufl. Stuttgart: Franckh, c1933 – 1r – 1 – us Wisconsin U Libr [830]

Durchfluege durch deutschland, die niederlande und frankreich / [Hess, J L von] – Hamburg, 1793-1797. 4v – 12mf – 9 – mf#HT-265 – ne IDC [914]

Durchflug eines humoristen durch deutschland, die schweitz und das suedliche franckreich / Raupach, Johann F – Breslau 1811 [mf ed Hildesheim 1995-98] – (= ser Fbc) – 1v on 3mf [ill] – 9 – €90.00 – 3-487-27763-8 – gw Olms [914]

Durchflusszytometrische untersuchungen von stosswelleninduzierten zellschaeden / Endl, Elmar – (mf ed 1994) – 2mf – 9 – €40.00 – 3-89349-897-4 – mf#DHS 897 – gw Frankfurter [530]

Durdent, Rene *see* Histoire de la convention nationale de france

Durell, Fletcher *see*
- Cooperation
- A new life in education

Durell, John Carlyon Vavasor *see* The historic church

Duren, Charles *see* Woman's place in religious meetings

Dürener zeitung 1875 – Dueren DE, 1957 2 nov-1959 30 jun [only local pgs] – 1 – (title varies: 3 jul 1889: general-anzeiger; 1896: duerener zeitung; 1946 as regional ed of: aachener volkszeitung, aachen) – gw Misc Inst [074]

Duret de Tavel *see*
- Calabria during a military residence of three years in a series of letters
- Sejour d'un officier francais en calabre

Dureteste, A *see* Cours de droit de l'indochine

Durey de Noinville, Jacques Bernard *see*
- Histoire du theatre de l'academie royale de musique en france
- Histoire du theatre de l'opera en france depuis l'etablissement de l'academie royale de musique, jusqu'a present

Durfee, Thomas *see* Some thoughts on the constitution of rhode island

Durga puja – Calcutta: Hindoo Patriot Press, 1871 [mf ed 1995] – (= ser Yale coll) – xxii/83p/lxx (ill) – 1 – 0-524-09006-8 – (notes & ill by prata'pachandra ghosha) – mf#1995-0006 – us ATLA [280]

Durgnat, Raymond *see* Strange case of alfred hitchcock

Durham and northumberland parish register society *see* Transactions of the durham and northumberland parish register society, 1898-1926

The durham book : being the first draft of the revision of the book of common prayer in 1661 / ed by Cuming, G J – London, 1961 – 9mf – 8 – €18.00 – ne Slangenburg [242]

Durham Central Labor Union *see*
- Carolina labor news
- Durham labor journal

Durham chronicle – Durham. England. -w. 1832-33; 1860-65. (7 reels) – 1 – uk British Libr Newspaper [072]

Durham Colored Primitive Baptist Association *see*
- Minutes of the durham colored primitive baptist association
- Minutes of the...annual session of the durham primitive baptist association

Durham county, uk history, topography and directory – 1894 – 10mf – 9 – NZ$40.00 – 0-908989-24-5 – (historical and descriptive sketches of the city, diocese, all wards, towns etc with lists of residents) – nz BAB [941]

[Durham-] durham news – CA. Feb 1969-Aug 1969 – 1r – 1 – $60.00 – mf#B02190 – us Library Micro [071]

Durham, Eunice Ribeiro *see* Assimilacao e mobilidade

Durham, Frank *see* Dubose heyward

Durham, J H *see* Carleton island in the revolution

Durham, John George Lambton, Earl of *see*
- Appendix (b) to report on the affairs of british north america
- Report on the affairs of british north america

Durham, John Pinckney *see* Baptist builders in louisiana

Durham, John Wyatt *see* Corals from the gulf of california and the north pacific coast of america

Durham labor journal / Durham Central Labor Union – 1955 jul 21-1956, 1957-62, 1963-1964 mar 12 – 5r – 1 – (cont by: carolina labor news [durham nc: 1964]) – mf#1223729 – us WHS [331]

Durham observations *see* Results of astronomical observations made at the observatory of the university, durham, from ...

Durham research review, 1950-80 – 55mf – 9 – mf#86655/5770 – uk Microform Academic [073]

Durham, Timothy L *see* Plasma free fatty acids at rest and exhaustion following theobromine ingestion

Durham university act 1861, report on the... 1862 : command n3173, 5709 and 5704-i – 1r – 1 – (with app. filmed with: higher education in london, report on the advancement of... (selbourne commission), 1889) – mf#96692 – uk Microform Academic [378]

Durham university journal – Durham, England 1968-95 – 1,5,9 – ISSN: 0012-7280 – mf#3186 – us UMI ProQuest [378]

Durham's heyward *see* Dubose heyward

Durharts baptist church. jefferson county. louisville, georgia : church records – Louisville. Jefferson County. Georgia.Durharts Baptist Church – 1 – us Southern Baptist [242]

Durieu de Maisonneuve, M C *see*
- Exploration scientifique de l'algerie

Durieux, Andre *see* Probleme juridique des dettes du congo belge et l'etat du congo

Durkee, Robert Peter *see* Dialogical preaching in the local church

Durkheim, Emile *see*
- De la division du travail social
- The elementary forms of the religious life

Durlab Singh *see* The sentinel of the east

Durlacher tageblatt *see* Durlacher wochenblatt

Durlacher wochenblatt – Karlsruhe DE, 1831-43, 1845-47 [gaps], 1850-1943 28 feb, 1949 15 jun-1964 – 1 – (title varies: 1 apr 1920: durlacher tageblatt. incl suppl: soweit der turmberg gruesst 1950-64) – gw Misc Inst [074]

Durm, Josef et al *see* Handbuch der architektur

Durnford and east's reports : term reports in the court of king's bench / Durnford, Charles & East, Edward H – v1-8. 1785-1800. London: J Butterworth, 1817 (all publ) – (= ser Durnford And East's Term Reports) – 70mf – 9 – $105.00 – (new ed with references to subsequent cases. also called: durnford and east's term reports) – mf#LLMC 84-767 – us LLMC [324]

Durnford, Charles *see* Durnford and east's reports

Durnovo, M N *see* Rech predsedatelia rybinskogo otdela vserossiiskogo natsionalnogo soiuza m n durnovo

Durocher, Georges *see* Bio-bibliographie du reverend pere paul-emile breton

Duroiselle, Charles *see*
- The ananda temple at pagan
- Jinacarita

Duron, Jorge Fidel *see*
- Indice de la bibliografia hondurena
- Ultimos dias de francisco morazan

Duron Y Gamero, Romulo Ernesto *see*
- Bosquejo historico de honduras
- Bosquejo historico de honduras, 1502 a 1921
- Honduras literaria
- Jose justo milla
- Limites de nicaragua

DuRoullet, Francois Louis Gaud Lebland, marquis *see* Lettres sur les drames-opera

DuRozoir, Charles *see* Description geographique, historique, militaire et routiere de l'espagne

Durrant, Earlene *see*
- Effects of adhesive spray and prewrap on taped ankle inversion before and after exercise
- Seasonal changes in selected physiological variables of female basketball players

Durrell, Gerald Malcolm *see* Overloaded ark

Durrett, M *see* Infared interactance

Durruti un anarquista integro / Gilabert, A G – Barcelona, 193? Fiche W916. (Blodgett Collection of Spanish Civil War Pamphlets) – 9 – us Harvard [946]

Durry, Marie Jeanne *see* Flaubert et ses projets inedits

Dur's elsass : humoristisch-satirisch wucheblaettle – Muelhausen / Elsass (Mulhouse F), 1907 2 oct-1914 25 jul – 1 – fr ACRPP [870]

Dursch, Georg Martin *see* Ghatakarparam oder das zerbrochene gefaess

Dursley, berkeley & sharpness gazette & west gloucestershire advertiser – [SW England] Gloucestershire oct 1878-dec 1950 [mf ed 2003] – 66r – 1 – (cont as: dursley, berkeley and sharpness gazette and wotton-under-edge advertiser [jan 1881-dec 1916]; dursley gazette and wotton-under-edge advertiser [1917]; dursley gazette and wotton-under-edge advertiser-dursley gazette [jan 1918-dec 1919]; dursley gazette [jan 1920-dec 1933]; dursley gazette-gloucestershire county gazette [1934]; gloucestershire county gazette [jan 1935-dec 1950]) – uk Newsplan [072]

Dursli der branntweinlaeuter *see* Die wassernot im emmenthal / fuenf maedchen / dursli der branntweinlaeuter

Durtschi, Shirley K *see* Emotions and cognitions of athletes competing in a high-risk sport

Duryas qas laksan r mtay slap col dau : ryan pralom lok manosancetana nyn kicckar pulis / Thu Dhan – Kracah: Ghun Thai S'an 2507 [1965] [mf ed 1990] – 1r with other items – 1 – (in khmer) – mf#mf-10289 seam reel 110/7 [S] – us CRL [959]

Dusaulx, Jean *see* Voyage a barege et dans les hautes pyrenees

Dusch – Goeteborg, 1899-1901 – 1r – 1 – sw Kungliga [078]

Duschak, Moritz *see*
- Die moral der evangelien und des talmud
- Schulgesetzgebung und methodik der alten israeliten

Dusenduewelswarf – Lunden DE, 1933 – 1 – gw Misc Inst [074]

Dusha kooperatsii / Armand, L M – 1917 – 1mf – 9 – mf#COR-5 – ne IDC [335]

Dushepoleznoe chtenie *see* Ezhemesiachnoe izdanie dukhovnogo soderzhaniia

Dushpastyr – New York: Zachary Orun, 1909-10 – us CRL [071]

The dusk of the gods : music-drama in three acts and a prelude = Goetterdaemmerung / Wagner, Richard – Boston: O Ditson, c1888 – 1r – 1 – (german and english) – us Wisconsin U Libr [790]

Dusmet de Arizcun, Xavier *see* Dos grandes aventureros espanoles del siglo 16

Dussaud, Rene *see*
- Histoire et religion des nosairais
- Introduction a l'histoire des religions
- Les monuments palestiniens et judaiques

Dussault, Joseph Daniel *see* Guide du jeune pianiste

Dussek, Johann Ladislaus *see*
- Alla tedesca
- Concerto for the grand or small piano forte with accompaniments
- Deux sonates pour le piano-forte
- Deux sonates pour le pianoforte avec accompagnement de violon et basse, oeuvre 34
- Dussek's 2o grand concerto in f for the piano-forte
- Grande sonate pour piano forte
- Partant pour la syrie
- Recueil d'airs connus varies, pour le piano...op 71
- Second concerto pour clavecin ou forte-piano, deux violons, alto, basse, cors, et hautbois ad libitum, oeuvre 14e
- Second grand concerto in f for the piano-forte
- Trois grandes sonates, pour piano...oeuvre 35
- Trois sonates pour clavecin ou forte-piano avec violon ad libitum...op 13
- Trois sonates pour le clavecin ou le forte-piano avec accompagnement de violon, oeuvre 10e

Dussek's 2o grand concerto in f for the piano-forte : with additional keys, arranged likewise for those without, op 27 / Dussek, Johann Ladislaus – London, Edinburgh: Corri, Dussek & Co [1795?] – 4mf – 9 – mf#fiche 1123 – us Sibley [780]

Dusseldorf / Stolz, Heinz – Leipzig, Germany. 1925 – 1r – us UF Libraries [914]

Düsseldorfer zeitung 1814 – Duesseldorf DE, 1848-49 – 4r – 1 – (filmed by other misc inst: 1814 1 jan & 1816 11 jun, 1817 jan-mar & 18 ap & 8 oct, 1822 20 jul, 1824-27, 1829-1832 mar, 1832 jul-1880, 1893-1900 jun, 1901-1903 jun, 1904-1926 7 may [132r]. with suppl: blaetter fuer scherz und ernst 1828-30, 1834-43, 1845-47, 1854-55 [7r]; das leben im bild 1924 n1-27 (gaps) [1r publ in berlin]; licht und schatten 1910-13 [2r]; roman-beilage 1906-1907 21 nov [1r]; von nah und fern [publ in stuttgart, fr oct 1812: illustrierte duesseldorfer zeitung, fr 1914: illustrierte weltschau] 1908-20 28 mar [5r]; welt und haus 1907 1dec-1914 30 jul [3r]; woechentliche unterhaltungs-beilage der duesseldorfer zeitung [fr 25 feb 1906: unterhaltungs-beilage der duesseldorfer zeitung] 1904-1907 24 nov [2r]; die zeit im bild 1924 n1-1926 n17 [1r publ in berlin]) – gw Misc Inst [074]

Dussieux, Louis *see*
- Le canada sous la domination francaise
- Geographie generale

Dust treatments for vegetable seeds / Tisdale, W B – Gainesville, FL. 1945 – 1r – us UF Libraries [630]

Dusty trails – [1974 jun?]-1980 jun – 1r – 1 – mf#483333 – us WHS [071]

Dutch activities in the east : seventeenth century: being a report on the records relating to the east in the state archives in the hague / ed by Ray, Nihar-ranjan – Calcutta: Book Emporium Ltd, 1945 – (= ser Samp: indian books) – us CRL [949]

Dutch and flemish school – (= ser Christie's pictorial archive: painting and graphic art) – 119mf – 9 – $890.00 – 0-907006-96-5 – (over 1300 artists, over 7000 reproductions) – uk Mindata [750]

Dutch and french bulb-culture in florida / Randall, G M – Deland, FL. 1926 – 1r – us UF Libraries [630]

The dutch at the north pole and the dutch in maine : a paper read...3d march, 1857 / Peyster, John Watts de – New York: New York Historical Society, 1857 – 1mf – 9 – mf#45388 – cn Canadiana [978]

The dutch boers and slavery in the trans-vaal republic : in a letter to r n fowler... / Chesson, Frederick William – London, 1869 – (= ser 19th c books on british colonization) – 1mf – 9 – mf#1.1.7088 – uk Chadwyck [960]

Dutch civil administrators [bestuursambtenaren] in the netherlands indies and netherlands new guinea, 1933-1962 : pts 2.1-2.6: papers of selected bestuursambtenaren – [mf ed 2007] – (= ser Dutch political conflict with the republic of indonesia, 1945-1949 2) – 725mf – 9 – €9930.00 – (witzh p/c & concordance) – mf#mmp120-125 – ne Moran [350]

Dutch communist daily newspaper : de waarheid, 1945-1990 – 2420mf – 9 – €7315.00 silver €3.30/mf (€6635.00 diazo €3.05/mf) – mf#m404 – ne MMF Publ [335]

Dutch documents relating to the gold coast and translations of letters and papers collected in the algemeen rijks archief (ara), state archives of the netherlands at the hague / Dantzig, A van – 1971 – 1 – us CRL [960]

Dutch East Indies. Koninklijk Nederlandsch-Indisch Leger *see* Handleiding voor de maleische taal

Dutch elm disease control / League of Wisconsin Municipalities – 1r – 1 – us WHS [634]

Dutch elm disease report – 1961-69 – 1r – 1 – (cont: dutch elm disease control program in wisconsin) – mf#367426 – us WHS [634]

Dutch etchers of the seventeenth century / Binyon, Robert Laurence – London 1895 – (= ser 19th c art & architecture) – 2mf – 9 – mf#4.2.378 – uk Chadwyck [760]

[Dutch flat-] forum – CA. 1875-78 [wkly] – 2r – 1 – $120.00 – mf#B02191 – us Library Micro [071]

[Dutch flat-] the placer times – CA. 1881-84 [wkly] – 1r – 1 – $110.00 – mf#B03130 – us Library Micro [071]

Dutch fork baptist church. richland county. south carolina : church records – 1958-72 – 1 – 5.13 – us Southern Baptist [242]

The dutch in malabar / Alexander, Padinjarethalakal Cherian; ed by Srinivasachariar, C S – Annamalainagar: Annamalai University, 1946 – (= ser Samp: indian books) – (foreword by c r reddy; int by ed) – us CRL [954]

The dutch press on microfiche – 9 – (catalogue available on request) – ne MMF Publ [074]

The dutch reformed church in south africa : with notices of the other denominations / M'Carter, John – Edinburgh: W & C Inglis, 1869 – 1mf – 9 – 0-7905-6304-5 – mf#1988-2304 – us ATLA [242]

The dutch republics of south africa : three letters to r n fowler...and charles buxton / Chesson, Frederick William – London, 1871 – (= ser 19th c books on british colonization) – 1mf – 9 – mf#1.1.7120 – uk Chadwyck [960]

Dutch song-books – 79 titles on 258mf – 9 – (dutch song books dating fr the 16th & 17th c. coll incl books ranging fr simple, inexpensive publ to highly illustrated works. coll based on bibl: nederlandsche liedboeken. lijst der in nederland tot het jaar 1800 uitgegeven liedboeken samegesteld onder leiding van d f scheurleer by d f scheurleer. the majority of the books come fr coll held by the royal library in the hague) – ne IDC [780]

Dutch theatre posters, 1853-1926 / Municipal Archives Rotterdam – 1997 – 386mf – 9 – €1260.00 – (with p/g & concordance in english) – mf#m448 – ne MMF Publ [790]

Dutch trade in asia, c1800-1835 : papers of hendrik doeff on japan and the east indies – [mf ed 2007] – 140mf – 9 – €1490.00 – (in dutch, also english, french, some german, japanese; with p/g & concordance) – mf#mmp129 – ne Moran [380]

Dutch trade in asia, pt 2 : papers of jan cock blomhoff on japan, c1817-1829 – [mf ed 2008] – ca 10mf – 9 – €165.00 – (printed p/guide & concordance based on: stukken afkomstig van jan cock blomhoff (bewerking 2005 door djk van vroa 1907, aanwinst 1907 xxxvi)) – mf#mmp137 – ne Moran [380]

[Dutch-creek] news – NV. 1 feb, mar, may 1907 [wkly] – 1r – 1 – $60.00 – mf#U044991 – us Library Micro [071]

Dutcher, Salem *see* Expressions of law and fact construed by the courts of georgia

Dutchess / Dutchess County Genealogical Society – 1973 jun-1980 spr, 1980 fall-1988 sum – 2r – 1 – mf#518814 – us WHS [929]

Dutchess County Genealogical Society *see* Dutchess

Dutchess. Presbytery (Pres. Church in the USA) *see* Minutes, 1762-1795

Dutchman / Pennsylvania Dutch Folklore Center – v6 n1-v7 n4 [1954 jun-1956 spr] – 1r – 1 – (cont: pennsylvania dutchman; cont by: pennsylvania dutchman) – mf#543668 – us WHS [390]

Duterte, Vicente *see* Ang palad, palad gayud

Duthie, Enid Lowry *see* L'influence du symbolisme francais dans le renouveau poetique de l'allemagne

Duties and encouragements of the christian ministry / Marsh, William – London, England. 1849 – 1r – us UF Libraries [240]

Duties and rewards of the christian minister / Grinfield, Thomas – London, England. 1832 – 1r – us UF Libraries [240]

The duties of christianity : theoretically and practically considered / Jackson, Thomas – London: John Mason, 1857 – 2mf – 9 – 0-524-07753-3 – (incl bibl ref) – mf#1991-3321 – us ATLA [240]

The duties of churches to their pastors : an essay / Wilson, Franklin – Charleston SC: Southern Baptist Publ Soc 1853 [mf ed 1994] – 1mf – 9 – 0-524-08884-5 – mf#1993-3348 – us ATLA [240]

The duties of educated young men in british america : being the annual university lecture of mcgill university, montreal, session 1863-4 / Dawson, John William – Montreal?: J Lovell, 1863 – 1mff – 9 – mf#23061 – cn Canadiana [378]

The duties of judge advocates... / Hughes, R M – London: Smith, Elder & Co, 1845 – 3mf – 9 – $4.50 – mf#LLMC 89-035 – us LLMC [340]

The duties of man, and other essays = Essays. Selections / Mazzini, Giuseppe – London: JM Dent; New York: EP Dutton, [1912?] – (= ser Everyman's library) – 1mf – 9 – 0-7905-9032-8 – (in english) – mf#1989-2257 – us ATLA [170]

Duties of sheriffs and constables, as defined by the laws, and interpreted by the supreme court, of the state of california / Harlow, William Sturtevant – San Francisco: Whitney, 1884. 549p. LL-83 – 1 – us L of C Photodup [347]

Duties of sheriffs and constables particularly under the practice in california, and the pacific states and territories / Harlow, William Sturtevant – 2d ed. San Francisco: Bancroft-Whitney, 1895. 588p. LL-82 – 1 – us L of C Photodup [340]

Duties of subjects / Hunter, Andrew – Edinburgh, Scotland. 1793 – 1r – us UF Libraries [240]

Duties of subjects and magistrates / Roberts, George – Monmouth, England. 1842 – 1r – us UF Libraries [240]

The duties of subjects to their rulers : with a special view to the present times: a sermon preached in the presbyterian church of scarborough... / George, James – Toronto?: W J Coates, 1838 – 1mf – 9 – mf#32392 – cn Canadiana [240]

Duties of the clergy / Heald, W M – Leeds, England. 1843 – 1r – us UF Libraries [240]
Duties of the deacons and priests in the church of england compared / Hale, William Hale – London, England. 1850 – 1r – us UF Libraries [241]
The duties of the heart = Hrdayah rila fararid al-qulub / Bachye, Rabbi [Bahya ben Joseph ibn Pakuda] – New York: E P Dutton 1909 [mf ed 1985] – (= ser Wisdom of the east series (new york, ny)) – 1mf – 9 – 0-8370-2139-1 – (trans with int by edwin collins) – mf#1985-0139 – us ATLA [270]
Duties of the individual to society / Gaskell, William – London, England. 1858 – 1r – us UF Libraries [240]
Duties of the poor / Travell, Ferdinando Tracy – London, England. 1836 – 1r – us UF Libraries [240]
Duties of the sick, stated and enforced / Secker, Thomas – London, England. 1821 – 1r – us UF Libraries [240]
Duties on trade at charleston, 1784-1789 – South Carolina Department of Archives and History, 1995 – 1r – 1 – $85.00 – (with guide) – mf#D3313 – us South C Archives [976]
Duties on trade at charleston, 1784-89 / South Carolina. Dept of Archives and History – 1r. M-6 – 1 – $75.00r – Out-of-state orders: us Scholarly Res – us South C Archives [380]
Duties, powers, and liability of national bank directors / A.S Pratt and Sons. Washington, DC – Washington, 1908 – 1 – mf#LL-1366 – us L of C Photodup [346]
Dutilleux, A *see* Histoire et cartulaire de l'abbaye demalbuisson
Dutilliet, Henri *see* Petit catechisme liturgique – catechisme du chant ecclesiastique
Dutoitspan herald and bultfontein advertiser – Kimberley: Josiah Angove. v1 n1 jun 14 1878-v7 n845 30 dec 1884 (3x/wk) [mf ed Cape Town: SA Library, 1982] – 1r – 1 – sa National [079]
Dutt, Manmatha Nath *see*
- The dharma sa'stra
- The garuda puranam
- Outlines of hindu metaphysics
- Prose english translation of agni puranam
- A prose english translation of mahanirvana tantram
- A prose english translation of vishnupuranam

Dutt, Meade Ervin *see* Jesus christ in human experience
Dutt, N K *see* The vedanta
Dutt, Nripendra Kumar *see*
- The aryanisation of india
- Origin and growth of caste in india

Dutt, Paramananda *see* Memoirs of moti lal ghose
Dutt, Rajani Palme *see*
- India to-day
- World politics, 1918-1936

Dutt, Romesh Chunder *see*
- The economic history of india under early british rule
- The great epics of ancient india
- Pratap singh, the last of the rajputs
- Sivaji

Dutt, Romesh Chunder et al *see* Land problems in india
Dutt, Sukumar *see*
- Early buddhist monachism, 600 bc-1000 bc
- Problem of indian nationality

Dutt, Surendra Nath *see* The life of benoyendra nath sen
Dutt, Toru *see* Ancient ballads and legends of hindustan
Dutto, Darren J *see* Leg spring model related to muscle activation, force, and kinematic patterns during endurance running to voluntary exhaustion
Dutton advance – Ontario Prov., Canada. Feb 1889- – 1 – cn Commonwealth Imaging [071]
Dutton enterprise – Ontario Prov., Canada. Dec 1881-1889 – 1 – cn Commonwealth Imaging [071]
Dutton, William Elliot *see* The eucharistic manuals of john and charles wesley
Duty and advantage of early rising / Wesley, John – London, England. 1816 – 1r – us UF Libraries [240]
Duty and conscience : addresses / King, Edward; ed by Randolph, Berkeley William – London: AR Mowbray; Milwaukee: Young Churchman, [1911?] – 1mf – 9 – 0-7905-9992-9 – mf#1989-1717 – us ATLA [240]
Duty and method of bearing good tidings to zion / Mccaul, Alexander – London, England. 1841 – 1r – us UF Libraries [939]
The duty and reward of propagating principles of religion and virtue exemplified in the history of abraham : a sermon preach'd before the trustees for establishing the colony of georgia in america / Burton, John – London: Printed by J March, 1733 (mf ed: Louisville [KY]: Lost Cause Press, 1974) – 3mf – 9 – mf#Sc Micro F-13628 – Located: NYPL – us Misc Inst [978]
The duty and the limitations of civil disobedience : a discourse / Bartlett, Samuel Colcord – Manchester, N.H.: Abbott, Jenks, 1853 – 1mf – 9 – 0-7905-6041-0 – mf#1988-2041 – us ATLA [240]
Duty, excellency, and pleasantness, of brotherly unity / Jamieson, John – Edinburgh, Scotland. 1819 – 1r – us UF Libraries [240]
Duty of a christian nation to her colonies and foreign dependencies / Ollivant, Alfred – Cambridge, England. 1850 – 1r – us UF Libraries [240]
Duty of attending week-day services in the church / Stebbing, Henry – London, England. 1840 – 1r – us UF Libraries [240]
Duty of being always ready / Skinner, William – Aberdeen, Scotland. 1839 – 1r – us UF Libraries [240]
Duty of british india in return for almight god's recent extraordin... / Wilson, Daniel – Calcutta, India. 1849 – 1r – us UF Libraries [240]
Duty of christians : in reference to their deceased ministers / Ryland, John – Bristol, England. 1805? – 1r – us UF Libraries [240]
Duty of christians to seek the salvation of israel – London, England. 18-- – 1r – us UF Libraries [240]
Duty of considering the example of departed good men / Ramsay, Edward Bannerman – Edinburgh, Scotland. 1830 – 1r – us UF Libraries [240]
Duty of contending earnestly for the faith once delivered to the sa... / Robertson, James – Edinburgh, Scotland. 1811 – 1r – us UF Libraries [240]
Duty of continued obedience to the church's law of custom in times / Scott, William – London, England. 1845 – 1r – us UF Libraries [240]
Duty of divinity students / Flint, Robert – Aberdeen, Scotland. 1861 – 1r – us UF Libraries [240]
Duty of english churchmen and the progress of the church in leeds / Hook, Walter Farquhar – London, England. 1851 – 1r – us UF Libraries [240]
Duty of family prayer / Blomfield, Charles James – London, England. 1845 – 1r – us UF Libraries [240]
Duty of hoping against hope / Keble, John – London, England. 1846 – 1r – us UF Libraries [240]
Duty of making known the gospel / Steere, A – London, England. 1875 – 1r – us UF Libraries [220]
Duty of ministers to be nursing fathers to the church and the duty... – London, England. 1796? – 1r – us UF Libraries [240]
Duty of paying custom, and the sinfulness of importing goods clande... – London, England. 1792 – 1r – us UF Libraries [240]
Duty of paying tribute enforced / Haldane, Robert – Edinburgh, Scotland. 1838 – 1r – us UF Libraries [240]
Duty of the church to her rulers / Paxton, George – Glasgow, Scotland. 1796 – 1r – us UF Libraries [240]
Duty of the clergy to enforce the frequent receiving of the sacrame... / Clapham, Samuel – London, England. 1806 – 1r – us UF Libraries [240]
The duty of the general assembly to all the churches under its care : a vindication of the minority in opposition to the resolutions on the state of the country / Hornblower, William Henry – Paterson: A Mead 1861 [mf ed 1992] – 1mf – 9 – 0-524-05544-0 – mf#1990-5148 – us ATLA [242]
The duty of the hour : extracts from pamphlet containing article from the sentinel (the orange and protestant advocate) being circulated among the protestant voters at the present election – [Kitchener, Ont?: Rittinger & Motz, 1908?] – 1mf – 9 – 0-665-97974-6 – mf#97974 – cn Canadiana [325]
Duval *see*
- Forme generale et particuliere de la convocation et de la tenue des assemblees nationales ou etats generaux de france, justifiee par pieces authentiques
- Recueil de pieces originales et authentiques concernant la tenue des etats generaux: d'orleans en 1560 sous charles 9; de blois en 1576 sous henri 3; de blois en 1588 sous henri 3; de paris en 1614 sous louis 13

Duval, A *see* La vie admirable de soeur marie de l'incarnation
Duval, Alexander *see* Maison a vendre
Duval, Alexandre *see*
- Aventure de saint-foix
- Edouard en ecosse
- Heritiers, ou, la naufrage
- Jeunesse de henri v
- La maison a vendre
- Projets de mariage, ou, les deux militaires
- Tyran domestique

Duval County Bridge Celebration Committee *see* Official program st johns river bridge celebr...
Duval county family welfare agency / Shepherd, Rose – s.l, s.l? 1935 – 1r – us UF Libraries [366]
Duval County (Fla) Council Of Social Agencies Of... *see* Membership directory
Duval, Georges *see*
- Monsieur daube
- Souvenirs de la terreur de 1788 a 1793
- Werther

Duval, Miles Percy *see* Cadiz to cathay
Duval, R *see* Anciennes litteratures chretiennes 2. la litterature syriaque
Duval, Rubens *see*
- La litterature syriaque
- Traite de grammaire syriaque

Duvalier, Francois *see* Face au peuple et a l'histoire
Duvall, Terry Glenn *see* Determining ministry priorities for macedonia baptist church during a transitional period
Duval-Thibault, Anna-Marie *see* Les deux testaments
Duvergier de Hauranne, Ernest *see* Huit mois en amerique
Duvernois, Henri *see* Nounouche
Duvernoy, Frederic *see*
- Deuxieme fantaisie, pour piano et cor ou violon
- Quatrieme fantaisie, pour le piano, cor ou violon
- Reveil de j j rousseau
- Songe de j j rousseau
- Troisieme fantaisie, pour piano, cor ou violon

Duvert, Felix-Auguste *see*
- Commissaire extraordinaire
- Homme blase
- Marchand de marrons
- Supplice de tantale

Duvert, M (Felix-Auguste) *see*
- Pont casse
- Sir jack, ou, qui est-ce qui veut se faire pendre?

Duveyrier, Henri *see*
- La confrerie musulmane de saidi mohammed ben ali es-senoausai et son domaine geographique
- La confrerie musulmane de sidi mohammed ben 'ali es senousi et son domaine geograohique en l'annee 1300 de l'hegire

Duveyrier, Honore Marie Nicolas *see* Histoire des premiers electeurs de paris en 1789
Duvillars, Pierre *see* L'erotisme au cinema
Dux christus : an outline study of japan / Griffis, William Elliott – New York: Publ for the Central Cttee on the United Study of Missions [by] Macmillan 1904 [mf ed 1986] – (= ser United study of missions) – 1mf – 9 – 0-8370-6501-1 – (incl bibl & ind) – mf#1986-0501 – us ATLA [950]
Dux-billiner zeitung – Dux (Duchcov CZ), 1939 2 oct-1940 30 sep, 1941-1944 30 sep – 1 – gw Misc Inst [077]
Duxbury 1644-1849 – Oxford MA [mf ed 1996] – (= ser Massachusetts vital record transcripts to 1850) – 1v on 11mf – 9 – 0-87623-245-4 – (mf 1t-2t: b,m,d 1644-1799. mf 3t: births & deaths 1733-85, 1835; marriages 1751-75, 1791, 1817, 1822. mf 3t-7t: births & deaths 1704-1867. mf 7t-8t: births 1645-1786. mf 8t. marriages 1644-1849. mf 8t-9t: out-of-town marriages 1682-1798. mf 9t: deaths 1652-1847, 1880. mf 10t: births 1843-49; marriages 1842-49. mf 11t: deaths 1843-49) – us Archive [978]
Duxbury 1661-1907 – Provo UT [mf ed 2005] – (= ser Massachusetts vital records) – 8v on 52mf – 9 – 0-87623-436-8 – (n1: duxbury church family records [1661-1897+]: mf1: birth & death ind [1665-1907]: a-s, marriages 1672-1844, a-b. mf2: birth & death ind [1665-1907] s-w, marriages 1677-1852, c-p. mf3: marriages 1667-1850, p-y, out-of-town marriages 1682-1798, births & deaths 1785-1886. mf4: 1763-1897. mf5: births & deaths 1714-1889+. mf6: births & deaths 1755-1888+. mf7: births & deaths 1695-1898+. mf8: births & deaths 1774-1888+. mf9: births & deaths 1674-1889+. mf10: births & deaths 1661-1888+; n2: duxbury church family records [1729-1971]: mf11: family records ind [1729-1971], a-z. mf12: deaths not in family records 1680-1889. mf13: births & deaths 1864-1940, 1952-72. mf14: births & deaths 1876-1971. mf15: births & deaths 1889-1972. mf16: births & deaths 1885-1972. mf17: births & deaths 1929-72. mf18: births & deaths 1946-72. mf19: births & deaths 1950-69. mf20: births & deaths 1958-72. mf21: births & deaths 1965-72; n3: duxbury church family records [1968-73]: mf22: family records ind [1968-73], a-u. mf23: family records ind [1968-73], v-z, births 1968-73. mf24: births 1971-73; duxbury index to births & deaths 1843-1955 in bks 1, 2, 3, 4, 5, 6: mf25: birth ind 1843-1955, a-f. mf26: birth ind 1843-1955, f-p. mf27: birth ind 1843-1955, p-z. mf28: death ind 1843-1955, a-e. mf29: death index 1843-1955, f-m. mf30: death ind 1843-1955, m-s. mf31: death ind 1843-1955, s-z; duxbury index to marriages & intentions 1843-1955 in bks 1, 2, 3 & 4: mf32: marriage intentions ind 1843-1958, a-h. mf33: marriage intentions ind 1843-1958, h-s. mf34: marriage intentions ind 1843-1958, s-y, b. mf35: marriage ind 1843-1955, a-c. mf35: marriage ind 1843-1955, c-h. mf36: marriage ind 1843-1955, h-r. mf37: marriage ind 1843-1955, r-z, l; bk 1: duxbury births, marriages, & deaths 1843-54: some of the pp display aging & are difficult to read mf38: births 1843-54, marriages 1843-48. mf39: marriages 1848-54, deaths 1843-54; bk 2: duxbury births, marriages, & deaths 1855-93: some of deaths on mf45 are hard to read mf40: births 1855-66. mf41: births 1867-89. mf42: births 1890-93, marriages 1855-68. mf43: marriages 1868-84. mf44: marriages 1884-93. mf45: deaths 1855-1859. mf46: deaths 1860-75. mf47: deaths 1875-89. mf48: deaths 1889-93; bk 3: duxbury births, marriages, & deaths 1894-1907: mf49: births 1894-1907, marriages 1894. mf50: marriages 1894-1907, deaths 1894-95. mf51: deaths 1895-1904. mf52: deaths 1904-07) – us Archive [978]
Duxer zeitung – Dux (Duchcov CZ), 1928 nov-1938 – 8r – 1 – gw Misc Inst [077]
Duxford, st peter & st John (2 parishes) 1599-1950 – 8mf – 9 – £10.00 – uk CambsFHS [929]
Dux-gong : titel und funktion des herzogs im frankenreich der merowinger und im china der zhou-dynastie / Holzinger, Regina – (mf ed 1995) – 2mf – 9 – €40.00 – 3-8267-2197-7 – mf#DHS 2197 – gw Frankfurter [900]
Duy Nasuan *see* Ekasar qang kar sahaprajajati
Duy Ryan *see* Kanjron cas nin cacak kmen r vivattan gamnit prajadhipateyy khmaer
Duyen anh *see* Thang gieng ngon nhu mot cap moi gan
Het duyfken in de steen-rotse : dat is, eene mede-lydende siele op de bittere passie iesu christi mediterende / Poirters, Adr – Antwerpen, 1657 – €15.00 – ne Slangenburg [241]
Het duyfken in de steen-rotse : dat is, eene mede-lydende siele op de bittere passie jesu christi mediterende... / [Poirters, Adrianus] – Antwerp: J Woons, [c1713] – 4mf – 9 – mf#O-3147 – ne IDC [090]
Het duyfken in de steen-rotse... / [Poirters, Adrianus] – Antwerp: C. Woons, 1665 – 5mf – 9 – mf#O-3254 – ne IDC [090]
Het duyfken in de steen-rotse... / [Poirters, Adrianus] – Antwerpen: A Bruers, [c1787] – 4mf – 9 – mf#O-3255 – ne IDC [090]
Duygu – Izmir. Sahib-i Imtiyaz: Hueseyin Rezmi; Sermuharriri: Haydar Ruesdi. n298. 21 mayis 1339 [1923] – (= ser O & t journals) – 1mf – 9 – $25.00 – us MEDOC [956]
Duy-Nguyen *see* Giot nang hong
Duytse lier : draayende veel van de nieuwste, deftige, en dartelende toonen / Luyken, Jan – Amsterdam: H. Bosch, 1729 – 2mf – 9 – mf#O-3237 – ne IDC [090]
Duytse lier... / Luyken, Jan – Amsterdam: Jacobus Wagenaar, 1671 – 2mf – 9 – mf#O-3110 – ne IDC [090]
Duytse lier... / Luyken, Jan – t'Amsterdam: Jan ten Houten, 1708 – 2mf – 9 – mf#O-3111 – ne IDC [090]
Dv droit des magistrats svr leurs svbiets... / [Beze, T] – [Heidelberg], 1578 – 1mf – 9 – mf#PFA-104 – ne IDC [240]
Dva goda raboty potrebitelskoi kooperatsii / Liubimov, I E – 1929 – 127p 2mf – 9 – mf#COR-331 – ne IDC [335]
Dva goda revoliutsii na ukraine : evoliutsiia i raskol bunda / Rafes, M G – 1920 – 2mf – 9 – mf#RPP-101 – ne IDC [325]
Dva goda v kambodzhe / Ardelian, Elena Lukinichna – Kharkov [Ukraine]: Prapor 1967 [mf ed 1989] – 1r with other items – 1 – mf#mf-10289 seam reel 001/13 [§] – us CRL [915]
Dva puti : k postanovke voprosa o vzaimootnosheniiakh mezhdu potrebitelskikh i selskokhoziaistvennykh kooperatsiei / Fishgendler, A M – 1923 – 47p 1mf – 9 – mf#COR-365 – ne IDC [335]
Dvadtsat' let sovetskoi vlasti : statisticheskii sbornik – M, 1937. 110p – 2mf – 9 – mf#RHS-19 – ne IDC [314]
Dvadtsatipiatiletie deiatel'nosti obshchestva vzaimnogo kredita s-peterburgskogo uezdnogo zemstva, 1871-1896 / Peterburgskoe Uezdnoe Zemskoe Obshchestvo Vzaimnogo Kredita – Spb, 1897 – 1mf – 9 – mf#REF-377 – ne IDC [332]
Dvan canna nhan man sami nra pum : [a novel] / Sin Kri, Mho pi Cha ra – Ran Kun: Tak lam ca pe thana 1957 [mf ed 1990] – 1r with other items – 1 – (in burmese) – mf#mf-10289 seam reel 193/1 [§] – us CRL [830]
Dvivedi, Durgaprasada *see* Vedic philosophy
Dvivedi, Manilal Nabhubhai *see* The imitation of srankara
Dvizhenie narodnykh natsionalistov *see* Zemshchina
Dvorak, Antonin *see* Cyprise, 10 milostnych pisni pro smyccovy kvartet, dle basni gustava pflegra moravskeho
Dvorak, Rudolf *see*
- Ein beitrag zur frage ueber die fremdwoerter im koraan
- Confucius und seine lehre
- Lao-tsi und seine lehre

Dvorianstvo i ego soslovnoe upravlenie za stolietie 1762-1855 godov / Korff, Sergei Aleksandrovich – C.-Peterburg, Tip. Trenke i Fiusno, 1906. 720 p. LL-4010 – 1 – us L of C Photodup [340]

Dvornik, F *see* Les slaves

Dvuglavyi orel – Kiev, Ukraine 1911 [mf ed Norman Ross] – 1 – mf#nrp-737 – us UMI ProQuest [077]

Dvukhnedelenoe izdanie : literaturno- i obshchestvenno-politicheskii zhurnal – Spb., 1906. v1-2 – 5mf – 9 – mf#R-3761 – ne IDC [077]

Dvukhnedelenyi illustrirovannyi zhurnal – M., 1912. v1-24 – 30mf – 9 – mf#R-4016 – ne IDC [077]

Dvukhnedelenyi khudozhestvenyi istoriko-literaturnyi zhurnal – M., 1913. nos 1-4 – 8mf – 9 – mf#R-4130 – ne IDC [077]

Dvukhnedelenyi nauchno-populiarnyi, obshchestvenno-politicheskii, ekonomicheskii i literaturno-khudozhestvennyii zhurnal – London. 1899+ (1) 1971+ (5) 1975+ (9) – 89mf – 9 – mf#1223 – ne IDC [077]

Dvukhnedelenyi zhurnal – New York. 1964-1967 (1) – 9mf – 9 – (missing: 1910(1, 3)) – mf#1808 – ne IDC [077]

Dvukhnedelenyi zhurnal : natsionalenye problemy – M., 1915. nos 1-4 – 6mf – 9 – mf#R-4137 – ne IDC [077]

Dvukhnedelenyi zhurnal : strakhovoe delo – Tvere, 1907-1916 – 206mf – 9 – (missing: 1915(6, 15, 21-23); 1916(19)) – mf#R-2318 – ne IDC [077]

Dvukhnedelenyi zhurnal, posviashchennyi voprosam byta i uslovii truda torgovo-promyshlennykh sluzhashchikh – Saratov, 1908. nos 1-4 – 4mf – 9 – mf#R-3987 – ne IDC [077]

Dvukhnedelnoe obozrenie / Vpered – London, 1876-1877. v1-49 – 24mf – 9 – mf#R-3452 – ne IDC [077]

Dvukhnedelnoe obozrenie, posviashchennoe voprosam bratskoi zhizni, kak ikh obiasnial liudiam khristos i kak napominaet teper l n tolstoi / ed by Chertkov, V G & Bulanzhe – Mundon (Essex), 1898. n1 – 1mf – 9 – mf#R-18004 – ne IDC [072]

Dvukhnedelnyi illiustrirovannyi literaturno-politicheskii zhurnal / Vseobshchaia biblioteka – Ithaca. 1842-1975 (1) – 106mf – 9 – mf#1985 – ne IDC [077]

Dvukhnedelnyi illiustrirovannyi voenno-literaturnyi zhurnal – Spb., 1914. v1(1-12) – 19mf – 9 – mf#R-1524 – ne IDC [077]

Dvukhnedelnyi zhurnal – M., 1915(1-5) – 3mf – 9 – (missing: 1915(4-5)) – mf#R-8141 – ne IDC [077]

Dwarf essex rape for winter forage / Scott, John M – Gainesville, FL. 1908 – 1r – us UF Libraries [630]

Dwarf survivals : and traditions as to pygmy races / Haliburton, Robert Grant – S.l: s.n, 1895? – 1mf – 9 – mf#05371 – cn Canadiana [573]

The dwarfs of mount atlas : statements of natives of morocco and of european residents there as to the existence of a dwarf race south of the great atlas / Haliburton, Robert Grant – London: D Nutt, 1891 – 1mf – 9 – mf#06389 – cn Canadiana [573]

Dwars door azie = Voyage d'exploration a travers l'asie / Deken, Constant de – Antwerpen: Clement Thibaut, 1902 [mf ed 1995] – (= ser Yale coll) – 430p (ill) – 1 – 0-524-09580-9 – (in dutch. trans fr french) – mf#1995-0580 – us ATLA [915]

The dwellers on the nile : or, chapters on the life, literature, history and customs of the ancient egyptians / Budge, Ernest Alfred Wallis – 2nd ed. London: Religious Tract Soc 1888 [mf ed 1989] – (= ser By-paths of bible knowledge 8) – 1mf – 9 – 0-7905-0619-X – (incl bibl ref & ind) – mf#1987-0619 – us ATLA [930]

Dwellings of the far south / Huss, Veronica E – s.l, s.l? 193-? – 1r – us UF Libraries [690]

Dwelly, Edward *see*
- Illustrated gaelic-english dictionary

Dwelshauvers-Dery, Victor Auguste Ernest *see* Histoire de la franc-maconnerie a liege avant 1830

Dwidjo Sewojo, M N G *see* De noodzakelijkheid van de instelling eener indische volksvertegenwoordiging met wetgevende macht

Dwight, Benjamin W *see* The higher christian education

"Dwight d eisenhower material" / Endacott, J Earl – 1 – us Kansas [920]

The dwight d eisenhower national security files, 1953-1960 – 2pt – (= ser National security file) – 1 – (pt1: subject files 30* isbn 1-55655-960-7 $5810. pt2: presidential files 24r* isbn 1-55655-961-5 $4380. with p/g) – us UPA [327]

Dwight David Eisenhower Army Medical Center *see* Examiner

Dwight doodles – Dwight, NE: Father Jerome Pokorny. 5v. v1 n1. may 1 1971-v5 n24. dec 15 1975 (semimthly) [mf ed with gaps filmed -1993] – 2r – 1 – (absorbed by: valparaiso hi-lites. publ in valparaiso ne, sep 1 1972-dec 15 1975) – us NE Hist [071]

Dwight doodles – Dwight, NE: Alfred Novacek. 12v. v1 n1. jan 15 1977-v12 n11 dec 15 1987 (mthly) [mf ed with gaps filmed - 1993] – 2r – 1 – (split from: valparaiso hi-lites) – us NE Hist [071]

Dwight, H G O *see*
- Christianity revived in the east
- Constantinople and its problems, its peoples, customs, religions and progress

Dwight, Harrison Gray Otis *see* Christianity revived in the east

Dwight, Henry Edwin *see* The life and writings of hon. vincent l. bradford

Dwight, John Sullivan *see* Dwight's journal of music

Dwight, Jonathan *see* Summer birds of prince edward island

Dwight l moody : impressions and facts / Drummond, Henry – New York: McClure, Phillips, 1900 – 1mf – 9 – 0-524-04012-5 – mf#1990-1184 – us ATLA [240]

Dwight l. moody : the man and his mission / Davis, George Thompson Brown et al – Chicago: Monarch, c1900 – 1mf – 9 – 0-7905-6052-6 – mf#1988-2052 – us ATLA [240]

Dwight l. moody materials concerning rutland baptist church / Mount Juliet. Tennessee. Rutland Baptist Church – 19p – 1 – 5.00 – (church records, 1821-1910, nov 1933-sept 1945. 616p. 29.64; 1) – us Southern Baptist [242]

Dwight, Mary Ann *see* Grecian and roman mythology

Dwight, Sereno Edwards *see*
- The hebrew wife
- Select discourses of sereno edwards dwight, pastor of park street church, boston, and president of hamilton college, in new york

Dwight, Theodore *see* Summer tours

Dwight, Timothy *see*
- The odore dwight woolsey
- Travels in new-england and new-york

Dwight, William Theodore *see* Select discourses of sereno edwards dwight, pastor of park street church, boston, and president of hamilton college, in new york

Dwight's american magazine : and family newspaper for the diffusion of useful knowledge and moral and religious principles – La Salle IL 1845-51 – 1 – mf#4154 – us UMI ProQuest [640]

Dwight's journal of music / ed by Dwight, John Sullivan – repr Boston. v1-41 n1-1051. 1852-81 – 11 – $350.00 set – (publ weekly then fortnightly) – us Univ Music [780]

Dwight's journal of music : a paper of art and literature – Killen TX 1852-81 – 1 – mf#3281 – us UMI ProQuest [780]

Dwight's journal of music : a paper of art and literature – v1-41. 1852-81 – 1 – us AMS Press [780]

Dwinell, Israel Edson *see* The higher criticism and a spent bible

Dwinger, Edwin Erich *see*
- Auf halbem wege
- Korsakoff
- Wir rufen deutschland
- Die zwoelf raeuber

Dwirnyk, J *see* Role de l'iconostase dans le culte divin

Dwivedi, Ram Awadh *see* Hindi literature

Dwoden, John *see* Outlines of the history of the theological literature of the church of england

Dworzecki, Mark *see* Yerushalayim de-lita bi-meri ube-sho'ah

Dwyer, Gregory B *see* Glycosylated hemoglobin and the oxygen kinetics in individuals with type 2 diabetes

Dyal, R S *see* Rate of decomposition of organic matters in soils of different degr...

Dyantyi, Vuyo Cedric *see* Analysing the understanding of the implementation of the curriculum 2005

Dyarchy in practice / Appadorai, Angadipuram – London; New York: Longmans Green & Co, 1937 – (= ser Samp: indian books) – (foreword by a b keith) – us CRL [954]

Dyatasarun al-dhay jah tazaynaws man al-dhashiyn al-arbah : seu, tatiani evangeliorum harmoniae arabice – Romae: Ex Typographia Polyglotta, SC de Propaganda Fide, 1888 – 1mf – 9 – 0-8370-1795-5 – mf#1987-6183 – us ATLA [220]

Dyaus asura, ahura mazda und die asuras : studien und versuche auf dem gebiete alt-indogermanischer religionsgeschichte / Bradke, P von – Halle, 1885 – €7.00 – ne Slangenburg [290]

Dybo, V A *see* Opyt sravneniia nostraticheskikh iazykov

Dyce, William *see*
- The national gallery
- Theory of the fine arts

Dyck, Johann Gottfried *see* Sechs wagen mit contrebande

Y dydd – [Wales] Gwynedd mehefin 1868-rhagfyr 1950 [mf ed 2004] – 97r – 1 – (missing: 1891) – uk Newsplan [072]

Dye, Eva Nicols *see* Bolenge

Dyea press – Dyea AL. 1898 apr 1 – 1r – 1 – mf#867443 – us WHS [071]

Dyea trail – Dyea AL. 1898 mar 26, apr 9, jun 25 – 1r – 1 – mf#867445 – us WHS [071]

Dyer, Alfred Saunders *see* Psalm-mosaics

Dyer, David *see* Tests of truth

Dyer, Helen S *see*
- A life for god in india
- Pandita ramabai
- Revival in india

Dyer, Isaac Watson *see* Maine corporation law

Dyer, John *see* Letters, official and private, from the rev dr carey

Dyer, Louis *see* Studies of the gods in greece at certain sanctuaries recently excavated

Dyer, Samuel *see* Dyer's new selection of sacred music

Dyer, Sidney *see*
- Dyer's psalmist
- The south western psalmist

Dyer, William H *see* "The oecumenical court"

Dyer's new selection of sacred music / Dyer, Samuel – 1834. 3rd ed – 1 – 9.31 – us Southern Baptist [242]

Dyer's psalmist / Dyer, Sidney – A collection of hymns and sacred songs. 1851 – 1 – us Southern Baptist [242]

Dyer's psalmist / Dyer, Sidney – Revised and corrected. 1853 – 1 – us Southern Baptist [242]

Dyes and pigments – Oxford, England 1980+ – 1,5,9 – ISSN: 0143-7208 – mf#42251 – us UMI ProQuest [660]

The dying god / Frazer, James George – London: Macmillan, 1911 – 1mf – 9 – 0-524-05846-6 – (incl bibl ref and ind) – mf#1990-3510 – us ATLA [200]

The dying indian's dream : a poem / Rand, Silas Tertius – Windsor, NS: C W Knowles, 1881 – 1mf – 9 – (with some additional latin poems) – mf#32271 – cn Canadiana [810]

The dying indian's dream *see* Rand and the micmacs

Dying postman / Gosse, P H – London, England. 18-- – 1r – us UF Libraries [240]

Dying scenes – London, England. 18-- – 1r – us UF Libraries [240]

Dying to sin / Stoughton, John – London, England. 1872? – 1r – us UF Libraries [240]

Dying token of affectionate remembrance of his people / Wing, J – Leicester, England. 18-- – 1r – us UF Libraries [240]

Dyk, Anna Margaretha van *see* The voices of women and young people who experienced domestic violence

Dyk, Braam van *see* Design of solvents for extractive distillation

Dyk citt kavi : [poems] / S Kabun et al – [Bhnam Ben]: Vappadharm 1985 [mf ed 1990] – 1r with other items – 1 – (in khmer) – mf#mf-10289 seam reel 129/14 [§] – us CRL [810]

Dyk jroh ramana bhaktra kanna khmaer loe : pralom lok knun manosancetana / Thu Dhan – Bhnam Ben: Qamat Pannagar [1965?] [mf ed 1990] – 2v in 1 on 1r with other items – 1 – (in khmer) – mf#mf-10289 seam reel 110/5 [§] – us CRL [959]

Dyk, Laetitia Arene van *see* Development of a conceptual model for organizational transformation with specific reference to absa group

Dyke, Charles E *see* Republican and democratic rule compared...

Dyke, Henry Jackson van *see* The variations of calvinism

Dyke, Henry Jackson van, Sr *see* Lectures and sermons of henry jackson van dyke

Dyke, Henry van *see*
- Essays in application
- The gospel for an age of doubt
- Little rivers

Dyke, Joseph Smith van *see* Theism and evolution

Dyke, Paul van *see* The age of the renascence

Dyke, Thomas *see* Travelling mems, during a tour through belgium, rhenish prussia, germany, switzerland, and france, in the summer and autumn of 1832

Dykes, James Oswald *see*
- Abraham, the friend of god
- From jerusalem to antioch
- The gospel according to st paul
- The law of the ten words
- Preaching christ crucified
- The relations of the kingdom to the world

Dykes, John Bacchus *see* Eucharistic truth and ritual

Dykes, Thomas *see* Proposed change in the mode of electing ministers of the church of...

Dykmans, M *see* Obituaire du monastere de groenendael dans la foret de soignes

Dymmer selv! : til selvpryvelse: samtiden anbefalet: anden raekke / Kierkegaard, Soeren – Kobenhavn: CA Reitzel, 1876 – 1mf – 9 – 0-7905-3788-5 – (himmelstrup) – mf#1989-0281 – us ATLA [240]

Dymond, Jonathan *see*
- Essays on the principles of morality
- An inquiry into the accordancy of war with the principles of christianity

Dyn : the review of modern art – Mexico. no. 1-6. Apr May 1942-Nov 1944 – 1 – us NY Public [700]

A dynamic faith / Jones, Rufus Matthew – 2nd ed. London: Headley; New York: Friends' Book & Tract Committee, 1902 [mf ed 1990] – 1mf – 9 – 0-7905-7597-3 – mf#1989-0822 – us ATLA [240]

Dynamic functional assessment of the lower extremity in the non-varsity athletic population / Groves, Michelle D – 1994 – 1mf – $4.00 – us Kinesology [612]

Dynamic maturity – Washington DC 1972-76 – 1,5,9 – (cont by: dynamic years) – ISSN: 0012-7388 – mf#7434.01 – us UMI ProQuest [618]

Dynamic science stories – New York. v1 n1-2. feb-may 1939 [all publ] – (= ser Science fiction periodicals, 1926-1978. series 1) – 1r – 1 – $95.00 – us UPA [830]

Dynamic years – Washington DC 1978-86 – 1,5,9 – (cont: dynamic maturity) – ISSN: 0148-799X – mf#7434,01 – us UMI ProQuest [618]

Dynamics of atmospheres and oceans – Oxford, England 1976+ – 1,5,9 – ISSN: 0377-0265 – mf#42182 – us UMI ProQuest [550]

Dynamics of comparative advantage and the resistance to free trade / Vollrath, Thomas L – Washington DC: US Dept of Agriculture, Economic Research Service [mf ed 1985] – (= ser Foreign agricultural economic report 214) – 9 – (with bibl) – us GPO [337]

Dynamics of meditation: (twelve interview-discourses on the various aspects of meditation) / Rajneesh, Bhagwan Shree; ed by Prem, Ma Ananda – 1st ed. Bombay: Jeevan Jagriti Kendra, 1972. Ma Yoga Laxmi, comp. 285p. 1 reel. 1190 – 1 – us Wisconsin U Libr [280]

The dynamics of morals : a sociopsychological theory of ethics / Mukerjee, Radhakamal – London: Macmillan & Co, [1950] – (= ser Samp: indian books) – (int by gardner murphy) – us CRL [170]

The dynamics of religion : an essay in english culture history / Robertson, John Mackinnon – London: University Press, 1897 – 1mf – 9 – 0-7905-6358-4 – (incl bibl ref) – mf#1988-2358 – us ATLA [210]

Die dynamik der grossstadt ins bild uebersetzen : zu den pariser bildern auguste chabauds 1907/08 / Schuerholz, Marietta Johanna – (mf ed 1998) – 4mf – 9 – €56.00 – 3-8267-2508-5 – mf#DHS 2508 – gw Frankfurter [750]

Dynamite – New York NY 1977-92 – 1,5,9 – mf#11547 – us UMI ProQuest [370]

Dynamometers and the measurement of power / Flather, John Joseph – New York, NY. 1907 – 1r – us UF Libraries [530]

Dynasties of mediaeval orissa / Misra, Binayak – Forward by Ramaprasad Chanda. Calcutta: K.N. Chatterji, 1933.111p – 1 – us Wisconsin U Libr [954]

The dynasts and the post-war age in poetry : a study in modern ideas / Chakravarty, Amiya Chandra – London: Oxford University Press, 1938 – (= ser Samp: indian books) – us CRL [420]

The dynasty of theodosius : or, eighty years' struggle with the barbarians. a series of lectures / Hodgkin, Thomas – Oxford: Clarendon Press, 1889 – 1mf – 9 – 0-7905-4923-9 – mf#1988-0923 – us ATLA [930]

Dynow, Zevi Elimelech *see* Bene yisakhar

Dyobouniotes, Konstantinos *see* Hippolyts schrift ueber die segnungen jakobs – hippolyts danielcommentar in handschrift no 573 des meteoronklosters

Dyocletianus leben / Buehel, Hans von; ed by Keller, Adelbert von – Quedlinburg, Leipzig: G Basse, 1841 [mf ed 1993] – (= ser Bibliothek der gesammten deutschen national-literatur von der aeltesten bis auf die neuere zeit sect1/22) – 64/212p – 1 – (incl bibl ref) – mf#8438 reel 5 – us Wisconsin U Libr [430]

Dyodekas emblematum sacrorum quorum consideratio accurata... / Saubert, J – Nurnberg: Durch Petrum Isselburger ins Kupffer gebracht, und beij Simon Halbmeijern gedruckt; zu finden beij Balthasaris Caijmoxen, 1625(-30). 4pts – 2mf – 9 – mf#0-1902 – ne IDC [090]

Dyondzo ya rimbewu / South Africa. Department of Health [Departement van Gesondheid] – Pretoria: Dept of Health [1979?] [mf ed Pretoria, RSA: State Library [199-]] – 22p [ill] on 1r with other items – 5 – mf#op 06756 r24 – us CRL [613]

Dyott, G M *see* Man hunting in the jungle

Dyrlund, Folmer *see* Tatere og natmandsfolk i danmark

Dyroff, A *see* D thomae aquinatis quaestiones disputatae de veritate. q 11 (fp13)

Dyroff, Adolf *see*
– Geschichte des pronomen reflexivum
– Ueber den existenzialbegriff

Dysmorphology and clinical genetics – Malden MA 1991-92 – 1,5,9 – ISSN: 0893-6633 – mf#18097 – us UMI ProQuest [575]

Dyson, C C *see* Madame de maintenon, her life and times 1635-1719

Dyson, Frank Watson *see* Observations of stellar parallax from photographs taken and measured at the royal observatory, greenwich

Dyson, Frank Watson et al *see* Observations made at the royal observatory, greenwich, in the year...in astronomy, magnetism and meteorology

Dyson, William Henry *see* Studies in christian mysticism

Dysphagia – Dordrecht, Netherlands 1986+ – 1,5,9 – ISSN: 0179-051X – mf#16988 – us UMI ProQuest [617]

Dyubhele, Noluntu Stella *see* Is the manufacturing sector an engine of growth in south africa?

Y dywysogaeth – Rhyl, Wales: John Morris 2 apr 1870-28 oct 1881 [mf jan-sep 1974] – 11v – 1 – (cont as: y llan [5 nov-3 dec 1881]; y llan a'r dywysogaeth [7 mar 1884-26 sep 1919, 19 oct 1928-27 may 1955]; y llan [3 jun 1955-]; fr 1884-85 & fr 1890-93 publ at merthyr tydfil; fr 1886-89 & fr 1894-98 at cardiff; fr 1899-1919 at lampeter; fr 1928-45 at caernarvon; fr 1946- at aberystwyth; wanting: oct-dec 1874) – uk Wales NatLib [072]

Y dywysogaeth – [Wales] LLGC 2 ebrill 1870-rhag 1950 [mf ed 2003] – 78r – 1 – (missing: 1874; cont as: y llan [5 tach-3 rhag 1881 & 21 maw-rhag 1884]; y llan a'r dywysogaeth [ion 1885-rhag 1919]; y llan and church news [ion 1920-rhag 1923]; y llan a'r dywysogaeth [ion 1924-rhag 1950]) – uk Newsplan [072]

Dz am dienstag : die bad doberaner heimatzeitung – Bad Doberan DE, 1962 17 apr-1967 30 mar – 2r – 1 – (title varies: 7 jan 1966: kuestenblick; publ in rostock) – gw Misc Inst [074]

Dz am sonntag – Duesseldorf DE, may 9 1926-feb 6 1927 – 1r – 1 – gw Misc Inst [074]

Dzallier d'Argenville, A N *see* Vies des fameux architectes [et sculpteurs] depuis la renaissance des arts...

Dzelzcelnieks trimda / Latvijas dzelzcelnieku centrs – n4,6-17 [1956, 1958-69] – 1r – 1 – mf#681791 – us WHS [071]

Dzerzhinets – (city unknown) 1944-45 [mf ed Norman Ross] – 1 – mf#nrp-97 – us UMI ProQuest [934]

Dzhanashvili, M G *see* Opisanie rukopisei tserkovnago muzeia dukhovenstva gruzinskoi eparkhii

Dzhangar – Moskva: Gos. izd-vo "Khudozh. litra" 1940 [mf ed Bloomington IN: Indiana Uni Lib, Preservation Dept 1984] – 1r – 1 – (the dzhangr tale, national epic of the kalmyk) – us Indiana Preservation [390]

Dzhangveladze, G A *see* Bankrotstvo antiproletarskikh partii v gruzii

Dzhaparidze, L S *see* Gruziia

Dzhavaxov, I A *see* Istorija cerkovnago razryva mezhdu gruziej i armeniej nachale 7 veka

Dzhezkazganskaia pravdv – Dzhezkazgan, Kazakhstan 1973-88 [mf ed Norman Ross] – 3r – 1 – mf#nrp-478 – us UMI ProQuest [077]

Dzhga-6 chasa vecher'ta – Sofia, Bulgaria. 15 Sept-Dec 1944 – 1r – 1 – us L of C Photodup [949]

Dziecko : organ centralnej organizacji opieki nad dziecmi zydowskimi w polsce – Warsaw, Poland 1932-33 [mf ed Norman Ross] – 1r – 1 – (with: sprawozdanie wydzialu towarzystwa biblioteki i czytelni publicznej "ezra" w krakowie: organ centralnej organizacji opieki nad dziecmi zydowskimi w polsce [krakow, poland]) – mf#nrp-2098 – us UMI ProQuest [939]

Dzieje kultury polskiej : napisal aleksander brueckner / Brueckner, Aleksander – Krakow: W L Anczyc Spolka 1931-46 [mf ed 1986] – 4v on 1r – 1 – (v4 ed by stanislaw kot & jan hulewicz; has imprint: krako'w, f pieczatkowski) – mf#1796 – us Wisconsin U Libr [943]

Dzieje polskiej mys'li politycznej w okresie porozbiorowym *see* Widoki i drogi rozwoju gospodarczego ziem polski

Dziennik baltycki – Gdansk, Poland. Oct 1945-1993 – 86r – 1 – us L of C Photodup [947]

Dziennik chicagoski = Chicago daily news – Chicago: Spolka Naklad. Wydawn. Polsk, 1890-[dec 15 1890-1942] – 1 – (in polish) – us CRL [071]

Dziennik literacki – Cracow. Poland. -w. 28 Mar 1947-31 Dec 1950. (2 reels) – 1 – uk British Libr Newspaper [947]

Dziennik literacki – L'viv, Poland 1861 [mf ed Norman Ross] – 1 – mf#nrp-980 – us UMI ProQuest [077]

Dziennik lodzki – Lodz, Poland. Jul-Sept 1945; Apr 1948-Aug 1953; Nov 1956-1992 – 72r – 1 – us L of C Photodup [943]

Dziennik ludowy – Chicago: [s.n.], mar 16 1907-apr 21 1925 – 45r – us CRL [071]

Dziennik ludowy = Polish people's daily – Chicago IL. 1907 mar 16/sep 23-1925 feb 2/apr 21 – 47r – 1 – (with gaps) – mf#851076 – us WHS [071]

Dziennik ludowy – Warsaw, Poland. 1962-89 – 63r – 1 – us L of C Photodup [943]

Dziennik ludowy – Warsaw. Poland. -d. 29 Aug 1945-Nov 1949. (17 reels) – 1 – uk British Libr Newspaper [943]

Dziennik narodowy = National polish daily – Chicago. apr 4 1908-jun 1909; jul 1913-1914; jul-dec 1915; sep 10 1917-jun 29 1918; jan 1920-sep 1923 – us CRL [071]

Dziennik podrozy do tatrow / Goszczy'nski, Seweryn – Petersburg 1853 [mf ed Hildesheim 1995-98] – 2mf – 9 – €60.00 – 3-487-29148-7 – gw Olms [947]

Dziennik polski – Cracow, Poland. -d. 22 June 1946-31 Oct 1951. 22 reels – 1 – uk British Libr Newspaper [943]

Dziennik polski – Dortmund DE), apr 3 1904-jun 2 1906 – 3r – 1 – gw Misc Inst [077]

Dziennik polski – Krakow, Poland. Apr 1945-1992 – 79r – 1 – us L of C Photodup [943]

Dziennik polski – London, UK. 1986- – 40+ r – 1 – uk British Libr Newspaper [072]

Dziennik polski = The polish daily – Detroit, MI: Polish American Pub Co, mar 1904-jul 27 1905; dec 1905-jun 19–?; 1913-jun 1923; 1924-jun 1941; sunday suppls for 1936-37 + for jul 1938-jun 1939 – 1 – us CRL [071]

Dziennik powszechny – Warsaw, Poland. Jan-Feb 1919 – 1r – 1 – us L of C Photodup [943]

Dziennik poznanski – Poznan, Poland. Mar-Apr 1923 – 1r – 1 – us L of C Photodup [943]

Dziennik zachodni – Katowice, Poland. Aug 1945; Oct 1947 (scattered issues); Oct 1949-1992 – 92r – 1 – us L of C Photodup [943]

Dziennik zjednoczenia – Chicago: Polish RCU of America, sep 1921-sep 1922; jan-jun 1927. (Country ed) – 7 – (issues for sep-dec 1921 filmed consecutively with: dziennik zjednoczenia (chicago: city ed)) – us CRL [071]

Dziennik zjednoczenia (city edition) – Chicago, IL: Polish RCU of America. City ed sep-dec 1921 – 80r – 1 – (filmed consecutively with: dziennik zjednoczenia (country ed) sep 1921-nov 1939) – us CRL [071]

Dziennik zwiakowy – Chicago. Jan 3 1911-July 1934; Apr 5 1935-Dec 31 1946. Incomplete – 1 – us NY Public [071]

Dziennik zwiazkowy – Chicago, IL: Polish National Alliance of US of NA, 1908-jan 1 1977 – 1 – us CRL [071]

Dziewicki, Michael Henry *see*
– Miscellanea philosophica
– Tractatus de apostasia
– Tractatus de blasphemia
– Tractatus de logica
– Tractatus de simonia

Dziko la nyasaland ndi anthu ace / Driessen, A – Bembeke, Malawi. 1938 – 1r – us UF Libraries [960]

Dzimbo sante – Chishawasha, Zimbabwe. 1930 – 1r – us UF Libraries [960]

Dzimbo sante – Chishawasha, Zimbabwe. 1951 – 1r – us UF Libraries [960]

Dzis i jutro – Warsaw, Poland. 10 Sep 1950-13 May 1956 – 6r – 1 – uk British Libr Newspaper [947]

Dzivhani, Makwarela David *see* The role of discipline in school and classroom management [microform]

E : the environmental magazine – Norwalk. 1990+ (1,5,9) – ISSN: 1046-8021 – mf#19273 – us UMI ProQuest [333]

E au akoanga no nga tumu tuatua i kitea i roto i te tutua na te atua = Theological lectures / Bogue, David – Cook Islands: London Missionary Soc, 1857 – 1r – 1 – (transl into raratongan by a buzacott) – mf#PMB Doc409 – at Pacific Mss [240]

E content – Wilton. 1999+ (1,5,9) – (cont: database) – ISSN: 1525-2531 – mf#14367,01 – us UMI ProQuest [000]

E di khrok ra cu nhan yan ma tuin mi ra khuin prannsum akkhara / Cam Sa on, U – Ran kun Mrui: u Mrat Thvan on, Umma on Ca pe 1974 [mf ed 1993] – on pt of 1r – 1 – mf#11052 r185 n9 – us Cornell [959]

E F *see* St basil and his rule

E F Schmidt Co *see* Reflections

E foi naquela noite de natal / Mendonca, Estevao De – Cuiaba, Brazil. 1969? – 1r – us UF Libraries [972]

E G kolbenheyers paracelsus-trilogie : eine metaphysik des deutschen menschen / Westhoff, Franz – Berlin: Junker und Duennhaupt, 1937 – 1 – (incl bibl ref) – us Wisconsin U Libr [430]

E h s mike *see* Echo news

E I wells correspondence 1861-1913 *see* E I wells papers

E I wells papers – 1861-1913 [mf ed 1981] – ca 145 items on 7mf – 9 – mf#51-180 – us South Carolina Historical [976]

E la casa un paradiso : predicazioni sulla famiglia cristiana tenute in una chiesa di montagna delle valli valdesi / Geymet, Enrico – Torre Pellice: Arti Grafiche "L'Alpina", 1943 – 1mf – 9 – 0-524-08105-0 – mf#1993-9011 – us ATLA [240]

E media professional – Wilton. 1997-1999 (1) 1997-1999 (5) 1997-1999 (9) – (cont: cd-rom professional. cont by: emedia) – ISSN: 1090-946X – mf#16703,02 – us UMI ProQuest [020]

E Mon, Buil *see* Gotama bhu ra rhan e buddhuppatti

E Mon, U *see* Mran ma ca ka pre nhan kabya kok nut khyak

E o *see* East oregonian

E o [East oregonian] – Pendleton OR: East Oregonian Pub Co [semiwkly] – 1 – (related to wkly ed: east oregonian (pendleton, or: weekly ed); daily ed: east oregonian (pendleton, or: daily evening ed)) – us Oregon Lib [071]

E permesso?? : ese nome permesso ce lo prendiamo – Sao Paulo, SP: [s.n.] 20 set 1896 – (= ser Ps 19) – mf#P18,02,50 – bl Biblioteca [440]

E pros romaious epistole : notes = St paul's epistle to the romans / Vaughan, Charles John – 6th ed. London: Macmillan, 1885 – 1mf – 9 – 0-8370-5624-1 – (incl ind of greek words) – mf#1985-3624 – us ATLA [227]

"E" reports for africa : from the united society for the propagation of the gospel, 1901-52 / United Society for the Propagation of the Gospel. Archives – 22r – 1 – mf#97360 – uk Microform Academic [220]

"E" reports for asia : from the united society for the propagation of the gospel, 1901-1952 – 63r – 1 – mf#97371 – uk Microform Academic [220]

E richard cross : a biographical sketch, with literary papers and religious and political addresses / Wilkinson, Marion – London: JM Dent, 1917 – 1mf – 9 – 0-524-07330-9 – mf#1991-3045 – us ATLA [920]

E S Sinegub et al *see* Otechestvennye vedomosti

E T A hoffmann / Bergengruen, Werner – Stuttgart: J G Cotta, 1944, c1939 [mf ed 1991] – (= ser Die dichter der deutschen) – 94p – 1 – mf#7483 – us Wisconsin U Libr [430]

E T A hoffmann : die drei reiche seiner gestaltenwelt / Willimczik, Kurt – Berlin: Junker und Duennhaupt, 1939 – 1 – us Wisconsin U Libr [430]

E T A hoffmann : lichnost i tvorchestvo / Ignatov, S S – Moskva: Tipografiia O L Somovoi, 1914 – 1 – (incl bibl ref) – us Wisconsin U Libr [430]

E T A hoffmann / Schaukal, Richard von – Berlin: Schuster & Loeffler, [1904] – 1r – 1 – (incl bibl ref (p.[99])) – us Wisconsin U Libr [920]

E T A hoffmanns elixiere des teufels und c. v. brentanos romanzen vom rosenkranz / Reiz, Elizabeth – Bonn 1920 – 1 – gw Mikropress [430]

E T A hoffmanns gespensterspiel / Escher, Karl – 2. Aufl. Berlin-Lichterfelde: E Runge, [19–?] – 1r – 1 – us Wisconsin U Libr [430]

E T A hoffmanns leben und werke : vom standpunkte eines irrenarztes / Klinke, Otto – Braunschweig: R Sattler, [1902?] – 1 – (incl bibl ref) – us Wisconsin U Libr [430]

E T A hoffmanns persoenlichkeit : anekdoten, schwaenke und charakterzuege aus dem leben des kammergerichtsrats, dichters und kapellmeisters ernst theodor amadeus hoffmann, nach mitteilungen seiner zeitgenossen aus den quellen zusammengetragen / Schollenheber, Wilhelm Heinrich – Muenchen: Verlag Parcus, 1922 – 1 – (incl bibl ref) – us Wisconsin U Libr [920]

E T A hoffmanns weltanschauung / Dahmen, Hans – Marburg a.L: N G Elwert, 1929 – 1r – 1 – (incl bibl ref) – us Wisconsin U Libr [430]

E von hartmann's philosophie des unbewussten / Ebrard, Johannes Heinrich August – Guetersloh: C Bertelsmann, 1876 – 1mf – 9 – 0-7905-7511-6 – (incl bibl ref) – mf#1989-0736 – us ATLA [190]

E voto dordraceno : toelichting op den heidelbergschen catechismus / Kuyper, Abraham – Amsterdam: JA Wormser, 1892-95 [mf ed 1990] – 4v on 6mf – 9 – 0-7905-7956-1 – mf#1989-1181 – us ATLA [240]

E w howe's monthly – Atchison KS. v11 n7, 10-v12 n10, 12-124 [1922 sep, dec-1923 dec, 1924 feb-1933 nov] – 1r – 1 – mf#853343 – us WHS [071]

E Ward & Co *see* The dress reform problem

EAA light plane world / EAA Ultralight Association – v5 n3-9 [1985 mar-sep], v6 n4, 9 [1986 apr, sep], v7 n1-4 [1987 jan-apr] – 1r – 1 – (cont: ultralight and the light plane; cont by: eaa experimenter) – mf#1266338 – us WHS [621]

EAA Ultralight Association *see* EAA light plane world

Eachard, Laurence *see* Dictionnaire geographique-portatif

Eaches, O P *see* Hebrews, james, and 1 and 2 peter

Eaches, Owen Philips *see* 1, 2 and 3 john, jude, and revelation

Eadie, Hazel Ballance *see* Lagooned in the virgin islands

Eadie, J I *see* An amharic reader

Eadie, John *see*
– An analytical concordance to the holy scriptures
– A commentary on the greek text of the epistle of paul to the colossians
– A commentary on the greek text of the epistle of paul to the ephesians
– A commentary on the greek text of the epistle of paul to the galatians
– A commentary on the greek text of the epistle of paul to the philippians
– A commentary on the greek text of the epistles of paul to the thessalonians
– The divine love
– Eadie's biblical cyclopaedia
– Paul the preacher

Eadie's biblical cyclopaedia : a dictionary of eastern antiquities, geography, natural history, sacred annals and biography, theology, and biblical literature illustrative of the old and new testaments = Biblical cyclopaedia / Eadie, John – new ed. London: Charles Griffin; Philadelphia: J B Lippincott, [1901?] – 2mf – 9 – 0-8370-1346-1 – mf#1987-6051 – us ATLA [220]

Eadmer *see* Eadmeri historia novorum in anglia (rs81)

Eadmeri historia novorum in anglia (rs81) : et opuscula duo de vita sancti anselmi et quibusdam miraculis ejus / Eadmer; ed by Rule, M – 1884 – (= ser The rolls series (rs)) – €19.00 – ne Slangenburg [931]

Eadon, W H and J A Auctioneers *see* Catalogue of the important collection of 360 modern paintings and water colour drawings

Eads, James Buchanan *see* Report on toronto harbour, ontario, 1882

Eagan, Marianne S *see* Kyphosis in active and sedentary postmenopausal women

Eager, George Boardman et al *see* The southern baptist pulpit

Eager, John Howard *see* Romanism in its home

Eagle – Brooklyn, NY. 1960-1963 (1) – mf#64913 – us UMI ProQuest [071]

Eagle – Bulter, PA. 1902-2000 (1) – mf#61768 – us UMI ProQuest [071]

Eagle – Butler, PA. 1873-1910 (1) – mf#68632 – us UMI ProQuest [071]

Eagle / Carl Vinson [Ship] – v4 n91-96, 104-122, 130, 144-147 [1986 sep 18-23, oct 12-30, nov 3, 24-27], v6 iss 7-9, 76, 119-126 [1988 jan 31-feb 2, jun 28, aug 18-25], v7 n14-27 [1989 may 14-27], v8 n12-37 [1990 feb 13-mar 16], v9 n15-16, 16, 23, 41, 46 [1991 apr 12-19, 26, sep 20, nov 15, dec 20; v20 n1-2 [1992 jan 3-10] – 1r – 1 – mf#1073902 – us WHS [071]

Eagle – Marinette WI. 1888 aug 18-1889 jun 22; 1889 jun 29-1890 nov 1; 1890 nov 8-1892 mar 26; 1892 apr 2-1893 jul 15; 1893 jul 22-1894 dec 1; 1894 dec 8-1896 may 30; 1896 jun 6-1897 nov 20; 1897 nov 27-1899 apr 22; 1899 apr 29-1900 jul 7; 1900 jul 14-1901 mar 9 – 10r – 1 – (cont: eagle [marinette wi: 1886]; cont by: weekly eagle) – mf#1107196 – us WHS [071]

Eagle – Marinette WI. 1888 jan 25-aug 11, 1886 oct 13-1888 jan 21 – 2r – 1 – (cont: marinette and peshtigo eagle; cont by: eagle [marinette wi: 1888]) – mf#1107195 – us WHS [071]

Eagle – North Bend, NE: Richard G and Vona V Van Cleef. 8v. v77 n5. nov 14 1974-v84 n52. sep 29 1982 (wkly) – 7r – 1 – (cont: north bend eagle. cont by: north bend eagle (1982)) – us NE Hist [071]

Eagle – Decatur, IN. 1867-1873 (1) – mf#62766 – us UMI ProQuest [071]

Eagle – East Providence, RI. 1882-1910 (1) – mf#66195 – us UMI ProQuest [071]

Eagle – Ekalaka, MT. 1909-1974 (1) – mf#64370 – us UMI ProQuest [071]

Eagle – Eldred, PA. 1888-1971 (1) – mf#65895 – us UMI ProQuest [071]

Eagle – Grafton, WV. 1884-1885 (1) – mf#67298 – us UMI ProQuest [071]

Eagle – South Sioux City, NE: F W Pace. 30v. v53 n33. jan 17 1930-v82 n31 dec 25 1958 (wkly) [mf ed 1930-58 (gaps) filmed 1974?] – 8r – 1 – (lacks: jan 23 1947, apr 17 1952, oct 3 1957. cont: south sioux city nebraska eagle) – us NE Hist [071]

Eagle – Oklahoma City, OK. 1953-1954 (1) – mf#65789 – us UMI ProQuest [071]

Eagle – Providence, RI. 1980-1985 (1) – mf#68463 – us UMI ProQuest [071]

Eagle – Reading, PA. 1868+ (1) – mf#68448 – us UMI ProQuest [071]

Eagle – Skamokawa, WA. 1899-1934 (1) – mf#67129 – us UMI ProQuest [071]

Eagle / Trumbull Co. Hubbard – jan 1985-dec 1994 [wkly] – 10r – 1 – mf#B34823-34832 – us Ohio Hist [071]

Eagle / Trumbull Co. Hubbard – v1 n1. oct 1966-dec 1984 [wkly] – 13r – 1 – mf#B3300-3312 – us Ohio Hist [071]

Eagle – White Cloud, MI. 1907-1973 (1) – mf#63886 – us UMI ProQuest [071]

Eagle – Wichita, KS. 1965-2000 (1) – mf#60477 – us UMI ProQuest [071]

Eagle – Augusta WI. 1899 mar 4/sep 30; 1899 oct/1901 mar 8-1914 jun 5/1915 oct 29 – 12r – 1 – (with gaps; cont: augusta eagle [augusta wi: 1874]; cont by: augusta eagle [augusta wi: 1915]) – mf#1044335 – us WHS [071]

Eagle *see* American eagle

The eagle – Chadron, NE: Students of Chadron State Normal School. v1 n1. sep 22 1920-(wkly) [mf ed with gaps filmed 1957-] – 1 – (some irregularities in numbering) – us NE Hist [071]

The eagle *see* Miscellaneous newspapers of teller county

Eagle and citizen times – Hawley, PA. 1967-1967 (1) – mf#65919 – us UMI ProQuest [071]

Eagle and county cork advertiser *see* Skibereen and west carbery eagle or south western advertiser

The eagle and lagos critic – Lagos, Nigeria. Mar 1883-Oct 1888 – 33ft – 1 – uk British Libr Newspaper [079]

Eagle beacon – Eagle, NE: A O Mayfield, 1899-v35 n8. may 11 1933 (wkly) [mf ed with gaps] – 7r – 1 – (cont by: nebraska beacon. publ in weeping water ne, jan 3 1929-33) – us NE Hist [071]

Eagle booster – Salem OR: [s.n.] – 1 – (began in 1935) – us Oregon Lib [071]

Eagle county miscellaneous newspapers – Denver, CO – 1r – 1 – (the pusher (jun 2 1894); the eagle county news (oct 27 1917, jun 14 1919, oct 2 1920); eagle county times (dec 31 1892)) – mf#MF Z99 Ea33 – us Colorado Hist [071]

The eagle county news *see* Eagle county miscellaneous newspapers

Eagle county times *see* Eagle county miscellaneous newspapers

Eagle eaglet – Eagle, NE: Interstate Newspaper Co, oct 1894-99// (wkly) [mf ed with gaps] – 2r – 1 – us NE Hist [071]

Eagle free press / Phoenix Indian Center – 1983 jan-1989 jan – 1r – 1 – mf#1055792 – us WHS [071]

Eagle herald – Marinette, WI. 1993-2000 (1) – mf#61937 – us UMI ProQuest [071]

Eagle leader / Fraternal Order of Eagles – 1974 nov-1978 jun, 1978 jul-1985 jun – 2r – 1 – mf#366588 – us WHS [071]

Eagle news – 1990 dec/jan, 3rd-4th qtr, 1991 2nd, 4th qtr – 1r – 1 – mf#1787344 – us WHS [071]

Eagle news – Poughkeepsie, NY. 1914-1942 (1) – mf#65174 – us UMI ProQuest [071]

Eagle of guatemala / Raine, Alice – New York, NY. 1947 – 1r – us UF Libraries [972]

Eagle point independent – Eagle Point OR: Eagle Point Independent, 1977-86 [wkly] – 1 – (cont by: upper rogue independent (1986-)) – us Oregon Lib [071]

Eagle quill – Eagle WI. 1898 jan 21-1904, 1905-1934, 1935-1943 oct 22, 1943 oct 29-1948, 1949-1952 aug 29 – 8r – 1 – mf#961934 – us WHS [071]

Eagle review – Alma, Arkansaw etc WI. 1977 apr 27-1979 apr 11 – 1r – 1 – mf#957377 – us WHS [071]

Eagle river democrat – Eagle River WI. 1893 apr 22-1895 may 27, 1895 jun 3-1896 aug 3 – 2r – 1 – (cont by: vilas county news) – mf#946339 – us WHS [071]

Eagle river review – Eagle River WI. 1890 jun 28/1891 dec 31-1926 jul 22/1927 jun 23 – 26r – 1 – (cont by: vilas county news; vilas county news-review [eagle river wi: 1927]) – mf#946349 – us WHS [071]

Eagle river shaft *see* Miscellaneous newspapers of summit county

Eagle river vindicator – Eagle River WI. 1887 jan 20-1890 jul 24 – 1r – 1 – mf#966917 – us WHS [071]

Eagle rock news herald – Los Angeles, CA. 1957-1957 (1) – mf#62178 – us UMI ProQuest [071]

Eagle rock sentinel – Los Angeles, CA. 1910-1968 (1) – mf#62179 – us UMI ProQuest [071]

Eagle standard – Fallon, NV. 1966-1968 (1) – mf#64748 – us UMI ProQuest [071]

Eagle times – Milton-Freewater OR: H E Judd, 1951- [wkly] [mf ed 1960] – 4r – 1 – (merger of: freewater times; milton eagle (1887-1951); cont by: milton-freewater valley herald (-1963)) – us Oregon Lib [071]

Eagle times *see* Freewater times (milton-freewater, or)

Eagle towncrier – Eagle WI. 1971 nov 5, 1973 aug 18, 1974 may 18, jul 13-aug 31 – 1r – 1 – mf#961953 – us WHS [071]

Eagle trade journal – Marinette, WI. 1936-1937 (1) – mf#67569 – us UMI ProQuest [071]

Eagle valley news – Richland OR: W L Flower, -1919 [wkly] [mf ed 1966-68] – 1r – 1 – us Oregon Lib [071]

Eagle, Walter *see* American negligence digest

Eagle watch / Rocky Mountain Arsenal [CO] – v3 iss 4,7,11-12 [1991 apr, jul, nov-dec], v4 iss 1-3,5,7,9 [1992 jan-mar, may, jul, sep], v5 iss 1 [1993 jan] – 1r – 1 – mf#2341341 – us WHS [071]

Eagle wing press – Naugatuck CT. 1981 nov/dec-1987 spr – 1r – 1 – (cont by: eagle [naugatuck ct]) – mf#1055794 – us WHS [071]

Eagle=gazette : [sesqui edition] / Fairfield Co. Lancaster – june 3, 1950 (400 pages 1800-1950) – 1r – 1 – mf#B2449 – us Ohio Hist [071]

Eagle's eye / Arizonans for National Security. Committee for National Security [AZ] – 1978 dec-1985 apr – 1r – 1 – mf#943720 – us WHS [071]

Eagle's eye / Brigham Young University – 1970 dec-1983 apr – 1r – 1 – mf#701758 – us WHS [071]

The eagle's eye – 1970-79 – (= ser American indian periodicals... 1) – 12mf – 9 – $105.00 – us UPA [305]

Eaglet – Eagle, NE: S S English, C W Hedges. 4v. v1 n1. sep 19 1890-v4 n50. aug 25 1894 (wkly) [mf ed with gaps] – 1r – 1 – (absorbed by: elmwood echo) – us NE Hist [071]

Eaglet / Polish Genealogical Society of Michigan – 1981 may-1988 sep – 1r – 1 – mf#1573053 – us WHS [929]

Eaglet times *see* [Templeton-] templeton times

Eaglin, James B *see*
- An evaluation of the probable impact of selected proposals for imposing mandatory minimum sentences in the federal courts
- The impact of the federal drug aftercare program
- The pre-argument conference program in the sixth circuit court of appeals
- A process-descriptive study of the drug aftercare program for drug-dependent federal offenders
- Sentencing federal offenders for crimes committed before november 1, 1987
- A validation and comparative evaluation of four predictive devices for classifying federal probation caseloads

Eales, Samuel John *see* Cantica canticorum

Ealing and acton gazette *see* Middlesex county times

Ealing and acton register – London UK, 26 may-22 sep 1877 – 1/4r – 1 – (incorp with: middlesex and surrey gazette) – uk British Libr Newspaper [072]

Ealing and chiswick guardian – London UK, 1986-22 dec 1988; 1989-12 dec 1990 – 8r – 1 – (aka: ealing borough guardian; guardian (ealing ed); 1987 master ng available under the same can number) – uk British Libr Newspaper [072]

Ealing borough guardian *see* Ealing and chiswick guardian

Ealing district recorder – [London & SE] BLNL 1986- – 1 – uk Newsplan [072]

Ealing gazette *see* Middlesex county times

Ealing gazette and west middlesex observer – London UK, 15 oct 1898-1910; 1912-29 sep 1923 – 24r – 1 – (fr sep 1923 incorp with: west middlesex gazette) – uk British Libr Newspaper [072]

Ealing guardian and county advertiser *see* Ealing guardian and middlesex advertiser

Ealing guardian and middlesex advertiser – London UK, 1852; 12 nov 1898-2 jun 1900 – 1r – 1 – (aka: ealing guardian and county advertiser) – uk British Libr Newspaper [072]

Ealing leader – London UK, 1986-91 – 29r – 1 – (aka: leader (ealing borough ed)) – uk British Libr Newspaper [072]

Eames, Wilberforce *see* Early new england catechisms

Ear and hearing – v1-17. 1975-96 – (= ser Journal Of The American Auditory Society) – 1,5,6,9 – $80.00r – (formerly: journal of the american auditory society. v1-5 1975-79 5r) – us Lippincott [610]

Ear, nose and throat journal – New York. 1976-2000 (1) 1976-2000 (5) 1976-2000 (9) – (cont: eye, ear, nose and throat monthly) – ISSN: 0145-5613 – mf#11129,01 – us UMI ProQuest [617]

Eardley, Culling Eardley *see* Englishman's thoughts on the scotch church

Eardley-Wilmot, S *see* Papers and addresses

Eardley-Wilmot, Sidney [comp] *see* Our journal in the pacific

Earl conrad/harriet tubman collection : from the holdings of the schomburg center for research in black culture, manuscripts, archives and rare books division: the new york public library, astor, lenox and tilden foundations – 1995 – 2r – 1 – $170.00 – (guide which covers all coll under "antebellum america and slavery" sold separately for $20.00 d3305.g2) – mf#D3305P06 – Dist. us Scholarly Res – us L of C Photodup [976]

Earl, Edward Curtis *see* The schoolhouse

Earl, G W *see* The eastern seas

Earl, George Windsor *see* Enterprise in tropical australia

Earl of aberdeen's correspondence with the rev dr chalmers / Gordon, George Hamilton – Edinburgh, Scotland. 1840 – 1r – us UF Libraries [240]

The earl of beaconsfield : with disraeli anecdotes never before published / Davin, Nicholas Flood – Toronto, Sydney NS: Belford, 1876 – 1mf – 9 – mf#24096 – cn Canadiana [920]

Earl rankin collection on cloze procedure / Rankin, Earl & Svobodny, Dolly – 1900-86 – 1000 titles on 250mf – 9 – (printed card indexes included. annual supplements) – us ATBI [370]

Earl, Stephen *see* Hills of the boasting woman

Earle, A B *see* Revival hymns

Earle, Absalom Backas *see* Bringing in sheaves

Earle, Alice Morse *see* The sabbath in puritan new england

Earle, Augustus *see* A narrative of a nine months' residence in new zealand, in 1827

Earle, John *see* Philology of the english tongue

Earle, William, Sir *see* The reunion of christendom in apostolic succession for the evangelization of the world

Earlestown and newton news – [NW England] Newton-le-Willows 1953-14 jul 1961 – 1 – (title change: newton & goldborne news [1981-84]) – uk MLA; uk Newsplan [072]

Earlewood baptist church – Richland Co, SC. 1644p – 1 – $13.05 – (scrapbooks and history 1939-feb 1992. wmu 1939-88. senior citizens 1976-91. minutes, 1939-45, 1947-49, 1952=1992. deacons' minutes 1962-68. church directory 1964-65. 1990 constitution and bylaws 1952) – mf#6703 – us Southern Baptist [242]

The earlier epistles of st paul : their motive and origin / Lake, Kirsopp – London: Rivingtons, 1911 – 2mf – 9 – 0-7905-1281-5 – (incl bibl ref and ind) – mf#1987-1281 – us ATLA [227]

The earlier pauline epistles : corinthians, galatians and thessalonians / ed by Bartlet, James Vernon – London: J M Dent; Philadelphia: J B Lippincott 1902 [mf ed 1989] – (= ser The temple bible) – 1mf – 9 – 0-7905-1802-3 – mf#1987-1802 – us ATLA [227]

The earlier prophecies of isaiah / Alexander, Joseph Addison – New-York: Wiley and Putnam, 1846 – 2mf – 9 – 0-8370-9521-2 – mf#1986-3521 – us ATLA [221]

The earliest cosmologies : the universe as pictured in thought by the ancient hebrews, babylonians, egyptians, greeks, iranians, and indo-aryans: a guidebook for beginners in the study of ancient literatures and religions / Warren, William Fairfield – New York: Eaton & Mains; Cincinnati: Jennings & Graham, c1909 – 1mf – 9 – 0-7905-0412-X – (incl bibl ref and indexes) – mf#1987-0412 – us ATLA [210]

The earliest english translations of buerger's lenore : a study in english and german romanticism / Emerson, Oliver Farrar – Cleveland: Western Reserve University Press, 1915 [mf ed 1989] – (= ser Western reserve studies 1/1) – 120p – 1 – mf#7095 – us Wisconsin U Libr [410]

The earliest english version of the fables of bidpai / "the morall philosophie of doni" by sir thomas north / ed by Jacobs, Joseph – London: D Nutt, 1888 [mf ed 1987] – lxxx/257/[1]p/pl (ill) – 1 – mf#1957 – us Wisconsin U Libr [390]

The earliest gospel : a historical study of the gospel according to mark / Menzies, Allan – London, New York: Macmillan, 1901 – 1mf – 9 – 0-8370-4389-1 – (incl ind of subjects and biblical passages cited) – mf#1985-2389 – us ATLA [226]

The earliest known coptic psalter : the text in the dialect of upper egypt / ed by Budge, Ernest Alfred Wallis, Sir – London: Kegan Paul, Trench, Truebner, 1898 – 1mf – 9 – 0-8370-1796-3 – mf#1987-6184 – us ATLA [220]

The earliest life of christ ever compiled from the four gospels : being the diatessaron of tatian, ca a.d. 160 / Hill, James Hamlyn – Edinburgh: T & T Clark, 1894 – 1mf – 9 – 0-7905-0164-3 – mf#1987-0164 – us ATLA [226]

The earliest sources for the life of jesus / Burkitt, Francis Crawford – Boston: Houghton Mifflin, 1910 – (= ser Modern Religious Problems) – 1mf – 9 – 0-7905-0622-X – mf#1987-0622 – us ATLA [220]

The earliest version of the babylonian deluge story and the temple library of nippur / Hilprecht, Hermann Vollrat – Philadelphia: University of Pennsylvania 1910 [mf ed 1986] – (= ser The babylonian expedition of the university of pennsylvania. series d: researches and treatises 5/1) – 1mf [ill] – 9 – 0-8370-7067-8 – (incl bibl ref) – mf#1986-1067 – us ATLA [470]

Early adolescents' knowledge of and attitudes toward hiv and aids / Blackwell, G F – 1991 – 2mf – 9 – $8.00 – us Kinesology [616]

Early alinari archives : art and architecture in italy – 122mf – 9 – $875.00 – 0-907006-54-X – (representative selection of important buildings, churches & monuments in towns by perhaps the finest art & architectural photographers in italy during the 19th c; over 7000 reproductions) – uk Mindata [770]

Early american – 1968-76 – (= ser American indian periodicals... 1) – 8mf – 9 – $95.00 – us UPA [305]

Early american churches / Embury, Aymar – Garden City NY: Doubleday, Page 1914 [mf ed 1989] – 1mf – 9 – 0-7905-4468-7 – mf#1988-0468 – us ATLA [720]

Early american herbaria : and related drawings from the british museum (natural history) / British Museum (Natural History) – [mf ed Chadwyck-Healey] – 5 herbaria on 4 colour + 13 b/w mf – 9,15 – (william bartram herbarium 2 col 3 b/w mf. mark catesby specimens in the samuel dale herbarium 2 b/w mf. john leonard riddell herbarium 4 b/w mf. thomas walter herbarium 2 b/w mf. william young herbarium 2 col 2 b/w mf. available separately or as single coll) – uk Chadwyck [580]

Early american history research reports from the colonial williamsburg foundation library / Colonial Williamsburg Foundation – [mf ed Chadwyck-Healey] – 933 reports on 1372mf – 9 – (with p/g) – uk Chadwyck [975]

Early american homes – Leesburg. 1996-2000 (1) 1996-2000 (5) 1996-2000 (9) – (cont: early american life. cont by: early american life) – ISSN: 1086-9948 – mf#6184,01 – us UMI ProQuest [640]

Early American Industries Association, Inc *see*
- Chronicle
- Chronicle of the early american industries association, inc

Early american life – Leesburg. 1971-1996 (1) 1970-1996 (5) 1977-1996 (9) – (cont by: early american homes) – ISSN: 0012-8155 – mf#6184 – us UMI ProQuest [640]

Early american life – Camp Hill. 2001+ (1) 2001+ (5) 2001+ (9) – (cont: early american homes) – mf#6184,02 – us UMI ProQuest [740]

Early american literature – Chapel Hill. 1986+ (1,5,9) – ISSN: 0012-8163 – mf#16329,01 – us UMI ProQuest [400]

Early american medical imprints, 1668-1820 – 105r – 1 – $11,025.00 – (based on robert b. austin's bibliography fr the national library of medicine. coll encompasses more than 1600 titles reflecting medical thought in america prior to 1821. includes printed guide) – mf#C39-22600 – us Primary [610]

Early american newsletter / Ad Hoc Committee on Indian Education et al – 1968-1985 feb – 1r – 1 – mf#618413 – us WHS [370]

Early american orderly books, 1748-1817 : from the new york historical society – 19r – 1 – (coll of 201 orderly bks fr the french and indian war to the end of the war of 1812) – mf#C39-27350 – us Primary [355]

Early american pamphlets (1796-1936) [at marietta college]... – 15r – 1 – mf#B27638-27652 – us Ohio Hist [073]

Early american philosophers / Jones, Adam Leroy – New York: Macmillan, 1898 [mf ed 1990] – (= ser Columbia university contributions to philosophy, psychology and education 24) – 1mf – 9 – 0-7905-6299-5 – (incl bibl ref) – mf#1988-2299 – us ATLA [190]

Early and central middle ages, c650-1200 : the manuscript record – 25r (coll) – 1 – us Primary [941]

The early and central middle ages, c650-1200 : the manuscript record – 2pt-coll – 25r – 1 – (coll of nearly 100 mss fr the 7th-12th centuries. pt1: mss from cambridge university library, sect a (mss dd-gg) 10r c39-27441. pt2: mss from cambridge university library, sect b (mss hh-mm, additional mss and the ely chapter of 974) 15r c39-27442. incl printed guide) – mf#C35-27440 – us Primary [090]

Early and late / Cox, John – London, England. 18-- – 1r – us UF Libraries [240]

The early annals of the english in bengal : the bengal public consultations for the first half of the eighteenth century / Wilson, Charles Robert – Calcutta: Asiatic Society, 1911- – (= ser Samp: indian books) – us CRL [954]

The early aryans in gujarata / Munshi, Kanaiyalal Maneklal – Bombay: University of Bombay, 1941 – (= ser Samp: indian books) – us CRL [930]

Early australian electoral rolls, vol 1 : new south wales 1903, western australia 1901, tasmania 1903 – 1 – A$150.00 – at Australian [350]

Early australian electoral rolls, vol 2 : victoria 1903, queensland 1903, south australia 1909 – 1 – A$155.00 – at Australian [350]

Early babylonian personal names from the published tablets of the so-called hammurabi dynasty (b.c. 2000) / Ranke, Hermann – Philadelphia: University of Pennsylvania, 1905 – 1mf – 9 – 0-8370-9103-9 – mf#1986-3103 – us ATLA [470]

The early baptists of philadelphia / Spencer, David – Philadelphia: W Syckelmoore, 1877 – 1mf – 9 – 0-524-04241-1 – mf#1990-5032 – us ATLA [242]

The early baptists of virginia : an address / Howell, Robert Boyte Crawford – Philadelphia: Press of the Society, 1857 – 1mf – 9 – 0-7905-6926-4 – mf#1988-2926 – us ATLA [242]

The early baptists of virginia / Howell, Robert Boyte Crawford – Philadelphia: Bible and Publication Society, [1876?] – 1mf – 9 – 0-524-04361-2 – mf#1990-5044 – us ATLA [242]

Early bible songs : with introduction on the nature and spirit of hebrew song / Drysdale, Alexander Hutton – [London]: Religious Tract Soc 1890 [mf ed 1985] – (= ser By-paths of bible knowledge 15) – 1mf – 9 – 0-8370-2982-1 – mf#1985-0982 – us ATLA [221]

Early bird – Arcanum, OH. 1994+ [1] – mf#69078 – us UMI ProQuest [071]

Early bird / Darke Co. Arcanum – may 1977-nov 1981 [wkly] – 7r – 1 – mf#B12994-13000 – us Ohio Hist [071]

Early bird series / Darke Co. Arcanum – jun 1969-apr 1977, nov 1981-dec 1993 [wkly] – 21r – 1 – mf#B33796-33816 – us Ohio Hist [071]

The early brahmanical system of gotra and pravara : a translation of the gotra-pravara-manjari of purusottama-pandita / Purusottama Pandita – Cambridge; New York: Cambridge University Press, 1953 – (= ser Samp: indian books) – (int by john brough) – us CRL [280]

Early british fiction : pre- 1750 – 53r – 1 – (based on william mcburney's checklist of english prose fiction, 1700-1739, and jerry begsley's check list of prose fiction published in england, 1740-1749. included are works of 34 women authors. with printed guide) – mf#C35-28210 – us Primary [830]

Early british relations with assam / Bhuyan, Suryya Kumar – Shillong: Assam Govt, 1928 – (= ser Samp: indian books) – us CRL [954]

Early buddhism / Davids, Thomas William Rhys – London: Constable 1914 [mf ed 1991] – (= ser Religions ancient and modern) – 1mf – 9 – 0-524-00828-0 – mf#1990-2074 – us ATLA [280]

Early buddhist monachism, 600 bc-1000 bc / Dutt, Sukumar – London: Kegan Paul, Trench, Trubner & Co; New York: EP Dutton & Co, 1924 – (= ser Samp: indian books) – us CRL [280]

Early buddhist scriptures : a selection / ed by Thomas, Edward J – London: Kegan Paul, Trench, Trubner & Co, 1935 – (= ser Samp: indian books) – us CRL [280]

Early canadian life – v1 n2-v4 n10 [1977 feb/mar-1980 oct] – 1r – 1 – mf#584046 – us WHS [071]

Early cape hottentots described in the writings of olfert dapper [1668] willem ten ryne [1686] and johannes gulielmus de grevenbroek [1695] [the] / ed by Schapera, Isaac – Cape Town: The Van Riebeeck Society 1933 – (= ser [Travel descriptions from south africa, 1711-1938]) – 4mf – 9 – (original texts, with trans into english by i schapera...& b farrington ...; int & notes by ed) – mf#zah-69 – ne IDC [307]

Early chapters of seneca history : jesuit missions in sonnontouan, 1656-1684 / Hawley, Charles – Auburn, NY?: Knapp, Peck & Thomson, 1884 – 1mf – 9 – mf#34454 – cn Canadiana [241]

Early childhood education journal – New York. 1995+(1,5,9) – (cont: day care and early education) – ISSN: 1082-3301 – mf#11177,01 – us UMI ProQuest [640]

Early Childhood Enhancement Project [Madison WI] *see* News from parents' place

Early childhood research quarterly – Norwood. 1998+ – 1,5,9 – ISSN: 0885-2006 – mf#25399 – us UMI ProQuest [370]

Early children's literature : the birmingham central library collection, 1538-1830 – 3pt – 1 – (pt1: 1538-1799 [367mf] £2150; pt2: 1800-20 [370mf] £2150; pt3: 1821-30 [348mf] £2150; with d/g) – uk Matthew [801]

Early chinese history : are the chinese classics forged? / Allen, Herbert J – London: SPCK; New York: ES Gorham, 1906 – 1mf – 9 – 0-524-01149-4 – mf#1990-2225 – us ATLA [951]

Early christian architecture in ireland / Stokes, Margaret MacNair – London: George Bell & Sons, 1878 – (= ser 19th c art & architecture) – 3mf – 9 – mf#4.1.50 – uk Chadwyck [720]

Early christian baptism and the creed : a study in ante-nicene theology / Chrehan, J – London, 1950 – 4mf – 8 – €11.00 – ne Slangenburg [240]

Early christian doctrine / Pullan, Leighton – 3rd. ed. London: Rivingtons, 1905 – (= ser Oxford church text books) – 1mf – 9 – 0-524-04850-9 – mf#1990-1342 – us ATLA [240]

Early christian ethics in the west : from clement to ambrose / Scullard, Herbert Hayes – London: Williams & Norgate, 1907 – 1mf – 9 – 0-8370-6371-X – (incl indof names and subjects) – mf#1986-0371 – us ATLA [230]

The early christian fathers : or, memorials of nine distinguished teachers of the christian faith during the first three centuries. including their testimony to the three-fold ministry of the church / Carmichael, William Miller – New-York: Alexander V Blake, 1844 – 1mf – 9 – 0-524-04011-7 – mf#1990-1183 – us ATLA [240]

The early christian martyrs and their persecutions / Herkless, John, Sir – London: Dent; Philadelphia: J.B. Lippincott, [1904?] – (= ser The Temple Series Of Bible Handbooks) – 1mf – 9 – 0-7905-5838-6 – mf#1988-1838 – us ATLA [240]

Early christian missions of ireland, scotland and england / Charles, Elizabeth Rundle – London: SPCK; New York: E & JB Young, 1893 – 1mf – 9 – 0-7905-4170-X – mf#1988-0170 – us ATLA [240]

Early christian numismatics : and other antiquarian tracts / King, Charles William – London: Bell & Daldy, 1873 – 1mf – 9 – 0-7905-5353-8 – mf#1988-1353 – us ATLA [930]

Early christian scotland, 400 to 1093 ad / Boyd, A K H – s.l, s.l? 18– – 1r – us UF Libraries [240]

Early christianity / Bainton, Roland Herbert – Princeton, NJ. 1969 – 1r – us UF Libraries [240]

Early christianity and paganism : a.d. 64 to the peace of the church in the fourth century / Spence-Jones, Henry Donald Maurice – New York: E P Dutton, [1901?] – 2mf – 9 – 0-7905-6449-1 – mf#1988-2449 – us ATLA [230]

Early christianity in arabia / Wright, Thomas – 9 – $10.00 – us IRC [240]

Early christianity outside the roman empire : two lectures delivered at trinity college, dublin / Burkitt, Francis Crawford – Cambridge: University Press; New York: Macmillan [distributor], 1899 – 1mf – 9 – 0-7905-4167-X – mf#1988-0167 – us ATLA [240]

The early christians in rome / Spence-Jones, Henry Donald Maurice – London: Methuen, 1910 – 2mf – 9 – 0-7905-6083-6 – mf#1988-2083 – us ATLA [240]

The early church / First Congregational Church, Emporia KS – 1857-1926, Additional church records through 1958 – 1 – us Kansas [240]

The early church : a history of christianity in the first six centuries / Duff, David; ed by Duff, David, jr – Edinburgh: T. & T. Clark, 1891 – 2mf – 9 – 0-7905-6586-2 – mf#1988-2586 – us ATLA [240]

The early church : a history of christianity in the first six centuries / Duff, David – Edingurgh: T. & T. Clark, 1891 – 2mf – us ATLA [240]

The early church / Horton, Robert F – London: T.C. & E.C. Jack, 1908 – (= ser Century Bible Handbooks) – 1mf – 9 – 0-7905-3201-8 – mf#1987-3201 – us ATLA [240]

The early church / Sheldon, Henry Clay – New York: Thomas Y Crowell, 1894 – 2mf – 9 – 0-524-03427-3 – mf#1990-0981 – us ATLA [240]

The early church from ignatius to augustine / Hodges, George – Boston: Houghton Mifflin, c1915 – 1mf – 9 – 0-524-02701-3 – mf#1990-0682 – us ATLA [240]

Early church history : to the death of constantine / Backhouse, Edward; ed by Tylor, Charles – 3rd ed. London: Simpkin, Marshall, Hamilton, Kent, 1892 [mf ed 1989] – 1mf – 9 – 0-7905-4319-2 – mf#1988-0319 – us ATLA [240]

Early church history to a.d. 313 / Gwatkin, Henry Melvill – 2nd ed. London: Macmillan, 1912 – 2mf – 9 – 0-7905-5699-5 – (incl bibl ref) – mf#1988-1699 – us ATLA [240]

The early church in the light of the monuments : a study in christian archaeology / Barnes, Arthur Stapylton – London; New York: Longmans, Green, 1913 – (= ser Westminster Library (London, England)) – 1mf – 9 – 0-7905-4126-2 – (incl bibl ref) – mf#1988-0126 – us ATLA [930]

The early church in the light of the monuments : a study in christian archaeology / Barnes, Arthur Stapylton – London; New York: Longmans, Green, 1913. (The Westminster library) – 1mf – us ATLA [240]

Early churches in palestine / Crowfoot, J W – 9 – $10.00 – us IRC [240]

The early churches of constantinopel : architecture and liturgy / Mathews, Thomas – University Park/London, 1971 – 6mf – 8 – €14.00 – ne Slangenburg [720]

Early clergy of pennsylvania and delaware / Hotchkin, Samuel Fitch – Philadelphia: PW Ziegler, 1890 [mf ed 1991] – 1mf – 9 – 0-524-00559-1 – mf#1990-0059 – us ATLA [242]

The early conflicts of christianity / Kip, William Ingraham – New York: D. Appleton, 1850, c1849 – 1mf – 9 – 0-7905-6002-X – mf#1988-2002 – us ATLA [240]

Early critical essays 1820-1822 / Cooper, James Fenimore – 9 – us Scholars Facs [840]

Early cuadro de costumbres in colombia / Duffey, Frank M – Chapel Hill, North Carolina. 1956 – 1r – us UF Libraries [972]

Early czech newspapers of texas / Institute of Texan Cultures. Library – Svobodan (La Grange) and Obzorn (Halletsville). 17 rolls – 1 – $400.00 – us TX Culture [071]

Early dawn – Oshkosh WI. 1877 jun 21, 1879 feb 13-1880 apr 9 – 2r – 1 – (cont by: reflector [oshkosh wi]) – mf#958831 – us WHS [071]

The early dawn – Bonthe, Sierra Leone. Jan 1885-Jun 1892 (imperfect) – 56ft – 1 – uk British Libr Newspaper [072]

Early day – Harrisburg, NE: Graves & Beard, 1889-1892// (wkly) [mf ed v4 n42. mar 11-jul 22 1892 (gaps) filmed 1986 – 1r – 1 – (cont: harrisburg gazette. merged with: labor wave to form: banner county news) – us NE Hist [071]

Early days and native ways in southern rhodesia / Jones, Neville – Bulawayo, Zimbabwe. 1944 – 1r – us UF Libraries [960]

Early days at york factory / Willson, Beckles – [Toronto?: s.n.], 1899 [mf ed 1982] – 1mf – 9 – 0-665-17839-5 – mf#17839 – cn Canadiana [971]

The early days of christianity / Farrar, Frederic William – New York: Funk & Wagnalls, 1883 – 2mf – 9 – 0-8370-6327-2 – (incl bibl ref and indexes) – mf#1986-0327 – us ATLA [220]

The early days of my episcopate / Kip, William Ingraham – New York: T. Whittaker, 1892 – 1mf – 9 – 0-7905-4986-7 – mf#1988-0986 – us ATLA [240]

The early days of thomas whittemore : an autobiography extending from a.d. 1800 to a.d. 1825 / Whittemore, Thomas – Boston: James M Usher, 1859, c1858 – 1mf – 9 – 0-7905-8214-7 – mf#1988-8097 – us ATLA [920]

Early development and parenting – Chichester. 1992-1998 (1,5,9) – (cont by: infant and child development) – ISSN: 1057-3593 – mf#19118 – us UMI ProQuest [150]

The early development of mohammedanism : lectures...may and june 1913 / Margoliouth, David Samuel – London: Williams & Norgate 1914 [mf ed 1991] – (= ser Hibbert lectures (london, england) 1913) – 1mf – 9 – 0-524-00932-5 – mf#1990-2155 – us ATLA [260]

Early dramas and romances : the robbers, fiesco, love and intrigue, demetrius, the ghost-seer, and the sport of destiny / Schiller, Friedrich – London: G Bell, 1917 [mf ed 1989] – (= ser Bohn's standard library. schiller's historical works) – xv/493p – 1 – (trans fr german) – mf#6555 – us Wisconsin U Libr [820]

Early drawings and illuminations : an introduction to the study of illustrated manuscripts: with a dictionary of subjects in the british museum / Birch, Walter de Gray – London: S Bagster, 1879 – 1mf – 9 – 0-7905-6461-0 – mf#1988-2461 – us ATLA [090]

Early eastern christianity : lecture for 1904, on the syriac-speaking church / Burkitt, Francis Crawford – London: John Murray 1904 [mf ed 1990] – (= ser St margaret's lectures 1904) – 1mf – 9 – 0-7905-6405-X – mf#1988-2405 – us ATLA [240]

Early, Eleanor *see*
– Lands of delight
– Ports of the sun

Early english books tract supplement : a fascinating look at england in the 16th and 17th centuries – 16th- and 17th c – 72r through unit 2 – 1 – (with p/g) – us UMI ProQuest [941]

The early english church / De Mille, James – S.l: s.n, 1877? – 1mf – 9 – mf#06872 – cn Canadiana [240]

The early english dissenters in the light of recent research (1550-1641) / Burrage, Champlin – Cambridge: University Press; New York: Putnam [distributor], 1912 – 2mf – 9 – 0-7905-4610-8 – (incl bibl ref) – mf#1988-0610 – us ATLA [240]

Early english newspapers : from the british museum, london and the bodleian library, oxford – ongoing coll 114 units ca 50r ea – 5465r – 1 – (from the colls of dr charles burney dating back to 1603 and rival collector john nicols in 1865) – mf#C39-28920 – us Primary [072]

Early english poetry, ballads and popular literature of the middle ages / Percy Society – v1-30 – 1 – $512.00 – mf#0446 – us Brook [810]

Early english printed books 1475-1640 / Cambridge University. Library – 4v – 1,9 – us AMS Press [010]

Early english text society extra series – v1-125. 1867-1920 – 469mf – 8 – mf#204 – ne IDC [420]

Early english text society original series – v1-128. 1864-1904 – 593mf – 8 – mf#1295 – ne IDC [420]

The early eucharist (a.d. 30-180) / Frankland, William Barrett – London: C J Clay; New York: Macmillan [distributor], 1902 – 1mf – 9 – 0-7905-3132-1 – mf#1987-3132 – us ATLA [240]

Early european banking in india : with some reflections on present conditions / Sinha, H – London: Macmillan and Co, 1927 – (= ser Samp: indian books) – us CRL [332]

The early fathers of the reformed church in the united states / Good, James Isaac – Reading, PA: D Miller, c1897 – 1mf – 9 – 0-524-07242-6 – mf#1991-2983 – us ATLA [242]

Early federal nominative reports – 1180mf – 9 – $1,770.00 coll – (individual titles also listed separately) – us LLMC [340]

Early flemish artists and their predecessors on the lower rhine / Conway, William Martin Conway, Baron – London 1887 – (= ser 19th c art & architecture) – 4mf – 9 – mf#4.2.27 – uk Chadwyck [750]

The early flemish painters / Crowe, Joseph Archer & Cavalcaselle, Giovanni Battista – London 1857 – (= ser 19th c art & architecture) – 5mf – 9 – mf#4.2.1265 – uk Chadwyck [750]

Early florentine woodcuts / Kristeller, Paul – London 1897 – (= ser 19th c art & architecture) – 5mf – 9 – mf#4.2.260 – uk Chadwyck [730]

Early florida citrus fruits in northern markets / Ferran, H R – Tallahassee, FL. 1934 – 1r – us UF Libraries [634]

Early florida pastimes / Seger, Alice – s.l, s.l? 193? – 1r – us UF Libraries [978]

Early friends and modern professors : in reply to strictures, by joseph john gurney / Martin, Henry – London: Edmund Fry, 1836 – 1mf – 9 – 0-524-07574-3 – mf#1991-3194 – us ATLA [240]

Early german art / Burlington Fine Arts Club. London – 1906 – 9 – uk Chadwyck [700]

Early hebrew orthography / Cross, Frank M – 1952 – 9 – $10.00 – us IRC [470]

Early hebrew story : its historical background / Peters, John Punnett – London: Williams & Norgate; New York: G P Putnam 1904 [mf ed 1989] – (= ser Crown theological library 7) – 1mf – 9 – 0-7905-2268-3 – (incl ind) – mf#1987-2268 – us ATLA [221]

The early heroes of islam / Salik, Saiyed Abdus – [Calcutta]: University of Calcutta, 1926 – (= ser Samp: indian books) – us CRL [260]

The early history of canadian banking, vol 1 : origin of the canadian banking system / Shortt, Adam – Toronto: Journal of the Canadian Bankers' Association, 1896 – 1mf – 9 – mf#13714 – cn Canadiana [332]

The early history of canadian banking, vol 4 : the first banks in lower canada / Shortt, Adam – Toronto: Journal of the Canadian Bankers' Association, 1897 [mf ed 1981] – 1mf – 9 – 0-665-13717-6 – mf#13717 – cn Canadiana [332]

The early history of canadian banking, vol 5 : the first banks in lower canada / Shortt, Adam – Toronto: Journal of the Canadian Bankers' Association, 1897 [mf ed 1981] – 1mf – 9 – 0-665-13718-4 – mf#13718 – cn Canadiana [332]

The early history of congregationalism in new jersey and the middle provinces / Brown, William Bryant – Boston: A Mudge, 1877 – 1mf – 9 – 0-524-02634-3 – mf#1990-4389 – us ATLA [242]

Early history of cuba, 1492-1586 / Wright, Irene Aloha – New York, NY. 1916 – 1r – us UF Libraries [972]

Early history of dorchester : and other parts of new brunswick / Milner, William Cochrane – [New Brunswick?: s.n, 1915?] – 1mf – 9 – 0-665-65262-3 – mf#65262 – cn Canadiana [929]

Early history of india / Ghosh, Nagendra Nath – Allahabad: Indian Press, 1948 – (= ser Samp: indian books) – us CRL [954]

The early history of india : from 600 b c to the muhammadan conquest, including the invasion of alexander the great / Smith, Vincent Arthur – 3rd rev enl ed. Oxford: Clarendon Press, 1914 [mf rev ed 1995] – (= ser Yale coll) – xii/512p (ill) – 1 – 0-524-09726-7 – mf#1995-0726 – us ATLA [930]

The early history of india from 600 bc to the muhammadan conquest : including the invasion of alexander the great / Smith, Vincent Arthur – Oxford: Clarendon Press, 1924 – (= ser Samp: indian books) – (rev by s m edwardes) – us CRL [930]

Early history of kamarupa : from the earliest times to the end of the sixteenth century / Barua, Kanaklal – Shillong: The Author, 1933 – (= ser Samp: indian books) – us CRL [954]

Early history of mobile baptists / Kennedy, Gladys – Unpub. mss. 192p – 1 – 6.72 – us Southern Baptist [242]

Early history of the athanasian creed : the results of some original research upon the subject, with an appendix containing four ancient commentaries... / Ommanney, George Druce Wynne – London: Rivingtons, 1880 – 1mf – 9 – 0-524-01123-0 – mf#1990-0337 – us ATLA [240]

Early history of the christian church from its foundation to the end of the fifth century = Histoire ancienne de l'eglise, tome 2 / Duchesne, Louis – London: John Murray, 1912 – 2mf – 9 – 0-7905-4631-0 – (incl bibl ref. in english) – mf#1988-0631 – us ATLA [240]

Early history of the christian church from its foundation to the end of the third century = Histoire ancienne de l'eglise, tome 1 / Duchesne, Louis – London: J. Murray, 1909 – 2mf – 9 – 0-7905-4413-X – (includes "note to second edition", and bibliographical references. in english) – mf#1988-0413 – us ATLA [240]

The early history of the church missionary society for africa and the east to the end of a.d. 1814 / Hole, Charles – London: Church Missionary Society, 1896 – 2mf – 9 – 0-7905-7051-3 – mf#1988-3051 – us ATLA [240]

Early history of the colony of victoria : from its discovery to its establishment as a self-governing province of the british empire / Labilliere, Francis Peter de – London, 1878 – (= ser 19th c british colonization) – 2v on 8mf – 9 – mf#1.1.5303 – uk Chadwyck [980]

Early history of the dekkan : down to the mahomedan conquest / Bhandarkar, Ramkrishna Gopal – Bombay: Printed at the Government Central Press, 1884 – (= ser Samp: indian books) – us CRL [930]

Early history of the disciples in the western reserve, ohio : with biographical sketches of the principal agents in their religious movement / Hayden, Amos Sutton – Cincinnati: Chase & Hall, 1875 – 2mf – 9 – 0-7905-4745-7 – mf#1988-0745 – us ATLA [240]

Early history of the federal supreme court / Muller, William Henry – Boston: Chipman, 1922. 117p. LL-1283 – 1 – us L of C Photodup [347]

The early history of the hebrews / Sayce, Archibald Henry – London: Rivingtons, 1897 – 2mf – 9 – 0-7905-0285-2 – (incl bibl ref and index) – mf#1987-0285 – us ATLA [939]

Early history of the vaisnava faith and movement in bengal : from sanskrit and bengali sources / De, Sushil Kumar – Calcutta: General Printers and Publ, 1942 – (= ser Samp: indian books) – us CRL [280]

Early history of vaishnavism in south india / Krishnaswamy Iyengar, Srinivasa – London; New York: Oxford University Press, 1920 – (= ser Samp: indian books) – us CRL [280]

Early human development – Amsterdam. 1977+ (1) 1977+ (5) 1987+ (9) – ISSN: 0378-3782 – mf#42126 – us UMI ProQuest [618]

Early ideals of righteousness : hebrew, greek, and roman / Kennett, Robert Hatch – Edinburgh: T & T Clark, 1910 [mf ed 1986] – 1mf – 9 – 0-8370-9957-9 – mf#1986-3957 – us ATLA [170]

Early impressions : or, evidences of the secret operations of the divine witness in the minds of children / Johnson, Jane – Philadelphia: TE Chapman, 1844 – 1mf – 9 – 0-524-07570-0 – mf#1991-3190 – us ATLA [240]

Early imprint publications on palestine, 1921-1939 – Chicago: The Middle Eastern Microfilm Project, 1993 – 1 – us CRL [956]

Early in nomine : a genesis of chamber music / Weidner, Robert Wright – U of Rochester 1960 [mf ed 19–] – 2v on 1r / 6mf – 1,9 – (v2 incl transcr of 38 selected in nomines of ferrabosco, mundy, parsley, taverner, tye, tallis & white; with bibl) – mf#film 1371 / fiche 118 – us Sibley [780]

Early indian religious thought / Vidyarthi, Pandeya Brahmeshwar – New Delhi, India. 1976 – 1r – us UF Libraries [280]

Early indian sculpture / Bachhofer, Ludwig – Paris: Pegasus Press, 1929 – (= ser Samp: indian books) – us CRL [730]

Early instrumental style of kurt weill / Luxner, Michael David – U of Rochester 1972 [mf ed 19–] – 4mf – 9 – mf#fiche 1146 – us Sibley [780]

Early irish laws and institutions / MacNeill, John – Dublin, Burns, Oates and Washbourne 1934 152 p. LL-2272 – 1 – us L of C Photodup [340]

Early iron age in malawi / Robinson, K R – Zomba, Malawi. 1969 – 1r – us UF Libraries [930]

Early israel and the surrounding nations / Sayce, Archibald Henry – New York: E R Herrick, 1899 [mf ed 1986] – 1mf – 9 – 0-8370-9739-8 – mf#1986-3739 – us ATLA [956]

Early ivories from samaria / Crowfoot, J W – PEF. 1938 – 9 – $10.00 – us IRC [930]

Early jewish colony in western guiana, 1658-1666 / Oppenheim, Samuel – New York, NY. 1907 ? – 1r – us UF Libraries [939]

Early latin hymnaries : an index of hymns in hymnaries before 1100 / Mearns, James – Cambridge: University Press, 1913 – 1mf – 9 – 0-7905-5014-8 – mf#1988-1014 – us ATLA [780]

Early laws of missouri pertaining to women : project of historical activities committee, 1966-1968 / Ingersoll, Mrs. Albert Converse – St. Louis National Society of Colonial Dames in the State of Missouri. 1969? 37p LL-285 – 1 – us L of C Photodup [340]

Early leaving : report of the central advisory council for education, 1954 – 2mf – 9 – mf#86961 – uk Microform Academic [324]

Early letters and classified papers, 1660-1740 *see* Collections from the royal society

Early letters of marcus dods, d.d : (late principal of new college, edinburgh) (1850-1864) = Correspondence. selections / Dods, Marcus – London: Hodder & Stoughton, [1910?] – 1mf – 9 – 0-7905-4350-8 – mf#1988-0350 – us ATLA [920]

The early life of abraham lincoln.. / Tarbell, Ida Minerva – New York: S.S. McClure, 1896. 240p. illus. ports. (McClure's Magazine Library, no. 3) – 1 – us Wisconsin U Libr [920]

Early life of george poindexter / Swearingen, Mack Buckley – Chicago, IL. 1934 – 1r – us UF Libraries [920]

The early life of jesus : sermons / Brooke, Stopford Augustus – London: David Stott, 1888 – 1mf – 9 – 0-7905-9151-0 – mf#1989-2376 – us ATLA [240]

The early life of jesus and new light on passion week / Whitman, Peleg Spencer – Philadelphia: Griffith & Rowland, 1914 – 1mf – 9 – 0-7905-2213-6 – (incl ind) – mf#1987-2213 – us ATLA [220]

Early life of mrs judson – London, England. 18– – 1r – us UF Libraries [240]

Early madrigals of alessandro striggio / Tadlock, Ray J – U of Rochester 1958 [mf ed 19–] – 2v on 1r – 1 – (with app & bibl) – mf#film 2091 – us Sibley [780]

Early man in zambia / Johnston, S – Lusaka, Zambia. 1970 – 1r – us UF Libraries [930]

Early marriages in geauga county [ohio] / Davis, Howard (Mrs) – 1968 – 1r – 1 – (typescript index, alphabetical by township, then by marriage. filmed by genealogical society of utah, 1974) – us Western Res [978]

Early methodism in the carolinas / Chreitzberg, Abel McKee – Nashville, TN: Pub House of the ME Church, South, 1897 – 1mf – 9 – 0-524-06989-1 – mf#1991-2842 – us ATLA [242]

Early methodism within the bounds of the old genesee conference from 1788 to 1828 : or, the first forty years of wesleyan evangelism in northern pennsylvania, central and western new york, and canada. containing sketches of interesting localities, exciting scenes, and prominent actors / Peck, George – New York: Carlton and Porter, 1860 – 2mf – 9 – 0-524-01746-8 – mf#1990-4138 – us ATLA [242]

Early methodist philanthropy / North, Eric McCoy – New York: Methodist Book Concern, c1914 – 1mf – 9 – 0-7905-5775-4 – (incl bibl ref) – mf#1988-1775 – us ATLA [242]

Early migrations : origin of the chinese race, philosophy of their early development, with an inquiry into the evidences of their american origin... / Brooks, Charles Wolcott – San Francisco: s.n, 1876 – 1mf – 9 – mf#14399 – cn Canadiana [572]

Early missions to and within the british islands / Hole, Charles – London: SPCK; New York: E & JB Young, [1888?] – 1mf – 9 – 0-7905-6480-7 – mf#1988-2480 – us ATLA [240]

Early moral and religious education : being a lecture delivered to the mechanics' institute and library association / Cook, John – Quebec?: Sinclair and Pooler, 1849 – 1mf – 9 – mf#52193 – cn Canadiana [230]

Early music – Oxford. 1973+ (1) 1976+ (5) 1976+ (9) – ISSN: 0306-1078 – mf#9852 – us UMI ProQuest [780]

Early music – [mf ed Marlborough 1996] – 2pt – 1 – (pt1: pembroke choir books & other music mss fr pembroke college, cambridge [3r] £285; pt2: music mss 1500-1793 fr national library of scotland [14r] £1300; with d/g) – uk Matthew [780]

Early music from low countries libraries – 3880mf – 9 – (pt1: concertos before 1820 (199 compositions) on 259mf; pt2: orchestral music before 1820 (372 compositions) on 427mf; pt3: church music 1750-1820 (303 compositions) on 541mf; pt4: vocal and instrumental tutors (398 works) on 811mf; pt5: historical organ collection : the baetz archive (ca 440 technical & decorative drawings) on 67mf; pt6: vocal music, 1650-1820 (311 compositions) on 609mf; pt7: keyboard music, 1620-1820 (449 compositions) on 326mf; pt8: music for solo instrument, 1600-1820 (404 compositions) on 361mf; pt9: 479mf) – ne IDC [780]

Early music from low countries libraries – 9 – (pt1: concertos before 1820 [259mf] €1755 m371; pt2: orchestral music before 1820 [427mf] €3295 [m374]; pt3: church music, 1750-1820 [541mf] €3680 [m377]; pt4: vocal & instrumental tutors [811mf] €5515 [m380]; pt5: historical organ collection: the baetz and witte archive, c1827-1902 (utrecht university) [107 b/w mf & 19mf of supporting materials] €730 [m383]; pt6: vocal music 1650-1820 [609mf] €4140 [m386]; pt7: keyboard music, 1620-1820 [326mf] €2265 [m387]; pt8: solo instrumental music 1620-1820 [361mf] €2455 [m388]; pt9: music for instrumental ensemble, 1680-1820 [479mf] €3260 [m389]; with p/g; previously "music from dutch libraries") – ne MMF Publ [780]

Early music history – Cambridge. 1990-1994 (1) – ISSN: 0261-1279 – mf#16527 – us UMI ProQuest [780]

Early new england catechisms : a bibliographical account of some catechisms published before the year 1800, for use in new england / Eames, Wilberforce – Worcester, Mass: C Hamilton, 1898 – 1mf – 9 – 0-524-00539-7 – (incl bibl ref) – mf#1990-0039 – us ATLA [240]

Early new england schools / Small, Walter Herbert – Boston: Ginn & Co, 1914 [mf ed 1990] – 1mf – 9 – 0-7905-6445-9 – (incl bibl ref) – mf#1988-2445 – us ATLA [370]

Early new hampshire baptist churches. new hampshire : church records – Reel 2 – Items 1-6 – 1 – (1. hopkinton baptist church society records. 1794-1864. 2. londonerry baptist church. 1799-1900 (2 vols.). 3. main and new hampshire quarterly meetings. 1783-1792. 4. marlow baptist church. 1777-1807, 1859-1900 (2 vols.). 5. mason baptist church. 1786-1836, original ms and typescripts. 6. meridith baptist church. 1779-1829, vol. 1) – us Southern Baptist [242]

Early new hampshire baptist churches. new hampshire : church records – Reel 3 – Items 1 and 2 – 1 – (1. meredity baptist church. 1823-44 (2 vols.). 2. new durham quarterly meeting. 1792-1801, vol. 1; 1801-1807, vol. 2; record of quarterly meeting. vol. 3, 1809; vol. 4, 1832-1857) – us Southern Baptist [242]

Early new hampshire baptist churches. new hampshire : church records – Reel 4 – Items 1-5 – 1 – (1. new durham quarterly meetings. 1857-1889; treasurer's report. 1868-1874. 2. new durham elders conference records. 1801-13, 1841-48 (2 vols.). 3. ministers conference records. 1843-1865, 1870-1885 (2 vols.). 4. new hampshire ministers conference of yearly meetings. 1884-1917. 5. northwood baptist church. 1779-1829) – us Southern Baptist [242]

Early new hampshire baptist churches. new hampshire : church records – Reel 5 – Items 1-7 – 1 – (1. sanbornton first baptist church. 1793-1848. 2. sandwich quarterly meetings minister's meeting. 1845-1911. 3. diary: curtis, silas. "a tour among freedmen" (in va, sc, nc, 1865). 4. white mountain quarterly meetings. 1842-1871. 5. seaman, job. papers. 1762-1820. 6. diary no. 1, original journal of elder job seamans. 1774 (breaks april 22, 1778-june 2, 1785; february 8, 1791-april 24, 1794; april 14, 1811-dec. 8, 1814). 7. diary no. 2. march 1802 – jan. 9, 1810) – us Southern Baptist [242]

Early new hampshire baptist churches. new hampshire : church records – 6 reels – 1 – us Southern Baptist [242]

Early new zealand / Sherrin & Wallace – 6mf – 9 – NZ$24.00 – 0-908797-37-0 – (earliest times to 1845. lists early european settlers) – mf#NZNB S675 – nz BAB [980]

Early newspapers / Allen Co. Lima – (1856-1900) – 10r – 1 – (enquire for titles and dates) – mf#B2439-2448 – us Ohio Hist [071]

Early newspapers / Delaware Co. Delaware – (oct 1821-dec 1857) [wkly] – 5r – 1 – (ask for titles) – mf#B1479-1483 – us Ohio Hist [071]

Early newspapers / Harrison Co. Cadiz – (1821-1932) scattered [wkly] – 1r – 1 – mf#B1264 – us Ohio Hist [071]

Early nineteenth century manuscripts from kumasi, ghana / Kongelige Bibliotek. Copenhagen – Mss. Orientalisk Samling Cod. Arab. 302 – 1 – us CRL [090]

Early old testament narratives : thirty-six lessons / Pulsford, William Hanson – Boston: Unitarian Sunday-School Society, c1893 – 1mf – 9 – 0-524-05688-9 – mf#1992-0538 – us ATLA [221]

The early persecutions of the christians / Canfield, Leon Hardy – New York: Columbia University: Longmans, Green [distributor], 1913 – (= ser Studies in History, Economics and Public Law) – 1mf – 9 – 0-7905-4194-7 – (incl bibl ref) – mf#1988-0194 – us ATLA [240]

Early photography books – Helios – 17r – 1 – $1845.00 – us UPA [770]

The early poetry of israel in its physical and social origins / Smith, George Adam – London: publ...by OUP, 1912 [mf ed 1988] – 1mf – 9 – 0-7905-0340-9 – (incl bibl ref & ind) – mf#1987-0340 – us ATLA [221]

Early prayer – London, England. 1856 – 1r – us UF Libraries [240]

Early presbyterian missions in the colonies and states / Tadlock, James Doak – Richmond, Va: Presbyterian Committee of Publication, c1896 – 1mf – 9 – 0-524-01756-5 – mf#1990-4148 – us ATLA [242]

Early presbyterianism in maryland / McIlvain, James William – [Baltimore MD?: s.n. 1890?] [mf ed 1990] – (= ser Johns hopkins university studies in historical and political science 8/3) – 1mf – 9 – 0-7905-6870-5 – (incl bibl ref) – mf#1988-2870 – us ATLA [242]

Early prevalence of monotheistic beliefs / Rawlinson, George – London, England. 1883? – 1r – us UF Libraries [240]

Early printed books on religion from colonial spanish america – 1543/4-ca 1800 – 406 titles on 1038mf – 9 – €8045.00 – (catecismos: 16 titles on 69mf [$742]; confesionarios: 13 titles on 31mf [$333]; doctrina cristiana, obras de devocion: 29 titles on 99mf [$1064]; himnos, villancicos: 3 titles on 5mf [$54]; manuales de rito: 16 titles on 43mf [$462]; hagiografías, milagros: 55 titles on 141mf [$1516]; biografías de religiosos: 30 titles on 84mf [$903]; historia eclesiastica: 5 titles on 24mf [$258]; ordenes religiosos, etc: 85 titles on 243mf [$2612]; inquisicion: 10 titles on 20mf [$215]; teología culta, derecho canonico: 12 titles on 33mf [$355]; mística y meditacion: 6 titles on 19mf [$204]; sermones en castellano: 26 titles on 31mf [$333]; sermones en lenguas amerindias y bilingues: 3 titles on 9mf [$97]; arte de predicar: 1 title on 2mf [$22]; evangelizacion: 5 titles on 23mf [$247]; iglesias, catedrales: 33 titles on 36mf [$387]; colegios religiosos: 5 titles on 5mf [$54]; papas (cartas apostolicas, etc): 5 titles on 5mf [$54]; arzobispos: 18 titles on 42mf [$452]; obispos: 13 titles on 32mf [$344]; concilios y sinodos: 6 titles on 22mf [$237]; miscelanea: 11 titles on 20mf [$215]) – ne IDC [241]

Early printed manuscript music / Westminster Abbey. Library – 32r – 1 – £1200.00 – (mostly italian of the 17th century, including sacred works by colona, foggia, gratiani and others. the mss music is mostly from the 18th century) – mf#WAM – uk World [780]

The early printed music collection / Christ Church. Oxford – 16th-17th C – 76r – 1 – £3,650.00 – (based on "catalogue of printed music" ed by aloys hiff, 1919. contents list available) – mf#XCM – uk World [780]

The early progress of the gospel : in eight sermons / Humphry, William Gilson – London: John W Parker, 1851 – 1mf – 9 – 0-7905-0374-3 – (incl bibl ref) – mf#1987-0374 – us ATLA [240]

Early promoted : a memoir of the rev william spiller cox. compiled by his father / [Cox, E W] – London, [1898] – 3mf – 9 – mf#HTM-45 – ne IDC [920]

Early prose writing / Lowell, James Russell – London, England. 1902 – 1r – us UF Libraries [420]

Early pupils of the spirit : the ethical development of the prophets of israel; and, what of samuel? / Whiton, James Morris – London: James Clarke, 1896 – 1mf – 9 – 0-8370-5828-7 – mf#1985-3828 – us ATLA [220]

Early quaker writings. 1st series 1650-1750 *see* Early quaker writings 17th-18th centuries

Early quaker writings. 2nd series 17th century *see* Early quaker writings 17th-18th centuries

Early quaker writings 17th-18th centuries / Friends House Library. The Religious Society of Friends – 35r – 1 – £1630.00 – (two series of major early quaker works. includes george fox, william penn, george bishop, thomas lason and many others. series 1: 1650-1750 25r £1150 eqw. series 2: 17th century 10r £480 eqs) – mf#EQWIS – uk World [243]

Early rare british film-makers' catalogues : 1896-1913 – 8r – 1 – £400.00 – mf#EFM – uk World [790]

Early rare photographic books : series a: northwestern museum of science and industry collection – 11r – 1 – £480.00 – mf#NPW – uk World [790]

Early rare photographic collections : pt a: photographs – pt b: register / Victoria and Albert Museum – 2pts – 24r (17 col 7 b/w) – 1,14 – £2150.00 coll – mf#VAA – uk World [770]

Early records of furnival's inn / Bland, Desmond S – New Castle upon Tyne: King's College, 1957 – 1mf – 9 – $1.50 – (edited from a middle temple manuscript) – mf#LLMC 84-276 – us LLMC [340]

Early records of upper east tennessee – 1987 jun-1990 oct [v1-4] – 1r – 1 – mf#1848270 – us WHS [978]

The early relation and separation of baptists and disciples / Gates, Errett – Chicago: Christian Century Co., 1904 – 1mf – 9 – 0-7905-5209-4 – mf#1988-1209 – us ATLA [240]

The early religion of israel : as set forth by biblical writers and by modern critical historians / Robertson, James – New York: Anson D F Randolph; Edinburgh: William Blackwood 1892 [mf ed 1986] – (= ser The baird lecture 1889) – 2mf – 9 – 0-8370-9982-X – (incl bibl ref & ind) – mf#1986-3982 – us ATLA [270]

The early religion of israel / Paton, Lewis Bayles – Boston: Houghton Mifflin, 1910 – (= ser Modern Religious Problems) – 1mf – 9 – 0-7905-1674-8 – mf#1987-1674 – us ATLA [270]

The early religious customs of new england : an address at the two hundredth anniversary of the building of the meeting-house in hingham, mass., august 8, 1881 / Young, Edward James – Cambridge: J Wilson, 1882 – 1mf – 9 – 0-524-01144-3 – mf#1990-0358 – us ATLA [240]

Early religious education : considered as the divinely appointed way to the regenerate life / Eliot, William Greenleaf – Boston: American Unitarian Association, 1868, c1855 – 1mf – 9 – 0-8370-7938-1 – mf#1986-1938 – us ATLA [240]

Early religious history of / Barr, John – 1852 – 1 – $50.00 – us Presbyterian [920]

Early religious history of maryland : maryland not a roman catholic colony. religious toleration not an act of roman catholic legislation / Brown, Benjamin F – Baltimore: Innes & Co, 1876 – 1mf – 9 – 0-524-04950-5 – mf#1990-1353 – us ATLA [241]

Early religious poetry of persia / Moulton, James Hope – Cambridge: University Press 1911 [mf ed 1991] – (= ser The cambridge manuals of science and literature) – 1mf – 9 – 0-524-00940-6 – (incl bibl ref) – mf#1990-2163 – us ATLA [290]

Early religious poetry of the hebrews / King, Edward George – Cambridge: University Press; New York: G P Putnam 1911 [mf ed 1989] – (= ser The cambridge manuals of science and literature) – 9 – 0-7905-1179-7 – (incl ind) – mf#1987-1179 – us ATLA [270]

The early roman episcopate to a.d. 384 / Beet, William Ernest – 1st ed. London: Charles H. Kelly, 1913 – 1mf – 9 – 0-7905-4078-9 – (incl bibl ref) – mf#1988-0078 – us ATLA [240]

Early roman-catholic missions to india : with sketches of jesuitism, hindu philosophy, and the christianity of the ancient indo-syrian church of malabar: an historical essay / Tinling, James Forbes Bisset – London: S W Partridge, 1871 [mf ed 1995] – (= ser Yale coll) – 103p – 1 – 0-524-09109-9 – mf#1995-0109 – us ATLA [241]

Early saint john methodism and history of centenary methodist church, saint john, nb : a jubilee souvenir / ed by Henderson, George A – Saint John, NB: G E Day, 1890 [mf ed 1980] – 3mf – 9 – (incl bibl ref) – mf#06985 – cn Canadiana [242]

The early schools of methodism / ed by Cummings, Anson Watson – New York: Phillips & Hunt; Cincinnati: Cranston & Stowe, 1886 – 1mf – 9 – 0-8370-7620-X – mf#1986-1620 – us ATLA [240]

Early schwenckfelder ministers in pennsylvania – Norristown PA: Board of Pub of the Schwenckfelder Church, 1941 [mf ed 2003] – (= ser Schwenckfeldiana 1/2) – 1r – 1 – (in english. incl trans fr german sources. incl bibl) – mf#2003-s008b – us ATLA [242]

Early science fiction novels / ed by Clareson, Thomas D – Greenwood Press – 99 titles on 382mf (24:1) – 9 – us UPA [830]

The early scottish church : the ecclesiastical history of scotland from the first to the twelfth century / Maclauchlan, Thomas – Edinburgh: T & T Clark, 1865 – 1mf – 9 – 0-7905-6655-9 – mf#1988-2655 – us ATLA [240]

Early scottish metrical tales – London: Hamilton, Adams 1889 [mf ed Bloomington IN: Indiana Uni Lib, Preservation Dept 1984] – 1r – 1 – us Indiana Preservation [810]

Early settlers of the bahama islands / Bethell, Arnold Talbot – Holt, England. 1930 – 1r – us UF Libraries [972]

Early settlers of the bahamas and colonists of nor... / Bethell, Arnold Talbot – Holt, England. 1937 – 1r – us UF Libraries [972]

Early Sites Research Society *see* Bulletin of the early...

Early sources of english unitarian christianity = Des origines du christianisme unitaire chez les anglais / Bonet-Maury, Gaston – London: British & Foreign Unitarian Association, 1884 – 1mf – 9 – 0-7905-4379-6 – (incl bibl ref. in english) – mf#1988-0379 – us ATLA [243]

The early spread of religous ideas : especially in the far east / Edkins, Joseph – New York: Fleming H Revell [1893?] [mf ed 1989] – (= ser By-paths of bible knowledge 19) – 1mf – 9 – 0-7905-1596-2 – (incl ind) – mf#1987-1596 – us ATLA [200]

Early study of nigerian languages / Hair, Paul Edward Hedley – London, England. 1967 – 1r – us UF Libraries [470]

Early texas newspapers / Institute of Texan Cultures. Library – Issues of The Weekly Telegraph (Houston), The Indianola Bulletin and The Texian Advocate (Victoria). Covers 1846-60. 3 rolls – 1 – $70.00 – us TX Culture [071]

The early trading companies of new france : a contribution to the history of commerce and discovery in north america / Biggar, Henry Percival – [Toronto]: University of Toronto Library, 1901 – (= ser University of toronto studies in history) – 4mf – 9 – 0-665-73647-9 – mf#73647 – cn Canadiana [380]

The early traditions of genesis / Gordon, Alexander Reid – Edinburgh: T & T Clark, 1907 – 1mf – 9 – 0-8370-3343-8 – (incl ind and bibliography) – mf#1985-1343 – us ATLA [221]

Early travel accounts by women, and women's experiences in india, africa *see* Colonial discourses, series 1

Early travels in india, 1583-1619 / ed by Foster, William – London; New York: Oxford University Press, 1921 – (= ser Samp: indian books) – us CRL [915]

Early travels in palestine : comprising the narratives of arculf, willibald, bernard... / ed by Wright, Thomas – London: Henry G Bohn, 1848 – 2mf – 9 – 0-7905-0538-X – (incl ind) – mf#1987-0538 – us ATLA [930]

Early trumbull county store ledgers – Warren, Trumbull, OH. 1809; 1840-48 – 1r – 1 – (these ledgers record accounts of early residents of warren, trumbull county, and painesville, then part of geauga county, at unidentified general stores) – us Western Res [978]

The early tudors : henry 7, henry 8 / Moberly, Charles Edward – New York: Scribner, 1887 – (= ser Epochs Of Modern History) – 1mf – 9 – 0-7905-5377-5 – (incl bibl ref) – mf#1988-1377 – us ATLA [941]

An early victorian railway station / Hutton, G H – 1953 – 6mf – 7 – mf#86520 – uk Microform Academic [941]

Early voyages and travels to russia and persia, by [him] and other englishmen... / Jenkinson, A – London: The Hakluyt Society, 1886. 2v – 4mf – 9 – (missing: v1) – mf#AR-2046 – ne IDC [915]

Early voyages to america : a paper read before the rhode island historical society / Baxter, James Phinney – Providence RI: Printed for the Society, 1889 – 1mf – 9 – mf#05937 – cn Canadiana [917]

Early welsh script / Lindsay, Wallace Martin – Oxford: J. Parker, 1912.64p. illus – 1 – us Wisconsin U Libr [000]

The early witnesses : or, piety and preaching of the middle ages / ed by Thompson, Joseph Parrish – New York: A D F Randolph, 1857 [mf ed 1990] – 1mf – 9 – 0-7905-6370-3 – mf#1988-2370 – us ATLA [240]

Early women authors : from early british fiction – 24r – 1 – (coll drawn from early british fiction: pre-1750 collection) – mf#C36-28211 – us Primary [420]

Early women's journals *see* Women advising women

The early work of aubrey beardsley / Beardsley, Aubrey Vincent – London 1899 – (= ser 19th c art & architecture) – 7mf – 9 – mf#4.2.1722 – uk Chadwyck [740]

Early years – Westport. 1971-1987 (1) 1972-1987 (5) 1975-1987 (9) – (cont by: teaching pre k-8) – ISSN: 0094-6532 – mf#6709 – us UMI ProQuest [370]

The early years of an african trader : being an account of john holt who sailed for west africa on 23rd jun 1862 150=london: privately publ for john holt & co(liverpool) by n neame, 1962 – us CRL [916]

The early years of christianity / Pressense, Edmond de – New York: Nelson & Phillips, [pref. 1872]-1879. Chicago: Dep of Photodup, U of Chicago Lib, 1978 (1r); Evanston: American Theol Lib Assoc, 1984 (1r) – 1 – 0-8370-0751-8 – (incl bibl ref and ind) – mf#1984-T068 – us ATLA [240]

The early years of john calvin : a fragment, 1509-1536 / M'Crie, Thomas; ed by Ferguson, William – Edinburgh: D Douglas, 1880 – 1mf – 9 – 0-7905-6414-9 – (incl bibl ref) – mf#1988-2414 – us ATLA [242]

The early years of the late bishop hobart / McVickar, John – New York: Protestant Episcopal Press, 1834 – 1mf – 9 – 0-7905-5121-7 – mf#1988-1121 – us ATLA [240]

Early zoroastrianism : lectures...oxford and in london, feb to may 1912 / Moulton, James Hope – London: Williams & Norgate 1913 [mf ed 1992] – (= ser Hibbert lectures (london, england) 1912) – 2mf – 9 – 0-524-02219-4 – mf#1990-2893 – us ATLA [290]

Earnest and affectionate address to the jews – London, England. 1818 – 1r – us UF Libraries [939]

Earnest and affectionate address to the people called methodists – London, England. 1807 – 1r – us UF Libraries [242]

Earnest christianity – Toronto: Published for the proprietors at the Wesleyan Book Room, [1873-1876] – 9 – mf#P04328 – cn Canadiana [240]

Earnest christianity illustrated : or, selections from the journal of the rev james caughey: containing several of mr caughey's sermons...with a brief sketch of mr caughey's life – London, CW [Ont]: C H Brown, 1855 – 5mf – 9 – 0-665-89696-4 – mf#89696 – cn Canadiana [242]

Earnest, Edward K *see* Seasonal changes in selected physiological variables of female basketball players

Earnest exhortation to a frequent reception of the holy sacrament o... / Park, James Allan – London, England. 18– – 1r – us UF Libraries [240]

An earnest inquiry into the true scriptural organization of the churches of god in christ jesus / Smith, Butler Kennedy – Indianapolis, IN: Indianapolis Printing & Pub House, 1871 [mf ed 1992] – 1mf – 9 – 0-524-02933-4 – mf#1990-0749 – us ATLA [240]

Earnest, Joseph Brummell *see* The religious development of the negro in virginia

An earnest plea of laymen of the new school presbyterian and congregational churches of new york and brooklyn... : and other evangelical efforts for the salvation of our country and the conversion of the world – New York: EO Jenkins, 1856 [mf ed 1992] – (= ser Presbyterian coll) – 1mf – 9 – 0-524-04729-4 – mf#1991-2134 – us ATLA [242]

Earp, George Butler *see* What we did in australia

Earth – New York. 1970-1971 (1) 1970-1971 (5) (9) – ISSN: 0012-8201 – mf#5971 – us UMI ProQuest [073]

Earth – Waukesha. 1997-1998 (1,5,9) – ISSN: 1056-148X – mf#19875 – us UMI ProQuest [550]

Earth – Wheaton, Illinois. v1-3 n6. apr 1930-july 1932 – 1 – us NY Public [410]

Earth and high heaven / Brown, Gwethalyn Graham Erichsen – Philadelphia, PA. 1944 – 1r – us UF Libraries [960]

Earth and planetary science letters – Amsterdam. 1966+ (1) 1966+ (5) 1987+ (9) – ISSN: 0012-821X – mf#42127 – us UMI ProQuest [520]

The earth and the stars / Abbot, Charles Greeley – New York: D Van Nostrand & Co 1925 [mf ed 1998] – (= ser Library of modern sciences) – 1r [pl/ill] – 1 – (incl ind) – mf#film mas 28210 – us Harvard [520]

The earth and the word : or, geology for bible students / Pattison, Samuel Rowles – Philadelphia: Lindsay & Blakiston; New York: Stanford & Delisser, 1858 – 1mf – 9 – 0-7905-1559-8 – mf#1987-1559 – us ATLA [220]

Earth first! : the radical environmental journal – v3 n4-5 [1983 may 1-jun 21], v4 n3-v8 n2 [1984 feb 2-1987 dec 22]; 1988 feb 2-1990 mar 20, 1988 feb 2-1990 mar 20, copy 2 – 3r – 1 – (cont: earth first! newsletter; cont by: earth first! journal) – mf#1549319 – us WHS [333]

Earth first! journal – v10 n5-8 [1990 may 1-sep 22], v11 n1 [1990 nov 1] – 2r – 1 – (cont: earth first!; cont by: earth first! [missoula mt: 1991]) – mf#1818322 – us WHS [333]

Earth first! newsletter – v2 n2 [1981 dec 21], v2 n4, 6 [1982 mar 20, jun 21] – 1r – 1 – (cont by: earth first) – mf#1819050 – us WHS [333]

Earth island journal – San Francisco. 1993+ (1,5,9) – ISSN: 1041-0406 – mf#19200 – us UMI ProQuest [639]

Earth journal – Boulder. 1994-1994 (1,5,9) – (cont: buzzworm) – ISSN: 1073-5852 – mf#18477,01 – us UMI ProQuest [639]

Earth, moon, and planets – Dordrecht. 1989-1996 (1) 1990-1996 (5) 1990-1996 (9) – ISSN: 0167-9295 – mf#14746,02 – us UMI ProQuest [520]

Earth science – Falls Church. 1946-1990 (1) 1971-1990 (5) 1975-1990 (9) – ISSN: 0012-8228 – mf#1456 – us UMI ProQuest [550]

Earth surface processes – Chichester. 1976-1980 (1,5,9) – (cont by: earth surface processes and landforms: the journal of the british geomorphological research group) – ISSN: 0360-1269 – mf#10812 – us UMI ProQuest [550]

Earth surface processes and landforms : the journal of the british geomorphological research group – Chichester. 1981+ (1,5,9) – (cont: earth surface processes) – ISSN: 0197-9337 – mf#10812,01 – us UMI ProQuest [550]

Earthly suffering and heavenly glory : with other sermons / Boardman, Henry Augustus – Philadelphia: J B Lippincott, 1878 [mf ed 1990] – 1mf – 9 – 0-7905-3587-4 – mf#1989-0080 – us ATLA [242]

Earth-oriented applications of space technology – Oxford. 1981-1986 (1,5,9) – (cont by: space technology) – ISSN: 0277-4488 – mf#49281 – us UMI ProQuest [629]

Earthquake and fire scrapbook of san jose public library – San Francisco, CA. 1906 – 2r – 1 – $100.00 – mf#B40306 – us Library Micro [978]

Earthquake engineering and structural dynamics – Chichester. 1972+ (1,5,9) – ISSN: 0098-8847 – mf#10803 – us UMI ProQuest [550]

Earthquake information bulletin – Reston. 1972-1985 (1) 1972-1985 (5) 1976-1985 (9) – (cont by: earthquakes and volcanoes) – ISSN: 0046-0931 – mf#7350 – us UMI ProQuest [530]

Earthquake notes / Seismological Society of America – Cambridge. 1972-1984 (1) 1976-1984 (5) 1976-1984 (9) – ISSN: 0012-8287 – mf#8176 – us UMI ProQuest [550]

Earthquakes and the interior of the earth / Klotz, Otto – [Toronto?: s.n, 1908?] [mf ed 1995] – 1mf – 9 – 0-665-74750-0 – mf#74750 – cn Canadiana [550]

Earthquakes and volcanoes – Reston. 1986-1994 (1) 1986-1994 (5) 1986-1994 (9) – (cont: earthquake information bulletin) – ISSN: 0894-7163 – mf#7350,01 – us UMI ProQuest [530]

The earth's antiquity in harmony with the mosaic record of creation / Gray, James – London, 1849 – (= ser 19th c evolution & creation) – 3mf – 9 – mf#1.1.1668 – uk Chadwyck [577]

Earth's equality – v3 n3-4 – 1r – 1 – mf#5072166 – us WHS [333]

Earth's grandest river, the st lawrence, and the thousand islands : an unrivaled summer resort – Watertown, NY: Hungerford & Coates, 1895 – 2mf – 9 – mf#04165 – cn Canadiana [917]

Earths in the universe and their inhabitants / Swedenborg, Emanuel – London, England. 1855 – 1r – us UF Libraries [240]

Earth-science reviews – Amsterdam. 1966+ (1) 1966+ (5) 1987+ (9) – ISSN: 0012-8252 – mf#42252 – us UMI ProQuest [550]

Earthtimes – n1, 4 [1970 apr, jul] – 1r – 1 – mf#1583212 – us WHS [333]

Earth/w / Williams Co. Edgerton – jan 1971-dec 1975 [wkly] – 3r – 1 – mf#B29340-29342 – us Ohio Hist [071]

Eascom history, 1 october 1944-1 april 1945 / U.S. Army Air Forces. Eastern Command, Europe – 1945 – 1 – us L of C Photodup [947]

Easley first baptist church. easley, south carolina : church records – 1873-1972 – 1 – 50.40 – us Southern Baptist [242]

Eason, Joshua Lawrence *see* Diagnostic study of technical incorrectness in the writings of...

East – Singapore, 25 Nov 17-31 Dec 1953 – 8ft – 1 – uk British Libr Newspaper [072]

East – Tokyo. 1964+ (1) 1975+ (5) 1975+ (9) – ISSN: 0012-8295 – mf#9951 – us UMI ProQuest [950]

The east *see* Vostok

East aberdeenshire observer, peterhead, fraserburgh & general advertiser – Peterhead, Scotland 1 oct 1875-9 mar 1893 – 1 – (cont: buchan observer, peterhead, fraserburgh & general advertiser [16 jan 1863-24 sep 1875]; cont by: buchan observer & east aberdeenshire advertiser [14 mar 1893-20 sep 1988]) – uk British Libr Newspaper [072]

East africa *see* Political party, trade union and pressure group materials

East africa and its invaders : from earliest times to the death of seyyid said in 1856 / Coupland, R – Oxford, 1938 – 11mf – 8 – mf#A-295 – ne IDC [956]

East Africa High Commission *see* Official gazette

East africa journal – Nairobi. 1964-1972 (1) 1971-1972 (5) – ISSN: 0012-8309 – mf#2104 – us UMI ProQuest [301]

East african – Dar es Salaam, Tanzania. n87-269. 1996 jul-1999 – 12r – us UF Libraries [079]

East african chiefs / Richards, Audrey Isabel – New York, NY. 1960, c1959 – 1r – us UF Libraries [960]

East african protectorate labour commission, report on the... 1912-13 – 1r – 1 – (with int by ehrlich and a clayton) – mf#96639 – uk Microform Academic [960]
East African Protectorate. Native Labour Commission, 1912-13 *see* Evidence and report
East African Universities Social Science Conference *see* Annual conference proceedings
East alabama today – Columbus, GA. 1968-1987 (1) – mf#68134 – us UMI ProQuest [071]
East algoma : facts about a wonderfully rich country that is open to the home-seekers of the world – Sault Ste Marie, Ont: Sault Express, [189-?] [mf ed 1984] – 1mf – 9 – mf#32068 – cn Canadiana [917]
East and north riding chronicle and driffield express – [Yorkshire & Humberside] East Riding jul 1872-11 aug 1917 [mf ed 2003] – 21r – 1 – (missing: 1874, 1894, 1909-11; cont: driffield express [jul 1872-20 dec 1873]; cont as: east riding chronicle and driffield express [jan 1887-aug 1917]) – uk Newsplan [072]
East and west : essays and sketches / Fitch, Adelaide Paddock – Toronto: W Briggs, 1911 – 3mf – 9 – 0-665-97317-9 – mf#97317 – cn Canadiana [840]
East and west / Guenon, Rene – London: Luzac & Co, 1941 – (= ser Samp: indian books) – (trans by william massey) – us CRL [900]
East and west : the story of a missionary band / Tuck, Mary N – London: London Missionary Society [1900?] [mf ed 1995] – (= ser Yale coll) – 219p (ill) – 1 – 0-524-10034-9 – mf#1995-1034 – us ATLA [240]
East and west ham gazette – [London & SE] Newham LSL, Stratford 1888-92 [wkly] – 1 – (cont as: west ham herald & south essex gazette [1893-1900]) – uk Newsplan [072]
East and west in religion / Radhakrishnan, Sarvepalli – London: George Allen & Unwin, 1933 – (= ser Samp: indian books) – us CRL [230]
East angels, a novel / Woolson, Constance Fenimore – New York, London: Harper & Brothers, 1886. 591p – 1 – us Wisconsin U Libr [830]
East anglian : or, notes and queries on subjects connected with the counties of suffolk, cambridge, essex and norfolk – London. 1858-1910 – 1 – mf#4749 – us UMI ProQuest [941]
East anglian daily times : east edition – Ipswich, England. 1984- – 208+ r – 1 – uk British Libr Newspaper [072]
East anglian daily times : essex edition – Ipswich, England. 1986- – 157+ r – 1 – uk British Libr Newspaper [072]
East anglian daily times : a morning paper – Ipswich, Suffolk, 13 oct 1874-to date – 1 – (lacking: 10 feb 1888-sep 1889; jan, aug 1897; sep-dec 1898, 1911) – uk Newsplan [072]
East antrim times *see* Larne times
East ardsley constables' accounts, 1653-1692 – 1r – 1 – mf#293 – uk Microform Academic [941]
East Ascension Genealogical and Historical Society [Ascension Parish LA] *see* Jambalaya
East asia journal of theology – Singapore. 1983-1986 (1,5,9) – (cont by: asia journal of theology) – ISSN: 0217-3859 – mf#13363 – us UMI ProQuest [240]
East asia millions – Reading: Bradley at the Crown Press, 1965- [mf ed 2003] – 1r – 1 – (latest iss consulted v109 n6 [dec 1982/jan 83]; mf: v92-109 1965-1982/83. iss by china inland mission & overseas missionary fellowship, 1965-67; by overseas missionary fellowship, 1968-) – mf#2003-s098 – us ATLA [240]
East asia millions – Philadelphia. 1989-1993 (1) 1993-1993 (5) 1993-1993 (9) – ISSN: 0012-8406 – mf#15243,02 – us UMI ProQuest [240]
East asian executive reports – Washington. 1988+ (1,5,9) – ISSN: 0272-1589 – mf#16426 – us UMI ProQuest [337]
East avenue baptist church. springfield, missouri : church records – 1890-1965 – 1 – us Southern Baptist [242]
East bay bridge – v1 n1-9 [1974 jun 14-nov 7] – 1r – 1 – mf#1055817 – us WHS [071]
East bay jewish observer – Oakland, CA. 1978-82 – 1 – us AJPC [071]
East bay labor journal / Building Trades Council of Alameda County et al – 1933 aug 4/1938-1975 jul/1981 mar – 1r – 1 – mf#690884 – us WHS [331]
East bay window – Phenix, RI. 1971-1992 (1) – mf#66259 – us UMI ProQuest [071]
East bay women for peace newsletter – 1979 jan-1988 apr, sep – 1r – 1 – (cont: el cerrito-richmond women for peace newsletter; cont by: women for peace-east bay newsletter) – mf#1329859 – us WHS [071]
East belfast herald & post – Belfast, Ireland 17 jan 1991-8 oct 1998 – 1 – uk British Libr Newspaper [072]
East Bengal (Pakistan). Legislative Assembly *see* Assembly proceedings
The east bengal times – Dacca: Raj Kumar Bhattacherji [mar 11 1933-jul 22 1939] (wkly) – 3r – 1 – us CRL [079]
East bridgewater 1754-1900 – Oxford, MA (mf ed 1995) – (= ser Massachusetts vital records) – 71mf – 9 – 0-87623-378-7 – (mf town records 1769-1825. mf 3: marriages 1759-1823; vital records 1754-1857. mf 5-10: town records 1823-54. mf 6: marriage banns 1823-50. mf 7: marriages: 1818-43. mf 8: births 1769-1849. mf 9: deaths 1822-44. mf 11-18: town meetings 1823-46. mf 19-27: highways 1822-1920. mf 28-35: mortgages 1837-58. mf 36-37: intents index 1849-84. mf 38-41: intentions 1849-82. mf 42-45: intentions 1883-1908. mf 46-48: birth index 1769-1943. mf 49-51: marriages index 1844-1943. mf 52-54: death index 1844-1943. mf 55: births 1843-51. mf 56: marriages, deaths 1844-51. mf 57-60: births 1852-90. mf 61-63: marriages 1852-90. mf 64-67: deaths 1852-90. mf 68-69: deaths 1891-1903. mf 70: marriages 1891-1903. mf 71: births 1891-99) – us Archive [978]
East bridgewater 1769-1849 – Oxford, MA (mf ed 1996) – (= ser Massachusetts vital record transcripts to 1850) – 5mf – 9 – 0-87623-246-2 – (mf 1t-2t: marriage banns 1823-49. mf 2t: marriages 1811-43; births 1769-1849. mf 3t: deaths 1822-44; marriage intentions 1849. mf 3t-4t: births 1844-49. mf 5t: marriages & deaths 1844-49) – us Archive [978]
East central europe – Pittsburgh. 1974-1979 (1) 1975-1979 (5) 1975-1979 (9) – ISSN: 0094-3037 – mf#6969 – us UMI ProQuest [943]
East Central Wisconsin Regional Planning Commission *see* Transportation improvement program for the appleton and oshkosh urbanized areas
East China Christian Education Association *see* Ccea newsletter
East city news – nov 1980-dec 1989 – 1 – mf#11.37 – nz Nat Libr [079]
East city news advertiser – Auckland, NZ. jan-dec 1985; jan-dec 1989 – 1 – mf#11.37 – nz Nat Libr [079]
East Cleveland. Ohio. First Presbyterian Church *see* Church records, ms 1528
East clevelander / Cuyahoga Co. East Cleveland – aug 1933-aug 1934 [wkly] – 1r – 1 – mf#B30931 – us Ohio Hist [071]
East coast advocate – Titusville, FL. 1890 aug 22-1921 mar 26 – 11r – us UF Libraries [071]
East coast bays news – Auckland, NZ. jul 1983-dec 1986 – 3r – 1 – mf#11.50 – nz Nat Libr [079]
East coast express and inland counties observer – Wicklow, 26 sep 1936-27 aug 1938; 7 jan 1939-14 dec 1940; 17 oct 1942-18 dec 1943 – 6r – 1 – (cont as: east coast express and dublin county and township times [17 apr 1937-?27 aug 1938; cont as: bray tribune and east coast express [?jan 1939-14 dec 1940]; cont as: bray tribune and people's weekly [?1941-29 jun 1946]; cont as: people's weekly [13 jul 1946-20 mar 1948]; cont as: people's [3 apr 1948-30 sep 1950]; cont as: irish people illustrated [14 oct 1950-20 dec 1952]) – ie National [072]
East coast express [bray tribune] – Bray sep 1936-38 – 2r – 1 – ie National [072]
East coast mail and wairoa guardian – New Zealand, 3 Jan-28 Dec 1908 (very imperfect) – 2r – 1 – uk British Libr Newspaper [079]
East coast messenger – Daytona, FL. v3 n36. 1887 nov 10 – 1r – us UF Libraries [071]
East Contra Costa Historical Society *see* Meganos
East cornwall times – Launceston, England. 14 May 1859-1861; 1864-Jun 1866; 8, 29 Dec 1866. -w. 2 reels – 1 – uk British Libr Newspaper [072]
East cumberland news – [NE England] Cumbria 2 jun 1883-dec 1909 [mf ed 2003] – 34r – 1 – uk Newsplan [072]
East, Edward H *see*
– Durnford and east's reports
– East's reports
East, Edward Murray *see* Inbreeding and outbreeding; their genetic and sociological significance
East end journal / Cuyahoga Co. East Cleveland – v1 n1. feb 1921-apr 1922 [wkly] – 1r – 1 – mf#B30930 – us Ohio Hist [071]
East end news – London, UK. jan-may 1986 – 1/4r – 1 – uk British Libr Newspaper [072]
East end news & advertiser – London, England. 17 july 1869-26 apr 1963 [wkly] – 86r – 1 – (aka: east end news and london shipping chronicle) – uk British Libr Newspaper [072]
East end news and london shipping chronicle *see* East end news & advertiser
East end worker – London, UK. 23 oct, 8, 22 nov 1926 – 1r – 1 – uk British Libr Newspaper [072]
East europe – New York. 1952-1975 (1) 1971-1975 (5) – ISSN: 0012-8430 – mf#1055 – us UMI ProQuest [321]
East european jewish affairs – London. 1992+(1,5,9) – (cont: soviet jewish affairs) – ISSN: 1350-1674 – mf#11225,01 – us UMI ProQuest [939]
East european quarterly – Boulder. 1967+ (1) 1974+ (5) 1974+ (9) – ISSN: 0012-8449 – mf#10071 – us UMI ProQuest [943]
East feliciana patriot – Clinton LA. 1868 may 9 – 1r – 1 – mf#1044061 – us WHS [071]
East fife mail : incorporating the fife mail and leven mail – Leven: Strachan & Livingston Ltd 1966- (wkly) [mf ed 3 jan 1994-] – 1 – (subtitle varies; not publ: 12 mar 1980, 15 jul-12 aug 1981;cont: fife mail (leven, scotland); absorbed: east fife observer) – ISSN: 1354-6066 – uk Scotland NatLib [072]
East florida, 1764-69 : from the public record office, london – (= ser Naval office shipping lists; British records relating to america in microform) – 1r – 1 – mf#96528 – uk Microform Academic [975]
East florida banner – Ocala, FL. 1871 feb 4; 1876 jan-mar 4 – 1r – us UF Libraries [071]
East florida courier – Starke, FL. 1888 mar. 31, apr. 11, apr. 28 – 1r – us UF Libraries [071]
East florida herald – St Augustine FL. 1823 jan 4 – 1r – 1 – (cont by: florida herald) – mf#787012 – us WHS [071]
East florida records – 175r – 1 – $6,125.00 – Dist. us Scholarly Res – us L of C Photodup [978]
East florida seminary : ocala / Crow, Charles L – s.l, s.l? 193-? – 1r – us UF Libraries [978]
East galway democrat – Ballinasloe, Ireland. -w. 11 oct 1913-1921; 26 apr 1936-27 aug 1949 – 8 1/2r – 1 – uk British Libr Newspaper [072]
East galway democrat – Galway 1911-13, 1921-36 – 19r – 1 – ie National [072]
East grinstead courier – 1983-96 – 38r – 1 – uk British Libr Newspaper [072]
East grinstead observer – [London & SE] West Sussex, East Grinstead Lib 5 jan 1977-; BLNL 1897, 1977-82, 1986- – 1 – uk Newsplan [072]
East ham recorder – London, UK. 28 jun-27 dec 1912; 13 jun-26 dec 1913; 13 feb 1914-17 dec 1915; 12 may-29 dec 1916; 26 jan, 9 feb, 14, 21, 28 dec 1917; 4 jan-29 nov 1918; 1922; 2 jan-17 apr 1925 – 5r – 1 – uk British Libr Newspaper [072]
East Harlem Interfaith *see* Networking news
East Harlem Youth Employment Services *see* Right on!
East hatley 1580-1950 – (= ser Cambridgeshire parish register transcript) – 3mf – 9 – £3.75 – uk CambsFHS [929]
The east in prayer *see* The lord's supper / the east in prayer
East India Association *see* Minutes and other records of the east india association, 1812-1814 and 1829-1847
East india (census) : general report of the census of india, 1901 / India. Census Commissioner – London: Printed for H M Stationery Office, by Darling & Son, 1904 [mf ed 1995] – (= ser Yale coll; [Great britain. parliament papers by command 2047) – xxv/582p (ill) – 1 – 0-524-10115-9 – mf#1995-1115 – us ATLA [315]
East India Company *see* Selection of papers from the records at the east-india house relating to the revenue, police, and civil and criminal justice under the company's governments in india
East india company factory records : sources from the british library, london – 3pt – 1 – (pt1: china & japan [34r] £3200; pt2: china [34r] £3200; pt3: fort st george (madras) [25r] £2350; with d/g) – uk Matthew [380]
The east india company in eighteenth-century politics / Sutherland, Lucy Stuart – London; New York: Oxford University Press, 1952 – (= ser Samp: indian books) – us CRL [380]
East india parliamentary papers – London, Printed for H M Stationery Office by Eyre and Spottiswoode, 1909 – (annual lists and general index to the parlimentary papers relating to the east indies published during the years 1801-1907 inclusive) – us CRL [324]
The east india sketch-book : comprising an account of the present state of society in calcutta, bombay, etc / Smith, Elizabeth – London 1832 [mf ed Hildesheim 1995-98] – 2v on 4mf – 9 – €120.00 – 3-487-27473-6 – gw Olms [954]
The east india sketch-book – London 1833 [mf ed Hildesheim 1995-98] – 2v on 4mf – 9 – €120.00 – 3-487-27479-5 – gw Olms [380]
The east india vade-mecum : or, complete guide to gentlemen intended for the civil, military, or naval service of the hon. east india company / Williamson, Thomas – London 1810 – (= ser 19th c british colonization) – 2v on 12mf – 9 – mf#1.1.1640 – uk Chadwyck [390]
The east india vade-mecum : or, complete guide to gentlemen intended for the civil, military, or naval service of the hon east india company / Williamson, Thomas – London 1810 [mf ed Hildesheim 1995-98] – 2v on 7mf – 9 – €140.00 – 3-487-27533-3 – gw Olms [380]
East indian railway company : address of the chairman lieut-general richard strachey, at the 46th annual general meeting of proprietors, held on the 4th jul 1893...london / Strachey, Richard – London [1893] – (= ser 19th c british colonization) – 1mf – 9 – mf#1.1.4311 – uk Chadwyck [380]
East indians in the west indies / Niehoff, Arthur – Milwaukee, WI. 1960 – 1r – us UF Libraries [972]
East, John *see* Jubilee of the bible, october 4, 1835
East kent times and district advertiser – [London & SE] Kent Arts & Lib, Ramsgate Lib 1896-1932 – 1 – (cont as: east kent times & mail [17 sep 1932-31 dec 1980]) – uk Newsplan [072]
East kent times and mail – [London & SE] Kent Arts & Lib, Ramsgate Lib 17 sep 1932-31 dec 1980 – 1 – (cont: east kent times & district advertiser [11 mar 1896-12 sep 1932]) – uk Newsplan [072]
East kilbride news – Cambuslang: J Lithgow & Sons 1952- (wkly) [mf ed 7 jan 1994-] – 1 – (absorbed: cambuslang advertiser; iss between jan 3 1953-feb 21 1953 called "picture edition"; imprint varies; numbering established sep 20 1952? & initially carries enumeration of cambuslang advertiser, but fr mar 7 1953 new numbering sequence is adopted: 1st yr n11 (mar 7 1953)-; enumeration ceased with n1479 (sep 19 1980)) – ISSN: 1353-4335 – uk Scotland NatLib [072]
East lincoln news – Lincoln (Uni Place), NE: E A McNeil. 1v. v22 n26. oct 14 1926-v22 n27. oct 21 1926 (wkly) [mf ed 1976] – 1r – 1 – (cont: university place news (1913). cont by: east lincoln news and the university place news) – us NE Hist [071]
East lincoln news and the university place news – Lincoln (Uni. Place), NE: E A McNeil. 3v. v22 n28. oct 28 1926-v24 n28. oct 25 1928 (wkly) [mf ed with gaps filmed 1976] – 1r – 1 – (cont: east lincoln news. cont by: university place news) – us NE Hist [071]
East london blackshirt – London, UK. apr, aug 1953; oct 1953-may 1956 – 1/4r – 1 – uk British Libr Newspaper [072]
East london daily dispatch – East London, South Africa: East London Daily Dispatch Ltd, [-1924]. oct 1922-1924 – us CRL [079]
East london daily dispatch – East London, South Africa: East London Daily Dispatch Ltd, oct 1922-24 – 1 – us CRL [960]
East london daily dispatch [and frontier advertiser] *see* Daily dispatch
The east london daily dispatch and frontier advertiser – East London, South Africa: East London Daily Dispatch Ltd [jul 6 1920-jun 30 1922] (daily ex sun) – 3r – 1 – us CRL [072]
The east london evangelist *see* The christian mission magazine, 1870-78...
East london observer – [London & SE] Tower Hamlets LHLA 19 sep 1857-29 dec 1944 [wkly] – 1 – uk Newsplan [072]
East london press – London, UK. 11 aug 1883-1886 – 3r – 1 – uk British Libr Newspaper [072]
East london recorder – London, UK. 17 jan 1874 – 1/4r – 1 – uk British Libr Newspaper [072]
East london reporter – London, UK. 1888-89; jun 1891 – 1/4r – 1 – uk British Libr Newspaper [072]
East london standard – East London SA, 1891-1899 [mf ed Cape Town: SA Library 1982] – 9r – 1 – (frequency varies. first publ as: frontier standard and east london gazette, jan 1890- sep 1891. cont: frontier standard and east london gazette) – sa National [079]
East lothian courier – Haddington: D & J Croal Ltd 1971- (wkly) [mf ed 7 jan 1994-] – 1 – (cont: haddingtonshire courier (haddington, scotland: 1942)) – uk Scotland NatLib [072]
East lothian news – Dalkeith: Scottish County Press 1971- (wkly) [mf ed 1 jul 1994-] – 1 – uk Scotland NatLib [072]
East lothian oracle – [Scotland] East Lothian, Edinburgh: Lamp of Lothian Press 23 mar 1934 (wkly) [mf ed 2004] – 1r – 1 – (a weekly journal of truth, progress and impartiality) – uk Newsplan [072]
East Madison Community Center *see* Newsletter
East Madison Community Center (Madison, WI) *see* Emcc news and activities
East Madison Community Center [Madison, WI] *see* Footnotes
East manchester reporter – Manchester ALS oct 1976-jul 1978, mar 1979-8 aug 1980, 12 sep 1991-dec 1995 – 1 – uk MLA; uk Newsplan [072]

East meets west : original records of western traders, travellers, missionaries and diplomats to 1852 – 3pt – 1 – (pt1: the log book of william adams (1564-1620) & other mss & rare printed materials fr bodleian library, oxford [22r] £2050; pt2: papers of englebert kaempfer (1651-1716) & related sources fr british library, london [10r] £925; pt3: papers of john scattergood (1681-1723), isaac titsingh (1740?-1812), heinrich julius klaproth (1783-1835) & other early materials fr british library, london [12r] £1125; pt4: east india company: ship's logs, ledgers & receipt books, 1605-1701 fr british library, london; pt5: east india company: ship's logs, 1701-1851, fr national maritime museum, greenwich [15r] £1400; with d/g) – uk Matthew [910]

East midland geographer – Nottingham. 1954-1991 (1) 1978-1981 (5) 1978-1981 (9) – ISSN: 0012-8481 – mf#10470 – us UMI ProQuest [900]

East of fife record – Anstruther, Scotland, UK. 1870-26 Nov 1875; 1876-9 Aug 1917. -w. 29 reels – 1 – uk British Libr Newspaper [072]

East of the barrier : or, side lights on the manchuria mission / Graham, J Miller – New York: Fleming H Revell, 1902 [mf ed 1986] – 1mf – 9 – 0-8370-6500-3 – (incl ind) – mf#1986-0500 – us ATLA [240]

East of the barrier, or side lights on the manchuria mission / Graham, J Miller – Edinburgh, London, 1902 – 3mf – 9 – mf#HTM-68 – ne IDC [915]

The east of to-day and to-morrow / Potter, Henry Codman – New York: Century, 1902 [mf ed 1995] – (= ser Yale coll) – 190p – 1 – 0-524-09643-0 – mf#1995-0643 – us ATLA [915]

East oregon herald – Burns OR: D L Grace, 1887-96 [wkly] [mf ed 1971-72] – 3r – 1 – (merged with: burns times to form: times-herald (1896-1929)) – us Oregon Lib [071]

East oregon herald *see* Times-herald (burns, or)

East oregon ranger – John Day OR: A R Jones, 1930-31 [wkly] – 1 – (1930 incl newspaper publ by mt vernon high school students. cont: long creek ranger; cont by: john day valley ranger) – us Oregon Lib [071]

East oregonian – Pendleton, Umatilla Co, OR: East Oregonian Pub Co. mar 1 1888-feb 28 1889; mar 1 1900-feb 28 1997; jul 19-sep 26 1997; may 10-jun 30 1998; jul 1998-sep 30 1999 – 1 – (began in 1887? aka: e o and: pendleton east oregonian) – us Oregon Hist [071]

East oregonian (pendleton, or: daily evening ed) – Pendleton OR: East Oregonian Pub Co, 1888- [daily ex sun] – 1 – (related to: semiwkly ed: e o; wkly ed: east oregonian (pendleton, or: weekly ed)) – us Oregon Lib [071]

East oregonian (pendleton, or: weekly ed) – Pendleton OR: M P Bull [wkly] – 1 – (began with oct 16 1875. ceased in 1911?. related to semiwkly ed: e o; daily ed: east oregonian (pendleton, or: daily evening ed)) – us Oregon Lib [071]

East otago review – North Otago, NZ. mar 1979-oct 1980 – 1r – 1 – mf#82.4 – nz Nat Libr [079]

East Pakistan (Pakistan). Assembly *see* Assembly proceedings

[East palo alto and san francisco peninsula area-] peninsula metro reporter – CA. aug 1979-dec 1979 – 1r – 1 – $60.00 – mf#B02192 – us Library Micro [071]

East park baptist church – Ventura. 1972-1996 (1) 1972-1996 (5) 1976-1996 (9) – 1r – 1 – $55.89 – mf#6665 – us Southern Baptist [242]

East pensacola heights – Pensacola, FL. 1908? – 1r – us UF Libraries [978]

East rand express – Germiston, Benoni, Boksburg. SA. 1898-1936 – 67r – 1 – sa National [079]

East retford advertiser – Nottinghamshire, 1 jan 1854-31 dec 1859 (wkly) – 3r – 1 – (incorp with: retford worksop isle of axholme and gainsborough news (mthly to n17 may 1855) then cont weekly as: retford advertiser etc to 11 jul 1857 then: retford newark worksop and gainsbro' advertiser etc) – uk Newsplan [072]

East riding telegraph – Beverley, England 1 jan 1898-7 nov 1903 [mf may-dec 1895, 1898, 1900] – 1 – (incorp with: beverley guardian; cont: beverley & east riding telegraph [4 may 1895-26 dec 1896]) – uk British Libr Newspaper [072]

The east saint louis race riot of 1917 : from national archives and state of illinois – 8r – 1 – $1435.00 – 0-89093-742-7 – (with p/g) – us UPA [362]

East side / Cuyahoga Co. Cleveland – ns: jul 1980-aug 1995 – 5r – 1 – mf#B36432-36436 – us Ohio Hist [071]

East side baptist church : church records – Paragould, AK. 1884p. 1912-jul 1986 – 1 – $84.78 – (lacking: oct 1953-55. formerly second baptist church) – us Southern Baptist [242]

East side baptist church. fort smith, arkansas : church records – 1953-70 – 1 – us Southern Baptist [242]

East side herald – Milwaukee WI. 1964 apr 30-1974 apr, 1968 oct 31 – 2r – 1 – mf#1166402 – us WHS [071]

East side journal – Los Angeles, CA. 1935-1967 (1) – mf#62180 – us UMI ProQuest [071]

East side monthly – Providence, RI. 1974-1992 (1) – mf#66289 – us UMI ProQuest [071]

East side news – Madison WI. 1912 oct 12-19 – 1r – 1 – mf#916985 – us WHS [071]

East side news – Madison WI. 1926 jul 15/1929-1958/60, 1961-1963 dec 5 – 12r – 1 – mf#939097 – us WHS [071]

East sider / Security State Bank [Madison WI] – 1926 sep-1930 jun – 1r – 1 – mf#921827 – us WHS [332]

East St Louis Central Trades and Labor Union *see* Southern illinois labor tribune

East stroudsburg press – East Stroudsburg, PA. -w 1919-1923 – 13 – $25.00r – us IMR [071]

East suffolk mercury – Suffolk, 29 oct 1992-to date – 1 – (fr 21 sep 1995 titled mercury; formed by amalg of the felixstowe, ipswich & woodbridge mercury papers) – uk Newsplan [072]

East Tennessee Historical Society *see* Tennessee ancestors

East tennessee roots – v1 n1-v4 n4 [1984 mar-1987 dec] – 1r – 1 – mf#1055828 – us WHS [071]

East texas family records / East Texas Genealogical Society – [1977-1984 spr], 1984 sum-1988 win – 2r – 1 – mf#822606 – us WHS [929]

East Texas Genealogical Society *see* East texas family records

East texas labor voice / Smith County Labor Council [TX] – v1 n1-11 [1981 dec-1982 dec] – 1r – 1 – mf#1048056 – us WHS [331]

East Tilton, New Hampshire. East Tilton Free Will Baptist Church *see* Records

East timor question, 1975-2002 / ed by Jolliffe, Jill – [mf ed 1997-2000] – 1056mf – 9 – €7340.00 – (with p/g; suppls available separately: 1997 [22mf] €195, 1998 [44mf] €310, 1999 [77mf] €540, 2000 [58mf] €440, 2001-02 €1195) – mf#m442 – ne MMF Publ [959]

East toledo sun / Lucas Co. Toledo – jul 1975-sep 1982 [wkly] – 6r – 1 – mf#B34051-34056 – us Ohio Hist [071]

East troy gazette – East Troy WI. 1880 jan 21-1882 aug 9 – 1r – 1 – mf#961926 – us WHS [071]

East troy news – East Troy WI. 1931 sep 2-1933 mar 8, 1933 mar 15-1934 oct 10, 1934 oct 17-1936 may 13, 1936 may 20-1937 dec 15, 1937 dec 22-1939 dec 27, 1940-1965, 1966 jan 5-1967 jul 26, 1967 aug 2-1969 mar 12, 1969 mar 19-1970 oct 7, 1970 oct 14-1972 apr 26, 1972 may-1973 may, 1973 jun-dec, 1974-2001 oct/dec – 119r – 1 – mf#4322447 – us WHS [071]

East village other – 1968 jul 15-30, mar 23-1969 apr 9, 1969 apr 16-1971 aug 10 – 2r – 1 – mf#362668 – us WHS [071]

East village other – New York. 1965-1972 – 1 – ISSN: 0012-8562 – mf#3171 – us UMI ProQuest [073]

East washingtonian – Pomeroy, WA. 1884-1998 (1) – mf#69385 – us UMI ProQuest [071]

East west = Tung hsi pao – 1974 feb 27, 1975 jul 9-1976 oct 27, 1976 nov 3-1978 feb 1, 1978 feb 8-1979 aug 22, 1979 sep 5-1980, 1981-1984 jan-dec 19, 1985-87 – 11r – 1 – mf#29595 – us WHS [071]

East west digest – London. 1976-1978 (1) 1976-1978 (5) 1976-1978 (9) – ISSN: 0012-8627 – mf#10471 – us UMI ProQuest [327]

East west journal – Brookline Village. 1980-86 (1,5,9) – (cont by: eastwest) – ISSN: 0191-3700 – mf#12433,02 – us UMI ProQuest [073]

East Winslow, Maine. East Winslow Baptist Church *see* Records

Eastbourne chronicle – England, Oct 1865-Dec 1907; 1910-12; 1950 – 36r – 1 – (lacking: aug-dec 1896) – uk British Libr Newspaper [072]

Eastbourne gazette – Eastbourne, England. Feb 1862-Sep 1912; 1913-27; 1950; 1986- – 80+ r – 1 – (lacking: 1897) – uk British Libr Newspaper [072]

Eastbourne herald – 1950; 1986-96 – 50r – 1 – uk British Libr Newspaper [072]

Eastbourne sun – 1921-2 jun 1923 – 1r – 1 – mf#49.4 – nz Nat Libr [079]

Eastburn, Manton *see* The annual sermon before the american sunday-school union

Eastbury illustrated, by elevations, plans, sections, views : and other delineations / Clarke, Thomas Hutchings – London 1834 – (= ser 19th c art & architecture) – 1mf – 9 – mf#4.1.312 – uk Chadwyck [720]

Easter greeting 1891 : reformed episcopal church, victoria, bc, financial statement to march 15th, 1891 – Victoria, BC?: s.n, 1891? – 1mf – 9 – mf#14684 – cn Canadiana [242]

Easter in heaven / Weninger, Francis Xavier – New York: D & J Sadlier, 1863 – 1mf – 9 – 0-8370-6857-6 – (subsequently issued in german under title: ostern im himmel) – mf#1986-0857 – us ATLA [240]

Easter School In Agricultural Science (3d : 1956... *see* Growth of leaves

Easter seal communicator – Chicago. 1973-1980 (1) 1980-1980 (5) 1980-1980 (9) – mf#7967 – us UMI ProQuest [613]

Easter seal news of wisconsin / Easter Seal Society for Crippled Children and Adults of Wisconsin – v29 n1-v44 n3 [1968 mar-1984 oct] – 1r – 1 – (cont: wisconsin easter seal news; cont by: easter seal network news) – mf#659305 – us WHS [360]

Easter Seal Society for Crippled Children and Adults of Wisconsin *see* Easter seal news of wisconsin

The easter sermons of st augustine / Weller, P T – Washington, DC. The Catholic University of America Studies in Sacred Theology, 1955 – 3mf – 9 – €7.00 – ne Slangenburg [240]

The easter song; being the first epic of christendom. / Sedulius, fifth century – Introd., verse-trans., and appendices by George Sigerson. Dublin: The Talbot Press, 1922. viii,269p – 1 – us Wisconsin U Libr [240]

Eastern africa history conference on language and culture in eastern africa papers : sponsored by the ministry of culture and social services and goethe institut, nyeri, 17-20 sep 1980 – Nairobi: University of Nairobi, Dept of History, 1980 – us CRL [470]

Eastern africa journal of rural development – Kampala. 1968-1982 (1) 1971-1982 (5) 1974-1982 (9) – ISSN: 0377-7103 – mf#6215 – us UMI ProQuest [338]

Eastern africa to-day / Joelson, Ferdinand Stephen – London, England. 1928 – 1r – us UF Libraries [960]

Eastern argus – Portland ME. 1810 jan-dec – 1r – 1 – (cont: standard [portland me]; cont by: weekly eastern argus) – mf#780678 – us WHS [071]

Eastern argus – Portland, ME: Day & Willis, sept 8 1803-dec 28 1824 – 7r – 1 – us CRL [071]

Eastern argus and borough of hackney times – Tower Hamlets, England 28 may 1881-14 sep 1912 [mf 1904] – 1 – (discontinued; cont: eastern argus & borough of hackney liberal [8 dec 1877-21 may 1881]) – uk British Libr Newspaper [072]

Eastern argus and shettleston and district weekly news – Glasgow, Scotland 12 dec 1914-14 may 1921 – 1 – (cont: eastern argus & shettleston & tollcross weekly news [18 oct 1913-5 dec 1914]) – uk British Libr Newspaper [072]

Eastern argus and shettleston and tollcross weekly news – [Scotland] Shettleston: A H Burnett 18 oct 1913-14 may 1921 (wkly) [mf ed 2003] – 7r – 1 – (cont: glasgow eastern argus; cont by: eastern argus and shettleston and district weekly news [jan 1915-may 1921]) – uk Newsplan [072]

Eastern arrow – 1983 apr-1991 dec – 1r – 1 – mf#1055832 – us WHS [071]

Eastern asia : a history, being the second edition of a brief history of eastern asia / Hannah, Ian Campbell – London; Leipsic: T Fisher Unwin, 1911 [mf ed 1995] – (= ser Yale coll) – 327p – 1 – 0-524-09236-2 – mf#1995-0236 – us ATLA [950]

Eastern Band of Cherokee Indians *see* Cherokee one feather

Eastern baptist church. eastern association. north carolina : church records – 1869-86 – 1 – 5.68 – us Southern Baptist [242]

Eastern bay of plenty picture news and kawerau gazette – mar 1980-dec 1984 – 20r – 1 – mf#16.22 – nz Nat Libr [079]

Eastern bay of plenty picture news and kawerau gazette *see* Kawerau and eastern bay news gazette

Eastern bengal ballads, mymensing : ratanu lahiri research fellowship lectures in two parts / Sen, Dineshchandra [comp] – Calcutta: The University, 1923 – (= ser Samp: indian books) – (foreword by lawrence john lumley dundas) – us CRL [780]

Eastern Board of Cherokee Indians *see* Cherokee one feather

Eastern buddhist *see* The cultural east

Eastern buddhist... : devoted to the study of mahayana buddhism – v1-8 n4. may 1921-aug 1958 [complete] – 2r – 1 – (filmed with: the cultural east) – ISSN: 0012-8708 – mf#ATLA S0068A – us ATLA [280]

The eastern calukyas / Ganguly, Dhirendra Chandra – Benares: DC Ganguly, 1937 – (= ser Samp: indian books) – us CRL [954]

Eastern Caribbean Conservation Conference, 1st *see* Conservation in the eastern caribbean

Eastern catholic life – Passaic, NJ: Eastern Catholic Press Assoc, nov 7 1965-1974 – 1 – us CRL [241]

Eastern cherokee applications of the united states court of claims, 1906-1909 / U.S. Court of Claims – (= ser Records relating to census rolls and other enrollments) – 348r – 1 – (with printed guide) – mf#M1104 – us Nat Archives [340]

Eastern chronicle – New Glasgow, NS. 1866-73 – 3r – 1 – ISSN: 0844-4374 – cn Library Assoc [071]

Eastern churches broadstreet – [complete] – 1r – 1 – mf#ATLA S0713A – us ATLA [243]

Eastern churches quarterly – Ramsgate. 1950-1964 (1) – mf#682 – us UMI ProQuest [240]

Eastern churches review – 1(1966)-5(1973) – 43mf – 9 – €82.00 – ne Slangenburg [243]

Eastern churches review – Oxford. 1976-1978 (1,5,9) – ISSN: 0012-8740 – mf#11440 – us UMI ProQuest [240]

Eastern clackamas news – Estacada OR: R M Standish, 1916-28 [wkly] – 1 – (cont: estacada progress (1908-16); cont by: clackamas county news (1928-57). 1916-18 incl newspaper pub by estacada high school students) – us Oregon Lib [071]

Eastern clay : fourteen stories / Gracias, Louis – Calcutta: L Gracias, 1948 – (= ser Samp: indian books) – us CRL [830]

Eastern Conference of Service Employees Unions *see* Eastern journal

Eastern counties advertiser and ilford gazette – London, UK. – 46r – 1 – (aka: eastern counties times and ilford gazette; eastern counties times and south essex recorder; eastern counties times and barking recorder) – uk British Libr Newspaper [072]

Eastern counties daily press – Norwich, England. 1870-74; 1924-30; 1935-37; 1979-81 – 123r – 1 – uk British Libr Newspaper [072]

Eastern counties herald – [Yorkshire & Humberside] Hull jul 1838-feb 1884 [mf ed 2003] – 37r – 1 – (cont: eastern counties herald, hull and general advertiser [jul 1838-dec 1850]; cont by: hull & eastern counties herald [jan 1862-feb 1884]) – uk Newsplan [072]

Eastern counties times and barking recorder *see* Eastern counties advertiser and ilford gazette

Eastern counties times and ilford gazette *see* Eastern counties advertiser and ilford gazette

Eastern counties times and south essex recorder *see* Eastern counties advertiser and ilford gazette

Eastern counties' times & south essex recorder – Redbridge, England 22 jun 1906-28 feb 1913 [mf 1896,1912] – 1 – (cont: eastern counties' times, & ilford gazette [6 oct 1893-5 apr 1902]; cont by: eastern counties' times & barking recorder [7 mar 1913-25 jul 1935]) – uk British Libr Newspaper [072]

Eastern courier – Auckland, NZ. jan 1977-jun 1989 – 46r – 1 – mf#11.32 – nz Nat Libr [079]

Eastern courier – George Town. Malaysia. -w. 6 Apr 1929-31 May 1930. (2 reels) – 1 – uk British Libr Newspaper [072]

Eastern customs in bible lands / Tristram, Henry Baker – 2nd ed. London:Hodder and Stoughton, 1894 – 1mf – 9 – 0-8370-5575-X – (incl indes) – mf#1985-3575 – us ATLA [220]

Eastern daily press *see* Eastern counties daily press

Eastern democrat – Eastport, ME: John Bent, may 1832-apr 1841 – 1r – 1 – us CRL [071]

Eastern economic journal – Bloomsburg. 1992+ (1,5,9) – ISSN: 0094-5056 – mf#18894 – us UMI ProQuest [330]

Eastern economist – New Delhi. 1970-1984 (1) 1970-1984 (5) 1975-1984 (9) – ISSN: 0012-8767 – mf#6392 – us UMI ProQuest [330]

The eastern era – St Thomas, Ont: Wrigley & Grayson, [1888] – 9 – ISSN: 1190-7258 – mf#P04114 – cn Canadiana [420]

Eastern european economics – Armonk. 1988-1996 (1,5,9) – ISSN: 0012-8775 – mf#16885 – us UMI ProQuest [330]

Eastern european politics and societies – Berkeley. 1987+ (1,5,9) – ISSN: 0888-3254 – mf#15675 – us UMI ProQuest [321]

Eastern european review – London. 10-31 may 1902 [wkly] – 20ft – 1 – uk British Libr Newspaper [947]

Eastern freeman – Ellsworth, ME: John Clark & Co, apr 22 1853-jul 28 1854 – 1 – us CRL [071]

The eastern frontier of british india, 1784-1926 / Bane, Anil Chandra – Calcutta: A Mukherjee & Co, 1946 – (= ser Samp: indian books) – us CRL [954]

Eastern fruit on western dishes; the morals of abou ben adhem / Locke, David Ross – Boston: Lee and Shepard; New York: Lee, Shepard and Dillingham, 1875 – 1 – us Wisconsin U Libr [830]

Eastern herald – Palatka, FL. 1875 sep 9-nov 6 – 1r – us UF Libraries [071]
The eastern hills chronicle – Shillong, India: The Khasi-Jaintia Press, sep 9 1960-nov 8 1961 – 1r – 1 – us CRL [079]
The eastern hills chronicle : a weekly paper – Shillong: The Khasi-Jaintia Press (wkly) [mf ed Chicago IL: Dept of Photoduplication, The University of Chicago Library [for the SEAsian MF Project at CRL] – 1r – 1 – mf#mf-10017 seam – us CRL [079]
Eastern hills journal series / Hamilton Co. Cincinnati – 3/1971-6/1984,6/1986-3-13/1991 [daily] – 15r – 1 – mf#B36099-36113 – us Ohio Hist [071]
Eastern horizon – Hong Kong. 1960-1981 (1) 1974-1981 (5) 1976-1981 (9) – ISSN: 0012-8813 – mf#10254 – us UMI ProQuest [073]
Eastern indian school of medieval sculpture / Banerji, Rakhal Das – Delhi: Manager of Publications, 1933 – (= ser Samp: indian books) – us CRL [730]
Eastern journal / Eastern Conference of Service Employees Unions – v1 n1-v3 n3 [1977 oct-1979 oct] – 1r – 1 – mf#641265 – us WHS [331]
Eastern laborer – Philadelphia PA. v1 n1-v2 n11 [1907 mar 23-may 4, 18, jun 8-nov 2, 16-30] – 1r – 1 – mf#868697 – us WHS [331]
Eastern law reporter / Canada. General – v1-14. 1906-14 (all publ) – 95mf – 9 – $142.00 – mf#LLMC 81-004 – us LLMC [340]
The eastern libyans / Bates, O – London, 1914 – 1mf2 – 8 – mf#A-676 – ne IDC [956]
Eastern lights : a brief account of some phases of life, thought and mysticism in india / Sircar, Mahendranath – Calcutta: Arya Pub House, 1935 – (= ser Samp: indian books) – us CRL [280]
Eastern magazine – Bangor. 1835-1836 – 1 – mf#3977 – us UMI ProQuest [073]
Eastern manners illustrative of the old testament history / Jamieson, Robert – Philadelphia: Presbyterian Board of Publ, [1838?] – 1mf – 9 – 0-8370-9797-5 – mf#1986-3797 – us ATLA [956]
Eastern mercury – Waltham Forest, England. 22 nov 1887-12 aug 1936 – n1-2718 – 1 – (aka: eastern mercury and leyton, leytonstone, woodford and chingford post; leyton, leytonstone, wanstead & eastern mercury, woodford & chingford post) – uk British Libr Newspaper [072]
Eastern mercury and leyton, leytonstone, woodford and chingford post *see* Eastern mercury
Eastern mercury and walthamstow post (walthamstow ed) – London, UK. 1950-10 may 1962 – 12r – 1 – (aka: walthamstow post) – uk British Libr Newspaper [072]
Eastern missions from a soldier's standpoint / Scott-Moncrieff, George Kenneth – London: Religious Tract Society, 1907 – 1mf – 9 – 0-7905-6784-9 – mf#1988-2784 – us ATLA [240]
Eastern monachism : an account of the origin, laws, discipline, sacred writings, mysterious rites, religious ceremonies, and present circumstances of the order of mendicants founded by gotama budha / Hardy, Robert Spence – London: Partridge and Oakey, 1850 – 2mf – 9 – 0-524-02426-X – mf#1990-3010 – us ATLA [280]
Eastern montana clarion – Ryegate, MT. 1935-1974 (1) – mf#64635 – us UMI ProQuest [071]
Eastern morning news and hull advertiser – [Yorkshire & Humberside] Hull 26 jan 1864-8 nov 1929 [mf ed 2003] – 240r – 1 – (missing: 1872; cont: eastern morning news [jan 1864-dec 1868]) – uk Newsplan [072]
Eastern Navajo Council *see* Resolutions of district councils
Eastern Nebraska Genealogical Society *see* Roots and leaves
Eastern news – Singapore. -d. 1 Jul 1940-31 Jul 1941. (12 reels) – 1 – uk British Libr Newspaper [079]
The eastern nigeria guardian – Port Harcourt, Eastern Region, Nigeria. Jan 27, 29-31, Feb 1-2, 6, 8-10, 12, 15-17, 19, 21-24, 26-28, Mar 9, 11-12, 20-21, 26-29 1940 – 1 – us NY Public [960]
Eastern North Carolina Genealogical Society *see* Quarterly review of the eastern north carolina genealogical society
The eastern observer – Homestead, PA: The Eastern Observer. v1 n1-v2 n22. jan 4 1942-nov 21 1943 – 1 – us CRL [073]
Eastern oregon news – Baker OR: Ryder Bros, -1939 [wkly] – 1 – (absorbed by: record courier (haines, or)) – us Oregon Lib [071]
Eastern oregon observer – Ontario OR: Elmo E Smith, -1947 [semiwkly] – 1 – (began in 1936. merged with: ontario argus (-1947) to form: ontario argus-observer (1947-70)) – us Oregon Lib [071]
Eastern oregon observer – Ontario OR: E E Smith [mf erd 1983] – (= ser Newspapers in microfilm: united states, 1948-1972) – v2 n35-56 [jun 28-sep 9 1938] [incomplete] on 1r – 1 – (merger of: ontario argus, and ontario argus-observer) – us UW Libraries [071]
Eastern oregon observer *see* Ontario argus
Eastern oregon republican – Union OR: Eastern Oregon Pub Co, -1891 [wkly] – 1 – (began in 1888; cont by: weekly eastern oregon republican (1891-94)) – us Oregon Lib [071]
Eastern oregon review – LaGrande OR: C J Shorb, -1980 [wkly] – 1 – (merged with: elgin recorder (-1980) to form: union county review-recorder (1980-)) – us Oregon Lib [071]
Eastern oregon review *see*
– Elgin recorder
– Union county review-recorder
Eastern oregon weekly tribune – Pendleton OR: M H Abbott, 1874-75 [wkly] – 1 – (cont by: oregon weekly tribune (1875-77)) – us Oregon Lib [071]
Eastern Orthodox Church. Russian Synod *see* Vsepoddannishii otchet ober-prokurora
Eastern outlook – Enugu, Nigeria. 7 jan 1954-15 dec 1955; 3 jan 1957-14 mar 1966 – 21r – 1 – (aka: nigerian outlook. imperfect) – uk British Libr Newspaper [079]
Eastern post – London. -w. 18 oct 1868-26 oct 1938 70r – 1 – uk British Libr Newspaper [072]
Eastern press – [Scotland] Glasgow: J Cossar 11 jun 1890-18 mar 1891 (wkly) [mf ed 2004] – 1r – 1 – uk Newsplan [072]
Eastern progress – Richmond, KY. 1986-2000 (1) – mf#63486 – us UMI ProQuest [071]
Eastern proverbs and emblems illustrating old truths – New York: Funk and Wagnalls, [1881?] – 1mf – 9 – 0-524-00928-7 – mf#1990-2151 – us ATLA [470]
The eastern province herald – Port Elizabeth SA, 1845-1981 – 556r – 1,16 – (1 jan 1982-31 dec 1999. title varies: eastern province news, 1850-53, port elizabeth herald, 1950-53) – sa National [072]
Eastern province news *see* The eastern province herald
The eastern question in its anglo-indian aspect : a paper read...on wednesday, may 16 1877 / Long, James – London, 1877 – (= ser 19th c british colonization) – 1mf – 9 – mf#1.1.2089 – uk Chadwyck [330]
Eastern reflector – Greenville, NC. 1882-1887 (1) – mf#65311 – us UMI ProQuest [071]
The eastern reporter – Albany: Wm Gould Jr & Co. v1-11. 1884-87 (all publ) – 102mf – 9 – $153.00 – (covers me, nh, vt, ma, ri, ct, ny and pa) – mf#LLMC 95-116 – us LLMC [340]
Eastern review / spectator / Franklin Co. Columbus – (sep 1958-sep 69), may 71-feb 1974 [wkly] – (= ser Spectator) – 15r – 1 – (title changes) – mf#B6690-6704 – us Ohio Hist [071]
The eastern seas : on voyages and adventures in the indian archipelago, in 1832-33-34,... / Earl, G W – London, 1837 – 6mf – 9 – mf#SE-20164 – ne IDC [915]
Eastern shore herald – Eastville, VA. 1904-1949 (1) – mf#66701 – us UMI ProQuest [071]
Eastern shore news – Accomac, VA. 1987-2000 (1) – mf#66786 – us UMI ProQuest [071]
Eastern shore town and country post – Accomac, VA. 1976-1977 (1) – mf#66660 – us UMI ProQuest [071]
Eastern shore whig and people's advocate – Easton, MD. 1828-1841 (1) – mf#63602 – us UMI ProQuest [071]
Eastern sotho / Ziervogel, D – Pretoria, South Africa. 1954 – 1r – us UF Libraries [960]
Eastern star – Machias, (East Falls), ME: J O Balch, dec 1823-nov 23 1824 – 1r – 1 – (filmed with: eastern star & washington advertiser) – us CRL [071]
Eastern star – Grahamstown, Johannesburg SA, 6 jan 1871-29 mar 1889 – 15r – 1 – mf#MS00259 – sa National [079]
Eastern star and washington advertiser – Machias East Falls, ME: Jeremiah O Balch, dec 2 1824-jul 21 1825 – 1r – (filmed with: eastern star (machias, me.)) – us CRL [071]
Eastern suburbs news – Wellington, NZ. aug 1978-dec 1987 – 10r – 1 – mf#41.19 – nz Nat Libr [079]
Eastern sun – Birmingham, AL. 1956-1962 (1) – mf#61985 – us UMI ProQuest [071]
Eastern sun – Kuala Lumpur, Malaysia. 1966-1971 (1) – mf#67802 – us UMI ProQuest [079]
Eastern telegraph – Dungog, apr 1912-nov 1922 – 3r – 9 – at Pascoe [079]
Eastern times – Cuttack, India. 1962-Oct 1966 – 10r – 1 – us L of C Photodup [079]
Eastern times & tower hamlets gazette – London, England. -w. Dec 1859-13 feb 1864. 2r – 1 – uk British Libr Newspaper [072]
Eastern world – London. 1947-1971 (1) – ISSN: 0012-8961 – mf#6703 – us UMI ProQuest [320]
Eastern world : a weekly journal for law, commerce, politics, literature, 1899-1908 – Yokohama, Japan – (= ser Asian journals) – 5r – 1 – £475.00 – (with d/g) – uk Matthew [950]
Eastern world – Yokohama. Japan. -w. 4 Feb 1899-14 Nov 1908. (5 reels) – 1 – uk British Libr Newspaper [072]
Eastern world, 1947-62 – 11r – 1 – mf#513 – uk Microform Academic [073]
Easterner – Muncie, IN. 1922-1940 (1) – mf#62909 – us UMI ProQuest [071]
Eastham, Cheshire, Eng (Parish) *see* Parish registers of eastham, cheshire, from ad 1598 to 1700
Easthampton 1785-1892 – Oxford, MA (mf ed 1987) – (= ser Massachusetts vital records) – 32mf – 9 – 0-87623-013-3 – (mf 1: records 1785-1819 v1. mf 2: records 1822-45 v2. mf 3-4: b,m,d 1844-64. mf 5-6: index to b,m,d 1787-1864. mf 7-12: b,m,d 1785-1850. mf 13-16: births 1865-92 v2. mf 17-19: marriages 1858-92 v2. mf 20-24: deaths 1858-92 v2. mf 25-27: index to births 1844-92. mf 28-30: index to marriages 1789-1892. mf 31-32: index to deaths 1845-92) – us Archive [978]
Eastlake, C L *see* Materials for a history of oil painting
Eastlake, Charles Lock *see*
– Contributions to the literature of the fine arts
– Materials for a history of oil painting
– The national gallery
Eastlake, Charles Locke *see*
– Hints on household taste in furniture, upholstery
– A history of the gothic revival
– Notes on the principal pictures in the louvre
– Notes on the principal pictures in the old pinakothek at munich
– Notes on the principal pictures in the royal gallery...at venice
– Notes on the...pictures in the brera gallery at milan
Eastlake, Elizabeth (Rigby) *see*
– Five great painters
– Life of john gibson, r a, sculptor
Eastlake, Elizabeth (Rigby) et al *see* Examples of decorative wrought ironwork of the 17th and 18th centuries
Eastland baptist church. nashville, tennessee : church records – 1911-61 – 1 – 53.73 – us Southern Baptist [242]
Eastleigh weekly news and gazette – England.1895-1900. -w. 6 reels – 1 – uk British Libr Newspaper [072]
Eastleigh weekly news and hants gazette – England, Oct 1895-96; 1897-1904; Feb 1905 – 80+ r – 1 – uk British Libr Newspaper [072]
Eastman, Barrett *see* Paris, 1900
Eastman, Edith V *see* Musical education and musical art
Eastman, George Herbert *see* Papers, 1913-1969
Eastman, George Washington *see*
– A practical system of book-keeping by single and double entry
Eastman Kodak Co. Research Laboratories *see* Abridged scientific publications from the kodak research laboratories
Eastman organic chemical bulletin – Rochester. 1927-1973 (1) 1970-1973 (5) – ISSN: 0096-221X – mf#1553 – us UMI ProQuest [540]
Eastman, P M *see* Robert raikes and the northamptonshire sunday schools
Eastman, Theophilus *see* Connexion between christian benevolence and spiritual propserity
Easton 1693-1900 – Oxford, MA (mf ed 1992) – (= ser Massachusetts vital records) – 63mf – 9 – 0-87623-131-8 – (mf 1-8: vital records 1693-1889. mf 9-13: vital records 1693-1813. mf 14-16: vital records 1773-1854. mf 17-19: town records 1732-74. mf 20-28: town records 1766-1816. mf 29-35: town records 1816-40. mf 36-38: intentions 1836-69. mf 39-41: birth index 1843-1912. mf 42-43: marriage index 1843-1913. mf 44-45: death index 1843-1912. mf 46-47: births 1843-62. mf 48: marriages 1843-54. mf 49: deaths 1843-62. mf 50-54: births 1863-1900. mf 55-58: deaths 1863-1900. mf 59-63: marriages 1854-1900) – us Archive [978]
Easton, Burton Scott *see* Recent work of the church on the data of the synoptic gospels
Easton centinel – Easton, PA., 1833-1836 – 13 – $25.00r – us IMR [071]
Easton, Peter Zaccheus *see* Does woman represent god?
Eastport sentinel – Eastport, ME: B Folsom, [jul 26 1823?-aug 1 1832; 1853-68; dec 23 1869-jul 1953; jan-may 1954] – 1 – us CRL [071]
Eastport sentinel, and passamaquody advertiser – Eastport, ME: Benjamin Folsom, [1819-jul 19 1823?] – us CRL [071]
East's reports : reports of cases argued and determined in the court of king's bench / East, Edward H – v1-16. 1800-12. London: J Butterworth & Son, 1805-18 (all publ) – 89mf – 9 – $133.00 – mf#LLMC 95-293 – us LLMC [324]
Eastside news – Portland OR: [s.n.] -1907 [daily ex sun] – 1 – (began in 1906; cont by: daily news (1907-12)) – us Oregon Lib [071]
Eastside sun / Lucas Co. Toledo – jan 1921-nov 1925 [wkly] – 2r – 1 – mf#B34057-34058 – us Ohio Hist [071]
Eastward. Presbytery (Pres. Church in the USA) *see* Minutes, 1771-92
Eastwest – Brookline Village. 1986-1991 (1,5,9) – (cont: east west journal; cont by: eastwest natural health) – ISSN: 0888-1375 – mf#12433,03 – us UMI ProQuest [073]
Eastwest natural health – Brookline Village. 1992-92 (1,5,9) – (cont by: natural health. cont: eastwest) – ISSN: 1061-4664 – mf#12433,04 – us UMI ProQuest [073]
East-west outlook / American Committee on East-West Accord – v5 n1-v9 n3 [1982 jan-1986 may] – 1r – 1 – (cont: just for the press) – mf#1265904 – us WHS [071]
Eastwick, Edward B *see* Venezuela
Eastwood and kimberley advertiser – Eastwood, Nottinghamshire, may 1895-1897, 1899-to date – 1 – uk Newsplan [072]
Eastwood baptist church. bowling green, kentucky : church records – 1953-79. 2060p – 1 – 92.70 – us Southern Baptist [242]
Easum, Chester Verne *see* The americanization of carl schurz
Easy – 1978 jan, 1981 aug – 2r – 1 – mf#4853329 – us WHS [071]
Easy conversations of english and japanese for those who learn the english language – Tokei [Tokyo]: printed by Matsmoto, 1872 – (= ser 19th c books on linguistics) – 4mf – 9 – mf#2.1.8 – uk Chadwyck [400]
An easy grammar of natural and experimental philosophy : for the use of schools / Phillips, Richard (David Blair pseud) – new ed. London: printed for Richard Phillips, 1808 – (= ser 19th c children's literature) – 3mf – 9 – mf#6.1.21 – uk Chadwyck [100]
An easy grammar of the primaeval language : commonly called hebrew, entitled orah mishor or, the "straight path" to real knowledge, fully exemplified by instructive and elegant extracts / Bolaffey, Hayim Victa – London: printed for Hatchard, & G & W B Whittaker, 1820 – (= ser 19th c books on linguistics) – 6mf – 9 – mf#2.1.15 – uk Chadwyck [470]
The easy instructor : or, a new method of teaching sacred harmony. / Little, William & Smith, William – 1798 – 1 – 5.00 – us Southern Baptist [242]
Easy lessons in general geography, with maps and illustrations : being introductory to "lovell's general geography" / Hodgins, John George – Montreal: printed & publ by J Lovell, 1863 [mf ed 1984] – 1mf – 9 – 0-665-45103-2 – (original iss in ser: lovell's series of school-books) – mf#45103 – cn Canadiana [912]
An easy mode of teaching the rudiments of latin grammar to beginners / Robertson, Thomas Jaffray – Montreal: J Lovell, 1861 [mf ed 1993] – 1mf – 9 – 0-665-91751-1 – mf#91751 – cn Canadiana [450]
Easy reader – Hermosa Beach, CA. 1996-1996 (1) – mf#62168 – us UMI ProQuest [071]
An easy walk through the british museum / Fagan, Louis Alexander – London 1891 – (= ser 19th c art & architecture) – 2mf – 9 – mf#4.2.826 – uk Chadwyck [060]
An easy way to use the psalms / Smith, Joseph Oswald – [s.l]: Ampleforth Abbey [1911?] [mf ed 1993] – 1mf – 9 – 0-524-05794-X – mf#1992-0621 – us ATLA [221]
Easy-english for natives in rhodesia / Mayr, F – Mariannhill, South Africa. 1928 – 1r – us UF Libraries [420]
Eating and dieting behaviors and weight concerns among ncaa division 1 women swimmers / Popovich, Angela M – 1998 – 1mf – 9 – $4.00 – mf#PE 3935 – us Kinesology [617]
Eating behaviors – New York, 2000+ [1,5,9] – ISSN: 1471-0153 – mf#42834 – us UMI ProQuest [150]
Eating disorder symptomatology in a male athletic population / Jewell, Elizabeth A – 1997 – 1mf – 9 – $4.00 – mf#PSY 1990 – us Kinesology [150]
Eating disorders among athletes : public policy to promote social and individual behavioral change / Clary, J M – 1992 – 1mf – 9 – $4.00 – us Kinesology [150]
Eating the tract – London, England. 18– – 1r – us UF Libraries [240]
Eaton, Abbie Fiske *see* Das spielmannskind / der stumme ratsherr
Eaton, Amasa M *see* Roger williams, the founder of providence the pioneer of religious liberty
Eaton, Arthur Wentworth Hamilton *see*
– The church of england in nova scotia and the tory clergy of the revolution
– The heart of the creeds
Eaton Bray – (= ser Bedfordshire parish register series) – 3mf – 9 – £7.50 – uk BedsFHS [929]
Eaton, Charlotte *see* Rome in the nineteenth century

Eaton (congleton), christ church – [Macclesfield Ferrets] – (= ser Cheshire monumental inscriptions) – 1mf – 9 – £2.50 – mf#80 – uk CheshireFHS [929]
Eaton, E K *see*
– Eaton's series of national and popular songs, for small military brass bands, from five to twelve instruments. op. 13
– Twelve pieces of harmony for...military brass bands of seventeen instruments
Eaton, Isabel *see* The philadelphia negro
Eaton, Jeanette *see* Gandhi, fighter without a sword
Eaton, John *see* The education of our girls
Eaton, John Richard Turner *see* The permanence of christianity
Eaton, John van *see* Expository and practical lectures on haggai and zechariah
Eaton, John W *see* Penuel
Eaton, Samuel J M *see* History of the presbytery of erie
Eaton, Samuel John Mills *see* History of the presbytery of erie
Eaton, Scott W *see* Analyzing computer applications in national collegiate athletic association's men's basketball programs
Eaton, Sherburne Blake *see* A discussion of the constitutionality of the act of congress of march 2, 1867, authorizing the seizure of books and papers for alleged frauds upon the revenue.
Eaton socon – (= ser Bedfordshire parish register series) – 2pt on 7mf – 9 – £16.00 – uk BedsFHS [929]
Eaton socon monumental inscriptions – St Neots Local HS 1980 – (= ser Bedfordshire parish register series) – 2mf – 9 – £3.00 – uk BedsFHS [929]
Eaton, Thomas Treadwell *see*
– The bible on women's public speaking
– Biographical materials, correspondence, sermons
Eaton's baptist church. south yadkin association. north carolina : church records – 1772-1902 – 1 – us Southern Baptist [242]
Eaton's series of national and popular songs, for small military brass bands, from five to twelve instruments. op. 13 / Eaton, E K – Boston: Henry Tolman, 1853. Includes: "Comin' thro' the Rye" and "Rule Britannia" in parts. MUSIC 1983, Item 1 – 1 – us L of C Photodup [780]
Eau claire advocate – Eau Claire WI. 1937 nov 11-1938 oct 20 – 1r – 1 – (cont: eau claire advocate and the chippewa valley commonwealth advocate; cont by: eau claire news) – mf#962634 – us WHS [071]
Eau claire argus – Eau Claire WI. 1879 jul 24 – 1r – 1 – mf#875207 – us WHS [071]
Eau claire [city directory : listing] – 1882, 1885, 1889-90, 1893-94 – 4r – 1 – mf#2913295 – us WHS [978]
Eau claire city telegraph – Eau Claire WI. 1858 jul 10 [v2 n3] – 1r – 1 – (cont: eau claire times; cont by: eau claire weekly free press) – mf#875332 – us WHS [071]
Eau claire county herald – Eau Claire WI. 1957 apr 4-1958 mar 13 – 1r – 1 – (cont: fall creek tribune) – mf#964808 – us WHS [071]
Eau claire county journal – Fall Creek WI. 1914 sep 4-1915 dec 17, 1915 dec 24-1916 dec 15 – 2r – 1 – (cont: fall creek cultivator; cont by: fall creek journal [fall creek wi: 1916]) – mf#1044379 – us WHS [071]
Eau claire county union – Augusta, Fairchild, Fall Creek WI. 1919 dec 5-1920 sep 10, 1920 sep 17-1921 dec 31, 1922 jan 1-1923 apr 13, 1923 apr 20-1924 sep 30, 1924 oct 1-1925 oct 8, 1925 oct 15-1926 dec 23, 1926 dec 30-1927 jan 6 – 7r – 1 – (cont: cooperative news-budget; augusta times; cont by: augusta eagle times; fairchild observer [fairchild wi: 1927]; fall creek journal [fall creek wi: 1927]) – mf#1044321 – us WHS [071]
Eau claire daily argus – Eau Claire WI. 1880 may 5, jun 15, 17-18, oct 6, 1881 jan 29 – 1r – 1 – mf#875337 – us WHS [071]
Eau claire daily news – Eau Claire WI. 1884 aug 30-nov 5 – 1r – 1 – mf#961935 – us WHS [071]
Eau claire herald – Eau Claire WI. 1862 mar 6-15, may 24, oct 1 – 1r – 1 – mf#962641 – us WHS [071]
Eau claire leader – Eau Claire WI. 1893 nov/1894 jan-1952 jan/feb – 272r – 1 – (with gaps; cont by: daily leader [eau claire, wi]; eau claire leader-telegram) – mf#967963 – us WHS [071]
Eau claire news – Eau Claire WI. 1938 oct 27-1940 jan 4 – 1r – 1 – (cont: eau claire advocate) – mf#962637 – us WHS [071]
Eau claire news – Eau Claire WI. 1875 may 15-1879 jan 18, 1879 jan 25-1882 oct 28, 1882 nov 4-dec 30, 1883 jul 21-1886 dec 25, 1890-92 – 5r – 1 – mf#967956 – us WHS [071]
Eau claire times – Eau Claire WI. 1857 may 23, aug 18-1858 jan 2 – 1r – 1 – mf#962646 – us WHS [071]
Eau claire und chippewa falls herold – Chippewa Falls, Eau Claire WI. 1890 dec 18-1892 jun 2, 1892 jun 9-sep 15 – 2r – 1 – (cont by: herold [eau claire, wi]) – mf#998630 – us WHS [071]
Eau claire weekly free press – Eau Claire WI. 1889 jan 3-1889, 1890 jan 14-jun 19, 1890 jan 7-jun 19 – 3r – 1 – (cont: eau claire free press [eau claire, wi: 1882]; cont by: weekly free press) – mf#999230 – us WHS [071]
Eau claire weekly leader – Eau Claire WI. 1895 mar 30-dec 28, 1896 jan 4-1897 jul 10 – 2r – 1 – (cont by: weekly leader) – mf#967965 – us WHS [071]
Eau claire weekly leader – Eau Claire WI. 1902 oct 11-1903 oct 31, 1903 nov 7-1904 nov 19, 1904 nov 26-1905 jun 24 – 3r – 1 – (cont: weekly leader) – mf#967970 – us WHS [071]
Eau claire weekly leader – Eau Claire WI. 1889 may 16-dec 30, 1890, 1891 jan-nov 9 – 3r – 1 – mf#967961 – us WHS [071]
Eau claire weekly sentinel – Eau Claire WI. 1884 apr 30 – 1r – 1 – mf#962644 – us WHS [071]
Eau de javelle / Gabriel, M – Paris, France. 1852 – 1r – us UF Libraries [440]
Eau gallie : the harbor city – Eau Gallie, FL. 1930? – 1r – us UF Libraries [978]
Eayrs, George *see* Richard baxter and the revival of preaching and pastoral service
Eban, Abba *see* Israel's position on the jordan canal project
Ebano – Malabo, Equatorial Guinea. Aug 3 1980-Jne 22 1991 – 1r – 1 – us L of C Photodup [079]
Ebano / Ordonez Arguello, Alberto – San Salvador, El Salvador. 1954 – 1r – us UF Libraries [972]
Ebauches / Price-Mars, Jean – Port-Au-Prince, Haiti. 1961 – 1r – us UF Libraries [972]
Ebb un flot : glueck un not / Lau, Fritz – Hamburg: M Glogau, 1921 – 1r – 1 – us Wisconsin U Libr [830]
Ebba news – New York. 1976-1976 (1) 1976-1976 (5) 1976-1976 (9) – ISSN: 0012-7485 – mf#10266 – us UMI ProQuest [590]
Ebbe und flut : ein hansischer roman deutscher zeitwende / Schupp, Johannes Martin – Muenchen: F Eher, 1942, c1938 – 1r – 1 – us Wisconsin U Libr [830]
Ebbe und fluth : gesammelte lyrische dichtungen: und ingurtha: trauerspiel in fuenf akten / Zuendt, Ernst Anton Joseph – Milwaukee, WI: Freidenker, 1894 – 1r – 1 – us Wisconsin U Libr [800]
Ebbecke, Dirk *see* Zur bedeutung der sozialperspektivitaet in der marktpsychologischen imageforschung
Ebbinghaus, Ernst A *see* Daz buoch von dem uebeln wibe
Ebbtide / Frederica Academy – v1 n1-9 [1978 spring-1981 jun] – 1r – 1 – mf#669859 – us WHS [071]
Die ebed jahwe-lieder in jesaja 40 ff : ein beitrag zur deuterojesaja-kritik / Staerk, Willy – Leipzig: JC Hinrichs, 1913 – (= ser Beitraege zur wissenschaft vom alten testament) – 1mf – 9 – 0-524-06343-5 – mf#1992-0881 – us ATLA [221]
Ebel, J G *see* Anleitung auf die nuetzlichste und genussvollste art in der schweitz zu reisen
Ebel, Johann *see* Manuel du voyageur en suisse
Ebel, Johann G *see* Anleitung, auf die nuetzlichste und genussvollste art die schweiz zu bereisen
Ebeling, Adolph *see* Lebende bilder aus dem modernen paris
Ebeling, Christoph Daniel *see*
– Amerikanische bibliothek
– Neue sammlungen von reisebeschreibungen
Ebeling, E *see*
– Assyrische rechtsurkunden
– Aus dem tagewerk eines assyrischen zauberpriesters
– Die babylonische fabel und ihre bedeutung fuer die literaturgeschichte
– Liebeszauber im alten orient
– Quellen zur kenntnis der babylonischen religion
Ebeling, Erich *see* Neubabylonische briefe aus ukruk.
Ebendorfer, Thomas *see* Chronica austriae (mgh6:13.bd)
Ebenezer : or, divine deliverances in china / Glover, Robert – New York: Alliance Press [1905] [mf ed 1995] – (= ser Yale coll) – 120p – 1 – 0-524-09468-3 – mf#1995-0468 – us ATLA [210]
Ebenezer baptist church : church minutes – Jonesboro, LA. 1456p. 1849-1997 – 1 – $65.52 – mf#6965 – us Southern Baptist [242]
Ebenezer baptist church – Silver Spring. 1971-1974 (1) 1971-1974 (5) (9) – 1r – 1 – $16.29 – mf#6490 – us Southern Baptist [242]
Ebenezer baptist church. aurora, indiana : church records – 1822-59 – 1 – 9.54 – us Southern Baptist [242]
Ebenezer baptist church. edgefield, south carolina : church records – 1973-83.Incomplete, 126p – 1 – 5.67 – us Southern Baptist [242]
Ebenezer baptist church. lincoln county. missouri : church records – Extinct.1915-23. 38p – 1 – 5.00 – us Southern Baptist [242]
Ebenezer grapevine – Providence, RI. 1980-1984 (1) – mf#68199 – us UMI ProQuest [071]
The ebenezer hazard collection – 8r – 1 – $280.00 – Dist. us Scholarly Res – us L of C Photodup [975]
Ebenezer Missionary Baptist Church *see* Reminder
Ebenhaezer : herdenking van hat vijftig-jarig bestaan van de christelijke gereformeerde gemeente, 14th street, chicago, 1867-1917 – [S.l.: s.n., 1917?] – 1mf – 9 – 0-524-06611-6 – mf#1991-2666 – us ATLA [240]
Das ebenhoech : geschichten von bauern und ihrem anhang / Huggenberger, Alfred – Frauenfeld: Huber 1919, c1911 [mf ed 1995] – 1r – 1 – (filmed with: daniel pfund) – mf#3884p – us Wisconsin U Libr [390]
Eber, P *see*
– Catechismuspredigten
– Pia et in verbo dei fvndata assertio, declaratio et confessio d pavli eberi de sacratissima coena
– Postilla
– Vom heiligen sacrament des leibs vnd bluts vnsers herren iesv christi unterricht vnd bekentnis
Eberhard, Christian August Gottlob *see* Hanchen und die kuechlein
Eberhard, Johann August *see*
– Philosophisches archiv
– Philosophisches magazin
Eberhard, Oscar *see* Bauernaufstand vom jahre 1381 in der englischen poesie
Eberhard, Otto *see* Der katechismus als paedagogisches problem
Eberhard von groote : ein beitrag zur geschichte der romantik am rhein / Giesen, Adolf – Gladbach-Rheydt, 1929 (mf ed 1992) – 1mf – 9 – €24.00 – 3-89349-019-1 – mf#DHS-AR 3 – gw Frankfurter [943]
Eberhardt, Jacqueline *see* A survey of family conditions with special reference to housing needs, orlando township, johannesburg
Eberhart, Jean M *see* Validity of the astrand-rhyming nomogram for moderately active adult females
Eberharter, Andreas *see*
– Das ehe- und familienrecht der hebrer
– Der kanon des alten testaments zur zeit des ben sira
Eberle, Josef *see* Gold am pazifik
Eberlein, Gerhard *see* Vortraege
Eberlein, Karl *see* Gedichte und gedanken
Eberlin, Elie *see* Juifs d'aujourd'hui
Eberlin, Johann Ernst *see* 115 versetten und cadenzen fuer die orgel in den gewoehnlichen 8 kirchen tonarten...ch 1
Ebermayer, Erich *see*
– Evil genius
– Kampf um odilienberg
Ebers, Carl Friedrich *see* Sarsena
Ebers, Fritz *see* Das grabbe-buch
Ebers, G *see* Durch gosen zum sinai
Ebers Georg *see* Josua
Ebers, Georg *see*
– Eine aegyptische koenigstochter
– Arachne
– Barbara blomberg
– Drei maerchen fuer alt und jung
– Eine frage
– Die frau buergemeisterin
– Die frau buergermeisterin
– Georg ebers gesammelte werke
– Die geschichte meines lebens
– Die gred
– Homo sum
– Im blauen hecht
– Im schmiedefeuer
– Josua
– Der kaiser
– Kleopatra
– Die nilbraut
– Per aspera
– A question
– Die schwestern
– Serapis
– The story of my life
– Uarda
– Ein wort
Ebers, George *see* Richard lepsius
Ebersbacher zeitung *see* Unterer filstal- und schurwaldbote 1906
Ebersberger anzeiger – Ebersberg, Obbay DE, 1886 2 dec-1897, 1899-1921 [gaps], 1938-1945 29 apr, 1949 1 oct-30 nov – 23r – 1 – (title varies: 1922-37: der oberbayer. with suppl: amtsblatt fuer den amtsbezirk ebersberg 1894-96, 1899, 1901-03) – gw Misc Inst [074]
Ebersold, Walter *see* Tell
Ebersole, Ezra Christian *see* The courts and legal profession of iowa.
Ebersole, Joseph L *see* Discovery problems in civil cases
Eberswalder kreisrundschau – Eberswalde DE, 1963 6 oct-1966 15 feb – 1r – 1 – gw Misc Inst [074]
Ebert, Adolf *see* Allgemeine geschichte der literatur des mittelalters im abendlande
Ebert, Joh Jacob *see* Der philosoph fuer jedermann
Ebert, Johannes *see* Sein und sollen des menschen bei immanuel hermann fichte
Ebert, Justus *see* American industrial evolution from the frontier to the factory
Ebey, Adam *see* The house that jack is building
Ebeye voice – v2 n4-5 [1968 feb 12, 1968-mar 4], n9 [1968 may 6] – 1r – 1 – mf#665900 – us WHS [071]
Ebhardt, Rolf *see* Hebbel als novellist
Ebn – Manhasset, 2001+ [1,5,9] – mf#19187,01 – us UMI ProQuest [621]
Ebner, A *see* Quellen und forschungen zur geschichte und kunstgeschichte des missale romanum im mittelater
Ebner, Adalbert *see* Quellen und forschungen zur geschichte und kunstgeschichte des missale romanum im mittelalter
Ebner, J *see* Die erkenntnislehre richards von st viktor
Ebner, Johann *see* Reise nach sued-afrika und darstellung meiner waehrend acht jahren daselbst als missionair unter den hottentotten gemachten erfahrungen...
Ebner, Theodor *see* Max eyth
Ebner-eschenbach / Reuter, Gabriele – Berlin: Schuster & Loeffler, [1904?] (mf ed 1990) – 1r – 1 – (filmed with: marie von ebner-eschenbach nach ihren werken geschildert) – us Wisconsin U Libr [430]
Ebner-Eschenbach, Marie von *see*
– Alte schule
– Altweibersommer
– Die arme kleine; stille welt
– Aus spaetherbsttagen
– Ausgewaehlte erzaehlungen
– Bozena
– Ein buch, das gern ein volksbuch werden moechte
– Dorf- und schlossgeschichten; neue dorf- und schlossgeschichten; zwei komtessen
– Drei novellen
– Erzaehlungen
– Das gemeindekind
– Genrebilder
– Glaubenslos?
– Glaubenslos?; unsuehnbar
– Lotti, die uhrmacherin
– Lotti, die uhrmacherin; agave; margarete
– Meine erinnerungen an grillparzer
– Neue dorf- und schlossgeschichten
– Parabeln und maerchen; gedichte; aphorismen; prinzessin leiladin; hirzepinzchen; erzaehlungen
– Saemtliche werke
– Das schaedliche / die totenwacht
– Die unbesiegbare macht; rittmeister brand
– Zwei comtessen
Ebone singles – 1997 may – 1r – 1 – mf#4109784 – us WHS [071]
Ebony – Chicago. 1945+ (1) 1968+ (5) 1960+ (9) – ISSN: 0012-9011 – mf#977 – us UMI ProQuest [978]
Ebony fashion fair magazine – 1994/1995 – 1r – 1 – mf#4027477 – us WHS [640]
Ebony jr! – Chicago. 1973-1985 (1) 1977-1985 (5) 1977-1985 (9) – ISSN: 0091-8660 – mf#8799 – us UMI ProQuest [370]
Ebony man: em – New York. 1996-1998 – 1,5,9 – ISSN: 0884-4879 – mf#20605 – us UMI ProQuest [305]
Ebrard, Friedrich Clemens *see* Die franzoesisch-reformierte gemeinde in frankfurt am main, 1554-1904
Ebrard, Johannes Heinrich August *see*
– Apologetics
– Biblical commentary on the epistle to the hebrews
– Bonifatius
– Christian ernst von brandenburg-baireuth
– Christliche dogmatik
– Das dogma vom heiligen abendmahl und seine geschichte
– E von hartmann's philosophie des unbewussten
– The gospel history
– Handbuch der christlichen kirchen- und dogmen-geschichte fuer prediger und studirende
– Die lehre von der stellvertretenden genugthuung
– Die praedestinationsfrage aufs neue betrachtet
Gli ebrei in libia, usi e costumi / Ha-Cohen, Mordecai – Roma, Italy. 1927 – 1r – us UF Libraries [939]
Ebright, Homer Kingsley *see* The petrine epistles
Ebstein *see* Dr. martin luthers krankheiten und deren einfluss auf seinen koerperlichen und geistigen zustand
Ebstein, Erich *see* Gottfried august buerger und philippine gatterer
Ebue-l'kemal *see* The divan project
Eburnea – Abidjan: Agence ivoirienne de presse. n1-10. apr 1967-jan/feb 1968; n12-61. apr 1968-jul 1972; n63-77. sep 1972-feb 1974; n79-124. jul 1974-1978] – 5r – 1 – us CRL [079]

Eby, Enoch et al *see* Hand-book of the general missionary and tract committee of the german baptist brethren church

Eby, Ezra E *see*
– A biographical history of waterloo township and other townships of the county, vol 1
– A biographical history of waterloo township and other townships of the county, vol 2
– A biographical history of waterloo township and other townships of the county, vols 1 and 2

Eby, Herbert Oscar *see* Extract of the district of columbia code

Ebyafayo by'obusiramu mu uganda : manuscript translations / Kulumba, Ali, Sheikh – [Kampala, Uganda: s.n, 1962 or 1963] – 1r – 1 – us CRL [470]

Ec and m : electrical construction and maintenance – Overland Park. 1981+ (1) 1981+ (5) 1981+ (9) – (cont: electrical construction and maintenance) – ISSN: 1082-295X – mf#366,01 – us UMI ProQuest [621]

Ec kiaa seksi penprop / Gelora KIAA – Djakarta, 1964-1965 – 3mf – 9 – mf#SE-1485 – ne IDC [950]

Eca : critico e humoristico – Florianopolis, SC. 01 jan 1932 – (= ser Ps 19) – bl Biblioteca [079]

Eca de queiroz. obras – Porto. v1-15. 1946-48 – 1 – $162.00 – mf#0191 – us Brook [440]

Ecad-entwurfsmanagement / Schuermann, Bernd – (mf ed 1999) – 3mf – 9 – €49.00 – 3-8267-2629-4 – mf#DHS 2629 – gw Frankfurter [510]

Ecbasis cuiusdam captivi per tropologiam (mgh7:24.bd) – 1935 – (= ser Monumenta germaniae historica 7: scriptores rerum germanicarum in usum scholarum (mgh7)) – €5.00 – ne Slangenburg [240]

Ecce ancilla domini : mary the mother of our lord: studies in the christian ideal of womanhood / Charles, Elizabeth Rundle – London: Society for Promoting Christian Knowledge, 1894 – 1mf – 9 – 0-8370-2632-6 – mf#1985-0632 – us ATLA [240]

Ecce deus : essays on the life and doctrine of jesus christ / Parker, Joseph – Boston: Roberts, 1867 [mf ed 1985] – 1mf – 9 – 0-8370-4666-1 – mf#1985-2666 – us ATLA [240]

Ecce deus : studies of primitive christianity / Smith, William Benjamin – London: Rationalist Press Association [by] Watts, 1912 – 1mf – 9 – 0-7905-5317-1 – mf#1988-1317 – us ATLA [240]

Ecce filius : or, the gospel of truth and grace by positive manifestation / Swinney, James Oswald – Chicago: Fleming H Revell, 1894 – 1mf – 9 – 0-8370-2208-8 – mf#1985-0208 – us ATLA [240]

Ecce home : ofte oogen-salve / Teelinck, W – Dordrecht, 1646 – 3mf – 9 – mf#PBA-324 – ne IDC [240]

Ecce homo : a critique on behalf of the cause of free enquiry and free expression – Ramsgate: Thomas Scott, 1866 [mf ed 1985] – 1mf – 9 – 0-8370-4793-5 – mf#1985-2793 – us ATLA [240]

Ecce homo / Gladstone, W E – London: Strahan, 1868 – 1mf – 9 – 0-8370-3305-5 – mf#1985-1305 – us ATLA [240]

Ecce homo : a survey of the life and work of jesus christ / Seeley, John Robert – Boston: Roberts, 1890 – 1mf – 9 – 0-8370-5211-4 – mf#1985-3211 – us ATLA [240]

Ecce pericles / Arevalo Martinez, Rafael – Guatemala, 1945 – 1r – us UF Libraries [972]

Ecce regnum : or an inquiry into the nature and a revelation of the glory of the kingdom of god, according to the scriptures / Josslyn, William R – New York: Wm B Mucklow, 1877 [mf ed 1985] – 1mf – 9 – 0-8370-3804-9 – mf#1985-1804 – us ATLA [240]

Ecce spiritus : a statement of the spiritual principle of jesus as the law of life / Hayward, Edward Farwell – Boston: George H Ellis, 1881 [mf ed 1985] – 1mf – 9 – 0-8370-4526-6 – mf#1985-2526 – us ATLA [240]

Ecce unitas : or, a plea for christian unity: in which its true principles and basis are considered / [Ralston, Thomas Neely] – Cincinnati: Hitchcock & Walden, 1875 [mf ed 1991] – 1mf – 9 – 0-7905-8559-6 – mf#1989-1784 – us ATLA [240]

Ecce venit : behold he cometh / Gordon, Adoniram Judson – New York: Fleming H Revell, c1889 [mf ed 1988] – 1mf – 9 – 0-7905-0258-5 – (incl bibl ref) – mf#1987-0258 – us ATLA [220]

Eccentric kinetic chain exercise as a conservative means of functionally rehabilitating chronic isolated posterior cruciate ligament insufficiency / MacLean, Christopher L – University of British Columbia, 1995 – 2mf – 9 – $8.00 – mf#PE3608 – us Kinesology [617]

Eccentric peak torque and maximal repetition work percentages of the dominant external rotators in college division 1 baseball players / Geisler, S A – 1991 – 1mf – 9 – $4.00 – us Kinesology [790]

Eccles and patricroft journal – [NW England] Eccles 31 jan 1874-3 nov 1977 – 1 – (title change: eccles journal [10 nov 1977-11 oct 1984]; eccles & irlam journal [18 oct 1984-27 nov 1986]; eccles journal [4 dec 1986-10 mar 1988]) – uk MLA; uk Newsplan [072]

Ecclesia : church problems considered in a series of essays / Stoughton, John et al; ed by Reynolds, Henry Robert – London: Hodder & Stoughton 1870 [mf ed 1991] – 2mf – 9 – 0-524-00594-X – mf#1990-0094 – us ATLA [240]

Ecclesia discens : the church's lesson from the age / Peile, James Hamilton Francis – London; New York: Longmans, Green, 1909 – 1mf – 9 – 0-7905-8553-7 – mf#1989-1778 – us ATLA [240]

Ecclesia lutherana : a brief survey of the evangelical lutheran church / Seiss, Joseph Augustus – Philadelphia: Lutheran Bookstore, 1867 – 1mf – 9 – 0-524-00975-9 – mf#1990-4033 – us ATLA [242]

The ecclesia monthly *see* Shen chao (ccs)

[Ecclesia, S ab] *see* Flos florum

Ecclesia the church, bible class lectures / Carroll, BH – 1903 – 1 – 5.00 – us Southern Baptist [242]

Ecclesia, the church of christ : a planned series of papers / Dolan, Gilbert et al; ed by Mathew, Arnold Harris – London: Burns and Oates, [190-?] – 1mf – 9 – 0-524-08337-1 – mf#1993-2027 – us ATLA [240]

Ecclesiae scholaeque tigurinae, de iisdem thesibus [zanchii] iudicium / Bullinger, Heinrich – 1mf – 9 – mf#PBU-264 – ne IDC [240]

Ecclesianthem : or, a song of the brethren. embracing their history and doctrine / Heckler, James Y – Lansdale, Pa: AK Thomas, 1883 – 1mf – 9 – 0-524-03262-9 – mf#1990-4665 – us ATLA [242]

Ecclesias evangelicas neqve haereticas neqve schismaticas...esse...apodixis / Bullinger, Heinrich – [Tigvri, Andrea Gesner f. et Rodolph Vuysenbach], 1552 – 2mf – 9 – mf#PBU-172 – ne IDC [240]

L'ecclesiaste / Podechard, Emmanuel – Paris: Victor Lecoffre 1912 [mf ed 1989] – (= ser Etudes bibliques) – 2mf – 9 – 0-7905-2422-8 – (in french & hebrew) – mf#1987-2422 – us ATLA [221]

L'ecclesiaste : traduit de l'hebreu avec une etude sur l'age et le caractere du livre / Renan, Ernest – Paris: Calmann Levy, 1882 [mf ed 1993] – 1mf – 9 – 0-524-05595-5 – (in french) – mf#1992-0450 – us ATLA [221]

Ecclesiastes : an introduction to the book: an exegetical analysis and a translation with notes / Tyler, Thomas – new ed. London: D Nutt, 1899 [mf ed 2003] – 1r – 1 – (incl bibl ref) – mf#b00659 – us ATLA [221]

Ecclesiastes : a new translation with notes explanatory, illustrative, and critical / Coleman, John Noble – 2nd rev enl ed. Edinburgh: Andrew Elliot, 1867 – 2mf – 9 – 0-8370-2708-X – mf#1985-0708 – us ATLA [221]

Ecclesiastes : or, the preacher / Streane, Annesley William – London: Methuen 1899 [mf ed 1989] – (= ser The churchman's bible) – 1mf – 9 – 0-7905-2092-3 – mf#1987-2092 – us ATLA [221]

Ecclesiastes : a study / Erdman, William Jacob – Philadelphia:[s.n.], c1895 – 1mf – 9 – 0-8370-3068-4 – mf#1985-1068 – us ATLA [221]

Ecclesiastes : words of koheleth son of david, king in jerusalem, translated anew, divided according to their logical cleavage,... / Genung, John Franklin – Boston: Houghton, Mifflin, 1904 – 1mf – 9 – 0-8370-3253-9 – mf#1985-1253 – us ATLA [221]

Ecclesiastes, or, koheleth / Zoeckler, Otto; ed by Lewis, Tayler – amer ed. New York: Charles Scribner, 1870 [mf ed 1986] – (= ser A commentary on the holy scriptures. old testament 10/2) – 1mf – 9 – 0-8370-6157-1 – (ann and int by ed. trans by william wells) – mf#1986-0157 – us ATLA [221]

Ecclesiastica : a triplet of old sermons / Henson, Hensley – [London?]: [s.n.], 1910 (London: Hugh Rees) – 1mf – 9 – 0-7905-7170-6 – mf#1988-3170 – us ATLA [240]

The ecclesiastical and missionary record for the presbyterian church of canada – Hamilton [Ont]: J Webster, [1844-1861] – 9 – mf#P04399 – cn Canadiana [242]

Ecclesiastical antiquities of london and its suburbs / Wood, Alexander – London: Burns and Oates, 1874 – 1mf – 9 – 0-524-01901-0 – mf#1990-0528 – us ATLA [240]

The ecclesiastical architecture of ireland to the close of the 12th century / Brash, R R – Dublin, 1875 – €14.00 – ne Slangenburg [720]

The ecclesiastical architecture of italy / Knight, Henry Gally – London 1842-44 – (= ser 19th c art & architecture) – 6mf – 9 – mf#4.2.1209 – uk Chadwyck [720]

Ecclesiastical art in germany during the middle ages = Vorschule zum studium der kirchlichen kunst des deutschen mittelalters / Luebke, Wilhelm – 4th ed. Edinburgh: Thomas C Jack; London: Simpkin, Marshall, 1877 [mf ed 1991] – 1mf – 9 – 0-524-00767-5 – (english trans fr 5th german ed by l a wheatley. with app) – mf#1990-0199 – us ATLA [720]

Ecclesiastical authority in england : church court records c1400-c1660 – 70r coll – 1 – (series 1: the church court records of ely 32r – pt 1: instance act books and court papers, 1374-1640 19r cl999-28641. pt 2: office act books and formularies, 1469-1639 13r cl999-28642. series 2: the church court records of chichester 38r – pt 1: instance act books, 1506-1696; deposition books, 1557-1694; and taxation books, 1606-1607 19r cl999-28643. pt 2: detection books, 1538-1700; excommunication papers, 1612-1665; churchwarden's presentments, 1573-1698; and other papers 19r cl999-28644) – mf#CL999-28640 – us Primary [240]

An ecclesiastical catechism of the presbyterian church : for the use of families, bible-classes, and private members – 6th rev ed. Richmond: Presbyterian Cttee of Publ, c1868 [mf ed 1986] – 1mf – 9 – 0-8370-8779-1 – (incl bibl ref) – mf#1986-2779 – us ATLA [242]

The ecclesiastical class book : or, history of the church: from the birth of christ to the present time / Goodrich, Charles Augustus – [rev ed]. New York: F J Huntington, 1839, c1835 – 1mf – 9 – 0-8370-7146-1 – mf#1986-1146 – us ATLA [240]

Ecclesiastical curiosities / Tyack, George Smith et al; ed by Andrews, William – London: W Andrews, 1899 – 1mf – 9 – 0-524-03750-7 – mf#1990-1097 – us ATLA [240]

An ecclesiastical dictionary : explanatory of the history, antiquities, heresies, sects, and religious denominations of the christian church / Farrar, John – London: Wesleyan Conference Office, 1864 [mf ed 1991] – 1mf – 9 – 0-524-01460-4 – mf#1990-0409 – us ATLA [052]

Ecclesiastical establishments not lawful / Marshall, Andrew – Glasgow, Scotland. 1837 – 1r – us UF Libraries [240]

Ecclesiastical gazette – Kingston [Ont]: Diocese of Ontario, [1874-18–?] – 9 – mf#P04258 – cn Canadiana [242]

An ecclesiastical history from the 1st to the 13th century / Butler, Clement Moore – Philadelphia: M'Calla & Stavely, 1868 [mf ed 1991] – 2mf – 9 – 0-524-00627-X – mf#1990-0127 – us ATLA [240]

An ecclesiastical history from the 13th to the 19th century / Butler, Clement Moore – Philadelphia: Claxton, Remsen & Haffelfinger, 1872 [mf ed 1991] – 2mf – 9 – 0-524-00628-8 – mf#1990-0128 – us ATLA [240]

An ecclesiastical history from the creation to the 18th century, a d / Bourne, Hugh [comp] – London: W Lister, 1865 [mf ed 1993] – 2mf – 9 – 0-524-06867-4 – (originally publ in the primitive methodist magazine fr 1825-42 incl; rev & abr by william antliff) – mf#1990-5286 – us ATLA [240]

The ecclesiastical history of eusebius in syriac... : with a collation of the ancient armenian version by adalbert merx / Wright, W – Cambridge, 1898 – 5mf – 9 – mf#AR-1671 – ne IDC [956]

An ecclesiastical history of ireland : from the introduction of christianity into that country to the year 1829 / Brenan, Michael John – rev ed. Dublin: James Duffy, 1864 [mf ed 1986] – 2mf – 9 – 0-8370-6025-7 – (incl bibl ref) – mf#1986-0025 – us ATLA [240]

The ecclesiastical history of ireland : from the earliest period to the present times / Killen, William Dool – London: Macmillan, 1875 – 3mf – 9 – 0-7905-7055-6 – (incl bibl ref) – mf#1988-3055 – us ATLA [240]

The ecclesiastical history of new england : comprising not only religious, but also moral, and other relations / Felt, Joseph Barlow – Boston: Congregational Library Association, 1855-1862 – 4mf – 9 – 0-7905-8263-5 – (incl bibl ref) – mf#1988-6141 – us ATLA [240]

The ecclesiastical history of new england / Mather, Cotton – v. 1 and 2. 1620-98 – 1 – 46.20 – us Southern Baptist [242]

Ecclesiastical history of newfoundland / Howley, Michael Francis – Boston: Doyle and Whittle, 1888, c1887 – 1mf – 9 – 0-8370-6984-X – mf#1986-0984 – us ATLA [240]

An ecclesiastical history of scotland : from the introduction of christianity to the present time / Grub, George – Edinburgh: Edmonston & Douglas, 1861 [mf ed 1990] – 4v on 4mf – 9 – 0-7905-4857-7 – (incl bibl ref) – mf#1988-0857 – us ATLA [240]

The ecclesiastical history of socrates, surnamed scholasticus, or the advocate : comprising a history of the church in seven books, from the accession of constantine, a.d. 305, to the 38th year of theodosius 2, including a period of 140 years = Ecclesiastical history / Socrates – London: Henry G Bohn, 1853 – (= ser Bohn's ecclesiastical library) – 2mf – 9 – 0-524-00652-0 – (in english) – mf#1990-0152 – us ATLA [240]

The ecclesiastical history of sozomen : comprising a history of the church from a.d. 324 to a.d. 440 – the ecclesiastical history of philostorgius = Ekklesiastike historia / Sozomen & Philostorgius – London: H G Bohn, 1855 – (= ser Bohn's ecclesiastical library) – 2mf – 9 – 0-7905-6568-4 – (in english) – mf#1988-2568 – us ATLA [240]

Ecclesiastical index : with the rectories, vicarages, perpetual and impropriate curacies, arranged alphabetically / ed by Knox, Robert – Dublin, 1839 – (= ser 19th c ireland) – 3mf – 9 – mf#1.1.8676 – uk Chadwyck [240]

Ecclesiastical institutions : being part 6 of the principles of sociology / Spencer, Herbert – London: Williams and Norgate, 1885 – 1mf – 9 – 0-524-06231-5 – (incl bibl ref) – mf#1991-0024 – us ATLA [200]

Ecclesiastical jurisdiction : a sketch of its origin and early progress / Edwards, Edwin – London, Benning, 1853. 176 p. LL-76 – 1 – us L of C Photodup [340]

Ecclesiastical law and rules of evidence : with special reference to the jurisprudence of the methodist episcopal church / Henry, William J & Harris, William Logan – rev ed. Cincinnati: Cranston and Stowe, 1885 – 2mf – 9 – 0-524-03445-1 – (incl bibl ref and ind) – mf#1990-4705 – us ATLA [242]

Ecclesiastical law in the state of new york / Hoffman, Murray – New York: Pott and Amery, 1868. 346p. LL-1305 – 1 – us L of C Photodup [340]

The ecclesiastical law of the church of england / Phillimore, Robert, Sir; ed by Phillimore, Walter George Frank, Sir – 2nd ed. London: Sweet and Maxwell, 1895 – 5mf – 9 – 0-7905-8145-0 – (incl bibl ref) – mf#1988-6092 – us ATLA [241]

Ecclesiastical manual : or, scriptural church government stated and defended / Lee, Luther – New York: publ at the Wesleyan Methodist Book Room, 1850 [mf ed 1984] – 3mf – 9 – 0-8370-0779-8 – (incl ind) – mf#1984-4147 – us ATLA [242]

Ecclesiastical metal work of the middle ages / Arundel Society, London – London 1868 – (= ser 19th c art & architecture) – 1mf – 9 – mf#4.2.1622 – uk Chadwyck [730]

The ecclesiastical or deutero-canonical books of the old testament commonly called the apocrypha / ed by Ball, Charles James – London, New York: Eyre and Spottiswoode, [1892?] – 3mf – 9 – 0-8370-1886-2 – mf#1987-6273 – us ATLA [221]

Ecclesiastical polity : the government and communion practised by the congregational churches in the united states of america, which were represented by elders and messengers in a national council at boston, a.d. 1865 – Boston: Congregational Pub Society, 1872 – 1mf – 9 – 0-524-07580-8 – mf#1991-3200 – us ATLA [242]

The ecclesiastical polity of the new testament : a study for the present crisis in the church of england / Jacob, George Andrew – London: Strahan, 1871 – 1mf – 9 – 0-524-04577-1 – mf#1992-0165 – us ATLA [241]

Ecclesiastical reminiscences of the united states / Waylen, Edward – New York: Wiley and Putnam, 1846 – 2mf – 9 – 0-524-06968-9 – (incl bibl ref) – mf#1990-5332 – us ATLA [917]

Ecclesiastical researches / Robinson, Robert – 1972 – 1 – us Southern Baptist [242]

Ecclesiastical topography : a collection of one hundred views of churches...of london / Woodburn, Samuel – London 1807 [i.e. 1810] – (= ser 19th c art & architecture) – 5mf – 9 – mf#4.1.308 – uk Chadwyck [720]

Ecclesiastical vestments : their development and history / Macalister, Robert Alexander Stewart – London: E Stock 1896 [mf ed 1990] – (= ser The camden library) – 1mf – 9 – 0-7905-5005-9 – (incl bibl ref) – mf#1988-1005 – us ATLA [260]

Ecclesiastici concentus canendi 1, 2, 3 & 4 vocibus... / Balbi, Luigi (Aloysius) – 1606 – (= ser Mssa) – 4mf – 9 – €60.00 – mfchl 172 – gw Fischer [780]

O ecclesiastico : periodico dedicado aos interesses da religiao – Maranhao: Typ Maranhense, 01 out 1852-set 1857; dez 1860; jan-set, dez 1861; jan-30 set 1862 – (= ser Ps 19) – mf#P17,02,55 – bl Biblioteca [200]

Ecclesiasticus : the greek text of codex 248 / ed by Hart, John Henry Arthur – Cambridge: University Press, 1909 [mf ed 1991] – 4mf – 9 – 0-8370-1979-6 – (text in greek. comm in english) – mf#1987-6366 – us ATLA [221]

Ecclesiasticus / ed by Schmidt, Nathaniel – London: J M Dent 1903 [mf ed 1993] – (= ser The temple bible) – 1mf – 9 – 0-524-06034-7 – mf#1992-0747 – us ATLA [221]

Ecclesiasticus (39, 12-49, 16) : ope artis criticae et metricae in formam originalem redactus / Schloegl, Nivard – Vindobonae [Vienna]: Mayer et Sociis, 1901 – 1mf – 9 – 0-8370-5096-0 – mf#1985-3096 – us ATLA [221]

L'ecclesiastique : ou, la sagesse de jesus, fils de sira / ed by Levi, Israel – Paris: E Leroux, 1898-1901 [mf ed 1990] – (= ser Bibliotheque de l'ecole des hautes etudes. sciences religieuses 10) – 5mf – 9 – 0-8370-1855-2 – (trans fr hebrew into french and comm by ed. incl bibl ref) – mf#1987-6242 – us ATLA [221]

Ecclesiastiques du Seminaire de St-Sulpice de Montreal *see*
– Reponse a une adresse de l'assemblee legislative du 20 septembre 1852
– Reponse a une adresse de l'assemblee legislative du vingt septembre 1852
– Return to an address to the governor general
– Statement of the affairs of the corporation of the ecclesiastics of the seminary of st sulpice, montreal

Les ecclesiastiques et les royalistes francais refugies au canada a l'epoque de la revolution, 1791-1802 / Dionne, Narcisse Eutrope – Quebec: [s.n.], 1905 [mf ed 1985] – 5mf – 9 – mf#SEM105P504 – cn Bibl Nat [241]

Ecclesine, Joseph B *see* A compendium of the laws and decisions relating to mobs, riots, invasion, civil commotion, insurrection, &c., as affecting fire insurance companies in the united states

Ecclesiography : or, the biblical church analytically delineated / Manly, John G – London: Partridge & Oakey, 1852 [mf ed 1994] – 5mf – 9 – 0-665-94720-8 – mf#94720 – cn Canadiana [240]

Ecclesiological essays / Legg, John Wickham – London: Alexander Moring 1905 [mf ed 1993] – (= ser The library of liturgiology and ecclesiology for english readers 7) – 1mf – 9 – 0-524-07018-0 – (incl bibl ref) – mf#1991-2871 – us ATLA [241]

Ecclesiological notes on the isle of man, ross, sutherland and the orkneys : or, a summer pilgrimage to s maughold and s magnus / Neale, John Mason – London: Joseph Masters, 1848 [mf ed 1990] – 1mf – 9 – 0-7905-6603-6 – mf#1988-2603 – us ATLA [720]

Ecclesiology / Dargan, Edwin Charles – 1897 – 1 – us Southern Baptist [242]

Ecclesiology : a fresh inquiry as to the fundamental idea and constitution of the new testament church / Fish, E J – New York: Authors' Publishing Co., 1875 – 1mf – 9 – 0-7905-3337-5 – mf#1987-3337 – us ATLA [240]

Ecclesiology : a treatise on the church and kingdom of god on earth / Morris, Edward Dafydd – New York: Scribner, 1885 – 1mf – 9 – 0-7905-9525-7 – mf#1989-1230 – us ATLA [240]

eCFO – Boston. 2001+ (1,5,9) – mf#31862 – us UMI ProQuest [650]

Ech, H van *see* Biological aspects that influence the domestication potential of englerophytum natalense

Echange [revue linneenne] : organe des naturalistes de la region lyonnaise – Lyon 1885-1943/45 155=n1-500 on 119mf – 9 – mf#2452c/2 – ne IDC [590]

Echanges economiques et relations sociales dans deux communautes villageoises de coree / Park, Song Yong – 2mf – 9 – (10417) – fr Atelier National [330]

Echanove Trujillo, Carlos Alberto *see* Santeria cubana

Echappe de la potence : souvenirs d'un prisonnier d'etat canadien en 1838 / Poutre, Felix – [Montreal: s.n.] 1862 [mf ed 1983] – 1mf – 9 – 0-665-43088-4 – mf#43088 – cn Canadiana [971]

Echard, J *see*
– Scriptores ordinis praedicatorum

Echavarria, Colon *see*
– Epopeya de la raza
– Soldado de san cristobal
– Soneto en la danzas de juan morel campos

Echavarria Olozaga, Felipe *see*
– Historia de una monstruosa farsa
– Proceso del gobierno del 13 de junio contra felipe

Echaz-bote : pfullinger stadtanzeiger – Pfullingen DE, 1950 17 may-1958 – 20r – 1 – (bezirksausgabe von reutlinger generalanzeiger) – gw Misc Inst [074]

Echelle de vocabulaire et d'orthographe partie de l'eleve / Vinette, Roland – Montreal: editions Centre de psychologie et de pedagogie, [entre 1953 et 1956] (mf ed 1994) – 2mf – 9 – mf#SEM105P2142 – cn Bibl Nat [440]

Echelon – v1 n1-v2 n6, v3 n1-2, v4 n1, [1979 jan/feb-1980 nov/dec, 1981 may-oct, 1982 apr/jun] – 1r – 1 – mf#625955 – us WHS [071]

Echeverri, Camilo Antonio *see* Obras completas de camilo antonio echeverri

Echeverri, Elio Fabio *see* Colombia a la mano

Echeverri Mejia, Oscar *see* 21 [i.e. veintiun] anos de poesia colombiana, 1942-1963

Echeverria, Amilcar *see* Antologia de prosistas guatemaltecos

Echeverria, Aquileo J *see* Concherias, romances, epigramas y otras poemas

Echeverria Barrera, Romeo Amilcar *see* Estudio acerca de una antologia de prosistas guate

Echeverria, Frederico de *see* Spain in flames

Echeverria Loria, Arturo *see*
– Juan rafael mora
– Poesias

Echeverria Rodriguez, Roberto *see* Golgotas

Echeverria, Ventura *see* Mostaza de semilla

Echeverria y Elgueza, Santiago Joseph de *see* Nos dr d santiago joseph de echeverria y elguezua

Echevers, Malin De *see* Galope de astros

Echeverz Azlor Espinal y Valdivielso, Pedro Ignacio de *see* Relacion historica de la fundacion de este convento de neustra senora del pilar

Echeverz, Francisco Miguel de *see* Novena del glorioso san ramon nonnato del real, y militar orden de n tra sra de la merced, redencion de cautivos

Echevez, Eliseo *see* Luz en la sombra

L'echo : journal scientifique, litteraire, industriel et agricole de la ville et de l'arrondissement de Castelnaudary. – Castelnaudary. mai 1844-nov 1848, 23 juil 1850-10 mars 1852 – 1 – fr ACRPP [073]

L'echo : organe de l'union st joseph de st hyacinthe – Saint-Hyacinthe, Quebec: B de LaBruere, [1891-1915?] [mf ed v1 n1 19 mars 1891-v1 n41 31 dec 1891; v1 n43 14 janv 1892-v2 n52 16 fevr 1893] – 9 – mf#P05053 – cn Canadiana [917]

L'echo : revue des theatres, de la litterature et des arts – Paris, 1839-11 nov 1844; 1er juil-27 dec 1846; 21 24 janv 1847; 1er janv-9 mai 1848; 31 janv, 25 sept 1851 – 1 – fr ACRPP [790]

Das echo – Ilberstedt DE, 1956 6 apr-1960 23 sep – 1r – 1 – (title varies: 1960: das sozialistische echo) – gw Misc Inst [074]

Das echo – v9-12. 1906-09 [gaps] – (= ser Mennonite serials coll) – 1r – 1 – mf#ATLA 1994-S015 – us ATLA [242]

Echo – 1968-1969, 1971 aug-1984 sum – 2r – 1 – mf#410607 – us WHS [071]

Echo – Beaver Dam WI. v1 n1-156 [1930 jul 1-1931 jan 1] – 1r – 1 – mf#1062044 – us WHS [071]

Echo – Boston MA. 1877 jul 15, oct 21 – 1r – 1 – mf#871993 – us WHS [071]

Echo / Butler Co. Fairfield – dec 1970-dec 1974, apr 1977-may 1990 [wkly] – 12r – 1 – mf#B35039-35050 – us Ohio Hist [071]

Echo – Cascade, MT. 1912-1916 (1) – mf#64311 – us UMI ProQuest [071]

Echo – Cle Elum, WA. 1907-1922 (1) – mf#69230 – us UMI ProQuest [071]

Echo – [Scotland] Glasgow 9 jan-dec 1893 [mf ed 2003] – 2r – 1 – (cont as: glasgow echo [1 sep-dec 1893]) – uk Newsplan [072]

Echo – v19 n11-v20 n12 [1983 sep-1984 sep], v21 n10, [1985 jul], v22 n3 [1985 dec], v22 n4-5, 10-11 [1986 jan-feb, jul-aug], v23 n1-2 [1986 oct-nov], v23 n6-7, 9 [1987 mar-apr, jun], v24 n1 [1987 oct] – 1r – 1 – (continued by: tower echo) – mf#1546699 – us WHS [071]

Echo / Cuyahoga Co. Cleveland – may 1911-4/17, 12/18-apr 1920 [wkly] – 3r – 1 – (in german) – mf#B3314-3316 – us Ohio Hist [335]

Echo : deutsche warte in bayern (fdp) – Nuernberg DE, 1946-49 – 1 – gw Misc Inst [325]

Echo – Dryden, NY. 1889-1890 (1) – mf#68703 – us UMI ProQuest [071]

Echo / Guernsey Co. Cumberland – jan 1898-feb 1899 (all damaged pages) [wkly] – 1r – 1 – mf#B30339 – us Ohio Hist [071]

Echo / Guernsey Co. Cumberland – mar 1899-jul 1952 (poor quality film) [wkly] – 13r – 1 – mf#B1465-1477 – us Ohio Hist [071]

Echo / Guernsey Co. Cumberland – (sep 1893-apr 1895) scattered [wkly] – 1r – 1 – mf#B6809 – us Ohio Hist [071]

Echo – Haverhill, England. 1888, 1890- – 3 iss on 57+ r – 1 – uk British Libr Newspaper [072]

Echo – Hysham, MT. 1919-1974 (1) – mf#64487 – us UMI ProQuest [071]

Echo – Leavenworth, WA. 1915-1983 (1) – mf#67985 – us UMI ProQuest [071]

Echo / Milwaukee State Teachers College – ?-1951 may 23 – 1r – 1 – (cont: echo weekly; cont by: wisconsin state times) – mf#601689 – us WHS [370]

Echo – Moundsville, WV. 1891-1929 (1) – mf#67393 – us UMI ProQuest [071]

Echo – Olympia, WA. 1868-1877 (1) – mf#67053 – us UMI ProQuest [071]

Echo – Providence, RI. 1970-1992 (1) – mf#66290 – us UMI ProQuest [071]

Echo / Sandusky Co. Greenspring – sep 1901-apr 1921 [wkly] – 5r – 1 – mf#B1697-1701 – us Ohio Hist [071]

Echo / Seneca Co. Green Springs – may 1921-dec 1971, jan 1977-nov 1979 [wkly] – 23r – 1 – mf#B31781-31803 – us Ohio Hist [071]

Echo / Seneca Co. Greenspring – sep 1901-apr 1921 [wkly] – 5r – 1 – mf#B1697-1701 – us Ohio Hist [071]

Echo – Sydney, Australia. 5, 12, 26 jul 1879; 9 aug 1881; 12 feb 1883; 2 jan 1885-aug 1892 – 42 1/2r – 1 – uk British Libr Newspaper [072]

Echo – Sydney, jan 1878-dec 1879, jul 1881-jul 1893 – 30r – A$1850.07 vesicular A$2015.07 silver – at Pascoe [079]

Echo – Troy, MT. 1914-1927 (1) – mf#64670 – us UMI ProQuest [071]

Echo / Ute Mountain Ute Tribe – 1982 jan-1987 dec/1988 jan – 1r – 1 – (cont: echo (towaoc, co)) – mf#941673 – us WHS [307]

Echo – Westerly, RI. 1855-1856 (1) – mf#66428 – us UMI ProQuest [071]

Echo – Woodlake, CA. 1913-1954 (1) – mf#62306 – us UMI ProQuest [071]

Echo *see* Warnow echo

The echo – Accra, Ghana. Oct 1937-Sep 1938; Jan-Feb 1939; Oct 1950-Dec 1951; Jul 1952-Dec 1953 (imperfect) – 11r – 1 – uk British Libr Newspaper [072]

The echo – Beaver Crossing, NE: J H Waterman, jan 10 1913-13// (wkly) [mf ed v1 n2. jan 17 1913-dec 13 1913] – 1r – 1 – us NE Hist [071]

The echo – Hastings, NE: I M Augustine. v1 n1. apr 1892- (mthly) [mf ed apr 1892] – 1r – 1 – us NE Hist [071]

The echo – London, 8 Dec 1868-31 July 1905 – 76r – 1 – uk British Libr Newspaper [072]

The echo / Sellar, Thomas – [Montreal?: s.n, 1866?] [mf ed 1984] – 1mf – 9 – 0-665-45721-9 – mf#45721 – cn Canadiana [242]

The echo – Sydney, Australia. 2 Jan 1885-22 Jul 1893 – 48r – 1 – uk British Libr Newspaper [079]

The echo *see*
– Gunnison county miscellaneous newspapers

Echo academico : publicacao academico litteraria – Rio de Janeiro, RJ. 08 jun 1872 – (= ser Ps 19) – mf#P19A,04,37 – bl Biblioteca [440]

Echo [accrington] – [North West] Accrington Lib 16 dec 1884-30 aug 1887 – 1 – uk MLA; uk Newsplan [072]

Das echo am memelufer – Tilsit [Sowjetsk RUS], 1841 1 apr-1842 29 dec, 1845, 1851-52, 1894 1 may-31 aug, 1914 27 aug-17 sep, 1939 1 oct-1944 30 jun – 1 – (title varies: 1848: echo am memelufer, 1 oct 1860: tilsiter zeitung, 1 sep 1937: memelwacht) – gw Misc Inst [074]

O echo americano : jornal hebdomadario, politico, literario e noticioso – Rio de Janeiro, RJ: Typ de N Lobo Vianna & Filhos, 25 mar 1860 – (= ser Ps 19) – mf#P25,03,08 n09 – bl Biblioteca [079]

Echo and north wexford and general advertiser *see* Echo and south leinster advertiser

The echo and protestant episcopal recorder *see* The echo

Echo and south leinster advertiser – Enniscorthy, Ireland. 1987-11 mar 1988; 4 oct 1991-92 – 7r – 1 – (aka: echo and north wexford and general advertiser) – uk British Libr Newspaper [072]

Echo and sports echo – jul-dec 1993; jan 1-15 1994; jan 17-apr 30 1994; may 2-dec 1994; jan 2-apr 15 1995; apr 18-29 1995; may-dec 30 1995; 1996 – 82r – 1 – (aka: sunderland echo) – uk British Libr Newspaper [072]

L'echo annamite : ogane de defense des interets francoannamites. – Saigon. 1920-avr 1931 – 1 – fr ACRPP [325]

Echo aus der katorga : notschrei an die menschheit: sammlung authentischer briefe aus den russischen gefaengnissen / Wicher, Stanislaus [comp] – Zuerich: Buchh des Schweizer Gruetlivereins, 1914 [mf ed 2004] – 1r – 1 – (filmed with: slavianskaia problema srednei evropy / g g khristiani (1919)) – us Wisconsin U Libr [365]

O echo caxiense – Caxias, MA: Typ Imperial, jan 1852 – (= ser Ps 19) – mf#P17,01,45 – bl Biblioteca [079]

Echo da camara dos deputados – Rio de Janeiro, RJ: Typ de Gueffier & C, 19 maio-28 ago 1832 – (= ser Ps 19) – mf#P02,04,33 – bl Biblioteca [320]

Echo da juventude : propriedade da sociedade progresso – Rio de Janeiro, RJ: Typ Americana de Jose Soares de Pinho, 03 maio-03 jun 1861 – (= ser Ps 19) – bl Biblioteca [079]

Echo da juventude : publicacao dedicada a literatura – Sao Luis, MA: Typ B de Mattos, 11 dez 1864-21 maio 1865 – (= ser Ps 19) – mf#P17,02,40 – bl Biblioteca [079]

O echo da rasao – Barbacena, MG: Typ da Sociedade Typographica, 12-19 dez 1840; fev, jun-13 jul 1842 – (= ser Ps 19) – mf#P17,02,60 – bl Biblioteca [972]

O echo da verdade – Maranhao: Typ do Progresso, 26 mar 1860 – (= ser Ps 19) – bl Biblioteca [079]

L'echo d'alger – Alger, Algeria: s.n, 1956-apr 25 1961 – 26r – 1 – us CRL [079]

L'echo d'alger – Alger. Journal republicain du matin. Dir. E. Bailac. 1919-39 – 1 – fr ACRPP [079]

Echo d'alger – Algiers – 16 1/2r – 1 – uk British Libr Newspaper [072]

Echo de belgique (de stem uit belgie) – London, UK. 25 sep 1914-4 feb 1916 – 1 – (aka: stem uit belgie, 11 feb 1916-21 feb 1919) – uk British Libr Newspaper [074]

L'echo de bulgarie – Sofia, Bulgaria 24 jan, 5,14 feb, 18 mar, 24 jul, 8 aug 1916-30 sep 1918; 2 dec 1918-10 jul 1919; 20 jun 1922-30 jan 1923 (imperfect) – 1 – uk British Libr Newspaper [077]

L'echo de bulgarie – Sofia, Bulgaria. jun 1920-feb 1921 – 1r – 1 – us L of C Photodup [077]

L'echo de chine : journal des interets francais en extreme-orient – Chang-hai. 1903-sept 1919 [wkly] – 1 – fr ACRPP [073]

Echo de france – London, UK. 15 Aug-9 Dec 1914 – 1 – uk British Libr Newspaper [072]

Echo de france – London, UK. 20 Sept 1922-14 Jul 1926 – 1 – uk British Libr Newspaper [072]

Echo de la bourse – Brussels Belgium, 16 feb-20 mar 1941; apr 1942-1943; 2 oct 1944-30 jul 1945 – 3 1/2r – 1 – uk British Libr Newspaper [074]

L'echo de la chanson : ou nouveau recueil de poesies, romances, vaudevilles, etc etc – Montreal?: s.n, 1843 – 2mf – 9 – (incl ind) – mf#49046 – cn Canadiana [780]

L'echo de la fabrique : journal industriel de lyon et du departement du rhone – Lyon, oct 1831-janv 1834 – 1 – fr ACRPP [338]

Echo de la litterature et des beaux-arts dans les deux mondes – Paris. 1840-48 – 1 – fr ACRPP [800]

Echo de la tamise – London, UK. 6-13 May 1858 – 1 – uk British Libr Newspaper [072]

Echo de londres et de grande-bretagne – London, UK. 15 May 1933-28 May 1940 – 1 – uk British Libr Newspaper [072]

Echo de l'orient : journal de smyrne – Smyrne. juil 1841-juin 1846 – 1 – (et collection dedeyan. a partir du 11 aout 1846 fusionne avec le journal de constantinople pour former: journal de constantinople. echo de l'orient) – fr ACRPP [950]

Echo de minas : folha catholica, politica, litteraria e noticiosa – Ouro Preto, MG. 07 mar 1873 – (= ser Ps 19) – bl Biblioteca [079]

L'echo de nancy – Sigmaringen. nov 1944-fevr 1945 – 1 – fr ACRPP [073]

Echo de nancy – Nancy, France. 19 feb 1943-23 aug 1944 – 1r – 1 – uk British Libr Newspaper [072]

L'echo de paris – Paris: Simond, jan-sep 14 1934; apr 27-29, may 1-15, jun 24-dec 1935 – 8r – 1 – us CRL [074]

l'echo de paris – Paris. 12 mars 1884-27 mars 1938 – 1 – (le 28 mars 1938 fusionne avec: le jour) – fr ACRPP [073]

L'echo de paris litteraire illustre – Paris. n2-79. fevr 1892-aout 1893 – 1 – fr ACRPP [073]

L'echo de selestat *see* Elsaesser volkszeitung

Echo de selestat – Selestat, France. 1919-23, 1925, 1928-8 mai 1940 – 1 – fr ACRPP [074]

L'echo de stan – Stanleyville: [s.n.], jan 1959-feb 20 1960 – (Issues filmed with: Bartlett, Robert E: Collection of African newspapers) – us CRL [079]

L'echo de st-cesaire – St Cesaire: College de St-Cesaire. v1 n1 1 mars 1931-v1 n5 1 dec 1931? // (mthly, irreg) [mf ed 1984] – 1r – 1 – mf#SEM35P196 – cn Bibl Nat [073]

L'echo de terrebonne – Terrebonne, IL. v1 n1 27 janv 1917-v2 n11 21 juin 1921 (mthly) [mf ed 1973] – 1r – 1 – (interrupted: janv 1918-juin 1920; special iss: juin 1918) – cn Bibl Nat [073]

Echo der gegenwart *see* Aachener anzeiger 1848

Echo der heimat – Linz, Austria. 26 jun 1946-12 feb 1948 – 1r – 1 – uk British Libr Newspaper [072]

Echo der mark *see* Iserlohner anzeiger

Echo der woche – Duesseldorf DE, sep 10 1893-dec 29 1895 – 1r – 1 – gw Misc Inst [074]

Echo der woche / a – Muenchen DE, 1947 7 feb-1950 22 sep [gaps], 1952 17 jan-17 may – 3r – 1 – (ed for bavaria) – gw Misc Inst [074]

Echo der woche / b? – Muenchen DE, 1948 10 jan-sep, 1949-1950 15 dec – 3r – 1 – gw Mikrofilm [074]

Echo der zeit *see* Katholischer beobachter / r

L'echo des coeurs : poeme declame aux noces d'or du cardinal taschereau / Gingras, Apollinaire – [Quebec?: s.n, 1892?] – 1mf – 9 – 0-665-91504-7 – mf#91504 – cn Canadiana [440]

L'echo des instituteurs : organe de leurs sentiments et de leurs interets, ouvert a tous leurs voeux et a toutes leurs reclamations – Paris, 1845-juin 1850 – 1 – fr ACRPP [330]

L'echo des jeunes : revue eclectique – Ste-Cunegonde de Montreal. v1 n1 nov 1891- (mthly) [mf ed 1983 v1 n1,12; v2 n15,21] – 1r – 5 – (ceased 189-?) – mf#SEM16P327 – cn Bibl Nat [073]
L'echo des laboratoires : bulletin de l'association des chefs de travaux et preparateurs des facultes des sciences – Paris. n6-19. juin 1910-oct 1913 – 1 – fr ACRPP [500]
L'echo des salons de paris depuis la restauration : ou recueil d'anecdotes sur l'ex-empereur buonaparte... / Verneur, Jacques – Paris 1814/15 [mf ed Hildesheim 1995-98] – 3v on 7mf – 9 – €140.00 – ISBN-10: 3-487-26337-8 – ISBN-13: 978-3-487-26337-3 – gw Olms [944]
L'echo des syndicats agricoles – Lille, nov 1904-aout 1944 – 1 – fr ACRPP [630]
L'echo des theatres, des arts, de la litterature – Paris, oct 1861-avr 1862 – 1 – fr ACRPP [700]
L'echo d'haiti – Port-au-Prince: Imp Vve J Chenet, jul 3-31, aug 14-sep 11, sep 25-oct 2, oct 23-nov 20 1894; may 21-jul 9 1895; jan 7-jun 7, jun 30 1899 – 4r – 1 – us CRL [079]
O echo do globo : jornal noticioso e commercial – Rio de Janeiro, RJ: Typ Primeiro de Janeiro, 30 maio-03 jul 1880 – (= ser Ps 19) – mf#P18A,01,23 – bl Biblioteca [380]
Echo do imperio : jornal do commercio, lavoura, industria e litteratura – Rio de Janeiro, RJ. 07-19 jun 1884 – (= ser Ps 19) – bl Biblioteca [073]
O echo do povo – Hongkong: J J da Silva e Souza, may 15, jul 17, 31, sep 25, oct 30, 1859; apr 27 1862 – 1r – us CRL [079]
O echo do rio : jornal politico e litterario – Rio de Janeiro, RJ: Typ Imparcial de Francisco de Paula Brito, 02 ago 1843-02 mar 1844 – (= ser Ps 19) – mf#P03A,03,13 – bl Biblioteca [073]
Echo do rio s francisco – Barra, BA: Typ do Echo do Rio S Francisco, 13 maio-jun, 05 ago 1877 – (= ser Ps 19) – mf#P18B,02,53 – bl Biblioteca [321]
L'echo d'oran : journal quotidien du matin – Oran, 1907-juin 1940; mars-nov 1943; fevr 1946-56 – 1 – fr ACRPP [079]
Echo d'oran – Algeria. 17 feb 1943-27 sep 1944 – 2r – 1 – uk British Libr Newspaper [072]
O echo dos andes – Manaus, AM. 16 nov 1882 – (= ser Ps 19) – mf#P11B,06,28 – bl Biblioteca [079]
O echo dos artistas : jornal litterario, critico e recreativo – Rio de Janeiro RJ: Typ de Domingos Luiz dos Santos, 02 jun-17 nov 1861 – (= ser Ps 19) – mf#P04A,04,01 n01 – bl Biblioteca [079]
O echo dos artistas – Vitoria, ES: Typ do Echo dos Artistas, 13 jan 1878 – (= ser Ps 19) – mf#P11B,05,13 – bl Biblioteca [321]
L'echo du calvaire : ou l'association du chemin de la croix perpetuel / Provancher, Leon – Quebec?: s.n, 1883 – 1mf – 9 – mf#12224 – cn Canadiana [366]
L'echo du college : organe de l'association des anciens eleves du college de levis – Levis: [s.n.] v1 n1 27 sep 1921- (daily) [mf ed 1991-] – 1 – (cont by: echo (levis, quebec); ceased 1962?) – mf#SEM35P352 – cn Bibl Nat [370]
L'echo du college de levis – Levis: [le College] v68 n1 dec 1988- [mf ed 1991-] – 9 – (cont: echo (levis, quebec)) – mf#SEM105P1358 – cn Bibl Nat [370]
Echo du commerce – New Orleans LA. 1808 sep 28 – 1r – 1 – mf#861616 – us WHS [380]
L'echo du kivu – Bukavu, may 11-nov 6 1959 – (Issues filmed with: Bartlett, Robert E: Collection of African newspapers) – us CRL [079]
L'echo du midi : journal politique, religieux et litteraire de la haute-garonne – Toulouse. 1821-janv 1829 – 1 – fr ACRPP [073]
L'echo du nord : politique, litteraire, industriel et commercial – Lille. 1902, 1937 – 1 – fr ACRPP [073]
Echo du nord – Lille, France. 2 feb-19 mar, 11, 12 jun 1940; 14 feb 1941-1942 – 4 1/4r – 1 – (aka: grand echo du nord de la france) – uk British Libr Newspaper [074]
L'echo du parlement – Brussels, Belgium 1 jan 1884-30 sep 1885 – 4r – 1 – uk British Libr Newspaper [074]
Echo du parlement – Brussels Belgium. 1 jan 1884-30 sep 1885 – 3 1/2r – 1 – uk British Libr Newspaper [074]
L'echo du peuple – Paris: Impr de Bureau et Comp, apr 9 1848 – us CRL [074]
L'echo du peuple – Paris: Maulde et Renou. [n1-2. undated-jun 8 1848] – us CRL [074]
L'echo du rhin : journal des interets moraux et materiels des classes ouvrieres et agricoles – Strasbourg. janv-avr 1848 – 1 – fr ACRPP [073]
L'echo du rhin – Mayence. Premier quotidien francais des pays Rhenans. 1920-21. mq no. 11-14, 24, 35-39, 210-240, 353, 363, 638 – 1 – fr ACRPP [074]
L'echo du soir – Paris: Imp Serriere et Co, apr 27-28, 30, may 1 1871 – (Filmed as pt of: Commune de Paris newspapers) – us CRL [074]
L'echo du tonkin – [Hanoi?: s.n. 1897 jan 1-apr 10, apr 17-dec 29, 1898 jan 1-29, mar 12-22, may 4-28, 1899 jan 7-28, feb 4-25, apr 1-jun 14, jun 21-28, nov 4-dec 6, dec 16-23 – 1r – 1 – mf#mf-4105 seam – us CRL [079]
Echo [enniscorthy] – Wexford 1907-09, 1913-15, 1927-85 – 76r – 1 – ie National [072]
L'echo francais : journal politique et litteraire, du commerce, des sciences, arts, tribunaux, theatres, modes – Paris. 10 janv 1829-6 fevr 1847 – 1 – fr ACRPP [073]
L'echo francais : revue des journaux de France – Rio de Janeiro, RJ: Imp Parisiense, 21 dez 1849 – (= ser Ps 19) – mf#P26,04,60 – bl Biblioteca [073]
Echo francais de londres – London, UK. 2 Dec 1837-20 Jan 1838 – 1 – uk British Libr Newspaper [072]
L'echo indigene : organe hebdomadaire de la defense des interets des musulmans algeriens – Constantine. n23-64. 1934 [wkly] – 1 – fr ACRPP [325]
Echo indigene – 1934 – 1 – us CRL [073]
Echo juvenil : orgao litterario e chistoso da sociedade fraternidade juvenil – Natal, RN: Typ do Conservador, 19 ago 1883; 24 abr 1884 – (= ser Ps 19) – bl Biblioteca [079]
Echo krakowskie – Krakow, Poland. May 1948-May 1949; 1953-Jan 1960. 12 reels – 1 – us L of C Photodup [943]
L'echo (levis, quebec) : organe de l'association des anciens du college de levis – Levis: [s.n.], [ca 1962]-v67 n2 printemps 1988 [mf ed 1991] – 9 – (cont: echo du college; cont by: echo du college de levis) – mf#SEM105P1359 – cn Bibl Nat [378]
O echo litterario : periodico instructivo – Recife, PE: Typ Correio do Recife, 10 maio-15 ago 1875 – (= ser Ps 19) – bl Biblioteca [410]
Echo macaense : pao tsung hai – Macau: Francisco H Fernandez, jul 18, oct 10 1893; may 8, 22 1898 – 1r – 1 – us CRL [079]
Echo maragogipano : orgao noticioso, litterario, agricola e commercial – Maragogipe, BA: [s.n.] 10 abr"jul,out 1884; 08 set 1886 – (= ser Ps 19) – mf#P11,02,16 – bl Biblioteca [079]
Echo miguelino – Natal, RN: Typ Independente, 29 set 1874 – (= ser Ps 19) – bl Biblioteca [079]
O echo nacional : pamphleto politico por dous velhos parlamentares – Rio de Janeiro, RJ: Typ Cosmopolita, 21 jan 1882 – (= ser Ps 19) – mf#P17,01,151 – bl Biblioteca [320]
L'echo national – Paris. 10 janv 1922-15 mai 1924 – 1 – fr ACRPP [073]
L'echo national – Paris: Imp Centrale de Napoleon Chaix et cie, sep 1848 – us CRL [074]
Echo news – Echo OR: W H Crary, -1942 [wkly] – 1 – (1921-23, 1927-31 incl newspaper pub by echo high school students) – us Oregon Lib [071]
Echo news – Echo, OR: W H Crary. v4 n27-v29 n47. feb 2 1917-may 29 1942 – 1 – (aka: tortoise, and e h s mike; 1921-23, 1927-31 incl newspaper publ by echo high school students) – us Oregon Hist [071]
The echo of clairvaux – Tracadie, NS: A T McInnes, [1880-18-?] – 9 – mf#P04292 – cn Canadiana [241]
Echo of niagara *see* Niagara peninsula newspapers, pt 1
Echo of the teacher – New York. N.Y. 1915 – 1 – us AJPC [071]
O echo pernambucano : periodico nacional, politico e noticioso – Pernambuco: Typ Voz do Brasil, 08 out-nov 1850; jan-11 jul 1851 – (= ser Ps 19) – mf#P19,3,38 – bl Biblioteca [320]
Echo pilot – Greencastle, PA. -w 1895-1912 – 13 – $25.00r – us IMR [071]
Das echo, post und beobachter – Chicago: Beobachter Pub Co, 1918-apr 1920 – 2r – us CRL [071]
Echo register – Echo OR: Umatilla Pub Co, -1909 [wkly] – 1 – (cont by: stanfield standard and echo register (1909)) – us Oregon Lib [071]
Echo rossii – Stockholm, Sweden. 1918-19 – 1r – 1 – sw Kungliga [078]
Echo sant'amarense : jornal politico, commercial e agricola – Santo Amaro, BA: [s.n.] 08 out 1881; 09 out 1886 – (= ser Ps 19) – mf#P11,02,18 – bl Biblioteca [073]
L'echo sioniste – Paris. v1-6 n9. sep 1899-sep 15 1905 – 1 – us NY Public [074]
Echo suburbano – Rio de Janeiro, RJ. 24 abr-31 dez 1911 – (= ser Ps 19) – mf#DIPER – bl Biblioteca [079]
O echo suburbano – Rio de Janeiro, RJ. 03 ago-26 out 1901 – (= ser Ps 19) – mf#DIPER – bl Biblioteca [079]
Echo sud-africain – London and Paris. 4 Oct 1894-10 Dec 1896 – 1 – uk British Libr Newspaper [072]
L'echo universel – Paris.10 dec 1868; 25 nov-9 dec 1869; 20 janv-4 aout 1870, 26 sept 1871-26 juil 1874; 2 fevr 1875-31 dec 1876; 1 juil-1 sept 1877 – 1 – (journal politique, litteraire, agricole et financier) – fr ACRPP [073]
Echo von elsass-lothringen – Strassburg (Strasbourg F), 1885-1886 1 feb – 1r – 1 – gw Misc Inst [074]
Echo von new orleans – New Orleans, LA. 1870-1870 (1) – mf#63511 – us UMI ProQuest [071]
Echo weekly / Milwaukee State Teachers College – 1937 dec 8-? – 1r – 1 – (continued by: echo) – mf#700003 – us WHS [370]
Echo z afryki : pismo miesieczne ilustrowane dla popierania zniesienia niewolnictwa i dla rozszerzenia misji katolickich w afryce – Krakow: Sodalicja Sw Piotra Klawera [mthly] [mf ed 2005] – v2-57 n2 (1894-feb 1949) [gaps] on 5r – 1 – (began in 1983; publ: krakow 1894-1914; salzburg: sodalicja sw piotra klawera 1915; krakow: sodalicja sw piotra klawera 1916-30; krosno: sodalicja klawerjanska 1931-grudz 1939; st louis mar 1940-1949; some vols lacking; some pgs damaged) – mf#2005c-s076 – us ATLA [241]
Echoes – Columbus. 1976-1995 (1) 1976-1995 (9) 1976-1995 (9) – ISSN: 0012-933X – mf#9811 – us UMI ProQuest [073]
Echoes / Imperial Order of the Daughters of the Empire – Toronto, Canada. mar 1911-dec 1913; mar 1914-dec 1916; mar 1917-dec 1918; mar 1919-dec 1921; mar-dec 1922; mar 1923-dec 1926; mar 1927-dec 1930; mar 1931-dec 1934; mar 1935-dec 1938; 1941-1951 – 11 1/2r – 1 – uk British Libr Newspaper [360]
Echoes from east and west : to which are added stray notes of mine own / Datta, Roby – Cambridge: Galloway and Porter, 1909 – (= ser Samp: indian books) – us CRL [880]
Echoes from edinburgh, 1910 : an account and interpretation of the world missionary conference / Gairdner, William Henry Temple – Author's ed. New York: Fleming H Revell, [1910?] – 1mf – 9 – 0-8370-6495-3 – mf#1986-0495 – us ATLA [240]
Echoes from hell : or, light after darkness... / Givens, Nick K – St Louis, MO: Columbia Book Concern, c1904 [mf ed 1986] – 1mf – 9 – 0-8370-8509-8 – mf#1986-2509 – us ATLA [241]
Echoes from mist-land : or, the nibelungen lay: revealed to lovers of romance and chivalry by auber forestier – 2nd ed. Chicago: S C Griggs; London: Truebner, 1889, c1887 [mf ed 1996] – liv/218p – 1 – (english prose trans of nibelungenlied) – mf#9576 – us Wisconsin U Libr [390]
Echoes from old calcutta : being chiefly reminiscences of the days of warren hastings, francis, and impey / Busteed, Henry Elmsley – Calcutta: Thacker, Spink and Co; London: W Thacker and Co, 1897 – (= ser Samp: indian books) – us CRL [954]
Echoes from palestine / Mendenhall, James William – Cincinnati: Walden & Stowe, 1883 [mf ed 1984] – 8mf – 9 – 0-8370-0792-5 – (incl bibl ref & ind) – mf#1984-4137 – us ATLA [915]
Echoes from the backwoods : or, scenes of transatlantic life / Levinge, Richard George Augustus – London: J & D A Darling, 1849 – 7mf – 9 – mf#45515 – cn Canadiana [917]
Echoes from the backwoods : or, sketches of transatlantic life / Levinge, Richard George Augustus – London: J & D A Darling, 1849 [mf ed 1984] – 7mf – 9 – 0-665-45515-1 – mf#45540 – cn Canadiana [917]
Echoes from the fleeting years / Gardner, George W – 1851-1923 – 1 – 5.53 – us Southern Baptist [242]
Echoes from the orient : a broad outline of theosophical doctrines / Judge, William Quan – New York: Path, 1890 – 1mf – 9 – 0-524-07074-1 – mf#1991-0056 – us ATLA [210]
Echoes from vagabondia / Carman, Bliss – Boston: Small, Maynard, 1912 – 1mf – 9 – 0-665-77956-9 – mf#77956 – cn Canadiana [810]
Echoes in the valley – White River Junction, VT. 1984-1986 (1) – mf#64785 – us UMI ProQuest [071]
Echoes of 1916 : a message to the preachers and elders of the church of the brethren / Lepley, Daniel F – [S.l.: s.n., 1916?] – 1mf – 9 – 0-524-05548-3 – mf#1990-5152 – us ATLA [242]
Echoes of the new creation : messages of the cross, the resurrection and the coming glory / Simpson, Albert B – Brooklyn, NY: Christian Alliance, c1903 [mf ed 1992] – (= ser Christian & missionary alliance coll) – 2mf – 9 – 0-524-03329-3 – mf#1990-4275 – us ATLA [240]
Echols collection on southeast asia *see* Western books on asia
Echols, J M *see* A checklist of indonesian serials in the cornell university library (1945-1970)
Les echos – Bamako, Mali: Impr EDIM. [n6-may 26/jun 9 1989-] – us CRL [079]
Echos de la semaine – Commune de Kinshasa: Ekatou MC, feb 1960 – us CRL [079]
Les echos de la vallee de munster – Munster.Juin 1865-mars 1869 – 1 – fr ACRPP [073]
Echos do povo – Rio de Janeiro, RJ: Typ da Luz, 1872 – (= ser Ps 19) – mf#P17,01,158 – bl Biblioteca [321]
Echos d'orient – 1(1897)-39(1940) – 9 – €744.00 – ne Slangenburg [950]
Echos d'orient – Paris, 1897/1898-1941/1942 – 348mf – 9 – mf#H-2505 – ne IDC [915]
Echos d'orient : revue d'histoire, de geographie et de liturgie orientales – Bucharest: Institut francais d'etudes byzantines, 1897-1942 [mf ed 2001] – (= ser Christianity's encounter with world religions, 1850-1950) – 9r – 1 – (in french) – mf#2001-s189 – us ATLA [956]
Echos forestiers : conference donnee par m j c chapais devant la reunion annuelle de l'association des ingenieurs forestiers de quebec tenue a quebec, le 8 janvier 1918 – Quebec: [s.n.], 1918 – 1mf – 9 – 0-659-92113-8 – mf#9-92113 – cn Canadiana [634]
Echos heroi-comiques du naufrage des anglais sur l'isle-aux-oeufs en 1711 / Hugolin, pere – Quebec: [s.n.], 1910 ([Quebec]: [Imp de l'Evenement]) (mf ed 1991) – 1mf – 9 – mf#SEM105P1473 – cn Bibl Nat [780]
Echos vedettes – Montreal: Publ Quebecor. v1 n1 26 janv 1963- [mf ed 1973-] – 1 – (suppl: tele-programme; has suppl: tele) – mf#SEM35P101 – cn Bibl Nat [073]
Echos weder-klanck passende op den gheestelycken wecker tot godtvruchtighe oeffeningen... / Bie, C de – Brussel: Cl. Schoevaerts, 1706 – 4mf – 9 – mf#O-149 – ne IDC [090]
Die echte biblisch-hebraeische metrik : mit grammatischen vorstudien / Schloegl, Nivard – Freiburg im Breisgau; St Louis, MO: Herder, 1912 – (= ser Biblische studien) – 1mf – 9 – 0-7905-2426-0 – mf#1987-2426 – us ATLA [470]
Die echtheit der biloamsprueche, num. 22-24 / Wobersin, Franz – Guetersloh: C Bertelsmann, 1900 – 1mf – 9 – 0-8370-9349-X – mf#1986-3349 – us ATLA [220]
Die echtheit der ignatianischen briefe : mit einer literarischen beilage, die alte lateinische uebersetzung der usher'schen sammlung der ignatiusbriefe und des polykarpbriefes / Funk, Franz Xaver von – Tuebingen: H Laupp, 1883 [mf ed 1990] – 1mf – 9 – 0-7905-5824-6 – (in german or latin. incl bibl ref) – mf#1988-1824 – us ATLA [240]
Die echtheit des zweiten briefes petri / Grosch, Hermann – Berlin: H Grosch, 1889 – 1mf – 9 – 0-8370-9627-8 – (in german and greek) – mf#1986-3627 – us ATLA [227]
Die echtheit des zweiten thessalonicherbriefs / Wrede, William – Leipzig: J C Hinrichs, 1903 – (= ser Tugal) – 1mf – 9 – 0-7905-1739-6 – (incl bibl ref and ind) – mf#1987-1739 – us ATLA [227]
Die echtheit des zweiten thessalonicherbriefs / Wrede, William – Leipzig, 1903 – (= ser Tugal 2-24/2) – 2mf – 9 – €5.00 – ne Slangenburg [227]
Echtheit, hauptbegriff und gedankengang der messianischen weissagung, jes. 9, 1-6 / Caspari, Wilhelm – Guetersloh: C Bertelsmann, 1908 – (= ser Beitraege zur foerderung christlicher theologie) – 1mf – 9 – 0-524-06041-X – (incl bibl ref) – mf#1992-0754 – us ATLA [220]
Eck, Ernst *see*
- Die stellung des erben, dessen rechte und verpflichtungen in dem entwurfe eines buergerlichen gesetzbuches fuer das deutsche reich
- Vortraege ueber das recht des buergerlichen gesetzbuchs

Eck, Herbert Vincent Shortgrave *see*
- The incarnation
- Sin

Eck, J et al *see* Appellation fuer die 12. ort einer lobl. eydtgnoschafft wider die vermeinte disputation zu bern gehalten
Eck, Johann *see*
- Apologia pro reverendis et illvstris principibvs catholicis
- Replica ioan eckii adversvs scripta secunda buceri apostatae super actis ratis ponae

Eck, Johann Georg *see* Leipziger gelehrtes tagebuch
Eck news / Eckankar (Organization) – v1 n2-v3 n2 [1981 oct/dec-1983 spr] – 1r – 1 – mf#670069 – us WHS [071]
Eck, Samuel *see*
- David friedrich strauss
- Johann calvin

Eck segge man bloss : schwaenke und geschichten / Henze, Wilhelm – Hannover: F Gersbach, 1931 – 1r – 1 – us Wisconsin U Libr [830]
Eckankar of Southern Nevada *see* Inner-view of las vegas
Eckankar (Organization) *see* Eck news

Eckard, L W *see* Historical sketches of the missions in japan, korea
Eckardt, Andre *see* History of korean art
Eckardt, Ludwig *see*
- Schiller's jugenddramen
- Wander-vortraege aus kunst und geschichte
Eckart, Dietrich *see*
- Dietrich eckart
- Familienvaeter
- Lorenzaccio
Eckart, Rudolf *see* Die frauengestalten der heiligen schrift in der dichtung
Eckart-Helm, Martina *see* Die blaue mauritius
Eckdall, Ella *see* Major general frederick funston ("1948")
Ecke, Gustav *see*
- Die evangelischen landeskirchen deutschlands im neunzehnten jahrhundert
- Die theologische schule albrecht ritschls
- Unverrueckbare grenzsteine
Ecke, Karl *see* Schwenckfeld, luther und der gedanke einer apostolischen reformation
Eckelmann, Ernst Otto *see* Schillers einfluss auf die jugenddramen hebbels
Eckenbrecker, Margarethe Hopfer Von *see* Was afrika mir gab und nahm
Eckenstein, Lina *see* Woman under monasticism
Ecker, Jakob *see* Moko oa bibele likolong le mahae
Ecker, KR *see* Aerobic and anaerobic performance measures in active and inactive young and middle-aged males
Eckermann : schauspiel in vier akten / Lissauer, Ernst – Berlin: Oesterheld, 1921 – 1r – 1 – us Wisconsin U Libr [820]
Eckermann, Johann Peter *see*
- Conversaciones con goethe en los ultimos anos de su vida
- Conversations with eckermann
- Gespraeche mit goethe in den letzten jahren seines lebens
Eckernfoerder zeitung – Eckernfoerde DE, jan 19-dec 31 1853, 1855-65, jan 12 1867-88, feb 4 1890-may 9 1945, oct 22 1949-1990 – 1 – (filmed by other misc inst: 1980 1 mar-[ca 5r/yr]) – gw Misc Inst [074]
Eckert auf grossfahrt : fahrtenerlebnisse eines hitlerjungen / Jank, Martin – 3. Druck. Berlin: Junge Generation Verlag, [194-?] – 1r – 1 – us Wisconsin U Libr [430]
Eckert, Britta *see* Natural resource management and local knowledge in transition
Eckert, Eduard Emil *see*
- La franc-maconnerie dans sa veritable signification
- Magazin der beweisfuehrung fuer verurtheilung des freimaurer-ordens
Eckert, Georg Heinrich *see* Goethes urteile ueber shakespeare aus seiner persoenlichkeit erklaert
Eckerth, W *see* Das waltherlied
Eckhard, J Georg *see* Monatlicher auszug aus allerhand
Eckhart, Ferenc *see* Short history of the hungarian people
Eckhart, Johannes *see*
- Eine lateinische rechtfertigungsschrift des meister eckhart
- Meister eckehart
- Meister eckeharts lehre von goettlichen und geschoepflichen sein
Eckhart, Meister *see*
- Meister eckhart und seine juenger
- Meister eckharts reden der unterscheidung
- Meister eckhart's sermons
Eckhart, Meister et al *see* Texte aus der deutschen mystik des 14. und 15. jahrhunderts
Eckhart von Hochheim *see*
- Meister eckehart
- Meister eckeharts lehre von goettlichen und geschoepflichen sein
Das eckige dornach : heerichs hombroicher bauten und die kunstlehre matares / Lippka, Regine – (mf ed 1997) – 2mf – 9 – €40.00 – 3-8267-2483-6 – mf#DHS 2483 – gw Frankfurter [720]
Eckl, C *see* Attitudes towards physical activity among american and german senior citizens
Eckman, F M *see* Our first decade in china, 1905-1915
Eckman, G P *see* Studies in the gospel of john
Eckmann, Heinrich *see*
- Das bluehende leben
- Eira und der gefangene
- Der stein im acker
- Das weib und die mutter
Eckstein, Ernst *see*
- Die bildschnitzer von weilburg
- Murillo
- Nero
- Die schoene von milet
Eckstein, Ludwig *see* Die sprache der menschlichen leibeserscheinung
L'eclair – Paris. 27 oct 1888, 15 mars 1889-28 janv 1926. Fait suite a: Le Peuple. 2 dec 1888-14 mars 1889. Le 29 janv 1926 absorbe par: L'Avenir de Paris – 1 – fr ACRPP [074]
L'eclair – Port-au-Prince: [s.n., 1889-]. [1ere annee, n1-2eme annee, n4. 8 fevr 1889-14 mars 1890] – 2r – 9 – us CRL [079]
L'eclair – Paris. juin 1852-53 – 1 – (revue hebdomadaire de la litterature, des sciences et des arts paraissant tous les samedis.) – fr ACRPP [073]
Eclair – Montpellier, France. 18 feb-22 aug 1941 – 1/2r – 1 – uk British Libr Newspaper [072]
Eclair / Planard, Eugene De – Paris, France. 1839 – 1r – us UF Libraries [440]
Eclaircissemens des antiquites de la ville de nismes / Caumette, Ch – Nismes 1766 [mf ed Hildesheim 1995-98] – 1v on 1mf [ill] – 9 – €40.00 – 3-487-25910-9 – gw Olms [944]
Eclaircissemens historiques en reponse aux calomnies dont les protestans du gard sont l'objet : et precis des agitations et des troubles de ce departement, depuis 1790 jusqu'a nos jours / Lauze de Peret, Pierre J – Paris 1818 [mf ed Hildesheim 1995-98] – 3v on 5mf – 9 – €100.00 – 3-487-26321-1 – gw Olms [944]
Eclaircissements des controverses salmuriennes... / Moulin, P du – Leyden, 1648 – 4mf – 9 – mf#PRS-145 – ne IDC [240]
Eclaircissements sur les cartes du tong-king : lettres edifiantes et curieuses... – Paris, 1780-1783. v16 – 1mf – 9 – mf#HT-571 – ne IDC [915]
Eclaircissements tires des deux lettres concernant l'ambassade des hollandais...la chine en 1655 : prevost d'exiles, a f histoire generale des voyages... – Paris, 1749-1761. v19 – 1mf – 9 – mf#HT-678 – ne IDC [915]
L'eclaireur : journal democratique quotidien de saint-etienne – Saint-Etienne, 1869-28 janv 1872 – 1 – fr ACRPP [074]
L'eclaireur de l'ain – Yonnax. 30 sept 1894-aout 1939; sept 1944-juin 1951 – 1 – (subtitle varies) – fr ACRPP [073]
Eclaireur de nice et de sud est – Nice, France. 16 feb, 16 apr, 31 jul 1941; aug 1941-1 jan 1942; 10, 11 jun 1944 – 2 1/2r – 1 – uk British Libr Newspaper [072]
L'eclaireur du peuple – no. 1-19. Paris. aout 1797 – 1 – fr ACRPP [073]
L'eclaireur du peuple : ou, le defenseur de 24 millions d'opprimes – no. 1-7. Paris. mars-avr 1796 – 1 – fr ACRPP [073]
L'eclaireur haytien : ou, le parfait patriote – Port-au-Prince: F Darfour, [n4-n5. 27 aout-8 sep 1818] – 1r – 1 – us CRL [079]
The eclectic almanac for the year 1839 / Armstrong, John – 1 – us Kansas [030]
Eclectic chinese-japanese-english dictionary : of eight thousand selected chinese characters, including an introduction to the study of these characters as used in japan / Gring, Ambrose Daniel [comp] – Yokohama: Kelly; Hong Kong: Kelly & Walsh, 1884 [mf ed 1995] – (= ser Yale coll) – clxvii, 650p – 1 – 0-524-09401-2 – (with app) – mf#1995-0401 – us ATLA [040]
Eclectic magazine of foreign literature – Boston. 1844-1907 (1) – mf#4445 – us UMI ProQuest [410]
Eclectic museum of foreign literature, music and art – New York. 1843-1844 (1) – mf#4567 – us UMI ProQuest [190]
Eclectic review – London. 1805-1868 (1) – mf#4240 – us UMI ProQuest [941]
Eclesiastiche sinfonie dette canzoni in aria francese... / Banchieri, Adriano – 1607 – (= ser Mssa) – 2mf – 9 – €35.00 – mfchl 179 – gw Fischer [780]
Eclesiastico...santa...san gabriel / Santano de Membrio, Juan – 1719 – 9 – sp Bibl Santa Ana [240]
L'eclipse – Paris. 1868-juin 1876 – 1 – (puis politique et financiere. puis revue comique illustree) – fr ACRPP [073]
Eclipse / National Association for Black Veterans – 1974 may-nov, 1977 jan/feb-may/jun-1989 jan-dec/1990 jan, feb-mar, may-jun/jul, sep; 1991 jan-dec, 1992 jan-nov, 1993 jan-feb, may, jul, sep-oct, 1994 feb-apr/may, oct/nov-dec, 1995-97 – 2r – 1 – mf#1111329 – us WHS [360]
Eclipse – Paris, France. 26 jun 1868-18 sep 1870; jun 1871-31 dec 1896; 7 jan-30 dec 1897 – 13 1/2r – 1 – uk British Libr Newspaper [072]
The eclipse of faith : or, a visit to a religious sceptic / Rogers, Henry – 4th ed. Boston: Crosby, Nichols, 1853 [mf ed 1989] – 2mf – 9 – 0-7905-2869-X – mf#1987-2869 – us ATLA [230]
Eclogae geologicae helvetiae – Basel. 1989-1991 (1) – ISSN: 0012-9402 – mf#13942 – us UMI ProQuest [550]
Eclogae historicorum de rebus byzantinis (cbh1,3) / ed by Labbe, Ph – Parisiis,1648 – (= ser Corpus byzantinae historiae (cbh)) – €12.00 – ne Slangenburg [243]
The eclogues of mantuan, 1448-1516 / Mantuanus Spagnuoli, Baptista – 1567 – 9 – us Scholars Facs [450]
Ecn – Highlands Ranch. 1999+ (1) 1999+ (5) 1999+ (9) – ISSN: 1523-3081 – mf#1418,02 – us UMI ProQuest [621]
Eco – London, UK. 23, 30 Nov 1895 – 1 – uk British Libr Newspaper [072]
L'eco coloniale della new england – Springfield, MA: Itala Print & Pub Co, dec 1917-oct 1919 – 1r – us CRL [071]
Yr eco cymraeg – [Wales] Isle of Anglesey 14 oct 1899-5 sep 1914 [mf ed 2003] – 8r – 1 – uk Newsplan [072]
Eco da voz portugueza por terras de santa cruz – Rio de Janeiro, RJ: Typ de M A da Silva Lima, 01 ago-15 set 1847 – (= ser Ps 19) – mf#P15,01,69 – bl Biblioteca [321]
L'eco d'america – Providence, RI. 1941-1942 (1) – mf#66333 – us UMI ProQuest [071]
Eco de ambos mundos – London, UK. 17 May 1873-26 Mar 1874 – 1 – uk British Libr Newspaper [072]
Eco de extremadura – Badajoz.1874-76 – 9 – sp Bibl Santa Ana [079]
Eco de extremadura – Caceres.1860-61 – 9 – sp Bibl Santa Ana [074]
El eco de extremadura – Caceres, 1860-1861 – 5 – sp Bibl Santa Ana [073]
El eco de extremadura – La Habana, 1892 y 1 n° de 1894 – 5 – sp Bibl Santa Ana [073]
El eco de filipinas – Manila, Philippine Islands. 1 sep 1890-19 oct 1891 [daily] – 3r – 1 – uk British Libr Newspaper [079]
Eco de fregenal – Fregenal de la Sierra. 1880-81 – 9 – sp Bibl Santa Ana [079]
Eco de fregenal.homenaje a arias montanomy bravo murillo – 1881 – 9 – sp Bibl Santa Ana [946]
Eco de la montana – Caceres.1898-99. No. sueltos – 1.50f – 9 – sp Bibl Santa Ana [074]
El eco de la montana – Caceres, 1894 y 1899. 2 numeros – 5 – sp Bibl Santa Ana [073]
El eco de la montana – Caceres, 1898-1899 – 5 – sp Bibl Santa Ana [073]
El eco de la opinion – Santo Domingo, Dominican Republic. 1879-1897 (1) – mf#67688 – us UMI ProQuest [079]
El eco de los barros – Villafranca de los Barros, 1895. 1 numero – 5 – sp Bibl Santa Ana [073]
Eco de plasencia – Plasencia. 1895-96. No. sueltos – 9 – sp Bibl Santa Ana [079]
El eco de trujillo – Trujillo, 1908-1911 – 5 – sp Bibl Santa Ana [073]
El eco del pueblo – Zafra, 1918. 1 numero – 5 – sp Bibl Santa Ana [073]
L'eco del rhode island – Providence, RI. 1897-1930 (1) – mf#66334 – us UMI ProQuest [071]
L'eco delle valli valdesi *see* Riforma
L'eco d'italia – New York. jan 1862-nov 1894. (Not collated) (incomplete) – 1 – us NY Public [073]
L'eco d'italia : periodico notizioso e commercial – Rio de Janeiro, RJ. 14 set-10 out 1879 – (= ser Ps 19) – mf#P19A,04,12 – bl Biblioteca [079]
L'eco d'italia : rivista italo-americana – New york, jan 9 1896-dec 31 1896 [triwkly] – 1 – us NY Public [071]
Eco d'italia *see* Cronaca
El eco extremeno – Plasencia, 1906. 1 numero – 5 – sp Bibl Santa Ana [073]
Eco mexicano – Los Angeles CA. v2 n87 [1886 apr 16] – 1r – 1 – mf#865849 – us WHS [071]
Eco-bulletin / Wisconsin's Environmental Decade – 1972 feb 14, apr 3, oct/nov, 1973 apr 16, may 14 -28, jun 25-jul 23, 1974 jun 10, 1977 jul 25, 1978 apr, 1979 jan-1980 oct, 1981 dec-1983 mar, may-1984 jun – 1r – 1 – (continued by: second decade) – mf#933809 – us WHS [333]
L'ecole chretienne de seville sous la monarchie des visigoths : recherches pour servir a l'histoire de la civilisation chretienne chez les barbares / Bourret, Joseph-Christian Ernest – Paris: Charles Douniol, 1855 [mf ed 1993] – 1mf – 9 – 0-524-06348-6 – (incl bibl ref) – mf#1990-1531 – us ATLA [946]
L'ecole de dieu : pedagogy and rhetoric in calvin's interpretation of deuteronomy / Blacketer, Raymond Andrew – Grand Rapids MI: Calvin Theological Seminary, 1998 [mf ed 1999] – 1r – 1 – $130.00 – mf#1999-B002 – us ATLA [242]
Ecole de l'homme et du citoyen : journal hebdomadaire – Paris, France. 29 oct, 29 nov 1870 – 1 – (incl app: republique ou monarchie) – mf#m.misc.268 – uk British Libr Newspaper [074]
Ecole de l'homme et du citoyen – n1-2. Paris. 29 oct-26 nov 1870 [wkly] – 1 – fr ACRPP [073]
L'ecole de medecine et de chirurgie de montreal, faculte de medecine de l'universite-victoria : et la soumission aux superieurs ecclesiastiques / Amicus – Montreal?: s.n, 1879 – 1mf – 9 – mf#01650 – cn Canadiana [610]
Ecole de service social affiliee a la Faculte des sciences sociales *see*
- Service social
Ecole des contribuables / Verneuil, Louis – Paris, France. 1934 – 1r – us UF Libraries [440]
Ecole des journalistes / Girardin, Emile De – Paris, France. 1835? – 1r – us UF Libraries [440]
Ecole des vieillards / Delavigne, Casimir – Paris, France. 182-? – 1r – us UF Libraries [440]
Ecole des vieillards / Delavigne, Casimir – Paris, France. 1824 – 1r – us UF Libraries [440]
L'ecole du monde : ou instruction d'un pere a son fils, touchant la maniere dont il faut vivre dans le monde / Noble, Eustache le – 3e ed. Paris: Martin Jouvenel. 3v. 1700 – 1r – 5 – mf#SEM16P64 – cn Bibl Nat [305]
L'ecole du pur amour de dieu dans la vie...de armelle nicolas... / [Poiret, D] – Cologne, 1704 – 10mf – 9 – mf#PPE-218 – ne IDC [240]
L'ecole emancipee : Revue pedagogique hebdomadaire – Paris. oct 1910-40 – 1 – fr ACRPP [370]
Ecole francaise d'athenes : bulletin de correspondance hellenique – Paris, 1877-1946. v1-70 – 470mf – 9 – mf#NE-335c – ne IDC [930]
Ecole Francaise d'Extreme-Orient *see* Bulletin
L'ecole guineenne / Toure, Ahmed Sekou – Conakry: Imp Nationale "Patrice Lumumba", 1968 – us CRL [079]
Ecole Libre Des Sciences Politiques (Paris, France) *see* Elie halevy, 1870-1937
L'ecole militaire de quebec / Lusignan, Alphonse – Montreal?: s.n, 1864 – 1mf – 9 – mf#35225 – cn Canadiana [355]
Ecole polytechnique de montreal : rapport du principal a l'honorable surintendant de l'instruction publique / Archambault, Urgel Eugene – Montreal?: Gazette, 1881 – 1mf – 9 – mf#05214 – cn Canadiana [378]
Ecole polytechnique [Montreal, Quebec] *see* Bulletin de l'ecole polytechnique de montreal
Ecole populaire de cooperation : la region, le recrutement, l'ecole en marche, resultats et conclusions / Godbout, Leopold – Quebec: Conseil superieur de la cooperation, [1944?] (mf ed 1994) – (= ser Bibliotheque populaire de cooperation) – 1mf – 9 – mf#SEM105P2063 – cn Bibl Nat [302]
L'ecole pour la vie / Toure, Ahmed Sekou – Conakry, Republique de Guinee: Bureau de presse de la presidence de la republique, 1976 – us CRL [079]
Ecole pratique des hautes etudes [France]. Centre d'histoire et civilisations de la Peninsule indochinoise. Groupe de recherches cam *see* Essai de transliteration raisonnee du cam
L'ecole primaire : journal d'education et d'instruction – Levis [Quebec]: Mercier & Cie, 1880 [mf ed 1re annee n1 1 janv 1880-1re annee n20 15 dec 1880] – 9 – mf#P05063 – cn Canadiana [370]
L'ecole publique = The public school – Montreal: Commission des ecoles catholiques de Montreal, Office des relations publiques. v1 n[1] oct 1969-v7 n4 juin 1976 [mf ed 1973-77] – 1r – 1 – mf#SEM35P20 – cn Bibl Nat [373]
L'ecole publique *see* The public school
L'ecole rurale – [Quebec?: s.n, 1904-19–?] [mf ed 1re annee n1 sept 1904-1re annee n10 juin 1905] – 9 – mf#P05123 – cn Canadiana [370]
Les ecoles d'agriculture de la province de quebec vengees : reponse a une "etude sur l'education agricole" de l'hon louis beaubien / Directeur de l'Ecole d'agriculture de Ste-Anne – Ste-Anne de la Pocatiere Quebec: F H Proulx, 1877 – 1mf – 9 – mf#12190 – cn Canadiana [378]
Ecoles de port-au-prince / Lherisson, Leonidas Caroux – Port-Au-Prince, Haiti. 1895 – 1r – us UF Libraries [972]
Les ecoles du manitoba : la question du jour traitee par un avocat constitutionnel / Fitzpatrick, Charles – S.l: s.n, 1896? – 1mf – 9 – mf#30202 – cn Canadiana [370]
Les ecoles episcopales et monastiques en occident (afm26) : avant les universites (768-1180) / Maitre, L – 1924 – (= ser Archives de la france monastique (afm)) – €12.00 – ne Slangenburg [378]
Les ecoles et l'enseignement de la theologie pendant la premiere moitie du 12e siecle / Robert, Gabriel – Paris: V. Lecoffre, 1909 – (= ser Etudes d'histoire des dogmes et d'ancienne litterature ecclesiastique) – 1mf – 9 – 0-7905-6673-7 – (incl bibl ref) – mf#1988-2673 – us ATLA [377]
Les ecoles primaires et l'enseignement obligatoire : texte de la conference donnee, samedi, au club de reforme / Dandurand, Raoul – [Quebec (Province)?: s.n, 1918?] – 1mf – 9 – 0-665-97214-8 – mf#97214 – cn Canadiana [370]
Les ecoles primaires et les ecoles normales en france, en suisse et en belgique : rapport presente au surintendant de l'instruction publique et aux membres du comite catholique / Magnan, Charles-Joseph – Quebec: [s.n.], 1909 – 4mf – 9 – 0-665-73276-7 – (incl bibl ref) – mf#73276 – cn Canadiana [370]

L'ecolier annamite : organe qui defend les interets des annamites et veille a l'avenir de leur pays – Saigon. nov 1924 – 1 – fr ACRPP [325]
Un ecolier du dix-septieme siecle : ou, l'ideal de l'education jesuitique / Reuss, Rodolph – Dole: L Bernin, 1901 [mf ed 1986] – 1mf – 9 – 0-8370-8610-8 – (in french. incl bibl ref) – mf#1986-2610 – us ATLA [377]
Ecological abstracts – Norwich. 1984-1990 (1,5,9) – ISSN: 0305-196X – mf#42591 – us UMI ProQuest [574]
Ecological applications – Washington. 1991-2000 (1,5,9) – ISSN: 1051-0761 – mf#18050 – us UMI ProQuest [574]
Ecological characterization of the caloosahatchee river / Drew, Richard D – Metaire, LA. 1985 – 1r – us UF Libraries [574]
Ecological characterization of the florida panhandle / Wolfe, Steven H – Washington, DC. 1988 – 1r – us UF Libraries [574]
Ecological characterization of the florida springs coast – Washington, DC. 1990 – 1r – us UF Libraries [574]
Ecological characterization of the lower everglades, florida bay / Schomer, N Scott – Washington, DC. 1982 – 1r – us UF Libraries [574]
Ecological characterization of the tampa bay watershed – Washington, DC. 1990 – 1r – us UF Libraries [574]
Ecological economics – Amsterdam. 1989+ (1,5,9) – ISSN: 0921-8009 – mf#42477 – us UMI ProQuest [333]
Ecological entomology – Oxford. 1980+ (1,5,9) – ISSN: 0307-6946 – mf#15522 – us UMI ProQuest [590]
Ecological indicators – Amsterdam. 2001+ (1,5,9) – ISSN: 1470-160X – mf#42857 – us UMI ProQuest [574]
Ecological issues on reintroducing wolves into yellowstone national park / ed by Cook, Robert S – [Denver CO]: Dept of the Interior, National Park Service, 1993 [mf ed 1995] – 4mf – 9 – (incl bibl ref) – us GPO [639]
Ecological modelling – Amsterdam. 1975+ (1) 1975+ (5) 1987+ (9) – ISSN: 0304-3800 – mf#42128 – us UMI ProQuest [574]
Ecological monographs – Durham NC 1931-43 – v1-13 on 138mf – 9 – mf#2226/2 – ne IDC [574]
Ecological research – Sakura-mura. 1992-1996 (1,5,9) – ISSN: 0912-3814 – mf#19107 – us UMI ProQuest [574]
Ecological restoration – Madison, 2000+ [1,5,9] – mf#14940,02 – us UMI ProQuest [574]
Ecological Society of America *see* Bulletin of the ecological society of america
Ecological survey of isle royale, lake superior / Adams, Charles Christopher – Lansing, MI. 1909 – 1r – us UF Libraries [574]
Ecologist – London. 1979+ (1) 1979+ (5) 1979+ (9) – ISSN: 0261-3131 – mf#10186,02 – us UMI ProQuest [574]
Ecologist – Wadebridge. 1975-1977 (1) 1976-1977 (5) 1976-1977 (9) – ISSN: 0012-9631 – mf#10186 – us UMI ProQuest [574]
Eco-logos – Denver. 1972-1979 (1) 1975-1979 (5) 1975-1979 (9) – mf#9714 – us UMI ProQuest [400]
Ecology – Brooklyn. 1970-2000 (1) 1920-2000 (5) 1975-2000 (9) – ISSN: 0012-9658 – mf#6112 – us UMI ProQuest [574]
Ecology – Lancaster 1920-46 – v1-27 on 259mf – 9 – mf#5525c/2 – ne IDC [574]
Ecology and habitat protection needs of the southeastern / Stys, Beth – Tallahassee, FL. 1993 – 1r – us UF Libraries [574]
Ecology center newsletter – Berkeley. 1971-1990 (1) 1987-1988 (5) 1987-1988 (9) – mf#8190 – us UMI ProQuest [574]
Ecology law quarterly – Berkeley. 1988+ (1,5,9) – ISSN: 0046-1121 – mf#15676 – us UMI ProQuest [340]
Ecology law quarterly – University of California at Berkeley. v1-26. 1971-2000 + ind 1-10. 1971-83 – 5,6,9 – $694.00 set – (v1-12 1971-85 + ind on reel $208. v13-26 1986-2000 on mf $486) – ISSN: 0046-1121 – mf#102561 – us Hein [346]
The ecology of cyclops in south-west nigeria and their relation to the occurrence of dracunculus medinensis, the guinea-worm, in that region / Onabamiro, Sanya Dojo – 1951 – us CRL [574]
Ecology of disease – Oxford. 1982-1983 (1,5,9) – ISSN: 0278-4300 – mf#49398 – us UMI ProQuest [574]
Ecology of hydric hammocks / Vince, Susan W – Washington, DC. 1989 – 1r – us UF Libraries [574]
Ecology of tampa bay, florida–an estuarine profile / Lewis, Roy R – Washington, DC. 1988 – 1r – us UF Libraries [574]
Ecology of the florida sandhill crane / Stys, Beth – Tallahassee, FL. 1997 – 1r – us UF Libraries [574]
Ecology of the seagrasses of south florida / Zieman, Joseph C – Washington, DC. 1982 – 1r – us UF Libraries [574]
Ecology of the south florida coral reefs / Jaap, Walter C – Washington, DC. 1984 – 1r – us UF Libraries [574]
Ecology today – Mystic. 1971-1972 (1) – ISSN: 0012-9666 – mf#6603 – us UMI ProQuest [574]
Eco-news – New York. 1971-1979 (1) 1977-1979 (5) 1977-1979 (9) – ISSN: 0163-5301 – mf#9113 – us UMI ProQuest [574]
Econometrica – Evanston. 1933+ (1,5,9) – ISSN: 0012-9682 – mf#12434 – us UMI ProQuest [330]
Economia agraria colombiana / Londono Mejia, Carlos Mario – Madrid, Spain. 1965 – 1r – us UF Libraries [333]
Economia brasileira e o mundo moderno / Bastos, Humberto – Sao Paulo, Brazil. 1948 – 1r – us UF Libraries [330]
Economia brasileira no alvorecer do seculo 19 / Brito, Rodrigues De – Salvador, Brazil. 1960 – 1r – us UF Libraries [972]
Economia colonial de venezuela / Arcila Farias, Eduardo – Mexico City? Mexico. 1946 – 1r – us UF Libraries [972]
Economia do petroleo / Instituto Brasileiro De Petroleo – Rio de Janeiro, Brazil. 1959 – 1r – us UF Libraries [972]
Economia do sisal / Banco De Angola Gabinete De Estudos Economicos – Lisboa, Portugal. 1966? – 1r – us UF Libraries [960]
Economia haitiana y su via de desarrollo / Pierre-Charles, Gerard – Mexico City? Mexico. 1965 – 1r – us UF Libraries [972]
Economia minera y petrolera / Balestrini C, Cesar – Caracas, Venezuela. 1959 – 1r – us UF Libraries [972]
Economia mundial – Madrid. Spain. -w. 9 Aug 1941-29 Dec 1945. (9 reels) – 1 – uk British Libr Newspaper [330]
Economia paulista no seculo 18 / Ellis Junior, Alfredo – Sao Paulo, Brazil. 1950 – 1r – us UF Libraries [972]
Economia y cultura en la historia de colombia / Nieto Arteta, Luis Eduardo – Bogota, Colombia. 1942 – 1r – us UF Libraries [972]
Economic advisor – 1983 feb 28, may/jun-1985 jan/feb – 1r – 1 – (cont: current events & bible prophecy newsletter) – mf#1130943 – us WHS [330]
An economic analysis of usda erosion control programs : a new perspective – Washington DC: US Dept of Agriculture, Economic Research Service, 1986 [mf ed 1986] – (= ser Agricultural economic report 560) – 9 – (with bibl) – us GPO [630]
Economic and financial prospects – Basel. 1989-1996 (1) – (cont: prospects; business news survey) – ISSN: 0256-3525 – mf#11496,01 – us UMI ProQuest [338]
Economic and financial review / Federal Reserve Bank of Dallas – Dallas. 1999+ (1,5,9) – ISSN: 1526-3940 – mf#29197 – us UMI ProQuest [332]
Economic and financial survey of angola, 1960-1965 / Banco De Angola, Lisbon Departamento De Estudos Economicos – Lisboa, Portugal. 1966? – 1r – us UF Libraries [960]
Economic and political weekly – Bombay: Sameeksha Trust, [v6-15. 1971-1980] – 20r – us CRL [300]
Economic and social investigations in england since 1833 : transactions of the manchester statistical society – 186mf – 1 – us Primary [330]
Economic and social investigations in ireland : transactions of the dublin social inquiry and statistical society, 1847-1919 – 4r – 1 – us Primary [330]
Economic and technical feasibility of increased ma / Agri Research, Inc – Manhattan, KS. 1964 – 1r – us UF Libraries [972]
Economic annals of bengal / Sinha, J C – London: Macmillan and Co, 1927 – (= ser Samp: indian books) – us CRL [339]
Economic aspect of the indian rice export trade / Latif, S A – Calcutta: Das Gupta & Co, [1923] – (= ser Samp: indian books) – us CRL [380]
Economic aspects of cane sugar production / Maxwell, Francis – London, England. 1927 – 1r – us UF Libraries [338]
The economic aspects of the history of the civilization of japan / Takekoshi, Yosaburo – New York: Macmillan, 1930. 3v – 1 – us Wisconsin U Libr [330]
Economic Associates *see* Industrial supply and distribution in puerto rico
The economic benefits of a sporting event to a community / Donovan, Maura E – 1998 – 2mf – 9 – $8.00 – mf#PE 3963 – us Kinesology [650]
Economic botany – New York. 1947+ [1]; 1971+ [5]; 1977+ [9] – ISSN: 0013-0001 – mf#957 – us UMI ProQuest [580]
Economic bulletin – Accra: Economic Society of Ghana. [v1 n10-v6 n4. oct 1957-62] – us CRL [330]
Economic bulletin for asia and the far east – New York. 1950-1974 (1) 1971-1973 (5) 1972-1972 (9) – (cont by: economic bulletin for asia and the pacific) – ISSN: 0424-2653 – mf#2509 – us UMI ProQuest [332]
Economic bulletin for asia and the pacific – New York. 1974-1993 (1) 1975-1993 (5) 1975-1993 (9) – (cont: economic bulletin for asia and the far east. cont by: asia-pacific development journal) – mf#2509,01 – us UMI ProQuest [332]
Economic bulletin of ghana – Accra: Economic Society of Ghana. [v7 n1. 1963] – us CRL [330]
The economic condition of canada and her trade policy / Cartwright, Richard – Kingston, Ont: [s.n.], 1892? – 1mf – 9 – mf#02549 – cn Canadiana [330]
Economic conditions in india / Padmanabha Pillai, Purushottama – London: George Routledge and Sons, 1925 – (= ser Samp: indian books) – us CRL [330]
Economic conditions in sind, 1592-1843 / Chablani, S P – Bombay: Orient Longmans Ltd, 1951 – (= ser Samp: indian books) – us CRL [339]
Economic cooperation among the negroes of georgia : report...with the proceedings of the 22nd annual conference for the study of negro problems...atlanta university...may the 28th 1917 / ed by Brown, Thomas I – Atlanta, GA: Atlanta University Press, 1917 (mf ed 1987) – 1r – 1 – us NY Public [305]
Economic daily news – jan 1 1987-dec 31 1998 – 112r – 1 – $80.00 – ch Transmission [079]
Economic data review / Democratic Policy Committee – n129-141, 146-148, 150-151, 153-162 [1984 mar 16-jun 8, jul 13-27, aug 10-24, sep 7-dec 14], n1-5, 7-8, 10-45 [1985 jan 4-feb 1, 15-22, mar 8-dec 6], n1-15, 17-34 [1986 jan 10-apr 18, may 2-dec 5], n1-7 [1987 jan 9-feb 27] – 1r – 1 – mf#1231672 – us WHS [330]
Economic democrat : a publication of the campaign for economic democracy / Campaign for Economic Democracy (CA) – 1980 oct-1986 jun – 1r – 1 – (cont: ced news; cont by: campaign california report) – mf#1082884 – us WHS [330]
Economic development : snapshots of world-movements in commerce, economic legislation, industrialism, and technical education / Sarkar, Benoy Kumar – Madras: BG Paul & Co, 1926 – (= ser Samp: indian books) – us CRL [338]
Economic development and cultural change – Chicago. 1952+ (1) 1970+ (5) 1977+ (9) – ISSN: 0013-0079 – mf#1400 – us UMI ProQuest [338]
Economic development in africa / Nyasaland Economic Symposium, Blantyre, Nyasaland, 1962 – Oxford, England. 1966 – 1r – us UF Libraries [338]
Economic development in brunei, hong kong, malaysia, singapore, south korea and taiwan *see* Asian economic history series, series 2
Economic development of colombia – Geneva, Switzerland. 1957 – 1r – us UF Libraries [338]
Economic development of guatemala / World Bank – Baltimore, MD. 1951 – 1r – us UF Libraries [338]
The economic development of india / Anstey, Vera – London; New York: Longmans, Green and Co, 1949 – (= ser Samp: indian books) – us CRL [330]
Economic development of the transkei – Ft Hare, South Africa. 1969? – 1r – us UF Libraries [338]
Economic development quarterly – Thousand Oaks. 1987+ (1,5,9) – ISSN: 0891-2424 – mf#17051 – us UMI ProQuest [330]
Economic development review – Park Ridge. 1983+ (1,5,9) – ISSN: 0742-3713 – mf#13622 – us UMI ProQuest [338]
Economic developments in brazil, 1949-1950 / Pan American Union Division Of Economic Research – Washington, DC. 1950 – 1r – us UF Libraries [338]
Economic digest – London. 1950-1954 (1) – mf#558 – us UMI ProQuest [332]
The economic effects of government assistance in commercial resort development : a case study of french lick, indiana / Isogawa, Hiroaki & Theobald, William F – 1990 – 2mf – 9 – $8.00 – us Kinesology [790]
Economic effects of irrigation : report of a survey of the direct and indirect benefits of the godavari pravara canals / Gadgil, Dhananjaya Ramchandra – Poona: Gokhale Institute of Politics and Economics, 1948 – (= ser Samp: indian books) – us CRL [333]
Economic entomology / Murray, Andrew – London, England. 1877 – 1r – us UF Libraries [590]
Economic facts – Nanking and Chengtu, China. Jun 1936-Apr 1946 – 1 – us Chinese Res [330]
Economic fluctuations in south africa, 1910-1949 / Du Plessis, Johannes Christiaan – Stellenbosch, South Africa. 1951 – 1r – us UF Libraries [330]
Economic forum : the minority business review – 1993 jan 15-dec, 1995 jan 20-1998 dec 18 – 2r – 1 – (cont: I i courier's economic forum) – mf#2684202 – us WHS [338]
Economic framework of south africa / Hurwitz, Nathan – Pietermaritzburg, South Africa. 1962 – 1r – us UF Libraries [330]
Economic geography – Worcester. 1925+ (1) 1969+ (5) 1975+ (9) – ISSN: 0013-0095 – mf#965 – us UMI ProQuest [900]
Economic geography of the transvaal / Williams, Owen – Aberystwyth, Wales, 1950 – us CRL [330]
Economic geology and the bulletin of the society of economic geologists / Society of Economic Geologists – El Paso. 1905+ [1,5,9] – ISSN: 0361-0128 – mf#1068 – us UMI ProQuest [550]
Economic growth and stability in a developing economy / Franzsen, D G – Pretoria, South Africa. 1960 – 1r – us UF Libraries [338]
The economic history of india, 1600-1800 / Mukerjee, Radhakamal – London; New York: Longmans, Green & Co, [1945] – (= ser Samp: indian books) – us CRL [330]
The economic history of india under early british rule : from the rise of the british power in 1757 to the accession of queen victoria in 1837 / Dutt, Romesh Chunder – London: Keagan Paul, Trench, Trubner & Co, [194-?] – (= ser Samp: indian books) – us CRL [330]
The economic history of liberia / Brown, George William – Washington, DC, Associated Publishers [c1941] – us CRL [330]
Economic history of the bombay, deccan, and karnatak, 1818-1868 170=foreword by dr gadgil / Choksey, Rustom Dinshaw – Poona: RD Choksey, 1945 – (= ser Samp: indian books) – us CRL [330]
Economic history review – Oxford. 1988+ (1,5,9) – ISSN: 0013-0117 – mf#17388 – us UMI ProQuest [330]
The economic impact of dean e. smith activities center events on chapel hill, north carolina / Applegate, Michael T & Mueller, Frederick O – 1993 – 2mf – $8.00 – us Kinesology [330]
Economic impact of scientific and technical change *see* Business and financial papers, 1780-1939
Economic impact on under-developed societies / Frankel, Sally Herbert – Cambridge, MA. 1953 – 1r – us UF Libraries [339]
Economic indicators – Washington. 1948+ [1]; 1968+ [5]; 1975+ [9] – ISSN: 0013-0125 – mf#1435 – us UMI ProQuest [332]
Economic inquiry – Huntington Beach. 1974+ (1,5,9) – (cont: western economic journal) – ISSN: 0095-2583 – mf#11792,01 – us UMI ProQuest [338]
Economic intelligence / United States of America. Chamber of Commerce – 1948 aug-1962 feb – 1r – 1 – mf#1055853 – us WHS [338]
Economic Intelligence Unit *see* Country reports from the eiu on microfiche
Economic issues : studies and issue briefs of the congressional research service, 1976-1982 / U.S. Congressional Research Service – 10r – 9 – $1935.00 – 0-89093-508-4 – (with p/g) – us UPA [330]
Economic journal : the quarterly journal of the royal economic society / Royal Economic Society (Great Britain) – London. 1891+ (1) 1971+ (5) 1975+ (9) – ISSN: 0013-0133 – mf#1282 – us UMI ProQuest [338]
Economic journal – v1-3. 1968-70. Colombo, Sri Lanka. -w – 1 – us Wisconsin U Libr [330]
Economic life in the vijayanagar empire / Mahalingam, T V – Madras: University of Madras, 1951 – (= ser Samp: indian books) – us CRL [954]
Economic literature, 1851-1900 : publications from the seligman collection at columbia university – (mf ed 2001) – 50r units – us Primary [330]
Economic modelling – Amsterdam. 1984-1995 (1,5,9) – ISSN: 0264-9993 – mf#17224 – us UMI ProQuest [330]
Economic morals : four lectures / Richmond, Wilfrid – London: WH Allen, 1890 – 1mf – 9 – 0-524-02867-2 – mf#1990-0724 – us ATLA [330]
Economic notes / Labor Research Association (US) – v52-v54 [1984 jan-1986 may] – 1r – 1 – (cont: labor notes; cont by: lra's economic notes) – mf#1326836 – us WHS [331]
Economic observer – Karachi, Pakistan. -w. Jan 1951-Feb 1956. 2 reels – 1 – uk British Libr Newspaper [072]
The economic organisation of agricultural production in west africa / La-Anyane, Seth – London, 1951 – us CRL [338]
Economic organisation of yam marketing in ghana / Nyanteng, V K – Legon, 1969 – us CRL [650]

Economic origins of jeffersonian democracy / Beard, Charles A – New York: Macmillan, 1915 – 6mf – 9 – $9.00 – mf#LLMC 95-074 – us LLMC [323]
Economic outlook usa – Ann Arbor. 1974-1989 (1) 1974-1989 (5) 1975-1989 (9) – ISSN: 0095-3830 – mf#2965 – us UMI ProQuest [332]
Economic perspectives – Chicago. 1977-1982 (1,5,9) – (cont: business conditions) – ISSN: 0164-0682 – mf#11375 – us UMI ProQuest [332]
Economic perspectives – Chicago. 1977-1982 [1,5,9] – (cont by: frb chicago economic perspectives) – ISSN: 0164-0682 – mf#11375 – us UMI ProQuest [332]
Economic perspectives : a review from the federal reserve bank of chicago – Chicago. 1989+ (1,5,9) – (cont: frb chicago economic perspectives) – ISSN: 1048-115X – mf#11375,02 – us UMI ProQuest [332]
Economic Planning Seminar Of The Commonwealth Of P... *see* Proceedings
Economic policy – Cambridge. 1989-1996 (1) – ISSN: 0266-4658 – mf#16529 – us UMI ProQuest [330]
Economic policy and programme for post-war india / Sarker, Nalini Ranjan – [Patna]: Patna University, 1945 – (= ser Samp: indian books) – us CRL [339]
Economic policy review / Federal Reserve Bank of New York – New York. 1995+ (1,5,9) – (cont: federal reserve bank of new york quarterly review) – mf#21279 – us UMI ProQuest [332]
Economic policy-making and development in brazil / Leff, Nathaniel H – New York, NY. 1968 – 1r – us UF Libraries [339]
Economic politica e economia brasileira / Graca, Arnobio – Sao Paulo, Brazil. 1962 – 1r – us UF Libraries [330]
Economic problems of modern india / ed by Mukerjee, Radhakamal – London: Macmillan and Co, 1939-1941 – (= ser Samp: indian books) – us CRL [339]
Economic quarterly federal reserve bank of richmond / Federal Reserve Bank of Richmond – Richmond. 1993+ (1) 1993+ (5) 1993+ (9) – (cont: economic review federal reserve bank of richmond) – ISSN: 1069-7225 – mf#5080,02 – us UMI ProQuest [332]
Economic reconstruction of india : a study in economic planning / Sen, Khagendra Nath – Calcutta: University of Calcutta, 1939 – (= ser Samp: indian books) – (foreword by pandit jawaharlal nehru) – us CRL [339]
Economic record – East Hawthorn. 1950+ (1) 1976+ (5) 1976+ (9) – ISSN: 0013-0249 – mf#11326 – us UMI ProQuest [330]
Economic report of the president : together with the annual report of the president / U.S. President – 1947-77 – 9 – $240.00 – mf#0654 – us Brook [330]
Economic report of the president transmitted to the congress : together with the annual report of the council of economic advisers – Washington. 1947+ [1]; 1971+ [5]; 1975+ [9] – ISSN: 0193-1180 – mf#2581 – us UMI ProQuest [332]
Economic review / Federal Reserve Bank of Atlanta – Atlanta. 1977+ (1) 1977+ (5) 1977+ (9) – (cont: monthly review federal reserve bank of atlanta) – ISSN: 0732-1813 – mf#5335,01 – us UMI ProQuest [332]
Economic review / Federal Reserve Bank of Dallas – Dallas. 1986-1999 (1) 1986-1999 (5) 1986-1999 (9) – ISSN: 0732-1414 – mf#15082 – us UMI ProQuest [332]
Economic review – London. 1891-1914 (1) – mf#2882 – us UMI ProQuest [330]
Economic review, 1959-1988 – National Institute of Economic and Social Research, London – 272mf – 9 – mf#85896 – uk Microform Academic [330]
Economic review federal reserve bank of cleveland – Cleveland. 1988+ (1,5,9) – ISSN: 0013-0281 – mf#15740 – us UMI ProQuest [332]
Economic review federal reserve bank of kansas city / Federal Reserve Bank of Kansas City – Kansas City. 1978+ (1) 1978+ (5) 1978+ (9) – (cont: federal reserve bank of kansas city monthly review) – ISSN: 0161-2387 – mf#8798,01 – us UMI ProQuest [332]
Economic review federal reserve bank of richmond / Federal Reserve Bank of Richmond – Richmond. 1974-1992 (1) 1974-1992 (5) 1976-1992 (9) – (cont: monthly review federal reserve bank of richmond. cont by: economic quarterly federal reserve bank of richmond) – ISSN: 0094-6893 – mf#5080,01 – us UMI ProQuest [332]
Economic review federal reserve bank of san francisco / Federal Reserve Bank of San Francisco – San Francisco. 1975+ (1) 1975+ (5) 1976+ (9) – (cont: business review federal reserve bank of san francisco) – ISSN: 0363-0021 – mf#335,02 – us UMI ProQuest [332]
The economic review of indonesia / ed by Ministries of Economic Affairs and Agriculture – Djakarta, 1947-1953 – 30mf – 9 – mf#SE-280 – ne IDC [330]
Economic review of the bank negara indonesia – Djakarta, 1966-1972 – 32mf – 9 – (missing: 1967(8)) – mf#SE-1334 – ne IDC [332]
The economic revolution of india and the public works policy / Connell, Arthur Knatchbull – London, 1883 – (= ser 19th c books on british colonization) – 3mf – 9 – mf#1.1.4004 – uk Chadwyck [330]
Economic section circular / Commercial Advisory Foundation in Indonesia – Djakarta, [1960]-1963. nos 1-2059 – 25mf – 9 – (missing: several nos) – mf#SE-1387 – ne IDC [330]
Economic section circular cr 1-4 / Commercial Advisory Foundation in Indonesia – Djakarta, 1963-1964 – 1mf – 9 – mf#SE-277 – ne IDC [330]
Economic section circular e / Commercial Advisory Foundation in Indonesia – Djakarta, 1964-1969 – 45mf – 9 – (missing: 1964(10-12); 1964/65(16)) – mf#SE-278 – ne IDC [330]
Economic section circular fr / Commercial Advisory Foundation in Indonesia – Djakarta, 1969(1-7) – 2mf – 9 – mf#SE-1388 – ne IDC [330]
Economic section circular ta / Commercial Advisory Foundation in Indonesia – Djakarta, 1963-1964 – 3mf – 9 – (missing: 1963(1-36); 1964(45-end)) – mf#SE-681 – ne IDC [330]
Economic society bulletin – Accra: Economic Society of Ghana. [v1 n1-9. jan-sep 1957] – us CRL [330]
The economic status of black women : an exploratory investigation / U S Commission on Civil Rights – Washington: GPO 1990 – 2mf – 9 – $3.00 – mf#llmc94-334 – us LLMC [305]
The economic status of black women : an exploratory investigation / U.S. Commission on Civil Rights – Washington: GPO, 1990. LLMC 94-334 – 2mf – 9 – $3.00 – us LLMC [300]
Economic studies – New York. 1896-1899 (1) – mf#2883 – us UMI ProQuest [330]
Economic study of 249 dairy farms in florida / Mckinley, Bruce – Gainesville, FL. 1932 – 1r – us UF Libraries [636]
Economic study of absentee ownership of citrus properties in florida / Hawthorne, H W – Gainesville, FL. 1935 – 1r – us UF Libraries [634]
Economic study of celery marketing / Brunk, Max E – Gainesville, FL. 1948 – 1r – us UF Libraries [634]
Economic study of depreciation of farm machinery on one hundred thi... / Woodruff, H Toliver – s.l, s.l? 1929 – 1r – us UF Libraries [630]
Economic study of farming in the plant city area / Zentgraf, Robert Louis – s.l, s.l? 1929 – 1r – us UF Libraries [630]
Economic study of potato farming in the hastings area for the crop year / Mckinley, Bruce – Gainesville, FL. 1928 – 1r – us UF Libraries [630]
Economic study of the lake hamilton citrus growers' association / Farun, Fred Nagib – s.l, s.l, s.l? 1934 – 1r – us UF Libraries [634]
Economic study of twenty-five large citrus groves in central florida / Smith, Herbert A – s.l, s.l? 1942 – 1r – us UF Libraries [634]
Economic survey / Gosudarstvennyi Bank. SSSR – v.1-7. Jul 1926-May 1932. 9 reels – 1 – us L of C Photodup [330]
Economic survey of commercial african farming among the sala... / Rees, Am Morgan – Lusaka, Zambia. 1955 – 1r – us UF Libraries [630]
Economic survey of sierra leone / Jack, D T – Freetown: Govt Printing Office, 1958 – us CRL [330]
Economic survey of zimbabwe rhodesia 1978-1979 / Zimbabwe. Ministry of Finance – Salisbury – (= ser Economic Survey of Rhodesia) – 2mf – 9 – (preceded by: economic survey of rhodesia) – mf#FS-360 – ne IDC [330]
Economic theory – Berlin. 1991-1991 (1) – ISSN: 0938-2259 – mf#18356 – us UMI ProQuest [330]
Economic times – Bombay, India. 1962-Aug 1977; 1978 – 57r – 1 – us L of C Photodup [079]
The economic transition in india / Morison, Theodore – London: John Murray, 1916 – (= ser Samp: indian books) – us CRL [330]
Economic trends – Norwich. 1974-1992 (1) 1976-1992 (5) 1976-1992 (9) – ISSN: 0013-0400 – mf#9882 – us UMI ProQuest [330]
Economic trends and outlook / American Federation of Labor and Congress of Industrial Organizations – 1956 apr-1958, 1959-1961 mar – 2r – 1 – mf#1111337 – us WHS [331]
Economic use of tractors in florida for 1929 / Rogers, Frazier – s.l, s.l? 1930 – 1r – us UF Libraries [630]
Economic week – New York. 1974-1985 (1) 1974-1985 (5) 1976-1985 (9) – mf#10128 – us UMI ProQuest [332]
Economic x-ray – Singapore. 7 jun 1937-30 sep 1941 (wkly) – 3r – 1 – uk British Libr Newspaper [330]
Economica – London. 1989+ (1,5,9) – ISSN: 0013-0427 – mf#15741 – us UMI ProQuest [330]
Economica / Senat Mahasiswa Fakultas Ekononi UO – Djakarta, 1966-1971 – 3mf – 9 – (missing: 1966-1969; 1970/1971(jan-sep)) – mf#SE-1473 – ne IDC [330]
Economics – Hassocks. 1975-1992 (1) 1975-1992 (5) 1975-1992 (9) – (cont by: economics and business education) – ISSN: 0300-4287 – mf#10794 – us UMI ProQuest [370]
Economics and business education – Hassocks. 1993-1996 (1,5,9) – (cont: economics. cont by: teaching business and economics) – ISSN: 0969-2509 – mf#20182 – us UMI ProQuest [330]
Economics and human behaviour / Florence, Philip Sargant – New York, NY. 1927 – 1r – us UF Libraries [330]
Economics and philosophy – Cambridge. 1989-1996 (1) 1991-1991 (5) 1991-1991 (9) – ISSN: 0266-2671 – mf#16530 – us UMI ProQuest [330]
Economics and politics – Oxford. 1989+ (1,5,9) – ISSN: 0954-1985 – mf#17389 – us UMI ProQuest [330]
Economics aspects of the war – [Kingston, Ont?]: Queen's University, c1916 – 2mf – 9 – 0-659-91666-5 – mf#9-91666 – cn Canadiana [933]
Economics bulletin. queen's university *see* Economics aspects of the war
Economics letters – Amsterdam. 1978+ (1) 1978+ (5) 1987+ (9) – ISSN: 0165-1765 – mf#42129 – us UMI ProQuest [330]
Economics of agriculture in a savannah village / Haswell, Margaret Rosary – London, England. 1953 – 1r – us UF Libraries [630]
The economics of discrimination / Turgeon, Lynn – Budapest: Center for Afro-Asian Research of the Hungarian Academy of Sciences, 1973 – 1 – mf#Sc Micro R-3611 n9 – Located: NYPL – us Misc Inst [305]
Economics of education review – Elmsford. 1984+ – 1,5,9 – ISSN: 0272-7757 – mf#49473 – us UMI ProQuest [370]
The economics of jesus : or, work and wages in the kingdom of god / Griffith-Jones, Ebenezer – Cincinnati: Jennings and Graham, [1905] – 1mf – 9 – 0-8370-3396-9 – mf#1985-1396 – us ATLA [230]
Economics of khaddar / Gregg, Richard Bartlett – Madras: S Ganesan, 1928 – (= ser Samp: indian books) – us CRL [338]
Economics of khadi / Gandhi, Mahatma – Ahmedabad: Navajivan Press, 1941 – (= ser Samp: indian books) – us CRL [338]
Economics of planning – Oslo. 1990-1991 (1,5,9) – ISSN: 0013-0451 – mf#18601,01 – us UMI ProQuest [332]
The economics of war / Fortier, Adelard – [Montreal?: s.n.] 1917 [mf ed 1994] – 1mf – 9 – 0-665-72128-5 – mf#72128 – cn Canadiana [355]
L'economie du cambodge / Prud'homme, Remy – Paris: Presses universitaires de France 1969 [mf ed 1989] – (= ser Tiers monde) – 1r with other items – 1 – (with bibl) – mf#mf-10289 seam reel 027/08 [§] – us CRL [330]
Economie et finances de saint-domingue / Trouillot, Henock – Port-Au-Prince, Haiti. 1965 – 1r – us UF Libraries [332]
Economie et humanisme – Lyon. Red. en chef, R. Delprat.1942-85 – 1 – fr ACRPP [073]
Economie haitienne / Moral, Paul – Port-Au-Prince, Haiti. 1959 – 1r – us UF Libraries [330]
Economisch en sociaal tijdschrift – Antwerpen. 1977-1992 (1) 1977-1980 (5) 1977-1980 (9) – ISSN: 0013-0575 – mf#11376 – us UMI ProQuest [300]
Economisch weekblad voor nederlandsch-indie – Batavia, 1932-51 – (= ser Korte Berichten voor Landbouw, Nijverheid en Handel) – 1310mf – 9 – (preceded by: korte berichten voor landbouw, nijverheid en handel. buitenzorg, 1910-32 v1-22) – mf#SE-28 – ne IDC [330]
economische opstellen uit de inheemsche pers *see* 1937 (1 jan-dec)
Economist – Aurora, IL. 1940-1957 (1) – mf#62499 – us UMI ProQuest [071]
Economist – Leiden. 1987-1996 (1,5,9) – ISSN: 0013-063X – mf#14484 – us UMI ProQuest [330]
Economist – London. 1843+ (1) 1969+ (5) 1971+ (9) – ISSN: 0013-0613 – mf#1011 – us UMI ProQuest [332]
Economist : a periodical paper explanatory of the new system of society projected by robert owen – v1-2. 1821-22 [all publ] – (= ser Radical periodicals of great britain, 1794-1914. period 1) – 10mf – 9 – $115.00 – us UPA [330]
Economist – Toronto, Canada. apr 1914-nov 1918; jan 1919-dec 1921 – 3r – 1 – uk British Libr Newspaper [071]
Economist – Vandergrift, PA. 1927-1927 (1) – mf#66111 – us UMI ProQuest [071]
The economist – London. -w. 1843-49. (12 reels) – 1 – uk British Libr Newspaper [330]
The economist – Toronto, Canada. Apr 1914-Dec 1921.-w. 2mqn reels – 1 – uk British Libr Newspaper [072]
The economist – Toronto: Economist Print and Pub Co, [1897?-19–] – 9 – mf#P05062 – cn Canadiana [360]
Economist newspapers – Chicago, IL. 1906-1924 (1) – mf#62551 – us UMI ProQuest [071]
Economist newspapers – Chicago, IL. 1907-1968 (1) – mf#62558 – us UMI ProQuest [071]
Economist newspapers – Chicago, IL. 1913-1967 (1) – mf#62545 – us UMI ProQuest [071]
Economist newspapers – Chicago, IL. 1915-1923 (1) – mf#62542 – us UMI ProQuest [071]
Economist newspapers – Chicago, IL. 1920-1923 (1) – mf#62556 – us UMI ProQuest [071]
Economist newspapers – Chicago, IL. 1924-1925 (1) – mf#62548 – us UMI ProQuest [071]
Economist newspapers – Chicago, IL. 1924-1931 (1) – mf#62563 – us UMI ProQuest [071]
Economist newspapers – Chicago, IL. 1924-1987 (1) – mf#67994 – us UMI ProQuest [071]
Economist newspapers – Chicago, IL. 1929-1967 (1) – mf#62560 – us UMI ProQuest [071]
Economist newspapers – Chicago, IL. 1930-1932 (1) – mf#62564 – us UMI ProQuest [071]
Economist newspapers – Chicago, IL. 1932-1947 (1) – mf#62543 – us UMI ProQuest [071]
Economist newspapers – Chicago, IL. 1935-1966 (1) – mf#62553 – us UMI ProQuest [071]
Economist newspapers – Chicago, IL. 1936-1972 (1) – mf#62561 – us UMI ProQuest [071]
Economist newspapers – Chicago, IL. 1953-1965 (1) – mf#62544 – us UMI ProQuest [071]
Economist newspapers – Chicago, IL. 1958-1967 (1) – mf#62546 – us UMI ProQuest [071]
Economist newspapers – Chicago, IL. 1959-1965 (1) – mf#62557 – us UMI ProQuest [071]
Economist newspapers – Chicago, IL. 1962-1968 (1) – mf#62566 – us UMI ProQuest [071]
Economist newspapers – Chicago, IL. 1969-1970 (1) – mf#62555 – us UMI ProQuest [071]
Economist newspapers – Chicago, IL. 1969-1970 (1) – mf#62547 – us UMI ProQuest [071]
Economist newspapers – Chicago, IL. 1969-1985 (1) – mf#62550 – us UMI ProQuest [071]
Economist newspapers – Chicago, IL. 1969-1987 (1) – mf#62559 – us UMI ProQuest [071]
Economist newspapers – Chicago, IL. 1969-1988 (1) – mf#62565 – us UMI ProQuest [071]
Economist newspapers – Chicago, IL. 1971 (1) – mf#62549 – us UMI ProQuest [071]
Economist newspapers – Chicago, IL. 1986-1988 (1) – mf#68264 – us UMI ProQuest [071]
Economist newspapers – Chicago, IL. 1990-1997 (1) – mf#68263 – us UMI ProQuest [071]
Economist newspapers – Chicago, IL. 1993-1994 (1) – mf#69005 – us UMI ProQuest [071]
Economist newspapers – Chicago, IL. 1995-1999 (1) – mf#68267 – us UMI ProQuest [071]
Economist para america latina – London. 1967-1970 (1) – mf#6026 – us UMI ProQuest [338]
El economista – (El Economista confidencial). Madrid. Spain. -w. 18 Oct 1958-31 Dec 1960. (Imperfect). (5 reels) – 1 – uk British Libr Newspaper [330]
L'economiste canadien-francais : organe officiel de l'union franco-canadienne – Montreal: [s.n., 1900?-19–?] [mf ed v1 n7 dec 1900] – 9 – mf#P05128 – cn Canadiana [330]
L'economiste europeen – Paris. v1-3. 1892-jun 1952 – 1 – us NY Public [330]
L'economiste haitien – Port-au-Prince: Francois Dalencour. [v1 n2. jan/feb 1940] – 1r – 9 – us CRL [079]
Economists at home and abroad / Iyengar, S Kesava – Hyderabad: Indian Institute of Economics, 1953 – (= ser Samp: indian books) – us CRL [330]
Economists' papers – [mf ed Marlborough 1991] – 3 ser – 1 – (ser 1: papers of william stanley jevons 1835-82 fr john rylands university library of manchester [25r] £2300; series 2: diaries of john neville keynes 1864-1917 fr cambridge university library [12r] £1125; series 3: papers of carl menger, 1840-1921 fr william r perkins library, duke university 2pts – pt1: notebooks, notes on economic principles & notes on money [21r] £1950, pt2: lectures, notes on methodology, correspondence, biographical materials, miscellanea & printed matter (incl the ann grundsaetze) [21r] £1950; with d/g) – uk Matthew [330]
Economy of a south indian temple / Ramakrishna Ayyar, Viravanallur Gopalier – Annamalainagar, India. 1946 – 1r – us UF Libraries [280]
Economy of brazil / Ellis, Howard Sylvester – Berkeley, CA. 1969 – 1r – us UF Libraries [330]

The economy of cambodia and its problems with industrialization = L'economie du cambodge et ses problemes d'industrialisation / Khieu Samphan – Universite de Paris 1959 [mf ed 1989] – 1r with other items – 1 – mf#mf-10289 seam reel 024/04 [§] – us CRL [338]

The economy of permanence : a quest for a social order based on non-violence / Kumarappa, Joseph Cornelius – Wardha, CP: All India Village Industries Association, 1946- – (= ser Samp: indian books) – (foreword by m k gandhi) – us CRL [301]

The economy of the central barotse plain / Gluckman, M – (= ser Institute for social research, university of zambia. papers 7) – 4mf – 7 – (4mf) – mf#363/6 – uk Microform Academic [960]

Ecorres, Charles des *see* Au pays des etapes

Ecos de amor y dolor / Fabricio Diaz, Francisco – Habana, Cuba. 1952 – 1r – us UF Libraries [972]

Ecos de los andes / Samper, Jose Maria – Paris, France. 1860 – 1r – us UF Libraries [972]

Ecos de silencio; poesia y cuentos / Duran Castillo, Benito – Habana, Cuba. 1956 – 1r – us UF Libraries [972]

Ecos de una guerra a muerte / Justiz Y Del Valle, Tomas Juan De – Habana, Cuba. 1941 – 1r – us UF Libraries [972]

Ecos del continente – Miami, FL. 1980 nov 01-1991 jan 01 – 1r – us UF Libraries [071]

Ecos del paraiso / Sanchez-Arjona, Vicente – Sevilla: Imp. Alvarez, Tomo 1. 1955 – 1 – sp Bibl Santa Ana [810]

Ecos del paraiso / Sanchez-Arjona, Vicente – Sevilla: Imp. Carlos Acuna, Tomo 3. 1955 – 1 – sp Bibl Santa Ana [810]

Ecos del paraiso / Sanchez-Arjona, Vicente – Sevilla: Imprenta Alvarez, Tomo 2. 1955 – 1 – sp Bibl Santa Ana [810]

Ecos perdidos / Diaz Perez, Nicolas – Poesias varias. 1881 – 9 – sp Bibl Santa Ana [810]

Ecotass : [deutsche ausgabe] – Paris. 1984-1988 (1) 1984-1988 (5) 1986-1988 (9) – ISSN: 0733-5997 – mf#49416 – us UMI ProQuest [332]

Ecotass : [edition francaise] – Paris. 1984-1990 (1) 1984-1990 (5) 1986-1990 (9) – ISSN: 0733-5970 – mf#49414 – us UMI ProQuest [332]

Ecotass : english edition – Oxford. 1984-1993 (1) 1984-1993 (5) 1987-1993 (9) – ISSN: 0733-5989 – mf#49415 – us UMI ProQuest [332]

Ecotass : [italian ed] – Elmsford. 1984-1990 (1) 1983-1990 (5,9) – ISSN: 0736-8429 – mf#49453 – us UMI ProQuest [332]

L'ecouteur aux portes – Paris. n1-2. oct 1789 – 1 – fr ACRPP [073]

Ecrin / Duport, Paul – Paris, France. 1843 – 1r – us UF Libraries [440]

Ecrin d'amour familial : details historiques au sujet d'une famille, comme il y en a tant d'autres au canada qui devraient avoir leur histoire / Beaubien, Charles Philippe – Montreal: Arbour & Dupont, 1914 – 4mf – 9 – 0-665-71645-1 – (incl bibl ref and ind) – mf#71645 – cn Canadiana [929]

Ecrin de la jeunesse – Montreal: Librairie Saint-Joseph, Cadieux & Derome, 1885 [mf ed 1979] – (= ser Bibliotheque religieuse et nationale.serie petit in 12) – 2mf – 9 – mf#SEM105P29 – cn Bibl Nat [241]

Ecriteaux : ou, rene le sage a la foire saint-germa / Barre, M – Paris, France. 1806 – 1r – us UF Libraries [440]

Les ecrits de monseigneur arthur robert : bibliographie analytique / Trottier, Guy N – 1955 [mf ed 1978] – (= ser Bibliographies du cours...1947-66) – 3mf – 9 – (with ind; pref by alexandre robert) – mf#SEM105P4 – cn Bibl Nat [241]

Les ecrits de saint paul – Paris: F Rieder 1926-28 [mf ed 1987] – 4v on 2r – 1 – (trans, int & ann by henri delafosse. filmed together: n1: freeing the human mind / harry elmer barnes [1931] & other titles) – mf#1889 – us Wisconsin U Libr [225]

Ecrits des cures de paris contra la politique et la morale des jesuites (1656) : avec une etude sur la querelle du laxisme / Recaulde, l de – Paris: Editions et librairie, 1921. Chicago: Dep of Photodup, U of Chicago Lib, 1972 (1r); Evanston: American Theol Lib Assoc, 1984 (1r) – 1 – 0-8370-1497-2 – (incl bibl ref) – mf#1984-B061 – us ATLA [241]

Les ecrits du docteur jean-baptiste meilleur : premier surintendant de l'instruction publique du bas-canada, 1843-1855, et fondateur du college de l'assomption, 1832 / Olivier, Rejean – Joliette: edition privee, 1993 [mf ed 1996] – (= ser Bibliographies du cours...1947-66) – 6mf – 9 – (with ind) – mf#SEM105P2659 – cn Bibl Nat [370]

Les ecrits francais – Dir. L. de Monti de Reze, M. Bresil, L. de Gonzague, Frick. no. spec. de nov 1913, no. 1-5. Paris. dec 1913-avr 1914 – 1 – fr ACRPP [800]

Les ecrits nouveaux – Paris. nov 1917-22. devenu: La Revue europeenne voir a ce titre – 1 – fr ACRPP [800]

Ecrits pour l'art – Fond. Gaston Dubedat. Paris. 1887-92, avr 1905 – 1 – fr ACRPP [700]

Ecrits pour l'art – Paris, 1887 [mf ed Chadwyck-Healey] – (= ser Art periodicals on microform) – 1r – 1 – uk Chadwyck [700]

Ecrits sur la bienheureuse marguerite bourgeoys, 1945-1962 : bibliographie analytique / Sainte-Zelia, soeur – 1962 [i.e. 1963] (mf ed 1978) – (= ser Bibliographies du cours...1947-66) – 3mf – 9 – (with ind; pref by soeur sainte-mechtilde-du-saint-sacrement) – mf#SEM105P4 – cn Bibl Nat [241]

Les ecritures coneiformes : expose des travaux qui ont prepare la lecture et l'interpretation... / Menant, Joachim – Paris: Benjamin Duprat, 1860 – 1mf – 9 – 0-8370-7651-X – (incl bibl ref) – mf#1986-1651 – us ATLA [470]

Les ecritures manicheennes / Alfaric, Prosper – Paris: Emile Nourry, 1918-19 [mf ed 1993] – 2v on 1mf – 9 – 0-524-08400-9 – (incl bibl ref) – mf#1993-4010 – us ATLA [290]

Ecroyd, William Farrer *see* The policy of self help

Ecstasy : the release of the soul from the body / Crookall, Robert – 1st ed. Moradabad: Darshana International, 1973 – us CRL [140]

Ectj : educational communication and technology – Washington. 1978-1988 (1,5,9) – (cont: av communication review) – ISSN: 0148-5806 – mf#1466,01 – us UMI ProQuest [370]

Ecuador *see*
- Gaceta del gobierno del ecuador
- Registro oficial

Ecuador. Direccion de Estadistica y Censos *see* Sintesis estadistica del ecuador 1955/1960-1955/1962

Ecuador. Direccion General de Estadistica *see* Ecuador en cifras 1938-1942

Ecuador en cifras 1938-1942 / Ecuador. Direccion General de Estadistica – (= ser Latin american & caribbean...1821-1982) – 6mf – 9 – uk Chadwyck [318]

Ecuador. Instituto Nacional de Estadistica *see* Anuario de estadistica 1963/1968-1966/1971

Ecuador. Instituto Nacional de Estadistica *see* Serie estadistica 1967/1972-1968/1973, 1974-1976

Ecuador. Ministerio de Gobiern *see* Informe...

Ecuador. Ministerio de Gobierno *see* Informe

Ecuador. Ministerio de Gobierno y Prevision Social *see* Informe...

Ecuador. Ministerio de Hacienda *see*
- Esposicion que dirije al congreso del ecuador en...el ministro de estado en el despacho de hacienda
- Esposicion que el ministro de hacienda del ecuador presenta a las camaras lejislativas reunidas en...
- Exposicion del ministro de hacienda a las camaras legislativas de...
- Informe a la nacion
- Informe anual del ministro de hacienda y credito publico
- Informe del ministro de hacienda...
- Informe del ministro de hacienda a la nacion
- Informe del ministro de hacienda y credito publico
- Informe del ministro de hacienda...a la h asamblea nacional refutando el presentado por...
- Informe del senor ministro de hacienda y credito publico al h congreso nacional
- Informe que el ministro de hacienda, credito publico, bancos, minas, comercio y marcas de fabrica presenta a la nacion
- Informe que...ministro de hacienda, credito publico, etc, presenta a la nacion en...
- Informe...presenta a la nacion
- Informe...presenta a la nacion y a sus representantes al congreso de...
- Memoria...
- Memoria...al congreso constitucional de...

Ecuador. Ministerio de Hacienda. *see* Informe del subsecretario de hacienda a la convencion nacional de...

Ecuador. Ministerio de lo Interior y Relaciones Exteriores *see*
- Esposicion...
- Exposicion...
- Informe...
- Memoria del subsecretario...a la convencion nacional de...

Ecuador Ministerio De Relaciones Exteriores *see* Posicion del ecuador en el conflicto colombo-perua

Ecuador. Ministerio de Relaciones Exteriores *see*
- Informe a la nacion...
- Informe del ministro de relaciones exteriores al congreso ordinario de...

Ecuador. Ministerio del Interior, Relaciones Esteriores e Instruccion Publica *see* Esposicion...dirijida a las camaras lejislativas del ecuador en...

Ecuador. Ministerio del Tesoro *see*
- Informe a la nacion
- Informe...presenta a la nacion

Ecue-yamba-o! / Carpentier, Alejo – Madrid, Spain. 1933 – 1r – us UF Libraries [972]

Ecumenical Center for Draft Counseling *see* Wisconsin draft counselor's newsletter

The ecumenical councils / Dubose, William Porcher – 3rd ed. New York: Scribner, 1900, c1897 – (= ser Ten Epochs of Church History) – 1mf – 9 – 0-524-00631-8 – mf#1990-0131 – us ATLA [240]

Ecumenical missionary conference, new york, 1900 : report of the ecumenical conference on foreign missions, held in carnegie hall and neighboring churches, april 21 to may 1 – New York: American Tract Society; London: Religious Tract Society, c1900 – 4mf – 9 – 0-8370-6112-1 – (includes appendix and index) – mf#1986-0112 – us ATLA [240]

Ecumenical press service : papers – 1946-76; 1984-91 – 1 – (missing pp) – mf#atla s0581 – us ATLA [240]

The ecumenical reconstruction of christianity: an exposition and evaluation of the christology of john hick / Miles, S Daniel – 1982 – 1 – 5.00 – us Southern Baptist [242]

Ecumenical review – Geneva. 1948+ (1) 1970+ (5) 1975+ (9) – ISSN: 0013-0796 – mf#1003 – us UMI ProQuest [240]

Ecumenical trends – Garrison. 1972+ (1) 1976+ (5) 1976+ (9) – ISSN: 0360-9073 – mf#9341 – us UMI ProQuest [240]

Ecumenism – Montreal. 1986+ (1,5,9) – ISSN: 0383-431X – mf#15163,02 – us UMI ProQuest [240]

Ecumenist – New York. 1962-1992 (1) 1971-1992 (5) 1976-1992 (9) – ISSN: 0013-080X – mf#6848 – us UMI ProQuest [240]

Ed l huntley's panegyric on the jews / Huntley, Ed L – Chicago: Ed L Huntley, [1891] – 1mf – 9 – 0-8370-3699-2 – mf#1985-1699 – us ATLA [270]

Edad pre-escolar / Barrera Moncada, Gabriel – Caracas, Venezuela. 1954 – 1r – us UF Libraries [972]

Edcentric – Eugene. 1970-1979 (1) 1974-1979 (5) 1975-1979 (9) – ISSN: 0046-1245 – mf#7172 – us UMI ProQuest [370]

Edd hayes' black college sports reports – 1991 nov, 1993 mar, sep – 1r – 1 – mf#4713484 – us WHS [790]

Edda Saemundar *see* Poetic edda

Eddins, A H *see*
- Brown rot of irish potatoes and its control
- Corn diseases in florida
- Potato diseases in florida

Eddins, William C, Jr *see* A comparison of bone mineral density between active and nonactive men with spinal cord injuries

Eddis, William *see* Letters from america, historical and descriptive

Eddowes's journal – Shrewsbury, England. 1849-52.-w. 2 reels – 1 – uk British Libr Newspaper [072]

Eddy, Daniel Clarke *see*
- Heroines of the missionary enterprise
- The memorial sermon preached in the baldwin-place meeting house on the last sabbath of its occupancy by the second baptist church
- Roger williams and the baptists
- A sketch of adoniram judson, dd, the burman apostle

Eddy, Elizabeth Ann (Berryman) *see* Papers

Eddy, George Sherwood *see* The new era in asia

Eddy, Mary Baker *see*
- Manual of the mother church, the first church of christ, scientist, in boston, massachusetts
- Miscellaneous writings, 1883-1896
- Pulpit and press
- Retrospection and introspection
- Science and health

Eddy, Richard *see*
- A history of the unitarians and the universalists in the united states
- Universalism in america

Eddy, Sherwood *see*
- Facing the crisis
- India awakening
- Japan and india

Eddystone : [a novel] / Jensen, Wilhelm – 2. aufl. Berlin: Paetel, 1894 [mf ed 1991] – 200p – 1 – mf#7502 – us Wisconsin U Libr [830]

The eddyville enterprise – Eddyville, NE: W C Bryner, 1906-20// (wkly) [mf ed 1911-20 (gaps)] – 2r – 1 – (some irregularities in numbering) – us NE Hist [071]

Eddyville first baptist church. eddyville, kentucky : church records – 1929-89. 1095p – 1 – 49.28 – us Southern Baptist [242]

Edeb yahu – Istanbul, 1908-09. Sahib-i Imtiyaz: Mehmed Remzi; Mueduer-i Mes'ul: Taslizade Hasan Ruesdue, Osman Nuri. n16. 17 kanunisani 1324 [1908] – (= ser O & t journals) – 1mf – 9 – $25.00 – us MEDOC [956]

Edebiyat-i umumiye mecmuasi – Istanbul: Kanaat Kuetuephanesi ve Matbaasi, 1916-19; Mueduer-i Mes'ul: Mehmed Celaleddin, Mueduer: Giridi Amed Saki. n1,2,4-20,22,25-32,38,57,59-65,67-68,70-71,76,109. 22 tesrinievvel 1332 [1916]-1 mart 1919 – (= ser O & t journals) – 13mf – 9 – $250.00 – us MEDOC [956]

Edel, May Mandelbaum *see* Papers, 1890-1939

Edelen family newsletter – v1-2 n5 [1972 jan-1973 sep] – 1r – 1 – mf#366593 – us WHS [929]

Edelhaeuser, Friedrich *see* Aufbau und evaluation eines messplatzes zur quantifizierenden erfassung von spastik

Edelmann, J F *see* Six sonates pour le clavecin avec accompagnement d'un violon ad libitum... oeuvre 1er

Edelmann, Mordecai Isaac *see* Mishle ha-talmud

Edelstein, Charles Louis *see* Role of cysteine proteases in hypoxia-induced renal proximal tubular injury

Der edelstein (cima4) : farbmikrofiche-edition der handschrift basel, oeffentliche bibliothek der universitaet basel, hs a n 3 17 / Boner, Ulrich – (mf ed 1987) – (= ser Codices illuminati medii aevi (cima) 4) – 27p on 3 color mf – 15 – €185.00 – 3-89219-004-6 – (int by klaus grubmueller. description by ulrike bodemann) – gw Lengenfelder [090]

Der edelstein / des teufels netz / sibyllenweissagung [cima7] : mikrofiche-edition der handschrift augsburg, universitaetsbibliothek, cod. i.3.2°3 / Boner, Ulrich – [mf ed 1987] – (= ser Codices illuminati medii aevi [cima] 7) – 3 color+4 b/w mf – 15,9 – €240.00 – 3-89219-007-0 – (description by ulrike bodemann) – gw Lengenfelder [090]

Eden – (New York). 1924-25 – 1 – us AJPC [830]

Eden, Charles Henry *see* China, historical and descriptive

Eden, Emily *see* Letters from india

Eden lost and won : studies of the early history and final destiny of man as taught in nature and revelation / Dawson, John William – New York: Fleming H Revell, 1896 – 1mf – 9 – 0-8370-2850-7 – mf#1985-0850 – us ATLA [220]

Eden, Michael John *see* Savanna ecosystem–northern rupununi, british guia...

Eden of the south / Webber, Carl – New York, NY. 1883 – 1r – us UF Libraries [978]

Eden, Robert *see*
- Church of scotland
- National church of england

The eden tableau, or, bible object-teaching : a study / Beecher, Charles – Boston: Lee and Shepard, 1880 – 1mf – 9 – 0-8370-2233-9 – mf#1985-0233 – us ATLA [220]

Eden underground news service – n1-45 [1976 apr 19-1977 dec 22] – 1r – 1 – mf#366600 – us WHS [071]

Eden union *see* Candelo / eden union / southern auckland advocate

Eden versus whistler the baronet and the butterfly – Paris [1899] – (= ser 19th c art & architecture) – 2mf – 9 – mf#4.2.744 – uk Chadwyck [750]

Eden, W *see* The history of new holland

Eden's land and garden with their marks yet to be seen / West, Landon – Pleasant Hill, Miami Country, Ohio: [s.n.], 1908 – 1mf – 9 – 0-524-04720-0 – mf#1990-5072 – us ATLA [220]

Eder ha-yekar – Tel-Aviv, Israel. 1947 – 1r – us UF Libraries [939]

Ederberg, B *see* Lahkusud eestis

Edersheim, Alfred *see*
- Elisha the prophet
- The life and times of jesus the messiah
- Prophecy and history in relation to the messiah
- Sketches of jewish social life in the days of christ
- The temple, it's [sic] ministry and services as they were at the time of jesus christ
- Tohu-va-vohu

Edersheim, Elisa Williamina *see* The laws and polity of the jews

Eder-Stein, Irmtraut *see* Reichskunstwart (bestand r 32) bd 16

Die edessenische abgar-sage / Lipsius, Richard Adelbert – Braunschweig: C.A. Schwetschke, 1880 – 1mf – 9 – 0-7905-5371-6 – (incl bibl ref) – mf#1988-1371 – us ATLA [470]

Edfeldt, Hans *see* Om bevisen foer guds verklighet

Edgar : dramma lirico in tre atti / Puccini, Giacomo – Milano: Ricordi 1944 [mf ed 1994] – 1r – 1 – (italian words) – mf#pres. film 133 – us Sibley [780]

Edgar allan poe : how to know him / Smith, Charles Alphonso – Garden City, NY. 1921 – 1r – us UF Libraries [420]

Edgar allan poe / Ingram, John Henry – London, England. v1-2. 1880 – 1r – us UF Libraries [420]

Edgar allan poe – New York, NY. 1945 – 1r – us UF Libraries [420]

Edgar, Andrew *see*
- The bibles of england
- Bibles of england
- Old church life in scotland
- Old church life in scotland. second series

Edgar, dramma lirico in quattro atti di ferdinando fontana : musica di giacomo puccini. teatro alla scala, carnevale-quaresima 1888-89 / Puccini, Giacomo – Milano: R Stabilimento Tito di Gio Ricordi...& c [1889] [mf ed 19–] – 1r – 1 – (libretto for the original production[1889] incl add & corr by puccini) – mf#pres. film 34 – us Sibley [780]
Edgar enterprise – Edgar WI. 1901 may 14 – 1r – 1 – mf#875426 – us WHS [071]
Edgar, Frank *see* Litafi na tatsuniyoyi na hausa
The edgar index – Edgar, NE: Osborne Sisters, sep 1898-jan 1900// (wkly) [mf ed -1899 (gaps)] – 1r – 1 – us NE Hist [071]
Edgar, James *see* The herald of zion
Edgar, James David *see*
– Canada and its capital
– Inaugural address delivered by j d edgar, esq, president of the ontario literary society, february 5th, 1863
– The insolvent act of 1864
– "Loyalty", "independence", and "veiled treason", defined
– A manual for oil men and dealers in land
– Speech of mr j d edgar, mp in the house of commons, july 3rd, 1894
Edgar, John *see* Limitations of liberty
Edgar, John G *see* War of the roses
Edgar, John Henry *see* A theological understanding of anger within grief
Edgar, Lewis M *see* A history of the primitive baptists in the western districts of tennessee and kentucky
Edgar news – Edgar WI. v19 n24-52 [1927 feb 24-sep 23]-1944 mar 10-jul 28 – 11r – 1 – mf#961936 – us WHS [071]
Edgar newsletter – v2 n1-5 [1973 jan/feb-sep/oct] – 1r – 1 – mf#1055859 – us WHS [071]
Edgar, oder, vom atheismus zur vollen wahrheit / Hammerstein, Ludwig von – Trier: Paulinus-Druckerei, 1886 – 1mf – 9 – 0-8370-7151-8 – (incl ind) – mf#1986-1151 – us ATLA [240]
Edgar post – Edgar, NE: James McNally. -v27 n14. apr 19 1921 (semiwkly) [mf ed 1895-1921 (gaps)] – 9r – 1 – (cont: post world. absorbed by: edgar sun (1914)) – us NE Hist [071]
Edgar quinet : his early life and writings / Heath, Richard – London: Truebner 1881 [mf ed 1990] – (= ser The english and foreign philosophical library 14) – 2mf – 9 – 0-7905-6179-4 – mf#1988-2179 – us ATLA [190]
Edgar, Robert *see* South african police and justice department files, 1916-26
Edgar, Robert McCheyne *see*
– Does god answer prayer?
– The genius of protestantism
Edgar sun – Edgar, NE: Asa D Scott. v15 n1. jan 2 1914-v71 n53. dec 30 1976 (wkly) [mf ed 1914-56,1958-76 (gaps) filmed [1972?]-1977 – 19r – 1 – (cont: sun. absorbed: edgar post. absorbed by: clay county sun. some irregularities in numbering) – us NE Hist [071]
Edgar weekly review – Edgar WI. 1954 jul 1-1957, 1958-62, 1963-1965 jan 28 – 3r – 1 – (continued by: record review) – mf#964266 – us WHS [071]
Edgar weekly times – Edgar, NE: H G Lyon, 1885 (wkly) [mf ed 12th yr n14. jul 25 1890 filmed [1973]] – 1r – 1 – (cont: edgar review) – us NE Hist [071]
Edgarton, S C *see* The floral fortune-teller
Edgartown 1651-1900 – Oxford, MA (mf ed 1994) – (= ser Massachusetts vital records) – 55mf – 9 – 0-87623-191-1 – (mf 1-7: town & vital 1651-1784. mf 8-14: town records 1790-1832. mf 8,11-14: marriages 1782-1838. mf 9-14: intentions 1800-38. mf 8,12-14: vitals 1718-1840. mf 15-20: town records 1832-51. mf 21-28: town records 1839-70. mf 21: deaths 1839-66. mf 21,24: marriages 1837-43, 1851. mf 21-28: intentions 1838-71. mf 23: births 1841-44. mf 25-28: dog records 1858-71. mf 29-32: birth index 1663-1963. mf 33-34: marriage index 1663-1843. mf 35-39: marriage index 1699-1799; marriage index 1844-1979. mf 40-43: death index 1663-1949. mf 44-45: births 1843-76. mf 45-46: marriages 1844-52. mf 46: deaths 1845-59. mf 47: births 1876-90. mf 48-49: marriages 1853-91. mf 49: out-of-town mariages 1699-1799. mf 50-53: deaths 1860-1907. mf 54: marriages 1891-1902. mf 55: births 1891-1901) – us Archive [978]
Edgartown 1656-1849 – Oxford, MA (mf ed 1996) – (= ser Massachusetts vital record transcripts to 1850) – 6mf – 9 – 0-87623-247-0 – (mf 1t: vital records 1656-1818. mf 1t-3t: births & deaths 1718-1840. mf 2t-3t: marriages 1754-1831; marriage intentions 1810-29. mf 4t-5t: intentions & marriages 1829-49; deaths 1832-49; vital records 1782-1851. mf 5t-6t: births 1843-49 mf 6t: marriages & deaths 1843-49) – us Archive [978]
Edgcumbe, E[Dward] R[Obert] Pearce *see* Zephyrus
Edge city – v1 n1-3 [1970 mar-jun] – 1r – 1 – mf#714266 – us WHS [071]
Edge, glades county, florida / Huss, Veronica E – s.l, s.l? 193-? – 1r – us UF Libraries [978]
Edge of the jungle / Beebe, William – Garden City, NY. 1927 – 1r – us UF Libraries [972]
Edge of the sea / Carson, Rachel – Boston, MA. 1955 – 1r – us UF Libraries [550]
Edge, William John *see* Letter to the right hon lord ashley
Edgefield baptist church. nashville, tennessee : church records – 1903-79 – 1 – us Southern Baptist [242]
Edgefield first baptist church. edgefield, south carolina : church records – 1854-1941. 872p – 1 – us Southern Baptist [242]
Edger, Lilian *see* The elements of theosophy
The edgerton bible case : the decision of the supreme court of wisconsin / Blaisdell, James Joshua – [s.l: s.n, 1890?] [mf ed 1986] – 1mf – 9 – 0-8370-9602-2 – mf#1986-3602 – us ATLA [377]
Edgerton, Franklin *see*
– Sanskrit historical phonology
– Vikrama's adventure
Edgerton independent – Edgerton WI. 1876: jan 7, may 19-1877 apr 6 – 1r – 1 – (continued by: wisconsin tobacco reporter) – mf#1108661 – us WHS [071]
Edgerton reporter – Edgerton WI. 1950 oct 19/1951-1998 jul/dec – 63r – 1 – (with gaps; cont: wisconsin tobacco reporter) – mf#1108667 – us WHS [071]
Edgerton union – Edgerton WI. 1866 jun 7, 28, jul 12 – 1r – 1 – mf#875413 – us WHS [071]
Edgerton, W F *see* The thutmosid succession
Edgeways and the saint : poems and a farce / Chattopadhyaya, Harindranath – Bombay: Nalanda Publications, [1946] – (= ser Samp: indian books) – us CRL [810]
Edgewood baptist church. colleton county. walterboro, south carolina : church records – 1959-72 – 1 – us Southern Baptist [242]
Edgewood baptist church. hopkinsville, kentucky : church records – 2Jul 1958-80 – 1 – us Southern Baptist [242]
Edgewood echo – Peru IL. 1886 may 14-jul 1 – 1r – 1 – mf#873755 – us WHS [071]
Edgewood this week / Butler Co. Trenton – aug 1982-jan 1986 [wkly] – 2r – 1 – mf#B35114-35115 – us Ohio Hist [071]
Edgewood this week / Butler Co. Trenton – may 20-sep 30 1986 [wkly] – 1r – 1 – mf#B5522 – us Ohio Hist [071]
Edgeworth de Firmont, Henry Essex *see* Relation de la mort de louis 16
Edgeworth, Francis Ysidro *see* Currency and finance in time of war
Edgeworth, Maria *see*
– Lettres intimes de maria edgeworth pendant ses voyages
– Women, education and literature
Edgeworth, Michael Pakenham *see* India in the age of empire
Edghill, Ernest Arthur *see*
– An enquiry into the evidential value of prophecy
– Faith and fact
– The revelation of the son of god
– The spirit of power as seen in the christian church of the second century
Edgware advertiser – [London & SE] BLNL 1986- – 1 – uk Newsplan [072]
Edgware and district journal – London UK, 28 dec 1945 – 1/4r – 1 – uk British Libr Newspaper [072]
Edgware and district post *see* Local (edgware edt)
Edgware and kingsbury recorder – London UK, 6 feb-27 mar, 8 may-11 dec 1986; 12 jan, 9, 2 apr, 9, 25 jun, 16, 23 jul, 15, 22, 29 oct 1987; 5 nov 1987-88; 12 jan-20 dec 1989; 4 jan-20 sep 1990 – 5 1/2r – 1 – uk British Libr Newspaper [072]
Edgware and mill hill times – 1986-97; sep-dec 1998; jan-jun 1999 80r – 1 – uk British Libr Newspaper [072]
Edgware local *see* Local (edgware edt)
Edgware local advertiser – Barnet, England. 10 jan 1985-27 mar 1986 – n39-102 – 1 – (cont: edgware and district local advertiser. cont as: edgware advertiser) – uk British Libr Newspaper [072]
Edgware post – London UK, 22 sep-22 dec 1988; 5 jan-10 aug 1989 – 1 1/2r – 1 – uk British Libr Newspaper [072]
The edgware reporter, stanmore and elstree chronicle – London. 29 mar 1890-31 mar 1894 [wkly] – n198-406 – 1 – (discontinued) – uk British Libr Newspaper [072]
Edhem, Ibrahim *see* Tuerkiye'nin sihhi-i ictimai cografyasi. bayazit vilayeti
Edib-i muhterem merhum ziya pasa'nin ruhyasi / Pasa, Ziya – Dersaadet: Kasbar Matbaasi, 1326 [1910] – (= ser Ottoman literature, writers and the arts) – 1mf – 9 – $25.00 – us MEDOC [470]
Edicion homenaje en conmemoracion de la... / Universidad De Santo Domingo – Ciudad Trujillo, Dominican Republic. 1942 – 1r – us UF Libraries [378]
Edicion tridentina del manual toledano y su incorporacion al ritual romano / Garcia, Alf I – 2mf – 8 – €5.00 – ne Slangenburg [240]
Das edict des antoninus pius; eine bisher nicht erkannte schrift novatian's vom jahre 249/50 : "cyprian", de laude martyrii / Harnack, Adolf von – Leipzig: J C Hinrichs, 1895 – (= ser Tugal) – 1mf – 9 – 0-7905-1764-7 – (incl bibl ref) – mf#1987-1764 – us ATLA [240]
Das edict des antonius pius / Harnack, Adolf von – Leipzig, 1895 – (= ser Tugal 1-13/4a) – 1mf – 9 – €3.00 – ne Slangenburg [240]
Edicto para la santa visita : contiene notables puntos de disciplina eclesiastica... / Perez Calama, Joseph – en Quito: imprenta de Raymundo, ano de 1791 – (= ser Books on religion...1543/44-c1800: obispos: obispos de quito) – 1mf – 9 – mf#crl-466 – ne IDC [241]
Edicto pastoral del ilustrisimo senor d d joseph perez calama, obispo de quito : sobre los dos puntos siguientes... / Perez Calama, Joseph – en Quito: impr de Raymundo de Salazar, ano de 1791 – (= ser Books on religion...1543/44-c1800: obispos: obispos de quito) – 1mf – 9 – mf#crl-393 – ne IDC [241]
Edicts, ordinances, declarations and decrees relative to the seigniorial tenure : required by an address of the legislative assembly, 1851 = Edits, ordonnances, declarations et arrets relatifs a la tenure seigneuriale... – Quebec: printed by E R Frechette, 1852 [mf ed 1983] – 4mf – 9 – mf#SEM105P250 – cn Bibl Nat [348]
Edictus ceteraeque langobardorum leges (mgh leges 4:2.bd) : cum constitutionibus et pactis principum. principum beneventanorm – 1869 – (= ser Monumenta germaniae historica leges 4. fontes iuris germanici antiqui in usum scholarum separatim editi (mgh leges 4)) – €11.00 – ne Slangenburg [342]
Ediderunt societatas orientales batava... / Acta Orientalia – Lugduni Batavorum, 1923-1943. v1-19 – 123mf – 8 – mf#CH-101c – ne IDC [956]
La edificante aventura de garin / Oteyza, Luis de – Madrid: La Novela Mundial, 1927 – sp Bibl Santa Ana [940]
Edification : a sermon preached in the church of s alban, the martyr, ottawa, on sunday, january 2, 1876 / Bedford-Jones, T – Ottawa: The Citizen Print & Pub Co, 1876 – 1mf – 9 – mf#10252 – cn Canadiana [240]
Les edifices religieux de la vieille geneve / Archinard, Andre – Geneve: Joel Cherbuliez, 1864 – 1mf – 9 – 0-524-03510-5 – (incl bibl ref) – mf#1990-1015 – us ATLA [240]
Edifying and curious letters of some missioners, of the society of jesus, from foreign missions – n p, 1707 – 3mf – 9 – mf#HT-565 – ne IDC [910]
Edil, Yehudah Leyb *see* Afike yehudah
Edinaia rossia : ezhedn natsional -progressiv gaz: izd rus nats kom g baku / ed by Podshibiakin, M F – Baku: [s n] 1919 [1918]-] – (= ser Asn 1-3) – n20-21 [1919] item 133, on reel n28 – 1 – mf#asn-1 133 – ne IDC [077]
Edinaia rus' : ezhedn natsional -progressiv gaz / ed by Mogilevskii, F – Odessa: T-vo na paiakh "Edinaia Rus'" 1919 [1919 13 [26] avg-1920 [?]] – (= ser Asn 1-3) – n1-109 [1919] [gaps] item 134, on reel n28 – 1 – mf#asn-1 134 – ne IDC [077]
Edinboro independent – Edinboro, PA. -w 1889-1912 – 13 – $25.00r – us IMR [071]
Edinburgh advertiser – Scotland, UK. 1770-71. -w. 2 reels – 1 – uk British Libr Newspaper [072]
Edinburgh annual register – Edinburgh. 1808-1826 (1) – mf#4182 – us UMI ProQuest [941]
Edinburgh catholic herald – [Scotland] Edinburgh: Scottish Observer & Herald Ltd 1 aug 1936-30 dec 1938 (wkly) [mf ed 2004] – 5r – 1 – (ceased with n2383 publ jan 6 1939; absorbed by: scottish catholic herald) – uk Newsplan [241]
Edinburgh citizen – [Scotland] Portobello, Edinburgh: T Adams & Sons 8 jan 1897-sep 1939 (wkly) [mf ed 2003] – 38r – 1 – (formed by union of: portobello advertiser, midlothian journal, and dalkeith, leith, musselburgh and penicuik weekly news and: edinburgh citizen; cont by: portobello advertiser and edinburgh citizen [jan-sep 1939]) – uk Newsplan [072]
Edinburgh citizen – Scotland. -w. 1920. 1 reel – 1 – uk British Libr Newspaper [072]
Edinburgh citizen & portobello advertiser – Portobello, Edinburgh: T Adams & Sons 1897-1938 (wkly) [mf ed 2003] – 35r – 1 – (formed by union of: portobello advertiser, midlothian journal, and dalkeith, leith, musselburgh and penicuik weekly news and: edinburgh citizen; cont by: portobello advertiser and edinburgh citizen) – uk Newsplan; uk Scotland NatLib [072]
Edinburgh co-operative citizen – [Scotland] London: National Co-Operative Publ Society oct 1934 [mf ed 2004] – 1v on 1r – 1 – uk Newsplan [334]
Edinburgh courant [edinburgh, scotland : 1705] – [Scotland] Edinburgh: J Watson 14-19 feb to 25-27 sep 1706, 12-14 mar 1707-17-20 mar 1710 [mf ed 2004] – 706v on 2r – 1 – uk Newsplan [072]
Edinburgh courant [edinburgh, scotland : 1855] – [Scotland] Edinburgh: printed & publ...by W Veitch 2 jul 1855-30 dec 1859 (3times/wk) [mf ed 2004] – 704v on 4r – 1 – (publ as suppl to the edinburgh evening courant on the days on wh the latter did not appear) – uk Newsplan [072]
Edinburgh daily stock and share list published under the authority of the committee of the stock exchange – [Scotland] Edinburgh 5 dec 1868, 10 jun 1881, 25 nov 1890-dec 1950 (daily) [mf ed 2003] – 154r – 1 – (publ suspended jul 30 1914-jan 4 1915; cont: edinburgh daily stock and share list national emergency edition published under the authority of the committee of the stock exchange [jul 1940-dec 1950]) – uk Newsplan [332]
Edinburgh directories, 1773-1975 : from the edinburgh central library – 146r – 1 – mf#97598 – uk Microform Academic [914]
Edinburgh evening courant – [Scotland] Edinburgh: printed by J M'Euen, W Brown & J Mosman 14 may 1705-6 feb 1886 (3times/wk) [mf ed 2004] – 164r – 1 – (absorbed: edinburgh advertiser (edinburgh, scotland : 1764); edinburgh evening post and scottish record; cont by: edinburgh courant [dec 1871-feb 1886]) – uk Newsplan [072]
Edinburgh evening courant – Scotland. 1857-59.-d. 3 reels – 1 – uk British Libr Newspaper [072]
The edinburgh evening courant – Edinburgh: James McEuen, 1727 – 1r – 1 – us CRL [072]
Edinburgh evening dispatch – [Scotland] Edinburgh: J Ritchie 4 jan 1898-dec 1950 (daily ex sun) [mf ed 2004] – 305r – 1 – (cont by: evening dispatch (edinburgh, scotland) [12 dec 1921-dec 1950]) – uk Newsplan [072]
Edinburgh evening dispatch – Scotland, UK. 1888; 1890. -w. 4 reels – 1 – uk British Libr Newspaper [072]
Edinburgh evening news – Edinburgh: Scotsman Publ 1995- (daily ex sun) – 1 – (cont: evening news (edinburgh, scotland)) – ISSN: 0307-5761 – uk Scotland NatLib [072]
Edinburgh evening post and scottish standard – [Scotland] Edinburgh: A Cannon 3 jan 1844-oct 1861 (twice/wk) [mf ed 2003] – 17r – 1 – (missing: 1860; continued by: edinburgh evening post & scottish record [jan 1845-oct 1861]) – uk Newsplan [072]
Edinburgh evening telephone – [Scotland] Edinburgh: printed...by J Miller & Sons 1 nov 1878-31 jan 1879 (daily) [mf ed 2003] – 79v on 1r – 1 – uk Newsplan [072]
Edinburgh free press – [Scotland] Edinburgh: E M Land 3-31 jan 1852 (wkly) [mf ed 2003] – 5v on 1r – 1 – uk Newsplan [072]
Edinburgh gazetteer – [Scotland] Edinburgh: printed by W G Moffat 16 nov 1792-jan 1794 [mf ed 2004] – 87v on 1r – 1 – (imprint varies) – uk Newsplan [072]
The edinburgh gazetteer : or geographical dictionary containing a description of the various countries, kingdoms, states, cities, towns, mountains, etc of the world... – Edinburgh 1822 [mf ed Hildesheim 1995-98] – 6v on 30mf – 9 – €300.00 – 3-487-29896-1 – (ill by aaron arrowsmith) – gw Olms [059]
Edinburgh guardian – [Scotland] Edinburgh: J W Finlay 16 apr 1853-16 jun 1855 (wkly) [mf ed 2003] – 114v on 4r – 1 – (cont by: daily express (edinburgh, scotland)) – uk Newsplan [072]
Edinburgh herald – [Scotland] Edinburgh: printed by Stewart, Ruthven & Co 15 mar 1790-aug 1806 (3times/wk) [mf ed 2004] – 18r – 1 – (cont by: edinburgh herald and chronicle [jan 1797-aug 1806]; merged with: patriot's weekly chronicle to form: herald and chronicle; ceased n978 (jun 11 1796)) – uk Newsplan [072]
Edinburgh herald & post (1994) – Edinburgh: Scotsman Communications 1994-2001 (wkly) [mf ed 2 may 1996-] – 1 – (cont: herald & post; cont by: herald & post (edinburgh, scotland: 2001)) – uk Scotland NatLib [072]
Edinburgh. International Exhibition of Industry, Science and Art, 1886 *see* Memorial catalogue of the french and dutch loan collection
Edinburgh law journal – v1-2. 1831-37 – 13mf – 9 – $19.50 – mf#LLMC 84-460 – us LLMC [340]
Edinburgh & leith advertiser – [Scotland] Edinburgh: J Gray 30 apr 1825-5 aug 1826 (wkly) [mf ed 2003] – 63v on 1r – 1 – uk Newsplan [072]
Edinburgh literary journal : or weekly register of criticism and belles lettres – Edinburgh. 1828-1832 (1) – mf#5313 – us UMI ProQuest [420]

Edinburgh magazine – Edinburgh. 1758-1762 (1) – mf#5314 – us UMI ProQuest [420]
Edinburgh magazine : or literary miscellany – Edinburgh. 1785-1803 (1) – mf#5943 – us UMI ProQuest [420]
Edinburgh magazine and literary miscellany – Edinburgh. 1739-1826 (1) – mf#4241 – us UMI ProQuest [420]
Edinburgh magazine and review – Edinburgh. 1773-1776 – 1 – mf#5315 – us UMI ProQuest [073]
Edinburgh mathematical society proceedings – Oxford. 1953-1975 (1) – ISSN: 0013-0915 – mf#1008 – us UMI ProQuest [510]
Edinburgh Medical Missionary Society *see* Quarterly paper
Edinburgh monthly review – Edinburgh. 1819-1821 (1) – mf#4189 – us UMI ProQuest [610]
Edinburgh new philosophical journal : exhibiting a view of the progressive discoveries and improvements in the sciences and the arts – Edinburgh. 1826-1864 (1) – mf#2792 – us UMI ProQuest [500]
Edinburgh observer – [Scotland] Edinburgh: M Anderson 27 may 1822-jun 1845 (twice/wk) [mf ed 2003] – 26r – 1 – (absorbed: edinburgh star (edinburgh, scotland : 1808); new north briton (edinburgh, scotland : 1832); cont by: edinburgh observer and churchman's family gazette [jul 1844-jun 1845]) – uk Newsplan [072]
Edinburgh observer : or town and country magazine – [Scotland] Edinburgh: Ruthvens, Printer 13 sep 1817-7 mar 1818 (fortnightly) [mf ed 2004] – 1r – 1 – uk Newsplan [072]
Edinburgh observer etc – Scotland, UK. May-Dec 1832. -w. 1/2 reel – 1 – uk British Libr Newspaper [072]
Edinburgh philosophical journal – Edinburgh. 1819-1826 (1) – mf#2775 – us UMI ProQuest [100]
Edinburgh property review and investment circular – [Scotland] Edinburgh: Edinburgh Publ Co 22 mar 1879-30 oct 1880 (wkly) [mf ed 2003] – 2v on 3r – 1 – uk Newsplan [333]
Edinburgh review : critical journal – Edinburgh; London. 1802-1910 (1) – mf#5712 – us UMI ProQuest [410]
Edinburgh review – Edinburgh. 1755-1756 (1) – mf#4242 – us UMI ProQuest [420]
Edinburgh review – v1-250 1802-1929 – 1 – $1,016.00 – us L of C Photodup [420]
Edinburgh review and dr strauss / Wheelwright, George – London, England. 1873 – 1r – us UF Libraries [240]
The edinburgh review and the affghan war / Urquhart, D – London, 1843 – 1mf – 9 – mf#ILM-2700 – ne IDC [956]
Edinburgh Royal Medical Society *see* Dissertations read to the edinburgh royal medical society, 1750-1970
Edinburgh saturday post – Scotland. -w. 12 May 1827-3 May 1828. (36 ft) – 1 – uk British Libr Newspaper [072]
Edinburgh star [Edinburgh, Scotland : 1808] – [Scotland] Edinburgh: printed by A & J Aikman 16 sep 1808-30 dec 1826 (2iss/wk) [mf ed 2004] – 14r – 1 – (absorbed: northern reporter and: edinburgh and leith advertiser [apr 1825-aug 1826]; absorbed by: edinburgh observer [may 1822-apr 1844]) – uk Newsplan [072]
Edinburgh trades council, 1859-1951 – (= ser Labour party in britain, origins and development at local level. series 1) – 18r – 1 – (int by Ian macdougall) – mf#97147 – uk Microform Academic [331]
Edinburgh weekly chronicle and scottish pilot – [Scotland] Edinburgh: W Cross 6 jan 1844-25 mar 1848 (wkly) [mf ed 2003] – 3r – 1 – (cont: edinburgh weekly chronicle and literary journal; cont by: edinburgh news and literary chronicle) – uk Newsplan [072]
Edinburgh weekly journal – [Scotland] Edinburgh: [s.n.] 2 dec 1756-31 may 1769 & 7 aug 1765-11 oct 1769 (wkly) [mf ed 2004] – 1r – 1 – (began in dec? 1756; publ originally by william auld who was joined in the enterprise fr 1765-1771 by william smellie) – uk Newsplan [072]
Edinburgh weekly magazine – Edinburgh. 1768-1784 (1) – mf#4243 – us UMI ProQuest [070]
Edinburgh weekly register – [Scotland] Edinburgh: printed by D Buchanan 30 nov 1808-27 dec 1809 (wkly) [mf ed 2003] – 1v on 1r – 1 – uk Newsplan [072]
Edinburgh weekly review – [Scotland] Edinburgh: J Reid 10 may 1862-jun 1886 (wkly) [mf ed 2003] – 33r – 1 – (absorbed: week (edinburgh, scotland : 1862); cont by: weekly review [jan 1866-dec 1875]; weekly review and reformer [jan 1876-dec 1877]; scottish reformer and weekly review [jan 1878-jun 1886]) – uk Newsplan [072]
Edinenie : ponedel'nichnaia i posleprazdnichnaia vnepartijnaia gazeta – Tbilisi, Georgia, 1918-19 – 2r – 1 – us UMI ProQuest [077]
Edinensis *see* Sunday railway travelling
Eding's digest of hawaii supreme court dec / Edings, W S – Honolulu: Bulletin Publ. Co, 1903 – (= ser Hawaii Supreme Court Reports) – 6mf – 9 – $9.00 – (covers v1-14 1847-1903) – mf#LLMC 90-001 – us LLMC [347]
Edings, W S *see* Eding's digest of hawaii supreme court dec
Edinost – Chicago IL, 1919-23* – 1r – 1 – (slovenian newspaper) – us IHRC [071]
Edinost – Chicago IL, 1920-25 – 5r – 1 – (slovenian newspaper) – us IHRC [071]
Edinost – Pittsburgh PA, 1911* – 1r – 1 – (slovenian newspaper) – us IHRC [071]
Edinost – Toronto, Ontario, 1942* – 1r – 1 – (slovenian newspaper) – us IHRC [071]
Edinstvo : gaz bespartiin, obshchestv -polit i koop / ed by Charnotskii, I A – Petropavlovsk [Akmol obl]: Pravl Raion soiuza step koop soiuzov 1918-19 [1918 [?]-1919 1 avg – (= ser Asn 1-3) – n13 [1918]-n312 [1919] [gaps] item 135, on reel n28,29 – 1 – mf#asn-1 135 – ne IDC [077]
Edinstvo : marksistkaia rabochaia gazeta, izdavaemaia pri blizhajshem uchastii g v plekhanova – St Petersburg, Russia, 1914 – 1r – 1 – us UMI ProQuest [077]
Edip, Esref *see* Sebil uer-resad
Edirne – (= ser Vilayet salnames) – 9 – (1300 [1883] def'a 9 4mf $60; 1305 [1888] def'a 14 4mf $325; 1308 [1891] def'a 17 3mf $55; 1309 [1892] def'a 18 11mf $180; 1310 [1893] def'a 19 12mf $195; 1313m [1897] def'a 23; 1314m [1898] def'a 24 5mf $75; 1319m [1903] def'a 28 20mf $320) – us MEDOC [956]
Edirne, Enis Dede *see* The divan project
Edirne rahnuemasi : (tarihce: 763-1337 hicri seneleri) / Osman, Tosyali Rifat – Edirne: Vilayet Matbaasi, 1920 – (= ser Ottoman histories and historical sources) – 2mf – 9 – $40.00 – us MEDOC [956]
Edis, Robert William *see*
- Decoration and furniture of town houses
- Healthy furniture and decoration

Edisi bahasa Indonesia *see* Ekonomi indonesia
Edison and ford in fort myers / Crowe, F Hilton – s.l, s.l? 1936 – 1r – us UF Libraries [978]
The edison collection of american sheet music – the antebellum scores : from the edison collection at the university of michigan, ann arbor – pre-1861 – ca 160r 20r per unit – 1 – (faetures works publ up to 1861. includes rare first editions of stephen foster to railroad ballads. coll accompanied by new title, composer, subject listing catalogue prepared by the university of michigan) – us Primary [780]
Edison echo – Edison, NE: Ronald R Furse. 22v. v1 n1. jan 2 1925-v22 n48. nov 28 1946 (wkly) [mf ed with gaps] – 5r – 1 – (absorbed by: public mirror (arapahoe ne)) – us NE Hist [071]
Edison Electric Institute *see* Eei bulletin
Edison kinetogram – Berlin DE, 1912 17 jan – 1 – gw Mikrofilm [790]
The edison news – Edison, NE: C E Reed (wkly) [mf ed 1911-14 (gaps) filmed [1972]] – 1r – 1 – us NE Hist [071]
Edison record – Edison, NE: H M Call. 2v. v1 n1. sep 4 1915-v2 n21. jan 19 1917 (wkly) [mf ed with gaps] – 1r – 1 – (absorbed by: public mirror (arapahoe ne)) – us NE Hist [071]
Edison, Thomas Alva *see* Thomas a edison papers
Edisto island church. south carolina : church records – 1860 – 1 – 5.00 – us Southern Baptist [242]
L'edit de calliste : etude sur les origines de la penitence chretienne / Ales, Adhemar d' – 2e ed. Paris: G Beauchesne, 1914 [mf ed 1990] – (= ser Bibliotheque de theologie historique) – 2mf – 9 – 0-7905-3627-7 – (in french. incl bibl ref) – mf#1989-0120 – us ATLA [240]
Edith, oder, die schlacht bei hastings : ein trauerspiel / Wesendonk, Mathilde – Stuttgart: G J Goeschen 1872 – 2mf – 9 – mf#mw-3 – ne IDC [820]
Edition of four sonatas and two sets of variations for piano by alexander reinagle / Reinagle, Alexander; ed by Hopkins, Robert Elliott – U of Rochester 1959 [mf ed 19–] – 2v on 7mf – 9 – mf#fiche 245, 536 – us Sibley [780]
Edition of the stabat mater and an analysis of the stabat mater by luigi boccherini / Grubb, Cassel William – U of Rochester 1949 [mf ed 19–] – 2v on 5mf – 9 – mf#fiche202 – us Sibley [780]
Editions of the bible and parts thereof in english : from the year 1505 to 1850: with an appendix containing specimens of translations, and bibliographical descriptions / Cotton, Henry – 2nd corr enl ed. Oxford: University Press, 1852 – 2mf – 9 – 0-7905-0182-1 – (incl ind) – mf#1987-0182 – us ATLA [220]
Editor and publisher – New York. 1901+ (1) 1971+ (5) 1975+ (9) – ISSN: 0013-094X – mf#5795 – us UMI ProQuest [070]
Editor and publisher international year book – New York. 1980+ (1,5,9) – ISSN: 0424-4923 – mf#12823 – us UMI ProQuest [070]
Editor and publisher market guide – New York. 1985+ (1,5,9) – mf#12824 – us UMI ProQuest [070]
Editor looks back / Green, George Alfred Lawrence – Cape Town, South Africa. 1947 – 1r – us UF Libraries [070]
The editor-bishop, linus parker : his life and writings / Galloway, Charles B – Nashville, Tenn.: Southern Methodist Publishing House, 1886 – 1mf – us ATLA [240]
The editor-bishop, linus parker : his life and writings / Galloway, Charles Betts – Nashville, Tenn: Southern Methodist Pub House, 1886 – 1mf – 9 – 0-7905-6610-9 – mf#1988-2610 – us ATLA [240]
Editorial from the daily mail of tuesday, february 5th, 1889 – Toronto?: Daily Mail, 1889 – 1mf – 9 – mf#02077 – cn Canadiana [241]
Editorial research reports – Washington. 1955-1986 (1) 1971-1986 (5) 1976-1986 (9) – (cont by: congressional quarterly's editorial research reports) – ISSN: 0013-0958 – mf#971 – us UMI ProQuest [320]
Editorial Sanchez Rodrigo *see* Lecturas para la juventud
Editoriales del neo-granadino / Ancizar, Manuel – Bogota, Colombia. 1936 – 1r – us UF Libraries [972]
Editorials from djakarta press translations of significant editorials from "harian rakjat, bintang timur, duta masjarakat, warta bhakti, berita indonesia" / US Information Service – Djakarta, 1961(jun 27)-1965(jan 19) – 15mf – 9 – (missing: 1962(aug-sep); 1963(jan-may 27, jun-jul); 1964(oct, nov 22-dec)) – mf#SE-534 – ne IDC [959]
Editorials from djakarta press translations of significant editorials from "merdeka, warta berita, sinar harapan, suluh indonesia, semesta, garuda" / US Information Service – Djakarta, 1961(jun 22)-1965(jan 28) – 36mf – 9 – (missing: 1963(jan-may 9, jun 1-23, sep 6-30, nov 1-19); 1964(jan 4-26, jan 28-feb 3, feb 5-12, feb 19-jul 1, jul 3-aug 13, aug 26-31, sep 18-29, nov-dec 31); 1965(jan 1-27)) – mf#SE-533 – ne IDC [959]
Editorials on file / ed by Trager, Oliver – 1970-92 – 9 – $700.00 – (1970-80. $298.00; 1981-82. $99.00; 1983-84. $99.00; 1985-86. $99.00; 1987-88. $99.00; 1989-90. $99.00; 1991-92. $99.00) – us Facts [070]
Editors Release Service *see* Washington crap report
Les edits et ordonnances royaux et le conseil superieur de quebec / Bellefeuille, Edouard Lefebvre de – [s.l: s.n, 1869?] [mf ed 1984] – 1mf – 9 – 0-665-10530-4 – (incl bibl ref) – mf#10530 – cn Canadiana [340]
Edits, ordonnances, declarations et arrets relatifs a la tenure seigneuriale : demandes par une adresse de l'assemblee legislative, 1851 = Edicts, ordinances, declarations and decrees relative to the seigniorial tenure... 1851 – Quebec: Impr de E R Frechette...1852 [mf ed 1983] – 4mf – 9 – mf#SEM105P248 – cn Bibl Nat [348]
Edits, ordonnances royaux, declarations et arrets du conseil d'etat du roi : concernant le canada – Quebec: Impr par P E Desbarats..1803 [mf ed 1985] – 2v on 1mf – 9 – 0-665-40522-7 – mf#40522 – cn Canadiana [348]
Edits, ordonnances royaux, declarations et arrets du conseil d'etat du roi concernant le canada / Nouvelle-France – Quebec: E R Frechette, 1854-1856 [mf ed 1977] – 1r – 5 – mf#SEM16P303 – cn Bibl Nat [348]
Edits, ordonnances royaux, declarations et arrets du conseil d'etat du roi concernant le canada / Nouvelle-France. Conseil superieur de Quebec – Quebec: E R Frechette. 3v. 1854-1856 [mf ed 1982] – 22mf – 9 – mf#SEM105P83 – cn Bibl Nat [348]
Edits, ordonnances royaux, declarations et arrets du conseil d'etat du roi, concernant le canada : mis en ordre chronologique et publies par ordre de son excellence sir robert shore milnes... / Nouvelle-France – Quebec: Impr par P E Desbarats...1803-1806 [mf ed 1982] – 1r – 1 – mf#SEM35P177 – cn Bibl Nat [348]
Edkins, C E *see* The ability of undergraduate physical education majors to verbally identify and visually discriminate critical elements of select sport skills
Edkins, J R *see* Chinese scenes and people
Edkins, Jane R *see* Chinese scenes and people
Edkins, Jane Rowbotham (Stobbs) *see* Chinese scenes and people
Edkins, Joseph *see*
- Ancient symbolism among the chinese
- China's place in philology
- Chinese buddhism
- Chinese scenes and people
- The early spread of religous ideas
- Religion in china

Edlefsen, Blaine *see* Symbolization and articulation of oboe tones
Edlund, Larry L *see* Effects of a swimming program on cystic fibrosis children
Edmands, John *see* The evolution of congregationalism
Edmiston, Paula A *see* The influence of participation in a sports training program on the self-concepts of the educable mentally retarded attending a one-week special olympics sports camp
Edmond et caroline : ou, la lettre et la reponse / Kreube, Frederic – Paris, France. 1819 – 1r – us UF Libraries [440]
Edmond ronayne over vrijmetselarij : drie lezingen. gehouden to grand rapids, mich... / Ronayne, Edmond – Grand Rapids, MI: De Standaard Drukkerij, [1880?] – 1mf – 9 – 0-524-06662-0 – mf#1991-2717 – us ATLA [240]
Edmond's select cases : unreported appeals / New York. (State) – v1-2. 1831-50 (all publ) – 13mf – 9 – $19.50 – (a pre-nrs title) – mf#LLMC 80-003 – us LLMC [340]
Edmonds, T *see* Scriptural representation of the abolition of the fourth command...
Edmonds, William Donald *see* The newspaper press in british west africa, 1918-1939
Edmondson, George W [comp] *see* From epworth to london with john wesley
Edmondston, Arthur *see* A view of the ancient and present state of the zetland islands
Edmonstone, Archibald *see* A journey to two of the oases of upper egypt
Edmonton and tottenham weekly guardian – London, UK. 1884-11 jun 1886; feb 1888-22 dec 1893; 1894-nov 1906 – 7 1/2r – 1 – uk British Libr Newspaper [072]
Edmonton bulletin – Edmonton, AB: Bulletin Pub Co, 1880-1906 – 25r – 1 – ISSN: 0845-3462 – cn Library Assoc [071]
Edmonton bulletin – Edmonton, Alberta, CN. jan 1907-jan 1951 – 188r – 1 – cn Commonwealth Imaging [071]
Edmonton bulletin – Canada. 28 sep 1906-8 jan 1912; 9 mar 1914 – 9 1/2r – 1 – (wanting 1913; 1914 imperfect) – uk British Libr Newspaper [971]
Edmonton capital – Edmonton, Alberta, CN. jan 1910-dec 1914 – 15r – 1 – cn Commonwealth Imaging [071]
Edmonton daily bulletin – Canada. 26 mar, 15 sep 1910; 12 jun 1911; 7 mar 1914-8 nov 1915 (Imperfect) – 10r – 1 – uk British Libr Newspaper [971]
Edmonton gazette – London, UK. 1984-31 Jul 1986. -w.6 1/2 reels – 1 – (incorp with the gazette from jul 1986) – uk British Libr Newspaper [072]
Edmonton journal – Edmonton, Alberta, CN. 1903- – 36r/r – 1 – Can$2660.00 silver Can$2500.00 vesicular – cn Commonwealth Imaging [073]
Edmonton native news / Canadian Native Friendship Center – 1972-79 – (= ser American indian periodicals... 2) – 4mf – 9 – $95.00 – us UPA [305]
Edmonton native news / Canadian Native Friendship Centre (Edmonton AB) – 1973-1983 mar – 1r – 1 – (cont: canadian native friendship centre : [newsletter]; cont by: newsletter (canadian native friendship centre, edmonton, ab)) – mf#592288 – us WHS [305]
Edmonton people's weekly – Edmonton, Alberta, CN. sept 1944-dec 1952 – 2r – 1 – cn Commonwealth Imaging [071]
Edmonton saturday news – Edmonton, Alberta, CN. dec 1905-dec 1912 – 3r – 1 – cn Commonwealth Imaging [071]
The edmonton ukranian news – Edmonton, Alberta, CN. jan 1928-dec 1971 – 16r – 1 – (in ukranian) – cn Commonwealth Imaging [071]
Edmund and margaret – London, England. 1828 – 1r – us UF Libraries [240]
Edmund campion : a biography / Simpson, Richard – new ed. London: J Hodges, 1896 [mf ed 1990] – (= ser Catholic standard library) – 2mf – 9 – 0-7905-8249-X – (1st ed publ in 1867. incl bibl ref) – mf#1988-8112 – us ATLA [241]
Edmunds, Albert Joseph *see*
- Buddhist and christian gospels
- Buddhist texts quoted as scripture by the gospel of john

Edmunds, Charles Keyser *see* Modern education in china
Edmunds, George Franklin *see* Canadian reciprocity treaty
Edmundson, George *see* The church in rome in the first century
Edn – Boston. 1962+ [1]; 1967+ [5,9] – ISSN: 0012-7515 – mf#1497 – us UMI ProQuest [621]
Edo bakufu kankobutsu shusei : collection of publications under the edo shogunate / ed by Kawase, Kazuma et al – 1599-1868 – 1771v on 120r – 1 – Y1284,000 – (with 350p guide Y12,000. in japanese) – ja Yushodo [950]

Edo bungaku sokan : collection of edo literature. essential illustrated novelistic and dramatic works. in the holdings of the daitokyu memorial library, tokyo / Kawase, Kazuma [comp] – 1530pts on 90r – 1 – Y858,000 – (with 72p guide ed by comp. in japanese) – ja Yushodo [480]

E-doc : your guide to technologies driving e-business – Silver Spring, 2000+ [1,5,9] – mf#29743 – us UMI ProQuest [000]

Edojidai ryuzu jibiki daishusei : collection of popular japanese language dictionaries in the edo period. in the holdings of the national diet library, tokyo – 338bks on 94r – 1 – Y1,135,000 – (with 80p guide. in japanese) – ja Yushodo [480]

Edouard, Emmanuel *see*
– Essai sur la politique interieure d'haiti
– Pantheon haitien
– Republique d'haiti 'a l'apotheose de victor hugo
– Rimes haitiennes
– Solution de la crise industrielle francaise

Edouard en ecosse : ou, la nuit d'un proscrit / Duval, Alexandre – Paris, France. 1814 – 1r – us UF Libraries [440]

Edp analyzer – Vista. 1963-1987 (1) 1971-1987 (5) 1975-1987 (9) – (cont by: i/s analyzer) – ISSN: 0012-7523 – mf#1618 – us UMI ProQuest [000]

Edp performance review – Phoenix. 1986-1989 (1,5,9) – (cont by: capacity management review) – ISSN: 0091-7206 – mf#15753 – us UMI ProQuest [650]

Edp weekly – Washington. 1980-1982 (1,5,9) – (cont by: computer age edp weekly) – ISSN: 0012-7558 – mf#12413 – us UMI ProQuest [000]

Edp weekly – Springfield. 1996-1998 (1,5,9) – (cont: computer age edp weekly) – mf#12413,02 – us UMI ProQuest [000]

Edp weekly – Springfield. 1999+ (1) – mf#12413,04 – us UMI ProQuest [000]

Edp weekly's it monitor – Springfield, 1998-1998 [1,5,9] – (cont: edp weekly. cont by: edp weekly) – mf#12413,03 – us UMI ProQuest [000]

Edsall, Robert Spencer *see* Relation between growth and yield of grapefruit trees as affected b...

Edschmid, Kasimir *see*
– Erika
– Feine leute, oder, die grossen dieser erde
– Das gute recht
– Hallo welt!
– Lord byron
– Das rasende leben
– Die sechs muendungen
– Sport um gagaly
– Timur

Edson, David Orr, M D *see* Getting what we want; how to apply psychoanalysis to your own problems

Eduard daniel clarke's, professors zu cambridge, reise durch russland und die tartarei in den jahren 1800-1801 – Weimar 1817 [mf ed Hildesheim 1995-98] – 1v on 4mf – 9 – €120.00 – 3-487-26517-6 – gw Olms [910]

Eduard moerike / Meyer, Herbert – Stuttgart: J F Steinkopf, 1950 – 1r – 1 – (incl bibl ref) – us Wisconsin U Libr [430]

Eduard moerike und klara neuffer : neue untersuchungen / Camerer, W – Marbach a.N: A Remppis 1908 [mf ed 1990] – 1r – 1 – (filmed with: uber die galgenlieder / christian morgenstern) – mf#2838p – us Wisconsin U Libr [430]

Eduard moerikes kuenstlerisches selbstverstaendnis. im spiegel seiner gedichte "die elemente", "goetthiche reminiszenz" und "neue liebe" / Aley, Peter – Frankfurt a.M., 1970 – 3mf – 9 – 3-89349-661-0 – gw Frankfurter [430]

Eduard reuss' briefwechsel mit seinem schueler und freunde karl heinrich graf : zur hundertjahrfeier seiner geburt = Correspondence. selections / Reuss, Eduard & Graf, Karl Heinrich; ed by Budde, Karl & Holtzmann, Heinrich Julius – Giessen: J Ricker, 1904 – 2mf – 9 – 0-524-05330-8 – mf#1990-1448 – us ATLA [943]

Eduard von bauernfelds gesammelte aufsaetze / ed by Hock, Stefan – Wien: Literarischer Verein, 1905 [mf ed 1993] – (= ser Schriften des literarischen vereins in wien 4) – xii/391p – 1 – (incl bibl ref and ind) – mf#8308 reel 1 – us Wisconsin U Libr [802]

Eduard von hartmann's religion der zukunft in ihrer selbstzersetzung / Heman, Carl Friedrich – Leipzig: JC Hinrichs, 1875 – 1mf – 9 – 0-7905-9959-7 – mf#1989-1684 – us ATLA [200]

Eduardo blanco : creador de la novela venezolana / Barnola, Pedro Pablo – Bogota, Colombia. 1954 – 1r – us UF Libraries [440]

Eduardo de Noronha. Lisboa: Imprensa Nacional, 1895 *see* O districto de lourenco marques e a africa do sul

Eduardo posada : secretaire perpetuel de l'academie nationale...colombie / Bayle, Constantino – Madrid: Razon y Fe, 1926 – 1 – sp Bibl Santa Ana [946]

Educacao popular : alfabetizacao e primeiras contas: experiencias na elaboracao de material didactico para adultos – Sao Paulo: Centro Ecumenico de Documentacao e Informacao, 1984 – us CRL [972]

Educacao publica em s paulo / Azevedo, Fernando De – Sao Paulo, Brazil. 1937 – 1r – us UF Libraries [370]

Educacao superior no brasil / Campos, Ernesto De Souza – Rio de Janeiro, Brazil. 1940 – 1r – us UF Libraries [378]

Educacion comercial en centro america / Haines, Peter George – Guatemala, 1964 – 1r – us UF Libraries [370]

Educacion de la mujer en america / Bayle, Constantino – Madrid: Razon y Fe, 1941 – 1 – sp Bibl Santa Ana [370]

Educacion en colombia / Bernal Escobar, Alejandro – Louvain, Belgium. 1965 – 1r – us UF Libraries [370]

Educacion en los estados unidos / Larrea, Julio C – Quito, Ecuador. 1960 – 1r – us UF Libraries [370]

Educacion guatemalteca / Guatemala Ministerio De Educacion Publica – Guatemala, 1962 – 1r – us UF Libraries [370]

La educacion integral / Bejarano y Sanchez, Eloy – 1898 – 9 – sp Bibl Santa Ana [370]

La educacion medica integral / Bejarano y Sanchez, Eloy – Madrid: Establecimiento Tipografico de J.A. Garcia, 1902 – 1 – sp Bibl Santa Ana [378]

Educacion moral / Alvarez Suarez, Augustin Enrique – Buenos Aires, Argentina. 1917 – 1r – us UF Libraries [370]

Educacion rural en las villas – Santa Clara, Cuba. 1959 – 1r – us UF Libraries [370]

La educacion y la justicia en los anos : memoria remitida al h congreso de la nacion por el ministro de educacion y justicia – Asuncion: Impr Nacional, 1932/1933-1934 – 1r – us CRL [370]

Educate – Philadelphia. 1969-1971 (1) 1969-1971 (5) (9) – ISSN: 0013-1121 – mf#5138 – us UMI ProQuest [370]

Educated women : the substance of an address delivered before the delta sigma society of mcgill university, december 1889 / Dawson, John William – Montreal?: s.n, 1889? – 1mf – 9 – mf#03665 – cn Canadiana [376]

Educating children : early and middle years – Washington. 1974-1976 (1) 1974-1976 (5) 1974-1976 (9) – mf#9139 – us UMI ProQuest [370]

Educating for a better world : now! – Los Angeles Athletic Club, 1984 – (= ser United states olympic academy 8) – 3mf – 9 – $12.00 – us Kinesology [790]

Educating in faith / Catholic Negro-American Mission Board – v1 n1-14 [1981 sum-1988 fall] – 1r – 1 – (cont: educating in faith) – mf#2302034 – us WHS [241]

Educating through dance : a multicultural theoretical framework / Staley, Kimberly T – Texas Woman's University, 1993 – 3mf – 9 – $12.00 – mf#PE 3672 – us Kinesology [790]

L'education : intellectuelle et morale / Compayre, Gabriel – Paris: Librairie Classique Paul Delaplane [1908?] [mf ed 1986] – 2mf – 9 – 0-8370-7686-2 – (in french. incl bibl) – mf#1986-1686 – us ATLA [370]

Education – Chula Vista. 1880+ (1) 1968+ (5) 1975+ (9) – ISSN: 0013-1172 – mf#13 – us UMI ProQuest [370]

Education : a framework for expansion, 1972. command n5174 – 1mf – 9 – mf#87032 – uk Microform Academic [324]

Education – Harlow. 1974-1994 (1) 1974-1994 (5) 1976-1994 (9) – ISSN: 0013-1164 – mf#8676 – us UMI ProQuest [370]

Education : a journal of reputation / Negro Needs Society – New York. v1-2 n4. 1935-36 [all publ] – (= ser Black journals, series 1) – 2mf – 9 – $45.00 – us UPA [370]

Education : pinellas county / Hunter, C M – s.l, s.l? 1936 – 1r – us UF Libraries [370]

Education / Scoville, Dorothy R – s.l, s.l? 1936 – 1r – us UF Libraries [370]

Education : tampa / Muse, Viola B – s.l, s.l? 193-? – 1r – us UF Libraries [370]

Education abstracts : [english edition] – Paris. 1949-1964 – 1 – mf#1549 – us UMI ProQuest [370]

Education abstracts – Washington. 1972-1977 (1) 1972-1977 (5) 1976-1977 (9) – ISSN: 0013-1210 – mf#6391 – us UMI ProQuest [370]

Education act, 1944 – 2mf – 9 – mf#86955 – uk Microform Academic [324]

Education acts in england and wales, 1870-1918 – 1r – 1 – mf#96736 – uk Microform Academic [324]

Education africaine – v24, n88-n.s. n48. 1935-58 – 1 – us CRL [370]

Education and citizenship in india / Alston, Leonard – London; New York: Longmans, Green, and Co, 1910 – (= ser Samp: indian books) – us CRL [370]

Education and computing – Amsterdam. 1991-1991 – 1,5,9 – ISSN: 0167-9287 – mf#42570 – us UMI ProQuest [370]

Education and culture ministry of education and culture / Indonesia. Kementerian pendidikan, pengadjaran dan kebudajaan – Djakarta, 1950(1-3), 1951-1957(1-9) – 10mf – 9 – mf#SE-477 – ne IDC [370]

Education and ethics = Questions de morale et d'education / Boutroux, Emile – New York: Macmillan, 1913 – 1mf – 9 – 0-7905-3546-7 – (in english) – mf#1989-0039 – us ATLA [170]

Education and job training : preparing for the 21st century workforce: hearing...house of representatives, 107th congress, 2nd session... angola, indiana, mar 22 2002 / United States. Congress. House. Committee on Education and the Workforce. Subcommittee on 21st Century Competitiveness – Washington: US GPO 2002 [mf ed 2002] – 2mf – 9 – 0-16-068970-8 – (incl ind) – us GPO [331]

Education and law journal *see* Brigham young university education and law journal

Education and life : an address delivered at the opening of the 32nd session of queen's university, kingston, canada / Watson, John – [s.l.]: Alma Mater Society of Queen's University, [1873?] [mf ed 1984] – 1mf – 9 – 0-665-32332-8 – mf#32332 – cn Canadiana [378]

Education and pictou academy / Anderson, William James – Pictou, NS?: s.n, 1850? – 1mf – 9 – mf#67252 – cn Canadiana [370]

The education and problems of the protestant ministry / Hill, David Spence – Worcester, Mass: Clark University Press, 1908 – 1mf – 9 – 0-524-07523-9 – (incl bibl ref) – mf#1991-3153 – us ATLA [242]

Education and social amelioration of women in pre-mutiny india / Datta, Kalikinkar – Patna: Patna Law Press, [1936?] – (= ser Samp: indian books) – us CRL [305]

Education and social issues : a history of education derivative – (= ser History of education) – 1287mf – 1 – us Primary [370]

Education and statesmanship in india : 1797 to 1910 / James, Henry Rosher – London; New York: Longmans, Green, and Co, 1911 – (= ser Samp: indian books) – us CRL [370]

Education and the future of religion : a sermon preached in the church of the gesu, rome, march 21, 1900, for the benefit of a free night-school / Spalding, John Lancaster – Notre Dame, Ind, USA: Ave Maria Press, [1900?] – 1mf – 9 – 0-8370-7831-8 – mf#1986-1831 – us ATLA [240]

Education and the higher life / Spalding, John Lancaster – 5th ed. Chicago: AC McClurg, 1897, c1890 – 1mf – 9 – 0-7905-6569-2 – mf#1988-2569 – us ATLA [370]

Education and training – London. 1970-1995 (1) 1971-1995 (5) 1975-1995 (9) – ISSN: 0040-0912 – mf#5814 – us UMI ProQuest [600]

Education and training in mental retardation – Reston. 1987-1993 – 1,5,9 – (cont: education and training of the mentally retarded. cont by: education and training in mental retardation and developmental disabilities) – ISSN: 1042-9859 – mf#12851,01 – us UMI ProQuest [370]

Education and training in mental retardation and developmental disabilities – Reston. 1994+ – 1,5,9 – (cont: education and training in mental retardation) – ISSN: 1079-3917 – mf#12851,02 – us UMI ProQuest [370]

Education and training of the mentally retarded – Reston. 1966-1986 (1) 1966-1986 (5) 1966-1986 (9) – (cont by: education and training in mental retardation) – ISSN: 0013-1237 – mf#12851 – us UMI ProQuest [370]

Education and treatment of children – Pittsburgh. 1990+ – 1,5,9 – ISSN: 0748-8491 – mf#17463 – us UMI ProQuest [370]

Education and urban society – Thousand Oaks. 1968+ (1) 1975+ (5) 1975+ (9) – ISSN: 0013-1245 – mf#10959 – us UMI ProQuest [306]

Education Association of China *see* Education association of china

Education association of china : monthly bulletin / Education Association of China – n16. 1908 [complete] – 1r – 1 – mf#ATLA S0703B – us ATLA [370]

Education Association of Fukien Province *see* Education association of fukien province

Education association of fukien province : journal / Education Association of Fukien Province – v1 n1-6. 1906-11 [complete] – 1r – 1 – mf#ATLA S0701C – us ATLA [370]

Education, bas-canada : reponse a une adresse de l'assemblee legislative du 9 avril 1853 demandant copie de tous les rapports, presentations et suggestions que les inspecteurs d'ecoles ont pu faire ou adresser... / Canada (Province). Surintendant de l'education pour le bas-Canada – Quebec: Impr Louis Perrault, 1853 [mf ed 1983] – 6mf – 9 – mf#SEM105P324 – cn Bibl Nat [370]

Education beyond apartheid : report – Johannesburg, 1971 – us CRL [370]

Education canada – Toronto. 1969+ (1) 1972+ (5) 1976+ (9) – ISSN: 0013-1253 – mf#7054 – us UMI ProQuest [370]

L'education catholique et le canada francais = Roman catholic education and french canada / Lussier, Irenee – Toronto: W J Gage, 1960 [mf ed 1992] – (= ser Quance lectures in canadian education) – 2mf – 9 – mf#SEM105P1573 – cn Bibl Nat [377]

L'education chretienne de la democratie : essai d'apologetique sociale / Calippe, Charles – Paris: Librairie Bloud, 1908 [mf ed 1986] – (= ser Questions de sociologie) – 1mf – 9 – 0-8370-8488-1 – (in french. incl bibl ref) – mf#1986-2488 – us ATLA [241]

Education daily – Gaithersburg. 1968+ (1) 1974+ (5) 1974+ (9) – ISSN: 0013-1261 – mf#10087 – us UMI ProQuest [370]

l'education dans la province de quebec : conference donnee au club assiniboia de regina, le 25 octobre 1916 / Mathieu, Olivier-Elzear – [Prince-Albert, Sask?: s.n, 1916?] – 1mf – 9 – 0-665-75189-3 – mf#75189 – cn Canadiana [370]

The education demanded by the people of the u. states : a discourse. delivered at union college, schenectady... / Wayland, Francis – Boston: Phillips, Sampson, 1855 – 1mf – 9 – 0-524-08694-X – mf#1993-3219 – us ATLA [370]

Education des femmes en haiti / Bouchereau, Madeleine G Sylvain – Port-Au-Prince, Haiti. 1944 – 1r – us UF Libraries [376]

L'education des sentiments / Thomas, P-Felix – 5e rev ed. Paris: Felix Alcan, 1910 [mf ed 1986] – (= ser Bibliotheque de philosophie contemporaine) – 1mf – 9 – 0-8370-7994-2 – (in french. incl bibl ref) – mf#1986-1994 – us ATLA [170]

Education dialogue – 1972 mar – 1r – 1 – mf#639823 – us WHS [071]

Education digest – Ann Arbor. 1935+ (1) 1968+ (5) 1970+ (9) – ISSN: 0013-127X – mf#218 – us UMI ProQuest [370]

L'education du caractere / Gillet, Martin Stanislaus – nouv ed. Paris: Desclee, De Brouwer, 1910 [mf ed 1986] – 1mf – 9 – 0-8370-8740-6 – (in french) – mf#1986-2740 – us ATLA [230]

Education enquiry : abstract of the answers and returns, 1835. command n62 – 16mf – 9 – mf#87118 – uk Microform Academic [941]

Education equipment – Tonbridge. 1977-1978 (1) 1978-1978 (5) 1978-1978 (9) – ISSN: 0013-1296 – mf#10651 – us UMI ProQuest [370]

Education et developpement – no. 1-81. Paris. oct 1964-72 – 5 – fr ACRPP [370]

Education for a new world / Montessori, Maria – Madras, India: Kalakshetra, 1948 – (= ser Samp: indian books) – us CRL [370]

Education for africans in tanganyika / George, Betty Grace (Stein) – Washington, DC. 1960 – 1r – us UF Libraries [370]

Education for girls and women in upper south carolina prior to 1890 : with related miscellaneous articles. a compilation by mrs henry towles crigler / Crigler, Sara Gossett – Greenville SC: [s.n] [mf ed Spartanburg SC: The Reprint Co Publ [1980?] – 4mf – 9 – mf#51-509 – us South Carolina Historical [376]

Education for life / Armstrong, Samuel Chapman – [S.l.: s.n., 1913?] – 1mf – 9 – 0-524-04006-0 – mf#1990-1178 – us ATLA [370]

Education for life : mass education / Kumarappa, Joseph Cornelius – Rajahmundry: Hindustan Pub Co, 1937 – (= ser Samp: indian books) – us CRL [370]

Education for teaching – London. 1974-1976 (1) 1974-1976 (5) 1974-1976 (9) – ISSN: 0013-1326 – mf#8910 – us UMI ProQuest [370]

Education for the disadvantaged child – Edmonton. 1976-1978 – 1,5,9 – ISSN: 0315-6621 – mf#11027 – us UMI ProQuest [370]

Education for the indian : fancy and reason on the subject; contract schools and non-sectarianism in indian education / Palladino, Lawrence Benedict – New York: Benziger Bros, 1892 – 1mf – 9 – 0-524-01124-9 – mf#1990-0338 – us ATLA [370]

Education for victory – Washington. 1942-1945 – 1 – mf#5182 – us UMI ProQuest [370]

Education forward – 1983 aug-1986 dec, 1987 jan-1990 may – 2r – 1 – (cont: newsletter - wisconsin department of public instruction, wisconsin. dept of public instruction) – mf#669600 – us WHS [370]

Education in canada : institution for the education of the youth of canada generally, and the most promising youth of the recently converted indian tribes, as teachers to their aboriginal countrymen... / Ryerson, Egerton – Leeds, England?: s.n, 1836? (Leeds England: R Inchbold). – 1mf – 9 – mf#40622 – cn Canadiana [370]

Education in chemistry – Cambridge. 1964+ (1) 1976+ (5) 1976+ (9) – ISSN: 0013-1350 – mf#5217 – us UMI ProQuest [540]

Education in france – New York. 1957-1970 – 1 – ISSN: 0424-5458 – mf#1514 – us UMI ProQuest [370]

Education in haiti / Cook, Mercer – Washington, DC. 1948 – 1r – us UF Libraries [370]

Education in india : a letter from the ex-principal of an indian government college to his appointed successor / Arnold, Edwin – London, 1860 – (= ser 19th c books on british colonization) – 1mf – 9 – mf#1.1.3583 – uk Chadwyck [370]

Education in india : a study of the lower ganges valley in modern times / Zellner, Aubrey Albert – New York: Bookman Associates, c1951 – (= ser Samp: indian books) – us CRL [370]

Education in india : today and tomorrow / Mukherji, S N – Baroda: Acharya Book Depot, 1950 – (= ser Samp: indian books) – us CRL [370]

Education in modern india : a brief review / Basu, Anathnath – Calcutta: Orient Book Co, [1945] – (= ser Samp: indian books) – us CRL [370]

Education in relation to the christianisation of national life : with supplement, presentation and discussion of the report in the conference on 17th june 1910 together with the discussion on christian literature – Edinburgh: Publ for the World Missionary Conference by Oliphant, Anderson & Ferrier; New York: Fleming H Revell, [1910?] – 2mf – 9 – 0-8370-6475-9 – (incl indes) – mf#1986-0475 – us ATLA [230]

Education in science – Hatfield. 1977-1996 (1,5,9) – ISSN: 0013-1377 – mf#11356 – us UMI ProQuest [500]

Education in science, 1962-1974 : the bulletin of the association for science education – 1r – 1 – mf#96879 – uk Microform Academic [370]

Education in the republic of haiti / Dale, George Allan – Washington, DC. 1959 – 1r – us UF Libraries [370]

Education in the society of friends : past, present, and prospective / Parrish, Edward – Philadelphia: J B Lippincott, 1865 – 1mf – 9 – 0-8370-8542-X – mf#1986-2542 – us ATLA [370]

Education in the two andovers : address... tuesday, sep 2nd 1856 / Fuller, Samuel – Andover: WF Draper, 1856 [mf ed 1993] – 1mf – 9 – 0-524-08362-2 – mf#1993-3062 – us ATLA [370]

Education journal – Toronto, ON. 1887-97 – 3r – 1 – cn Library Assoc [370]

Education law bulletin – Cambridge. 1977-1988 (1) 1977-1988 (5) 1977-1988 (9) – mf#11623 – us UMI ProQuest [344]

Education laws and regulations of the virgin islands / Virgin Islands Of The United States Laws, Statute – Orford, NH. 1965 – 1r – us UF Libraries [370]

Education libraries – Boston. 1986+ (1,5,9) – ISSN: 0148-1061 – mf#15322,01 – us UMI ProQuest [020]

Education libraries bulletin – Leicester. 1958-1988 (1) 1972-1988 (5) 1975-1988 (9) – (cont by: education libraries journal) – ISSN: 0013-1407 – mf#6975 – us UMI ProQuest [020]

Education libraries journal – Leicester. 1989-1996 (1) 1989-1996 (5) 1989-1996 (9) – (cont: education libraries bulletin) – ISSN: 0957-9575 – mf#6975,01 – us UMI ProQuest [020]

An education management perspective on gender inequities in science education : in a free state primary school / Makate, Paulina Pulane – Pretoria: Vista University 2001 [mf ed 2002] – 2mf – 9 – (incl bibl ref) – mf#mfm15328 – sa Unisa [370]

L'education morale des le berceau / Perez, Bernard – 4e rev ed. Paris: Felix Alcan, 1901 [mf ed 1986] – 1mf – 9 – 0-8370-7900-4 – (in french. incl bibl ref) – mf#1986-1900 – us ATLA [150]

Education, no 640 : new smyrna / Sweett, Zelia Wilson – s.l, s.l? 1936 – 1r – us UF Libraries [370]

"The education of a kansan" / Klingberg, Fran J – 1883-1911 – 1 – us Kansas [978]

The education of a ministry, the proper work and care of the churches : a discourse / Shedd, William Greenough Thayer – Boston: T R Marvin 1855 [mf ed 1991] – 1mf – 9 – 0-7905-9890-6 – mf#1989-1615 – us ATLA [240]

The education of christ : hill-side reveries / Ramsay, William Mitchell – New York: G P Putnam, 1902 – 1mf – 9 – 0-8370-4835-4 – mf#1985-2835 – us ATLA [240]

The education of deaf children : the possible place of finger spelling and signing, 1968 – 2mf – 9 – mf#87023 – uk Microform Academic [324]

The education of india : a study of british educational policy in india, 1835-1920, and of its bearing on national life and problems in india to-day / Mayhew, Arthur – London: Faber and Gwyer, 1926 – (= ser Samp: indian books) – us CRL [370]

Education of life *see* Jen te chiao yu (ccm153)

The education of our girls : an address, delivered at tilden ladies' seminary, west lebanon, n.h. june 21, 1877 / Eaton, John – New York: Wm B Folger, 1877 – 1mf – 9 – 0-8370-7787-7 – mf#1986-1787 – us ATLA [376]

Education of pauper and destitute children in the 19th c : a collection of 11 theses / Marsden, W E [comp] – 42mf – 9 – mf#87091 – uk Microform Academic [080]

Education of the adolescent : report of the consultative committee (hadow report), 1927 – 4mf – 9 – mf#86949 – uk Microform Academic [324]

The education of the artist / Chesneau, Ernest Alfred – London 1886 – (= ser 19th c art & architecture) – 4mf – 9 – mf#4.2.1227 – uk Chadwyck [700]

Education of the central nervous system / Halleck, Reuben Post – New York, NY. 1896 – 1r – us UF Libraries [611]

Education of the handicapped : a history of education derivative – (= ser History of education) – 385mf – 1 – us Primary [370]

The education of the negro prior to 1861 : a history of the education of the colored people of the united states from the beginning of slavery to the civil war / Woodson, Carter Godwin – New York: G P Putnam, 1915 [mf ed 1990] – 2mf – 9 – 0-7905-7097-1 – (incl bibl ref) – mf#1988-3097 – us ATLA [976]

Education of the spanish speaking child : in the five southwestern states – 1933 – 1r – 1 – $50.00 – mf#C63011 – us Library Micro [370]

Education of the visually handicapped – Washington. 1969-1988 (1) 1973-1988 (5) 1975-1988 (9) – (cont: international journal for the education of the blind. cont by: re:view) – ISSN: 0013-1458 – mf#8567 – us UMI ProQuest [360]

The education of the will : the theory and practice of self-culture = Education de la volonte / Payot, Jules – 3rd amer ed. New York: Funk & Wagnalls, 1911 [mf ed 1991] – 2mf – 9 – 0-7905-8550-2 – (trans fr french into english by smith ely jelliffe) – mf#1989-1775 – us ATLA [150]

The education of the women of india / Cowan, Minna Galbraith – Edinburgh: Oliphant, Anderson & Ferrier, [1912?] – 1mf – 9 – 0-7905-4267-6 – (incl bibl ref) – mf#1988-0267 – us ATLA [376]

The education of woman / Gerhart, Emanuel Vogel – Lancaster, Pa: Daily Express, 1864 – 1mf – 9 – 0-8370-7792-3 – mf#1986-1792 – us ATLA [376]

Education of women : a history of education derivative – (= ser History of education) – 1200mf – 1 – us Primary [376]

The education of women : from the history of education collection – (= ser The history of education coll) – 1148mf – 9 – (coll documents educational theories and practices from the 15th century through 1917 and provides a perspective on women's rights and their involvement in the teaching profession. coll filmed from holdings of the milbank memorial library of teachers college, columbia university, the harvard university library and other institutions) – mf#C36-27891 – us Primary [376]

The education of women in china / Burton, Margaret E – New York: Fleming H Revell, c1911 – 1mf – 9 – 0-7905-4109-2 – (incl bibl ref) – mf#1988-0109 – us ATLA [376]

The education of women in japan / Burton, Margaret E – New York: Fleming H Revell, c1914 – 1mf – 9 – 0-7905-4110-6 – (incl bibl ref) – mf#1988-0110 – us ATLA [376]

L'education ou la grande question sociale du jour : recueil de documents propres a eclairer les gens de bonne foi: mai 1886 – Montreal?: s.n, 1886 – 3mf – 9 – mf#25035 – cn Canadiana [370]

Education policy bulletin – Lancaster. 1979-1980 – 1,5,9 – ISSN: 0305-9847 – mf#10997,01 – us UMI ProQuest [370]

Education, politics, and war / Radhakrishnan, Sarvepalli – Poona, India: International Book Service, 1944 – (= ser Samp: indian books) – us CRL [954]

The education problem in india / Banerjee, Gooroodass – Calcutta: SK Lahiri and Co, 1914 – (= ser Samp: indian books) – us CRL [370]

Education recaps – Princeton. 1966-1978 (1) 1970-1978 (5) – ISSN: 0013-1504 – mf#2729 – us UMI ProQuest [370]

Education Resource Associates *see* Bern

Education review – St John, Fredricton, NB. 1887-1931 – 8r – 1 – cn Library Assoc [370]

Education statistics for the united kingdom 1967-1975 – [mf ed Chadwyck-Healey] – (= ser British government publications...1801-1977) – 16mf – 9 – uk Chadwyck [314]

Education summary – Waterford. 1948-1973 (1) 1972-1973 (5) (9) – ISSN: 0013-1520 – mf#6781 – us UMI ProQuest [370]

Education, to whom does it belong? / Bouquillon, Thomas – 2nd ed. Baltimore: John Murphy, 1892 – 1mf – 9 – 0-524-07510-7 – (incl bibl ref) – mf#1991-3140 – us ATLA [370]

Education today – Toronto. 1989+ – 1,5,9 – (cont: ontario education) – ISSN: 0843-5081 – mf#17268 – us UMI ProQuest [370]

L'education traditionelle a travers les chants et les poemes / Dione, Salif – 1982 – us CRL [370]

Education update / AFL-CIO – 1978 jun-1989 may, 1989 july-1994 nov – 2r – 1 – mf#1708761 – us WHS [331]

Education usa – Gaithersburg. 1958-1991 (1) 1970-1991 (5) 1976-1991 (9) – ISSN: 0013-1571 – mf#2226 – us UMI ProQuest [370]

Education week – Washington. 1981+ – 1,5,9 – ISSN: 0277-4232 – mf#13364 – us UMI ProQuest [370]

Educational : word lessons / Baillairge, Charles P Florent – [Quebec?: s.n, 19–?] – 1mf – 9 – 0-665-98306-9 – mf#98306 – cn Canadiana [370]

Educational achievement and black-white inequality / Jacobson, Jonathan et al – Washington DC: National Center for Education Statistics, US Dept of Education... – 2mf – 9 – (incl bibl ref) – us GPO [370]

Educational administration – London. 1977-1981 – 1,5,9 – (cont by: educational management and administration) – ISSN: 0305-7496 – mf#11557 – us UMI ProQuest [370]

Educational administration abstracts – Columbus. 1966+ (1) 1966+ (5) 1966+ (9) – ISSN: 0013-1601 – mf#6393 – us UMI ProQuest [370]

Educational administration and supervision – v1-46. 1915-60 – 1 – us AMS Press [370]

Educational administration quarterly – Thousand Oaks. 1971+ (1) 1965+ (5) 1975+ (9) – ISSN: 0013-161X – mf#6271 – us UMI ProQuest [370]

Educational and Cooperative Union of America et al *see* Iowa union farmer

Educational and psychological measurement (epm) – Durham. 1941+ [1,5]; 1975+ [9] – ISSN: 0013-1644 – mf#1487 – us UMI ProQuest [150]

Educational and psychological research – Hattiesburg. 1981-1988 (1) 1981-1988 (5) 1981-1988 (9) – ISSN: 0279-0688 – mf#12677 – us UMI ProQuest [370]

Educational and training technology international (etti) – London. 1989-1994 – 1,5,9 – (cont: programmed learning and educational technology. cont by: innovations in education and training international) – ISSN: 0954-7304 – mf#11223,01 – us UMI ProQuest [370]

Educational assessment – Mahwah, 1998+ – 1,5,9 – ISSN: 1062-7197 – mf#25216 – us UMI ProQuest [370]

The educational background of the gifted indian pre-school child / Jaggan, Vijay Aheer Jaggan – Uni of South Africa 2000 [mf ed Johannesburg 2000] – 5mf – 9 – mf#mfm14912 – sa Unisa [370]

Educational broadcasting review – Washington. 1967-1973 (1) 1971-1973 (5) – ISSN: 0013-1660 – mf#2558 – us UMI ProQuest [380]

Educational centers : daytona beach / Goebel, Rubye K – s.l, s.l? 1936 – 1r – us UF Libraries [370]

The educational circular – [S.l: s.n, 1876?-18–] – 9 – (incl ind) – mf#P04010 – cn Canadiana [370]

The educational conquest of the far east / Lewis, Robert Ellsworth – New York: FH Revell, c1903 – 1mf – 9 – 0-524-03586-5 – (incl bibl ref) – mf#1990-1046 – us ATLA [370]

Educational controversies in india : the cultural conquest of india under british imperialism / Boman-Behram, B K – Bombay: DB Taraporevala Sons & Co, [1943] – (= ser Samp: indian books) – us CRL [954]

Educational development – Bishop's Stortford. 1974-1975 (1) 1974-1975 (5) (9) – ISSN: 0013-1695 – mf#8912 – us UMI ProQuest [700]

Educational digest – Gormley. 1975-1993 (1) 1975-1993 (5) 1975-1993 (9) – ISSN: 0046-1482 – mf#10780 – us UMI ProQuest [370]

Educational documentation and information : bulletin of the international bureau of education / International Bureau of Education – Paris. 1950-1984 (1) 1971-1984 (5) 1976-1984 (9) – (cont by: bulletin of the international bureau of education) – ISSN: 0303-3899 – mf#461 – us UMI ProQuest [370]

Educational evaluation and policy analysis – Washington. 1979+ – 1,5,9 – ISSN: 0162-3737 – mf#11834 – us UMI ProQuest [370]

Educational executives' overview – New York. 1960-1963 – 1 – ISSN: 0424-575X – mf#5536 – us UMI ProQuest [370]

Educational facility planner – Columbus. 1989-1995 (1) 1989-1995 (5) 1989-1995 (9) – (cont: cefp journal) – mf#10304,01 – us UMI ProQuest [370]

Educational Film Library Association *see* Efla evaluations

Educational formations of the jesuits... / Jacobsen, Jerome V – Madrid: Razon y Fe, 1940 – sp Bibl Santa Ana [241]

Educational forum – Indianapolis. 1936+ (1) 1968+ (5) 1975+ (9) – ISSN: 0013-1725 – mf#1049 – us UMI ProQuest [370]

Educational foundations – Ann Arbor. 1986+ – 1,5,9 – ISSN: 1047-8248 – mf#17147 – us UMI ProQuest [370]

Educational gerontology – New York. 1976+ (1,5,9) – ISSN: 0360-1277 – mf#11135 – us UMI ProQuest [618]

Educational handwork / Kidner, Thomas Bessill – Toronto: Educational Book Co, c1910 [mf ed 1996] – 2mf – 9 – 0-665-80967-0 – mf#80967 – cn Canadiana [331]

Educational horizons – Bloomington. 1921+ (1) 1972+ (5) 1972+ (9) – ISSN: 0013-175X – mf#8224 – us UMI ProQuest [370]

The educational ideal in the ministry / Faunce, William Herbert Perry – New York: Macmillan, 1908 – (= ser Lyman Beecher Lectures) – 1mf – 9 – 0-7905-3739-7 – mf#1989-0232 – us ATLA [240]

The educational ideal in the ministry : the lyman beecher lectures at yale university in the year 1908 / Faunce, William Herbert Perry – New York: Macmillan, 1908 – 1mf – us ATLA [240]

Educational ideas and institutions in ancient india / Sarkar, Subimal Chandra – [Patna: Patna College, 1928] – (= ser Samp: indian books) – us CRL [370]

The educational journal *see* The educational weekly

Educational journal of virginia – Richmond, VA. 1869-1891 (1) – mf#66817 – us UMI ProQuest [071]

Educational journal of western canada – Brandon [Man: s.n, 1899-1903] – 9 – mf#P04401 – cn Canadiana [370]

Educational leader / Church of Our Lord Jesus Christ of the Apostolic Faith – 1979 – 1r – 1 – mf#4024258 – us WHS [242]

Educational leadership – Alexandria. 1943+ [1]; 1968+ [5]; 1975+ [9] – ISSN: 0013-1784 – mf#1454 – us UMI ProQuest [370]

Educational management abstracts – Abingdon. 2000+ – 1 – ISSN: 1467-582X – mf#20975,01 – us UMI ProQuest [370]

Educational management and administration – London. 1982-1995 – 1,5,9 – (cont: educational administration) – ISSN: 0263-211X – mf#11557,01 – us UMI ProQuest [370]

Educational measurement, issues and practice – Washington. 1989+ – 1,5,9 – ISSN: 0731-1745 – mf#17157 – us UMI ProQuest [370]

Educational media – Toronto. 1969-1971 (1) 1969-1971 (5) (9) – ISSN: 0013-1814 – mf#6031 – us UMI ProQuest [370]

Educational media international – London. 1973+ (1) 1975+ (5) 1975+ (9) – ISSN: 0952-3987 – mf#8692 – us UMI ProQuest [370]

Educational method : a journal of progressive public schools – Washington. 1921-1943 – 1 – mf#2209 – us UMI ProQuest [370]

Educational missions / Barton, James L – New York: Student Volunteer Movement for Foreign Missions, 1913 – 1mf – 9 – 0-7905-4329-X – (incl bibl ref) – mf#1988-0329 – us ATLA [240]

Educational missions in india : revised special report of the committee for the propagation of the gospel in foreign parts, especially in india, to the general assembly of the church of scotland / M'Murtrie, John – [s.l.: Printed by W. Blackwood, 1890?]. Chicago: Dep of Photodup, U of Chicago Lib, 1971 (1r); Evanston: American Theol Lib Assoc, 1984 (1r) – 1 – 0-8370-0466-7 – (incl ind) – mf#1984-B245 – us ATLA [240]

The educational museum and school of art and design for upper canada : with a plan of the english educational museum, etc / Canada (Province). Dept of Public Instruction for Upper Canada – Toronto: printed by Lovell [sic] & Gibson, 1858 [mf ed 1983] – 1mf – 9 – mf#SEM105P334 – cn Bibl Nat [378]

Educational music magazine – Chicago. 1949-1957 – 1 – mf#282 – us UMI ProQuest [370]

Educational news : a weekly record and review – [Scotland] Edinburgh: Educational Institute of Scotland jan 1876-apr 1918 (wkly) [mf ed 2004] – 90r – 1 – (missing: 1882; cont: scottish educational journal (edinburgh, scotland : 1856); cont by: scottish educational journal [may 1918-dec 1950]) – uk Newsplan [370]

Educational news bulletin – 1908 nov11-1913 feb 23, 1913 mar 24-1921 jun – 1r – 1 – mf#481970 – us WHS [370]

Educational performance of athletes and nonathletes in two mississippi rural high schools / Jefferson, Ceroy – 1999 – 1mf – 9 – $4.00 – mf#PSY 2089 – us Kinesology [306]

Educational perspectives – Honolulu. 1962+ (1) 1976+ (5) 1976+ (9) – ISSN: 0013-1849 – mf#8581 – us UMI ProQuest [370]
Educational philosophy and theory – Abingdon. 1977+ – 1,5,9 – ISSN: 0013-1857 – mf#11558 – us UMI ProQuest [370]
The educational philosophy of mahatma gandhi / Patel, M S – Ahmedabad: Navajivan Pub House, 1953 – (= ser Samp: indian books) – (foreword by hansa mehta) – us CRL [370]
The educational philosophy of national socialism / Kneller, George Frederick – New Haven: Yale University Press, 1941. viii,299p. Bibliography – 1 – us Wisconsin U Libr [370]
Educational policy – Los Altos. 1987+ – 1,5,9 – ISSN: 0895-9048 – mf#16647 – us UMI ProQuest [370]
The educational psychological effect of the cochlear implant on the hearing-impaired child's family / Bezuidenhout, Elsie Petronella – Uni of South Africa 2001 [mf ed Johannesburg 2001] – 4mf [ill] – 9 – (incl bibl ref) – mf#mfm14822 – sa Unisa [150]
An educational psychological perspective on partner roles in heterosexual marriages / Phetla, Rabi Joseph – Uni of South Africa 2000 [mf ed Johannesburg 2000] – 4mf – 9 – (incl bibl ref) – mf#mfm15055 – sa Unisa [150]
Educational psychologist – Hillsdale. 1963+ (1,5,9) – ISSN: 0046-1520 – mf#12435 – us UMI ProQuest [150]
Educational psychology – Dorchester-on-Thames. 1994-1996 (1,5,9) – ISSN: 0144-3410 – mf#20932 – us UMI ProQuest [150]
Educational psychology review – New York. 1989+ (1,5,9) – ISSN: 1040-726X – mf#17662 – us UMI ProQuest [150]
Educational reconstruction – Bombay: Vora & Co, Publishers, 1938 – (= ser Samp: indian books) – us CRL [370]
Educational reconstruction / Kuppuswamy, Bangalore – Mysore: Sri Kantha Business Syndicate, 1949 – (= ser Samp: indian books) – us CRL [370]
Educational reconstruction, 1943 : command n6458 – 1mf – 9 – mf#86954 – uk Microform Academic [324]
Educational record – Washington. 1920-1997 (1) 1969-1997 (5) 1975-1997 (9) – (cont by: presidency) – ISSN: 0013-1873 – mf#1020 – us UMI ProQuest [370]
The educational record of the province of quebec – Montreal: Protestant Committee of the Board of Education, [1881?-1965] – 9 – mf#P04834 – cn Canadiana [370]
Educational reform in japan, 1945-1952 – 2pt – (= ser The occupation of japan) – 9 – (pt1 461mf isbn 0-88692-199-6 $4280; guide only $290. pt2 562mf isbn 1-55655-678-0 $6620; guide only $605) – us UPA [370]
Educational research bulletin – Columbus. 1922-1961 – 1 – mf#848 – us UMI ProQuest [370]
Educational research quarterly – West Monroe. 1976+ – 1,5,9 – ISSN: 0196-5042 – mf#10991 – us UMI ProQuest [370]
Educational researcher – Washington. 1964+ (1) 1977+ (5) 1977+ (9) – ISSN: 0013-189X – mf#6185 – us UMI ProQuest [370]
Educational resources and techniques – Tyler. 1960-1992 (1) 1972-1992 (5) 1975-1992 (9) – ISSN: 0424-5997 – mf#5999 – us UMI ProQuest [370]
Educational review – Birmingham. 1994-1996 – 1,5,9 – ISSN: 0013-1911 – mf#20933 – us UMI ProQuest [370]
Educational review – Garden City. 1891-1928 – 1 – ISSN: 0190-4191 – mf#2884 – us UMI ProQuest [370]
The educational review – Shanghai: China Christian Educational Assoc. v1-30 n4. may 1907-nov 1938 (frequency varies) [all publ] – (= ser Missionary periodicals from the china mainland) – 6r – 1 – $920.00 – (title varies) – us UPA [242]
Educational Services [Washington DC] *see* Cambodian
The educational situation / Dewey, John – Chicago: The University of Chicago Press, 1902 [mf ed 1970] – (= ser University of chicago. contributions to education 3; Library of american civilization 14968) – 1mf – 9 – us Chicago U Pr [370]
Educational statistics in the north-western state, nigeria / Planning and Administrative Division, Ministry of Education and Community Development – Sokoto, North-Western State: The Division, 1968-72 – us CRL [370]
Educational studies – Dorchester-on-Thames. 1993-1996 – 1,5,9 – ISSN: 0305-5698 – mf#20934 – us UMI ProQuest [370]
Educational studies – Ypsilanti. 1970+ (1) 1976+ (5) 1976+ (9) – ISSN: 0013-1946 – mf#10412 – us UMI ProQuest [370]
Educational studies in mathematics – Dordrecht. 1984+ (1,5,9) – ISSN: 0013-1954 – mf#14745 – us UMI ProQuest [510]
The educational system of the ancient hindus / Das, Santosh Kumar – Calcutta: [sn], 1930 (Calcutta: Mitra Press) – (= ser Samp: indian books) – us CRL [377]
The educational systems of the puritans and jesuits compared : a premium essay written for "the society for the promotion of collegiate and theological education at the west" / Porter, Noah – New York: M W Wood, 1851 – 1mf – 9 – 0-8370-7657-9 – (incl bibl ref) – mf#1986-1657 – us ATLA [377]
Educational technology, research and development – Washington. 1989+ – 1,5,9 – ISSN: 1042-1629 – mf#16914 – us UMI ProQuest [370]
Educational theatre journal – Washington. 1949-1978 (1) 1969-1978 (5) 1975-1978 (9) – (cont by: theatre journal) – ISSN: 0013-1989 – mf#1070 – us UMI ProQuest [790]
Educational theory – Urbana. 1951+ (1) 1978+ (5) 1978+ (9) – ISSN: 0013-2004 – mf#7203 – us UMI ProQuest [370]
Educational thoughts for the diamond jubilee year : inaugural address delivered in convocation hall, manitoba college, winnipeg, november 19th, 1897 / Bryce, George – Winnipeg?: s.n, 1897? – 1mf – 9 – mf#02850 – cn Canadiana [378]
Educational thoughts for the diamond jubilee year : inaugural address delivered in convocation hall, manitoba college, winnipeg, november 19th, 1897 / Bryce, George – [Winnipeg?: s.n, 1897?] [mf ed 1980] – 1mf – 9 – 0-665-02850-4 – mf#02850 – cn Canadiana [378]
The educational weekly – Toronto: Grip Print and Pub Co, [1884?-1887] – 9 – (merged with: canada school journal to become: the educational journal) – mf#P04915 – cn Canadiana [370]
Educationalist – Brighton, CW [Ont]: H Spencer, [1860-18– or 19–] – 9 – mf#P04412 – cn Canadiana [370]
Educationist – Indianapolis. 1873-1875 – 1 – mf#5039 – us UMI ProQuest [370]
Educator / Air Force Institute of Technology (US) – 1980 jul-1988 apr, fall – 1r – 1 – mf#1897880 – us WHS [629]
Educator – 1969 jun-1972 jan – 1r – 1 – (continued by: national educator) – mf#2870370 – us WHS [370]
Educator – Easton. 1838-1839 – 1 – mf#4615 – us UMI ProQuest [370]
Educator : official voice of local 1862, i.a.m.a.w. / International Association of Machinists and Aerospace Workers – v15 n2-v24 n2 [1981 feb-1990 may] – 1r – 1 – mf#1074437 – us WHS [331]
The educator – London [Ont.]: Jones & Co, [1868-18–?] – 9 – mf#P04406 – cn Canadiana [370]
The educator : a monthly illustrated magazine designed to promote the cause of education among the colored population of the united states – Baltimore MD: Industrial Dept of the Centenary Biblical Institute, oct 1886-sep 1888 [mf ed 2005] – 2v on 1r – 1 – mf#2005-S058 – us ATLA [370]
Educators' perceptions of outcomes-based education : with reference to the foundations phase / Salie, Durriyah – Pretoria: Vista University 2002 [mf ed 2002] – 2mf – 9 – (incl bibl ref) – mf#mfm15205 – sa Unisa [370]
The educator's role in a multicultural classroom : with reference to primary schools in pretoria / Setshedi, Justinus Rankgakgata – Pretoria: Vista University 2003 [mf ed 2003] – 2mf – 9 – (incl bibl ref) – mf#mfm15202 – sa Unisa [370]
Educause review – Boulder. 2000+ (1) 2000+ (5) 2000+ (9) – (cont: educom review) – ISSN: 1527-6619 – mf#10440,03 – us UMI ProQuest [378]
Educom – Princeton. 1974-1983 (1) 1974-1983 (5) 1974-1983 (9) – (cont by: educom bulletin) – ISSN: 0424-6268 – mf#10440 – us UMI ProQuest [378]
Educom bulletin – Princeton. 1984-1988 (1) 1984-1988 (5) 1984-1988 (9) – (cont: educom. cont by: educom review) – ISSN: 1045-9154 – mf#10440,01 – us UMI ProQuest [370]
Educom review – Washington. 1989-1999 (1) 1989-1999 (5) 1989-1999 (9) – (cont: educom bulletin. cont by: educause review) – ISSN: 1045-9146 – mf#10440,02 – us UMI ProQuest [378]
Edularios de la monarquia espanola de margarita, nueva andalucia...tomo 1 y 2. caracas, 1967 / Otte, Enrique – Madrid: Graf. Calleja, 1968 – 1 – sp Bibl Santa Ana [946]
Eduquemos a las madres / Rodriguez Pedreira, Jose – Plasencia: Imp. Gabriel y Galan, 1959 – 1 – sp Bibl Santa Ana [946]
Edvard griegs briefwechsel / ed by Oelmann, Klaus Henning – 9 – (v1: die briefe max abrahams an edvard grieg (mf ed 1994) 2mf €40 isbn: 3-8267-2018-0 dhs 2018) – gw Frankfurter [780]
Edward 6 and the book of common prayer : an examination into its origin and early history, with an appendix of unpublished documents / Gasquet, Francis Aidan & Bishop, Edmund – London: J. Hodges, 1890 – 2mf – 9 – 0-7905-5883-1 – (incl bibl ref) – mf#1988-1883 – us ATLA [240]
Edward 6th and the book of common prayer / Gasquet, Abbot & Bishop, E – London, 1948 – 4mf – 8 – €11.00 – ne Slangenburg [242]
The edward barnsley drawings collection : furniture and interior design – 7r – 1 – £380.00 – (incl printed guide) – mf#EBT – uk World [740]
Edward bickersteth : missionary bishop in japan / Bickersteth, Marion (forsyth) – Tokyo: Kyo Bun Kwan, 1914 [mf ed 1995] – (= ser Yale coll) – x/187p/iii (ill) – 1 – 0-524-10188-4 – (pref note by comp) – mf#1995-1188 – us ATLA [920]
Edward, Brother *see* Plenty how-do from africa
Edward burne-jones / Bell, Malcolm – London 1892 – (= ser 19th c art & architecture) – 4mf – 9 – mf#4.2.747 – uk Chadwyck [750]
Edward drumgoole papers / Drumgoole, Edward – 1770-1871. University of North Carolina Library. Guide – 1 – $90.00 – us CIS [920]
The edward everett papers, 1675-1930 – [mf ed 1972] – 70r – 1 – (with p/g) – us MA Hist [975]
Edward gayer andrews : a bishop of the methodist episcopal church / McConnell, Francis John – New York: Eaton & Mains, c1909 – 1mf – 9 – 0-524-06771-6 – mf#1991-2778 – us ATLA [242]
Edward, Georg *see* Balladen und lieder
Edward harold browne, d.d : lord bishop of winchester and prelate of the most noble order of the garter / Kitchin, George William – London: J. Murray, 1895 – 2mf – 9 – 0-7905-8085-3 – mf#1988-8021 – us ATLA [240]
Edward, Henry *see* Pastoral letter to the clergy and laity of the diocese of westminst...
Edward Hunt & Co. Bristol, England *see* City of bristol, newport and welch towns directory
Edward irving : man, preacher, prophet / Root, Jean C – 1912 – 1 – $50.00 – us Presbyterian [240]
Edward irving and his circle, including some consideration of the "tongues" movement in the light of modern psychology / Drummond, A L – 9 – $12.00 – us IRC [240]
Edward lawrence scull : a brief memoir, with extracts from his letters and journals / Thomas, Allen Clapp – Cambridge: Riverside Press, 1891 – 1mf – 9 – 0-524-06027-4 – mf#1991-2387 – us ATLA [240]
Edward S Joynes *see* Der zerbrochene krug
Edward white, his life and work / Freer, Frederick Ash – London: Elliot Stock, 1902 – 1mf – 9 – 0-524-07749-5 – mf#1991-3317 – us ATLA [920]
Edward woodruff vs north bloomfield gravel mining co controversial hydraulic mining case near mary – 10r – 1 – $500.00 – (with 16mm ind) – mf#B7006 – us Library Micro [574]
Edward woodruff vs north bloomfield gravel mining company : controversial hydraulic mining case near marysville, california – 10r – 5 – $500.00 – mf#B50530 – us Library Micro [346]
Edward youngs gedanken ueber die originalwerke : in einem schreiben an samuel richardson = Conjectures on original composition / ed by Jahn, Kurt – Bonn: A Marcus & E Weber 1910 – (= ser Kleine texte fuer theologische und philologische vorlesungen und uebungen 60) – 1mf – 9 – 0-524-04690-5 – (repr of earliest german trans, leipzig 1760 by h e von teubern) – mf#1990-1317 – us ATLA [410]
Edwardes, Marian *see* A dictionary of non-classical mythology
Edwardes, Stephen Meredyth *see*
- Babur
- Mughal rule in india

Edwardian inventories for bedfordshire / ed by Eeles, Francis Carolus – London: Longmans, Green 1905 [mf ed Wakefield: EP Microform Ltd 1971] – (= ser Alcuin club colls 6) – n6 on reel – 1 – (inventories for 1552; ed fr transcr by j e brown) – uk Microform Academic [941]
Edwardian inventories of church goods *see* Documents towards a history of reformation in cornwall
Edwards *see* Citrus
Edwards academy, greeneville, tn executive committee minutes, 1877-1881 – 1r – 1 – $35.00 – mf$-47 – us Commission [378]
Edwards, Alfred George *see* Landmarks in the history of the welsh church
Edwards, Amelia Ann Blanford *see* Thousand miles up the nile
Edwards, Bela Bates *see*
- The missionary gazetteer
- Selections from german literature
- Writings of professor b. b. edwards

Edwards, Bryan *see* The history, civil and commercial, of the british colonies in the west indies
Edwards, C *see* Han-mu-la-pi fa tien
Edward's chancery appeals reports / New York. (State) – v1-4. 1831-50 (all publ) – 33mf – 9 – $49.50 – (a pre-nrs title) – mf#LLMC 80-021 – us LLMC [340]
Edwards, Charles *see* A practical treatise on parties to bills and other pleadings in chancery: with precedents
Edwards, Charles Lincoln *see* Bahama songs and stories
Edwards, Clark *see* Us agriculture's potential to supply world food markets
Edwards, David *see*
- Medical criticism
- Pulpit criticism

Edwards, Dennis [comp] *see* Cape town guide [the]
Edward's dream : the philosophy of a humorist = Eduards traum / Busch, Wilhelm; ed by Carus, Paul – Chicago: The Open Court Pub Co, 1909 [mf ed 1989] – 74p – 1 – (trans fr the german of wilhelm busch by ed) – mf#7096 – us Wisconsin U Libr [890]
Edwards, E *see* Liber monasterii de hyda (rs45)
Edwards, E H *see*
- Fire and sword in shansi

Edwards, Edward *see*
- The administrative economy of the fine arts in england
- Anecdotes of printers who have resided or been born in england
- The napoleon medals

Edwards, Edwin *see* Ecclesiastical jurisdiction
Edwards, Eliezer *see* Words, facts, and phrases
Edwards flyers – Edwards Air Force Base, CA. 1964-1966 (1) – mf#61966 – us UMI ProQuest [071]
Edwards, George *see*
- Advertisement
- A certain way to save our country
- The discovery of the true and natural era of mankind
- Effectual means of relieving the exigencies and grievances of the times
- An explanatory address, and vindication, to the legislature
- The five practical plans
- The golden age
- Humble and explanatory memorial of dr george edwards
- Humble petitions, etc
- The income tax fathered
- A letter addressed to the different orders of the united kingdom
- The means of saving our country
- No 3
- The pioneer work of the presbyterian church in montana
- A plain practical plan
- The plan and documents, pt 1
- The practical system of human economy
- Radical means of counteracting the present scarcity, and preventing famine in future
- Reasons why a true or genuine system of public and private welfare
- The royal redress of the times
- A short view of a work, entitled "nature's policy for man and nations"
- Summary means
- The whole and sole cause of our present critical situation

Edwards, Gus C *see* Legal laughs
Edwards, J *see* Reminiscences of the early life and missionary labours...
Edwards, J F *see* The life and teaching of tukaram
Edwards, John *see*
- The inquisitions
- The scripture-doctrine of the five points

Edwards, John Baker *see*
- On the water supply of montreal and its suburbs
- On trichina spiralis
- Report of a meeting of the montreal natural history society

Edwards, John Hugh *see* David lloyd george
Edwards, Jonathan *see* Sermons, 1766-1800
Edwards, Joseph Plimsoll *see*
- The history of freemasonry in nova scotia
- Louisbourg
- The public records of nova scotia

[Edwards, Martin Luther] *see* The bible and reason against atheism
Edwards, Morgan *see*
- The customs of primitive churches
- Manuscript sermons
- Materials toward a history of baptists in prov. of penn., r.i., n.j., del., md., va., n.c., s.c., and georgia
- Materials towards a history of the baptists in delaware state

Edwards, Oliver *see* Englische dichtung aus goethes zeitalter im licht deutscher kunstlehre
Edwards' philatelic press list [and] advertiser of philatelists' supplies – Montreal: J. Edwards, [1896?-1898] – 9 – ISSN: 1190-7371 – mf#P04544 – cn Canadiana [760]

EFFECT

Edwards, S J Celestine *see* From slavery to a bishopric
Edwards, Smalley et al *see* The atonement
Edwards, Thomas *see* Twelve favourite new country dances for the violin, harp or pianoforte
Edwards, Thomas Charles *see*
- A commentary on the first epistle to the corinthians
- The god-man
Edwards, Thomas, jr *see* Quiz book: questions and answers on the subject of bailments and carriers.
Edwards, William *see* Four centuries of nonconformist disabilities, 1509-1912
Edwards, William Cameron *see* 1868-1918
Edwards, William Henry *see* Shaksper not shakespeare
Edwards, William Seymour *see* On the mexican highlands
Edwin arnold as poetizer and as paganizer : containing an examination of the "light of asia" for its literature and for its buddhism / Wilkinson, William Cleaver – New York: Funk & Wagnalls 1884 [mf ed 1993] – (= ser Standard library 131) – 3mf – 9 – 0-524-07927-7 – mf#1991-3472 – us ATLA [280]
Edwin octavius tregelles, civil engineer and minister of the gospel / Tregelles, Edwin Octavus; ed by Fox, Sarah E – London: Hodder and Stoughton, 1892 – 2mf – 9 – 0-524-07767-3 – mf#1991-3335 – us ATLA [240]
Edwin von manteuffel als quelle zur geschichte friedrich wilhelms 4 / Schmitz, Elisabeth – Muenchen, Berlin, 1921 (mf 1992) – 1mf – 9 – €24.00 – 3-89349-057-4 – mf#DHS-AR 19 – gw Frankfurter [943]
Edwins, A W et al *see* Our first decade in china, 1905-1915
Edworth – (= ser Bedfordshire parish register series) – 1mf – 9 – £3.00 – uk BedsFHS [929]
Edworth, st george monumental inscriptions – Bedfordshire Family HS 1996 – (= ser Bedfordshire parish register series) – 1mf – 9 – £1.25 – uk BedsFHS [929]
Edzardi, Anton *see* Untersuchungen ueber koenig rother
Ee : systems engineering today – Radnor. 1942-1974 (1) 1971-1972 (5) – ISSN: 0090-5356 – mf#1040 – us UMI ProQuest [621]
Eeden, Guy Van *see* Crime of being white
Eeden, Jeanne van *see* The representation of mythical africa at the lost city
Eeg-emg : zeitschrift fuer elektroenzephalographie elektromyographie und verwandte gebiete – Stuttgart. 1975-1976 (1) 1975-1976 (5) 1975-1976 (9) – ISSN: 0012-7590 – mf#10158 – us UMI ProQuest [610]
Eei bulletin / Edison Electric Institute – New York. 1933-1974 (1) – ISSN: 0012-7604 – mf#8531 – us UMI ProQuest [621]
Eek, Dirk van *see* Napoleon im spiegel der goetheschen und der heineschen dichtung
Eekhof, Albert *see* De hervormde kerk in noord-amerika (1624-1664)
The eel ground times – Eel Ground [NB]: S B W Francis, [1879] – 9 – mf#P04928 – cn Canadiana [073]
Eel river weekly – South Whitley, IN. 1934-1935 (1) – mf#62979 – us UMI ProQuest [071]
Eeles, Francis Carolus *see*
- Edwardian inventories for bedfordshire
- Traditional ceremonial and customs connected with the scottish liturgy
Eells, Myron *see*
- Father eells
- A history of indian missions of the pacific coast
- History of the congregational association of oregon and washington territory, the home missionary society of oregon and adjoining territories, and the northwestern association of congregational ministers
- Marcus whitman, pathfinder and patriot
- The relations of the congregational colleges to the congregational churches
- Ten years of missionary work
De eendracht : het hollandsche orgaan voor de oostelijke provincie der kaap kolonie en voor de aangrenzende districten van den oranje vrijstaat – Aliwal north SA, oct 1885-dec 24 1886 (wkly) [mf ed Cape Town: SA library 1986] – 1r – 1 – mf#MS00379 – sa National [079]
Eenige bybelse figuuren gelykenissen en zinnebeelden / Luyken, Jan – n.p, n.d. – 1mf – 9 – mf#0-3112 – ne IDC [090]
Eenige schetsen voor eene geschiedenis van de trekboeren, thans bekend ander de naam van de gereformeerde gemeente te st januario, humpata, distrikt mossamedes, provincie angola, op de west kust van zuid-afrika – Amsterdam, Pretoria: Hoveker & Wormser, 1897 – 1 – us CRL [960]
Eenvoudige japansche spraakkunst : erste deel / Moosdijk, P van der – erste druk. Batavia: N V Boekhandel Visser and Co, 2602 (apr 1942) – 112p 2mf – 9 – mf#SE-2002 mf112-113 – ne IDC [480]
Eenzaam buitenleven met aantekeningen en zinnebeelden verrykt / Sluiter, W – t'Amsterdam: Jacob van Royen, 1717 – 4mf – 9 – mf#0-431 – ne IDC [090]
Eeo spotlight – Washington. 1974-1978 (1) 1974-1978 (5) 1974-1978 (9) – (cont by: spotlight) – ISSN: 0190-2326 – mf#9154,01 – us UMI ProQuest [331]
Eeo today – New York. 1978-1982 (1,5,9) – (cont by: employment relations today) – ISSN: 0362-5818 – mf#11881 – us UMI ProQuest [331]
Eeq : exceptional education quarterly – Rockville. 1980-1983 – 1,5,9 – ISSN: 0196-6960 – mf#12729 – us UMI ProQuest [370]
Eerdmans, Bernadus Dirk *see*
- Das buch exodus
- Das buch leviticus
- Die komposition der genesis
- Die vorgeschichte israels
Eerdmans, Bernardus Dirk *see* Het roomsche gevaar
Eerlycke tytkorting : bestaende in verscheyde rymen / Krul, J H – Haerlem: Hendrick van Marcke end Theunis Jansen, 1634 – 9mf – 9 – mf#0-3102 – ne IDC [090]
Eerlycke tytkorting : bestaende in verscheyde rymen / Krul, J H – t'Amsterdam: Cornelis van Breugel, 1635 [1640] – 1mf – 9 – mf#0-3101 – ne IDC [090]
El eersheh / Griffith, F L – London, 1893-1894. 2pts – 6mf – 9 – mf#NE-20396 – ne IDC [956]
Eerste boekjaar der indische partij 1912 : samengesteld...enz / Ham, J G van – Bandoeng, 1913 – (= ser Publicaties der Indische Partij 7) – 2mf – 8 – mf#SE-1427 – ne IDC [959]
De eerste christelijke gereformeerde gemeente, 1867-1917, muskegon, michigan – Muskegon, MI: [s.n, 1917] [mf ed 1993] – (= ser Reformed church coll) – 24p on 1mf – 9 – 0-524-06612-4 – mf#1991-2667 – us ATLA [242]
Eerste deel der bouw-kunst, ofte grondige bewijs-redenen... / Vermaarsch, J – Leyden, 1664 – 3mf – 9 – mf#0A-78 – ne IDC [720]
Eerste onderzoekingen met den mikrometer van airy : volbragt op het observatorium der hoogeschool te leiden / Kaiser, Frederik – [Amsterdam: C G van der Post 1857] [mf ed 1998] – 1r – 1 – mf#film mas 28292 – us Harvard [520]
De eerste vier verzen van den zestienden psalm; nog eens, de eerste vier verzen van psalm 16 / Wildeboer, Gerrit – Groningen: [s.n.], 1891-[1893] [mf ed 1993] – 12/[4]p on 1mf – 9 – 0-524-08613-3 – (incl bibl ref) – mf#1993-0048 – us ATLA [221]
Eesti baptismi ajalugu: i. arkamise aeg / Tuttar, H & Dahl, H V – Estonian Baptist History: 1. Revival Time. Tallinn: Publishing House of the Estonian Baptist Churches, 1929. 134p – 1 – 5.36 – us Southern Baptist [242]
Eesti haal – London, England. -m. Dec 1947-Dec 1952. 1 reel – 1 – uk British Libr Newspaper [072]
Eesti haal – London, UK. Jan 1953- – 1 – uk British Libr Newspaper [072]
Eesti paevaleht – Stockholm, Sweden. 1979- – 1 – sw Kungliga [078]
Eesti post – Geislingen a.d. Steige DE, 1948 1 sep-1952 13 aug – 2r – 1 – uk British Libr Newspaper [074]
Eesti raamatute uldnimestik – Tartu, Estonia. 1937-1939 – 1 – us NY Public [010]
Eesti rada – Augsburg, DE. 4 nov 1947-aug 1952 – 1 – uk British Libr Newspaper [947]
De eeuwige cirkel leven en strijd van de indianen en marrons in suriname / Assid Door – Den Haag, Netherlands: Van der Laan 1946 [mf ed 1990] – 1r – 1 – us UF Libraries [972]
Het eeuwigh leven / Hertogenbosch, Ioannes Evangelista van 'S – Tot Loven, 1643 – 4mf – 8 – €11.00 – ne Slangenburg [240]
Efail isaf, tabernacle, monumental inscriptions – 1mf – 9 – £1.25 – uk Glamorgan FHS [929]
Efemerides burgalescas / Albarellos, Juan – Madrid: Archivo Ibero Americano, 1964 – 1 – sp Bibl Santa Ana [946]
Efemerides de mompos : (con licencia eclesiastica) / Rodriguez Hontiyuelo, Mariano – Cartagena, Colombia. 1935 – 1r – us UF Libraries [972]
Efemerides para escribir la historia / Pardo, Jose Joaquin – Guatemala, 1944 – 1r – us UF Libraries [972]
Efendi, Ahmed *see* The divan project
Efendi, Ahmed Asim *see* Asim tarihi
Efendi, A'ma Yusuf Garibi *see* The divan project
Efendi, Esrar Dede *see* The divan project
Efendi, Halat *see* The divan project
Efendi, Hanyevi Sefik *see* The divan project
Efendi, Hayrullah *see* Tarih-i devlet-i aliyye-i 'osmaniyye
Efendi, Hazik *see* The divan project
Efendi, Hudayi Aziz Mahmud *see* The divan project
Efendi, Kaygulu *see* The divan project
Efendi, Muestak *see* The divan project
Efendi, Munse'at-i Rif'at *see* The divan project
Efendi, Mustafa *see* Tarih-i nefis
Efendi, Nes'et *see* The divan project
Efendi, Rasit *see* Tarih-i rasit
Efendi, Selaniki Mustafa *see* Tarih-i selaniki
Efendi,Vecdi *see* The divan project
Efermerides navais brasileiras / Vasconcelos, Alberto – Rio de Janeiro, Brazil. 1961 – 1r – us UF Libraries [972]
Efes dammaim : a series of conversations at jerusalem between a patriarch of the greek church and a chief rabbi of the jews, concerning the malicious charge against the jews of using christian blood = Efes damim / Levinsohn, Isaac Baer – London: Longman, Brown, Green, and Longmans, 1841 – 1mf – 9 – 0-7905-0096-5 – mf#1987-0096 – us ATLA [939]
Effacing the 'god-king' : internal developments in cambodia since march 1970 / Osborne, Milton E – [n.p. 1971?] [mf ed 1989] – 1r with other items – 1 – mf#mf-10289 seam reel 016/13 [§] – us CRL [959]
L'effect de la guerre sur nos methodes d'elevage et d'agriculture / Barre, Stanislas Morrier – [Quebec?: Imp[r] l'Action sociale], 1917 – 1mf – 9 – 0-665-66846-5 – mf#66846 – cn Canadiana [630]
The effect of a 10-week stress management course on self-reported stress-related physical and psychological symptoms / Roberts, Renee – 1997 – 2mf – 9 – $8.00 – mf#PSY 1986 – us Kinesology [612]
The effect of a 12-week resistive training program in the home using the body bar on dynamic and absolute strength in middle-age women / Mortell, Rosemarie – 1992 – 1mf – $4.00 – us Kinesology [612]
The effect of a 12-week resistive training program on the blood lipid levels of previously sedentary adult women / Martin, James R – 1994 – 2mf – 9 – $8.00 – mf#PH 1589 – us Kinesology [612]
The effect of a 50-km ultramarathon on vitamin b6 metabolism and plasma and urinary urea nitrogen / Grediagin, Ann – 2000 – 222p on 3mf – 9 – $15.00 – mf#PH 1727 – us Kinesology [612]
The effect of a 60-minute duration exercise, at the intensities of the lactate and the individual anaerobic thresholds, on the cardiovascular drift / Buhre, U T – 1992 – 2mf – 9 – $8.00 – us Kinesology [613]
The effect of a circuit weight training program followed by a detraining period on saliva cortisol and testosterone in males / Mazzocca, Augustus D – 1989 – 164p 2mf – 9 – $8.00 – us Kinesology [612]
The effect of a lifetime of physical activity on the quantity of bone in the canine / Fedler, Joan M & Maynard, Jerry A – 1989 – 1mf – 9 – $4.00 – us Kinesology [613]
The effect of a motor development program on preschool children's motor skills / Bargen, Melinda – 2000 – 69p on 1mf – 9 – $5.00 – mf#PSY 2148 – us Kinesology [612]
The effect of a nutrition intervention on body weight and body composition of hiv-infected individuals at risk for wasting syndrome / Timpel, Tara J – 1999 – 1mf – 9 – $4.00 – mf#HE 629 – us Kinesology [790]
The effect of a physical activity intervention based on the transtheoretical model in changing physical-activity-related behavior on low-income elderly volunteers / Braatz, Janelle S – 1997 – 3mf – 9 – $12.00 – mf#HE 630 – us Kinesology [614]
Effect of a psychological skills training program on competition anxiety and performance of selected national youth sports program campers / Cox, Kimberley A – Temple University, 1995 – 2mf – 9 – $8.00 – mf#PSY1842 – us Kinesology [150]
Effect of a rehabilitation programme on anxiety, depression and coping strategies in alcoholics / Soobedar, Latifa – Stellenbosch: U of Stellenbosch 1998 [mf ed 1998] – 1mf – 9 – mf#mf.1277 – sa Stellenbosch [616]
Effect of a softshell prophylactic ankle stabilizer on performance in events involving speed, agility, and vertical jump during long-term use / Locke, Alison B – Temple University, 1996 – 1mf – 9 – mf#PE 3662 – us Kinesology [617]
The effect of a study of the biblical concept of church on establishing long range planning goals / Duncan, Daniel Wendell – 1982 – 1 – 7.52 – us Southern Baptist [242]
The effect of a weight training program on the bone density of women aged 40-50 years / Dornemann, Timothy M – 1994 – 1mf – $4.00 – us Kinesology [612]
The effect of a wilderness therapy program on youth-at-risk, as measured by locus of control and self-concept / Anderson, Amy – Brigham Young University, 1995 – 1mf – 9 – mf#PSY 1879 – us Kinesology [150]
The effect of active recovery on the post-exercise diffusion capacity / Chen, Kevin Y – 1998 – 1mf – 9 – $4.00 – mf#PH 1614 – us Kinesology [612]
Effect of acupuncture tens on second degree ankle sprains / Javens, J A – 1988 – 1mf – 9 – $4.00 – us Kinesology [790]
The effect of acute exercise on state anxiety and acoustic startle eyeblink response in physically active and inactive men / Tieman, James G & Dishman, Rodney K – 1990 – 2mf – 9 – $8.00 – us Kinesology [150]
The effect of adhesive ankle strapping upon isokinetic strength as measured by use of the biodex dynamometer / Wennerberg, D K – 1989 – 2mf – 9 – $8.00 – us Kinesology [613]
The effect of aerobic exercise on the rate of protein catabolism during a seventy-two hour fast / Sourisseau, George E – 1981 – 1mf – 9 – $4.00 – us Kinesology [790]
Effect of age and thirty minutes of exercise on prostacyclin thromboxane a2 ratios and circulating concentrations of prostacyclin and thromboxane a2 / Todd, M K – 1990 – 2mf – 9 – $8.00 – us Kinesology [612]
Effect of age on reaction and movement times in girls and women / Marranca, Harriett A & Gench, Barbara E – 1992 – 2mf – $8.00 – us Kinesology [150]
The effect of aids on the attitudes of secondary school learners towards sexual behaviour / Mothabeng, Mampheletso Bernice – Pretoria: Vista University 2002 [mf ed 2002] – 3mf – 9 – (incl bibl ref) – mf#mfm15248 – sa Unisa [373]
The effect of altering speed of backward movement of the trunk on anticipatory postural adjustments / Weissblueth, Eyal & Cole, Kelly J – 1991 – 2mf – 9 – $8.00 – us Kinesology [612]
Effect of an active attentional strategy on running economy of low economical runners / Smith, Alan L & Gill, Diane L – 1993 – 2mf – 9 – $8.00 – us Kinesology [150]
Effect of an exercise program on quality of life of women with fibromyalgia / Gandhi, Namita – 2000 – 1mf – 9 – $4.00 – mf#HE 657 – us Kinesology [616]
The effect of an incentive-based wellness challenge program on physical fitness in industrial workers / Cox, Kelly M – 1997 – 2mf – 9 – $8.00 – mf#HE 588 – us Kinesology [612]
Effect of an interactive multimedia computer tutorial on students' understanding of ballet allegro terminology / Fisher-Stitt, Norma S – Temple, University, 1996 – 2mf – 9 – $8.00 – mf#PE 3640 – us Kinesology [790]
The effect of an ncaa division 1 wrestling season on selected physiological variables / Schultz, Mark P – 1997 – 1mf – 9 – $4.00 – mf#PH 1605 – us Kinesology [612]
The effect of angle, velocity, and rotation of incidence on the angle deviation of rebounding tennis balls / Smith, James F – 1988 – 153p 2mf – 9 – $8.00 – us Kinesology [790]
The effect of anterior cruciate ligament reconstruction on ground reaction forces during locomotion / Simenz, Christopher J – 1999 – 1mf – 9 – $4.00 – mf#PE 3992 – us Kinesology [611]
The effect of arm and leg versus legs alone exercise on the stairmaster 4000pt in females / Belford, Michele L – University of Wisconsin-La Crosse, 1995 – 1mf – 9 – mf#PH 1486 – us Kinesology [612]
The effect of arm movement on the biomechanics of standing up / Carr, J H – 1991 – 2mf – 9 – $8.00 – us Kinesology [790]
The effect of athletic participation on school discipline / Hudson, Scott B – 1999 – 1mf – 9 – $4.00 – mf#PSY 2090 – us Kinesology [150]
Effect of batting stance on ground reaction forces, bat velocity, and response time / LaBranche, Matthew – 1994 – 1mf – $4.00 – us Kinesology [612]
The effect of bench height on heart rateof college-age women of short and tall stature / Wright, Susan K – 1999 – 72p on 1mf – 9 – $5.00 – mf#PH 1697 – us Kinesology [612]
The effect of bicycle crank arm length on oxygen consumption at a constant workload and cadence / Morris, D M – 1992 – 1mf – 9 – $4.00 – us Kinesology [612]
The effect of body position on spinal cord injured swimmers / Malone, KN – 1990 – 1mf – 9 – $4.00 – us Kinesology [790]
The effect of breathing technique on blood pressure response to weight lifting / Linsenbardt, S – 1989 – 1mf – 9 – $4.00 – us Kinesology [612]
The effect of bromelain on recovery from exercise-induced skeletal muscle injury / Walker, J A – 1990 – 1mf – 9 – $4.00 – us Kinesology [615]

The effect of cadence on aerobic and anaerobic contributions to the total energy requirements of cycling at constant power output / Reimer, Brad W & Sanderson, David – 1991 – 2mf – 9 – $8.00 – us Kinesology [613]

The effect of caffeine and ephedrine on strength, power, and quickness / Putnam, Shawn R – 2000 – 134p on 2mf – 9 – $10.00 – mf#PH 1718 – us Kinesology [612]

The effect of calcium supplementation on blood pressure and hemodynamic variables in hypertensive males / Shaeffer, Kristen L – Springfield College, 1994 – 2mf – 9 – $8.00 – mf#PH1475 – us Kinesology [612]

Effect of calcium-deficient roughages upon mild production and welfare of dairy cows / Becker, R B – Gainesville, FL. 1933 – 1r – us UF Libraries [636]

The effect of carbohydrate-electrolyte ingestion on sprint performance following high intensity running in males / Robinson, Ellyn M – 1999 – 2mf – 9 – $8.00 – mf#PH 1658 – us Kinesology [612]

The effect of carbonated solutions on gastric emptying during prolonged cycling / Beard, Glenn C – 1990 – 77p 1mf – 9 – $4.00 – us Kinesology [612]

Effect of cd-rom enhanced lectures on substance abuse test scores / George, Joelle – 1997 – 1mf – 9 – $4.00 – mf#HE 608 – us Kinesology [360]

Effect of chronic cocaine on selected physiological responses during rest and exercise in rats / Kelly, K Patrick – 1993 – 2mf – $8.00 – us Kinesology [619]

Effect of chronic ethanol consumption and moderate intensity endurance training on murine plasma corticosterone concentration / Sipp, T L – 1992 – 1mf – 9 – $4.00 – us Kinesology [612]

The effect of chronic exercise stress on hippocampal glucocorticoid and serotonin 1a receptors / Jones, T B – 1998 – 2mf – 9 – $8.00 – mf#PH 1623 – us Kinesology [612]

The effect of cocaine on muscle carbohydrate metabolism and endurance during high intensity exercise in rats / Braiden, Russell W – 1993 – 1mf – $4.00 – us Kinesology [619]

The effect of cold water baths on post treatment leg electrical activity and isometric strength / Schroeder, Christine – 1981 – 1mf – 9 – $4.00 – us Kinesology [790]

The effect of computer technology in the sports information offices of the mid-american conference / Taylor, Chris – 1998 – 1mf – 9 – $4.00 – mf#PE 3887 – us Kinesology [790]

Effect of copper sulfate and potassium arsenate on the accumulatio... / Camp, John Perlin – s.l, s.l? 1927 – 1r – us UF Libraries [630]

The effect of core stabilization training on function performance in swimming / Scibek, Jason S – 1999 – 1mf – 9 – $4.00 – mf#PE 3945 – us Kinesology [611]

The effect of cranklength on oxygen consumption when cycling at a constant work rate / Carmichael, J Kevin – 1981 – 1mf – 9 – $4.00 – us Kinesology [790]

The effect of daily bean ingestion on the lipid profiles of normocholesterolemic college students / Damson, R L – 1991 – 1mf – 9 – $4.00 – us Kinesology [612]

The effect of delayed onset muscle soreness on selected responses to endurance exercise / Lawrence, Kristen E – 1999 – 1mf – 9 – $4.00 – mf#PE 3938 – us Kinesology [617]

The effect of demographic and sport related factors on motivational orientation / Martens, Matthew P – 1997 – 2mf – 9 – $8.00 – mf#PSY 1951 – us Kinesology [150]

Effect of dietary education, exercise, and two dietary programs on women's body composition, caloric intake, dietary composition, and subjective feelings concerning the programs / Peugnet, Jeffrey C – Brigham Young University, 1995 – 2mf – 9 – $8.00 – mf#PH 1503 – us Kinesology [612]

Effect of different cooling methods on thermoregulation following intermittent anaerobic exercise in the heat / Schiller, Eric R – 1996 – 1mf – 9 – $4.00 – mf#PH 1606 – us Kinesology [612]

The effect of different interval durations on measures of exercise intensity / Hrovatin, Lauri A – 1999 – 1mf – 9 – $4.00 – mf#PE 3988 – us Kinesology [790]

The effect of different interval magnitudes on measures of exercise intensity / Florhaug, Jessica A – 1999 – 1mf – 9 – $4.00 – mf#PH 1668 – us Kinesology [612]

The effect of distraction during cycle ergometry : on ratings of preceived exertion and affect scores in overweight individuals / Williams, Lauren H – 2000 – 75p on 1mf – 9 – $5.00 – mf#PSY 2147 – us Kinesology [150]

The effect of estrogen status on muscle tissue damage in women following an eccentric exercise bout / Styers, Anna – 1999 – 1mf – 9 – $4.00 – mf#PE 4022 – us Kinesology [612]

The effect of exercise and alcohol on perceived exertion, blood lactate, heart rate and thermoregulation in a hot environment / Landry, Jennifer A – 1995 – 2mf – 9 – $8.00 – mf#PH 1557 – us Kinesology [612]

Effect of exercise intensity and duration on postexercise metabolism in obese adults / Creel, David B – Indiana University, 1995 – 2mf – 9 – $8.00 – mf#PH1455 – us Kinesology [612]

The effect of exercise intensity on the extent of and recovery from fatigue of long duration / Stefke, Elmar J & Lehman, Steven L – 1993 – 1mf – $4.00 – us Kinesology [617]

Effect of exercise of moderate duration and intensity upon rate-pressure product in women / Pinkerton, Jana L – 1998 – 1mf – 9 – $4.00 – mf#PE 3926 – us Kinesology [617]

Effect of exercise on bone mineral density of the forearm in premenarcheal girls / Anderson, Francine M – 1999 – 198p on 3mf – 9 – $15.00 – mf#PE 4207 – us Kinesology [612]

The effect of exercise on glyceraldehyde-3-phosphate dehydrogenase and superoxide dismutase activities in the post-ischemic heart / Gow, Andrew J – 1994 – 1mf – $4.00 – us Kinesology [615]

The effect of exercise training on fasting blood glucose levels in adolescents / Bauman, Mara J – 1998 – 1mf – 9 – $4.00 – mf#PH 1628 – us Kinesology [612]

Effect of exercise training on the function of the rat myocardium during reperfusion following global ischemia / Libonati, Joseph R & Paolone, Albert M – 1993 – 2mf – $8.00 – us Kinesology [612]

The effect of exogenous recombinant porcine somatotropin on pig common calcanean tendon biochemistry / Choy, Valerie E & Vailas, Arthur C – 1990 – 1mf – 9 – $4.00 – us Kinesology [612]

Effect of fatigue on open kinetic chain proprioception and closed kinetic chain neuromuscular control / Myers, Joseph B – 1998 – 2mf – 9 – $8.00 – mf#PSY 2020 – us Kinesology [612]

The effect of fatigue on postural stability and neuropsychological function / Crowell, Dean H – 2000 – 117p on 2 2mf – 9 – $10.00 – mf#PSY 2146 – us Kinesology [150]

Effect of fertilizer on growth and composition of carpet and other grasses / Blaser, R E – Gainesville, FL. 1943 – 1r – us UF Libraries [630]

Effect of flexibility exercises on range of motion and physical performance of developmentally disabled adults / Gbenedio, Nelson A – 1999 – 304p on 4mf – 9 – $20.00 – mf#PE 4194 – us Kinesology [613]

The effect of foot landing position on foot mechanics during gait / Jiang, Peixing – 1996 – 2mf – 9 – $8.00 – mf#PE 3822 – us Kinesology [612]

Effect of footing shape on foundation vibrations / Chlawson, James William – s.l, s.l? 1959 – 1r – us UF Libraries [500]

The effect of free agency on player loyalty in major league baseball / Montgomery, Daron – 1998 – 1mf – 9 – $4.00 – mf#PE 3854 – us Kinesology [790]

Effect of frequent cutting and nitrate fertilization on the growth behavior / Leukel, W A – Gainesville, FL. 1934 – 1r – us UF Libraries [630]

Effect of frequent fires on chemical composition of forest soils in the longleaf pine region / Heyward, Frank – Gainesville, FL. 1934 – 1r – us UF Libraries [630]

The effect of half-time warm-up procedures upon injuries to high school varsity football players / Howat, Kenneth J – University of North Carolina at Chapel Hill, 1995 – 1mf – 9 – $4.00 – mf#PE3600 – us Kinesology [617]

Effect of hardiness training on math and science grades : in economically and/or educationally disadvantages junior high and high school students / Shoemaker, Mindy – 1997 – 1mf – 9 – $4.00 – mf#PSY 1997 – us Kinesology [373]

Effect of heart rate deceleration biofeedback training on golf putting performance / Damarjian, Nicole M & Crews, Debra J – 1992 – 1mf – 9 – $4.00 – us Kinesology [150]

The effect of heat and cold on ankle stability / Keenan, Karen A – University of North Carolina at Chapel Hill, 1995 – 1mf – 9 – $4.00 – mf#PE3604 – us Kinesology [617]

The effect of high energy insoles on vertical jump performance / Rauch, Ursula – 1997 – 1mf – 9 – $4.00 – mf#PE 3806 – us Kinesology [612]

Effect of hydrocyanic acid gas fumigation on the subsequent growth / Wilmot, Royal James – s.l, s.l? 1932 – 1r – us UF Libraries [630]

The effect of intermittent hyperbaric oxygenation on short term recovery from grade 2 medial collateral ligament injuries / Soolsma, Serge J – University of British Columbia, 1996 – 2mf – 9 – $8.00 – mf#PE 3671 – us Kinesology [617]

Effect of intradialytic exercise on urea kinetics / Leung, Raymond W M – 1999 – 2mf – 9 – $8.00 – mf#PH 1656 – us Kinesology [612]

The effect of keyboard design on finger, forearm, and shoulder muscle activity / Stone, Corey W – 2000 – 129p on 2mf – 9 – $10.00 – mf#PE 4115 – us Kinesology [612]

The effect of lactic acid on fat mobilization and utilization in trained subjects / Vega des Jesus, Ramon – 1988 – 84p 1mf – 9 – $4.00 – us Kinesology [612]

The effect of mass on the kinematics of steady state wheelchair propulsion in adults and children with spinal cord injury / Bednarczyk, Janet H & Sanderson, David – 1993 – 1mf – 9 – $4.00 – us Kinesology [612]

The effect of menstrual cycle phase on diffusing capacity of the lung / Bacon, Catherine Jane – 1997 – 2mf – 9 – $8.00 – mf#PH 1585 – us Kinesology [612]

Effect of mental imagery of a motor task on the hoffmann reflex / Hale, Brendon S – 1998 – 1mf – 9 – $4.00 – mf#PSY 2124 – us Kinesology [612]

Effect of method of rearing sc white leghorn chicks upon rate of growth, feed / Mehrhof, N R – Gainesville, FL. 1943 – 1r – us UF Libraries [636]

The effect of moderate exercise on lipid profiles in a healthy college age population / Calnin, R J – 1991 – 2mf – 9 – $8.00 – us Kinesology [613]

The effect of modified pnf trunk strengthening on functional performance in female rowers / Galilee-Belfer, Adam – 1999 – 1mf – 9 – $4.00 – mf#PE 3948 – us Kinesology [612]

Effect of mulch and chemical treatments on microbiological action i... / Batista Y Cuba, Juan Wilfredo – s.l, s.l? 1943 – 1r – us UF Libraries [630]

The effect of muscle ischemia on sarcoplasmic reticulum ca2+-atpase function : an in situ rat model / Stavrianeas, Stasinos – University of Oregon, 1995 – 1mf – 9 – $4.00 – mf#PH1480 – us Kinesology [612]

Effect of muscle length on motor unit firing behavior in human tibialis anterior muscle / Vander Linde, Darl W & Kukulka, Carl G – 1989 – 2mf – 9 – $8.00 – us Kinesology [612]

Effect of music programming on walking velocity / Zilonka, Elaine M – 1999 – 1mf – 9 – $4.00 – mf#PH 1674 – us Kinesology [790]

The effect of mutual choice placement on the satisfaction of student and cooperating teachers in physical education / Johnson, Susan M – 1982 – 2mf – 9 – $8.00 – us Kinesology [790]

Effect of oil sprays on the transpiration of citrus / Merrin, George Alfred – s.l, s.l? 1929 – 1r – us UF Libraries [634]

The effect of open and closed kinetic chain strength training on change in vertical jump height / Oates, Deniece D – 1997 – 1mf – 9 – $4.00 – mf#PE 3810 – us Kinesology [612]

The effect of oral smokeless tobacco on the cardiovascular and metabolic responses in humans during rest and exercise / Van Duser, Bruce L & Chevrette, John M – 1991 – 2mf – 9 – $8.00 – us Kinesology [613]

The effect of orthotic correction on walking and running efficiency in subjects with excessive pronation / Comeau-Stender, Susan M – 1997 – 1mf – 9 – $4.00 – mf#PE 3791 – us Kinesology [612]

Effect of peer group presence on the gross motor performance of young children / Bates, MK – 1990 – 2mf – 9 – $8.00 – us Kinesology [150]

The effect of perception of performance outcomes on mood following exercise / Jennings, A – 1990 – 2mf – 9 – $8.00 – us Kinesology [150]

The effect of phase 2 cardic rehabilitation on self-efficacy and quality of life / Doyle, Mike N – 2000 – 38p on 1mf – 9 – $5.00 – mf#HE 679 – us Kinesology [617]

The effect of placement site on isometric force measurements with the nicholas manual muscle tester in college women / Templeton, Charles L – 1995 – 1mf – 9 – $4.00 – mf#PE 3777 – us Kinesology [612]

The effect of planned exercise as a disinhibitor of dietary restraint : an investigation of perceived control and resultant affect / Hart, Elizabeth A – 1994 – 3mf – $12.00 – us Kinesology [150]

Effect of practice schedule variation on the acquisition, retention, and transfer of an applied motor skill by children with and without mental retardation / Sutlive, Vinson H, 3rd – Indiana University, 1995 – 3mf – 9 – $12.00 – mf#PSY1868 – us Kinesology [611]

The effect of pre-competitive practice on the activation and mood of high school football players / Thomas, PR – 1991 – 1mf – 9 – $4.00 – us Kinesology [150]

The effect of prior aerobic exercise upon single session strength performance / Snyder, Robert – 1999 – 55p on 1mf – 9 – $5.00 – mf#PH 1696 – us Kinesology [612]

Effect of proprioceptive neuromuscular facilitation stretch techniques in trained and untrained older adults / Ferber, Reed – 1998 – 2mf – 9 – $8.00 – mf#PE 3811 – us Kinesology [612]

The effect of rapid weight loss/weight gain on muscular power of intercollegiate wrestlers / Vorhis, Phillip E – 1994 – 1mf – $4.00 – us Kinesology [612]

The effect of refinement and teacher feedback on female junior high school students' volleyball practice success and achievement / Pellett, Tracy L – 1993 – 2mf – $8.00 – us Kinesology [376]

The effect of regularly scheduled daily supervisor verbal feedback on use of personal protective equipment / Vink, Marc P & Legos, Patricia M – 1993 – 2mf – $8.00 – us Kinesology [612]

The effect of relaxation training on sport climbing performance of college students / Fraser, Robert G – 1998 – 1mf – 9 – $4.00 – mf#PSY 2058 – us Kinesology [790]

The effect of resistance training on resting blood pressure in hypertensive women / Lynes, Liliana K – 1994 – 3mf – $12.00 – us Kinesology [612]

The effect of running speed and turning direction on lower extremity joint moment / Lee, Ki-Kwang – 1999 – 2mf – 9 – $8.00 – mf#PE 3976 – us Kinesology [611]

The effect of salbutamol on performance in elite non-asthmatic athletes / Meeuwisse, Willem H & McKenzie, Donald C – 1990 – 1mf – 9 – $4.00 – us Kinesology [613]

The effect of seat-tube angle variation on cardiorespiratory responses during submaximal bicycling / Heil, DP – 1992 – 2mf – 9 – $8.00 – us Kinesology [612]

Effect of seed-potato treatment on yield and rhizoctonisis in florida from 1924 to 1929 / Gratz, L O – Gainesville, FL. 1930 – 1r – us UF Libraries [630]

The effect of selected buffering agents on performance in the competitive 1600 meter run / Avedisian, Lori-Ann – Oregon State University, 1995 – 1mf – 9 – mf#PH 1484 – us Kinesology [612]

The effect of semiconductor tapes in reduction of chronic pain / Swalberg, Mary – 1996 – 1mf – 9 – $4.00 – mf#PE 3792 – us Kinesology [617]

The effect of seven weeks of training on the dietary intake and skinfolds of a woman 82 years of age / Shimidzu, Linda K – 1991 – 1mf – 9 – $4.00 – us Kinesology [613]

The effect of short term emg biofeedback on neck muscle relaxation for rotary pursuit performance / Li, J – 1990 – 2mf – 9 – $8.00 – us Kinesology [612]

The effect of short-term slide board training on lower extremity lateral movement / Utsumi, Toshio – University of North Carolina at Chapel Hill, 1995 – 1mf – 9 – $4.00 – mf#PH1482 – us Kinesology [611]

Effect of slide board training as a component of pre-season conditioning on concentric and eccentric quadriceps peak torque, vertical jump height, and agility / Thomas, Tammy R – 1994 – 1mf – $4.00 – us Kinesology [612]

Effect of sodium and water intake on plasma aldosterone during prolonged exercise in warm environment / Shi, Xiaocai & Costill, David L – 1990 – 1mf – 9 – $4.00 – us Kinesology [612]

The effect of sodium citrate ingestion on 1600 meter running performance / Guerra, Arthur – Oregon State University, 1995 – 1mf – 9 – mf#PH 1494 – us Kinesology [612]

Effect of soil reaction on the assimilation of certain primary nutr... / Henderson, J R – s.l, s.l? 1934 – 1r – us UF Libraries [630]

Effect of soil type on the nitrification of dried blood and ammoniu... / Wooten, Robert B – s.l, s.l? 1931 – 1r – us UF Libraries [630]

Effect of soil types, soil pasteurisation and inoculation on the growth of colophospermum mopane families / Hangula, Rusta J K – Stellenbosch: U of Stellenbosch 1998 [mf ed 1998] – 3mf – 9 – mf#mf.1355 – sa Stellenbosch [634]

The effect of speed and treadmill compliance on oxygen consumption during walking / Nelson, Jo A – 1999 – 2mf – 9 – $8.00 – mf#PH 1688 – us Kinesology [612]

Effect of sprint training upon sarcoplasmic reticulum ca2+ atpase and na+-ca2+ exchanger mrna expression in rat myocardium / Gow, Andrew J – Temple University, 1995 – 2mf – 9 – $8.00 – mf#PH1460 – us Kinesology [612]

Effect of stilbestrol on udder development, pelvic changes, lactation and reproduction – Gainesville, FL. 1948 – 1r – us UF Libraries [636]

The effect of strengthening external hip rotators on abnormal pronation of the subtalar joint / Stein, Tamara – 1999 – 1mf – 9 – $4.00 – mf#PE 4021 – us Kinesology [612]

The effect of submaximal exercise in a neutral or hot-humid environment on recovery hemodynamics in men and women / Fisher, Michele M – Springfield College, 1995 – 3mf – 9 – $12.00 – mf#PH1458 – us Kinesology [612]

Effect of substituted cations in the soil complex on the decomposit... / Whitehead, Thomas – s.l, s.l? 1941 – 1r – us UF Libraries [630]

The effect of substrate utilization, manipulated by nicotinic acid, on excess postexercise oxygen consumption / Trost, Stewart G – Oregon State University, 1994 – 2mf – 9 – $8.00 – mf#PH 1513 – us Kinesology [612]

The effect of surface electomyography visual biofeedback on the ability to minimize mid-trapezius muscle activity during an arm flexion task in females / Hillenmayer, Dawn M – 1999 – 1mf – 9 – $4.00 – mf#PE 4059 – us Kinesology [617]

The effect of swim training on plasma somatomedin-c levels in 8- to 10-year-old children / Counts, Charlene L M & Ben-Ezra, Victor – 1991 – 1mf – 9 – $4.00 – us Kinesology [612]

The effect of tactile and whole/part drill on the acquisition of opposition in a successful basketball lay-up / Carlson, PD – 1991 – 2mf – 9 – $8.00 – us Kinesology [150]

The effect of t'ai chi ch'uan upon selected fitness components of older women / Inamura, Chikako – 1999 – 68p on1mf – 9 – $5.00 – mf#PSY 2127 – us Kinesology [613]

Effect of the achilles tendon adhesive taping and pro m-p achilles strap on eccentric plantar flexion peak torque / Morales, Alan D – 1994 – 1mf – $4.00 – us Kinesology [617]

Effect of the aircast on functional movements using the biotran / Rockhill, Bryan H – 1994 – 1mf – $4.00 – us Kinesology [617]

The effect of the airstirrup and a conventional method of strapping the ankle on agility and vertical jump performance / Brassard, Marc F – 1988 – 97p 1mf – 9 – $4.00 – us Kinesology [617]

The effect of the education of third world women on family health : a chinese example / Hallmann, Jayne E & Hill, J Stanley – 1992 – 2mf – $8.00 – us Kinesology [613]

The effect of the glacial epoch upon the distribution of insects in north america / Grote, Augustus Radcliffe – S.l: Salem Press, 1876? – 1mf – 9 – mf#32882 – cn Canadiana [590]

The effect of the lead leg plant on factors affecting distance on kickoffs in football / Snowden, Steven R & Dowell, Linus J – 1991 – 2mf – 9 – $8.00 – us Kinesology [612]

The effect of the menstrual cycle on bioimpedance reliability / Larson, Lois A & Porcari, John P – 1993 – 1mf – $4.00 – us Kinesology [612]

The effect of the paradoxical intervention of symptom prescription on state anxiety levels and performance in young competitive swimmers / Greenberg, Doreen L – 1994 – 2mf – $8.00 – us Kinesology [150]

The effect of the presence of the coach on pain perception and pain tolerance of athletes / Coutu, Debra L – Springfield College, 1995 – 2mf – 9 – $8.00 – mf#PSY1841 – us Kinesology [150]

The effect of the reciprocal approach in teaching on the process of self-discovery for beginning modern dance students at the secondary level / Blomquist, Melinda E – 1998 – 2mf – 9 – $8.00 – mf#PE 3912 – us Kinesology [790]

Effect of three shoulder exercise programs on strength, proprioception, neuromuscular control, and functional performance / Padua, Darin A – 1998 – 2mf – 9 – $8.00 – mf#PSY 2024 – us Kinesology [612]

The effect of three training methods on the teaching preparation of counselor-teachers in a resident environmental education program / Johnson, Susan L – 1982 – 2mf – 9 – $8.00 – us Kinesology [790]

The effect of time of season on the athletic identity in collegiate swimmers / Antshel, Kevin M – 1994 – 2mf – $8.00 – us Kinesology [150]

Effect of time of turning and method of supplementing green manures / Bedsole, Malcolm R – s.l, s.l? 1930 – 1r – us UF Libraries [630]

Effect of timing of upper body cycling exercise on the recovery from delayed-onset muscle soreness / Rescino, Mark H – 1999 – 2mf – 9 – $8.00 – mf#PE 3919 – us Kinesology [617]

The effect of toe and plantar flexor strength training on vertical jump performance of folk dancers / Meiners, Earlet P – 1991 – 1mf – $4.00 – us Kinesology [612]

The effect of trained hearing peer tutors on the physical activity levels of deaf students in inclusive elementary school physical education classes / Lieberman, Lauren J – Oregon State University, 1996 – 2mf – 9 – $8.00 – mf#PE 3661 – us Kinesology [370]

Effect of training frequency on cervical rotation strength / DeFilippo, G J – 1991 – 2mf – 9 – $8.00 – us Kinesology [612]

Effect of training on lactate utilization by rat muscle mitochondria / Murakami, Joan R – 1989 – 59p 1mf – 9 – $4.00 – us Kinesology [612]

The effect of training status on resting metabolic rate and substrate utilization in women / Bowden, Victoria L – 1997 – 1mf – 9 – $4.00 – mf#PH 1596 – us Kinesology [612]

The effect of transverse pedal spacing on cycling efficiency / Carling, Jon & Fisher, A Garth – 1992 – 1mf – $4.00 – us Kinesology [613]

The effect of treadmill compliance and foot type electromyography of lower extremity muscles during running / Backmann, Christine K – 1997 – 2mf – 9 – $8.00 – mf#PE 3937 – us Kinesology [612]

The effect of two types of plyometric training in improving vertical jump ability in female college soccer players / Villarreal, Jose & Considine, William J – 1992 – 2mf – $8.00 – us Kinesology [612]

The effect of ultrasound on temperature rise in the preheated triceps surae muscle group / Harris, Shane T – Brigham Young University, 1994 – 1mf – 9 – mf#PE 3650 – us Kinesology [617]

The effect of upper body excerise on secondary lymphedema following breast cancer treatment / Kalda, Andrea L – 1999 – 69p on 1mf – 9 – $5.00 – mf#HE 672 – us Kinesology [617]

Effect of various factors upon the ascorbic acid content of some florida-grown mangos / Mustard, Margaret J – Gainesville, FL. 1945 – 1r – us UF Libraries [630]

The effect of various lifting intensities in release of human growth hormone / Kang, H -Y – 1990 – 1mf – 9 – $4.00 – us Kinesology [612]

Effect of varying amounts of nitrogen and potassium on the yield an... / Miles, Ivan Ernest – s.l, s.l? 1931 – 1r – us UF Libraries [630]

The effect of varying treadmill surface compliance on oxygen uptake during running / Leahy, Guy D – 1996 – 1mf – 9 – $4.00 – mf#PH 1662 – us Kinesology [612]

Effect of visual feedback and verbal encouragement on eccentric quadriceps and hamstrings peak torque of males and females / Lukasiewicz, William C – 1997 – 1mf – 9 – $4.00 – mf#PSY 2103 – us Kinesology [612]

Effect of vitamin e supplementation on delayed-onset muscle soreness / Blackwell, Ryan – 1997 – 1mf – 9 – $4.00 – mf#PH 1632 – us Kinesology [615]

Effect of water running and cycling on vo2max and 2-mile performance / Eyestone, Edward D & Fisher, A Garth – 1990 – 1mf – $4.00 – us Kinesology [617]

The effect of wrist weight on the hemodynamic response to exercise in coronary artery disease / Kaplan, Linda & Paolone, Vincent J – 1993 – 2mf – $8.00 – us Kinesology [612]

The effect of yoga and relaxation techniques on outcome variables associated with osteoarthritis of the hands and finger joints / Garfinkel, Marian S & Levy, Marvin R – 1992 – 2mf – 9 – $8.00 – us Kinesology [613]

Effective clinical practice (ecp) – Philadelphia, 1998+ [1,5,9] – ISSN: 1099-8128 – mf#28821 – us UMI ProQuest [616]

Effective workers in needy fields / McDowell, William Fraser et al – New York: Student Volunteer Movement for Foreign Missions, 1902 – 1mf – 9 – 0-524-08519-6 – (incl bibl ref) – mf#1993-1049 – us ATLA [240]

Effectiveness of a minimal physician delivered stage-based intervention regarding readiness to change specific to physical activity / Monahan, Bridget – 1999 – 1mf – 9 – $4.00 – mf#HE 652 – us Kinesology [360]

Effectiveness of a walking club and a self-directed physical activity program in increasing moderate intensity physical activity among african-american females / Rogers, Tecora M – 1997 – 4mf – 9 – $16.00 – mf#HE 645 – us Kinesology [613]

The effectiveness of acupressure in the treatment of primary dysmenorrhea / Fontenot, M E – 1989 – 1mf – 9 – $4.00 – us Kinesology [612]

Effectiveness of an abdominal training protcol on an unstable surface / Brovender, Samuel J – 2001 – 75p on 1mf – 9 – $5.00 – mf#PSY 2165 – us Kinesology [613]

The effectiveness of behavioral contracts in promoting the maintenance of cancer risk reduction behavior while utilized in a college cancer avoidance course / Burkley, Renee L – 1990 – 69p on 1mf – 9 – $4.00 – us Kinesology [150]

Effectiveness of exercise versus exercise plus tape in the management of females with patellofemoral pain / Froehling, Lori A – University of Wisconsin-La Crosse, 1996 – 1mf – 9 – mf#PE 3642 – us Kinesology [617]

Effectiveness of fundraising techniques for collegiate women's and olympic sports' facilities / James, W R – 1998 – 1mf – 9 – $4.00 – mf#PE 3960 – us Kinesology [790]

Effectiveness of human resource management strategies in the engineering studies department in a technical college : a case study / Nconco, Fezile Wilfred – Pretoria: Vista University 2002 [mf ed 2002] – 4mf – 9 – (incl bibl) – mf#mfm15198 – sa Unisa [650]

The effectiveness of individualized mental training program on attentional styles, competitive trait anxiety and performance of female softball players / Ethridge, M Kriss – 1997 – 1mf – 9 – $4.00 – mf#PSY 1942 – us Kinesology [790]

The effectiveness of job specific training on the work performance of female student nurses / McCannon, Robin K & Miller, Marilyn K – 1993 – 2mf – $8.00 – us Kinesology [376]

The effectiveness of microcurent electrical nerve stimulation (m.e.n.s.) in the treatment of post acute lymphedema in ankle injuries / Galley, Suzi-Lyn – 1994 – 1mf – $4.00 – us Kinesology [617]

The effectiveness of repeated submaximal concentric exercise and heated whirlpool in the treatment of delayed onset muscular soreness / Miller, MK – 1991 – 1mf – 9 – $4.00 – us Kinesology [613]

Effectiveness of selected components in behavioral weight-loss interventions : a meta-analysis / Wood, Nadine MS & Donatelle, Rebecca J – 1992 – 2mf – 9 – $8.00 – us Kinesology [150]

The effectiveness of static magnetic therapy on clinically induced delayed onset muscle soreness / Royle, Nancy L – 1999 – 1mf – 9 – $4.00 – mf#PE 4024 – us Kinesology [617]

Effectiveness of the breathe right nasal strip in collegiate middle and long distance runners / Roehl, Matthew J – 1997 – 1mf – 9 – $4.00 – mf#PH 1600 – us Kinesology [612]

The effectiveness of the cascade model in the training of educators for implementing outcomes-based education / April, Ewart Zolile – Pretoria: Vista University 2001 [mf ed 2001] – 4mf – 9 – (incl bibl ref) – mf#mfm15223 – sa Unisa [370]

The effectiveness of the hinged golf club as a training aid to develop consistency in novice golfers / Dexter-Fogarty, Tracey – Springfield College, 1995 – 2mf – 9 – $8.00 – mf#PE3589 – us Kinesology [612]

Effectiveness of the schoollunch in improving the nutritional status of school children – Gainesville, FL. 1946 – 1r – us UF Libraries [613]

The effectiveness of the stages of change model and experimental exercise prescriptions in increasing female adults' physical activity and exercise behavior / Cardinal, Bradley J & Sachs, Michael L – 1993 – 3mf – 9 – $12.00 – us Kinesology [150]

The effects of 6-week and 12-week rehabilitation programs on the depression level of cardiac patients / Hazavehei, Seyyed M M – Texas Woman's University, 1993 – 2mf – 9 – $8.00 – mf#PSY 1889 – us Kinesology [150]

The effects of 15 weeks of resistive training with chromium supplementation : upon muscle strength, body composition, and urinary chromium excretion in untrained college-aged female / Henry, Dahlia – 2000 – 93p on 1mf – 9 – $5.00 – mf#PE 4170 – us Kinesology [612]

The effects of 90 days of km supplementation on aerobic capacity and general well-being of healthy adults / Pugliese, Ari – University of Wisconsin-La Crosse, 1996 – 1mf – 9 – $4.00 – mf#PH1472 – us Kinesology [613]

The effects of a 6-week stretching program, using flex bands, on the low back and hamstring flexibility of cardiac rehabilitation patients / Robertson, Sara L – 1997 – 1mf – 9 – $4.00 – mf#PH 1599 – us Kinesology [612]

Effects of a 30-minute walk on ground reaction forces : during walking with an external load / Cardillo, Cheryl M – 2000 – 79p on 1mf – 9 – $5.00 – mf#PE 4129 – us Kinesology [612]

The effects of a carbohydrate-electrolyte replacement drink taken during high intensity exercise on sprint capacity at the end of exercise / Ball, Thomas C – 1994 – 2mf – $8.00 – us Kinesology [612]

Effects of a chair exercise program (sit and be fit tm) for older adults : on functional health-related components of fitness / Kinkade-Schall, Kristi L – 2000 – 60p on 1mf – 9 – $5.00 – mf#HE 668 – us Kinesology [613]

Effects of a competitive season on body composition in female intercollegiate athletes / Williams, Salena – University of Wisconsin-La Crosse, 1995 – 1mf – 9 – mf#PE 3679 – us Kinesology [617]

Effects of a creative dance program on the perceptual motor performance of trainable mentally retarded children / Barnes, Carolyn M – 1978 – 1mf – 9 – $4.00 – us Kinesology [790]

The effects of a crosstraining program on strength development / Barton, Andrew R – 1996 – 2mf – 9 – $8.00 – mf#PH 1546 – us Kinesology [612]

The effects of a different arm swing on vertical jump and toe-touch jump performance / Munkasy, B A – 1990 – 2mf – 9 – $8.00 – us Kinesology [790]

The effects of a farm youth hearing study on parental hearing protection knowledge, attitudes and behavior / Knobloch, Mary Jo – 1996 – 1mf – 9 – $4.00 – mf#HE 593 – us Kinesology [362]

Effects of a flexibility exercise program upon perceived lower back pain / Rough, Lynn – 1999 – 1mf – 9 – $4.00 – mf#PE 3981 – us Kinesology [617]

The effects of a flexibility training program on flexibility test scores in elementary school children / Dulaney, N M – 1991 – 1mf – 9 – $4.00 – us Kinesology [790]

The effects of a friendship enhancement program for individuals with development disabilities / Lyons, Rebecca A – 1999 – 2mf – 9 – $8.00 – mf#RC 536 – us Kinesology [612]

The effects of a group-oriented contingency management system on behaviorally disordered students in physical education / Vogler, E Williams – 1980 – 2mf – 9 – $8.00 – us Kinesology [790]

The effects of a health related physical fitness curriculum on selected fitness variables / Waite, Terence M – 1988 – 86p 1mf – 9 – $4.00 – us Kinesology [613]

The effects of a leisure activity visitation training program on the visitor's perceived satisfaction of visits with individuals with dementia related diseases including alzheimer's disease / Waskiewicz, Becky A – 1994 – 1mf – $4.00 – us Kinesology [790]

Effects of a lower limb strength training program on balance measures in men with mental retardation / Suomi, Rory & Surburg, Paul R – 1991 – 4mf – 9 – $16.00 – us Kinesology [612]

The effects of a miniumum impact camping slide-tape program on wilderness visitors' awareness of minimum impact camping / Anderson, Lynn S – 1981 – 2mf – 9 – $8.00 – us Kinesology [790]

The effects of a modern football uniform on thermoregulation / Ross, John L – 1999 – 1mf – 9 – $4.00 – mf#PH 1678 – us Kinesology [612]

The effects of a modified ball in developing the volleyball pass and set for high school students / Weidner, Julie A – 1998 – 1mf – 9 – $4.00 – mf#PE 3904 – us Kinesology [790]

Effects of a multimedia performance principle training program on correct analysis and diagnosis of throwlike movements / Williams, Emyr W – Ohio State University, 1995 – 2mf – 9 – $12.00 – mf#PE3623 – us Kinesology [790]

Effects of a partnering class on dancers' muscular strength, flexibility, and body composition / Vetter, Rheba E – 2000 – 3mf – 9 – $12.00 – mf#PH 1691 – us Kinesology [612]

The effects of a "prescriptive individualized program" and a "nonprescriptive group task program" on fundamental motor pattern and ability acquisition, self-concept, and socialization skills of kindergarten children / Moyer, Steve W – 1981 – 2mf – 9 – $8.00 – us Kinesology [790]

The effects of a project learning tree workshop on pre-service teachers' attitudes toward teaching environmental education / Kunz, Dorothea E – 1989 – 114p on 2mf – 9 – $8.00 – us Kinesology [370]

Effects of a proximal provocation on carpal tunnel syndrome / Farrell, Kevin P – 1998 – 4mf – 9 – $16.00 – mf#PE 3910 – us Kinesology [612]

Effects of a semirigid and a softshell prophylactic ankle stabilizer on performance / Macpherson, Kevin – 1994 – 1mf – $4.00 – us Kinesology [617]

The effects of a six week, 11 hour ropes course unit on the attitudes towards physical activity of high school students with behavior disorders / Lee, Jeff – 1999 – 1mf – 9 – $4.00 – mf#PSY 2098 – us Kinesology [373]

The effects of a six-month exercise maintenance program on the cardiovascular fitness levels of participants / Carney, Deborah A – 1981 – 2mf – 9 – $8.00 – us Kinesology [790]

The effects of a stair climbing program on leg strength, flexibility and functional mobility in men and women aged 76 to 86 years / Creviston, Todd A – 1996 – 1mf – 9 – $4.00 – mf#PH 1548 – us Kinesology [612]

The effects of a strength training program on the body image, self-concept, and dynamic strength of seventh grade girls / Lucas, Jason – 1994 – 1mf – $4.00 – us Kinesology [150]

Effects of a swimming program on cystic fibrosis children / Edlund, Larry L – 1980 – 1mf – 9 – $4.00 – us Kinesology [790]

The effects of a ten-week step aerobic training program on aerobic capacity of college-aged females / Chapek, Constance L & Porcari, John P – 1992 – 1mf – $4.00 – us Kinesology [613]

The effects of a ten-week step aerobic training program on the body composition of college-aged women / Huntley, Elizabeth A & Porcari, John P – 1992 – 1mf – $4.00 – us Kinesology [612]

The effects of a therapeutic horseback riding experience : on selected behavioral and psychological factors of ambulatory adults diagnosed with multiple sclerosis / Patterson, Tara S – 2000 – 86p on 1mf – 9 – $5.00 – mf#HE 681 – us Kinesology [150]

The effects of a torso strengthening and rotational power program versus strength training on angular hip, angular shoulder, and linear bat head velocity in male college baseball players / Lund, Robin J – 1997 – 1mf – 9 – $4.00 – mf#PE 3893 – us Kinesology [612]

The effects of a treatment program for chronic pain patients using enjoyable imagery with biofeedback induced relaxation / Mckee, Patrick J – 1981 – 2mf – 9 – $8.00 – us Kinesology [615]

The effects of a type and interest-based career exploration program on the career maturity and goal stability of collegiate student-athletes / Ludwig, Martha M – 1993 – 4mf – 9 – $16.00 – mf#PE 3944 – us Kinesology [150]

The effects of a verbalized preperformance routine on free-show shooting accuracy of adult basketball participants / Oliver, Jon A – 1998 – 1mf – 9 – $4.00 – mf#PSY 2065 – us Kinesology [612]

The effects of a water exercise program on the manifestations of fibromyalgia / Westfall, Jacquelyn K – 1999 – 97p on 1mf – 9 – $5.00 – mf#HE 664 – us Kinesology [617]

The effects of a weight training course on stress levels and locus of control in college females / Fuller, Tamela G & Rhodes, Ronald L – 1992 – 1mf – $4.00 – us Kinesology [150]

The effects of accupressure therapy on exercise induced delayed onset muscle soreness and muscle function / Charles-Liscombe, Robert S – 1997 – 2mf – 9 – $8.00 – mf#PE 3865 – us Kinesology [617]

The effects of acquaintance rape prevention programming on male athletes' sexual and dating attitudes / Andersen, Steven J & Dosch, Margaret – 1992 – 1mf – $4.00 – us Kinesology [150]

The effects of acute and chronic exercise on serum potassium in hemodialysis patients / Carney, Colleen M – 1999 – 2mf – 9 – $8.00 – mf#PH 1676 – us Kinesology [612]

The effects of acute dietary creatine supplementation on power output indices and blood lactate concentrations during high-intensity intermittent cycling exercise / Capriotti, Paul V – 1998 – 2mf – 9 – $8.00 – mf#PH 1665 – us Kinesology [612]

The effects of acute exercise of varying intensities on subjects with type 1 diabetes mellitus / Dauley, Patricia A – 1993 – 2mf – $8.00 – us Kinesology [615]

Effects of acute exercise on children with attention deficit-hyperactivity disorder / Tantillo, Mary – 1996 – 2mf – 9 – $8.00 – mf#PSY 2005 – us Kinesology [616]

Effects of acute resistive exercise on the resting metabolic rate of women / Brady, Christine P – Temple University, 1996 – 1mf – 9 – mf#PH 1488 – us Kinesology [612]

Effects of adhesive spray and prewrap on taped ankle inversion before and after exercise / Keetch, Anita & Durrant, Earlene – 1992 – 1mf – $4.00 – us Kinesology [617]

The effects of aerobic dance exercise and nutrition intervention in cholesterol levels / Vetro, VL – 1990 – 2mf – 9 – $8.00 – us Kinesology [612]

Effects of aerobic exercise on the lipid profile levels of patients with moderate to severe burn injury / Bacon Hilda – 1994 – 1mf – $4.00 – us Kinesology [615]

The effects of age and endurance training on rat adrenal tissue : a morphological analysis and determination of catecholamine content / Schmidt, Kathryn & Stanley, William C – 1990 – 1mf – 9 – $4.00 – us Kinesology [612]

Effects of age and ethanol on thermoregulatory responses of men to a cold air stress / Hopkins, Ruth A – Temple University, 1995 – 2mf – 9 – $8.00 – mf#PH1464 – us Kinesology [612]

Effects of age and gender on functional rotation and lateral flexion of the back and the neck / Netzer, Ofra & Payne, V Gregory – 1992 – 1mf – 9 – $4.00 – us Kinesology [612]

Effects of age, velocity, and added mass on postural adjustments associated with a rapid armraising movement / Manchester, Diane L & Woollacott, Marjorie H – 1990 – 3mf – 9 – $12.00 – us Kinesology [612]

The effects of alcohol upon the human system : an essay upon the cause, nature and treatment of alcoholism / Watkins, Thomas C – [Hamilton, Ont?: s.n, 189-?] [mf ed 1994 – 1mf – 9 – 0-665-94627-9 – mf#94627 – cn Canadiana [615]

Effects of (alpha)-adrenergic stimulation on sr ca2+ atpase and na/ca mrna expression in cultured neonatal rat ventricular myocytes / Toaldo, Gina-Lee – Temple University, 1995 – 1mf – 9 – $4.00 – mf#PH1481 – us Kinesology [612]

The effects of alpha-tocopherol on metabolic determinations in graded exercise / Keroack, Christopher R & Paolone, Vincent J – 1992 – 1mf – $4.00 – us Kinesology [612]

The effects of amino acid supplementation on endurance performance / Laporte, Rebecca J & Paolone, Vincent J – 1993 – 2mf – $8.00 – us Kinesology [612]

The effects of an application of sunscreen on selected physiological variables during exercise in the heat / Connolly, Declan A – Oregon State University, 1995 – 2mf – 9 – $8.00 – mf#PH 1490 – us Kinesology [612]

The effects of an evaluative audience upon college males' self-efficacy, perceived ability, anxiety, and learning of a novel motor task / Simensky, Steven G & Ewing, Martha E – 1991 – 2mf – 9 – $8.00 – us Kinesology [150]

The effects of an induced internal and external attentional focus upon upper body strength / Hein, Erica J & Pein, Richard L – 1993 – 1mf – $4.00 – us Kinesology [150]

Effects of an intercollegiate sport season on selected personality traits and mental preparation skills / Drake, Benjamin C – 1997 – 1mf – 9 – $4.00 – mf#PSY 1983 – us Kinesology [150]

The effects of an interdependent group-oriented contingency on middle school students' physical activity levels during physical education / Schuldheisz, Joel M – 1998 – 3mf – 9 – $12.00 – mf#PE 3889 – us Kinesology [790]

Effects of an interval training dance class on select cardiovascular variables / Christensen, Kimberly M – University of Oregon, 1994 – 1mf – 9 – $4.00 – mf#PE3585 – us Kinesology [790]

The effects of ankle sprains and external support on muscle onset latencies / Van den Eikhof, Victoria E – University of Oregon, 1996 – 1mf – 9 – mf#PE 3676 – us Kinesology [617]

Effects of ankle taping on the neuromuscular regulation of impact forces at heel strike / Pettitt, Robert W – 1998 – 1mf – 9 – $4.00 – mf#PE 3903 – us Kinesology [617]

Effects of apartheid on education, science, culture / Unesco – New York, NY. 1967 – 1r – us UF Libraries [322]

Effects of aquatic simulated and dry land plyometrics on vertical jump height / Stemm, John D – 1993 – 1mf – $4.00 – us Kinesology [574]

Effects of arch support on changes in arch height, vertical ground reaction force and center of pressure under different foot positions while loading and demonstrated by contact bone-on-bone forces / Chen, Shing-Jye – 2000 – 1mf – 9 – $4.00 – mf#PE 4077 – us Kinesology [611]

The effects of athletic training on bone mineral density in female collegiate gymnasts / Nichols, David L & Sanborn, Charlotte F – 1992 – 2mf – $8.00 – us Kinesology [612]

The effects of attentional focus and trait anxiety between starting and nonstarting division 1 basketball players / Braithwaite, Rock – 1998 – 1mf – 9 – $4.00 – mf#PSY 2056 – us Kinesology [790]

The effects of auditory biofeedback on the accuracy of the tennis volley / Holcombe, Robert A & Lewis, Kathryn – 1991 – 2mf – $8.00 – us Kinesology [150]

The effects of balance training on the segmental reflex system of elderly subjects / Mynark, Richard G – 1999 – 3mf – 9 – $12.00 – mf#PSY 2075 – us Kinesology [612]

The effects of body segment length and head position upon sit and reach flexibility performance / Tardie, Gregory B & Pechar, Gary S – 1992 – 1mf – 9 – $4.00 – us Kinesology [612]

The effects of caffeine ingestion on heart rate, blood pressure, and physical work capacity at submaximal levels in 15 caffeine habituated non-athletic male subjects / Guzolik, Gerald L – 1997 – 1mf – 9 – $4.00 – mf#PH 1633 – us Kinesology [612]

Effects of caffeine on central on peripheral hemodynamics at rest and during exercise / Baruch, Amy R – Springfield College, 1994 – 2mf – 9 – $8.00 – mf#PH1452 – us Kinesology [612]

Effects of caffeine on sprint performance / Sweenor, Kymberlie A – 1998 – 2mf – 9 – $8.00 – mf#PH 1612 – us Kinesology [790]

Effects of caloric restriction and resistive exercise on the resting energy expenditure of weight-reduced obese women / Stopford, Jane L & Kendrick, Zebulon V – 1992 – 2mf – 9 – $8.00 – us Kinesology [612]

Effects of cannabis, tobacco, methaqualone smoking on the oral environment / Darling, Mark Roger – U of the Western Cape 1989 [mf ed S.l: s.n. 1989] – 2mf [ill] – 9 – (summary in afrikaans & english; incl bibl) – sa Misc Inst [616]

Effects of carbohydrate supplementation on immune function with long endurance running and cycling / Blodgett, Andrew D – 1998 – 1mf – 9 – $4.00 – mf#PE 4083 – us Kinesology [611]

The effects of cardiac rehabilitation on coronary heart disease risk factors in post myocardial infarction patients / Cobham, H W – 1991 – 2mf – 9 – $8.00 – us Kinesology [612]

The effects of case methods on pre[-]service physical education teachers' value orientations / Timken, Gay L – 2000 – 2mf – 9 – $8.00 – mf#PE 4086 – us Kinesology [370]

Effects of certain environmental factors on germination of florida cigar-wrapper tobacco seeds / Kincaid, Randall R – Gainesville, FL. 1935 – 1r – us UF Libraries [630]

Effects of change in inputs in policy-making for the south african public service / Ababio, Ernest Peprah – Uni of South Africa 2000 [mf ed Johannesburg 2000] – 6mf – 9 – (incl bibl ref) – mf#mfm15039 – sa Unisa [350]

The effects of chromium supplementation and a low carbohydrate diet on high-intensity endurance performance / Kocher, Pamela L – 1994 – 1mf – $4.00 – us Kinesology [612]

Effects of coach interactions on college soccer players' behavior and perception / Cardinal, Jeffrey S – 1998 – 2mf – 9 – $8.00 – mf#PSY 2015 – us Kinesology [150]

Effects of cocaine on glucagon and insulin in exercising rats / Mitchell, James A – Brigham Young University, 1995 – 1mf – 9 – mf#PH 1501 – us Kinesology [615]

The effects of contextual interference and three levels of difficulty on the acquisition, retention, and transfer of hockey striking skills by second grade children / Halliday, Nancy & Goldberger, Michael – 1992 – 2mf – 9 – $8.00 – us Kinesology [150]

Effects of contextual interference on initial learning of tennis groundstrokes / Smithee, Larry L – Brigham Young University, 1994 – 1mf – 9 – mf#PSY 1901 – us Kinesology [150]

The effects of continuous and intermittent exercise bouts of equal work output on post-exercise energy expenditure / Ziegenfuss, T N – 1991 – 1mf – 9 – $4.00 – us Kinesology [612]

The effects of cooperative and individualistic goal structures on the learning domains of beginning tennis students / Brown, Joseph D – 1988 – 92p 1mf – 9 – $4.00 – us Kinesology [150]

The effects of cooperative games on classroom cohesion / Ringgenberg, Scott W – 1998 – 1mf – 9 – $4.00 – mf#PSY 2057 – us Kinesology [150]

The effects of coping skills on physiological and cognitive adaptation to a high risk activity / Lewis, Debra A – 1982 – 1mf – 9 – $4.00 – us Kinesology [616]

The effects of couple communication training on marital perceptions / Griffith, William Herbert – 1981 – 1 – $5.20 – us Southern Baptist [306]

The effects of creatine : on handgrip dynamometer maximal contraction and submaximal contraction / Martin, Bryant R – 2000 – 56p on 1mf – 9 – $6.00 – mf#PH 1700 – us Kinesology [612]

The effects of creative dance on movement creativity in third grade children / Funk, Wendy W – Brigham Young University, 1995 – 2mf – 9 – $8.00 – mf#PE 3644 – us Kinesology [790]

The effects of cross and self fertilisation in the vegetable kingdom / Darwin, Charles Robert – London, 1876 – (= ser 19th c evolution & creation) – 6mf – 9 – mf#1.1.4258 – uk Chadwyck [574]

The effects of decadron phonophoresis on serum levels of dexamethasone sodium phosphate / Darrow, Heather – 1998 – 1mf – 9 – $4.00 – mf#PE 3918 – us Kinesology [615]

Effects of deep water and treadmill running on oxygen uptake and energy expenditure in seasonally trained cross country runners / DeMaere, Jodi Michelle – 1996 – 1mf – 9 – $4.00 – mf#PH 1590 – us Kinesology [612]

The effects of dehydration and temperature on movement and reaction time in college age males / Whittle, R C – 1991 – 1mf – 9 – $4.00 – us Kinesology [613]

Effects of dehydration on ratings of perceived exertion at the lactate and ventilatory thresholds / Dengel, DR – 1990 – 2mf – 9 – $8.00 – us Kinesology [150]

The effects of diet and exercise of varying intensities on the body composition of adult women / Bradley, Carolyn G – 1980 – 2mf – 9 – $8.00 – us Kinesology [790]

The effects of dietary carbohydrates on resting metabolic rate / Jewell, David A – 1998 – 1mf – 9 – $4.00 – mf#PH 1629 – us Kinesology [612]

Effects of different exercise promotion strategies and stage of exercise on reported physical activity, self-motivation, and stages of exercise in worksite employees / Cash, Tamra L – 1997 – 3mf – 9 – $12.00 – mf#PSY 2101 – us Kinesology [150]

The effects of different resistances on peak power during the wingate anaerobic test / Hermina, Waldemar – 1999 – 1mf – 9 – $4.00 – mf#PH 1653 – us Kinesology [612]

Effects of elevated muscle temperature on exercise-induced muscle sympathetic nerve activity / Gracey, Kathryn H – 1997 – 1mf – 9 – $4.00 – mf#PH 1567 – us Kinesology [612]

The effects of endurance exercise on metabolic water production and plasma volume changes / Pivarnik, James M – 1982 – 1mf – 9 – $4.00 – us Kinesology [790]

The effects of enhanced eccentric training on improvement of strength / Follenius, Christopher & Stopka, Christine – 1993 – 1mf – 9 – $4.00 – us Kinesology [613]

Effects of environmental treatments on the emergence of aquatic locomotor behaviors / Sullivan, Ann-Catherine – 1997 – 2mf – 9 – $8.00 – mf#PSY 2106 – us Kinesology [612]

The effects of ergogenic aid supplementation on the sprint capacity of male cyclists / Hansen, Christopher A – 1999 – 1mf – 9 – $4.00 – mf#PE 3942 – us Kinesology [617]

Effects of ethanol on thermoregulatory responses during cold air exposure in male and female subjects / Seo, Chungjin – Temple University, 1996 – 2mf – 9 – $8.00 – mf#PH 1506 – us Kinesology [612]

Effects of exercise and vitamin b12 supplementation on the depression scale scores of a wheelchair confined population / Dalton, Richard B – 1980 – 2mf – 9 – $8.00 – us Kinesology [616]

Effects of exercise intensity on glucose tolerance and insulin sensitivity / Shriver, Timothy C – Iowa State University, 1993 – 1mf – 9 – $4.00 – mf#PH1477 – us Kinesology [612]

The effects of exercise mode on postexercise oxygen consumption, urinary nitrogen, and fat utilization / Kolkhorst, FW – 1990 – 2mf – 9 – $8.00 – us Kinesology [612]

The effects of exercise on blood volume during dialysis : in patients with end stage renal disease / Hall, Christopher K – 2000 – 118p on 2mf – 9 – $10.00 – mf#PH 1709 – us Kinesology [617]

The effects of exercise on bone mineral density in postmenopausal women : a meta-analysis / Smith, Dana M – 1999 – 1mf – 9 – $4.00 – mf#PE 3993 – us Kinesology [612]

The effects of exercise on individuals with down syndrome / Shaffer, Heather – 1998 – 1mf – 9 – $4.00 – mf#PH 1647 – us Kinesology [612]

The effects of exercise on low-density-liporotein-receptor mediated clearance and atherosclerotic lesions in dietary induced hyperlipidemic nzw rabbits / Pujol, Thomas J & Westerfield, R Carl – 1991 – 1mf – 9 – $4.00 – us Kinesology [612]

The effects of exercise on myocardial capillary bed and connective tissue in senescent rats / Finch, Merry B – 1982 – 1mf – 9 – $4.00 – us Kinesology [790]

The effects of exercise on premenstrual syndrome and progesterone concentrations / Anzelc-Spesia, Meredith L – 1997 – 2mf – 9 – $8.00 – mf#PH 1615 – us Kinesology [615]

The effects of exercise on stress and functional abilities in community dwelling elderly / Sergent, Evelyn H – Purdue University, 1995 – 1mf – 9 – $4.00 – mf#PSY1863 – us Kinesology [150]

The effects of exercise on the strength of the low back / Knecht, John & Prentice, William E – 1992 – 1mf – 9 – $4.00 – us Kinesology [617]

The effects of exercise on weight loss, fat loss and circumference changes / Choffletti, Caryn E – 1997 – 2mf – 9 – $8.00 – mf#PH 1547 – us Kinesology [613]

The effects of exercise training and severe caloric restriction on lean-body mass in the obese / Leutholtz, Brian C & Heusner, William – 1991 – 1mf – $4.00 – us Kinesology [612]

The effects of family participation in an outdoor adventure program / Kugath, Steven D – 1997 – 3mf – 9 – $12.00 – mf#RC 534 – us Kinesology [790]

Effects of fatigue on mechanical and muscular components of performance during drop landings / James, C Roger et al – 1991 – 2mf – 9 – $8.00 – us Kinesology [612]

Effects of fatigue on shock attenuation during running / Mercer, John A – 1999 – 2mf – 9 – $8.00 – mf#PE 3920 – us Kinesology [612]

The effects of fixed and hinged ankle foot orthoses on gait myoelectric activity and standing joint alignment in children with cerebral palsy / Lough, Loretta K & Soderberg, Gary L – 1990 – 3mf – 9 – $12.00 – us Kinesology [612]

The effects of four consecutive days of acute exercise on macrophage antigen presentation / Ceddia, Michael A – 1999 – 2mf – 9 – $8.00 – mf#PE 1654 – us Kinesology [612]

The effects of freedom of choice on the participants in a leisure education program / Richard, Anne B – 1989 – 151p on 2mf – 9 – $8.00 – us Kinesology [370]

Effects of freezing temperatures on sugarcane in the florida everglades / Bourne, B A – Gainesville, FL. 1935 – 1r – us UF Libraries [630]

The effects of functional isometric weight training in conjunction with dynamic weight training on two bench press measurement tests / Johnston, David L – University of Wisconsin-La Crosse, 1995 – 2mf – 9 – $8.00 – mf#PH 1498 – us Kinesology [612]

The effects of galvanic current and ice on muscle temperature / Grutzner, Sally J – 1997 – 1mf – 9 – $4.00 – mf#PE 3834 – us Kinesology [612]

The effects of game stress situations on the heart rates of selected high school football coaches / Delashmit, S J – 1991 – 1mf – 9 – $4.00 – us Kinesology [150]

The effects of gender on alt-pe motor in junior high physical education / Woerfel, Laurie A & Wurzer, David J – 1991 – 2mf – 9 – $8.00 – us Kinesology [790]

The effects of glasnost and perestroika on the soviet sport system / Kim, Jong-Il. & DePauw, Karen P – 1993 – 2mf – 9 – $8.00 – us Kinesology [790]

The effects of gloves : on grip strength and three-point pinch in adults / Rock, Kim M – 2000 – 46p on 1mf – 9 – $5.00 – mf#PE 4126 – us Kinesology [612]

The effects of goal setting and imagery training programs on the free-throw performance of female basketball players / Lerner, Bart S – West Virginia University, 1995 – 2mf – 9 – $8.00 – mf#PSY1850 – us Kinesology [150]

The effects of goal setting on performance enhancement in a competitive athletic setting / Stitcher, Thomas P – 1989 – 85p 1mf – 9 – $4.00 – us Kinesology [790]

The effects of graded treadmill running on foot and ankle kinematics in recreational runners / Wasielewski, Noah J – 1999 – 128p on 21mf – 9 – $10.00 – mf#PE 4180 – us Kinesology [612]

The effects of group process and sport imagery on the sport experience of high school athletes / Sankar, Dan – 1997 – 1mf – 9 – $4.00 – mf#PSY 1995 – us Kinesology [302]

The effects of hand cooling during strenuous exercise on metabolic and cardiorespiratory function / Vanheest, Jaci L – 1988 – 87p 1mf – 9 – $4.00 – us Kinesology [612]

The effects of hang board exercise on grip strength and climbing performance in college age male indoor rock climbers / Jurrens, Jay D – 1997 – 1mf – 9 – $4.00 – mf#PH 1571 – us Kinesology [612]

The effects of hangboard exercise on climbing performance and grip strength in college age female indoor rock climbers / Kingsley, Angie M – 1997 – 1mf – 9 – $4.00 – mf#PH 1576 – us Kinesology [612]

The effects of heat and ice on hamstring flexibility : utilizing proprioceptive neuromuscluar facilitation stretching / Lumpkin, Kelly J – 1999 – 60p on 1mf – 9 – $5.00 – mf#PE 4096 – us Kinesology [617]

The effects of high and low intensity eccentric exercise on muscle soreness and strength / Scharnhorst, R L – 1991 – 2mf – 9 – $8.00 – us Kinesology [790]

The effects of high impact exercise versus low impact exercise on bone density in postmenopausal women / Grove, C A – 1990 – 4mf – 9 – $16.00 – us Kinesology [613]

The effects of high resistances on peak power output and total mechanical work during short-duration high intensity exercise in the elite female athlete / Sidner, Aaron B – 1998 – 1mf – 9 – $4.00 – mf#PH 1624 – us Kinesology [612]

Effects of high school concepts-based physical education on student behavior, knowledge, and motivation / Mickelson, Connie L & Roundy, Elmo S – 1992 – 2mf – $8.00 – us Kinesology [790]

The effects of high spatial constraints in determining the nature of the speed-accuracy trade-off in aimed hand movements / Kim, Kyoung N – 1988 – 39p 1mf – 9 – $4.00 – us Kinesology [790]

Effects of high versus low glycemic index-rated carbohydrate foods on exercise performance and fat / Andrews, Steven J – 1998 – 1mf – 9 – $4.00 – mf#PH 1660 – us Kinesology [612]

The effects of high volume resistance training on lipid profiles and insulin sensitivity / Hair, Christopher Heath – 1997 – 1mf – 9 – $4.00 – mf#PH 1569 – us Kinesology [612]

The effects of high-volt pulsed current electrical stimulation on delayed onset muscle scoreness / Butterfield, David L – 1996 – 1mf – 9 – $4.00 – mf#PE 3798 – us Kinesology [617]

The effects of hip position and angular velocity on quadriceps and hamstring eccentric peak torque / Hopkins, Joe R & Sitler, Michael R – 1992 – 1mf – 9 – $4.00 – us Kinesology [612]

The effects of hormone replacement therapy and active lifestyle on immune function in postmenopausal women / Hough, Holly J – 1998 – 2mf – 9 – $8.00 – mf#PH 1631 – us Kinesology [612]

The effects of hydration status and blood glucose on mental performance during extended exercise in the heat / Puchkoff, Julie E – 1997 – 1mf – 9 – $4.00 – mf#PH 1602 – us Kinesology [612]

The effects of imaginary maximal muscle contraction training on the voluntary neural dirve to muscle / Yue, Guang H & Cole, Kelly J – 1990 – 2mf – 9 – $8.00 – us Kinesology [150]

Effects of imitative learning on the acquisition of rotary pursuit skill by educable mentally retarded boys / Nierengarten, Mark E – 1982 – 2mf – 9 – $8.00 – us Kinesology [616]

The effects of impact plus resistance training on the musculoskeletal system in premenopausal women / Winters, Kerri M – 2000 – 2mf – 9 – $8.00 – mf#PE 4087 – us Kinesology [617]

Effects of incremental versus constant-load exercise upon selected visual parameters in college-aged males and females / Larouere, Brian – 1998 – 1mf – 9 – $4.00 – mf#PSY 2097 – us Kinesology [612]

Effects of individual leisure counseling on perceived freedom in leisure, perceived self-efficacy, depression, and abstinence of adults in a residential program for substance / Collins, G Colleen – 1997 – 3mf – 9 – $12.00 – mf#RC 513 – us Kinesology [790]

The effects of ingesting a carbohydrate electrolyte solution on cycling performance in women / Osterkamp, Christine M – 1999 – 1mf – 9 – $4.00 – mf#PH 1679 – us Kinesology [612]

Effects of ingesting protein with various forms of carbohydrate following resistance exercise on substrate availability and markers of catabolism / Lundberg, Jennifer L – 2000 – 1mf – 9 – $4.00 – mf#PE 4071 – us Kinesology [612]

Effects of instruction on the analytical proficiency of physical education majors in fundamental sport skills analysis / Gangstead, Sandra K – 1982 – 2mf – 9 – $8.00 – us Kinesology [790]

The effects of instructions and movement reversals on the accuracy and kinematics of a rapid sequential tapping movement / Song, S – 1992 – 1mf – 9 – $4.00 – us Kinesology [150]

The effects of integrating geometry into physical education / Bastasch, Jeanne D – 1999 – 1mf – 9 – $4.00 – mf#PE 3987 – us Kinesology [370]

The effects of integration in physical education on the motor performance and perceived competence characteristics of educable mentally retarded and nonhandicapped children / Smith, S D – 1989 – 3mf – 9 – $12.00 – us Kinesology [150]

The effects of intermittent compression and cold on edema in postacute ankle sprains / Brewer, K D – 1990 – 1mf – 9 – $4.00 – us Kinesology [616]

The effects of intermittent compression on edema in post-acute ankle sprains / Rucinski, Terri J – 1989 – 47p 1mf – 9 – $4.00 – us Kinesology [617]

The effects of intermittent hyperbaric oxygen on pain perception an eccentric strength in a human injury model / Staples, James R – University of British Columbia, 1996 – 2mf – 9 – $8.00 – mf#PE 3673 – us Kinesology [617]

The effects of internal and external imagery on muscular and ocular concomitants / Hale, Bruce D – 1981 – 2mf – 9 – $8.00 – us Kinesology [616]

The effects of internet-based instructional lesson planning on teacher trainee performance / Brown, Seth E – 1999 – 2mf – 9 – $8.00 – mf#PE 3968 – us Kinesology [370]

Effects of irrigation with sewage effluent on the yields and establishment of napier grass / Stokes, W E – Gainesville, FL. 1930 – 1r – us UF Libraries [630]

Effects of leg exercise and insulin injection sites on blood glucose in persons with insulin dependent diabetes mellitus (iddm) / Gagalis, Zisis – 1992 – 1mf – $4.00 – us Kinesology [616]

The effects of leisure education on leisure satisfaction, leisure participation, and self-confidence for individuals with brain injuries / Prvu, Janet A – 1998 – 2mf – 9 – $8.00 – mf#HE 640 – us Kinesology [370]

The effects of leisure education on life satisfaction and leisure satisfaction among japanese american older adults / Shimura, Kenichi & Gushiken, Thomas – 1993 – 2mf – $8.00 – us Kinesology [790]

Effects of leukocytes on equine satellite cell proliferation / Watanabe, Kaori – 2000 – 82p on 1mf – 9 – $5.00 – mf#PH 1724 – us Kinesology [611]

Effects of limited and expanded rest intervals on the navy physical readiness test / Gray, John G – 1998 – 1mf – 9 – $4.00 – mf#PH 1669 – us Kinesology [612]

Effects of lower limb dominance on dynamic postural stability / Ross, Scott E – 2000 – 81p on 1mf – 9 – $5.00 – mf#PE 4111 – us Kinesology [617]

Effects of low-intensity, pain-free exercise on muscle metabolism in patients with peripheral vascular disease evaluated by 31p-nmr spectroscopy / Marburger, Lorri K & Stopka, Christine – 1992 – 2mf – 9 – $8.00 – us Kinesology [612]

The effects of magnetic therapy on physiological strength / Bottesch, Jessica M – 1999 – 1mf – 9 – $4.00 – mf#PE 4020 – us Kinesology [611]

The effects of menstrual cycle phase on competitive swimming performance / Rogers, Mary Jane L – University of North Carolina at Chapel Hill, 1994 – 1mf – 9 – $4.00 – mf#PSY1859 – us Kinesology [150]

The effects of migrant labour on the family system / Mazibuko, Ronald Patrick – Uni of South Africa 2000 [mf ed Johannesburg 2000] – 5mf – 9 – (incl bibl ref) – mf#mfm14819 – sa Unisa [306]

The effects of music on patients in a cardiac rehabilitation program / Holstein, Robyn E – 2000 – 41p on 1mf – 9 – $5.00 – mf#HE 690 – us Kinesology [617]

The effects of music on psychophysiological stress responses to graded exercise / Bronwley, K A – 1991 – 1mf – 9 – $4.00 – us Kinesology [790]

Effects of nutritional intervention on blood glucose levels in women (age 73-85) / Warren, John R & Shier, Nathan W – 1992 – 3mf – 9 – $12.00 – us Kinesology [613]

The effects of opioid receptor antagonism on plasma catecholamines and fat metabolism during prolonged exercise above or below lactate threshold in males / Hikoi, Hirotaka – 1999 – 2mf – 9 – $8.00 – mf#PH 1663 – us Kinesology [612]

The effects of oral contraception and hypoxia on respiratory parameters during graded exercise / Engelhard-Colton, Nancy – 1989 – 123p 2mf – 9 – $8.00 – us Kinesology [612]

Effects of overuse injury proneness and task difficulty on joint kinetic variability during landing / James, Charles R – University of Oregon, 1996 – 3mf – 9 – $12.00 – mf#PE 3654 – us Kinesology [612]

The effects of participant belaying on self efficacy of college students in indoor rock climbing / Zmudy, Mark – 1999 – 1mf – 9 – $4.00 – mf#PSY 2100 – us Kinesology [150]

The effects of participation in a leisure education program upon the lesiure behavior of mentally retarded adults / Marshall, Katharine R – 1982 – 1mf – 9 – $4.00 – us Kinesology [616]

The effects of perceived directors' leadership behaviors and selected demographic variables on physical education instructors' job satisfaction / Yang, Chih-hsien – 1994 – 2mf – $8.00 – us Kinesology [150]

Effects of perceived quality of life between coronary artery bypass graft and heart transplantation patients with regard to cardiac rehabilitation / Hunt, L E – 1991 – 2mf – 9 – $8.00 – us Kinesology [150]

The effects of perceived risk, risk-taking behaviors, and body size on injury in youth sport / Kontos, Anthony P – 2000 – 128p on 2mf – 9 – $10.00 – mf#PSY 2133 – us Kinesology [150]

The effects of pilates-based training on balance and gait in an elderly population / Hall, David W – 1998 – 55p on 1mf – 9 – $5.00 – mf#PSY 2131 – us Kinesology [612]

The effects of plyometric training on sprinting performance of collegiate males / Curley, Jeffrey J – University of North Carolina at Chapel Hill, 1995 – 1mf – 9 – $4.00 – mf#PH1456 – us Kinesology [611]

Effects of positive reinforcement on influencing grip strength performance of college-aged females / Putnam, Kelly – Texas Woman's University, 1994 – 2mf – 9 – $8.00 – mf#PSY 1898 – us Kinesology [150]

Effects of posture specific therapeutic exercise : on chronic back pain and disability / Brinton, Maria – 1999 – 93p on 1mf – 9 – $5.00 – mf#PE 4114 – us Kinesology [617]

The effects of pre-exercise consumption of low and high glycemic index carbohydrate foods on endurance running performance / Nagae, Sarah E – 1998 – 1mf – 9 – $4.00 – mf#PH 1659 – us Kinesology [612]

Effects of prolonged exercise on leptin : changes during exercise and recovery / Zafeiridis, Andreas – 1997 – 2mf – 9 – $8.00 – mf#PH 1581 – us Kinesology [612]

The effects of prototypic examples and video replay on adolescent girls' acquisition of basic field hockey skills / Russell, Diane & Sinclair, Gary D – 1991 – 2mf – 9 – $8.00 – us Kinesology [150]

Effects of prudence on the temporal and spiritual welfare of man / Bunbury, Robert Shirley – London, England. 1841 – 1r – us UF Libraries [240]

The effects of pubertal status on energy expenditure during cycling / Polzien, Kristen M – 2000 – 58p on 1mf – 9 – $5.00 – mf#PE 4152 – us Kinesology [612]

Effects of rational behavior training on attitudes of rehabilitation support personnel / Hooge, N C – 1991 – 1mf – 9 – $4.00 – us Kinesology [150]

The effects of recovery time on throwing velocity and accuracy of college baseball pitchers / Hendrickson, William R – 1993 – 1mf – $4.00 – us Kinesology [150]

The effects of redeployment on the recruitment and selection of teachers in secondary schools in port elizabeth / Wyk, Heidi van – Vista University 2000 [mf ed Johannesburg 2000] – 2mf – 9 – (incl bibl ref) – mf#mfm14759 – sa Unisa [373]

Effects of relationship status, setting and sex of perpetrator on college student evaluations of dating violence / Bethke, T – 1990 – 1mf – 9 – $4.00 – us Kinesology [360]

The effects of resistance exercise on lower extremity power in women with multiple sclerosis / Summers, Louisa – 2000 – 2mf – 9 – $8.00 – mf#HE 659 – us Kinesology [616]

The effects of resistance exercise on peripheral blood cytokine production in women ages 65-79 / Teranishi, Cheri T – 2000 – 109p on 2mf – 9 – $10.00 – mf#PH 1713 – us Kinesology [612]

The effects of resistance training during early cardiac rehabilitation (phase 9) on strength and body composition / Potvin, Andre N – 1988 – 2mf – 9 – $8.00 – mf#HE 624 – us Kinesology [613]

The effects of resistance training on fracture risk and psychological variables in postmenopausal women / Shaw, Janet M – Oregon State University, 1995 – 2mf – 9 – $8.00 – mf#PH 1507 – us Kinesology [612]

Effects of resistance training on ground reaction forces : during gait termination in older adults / Niemann-Carr, Nicole J – 2000 – 106p on 2mf – 9 – $10.00 – mf#PE 4116 – us Kinesology [612]

The effects of retroactive inhibition and contextual interference on learning a motor task / Liu, X. – 1991 – 2mf – 9 – $8.00 – us Kinesology [150]

The effects of rewards on intrinsic motivations of exercisers and nonexercisers / Tally, Elizabeth – 1993 – 1mf – $4.00 – us Kinesology [150]

The effects of road surface pitch on the subtalar joint while running at selected velocities / Pankey, Robert B – 1988 – 107p 2mf – 9 – $8.00 – us Kinesology [612]

The effects of running with a functional knee brace on lower extremity joint moments of force in anterior cruciate ligament injured subjects / Hunter, P B – 1990 – 2mf – 9 – $8.00 – us Kinesology [790]

Effects of same-day strength training : on selected physiological variables in female collegiate basketball players / Woolstenhulme, Mandy – 2000 – 56p on 1mf – 9 – $5.00 – mf#PH 1699 – us Kinesology [612]

Effects of same-day strength training on shooting skills of female collegiate basketball players / Kerbs, Brooke – 2000 – 1fm – 9 – $4.00 – mf#PE 4069 – us Kinesology [790]

Effects of same-sex and coeducational physical education on perceptions of self-confidence and class environment / Lirgg, Cathy D & Feltz, Deborah D – 1991 – 3mf – 9 – $12.00 – us Kinesology [150]

Effects of seat and back rest inclination on wheelchair propulsion of individuals with spastic cerebral palsy / Skaggs, Steve O – 1995 – 2mf – 9 – $8.00 – mf#PSY 1999 – us Kinesology [616]

Effects of selected curriculum materials and teaching experience on the preactive planning of physical educators / Ballat, Paul C – Temple University, 1995 – 3mf – 9 – $12.00 – mf#PE3582 – us Kinesology [370]

Effects of self-help stress management and parental training on the adjustment of single divorced mothers / Loots, J S – U of the Western Cape 1991 [mf ed S.l: s.n.] 1991 – 6mf – 9 – (summary in afrikaans & english) – sa Misc Inst [150]

The effects of self-talk on batting performance / Hamel, J M – 1991 – 2mf – 9 – $8.00 – us Kinesology [150]

Effects of sensory balance training in older adults / Hu, M – 2mf – 9 – $8.00 – us Kinesology [150]

The effects of short-term exercise on lipid and lipoprotein metabolism in obese males with abnormal glucose / Denton, Julia C – 1997 – 2mf – 9 – $8.00 – mf#PH 1593 – us Kinesology [612]

The effects of single versus multiple measures of biofeedback on basketball free throw shooting performance / Kavussanu, Maria et al – 1992 – 2mf – 9 – $8.00 – us Kinesology [150]

The effects of size and weight on basketball free throw performance : a biomechanical analysis of unskilled college women / Wilkerson, Bethany A – 1989 – 48p 1mf – 9 – $4.00 – us Kinesology [612]

Effects of slide board training : on the lateral movement of college-aged football players / Petersen, Tianna S – 2000 – 97p on 1mf – 9 – $5.00 – mf#PE 4140 – us Kinesology [612]

Effects of social environment on feeling states and self-efficacy in a group exercise class / Elfering, Melissa – 1998 – 1mf – 9 – $4.00 – mf#PSY 2036 – us Kinesology [150]

The effects of social physique anxiety, gender, age, and depression on perceived exercise behavior / Lantz, C D – 1991 – 2mf – 9 – $8.00 – us Kinesology [150]

The effects of social support on men's exercise-related cardiovascular reactivity / Hollander, Daniel B – 1998 – 142p on 3mf – 9 – $15.00 – mf#PSY 2176 – us Kinesology [150]

Effects of sodium bicarbonate loading on running time to exhaustion in male and female runners / Pardo, Javier – 1996 – 2mf – 9 – $8.00 – mf#PH 1559 – us Kinesology [612]

The effects of solid and liquid carbohydrate feedings on high intensity intermittent exercise performance / Walton, Peter T – University of British Columbia, 1996 – 1mf – 9 – $4.00 – mf#PH1483 – us Kinesology [612]

The effects of spatting and ankle taping on inversion before and after exercise / Pederson, Troy S – Brigham Young University, 1995 – 1mf – 9 – mf#PE 3665 – us Kinesology [617]

The effects of sport specificity on the utilization of stored elastic energy during a drop jump / Schiralli, Beth – 1998 – 1mf – 9 – $4.00 – mf#PE 3986 – us Kinesology [611]

The effects of sports massage upon subsequent quadricep force output, power, and total work / Kennard, Barbara A – 1998 – 1mf – 9 – $4.00 – mf#PE 3980 – us Kinesology [612]

The effects of stage-matched intervention on physical activity and coronary heart disease risk factors in women / Michalowski, Jenna R – 1999 – 1mf – 9 – $4.00 – mf#HE 631 – us Kinesology [614]

The effects of stage-matched intervention on the stages of change and exercise self-efficacy / Harder, Meghan – 1999 – 1mf – 9 – $4.00 – mf#PSY 2102 – us Kinesology [150]

Effects of static and hold-relax stretching on hamstring range of motion using the flexability le1000 / Gribble, Phillip A – 1998 – 1mf – 9 – $4.00 – mf#PH 1618 – us Kinesology [612]

The effects of static stretching on peak power and peak velocity during the benchpress / McLellan, Ernst W – 2000 – 33 on 1mf – 9 – $5.00 – mf#PE 4100 – us Kinesology [612]

Effects of statically performed toe touch stretches on torque production of the hamstring and quadriceps muscle groups / Thigpen, Lydia K – 1988 – 181p 2mf – 9 – $8.00 – us Kinesology [617]

Effects of step height variation on knee joint moments of force during lateral step-up exercises / Rauch, Mignone – Texas Woman's University, 1994 – 2mf – 9 – $8.00 – mf#PE 3667 – us Kinesology [612]

The effects of stick length on the kinematics of maximal velocity throwing in lacrosse / Stevenson, John R – 1983 – 2mf – 9 – $8.00 – us Kinesology [790]

Effects of strength training on muscle mass and musculoskeletal injury in middle aged and older men / Redmond, R A – 1991 – 2mf – 9 – $8.00 – us Kinesology [612]

The effects of stretching, ice massage, and rest an anterior shin pain / Wilson, Natalie – 1997 – 1mf – 9 – $4.00 – mf#PE 3876 – us Kinesology [615]

Effects of submaximal exercise on the mood of female bulimics / Glazer, A R – 1991 – 1mf – 9 – $4.00 – us Kinesology [150]

Effects of subtitle b of s. 1766 to the public utility holding company act : hearing...united states senate, 107th congress, 2nd session to examine the effects of...amendments to the public utility holding company act, on energy markets and energy consumers, feb 6 2002 / United States. Congress. Senate. Committee on Energy and Natural Resources – Washington: US GPO 2002 [mf ed 2002] – 1mf – 9 – (incl bibl ref) – us GPO [343]

The effects of success and failure on casual attributions among scholastic wrestlers / Scott, David – 1982 – 1mf – 9 – $4.00 – us Kinesology [150]

Effects of summer cover crops on crop yields and on the soil / Stokes, W E – Gainesville, FL. 1936 – 1r – us UF Libraries [630]

The effects of supervised cardiac rehabilitation on selected coronary artery disease risk factors following coronary artery bypass graft surgery / Goebel, BM – 1991 – 2mf – 9 – $8.00 – us Kinesology [612]

Effects of supervisory profiling on targeted feedback behaviors of preservice physical educators / Smith, John O & Sinclair, Gary D – 1991 – 2mf – 9 – $8.00 – us Kinesology [150]

The effects of surface type on plantar pressure distribution and running kinematics / Killgore, Garry L – 1989 – 143p 2mf – 9 – $8.00 – us Kinesology [612]

The effects of sustained heavy exercise on the development of pulmonary interstitial edema in trained male cyclists / O'Hare, Turlough J – 1998 – 1mf – 9 – $4.00 – mf#PH 1616 – us Kinesology [612]

The effects of television viewing on the self-regulation of exercise intensity / Viteri, Jacqueline E – 1994 – 2mf – $8.00 – us Kinesology [150]

The effects of the cross walk#zy's resistive arm poles on the metabolic costs of treadmill walking / Foley, Thomas S – 1994 – 1mf – $4.00 – us Kinesology [612]

The effects of the donjoy defiance knee brace on functional performance measures in females with acl reconstructions / Piland, Scotty G – 1998 – 1mf – 9 – $4.00 – mf#PE 3870 – us Kinesology [617]

Effects of the ejectment – London, England. 18-- – 1r – us UF Libraries [240]

The effects of the kids' connection program on sixth graders' drug knowledge and self-concept / Stumbaugh, T A – 1991 – 1mf – 9 – $4.00 – us Kinesology [150]

Effects of the menstrual cycle and oral contraceptives on athletic performance / Lebrun, Constance MT & McKenzie, Donald C – 1991 – 2mf – 9 – $8.00 – us Kinesology [613]

The effects of the menstrual cycle on exercise performance / McCracken, M A – 1990 – 1mf – 9 – $4.00 – us Kinesology [612]

Effects of the menstrual cycle phases on the energy intake and expenditure in physically active and inactive women / Holliman, Susan C – 1993 – 3mf – $12.00 – us Kinesology [612]

The effects of the new tariff on the upper canada trade – [Toronto?: s.n, 1859?] [mf ed 1984] – 1mf – 9 – 0-665-46027-9 – mf#46027 – cn Canadiana [380]

The effects of the strength shoe on vertical jump performance in male collegiate basketball players / Cody, SM – 1989 – 1mf – 9 – $4.00 – us Kinesology [790]

Effects of the use of two different teaching styles on motor skill acquisition of fifth-grade students / Moore, Robert E – East Texas State University, 1996 – 1mf – 9 – mf#PSY 1894 – us Kinesology [150]

Effects of thick-bar resistance training on strength measures in experienced weightlifters / Kruger, Matthew J – 1999 – 1mf – 9 – $4.00 – mf#PE 3984 – us Kinesology [611]

The effects of three different ankle training programs on functional stability and single limb stance / Malley, Cody – 1998 – 1mf – 9 – $4.00 – mf#PSY 2021 – us Kinesology [612]

Effects of three different hyperhydration strategies on cardiovascular and thermoregulatory resonses, blood volume and running performance / Collins, Michael G – 1999 – 2mf – 9 – $8.00 – mf#PH 1689 – us Kinesology [612]

The effects of three selected training programs on shoulder external rotation strength, flexibility, and throwing velocity in collegiate baseball players / Ploeger, Robin – 1993 – 2mf – $8.00 – us Kinesology [617]

Effects of thymopentin on the responses of hypothalamic-pituitary-adrenal axis to a high intensity dynamic exercise protocol / Golan, Ron & Kendrick, Zebulon V – 1993 – 2mf – 9 – $8.00 – us Kinesology [613]

The effects of toys, prompts, and flotation devices on the learning of water orientation skills : for preschoolers with or without developmental delay / Clawson, Cindy A – 1999 – 82p on 1mf – 9 – $5.00 – mf#PE 4168 – us Kinesology [150]

The effects of trade liberalisation on south african agriculture : with specific refe[re]nce to fertiliser and seed / Breitenbach, Marthinus Christofel – Vista University 2000 [mf ed Johannesburg 2000] – 5mf – 9 – (incl bibl ref) – mf#mfm14799 – sa Unisa [380]

The effects of training and detraining on corticosterone rhythms and dietary fat selection in the osborne-mendel rat / Schlabach, G A – 1991 – 2mf – 9 – $8.00 – us Kinesology [590]

Effects of training in strength shoes$_{(tm)}$ on speed, jumping ability, and calf girth / Pethan, Scott M – 1993 – 1mf – $4.00 – us Kinesology [612]

Effects of training on resting blood pressure in men at risk for coronary heart disease : strength vs aerobic exercise training / Dawson, P K – 1990 – 2mf – 9 – $8.00 – us Kinesology [612]

Effects of training utilizing two isotonic weight resisted exercises : on modified vertical jump performance / Troczynski, Les B – 1999 – 261p on 3mf – 9 – $15.00 – mf#PE4142 – us Kinesology [612]

The effects of twelve weeks of walking or exerstriding on upper body muscular strength and endurance / Karawan, Ariel & Porcari, John P – 1992 – 2mf – $8.00 – us Kinesology [612]

The effects of two aerobic fitness levels on excess post-exercise oxygen consumption in young adults / Short, Kevin R – 1994 – 1mf – $4.00 – us Kinesology [612]

The effects of two attentional training packages on self-efficacy, state anxiety, perceived workload, and task performance / Wilson, Casey D – 1999 – 1mf – 9 – $4.00 – mf#PSY 2066 – us Kinesology [790]

The effects of two educational processes on energy, nutrient, and food group intakes of sedentary, overweight women who are consuming self-help, low-fat, ad libitum diets / Jensen, J Keith – 1997 – 2mf – 9 – $8.00 – mf#HE 614 – us Kinesology [614]

The effects of two instructional conditions on sport skill specific analytic proficiency of physical education majors / Leis, Hans J & Gangstead, Sandra K – 1993 – 1mf – 9 – $4.00 – us Kinesology [790]

Effects of two resistance training protocols on insulin-like growth factors, muscle strength, and bone mass in older adults / Maddalozzo, Gianni F – 1999 – 2mf – 9 – $8.00 – mf#PE 3971 – us Kinesology [612]

The effects of varied rest interval lengths on depth jump performance / Read, M Michael – 1997 – 1mf – 9 – $4.00 – mf#PE 3805 – us Kinesology [612]

The effects of various exercise modalities on serum cholesterol and triglyceride concentrations / Crowder, Todd A & Roberts, John A – 1989 – 3mf – 9 – $12.00 – us Kinesology [612]

Effects of varying levels of fatigue on the rate of force development in females / Ewing, John L, Jr – 1982 – 2mf – 9 – $8.00 – us Kinesology [790]

The effects of video-computerized feedback on competitive state anxiety, self-efficacy, effort, and baseball hitting-task performance / Leslie, P J – 1998 – 2mf – 9 – $8.00 – mf#PSY 2037 – us Kinesology [790]

The effects of visual imagery ability combined with visual mental practice techniques upon motor performance / Hoffman, Diana M – 1980 – 1mf – 9 – $4.00 – us Kinesology [612]

Effects of visual training on visual pursuit, catching, and attentiveness : a case study / Shimakawa, Tsuguyo & Fisher, Janet M – 1992 – 2mf – 9 – $8.00 – us Kinesology [150]

The effects of visual-verbal modeling on the form and outcome of basketball shooting in beginners / Thomas, Milton B – 1998 – 1mf – 9 – $4.00 – mf#PE 4015 – us Kinesology [790]

Effects of warm-up prior to eccentric exercise : on indirect markers of muscle damage / Evans, Rachel – 2000 – 133p on 2mf – 9 – $10.00 – mf#PE 4132 – us Kinesology [617]

The effects of weight loss on plasma cholesterol and lipoproteins among female participants in a residential wellness program / Teague, S L – 1991 – 2mf – 9 – $8.00 – us Kinesology [613]

The effects on extracurricular participation of academic achievement, self-concept, and locus of control among high school students / Johnson, Scott R – 2000 – 224p on 3mf – 9 – $15.00 – mf#PSY 2169 – us Kinesology [150]

Effectual means of relieving the exigencies and grievances of the times : or of introducing the new and happy era of mankind... / Edwards, George – London: printed by W Pople, 1814 – (= ser 19th c economics) – 1mf – 9 – mf#1.1.90 – uk Chadwyck [339]

Effectual remedy to the disputes presently existing in the associat... / Taylor, William – Glasgow, Scotland. 1799 – 1r – us UF Libraries [240]

Die effek van 'n prestasierapport in liggaamlike opvoeding vir seuns / Malan, Jan H – Stellenbosch: U van Stellenbosch 1982 [mf ed 1982] – 3mf – 9 – mf#af.2005-567 – sa Stellenbosch [370]

Effekte von protease-inhibitoren auf das wachstum periodontaler bakterien und elastase / Weist, Torsten & Ryll, Diana – (mf ed 1999) – 1mf – 9 – €30.00 – 3-8267-2600-6 – mf#DHS 2600 – gw Frankfurter [617]

Les effets de la loi quebecoise interdisant la publicite destinee aux enfants rapport / Comite federal-provincial sur la publicite destinee aux enfants (Canada) – [Ottawa]: Ministere des communications du Canada; [Quebec]: Ministere des communications du Quebec, 1985 [mf ed 1996] – 2mf – 9 – mf#SEM105P2719 – cn Bibl Nat [340]

Effetti d'amore, canzonette a quatro voci / Scaletta, Orazio – 1595 – (= ser Mssa) – 2mf – 9 – €35.00 – mfchl 403 – gw Fischer [780]

Efficacite au debusquage des billots et au maniement des chevaux / Koroleff, Alexander – Montreal: Section forestiere, l'Association canadienne de la pulpe et du papier, 1942 [mf ed 1994] – 1mf – 9 – mf#SEM10P2141 – cn Bibl Nat [634]

Efficacy cognitions, intrinsic motivation, and exercise behavior / Oman, R F – 1989 – 2mf – 9 – $8.00 – us Kinesology [150]

Efficacy of prayer / Foreign Chaplain – London, England. 1873 – 1r – us UF Libraries [240]

The efficacy of prayer : being the donnellan lectures for the year 1877 / Jellett, John Hewitt – 3rd ed. Dublin: Hodges, Foster, and Figgis; London: Macmillan, 1880 – 1mf – 9 – 0-7905-1207-6 – (incl bibl ref) – mf#1987-1207 – us ATLA [240]

Efficacy of prayer in relation to the divine judgments / Veitch, James – Edinburgh, Scotland. 1865 – 1r – us UF Libraries [240]

The efficacy of topical ibuprofen in an inflammatory model : delayed onset muscle soreness / Mack, Rana L – University of British Columbia, 1995 – 1mf – 9 – $4.00 – mf#PE3609 – us Kinesology [617]

The efficacy of water displacement as a potential tool for assessing total body composition / Friedman, Amy A – 1999 – 1mf – 9 – $4.00 – mf#PE 3925 – us Kinesology [617]

Efficiency : a study of the why and how of adult class work / Pounds, John Edward – St Louis, MO: Christian Board of Publication, c1912 – 1mf – 9 – 0-524-06655-8 – mf#1991-2710 – us ATLA [374]

Efficient religion / Andrews, George Arthur – New York: Hodder & Stoughton, c1912 – 1mf – 9 – 0-7905-7676-7 – mf#1989-0901 – us ATLA [240]

Effie vernon : or, life and its lessons / Addison, Julia – London: E Marlborough & Co, 1861 – (= ser 19th c women writers) – 5mf – 9 – mf#5.1.36 – uk Chadwyck [830]

Effinger, John R *see* Unitarianism, its history and position

Effingham County Genealogical Society *see* Crossroad trails

Die effizienz eines individuellen intensiv-prophylaxe-programms bei koerperbehinderten patienten mit spastischer zerebralparese : eine klinisch kontrollierte interventionsstudie / Hofmann, Eva – (mf ed 1996) – 1mf – 9 – €30.00 – 3-8267-2362-7 – mf#DHS 2362 – gw Frankfurter [617]

Die effizienz eines individuellen intensiv-prophylaxeprogramms bei patienten mit morbus-down-syndrom / Otten, Ursula – (mf ed 1998) – 1mf – 9 – €30.00 – 3-8267-2506-9 – mf#DHS 2506 – gw Frankfurter [617]

Effner, Ute Antonie *see* Photoelektronenspektroskopie an alkalimetall/si-grenzflaechen mit synchrotronstrahlung

L'effort – Lyon. aout 1940-aout 1944 – 1 – fr ACRPP [073]

L'effort – Port-au-Prince, Haiti: [s.n., 1902-]. [1ere annee, n1-n28. 28 mars-22 juil 1902] – 2 sheets – 9 – us CRL [079]

L'effort – Poitiers puis Paris. juin 1910-juin 1914 – 1 – (puis l'effort libre.) – fr ACRPP [073]

L'effort : revue federale de litterature, de sociologie et d'art – Paris. n10-19. 1900 – 1 – fr ACRPP [073]

L'effort *see* Voices from wartime france, 1939-45

Effort / Jeremie – Port-Au-Prince, Haiti. 1905 – 1r – us UF Libraries [972]

L'effort culturel du peuple espagnol en armes / Spain. Ministerio de Instruccion Publica y Bellas Artes – [n. p.] 1937 – 9 – mf#w852 – us Harvard [946]

Effort du gouvernement dans le domaine de l'educat... – Port-Au-Prince, Haiti. v1-2. 1956 – 1r – us UF Libraries [370]

L'effort indochinois – Hanoi: Impr Trang-Bac Tan-Vau, jun 13 1939-jun 6 1941 – 1r – 1 – mf#mf-11768 seam – us CRL [079]

L'effort libre *see* L'effort

Les efforts de la liberte et du patriotisme contre le despotisme, du sr de maupeou chancelier de france : ou recueil des ecrits patriotiques publies pour maintenir l'ancien gouvernement francais – Londres 1772-73 [mf ed Hildesheim 1995-98] – 4v on 14mf – 9 – €140.00 – 3-487-26201-0 – gw Olms [944]

Efforts et resultats / Vincent, Stenio – Port-Au-Prince, Haiti. 1938 – 1r – us UF Libraries [972]

Effront, Jean *see* Enzymes and their applications

Effrontes / Augier, Emile – Paris, France. 1861 – 1r – us UF Libraries [440]

Efimovich, L G *see* Poltavskii den'

Efiopskie rukopisi v s-peterburge / Turaev, B – Spb, 1906 – 3mf – 9 – mf#R-10778 – ne IDC [956]

Efl gazette – Oxford. 1985-1987 (1) – ISSN: 0732-5819 – mf#49412 – us UMI ProQuest [420]

Efla evaluations / Educational Film Library Association – New York. 1973-1987 (1,5,9) – (cont by: afva evaluations) – ISSN: 0146-3152 – mf#9916 – us UMI ProQuest [790]

EFO collector – v1 n1-v11 n3 [1978 jun-1988 dec] – 1r – 1 – mf#1055785 – us WHS [071]

Efremov, P A *see*
- Trutene, 1769-1770
- Zhivopisets. 1772-1773

Efsus / Hanim, Nigar – Istanbul: Ahtar Matbaasi, 1308-9 [1891-2] – (= ser Ottoman literature, writers and the arts) – 3mf – 9 – $55.00 – us MEDOC [470]

Efta bulletin – Geneva. 1977-1992 (1,5,9) – ISSN: 0012-7655 – mf#11656 – us UMI ProQuest [338]

Efta news/efta bulletin : [english ed] – Geneva. 1993-1994 (1,5,9) – mf#20356 – us UMI ProQuest [338]

Efterretninger om geistlige embeder i norge / Boeck, Thorvald Olaf – Christiania: J Dybwad, [187-] [mf ed 1986] – 660p – 1 – mf#7432 – us Wisconsin U Libr [240]

Efterretninger om groenland : udbragne af en journal holden fra 1721-1788 / Egede, P – Kobenhavn, [1788] – 7mf – 9 – mf#N-198 – ne IDC [919]

L'egalitaire : journal de l'organisation sociale – n1-2. Paris. mai-juin 1840 – 1 – fr ACRPP [325]

L'egalitaire / Parti socialiste (SFIO) – Brest, mars 1907-sep 1908 – 1 – fr ACRPP [335]

Egalitatea – Bucharest, Romania. -w. 15 April 1890-25 Dec 1892. 1 reel – 1 – uk British Libr Newspaper [949]

L'egalite – Cap-Haitien: St-Cap Louis Blot, feb 15 1881-jun 12 1883 – 2 sheets – 9 – us CRL [079]

L'egalite : journal de l'association internationale des travailleurs de la suisse romande – Geneva. v1-4 n23. 1868-72 – (= ser Important periodicals of italian and international socialism, 1868-1917) – 1r – 1 – $150.00 – us UPA [335]

L'egalite : journal republicain socialiste. – Paris. 8 fevr 1889-7 oct 1891 – 1 – fr ACRPP [335]

L'egalite – Paris: Dondey-Dupre, apr-may 1849 – us CRL [074]

L'egalite de roubaix-tourcoing – Roubaix. fevr 1896-juin 1914 – 1 – fr ACRPP [073]

Egalite des hommes, des peuples, des races – Alger. Dir, Ferhat Abbas. no.37-133. aout 1946-juil 1948 – 1 – fr ACRPP [322]

Egan, Pierce *see* Walks through bath

Egan, Thomas J *see* History of the halifax volunteer battalion and volunteer companies, 1859-1887

Egana, Manuel R *see* Tres decadas de produccion petrolera

Egar, John Hodson *see* The threefold grace of the holy trinity

Egas, Eugenio *see* Galeria dos presidentes de sao paulo

Egbert, James *see* Alexander campbell and christian liberty

Eg-binnenmarkt und entwicklung von politikfeldern / Knaepper, Matthias – (mf ed 1993) – 3mf – 9 – €49.00 – 3-89349-686-6 – mf#DHS 686 – gw Frankfurter [327]

Ege, Ernst *see* Helmbrecht

Egede, H *see*
- Description et histoire naturelle du groenland
- Det gamle groenlands nye perlustration

Egede, Hans *see* Description et histoire naturelle du groenland

Egede, P *see*
- Efterretninger om groenland
- Omstaendelig og udfoerlig relation angaaende den gronlandske missions begyndelse og fortsaettelse

Egel, Karl Georg *see* Das lied der matrosen

Egelhaaf, Gottlob *see*
- Gustav adolf in deutschland, 1630-1632
- Landgraf philipp von hessen / m butzers bedeutung fuer das kirchliche leben in hessen

Egerer fronleichnamsspiel / ed by Milchsack, Gustav – Stuttgart: Litterarischer Verein, 1881 (Tuebingen: L F Fuess) – (incl bibl ref) – us Wisconsin U Libr [430]

Egerer fronleichnamsspiel / ed by Milchsack, Gustav – Stuttgart: Litterarischer Verein in Stuttgart, 1881 (Tuebingen: L F Fues) [mf ed 1993] – (= ser Blvs 156) – 364p – 1 – (incl bibl ref) – mf#8470 reel 32 – us Wisconsin U Libr [790]

Egerer passionsspiel *see* Egerer fronleichnamsspiel

Egerer tagblatt egerer zeitung – Cheb, Czechoslovakia. Jul-Aug 1938 – 1r – 1 – us L of C Photodup [077]

Das egerer urgichtenbuch : 1543-1579 / ed by Skala, Emil – Berlin: Akademie-Verlag, 1972 [mf ed 1994] – (= ser Deutsche texte des mittelalters 67) – li/175p/1pl – 1 – (incl bibl ref and ind) – mf#8623 reel 19 – us Wisconsin U Libr [340]

Egerer zeitung – Eger (Cheb CZ), 1923 3 jan-1938 – 33r – 1 – gw Misc Inst [077]

Egerland – Eger (Cheb CZ), 1928 nov-1933 – 6r – 1 – gw Misc Inst [077]

Egerton, F Clement C *see* Angola in perspective

Egerton, Fred, mrs *see* Admiral of the fleet, sir geoffrey phipps hornby gcb

Egerton, Hakluyt *see*
- England and rome
- Is the new theology christian?
- Liberal theology and the ground of faith

Egerton, Hugh Edward *see* Canadian constitutional development

An egg check list of north american birds : giving accurate descriptions of the color and size of the eggs, and locations of the nests of the land and water birds of north america / Davie, Oliver – Columbus OH: Hann & Adair, 1885 – 1mf – 9 – mf#27814 – cn Canadiana [590]

Egg industry – Mount Morris. 1987+ (1,5,9) – (cont: poultry tribune) – ISSN: 0896-2804 – mf#11516,01 – us UMI ProQuest [630]

Eggebrecht, Axel *see* Goethe, schiller

Eggenburger zeitung – Eggenburg, Austria. 10 oct 1946-12 feb 1948 – 1r – 1 – uk British Libr Newspaper [072]

Egger, Franz *see* Absolute oder relative wahrheit der heiligen schrift?

Egger, Victor *see* La parole interieure: essai de psycologie descriptive

Eggers, Chr U D von *see*
- Deutsches gemeinnuetziges magazin
- Neues deutsches magazin

Eggers, Christian U. von *see* Reise durch franken, baiern, oesterreich, preussen und sachsen

Eggers, Hans *see*
- Der althochdeutsche isidor
- Zwei psalter

Eggers, Kurt *see*
- Deutsches bekenntnis
- Die geburt des jahrtausends
- Das grosse wandern
- Hutten
- Der junge hutten
- Schicksalsbrueder
- Der tanz aus der reihe
- Tausend jahre kakelduett
- Vater aller dinge
- Von der freiheit des kriegers

Eggersmann, Christian *see* Beeinflussung von sensorischer reizschwelle und urodynamischen parametern durch lidocain-haltiges gleitgel fuer topische anwendung in der urethra

Eggert, H *see* Die concurrenz fuer entwuerfe zum neuen reichstagsgebaeude

Eggert, KE *see* First metatarsophalangeal joint range of motion as a factor in turf toe injuries

Eggert-Windegg, Walther *see* Briefe

Eggington, st michael and all angels monumental inscriptions – Bedfordshire Family HS 1978 – (= ser Bedfordshire parish register series) – 1mf – 9 – £1.25 – uk BedsFHS [929]

Eggler, Jurg *see* Iconographic motifs from palestine/israel and daniel 7:2-14

Eggleston, Edward *see*
- The beginners of a nation
- The circuit rider
- The schoolmaster's stories, for boys and girls

Eggleston, George Cary *see* Dorothy south: a love story of virginia just before the war

Egharevba, Jacob U *see*
- Benin games and sports
- Benin law and custom
- The city of benin
- Concise lives of famous iyases of benin
- The origin of benin
- Some tribal gods of southern nigeria

Eginitis, Demetrios *see* Annales de l'observatoire national d'athenes

Egiptio, A *see* Avisi particulari

Egit, Jacob *see* Tsu a nay lebn

Eglentiers poetens borst-weringh / Rodenburgh, Th – t'Amsterdam: Paulus van Ravesteyn, voor JE Cloppenburgh, 1619 – 6mf – 9 – mf#O-735 – ne IDC [090]

Egle-tal / Preil, Joshua Joseph – Warsaw, Poland. 1898 – 1r – us UF Libraries [939]

Egli, E *see*
- Actensammlung zur geschichte der zuercher reformation in den jahren 1519-1533
- Analecta reformatoria
- Heinrich bullingers diarium
- Die reformation im bezirke affolter
- Die schlacht von cappel 1531
- Schweizerische reformationsgeschichte
- Die st. galler taeufer
- Die zuericher wiedertaeufer zur reformationszeit
- Zwingli's tod nach seiner bedeutung fuer kirche und vaterland

Egli, Emil *see*
- Analecta reformatoria
- Die zuericher wiedertaeufer zur reformationszeit

Eglineagle – 1981 may-1982, 1983 jan-nov 4, 1984 feb 17-1985 jul, 1985 aug-1986 may, 1986 jun-1987 jun, 1987 jul-1988 jul, 1988 aug-1989 apr, 1989 may-dec – 8r – 1 – mf#646852 – us WHS [071]

L'eglise a l'epoque du concile de trente (he17) – Paris, 1948 – (= ser Histoire de l'eglise (he)) – €25.00 – ne Slangenburg [241]

L'eglise anglicane avant la reforme, abrege d'histoire ecclesiastique : en trois parties / Benoit, Henry E – Montreal: [s.n, 1900?] – 2mf – 9 – 0-665-91628-0 – (int by h m m hackett) – mf#91628 – cn Canadiana [242]

L'eglise au bresil pendant l'empire et pendant la republique / Badaro, F – Roma: Stabilimento Bontempelli, 1895 [mf ed 1990] – 1mf – 9 – 0-7905-7041-6 – (in french) – mf#1988-3041 – us ATLA [230]

L'eglise au pouvoir des laiques (888-1057) (he7) – Paris, 1940 – (= ser Histoire de l'eglise (he)) – €27.00 – ne Slangenburg [240]

L'eglise au temps du grand schisme et de la crise conciliaire (1378-1449) (he14) – Paris, 1962-64 – (= ser Histoire de l'eglise (he)) – €61.00 – ne Slangenburg [240]

L'eglise byzantine de 527 a 847 / Pargoire, R P J – Paris, 1905 – 5mf – 9 – mf#H-2940 – ne IDC [243]

L'eglise catholique au canada : precis historique et statistique publie en 1909 a l'occasion du premier concile plenier de quebec – Quebec: Editions de l'Action sociale catholique, 1914 – 2mf – 9 – 0-665-73544-8 – mf#73544 – cn Canadiana [241]

Eglise catholique. Diocese de Quebec *see* – Catechisme a l'usage du diocese de quebec

Eglise catholique. Diocese de Quebec Eveque *see* Lettre circulaire a messieurs les cures du district de quebec

L'eglise catholique et la liberte aux etats-unis / Meaux, Marie Camille Alfred, Vicomte de – 2. ed. Paris: Victor Lecoffre 1893 [mf ed 1992] – 1mf – 9 – 0-524-03851-1 – (in french) – mf#1990-4898 – us ATLA [241]

L'eglise catholique et les protestants / Romain, Georges – Paris: Librairie Bloud & Barral, 1900 [mf ed 1986] – 1mf – 9 – 0-8370-8784-8 – (in french. incl bibl ref) – mf#1986-2784 – us ATLA [230]

L'eglise catholique et l'etat sous la troisieme republique (1870-1906) / Debidour, Antonin – Paris: F Alcan. 2v. 1906-09 [mf ed 1990] – (= ser Bibliotheque d'histoire contemporaine) – 3mf – 9 – 0-7905-6986-8 – (in french. incl bibl ref) – mf#1988-2986 – us ATLA [230]

Eglise catholique, la renaissance, le protestantisme *see* The catholic church, the renaissance and protestantism

Eglise Catholique. Province de Quebec *see* Le catechisme des provinces ecclesiastiques de quebec, montreal, ottawa

L'eglise chretienne au temps de saint ignace d'antioche / Genouillac, Henri de – Paris: Beauchesne, 1907 [mf ed 1992] – 1mf – 9 – 0-524-03583-0 – (in french. incl bibl ref) – mf#1990-1043 – us ATLA [240]

Eglise d'Angleterre en Canada Diocese of Huron. Synod *see* Constitution, rules and canons of the incorporated synod of the diocese of huron

Eglise d'Angleterre en Canada Diocese of Montreal. Synod *see*
- Constitution, rules of order, by-laws, rules and canons of the synod of the diocese of montreal

Eglise d'Angleterre en Canada Diocese of Nova Scotia. Diocesan Synod *see* Constitution, canons, rules and regulations of the diocesan synod of nova scotia

Eglise d'Angleterre en Canada. Diocese of Ontario *see* Canons of the synod of the diocese of ontario and of the provincial synod of canada

Eglise d'Angleterre en Canada. Province du Canada *see* Constitution, rules of order, canons, etc of the synod of the province of "canada"

Eglise d'Angleterre en Canada Province of Canada *see* Constitution, rules of order, canons etc of the synod of the province of "canada"

L'eglise d'apres calvin / Farsat, Henri – Geneve: Impr Ziegler, 1874 [mf ed 1993] – (= ser Presbyterian coll) – 1mf – 9 – 0-524-06536-5 – (in french) – mf#1991-2620 – us ATLA [242]

L'eglise d'apres l'institution chretienne de jean calvin / Daulte, Henri – Lausanne: Georges Bridel, 1885 [mf ed 1993] – (= ser Presbyterian coll) – 2mf – 9 – 0-524-07409-7 – mf#1991-3069 – us ATLA [242]

L'eglise de berne : ses adversaires et ses defenseurs / Mestral, Armand de – Lausanne: Georges Bridel, 1866 [mf ed 1991] – 1mf – 9 – 0-524-00770-5 – (in french) – mf#1990-0202 – us ATLA [240]

L'eglise de calvin a strasbourg (1538-1541) / Berton, Eugene – Montauban: Macabiau-Vidallet, 1881 [mf ed 1992] – 1mf – 9 – 0-524-03571-7 – (in french. incl bibl ref) – mf#1990-1031 – us ATLA [242]

L'eglise de france sous la troisieme republique / Lecanuet, Edouard – nouv rev corr ed. Paris: J de Gigord, 1910 [mf ed 1991] – 2v on 3mf – 9 – 0-524-00569-9 – (in french. incl bibl ref) – mf#1990-0069 – us ATLA [944]

L'eglise de geneve, 1555-1909 : esquisse historique de son organisation, suivie de ses diverses constitutions, de la liste de ses pasteurs et professeurs, et d'une table biographique / Heyer, Henri – Geneve: A Jullien, 1909 [mf ed 1992] – 2mf – 9 – 0-524-04191-1 – (in french) – mf#1990-1230 – us ATLA [242]

L'eglise de paris pendant la revolution francaise, 1789-1801 / Delarc, Odon – Paris: Desclee, de Brouwer, [1895-1897?] [mf ed 1990] – 3v on 4mf – 9 – 0-7905-7045-9 – (in french. incl bibl ref) – mf#1988-3045 – us ATLA [241]

L'eglise de rome : reponse du reverend charles chiniquy au rev j m bbuyere [sic], grand-vicaire de london, ontario... / Chiniquy, Charles – Montreal: Impr de l'Aurore, 1870 – 1mf – 9 – mf#SEM105P45 – cn Bibl Nat [241]

L'eglise de russie / Boissard, L – Paris: Joel Cherbuliez, 1867 [mf ed 1986] – 2v on 4mf – 9 – 0-8370-7848-2 – (in french. incl bibl ref) – mf#1986-1848 – us ATLA [240]

Eglise de st francois d'assise / Desmazures, Adam Charles Gustave – Montreal: s.n, 1870 – 1mf – 9 – mf#03904 – cn Canadiana [720]

L'eglise en notre temps / Fertin, Pierre – [Port-au-Prince: Impr La Phalange] 1963 [mf ed 1963] – 2mf – 9 – (incl bibl) – mf#Sc Micro F-42 – us NY Public [240]

L'eglise et la critique / Mignot, Eudoxe Irenee Edouard – 2e ed. Paris: Librairie Victor Lecoffre, 1910 [mf ed 1986] – 1mf – 9 – 0-8370-8841-0 – mf#1986-2841 – us ATLA [241]

L'eglise et la critique biblique (ancien testament) / Brucker, Joseph – Paris: P Lethielleux, [1907?] [mf ed 1986] – (= ser Questions d'ecriture sainte) – 1mf – 9 – 0-8370-6888-6 – (in french. incl bibl ref, app and ind) – mf#1986-0888 – us ATLA [241]

L'eglise et la remission des peches aux premiere siecles / Galtier, P – Paris, 1932 – 9mf – 8 – €18.00 – ne Slangenburg [240]

L'eglise et la renaissance (1449-1517) (he15) – (= ser Histoire de l'eglise (he)) – €19.00 – ne Slangenburg [941]

L'eglise et la science : precis historique / Francais, J – Paris: Librairie critique, 1908 [mf ed 1986] – 1mf – 9 – 0-8370-8669-8 – (in french. incl bibl ref) – mf#1986-2669 – us ATLA [210]

L'eglise et la sorcellerie : preecis historique, suivi des documents officiels, des textes principaux, et d'un proces inedit / Francais, J – Paris: E Nourry, 1910 [mf ed 1990] – 1mf – 9 – 0-7905-5989-7 – (in french) – mf#1988-1989 – us ATLA [130]

L'eglise et le progres du monde / Devas, Charles Stanton – Paris: Victor Lecoffre, 1909 [mf ed 1986] – 1mf – 9 – 0-8370-7213-1 – (trans fr english into french by j-d folghera. incl bibl ref) – mf#1986-1213 – us ATLA [241]

L'eglise et l'enseignement populaire : sous l'ancien regime / Allain, Ernest – Paris: Bloud, [18–?] [mf ed 1986] – (= ser Science et religion) – 1mf – 9 – 0-8370-7521-1 – (in french) – mf#1986-1521 – us ATLA [377]

L'eglise et les campagnes au moyen age / Prevost, Gustave Amable – Paris: H Champion, 1892 [mf ed 1990] – 1mf – 9 – 0-7905-7188-9 – (in french. incl bibl ref) – mf#1988-3188 – us ATLA [240]

L'eglise et l'etat au canada apres la conquete du pays par les anglais : mgr briand et les gouverneurs de son temps / Gosselin, Auguste – [Evreux, France?: s.n.], 1916 – 1mf – 9 – 0-665-74320-3 – mf#74320 – cn Canadiana [230]

L'eglise et l'etat dans la seconde moitie du 3e siecle (249-284) / Aube, Benjamin – Paris: E Perrin, 1885 [mf ed 1990] – 2mf – 9 – 0-7905-6221-9 – (in french, latin and greek. incl bibl ref) – mf#1988-2221 – us ATLA [230]

L'eglise et l'etat en france / Desdevises du Dezert, Georges – Paris: Societe francaise d'impr et de librairie, 1907-08 [mf ed 1990] – 2v on 2mf – 9 – 0-7905-6988-4 – (in french) – mf#1988-2988 – us ATLA [241]

L'eglise et l'etat sous la monarchie de juillet / Thureau-Dangin, Paul – Paris: E Plon, 1880 [mf ed 1992] – 2mf – 9 – 0-524-03664-0 – (in french. incl bibl ref) – mf#1990-1092 – us ATLA [241]

L'eglise et l'orient au moyen age : les croisades / Brehier, Louis – 3. ed. Paris: J Gabalda, 1911 [mf ed 1991] – (= ser Bibliotheque de l'enseignement de l'histoire ecclesiastique) – 1mf – 9 – 0-524-01102-8 – (in french) – mf#1990-0316 – us ATLA [931]

Eglise et theologie – Ottawa. 1975-1999 (1) 1976-1999 (5) 1976-1999 (9) – (cont by: theoforum) – ISSN: 0013-2349 – mf#9915 – us UMI ProQuest [240]

L'eglise evangelique reformee de florence : depuis son origine jusqu' a nos jours: notice historique d'apres les sources originales / Andre, Louis Edouard Tony – Florence: Impr et librairie claudienne, 1899 [mf ed 1992] – 1mf – 9 – 0-524-03210-6 – (incl bibl ref) – mf#1990-0838 – us ATLA [242]

L'eglise francaise de strasbourg au seizieme siecle : apres des documents inedits / Erichson, Alfred – Paris: Fischbacher, 1886 [mf ed 1992] – 1mf – 9 – 0-524-03580-6 – (in french) – mf#1990-1040 – us ATLA [242]

L'eglise georgienne : des origines jusqu'a nos jours / Tamarati, Michel – Rome: Societe typographico-editrice romaine, 1910 [mf ed 1986] – 2mf – 9 – 0-8370-7512-2 – (incl ind) – mf#1986-1512 – us ATLA [243]

L'eglise naissante et le catholicisme / Batiffol, Pierre – Paris: J Gabalda, 1909 [mf ed 1986] – 2mf – 9 – 0-8370-9923-4 – (in french. incl bibl ref) – mf#1986-3923 – us ATLA [240]

Eglise Nationale Protestante de Geneve *see* Eglise nationale protestante de geneve

Eglise nationale protestante de geneve : memorial des seances du consistoire / Eglise Nationale Protestante de Geneve – v1-119. 1873-1991 – Inquire – 1 – mf#ATLA S0337 – us ATLA [242]

L'eglise orthodoxe russe : organisation, dogmes, heresies (doukhoborstes et molokanes) / Laflamme, Joseph Clovis Kemner – Quebec: impr de L-J Demers & Freres...1901 [mf ed 1985] – (= ser Deux conferences donnees) – 1mf – 9 – mf#SEM105P464 – cn Bibl Nat [243]

L'eglise orthodoxe russe : organisation, dogmes, heresies, doukhoborstes et molokanes...quebec, 1900-1901 / Laflamme, Joseph Clovis Kemler – [Quebec?: s.n.] 1901 [mf ed 1994] – 1mf – 9 – 0-665-73268-6 – mf#73268 – cn Canadiana [243]

L'eglise primitive (he1) – Paris, 1934 – (= ser Histoire de l'eglise (he)) – €23.00 – ne Slangenburg [941]

Une eglise reformee au 17 siecle : ou, histoire de l'eglise wallonne de hanau. depuis sa fondation jusqu'a l'arrivee dans son sein des refugies francais / Leclercq, J B – Hanau: Imprimerie des orphelins, 1868 – 1mf – 9 – 0-524-02006-X – (incl bibl ref) – mf#1990-0551 – us ATLA [240]

L'eglise reformee de paris sous henri 4 : rapports de l'eglise et de l'etat, vie publique et privee des protestants, leur part dans l'histoire de la capitale, le mouvement des idees, les arts, la societe, le commerce / Pannier, Jacques – Paris: Honore Champion, 1911 [mf ed 1992] – (= ser Presbyterian coll) – 2mf – 9 – 0-524-03940-2 – (incl bibl ref) – mf#1990-4934 – us ATLA [242]

L'eglise romaine et les origines de la renaissance / Guiraud, Jean – 4e ed. Paris: V Lecoffre, 1909 [mf ed 1990] – (= ser Bibliotheque de l'enseignement de l'histoire ecclesiastique) – 1mf – 9 – 0-7905-5892-0 – (in french. incl bibl ref) – mf#1988-1892 – us ATLA [241]

L'eglise russe et l'eglise catholique : lettres du r p rozaven de la compagnie de jesus / Gagarin, Jean [comp] – nouv ed. Paris: E Plon, 1876 [mf ed 1986] – 1mf – 9 – 0-8370-7188-7 – (in french) – mf#1986-1188 – us ATLA [230]

L'eglise selon l'evangile / Gasparin, Agenor, comte de – Paris: M Levy, 1878-79 [mf ed 1990] – (= ser Coll michel levy) – 2v on 2mf – 9 – 0-7905-3440-1 – (in french) – mf#1987-3440 – us ATLA [240]

L'eglise sous la croix pendant la domination espagnole : chronique de l'eglise reformee de lille / Frossard, Charles Louis – Paris: Grassart, 1857 [mf ed 1992] – 1mf – 9 – 0-524-02003-5 – (in french) – mf#1990-0548 – us ATLA [242]

Eglise unie d'Angleterre et d'Irlande Province of Canada. Provincial Synod *see* Provincial synod of the united church of england and ireland in canada

Les eglises de jerusalem : la discipline et la liturgie au 4e siecle / Cabrol, Fernand – Paris: H Oudin, 1895 – 1mf – 9 – 0-524-00519-2 – (incl bibl ref) – mf#1990-0019 – us ATLA [240]

Les eglises de jeruzalem : la discipline et la liturgie au 4th siecle / Cabrol, F – Paris, 1895 – 4mf – 8 – €11.00 – ne Slangenburg [243]

Les eglises du refuge en angleterre / Schickler, Fernand de – Paris: Fischbacher, 1892 – 4mf – 9 – 0-7905-7143-9 – mf#1988-3143 – us ATLA [240]

Les eglises orientales et le saint-siege / Lamy, Thomas Joseph – Bruxelles: Societe belge de librairie, 1895 – 1mf – 9 – 0-8370-7806-7 – mf#1986-1806 – us ATLA [240]

Eglises reformees de france : cinquante ans de souvenirs religieux et ecclesiastiques, 1830-1880 / Pedezert, Jean – Paris: Librairie Fischbacher, 1896 – 6mf – 9 – 0-524-07360-0 – mf#1990-5397 – us ATLA [240]

Eglises separees / Duchesne, Louis – 2e ed. Paris: Albert Fontemoing, 1905 – 1mf – 9 – 0-8370-7625-0 – (incl bibl ref) – mf#1986-1625 – us ATLA [240]

Eglitis, Anslavs *see* Svabu kaprico

Egloffstein, Hermann, Freiherr von *see*
– Alt-weimars abend
– Carl august im niederlaendischen feldzug 1814

Egloffstein, Hermann, Freiherr von und zu *see* Carl august im niederlaendischen feldzug 1814

Eglogas del pastor de extremadura / Rocha, Manuel de la – 1821 – 9 – sp Bibl Santa Ana [810]

Las eglogas y georgicas...virgilio / Mesa, Cristobal – 1793 – 9 – sp Bibl Santa Ana [450]

Eglwys Brewis, glamorgan, parish church of st brewis : baptisms 1722-1939, burials 1721-1935, marriages 1724-1860; & gileston, st giles, baptisms 1701-1940, burials 1702-1946, marriages 1701-1836 – 1mf – 9 – £1.25 – uk Glamorgan FHS [929]

Eglwys brewis, st brewis; bethesda'r fro congregational, monumental inscriptions – 1mf – 9 – £1.25 – uk Glamorgan FHS [929]

Eglwysilan, glamorgan, parish church of st ilan : baptisms 1679-1900, burials 1694-1891, marriages 1695-1837 – 4mf – 9 – £5.00 – uk Glamorgan FHS [929]

Eglwysilan, st ilan, monumental inscriptions – 1mf – 9 – £1.25 – uk Glamorgan FHS [929]

Egmont star – Hawera. New Zealand. -d. 15 Jul 1899-26 Sep 1914. (Imperfect). (29 reels) – 1 – uk British Libr Newspaper [079]

Egner, Fritz *see* Der dichterische essay

The ego and its place in the world / Shaw, Charles Gray – London: G Allen, 1913 – 2mf – 9 – 0-7905-7468-3 – mf#1989-0693 – us ATLA [100]

The ego and the mechanisms of defence = Das ich und die abwehrmechanismen / Freud, Anna – London: Hogarth Press, 1937 [mf ed 1993] – (= ser The international psycho-analytical library 30) – 1mf – 9 – 0-524-08103-4 – (english by cecil barnes) – mf#1993-9009 – us ATLA [150]

Ego documents from the netherlands, 16th century-1814 : pt 1: manuscript travel journals in languages other then dutch, 16th century-1814 / ed by Dekker, R M – 335mf – 9 – €1785.00 – (p/g in english) – mf#m428 – ne MMF Publ [914]

Ego, Michael M *see* Leisure preference patterns of second-generation japanese-americans of selected cities in the united states

The egoist: an individual review – v. 1-6, no. 5. 1914-19 – 1 – us L of C Photodup [410]

Egorova, N F *see* Delo [odessa: 1919]

Egozcue, J *see* Memorias de la comision del mapa geologico de espana. memoria geologica-minera de la provincia de caceres

Egregia cantio... / Kerle, Jacobus de – 1574 – (= ser Mssa) – 1mf – 9 – €20.00 – mfchl 271 – gw Fischer [780]

La egregia figura de carlos de yuste. (las postrimerias de su vida y su muerte ejemplar) / Gutierrez Macias, Valeriano – Badajoz: Dip. Provincial, 1958. Sep. REE – sp Bibl Santa Ana [920]

Egremont (ashfield) 1838-1897 – Oxford, MA (mf ed 1988) – (= ser Massachusetts vital records) – 16mf – 9 – 0-87623-044-3 – (mf 1: town & vital records 1834-43. mf 2: town & vital records 1841-44. mf 3: town & vital records 1844-48. mf 4: town & vital records 1847-53. mf 5: town & vital records 1852-53. mf 6-7: index to births 1845-1987. mf 8-9: index to marriages 1845-1987. mf 10-11: index to deaths 1845-1987. mf 12: marriage intentions 1846-83. mf 13: births 1844-61. mf 14: marriages & deaths 1844-61. mf 15: births 1862-97; marriages 1862-97. mf 16: marriages 1880-97; deaths 1861-97) – us Archive [978]

Egremont (liscard), st john – (= ser Cheshire monumental inscriptions) – 1mf – 9 – £2.50 – mf#187 – uk CheshireFHS [929]

Eguia, F *see* Papel o escrito...sobre las bebidas heladas...

Eguiara y Eguren, Juan Jose de *see*
– Maria santissima pintandose milagrosamente en su bellissima imagen de guadalupe de mexico
– Praelecto theologica
– Relectio exponens vigessimam sextam distinctionem libri tertij sententiarum magistri
– Vida del venerable padre don pedro de arellan, y sossa, sacerdote

Eguibar y Muniz, Juan Jose de *see* Zalamea de la serena (badajoz) jamas fue "ilipa"

Egville, J-H d' *see*
– Le mariage mexicain, divertissement ballet
– Telemaque

Egvilly, A d' *see* Memoires historiques et politiques de 1820 a 1830

Egyenloseg – Budapest. 1903-16 – 1 – us L of C Photodup [073]

Egypt – A monthly record of Egyptian and Near East news. London. -m. Mar 1911-Feb 1913. (22 ft) – 1 – uk British Libr Newspaper [960]

Egypt / Clement, Clara Erskine – Boston, MA. 1880 – 1r – us UF Libraries [960]

Egypt : in translations / Breasted, James Henry et al – New York: Parke, Austin & Lipscomb c1917 [mf ed 1992] – (= ser The sacred books and early literature of the east 2) – 2mf – 9 – 0-524-04428-7 – (trans & bibl by james h breasted; incl bibl ref) – mf#1991-0002 – us ATLA [470]

Egypt : internal affairs and foreign affairs, 1945-jan 1963 / U.S. State Dept – (= ser Confidential u s state department central files) – 1 – $20,780.00 coll – (internal affairs & foreign affairs, 1945-49 19r isbn 0-89093-648-X $3675. 1950-54 38r isbn 0-89093-649-8 $7370. internal affairs, 1955-59 30r isbn 1-55655-144-4 $5810. foreign affairs, 1955-59 7r isbn 1-55655-145-2 $1340. egypt/united arab republic: internal affairs & foreign affairs, 1960-jan 1963 19r isbn 1-55655-807-4 $3675. with p/g) – us UPA [327]

Egypt / Waters, Clara Erskine Clement – rev and enl. Chicago: Werner, 1895 – 1mf – 9 – 0-524-04599-2 – mf#1992-0187 – us ATLA [930]

Egypt *see*
– Al-jaridah al-rasmiyah. (official gazette)
– Al-waqai al-misriya

Egypt and babylon from sacred and profane sources / Rawlinson, George – New York: Charles Scribner, 1885 [mf ed 1988] – 1mf – 9 – 0-7905-0205-4 – (incl bibl ref) – mf#1987-0205 – us ATLA [930]

Egypt and israel / Petrie, W M Flinders – London, England. 1911 – 1r – us UF Libraries [327]

Egypt and israel / Petrie, William Matthew Flinders – London: SPCK; New York: E S Gorham, 1911 – 1mf – 9 – 0-7905-1833-3 – (incl ind) – mf#1987-1833 – us ATLA [220]

Egypt, and mohammed ali : or, travels in the valley of the nile / Saint John, James A – London 1834 [mf ed Hildesheim 1995-98] – 2v on 8mf – 9 – €160.00 – 3-487-27368-3 – gw Olms [916]

Egypt and syria / Dawson, John William – London, England. 1892 – 1r – us UF Libraries [327]

Egypt and the books of moses : or, the books of moses illustrated by the monuments of egypt = Buecher mose's und aegypten / Hengstenberg, Ernst Wilhelm – Edinburgh: Thomas Clark, 1845 – 1mf – 9 – 0-8370-9392-9 – (incl bibl ref. in english) – mf#1986-3392 – us ATLA [221]

Egypt and the christian crusade / Watson, Charles Roger – Philadelphia, PA: Board of Foreign Missions of the United Presbyterian Church of N A, c1907 – 1mf – 9 – 0-8370-6447-3 – (incl ind) – mf#1986-0447 – us ATLA [240]

Egypt and the egyptian question / Wallace, Donald Mackenzie – London 1883 – (= ser 19th c british colonization) – 6mf – 9 – mf#1.1.9743 – uk Chadwyck [320]

Egypt and the pentateuch : an address to the members of the open air mission / Cooper, William Ricketts – London: Samuel Bagster, [1875] – 1mf – 9 – 0-8370-2741-1 – mf#1985-0741 – us ATLA [221]

Egypt baptist church. millington, tennessee : church records – 1840-1961 – 1 – 51.12 – us Southern Baptist [242]

Egypt, cyprus and asiatic-turkey / Farley, James Lewis – London: Truebner & Co., 1878. xvi,270p – 1 – us Wisconsin U Libr [915]

Egypt for the egyptians : a retrospect and a prospect – London, 1880 – (= ser 19th c books on british colonization) – 3mf – 9 – mf#1.1.9746 – uk Chadwyck [960]

Egypt, india, and the colonies / Fitzgerald, William Forster Vesey. – London, 1870 – (= ser 19th c books on british colonization) – 3mf – 9 – mf#1.1.3701 – uk Chadwyck [337]

Egypt. Laws, Statutes, etc *see* La legislation en matiere immobiliere en egypte; recueil des lois, reglements et instructions administratives relatifs a la propriete immobiliere

Egypt. Maslahat al-Ihsa wa-al-Ta'dad *see*
– Annuaire statistique
– Annuaire statistique 1901-1959
– Statistical returns 1881-1897

Egypt. Maslahat al-Ihsa wa-al-Tadad *see* Recensement general de l'egypte

Egypt, nubia, and abyssinia / Conder, Josiah – London 1827 [mf ed Hildesheim 1995-98] – 2v on 6mf – 9 – €120.00 – 3-487-27383-7 – gw Olms [916]

Egypt, nubia and ethiopia illustrated by one hundred stereoscopic photographs taken by francis frith for messers : negretti and zambra; with descriptions and numerous wood engravings by joseph bonomi; and notes by samuel sharpe – London: Smith, Elder, 1862 – (filmed with: christianity, islam and the negro race/e w blyden) – us CRL [960]

The egypt of the hebrews and herodotos / Sayce, Archibald Henry – London: Rivington, Percival, 1896 – 1mf – 9 – 0-7905-0286-0 – (incl bibl ref and index) – mf#1987-0286 – us ATLA [960]

Egypt, palestine, and phoenicia : a visit to sacred lands / Bovet, Felix – London: Hodder & Stoughton 1882 [mf ed 1989] – 1mf [ill] – 9 – 0-7905-2885-1 – (trans by w h lyttelton; biogr sketch of aut by f godet) – mf#1987-2885 – us ATLA [915]

Egypt past and present / Adams, William Henry Davenport – London; New York: T Nelson, 1894 – 1mf – 9 – 0-524-00620-2 – mf#1990-0120 – us ATLA [916]

Egypt under the pharaohs : a history derived entirely from the monuments = Geschichte aegypten's unter den pharaonen / Brugsch, Heinrich Karl – new rev condensed ed. London: J Murray, 1891 [mf ed 1992] – 2mf – 9 – 0-524-02200-3 – (in english) – mf#1990-2874 – us ATLA [930]

L'egypte et la syrie : ou moeurs, usages, costumes et monumens des egyptiens, des arabes et des syriens; precede d'un precis historique; accompagne de notes et eclaircissemens... / Breton de LaMartiniere, Jean – Paris 1814 [mf ed Hildesheim 1995-98] – 6v on 12mf – 9 – €120.00 – 3-487-27378-0 – gw Olms [956]

Egypte's internationaal statuut... / Houten, H R van – 's-Gravenhage, 1930 – 2mf – 9 – mf#ILM-1940 – ne IDC [956]

Egyptian belief and modern thought / Bonwick, James – London: C Kegan Paul, 1878 – 2mf – 9 – 0-524-02196-1 – mf#1990-2870 – us ATLA [290]

Egyptian birds / Whymper, C – London, 1909 – 8mf – 8 – mf#Z-1960 – ne IDC [590]

Egyptian bondage / Wackerbarth, Francis Diedrich – London, England. 1842 – 1r – us UF Libraries [240]

Egyptian ceramic art / Wallis, Henry – London [1900] – (= ser 19th c art & architecture) – 2mf – 9 – mf#4.2.1182 – uk Chadwyck [730]

The Egyptian church / Dowling, Theodore Edward – London: Cope & Fenwick, [1909?] – 1mf – 9 – 0-7905-4408-3 – mf#1988-0408 – us ATLA [240]

The egyptian coffin texts, vol 2 : texts of spells 76-163 / Buck, Adriaan de – 1938 – 9 – $12.00f – 0-226-07946-5 – us Oriental [930]

The egyptian coffin texts, vol 6 : texts of spells 472-786 / Buck, Adriaan de – 1956 – 9 – $18.00f – 0-226-07944-9 – us Oriental [930]

The egyptian conception of immortality / Reisner, George Andrew – Boston: Houghton Mifflin, c1912 – (= ser The Ingersoll Lecture) – 1mf – 9 – 0-7905-8565-0 – mf#1989-1790 – us ATLA [290]

Egyptian daily post – Cairo. Egypt. -d. 29 apr 1909-26 nov 1910 – 8 1/2r – 1 – uk British Libr Newspaper [079]

Egyptian gazette – Alexandria, Cairo, Egypt. -d. Aug 1882; feb 1884; 1893-1938; mar 1939-mar 1940; 1941-feb 1944; 1945-sep 1952; 1958; 1959, aug 1960-1965 – 227r – 1 – uk British Libr Newspaper [072]

The egyptian gazette – Alexandia: [s.n.], 1952-66 – 1 – us CRL [079]

The egyptian heaven and hell / Budge, Ernest Alfred Wallis – Chicago: Open Court; London: Kegan Paul, Trench, Truebner 1906 [mf ed 1989] – (= ser Books on egypt and chaldaea 20-22) – 3v on 2mf – 9 – 0-8370-1178-7 – mf#1987-6014 – us ATLA [290]

Egyptian ideas of the future life / Budge, Ernest Alfred Wallis – 2nd ed. London: Kegan Paul, Trench, Truebner 1900 [mf ed 1989] – (= ser Books on egypt and chaldaea 1) – 1mf – 9 – 0-8370-1179-5 – mf#1987-6015 – us ATLA [290]

Egyptian ideas of the future life / Budge, Ernest Alfred Wallis – London, 1899 – (= ser Books on egypt and chaldaea) – 3mf – 9 – (books on egypt and chaldaea. v1) – mf#NE-20019 – ne IDC [956]

Egyptian inscriptions from the british museum and other sources / Sharpe, S – London, 1837-1855. 3v – 14mf – 9 – mf#NE-455 – ne IDC [930]

Egyptian letters to the dead : mainly from the old and middle kingdoms / ed by Gardiner, A H & Sethe, K – London, 1928 – 3mf – 9 – mf#NE-463 – ne IDC [930]

Egyptian magic / Budge, Ernest Alfred Wallis – London: Kegan Paul, Trench, Truebner 1899 [mf ed 1989] – (= ser Books on egypt and chaldaea 2) – 1mf – 9 – 0-8370-1181-7 – (incl bibl ref) – mf#1987-6016 – us ATLA [290]

Egyptian mail – Cairo. Egypt. -w. 7 Mar 1916-30 Jun 1922; 5 oct 1930; 21 jan 1936; 12 may 1937; 1943-1945 (very imperfect); 8 may 1946-27 sep 1952; 4 jan 1958-31 dec 1960; 1961-1965; 16 apr 1966-10 feb 1968 (imperfect) – 13r – 1 – (aka: natal daily news; baraza and egyptian gazette) – uk British Libr Newspaper [072]

Egyptian mail – Cairo: Societe oriental de publicite, 1952-55; 1955-66 – (filmed consecutively with: egyptian gazette) – us CRL [380]

Egyptian poetry : from its renaissance to the present time / Megally, S – n.p, 1974 – 2mf – 9 – mf#NE-385 – ne IDC [470]

Egyptian press extracts – [Cairo?]: [s.n.], feb 5, 1941-jan 1943; jan 19 1945-apr 19 1946 – us CRL [079]

Egyptian religion – New York: Alma Egan Hyatt Foundation, 1933-36 [mf ed 2000] – (= ser Christianity's encounter with world religions, 1850-1950) – 1r – 1 – (in english, french or german) – mf#2000-s002 – us ATLA [290]

Egyptian religion / Lieblein, Jens Daniel Carolus – Leipzig: IC Hinrichs, 1884 [mf ed 1991] – 1mf – 9 – 0-524-01840-5 – (incl critique of hibbert lectures for 1879 delivered by peter le page renouf) – mf#1990-2675 – us ATLA [290]

Egyptian republic – Centralia, IL. 1859-1861 (1) – mf#62527 – us UMI ProQuest [071]

The egyptian saudaan : its history and monuments / Budge, Ernest Alfred Wallis – Philadelphia: JB Lippincott, 1907 – 4mf – 9 – 0-7905-7207-9 – mf#1988-3207 – us ATLA [960]

Egyptian standard (daily edition) – Cairo, Egypt. 20 aug 1907-24 jan 1908 – 1r – 1 – uk British Libr Newspaper [079]

Egyptian tomb steles and offering stones of the museum of anthropology and ethnology of the university of california / Lutz, H F – Leipzig, 1927 – 4mf – 8 – mf#H-241 – ne IDC [930]

Egyptian trade journal and sudan gazette – London, UK. Oct 1906. -irr. 8 feet – 1 – uk British Libr Newspaper [072]

L'egyptienne : revue mensuelle politique, feminisme, sociologie – Cairo. v1-16 n1-164. feb 1925-apr 1940 [mnthly] – (= ser Arabic journals and periodicals) – 5r – 1 – $950.00 – (in french. some iss missing) – us MEDOC [073]

De egyptische kerk : een en ander over de kopten / Vlieger, A de – Middelburg: K le Cointre, 1896 [mf ed 1986] – 44p on 1mf – 9 – 0-8370-7837-7 – (incl bibl ref) – mf#1986-1837 – us ATLA [243]

Egypt's past, present and future / Howell, Joseph Morton – Dayton, OH: Service Publishing Company, 1929. xi,378p. plates, ports – 1 – us Wisconsin U Libr [956]

Egyseges magyarsag = United hungarians – Niagara Falls, ON. v1-3 n26. oct 9 1959-jul 1 1961// – 2r – 1 – Can$175.00 – (strongly anti-communist hungarian-language paper) – cn McLaren [947]

Das ehbuechlin / Alber, E – np, [1539] – 1mfmf – 9 – mf#TH-1 mf 8 – ne IDC [242]

Ehe die spur sich verliert / Langenbucher, Erich – Berlin: Junge Generation Verlag, [1942] – 1r – 1 – us Wisconsin U Libr [830]

Das ehe- und familienrecht der hebrer : mit ruecksicht auf die ethnologische forschung / Eberharter, Andreas – Muenster i W: Aschendorff, 1914 [mf ed 1989] – (= ser Alttestamentliche abhandlungen 5/1-2) – 1mf – 9 – 0-7905-1937-2 – (incl ind) – mf#1987-1937 – us ATLA [270]

Ehe und Jungfraeulichkeit im Neuen Testament / Fischer, Joseph – 1.& 2. aufl. Muenster in Westfalen: Aschendorff 1919 [mf ed 2005] – (= ser Biblische zeitfragen; 9. folge 3/4) – 1r with other items – 1 – 0-524-10544-8 – mf1074b – us ATLA [225]

Das eheliche und unverehelichte leben der ersten christen, nach ihren eigenen zeugnissen und exempeln / Arnold, Gottfried – Franckfurt: T. Fritsch, 1702. Chicago: Dep of Photodup, U of Chicago Lib, 1975 (1r); Evanston: American Theol Lib Assoc, 1984 (1r) – 1 – 0-8370-0467-5 – mf#1984-B261 – us ATLA [240]

Ehelolf, H *see* Keilschrifturkunden aus boghazkoey

Das eherne gesetz : ein buch fuer die kommenden / Beumelburg, Werner – 11.-20. tausend. Oldenburg i O: G Stalling c1934 [mf ed 1989] – 1r – 1 – (filmed with: der feigling & other titles) – mf#7017 – us Wisconsin U Libr [830]

Das eherne gesetz : die dichtungen georg buechners – 1. aufl. Berlin: Verlag der Nation 1950 [mf ed 1993] – (= ser Nationales erbe 1) – 1r – 1 – (filmed with: the plays of georg buechner / trans & int by geoffrey dunlop) – mf#8526 – us Wisconsin U Libr [810]

Ehespiegel : das ist, alles was vom heyligen ehestande nuetzliches, noetiges, vnd troestliches mag gesagt werden in sibentzig brautpredigten: zusammen verfasset / Spangenberg, C – Strassburg, 1561 – 7mf – 9 – mf#TH-1 mf 1401-1407 – ne IDC [242]

Ehespiegel mathesij / Mathesius, J – Leipzig, 1591 – 7mf – 9 – mf#TH-1 mf 963-969 – ne IDC [242]

Ehinger tagblatt – Ehingen / Donau DE, 1975-– 114r until 1990 – 1 – (bezirksausgabe von suedwest-presse, ulm) – gw Misc Inst [074]

Ehitus ja arhitektuur – Tallinn: [s.n.] n2. 1978 – us CRL [079]

Ehmer, Wilhelm *see* Der flammende pfeil

Ehnasya / Flinders Petrie, W M – London, 1905 – (= ser Mees 26) – 10mf – 8 – €19.00 – ne Slangenburg [930]

Ehnasya / Petrie, William Matthew Flinders – 1905 – 9 – $10.00 – us IRC [930]

Ehnasya, 1904 / Petrie, W M – London, 1905 – 3mf – 9 – mf#NE-20352 – ne IDC [956]

Ehni, Jacques *see* Die urspruengliche gottheit des vedischen yama

Ehp *see* Environmental health perspectives (ehp)

[Ehrbaren stadt braunschweig christenliche ordnung] der erbarn stadt braunschwyg christenliche ordenung, zu dienst dem heiligen evangelio christlicher lieb, zucht, fride und eynigkeit : auch darunter vil christlicher lere / Bugenhagen, Johannes – Nuernberg: Paypus 1531 – (= ser Hqab. literatur des 16. jahrh.) – 3mf – 9 – €40.00 – mf#1531 – gw Fischer [780]

Ehrenberg, Victor *see* Rechtsgeleerd advies van prof dr v ehrenberg in zake de zuid

Ehrenburg, Ilia Grigorevich *see*
– Estampas de espana
– Not intervention...conquest

Ehrenfeld, Alexander *see* Die letzte stunde

Ehrenfels, Omar Rolf Leopold Werner, Freiherr von *see* Kadar of cochin

Ehrenfried, Sabine *see* Die untersuchung der semantischen dimension von fachlichkeit an texten der historiographie und politik

Ehren-gebu oesterreichischer helden-tugenden : mit welchen weilandt der durchleuechtigste fuerst, und herr, herr ferdinandus carolus ertzhertzog zu oesterreich, etc in lebenszeiten herrlich gezieret ware / Bidermann, E – Ynsspruġg: Bey Hieronymo Paur, [1663] – 1mf – 9 – mf#O-1807 – ne IDC [943]

Ehrenhaus, Martin *see* Die operndichtung der deutschen romantik

Ehren-rangliste des ehemaligen deutschen heeres.. / Bund Deutscher Offizier – Berlin: E.S. Mittler, 1926. xvii,1275p – 1 – us Wisconsin U Libr [943]

Ehrenreich, Paul Max Alexander *see* Contribuicoes para a etnologia do brasil

Ehrensberger, H *see* Libri liturgici bibliothecae apostolicae vaticanae manu scripti

Ehrenstein, Albert *see*
– Die gedichte von albert ehrenstein
– Menschen und affen
– Der selbstmord eines katers
– Tubutsch

Ehrenstein, August *see* Chronik des abenteuerlichen, wundervollen und seltsamen in den schicksalen beruehmter reisenden

Ehrentheil, Moritz *see* Judisches familien-buch

Ehrentreu, Ernst *see* Untersuchungen uber die massora

Ehrentreu, Heinrich *see* Or ha-emet

Ehret, Joseph *see* Vokieciu literaturos istorija

Ehrhard, A *see*
– Die altchristliche literatur und ihre erforschung seit 1880. allgemeine uebersicht und erster literaturbericht
– Die altchristliche literatur und ihre erforschung von 1884-1900

Ehrhard, Albert *see*
– Die altchristliche literatur und ihre erforschung seit 1880
– Die altchristliche litteratur und ihre erforschung von 1884-1900
– Der katholizismus und das zwanzigste jahrhundert im lichte der kirchlichen entwicklung der neuzeit
– Das mittelalter und seine kirchliche entwicklung
– Das religioese leben in der katholischen kirche

Ehrhard, Jean *see* Communaute ou secession?

Ehrhardt, Albert *see* Ueberlieferung und bestand der hagiographischen und homiletischen literatur der griechischen kirche

Ehrhardt, Ingrid *see* Erich a schelling (1904-1986)

Ehrhardt, Lucien Andre *see* Hacienda publica en el salvador

Ehrhardt, Rita *see* Verstaendlichkeit von fachtexten

Ehrhardt, Traugott *see* Die geschichte der festung koenigsberg/pr., 1257-1945

Ehrismann, Gustav *see*
– Der renner
– Rudolf von ems weltchronik

Ehrke, Hans *see*
– Gewappnetes herz
– Makedonka

Ehrle, F *see* Historia bibliothecae romanorum pontificum

Ehrle, Franz *see*
– Bibliothektechnisches aus der vatikana
– Martin de alpartils chronica actitatorum temporibus domini benedicti 13. band 1, einleitung, text der chronik, anhang ungedruckter aktenstuecke
– Der sentenzenkommentar peters von candia
– Der sentenzenkommentar peters von candia, des pisaner papstes alexanders 5

Ehrler, Hans Heinrich *see*
– Briefe aus meinem kloster
– Briefe vom land
– Bruder hermanns klause
– Die drei begegnungen des baumeisters wilhelm
– Elisabeths opferung
– Die frist
– Fruehlings-lieder
– Gedichte
– Das gesetz der liebe
– Gesicht und antlitz
– Der hof des patrizierhauses
– Die lichter schwinden im licht
– Die liebe leidet keinen tod
– Meine fahrt nach berlin
– Mit dem herzen gedacht
– Der morgen
– Die reise in die heimat
– Die reise ins pfarrhaus
– Unter dem abendstern
– Wolfgang

Ehrlich, Cyril *see* The uganda company, limited

Ehrlich, Eugen *see* Das zwingende und nichtzwingende recht im buergerlichen gesetzbuch fuer das deutsche reich

Ehrlich, Lea et al *see* Mount st helens newspaper collection index, 1980-1981

Die ehrliche frau nebst harlequins hochzeit- und kindbetterinschmaus. der ehrlichen frau schlampampe krankheit und tod : lustspiele / Reuter, Christian; ed by Ellinger, Georg – Halle: Max Niemeyer, 1890 – (incl bibl ref) – us Wisconsin U Libr [430]

Ehrlicher, Fritz *see* Untersuchung zum hypalgetischen effekt der transkutanen elektrischen nervenstimulation auf die schmerzrezeption und schmerzperzeption

Ehrmann, Eliezer L *see* Arbeitsplan fur chanukka

Ehrmann, Theophil F *see*
– Beitraege zur laender- und staatenkunde der tartarei
– Neueste beitraege zur kunde von indien

Ehrt, Carl *see* Abfassungszeit und abschluss des psalters zur pruefung der frage nach makkabaeerpsalmen

Ehstnische sprachlehre fuer beide hauptdialekte, den revalschen und doerptschen : nebst einem vollstaendigen woerterbuch / ed by Hupel, August Wilhelm – Riga; Leipzig: Hartknoch 1780 – (= ser Whsb) – 6mf – 9 – €70.00 – mf#Hu 207 – gw Fischer [430]

Ehwald, R *see*
– Adhelmi opera
– Emil brauns briefwechsel

Ei Lwin, Sitkaing *see* Jo thvan on nhan mui sita

Eibenschuetz, S *see* Illustrierte gemeinde-zeitung

Eich, Hedwig *see*
– Die koenigsfanfare

Eichas, Tyler M *see* Relationships among perceived leadership styles, member satisfaction and team cohesion in high school basketball teams

Eichberg, Oskar *see*
– Parsifal
– Richard wagners sinfonie in c-dur

Eichelberger, Robert L *see* Japan and america, c1930-1955 – the pacific war and the occupation of japan, series 1

Die eichen europa's und des orients / Kotschy, C G T – Wien, Olmuez, 1862 – 4mf – 9 – mf#8394 – ne IDC [956]

Eichendorff, Hermann, Freiherr von *see* Joseph freiherr von eichendorff

Eichendorff, Joseph, Freiherr von *see*
– Eichendorff-lese
– Fruhlingsnacht [f von eichendorff] r schumann, op 39, no 12
– Gedichte von joseph freiherrn von eichendorff
– Joseph freiherrn v. eichendorffs werke
– Joseph und wilhelm eichendorffs jugendgedichte
– Mein herz still in sich singet

Eichendorff-lese : aus den romanen, novellen und gedichten des grossen romantikers / Eichendorff, Joseph, Freiherr von; ed by Hayduk, Alfons – Prag: Noebe & Co, 1944 – 1r – 1 – us Wisconsin U Libr [800]

Eichendorffs erlebnis und gestaltung der sinnenwelt / Wehrli, Rene – Frauenfeld; Leipzig: Hube & Co, 1938 – 1r – 1 – us Wisconsin U Libr [430]

Eichendorffs historische trauerspiele : eine studie / Erdmann, Julius – Halle (Saale): M Niemeyer, 1908 – 1r – 1 – us Wisconsin U Libr [430]

Eichendorffs jugenddichtungen / Hoeber, Eduard – Berlin: Vogt, 1894 – 1r – 1 – us Wisconsin U Libr [430]

Eichendorffs lyrik : eine studie zur analyse ihrer stoff- und motivkreise / Fassbinder, Franz – Koeln: J P Bachem, 1911 – 1r – 1 – (incl bibl ref and index) – us Wisconsin U Libr [430]

Eichendorffs menschengestaltung / Riepe, Christian – Berlin: Junker und Duennhaupt, 1941 – 1 – (incl bibl ref) – us Wisconsin U Libr [430]

Eichendorffs verhaeltnis zur religion / Wettig, Heinrich Marcellus – Mainz: [s.n.], 1921 – 1r – 1 – us Wisconsin U Libr [430]

Eichendorffs weltbild / Jakubczyk, Karl – 2. Aufl. Habelschwerdt: Frankes Buchhandlung, 1924 – 1r – 1 – us Wisconsin U Libr [430]

Eichentopf, Hans *see* The odor storms erzaehlungskunst in ihrer entwicklung

Eichhoff, Johann Peter *see* Materialien zur geistlichen und weltlichen statistick des niederrheinischen und westphaelischen kreises und der angraenzenden laender nebst nachrichten zum behuf ihrer aeltern geschichte

Eichholz, Johann *see*
– Darstellungen aus der schweitz
– Neue briefe ueber italien

Eichhorn, Albert *see* Das abendmahl in neuen testament

Eichhorn, Franz *see* In der grunen holle

Eichhorn, Johann Gottfried *see*
– Allgemeine bibliothek der biblischen litteratur
– Introduction to the study of the old testament

Eichler, George Augustus *see* Studies in student leadership

Eichler, P A *see* Die dschinn, teufel und engel im koran

Eichmann / Clarke, Comer – New York, NY. 1960 – 1r – us UF Libraries [939]

Eichmann, E *see* Weihe und kroenung des papstes im mittelalter

Eichmann, F *see* Die reformen des osmanischen reiches
Der eichmann-prozess in der deutschen oeffentlichen meinung : eine dokumentensammlung / Lamm, Hans – Frankfurt am Main: Ner-Tamid-Verlag, 1961 (mf ed 1995) – 1r – 1 – (incl bibl ref) – mf#*ZP-1485 – us NY Public [340]
Eichner, Ernst *see* Six quatuors pour une flute, violon, alto et basse, oeuvre 4
Eicholtz, W H and Sons *see* Records
Eichsfelder heimatzeitung – Heiligenstadt, Thuer DE, 1966 8 sep-1969 9 oct – 1r – 1 – (covers districts heilgenstadt & worbis) – gw Misc Inst [074]
Eichsfelder tageblatt *see* Mitteldeutsche allgemeine [main ed]
Eichstaedt, Heinrich Karl Abrah *see* Jenaische allgemeine literatur-zeitung
Eichstaetter kurier – Eichstaett DE, 1983 1 jun- – 9r/yr – 1 – (ba v. donau-kurier, ingolstadt) – gw Misc Inst [074]
Eichthal, Rudolf von *see* Die wunderkur
Eick, Eugen *see* Tagebuch des letzten abtes zu liesborn carolus von kerssenbrock (1750-1828)
Eicken, Heinrich von *see* Geschichte und system der mittelalterlichen weltanschauung
Eickenroth, Manuela P *see* Theaterunternehmen zwischen kunst, kommerz und politik
Eickhorst, William *see* Dekadenz in der neueren deutschen prosadichtung
Eickstedt, Valentin von *see* Epitome annalium pomeraniae
Der eid bei den semiten : in seinem verhaeltnis zu verwandten erscheinungen sowie die stellung des eides im islam / Pedersen, Johannes – Strassburg: K J Truebner, 1914 – (= ser Studien zur geschichte und kultur des islamischen orients) – 1mf – 9 – 0-7905-3157-7 – (incl bibl ref) – mf#1987-3157 – us ATLA [270]
Eid bleibt eid : 2 novellen / Berglar-Schroeer, Paul – feldpostausg. Dresden: H B Schulze, 1943 [mf ed 1989] – 112p – 1 – mf#7010 – us Wisconsin U Libr [830]
Eid gegen den modernismus : gutachten ueber den durch das paepstliche motu proprio "sacrorum antistitum" vom 1. september 1910 fuer den katholischen klerus vorgeschriebenen / Kiefl, Franz Xaver – Kempten: J Koesel, 1912 – 1mf – 9 – 0-524-04495-3 – mf#1990-1257 – us ATLA [241]
Der eid im alten testament : vom standpunkte der vergleichenden religionsgeschichte aus / Happel, Julius – Leipzig: Wilhelm Friedrich, [1893] – 1mf – 9 – 0-8370-3467-1 – mf#1985-1467 – us ATLA [221]
Der eid wider den modernismus und die theologische wissenschaft / Mausbach, Joseph – Koeln: J P Bachem, 1911 – 1mf – 9 – 0-8370-8455-5 – (incl bibl ref) – mf#1986-2455 – us ATLA [240]
Eide, I *see* Pochemu germanskii fashizm usilivaet opasnost' voiny
Eidelberg, Paul *see* Sadat's strategy
Eiderstedter anzeigenblatt – Garding DE, 1949 1 mar-30 sep – 1r – 1 – gw Misc Inst [074]
Eiderstedter nachrichten – Garding DE, 1864 5 may-1874, 1876-1945 25 may – 1 – gw Misc Inst [074]
Eiderstedter wochenblatt *see* Der dittmarser und eiderstedter bote
Eidgenoessische Sternwarte Zuerich *see* Publikationen der eidgenoessischen sternwarte in zuerich
Eidlitz, Walther *see* Unknown india
Eidos – London, 1950 [mf ed Chadwyck-Healey] – (= ser Art periodicals on microform) – 1r – 1 – uk Chadwyck [760]
Eidylivm de foelici et christiana profectione... caroli a lotharingia...ad sacrum bellum in turcos susceptum 1572 – Parisiis, 1572 – 1mf – 9 – mf#H-8192 – ne IDC [956]
Eifeler nachrichten – Monschau DE, 1978 1 sep-1991 – ca 8r/yr – 1 – (bezirksausgabe von aachener nachrichten) – gw Misc Inst [074]
Eifeler volkszeitung – Schleiden DE, 1957 2 nov-1959 30 jun – 1 – (bezirksausgabe von aachener volkszeitung, aachen) – gw Misc Inst [074]
Eifelsagen, lieder und gedichte / Zirbes, Peter – Coblenz: P Zirbes 1891 [mf ed Bloomington IN: Indiana Uni Lib, Preservation Dept 1984] – 275p on 1r – 1 – us Indiana Preservation [390]
Eiffe, Peter Ernst *see* Seemannsgarn
The eiffel tower / Tissandier, Gaston – London 1889 – (= ser 19th c art & architecture) – 2mf – 9 – mf#4.2.910 – uk Chadwyck [720]
Yr eifion – [Wales] University of Wales-Bangor ion 1856-rhag 1859 [mf ed 2003] – 4r – 1 – (cont by: yr arweinydd [ion 1857-rhag 1859]) – uk Newsplan [072]
Die eigenart der alttestamentlichen religion : eine akademische antrittsrede / Bertholet, Alfred – Tuebingen: JCB Mohr, 1913 – 1mf – 9 – 0-524-02940-7 – mf#1990-3152 – us ATLA [270]
Die eigenart der biblischen religion / Orelli, Conrad von – Gr Lichterfelde-Berlin: E Runge 1906 [mf ed 1993] – (= ser Biblische zeit- und streitfragen 2/12) – 1mf – 9 – 0-524-05818-0 – mf#1992-0645 – us ATLA [220]
Eigenart der biblischen religion = The peculiarity of the religion of the bible / Orelli, Conrad v – New York: Eaton & Mains; Cincinnati: Jennings & Graham, c1908 – 1mf – 9 – 0-8370-4631-9 – (in english. incl bibl ref) – mf#1985-2631 – us ATLA [220]
Das eigenartige des christentums als religion / Noesgen, Karl Friedrich – Halle (Saale): Richard Muehlmann (Max Grosse), 1902 – 1mf – 9 – 0-8370-4597-5 – mf#1985-2597 – us ATLA [240]
Eigenbrodt, Wolrad *see* Hagedorn und die erzaehlung in reimversen
Der eigene – Berlin DE, 1906-1930/33 n9 – 4r – 1 – gw Misc Inst [074]
Das eigene verschulden ein verschulden gegen sich selbst oder gegen dritte / Schumann, Wolfgang – Marburg, 1931 (mf ed 1994) – 1mf – 9 – €24.00 – 3-8267-3066-6 – mf#DHS-AR 3066 – gw Frankfurter [340]
Eigene wege 1960 / Betts, Peter John et al – Bern: Sinwel-Verlag, 1960 [mf ed 1993] – 1v (ill) – 1 – mf#8310 – us Wisconsin U Libr [430]
Das eigentliche ist unsichtbar : eine biographische annaeherung an den schriftsteller felix hartlaub / Marose, Monika – (mf ed 2001) – 5mf – 9 – €59.00 – 3-8267-2750-9 – mf#DHS 2750 – gw Frankfurter [430]
Eight books of caesar's gallic war / Harper, William Rainey – New York, NY. 1891 – 1r – us UF Libraries [025]
Eight charges delivered at so many several general sessions... : held at charles town...in the years 1703, 1704, 1705, 1706, 1707 / Hogue, L Lynn – Uni of Tennessee, 1972 [mf ed Spartanburg SC: Reprint Co, [1984?] – 5mf – 9 – mf#51-502/503 – us South Carolina Historical [347]
Eight days with the spiritualists / Gillingham, James – Chard, England. 1872 – 1r – us UF Libraries [240]
Eight essays on various subjects / Maitland, Samuel Roffey – London: Francis & John Rivington, 1852 – 1mf – 9 – 0-7905-0139-2 – (incl bibl ref and index) – mf#1987-0139 – us ATLA [240]
Eight hour herald / Illinois State Federation of Labor – v1 n18, v4 n87-v7 n204 [1892 dec 10, 1895 oct 19-1898 jan 13] – 1r – 1 – mf#1111352 – us WHS [331]
Eight hour miller / International Union of Flour and Cereal Mill Employees – 1903-04, 1909-10 – (= ser Labor union periodicals, pt 3: food and agricultural industries) – 1r – 1 – $210.00 – 1-55655-609-8 – us UPA [660]
Eight hour miller : official journal... / International Union of Flour and Cereal Mill Employees – 1903 jun, oct, 1905 feb,sep, 1906 apr-jun, 1909 mar-jun – 1r – 1 – mf#4934995 – us WHS [331]
The eight leading churches : their history and teaching / Berry, George Keys – Portland OR: G K Berry c1914 [mf ed 1992] – 1mf [ill] – 9 – 0-524-02697-1 – (incl bibl ref) – mf#1990-0678 – us ATLA [240]
Eight lectures on miracles : preached...in the year 1865 on the foundation of the late rev. john bampton, m.a., canon of salisbury / Mozley, James Bowling – 2nd ed. London: Rivingtons, 1867 – 1mf – 9 – 0-8370-4517-7 – (incl bibl ref) – mf#1985-2517 – us ATLA [210]
Eight o'clock – Auckland, NZ. jan 1975-dec 1978; jan 1981-dec 1984 – 29r – 1 – mf#11.21 – nz Nat Libr [079]
Eight sermons on christian union / Evans, J H – London, England. 1843? – 1r – us UF Libraries [240]
Eight silver pattern books : from birmingham city library / Boulton, Mathew – 2r – 1 – mf#96387 – uk Microform Academic [730]
Eight views of baptism : or, internal evidences of adult baptism, being a review of "the baptized child" / Hague, William – Boston: Gould, Kendall & Lincoln, 1836 [mf ed 1984] – 1mf – 9 – 0-8370-0980-4 – (incl bibl ref) – mf#1984-4336 – us ATLA [240]
Eight years in the toils : sketches from a gambler's life / Andrews, John D – Butte City: Joseph Andrews, [c1890] (mf ed 19–) – 1r – 1 – mf#*ZH-348 – us NY Public [920]
Eighteen centuries of the church in england / Hore, Alexander Hugh – Oxford: Parker, 1881 – 2mf – 9 – 0-8370-9875-0 – (incl ind) – mf#1986-3875 – us ATLA [240]
Eighteen centuries of the orthodox greek church / Hore, Alexander Hugh – London: James Parker, 1899 – 2mf – 9 – 0-7905-4925-5 – mf#1988-0925 – us ATLA [243]
Eighteen hundred and eleven : a poem / Barbauld, Anna Letitia – London: printed for J Johnson & Co, 1812 – (= ser 19th c women writers) – 1mf – 9 – mf#5.1.1 – uk Chadwyck [810]
Eighteen marches & c : for two violins (flutes or hautboys), two french horns, ad libitum, and a bass / Key, Joseph – London: printed for Messrs Thompson...[179-] [mf ed 1989] – 1r – 1 – mf#pres. film 50 – us Sibley [780]
Eighteen months in india, 1936-1937 : being further essays and writings / Nehru, Jawaharlal – Allahabad: Kitabistan, 1938 – (= ser Samp: indian books) – us CRL [954]
Eighteen months in jamaica : with recollections of the late rebellion / Foulks, Theodore – London 1833 [mf ed Hildesheim 1995-98] – 1v on 1mf – 9 – €40.00 – 3-487-26928-7 – gw Olms [918]
Eighteen unratified indian treaties in califronia – 1851-53 – 1r – 1 – $50.00 – mf#C63019 – us Library Micro [978]
Eighteen years in uganda & east africa / Tucker, Alfred Robert – London: E Arnold, 1908 – 3mf – 9 – 0-7905-7087-4 – mf#1988-3087 – us ATLA [916]
Eighteen years on lake bangweulu / Hughes, J E – London: Field House, [1933] – 1 – us CRL [960]
Eighteen years on the gold coast : from the royal commonwealth society library / Cruikshank, B – 1853 – 12mf – 7 – mf#2985 – uk Microform Academic [900]
Eighteen years on the gold coast of africa : including an account of the native tribes, and their intercourse with europeans / Cruickshank, B – London: Hurst and Blackett, 1853. 2v – 13mf – 9 – mf#A-298 – ne IDC [916]
The eighteenth century collection – ongoing (through unit 295 ca 35r per unit) – ca 10,325r total – 1 – (ongoing project based upon the eighteenth century short title catalogue (estc) holdings as well as those from over 1500 university, private and public libraries worldwide. a variety of materials is included – from books and broadsides, bibles, tract boks and sermons to printed ephemera, all providing a diverse coll of material on the enlightenment in great britain between 1701 and 1800. subject breakouts available: law 630r c39-28310. religion and philosophy 2494r c39-28320. social science 1058r c39-28330. history and geography 1756r c39-28340. literature and language 2900r c39-28350. fine arts, music, art and architecture 124r c39-28360. science, technology and medicine 1044r c39-28370. general reference and misc 319r c39-28380) – mf#C39-28300 – us Primary [941]
The eighteenth century collection : specialist literature subsets – 638r – 1 – (joseph addison 14r. almanacs and advertising of the 18th century 6r. george berkley 4r. birth of the gothic novel 18r representing works of horace walpole, clara reeve, anne radcliffe, matthew gregory lewis and william beckford. william blackstone 8r. edmund burke 14r. daniel defoe 27r. john dryden 9r. 18th century drama 28r representing the works of joseph addison, william congreve, john gray, oliver goldsmith, george lillo, nicholas rowe, and richard brinsley sheridan. 18th editions of shakespeare 30r. henry fielding 35r. the french revolution 11r. edward gibbon 7r. oliver goldsmith 24r. david hume 30r. johnson and boswell 60r. thomas paine 5r. poets of the mid and late 18th century 35r representing works of mark akenside, robert burns, thomas chatterton, william collins, william cowper, george crabbe, thomas gray, james macpherson, christopher smart and james thompson. alexander pope 47r. rise of methodism 26r. samuel richardson 18r. tobias smollett 25r. laurence stern 29r. jonathan swift 32r. women writers of the 18th century 26r representing works of fanny burney, elizabeth inchbald, sarah fielding, charlotte lennox, ann radcliffe, clara reeve, mary wollstonecraft, charlotte smith and anne finch) – us Primary [420]
Eighteenth century english provincial newspapers : from the british library, london – 4 series – 125r – 1 – (series 1: bath newspapers 31r c39-16201 – pt1: the bath journal and other papers 15r c39-16202; pt2: the bath chronicle and other papers 16r c39-16203. series 2: derby newspapers 20r c39-16204. series 3: ipswich newspapers 32r c39-16205 – pt1: the ipswich journal or the weekly mercury 1720-31, the ipswich gazette 1732-37, and the ipswich journal 1739-66 16r c39-16206; pt2: the ipswich journal 1767-1800 16r c39-16207. series 4: newcastle-upon-tyne newspapers 42r c39-16208 – pt1: the newcastle gazette or northen courant 1710-12, the newcastle courant 1711-1800, the newcastle weekly mercury 1722-23, the newcastle country journal or impartial intelligencer 1734-38 20r c39-16209; pt2: newcastle chronicle or general weekly advertiser 1764-1800, newcastle advertiser 1788-1800, newcastle gazette 1744-52, newcastle intelligencer 1755-59, newcastle journal 1739-88 22r c39-16210. incl list of contents on reel for each series) – mf#C39-16200 – us Primary [072]
Eighteenth century english romantic poetry / Partridge, Eric – Paris, France. 1924 – 1r – us UF Libraries [420]
Eighteenth century journals : from the hope collection at the bodleian library, oxford – 20r – 1 – £1800.00 – (incl: the actor, anti-theatre, the bee reviv'd, the covent garden chronicle, the eaton chronicle, the free briton, the microcosm, pig's meat, the rhapsodist, the spy at oxford & cambridge, towntalk, the tribune, the watchman, the world & 62 other titles; with d/g) – uk Matthew [073]
Eighteenth century law : from the eighteenth century collection – ongoing – us Primary [340]
Eighteenth century life – Pittsburgh. 1989+ (1,5,9) – ISSN: 0098-2601 – mf#18020 – us UMI ProQuest [975]
Eighteenth century nonconformity / Colligan, James Hay – London; New York: Longmans, Green, 1915 – 1mf – 9 – 0-7905-4620-5 – mf#1988-0620 – us ATLA [240]
Eighteenth century short title catalogue / British Library – [mf ed London: The British Library 1990] – 220mf – 9 – (incl bibl) – uk British Library [010]
Eighteenth-century english organ voluntary : a critical and analytical study with a new edition of four voluntaries by henry heron / Godding, Marc Antone – U of Rochester 1971 [mf ed 19–] – 4mf – 9 – mf#fiche 529 – us Sibley [780]
Eighth census of the united states, 1860 / U.S. Bureau of the Census – (= ser 1860 census schedules) – 1438r – 1 – mf#M653 – us Nat Archives [317]
Eighth census of the united states for the northern district of halifax county, virginia, 1860 : schedules of free inhabitants, slave inhabitants, mortality, agriculture, industry, and social statistics – 1r – 1 – mf#M1808 – us Nat Archives [317]
The eighth of december, 1854 : some account of the definition of the immaculate conception of the most blessed mother of god = Ineffabilis deus – London: T Jones, 1854 – 1mf – 9 – 0-8370-7776-1 – (in english and latin) – mf#1986-1776 – us ATLA [240]
Eighth region auto worker labor news / International Union, United Automobile Workers – v10n9-v15 n6 [1949 feb-1954 feb] – 1r – 1 – mf#1055889 – us WHS [331]
The eight-hour law. / National Eight-Hour Delegation – Washington: Darby, 1880. 32p. LL-2351 – 1 – us L of C Photodup [340]
Eightrock : music, dance, film, literature – v2 n3 [1993], 1994 – 1r – 1 – mf#2901394 – us WHS [073]
Eighty second field hospital, usar – v12 n3 [1983 dec], v12 n10 [1984 aug] – 1r – 1 – mf#1131953 – us WHS [360]
Eigyo hokokusho shusei : collected annual reports of major companies in japan, 1872-1945 – 1 – (1st ser: 917 companies [400r] 88p ind [y3576,000]; 2nd ser: (suppl to 1st) 156 companies newly collected 304 companies [110r] 40p ind [y1093,000]; 3rd ser: 831 companies [120r] 48p ind [y1109,000]; 4th ser: 3057 companies [480r] 136p ind [y4400,000]; 5th ser: 5733 companies (suppl to former ser 2447 companies) newly collected 3286 companies [852r] 200p ind [y10,224,000]; 6th ser: 1006 companies: newly collected 183 companies (suppl to former series 823 companies) [180r] 48p ind [y2520,000]) – ja Yushodo [338]
The eiheiji : a brief account of the monastery, with the short history of the soto sect, a biographical sketch of the founder etc etc – [Japan: s.n, 191-] [mf ed 1995] – (= ser Yale coll) – 34p – 1 – 0-524-09299-0 – mf#1995-0299 – us ATLA [280]
Eileen haddon collection of southern rhodesia archives, manuscripts, and documents – Chicago, University of Chicago, Photodup Dept, 1972 – 35r – us CRL [324]
Eilenburger nachrichten – Eilenburg DE, 1993 2 jan-2 apr – 1 – gw Misc Inst [074]
Eilenburger neueste nachrichten – Eilenburg DE, 1914-19 – 7r – 1 – gw Misc Inst [074]
Eilenburger Wochenblatt – Eilenburg DE, 1836-39, 1841-65 – 7r – 1 – gw Misc Inst [074]
Ein dem untergang naher aramaeer war mein vater... : das "kleine geschichtliche credo" und seine wirkungsgeschichte im midrasch der pessach-haggada / Homolka, Walter – (mf ed 1993) – 1mf – 9 – €37.50 – 3-89349-708-0 – mf#DHS 708 – gw Frankfurter [270]
Einander : oden, lieder, gestalten / Werfel, Franz – Muenchen: K Wolff, 1923 – 1r – 1 – us Wisconsin U Libr [800]
Einbecker morgenpost – Einbeck DE, 1988- – 5r/yr – 1 – gw Misc Inst [074]
Einblicke in das sprachliche der semitischen urzeit betreffend die entstehungsweise der meisten hebraeischen wortstaemme / Herzfeld, Levi – Hannover: Hahn, 1883 – 1mf – 9 – 0-8370-9157-8 – (incl ind) – mf#1986-3157 – us ATLA [470]

Einblicke in england und london im jahre 1818 : ausfuehrliche bearbeitung der in oeffentlichen blaettern schon mitgetheilten bruchstuecke / Bornemann, Johann W – Berlin 1819 [mf ed Hildesheim 1995-98] – (= ser Fbc) – 2mf – 9 – €60.00 – 3-487-28819-2 – gw Olms [941]

Einding, Rudolf Georg *see* Der opfergang

Einer baut einen dom : freiheitsgedichte / Holzapfel, Karl Maria – Berlin: W Heyer, 1934 – 1 – us Wisconsin U Libr [810]

Einer mutter sohn : roman / Viebig, Clara – Berlin: E Fleischel, 1906 – 1r – 1 – us Wisconsin U Libr [830]

Eines kriegsknechts abenteuer / Schuecking, Levin – Berlin: Carl Flemming und C T Wiskott, c1922 – 1r – 1 – us Wisconsin U Libr [830]

Das einfache leben : roman / Wiechert, Ernst Emil – Muenchen: A Langen/G Mueller 1939 [mf ed 1991] – 1r – 1 – (filmed with: der exote / ernst wiechert & other titles) – mf#3043p – us Wisconsin U Libr [830]

Einfluesse auf das fruehwerk jakob steinhardts : zur geistesgeschichtlichen verortung eines juedischen expressionisten mit einem ausblick auf sein gesamtwerk / Kaufmann, Dorothee – (mf ed 2000) – 5mf – 9 – €59.00 – 3-8267-2719-3 – mf#DHS 2719 – gw Frankfurter [700]

Die einfluesse der antike in wielands hermann : beitrag zur entwicklungs-geschichte der deutschen literatur im 18. jahrhundert / Doell, M – Muenchen: M Schnidtmann, 1897 – 1r – 1 – (incl bibl ref) – us Wisconsin U Libr [430]

Einfluesse impliziter eignungstheorien auf die beobachtungsgenauigkeit in assessment-centern : eine feldstudie bei der allianz versicherungs-ag, zn koeln / Blankenburg, Roland & Gambla, Michael – (mf ed 1994) – 2mf – 9 – €40.00 – 3-89349-882-6 – mf#DHS 882 – gw Frankfurter [150]

Der einfluss babyloniens auf das verstaendnis des alten testamentes / Jeremias, Alfred – Berlin: Edwin Runge 1908 [mf ed 1989] – (= ser Biblische zeit- und streitfragen 4/2) – 1mf – 9 – 0-7905-0581-9 – (incl bibl ref) – mf#1987-0581 – us ATLA [221]

Der einfluss der bibelkritik auf das christliche glaubensleben : vortrag / Stave, Erik – Tuebingen: J C B Mohr, 1903 – (= ser Sammlung Gemeinverstaendlicher Vortraege Und Schriften Aus Dem Gebiet Der Theologie Und Religionsgeschichte) – 1mf – 9 – 0-7905-2086-9 – mf#1987-2086 – us ATLA [220]

Einfluss der englischen philosophen seit bacon auf die deutsche philosophie des 18. jahrhunderts : von der koeniglich preussischen akademie der wissenschaften mit einem preise ausgezeichnete untersuchung / Zart, Gustav – Berlin: F Duemmler, 1881 – 1mf – 9 – 0-524-00670-9 – mf#1990-0170 – us ATLA [190]

Einfluss der grieschischen skepsis auf die entwicklung / Horovitz, Saul – Breslau, Germany. 1915 – 1r – us UF Libraries [939]

Der einfluss der mysterienreligionen auf das aelteste christentum / Clemen, Carl – Giessen: A Toepelmann, 1913 – (= ser Religionsgeschichtliche Versuche und Vorarbeiten) – 1mf – 9 – 0-7905-4255-2 – (incl bibl ref) – mf#1988-0255 – us ATLA [230]

Der einfluss der philosophie charles bonnets auf friedrich heinrich jacobi / Isenberg, Karl – Borna-Leipzig: R Noske 1906 [mf ed 1990] – 1r – 1 – (incl bibl ref. filmed with: tiefgluth / pedro ilgen) – mf#2738p – us Wisconsin U Libr [430]

Der einfluss der protestantischen schulphilosophie auf die orthodox-lutherische dogmatik / Weber, Emil – Leipzig: A Deichert, 1908 – 1mf – 9 – 0-8370-8798-8 – (incl bibl ref) – mf#1986-2798 – us ATLA [242]

Der einfluss der reformirten kirche auf preussens groesse / Zahn, Adolf – Halle: Richard Muehlmann, 1871 – 1mf – 9 – 0-524-07215-9 – mf#1990-5373 – us ATLA [242]

Der einfluss des humanismus und der reformation auf das gleichzeitige erziehungs- und schulwesen : bis in die ersten jahrzehnte nach melanchthons tod / Roth, Friedrich – Halle, Verein fuer Reformationsgeschichte, 1898 – (= ser Schriften des vereins fuer reformationsgeschichte 15. jahrg/60) – 1mf – us ATLA [370]

Der einfluss des humanismus und der reformation auf das gleichzeitige erziehungs- und schulwesen : bis in die ersten jahrzehnte nach melanchthons tod / Roth, Friedrich – Halle: Verein fuer Reformationsgeschichte 1898 [mf ed 1990] – (= ser Schriften des vereins fuer reformationsgeschichte 15/60) – 1mf – 9 – 0-7905-4898-4 – (incl bibl ref) – mf#1988-0898 – us ATLA [370]

Die einfluss des islaam auf das haeusliche, sociale und politische leben seiner bekenner : eine culturgeschichtliche studie / Pischon, Carl Nathanael – Leipzig: FA Brockhaus, 1881 – 1mf – 9 – 0-524-01979-7 – mf#1990-2770 – us ATLA [260]

Der einfluss des supreme court auf die politik der u.s.a. von 1789 bis zum ende des zweiten weltkriegs. / Jordan, Horst W – Mainz? 1952? LL-389 – 1 – us L of C Photodup [347]

Der einfluss einer hiv-1-infektion humaner monozyten/makrophagen auf die genexpression immunregulatorischer proteine in vitro / Glienke, Wolfgang – (mf ed 1996) – 2mf – 9 – €40.00 – 3-8267-2366-X – mf#DHS 2366 – gw Frankfurter [540]

Der einfluss eines individuellen intensiv-prophylaxeprogramms auf die parodontale gesundheit bei alterspatienten : eine klinisch kontrollierte interventionsstudie / Poettker, Nina – 2000 – 2mf – 9 – 3-8267-2677-4 – mf#DHS 2677 – gw Frankfurter [617]

Der einfluss eines neuen 30 mm ballonkatheters auf coronararteriendissektionen bei erkutanter transluminaler coronarangioplastie im vergleich zum konventionellen 20 mm ballonkatheter / Olschner, Gerhard Konrad – (mf ed 1995) – 1mf – 9 – €30.00 – 3-8267-2207-8 – mf#DHS 2207 – gw Frankfurter [617]

Der einfluss philos auf die aelteste christliche exegese (barnabas, justin und clemens von alexandria) : ein beitrag zur geschichte der allegorisch-mystischen schriftauslegung im christlichen altertum / Heinisch, Paul – Muenster i.W: Aschendorff, 1908 [mf ed 1989] – (= ser Alttestamentliche abhandlungen 1/1-2) – 1mf – 9 – 0-7905-0896-6 – (incl ind. in german & greek) – mf#1987-0896 – us ATLA [221]

Der einfluss portugals bei der wahl pius 6 / Harder, Ernst – Koenigsberg: Hartung, [1882?] – 1mf – 9 – 0-8370-8578-0 – mf#1986-2578 – us ATLA [240]

Der einfluss von angstneigung und falscher physiologischer rueckmeldung auf die kontingente negative variation / Haensel, Frank – (mf ed 1991) – 1mf – 9 – €49.00 – 3-89349-400-6 – mf#DHS 400 – gw Frankfurter [150]

Der einfluss von goethes wilhelm meister auf das drama der romantiker / Wendriner, Karl Georg – Leipzig, 1909 – 2mf – 9 – 3-89349-317-4 – gw Frankfurter [430]

Einfluss von kulturbedingungen auf die physiologie vaskulaerer endothelialer zellen in vitro / Schrimpf, Gangolf – (mf ed 1996) – 2mf – 9 – €40.00 – 3-8267-2300-7 – mf#DHS 2300 – gw Frankfurter [540]

Der einfluss von olsalazin auf den gastrointestinalen transit : untersuchungen mit dem metalldetektor / Schwarz, Gunther – Mainz: Gardez, 1993 (mf ed 1996) – (= ser Neue medizinische Bibliothek 2) – 1mf – 9 – €24.00 – 3-8267-9655-1 – mf#DHS 9655 – gw Frankfurter [610]

Einfluss von pulsoximetrie und kapnometrie auf die sicherheit beatmeter patienten bei intensivverlegungen / Marx, Gernot – (mf ed 1996) – 2mf – 9 – €40.00 – mf#DHS 2326 – gw Frankfurter [617]

Einfluss von wasserdampf auf den ablauf der heterogen katalysierten oxidativen kupplung von methan / Jankowski, Joachim – (mf ed 1992) – 1mf – 9 – €37.50 – 3-89349-550-9 – mf#DHS 550 – gw Frankfurter [540]

Der einfluss wilhelm meisters auf den roman der romantiker / Donner, Joakim Otto Evert – Berlin: R Heinrich 1893 [mf ed 1990] – 1r – 1 – (incl bibl ref. filmed with: goethes tasso / kuno fischer) – mf#7366 – us Wisconsin U Libr [430]

Die einflussnahme der amerikanischen besatzungsmacht auf die berliner kulturpolitik in den nachkriegsjahren, 1945-1947 / Kanzler, Melanie – (mf ed 1992) – 2mf – 9 – €49.00 – 3-89349-508-8 – mf#DHS 508 – gw Frankfurter [327]

Einfuehrung der reformation in die kurmark brandenburg durch joachim 2 / Steinmueller, Paul – Halle: Verein fuer Reformationsgeschichte 1903 [mf ed 1990] – (= ser Schriften des vereins fuer reformationsgeschichte 20/76) – 1mf – 9 – 0-7905-5136-5 – (incl bibl ref) – mf#1988-1136 – us ATLA [943]

Die einfuehrung der reformation in hamburg / Sillem, Carl Hieronymus Wilhelm – Halle: Verein fuer Reformationsgeschichte, 1886 – (= ser [Schriften des Vereins fuer Reformationsgeschichte]) – 1mf – 9 – 0-7905-4712-0 – (incl bibl ref) – mf#1988-0712 – us ATLA [943]

Die einfuehrung der reformation in rostock / Vorberg, Axel – Halle: Verein fuer Reformationsgeschichte 1897 [mf ed 1990] – (= ser Schriften des vereins fuer reformationsgeschichte) – 1mf – 9 – 0-7905-4839-9 – mf#1988-0839 – us ATLA [242]

Einfuehrung in das roemische brevier / Lietzmann, Hans – Bonn: A Marcus und E Weber, 1917 – (= ser Kleine texte fuer vorlesungen und uebungen) – 1mf – 9 – 0-524-06591-8 – mf#1990-5257 – us ATLA [240]

Einfuehrung in das theologische studium / Wernle, Paul – Tuebingen: JCB Mohr, 1908 – 2mf – 9 – 0-7905-3624-2 – mf#1989-0117 – us ATLA [240]

Einfuehrung in die astrophysik / Waldmeier, Max – Basel: Birkhaeuser 1948 [mf ed 2005] – (= ser Lehrbuecher und monographien aus dem gebiete der exakten wissenschaften; 18. astronomisch-geophysikalische reihe 3) – 1r – 1 – (incl ind) – mf#film mas 37000 – us Harvard [520]

Einfuehrung in die deutsche literatur : dichtungen in poesie und prosa erlaeutert fuer schule und haus, zugleich eine geschichte der deutschen literatur von den aeltesten zeiten bis zur gegenwart / ed by Meyer, Johannes – Berlin: Gerdes & Hoedel, 1905-1913 – 1 – (incl bibl ref and index) – us Wisconsin U Libr [430]

Einfuehrung in die evangelische missionskunde : im anschluss an die basler mission / Bornemann, Wilhelm – Tuebingen: J C B Mohr, 1902 – 1mf – 9 – 0-7905-5574-3 – (incl bibl ref) – mf#1988-1574 – us ATLA [242]

Einfuehrung in die geschichte der theologischen literatur der fruehscholastik / Landgraf, Arthur M – Regensburg, 1948 – 3mf – 8 – €7.00 – ne Slangenburg [240]

Einfuehrung in die hoehere geisteskultur des islam / Horten, Max – Bonn: F Cohen, 1914 – 1mf – 9 – 0-524-01554-6 – (incl bibl ref) – mf#1990-2508 – us ATLA [260]

Einfuehrung in die weltliteratur : (von den aeltesten zeiten bis zur gegenwart) im anschluss an das leben und schaffen goethes / Bartels, Adolf – Muenchen: G D W Callwey 1913 [mf ed 1999] – 3v on 2r – 1 – (incl bibl ref & ind) – mf#10154 – us Wisconsin U Libr [410]

Einfuehrung in goethes "faust" / Petsch, Robert – 2., durchges Aufl. Hamburg: Broschek, [1941] – 1r – 1 – (incl bibl ref) – us Wisconsin U Libr [430]

Einfuehrung in goethes faust / Lienhard, Friedrich – Leipzig: Quelle & Meyer, 1913 – 1r – 1 – us Wisconsin U Libr [430]

Einfuehrung in goethes faust / Petsch, Robert – [Prag: Deutscher Verein zur Verbreitung gemeinnuetziger Kenntnisse, 1910] – 1r – 1 – (incl bibl ref) – us Wisconsin U Libr [430]

Einfuehrung in richard wagners werke und schriften / Pfordten, Hermann Ludwig, Freiherr von der – 2. Aufl. Bielefeld: Velhagen & Klasing, 1921 – 1r – 1 – (incl bibl ref) – us Wisconsin U Libr [780]

Einfuehrung in theorie, geschichte und funktion der ddr-literatur / ed by Schmitt, Hans-Juergen – Stuttgart: J B Metzler, c1975 – 1r – 1 – (incl bibl ref) – us Wisconsin U Libr [430]

Das einfuehrungsgesetz vom 18. august 1896 / Niedner, Alexander – 2. umgearb verm Aufl. Berlin: C Heymann, 1901 – (= ser Civil law 3 coll; Kommentar zum buergerlichen gesetzbuche und seinen nebengesetzen) – 6mf – 9 – (incl bibl ref and index) – mf#LLMC 96-560 – us LLMC [340]

Einführung in goethe's meisterwerke; selections from goethe's poetical and prose works : with copious biographical, literary, critical and explanatory notes, a vocabulary of difficult words and an introduction containing a life of goethe, for school and home by dr wilhelm bernhardt / ed by Bernhardt, Wilhelm – Boston: D C Heath & Co, 1896 [mf ed 1994] – xii275p – 1 – mf#8637 – us Wisconsin U Libr [430]

Der eingang des johannesevangeliums (kapitel 1, 6-18) : in meditationen / Philippi, Friedrich Adolph – Stuttgart: Samuel Gottlieb Liesching, 1866 – 1mf – 9 – 0-8370-4730-7 – mf#1985-2730 – us ATLA [225]

Die eingangsbuecher des parzival und das gesamtwerk / Cucuel, Ernst – Frankfurt: Diesterweg, 1937 – 1r – 1 – (incl bibl ref) – us Wisconsin U Libr [390]

Die eingeborenen sued-afrika's : atlas, enthaltend dreissig tafeln recentypen: 60 portraits...nach original-photographien des verfassers in kupfer radirt von professor hugo buerckner / Fritsch, Gustav – Breslau: Ferdinand Hirt 1872 – (= ser [Travel descriptions from south africa, 1711-1938]) – 2mf – 9 – mf#zah-48 – ne IDC [916]

Die eingeborenen sued-afrika's : ethnographisch und anatomisch beschrieben / Fritsch, Gustav – Breslau: Ferdinand Hirt 1872 – (= ser [Travel descriptions from south africa, 1711-1938]) – 7mf – 9 – mf#zah-47 – ne IDC [916]

Eingriffsqualitaet und rechtliche regelung polizeilicher videoaufnahmen / Jendro, Frank – (mf ed 1992) – 3mf – 9 – €49.00 – 3-89349-590-8 – mf#DHS 590 – gw Frankfurter [340]

Einhaelligkeit der dienern der kirhen zuo zuerich vnd herren joannis caluinj / Bullinger, Heinrich – [Zuerich, Ruodolff Wyssenbach, 1551] – 1mf – 9 – mf#PBU-262 – ne IDC [240]

Einhard *see*
– Einhard's life of charlemagne
– Life of charlemagne

Einhardi vita karoli magni (mgh7:25.bd) – 1911 – (= ser Monumenta germaniae historica 7: scriptores rerum germanicarum in usum scholarum (mgh7)) – €5.00 – ne Slangenburg [240]

Einhard's life of charlemagne : the latin text = Vita karoli magni imperatoris / Einhard; ed by Garrod, Heathcote William & Mowat, Robert Balmain – Oxford: Clarendon Press, 1915 – 1mf – 9 – 0-524-00634-2 – mf#1990-0134 – us ATLA [940]

Der einheimische klerus in den heidenlaendern / Huonder, Anton – Freiburg i.B.; St. Louis, Mo.: Herder, 1909 – (= ser Missions-Bibliothek) – 1mf – 9 – 0-7905-6186-7 – (incl bibl ref) – mf#1988-2186 – us ATLA [240]

Die einheit – Berlin DE, 1926 1 feb-1929 10 jun [gaps] – 4r – 1 – gw Misc Inst [074]

Einheit – Bitterfeld DE, 1958 6 oct-1968 26 mar [gaps] – 3r – 1 – gw Misc Inst [074]

Einheit – Zeitz DE, 1948 18 dec-1991 12 dec – 14r – 1 – (hydriewerk zeitz. title varies: n25 1990: hyzet) – gw Misc Inst [074]

Einheit / Zentralkomitee der Sozialistischen Einheitspartei Deutschlands. Berlin – v1-37. 1946-82 (incomplete) [mnthly] – 1 – us Wisconsin U Libr [325]

Einheit *see* Young czechoslovakia

Die einheit der biblischen urgeschichte (1 mos 1-3) : und die uebereinstimmung des schoepfungsberichtes mit den naturverhaeltnissen der erde: nachgewiesen mit beziehung auf die ansichten dr. delitzsch's, dr. hoelemann's, und dr. keil's / Keerl, Philipp Friedrich – Basel: Bahnmaier (C Detloss), 1863 – 1mf – 9 – 0-8370-3861-8 – (incl bibl ref) – mf#1985-1861 – us ATLA [221]

Die einheit der genesis : ein beitrag zur kritik und exegese der genesis / Kurtz, Johann Heinrich – Berlin: J A Wohlgemuth, 1846 – 1mf – 9 – 0-7905-3455-X – (includes bibliographic references) – mf#1987-3455 – us ATLA [221]

Einheit der weltbewegung *see* Weltfront gegen imperialistischen krieg und faschismus

Einheit der weltbewegung gegen imperialistischen krieg und faschismus : halbmonatsorgan des weltkomitees zum kampf gegen imperialistischen krieg und faschismus – Paris (F), 1935 feb-jul – 1r – 1 – (only 1935 n1: einheit; publ: weltfront...paris) – gw Misc Inst [320]

Einheit fuer hilfe und verteidigung : zeitschrift der internationalen solidaritaetsbewegung – Paris (F), 1936 jan-1938 may – 1r – 1 – gw Misc Inst [320]

Der einheitliche festellungsbescheid / Stengel, Karl – Leipzig, 1940 (mf ed 1994) – 1mf – 9 – €24.00 – 3-8267-3001-1 – mf#DHS-AR 3001 – gw Frankfurter [340]

Einheitliches religionsbuch : enthaltend biblische geschichte, kirchengeschichte, katechismus mit erlaeuterungen und kirchenlieder / Zuck, Otto – Dresden: Gerhard Kuehtmann, 1896 – 1mf – 9 – 0-8370-8079-7 – mf#1986-2079 – us ATLA [240]

Die einheitlichkeit der paulinischen briefe : an der hand der bisher mit bezug auf sie [i.e. die] aufgestellten interpolations- und compilations-hypothesen / Clemen, Carl – Goettingen: Vandenhoeck und Ruprecht, 1894 – 1mf – 9 – 0-8370-2684-9 – (incl ind of names) – mf#1985-0684 – us ATLA [227]

Die einheitlichkeit des buches daniel / Gall, August, Freiherr von – Giessen: [s.n.], 1895 – 1mf – 9 – 0-7905-3439-8 – mf#1987-3439 – us ATLA [221]

Die einheitsfront – Berlin-Bohnsdorf DE, 1922-26 – 2r – 1 – gw Misc Inst [074]

Die einheitsfront : kampforgan gegen den faschismus – New York NY (USA), 1934 aug – 1r – 1 – (only publ once) – gw Misc Inst [320]

Die einheitslehre der goettlichen trinitaet nach der kirchlichen tradition bewiesen und gegen die irrlehren / Oischinger, Johann Nepomuk Paul – Muenchen: J J Lentner (G Stahl), 1862 – 1mf – 9 – 0-8370-3936-3 – (incl bibl ref) – mf#1985-1936 – us ATLA [240]

Die einheitslehre (monismus) als religion : eine studie / Bulova, Josef Ad – 2. aufl [s.l: J Bulova, 1897? (Stuttgart: Hoffmann) – 1mf – 9 – 0-7905-3702-8 – mf#1989-0195 – us ATLA [210]

Einhorn, David *see*
– Ausgewaehlte predigten und reden
– Onheyb

Einige bemerkungen zu adolf harnacks pruefung der geschichte des neutestamentlichen kanons (1.band. 1.haelfte) / Zahn, Theodor – Erlangen: A Deichert, 1889 – 1mf – 9 – 0-8370-9354-6 – (incl bibl ref) – mf#1986-3354 – us ATLA [225]

Einige erinnerungen von meinen reisen in russland, der tuerkei und italien : zur unterhaltung fuer alle leser, besonders fuer das weibliche geschlecht – Augsburg 1831 [mf ed Hildesheim 1995-98] – 1mf – 9 – €40.00 – 3-487-29525-3 – gw Olms [910]

Einige gedanken zur ausdifferenzierung von staat und recht / Jansen, Brigitte E S – (mf ed 1995) – 1mf – 9 – €30.00 – 3-8267-2111-X – mf#DHS 2111 – gw Frankfurter [370]

Einige mitteilungen ueber seinen diesjaehrigen besuch in der colonia eritrea / Schweinfurth, G – Berlin, 1892 – 1mf – 9 – mf#13069 – ne IDC [956]

Einige notizen ueber bonny an der kuste von guinea : seine sprache und seine bewohner – Goettingen: In der Dieterichschen Univ.-Buchdruckerei, 1848 – 1 – (with: observations..by richard wharton) – us CRL [960]

Einige worte uber die inuit (eskimo) des smith-sundes : nebst bemerkungen uber inuit-schadel / Bessels, Emil – S.l: s.n, 18–? – 1mf – 9 – mf#16948 – cn Canadiana [305]

Die einigkeit – Berlin DE, 1918 14 dec-1923 n25, 1924-27, 1931 24 jan, 12 dec – 4r – 1 – (title varies: 1914?: mitteilungsblatt der geschaeftskommission der freien vereinigung deutscher gewerkschaften; 1915?: rundschreiben an die vorstaende und mitglieder aller der freien vereinigung deutscher gewerkschaften angeschlossenen vereine; 1918?: der syndikalist. incl suppl: die junge menschheit 1921 n3-12) – gw Misc Inst [331]

Einigkeit *see* Eynigkeyt

Einigung – Bonn DE, 1958, nov-1960 jul, 1960 nov-1966, 1971 n4 – 1r – 1 – (title varies: 1 nov 1960: initiative) – gw Misc Inst [074]

Einkehr : neue gedichte / Anacker, Heinrich – Muenchen: F Eher, 1934 [mf ed 1988] – 174p – 1 – mf#6939 n11 – us Wisconsin U Libr [810]

Einkehr : neue gedichte / Beck, Friedrich – Wien: A Beck, 1931 [mf ed 1989] – 151p – 1 – mf#7002 – us Wisconsin U Libr [810]

Einkehr bei josef hofmiller – Lindau: J Thorbecke 1948 [mf ed 1990] – 1r – 1 – (incl bibl ref. filmed with: gestern / hugo von hofmannsthal) – mf#2729p – us Wisconsin U Libr [943]

Einleitende untersuchungen und commentar ueber die briefe *see* Commentary on the epistles of st john

Einleitung in das alte testament / Bleek, Friedrich; ed by Bleek, Johannes Friedrich & Kamphausen, Adolf – 4. aufl. Berlin: G Reimer, 1878 [mf ed 1991] – (= ser Einleitung in die heilige schrift 1) – 2mf – 9 – 0-8370-1962-1 – (incl bibl ref & ind) – mf#1987-6349 – us ATLA [221]

Einleitung in das alte testament : einschliesslich apokryphen und pseudepigraphen / Strack, Hermann Leberecht – 6. neubearb aufl. Muenchen: C H Beck (Oskar Beck), 1906 [mf ed 1985] – 1mf – 9 – 0-8370-5440-0 – (incl bibl) – mf#1985-3440 – us ATLA [221]

Einleitung in das alte testament : mit einschluss der apocryphen und der pseudepigraphen alten testaments / Koenig, Eduard – Bonn: Eduard Weber's Verlag (Julius Flittner) 1893 [mf ed 1985] – (= ser Sammlung theologischer handbuecher 2/1) – 2mf – 9 – 0-8370-3965-7 – (incl app & ind) – mf#1985-1965 – us ATLA [221]

Einleitung in das alte testament / Sellin, Ernst – 3. neubearb. aufl. Leipzig: Quelle und Meyer, 1920. Chicago: Dep of Photodup, U of Chicago Lib, 1970 (1r); Evanston: American Theol Lib Assoc, 1984 (1r) – (= ser Evangelisch-theologische Bibliothek) – 1 – 0-8370-0443-8 – (includes bibliographies and index) – mf#1984-B137 – us ATLA [221]

Einleitung in das neue testament / Barth, Fritz – Guetersloh: C Bertelsmann, 1908 – 2mf – 9 – 0-7905-0857-5 – (incl bibl ref and indexes) – mf#1987-0857 – us ATLA [225]

Einleitung in das neue testament / Belser, Johannes Evangelist – Freiburg i B: Herder, 1905 – 3mf – 9 – 0-524-05594-7 – (incl bibl ref) – mf#1992-0449 – us ATLA [225]

Einleitung in das neue testament / Bleek, Friedrich – 4. aufl. Berlin: G Reimer, 1886 – (= ser Einleitung in die Heilige Schrift) – 3mf – 9 – 0-7905-8298-8 – mf#1987-6403 – us ATLA [225]

Einleitung in das neue testament / Juelicher, Adolf – 1. und 2. aufl. Freiburg i.B: J C B Mohr (Paul Siebeck), 1894 – 1mf – 9 – 0-8370-3813-8 – (incl ind) – mf#1985-1813 – us ATLA [225]

Einleitung in das neue testament / Schaefer, Aloys – Paderborn: Ferdinand Schoeningh, 1898 – (= ser Wissenschaftliche Handbibliothek) – 1mf – 9 – 0-7905-2056-7 – (incl bibl ref) – mf#1987-2056 – us ATLA [225]

Einleitung in das neue testament / Zahn, Theodor – Leipzig: Deichert, 1897-1899 – 11mf – 9 – 0-8370-1975-3 – (incl bibl ref) – mf#1987-6362 – us ATLA [220]

Einleitung in das neue testament *see* Introduction to the new testament

Einleitung in das nibelungenlied / Muth, Richard Von – Paderborn, Germany. 1877 – 1r – us UF Libraries [780]

Einleitung in das system der christlichen lehre, oder, propaedeutische entwicklung der christlichen lehrwissenschaft : ein versuch / Beck, Johann Tobias – 2. verm Aufl. Stuttgart: JF Steinkopf, 1870 – 1mf – 9 – 0-7905-9134-0 – mf#1989-2359 – us ATLA [240]

Einleitung in den codex napoleon / Seidensticker, Johann Anton Ludwig – Tuebingen, 1908 (mf ed 1994) – 6mf – 9 – €99.00 – 3-8267-3072-0 – mf#DHS-AR 3072 – gw Frankfurter [944]

Einleitung in den hexateuch / Holzinger, Heinrich – Freiburg i.B.: J C B Mohr (Paul Siebeck), 1893 – 2mf – 9 – 0-7905-1068-5 – mf#1987-1068 – us ATLA [221]

Einleitung in den talmud / Strack, Hermann Leberecht – 4., neubearbeitete Aufl. Leipzig: J.C. Hinrichs, 1908 – (= ser Schriften Des Institutum Judaicum In Berlin) – 1mf – 9 – 0-8370-5441-9 – (incl bibliographies and ind of words and names) – mf#1985-3441 – us ATLA [270]

Einleitung in die buecher des alten testamentes / Baudissin, Wolf Wilhelm, Graf – Leipzig: S Hirzel, 1901 – 2mf – 9 – 0-8370-9442-9 – (includes bibliographies and indexes) – mf#1986-3442 – us ATLA [221]

Einleitung in die christliche ethik / Weiss, Hermann – Freiburg i B: JCB Mohr, 1889 – (= ser Sammlung theologischer Lehrbuecher) – 1mf – 9 – 0-7905-8969-9 – (incl bibl ref) – mf#1989-2194 – us ATLA [170]

Einleitung in die dogmengeschichte / Kliefoth, Theodor – Parchim: DC Hinstorff, 1839 – 1mf – 9 – 0-524-00050-6 – (incl bibl ref) – mf#1989-2750 – us ATLA [240]

Einleitung in die drei ersten evangelien / Wellhausen, Julius – Berlin: Georg Reimer, 1905 – 1mf – 9 – 0-8370-9594-8 – mf#1986-3594 – us ATLA [220]

Einleitung in die ethik / Stange, Carl – Leipzig: Dieterich, 1901 – 2mf – 9 – 0-8370-6415-5 – mf#1986-0415 – us ATLA [170]

Einleitung in die evangelische dogmatik / Lobstein, Paul – Freiburg i. B.: J.C.B. Mohr, 1897 – 1r – 1 – 0-8370-0553-1 – mf#1984-6062 – us ATLA [242]

Einleitung in die evangelische dogmatik *see* Collected works

Einleitung in die goettlichen buecher des alten bundes *see* Introduction to the old testament

Einleitung in die heilige schrift : zweiter theil: einleitung in das neue testament / Bleek, Friedrich – Berlin, 1862 – 14mf – 8 – €27.00 – ne Slangenburg [220]

Einleitung in die heilige schrift, alten und neuen testaments / Kaulen, Franz – 2. verb aufl. Freiburg i B: Herder 1884 [mf ed 1994] – (= ser Theologische bibliothek) – 6mf – 9 – 0-524-08608-7 – mf#1993-0043 – us ATLA [220]

Einleitung in die kanonischen buecher des alten testaments / Cornill, Carl Heinrich – 6. neubearb aufl. Tuebingen: J C B Mohr (Paul Siebeck), 1908 [mf ed 1986] – (= ser Grundriss der theologischen wissenschaften 2/1) – 1mf – 9 – 0-8370-9692-8 – (incl bibl ref & ind) – mf#1986-3692 – us ATLA [221]

Einleitung in die monumentale theologie / Piper, Ferdinand – Gotha: Rud. Besser, 1867 – 3mf – 9 – 0-7905-8045-4 – (incl bibl ref) – mf#1988-6026 – us ATLA [240]

Einleitung in die neugriechische grammatik / Chatzidakis, G N – Leipzig: Breitkopf & Haertel, 1892 – (= ser Bibliothek indogermanischer grammatiken) – 1mf – 9 – 0-8370-1190-6 – mf#1987-6020 – us ATLA [450]

Einleitung in die philosophie : mit zugrundlegung von schleiermachers dialektik / Brodbeck, Adolf – Tuebingen: LF Fues, 1881 – 1mf – 9 – 0-524-00431-5 – mf#1989-3131 – us ATLA [100]

Einleitung ins alte testament. selections *see* Introduction to the study of the old testament

Einleitung ins neue testament : aus schleiermacher's handschriftlichen nachlasse und nachgeschriebenen vorlesungen / Schleiermacher, Friedrich [Ernst Daniel]; ed by Wolde, Georg – Berlin: G Reimer 1845 [mf ed 1991] – 2mf – 9 – 0-524-00332-7 – (pref by friedrich luecke; incl bibl ref) – mf#1989-3032 – us ATLA [225]

Einleitung zum kriegs-process : worinnen von der kriegs-jurisdiction und wem dieselbe zustehe / denen personen und sachen / welche unter die kriegs-jurisdiction gedhren... / Ludovici, Jacob Friedrich – 8th ed. Halle: Wasenhaus, 1737 – 3mf – 9 – $4.50 – mf#LLMC 89-020 – us LLMC [355]

Einsame wanderungen in der schweiz im jahr 1809 / Uklanski, Carl T von – Berlin 1810 [mf ed Hildesheim 1995-98] – (= ser Fbc) – 387p on 3mf – 9 – €90.00 – 3-487-29366-8 – gw Olms [914]

Einsatzbedingungen der ostarbeiter sowie der sowjetrussischen kriegsgefangene *see* Nsdap (national socialist german workers party) nazi publications

Einsetzung eines koenigs : roman / Zweig, Arnold – Amsterdam: Querido, 1937 – 1r – 1 – (completes the triptych, trilogie des uebergangs; bks 1 and 2 of the series are respectively: erziehung vor verdun and der streit um dem sergeanten grischa) – us Wisconsin U Libr [830]

Der einsiedler – Koenigsberg (Kaliningrad RUS), 1740-41 – 1 – gw Misc Inst [077]

Der einsiedler am starnberger see : historischer roman / Frankenburg, Robert – Dresden: R H Dietrich [188-?] [mf ed 1989] – 1r – 1 – mf#7260 – us Wisconsin U Libr [830]

Einsiedler und genosse : soziale gedichte nebst einem vorspiel von bruno wille / Wille, Bruno – Berlin: Freie Verlags-Anstalt [1890?] [mf ed 1991] – 1r – 1 – (filmed with: prisoner halm / karl wilke) – mf#3054p – us Wisconsin U Libr [810]

Einsprueche : multidisziplinaere beitraege zur frauenforschung / ed by Volland, Gerlinde – 3-8267-9707-8 – mf#DHS 9707 – gw Frankfurter [305]

Einst auf der lorettohoehe : aufzeichnungen des leutnants bruckner / Goltz, Joachim, Freiherr von der – Muenchen: A Langen, G Mueller, c1934 – 1r – 1 – us Wisconsin U Libr [943]

Einstein theory of relativity / Lieber, Lillian Rosanoff – Lancaster, PA. 1936 – 1r – us UF Libraries [530]

Eintracht – Chicago IL (USA), 1927-1939 22 [gaps 2r], 1972/73-1985 7 sep, 2002 5 jan-31 aug [1r] – 1 – gw Misc Inst [071]

Eintracht – Skokie IL (USA), 1927-39 [2r], 1972– – 1 – gw Misc Inst [071]

Die einwanderung der israelitischen staemme in kanaan : historisch-kritische untersuchungen / Steuernagel, Carl – Berlin: C A Schwetschke und Sohn, 1901 (mf ed 1995) – 1r – 1 – (incl bibl ref and ind) – mf#*ZP-1486 – us NY Public [221]

Die einwanderung der israelitischen staemme in kanaan : historisch-kritische untersuchungen / Steuernagel, Carl – Berlin: C A Schwetschke, 1901 – 1mf – 9 – 0-7905-0391-3 – (incl bibl ref and indexes) – mf#1987-0391 – us ATLA [939]

Die einwirkung des buergerlichen gesetzbuchs auf zuvor entstandene rechtsverhaeltnisse : eine darstellung der fragen der uebergangszeit / Habicht, Hermann – 3. verb u verm Aufl. Jena: G Fischer, 1901 – (= ser Civil law 3 coll; Abhandlung zum privatrecht und civilprozess des deutschen reiches) – 9mf – 9 – (incl bibl ref) – mf#LLMC 96-518 – us LLMC [346]

Die einwirkung des christenthums auf die althochdeutsche sprache : ein beitrag zur geschichte der deutschen kirche / Raumer, Rudolf von – Stuttgart: SG Liesching, 1845 – 2mf – 9 – 0-7905-6312-6 – (incl bibl ref) – mf#1988-2312 – us ATLA [430]

Die einwirkungen der reformation auf die organisation und besetzung des reichskammergerichts / Broehmer, Heinrich – Heidelberg, c1930 (mf ed 1995) – 1mf – 9 – €24.00 – 3-8267-3168-9 – mf#DHS 3168 – gw Frankfurter [340]

Einzig, Paul *see* Fa-hsi-ssu chu i chih ching chi chi ch'u

Der einzige und sein eigentum / Stirner, Max – Berlin, 1924 – 5mf – 8 – €12.00 – ne Slangenburg [100]

Der einzige und seine liebe / Kroeger, Timm – Hamburg: A Janssen 1905 [mf ed 1990] – 1r – 1 – (filmed with: kotzebue in england / walter sellier) – mf#2776p – us Wisconsin U Libr [830]

Der einzige weg : zeitschrift fuer die vierte internationale – Zuerich, Fraumuenster (CH), 1937 dec-1939 may [gaps] – 1 – gw Misc Inst [335]

Eir news – 1987 apr 28-may 12 – 1r – 1 – (cont: new solidarity (new york, ny); cont by: new federalist) – mf#1270793 – us WHS [071]

Eira und der gefangene : roman / Eckmann, Heinrich – taschenausg. Braunschweig: G Westermann, c1935 [mf ed 1989] – 394p – 1 – mf#7202 – us Wisconsin U Libr [830]

Eire-ireland – St. Paul. 1965-1973 (1) – ISSN: 0013-2683 – mf#10118 – us UMI ProQuest [000]

An eirenic itinerary : impressions of our tour, with addresses and papers on the unity of christian churches / McBee, Silas – New York: Longmans, Green, 1911 [mf ed 1990] – 1mf – 9 – 0-7905-5115-2 – mf#1988-1115 – us ATLA [240]

An eirenicon : in a letter to the author of "the christian year" / Pusey, Edward Bouverie – Oxford: John Henry & James Parker [dist]: Rivingtons [dist] 1865 [mf ed 1986] – 1mf – 9 – 0-8370-9102-0 – (incl bibl ref) – mf#1986-3102 – us ATLA [242]

Eirenicum de pace ecclesiae catholicae / Du Jon (Junius), F – Lugduni Batavorum, 1593 – 4mf – 9 – mf#PRS-140 – ne IDC [241]

Der eisbaer – Berlin DE, 1918-19 [gaps] – 1 – gw Mikrofilm [073]

Der eisbaer – Wien (A), 1919 1 aug-1920 1 oct – 1r – 1 – gw Mikrofilm [074]

Eiselein, Joseph *see* Saemtliche werke

Eiselen, F *see* Goethes paedagogik

Eiselen, F C *see*
- The christian view of the old testament
- Sidon

Eiselen, Frederick Carl *see*
- The minor prophets
- The worker and his bible

Eiselen, Werner *see* Initiation rites of the bamasemola

Eisen, Gustavus A *see* Ancient oriental cylinder and other seals

Eisen, Heinrich *see* Die verlorene kompanie

Das eisen im feuer : roman / Viebig, Clara – 14. aufl. Berlin: E Fleischel 1913 [mf ed 1989] – 1r – 1 – (filmed with: dilettanten des lebens) – mf#7154 – us Wisconsin U Libr [830]

Eisenacher tagespost *see* Thueringische landeszeitung

Eisenacher Versammlung zur Besprechung der Socialen Frage, (1872) *see* Verhandlungen...am 6 und 7 october 1872

Eisenbach, Artur *see* Remilitarizatsye in mayrev-daytshland

Eisenbahn-journal – (Hamburg-) Altona DE, 1835-37 – 1 – gw Misc Inst [380]

Eisenbahn-journal und national-magazin fuer die fortschritte im handel – Amberg/Oberpf DE, 1835-37 – 1 – gw Misc Inst [380]

Eisenbahn-zeitung – Strassburg (Strasbourg F), 1893 9 jul-1898 29 aug [gaps] – 1 – (also a theatre & concert paper for strassburg) – fr ACRPP [790]

Eisenbahn-zeitung – Luebeck DE, 1865 6 jun-1923 15 sep – 116r – 1 – (title varies: 30 aug 1900: luebecker nachrichten; 1 apr 1919: luebecker vorstadt-zeitung; 1 apr 1921: luebecker neueste nachrichten [until 5 jun 1865 in hamburg-bergedorf]) – gw Mikrofilm [380]

Eisenberg, Helen *see* Pleasure chest

Eisenbeth, Maurice *see* Juifs de l'afrique du nord

Eisenhofer, L *see* Procopius von gaza

Eisenhofer, Ludwig *see* Procopius von gaza

Eisenhower, Dwight D *see* The diaries of dwight d eisenhower, 1953-1961

Eisenlohr, August *see* Ein altbabylonischer felderplan

Eisenmann, Joey C *see* Blood lipids and peak oxygen consumption in young distance runners

Eisenring – Temeschburg (Timisoara RO), 1924 16 mar-1929 25 oct – 1r – 1 – (lacking: 1924) – gw Misc Inst [077]

Eisenring, Carl Jacob *see* Die wahre union und die zwinglifeier

Eisenstadt, Benzion *see* Rabane mins k va-hakhameha

Eisenstadter, Meir *see* Imre yosher

Eisenstein, Judah David *see* Otsar ma'amare hazal konkordantsya le-ma'amarinm pitgamin

Eisentraut, Englehard *see* Studien zur apostelgeschichte

Eisenzeitliche keramik aus galilaa : von der forschungsgeschichte zu einer neuen klassifikation / Borgonon, Helena Pastor – (mf ed 2000) – 5mf – 9 – €59.00 – 3-8267-2736-3 – mf#DHS 2736 – gw Frankfurter [930]

Eiserne blaetter – Berlin DE, 1931 4 jan-1939 – 6r – 1 – gw Misc Inst [074]

Eiserne blaetter : wochenschrift fuer deutsche politik und kultur – v1-21. 1919-39 – 19r – 1 – (lacks v12-13. v16 & some pp) – mf#atla s0047 – us ATLA [943]

Das eiserne Jahr : roman / Bloem, Walter – Leipzig: Grethlein, c1910 [mf ed 1989] – 449p – 1 – mf#7032 – us Wisconsin U Libr [830]

Das eiserne jahr : heer wider heer; volk wider volk; die schmiede der zukunft: die kriegsroman-trilogie von siebzig-einundsiebzig / Bloem, Walter – volksausg. Berlin: Globus Verlag, c1940 [mf ed 1989] – 638p – 1 – mf#7032 – us Wisconsin U Libr [830]

Eiserne sonette / Winckler, Josef – Leipzig: Insel-Verlag, [191-?] – 1r – 1 – (subsequently publ as: eiserne welt) – us Wisconsin U Libr [810]

Eiserne welt *see* Eiserne sonette

Der eislebische christliche ritter : ein reformationsspiel / Rinckhart, Martin; ed by Mueller, Carl – Halle: Max Niemeyer 1883 [mf ed 1993] – (= ser Neudrucke deutscher literaturwerke des 16. und 17. jahrhunderts 53-54) – 11r – 1 – mf#3387p – us Wisconsin U Libr [820]

Eisler, Leopold *see* Beitraege zur rabbinischen sprach- und alterthumskunde

Eisler, Moritz *see* Vorlesungen uber die judischen philosophen des mittelalters

Eisler, Robert *see* Weltenmantel und himmelszelt

Eisler, Rudolf *see*
- Geschichte des monismus
- Der zweck

Eisner, Kurt *see*
- Schuld und suehne
- Treibende kraefte

Eisner, Sigmund *see* Tale of wonder

Eissfeldt, Otto *see*
- Erstlinge und zehnten im alten testament
- Der maschal im alten testament

Eitel, Ernest John *see*
- Buddhism
- Hand-book of chinese buddhism

Eitelberger von Edelberg, R *see* Aretino

Either great destruction of human life and final monarchy : or glorious resurrection for peace and harmony and universal liberty / Smolnikar, Andrew B – Columbus, [OH]: Ohio State Journal Co, 1856 – 1r – 1 – us Western Res [100]

Eitle, Hermann *see* Die unterordnung der saetze bei chaucer

Eitner, Gustav *see*
- Friedrichs von logau saemmtliche sinngedichte

Eitner, Rob *see* Musik-beilagen zu den gedichten des koenigsberger dichterkreises

Eitner, Robert *see* Publikationen aelterer praktischer und theoretischer musikwerke

Eitzen, P von *see*
- Admonitio de praecipvis capitibvs controversiarvm de coena domini
- Rechte vnd ware meinung vnd verstand goettlicher schrifft vnd der augspurgischen bekandtnus

Eiu [!] chr[i]stenlich widerfechtug / Jud, L – Zuerich, Johann Hager, 1524 – 1mf – 9 – mf#PBU-275 – ne IDC [240]

Eiusdem aenigmatum libellus / Junius, H – Ed 3. Antverpiae: Ex officina Christophori Plantini, 1569 – 2mf – 9 – mf#O-1454 – ne IDC [090]

Eive, Gloria *see* Manuscript collection of 18th century italian manuscripts in the university of california – berkeley music library

Eizalde, Bernardo *see* Intendencia de extremadura: circular

Eizenhofer, L *see* Das prager sakramentar (tab38-42)

Ejc *see* European journal of cancer, pt b

The ejected of 1662 in cumberland and westmoreland : their predecessors and successors / Nightingale, Benjamin – Manchester: University Press, 1911 – (= ser Publications of the University of Manchester) – 4mf – 9 – 0-7905-8112-4 – (incl bibl ref) – mf#1988-6074 – us ATLA [240]

Ejemplos de ortografia espanola... / Segura de la Garmilla, Ramon & Ganan Gonzalez, Felix – Madrid: Razon y Fe, 1927 – 1 – sp Bibl Santa Ana [440]

Ejercicio de 1931. presupuesto ordinario formado para el referido ejercicio por la comision permanente y aprobado por el ayuntamiente pleno y e ilmo. sr.delegado de hacienda / Badajoz – Badajoz: Tipografia Espanola, S.A. – sp Bibl Santa Ana [946]

Ejercicio facil e importante para el trance de la agonia – Manila: La Industrial 1882 [mf ed Bloomington IN: Indiana Uni Lib, Preservation Dept 1984] – 1r – 1 – us Indiana Preservation [241]

Ejercicio social del ano 1912. memoria y balance leidos y aprobados en junta general de socios...1913 / Caja General Frexnense – Badajoz: Tip. Lib. y Enc.Uceda Hermanos, 1913 – sp Bibl Santa Ana [946]

Ejercicios de lenguaje y gramatica elemental / Mendez Pereira, Octavio – Boston, MA. v1-2. 1921 – 1r – us UF Libraries [972]

El ejercito aleman tal como es : diarios de oficiales y soldados alemanes, hechos prisioneros o caidos en el frente ruso – Mexico: editorial "el libro libre," 1944 – 1 – (int by bodo uhse) – us Wisconsin U Libr [972]

Ejercito de extremadura en 1644. competencias de jurisdiccion / Ruiz Garcia, Felix – Badajoz: Imprenta Diputacion Provincial, 1971 – sp Bibl Santa Ana [340]

El ejercito de la monarquia y el ejercito de la republica / Vidal, Fabian – Barcelona: Ed Espanolas [between 1930-1940?] [mf ed 1977] – (= ser Blodgett coll) – 1mf – 9 – mf#w1247 – us Harvard [946]

El ejercito en la sociodad contemporanea... : conferencia pronunciada en el istituto barbara de braganza de badajoz el 11 de abril de 1969 / Moro Cardenas, Ezequiel – Badajoz: imprenta doncel, 1969 – 1 – sp Bibl Santa Ana [946]

Ejercito popular unido, ejercito de la victoria : texto del informe...comite central del partido comunista de espana, celebrado en valencia de dia 13 de noviembre de 1937 / Ibarruri, Dolores – Madrid: El Partido 1938 [mf ed 1977] – (= ser Blodgett coll) – 1mf – 9 – mf#w953 – us Harvard [946]

Un ejercito popular y democratico al servicio del pueblo / Spain. Ministerio de Defensa Nacional – Barcelona, 19?? Fiche W855. (Blodgett Collection of Spanish Civil War Pamphlets) – 9 – us Harvard [946]

Ekari cu phura : [a novel] / Cin Tan, Takkasuil – Ran Kun: Siha ratana pum nhip tuik 1978- [mf ed 1990] – [ill] 1r with other items – 1 – (in burmese) – mf#mf-10289 seam reel 150/2 [§] – us CRL [830]

Ekasar qang kar sahaprajajati : bi qatit dau qanagat / Duy Nasuan – Bhnam Ben: Ron Bumb Qapsara [1971?] [mf ed 1990] – 1r with other items – 1 – (in khmer) – mf#mf-10289 seam reel 126/3 [§] – us CRL [341]

Ekaterina konstantinovna breshkovskaia / Kovarskii, B – 1917 – (= ser Avtobiografiia babushki russkoi revoliutsii) – 1mf – 9 – mf#RPP-231 – ne IDC [325]

Ekaterinoslav – Dnepropetrovsk, 1917-18 – 1 – us UMI ProQuest [077]

Ekaterinoslav. Sovet rk i kd *see* Izvestiia ekaterinoslavskogo soveta rabochikh i soldatskikh deputatov

Ekaterinoslavskie gubernskie vedomosti – Dnepropetrovsk, 1849-1918 – 1 – us UMI ProQuest [077]

Ekaterinoslavskii vestnik / ed by Kazetskii, N N – Taganrog [Obl voiska Don]: [s n] 1919 [1918 [20 apr]-1919 [?]] – (= ser Asn 1-3) – n68 [1919] item 137, on reel n29 – 1 – mf#asn-1 137 – ne IDC [077]

Ekeberg, Carl *see* Precis historique de l'economie rurale des chinois

Ekell archives – v2 n2-4 [1988 spr-fall] – 1r – 1 – mf#1569850 – us WHS [071]

Ekendahl, D G von *see* Napoleons ansichten von der gottheit jesu, sowie von religion, priestern und krichenthum, protestantismus und katholicismus

Ekhaya lesikolo – Salisbury, Zimbabwe. 1967 – 1r – us UF Libraries [960]

Ekhaya lesikolo – Salisbury, Zimbabwe. 1970 – 1r – us UF Libraries [960]

Ekho : gaz bespartiin, demokrat / ed by Zavinskii, E T – Vladivostok [Primor obl]: G Perkins, izd angl sektsii George Highfield 1919-20 [1919 26 fevr-[1921] [?] – (= ser Asn 1-3) – n1 [1918]-n195 [1919] [gaps] item 444, on reel n86, 87 – 1 – mf#asn-1 444 – ne IDC [077]

Ekho – St Petersburg, 1906 – 1 – us UMI ProQuest [077]

Ekho kavkaza – Vladikavkaz, 1906 – 1 – (reel contains short runs of multiple titles. for complete listing of titles on a reel, please inquire) – us UMI ProQuest [077]

Ekho litvy – Vilna, USSR. 1955-1990 (1) – mf#61062 – us UMI ProQuest [077]

Ekiken jikkun *see* The way of contentment

Ekinci – Baku, 1875-77 – 1r – 1 – (on single reel with: musavat) – us UMI ProQuest [077]

Ekinci *see* Musavat

Ekistics : the problems and science of human settlements – Athens. 1955+ [1]; 1970+ [5]; 1975+ [9] – ISSN: 0013-2942 – mf#1953 – us UMI ProQuest [710]

Ekitabo kye kika nsenene. : manuscript translations / Kagwa, Apolo – [Kampala, Uganda: s.n, 1962 or 1963] – 1r – 1 – us CRL [470]

Ekitabo ky'ekika kya nsenene = The history of the grasshopper clan / Kagwa, Apolo – Mengo, Uganda: A K Press, [1905?] – 1 – us CRL [960]

Ekkard, Fr *see* Der reisende

Ekkard, Friedrich *see* Allgemeines register ueber die goettingischen gelehrten anzeigen von 1753 bis 1782

Ekkehard : audifax und hadumoth / Scheffel, Joseph Viktor von; ed by Handschin, Charles Hart & Luebke, William F – New York: American Book Company, c1911 – 1r – 1 – us Wisconsin U Libr [430]

Ekkehard : eine geschichte aus dem zehnten jahrhundert / Scheffel, Joseph Viktor von – 14., vom Verfasser durchgesehene Aufl., Stuttgart: J B Metzler, 1875 – 1r – 1 – us Wisconsin U Libr [830]

Ekkehard : eine geschichte aus dem zehnten jahrhundert / Scheffel, Joseph Viktor von – Stuttgart: Adolf Bonz, 1909 – 1r – 1 – (incl bibl ref) – us Wisconsin U Libr [830]

Ekkehard : a tale of the 10th century / Scheffel, Joseph Viktor von – London: J M Dent; New York: E P Dutton, [1911] – 1 – (incl bibl ref) – us Wisconsin U Libr [430]

Ekkehard : a tale of the tenth century / Scheffel, Joseph Viktor von – New York: T Y Crowell, 1895 – 1r – 1 – (includes biographical sketch of author, by nathan haskell dole) – us Wisconsin U Libr [430]

Ekkehard, Friedrich *see* Sturmgeschlecht

Ekklesia – Athens. Greece. -w. 25 Jul 1928-8 Mar 1941, 1 Jan 1947-15 Dec 1949. (Very imperfect). (4 reels) – 1 – uk British Libr Newspaper [072]

Ekklesia – Buenos Aires: Concilio Argentino de la Federacion Luterana Mundial [v1 n1-v11 n26 (sep 1957-agosto 1967)] (irreg) – 1r – 1 – us CRL [242]

Ekklesiastike aletheia – Istanbul, Turkey. In Greek. -w. 15 May 1885-30 Dec 1897; 11 Jan 1920-29 Sept 1923. 9 reels – 1 – uk British Libr Newspaper [949]

Eklund, Johan Alfred *see* Nirvana

Ekman, Erik Jakob *see* Pauli bref till efesierne

Ekomicheskii byt krest'ian saratovskogo, kuznetskogo uezdov saratovskoi gubernii opyt issledovaniia fizicheskikh, ekonomicheskikh i tekhnicheskikh uslovii / Smirnov, N – M, 1884 – 2mf – 8 – mf#RZ-132 – ne IDC [314]

Ekonom – Czech Republic, 1999- – 4r per y standing order – 1 – (1991 (oct, dec only) 1r. 1992-98 4r per y) – us UMI ProQuest [077]

Ekonom – Spb., 1841-1853 – 162mf – 9 – (missing: 1841-1843; 1849(52); 1850(28, 52); 1851(52); 1852(51-52); 1853(52)) – mf#R-3400 – ne IDC [077]

Ekonomi / Ikatan Sardjana Ekonomi – Djakarta, 1959-1970 – 26mf – 9 – (missing: 1961(4); 1962-1964; 1966-1969) – mf#SE-282 – ne IDC [330]

Ekonomi – Nicosia, Cyprus. Feb 20 1979-1991 – 7r – 1 – (scattered issues lacking) – us L of C Photodup [079]

Ekonomi buyuk gazete – Istanbul: Aydinlik Basimevi, jun 23 1949-mar 1952; aug 13 1953 – 9r – 1 – us CRL [330]

Ekonomi dan industri / Biro Statistik dan Dokumentasi, Departemen Perindustrian Rakjat – Djakarta, 1958-1964(3) – 40mf – 9 – (missing: 1959(7-12); 1960(9-12); 1961(2, 5-12); 1962(4-12)) – mf#SE-283 – ne IDC [330]

Ekonomi dan masjarakat – Djakarta, 1959-1964 – 15mf – 9 – (missing: 1959, v1(4); 1961-1963, v3-4) – mf#SE-285 – ne IDC [330]

Ekonomi gazetesi – Istanbul: Aydinlik Basimevi, [-1949]. [nov 1948-jun 22 1949] – 1r – 1 – us CRL [330]

Ekonomi gazetesi – Istanbul: Ekonomi Matbaasi, nov 5 1953-nov 29 1954 – 2r – us CRL [330]

Ekonomi indonesia / ed by Edisi bahasa Indonesia – Djakarta, 1969-1972 – 244mf – 9 – (missing: 1969-1970, v1-2(1-208); 1972, v5(576)) – mf#SE-1474 – ne IDC [330]

Ekonomi indonesia – Djakarta, 1970(aug)-1972 – 80mf – 9 – (missing: 1971, v1(142); 1972, v2(341)) – mf#SE-1475 – ne IDC [330]

Ekonomi, keuangan dan bank / Bank Indonesia – Djakarta, 1964-1966 – 23mf – 9 – mf#SE-286 – ne IDC [332]

Ekonomi luar negeri / Indonesia. Kementerian perekonomian – Djakarta, 1954-1955 – 24mf – 9 – (missing: 1954(1, 2, 24-end); 1955(1-3, 19)) – mf#SE-290 – ne IDC [330]

Ekonomi nasional – Jakarta, Indonesia. 1963-1965 (1) – mf#67830 – us UMI ProQuest [079]

Ekonomicheskaia chast' / Materialy po otsenke zemel' Nizhegorodskoi gubernii – Nizhnii-Novgorod, 1897-1900. v1-14(2) – 114mf – 8 – (missing: v1, v13) – mf#RZ-67 – ne IDC [314]

Ekonomicheskaia gazeta – Moscow, Russia. 1985-1989 – 5r – (gaps) – us UF Libraries [077]

Ekonomicheskaia gazeta – Moscow, 1986-89 – 5r – 1 – us UMI ProQuest [077]

Ekonomicheskaia politika sssr : uchebnik dlia sovparzhkol i marksiszko-leninkikh kruzhkov... / ed by Bokhanovskii, B – [Leningrad]: Ogiz Priboi, 1931 [mf ed 2002] – 1r – 1 – (filmed with: moskva v oktiabre /...pod redakziei n ovsiannikova (1919) and: hongrie / de j duckerz (1888). incl bibl ref) – mf#5215 – us Wisconsin U Libr [330]

Ekonomicheskaia priroda kooperativov i ikh klassifikatsiia / Tugan-Baranovskii, M I – 1914 – 127p 2mf – 9 – mf#COR-127 – ne IDC [335]

Ekonomicheskaia zhizn' – 6Nov 1918-13 Jun 1941. 1920 wanting – 1 – 485.00 – us L of C Photodup [947]

Ekonomicheskaia zhizn' – Moscow, 1923-30 – 58r – 1 – us UMI ProQuest [077]

Ekonomicheskie osnovy khristianskikh prasdnikov / Lippert, Julius – Moskva: Gos izd-vo, 1925 [mf ed 2002] – 1r – 1 – (filmed with: skazaniia ob antikhristie v slavianskikh perevodakh s zamiechaniiami o slavianskikh perevodakh tvorenii sv ippolita (1874)) – mf#5225 – us Wisconsin U Libr [390]

Ekonomicheskie zapiski – Spb., 1853-1862 – 134mf – 9 – (missing: 1853(1-52); 1860(6-9); 1862(48)) – mf#R-3402 – ne IDC [077]

Ekonomicheskii biulleten / Kon'iunkturnyi Institut. Moscow – 1922-28. Scattered issues missing – 1 – us L of C Photodup [330]

Ekonomicheskii biulleten koniunkturnogo instituta pri petrovskoi selsko-khoziaistvennoi akademii tssu sssr / ed by Kondratev, N D – 1922-1929(9) – 76mf – 9 – (missing:1928(2-4,11-12)) – mf#RHS-1 – ne IDC [335]

Ekonomicheskii magazin, selsko-khoziaistvennyi zhurnal, izdavavshiisia v 1780-1789 gg bibliograficheskoe opisanie / Neustroev, A N – Spb., 1874 – 1mf – 9 – mf#R-5656 – ne IDC [077]

Ekonomicheskii zhurnal – Spb., 1885-93(11) – 142mf – 9 – mf#COR-705 – ne IDC [077]

Ekonomicheskoe polozhenie rossii nakanune velikoi oktiabr'skoi sotsialisticheskoi revoliutsii, mart-oktiabr' 1917 : dokumenty i materialy / ed by Sidorov, A L et al – M, L, 1957. 2v – 24mf – 9 – mf#REF-146 – ne IDC [332]

Ekonomicheskoe voz – M., 1915. nos 1-11 – 11mf – 9 – mf#R-3979 – ne IDC [077]

Ekonomicheskoe vozrozhdenie *see* Ezhemesiachnyi zhurnal

Ekonomicheskoye Obozreniye – Moscow. Jan. 1929-Mar. 1930 – 1 – us NY Public [947]

'N ekonomies-antropologiese studie van die abakwamkhwanazi van natal / Erasmus, J C – Stellenbosch: U van Stellenbosch 1978 – 3mf – 9 – (in afrikaans) – mf#mf.155 – sa Stellenbosch [305]

Ekonomika i zhizn – Moscow, Russia. n1-8459. 1991-1993 jun – 5r – (Gaps) – us UF Libraries [077]

Ekonomika i zhizn' – Russia, 1999- – 5r per y – 1 – (backfile through 1998 $85/r) – us UMI ProQuest [077]

Ekonomika Sel'skogo Khoziaistva – 1925-62. (Scattered issues lacking) – 1 – 480.00 – us L of C Photodup [630]

Ekonomist rossii – 1909-12 – 46mf – 9 – $300.00 – us UMI ProQuest [330]

Ekonomski Dnevnik – Belgrade, Yugoslavia. Jan-Jul 1952 – 1r – 1 – us L of C Photodup [949]

Ekphrasis tes hagias sophias / Antoniades, M – Athen/Leipzig. v1-3. 1907-09 – 3v – €113.00 – ne Slangenburg [243]

Ekran – Moscow, 1921-22 – 15mf – 9 – us UMI ProQuest [790]

Ekrem, Ali *see* Lisan-i osmani

Ekrem, Ali [Bolayir] *see* Ordunun defteri

Ekrem, Recaizade Mahmut *see* Pejmurde

Eksjoeposten – Eksjoe, Sweden. 1894-95 – 1r – 1 – sw Kungliga [078]

Ekskuzovich, Nikolai, defendent *see* Protsess" ekskuzovicha, fikhgendlera i daina zasiedanie ugolovnogo departamenta odesskoi sudebnoi palaty, s" uchastiem" prisiazhnykh" zasiedatelei. stenograficheskii otchet" i izd. a. s. karfunkelia

'N eksplorasie van paradokse en die nut daarvan vir die gesinsterapie praktyk / Verster, Barbara C – Uni of South Africa 2000 [mf ed Johannesburg 2000] – 3mf – 9 – (incl bibl ref; text in afrikanns) – mf#mfm14930 – sa Unisa [300]

Ekspor = Export – Djakarta, Biro Pusat Statistik. 1957, 1959-63. (incomplete) – 1 – (in indonesian; tables of contents and explan. remarks also in english) – us Wisconsin U Libr [380]

Eksport promyslovoi kooperatsii / Zhabin, A I – 1929 – 19p 1mf – 9 – mf#COR-422 – ne IDC [335]

Ekspres / Aksi Press – Djakarta, 1970-1972. v1-3 – 115mf – 9 – (missing: 1970, v1(5, 8, 12); 1971, v2(54, 56, 67, 70, 76); 1972, v3(85, 87, 88, 91, 93-97, 101, 102)) – mf#SE-1476 – ne IDC [959]

Ekspres = Express – Chicago, IL. 1936 – 1 – us AJPC [071]

Ekstra bladet – Copenhagen, Denmark. 1945 – 2r – 1 – uk British Libr Newspaper [074]

Ekstrennyi biulleten' poslednikh izvestii gazety "trud"; telegrammy / ed by Kartsov, V – Minusinsk [Enis gub]: [s n] 1918-19 – (= ser Asn 1-3) – 10 nenum vyp [1918], 2 nenum vyp [1919] item 443, on reel n86 – 1 – (suppl of: trud [asn-1 420]) – mf#asn-1 443 – ne IDC [077]

Ekvall, David P *see* Outposts

Ekwelie, Sylvanus Ajani *see* The content of broadcasting in nigeria

Ekwensi, Cyprian *see* Lokotown and other stories

[El centro] el centro progress – CA. 1913-1922 – 18r – 1 – $1080.00 – mf#C03206 – us Library Micro [071]

[El centro-] el centro star – CA. 1908 – 1r – 1 – $60.00 – mf#C03207 – us Library Micro [071]

[El cerrito-] el cerrito journal – CA. 1950-51 – 2r – 1 – $120.00 – mf#B02197 – us Library Micro [073]

El chicano – Colton, CA. 1968- – 39+ r – 1 – $1950.00 ($90.00y) – mf#R04017 – us Library Micro [071]

El cojo ilustrado – 1892-1915 – 1 – us L of C Photodup [073]

El dorado *see* Amador/el dorado

El dorado, arkansas *see* El dorado second baptist church. el dorado, arkansas

[El dorado-] el dorado canyon miner – NV. 1917 – 1r – 1 – $60.00 – mf#U04492 – us Library Micro [622]

El dorado fantasma / ed by Bayle, Constantino – Madrid: Razon y Fe, 1931 – 1 – sp Bibl Santa Ana [440]

El dorado reporter – Placerville, CA. 1912-1914 (1) – mf#62232 – us UMI ProQuest [071]

El dorado second baptist church. el dorado, arkansas : church records – 1923-75 – 1 – $119.97 – us Southern Baptist [242]

El dorado springs first baptist church. el dorado springs, missouri : church records – 1882-feb 1967 – 1 – $49.23 – us Southern Baptist [242]
El dorado springs, missouri *see* El dorado springs first baptist church. el dorado springs, missouri
El dorado/amador – 1928-38; 1992- – (= ser California telephone directory coll) – 13+ r – 1 – $650.00 – mf#P00004 – us Library Micro [917]
El iris – Badajoz, 1889-1890 – 5 – sp Bibl Santa Ana [073]
El Maliki, Abderrahmane *see* L'exode rural au maroc
[El monte-] el monte herald and press – CA. 1977-89 – 25r – 1 – $1500.00 – mf#R02198 – us Library Micro [071]
[El monte-] el monte herald and valley herald – CA. 1959-1960 – 2r – 1 – $120.00 – mf#H04008 – us Library Micro [071]
El Monte Historical Society [CA] *see* Landmark
El moro monitor *see* Miscellaneous newspapers of las animas county, reel 1
El paso county miscellaneous newspapers – Denver, CO (mf ed 1991) – 1r – 1 – (the fountain valley news (may 9-23 1958); fountain valley news (feb 24 1961-dec 28 1962)) – mf#MF Z99 El69f – us Colorado Hist [071]
El paso county miscellaneous newspapers, reel 1 – Denver, CO (mf ed 1991) – 1r – 1 – (westside town and country weekly times (aug 18 1971-feb 14 1973); westside times (feb 21-apr 25 1973)) – mf#MF Z99 El69c Reel 1 – us Colorado Hist [071]
El paso county miscellaneous newspapers, reel 2 – Denver, CO (mf ed 1991) – 1r – 1 – (calhan news & ramah record (apr 12 1928); colorado city iris (may 23 1890); colorado city journal (nov 28 1861); colorado mountaineer (oct 21 1875-sep 6 1876); colorado springs farm news (mar 29 1935); colorado springs independent (aug 2 1934); colorado springs minority press (jun 30 1982-jul 18 1983); colorado springs observer (sep 12-29 1926); colorado springs sentinel (aug 28-oct 23 1969); colorado state republic (jul 9 1885); the colorado voice (jun 18 1948-aug 26 1949); el paso county courier (jul 15-29 1965); the magnet (jun 9 1880, apr 20 1881); the mining investor (sep 23 1907); the new west (dec 1878); the plain dealer (oct 10 1894); public opinion (oct 17 1914); queen bee (dates unknown); ute pass weekly news (jul 7-sep 8 1922); the voice of colorado (apr 17 1936); the weekly times journal (mar 1-may 17 1974); the wellspring (jan 1978-dec 1979); falcon herald (nov 21 1888); fountain herald (sep 17-oct 8 1937); the cheyenne news & ivywild times (jun 13 1913); the manitou item (may 27 1882); pike's peak news (1891-97); pike's peak daily news (aug 28 1900-jul 9 1934); the monument mentor (sep 13 1879); the columbine herald (oct 25 1957-dec 12 1958); west creek times (mar 20 1896)) – mf#MF Z99 El69c Reel 2 – us Colorado Hist [071]
El paso county miscellaneous newspapers, reel 3 – Denver, CO (mf ed 1991) – 1r – 1 – (the weekly telegraph (jan 4-may 24, dec 6 1901-dec 26 1902); colorado telegraph semi-weekly edition (may 28-jun 7 1901); colorado telegraph weekly edition (jun 14-nov 29 1901)) – mf#MF Z99 El69c Reel 3 – us Colorado Hist [071]
El paso del guadiana / Crespo, Pedro – Madrid: sala editorial, 1977 – 1 – sp Bibl Santa Ana [946]
El salvador : the making of us policy, 1977-1984 – [mf ed Chadwyck-Healey] – (= ser National security archive, washington dc: the making of us policy) – 870mf – 9 – (with 2v p/g & ind) – uk Chadwyck [327]
El salvador : pais en marcha ascendente / El Salvador Secretaria De Informacion – San Salvador, El Salvador, 1953 – 1r – us UF Libraries [972]
El salvador : tierra de realidad y esperanza / Estrella De Centroamerica – San Salvador, El Salvador, 1949 – 1r – us UF Libraries [972]
El salvador 2 : war, peace, and human rights, 1980-1994 – [mf ed Chadwyck-Healey] – (= ser National security archive, washington dc: the making of us policy) – 220mf – 9 – (with p/g & ind) – uk Chadwyck [327]
El salvador al dia – San Salvador. 10 dec 1954-16 sep 1955 – 1/4r – 1 – uk British Libr Newspaper [079]
El salvador de hoy / Gonzalez Ruiz, Ricardo – San Salvador, El Salvador, 1952 – 1r – us UF Libraries [972]
El Salvador Direccion General De Estadistica *see* Republica de el salvador
El Salvador. Direccion General de Estadistica *see* Anuario estadistico 1911-1965
[El segundo-] el segundo tribune – CA. 1965-1972 – 5r – 1 – $300.00 – mf#H04010 – us Library Micro [071]
Elam baptist church – Jones Co, GA – 1 – $52.38 – (minutes 1874-1922, 1928-82, 1977 church membership record. minutes apr 1982-dec 1998. 1874-1898 original records very light) – mf#6875 – us Southern Baptist [242]
Elam, Charles *see* Winds of doctrine
Die el-amarna-tafeln : mit einleitung und erlaeuterungen / ed by Knudtzon, Joergen Alexander – Leipzig: JC Hinrichs, 1915 – (= ser Vorderasiatische Bibliothek) – 4mf – 9 – 0-524-08513-7 – mf#1993-0038 – us ATLA [930]
L'elan – Paris. n1-9. avr 1915-fevr 1916 – 1 – fr ACRPP [073]
L'elan – Paris. v1-10. Apr 1915-1916 – 1 – us NY Public [073]
Elan – Stendal DE, 1976-90 [gaps] – 3r – 1 – (erdoel-erdgas) – gw Misc Inst [074]
Elancee – 1983 aug – 1r – 1 – (cont: elan (new york, ny)) – mf#4722305 – us WHS [071]
Elastomerics – New York. 1977-1992 (1) 1977-1992 (5) 1977-1992 (9) – (Cont: Rubber age) – ISSN: 0146-0706 – mf#144,01 – us UMI ProQuest [670]
Elath, Eliahu *see* Ukhluse kikar ha-yarden ve-hayehem
Elb, Max *see* Zur hundert jahr-feier des kranken-unterstutzungs-instituts
Die elbaue *see* Koetzschenbrodaer zeitung
Elbe, A von der *see* Chronika eines fahrenden schuelers
Elbe-elster rundschau – Bad Liebenwerda DE, 1992- – 67r/yr – 1 – (bezirksausgabe von lausitzer rundschau, cottbus; covers: bad liebenwerda & herzberg, fr 1993 only bad liebenwerda) – gw Misc Inst [074]
Elbe-elster rundschau – Jessen DE, 1992- – 5r/yr – 1 – (bezirksausgabe von lausitzer rundschau, cottbus; covers jessen & wittenberg) – gw Misc Inst [074]
Elbe-elster rundschau – Herzberg, Elster DE, 1993- – 2r/yr – 1 – (bezirksausgabe von lausitzer rundschau, cottbus; earlier ed s.u. bad liebenwerda) – gw Misc Inst [074]
Elbe-jeetzel-zeitung – Luechow (Wendland) DE, 1977- – ca 6r/yr – 1 – (incl suppl: 100 jahre luechower heimatzeitung 1854-54, jubilee-ed 1954 4/5 dec [1r]) – gw Misc Inst [074]
Elberfelder zeitung *see* Provinzial-zeitung
Elbers, Gerald W *see* Scientific revolution
Elbetal-zeitung *see* Aussiger tagblatt
Elbing, Ulrich *see* Autoaggression und pathologische informationsverarbeitung bei geistigbehinderten mit autistischen zuegen
Elbinger anzeigen *see* Koeniglich (genehmigte) west-preussische elbingsche zeitung von staats- und gelehrten sachen
Elbinger morgenblatt – Elbing (Elblag PL), 1848 1 may-1849 30 mar 30 – 1r – 1 – gw Misc Inst [077]
Elbit-rundblick – Wittenberg DE, 1961 7 may-1991 18 feb [gaps] – 5r – 1 – (gummiwerke piesteritz) – gw Misc Inst [670]
Elbogen, Ismar *see* Der juedische gottesdienst in seiner geschichtlichen entwicklung
Elbogener zeitung – Elbogen (Loket CZ), 1938 27 aug-31 dec – 1r – 1 – gw Misc Inst [077]
'Elbonah shel torah / Feigensohn, Samuel Shraga – Berlin, Germany. 1928/29 – 1r – 1 – us UF Libraries [074]
Elbow drums / Calgary Indian Friendship Centre – 1969 jan-mar, may, aug-dec, 1970 apr, sep, 1971 apr – 1r – 1 – mf#1055905 – us WHS [305]
Elbridge community baptist church. elbridge, new york : church records – 1813-1966 – 1 – (formerly first baptist church) – us Southern Baptist [242]
Elbridge gerry papers 1744-1895 – [mf ed 1988] – 7r – 1 – (with p/g) – us MA Hist [975]
Elbthal-morgen-zeitung – Dresden DE, 1898-1905 30 sep – 12r – 1 – gw Misc Inst [074]
El-Busaidy, Hamed Bin Saleh *see* Ndoa na talaka
Elbwart, Wilhelm, Edler von *see* Stadt im sommerwind
El-carmel – 1920-1934 – 8r – 1 – mf#J-91-205 – ne IDC [956]
Elchasai : Ein Religionsstifter Und Sein Werk / Brandt, Wilhelm – Leipzig: J C Hinrichs, 1912 – 1mf – 9 – 0-7905-1573-3 – (Incl bibl ref and indexes) – mf#1987-1573 – us ATLA [210]
Eldad ha-dani – Pressburg, Czechoslovakia. 1891 – 1r – us UF Libraries [077]
Elder edda of saemund sigfusson : and, the younger edda of snorre sturleson – London: Norroena Society, 1907 – (= ser Anglo saxon classics) – 4mf – 9 – 0-524-08190-5 – mf#1991-0303 – us ATLA [430]
Elder, Elinor Goertz *see* Carl philipp emanuel bach's concept of the free fantasia
Elder Family *see* Letters and scrapbook
Elder lott cary biography / Taylor, J B – 1837 – 1 – $5.00 – us Southern Baptist [242]
Elder, William *see*
– Biography of elisha kent kane
– Infant sprinkling
– Reasons for relinquishing the principles of adult baptism and embracing those of infant baptism
– The university, mediaeval and modern
Elderberry times – Old Town, ME. jan-oct 1974 – 1 – us CRL [071]
Elders' journal – Kirtland, Ohio etc. v1 n1-4. oct 1837-aug 1838 – 1 – us NY Public [242]
Elder's thoughts on union / Dickson, David – Edinburgh, Scotland. 1870 – 1r – us UF Libraries [242]
Eldersveld, S *see* De weldadigheid gods en de eerste christelijk gereformeerde kerk, kalamazoo, mich, 1869-1912
Eldin, F *see* Haiti
Eldorado of the ancients / Peters, Karl – New York, NY. 1969 – 1r – us UF Libraries [420]
The eldorado of the ancients / Peters, Karl – London: C.A. Pearson, 1902. x,447p. illus. 2 fold. maps – 1 – us Wisconsin U Libr [930]
Eldridge Cleaver Crusades *see* Crusader
Eldridge, Paul *see* And thou shalt teach them
[Eldridge-] the eldridge gazette – CA. 1975 – 3r – 1 – $180.00 (subs $50/y) – mf#B06024 – us Library Micro [071]
Eldridge, William B *see*
– The district court executive pilot program
– The second circuit sentencing study
Eleanor leslie : a memoir / Stone, Jean Mary – London: Art and Book Co, 1898 – 1mf – 9 – 0-8370-7023-6 – (incl ind) – mf#1986-1023 – us ATLA [920]
E-learning – Cleveland. 2000+ (1,5,9) – ISSN: 1530-6399 – mf#31572 – us UMI ProQuest [000]
Eleccion al mejor deportista provincial de badajoz-1970 / Junta Provincial de Educacion Fisica y Deportes – Badajoz: graf. nemesio jimenez, 1971 – sp Bibl Santa Ana [946]
Elecciones de 1964 / El Salvador Presidencia Departamento De Relacion – San Salvador, El Salvador. 1964 – 1r – us UF Libraries [972]
Las elecciones in merida – 1881 – 9 – sp Bibl Santa Ana [946]
Elected or appointed officials? : a paper submitted to the american academy of political and social science / Bourinot, John George – Philadelphia: American Academy of Political and Social Science, 1895? – 1mf – 9 – mf#00228 – cn Canadiana [350]
L'electeur – Paris. 5 juin 1868-30 mars 1871 – 1 – (puis l' electeur libre.) – fr ACRPP [073]
L' electeur libre *see* L'electeur
Electeurs de la province de quebec – Quebec: s.n, 1878 – 2mf – 9 – mf#03183 – cn Canadiana [325]
Electeurs de la province de quebec! : prenez et lisez! – [Quebec?: Belleau], 1891 – 2mf – 9 – 0-665-11169-X – mf#11169 – cn Canadiana [325]
Election and conversion : A Frank Discussion of Dr. Pieper's Book on "Conversion and Election" / Keyser, Leander Sylvester – Burlington, Iowa: German Literary Board, 1914 – 1mf – 9 – 0-7905-9991-0 – mf#1989-1716 – us ATLA [240]
Election and service / Peake, Arthur Samuel – London: Hodder & Stoughton 1908 [mf ed 1989] – (= ser Aids to the devotional study of scripture 2) – 1mf – 9 – 0-7905-1776-0 – mf#1987-1776 – us ATLA [225]
Election archives – New Delhi. 1970-1973 (1) – ISSN: 0046-1644 – mf#7945 – us UMI ProQuest [325]
Election case law : a summary of judicial precedent on election issues other than campaign financing – 1989; 1993 – 7mf – 9 – $10.50 – mf#LLMC 95-026 – us LLMC [342]
The election, confirmation and homage of bishops of the church of england : a paper...dec 6 1899 / Browne, George Forrest – London: SPCK 1900 [mf ed 1993] – (= ser Church historical society (series) 60) – 1mf – 9 – 0-524-05532-7 – (incl ind) – mf#1990-5136 – us ATLA [242]
Election des conseillers de ville : quartier st roch... – [Quebec?: s.n, 1856?] [mf ed 1984] – 1mf – 9 – 0-665-17040-8 – mf#17040 – cn Canadiana [350]
Election du comte northumberland / Laterriere, Pierre de Sales – S.l: s.n, 1820? – 1mf – 9 – mf#21082 – cn Canadiana [325]
L'election du quartier-est de montreal : contenant l'adresse de joseph papineau, ecr aux electeurs / Papineau, Joseph – A Montreal: 1810 – 1mf – 9 – mf#63426 – cn Canadiana [325]
The election law. constitutionality thereof maintained. an opinion of one of the most eminent constitutional lawyers in the united states on the election law of mississippi – n.p., 1871? 10 p. LL-649 – 1 – us L of C Photodup [342]
Election manifesto : 1967 / Indian National Congress – New Delhi: N Balakrishnan, [1967?] – us CRL [325]
Election manifesto : 1967 / Sangh, Bharatiya Jana – [s.l.]: The Sangh, [1967?] (Delhi: Arjun Press) – us CRL [325]
Election manifesto – New Delhi: [s.n., 1967?] – us CRL [325]
Election manifesto : samyukta socialist party – New Delhi: The Party, [1967?] – us CRL [325]
Election manifesto : swatantra party election manifesto – [Bombay: Printed at Inland Printers, 1967?] – us CRL [325]
Election manifesto and immediate programme : fourth general election, 1967 / Bangla Congress – New Delhi: [s.n.] 1967? – us CRL [325]
Election manifesto of akhil bharat hindu mahasabha – New Delhi: The Mahasabha, 1966 – us CRL [325]
Election manifesto of the communist party of india – New Delhi: D P Sinha, 1966 – us CRL [325]
Election manifesto of the communist party of india (marxist) – New Delhi: The Party, 1967 – us CRL [325]
Election manifesto of the republican party of india, 1967 / ed by Khobragade, B D – Chanda, 1967 – us CRL [325]
Election manifesto of the tamilnad toilers welfare party, madras state : Released oct 9, 1966, at the conference of the party held in madras city – Madras: The Party, 1966 – us CRL [325]
The election of grace / Taylor, William MacKergo – London: Hodder & Stoughton, 1868 [mf ed 1985] – 1mf – 9 – 0-8370-2198-7 – mf#1985-0198 – us ATLA [240]
Election programmes in new zealand politics – 1911-1996 – mf#ZB 32 – nz Nat Libr [325]
Election returns, 1858-1962 / Minnesota. Secretary of State – 83r – 1 – $30.00r – us Minn Hist [325]
Election scratch sheet / International Brotherhood of Teamsters, Chauffeurs, Warehousemen and Helpers of America – n1-n4 – 1r – 1 – mf#634173 – us WHS [331]
Elections de 1881 : situation politique et administrative de la province de quebec – Montreal: La Patrie, 1881 – 1mf – 9 – 0-665-04261-2 – mf#04261 – cn Canadiana [325]
Elections de 1887 : le vraie question – Quebec?: C Darveau, 1887 – 1mf – 9 – mf#30159 – cn Canadiana [325]
Elections de 1958 / France – D'apres la presse regionale. Dossiers de Presse – 1 – fr ACRPP [944]
Les elections episcopales dans l'eglise de france du 9e au 12e siecle : etude sur la decadence du principe electif, 814-1150 / Imbart de La Tour, Pierre – Paris: Hachette, 1891 [mf ed 1990] – 6mf – 9 – 0-7905-4871-2 – (in french) – mf#1988-0871 – us ATLA [240]
Elections generales de 1900 : conseils pratiques pour l'organisation, qualification des electeurs, instructions aux agents pour la province de quebec – [Montreal?: s.n, 1900?] – 1mf – 9 – 0-665-91667-1 – mf#91667 – cn Canadiana [325]
Elections legislatives de 1908 / Janvier, Louis Joseph – Port-Au-Prince, Haiti. 1908 – 1r – us UF Libraries [323]
Elections presidentielles et legislatives du 27 dec 1974 – Conakry: Impr national P Lumumba, 1975 – us CRL [325]
Elective course of lectures in systematic theology / Curtis, Olin Alfred – Madison, NJ: Drew Theological Seminary, 1901 – 1mf – 9 – 0-8370-2792-6 – mf#1985-0792 – us ATLA [240]
The electoral act for van diemen's land – Launceston, 1851 – (= ser 19th c books on british colonization) – 1mf – 9 – mf#1.1.7004 – uk Chadwyck [325]
Electoral campaign speeches, oct 26-nov 5 1959 – Tunis: Secretariate of State for Information Publications, [1959?] – us CRL [325]
The electoral government of greater britain : a suggestion. first article: proposed referendum senates for the parliaments of great britain and ireland. second article: proposed supreme britannic senate or political assembly / Thwaite, Benjamin Howarth – [London 1895] – (= ser 19th c british colonization) – 1mf – 9 – mf#1.1.3809 – uk Chadwyck [323]
Electoral regulations : 1977 / Nigeria – n.p., Nigeria. 1977 – 1r – us UF Libraries [325]
Electoral studies – Kidlington. 1982+ (1,5,9) – ISSN: 0261-3794 – mf#17225 – us UMI ProQuest [325]
Electors of ottawa : you will soon be called upon to elect your representatives for the house of commons / Monk, Henry Wentworth – [Ottawa?: s.n. 1887?] [mf ed 1987] – 1mf – 9 – 0-665-52111-1 – mf#52111 – cn Canadiana [325]
Electorum symbolorum et parabolarum historicarum syntagmata : ex Horo, Clemente, Epiphanio et aliis cum notis et observationibus / Caussin, N – Parisiis: Sumptibus Romani de Beauvais, 1618 – 11mf – 9 – mf#O-55 – ne IDC [090]
Electra – Amsterdam, Netherlands. -w. May 1899-Nov 1902. 2 reels – 1 – uk British Libr Newspaper [949]
Electra : orgam da liga anti clerical paranaense – Curitiba, PR: Typ Imp Paranaense, ago 1901-ago 1903 – (= ser Ps 19) – mf#P16,02,92 – bl Biblioteca [079]

Electre / Poizat, Alfred – Paris, France. 1907 – 1r – us UF Libraries [440]
Electric light and power – Tulsa. 1979+ (1,5,9) – ISSN: 0013-4120 – mf#12414 – us UMI ProQuest [621]
Electric machines and electromechanics – Washington. 1976-1982 (1,5,9) – (Cont by: Electric machines and power systems) – ISSN: 0361-6967 – mf#11136 – us UMI ProQuest [621]
Electric machines and power systems – Washington. 1983-1998 (1,5,9) – (Cont: Electric machines and electromechanics) – ISSN: 0731-356X – mf#11136,01 – us UMI ProQuest [621]
Electric messages from japan *see* Oms (oriental missionary standard) outreach
Electric perspectives – Washington. 1981+ (1,5,9) – ISSN: 0364-474X – mf#12929 – us UMI ProQuest [621]
Electric power components and systems – Philadelphia, 2001+ [1,5,9] – (cont: electric machines and power systems) – ISSN: 1532-5008 – mf#11136,02 – us UMI ProQuest [621]
Electric power statistics – Washington. 1973-1978 (1) 1973-1978 (5) 1973-1978 (9) – ISSN: 0013-4139 – mf#6831 – us UMI ProQuest [333]
Electric power systems research – Lausanne. 1977-1994 (1) 1977-1994 (5) 1987-1994 (9) – ISSN: 0378-7796 – mf#42130 – us UMI ProQuest [621]
Electric press – [NW England] Stalybridge Lib dec 1909-jan 1910 – 1 – uk MLA; uk Newsplan [072]
Electric technology = Elektrichestvo – Oxford. 1990-1990 (1,5,9) – (Cont: Electric technology USSR. Cont by: Electrical technology) – ISSN: 0013-4155 – mf#49062,01 – us UMI ProQuest [621]
Electric technology ussr = Elektrichestvo – Oxford. 1958-1990 (1,5) 1957-1990 (9) – (cont by: electric technology) – ISSN: 0013-4155 – mf#49062 – us UMI ProQuest [621]
Electrica berlanguena S.A. Berlanga, Badajoz *see* Memoria y balance general en 30 de junio de 1938. 37 ejercicio social. leida y aprobada...1938
Electrical business – Evanston. 1979-1982 (1,5,9) – ISSN: 0162-8534 – mf#12571 – us UMI ProQuest [621]
Electrical communication : english ed – Paris. 1922-1994 (1) 1970-1994 (5) 1975-1994 (9) – ISSN: 1242-0565 – mf#4 – us UMI ProQuest [380]
Electrical construction and maintenance – New York. 1920-1980 (1) 1970-1980 (5) 1977-1980 (9) – (Cont by: EC and M : electrical construction and maintenance) – ISSN: 0013-4260 – mf#366 – us UMI ProQuest [621]
Electrical consultant – Cos Cob. 1960-1986 [1]; 1970-1986 [5]; 1975-1986 [9] – (cont by: electrical systems design) – ISSN: 0361-4972 – mf#1892 – us UMI ProQuest [621]
Electrical design and mfg – Libertyville. 1992-1995 (1,5,9) – (Cont: Electrical manufacturing) – ISSN: 1065-7436 – mf#16958,01 – us UMI ProQuest [670]
Electrical distribution – Northfleet. 1960-1969 (1) – ISSN: 0422-8693 – mf#2994 – us UMI ProQuest [621]
Electrical energy management – Cos Cob. 1981-1982 (1) 1981-1982 (5) 1981-1982 (9) – ISSN: 0194-4746 – mf#12537,01 – us UMI ProQuest [621]
Electrical engineer – Chippendale. 1969-1995 (1) 1970-1995 (5) 1976-1995 (9) – ISSN: 0013-4309 – mf#3433 – us UMI ProQuest [621]
Electrical engineer – Johannesburg. 1980-1983 (1,5,9) – mf#12147 – us UMI ProQuest [621]
Electrical engineering – New York. 1954-1963 (1) – ISSN: 0095-9197 – mf#871 – us UMI ProQuest [621]
Electrical engineering in japan – Washington. 1984-1995 (1,5,9) – ISSN: 0424-7760 – mf#14347 – us UMI ProQuest [621]
Electrical engineering measurements for commercial... / Parr, George Dudley Aspinall – London, England. 1903 – 1r – us UF Libraries [621]
The electrical experimenter – New York: [s.n., -1920]. [v6, n1-5 may-sep 1918; n61-65; v6 n7-12 nov 1918-apr 1919; n67-72] – 1r – us CRL [071]
Electrical machinery / Croft, Terrell – New York, NY. 1917 – 1r – 1 – us UF Libraries [621]
Electrical manufacturing – Libertyville. 1987-1992 (1,5,9) – (Cont by: Electrical design and mfg) – ISSN: 0895-3716 – mf#16958 – us UMI ProQuest [670]
Electrical, mechanical and milling news – Toronto: C H Mortimer, [1889-1890] – 9 – mf#P06115 – cn Canadiana [621]
Electrical news *see* Canadian electrical news and engineering journal [electrical news]
Electrical power engineer, 1920-59 : the official journal of the electrical power engineers association – v1-41 – 32r – 1 – mf#531 – uk Microform Academic [621]
Electrical practice – New York. 1974-1977 (1) 1974-1977 (5) 1974-1977 (9) – ISSN: 0094-9434 – mf#10688 – us UMI ProQuest [621]
Electrical progress and monthly register – London, UK. nov-dec 1906 [mthly] – 8ft – 1 – uk British Libr Newspaper [621]
Electrical review – London. 1872-1989 [1]; 1958-1989 [5,9] – ISSN: 0013-4384 – mf#1225 – us UMI ProQuest [621]
Electrical south – Atlanta. 1938-1978 (1) 1970-1978 (5) 1973-1978 (9) – ISSN: 0013-4392 – mf#351 – us UMI ProQuest [621]
Electrical systems design – Cos Cob. 1986-1990 (1) 1986-1990 (5) 1986-1990 (9) – (cont: electrical consultant) – ISSN: 0899-6083 – mf#1892,01 – us UMI ProQuest [621]
Electrical technology – Oxford. 1991+ (1,5,9) – (Cont: Electric technology = Elektrichestvo) – ISSN: 0965-5433 – mf#49062,02 – us UMI ProQuest [621]
Electrical times – London. 1960-1988 (1) 1972-1988 (5) 1984-1988 (9) – ISSN: 0013-4414 – mf#459 – us UMI ProQuest [621]
Electrical times e ingeniero industrial – London, England. 28 jan 1915 – 1 – (cont as: ingeniero industrial, apr 1915-aug 1917) – uk British Libr Newspaper [621]
Electrical union world / International Brotherhood of Electrical Workers, Local 3 – 1960 jan 1-1961 dec 15, 1975-81, 1982-86 – 3r – 1 – mf#1055917 – us WHS [331]
Electrical union world – New York, NY. 1959-1969 (1) – mf#65070 – us UMI ProQuest [071]
Electrical west – San Francisco. 1895-1970 (1) – ISSN: 0095-9219 – mf#31 – us UMI ProQuest [621]
Electrical wholesaling – Chicago. 1920-1996 (1) 1967-1981 (5) 1967-1981 (9) – ISSN: 0013-4430 – mf#368 – us UMI ProQuest [621]
Electrical world – New York. 1883+ (1) 1965+ (5) 1975+ (9) – ISSN: 0013-4457 – mf#35 – us UMI ProQuest [621]
Electricite – Paris, France. 5 jan-aug 1876; 5 jul 1878-20 dec 1879; 1880-9 aug 1894 – 8 1/2r – 1 – uk British Libr Newspaper [072]
Electricity on the farm – New York. 1927-1974 (1) 1971-1973 (5) – ISSN: 0013-4554 – mf#1968 – us UMI ProQuest [621]
Electrificacao rural no nordeste / Banco do Nordeste do Brasil Escritorio Tecnico de... – Fortaleza, Brazil. 1959 – 1r – us UF Libraries [338]
Electri-onics – Libertyville. 1983-1987 (1) 1983-1987 (5) 1983-1987 (9) – (Cont: Insulation/circuits) – ISSN: 0745-4309 – mf#12329,03 – us UMI ProQuest [621]
Electro optics – Des Plaines. 1983-1983 (1) 1983-1983 (5) 1983-1983 (9) – (Cont: Electro-optical systems design) – ISSN: 0745-5003 – mf#5922,01 – us UMI ProQuest [621]
Electrochemical Society *see*
– Journal of the electrochemical society
– Transactions of the electrochemical society
Electrochemical society extended abstracts – Pennington. 1955-1995 (1) 1971-1995 (5) 1974-1995 (9) – (cont by: electrochemical society meeting abstracts) – ISSN: 0160-4619 – mf#2299 – us UMI ProQuest [540]
Electrochemical society meeting abstracts – Pennington. 1996-1996 (1) 1996-1996 (5) 1996-1996 (9) – (cont: electrochemical society extended abstracts) – ISSN: 1091-8213 – mf#2299,01 – us UMI ProQuest [540]
Electrochemical society meetings – Pennington, 1951-66 (1) – (cont: american electro-chemical society meetings) – mf#13067,01 – us UMI ProQuest [540]
Electrochemical technology – Princeton. 1963-1968 (1) – ISSN: 0424-8090 – mf#1606 – us UMI ProQuest [621]
Electrochemistry communications – 1999+ (1,5,9) – ISSN: 1388-2481 – mf#42858 – us UMI ProQuest [540]
Electrochimica acta – Oxford. 1959+ (1,5,9) – ISSN: 0013-4686 – mf#49063 – us UMI ProQuest [540]
Electroencephalography and clinical neurophysiology – Limerick. 1949-1998 (1) 1949-1998 (5) 1987-1998 (9) – (Cont by: Clinical neurophysiology) – ISSN: 0013-4694 – mf#42253 – us UMI ProQuest [616]
Electrolysis of organic compounds / Kolbe, Hermann – Edinburgh, Scotland. 1900 – 1r – us UF Libraries [540]
Electromagnetics – Washington. 1988-1996 (1,5,9) – ISSN: 0272-6343 – mf#14241 – us UMI ProQuest [550]
Electromechanical design – Boston. 1957-1975 (1) 1971-1975 (5) – ISSN: 0013-4716 – mf#1529 – us UMI ProQuest [621]
Electromedical Devices and Radiological Health, 21st anniversary, 1973-1994 / South Africa. Directorate: Electromedical Devices and Radiological Health – Pretoria: Dept of Health [1994] [mf ed Pretoria, RSA: State Library [199-]] – 37p [ill] on 1r with other items – 5 – mf#op 11922 r24 – us CRL [360]
An electromyographic analysis of selected abdominal exercises / Seamons, Todd D – 1997 – 2mf – 9 – $8.00 – mf#PH 1607 – us Kinesology [612]
Electromyographic changes during intense isokinetic strength training / Lamack, Daniel D – Iowa State University, 1993 – 1mf – 9 – $4.00 – mf#PH1467 – us Kinesology [612]
An electromyographic comparison of abdominal exercises on selected commercially available equipment / Leung, Wai M – 1997 – 2mf – 9 – $8.00 – mf#PH 1558 – us Kinesology [612]
An electromyographic comparison of seated and standing up-hill cycling / Griffith, Gareth E – 1997 – 2mf – 9 – $8.00 – mf#PE 3757 – us Kinesology [612]
An electromyographic investigation of four elastic tubing closed kinetic chain exercises after acl reconstruction / Metzger, Kimbie – 1996 – 2mf – 9 – $8.00 – mf#PE 3814 – us Kinesology [617]
Electron microscopy reviews – Elmsford. 1988-1992 (1,5,9) – ISSN: 0892-0354 – mf#49519 – us UMI ProQuest [578]
Electronic and computer music : an annotated bibliography of writings in english / Bahler, Peter Benjamin – U of Rochester 1966 [mf ed 19–] – 1r / 3mf – 1,9 – mf#film 752 / fiche445 – us Sibley [780]
Electronic business – Boston. 1984-1993 (1,5,9) – (Cont by: Electronic business buyer) – ISSN: 0163-6197 – mf#14878 – us UMI ProQuest [621]
Electronic business – Highlands Ranch. 1997+ (1,5,9) – (Cont: Electronic business today) – ISSN: 1097-4881 – mf#14878,03 – us UMI ProQuest [621]
Electronic business buyer – Highlands Ranch. 1993-1995 (1,5,9) – (Cont: Electronic business. Cont by: Electronic business today) – ISSN: 1073-1059 – mf#14878,01 – us UMI ProQuest [621]
Electronic business today – Highlands Ranch. 1995-1997 (1,5,9) – (Cont: Electronic business buyer. Cont by: Electronic business) – ISSN: 1085-8288 – mf#14878,02 – us UMI ProQuest [621]
Electronic design – Cleveland. 1953+ (1) 1965+ (5) 1974+ (9) – ISSN: 0013-4872 – mf#1483 – us UMI ProQuest [621]
Electronic engineering – London. 1941+ [1]; 1971+ [5]; 1976+ [9] – ISSN: 0013-4902 – mf#1204 – us UMI ProQuest [621]
Electronic engineering design – London. 2002+ (1) – (cont: Electronic engineering) – mf#1204,01 – us UMI ProQuest [621]
Electronic engineering times – Manhasset. 1991+ (1,5,9) – ISSN: 0192-1541 – mf#14472 – us UMI ProQuest [621]
Electronic imaging – Boston. 1982-1985 (1,5,9) – ISSN: 0737-6553 – mf#13097 – us UMI ProQuest [621]
Electronic instrument digest – Chicago. 1970-1972 [1]; 1965-1972 [5] – ISSN: 0013-4929 – mf#5884 – us UMI ProQuest [621]
Electronic learning – New York. 1983-1996 – 1,5,9 – ISSN: 0278-3258 – mf#14114 – us UMI ProQuest [370]
Electronic library – Bradford. 1989+ (1,5,9) – ISSN: 0264-0473 – mf#17523 – us UMI ProQuest [020]
Electronic manufacturing – Libertyville. 1987-1989 (1) 1987-1989 (5) 1987-1989 (9) – (Cont by: Contract and captive electronic manufacturing and printed circuit production) – ISSN: 0895-3708 – mf#16957 – us UMI ProQuest [621]
Electronic media – Chicago. 1982+ (1,5,9) – (cont: advertising age electronic media edition) – ISSN: 0745-0311 – mf#13840,01 – us UMI ProQuest [380]
Electronic media coverage of federal civil proceedings : an evaluation of the pilot program in six district courts and two courts of appeals / Johnson, Molly T et al – 1994 – 1mf – 9 – $1.50 – mf#llmc99-031 – us LLMC [347]
Electronic music review – Leicester. 1967-1968 (1) – ISSN: 0424-8260 – mf#5915 – us UMI ProQuest [780]
Electronic news – New York. 1988+ (1,5,9) – ISSN: 1061-6624 – mf#17276 – us UMI ProQuest [621]
Electronic packaging and production – Newton. 1970+ (1) 1971+ (5) 1972+ (9) – ISSN: 0013-4945 – mf#5883 – us UMI ProQuest [621]
Electronic products – Garden City. 1960+ (1) 1971+ (5) 1974+ (9) – ISSN: 0013-4953 – mf#1552 – us UMI ProQuest [621]
Electronic progress – Lexington. 1956-1992 (1) 1974-1992 (5) 1976-1992 (9) – ISSN: 0013-4961 – mf#9484 – us UMI ProQuest [621]
Electronic publishing – Chichester. 1989-1995 (1,5,9) – ISSN: 0894-3982 – mf#17045 – us UMI ProQuest [070]
Electronic publishing abstracts – Oxford. 1984-1988 (1) 1983-1988 (5) 1984-1988 (9) – ISSN: 0739-2907 – mf#49483 – us UMI ProQuest [070]
Electronic publishing and printing – Chicago. 1989-1990 (1,5,9) – (Cont: EP and P. Cont by: Computer publishing magazine) – ISSN: 1044-0852 – mf#15283,01 – us UMI ProQuest [070]
Electronic servicing – Overland Park. 1957-1981 (1) 1971-1981 (5) 1976-1981 (9) – ISSN: 0013-497X – mf#1097 – us UMI ProQuest [380]
Electronic servicing and technology – Hicksville. 1981+ (1,5,9) – ISSN: 0278-9922 – mf#12980 – us UMI ProQuest [380]
Electronic systems technology and design – Tulsa, 1999-1999 [1,5,9] – (cont: computer design) – ISSN: 1524-1238 – mf#8508,01 – us UMI ProQuest [621]
Electronic warfare – Palo Alto. 1975-1977 (1) 1975-1977 (5) 1975-1977 (9) – (Cont by: Electronic warfare defense electronics) – ISSN: 0363-258X – mf#10742 – us UMI ProQuest [621]
Electronic warfare defense electronics – Palo Alto. 1977-1979 (1,5,9) – (Cont: Electronic warfare) – ISSN: 0164-3363 – mf#10742,01 – us UMI ProQuest [621]
Electronics – New York. 1930-1984 (1) 1965-1984 (5) 1970-1984 (9) – (cont by: ElectronicsWeek) – ISSN: 0013-5070 – mf#26 – us UMI ProQuest [621]
Electronics – Cleveland. 1985-1995 (1,5,9) – (cont: ElectronicsWeek) – ISSN: 0883-4989 – mf#26,02 – us UMI ProQuest [621]
Electronics and communications in japan – New York. 1984-1984 (1,5,9) – ISSN: 0424-8368 – mf#14348 – us UMI ProQuest [621]
Electronics and communications in japan, pt 1 : communications – New York. 1985-1994 (1,5,9) – ISSN: 8756-6621 – mf#15471 – us UMI ProQuest [380]
Electronics and communications in japan, pt 2 : electronics – New York. 1985-1994 (1,5,9) – ISSN: 8756-663X – mf#15472 – us UMI ProQuest [621]
Electronics and communications in japan, pt 3 : fundamental electronic science – New York. 1989-1994 (1,5,9) – ISSN: 1042-0967 – mf#18185 – us UMI ProQuest [380]
Electronics and power – Stevenage. 1955-1987 (1) 1975-1987 (5) 1975-1987 (9) – (Cont by: IEE review) – ISSN: 0013-5127 – mf#10673 – us UMI ProQuest [621]
Electronics and wireless world – London. 1983-1989 (1) 1983-1989 (5) 1983-1989 (9) – (cont: Wireless world. cont by: Electronics world + wireless world) – ISSN: 0266-3244 – mf#661,01 – us UMI ProQuest [380]
Electronics illustrated – New York. 1958-1972 (1) 1971-1972 (5) 1971-1972 (9) – ISSN: 0013-5178 – mf#1748 – us UMI ProQuest [621]
Electronics letters – Stevenage. 1965+ (1) 1975+ (5) 1983+ (9) – ISSN: 0013-5194 – mf#10674 – us UMI ProQuest [621]
Electronics now – Farmingdale. 1993-1999 (1) 1993-1999 (5) 1993-1999 (9) – (Cont: Radio-electronics) – ISSN: 1067-9294 – mf#204,01 – us UMI ProQuest [380]
Electronics purchasing – Newton. 1986-1993 (1,5,9) – ISSN: 0889-0196 – mf#14879 – us UMI ProQuest [650]
Electronics test – San Francisco. 1983-1990 (1) 1983-1990 (5) 1983-1990 (9) – ISSN: 0164-9620 – mf#13089 – us UMI ProQuest [621]
Electronics weekly – Surrey. 1968-1988 (1) 1984-1988 (5) 1984-1988 (9) – ISSN: 0013-5224 – mf#3145 – us UMI ProQuest [621]
Electronics world – Sutton. 1996+ (1,5,9) – mf#661,03 – us UMI ProQuest [380]
Electronics world + wireless world – Sutton. 1989-1996 (1) 1989-1996 (5) 1989-1996 (9) – (Cont: Electronics and wireless world) – ISSN: 0959-8332 – mf#661,02 – us UMI ProQuest [380]
ElectronicsWeek – New York. 1984-1985 (1,5,9) – (cont: Electronics. cont by: Electronics) – ISSN: 0748-3252 – mf#26,01 – us UMI ProQuest [621]
Electro-optical systems design – Chicago. 1969-1982 (1) 1971-1982 (5) 1971-1982 (9) – (Cont by: Electro optics) – ISSN: 0424-8457 – mf#5922 – us UMI ProQuest [621]
Electro-technology newsletter – Beverly Shores. 1928-1976 (1) 1965-1976 (5) – ISSN: 0146-3667 – mf#1010 – us UMI ProQuest [621]
The electrothermic production of iron and steel / Stansfield, Alfred – [S.l: s.n, 1904?] [mf ed 1991] – 1mf – 9 – 0-665-99522-9 – mf#99522 – cn Canadiana [660]
Elefantes blancos, paginas doradas. nuevas noticias de indochina / Elias de Tejada Spinola, Francisco – Barcelona: Talleres Graficos Rafael Salva, 1957 – sp Bibl Santa Ana [590]
Elefherotypia – 1990- – 1 – enquire for prices – (yrly reel count varies) – us UMI ProQuest [079]
Elegancias : revista mensual ilustrada artistica, literaria, modas y actualidades – Paris. n1-46. mai 1911-aout 1914 – 1 – (mq no. 2, 5. en 1913-14 contient des suppl., n1-13) – fr ACRPP [073]

Elegant : the magazine for fashionable living – 1964 nov-1965 feb,apr-may, jul-sep, 1967 mar – 1r – 1 – mf#4879176 – us WHS [071]

Elegant extracts for the german flute or violin selected from the most favorite songs etc : sung in the theatres and public places – Baltimore: J Carr 1794-98 – 1 – us L of C Photodup [780]

Elegant, Robert S *see* Dragon's seed

O elegante : jornal litterario, critico e humoristico – Florianopolis, SC. 12 ago, dez 1923; 18 jan 1925 – (= ser Ps 19) – mf#UFSC/BPESC – bl Biblioteca [073]

Die elegante welt – Berlin DE, 1912-43 [gaps] – 1 – gw Misc Inst [074]

ElegantiolaeEutropius *see* In orationes quasdam ciceronis...

Elegantissimorum emblematum corpusculum Latinis Belgicisque versibus elucidatum – Lugduni Batavorum: Ex chalcographia Petri vander Aa, 1696 – 1mf – 9 – mf#O-3065 – ne IDC [090]

Eleggua project newsletter – 1996 jan, 1997 sep – 1r – 1 – mf#5266144 – us WHS [071]

Elegia a avaro barba / Mongo – Havana, Cuba – 1r – 1 – us UF Libraries [972]

Elegia a jesus menendez / Guillen, Nicolas – La Habana, Cuba. 1978 – 1r – us UF Libraries [972]

Elegias de varones ilustres de indias / Castellanos, Juan De – Bogota, Colombia. v1-4. 1955 – 2r – us UF Libraries [972]

Elegias en la viva muerte de enrique munoz meany / Ovalle Lopez, Werner – Guatemala, 1961 – 1r – us UF Libraries [972]

Elegie, september 1823 : goethes reinschrift mit urikens von levetzow brief an goethe und ihrem jugendbildnis / Goethe, Johann Wolfgang von; ed by Suphan, Bernhard – Weimar: Goethe-Gesellschaft, 1900 [mf ed 1993] – (= ser Schriften der goethe-gesellschaft 15) – 19p/2lea/10pl – 1 – (incl bibl ref) – mf#8657 reel 4 – us Wisconsin U Libr [920]

Elegy : on the death of the rev james spencer, ma – S.l: s.n, 1863? – 1mf – 9 – mf#41284 – cn Canadiana [080]

Elektra : Tragedy in one act / Strauss, Richard & Hofmannsthal, Hugo von – Berlin: A Fuerstner, c1910 – 1r – 1 – us Wisconsin U Libr [790]

Elektra : tragoedie in einem aufzuge / Strauss, Richard & Hofmannsthal, Hugo von – Berlin: A Fuerstner, c1908 – 1r – 1 – us Wisconsin U Libr [780]

Elektrische bahnen : zentralblatt fuer elektrischen zugbetrieb und alle arten von triebfahrzeugen mit elektrischem antrieb – Berlin. 1973-1980 (1) 1975-1980 (5) 1975-1980 (9) – ISSN: 0013-5437 – mf#9074 – us UMI ProQuest [380]

Elektrische messungen an gestaeubten mosi$_2$-schichten / Lippert, Gunther – (mf ed 1995) – 1mf – 9 – €30.00 – 3-8267-2116-0 – mf#DHS 2116 – gw Frankfurter [621]

Elektrochemische thermospray-massenspektrometrie : on-line-methode zur aufklaerung von elektrochemischen reaktionen in sauren elektrolytloesungen sowie von prozessen in membranen / Stassen, Ingo – (mf ed 1995) – 2mf – 9 – €40.00 – 3-8267-2264-7 – mf#DHS 2264 – gw Frankfurter [540]

Elektrokinetische untersuchungen von aerosiloberflaechen und aerosil-triarylmetylhalogenid-adsorbaten in organischen medien / Simon, Frank – (mf ed 1993) – 1mf – 9 – €30.00 – 3-89349-768-4 – mf#DHS 768 – gw Frankfurter [540]

Elektronenmikroskopische untersuchungen zur stereoselektiven bildung mizellarer lipidfasern aus n-alkylaldonamiden / Boettcher, Christoph – (mf ed 1992) – 2mf – 9 – €49.00 – 3-89349-458-8 – mf#DHS 458 – gw Frankfurter [540]

Elektronenspektroskopische untersuchung von alkan- und alkanthiolfilmen auf festkoerperflaechen / Heinz, Bertram – (mf ed 1998) – 3mf – 9 – €49.00 – 3-8267-2531-X – mf#DHS 2531 – gw Frankfurter [530]

Elektrophysiologische und haemodynamische effekte von magnesium auf spaete reperfusionsarrhythmien bei akutem myokardinfarkt / Ketteler, Thomas – (mf ed 1999) – 3mf – 9 – €49.00 – 3-8267-2627-8 – mf#DHS 2627 – gw Frankfurter [616]

Elektrophysiologische untersuchungen zur verarbeitung grammatischer information in der finnischen sprache / Muente, Ava Sinikka – (mf ed 1995) – 1mf – 9 – €30.00 – 3-8267-2229-9 – mf#DHS 2229 – gw Frankfurter [612]

Elektrotechnischer anzeiger – Berlin DE, 1892-1909 – 25r – 1 – uk British Libr Newspaper [621]

Elektrotechnisches echo – Magdeburg DE, 1890-93, 1896-1904 [wkly] – 6r – 1 – uk British Libr Newspaper [600]

Elele hawaii – Honolulu HI. 1852 mar 24 – 1r – 1 – (cont: nonanona) – mf#634416 – us WHS [071]

Elelin / Rojas, Ricardo – Buenos Aires, Argentina. 1929 – 1r – us UF Libraries [972]

Elemens de chymie-pratique : contenant la description des operations fondamentales de la chymie, avec des explications et des remarques sur chaque operation / Macquer, Pierre Joseph – Paris: J T Herissant, 1751 – 1 – us Wisconsin U Libr [540]

Elemens de la grammaire francaise / Lhomond, Charles-Francois – 1re ed. Quebec: J Neilson, 1800 [mf ed 1971] – 1r – 5 – mf#SEM16P57 – cn Bibl Nat [440]

Elemens de la grammaire francaise / Lhomond, Charles-Francois – 2e ed. Quebec: J Neilson, 1810 [mf ed 1971] – 1r – 5 – mf#SEM16P58 – cn Bibl Nat [440]

Elemens de la grammaire francaise / Lhomond, Charles-Francois – 3e ed. Quebec: J Neilson, 1819 [mf ed 1971] – 1r – 5 – mf#SEM16P59 – cn Bibl Nat [440]

Elemens de la grammaire francaise / Lhomond, Charles-Francois – Montreal: James Brown, 1820 [mf ed 1971] – 1r – 5 – mf#SEM16P61 – cn Bibl Nat [440]

Elemens de la grammaire francaise / Lhomond, Charles-Francois – Montreal: Jh Victor Delorme, 1817 [mf ed 1971] – 1r – 5 – mf#SEM16P60 – cn Bibl Nat [440]

Elemens de la grammaire japonaise / Rodrigues, Joao – Paris: Donday-Dupre 1825 – (= ser Whsb) – 2mf – 9 – €30.00 – (transduits du portugais sur le manuscrit de la bibliotheque du roi...par m c landresse; precedes d'une explication des syllabaires japonais, et de deux plaches contenant les signes de ces syllabaires, par m abel-remusat) – mf#Hu 329 – gw Fischer [480]

Elemens de la grammaire latine / Lhomond, Charles-Francois – Montreal: Roy & Bennett, 1797 [mf ed 1971] – 1r – 5 – mf#SEM16P62 – cn Bibl Nat [450]

Elemens de la grammaire latine, a l'usage des colleges / Lhomond, Charles-Francois – nouv ed. Quebec: J Neilson, 1813 [mf ed 1971] – 1r – 5 – mf#SEM16P174 – cn Bibl Nat [450]

Elemens de la grammaire latine, a l'usage des colleges : methode / Lhomond, Charles-Francois – Montreal: Louis Roy, 1796 [mf ed 1971] – 1r – 5 – mf#SEM16P173 – cn Bibl Nat [450]

Elemens de la grammaire latine, a l'usage des colleges : syntaxe / Lhomond, Charles-Francois – Montreal: Louis Roy, 1796 [mf ed 1971] – 1r – 5 – mf#SEM16P172 – cn Bibl Nat [450]

Elemens de la grammaire mandchoue / Gabelentz, Hans Conon von der – Altenbourg: Comptoir de la litterature 1832 – (= ser Whsb) – 2mf – 9 – €30.00 – mf#Hu 302 – gw Fischer [480]

Elemens de la langue des celtes gomerites ou bretons : introduction a cette langue et par elle a celles de tous les peuples connus / Brigant, Jacques le – Strasbourg: Lorenz & Schouler 1779 – (= ser Whsb) – 1mf – 9 – €20.00 – mf#Hu 108 – gw Fischer [490]

Elemens de l'architecture navale : ou traite pratique de la construction des vaisseaux / Duhamel du Monceau, Henri Louis – Paris: Chez Charles Antoine Jombert, 1752 [mf ed 1983] – 7mf – 9 – mf#SEM105P291 – cn Bibl Nat [623]

Elemens de musique theorique et pratique : suivant les principes de m rameau / Alembert, Jean Le Rond d' – nouv ed. A Lyon: Chez Jean-Marie Bruyset...1762 [mf ed 19–] – 6mf – 9 – mf#fiche325 – us Sibley [780]

Elementa linguae chaldaicae : quibus accedit series patriarcharum chaldaeorum / Guriel, Joseph – Romae: S Congregationis de Propaganda Fide, 1860 – 1mf – 9 – 0-524-07652-9 – mf#1992-1093 – us ATLA [470]

Elementa linguae daco-romanae sive valachicae : emendata, faciltiata, et in meliorem ordinem redacta / Sincai, Gheorghe – Budae: Typis Regiae Universitatis Pestanae 1805 – (= ser Whsb) – 2mf – 9 – €30.00 – mf#Hu 193 – gw Fischer [450]

Elementa musica : of niew licht tot het welverstaan van de musiec en de bascontinuo / Blankenburg, Quirinus van – 's Gravenhage: L Berkoske 1739 [mf ed 19–] – 2pt in 1v on 6mf – 9 – mf#fiche 365 – us Sibley [780]

Elementarlehre der syrischen sprache = uhlemann's syriac grammar / Uhlemann, Friedrich – 2nd ed. New York: D Appleton, 1875, c1855 – 2mf – 9 – 0-8370-7746-X – (grammar and exercises in english; readings in syriac) – mf#1986-1746 – us ATLA [470]

Elementary advice to the body-politic, on the subject of taxation : by a state physician, who can administer to a mind diseased – London: John Hatchard & Son, 1823 – (= ser 19th c economics) – 1mf – 1 – mf#1.1.283 – uk Chadwyck [336]

Elementary agriculture and nature study / Brittain, John – Toronto: Educational Book Co, [909?] – 4mf – 9 – 0-665-98104-X – (with suppl chaps on: "the physics of some common tools" by carleton j lynde; "fruit growing in new brunswick" by w w hubbard and "common weeds of new brunswick" by d wiley hamilton) – mf#98104 – cn Canadiana [630]

Elementary anatomy and physiology, first aid, elementary hygiene : a preliminary handbook for nurses in mental hospitals and mental defective institutions = Elementêre anatomie en fisiologie, noodhulp by ongelukke, elementere gesondheidsleer / South Africa. Department of Health [Departement van Gesondheid] – rev ed. [Pretoria]: Govt Printer 1955 [mf ed Pretoria, RSA: State Library [199-]] – 203p [ill] on 1r with other items – 5 – (also available in afrikaans) – mf#r26 – us CRL [362]

An elementary and practical grammar of the galla or oromo language / Hodson, Arnold Wienholt & Walker, Craven H – London: Society fro Promoting Christian Knowledge, 1922 – 1 – us CRL [490]

Elementary art : or, the use of the lead pencil / Harding, James Duffield – London 1834 – (= ser 19th c art & architecture) – 4mf – 9 – mf#4.2.399 – uk Chadwyck [700]

Elementary chemistry for high schools / Evans, Nevil Norton – Toronto: Educational Book Co, c1914 [mf ed 1998] – 3mf – 9 – 0-665-99132-0 – mf#99132 – cn Canadiana [540]

An elementary course of biblical theology = Doctrinae christianae par theoretica / Storr, Gottlob Christian – 2nd ed. Andover: Gould & Newman, New York: Griffin, Wilcox, 1836 [mf ed 1988] – 2mf – 9 – 0-7905-0160-0 – (in english & greek trans by samuel simon schmucker; trans by karl christian flatt, 1803 under title: lehrbuch der christlichen dogmatik; schmucker trans this into english, 1826 in 2v, the 2nd ed is condensed fr the 1st; incl bibl ref & ind) – mf#1987-0160 – us ATLA [225]

Elementary decoration...of dwelling-houses / Facey, James William – London 1882 – (= ser 19th c art & architecture) – 2mf – 9 – mf#4.2.55 – uk Chadwyck [740]

Elementary electronics – New York. 1973-1981 (1) 1973-1981 (5) 1973-1981 (9) – (Cont by: Science and electronics) – ISSN: 0013-595X – mf#8371 – us UMI ProQuest [621]

Elementary english – Urbana. 1924-1975 (1) 1969-1975 (5) 1975-1975 (9) – (cont by: language arts) – ISSN: 0013-5968 – mf#916 – us UMI ProQuest [370]

Elementary english grammar / Latham, Robert Gordon – Cambridge, England. 1854 – 1r – us UF Libraries [420]

The elementary forms of the religious life : a study in religious sociology = Formes elementaire de la vie religieuse / Durkheim, Emile – London: G. Allen & Unwin; New York: Macmillan, [1915?] – 2mf – 9 – 0-7905-6287-1 – (incl bibl ref. in english) – mf#1988-2287 – us ATLA [301]

The elementary geography of canada : for the use of schools / Borthwick, John Douglas – Montreal: J B Rolland, 1871 – 1mf – 9 – 0-665-05863-2 – mf#05863 – cn Canadiana [917]

Elementary grammar : with full syllabary and progressive reading book of the assyrian language in the cuneiform type / Sayce, Archibald Henry – London: Samuel Bagster, [1875?] [mf ed 1986] – (= ser Archaic classics) – 1mf – 9 – 0-8370-8470-9 – mf#1986-2470 – us ATLA [470]

Elementary grammar of cibemba / Sims, George W – Ft Rosebury, Zaire. 1959 – 1r – us UF Libraries [470]

Elementary grammar of the thonga-shangaan language / Junod, Henri Alexandre – Lausanne, Switzerland. 1932 – 1r – us UF Libraries [470]

An elementary hebrew grammar : with reading and writing lessons and vocabularies / Green, William Henry – new corr ed. New York: John Wiley; London: Chapman & Hall, 1898, c1871 [mf ed 1986] – 1mf – 9 – 0-8370-9149-7 – (contains hebrew-english & english-hebrew vocabularies) – mf#1986-3149 – us ATLA [470]

Elementary illustrations of the celestial mechanics of laplace / Laplace, Pierre Simon, marquis de – London: J Murray 1832 [mf ed 1998] – 1r [ill] – 1 – (trans by thomas young) – mf#film mas 28229 – us Harvard [520]

Elementary instruction in the art of illuminating and missal painting on vellum / De Lara, D Laurent – 2nd ed. London [1857] – (= ser 19th c art & architecture) – 1mf – 9 – mf#4.2.1733 – uk Chadwyck [740]

Elementary law / Robinson, William Callyhan – Boston: Little, Brown, 1882 – 379p – 1 – mf#LL-1246 – us L of C Photodup [340]

Elementary lessons in english for home and school use / Whitney, William Dwight & Knox-Heath, Nelly Lloyd; ed by MacCabe, John Alexander – Toronto, Winnipeg: W J Gage, 1883? – (= ser W j gage and co's language series) – 3mf – 9 – mf#25759 – cn Canadiana [420]

Elementary messenger – 1920-39 – 1 – us Southern Baptist [242]

Elementary questions : for the use of children – London, England. 1820 – 1r – 1 – us UF Libraries [240]

Elementary school guidance and counseling – Alexandria. 1967-1997 (1) 1971-1997 (5) 1975-1997 (9) – ISSN: 0013-5976 – mf#3176 – us UMI ProQuest [370]

Elementary school journal – Chicago. 1900+ (1) 1966+ (5) 1977+ (9) – ISSN: 0013-5984 – mf#138 – us UMI ProQuest [370]

Elementary tonga grammar : with exercises and key / Collins, B – Lusaka, Zambia. 1958 – 1r – 1 – us UF Libraries [490]

An elementary treatise on algebra / Bridge, Bewick – New York, Montreal: D & J Sadlier, 1876 – 2mf – 9 – 0-665-94527-2 – mf#94527 – cn Canadiana [510]

An elementary treatise on mechanics, pt 1 : statics / Cherriman, John Bradford – Toronto: Copp, Clark, 1870 – 2mf – 9 – mf#32017 – cn Canadiana [621]

An elementary treatise on mechanics, pt 2 : dynamics of a particle / Cherriman, John Bradford – Toronto: Copp, Clark, 1877 – 3mf – 9 – mf#08576 – cn Canadiana [621]

An elementary treatise on the american law of real property / Tiedeman, Christopher Gustavus – St. Louis, Thomas, 1885. 785 p. LL-1025 – 1 – us L of C Photodup [346]

Elementary treatise on the differential calculus founded / Rice, John Minot – New York, NY. 1877 – 1r – us UF Libraries [510]

Elemente : Drei Einakter / Wolf, Friedrich – Ludwigsburg (Wuerttemberg): Chronos Verlag, 1922 – 1r – 1 – us Wisconsin U Libr [820]

Elemente der mathematik = Revue de mathematiques elementaires – Basel. 1992-1996 (1) – ISSN: 0013-6018 – mf#13943 – us UMI ProQuest [510]

Elemente deskriptiver und inferentieller statistik und ihre vorlaeufer / Kostrzewa, Frank – (mf ed 1993) – 1mf – 9 – €37.50 – 3-89349-696-3 – mf#DHS 696 – gw Frankfurter [430]

Elementere anatomie en fisiologie, noodhulp by ongelukke, elementere gesondheidsleer : 'n voorbereidingsleerboek vir verpleegsters in hospitale vir sielsiekes en inrigtings vir swaksinniges = Elementary anatomy and physiology, first aid, elementary hygiene / South Africa. Department of Health [Departement van Gesondheid] – hersiene druk. [Pretoria]: Dept van Gesondheid 1956 [mf ed Pretoria, RSA: State Library [199-]] – 206p [ill] on 1r with other items – 5 – (also available in english) – mf#op 00654 r26 – us CRL [610]

Elementi di lingua etrusca / Pallottino, Massimo – Firenze, Italy. 1936 – 1r – us UF Libraries [440]

Elementi grammaticali del caldeo biblico e del dialetto talmudico babilonese = Grammar of the biblical chaldaic language and the talmud babli idioms / Luzzatto, Samuel David – New York: John Wiley, 1876 – 1mf – 9 – 0-8370-7083-X – (in english) – mf#1986-1083 – us ATLA [470]

Elementi grammaticali del caldeo biblico e del dialetto talmudico babilonese / Luzzatto, Samuel David – Padova: A Bianchi, 1865 – 1mf – 9 – 0-8370-9167-5 – mf#1986-3167 – us ATLA [470]

Elementi scientifici di etica civile e diritto / Augias, Carlo – Ancona: tip del commercio, 1878 – 275p – 1 – mf#LL-4027 – us L of C Photodup [340]

Elementi teorico-practici de musica by francesco galeazzi : an annotated english translation and study of vol 1. presented by angelo frascarelli – U of Rochester 1968 [mf ed 19–] – 12mf – 9 – mf#fiche 201 – us Sibley [780]

Elemento afronegroide en el espanol de puerto rico / Alvarez Nazario, Manuel – San Juan, Puerto Rico. 1961 – 1r – 1 – us UF Libraries [972]

Elemento italiano na formacao do brazil / Pettinati, Francesco – Sao Paulo, Brazil. 1939 – 1r – us UF Libraries [972]

O elemento negro: historia, folklore, linguistica / Ribeiro, Joao – Rio de Janeiro, 193-. 237p – 1 – us Wisconsin U Libr [305]

Elementos de aritmetica / Garcia, Juan Justo – 1782 – 9 – sp Bibl Santa Ana [510]

Elementos de aritmetica, algebra y geometria / Garcia, Juan Justo – Madrid: Joaquin Ibarra, 1782 – 1 – sp Bibl Santa Ana [510]

Elementos de aritmetica, tomo 1 / Garcia, Juan Justo – 1801 – 9 – (tomo 1 1801. tomo 1 1821. tomo 2 1822) – sp Bibl Santa Ana [510]

Elementos de derecho administrativo con aplicacion / Troncoso De La Concha, Manuel De Jesus – Ciudad Trujillo, Dominican Republic. 1938 – 1r – us UF Libraries [350]

Elementos de derecho civil y penal de costa rica / Jimenez, Salvador – San Jose, Costa Rica. v1-2. 1876 – 1r – us UF Libraries [350]

Elementos de derecho electoral : desarrollados conforme al programa y explicaciones del profesor titular de la asignatura, en la univerisdad de la habana / Lancis y Sanchez, Antonio – Habana Publicaciones Universitarias, 1954 – 164p – 1 – mf#LL-8024 – us L of C Photodup [325]

Elementos de filosofia / Prisco, Jose – 1884 – 2v – 9 – (trans by gabino tejado) – sp Bibl Santa Ana [190]

Elementos de filosofia moral / Romero de Castillay Perosso, Francisco – 1893 – 9 – sp Bibl Santa Ana [170]

Elementos de folk-lore musical brasiliero / Vale, Flausino Rodrigues – Sao Paulo, Brazil. 1936 – 1r – 1 – us UF Libraries [780]

Elementos de geografia de cuba / Marrero, Levi – Habana, Cuba. 1946 – 1r – 1 – us UF Libraries [918]

Elementos de geografia general y de colombia / Sanchez Eusse, Hernando – Medellin, Colombia. 1965 – 1r – 1 – us UF Libraries [918]

Elementos de geometria y fisica experimental... / Ameller, C – Cadiz, 1788 – 5mf – 9 – sp Cultura [510]

Elementos de gramatica / Servan, Juan – 1882 – 9 – sp Bibl Santa Ana [440]

Elementos de gramatica de la lengua keshua / Berrios, Jose David – La Paz, Bolivia. 1919 – 1r – us UF Libraries [490]

Elementos de gramatica quioca / Santos, Eduardo Dos – Lisboa, Portugal. 1962 – 1r – us UF Libraries [440]

Elementos de grammatica tetense : lingua chi-nyungue / Courtois, Victor Joseph – Nova ed. Coimbra: Impr. da Universidade, 1899, cover 1900. Chicago: Dep of Photodup, U of Chicago Lib, 1972 (1r) Evanston: American Theol Lib Assoc, 1984 (1r) – 1 – 0-8370-0097-1 – mf#1984-B304 – us ATLA [470]

Elementos de historia de costa rica / Montero Barrantes, Francisco – San Jose, Costa Rica. v1-2. 1892-94 – 1r – 1 – us UF Libraries [972]

Elementos de historia de honduras / Salgado, Felix – Tegucigalpa, Mexico. 1945 – 1r – us UF Libraries [972]

Elementos de historia universal / Villanueva y Canedo, Luis – 1845 – 9 – (tomo 2 1846) – sp Bibl Santa Ana [900]

Elementos de logica / Romero de Castilla, Tomas – 1886 – 9 – sp Bibl Santa Ana [160]

Elementos de matematicas o...introduccion a la fisica experimental / Cibat, A – Barcelona, SA – 4mf – 9 – sp Cultura [510]

Elementos de psicologia / Perez Enciso, Guillermo – Caracas, Venezuela. 1955 – 1r – 1 – us UF Libraries [150]

Elementos de psicologia / Romero de Castilla, Tomas – 1876 – 9 – sp Bibl Santa Ana [150]

Elementos de religion. la doctrina de nuestro senor jesucristo / Garcia Garcia, Casimiro – Caceres: tip extremadura, 1940 – 1 – sp Bibl Santa Ana [240]

Elementos de trigonometria rectilinea / Luna y Gomez, Sergio – Sevilla: Est tip angel saavedra, 1908 – 1 – sp Bibl Santa Ana [510]

Elementos de...logica / Garcia, Juan Justo – 1821 – 9 – sp Bibl Santa Ana [160]

Elementos geograficos en la economia cubana / Marrereo Y Artiles, Levi – Habana, Cuba. 1949 – 1r – 1 – us UF Libraries [330]

Elementos griegos y latinos que entran en la composicion de numerosos tecnicismos espanoles, franceses e ingleses : madrid, 1929 / Bayle, Constantino & Ramos Yebes, Jose M – Madrid: Razon y Fe, 1930 – 1 – sp Bibl Santa Ana [400]

Elementos para um diccionario chorographico da provincia... / Lapa, Joaquim Jose – Lisboa, Portugal. 1889 – 1r – 1 – us UF Libraries [960]

Elementos preliminares para poder formar un systema de gobierno del hospicio general / Anzano, T – Madrid, 1778 – 4mf – 9 – sp Cultura [360]

Elementos terrestres / Odio, Eunice – Guatemala, 1948 – 1r – 1 – us UF Libraries [550]

Elementos...gramatica castellana / Lemus y Rubio, Pedro – 1897 – 9 – sp Bibl Santa Ana [440]

Elements – Washington. 1974-1979 (1) 1974-1979 (5) 1979-1979 (9) – mf#10460 – us UMI ProQuest [320]

Elements d'archeologie chretienne / Marucchi, Orazio – Paris: Desclee, Lefebvre, 1899-1902 – 4mf – 9 – 0-524-00643-1 – (Incl bibl ref) – mf#1990-0143 – us ATLA [930]

Elements d'archeologie chretienne / Reusens, Edmond – 2e rev et considerablement augm ed. Aix-la-Chapelle: Rudolf Barth, 1885-1886 – 3mf – 9 – 0-524-03419-2 – mf#1990-0973 – us ATLA [930]

Elements d'archéologie nationale, procedes d'une histoire de l'art monumental chez les anciens / Batissier, L – Paris, 1843 – 7mf – 9 – mf#OA-133 – ne IDC [720]

Elements de botanique et de physiologie vegetale : suivis d'une petite flore simple et facile pour aider a decouvrir les noms des plantes les plus communes du canada / Brunet, Ovide – Quebec: P. Delisle, 1870 – 2mf – 9 – (incl ind) – mf#00299 – cn Canadiana [580]

Elements de droit constitutionnel / Dorsainvil, Jean Baptiste – Paris, France. 1912 – 1r – 1 – us UF Libraries [342]

Elements de grammaire bega / Meeussen, A E – Tervuren, Belgium. 1960 – 1r – 1 – us UF Libraries [490]

Elements de la grammaire de la langue romane, avant l'an 1000 : precedes de recherches sur l'origine et la formation de cette langue / Raynouard, Francois-Just-Marie – Paris: Didot 1816 – (= ser Whsb) – 2mf – 9 – €30.00 – mf#Hu 068 – gw Fischer [410]

Elements de la grammaire francaise / Lhomond, C F – nouv augm ed. Montreal: E R Fabre, 1844 [mf ed 1985] – 1mf – 9 – 0-665-01726-X – (with app) – mf#01726 – cn Canadiana [440]

Elements de la grammaire turke : a l'usage des eleves de l'ecole royale et speciale des langues orientales vivantes / Jaubert, Pierre Amedee Emilien Probe – Paris: Impr royale 1823 – (= ser Whsb) – 3mf – 9 – €40.00 – mf#Hu 217 – gw Fischer [460]

Elements de langue peule du nord-cameroun / Dauzats, Andre – 2e ed. Albi, France: Impr Albigeoise, 1944 – 1 – us CRL [470]

Elements d'instruction morale et civique : degres moyen et superieur: la famille et l'ecole, la societe et la patrie, la nature humaine et la morale, la societe politique / Compayre, Gabriel – nouv ed. Paris: Librairie Classique Paul Delaplane, [ca 1882] – 1mf – 9 – 0-8370-7687-0 – mf#1986-1687 – us ATLA [370]

Elements in baptist development : a study of denominational contributions to national life, christian ideals and world movements / King, Henry Melville et al; ed by Boone, Ilsley – Boston: Backus Historical Society, 1913 – 1mf – 9 – 0-7905-5690-1 – mf#1988-1690 – us ATLA [242]

Elements in luvale beliefs and rituals / White, C M N – (= ser Institute for social research, university of zambia. papers 32) – 3mf – 7 – mf#4734 – uk Microform Academic [306]

Elements in luvale beliefs and rituals / White, C M N – Manchester, England. 1969 – 1r – 1 – us UF Libraries [306]

Elements of anatomy : designed for the use of students in the fine arts / Sharpe, James Birch – London 1818 – (= ser 19th c art & architecture) – 1mf – 9 – mf#4.2.1663 – uk Chadwyck [700]

Elements of architectural criticism for the use of students / Gwilt, Joseph – London 1837 – (= ser 19th c art & architecture) – 2mf – 9 – mf#4.1.15 – uk Chadwyck [720]

Elements of architectural criticism for the use of students, amateurs, and reviewers / Gwilt, Joseph – London: John Williams, 1837 – (= ser 19th c visual arts & architecture) – 2mf – 9 – mf#4.1.15 – uk Chadwyck [720]

Elements of art, a poem : in six cantos / Shee, Martin Archer – London 1809 – (= ser 19th c art & architecture) – 5mf – 9 – mf#4.2.1066 – uk Chadwyck [700]

Elements of buddhist iconography / Coomaraswamy, Ananda Kentish – Cambridge, Massachusetts: Harvard University Press, 1935 – (= ser Samp: indian books) – us CRL [700]

The elements of business law / Huffcut, Wilson; ed by Bogert, George Gleason – Boston: Ginn & Co, 1917 – 4mf – 9 – $6.00 – mf#LLMC 94-270 – us LLMC [346]

The elements of case management / Schwarzer, William – Washington: FJC, 1991 – 1mf – 9 – $1.50 – mf#LLMC 95-384 – us LLMC [340]

Elements of criticism / Kames, Henry Jones – New York, NY. 1855 – 1r – us UF Libraries [410]

Elements of divine truth : a series of lectures on christian theology to sabbath-school teachers / Symington, Andrew – Edinburgh: Johnstone & Hunter, 1854 [mf ed 1986] – 2mf – 9 – 0-8370-6527-5 – (incl ind) – mf#1986-0527 – us ATLA [242]

Elements of divinity : or, a concise and comprehensive view of bible theology / Ralston, Thomas Neely; ed by Summers, Thomas Osmond – Nashville, Tenn: AH Redford for the ME Church, South, 1878 [mf ed 1991] – 3mf – 9 – 0-7905-9447-1 – (1st printed 1854) – mf#1989-2672 – us ATLA [240]

Elements of divinity : A Series of Lectures on Biblical Science, Theology, Church History, and Homiletics / Smith, George – Nashville, Tenn: Southern Methodist Pub House, 1881 – 2mf – 9 – 0-524-00135-9 – mf#1989-2835 – us ATLA [240]

Elements of ethics / Davis, Noah Knowles – New York: Silver, Burdett, c1907 – 1mf – 9 – 0-8370-6106-7 – (incl bibl ref and ind) – mf#1986-0106 – us ATLA [170]

The elements of ethics : an introduction to moral philosophy / Muirhead, John Henry – New York: Charles Scribner, 1892 – 1mf – 9 – 0-8370-6224-1 – mf#1986-0224 – us ATLA [170]

Elements of gaelic grammar / Gillies, Hugh Cameron – London, England. 1902 – 1r – 1 – us UF Libraries [490]

Elements of general and christian theology / Townsend, Luther Tracy – New York: Nelson & Phillips; Cincinnatti: Hitchcock & Walden, 1879 – 1mf – 9 – 0-8370-5656-X – mf#1985-3656 – us ATLA [240]

Elements of handicraft and design / Benson, William Arthur Smith – London 1893 – (= ser 19th c art & architecture) – 2mf – 9 – mf#4.2.32 – uk Chadwyck [740]

Elements of hebrew syntax by an inductive method / Harper, William Rainey – New York: Charles Scribner, 1888 – 1mf – 9 – 0-8370-9241-8 – (incl ind of hebrew words) – mf#1986-3241 – us ATLA [470]

Elements of hindi and braj bhakha grammar / Ballantyne, James Robert – 2nd ed. London: Treubner, 1868 [mf ed 1995] – (= ser Yale coll) – 38p – 1 – 0-524-09050-5 – mf#1995-0050 – us ATLA [490]

Elements of hindu culture and sanskrit civilization / Acharya, Prasanna Kumar – Lahore: Mehar Chand Lacchman Das, 1939 – (= ser Samp: indian books) – us CRL [280]

Elements of human genetics : professional educational material / South Africa. Genetics Division of the Department of Health – Pretoria: Dept of Health 1975 [mf ed Pretoria, RSA: State Library [199-]] – 128p [ill] on 1r with other items – 5 – mf#op 07142 r23 – us CRL [575]

The elements of international law : with an account of its origin, sources and historical development / Davis, George B – N.Y./London: Harper & Bros, 1900 – 7mf – 9 – $10.50 – mf#LLMC 92-188 – us LLMC [341]

Elements of jurisprudence – London: T Payne & Son, 1783 – 2mf – 9 – $3.00 – mf#LLMC 95-170 – us LLMC [340]

The elements of jurisprudence / Holland, Thomas E – 6th ed. Oxford: Clarendon Press, 1893 – 5mf – 9 – $7.50 – mf#LLMC 95-177 – us LLMC [340]

The elements of jurisprudence / Holland, Thomas E – 9th ed. N.Y./London: Oxford Univ Pr, Am.Branch/Henry Frowde, 1900 – 5mf – 9 – $7.50 – mf#LLMC 95-176 – us LLMC [340]

The elements of law: being a comprehensive summary of american civil jurisprudence / Hilliard, Francis – Boston: Hilliard, Gray, 1835. 345p. LL-1077 – 1 – us L of C Photodup [346]

Elements of life insurance / Dawson, Miles M – 3rd ed. Chicago, New York: The Spectator Co, 1911 – 2mf – 9 – $3.00 – mf#LLMC 92-177 – us LLMC [368]

Elements of logic = Logica / Balmes, Jaime Luciano – New York: P O'Shea, 1873 – 1mf – 9 – 0-7905-9130-8 – (In English) – mf#1989-2355 – us ATLA [160]

The elements of mercantile law / Parsons, Theophilus – 2d ed. Boston: Little, Brown, 1862. 684p. LL-1007 – 1 – us L of C Photodup [346]

The elements of moral science / Dagg, John Leadley – New York: Sheldon, 1860 – 1mf – 9 – 0-7905-8776-9 – mf#1989-2001 – us ATLA [170]

The elements of moral science / Wayland, Francis – rev and improved ed. New York: Sheldon, c1865 – 1mf – 9 – 0-524-08852-7 – mf#1993-2137 – us ATLA [170]

Elements of nyanja for english-speaking students / Price, Thomas – Blantyre, Malawi. 1941, 1943, 1953 – 3r – 1 – us UF Libraries [470]

The elements of pain and conflict in human life considered from a christian point of view / Sorley, William Ritchie et al – Cambridge: University Press, 1916 – 1mf – 9 – 0-7905-9670-9 – mf#1989-1395 – us ATLA [210]

Elements of physiophilosophy / Oken, Lorenz – London, 1847 – (= ser 19th c evolution & creation) – 8mf – 9 – (fr german by alfred tulk) – mf#1.1.10897 – uk Chadwyck [573]

The elements of picturesque scenery : or studies of nature made in travel / Twining, Henry – London 1853-65 – (= ser 19th c art & architecture) – 10mf – 9 – mf#4.2.1595 – uk Chadwyck [700]

Elements of political economy : or, how individuals and a country become rich / Ryerson, Egerton – Toronto: Copp, Clark, 1877 – 2mf – 9 – mf#12795 – cn Canadiana [330]

Elements of politics / Sidgwick, Henry – London, England. 1919 – 1r – 1 – us UF Libraries [320]

Elements of practical radio mechanics / Marshall, Samuel Louis – New York, NY. 1945 – 1r – us UF Libraries [621]

Elements of prophecy / Kelly, William – London: G Morrish, 1876 – 1mf – 9 – 0-7905-1209-2 – mf#1987-1209 – us ATLA [220]

Elements of religion / Jacobs, Henry Eyster – Philadelphia: Board of Publication of the General Council of the Evangelical Lutheran Church in North America, 1913, c1898 – 1mf – 9 – 0-7905-7347-4 – mf#1989-0572 – us ATLA [240]

Elements of religious pedagogy *see* – Tsung chiao chiao hsueh fa ta kang

The elements of remembrance and celebration in christian worship and education / Oswalt, Lynn T – 1981 – 1 – 8.56 – us Southern Baptist [242]

The elements of rhetoric / De Mille, James – New York: Harper, 1878 – 7mf – 9 – (incl ind) – mf#06019 – cn Canadiana [420]

Elements of right and of the law : to which is added an historical and critical essay upon the several theories of jurisprudence / Smith, George H – 2nd ed. Chicago: Callaghan, 1887 – 5mf – 9 – $7.50 – mf#LLMC 95-210 – us LLMC [340]

Elements of shona (zezuru dialect) / Fortune, George – London, England. 1957 – 1r – 1 – us UF Libraries [470]

Elements of southern sotho / Paroz, R A – Morija, Zimbabwe. 1946 – 1r – 1 – us UF Libraries [470]

Elements of structures / Hool, George Albert – New York, NY. 1912 – 1r – 1 – us UF Libraries [500]

Elements of syriac grammar : by an inductive method / Wilson, Robert Dick – New York: Charles Scribner, 1891 – 1mf – 9 – 0-8370-7676-5 – (incl ind) – mf#1986-1676 – us ATLA [470]

[Elements of telugu grammar] / Papayya Sastri, B – Anakapalle: SVRV Press, 1906 – 1 – us CRL [490]

The elements of texas pleading / Roberts, Oran Milo – Austin: Jones, 1890. 83 2p. LL-1397 – 1 – us L of C Photodup [340]

The elements of the christian religion / Blomgren, Carl August – Rock Island, IL: Augustana Book Concern, c1907 – 1mf – 9 – 0-524-06384-2 – (incl bibl ref) – mf#1991-2506 – us ATLA [240]

The elements of the gospel harmony : with a catena on inspiration, from the writings of the ante-nicene fathers / Westcott, Brooke Foss – Cambridge: Macmillan, 1851 – 1mf – 9 – 0-7905-2209-8 – mf#1987-2209 – us ATLA [220]

Elements of the law of bailments and carriers : including pledge and pawn and innkeepers / Van Zile, Philip Taylor – Chicago: Callaghan, 1908 – 856p – 1 – mf#LL-1510 – us L of C Photodup [340]

Elements of the law of damages / Sedgwick, Arthur George – 2d ed. Boston: Little, Brown, 1909 – 368p – 1 – mf#LL-1017 – us L of C Photodup [346]

Elements of the law of negotiable contracts / Johnson, Elias Finley – Ann Arbor, MI: Wahr, 1898 – 707p – 1 – mf#LL-435 – us L of C Photodup [346]

Elements of the law of partnership / Mechem, Floyd Russell – 2nd ed. Chicago: Callaghan, 1920 – 501p – 1 – mf#LL-645 – us L of C Photodup [346]

Elements of the science of religion = Inleiding tot de godsdienst wetenschap / Tiele, Cornelis Petrus – Edinburgh: William Blackwood, 1897-1899 – (= ser Gifford lectures) – 2mf – 9 – 0-524-01516-3 – (In English) – mf#1990-2492 – us ATLA [200]

Elements of theology : or, an exposition of the divine origin, doctrines, morals and institutions of christianity / Lee, Luther – 4th ed. Syracuse, NY: Wesleyan Book Room, 1865 [mf ed 1992] – (= ser Methodist coll) – 2mf – 9 – 0-524-04436-8 – mf#1991-2101 – us ATLA [240]

Elements of theology natural and revealed / Fairchild, James Harris – Oberlin, O[hio]: Edward J Goodrich, c1892 – 1mf – 9 – 0-8370-3793-X – (incl ind) – mf#1985-1793 – us ATLA [210]

The elements of theosophy / Edger, Lilian – London: Theosophical Pub Society, 1907 – 1mf – 9 – 0-524-01279-2 – mf#1990-2315 – us ATLA [100]

The elements of torts / Cooley, Thomas McIntyre – Chicago, Callaghan, 1895. 335 p. LL-608 – 1 – us L of C Photodup [340]

Elements of Unity / Dow, William – Edinburgh, Scotland. 1865 – 1r – us UF Libraries [240]

ELEMENTS

Elements pour une etude du systeme adverbial du francais contemporain / Lenepveu, Veronique – 1mf – 9 – mf#10295 – fr Atelier National [440]
Elenchus vegetabilium et animalium... / Kramer, W H – Nashville. 1969-1972 (1) – 5mf – 9 – mf#10576 – ne IDC [590]
Elencos y discursos academicos / Luz Y Caballero, Jose De La – Habana, Cuba. 1950 – 1r – us UF Libraries [370]
Das elend der aufklaerung : ueber ein dilemma in deutschland / Vormweg, Heinrich – Darmstadt: Luchterhand, c1984 [mf ed 1993] – (= ser Sammlung luchterhand 524) – 132p – 1 – (incl bibl ref) – mf#8262 – us Wisconsin U Libr [080]
Das elend des polyphem : zum thema der subjektivitaet bei thomas bernhard, peter handke, wolfgang koeppen und botho strauss / Hofe, Gerhard vom & Pfaff, Peter – Koenigstein: Athenaeum, 1980 [mf ed 1993] – 137p – 1 – (incl bibl ref) – mf#8272 – us Wisconsin U Libr [430]
Elend und groesse unserer tage : anekdoten, 1933-1947 / Weiskopf, Franz Carl – Berlin: Dietz, 1950 – 1r – 1 – us Wisconsin U Libr [880]
Elenev, IU A *see*
– Eniseiskii vestnik
– Rodina [cheliabinsk: 1917-1919]
– Utro sibiri
Elephant – New York. 1848-1848 – 1 – mf#3728 – us UMI ProQuest [073]
Die elephantiner papyri und die buecher esra-nehemja : mit einem supplement zu meiner erklaerung der hebraeischen eigennamen / Jahn, Gustav – Leiden: E J Brill, 1913 – 1mf – 9 – 0-7905-2112-1 – mf#1987-2112 – us ATLA [930]
Eler, Andre-Frederic *see*
– Trois quatuors pour flute, violon, alto et basse, oeuvre 7
– Trois trios pour deux violons et violoncelle
Elert, Werner *see*
– Jacob boehmes deutsches christentum
– Die voluntaristische mystik jacob boehmes
Elet es irodalom – Budapest, 1979-1982 – 4r – 1 – gw Mikropress [949]
Elets. sovet rk i kd *see* Izvestiia eletskogo soveta rabochikh, soldatskikh i krest'ianskikh deputatov
L'elettricista : rivista mensile di elettrotecnica – Rome, Italy 1 jan 1902-15 dec 1909 – 1 – uk British Libr Newspaper [072]
Das eleusische fest urspruenglich identisch mit dem laubhuettenfest der juden / Haury, Jakob – Muenchen: J Lindauer, 1914 – 1mf – 9 – 0-524-01551-1 – (incl bibl ref) – mf#1990-2505 – us ATLA [250]
Eleutheria – Athens. Greece. -d. 19 Nov 1944-Dec 1965. (1944-46 very imperfect). (76 reels) – 1 – uk British Libr Newspaper [949]
Eleutheron bema – Athens, Greece. 1 jan-31 dec 1928; 1 jan 1938-22 april 1941 [daily] – 18r – 1 – (lacking jul 1939) – uk British Libr Newspaper [074]
Eleva gazette – Eleva WI. 1921 jul 14-1922 nov30, 1922 dec 7-1923 aug 2 – 2r – 1 – mf#962670 – us WHS [071]
L'elevage du cheval en canada / Duchene, John D – Quebec: impr Darveau, Jos. Beauchamp...1901 [mf ed 1985] – 2mf – 9 – mf#SEM105P471 – cn Bibl Nat [636]
Elevage in afrique occidentale francaise / Doutressoulle, Georges – Paris, France. 1947 – 1r – 1 – us UF Libraries [440]
Elevations a dieu : ou, ecole de l'amour divin / Carafa, Vincent – Paris: Regis Ruffet, 1863 – 1mf – 9 – 0-8370-7451-7 – mf#1986-1451 – us ATLA [240]
Elevations et motets a 2. et 3. voix, et a voix seule, deux dessus de violon, ou deux flutes : avec la basse-continue / Brossard, Sebastien de – Paris: Christophe Ballard 1699 [mf ed 19–] – 2mf – 9 – mf#fiche 979 – us Sibley [780]
Elevations poetiques / Burque, Francois-Xavier – Quebec: J-P Garneau de la Librairie Garneau: Impr Ernest Tremblay. 2v [1923?] (mf ed 1992) – 7mf – 9 – mf#SEM105P1511 – cn Bibl Nat [440]
[Eleven letters on free trade vs protection which appeared in the canadian illustrated news] / Dewart, William – [S.l: s.n, 1875?] – 1mf – 9 – 0-665-29885-4 – mf#29885 – cn Canadiana [380]
Eleven plates representing works of indian sculpture : chiefly in english collections – London: Probsthain & Co, [1911] – (= ser Samp: indian books) – us CRL [730]
Eleven points river, shannon county / Baptist Associations. Missouri – 1973-87, 1981-87 – 596p – 1 – us Southern Baptist [242]
Eleven years in central south africa / Thomas, Thomas Morgan – Bulawayo, Zimbabwe. 1970 – 1r – 1 – us UF Libraries [960]
Eleven years in ceylon : comprising sketches of the field sports and natural history of that colony, and an account of its history and antiquities / Forbes, Jonathan – London. 2v. 1840 – (= ser 19th c books on british colonization) – 9mf – 9 – mf#1.1.5708 – uk Chadwyck [954]
Eleventh census of the united states, 1890 / U.S. Bureau of the Census – (= ser 1890 census schedules) – 3r – 1 – mf#M407 – us Nat Archives [317]
Eleventh hour – London, England. 18– – 1r – 1 – us UF Libraries [240]
Eleventh hour emergency bulletin – London, England. mar 1934-17 jul 1935 – 1r – 1 – uk British Libr Newspaper [073]
Eleventh hour messenger / Wesleyan Holiness Association of Churches (US) – 1977 sep-1984 – 1r – 1 – mf#966382 – us WHS [242]
Eleventh-hour laborers : a series of articles from the watchword / Chapell, Frederick Leonard – South Nyack NY: Christian Alliance Pub Co, [1899?] [mf ed 1992] – (= ser Alliance colportage library 1/7; Christian and missionary alliance coll) – 2mf – 9 – 0-524-02246-1 – mf#1990-4253 – us ATLA [240]
Eleves ensemble / Fournier, Narcisse – Paris, France. 1848 – 1r – 1 – us UF Libraries [440]
Elf jahre freimaurer! / Daiber, Albert Ludwig – Stuttgart: Strecker & Schroeder 1906 – 1mf – 9 – mf#vrl-199 – ne IDC [366]
Elf jahre gouverneur in deutsch-suedwestafrika / Leutwein, Theodor – 3. aufl. Berlin: E S Mittler, 1908 – 1 – us CRL [920]
Elf preussische offiziere : novelle / Paulus, Helmut – Dresden: W Heyne, 1941 – 1r – 1 – us Wisconsin U Libr [830]
Elf sendschreiben an den heiligen vater in rom / Wieczorek, Rudolph – New York: Aug W Steinhaus, 1870 [mf ed 1986] – 1mf – 9 – 0-8370-8075-4 – mf#1986-2075 – us ATLA [241]
Elfe, Thomas *see* Account book
Elfering, Melissa *see* Effects of social environment on feeling states and self-efficacy in a group exercise class
Les elfes : ballet-fantastique en trois actes de mm. de saint-georges et mazilier [pseud] / Saint-Georges, Henri – Paris: V Jonas, ed-libraire de l'Opera, 1856 – 1 – mf#*ZBD-*MGTZ pv4-Res – Located: NYPL – us Misc Inst [790]
El'fimov, P P *see* Sibirskii put'
Elf-Lords of Rivendell (Whitewater, WI) *see* G
Elford, Frederic C *see*
– Farm poultry
– Poultry-keeping in town and country
Elfriede : eine erzaehlung / Taylor, George – Leipzig: S Hirzel, 1885 [mf ed 1994] – 371p – 1 – mf#8749 – us Wisconsin U Libr [880]
Elfriede : schauspiel in drei acten / Anzengruber, Ludwig – Wien: L Rosner, 1873 [mf ed 1993] – 41p – 1 – mf#8459 – us Wisconsin U Libr [820]
Elfsborgs lans allehanda – Vanersborg, Sweden. 1984- – 1 – sw Kungliga [078]
Elfsborgs lans annonsblad – Skara, Vanersborg, Sweden. 1886-1984 – 1 – sw Kungliga [078]
Elfsborgs lans tidning – Alingsas, 1892-1978 – 200r – 1 – sw Kungliga [078]
Elfsborgs lans tidning *see* Alingsas tidning
The elgar diaries, letters and manuscripts : from birmingham university library, uk – 1889-1939 – (= ser Twentieth Century Composers) – 17r – 1 – us Primary [780]
Elger, W den *see*
– Zinne-beelden der liefde
– Zinnebeelden der liefde
– Zinne-beelden der liefde
Elgin and morayshire courier – Scotland, jun 1849-dec 1851; jan 1855-dec 1860; jan 1869-jun 1874 [wkly] – 8r – 1 – uk British Libr Newspaper [072]
Elgin courant – Scotland. -w. 1845, 1866, 1869, 1873, 1875-79 – 9r – 1 – uk British Libr Newspaper [072]
Elgin courier – Scotland. -w. 13 July 1827-3 July 1829; April 1845-Dec 1847; Jan-May 1849. I 1 2 reels – 1 – uk British Libr Newspaper [072]
Elgin daily courier – Elgin, IL. 1884-1925 (1) – mf#69621 – us UMI ProQuest [071]
Elgin daily news – Elgin, IL. 1876-1925 (1) – mf#69622 – us UMI ProQuest [071]
Elgin news – Elgin OR: Special Advertising, [wkly] – 1 – us Oregon Lib [071]
Elgin recorder – Elgin OR: Recorder Pub Co, -1980 [wkly] – 1 – (merged with: eastern oregon review (-1980) to form: union county review-recorder (1980-)) – us Oregon Lib [071]
Elgin recorder *see*
– Eastern oregon review
– Union county review-recorder
Elgin register – Elgin, NE: Ernest S Scofield. 2v. v1 n1. dec 24 1903-v2 n17. apr 13 1905 (wkly) [mf ed feb 16-apr 6 1905 (gaps) filmed 1958] – 1r – 1 – (cont by: neligh register) – us NE Hist [071]
The elgin review – Elgin, NE: Ernest C Scofield. v1 n1. jan 1 1897)- (wkly) – 1 – (some irregularities in numbering) – us NE Hist [071]
Elguero, Francisco *see* Museo intelectual. vanguardia. mexico, 1928
Elh – Baltimore. 1934+ (1) 1971+ (5) 1975+ (9) – ISSN: 0013-8304 – mf#2098 – us UMI ProQuest [420]
El-hack : organe de defense des interets musulmans – 2e. Oran. n1-46. 1911-12 [wkly] – (= ser Hack) – 1 – (journal politique) – fr ACRPP [320]
El-Hage Rahmat-Ullah Effendi de Dehli *see* Idh-har-haqq
Elhanan Bet Isaac *see* Tosafot 'al masekkhet 'avodah zarah
Elhorst, Hendrik Jan *see*
– De profetie van amos
– De profetie van micha
Eli : an oratorio / Costa, Michael – Boston: O Ditson [1858?] [mf ed 1993] – 1r – 1 – (words selected & written by william bartholomew; 1st performed at the birmingham musical festival, aug 29th 1855) – mf#pres. film 128 – us Sibley [780]
Eli and sybil jones : their life and work / Jones, Rufus Matthew – Philadelphia: H T Coates, c1889 – 1mf – 9 – 0-7905-5348-1 – mf#1988-1348 – us ATLA [920]
Eli, samuel, and saul : a transition chapter in israelitish history / Salmond, Charles Adamson – Edinburgh: T & T Clark; London: Simpkin, Marshall, Hamilton, Kent, [1904?] – (= ser Bible class primers) – 1mf – 9 – 0-7905-0154-6 – mf#1987-0154 – us ATLA [221]
Eli trembling for the ark of god / Mackenzie, William Bell – London, England. 1866 – 1r – 1 – us UF Libraries [240]
Elia, Paschal d' *see* Chung-kuo tien chu chiao chuan chiao shih (ccm121)
Elia, Silvio Edmundo *see* Problema da lingua brasileira
Eliade, Mircea *see*
– Images et symboles: essais sur le symbolisme magico-religieux
– Le mythe de l'eternel retour
Elias 1, Patriarch of the Nestorians *see* Tvrts mml svry
Elias bar Shinaya *see* A treatise on syriac grammar
Elias, de San Juan Baptista *see* Compendio de las excelencias de la bulla de la sancta cruzada en lengua mexicana
Elias de Tejada, F *see*
– El hegelismo juridico espanol
– El racismo
Elias de tejada, f las doctrinas politicas de la edad media : madrid, 1946 / Iturrios, J – Madrid: Razon y Fe, 1947 – 1 – sp Bibl Santa Ana [946]
Elias De Tejada, Francisco *see* Pensamiento politico de los fundadores de nueva gr...
Elias de Tejada, Francisco *see*
– Las doctrinas politicas de eugenio maria de hostos
– Die geburtsstunde...de heyelte
– El reino de galicia, tomo 1
– Sociologia del africa negra
Elias de Tejada Spinola, Francisco *see*
– Las doctrinas politicas en portugal (edad media)
– Elefantes blancos, paginas doradas. nuevas noticias de indochina
– Encrucijada juridica de la costa de marfil
– La filosofia juridica del profesor de asis garrote
– La filosofia juridica en la espana actual
– Geronimo castillo de bovadilla
– El hegelismo juridico espanol
– Ideas politicas de angel ganivet
– La monarquia tradicional
Elias de tejada spinola, francisco. la tradicion gallega. prologo de r otero pedrayo : madrid, 1944 / Marquez, Gabino – Madrid: Razon y Fe, 1945 – 1 – sp Bibl Santa Ana [946]
Elias Garcia, A *see* As moedas visigodas da lusitania
Elias, jahve, und baal / Gunkel, Hermann – Tuebingen: J C B Mohr 1906, c1905 [mf ed 1986] – (= ser Religionsgeschichtliche volksbuecher fuer die deutsche christliche gegenwart 2/8) – 1mf – 9 – 0-8370-9389-9 – mf#1986-3389 – us ATLA [221]
Elias Perez, Alberto *see* El despido del trabajador (comentarios del decreto de 26 de octubre de 1956)
Elias und die religioesen verhaeltnisse seiner zeit / Sanda, Albert – 1. & 2. aufl. Muenster i W: Aschendorff 1914 [mf ed 1992] – (= ser Biblische zeitfragen 7/1-2) – 1mf – 9 – 0-524-04111-3 – mf#1992-0069 – us ATLA [221]
Eliasberg, Alexander *see* Sagen polnischer juden
Eliasberg, Alexander et al *see* Neue juedische monatshefte
Eliashevich, I Ya *see* S krestom i evangeliem protiv kolkhozov
Eliav, Mordechai *see* Zeh ha-yom
Elie goulet de la societe des ecrivains canadiens : bio-bibliographie analytique / Allen, Marie-B – 1959 [mf ed 1978] – (= ser Bibliographies du cours...1947-66) – 3mf – 9 – (with ind; pref by pere Hilaire de la Perade) – mf#SEM105P4 – cn Bibl Nat [410]
Elie halevy, 1870-1937 / Ecole Libre Des Sciences Politiques (Paris, France) – Paris, France. 1939? – 1r – 1 – us UF Libraries [944]
Elie lescot – New York, NY. 1944 – 1r – 1 – us UF Libraries [972]
Elie, Louis E *see*
– Histoire d'haiti
– President boyer et l'empereur
Eliet, Edouard *see* Langues spontanees, dites commerciales, du congo
Eli'ezer ben-yehudah – Yerushalayim, Israel. 1924 – 1r – 1 – us UF Libraries [939]
Eliezer, Of Beaugency *see* Perush 'al yehezkel ve-tere 'asar
The eligibility of women not a scriptural question – [New York: Hunt & Eaton, 189l] Beltsville, Md: NCR Corp, 1978 (1mf); Evanston: American Theol Lib Assoc, 1984 (1mf) – (= ser Women & the church in america) – 9 – 0-8370-1611-8 – mf#1984-2012 – us ATLA [240]
Eliiezer ben-yehudah / Klausner, Joseph – Tel-Aviv, Israel. 1939 – 1r – 1 – us UF Libraries [939]
Elijah Ben Solomon *see* Sefer minhat eliyahu
Elijah fed by ravens / Sibly, Manoah – London, England. 1796 – 1r – 1 – us UF Libraries [240]
Elijah, his life and times / Milligan, William – New York: Fleming H Revell, [189-?] – 1mf – 9 – 0-8370-9969-2 – (incl bibl ref) – mf#1986-3969 – us ATLA [221]
Elijah, the favored man : a life and its lessons for to-day / Patterson, Robert M – Philadelphia: Presbyterian Board of Publ, c1880 – 1mf – 9 – 0-7905-1558-X – (incl bibl ref) – mf#1987-1558 – us ATLA [221]
Elijah the prophet / Royer, Galen Brown – Elgin, IL: Brethren Pub House, 1905 – 1mf – 9 – 0-524-04236-5 – mf#1990-5027 – us ATLA [221]
Elijah the prophet / Taylor, William Mackergo – New York: Harper, c1875 – 1mf – 9 – 0-7905-1019-7 – (incl bibl ref and ind) – mf#1987-1019 – us ATLA [221]
Elijah wadsworth family papers, 1792-1868 / Wadsworth, Elijah – [mf ed 1980] – 2r – 1 – (incl a 7p guide. correspondence, agreements, & deeds of the wadsworth family, early settlers in the reserve, & military papers of the 4th division, ohio militia, commanded by wadsworth, 1804-1813) – mf#ms2729 – us Western Res [978]
Elim : or, hymns of holy refreshment / ed by Huntington, Frederic Dan – Boston: E P Dutton, 1865 [mf ed 1984] – 4mf – 9 – 0-8370-0817-4 – (incl ind) – mf#1984-4165 – us ATLA [810]
Elim baptist church. macon, georgia : church records – 1843-68, 1926-63 – 1 – us Southern Baptist [242]
Elima – Kinshasa: Essolomwa-Nkoy Ea Linganga, [dec 31 1972/jan 1/2 1973-1977] – 30r – 1 – us CRL [079]
Eliminate administrative discharges in lieu of court martial : guidance for plea agreements in military courts is needed; report to the congress by the comptroller general of the u.s. – Washington: General Accounting Office, Apr 28 1978 (FPCD-77-47) – 1mf – 9 – $1.50 – mf#LLMC 96-071 – us LLMC [355]
Elimu Center for African Arts et al *see* Malu Kai
Eliodoro Valle, Rafael *see* El convento de tepotzotlan...1924
Eliot, Andrew *see* A sermon preached october 25th, 1759
Eliot, Charles *see* Hinduism and buddhism
Eliot, Charles Norton Edgecumbe *see* Turkey in europe, by odysseus pseud
Eliot, Charles William et al *see* The religion of the future, and other essays
Eliot, Charlotte *see* William greenleaf eliot
Eliot, George *see*
– Adam bede
– Felix Holt
– Impressions of theophrastus such
– Nineteenth century literary manuscripts
– Scenes of clerical life
– Works of george eliot
Eliot, Ida Mitchell et al *see* Caterpillars and their moths
Eliot, John *see*
– Eliot's brief narrative
– The indian primer
Eliot memorial : sketches historical and biographical of the eliot church and society, boston / Thompson, Augustus Charles – Boston: Pilgrim Press c1900 [mf ed 1990] – 2mf – 9 – 0-7905-8158-2 – mf#1988-6105 – us ATLA [240]
Eliot ness papers *see* Ness, eliot, papers, ms 3699
Eliot, Samuel Atkins *see* Social classes in a republic
Eliot, Simon *see* Publishers' circular 1837-1900
Eliot, T S *see* Dante
Eliot, William *see* Parish church of aston-juxta-birmingham
Eliot, William Greenleaf *see*
– A discourse
– Discourses on the unity of god, and other subjects
– Early religious education
– Lectures to young women

Eliot's brief narrative : brief narrative of the progress of the gospel amongst the indians in new-england, in the year 1670 / Eliot, John – [Boston: Directors of the Old South Work, 1896?] – 1mf – 9 – 0-524-04130-X – mf#1990-1200 – us ATLA [240]

Eliovson, Sima *see* South africa

Elisabeth of Romania, Queen *see* Aus carmen sylva's leben

Elisabeth, Queen *see*
- Es klopft
- From memory's shrine
- Jehovah
- Songs of toil

Elisabeth seton / Conan, Laure – Montreal: la Cie de publ de la Revue canadienne, 1903 [mf ed 1988] – 2mf – 9 – mf#SEM105P914 – cn Bibl Nat [241]

Elisabeth seton und das entstehen der katholischen kirche in den vereinigten staaten, pt 2 / Barberey, Helene, Freifrau von – Muenster, 1873 (mf ed 1993) – 2mf – 9 – €31.00 – 3-89349-362-X – mf#DHS-AR 362 – gw Frankfurter [241]

Elisabeth seton und das entstehen der katholischen kirche in den vereinigten staaten, pt1 / Barberey, Helene, Freifrau von – Muenster, 1873 (mf ed 1993) – 2mf – 9 – €31.00 – 3-89349-361-1 – mf#DHS-AR 362 – gw Frankfurter [241]

Elisabeth von Ansbach-Bayreuth, Markgraefin [Elisabeth Craven] *see* Voyage de milady craven a constantinople

Elisabeth von england : schauspiel / Bruckner, Ferdinand – 8. & 9. aufl. Berlin: S Fischer 1932 [mf ed 1991] – 1r – 1 – (filmed with: hinter der maske / armin gimmertahl) – mf#2909p – us Wisconsin U Libr [820]

Elisabeth-de-la-Trinite, soeur *see* Bibliographie analytique de l'oeuvre de monseigneur albert tessier...

Elisabeths opferung : novellen / Ehrler, Hans Heinrich – 2. aufl. Stuttgart: Greiner & Pfeiffer [1924?] [mf ed 1989] – 1r – 1 – (filmed with: menschen und affen / albert ehrenstein) – mf#7207 – us Wisconsin U Libr [830]

Elisavetpol'skij garnizonnyj sovet soldatskikh deputatov *see* Izvestiia soveta soldatskikh deputatov elisavetpol'skogo garnizona

Elischa ben abuja-acher / Back, Samuel – Frankfurt am Main, Germany. 1891 – 1r – 1 – us UF Libraries [939]

Elise ruediger geb von hohenhausen : ein bild ihres lebens und schaffens / Esche, Annelinde – Emsdetten, 1939 (mf ed 1992) – 1mf – 9 – €24.00 – 3-89349-004-3 – mf#DHS-AR 7 – gw Frankfurter [430]

Elise von hohenhausen : eine westfaelische dichterin und uebersetzerin / Hackenberg, Fritz – Muenster, 1913-15 (mf ed 1992) – 1mf – 9 – 3-89349-005-1 – mf#DHS-AR 8 – gw Frankfurter [430]

Elise von hohenhausen, geb von ochs (1789-1857) : zum forschungsstand / Haensel-Hohenhausen, Markus – (mf ed 1992) – 2mf – 9 – €49.00 – 3-89349-530-4 – mf#DHS 530 – gw Frankfurter [430]

Elisha Mitchell Scientific Society, Chapel Hill *see* Nc journal of the elisha mitchell scientific society

Elisha the prophet : A Type of Christ / Edersheim, Alfred – London: William Hunt, 1873 – 1mf – 9 – 0-7905-2406-6 – mf#1987-2406 – us ATLA [221]

Elisha's tribute to the memory of elijah / Graham, John – Ayr, Scotland. 1853 – 1r – 1 – us UF Libraries [221]

Elite athletes in flow : the psychology of optimal sport experience / Jackson, Susan A & Gould, Daniel – 1992 – 3mf – 9 – $12.00 – us Kinesology [150]

Elite directory of vancouver – Vancouver: Thomson Sta[tionery] Co c1908 [mf ed 1995] – 2mf – 9 – 0-665-75463-9 – mf#75463 – cn UBC Preservation [917]

Les elites khmeres – [Phnom-Penh: Universite bouddhique Preah Sihanouk Raj 1965?] [mf ed 1989] – (= ser Culture et civilisation 8) – 1r with other items – 1 – mf#mf-10289 seam reel 009/07 [§] – us CRL [959]

Elites of barotseland, 1878-1969 / Caplan, Gerald L – Berkeley, CA. 1970 – 1r – 1 – us UF Libraries [960]

Elitros / Albis, Victor H – Bogota, Colombia. 1952 – 1r – us UF Libraries [972]

Eliyahu, dan, menasheh – Kuskus-Tiv'on? Israel. 1939? – 1r – 1 – us UF Libraries [939]

Elizabeth *see* In the mountains

Elizabeth 1 and the english parliament / Walter, Beate – (mf ed 1994) – 1mf – 9 – €30.00 – 3-8267-2000-8 – mf#DHS 2000 – gw Frankfurter [941]

Elizabeth aubrey – London, England. 18– – 1r – 1 – us UF Libraries [240]

Elizabeth baptist church. shelby, north carolina : church records – 1910-54. bulletins. 1955-63 – 1 – $47.79 – us Southern Baptist [242]

Elizabeth, Charlotte *see* Israel's ordinances

Elizabeth [city directory : listing] – 1905 – 1r – 1 – mf#3072956 – us WHS [978]

Elizabeth colenso : her work for the melanesian mission / Swabey, Frances Edith – 1956-1959 – 1r – 1 – (incl ind) – mf#pmb560 – at Pacific Mss [240]

Elizabeth fry / Ashby, Irene M – London: Edward Hicks, Jr, 1892 – 1mf – 9 – 0-524-07506-9 – mf#1991-3136 – us ATLA [365]

Elizabeth fry / Lewis, Georgina King – 3rd ed. London: Headley, [1909?] – 1mf – 9 – 0-524-01000-5 – mf#1990-0277 – us ATLA [365]

Elizabeth herald – Elizabeth, PA. 1889-1912 [wkly] – 13 – $25.00r – us IMR [071]

Elizabeth musande wokuthuringen – Gwelo, Zimbabwe. 1959 – 1r – 1 – us UF Libraries [960]

Elizabeth, Queen, consort of Frederick 1, King of Bohemia *see* Briefe der elisabeth stuart, koenigin von boehmen

Elizabeth rudder / Yates, William – London, England. 18– – 1r – 1 – us UF Libraries [240]

Elizabeth von Brandenburg *see* Aus nacht zum licht

Elizabethan bishops and the civil power – London, England. 1897 – 1r – 1 – us UF Libraries [240]

The elizabethan bishops and the civil power : statute 8 eliz. c. 1. (a.d. 1565-6) – London: SPCK 1897 [mf ed 1993] – (= ser Church historical society (series) 22) – 1mf – 9 – 0-524-07197-7 – mf#1990-5355 – us ATLA [242]

The elizabethan clergy and the settlement of religion, 1558-1564 / Gee, Henry – Oxford: Clarendon Press, 1898 – 1mf – 9 – 0-7905-5144-6 – mf#1988-1144 – us ATLA [240]

Elizabethan demonology : An Essay in Illustration of the Belief in the Existence of Devils, and the Powers Possessed by Them, as it was Generally Held during the Period of the Reformation, and the Times Immediately Succeeding, with special Reference to Shakspere and his Works / Spalding, Thomas Alfred – London: Chatto and Windus, 1880 – 1mf – 9 – 0-524-02374-3 – mf#1990-2985 – us ATLA [210]

Elizabethan ireland and the settlement of ulster : the carew papers at lambeth palace library / Lambeth Palace Library – 1574-1616 – 15r – 1 – £720.00 – mf#CAR – uk World [941]

Elizabethan part-song books : from carlisle cathedral library – 2v – 1r – 1 – mf#96028 – uk Microform Academic [780]

The elizabethan prayer-book and ornaments : with an appendix of documents / Gee, Henry – London, New York: Macmillan, 1902 – 1mf – 9 – 0-7905-5145-4 – (incl bibl ref) – mf#1988-1145 – us ATLA [240]

The elizabethan religious settlement : a study of contemporary documents / Birt, Henry Norbert – London: G. Bell, 1907 – 2mf – 9 – 0-7905-5631-6 – (incl bibl ref) – mf#1988-1631 – us ATLA [240]

Elizabethton first baptist church. elizabethton, tennessee : church records – 1842-1943 – 1 – us Southern Baptist [242]

Elizabethtown baptist church. bladen association. north carolina : church records – 1903-24 – 1 – us Southern Baptist [242]

Elizabethtown chronicle – Elizabethtown, PA. -w 1928-1983 – 13 – $25.00r – us IMR [071]

Elizalde Ita y Parra, Jose Mariano Gregorio de *see* Dia festivo proprio para el culto, y rezo del senor san joachin el veinte de marzo de cada ano

Elizondo Arce, Hernan *see* Memorias de un pobre diablo

Elk county gazette – St Mary's, PA. 1873-1915 (1) – mf#66087 – us UMI ProQuest [071]

Elk County Historical Society (PA) *see* Elk horn

Elk creek baptist church. stewart county. cumberland city, tennessee : church records – feb 1961-aug 1967 – 1 – us Southern Baptist [242]

Elk creek baptist church. stewart county. cumberland city, tennessee : church records – feb 1961-aug 1967 – 1 – us Southern Baptist [242]

The elk creek herald – Elk Creek, NE: N H Libby. 16v. 1894-v16 n42. aug 18 1910 (wkly) [mf ed 1895-1910 (gaps)] – 3r – 1 – us NE Hist [071]

Elk democrat – Ridgway, PA. -w 1869-1912 – 13 – $25.00r – us IMR [071]

[Elk grove-] elk grove citizen – CA. 1909-1912; 1947-1965; 1968 – 64r – 1 – $3840.00 (subs $150/y) – mf#B05030 – us Library Micro [071]

Elk horn / Elk County Historical Society (PA) – 1969 win-1982 spr – 1r – 1 – mf#618179 – us WHS [978]

Elkan, Hugo *see* Die gesta innocentii 3. im verhaeltniss zu den regesten desselben papstes

Elkhart Lake Area Chamber of Commerce *see* Depot dispatch

Elkhart lake herald – Elkhart Lake WI. 1906 jul 10 – 1r – 1 – mf#962668 – us WHS [071]

Elkhart lake libelle – Elkhart Lake WI. 1901 aug 17 – 1r – 1 – mf#962667 – us WHS [071]

Elkhorn exchange – Elkhorn, NE: Jeffries & Goodhard, 1891 (wkly) [mf ed v2 n32. nov 25 1892-jul 19 1918 (gaps) filmed 1979] – 2r – 1 – (merged with: waterloo gazette, millard courier and: bennington herald to form: douglas county gazette. publ in waterloo jan 18 1907-. called sub-ed and later assoc newspaper of: waterloo gazette 1901?-jan 5 1934) – us NE Hist [071]

Elkhorn independent – Elkhorn WI. 1855 jun 1-1861 jan 31, 1861-64, 1865-1868 jun 3 – 3r – 1 – (cont: walworth county independent [elkhorn, wi: 1854]; walworth county reporter; geneva express; cont by: walworth county independent [elkhorn, wi: 1868]) – mf#1139561 – us WHS [071]

Elkhorn independent – Elkhorn WI. 1892 feb 4/dec-1999 oct-dec [gaps] – 157r – 1 – (cont: walworth county independent (elkhorn, wi: 1868)) – mf#1139567 – us WHS [071]

Elkhorn liberal – Elkhorn WI. 1875 jul 17-1876 jan 7 – 1r – 1 – mf#962657 – us WHS [071]

Elkhorn pen and plow – Oakdale, NE: K P McCormick and Sarah E Taylor. 7v. v3 n1. apr 17 1879-v7 n27. oct 18 1883 (wkly) [mf ed with gaps] – 3r – 1 – (cont: oakdale pen and plow. cont by: oakdale journal (1883). v5 n10 jun 16 1881-v7 n27 oct 18 1883 called also whole n218-399) – us NE Hist [071]

Elkhorn valley mirror – Norfolk, NE: D McIntosh. -v2 n13. may 12 1927 (wkly) [mf ed v1 n13. may 13 1926)-may 12 1927] – 1r – 1 – (merged with: madison star-mail (1923) to form: madison star-mail and elkhorn valley mirror consolidated) – us NE Hist [071]

Elkhorn valley news – Norfolk, NE: Norton & Sprecher, 1881-88// (wkly) [mf ed 188488 (gaps)] – 1r – 1 – (cont by: norfolk weekly news) – us NE Hist [071]

Elkin, William Lewis *see*
- Transactions of the astronomical observatory of yale university

Elkington & Co *see* The manufactures of elkington and co

Elkins family exchange newsletter – 1982 may-1985 feb – 1r – 1 – mf#1130727 – us WHS [929]

Elkins, Frank *see* The complete ski guide.

Elkins, Stephen Benton *see* Address delivered before the alumni association of the university of the state of missouri

Elkinton, Joseph Scotton *see* Selections from the diary and correspondence of joseph s. elkinton, 1830-1905

Elko baptist church. elko, south carolina : church records – 1896-1971 – 1 – us Southern Baptist [242]

[Elko-] daily argonaut – NV. 1897; 1898-99 [daily] – 2r – 1 – $120.00 – mf#U04493 – us Library Micro [071]

[Elko-] elko daily free press – NV. 1876-77; 1883-90; 1892-1918; 1928- [daily] – 162r – 1 – $9720.00 (subs $300y) – mf#UN04494 – us Library Micro [071]

[Elko-] enterprise – NV. 1916-17 [wkly] – 1r – 1 – $60.00 – mf#U04497 – us Library Micro [071]

[Elko-] independent – NV. 1869-73; 1875-1914; 1923; 1929; 1911-42 (scats) [wkly] – 13r – 1 – $780.00 – mf#U04500 – us Library Micro [071]

[Elko-] nevada silver tidings – NV. 1897; jan-jul 1899 [wkly; biwkly] – 2r – 1 – $120.00 – mf#U04498 – us Library Micro [071]

[Elko-] northeastern nevada historical society quarterly – NV. 1970- – 3r – 1 – $180.00 (subs $50y) – mf#U04838 – us Library Micro [073]

[Elko-] the telegram – MI – 1r – 1 – $110.00 – mf#U04499 – us Library Micro [071]

[Elko-] the weekly tidings – NV. 1898 – 1r – 1 – $60.00 – mf#U04502 – us Library Micro [071]

[Elko-] weekly elko independent – NV. 1869-71, 1975-79, 1887-99, 1908-18, 1924, 1927-79, 1982- [daily] – 78r – 1 – $4680.00 (subs $60y) – (aka: daily independent) – mf#U04496 – us Library Micro [071]

[Elko-] weekly post – NV. 1876-77 – 1r – 1 – $60.00 – mf#U04501 – us Library Micro [071]

Elk's gulch gazette / Benevolent and Protective Order of Elks – v1 n1 [1919 feb 16] – 1r – 1 – mf#655055 – us WHS [071]

Elkton baptist church. elkton, kentucky : church records – 1825-1909 – 1 – us Southern Baptist [242]

Ellbogen, Ismar *see* Aus dem leben der juden deutschlands im mittelalter

Ellenberger, David Frederic *see* Documentary history of the basotho [lesotho]

Ellenberger, Heinrich *see* Tsel ve-or

Ellenberger, J S *see* Legislative history of the securities act of 1933 and securities exchange act of 1934

Ellenberger, Victor *see*
- Century of mission work in basutoland
- Sur les hauts-plateaux du lessouto

[Ellendale-] the ellendale star – NV. 1909 – 1r – 1 – $60.00 – mf#U04839 – us Library Micro [071]

Ellenikos aster = The greek star – Chicago IL. [1915 apr 16-1929 aug 23] – 1r – 1 – mf#872596 – us WHS [071]

Ellerbek, Soren Anton *see* Antung ved jalufloden

Ellerd, Andria *see* Variables related to knowledge levels of aging and planning for future aging of texas high school graduates

Ellerker, R *see* Coleccion de los mas preciosos adelantamientos de la medicina...

Ellermann, Heinrich *see* Das gedicht (mme5)

Ellery queen's mystery magazine – New York. 1973+ (1) 1973+ (5) 1974+ (9) – ISSN: 1054-8122 – mf#8351 – us UMI ProQuest [420]

Ellery, Robert Lewis John *see* Report of the board of visitors to the observatory

Ellesby, James *see* Caution against ill company

Ellesmere, Francis Egerton, Earl of *see* Essays on history, biography, geography, engineering, etc

Ellesmere guardian – 1891-99; 1904-06; apr-dec 1913; 1915-22; 1925-45; jul 1946-apr 1983 – 1 – (1900-03, 1907-mar 1913; 1923-24 unavailable) – mf#70.2 – nz Nat Libr [079]

Ellesmere port news and advertiser – [NW England] Ellesmere Port Lib 1964-69, feb 1969-1975 – 1 – (cont: ellesmere port advertiser) – uk MLA; uk Newsplan [072]

Ellesmere port pioneer – [NW England] Ellesmere Port 9 jan-2 jul 1920, 1921-22, 3 aug-28 dec 1923, 4 jul-24 dec 1924, 1925-25 nov 1927, 1928-59, 6 may 1960 – 1 – uk MLA; uk Newsplan [072]

Ellesmere port weekly news, whitby & sutton reporter – [NW England] Ellesmere Port, Cheshire Record Off 26 mar-2 dec 1909 – 1 – uk MLA; uk Newsplan [072]

Ellet, Elizabeth Fries (Lummis) *see* Women artists in all ages and countries

Ellice, Edward *see* Les communications de mercator

Ellicott, Charles J *see* Critical and grammatical commentary on st paul's epistle to the ephesians

Ellicott, Charles John *see*
- A critical and grammatical commentary on st paul's epistle to the ephesians
- A critical and grammatical commentary on the pastoral epistles
- Epistles of st peter, st john, and st jude
- The gospel according to st luke
- Historical lectures on the life of our lord jesus christ
- Modern unbelief
- A new testament commentary for english readers
- The revisers and the greek text of the new testament
- The second epistle to the corinthians
- The third book of moses

Ellies Du Pin, Lud *see* Opera omnia

Elligen, J *see* The terrible deeds of george l shaftesbury

Elliger, W *see* Die stellung der alten christen zu den bildern in den ersten vier jahrhunderten

Ellil in sumer und akkad / Noetscher, F – Hannover, 1927 – 2mf – 9 – mf#NE-408 – ne IDC [956]

Elling news *see* Esmond leader and elling news

Ellinger, Georg *see*
- Angelus silesius saemtliche poetische werke
- Cherubinischer wandersmann
- Die ehrliche frau nebst harlequins hochzeit- und kindbetterinschmaus. der ehrlichen frau schlampampe krankheit und tod
- Heilige seelenlust
- Philipp melanchthon

Ellingson, Lyndall A *see* Breast self-examination, the health belief model and sexual orientation in women

Ellingson, Susan M *see* The examination of the validity of the tarskij equation for predicting vo2 max in an active female population

Ellingwood, Albert R *see* Departmental cooperation in state government

Ellinor : oder, Traeumen und Erwachen. Phantastisches Ballet in 3 Akten und 6 Bildern. Musik von Hertel / Taglioni, Paul – Berlin: Eigenthum von P Taglioni [1860?] – 1 – mf#*ZBD-*MGTZ pv1-Res – Located: NYPL – us Misc Inst [790]

Ellinwood, DeWitt *see*
- [Papers, 1913?-1924?]
- Papers, 1914-1919
- Papers, 1914-1920
- Papers, 1916-1925
- Papers, 1917?-1957
- Selections from the home political files, 1915-1919 of the national archives of india
- Selections from the home political files july 1914-1916, 1918, and the army department files 1914-may 1919

Ellinwood, Frank Field *see*
- The great conquest
- Oriental religions and christianity
- Questions and phases of modern missions

Ellinwood, Leonard *see*
- Bibliography of american hymnals
- Dictionary of american hymnology

Ellinwood, Mary Gridley *see* Frank field ellinwood
Elliot, Charles Burke *see* An outline of the law of insurance
Elliot, Daniel Giraud *see* The wild fowl of the united states and british possessions
Elliot, Edward *see* Biographical story of the constitution
Elliot, George *see* God is spirit, god is love
Elliot, Henry Miers *see*
- The history of india, as told by its own historians
- Memoirs on the history, folk-lore, and distribution of the races of the north western provinces of india..

Elliot, Hugh Samuel Roger *see* Modern science and the illusions of professor bergson
Elliot, James Rupert *see* The trade relations of the farmers of nova scotia
Elliot, Jonathan *see* The debates in the several state conventions on the adoption of the federal constitution, as recommended by the general convention at philadelphia, in 1787
Elliot, Robert Henry *see*
- The experience of a planter in the jungles of mysore
- Gold, sport, and coffee planting in mysore

Elliot, William *see* The midnight cry
Elliott, Aubrey *see* Magic world of the xhosa
Elliott, Charles *see*
- The bible and slavery
- Christus mediator
- General introduction to the prophetic writings of the old testament
- History of the great secession from the methodist episcopal church in the year 1845
- History of the great secession from the methodist episcopal church in the years 1845
- Indian missionary reminiscences, principally of the wyandot nation
- Lettres from the north of europe
- Sinfulness of american slavery
- South-western methodism
- A treatise on the inspiration of the holy scriptures
- A vindication of the mosaic authorship of the pentateuch

Elliott, Charles John *see* Some strictures on a book entitled the communicant's manual, with two prefaces, by the rev. e. king ...
Elliott, Charles Wyllys *see* The new england history
Elliott Coues *see* Key to north american birds
Elliott, E N *see* Cotton is king, and pro-slavery arguments
Elliott, Edward B *see* Christian's view of the cause and remedy of the present national di...
Elliott, Edward Bishop *see*
- Apocalypsis alfordiana
- Horae apocalypticae

Elliott empire – v1 n1-v2 n4 [1978 mar-1979 dec] – 1r – 1 – mf#639187 – us WHS [071]
Elliott, Ernest Eugene *see*
- Making good in the local church
- The problem of lay leadership

Elliott family quarterly – v1 n1-v3 n4 [1985 mar-1987 dec] – 1r – 1 – mf#1336482 – us WHS [929]
Elliott, George et al *see* Social ministry
Elliott, John Frederick *see* An essay shewing the expediency of emigration, under certain circumstances
Elliott, John H *see* Shall the name be changed?
Elliott, Juliet Georgiana *see* Miss elliott's accounts, 1844-1874
Elliott, Mary (Belson *see* Little lessons for little folks
Elliott, Mary (Belson) *see* The orphan boy
Elliott, Mary (Belson) et al *see* Innocent poetry for infant minds
Elliott, R *see* Views in india, china, and on the shores of the red sea
Elliott, R G *see* Letters
Elliott, Richard *see* Eternal realities considered as forming the principle of missionary
Elliott, Robert *see* Robert elliott's poems
Elliott, Walter *see*
- The life of father hecker
- The life of jesus christ

Elliott, William Allan *see* Notes for a sindebele dictionary and grammar
Elliott-Binns, L E *see* Galilean christianity
Ellis *see* Batavia in post-war days
Ellis, Aaron *see* Bible vs tradition
Ellis, Alfred Burdon *see*
- The ewe-speaking peoples of the slave coast of west africa
- History of the first west india regiment
- The tshi-speaking peoples of the gold coast of west africa

Ellis, Brabazon *see*
- Doctrine of the church of england as contrasted with the church of...
- Dr hook's test of controversy examined

Ellis, Carleton *see* Soilless growth of plants
Ellis, Charles Mayo *see*
- An essay on transcendentalism
- The history of roxbury town

[Ellis, Charles Mayo] *see* An essay on transcendentalism
Ellis County. Kansas *see* Register of marriages, births and deaths in ellis county
Ellis County. Kansas. Big Creek Township. Justice of the Peace *see* Criminal and civil dockets
Ellis, Edward Sylvester *see*
- Among the esquimaux
- Fire, snow and water
- A hunt on snow-shoes
- The last struggle
- Red jacket
- Tecumseh, chief of the shawanoes
- The young gold seekers of the klondike

Ellis, Edwin John *see* Real blake
Ellis, F H *see* Character forming in school
Ellis, G *see* Memoir of a map of the countries comprehended between the black sea and the caspian
Ellis, George Edward *see*
- An address delivered in the first church, salem
- A half-century of the unitarian controversy
- Life and religion of the hindoos
- Memoir of jared sparks, ll.d
- The puritan age and rule in the colony of the massachusetts bay, 1629-1685

Ellis, Griffith Ogden *see*
- Blackstone quizzer b.
- Quizzer no. 2...being questions and answers on criminal law.
- Quizzer no. 10...being questions and answers on bills, notes, and cheques.

Ellis, H *see*
- Chronica
- Journal of the proceedings of the late embassy to china

Ellis, Harriet Warner *see* Our eastern sisters and their missionary helpers
Ellis, Henry *see* Journal of the proceedings of the late embassy to china
Ellis, Howard Sylvester *see* Economy of brazil
Ellis island, 1900-1933 – (= ser Records of the immigration and naturalization service, series a: subject correspondence files 3; Research colls in american immigration) – 18r – 1 – $3485.00 – 1-55655-541-5 – (with p/g) – us UPA [324]
Ellis, J *see* The natural history of many curious and uncommon zoophytes,...
Ellis, J E *see* Life of william ellis, missionary to the south seas and to madagascar...
Ellis, J J *see*
- Life story of george whitefield
- Life story of william carey

Ellis, John *see*
- A reply to "the academy's" review of "the wine question in the light of the new dispensation"
- Unglaube und offenbarung
- The wine question in the light of the new dispensation

Ellis, John Breckenridge *see* Fran
Ellis, Joseph J *see* From darkness to light
Ellis Junior, Alfredo *see*
- Bandeirismo paulista e o recuo do meridiano
- Capitulos da historia social de s paulo
- Confederacao ou separacao
- Economia paulista no seculo 18
- Feijo e a primeira metade do seculo 19
- Meio seculo de bandeirismo
- Nossa guerra
- Primeiros troncos paulistas e o cruzamento euro-am...

Ellis, Manfred Maria *see* Deutsche schriften
Ellis, Marjorie K *see* Comparison of balance and maximal oxygen consumption among hearing, congenital non-hearing and acquired non-hearing female intercollegiate athletes
Ellis, Mina Benson Hubbard *see* A woman's way through unknown labrador
Ellis, Philip William *see* Discours de m p w ellis
Ellis, Richard *see* The refugees in catalonia
Ellis, Sumner *see* Life of edwin h. chapin, d.d
Ellis, Thomas F *see*
- Adolphus and ellis' reports
- Adolphus and ellis' reports, new series

Ellis, W *see*
- The martyr church
- Narrative of a tour through hawaii, or owhyhee
- Polynesian researches during a residence of nearly six years in the south sea islands...
- Three visits to madagascar during the years 1853-1854-1856

Ellis, Wade Hampton *see*
- Lectures on private corporations
- One way to restrict monopoly

Ellis, William *see*
- The american mission in the sandwich islands
- History of madagascar...the progress of the christian mission established in 1818
- Journal of three voyages along the coast of china, in 1831, 1832 and 1833
- The life and correspondence of william and alice ellis, of airton
- The martyr church
- Narrative of a tour through hawaii, or owhyhee
- New britain
- Polynesian researches
- Three visits to madagascar during the years 1853, 1854-1856...

Ellis, William Hodgson *see* Wayside weeds
Ellis, William Patterson *see* Liber albus civitatis oxoniensis
Ellis, William S *see*
- High school chemistry
- Introductory chemistry
- A report on elementary technical education for ontario

Ellis, William Smith *see* The antiquities of heraldry...from literature, coins, gems, vases, and other monuments of pre-christian and mediaeval times.
Ellis, William Thomas *see*
- "Billy" sunday, the man and his message
- Men and missions

Ellison, Randall Erskine *see* An english-kanuri sentence book
Ellitt, Simon Bolivar *see* Important timber trees of the united states
Ellman, John *see* A letter on the corn laws
Ellon advertiser – Turiff: W Peters & Son Ltd 1979- (wkly) [mf ed 1 jul 1994-] – 1 – (cont: ellon and district advertiser) – uk Scotland NatLib [072]
Ellon times & east gordon advertiser – Ellon: Ellon Times 1990- (wkly) [mf ed 7 jul 1994-] – 1 – uk Scotland NatLib [072]
Ellrich, August *see* Die ungarn wie sie sind
Ells, Robert Wheelock *see*
- Ancient channels of the ottawa river
- Bulletin on graphite
- Marl deposits in ontario, quebec, new brunswick and nova scotia
- Notes on recent sedimentary formations on the bay of fundy coast
- On the geology of the ottawa and parry sound railway
- Palaeozoic outliers in the ottawa river basin
- The physical features and geology of the route of the proposed ottawa canal between the st lawrence river and lake huron
- Rapport sur la geologie de l'interieur de la peninsule de gaspe et d'une partie de l'ile du prince-edouard
- Rapport sur les formations geologiques de l'est des comtes d'albert et westmoreland, nouveau-brunswick et de certaines parties des comtes de cumberland et colchester, nouvelle-ecosse
- Recent conclusions in quebec geology
- Report on a portion of the province of quebec
- Report on the geological formations of eastern albert and westmoreland counties, new brunswick, and of portions of cumberland and colchester counties, nova scotia
- Report on the geology of a portion of the eastern townships
- Report on the geology of northern new brunswick
- Report on the mineral resources of the province of quebec

Ells, Sydney Clarke *see* Report on james bay surveys exploration
Ellsworth american – Ellsworth, ME: Wm H Chaney, 1855-1920 – 1 – us CRL [071]
Ellsworth County. Kansas. District Court *see*
- Records

Ellsworth, Henry William *see* The papers of henry william ellsworth, 1845-1849
Ellsworth herald – Ellsworth, ME: Couillard & Hilton, oct 17 1851-dec 1854 – 1r – 1 – us CRL [071]
Ellsworth. Kansas *see* Records
Ellsworth record – Ellsworth WI. 1900 jan 11/1904-1982 jan 7/nov 11 [gaps] – 27r – 1 – (cont: ellsworth record and ellsworth gleaner) – mf#877754 – us WHS [071]
Ellsworth record and ellsworth gleaner – Ellsworth WI. [1894 oct 26-1899 jan 5], 1899 jan 12-1900 jan 4 – 2r – 1 – (cont: ellsworth gleaner; cont by: ellsworth record) – mf#877819 – us WHS [071]
Ellwood, Charles Abram *see* Man's social destiny in the light of science
Ellwood, Thomas *see*
- Epistle to friends
- The history of the life of thomas ellwood

Elm grove baptist church (formerly: jonathan creek). murray, kentucky : church records – 1846-1977 – 1 – us Southern Baptist [242]
Elm grove elm leaves – Elm Grove WI. 1980 jan 24/jun-2003 nov-dec [gaps] – 103r – 1 – (cont: elm leaves) – mf#1004370 – us WHS [071]
Elm leaves – Brookfield, Elm Grove WI. 1946 feb 1/1947-1980 jan 1/17 [gaps] – 42r – 1 – (continued by: elm grove elm leaves) – mf#1004368 – us WHS [071]
El'maqsad (vies des saintes du rif) / Al-Badisi, "Abd al-Haqq ibn Ismail" – Paris, 1926 – 1 – (ann by g s colin) – us CRL [260]
Elmcreek beacon – Elmcreek, NE: E C Krewson. 76v. v1 n1. jun 10 1898-76th yr n28. sep 27 1973 (wkly) [mf ed with gaps filmed - [1990]] – 20r – 1 – (merged with: overton observer to form: beacon-observer) – us NE Hist [071]
Elmes, James *see*
- The arts and artists...of the schools of painting, sculpture and architecture
- A general and bibliographical dictionary of the fine arts
- Lectures on architecture
- Memoirs of the life and works of sir christopher wren
- Metropolitan improvements

Elmhirst, Philip *see* Occurrences during a six months' residence in the province of calabria ulteriore, in the kingdom of naples
Elmira independent – Ontario, CN. 1986- – 1r/y – 1 – Can$93.00 – cn Commonwealth Imaging [071]
Elmo *see* Mirth for the million
Elmore, H M *see* H m elmore's britischen schiffskapitaen's, vermischte nachrichten von verschiedenen gegenden, inseln und handelsplaetzen in asien
Elmore, Wilber Theodore *see* Dravidian gods in modern hinduism
El-moutakid : independent, politique, critique et moral – Constantine, 1925 – (= ser Moutakid) – 1 – (in arabic) – fr ACRPP [073]
Elmshorner anzeiger – Elmshorn DE, 1960 23 jan 23-1965 – 1r – 1 – gw Misc Inst [074]
Elmshorner nachrichten – Elmshorn DE, 1978 1 sep- – ca 5r/yr – 1 – gw Misc Inst [074]
Elmshorner nachrichten *see* Pinneberger kreisblatt
Elmslie, W A L *see*
- The books of chronicles
- The mishna on idolatry aboda zara

Elmslie, Walter Angus *see* Among the wild ngoni
Elmslie, William Alexander Leslie *see* The mishna on idolatry
Elm-tree on the mall / France, Anatole – New York, NY. 1922 – 1r – 1 – us UF Libraries [071]
Elmwood argus – Elmwood WI. [1938 oct 27/1951], 1952-65, 1966 jan 6-1968 oct 24, 1968 oct 31-1970 nov 12, 1970 nov 19-1973 jun 28, 1973 jul 5-1975 dec 25, 1976-95 – 21r – 1 – (continued by: spring valley sun (spring valley, pierce and saint croix counties, wi: 1952); spring valley elmwood sun-argus) – mf#1001432 – us WHS [071]
Elmwood echo – Elmwood, NE: A W Mayfield, nov 1886-v10 n [33] jul 3 1896 (wkly) [mf ed v7 n313. nov 18 1892-96 (gaps)] – 1r – 1 – (merged with: elmwood leader to form: elmwood leader-echo) – us NE Hist [071]
Elmwood leader – Elmwood, NE: H D Barr, sep 1891-v5 n[44] jul 3 1896 (wkly) [mf ed 1892-96 (gaps)] – 2r – 1 – (merged with elmwood echo to form elmwood leader-echo) – us NE Hist [071]
Elmwood leader-echo – Elmwood, NE: J A Clements. v5 n[45] jul 10 1896-v67 n3. sep 24 1953 (wkly) [mf ed with gaps)] – 24r – 1 – (formed by the union of: elmwood leader and: elmwood echo. absorbed by: plattsmouth journal. issues for jul 10 1896-sep 5 1924 called v5 n[45]-v34 n17 cont the numbering of elmwood leader. iss for sep 12 1924-1953 called v38 n18-v67 cont the numbering of elmwood echo) – us NE Hist [071]
Elmwood sector – Providence, RI. 1924-1925 (1) – mf#66291 – us UMI ProQuest [071]
Elnecave, Nissim *see* Problema de la identidad judia
La elocuencia militar / Barado, Francisco – 1878 – 9 – sp Bibl Santa Ana [946]
Eloesser, Arthur *see*
- Heinrich v. kleist
- Modern german literature

Eloge de charles 5, roi de france : discours qui a remporte le prix de l'academie francoise en 1767 / LaHarpe, Jean-Francois de – Paris 1767 [mf ed Hildesheim 1995-98] – 1v on 1mf – 9 – €40.00 – 3-487-26303-3 – gw Olms [944]
Eloge de henri le grand, roi de france et de navarre / Sapt, ... de – Lyon 1768 [mf ed Hildesheim 1995-98] – 1v on 1mf [ill] – 9 – €40.00 – 3-487-26109-X – gw Olms [944]
Eloge de la ville de moukden et de ses environs : poeme compose par kien-long, empereur de la chine & de la tartarie, actuellement regnant / Amiot, J J M – Paris: N M Tilliard, 1770 – 5mf – 9 – mf#HT-634 – ne IDC [915]
Eloge historique de mr rameau : compositeur de la musique du cabinet du roi, associe de l'academie des sciences, arts & belles-lettres de dijon / Maret, Hugues – Dijon: Causse 1766 [mf ed 19–] – 2mf – 9 – mf#fiche 520 – us Sibley [780]
Eloges de plusieurs personnes o s b / Blemur, R M J de – Paris, 1679 – €35.00 – ne Slangenburg [241]
Eloges et discours sur la triomphante reception du roy en sa ville de Paris... / [Machault, J B de] – Paris: Pierre Recolet, 1629 – 7mf – 9 – mf#O-62 – ne IDC [090]
Eloges historiques de charles 5, et de henri 4, rois de france / Villette, Charles M de – Amsterdam [i e Paris] 1772 [mf ed Hildesheim 1995-98] – 1v on 2mf [ill] – 9 – €60.00 – 3-487-26302-5 – gw Olms [944]
Elogia virorum literis et sapientia illustrium : ad vivum expressis imaginibus exornata, Vol 1 / Tomasini, G F – Patavii: Ex typographia Sebastiani Sardi, 1644 – 5mf – 9 – mf#O-1364 – ne IDC [090]

Elogio de eugenio hermoso : discurso del academico electo y contestacion del excmo sr. Enrique lafuente ferrari / Mosquera Gomez, Luis – Madrid: Blass S.A., 1964 – 1 – sp Bibl Santa Ana [946]
Elogio de los fundadores / Carbonell, Miguel Angel – Habana, Cuba. 1939 – 1r – 1 – us UF Libraries [972]
Elogio de los padres y hermanos de la compania de jesus : muertos por cristo en espana, 1936-39 / Lerida, Felipe – Buenos Aires: [s.n.] 1939 [mf ed 1977] – (= ser Blodgett coll) – 1mf – 9 – mf#w995 – us Harvard [946]
Elogio del dr enrique jose varona y pera / Dihigo, Juan Miguel – Habana, Cuba. 1935 – 1r – 1 – us UF Libraries [972]
Elogio del dr francisco de p coronado y alvaro / Santovenia Y Echaide, Emeterio Santiago – Habana, Cuba. 1948 – 1r – 1 – us UF Libraries [972]
Elogio del dr jose a rodriguez garcia / Dihigo, Juan Miguel – Habana, Cuba. 1935 – 1r – 1 – us UF Libraries [972]
Elogio del dr mario garcia kohly / Dihigo, Juan Miguel – Habana, Cuba. 1937 – 1r – 1 – us UF Libraries [972]
Elogio del sr nestor leonelo carbonell / Justiz y Del Valle, Tomas Juan De – Habana, Cuba. 1946 – 1r – 1 – us UF Libraries [972]
Elogio historico del...doctor joseph cervi... / Ortega, J – Madrid, 1748 – 1mf – 9 – sp Cultura [610]
Elogio poetico...a...personas...de extremadura / Salas, Francisco Gregorio de – Madrid: Andres Ramirez, 1773 – 1 – sp Bibl Santa Ana [946]
Elogio...antonio mendes correia / Castro, Jose de – Madrid: Archivo Ibero Americano, 1965 – 1 – sp Bibl Santa Ana [946]
Elogio...jose moreno nieto / Torres Aguilar-Amat, Salvador – 1882 – 9 – sp Bibl Santa Ana [810]
Elogios de los cinco principios / Mogroveio de Cerda, Ivan – 1636 – 9 – sp Bibl Santa Ana [810]
Elogios en loor...don jaime...don fernando cortes / Lasso de la Vega Cotino, Garbriel – 1601 – 9 – sp Bibl Santa Ana [810]
Elogios poeticos / Salas, Francisco Gregorio de – 1773 – 9 – sp Bibl Santa Ana [810]
Elohim ausserhalb des pentateuch : Grundlegung zu einer Untersuchung ueber die Gottesnamen im Pentateuch / Baumgaertel, Friedrich – Leipzig: J C Hinrichs, 1914 – (= ser Beitraege zur wissenschaft vom alten testament) – 1mf – 9 – 0-7905-2520-8 – mf#1987-2520 – us ATLA [221]
The elohim revealed in the creation and redemption of man / Baird, Samuel John – Philadelphia: Parry & McMillan, 1860, c1859 [mf ed 1989] – 2mf – 9 – 0-7905-0852-4 – (incl bibl ref & ind) – mf#1987-0852 – us ATLA [240]
The elohistic and jehovistic theory minutely examined : with some remarks on scripture and science / Biley, Edward – London: Bell and Daldy, 1865 – 1mf – 9 – 0-7905-3009-0 – mf#1987-3009 – us ATLA [221]
Elola, Jose *see* El credo y la razon
Eloquence a virtue : or, outlines of a systematic rhetoric = Die beredsamkeit eine tugend / Theremin, Franz – rev ed. Andover: Warren F Draper, 1859 [mf ed 1991] – 1mf – 9 – 0-524-00349-1 – (english trans fr german by william g t shedd. with int essay) – mf#1989-3049 – us ATLA [400]
Eloquence de la chaire / Boucher, Edouard – Lille, France. 1894 – 1r – 1 – us UF Libraries [025]
Eloquencia forense / Roxo de Flores, Felipe – 1793 – 9 – sp Bibl Santa Ana [340]
Elora backwoodsman – Elora CW. v1-7 n11. apr 3 1852-jul 28 1858// (wkly) – 1r – 1 – Can$110.00 – cn McLaren [071]
Elordury, E *see* Gomez monsegu c.p., bernardo y elias de tejada, f. la riqueza espiritual de espana
Elore – New York: "Elore" Publishing Association, dec 10 1917-21 – 8r – 1 – us CRL [071]
Elore forward / Hungarian Socialist Federation of America – v5 n43-v7 n13 [1909 oct 23-1911 mar 24] – 1r – 1 – mf#622801 – us WHS [335]
El-ouma – Paris. n28, 58-59, 61-71. dec 1934-avr 1939 – (= ser Ouma) – 1 – fr ACRPP [073]
Eloy, F J Nicholas *see* Dictionnaire historique de la medecine ancienne et moderne (ael3/19)
Elphinstone, Howard Graham *see* Road to swahili
Elphinstone, Mountstuart *see*
– An account of the kingdom of caubul
– Geschichte der englischen gesandtschaft an den hof von kabul, im jahre 1808
– The rise of the british power in the east
– Tableau du royaume de caboul, et de ses dependances, dans la perse, la tartarie et l'inde
Elpidin, M K *see* Podpolnoe slovo
Elrington, Charles K Richard *see* Apostolical succession
Elroy chronicle – Elroy WI. 1890 mar 19-1891 dec 30, 1892 jan 6-nov 30 – 2r – 1 – mf#1125705 – us WHS [071]
Elroy head light – Elroy WI. 1874 sep 10, 1875 feb 4-nov 18, 1875 nov 25-1876 may 11 – 3r – 1 – mf#967949 – us WHS [071]
Elroy leader – Elroy WI. 1898 sep-1903, [1904-08], 1909-12, [1913-18], 1919-1922 may 4 – 5r – 1 – (continued by: elroy tribune; elroy tribune; elroy leader-tribune) – mf#1046026 – us WHS [071]
Elroy leader=tribune – Elroy WI. 1922 may 11-1924, 1925-27, 1928-1930 may 22, 1931-33, 1934-1937 jun, 1937 jul-1940, 1941 may 15,22, nov 13, 1942 jun 18,25, oct 1, dec, 1943 nov 5-1946, 1947-64, 1965 jan 7 – 14r – 1 – (cont: elroy leader; elroy tribune; cont by: elroy-kendall-wilton tribune-keystone) – mf#1046030 – us WHS [071]
Elroy tribune – Elroy WI. 1899 may 4-1900, 1901-17, 1918-1920, 1921 mar 3,17,24, apr 28, sep 29, nov 3 – 7r – 1 – (cont: tribune (elroy, wi)) – mf#968074 – us WHS [071]
Elroy tribune – Elroy WI. 1881 nov16-1885, 1886-88, 1889-92, 1893-96, 1897-1898 jan 20 – 5r – 1 – (continued by: tribune (elroy, wi)) – mf#968076 – us WHS [071]
Elroy union – Elroy WI. 1873 jan 11-1874 jan 8 – 1r – 1 – mf#929416 – us WHS [071]
Elroy-kendall-wilton tribune-keystone – Elroy, Kendall, Wilton WI. 1965 jan 14-1966 dec 29, 1967 jan 5-1969 feb 28, 1969 mar 6-1971 mar 4, 1971 mar 11-1972 dec 28, 1973-77, 1977 oct 6-1978 jun 28, 1978 jul 6-1979 mar 29, 1979 apr 5-dec 27, 1980, 1981, 1982 jan-mar 4 – 14r – 1 – (cont: kendall and wilton keystone; elroy leader=tribune; cont by: tribune keystone) – mf#1046035 – us WHS [071]
Els, Hans van *see* Grabbe als kritiker
Der elsaesser – Strassburg (Strasbourg F), 1885-1904, 1906-18 [gaps] – 53r – 1 – (with app: l'alsacien until 1893) – gw Misc Inst [074]
Der elsaesser bauer – Strassburg (Strasbourg F), 1921 3 feb-1929 23 sep, 1934-35 [gaps] – 1 – fr ACRPP [630]
Die elsaesser hausfrau : la menagere alsacienne – Strassburg (Strasbourg F), 1922-1929 20 sep [gaps] – 1 – fr ACRPP [640]
Elsaesser jornal *see* Niederrheinischer kurier
Elsaesser kurier – Colmar / Elsass (F), 1897 2 may-1917, 1920-1940 15 may – 1 – fr ACRPP [074]
Elsaesser tagblatt – Colmar / Elsass (F), 1890, 1892-1918 – 1 – (: with suppl: elsaesser erzaehler 1892-1913 [gaps]; landwirtschaftliches wochenblatt 1891-1913 [gaps]) – fr ACRPP [074]
Elsaesser volksblatt – Colmar / Elsass (F), 1913 apr-dec, 1915-1916 14 dec – 1 – (predecessor: molsheimer kreisblatt) – fr ACRPP [074]
Elsaesser volkszeitung – Schlettstadt (Selestat F), 1899-1918 [gaps], 19191-23, 1925, 1928-1940 8 may – 1 – (title varies: n99 1907: schlettstadter volksblatt; after 1st world war: l'echo de selestat) – fr ACRPP [074]
Elsaessische nachrichten : amtliche bekanntmachungen fuer den kreis schlettstadt – Schlettstadt (Selestat F), 1879-80, 1883-1901 28 sep – 1 – fr ACRPP [074]
Elsaessische stammeskunde / Bouchholtz, Fritz [comp] – Jena: E Diederichs Verlag 1944 [mf ed 1993] – (= ser Stammeskunde deutscher landschaften) – 1r – 1 – (incl bibl ref & ind; filmed with: die deutschen volksbuecher / paul ernst) – mf#8300 – us Wisconsin U Libr [390]
Elsaessische volks- und handelszeitung *see* L'alsacien / elsaessische volks- und handelszeitung
Der elsaessische volksbote – Rixheim (F), 1869 n24-1870 n43 [gaps] – 1r – 1 – gw Misc Inst [074]
Der elsaessische volksbote – Strassburg (Strasbourg F), 1899 oct-1906 feb [gaps] – 1 – fr ACRPP [074]
Elsaessische volkszeitung und colmarer anzeiger *see* L'alsacien / elsaessische volks- und handelszeitung
Elsaessischer anzeiger = Affiches alsaciennes – Colmar / Elsass (F), 1876-80 – 2r – 1 – gw Misc Inst [074]
Elsaessisches sonntagsblatt fuer unterhaltung und belehrung – Strassburg (Strasbourg F), 1929-1939 27 aug [gaps] – 1 – fr ACRPP [074]
Elsaessisches volksblatt – Strassburg (Strasbourg F), 1868-82 [gaps] – 3r – 1 – gw Misc Inst [074]
Elsam, Richard *see* An essay on rural architecture
Das elsass : neue historisch-topographische beschreibung der beiden rhein-departemente / Aufschlager, Johann F – Strasburg 1825 [mf ed Hildesheim 1995-98] – 2v on 6mf – 9 – €120.00 – 3-487-29704-3 – gw Olms [914]
Das elsass und die erneuerung des katholischen lebens in deutschland von 1814 bis 1848 / Schnuetgen, Alexander – Strassburg, 1913 (mf ed 1992) – 1mf – 9 – 3-89349-157-0 – mf#DHS-AR 43 – gw Frankfurter [241]
Elsass-lothringen : naturansichten und lebensbilder / Noe, Heinrich – Glogau 1872 [mf ed Hildesheim: 1995-98] – (= ser Fbc) – viii/275p on 2mf [ill] – 9 – €60.00 – 3-487-29489-3 – gw Olms [914]
Der elsass-lothringer – Colmar / Elsass (F), 1913-1914 n177 – 1r – 1 – gw Misc Inst [074]
Elsass-lothringer zeitung (elz) – Strassburg (Strasbourg F), 1932 1 oct-31 dec, 1933 2 jan-1937 30 apr, 1938 1 apr-1939 27 aug, 1940 2 jan-8 jun – 1 – gw Misc Inst [074]
Die elsass-lothringische volkspartei – Colmar / Elsass (F), 1896 mar-1912 [gaps] – 1 – fr ACRPP [074]
Elsass-lothringisches landesprivatrecht / Kisch, Wilhelm – Halle (Saale): Waisenhaus, 1905 – (= ser Civil law 3 coll; Das buergerliche recht des deutschen reiches und preussens) – 11mf – 9 – (incl bibl ref and ind) – mf#LLMC 96-574 – us LLMC [346]
Elsass-lothringisches morgenblatt – Muelhausen / Elsass (Mulhouse F), 1901-04 [gaps] – 5r – 1 – gw Misc Inst [074]
Elsbach, A C *see* Der lebensgehalt der wissenschaften
Else von der tanne : oder, das glueck domini friedemann leutenbachers, armen dieners am wort gottes zu wallrode im elend: erzaehlung / Raabe, Wilhelm Karl – Leipzig: P Reclam, 1943 – 1r – 1 – us Wisconsin U Libr [830]
Elsee, Charles *see*
– Neoplatonism in relation to christianity
Elsevier, Abraham *see* The elsevier republics
Elsevier, Bonaventure *see* The elsevier republics
The elsevier republics / Elsevier, Bonaventure & Elsevier, Abraham – 106mf (16:1-18:1) – 9 – $1845.00 – (complete text of all 35 original republics publ in latin fr 1625-50. with p/g in english. int by daniel traister) – us UPA [900]
Elsholtz, Franz von *see* Ansichten und umrisse aus den reise-mappen zweier freunde
El'shtein, N S *see* Berdianskaia zhizn'
Elsie dinsmore / Finley, Martha – Akron, OH. 1943 – 1r – 1 – us UF Libraries [025]
The elsie leader – Elsie, NE: I J Howe, 1894-v8 n27. apr 11 1902// (wkly) [mf ed 1895-1902 (gaps)] – 1r – 1 – (cont by: leader (madrid ne). issues for jan 7-21 1897 incorrectly dated jan 7-21 1896) – us NE Hist [071]
El-siglo – Santiago, juil 1970-sept 1973 – (= ser Siglo) – 1 – fr ACRPP [073]
Elsmore Council. Kansas *see* Lodge records
Elsner, Richard *see*
– Die deutsche dichtung 1936-1937
– Idylle in bauerbach
Elson, Louis Charles *see* The history of american music
Elst, Ferdinand Vander *see* Katanga
Elster, Ernst *see*
– Beitraege zur deutschen literaturwissenschaft
– Friedrich gottlieb klopstock
– Heinrich heines buch der lieder
Elster, Hanns Martin *see* Ausgewaehlte werke
Elstow – (= ser Bedfordshire parish register series) – 2mf – 9 – £5.00 – uk BedsFHS [929]
Elstraer zeitung – Elstra DE, 1910 22 dec-1937 – 34r – 1 – gw Misc Inst [074]
Elstub, W *see* Memorial service
Elsum, J *see* Epigrams upon the paintings of the most eminent masters, ancient and modern...
Elsworth 1538-1950 – 8mf – 9 – £10.00 – uk CambsFHS [929]
Elt : the magazine of equipment leasing and finance – Arlington. 2000+ (1,5,9) – mf#20236,01 – us UMI ProQuest [620]
Elt documents – Oxford. 1986-1986 (1,5) 1985-1986 (9) – ISSN: 0736-2048 – mf#49484 – us UMI ProQuest [420]
Elt journal – Oxford. 1981+ (1) 1981+ (5) 1981+ (9) – (cont: english language teaching journal: elt) – ISSN: 0951-0893 – mf#1389,02 – us UMI ProQuest [420]
Elten, land und leute : eine chronik vergangener zeiten / Gies, L – Cleve, 1951 – €11.00 – ne Slangenburg [943]
Eltere yidishe literatur / Stiff, Nahum – Kiev, Ukraine. 1929 – 1r – 1 – us UF Libraries [470]
Das elternhaus : briefe grosser deutscher / Roch, Herbert [comp] – Berlin: P Neff 1943 [mf ed 1993] – 1r – 1 – (filmed with: die schoensten novellen unserer romantik / walter von malo [comp]) – mf#3392p – us Wisconsin U Libr [860]
Eltham and district times – London, UK. 1905-18 dec 1975; 1976-sep 1978; 19 oct-30 nov 1978; 15 feb 1979-11 sep 1980; 9 oct 1980-23 oct 1986; nov 1986-1991; 9 jan-3 dec 1992; 28 jan 1993-jun 1998 – 228 1/2r – 1 – (aka: eltham and kentish times; eltham times; eltham blackheath and greenwich times+eltham & greenwich times) – uk British Libr Newspaper [072]
Eltham and greenwich times *see* Eltham and district times
Eltham and kentish times *see* Eltham and district times
Eltham and sidcup news shopper – London UK, 1986; 1988-92; 1994-96; 8 jan-dec 1997 – 41 1/4r – 1 – (aka: news shopper (eltham and sidcup)) – uk British Libr Newspaper [072]
Eltham blackheath and greenwich times *see* Eltham and district times
Eltham times *see* Eltham and district times
Eltisley 1599-1900 – (= ser Cambridgeshire parish register transcript) – 3mf – 9 – £3.75 – uk CambsFHS [929]
Elton first baptist church. elton, louisiana : church records – 1910-sep 1991 – 1 – $90.14 – us Southern Baptist [242]
Eltz-Hoffmann, Lieselotte von *see* Adalbert stifter und wien
Elu ja sonage = With life and word / Marley, K L – Toronto: Estonian Free Church Publishing House. Publ. No. 6295 d. One item of four on a reel. Biographies of Evangelical Covenant religious workers. 71p – 1 – us Southern Baptist [242]
Elucidatio musicae choralis : Das ist: gruendlich und wahre erlaeuterung oder unterweisung wie die edle und uralte choral-music fundamentaliter...erlehrnet werden / Samber, Johann Baptist – Salzburg: gedruckt bey J J Mayr 1710 [mf ed 19–] – 4mf – 9 – mf#fiche 895 – us Sibley [780]
An elucidation of the principles of english architecture / Kendall, John – London 1818 – (= ser 19th c art & architecture) – 1mf – 9 – mf#4.2.406 – uk Chadwyck [720]
Elucidationes de potestate papae / Caceres, Diego de – 1642 – 9 – sp Bibl Santa Ana [240]
Elucidationes in omnia sanctorum apostolorum scripta / Arias Montano, Benito – 1588 – 9 – sp Bibl Santa Ana [240]
Elucidationes in quator evangelia metthaei, marci, lucae, iohannis... / Arias Montano, Benito – Antuerpiae: officina christophori plantini, 1575 – sp Bibl Santa Ana [225]
Las "elucidationesin evangelia" de benito arias montano / Garcia Garcia, Rafael – Malaga: Revista Espanola de Estudios Biblicos, 1928 – 1 – sp Bibl Santa Ana [780]
Elucidatorium ecclesiasticum ad officium ecclesiae pertinentia planius exponens in quattuor libros completens / Clichtove, J – Basileae, 1517 – 9mf – 9 – mf#CA-78 – ne IDC [240]
Elucidatorium ecclesiasticum ad officium ecclesiae pertinentia planius exponens in quatuor libros completens / [Clichtove, J] – Parisiis, 1540 – 9mf – 9 – mf#CA-77 – ne IDC [240]
Elucubratio de dogmatica romani pontificis infallibilitate eiusque definibilitate / Cardoni, Giuseppe – Romae: typis civilitatis catholicae, 1870 – 1mf – 9 – 0-8370-8490-3 – (incl bibl ref and ind) – mf#1986-2490 – us ATLA [240]
Eluttu – Madras: C S Chellappa. [n1-99. 1959-mar 1967] – 1r – us CRL [079]
Elvehjem echo / Elvehjem Neighborhood Association – 1969 dec 5-1978 dec, 1979-88 – 2r – 1 – mf#396795 – us WHS [071]
Elvehjem Neighborhood Association *see* Elvehjem echo
Elven, Cornelius *see* Is thy heart right?
Elvenich, Peter Joseph *see*
– Der papst und die wissenschaft
– Der unfehlbare papst. erster vortrag
Elvert, Christian *see* Geschichte der juden in mahran und oesterr-schlesien
Elvira / Vincenzi, Moises – San Jose, Costa Rica. 1940 – 1r – us UF Libraries [972]
Elviro Meseguer, Francisco *see*
– Discursos
– Pregon de la semana santa cacerena
– Torrijos y la eucaristia
Elwang, William Wilson *see* The social function of religious belief
Elwell, Joseph Browne *see* Practical bridge
Elwenspoek, Curt *see* Die roten lotosblueten
Elwin, E F *see* Indian jottings from ten years' experience in and around poona city
Elwin, Edward Fenton *see*
– Indian jottings
– Thirty-four years in poona city
– Thirty-nine years in bombay city
Elwin, Verrier *see*
– The agaria
– The baiga
– The dawn of indian freedom
– Folk-songs of chhattisgarh
– Folk-songs of the maikal hills
– Folk-tales of mahakoshal
– India's north-east frontier in the nineteenth century
– Maria murder and suicide
– The muria and their ghotul
– Myths of middle india
– The tribal art of middle india
– Truth about india
Elwin, Warwick *see* The minister of baptism

Elwood, Anne *see* Narrative of a journey overland from england, by the continent of europe, egypt, and the red sea to india
The elwood bulletin – Elwood, NE: Harry E Moore, aug 6 1896 (wkly) [mf ed 1896-1902,1908- (gaps)] – 1 – (publ as: the bulletin oct 29-dec 1896) – us NE Hist [071]
The elwood republican – Elwood, NE: H R Johnson, 1893 (wkly) [mf ed v3 n41. oct 12 1895-1897 (gaps) filmed 1979] – 1r – 1 – (absorbed: gosper county citizen. issues for -v4 n43 also called whole n199. v4 n46-v5 n36 also called v12 n47-v14 n37 cont the numbering designation of the gosper county citizen. issues for sep 9-oct 14 1897 called v1 n1-v1 n6. issues for oct 21 1897- called v14 n44-) – us NE Hist [071]
Elworth, st peter – (= ser Cheshire monumental inscriptions) – 2mf – 9 – £4.00 – mf#200 – uk CheshireFHS [929]
Elworthy, Frederic Thomas *see* The evil eye
Ely *see* Herman family, 1775-1852
Ely cathedral 1690-1974 – (= ser Cambridgeshire parish register transcript) – 1mf – 9 – £1.25 – uk CambsFHS [929]
Ely cemetery burial book 1855-1950 – 15mf – 9 – £18.75 – uk CambsFHS [929]
Ely: chettisham 1599-1946 – (= ser Cambridgeshire parish register transcript) – 1mf – 9 – £1.25 – uk CambsFHS [929]
[Ely-] ely daily times – NV. 1920- [daily] – 164r – 1 – $9840.00 (subs $140y) – mf#UN04503 – us Library Micro [071]
Ely heritage – 1984 sum-1990 win – 1r – 1 – mf#1789273 – us WHS [071]
Ely holy trinity, index 1559-1881 – (= ser Cambridgeshire parish register transcript) – 4mf – 9 – £5.00 – (baptisms 1559-1877 [11mf] £14.75; burials 1559-1863 [9mf] £11.25; marriages & banns 1559-1881 [13mf] £14.75) – uk CambsFHS [929]
Ely, John *see* Review of nonconformity
Ely lectures on the revised version of the new testament : with an appendix containing the chief textual changes / Kennedy, Benjamin Hall – London: Richard Bentley, 1882 – 1mf – 9 – 0-8370-3878-2 – (incl app in defence of trans, on reasons for the need of the revised version, and on textual corrections in the revised version) – mf#1985-1878 – us ATLA [225]
[Ely-] mining expositor – NV. feb-dec 1908; 1913-14 [daily] – 3r – 1 – $180.00 – mf#U04504 – us Library Micro [071]
[Ely-] mining record – NV. 1905-11, 1916-79 [wkly] – 39r – 1 – $2340.00 – (aka: ely record) – mf#UN04505 – us Library Micro [071]
Ely record *see* [Ely-] mining record
Ely, Richard T *see* Wisconsin progressives
Ely, Richard Theodore *see*
– Social aspects of christianity
– Social aspects of christianity, and other essays
– The social law of service
Ely, Richard Theodore et al *see* The labor problem
Ely, Seth *see* Sacred music
Ely st mary's 1559-1880 – (= ser Cambridgeshire parish register transcript) – 15mf – 9 – £18.75 – uk CambsFHS [929]
The ely volume : or, the contributions of our foreign missions to science and human well-being / Laurie, Thomas – Boston: American Board of Commissioners for Foreign Missions, 1881 – 2mf – 9 – 0-8370-7229-8 – (incl ind) – mf#1986-1229 – us ATLA [240]
[Ely-] weekly mining expositor – NV. 1907-15 [wkly] – 8r – 1 – $480.00 – mf#U04506 – us Library Micro [071]
[Ely-] white pine news – NV. feb 1867; may-aug 1870; july-sep 1872; 1881-1923 [wkly] – 23r – 1 – $1380.00 – mf#U04507 – us Library Micro [071]
[Ely-] white pine suffragist – NV. 31 oct 1914 – 1r – 1 – $60.00 – mf#U04508 – us Library Micro [322]
Ely, William D *see* Keyhole for roger williams' key
Elyot, Thomas *see*
– Bibliotheca eliotae: eliotis librarie
– The castel of helth
Elyria. Ohio. First Baptist Church, Women's Home Mission Society *see* Church records, ms 788
Elyria. Ohio. Presbytery *see*
– Presbyteries of elyria and lorain records, 1836-1863
– Presbytery of elyria record book, 1864-1866
Elysius jucundarum quaestionum campus, omnium literarum amoenissima varietate... / Reyes Franco, G – Francfurt, 1670 – 24mf – 9 – sp Cultura [450]
Elze, Karl *see* Gedichte
Elze, Theodor [comp] *see* Primus trubers briefe
Em thi truong hoc anh thi truong doi / Truc Quan – Saigon: Thuy Chung 1974 [mf ed 1992] – on pt of 1r – 1 – mf#11052 r377 n5 – us Cornell [959]
Em torno de alguns tumulos afro-cristaos de uma area africana / Freyre, Gilberto – Salvador, Brazil. 1959 – 1r – 1 – us UF Libraries [960]
Die emanation der motivik auf die thematik des musicals cats / Petri, Hasso Gottfried – (mf ed 2000) – 3mf – 9 – €49.00 – 3-8267-2731-2 – mf#DHS 2731 – gw Frankfurter [780]
La emancipacion de america y su reflejo en la cultura espanola. madrid, 1944 / Fernandez Almagro, Melchor – Madrid: Razon y Fe, 1947 – 1 – sp Bibl Santa Ana [972]
Emancipacion de hispanoamerica / Amunategui Y Solar, Domingo – Santo Domingo, Chile. 1936 – 1r – 1 – us UF Libraries [972]
O emancipador : o germinal / Marques, Lourenco – Special issues for jun 26, jul 15,12,19,26, aug 2,9,16,23,30, sep 13,20,27, oct 14 1926; oct 25, 1926-jul 1919/37 – 1r – us CRL [070]
L'emancipation – Lyon. no.1-18,22. oct-nov 1880 – 1 – fr ACRPP [073]
L'emancipation : organe central de l'unite totale des travailleurs. ed. nationale. – Saint-Denis. 7 no. nov 1934-juin 1936 – 1 – (elements repris par: l' emancipation nationale) – fr ACRPP [325]
L'emancipation – Saint-Denis. 32 no. mars 1902-1939 – 1 – fr ACRPP [073]
L'emancipation – Toulouse. aout 1838-mars 1839, 2 oct 1844, 1er mai 1845, 11 nov 1846, 10 oct 1847-52 – 1 – fr ACRPP [073]
Emancipation / Anarchist Association of the Americas – v3 iss 1=10-v5, iss 6=37 [1980 jan 19-1982 jul] – 1r – 1 – (cont: anarchy times) – mf#670072 – us WHS [320]
Emancipation – Leopoldville: M A Nguvulu, apr 6-27 1960, jan 1 1961 – us CRL [079]
L'emancipation nationale – Paris puis Marseille. juil 1936-aout 1944 – 1 – fr ACRPP [073]
Emancipation nationale – Marseilles, France. 13 jun 1942; 1 jun 1944 – 1r – 1 – uk British Libr Newspaper [074]
L' emancipation nationale *see* L'emancipation
The emancipation of massachusetts / Adams, Brooks – Boston: Houghton, Mifflin, 1899, c1886 – 1mf – 9 – 0-7905-4006-1 – (incl bibl ref) – mf#1988-0006 – us ATLA [975]
L'emancipation sexuelle de la femme / Pelletier, Madeleine – Paris: Giard et Briere, 1911 – (= ser Les femmes [coll]) – 1mf – 9 – mf#10443 – fr Bibl Nationale [305]
Emancipator – Jonesborough. 1820-1820 (1) – mf#3087 – us UMI ProQuest [976]
Emancipator / Socialist Labor Party – Cleveland OH. 1894 jul 7, 14-aug 25, sep 22, 29 – 1r – 1 – mf#869128 – us WHS [335]
Emancipator and republican – Boston, MA. v8 n23-9 n35. 5 oct 1843-25 dec 1844 – 1 – us NY Public [071]
Emancipator and republican – Boston MA. 1833 may 18-1837 dec 28, 1838 jan 4-1841 jun 24, 1841 jul 1-1844 aug 7, 1844 aug 14-1847 dec 1, 1847 dec 8-1850 dec 26 – 5r – 1 – (cont: emancipator and free soil press; boston weekly republican; cont by: weekly chronotype; commonwealth and emancipator) – mf#869004 – us WHS [071]
Emanu-el – San Francisco, CA. 1897-1949; 1958-67 – 1 – us AJPC [071]
Emanuel, Charles Herbert Lewis *see* Century and a half of jewish history
Emanuel geibel : erster theil / Goedeke, Karl – Stuttgart: J G Cotta, 1869 (mf ed 1990) – 1 – (no more publ) – us Wisconsin U Libr [430]
Emanuel geibel : ein gedenkbuch / Holz, Arno; ed by Holz, Arno – Leipzig: O Parrisins, 1884 (mf ed 1990) – 1 – us Wisconsin U Libr [430]
Emanuel geibel als uebersetzer und nachahmer englischer dichtungen / Volkenborn, Heinrich – Muenster: Theissing, 1910 (mf ed 1990) – 1 – (incl bibl ref) – us Wisconsin U Libr [430]
Emanuel geibel, saenger der liebe, herold des reiches : ein deutsches dichterleben / Gaedertz, Karl Theodor – Leipzig: G Wigand, 1897 (mf ed 1990) – 1r – 1 – us Wisconsin U Libr [430]
Emanuel geibel's briefe an karl freiherrn von der malsburg und mitglieder seiner familie / Geibel, Emanuel; ed by Duncker, Albert – Berlin: Paetel, 1885 (mf ed 1990) – 1 – us Wisconsin U Libr [860]
Emanuel geibels gesammelte werke : in acht baenden = Works – Stuttgart: J G Cotta, 1883 (mf ed 1990) – 8v in 4 – 1 – us Wisconsin U Libr [802]
Emanuel geibels leben, werke und bedeutung fuer das deutsche volk / Leimbach, Karl Ludwig – 2. verm neubearb aufl. Wolfenbuettel: J Zwissler, 1894 (mf ed 1990) – 1 – us Wisconsin U Libr [430]
Emanuel geibels lyrik : auf ihre deutschen vorbilder geprueft / Stichternath, Friedrich – Muenster i/W: F Coppenrath, 1911 (mf ed 1990) – 1 – (incl bibl ref) – us Wisconsin U Libr [430]
Emanuel greenwald, pastor and doctor of divinity : footprints of his life, together with his earliest extant and latest discourses / Haupt, C Elvin [comp] – Lancaster, PA: G L Fon Dersmith, 1889 [mf ed 1993] – 181p/1pl on 1mf – 9 – 0-524-08380-0 – mf#1993-3080 – us ATLA [920]
Emanuel, Nathan H *see* Influence of ancient hebrew music on gregorian chant
Emanuel swedenborg : as a man of science / Fernald, Woodbury Melcher – Boston: Otis Clapp, 1860 [mf ed 1984] – 7mf – 9 – 0-8370-0929-4 – (incl bibl ref and app) – mf#1984-4296 – us ATLA [500]
Emanuel, W V *see* The naval side of the spanish war
Emanzipation der juden in anhalt-dessau / Horwitz, Ludwig – Dessau, Germany. no date – 1r – 1 – us UF Libraries [939]
Emard, Joseph-Medard *see*
– L'agriculture
– Allocution prononcee a l'ouverture du congres de l'enseignement secondaire tenu a quebec, juin 1914
– Au congres eucharistique de malte
– Au jeudi saint
– Au jour de l'an
– La benediction abbatiale
– Le bon pasteur
– Le code de droit canonique
– Le congres eucharistique de montreal
– De l'influence eucharistique sur l'apostolat des premiers missionnaires au canada
– L'episcopat, son origine et son oeuvre
– La guerre
– Messages
– La pentecote
– Le pretre-soldat
– Saint pierre
– La succession apostolique
– Les tendresses du sacre-coeur de jesus
Emard, M R *see* Religion and leisure
La embajada del marques de cogolludo a roma en 1687 y el duque de medinaceli y la giorgina. madrid, 1929 / Villaurrutia, W R – Madrid: Razon y Fe, 1930 – 1 – sp Bibl Santa Ana [946]
Embajada del obispo de cartagena fr antonio trejo pidiendo al papa la definicion de la inmaculada / Pou Marti, Jose – Archivo Ibero Americano, 1932 – 1 – sp Bibl Santa Ana [240]
Las embajadas de don juan antonio de vera y zuniga en italia : conferencia / Garcia Arias, Luis – Madrid: Graf Valera, 1950 – 1 – sp Bibl Santa Ana [946]
L'emballement : poeme antiimperialiste / Gingras, Apollinaire – [Quebec (Province): s.n.], c1920 – 1mf – 9 – 0-665-71508-0 – mf#71508 – cn Canadiana [810]
The embargo / Bryant, William C – 1808-1809 – 9 – 5.00 – us Scholars Facs [830]
The embassy of john van campen and constantine noble to sing la mong, vice roy of fo-kyen : the embassy of 1662 / Montanus, A – London, 1745-1747. v3 – 2mf – 9 – mf#A-271 – ne IDC [910]
The embassy of peter de goyer and jacob de keyzer from the dutch east india company to the emperor of china, in 1655... / Nieuhof, J – London, 1745-1747. v3 – 3mf – 9 – mf#A-271 – ne IDC [915]
The embassy of shah rakh : son of tamerlan, and other princes, to the emperor of katay, or china – London. v4. 1745-1747 – 1mf – 9 – mf#A-271 – ne IDC [915]
The embassy of sir thomas roe to india, 1615-19 : as narrated in his journal and correspondence / ed by Foster, William – London: Oxford University Press, 1926 – (= ser Samp: indian books) – us CRL [954]
The embassy of the lord van hoorn to kang hi, emperor of china and eastern tartary : the embassy of 1664 / Montanus, A – London, 1745-1747. v3 – 2mf – 9 – mf#A-271 – ne IDC [915]
Embden, Ludwig von *see* The family life of heinrich heine
Embedded systems programming – San Francisco. 1988-1994 (1,5,9) – ISSN: 1040-3272 – mf#17140 – us UMI ProQuest [000]
Embel, Franz *see* Fussreise von wien nach dem schneeberge
Ember, A *see* Oriental studies published in commemoration of the fortieth anniversary of paul haupt as director of the john hopkins university
Emberson, Alfred *see* All about victoria, british columbia
Emberson, Frederick C *see*
– Are we immortal?
– The art of teaching
– Hash (wholesale to boarding houses)
– The yarn of the love sick parsee
Emblem / Marine Corps Logistics Base (Albany, GA) – Albany GA. v26 n22-41 [1981 jun 5-oct 16], v27 n18-? [1982 may-dec], 1983, 1984, 1985-86, 1987-1988 aug, 1989, 1990, – 7r – 1 – mf#966679 – us WHS [071]
Emblem – Middletown, OH. 1851-1853 (1) – mf#65585 – us UMI ProQuest [071]
Emblem topics / Supreme Emblem Club of the United States of America – 1977 sep-1984 jun, 1984 sep-1988 dec – 2r – 1 – mf#716755 – us WHS [929]
Emblema sacrum ex apocal loh theologi : de quo praeside Iesu Christo... / Wirz, J – Tiguri: Ex officina Bodmeriana, 1631 – 1mf – 9 – mf#O-1471 – ne IDC [090]
Los emblemas de alciato : traducidos en rhimas espanolas anadidos de figuras y de nuevos emblemas en la tercera parte de la obra / Alciato, Andrea – Lyons: M Bonhomme, 1549 – 3mf – 9 – mf#O-1476 – ne IDC [090]
Emblemas morales de don iuan de horozco y covarruvias arcediano de cuellar en la santa yglesia de segovia : dedicadas a la buena memoria del presidente don diego de covarruvias y leyva su tio... – Caragoca: Alonso Rodriguez, 1604 – 11mf – 9 – mf#O-638 – ne IDC [090]
Emblemas morales de don iuan de horozco y covarruvias arcediano de cuellar en la santa yglesia de segovia : dedicadas a la buena memoria del presidente don diego de covarruvias y leyva su tio... – Segovia: Impresso Iuan de la Cuesta, 1591 – 8mf – 9 – mf#O-10 – ne IDC [090]
Emblemas morales de don sebastian de covarrubias orozco : capellan del rey n s maestrescuela... / Covarrubias Orozco, S de – Madrid: Luis Sanchez, 1610 – 11mf – 9 – mf#O-214 – ne IDC [090]
Emblemas moralizadas... / Soto, Hernando de – Madrid: Por les herederos de Ian Iniguez de Lequerica, 1599 – 4mf – 9 – mf#O-1908 – ne IDC [090]
Emblemas nacionales / Galvez G, Maria Albertina – Guatemala, 1958 – 1r – 1 – us UF Libraries [972]
Emblemata : cum aliquot nummis antiqui operis... / Sambucus, J – Antverpiae: Ex officina Christophori Plantini, 1564 – 3mf – 9 – mf#O-743 – ne IDC [090]
Emblemata : cum claudii minois divionensis ad eadem commentariis / Alciato, Andrea – Lugduni Batavorum: Ex officina Plantiniana, apud Franciscum Raphelengium, 1591 – 8mf – 9 – mf#O-3028 – ne IDC [090]
Emblemata : eiusdem aenigmatum libellus ad d arnoldum rosenbergum / Junius, H – Antverpiae: Ex officina Christophori Plantini, 1585 – 2mf – 9 – mf#O-1949 – ne IDC [090]
Emblemata : eiusdem aenigmatum libellus. cum nova et emblematum et aenigmatum appendice / Junius, H – Lugduni Batavorum: Ex officina Plantiniana, apud Franciscum Raphelengium, 1596 – 2mf – 9 – mf#O-18 – ne IDC [090]
Emblemata : emblemes chrestienes et morales, sinne-beelden streckende tot christelicke bedenckinghe ende leere der zedicheyt... / Heyns, Z – Rotterdam: Pieter van Waesberge, 1625 – 9mf – 9 – mf#O-291 – ne IDC [090]
Emblemata : from the british library copy (11408 aaa 43) of the edition of mathias bonhomme, lyons, 1550 / Alciato, Andrea – 1r – 1 – mf#97086 – uk Microform Academic [760]
Emblemata : sive loca quadam ex Adami Adami... / Meyern, J G – Ratisbonae: Typis R"dlmayerianis, 1760 – 2mf – 9 – mf#ILM-971 – ne IDC [090]
Emblemata... / Sambucus, J – Antverpiae: Apud Christophorum Plantinum, 1584 – 3mf – 9 – mf#O-1269 – ne IDC [090]
Emblemata... / Sambucus, J – Tertio ed. Antverpiae: Ex officina Christophori Plantini, 1569 – 4mf – 9 – mf#O-1466 – ne IDC [090]
Emblemata a jano jac boissardo vesuntino delineata sunt / Lebey de Batilly, D – Francofurti ad Moenum, 1596 – 2mf – 9 – mf#O-82 – ne IDC [090]
Emblemata, ad d arnoldum cobelium : eiusdem aenigmatum libellus, ad d arnoldum rosenbergum / Junius, H – Antverpiae: ex officina christophori plantini, 1565 – 2mf – 9 – mf#O-3098 – ne IDC [090]
Emblemata adriani iunii medici : overgheset in nederlandsche tale deur m a gilles / Junius, H – Antwerp: Plantin, 1575 – 1mf – 9 – mf#O-812 – ne IDC [090]
Emblemata afbeeldinghen amatoria van minne : Emblemes d'amour / [Hooft, P C] – Amsterdam: Willem lanszoon, 1618 – 2mf – 9 – mf#P-876 – ne IDC [090]
Emblemata amatoria : afbeeldinghen van minne. emblemes d'amour / [Hooft, P C] – t'Amsterdam: Willem lanszoon, 1618 – 2mf – 9 – mf#O-3085 – ne IDC [090]
Emblemata amatoria = Emblems of love / Ayres, Ph. – London: Sold by R. Bently in Covent Garden; S. Tidmarch at the Kings head in Cornhill, 1683 – 1mf – 9 – (In four languages. Dedicated to the ladys) – mf#O-1230 – ne IDC [090]
Emblemata amatoria : iam demum emendata / [Heinsius, D] – [Amstelredam: D Pietersz, 1608] – 1mf – 9 – mf#O-3176 – ne IDC [090]
Emblemata amatoria : iam demum emendata / [Heinsius, D] – [Amsterdam: D P Pers, c1605] – 1mf – 9 – mf#O-3178 – ne IDC [090]
Emblemata amatoria : iam demum emendata. / [Heinsius, D] – [Amsterdam: D Pietersz, 1612] – 1mf – 9 – mf#O-3177 – ne IDC [090]

Emblemata amatoria : iam demum emendata / [Timmermans, I A] – [Amsterdam, c1608-12] – 1mf – 9 – mf#O-3181 – ne IDC [090]

Emblemata amatoria *see* Emblemes d'amour en quatre langue (sic)

Emblemata amatoria georgii camerarii – Venetiis: Sumpt P P Tozzii, [1627] – 3mf – 9 – mf#O-69 – ne IDC [090]

Emblemata amoris...studio et opera rapheelis custodis... / [Vaenius, O] – Augustae vindolicorum: [gedruckt durch luca schultes; in verlegung raphaelis custodis kupfferstechers], 1622 [1623] – 2mf – 9 – mf#O-1450 – ne IDC [090]

Emblemata andreae alciati iurisconsulti clarissimi / Alciato, Andrea – Lugduni: Apud Gulielmum Rouillium, 1548 – 2mf – 9 – mf#O-1474 – ne IDC [090]

Emblemata andreae alciati...imaginibusque... illustrata / Alciato, Andrea – Francofurti ad Moenum: Apud Georgium Coruinum, sumptibus Sigismundi Feyerabendt etc Simonis Huteri, 1567 – 7mf – 9 – mf#O-1445 – ne IDC [090]

Emblemata anniversaria academiae altorfinae studiorum iuventutis exercitandorum causa proposita et variorum orationibus exposita / Academia Altorfina – Norimbergae: impensis levini hulsij, 1597 – 4mf – 9 – mf#O-535 – ne IDC [090]

Emblemata anniversaria academiae noribergensis, quae est altorffii / Academia Altorfina – Nurembergae: per abr wagenman, 1617 – 11mf – 9 – mf#O-536 – ne IDC [090]

Emblemata Augustissimi imperatoris Josephi 1 *see* Supremis honoribus affixa

...Emblemata centum : regio politica aeneis laminis affabre caelata... / Solorzano Pereyra, J de – [Matriti: D Garcia Morras, 1653] – 26mf – 9 – mf#O-759 – ne IDC [090]

Emblemata centum, regio politica / Solorzano Pereyra, J de – [Madrid: D Garcia Morras, 1651] – 17mf – 9 – mf#O-1469 – ne IDC [090]

Emblemata cum privilegijs / Maccio, P – [Bononiae: Clemens Ferronius...excudebat, 1628] – 5mf – 9 – mf#O-681 – ne IDC [090]

Emblemata d a alciati denuo ab ipso autore recognita... / Alciato, Andrea – Lugduni: Apud Guliel. Rovilium, 1550 – 3mf – 9 – mf#O-112 – ne IDC [090]

Emblemata et epigrammata miscellanea selecta ex stromatis peripateticis Antonii Fayi / La Faye, A de – Genevae: Apud Petrum & Iacobum Chouet, 1610 – 4mf – 9 – mf#O-20 – ne IDC [090]

Emblemata ethico-politica carmine explicata : ad serenissimum principem Leopoldum Wilhelmum... / Kreihing, J – Antverpiae: Apud Iacobum Meursium, 1661 – 3mf – 9 – mf#O-325 – ne IDC [090]

Emblemata florentii schoonhovii i c goudani... / Schoonhovius, F – ed 3. Amstelodami: Joannem Janssonium, 1635 – 3mf – 9 – mf#O-1270 – ne IDC [090]

Emblemata florentii schoonhovii i c goudani... / Schoonhovius, F – ed 4. Amstelodami: Joannem Janssonium, 1648 – 3mf – 9 – mf#O-1467 – ne IDC [090]

Emblemata florentii schoonhovii i c goudani... / Schoonhovius, F – Goudae: Apud Andream Burier, 1618 – 5mf – 9 – mf#O-428 – ne IDC [090]

Emblemata heroica : of de medalische sinnebeelden der ses en dertig graaven van Holland... / Smids, L – Leyden: Dirk Haak, 1714 – 3mf – 9 – mf#O-432 – ne IDC [090]

Emblemata heroica... / Smids, L – Amsterdam: Johannes Oosterwyk en Hendrick vande Gaete, 1712 – 2mf – 9 – mf#O-3165 – ne IDC [090]

Emblemata horatiana : imaginibus in aes incisis atque latino, germanico, gallico et belgico carmine illustrata / Vaenius, O – Amstelaedami: Apud Henricum Wetstenium, 1684 – 3mf – 9 – mf#O-3188 – ne IDC [090]

Emblemata i sambuci : in nederlantsche tale ghetrouwelick overgheset – t'Antwerpen: Christoffel Plantyn, 1566 – 3mf – 9 – mf#O-3162 – ne IDC [090]

Emblemata Iosephina cum eulogijs opera r.p.d... / Stengel, C – Augustae Vindelicorum: Typis Veronicae Apergerin, 1658 – 1mf – 9 – mf#O-1911 – ne IDC [090]

Emblemata moralia : scripta quondam Hispanice a Johanne de Boria, latinitate autem donata a LCCP / Boria, J de – Berolini: Sumptibus Johann. Michael. Rudigeri, 1697 – 3mf – 9 – mf#O-6 – ne IDC [090]

Emblemata moralia et bellica : nunc recens in lucem edita / Bruck, J – Argentorati: per iacobum ab heyden iconographum, 1615 – 2mf – 9 – mf#O-82 – ne IDC [090]

Emblemata moralia et oeconomica... / Lubbaeus, R – Arnhemi: Apud Ioannem Iansonium, 1609 – 1mf – 9 – mf#O-585 – ne IDC [090]

Emblemata moralia nova : das ist: achtzig sinnreiche nachdenkliche Figuren auss heyliger Schrifft in Kupfferstuecken fuergestellet... / Cramer, D – Franckfurt am Mayn, 1630 – 3mf – 9 – mf#O-588 – ne IDC [090]

Emblemata nicolai reusneri ic partim ethica... – Francofort ad Moenum: per ioannem feyerabendt, impensis sigusmundi feyerabendij, 1581 – 4mf – 9 – mf#O-730 – ne IDC [090]

Emblemata nobilitati et vulgo scitu digna : singulis historijs symbola adscripta et elegantis versiis historiam explicantes / Bry, J Th de – Franco[furti] ad M[oenum], 1592 – 2mf – 9 – mf#O-181 – ne IDC [090]

Emblemata nobilitatis : stamm- und wappenbuch von theodor de bry / ed by Warnecke, F – Berlin: J A Stargardt, 1894 – 3mf – 9 – mf#O-1530 – ne IDC [929]

Emblemata nova : das ist new Bilderbuch: darinnen durch sonderliche Figuren der jetzigen Welt Lauff und Wesen verdeckter Weise abgemahlet... / Friedrich, A – Francoforti: Apud Lucam Iennis, 1617 – 3mf – 9 – mf#O-1245 – ne IDC [090]

Emblemata of zinnewerck : voorghestelt in beelden, ghedichten en breeder uijt-legginghen tot uijtdrukinghe... / Brune, J de – t'Amsterdam: Jan Jacobsz Schipper, 1661 – 7mf – 9 – mf#O-565 – ne IDC [090]

Emblemata of zinnewerck : zelfde als voorgaande / Brune, J de – t'Amsterdam: Ian Evertsen Kloppenburch, 1624 – 7mf – 9 – mf#O-180 – ne IDC [090]

Emblemata ofte sinnebeelden... / Zevecotius, J – Amsterdam: J Janssonius, 1638 – 4mf – 9 – mf#O-3269 – ne IDC [090]

Emblemata physico-ethica : hoc est naturae morum moderatricis picta praecepta / Taurellus, N – Noribergae: In Bibliopolio Simonis Halbmayeri, 1617 – 3mf – 9 – mf#O-777 – ne IDC [090]

Emblemata politica : accedunt dissertationes politicae de romanorum imperio... / Boxhorn, M Z – Amstelodami: Apud Joannem Janssonium, 1651 – 3mf – 9 – mf#O-563 – ne IDC [090]

Emblemata politica : quibus ea, quae principatum spectant...opus novum / Bruck, J – Argentinae: apud jacobum ab heyden, colonae: apud abrahamum hogenberg, 1618 – 4mf – 9 – mf#O-8 – ne IDC [090]

Emblemata politica, et orationes / Boxhorn, M Z – Amstelodami: Ex offina Johannis Janssoni, 1635 – 3mf – 9 – mf#O-3039 – ne IDC [090]

Emblemata politica in aula magna curiae noribergensis depicta / Isselburg, P – ed 2. Nuernberg: In Verlegung Wolff Endters, 1640 – 1mf – 9 – mf#O-31 – ne IDC [090]

Emblemata politica in aula magna curiae noribergensis depicta : quae sacra virtutum suggerunt monita prudenter administrandi fortiterque defendendi republicam / Isselburg, P – [Nuremberg], 1617 – 1mf – 9 – mf#O-05 – ne IDC [090]

Emblemata pro toga et sago / Bruck, J – Norimbergae: Apud Pauli Fuerstii, n.d. (end of the 17th c) – 1mf – 9 – mf#O-1235 – ne IDC [090]

Emblemata sacra : dat is, eenighe geestelicke sinnebeelden met nieuwe ghedichten... / H(ulsius), B – n.p, 1631 – 3mf – 9 – mf#O-312 – ne IDC [090]

Emblemata sacra : das ist gottliche andachten, voller flammender begierden einer buszfertigen, geheiligten und liebreichen seelen / Hoburg, C – Amsterdam: Henrico Betkio; Franckfurt: Christoffel le Blon, 1661 – 3mf – 9 – mf#O-3226 – ne IDC [090]

Emblemata saecularia : mira et iucunda varietate saeculi huius mores ita exprimentia... / Bry, J Th de & Bry, J I de – Francoforti, 1596 – 2mf – 9 – mf#O-61 – ne IDC [090]

Emblemata secularia : mira et iucunda varietate seculi huius nores ita exprimentia... / Bry, J Th de – Oppenhemii: Typis Hieronymi Galleri, [1611] – 4mf – 9 – mf#O-62 – ne IDC [090]

Emblemata selectiora *see* Typis elegantissimis expressa

Emblemata sive symbola a principibus, viris ecclesiasticis, ac militaribus, aliisque usurpanda / Vaenius, O – Bruxellae: Ex officina Huberti Antonii, 1624 – 1mf – 9 – mf#O-791 – ne IDC [090]

Emblemata v c andreae alciati mediolanensis iurisconsulti / Alciato, Andrea – Lugduni Batavorum: Ex officina Christophori Plantini, 1584 – 5mf – 9 – mf#O-3205 – ne IDC [090]

Emblemata v c andreae alciati mediolanensis iurisconsulti / Alciato, Andrea – Lugduni Batavorum: Ex officina Plantiniana, Apud Franciscum Raphelengium, 1591 – 5mf – 9 – mf#O-1291 – ne IDC [090]

Emblemata...cum imaginibus plerisque restitutis ad mentem auctoris / Alciato, Andrea – Patavij: Apud Pet. Paulum Tozzium, 1618 – 5mf – 9 – mf#O-1293 – ne IDC [090]

Emblematische gemueths-vergnuegung : bey betrachtung 715 der curieusten und ergaezlichsten sinnbildern... / [Offelen, H] – Augspurg: bey lorentz kroninger und goebels seel erben, 1693 – 2mf – 9 – mf#O-1453 – ne IDC [090]

Emblematum christianorum centuria : cum corundem Latina interpretatione / Montenay, G de – Tiguri: Apud Christophorum Froschouerum, 1584 – 3mf – 9 – mf#O-61 – ne IDC [090]

Emblematum ethico-politicorum centuria... : editio ultima / Zincgreff, J W – Heidelbergae: apud clementem ammonium, 1666 – 3mf – 9 – mf#O-1275 – ne IDC [090]

Emblematum ethico-politicorum centuria iulii guilielmi zincgrafii / Zincgreff, J W – Franckfurt am Mayn: Verlegts Thomas Michael Goetz, 1698 – 3mf – 9 – mf#O-1366 – ne IDC [090]

Emblematum liber / Alciato, Andrea – [Augustae Vindelicorum], 1531 – 2mf – 9 – mf#O-1229 – ne IDC [090]

Emblematum liber : Ipsa emblemata ab auctore delineata: a Theodoro de Bry sculpta, & nunc recens in lucem edita / Boissard, J J – Francofurti ad Moenum, 1593 – 2mf – 9 – mf#O-556 – ne IDC [090]

Emblematum liber divo matthiae, romanorum imperatori augustissimo... / Westhovius, W – Ratisbonae: Sub incude Typographica Matthiae Myll, 1613 – 1mf – 9 – mf#O-1978 – ne IDC [090]

Emblematum sacra : Hoc. est, decades quinque emblematum ex sacra scriptura... / Cramer, D & Bachman, C – Francofurti: Sumptibus Lucae Jennisl, 1624 – 6mf – 9 – mf#O-1338 – ne IDC [090]

Emblematum sacrorum et civilium miscellaneorum sylloge prior (-posterior) / Bornitz, J – Heidelbergae: Cl. Ammonius, 1659 2pts – 3mf – 9 – mf#O-76 – ne IDC [090]

Emblematum sacrorum quorum consideratio accurata... / Saubert, J – Nuernberg, 1625 – 2mf – 9 – mf#O-746 – ne IDC [090]

Emblemes : ou devises Chrestiennes, composees par damoiselle Georgette de Montenay / Montenay, G de – Lyon: Jean Marcorelle, 1571 – 3mf – 9 – mf#O-375 – ne IDC [090]

Emblemes and epigrames...a d 1600 / Thynne, Francis; ed by Furnivall, F J – London: publ for the early english text society by n truebner, 1876 – 2mf – 9 – mf#O-02 – ne IDC [090]

Les emblemes d'alciat / Duplessis, G – Paris: J Rouam, 1884 – 2mf – 9 – mf#O-1198 – ne IDC [090]

Emblemes d'amour en quatre langue (sic) / Emblemata amatoria – Londe: l'Amoureux, 1690 – 1mf – 9 – mf#O-600 – ne IDC [090]

Les emblemes de l'amour humain... / Vaenius, O – Brusselles: Francois Foppens, 1667 – 4mf – 9 – mf#O-1273 – ne IDC [090]

Emblemes divers : repr, sentez dans 140 figures en tailledouce / Baudoin, J – Paris: Loyson, 1659-60 – 11mf – 9 – mf#O-1232 – ne IDC [090]

Emblemes nouveaux : esquels le cours de ce monde est depeint et represente... / Friedrich, A – Francoforti: Apud Iacobum de Zetter, 1617 – 2mf – 9 – mf#O-1852 – ne IDC [090]

Emblemes ou devises chretiennes / [Philotheus] – Utrecht: Antoine Schouten, 1697 – 3mf – 9 – mf#O-603 – ne IDC [090]

Emblemes royales...louis le grand / Martinet, J – Paris: Claude Barbin, 1673 – 2mf – 9 – mf#O-364 – ne IDC [090]

Les emblemes...mis en rime francoyse / Alciato, Andrea – n.p, n.d. – 2mf – 9 – mf#O-1472 – ne IDC [090]

Emblems for the improvement and entertainment of youth – London: R Ware, 1755 – 3mf – 9 – mf#O-1452 – ne IDC [090]

Emblems from and for the factory / Richardson, J – London, England. 1851 – 1r – 1 – us UF Libraries [240]

Emblems of saints : By Which They Are Distinguished In Works Of Art / Husenbeth, Frederick Charles; ed by Jessopp, Augustus – 3rd ed. Norwich: Printed for the Norfolk and Norwich Archaeological Society by A.H. Goose, 1882 – 2mf – 9 – 0-7905-8107-8 – mf#1988-6069 – us ATLA [700]

Emblems of the holy spirit / Simpson, Albert B – Nyack: Christian Alliance Pub Co, c1895 [mf ed 1992] – (= ser Christian & missionary alliance coll) – 1mf – 9 – 0-524-03740-X – mf#1990-4845 – us ATLA [220]

Embo journal – Oxford. 1986+ (1,5,9) – ISSN: 0261-4189 – mf#16449 – us UMI ProQuest [574]

Emboscada a morfeo / Jesus Castro, Tomas de – Madrid, Spain. 1964 – 1r – 1 – us UF Libraries [972]

Embracing leer and leven : the theology of simon oomius in the context of nadere reformatie orthodoxy / Schuringa, Gregory D – Calvin Theological Seminary 2003 [mf ed 2004] – 1r – 1 – 0-524-10506-5 – mf#d00008 – us ATLA [242]

Embree, Beatrice *see* The girls of miss clevelands'

Embroiderers' Guild of America *see* Madison area chapter, embroiderers' guild of america

Embroidery : its history, beauty, and utility, with plain instructions to learners / Wilcockson, Emma Elizabeth – London: Darton & Co, [1857] – (= ser 19th c art & architecture) – 1mf – 9 – mf#4.1.27 – uk Chadwyck [740]

Embrujo del microfono / Moreno, Magda – Medellin, Colombia. 1948 – 1r – 1 – us UF Libraries [972]

Embury, Aymar *see* Early american churches

Emcc news and activities / East Madison Community Center (Madison, WI) – 1980 jul-sep – 1r – 1 – (cont: newsletter (east madison community center (madison, wi)); cont by: footnotes (madison, wi)) – mf#1830318 – us WHS [360]

Emden, Jacob *see* Megilat sefer

Emder volksblatt – Emden DE, 1848 15 may-1849 30 sep – 1r – 1 – gw Misc Inst [074]

Emder zeitung *see* Rhein-ems-zeitung

Emedia – Wilton. 1999-99 (1,5,9) – (cont: e media professional; cont by: emedia magazine) – ISSN: 1525-4658 – mf#16703,03 – us UMI ProQuest [380]

Emedia – Wilton. 2002+ (1,5,9) – ISSN: 1525-4658 – mf#16703,05 – us UMI ProQuest [380]

Emedia magazine – Wilton. 2001+ (1) – (cont: emedia) – ISSN: 1529-7306 – mf#16703,04 – us UMI ProQuest [380]

Emel mecmuasi – Bazargic, RM. Mesul Mueduerue: Muestecib H Fazil. n1. 1 kanunisani 1930-6,8,11-17,19-22,28,33-36,40-41. 1 eyluel 1931; 3 sene n1. 1 kanunisani 1932, 2,5,10,11; 4 sene n3. 1 mart 1933, 8-12; 5 sene n2. subat 1934, 3,5,10,11; 6 sene n5. mayis 1935 – (= ser O & t journals) – 16mf – 9 – $265.00 – us MEDOC [956]

Emel'ianov, A I *see* Novyi put' [kustanai: 1918-1919]

Emelina / Dario, Ruben – Paris, France. 1927 – 1r – us UF Libraries [972]

Emendas a constituicao de 1946 / Brazil. Congresso Nacional – Brasilia, Brazil. 1970 – n11 on 1r – 1 – us UF Libraries [342]

Emendationen zu stellen des neuen testaments / Koennecke, Clemens – Guetersloh: C Bertelsmann, 1908 – (= ser Beitraege zur foerderung christlicher theologie) – 1mf – 9 – 0-524-06212-9 – mf#1992-0850 – us ATLA [225]

Emendationes et adnotationes ad tertulliani apologeticum (fp12) / ed by Rauschen, G – 1919 – (= ser Florilegium patristicum (fp)) – €5.00 – ne Slangenburg [240]

Emendationes in plerosque sacrae scripturae veteris testamenti libros : Secundum Veterum Versiones Nec Non Auxiliis Criticis Caeteris Adhibitis / Graetz, Heinrich; ed by Bacher, Wilhelm – Breslau [Wroclaw]: Schlesische Buchdr, 1892-1894 – 2mf – 9 – 0-524-07603-0 – mf#1992-1087 – us ATLA [220]

Emerald – Baltimore. 1810-1811 (1) – mf#3738 – us UMI ProQuest [920]

Emerald : or Miscellany of literature, containing sketches of the manners, principles and amusements of the age – Boston. 1806-1808 (1) – mf#3572 – us UMI ProQuest [790]

Emerald and baltimore literary gazette – Baltimore. 1828-1849 (1) – mf#4369 – us UMI ProQuest [420]

Emerald city chronicle – v1 n1-v2 n14 [1977 oct 11/25-1978 jul 11/25], v3 n1-2, [1979 jan 19-feb 2] – 1r – 1 – mf#382318 – us WHS [071]

Emerald city flyer – n3-4 [1979 mar 2/16-16/30] – 1r – 1 – mf#937128 – us WHS [071]

Emerald empire news – Eugene OR: C E Darling, 1961-62 [daily ex weekends] – 1 – us Oregon Lib [071]

l'emeraude, morceaux choisis de litterature moderne – Paris: Urbain Canel et Ad Guyot, 1832 [mf ed 1986] – (= ser Annales romantiques) – 219p – 1 – mf#9635 – us Wisconsin U Libr [800]

Emerge – 1989 oct-1990 nov, 1991 jan-nov – 2r – 1 – (continued by: savoy (new york, ny)) – mf#1711087 – us WHS [071]

Emergence : arts magazine of the n c cultural arts coalition, inc / North Carolina Cultural Arts Coalition – v2 [1980] – 1r – 1 – mf#4851574 – us WHS [700]

Emergence '76 : official publication... / American Revolution Bicentennial Commission of Texas – v2 n1-v3 n1 [1974 win-1975 spr] – 1r – 1 – (cont: bicentennial in texas; bicentennial in texas; cont by: bicentennial in texas (1976)) – mf#936314 – us WHS [975]

Emergency – Torrance. 1985-1998 (1,5,9) – ISSN: 0162-5942 – mf#14425,01 – us UMI ProQuest [610]

Emergency care quarterly – Gaithersburg. 1985-1991 (1,5,9) – ISSN: 8755-8467 – mf#14922 – us UMI ProQuest [610]

The emergency in china / Pott Francis Lister Hawks – New York: Missionary Education Movt of the US & Canada 1913 [mf ed 1995] – (= ser Yale coll) – 1r – 1 – 0-524-09563-9 – mf#1995-0563 – us ATLA [951]

Emergency legislation of the us 1775-1918 : dealing with the control and taking of private property for the public use, benefit, or welfare; presidential proclamations and executive orders thereunder, to and including jan 31, 1918; to which is added a reprint of analogous legislation since 1775 / ed by Clark, J Reuben, Jr – Washington: GPO, 1918 (all publ) – 12mf – 9 – $18.00 – mf#llmc 94-357 – us LLMC [342]

Emergency librarian – Seattle. 1980-1997 (1) 1980-1997 (5) 1980-1997 (9) – (Cont by: Teacher librarian) – ISSN: 0315-8888 – mf#12834 – us UMI ProQuest [020]

Emergency medicine – New York. 1969+ (1) 1969+ (5) 1969+ (9) – ISSN: 0013-6654 – mf#9896 – us UMI ProQuest [610]

Emergency medicine clinics of north america – Philadelphia. 1983+ (1,5,9) – ISSN: 0733-8627 – mf#13381 – us UMI ProQuest [610]

Emergency medicine journal – emj – London, 2001+ [1,5,9] – (cont: journal of accident and emergency medicine) – ISSN: 1472-0205 – mf#15504,02 – us UMI ProQuest [610]

Emergency planning digest – Ottawa. 1979-1986(1,5,9) – (Cont by: Emergency preparedness digest) – ISSN: 0317-3518 – mf#12148 – us UMI ProQuest [360]

Emergency preparedness digest – Ottawa. 1987-1993 (1,5,9) – (Cont: Emergency planning digest) – ISSN: 0837-5771 – mf#12148,01 – us UMI ProQuest [360]

Emergency press – Dearborn Heights. 1973-1973 (1) 1973-1973 (5) (9) – mf#7022 – us UMI ProQuest [130]

Emergency radiology : a journal of practical imaging – v1-3. 1994-1996 – 1 – $80.00r – us Lippincott [616]

Emergency reporter – [NW England] Ashton-under-Lyne, Stalybridge Lib 26 jun-31 jul 1959 – 1 – uk MLA; uk Newsplan [072]

Emergency treatment of sports injuries = Noodbehandeling van sportbeserings / Velden, D P van – Pretoria: The Dept 1986 [mf ed Pretoria, RSA: State Library [199-]] – (= ser Keep well: teachers' guide 5) – 18p [ill] on 1r with other items – 5 – (in english & afrikaans) – mf#op 08471 r26 – us CRL [362]

Emerging colombia / Hunter, John Merlin – Washington, DC. 1962 – 1r – 1 – us UF Libraries [972]

Emerging markets, finance and trade – Armonk. 2002+ (1,5,9) – ISSN: 1540-496X – mf#16894,03 – us UMI ProQuest [380]

Emerging themes of african history / International Congress of African Historians – Nairobi, Kenya. 1968 – 1r – 1 – us UF Libraries [960]

Emerging trends – Princeton. 1988+ (1,5,9) – ISSN: 0889-8936 – mf#15287 – us UMI ProQuest [240]

L'emerillon : organe officiel de l'ordre des commandeurs de jacques-cartier – Ottawa: [s.n.] V1 n1 janv 1930-1965?// [mf ed 1976] – 4r – 5 – (interrupted: dec 1944-oct 1947) – mf#SEM16P256 – cn Bibl Nat [073]

Emerita augusta. apuntes monograficos acerca de la catedral metropolitana de santa maria jerusalen / Gonzalez y Gomez de Soto, Juan Jose – Merida, 1903 – 1 – sp Bibl Santa Ana [240]

Emerson : a lecture / Birrell, Augustine – London: P Green, 1903 [mf ed 1992] – (= ser Essex hall lecture 1903; Unitarian/universalist coll) – 1mf – 9 – 0-524-02245-3 – mf#1990-4252 – us ATLA [080]

Emerson and his friends / Sunderland, Jabez Thomas – Calcutta: R Chatterjee, 1941 – (= ser Samp: indian books) – us CRL [920]

Emerson, Brown, 1778-1872 *see* The causes and effects of war

Emerson, Caleb *see* The caleb emerson family papers, 1795-1905

The emerson crescent – Emerson, NE: I C Trumbauer, nov 1903 (wkly) [mf ed 1904-05 (gaps)] – 1r – 1 – us NE Hist [071]

Emerson, Edward Randolph *see* A lay thesis on bible wines

Emerson, Edward Waldo *see* Emerson in concord

Emerson enterprise – Emerson, NE: W F Bancroft, 1892 (wkly) [mf ed -1914 (gaps)] – 9r – 1 – (cont by: emerson tri-county press) – us NE Hist [071]

Emerson, George Homer *see*
- The bible and modern thought
- The doctrine of probation examined
- Life of alonzo ames miner, s.t.d., ll.d
- Memoir of ebenezer fisher, d.

Emerson in concord : a memoir / Emerson, Edward Waldo – Boston: Houghton, Mifflin, 1889 – 1mf – 9 – 0-524-01083-8 – mf#1990-4048 – us ATLA [420]

Emerson, Oliver Farrar *see* The earliest english translations of buerger's lenore

Emerson, Ralph Waldo *see*
- An address delivered before the senior class
- The collection of ralph waldo emerson, 1822-1903
- The conduct of life
- The correspondence of thomas carlyle and ralph waldo emerson, 1834-1872
- Letters and social aims
- Natural history of intellect
- Nature
- Society and solitude
- Works of ralph waldo emerson

The emerson tri-county press – Emerson, NE: M R Blakee. -v94 n31. jul 30 1985 (wkly) [mf ed 1934-85 (gaps) filmed 1974-] – 18r – 1 – (cont: emerson enterprise. absorbed by: nebraska journal-leader) – us NE Hist [071]

Emerson, Wilimena H *see* The descendants of john eliot from 1598-1905

Emerson, William Canfield *see* Seminoles

Emerson's magazine and putnam's monthly – 1858 jan-jun – 1r – 1 – mf#2697708 – us WHS [071]

Emerson's magazine and putnam's monthly – New York. 1854-1858 – 1 – mf#5316 – us UMI ProQuest [073]

Emerton, Ephraim *see*
- Desiderius erasmus of rotterdam
- An introduction to the study of the middle ages
- Mediaeval europe
- Unitarian thought

Emery, James Augustin *see* Combination and social progress

Emery, Louis *see* Introduction a l'etude de la theologie protestante

Emery, Michael S *see* Exercise induced hypoxemia as a determinant of maximal aerobic capacity

Emery-Coderre, Joseph *see* Jurisprudence medicale

Emery's journal of agriculture – v2 n1-14 [1858 jul 1-sep 30] – 1r – 1 – (continued by: prairie farmer (chicago, il: 1843)) – mf#574019 – us WHS [630]

Der emes – (New York). 1921 – 1 – us AJPC [939]

Emes – Moscow. 1921-1935. (incomplete) – 1 – us NY Public [073]

Emet ve-emunah / Mirkin, Katriel Zevi – St Petersburg, Russia. 1905 – 1r – 1 – us UF Libraries [939]

Emeth – London, UK. 7 Jun 1931 – 1 – uk British Libr Newspaper [072]

The emeth (truth) – Boston. Die Wahrheit. 1895-96 – 1 – us AJPC [335]

Les emeutiers : les deux lundis / Levy, Armand & Valleton, Henri – Paris: Impr Lacour et cie, [1849?] – us CRL [074]

An emg study of four elastic tubing closed kinetic chain exercises : a preliminary study / Bachler, Levi R – Brigham Young University, 1994 – 1mf – 9 – mf#PE 3627 – us Kinesology [617]

The emi pathe film library catalogue : 75 years of newsreels, 1896-1970 – 68r – 1 – £2250.00 – (part 1: subject and personality ind part 2: locations ind) – mf#EMI – uk World [790]

Emig, William Harrison *see* Stain technique

La emigracion en baleares y canarias / Diaz Perez, Nicolas – 1882 – 9 – sp Bibl Santa Ana [946]

La emigracion extremena a indias : siglo 16. fichero documental / Rubio, Angel – Santiago de Chile: Imprenta Universitaria, 1948 – 1 – sp Bibl Santa Ana [946]

The emigrant : and other poems / McLachlan, Alexander – Toronto: Rollo & Adam, 1861 [mf ed 1984] – 3mf – 9 – 0-665-46136-4 – mf#46136 – cn Canadiana [810]

The emigrant / Head, Francis Bond – London: J Murray, 1846 [mf ed 1984] – 5mf – 9 – 0-665-45546-1 – mf#45546 – cn Canadiana [971]

The emigrant and sportsman in canada : some experiences of an old country settler. with sketches of canadian life, sporting adventures, and observations on the forests and fauna / Rowan, John J – London 1876 – (= ser 19th c british colonization) – 5mf – 9 – mf#1.1.7315 – uk Chadwyck [790]

Emigrant savings bank records, 1841-1945 – 23r – 1 – $2645.00 – (guide sold separately: d3479.g $15) – mf#D3479 – us NY Public [332]

Emigrants : a selection of 17 guides and pamphlets. from the british library and other libraries – (= ser BRRAM series) – 2r – 1 – £134 / $268 – (int by charlotte erickson) – mf#r95799 – uk Microform Academic [304]

The emigrants / Imlay, Gilbert & Wollstonecraft, Mary – 1793 – 9 – us Scholars Facs [830]

The emigrant's guide to australia : with a memoir of mrs chisholm / Mackenzie, Eneas – London [1853] – (= ser 19th c british colonization) – 2mf – 9 – mf#1.1.3544 – uk Chadwyck [919]

The emigrant's guide to new brunswick, british north america / Atkinson, Christopher William – Berwick-upon-Tweed, England?: s.n, 1842 – 2mf – 9 – mf#28523 – cn Canadiana [917]

The emigrant's guide to upper canada : or, sketches of the present state of that province; collected from a residence therein during the years 1817, 1818, 1819; interspersed with reflections / Stuart, Charles – London 1820 [mf ed Hildesheim 1995-98] – 1v on 2mf – 9 – €60.00 – 3-487-27095-1 – gw Olms [917]

The emigrant's introduction to an acquaintance with the british american colonies : and the present condition and prospects of the colonists / Hill, S S – London: Parbury, 1837 – 4mf – 9 – mf#37257 – cn Canadiana [971]

The emigrant's note book and guide : with recollections of upper and lower canada, during the late war / Morgan, J C – London 1824 [mf ed Hildesheim 1995-98] – 1v on 3mf – 9 – €90.00 – 3-487-27087-0 – gw Olms [920]

L'emigration : quelques conseils aux emigrants / Vekeman, Gustave – Sherbrooke, Quebec?: s.n, 1884 – 1mf – 9 – mf#25440 – cn Canadiana [320]

Emigration : considered chiefly in reference to the practicability and expediency of importing and of settling throughout the territory of new south wales, a numerous, industrious and virtuous agricultural population / Lang, John Dunmore – Sydney, 1833 – (= ser 19th c british colonization) – 1mf – 9 – mf#1.1.3501 – uk Chadwyck [304]

Emigration : free, assisted, and full-paying passages. together with the conditions for obtaining free land grants, rules for emigration clubs etc / Bate, John – [London], [1869] – (= ser 19th c books on british colonization) – 1mf – 9 – (publ under the authority of the national emigration aid society) – mf#1.1.9000 – uk Chadwyck [304]

Emigration : its advantages to great britain and her colonies. together with a detailed plan for the formation of the proposed railway between halifax and quebec, by means of colonization / MacDougall, Patrick Leonard – London 1848 – (= ser 19th c british colonization) – 1mf – 9 – mf#1.1.523 – uk Chadwyck [304]

Emigration : papers relative to emigration to the british provinces in north america [in coninuation of the papers presented dec 1847] – London: printed by W Clowes for HMSO 1848 [mf ed 1987] – 1mf – 9 – 0-665-63351-3 – mf#63351 – cn Canadiana [320]

Emigration : where to go, and who should go. new zealand and australia (as emigration fields) in contrast with the united states and canada. canterbury and the diggings / Hursthouse, Charles Flinders – London, [1852?] – (= ser 19th c british colonization) – 2mf – 9 – mf#1.1.5631 – uk Chadwyck [304]

Emigration : Who should emigrate. How to emigrate. And where to emigrate / Aspdin, James – (= ser 19th c books on british colonization) – 1mf – 9 – mf#1.1.4697 – uk Chadwyck [304]

Emigration. / Maconochie, Alexander – London. 1848 – 1 – us CRL [320]

Emigration and immigration : a study in social science / Mayo-Smith, Richmond – London 1890 – (= ser 19th c british colonization) – 4mf – 9 – mf#1.1.5942 – uk Chadwyck [304]

Emigration and superabundant population considered, in a letter to lord ashley : by amicus populi – London 1848 – (= ser 19th c british colonization) – 1mf – 9 – mf#1.1.6779 – uk Chadwyck [304]

Emigration en canada : description du pays: ses avantages : la terre promise du cultivateur / Bodard, Auguste – S.l: s.n, 1891? – 1mf – 9 – mf#30025 – cn Canadiana [304]

Emigration fields : north america, the cape, australia, and new zealand, describing these countries, and giving a comparative view of the advantages they present to british settlers / Matthew, Patrick – Edinburgh: A & C Black, 1839 [mf ed 1983] – 4mf – 9 – 0-665-38230-8 – mf#38230 – cn Canadiana [320]

Emigration from india : the export of coolies, and other labourers, to mauritius – London, 1842 – (= ser 19th c books on british colonization) – 1mf – 9 – mf#1.1.987 – uk Chadwyck [331]

Emigration from the british islands : considered with regard to its bearing and influence upon the interests and prosperity of great britain – London, 1862 – (= ser 19th c books on british colonization) – 1mf – 9 – mf#1.1.9512 – uk Chadwyck [304]

Emigration from the british islands : considered with regard to its bearing and influence upon the interests and prosperity of great britain – London: J Ridgeway, 1862 [mf ed 1983] – 1mf – 9 – 0-665-44468-0 – mf#44468 – cn Canadiana [320]

Emigration gazette – London, 23 Oct 1841-13 May 1843 – 1r – 1 – uk British Libr Newspaper [072]

Emigration, land and railway frauds : the "colonists' handbook", canada, 1882 / Hind, Henry Youle – [Windsor, NS?: s.n, 1882?] [mf ed 1994] – 1mf – 9 – 0-665-94578-7 – mf#94578 – cn Canadiana [364]

The emigration of gentlemen's sons to the united states and canada / Bradley, Arthur Granville – Westminster England: Women's Printing Society, 18-- – 1mf – 9 – mf#00981 – cn Canadiana [971]

Emigration to british north america under the early passenger acts, 1803-1842 / Walpole, Kathleen A – London University 1929 – (= ser BRRAM theses coll) – 1r – 1 – £57 / $114 – mf#r19098 – uk Microform Academic [304]

Emil brauns briefwechsel : mit den bruedern grimm und joseph von lassberg / ed by Ehwald, R – Gotha: F A Perthes, 1891 [mf ed 1989] – xii/169p – 1 – (incl bibl and ind) – mf#7063 – us Wisconsin U Libr [860]

Emil goetts vermaechtnis / Droop, Fritz – Konstanz a.B.: Reuss & Itta, 1917 – 1 – us Wisconsin U Libr [430]

The emil j gumbel collection : political papers of an anti-nazi scholar in weimar and exile, 1914-1966 – 8r – 1 – $1375.00 – 1-55655-212-2 – (with p/g. filmed fr holdings of the leo baeck institute, new york city) – us UPA [320]

Emil kuhs kritische und literarhistorische aufsaetze, 1863-1876 / ed by Schaer, Alfred – Wien: Literarischer Verein, 1910 – xvi/457p – 1 – us Wisconsin U Libr [840]

Emile, or, concerning education : extracts: containing the principal elements of pedagogy found in the first three books / Rousseau, Jean-Jacques – Boston: D C Heath, 1898, c1893 – 1mf – 9 – 0-8370-7664-1 – mf#1986-1664 – us ATLA [370]

Emile zola : principes et caracteres generaux de son oeuvre / Robert, Guy – Paris, France. 1952 – 1r – 1 – us UF Libraries [440]

Emilia / Arvelo, Teresa – Guatemala, 1961 – 1r – us UF Libraries [972]

Emilia galotti / Lessing, Gotthold Ephraim; ed by Gast, E R – Goth [Germany]: F A Perthes, 1886 – 1r – 1 – (incl bibl ref) – us Wisconsin U Libr [820]

Emily c judson : a memorial / Wyeth, Walter Newton – Philadelphia: the aut, 1890 [mf ed 1984] – (= ser Women & the church in america 140; Missionary memorials 3) – 1mf – 9 – 0-8370-1453-0 – mf#1984-2140 – us ATLA [240]

Emily, Jules *see* Mission marchand

Emily n blair family papers *see* Blair, emily n, family papers, ms 4342

Emily newell blair family papers, 1785-1972 / Blair, Emily Newell – [mf ed 1988] – 17r – 1 – mf#ms4342 – us Western Res [305]

Emin pascha : ein deutscher forscher und kaempfer im innern afrikas / Staby, Ludwig – Stuttgart 1890 [mf ed Hildesheim 1995-98] – 1v on 2mf [ill] – 9 – €60.00 – 3-487-27315-2 – gw Olms [960]

Emin pascha : eine sammlung von reisebriefen und berichten dr emin-pascha's aus den ehemals aegyptischen aequatorialprovinzen und deren grenzlaendern / ed by Schweinfurth, Georg – Leipzig 1888 [mf ed Hildesheim 1995-98] – 1v on 4mf [ill] – 9 – €120.00 – 3-487-27313-6 – gw Olms [910]

Emin pasha and the rebellion at the equator : a story of nine months' experiences in the last of the soudan provinces... / Jephson, A J Mounteney – London, 1890 – 7mf – 9 – mf#HT-95 – ne IDC [918]

Emin, Resulzade Mehmet *see* Azerbaycan cumhuriyet keyfiyeti tesekkuelue ve simdiki vaziyeti

Eminent indians on indian politics : with sketches of their lives, portraits and speeches... / ed by Parekh, Chunilal Lalubhai – Bombay 1892 – (= ser 19th c british colonization) – 7mf – 9 – mf#1.1.6224 – uk Chadwyck [954]

Eminent missionary women / Gracey, J T [Mrs] – New York: Eaton & Mains; Cincinnati: Curts & Jennings, 1898 [mf ed 1977] – 3mf – 9 – 0-8370-0186-2 – mf#1984-2109 – us ATLA [305]

Eminent piety essential to eminent usefulness / Reed, Andrew – London, England. 1833 – 1r – 1 – us UF Libraries [240]

L'emirat des trarzas / Marty, Paul – Paris: E Leroux, 1919 – 1 – us CRL [960]

Emk quarterly – Staten Island. 1969-1972 (1) – ISSN: 0012-7752 – mf#5989 – us UMI ProQuest [920]

Emlekfuzet a satoraljaujhelyi aut orth izr hitkozsegi – Satoraljaujhely, Hungary. 1912 – 1r – 1 – us UF Libraries [939]

Emlekfuzete a magyar zsidosag egyenjogositasanak 5o evfordulojara – Budapest, Hungary. 1917 – 1r – 1 – us UF Libraries [939]

Emlekkoenyve / Federation of European National Societies of the Theosophical Society. Congress – Budapest: Magyar Teozofiai Tarsulat, 1929 [irreg] [mf ed 2004] – 1v on 1r – 1 – (mf: 9th [1929]. no transactions publ at 10th & 11th congresses. film incl earlier & later titles: transactions of the...annual congress of the federation of european sections of the theosophical society; transactions of the...congress of the federation of european national societies of the theosophical society; federation of european national societies of the theosophical society, congreso de barcelona; and: history of the efts summary) – mf#2003-s124d – us ATLA [290]

Emlekkonyv dr kiss arnold, budai vezeto forabbi, hetvenedik szulet – Budapest, Hungary. 1939 – 1r – 1 – us UF Libraries [939]

Emma / Hodgkins, B – London, England. 18– – 1r – us UF Libraries [240]

Emma : Ou, Un Ange Gardien / Laya, Leon – Paris, France. 1844 – 1r – us UF Libraries [440]

Emma Goldman Clinic for Women *see*
– Emma's periodical rag
– Newsletter

Emma goldman papers : span cultures and continents, tracing the origins of significant social movements / University of California, Berkeley. The Emma Goldman Project – [mf ed Chadwyck-Healey] – 69r – 13 – (with p/g ed by candace falk. foreword by leon f litwack) – uk Chadwyck [305]

Emmanuel baptist church. calinville, illinois : church records – 1921-79 – 1 – us Southern Baptist [242]

Emmanuel baptist church. jefferson city, tennessee : church records – 1924-65 – 1 – us Southern Baptist [242]

Emmanuel Lutheran Church, Hoisington, KS *see* Records

The emmanuel movement in a new england town : a systematic account of experiments and reflections designed to determine the proper relationship between the minister and the doctor in the light of modern needs / Powell, Lyman Pierson – New York: G P Putnam, 1909 [mf ed 1990] – 1mf – 9 – 0-7905-5618-9 – (incl bibl ref) – mf#1988-1618 – us ATLA [241]

Emmanuelis barradas s i tractatus tres historico-geographici / Barradas, M – Romae: C de Luigi, 1906 – 5mf – 9 – mf#SEP-13 – ne IDC [910]

Emmanuelis...in communi patriae plausu celebrata et demississime dicata... / Fama prognostica ad cunas serenissimi principis Maximiliani – [Munich]: Typis Lucae Straubii, 1662 – 1mf – 9 – mf#O-1570 – ne IDC [090]

Emma's periodical rag / Emma Goldman Clinic for Women – 1979 oct-1981 fall – 1r – 1 – (cont: newsletter (emma goldman clinic for women)) – mf#672686 – us WHS [305]

L'emmaus di s luca / Bazzocchini, Benvenuto – Roma: Frederico Pustet, 1906 [mf ed 1993] – 1mf – 9 – 0-524-07333-3 – (in italian) – mf#1992-1064 – us ATLA [225]

Emmel, Hildegard *see* Masken in volkstuemlichen deutschen spielen

Emmel, M W *see*
– Etiology of fowl paralysis, leukemia and allied conditions in animals
– Field experiments in the use of sulfur to control lice, fleas and mites of chickens
– Toxic principle of the tung tree

Emmel, M W (Mark Wirth) *see* "Swollen joints" in range calves

Emmendingen als schauplatz von goethes hermann und dorothea / Hagen, Rosa – Emmendingen: Doelter, 1912 – 1r – 1 – us Wisconsin U Libr [430]

Der emmentaler bauer bei jeremias gotthelf : ein beitrag zur baeuerlichen ethik / Barthel, Helene – Muenster in Westf: Verlag der Aschendorffschen Verlagsbuchhandlung, 1931 [mf ed 1989] – (= ser Veroeffentlichungen der volkskundlichen kommission des provinzialinstituts fuer westfaelische landes- und volkskunde. 1. reihe 3) – vii/147p – 1 – (incl bibl ref) – mf#7028 – us Wisconsin U Libr [430]

Emmerich, Anna Katharina *see* Leben der heil. jungfrau maria

Emmerich, Klaus Dieter *see* Primaere mechanische infarktgefaessrekanalisation im akuten myokardinfarkt

Emmerich, Kristin *see* Kephalometrische untersuchung der skelettalen und dentalen veraenderungen mit der elasto-headgear-apparatur

Emmert, David *see*
– After twenty-five years
– Reminiscences of juniata college

Emmet county republican – Estherville, IA. 1887-1902 (1) – mf#63203 – us UMI ProQuest [071]

Emmet, Cyril William *see* St paul's epistle to the galatians

Emmet, John *see* Religious art

Emmius, U *see*
– Den david-jorischen gheest in leven ende leere
– Ein grundlick bericht van der lere und dem geist des ertzketters david joris

Emmius, [U] *see*
– Grondelicke onderrichtinghe
– Guilhelmus ludovicus comes nassovius...
– Mensonis altingil vita...

El emmo cardenal goma / Bayle, Constantino – Madrid: Razon y Fe, 1940 – sp Bibl Santa Ana [240]

Emmons, Michael L *see* Affirmez vous!

Emmons, Nathanael *see* The cambridge platform of church discipline

Emmott, Elizabeth B *see* Loving service

Emmott, Elizabeth Braithwaite *see*
– Loving service
– The story of quakerism

Emniyet – Filibe, 1896-97. Sahib-i Imtiyaz ve Mueduer-i Mes'ul: Emin Tevfik; Mueduer ve Sermuharriri: Selaniki Hilmi. n1-72. 6 mayis 1312-12 temmuz 1313 [1896-97] – (= ser 0 & t journals) – 3mf – 9 – $95.00 – us MEDOC [956]

Emo national digest / Canada. Emergency Measures Organization – Ottawa. 1961-1974 (1) 1972-1973 (5) (9) – ISSN: 0012-7787 – mf#6807 – us UMI ProQuest [350]

Emoan, Max *see* Monografie der reproductiven phaenomene

Emory international law review – v1-14. 1986-2000 – 9 – $293.00 set – (Title varies: v1-3 1986-89 as emory journal of international dispute resolution) – ISSN: 1052-2840 – mf#110491 – us Hein [341]

Emory, John *see*
– A defence of "our fathers" and of the original organization of the methodist episcopal church against the rev alexander m'caine and others
– The episcopal controversy reviewed

Emory journal of international dispute resolution *see* Emory international law review

Emory law journal – Atlanta. 1974+ (1) 1974+ (5) 1976+ (9) – (Cont: Journal of public law) – ISSN: 0094-4076 – mf#3498,01 – us UMI ProQuest [340]

Emory law journal – v1-49. 1952-2000 – 5,6,9 – $975.00 set – (v1-33 1952-84 on reel or mf $484. v34-49 1985-2000 on mf $491. title varies: v1-22 1952-73 as journal of public law) – ISSN: 0094-4076 – mf#102581 – us Hein [343]

Emory, Robert *see*
– The episcopal controversy reviewed
– History of the discipline of the methodist episcopal church

Emory university quarterly – Atlanta. 1945-1967 – 1 – ISSN: 0884-4844 – mf#298 – us UMI ProQuest [378]

Emory, W H *see* Report on the united states and mexican boundary survey

Emotional and behavioural problems adolescent learners experience without proper pedagocial guidance and assistance : with reference to temba district in the north west province / Kadiege, Keti Kolobe Herman – Pretoria: Vista University 2003 [mf ed 2003] – 2mf – 9 – (incl bibl ref) – mf#mfm15314 – sa Unisa [362]

Die emotionale einstellung agoraphobischer patienten und ihrer partner als praediktor fuer den therapieerfolg / Rodde, Sibyll – (mf ed 1998) – 2mf – 9 – €40.00 – 3-8267-2555-7 – mf#DHS 2555 – gw Frankfurter [150]

The emotions / McCosh, James – New York: Scribner, 1880 – 1mf – 9 – 0-7905-7530-2 – mf#1989-0755 – us ATLA [150]

Emotions and cognitions of athletes competing in a high-risk sport / Durtschi, Shirley K – 1998 – 324p on 4mf – 9 – $20.00 – mf#PSY 2175 – us Kinesology [150]

The emotions and the will / Bain, Alexander – 4th ed London: Longmans, Green, 1899 – 2mf – 9 – 0-7905-3528-9 – mf#1989-0021 – us ATLA [100]

Empac! : newsletter of ethnic millions political action / Ethnic Millions Political Action Committee – n7-n12 [1976 mar-1977 jun] – 1r – 1 – (cont: new america) – mf#638843 – us WHS [322]

Empadronamiento de moriscos de granada (anno 1573-1595, 1610, 1733) – Cordoba – 1r – 5,6 – sp Cultura [946]

Empangan darah : satoe lelakon tertarik dari perang besar taon 1906 / Tan, Boen Soan – Soerabaia: Tan's Drukkerij, 1935 [mf ed 1998] – (= ser Penghidoepan 130) – 1r – 1 – (coll as pt of the colloquial malay collection. filmed with: multi-millionair / ong khing han) – mf#10002 – us Wisconsin U Libr [830]

Emparons-nous de l'industrie / Bouchette, Errol – Ottawa: Impr generale, 1901 [mf ed 1985] – 1mf – 9 – mf#SEM105P466 – cn Bibl Nat [338]

Empedocle : ou, le philosophe de l'amour et de la hain / Brun, Jean – Paris: Seghers 1966 [mf ed 1985] – (= ser Philosophes de tous les temps 27) – 1r – 1 – (incl bibl) – mf#1511 – us Wisconsin U Libr [180]

El emperador carlos 5 / Mignet, Mr – 1855 – 9 – sp Bibl Santa Ana [920]

El emperador carlos 5...yuste / Mignet, Mr; ed by Lobo, Miguel – 1855 – 9 – sp Bibl Santa Ana [946]

El emperador d pedro 2 y el instituto historico (5) / Affonse Celso de Assis Figueiredo – Buenos Aires: [imprenta mercatali] 1938 [mf ed 2000] – (= ser Biblioteca de autores brasilenos traducidos al castellano) – 1r – 1 – mf#*Z-9268 – us NY Public [972]

L'empereur almamy samori toure : grand administrateur et grand stratege – Conakry: Imprimerie nationale, 1971 – us CRL [920]

The emperor / Payne, Robert – London: William Heinemann Ltd, 1953 – (= ser Samp: indian books) – us CRL [954]

The emperor akbar, a contribution towards the history of india in the 16th century / Noer, Graf Friedrich Christian Karl August von – Trans. and partly rev. by Annette S. Beveridge. Calcutta: Thacker, Spink & Co; London: Truebner, 1890. 2v – 1 – us Wisconsin U Libr [920]

Emperor alexander – Glasgow? Scotland. 18– – 1r – 1 – us UF Libraries [240]

The emperor hadrian : a picture of the graeco-roman world in his time = Geschichte des roemischen kaisers hadrian und seiner zeit / Gregorovius, Ferdinand – London, New York: Macmillan, 1898 – 1mf – 9 – 0-7905-5228-0 – (incl bibl ref. in english) – mf#1988-1228 – us ATLA [930]

The emperor julian and his generation : an historical picture = Ueber den kayser julianus und sein zeitalter / Neander, August – London: J.W. Parker, 1850 – 1mf – 9 – 0-7905-5957-9 – (incl bibl ref. in english) – mf#1988-1957 – us ATLA [930]

Empey, Arthur Guy *see* "Over the top"

Empey, Michael D *see* An investigation of the career mobility patterns of national football league head coaches

Empfindsame reise nach schilda / [Rebmann, A G F] – Leipzig, 1793 – 3mf – 9 – mf#HT-218 – ne IDC [914]

Die empfindsamen in darmstadt : studien ueber maenner und frauen aus der wertherzeit / Tornius, Valerian – Leipzig: Klinkhardt & Biermann, [1910] – 1r – 1 – (incl bibl ref) – us Wisconsin U Libr [430]

Ein empfindsamer besuch im invaliden-hotel zu paris : nebst historischen notizen ueber dessen entstehung, fortgang und gegenwaertigen zustand – Berlin 1855 [mf ed Hildesheim 1995-98] – 1v on 1mf – 9 – €40.00 – 3-487-25926-5 – gw Olms [914]

Emphasis on faith and living / Missionary Church in the United States and Canada – 1979 jan 15-1982 jun, 1982 jul-1985 – 2r – 1 – (cont: gospel banner (goshen, in); missionary banner (elkhart, in); emphasis! (fort wayne, in: 1967); cont by: missionary church today) – mf#609566 – us WHS [240]

The emphasised bible : a new translation, designed to set forth the exact meaning, the proper terminology and the graphic style of the sacred originals / Rotherham, Joseph Bryant – London: H R Allenson, 1902 – 12mf – 9 – 0-8370-1887-0 – mf#1987-6274 – us ATLA [220]

Empire / Fabian Society (Great Britain). Colonial Bureau – London: Fabian Society 1938-72 – 53mf – 9 – (1949-72 called: venture; suspended: 1939-41) – uk Microform Academic [335]

Empire – Toronto. v1-8 n2215. dec 27 1887-feb 6 1895// (daily) – 22r – 1 – Can$2420.00 – (founded as the new organ of macdonald conservatism; before it was absorbed by the mail in 1895, it was the most influential conservative daily in canada) – cn McLaren [071]

Empire – Juneau, AK. 1970-2000 (1) – mf#60406 – us UMI ProQuest [071]

Empire – New Orleans LA. 1876 nov 26 – 1r – 1 – mf#861949 – us WHS [071]

Empire / New York State Legislative Institute – v4 n5-v5 n5 [1978 oct/nov-1979 oct/nov] – 1r – 1 – (cont: empire state report; cont by: empire state report) – mf#577310 – us WHS [320]

Empire – Sydney, 1850-75 – 45r – 1 – A$1732.50 vesicular A$1980.0 silver – at Pascoe [079]

Empire – Toronto, Canada. 12 jan 1888-6 feb 1895 (wanting jun-nov 1890) – 27 1/2r – 1 – uk British Libr Newspaper [071]

The empire – London, England. Jan 1854-Jun 1856 – 3 1/2r – 1 – uk British Libr Newspaper [072]

The empire – Toronto, Canada. -w, -d. 12 Jan 1888-6 Feb 1895. Lacking June-Nov 1890. 27 1 2 reels – 1 – uk British Libr Newspaper [072]

Empire and colonial administration : the papers of lachlan macquarie (1762-1824) and family from the state library of new south wales – 8r – 1 – £750.00 – (with d/g) – uk Matthew [320]

Empire and commonwealth : archives of the royal commonwealth society from cambridge university library – 2pt – 1 – (pt1: colour question in imperial policy c1830-1939 [25r] £2350; pt2: imperial & commonwealth conferences 1887-1955 [18r] £1700; with d/g) – uk Matthew [900]

Empire builder – Coos Bay OR: J F Kutch, 1966-75 [wkly] – 1 – (cont: empire charleston builder; cont by: builder (coos bay, or)) – us Oregon Lib [071]

Empire builder (coos bay, or) – Coos Bay OR: W N & M E Grannell, 1977-78 [wkly] – 1 – (cont: coos bay empire builder; cont by: bay reporter) – us Oregon Lib [071]

Empire charleston builder – Empire OR: C S McDonald, 1953-66 [wkly] – 1 – (cont by: empire builder (1966-75). place of publ moved to coos bay feb 11 1965) – us Oregon Lib [071]

L'empire chinois : le bouddhisme en chine et au thibet / Lamairesse – Paris: E Flammarion [1893?] [mf ed 1992] – 1mf – 9 – 0-524-04986-6 – (in french) – mf#1990-3444 – us ATLA [280]

L'empire chinois : faisant suite a l'ouvrage intitule souvenirs d'un voyage dans la tartarie et le thibet / Huc, evariste – Paris 1854 [mf ed Hildesheim 1995-98] – 2v on 6mf – 9 – €120.00 – 3-487-27624-0 – gw Olms [915]

Empire cotton growing review, 1924-58 – v1-35 – 10r – 1 – mf#146 – uk Microform Academic [630]

Empire Cream Separator Co *see* Empire milking machines

Empire du bresil / Brazil. Commissao Brazileira na Exposicao Universa... – Rio de Janeiro, Brazil. 1873 – 1r – 1 – us UF Libraries [972]

Empire du bresil / Roy, J-J-E – Tours, France. 1861 – 1r – 1 – us UF Libraries [972]

Empire in asia : how we came by it: a book of confessions / McCullagh Torrens, William – Allahabad: LM Basu, 1938 – (= ser Samp: indian books) – us CRL [950]

Empire in brazil / Haring, Clarence Henry – Cambridge, MA. 1958 – 1r – 1 – us UF Libraries [972]

The empire in india : letters from madras and other places / Bell, Evans – London, 1864 – (= ser 19th c books on british colonization) – 5mf – 9 – mf#1.1.6071 – uk Chadwyck [954]

Empire journal of experimental agriculture – Oxford. 1933-1964 (1) – mf#1244 – us UMI ProQuest [630]

Empire, jun 1938-jan 1949/venture, feb 1949-aug 1972/third world, sept 1972-jun 1975 – Fabian Society. 1938-jun 1975 – 59mf – 9 – mf#86989 – uk Microform Academic [073]

L'empire khmer : histoire et documents / Maspero, Georges – Phnom-Penh: Impr du Protectorat 1904 [mf ed 1989] – 1r with other items – 1 – (with bibl footnotes) – mf#mf-10289 seam reel 006/02 [§] – us CRL [930]

L'empire liberal : etudes, recits, souvenirs / Ollivier, Emile – Paris. v1-17. 1899-1915 – 1 – $180.00 – (in french) – mf#0435 – us Brook [944]

Empire milking machines / Empire Cream Separator Co – Bloomfield NJ; Montreal: Empire Cream Separator Co [1918?] [mf ed 1994] – 1mf – 9 – 0-665-72778-X – mf#72778 – cn Canadiana [630]

The empire of christ : being a study of the missionary enterprise in the light of modern religious thought / Lucas, Bernard – London: Macmillan, 1907 – 1mf – 9 – 0-8370-6208-X – mf#1986-0208 – us ATLA [240]

The empire of the nabobs : a short history of british india / Hutchinson, Lester – London: George Allen and Unwin Ltd, 1937 – (= ser Samp: indian books) – us CRL [954]

The empire of the ptolemies / Mahaffy, John Pentland – London: Macmillan, 1895 – 2mf – 9 – 0-524-04470-8 – (incl bibl ref) – mf#1992-0139 – us ATLA [930]

L'empire peul du macina / Ba, Amadou Hampate – 1955 – 1 – us CRL [960]

Empire press – Waterville, WA. 1921-1983 (1) – mf#67178 – us UMI ProQuest [071]

Empire star – Rochester, NY. 1948-1960 (1) – mf#65188 – us UMI ProQuest [071]

Empire state report – Albany. 1974+ (1,5,9) – ISSN: 0747-0711 – mf#11453 – us UMI ProQuest [350]

Empire state report / New York State Legislative Institute – v5 n6-v7 n28 [1979 dec-1981 jul 20] – 1r – 1 – (cont: empire) – mf#577311 – us WHS [323]

Empire state report / New York State Legislative Institute – v1 n1-v4 n4 [1978 jan/feb-sep 15] – 1r – 1 – mf#578273 – us WHS [320]

Empire true fissure *see* Clear creek county miscellaneous newspapers

Empire writes back – ca 10r [mf ed summer 2003] – 1 – $1300.00 – (pt1: indian views on britain & empire, 1810-1915, fr british library, london [10r] £950; pt2: black & asian visitors to britain, 1734-1942 [16r] £1500; with d/g) – uk Matthew [305]

Empires and emperors of russia, china, korea, and japan.notes and recollections / Vay, Peter – New York: Dutton, 1906. xxxii, 399p. illus – 1 – us Wisconsin U Libr [950]

Empires of the veld : being fragments of the unwritten history of the two late boer republics, with other papers for the most part descriptive of the life and character of the people / Kok, K J de – Durban: J C Juta, 1904 – 1 – us CRL [960]

Empirical psychology : or, the science of mind from experience / Hickok, Laurens Perseus – rev ed. Boston: Ginn, Heath, 1884, c1882 – 1mf – 9 – 0-7905-3954-3 – mf#1989-0447 – us ATLA [150]

Empirical research in theatre – Bowling Green. 1981-1981 (1,5,9) – ISSN: 0361-2767 – mf#12026 – us UMI ProQuest [790]

Empirical study of class actions in four federal district courts : final report... / Willging, Thomas E et al – 1996 – 3mf – 9 – $4.50 – mf#llmc99-024 – us LLMC [347]

An empirical study of rule 11 sanctions / Kassin, Saul M – Washington: FJC, 1985 – 1mf – 9 – $1.50 – mf#LLMC 95-324 – us LLMC [340]

Empirical study of the development of factory shops in the clothing industry in the cape peninsula / Visser, Dirk Jacobus – U of the Western Cape 1991 [mf ed S.l: s.n.] 1991 – 3mf – 9 – (summary in afrikaans & english; incl bibl) – sa Misc Inst [380]

Empirismus und skepsis in dav hume's philosophie : als abschliessender zersetzung der englischen erkenntnisslehre, moral und religionswissenschaft / Pfleiderer, Edmund – Berlin: G Reimer, 1874 – 2mf – 9 – 0-7905-8718-1 – mf#1989-1943 – us ATLA [190]

L'emplacement du baucher de michel servet / Doumergue, Emile – Geneve: A Jullien, 1903 [mf ed 1992] – 1mf – 9 – 0-524-03578-4 – (in french) – mf#1990-1038 – us ATLA [930]

L'emplacement du fort de dollard des ormeaux : etude reproduite de la revue d'histoire de l'amerique francaise / Morin, Victor – Montreal: [s.n.] 1952 [mf ed 1987] – 1mf – 9 – mf#SEM105P764 – cn Bibl Nat [971]

L'employe cegetiste des mines – Lens. n1-5, 10-12, 15-16, 18, 20, 22-24, 26. mai 1937-juin 1939 – 1 – fr ACRPP [622]

Employee assistance quarterly / ed by McClellan, Keith – v1- 1985- – 1, 9 ($250.00 in US $350.00 outside hardcopy subsc) – us Haworth [360]

Employee benefit plan review – Chicago. 1973+ (1,5,9) – ISSN: 0013-6808 – mf#9071 – us UMI ProQuest [331]

Employee benefits journal – Brookfield. 1985+ (1,5,9) – ISSN: 0361-4050 – mf#15123 – us UMI ProQuest [331]

Employee benefits report – Boston. 1989-1993 (1) – ISSN: 0884-478X – mf#11837,02 – us UMI ProQuest [650]

Employee health and fitness – Atlanta. 1983+ (1,5,9) – ISSN: 0199-6304 – mf#12278 – us UMI ProQuest [331]

Employee news / Sta-Rite Industries, Inc – v1 n1-v9 [i.e. 8] n4 [1976 spr-1983 fall] – 1r – 1 – (cont: sta-rite conveyor; cont by: today (delavan, wi)) – mf#1476971 – us WHS [331]

Employee perceptions of share ownership schemes : an empirical study / Mazibuko, Noxolo Ellen – Pretoria: Vista University 2000 [mf ed 2000] – 6mf – 9 – (incl bibl ref) – mf#mfm15335 – sa Unisa [306]

Employee press – v4 n2-v13 n11 [1970 mar-1979 dec] – 1r – 1 – (cont: uc clerical, technical, and professional employee press; cont by: afscme 1695 union news) – mf#637178 – us WHS [331]

Employee relations – Bradford. 1992-1995 (1,5,9) – ISSN: 0142-5455 – mf#15743 – us UMI ProQuest [331]

Employee relations law journal – New York. 1975+(1,5,9) – ISSN: 0098-8898 – mf#11882 – us UMI ProQuest [344]

Employee relations law journal – v1-26. 1976-2001 – 5,6,9 – $1048.00 set – (v1-10 1976-85 on reel $245. v11-26 1985-2001 on mf $803) – ISSN: 0098-8898 – mf#102591 – us Hein [331]

Employee responsibilities and rights journal – New York. 1988-1996 (1,5,9) – ISSN: 0892-7545 – mf#17663 – us UMI ProQuest [331]

Employee services management – Oak Brook. 1981-1999 (1) 1981-1999 (5) 1981-1999 (9) – (Cont: Recreation management) – ISSN: 0744-3676 – mf#7200,01 – us UMI ProQuest [790]

Employee's advocate / American Federation of Government Employees – 1979 jun 8-1986 apr – 1r – 1 – mf#1477250 – us WHS [331]

Employees' compensation appeals board decisions – v1-39. 1946-88 – 503mf – 9 – $754.00 – (ind/digest for v1-34) – mf#llmc 81-219 – us LLMC [344]

Employees' Mutual Benefit Association [Milwaukee WI] *see* Rail and wire

Employees' newsletter / Loewi and Co – v3 iss 4 [1977 win] – 1r – 1 – (continued by: insider (milwaukee, wi)) – mf#638835 – us WHS [331]

Employees organize = Empleados organizan / Service Employees International Union – v9 [i.e. 4] n10-v9 n6 [1973 may 21-1978 aug 8] – 1r – 1 – mf#615664 – us WHS [331]

Employers' and public accountants' attitudes towards employee reporting in south africa / Stainbank, Lesley June – Uni of South Africa 2000 [mf ed Johannesburg 2000] – 7mf – 9 – (incl bibl ref) – mf#mfm14830 – sa Unisa [650]

Employers' liability / Osgood, William Newton – Boston: Hodges, 1891 – 30p – 1 – mf#LL-1379 – us L of C Photodup [344]

The employers' liability act, 1880 : and the workmen's compensation act, 1906: with the statutes relating to and cases decided on the previous workmen's compensation acts in england, scotland, and ireland... / Ruegg, Alfred Henry – London: Butterworth; Toronto: Canada Law Book Co, 1910 – 13mf – 9 – 0-659-91990-7 – mf#91990 – cn Canadiana [344]

The employers' liability acts and the assumption of risks in new york, massachusetts, indiana, alabama, colorado, and england / Dresser, Frank Farnum – St. Paul, Keefe-Davidson, 1902-08. 2 v. LL-258 – 1 – us L of C Photodup [344]

Employers' liability, workmen's compensation and liability insurance. / Conner, Jeremiah Frederick – Chicago, Spectator 1916 262 p. LL-699 – 1 – us L of C Photodup [344]

Employment and earnings – Washington. 1954+ [1]; 1970+ [5]; 1975+ [9] – ISSN: 0013-6840 – mf#1710 – us UMI ProQuest [331]

Employment equity in canadian newspaper sports journalism : a comparative study of the work experiences of women and men sports reporters / Depatie, Caroline – 1997 – 2mf – 9 – $8.00 – mf#PE 3787 – us Kinesology [070]

Employment information bulletin / American Historical Association – 1976 oct-1982 apr – 1r – 1 – (cont: professional register bulletin) – mf#275764 – us WHS [331]

The employment of indians in the war of 1812 / Cruikshank, Ernest Alexander – Washington: GPO, 1896 – 1mf – 9 – mf#03635 – cn Canadiana [971]

Employment relations today – New York. 1983+ (1,5,9) – (Cont: EEO today) – ISSN: 0745-7790 – mf#11881,01 – us UMI ProQuest [331]

Employment security in indiana / Indiana. Employment Security Board – n1-42. 1936-77 – (= ser Harvard law school library coll) – 44mf – 9 – us Harvard Law [344]

Employment security review – Washington. 1934-1963 (1) – mf#1436 – us UMI ProQuest [331]

Employment service review – v1-18. 1934-51 – (= ser Employment Security Service) – 9 – (Began as: Employment Security Service) – mf#LLMC 84-351 – us LLMC [331]

Employment service review – Washington. 1964-1968 (1) – ISSN: 0424-9380 – mf#1660 – us UMI ProQuest [331]

Die empoerung – Berlin DE, 1922 n1-4 – 1 – gw Misc Inst [074]

Emporia Evangelical United Brethren Mission *see* Records

Emporia gazette – Emporia KS. 1917 jan 1-apr 9, 1917 apr 10-jul 12, 1917 jul 13-oct 13, 1917 oct 15-1918 jan 11, 1918 apr 25-aug 6, 1918 aug 7-nov 20, 1918 jan 12-1918 apr 24, 1918 nov 21-1919 feb 28, 1919 mar 1-may 29, 1919 may 30-aug 28, 1919 aug 29-1920 jan 2 – 11r – 1 – mf#845949 – us WHS [071]

Emporia. Kansas. St. Andrew's Episcopal Church *see* Records

Emporio italiano – London, UK. Mar-Jul 1857 – 1 – uk British Libr Newspaper [072]

Emporium : rivista mensile illustrata d'arte letteratura scienze e varieta – v1-84. 1895-1936 – 9 – $1188.00 – mf#0192 – us Brook [700]

Emporium Independent – Emporium, PA. -w 1904-1912 – 13 – $25.00r – us IMR [071]

Emporium of arts and science – Philadelphia. 1812-1814 (1) – mf#3739 – us UMI ProQuest [700]

Emporo italiano – London, UK. 31 Mar-1 Jul 1857. -w. 18 feet – 1 – uk British Libr Newspaper [072]

Empowering young people through narrative / Steyn, Lynette – Uni of South Africa 2001 [mf ed Johannesburg 2001] – 3mf – 9 – (incl bibl ref) – mf#mfm15012 – sa Unisa [240]

Empreintes – n1-11. Bruxelles. 1946-52 – 1 – fr ACRPP [073]

Empresa Ber-Maq *see* Guia de espectaculos. feria y fiestas agosto 1974

Una empresa el siglo 18 : los navios de la ilustracion... / Basterra, Ramon de – Madrid: Razon y Fe, 1926 – 1 – sp Bibl Santa Ana [355]

Empresas espirituales y morales... / Villava, J F de – Baeca: Fernando Diaz de Montoya, 1613 – 9mf – 9 – mf#0-1968 – ne IDC [090]

Empresas morales : compuestas por el exellentissimo senor, don Juan de Borja... / Boria, J de – Brusselas: Por Francisco Foppens, 1680 – 6mf – 9 – mf#0-569 – ne IDC [090]

Empress express – Alberta, CN. jan 1913-dec 1936 – 6r – 1 – cn Commonwealth Imaging [071]

Empros – Athens, Greece. -d. 1 Oct 1917-3 Feb 1919. Imperfect. 3 reels – 1 – uk British Libr Newspaper [949]

Emprunt de trois millions de piastres / Michel, Antoine – Port-Au-Prince, Haiti. 1934 – 1r – 1 – us UF Libraries [972]

Les emprunts de la bible hebraique au grec et au latin / Vernes, Maurice – Paris: Ernest Leroux, 1914 – (= ser Bibliotheque de l'ecole des hautes etudes) – 1mf – 9 – 0-524-05703-6 – mf#1992-0553 – us ATLA [470]

Emrah *see* The divan project

Emre,Yunus *see* The divan project

Emri, Murad *see* The divan project

Emrich, George *see* Thomas jefferson's march

Emrich, Hermann *see* Goethes intuition

Emrich, Wilhelm *see* Paulus im drama

Ems, Rudolf von *see*
– Alexander
– Der guote gerhart
– Rudolf von ems weltchronik
– Rudolfs von ems willehalm von orleans
– Weltchronik

Emscherzeitung *see* Der reichsfreund

Emsdettener volkszeitung, emsdettener tageblatt – Emsdetten DE, 1953-70 [gaps] – 1 – gw Misc Inst [074]

Emslaendische rundschau : edition lingen, meppen, nordhorn – Meppen DE, 1953-1961 28 oct [only local ed] – 1 – (bezirksausgabe von westfaelische rundschau, dortmund) – gw Misc Inst [074]

Ems-zeitung – Papenburg DE, 1987- – 7r/yr – 1 – (bezirksausgabe von neue osnabruecker zeitung) – gw Misc Inst [074]

Emt journal – St Louis. 1977-1981 (1,5,9) – ISSN: 0147-5851 – mf#11874 – us UMI ProQuest [610]

Emu : a quarterly magazine to popularize the study and protection of native birds – Melbourne 1901/02-1945/46 – v1-45 on 323mf – 9 – mf#z-2018c/2 – ne IDC [590]

Emu be tacifi ilan be hafukiyara manju gisun-i buleku bithe = Yi xue san shang qing wen jian / Juwentu – Jingdu: Ying hua tang Xu shi shu fang, Qianlong bing yin [1746] [mf ed 1966] – (= ser Tenri coll of manchu-books in manchu-characters. series 1, linguistics 50; Mango bunkenshu. 1, gogaku hen) – 4v on 1r – 1 – (in manchu and chinese) – ja Yushodo [480]

Emunah bi-shete rashuyot ve-rishumeha be-sifrut yisra'el / Rubin, Salomon – Podgorze-Krakow, Poland. 1908 – 1r – 1 – us UF Libraries [939]

Emunat ha-tehiyah / Potschowsky, Moses Nathaneel – Berdichev, Ukraine. 1896 – 1r – 1 – us UF Libraries [939]

Emunot veha-de'ot / Sa'adia Ben Joseph – Kraka, Poland. 1880 – 1r – 1 – us UF Libraries [939]

En / Yanes, Miguel – Habana, Cuba. 1935 – 1r – us UF Libraries [972]

En alas del deseo / Prosperi, Ramon F – Panama, 1927 – 1r – 1 – us UF Libraries [972]

En america meridional / Maseras, Alfonso – Barcelona, Spain. 1922 – 1r – 1 – us UF Libraries [972]

Een en ander uit den eersten tijd der medische zending te moekden / Wartena, A J – [Rotterdam]: J M Bredee, 1917 [mf ed 1995] – (= ser Yale coll; Lichtstralen op den akker der wereld [23. jaarg 1917] 2) – 28p (ill) – 1 – 0-524-09705-4 – (in dutch) – mf#1995-0705 – us ATLA [610]

En avant...marche! : grande revue en 3 actes et 10 tableaux representee pour la premiere fois au theatre national francais de Montreal (direction G Gauvreau) le 21 decembre 1914 / Christe, Pierre – [Quebec (Province): s.n, 1914?] (mf ed 1994) – 1mf – 9 – mf#SEM105P2187 – cn Bibl Nat [790]

En butinant : scenes et croquis de mongolie / Oost, Joseph van – Chang-hai: Impr de la Mission Catholique, 1917 [mf ed 1995] – (= ser Yale coll) – vi/157p (ill) – 1 – 0-524-10035-7 – (in french) – mf#1995-1035 – us ATLA [241]

En camisa rosa. las sonatas del otono / Trigo, Felipe – Madrid: Renacimiento, 1921 – sp Bibl Santa Ana [810]

En canot de papier de quebec au golfe du mexique : 2,500 milles a l'aviron / Bishop, Nathaniel Holmes – Paris: E Plon, 1879 – 5mf – 9 – (also available in english) – mf#00140 – cn Canadiana [917]

En chaland sous les tropiques / Neufville, Gilbert de – Paris: B Grasset, 1926 – 1 – us CRL [960]

En chaland sous les tropiques / Neufville, Gilbert De – Paris, France. 1926 – 1r – 1 – us UF Libraries [025]

En chine au tche-ly sud-est : une mission d'apres les missionnaires / Leroy, Henri Joseph – [Lille]: Desclee, de Brouwer, 1900 [mf ed 1995] – (= ser Yale coll) – xl/458p (ill) – 1 – 0-524-10230-9 – (in french) – mf#1996-1230 – us ATLA [241]

En claro / Arrufat, Anton – Habana, Cuba. 1962 – 1r – 1 – us UF Libraries [972]

En cour d'appel : john fraser et al, appellants sic et john munro et al, intimes... – S.l: s.n, 1820? – 1mf – 9 – mf#61579 – cn Canadiana [346]

En defensa de los vascos – Santiago de Chile, 1937 – (= ser Blodgett coll) – 9 – mf#fiche w858 – us Harvard [946]

En dehors *see* Dehors

En el 4th centenario del doctor arias montano. la version metricolatina del salterio hebraico / Diego, Sandalio – Madrid: reeb, 1928 – 1 – sp Bibl Santa Ana [780]

En el a no de enero / Soler Puig, Jose – La Habana, Cuba. 1963 – 1r – 1 – us UF Libraries [972]

En el alcazar de la reina. antologia poetica guadalupense : caceres, 1967 / Corredor, Antonio – Madrid: Graf Calleja, 1968 – 1 – sp Bibl Santa Ana [810]

En el amanecer de una nueva era / Garcia Bauer, Carlos – Guatemala, 1951 – 1r – 1 – us UF Libraries [972]

En el cafetal / Malpica La Barca, Domingo – Habana, Cuba. 1890 – 1r – 1 – us UF Libraries [972]

En el camino / Piedra-Bueno, Andres De – La Habana, Cuba. 1926 – 1r – 1 – us UF Libraries [972]

En el centenario de ayestaran / Gay-Calbo, Enrique – Habana, Cuba. 1945 – 1r – 1 – us UF Libraries [972]

En el cuarto de mi mujer / Hurtado, Antonio – 1866 – 9 – sp Bibl Santa Ana [830]

En el darien / Colombia Ministerio De Guerra – Bogota, Colombia. 1910 – 1r – 1 – us UF Libraries [972]

En el n° 2 centenario del nacimiento del principe de la paz (1767-1967). badajoz y su hijo godoy, el mas ilustre y calumniado de los 302 ilustres badajocenses / Lopez, Benigno – Badajoz: Imp. INCA, s.a. – 1 – sp Bibl Santa Ana [946]

En el nombre del padre / Arce, Manuel Jose – Guatemala, 1955 – 1r – 1 – us UF Libraries [972]

En el pais de los eternos hielos / Bayle, Constantino & Segundo, Llorente – Madrid: Razon y Fe, 1941 – sp Bibl Santa Ana [240]

En el portal de belen / Sanchez Loro, Domingo – Caceres: publicaciones del departamento de seminarios de la jefatura provincial del movimiento, 1953 – 1 – sp Bibl Santa Ana [946]

En el pueblo dormido / Vargas Zuniga, Rodrigo – Madrid: Imp. Hispano-Africana, 1915 – 1 – sp Bibl Santa Ana [946]

En el reino de la frivolidad / Gomez Carrillo, Enrique – Madrid, Spain. 1923 – 1r – 1 – us UF Libraries [972]

En el remoto cigango. jornadas japonesas / Oteyza, Luis de – Madrid: Editorial Pueyo, S.L. 1927 – sp Bibl Santa Ana [946]

En el salon de los virreyes / Escobar Camargo, Antonio – Bogota, Colombia. 1957 – 1r – 1 – us UF Libraries [972]

En el templo de apolo / Madriz, Ernesto – Leon, Nicaragua. 1949? – 1r – 1 – us UF Libraries [972]

En el umbral del misterio / Roso de Luna, Mario – Madrid: Editorial Pueyo, 1921 – 1 – sp Bibl Santa Ana [240]

En episode i soeren kierkegaards ungdomsliv / Heiberg, Peter Andreas – Kobenhavn: Gyldendal, 1912 – (= ser Kierkegaard-studier) – 1mf – 9 – 0-524-00439-0 – mf#1989-3139 – us ATLA [190]

En este pais! / Urbaneja Achelpohl, Luis Manuel – Caracas, Venezuela. 1950 – 1r – 1 – us UF Libraries [972]

En familia : poesias / Sanchez-Arjona, Vicente – Sevilla: establecimiento tipografico juan mejias, 1930 – 1 – sp Bibl Santa Ana [810]

En familia : versos viejos / Gomez-Bravo, Vicente – Badajoz: Arqueros, 1952 – sp Bibl Santa Ana [946]

En fyrstes fald : cambodia og nixon-doktrinen / Svejstrup, Paul – Kobenhavn [Denmark]: Rhodos [1971] [mf ed 1989] – (= ser Politisk analyse) – 1r with other items – 1 – (incl bibl ref) – mf#mf-10289 seam reel 020/05 [§] – us CRL [327]

En guinee – Paris: H Le Soudier, 1895 – 1 – us CRL [960]

En hydravion au-dessus du continent noir / Bernard, Marc – Paris: B Grasset, 1927 – 1 – us CRL [550]

En la camara / Uribe Echeverri, Carlos – Bogota, Colombia. 1926 – 1r – 1 – us UF Libraries [972]

En la catedral de toledo : horas de luz...toledo, 1928 / Segura Saenz, Pedro; ed by Bayle, Constantino – Madrid: Razon y Fe, 1928 – 9 – sp Bibl Santa Ana [720]
En la contienda / Cuevas Zequeira, Sergio – Habana, Cuba. 1901 – 1r – 1 – us UF Libraries [972]
En la cumbre se pierden los caminos / Sosa, Julio Bautista – Panama, 1957 – 1r – 1 – us UF Libraries [972]
En la espana leal ha nacido un ejercito / Bates, Ralph – Mexico, 193? – (= ser Blodgett coll) – 9 – mf#fiche w750 – us Harvard [946]
En la noche de mundo / Rodriguez Acosta, Ofelia – Habana, Cuba. 1940 – 1r – 1 – us UF Libraries [972]
En la paz de la aldea / Richard Lavalle, Enrique – Buenos Aires, Argentina. 1913? – 1r – 1 – us UF Libraries [972]
En la raiz que sangra, versos / Ona, Gines De – Habana, Cuba. 1960 – 1r – 1 – us UF Libraries [972]
En la ruta de los libertadores / Marrero Aristy, Ramon – Ciudad Trujillo, Dominican Republic. 1943 – 1r – 1 – us UF Libraries [972]
En la sesion del 23 octubre...vacantes...fueron elegidos...academicos de numero...duque de t'serclares... / Fita, Fidel – Madrid: Est. Imp. Fortanet, 1908 – 1 – sp Bibl Santa Ana [946]
En la sombra / Hurtado, Antonio – 1870 – 9 – sp Bibl Santa Ana [830]
En las bancas del foro / Sanabria Campos, Jose Antonio – EL Salvador, El Salvador. 1957? – 1r – 1 – us UF Libraries [972]
En las lomas de el purial / Juarez Fernandez, Bel – Habana, Cuba. 1962 – 1r – 1 – us UF Libraries [972]
En las tierras del oro del imperio del sol : madrid, 1945 / Real, Cristobal – Madrid: Razon y Fe, 1947 – 1 – sp Bibl Santa Ana [946]
En las zarzas del oreb / Vargas Vila, Jose Maria – Paris, France. 1913 – 1r – 1 – us UF Libraries [972]
En literair anmeldelse : to tidsaldre, novelle af forfatteren til "en hverdags-historie" / Kierkegaard, Soeren – Kobenhavn: C A Reitzel, 1846 – (= ser Himmelstrup) – 1mf – 9 – 0-7905-3795-8 – mf#1989-0288 – us ATLA [190]
En los andamios / Trigo, Felipe – Madrid: Renacimiento, s.a. – sp Bibl Santa Ana [946]
En los caminos de la libertad / Albornoz, Alvaro de – Brooklyn, NY. 1939 – (= ser Blodgett coll) – 9 – mf#fiche w706 – us Harvard [946]
En los matrimonios rotos que hacemos con los hijos? / Aradillas Agudo, Antonio – Madrid: Ediciones Maisal, S.A., 1977 – 1 – sp Bibl Santa Ana [306]
En marcha – Miami, FL. 1981 nov-1984 jun – 1r – 1 – (missing: 1982 feb, may, oct, dec; 1983 mar, may, jul-aug, oct-dec; 1984 jan-apr) – us UF Libraries [071]
En marcha hacia la victoria / Carrillo, Santiago – Valencia? 1937 – (= ser Blodgett coll) – 9 – mf#fiche w775 – us Harvard [946]
En marge de la legende doree / Saintyves, Pierre – Paris: E Nourry 1930 [mf ed Bloomington IN: Indiana Uni Lib, Preservation Dept 1984] – viii 596p on 1r [ill] – 1 – us Indiana Preservation [390]
En marge d'une confederation economique inter-anti... / Coen, Edwidg – Port-Au-Prince, Haiti. 195- – 1r – 1 – us UF Libraries [972]
En memoria de ramon / Sociedad Economica de Amigos del Pais de Badajoz – Badajoz: imprenta de la diputacion provincial, 1963 – sp Bibl Santa Ana [330]
En mi barrio / Pineiro, Abelardo – Habana, Cuba. 1962 – 1r – 1 – us UF Libraries [972]
En notas de bibliografia franciscana / Lopez, Atanasio & Moreno de Robles, Francisco – Archivo Ibero Americano, 1926 – 1 – sp Bibl Santa Ana [241]
En oceanie : voyage autour du monde en 365 jours 1884-1885 / Cotteau, E – Paris, 1895 – 5mf – 9 – mf#HT-36 – ne IDC [919]
En plein confict / Dubruel, Marc; ed by Bayle, Constantino – Madrid: Razon y Fe, 1928 – 9 – sp Bibl Santa Ana [972]
En plena polemica... / Delgado, P J – Madrid: Razon y Fe, 1927 – 1 – sp Bibl Santa Ana [999]
En pos de la felicidad / Guiral Moreno, Mario – Habana, Cuba. 1920 – 1r – 1 – us UF Libraries [972]
En pro y en contra (criticas) / Gonzalez Serrano, Urbano – Madrid: Libreria de Victoriano Suarez, S.A. – 1 – sp Bibl Santa Ana [946]
En quoi la langue esquimaude differe-t-elle grammaticalement des autres langues de l'amerique du nord? / Adam, Lucien – Copenhague?: Thiele, 1884 [mf ed 1984] – 1mf – 9 – 0-665-05096-8 – (incl bibl ref) – mf#05096 – cn Canadiana [490]
En reponse aux assertions de l'historien / Thoby, Perceval – Port-Au-Prince, Haiti. 1939 – 1r – 1 – us UF Libraries [972]
En resa til norra america... / Kalm, P – Stockholm. 3v. 1753-1761 – 18mf – 9 – mf#1000 – ne IDC [917]
En route pour la mer glaciale / Petitot, Emile – Paris: Letouzey et Ane, [1887?] [mf ed 1982] – 5mf – 9 – 0-665-30446-3 – (incl bibl ref) – mf#30446 – cn Canadiana [917]
En route pour la mer glaciale / Petitot, Emile Fortune Stanislas Joseph – Paris: Letouzey et Ane, 1887? – 5mf – 9 – (incl bibl ref) – mf#30446 – cn Canadiana [917]
En route pour le canada / Bodard, Auguste – Montreal?: s.n, 1893? – 1mf – 9 – mf#26226 – cn Canadiana [304]
En route to the klondike, pt 1 : a series of photographic views of the picturesque land of gold and glaciers / Roche, Frank La – Chicago, New York: W B Conkey, 1898? – (= ser People's series) – pt1 on 1mf – 9 – mf#17282 – cn Canadiana [917]
En route to the klondike, pt 2 : a series of photographic views / Roche, Frank La – Chicago, New York: W B Conkey, 1898? – pt2 on 1mf – 9 – mf#17283 – cn Canadiana [917]
En route to the klondike, pt 3 : a series of photographic views / Roche, Frank La – Chicago, New York: W B Conkey, 1898? – pt3 on 1mf – 9 – mf#17284 – cn Canadiana [917]
En route to the klondike, pt 4 : a series of photographic views / Roche, Frank La – Chicago, New York: W B Conkey, 1898? – pt4 on 1mf – 9 – mf#17285 – cn Canadiana [917]
En route to the klondike, pt 5 : a series of photographic views / Roche, Frank La – Chicago, New York: W B Conkey, 1898? – pt5 on 1mf – 9 – mf#17286 – cn Canadiana [917]
En route to the klondike, pt 6 : a series of photographic views / Roche, Frank La – Chicago, New York: W B Conkey, 1898? – pt6 on 1mf – 9 – mf#17287 – cn Canadiana [917]
En route to the klondike, pts 1-6 : a series of photographic views of the picturesque land of gold and glaciers – Chicago, New York: W B Conkey, 1898? – 6v on 1mf – 9 – mf#17281 – cn Canadiana [917]
En sahara a travers le pays des maures nomades / Donnet, Gaston – Paris: H May, [1901] – 1 – us CRL [916]
En sampan sur les lacs du cambodge et a angkor / Gas-Faucher, F – Marseille: Barlatier 1922 [mf ed 1989] – 1r with other items – 1 – (notes d'un touriste) – mf#mf-10289 seam reel 004/01 [§] – us CRL [915]
'En shim'on / Finkelstein, Simon Isaac – New York, USA. v1-2. 1935 – 1r – 1 – us UF Libraries [939]
En soledad de amor herido / Inchaustegui Cabral, Hector – Santiago, Dominican Republic. 1943 – 1r – 1 – us UF Libraries [972]
En temps de guerre : recueil d'extraits de journaux, de documents diplomatiques, etc / Squair, John [comp] – Toronto: Copp, Clark, 1916 [mf ed 1996] – 2mf – 9 – 0-665-78446-5 – mf#78446 – cn Canadiana [933]
En tres y dos / Fornet, Ambrosio – Habana, Cuba. 1964 – 1r – 1 – us UF Libraries [972]
En una ciudad llamada san juan / Marques, Rene – Habana, Cuba. 1962 – 1r – 1 – us UF Libraries [972]
En una ciudad llamada san juan / Marques, Rene – Mexico City? Mexico. 1960 – 1r – 1 – us UF Libraries [972]
En una silla de ruedas / Lyra, Carmen – San Salvador, El Salvador. 1960 – 1r – 1 – us UF Libraries [972]
En verite / Taylor, Raynor – London: printed for aut...[c179-?] [mf ed 19–] – 1mf – 9 – mf#fiche 952 – us Sibley [780]
En veteran : nogle blade af chinas missionshistorie / Morthensen, Eilert – Kobenhavn: Danske Missionsselskab, 1914 [mf ed 1995] – (= ser Yale coll; Missions-bibliotheket, aarg. 1914 4) – 59p (ill) – 1 – 0-524-09796-8 – (in danish) – mf#1995-0796 – us ATLA [240]
En viaje, 1881-1882 / Cane, Miguel – Buenos Aires, Argentina. 1940 – 1r – 1 – us UF Libraries [972]
En villanueva de la sierra tuvo su origen la "fiesta del arbol" / Gutierrez Macias, Valeriano – Badajoz: imp dip provincial, 1968 – sp Bibl Santa Ana [390]
En yu (ccs) : t'uan ch'i yueh k'an = Christian fellowship monthly – Beijing. n1-7 1948; v3 n6-7 1951 [complete] [mf ed 198?] – (= ser Chinese christian serials coll) – 1 – (formed by merger of: en yu (1947) and: t'uan ch'i yueh k'an. suspended probably after n7 1948 for a time) – mf0296hb – us ATLA [240]
En yu (ccs) = Friend in god – Beijing. n1 jun 1947?-n8 dec 1947 [complete] [mf ed 198?] – (= ser Chinese christian serials coll) – 1 – (merged with: t'uan ch'i yueh k'an to form: en yu (1948)) – mf0296ha – us ATLA [240]
Enakievskij sovet rk i kd *see* Izvestiia enakievskogo soveta rabochikh i soldatskikh deputatov
Enakopravnost / Cuyahoga Co. Cleveland – dec 1942-mar 1957 [daily] – 16r – 1 – (In Slovenian) – mf#B30732-30747 – us Ohio Hist [071]
Enakopravnost – Cleveland, OH, jan 2-dec 21 1925 – 2r – 1 – (Daily independent slovenian language newspaper) – us Western Res [071]
Enakopravnost – Cleveland OH, 1919, 1940-43, 1949, 1955-56* – 1r – 1 – (slovenian newspaper) – us IHRC [071]
Enamorado Cuesta, Jose *see* Salve hispania
Enamorado-Cuesta, Jose *see*
– Estampas del vivac
– Princesa y el oso blanco
Enantioselektive reduktion von ketonen mit neuen nad(H)-abhaengigen oxidoreduktasen / Zelinski, Thomas – (mf ed 1996) – 2mf – 9 – €40.00 – 3-8267-2337-6 – mf#DHS 2337 – gw Frankfurter [540]
Enaratio priorum capitum evangelii johannis... / Pezelius, C – Neustadii, [1586] – 7mf – 9 – mf#PBA-290 – ne IDC [221]
Enardo and rosael / Tapia Y Rivera, Alejandro – New York, NY. 1952 – 1r – 1 – us UF Libraries [972]
Enarem *see* Ang pulahan
Enarratio epistolae pavli, scriptae ad philippenses / Major G – [Wittenberg], 1559 – 4mf – 9 – mf#TH-1 mf 920-923 – ne IDC [242]
Enarratio in evangelium matthaei... / Oecolampadius, J – Basileae, [Cratander], 1536 – 6mf – 9 – mf#PBU-393 – ne IDC [225]
Enarratio psalmi sexagesimi octavi / Major, G – Lipsiae, 1551 – 2mf – 9 – mf#TH-1 mf 924-925 – ne IDC [242]
Enarrationes in quinque priora capita libri geneseos : et alii tractatus / Politi, Ambr Cathar – Romae, 1552 – 14mf – 8 – €27.00 – ne Slangenburg [220]
Enarrationis psalmorum davidis / Moeller, Heinrich – Novissima editio, prioribus emendatior. Genevae: Apud Petrum & Iacobum Chovet, 1610. Chicago: Dep of Photodup, U of Chicago Lib, 1974 (1r); Evanston: American Theol Lib Assoc, 1984 (1r) – 1 – 0-8370-0015-7 – (Incl ind) – mf#1984-B400 – us ATLA [221]
Enault, Louis *see* Goethe og werther
Enbaev, A M *see* Kustarnaia promyshlennost i promyslovaia kooperatsiia v natsionalnykh respublikakh i oblastiakh sssr
Encantos do oeste / Couto De Magalhaes, Agenor – Rio de Janeiro, Brazil. 1945 – 1r – 1 – us UF Libraries [972]
Encarnacion-Garcia, Haydee *see* Sociocultural differences in eating[-]disordered behaviors and body image perception
Encephale – Paris. 1978-1980 (1,5,9) – ISSN: 0013-7006 – mf#11722 – us UMI ProQuest [616]
L'enchaine / Comite Regional de l'Oranie du Parti Communiste – no. 1. Oran. aout 1932 – 1 – fr ACRPP [335]
The enchantment of art : as part of the enchantment of experience: essays / Phillips, Duncan – New York: J Lane; Toronto: Bell & Cockburn, 1914 – 4mf – 9 – 0-659-91873-0 – mf#9-91873 – cn Canadiana [700]
Encheiridion i dogmatik jaemte dogmhistoriska anmaerkningar / Lindberg, Conrad Emil – Rock Island, IL: Lutheran Augustana Book Concern, 1898 – 1mf – 9 – 0-524-05014-7 – mf#1991-2184 – us ATLA [240]
Enchiridion : The Small Catechism of Dr. Martin Luther = Kleine katechismus / Luther, Martin – 3rd ed. Reading, Pa: Pilger Book Store, [186-?] – 1mf – 9 – 0-524-04380-9 – (In English) – mf#1991-2084 – us ATLA [242]
Enchiridion : oder handbuechlein fuer die, so jres glaubens vnd der seligkeit halben geistliche anfechtung haben / Waldner, W – np, 1566 – 2mf – 9 – mf#TH-1 mf 1467-1468 – ne IDC [242]
Enchiridion canonico-morale de confessario ad inhonesta & turpia solicitante : nec non de decretis & constitutionibus pontificiis ad hoc nefarium crimen exterminandum emanatis / Vilaplana, Hermenegildo de – Ed secunda locupletior in paucis – Mexici:...Bibliothecae Mexicanae destinata 1765 – (= ser Books on religion...1543/44-c1800: confesionarios) – 3mf – 9 – mf#crl-28 – ne IDC [241]
Enchiridion controversiarum / Vorstius, C – Hannoviae, 1608 – 2mf – 9 – mf#PBA-338 – ne IDC [240]
Enchiridion controversiarum praecipuarum... / Coster, F – Coloniae Agrippinae, 1593 – 8mf – 9 – mf#CA-47 – ne IDC [180]
Enchiridion d timothei kirchneri jn welchem die fuernembsten hauptstueck der christlichen lehre durch frag vnd antwort auss gottes wort gruendtlich erklaeret / Kirchner, T – Heydelberg, 1584 – 10mf – 9 – mf#TH-1 mf 817-826 – ne IDC [242]
Enchiridion musicae mensuralis / Rhaw, Georg – 6th ed, [Wittenberg: Georg Rhaw 1536] [mf ed 19–] – 1r – 1 – mf#film 407 – us Sibley [780]
Enchiridion musicae mensuralis [1520] / Rhau, Georg – (= ser Mssa) – 2mf – 9 – €35.00 – mfchl 68 – gw Fischer [780]
Enchiridion musicae mensuralis [1546] / Rhau, Georg – (= ser Mssa) – 1mf – 9 – €20.00 – mfchl 68a – gw Fischer [780]
Enchiridion mvsicae mensvralis [pt 2] / Rhaw, Georg – [Lipsiae?] 1531 [mf ed 19–] – 1mf – 9 – mf#fiche 928 – us Sibley [780]
Enchiridion o manual instrumento de salud contra el morbo articular que llaman gota... / Gomez Miedes, B – Zaragoza, 1589 – 4mf – 9 – sp Cultura [610]
Enchiridion oder handbuchlin eins waren christenlichen vn strytbarlichen lebens... / Erasmus – Basel, [Valentin Curio], 1521 – 3mf – 9 – mf#PBU-537 – ne IDC [240]
Enchiridion rome : or, manual of detached remarks on...ancient and modern rome / Weston, Stephen – London 1819 – (= ser 19th c art & architecture) – 3mf – 9 – mf#4.1.447 – uk Chadwyck [930]
Enchiridion sive manuale confessariorum et poenitentium / Martinus ab Azpilcueta Navarrus – Antverpiae, 1575 – 18mf – 8 – €35.00 – ne Slangenburg [240]
Enchiridion symbolorum et definitionum : Quae in Rebus Fidei et Morum a Conciliis Oecumenicis et Summis Pontificibus Emanarunt in Auditorum Usum / ed by Denzinger, Heinrich – Wirceburgi [Wuerzburg]: Stahel, 1854 – 1mf – 9 – 0-7905-9184-7 – mf#1989-2409 – us ATLA [240]
Enchiridion utriusque musicae practicae [1520] / Rhau, Georg – (= ser Mssa) – 1mf – 9 – €20.00 – mfchl 67 – gw Fischer [780]
Enchiridion utriusque musicae practicae [1546] / Rhau, Georg – (= ser Mssa) – 1mf – 9 – €20.00 – mfchl 67a – gw Fischer [780]
Enchiridion veteris et novi testamenti : ...Handbuechlein dess Alten und Neuwen Testaments... – Franckfurt am Mayn: Apud Paulum Reffeler, impensis Sigismundi Feyerabent, 1573 – 10mf – 9 – mf#O-81 – ne IDC [090]
Enchiridion vtrivsqve mvsicae practi cx a georgio rhauo ex variis musicorum libris : pro pueris in schola vitebergensi congestum / Rhaw, Georg – [Vitebergae: apud Georgivm Rhav] 1546 [mf ed 19–] – 2pt in 1v, 2mf – 9 – mf#fiche 927 – us Sibley [780]
Enchiridium ethicum : ex aristotele olim collectum / Puteanus, Erycius – Lovanii: Typis Philippi Dormalii 1620 – (= ser Ethics in the early modern period) – 1mf – 9 – mf#pl-108 – ne IDC [170]
Enchiridium religionis reformatae / Walaeus, A – Ed 2. Lugduni Batavorum, 1660 – 6mf – 9 – mf#PBA-396 – ne IDC [240]
Enchiridon : ov, brief recveil dv droict escrit: garde et observe ov abroge en france / Imbert, Jean – Cologny: I Arnavld, 1615 – 1 – (rev corr aug & add by m p gvenois) – us Wisconsin U Libr [944]
Las enciclicas : rerum novarum, quadragesimo anno, divini redemptoris contra el comunismo y divini illius magistri sobre la educacion cristiana, al alcance de todos / Marquez, Gabino – Madrid: Apostolado de la Prensa, S A, 2nd ed 1941 – 1 – sp Bibl Santa Ana [946]
Las enciclicas : rerum novarum, quadragesimo anno y divini redemptoris contra el comunismo, al alcance de todos / Marquez, Gabino – Toledo: Ed Catolica Toledana, 1938 – 1 – sp Bibl Santa Ana [946]
Enciclopedia colombiana / Castro, Salomon G – Bogota, Colombia. 1929 – 1r – 1 – us UF Libraries [972]
Enciclopedia dos municipios brasileiros – Rio de Janeiro, 1957 – 36v – 9 – us Brook [972]
Enciclopedia dos municipios brasileiros – Rio de Janeiro, Brazil. v1-35. 1957-1963 – 9r – 1 – us UF Libraries [972]
Enciclopedia metodica criticoragionata delle belle arti / Zani, Pietro – Parma. v1-19. pt2 v1-9. 1817-24 – 9 – us Brook [700]
Encina, Francisco Antonio *see*
– Independencia de nueva granada y venezuela
– Primera republica de venezuela
Encina y la Carrera, Juan Ignacio de la *see* Por los...monasterios de...
Encinitas coast dispatch *see* [Encinitas-] the progress
[Encinitas-] encinitas coast dispatch – CA. 1925- – 138r – 1 – $8280.00 (subs $100/y) – mf#HC02201 – us Library Micro [071]

[Encinitas-] the progress – CA. 1925-1926, 1927-1982 – 51r – 1 – $3060.00 – (aka: encinitas coast dispatch) – mf#H03211 – us Library Micro [071]

[Encinitas/del mar-] blade-citizen – CA. 1982 – 144r – 1 – $8640.00 (subs $600/y) – mf#H04042 – us Library Micro [071]

Enclitic / Center for Media Literacy. University of Minnesota – v1 1-v4:2 [1977 spr-1980 fall], v5 1-v10:2 [1981 spr-1988 fall] – 2r – 1 – mf#1552752 – us WHS [302]

Encobra / National Coalition of Blacks for Reparations in America – summer 1994, 1995, 1996, 1997 – 1r – 1 – mf#3995651 – us WHS [305]

Encomendero / Gavidia, Francisco – San Salvador, El Salvador. 1960 – 1r – us UF Libraries [972]

Encomiendas, tomo 1 : caracas, 1927 / Davila, Vicente – Madrid: Razon y Fe, 1930 – 1 – sp Bibl Santa Ana [946]

Encomium moriae : stulticiae / Erasmus, D – Lausanne, Basel: Froben, 1515 – 5mf – 9 – mf#0-247 – ne IDC [700]

Encomium musicae vocalis et instrumentalis : das ist: eine christliche predigt vom lob der lieben musicae, wie man gott nicht allein mit lebendiger stimme, sondern auch mit instrumenten loben solle / Friedrich, Martinus – Jena: T Steinmann 1610 [mf ed 19–] – 1mf – 9 – mf#fiche 4 – us Sibley [780]

Encontro com o tempo / Bastos, Joaquim Justino Alves – Porto Alegre, Brazil. 1965 – 1r – 1 – us UF Libraries [972]

Encontro De Brasilia, 1970 *see* Documentacao, 27 a 31 de julho de 1970

Encontro De Geologos (1st: 1966: Porto Alegre, Brazil) *see* Anais

Encore – New York. 1972-1974 (1) 1973-1974 (5) 1973-1974 (9) – (Cont by: Encore American and worldwide news) – ISSN: 0046-1954 – mf#7290 – us UMI ProQuest [305]

Encore – 1983 jun-1989 dec – 1r – 1 – (cont: carnival glass encore) – mf#1803636 – us WHS [071]

Encore – New York. 1972-1995 (1) 1975-1995 (5) 1975-1995 (9) – ISSN: 0071-0164 – mf#6927 – us UMI ProQuest [400]

Encore american and worldwide news – New York. 1975-1982 (1) 1976-1982 (5) 1976-1982 (9) – (cont: encore) – ISSN: 0161-6536 – mf#7290,01 – us UMI ProQuest [071]

Encore american and worldwide news – 1977 apr 4 – 1r – 1 – (cont: encore (new york, ny: 1972)) – mf#270389 – us WHS [071]

Encore deux annees : ou 1832 et 1833 episodes / Jailly, Hector de – Paris 1834 [mf ed Hildesheim 1995-98] – 1v on 3mf [ill] – 9 – €90.00 – 3-487-26074-3 – gw Olms [944]

Encore un pourceaugnac / Scribe, Eugene – Paris, France. 1817 – 1r – 1 – us UF Libraries [440]

Encounter – Indianapolis. 1940+ (1) 1969+ (5) 1975+ (9) – ISSN: 0013-7081 – mf#1556 – us UMI ProQuest [073]

Encounter – London. 1989-1990 – 1,5,9 – ISSN: 0013-7073 – mf#17722 – us UMI ProQuest [073]

Encounter, 1953-84 – 37r – 1 – mf#179 – uk Microform Academic [073]

O encouracado – Bahia: [s.n.] 12 ago 1889 – (= ser Ps 19) – mf#P18B,02,05 – bl Biblioteca [321]

Encouragement for babes in the church / Sibly, Manoah – London, England. 1796 – 1r – 1 – us UF Libraries [240]

Encouragement to parents – London, England. 18– – 1r – 1 – us UF Libraries [240]

Encouragement to perseverance and holy importunity in prayer / Harris, Robert – London, England. 1841 – 1r – 1 – us UF Libraries [240]

Encouragement to the faithful ministers of christ / Davies, John – London, England. 1805 – 1r – 1 – us UF Libraries [240]

Encouragements of ordination / Stanley, Arthur Penrhyn – Oxford, England. 1864 – 1r – 1 – us UF Libraries [240]

Encrucijada juridica de la costa de marfil / Elias de Tejada Spinola, Francisco – Sevilla: publicaciones de la universidad de sevilla, 1974 – 1 – sp Bibl Santa Ana [340]

Encuentro femenil – v1 n1-2 [1973 spr-1973] – 1r – 1 – mf#564074 – us WHS [071]

Encuesta continental sobre el control de la inflac... / Inter-American Council Of Commerce And Production – Montevideo, Uruguay. 1945 – 1r – 1 – us UF Libraries [972]

Encuesta sobre la cultura de los ladinos en guatemala / Adams, Richard Newbold – Guatemala, 1964 – 1r – 1 – us UF Libraries [972]

The encyclical and modernist theology / Lebreton, J – London: Catholic Truth Society, 1908 – 1mf – 9 – 0-8370-8123-8 – (incl bibl ref) – mf#1986-2123 – us ATLA [240]

Encyclical letter / Pius 9, Pope – London, England. 1847 – 1r – 1 – us UF Libraries [241]

Encyclical letter of our holy father pope pius 9, ordering prayers and announcing new jubilee : to all the patriarchs, primates, archbishops, and bishops of the catholic world – S.l: s.n, 1851? – 1mf – 9 – mf#43402 – cn Canadiana [241]

Encyclical letter of our most holy lord pius the ninth – London, England. 1847? – 1r – 1 – us UF Libraries [241]

Encyclical letter of pope leo the 12th – Dublin, Ireland. 1824 – 1r – 1 – us UF Libraries [241]

The encyclical letter of pope pius 9. on the immaculate conception – London: James Miller, [ca 1849] – 1mf – 9 – 0-8370-7925-X – mf#1986-1915 – us ATLA [240]

The encyclical on "modernism" / Smith, Sydney Fenn – London: Catholic Truth Society, [ca 1907] – 1mf – 9 – 0-8370-7988-8 – mf#1986-1988 – us ATLA [240]

Encyclique acerbo nimis sur l'enseignement de la doctrine chretienne *see* Encyclique vix pervenit sur les contrats, 1er novembre 1745

L'encyclique rerum novarum "sur la condition des ouvriers" / Guerin, M – Montreal: Secretariat General de l'ACJC, 1920 – 1mf – 9 – 0-665-97224-5 – mf#97224 – cn Canadiana [305]

Encyclique vix pervenit sur les contrats, 1er novembre 1745 / Benoit 14, Pope – Montreal: ecole sociale populaire, [1946?] (mf ed 1994) – 1mf – 9 – (filmed with: encyclique acerbo nimis sur l'enseignement de la doctrine chretienne, 15 avril 1905 by pope pius 10) – mf#SEM105P2140 – cn Bibl Nat [241]

Encyclopaedia biblica / Cheyne, Thomas Kelly – London, England. v1-4. 1903 – 1r – us UF Libraries [220]

Encyclopaedia biblica : a critical dictionary of the literary, political and religious history, the archaeology, geography, and natural history of the bible / ed by Cheyne, Thomas Kelly & Black, John Sutherland – new ed. New York: Macmillan. 1v. 1914 – 27mf – 9 – 0-8370-1990-7 – mf#1987-6377 – us ATLA [052]

An encyclopaedia of architecture : historical, theoretical / Gwilt, Joseph – new ed. London 1867 – (= ser 19th c art & architecture) – 15mf – 9 – mf#4.2.305 – uk Chadwyck [720]

An encyclopaedia of architecture : historical, theoretical and practical... / Gwilt, J – London, 1888 – 25mf – 9 – mf#O-282 – ne IDC [720]

An encyclopaedia of cottage, farm, and villa architecture and furniture : containing numerous designs for dwellings, from the cottage to the villa, including farm houses, farmeries, and other agricultural buildings... / Loudon, John Claudius – new ed. London: Longman, Rees, Orme...1836 – (= ser 19th c art & architecture) – 12mf – 9 – mf#4.1.156 – uk Chadwyck [720]

Encyclopaedia of evidence / ed by Camp, Edgar W et al – Los Angeles: L D Powell Co. v1-14+1st, 2nd suppl vols. 1902-09 (all publ) – 161mf – 9 – $241.00 – mf#LLMC 82-510 – us LLMC [347]

Encyclopaedia of forms and precedents / ed by McConnel, W H & Mack, William – Northport, NY: Ed Thompson. v1-18. 1896-1904 (all publ) – 208mf – 9 – $312.00 – mf#LLMC 82-511 – us LLMC [347]

Encyclopaedia of forms and precedents for pleading and practice : at common law, in equity, and under the various codes and practice acts – Northport, NY: Cockcroft, 1896-1904 – 18v – 1 – mf#LL-898 – us L of C Photodup [347]

Encyclopaedia of heraldry / Burke, J B & Burke, J B – 1r – 1 – mf#2154 – uk Microform Academic [929]

The encyclopaedia of missions : descriptive, historical, biographical, statistical / ed by Bliss, Edwin Munsell – New York: Funk & Wagnalls, c1891 [mf ed 1986] – 4mf – 9 – 0-8370-7124-0 – (incl ind) – mf#1986-1124 – us ATLA [030]

Encyclopaedia of pleading and practice / ed by McKinney, William M – Northport, NY: Ed Thompson. v1-23+1st, 2nd suppls. 1895-1909 (all publ) – 340mf – 9 – $510.00 – mf#LLMC 82-512 – us LLMC [347]

The encyclopaedia of pleading and practice : under the codes and practice acts, at common law, in equity and in criminal cases – Northport, N.Y., Thompson, 1895-1902. 23 v – 1 – (suppl: northport, ny 1903-09 4v ll-638) – us L of C Photodup [345]

Encyclopaedia of the presbyterian church in the united states of america : including the northern and southern assemblies / ed by Nevin, Alfred et al – Philadelphia: Presbyterian Encyclopaedia Pub Co, 1884 – 3mf – 9 – 0-524-04181-4 – mf#1990-4985 – us ATLA [242]

Encyclopaedia of theology = Theologik / Raebiger, Julius Ferdinand – Edinburgh: T & T Clark 1884-85 [mf ed 1991] – (= ser Clark's foreign theological library. new series 20,21) – 2v on 2mf – 9 – 0-7905-9076-X – (incl bibl ref; trans with additions by john macpherson) – mf#1989-2301 – us ATLA [200]

An encyclopaedia on the evidences : or, masterpieces of many minds / Monser, John Waterhaus – St Louis: John Burns, 1880 [mf ed 1992] – 2mf – 9 – 0-524-05410-X – mf#1992-0420 – us ATLA [240]

The encyclopaedia sinica / ed by Couling, Samuel – London: Oxford University Press, 1917 – 7mf – 9 – 0-524-08188-3 – mf#1991-0301 – us ATLA [030]

Encyclopaedie der deutschen national-literatur : oder, biographisch-kritisches lexikon der deutschen dichter und prosaisten seit den fruehesten zeiten: nebst proben aus ihren werken / Wolff, Oskar Ludwig Bernhard – Leipzig: C Wigand, 1835-42 – 1r – 1 – us Wisconsin U Libr [430]

Encyclopaedie der heilige godgeleerdheid / Kuyper, Abraham – Amsterdam: JA Wormser, 1894 [mf ed 1990] – 3v on 4mf – 9 – 0-7905-7957-X – mf#1989-1182 – us ATLA [240]

Encyclopaedie der theologie / Hofmann, Johann Christian Konrad von; ed by Bestmann, Hugo Johannes – Noerdlingen: CH Beck, 1879 – 1mf – 9 – 0-8370-3622-4 – mf#1985-1622 – us ATLA [240]

Encyclopaedisches woerterbuch der medicinischen wissenschaften (ael3/13) / ed by Graefe, C F von & Hufeland, C W – Berlin 1828-49 [mf ed 1994] – (= ser Archiv der europaeischen lexikographie: fach-enzyklopaedien) – 37v on 151mf – 9 – €1460.00 – 3-89131-181-8 – (int by michael stolberg) – gw Fischer [610]

An encyclopaedist of the dark ages : isidore of seville / Brehaut, Ernest – New York: Columbia University: Longmans, Green, agents, 1912 [mf ed 1989] – (= ser Studies in history, economics and public law 48/1) – 1mf – 9 – 0-7905-4542-X – (incl bibl ref) – mf#1988-0542 – us ATLA [931]

Encyclopedia judaica : das judentum in geschichte und gegenwart – Berlin. bd1-10 (Aach-Lyra). 1928-34 – 168mf – 9 – €320.00 – ne Slangenburg [939]

Encyclopedia of american quaker genealogy – v2 – 1r – 1 – mf#1058243 – us WHS [243]

An encyclopedia of canadian biography : containing brief sketches and half-tone engravings of prominent business and professional men identified with the sovereign life assurance company of canada – [Toronto?: Sovereign Life Assurance Co, 1906?] [mf ed 1994] – 1mf – 9 – 0-665-73381-X – mf#73381 – cn Canadiana [338]

An encyclopedia of law and forms / Spalding, Hugh Mortimer – Philadelphia, Ziegler, 1880. 676 p. LL-1004 – 1 – us L of C Photodup [340]

The encyclopedia of ornament / Shaw, Henry – London 1842 – (= ser 19th c art & architecture) – 3mf – 9 – mf#4.2.1338 – uk Chadwyck [740]

Encyclopedia of religious knowledge / Brown, J Newton – v1-2. 1835 – 1 – us Southern Baptist [240]

Encyclopedia of southern baptists – v4 – 1 – us Southern Baptist [242]

Encyclopedia of southern baptists – 5 – (original and ed mss alphabetically arr) – us Southern Baptist [242]

Encyclopedia of the laws of england : being a new abridgement by the most eminent legal authorities – London/Edinburgh: Sweet & Maxwell/William Green & Sons. v1-12. 1897-1903 (all publ) – 84mf – 9 – $126.00 – (incl suppl and ind vol) – mf#LLMC 84-800 – us LLMC [342]

Encyclopedia of the us supreme court reports – Charlottesville, Michie. v1-13. 1908-23 (all publ) – (= ser United states supreme court reports) – 154mf – 9 – $231.00 – (covers the us reports v1-259) – mf#LLMC 90-021 – us LLMC [348]

Encyclopedias of artists from 17th to early 19th century = Kuenstlerlexika des 17. bis fruehen 19. jahrhunderts / ed by Schuette, Ulrich – [mf ed 2001] – 180mf (1:24) – 9 – diazo €2200.00 (silver €2600 isbn: 978-3-598-34552-4) – ISBN-10: 3-598-34551-8 – ISBN-13: 978-3-598-34551-7 – gw Saur [700]

Encyclopedie : ou dictionnaire raisonne des sciences... / ed by Diderot, M & d'Alembert, M – Paris 1751-80, 1751-65. v1-17 – 1317mf – 8 – (suppl 1-4 [1776-77]; suppl 1-2 [1780]; recueil des planches 1-12 [1762-77]) – mf#5437/2 – ne IDC [055]

Encyclopedie : ou dictionnaire raisonne des sciences, des arts et des metiers / Diderot, Denis & Alembert, J le Rond d' – Paris. v1-17 + suppl 1-4 + planches 1-12. 1751-1777 – 9 – €1333.00 – ne Slangenburg [030]

Encyclopedie (ael1/7) : ou dictionnaire universel raisonne des connoissances humaine / Fortune-Barthelemy de Felice – Yverdon 1770-80 [mf ed 1993] – (= ser Archiv der europaeischen lexikographie, abt 1: enzyklopaedien) – 42v+6 suppl vols on 257mf – 9 – €2020.00 – 3-89131-069-2 – gw Fischer [030]

Encyclopedie (ael1/34) : ou dictionnaire raisonne des sciences, des arts et des metiers / ed by Diderot, M & d'Alembert, M – Paris 1751-80 [mf ed 1996] – (= ser Archiv der europaeischen lexikographie, abt 1: enzyklopaedien) – 35v on 180mf – 9 – €1740.00 – 3-89131-224-5 – gw Fischer [030]

Encyclopedie coloniale et maritime – Paris, sep 1950-avr 1957 – 1 – (devenu: encyclopedie mensuelle d'outre-mer) – fr ACRPP [944]

Encyclopedie de la musique et dictionnaire du conservatoire / Lavignac, Albert – Paris. v1-2. 1920-31 – 1 – $216.00 – mf#0318 – us Brook [780]

Encyclopedie der christelijke theologie / Doedes, Jacobus Izaak – 2. verm uit. Utrecht: Kemink & Zoon, 1883 – 1mf – 9 – 0-8370-2944-9 – mf#1985-0944 – us ATLA [240]

Encyclopedie der evangelischen kirchenmusik / Kummerle, Salomon – 1885-95 – 1 – us L of C Photodup [780]

Encyclopedie des voyages : contenant l'abrege historique des moeurs, usages, habitudes domestiques... / Grasset S Sauveur – [Paris]: ...chez Deroy...1796 [mf ed 1986] – 5v on 1mf – 9 – 0-665-48962-5 – mf#48962 – cn Canadiana [390]

Encyclopedie des voyages : contenant l'abrege historique des moeurs, usages, habitudes domestiques... / Grasset-Saint-Saveur, J – [Paris], 1796. 5v – 47mf – 9 – mf#A-309 – ne IDC [910]

Encyclopedie mensuelle d'outre-mer *see* Encyclopedie coloniale et maritime

Encyclopedie mensuelle vivante, ouverte et libre *see* L'esprit francais

Encyclopedie methodique 1782-1832 – Paris: Panckoucke, Charles-Joseph [mf ed 2004] – (= ser AEL 1/50) – 206v on 1516mf – 9 – €7800 diazo €9360 silver – 3-89131-453-1 – (vols may be purchased individually; with accompanying vol: harald fischer: die encyclopedie methodique. zum inhalt und aufbau des werkes 140p [2004] isbn 3-89131-414-0 €42) – gw Fischer [030]

Encyclopedie methodique (ael3/12) : medecine / Azyr, Felix Vicq d' et al – ParisLiege) 1787-1830 [mf ed 1996] – (= ser Archiv der europaeischen lexikographie: fach-enzyklopaedien) – 13v on 55mf – 9 – €700.00 – 3-89131-180-X – (int by michael stolberg) – gw Fischer [610]

Encyclopedie moderne (ael1/40) : dictionnaire abrege des sciences, des lettres, des arts, de l'industrie, de l'agriculture et du commerce / ed by Renier, Leon – nouv ed. Paris 1860-83 [mf ed 1998] – (= ser Archiv der europaeischen lexikographie, abt 1: enzyklopaedien) – 26v+12 suppl vols+5 vols pl on 120mf – 9 – €1000.00 – 3-89131-314-4 – gw Fischer [030]

Encyclopedie oeconomique (ael1/27) : ou systeme general, 1. d'oeconomie rustique, 2. d'oeconomie domestique et 3. d'oeconomie politique – Yverdon 1770-71 [mf ed 1994] – (= ser Archiv der europaeischen lexikographie, abt 1: enzyklopaedien) – 16v on 35mf – 9 – €680.00 – 3-89131-205-9 – gw Fischer [330]

Encyclopedique ou universel *see* Journal encyclopedique

Encyklika und syllabus vom 8. dezember 1864 : als ein beitrag zum verstaendnis der kirchlichen lage der gegenwart fuer evangelische christen / Roenneke, K – Guetersloh: C Bertelsmann, 1891 – 1mf – 9 – 0-8370-8088-6 – (incl ind) – mf#1986-2088 – us ATLA [240]

Encyklopaedie der gesammten musikalischen wissenschaften : oder universal-lexikon der tonkunst – Stuttgart. 6v+suppl. 1835-42 – 1 – $204.00 – mf#0193 – us Brook [780]

Encyklopaedie der theologischen wissenschaften / Rosenkranz, Karl – 2. gaenzlich umgearb aufl. Halle: C A Schwetschke, 1845 – 1mf – 9 – 0-7905-9856-6 – mf#1989-1581 – us ATLA [200]

Encyklopaedie der theologischen wissenschaften nebst methodenlehre : zu akademischen vorlesungen und zum selbststudium / Krieg, Cornelius – Freiburg im Breisgau; St Louis, MO: Herder, 1899 – 1mf – 9 – 0-8370-4001-9 – (incl bibl and ind) – mf#1985-2001 – us ATLA [200]

Encyklopaedie und methodologie der theologie / Kihn, Heinrich – Freiburg i B: Herder 1892 [mf ed 1993] – (= ser Theologische bibliothek) – 2mf – 9 – 0-524-08451-3 – (incl ind) – mf#1993-2056 – us ATLA [240]

Encyklopaedisches journal / ed by Dohm, Chr Wilh von – Cleve, Duesseldorf 1774 – (= ser Dz) – 13st on 8mf – 9 – €2460.00 – mf#k/n303 – gw Olms [074]

Encyklopaedisches woerterbuch der wissenschaften, kuenste und gewerbe (ael1/6.1) – Altenburg 1824-36 [mf ed 1992] – (= ser Pierers enzyklopaedisches woerterbuch) – 26v + 6suppl vols on 197mf – 9 – €1080.00 – 3-89131-079-X – gw Fischer [030]

End not yet / Tayler, W Elfe – Bristol, England. 1859 – 1r – 1 – us UF Libraries [240]

"The end of controversy" controverted : a refutation of milner's "end of controversy," in a series of letters addressed to the most reverend francis patrick kenrick, roman catholic archbishop of baltimore / Hopkins, John Henry – New York: Pudney & Russell, 1855, c1854 [mf ed 1986] – 2v on 3mf – 9 – 0-8370-9070-9 – (incl bibl ref) – mf#1986-3070 – us ATLA [241]

End of our being / Dickson, David – Edinburgh, Scotland. 1827 – 1r – 1 – us UF Libraries [240]

The end of religious controversy : in a friendly correspondence between a religious society of protestants and a catholic divine / Milner, John – Baltimore: J Murphy, 1851 – 1mf – 9 – 0-8370-8282-X – (incl bibl ref) – mf#1986-2282 – us ATLA [240]

End of the curse / Hood, Edwin Paxton – London, England. 1869 – 1r – 1 – us UF Libraries [240]

End of the free-will controversy / Travis, Henry – London, England. 1875 – 1r – 1 – us UF Libraries [240]

The end of the irish parliament / Fisher, Joseph Robert – London: E. Arnold, 1911.xii,315p. With: Lehrbuch der mechanischen naturlehre by E.G. Fisher. 1 reel. 1292 – 1 – us Wisconsin U Libr [941]

The end of the law : being the warburton lectures given in lincoln's inn chapel during the years 1907-1911 / Glazebrook, Michael George – London: Rivingtons, 1911 – 1mf – 9 – 0-8370-9947-1 – mf#1986-3947 – us ATLA [240]

The end of the law : or, christ and buddhism / Gilmore, David Chandler & Smith, J F – Calcutta: Association Press [1914?] [mf ed 1991] – 1mf – 9 – 0-524-01440-X – mf#1990-2435 – us ATLA [230]

End of the rainbow / Valparaiso Realty Company – New Valparaiso, FL. 192- – 1r – 1 – us UF Libraries [978]

End of the world in 1867 – London, England. 18– – 1r – 1 – us UF Libraries [240]

End poverty paper *see* [Los angeles-] upton sinclair's epic news

Endacott, J Earl *see*
- D d eisenhower home and family
- "Dwight d eisenhower material"

Endang / Kentjana – Djakarta, 1953-1955 – 3mf – 9 – (missing: 1953, v1; 1954, v2(1, 3-12, 14-18)) – mf#SE-901 – ne IDC [959]

Endang *see* Almanak umum nasional

Das ende der eisernen schar : mit dem "polnischen tagebuch 1939" / Bodenreuth, Friedrich – Feldpostausg. Leipzig: Reclam, 1940 [mf ed 1991] – 75p – 1 – mf#7501 – us Wisconsin U Libr [880]

Ende der illusionen / Picht, Werner Robert Valentin – Berlin, Germany. 1941 – 1r – 1 – us UF Libraries [025]

Das ende der zeit / Guttmann, Bernhard – Freiburg im Breisgau: Zaehringer Verlag, 1948 (mf ed 1995) – 1r – 1 – mf#*ZP-1500 – us NY Public [230]

Ende gut, alles gut : erzaehlung aus dem ries / Meyr, Melchior – Bayreuth: Gauverlag Bayreuth, 1944 – 1r – 1 – us Wisconsin U Libr [430]

Ende, J van den *see* Michael servet, een der vele slachtoffers van den ketterjager kalvijn

Ende und anfang : ein lebensbuch / Muehlen, Hermynia zur – Berlin: S Fischer, c1929 – 1r – 1 – us Wisconsin U Libr [880]

The endeavor herald : for christ and the church – Toronto: Endeavor Herald Co, [1888?-189- or 19–] – 9 – mf#P06056 – cn Canadiana [240]

Endeavors after the christian life : discourses / Martineau, James – Boston: American Unitarian Assoc, 1881 [mf ed 1993] – (= ser Unitarian/ universalist coll) – 2mf – 9 – 0-524-07163-2 – mf#1991-2952 – us ATLA [243]

Endeavour : french edition – Elmsford. 1942-1976 (1) 1976-1976 (5) 1976-1976 (9) – mf#7713 – us UMI ProQuest [500]

Endeavour : german edition – Elmsford. 1942-1976 (1) 1976-1976 (5) 1976-1976 (9) – mf#7714 – us UMI ProQuest [500]

Endeavour – Oxford. 1977+ (1,5,9) – ISSN: 0160-9327 – mf#49280 – us UMI ProQuest [500]

Endeavour : spanish edition – Elmsford. 1942-1976 (1) 1976-1976 (5) 1976-1976 (9) – mf#7715 – us UMI ProQuest [500]

Endecasilabo castellano / Henriquez Urena, Pedro – Buenos Aires, Argentina. 1945 – 1r – 1 – us UF Libraries [972]

Endemann, Karl *see* Versuch einer grammatik des sotho

Enderbrock, D M *see* The parental obligation to care for the religious education of children

Enderby, Charles *see* Proposal for re-establishing the british southern whale fishery

Enderling, Paul *see* Die glocken von danzig

Enders, Barthol *see* Der begriff dogma entwickelt aus der entscheidung ueber die unbefleckte empfaengniss mariae

Enders, Carl *see* Festschrift fuer berthold litzmann zum 60. geburtstag 18.4.1917

Enders, Carl Friedrich *see* Gottfried keller

Enders, Ludwig *see*
- Aus dem kampf der schwaermer gegen luther
- Ausgewaehlte schriften
- Luther und emser
- Ein schoener dialogus von martino luther und der geschickten botschaft aus der hoelle

Enderuni,Vasif-i *see* The divan project

Das endinger judenspiel / ed by Amira, Karl von – Halle: Max Niemeyer 1883 [mf ed 1993] – (= ser Neudrucke deutscher literaturwerke des 16. und 17. jahrhunderts 41) – 11r – 1 – (incl bibl ref) – mf#3387p – us Wisconsin U Libr [430]

Endl, Elmar *see* Durchflusszytometrische untersuchungen von stosswelleninduzierten zellschaeden

Endless being : or, man made for eternity / Barlow, Joseph Lorenzo – New York: F H Revell, c1888 – 1mf – 9 – 0-7905-8635-5 – mf#1989-1860 – us ATLA [240]

The endless future : showing the probable connection between human probation and the endless universe that is to be / Cook, E Wake – Nashville, Tenn: Southern Methodist Pub House, 1885 – 1mf – 9 – 0-524-06481-4 – mf#1991-2581 – us ATLA [240]

The endless future of the human race : a letter to a friend / Henry, Caleb Sprague – New York: D Appleton, 1879 – 1mf – 9 – 0-524-04730-8 – (incl bibl ref) – mf#1991-2135 – us ATLA [240]

Endless punishment : In the very words of its advocates / Sawyer, Thomas Jefferson – Boston: Universalist Pub House, 1880 – 1mf – 9 – 0-524-04439-2 – (incl bibl ref) – mf#1991-2104 – us ATLA [240]

Endless punishment rejected : conversation between inquirer and expositor / Ballou, Adin – [S.l.: s.n., 1850?] (Hopedale, Mass: AG Spalding) – 1mf – 9 – 0-524-05563-7 – mf#1991-2297 – us ATLA [240]

Endlicher bericht abdiae praetorij von seiner lere in den artickeln, darin er von doctore andrea musculo auffs hefftigste angegriffen wird / Praetorius, A – [Wittemberg], 1563 – 5mf – 9 – mf#TH-1 mf 1281-1285 – ne IDC [242]

Die endlose strasse : ein frontstueck in vier bildern / Graff, Sigmund & Hintze, Carl Ernst; ed by Matthaesius, Friedrich – Bielefeld: Velhagen & Klasing, 1936 – 1r – 1 – us Wisconsin U Libr [820]

Der endlose wald : roman aus dem boehmerwald / Attenberger, Toni – 3. aufl. Frankfurt/Main: Breidenstein, 1940 [mf ed 1988] – 268p – 1 – mf#6968 – us Wisconsin U Libr [830]

Endocrine pathology – Cambridge. 1991-1994 (1,5,9) – ISSN: 1046-3976 – mf#18078 – us UMI ProQuest [616]

The endocrinologist – v1-6. 1991-96 – 1,5,6,9 – $95.00r – us Lippincott [616]

Endocrinology – Philadelphia. 1917-1977 (1) 1967-1977 (5) 1970-1977 (9) – ISSN: 0013-7227 – mf#2357 – us UMI ProQuest [616]

Endocrinology and metabolism clinics of north america – Philadelphia. 1987+ (1,5,9) – ISSN: 0889-8529 – mf#12718,01 – us UMI ProQuest [616]

Endocrinology index – Washington. 1968-1979 [1]; 1974-1979 [5,9] – ISSN: 0013-7235 – mf#6493 – us UMI ProQuest [616]

Endoluminale bestimmung der blutflussgeschwindigkeit bei der kathederbehandlung der peripheren arteriellen verschlusskrankheit : ein vergleich mit angiographie, intraarteriellen druckmessungen und klinischen befunden / Mackowski, Joanna Magdalena – (mf ed 1996) – 2mf – 9 – €40.00 – 3-8267-2387-2 – mf#DHS 2387 – gw Frankfurter [616]

Endore, S Guy *see*
- Babouk
- The sleepy lagoon mystery

Endoscopy – Stuttgart. 1975+ (1,5,9) – ISSN: 0013-726X – mf#10159 – us UMI ProQuest [617]

Endothelial selectins and pulmonary gas exchange in female aerobic athletes / Hunte, Garth S – 2000 – 156p on 2mf – 9 – $10.00 – mf#PH 1712 – us Kinesology [612]

Endowment of romanism in ireland though the "christian brothers"... / Kerr, Rev Dr – Edinburgh, Scotland. 18– – 1r – 1 – us UF Libraries [240]

The endowments and establishment of the church of england / Brewer, John Sherren; ed by Dibdin, Lewis Tonna – 2nd rev ed. London: John Murray, 1885 – 1mf – 9 – 0-7905-4437-7 – (incl bibl ref) – mf#1988-0437 – us ATLA [241]

Endowments of the church and their origin / Dorington, John Edward – London, England. 1884 – 1r – 1 – us UF Libraries [240]

Endres, Franz Carl *see* Symbolik von goethes faust

Endres, J A *see*
- Forschungen zur geschichte der fruehmittelalterlichen philosophie
- Petrus damiani und die weltliche wissenschaft

Endres, Norbert *see* Faechererweiterungen symmetrischer moduale und homotopiemengen von produktabbildungen auf sphaeren

Endres tuchers baumeisterbuch der stadt nuernberg (1464-1475) / ed by Lexer, Matthias – Stuttgart: Litterarischer Verein, 1862 [mf ed 1993] – (= ser Blvs 64) – xiv/387p – 1 – mf#8470 reel 13 – us Wisconsin U Libr [914]

Endress, Gerhard *see* Die arabischen uebersetzungen von aristoteles' schrift de caeolo

Endrikat, Fred *see* Liederliches und lyrisches

Ends are means : a critique of social values / Shelvankar, Krishnarao Shivarao – [London]: Lindsay Drummond Ltd, 1938 – (= ser Samp: indian books) – (int by H Levy) – us CRL [303]

Der endtchrist / Gwalther, R – Zuerich, Froschouer, 1546 – 2mf – 9 – mf#PBU-290 – ne IDC [240]

Enduran, Ludoix *see*
- La traite des negres
- La traite des negres; ou, deux marins au senegal

Endure hardness / Nicholson, William – Winchester, England. 1840? – 1r – 1 – us UF Libraries [240]

Enea silvio de' piccolomini, als papst pius der zweite : und sein zeitalter / Voigt, Georg – Berlin: G Reimer, 1856-1863 – 4mf – 9 – 0-7905-7157-9 – (incl bibl ref) – mf#1988-3157 – us ATLA [241]

Eneas-roman (cima2) : farbmikrofiche-edition der handschrift heidelberg, universitaetsbibliothek, cod.pal.germ.403 / Veldeke, Heinrich von – (mf ed 1987) – (= ser Codices illuminati medii aevi (cima) 2) – 27p on 6 color mf – 15 – €335.00 – 3-89219-002-X – (int by hans fromm) – gw Lengenfelder [090]

Eneas-roman (cima59) : farbmikrofiche-edition der handschrift wien, oesterreichische nationalbibliothek, cod 2861 / Veldeke, Heinrich von – (mf ed 2000) – (= ser Codices illuminati medii aevi (cima) 59) – 55p on 4 color mf – 15 – €280.00 – 3-89219-059-3 – (int & description by marcus schroeter) – gw Lengenfelder [090]

La eneida de... / Virgilio – Coria: Imp. de P. Evaristo Montero, 1873 – 1 – sp Bibl Santa Ana [946]

Eneide / Veldeke, Heinrich von; ed by Schieb, Gabriele & Frings, Theodor – Berlin: Akademie-Verlag, 1964-70 [mf ed 1993] – (= ser Deutsche texte des mittelalters 58-59, 62) – 3v – 1 – (incl bibl ref and ind) – mf#8623 reel 16-17 – us Wisconsin U Libr [810]

L'eneide di virgile traduite en vers francois / [Perrin, P] – Paris: Des caracteres de P Moreau, 1648 – 6mf – 9 – mf#O-1357 – ne IDC [090]

L'eneide di virgile traduite en vers francois / [Perrin, P] – Paris: Moreau, Pasle, 1648-58 – 11mf – 9 – mf#O-1936 – ne IDC [090]

Enelow, Hyman Gerson *see* Yahvism and other discourses

Enemy and the standard of defence / Symington, William – Glasgow, Scotland. 1852 – 1r – 1 – us UF Libraries [240]

The enemy side of the hill / U.S. Army. Historical Division – The 1945 background on interrogation of German commanders. 1949 – 1 – us L of C Photodup [943]

Enemy slain by prayer – London, England. 186-? – 1r – 1 – us UF Libraries [240]

Energie – Halle S DE, 1967 6 jan-1990 [gaps] – 4r – 1 – (energieversorgung halle) – gw Misc Inst [333]

Energy : a crisis in power / Holdren, John & Herrera, Philip – (= ser Sierra club books out of print) – 1r – 5 – $50.00 – mf#B70009 – us Library Micro [333]

Energy : efficient and final cause / McCosh, James – New York: Scribner 1883 [mf ed 1991] – 1mf – 9 – 0-7905-9805-1 – mf#1989-1530 – us ATLA [100]

Energy – Norwalk. 1980-1996 (1,5,9) – ISSN: 0149-9386 – mf#12364 – us UMI ProQuest [333]

Energy : official organ of the canadian philatelic press association – Berlin [Kitchener], Ont: Energy Pub Co, [1899-1901?] – 9 – (merged with: canada stamp sheet to become: canada stamp sheet and energy) – mf#P04551 – cn Canadiana [760]

Energy – Oxford. 1976+ (1,5,9) – ISSN: 0360-5442 – mf#49104 – us UMI ProQuest [530]

Energy abstracts 1994 – us Ei [333]

Energy abstracts 1995 – us Ei [333]

Energy abstracts 1996 – us Ei [333]

Energy action / Missouri Energy Foundation – v1 n1-v4 n1 [1981 mar/apr-1984 spr] – 1r – 1 – (cont: safe energy news) – mf#1207933 – us WHS [333]

Energy and buildings – Lausanne. 1977-1994 (1) 1977-1994 (5) 1987-1994 (9) – ISSN: 0378-7788 – mf#42158 – us UMI ProQuest [690]

Energy and Chemical Workers Union *see* Journal

Energy and labor / Cuningham, Granville Carlyle – S.l: s.n, 1891? – 1mf – 9 – (incl bibl ref) – mf#58821 – cn Canadiana [331]

Energy Awareness Center et al *see* Nuclear hazards

Energy balance and the components of total daily energy expenditure in endurance trained and untrained women / Beidleman, B A – 1991 – 3mf – 9 – $12.00 – us Kinesology [613]

Energy conversion and management – Oxford. 1961+ (1,5,9) – ISSN: 0196-8904 – mf#49064 – us UMI ProQuest [333]

Energy cost of walking with and without arm activity on the cross walk dual motion cross trainer / Knox, Kelly M – 1993 – 1mf – $4.00 – us Kinesology [612]

Energy cost of walking/jogging in a laminar flow resistance pool / Waldo, Brian R – 1997 – 1mf – 9 – $4.00 – mf#PH 1583 – us Kinesology [612]

The energy cost of women walking with and without hand weights while performing rhythmic arm movements / Zywicki, Scott S & Butts, Nancy Kay – 1992 – 1mf – $4.00 – us Kinesology [612]

Energy developments in japan – Chicago. 1978-1985 (1,5) – ISSN: 0161-8091 – mf#49454 – us UMI ProQuest [333]

Energy digest – Wheathampstead. 1973-1990 (1) 1972-1990 (5) 1975-1990 (9) – ISSN: 0367-1119 – mf#8662 – us UMI ProQuest [690]

Energy economics – Kidlington. 1979+ (1,5,9) – ISSN: 0140-9883 – mf#17226 – us UMI ProQuest [333]

Energy engineering : journal of the Association of Energy Engineers – Atlanta. 1979+ (1,5,9) – (Cont: Building systems design) – ISSN: 0199-8595 – mf#756,01 – us UMI ProQuest [690]

Energy expenditure and substrate utilization : in non-obese african-american women and caucasian women / Washinton, Sara B – 2000 – 56p on 1mf – 9 – $5.00 – mf#PH 1720 – us Kinesology [612]

Energy expenditure of step training vs low impact aerobics using three common movement patterns / Barry, Dawn M – Purdue University, 1995 – 1mf – 9 – mf#PH 1485 – us Kinesology [612]

The energy flow of the human being and the universe : tai ji philosophy as an artistic and philosophical foundation for the development of chinese contemporary dance / Yu, Jin-Wen – 1994 – 4mf – $16.00 – us Kinesology [306]

Energy information database / ed by Voight, R & Franklin, G – 1956-80 – 9 – $22400.00; $225.00t – (annual update $22125.00y) – us IRE [621]

Energy international – San Francisco. 1964-1980 (1) 1964-1980 (5) 1964-1980 (9) – ISSN: 0013-7529 – mf#9669 – us UMI ProQuest [333]

Energy journal – Cambridge. 1980+ (1,5,9) – ISSN: 0195-6574 – mf#12886 – us UMI ProQuest [333]

Energy law journal – v1-22. 1980-2001 – 9 – $468.00 set – ISSN: 0270-9163 – mf#105041 – us Hein [340]

Energy law journal – Washington. 1980-2001 (1) 1980-2001 (5) 1980-2001 (9) – ISSN: 0270-9163 – mf#12613 – us UMI ProQuest [333]

Energy management – Cleveland. 1981-1983 (1,5,9) – ISSN: 0199-5650 – mf#12651 – us UMI ProQuest [333]

Energy news digest / Citizens Energy Council – 1980 nov 24-1986 jan – 1r – 1 – (cont: energy news digest of nuclear hazards versus alternative energies; cont by: bulletin (citizens energy council)) – mf#1214368 – us WHS [333]

Energy news digest of nuclear hazards versus alternative energies / Citizens Energy Council – n229-230,231 [i.e. 232?], 234-236 [1980 jan 31-apr 17, jun 21, aug 28-oct 31] – 1r – 1 – (cont: nuclear hazards; cont by: energy news digest) – mf#1231043 – us WHS [333]

Energy pipelines and systems – Houston. 1974-1974 (1) 1974-1974 (5) (9) – ISSN: 0093-0512 – mf#9609 – us UMI ProQuest [333]

Energy policy – Kidlington. 1982+ (1,5,9) – ISSN: 0301-4215 – mf#13328 – us UMI ProQuest [333]

Energy progress – New York. 1981-1988 (1) 1981-1988 (5) 1981-1988 (9) – ISSN: 0278-4521 – mf#13368 – us UMI ProQuest [333]

ENERGY

Energy realities : rates of consumption, energy reserves, and future options: hearing...house of representatives, 107th congress, 1st session, may 3 2001 / United States. Congress. House. Committee on Science. Subcommittee on Energy – Washington: US GPO 2002 [mf ed 2002] – 4mf – 9 – (incl bibl ref) – us GPO [333]
Energy research abstracts – Oak Ridge. 1977-1992 (1,5,9) – (Cont: ERDA energy research abstracts) – ISSN: 0160-3604 – mf#11139,02 – us UMI ProQuest [333]
Energy sources – New York. 1975-1996 (1,5,9) – ISSN: 0090-8312 – mf#11018 – us UMI ProQuest [530]
Energy systems and policy – New York. 1974-1991 (1) 1974-1991 (5) 1974-1991 (9) – ISSN: 0090-8347 – mf#11081 – us UMI ProQuest [333]
Energy user news – Troy, 1998+ [1,5,9] – ISSN: 0162-9131 – mf#18824 – us UMI ProQuest [333]
Energyfiche : retrospective 1976-1993 – 9 – (access to key information identified by energy information abstracts in environment abstracts database) – us CIS [020]
Enesco, Georges *see*
– Suite pour piano, op 10
– Variations pour 2 pianos, op 5
L'enfamt d'argigente : suivi de: le grec et la nature / Festugiere, A J – Paris, 1950 – €7.00 – ne Slangenburg [110]
L'enfance bambara : approche psycho-culturale de trois phases pre-circoncisionelles en pays bambara / Couloubaly, Pascal Baba F – Dakar: IFAN, 1986 – us CRL [300]
L'enfance de suzette : livre de lecture courante a l'usage des jeunes filles / Halt, Marie Malezieux – Paris: P Delaplaine, 1892 – (= ser Les femmes [coll]) – 2mf – 9 – mf#8844 – fr Bibl Nationale [830]
Enfant du faubourg / Deslandes, Paulin – Paris, France. 1837 – 1r – 1 – us UF Libraries [440]
L'enfant mysterieux / Dick, Vinceslas-Eugene – Quebec: J A Langlais [1890?] [mf ed 1980] – 2v on 1mf – 9 – 0-665-05666-4 – mf#05666 – cn Canadiana [830]
Enfant truque / Natanson, Jacques J – Paris, France. 1931 c1922 – 1r – 1 – us UF Libraries [440]
Enfantillage / Melesville, M – Paris, France. 1844 – 1r – us UF Libraries [440]
Les enfants celebres / Chaumette, E J M – Limoges: E Ardant 1888? [mf ed 1988] – 1r – 1 – mf#2182 – us Wisconsin U Libr [920]
Les enfants de l'orpailleur / Montbrillant, A de – 2e ed. Quebec: J-A Langlais, [1888?] [mf ed 1984] – 2mf – 9 – 0-665-38143-3 – mf#38143 – cn Canadiana [830]
Enfermedad de centro-america / Mendieta, Salvador – Barcelona, Spain. v1-3. 1934 – 1r – 1 – us UF Libraries [972]
Enfermedades de los conquostadores / Figueroa Marroquin, Horacio – San Salvador, El Salvador. 1957 – 1r – 1 – us UF Libraries [972]
Enfermedades rojas del cerdo / Lopez Sanchez, Ernesto – Badajoz: La Alianza, 1934 – 1 – sp Bibl Santa Ana [946]
Enfield 1770-1890 – Oxford, MA (mf ed 1984) – (= ser Massachusetts vital records) – 80mf – 9 – 0-931248-73-6 – (Mf 1-3: B,M,Intents, Deaths 1783-1849. Mf 4-6: B,M,D 1844-54. Mf 7-9: B,M,D 1859-92. Mf 10-11: Marriage Intentions 1849-1905. Mf 12-28: Index to Births 1770-1938. Mf 29-45: Index to Marriage Intents 1816-1905. Mf 46-62: Index to Marriages 1816-1937. Mf 63-79: Index to Deaths 1802-1938. Mf 80: Index to Burials) – us Archive [978]
Enfield advertiser – London, UK. 1986-20 dec 1990; 1991; 1992 – 23r – 1 – uk British Libr Newspaper [072]
Enfield chronicle – Enfield, England 25 mar-4 aug 1899 [mf 25 mar 1898-16 mar 1900, may 1901-25 sep 1903] (mf ed 2003) – 1 – (cont by: enfield & edmonton chronicle & barnet herald [11 aug 1899-23 dec 1904]) – uk British Libr Newspaper [072]
Enfield & edmonton chronicle & barnet herald – Enfield, England 11 aug 1899-23 dec 1904 [mf 25 mar 1898-16 mar 1900, may 1901-25 sep 1903] – 1 – (cont: enfield chronicle [25 mar 1898-4 aug 1899]; discontinued) – uk British Libr Newspaper [072]
Enfield edmonton palmers green and southgate independent *see* Enfield independent
Enfield express – London, UK. 4 jan-1 nov 1889 – 1/2r – 1 – uk British Libr Newspaper [072]
Enfield gazette – [London & SE] Enfield Public Lib 1859-1974, 1976-84; BLNL 1984- – 1 – uk Newsplan [072]
Enfield gazette and observer *see* Meyers enfield observer and local general advertiser
Enfield independent – London, UK. 8 jan 1986-1993 – 39 1/2r – 1 – (aka: enfield edmonton palmers green and southgate independent) – uk British Libr Newspaper [072]
Enfield, Richard *see* On the duty of the educated and wealthy classes to sunday schools
Enfield town express – Enfield UK, 10 jul-18 dec 1992; 1993 – 2r – 1 – uk British Libr Newspaper [072]
Enfield Waltham Sunday School *see* Reports
Enfield weekly herald *see* Enfield weekly herald and enfield highway and ponders end advertiser
Enfield weekly herald and enfield highway and ponders end advertiser – London, UK. 1951 – 1/2r – 1 – (aka: enfield weekly herald) – uk British Libr Newspaper [072]
Enfield, William *see* New encyclopaedia
Enfins seuls! / Sablons, Albert – Paris, France. 1934 – 1r – 1 – us UF Libraries [440]
Enfoque metropolitano 3 – Miami, FL. 1994 jan 01-1999 jul 14 – 1r – 1 – (missing: 1998 jan 15-1999 jul 14) – us UF Libraries [071]
Enforcement journal – Venice. 1975-1991 (1) 1984-1991 (5) 1984-1991 (9) – ISSN: 0042-2347 – mf#10582 – us UMI ProQuest [360]
Enforcement of judgments and orders / Obi-Okoye, A – Enugu, Nigeria: Reveille 1973 – 58p – 1 – mf#LL-12047 – us L of C Photodup [340]
Engadin express and alpine post – Samaden (CH) 1 jun 1901-24 aug 1939 [wkly] – 56r – 1 – (in english & german; cont: alpine post" wh is entered under st moritz) – uk British Libr Newspaper [074]
Engadine district news – Engadine – 3r – A$227.39 vesicular A$243.89 silver – at Pascoe [079]
Engadine district news – Engadine, nov 1964-77 – 3r – A$115.50 vesicular A$132.00 silver – at Pascoe [079]
Engage – Washington. 1968-1972 (1) – ISSN: 0013-7618 – mf#7790 – us UMI ProQuest [240]
Engagement, alienation and self-discovery in the poetry of arthur kenneth nortje / Solomons, Abubakar – U of the Western Cape 1986 [mf ed S.l: s.n. 1986] – 2mf – 9 – (incl bibl) – sa Misc Inst [420]
Engagement in defence of the liberties of the church and people of... – Edinburgh, Scotland. 1840? – 1r – 1 – us UF Libraries [240]
Engage/social action – Washington. 1973-1974 (1) – (Cont by: ESA Engage/social action) – ISSN: 0090-3485 – mf#7147 – us UMI ProQuest [301]
Engaging in ministry with an ethnic minority local church / Lyght, Ernest Shaw – Princeton, NJ, 1979. Chicago: Dep of Photodup, U of Chicago Lib, 1979 (1r); Evanston: American Theol Lib Assoc, 1984 (1r) – 1 – 0-8370-1374-7 – mf#1984-T219 – us ATLA [240]
Engano de las razas / Ortiz, Fernando – Habana, Cuba. 1945 – 1r – 1 – us UF Libraries [972]
Engasser, Quirin *see* Der ursaecher
Engberg, Jon E *see* Piano trio
Engberg, Robert M *see* Notes on the chalcolithic and early bronze age pottery at megiddo
Engdahl, Richard *see* Beitraege zur kenntnis der byzantinischen liturgie
Enge, Kevin M *see* Habitat occurrence of florida's native amphibians and reptiles
Engel, A *see*
– Calvinischer betlersmantel darin angezeiget wird mit was kleider sie sich bekapen den schalck verbergen vnd zudecken koennen
– Thewrungs spiegel darinnen gewiesen wird woher thewrung koeme vnd warumb
Engel, Alexander *see* Protektion
Engel, Carl Dietrich Leonhard *see*
– Das engelsche volksschauspiel doctor johann faust als faelschung
– Zusammenstellung der faust-schriften vom 16. jahrhundert bis zur mitte 1884
Engel, Eduard *see* Fremdwoerterbuch
Engel, Georg Julius Leopold *see* Die last
Engel, Johann Christian von *see* Geschichte von halitsch und wladimir bis 1772..
Engel, Johann Jakob *see* Der philosoph fuer die welt
Engel, M R *see* Der kampf um roemer kapitel 7
Engel, Moritz *see* Die loesung der paradiesfrage
Engel, Sabine von *see* Wir haben dich gemeint
Engelbach, Georg *see* Religioese fragen
Engelberg : eine dichtung / Meyer, Conrad Ferdinand – 7. aufl. Leipzig: H Haessel, 1900 [mf ed 1990] – 112p – 1 – mf#7607 – us Wisconsin U Libr [810]
Engelbert, Frater *see* Ons eerste geschiedenisboekje
Engelbrecht, Arnold Raymond *see* Die verband tussen taalbeplanning en moedertaalonderrig met verwysing na suid-afrika as veeltalige land
Engelbrecht, Gert Hermias *see* Basiese elemente van 'n effektiewe jeugbediening
Engelbrecht, Ockert Michiel *see* Die finansiele posisie van gades na egskeiding met spesifieke verwysing na die "clean break"-beginsel
Engelgardt, M A *see* Printsipy trudovoi teorii
Engelgrave, H *see*
– Coelum empyreum
– Lux evangelica sub velum sacrorum emblematum recondita in anni dominicas...
Engelhard / Wuerzburg, Konrad von; ed by Gereke, Paul – Halle a. S: M Niemeyer, 1912 [mf ed 1993] – (= ser Altdeutsche textbibliothek n17) – xi/220p – 1 – (incl bibl ref) – mf#8193 reel 2 – us Wisconsin U Libr [810]
Engelhard, Karl *see* Friedrich hebbel als lyriker
Engelhard-Colton, Nancy *see* The effects of oral contraception and hypoxia on respiratory parameters during graded exercise
Engelhardt, B, baron von *see* Observations astronomiques
Engelhardt, D J *see* Richard von st victor und johannes rusbroek
Engelhardt, E v *see* Der herr sicher
Engelhardt, Lisa von *see* Ferdinand v wrangel und seine reise laengs der nordkueste von sibirien und auf dem eismeere
Engelhardt, M von *see* Das christentum justins des maertyrers
Engelhardt, Moritz von *see*
– Das christenthum justins des maertyrers
– Reise in die krym und den kaukasus
Engelhardt, Zephyrin *see*
– The franciscans in arizona
– The franciscans in california
– The holy man of santa clara
– Missionary work of the franciscans
– The missions and missionaires of california... santa barbara, 1929
– The missions and missionaries of california
Engelholms Tidning *see* Oresundsposten
Engelholms tidning – AEngelholm, 1867-1946 – 9 – sw Kungliga [078]
Engelholms tidning *see* Hoganaes tidning
Engelholmsposten – Aengelholm, 1897-1901 – 8r – 1 – sw Kungliga [078]
Engelke, Gerrit *see*
– Rhythmus des neuen europa
– Vermaechtnis
Engelkemper, Wilhelm *see*
– Heiligtum und opferstaetten in den gesetzen des pentateuch
– Die religionsphilosophische lehre saadja gaons ueber die hl schrift
Engelkes, Gustav Gerhard *see* Der kornett des koenigs
Engelmann, Arthur *see* Das alte und das neue buergerliche recht deutschlands
Engelmann, Julius *see* Taschenbuch fuer reisende durch deutschland und die angraenzenden laender
Engel-Mitscherlich, Hilde *see* Hebbel als dichter der frau
Engels, Friedrich *see*
– Die bakunisten an der arbeit
– Chia tsu ssu yu ts'ai ch'an chi kuo chia chih ch'i yuean
– Der deutsche bauerkrieg
– Die entwicklung des sozialismus
– Feuerbach, the roots of the socialist philosophy
– Osnovni zasady komunizmu
– Rozvytok sotsiializmu vid utopii do nauky
Engels, Johann *see* Denkwuerdigkeiten der natur und kunst, religion und geschichte, schiffahrt und handlung in den koeniglich preussischen niederrheinisch-westf provinzen
Engels, Johann Peter *see* Das normaldruckglaukom
Engel's kak literaturnyi kritik / Shiller, Frants Petrovich – Moskva: Gos izd-vo khudozhestvennoi lit-ry, 1933 [mf ed 2002] – 1r – 1 – (filmed with: k biografii adama mitskevicha v 1821-1829 godakh / fedor verzhbovskii [teodor wierzbowski], (1898). incl bibl ref) – mf#5239 – us Wisconsin U Libr [940]
Der engels- und teufelsglaube des apostels paulus / Kurze, Georg – Freiburg i B: Herder, 1915 [mf ed 1993] – 1mf – 9 – 0-524-05919-5 – (incl bibl ref) – mf#1992-0676 – us ATLA [225]
Das engelsche volksschauspiel doctor johann faust als faelschung / Bruinier, Johannes Weijgardus; ed by Engel, Carl Dietrich Leonhard – Halle a/S: M Niemeyer 1894 [mf ed 1990] – (= ser Faust vor goethe: untersuchungen) – 1r – 1 – (incl bibl ref. filmed with: "old-iniquity": der schluessel zu goethes "faust" / ottomar beta) – mf#7341 – us Wisconsin U Libr [790]
Das engelsche volksschauspiel doctor johann faust als faelschung / Bruinier, Johannes Weijgardus – Halle a/S: M Niemeyer 1894 [mf ed 1990] – 1r – 1 – (filmed with: 'old-iniquity': der schluessel zu goethes 'faust' / ottomar beta) – mf#7341 – us Wisconsin U Libr [790]
Engelsmann, Walter *see* Goethe und beethoven
Engelstoft, Christian Thorning *see* De confutatione latina
Der engelwirt : eine schwabengeschichte / Strauss, Emil – 56.-63. aufl. Berlin c1921 [mf ed 1985] – 1r – 1 – mf#1493 – us Wisconsin U Libr [390]
Engenheiro frances no brasil / Freyre, Gliberto – Rio de Janeiro, Brazil. 1940 – 1r – 1 – us UF Libraries [972]
Engenheiros E Economistas Consultores *see* Medio sao francisco
Engert, Horst *see* Die tragik der dem leben nicht gewachsenen innerlichkeit in den werken gerhart hauptmanns
Engert, Joseph *see*
– Der deismus in der religions- und offenbarungskritik des hermann samuel reimarus
– Der naturalistische monismus haeckels
Engert, Thaddaeus *see*
– Der betende gerechte der psalmen
– Der deutsche modernismus
– Die suenden der paepste im spiegel der geschichte
Engineer – Washington. 1976+ (1) 1976+ (5) 1976+ (9) – ISSN: 0046-1989 – mf#7910 – us UMI ProQuest [620]
Engineer update – v1 n3 [1977 jul], v2 n3-4 [1978 may-jun], v3 n6-v13 n12 [1979 aug-1989 dec] – 1r – 1 – mf#435205 – us WHS [071]
Engineering – London: Charles Robert Johnson, [v9. jan-jun 1895] – us CRL [620]
Engineering and boiler house review – London. 1899-1968 (1) – mf#1264 – us UMI ProQuest [621]
Engineering and mining journal – New York. 1866+ (1) 1970+ (5) 1975+ (9) – ISSN: 0095-8948 – mf#25 – us UMI ProQuest [622]
Engineering applications of artificial intelligence – Oxford. 1991-1996 (1,5,9) – ISSN: 0952-1976 – mf#49609 – us UMI ProQuest [000]
Engineering computations – Bradford. 2001+ (1,5,9) – ISSN: 0264-4401 – mf#31585 – us UMI ProQuest [621]
Engineering costs and production economics – Amsterdam. 1976-1990 (1) 1976-1990 (5) 1987-1990 (9) – (Cont by: International journal of production economics) – ISSN: 0167-188X – mf#42184 – us UMI ProQuest [620]
Engineering cybernetics – Silver Spring. 1977-1984 (1,5,9) – (Cont by: Soviet journal of computer and systems sciences) – ISSN: 0013-788X – mf#14349 – us UMI ProQuest [000]
Engineering d a polytechnische zeitung – Berlin, Germany 2 jan-25 dec 1875, 4 jan 1879-30 dec 1882, 5 jan-27 dec 1884 – 1 – (cont: allgemeine deutsche polytechnische zeitung [4 jan 1873-26 dec 1874]) – uk British Libr Newspaper [620]
Engineering design graphics journal – College Station. 1974+ (1) 1974+ (5) 1975+ (9) – ISSN: 0046-2012 – mf#9489 – us UMI ProQuest [620]
Engineering digest – New York. 1962-1965 [1,5,9] – ISSN: 0423-1376 – mf#1700 – us UMI ProQuest [620]
Engineering economist – Norcross. 1955+ (1) 1967+ (5) 1975+ (9) – ISSN: 0013-791X – mf#1185 – us UMI ProQuest [620]
Engineering education – Washington. 1910-1991 (1) 1970-1991 (5) 1976-1991 (9) – ISSN: 0022-0809 – mf#591 – us UMI ProQuest [378]
Engineering fracture mechanics – New York. 1968+ (1,5,9) – ISSN: 0013-7944 – mf#49065 – us UMI ProQuest [620]
Engineering geology – Amsterdam. 1965+ (1) 1965+ (5) 1986+ (9) – ISSN: 0013-7952 – mf#42193 – us UMI ProQuest [624]
Engineering in medicine – London. 1976-1988 (1,5,9) – (Cont by: Proceedings of the Institution of Mechanical Engineers Pt H, Journal of engineering in medicine) – ISSN: 0046-2039 – mf#11217 – us UMI ProQuest [610]
Engineering Index – v1 1884-1891 – 1 – us NY Public [620]
The engineering index annual – 1,5,6,13 – (1884-1969 $400y, $425y overseas; 1970-1980 $500y, $540y overseas; 1981 $560y, $600y overseas; 1982 $630y, $670y overseas; 1983 $680y, $720y overseas; 1984 $720y, $760y overseas; 1985-86 $745y, $790y overseas; 1987 $840y, $890y overseas; 1988 $910y, $960y overseas; 1989 $940y, $990y overseas; 1990 $1,210y, $1,250y overseas; 1991 $1,460y $1,560y overseas; 1992 $1,540y $1,640y overseas; 1993 $1,820y $1,980y overseas; 1994 $2,100y $2,200y overseas) – us Ei [620]
The engineering index annual 1995 – v94. 1996 – 1,5,6,13 – $2,270.00; $2,375.00 outside North America – (abstracts of the worldwide engineering literature on annual basis, organized by subject) – us Ei [620]
The engineering index combination – $3,490.00; $3,750.00 outside North America – (1996 monthly. 1995 annual) – us Ei [620]
The engineering index monthly 1996 – 1,5,6,13 – $2,645.00y $2,970.00y outside North America – us Ei [620]
Engineering issues – New York. 1973-1978 (1) 1974-1978 (5) 1974-1978 (9) – (Cont by: Issues in engineering) – ISSN: 0093-8343 – mf#8148 – us UMI ProQuest [620]
Engineering journal – Chicago. 1964+ (1) 1971+ (5) 1975+ (9) – ISSN: 0013-8029 – mf#6186 – us UMI ProQuest [624]

Engineering journal = Revue de l'ingenierie – Montreal. 1975-1983 (1) 1975-1982 (5) 1975-1982 (9) – ISSN: 0013-8010 – mf#10493 – us UMI ProQuest [620]
Engineering magazine – New York. 1891-1911 (1) – mf#2888 – us UMI ProQuest [620]
Engineering management international – Amsterdam. 1981-1988 (1,5,9) – (Cont by: Journal of engineering and technology management: JET-M) – ISSN: 0167-5419 – mf#42586 – us UMI ProQuest [650]
Engineering news – London. 1961-1968 (1) – ISSN: 0423-1503 – mf#3160 – us UMI ProQuest [620]
Engineering news – London. 1989-1993 (1) – ISSN: 0267-5145 – mf#16026 – us UMI ProQuest [620]
Engineering news-record – New York. 1874-1986 (1) 1965-1986 (5) 1970-1986 (9) – (cont by: enr) – ISSN: 0013-807X – mf#34 – us UMI ProQuest [624]
Engineering opportunities – London. 1975-1976 (1) – ISSN: 0046-2063 – mf#9975 – us UMI ProQuest [620]
Engineering outlook – n8 [1941 may] – 1r – 1 – (continued by: monthly outlook) – mf#1269014 – us WHS [071]
Engineering production – London. 1971-1972 (1) – ISSN: 0013-8053 – mf#5748 – us UMI ProQuest [620]
Engineering sciences and mechanics [aasms43] – 1983 – (= ser Aasms 1968) – 2mf – 9 – $10.00 – 0-87703-215-7 – (suppl to v50, advances) – us Univelt [629]
Engineering structures – Kidlington. 1989+ (1,5,9) – ISSN: 0141-0296 – mf#17227 – us UMI ProQuest [624]
Engineering technician in the news – El Paso. 1972-1973 (1) 1972-1972 (5) (9) – ISSN: 0013-8126 – mf#7743 – us UMI ProQuest [620]
Engineering thermodynamics / Rogers, Gordon Frederick Chrichton – London, England. 1957 – 1r – us UF Libraries [621]
Engineers' Association and Scientists' Association [Marconi] *see* Communique
Engineer's digest – Willow Grove. 1989-1992 (1) – ISSN: 0199-0101 – mf#15345 – us UMI ProQuest [620]
Engineers' digest : english edition – London. 1940-1982 (1) 1971-1982 (5) 1977-1982 (9) – ISSN: 0013-8169 – mf#1251 – us UMI ProQuest [620]
Engineers news / International Union of Operating Engineers – 1952 jan-1959 aug, 1959 sep-1967 may, 1967 jun-1974 jun, 1974 jul-1980 dec, 1981-84 – 5r – 1 – mf#367504 – us WHS [320]
Engineers' outlook / American Federation of Technical Engineers – 1949 dec-1959, 1960-68, 1969-1976 jun/jul – 5r – 1 – (cont: monthly outlook; cont by: outlook (washington, dc: 1976)) – mf#1240278 – us WHS [620]
Engineers' surveying instruments / Baker, Ira Osborn – New York, NY. 1892 – 1r – 1 – us UF Libraries [624]
Enginemen's press / Brotherhood of Locomotive Firemen and Enginemen – 1959 oct 16-1964, 1965 jan 1-1967 dec 22, 1968 jan 5-dec 20 – 3r – 1 – mf#1111392 – us WHS [380]
Enginews / Teledyne Total Motor et al – 1975 sep-1983 3rd qtr – 1r – 1 – mf#1099327 – us WHS [620]
Engl von Wagrain, F F T *see* Sapientia politica symbolica
Engla and seaxna scopas and boceras : anglosaxonum poetae atque scriptores prosaici, quorum partim integra opera, partim loca selecte / ed by Ettmueller, Ludwig – Quedlinburgii, Lipsiae: G Basse, 1850 [mf ed 1993] – (= ser Bibliothek der gesammten deutschen national-literatur von der aeltesten bis auf die neuere zeit sect1/28) – xxiv/304p – 1 – mf#8438 reel 6 – us Wisconsin U Libr [420]
England : The Fortress Of Christianity / Croly, George – London, England. 1837 – 1r – us UF Libraries [240]
England / Gairdner, James – London: SPCK; New York: Pott, Young, [1879?] – (= ser Early Chroniclers Of Europe) – 1mf – 9 – 0-7905-5937-4 – (Incl bibl ref) – mf#1988-1937 – us ATLA [941]
England / Raumer, Friedrich L von – Leipzig 1842 [mf ed Hildesheim 1995-98] – 3v on 12mf – 9 – €120.00 – 3-487-27978-9 – gw Olms [941]
England : a short history / Smith, Goldwin Albert – New York: Scribner, (1971) – xv/549p – 1 – us Wisconsin U Libr [941]
England Air Force Base [LA] *see* Tiger talk
England and america : a comparison of the social and political state of both nations / Wakefield, Edward G – London 1833 [mf ed Hildesheim 1995-98] – 2v on 4mf – 9 – €120.00 – 3-487-26747-0 – gw Olms [300]
England and canada : a summer tour between old and new westminster: with historical notes / Fleming, Sandford – London: S Low, Marston, Searle & Rivington, 1884 – 6mf – 9 – 0-665-90677-3 – (incl ind) – mf#90677 – cn Canadiana [910]
England and christendom / Manning, Henry Edward – London: Longmans, Green, 1867 – 1mf – 9 – 0-7905-5431-3 – mf#1988-1431 – us ATLA [240]
England and her colonies considered in relation to the aborigines : with a proposal for affording them medical relief / Aborigines Protection Society, London – 2nd ed. [London, 1842?] – (= ser 19th c books on british colonization) – 1mf – 9 – mf#1.1.3668 – uk Chadwyck [941]
The england and holland of the pilgrims / Dexter, Henry Martyn & Dextwer, Morton – Boston: Houghton, Mifflin, 1905 – 2mf – 9 – 0-7905-4556-X – (incl bibl ref) – mf#1988-0556 – us ATLA [975]
England and ireland : a counter-proposal / Booth, Charles – London, 1886 – (= ser 19th c ireland) – 1mf – 9 – mf#1.1.411 – uk Chadwyck [941]
England and ireland : a lecture delivered at montreal, december 17th, 1880 / Bray, Alfred James – Montreal: Printed for aut by John Lovell & Son, 1881 – 1mf – 9 – mf#36582 – cn Canadiana [941]
England and ireland / Mill, John Stuart – London, 1868 – (= ser 19th c ireland) – 1mf – 9 – mf#1.1.4232 – uk Chadwyck [941]
England and rome : a history of the relations between the papacy and the english state and church from the norman conquest to the revolution of 1688 / Ingram, T Dunbar – London; New York: Longmans, Green,1892 – 2mf – 9 – us ATLA [240]
England and rome : a history of the relations between the papacy and the english state and church from the norman conquest to the revolution of 1688 / Ingram, Thomas Dunbar – London; New York: Longmans, Green, 1892 – 2mf – 9 – 0-7905-4980-8 – (incl bibl ref) – mf#1988-0980 – us ATLA [941]
England and rome : a study in catholic assent / Egerton, Hakluyt – Leighton Buzzard: Faith Press, 1910 – 1mf – 9 – 0-524-03792-2 – mf#1990-4864 – us ATLA [240]
England and rome : three letters to a pervert / Burgon, John William – [rev enl ed] New York: E P Dutton, 1869 – 1mf – 9 – 0-8370-8487-3 – (incl bibl ref) – mf#1986-2487 – us ATLA [240]
England and russia in central asia : with two maps and appendices / Boulger, Demetrius Charles de Kavanagh – London: W H Allen 1879 [mf ed 1987] – 2v on 1r – 1 – mf#9254 – us Wisconsin U Libr [327]
England and south africa / Clotten, Francis Egon – London, 1891 – (= ser 19th c books on british colonization) – 1mf – 9 – mf#1.1.4706 – uk Chadwyck [337]
England and south africa / Gibbs, Edward J – London, 1889 – (= ser 19th c books on british colonization) – 2mf – 9 – mf#1.1.7474 – uk Chadwyck [327]
England and the english / Bulwer Lytton, Edward G – London 1833 [mf ed Hildesheim 1995-98] – 2v on 6mf – 9 – €120.00 – 3-487-28816-8 – gw Olms [941]
England and the holy see : an essay towards reunion / Jones, Spencer – London, New York: Longmans, Green, 1902 – 2mf – 9 – 0-7905-5287-6 – mf#1988-1287 – us ATLA [240]
England and the union *see* Union
England, canada and the great war by lieutenant-colonel l g desjardins : ex-member of the house of commons, ottawa, and of the legislative assembly, quebec: oct 1 1918 – [Canada: s.n, 1918?] [mf ed 1996] – 1mf – 9 – 0-665-78438-4 – mf#78438 – cn Canadiana [933]
England delineated in two volumes – London 1804 [mf ed Hildesheim 1995-98] – 2v on 4mf – 9 – €120.00 – 3-487-27969-X – gw Olms [914]
England described : being a concise delineation of every county in england and wales; with an account of its most important products... / Aikin, John – London 1818 [mf ed Hildesheim 1995-98] – (= ser Fbc) – 3mf – 9 – €90.00 – 3-487-28811-7 – gw Olms [914]
England. Exchequer Chamber *see* Meeson and welsby's reports
England expects every man to do his duty!!! / Mears, Thomas – Southampton, England. 1805? – 1r – 1 – us UF Libraries [941]
England, germany, and the transvaal, 1895-1902 / Penner, Cornelius D – Chicago, IL. 1937 – 1r – 1 – us UF Libraries [327]
England, her colonies and her enemies : how she may make the former protect her against the latter, and how make them sources of boundless wealth and power – London: J Ridgway, 1840 – 1mf – 9 – mf#21772 – cn Canadiana [941]
England in egypt / Milner, Alfred Milner, 1st viscount – London 1892 – (= ser 19th c british colonization) – 5mf – 9 – mf#1.1.3412 – uk Chadwyck [327]
England. Inns of Court *see*
– Acts of parliament and bench table orders of the inner temple
– Gesta greyorum...
England, John *see* The works of the right reverend john england, first bishop of charleston
England, Joyce *see* A comparison of the tensile strength of the umbilical cord of babies of smoking and non-smoking mothers
England, Kathleen M *see* Analysis of the instructional ecology in tutorial tennis settings
The england of the pacific : or new zealand as an english middle-class emigration-field. a lecture...together with a reprint of letters to the daily news on the english agricultural labourer in new zealand / Clayden, Arthur – London, 1879 – (= ser 19th c books on british colonization) – 1mf – 9 – mf#1.1.4966 – uk Chadwyck [980]
England, palestine, egypt and india : connected by a railway system. popularly explained / MacBean, S – London 1876 – (= ser 19th c british colonization) – 3mf – 9 – mf#1.1.8495 – uk Chadwyck [380]
England, scotland, and ireland : containing a description of the character, manners, customs, dress, diversions, and other peculiarities of the inhabitants of great britain / Pyne, William H – London 1827 [mf ed Hildesheim 1995-98] – 4v on 8mf – 9 – €160.00 – 3-487-28887-7 – gw Olms [914]
England, turkey, and russia / Croly, George – London, England. 1854 – 1r – 1 – us UF Libraries [327]
England two hundred years ago / Gillett, Ezra Hall – Philadelphia: Presbyterian Board of Publication, c1866 – 1mf – 9 – 0-7905-4589-6 – mf#1988-0589 – us ATLA [941]
England und italien / Archenholz, J W von – Leipzig, 1787 – 5v on 19mf – 9 – mf#HT-270 – ne IDC [914]
England und italien : [fuenf theile] / Archenholtz, Johann W von – Carlsruhe 1787 [mf ed Hildesheim 1995-98] – 5v on 11mf – 9 – €110.00 – 3-487-28830-3 – gw Olms [914]
England und italien – Leipzig DE, 1787 [gaps] – 1r – 1 – gw Misc Inst [940]
England und schottland : reisetagebuch / Lewald, Fanny – Braunschweig 1851/52 [mf ed Hildesheim 1995-98] – (= ser Fbc) – 2v on 8mf – 9 – €160.00 – 3-487-27979-7 – gw Olms [914]
England under protector somerset : an essay / Pollard, Albert Frederick – London: K Paul, Trench, Truebner, 1900 – 1mf – 9 – 0-7905-7183-8 – (incl bibl ref) – mf#1988-3183 – us ATLA [941]
England under the old religion : and other essays / Gasquet, Francis Aidan – London: G Bell; New York: Macmillan [distributor], 1912 – 1mf – 9 – 0-7905-5207-8 – mf#1988-1207 – us ATLA [941]
England versus rome : a brief hand-book of the roman catholic controversy for the use of members of the english church / Swete, Henry Barclay – London: Rivingtons, 1868 – 1mf – 9 – 0-7905-6839-X – (incl bibl ref) – mf#1988-2839 – us ATLA [241]
England, wales, irland und schottland : erinnerungen an natur und kunst aus einer reise in den jahren 1802 und 1803 / Goede, Christian A – Dresden 1804/05 [mf ed Hildesheim 1995-98] – 5v on 18mf – 9 – €180.00 – 3-487-28892-3 – gw Olms [914]
England's alternative / Wilton, W – Evesham, England. 1798 – 1r – 1 – us UF Libraries [941]
England's antiphon / MacDonald, George – [S.l.]: Lippincott: Macmillan, [1868?] – (= ser Sunday Library for Household Reading) – 1mf – 9 – 0-7905-7982-0 – mf#1989-1267 – us ATLA [420]
England's danger / Horton, Robert Forman – London: James Clarke, 1899 – 1mf – 9 – 0-8370-8436-9 – mf#1986-2436 – us ATLA [240]
England's duty to india in respect of the education and public employment of the native indians : containing extracts from public documents letters, minutes and speeches of british and anglo-indian statesmen, relating to indian affairs – Calcutta, 1889 – (= ser 19th c books on british colonization) – 1mf – 9 – mf#1.1.4916 – uk Chadwyck [954]
Englands einfluss auf die lehrdichtung Hallers / Wyplel, Ludwig – Wien: Im Selbstverlage des Verfassers, 1888 – 1r – 1 – us Wisconsin U Libr [410]
Englands einfluss auf georg rudolf weckherlin / Boehm, Wilhelm – Goettingen: Dieterich'sche Univ.-Buchdruckerei, 1893 [mf ed 1993] – 80p – 1 – (incl bibl ref) – mf#7780 – us Wisconsin U Libr [410]
"England's greatness" : anniversary sermon delivered to the members of st george's society of ottawa and the sons of england, st andrew's church, ottawa, april 23rd, 1899 / Herridge, William Thomas – Ottawa: s.n, 1899? – 1mf – 9 – mf#05558 – cn Canadiana [242]
England's mission to india : some impressions from a recent visit / Barry, Alfred – London: SPCK, 1895 [mf ed 1995] – (= *see* Yale coll) – 214p – 1 – 0-524-09340-7 – mf#1995-0340 – us ATLA [242]
Engle, George C *see* Parental involvement in the promotion of a culture of learning
Englekirk, John Eugene *see* Literatura norteamericana no brasil
Engler, Bruno *see* Die verwaltung der stadt muenster von den letzten zeiten der fuerstbischoeflichen bis zum ausgang der franzoesischen herrschaft 1802-1813
Englert, Winfried Philipp *see* Christus und buddha in ihrem himmlischen vorleben
Englesea-brook, primitive methodist chapel – (= ser Cheshire monumental inscriptions) – 1mf – 9 – £2.50 – mf#81 – uk CheshireFHS [929]
Englewood messenger *see* Arapahoe county miscellaneous newspapers
Englewood news *see* Arapahoe county miscellaneous newspapers
Englewood times – Chicago, IL. 1905-1928 (1) – mf#62552 – us UMI ProQuest [071]
Englich, Ulrich *see* Roentgenographische und spektroskopische untersuchungen an alkali-mangan(3)-verbindungen vom kryolith-typ
Das englische christenvolk und wir / Wurster, Paul – Tuebingen: Kloeres 1915 [mf ed 1987] – (= ser Durch kampf zum frieden 4) – 1r – 1 – mf#6840 – us Wisconsin U Libr [933]
Englische correspondenz – London, UK. aug 1850-nov 1851; 1852-aug 1856; feb 1860-1861 – 1 – uk British Libr Newspaper [072]
Englische dichtung aus goethes zeitalter im licht deutscher kunstlehre / Edwards, Oliver – Bonn a. Rh: L Roehrscheid, 1930 – 1r – 1 – (incl bibl ref) – us Wisconsin U Libr [410]
Englische fluechtlinge in zuerich... / Vetter, T – Zuerich, Orell Fuessli, 1893 – (= ser Neujahrsblatt) – 1mf – 9 – mf#PBU-440 – ne IDC [240]
Die englische fluechtlings-gemeinde in frankfurt am main 1554-1559 / Jung, Rudolf – Frankfurt a M: J Baer, 1910 – (= ser Frankfurter historische Forschungen) – 1mf – 9 – 0-524-04494-5 – mf#1990-1256 – us ATLA [943]
[Englische motetten und madrigale, 1520-1598] : aus der sammlung wagener, marburg. abschrift von carl dreher, karlsruhe – [18–] [mf ed 19–] – 1r – 1 – mf#pres. film 31 – us Sibley [780]
Englische rundschau – Koeln DE, 1959-1960 [gaps] – 1 – gw Misc Inst [074]
Die englische-franzosische friedensverhandlung : december 1799-januar 1800 / Bowman, Hervey Meyer – Leipzig?: O Schmidt, 1899 – 1mf – 9 – mf#24347 – cn Canadiana [940]
English – Oxford. 1936+ (1) 1972+ (5) 1973+ (9) – ISSN: 0013-8215 – mf#1257 – us UMI ProQuest [420]
The english / Maclear, George Frederick – London: Society for Promoting Christian Knowledge; New York: E & JB Young, [1878?] – (= ser Conversion of the West) – 1mf – 9 – 0-7905-4771-6 – (incl bibl ref) – mf#1988-0771 – us ATLA [240]
English Aboricultural Society *see* Transactions of the english aboricultural society
English altars from illuminated manuscripts / ed by Hope, William Henry St John – London 1899 – (= ser 19th c art & architecture) – 1mf – 9 – mf#4.2.1681 – uk Chadwyck [720]
English america, vol 1 : or, pictures of canadian places and people / Day, Samuel Phillips – London: T C Newby, 1864 [mf ed 1983] – 4mf – 9 – 0-665-44231-9 – mf#44231 – cn Canadiana [370]
English america, vol 2 : or, pictures of canadian places and people / Day, Samuel Phillips – London: T C Newby, 1864 [mf ed 1983] – 4mf – 9 – 0-665-44232-7 – mf#44232 – cn Canadiana [917]
English america, vols 1-2 : or, pictures of canadian places and people / Day, Samuel Phillips – London: T C Newby. 2v. 1864 – 1mf – 9 – 0-665-44230-0 – mf#44230 – cn Canadiana [971]
The english and american register – Berlin DE, 1887 5 nov-1889 28 dec – 1r – 1 – gw Misc Inst [074]
An english and burman vocabulary : preceded by a concise grammar, in which the burman definitions and words are accompanied with a pronunciation in the english character / Hough, George Henry – Serampore, 1825 – (= ser 19th c books on linguistics) – 5mf – 9 – mf#2.1.9 – uk Chadwyck [040]
English and burman vocabulary : preceded by a concise grammar, in which the burman definitions and words are accompanied with a pronunciation in the english character; designed to extend the colloquial use of the burman language / Hough, George H – Serampore: Mission press 1814 – (= ser Whsb) – 5mf – 9 – €60.00 – mf#Hu 286 – gw Fischer [480]

English and french manual of conservation = Manuel de conversation anglaise et francaise / Wright, Alexander – Montreal: Railway and Commercial Print Co, 1895 – 2mf – 9 – (text in english and french in parallel columns) – mf#28666 – cn Canadiana [410]

An english and japanese and japanese and english vocabulary / Medhurst, Walter Henry – Batavia, 1830 – (= ser 19th c books on linguistics) – 4mf – 9 – mf#2.1.36 – uk Chadwyck [040]

English and japanese and japanese and english vocabulary / Medhurst, Walter Henry – Batavia: Lithography 1830 – (= ser Whsb) – 4mf – 9 – €50.00 – mf#Hu 331 – gw Fischer [410]

English and latin : a manual of prose composition / Ogle, Marbury Bladen – New York, London: The Century Co, c1926 – (= ser The century college latin series) – 1r – 1 – mf#1263 – us Wisconsin U Libr [410]

English and scottish psalm and hymn tunes, 1543-1677 / Frost, Maurice – 1 – $25.90 – us Southern Baptist [780]

English and scottish silver spoons / How, G E P – London. v1-3. 1952-1957 – €7.00 – ne Slangenburg [730]

English and tamil first book – Madras: Church of Scotland Missionary Press, 1857 – 1 – us CRL [242]

English and telugu dictionary / Brown, Charles Philip – Madras: SPCK, 1895 – 1r – 1 – 0-8370-1479-4 – mf#1984-B020 – us ATLA [040]

English and tongan vocabulary : also a tongan and english vocabulary, with a list of idiomatic phrases; and tongan grammar / Baker, Shirley W – 1897 – 1r – 1 – mf#pmb doc470 – at Pacific Mss [040]

English and victorian pictures, drawings and watercolours – (= ser Christie's pictorial sales review) – 117mf – 9 – $950.00 – 0-907006-97-3 – (13,000 images) – uk Mindata [740]

English apologetic theology / Macran, Frederick Walter – London: Hodder and Stoughton, 1905 – (= ser Donnellan Lectures) – 1mf – 9 – 0-7905-7966-9 – mf#1989-1191 – us ATLA [240]

English applications to pay deposit 1864 *see* Northern territory land applications – various

English art : subject collections – (= ser Art exhibition catalogues on microfiche) – 504 catalogues on 613mf – 9 – £3220.00 – (individual titles not listed separately) – uk Chadwyck [700]

English art in 1884 / Blackburn, Henry – New York, 1885 – (= ser 19th c art & architecture) – 6mf – 9 – mf#4.2.1042 – uk Chadwyck [700]

The english augsburg confession of 1536 / ed by Jacobs, Henry Eyster – Philadelphia: Lutheran Publ Society: Publ for the Joint Committee, 1888 – 1mf – 9 – 0-8370-8623-X – mf#1986-2623 – us ATLA [240]

The english baby in india and how to rear it / Kingscote, Adeline Georgina Isabella – London, 1893 – (= ser 19th c british colonization) – 3mf – 9 – mf#1.1.2471 – uk Chadwyck [618]

English baptist reformation : from 1609 to 1641 a d / Lofton, George Augustus – Louisville, KY: Chas T Dearing, 1899 – 1mf – 9 – 0-524-03723-X – mf#1990-4828 – us ATLA [242]

English baptist reformation / Lofton, George A – 1899 – 284p – 1 – us Southern Baptist [242]

The english baptists : who they are, and what they have done / ed by Clifford, John – London: E Marlborough, 1881 – 1mf – 9 – 0-7905-5866-1 – mf#1988-1866 – us ATLA [242]

English benedictine calendars after a d 1100, vol 1-2 (hbs77,81) / Wormald, F – 1939, 1946 – (= ser Henry bradshaw society (hbs)) – 2v – 8 – €12.00 – (v1 3mf. v2 2mf) – ne Slangenburg [241]

English bible : extracts from the important english versions of the bible from wiclif's to the king james version – [Boston: Directors of the Old South Work, 1896?] – 1mf – 9 – 0-524-04127-X – mf#1990-1197 – us ATLA [220]

The english bible : an historical survey, from the dawn of english history to the present day / Payne, Julius D – London: Wells, Gardner, Darton, 1911 – 1mf – 9 – 0-8370-9975-7 – (incl ind) – mf#1986-3975 – us ATLA [220]

The english bible : a sketch of its history / Milligan, George – New York: A D F Randolph, 1895 – 1mf – 9 – 0-8370-4435-9 – mf#1985-2435 – us ATLA [220]

The english bible and our duty with regard to it : a plea for revision / Abbott, Thomas Kingsmill – 2nd ed. Dublin: Hodges, Foster, 1871 – 1mf – 9 – 0-8370-2030-1 – mf#1985-0030 – us ATLA [220]

English bible versions : a tercentenary memorial of the king james version, from the new york bible and common prayer book society established a d 1809 / Barker, Henry – New York: ltd ed iss for the Society by Edwin S Gorham, 1911 – 1mf – 9 – 0-7905-0248-8 – (incl ind) – mf#1987-0248 – us ATLA [220]

The english black monks of st benedict : a sketch of their history from the coming of st augustine to the present day / Taunton, Ethelred Luke – London: John C Nimmo; New York: Longmans, Green 1897 [mf ed 1991] – 2v on 2mf – 9 – 0-524-00791-8 – mf#1990-0223 – us ATLA [241]

The english booktrade, 1660-1853 : 156 titles relating to the early history of english publishing, bookselling and the struggle for copyright and the freedom of the press / ed by Park, Stephen – New York: Garland Publ Inc, 1985? – 1mf – 9 – $1.50 – mf#LLMC 91-078 – us LLMC [343]

The english bread-book for domestic use : adapted to families of every grade / Acton, Eliza – London: Longman, Brown, Green, Longmans & Roberts, 1857 (mf ed 1982) – 1r – 1 – mf#*ZU-221 – us NY Public [640]

English calendars before a d 1100 (hbs72) / Wormald, F – 1934 – (= ser Henry bradshaw society (hbs)) – 5mf – 8 – €12.00 – ne Slangenburg [390]

English caricaturists and graphic humourists of the nineteenth century / Everitt, Graham – London 1886 – (= ser 19th c art & architecture) – 7mf – 9 – mf#4.2.1190 – uk Chadwyck [740]

An english carmelite : the life of catharine burton, mother mary xaveria of the angels, of the english teresian convent at antwerp / Burton, Catharine – London: Burns & Oates, 1876 [mf ed 1986] – (= ser Quarterly series 18) – 1mf – 9 – 0-8370-6889-4 – mf#1986-0889 – us ATLA [241]

English cartoons and satirical prints, 1320-1832, in the british museum : the world's greatest collection of english satirical prints – [mf ed Chadwyck-Healey] – 32r – 1 – (coll of some 17,400 prints, the earliest dated 1320, the majority concentrated in the 2nd half of the 18th c & the 1st decades of the 19th. with: catalogue of political and personal satires ed by f g stephens and m d george [11r]) – uk Chadwyck [740]

The english cathedral of the nineteenth century / Beresford-Hope, Alexander James Beresford – London: John Murray, 1861 – (= ser 19th c art & architecture) – 4mf – 9 – mf#4.1.206 – uk Chadwyck [720]

English cathedrals illustrated / Bond, Francis – London 1899 – (= ser 19th c art & architecture) – 4mf – 9 – mf#4.2.1112 – uk Chadwyck [720]

The english catholic nonjurors of 1715 : being a summary of the register of their estates, with genealogical and other notes, and an appendix of unpublished documents in the public record office / ed by Estcourt, Edgar Edmund & Payne, John Orlebar – London: T Baker, 1900 – 1mf – 9 – 0-524-03338-2 – mf#1990-0919 – us ATLA [941]

English church / Manning, Henry Edward – London, England. 1835 – 1r – 1 – us UF Libraries [240]

The english church : from the accession of charles 1 to the death of anne (1625-1714) / Hutton, William Holden – London, New York: Macmillan, 1903 [mf ed 1990] – (= ser A history of the english church 6) – 1mf – 9 – 0-7905-4822-4 – (incl bibl ref) – mf#1988-0822 – us ATLA [240]

The english church : from the norman conquest to the accession of edward 1 (1066-1272) / Stephens, William Richard Wood – London, New York: Macmillan, 1909 [mf ed 1990] – (= ser A history of the english church 2) – 1mf – 9 – 0-7905-5971-4 – mf#1988-1971 – us ATLA [240]

The english church and its bishops, 1700-1800 / Abbey, Charles John – London: Longmans, Green, 1887 – 2mf – 9 – 0-7905-4420-2 – (incl bibl ref) – mf#1988-0420 – us ATLA [240]

The english church and the ministry of the reformed churches / Denny, Edward – London: SPCK 1900 [mf ed 1993] – (= ser Church historical society (series) 57) – 1mf – 9 – 0-524-05539-4 – mf#1990-5143 – us ATLA [242]

The english church and the reformation / Carter, Charles Sydney – London; New York: Longmans, Green, 1915 – 1mf – 9 – 0-524-00984-8 – (incl bibl ref) – mf#1990-0261 – us ATLA [242]

The english church from its foundation to the norman conquest (597-1066) / Hunt, William – London, New York: Macmillan, 1899 [mf ed 1990] – (= ser A history of the english church 1) – 2mf – 9 – 0-7905-6182-4 – (incl bibl ref) – mf#1988-2182 – us ATLA [240]

The english church from the accession of george 1 : to the end of the 18th century (1714-1800) / Overton, John Henry & relton, Frederic – London, New York: Macmillan, 1906 [mf ed 1990] – (= ser A history of the english church 7) – 1mf – 9 – 0-7905-7067-X – (incl bibl ref) – mf#1988-3067 – us ATLA [242]

English church furniture / Cox, John Charles & Harvey, Alfred – New York: E P Dutton, 1907 – (= ser The Antiquary's Books) – 2mf – 9 – 0-7905-5026-1 – mf#1988-1026 – us ATLA [740]

English church history from the death of archbishop parker to the death of king charles 1 : four lectures / Plummer, Alfred – Edinburgh: T & T Clark 1904 [mf ed 1990] – 1mf – 9 – 0-7905-5732-0 – (incl bibl ref) – mf#1988-1732 – us ATLA [240]

English church history from the death of charles 1 to the death of william 3 : four lectures / Plummer, Alfred – Edinburgh: T & T Clark, 1907 – 1mf – 9 – 0-7905-5791-6 – (incl bibl ref) – mf#1988-1791 – us ATLA [242]

English church history from the death of king henry 7 to the death of archbishop parker : four lectures / Plummer, Alfred – Edinburgh: T & T Clark 1905 [mf ed 1990] – 1mf – 9 – 0-7905-5554-9 – (incl bibl ref) – mf#1988-1554 – us ATLA [242]

The english church in other lands : or, the spiritual expansion of england / Tucker, Henry William – London, New York: Longmans, Green, 1891 – (= ser Epochs of Church History) – 1mf – 9 – 0-524-00794-2 – mf#1990-0226 – us ATLA [240]

The english church in the 14th and 15th centuries / Capes, William Wolfe – London, New York: Macmillan, 1900 [mf ed 1989] – (= ser A history of the english church 3) – 1mf – 9 – 0-7905-4195-5 – (incl bibl ref) – mf#1988-0195 – us ATLA [240]

The english church in the 16th century : from the accession of henry 8 to the death of mary / Gairdner, James – London: Macmillan, 1902 [mf ed 1992] – (= ser Anglican/episcopal coll) – 2mf – 9 – 0-524-02635-1 – (incl bibl ref) – mf#1990-4390 – us ATLA [242]

The english church in the 17th century / Carter, Charles Sydney – London, New York: Longmans, Green 1909 [mf ed 1990] – (= ser Anglican church handbooks) – 1mf – 9 – 0-7905-5382-1 – (incl bibl ref) – mf#1988-1382 – us ATLA [242]

The english church in the 18th century / Carter, Charles Sydney – London, New York: Longmans, Green 1910 [mf ed 1989] – (= ser Anglican church handbooks) – 1mf – 9 – 0-7905-4197-1 – (incl bibl ref) – mf#1988-0197 – us ATLA [242]

The english church in the 19th century / Warre Cornish, Francis – London: Macmillan, 1910 [mf ed 1990] – (= ser A history of the english church 8) – 2v on 2mf – 9 – 0-7905-7034-3 – (incl bibl ref) – mf#1988-3034 – us ATLA [242]

The english church in the eighteenth century / Abbey, Charles J – London: Longmans, Green, 1878 – 3mf – 9 – 0-7905-4360-5 – (incl bibl ref) – mf#1988-0360 – us ATLA [240]

The english church in the middles ages / Hunt, William – London: Longmans, Green, 1888 [mf ed 1990] – (= ser Epochs of church history) – 1mf – 9 – 0-7905-4760-0 – mf#1988-0760 – us ATLA [240]

The english church in the nineteenth century / Stock, Eugene – London, New York: Longmans, Green, 1910 – 1mf – 9 – 0-7905-5974-9 – mf#1988-1974 – us ATLA [240]

The english church in the nineteenth century (1800-1833) / Overton, John Henry – London, New York: Longmans, Green, 1894 – 1mf – 9 – 0-7905-5781-9 – (incl bibl ref) – mf#1988-1781 – us ATLA [240]

The english church in the reigns of elizabeth and james 1 (1558-1625) / Frere, Walter Howard – London, New York: Macmillan, c1904 [mf ed 1990] – (= ser A history of the english church 5) – 1mf – 9 – 0-7905-4641-8 – mf#1988-0641 – us ATLA [240]

English church life from the restoration to the tractarian movement / Legg, Wickham – London, New York: Longmans, Green, 1914 – 2mf – 9 – 0-7905-5104-7 – (incl bibl ref) – mf#1988-1104 – us ATLA [240]

The english church mission in corea : its faith and practice – London: A R Mowbray; Milwaukee: Young Churchman, 1917 [mf ed 1995] – (= ser Yale coll) – 80p/[8]pl (ill) – 1 – 0-524-09505-1 – (pref by the right rev bishop corfe) – mf#1995-0505 – us ATLA [240]

English church ways : Described To Russian Friends In Four Lectures / Frere, Walter Howard – Milwaukee: Young Churchman, 1914 – 1mf – 9 – 0-7905-5599-9 – mf#1988-1599 – us ATLA [240]

English circular – London, UK. 1841-44 [irregular] – 1/2 r – 1 – uk British Libr Newspaper [072]

English clandestine satire, 1660-1704 : popular culture, entertainment and information in the early modern period – 24r – 1 – £2250.00 – (with d/g) – uk Matthew [870]

The english classification tests / Orville, Glenn – Nebraska [c1930] (mf ed 1994) – 2mf – 9 – €31.00 – 3-8267-3089-5 – mf#DHS-AR 3089 – gw Frankfurter [370]

English clerical biographical dictionary *see* Crockford's clerical directory

English composition and rhetoric / Bain, Alexander – London, England. v1-2. 1893 – 1r – 1 – us UF Libraries [420]

English constitution : and other political essays / Bagehot, Walter – New York, NY. 1884 – 1r – 1 – us UF Libraries [323]

English contemporary art / Sizeranne, Robert de la – Westminster 1898 – (= ser 19th c art & architecture) – 4mf – 9 – mf#4.2.314 – uk Chadwyck [700]

English costume of the eighteenth century / Laver, James – London, England. 1931 – 1r – 1 – us UF Libraries [390]

English country dances : arranged for children's performance / ed by Kidson, Frank – London: J Curwen & Sons Ltd, c1914 – 1 – mf#*ZBD-*MGS – Located: NYPL – us Misc Inst [790]

The english cricketers' trip to canada and the united states / Lillywhite, Frederick, 1829-1866 – London: F Lillywhite, 1860 [mf ed 1983] – 2mf – 9 – 0-665-38217-0 – mf#38217 – cn Canadiana [790]

English daily bulletin – Aden: Aden News Agency, jul 20-30, aug 1-2, dec 22, 1987; mar 9-14, 16-26, apr 3-10 1988 – 1r – 1 – us CRL [079]

English dance and song – London. 1975+(1,5,9) – ISSN: 0013-8231 – mf#10536 – us UMI ProQuest [780]

English dioceses : a history of their limits from the earliest times to the present day / Hill, Geoffry – London: E Stock, 1900 – 1mf – 9 – 0-7905-4815-1 – mf#1988-0815 – us ATLA [240]

English earthenware / Church, Arthur Herbert – [London] 1884 – (= ser 19th c art & architecture) – 3mf – 9 – mf#4.2.1445 – uk Chadwyck [730]

English economic and political pamphlets – 1700-1870 – 1 – $960.00 – mf#0194 – us Brook [410]

English, Edmund Francis *see* Views of lansdown tower, bath...from drawings by willes maddox

English education – Urbana. 1969+ (1) 1969+ (5) 1969+ (9) – ISSN: 0007-8204 – mf#7204 – us UMI ProQuest [420]

English electric journal – Stafford. 1920-1968 (1) – mf#3311 – us UMI ProQuest [621]

English episcopal palaces : province of canterbury / Morewood, Caroline C et al; ed by Rait, Robert Sangster – London: Constable, 1910 – 1mf – 9 – 0-7905-5678-2 – mf#1988-1678 – us ATLA [720]

English episcopal palaces : Province of york / Niemeyer, N et al; ed by Rait, Robert Sangster – London: Constable, 1911 – 1mf – 9 – 0-7905-5679-0 – mf#1988-1679 – us ATLA [720]

English evangelical lutheran synod of the northwest : minutes – 1891-1962 [complete] – 3r – 1 – mf#ATLA S0053 – us ATLA [242]

The english factories in india : new series / Fawcett, Charles – Oxford: Clarendon Press, 1936-1955 – (= ser Samp: indian books) – us CRL [380]

English Folk Dance and Song Society *see* The dancing english

English for specific purposes – New York. 1980+ (1,5,9) – ISSN: 0889-4906 – mf#49413 – us UMI ProQuest [420]

English furniture, decoration, woodwork and allied arts / Strange, Thomas Arthur – London [1900] – (= ser 19th c art & architecture) – 4mf – 9 – mf#4.2.1513 – uk Chadwyck [740]

The english general baptists of the seventeenth century / Taylor, Adam – London: printed...by T Bore, 1818 [mf ed 1994] – (= ser The history of the english general baptists 1) – 6mf – 9 – 0-524-08817-9 – mf#1993-3309 – us ATLA [242]

English gentleman *see* The argus

English, George B *see* A narrative of the expedition to dongola and sennaar

English gift books and literary annuals, 1823-1857 – [mf ed Chadwyck-Healey] – 23 titles on 697mf – 9 – (with ind of contributors and " ind to the annuals 1820-1850" by andrew boyle 1967. titles of coll also listed individually) – uk Chadwyck [800]

English grammar in american high schools since 1900 / Gruen, Ferdinand Bernard – Washington DC, 1934 (mf ed 1994) – 4mf – 9 – €45.00 – 3-8267-3088-7 – mf#DHS-AR 3088 – gw Frankfurter [373]

English graphic satire : and its relation to different styles of painting, sculpture / Buss, Robert William – [London] 1874 – (= ser 19th c art & architecture) – 4mf – 9 – mf#4.2.928 – uk Chadwyck [740]

The english guide to the paris exhibition, 1878 / Great Britain. Royal Commission for the Paris Exhibition, 1878 – 6th ed. London [1878] – (= ser 19th c art & architecture) – 2mf – 9 – mf#4.2.900 – uk Chadwyck [700]

English historical pamphlets – 1st ser. c1819-1854 – 1 – us Brook [941]

English historical review – Harlow. 1886+ (1) 1971+ (5) 1976+ (9) – ISSN: 0013-8266 – mf#955 – us UMI ProQuest [941]

The english hymn : its development and use in worship / Benson, Louis F – New York: Hodder & Stoughton: G H Doran, c1915 – (= ser Stone Lectures) – 2mf – 9 – 0-7905-4430-X – (incl bibl ref) – mf#1988-0430 – us ATLA [780]

The english hymnal : with tunes – London: Oxford University Press, 1909 – 11mf – 9 – 0-524-06087-8 – mf#1991-2400 – us ATLA [780]

English hymns : their authors and history / Duffield, Samuel Willoughby – 3rd rev corr ed. New York: Funk & Wagnalls, 1888 – 2mf – 9 – 0-7905-5386-4 – mf#1988-1386 – us ATLA [780]

English hymns and their tunes in the 16th and 17th centuries / Parks, Edna D – 1957 – 1 – us Southern Baptist [242]

English hymn-tunes of the eighteenth century / Hasel, Haven Binford – U of Rochester 1946 [mf ed 19–] – 4mf – 9 – mf#fiche 206 – us Sibley [780]

English illuminated manuscripts / Thompson, Edward Maunde – London 1895 – (= ser 19th c art & architecture) – 2mf – 9 – mf#4.2.768 – uk Chadwyck [740]

English illustrated magazine – New York. 1883-1913 (1) – mf#2885 – us UMI ProQuest [700]

English illustration : 'the sixties': 1855-1870 / White, Gleeson – Westminster 1897 – (= ser 19th c art & architecture) – 6mf – 9 – mf#4.1.244 – uk Chadwyck [740]

The english in america : from the first english discoveries to the present day / Brownell, Henry Howard – Hartford, CT: Hurlbut, Kellogg, 1861 [mf ed 1983] – 7mf – 9 – 0-665-43306-9 – mf#43306 – cn Canadiana [970]

The english in america : the puritan colonies / Doyle, John Andrew – London: Longmans, Green, 1887 – 3mf – 9 – 0-524-02818-4 – (incl bibl ref) – mf#1990-4439 – us ATLA [975]

English in education – Sheffield. 1976+ (1,5,9) – ISSN: 0425-0494 – mf#11024 – us UMI ProQuest [420]

The english in india : a problem of politics / Marriott, John Arthur Ransome – London, New York: Oxford University Press, 1932 – (= ser Samp: indian books) – us CRL [954]

The english in ireland in the eighteenth century / Froude, James Anthony – New York: Scribner, Armstrong, 1873-1874 – 4mf – 9 – 0-7905-5655-3 – mf#1988-1655 – us ATLA [941]

English in new zealand – Aukland. 1978-1980 (1,5,9) – mf#11118 – us UMI ProQuest [420]

English in the west indies / Froude, James Anthony – New York, NY. 1888 – 1r – 1 – us UF Libraries [420]

The english in the west indies : or, the bow of ulysses / Froude, James Anthony – New York: Scribner, 1888 – 1mf – 9 – 0-7905-5394-5 – mf#1988-1394 – us ATLA [918]

English in western india : being the early history of the factory at surat, of bombay, and the subordinate factories on the western coast, from the earliest period until the commencement of the eighteenth century... / Anderson, Philip, d. 1857 – Bombay: Smith, Taylor, 1854 – (= ser Yale coll) – xii/191p – 1 – 0-524-10072-1 – mf#1995-1072 – us ATLA [954]

English intercourse with siam in the seventeenth century / Anderson, J – London, 1890 – 6mf – 9 – mf#SE-20161 – ne IDC [915]

English interference with irish industries / MacNeill, John Gordon Swift – [London], 1886 – (= ser 19th c ireland) – 2mf – 9 – mf#1.1.8373 – uk Chadwyck [338]

English journal – Urbana. 1912+ (1) 1967+ (5) 1970+ (9) – ISSN: 0013-8274 – mf#481 – us UMI ProQuest [420]

English labourer's chronicle – Leamington. England. -w. 1877-94. (6 reels) – 1 – uk British Libr Newspaper [072]

English language teaching – Oxford. 1946-1973 (1) 1970-1973 (5) 1971-1972 (9) – (Cont by: English language teaching journal: ELT) – ISSN: 0013-8290 – mf#1389 – us UMI ProQuest [420]

English language teaching journal (elt) – Oxford. 1973-1981 (1) 1973-1981 (5) 1976-1981 (9) – (cont by: elt journal. cont: english language teaching) – ISSN: 0013-8290 – mf#1389,01 – us UMI ProQuest [420]

English law and irish tenure / Gibbs, Frederick Waymouth – London, 1870 – (= ser 19th c ireland) – 2mf – 9 – mf#1.1.1864 – uk Chadwyck [340]

English leader – London, UK. 1866-67. -irr. 1 1/2 reels – 1 – uk British Libr Newspaper [072]

English life and manners in the later middle ages / Abram, Annie – London: G Routledge; New York: EP Dutton, 1913 [mf ed 1989] – 1mf – 9 – 0-7905-4303-6 – (incl bibl ref) – mf#1988-0303 – us ATLA [931]

English life of jesus, pt 2 : Comprising An Analysis Of The Career Of John The Baptist, And Of The Beginning Of The Public Ministry Of Jesus / Scott, Thomas – Ramsgate: T Scott, [c1866] – 1mf – 9 – 0-8370-5193-2 – mf#1985-3193 – us ATLA [220]

English literature in account with religion, 1800-1900 / Chapman, Edward Mortimer – Boston: Houghton Mifflin, 1910 – 2mf – 9 – 0-7905-7809-3 – mf#1989-1034 – us ATLA [420]

English literature in transition (1880-1920) – Greensboro. 1957+ (1) 1976+ (5) 1976+ (9) – ISSN: 0013-8339 – mf#10626 – us UMI ProQuest [420]

English lucian : or weekly discoveries of the witty intrigues, comical passages and remarkable transactions in town and country – London. 1698-1698 (1) – mf#5317 – us UMI ProQuest [870]

English lutheranism in the northwest / Trabert, George Henry – Philadelphia: General Council Publication House, 1914 – 1mf – 9 – 0-524-01098-6 – mf#1990-4063 – us ATLA [242]

English medieval embroidery / Hartshorne, Charles Henry – London 1848 – (= ser 19th c art & architecture) – 3mf – 9 – mf#4.2.576 – uk Chadwyck [740]

English men of science : their nature and nurture / Galton, Francis – London, 1874 – (= ser 19th c evolution & creation) – 3mf – 9 – mf#1.1.9544 – uk Chadwyck [500]

English mezzotint portraits / Burlington Fine Arts Club. London – 1902 – 9 – uk Chadwyck [760]

English minstrelsie / Baring-Gould, Sabine – Edinburgh, 1895 – 8v – 1 – us L of C Photodup [780]

English miracle plays, moralities, and interludes : specimens of the pre-elizabethan drama / ed by Pollard, Alfred William – 4th rev ed. Oxford: Clarendon Press, 1904 – ix/250p – 1 – us Wisconsin U Libr [820]

English monastic life / Gasquet, Francis Aidan, Cardinal – 2nd rev ed. London: Methuen, 1904 [mf ed 1990] – (= ser The antiquary's books) – 1mf – 9 – 0-7905-5208-6 – (1st ed 1904. incl bibl ref) – mf#1988-1208 – us ATLA [931]

English musical gazette – London. 1819. Drexel 279 no. 1-3. Jan.-Mar. 1819 – 1 – us NY Public [780]

English nonconformity / Tayler, John James – London, England. 1859 – 1r – 1 – us UF Libraries [240]

English nonconformity / Vaughan, Robert – London: Jackson, Walford and Hodder, 1862 – 2mf – 9 – 0-7905-6961-2 – mf#1988-2961 – us ATLA [240]

English opera : or the vocal musick in psyche... to which has been adjoyned the instrumental musick in the tempest / Locke, Matthew – London: printed by T Ratcliff & N Thompson for aut 1675 [mf ed 19–] – 2mf – 9 – (words by thomas shadwell) – mf#fiche 1099 – us Sibley [780]

English orders for consecrating churches in the 17th century (hbs41) / Wickham Legg, J – 1911 – (= ser Henry bradshaw society (hbs)) – 7mf – 8 – €15.00 – ne Slangenburg [240]

The english ordinal : its history, validity and catholicity / Walcott, Mackenzie Edward Charles – London: F & J Rivington 1851 [mf ed 1992] – 1mf – 9 – 0-524-03202-5 – (incl bibl ref) – mf#1990-4651 – us ATLA [242]

English painters of the present day / Atkinson, John Beavington – London 1871 – (= ser 19th c art & architecture) – 2mf – 9 – mf#4.2.103 – uk Chadwyck [750]

The english parish church : an account of the chief building types & of their materials during nine centuries / Cox, John Charles – London: BT Batsford; New York: Scribner, [1914?] – 1mf – 9 – 0-7905-7214-1 – mf#1988-3214 – us ATLA [720]

English pen artists of to-day / Harper, Charles George – London 1892 – (= ser 19th c art & architecture) – 5mf – 9 – mf#4.2.151 – uk Chadwyck [740]

English philosophers and schools of philosophy / Seth, James – London: JM Dent; New York: EP Dutton, 1912 – (= ser The Channels of English Literature) – 1mf – 9 – 0-7905-7370-9 – (Incl bibl ref) – mf#1989-0595 – us ATLA [100]

The english physitians guide or a holyguide; leading the way to know all things, past, present & to come / Heydon, John – London, 1662 – 1 – us Wisconsin U Libr [150]

English place-name society – v1-19. 1924-1943 – 153mf – 8 – mf#H-770c – ne IDC [420]

English poetry, 1750-1855 – 1pt – 1 – (pt 1: recollections, conversations & commonplace books of the reverend john mitford (1781-1859) fr british library, london [24r] £950; with d/g) – uk Matthew [420]

The english poets, lessing, rousseau : essays / Lowell, James Russell – London: W Scott; Toronto: W J Gage, 1888 – (= ser The camelot series) – 4mf – 9 – (with "an apology for a preface") – mf#29481 – cn Canadiana [410]

English political institutions / Marriott, John Arthur Ransome – Oxford, England. 1925 – 1r – 1 – us UF Libraries [941]

English poor laws, 1639-1890 : a unique collection of pamphlets and books on great britain's poor laws – [mf ed UMI] – 655mf – 9 – (with p/g) – us UMI ProQuest [344]

The english pre-raphaelite painters : their associates and successors / Bate, Percy H – London 1899 – (= ser 19th c art & architecture) – 4mf – 9 – mf#4.2.940 – uk Chadwyck [750]

The english presbyterian messenger – London: Hamilton, Adams & Co [mthly] [mf ed 2006] – v1 n19-33 (oct 1846-dec 1847); ns: v1-15 n188 (jan 1848-aug 1863) [gaps] on 2r – 1 – (began with v1 in may 1845; cont by: messenger & missionary record of the presbyterian church in england; imprint varies; several iss missing; some pgs damaged) – mf#2006c-s009 – us ATLA [242]

English, present and past / Aiken, Janet Rankin – New York, NY. 1930 – 1r – 1 – us UF Libraries [420]

English psalter *see* Selected illuminations from manuscripts in the fitzwilliam museum, cambridge

English psalters – 14th c – (= ser Holkham library manuscript books 23,24) – 1r – 1 – mf#96583 – uk Microform Academic [090]

English psychology / Ribot, Theodule Armand – Trans. from the French. New York: D. Appleton and Co., 1891. viii,328p – 1 – us Wisconsin U Libr [150]

English puritanism and its leaders : cromwell, milton, baxter, bunyan / Tulloch, John – Edinburgh: W Blackwood, 1861 – 2mf – 9 – 0-7905-9648-2 – (incl bibl ref) – mf#1989-1373 – us ATLA [941]

The english puritans / Brown, John – Cambridge: University Press; New York: Putnam 1910 [mf ed 1989] – (= ser The cambridge manuals of science and literature) – 1mf – 9 – 0-7905-4157-2 – (incl bibl ref) – mf#1988-0157 – us ATLA [243]

English puritarisme : containening the maine opinions of the rigidest sort of those that are called puritanes in the realme of England / Bradshaw, W – n.p, 1605 – 1mf – 9 – mf#PW-37 – ne IDC [243]

The english reader : or, pieces in prose and verse: selected from the best writers: designed to assist young persons to read with propriety and effect... – Montreal: C Bryson, 1841 [mf ed 1984] – 3mf – 9 – 0-665-39851-4 – mf#39851 – cn Canadiana [420]

English record – Albany. 1950+ (1) 1971+ (5) 1977+ (9) – ISSN: 0013-8363 – mf#1925 – us UMI ProQuest [420]

The english reformation : how it came about and why we should uphold it / Geikie, John Cunningham – [American ed.] New York: D. Appleton, 1879. Beltsville, Md: NCR Corp, 1978 (6mf); Evanston: American Theol Lib Assoc, 1984 (6mf) – 9 – 0-8370-1055-1 – (incl bibl ref and index) – mf#1984-4418 – us ATLA [242]

The english reformation / Hutton, William Holden – London: SPCK 1899 [mf ed 1993] – (= ser Church historical society (series) 53) – 1mf – 9 – 0-524-05505-X – mf#1990-1500 – us ATLA [242]

The english reformation and its consequences : four lectures / Collins, William Edward – 2nd rev ed. London: SPCK 1908 [mf ed 1993] – (= ser Church historical society (series)) – 1mf – 9 – 0-524-06350-8 – (with notes & app) – mf#1990-1533 – us ATLA [242]

The english reformation and puritanism : with other lectures and addresses / Hulbert, Eri Baker; ed by Wyant, Andrew Robert Elmer – Chicago: University of Chicago Press, 1908 [mf ed 1993] – 2mf – 9 – 0-524-08429-7 – mf#1993-1039 – us ATLA [242]

English reports – v1-176 + 2v table of cases – 1 – $2595.00 – mf#409150 – us Hein [340]

The english reports (full reprint) – v1-176. 1378-1865. Repr. Edinburgh/London: Wm Green & Sons/Stevens & Sons, 1900-32 – 2456mf – 9 – $3,685.00 – (includes 2 index vols for which w green & sons is the sole publisher) – us LLMC [340]

English republic – London, UK. 1851-55. -irr. 1 reel – 1 – uk British Libr Newspaper [072]

English retraced : or, remarks, critical and philological: founded on a comparison of the breeches bible with the english of the present day / Gurnhill, James – Cambridge [England]: H Wallis, 1862 – 1mf – 9 – 0-8370-3426-4 – mf#1985-1426 – us ATLA [420]

English review – London. 1844-1853 (1) – mf#4244 – us UMI ProQuest [420]

English review – London. 1908-1937 (1) – mf#2270 – us UMI ProQuest [420]

English review – London. v1-64. dec 1908-july 1937 – 1 – us NY Public [420]

English review magazine – London. 1948-1950 (1) – mf#2271 – us UMI ProQuest [420]

English review of literature, science, discoveries, inventions and practical controversies and contests – London. 1783-1796 (1) – mf#4245 – us UMI ProQuest [420]

The english revisers' greek text : shown to be unauthorized except by egyptian copies discarded by greeks and to be opposed to the historic text of all ages and churches / Samson, George Whitefield – Cambridge, MA: Moses King, c1882 – 1mf – 9 – 0-8370-5032-4 – mf#1985-3032 – us ATLA [220]

The english rite : being a synopsis of the sources and revisions of the book of common prayer, with an introduction and an appendix / Brightman, Frank Edward – London: Rivingtons, 1915 – 3mf – 9 – 0-524-03133-9 – (incl bibl ref) – mf#1990-4582 – us ATLA [240]

English rule and native opinion in india : from notes taken in 1870-1874 / Routledge, James – London 1878 – (= ser 19th c british colonization) – 4mf – 9 – mf#1.1.7514 – uk Chadwyck [320]

English ruling cases / Great Britain. Courts – 1307-1908 – 9r – 1 – $450.00 – us Trans-Media [347]

English ruling cases – v1-26. 1307-1894. Boston: Boston Book Co, 1902 (all publ) – 237mf – 9 – $355.00 – (publ as: ruling cases, subscription edition. arr, ann and ed by robert campbell et al with american notes by irivng browne) – mf#LLMC 81-406 – us LLMC [340]

The english school of painting / Chesneau, Ernest Alfred – London 1885 – (= ser 19th c art & architecture) – 5mf – 9 – mf#4.2.102 – uk Chadwyck [750]

English section, teacher training and education faculty, university of north sumatra – Medan, 1962-1963 – 3mf – 9 – (missing: 1962(1)) – mf#SE-762 – ne IDC [959]

English social movements / Woods, Robert Archey – New York: Scribner, 1891 – 1mf – 9 – 0-524-01878-2 – mf#1990-2713 – us ATLA [360]

English society at home / Du Maurier, George Louis Palmella Busson – London 1880 – (= ser 19th c art & architecture) – 2mf – 9 – mf#4.2.1468 – uk Chadwyck [306]

English sources of goethe's gretchen tragedy : a study on the life and fate of literary motives / Liljegren, Sten Bodvar – Lund: C W K Gleerup, 1937 [mf ed 1993] – (= ser Acta regiae societatis humaniorum litterarum lundensis 24) – 1 – (incl bibl ref) – mf#8605 – us Wisconsin U Libr [410]

English stage after the restoration, 1733-1822 : archives of the 18th and 19th century british theatres royal from the british library, london – 35r coll – 1 – (coll contains accounts and ledgers of revered theatrical institutions, giving greater insight into their operations, production methods and audiences) – mf#C35-12500 – us Primary [790]

English state trials, 1163-1858 / ed by Cobbett, William et al – 14r – 1 – $1820.00 – mf#S1847 – us Scholarly Res [347]

English studies in africa – Johannesburg. 1974+ (1) 1974+ (5) 1977+ (9) – ISSN: 0013-8398 – mf#9226 – us UMI ProQuest [420]

The english teacher / Narayan, R K – London: Eyre & Spottiswoode, 1945 – (= ser Samp: indian books) – us CRL [830]

English theories of public address / Sandford, William Phillips – Columbus, OH. 1938 – 1r – 1 – us UF Libraries [080]

English today – Cambridge. 1987+ (1,5,9) – ISSN: 0266-0784 – mf#16531 – us UMI ProQuest [420]

English topography : or, a series of historical and statistical descriptions of the several counties of england and wales / Nightingale, Joseph – London 1816 [mf ed Hildesheim 1995-98] – 1v on 3mf – 9 – €90.00 – 3-487-27550-3 – gw Olms [941]

An english translation of the satyarth prakash : (literally: expose of right sense (of vedic religion) of maharshi swami dayanand saraswati, being, a guide to vedic hermeneutics / Prasad, Durga – Lahore: Virjanand Press, 1908 – (= ser Samp: indian books) – us CRL [280]

An english translation with the sanskrit text of the tattva-kaumudai (saankhya) of vaachaspati misra = Sankhyatattvakaumudi / Vacaspatimisra – [Bombay?]: Pub for the Bombay Theosophical Publication Fund..., 1896 [mf ed 1993] – 1mf – 9 – 0-524-07304-X – (in english & sanskrit) – mf#1991-0090 – us ATLA [290]

English translations of politcial speeches in fiji / Norton, Robert – 1965-1968 – 1r – 1 – mf#pmb1228 – at Pacific Mss [320]

The english version of the new testament compared with king james' translation in use by all protestants / Simkins, William Washington – Pella, IA: Betzer & Gregoire, 1882 – 1mf – 9 – 0-8370-9180-2 – mf#1986-3180 – us ATLA [225]

English wayfaring life in the middle ages (14th century) = Anglais au moyen age / Jusserand, Jean Jules – 2nd ed. New York: GP Putnam, 1890 [mf ed 1993] – 2mf – 9 – 0-524-06353-2 – (english trans by lucy toulmin smith) – mf#1990-1536 – us ATLA [931]

English wayfaring life in the middle ages (14th century) / Jusserand, Jean Adrien Antoine Jules – London, England. 1901 – 1r – 1 – us UF Libraries [941]

English, William F *see* Evolution and the immanent god

English woman's review and drawing room journal – London, UK. 21 Mar 1857-1859. -f. 1 1/2 reels – 1 – uk British Libr Newspaper [072]

The english works of raja ram mohun roy / Bose, Eshan Chunder [comp]; ed by Ghose, Jogendra Chunder – Calcutta: Oriental Press, 1885-1887 – (= ser Samp: indian books) – us CRL [280]

The english works of raja rammohun roy / ed by Nag, Kalidas & Burman, Debajyoti – Calcutta: Sadharan Brahmo Samaj, 1945-1948 – (= ser Samp: indian books) – us CRL [280]

The english works of raja rammohun roy : with an english translation of "tuhfatul muwahhiddin" – Allahabad: The Panini Office, 1906 – (= ser Samp: indian books) – us CRL [280]

The english works of sir henry spelman, kt. / Spelman, Henry – London, Browne, Mears, Clay etc. 1723. 42, xxvi, 38, li-lxv, 67-194, 168, 4, 256, 24 p. LL-979 – 1 – us L of C Photodup [340]

The english works of wyclif hitherto unprinted / Wycliffe, John; ed by Matthew, Frederic David – London: publ for the Early English Text Society by Truebner, 1880 [mf ed 1990] – (= ser Early english text society (series) 74) – 2mf – 9 – 0-7905-8018-7 – (incl bibl ref) – mf#1988-8018 – us ATLA [240]

English-american / Gage, Thomas – London, England. 1928 – 1r – 1 – us UF Libraries [420]

English-bemba phrase book / Lewanika, Godwin A M – 3rd ed. London: Macmillan, 1955 – us CRL [040]

English-bemba phrase book – London, Macmillan, 1955 – us CRL [040]

English-bulu vocabulary – Elat, Cameroun. 192-? – 1r – 1 – us UF Libraries [040]

English-cinyanja dictionary – London, England. 195-? – 1r – 1 – us UF Libraries [040]

English-german literary influences : bibliography and survey / Price, Lawrence Marsden – Berkeley: University of California Press, 1919 – 1r – 1 – (incl bibl ref and ind) – us Wisconsin U Libr [410]

An english-hebrew lexicon : being a complete verbal index to gesenius' hebrew lexicon / Potter, Joseph Lewis – New York: Hurd & Houghton, 1877, c1872 [mf ed 1986] – 1mf – 9 – 0-8370-9498-4 – mf#1986-3498 – us ATLA [040]

An english-kanuri sentence book / Ellison, Randall Erskine – London: Publ on behalf of the Govt of Nigeria by the Crown Agents for the Colonies, 1937 – 1 – us CRL [490]

English-korean dictionary / Jones, George Heber – Tokyo Kyo Bun Kwan, [1916] [mf ed 1995] – (= ser Yale coll) – 1 – 0-524-09576-0 – mf#1995-0576 – us ATLA [040]

English-kwanyama dictionary / Tobias, George Wolfe Robert – Johannesburg, South Africa. 1954 – 1r – 1 – us UF Libraries [040]

English-lamba vocabulary / Doke, Clement Martyn – Johannesburg, South Africa. 1933 – 1r – 1 – us UF Libraries [040]

English-lamba vocabulary / Doke, Clement Martyn – Johannesburg, South Africa. 1963 – 1r – 1 – us UF Libraries [040]

English-language newspapers published in china – 307r – 1 – (the china news, taipei: apr-sep 1952, aug 1956-1961, 1963-1979, sep 1982-apr 1985 88r l9300123. china post, taipei, international air mail edition: 1963, oct 1964-mar 1970, oct 1970-sep 1973, 1974-1980 55r l9300124. china press, shanghai: 1916-1924, apr-may 1949 54r l9300125. north china star, tientsin: nov 1929- 1941 75r l9300126. peiping shronicle, peking: jun 1932-1942 32r l9300127. shanghai news, shanghai: jun 1950-1952 3r l9300128) – Dist. us Scholarly Res – us L of C Photodup [079]

English-lozi phrase book / Lewanika, Godwin A M – London: Macmillan, 1949 – 1 – us CRL [040]

English-lozi vocabulary / Burger, J P – Mongu, Zambia. 1960 – 1r – 1 – us UF Libraries [040]

Englishman : being the sequel of the Guardian – London. 1713-1715 (1) – mf#4812 – us UMI ProQuest [420]

Englishman – Calcutta, India. jan 1861-may 1931, jan 1932-mar 1934 [daily] – 433r – 1 – uk British Libr Newspaper [079]

The englishman – [London]: Printed for Staples Steare, jun 14, jul 13,23-30, aug 6-20 1768 – 1 – us CRL [073]

The englishman at home, his responsibilities and privileges / Porritt, Edward – New York: Thomas Y Crowell, 1893 – 5mf – 9 – $7.50 – (while the title implies a civil liberties thrust, this work is really an extensive and detailed description of the english legal and governmental structure at the turn of the century) – mf#LLMC 92-156 – us LLMC [340]

An englishman defends mother india : a complete constructive reply to "mother india" / Wood, Ernest – Madras: Ganesh & Co; New York: Tantrik Press, 1930 – (= ser Samp: indian books) – us CRL [920]

Englishman directed in the choice of his religion – London, England. 1821 – 1r – 1 – us UF Libraries [240]

Englishman directed in the choice of his religion / Weston, Edward – London, England. 1799 – 1r – 1 – us UF Libraries [240]

The englishman's greek concordance of the new testament : being an attempt at a verbal connexion between the greek and the english texts, including a concordance to the proper names, with indexes, greek-english and english-greek, and a concordance of various readings – 9th ed. London: Samuel Bagster, 1903 – 11mf – 9 – 0-524-06064-9 – mf#1992-0777 – us ATLA [225]

The englishman's hebrew and chaldee concordance of the old testament : being an attempt at a verbal connexion between the original and the english translation – London: Longman, Green, Brown, and Longmans, 1843 – 17mf – 9 – 0-524-06065-7 – mf#1992-0778 – us ATLA [221]

Englishman's magazine – London. 1831-1831 – 1 – mf#4246 – us UMI ProQuest [073]

Englishman's magazine – London. 1842-1843 – 1 – mf#4848 – us UMI ProQuest [073]

Englishman's overland mail – Calcutta, India. jan 1864-dec 1876 [wkly] – 16r – 1 – uk British Libr Newspaper [079]

The englishman's right *see* The canadian's right the same as the englishman's

Englishman's saturday evening journal – Calcutta, India. jan 1863-dec 1866, mar 1867-dec 1875 [lacking: jan, feb 1867] – 24r – 1 – uk British Libr Newspaper [079]

Englishman's thoughts on the scotch church / Eardley, Culling Eardley – London, England. 1841 – 1r – 1 – us UF Libraries [242]

Englishmen at home / Nandalala Ghosha – 1st ed. Calcutta, 1888 – (= ser 19th c british colonization) – 3mf – 9 – mf#1.1.9685 – uk Chadwyck [914]

Englishmen introduced to the free church of scotland / Agnew, David Carnegie A – Perth, Australia. 1851 – 1r – 1 – us UF Libraries [242]

English-sesuto vocabulary / Casalis, A – Morija, Zimbabwe. 1915 – 1r – 1 – us UF Libraries [040]

English-speaking missions in the congo independent state (1878-1908) / Slade, Ruth M – Brussel, 1959 – 11mf – 8 – €21.00 – ne Slangenburg [240]

English-swahili dictionary / Madan, Arthur Cornwallis – Oxford, England. 1902 – 1r – 1 – us UF Libraries [040]

English-swahili phrase-book / Universities' Mission To Central Africa – London, England. 1941 – 1r – 1 – us UF Libraries [040]

English-swahili vocabulary / Madan, Arthur Cornwallis – London, England. 1911 – 1r – 1 – us UF Libraries [040]

English-swahili vocabulary / Madan, Arthur Cornwallis – London, England. 1929 – 1r – 1 – us UF Libraries [040]

English-tagalog vocabulary / Villa, Jose – Manila: University Publ Co, 1946 – us CRL [040]

An english-telugu scientific dictionary / Holler, P – Rajahmundry: Printed at Vivekavardhani Press, 1900 – 1 – (filmed with: a small english-telugu dictionary) – us CRL [490]

English-tonga phrase book... – London: Macmillan, 1950 – 1 – us CRL [040]

English-tonga phrase-book for rhodesia (north and south) / Torrend, J – Mariannhill, South Africa. 1930 – 1r – 1 – us UF Libraries [040]

An english-tswa dictionary / Persson, J A – [Cleveland, Transvaal]: Inhambane Mission Press, 1928 – 1 – us CRL [040]

English-venda vocabulary / Marole, L T – Morija, Zimbabwe. 1954 – 1r – 1 – us UF Libraries [040]

English-vernacular dictionary of the bantu-botatwe dialects / Torrend, J – Natal, South Africa. 1931 – 1r – 1 – us UF Libraries [040]

The englishwoman in egypt : letters from cairo, written...in 1842, 1843 & 1844, with e w lane / [Poole, S] – London, 1844-1845. 3v – 9mf – 9 – mf#HT-115 – ne IDC [916]

The englishwoman in india : containing information for the use of ladies proceeding to, or residing in, the east indies, on the subjects of their outfit, furniture, housekeeping, the rearing of children, duties and wages of servants – London, 1864 – (= ser 19th c books on british colonization) – 3mf – 9 – mf#1.1.2801 – uk Chadwyck [640]

An englishwoman in india two hundred years ago *see* The pirates of malabar

Englishwoman's domestic magazine – London. 1852-1879 (1) – mf#2793 – us UMI ProQuest [640]

An englishwoman's twenty-five yaers in tropical africa : being the biography of gwen elen lewis, missionary to the cameroons and the congo / Hawker, George – London, New York: Hodder & Stoughton [1911?] [mf ed 1990] – 1mf – 9 – 0-7905-5939-0 – mf#1988-1939 – us ATLA [242]

Engman, Evt *see* Population growth in africa

Engravings and etchings of the principal statues, busts, bass-reliefs / Blundell, Henry – [London?] 1809 – (= ser 19th c art & architecture) – 6mf – 9 – mf#4.1.178 – uk Chadwyck [760]

Engravings from the pictures of the national gallery – London 1834 – (= ser 19th c art & architecture) – 5mf – 9 – mf#4.2.1080 – uk Chadwyck [760]

Engravings from the small and large books of designs / Blake, William – Copy A – 1col r – 14 – mf#C97275 – uk Microform Academic [810]

Engravings of ancient cathedrals, hotels de ville, and other public buildings of celebrity, in france, holland, germany, and italy : drawn on the spot, and engraved by john coney: with illustrative descriptions – London: Moon, Boys & Graves, 1832 – (= ser 19th c art & architecture) – 2mf – 9 – mf#4.1.183 – uk Chadwyck [720]

Engravings of the most remarkable of the sepulchral brasses in norfolk / Cotman, John Sell – London 1819 – (= ser 19th c art & architecture) – 8mf – 9 – mf#4.2.1443 – uk Chadwyck [760]

Engrossed bills and resolutions of the us senate, 1789-1817 / U.S. Senate – (= ser Records of the united states senate) – 5r – 1 – (with printed guide) – mf#M1260 – us Nat Archives [324]

Engstler, Otto Hans *see* Das leben des hartwig brueckner

Engstrom,R Todd *see* Avian communities in florida habitats

Enhancing child protection laws after the april 16, 2002 supreme court decision, ashcroft v free speech coalition : hearing...house of representatives, 107th congress, 2nd session, may 1 2002 / United States. Congress. House. Committee on the Judiciary. Subcommittee on Crime, Terrorism, and Homeland Security – Washington: US GPO 2002 [mf ed 2002] – 1mf – 9 – 0-16-068723-3 – us GPO [322]

Enhancing students' personal resources through narrative / Rapmund, Valerie Joan – Uni of South Africa 2000 [mf ed Johannesburg 2000] – 8mf [ill] – 9 – mf#mfm14927 – sa Unisa [400]

Eni va! – Ibadan, Nigeria. v1-7. 19– – 1r – 1 – us UF Libraries [960]

Enid peace and justice center newsletter – 1985 nov-1989 nov – 1r – 1 – mf#1802401 – us WHS [321]

L'enigma della genesi nel pensiero antico e moderno / Minocchi, Salvatore – Firenze: Enrico Ariani, 1908 [mf ed 1985] – 1mf – 9 – 0-8370-4444-8 – (in italian. incl bibl ref) – mf#1985-2444 – us ATLA [221]

L'enigme du macina / Ouane, Ibrahima Manadou – Monte-Carlo: Regain, [1952] – 1 – us CRL [960]

Eniseiskii kooperator – Krasnoiarsk, 1922-1924(5) – 19mf – 9 – (missing:1922(1-11)) – mf#COR-586 – ne IDC [335]

Eniseiskii vestnik : ezhedn polit i obshchestv gaz / ed by Elenev, IU A – Krasnoiarsk [Enis gub]: Enis gub uprava 1919 [1919 16 ianv-[dek] – (= ser Asn 1-3) – n1-231 [1919] item 138, on reel n29,30 – 1 – (suppl: telegrammy [asn-1 395]; cont: volia sibiri) – mf#asn-1 138 – ne IDC [077]

Enkele opmerkingen naar aanleiding van de enquete inzake : "de stem van Ambon" / Grondel, A H – Bussum, 1956 – 1mf – 8 – mf#SE-1589 – ne IDC [959]

Enkele vragen en antwoorden over den heiligen doop : Ter Terechtwijzing voor Degenen Die Nog Geen Belijdenis Hebben Gedaan / Noordewier, Jacob – Holland, MI: Holkeboer, [19–?] – 1mf – 9 – 0-524-06100-9 – mf#1991-2413 – us ATLA [240]

Enking, Ottomar *see* Gerhart hauptmanns "till eulenspiegel"

Enkoepings nyheter – Enkoeping, Motala. 1925-31 – 9 – sw Kungliga [078]

Enkoepings tidning – Enkoeping, 1880-85 – 3r – 1 – sw Kungliga [078]

Enkoepings tidning – Enkoeping, 1885-1918 – 20r – 1 – sw Kungliga [078]

Enkopings weckoblad – Enkoping, Sweden. 1856; 1863-79 – 5r – 1 – sw Kungliga [078]

Enkopingsposten – Enkoping, Sweden. 1879-1978 – 174r – 1 – sw Kungliga [078]

Enkopingsposten – Enkoping, Sweden. 1979- – 1 – sw Kungliga [078]

The enlarging conception of god / Youtz, Herbert Alden – New York: Macmillan, 1914 – 1mf – 9 – 0-7905-8753-X – mf#1989-1978 – us ATLA [210]

Enlightened conformists / Thorn, William – London, England. 18– – 1r – 1 – us UF Libraries [240]

Enlightened non-zoroastrians on mazdayasnism, the excellent religion / Motivala, Jehangir Jamshedji & Sahiar, Bahmanji Navrojji – Bombay: Mistry Printing Works, 1897-1899 – 1mf – 9 – 0-524-02218-6 – mf#1990-2892 – us ATLA [290]

Enlightener / Young Men's Christian Association (La Crosse, WI) – 1885 dec [v2 n12] – 1r – 1 – mf#1331219 – us WHS [240]

Enlist India for freedom! / Thompson, Edward John – London: Victor Gollancz Ltd, 1940 – (= ser Samp: indian books) – us CRL [954]

Enlisted times / Military Studies Center – iss 1-17 [1979 apr-1980 aug/sep] – 1r – 1 – mf#675931 – us WHS [355]

Enlow, C R *see* Lawns in florida

L'ennemie sociale : histoire documentee des faits et gestes de la franc-maconnerie de 1717 a 1890 en france, en belgique et en italie / Rosen, Paul – Paris: Bloud & Barral 1890 – 5mf – 9 – mf#vrl-158 – ne IDC [366]

Les ennemis de wagner : a propos des representations de lohengrin a l'eden-theatre / Verdun, Paul – Paris: Dupret 1887 – 1mf – 9 – mf#wa-110 – ne IDC [790]

Ennepethal-zeitung – Gevelsberg DE, 1949 15 oct-1980 – 149r – 1 – (title varies: 19 aug 1902: gevelsberger zeitung; 16 dec 1938: gevelsberger zeitung milsper-voerder zeitung; 15 oct 1949: gevelsberger zeitung; 2 jan 1958: gevelsberger zeitung. ennepetaler zeitung; fr 30 dec 1972 ba v. westdeutsche zeitung, wuppertal) – gw Mikrofilm [074]

Ennery, Adolphe D' *see*
- Cartouche
- Feu peterscott
- Memoires de deux jeunes mariees
- Premier jour de bonheur

Ennery, Jonas *see* Imre lev

Ennes, Antonio Jose *see* Providencias publicadas pelo commissario regio na providencia de mocambique

Ennes, Ernesto *see* Estudos sobre historia do brasil

Ennis chronicle – Ennis 1789, 1818-jan 1819 – 2r – 1 – ie National [072]

Ennis chronicle and clare advertiser – Ennis. Ireland. -sw. 5 Jan 1828-30 Nov 1831. (4 reels) – 1 – uk British Libr Newspaper [072]

Ennis gazette – Ennis, Ireland. 26 may 1813 – 1/4r – 1 – uk British Libr Newspaper [072]

Ennis, Luna May *see* Music in art

Enniscorthy guardian – Enniscorthy, Ireland. 4 may 1889-1896 – 5r – 1 – (cont as: guardian (wexford)) – uk British Libr Newspaper [072]

Enniscorthy guardian – Enniscorthy, dec 1904-1985 – 100r – 1 – ie National [072]

Enniscorthy news – Ireland, mar 1861-1868; 1871-84; 1886-91; 1894-96 [wkly] – 11r – 1 – uk British Libr Newspaper [072]

Enniscorthy news and county of wexford advertiser – Enniscorthy, Ireland. 2 mar 1861-68; 1870-96 – 13 1/2r – 1 – (aka: news and county of wexford advertiser) – uk British Libr Newspaper [072]

Enniscorthy recorder and gorey correspondent – Enniscorthy, Ireland. 1893-96; 1900 – 3 1/2r – 1 – uk British Libr Newspaper [072]

Enniskillen advertiser and north western counties gazette – Enniskillen, Ireland. 21 jul 1864-28 sep 1876 [wkly] – 5 1/2r – 1 – uk British Libr Newspaper [072]

Enniskillen chronicle and erne packet – Enniskillen, Ireland. 27 may 1813; 1824.may 1849; 23 aug 1849-4 nov 1850; 20 mar-25 dec 1851; 1852-1889; 2 mar 1890-27 jul 1893 [wkly] – 45 3/4r – 1 – (incorp with: impartial reporter 1893. aka: fermanagh mail and enniskillen chronicle) – uk British Libr Newspaper [072]

Enniskillen gazette and commercial advertiser – Enniskillen, Ireland. 2 mar 1854 – 1/4r – 1 – uk British Libr Newspaper [072]

Enniskillen sentinel – Enniskillen, Ireland. 25 jan, 1 feb, 29 mar, 5 apr 1877 – 1/4r – 1 – uk British Libr Newspaper [072]

Enniskillen watchman – Enniskillen, Ireland. 28 sep-26 oct 1848 [wkly] – 1/4r – 1 – uk British Libr Newspaper [072]

The enniskillener – Enniskillen. Ireland. -w. 13 Feb 1830-29 Dec 1836, 2 Jan-20 Feb 1840. (3 reels) – 1 – uk British Libr Newspaper [072]

Enniskillener or fermanagh constitution – Enniskillen, Ireland. 13 feb 1830-1836; 2 jan-20 feb 1840 – 2 1/2r – 1 – uk British Libr Newspaper [072]

Ennstaler kurier – Nezen, Austria. 4 nov 1945-23 mar 1947 – 1/2r – 1 – uk British Libr Newspaper [074]

Enoch : or, idris, and death / Davidson, Judson D – Napanee [ON]:...the Beaver Office, 1906 [mf ed 1999] – 1mf – 9 – 0-659-91223-6 – mf#9-91223 – cn Canadiana [810]

Enoch, frere *see* Le vrai franc-macon

Enoch, Samuel *see* Der treue zions-waechter

Enock, C Reginald *see* Panama canal

Enock, Charles Reginald *see* Tropics

Enon baptist church of christ (now first baptist church). huntsville, alabama : church records – jun 1809-apr 1861 – 1 – us Southern Baptist [242]

Enon baptist church. pickens county. easley, south carolina : church records – 1851-aug 1942 – 1 – us Southern Baptist [242]

Enon, Pennsylvania. Enon Baptist Church *see* Minutes

Enosis – Larnaca, Cyprus. 27 mar 1886-89 dec 1889; 4 jan 1890-10 jan 1893 – 2r – 1 – uk British Libr Newspaper [072]

Enquete cinematographique sur les steelbands de trinidad / Verba, Daniel – 1mf – 9 – mf#10453 – fr Atelier National [790]

Enquete coloniale dans l'afrique francaise occidentale et equatoriale sur l'organisation de la famille indigene, les fiancailles, le mariage : avec une esquisse generale des langues de l'afrique / Delafosse, Maurice – Paris: societe d'editions geographiques, maritimes et coloniales, 1930 – 1 – (et une esquisse ethnologique des principales populations de l'afriq. francaise equatoriale par le dr poutrin) – us CRL [306]

Enquete dans l'affaire du chemin de fer de la baie des chaleurs : rapports des commissaires, procedes de la commission, deposition des temoins, etc, etc / Quebec (Province). Commission royale d'enquete – Quebec: I Turcot, 1892 – 2mf – 9 – 0-665-93303-7 – (incl bibl ref) – mf#93303 – cn Canadiana [380]

L'enquete devant le comite des comptes publics de l'assemblee legislative dans l'affaire de la peinture et de la tapisserie – Quebec: [s.n.], [1888?] (mf ed 1987) – 1mf – 9 – mf#SEM105P565 – cn Bibl Nat [323]

Enquete parlementaire sur les actes du gouvernement de la defense nationale...depositions des temoins / France. Commission d'enquete sur les actes du gouvernement de la defense nationale – 6v. 1872-75 – 1 – us L of C Photodup [324]

Enquete sur la loi de la convention collective et son application : analyse des decisions des tribunaux / Lambert, Joseph A – Quebec: Comite...fevrier 1965 [mf ed 1979] – 1mf – 9 – mf#SEM105P8 – cn Bibl Nat [340]

Enquete sur la loi de la convention collective et son explication : analyse des decisions des tribunaux / Vaillancourt, Gerard – Quebec: Comite...1965 (i.e. 1964) [mf ed 1979] – 1mf – 9 – mf#SEM105P9 – cn Bibl Nat [340]

Enquete sur le crime organise (Quebec) *see* L'introduction frauduleuse de viande impropre sur le marche de la consommation humaine et la fraude en rapport avec la viande chevaline

Enquete sur les langues parless au senegal par les eleves / Wioland, Francois – Dakar, Senegal. 1965 – 1r – 1 – us UF Libraries [960]

Enquete sur les services de sante (1948), province de quebec – [Quebec: Ministere de la sante. 4v. 1951?] (mf ed 1993) – 1r – 1 – mf#SEM35P385 – cn Bibl Nat [360]

Enquete...sur les evenements du 21 mai 1832, a montreal / Bas-Canada. Parlement. Chambre d'assemblee – [Quebec (Province): s.n. 2v. 1833?] (mf ed 1982) – 6mf – 9 – mf#SEM105P110 – cn Bibl Nat [325]

Enquire within for information about manitoba and the boundless wheat fields of the new northwest : through which runs the canadian pacific railway / Begg, Alexander – [Liverpool?: s.n, 1883?] [mf ed 1981] – 1mf – 9 – mf#01087 – cn Canadiana [630]

Enquirer – Battle Creek, MI. 1994-2000 (1) – mf#61501 – us UMI ProQuest [071]

Enquirer : (city edition) – Cincinnati, OH. 1841-1920 (1) – mf#65412 – us UMI ProQuest [071]

Enquirer – Columbus, GA. 1855-1988 (1) – mf#60445 – us UMI ProQuest [071]

Enquirer – Oconto WI. 1881 jul 30-1886, 1887-1889 jul 11, 1889 jul 18-1893, 1894-97, 1898-1901 jul 15, 1905 jul 7-1906, 1907-09, 1910-1922 nov 10 – 8r – 1 – (continued by: oconto county reporter-enterprise) – mf#958186 – us WHS [071]

Enquirer – Georgetown, SC. 1880-1889 (1) – mf#66488 – us UMI ProQuest [071]

Enquirer / Hamilton Co. Cincinnati – jul 1-15 1937 [daily] – 1r – 1 – mf#B1260 – us Ohio Hist [071]

Enquirer / Hamilton Co. Cincinnati – oct 1906-jan 1907 [daily] – 3r – 1 – mf#B1261-1263 – us Ohio Hist [071]

Enquirer : (kentucky edition) – Cincinnati, OH. 1901-1965 (1) – mf#65413 – us UMI ProQuest [071]

Enquirer / Lucas Co. Oregon – feb 1981-mar 1982 [wkly] – 1r – 1 – mf#B13554 – us Ohio Hist [071]

Enquirer – Memphis, TN. 1847-1849 (1) – mf#66557 – us UMI ProQuest [071]

Enquirer – Memphis, TN. 1936-1940 (1) – mf#66558 – us UMI ProQuest [071]

Enquirer – Richmond, VA. 1861-1872 (1) – mf#61137 – us UMI ProQuest [071]

Enquirer / Vinton Co. McArthur – 1874-81, 1883 [wkly] – 4r – 1 – mf#B10803-10806 – us Ohio Hist [071]

Enquirer / Vinton Co. McArthur – feb 1873-dec 1873 [wkly] – 1r – 1 – mf#B148 – us Ohio Hist [071]

Enquirer *see* [Oakland-] oakland post enquirer

The enquirer – Blantyre: Commercial Development Graphic Ltd, aug 2/8-sep 20/26; oct 5-dec 15, 31 1993; jan-dec 1994; jan-feb 3/9, 17/23-aug 3/8, 25/31-dec 19 1995; jan-jun 19/25; jul 5/11-19/25 1996 – 1r – 1 – us CRL [079]

The enquirer – Quebec: W H Shadgett, 1821-1822 – 9 – (incl some text in french) – ISSN: 1190-691X – mf#P04145 – cn Canadiana [073]

The enquirer – Toronto: T J Hamilton, [1880?-19–?] – 9 – mf#P04267 – cn Canadiana [240]

Enquirer and message / Hamilton Co. Cincinnati – nov 1843-jan 1845 (poor quality) [daily] – 4r – 1 – mf#B1253-1256 – us Ohio Hist [071]

Enquirer plus edition – Columbus, GA. 1981-1987 (1) – mf#68247 – us UMI ProQuest [071]

Enquirer thru journal / Miami Co. Piqua – (1862-83) scattered [wkly] – 2r – 1 – mf#B8560-8561 – us Ohio Hist [071]

Enquirer's difficulties / Stock, John – London, England. 18– – 1r – 1 – us UF Libraries [240]

Enquiridion contra el morbo articular / Gomez Miedes, B – Madrid, 1731 – 4mf – 9 – sp Cultura [610]

Enquiry – v1-2 n2,3. 1942-1945 [all publ] – (= ser Radical periodicals in the united states, 1881-1960. series 1) – 5mf – 9 – $85.00 – us UPA [073]

An enquiry concerning the intellectual and moral faculties and literature of negroes : followed with an account of the life and works of 15 negroes and mulattoes distinguished in science, literature and the arts / Gregoire, H – Printed by T Kirk, 1810 – (filmed with: a tribute for the negro/w armistead) – us CRL [305]

An enquiry into the constitution, discipline, unity, and worship of the primitive church : which flourished within the first three hundred years after christ / King, Peter King, Lord – London: S Cornish, 1839 [mf ed 1992] – 1mf – 9 – 0-524-05149-6 – mf#1990-1405 – us ATLA [240]

An enquiry into the evidential value of prophecy / Edghill, Ernest Arthur – London, New York: Macmillan, 1906 [mf ed 1989] – (= ser Hulsean prize essay) – 2mf – 9 – 0-7905-1985-2 – (incl ind) – mf#1987-1985 – us ATLA [220]

An enquiry into the nature and effects of the paper credit of great britain / Thornton, Henry – London: J Hatchard & F & C Rivington, 1802 – (= ser 19th c economics) – 4mf – 9 – mf#1.1.109 – uk Chadwyck [332]

An enquiry into the objections against george psalmanaazaar of formosa... – London: Bernard Lintott, [1705] – 1mf – 9 – mf#HT-666 – ne IDC [910]

An enquiry into the obligations of christians : to use means for the conversion of the heathens / Carey, William – London, Hodder and Stoughton, 1891. Chicago: Dep of Photodup, U of Chicago Lib, 1974 (1r); Evanston: American Theol Lib Assoc, 1984 (1r) – 1 – 0-8370-0059-9 – mf#1984-6017 – us ATLA [240]

Enr – New York. 1987+ (1,5,9) – (cont: engineering news-record) – ISSN: 0891-9526 – mf#34,01 – us UMI ProQuest [624]

L'enrage – n1-12. 1968 – 1 – us AMS Press [073]

Enraght, R W *see* Real presence and holy scripture

Enrichis des figures : eeceuil des plusieurs enigmes, airs, devises, et medailles – Amsterdam: Jansson de Waesberge, 1684 – 3mf – 9 – mf#O-2012 – ne IDC [090]

Enrichment of private prayer / Moberly, Robert Campbell – London: SPCK, 1897 – 1mf – 9 – 0-524-07209-4 – mf#1990-5367 – us ATLA [240]

Enrico 4 : tragedia in tre atti / Pirandello, Luigi – 3. ed. Firenze: R Bemporad & Figlio, (c1923) – 1r – 1 – mf#1268 – us Wisconsin U Libr [820]

Enrico arnaud : pastore e duce de'valdesi, 1641-1721 / Comba, Emilio – Firenze: Claudiana, 1889 – 1mf – 9 – 0-7905-5520-4 – (incl bibl ref) – mf#1988-1520 – us ATLA [945]

Enrico heine nella letteratura italiana : avanti la "rivelazione" di t massarani / Bonardi, Carlo – Livorno: R Guisti, 1907 – viii/150p – 1 – (incl bibl ref) – mf#8770 – us Wisconsin U Libr [440]

Enrique a laguerre y su obra 'la resaca' / Morfi, Angelina – San Juan, Puerto Rico. 1964 – 1r – 1 – us UF Libraries [972]

Enrique abril, heroe / Felices, Jorge – San Juan, Puerto Rico. 1947 – 1r – 1 – us UF Libraries [972]

Enrique gomez carillo / Torres, Edelberto – Guatemala, 1956 – 1r – 1 – us UF Libraries [972]

Enrique gomez carrillo / Mendoza, Juan Manuel – Guatemala. v1-2. 1946 – 1r – 1 – us UF Libraries [972]

Enrique hernandez miyares y su poesia / Maderal, Luis – Marianao, Cuba. 1959 – 1r – 1 – us UF Libraries [972]

Enrique jose varona / Entralgo, Elias Jose – Habana, Cuba. 1937 – 1r – 1 – us UF Libraries [972]

Enrique jose varona / Vitier, Medardo – Habana, Cuba. 1949 – 1r – 1 – us UF Libraries [972]

Enrique pineyro, historiador / Cordova, Federico – Habana, Cuba. 1944 – 1r – 1 – us UF Libraries [972]

Enrique pineyro y la critica literaria / Bueno, Salvador – Habana, Cuba. 1957? – 1r – 1 – us UF Libraries [972]

Enriquecimiento sin causa / Fabrega P, Jorge – Panama, 1955 – 1r – 1 – us UF Libraries [972]

Enriques, Federigo *see* Problems of science

Enriquez Anselmo, Juan *see* Dos medicos extremenos de ayer

Enriquez, Carlos *see*
– Feria de guaicanama
– Tilin garcia

Enriquez, Jose T et al *see* Filipino language lexicon

Enriquillo / Galvan, Manuel De Jesus – Barcelona, Spain. 1909 – 1r – us UF Libraries [972]

Enriquillo / Galvan, Manuel De Jesus – Buenos Aires, Argentina. 1944 – 1r – us UF Libraries [972]

Enriquillo / Galvan, Manuel De Jesus – Ciudad Trujillo, Dominican Republic. 1955 – 1r – us UF Libraries [972]

Enriquillo : leyenda historica dominicana / Galvan, Manuel De Jesus – New York, NY. 1964 – 1r – 1 – us UF Libraries [972]

Enriquillo : leyenda historica dominicana / Galvan, Manuel De Jesus – Santo Domingo, Dominican Republic. 1962 – 1r – 1 – us UF Libraries [972]

Enrolled acts and resolutions of congress, 1893-1956 / U.S. Congress – (= ser General records of the united states government) – 139r – 1 – mf#M1326 – us Nat Archives [323]

Enrolled original acts and resolutions of the congress of the united states, 1789-1823 / U.S. Congress – (= ser General records of the united states government) – 17r – 1 – mf#M337 – us Nat Archives [323]

The enrollment and persistence of african-american doctoral students in physical education and related disciplines / King, Susan E & Anderson, William G – 1992 – 3mf – 9 – $12.00 – us Kinesology [790]

Enrollment cards for the five civilized tribes, 1898-1914 / U.S. Bureau of Indian Affairs – (= ser Records relating to census rolls and other enrollments) – 93r – 1 – (with printed guide) – mf#M1186 – us Nat Archives [317]

Enrollment report / Wisconsin. Board of Regents of State Colleges – 1954/55-1964/65 – 1r – 1 – (continued by: enrollment report (wisconsin state universities system)) – mf#530977 – us WHS [071]

Enrollment report / Wisconsin State Universities System – 1965/66-1971/72 – 1r – 1 – (cont: enrollment report (wisconsin. board of regents of state colleges)) – mf#530982 – us WHS [378]

L'enrolment des gamins de paris pour l'armee d'italie – Paris, [1848?] – us CRL [944]

The enron collapse and its implications for worker retirement security : hearings... house of representatives, 107th congress, 2nd session...washington dc, feb 6 2002 and feb 7 2002 / United States. Congress. House. Committee on Education and the Workforce – Washington: US GPO 2002 [mf ed 2002] – 4mf – 9 – 0-16-069012-9 – (incl bibl ref & ind) – us GPO [345]

O ensaio : orgao do gremio litterario ubirajara – Rio de Janeiro, RJ: Typ de Jose Dias de Oliveira, 03 jun-16 ago 1882 – (= ser Ps 19) – mf#P19A,04,29 – bl Biblioteca [440]

Ensaio critico sobre a viagem ao brasil em 1852 / Pascual, Antonio Diodoro De – Rio de Janeiro, Brazil. 1861 – 1r – 1 – us UF Libraries [972]

Ensaio de diccionario kimbundu-portuguez. / Cordeiro da Matta, J D – Lisbon: A M Pereira, 1893 – 1 – us CRL [040]

Ensaio de um estudo geografico da rede urbana de angola / Amaral, Ilidio Do – Lisboa, Portugal. 1962 – 1r – 1 – us UF Libraries [916]

Ensaio historico sobre a independencia / Marques, Xavier – Rio de Janeiro, Brazil. 1924 – 1r – 1 – us UF Libraries [972]

Ensaio juridico e litterario – Recife, PE: Typ Industrial, 01 maio-01 jun 1878 – (= ser Ps 19) – mf#P17,02,139 – bl Biblioteca [340]

Ensaio juvenil : orgao do clube juvenil – Campanha, MG: Typ do Monitor Mineiro, 30 ago-23 set 1889 – (= ser Ps 19) – mf#P17,02,79 – bl Biblioteca [079]

Ensaio litterario – Recife, PE: Typ do Correio da Tarde, 15 fev 1865 – (= ser Ps 19) – mf#P16,01,14 – bl Biblioteca [440]

Ensaio sobre a problematica dos transportes / Carvalho, Osvaldo Ferraro – Rio de Janeiro, Brazil. 1957 – 1r – 1 – us UF Libraries [380]

Ensaio sobre o parnasianismo brasileiro / Martins, Jose V De Pina – Coimbra, Portugal. 1945 – 1r – 1 – us UF Libraries [972]

Ensaios / Marques, Xavier – Rio de Janeiro, Brazil. 1944 – 1r – us UF Libraries [972]

Ensaios / Milliet, Sergio – Sao Paulo, Brazil. 1938 – 1r – us UF Libraries [972]

Ensaios : revista litteraria – Ouro Preto, MG: Typ d'O Jornal de Minas, dez 1890 – (= ser Ps 19) – mf#P17,2,90 – bl Biblioteca [440]

Ensaios : revista mensal scientifica e litteraria – Ouro Preto, MG: Typ Silva Cabral, mar-abr 1893; jan 1894 – (= ser Ps 19) – mf#P17,02,84 – bl Biblioteca [500]

Ensaios americanos / Freitas, Newton – Rio de Janeiro, Brazil. 1945 – 1r – 1 – us UF Libraries [972]

Ensaios biograficos / Gontijo De Carvalho, Antonio – Sao Paulo, Brazil. 1951 – 1r – 1 – us UF Libraries [920]

Ensaios brasilianos / Roquette-Pinto, Edgardo – Sao Paulo, Brazil. 1940 – 1r – 1 – us UF Libraries [972]

Ensaios de anthropologia brasiliana / Roquette-Pinto, Edgardo – Sao Paulo, Brazil. 1933 – 1r – 1 – us UF Libraries [972]

Ensaios de etnologia brasileira / Baldus, Herbert – Sao Paulo, Brazil. 1937 – 1r – 1 – us UF Libraries [972]

Ensaios de geographia linguistica / Castro, Eugenio De – Sao Paulo, Brazil. 1941 – 1r – 1 – us UF Libraries [440]

Ensaios de historia e critica / Araujo Jorge, Arthur Guimaraes de – Rio de Janeiro, Brazil. 1948 – 1r – 1 – us UF Libraries [972]

Ensaios de sciencia por diversos amadores – Rio de Janeiro, RJ: Typ Brown & Evaristo, mar, jul 1876 – (= ser Ps 19) – mf#P17,01,161 – bl Biblioteca [500]

Ensaios poeticos / Castro Sampaio, Manuel de – 1858 – 9 – sp Bibl Santa Ana [810]

Ensaios sul-americanos / Mesquita, Julio De – Sao Paulo, Brazil. 1956 – 1r – 1 – us UF Libraries [972]

Ensaistas brasileiros / Oliveira, Jose Osorio de – Lisboa, Portugal. 194-? – 1r – 1 – us UF Libraries [972]

Ensayistas colombianos / Hernandez De Alba, Guillermo – Buenos Aires, Argentina. 1957 – 1r – 1 – us UF Libraries [972]

Ensayistas contemporaneos / Lizaso, Felix – Habana, Cuba. 1938 – 1r – 1 – us UF Libraries [972]

Ensayo bibliografico sobre san pedro de alcantara / Recio Veganzones, Alejandro – Madrid, 1962 – 1 – sp Bibl Santa Ana [240]

Ensayo biografico batista / Acosta Rubio, Raoul – Habana, Cuba. 1943 – 1r – 1 – us UF Libraries [972]

Ensayo biografico de francisco morazan / Pineda M, Leonidas – Tegucigalpa, Mexico. 1944 – 1r – 1 – us UF Libraries [972]

Ensayo biografico del procer jose leon sandoval / Alvarez Lejarza, Emilio – Managua, Nicaragua. 1947 – 1r – 1 – us UF Libraries [972]

Ensayo biologico sobre hernando cortes, tomo 1 : oaxaca, 1933 / Gutirrrez, Solano – Madrid: Razon y Fe, 1935 – 1 – sp Bibl Santa Ana [920]

Ensayo critico y antologico acerca / Ayala Duarte, Crispin – Caracas, Venezuela. 1936 – 1r – 1 – us UF Libraries [972]
Ensayo de catalogo de los lugares de senorio temporal, los obispos de espana en la edad media / Perez-Villamil, Manuel – Madrid: Fortanet, 1916 – sp Bibl Santa Ana [240]
Ensayo de divulgacion cientifica sobre / Pieter, Heriberto – Ciudad Trujillo, Dominican Republic. 1949 – 1r – 1 – us UF Libraries [972]
Ensayo de haikai antillano / Benet Y Castellon, Eduardo – Dienfuegos, Cuba. 1957 – 1r – 1 – us UF Libraries [972]
Ensayo de historia americana / Gilii, Filippo Salvadore – Bogota, Colombia. 1955 – 1r – 1 – us UF Libraries [972]
Ensayo de medicina general o sea de filosofia medica... / Nieto y Serrano, M – Madrid, 1860 – 10mf – 9 – sp Cultura [610]
Ensayo de un registro del censo organizado... por el distrito de rosmblon en 1895 – Manila, 1896 – 6mf – 9 – sp Cultura [946]
Ensayo de una biblioteca espanola de libros raros y curiosos / Gallardo, B J – Madrid, 1963-1966 – 50mf – 9 – sp Cultura [020]
Ensayo en la generacion del treinta / Robles De Cardona, Mariana – San Juan, Puerto Rico. 1960 – 1r – 1 – us UF Libraries [972]
Ensayo historico de olivenza / Parra, Victoriano C – Badajoz: Tip. y Lib. Antonio Arqueros, 1909 – 1 – sp Bibl Santa Ana [946]
Ensayo historico sobre las tribus nonualcas y su c... / Dominquez Sosa, Julio Alberto – San Salvador, El Salvador. 1964 – 1r – 1 – us UF Libraries [972]
Ensayo historico-critico de las relaciones diploma / Perez Concha, Jorge – Quito, Ecuador. v1-2. 1961-1964 – 1r – 1 – us UF Libraries [972]
Ensayo para una teoria de extremadura / Becerro de Bengoa, Ricardo – Caceres: Imprenta Garcia Floriano, 1950 – sp Bibl Santa Ana [946]
Ensayo politico sobre el reino de la nueva espana / Bayle, Constantino & Humboldt, Alejandro de – Madrid: Razon y Fe, 1944.- 5v – 1 – sp Bibl Santa Ana [320]
Ensayo politico sobre la isla de cuba / Humboldt, Alexander Von – Habana, Cuba. 1960 – 1r – 1 – us UF Libraries [972]
Ensayo sobre el catolicismo, el liberalismo y el socialismo / Donoso Cortes, Juan Francisco – Buenos Aires: Editorial Americalee, 1943 – 1 – sp Bibl Santa Ana [320]
Ensayo sobre el catolicismo liberalismo / Donoso Cortes, Juan Francisco – 1851 – 9 – (ed 1 1880) – sp Bibl Santa Ana [240]
Ensayo sobre el destino / Masferrer, Alberto – San Salvador, El Salvador. 1963 – 1r – 1 – us UF Libraries [972]
Ensayo sobre el rio de la plata y la revolucion francesa / Cailler-Bois, Ricardo – Buenos Aires, 1929 – 1 – sp Bibl Santa Ana [946]
Ensayo sobre plantas usuales de costa rica / Pittier, Henri – San Jose, Costa Rica. 1957 – 1r – 1 – us UF Libraries [972]
Ensayo sobre virgilio / Ricci, Clemente – Buenos Aires, 1931; Madrid: Razon y Fe, 1931 – 1 – sp Bibl Santa Ana [450]
Ensayo teorico e historico sobre la generacion de los conocimientos humanos : traduccion de a. garcia moreno...tomo 1 / Tiberghien, G – Madrid: Imp. de Federico Escamez Centeno, s.a. – 9 – sp Bibl Santa Ana [190]
Ensayo...biblioteca...libros raros / Gallardo, Bartolome Jose – 1863-66 – 4v – 9 – sp Bibl Santa Ana [070]
Ensayos americanos (critica literaria) / Freitas, Newton – Buenos Aires, Argentina. 1942 – 1r – 1 – us UF Libraries [972]
Ensayos centroamericanos / Rey, Julio Adolfo – Santa Ana, El Salvador. 1955 – 1r – 1 – us UF Libraries [972]
Ensayos critos / Lozano Y Lozano, Juan – Bogota, Colombia. 1934 – 1r – 1 – us UF Libraries [972]
Ensayos de etice. duns escoto en extremadura 2 / Colegio San Antonio – Caceres: imp la minerva, 2nd parte 1956 – 1 – sp Bibl Santa Ana [946]
Ensayos de historia politica y diplomatica / Rivas, Angel Cesar – Madrid, Spain. 191- – 1r – 1 – us UF Libraries [972]
Ensayos de literatura cubana / Chacon y Calvo, Jose Maria – Madrid, Spain. 1922 – 1r – 1 – us UF Libraries [972]
Ensayos de poesia indigena en cuba / Varela, Jose Luis – Madrid, Spain. 1951 – 1r – 1 – us UF Libraries [972]
Ensayos escogidos / Picon-Salas, Mariano – Santiago, Chile. 1958 – 1r – 1 – us UF Libraries [972]
Ensayos historicos : publicacion de homenaje / Bayle, Constantino & Riverola, Rofolfo – Madrid: Razon y Fe, 1944 – 1 – sp Bibl Santa Ana [946]
Ensayos literarios / Arnes Luna, Alfredo – Madrid: Imp Giralda, 1948 – 1 – sp Bibl Santa Ana [440]
Ensayos literarios / Fernandez Spencer, Antonio – Ciudad Trujillo, Dominican Republic. 1960 – 1r – 1 – us UF Libraries [972]
Ensayos martianos / Marinello, Juan – Santa Clara, Cuba. 1961 – 1r – 1 – us UF Libraries [972]
Ensayos poeticos / Sanchez Arjona y Sanchez Arjona, Jose – 1872 – 9 – sp Bibl Santa Ana [810]
Ensayos sobre etnologia argentina / Cabrera, Pablo – Madrid: Razon y Fe, 1932 – 1 – sp Bibl Santa Ana [305]
Ensayos sobre planeacion / Currie, Lauchlin Bernard – Bogota, Colombia. 1965 – 1r – 1 – us UF Libraries [972]
Ensayos y apuntes / Esquenazi-Mayo, Roberto – Habana, Cuba. 1956 – 1r – 1 – us UF Libraries [972]
Ensayos y dialogos / Castellanos, Francisco Jose – Habana, Cuba. 1961 – 1r – 1 – us UF Libraries [972]
Ensayos y semblanzas / Mesa, Carlos E – Bogota, Colombia. 1956 – 1r – 1 – us UF Libraries [972]
Ensayos y textos elementales de historia / Sierra, Justo – Mexico City? Mexico. 1948 – 1r – 1 – us UF Libraries [972]
Ense, K A Varnhagen von *see* Tagebuecher von k a varnhagen von ense
Ense, Karl August Varnhagen von *see* K l von knebel's literarischer nachlass und briefwechsel
Les enseignants : le trait d'union et l'apostrophe du monde enseignant – Montreal: [s.n.] v1 n1 15 oct 1970- (mthly) [mf ed 1973-] – 1 – mf#SEM35P21 – cn Bibl Nat [370]
L'enseigne : son histoire, sa philosophie, ses particularites, les boutiques, les maisons, la rue, la reclame commerciale a lyon / Grand-Carteret, John – Grenoble: Librairie Dauphinoise; Moutiers: Librairie Savoyarde, 1902 [mf ed 1988] – 6mf – 9 – (with ind) – mf#SEM105P661 – cn Bibl Nat [944]
L'enseignement catholique dans la france contemporaine : etudes et discours / Baudrillart, Alfred – Paris: Bloud, 1910 [mf ed 1986] – 2mf – 9 – 0-8370-7606-4 – (in french. incl bibl ref and ind) – mf#1986-1606 – us ATLA [241]
L'enseignement de jesus / Batiffol, Pierre – Paris: Librairie Bloud, 1909 [mf ed 1993] – (= ser Bibliotheque de l'enseignement scripturaire) – 1mf – 9 – 0-524-05789-3 – (in french) – mf#1992-0616 – us ATLA [240]
Enseignement de l'histoire en haiti / Pressoir, Catts – Mexico City? Mexico. 1950 – 1r – 1 – us UF Libraries [972]
L'enseignement de saint paul dans les epitus de l'annee liturgique, 1933 / Soubigon, P – Madrid: Razon y Fe, 1934 – 1 – sp Bibl Santa Ana [240]
L'enseignement des apotres / Bovon, Jules – Lausanne: Georges Bridel, 1894 [mf ed 1989] – (= ser Etude sur l'oeuvre de la redemption 1/2) – 2mf – 9 – 0-7905-2462-7 – mf#1987-2462 – us ATLA [225]
L'enseignement des jesuites au canada : college sainte-marie de montreal / Bellay, A – S.l: s.n, 1891? – 1mf – 9 – mf#03563 – cn Canadiana [377]
L'enseignement des lettres classiques d'ausone a alcuin : introduction a l'histoire des ecoles carolingiennes / Roger, Maurice – Paris: Alphonse Picard, 1905 [mf ed 1986] – 2mf – 9 – 0-8370-8857-7 – (in french. incl ind) – mf#1986-2857 – us ATLA [931]
L'enseignement des sciences sociales : compte-rendu d'une journee d'etudes de professeurs des colleges classiques et de la faculte des sciences sociales...le 11 mars 1962 / Universite de Montreal. Faculte des sciences sociales, economiques et politiques – [Montreal: s.n, 1962] (mf ed 2001) – 9 – cn Bibl Nat [300]
L'enseignement du buddhisme des origines a nos jours / Huot Tath Vajirappano – [Phnom-Penh: Universite buddhique Preah Sihanouk Raj 1962?] [mf ed 1989] – (= ser Culture et civilisation khmeres 1) – 1r with other items – 1 – mf#mf-10289 seam reel 008/11 [§] – us CRL [280]
L'enseignement et l'education en republique de guinee / Toure, A S – Conakry: Impr nationale "Patrice Lumumba", 1972 – us CRL [370]
L'enseignement manuel de l'enfant dans l'ecole primaire / Dupuy, Paul – Montreal?: s.n, 1889 – 1mf – 9 – (en tete du titre: Enseignement pratique et technique) – mf#56035 – cn Canadiana [370]
Enseignement medico-social pour coloniaux / Habig, Jean-Marie – Bruxelles: Edition Universelle, 1946-48 – 1 – us CRL [960]
L'enseignement menager au canada francais : bibliographie analytique, 1955-1963 / St-Claude-Marie, soeur – 1964 [mf ed 1979] – (= ser Bibliographies du cours...1947-66) – 1mf – 9 – (with ind; pref by sister sainte-isabelle) – mf#SEM105P4 – cn Bibl Nat [339]
L'enseignement menager dans la province de quebec : reglements et programmes – Quebec: [Departement de l'instruction publicque], 1943 [mf ed 1993] – 2mf – 9 – mf#SEM105P1827 – cn Bibl Nat [350]
Enseignement pratique et technique : l'enseignement manuel de l'enfant dans l'ecole primaire / Dupuy, Paul – Montreal?: s.n, 1889 – 1mf – 9 – mf#51220 – cn Canadiana [370]
L'enseignement primaire – [Quebec: s.n., 1881-19–] – 9 – mf#P04117 – cn Canadiana [370]
Enseignement primaire et reformes scolaires : conference faite devant l'association des instituteurs de la circonscription de l'ecole normal jacques-cartier, le 26 mai 1893 / Robillard, L G – St-Jerome, Quebec?: s.n, 1893? – 1mf – 9 – mf#04776 – cn Canadiana [370]
L'enseignement qu'on appelle agricole et l'evolution des espaces sociaux a travers l'exemple de la basse-normandie / Lille, Jean – 2mf – 9 – fr Atelier National [307]
Les enseignements de la parole. spiritualisme chretien *see* La parole
Ensenanza de la historia en colombia / Aguilera, Miguel – Mexico City? Mexico. 1951 – 1r – 1 – us UF Libraries [972]
Ensenanza de la historia en cuba / Pan American Institute of Geography and History – Mexico City? Mexico. 1951 – 1r – 1 – us UF Libraries [972]
Ensenanza de la historia en puerto rico / Rivera, Antonio – Mexico City? Mexico. 1953 – 1r – 1 – us UF Libraries [972]
Ensenanza de la historia en venezuela / Vazquez, Pedro Tomas – Mexico City? Mexico. 1951 – 1r – 1 – us UF Libraries [972]
La ensenanza de lenguas civilizadas a los barbaros. un caso de teologia pastoral misionera / Bayle, Constantino – Madrid: Razon y Fe, 1933 – 1 – sp Bibl Santa Ana [240]
La ensenanza en barcelona a fines del siglo 18 / Oriol Moncanut, Ana Maria – Madrid: Archivo Ibero Americano, 1960 – 1 – sp Bibl Santa Ana [946]
Ensenanza...badajoz / Saa Maldonado, Manuel – 1873 – 9 – sp Bibl Santa Ana [370]
Ensenanzas de la campana de corea / Ruiz Novoa, Alberto – Bogota, Colombia. 1956 – 1r – 1 – us UF Libraries [972]
Ensenanzas de una revolucion / Ferrara, Orestes – Habana, Cuba. 1932 – 1r – 1 – us UF Libraries [972]
Ensenanzas y profecias / Tejera, Diego Vincente – Habana, Cuba. 1916 – 1r – 1 – us UF Libraries [972]
The ensign *see* Mataura ensign
Ensign, M R *see* Grading, packing and stowing florida produce
O ensino da historia no brasil – Mexico City?, Mexico. 1953 – 1r – us UF Libraries [972]
Den enskilde och kyrkan : foeredrag hallet vid studentmoetet i huskvarna 1909 / Soederblom, Nathan – Uppsala: L Norblad [1909?] [mf ed 1990] – (= ser Sveriges kristliga studentroerelses skriftserie 4) – 1mf – 9 – 0-7905-6320-7 – mf#1988-2320 – us ATLA [240]
Ensor, George *see* Addresses to the people of ireland
Entalpiia plavleniia solevykh evtektik / Cherneeva, L I et al – Moskva: in-t vysokikh temperatur an sssr, 1980 – us CRL [947]
Entalpiia, teploemkost, teplota i entropiia plavleniia nekotorykh tugoplavkikh metallov / Chekhovskoi, V I A – Moskva: in-t vysokikh temperatur an sssr, 1979 – us CRL [947]
Entartung / Nordau, Max Simon – Berlin: C. Duncker, 1892-93. 2v – 1 – us Wisconsin U Libr [430]
Entdeckte geheime correspondenz des exkaisers napoleon buonaparte von der insel st helena mit seinen freunden und anhaengern in europa und in deutschland insbesondere : enthaltend seine freymuethigen ansichten und meinungen ueber die politischen ereignisse, welche in europa seit seiner gefangennehmung statt gefunden haben und seine blicke in die zukunft – Heimburg u. Constanz [i.e. Sondersha 1817] [mf ed Hildesheim 1995-98] – 1v on 2mf – 9 – €60.00 – 3-487-26350-8 – gw Olms [940]
Die entdeckung der nilquellen : reisetagebuch / Speke, John H – Leipzig 1864 [mf ed Hildesheim 1995-98] – 2v on 6mf – 9 – €120.00 – 3-487-27361-6 – (trans fr english) – gw Olms [916]
Die entdeckung des rheingolds aus seinen wahren dekorationen / Wirth, Moritz – Leipzig: C Wild 1896 – 3mf – 9 – mf#wa-115 – ne IDC [780]
Entdeckungen ueber die theorie des klanges / Chladni, Ernst Florens Friedrich – Leipzig: Bey Weidmanns erben und Reich 1787 [mf ed 19–] – 2mf – 9 – mf#fiche 391 – us Sibley [780]
Entdeckungs-reise in die sued-see und nach der berings-strasse zur erforschung einer nordoestlichen durchfahrt : unternommen in den jahren 1815, 1816, 1817 und 1818... / Kotzebue, Otto von – Weimar 1821 [mf ed Hildesheim 1995-98] – 3v on 8mf – 9 – €160.00 – 3-487-26631-8 – gw Olms; ne IDC [590]
Entdeckungsreise nach australien : unternommen auf befehl sr maj des kaisers von frankreich und koenigs von italien mit den korvetten der geograph und der naturalist, und der goelette kasuarina in den jahren 1800 bis 1804 / Peron, Francois A – Weimar 1808 [mf ed Hildesheim 1995-98] – 1v on 3mf [ill] – 9 – €90.00 – 3-487-26550-8 – (other ed: weimar 1819 [4mf] isbn: 3-487-26513-3 €40) – gw Olms [919]
Entdeckungsreise nach australien : unternommen mit den corvetten, der geograph und der naturalist, und der goelette casuarina in den jahren 1800 bis 1804 / Peron, Francois – Weimar 1819 [mf ed Hildesheim 1995-98] – 1v on 4mf – 9 – €120.00 – ISBN-10: 3-487-26513-3 – ISBN-13: 978-3-487-26513-1 – gw Olms [919]
Entdeckungsreise nach der westkueste von korea und der grossen lutschu-insel / Hall, Basil – Weimar 1819 [mf ed Hildesheim 1995-98] – 1v on 2mf – 9 – €60.00 – 3-487-26509-5 – gw Olms [915]
Die entdeckungsreisen in nord- und mittel-afrika von richardson, overweg, barth und vogel / ed by Arenz, Karl – Leipzig: C B Lorck 1857 [mf ed Hildesheim 1995-98] – 1v on 2mf – 9 – €60.00 – 3-487-27386-1 – gw Olms; us CRL [916]
Enten-eller : et livs-fragment / Kierkegaard, Soeren; ed by Eremita, Victor – Kobenhavn: C A Reitzel, 1843 – (= ser Himmelstrup) – 2mf – 9 – 0-7905-3789-3 – mf#1989-0282 – us ATLA [190]
L'entente *see* Entente franco-musulmane
Entente – London, UK. 14 Jul-9 Oct 1920 – 1 – uk British Libr Newspaper [072]
Entente franco-musulmane : organe hebdomadaire d'union et de defense des interets des musulmans algeriens – Constantine, aout 1935-janv 1942 – 1 – (puis l'entente) – fr ACRPP [960]
Entering wedge – Durand WI. 1893 jul 6/1895 feb 14-1917 jul 5-1918 jul 18 [gaps] – 16r – 1 – (continued by: entering wedge and the pepin county courier) – mf#964795 – us WHS [071]
Entering wedge and the pepin county courier – Durand WI. 1918 may 2-jul 4 – 1r – 1 – (cont: entering wedge; cont by: courier-wedge) – mf#964797 – us WHS [071]
Enterprise – Barneveld, Ridgeway WI. 1889 jun 7-dec 6 – 1r – 1 – mf#966269 – us WHS [071]
Enterprise – BC, Canada. aug 1903-aug 1904 – 1 – cn Commonwealth Imaging [071]
Enterprise – Belle Vernon, PA. 1976-1981 (1) – mf#68102 – us UMI ProQuest [071]
Enterprise / Belmont Co. Barnesville – 1895-1910, 1923-32 [wkly] – 12r – 1 – mf#B922-933 – us Ohio Hist [071]
Enterprise / Belmont Co. Barnesville – 1933-75 [wkly] – 27r – 1 – mf#B6125-6151 – us Ohio Hist [071]
Enterprise / Belmont Co. Barnesville – jan 1976-dec 1984 [wkly] – 9r – 1 – mf#B16045-16053 – us Ohio Hist [071]
Enterprise / Belmont Co. Barnesville – jan 1985-dec 1988 [wkly] – 4r – 1 – mf#B34833-34836 – us Ohio Hist [071]
Enterprise – Benwood, WV. 1913-1919 (1) – mf#67208 – us UMI ProQuest [071]
Enterprise – Berwick, PA. 1954-1983 (1) – mf#65842 – us UMI ProQuest [071]
Enterprise – Boron, CA. 1954-1969 (1) – mf#62104 – us UMI ProQuest [071]
Enterprise – Boscobel WI. 1894 jan 3-1895 sep 28 – 1r – 1 – mf#963579 – us WHS [071]
Enterprise – Burgettstown, PA. 1982-1995 (1) – mf#65850 – us UMI ProQuest [071]
Enterprise – Cairo, WV. 1911-1913 (1) – mf#67227 – us UMI ProQuest [071]
Enterprise – Ceredo, WV. 1881-1885 (1) – mf#67231 – us UMI ProQuest [071]
Enterprise / Champaign Co. Saint Paris – sep 1878-jan 1879 [wkly] – 1r – 1 – mf#B11640 – us Ohio Hist [071]
Enterprise – Big Springs, NE: Alfred R Evans. v1 n32. sep 25 1952-v10 n25. aug 3 1962 (wkly) [mf ed with gaps] – 10v on 4r – 1 – (cont: big springs enterprise. absorbed by: keith county news (ogalalla ne 1897)) – us NE Hist [071]
Enterprise – Anselmo, NE: C O Anderson, 1906-v34 n44. jan 25 1940 (wkly) – 10r – 1 – (cont by: anselmo enterprise) – us Bell [071]
Enterprise – Pawnee City, NE: W F Wright, oct 17 1883// (wkly) [mf ed 1878-83 (gaps) filmed [1968]] – 2r – 1 – (cont by: pawnee press) – us NE Hist [071]
Enterprise – Stapleton, NE: H E Roush. 17v. v1 n1. jul 25 1912-v17 n42. apr 18 1929 (wkly) [mf ed with gaps filmed 1969?]] – 11r – 1 – (cont by: stapleton enterprise) – us NE Hist [071]
Enterprise – 1893 oct 26-1894 mar 15, 1894 mar 22-1895 oct 3, 1895 oct 10-1897 apr 22, 1897 apr 29-1898 oct 27, 1898 nov 3-1900 may 10, 1900 may 17-sep 6 – 6r – 1 – (cont: delavan enterprise (delavan, wi: 1878); cont by: delavan enterprise (delavan, wi: 1900)) – mf#1139460 – us WHS [071]

Enterprise – Evansville WI. 1910 nov 23-1911 jan 18 – 1r – 1 – (cont: enterprise and the tribune) – mf#961949 – us WHS [071]

Enterprise – Luck WI. 1975 nov 13-1976 jun 24, 1976 jul 1-1977 apr 28 – 2r – 1 – (cont: enterprise-herald (luck, wi); cont by: luck enterprise (luck, wi: 1976)) – mf#1125745 – us WHS [071]

Enterprise – Kennard, NE: L F Hilton. v16 n52. jan 24 1913)- (wkly) [mf ed with gaps] – 1 – (cont: kennard enterprise. absorbed: fort calhoun chronicle. publ in blair ne, sep 19 1913- . some irregularities in numbering) – us NE Hist [071]

Enterprise – Colby, Neillsville WI. 1877 mar 24-1878 may 4 – 1r – 1 – (cont: langlade enterprise) – mf#956594 – us WHS [071]

Enterprise – Sparta WI. 1885 jan 10-17 – 1r – 1 – (cont: monroe county democrat (sparta, wi: 1879); cont by: democrat enterprise) – mf#1044260 – us WHS [071]

Enterprise – North Platte, NE: W LaMunyon and J H Peake, 1873-v7 n48. dec 31 1874 (wkly) [mf ed with gaps] – 1r – 1 – (cont: north platte enterprise. cont by: north platte enterprise (1875). issues for nov 8 1873-dec 31 1874 called v6 n41-v7 n48) – us NE Hist [071]

Enterprise – Oconomowoc WI. 1900 oct 26-1901 mar 1 – 1r – 1 – (cont: oconomowoc republican; cont by: oconomowoc enterprise) – mf#1009821 – us WHS [071]

Enterprise – Evansville WI. 1881 oct 26-1883 may 29, 1882 feb 28-1888, 1889-1906, 1907-1908 may 15 – 6r – 1 – (cont: pudding-stick [evansville, wi]; cont by: enterprise and the tribune) – mf#961948 – us WHS [071]

Enterprise – Virginia, NE: W S Taylor, 1896 (wkly) [mf ed with gaps] – 1r – 1 – (cont: virginia enterprise (virginia ne)) – us NE Hist [071]

Enterprise – Luck WI. 1904 oct 14/1905 dec 28-1923 jun 7/1926 dec 30, 1924, 1925 [gaps] – 15r – 1 – (continued by: luck enterprise (luck, wi: 1926)) – mf#1125742 – us WHS [071]

Enterprise – Wittenberg WI. 1893 nov 16-1897 dec 31, 1898-1904, 1905 jan-13-1906 may 4 – 3r – 1 – (continued by: wittenberg enterprise) – mf#952005 – us WHS [071]

Enterprise / Cuyahoga Co. Berea – v1 n1. (5/1898-1905,1918-22,1925-7/1955) (damaged) [wkly] – 19r – 1 – mf#B34740-34758 – us Ohio Hist [071]

Enterprise – Dubuque, IA. 1901-1905 (1) – mf#68016 – us UMI ProQuest [071]

Enterprise – East Greenwich, RI. 1888-1889 (1) – mf#66190 – us UMI ProQuest [071]

Enterprise : for entrepreneurs – Sutton. 2002+ (1,5,9) – mf#32596 – us UMI ProQuest [338]

Enterprise – Harlem, MT. 1899-1926 (1) – mf#64435 – us UMI ProQuest [071]

Enterprise – High Point, NC. 1886-1888 (1) – mf#69224 – us UMI ProQuest [071]

Enterprise – High Point, NC. 1915-2000 (1) – mf#61674 – us UMI ProQuest [071]

Enterprise – Homestead, FL. 1914-1931 (1) – mf#61159 – us UMI ProQuest [071]

Enterprise – Hudson, OH. 1877-1881 (1) – mf#65529 – us UMI ProQuest [071]

Enterprise – Lakeview, MI. 1989-2000 (1) – mf#68555 – us UMI ProQuest [071]

Enterprise – Livingston, MT. 1999-1999 (1) – mf#61579 – us UMI ProQuest [071]

Enterprise / Lorain Co. Wellington – 1/6-9/21, 1876; 8/28-9/4, 1901 (all damaged) [wkly] – 1r – 1 – mf#B31711 – us Ohio Hist [071]

Enterprise / Lorain Co. Wellington – jan 1985-dec 1986 [wkly] – 2r – 1 – mf#B11249-11250 – us Ohio Hist [071]

Enterprise / Lorain Co. Wellington – jan 1988-dec 1990 [wkly] – 3r – 1 – mf#B31225-31227 – us Ohio Hist [071]

Enterprise / Lorain Co. Wellington – jan 1991-dec 1994 [wkly] – 3r – 1 – mf#B34735-34737 – us Ohio Hist [071]

Enterprise / Lorain Co. Wellington – jan-dec 1914 [wkly] – 1r – 1 – mf#B33024 – us Ohio Hist [071]

Enterprise / Lorain Co. Wellington – jan-dec 1984 [wkly] – 1r – 1 – mf#B25608 – us Ohio Hist [071]

Enterprise / Lorain Co. Wellington – jan-dec, 1987 [wkly] – 1r – 1 – mf#B29154 – us Ohio Hist [071]

Enterprise / Lorain Co. Wellington – v1 n1. (9/1867-5/86, 89-1983) [wkly, semiwkly, wkly] – 65r – 1 – mf#B13580-13644 – us Ohio Hist [071]

Enterprise – McComb, MS. 1931-1935 (1) – mf#64053 – us UMI ProQuest [071]

Enterprise – McComb, MS. 1935-1945 (1) – mf#64054 – us UMI ProQuest [071]

Enterprise – Mebane, NC. 1991-2000 (1) – mf#68810 – us UMI ProQuest [071]

Enterprise – Medical Lake, WA. 1922-1943 (1) – mf#69245 – us UMI ProQuest [071]

Enterprise / Medina Co. Wadsworth – may 1866-apr 1877 [wkly] – 3r – 1 – mf#B5995-5997 – us Ohio Hist [071]

Enterprise – Mullins, SC. 1914-1989 (1) – mf#66508 – us UMI ProQuest [071]

Enterprise / Muskingum Co. New Concord – aug 1880-jul 1882 [wkly] – 1r – 1 – mf#B4117 – us Ohio Hist [071]

Enterprise / National Association of Manufacturers (US) – 1977 feb-1979 dec, 1980-1982 mar, 1982 apr-1984 apr, 1984 may-1987 spr – 1r – 1 – (cont: nam reports) – mf#492859 – us WHS [338]

Enterprise – New Concord, OH. 1880-1966 (1) – mf#65601 – us UMI ProQuest [071]

Enterprise – Newport, RI. 1886-1897 (1) – mf#66220 – us UMI ProQuest [071]

Enterprise – Noblesville, IN. 1905-1909 (1) – mf#62928 – us UMI ProQuest [071]

Enterprise – Oshkosh WI. 1897 feb 1-1898 jun 15, 1897 sep 15, 1898 jun 16-1899 jan 31, 1899 apr 27 – 3r – 1 – mf#959538 – us WHS [071]

Enterprise – Paris. 1969-1974 (1) 1972-1974 (5) (9) – ISSN: 0013-9068 – mf#5170 – us UMI ProQuest [338]

Enterprise – Patoka, IL. 1885-1887 (1) – mf#62677 – us UMI ProQuest [071]

Enterprise / Richland Co. Butler – (1893-1907) scattered [wkly] – 3r – 1 – mf#B2925-2927 – us Ohio Hist [071]

Enterprise / Richland Co. Butler – (feb-nov 1905) [wkly] – 1r – 1 – mf#B10347 – us Ohio Hist [071]

Enterprise – Riverside, CA. 1891-1893 (1) – mf#62254 – us UMI ProQuest [071]

Enterprise / Sandusky Co. Clyde – (feb 1906-oct 1908) damaged material [wkly] – 1r – 1 – mf#B29745 – us Ohio Hist [071]

Enterprise – Sheridan, MT. 1904-1909 (1) – mf#64647 – us UMI ProQuest [071]

Enterprise – Shortsville, NY. 1883-1976 (1) – mf#68983 – us UMI ProQuest [071]

Enterprise – Sidney, NY. 1923-1927 (1) – mf#68988 – us UMI ProQuest [071]

Enterprise – South Hill, VA. 1989-2000 (1) – mf#61144 – us UMI ProQuest [071]

Enterprise – Spartanburg, SC. 1871-1880 (1) – mf#66520 – us UMI ProQuest [071]

Enterprise / Star Co. Wilmot – aug 1883-dec 1886 [wkly] – 1r – 1 – mf#B33870 – us Ohio Hist [071]

Enterprise / Trumbull Co. Hubbard – feb 1901-feb 1902 [wkly] – 1r – 1 – mf#B3438 – us Ohio Hist [071]

Enterprise / Vinton Co. Hamden – v1 n1. (1/1880-12/1883,1-12/1886) [wkly] – 2r – 1 – mf#B32134-32135 – us Ohio Hist [071]

Enterprise – Westerly, RI. 1867-1868 (1) – mf#66429 – us UMI ProQuest [071]

Enterprise – White Salmon, WA. 1916-1980 (1) – mf#67183 – us UMI ProQuest [071]

Enterprise – White Sulphur Springs, MT. 1907-1907 (1) – mf#69185 – us UMI ProQuest [071]

Enterprise – Williamson, WV. 1911-1913 (1) – mf#67524 – us UMI ProQuest [071]

Enterprise – Williamston, NC. 1986-2000 (1) – mf#68177 – us UMI ProQuest [071]

Enterprise – Willimantic, CT. 1877-1879 (1) – mf#62375 – us UMI ProQuest [071]

Enterprise – Winneconne WI. 1884, 1885 aug 28, oct 16 – 2r – 1 – us WHS [071]

The enterprise – Bennet, NE: A G Hammond (wkly) [mf ed v1 n43. oct 22 1908-oct 21 1910 (gaps)] – 1r – 1 – us NE Hist [071]

The enterprise – Omaha, NE: G F Franklin, jan 1893- (wkly) [mf ed aug 1895-feb 1911 (gaps) filmed in 1977] – 1r – 1 – us NE Hist [071]

The enterprise – Omaha, NE: G F Franklin. v3 n32. aug 10 1895 [mf ed 1947] – (= ser Negro Newspapers on Microfilm) – 1r – 1 – us L of C Photodup [071]

The enterprise – Swellendam SA, 1883-86 (wkly) [mf ed Cape Town: SA library 1986] – 1r – 1 – mf#MS00442 – sa National [079]

Enterprise and harpersville budget – Afton, NY. 1881-1969 (1) – mf#68515 – us UMI ProQuest [071]

Enterprise and journal – Beaumont, TX. 1880-1935 (1) – mf#66579 – us UMI ProQuest [071]

Enterprise and the tribune – Evansville WI. 1908 may 20-1910 nov 10 – 1r – 1 – (cont: enterprise (evansville, wi: 1881); cont by: enterprise (evansville, wi: 1910)) – mf#961948 – us WHS [071]

Enterprise chieftain – Enterprise OR: G P Cheney, 1938-43 [wkly] – 1 – (cont: enterprise record chieftain (1911-38). absorbed: joseph herald (1902-42); wallowa sun (1906-42); cont by: wallowa county chieftain (1943-)) – us Oregon Lib [071]

Enterprise chieftain – Enterprise, Wallowa County, OR: G P Cheney. 55th yr n22-59th yr n47. oct 6 1938-mar 25 1943 – 1 – (cont: enterprise record chieftain, joseph herald, and wallowa sun; cont by: wallowa county chieftain) – us Oregon Hist [071]

Enterprise chieftain *see* – Wallowa sun

Enterprise courier – Oregon City OR: E P Kaen & Walter W R May 1950-61 [daily ex mon & sat] – 1 – (merger of oregon city enterprise (oregon city, or: daily) and: banner-courier; cont by: new enterprise-courier) – us Oregon Lib [071]

Enterprise (edmonton edition) – Edmonds, WA. 1986+ (1) – mf#68343 – us UMI ProQuest [071]

Enterprise in tropical australia / Earl, George Windsor – London, 1846 – (= ser 19th c books on british colonization) – 2mf – 9 – mf#1.1.6338 – uk Chadwyck [980]

Enterprise journal – Beaumont, TX. 1936-2000 (1) – mf#61861 – us UMI ProQuest [071]

Enterprise journal – McComb, MS. 1945-1999 (1) – mf#61549 – us UMI ProQuest [071]

Enterprise messenger – Merna, NE: Orien B Winter. 42nd yr n46. mar 27 1947- (wkly) – (= ser Merna messenger; Anselmo enterprise) – 3r – 1 – (formed by the union of: merna messenger and: anselmo enterprise) – us Bell [071]

Enterprise / news / Wyandot Co. Nevada – sep 1906-dec 1932 [wkly] – 8r – 1 – (title changes) – mf#B4109-4116 – us Ohio Hist [071]

Enterprise news – Cambridge Springs, PA. 1900-1979 (1) – mf#65853 – us UMI ProQuest [071]

Enterprise news – Wittenberg WI. 1980 aug 28-dec 31, 1981, 1982 jan 1-apr 22 – 3r – 1 – (cont: wittenberg enterprise and birnamwood news (wittenberg, wi: 1971); cont by: wittenberg enterprise and birnamwood news (wittenberg, wi: 1982)) – mf#952093 – us WHS [071]

Enterprise newsletter / Fisher Institute – 1982 dec-1984 sep – 1r – 1 – mf#1130995 – us WHS [338]

Enterprise news-record – Enterprise OR: Enterprise Press, 1910-11 [semiwkly] – 1 – (cont: news-record (1907-10). merged with: wallowa county chieftain (1909-11) to form: enterprise record chieftain (1911-38)) – us Oregon Lib [071]

Enterprise (oregon city, or) – Oregon City OR: [s.n.] [wkly] – 1 – (cont: oregon city enterprise (oregon city, or: 1871); cont by: oregon city enterprise (oregon city, or: weekly)) – us Oregon Lib [071]

Enterprise (parkrose, or) – Parkrose OR: Margaret Thompson Hill, 1957-58 [wkly] – 1 – (cont: parkrose-east county enterprise (-1957); cont by: parkrose-east county enterprise (1958-62)) – us Oregon Lib [071]

Enterprise (parkrose, or: 1962) – Parkrose OR: Parkrose Enterprise, 1962-64 [wkly] – 1 – (cont: parkrose mid county enterprise. merged with: greater eastside news to form: greater enterprise news) – us Oregon Lib [071]

Enterprise record chieftain – Enterprise OR: Enterprise Press, 1911-38 [wkly] – 1 – (merger of: enterprise news-record; wallowa county chieftain (1909-11). absorbed: wallowa county reporter; cont by: enterprise chieftain (1938-43)) – us Oregon Lib [071]

Enterprise recorder – Enterprise, FL. 1908 jun 25-1909 aug 26 – 1r – us UF Libraries [071]

Enterprise (redmond, or) – Redmond OR: Douglas Mullarky, 1912- [wkly] – 1 – (ceased in 1914. absorbed by: redmond spokesman) – us Oregon Lib [071]

Enterprise (redmond, or) *see* Redmond spokesman

Enterprise series / Madison Co. London – 2/1872-75, 7/1876-12/1934 [wkly, semiwkly] – 43r – 1 – mf#B7974-8016 – us Ohio Hist [071]

Enterprise-courier *see* Banner-courier

Enterprise-herald – Luck WI. 1941 mar 6/1942 aug 27-1975 feb 2-nov 6 [gaps] – 17r – 1 – (cont: milltown herald (milltown, wi: 1926); luck enterprise (luck, wi: 1926)) – mf#1125744 – us WHS [071]

Enterprise-herald – Wausa, NE: Lynn & Knot (wkly) [mf ed 1895-1902 (gaps) filmed 1979-[1980]] – 3r – 1 – (cont: wausa enterprise) – us NE Hist [071]

Enterprise-leader – Deerfield WI. 1898 dec 29-1900 feb 23 – 1r – 1 – (cont: deerfield enterprise; leader (deerfield, wi)) – mf#936894 – us WHS [071]

Enterprise-messenger – Merna, NE: Orien B. Winter. 42nd yr n46. mar 27 1947- (wkly) [mf ed aug 19 1954-oct 30 1958 (gaps)] – 1r – 1 – (formed by the union of: merna messenger and: anselmo enterprise) – us NE Hist [071]

Enterpriser – Berea, OH. 1898-1955 (1) – mf#65383 – us UMI ProQuest [071]

Enterprise-review / Huron Co. Greenwich – oct 1949-oct 1965, 1971-feb 1982 [wkly] – 13r – 1 – mf#B23158-23170 – us Ohio Hist [071]

Entertainer : containing remarks on men, manners, religion and policy – London. 1717-1718 – 1 – mf#4247 – us UMI ProQuest [073]

Entertainment and sports law journal *see* University of miami entertainment and sports law review

Entertainment and sports lawyer – v1-4. 1994-98 – $59.00 set – (title varies: v1 (1994) as detroit college of law, entertainment and sports law forum. v2-3 as detroit college of law at michigan state university entertainment and sports law journal) – ISSN: 1079-4557 – mf#115911 – us Hein [346]

Entertainment and sports lawyer (aba) – v1-17. 1982-2000 – 9 – $187.00 set – ISSN: 0732-1880 – mf#112101 – us Hein [346]

Entertainment connection magazine – v8 iss 59 [1995 oct] – 1r – 1 – mf#3451782 – us WHS [790]

Entertainment design – New York. 1999+ (1) 1999+ (5) 1999+ (9) – (cont: tci) – ISSN: 1520-5150 – mf#2339,02 – us UMI ProQuest [790]

The entertainment of his most excellent majestie charles 2 in his passage through the city of london to his coronation / Ogilby, J – London: Tho. Roycroft, 1662 – 7mf – 9 – mf#O-381 – ne IDC [090]

Entertainment weekly – New York. 1990+ (1,5,9) – ISSN: 1049-0434 – mf#18585 – us UMI ProQuest [790]

Entfaltung der sozialwissenschaftlichen rationalitaet durch eine transklassische logik / Pusch, Fred – Dortmund: projekt vlg, 1992 (mf ed 1996) – 4mf – 3-8267-9705-1 – mf#DHS 9705 – gw Frankfurter [301]

Der entfesselte saeugling : eine komische geschichte fuer erwachsene / Vesper, Will – Muenchen: A Langen/G Mueller 1935 [mf ed 1991] – 1r – 1 – (filmed with: blumbergshof / siegfried von vegesack) – mf#2944p – us Wisconsin U Libr [830]

Entgiftung elektrophiler xenobiotika in koniferen durch konjugation mit glutathion und metabolismus der glutathion-konjugate / Schroeder, Peter – (mf ed 1998) – 2mf – 9 – €40.00 – 3-8267-2578-6 – mf#DHS 2578 – gw Frankfurter [574]

Die entgleisten / Frank, Leonhard – Berlin: R Hobbing, [1929] (mf ed 1990) – 1r – 1 – (filmed with: trenck) – us Wisconsin U Libr [790]

Die entgleisten / Frank, Leonhard – Berlin: R Hobbing, [1929] (mf ed 1990) – 1r – 1 – (filmed with: trenck) – us Wisconsin U Libr [790]

Die entgleisten / Frank, Leonhard – Berlin: R Hobbing (mf ed 1990) – 1r – 1 – (filmed with: trenck) – us Wisconsin U Libr [790]

Enthoven, Reginald Edward *see* The folklore of bombay

Enthuellungen der geheimnisse der freimaurerei : erklaerung saemmtlicher geheimer zeichen, charaktere...: eine apologie gegen die anfeindungen der ordensgegner / Backoffner, Rudolph [comp] – Berlin: Im Selbstverlag des Verfassers [1881?] – 2mf – 9 – mf#vrl-193 – ne IDC [366]

Enthusiasmus und bussgewalt beim griechischen moenchtum : eine studie zu symeon dem neuen theologen / Holl, Karl – Leipzig: J C Hinrichs, 1898 [mf ed 1990] – 1mf – 9 – 0-7905-5470-4 – (in german & greek. incl bibl ref) – mf#1988-1470 – us ATLA [243]

Enthusiasmus und bussgewalt beim griechischen moenchtum : eine studie zu symeon dem neuen theologen / Holl, Karl – Leipzig, 1898 – 6mf – 8 – €14.00 – ne Slangenburg [241]

The enthusiasts of port-royal / Rea, Lilian – New York: Scribner, 1912 – 1mf – 9 – 0-7905-6079-8 – (incl bibl ref) – mf#1988-2079 – us ATLA [240]

Entidad constructora benfica nuestra senora de la soledad. obispado de badajoz. estatutos sociales y de la comunidad de vecinos – Badajoz: Imprenta Espanola, 1968 – 1 – sp Bibl Santa Ana [946]

Entire absolution of the penitent : sermon 2 / Pusey, E B – Oxford, England. 1846 – 1r – 1 – us UF Libraries [240]

Entire commentary upon the whole epistle of the apostle paul to the ephesians / Baynes, Paul – Edinburgh: James Nichol, 1866 – 1mf – 9 – 0-8370-2219-3 – (Incl ind) – mf#1985-0219 – us ATLA [227]

Entire holiness : an essay / Wallace, John H – Auburn, NY: Wm J Moses, 1853 – 1mf – 9 – 0-524-07171-3 – mf#1991-2960 – us ATLA [240]

An entirely new and original military opera in three acts, entitled : leo, the royal cadet / Cameron, George Frederick – [S.l: s.n, 1889?] [mf ed 1980] – 1mf – 9 – 0-665-05123-9 – mf#05123 – cn Canadiana [790]

Die entjungferung der welt : ein goettlicher roman / Seeliger, Ewald Gerhard – Wien: Gloriette-Verlag c1923 [mf ed 1996] – 1r – 1 – (filmed with: am alltag vorbei / peter scher) – mf#4029p – us Wisconsin U Libr [830]

Entomolgy / Folsom, Justus Watson – Philadelphia, PA. 1906 – 1r – us UF Libraries [590]

Entomologia experimentalis et applicata – Amsterdam. 1991-1995 (1,5,9) – ISSN: 0013-8703 – mf#16781 – us UMI ProQuest [590]

Entomological magazine / ed by Newman, E – London etc 1833-38 – v1-5 on 49mf – 9 – mf#z-937/2 – ne IDC [590]
Entomological news und proceedings / Academy of National Sciences Philadelphia. Entomological Section – Philadelphia 1890-1920 – v1-31 on 241mf – 9 – mf#z-938/2 – ne IDC [590]
Entomological notes / Neal, James Clinton – Lake City, FL. 1890 – 1r – 1 – us UF Libraries [636]
Entomological review – Washington. 1959-1996 (1) 1959-1996 (5) 1959-1996 (9) – ISSN: 0013-8738 – mf#14350 – us UMI ProQuest [590]
Entomological Society of America *see*
– Bulletin of the entomological society of america
– Miscellaneous publications of the entomological society of america
Entomological society of america. annals *see* Annals of the entomological society of america
Entomological Society of British Columbia *see* Journal of the entomological society of british columbia
Entomological society of london. transactions – 1836-83 – 9 – $660.00 – (1884-1975 $3220 [0196]) – mf#0195 – us Brook [590]
Entomologische hefte – Frankfurt am Main 1803 – v1-2 on 5mf – 9 – mf#z-940/2 – ne IDC [590]
Entomologische mitteilungen / Schenkling, S & Schaufuss, C [comp] – Berlin 1912-28 – v1-17 on 109mf – 9 – mf#z-941/2 – ne IDC [590]
Entomologische monatsblatter / ed by Kraatz, G – Berlin 1876-80 – v1-2 on 7mf – 9 – mf#2436/2 – ne IDC [590]
Entomologische nachrichten / ed by Katter, E & Karsch, E – Quedlinburg 1875-1900 – v1-26 on 169mf – 9 – mf#z-942/2 – ne IDC [590]
Entomologischer Verein in Stettin *see* Linnaea entomologica
Entomologist – London 1840/1842-1920 – v1-53 on 359mf – 9 – mf#z-949/2 – ne IDC [590]
L'entomologiste genevois – Peney-Geneve 1889-[1890] – n1-10 on 4mf – 9 – mf#z-950/2 – ne IDC [590]
Entomologist's monthly magazine – London 1864/65-1920 – v1-56 on 343mf – 9 – mf#z-952/2 – ne IDC [590]
Entomologist's record and journal of variation / ed by Tutt, J W – London 1890-1908 – v1-18 on 140mf – 9 – mf#2453/2 – ne IDC [590]
The entomophthoreae of the united states / Thaxter, R – Boston, 1886-1893. v4(6, p133-201) – (= ser Memoirs Boston Society of Natural History) – 2mf – 9 – mf#Z-2249 – ne IDC [590]
Entomostraca seu insecta testacea... / Mueller, Otto Frederik – Lipsiae: Havniae: Sumtibus J G Muelleriani, 1785 – 134p/21pl (ill) – 1 – (incl ind) – us Wisconsin U Libr [590]
L'entr'acte : programme des spectacles – Paris, juil-dec 1832, 1850-53, 1855, 1858, janv-juin 1865, juin-dec 1871, janv-juin 1873 – 1 – fr ACRPP [790]
Entr'acte and limelight – London, 1881-85 – 1 – us L of C Photodup [790]
Entrada por las raices / Arrivi, Francisco – San Juan, Puerto Rico. 1964 – 1r – 1 – us UF Libraries [972]
Entraigo, Elias Jose *see* Cubania de fray candil
Entralgo, Elias Jose *see*
– America latina y su enrique jose varona
– Apologia de las 7 de la manana
– Cartas a luz caballero
– Diputados por cuba en las cortes de espana
– Enrique jose varona
– Genuina labor periodistica de enrique jose varona
– Ideario de varona en la filosofia social
– Insurreccion de los diez anos
– Universidad de berriel
Entrambasaguas, Joaquin de *see* Una familia de ingenios, los ramirez de prado
Entrambasaguas, joaquin de. la biblioteca en ramirez de prado : madrid, 1943 / Hornedo, R M – Madrid: Razon y Fe, 1945 – 1 – sp Bibl Santa Ana [020]
Entrambasaguas Pena, Joaquin *see* Poesias de dona catalina clara ramirez de guzman
The entrance – Essex, Ont: [s.n, 1894-189- or 19–] – 9 – mf#P04426 – cn Canadiana [370]
Entrance into the millennial kingdom / Govett, Robert – Norwich, England. 1883 – 1r – 1 – us UF Libraries [240]
Entre amor y musica / Dominguez Rodlan, Maria Luisa – Habana, Cuba. 1954 – 1r – 1 – us UF Libraries [972]
Entre cubanos / Ortiz, Fernando – Paris, France. 1913 – 1r – 1 – us UF Libraries [972]
Entre dos filos : managua, 1927 / Chamorro, Pedro Joaquin; ed by Bayle, Constantino – Madrid: Razon y Fe, 1929 – 9 – sp Bibl Santa Ana [972]
Entre dos siglos : el uruguay alrededor de 1800 / Falcao Espalter, Mario – Madrid: Razon y Fe, 1927 – 1 – sp Bibl Santa Ana [972]
Entre el deber y el derecho / Hurtado, Antonio – 1873 – 9 – sp Bibl Santa Ana [830]
Entre encajes / Gomez Carrillo, Enrique – Barcelona, Spain. 1905 – 1r – 1 – us UF Libraries [972]
Entre la piedra y la cruz : novela / Monteforte Toledo, Mario – Guatemala, 1948 – 1r – 1 – us UF Libraries [830]
Entre la selva de neon / Velasquez, Rolando – San Salvador, El Salvador. 1956 – 1r – 1 – us UF Libraries [972]
Entre le victoria, l'albert et l'edouard : ethnographie de la partie anglaise du vicariat de l'uganda: origines, histoire, religion, coutumes – Rennes, France: Impr Oberthur, 1920 – 1 – us CRL [305]
Entre mis cuatro paredes / Gonzalez Castell, Rafael – Montijo (Badajoz): Imp. Izquier do Hidalgo, 1934 – 1 – sp Bibl Santa Ana [946]
"Entre nous" / Beausoleil, Joseph Maxime – Montreal: s.n, 1897 – 1mf – 9 – mf#03530 – cn Canadiana [610]
Entre nous : causeries du samedi / Ledieu, Leon – [Québec?: s.n.], 1889 [mf ed 1980] – 3mf – 9 – mf#08671 – cn Canadiana [610]
Entreactos / Ramos, Jose Antonio – Habana, Cuba. 1913 – 1r – us UF Libraries [972]
Entrelineas – Kansas City. 1971-1976 (1) 1974-1976 (5) 1974-1976 (9) – ISSN: 0013-9017 – mf#9169 – us UMI ProQuest [305]
d'Entremont, Laura S *see* Development and validity of the teachers' attitude, comfort and training scale (tacts) on sexuality education
Entrepreneur / Grove City College (PA). Young Americans for Freedom – v1 n1-v3 n3 [1972 dec-1975 oct], v1 n1-v3 n3, [1975 dec-1978 may], v1 n1-6 [1979 nov-1981 may] – 1r – 1 – mf#625715 – us WHS [071]
Entrepreneur – Santa Monica. 1984+ (1,5,9) – ISSN: 0163-3341 – mf#14441,02 – us UMI ProQuest [650]
Entrepreneurship and regional development – London. 1991-1996 (1,5,9) – ISSN: 0898-5626 – mf#17307 – us UMI ProQuest [650]
Entrepreneurship, innovation, and change – New York. 1992-1996 (1) – ISSN: 1059-0137 – mf#19619 – us UMI ProQuest [650]
Entrepreneurship theory and practice: et&p – Waco. 1988+ (1,5,9) – (cont: american journal of small business) – ISSN: 1042-2587 – mf#13012,01 – us UMI ProQuest [650]
Entreprises internationales, transnationales et multinationales : bibliographie selective et annotee / Nadeau, Johan – [Quebec: Bibliotheque de l'Assemblee nationale, Division de la reference parlementaire], 1988 [mf ed 1994] – 1mf – 9 – (with ind) – mf#SEM105P2086 – cn Bibl Nat [338]
Entretenimientos poeticos / Fernandez Garcia, Manuel – 1877 – 9 – sp Bibl Santa Ana [810]
Entretien au peuple : un mal a combattre: la tuberculose / Gauvreau, Joseph – [Montreal] : Institut Bruchesi, 1912 – 1mf – 9 – 0-665-76567-3 – mf#76567 – cn Canadiana [616]
Entretien de jean pichu avec son sergeant au suject du [...] – Paris: [s.n.]. [n1-3 1849] – us CRL [920]
Entretien d'origene : avec heraclide et les eveques ses collegues sur le pere, le fils et l'ame / Origenes (Origen); ed by Scherer, Jean – Le Caire, 1949 – 7mf – 9 – €15.00 – ne Slangenburg [240]
Entretien sur les arts industriels / Desmazures, Adam Charles Gustave – Montreal: s.n, 1870 – 1mf – 9 – mf#03905 – cn Canadiana [740]
Les entretiens d'ariste et d'eugene / [Bouhours, D] – Paris: Seb. Mabre-Cramoisy, 1671 – 6mf – 9 – mf#0-1336 – ne IDC [090]
Les entretiens d'ariste et d'eugene : nouvelle edition... / [Bouhours, D] – Amsterdam: Estienne Roger, 1703 – 4mf – 9 – mf#0-3038 – ne IDC [090]
Les entretiens de l'autre monde sur ce qui se passe dans celui-ci : ou dialogues grotesques et pittoresquesentre feu louis 15, feu le prince de conti... – Londres [i e Hollande] 1784 [mf ed Hildesheim 1995-98] – 1v on 3mf – 9 – €90.00 – 3-487-26204-5 – gw Olms [944]
Entretiens des non-combattants durant la guerre 1914-1918 – Paris. 2e s., n1; 4e s., n 4 5. nov 1915-18 – 1 – fr ACRPP [933]
Entretiens d'ete de pontigny – Versailles. aout sept 1910, 1926-27 (I, IX-X) – 1 – fr ACRPP [073]
Entretiens d'un mois de marie / ed by Speelman, R P – Tournai: J Casterman, 1856 – 1mf – 9 – 0-8370-9182-9 – mf#1986-3182 – us ATLA [240]
Les entretiens idealistes : cahiers mensuels d'art et de philosophie – Paris. oct 1906-juil 1914 – 1 – fr ACRPP [073]
Entretiens politiques et litteraires – Paris. n1-57. mars 1890-93 – 1 – fr ACRPP [073]
Entretiens sur la demonstration catholique de la revelation chretienne / Dechamps, Victor Auguste – 3. ed. Paris: H Casterman 1861 [mf ed 1992] – 2mf – 9 – 0-524-03791-4 – (1st printed 1857) – mf#1990-4863 – us ATLA [241]
Entretiens sur la franc-maconnerie / Lessing, Gotthold Ephraim – [s.l: s.n] 1784 – 1mf – 9 – mf#vrl-70 – ne IDC [366]
Entretiens sur la grammaire / Sauveur, Lambert – New York, NY. 1879 – 1r – 1 – us UF Libraries [440]
Entretiens sur le bon usage de la liberte / Grenier, Jean – Paris, France. 1948 – 1r – 1 – us UF Libraries [960]
Entrez donc! : reponse aux objections qui retiennent hors de la societe de temperance / Hugolin, pere – [Montreal?: s.n, 1908?] [mf ed 1995] – 1mf – 9 – 0-665-74630-X – mf#74630 – cn Canadiana [170]
Entries in the family bible of john glen of glasgow – 1r – 1 – mf#3374 – uk Microform Academic [929]
Die entrueckten : drei erzaehlungen / Bernewitz, Elsa – feldpostausg. Muenchen: A Langen/G Mueller, c1943 [mf ed 1989] – (= ser Die kleine buecherei 63) – 60p – 1 – mf#7010 – us Wisconsin U Libr [880]
Die entscheidung – Berlin DE, 1932 9 oct-1933 19 mar – 1r – 1 – gw Mikrofilm [074]
Die entscheidung – Hamburg DE, 1920-21 – 1 – gw Misc Inst [074]
Entscheidung in modellen : die bedeutung der kognitionspsychologie fuer die entscheidungstheorie unter besonderer betrachtung der geltung der theorie mentaler modelle fuer die bounded rationality / Goessling, Tobias – (mf ed 1996) – 2mf – 9 – €30.00 – 3-8267-2291-4 – mf#DHS 2291 – gw Frankfurter [330]
Entscheidungen der film-pruefstelle – Muenchen DE, 1921 9 nov-1924 11 mar, 1924 12 may-1926, 1929 30 dec-1936, 1937 29 mar-24 dec, 1940-1944 2 dec, 1944 18 dec-1945 13 jan – 8r – 1 – (also as suppl to: deutscher reichsanzeiger und preussischer staatsanzeiger, berlin) – gw Mikrofilm [790]
Der entscheidungskampf – Essen DE, 1924 n1-64 – 1 – gw Misc Inst [074]
Die entscheidungsschlacht : und andere kriegsnovellen / Bleibtreu, Karl et al – Stuttgart: Die Lese Verlag [1914?] [mf ed 1995] – 1r – 1 – (filmed with: gravelotte / carl bleibtreu & other titles) – mf#3790p – us Wisconsin U Libr [830]
Entschiendene schulreform : abhandlungen zur erneuerung der deutschen erziehung / ed by Oestreich, Paul – Berlin: Oldenburg & Co 1922-28 [mf ed 1979] – 51v on 2r – 1 – (v25 in this set is not original series ed, but is 5th ed of the same work; v51 entitled: sammlung entschiendene schulreform) – mf#film mas c620 – us Harvard [370]
Entschladen, Frank *see* Intrazellulaere regulationsmechanismen der t-zell migration
Entsiklopedicheskii slovar' = The encyclopedic dictionary / ed by Andreev, I E et al – St Petersburg: Brockhaus & Efron Publishers. v1-43. 1890-1907 – 767mf – 9 – $4,500.00 – us UMI ProQuest [947]
Entsiklopediia bankovogo dela *see* Rukovodstvo dlia deiatelei i lits, pribegaiushchikh k uslugam bankov
Die entstehung der altkatholischen kirche : eine kirchen- und dogmengeschichtliche monographie / Ritschl, Albrecht – 2., durchgaengig neu ausgearb. Aufl. Bonn: A. Marcus, 1857. Chicago: Dep of Photodup, U of Chicago Lib, 1967 (1r); Evanston: American Theol Lib Assoc, 1984 (1r) – 1 – 0-8370-0450-0 – (incl bibl ref and ind) – mf#1984-B066 – us ATLA [241]
Die entstehung der apokalypse / Voelter, Daniel – 2. voellig neu gearb aufl. Freiburg i.B: J C B Mohr (Paul Siebeck), 1885 – 1mf – 9 – 0-8370-9332-5 – (incl bibl ref) – mf#1986-3332 – us ATLA [221]
Die entstehung der bibel / Zittel, Emil – 5. verb. Aufl. Leipzig: Philipp Reclam, [c1891] – 1mf – 9 – 0-8370-5967-4 – (incl bibl ref) – mf#1985-3967 – us ATLA [220]
Die entstehung der bischoeflichen fuerstenmacht / Hauck, Albert – Leipzig: A. Edelmann, 1891 – 1mf – 9 – 0-7905-6477-7 – (incl bibl ref) – mf#1988-2477 – us ATLA [943]
Die entstehung der eckermannschen gespraeche und ihre glaubwuerdigkeit : mit einem faksimile und einem anhang ungedruckter briefe von und an eckermann / Petersen, Julius – 2. verm verb aufl. Frankfurt am Main: M Diesterweg, 1925 [mf ed 1993] – (= ser Deutsche forschungen 2) – [6]/174p – 1 – (incl bibl ref) – mf#8023 reel 1 – us Wisconsin U Libr [430]
Die entstehung der evangelischen gottesdienstordnungen sueddeutschlands im zeitalter der reformation / Waldenmaier, Hermann – Leipzig: Verein fuer Reformationsgeschichte 1916 [mf ed 1991] – (= ser Schriften des vereins fuer reformationsgeschichte 34/125-126) – 1mf – 9 – 0-524-01593-7 – (incl bibl ref) – mf#1990-0459 – us ATLA [242]
Die entstehung der gotteslehre des aristoteles / Arnim, H von – Wien, 1931 – 2mf – 8 – €5.00 – ne Slangenburg [120]
Die entstehung der jobsiade / Dickerhoff, Hans – Muenster i.Westf.: Aschendorff, 1908 – 1r – 1 – us Wisconsin U Libr [430]
Die entstehung der kindertaufe im dritten jahrhundert n. chr. und die wiedereinfuehrung der biblischen taufe im siebzehnten jahrhundert n. chr : der kirchen- und weltgeschichte gemaess / Rauschenbusch, August – 2., sehr verm. Aufl. Hamburg: J.G. Oncken, 1898 – 1mf – 9 – 0-7905-5917-X – mf#1988-1917 – us ATLA [240]
Die entstehung der konziliaren theorie : zur geschichte des schismas und der kirchenpolitischen schriftsteller konrad von gelnhausen (1390) und heinrich von langenstein (1397) / Kneer, August – Roma: Filippo Cuggiani, 1893 – 1mf – 9 – 0-8370-8191-2 – (incl bibl ref) – mf#1986-2191 – us ATLA [240]
Die entstehung der lutherischen und der reformierten kirchenlehre : samt ihren innerprotestantischen gegensaetzen / Tschackert, Paul – Goettingen: Vandenhoeck and Ruprecht, 1910 – 2mf – 9 – 0-8370-8717-1 – (in german and latin. incl bibl ref and index) – mf#1986-2717 – us ATLA [242]
Die entstehung der modernen unterhaltungsliteratur : studien zum trivialroman des 18. jahrhunderts / ed by Greiner, Martin – Reinbek/Hamburg: Rowohlt, 1964 – 1r – 1 – (incl bibl ref and index) – us Wisconsin U Libr [430]
Die entstehung der neutestamentlichen hirtenbriefe : ein versuch / Hesse, F H – Halle a. S: C A Kaemmerer, 1889 – 1mf – 9 – 0-8370-3578-3 – (incl bibl ref) – mf#1985-1578 – us ATLA [225]
Die entstehung der paulinischen christologie / Brueckner, Martin – Strassburg: JH Ed Heitz (Heitz & Muendel), 1903 – 1mf – 9 – 0-8370-2491-9 – mf#1985-0491 – us ATLA [227]
Entstehung der perikopen der roemischen meszbuecher / Beissel, Stephan – Freiburg Brsg., 1907 – 4mf – 8 – €11.00 – ne Slangenburg [240]
Die entstehung der preussischen landeskirche unter der regierung koenig friedrich wilhelms des dritten : ein beitrag zur geschichte der kirchenbildung im deutschen protestantismus / Foerster, Erich – Tuebingen: JCB Mohr, 1905-1907 – 3mf – 9 – 0-524-02954-7 – (incl bibl ref) – mf#1990-4506 – us ATLA [242]
Die entstehung der reformatio ecclesiarum hassiae von 1526 / Friedrich, Julius – Giessen 1905 [mf ed 1994] – 2mf – 9 – mf#DHS-AR3009 – gw Frankfurter [240]
Die entstehung der schrift, die verschiedenen schriftsysteme und das schrifttum der nicht alfabetarisch schreibenden voelker / Wuttke, Heinrich – Leipzig: In Commission bei T D Weigel, 1877 – 2mf – 9 – 0-8370-9597-2 – mf#1986-3597 – us ATLA [400]
Die entstehung der schriften des neuen testaments : vortraege / Wrede, William – Tuebingen: J C B Mohr 1907 [mf ed 1993] – (= ser Lebensfragen 18) – 1mf – 9 – 0-524-05825-3 – mf#1992-0652 – us ATLA [225]
Die entstehung der speisesakramente = Till fragan om uppkomsten af sakramentala maltider / Reuterskioeld, Edgar – Heidelberg: C Winter, 1912 [mf ed 1992] – (= ser Religionswissenschaftliche bibliothek 4) – 1mf – 9 – 0-524-02039-6 – (german trans fr swedish by hans sperber. incl bibl ref) – mf#1990-2814 – us ATLA [200]
Die entstehung der weisheit salomos : ein beitrag zur geschichte des juedischen hellenismus / Focke, Friedrich – Goettingen: Vandenhoeck & Ruprecht, 1913 – 1mf – 9 – 0-7905-0887-7 – (incl bibl ref and index) – mf#1987-0887 – us ATLA [221]
Die entstehung der welt : eine kritische beleuchtung der angaben des alten testamentes gegenueber der wissenschaft / Jedlicska, Johann – Wien: H Hierhammer & H Geitner, 1903 – 1mf – 9 – 0-524-05809-1 – mf#1992-0636 – us ATLA [220]
Die entstehung des aeltesten schriftsystems : oder der ursprung der keilschriftzeichen / Delitzsch, Friedrich – Leipzig: J C Hinrichs, 1897 – 1mf – 9 – 0-7905-0077-9 – (incl ind) – mf#1987-0077 – us ATLA [470]
Die entstehung des alten testamentes / Staerk, Willy – Leipzig: G J Goeschen 1905 [mf ed 1992] – (= ser Sammlung goeschen) – 1mf – 9 – 0-8370-5358-7 – (incl bibl ref) – mf#1985-3358 – us ATLA [221]
Die entstehung des alten testaments : rede zur rektoratsfeier des jahres 1896 und zur einweihung der neuen basler universitaetsbibliothek am 6. november / Duhm, Bernhard – Freiburg i. B: J C B Mohr, 1897 – (= ser [Sammlung Gemeinverstaendlicher Vortraege Und Schriften Aus Dem Gebiet Der Theologie Und Religionsgeschichte]) – 1mf – 9 – 0-7905-3334-0 – mf#1987-3334 – us ATLA [221]
Die entstehung des christentums : neue beitraege zum christusproblem / Kalthoff, Albert – Leipzig: Eugen Diederichs, 1904 – 1mf – 9 – 0-8370-3839-1 – mf#1985-1839 – us ATLA [240]

Die entstehung des christentums / Pflueger, Paul – Leipzig: Adolf Buerdeke, 1910 – 1mf – 9 – 0-7905-0439-1 – mf#1987-0439 – us ATLA [220]

Die entstehung des christustypus in der abendlaendischen kunst / Hauck, Albert – [Heidelberg?: C Winters?, 1880?] – (= ser Sammlung von Vortraegen fuer das deutsche Volk) – 1mf – 9 – 0-524-03283-1 – mf#1990-0894 – us ATLA [700]

Die entstehung des deuteronomischen gesetzes : kritisch und biblisch-theologisch untersucht / Steuernagel, Carl – Halle (Saale): J Krause, 1896 – 1mf – 9 – 0-8370-9310-4 – mf#1986-3310 – us ATLA [221]

Die entstehung des glaubens an die auferstehung jesu : eine historisch-kritische untersuchung / Voelter, Daniel – Strassburg: J H Ed Heitz, 1910 – 1mf – 9 – 0-7905-0449-9 – mf#1987-0449 – us ATLA [240]

Die entstehung des gottesgedankens und der heilbringer / Breysig, Kurt – Berlin: G Bondi, 1905 – 1mf – 9 – 0-524-02417-0 – (incl bibl ref) – mf#1990-3001 – us ATLA [210]

Die entstehung des johannesevangeliums / Clemen, Carl – Halle a S: Max Niemeyer, 1912 – 2mf – 9 – 0-7905-0685-8 – (incl ind) – mf#1987-0685 – us ATLA [225]

Die entstehung des neuen testaments / Clemen, Carl – Leipzig: C J Goeschen 1906 [mf ed 1989] – (= ser Sammlung goeschen) – 1mf – 9 – 0-7905-1637-3 – (incl ind) – mf#1987-1637 – us ATLA [225]

Die entstehung des neuen testaments / Holtzmann, Heinrich Julius – Tuebingen: J C B Mohr (Paul Siebeck) 1906 [mf ed 1985] – (= ser Religionsgeschichtliche volksbuecher fuer die deutsche christliche gegenwart 1/11) – 1mf – 9 – 0-8370-3632-1 – mf#1985-1632 – us ATLA [225]

Die entstehung des neuen testaments / Krueger, Gustav – Freiburg i B: J C B Mohr, 1896 – 1mf – 9 – 0-524-04406-6 – mf#1992-0099 – us ATLA [225]

Die entstehung des neuen testaments und die wichtigsten folgen der neuen schoepfung / Harnack, Adolf von – Leipzig: J C Hinrichs, 1914 – 1mf – 9 – 0-7905-5406-2 – (incl bibl ref) – mf#1988-1406 – us ATLA [225]

Die entstehung des volkes israel : akademische rede zur feier des jahresfestes der grossherzoglich hessischen ludwigs-universitaet am 1. juli 1897 / Stade, Bernhard – Giessen: V Muenchow, 1897 – 1mf – 9 – 0-7905-3483-5 – mf#1987-3483 – us ATLA [939]

Entstehung und aufbau von gottfried kellers seldwyler novelle "kleider machen leute" / Wuest, Paul – Bonn: F Cohen, 1914 – 1 – (incl bibl ref) – us Wisconsin U Libr [430]

Die entstehung und der charakter unserer evangelien / Blass, Friedrich Wilhelm – Leipzig: A Deichert, 1907 – 1mf – 9 – 0-7905-9240-1 – mf#1989-2465 – us ATLA [220]

Die entstehung und fortbildung des luthertums und die kirchlichen bekenntnisschriften desselben von 1548-76 / Heppe, Heinrich – Cassel: J G Krieger, 1863 – 1mf – 9 – 0-8370-8747-3 – mf#1986-2747 – us ATLA [242]

Entstehung und geschichte des altaegyptischen goetterglaubens / Strauss und Torney, Victor von – Heidelberg: C Winter, 1891 – 1mf – 9 – 0-524-04538-0 – mf#1990-3372 – us ATLA [290]

Entstehung und herkunft der ionischen saeule / Luschan, Felix von – Leipzig: JC Hinrichs, 1912 [mf ed 1989] – (= ser Der alte orient 13/4) – 1mf – 9 – 0-7905-2025-7 – mf#1987-2025 – us ATLA [720]

Entstehungsgeschichte der kirchenslavischen sprache / Jagic, Vatroslav – neue erw ausg. Berlin: Weidmann, 1913 – 2mf – 9 – 0-7905-6754-7 – mf#1988-2754 – us ATLA [400]

Die entstehungsgeschichte des entwurfs eines buergerlichen gesetzbuches fuer das deutsche reich : in verbindung mit einer uebersicht der privatrechtlichen kodifikationsbestrebungen in deutschland / Vierhaus, Felix – Berlin: J Guttentag, 1888 – (= ser Civil law 3 coll; Beitraege zur erlaeuterung und beurtheilung des entwurfes eines buergerlichen gesetzbuches fuer das deutsche reich) – 1mf – 9 – mf#LLMC 96-503 – us LLMC [346]

Die entstehungsgeschichte des trienter rechtfertigungsdekretes : ein beitrag zur dogmengeschichte des reformationszeitalters / Hefner, Joseph – Paderborn: Ferdinand Schoeningh, 1909 – 2mf – 9 – 0-8370-8824-0 – mf#1986-2824 – us ATLA [240]

Die entstehungsgeschichte von hebbels moloch / Saedler, Heinrich – Bonn: H Ludwig, 1914 – 1r – 1 – (incl bibl ref) – us Wisconsin U Libr [430]

Die entstehungszeit von luther's geistlichen liedern / Achelis, Ernst Christian – Marburg: C.L. Pfeil, 1883 – 1mf – 9 – 0-7905-5440-2 – (incl bibl ref) – mf#1988-1440 – us ATLA [242]

Die entvolkerung des platten landes in pommern von 1890-1905 und ihre ursachen / Langerstein, Julius – Greifswald, 1912 – 1 – gw Mikropress [943]

Die entwickelung der evangelischen mission : im letzten jahrzehnt (1878-1888) / Burkhardt, Gustav Emil – Bielefeld: Velhagen & Klasing, 1890 – 1mf – 9 – 0-7905-6102-6 – (incl bibl ref) – mf#1988-2102 – us ATLA [242]

Die entwickelung der katholischen kirche im 19. jahrhundert : vortraege / Sell, Karl – Leipzig: J C B Mohr (Paul Siebeck), 1898 – 1mf – 9 – 0-8370-7908-X – (incl bibl ref) – mf#1986-1908 – us ATLA [241]

Die entwickelung des israelitischen prophetenthums / Maybaum, Siegmund – Berlin: Ferd. Duemmlers, 1883 – 1mf – 9 – 0-8370-4331-X – (incl bibl ref) – mf#1985-2331 – us ATLA [221]

Entwickelung des paulinischen lehrbegriffs / Daehne, August Ferdinand – Halle: C A Schwetschke, 1835 – 1mf – 9 – 0-8370-9609-X – (incl ind) – mf#1986-3609 – us ATLA [225]

Die entwickelung unserer orientpolitik / Rapp, Adolf – Tuebingen: Kloeres, 1916. 28p – 1 – us Wisconsin U Libr [943]

Entwickelungsgedanke und gotteserfahrung / Mueller, Adolf – Halle a S: Max Niemeyer, 1908 [mf ed 1985] – 1mf – 9 – 0-8370-4522-3 – mf#1985-2522 – us ATLA [230]

Entwickelungsgeschichte der absichtssaetze / Weber, Philipp – Wuerzburg: A Stuber, 1884-85 – (= ser Beitraege zur historischen syntax der griechischen sprache) – 1mf – 9 – 0-8370-1657-6 – (incl bibl ref) – mf#1987-6087 – us ATLA [450]

Entwickelungsgeschichte des substantivierten Infinitivs / Birklein, Franz – Wuerzburg: A. Stuber, 1888 – (= ser Beitraege zur historischen syntax der griechischen sprache) – 1mf – 9 – 0-8370-1403-4 – mf#1987-6062 – us ATLA [450]

Entwickelungsgeschichte von der lehre von der person christi *see* History of the development of the doctrine of the person of christ

Die entwicklung der alttestamentlichen gottesidee in vorexilischer zeit : historisch-kritische bedenken gegen moderne auffassungen / Moeller, Wilhelm – Guetersloh: Bertelsmann, 1903 – (= ser Beitraege zur foerderung christlicher theologie) – 1mf – 9 – 0-7905-9353-X – mf#1989-2578 – us ATLA [221]

Die entwicklung der amtshaftung in deutschland seit dem 19. jahrhundert / Gehre, Horst – Bonn, 1958 – 1 – gw Mikropress [943]

Die entwicklung der christlichen religion : innerhalb des neuen testaments / Clemen, Carl – Leipzig: G J Goeschen 1908 [mf ed 1985] – (= ser Sammlung goeschen) – 1mf – 9 – 0-8370-2685-7 – (incl ind) – mf#1985-0685 – us ATLA [225]

Die entwicklung der dermato-venerologie an der fakultaet/dem bereich medizin der karl-marx-universitaet von 1945 bis 1975 / Baumann, Simone – Leipzig, 1989 (mf ed 1994) – 2mf – 9 – €31.00 – 3-8267-2006-7 – mf#DHS-AR 2006 – gw Frankfurter [616]

Der entwicklung der deutschen kultur im spiegel des deutschen lehnworts / Seiler, Friedrich – Halle (Saale): Buchhandlung des Waisenhauses 1921-25 [mf ed 1979] – 8v in 6 on 1r – 1 – mf#film mas 8654 – us Harvard [430]

Die entwicklung der ehe / Achelis, Thomas – Berlin: E Felber 1893 [mf ed Bloomington IN: Indiana Uni Lib, Preservation Dept 1984] – 1r – 1 – us Indiana Preservation [390]

Die entwicklung der frage der buendniszugehoerigkeit eines wiedervereinigten deutschlands von der maueroeffnung bis zum treffen von michail gorbatschow und helmut kohl in schelesnowodsk... / Brenner, Stefan – (mf ed 1992) – 5mf – 9 – €62.50 – 3-89349-614-9 – mf#DHS 614 – gw Frankfurter [327]

Die entwicklung der hoheren schulbildung in indonesien von 1600-1941 / Oei-Tan Soey Nio – Bonn, 1965 – 1 – gw Mikropress [959]

Die entwicklung der lehre von der person christi im 19. jahrhundert / Guenther, Ernst – Tuebingen: JCB Mohr, 1911 – 2mf – 9 – 0-524-06044-4 – (incl bibl ref) – mf#1992-0757 – us ATLA [240]

Die entwicklung der novellistischen kompositionstechnik kleists bis zur meisterschaft : der findling. die verlobung in st domingo. das erdbeben in chili. die marquise von o., unter ausschluss des kohlhaas-fragmentes / Guenther, Kurt Martin – Altenburg: S Geibel, 1911 – 1r – 1 – (incl bibl ref) – us Wisconsin U Libr [430]

Die entwicklung der praemaxillia und maxilla bei feten mit lippen-kiefer-gaumen-spalten / Baric, Iva – (mf ed 1999) – 1mf – 9 – €30.00 – 3-8267-2667-7 – mf#DHS 2667 – gw Frankfurter [617]

Die entwicklung der protestantischen theologie in deutschland seit kant und in grossbritannien seit 1825 / Pfleiderer, Otto – Freiburg i.B.: J C B Mohr, 1891 – 2mf – 9 – 0-8370-8702-3 – (incl bibl ref) – mf#1986-2702 – us ATLA [242]

Entwicklung der protestantischen theologie in deutschland seit kant und in grossbritannien seit 1825 *see* The development of theology in germany since kant and its progress in great britain since 1825

Entwicklung der terminologie in der afrikanische sprache der xhosa (Suedafrika) / Drame, Anja – (mf ed 2001) – 105p – 9 – €40.00 – 3-8267-2767-3 – mf#DHS 2767 – gw Frankfurter [470]

Entwicklung der wahlen und politischen parteien in gross-dortmund / Graf, Hans – Hannover, Germany. 1958 – 1r – 1 – us UF Libraries [325]

Die entwicklung der werkzeuge / Kick, Friedrich – Prag: Verlag des deutschen Vereines zur Verbreitung gemeinnuetziger Kenntnisse, 187- – 1r – 1 – (incl bibl ref) – us Wisconsin U Libr [621]

Die entwicklung des aeltesten japanischen seelenlebens : nach seinen literarischen ausdrucksformen / Leo, Justus – Leipzig: R Voigtlaender, 1907 – (= ser Beitraege zur kultur- und universalgeschichte) – 1mf – 9 – 0-524-01783-2 – (incl bibl ref) – mf#1990-2631 – us ATLA [470]

Die entwicklung des arbeitshauses unter besonderer beruecksichtigung der reformatorischen bestrebungen und der vehaeltnisse in westfalen / Kipper, Karl – Goettingen 1933 (mf ed 1995) – 1mf – 9 – €24.00 – 3-8267-3164-6 – mf#DHS-AR 3164 – gw Frankfurter [943]

Die entwicklung des bildlichen ausdrucks in der prosa klemens brentanos / Poerner, Martin – [S.l: s.n.], 1911 [mf ed 1989] – 76p – 1 – mf#7086 – us Wisconsin U Libr [430]

Die entwicklung des christentums *see* The development of christianity

Die entwicklung des christentums zur universal-religion / Beth, Karl – Leipzig: Quelle & Meyer, 1913 – 1mf – 9 – 0-7905-5629-4 – (incl bibl ref) – mf#1988-1629 – us ATLA [240]

Die entwicklung des lyrischen stils bei detlev von liliencron / Assmann, Elisabeth – Koenigsberg: O Kuemmel, 1936 [mf ed 1991] – xvi/145p – 1 – (incl bibl ref) – mf#7594 – us Wisconsin U Libr [430]

Die entwicklung des religionsbegriffs bei schleiermacher / Huber, Eugen – Leipzig: Dieterich, 1901 – (= ser Studien zur Geschichte der Theologie und der Kirche) – 1mf – 9 – 0-7905-9000-X – mf#1989-2225 – us ATLA [200]

Die entwicklung des sozialismus / Engels, Friedrich – Hottingen-Zuerich, 1883 – 1 – gw Mikropress [335]

Entwicklung, durchfuehrung und evaluation eines kurses 'gegenseitige ganzkoerperuntersuchung von medizinstudierenden' zur schulung der praktischen fertigkeiten im koerperlichen untersuchen / Birkner, Thomas – (mf ed 1996) – 2mf – 9 – €40.00 – 3-8267-2296-5 – mf#DHS 2296 – gw Frankfurter [610]

Entwicklung einer antisense-strategie und wanderungsdynamik c-erbb-2/egfr ueberexprimierender brust-adenokarzinomzellinien in einer 3d-kollagen matrix / Dittmar, Thomas – (mf ed 1999) – 2mf – 9 – €40.00 – 3-8267-2615-4 – mf#DHS 2615 – gw Frankfurter [616]

Entwicklung einer dna-vakzine gegen das glykoprotein b von varizelle-zoster-virus / Krukenkamp, Christoph – (mf ed 1997) – 2mf – 9 – €40.00 – 3-8267-2415-1 – mf#DHS 2415 – gw Frankfurter [574]

Entwicklung eines biosensors fuer glukarat / Wunder, Uwe – (mf ed 1995) – 1mf – 9 – €30.00 – 3-8267-2267-1 – mf#DHS 2267 – gw Frankfurter [574]

Entwicklung eines kombinierten laser-elektroden-katheters zur av-knoten-koagulation bei tachykarden rhythmusstoerungen / Hug, Bernhard – (mf ed 1993) – 2mf – 9 – €49.00 – 3-89349-690-4 – mf#DHS 690 – gw Frankfurter [621]

Entwicklung eines neuen toc-messverfahrens auf basis ueberkritischer nassoxidation mit massenspektrometrischem nachweis / Schiller, Christian – (mf ed 1999) – 2mf – 9 – €40.00 – 3-8267-2652-9 – mf#DHS 2652 – gw Frankfurter [540]

Entwicklung eines nichtviralen, episomalen vektors fuer saeugetierzellen / Piechaczek, Christoph – (mf ed 1999) – 2mf – 9 – €40.00 – 3-8267-2666-9 – mf#DHS 2666 – gw Frankfurter [574]

Die entwicklung friedrich rueckerts bis 1810 und seine dichterischen anfaenge : mit benutzung seines handschriftlichen nachlasses dargestellt / Magon, Leopold – Muenster, 1914 (mf ed 1995) – 1mf – 9 – €24.00 – 3-8267-3116-6 – mf#DHS-AR 3116 – gw Frankfurter [430]

Entwicklung, implementierung und anwendung einer korrelationsmethode fuer frequenzabhaengige polarisierbarkeit : chemie im gardez, bd 1 / Haettig, Christof; ed by Hess, Bernd – Mainz: Gardez 1995 (mf ed 1996) – 2mf – 9 – €31.00 – 3-8267-9665-9 – mf#DHS 9665 – gw Frankfurter [540]

Entwicklung in einer monetaer gesteuerten weltwirtschaft : die verschuldungskrise der dritten welt aus der sicht einer kreditorientierten wirtschaftstheorie / Strecker, Otto A – (mf ed 1993) – 1mf – 9 – €37.50 – 3-89349-638-6 – mf#DHS 638 – gw Frankfurter [330]

"Entwicklung ist das zauberwort" : darwinistische naturverstaendnis im werk julius harts als baustein eines neuen naturalismus-paradigmas / Kaiser, Dagmar – Mainz: Gardez, 1995 (mf ed 1996) – (= ser Germanistik im gardez 3) – 4mf – 9 – €45.00 – 3-8267-9663-2 – mf#DHS 9663 – gw Frankfurter [430]

Entwicklung und erprobung eines computerunterstuetzten curriculums des grundlegenden chemieunterrichtes fuer die schwerpunktthemen "einfuehrung in die chemie" sowie "atombau und chemische bindung" / Koehler, Georg – (mf ed 1993) – 3mf – 9 – €49.00 – 3-89349-726-9 – mf#DHS 726 – gw Frankfurter [540]

Entwicklung und offenbarung / Simon, Theodor – Berlin: Trowitzsch, 1907 [mf ed 1990] – 1mf – 9 – 0-7905-3480-0 – (incl bibl ref) – mf#1987-3480 – us ATLA [210]

Der entwicklungsgedanke in schleiermachers glaubenslehre / Meyer, Albert – Borna-Leipzig: R Noske, 1910 – 1mf – 9 – 0-7905-9516-8 – (incl bibl ref) – mf#1989-1221 – us ATLA [240]

Der entwicklungsgedanke und das christentum / Beth, Karl – Berlin: Edwin Runge, 1909 – 1mf – 9 – 0-8370-2312-2 – (incl ind of names) – mf#1985-0312 – us ATLA [210]

Entwicklungsgeschichte der vorstellungen vom zustande nach dem tode : auf grund vergleichender religionsforschung / Spiess, Edmund – Jena: H Costenoble, 1877 [mf ed 1992] – 2mf – 9 – 0-524-02434-0 – (incl bibl ref) – mf#1990-3018 – us ATLA [230]

Die entwicklungsgeschichte des haeutigen labyrinthorgans bei menschlichen embryonen und feten anhand computergestuetzter dreidimensionaler rekonstruktionen / Lang, Thomas – (mf ed 2000) – 1mf – 9 – €30.00 – 3-8267-2706-1 – mf#DHS 2706 – gw Frankfurter [617]

Entwicklungsgeschichtliche goethe-kritik / Wolff, Eugen – Oldenburg: Schulze (R Schwartz), 1925 – 1r – 1 – (incl bibl ref) – us Wisconsin U Libr [430]

Der entwicklungspolitische runde tisch in der ddr und im vereinigten deutschland : ziele, arbeitsweise und ergebnisse einer aussergewoehnlichen institution / Belle, Manfred – (mf ed 1994) – 2mf – 9 – €40.00 – 3-89349-854-0 – mf#DHS 854 – gw Frankfurter [943]

Entwicklungstendenzen in der landwirtschaftlichen produktion nach der einfuehrung moderner reistechnologie : eine darstellung am beispiel von bangladesh / Hemrich, Guenter – (mf ed 1993) – 2mf – 9 – €40.00 – 3-89349-784-6 – mf#DHS 784 – gw Frankfurter [338]

Entwurf, aufbau und erprobung eines rastertunnel-messkopfes fuer den einsatz in einem rasterelektronenmikroskop / Foerster, Matthias – (mf ed 1995) – 2mf – 9 – €40.00 – 3-8267-2199-3 – mf#DHS 2199 – gw Frankfurter [530]

Entwurf des verfassungsgesetzes fuer die evangelische kirche des herzogthums oldenburg – Oldenburg: F Schmidt, 1849 – 1mf – 9 – 0-524-03898-8 – mf#1990-1157 – us ATLA [242]

Entwurf einer grundbuchordnung fuer das deutsche reich : kommission zur ausarbeitung des entwurfes eines buergerlichen gesetzbuches / Germany. Kommission zur Ausarbeitung des Entwurfes eines Buergerlichen Gesetzbuchs – Berlin: gedruckt in der Reichsdruckerei, 1883 – (= ser Civil law 3 coll) – 5mf – 9 – (incl bibl ref) – mf#LLMC 96-587 – us LLMC [346]

Entwurf einer historischen architektur, in abbildung unterschiedener beruehmten gebaeude, des alterthums und fremder voelcker... / Fischer von Erlach, J B – Wien, 1721 – 3mf – 9 – mf#OA-29 – ne IDC [720]

Entwurf eines ausfuehrungsgesetzes zum buergerlichen gesetzbuche nebst begruendung – Berlin: J Guttentag, 1899 – (= ser Civil law 3 coll) – 4mf – 9 – (The "entwurf" is contained in the 1st, and the "begruendung" in 2nd series of arabic pagination. 2nd pagination group is entitled: begruendung zu dem entwurf eines ausfuehrungsgesetzes zum buergerlichen gesetzbuche) – mf#LLMC 96-514 – us LLMC [346]

Entwurf eines buergerlichen gesetzbuches fuer das deutsche reich / Germany. Kommission zur Ausarbeitung des Entwurfes eines Buergerlichen Gesetzbuchs – Berlin: gedruckt in der Reichsdruckerei. 3v. 1880-82 – (= ser Civil law 3 coll) – 28mf – 9 – (incl bibl ref) – mf#LLMC 96-586 – us LLMC [346]

Entwurf eines buergerlichen gesetzbuches fuer das deutsche reich, erste berathung : erstes buch, allgemeiner theil, zweites buch, recht der schuldverhaeltnisse, drittes buch, sachenrecht / Germany. Kommission zur Ausarbeitung des Entwurfes eines Buergerlichen Gesetzbuchs – Berlin: gedruckt in der Reichsdruckerei, 1885 – (= ser Civil law 3 coll) – 4mf – 9 – (incl bibl ref) – mf#LLMC 96-588 – us LLMC [346]

Entwurf eines buergerlichen gesetzbuches fuer das deutsche reich, erste berathung : viertes buch, familienrecht / Germany. Kommission zur Ausarbeitung des Entwurfes eines Buergerlichen Gesetzbuchs – Berlin: gedruckt in der Reichsdruckerei, 1886 – (= ser Civil law 3 coll) – 2mf – 9 – (incl bibl ref) – mf#LLMC 96-589 – us LLMC [346]

Entwurf eines buergerlichen gesetzbuches fuer das deutsche reich, erste berathung : fuenftes buch, erbrecht / Germany. Kommission zur Ausarbeitung des Entwurfes eines Buergerlichen Gesetzbuchs – Berlin: gedruckt in der Reichsdruckerei, 1887 – (= ser Civil law 3 coll) – 2mf – 9 – (incl bibl ref) – mf#LLMC 96-590 – us LLMC [346]

Entwurf eines buergerlichen gesetzbuches fuer das deutsche reich, erste lesung : ausgearbeitet durch die von dem bundesrathe berufene kommission / Germany. Bundesrat – Berlin: Guttentag, 1888 – (= ser Civil law 3 coll) – 6mf – 9 – (incl bibl ref) – mf#LLMC 96-502 – us LLMC [346]

Entwurf eines buergerlichen gesetzbuches fuer das deutsche reich, erste lesung : ausgearbeitet von der in folge des beschlusses des bundesrathes vom 22. juni 1874 eingesetzten kommission – Berlin: gedruckt in der Reichsdruckerei, 1887 – (= ser Civil law 3 coll) – 7mf – 9 – (incl bibl ref) – mf#LLMC 96-591 – us LLMC [346]

Entwurf eines buergerlichen gesetzbuchs fuer das deutsche reich, zweite lesung : nach den beschluessen der redaktionskommission – Berlin: J Guttentag. 3v in 1. 1895 – (= ser Civil law 3 coll) – 8mf – 9 – (incl bibl ref) – mf#LLMC 96-505 – us LLMC [346]

Entwurf eines buergerlichen gesetzbuchs in der fassung der dem reichstag gemachten vorlage – Berlin: J Guttentag, 1896 – (= ser Civil law 3 coll) – 6mf – 9 – mf#LLMC 96-509 – us LLMC [346]

Der entwurf eines buergerlichen gesetzbuchs und das deutsche recht / Gierke, Otto Friedrich von – Veraend u verm Ausg. Leipzig: Duncker & Humblot, 1889 – (= ser Civil law 3 coll) – 7mf – 9 – (incl bibl ref) – mf#LLMC 96-504 – us LLMC [346]

Entwurf eines buergerlichen gesetzbuchs und eines zugehoerigen einfuehrungsgesetzes : sowie eines gesetzes, betreffend aenderungen des gerichtsverfassungsgesetzes, der civilprozessordnung, der konkursordnung und der einfuehrungsgesetze zur civilprozessordnung und zur konkursordnung – Berlin: J Guttentag, 1898 – (= ser Civil law 3 coll) – 6mf – 9 – mf#LLMC 96-511 – us LLMC [346]

Entwurf eines einfuehrungsgesetzes zum buergerlichen gesetzbuch : in der fassung der dem reichstag gemachten vorlage – Berlin: J Guttentag, 1896 – (= ser Civil law 3 coll) – 1mf – 9 – mf#LLMC 96-510 – us LLMC [346]

Entwurf eines einfuehrungsgesetzes zum buergerlichen gesetzbuche fuer das deutsche reich, erste lesung : nebst motiven / Germany. Kommission zur Ausarbeitung des Entwurfes eines Buergerlichen Gesetzbuchs – Berlin: J Guttentag, 1888 – (= ser Civil law 3 coll) – 4mf – 9 – mf#LLMC 96-501 – us LLMC [346]

Entwurf eines einfuehrungsgesetzes zum buergerlichen gesetzbuches fuer das deutsche reich, erste lesung : ausgearbeitet von der durch beschluss des bundesrathes vom 22. juni 1874 eingesetzten kommission – Berlin: gedruckt in der Reichsdruckerei, 1888 – (= ser Civil law 3 coll) – 1mf – 9 – mf#LLMC 96-592 – us LLMC [346]

Entwurf eines familienrechts fuer das deutsche reich / Germany. Kommission zur Ausarbeitung des Entwurfes eines Buergerlichen Gesetzbuchs – Berlin: Reichsdruckerei. v1-2. 1880 – (= ser Civil law 3 coll) – 38mf – 9 – (incl suppl: anlagen zu den motiven des entwurfs eines familienrechts fuer das deutsche reich. incl bibl ref) – mf#LLMC 96-539 – us LLMC [346]

Entwurf eines gesetzes fuer das deutsche reich : betreffend die zwangsvollstreckung in das unbewegliche vermoegen / Germany. Kommission zur Ausarbeitung des Entwurfes eines Buergerlichen Gesetzbuchs – Berlin: gedruckt in der Reichsdruckerei, 1888 – (= ser Civil law 3 coll) – 4mf – 9 – (incl bibl ref) – mf#LLMC 96-593 – us LLMC [346]

Entwurf eines rechtes der erbfolge fuer das deutsche reich : nebst dem entwurfe eines ausgegebenen einfuehrungsgesetzes / Germany. Kommission zur Ausarbeitung des Entwurfes eines Buergerlichen Gesetzbuchs – Berlin: Reichsdruckerei, 1879 – (= ser Civil law 3 coll) – 19mf – 9 – (incl suppl: begruendung des entwurfes eines rechtes der erbfolge entwurfes eines einfuehrungsgesetzes. incl bibl ref) – mf#LLMC 96-538 – us LLMC [346]

Entwurf und erprobung eines konzepts fuer die ltg in der lehrerbildung an paedagogischen hochschulen / Albrecht, Helmut – (mf ed 1992) – 3mf – 9 – €49.00 – 3-89349-626-2 – mf#DHS 626 – gw Frankfurter [370]

Entwurzelt : szenenfolge aus dem leben einer deutschen kolonie im kaukasus: zeit 1923-1935 / Walling, Hermine – Berlin: Volksbund fuer das Deutschtum im Ausland, 1938 – 1r – 1 – us Wisconsin U Libr [430]

Die entzuendliche aktivitaet und der eisengehalt der leber bei chronischer hepatitis b und c / Beinker, Nele Karen – (mf ed 1996) – 1mf – 9 – €30.00 – 3-8267-2379-1 – mf#DHS 2379 – gw Frankfurter [616]

Enukidze, Avel'Sofronovich *see* Doklad po konstitutsionnym voprosam na 7 s"ezd sovetov soiuza ssr. 5 fevralia 1935 g

Enuma elish *see* The seven tablets of creation

Enumeration des genres de plantes de la flore du canada : precedee des tableaux analytiques des familles et destinee aux eleves qui suivent le cours de botanique descriptive donne a l'universite laval / Brunet, Ovide – Quebec: G & G E Desbarats, 1864 – 1mf – 9 – mf#33333 – cn Canadiana [580]

Enumeration des plantes decouvertes : par les voyageurs dans les iles de la societe principalement dans celle de tahiti / Guillemin, J A – 18?– 1r – 1 – mf#pmb doc29 – at Pacific Mss [919]

Envar uel-'Astkin *see* The divan project

Envar-i vicdan – Trabzon: Ikbal Matbaasi, S Mirgovic Matbaasi, 1910-12. Mueessisi: Zeynelabidin; Sahib-i Imtiyaz: Ali Riza; Mueduer-i Mes'ul: Ali Osman. n88/105. 10 tesrinisani 1911 – (= ser O & t journals) – 1mf – 9 – $25.00 – us MEDOC [956]

Envar-i zeka – Istanbul: Mahmud bey Matbaasi, Matbaa-i Ebuezziya, 1883-85. Sahib-i Imtiyaz ve Muharriri: Mustafa Resid. n13-24. 1299-1300 [1883-84] – (= ser O & t journals) – 3mf – 9 – $75.00 – us MEDOC [956]

L'envers du decor – Montreal: Theatre du Nouveau-Monde. v1 no 1- nov 1968- [mf ed 1973] – 1r – 1 – mf#SEM35P22 – cn Bibl Nat [790]

Enviroaction : environmental digest of the national wildlife federation – Washington. 1991-1996 (1,5,9) – (cont: national wildlife federation's conservation) – mf#16507,01 – us UMI ProQuest [639]

Environment – Washington. 1958+ (1) 1970+ (5) 1970+ (9) – ISSN: 0013-9157 – mf#3043 – us UMI ProQuest [333]

Environment abstracts and envirofiche – 1975- – apply for prices – (microfiche and ind coll available: complete and conferences collection. available on: monthly subsc, quarterly with cdrom ind subsc, print ind and cdrom and magnetic tape formats) – us CIS [020]

Environment and behavior – Thousand Oaks. 1969+ (1) 1975+ (5) 1975+ (9) – ISSN: 0013-9165 – mf#10963 – us UMI ProQuest [301]

Environment and Natural Resources Policy Division of the Library of Congress *see* Toxic substances control act, 1976

Environment international – New York. 1978+ (1,5,9) – ISSN: 0160-4120 – mf#49298 – us UMI ProQuest [333]

Environment monthly – New York. 1972-1976 (1) 1972-1976 (5) 1976-1976 (9) – ISSN: 0013-919X – mf#6767 – us UMI ProQuest [333]

The environment of early christianity / Angus, Samuel – London: Duckworth 1914 [mf ed 1991] – (= ser Studies in theology) – 1mf – 9 – 0-524-00000-X – mf#1989-2700 – us ATLA [240]

Environment report – Washington. 1971-1996 (1) 1970-1996 (5) 1972-1996 (9) – ISSN: 0013-9203 – mf#6030 – us UMI ProQuest [333]

Environment today – Marietta. 1991-1995 (1,5,9) – ISSN: 1054-7517 – mf#18141 – us UMI ProQuest [333]

Environmental action – Tacoma Park. 1979-1996 (1,5,9) – ISSN: 0013-922X – mf#12215 – us UMI ProQuest [333]

Environmental administrative decisions : vols 1-7, march 1972 to march 1997 – Washington: GPO 1995- – 79mf – 9 – $118.00 – (v1-3 mar 1972-feb 1992, contain decisions of the administrator and the judicial officers. v4 & following contain decisions of the environmental appeals board. vols added as they become available) – mf#llmc97-300 – us LLMC [350]

Environmental affairs – Chestnut Hill. 1971-1977 (1) 1975-1977 (5,9) – (cont by: boston college environmental affairs law review) – ISSN: 0046-2225 – mf#10215 – us UMI ProQuest [333]

Environmental affairs *see* Boston college environmental affairs law review

Environmental and experimental botany – Oxford. 1961+ (1,5,9) – ISSN: 0098-8472 – mf#49066 – us UMI ProQuest [580]

Environmental and resource economics – Dordrecht. 1991-1996 (1,5,9) – ISSN: 0924-6460 – mf#18603 – us UMI ProQuest [333]

Environmental biology of fishes – The Hague. 1988+ (1,5,9) – ISSN: 0378-1909 – mf#16782 – us UMI ProQuest [590]

Environmental claims journal – New York. 1988-1990 (1,5,9) – ISSN: 1040-6026 – mf#16927 – us UMI ProQuest [333]

Environmental claims journal – v1-7. 1988-95 – 9 – $281.00 set – ISSN: 1040-6026 – mf#112381 – us Hein [344]

Environmental comment – Washington. 1973-1981 (1) 1975-1981 (5) 1976-1981 (9) – ISSN: 0149-6573 – mf#9122 – us UMI ProQuest [333]

Environmental conservation – Cambridge. 1997+ (1,5,9) – ISSN: 0376-8929 – mf#26172 – us UMI ProQuest [333]

Environmental control and agri-technology – 1976 – 9 – $20.00 – us Univelt [629]

Environmental control and safety management – Morristown. 1949-1971 (1) 1970-1971 (5) – ISSN: 0036-2514 – mf#378 – us UMI ProQuest [333]

Environmental control news for southern industry – Memphis. 1971-1980 (1) 1971-1980 (5) 1976-1980 (9) – ISSN: 0013-9238 – mf#7555 – us UMI ProQuest [333]

Environmental engineering – Bury St. Edmunds. 1988+ (1,5,9) – (Cont: Journal of the Society of Environmental Engineers) – ISSN: 0954-5824 – mf#17137 – us UMI ProQuest [628]

Environmental entomology – Lanham. 1972+ (1) 1972+ (5) 1976+ (9) – ISSN: 0046-225X – mf#6784 – us UMI ProQuest [574]

Environmental ethics – Denton. 1979+ (1,5,9) – ISSN: 0163-4275 – mf#12990 – us UMI ProQuest [333]

Environmental finance – v1-2. 1991-92 (all publ) – 9 – $40.00 set – mf#113351 – us Hein [336]

Environmental geology – New York. 1982-1983 (1) 1982-1983 (5) 1982-1983 (9) – (Cont by: Environmental geology and water sciences) – ISSN: 0099-0094 – mf#13163 – us UMI ProQuest [550]

Environmental geology – Berlin. 1993+ (1,5,9) – (Cont: Environmental geology and water sciences) – ISSN: 0943-0105 – mf#13163,02 – us UMI ProQuest [550]

Environmental geology and water sciences – New York. 1984-1992 (1,5,9) – (Cont: Environmental geology. Cont by: Environmental geology) – ISSN: 0177-5146 – mf#13163,01 – us UMI ProQuest [550]

Environmental health – London. 1967-1997 (1) 1972-1997 (5) 1972-1997 (9) – (Cont by: Environmental health journal: EHJ) – ISSN: 0013-9270 – mf#3456 – us UMI ProQuest [350]

Environmental health journal (ehj) – London. 1998+ (1) – (cont: environmental health) – mf#3456,01 – us UMI ProQuest [350]

Environmental health perspectives (ehp) – Research Triangle Park. 1979+ (1,5,9) – ISSN: 0091-6765 – mf#12113 – us UMI ProQuest [333]

Environmental history – Durham. 1996+ (1,5,9) – ISSN: 1084-5453 – mf#26276 – us UMI ProQuest [333]

Environmental history review – Newark. 1990-1995 (1,5,9) – (Cont: ER Environmental review) – ISSN: 1053-4180 – mf#12757,01 – us UMI ProQuest [333]

Environmental impact assessment review – New York. 1985+ (1) 1985+ (5) 1987+ (9) – ISSN: 0195-9255 – mf#42414 – us UMI ProQuest [333]

Environmental impact statement relative to proposed compact of free association : summary of the environmental impact statement scoping meeting, august 28 1980, between the u.s. and the governments of palau, the marshall islands and the federated states of micronesia / Micronesia. (U.S.) – Washington: office for micronesian status negotiations, 5 jan 1981 – (= ser Micronesia: evolution to separate political entities) – 1mf – 9 – $1.50 – mf#LLMC 82-100F title 9 – us LLMC [324]

Environmental law – Northwestern School of Law of Lewis & Clark. v1-27 (1970-97) – $468.00 set – ISSN: 0046-2276 – mf#102601 – us Hein [344]

Environmental law – Portland. 1970+ (1) 1973+ (5) 1973+ (9) – ISSN: 0046-2276 – mf#9812 – us UMI ProQuest [333]

Environmental law – v1-30. 1970-2000 – 9 – $631.00 – ISSN: 0046-2276 – mf#102601 – us Hein [344]

Environmental law in a developing country, botswana / Fink, Susan E – Uni of South Africa 2000 [mf ed Pretoria: UNISA 2000] – 2mf – 9 – (incl bibl ref) – mf#mfm14719 – sa Unisa [344]

Environmental law journal – v1 (1994) – 9 – $9.00 – mf#116541 – us Hein [344]

Environmental law journal – v1. 1994 – 9 – $15.00 – mf#116541 – us Hein [344]

Environmental law journal *see* New york university environmental law journal

Environmental management – Heidelberg. 1981+ (1,5,9) – ISSN: 0364-152X – mf#13164 – us UMI ProQuest [333]

Environmental manager – New York. 1991-1998 (1,5,9) – ISSN: 1043-786X – mf#18317 – us UMI ProQuest [333]

Environmental modelling and software : with environment data news – Southampton. 1997+ (1) – ISSN: 1364-8152 – mf#42694,01 – us UMI ProQuest [333]

Environmental monitoring and assessment – Dordrecht. 1984-1996 (1,5,9) – ISSN: 0167-6369 – mf#14747 – us UMI ProQuest [333]

Environmental Peace Action Workshop Coalition *see* Family voice

Environmental policy and law – IOS Press. v1-31. 1975-2001 – 9 – $313.00 set – mf#11748 – us Hein [344]

Environmental policy and law – Lausanne. 1995-1996 (1,5,9) – ISSN: 0378-777X – mf#21537 – us UMI ProQuest [333]

Environmental pollution – Barking. 1987+ (1,5,9) – ISSN: 0269-7491 – mf#42437 – us UMI ProQuest [333]

Environmental pollution series a : ecological and biological – London. 1970-1986 (1) 1970-1986 (5) 1972-1977 (9) – ISSN: 0143-1471 – mf#42256 – us UMI ProQuest [574]

Environmental pollution series b : chemical and physical – Essex. 1981-1986 (1,5,9) – ISSN: 0143-148X – mf#42257 – us UMI ProQuest [333]

Environmental practice news *see* William and mary environmental law and policy review

Environmental professional – Elmsford. 1989-1995 (1,5,9) – ISSN: 0191-5398 – mf#16353 – us UMI ProQuest [333]

Environmental progress – New York. 1982+ (1,5,9) – ISSN: 0278-4491 – mf#13369 – us UMI ProQuest [333]

Environmental psychology and nonverbal behavior – New York. 1976-1979 (1,5,9) – (Cont by: Journal of nonverbal behavior) – ISSN: 0361-3496 – mf#11178 – us UMI ProQuest [150]

Environmental quality : the annual report of the Council on Environmental Quality / Council on Environmental Quality – Washington. 1975-1992 (1) 1976-1980 (5) 1976-1980 (9) – ISSN: 0095-2044 – mf#6457 – us UMI ProQuest [333]

Environmental quality management – New York. 1996+ (1,5,9) – ISSN: 1088-1913 – mf#19152,01 – us UMI ProQuest [350]

Environmental quarterly – Little Neck. 1955-1970 (1) – ISSN: 0013-9343 – mf#2204 – us UMI ProQuest [333]

Environmental regulation and permitting – New York. 1996+ (1,5,9) – ISSN: 1083-6624 – mf#24844 – us UMI ProQuest [333]

Environmental Research Foundation *see* Rachel's hazardous waste news

Environmental Research Foundation (Princeton, NJ) *see* Hazardous waste news

Environmental science and technology – v1-1967- – 1,5,6,9 – us ACS [628]

Environmental toxicology – New York. 1999+ (1) – (Cont: Environmental toxicology and water quality) – ISSN: 1520-4081 – mf#18115,02 – us UMI ProQuest [333]

Environmental toxicology and chemistry – New York. 1995+ (1,5,9) – ISSN: 0730-7268 – mf#23087 – us UMI ProQuest [333]

Environmental toxicology and water quality – New York. 1991-1996 (1,5,9) – (Cont: Toxicity assessment. Cont by: Environmental toxicology) – ISSN: 1053-4725 – mf#18115,01 – us UMI ProQuest [333]

Environmetrics – West Sussex. 1991-1991 (1,5,9) – ISSN: 1180-4009 – mf#18537 – us UMI ProQuest [510]

Environs : environmental law and policy journal – v1-24. 1977-2001 – 9 – $175.00 set – mf#117381 – us Hein [344]

Environs of st petersburg : clearwater – s.l, s.l? 193-? – 1r – 1 – us UF Libraries [550]

Environs of st petersburg : new port richey – s.l, s.l? 193-? – 1r – 1 – us UF Libraries [550]

Environs of st petersburg : palm harbor – s.l, s.l? 193-? – 1r – 1 – us UF Libraries [550]

Environs of st petersburg : safety harbor – s.l, s.l? 193-? – 1r – 1 – us UF Libraries [550]

Environs of st petersburg : tampa shores – s.l, s.l? 193-? – 1r – 1 – us UF Libraries [550]

Envol – Leopoldville: Societe litteraire d'Afrique. [n26-49. 1957-1958] – us CRL [079]

Envoy – Pittsburgh. 1964-1992 [1]; 1974-1992 [5,9] – ISSN: 0013-9408 – mf#8109 – us UMI ProQuest [150]

Enzootic bronchopneumonia of dairy calves / Sanders, D A – Gainesville, FL. 1940 – 1r – 1 – us UF Libraries [636]

Der enztaeler – Neuenburg-Ulm DE, 1980-1992 30 sep – 67r – 1 – (bezirksausgabe von suedwest-presse, ulm) – gw Misc Inst [074]

Enzyklopaedische information im 19. jahrhundert : die ergaenzungswerke zum brockhaus konservationslexikon: "zeitgenossen" – "die gegenwart" – "unsere zeit" / ed by Seemann, Otmar – [mf ed 1994-95] – diazo €4290.00 (silver €4900 isbn: 978-3-598-32295-2) – ISBN-10: 3-598-32294-1 – ISBN-13: 978-3-598-32294-5 – (with ind) – gw Saur [030]

Enzyme – Basel. 1966-1974 (1) 1971-1973 (5) – ISSN: 0013-9432 – mf#2055 – us UMI ProQuest [612]

Enzyme and microbial technology – New York. 1979+ (1,5,9) – ISSN: 0141-0229 – mf#13329 – us UMI ProQuest [574]

Enzymes / Waksman, Selman A – Baltimore, MD. 1926 – 1r – us UF Libraries [574]

Enzymes and their applications / Effront, Jean – New York, NY. 1902 – 1r – 1 – us UF Libraries [574]

Enzymologische aspekte der homofermativen milchsaeuregaerung in mutans-streptokokken / Gansser, Georgine – (mf ed 1998) – 2mf – 9 – €40.00 – 3-8267-2565-4 – mf#DHS 2565 – gw Frankfurter [574]

Enzymology – Amsterdam. 1961-1981 (1) 1961-1981 (5) (9) – ISSN: 0005-2744 – mf#42172 – us UMI ProQuest [574]

Eola park / Harold, William G – s.l, s.l? 1936 – 1r – 1 – us UF Libraries [978]

Eom, Han J *see* Computer-aided recording and mathematical analysis of team performance in volleyball

Eon de Beaumont, Charles Genevieve Louis Auguste Andre Timo *see* Lettres, memoires et negociations particulieres du chevalier d'eon

Eos : an epic of the dawn and other poems / Davin, Nicholas Flood – Regina: Leader Co, 1889 – 2mf – 9 – mf#30129 – cn Canadiana [810]

Eos : a prairie dream, and other poems / Davin, Nicholas Flood – Ottawa?: Citizen Print and Pub Co, 1884 – 1mf – 9 – mf#30125 – cn Canadiana [810]

Eos : transactions / American Geophysical Union – 1920-93 [wkly] – 1,5,6,13 – (1920-29 $50. 1930-34 $50. 1935-58 $25y. 1959-78 $25y. 1971-76 $20y. 1979 $35. 1982 $45. 1992 v73 $190. 1993 v74 $205y. 1994 v75 $230. 1995 v76 $260) – us AGU [550]

Eothen : or, traces of travel brought home from the east / Kinglake, Alexander William – Toronto: Adam & Stevenson, 1871 – 3mf – 9 – mf#06720 – cn Canadiana [915]

Eoule, Rowland Edmund Prothero *see* Aggressive irreligion

Ep and p – Chicago. 1986-1988 (1,5,9) – (cont by: electronic publishing and printing) – ISSN: 0887-1876 – mf#15283 – us UMI ProQuest [070]

Ep news / European Parliament. Secretariat. Luxembourg – v1- 1979- – 1 – us Wisconsin U Libr [940]

Epa journal : a magazine on national and global environmental perspectives – Washington. v1-21 n2. jan 1975-oct/dec 1995 – 140mf – 9 – $210.00 – (lacks: v1 n2-3 & 5-7, v2 n1-6 & 8-10; vols added as they become available) – mf#llmc97-301 – us LLMC [339]

Epa journal / United States. Environmental Protection Agency – Washington. 1977-1995 (1,5,9) – ISSN: 0145-1189 – mf#11646 – us UMI ProQuest [333]

Epa pesticide label file / U.S. National Technical Information Service – [qrtly] – 9 – (information on registered pesticide labels) – us NTIS [630]

Epaitres catholiques, apocalypse : traduction et commentaire – Paris: Librairie Bloud, 1905 – (= ser La pensee chretienne) – 1mf – 9 – 0-524-06039-8 – mf#1992-0752 – us ATLA [221]

Epalda Guerrero, Juan *see* Resumen del estado...badajoz

Epalza Guerrero, Juan *see* Resumen...instituto provincial de segunda ensenanza de badajoz... 1874 a 1875...por el secretario accidental don...

Epargne – London, UK. 17 dec 1870-8 oct 1871 – 1 – (aka: nouvelle epargne, 15 oct-14 dec 1871) – uk British Libr Newspaper [072]

Epaves poetiques / veronica : drame en cinq actes / Frechette, Louis – Montreal: Beauchemin, 1908 – 4mf – 9 – 0-665-75560-0 – (incl english text) – mf#75560 – cn Canadiana [820]

Epee de jeanne d'arc : ou, les cino-demoiselles / Marechalle, Alexandre Marie – Paris, France. 1819 – 1r – 1 – us UF Libraries [440]

Ephemera – [NW England] Manchester, Stalybridge Lib 19 oct 1871 – 1 – uk MLA; uk Newsplan [072]

Ephemeriden der litteratur und des theaters / ed by Bertram, Christ August von – Berlin 1785-87 – (= ser Dz) – 6v on 18mf – 9 – €180.00 – mf#k/n4207 – gw Olms [790]

Ephemeriden der menschheit : oder bibliothek der sittenlehre und der politik / ed by Iselin, Isaak – Basel 1776-78, Leipzig 1780-84, 1786 – (= ser Dz) – 9jge[zu je 12st] on 85mf – 9 – (fr 1782ff ed by wilh gottl becker; 1776 st 2ff:...der politik und der gesetzgebung) – mf#k/n2726 – gw Olms [301]

Ephemerides – (= ser French economists of the 18th century) – 9 – $735.00 – us UPA [330]

Ephemerides : und volkslieder / Goethe, Johann Wolfgang von; ed by Martin, Ernst – Heilbronn: Henninger, 1883 [mf ed 1993] – (= ser Deutsche litteraturdenkmale des 18. und 19. jahrhunderts 14) – xx/47p – 1 – (german, french and latin text. int in german) – mf#8676 reel 2 – us Wisconsin U Libr [430]

Ephemerides du citoyen : ou chronique de l'esprit national – Paris. nov 1765-72, dec 1774-juin 1776, janv-mai 1788 – 1 – (puis ou bibliotheque raisonnee des sciences morales et politiques. devenu: nouvelles ephemerides economiques ou bibliotheque raisonnee de l'histoire, de la morale et de la politique.) – fr ACRPP [073]

Ephemerides historiques et politiques du regne de louis 18 depuis la restauration / Desmarais, Cyprien – Paris 1825 [mf ed Hildesheim 1995-98] – 1v on mf [ill] – 9 – €90.00 – 3-487-26341-6 – gw Olms [944]

Ephemerides khmeres : revu par l'oknha hora tipachak [yok] / Um-Pou. – Phnom-Penh: Editions de la Bibliotheque royale 1934 [mf ed 1990] – 1r with other items – 1 – (title & text in khmer; added t.p. in french) – mf#mf-10289 seam reel 120/9 [$] – us CRL [520]

Ephemerides liturgicae – 1(1887)-82(1968) – 1021mf – 9 – €1946.00 – ne Slangenburg [243]

Ephemeris – Athens, Greece. -d. 5-11 May, 22 Oct 1874-12 March 1880; 1 Jan 1886-31 Dec 1891; 1 Jan-30 June 1897. 18 reels – 1 – uk British Libr Newspaper [949]

Ephermerides ordinis cartusiensis / ed by Vasseur, L Le – Monstrolii. v1-5. 1890-93 – €111.00 – ne Slangenburg [241]

Der epheserbrief des apostels paulus / Belser, Johannes Evangelist – Freiburg i B, St Louis: Herder Verlagshandlung, 1908 – 1mf – 9 – 0-8370-2259-2 – (incl ind) – mf#1985-0259 – us ATLA [227]

Ephesian / Ephesus Seventh Day Adventist Church – 1982 apr – 1r – 1 – mf#5266196 – us WHS [242]

The ephesian gospel / Gardner, Percy – London: Williams and Norgate; New York: Putnam, 1915 – (= ser Crown Theological Library) – 1mf – 9 – 0-7905-3196-8 – mf#1987-3196 – us ATLA [226]

Ephesian studies : Expository Readings On The Epistle Of Saint Paul To The Ephesians / Moule, Handley Carr Glyn – New York: A C Armstrong, 1900 – 1mf – 9 – 0-8370-4505-3 – mf#1985-2505 – us ATLA [221]

Ephesiaques / Xenophon Of Ephesus – Paris, France. 1926 – 1r – us UF Libraries [450]

Ephesus and the temple of diana / Falkener, Edward – London: Day, 1862 – 1mf – 9 – 0-524-06461-X – (incl bibl ref) – mf#1992-0889 – us ATLA [930]

Ephesus baptist church. beulah, north carolina : church records – 1835-1919 – 1 – us Southern Baptist [242]

Ephesus baptist church. winchester, kentucky : church records – 1848-1979 – 1 – us Southern Baptist [242]

Ephesus Seventh Day Adventist Church *see* Ephesian

Ephialtes : Eine pathologisch-mythologische Abhandlung ueber die Alptraeume und Alpdaemonen des klassischen Altertums / Roscher, Wilhelm Heinrich – Leipzig: BG Teubner, 1900 – (= ser Abhandlungen der philologisch-historischen classe der koenigl saechsischen gesellschaft der wissenschaften) – 1mf – 9 – 0-524-02321-2 – (incl bibl ref) – mf#1990-2944 – us ATLA [250]

Ephphatha : or, the amelioration of the world: sermons / Farrar, Frederic William – London: Macmillan, 1880 – 1mf – 9 – 0-7905-9197-9 – mf#1989-2422 – us ATLA [240]

Ephraem der Syrer (Ephraem Syrus, Saint) *see*
- Ausgewaehlte reden und lieder / nisibenische hymnen, bd. 1 (bdk37 1.reihe)
- Hymnen gegen die irrlehrer, 2. bd (bdk61 1.reihe)
- Hymni et sermones
- Opera omnia, graece, syriace et latine
- Opera versio armenica
- Rabulae episcopi edesseni, balaei aliorumque

Ephraem Syrus, Saint *see* S ephraemi syri, rabulae episcopi edesseni, balaei aliorumque opera selecta

Ephraemius (cshb41) / ed by Bekkeri, Imm – Bonnae, 1840 – (= ser Corpus scriptorum historiae byzantinae (cshb)) – €17.00 – ne Slangenburg [243]

Ephraim brown papers *see* Brown, ephraim, papers, ms 1872

Ephraim, Charlotte *see* Wandel des griechenbildes im achtzehnten jahrhundert

Ephraim george squier – 15r – 1 – $525.00 – Dist. us Scholarly Res – us L of C Photodup [930]

Ephraim george squier papers, 1835-1872 – 4r – 1 – $450.00 – (with printed guide) – us UMI ProQuest [972]

Ephraim's quotations from the gospel (ts7/2) / ed by Burkitt, F C – 1901 – (= ser Texts and studies (ts)) – 2mf – 9 – €5.00 – ne Slangenburg [226]

Ephrata codex / Ephrata Community – 1746 – 1 – (die bitter gute, oder das gesang der einsamen turtel-taube, der christlichen kirche hier auf erden) – us L of C Photodup [780]

Ephrata Community *see*
- Ephrata codex
- Das lied der liederen, welches ist solomons
- Music book of the ephrata cloister
- Music for the hymns in turtel-taube
- Music for the zionitischer weyrauchs huegel
- Notes written in 1749 by the sisters of the sisterhood in the sisterhouse of the seven day baptists, at ephrata in lancaster county, pennsylvania
- Paradisches wunder-spiel welches sich in diesen letzten zeiten und tagen in denen abendlaendischen welt-theilen als ein vorspiel der neuen welt hervorgethan

Ephrem lougpre mystique franciscain / Barrado Manzano, Arcangel – Madrid: Graf Calleja, 1969 – 1 – sp Bibl Santa Ana [241]

Epia society digest – 1983 apr-1987 jun – 1r – 1 – (continued by: freedom newsdigest) – mf#1495271 – us WHS [071]

The epic – (Los Angeles). 1934 – 1 – us AJPC [073]

The epic fast / Nair, Pyarelal – Ahmedabad: (M.M. Bhatt), 1932. xii,325p. On t-p, New York: Universal Publishing Co., 1934. Includes articles by Gandhi and others – 1 – us Wisconsin U Libr [640]

The epic fast / Pyarelal – Ahmedabad: Mohanlal Maganlal Bhatt, 1932 – (= ser Samp: indian books) – us CRL [954]

Epic india : or, india as described in the mahabharata and the ramayana / Vaidya, Chintaman Vinayak – Bombay: Mrs R A Sagoon, 1907 – 2mf – 9 – 0-524-06233-1 – mf#1991-0026 – us ATLA [954]

Epic mythology / Hopkins, Edward Washburn – Strassburg: KJ Truebner, 1915 – (= ser Grundriss der indo-arischen philologie und altertumskunde) – 1mf – 9 – 0-524-01178-8 – (incl bibl ref) – mf#1990-2254 – us ATLA [280]

The epic of gilgamish / Langdon, Stephen – Philadelphia: University Museum, 1917 – (= ser Publications of the Babylonian Section) – 1mf – 9 – 0-524-07943-9 – mf#1991-0193 – us ATLA [470]

The epic of mount everest / Younghusband, Francis Edward – London: Edward Arnold & Co, 1931 – (= ser Samp: indian books) – us CRL [900]

The epic of the inner life : being the book of job / Genung, John Franklin – Boston: Houghton, Mifflin, 1900, c1891 – 1mf – 9 – 0-7905-0083-3 – mf#1987-0083 – us ATLA [220]

La epica juglaresca alemana del siglo 12 / Albrecht, Hellmuth F G – Tucuman: Universidad Nacional de Tucuman, Facultad de Filosofia y Letras, 1963 [mf ed 1993] – (= ser Cuadernos de humanitas 15) – 109p – 1 – (incl poems in german and spanish. incl bibl ref) – mf#8171 – us Wisconsin U Libr [430]

Epicedia in praematvrvm obitvm...ioannis stuckii iohan gvil stuckii tigurin, theologi f[iiii] / Stucki, J W – Haidelbergae, Ioannes Lancellotus, 1599 – 1mf – 9 – mf#PBU-644 – ne IDC [240]

Epicedion thomae sporeri... / Dietrich, Sixt – 1534 – (= ser Mssa) – 1mf – 9 – €20.00 – mfchl 207 – gw Fischer [780]

Epics, myths and legends of india : a comprehensive survey of the sacred lore of the hindus and buddhists / Thomas, Paul – Bombay: DB Taraporevala Sons, [19–] – (= ser Samp: indian books) – us CRL [390]

Epictetus *see*
- Discourses of epictetus
- The works of epictetus
- Works of epictetus

Epicure / ed by Usener, Hermanus – Lipsiaea, 1887 – €27.00 – ne Slangenburg [100]

Epicureanism / Wallace, William – London: SPCK; New York: E & JB Young, [190-?] [mf ed 1991] – (= ser Chief ancient philosophies) – 1mf – 9 – 0-524-00200-2 – (incl bibl ref) – mf#1989-2900 – us ATLA [180]

Epicurus / Taylor, Alfred Edward – London: Constable, 1911 [mf ed 1990] – (= ser Philosophies ancient and modern (london, england)) – 1mf – 9 – 0-7905-7475-6 – (incl bibl ref) – mf#1989-0700 – us ATLA [180]

Epidemia de tercianas...en varios pueblos de urgel...en 1785 / Balaguer, G – Barcelona, 1785 – 2mf – 9 – sp Cultura [614]

La epidemia de viruela / Huertas y Barrero, Francisco – 1886 – 9 – sp Bibl Santa Ana [946]

Epidemic infantile gastro-enteritis = Epidemiese gastroenteritis (suigelinge) / South Africa. Department of Health [Departement van Gesondheid] – [Pretoria: Dept of Health 1975?] [mf ed Pretoria, RSA: State Library [199-]] – 14p [ill] on 1r with other items – 5 – mf#op 06538 r25 – us CRL [618]

Epidemiologia : sive tractatus de peste ad regni sardiniae... / Angelerius, Q T – Madrid, 1598 – 4mf – 9 – sp Cultura [614]

An epidemiologic investigation of the relationship between religiosity, selected health behaviors, and blood pressure / Hixson, Karen A – University of North Carolina at Greensboro, 1996 – 2mf – 9 – $8.00 – mf#HE 565 – us Kinesology [613]

Epidemiologische untersuchung zu candida antikoerper- und antigen-titern im serum von probanden unterschiedlicher altersstufen / Stoltzenberg, Katharina – (mf ed 1998) – 1mf – 9 – €30.00 – 3-8267-2593-X – mf#DHS 2593 – gw Frankfurter [576]

Epidemiology – v3-5. 1992-1996 – 5r – 1,5,6,9 – $80.00r – us Lippincott [614]

Epidemiology and infection – Cambridge. 1987+ (1,5,9) – (Cont: Journal of hygiene) – ISSN: 0950-2688 – mf#12125,01 – us UMI ProQuest [614]

Epigrafes hebreos de bejar y salamanca / Fita, Fidel – Madrid: Tip Fortanet, 1907 – sp Bibl Santa Ana [946]

Epigrafia en cuba / Dihigo, Juan Miguel – Habana, Cuba. 1928 – 1r – 1 – us UF Libraries [972]

Epigrafia romana / Blazquez, Vidal – Madrid: Fortanet, 1920 – 1 – sp Bibl Santa Ana [946]

Epigrafia romana de extremadura : merida, guarena, torremejia, almendralejo, villafranca de los barros / Monsalud, Marques de – Madrid: Est. Tip. Fortanet, 1897 – sp Bibl Santa Ana [946]

Epigrafia romana de extremadura. marcas de alfareros y grafitos (villafranca de los barros) / Monsalud, Marques de – Madrid: Est. Tip. Fortanet, 1907 – 1 – sp Bibl Santa Ana [946]

Epigrafia romana de medina de las torres y fregenal de la sierra / Monsalud, Marques de – Madrid: Est. Tip. Fortanet, 1898 – 1 – sp Bibl Santa Ana [946]

Epigrafia romana de merida / Fita, Fidel & Rodriguez Villa, Antonio – Madrid: Tip. Fortanet, 1896 – sp Bibl Santa Ana [946]

Epigrafia romana de montanchez, rena, banos de la encina. santisteban del puerto, cartagena y cadiz / Fita, Fidel – Madrid: Tip. de Fortanet, 1901 – sp Bibl Santa Ana [946]

Epigrafia romana de zaragoza y extremadura. zaragoza, tarazona, almendralejo, valle de santa ana, jerez de los caballeros / Monsalud, Marques de – Madrid: Est. Tip. Fortanet, 1898 – 1 – sp Bibl Santa Ana [946]

Epigrafia romana, griega y visigotica de extremadura (merida, solana de los barros e italica) / Monsalud, Marques de – Madrid: Est. Tip. Fortanet, 1907 – 1 – sp Bibl Santa Ana [946]

Epigrafia romana y griega de la provincia de caceres. nuevas ilustraciones : caceres, plasenzuela, valdelacasa / Fita, Fidel – Madrid: Fortanet, 1917 – 1 – sp Bibl Santa Ana [946]

Epigrafia romana y visigotica / Fita, Fidel – Madrid: Tip. Fortanet, 1896 – sp Bibl Santa Ana [946]

Epigrafia romana y visigotica de extremadura (merida y barcarrota) / Monsalud, Marques de – Madrid: Fortanet, 1904 – sp Bibl Santa Ana [946]

Epigrafia romana y visigotica de extremadura y andalucia / Monsalud, Marques de – Madrid: Establec Tipografico de Fortanet, 1908 – 1 – sp Bibl Santa Ana [946]

Epigrafia romana y visigotica de garlitos, capilla, belalcazar y el guijo / Fita, Fidel – Madrid: Fortanet, 1912 – 1 – sp Bibl Santa Ana [946]

Epigrafia romana y visigotica de montemolin / Fita, Fidel & Hinojois, Marques de – Madrid: Tip. Fortanet, 1918 – sp Bibl Santa Ana [946]

Epigrafia romana y visigotica. poza de la sal. merida. alburquerque / Fita, Fidel – Madrid: Fortanet, 1915 – 1 – sp Bibl Santa Ana [946]

Epigrafia visigotica y romana de barcelona, merida, morente y bujalance / Fita, Fidel – Madrid: Fortanet, 1909 – sp Bibl Santa Ana [946]

Epigramas americanos / Diez Canedo, Enrique – Mexico, 1945 – 1 – sp Bibl Santa Ana [972]

Epigrammata *see* Metamorphoses...

Epigrammata, 1614 / Porter, Thomas – (= ser Holkham library manuscript books 436) – 1r – 1 – mf#97105 – uk Microform Academic [090]

Epigramme / Knortz, Karl – Lyck [Ostpreussen]: Emil Wiebe, 1878 – 1r – 1 – us Wisconsin U Libr [840]

Epigramme : nebst einer auswahl aus seinen uebrigen gedichten / Grob, Johann; ed by Lindqvist, Axel – Leipzig: K W Hiersemann, 1929 – (incl bibl ref and ind) – us Wisconsin U Libr [810]

Epigramme : nebst einer auswahl aus seinen uebrigen gedichten / Grob, Johann; ed by Lindqvist, Axel – Leipzig: K W Hiersemann, 1929 – 1 – (incl bibl ref and index) – us Wisconsin U Libr [810]

Epigramme und sprueche / Morgenstern, Christian – Muenchen: R Piper, 1920 – 1r – 1 – us Wisconsin U Libr [880]

Epigrams and excerpts selected and arranged by fred c mullinix / Lamm, Henry – East Aurora, NY. Roycrofters 1922 – 354p – 1 – mf#LL-319 – us L of C Photodup [340]

Epigrams upon the paintings of the most eminent masters, ancient and modern... / Elsum, J – London, 1700 – 2mf – 9 – mf#O-1179 – ne IDC [700]

Epigraphical echoes of kalidasa / Sivaramamurti, C – Madras: Thompson & Co, 1944 – (= ser Samp: indian books) – (foreword by KN Dikshit) – us CRL [730]

Epilegomena zu meiner wissenschaft der logischen idee : als replik gegen die kritik der herren michelet und lassalle / Rosenkranz, Karl – Koenigsberg: Borntraeger, 1862 – 1mf – 9 – 0-7905-9468-4 – mf#1989-2693 – us ATLA [160]

Epilepsia – Copenhagen. 1993-1995 (1,5,9) – ISSN: 0013-9580 – mf#18705 – us UMI ProQuest [616]

Epilepsy research – Amsterdam. 1989+ (1,5,9) – ISSN: 0920-1211 – mf#42509 – us UMI ProQuest [616]

Epilogus chronologicus... *see* The irish dominicans of the seventeenth century

Epimetheus / Murray, John Clark – Montreal: W F Brown; W Drysdale, 1897? – 1mf – 9 – mf#11214 – cn Canadiana [830]

Epinay, Adrien d' *see* Renseignements pour servir a l'histoire de l'ile de france jusqu'a l'annee 1810 inclusivement

Epiney-Burgard, G *see* Gerard grote (1340-1384)

Epinicion, christo cantatum ab ioanne calvino, calendis januarii, anno 1541 / Calvin, J – Genevae: Per Joannem Girardum, 1544 – 1mf – 9 – mf#CL-75 – ne IDC [242]

Epiphanius *see* Ad physiologum

Epiphanius (gcsej3a) / ed by Holl, K – (= ser Griechische christlichen schriftsteller der ersten jahr- hunderte (gcsej)) – (bd1: 1915 €17. bd2: 1922 €19. bd3: 1933 €18) – ne Slangenburg [240]

Epiphanius, Saint, Bishop of Constantia in Cyprus *see* Des heiligen epiphanius von salamis

Epiphanius von Salamis (Epiphanius of Constantia, Saint) *see* Der festgeankerte / anakephalaios / gegen die antikomarianiten (bdk38 1.reihe)

Epiphany – San Francisco. 1988-1995 (1,5,9) – ISSN: 0273-6969 – mf#15668 – us UMI ProQuest [240]

Epirotica *see* Historia politica et patriarchica (cshb47)

Les epis : poesies fugitives et petits poemes / Lemay, Pamphile – Montreal: J-A Guay, 1914 [mf ed 1995] – 3mf – 9 – 0-665-74850-7 – mf#74850 – cn Canadiana [810]

Die epische kunst und kunsttechnik ernst von wildenbruchs / Morisse, Anne-Marie – Bonn: C Georgi, 1912 – 1r – 1 – us Wisconsin U Libr [430]

Der epische stil von hermann und dorothea / Steckner, Hans – Halle (Saale): M Niemeyer 1927 [mf ed 1993] – (= ser Saechsische forschungsinstitute in leipzig. forschungsinstitut fuer neuere philologie. 1. altgermanistische abteilung 4) – 1r – 1 – (incl bibl ref) – mf#8591 – us Wisconsin U Libr [430]

Die epischen werke otto ludwigs und ihr verhaeltnis zu charles dickens / Lueder, Fritz – Leipzig: A Hoffmann, 1910 – 1r – 1 – us Wisconsin U Libr [410]

Episcopacy and unity : a historical inquiry into the relations between the church of england and the non-episcopal churches at home and abroad, from the reformation to the repeal of the occasional conformity act / Wilson, Henry Albert – London; New York: Longmans, Green, 1912 – 1mf – 9 – 0-7905-7266-4 – (incl bibl ref) – mf#1988-3266 – us ATLA [240]

Episcopacy exclusive / Beman, Nathan S S – 1856 – 1 – $50.00 – (letters to rev john hughes 1851) – us Presbyterian [240]

Episcopacy, tradition, and the sacraments / Fitzgerald, William – Dublin, Ireland. 1839 – 1r – 1 – us UF Libraries [240]

Episcopal and presbyterial government conjoyned : proposed as an expedient for the compremising of the differences, and preventing of those troubles about the matter of church-government / Us[s]her, J [Archbishop of Armagh] – London, 1679 – 1mf – 9 – mf#PW-55 – ne IDC [242]

Episcopal Church *see*
- Annual catalogue of kemper hall, kenosha, wisconsin
- Church times
- Dayspring
- Hawaiian church chronicle
- Iowa churchman
- Linkage
- Living church
- Milwaukee churchman
- Northwestern church
- Parish record
- Reporter: a periodical devoted to religion, law, legislation, and public events
- Southern wisconsin episcopal life

The episcopal church : its doctrine, its ministry, its discipline, its worship, and its sacraments / Hodges, George – New York: Thomas Whittaker, c1892 – 1mf – 9 – 0-8370-8678-7 – mf#1986-2678 – us ATLA [241]

The episcopal church : its teaching and worship / Griswold, Latta – New York: Morehouse-Gorham, c1917 – 1mf – 9 – 0-524-06248-X – mf#1990-5203 – us ATLA [240]

Episcopal Church. Commission of Home Missions to Colored People *see* Annual report of the commission of home missions to colored people

Episcopal Church. Commission on a Nation-wide Preaching Mission *see* A nation-wide preaching mission

Episcopal Church Diocese Of Florida *see*
- Diocese of florida annual council
- Journal of the annual convention
- Journal of the annual council
- Journal of the proceedings of the annual council
- Proceedings in organizing the diocese and journal

Episcopal Church. Diocese of Georgia. Council of Colored Churchmen *see*
- Journal of the...annual council of colored churchmen, diocese of georgia
- Journal of the...annual session of the council of colored churchmen in the diocese of georgia

Episcopal Church. Diocese of Milwaukee *see* Young churchman

Episcopal Church. Diocese of Tennessee. Convocation of the Colored Churchmen *see*
- Proceedings of the...annual convocation of the colored churchmen, diocese of tennessee
- Proceedings of the...convocation of the colored churchmen in the diocese of tennessee

Episcopal Church Docese Of Florida *see* Journal of the proceedings of the annual convention

Episcopal Church Foundation *see*
- Newsletter
- Serving the whole church everywhere

Episcopal Church In Scotland General Synod (1838) *see* Code of canons of the episcopal church in scotland

Episcopal Church In Scotland General Synod (1862-1863) *see* Code of canons of the episcopal church in scotland

The episcopal church of scotland : from the reformation to the revolution / Lawson, John Parker – Edinburgh: Gallie and Bayley, 1844 – 1mf – us ATLA [240]

The episcopal church of scotland : from the reformation to the revolution / Lawson, John Parker – Edinburgh: Gallie and Bayley, 1844 – 2mf – 9 – 0-7905-5299-X – mf#1988-1299 – us ATLA [242]

Episcopal Church. Office of the Indian Commission *see* Annual report of the indian commission to the domestic committee of the board of missions

Episcopal church record : st peter's episcopal church – s.l, s.l? 1937? – 1r – 1 – us UF Libraries [240]

The episcopal controversy reviewed / Emory, John; ed by Emory, Robert – New-York: T Mason & G Lane for the Methodist Episcopal Church, 1838 – 1mf – 9 – 0-7905-5087-3 – mf#1988-1087 – us ATLA [240]

Episcopal counsel upon ministerial duties / O'brien, James Thomas – Dublin, Ireland. 1853 – 1r – 1 – us UF Libraries [240]

Episcopal dioceses records – CA. Sacramento: 1849-1933; Orleans: 1912-1933; Benicia... – 6r – 1 – $300.00 – mf#C50013 – us Library Micro [240]

Episcopal elections : ancient and modern: a study in ecclesiastical polity / Dawson, Samuel Edward – Montreal: Dawson, 1877 – 1mf – 9 – mf#24165 – cn Canadiana [240]

Episcopal elections : a letter to the ven archdeacon whitaker, prolocutor of the provincial synod of canada by john travers lewis...bishop of ontario – Ottawa?: s.n, 1877? – 1mf – 9 – mf#24164 – cn Canadiana [240]

The episcopal invitation / Strong, Robert – Philadelphia: Presbyterian Publ Committee, c1866 – 1mf – 9 – 0-8370-8792-9 – mf#1986-2792 – us ATLA [230]

Episcopal magazine – Philadelphia. 1820-1821 (1) – mf#4447 – us UMI ProQuest [240]

Episcopal methodism as it was and is : or, an account of the origin, progress, doctrines...of the methodist episcopal church in the united states / Gorrie, Peter Douglass – Auburn: Derby & Miller 1852 [mf ed 1990] – 1mf – 9 – 0-7905-6469-6 – mf#1988-2469 – us ATLA [242]

Episcopal oath of allegiance to the pope – London, England. 18– – 1r – 1 – us UF Libraries [240]

Episcopal recorder – Philadelphia. 1823-1851 (1) – ISSN: 0013-9610 – mf#4834 – us UMI ProQuest [240]

The episcopal succession in england, scotland and ireland, a.d. 1400 to 1875 : with appointments to monasteries and extracts from consistorial acts taken from mss in public and private libraries in rome, florence, bologna, ravenna and paris / Brady, William Maziere – Rome: Tipografia della Pace, 1876-1877 – 4mf – 9 – 0-7905-5578-6 – mf#1988-1578 – us ATLA [240]

Episcopal teacher – 1986-88 – 1r – 1 – mf#ATLA S0443 – us ATLA [242]

Episcopal watchman – Hartford. 1827-1833 (1) – mf#3699 – us UMI ProQuest [240]

Episcopalian – v139 n1-v142 n3 [1974 jan-1977 mar], v142 n4-v146 n12 [1977 apr-1981 dec], v147 n1-v151 n11 [1982 jan-1986 nov], v151 n12-v155 n3 [1986 dec-1990 mar] – 4r – 1 – (cont: forth; cont by: episcopal life) – mf#154551 – us WHS [242]

An episcopalian demand for christian schools / McMillan, Thomas – [s.l: s.n, c1903] [mf ed 1986] – 1mf – 9 – 0-8370-7565-3 – mf#1986-1565 – us ATLA [377]

Episcopalians / Addison, Daniel Dulany – New York: Baker & Taylor, c1904 – 1mf – 9 – 0-524-02463-4 – mf#1990-4322 – us ATLA [242]

L'episcopat, son origine et son oeuvre : allocution prononcee le 26 sep 1909; la femme chretienne, sa mission sociale: sermon preche le 13 oct 1909 / Emard, Joseph-Medard – Valleyfield [Quebec: s.n, 1909?] [mf ed 1995] – 1mf – 9 – 0-665-74240-1 – mf#74240 – cn Canadiana [305]

The episcopate of charles wordsworth : bishop of st. andrews, dunkeld, and dunblane, 1853-1892 / Wordsworth, John – London; New York: Longmans, Green, 1899 – 1mf – 9 – 0-7905-8216-3 – mf#1988-8099 – us ATLA [240]

Episcopate with two voices / Denison, George Anthony – Oxford, England. 1874 – 1r – 1 – us UF Libraries [240]

Episcopi carpentoracti s r e card epistolarum libri sexdecim / Sadoleti, Iac (Jacopo Sadoleto) – Col. Agrippinae, 1580 – 7mf – 8 – €15.00 – ne Slangenburg [227]

Episcopius, J *see* Paradigmata graphices variorum artificium ex formis nicolai visscher

Episcopius, S *see* Opera theologica

Episcopologio cauriense / Orti Belmonte, Miguel Angel – Caceres: dip. prov. de caceres, servicios culturales, 1958. col. estudios extremenos – 1 – sp Bibl Santa Ana [240]

Un episode de la lutte fratricide : deux mois de bombardement – Paris, 1938? Fiche W859. (Blodgett Collection of Spanish Civil War Pamphlets) – 9 – us Harvard [946]

Un episode de l'epopee du kajoor : la battaille de dekhele / Wade, Magatte – 1980 – us CRL [960]

Un episode de l'histoire de la dime au canada 1705-1707 / Gosselin, Auguste – Ottawa: J Hope, 1903 – 1mf – 9 – 0-665-98277-1 – mf#98277 – cn Canadiana [241]

L'episode de l'ile de sable / Cazes, Paul de – [S.l: s.n.], 1892? – 1mf – 9 – mf#06958 – cn Canadiana [971]

Episode de l'independence d'haiti / Latortue, Paul Emile – Port-Au-Prince, Haiti. 1896 – 1r – 1 – us UF Libraries [972]

The episode of the quarrel between titania and oberon : from shakespeare's a midsummer night's dream: specially arranged for representation with the mendelssohn music by f a dixon – Ottawa: J Durie, 1898 – 1mf – 9 – mf#13534 – cn Canadiana [420]

Episoden aus einer reise nach paris im sommer 1809 / Storck, Adam – Duisburg [u.a.] 1810 [mf ed 1995-98] – (= ser Fbc) – 330p on 2mf [ill] – 9 – €60.00 – 3-487-29679-9 – gw Olms [914]

Episodio de la guerra de la independencia / Requesens, Francisco de – Madrid: Tip. Fortanet, 1889 – 1 – sp Bibl Santa Ana [946]

Episodio de la...independencia / Requesens, Francisco de – 1889 – 9 – sp Bibl Santa Ana [946]

Episodios / Ramirez Moreno, Augusto – Bogota, Colombia. 1930 – 1r – us UF Libraries [972]

Episodios de la guerra de 1899 a 1903 / Arbelaez, Tulio – Bogota, Colombia. 1936 – 1r – 1 – us UF Libraries [972]

Episodios de la revolucion cubana / Cruz, Carlos Manuel De La – Habana, Cuba. 1890 – 1r – 1 – us UF Libraries [972]

Episodios de la revolucion cubana / Cruz, Carlos Manuel De La – Habana, Cuba. 1911 – 1r – 1 – us UF Libraries [972]

Episodios historicos do brasil / Moniz, Heitor – Rio de Janeiro, Brazil. 1942 – 1r – 1 – us UF Libraries [972]

Die epistel des heiligen jakobus / Belser, Johannes Evangelist – Freiburg i B, St Louis, MO: Herder, 1909 – 1mf – 9 – 0-8370-9603-0 – (incl bibl ref and ind) – mf#1986-3603 – us ATLA [227]

Epistel oder sandtbrief huldrych zuinglis von des herren nachtmahl / Zwingli, H – Zuerich, Johannes Hager, 1525 – 1mf – 9 – mf#PBU-514 – ne IDC [242]

Die epistel s paulj an titum / Spangenberg, C – Strassburg, 1564 – 3mf – 9 – mf#TH-1 mf 1408-1410 – ne IDC [242]

Epistelen till de romare : till uppbyggelse i tron och gudaktigheten / Rosenius, Carl Olof – Stockholm: AL Norman, 1867-1868 – 3mf – 9 – 0-524-05267-0 – mf#1991-2259 – us ATLA [227]

Episteln / Penzoldt, Ernst – Berlin: Suhrkamp 1942 [mf ed 1996] – 1r – 1 – (filmed with: lohmer lesebuch / rudolf paulsen) – mf#3981p – us Wisconsin U Libr [860]

Episteme – Romano. 1973-1973 (1) – ISSN: 0013-9637 – mf#8915 – us UMI ProQuest [900]

The epistle general of james = Brief des jakobus / Lange, Johann Peter & Oosterzee, Johannes Jacobus – 5th ed. New York: Charles Scribner, c1867 – (= ser Theologisch-homiletisches Bibelwerk) – 1mf – 9 – 0-8370-6747-2 – (includes bibliographies. in english) – mf#1986-0747 – us ATLA [227]

The epistle general of jude = Der brief judae / Fronmueller, G F C; ed by Mombert, Jacob Isidor – New York: Charles Scribner, c1867 [mf ed 1986] – (= ser A commentary on the holy scriptures. new testament 9/4) – 1mf – 9 – 0-8370-6737-5 – (trans with additions by ed) – mf#1986-0737 – us ATLA [227]

The epistle of james : practically explained = Brief jakobi / Neander, August – New-York: Sheldon, c1852 – 1mf – 9 – 0-7905-0108-2 – mf#1987-0108 – us ATLA [227]

The epistle of james and other discourses / Dale, R W – London: Hodder and Stoughton, 1895 – 1mf – 9 – 0-7905-1510-5 – mf#1987-1510 – us ATLA [227]

The epistle of our lord to the seven churches of asia / Dods, Marcus – Edinburgh: John Maclaren 1867 [mf ed 1985] – 1mf – 9 – 0-8370-2936-8 – mf#1985-0936 – us ATLA [227]

The epistle of paul the apostle to the ephesians – Cambridge: University Press, 1914 – (= ser Cambridge Greek Testament For Schools And Colleges) – 1mf – 9 – 0-7905-2190-3 – (incl ind) – mf#1987-2190 – us ATLA [227]

The epistle of paul the apostle to the ephesians / Whitaker, George Herbert – London: Methuen 1902 [mf ed 1989] – (= ser The churchman's bible) – 1mf – 9 – 0-7905-2212-8 – mf#1987-2212 – us ATLA [227]

The epistle of paul the apostle to the galatians / Robinson, A W – London: Methuen; Boston: L C Page 1900 [mf ed 1989] – (= ser The churchman's bible) – 1mf – 9 – 0-7905-3104-6 – mf#1987-3104 – us ATLA [227]

The epistle of paul the apostle to the romans : notes, comments, maps, and illustrations / Abbott, Lyman – New York: A S Barnes, [1888] – 1mf – 9 – 0-8370-2019-0 – (series title on binding: abbott's commentary) – mf#1985-0019 – us ATLA [227]

The epistle of paul the apostle to the romans : with introduction and notes / Moule, Handley Carr Glyn – stereotyped ed. Cambridge: University Press, 1896 – 1mf – 9 – 0-8370-4506-1 – (incl ind) – mf#1985-2506 – us ATLA [227]

The epistle of paul the apostle to the romans : with notes, comments, maps and illustrations / Abbott, Lyman – New York: A S Barnes, c1888. Chicago: Dep of Photodup, U of Chicago Lib, 1973 (1r); Evanston: American Theol Lib Assoc, 1984 (1r) – 1 – 0-8370-0520-5 – (incl bibl ref) – mf#1984-B371 – us ATLA [227]

The epistle of paul to philemon : a theological and homiletic commentary = Der brief an philemon / Oosterzee, Johannes Jacobus van – New York: Charles Scribner, c1868 [mf ed 1986] – (= ser A commentary on the holy scriptures. new testament 8/4) – 1mf – 9 – 0-8370-6766-9 – (english trans fr german with additions by horatio balch hackett) – mf#1986-0766 – us ATLA [227]

The epistle of paul to the churches of galatia / Macgregor, James – Edinburgh: T & T Clark 1879 [mf ed 1985] – (= ser Handbook for bible classes) – 1mf – 9 – 0-8370-4231-3 – mf#1985-2231 – us ATLA [227]

The epistle of paul to the colossians = Der brief st pauli an die kolosser / Riddle, Matthew Brown – New York: Scribner, Armstrong, 1874, c1870 [mf ed 1985] – (= ser A commentary on the holy scriptures. new testament 7/3) – 1mf – 9 – 0-8370-4705-6 – (trans fr german by matthew brown riddle) – mf#1985-2705 – us ATLA [227]

The epistle of paul to the ephesians = Der brief st pauli an die epheser / Braune, Karl – New York: Scribner, Armstrong, 1874, c1870 [mf ed 1985] – (= ser A commentary on the holy scriptures. new testament 7/2) – 1mf – 9 – 0-8370-4691-2 – (trans fr germa by matthew brown riddle. incl bibl ref) – mf#1985-2691 – us ATLA [227]

The epistle of paul to the ephesians : with introduction and notes / Candlish, James S – Edinburgh: T & T Clark, 1895 – 1mf – 9 – 0-8370-2577-X – mf#1985-0577 – us ATLA [227]

The epistle of paul to the galatians = Der brief pauli an die galater / Schmoller, Otto; ed by Riddle, Matthew Brown – New York: Scribner, Armstrong, 1874 [mf ed 1985] – (= ser A commentary on the holy scriptures. new testament 7) – 1mf – 9 – 0-8370-5121-5 – (trans fr german by charles casey starbuck) – mf#1985-3121 – us ATLA [227]

The epistle of paul to the galatians / Duncan, George – Harper. 1934 – 9 – $10.00 – us IRC [240]

The epistle of paul to the philippians = Der brief st pauli an die philipper / Braune, Karl – New York: Scribner, Armstrong, 1874, c1870 [mf ed 1985] – (= ser A commentary on the holy scriptures. new testament 7/4) – 1mf – 9 – 0-8370-4702-1 – (trans fr german by horatio balch hackett. with additions) – mf#1985-2702 – us ATLA [227]

The epistle of paul to the romans = Der brief pauli an der roemer / Lange, Johann Peter; ed by Schaff, Philip & Riddle, Matthew Brown – New York: Charles Scribner, c1869 [mf ed 1986] – (= ser A commentary on the holy scriptures. new testament 5) – 2mf – 9 – 0-8370-6748-0 – (trans fr german by john fletcher hurst) – mf#1986-0748 – us ATLA [227]

The epistle of paul to the romans / Dodd, C H – 1932 – 9 – $10.00 – us IRC [240]

The epistle of paul to titus = Die pastoralbriefe / Oosterzee, Johannes Jacobus van – New York: Charles Scribner, c1868 – (= ser A commentary on the holy scriptures. new testament 8/3) – 1mf – 9 – 0-8370-6767-7 – (english trans fr german with additions by george edward day) – mf#1986-0767 – us ATLA [227]

The epistle of priesthood : studies in the epistle to the hebrews / Nairne, Alexander – Edinburgh: T & T Clark; New York: Charles Scribner [distributor], 1913 – 2mf – 9 – 0-7905-1255-6 – (incl bibl ref and ind) – mf#1987-1255 – us ATLA [227]

The epistle of psenosiris : an original document from the diocletian persecution (papyrus 713 brit. mus.) = Epistle of psenosiris / Psenosiris; ed by Deissmann, Gustav Adolf – London: Adam and Charles Black, 1902 – 1mf – 9 – 0-524-05515-7 – (in english and greek) – mf#1990-1510 – us ATLA [240]

The epistle of st james : the greek text with introduction, commentary as far as chapter 4, verse 7, and additional notes / Hort, Fenton John Anthony – London: Macmillan, 1909 – 1mf – 9 – 0-8370-3661-5 – (incl indof greek and hebrew words and of subjects) – mf#1985-1661 – us ATLA [227]

The epistle of st james : the greek text with introduction, notes and comments / Mayor, Joseph Bickersteth – London; New York: Macmillan, 1892 – 2mf – 9 – 0-8370-9719-3 – (incl bibl ref and indexes) – mf#1986-3719 – us ATLA [227]

The epistle of st james, 1 1-4.7 / Hort, Fenton John Anthony – 1909 – 9 – $10.00 – us IRC [240]

The epistle of st james: with an introduction and notes / Knowling, Richard John – London: Methuen, 1904. 1 fiche – 9 – us ATLA [240]

The epistle of st jude and the second epistle of st peter / Mayor, Joseph Bickersteth – London: Macmillan, 1907.1 fiche. 8370-4333-6 – 9 – us ATLA [240]

The epistle of st paul to the romans / Moule, Handley Carr Glyn – 5th ed. New York: A C Armstrong, 1901 – 2mf – 9 – 0-8370-4507-X – mf#1985-2507 – us ATLA [227]

The epistle of the apostle paul to romans : a new translation, with notes / Godwin, John Henry – London: Hodder & Stoughton, 1873 – 1mf – 9 – 0-8370-3325-X – mf#1985-1325 – us ATLA [227]

An epistle to all buddhists throughout the world / Richard, Timothy – [Shanghai?: s.n, 1916?] [mf ed 1992] – 1mf – 9 – 0-524-03124-X – (in english & chinese) – mf#1990-3177 – us ATLA [280]

Epistle to friends / Ellwood, Thomas – Manchester, England. 18– – 1r – 1 – us UF Libraries [240]

An epistle to the clergy of the southern states / Grimke, Sarah Moore – [New York: s.n, 1836] [mf ed 1984] – (= ser Women & the church in america 18) – 1mf – 9 – 0-8370-0233-8 – mf#1984-2018 – us ATLA [976]

The epistle to the colossians : analysis and examination notes / Garrod, George Watts – London, New York: Macmillan, 1898 – 1mf – 9 – 0-7905-0026-4 – (incl bibl ref and index) – mf#1987-0026 – us ATLA [227]

The epistle to the ephesians : from notes of readings / Bellett, John Gifford – London: Robert L Allan, 1871 – 1mf – 9 – 0-8370-2257-6 – mf#1985-0257 – us ATLA [227]

The epistle to the ephesians : its doctrine and ethics / Dale, Robert William – London: Hodder and Stoughton, 1883 – 2mf – 9 – 0-7905-0075-2 – (incl bibl ref) – mf#1987-0075 – us ATLA [227]

The epistle to the ephesians / Parker, Joseph – New York: A C Armstrong, 1905 – 1mf – 9 – 0-8370-4668-8 – mf#1985-2668 – us ATLA [227]

The epistle to the ephesians : with introduction and notes / Moule, Handley Carr Glyn – Stereotyped ed. Cambridge: University Press; New York: Macmillan [distributor], 1886 – 1mf – 9 – 0-8370-6823-1 – (incl bibl ref and index) – mf#1986-0823 – us ATLA [227]

The epistle to the galatians : an essay on its destination and date: with an appendix on the visit to jerusalem recorded in chapter 2: being an enlargement of the norrisian prize essay for 1898 on "the locality of the churches of galatia" / Askwith, Edward Harrison – London, New York: Macmillan, 1899 – 1mf – 9 – 0-8370-9921-8 – (incl bibl ref) – mf#1986-3921 – us ATLA [227]

The epistle to the galatians / Findlay, George Gillanders – New York: A.C. Armstrong, [1889] – 2mf – 9 – 0-8370-2414-5 – mf#1985-0414 – us ATLA [227]

The epistle to the galatians / ed by Perowne, Edward Henry – sereotyped ed. Cambridge: University Press, 1890 – (= ser The Cambridge Bible for Schools and Colleges) – 1mf – 9 – 0-8370-6832-0 – mf#1986-0832 – us ATLA [227]

The epistle to the galatians : with an introduction, explanatory notes, practical thoughts, and prayers, for private and family use / Headland, Edward – London: Hatchard, 1866 – 1mf – 9 – 0-7905-3256-5 – mf#1987-3256 – us ATLA [227]

The epistle to the galatians, in greek and english : with an analysis and exegetical commentary / Turner, Samuel Hulbeart – New York: Dana and Co, 1856 – 1mf – 9 – 0-524-06058-4 – mf#1992-0771 – us ATLA [227]

Epistle to the hebrews : the harklean version, chapters 11, 28 – 13, 25 / ed by Bensly, Robert L – Cambridge, 1889 – €3.00 – ne Slangenburg [227]

The epistle to the hebrews : being the substance of three lectures delivered in the chapel of the honourable society of lincoln's inn... / Maurice, Frederick Denison – London: J W Parker, 1846 – 1mf – 9 – 0-7905-1358-7 – mf#1987-1358 – us ATLA [227]

The epistle to the hebrews = Der brief an die hebraeer / Moll, Carl Bernhard – New York: Charles Scribner, c1868 [mf ed 1986] – (= ser A commentary on the holy scriptures. new testament 8/5) – 1mf – 9 – 0-8370-6763-4 – (trans fr german with asahel clark kendrick) – mf#1986-0763 – us ATLA [227]

The epistle to the hebrews : an exposition / Saphir, Adolph – 3rd american ed. New York: CC Cook, [1902?] – 3mf – 9 – 0-524-05064-3 – mf#1992-0317 – us ATLA [227]

The epistle to the hebrews / Goodspeed, Edgar Johnson – New York: Macmillan, 1908 – 1mf – 9 – 0-8370-3338-1 – (incl indes) – mf#1985-1338 – us ATLA [227]

The epistle to the hebrews : the greek text with notes and essays / Westcott, Brooke Foss – 3d ed. London; New York: Macmillan, 1903. Beltsville, Md: NCR Corp, 1978 (7mf); Evanston: American Theol Lib Assoc, 1984 (7mf) – (= ser Biblical crit – us & gb) – 9 – 0-8370-0675-9 – (incl ind) – mf#1984-1098 – us ATLA [227]

The epistle to the hebrews : with notes = Pros hebraious / Vaughan, Charles John – London; New York: Macmillan, 1890 – 1mf – 9 – 0-8370-9910-2 – (discussion in english and greek; text in greek. incl ind of greek words) – mf#1986-3910 – us ATLA [227]

The epistle to the hebrews / Rendall, Frederic – London, New York: Macmillan, 1888 – 1mf – 9 – 0-7905-0212-7 – (incl ind) – mf#1987-0212 – us ATLA [227]

The epistle to the hebrews : the sirst apology for christianity: an exegetical study / Bruce, Alexander Balmain – New York City: Charles Scribner, 1899 [mf ed 1989] – 2mf – 9 – 0-7905-0677-7 – mf#1987-0677 – us ATLA [227]

The epistle to the hebrews / Wickham, E C – London: Methuen, 1910 – (= ser Westminster Commentaries) – 1mf – 9 – 0-7905-2218-7 – (incl ind) – mf#1987-2218 – us ATLA [227]

The epistle to the hebrews : with introduction and notes / Davidson, Andrew Bruce – Edinburgh: T & T Clark, [1887?] – 1mf – 9 – 0-8370-2834-5 – mf#1985-0834 – us ATLA [227]

The epistle to the hebrews : with notes, critical, explanatory and practical / Cowles, Henry – New York: D Appleton, 1878 – 1mf – 9 – 0-8370-2752-7 – mf#1985-0752 – us ATLA [227]

The epistle to the hebrews compared with the old testament / Newton, Adelaide Leaper – 5th ed. New York: R Carter, 1867 – 1mf – 9 – 0-524-04915-7 – mf#1992-0258 – us ATLA [227]

The epistle to the philippians : with introduction and notes / Moule, Handley Carr Glyn – stereotyped ed. Cambridge: University Press, 1899 – 1mf – 9 – 0-8370-4508-8 – (incl ind) – mf#1985-2508 – us ATLA [227]

The epistle to the romans : a commentary, logical and historical / Stifler, James M – New York: Fleming H Revell, 1897 – 1mf – 9 – 0-8370-5419-2 – mf#1985-3419 – us ATLA [227]

The epistle to the romans = Roemerbrief / Barth, Karl – London: OUP, 1950 [mf ed 2004] – (= ser Modern german commentaries on the epistle to the romans) – 1r – 1 – 0-524-10392-5 – (german original publ in 1918. trans fr 6th ed by edwyn clement hoskyns; new pref by aut. incl ind) – mf#b00719 – us ATLA [227]

The epistle to the romans : with introduction and notes / Brown, David – Edinburgh: T & T Clark, [18–] – 1mf – 9 – 0-8370-2476-5 – mf#1985-0476 – us ATLA [227]

The epistle to the romans : with notes critical and practical / Sadler, Michael Ferrebee – 2nd rev ed. London: George Bell, 1889 – 1mf – 9 – 0-8370-5020-0 – mf#1985-3020 – us ATLA [227]

The epistle to timothy and the woman question / Schmauk, Theodore Emanuel – [Philadelphia: Evangelical Lutheran Theological Seminary, 1899] Beltsville, MD: NCR Corp, 1978 (1mf); Evanston: American Theol Lib Assoc, 1984 (1mf) – (= ser Women & the church in america) – mf#1984-6256 – us ATLA [240]

The epistles and gospels of the sundays throughout the year : with notes, critical and explanatory / M'Carthy, Daniel – Dublin: James Duffy; London: Burns, Lambert, and Oates, 1868 – 2mf – 9 – 0-8370-9402-X – (incl indes) – mf#1986-3402 – us ATLA [227]

The epistles general of john = Die drei briefe des apostels johannes / Braune, Karl – New York: Charles Scribner, c1867 [mf ed 1986] – (= ser A commentary on the holy scriptures. new testament 9/3) – 1mf – 9 – 0-8370-6724-3 – (trans fr german by jacob isidor mombert. incl bibl) – mf#1986-0724 – us ATLA [227]

The epistles general of peter = Die briefe petri / Fronmueller, Petri – New York: Scribner, Armstrong, c1867 [mf ed 1986] – (= ser A commentary on the holy scriptures. new testament 9/2) – 1mf – 9 – 0-8370-6738-3 – (incl bibl) – mf#1986-0738 – us ATLA [227]

The epistles o' hugh airlie i.e. j kerr lawson : (formely o' scotland, presently conneckit wi' tam tamson's warehoose in toronto – Toronto: Grip Print & Pub Co, 1888 – 2mf – 9 – (ill by john wilson bengough) – mf#08952 – cn Canadiana [860]

The epistles of paul the apostle : a sketch of their origin and contents / Findlay, George Gillanders – New York: Wilbur B Ketcham, [1892] – 1mf – 9 – 0-8370-3129-X – mf#1985-1129 – us ATLA [227]

The epistles of paul the apostle to the colossians and to philemon / Knight, Henry Joseph Corbett – London: Methuen 1907 [mf ed 1989] – 1mf – 9 – 0-7905-2241-1 – mf#1987-2241 – us ATLA [227]

The epistles of paul the apostle to the thessalonians, corinthians, galatians, romans and philippians / Drummond, James – New York: G P Putnam, 1899 – 1mf – 9 – 0-8370-2976-7 – mf#1985-0976 – us ATLA [227]

The epistles of paul to the corinthians, galatians, ephesians, philippians, colossians, thessalonians, timothy, titus and philemon – With introd. and commentary by Abiel Abbot Livermore. Boston: Lockwood, Brooks, 1881. 308p – 1 – us Wisconsin U Libr [240]

The epistles of s john – Cambridge: University Press, 1886 – (= ser Cambridge Greek Testament for Schools and Colleges) – 1mf – 9 – 0-524-06853-4 – mf#1992-0995 – us ATLA [227]

The epistles of s paul from the codex laudianus : (l [wordsworth's o2]) numbered laud. lat. 108 in the bodleian library at oxford / ed by Buchanan, Edgar Simmons – London: Heath Cranton & Ouseley, 1914 – (= ser Sacred Latin Texts) – 1mf – 9 – 0-8370-1888-9 – mf#1987-6275 – us ATLA [227]

The epistles of saint paul to the thessalonians, galatians and romans : essays and dissertations / Jowett, Benjamin; ed by Campbell, Lewis – London: John Murray, 1894 – 1mf – 9 – 0-8370-9552-2 – mf#1986-3552 – us ATLA [227]

Epistles of ss clement of rome and barnabas and the shepherd of hermas – London: Griffith Farran, Browne, [1888?] [mf ed 1991] – (= ser Ancient and modern library of theological literature; Apostolic fathers 1) – 1mf – 9 – 0-7905-8642-8 – (with int) – mf#1989-1867 – us ATLA [240]

Epistles of st ignatius and st polycarp : with introductory preface comprising a history of the christian church in the 2nd century – London: Griffith Farran, 1889 [mf ed 1991] – (= ser Ancient and modern library of theological literature; Apostolic fathers 2) – 1mf – 9 – 0-7905-8665-7 – mf#1989-1890 – us ATLA [240]

The epistles of st john : a series of lectures on christian ethics / Maurice, Frederick Denison – London, New York: Macmillan, 1893 – 1mf – 9 – 0-7905-3149-6 – mf#1987-3149 – us ATLA [227]

The epistles of st john : twenty-one discourses / Alexander, William – NY: A C Armstrong, 1899 – 1mf – 9 – 0-8370-2064-6 – (with greek text, comparative versions, and notes chiefly exegetical) – mf#1985-0064 – us ATLA [227]

The epistles of st john / Westcott, Brooke Foss – 1883 – 9 – $15.00 – us IRC [240]

The epistles of st john : with introduction and appendices / Plummer, Alfred – stereotyped ed. Cambridge:University Press, 1883 – 1mf – 9 – 0-8370-5396-X – (includes general index) – mf#1985-3396 – us ATLA [227]

The epistles of st paul to the colossians and philemon / Maclaren, Alexander – New York: Armstrong, 1901. vii,493p – 1 – us Wisconsin U Libr [243]

The epistles of st paul to the colossians, thessalonians, and timothy : with notes critical and practical / Sadler, Michael Ferrebee – 2nd ed. London, New York: George Bell, 1893 – 1mf – 9 – 0-8370-5021-9 – mf#1985-3021 – us ATLA [227]

The epistles of st paul to the corinthians : with critical notes and dissertations / Stanley, Arthur Penrhyn – 5th ed. London: John Murray, 1882. Beltsville, Md: NCR Corp, 1978 (7mf); Evanston: American Theol Lib Assoc, 1984 (7mf) – (= ser Biblical crit – us & gb) – 9 – 0-8370-0225-7 – mf#1984-1040 – us ATLA [227]

Epistles of st paul to the ephesians, colossians, and philemon / Davies, John Llewelyn – London: Macmillan, 1866 – 1mf – 9 – us ATLA [227]

The epistles of st paul to the ephesians, the colossians, and philemon : with introductions and notes / Davies, John Llewelyn – London: Macmillan, 1866 – 1mf – 9 – 0-8370-2841-8 – (with int & notes) – mf#1985-0841 – us ATLA [227]

The epistles of st paul to the galatians, ephesians, and philippians : with notes critical and practical / Sadler, Michael Ferrebee – 2nd ed. London, New York: George Bell, 1892 – 1mf – 9 – 0-8370-5022-7 – mf#1985-3022 – us ATLA [227]

The epistles of st paul to the thessalonians, galatians and romans : translation and commentary / Jowett, Benjamin; ed by Campbell, Lewis – 3rd ed. London: John Murray, 1894 – 2mf – 9 – 0-8370-9553-0 – mf#1986-3553 – us ATLA [227]

The epistles of st paul to the thessalonians, galatians, romans : with critical notes and dissertations / Jowett, Benjamin – 2d ed. London: John Murray, 1859. Beltsville, Md: NCR Corp, 1978 (13mf); Evanston: American Theol Lib Assoc, 1984 (13mf) – (= ser Biblical crit – us & gb) – 9 – 0-8370-0197-8 – mf#1984-1019 – us ATLA [227]

The epistles of st paul to titus, philemon, and the hebrews : with notes critical and practical / Sadler, Michael Ferrebee – 2nd ed. London, New York: George Bell, 1893 – 1mf – 9 – 0-8370-5023-5 – mf#1985-3023 – us ATLA [227]

The epistles of st paul written after he became a prisoner : arranged in the probable chronological order, viz ephesians, colossians, philemon...with explanatory notes / Boise, James Robinson – New York: D Appleton, 1887 [mf ed 1985] – 1mf – 9 – 0-8370-2406-4 – mf#1985-0406 – us ATLA [227]

The epistles of st peter / Jowett, John Henry – New York: A C Armstrong, 1906 – 1mf – 9 – 0-8370-3808-1 – mf#1985-1808 – us ATLA [227]

Epistles of st peter, st john, and st jude : with commentaries / Mason, Arthur James et al; ed by Ellicott, Charles John – London: Cassell, Petter, Galpin, [18-?] – (= ser New testament commentary) – 1mf – 9 – 0-524-05100-3 – mf#1992-0321 – us ATLA [227]

The epistles of the new testament : an attempt to present them in current and popular idiom / Hayman, Henry – London: A and C Black, 1900 – 2mf – 9 – 0-8370-1952-4 – mf#1987-6339 – us ATLA [227]

Epistles on women / Aikin, Lucy – London. 1810 – 1 – us CRL [840]

The epistles to the colossians and the philemon : with introduction and notes / Moule, Handley Carr Glyn – stereotyped ed. Cambridge: University Press, 1898 – 1mf – 9 – 0-8370-4510-X – (incl bibl ref, appendixes and ind) – mf#1985-2510 – us ATLA [227]

The epistles to the colossians and to the ephesians / Alexander, Gross – New York: Macmillan, 1910 – (= ser The bible for home and school) – 1mf – 9 – 0-7905-3064-3 – (incl bibl ref) – mf#1987-3064 – us ATLA [227]

The epistles to the hebrews, colossians, ephesians, and philemon, the pastoral epistles, the epistles of james, peter, and jude : together with a sketch of the history of the canon of the new testament / Cone, Orello – New York: G P Putnam, 1901 – 1mf – 9 – 0-7905-0005-1 – (incl bibl ref) – mf#1987-0005 – us ATLA [227]

The epistles to the thessalonians / Denney, James – New York: A C Armstrong, 1892 – 1mf – 9 – 0-8370-2890-6 – mf#1985-0890 – us ATLA [227]

The epistles to the thessalonians : with an introduction, explanatory notes, practical thoughts, and prayers, for private and family use / Headland, Edward – London: Hatchard, 1863 – 1mf – 9 – 0-7905-3257-3 – mf#1987-3257 – us ATLA [227]

The epistles to timothy and titus : with introduction and notes / Humphreys, A E – stereotyped ed. Cambridge: University Press, 1897 – 1mf – 9 – 0-8370-3692-5 – (includes appendix & index) – mf#1985-1692 – us ATLA [227]

Epistola ad ciceronem *see* Epistola ad quintum fratrem...

Epistola ad quintum fratrem... / Cicero, Marcus Tullius – 15th c – (= ser Holkham library manuscript books 379,123) – 1r – 1 – (filmed with: epistola ad ciceronem by quintus. oeconomica by aristoteles. opuscula et epistolae by s ambrosius) – mf#96728 – uk Microform Academic [450]

Epistola ad senatum populumque genevensem, qua in obedientiam romani pontificis eos reducere conatur. joannis calvini responsio / Sadoleto, J & Calvin, J – Argentorati: Per Wendelinum Rihelium, 1539 – 2mf – 9 – mf#CL-18 – ne IDC [227]

Epistola ad trajanum ex plutarcho *see* Epistolae, libri 9

Epistola beati pauli apostoli ad romanos / Agus, Joseph – Ratisbonae: Sumptibus, chartis et typis Friderici Pustet, 1888 – 2mf – 9 – 0-524-06567-5 – mf#1992-0910 – us ATLA [227]

Epistola constantinolopi recens scripta. de praesenti turcici imperij statu, and gubernatoribus praecipuis, and de bello persico / Billerbeg, F de – n.p, 1582 – 1mf – 9 – mf#H-8201 – ne IDC [956]

Epistola d pavli ad galatas cvm commentario / Corner, C – Heidelbergae, 1583 – 5mf – 9 – mf#TH-1 mf 339-343 – ne IDC [242]

Epistola de laudibus paetriae nostrae / Carrillo Chumacero, Fernando – Majera, 1617 – 1 – sp Bibl Santa Ana [240]

Epistola exhortatoria en orden a que los predicadores evangelicos no priven de la doctrina a las almas en los sermones de fiestas / Barcia y Zambrana, Jose de – en la Puebla: impr de Diego Fernandez de Leon...ano de 1693 – (= ser Books on religion...1543/44-c1800: arte de predicar) – 2mf – 9 – mf#crl-335 – ne IDC [241]

Epistola joannis calvini, qua fidem admonitionis ab eo nuper editae, apud polonos confirmat / Calvin, J – Genevae: ex officina francisci perrini, 1563 – 1mf – 9 – mf#CL-40 – ne IDC [242]

Epistola religiosa...ceferino gonzalez... / Barrantes Moreno, Vicente – 1873 – 9 – sp Bibl Santa Ana [240]

Epistolae : formae tplila 55 / Gemblacensis, Guibertus – 1989 – (= ser ILL – ser a; Cccm 66-66a) – 15mf+196p – 9 – €80.00 – 2-503-63662-4 – be Brepols [227]

Epistolae / Hieronymus, Saint – 15th c – (= ser Holkham library manuscript books 126) – 1r – 1 – mf#96530 – uk Microform Academic [227]

Epistolae / Hieronymus, Saint – 15th c – (= ser Holkham library manuscript books 124) – 1r – 1 – mf#97102 – uk Microform Academic [227]

Epistolae / Hieronymus, Saint – 15th c – (= ser Holkham library manuscript books 125) – 1r – 1 – mf#2724 – uk Microform Academic [227]

Epistolae / Petrarch [Francesco Petrarca] – (= ser Holkham library manuscript books 428) – 1r – 1 – mf#97387 – uk Microform Academic [860]

Epistolae / Petrus Blessensis (Petrus von Blois) – Bruxellis: apud Fratres Communis vitae, c1480 – €37.00 – ne Slangenburg [227]

Epistolae : praestantium ac eruditorum virorum... quae a j arminio, c vorstio, s episcopio, h grotio, c barlaeo, conscriptae sunt / Vorstius, C – ed 2. Amstelaedami, 1684 – 18mf – 9 – mf#PBA-340 – ne IDC [227]

Epistolae / Xavier, Francis, Saint – Hongkong: typis Societatis missionum ad exteros, 1888-90 [mf ed 1995] – (= ser Yale coll) – 2v – 1 – 0-524-09638-4 – (in latin) – mf#1995-0638 – us ATLA [241]

Epistolae *see*
– Historia...
– Res gestae alexandri magni

Epistolae 3 de recta et legitima ecclesiarum bene instituendarum ratione ac modo / Lasco, J – Basilae, [1556] – 4mf – 9 – mf#PBA-222 – ne IDC [240]

Epistolae ab ecclesiae helveticae reformatoribus vel ad eos scriptae / ed by Fuesslin, J C – Zuerich, Heidegger, 1742 – 6mf – 9 – mf#PBU-422 – ne IDC [242]

Epistolae ad henr bullingerum / Lasco, J; ed by Gerses, D – Groningen/Bremen: corn barlinkhof & g w rump, 1754. v4(1) – (= ser Scrinium antiquarium) – 1mf – 9 – mf#PBU-477 – ne IDC [242]

Epistolae catholicae breviter explicatae : ad usum seminariorum et cleri / Steenkiste, J-A van – Brugis: Beyaert-Defoort, 1876 – 1mf – 9 – 0-524-07660-X – mf#1992-1101 – us ATLA [227]

Epistolae, declamationes, epistolae mutuae *see* Divinae institutiones...

Epistolae duae : quarum prima Adriani Pauli, altera J. de L. responsoria / Labadie, Jean de – n.p, 1672 – 1mf – 9 – mf#PPE-192 – ne IDC [240]

Epistolae dvae ad ecclesias polonicas / Bullinger, Heinrich – Tigvri, Christoph Froschouer, 1561 – 1mf – 9 – mf#PBU-219 – ne IDC [240]

Epistolae et poemata / Boxhorn, M Z – Amstelodami: Ex officina Caspari Commelini, 1662 – 6mf – 9 – mf#O-3209 – ne IDC [090]

...Epistolae familiares / Symmachus, Q A – Argentoraci, 1510 – 2mf – 9 – mf#H-8224 – ne IDC [956]

Epistolae karolini aevi, tom 2 (mgh epistolae 1:4.bd) – 1895 – (= ser Monumenta germaniae historica epistolae. 1. epistolae in quarto (mgh epistolae 1)) – €32.00 – ne Slangenburg [227]

Epistolae karolini aevi, tom 3 (mgh epistolae 1:5.bd) – 1898-1899 – (= ser Monumenta germaniae historica epistolae. 1. epistolae in quarto (mgh epistolae 1)) – €35.00 – ne Slangenburg [227]

Epistolae karolini aevi, tom 4 (mgh epistolae 1:6.bd) – 1902-1925 – (= ser Monumenta germaniae historica epistolae. 1. epistolae in quarto (mgh epistolae 1)) – €40.00 – ne Slangenburg [227]

Epistolae karolini aevi, tom 5 (mgh epistolae 1:7.bd) – 1912-1928 – (= ser Monumenta germaniae historica epistolae. 1. epistolae in quarto (mgh epistolae 1)) – €25.00 – ne Slangenburg [227]

Epistolae karolini aevi, tom 6 fasc 1 (mgh epistolae 1:8.bd) : hincmari archiepiscopi remensis epistolarum pars 1 – 1939 – (= ser Monumenta germaniae historica epistolae. 1. epistolae in quarto (mgh epistolae 1)) – €14.00 – ne Slangenburg [227]

Epistolae, libri 9 / Pliny The Younger [Gaius Plinius Caecilius Secundus] – 15th c – (= ser Holkham library manuscript books 396) – 1r – 1 – (filmed with: livius: orationes hannibalis et scipionis / epistola ad trajanum ex plutarcho) – mf#95968 – uk Microform Academic [450]

Epistolae merowingici et karolini aevi, tom 1 (mgh epistolae 1:3.bd) – 1892 – (= ser Monumenta germaniae historica epistolae. 1. epistolae in quarto (mgh epistolae 1)) – €31.00 – ne Slangenburg [227]

Epistolae nomine ssmi domini benedicti 15 : humaniter missae ab petro gasparri coetui virorum delectorum... – [S.l: s.n, 1915?] – 1mf – 9 – 0-524-03226-2 – mf#1990-0854 – us ATLA [240]

Epistolae obscurorum virorum : the latin text with an english rendering, notes, and an historical introduction / Crotus Rubeanus – London: Chatto & Windus, 1909 – 2mf – 9 – 0-7905-4336-2 – (incl bibl ref. in english and latin) – mf#1988-0336 – us ATLA [227]

Epistolae principvm, rervmpvblicarvm, ac sapientvm virorvm : ex antiquis and recentioribus, tam graecis, quam latinis, historijs and annalibus collectae / Donzellini, G – Venetiis, 1574 – 5mf – 9 – mf#H-8197 – ne IDC [956]

Epistolae qvaedam ioachimi morlin doctoris theologiae ad d andream osiandrum, et responsiones / Moerlin, J – np, 1551 – 1mf – 9 – mf#TH-1 mf 1438 – ne IDC [242]

Epistolae reverendi, danielis hoffmanni de libro concordiae / Hoffmann, D – Ienae, 1597 – 1mf – 9mf – mf#TH-1 mf 700 – ne IDC [242]

Epistolae theologicae / Beza, Theodor de – Geneve, E Vignon, 1573 – 5mf – 9 – mf#PFA-103 – ne IDC [240]

Epistolae tigurinae – Cantabrigiae, J Gul Parker, 1848 – 6mf – 9 – mf#PBU-430 – ne IDC [240]

Epistolario... / Muratori, Lodovico A; ed by Curato da Matteo Campori – Plodena. v1-14. 1901 – 1 – $240.00 – mf#0381 – us Brook [440]

Epistolario de heroes / Cabrales, Gonzalo – Habana, Cuba. 1922 – 1r – 1 – us UF Libraries [972]

Epistolario de menendez pelayo con jose lopez prudencio (1902-1910) / Rodriguez Monino, Antonio – Badajoz: imprenta de la diputacion provincial, 1958 – sp Bibl Santa Ana [946]

Epistolario de nueva espana... / Paso y Troncoso, Francisco del – Madrid: Razon y Fe, 1940 – sp Bibl Santa Ana [240]

Epistolario historico / Jimenez Malaret, Rene – San Juan, Puerto Rico. 1953 – 1r – 1 – us UF Libraries [972]

Epistolario poetico completo : noticia preliminar por a rodriguez-monino / Aldana, Francisco de – Badajoz: Dip. Provincial, 1946 – 1 – sp Bibl Santa Ana [440]

Epistolario y otros poemas / Lazaro, Angel – Habana, Cuba. 1952 – 1r – 1 – us UF Libraries [972]

Epistolarum fasciculus : scrinium antiquarium, v4 (1) / Bullinger, Heinrich; ed by Gerdes, D – Groningen/Bremen: Corn Barlinkhof & G W Rump, 1754 – 1mf – 9 – mf#PBU-423 – ne IDC [240]

Epistolarum libri duo... / Zanchi, G – Genevae, 1613 – 5mf – 9 – mf#PBU-704 – ne IDC [240]

The epistolary literature of the assyrians and babylonians / Johnston, Christopher – 1898 – 1mf – 9 – 0-8370-7467-3 – (text in english and akkadian; commentary in english) – mf#1986-1467 – us ATLA [470]

Epistolas a mi amigo aristos telasca / Cestero Burgos, Tulio A – Ciudad Trujillo, Dominican Republic. 1949 – 1r – 1 – us UF Libraries [972]

Epistolas familiares / Guevara, Antonio – 9 – sp Bibl Santa Ana [240]

Epistole thurci... – Lugduni, 1520 – 1mf – 9 – mf#H-8136 – ne IDC [956]

Epistolica quaestio de vitae termino...fatali, an mobili? / Beverovicius, J – ed 2a. Lugd. Batavorum, 1636 – 11mf – 8 – €21.00 – ne Slangenburg [230]

Epistoloe obscurorum virorum – London, England. 1925 – 1r – 1 – us UF Libraries [025]

Epistolographoi hellenikoi – Paris, France. 1873 – 1r – 1 – us UF Libraries [960]

Epistre de jaques sadolet cardinal, envoyee au senat et peuple de geneve : par laquelle il tasche les reduire soubz la puissance de l'evesque de romme / Sadoleto, J & Calvin, J – Geneve: Par Michel Du Bois, 1540 – 2mf – 9 – mf#CL-45 – ne IDC [240]

L'epistre d'othea (cima31) : farbmikrofiche-edition der handschrift erlangen-nuernberg, universitaetsbibliothek, ms 2361 / Pizan, Christine de – (mf ed 1996) – (= ser Codices illuminati medii aevi (cima) 31) – 98p on 5 color mf – 15 – €320.00 – 3-89219-031-3 – (int by helga lengenfelder) – gw Lengenfelder [090]

Epistre envoye aux reliques de la dissipation horrible de l'antechrist / Farel, Guillaume – Geneve, 1544 – 1mf – 9 – mf#PFA-157 – ne IDC [240]

Epistre envoyée au duc de lorraine / Farel, Guillaume – [Geneve], Jean Girard, 1543 – 2mf – 9 – mf#PFA-153 – ne IDC [240]

Epistre envoyee aux fideles conversans entre les chrestiens papistiques / Viret, P – n.p, 1543 – 2mf – 9 – mf#PFA-190 – ne IDC [240]

Epistre exhortatoire...tous ceux qui ont congnoissance de l'evangile / Farel, Guillaume – [Geneve], 1544 – 1mf – 9 – mf#PFA-158 – ne IDC [240]

Epistula ad romanos : secundum editionem sancti hieronymi – Oxonii: E Typographeo Clarendoniano, 1913 – (= ser Novum testamentum domini nostri iesu christi latine) – 2mf – 9 – 0-8370-1856-0 – mf#1987-6243 – us ATLA [227]

Epistulae et chartae ad historiam primi belli sacri spectantes... : die kreuzzugsbriefe aus den jahren 1088-1100, eine quellensammlung... / Hagenmeyer, H – Innsbruck, 1901 – 6mf – 9 – mf#H-3090 – ne IDC [956]

Epistulae pauli et catholicae fere integrae : Ex libro porphyrii episcopi palimpsesto saeculi octavi vel noni / ed by Tischendorf, Constantin von – Lipsiae: J C Hinrichs, 1865 [mf ed 1986] – 4mf – 9 – 0-8370-9429-1 – mf#1986-3429 – us ATLA [227]

Epistulae pauli et catholicae palimpsestae (msi5) / ed by Tischendorf, G F C – Lipsiae, 1865 – (= ser Monumenta sacra inedita. nova collectio v5) – €56.00 – ne Slangenburg [227]

Epitalamio del prieto trinidad / Sender, Ramon Jose – Mexico City? Mexico. 1942 – 1r – 1 – us UF Libraries [972]

Epitaph – La Farge WI. 1972 jan 12-1972 dec 27, 1973 jan 3-1974 jan 30, 1974 feb 20-1975 aug 14 – 3r – 1 – (continued by: la farge epitaph) – mf#1045276 – us WHS [071]

Epitaph-news – La Farge, Viola WI. 1982 jul 1/1983, 1984-1997 jan/dec – 18r – 1 – (cont: la farge epitaph (la farge, wi: 1982); viola news (viola, wi: 1914)) – mf#1045304 – us WHS [071]

Epitaphs of the catacombs : or, christian inscriptions in rome during the first four centuries / Northcote, James Spencer – London: Longmans, Green, 1878 – 1mf – 9 – 0-7905-7066-1 – (incl bibl ref) – mf#1988-3066 – us ATLA [930]

Epithalamium symbolicum conjugibus porphyrogenitis : serenissimo potentissimo Ferdinando 3... – Graecii in Styriis: Typis Ernesti Widmanstadii, 1631 – 3mf – 9 – mf#O-1989 – ne IDC [090]

Epitoma vaticana ex apollodori bibliotheca / Apollodorus – Lipsiae, Germany. 1891 – 1r – 1 – us UF Libraries [450]

Epitomae historiarum libri 18 (cshb50) / Ioannes Zonarad – Bonnae, 1897 – (= ser Corpus scriptorum historiae byzantinae (cshb)) – €31.00 – ne Slangenburg [243]

Epitome annalium pomeraniae : kurtzer bericht von belegenheit des landes stettin pommern wie dieselbe von etzlichen jahren / Eickstedt, Valentin von – 1600-1799 – 1r – 1 – (copy of/and/or trans of: epitome annalium pomeraniae, publ eventually in 1728 in greifswald) – us Wisconsin U Libr [943]

Epitome bibliothecae conradi gesneri / Simler, J – Zuerich, Christoph Froschauer, 1555 – 5mf – 9 – mf#PBU-409 – ne IDC [240]

Epitome breve de la vita y muerte del illustrissimo dotor don bernardo de almansa : criollo de la ciudad de lima, tesorero de la ciudad de cartagena, arcediano de la plata, inquisidor de logron[n]o y de toledo... / Solis y Valenzuela, Pedro de – en Lima: Por Pedro de Cabrera...ano de 1646 – (= ser Books on religion...1543/44-c1800: biografias de religiosos) – 2mf – 9 – mf#crl-132 – ne IDC [241]

Epitome colloqvii montisbelgartensis inter d iacobvm andreae, et d theodorum bezam / Andreae d A, J – Tvbingae, 1588 – 1mf – 9 – mf#TH-1 mf 20 – ne IDC [242]

Epitome de la vida y hechos del invicto emperador carlos 5 / Vera y Figueroa, Juan Antonio – Madrid: Juan Sanchez, 1649 – 1 – sp Bibl Santa Ana [920]

Epitome de...milagros...pedro de alcantara / Manzanares, Blas de – 1824 – 9 – sp Bibl Santa Ana [830]

Epitome emblematum panegyricorum academiae altorfinae : studiosae iuventute proposita – Noribergae: Impensis Levini Hulsii, 1602 – 3mf – 9 – mf#O-12 – ne IDC [090]

Epitome exegeticae biblicae / Hetzenauer, Michael – Oeniponte [Innsbruck]: Sumptibus Librariae Academicae Wagnerianа, 1903 – (= ser Studium biblicum novi testamenti catholicum) – 1mf – 9 – 0-524-07335-X – mf#1992-1066 – us ATLA [220]

Epitome historiae sacrae : ad usum tyronum linguae latinae / Lhomond, Charles-Francois – Quebec: apud Joannem Neilson, 1803 [mf ed 1976] – 1r – 5 – mf#SEM16P66 – cn Bibl Nat [450]

Epitome historial...juan de la puebla / Tirado, Juan – 1724 – 9 – sp Bibl Santa Ana [946]

Epitome historico de merida / Gonzalez y Gomez de Soto, Juan Jose – Merida: tip juan f rivera silva, 1906 – 1 – sp Bibl Santa Ana [946]

Epitome historico...fregenal / Sanchez Cid, Antonio – 1843 – 9 – sp Bibl Santa Ana [946]

Epitome in evangelia et epistolas in usum ministrorum ecclesiae / Hedion, C – Strasbourg, 1537 – 3mf – 9 – mf#PPE-109 – ne IDC [240]
Epitome livii see Opera...
Epitome of greek grammar / Strong, James – [S.l.: s.n.], c1856 (New York: John F Trow) – 1mf – 9 – 0-8370-9188-8 – mf#1986-3188 – us ATLA [450]
An epitome of jainism : being a critical study of its metaphysics, ethics, and history etc in relation to modern thought / Nahar, Puran Chand & Ghosh, Krishnachandra – Calcutta: H Duby, 1917 [mf ed 1992] – 1v on 2mf – 9 – 0-524-04863-0 – mf#1990-3425 – us ATLA [180]
An epitome of leading common law cases : with some short notes thereon...smith's leading cases / Indermaur, John – 5th ed. Boston, Soule and Bugbee, 1883. 107 p. LL-420 – 1 – us L of C Photodup [346]
Epitome of rev dr erick pontoppidan's explanation of martin luther's small catechism – Chicago: J Anderson, 1900 – 1mf – 9 – 0-524-00308-4 – (incl english trans of luther's kleine katechismus) – mf#1989-3008 – us ATLA [242]
Epitome of the patent laws in canada and united states / Fetherstonhaugh, Edward J – Montreal: Fetherstonhaugh & Blackmore [1905?] [mf ed 1996] – 1mf – 9 – 0-665-79788-5 – mf#79788 – cn Canadiana [346]
Epitome pastoralis ad usum cleri in statibus foederatis americae / Weninger, Francis Xavier – Buffalone: C Wieckmann & S Brandt, 1855 – 1mf – 9 – 0-8370-6789-8 – mf#1986-0789 – us ATLA [240]
Epitome rerum ab ioanne et alexio comnenis gestarum (cshb26) / Joannis Cinnami – Bonnae, 1836 – (= ser Corpus scriptorum historiae byzantinae (cshb)) – €25.00 – (ad fidem cod vat rec aug meineke. Nicephori bryennii commentarii. Rec aug meineke) – ne Slangenburg [243]
Epitome theologiae christianae / Abaelardus, Petrus (Abelard, Peter); ed by Rheinwald, F H – Berolini, 1835 – 3mf – 8 – €7.00 – ne Slangenburg [240]
Epitome thesavri antiqvitatvm : hoc est, impp rom orientalium et occidentalium iconum ex antiquis numismatibus quam fidelissime deliniatarum / Strada, Jacobus de – Lugduni, 1553 – 1 – us Wisconsin U Libr [450]
Epitome trium terrae partium... / Vadian, J – Tigvri, Christoph Froschover, 1534 – 4mf – 9 – mf#PBU-400 – ne IDC [240]
Epitome...carlos 5 / Vera y Figueroa, Juan Antonio – 1622 – 9 – sp Bibl Santa Ana [946]
Epitre a m prendergast : apres avoir lu "un soir d'automne" / Chauveau, Pierre J O – S.l: s.n, 1881? – 1mf – 9 – mf#01106 – cn Canadiana [360]
Epitres et evangiles des dimanches et fetes de l'annee : precedes des prieres durant la sainte messe et des vepres et complies du dimanche – Montreal: E R Fabre, 1846 [mf ed 1991] – 4mf – 9 – 0-665-90556-4 – mf#90556 – cn Canadiana [241]
L'epoca – Rome, Italy 30 dec 1917-9 jul 1919 (imperfect) – 1 – uk British Libr Newspaper [074]
A epoca : orgam dos interesses da republica – Manaus, AM: 5 dez 1889 – (= ser Ps 19) – mf#P11B,06,36 – bl Biblioteca [079]
Epoca – London, UK. 4 Jan-15 Dec 1842 – 1 – uk British Libr Newspaper [072]
La epoca – Madrid. Spain. -d. 2 Jan 1870-22 May 1874, 1 Jan 1875-11 Jul 1936 – 223r – 1 – uk British Libr Newspaper [074]
La epoca – San Antonio, mar 1918-jun 7, 1931 – 14r – us CRL [079]
La epoca – Santiago, Chile. Mar 1987-Feb 1990 – 39r – 1 – us L of C Photodup [079]
La epoca – Madrid, Spain. Jul 1909-10 Jun 1936 – 54r – 1 – (some issues lacking) – us L of C Photodup [074]
La epoca – Tegucigalpa, Honduras: F Z Duron, 1956-jan 18 1958 – 5r – 1 – us CRL [079]
Epoca antigua y de la conquista / Barberena, Santiago Ignacio – San Salvador, El Salvador. 1914 – 1r – 1 – us UF Libraries [972]
Epoca colonial / Barberena, Santiago Ignacio – San Salvador, El Salvador. 1917 – 1r – 1 – us UF Libraries [972]
Epoch – Ithaca. 1955+ (1) 1964+ (5) 1964+ (9) – ISSN: 0145-1391 – mf#1027 – us UMI ProQuest [400]
An epoch in printing : being the first matter set on the first linotype machine manufactured in canada / Faustus – Montreal: Linotype Co, [1892] [mf ed 1980] – 1mf – 9 – 0-665-03077-0 – mf#03077 – cn Canadiana [680]
Epoch makers of modern missions / McLean, Archibald – New York: F H Revell, c1912 – 1mf – 9 – 0-7905-5120-9 – mf#1988-1120 – us ATLA [240]
The epoch of creation : the scripture doctrine contrasted with the geological theory / Lord, Eleazar – New York: Scribner, 1851 – 1mf – 9 – 0-524-05988-8 – mf#1992-0725 – us ATLA [220]
Die epochen der kirchlichen geschichtschreibung / Baur, Ferdinand Christian – Tuebingen: Fues, 1852 – 1mf – 9 – 0-7905-4020-7 – mf#1988-0020 – us ATLA [240]
Epochs in baptist history : read before the baptist ministers' and laymen's union of kansas city and vicinity... / Griffith, Elmer Cummings – Liberty, MO: Advance, [1908?] – 1mf – 9 – 0-524-06619-1 – mf#1991-2674 – us ATLA [242]
Epochs in buddhist history : the haskell lectures, 1921 / Saunders, Kenneth James – Chicago, IL: University of Chicago Press, 1924 – (= ser Samp: indian books) – us CRL [280]
Epochs in church history : and other essays / Washburn, Edward Abiel; ed by Tiffany, Charles Comfort – New York: EP Dutton, 1883 – 1mf – 9 – 0-524-04023-0 – mf#1990-1195 – us ATLA [240]
Epochs in the life of jesus : a study of development and struggle in the messiah's work / Robertson, A T – New York: Charles Scribner, 1907 [mf ed 1988] – 1mf – 9 – 0-7905-0220-8 – (incl ind) – mf#1987-0220 – us ATLA [240]
Epochs in the life of paul : a study of development in paul's career / Robertson, A T – New York: Charles Scribner, 1909 – 1mf – 9 – 0-8370-4920-2 – (incl ind) – mf#1985-2920 – us ATLA [920]
Epochs of the arts / Hoare, Prince – London 1813 – (= ser 19th c art & architecture) – 5mf – 9 – mf#4.2.288 – uk Chadwyck [700]
Epokha bielinskago : obshchii ocerk: iz lektsii, chitannykh v petrogradskom universitetie / Vengerov, S A – Izd. 2-e. [Petrograd]: Kn-vo "Prometei" N N Mikhailova, 1915 – 1 – us CRL [947]
L'epopee canadienne : histoire du canada: cahier d'exercices sur le manuel de 6e et de 7e annee / Brisebois, Raymond – Montreal: la Librairie des ecoles, les freres des ecoles chretiennes, [1961?] (mf ed 1992) – 2mf – 9 – (with ind) – mf#SEM105P1694 – cn Bibl Nat [971]
L'epopee canadienne : histoire du canada: cahier d'exercices sur le manuel de 6e et de 7e annee / Brisebois, Raymond – Montreal: Lidec inc, [1961?] (mf ed 1992) – 2mf – 9 – (with ind) – mf#SEM105P1691 – cn Bibl Nat [971]
L'epopee canadienne : histoire du canada: cahier d'exercices sur le manuel de 6e et de 7e annee: [guide du professeur] / Brisebois, Raymond – [Montreal]: la Librairie des ecoles, les freres des ecoles chretiennes, [1961?] (mf ed 1992) – 2mf – 9 – (with ind) – mf#SEM105P1692 – cn Bibl Nat [971]
L'epopee de samba gueladiegui : etude d'une version inedite / Ly, Amadou – 1978 – us CRL [470]
L'epopee d'el hadj omar : approche litteraire et historique / Dieng, M Samba – 1984 – us CRL [470]
Une epopee zarma : wangougna issa korombeize modi, ou, issa koygolo "mere de la science de la guerre" / Mahamane, Tandina Ousmane – 1984 – us CRL [470]
Epopeia – Moscow-Berlin. n1-4. 1922-1923. – 1 – us NY Public [073]
Epopeia de antonio joao / Mello, Raul Silveira De – Rio de Janeiro, Brazil. 1969 – 1r – 1 – us UF Libraries [972]
Epopeya de la raza / Echavarria, Colon – San Juan, Puerto Rico. 1946 – 1r – 1 – us UF Libraries [972]
Epopeya de marti desda paula hasta dos rios / Casals Llorente, Jorge – Matanzas, Cuba. 1953 – 1r – 1 – us UF Libraries [972]
Epopeya del moncada / Rodriguez Santos, Justo – Habana, Cuba. 1963 – 1r – 1 – us UF Libraries [972]
L'epoque : journal complet et universel – Paris. 1846 – 1 – fr ACRPP [073]
L'epoque – Journal politique et litteraire. Red. en chef E. Feydeau. Paris. 9 mars-21 sept 1865, 25 mai-1er dec 1868, 28 mars-25 juil 1869 – 1 – fr ACRPP [944]
L'epoque – Paris. n277-386. 12 mars-30 juin 1938 – 1 – fr ACRPP [073]
Epoque – Paris, France. 9 may-6 dec 1945 – 1/2r – 1 – uk British Libr Newspaper [072]
L'epoque carolingienne (he6) – Paris, 1937 – (= ser Histoire de l'eglise (he)) – €25.00 – ne Slangenburg [931]
Epoque de 1815 : ou choix de propositions, lois, rapports discutes a la chambre des deputes... / Guerard, Emile – Paris 1821 [mf ed Hildesheim 1995-98] – 1v on 3mf – 9 – €90.00 – ISBN-10: 3-487-26325-4 – ISBN-13: 978-3-487-26325-0 – gw Olms [350]
L'epoque moderne : journal politique, litteraire, industriel, commercial et financier du credit europeen – London, England 2 jun-1 dec 1877 – 1 – (discontinued) – uk British Libr Newspaper [300]
Epoque ou nous vivons / Capek, Karel – Paris, France. 1939 – 1r – 1 – us UF Libraries [440]
Epoux avant le mariage / Desaugiers, Marc-Antoine – Paris, France. 1808 – 1r – 1 – us UF Libraries [440]
Eppelbaum, Menahem Baerush see
– Oyfbroyz
– Zeydns shkia
Eppendorf-winterhuder nachrichten – Hamburg DE, 1883-1905 – 24r – 1 – gw Misc Inst [074]
Epping and district times – Eastwood, 1923; 1946 – 2r – A$133.01 vesicular A$144.01 silver – (aka: northern district times) – at Pascoe [079]
Epping, Joseph see Astronomisches aus babylon
Eppinger zeitung see
– Heilbronner stimme
– Neue eppinger zeitung
Eppler, Christoph Friedrich see
– Geschichte der gruendung der armenisch-evangelischen gemeinde in schamachi
– Karl rudolf hagenbach
Eppler, Paul see Geschichte der basler mission, 1815-1899
L'epreuve : Revue d'art mensuelle – IV, n.s., no. 3-5; VI, n.s., no. 19-21. Paris. janv-mars 1903, sept-nov 1904 – 1 – fr ACRPP [700]
Epshtayn, Yehoshu'a Ben Nahman see Keren yehoshu'a...
Epsitolae binae de virginitate, syriace / Clemens Romanus; ed by Beelen, J Th – Lovanii, 1856 – €23.00 – ne Slangenburg [227]
Epsom observer & mid surrey county chronicle – [London & SE] Sutton 6 feb 1903-dec 1907 [mf ed 2003] – 5r – 1 – uk Newsplan [072]
Epstein, A L see Juridical techniques and the judicial process
Epstein, A L (Arnold Leonard) see Politics in an urban african community
Epstein, Jehudo see Mein weg von ost nacht west
Epstein, Kalonymus Kalman see Ma'or va-shemesh 'al hamishah humshe torah
The epworth singers and other poets of methodism / Christophers, Samuel Woolcock – New York: A. D. F. Randolph, [pref. 1874]. Chicago: Dep of Photodup, U of Chicago Lib, 1972 (1r); Evanston: American Theol Lib Assoc, 1984 (1r) – 1 – 0-8370-0098-X – (incl ind) – mf#1984-B303 – us ATLA [810]
Eq : educause quarterly – Boulder. 2000+ – 1,5,9 – (cont: cause/effect) – ISSN: 1528-5324 – mf#12570,01 – us UMI ProQuest [370]
Equador – Manaus, AM. 01 jan-20 maio 1888 – (= ser Ps 19) – mf#P11B,06,37 – bl Biblioteca [079]
Equador : revista dos interesses publicos – Alemquer, PA. 17 nov 1888 – (= ser Ps 19) – bl Biblioteca [073]
O equador / Thome, S – jul 1926-sep 1927 – us CRL [079]
Equal employment opportunity commission administrative history see Civil rights during the johnson administration, 1963-1969
Equal justice / Communist Party. USA International Labor Defense – v1-16 n3. 1926-42 [all publ] – (= ser Radical periodicals in the united states, 1881-1960. series 1) – 37mf – 9 – $335.00 – us UPA [335]
Equal opportunities international – Patrington. 1992-1995 (1,5,9) – ISSN: 0261-0159 – mf#16488 – us UMI ProQuest [331]
Equal opportunity forum – 1977 oct-1979 dec, 1980-1982 may – 2r – 1 – mf#513616 – us WHS [322]
Equal opportunity forum – Venice. 1979-1982 (1,5,9) – ISSN: 0192-1533 – mf#12235 – us UMI ProQuest [331]
Equal rights : independent feminist weekly – ser 1 v1 n1-9; ser 2 v1-2 n43. 1935-36 [all publ] – (= ser Periodicals on women and women's rights, series 2) – 1r – 1 – $140.00 – us UPA [322]
Equal rights – v. 1-40. 1923-54 – 1 – us L of C Photodup [360]
Equal Rights Association for the Province of Ontario see Address by the provincial council to the people of ontario
Equality / Bellamy, Edward – New York: D Appleton, 1897 – 1mf – 9 – 0-7905-7321-0 – (Sequel to "Looking backward, 2000-1887") – mf#1989-0546 – us ATLA [335]
Equality – New York. N.Y. 1939-40 – 1 – us AJPC [071]
Equality – The American Way: Equal Rights and Equal Opportunity for All. New York. v. 1-2 no. 10. May 1939-Oct Nov 1940 – 1 – us NY Public [977]
Equality now! : monthly newsletter...madison chapter / National Organization for Women – 1973 jul-1984 dec – 1r – 1 – (cont: do it now (madison, wi)) – mf#998809 – us WHS [322]
Equality of jew and gentile in the new testament dispensation / Mccaul, Alexander – London, England. 1838 – 1r – us UF Libraries [225]
Equalization of the duty on coals : a short address, shewing the impolicy, the partiality, and inhumanity of the present local duties on coals – London: printed by Evans & Ruffy, 1817 – (= ser 19th c economics) – 1mf – 9 – mf#1.1.218 – uk Chadwyck [336]
Equalop / Planners for Equal Opportunity – v2 n1-v5 n1, 2, 3 [1968 spr-1971 win] – 1r – 1 – mf#1111411 – us WHS [322]
Equals – London, Race Relations Board. n1-14; apr may 1975-jun jul 1977 [bimnthly] – 1 – us Wisconsin U Libr [322]
Equatorial america / Ballou, Maturin Murray – Boston, MA. 1892 – 1r – 1 – us UF Libraries [972]
equilibre – 1991 sep 18/25-nov 27/dec 4 – 1r – 1 – mf#4695761 – us WHS [071]
Equilibrios poeticos de un! desequilibrado! : tomo 1-4 / Sanchez-Arjona, Vicente – Sevilla: Imp Alvarez, 1956 – 1 – sp Bibl Santa Ana [810]
Equiloecq, F V see Essai sur la litterature merveilleuse des noirs
Equine events – 1978 jul-1980 dec [v3 n6-v5 n12], 1981 jan-1983 jan [v6 n1-v8 n1] – 2r – 1 – mf#3862808 – us WHS [790]
Equine glanders and its eradication / Dawson, Charles F – Lake City, FL. 1905 – 1r – 1 – us UF Libraries [636]
Equine practice – Santa Barbara. 1979-1995 (1) 1979-1995 (5) 1979-1995 (9) – ISSN: 0162-8941 – mf#12428 – us UMI ProQuest [636]
Equinoxe de janviers / Marres, Jacques – Bruxelles, Belgium. 1959 – 1r – 1 – us UF Libraries [960]
Equipment management (em) – Lincolnwood. 1988-1991 (1,5,9) – ISSN: 0733-3056 – mf#16397,02 – us UMI ProQuest [650]
Equipping museums and libraries for the 21st century : hearing...house of representatives, 107th congress, 2nd session...washington dc, feb 14 2002 / United States. Congress. House. Committee on Education and the Workforce. Subcommittee on Select Education – Washington: US GPO 2002 [mf ed 2002] – 1mf – 9 – 0-16-068869-8 – (incl ind) – us GPO [020]
Equipping the laity as worship leaders in the ministry of big bethel baptist church / Dawson, Walter Robert – 1982 – 1 – $5.00 – us Southern Baptist [242]
Equitable Fraternal Union [Neenah, WI] see Friend and guide
Equitable Fraternal Union [Neenah, WI] et al see Friend and guide, the messenger
Equitable reserve guide – 1942 sep-1952 dec, 1953 jan-1961 dec, 1962 jan-1975 jun, 1975 jul-1986 may – 4r – 1 – (cont: friend and guide, the messenger; cont by: guide (equitable reserve association)) – mf#1095281 – us WHS [071]
Equitas see Exposition of the case of lieutenant-colonel bouchette, surveyor-general
Equity / Cook, Forrest – Los Angeles 1938 112 p. LL-360 – 1 – us L of C Photodup [342]
Equity / Institute for Global Education – n1-41 [1980 oct-1988 feb] – 1r – 1 – mf#1321416 – us WHS [071]
Equity : its principles in procedure, codes and practice acts / Hughes, William Taylor – St Louis: Central Law Journal Co, 1911 – 1 – mf#LL-86 – us L of C Photodup [347]
The equity – Shawville (Quebec). v1 n1 jun 7 1883- (wkly) [mf ed 1973-] – 1 – mf#SEM35P100 – cn Bibl Nat [073]
Equity as applied in the state and federal courts in texas and other states / Simkins, William Stewart – 2nd ed. Kansas City, MO: Vernon, 1911 – 1002p – 1 – mf#LL-1333 – us L of C Photodup [347]
Equity case files from the western district court of texas at el paso relating to the chinese exclusion acts, 1892-1915 / U.S. District Court – (= ser Records of district courts of the united states) – 34r – 1 – (with printed guide) – mf#M1610 – us Nat Archives [347]
Equity case files of the us circuit court for the southern district of new york,1791-1846 / U.S. Circuit and District Courts – (= ser Records of district courts of the united states) – 23r – 1 – (with printed guide) – mf#M884 – us Nat Archives [347]
Equity co-op news / Equity Cooperative Livestock Sales Association – 1935 aug-1936 mar – 1r – 1 – (cont: equity news; cont by: equity co-operative livestock news) – mf#3396812 – us WHS [636]
Equity co-operative livestock news / Equity Cooperative Livestock Sales Association – 1936 apr-1942 dec – 1r – 1 – (cont: equity co-op news; cont by: equity co-operative livestock sales association news) – mf#3396813 – us WHS [636]
Equity Cooperative Livestock Sales Association see
– Equity co-op news
– Equity co-operative livestock news
– Equity news

Equity jurisdiction, pleading and practice in maine / Whitehouse, Robert Treat – Portland: Loring, Short & Harmon, 1900 – 949p – 1 – mf#LL-1647 – us L of C Photodup [347]

Equity jurisprudence / Steele, Sherman – New York: Prentice-Hall, 1927 – 897p – 1 – mf#LL-1071 – us L of C Photodup [342]

Equity news / Actors' Equity Association – 1972-1979, 1980 jan-1994 dec – 2r – 1 – (cont: equity (new york, ny)) – mf#1385605 – us WHS [790]

Equity news / American Society of Equity – 1913 may 10-dec, 1914-1918 feb 1 – 2r – 1 – (cont: wisconsin equity news (1908); cont by: wisconsin equity news (1919)) – mf#3416323 – us WHS [630]

Equity news / Equity Cooperative Livestock Sales Association – 1935 jan-jul – 1r – 1 – (continued by: equity co-op news) – mf#3396811 – us WHS [636]

Equity news : the official paper / American Society of Equity – 1920 jan 7-1921, 1922-25, 1926-29, 1930-32, 1933-1934 jun 1 – 5r – 1 – (cont: wisconsin equity news (1919); cont by: equity union farmer) – mf#3416341 – us WHS [630]

Equity newsletter / American Friends Service Committee – n4 [1971 dec 12], v4 n1 [1972 feb 18] – 1r – 1 – mf#722005 – us WHS [071]

Equity practice in the united states circuit courts / Shiras, Oliver Perry – 2nd ed. Chicago: Callaghan, 1898 – 226p – 1 – mf#LL-1134 – us L of C Photodup [347]

Equity precedents / Curtis, Charles Ticknor – 4th ed. Boston: Little, Brown, 1869 – 596p – 1 – mf#LL-555 – us L of C Photodup [347]

Equity precedents : supplementary to mr justice story's treatise on equity pleadings / Curtis, George Ticknor – Boston: Little, Brown, 1850 – 562p – 1 – mf#LL-557 – us L of C Photodup [347]

Equity procedure : embodying the principles of pleading and practice applicable to courts of equity...in virginia and west virginia... / Hogg, Charles Edgar – Cincinnati: Anderson Co, 1921 – 2v – 1 – mf#LL-16206 – us L of C Photodup [347]

Equity records of the us circuit court for the eastern district of pennsylvania, 1790-1847 / U.S. Circuit and District Courts – (= ser Records of district courts of the united states) – 23r – 1 – (with printed guide) – mf#M985 – us Nat Archives [347]

Equity union farmer / National Farmers' Union [US] – 1934 oct – 1r – 1 – (cont: equity news [wisconsin edition]; cont by: farmers' equity union news) – mf#3416352 – us WHS [630]

Equus – Harrisburg. 1979+ (1,5,9) – ISSN: 0149-0672 – mf#12057 – us UMI ProQuest [636]

Er biblen guds ord? : kritiske betragtninger over de vigtigste afsnit af det gamle og nye testamante / Johnson, N S – Sioux Falls, Dakota: Forfatterens Forlag, 1887 – 3mf – 9 – 0-524-07887-4 – mf#1991-3432 – us ATLA [220]

ER environmental review – Pittsburgh. 1976-1989 (1) 1976-1989 (5) 1976-1989 (9) – (cont by: environmental history review) – ISSN: 0147-2496 – mf#12757 – us UMI ProQuest [333]

Er hilft uns frei aus aller not : erlebnisberichte aus den septembertagen 1939 / ed by Kammel, Richard – Posen: Lutherverlag, 1940 – 1mf – 9 – 0-524-08113-1 – mf#1993-9019 – us ATLA [240]

Er werd een stad geboren / Faber, G H von – Soerabaja, 1953 – 5mf – 9 – mf#SE-1424 – ne IDC [959]

Era – Fremantle, Australia. 2 feb 1867-26 dec 1868; 1869-3 jul 1886 – 12r – 1 – uk British Libr Newspaper [072]

Era – Lake Mills WI. 1875 july 21 – 1r – 1 – mf#957246 – us WHS [071]

Era – New Orleans, LA. 1863-1864 (1) – mf#63535 – us UMI ProQuest [071]

The era – London. -w. 1865-90. (68 reels) – 1 – uk British Libr Newspaper [072]

Era almanack and annual : the dramatic and musical – London. 1868-1919 (1) – mf#5318 – us UMI ProQuest [790]

Era dispatch / quiver / news / dispatch / Champaign Co. Saint Paris – (1884-1953) gap fillers [wkly] – 3r – 1 – (title changes) – mf#B12109-12111 – us Ohio Hist [071]

Era magazine : an illustrated monthly – v13 [1904 jan-jun] – 1r – 1 – (cont: era [philadelphia, pa: 1901]; cont by: new england magazine) – mf#2892072 – us WHS [071]

L'era nuova – Paterson NJ, 1908-19 – 3r – 1 – (italian newspaper) – us IHRC [071]

The era of expressionism = Expressionismus / ed by Raabe, Paul – London: Calder & Boyars, c1974 [mf ed 1993] – (= ser German expressionism) – 423p – 1 – (tran by j m ritchie. incl bibl ref and ind. ann by ed) – mf#8281 – us Wisconsin U Libr [430]

Era of progress – Pepin WI. 1861 dec 25 – 1r – 1 – mf#960866 – us WHS [071]

Era rossii / Narodnaia Natsional'naia Partiia – Moscow, Russia. n1(ian 1994)-n2(mai.1994), n4(sent 1994)-n7(ianv 1995), spetsvyp n1(apr 1995), n9(apr 1995)-n11(iiun 1995), n13(dek 1995)-n19(iiul 1996), n21(sep 1996), n24(ian 1997)-n31(avg 1998) – 1 – mf#mf-12248 (reel 4) – us CRL [077]

Era series / Columbiana Co. Salem – dec 1873-feb 1887, feb 1889-1918 [wkly, daily, semiwkly] – 20r – 1 – mf#B4314-4333 – us Ohio Hist [071]

Eraclius : deutsches und franz"sisches gedicht des 12. jahrhunderts... / Otto; ed by Massmann, H F – Quedlinburg; Leipzig: G Basse, 1842 – us Wisconsin U Libr [430]

Eraskine, T *see* Remarks on the internal evidence for the truth of revealed

Erasmus / Finch, A Elley – London, England. 1875 – 1r – us UF Libraries [240]

Erasmus : onderzoek naar zijne theologie en zijn godsdienstig gemoedsbestaan / Lindeboom, Johannes – Leiden: AH Adriani, 1909 – 1mf – 9 – 0-7905-7010-6 – (incl bibl ref) – mf#1988-3010 – us ATLA [240]

Erasmus *see*
- Enchiridion oder handbuchlin eins waren christenlichen vn strytbarlichen lebens...
- Ein expostulation oder klag jhesu zu dem menschen, der vss eygnem mutwill verdampt wuert...
- Ein fast nutzlich vslegung des ersten psalmen...
- Ein klag des frydes...
- Paraphrases zu tuetsch, die epistlen sancti pauli...
- Paraphrasis

Erasmus and other essays / Dods, Marcus – London: Hodder & Stoughton, 1891 – 1mf – 9 – 0-7905-5983-8 – mf#1988-1983 – us ATLA [240]

Erasmus, D *see* Encomium moriae

Erasmus, Desiderius *see*
- The complaint of peace
- In praise of folly
- Novvm instrumentu omne
- Opera omnia
- Opus epistolarum
- Praise of folly
- Proverbs or adages, gathered out of the "chiliades" and englished by richard taverner

Erasmus, J C *see* 'N ekonomies-antropologiese studie van die abakwamkhwanazi van natal

Erasmus, R G *see* Strategic human resource management and organizational performance

Erasmus, Rudolf *see* Verslag van die kommissie van ondersoek na bedryfsgesondheid

Erastus, Thomas *see* Dispvtationvm de medicina nova philippi paracelsi pars prima: in qva, qvae de remediis..

Erath, A *see* Augustus velleris aurei ordo per emblemata, ectheses politicas et historiam demonstratus

Eravati kam nam bhe mha lvam te ma chui kra to pa nhan a khra vatthu tui mya : [short stories] / Myui Su Ra – Ran Kun: Ca pe Biman 1987 [mf ed 1990] – (= ser Ca pe Biman thut pann su lak cvai ca can) – 1r with other items – 1 – (in burmese) – mf#mf-10289 seam reel 166/3 [§] – us CRL [830]

Erazo, Salvador L *see* Parnaso salvadoreno

Erb, Frank Otis *see* The development of the young people's movement

Erb, William Harvey *see* Dr. nevin's theology

Erbach, Albrecht *see* Bemerkungen auf einer reise durch einen theil der schweiz und einige ihrer naechsten umgebungengeschrieben im bluethen-monath

Erbach, Christian *see*
- Modi sacri sive cantus musici...
- Modorum sacrorum sive cantionum...liber secundus

Erbacher zeitung – Erbach / Donau DE, 1902-13 – 6r – 1 – gw Misc Inst [074]

Erbach-Erbach, Eberhard zu *see* Wandertage eines deutschen touristen im strom- und kuestengebiet des orinoko

Erbach-Erbach, Ernst zu *see* Reisebriefe aus amerika

Der erbarzt – Berlin DE, 1934 n1-7, 1938 n8 – 1 – gw Misc Inst [575]

Das erbe : ein buch gedanken, bilder und gestalten / ed by Klein, Timotheus – Muenchen: R Piper 1921 [mf ed 1993] – 1r [ill] – 1 – (filmed with: sputnik contra bombe / gerhard wolf [ed]) – mf#3336p – us Wisconsin U Libr [800]

Das erbe der uraniden : roman / Dominik, Hans – Berlin: Scherl, 1943 [mf ed 1989] – 340p – 1 – mf#7181 – us Wisconsin U Libr [830]

Das erbe des ostdeutschen volksgesanges / Salmen, Walter – Wuerzburg: Holzner-Verlag 1956 [mf ed Bloomington IN: Indiana Uni Lib, Preservation Dept 1984] – 1r – 1 – us Indiana Preservation [390]

Das erbe des wucherers : original-volksstueck mit gesang in fuenf acten / Costa, Carl – [Wien?: s.n. 18–?] [mf ed 1993] – 1r – 1 – mf#8538 – us Wisconsin U Libr [820]

Erbe und gegenwart : eine anthologie zur schoenen literatur / ed by Hoefer, Karlheinz et al – rev ed. Leipzig: Fachbuchverlag, 1962 – 1 – us Wisconsin U Libr [800]

Erbe und gegenwart : eine auswahl aus der deutschen literatur / ed by Baer, Heinz et al – Leipzig: Fachbuchverlag, 1959 – 1 – us Wisconsin U Libr [800]

Erbe und tradition in der literatur / Dahnke, Hans-Dietrich – Leipzig: VEB Bibliographisches Institut, 1977 – 1r – 1 – (incl bibl ref) – us Wisconsin U Libr [430]

Das erbe wolgasts : ein querschnitt durch die heutige jugendschriftenfrage / Fronemann, Wilhelm – Langensalza: J Beltz, 1927 – 1 – us CRL [890]

Erbes, C *see* Die todestage der apostel paulus und petrus

Erbes, Carl *see* Die todestage der apostel paulus und petrus

Erblaendische staatsanzeigen / ed by Luca, Ignaz de – Wien 1785 – (= ser Dz. historisch-politische abt) – 1v[=2pt] on 6mf – 9 – €120.00 – mf#k/n1692 – gw Olms [074]

Erbrecht / Crome, Carl – Tuebingen: J C B Mohr, 1912 – (= ser Civil law 3 coll; System des deutschen buergerlichen rechts) – 9mf – 9 – (Incl bibl ref and index) – mf#LLMC 96-550 – us LLMC [348]

Erbrecht / Leonhard, Franz – 2., vollst neu bearb Aufl. Berlin: C Heymann, 1912 – (= ser Civil law 3 coll; Kommentar zum buergerlichen gesetzbuche und seinen nebengesetzen) – 7mf – 9 – (Incl bibl ref and index) – mf#LLMC 96-559 – us LLMC [346]

Erbt, Wilhelm *see*
- Deutsche einsamkeiten
- Handbuch zum alten testament
- Israel und juda
- Jeremia und seine zeit
- Das markusevangelium
- Die purimsage in der bibel
- Die sicherstellung des monotheismus durch die gesetzgebung im vorexilischen juda
- Von jerusalem nach rom

Ercilla – Santiago, Chile: Prensas de la editorial Ercilla, SA. v3 n112-v10 n488 jul 1937-sep 5 1944 8r. v10 n489-v13 n632 sep 12 1944-jun 10 1947 4r – 12r – 1 – us CRL [079]

Ercilla y Zuniga, Alonso de *see* La araucana

Erck, Wentworth *see* The land question

Erd / Korn, Rachel H – Warsaw, Poland. 1936 – 1r – us UF Libraries [939]

Erd- oder feuerbestattung : der biblische brauch auf ethnographischem hintergrund / Caspari, Wilhelm – Berlin: Edwin Runge 1914 [mf ed 1989] – (= ser Biblische zeit- und streitfragen 9/10) – 1mf – 9 – 0-7905-1631-4 – mf#1987-1631 – us ATLA [220]

Erda energy research abstracts / United States Energy Research and Development Administration – Oak Ridge. 1976-1977 (1,5,9) – (cont: erda research abstracts; cont by: energy research abstracts) – ISSN: 0361-9869 – mf#11139,01 – us UMI ProQuest [333]

Erda research abstracts / United States Energy Research and Development Administration – Oak Ridge. 1976-1976 (1,5,9) – (cont by: erda energy research abstracts) – ISSN: 0361-9877 – mf#11139 – us UMI ProQuest [333]

Erdachte briefe / Eschmann, Ernst Wilhelm – Baden-Baden: H Buehler, 1946 (mf ed 1990) – 1 – us Wisconsin U Libr [860]

Erdachte gespraeche / Ernst, Paul – Frontbuchhandelsausg. Muenchen: A Langen/ G Mueller, 1944 – 1 – us Wisconsin U Libr [830]

Der erdball und seine naturwunder : populaires handbuch der physischen geographie / Vollmer, Carl G – Berlin 1855 [mf ed Hildesheim 1995-98] – 4mf – 9 – €120.00 – 3-487-29978-X – gw Olms [910]

Erdbeschreibung des kurfuerstenthums hessen : nach der neuesten staatseintheilung abgefasst und zum gebrauch fuer buerger- und volksschulen eingerichtet / Wiegand, Conrad – Cassel 1825 [mf ed Hildesheim 1995-98] – 2mf – 9 – €60.00 – 3-487-29497-4 – gw Olms [914]

Erdbrink, Gerhard Rudolf *see* Geutzlaff, de apostel der chinezen

Die erde : neue dichtungen / Bonsels, Waldemar et al – Muenchen-Schwabing: E W Bonsels 1905-06 [mf ed 1993] – 1r – 1 – (filmed with: bluethen und perlen / p f l warns [ed]) – mf#3344p – us Wisconsin U Libr [810]

Erde, D *see* Mensheviki

Die erde und ihre bewohner : ein hand- und lesebuch fuer alle staende / Hoffmann, Karl F – Stuttgart 1835 [mf ed Hildesheim 1995-98] – 5mf – 9 – €100.00 – 3-487-29953-4 – gw Olms [910]

Erdgeist : tragoedie in vier aufzuegen / Wedekind, Frank – Muenchen: G Mueller 1920 [mf ed 1996] – 1r – 1 – (filmed with: die ungleichen schalen / jakob wassermann) – mf#4056p – us Wisconsin U Libr [820]

Erdkraft : roman / Dreyer, Max – 3. aufl. Muenchen: F Eher, 1944 [mf ed 1989] – 371p – 1 – mf#7185 – us Wisconsin U Libr [830]

Die erdkunde im verhaeltniss zur natur und zur geschichte des menschen / Ritter, Carl – Berlin 1832-59 [mf ed Hildesheim 1995-98] – 21v on 137mf – 9 – €822.00 – 3-487-29999-2 – gw Olms [900]

Erdman, Charles Rosenbury *see* The gospel of john

Erdman, William Jacob *see*
- Ecclesiastes
- The symbolic structure of the gospel according to john

Erdmann, C *see* Studien zur briefliteratur deutschlands im 11. jahrhundert (mgh schriften:1.bd)

Erdmann, David *see*
- The books of samuel
- Luther und seine beziehungen zu schlesien, insbesondere zu breslau

Erdmann, Johann F *see* Beitraege zur kenntniss des innern von russland

Erdmann, Julius *see* Eichendorffs historische trauerspiele

Erdmann, Oskar *see* Otfrids evangelienbuch

Erdmann, Stephan *see* Jenseits des rationalitaetsprinzips

Der erdoelpionier *see* Der bohrkumpel

Erdues, K *see* Zwingli 67 tetele...

Erdumsegelung der koenigl schwedischen fregatte eugenie : in den jahren 1851 bis 1853 ausgefuehrt unter dem befehl des commandeur-capitains c a virgin behufs anknuepfung politischer und commercieller beziehung und wissenschaftlicher beobachtungen und entdeckungen / Etzel, Anton von – Berlin 1856 [mf ed Hildesheim 1995-98] – 2v on 4mf – 9 – €120.00 – ISBN-10: 3-487-26613-X – ISBN-13: 978-3-487-26613-8 – gw Olms [910]

L'ere nigerienne : [essai d'epopee anectotique sur l'histoire de l'ouest-africain francais] / Perron, Michel – Paris: Editions de la Pensee latine, 1926 – 1 – us CRL [960]

L'ere nouvelle : grand organe quotidien de l'entente des gauches – Paris. 27 dec 1919-juin 1940 – 1 – fr ACRPP [335]

l'ere nouvelle – Paris. juil 1893-nov 1894 – 1 – (revue mensuelle de socialisme scientifique) – fr ACRPP [073]

Ere nouvelle – 1992 oct 30/nov 17, dec 20/1992 jan 5 – 1r – 1 – mf#4695770 – us WHS [071]

Erec / Aue, Hartmann von der; ed by Leitzmann, Albert – Halle/Saale: M Niemeyer, 1939 [mf ed 1993] – (= ser Altdeutsche textbibliothek n39) – xxxvi/262p – 1 – mf#8193 reel 4 – us Wisconsin U Libr [830]

The erechtheion at athens : fragments of athenian architecture / Inwood, Henry William – London 1831 – (= ser 19th c art & architecture) – 8mf – 9 – mf#4.2.1610 – uk Chadwyck [720]

Erection of french river bridge : canadian pacific railway / Monsarrat, Charles Nicholas – [S.l: s.n, 1908?] [mf ed 1991] – 1mf – 9 – 0-665-99508-3 – mf#99508 – cn Canadiana [624]

Erekhe ha-noutariukin = Lexikon der abbreviaturen / Haendler, G H – Frankfurt (Main): J Kauffmann, 1897 – 1mf – 9 – 0-8370-7153-4 – mf#1986-1153 – us ATLA [470]

Erem, Moshe *see* Sionismo ante el juicio internacional

...L'eremita, la carcere, e l' diporto : opera nella quale si contengano nouelle, and altre cose morali... / Granucci, N – Lucca, 1569 – 4mf – 9 – mf#H-8309 – ne IDC [956]

Eremita, Victor *see* Enten-eller

Eremitenschule in altbayern / Heigenmooser, J – Berlin, 1903 – €7.00 – ne Slangenburg [241]

Eremites et reclus : etudes sur d'anciennes formes de vie religieuses / Gougaud, L – Liguge, 1928 – 3mf – 8 – €7.00 – ne Slangenburg [210]

Gli eretici d'italia : discorsi storici / Cantu, Cesare – Torino: Unione tipografico-editrice, 1865-1866 – 4mf – 9 – 0-524-08266-9 – (incl bibl ref) – mf#1993-3021 – us ATLA [240]

Erets avotenu / Levontin, Zalman David – Tel-Aviv, Israel. v1-3. 1923 – 1r – 1 – us UF Libraries [939]

Erets rusyah u-melo'ah / Levinsohn, Joshua – Vilna, Lithuania. 1868 – 1r – 1 – us UF Libraries [025]

Erets yisra'el / Morpurgo, Luciano – Rome, Italy. 1930 – 1r – 1 – us UF Libraries [939]

Erets yisrael / Ben-Yehuda, Eliezer – Jerusalem, Israel. 1883 – 1r – 1 – us UF Libraries [939]

Erevan – Sofia, Bulgaria. Jul 1952-May 1958 (scattered issues) – 1r – 1 – us L of C Photodup [949]

Erez israel : jahrbuch des keren kajemeth lejisrael (juedischer nationalfonds) – Berlin: Jewish National Fund. v1. 1921-23 – (= ser German-jewish periodicals...1768-1945, pt 3) – 1r – 1 – $165.00 – mf#B70 – us UPA [939]

Erez yisrael un di idishe arbeiterschaft / Zhitlowsky, Chaim – New York, NY. 1918 – 1r – 1 – us UF Libraries [939]

Erez-yisrael we-suriyah ha-deromith / Press, Jesaias – Wien, Austria. 1921 – 1r – 1 – us UF Libraries [939]

Erfa 1840 – Euskirchen DE, 1840 5 jul-1849, 1851-1863 26 dec – 1 – (with gaps. title varies: 9 jan 1842: intelligenzblatt fuer die kreise euskirchen, rheinbach und ahrweiler; 11 feb 1844: erfa; 2 jun 1844: kreis-intelligenzblatt fuer euskirchen, rheinbach und ahrweiler; 6 nov 1847: intelligenzblatt fuer den kreis euskirchen und den kreis rheinbach; 1 jan 1848: intelligenzblatt fuer die kreise rheinbach und euskirchen; 17 dec 1851: kreis-intelligenzblatt fuer rheinbach und euskirchen; 3 jul 1852: kreis-intelligenzblatt fuer euskirchen und rheinbach; 20 sep 1856: kreis-intelligenzblatt fuer euskirchen und rheinbach) – gw Misc Inst [074]

Die erfahrung in platons ideenlehre : die idee als gestalt der erfahrung / Joannou, Petros-Perikles – Muenchen, 1936 (mf ed 1992) – 1mf – 9 – €24.00 – 3-89349-056-6 – mf#DHS-AR 18 – gw Frankfurter [180]

Erfahrungen und widersprueche : versuche ueber literatur / Fuehmann, Franz – 1. aufl. Rostock: Hinstorff, 1975 [mf ed 1992] – 222p – 1 – mf#8253 – us Wisconsin U Libr [430]

Der erfahrungsbeweis fuer die wahrheit des christenthums / Wendt, Hans Hinrich – Goettingen: Vandenhoeck & Ruprecht, 1897 [mf ed 1985] – 1mf – 9 – 0-8370-5784-1 – mf#1985-3784 – us ATLA [240]

Erfassung und beurteilung geobotanischer daten : kritische betrachtung ausgewaehlter methoden / Schleiter, Ingrid M – (mf ed 1995) – 3mf – 9 – €49.00 – 3-8267-2106-3 – mf#DHS 2106 – gw Frankfurter [574]

Der erfolg der werbung : ansaetze zur messung des oekonomischen werbeerfolgs / Feller, Dirk – (mf ed 1997) – 1mf – 9 – €30.00 – 3-8267-2432-1 – mf#DHS 2432 – gw Frankfurter [650]

Die erforschung afrikas. / Hassert, Kurt – Leipzig: W. Goldmann, 1943. 259p. maps. Bibliog – 1 – us Wisconsin U Libr [960]

Der erft-bote – Bedburg 1950 2 mar-17 jun [mf ed 2004] – 1r – 1 – (filmed by misc inst: 1941 1 jul-31 dec [1r] with suppl) – gw Mikrofilm; gw Misc Inst [074]

Erfte indianer : die dem christoph columbus verkommen – s.l, s.l? 1755 – 1r – 1 – us UF Libraries [975]

Erfttal-nachrichten – [Erftstadt-]Oberliblar 1950 3 mar-1951 27 apr [mf ed 2004] – 1r – 1 – gw Mikrofilm [074]

Erfurt, Ebernant von *see* Heinrich und kunigunde

Erfurter wochenzeitung – Erfurt DE, 1966 21 sep-1970 20 may – 1r – 1 – gw Misc Inst [074]

Erfurth, Fritz *see*
- Die "deutschen sagen" der brueder grimm
- Die deutschen sagen der brueder grimm

Erfurtische gelehrte nachrichten – Erfurt DE, 1754-79 – 10r – 1 – (title varies: 1769: erfurtische gelehrte zeitungen; 1789: erfurtische gelehrte zeitung [...]) – gw Misc Inst [074]

Erfurtische gelehrte zeitung / ed by Froriep, Just Friedrich – Erfurt 1780-96 – (= ser Dz) – 17jge on 51mf – 9 – €510.00 – mf#k/n348 – gw Olms [430]

Erfurtische gelehrte zeitungen *see* Erfurtische gelehrte nachrichten

Ergaenzungen des allgemeinen landrechts fuer die preussischen staaten : enthaltend eine vollstaendige zusammenstellung aller noch geltenden, das allgemeine landrecht abaenderden,... / ed by Strombeck, Friedrich Heinrich von – 3. verm verb aufl. Leipzig: F A Brockhaus. v1-3. 1829 – (= ser Civil law 3 coll) – 35mf – 9 – mf#LLMC 96-542 – us LLMC [342]

Ergaenzungen zu moehler's symbolik : aus dessen schrift, neue untersuchungen der lehrgegensaetze zwischen den katholiken und protestanten / ed by Raich, Johann Michael – Mainz: Fl Kupferberg, 1889 [mf ed 1986] – 1mf – 9 – 0-8370-7255-7 – (incl bibl ref) – mf#1986-1255 – us ATLA [241]

Ergaenzungs-conversationslexikon (ael1/19) / ed by Steger, Franz – Leipzig, Leipzig/ Meissen 845-59 [mf ed 1993] – (= ser Archiv der europaeischen lexikographie, abt 1: enzyklopaedien) – 14v on 56mf – 9 – €590.00 – 3-89131-107-9 – gw Fischer [030]

Erganzungsheft ... : der mitteilungen aus den deutschen schutzgebieten – Berlin: E S Mittler. [n1-13. 1908-1917] – 2r – us CRL [074]

Ergebnisdarstellung einer bem-berechnung mit cad-system icem ddn / Fischer, Thomas – (mf ed 1996) – 2mf – 9 – €40.00 – 3-8267-2298-1 – mf#DHS 2298 – gw Frankfurter [621]

Ergebnisse der 21. jahrestagung des arbeitskreises 'deutsche literatur des mittelalters' / Ernst-Moritz-Arndt-Universitaet Greifswald. Sektion Germanistik, Kunst- und Musikwissenschaft – Greifswald: Ernst-Moritz-Arndt-Universitaet Greifswald, 1989 [mf ed 1993] – (= ser Wissenschaftliche beitraege der ernst-moritz-arndt-universitaet greifswald. deutsche literatur des mittelalters 4) – 236p – 1 – (incl bibl ref) – mf#8161 – us Wisconsin U Libr [430]

Ergebnisse der 22. und 23. jahrestagung des arbeitskreises deutsche literatur des mittelalters / Ernst-Moritz-Arndt-Universitaet Greifswald. Institut für Deutsche Philologie – Greifswald: Ernst-Moritz-Arndt-Universitaet, Institut fuer Deutsche Philologie, 1990 [mf ed 1993] – (= ser Wissenschaftliche beitraege der ernst-moritz-arndt-universitaet greifswald. deutsche literatur des mittelalters 6) – 196p – 1 – (incl bibl ref) – mf#8161 – us Wisconsin U Libr [430]

Die ergebnisse der protestantischen mission in vorderindien : mit besonderer beruecksichtigung der leistungen der evangelischen missionsgesellschaft in basel / Schweizer, R – Bern: Karl H Mann, 1868 [mf ed 1995] – (= ser Yale coll) – viii/223p – 1 – 0-524-09115-3 – (in german) – mf#1995-0115 – us ATLA [242]

Ergebnisse einer bereisung des gebiets zwischen okawango... / Seiner, F – Rome. 1968-1996 (1,5,9) – 19mf – 9 – mf#6126 – ne IDC [910]

Ergebnisse meiner naturhistorisch-oeconomischen reisen / Bronn, Heinrich – Heidelberg [u a] 1826-[1831] [mf ed Hildesheim 1995-98] – 2v on 9mf – 9 – €180.00 – 3-487-27773-5 – gw Olms [910]

Ergonomics – London. 1988+ (1,5,9) – ISSN: 0014-0139 – mf#17316 – us UMI ProQuest [620]

Ergonomics abstracts – London. 1990-1995 (1,5,9) – ISSN: 0046-2446 – mf#17317 – us UMI ProQuest [620]

Ergonomische behandlungskonzepte in der zahnaerztlichen propaedeutik : eine qualitative und quantitative analyse / Klenke, Carsten – 2000 – 2mf – 9 – 3-8267-2685-5 – mf#DHS 2685 – gw Frankfurter [617]

Ergriffenes dasein : deutsche lyrik, 1900-50 / Holthusen, Hans Egon & Kemp, Friedhelm – Ebenhausen/Muenchen: W Langewiesche-Brandt, 1953 – 1r – 1 – (incl bibl ref and ind) – us Wisconsin U Libr [430]

Ergriffenes dasein : deutsche lyrik des 20. jahrhunderts / Holthusen, Hans Egon & Kemp, Friedhelm – Ebenhausen/Muenchen: Langweische-Brandt, 1957, c1953 – 1r – 1 – (incl bibl ref and ind) – us Wisconsin U Libr [430]

Erh ch'eng yen chiu / Kuan, Tao-chung – Shang-hai: Chung-hua shu chu, Min kuo 26 [1937] – (= ser P-k&k period) – us CRL [951]

Erh chiu / Ch'in, Shou-ou – Shang-hai: T'ai p'ing shu chu, 1944 – (= ser P-k&k period) – us CRL [480]

Erh nien lai chih nan-hui chien she – [China: Chiang-su Nan-hui hsien cheng fu chien she k'o], 1935 – (= ser P-k&k period) – us CRL [330]

Erh nu ying hsiung : ssu mu chu / Sun, I-chien – [China]: Ch'ing nien shu tien, Min kuo 31 [1942] – (= ser P-k&k period) – us CRL [951]

Erh shih chiu nien tu kung li ko yuan hsiao t'ung i chao sheng pao kao – [China: sn, 1940] – (= ser P-k&k period) – us CRL [350]

Erh shih i nien tu ch'uan kuo kao teng chiao yu t'ung chi – Shang-hai: Shang wu yin shu kuan, Min kuo 24 [1935] – (= ser P-k&k period) – us CRL [370]

Erh shih jen so hsuan tuan p'ien chia tso chi / ed by Chao, Chia-pi chi – Shang-hai: Liang yu t'u shu kung ssu, 1937 – (= ser P-k&k period) – us CRL [480]

Erh shih pa nien tu nan-ching shih cheng kai k'uang – Nan-ching: Nan-ching t'e pieh shih cheng fu mi shu ch'u, Min kuo 28 [1939] – (= ser P-k&k period) – us CRL [951]

Erh shih san nien tu chung-kuo wen i nien chien / Yang, Chin-hao pien – Shang-hai: Pei hsin shu chu, 1936 – (= ser P-k&k period) – us CRL [480]

Erh shih ssu nien chih lin-an : lin-an wu su pao min kuo erh shih wu nien yuan tan t'e k'an – [China: sn], 1936 – (= ser P-k&k period) – us CRL [951]

Erh shih ssu nien tu chin yen nien pao – [China]: Chun shih wei yuan hui chin yen tsung hui, Min kuo 25 [1936] – (= ser P-k&k period) – us CRL [360]

erh shih wu nien tu / China Chiao t'ung pu – [China: Chiao t'ung pu, 1936] – (= ser P-k&k period) – us CRL [380]

Erh t'ung chiao yu shih chi wen t'i / Sun, Yu – Shang-hai: Shang wu yin shu kuan, [1935] – (= ser P-k&k period) – us CRL [370]

Erh t'ung chih yu ti erh chi – Shang-hai: Shen pao kuan, Min kuo 25 [1936] – (= ser P-k&k period) – us CRL [370]

Erh t'ung hsin li hsueh chi ch'i ying yung / Hsiao, Hsiao-jung – Shang-hai: Shang wu yin shu kuan, [1935] – (= ser P-k&k period) – us CRL [150]

Erh t'ung kuan li fa (ccm3) = Principles in child training / Barbour, Dorothy Dickinson – 2nd ed. Shanghai. 2v. 1932 [mf ed 198?] – (= ser Ccm 3) – 1 – mf#1984-b500 – us ATLA [240]

Erh t'ung sheng huo / Chu, Chao-ts'ui – Shang-hai: Shih chieh shu chu, 1933 – (= ser P-k&k period) – us CRL [305]

Erh t'ung tzu chih chih tao shu / Li, K'ang-fu – Shang-hai: Shih chieh shu chu, Min kuo 21 [1932] – (= ser P-k&k period) – us CRL [370]

Erh t'ung wen hsueh hsiao lun / Chou, Tso-jen – Shang-hai: Erh t'ung shu chu, Min kuo 21 [1932] – (= ser P-k&k period) – us CRL [390]

Erh t'ung yu ch'eng jen ch'ang yung tzu hui chih tiao ch'a chi pi chiao / Tu, Tso-chou – [China]: Hsia-men ta hsueh, Min kuo 22 [1933] – (= ser P-k&k period) – us CRL [480]

Erh tz'u shih chieh ta chan chung chih mei-kuo ti wai chiao cheng ts'e / Hsieh, Jen-chao – [Ch'ung-ch'ing: Wu shih nien tai ch'u pan she, 1942] – (= ser P-k&k period) – us CRL [337]

Erh tz'u ta chan hsin chan shu / Wintringham, Tom – Ch'ung-ch'ing: Shih tai shu chu, Min kuo 30 [1941] – (= ser P-k&k period) – us CRL [951]

Erhard, Christian Daniel *see* Amalthea

Erhardi weigelii sacrae caesar : philosophia mathematica: theologia naturalis solida: per singulas scientias continuata: universae artis inveniendi prima stamina complectens / Weigel, Erhard – Jenae: Sumptibus Matth. Brickneri, bibliopolae Jen. & Helmstad., Typis Pauli Ehrichii, 1693 – 1 – us Wisconsin U Libr [100]

Erh-ch'i *see* Lao pai hsing tsen yang k'ang jih

Die erhebung europas gegen napoleon 1 : drei vorlesungen, gehalten zu muenchen am 24, 27 und 30 maerz 1860 / Sybel, Heinrich von – Muenchen 1860 [mf ed Hildesheim 1995-98] – 1v on 1mf – 9 – €40.00 – 3-487-26388-2 – gw Olms [940]

Die erhebung preussens im jahre 1813 und die rekonstruktion des staates / Ranke, Leopold von; ed by Kaemmel, Otto – Leipzig um 1900 (mf ed 1992) – 2mf – 9 – €24.00 – 3-89349-115-5 – mf#DHS-AR 84 – gw Frankfurter [943]

Erhebungen – Luebeck DE, 1809 – 1r – 1 – gw Misc Inst [074]

Erhebungsinstrumentarium zu akzeptanzproblemen beim einsatz innovativer informationstechnologie im buero- und verwaltungsbereich analysiert am beispiel eines industriebetriebes / Wagner, Albert – Nuernberg, 1983 (mf ed 1994) – 2mf – 9 – €19.00 – 3-89349-904-0 – mf#DHS-AR 904 – gw Frankfurter [650]

Erheiterungen – 1811-27 [mf ed 1997] – (= ser Die zeitschriften des august von kotzebue) – 102mf – 9 – €1190.00 – 3-89131-234-2 – gw Fischer [430]

Erholungsstunden – 1883 jan 4-1885 dec 24, 1887 jan 6-1889, 1890 jan 4-1891 dec 19, 1893 dec 23-1896 dec 26, 1897 jan 2-1899 dec 30, 1900 jan 13-1903 dec 5 – 6r – 1 – mf#1223728 – us WHS [071]

Erich a schelling (1904-1986) : ein architekt zwischen traditionalismus und moderne / Ehrhardt, Ingrid – (mf ed 1999) – 5mf – 9 – €59.00 – 3-8267-2637-5 – mf#DHS 2637 – gw Frankfurter [720]

Erich, Herbert *see* Die schoene von milet

Erichson, A *see*
- Ulrich zwingli und die elsaessischen reformatoren
- Zwingli's tod und dessen beurtheilung durch zeitgenossen

Erichson, Alfred *see*
- Bibliographia calviniana
- Die calvinische und die altstrassburgische gottesdienstordnung
- L'eglise francaise de strasbourg au seizieme siecle

Ericksen, Ephraim Gordon *see* Africa company town

Erickson, Jeff D A *see* Physiological responses to recreational snowshoeing in females

Ericson, Jack T *see* Oneida community

The ericson journal – Ericson, NE: A C Bell, 1912 (wkly) [mf ed 1914-39 (gaps)] – 7r – 1 – (issues for apr 13, 27 1922 called v10 n17) – us NE Hist [071]

Ericsson s [sic] caloric engine : manufactured by charles pierson, niagara, cw / Pierson, Charles – [Toronto?: s.n] 1860 [mf ed 1987] – 1mf – 9 – 0-665-34173-3 – mf#34173 – cn Canadiana [621]

Erie Co. Sandusky *see*
- Erie county news
- Weekly journal

Erie Co. Vermillion *see* Bugle

Erie county news / Erie Co. Sandusky – v1 n1. may 1863-feb 1865 [wkly] – 1r – 1 – mf#B5537 – us Ohio Hist [071]

Erie county reporter – Huron, OH. 1880-1972 (1) – mf#65541 – us UMI ProQuest [071]

Erie herald *see* Miscellaneous newspapers of weld county

Erie. Presbytery (Pres. Ch. in the USA New School) *see* Minutes, 1838-1870

Erie. Presbytery (Pres. Church in the USA) *see* Minutes, 1802-1924

Erie review *see* Miscellaneous newspapers of weld county

Erie Society for Genealogical Research *see* Keystone kuzzins

Erie tageblatt – Erie, PA (USA), 1920 1 oct-1921, 1924-1929 30 sep, 1930-31 [gaps], 1933 6 jan-1934 16 mar – 18r – 1 – gw Misc Inst [071]

Erika : erzaehlung / Edschmid, Kasimir – Berlin: P Zsolnay, 1938 – 1r – 1 – us Wisconsin U Libr [830]

Erinensis [Walter Cavendish Crofton] *see* Brief sketch of the life of charles, baron metcalfe, of fernhill, in berkshire...

Erinnerungen / Alexis, Willibald; ed by Ewerts, Max – Berlin, 1900 (mf ed 1992) – 3mf – 9 – €24.00 – 3-89349-105-8 – mf#DHS-AR 46 – gw Frankfurter [880]

Erinnerungen / Matthisson, Friedrich von – Zuerich 1810-16 [mf ed Hildesheim 1995-98] – 9v on 15mf – 9 – €150.00 – 3-487-27735-2 – gw Olms [800]

Erinnerungen / Thoma, Ludwig – Muenchen: R Piper, c1947 – 1r – 1 – us Wisconsin U Libr [943]

Erinnerungen an anzengruber / Rosner, L – Leipzig: J Klinkhardt, 1891 – 1r – 1 – us Wisconsin U Libr [430]

Erinnerungen an franz grillparzer : fragmente aus tagebuchblaettern / Wartenegg, Wilhelm von – Wien: C Konegen, 1901 – 1r – 1 – us Wisconsin U Libr [920]

Erinnerungen an friedrich hebbel / Kulke, Eduard – Wien: C Konegen, 1878 – 1r – 1 – us Wisconsin U Libr [920]

Erinnerungen an friedrich nietzsche / Deussen, Paul – Leipzig: F A Brockhaus, 1901 – 1r – 1 – (incl ind) – us Wisconsin U Libr [190]

Erinnerungen an gottfried keller / Frey, Adolf – 2. erw aufl. Leipzig: H Haessel, 1893 – 1r – 1 – us Wisconsin U Libr [920]

Erinnerungen an gustav nachtigal / Berlin, Dorothea – Berlin 1887 [mf ed Hildesheim 1995-98] – 1v on 2mf – 9 – €60.00 – 3-487-27321-7 – gw Olms [916]

Erinnerungen an italien : besonders an rom / Kahlert, August – Breslau 1843 [mf ed Hildesheim 1995-98] – 3mf – 9 – €90.00 – 3-487-29221-1 – gw Olms [880]

Erinnerungen an italien : in briefen / Pannasch, Anton – Wien 1826 [mf ed Hildesheim 1995-98] – 2mf – 9 – €60.00 – 3-487-29277-7 – gw Olms [880]

Erinnerungen an richard wagner = Souvenirs sur richard wagner / Schure, Edouard – Leipzig: Breitkopf & Haertel 1900 – 1mf – 9 – (trans fr french by fritz ehrenberg) – mf#wa-101 – ne IDC [780]

Erinnerungen an richard wagner : ein vortrag gehalten am 13. april 1883 im wissenschaftlichen club zu wien / Wolzogen, Hans von; ed by Wiener Akademischen Wagner-Verein – Wien: Konegen, 1883 – 1mf – 9 – mf#wa-121 – ne IDC [780]

Erinnerungen an stefan george : mit einer bibliographie / Bondi, Georg – Berlin: G Bondi, 1934 [mf ed 1990] – 31/[1]p – 1 – mf#7294 – us Wisconsin U Libr [920]

Erinnerungen aus aegypten und kleinasien / Prokesch-Osten, Anton – Wien 1829-1831 [mf ed Hildesheim 1995-98] – 3v on 9mf – 9 – €180.00 – ISBN-10: 3-487-26709-8 – ISBN-13: 978-3-487-26709-8 – gw Olms [880]

Erinnerungen aus dem leben eines deutschen in paris / Depping, Georges – Leipzig 1832 [mf ed Hildesheim 1995-98] – 4mf – 9 – €120.00 – 3-487-29672-1 – gw Olms [914]

Erinnerungen aus dem leben eines ostindischer missionaers – Halle: Julius Fricke [mf ed 1995] – (= ser Yale coll) – vi/470p – 1 – 0-524-10153-1 – (in german) – mf#1995-1153 – us ATLA [920]

Erinnerungen aus der suedafrikanischen mission *see* Reminiscences of the south african mission

Erinnerungen aus italien, england und amerika / Chateaubriand, Francois R de – Dresden 1816 [mf ed Hildesheim 1995-98] – 1v on 2mf – 9 – €60.00 – 3-487-27336-5 – gw Olms [880]

Erinnerungen aus meinem leben / Freytag, Gustav – Leipzig: S Hirzel, 1887(mf ed 1990) – 1 – us Wisconsin U Libr [430]

Erinnerungen aus paris : 1817-1848. / Leo, Sophie – Berlin 1851 [mf ed Hildesheim 1995-98] – 2mf – 9 – €60.00 – 3-487-29680-2 – gw Olms [880]

Erinnerungen aus paris : im Jahr 1831. / Seybold, Friedrich – Stuttgart 1832 [mf ed Hildesheim 1995-98] – 2mf – 9 – €60.00 – 3-487-29678-0 – gw Olms [880]

Erinnerungen aus paris im jahre 1804 [achtzehnhundertvier] / Kotzebue, August von – Berlin 1804 [mf ed Hildesheim 1995-98] – 4mf – 9 – €120.00 – 3-487-29688-8 – gw Olms [880]

Erinnerungen aus suedeuropa : geschichtliche, topographische und literarische mittheilungen aus italien, dem suedlichen frankreich, spanien und portugal / Bellermann, Christian F – Berlin 1851 [mf ed Hildesheim 1995-98] – (= ser Fbc) – 1v on 2mf [ill] – 9 – €60.00 – 3-487-27781-6 – gw Olms [914]

Erinnerungen eines alten lutheraners / Hammerstein, Ludwig von – Freiburg i B: Herder, 1882 – 1mf – 9 – 0-8370-6666-2 – mf#1986-0666 – us ATLA [242]

Erinnerungen vom journalisten zum historiker der deutschen arbeiterbewegung / Mayer, Gustav – Zurich, 1949 – 1 – gw Mikropress [331]

Erinnerungen von einer reise in den jahren 1803, 1804 und 1805 / Schopenhauer, Johanna – Rudolstadt 1813-17 [mf ed Hildesheim 1995-98] – 6v on 8mf – 9 – €160.00 – 3-487-27790-5 – gw Olms [910]

Erinnerungen von einer reise nach st petersburg im jahre 1814 / Schlippenbach, Ulrich von – Hamburg 1818 [mf ed Hildesheim 1995-98] – 2v on 4mf – 9 – €120.00 – 3-487-29007-3 – gw Olms [880]

Erinnerungsblaetter deutscher regimenter : die anteilnahme der truppenteile der ehemaligen deutschen armee am weltkriege – Oldenburg i.O: G Stalling v30-371 1922-39 [mf ed 1979] – 21r – 1 – (bearbeitet unter benutzung der amtlichen kriegstagebuecher) – mf#film mas c678 – us Harvard [355]

Erinnerungs-blaetter von einer reise nach paris im sommer 1811 / Halem, Gerhard A von – Hamburg 1813 [mf ed Hildesheim 1995-98] – 2mf – 9 – €60.00 – 3-487-29689-6 – gw Olms [914]

Erinnerungs-blaetter zur einweihungsfeier des zwingli-denkmals in zuerich – Zuerich, 1885 – 2pts on 2mf – 9 – mf#ZWI-31 – ne IDC [242]

Erinnerungs-skizzen aus russland, der tuerkei und griechenland : entworfen waehrend des aufenthalts in jenen laendern in den jahren 1833 und 1834 / Tietz, Friedrich – Coburg [u a] 1836 [mf ed Hildesheim 1995-98] – 2v on 4mf – 9 – €120.00 – 3-487-27711-5 – gw Olms [880]

Eritassard hayastan – Boston: [s.n.], dec 12 1917-1921; jan 14-apr 29 1922; aug 7 1922-oct 13 1923; jan 6, feb 17 1934; sep 21 1948-aug 19 1949; aug 23 1949-may 16 1952 – 7r – 1 – us CRL [071]

Eritassard hayastan – Providence, RI. 1915-1916 (1) – mf#66292 – us UMI ProQuest [071]

Erith and crayford times – London UK, jan-16 oct 1986; nov 1986-87; 1989-92; 25 feb-dec 1993 – 29r – 1 – uk British Libr Newspaper [072]

Erith, belvedere & district free press – [London & SE] Bexley 15 nov 1930-30 oct 1937 [mf ed 2003] – 7r – 1 – uk Newsplan [072]

Erith observer *see* Erith times belvedere and abbey wood chronicle and general district advertiser etc

Erith times belvedere and abbey wood chronicle and general district advertiser etc – London UK, 1889; 1896 – 2r – 1 – (after 26 dec 1919 incorp with: erith observer) – uk British Libr Newspaper [072]

"Eritis sicut deus" : ein beitrag zur geschichte des religioesen romans / Dobbriner, Paul – Lucka, 1913 (mf ed 1993) – 2mf – 9 – €31.00 – 3-89349-345-X – mf#DHS-AR 198 – gw Frankfurter [410]

Eritrean daily news – Asmara, Ethiopia. 28 feb-4 nov 1947 – 1r – 1 – uk British Libr Newspaper [079]

Eritrean weekly news – Asmara, Ethiopia. 30 aug 1945-1950 – 2r – 1 – uk British Libr Newspaper [079]

Eriugena, Ioannes Scotus *see* De divina praedestinatione

Erixon, Sigurd *see* Byggnadskultur

Erkelenzer volkszeitung 1957 – Erkelenz DE, 1957 2 nov-1959 30 jun – 1 – (regional ed of aachener volkszeitung, aachen) – gw Misc Inst [074]

Erkennen und wissen nach gregor von rimini / Wuersdoerfer, J – 1917 – (= ser Bgphma 20/1) – €7.00 – ne Slangenburg [100]

Erkenning van ambon / ed by Bureau Zuid-Molukken – Es-Gravenhage. n1. 1950 – 1mf – 9 – mf#SE-1592 – ne IDC [959]

Erkenntnis – Dordrecht. 1989-1996 (1,5,9) – ISSN: 0165-0106 – mf#14748,02 – us UMI ProQuest [100]

Die erkenntnislehre anselms von canterbury / Fischer, J – 1911 – (= ser Bgphma 10/3) – €5.00 – ne Slangenburg [140]

Die erkenntnislehre bonaventuras / Luyckx, B A – 1923 – (= ser Bgphma 23/3-4) – €14.00 – ne Slangenburg [100]

Die erkenntnislehre des wilhelm von auvergne / Baumgartner, Matthias – Muenster: Aschendorff, 1893 – (= ser Beitraege zur geschichte der philosophie des mittelalters) – 1mf – 9 – 0-524-00247-9 – (incl bibl ref) – mf#1989-2947 – us ATLA [120]

Die erkenntnislehre des wilhelm von auvergne / Baumgartner, Matthias – Muenster, 1893 – (= ser Bgphma 2/1) – 2mf – 8 – €5.00 – ne Slangenburg [100]

Die erkenntnislehre richards von st viktor / Ebner, J – 1917 – (= ser Bgphma 19/4) – €7.00 – ne Slangenburg [241]

Die erkenntnislehre s a kierkegaards : eine wuerdigung seiner verfasserwirksamkeit vom zentralen gesichtspunkte aus / Slotty, Martin – Cassel: Pillardy & Augustin, 1915 – 1mf – 9 – 0-524-00133-2 – mf#1989-2833 – us ATLA [120]

Die erkenntniss des christenthumes vom naturwissenschaftlichen standpuncte : ein beitrag zur dogmatischen reform der protestantischen kirche / Bonorden, Hermann Friedrich – Leipzig: Siegismund & Volkening, 1876 – 1mf – 9 – 0-8370-2413-7 – mf#1985-0413 – us ATLA [210]

Die erkenntnistheoretische bedeutung des gefuehlsmaessigen erfassens bei schleichermeier / Hammer, Anton – Freiburg, 1934 (mf ed 1992) – 1mf – 9 – €24.00 – 3-89349-055-8 – mf#DHS-AR 17 – gw Frankfurter [110]

Die erkenntnistheoretischen und metaphysischen grundlagen der dogmatischen systeme von a.e. biedermann und r.a. lipsius / Fleisch, Urban – Berlin: C.A. Schwetschke, 1901 – 1mf – 9 – 0-8370-3153-2 – (incl bibl ref) – mf#1985-1153 – us ATLA [140]

Erkhardt, S L *see* Volia naroda

Erkhe ruah ve-sifrut / Benari, Nahum – Tel-Aviv, Israel. 1953 – 1r – 1 – us UF Libraries [939]

Erklaerung der briefe an die ephesier, philipper, kolosser : und des ersten briefes an die thessalonicher / Bisping, August – Muenster: Aschendorff, 1855 – 1mf – 9 – 0-524-07174-8 – mf#1992-1044 – us ATLA [227]

Erklaerung der briefe petri / Beck, Johann Tobias; ed by Lindenmeyer, Julius – Guetersloh: C Bertelsmann, 1896 – 1mf – 9 – 0-8370-2226-6 – mf#1985-0226 – us ATLA [240]

Erklaerung der glaubensartikel und hauptlehren der methodistenkirche / Sulzberger, Arnold – Bremen: Verlag des Tractathauses, [1880?] – 1mf – 9 – 0-7905-8925-7 – mf#1989-2150 – us ATLA [242]

Erklaerung der historie des leidens und sterbens unsers herrn christi jesu : nach den vier evangelisten also angestellet dass wir dadurch zur erkenntnis der liebe christi erwecket werden und am innerlichen menschen seliglich zunehmen moegen / Gerhard, Johann – Berlin: Gustav Schlawitz, 1868 – 1mf – 9 – 0-8370-3254-7 – mf#1985-1254 – us ATLA [240]

Erklaerung der offenbarung johannes, cap 1-12 / Beck, Johann Tobias; ed by Lindenmeyer, Julius – Guetersloh: C Bertelsmann, 1884 – 1mf – 9 – 0-8370-2227-4 – mf#1985-0227 – us ATLA [225]

Erklaerung der propheten micha und joel : nebst einer einleitung in die prophetie / Beck, Johann Tobias; ed by Lindenmeyer, Julius – Guetersloh: C Bertelsmann, 1898 – 1mf – 9 – 0-8370-2228-2 – mf#1985-0228 – us ATLA [221]

Erklaerung der zwei briefe an die thessalonicher und des briefes an die galater / Schaefer, Aloys – Muenster i. W: Aschendorff, 1890 – 1mf – 9 – 0-8370-5065-0 – (includes bibliographies) – mf#1985-3065 – us ATLA [227]

Erklaerung des barnabasbriefes : ein anhang zu de wette's exegetischen handbuch zum neuen testament / Mueller, Johann Georg – Leipzig: S Hirzel, 1869 – 1mf – 9 – 0-8370-9568-9 – (incl ind of greek words) – mf#1986-3568 – us ATLA [227]

Erklaerung des briefes an die hebraeer / Bisping, August – Muenster: Aschendorff, 1854 – 1mf – 9 – 0-524-07176-4 – mf#1992-1046 – us ATLA [227]

Erklaerung des briefes an die roemer / Bisping, August – 2. verb verm aufl. Muenster: Aschendorff, 1860 – 1mf – 9 – 0-524-07177-2 – mf#1992-1047 – us ATLA [227]

Erklaerung des briefes an die roemer / Schaefer, Aloys – Munster i W: Aschendorff, 1891 – 1mf – 9 – 0-524-06801-1 – mf#1992-0964 – us ATLA [227]

Erklaerung des buchs baruch / Reusch, Franz Heinrich – Freiburg i.B.: Herder, 1853 – 1mf – 9 – 0-7905-0320-4 – (incl bibl ref) – mf#1987-0320 – us ATLA [221]

Erklaerung des ersten briefes an die korinther / Bisping, August – Muenster: Aschendorff, 1855 – 1mf – 9 – 0-524-07175-6 – (incl bibl ref) – mf#1992-1045 – us ATLA [227]

Erklaerung des ersten briefes an die korinther / Schaefer, Aloys – Munster i W: Aschendorff, 1903 – 1mf – 9 – 0-524-05694-3 – (incl bibl ref) – mf#1992-0544 – us ATLA [227]

Erklaerung des hebraerbriefes / Schaefer, Aloys – Munster i W: Aschendorff, 1893 – 1mf – 9 – 0-524-05695-1 – (incl bibl ref) – mf#1992-0545 – us ATLA [227]

Erklaerung des propheten isaias / Knabenbauer, Joseph – Freiburg i B: Herder, 1881 – 2mf – 9 – 0-524-05915-2 – (incl bibl ref) – mf#1992-0672 – us ATLA [221]

Erklaerung des zweiten briefes an die korinther / Schaefer, Aloys – Munster i W: Aschendorff, 1903 – 1mf – 9 – 0-524-05696-X – (incl bibl ref) – mf#1992-0546 – us ATLA [227]

Erklaerung des zweiten briefes an die korinther, und des briefes an die galater / Bisping, August – Muenster: Aschendorff, 1857 – 1mf – 9 – 0-524-07178-0 – mf#1992-1048 – us ATLA [227]

Erklaerung des zweiten briefes an die thessalonicher, der drei pastoralbriefe und des briefs an philemon / Bisping, August – Muenster: Aschendorff, 1858 – 1mf – 9 – 0-524-07179-9 – mf#1992-1049 – us ATLA [227]

Erklaerungsfunktionalitaet wissensbasierter systeme : theoretische und empirische untersuchungen zur entwicklung von expertensystemen und der transformation von arbeit durch den einsatz lernfoerderlicher technologien / Kozok, Barbara – (mf ed 1998) – 4mf – 9 – €56.00 – 3-8267-2498-4 – mf#DHS 2498 – gw Frankfurter [320]

Erklarung der propheten nahum und zephanja : nebst einem prophetischen totalbild der zukunft / Beck, Johann Tobias; ed by Gutscher, H & Lindenmeyer, Julius – Guetersloh: Bertelsmann, 1899 – 1mf – 9 – 0-8370-2229-0 – mf#1985-0229 – us ATLA [221]

Erkow ap'e tsap'e / Mankowni, N L – 1965 – 9 – $25.00 – us Scholars Facs [490]

Erlaeuterung der babylonischen keilinschriften aus behistun / Grotefend, Georg Friedrich – Goettingen: Dieterich, 1853 – (= ser Abhandlungen der koeniglichen gesellschaft der wissenschaften zu goettingen) – 1mf – 9 – 0-8370-7689-7 – mf#1986-1689 – us ATLA [470]

Erlaeuterung der inschrift aus den oberzimmern in nimrud / Grotefend, Georg Friedrich – [s.l: s.n, 1853?] [mf ed 1986] – 1mf – 9 – 0-8370-7700-1 – mf#1986-1700 – us ATLA [490]

Erlaeuterung der keilinschriften babylonischer backsteine : mit einigen andern zugaben und einer steindrucktafel / Grotefend, Georg Friedrich – Hannover: Hahn, 1852 – 1mf – 9 – 0-8370-7701-X – mf#1986-1701 – us ATLA [470]

Erlaeuterung einer inschrift des letzten assyrisch-babylonischen koenigs aus nimrud : mit drei andern zugaben und einer steindrucktafel / Grotefend, Georg Friedrich – Hannover: Hahn, 1853 – 1mf – 9 – 0-8370-7702-8 – mf#1986-1702 – us ATLA [470]

Erlaeuterung zweier ausschreiben des koeniges nebukadnezar in einfacher babylonischer keilschrift / Grotefend, Georg Friedrich – Goettingen: Dieterich, 1853 – (= ser Abhandlungen der koeniglichen gesellschaft der wissenschaften zu goettingen) – 1mf – 9 – 0-8370-7723-0 – mf#1986-1723 – us ATLA [470]

Erlaeuterungen ausgewaehlter werke goethes : fuer die obersten klassen hoeherer lehranstalten sowie zum selbstunterricht / Klaucke, Paul – Berlin: W. Weber. 3v. 1887 (mf ed 1990) – 1 – us Wisconsin U Libr [430]

Erlaeuterungen der theoretischen und praktischen philosophie nach herrn feders ordnung / Tittel, Gottlob August – Frankfurt am Main: Bei Johann Gottlieb Garbe 1785 – (= ser Ethics in the early modern period) – 4mf – 9 – mf#pl-143 – ne IDC [170]

Erlaeuterungen und aufsaetze zur einfuehrung in goethes faust fuer lehrer und den gebildeten / Buurman, Ulrich – Leipzig: Renger 1901 [mf ed 1990] – 1r – 1 – (incl bibl ref; filmed with: a passage in the night / sholem asch) – mf#7342 – us Wisconsin U Libr [430]

Erlaeuterungen zu dunkeln stellen im buche hiob / Richter, Georg – Leipzig: JC Hinrichs, 1912 – (= ser Beitraege zur wissenschaft vom alten testament) – 1mf – 9 – 0-524-06680-9 – mf#1992-0933 – us ATLA [221]

Erlaeuterungen zu dunkeln stellen in den kleinen propheten / Richter, Georg – Guetersloh: C Bertelsmann, 1914 – (= ser Beitraege zur foerderung christlicher theologie) – 1mf – 9 – 0-524-06742-2 – mf#1992-0945 – us ATLA [221]

Erlaeuterungen zu goethes egmont fuer schule und haus / ed by Hoffmann, Professor Dr – Leipzig: H Beyer, [19–?] (mf ed 1990) – 1 – us Wisconsin U Libr [430]

Erlaeuterungen zu goethe's 'faust' : 1. [und] 2. teil / Bischoff, Erich – Leipzig: H Breyer, [19–?] [mf ed 1990] – 2v – 1 – (incl bibl ref) – mf#7341 – us Wisconsin U Libr [430]

Erlaeuterungen zu goethes werken / Duentzer, Heinrich – Leipzig: E Wartig, 1886 [mf ed 1989] – (= ser Erlaeuterungen zu den deutschen klassikern 1) – 38v in 11 – 1 – mf#6988 – us Wisconsin U Libr [410]

Erlaeuterungen zu goethes werken fuer schulgebrauch und selbststudium als litteraturkundliches repetitorium / Rothe, B – Breslau: G Sperber, 1897 – 1 – us Wisconsin U Libr [430]

Erlaeuterungen zu kant's religion innerhalb der grenzen der blossen vernunft / Kirchmann, Julius Hermann von – Berlin: L Heimann, 1869 – (= ser Philosophische bibliothek) – 1mf – 9 – 0-7905-9399-8 – mf#1989-2624 – us ATLA [200]

Erlaeuterungen zu lessing's hamburgischer dramaturgie / Bischoff, Erich – Leipzig: H Beyer, [18–?] [mf ed 1992] – 168p – 1 – mf#7589 – us Wisconsin U Libr [430]

Erlaeuterungen zu nietzsches zarathustra / Messer, August – Stuttgart: Strecker und Schroeder, 1922 – 1 – us Wisconsin U Libr [190]

Erlaeuterungen zu wagners tristan und isolde von karl heckel / Heckel, Karl – Mannheim: Heckel 1893 – 1mf – 9 – mf#wa-42 – ne IDC [780]

Erlaftal bote – Scheibbs, Austria. 27 jul 1946-7 feb 1948 – 1/2r – 1 – uk British Libr Newspaper [074]

Erlanger nachrichten *see* Erlanger tagblatt 1858

Erlanger tagblatt 1858 – Erlangen DE, 1977- – ca 14r/yr – 1 – (title varies: 2 jan 1981: erlanger nachrichten) – gw Misc Inst [074]

Erlangische gelehrte anmerkungen und nachrichten – Erlangen 1770-87 – (= ser Dz) – jg25-42 on 148mf – 9 – €888.00 – mf#k/n274 – gw Olms [430]

Erlangische gelehrte anmerkungen und nachrichten *see* Compendium historiae literariae novissimae

Erlangische gelehrte zeitung *see* Compendium historiae literariae novissimae

Das erlebnis und die dichtung : lessing, goethe, novalis, hoelderlin / Dilthey, Wilhelm – 6. aufl. Leipzig: B G Teubner, 1919 [mf ed 1993] – 476p (ill) – 1 – mf#8232 – us Wisconsin U Libr [430]

Erlebnisse eines schuldenbauers / Gotthelf, Jeremias [Albert Bitzius]; ed by Hunziker, Rudolf & Baehler, Eduard – Erlenbach, Zuerich: Eugen Rentsch Verlag, 1924 [mf ed 1993] – (= ser Saemtliche werke in 24 baenden 14) – 497p – 1 – (incl bibl ref) – mf#8522 reel 4 – us Wisconsin U Libr [830]

Erlebnisse in abessinien in den jahren 1858-1868 / Waldmeier, T – Basel, 1869 – 2mf – 9 – mf#HT-154 – ne IDC [916]

Erlebtes, 1862-1901 / Meinecke, Friedrich – Leipzig: Koehler & Amelang, 1941. 224p – 1 – us Wisconsin U Libr [920]

Die erleichterungen der schammaiten und die erschwerungen der hilleliten : ein beitrag zur entwicklungsgeschichte der halachah / Schwarz, Adolf – Wien: Isr-Theol Lehranstalt, 1893 – 1mf – 9 – 0-8370-5182-7 – (incl bibl ref) – mf#1985-3182 – us ATLA [270]

Erlenvein, A A *see* Narodnyia skazki

Erler, Fritz *see*
- Soll deutschland rusten? die spd zum wehrbeitrag
- Wehr- und aussenpolitik im gespaltenen deutschland. referat auf dem spd-parteitag 1958 in stuttgart

Erler, Otto *see*
- Der galgenstrick
- Die gewissenhaften
- Struensee
- Die tragischen probleme des struensee-stoffes

Erlernte hilflosigkeit : experimentelle induktion emotionaler veraenderungen und kognitiver interferenzen durch nicht-kontingente lernbedingungen / Finzer, Michael – (mf ed 1994) – 2mf – 9 – €40.00 – 3-89349-855-9 – mf#DHS 855 – gw Frankfurter [150]

Erleuterung der egyptischen altertuemer / Semler, J S – Breslau, Leipzig, 1748 – 4mf – 9 – mf#VR-11.73 – ne IDC [956]

Erlikh, Yisrael *see* Rabi mendele mi-kotsk

Erloeserin : ein hetaerengespraech / Brod, Max – Berlin: E Rowohlt 1921 [mf ed 1989] – 1r – 1 – (filmed with: ein gelegenheitsgedicht von brockes / friedrich gundolf) – mf#7089 – us Wisconsin U Libr [880]

Die erloesung / ed by Bartsch, Karl – Quedlinburg, Leipzig: G Basse, 1858 [mf ed 1993] – (= ser Bibliothek der gesammten deutschen national-literatur von der aeltesten bis auf die neuere zeit sect1/37) – lxx/381p – 1 – (incl bibl ref and ind) – mf#8438 reel 8 – us Wisconsin U Libr [810]

Erloesung / Herrmann, R – Tuebingen: J C B Mohr 1905 [mf ed 1991] – 1mf – 9 – 0-524-00895-7 – (incl bibl ref) – mf#1990-2118 – us ATLA [230]

Erm – Morgantown. 1976-1979 – 1,5,9 – ISSN: 0572-3698 – mf#11029 – us UMI ProQuest [370]
Erman, A *see*
- Aegyptische chrestomathie
- Aegyptische grammatik
- Die aegyptische religion
- Aegyptisches glossar
- Ausfuehrliches verzeichnis der aegyptischen altertuemer und gipsabguesse
- Ein denkmal memphitischer theologie
- Gespraech eines lebensmueden mit seiner seele
- Die hieroglyphen
- Hymnen an das diadem der pharaonen
- Kurzer arbriss der aegyptischen grammatik
- Die literatur der aegypter
- Die maerchen des papyrus westcar
- Neuaegyptische grammatik
- Zaubersprueche fuer mutter und kind

Erman, Adolf *see*
- Handbook of egyptian religion
- Life in ancient egypt
- Reise um die erde durch nord-asien und die beiden oceane in den jahren 1828, 1829 und 1830

Erman, G *see*
- Deutschland im jahre 2000
- Die moderne gesellschaft, ihre geselligkeit und ihre moral

Erman, Wilhelm *see*
- Bibliographie der deutschen universitaeten

Ermanskii, A K *see* Nashi blizhaishie trebovaniia i konechnaia tsel
Ermatinger, Charles Oakes *see*
- Canadian franchise and election laws
- Record of the celebration of the centenary of the talbot settlement
- The talbot regime

Ermatinger, Emil *see*
- Deutsche dichter, 1700-1900
- Die deutsche lyrik in ihrer geschichtlichen entwicklung
- Die weltanschauung des jungen wieland
- Wieland und die schweiz

L'ermitage : revue mensuelle artistique et litteraire. – Paris. avr 1890-juin 1906 – 1 – fr ACRPP [073]
Las ermitas de cordoba / Aragon Fernandez, Antonio – Madrid: Razon y Fe, 1927 – 1 – sp Bibl Santa Ana [946]
Ermitazh = Hermitage – Moscow. n1-15. may 1922-aug 1922 – 5mf – 9 – (cont: vestnik teatra) – us UMI ProQuest [790]
L'ermite au palais : moeurs indiciaires du dix-neuvieme siecle / Saint-Hilaire, Emile Marco de – Paris 1832 [mf ed Hildesheim 1995-98] – 2v on 4mf – 9 – €120.00 – ISBN-10: 3-487-25862-5 – ISBN-13: 978-3-487-25862-1 – gw Olms [944]
L'ermite toulonnais faisant suite a l'ermite en province de m de jouy / Bellue, Pierre – Paris 1828 [mf ed Hildesheim 1995-98] – 3mf – 9 – €90.00 – 3-487-29763-9 – gw Olms [914]
Les ermites de la bigorre / Laforgue, E – Lourdes, 1923 – €3.00 – ne Slangenburg [241]
Ermittlung von grundlagen zur ultrafiltration / Driessen, Helmut – Manuskript, 1977 (mf ed 1993) – 1mf – 9 – €24.00 – 3-89349-641-6 – mf#DHS 641 – gw Frankfurter [621]
Ermlaendische zeitung – Braunsberg (Braniewo PL), 1897, 1898 1 jul-1899, 1901, 1902 1 jul-31 dec, 1904-05 – 9r – 1 – gw Misc Inst [077]
Ermoni, Vincent *see*
- L'agape dans l'eglise primitive
- Le careme
- Histoire du credo
- Les origines de l'episcopat
- Saint jean damascene

Die ernaehrungsbedingten mangelkrankheiten der erwachsenen feldarbeitersklaven im antebellum sueden der usa. 1810-1860 : eine revision der von fogel/engermann, savitt und gibbs et al berechneten naehr- und mineralstoffwerte der sklavennahrung anhand von 422 quellen aus 10 suedstaaten / Bernhagen, Joerg – (mf ed 1997) – 5mf – 9 – €59.00 – 3-8267-2425-9 – mf#DHS 2425 – gw Frankfurter [976]
Ernest and ida : or christmas at montagu house / Armstrong, Jessie F – London: Houlston & Sons, 1888 – (= ser 19th c children's literature) – 2mf – 9 – mf#6.1.40 – uk Chadwyck [830]
Ernest bloch's sacred service / Fulton, Alvin W – U of Rochester 1953 [mf ed 19–] – 2mf – 9 – mf#fiche 759, 995 – us Sibley [780]
Ernest maltravers / Lytton, Edward Bulwer Lytton, Baron – Boston, MA. 189- – 1r – 1 – us UF Libraries [025]
Ernest oppenheimer and the economic development of southern africa / Gregory, Theodor Emanuel Gugenheim – Cape Town, South Africa. 1962 – 1r – 1 – us UF Libraries [338]
Ernest renan / Barry, William Francis – London: Hodder & Stoughton, 1905 – (= ser Literary lives) – 1mf – 9 – 0-7905-4325-7 – mf#1988-0325 – us ATLA [140]
Ernesti, Heinrich Friedrich Theodor Ludwig *see*
- Die theorie vom ursprunge der suende aus der sinnlichkeit
- Die theorie vom ursprunge der suende aus vorzeitlicher selbstentscheidung

Ernesti prueferi elementa metagrammatices / Pruefer, Carl E – Berolini: Pauli 1830 – (= ser Whsb) – 4mf – 9 – €50.00 – mf#Hu 015 – gw Fischer [410]
Ernestine *see* Palm room ballads
Ernesto cardenal : una mitica aventura poetica / Hernandez, Antonio Angel Delgado – (mf ed 1998) – 3mf – 9 – €49.00 – 3-8267-2530-1 – mf#DHS 2530 – gw Frankfurter [440]
Ernesto pinto. el santo del siglo / Bayle, Constantino – Madrid: Razon y Fe, 1941 – sp Bibl Santa Ana [240]
Ernestus, Johann August *see*
- Principles of biblical interpretation
- Tria symbola oecumenica, augustanam confessionem et apologiam ejus

Erneuerung der heiligen mission in zwei abtheilungen : von den zwei fahnen, die fahne christi und die des luzifer, die merkmale der kirche reflectirt im charakter ihrer kinder / Weninger, Francis Xavier – Cincinnati: [s.n.], 1885 – 2mf – 9 – 0-8370-6790-1 – mf#1986-0790 – us ATLA [240]
Die erneuerung des paulinischen christentums durch luther dekanatsrede gehalten am 31. oktober 1902 in wien / Feine, Paul – Leipzig: J C Hinrichs, 1903 – 1mf – 9 – 0-8370-5991-7 – mf#1985-3991 – us ATLA [225]
Erneuerung mit sachsenstimme *see* Sachsenstimme
Ernle, Rowland Edmund Prothero *see* The psalms in human life
Ernouf, Alfred Auguste *see* Histoire de trois ouvriers francais: richard lenoir, abraham louis breguet, michel brezin
Ernsberger, C S *see* A history of the wittenberg synod of the general synod of the evangelical lutheran church, 1847-1916
Ernst, Alfred *see*
- L'art de richard wagner
- Richard wagner et le drame contemporain

Ernst and Ernst *see* Budget control
Ernst, Arthur *see* Ontario chronicle
Ernst der bekenner : herzog von braunschweig und lueneburg / Wrede, Adolf – Halle: Verein fuer Reformationsgeschichte, 1888 – (= ser [Schriften des vereins fuer reformationsgeschichte]) – 1mf – 9 – 0-7905-4779-1 – (incl bibl ref) – mf#1988-0779 – us ATLA [240]
Ernst, Ferdinand *see* Bemerkungen auf einer reise durch das innere der vereinigten staaten von nord-amerika im jahre 1819
Ernst freiherrn von feuchtersleben's saemmtliche werke : mit ausschluss der rein medizinischen = Works / ed by Hebbel, Friedrich – Wien: C Gerold. 7v. 1851-53 (mf ed 1990) – 1 – us Wisconsin U Libr [802]
Ernst fries (1801-1833) : studien zu seinen landschaftszeichnungen / Bott, Elisabeth – Heidelberg, 1978 – 3mf – 9 – 3-89349-761-7 – gw Frankfurter [740]
Ernst, Fritz *see* Essais
Ernst, Fritz [comp] *see* Schriften
Ernst haeckel, der monistische philosoph : eine kritische antwort auf seine "weltraethsel" / Hoenigswald, Richard – Leipzig: E Avenarius, 1900 – 1mf – 9 – 0-7905-7304-0 – (incl bibl ref) – mf#1989-0529 – us ATLA [190]
Ernst hardt und die neuromantik : ein mahnruf an die gegenwart / Schumann, Harry – Loetzen: P Kuehnel, 1913 – 1 – us Wisconsin U Libr [430]
Ernst herzog / ed by Bartsch, Karl – Wien: W Braunmueller, 1869 – 1 – us Wisconsin U Libr [430]
Ernst, Johann *see*
- Cyrpian und das papsttum
- Die ketzertaufangelegenheit in der altchristlichen kirche nach cyprian

Ernst juenger : ein leben im umbruch der zeit / Mueller, Wulf Dieter – Berlin: Junker und Duennhaupt, 1934 – 1 – us Wisconsin U Libr [920]
Ernst juenger : mensch und werk / Becher, Hubert – Warendorf, Westfalen: J Schnell, 1949 [mf ed 1991] – (= ser Gestalt und werk 2) – 110p – 1 – (incl bibl ref) – mf#7503 – us Wisconsin U Libr [430]
Ernst juenger : die wandlung eines deutschen dichters und patrioten / Paetel, Karl Otto – New York: F Krause, 1946 – 1 – (incl bibl ref) – us Wisconsin U Libr [430]
Ernst juenger und das schicksal des menschen / Nebel, Gerhard – Wuppertal: Marees-Verlag, 1948 – 1 – us Wisconsin U Libr [430]
Ernst kochs "prinz rosa-stramin" : ein beitrag zur hessischen literaturgeschichte / Froeb, Hermann – Marburg a.L: N G Elwert, 1925 – 1 – (incl bibl ref) – us Wisconsin U Libr [430]
Ernst moritz arndt : Deutsche volkwerdung: sein politisches vermaechtnis an die deutsche gegenwart; kernstellen aus seinem schriften und briefen / ed by Petersen, Carl & Ruth, Paul Hermann – Breslau: Hirt, [1934] [mf ed 1988] – (= ser Hirts deutsche sammlung. literarische abt gruppe 9: gedankliche prosa 12) – 160p – 1 – mf#6954 – us Wisconsin U Libr [800]
Ernst moritz arndt : ein lebensbild / Muesebeck, Ernst – Gotha: F A Perthes, 1914- – 1 – (incl bibl ref) – us Wisconsin U Libr [943]
Ernst moritz arndt : der vorkaempfer fuer einheit und demokratie / Scurla, Herbert – Berlin: Kongress-Verlag, 1952 [mf ed 1993] – 169p/4pl (ill) – 1 – mf#8459 – us Wisconsin U Libr [920]
Ernst moritz arndt : der weg eines deutschen mannes / Heine, Gerhard – Leipzig: L Klotz c1939 [mf ed 1988] – 1r – 1 – (filmed with: ludwig anzengruber / sigismund friedmann) – mf#6954 – us Wisconsin U Libr [943]
Ernst moritz arndt in schweden : neue beitraege zum verstaendnis seines lebens und dichtens / Guelzow, Erich – Greifswald: L Bamberg, 1920 – 1 – us Wisconsin U Libr [943]
Ernst moritz arndts briefe an eine freundin / ed by Guelzow, Erich – Stuttgart: J G Cotta, 1928 [mf ed 1988] – 240p – 1 – mf#6954 – us Wisconsin U Libr [920]
Ernst moritz arndts fragmente ueber menschenbildung in ihrer paedagogischen bedeutung / Koelle, Conrad – Langensalza: H Beyer, 1916 – 1 – us Wisconsin U Libr [943]
Ernst moritz arndt's reise durch schweden im jahr 1804 – Berlin 1806 [mf ed Hildesheim 1995-98] – 4v on 8mf – 9 – €160.00 – 3-487-28946-6 – gw Olms [914]
Ernst moritz arndt's saemmtliche werke – Leipzig: K R Vogelsberg, 1892 [mf ed 1988] – 14v on 2r – 1 – mf#6952 – us Wisconsin U Libr [802]
Ernst, Otto *see* Nietzsche der falsche prophet
Ernst, Paul *see*
- Die deutschen volksbuecher
- Erdachte gespraeche
- Gesammelte werke
- Das kaiserbuch
- Das leben ein gleichnis
- Leo tolstoi und der slavische roman
- Manfred und beatrice
- Eine nacht in florenz
- Sechs geschichten

Ernst penzoldt und das theater / Rahn, Konstanze – Frankfurt a.M. 1976 (mf ed 1993) – 2mf – 9 – €31.00 – 3-89349-647-5 – mf#DHS 647 – gw Frankfurter [790]
Ernst simmels psychoanalytische klinik "sanatorium schloss tegel gmbh" (1927-1983) : beitrag zur wissenschaftsgeschichte einer pyschoanalytischen psychosomatik / Schultz-Venrath, Ulrich – (mf ed 1995) – 3mf – 9 – €49.00 – 3-8267-2081-4 – mf#DHS 2081 – gw Frankfurter [616]
Ernst, Stacey L *see* Prediction of injury in high school volleyball players with perceived leadership behavior of the coach
Ernst theodor amadeus hoffmann / Kroll, Erwin – Leipzig; Breitkopf & Haertel, 1923 – 1 – (incl bibl ref) – us Wisconsin U Libr [430]
Ernst theodor amadeus hoffmann : lebensschicksal eines seltsamen mannes / Krieger, Erhard – Kitzingen/Main: Holzner-Verlag [1952] – 1 – (incl bibl ref) – us Wisconsin U Libr [430]
Ernst toller : eine studie / Signer, Paul – Berlin: Verlag Landsberg, 1924 – 1 – us Wisconsin U Libr [430]
Ernst troeltsch : eine kritische zeitstudie / Kaftan, Theodor – Schleswig: J Bergas, 1912 – 1mf – 9 – 0-7905-7655-4 – mf#1989-0880 – us ATLA [190]
Ernst, U *see* Geschichte des zuercherischen schulwesens...
Ernst, Ulrich *see* Geschichte des zuercherischen schulwesens
Ernst von wildenbruch : ernstes und heiteres aus seinem leben / Duncker, Dora – Berlin: H Paetel, 1909 – 1 – us Wisconsin U Libr [430]
Ernst von wildenbruch / Litzmann, Berthold – Berlin: G Grote. 2v. 1913-16 – 1 – (incl bibl ref and ind) – us Wisconsin U Libr [430]
Ernst von wildenbruchs dramatische technik / Mannes, Ulrich – Jena: [Universitaet Jena], 1934 – 1 – (incl bibl ref) – us Wisconsin U Libr [430]
Ernst zahn : das werk und der dichter / Spiero, Heinrich – Stuttgart: Deutsche Verlags-Anstalt, 1927 – 1 – (incl bibl ref) – us Wisconsin U Libr [430]
Ernst zahns gesammelte werke : erste serie / Zahn, Ernst – Stuttgart: Deutsche Verlags-Anstalt, [192-?] – 1 – us Wisconsin U Libr [802]
Ernste blicke in den wahn der modernen kritik des alten testamentes / Zahn, Adolf – Guetersloh: C Bertelsmann, 1893 – 1mf – 9 – 0-8370-9758-4 – mf#1986-3758 – us ATLA [221]
Ernste blicke in den wahn der modernen kritik des alten testamentes. neue folge / Zahn, Adolf – Guetersloh: C Bertelsmann, 1894 – 1mf – 9 – 0-8370-5941-0 – (incl bibl ref) – mf#1985-3941 – us ATLA [221]
Die ernsthaften toren : novellen / Ulitz, Arnold – Muenchen: A Langen 1922 [mf ed 1996] – 1r – 1 – (filmed with: leutnant bertram / bodo uhse) – mf#4049p – us Wisconsin U Libr [830]
Ernsting, Arthur Conrad *see* Nucleus totius medicinae (ael3/11)
Ernstinger, Hans Georg *see* Hans georg ernstingers raisbuch
Ernst-Moritz-Arndt-Universitaet *see*
- Studien zur literatur des spaetmittelalters
- Zur gesellschaftlichen funktionalitaet mittelalterlicher deutscher literatur

Ernst-Moritz-Arndt-Universitaet Greifswald. Institut für Deutsche Philologie *see* Ergebnisse der 22. und 23. jahrestagung des arbeitskreises deutsche literatur des mittelalters
Ernst-Moritz-Arndt-Universitaet Greifswald. Sektion Germanistik, Kunst- und Musikwissenschaft *see* Ergebnisse der 21. jahrestagung des arbeitskreises 'deutsche literatur des mittelalters'
Erntekranz : gewunden aus den evangelien-perikopen des kirchenjahrs / Liefeld, Friedrich Wilhelm Albert – Milwaukee, WI: G. Brunder. 2v in 1. 1881 (mf ed 1990) – 1 – us Wisconsin U Libr [810]
Un eroe dell'ala rivoluzionaria italiana, giordano viezzoli – Paris, 1936? Fiche W 860. (Blodgett Collection of Spanish Civil War Pamphlets) – 9 – us Harvard [946]
Erokhin, N V *see* Cherez kooperatsiiu k elektrifikatsii
Eros und die evangelien : aus den notizen eines vagabunden / Bonsels, Waldemar – Frankfurt/M: Ruetten & Loening, 1921 [mf ed 1989] – 213p – 1 – mf#7051 – us Wisconsin U Libr [830]
Erosion of the rule of law in south africa – Geneva, Switzerland. 1968 – 1r – 1 – us UF Libraries [960]
EroSpirit – v1 n1-v5 n2 [1986 jan-1990 may/jun] – 1r – 1 – mf#1549705 – us WHS [071]
Erote ed anterote torneo celebrato dall'altezza serenissima elettorale di massimiliano emanuele... : con la serenissima elettrice maria antonia... / [Terzago, V.] – In Monaco: Per Giovanni lecklino, 1686 – 1mf – 9 – mf#0-1955 – ne IDC [090]
Erotemata musicae practicae : ex probatissimis quibusque hujus divinae & dulcissimae artis scriptoribus...breviter selecta... / Lossius, Lucas – Noribergae: Gerlach 1590 – (= ser Hqab. literatur des 16. jahrh.) – 2mf – 9 – €30.00 – mf#1590a – gw Fischer [780]
Erotemata musicae practicae / Lossius, Lucas – 1565 – (= ser Mssa) – 2mf – 9 – €35.00 – mfchl 62 – gw Fischer [780]
Erotemata musices practicae / Wilflingseder, Ambrosius – 1563 – (= ser Mssa) – 9 – €70.00 – mfchl 71 – gw Fischer [780]
Erotematum musicae libri duo / Beurhusius, Friedrich – 1580 – (= ser Mssa) – lost – 9 – mfchl 50 – gw Fischer [780]
Erotematum musicae libri duo : ex optimis huius artis scriptoribus vera perspicuaque methodo descripti / Beurhaus, Friedrich – Noribergae: Gerlach, Montanus 1580 [mf ed 1978?] – 2mf – 9 – mf#fiche 361 – us Sibley [780]
L'erotisme au cinema : pin-up / Duvillars, Pierre – Paris: edition du 20e siecle, 1951 – (= ser Les femmes [coll]) – 1mf – 9 – mf#8069 – fr Bibl Nationale [790]
Errandonea, Ignacio *see*
- Donoso cortes, juan. obras completas recopiladas por...2 vol. madrid, 1946
- Estudios clasicos. morfologia griega. sintaxis griega del p. santiago morillos j
- Nuevo salterio latino-espanol. version latina promulgada por s.s. pio 12...
- Poesia cristiana. antologia de poesia romano-cristiana y latino medieval (siglos 4-15) toledo, 1946
- Santos coco, francisco. la pronunciacion del latin. badajoz, 1929
- Vida y obras de don juan pablo forner y segarra, madrid, 1944

Errante, Vincenzo *see* Il mito di faust
Errard, J *see* Fortificatio, das ist kuenstliche und wolgegruendte demonstration
Errata de l'Essai sur la musique ancienne et moderne : ou lettre a l'auteur de cet essai / Latour de Franqueville, Mme de – [Paris?] 1780 [mf ed 19–] – 2mf – 9 – mf#fiche 1243 – us Sibley [780]
Errazuriz, Crescente *see* Pedro de valdivia
Errett, Isaac *see*
- Evenings with the bible
- Fifty-nine years of history
- Life and writings of george edward flower
- Linsey-woolsey
- Our position
- The querists' drawer
- Talks to bereans

Die errettung des ruhrgebiets (1918-1920) – Darstellungen aus den nachkriegskampfen deutscher truppen und freikorps. 9. Bd. Berlin 1943 – 1 – gw Mikropress [943]

Erreur revolutionnaire et notre etat social / Magloire, Auguste – Port-Au-Prince, Haiti. 1909 – 1r – 1 – us UF Libraries [972]

Errington, George *see* The irish land question

Erro y Aspiroz, Juan B *see*
– Alfabeto de la lengua primitiva de espana
– El mundo primitivo o examen filosofico de la antigueedad y cultura de la nacion bascongada

Erromanga : the martyr isle / Robertson, H A; ed by Fraser, John – New York: A C Armstrong; London: Hodder and Stoughton, 1902 – 2mf – 9 – 0-8370-6351-5 – mf#1986-0351 – us ATLA [240]

Erroneous statements concerning atonement and its results considered / Newton, Benjamin Wills – London, England. 1877 – 1r – 1 – us UF Libraries [240]

Error detected and fiction rebuked / Maddock, Theophilus – London, England. 1794 – 1r – 1 – us UF Libraries [240]

The error of modern missouri : its inception, development, and refutation / ed by Schodde, George Henry – Columbus, Ohio: Lutheran Book Concern, 1897 – 2mf – 9 – 0-7905-7145-5 – mf#1988-3145 – us ATLA [240]

Errores actuales que se hallan extendidos en espana causando gravisimos estragos en el pueblo catolico... / Marquez, Gabino – Madrid: Razon y Fe, 1935 – 1 – sp Bibl Santa Ana [240]

Los errores de nuestro tiempo / Donoso Cortes, Juan Francisco – Madrid: Feria Nacional del Libro, 1955 – sp Bibl Santa Ana [946]

Errores del diccionario de madrid / Malaret, Augusto – San Juan, Puerto Rico. 1936 – 1r – 1 – us UF Libraries [440]

Errores modernos, expuestos y refutados por... : con un apendice sobre la nueva bula de la santa cruzada / Marquez, Gabino – Jerez de la Frontera: Tipolitografia de Salido Hermanos, 1917 – 1 – sp Bibl Santa Ana [946]

Errores y omisiones de la obra "bibliografia del general... / Salas, Carlos I – Buenos Aires, Argentina. 1912 – 1r – 1 – us UF Libraries [972]

Error's chains : How forged and broken. a complete, graphic, and comparative history of the many strange beliefs, superstitious practices... of mankind throughout the world... / Dobbins, Frank Stockton – New York: Standard Pub House, 1883 – 2mf – 9 – 0-524-05841-5 – mf#1990-3505 – us ATLA [200]

Errors in criminal proceedings in all states and territories and federal courts where judgments have been affirmed / Walker, William Slee – Cincinnati, Anderson, 1916 – 550p – 1 – mf#LL-1551 – us L of C Photodup [345]

Errors of campbellism : being a review of all the fundamental errors of the system of faith and church polity of the denomination founded by alexander campbell / Stuart, T McK – Cincinnati: Jennings and Graham; New York: Eaton and Mains, c1890 – 1mf – 9 – 0-7905-9692-X – mf#1989-1417 – us ATLA [240]

The errors of hopkinsianism detected and refuted : in six letters to the rev s williston, pastor of the presbyterian church in durham, n.y / Bangs, Nathan – New York: Printed for the author, 1815. Beltsville, MD: NCR Corp, 1978 (4mf); Evanston: American Theol Lib Assoc, 1984 (4mf) – 9 – 0-8370-0913-8 – mf#1984-4259 – us ATLA [243]

The errors of the plymouth brethren / Carmichael, James – Montreal: W Drysdale, 1888 – 1mf – 9 – mf#36585 – cn Canadiana [242]

Der ersatz der religion durch vollkommeneres und die abstreifung alles asiatismus / Duehring, Eugen Karl – 3. umgearb aufl. Leipzig: Theod Thomas, 1906 – 1mf – 9 – 0-8370-2986-4 – (incl app partly containing other works by the author) – mf#1985-0986 – us ATLA [270]

Ersatzversuche fuer das biblische christusbild / Rohr, Ignaz – 2. aufl. Muenster i W: Aschendorff 1908 [mf ed 1992] – (= ser Biblische zeitfragen) – 1mf – 9 – 0-524-05629-3 – (incl bibl ref) – mf#1992-0484 – us ATLA [220]

Ersch, Johann S *see* Allgemeine encyclopaedie der wissenschaften und kuenste

Ersch, Johann Samuel *see*
– Allgemeine encyclopaedie der wissenschaften und kuenste
– Allgemeine literatur-zeitung

Die erschaffung der welt und der menschen : und deren geschichte bis nach der suendfluth / Westermayer, Anton – Schaffhausen: Friedr Hurter, 1861 [mf ed 1993] – (= ser Das alte testament und seine bedeutung 1) – 2mf – 9 – 0-524-06223-4 – mf#1992-0861 – us ATLA [221]

Die erschuetterung des optimismus durch das erdbeben von lissabon 1755 : was ist heute die religioese aufgabe der universitaeten? / Luetgert, Wilhelm & Schlatter, Adolf von – Guetersloh: C Bertelsmann, 1901 – (= ser Beitraege zur foerderung christlicher theologie) – 1mf – 9 – 0-7905-9317-3 – mf#1989-2542 – us ATLA [210]

Ershov, A *see* Komsomol i kooperatsiia

Erskine church echoes – Hamilton, Ont.: The Church, [188- or 189–189- or 19–] – 9 – mf#P05069 – cn Canadiana [242]

Erskine dale, pioneer / Fox, John – Toronto: G J McLeod, 1920 [mf ed 1995] – 4mf – 9 – 0-665-74268-1 – (ill by f c yohn) – mf#74268 – cn Canadiana [830]

Erskine, Ebenezer *see* Plant of renown

Erskine, James St. Clair *see* Remarks on the report of the faculty of advocates, appointed to consider the provisions of the bill for the better regulating of the process of the courts of law in scotland

Erskine of linlathen : selections and biography / Henderson, Henry F – Edinburgh, London: Oliphant, Anderson & Ferrier, 1899 [mf ed 1984] – 4mf – 9 – 0-8370-0166-8 – (incl ind) – mf#1984-0034 – us ATLA [242]

Erskine, Payne *see* Joyful heatherby

Erskine presbyterian church, hamilton, canada : semi-jubilee, 1880-1905: brief histories of the church, its ministers and organizations – [Hamilton ON?: s.n, 1905? [mf ed 1994] – 1mf – 9 – 0-665-72767-4 – mf#72767 – cn Canadiana [242]

Erskine, Thomas *see*
– The brazen serpent
– The doctrine of election
– Remarks on the internal evidence for the truth of revealed religion

Erskine, Thomas, Sir *see* The supernatural gifts of the spirit

The erskines / Macewen, Alexander Robertson – Edinburgh: Oliphant, Anderson & Ferrier, 1900 – (= ser Famous Scots Series) – 1mf – 9 – 0-7905-5187-X – mf#1988-1187 – us ATLA [920]

Das erst capitel des propheten jeheskiels... : von dem ampt der oberen vnd der vnderthonen / Oecolampadius, J – [Basel, Andreas Cratander, 1527] – 1mf – 9 – mf#PBU-377 – ne IDC [240]

Erst mensch, dann christ und so ein ganzer mensch / Weiss, Albert Maria – Freiburg im Breisgau; St. Louis, MO: Herder, 1878 [mf ed 1986] – (= ser Apologie des christenthums vom standpunkte der sittenlehre 1) – 2mf – 9 – 0-8370-7035-X – (incl bibl ref) – mf#1986-1035 – us ATLA [241]

Der erste brief johannis : in berichtigter lutherscher uebersetzung / Neander, August – Berlin: Wiegandt und Grieben, 1851 – 1mf – 9 – 0-8370-9571-9 – mf#1986-3571 – us ATLA [227]

Der erste brief johannis / Rothe, Richard; ed by Muehlhaeuser, R – Wittenberg: Hermann Roelling, 1878 – 1mf – 9 – 0-524-06859-3 – mf#1992-1001 – us ATLA [227]

Erste brief johannis = The first epistle of john / Neander, August – New York: Lewis Colby, 1852 – 1mf – 9 – 0-8370-9643-X – (In English) – mf#1986-3643 – us ATLA [227]

Erste brief johannis in predigten *see* Commentary on the first epistle of st john

Der erste brief pauli an die korinther *see* The first epistle of paul to the corinthians

Der erste brief petri / Schott, Theodor – Erlangen: Andreas Deichert, 1861 – 1mf – 9 – 0-8370-9655-3 – (incl bibl ref) – mf#1986-3655 – us ATLA [227]

Das erste buch der tora : genesis uebersetzt und erklaert / Jacob, B – Berlin, 1934 – €84.00 – ne Slangenburg [221]

Der erste clemensbrief = First epistle of clement to the corinthians / Clement 1, Pope; ed by Knopf, Rudolf – Leipzig: J C Hinrichs, 1899 – (= ser Tugal) – 1mf – 9 – 0-7905-4028-2 – mf#1988-0028 – us ATLA [240]

Der erste clemensbrief / Knopf, R – Leipzig, 1899 – (= ser Tugal 2-20/1) – 3mf – 9 – €7.00 – ne Slangenburg [227]

Der erste clemensbrief in altkopischer uebersetzung = First epistle of clement to the corinthians – Leipzig: J C Hinrichs, 1908 – (= ser Tugal) – 1mf – 9 – 0-7905-1809-0 – (incl ind; in coptic) – mf#1987-1809 – us ATLA [227]

Der erste clemensbrief in altkoptischer uebersetzung / Schmidt, Carl – Leipzig, 1908 – (= ser Tugal 3-32/1) – 3mf – 9 – €7.00 – ne Slangenburg [227]

Die erste deutsche bibel / ed by Kurrelmeyer, W – Stuttgart: Litterarischer Verein, 1904-15 (Tuebingen: H Laupp, Jr) [mf ed 1993] – (= ser Blvs 234, 238, 243, 246, 249, 251, 254, 258, 259, 266) – 10v – 1 – (middle high german text) – mf#8470 reels 48-54 – us Wisconsin U Libr [220]

Die erste deutsche bibel / ed by Kurrelmeyer, W – Stuttgart: Litterarischer Verein. 10v. 1904-15 (Tuebingen: H Laupp, Jr) – us Wisconsin U Libr [220]

Die erste epistel paulj an timotheum / Spangenberg, C – Strassburg, 1564 – 5mf – 9 – mf#TH-1 mf 1411-1415 – ne IDC [242]

Die erste erhebung der bergarbeiter *see* 1889

Der erste evangelische gottesdienst in strassburg : vortrag. gehalten im evangelischen vereinshause zu strassburg... / Smend, Julius – Strassburg: JH Ed Heitz, 1897 – 1mf – 9 – 0-524-02285-2 – mf#1990-0590 – us ATLA [242]

Erste frage : kann man noch mensch sein, ohne christ zu sein? = Solution de grands problemes / Martinet, Antoine, abbe – Tuttlingen: E L Kling, 1858 [mf ed 1986] – (= ser Loesung grosser fragen 1) – 1mf – 9 – 0-8370-7309-X – (incl bibl ref. german trans fr french by anton weiskopf) – mf#1986-1309 – us ATLA [241]

Erste internationale film-zeitung 1909 : parlamentsausgabe – Berlin DE, 1909-1915 3 jul [gaps], 1918 17 aug-1920 11 dec – 8r – 1 – (filmed with suppls: der kinematographen-operateur, pathe-woche, das filmrecht) – gw Mikrofilm [790]

Erste internationale kinematographen-zeitung – Hamburg DE, 1908 25 mar – 1 – gw Mikrofilm [790]

Der erste korintherbrief / Weiss, Johannes – 9. aufl. Goettingen: Vandenhoeck & Ruprecht, 1910 – (= ser Kritisch exegetischer Kommentar ueber das Neue Testament) – 1mf – 9 – 0-7905-3177-1 – (incl bibl ref) – mf#1987-3177 – us ATLA [227]

Erste liebe : roman aus der jugendzeit / Wehner, Josef Magnus – Hamburg: Hanseatische Verlagsanstalt, c1941 – 1 – us Wisconsin U Libr [830]

Der erste petrusbrief : seine entstehung und stellung in der geschichte des urchristentums / Voelter, Daniel – Strassburg: J H Ed Heitz, 1906 – 1mf – 9 – 0-8370-9333-3 – (incl bibl ref) – mf#1986-3333 – us ATLA [227]

Der erste petrusbrief und die neuere kritik / Weiss, Bernhard – Gr Lichterfelde-Berlin: Edwin Runge 1906 [mf ed 1993] – (= ser Biblische zeit- und streitfragen 2/9) – 1mf – 9 – 0-524-05944-6 – mf#1992-0701 – us ATLA [227]

Das erste pontificalschreiben des apostelfuersten petrus : wissenschaftliche und praktische auslegung des ersten briefes des heil. petrus im geiste der kirche und im hinblick auf den geist der zeit: eine festschrift zur erinnerung an das fuenfundzwanzigjaehrige papst-jubilaeum des heiligen vaters pius 9 / Hundhausen, Ludwig Joseph – Mainz: Franz Kirchheim, 1873 – 2mf – 9 – 0-7905-1069-3 – (in german and greek. incl ind) – mf#1987-1069 – us ATLA [227]

Die erste schlacht, vom werden und von den ersten kaempfen des bataillons edgar andre / Uhse, Bodo – 2Auf. Strasbourg, 1938. Fiche W1235. (Blodgett Collection of Spanish Civil War Pamphlets) – 9 – us Harvard [946]

Das erste sendschreiben des apostel paulus an die korinthier / Heinrici, Carl Friedrich Georg – Berlin: Wilhelm Hertz, 1880 – 2mf – 9 – 0-7905-0037-X – mf#1987-0037 – us ATLA [227]

[Erste] sinfonie, (es-dur), op 28 : [1. allegro maestoso. 2. scherzo. 3. grave und finale] fuer grosses orchester / Bruch, Max – Leipzig: C F W Siegel [187-] [mf ed 1995] – 1r – 1 – mf#pres. film 140 – us Sibley [780]

Die erste stunde nach dem tode : eine gespenstergeschichte / Brod, Max – Leipzig: K Wolff 1916 [mf ed 1989] – 1r – 1 – (filmed with: ein gelegenheitsgedicht von brockes / friedrich gundolf) – mf#7089 – us Wisconsin U Libr [830]

Der erste theil newer teutscher lieder mit 5 stimmen / Lasso, Orlando di – 1576 – (= ser Mssa) – 2mf – 9 – €60.00 – mfchl 295 – gw Fischer [780]

Der erste thessalonicherbrief / Schmidt, Paul Wilhelm – Berlin: Georg Reimer, 1885 – 1mf – 9 – 0-8370-5114-2 – mf#1985-3114 – us ATLA [227]

Die erste und zweite fassung von goethes "wanderjahren" / Bimler, Kurt – Beuthen, O-S: M Immerwahr, 1907 [mf ed 1990] – 85p – 1 – (incl bibl ref) – mf#7371 – us Wisconsin U Libr [430]

Der erste und zweite petrusbrief und der judasbrief / Wohlenberg, Gustav – 1. u 2. aufl. Leipzig: A Deichert, 1915, c1914 – 1mf – 9 – 0-8370-9438-0 – mf#1986-3438 – us ATLA [227]

Erste, zweite und dritte berathung des entwurfs eines buergerlichen gesetzbuchs im reichstage / Germany. Reichstag – Berlin: J Guttentag, 1896 – (= ser Civil law 3 coll) – 10mf – 9 – mf#LLMC 96-508 – us LLMC [346]

Die ersten ausgaben von grimmelshausens simplicissimus : eine kritische untersuchung / Borcherdt, Hans Heinrich – Muenchen: H Stobbe, 1921 [mf ed 1990] – (= ser Einzelschriften zur buecher- und handschriftenkunde 1) – 64p – 1 – (incl bibl ref) – mf#7423 – us Wisconsin U Libr [430]

Die ersten bnovellen otto ludwigs und ihr verhaeltnis zu ludwig tieck / Greiner, Wilhelm – Poessneck i.Th.: B Feigenspan, 1903 – 1r – 1 – (incl bibl ref) – us Wisconsin U Libr [430]

Die ersten buecher stefan georges : eine annaeherung an das werk / Lachmann, Eduard – Berlin: G Bondi, 1933 (mf ed 1990) – 1r – 1 – (filmed with: das werk georges) – us Wisconsin U Libr [430]

Die ersten deutschen zeitungen / ed by Weller, Emil Ottokar – Stuttgart: Litterarischer Verein, 1872 (Tuebingen: H Laupp) [mf ed 1993] – (= ser Blvs 111) – 383p – 1 – mf#8470 reel 23 – us Wisconsin U Libr [010]

Die ersten deutschen zeitungen / Weller, Emil Ottokar; ed by Weller, Emil – Stuttgart: Litterarischer Verein, 1872 (Tuebingen: H Laupp) [mf ed 1993] – (= ser Blvs 109) – 383p – 1 – mf#8470 reel 23 – us Wisconsin U Libr [074]

Die ersten jahre der kirche calvins, 1541-1546 / Cornelius, Carl Adolf – Muenchen: Verlag der K Akademie, 1895 – 1mf – 9 – 0-524-02792-7 – (incl bibl ref) – mf#1990-0696 – us ATLA [242]

Die ersten poetischen versuche hamerlings : zur geschichte seines zwettler aufenthalts / Rabenlechner, Michael Maria – Hamburg: Verlagsanstalt und Druckerei A.-G. (vormals J F Richter), 1896 – 1r – 1 – us Wisconsin U Libr [430]

Erster [-dritter] theyl biblischer gebett / Spangenberg, C – np, 1582-1583. 3v – 13mf – 9 – mf#TH-1 mf 1632-1644 – ne IDC [242]

Erstes erlebnis : vier geschichten aus kinderland / Zweig, Stefan – Leipzig: Insel-Verlag, 1919 – 228p – 1 – mf#7981 – us Wisconsin U Libr [830]

Erstes trio, in f dur, pianoforte, violine und violoncello / Bargiel, Woldemar – neue rev ausg. Leipzig: F E C Leuckart [between 1870-73] [mf ed 1992] – 1r – 1 – mf#pres. film 123 – us Sibley [780]

Erstlinge und zehnten im alten testament : ein beitrag zur geschichte des israelitisch-juedischen kultus / Eissfeldt, Otto – Leipzig: J C Hinrichs, 1917 – (= ser Beitraege zur wissenschaft vom alten testament) – 1mf – 9 – 0-524-02421-9 – (incl bibl ref) – mf#1990-3005 – us ATLA [221]

Die erstlingsnovellen heinrich von kleist / Davidts, Hermann – Berlin, 1913 (mf ed 1995) – 1mf – 9 – €24.00 – 3-8267-3146-8 – mf#DHS-AR 3146 – gw Frankfurter [430]

Ertl, Emil *see*
– Meisternovellen

Ertl, Ernst *see* Werkmeister im "paradies"

Der ertrag der ausgrabungen im orient fuer die erkenntnis der entwicklung der religion israels / Sellin, Ernst – Leipzig: A Deichert, 1905 – 1mf – 9 – 0-7905-0440-5 – (incl bibl ref) – mf#1987-0440 – us ATLA [270]

Erttmann, Paul Oskar *see* Die otenkamps

Ertugrul – Bursa. Mueduer-i Mes'ul: Ismail Hakki, Ahmed Refik; Sermuharriri: Hakki Baha, Ziya Sakir. n461. 27 subat 1919, 463-464, 551. 6 mayis 1920 – (= ser O & t journals) – 1mf – 9 – $25.00 – us MEDOC [956]

Erudicion evangelica...santa oracion / Jimenez, Antonio – 1627 – 9 – sp Bibl Santa Ana [240]

La erudicion extremena y la academia de la historia / Rodriguez Monino, Antonio – Badajoz, 1946 – 1 – sp Bibl Santa Ana [946]

Eruvin : or, miscellaneous essays on subjects connected with the nature, history, and destiny of man / Maitland, Samuel Roffey – 2nd ed. London: Francis & John Rivington, 1850 – 1mf – 9 – 0-7905-0140-6 – (incl bibl ref and ind) – mf#1987-0140 – us ATLA [110]

Ervaringen gedurende mijn twaalfjarig zendingsleven / Wiersma, J N – Rotterdam: D J P Storm Lotz, 1876 [mf ed 1995] – (= ser Yale coll) – 255p – 1 – 0-524-09794-1 – (in dutch) – mf#1995-0794 – us ATLA [959]

Ervin, James R *see* An assessment of the marketing and promotions of women's lacrosse in ncaa division 1

Erving 1844-1906 – Oxford, MA (mf ed 1983) – (= ser Massachusetts vital records) – 6mf – 9 – 0-931248-54-X – (Mf 1: Index: B,M,D 1844-1906. Mf 2: Births 1844-80. Mf 3: Births 1881-1905. Mf 4: Marriages 1844-95. Mf 5: Marriages 1896-1906; Deaths 1845-67. Mf 6: Deaths 1868-1906) – us Archive [978]

Das erwachen des deutschen nationalbewusstseins in der preussischen judenheit (von moses mendelssohn bis zum beginn der reaktion) : ein geistesgeschichtlicher beitrag zur emanzipationsgeschichte der deutschen juden / Offenburg, Benno – Hamburg, 1933 (mf ed 1996) – (= ser Monographien zur wissenschaft des judentums) – 1mf – 9 – €24.00 – 3-8267-3189-1 – mf#DHS 70001 – gw Frankfurter [939]

Erwaegungen fuer die bischoefe des concilium's ueber die frage der paepstlichen unfehlbarkeit / Doellinger, Johann Joseph Ignaz von – Regensburg: G J Manz, 1869 – 1mf – 9 – 0-8370-7934-9 – (incl bibl ref) – mf#1986-1934 – us ATLA [240]

Die erwaehlung israels in der wueste / Bach, Robert – [1951?] Chicago: Dep of Photodup, U of Chicago Lib, 1963 (1r); Evanston: American Theol Lib Assoc, 1984 (1r) – 1 – 0-8370-0427-6 – mf#1984-B012 – us ATLA [221]

Erwartung und angebot : studien zum gegenwaertigen verhaeltnis von literatur und gesellschaft in der ddr / Kaufmann, Eva – Berlin: Akademie-Verlag, 1976, c1975 [mf ed 1992] – (= ser Literatur und gesellschaft) – 237p – 1 – (incl bibl ref and ind) – us Wisconsin U Libr [430]

Erweis der echtheit und glaubwuerdigkeit des pentateuch fuer die wissenschaft / Rupprecht, Eduard – Guetersloh: C Bertelsmann, 1896-1897 – 3mf – 9 – 0-7905-3053-8 – mf#1987-3053 – us ATLA [221]

Erweiterte b-bild-diagnostik in der mammasonographie mittels texturanalyse und speckle-muster-reduktion / Bader, Werner – (mf ed 1999) – 3mf – 9 – €49.00 – 3-8267-2654-5 – mf#DHS 2654 – gw Frankfurter [618]

Erweiterungen und aenderungen im vierten evangelium / Wellhausen, Julius – Berlin: Georg Reimer, 1907 – 1mf – 9 – 0-8370-5770-1 – (incl bibl ref) – mf#1985-3770 – us ATLA [226]

Erwin first baptist church (formerly indian creek). erwin, tennessee : church records – 1822-1970 – 1 – $93.42 – us Southern Baptist [242]

Erwin, Frank Alexander *see* A summary of torts

Erwin progress – Painted Post, NY. 1972-1973 (1) – mf#65155 – us UMI ProQuest [071]

Erwin, Theodor *see* Temas alemaes

Eryci puteani bruma : chimonopaegnion, de laudibus hiemis, ut ea potissimum apud belgas... / Puteanus, E – Monaci: [Ex formis Annae Bergiae viduae. Apud Raphaelem Sadelerum], 1619 – 1mf – 9 – mf#O-1945 – ne IDC [090]

Erynnerung was denen, so sich ynn ehestand begeben, zu bedencken sey / Menius, J – Wittemberg, 1528 – 1mf – 9 – mf#TH-1 mf 1163 – ne IDC [242]

Der erythroide anionenaustauscher ae1 : struktuelle untersuchungen durch mutagenese der murinen aei-cdna und bestimmung der kopienzahl der humanen ae1-mrna waehrend der erythroiden differenzierung / Koenig, Joerg Udo – (mf ed 1995) – 2mf – 9 – €40.00 – 3-8267-2094-6 – mf#DHS 2094 – gw Frankfurter [575]

Erzaehlende schriften / Holtei, Karl von – Breslau: Edward Trewendt 1861-74 [1861-68] [mf ed 1979] – 3r – 1 – mf#film mas c473 – us Harvard [830]

Der erzaehler *see* Duesseldorfer merkur

Der erzaehler an der saale *see* Hoefer intelligenz-blatt

Der erzaehler an der spree – Bautzen DE, 1842 29 apr-1851 26 apr – 4r – 1 – gw Misc Inst [074]

Die erzaehlung des hexateuch : auf ihre quellen untersucht / Smend, Rudolf – Berlin, G. Reimer, 1912 – 1mf – 9 – 0-7905-3229-8 – mf#1987-3229 – us ATLA [221]

Erzaehlung des russischen flott-captains rikord von seiner fahrt nach den japanischen kuesten in den jahren 1812 und 1813 und von seinen unterhandlungen mit den japanern : gedruckt auf allerhoechsten befehl st petersburg 1816 / Rikord, Petr – Leipzig 1817 [mf ed Hildesheim 1995-98] – 1v on 2mf – 9 – €60.00 – 3-487-27537-6 – gw Olms [880]

Erzaehlung des sempacher krieges... / Bullinger, Heinrich; ed by Geilfus, G – Winterthur, Ziegler, 1865) – 1mf – 9 – mf#PBU-484 – ne IDC [240]

Erzaehlungen / Arnim, Ludwig Achim, Freiherr von – Berlin: Union Verlag, 1957 [mf ed 1993] – (= ser Perlenkette 25) – 217p (ill) – 1 – (aft by hans krey) – mf#8464 – us Wisconsin U Libr [880]

Erzaehlungen / Bosshart, Jakob – Leipzig: H Haessel, 1921-22 [mf ed 1989] – 4v – 1 – mf#7055 – us Wisconsin U Libr [830]

Erzaehlungen / Ebner-Eschenbach, Marie von – 2nd rev ed. Stuttgart: Cotta, 1879 [mf ed 1993] – 167p – 1 – mf#8572 – us Wisconsin U Libr [830]

Erzaehlungen / Kleist, Heinrich von – Leipzig: Bibliographisches Institut, [1905?] [mf ed 1995] – (= ser Meyers volksbuecher) – 151p – 1 – mf#8787 – us Wisconsin U Libr [830]

Erzaehlungen / Stifter, Adalbert; ed by Aprent, Johannes – Osnabrueck: Bernhard Wehberg, 1903 [mf ed 1995] – 351p – 1 – mf#8886 – us Wisconsin U Libr [830]

Erzaehlungen aus altdeutschen handschriften / Keller, Adelbert von [comp] – Stuttgart: Literarischer Verein, 1855 [mf ed 1993] – (= ser Blvs 35) – 712p – 1 – (rhymed tales dating fr 15th c) – mf#8470 reel 8 – us Wisconsin U Libr [390]

Die erzaehlungen eduard von keyserlings : ein beitrag zur deutschen literaturgeschichte / Knoop, Kaete – Marburg a.L.: N G Elwert, 1929 – 1r – 1 – (incl bibl ref (3rd-4th prelim. leaves)) – us Wisconsin U Libr [430]

Erzaehlungen fuer die jugend – St Louis, MO: Lutherischer Concordia-Verlag, [18–?] [mf ed 1993] – 1r – 1 – mf#8364 – us Wisconsin U Libr [830]

Erzaehlungen und novellen / Hebbel, Friedrich – Pesth: G Heckenast, 1855 [mf ed 1990] – 154p – 1 – mf#7449 – us Wisconsin U Libr [830]

Erzaehlungen und schwaenke / ed by Lambel, Hans – Leipzig: F A Brockhaus, 1872 [mf ed 1993] – (= ser Deutsche classiker des mittelalters 12) – xiv/358p – 1 – (incl bibl ref and ind) – mf#8189 reel 2 – us Wisconsin U Libr [800]

Erzaehlungseingaenge in der deutschen literatur / Leib, Fritz – Giessen, 1913 (mf ed 1995) – 2mf – 9 – €31.00 – 3-8267-3109-3 – mf#DHS-AR 3109 – gw Frankfurter [430]

Die erzaehlungstechnik viktor scheffels / Grebe, Walter – Barmen-Wichlingh: Montanus & Ehrenstein, 1919 – 1r – 1 – (incl bibl ref) – us Wisconsin U Libr [430]

Erzahlungen und novellen *see* Hebbels herodes und mariamne

Der erzbeter : und drei andere legenden der bosheit / Kremer, Hannes – 5.aufl. Muenchen: F Eher, 1943 [mf ed 1992] – (= ser Soldaten-kameraden! v16) – 50p (ill) – 1 – mf#7524 – us Wisconsin U Libr [390]

Erzbischof albrecht 2. von magdeburg / Schmidt, Hermann – 1880 [mf ed 1990] – 1mf – 9 – 0-7905-6781-4 – (incl bibl ref) – mf#1988-2781 – us ATLA [241]

Erzbischof bruno von trier : ein beitrag zur geschichte der geistigen stroemungen im investiturstreit / Schlechte, Horst – Leipzig, 1934 (mf ed 1993) – 1mf – 9 – €24.00 – 3-89349-274-7 – mf#DHS-AR 131 – gw Frankfurter [241]

Der erzbischof von koeln johannes cardinal von geissel und seine zeit / Baudri – Koeln, 1881 (mf ed 1993) – 2mf – 9 – €31.00 – 3-89349-359-X – mf#DHS-AR 359 – gw Frankfurter [240]

Erzeugung von phototaktischen verhaltensvarianten aus den mutanten d1, km1 und flx15 des archaebakteriums halobakterium salinarium unter verwendung der mutagene ethylmethansulfonat und n-methyl-n'-nitro-n-nitrosoguanidin / Beermann, Kerstin – (mf ed 1997) – 2mf – 9 – €40.00 – 3-8267-2407-0 – mf#DHS 2407 – gw Frankfurter [574]

Erzgebirgisches nachrichts- und anzeigeblatt *see* Nuetzliches und unterhaltendes marienberger wochenblatt fuer alle staende

Der erzieherische gehalt in j j breitingers "critischer dichtkunst" *see* Gockel, hinkel und gackeleia

Der erzieherische gehalt in j j breitingers 'critischer dichtkunst' : abhandlung... / Braeker, Jakob – St Gallen [Switzerland]: H Tschudy 1950 [mf ed 1993] – 1r – 1 – (incl bibl ref) – mf#8525 – us Wisconsin U Libr [430]

Die erziehung der deutschen jungmannschaft im reichsarbeitsdienst *see* Nsdap (national socialist german workers party) nazi publications

Die erziehung des weiblichen geschlechts in indien und anderen heidenlaendern : ein aufruf an die christlichen frauen deutschlands und der schweiz / Hoffmann, Wilhelm – 3. gaenzlich umgearb Aufl. Heidelberg: Winter, 1853 – 1mf – 9 – 0-524-02085-X – (incl bibl ref) – mf#1990-2849 – us ATLA [360]

Die erziehung in der religion jesu im unterschiede zu der im dogmatischen christentume : ein beitrag zur abhilfe eines unertraeglichen notstandes in unserer jugenderziehung / Lietz, Hermann – Langensalza: Hermann Beyer, 1896 – 1mf – 9 – 0-8370-7958-6 – (incl bibl ref) – mf#1986-1958 – us ATLA [240]

Die erziehung zur ehe : eine satire / Hartleben, Otto Erich – Berlin: S Fischer 1893 [mf ed 1986] – 1r – 1 – (with: friends, society of / lower, t) – mf#1669 – us Wisconsin U Libr [870]

Erziehungs-blatter : organdes deutsch-amerikanischen lehrerbundes – 5. jahrg, neue folge, 2. ed, 19. heft [juli 1875], 16. jahrg, neue folge, 13. bd, 6. heft=186? [marz 1886] – 1r – 1 – mf#1003074 – us WHS [370]

Erziehungs-blatter fur schule und haus : organ des deutsch-amerikanischen lehrerbundes – Milwaukee: Hailmann and Dorflinger, 1875-1899. [ns v2 n9-v9 n12 jun 1875-sep 1882] – us CRL [071]

Der erziehungsgedanke im jugendstrafrecht : eine empirische analyse / Neus, Alexandra – (mf ed 1997) – 3mf – 9 – €49.00 – mf#DHS 2941 – gw Frankfurter [345]

Der erzketzer : ein roman vom leiden des wahrhaftigen / Wolzogen, Ernst von – 2. aufl. Berlin: F Fontane 1910 [mf ed 1992] – 2v on 1r – 1 – (filmed with: faust / f marlow [ludwig hermann wolfram]) – mf#3062p – us Wisconsin U Libr [830]

Erzstufen novellen und erzaehlungen / Schmid, Herman – 2.aufl. Leipzig: Keil [18–?] – (= ser Gesammelte schriften. volks- und familien-ausgabe 3) – 1 – (bound with: tannengruen and am kamin) – mf#film mas c438 – us Harvard [800]

Erzurum – (= ser Vilayet salnames) – 9 – (1304H [1887] def'a 10 5mf $75; 1315H/1313M [1897] def'a 13 4mf $190) – us MEDOC [956]

Es espana otra china? / Linebarger, Paul Myron Anthony – [Barcelona: s.n. 1931] [mf ed 1977] – (= ser Blodgett coll) – 1mf – 9 – mf#w999 – us Harvard [946]

Es geht ein pfluger ubers land / Wiechert, Ernst Emil – Muenchen, Germany. 1951 – 1r – 1 – us UF Libraries [430]

Es ist ein bann unter dir, israel : ein wort gegen den ueblichen gebrauch und die herkoemmliche stellung der apokryphen in der evangelischen kirche an alle evangelische christen / Wild, Friedrich Karl – Noerdlingen: C H Beck, 1854 – 1mf – 9 – 0-7905-0358-1 – (incl bibl ref) – mf#1987-0358 – us ATLA [220]

Es ist zeit : roman / Flake, Otto – Berlin: S Fischer, 1929 (mf ed 1990) – 1r – 1 – us Wisconsin U Libr [830]

Es klingt wie eine sage : zrzaehlungen aus alten chroniken / Storm, Theodor; ed by Langenbucher, Hellmuth – Bayreuth: Der Gauverlag, 1944 – 1r – 1 – us Wisconsin U Libr [390]

Es klopft / Elisabeth, Queen – 3. aufl. Regensburg: W Wunderling, 1887 (mf ed 1990) – 1 – us Wisconsin U Libr [430]

ES magazine – 1987 nov – 1r – 1 – mf#4851570 – us WHS [071]

Es muss tag werden – Muenchen DE, 1848 6 dec-1849 22 jan – 1r – 1 – gw Misc Inst [074]

Es navidad / Espada Marrero, J – Puerto Rico? 1959 – 1r – 1 – us UF Libraries [972]

Es necesario / Oraa, Francisco De – Habana, Cuba. 1964 – 1r – 1 – us UF Libraries [972]

Es reiten die chungusen : kaempfe mit mandschurischen bahnraeubern / Boenisch, Hermann Friedrich – Berlin: Zsolnay, K H Bischoff, 1942 [mf ed 1989] – 306p (ill) – 1 – mf#7050 – us Wisconsin U Libr [830]

Es reiten die wilden jaeger : roman / Burre, Paul – Jena: E Diederichs, 1943 [mf ed 1989] – 308p – 1 – mf#7096 – us Wisconsin U Libr [830]

Es shtarbt a shtetl / Weissbrod, Abraham – Munich, Germany. 1948 – 1r – 1 – us UF Libraries [939]

Es tiempo / Movimiento Estudiantil Chicano de Aztlan (Los Altos Hills, CA) – v1 iss1, 3 [1971 jan, sep 16] – 1r – 1 – mf#1269389 – us WHS [071]

Es war : roman / Sudermann, Hermann – 33. Aufl. Stuttgart: J G Cotta, 1902 – 1r – 1 – us Wisconsin U Libr [830]

Es war einmal : modern fairy tales for beginners in german / Baumbach, Rudolf & Wildenbruch, Ernst von – New York: American Book Co, c1893 [mf ed 1995] – 174p – 1 – (with english notes and a german-english vocabulary by wilhelm bernhardt) – mf#8972 – us Wisconsin U Libr [390]

Es war in einer sommernacht : novelle / Gerstner, Hermann – Muenchen: Zentralverlag der NSDAP, F Eher 1944 [mf ed 1990] – 1r – 1 – (filmed with: die regulatoren in arkansas / friedrich gerstacker) – mf#2609p – us Wisconsin U Libr [830]

Es werde licht : poesien / Jacoby, Leopold – 4. aufl. Muenchen: M Ernst, 1893 – 1r – 1 – us Wisconsin U Libr [810]

Es wird zeit : [poems] / Becher, Johannes Robert – Moskau: Verlagsgenossenschaft Auslaendischer Arbeiter in der UdSSR, 1933 [mf ed 1989] – 77p – 1 – mf#6994 – us Wisconsin U Libr [810]

Esa engage/social action – Washington. 1975-1987 (1,5,9) – (cont: engage/social action; cont by: christian social action) – ISSN: 0164-5528 – mf#7147,01 – us UMI ProQuest [360]

Es'ad *see* The divan project

Esad *see* Banet suad serhi

Esanzo, chants pour mon pays : poemes / Bolamba, Antoine Roger – Paris: Presence africaine, [1955] – 1 – us CRL [960]

Es'ar *see* The divan project

Esarhaddon, King of Assyria *see* Unpublished inscriptions of esarhaddon

Esatiz-i elhan / Yekta, Rauf – Istanbul: Mahmud Bey Matbaasi; Evkaf Islamiye, 1318-41 [1903-25] – (= ser Ottoman literature, writers and the arts) – 3mf – 9 – $55.00 – us MEDOC [470]

Esau : ou, la royaute populaire – Paris 1848 – us CRL [074]

Esau [pseud] *see* Herrin und sklave nach sacher masoch

Esbatiment moral, des animaux : louez le seigneur vous qui estes de la terre, dragons... / [Heyns, P] – Anvers: Gerard Smits, [1578] – 3mf – 9 – mf#O-3082 – ne IDC [090]

Esber / Hamid, Abduelhak – Istanbul, 1922 – (= ser Ottoman literature, writers and the arts) – 3mf – 9 – $55.00 – us MEDOC [470]

O esboco : periodico semanal, literario e recreativo – Rio de Janeiro, RJ: Typ H Lombaerts & C, 06 jan-04 maio 1889 – (= ser Ps 19) – mf#P17,01,138 – bl Biblioteca [410]

Esboco gramatical e vocabulario da lingua dos indi... / Rondon, Candido Mariano Da Silva – Rio de Janeiro, Brazil. 1948 – 1r – 1 – us UF Libraries [490]

Esbozo de un tema / Rodriguez Bou, Ismael – San Juan, Puerto Rico. 1965 – 1r – 1 – us UF Libraries [972]

Esbozo de una historia de las ideas en el brasil / Cruz Costa, Joao – Mexico City? Mexico. 1957 – 1r – 1 – us UF Libraries [972]

Esbozo de una politica agricola para honduras / Universidad Nacional Autonoma De Honduras Institut – Tegucigalpa, Mexico. 1964 – 1r – 1 – us UF Libraries [972]

Escadron volant de la reine / Dumanoir, Philippe – Paris, France. 1845 – 1r – 1 – us UF Libraries [440]

Escagedo Salmon, Mateo *see* La biblioteca del camarista de castilla d fernando jose de velasco y ceballos

Escalante, Aquiles *see* Geografia del atlantico

Escalante, B de *see* An account of the empire of china...

Escalante de Mendoza, J *see* Ytinerario de navegacion de los mares y tierras occidentales...

Escalante, Gumersindo *see* Manual de misionologia. vitoria, 1933

Escallon, Rafael *see* Doctrinas de la procuraduria general de la nacion

Escalofon de capataces y camineros de la misma en 31 de diciembre de 1959 / Jefatura de obras publicas de la provincia de Badajoz – Badajoz: Imp. Arqueros, 1960 – sp Bibl Santa Ana [946]

[Escalon-] escalon times – CA. 1927-58; 1968- – 45r – 1 – $2700.00 (subs $90/y) – mf#BC02202 – us Library Micro [071]

Escambia county historical quarterly – v3 n1 [1975 mar], v5 n2/3-? [1977 jun-1985] – 1r – 1 – (continued by: escambia county historical journal) – mf#1101440 – us WHS [071]

Escambia county history / Hargis, Modeste – s.l, s.l? 1936 – 1r – us UF Libraries [978]

Escambia county place-names / Hargis, Modeste – s.l, s.l? 1937 – 1r – us UF Libraries [978]

Escamps, Henri d' *see* Histoire et geographie de madagascar

O escandalo – Sabara, MG. 08 ago 1891 – (= ser Ps 19) – mf#P17,02,80 – bl Biblioteca [323]

Escapade / Trarieux, Gabriel – Paris, France. 1913 – 1r – us UF Libraries [440]

Escape *see* Gay life

Escape for thy life / Miller, John C – London, England. 18– – 1r – 1 – us UF Libraries [240]

Escape to the tropics / Holdridge, Desmond – New York, NY. 1937 – 1r – 1 – us UF Libraries [972]

Escape to wisconsin : newsletter of the wisconsin division of tourism – v1 n1-v6 n1 [1982 apr-1988 jan] – 1r – 1 – (continued by: wisconsin – you're among friends) – mf#611560 – us WHS [338]

Escaped from the gallows : souvenirs of a canadian state prisoner in 1838 / Bois, Louis-Edouard – Montreal: Beauchemin & Valois, 1885 – 2mf – 9 – (trans fr french) – mf#12133 – cn Canadiana [880]

Escapes – 1991 spr/sum – 1r – 1 – mf#2343966 – us WHS [071]

Escarabajo-vampiro : o el juicio de un dictador / Ponce De Avalos, Reynaldo – Tegucigalpa, Mexico. 1959 – 1r – 1 – us UF Libraries [972]

Escaramuzas en la frontera cacerena : con ocasion de las guerras por la independencia de portugal / Velo Nieto, Gervasio – Madrid: Imp F Martinez, 1952 – sp Bibl Santa Ana [946]

Escarceos de toponimia extremena / Garcia de Diego, Jose A – Badajoz: Dip. Provincial, 1975 – 1 – sp Bibl Santa Ana [946]

Escarceos historicos / Naranjo Martinez, Enrique – Bogota, Colombia. 1956 – 1r – 1 – us UF Libraries [972]

Escarceos literarios / Guardia Quiros, Victor – San Jose, Costa Rica. 1938 – 1r – 1 – us UF Libraries [972]
Escardo, Rolando Tomas *see* Rafagas
L'escarmouche – Paris. n1-3. 12 nov 1893-16 mars 1894 – 1 – fr ACRPP [073]
La escena espanola...teatro / Garcia de la Huerta, Vicente – 1786 – 9 – sp Bibl Santa Ana [440]
Eschallier, J C *see* Essai sur l'habitat sedentaire traditionnel au sahara algerien
Eschantillon des principaux paradoxes de la papaute / Rivet, A – La Rochelle, 1603 – 3mf – 9 – mf#CA-149 – ne IDC [240]
Die eschatologischen aussagen jesu in den synoptischen evangelien / Haupt, Erich – Berlin: Reuther & Reichard, 1895 – 1mf – 9 – 0-8370-3525-2 – (incl list of biblical texts cited) – mf#1985-1525 – us ATLA [220]
The eschatology of jesus / Jackson, Henry Latimer – London; New York: Macmillan, 1913 – 1mf – 9 – 0-7905-1116-9 – (incl bibl ref) – mf#1987-1116 – us ATLA [240]
The eschatology of jesus : or, the kingdom come and coming: a brief study of our lord's apocalyptic language in the synoptic gospels... / Muirhead, Lewis A – London: Andrew Melrose, 1904 – 1mf – 9 – 0-8370-4535-5 – (incl bibl ref) – mf#1985-2535 – us ATLA [220]
The eschatology of the gospels / Dobschtz, Ernst von – London: Hodder & Stoughton, 1910 [mf ed 1989] – 1mf – 9 – 0-7905-0703-X – (incl bibl ref) – mf#1987-0703 – us ATLA [226]
Eschbach, Alphons *see* Desputationes physiologico-theologicae
Eschbach, E R *see* Historical sketch of evangelical reformed church, frederick, maryland
Esche, Annelinde *see* Elise ruediger geb von hohenhausen
Eschelbach, Hans *see* Ueber die poetischen bearbeitungen der sage vom ewigen juden
Eschelbacher, Joseph *see*
– Das judentum und das wesen des christentums
– Yahadut u-mahut ha-notsriyut
Eschenbach, A *see* Die lehren des bergwerksstreikes vom mai 1889
Eschenbach, Wolfram von *see*
– Parzival
– Die werke wolframs von eschenbach
– Willehalm
– Wolfram's von eschenbach parzival und titurel
Eschenburg, Joh Joachim *see*
– Braunschweigisches magazin
– Brittisches museum fuer die deutschen
Eschenburg, Theodor *see* Das kaiserreich am scheideweg
Escher, Andreas *see* Die rechtsstellung des verwaltungsrates nach dem aktienrecht der vereinigten staaten von nordamerika
Escher, H *see* Die glaubensparteien in der eidgenossenschaft und ihre beziehungen zum ausland, vornehmlich zum hause habsburg und zu den deutschen protestanten 1527-1531
Escher, Karl *see* E T A hoffmanns gespensterspiel
Escherich, G *see* Im lande des negus
Eschmann, Ernst Wilhelm *see* Erdachte briefe
Eschmann, Johann Carl *see* Guide du jeune pianiste
Eschment, Ulrich-Alexander *see* Benjamin britten
Eschner, Max *see* Die deutschen schutzgebiete in afrika
Eschwege, Wilhelm L von *see* Journal von brasilien
Eschweger kreisblatt *see* Eschweger tageblatt
Eschweger tageblatt – Eschwege DE, 1878 12 jan-1882, 1884-1930, 1932-39, 1940 1 jul-1944 – 104r – 1 – (with gaps. filmed with suppl. title varies: 1889: eschweger tageblatt und kreisblatt; 20 sep 1902: eschweger tageblatt; 1934: eschweger tageblatt fuer kurhessen) – gw Misc Inst [074]
Eschweger tageblatt fuer kurhessen *see* Eschweger tageblatt
Eschweger tageblatt und kreisblatt *see* Eschweger tageblatt
Eschweger zeitung *see* Fulda-werra-zeitung
Eschweiler nachrichten – Eschweiler DE, 1988- – 8r/yr – 1 – (regional ed of aachener nachrichten) – gw Misc Inst [074]
L'esclavage chez les anciens hebreux : etude d'archeologie biblique / Andre, Louis Edouard Tony – Paris: Librairie Fischbacher, 1892 [mf ed 1989] – 1mf – 9 – 0-7905-3066-X – (in french. incl bibl ref) – mf#1987-3066 – us ATLA [939]
L'esclavage dans l'antiquite et son abolition par le christianisme / Desbarats, George Edouard – S.l: s.n, 1858? – 1mf – 9 – mf#22699 – cn Canadiana [230]
Esclaves / Ryner, Han – Constans-Honorine (Seine-et-Oise), France. 1925 – 1r – us UF Libraries [440]
Les esclaves chretiens : depuis les premiers temps de l'eglise jusqu' a la fin de la domination romaine en occident / Allard, Paul – 5e ed. entierement refondue. Paris: V. Lecoffre, 1914 – 1mf – 9 – 0-7905-4061-4 – (incl bibl ref) – mf#1988-0061 – us ATLA [930]
Esclaves, serfs et mainmortables / Allard, Paul – Nouv. ed., rev. et augm. Bruxelles: A. Vromant; Paris: Sanard & Derangeon, 1894 – 1mf – 9 – 0-7905-6520-X – (Incl bibl ref) – mf#1988-2520 – us ATLA [940]
La escoba – Valencia de Alcantara, 1911. 2 numeros – 5 – sp Bibl Santa Ana [073]
Escobar Camargo, Antonio *see* En el salon de los virreyes
Escobar, Diego Antonio de *see* Sermon epidictico, que en las honras, que de orden de n m r p f fernando alonso gonzalez...
Escobar, Felipe J *see* Legado de los proceres
Escobar, Francisco *see* La vision de san alonso rodriguez pintada por francisco de zurbaran en 1630
Escobar, L *see* Las quatrocientas respuestas a otras tantas preguntas
Escobar, M *see* Tratado...de la esencia, causas y curaciones de los bubones y carbuncos...
Escobar, Matias de *see* Voces de triton sonoro
Escobar, Paulo Emilio *see* Ferrocarriles de colombia en 1925-26
Escobar Prieto, Eugenio *see*
– Antiguedad y limites del obispado de coria
– El castillo de piedrabuena
– Oracion funebre que en el 4th centenario de la muerte dela reina catolica pronuncio en la iglesia catolica de santa maria de caceres...
Escobar Uribe, Arturo *see* Rezadores y ayudados
Escobar Velado, Oswaldo *see* Christoamerica
Escofet, Jose *see*
– Francisco de pizarro o el pais del oro
– Hernando cortes o la conquista de mejico
Escoiquiz, Juan de *see* Tratado de las obligaciones del hombre
A escola – Maceio, AL: Typ de Amintas de Mendonca, 8 abr, ago 1883; 8 abr, 8 jun 1884; 10 maio 1885 – (= ser Ps 19) – 1,5,6 – mf#P18B,01,07 – bl Biblioteca [079]
A escola : revista scientifica, litteraria e noticiosa – Bahia: Imprensa Economica, 10 out 1880 – (= ser Ps 19) – 1,5,6 – bl Biblioteca [079]
A escola nacional de musica e as pesquisas de folclore musical do brasil – Rio de Janeiro: Centro de Pesquisas Folcloricas 1944 [mf ed Bloomington IN: Indiana Uni Lib, Preservation Dept 1984] – 43p on 1r [ill] – 1 – us Indiana Preservation [390]
Escola secund aria numa sociedade em mundanca / Pereira, Joao Baptista Borges – Sao Paulo, Brazil. 1969 – 1r – 1 – us UF Libraries [972]
El escolar extremeno – Badajoz, 1896 y 1897 – 5 – sp Bibl Santa Ana [073]
Escombros / Lugo Lovaton, Ramon – Ciudad Trujillo, Dominican Republic. 1955 – 1r – us UF Libraries [972]
Escompto Bank NV, Djakarta *see* Report
[Escondido-] escondido times – CA. 1893-94; 1907-08 – 4r – 1 – $240.00 – mf#RC02203 – us Library Micro [071]
Escondido Genealogical Society *see*
– Hidden valley journal
– Hidden valley newsletter
– Hidden valley quarterly
[Escondido-] times advocate – CA. 1912- – 838r – 1 – $50,280.00 (subs $1440/y) – mf#RC02204 – us Library Micro [071]
[Escondido-] weekly times advocate – CA. 1909-1960 – 30r – 1 – $1800.00 – mf#R03213 – us Library Micro [071]
Escontria, Alfredo *see* Breve estudio de la obra y personalidad del escultor y arquitecto don manuel tolsa. mexico
Escott, Bickham S *see* Letter to the farmers
Escova de pitanguy : orgam critico – Pitangui, MG. 23 set 1883 – (= ser Ps 19) – mf#P17,02,110 – bl Biblioteca [079]
Escovar Ballesteros, Salvador *see* Reportaje sobre el primer congreso pedagogico cent...
Escragnolle Doria, Luiz Gastao De *see* Memoria historica
Escragnolle Taunay, Affonso De *see*
– Leonor de avila
– No rio de janeiro de dom pedro 2
Escrava Isaura / Guimaraes, Bernardo – Rio de Janeiro, Brazil. 1941 – 1r – us UF Libraries [972]
Escravidao africana no brasil / Moraes Filho, Evaristo De – Sao Paulo, Brazil. 1933 – 1r – 1 – us UF Libraries [972]
Escriptura y fundacion del convento de jesus de merida – 1 – sp Bibl Santa Ana [946]
Escritas literarios de rufino jose cuervo / Bayony Posada, Nicolas – Madrid: Razon y Fe, 1940 – 1 – sp Bibl Santa Ana [440]
Escrito de expresion de agravios de d jose moreno / Cortina, Jose Antonio – Habana, Cuba. 1875 – 1r – 1 – us UF Libraries [972]
Escrito de memoria / Vallenilla Lanz, Laureano – Mexico City? Mexico. 1961 – 1r – 1 – us UF Libraries [972]
Escrito de replica en los autos / Cruz, Carlos Manuel De La – Habana, Cuba. 1916 – 1r – 1 – us UF Libraries [972]
Escrito en derecho...duque de villahermosa – 1831 – 9 – sp Bibl Santa Ana [946]
Escrito en vista de pruebas / Rodriguez Valdes, Manuel M – 1851 – 9 – sp Bibl Santa Ana [440]
Escrito y cantado, 1954-1959 / Vitier, Cintio – Habana, Cuba. 1959 – 1r – 1 – us UF Libraries [972]
Escritoir : or, masonic and miscellaneous album – Albany. 1826-1827 – 1 – mf#3764 – us UMI ProQuest [073]
Las escritoras espanolas / Nelken, Margarita – Barcelona, 1930; Madrid: Razon y Fe, 1931 – 1 – sp Bibl Santa Ana [440]
Escritores de costa rica / Abreu Gomez, Ermilo – Washington, DC. 1950 – 1r – 1 – us UF Libraries [972]
Escritores de costa rica / Sotela, Rogelio – San Jose, Costa Rica. 1942 – 1r – 1 – us UF Libraries [972]
Escritores de hispanoamerica / Ragucci, Rodolfo M – Buenos Aires, Argentina. 1961 – 1r – 1 – us UF Libraries [972]
Escritores espanoles : carolina coronado / Ossorio, Bernard M – 1889 – 9 – sp Bibl Santa Ana [440]
Escritos / Arosemena, Pablo – Panama, v1-2. 1930 – 1r – us UF Libraries [972]
Escritos / Rueda Vargas, Tomas – Bogota, Colombia. v1-3. 1963 – 1r – us UF Libraries [972]
Escritos / Suarez, Marco Fidel – Bogota, Colombia. 1935 – 1r – us UF Libraries [972]
Escritos anejos. serios y humuristicos / Doncel y Ordaz, Jose – Badajoz: Uceda Hermanos, 1906 – 1 – sp Bibl Santa Ana [800]
Escritos de domingo del monte / Delmonte Y Aponte, Domingo – Habana, Cuba. v1-2. 1929 – 1r – 1 – us UF Libraries [972]
Escritos de luperon / Luperon, Gregorio – Ciudad Trujillo, Dominican Republic. 1941 – 1r – 1 – us UF Libraries [972]
Escritos e discursos literarios / Nabuco, Joaquim – Sao Paulo, Brazil. 1949 – 1r – 1 – us UF Libraries [972]
Escritos escogidos / Suarez, Marco Fidel – Bogota, Colombia. 1952 – 1r – 1 – us UF Libraries [972]
Escritos ineditos de ruben dario – New York, NY. 1938 – 1r – 1 – us UF Libraries [440]
Escritos literarios / Luz y Caballero, Jose de la – Habana, Cuba. 1946 – 1r – 1 – us UF Libraries [972]
Escritos literarios y cientificos / Cagigal, Juan Manuel – Caracas, 1930; Madrid: Razon y Fe, 1931 – 1 – sp Bibl Santa Ana [440]
Escritos politico-economicos / Samper, Miguel – Bogota, Colombia. v1-2. 1925 – 1r – 1 – us UF Libraries [330]
Escritos politicos / Arevalo, Juan Jose – Guatemala, 1945-46 – 2r – 1 – us UF Libraries [320]
Escritos selectos / Rosa, Ramon – Buenos Aires, Argentina. 1957 – 1r – us UF Libraries [972]
Escritos varios / Martinez Silva, Carlos – Bogota, Colombia. 1954 – 1r – 1 – us UF Libraries [972]
La escritura ogneica en extremadura / Roso de Luna, Mario – Madrid: Fortanet, 1904. B.R.A.H. 44 y 45, 1904, pp. 357-359 y 352-353 – sp Bibl Santa Ana [946]
Escrituras completas. revelaciones de antano / Picon La Res, Eduardo – Madrid: Razon y Fe, 1940 – 1 – sp Bibl Santa Ana [946]
Las escrituras...tapia / Tapia – v2-6, v8 1620-26 – 9 – sp Bibl Santa Ana [946]
Escrutinio phisico medico de un peregrino especifico de las calenturas intermitentes / Munoz y Peralta, M – Sevilla, 1699 – 2mf – 9 – sp Cultura [610]
Escrutinio sociologico de la historia colombiana / Lopez De Mesa, Luis – Bogota, Colombia. 1956 – 1r – 1 – us UF Libraries [972]
Escuadra del almirante cervera / Risco, Alberto – Madrid, Spain. 1920 – 1r – 1 – us UF Libraries [972]
O escudeiro baptista – Campos, RJ. 01 jan-dez 1909 – (= ser Ps 19) – mf#DIPER – bl Biblioteca [079]
Escuder, Ricardo *see* El pericon
Escudero Gonzalez, Jose *see*
– La catastrofe de barcelona
– La tragedia de ribadelago
El escudo – 1952-57. 542p – 1 – us Southern Baptist [242]
Escudo de armas de mexico : celestial proteccion de esta nobilissima ciudad, de la nueva-espana, y de casi todo el nuevo mundo, maria santissima...arzobispal el ano de 1531... / Cabrera y Quintero, Cayetano – Impr en Mexico: Por la Viuda de Joseph Bernardo de Hogal...ano de 1746 – (= ser Books on religion...1543/44-c1800: milagros y culto de la virgen) – 7mf – 9 – mf#crl-89 – ne IDC [241]
Escudo de las indulgencias de la religion...de s francisco / Almendralejo, Pedro de – 1699 – 9 – sp Bibl Santa Ana [240]
El escudo de merida y su origen romano / Alvarez Saenz de Buruaga, Jose – Madrid: revista de archivos, bibliotecas y museos, 1954 – 1 – sp Bibl Santa Ana [946]
Escudo del estado espanol / Spain. Ministerio del Interior – n.p, 1938 – (= ser Blodgett coll) – 9 – mf#fiche w861 – us Harvard [946]
Escudo oficial del municipo de la habana / Garcia Ensenat, Ezequiel – Habana, Cuba. 1943 – 1r – 1 – us UF Libraries [972]
Escuela de ciencias economicas y sociales / Yglesias R, Eduardo – San Jose, Costa Rica. 1953 – 1r – 1 – us UF Libraries [300]
La escuela de la amistad o el filosofo enamorado / Forner Segarra, Juan Pablo – 1796 – 9 – sp Bibl Santa Ana [830]
La escuela de medicina de guadalupe / Colegio Oficial de Medicos de la Provincia de Caceres – Caceres: Imprenta de la Viuda de Garcia Floriano, 1952 – sp Bibl Santa Ana [378]
Escuela del buen amor / Sierra Berdecia, Fernando – San Juan, Puerto Rico. 1963 – 1r – 1 – us UF Libraries [972]
Escuela Diocesana de Catequetica *see* Memoria del curso 1966
Escuela elemantal de trabajo y de capataces agricolas de caceres : memoria resumen de la actuacion...entre los cursos 1935-36 y 1943-44... / Galan Saval, Ricardo – Caceres: imp garcia floriano cumbreno, 1945 – sp Bibl Santa Ana [360]
Escuela Graduada Jose Luis Cotallo *see* Estatutos de la asociacion de padres de familia y amigos de la escuela
Escuela luminosa / Zamora, Victor – Habana, Cuba. 1957 – 1r – 1 – us UF Libraries [972]
Escuela Oficial de Maestria *see* Escuela oficial de maestria industrial de badajoz
Escuela oficial de maestria industrial de badajoz / Escuela Oficial de Maestria – Badajoz: Tipgrafia Clasica, 1965 – sp Bibl Santa Ana [370]
Escuela rural / Gonzalez De Padrino, Flor – Caracas, Venezuela. 1947 – 1r – 1 – us UF Libraries [972]
Escuela secundaria guatemalteca, problemas y soluc... / Canbronero Salazar, Miguel Angel – Guatemala, 1961 – 1r – 1 – us UF Libraries [972]
Escuela Superior De Administracion Publica America *see* Hacia una integracion metropolitana de san jose
Escuelas de Maria Santisima *see* Constituciones
Escuelas parroquiales del sagrado corazon : olivenza (badajoz) 1969 – Olivenza (Badajoz): Tip. Martinez Rengifo, 1970 – 1 – sp Bibl Santa Ana [377]
Escuelas practicas de agricultura / Mexico Departamento De Ensenanza Agricola – Mexico City? Mexico. 1946 – 1r – 1 – us UF Libraries [630]
Escultor de la sombra / Arrivi, Francisco – San Juan, Puerto Rico. 1965 – 1r – 1 – us UF Libraries [972]
El escultor extremeno juan de avalos / Segura Otano, Enrique – Badajoz: imp diputacion provincial, 1958 – sp Bibl Santa Ana [946]
Escultura en el ecuador / Navarro, Jose Gabriel – Madrid, Spain. 1929 – 1r – 1 – us UF Libraries [972]
La escultura en el ecuador, 1929 / Navarro, Jose Gabriel – Madrid: Razon y Fe, 1931 – 1 – sp Bibl Santa Ana [730]
Esculturas protohistoricas de la peninsula hispanica / Paredes Guillen, Vicente – Caceres: tip enc y lib jimenez, 1902 – 1 – sp Bibl Santa Ana [930]
Esd : The electronic system design magazine – Boston. 1987-1989 (1,5,9) – (cont: digital design) – ISSN: 0893-2565 – mf#13088,01 – us UMI ProQuest [000]
Esdaile, James *see* Civil and religious institutions necessarily and inseparably connec...
Esdalls newsletter – Dublin, Ireland. 16 feb 1746-24 jun 1752 (imperfect) – 1/2r – 1 – (10 jul 1747 not filmed) – uk British Libr Newspaper [072]
Esdras : in esdrae librvm...item de vita & obitu eiusdem narratio, scripta...lo guilelmo stuckio / Wolf, J – Tigvri: Christoph Froschauer, 1584 – 6mf – 9 – mf#PBU-659 – ne IDC [240]
La esencial heterogeneidad del ser en antonio machado / Frutos Cortes, Eugenio – Madrid, 1959. Sep. Rev. Filosofia, tomo XVIII, no 69-70 – sp Bibl Santa Ana [946]
Esequie del divino michelagnolo buonarroti : celebrate in firenze dall'accademia de pittori, scultori, e architettori – Firenze: Appresso i Giunti, 1564 – 1mf – 9 – mf#O-1990 – ne IDC [090]
Esequie del serenissimo principe francesco : celebrate in fiorenza dal serenissimo fernandino il granduca di toscana suo fratello... / Cavalcanti, A – Fiorenza: Gio Batista Landini, 1634 – 1mf – 9 – mf#O-1540 – ne IDC [090]
Esequie della maesta christianiss : di luigi 13 il giusto re di francia e di navarra, celebrate in firenze dall' altezza serenis... / Dati, C R – Firenze: Nella stamperia di SAS, 1644 – 2mf – 9 – mf#O-1830 – ne IDC [090]

Esequie dell'ill mo & ecc mo principe don francesco medici celebrate dal ser mo don cosimo 2, gran duca di toscana 4 / Adimari, A – Firenze: per gio donato e benardino giunti e compagni, 1614 – 1mf – 9 – mf#O-1822 – ne IDC [090]
Esequie fatte in padoua al gran prior di lombardia f agostino forzadura... / Malsucio, R – [Padova, 1664] – 2mf – 9 – mf#O-1097 – ne IDC [090]
'Eser shenot redifot / Tsentsiper, Aryeh Leib – Tel-Aviv, Israel. 1930 – 1r – 1 – us UF Libraries [939]
Eseranto / O'conner, John Charles – New York, NY. 1907 – 1r – us UF Libraries [025]
Eser-i eslaftan heft meclis / Ali, Mustafa bin Ahmet – Dersaadet [Istanbul]: Ikdam Matbaasi, 1316 [1900] – (= ser Ottoman histories and historical sources) – 1mf – 9 – $25.00 – us MEDOC [956]
Esery : partiia sotsialistov-revoliutsionerov / Chernomordik, S – Kharkov, 1930 – 56p 1mf – 9 – mf#RPP-259 – ne IDC [325]
La esfera / Sender, Ramon Jose – Buenos Aires. 1947 – 1 – us CRL [830]
Esfinge / Carrion, Miguel De – Habana, Cuba. 1961 – 1r – us UF Libraries [972]
Esfuerzo de mexico por la independencia de cuba / Chavez Orozco, Luis – Mexico City? Mexico. 1930 – 1r – 1 – us UF Libraries [972]
Esguerra Camargo, Luis *see* Introduccion al estudio del problema immigratorio
Eshcol / Humphrey, Simon James – New York: Fleming H Revell, c1893 – 1mf – 9 – 0-8370-6667-0 – mf#1986-0667 – us ATLA [240]
Eshelman, Matthew Mays *see*
– A history of the church of the brethren
– The history of the danish mission
– A model life
– Non-conformity to the world
– The open way into the book of revelation
– Two sticks
Eshet, Dan *see* Life, liberty and leisure
Esipov, G *see* Raskolnichi dela 18 stoletiia, izvlechennyia iz del preobrazhenskogo prikaza i tainoi rozysknykh del kantseliarii
Esipov, G V *see* Sobranie dokumentov po delu tsarevicha alekseia petrovicha, vnov naidennykh...
Eskelund, Karl *see*
– Drums in bahia
– While god slept
Eskilstuna allehanda – Eskilstuna, Sweden. 1844-69 – 9r – 1 – sw Kungliga [078]
Eskilstuna tidning – Eskilstuna, Sweden. 1837-38 – 1r – 1 – (1867-93 15r) – sw Kungliga [078]
Eskilstuna weckoblad – Eskilstuna, Sweden. 1840-41 – 1r – 1 – sw Kungliga [078]
Eskilstunakorrespondenten – Eskilstuna, 1865-66 – 9 – sw Kungliga [078]
Eskilstunakuriren – Eskilstuna, Sweden. 1979- – 1 – (strengnas tidning) – sw Kungliga [078]
Eskilstunakuriren – Eskilstuna, Sweden. 1890-1978 – 493r – 1 – (Strengnas Tidning, 1964-78) – sw Kungliga [078]
Eskilstunaposten – Eskilstuna, Sweden. 1883-1893 – 7r – 1 – sw Kungliga [078]
Eskilstunatidningen – Eskilstuna, Sweden. 1895-1898 – 3r – 1 – sw Kungliga [078]
Eskimoliv / Nansen, F – Kristiania, 1891 – 6mf – 9 – mf#N-325 – ne IDC [919]
Eskual herria – Los Angeles. 1893-98 – 1 – fr ACRPP [073]
Eslava, Hilarion *see* Museo organico espanol
Eslovs tidning *see* Nordvastra skanes tidningar engelholms tidning
Esm magazine – Oak Brook, 2000+ [1,5,9] – (cont: employee services management) – mf#7200,02 – us UMI ProQuest [790]
Esmark, Jens *see* Reise von christiania nach drontheim durch oesterdalen und zurueck ueber dovre
Esmeralda : a grand ballet in one act and five tableaux / Perrot, Jules – New York: W Corbyn, 1856 – 1 – (Transl fr the French) – mf#*ZBD-*MGTZ pv4-Res – Located: NYPL – us Misc Inst [790]
Esmeralda : ou, notre dame de paris. a dramatic ballet pantomime in two acts and five tableaux. first performed in new york, at the park theatre, on the 18th sep 1848, by the french ballet company of h monplaisir / Perrot, Jules – New York: [Sold by Mr. Corbyn's Dramatic Agency] 1848 – 1 – (in french and english in parallel clms) – mf#*ZBD-*MGTZ pv2-Res – Located: NYPL – us Misc Inst [790]
Esmeralda star : independent california newspaper – Aurora, CA. v1 n1. may 20 1862 – 1r – 1 – us Western Res [071]
Esmond bee (1902) – Esmond, ND: Allison Bros, 1902; -v20 n14 sep 27 1919 (wkly) – 1 – (cont: goa bee; absorbed by: benson county farmers press) – mf#01701-01705 – us North Dakota [071]
Esmond bee (1945) – Esmond, ND: C L Jensen, 1945? -v2 n12 dec 30 1946 (wkly) – 1 – (merged with: maddock standard to form: standard (maddock, nd)) – mf#01705 – us North Dakota [071]
Esmond leader *see* Esmond leader and elling news
Esmond leader and elling news – Esmond, Benson Co, ND: E E Saunders, 1904; Leader v1 n42 = News v3 n22 jan 25 1905 (wkly) – 1 – (formed by the union of: esmond leader and: elling news. absorbed by: oberon reporter. missing: 1904 sep 22, nov 3, dec 8,22) – mf#06297 – us North Dakota [071]
Eso y mas / Salarrue – San Salvador, El Salvador. 1962 – 1r – 1 – us UF Libraries [972]
Esoteria y fervor populares de puerto rico / Garrido, Pablo – Madrid, Spain. 1952 – 1r – 1 – us UF Libraries [972]
Esoteric basis of christianity / Kingsland, William – London, England. 1891 – 1r – 1 – us UF Libraries [240]
Esoteric christianity : or, the lesser mysteries / Besant, Annie Wood – New York: John Lane, 1910 [mf ed 1991] – 1mf – 9 – 0-524-01042-0 – mf#1990-2190 – us ATLA [230]
L'esoterisme de hebbel / Bastier, Paul – Paris: E Larose, 1910 [mf ed 1990] – 70p – 1 – mf#7449 – us Wisconsin U Libr [430]
Espace commun portugais / Massart, Jean Jacques – Bruxelles, Belgium. 1969 – 1r – 1 – us UF Libraries [946]
Espace geographique – Paris. 1977-1993 (1) 1977-1980 (5) 1977-1980 (9) – ISSN: 0046-2497 – mf#11723 – us UMI ProQuest [900]
Espada Marrero, J *see*
– Es navidad
– Hijo prodigo y otros poemas
Espada Rodriguez, Jose *see* Canto a los argonautas y otros poemas
Espada y otras narraciones / Salarrue – San Salvador, El Salvador. 1960 – 1r – 1 – us UF Libraries [972]
L'espagne accuse : discours d'alvarez del vayo, ministre des affaires etrangeres d'espagne, a la societe des nations... / Alvarez del Vayo, Julio – Paris: Le Comite franco-espagnol [1936] [mf ed 1980] – (= ser Blodgett coll) – 1mf – 9 – (pref by jean zyromski) – mf#w1507 – us Harvard [341]
L'espagne chretienne / Leclercq, Henri – 2e. ed. Paris: V Lecoffre, 1906 [mf ed 1990] – (= ser Bibliotheque de l'enseignement de l'histoire ecclesiastique) – 1mf – 9 – 0-7905-5369-4 – (in french) – mf#1988-1369 – us ATLA [240]
L'espagne de franco : synthese de trois conferences donnees du 17 janvier au 10 fevrier 1938 / Joubert, Henri – Paris, 1938 – (= ser Blodgett coll) – 9 – mf#fiche w976 – us Harvard [946]
L'espagne du sud : l'extremadoure / Sermet, Jean – Paris: B Arthaud, 1953 – 1 – sp Bibl Santa Ana [946]
L'espagne et la paix – Paris, 1937? – (= ser Blodgett coll) – 9 – mf#fiche w 965 – us Harvard [946]
L'espagne et le portugal : ou moeurs, usages et costumes des habitans de ces royaumes precede d'un precis historique / Breton de LaMartiniere, Jean – Paris 1815 [mf ed Hildesheim 1995-98] – 6v on 12mf – 9 – €120.00 – 3-487-29879-1 – gw Olms [914]
Espaklaria – (New York). 1907 – 1 – us AJPC [073]
Espaldas a si mismo / Bosch, Juan – Ciudad Trujillo, Dominican Republic. 1942 – 1r – 1 – us UF Libraries [972]
Espana : la abortada republica / ed by Bayle, Constantino – Madrid: Razon y Fe, 1931 – 1 – sp Bibl Santa Ana [946]
Espana : los anos vitales / Bolin, Luis – Madrid: Graf. Calleja, 1967 – 1 – (pref by arthur bryant) – sp Bibl Santa Ana [946]
Espana / Guillen, Nicolas – Mexico City? Mexico. 1937 – 1r – us UF Libraries [946]
Espana – Madrid, Spain. -w. 7 Sept 1916-31 July 1919. 2 reels – 1 – uk British Libr Newspaper [072]
Espana : primera asamblea de accion catolica / ed by Bayle, Constantino – Madrid: Razon y Fe, 1930 – 1 – sp Bibl Santa Ana [241]
Espana – Tangier. 9 sep 1940-23 may 1945 – 7r – 1 – uk British Libr Newspaper [072]
Espana 1958 : 1st festival hispanoamericano. caceres / Festival de Folklore Hispanoamericano 1, 1958 – Madrid: Graf. Nilo, 1958 – 1 – sp Bibl Santa Ana [390]
Espana ante el hemisferio de occidente, tomo 2 : colombia, venezuela, ecuador, peru, bolivia, chile, argentina / Lazurtegui, Julio de – Madrid: Razon y Fe, 1926 – 1 – sp Bibl Santa Ana [972]
Espana ante la independencia de los estados unidos por el dr juan f... / ed by Bayle, Constantino – Madrid: Razon y Fe, 1926 – 1 – sp Bibl Santa Ana [946]
Espana antes y despues de la venida del directorio / Luna Rancel, Felipe – Jerez de los Caballeros, Competidora. 1924 – 1 – sp Bibl Santa Ana [946]
Espana economica y financiera – Madrid, Spain. 3 jan 1942-8 sep 1945 [wkly] – 4r – 1 – (imperfect) – uk British Libr Newspaper [330]
Espana en america : dos veladas literarias celebradas en la ciudad de burgos 1892-1893, con motivo del cuarto centenario de colon / Gomez Bravo, Vicente – Cadiz: Est. Ceron, 1943 – 1 – sp Bibl Santa Ana [440]
Espana en el congreso de viena segun la correspondencia oficial de d pedro gomez labrador / Villaurrutia, W R – Madrid: tip rev archiv, bibliot y mus, 1907 – 1 – sp Bibl Santa Ana [946]
Espana en el siglo 20 : galeria de personalidades ilustres – Madrid: 1st part. s.i., s.a. – 1 – sp Bibl Santa Ana [946]
Espana en indias / Bayle, Constantino – Madrid: Editroa Nacional, 1942 – 1 – sp Bibl Santa Ana [972]
Espana en indias / Perez Bustamante, Ciriaco & Bayle, Constantino – Madrid: Revista de Indias, 1940 – 1 – sp Bibl Santa Ana [972]
Espana en sus gloriosas jornadas de julio y agosto de 1936 / Saenz, Vicente – San Jose, Costa Rica: La Tribuna 1936 [mf ed 1977] – (= ser Blodgett coll) – 1mf – 9 – mf#w1154 – us Harvard [946]
Espana en trento. 2. el concilio de trento en las indias espanolas / ed by Bayle, Constantino – Madrid: Razon y Fe, 1945 – 1 – sp Bibl Santa Ana [946]
Espana evangelica – 1984-89 [complete] – Inquire – 1 – (cont: carts circular) – mf#ATLA S0365 – us ATLA [242]
La espana franquista, satelite de hitler / Ibarruri, Dolores – Toulouse? 194? Fiche W954. (Blodgett Collection of Spanish Civil War Pamphlets) – 9 – us Harvard [946]
Espana; impresiones y reflejos / Jerrold, Douglas – Salamanca, 1937 – (= ser Blodgett coll) – 9 – mf#fiche w969 – us Harvard [946]
Espana, Leyes Decretos... *see* Reglamento de pastor hierbas y rastrojeras
Espana, Leyes, Decretos... *see* Ley de 15 de julio de 1952 sobre explotaciones agrarias
Espana. Leyes, decretos, etc *see* Codigo de la circulacion
Espana libro – Brooklyn. v.15-38. 1953-76 – 1 – us L of C Photodup [946]
Espana moscovita y sus consecuencias / Aristeguieta Silva, F – [Brussels: s.n. 1938] – (= ser Blodgett coll) – 1v – 9 – mf#w726 – us Harvard [946]
La espana oriental – Manila. Philippine Islands. -w. 7 Apr 1889-30 Jun 1890. (Imperfect). (1 reel) – 1 – ((edicion hispano-tagalog). 4 jul-26 dec 1889. (imperfect). (33 ft)) – uk British Libr Newspaper [072]
Espana roja – n.p., 193? Fiche W 1506. (Blodgett Collection of Spanish Civil War Pamphlets) – 9 – us Harvard [946]
Espana sagrada – Madrid, 1784 v34; 1826 v44; 1832, v45 – 30mf – 8 – mf#1570 – ne IDC [240]
Espana sagrada : theatro geogr-hist de la iglesia de espana / Florez, Henrique – Madrid. 1(1754)-52(1946) – 440mf – 9 – €852.00 – ne Slangenburg [440]
Espana sagrada...tomos 1-51 / Florez, E et al – Madrid, 1745-1879 – 425mf – 9 – sp Cultura [946]
Espana vista otra vez / Noel, Martin – Madrid: Edit. Espana, 1929 – 1 – sp Bibl Santa Ana [946]
Espana y algunos espanoles / Galvez, Manuel – Buenos Aires: Editorial Huarpes, 1945 – 1 – us Wisconsin U Libr [946]
Espana y america : revista quincenal de religion, ciencia, literatura y arte – Madrid. ano 1-25. 1903-27 – 1 – us NY Public [073]
Espana y colon : madrid, 1935 / Real, Cristobal – Madrid: Razon y Fe, 1936 – 1 – sp Bibl Santa Ana [946]
Espana y cuba / Spain Ministerio de Ultramar – Madrid, Spain. 1896 – 1r – 1 – us UF Libraries [327]
Espana y el clero indigena de america / ed by Bayle, Constantino – Madrid: Razon y Fe, 1931 – 1 – sp Bibl Santa Ana [946]
Espana y francisco franco : 25 aniversario de la exaltacion a la jefatura del estado / Delegacion Nacional de Prensa, Propaganda y Radio – Badajoz: Graf Extremena, 1961 – sp Bibl Santa Ana [946]
Espana y franco / Gimenez Caballero, Ernesto – Cegama, 1938 – (= ser Blodgett coll) – 9 – mf#fiche w917 – us Harvard [946]
Espana y la guerra imperialista / Diaz, Jose – Mexico, 1939 – (= ser Blodgett coll) – 9 – mf#fiche w833 – us Harvard [946]
Espana y sus hombres. resena historico-biografica de sus principales personalidades – Madrid: Tip. Julian Frances, 1916 – 1 – sp Bibl Santa Ana [920]
Espana-paris : organo del centre hispano-americain espana-paris – Paris. n1-14. 14 avr-21 juil 1900 – 1 – (lacking: n13) – fr ACRPP [073]
Espanol – Madrid. Spain. -w. 31 Oct 1942-29 Dec 1945. (Imperfect). (6 reels) – 1 – uk British Libr Newspaper [072]
Un espanol al servicio de la santa sede. don juan de carvajal, cardenal de sant'angelo, lugado en alemania y hungria (1.399-1.469) / Gomez Canedo, Luis – Madrid: C.S.I.C. Instituto Jeronimo Zurita, 1947 – 1 – sp Bibl Santa Ana [240]
Espanol – castellano : die namen des spanischen / Diederichs, Joern – (mf ed 1993) – 3mf – 9 – €49.00 – 3-89349-676-9 – mf#DHS 676 – gw Frankfurter [440]
Espanol de ambos mundos – London, UK. 7 aug 1860-10 jan 1862 – 1 – uk British Libr Newspaper [072]
El espanol de jalisco : madrid, 1967 / Cardenas, Daniel – Madrid: graf calleja, 1968 – 1 – sp Bibl Santa Ana [946]
Espanol en la espanola / Garcia Rodriguez, Jose Maria – Ciudad Trujillo, Dominican Republic. 1947 – 1r – 1 – us UF Libraries [972]
Espanol en mejico, los estados unidos, y / Henriquez Urena, Pedro – Buenos Aires, Argentina. 1938 – 1r – 1 – us UF Libraries [972]
Espanol hablado en santander / Florez, Luis – Bogota, Colombia. 1965 – 1r – 1 – us UF Libraries [972]
Espanol, Juan *see* Quienes son los responsables?
Espanola, saint domingue, haiti / Pierre Audain, Julio J – Mexico City? Mexico. 1961 – 1r – 1 – us UF Libraries [972]
Espanoles e ingleses en america durante el siglo 17. el conde de gandomar y su intervencion en el proceso, prision y muerte de sir walter raleigh : santiago de compostela, 1928 / Perez Bustamante, Ciriaco – Madrid: Razon y Fe, 1929 – 1 – sp Bibl Santa Ana [972]
Espanoles fuera de espana / Figueroa y Melgar, Alfonso de – Badajoz: imp dip provincial, 1973 – sp Bibl Santa Ana [946]
Los espanoles y magallanes en la expedicion del estrecho / Bayle, Constantino – Madrid: Razon y Fe, 1921 – 1 – sp Bibl Santa Ana [946]
Espanolidad literaria de jose marti / Marinello, Juan – Habana, Cuba. 1942 – 1r – 1 – us UF Libraries [972]
Espanolismo y antiespanolismo en la america hispana. la poblacion hispanoamericana a partir de la independencia : madrid, 1945 / Baron Castro, Rodolfo – Madrid: Razon y Fe, 1947 – 1 – sp Bibl Santa Ana [946]
Esparragalejo. Ayuntamiento *see* Revista de los esparrenses, no 5 dedicada a la 2nd semana cultural y fiesta del emigrante. agosto 1980
Espartero : historia / Florez, Jose Segundo – 1844-45 – 4v – 9 – sp Bibl Santa Ana [946]
Espectaculos de la semana – Miami, FL. 1975 feb 6-1973 may 25 [gaps] – 3r – 1 – us UF Libraries [071]
El espectador – 1937-60.Incomplete. Bilingual weekly.6 reels – 1 – us Stanford [071]
El espectador – Bogota, Columbia. 1966-1988 (1) – mf#61960 – us UMI ProQuest [079]
Espectador – Bogota, Colombia. 1989 apr 01-2000 jan 31 – 159r – (Gaps) – us UF Libraries [079]
O espectador – Rio de Janeiro, RJ: Typ Camoes, 28 set, nov-dez 1881; jun 1882-25 out 1885 – (= ser Ps 19) – mf#P19A,04,17 – bl Biblioteca [790]
O espectador da america do sul – Rio de Janeiro, RJ: Typ de Quirino & Irmao, 16 jul 1863-30 jun 1864 – (= ser Ps 19) – mf#P18A,01,21 – bl Biblioteca [079]
Espejismo de la selva / Berti, Jose – Caracas, Venezuela. 1947 – 1r – 1 – us UF Libraries [972]
Espejo, Christobal *see* Las antiguas ferias de medina del campo
Espejo crystalino de paciencia y viva imagen de christo crucificado en la admirable vida y virtudes de la venerable madre sor maria ynes de los dolores / Mora, Juan Antonio de – en Mexico:...Viuda de Miguel de Rivera Calderon, ano de 1729 – (= ser Books on religion...1543/44-c1800: biografias de religiosos) – 4mf – 9 – mf#crl-139 – ne IDC [241]
Espejo de conciencia que trata de todos los estados – 1525 – 9 – sp Bibl Santa Ana [946]
El espejo de la muerte : en que se notan los medios de prepararse para morir... / Bundeto, C – Antwerp: Gallet, 1700 – 2mf – 9 – mf#O-2059 – ne IDC [090]
Espejo de paciencia / Balboa Troya y Quesada, Silvestre de – Habana, Cuba. 1960, 1962 – 2r – 1 – us UF Libraries [972]
Espejo de principes y cavalleros...(tercera parte de -) / [Ortunez de Calahorra, D] – Alcala: por iuan de lequerica, 1587 – 8mf – 9 – mf#O-73 – ne IDC [090]

Espejo divino en lengua mexicana en que pueden verse los padres y tomar documento para acertar a doctrinar bien a sus hijos, y aficionallos alas virtudes / Mijangos, Juan de – en Mexico: impr Diego Lopez Danalos 1607 – (= ser Books on religion...1543/44-c1800: doctrina cristiana, obras de devocion) – 6mf – 9 – mf#crl-42 – ne IDC [241]
Espeleologia, cursillo dictado / Nunez Jimenez, Antonio – Havana, Cuba. 1949 – 1r – 1 – us UF Libraries [972]
Espelho – London, UK. Sep 1914-15 Jun 1919 – 1 – uk British Libr Newspaper [072]
O espelho – Rio de Janeiro, RJ: Imprensa Nacional, 01 out 1821-27 jun 1823 – (= ser Ps 19) – mf#P01B,05,16 – bl Biblioteca [320]
O espelho da justica – Rio de Janeiro: Typ de Thomas B Hunt, 01 dez 1830-03 jun 1831 – (= ser Ps 19) – mf#P02,04,15 – bl Biblioteca [320]
O espelho diamantino : periodico de politica, litteratura, bellas artes, theatro e modas – Rio de Janeiro, RJ: Typ de Plancher-Seignot, 01 out 1827-28 abr 1828 – (= ser Ps 19) – mf#P01,03,17 – bl Biblioteca [079]
Espelho dos livros – Rio de Janeiro, Brazil. 1936 – 1r – 1 – us UF Libraries [972]
Espelkamper nachrichten – Rahden DE, 1858 may-1966 – 1 – gw Misc Inst [074]
Espenberger, J N *see* Die philosophie des petrus lombardus
d'Espence, C *see* Opera omnia quae superstes adhuc editit
Espendez Navarro, Juan *see* Caserio del carmen
Esper, E J C *see* Die pflanzenthiere in abbildungen nach der natur...
Espera / Dominguez, Franklin – Ciudad Trujillo, Dominican Republic. 1959 – 1r – us UF Libraries [972]
A esperanca : orgao litterario, recreativo e noticioso – Florianopolis, SC, 7 out 1907 – (= ser Ps 19) – mf#UFSC/BPESC – bl Biblioteca [079]
Esperanca : periodico litterario e critico – Manaus, AM: Typ do Commercio do Amazonas, 21 jan 1877 – (= ser Ps 19) – mf#P11B,06,29 – bl Biblioteca [440]
L'esperance – Central Falls, RI. 1891-1899 (1) – mf#66180 – us UMI ProQuest [071]
L'esperance – Paris. 9e annee, no. 1-24. 7 janv-16 dec 1847 – 1 – fr ACRPP [073]
Esperances / Bilhaud, Paul – Paris, France. 1885 – 1r – us UF Libraries [440]
La esperanza en dios carta patoral / Perez Munoz, Adolfo – Badajoz: Tip. y Encuadern. de Uceda Hnos, 1917 – 1 – sp Bibl Santa Ana [240]
Espias-Sanchez, Manuel *see* La familia en directo
Espigando en! mi hereda! : tomo 1 / Sanchez-Arjona, Vicente – Sevilla: Imp Zambrano, 1952 – 1 – sp Bibl Santa Ana [810]
Espigando en! mi hereda! : tomo 2 / Sanchez-Arjona, Vicente – Sevilla: Imp Zambrano, 1953 – 1 – sp Bibl Santa Ana [810]
Espigando en! mi hereda! : tomo 3 / Sanchez-Arjona, Vicente – Sevilla: Imp Zambrano, 1955 – 1 – sp Bibl Santa Ana [810]
Espigando en! mi hereda! : tomo 4 / Sanchez-Arjona, Vicente – Sevilla: Imp Zambrano, 1955 – 1 – sp Bibl Santa Ana [810]
Espigas al sol / Weber, Delia – Ciudad Trujillo, Dominican Republic. 1959 – 1r – 1 – us UF Libraries [972]
Espigas intelectuales / Nieto Rojas, Jose Maria – Bogota, Colombia. 1946 – 1r – 1 – us UF Libraries [972]
Espil, Alberto *see* Revolucion de 1893 y don julio a costa, gobernador...
Espinar, Jaime *see* Noviembre de madrid
Espindola, Nicolas de *see* Exercicios de desagravios de christo senor nuestro en la cruz
Espine, J *see* Opuscules theologiques
Espiney, Francois d' (Francois de Sales, Saint) *see* Souhaits de bonne annee
Espino, Alfredo *see* Jicaras tristes
Espino de Caceres, Diego *see*
– Speculum testamentorum...
– Speculum testamentorum; sive..
Espino, Miguel Angel *see*
– Como cantan alla
– Hombres contra la muerte
– Trenes
Espinosa, Aurelio Macedonio *see*
– El romancero espanol
– Studies in new mexican spanish
Espinosa, Ciro *see* Indagacion y critica
Espinosa, Francisco *see*
– Cien de las mejores poesias liricas salvadorenas
– Folk-lore savladoreno
Espinosa, Isidro Felix *see* Cronica de la provincia franciscana de los apostoles san pedro y san pablo de michoacan. 2nd ed. mexico, 1946
Espinosa, Isidro Felix de *see*
– Chronica apostolica
– Nuevas empressas del peregrino americano septentrional atlante
– El peregrino septentrional atlante
Espinosa, J Manuel *see* First expedition of vargas into new mexico, 1692
Espinosa, Jose Maria *see*
– Memorias de un abanderado
Espion / Halevy, Leon – Bruxelles, Belgium. 1829 – 1r – us UF Libraries [440]
L'espionne boche : drame militaire canadien / Lemay, Joseph Henri – Sherbrooke [Quebec]: Cie de publication de "La Tribune", 1916 [mf ed 1998] – 2mf – 9 – 0-665-65385-9 – mf#65385 – cn Canadiana [820]
O espiritismo : orgao dedicado ao estudo – Rio de Janeiro, RJ. 22 out-nov 1881 – (= ser Ps 19) – mf#P19A,04,22 – bl Biblioteca [130]
Espiritismo no brazil / Ribeiro, Leonidio – Sao Paulo, Brazil. 1931 – 1r – 1 – us UF Libraries [972]
Espirito da sociedade colonial / Calmon, Pedro – Sao Paulo, Brazil. 1935 – 1r – 1 – us UF Libraries [972]
O espirito santense : jornal politico, scientifico, litterario e noticioso – Vitoria, ES: Typ do Espirito Santense, 08 set 1870-dez 1875; jan-maio, jul-dez 1876; jan 1877-dez 1879; jun 1880-14 jun 1889 – (= ser Ps 19) – mf#P11B,05,12 – bl Biblioteca [073]
Espirito Santo (Brazil) Governor *see* Relatorios dos presidentes, 1a republica, 1892-1930
Espirito Santo (Brazil) President *see* Relatorios dos presidentes, epoca do imperio, 1833-1888
El espiritu de santa teresa y el de san ignacio / ed by Bayle, Constantino – Madrid: razon y fe, 1922 y 1923 – 1 – sp Bibl Santa Ana [240]
El espiritu del siglo / Juras Reales, Baron de – 1833 – 9 – sp Bibl Santa Ana [946]
El espiritu genuino de falange espanola. es catolico? / Bayle, Constantino – Madrid: Razon y Fe, 1937 – 1 – sp Bibl Santa Ana [241]
Espiritu y camino de hispanoamerica / Frankl, Victor – Bogota, Colombia. 1953 – 1r – 1 – us UF Libraries [972]
El espiritu y el apostolado de sor maria josefa rosello... / Noberasco, Felipe – Madrid: Razon y Fe, 1926 – 1 – sp Bibl Santa Ana [241]
Espiritualidad y civilizacion / Mora Y Varona, Gaston – Habana, Cuba. 1938 – 1r – 1 – us UF Libraries [972]
Epistolario de nueva espana 1505-1812 recopilado por francisco del paso y troncoso, tomo 16, mejico, 1942 / Bayle, Constantino – Madrid: Razon y Fe, 1944 – 1 – sp Bibl Santa Ana [946]
L'esplorazione del guiba : viaggio di scoperta nel cuore dell'africa, eseguito sotto gli auspici della societa geografica italiana / Bottego, Vittorio – Roma: Societa editrice nazionale, [1900] – 1 – us CRL [916]
L'espoir de nice et du sud-est – Nice. 1960-72 – 1 – fr ACRPP [073]
O esporte : semanario ilustrado – Florianopolis, SC. 05 maio, 23 jul 1927 – (= ser Ps 19) – mf#UFSC/BPESC – bl Biblioteca [079]
La esposa de donoso cortes / Munoz de San Pedro, Miguel – Badajoz, 1953. Sep. de la Revista de Rstudios Extremenos – sp Bibl Santa Ana [946]
Esposicion... / Colombia. Ministerio de Relaciones Exteriores – Bogota: Impr del Neo-granadino, [1856-1858] (annual) – 1r – 1 – us CRL [972]
Esposicion... / Colombia. Secretaria de lo Interior i Relaciones Exteriores – Bogota: Impr de la Nacion, -1866] [1865-1866] (annual) – 1r – 1 – us CRL [972]
Esposicion... / Ecuador. Ministerio de lo Interior y Relaciones Exteriores – Quito: Impr Nacional [1863,1867,1871,1875] (annual) – 1r – 1 – us CRL [972]
Esposicion que dirije al congreso del ecuador en...el ministro de estado en el despacho de hacienda / Ecuador. Ministerio de Hacienda – Quito: impr de joaquin teran, [1846] (annual) – 1r – 1 – us CRL [336]
Esposicion que el ministro de hacienda del ecuador presenta a las camaras lejislativas reunidas en... / Ecuador. Ministerio de Hacienda – Quito: Impr de Bermeo [1849, 1854-1855, 1857] (annual) – 1r – 1 – us CRL [336]
Esposicion que el secretario de estado en el despacho de lo interior de la nueva granada presenta al congreso constitucional de... / Colombia. Secretaria del Interior – Bogota: Impr de J A Cualla [1845] (annual) – 1r – 1 – us CRL [972]
Esposicion que el secretario de estado en el despacho de relaciones esteriores de la republica de Colombia hace al congreso de...sobre los negocios de su departamento / Colombia. Ministerio de Relaciones Exteriores – Bogota: Impr de Pedro Cubides [1827] (annual) – 1r – 1 – us CRL [972]
Esposicion que el secretario de estado en el despacho del interior de la republica de Colombia hizo al congreso de...sobre los negocios de su departamento / Colombia. Secretaria del Interior – Bogota: Impr de la Republica [1824] (annual) – 1r – 1 – us CRL [972]
Esposicion que presenta al congreso constitucional de [...] / el Ministro Secretario de Estado en el Departamento del Interior y Relaciones Esteriores – Sucre: impr de la libertad, 1840 – 1r – 1 – us CRL [323]
Esposicion que presenta en bolivia el ministro de estado en el despacho del interior a la convencion nacional en [...] – Chuquisaca: impr de beeche y cia, 1843 – 1r – 1 – us CRL [320]
Esposicion...dirijida a las camaras lejislativas del ecuador en... / Ecuador. Ministerio del Interior, Relaciones Esteriores e Instruccion Publica – Quito: Impr de Bermeo, 1856-[1856-1858] (annual) – 1r – 1 – us CRL [972]
Esposizione della divina commedia di dante-alighieri / Trucchi, Ernesto – Milano: A Montaldi & C, (1943) – 3v – 1 – us Wisconsin U Libr [440]
Esposocion(sic) que dirige a las cortes constituyentes en defensa de su padre don manuel godoy... / Chinchon, Condesa de – Madrid: imp f andres y comp, 1855 – 1 – sp Bibl Santa Ana [946]
Espresso – [Italy], 1955- – 4r per y – 5 – enquire for prices – us UMI ProQuest [074]
Espresso – Rome, Italy. 1955+ (1) 1977+ (5) 1977+ (9) – ISSN: 0423-4243 – mf#9004 – us UMI ProQuest [073]
L'esprit – Paris. n1-2. mai 1926-janv 1927 – 1 – fr ACRPP [073]
Esprit : revue internationale – Paris. oct 1932-juil 1941, dec 1944-1956 – 1 – fr ACRPP [073]
Esprit createur – Baton Rouge. 1961-1966 (1) – ISSN: 0014-0767 – mf#5822 – us UMI ProQuest [400]
Esprit d'alexandre vinet : pensees et reflexions, extraites de tous ses ouvrages et de quelques manuscrits inedits – Paris: Joel Cherbuliez, 1861 [mf ed 1993] – 2v on 3mf – 9 – 0-524-08660-5 – (incl bibl ref) – mf#1993-2120 – us ATLA [242]
L'esprit de la fronde : ou histoire politique et militaire des troubles de france pendant la minorite de louis 14 / Mailly, Jean – Paris 1772-1773 [mf ed Hildesheim 1995-98] – 5v on 37mf – 9 – €370 – ISBN-10: 3-487-26094-8 – ISBN-13: 978-3-487-26094-5 – gw Olms [323]
L'esprit de la gaule / Reynaud, Jean – Paris: Furne, Jouvet, 1866 [mf ed 1992] – 1mf – 9 – 0-524-02040-X – (in french) – mf#1990-2815 – us ATLA [200]
L'esprit de la ligue : ou histoire politique des troubles de france, pendant les 16e et 17e siecles / Anquetil, Louis P – Paris 1767 [mf ed Hildesheim 1995-98] – 3v on 8mf – 9 – €160.00 – 3-487-26098-0 – gw Olms [944]
L'esprit de mr arnaud : tire de sa conduite, des ecrits de luy & de ses disciples, particuliere-ment de l'apologie pour les catholiques – Deventer. v1-2. 1684 – €36.00 – ne Slangenburg [241]
Esprit de parti / Bert, Pierre Nicolas – Paris, France. 1818 – 1r – 1 – us UF Libraries [440]
L'esprit des cours de l'europe – La Haye puis Amsterdam. juin 1699-avr 1710 (I-XIX) – 1 – fr ACRPP [073]
L'esprit d'henri 4 : ou anecdotes les plus interessantes, traits sublimes, reparties ingenieuses, & quelques lettres de ce prince / Prault, Louis – Paris 1770 [mf ed Hildesheim 1995-98] – 1v on 3mf – 9 – €90.00 – ISBN-10: 3-487-26105-7 – ISBN-13: 978-3-487-26105-8 – gw Olms [944]
L'esprit francais – Paris.n1-87. oct 1929-33 – apply for price – (puis encyclopedie mensuelle vivante, ouverte et libre.) – fr ACRPP [073]
L'esprit nouveau – Paris.n1-23. janv-juin 1867 – 1 – fr ACRPP [073]
L'esprit nouveau : revue internationale illustree de l'activite contemporaine – Paris. n1-29. 1920-25 – 1 – fr ACRPP [073]
L'esprit public – Paris. 16 fevr 1862-1er juil 1864 – 1 – fr ACRPP [073]
Esprits de la vie a madagascar / Faublee, Jacques – Paris, France. 1954 – 1r – 1 – us UF Libraries [960]
Espronceda / Blanco Garcia, Francisco – Madrid: Saenz de Jubera, 1909 – sp Bibl Santa Ana [440]
Espronceda. ilustraciones biograficas y criticas / Alonso Cortes, Narciso – Valladolid: Libreria Santaren, 1942 – sp Bibl Santa Ana [920]
Espronceda, Jose de *see*
– A salamanca diak
– Blanca de borbon
– De gibraltar a lisboa
– El diablo mundo
– El estudiante de salamanca
– Ni el tio ni el sobrino
– Obras poeticas
– Obras poeticas completas
– Obras poeticas de
– Obras poeticas de d..., ordenadas y anotadas por j.e. hartzenbusch a saber
– Obras poeticas de.... precedidas de la biografia del autor
– Obras poeticas y escritos en prosa
– Obras poeticas y escritos en prosa. coleccion completa...ordenada por don patricio de la esnosura
– Paginas olvidadas
– Paginas olvidadas de...
– Poesias
– Poesias elegidas
– Sancho saldana
– Sancho saldana o el castellano de cuellar
– Las tres reinas
Espronceda su epoca su vida y sus obras / Cascales Munoz, Jose – Madrid: Biblioteca Hispania, 1914 – 1 – sp Bibl Santa Ana [946]
Espronceda...obras / Rodriguez Solis, Enrique – 1884 – 9 – sp Bibl Santa Ana [890]
Espuelazo / New England Farm Workers' Council – v3 n10-v8 n4 [1974 jun-1980 sep/dec] – 1r – 1 – mf#367506 – us WHS [331]
Espumas flutuantes / Alves, Castro – Rio de Janeiro, Brazil. 1947 – 1r – 1 – us UF Libraries [972]
Esputa, J *see* The tuneer's polka
Esq – Pullman. 1972+ (1,5,9) – ISSN: 0093-8297 – mf#17356,01 – us UMI ProQuest [400]
Esquema historico de las letras en cuba / Fernandez De Castro, Jose Antonio – Habana, Cuba. 1949 – 1r – 1 – us UF Libraries [972]
Esquema sobre los factores alogenos de la poblacio / Martin, Juan Luis – Habana, Cuba. 1944 – 1r – 1 – us UF Libraries [972]
Esquenazi-Mayo, Roberto *see* Ensayos y apuntes
Esquer, A *see* Essai sur les castes dans l'inde
Esquila / Crusco, Romualdo – Habana, Cuba. 1942 – 1r – us UF Libraries [972]
Esquire – Chicago. 1933-1978 (1) 1969-1978 (5) 1960-1978 (9) – (cont by: esquire fortnightly) – ISSN: 0014-0791 – mf#2318 – us UMI ProQuest [073]
Esquire – New York. 1979+ (1,5,9) – (cont: esquire fortnightly) – ISSN: 0194-9535 – mf#2318,02 – us UMI ProQuest [740]
Esquire fortnightly – New York. 1978-1979 (1) 1978-1979 (5) 1978-1979 (9) – (cont: esquire. cont by: esquire) – ISSN: 0884-5220 – mf#2318,01 – us UMI ProQuest [073]
Esquire good grooming guide – New York. 1966-1968 (1) – ISSN: 0014-0805 – mf#6761 – us UMI ProQuest [640]
Esquiros, Alphonse *see*
– L'accusateur public
– Les vierges folles
Esquisse bio-bibliographique de monsieur le notaire leonidas bachand : president de l'alliance francaise de sherbrooke / Desjardins, soeur – 1962 [mf ed 1978] – (= ser Bibliographies du cours...1947-66) – 1mf – 9 – (with ind; pref by Maurice O'Bready) – mf#SEM105P4 – cn Bibl Nat [920]
Esquisse biographique de sir george-etienne cartier / David, Laurent-Olivier – Montreal?: s.n, 1873 – 1mf – 9 – mf#03645 – cn Canadiana [920]
Esquisse de la langue holoholo / Coupez, A – Tervuren, Belgium. 1955 – 1r – 1 – us UF Libraries [470]
Esquisse de la langue ombo / Meeussen, A E – Tervuren, Belgium. 1952 – 1r – 1 – us UF Libraries [470]
Esquisse de la vie et des travaux apostoliques de sa grandeur mgr fr xavier de laval-montmorency : premier eveque de quebec: suivie de l'eloge funebre du prelat / Bois, Louis-Edouard – Quebec?: A Cote, 1845 – 2mf – 9 – mf#32940 – cn Canadiana [241]
Esquisse de l'histoire du bresil / Silva Paranhos, Jose Maria Da, Junior – Rio de Janeiro, Brazil. 1958 – 1r – 1 – us UF Libraries [972]
Esquisse de sociologie haitienne / Baguidy, Joseph D – Port-Au-Prince, Haiti. 1946 – 1r – 1 – us UF Libraries [301]
Esquisse du commerce de pelleteries des anglois, dans l'amerique septentrionale : avec des observations relatives a la compagnie du nord-ouest de Montreal / Selkirk, Thomas Douglas, 5th Earl of – Montreal: James Brown, 1819 [mf ed 1971] – 1r – 5 – mf#SEM16P82 – cn Bibl Nat [380]
Esquisse d'une philosophie de la religion d'apr es la psychologie et l'histoire *see* Outlines of a philosophy of religion based on psychology and history
Esquisse d'une philosophie de la religion d'apres la psychologie et l'histoire / Sabatier, Auguste – 7e ed. Paris: Fischbacher, 1903 – 1mf – 9 – 0-8370-5014-6 – (incl name ind) – mf#1985-3014 – us ATLA [100]

Esquisse geologique du canada : pour servir a la preparation d'un chronographe geologique du canada et des autres parties de l'amerique septentrionale britannique / Ami, Henry Marc – Quebec: Le Naturaliste canadien, 1902 – 1mf – 9 – 0-665-73485-9 – mf#73485 – cn Canadiana [550]

Esquisse geologique du canada : pour servir a l'intelligence de la carte geologique et de la collection des mineraux economiques envoyees a l'exposition universelle de paris, 1855 / Logan, William Edmond & Hunt, Thomas Sterry – Paris: Hector Bossange et fils, 1855 [mf ed 1983] – 2mf – 9 – mf#SEM105P227 – cn Bibl Nat [550]

Esquisse morale et politique des etats-unis de l'amerique du nord / Murat, Achille – Paris 1832 [mf ed Hildesheim 1995-98] – 1v on 3mf – 9 – €90.00 – 3-487-27175-3 – gw Olms [975]

Esquisse sur richard wagner / Grandmougin, Charles Jean – Paris: Durand, Schoenewerk [1873] – 1mf – 9 – mf#wa-35 – ne IDC [780]

Esquisses allemandes / Jolivet, A et al – Paris: Aubier, [1942] – 1 – us Wisconsin U Libr [430]

Esquisses biographiques : 1795-1855 Jean-Joseph Girouard, l'ancien depute du comte du Lac des Deux-Montagnes, un des prisonniers politiques de 1837-38 / Baillairge, George Frederick – Joliette, PQ: Bureaux de bon combat, du couvent et de la famille, 1893 – 3mf – 9 – mf#06816 – cn Canadiana [929]

Esquisses des moeurs turques au 19e siecle : ou scenes populaires, usages religieux, ceremonies publiques, vie interieure, habitudes sociales, idees politiques des mahometans; en forme de dialogues / Palaiologos, Gregorios – Paris 1827 [mf ed Hildesheim 1995-98] – 3mf – 9 – €90.00 – 3-487-29130-4 – gw Olms [956]

Esquisses havanaises / Vaudoyer, Jean Louis – Paris, France. 1930 – 1r – 1 – us UF Libraries [972]

Esquisses historiques : ou coup d'oeil rapide jete sur quinze annees de notre histoire nationale / Simonot, J F – Paris 1823 [mf ed Hildesheim 1995-98] – 2v on 4mf – 9 – €120.00 – ISBN-10: 3-487-26375-0 – ISBN-13: 978-3-487-26375-5 – gw Olms [944]

Esquivel, Antonio *see* Acusado a la inquisicion

Esquivel Obregon, Toribio *see*
– Apuntes para la historia del derecho de mejico
– Hernando cortes y el derechos internacional en el siglo 16

Esr magazine *see* Empire state report

Esra, nehemia und esther / Siegfried, Carl – Goettingen: Vandenhoeck & Ruprecht, 1901 – (= ser Handkommentar zum alten testament) – 1mf – 9 – 0-8370-9503-4 – (Includes bibliographies) – mf#1986-3503 – us ATLA [221]

Esref – Istanbul: Birinci sene aded 1-52. 25 safer 1327-21 safer 1328 [18 mar 1909-5 mar 1910] – (= ser O & t journals) – 7mf – 9 – $110.00 – (missing iss n52 otherwise complete) – us MEDOC [956]

Esref, Kulliyat-i Sa'ir *see* The divan project

Esref, Mehmet *see* Tarih-i umumi ve osmani atlasi

Esref us- Su'ara *see* The divan project

Esrim ve-arbaah sifre ha-kodesh : meduyakim hetev al pi ha-masorah ve-al pi depusim rishonim, im hilufim ve-hagahot min kitve yad atikim ve-targumim yeshanim – [London: s.n.], 1894 (Vienna: Carl Fromme) – 5mf – 9 – 0-8370-1797-1 – mf#1987-6185 – us ATLA [221]

L'essai – Montreal: [s.n, 1894-1895?] [mf ed 1re annee n1 1 dec 1894, 1re annee n2 15 dec 1894-1re annee n3 1 janv 1895] – 9 – ISSN: 1190-7622 – mf#P04138 – cn Canadiana [440]

Essai bibliographique : histoire et geographie, manuels pour les eleves de langue francaise, approuves par le comite catholique du conseil de l'instruction publique de la province de quebec, 1950-1959 / Despins, Simonne – 1962 [mf ed 1978] – (= ser Bibliographies du cours...1947-66) – 1mf – 9 – mf#SEM105P4 – cn Bibl Nat [900]

Essai bibliographique : la religion et le francais: manuels pour les eleves de langue francaise, approuves par le comite catholique du conseil de l'instruction publique de la province de quebec, 1950-1959 / Robert, frere – 1961 [mf ed 1978] – (= ser Bibliographies du cours...1947-66) – 2mf – 9 – (pref by Charles Bilodeau) – mf#SEM105P4 – cn Bibl Nat [440]

Essai critique sur l'histoire de charles 7, d'agnes sorelle et de jeanne d'arc / Delort, Joseph – Paris 1824 [mf ed Hildesheim 1995-98] – 296 S. 2mf – 9 – €60.00 – ISBN-10: 3-487-26224-X – ISBN-13: 978-3-487-26224-6 – gw Olms [944]

Essai de bibliographie canadienne, vol 1 : inventaire d'une bibliotheque comprenant imprimes, manuscrits, estampes, etc relatifs a l'histoire du canada et des pays adjacents avec des notes bibliographiques / Gagnon, Phileas – Quebec: impr pour l'auteur, 1895 [mf ed 1983] – 8mf – 9 – 0-665-03756-2 – mf#03756 – cn Canadiana [971]

Essai de bibliographie canadienne, vol 2 : inventaire d'une bibliothque comprenant imprimes, manuscrits, estampes, etc relatifs a l'histoire du canada... / Gagnon, Phileas – Quebec: Cite de Montreal, 1913 – v2 on 6mf – 9 – mf#03757 – cn Canadiana [971]

Essai de bibliographie canadienne, vols 1-2 : inventaire d'une bibliotheque comprenant imprimes, manuscrits, estampes, etc relatifs a l'histoire du canada et des pays adjacents avec des notes bibliographiques / Gagnon, Phileas – Quebec: impr pour l'auteur. 2v. 1895-1913 – 1mf – 9 – 0-665-03755-4 – mf#03755 – cn Canadiana [971]

Essai de bibliographie sur le regime municipal dans la province de quebec / Drapeau, Julien – 1955 [mf ed 1978] – (= ser Bibliographies du cours...1947-66) – 2mf – 9 – (with ind; pref by mtre Jean-Louis Doucet) – mf#SEM105P4 – cn Bibl Nat [350]

Essai de bio-bibliographie / Marie-Clement, frere – 1964 [mf ed 1979] – (= ser Bibliographies du cours...1947-66) – 2mf – 9 – (with ind; pref by frere Gaetan; ill by frere Jean-Vital) – mf#SEM105P4 – cn Bibl Nat [241]

Essai de bio-bibliographie de monsieur louis-gerard "gerry" gosselin : avocat, ecrivain, journaliste / Sainte-Marie-de-l'Ange-Gardien, soeur – 1962 [mf ed 1978] – (= ser Bibliographies du cours...1947-66) – 2mf – 9 – (with ind; pref by soeur sainte-marie-de-la presentation) – mf#SEM105P4 – cn Bibl Nat [920]

Essai de bio-bibliographie sur cecile rouleau : educatrice / Roy-Tessier, Antoinette – 1964 [mf ed 1979] – (= ser Bibliographies du cours...1947-66) – 1mf – 9 – (with ind) – mf#SEM105P4 – cn Bibl Nat [370]

'Essai de grammaire' and 'dictionnaire fidjien-francais' / Mathieu, C et al [comp?] – 1854 – 1r – 1 – mf#pmb451 – at Pacific Mss [490]

Essai de grammaire malgache / Montagne, Lucien – Paris: Societe d'editions, geographiques, maritimes et coloniales, 1931 – 1 – us CRL [490]

Essai de manuel de la langue agni, parlee dans la moitie orientale de la cote d'ivoire – Paris: Libraire africaine et coloniale, 1900 – 1 – us CRL [490]

Essai de methodologie des sciences theologiques / Vaucher, Edouard – Paris: Jules Claye, 1878 – 1mf – 9 – 0-8370-5621-7 – mf#1985-3621 – us ATLA [200]

Essai de statistique de l'ile bourbon : consideree dans sa topographie, sa population, son agriculture, son commerce, etc; ouvrage couronne en 1828 par l'academie royale des sciences; suivi d'un projet de colonisation de l'interieur de cette ile / Thomas, Pierre – Paris 1828 [mf ed Hildesheim 1995-98] – 2v on 5mf – 9 – €100.00 – 3-487-27237-7 – gw Olms [314]

Essai de transliteration raisonnee du cam / Ecole pratique des hautes etudes [France]. Centre d'histoire et civilisations de la Peninsule indochinoise. Groupe de recherches cam – [s.l: s.n. 196-?] [mf ed 1989] – 1r with other items – 1 – (with bibl footnotes) – mf#mf-10289 seam reel 023/12 [§] – us CRL [480]

Essai d'une bibliographie sur la question d'orient : orient europeen 1821-1897 / Bengesco, G – Bruxelles, Paris, 1897 – 4mf – 9 – mf#AR-1707 – ne IDC [956]

Essai d'une introduction a la dogmatique protestante / Lobstein, Paul – Paris: Librairie Fischbacher, 1896 [mf ed 1985] – 1mf – 9 – 0-8370-4155-4 – (incl bibl ref) – mf#1985-2155 – us ATLA [242]

Essai d'une introduction a la dogmatique protestante *see*
– Collected works
– Introduction to protestant dogmatics

Essai historique sur la louisiane / Gayarre, Charles – [Nouvelle-Orleans?: s.n.] 2v. 1830 [mf ed 1985] – 2v on 1mf – 9 – 0-665-49265-0 – mf#49265 – cn Canadiana [978]

Essai historique sur les facultes de theologie de saumur et de sedan / Auziere, Louis – 1836 – 1mf – 9 – 0-524-04007-9 – mf#1990-1179 – us ATLA [240]

Essai politique sur le royaume de la nouvelle-espagne / Humboldt, Alexander von – Paris 1811 [mf ed Hildesheim 1995-98] – 5v on 17mf – 9 – €170.00 – ISBN-10: 3-487-26989-9 – ISBN-13: 978-3-487-26989-4 – gw Olms [972]

Essai statistique sur le royaume de portugal et d'algarve : compare aux autres etats de l'europeet suivi d'un coup d'oeil sur l'etat actuel des sciences, des lettres et des beaux-arts parmi les portugais des deux hemispheres / Balbi, Adriano – Paris 1822 [mf ed Hildesheim 1995-98] – 2v on 8mf – 9 – €160.00 – 3-487-29830-9 – gw Olms [946]

Essai sur jerome savonarole : d'apres sa predication / Manen, G – Montauban: J Granie, 1897 – 1mf – 9 – 0-524-08518-8 – mf#1993-1048 – us ATLA [240]

Essai sur la condition de la femme au siam / Duplatre, Louis – Lyon: Rey, 1922 – (= ser Les femmes [coll]) – 2mf – 9 – mf#8472 – fr Bibl Nationale [305]

Essai sur la constitution geologique de... / Cappelle, Herman Van – Baarn, Surinam. 1907 – 1r – 1 – us UF Libraries [550]

Essai sur la democratie au cambodge / Preschez, Philippe – [Paris: Centre d'etude des relations internationales] 1961 [mf ed 1989] – (= ser Publications du centre d'etude des relations internationales. ser c: recherches 4) – 1r with other items – 1 – (incl bibl) – mf#mf-10289 seam reel 020/13 [§] – us CRL [320]

Essai sur la doctrine socinienne / Amphoux, Henri – Strasbourg: Imprimerie de Veuve Berger-Levrault, 1850 – 1mf – 9 – 0-7905-6580-3 – mf#1988-2580 – us ATLA [240]

Essai sur la legende du buddha : son caractere et ses origines / Senart, Emile – 2nd ed. Paris: Ernst Leroux, 1882 [mf ed 1995] – (= ser Yale coll) – xxxv/496p – 1 – 0-524-09715-1 – (in french. 1st publ in: journal asiatique, 1873-75) – mf#1995-0715 – us ATLA [280]

Essai sur la litterature merveilleuse des noirs : suivi de contes indigenes de l'ouest-africain francaise / Equiloecq, F V – Paris: E Leroux, 1913-16 – 1 – us CRL [490]

Essai sur la manifestation des convictions religieuses et sur la separation de l'eglise et de l'etat *see* An essay on the profession of personal religious conviction and upon the separation of church and state

Essai sur la musique ancienne et moderne / Laborde, Jean-Benjamin – Paris: Impr de P-D Pierres & ...chez E Onfroy 1780 [mf ed 19–] – 4v on 47mf – 9 – mf#fiche 588, 749 – us Sibley [780]

Essai sur la pedagogie de leibniz / Vernay, Joseph – Heidelberg, 1914 (mf ed 1993) – 2mf – 9 – €31.00 – 3-89349-310-7 – mf#DHS-AR 166 – gw Frankfurter [430]

Essai sur la physionomie des serpens / Schlegel, H – La Haye, 1837 – 9mf – 9 – mf#Z-2241 – ne IDC [590]

Essai sur la polemique et la philosophie de saint clement d'alexandrie / Hebert-Duperron, Victor – Caen: A Hardel, 1855 – 1mf – 9 – 0-524-05321-9 – (incl bibl ref) – mf#1990-1439 – us ATLA [240]

Essai sur la politique interieure d'haiti / Edouard, Emmanuel – Paris, France. 1890 – 1r – 1 – us UF Libraries [972]

Essai sur la propogation de l'alphabet phenicien dans l'ancien monde / Lenormant, Francois – Paris: Maisonneuve. 2v. 1872-75 – 2mf – 9 – 0-8370-9076-8 – (incl bibl ref) – mf#1986-3076 – us ATLA [470]

Essai sur la psychologie des actions humaines d'apres les systemes d'Aristote et de saint Thomas d'Aquin / Lecoultre, Henri – Paris: Fischbacher, 1883 – 1mf – 9 – 0-7905-8824-2 – (Incl bibl ref) – mf#1989-2049 – us ATLA [150]

Essai sur la statistique de la population francaise / Angeville, A d' – Paris, 1836 – 1r – 1 – mf#96873 – uk Microform Academic [944]

Essai sur la vie et la doctrine de saint-martin, le philosophe inconnu / Caro, Elme – Paris: Hachette, 1852 – 1mf – 9 – 0-7905-7562-0 – mf#1989-0787 – us ATLA [140]

Essai sur l'accompagnement du clavecin : pour parvenir facilement & en peu de tems a accompagner... / Clement, Charles Francois – Paris: C Ballard 1758 [mf ed 19–] – 1mf – 9 – mf#fiche 392 – us Sibley [780]

Essai sur l'architecture / Laugier, M A – nouv ed. Paris, 1755 – 4mf – 9 – mf#OA-66 – ne IDC [720]

Essai sur le behaisme *see* The universal religion, bahaism

Essai sur le canal de suez : droit et politique / Moussa, A – Paris, 1935 – 2mf – 9 – mf#ILM-2332 – ne IDC [956]

Essai sur le chameau au sahara occidental / Monteil, Vincent – Saint-Louis du Senegal, 1952 – 1 – us CRL [636]

Essai sur le commerce de russie : avec l'histoire de ses decouvertes / Marbault – Amsterdam: [s.n.] 1777 [mf ed 1984] – 4mf – 9 – 0-665-46130-5 – mf#46130 – cn Canadiana [380]

Essai sur le fondement metaphysique de la morale / Rauh, Frederic – Paris: Felix Alcan, 1891 – 1mf – 9 – 0-8370-6340-X – mf#1986-0340 – us ATLA [170]

Essai sur le gnosticisme egyptien : ses developpements et son origine egyptienne / Amelineau, Emile – Paris: E Leroux 1887 [mf ed 1990] – (= ser Annales du musee guimet 14) – 4mf [ill] – 9 – 0-7905-8259-7 – (incl bibl ref) – mf#1988-6137 – us ATLA [290]

Essai sur le libre arbitre : sa theorie et son histoire / Fonsegrive, George – 2e ed. Paris: Felix Alcan, 1896 – 2mf – 9 – 0-8370-6184-9 – (incl bibl ref) – mf#1986-0184 – us ATLA [210]

Essai sur le mysticisme speculatif en allemagne au quatorzieme siecle / Delacroix, Henri – Paris: F Alcan, 1900 – 1mf – 9 – 0-7905-7438-1 – (incl bibl ref) – mf#1989-0663 – us ATLA [180]

Essai sur le pali : ou langue sacree de la presqu'ile au-dela du gange: avec six planches lithographiees, et la notice des manuscrits de la bibliotheque du roi – Paris: Dondey-Dupre 1826 – (= ser Whsb) – 3mf – 9 – €40.00 – (incl: observations grammaticales sur quelques passages de l'essai sur le pali de mm e burnouf et lassen / e burnouf) – mf#Hu 267 – gw Fischer [490]

Essai sur le patois lorrain des environs du comte du ban de la roche, fief royal d'alsace / Oberlin, Jeremias Jacob – Strasbourg: Stein 1775 – (= ser Whsb) – 4mf – 9 – €50.00 – mf#Hu 083 – gw Fischer [440]

Essai sur l'ecclesiologie de zwingle / Bachofen, Charles – Geneve: Rivera & Dubois, 1890 – 1mf – 9 – 0-7905-5620-0 – mf#1988-1620 – us ATLA [242]

Essai sur les castes dans l'inde / Esquer, A – Pondichery: A. Saligny, 1870. 500p. Bibliographic footnotes – 1 – us Wisconsin U Libr [954]

Essai sur les doctrines sociales et politiques de taki-d-din-ahmad b. taimiya / Laoust, Henri – Le Caire, 1939 – 13mf – 8 – €25.00 – ne Slangenburg [260]

Essai sur les eglises romanes et roma-bysantines du departement du Puy-de-Dome / Mallay, A G – Moulins, 1838-1841 – 5mf – 9 – mf#OA-132 – ne IDC [720]

Essai sur les hallucinations conscientes / Bessonnet, Rene – Paris: H Jouve 1898 [mf ed Bloomington IN: Indiana Uni Lib, Preservation Dept 1984] – 1r – 1 – us Indiana Preservation [390]

Essai sur les invasions maritimes des normands dans les gaules : suivi d'un apercu des effets que les etablissemens des hommes du nord ont eus sur la langue... / Capefigue, Jean Baptiste Honore Raymond – Paris 1823 [mf ed Hildesheim 1995-98] – 1v on 3mf – 9 – €90.00 – ISBN-10: 3-487-25954-0 – ISBN-13: 978-3-487-25954-3 – gw Olms [930]

Essai sur les lettres de change et les billets promissoires / Girouard, Desire – Montreal: J Lovell, 1860 – 3mf – 9 – (incl bibl ref) – mf#41549 – cn Canadiana [332]

Essai sur les moeurs du tems / Reboul, Robert – Londres [u.a. 1768 [mf ed Hildesheim 1995-98] – 1v on 4mf – 9 – €120.00 – ISBN-10: 3-487-25886-2 – ISBN-13: 978-3-487-25886-7 – gw Olms [390]

Essai sur les motazelites : les rationalistes de l'islaam / Galland, Henri – Paris: Librairie orientale and americaine, [1906?] – 1mf – 9 – 0-524-01282-2 – (incl bibl ref) – mf#1990-2318 – us ATLA [260]

Essai sur les origines des partis saduceen et pharisien... / Montet, Edouard Louis – Vienne: Adolphe Holzhausen, 1883 – 1mf – 9 – 0-8370-4472-3 – (incl bibl ref) – mf#1985-2472 – us ATLA [270]

Essai sur les origines du romancero / Foulche-Delbosc, Raymond – Paris: [s n] 1912 [mf ed Bloomington IN: Indiana Uni Lib, Preservation Dept 1984] – 47p on 1r – 1 – us Indiana Preservation [390]

Essai sur les principes regissant l'administration de la justice aux indes orientales hollandaises surtout dans les iles de java et de madoura et leur application / Winckel, Christiaan Philip Karel – Samarang: van Dorp, 1880 – 315p – 1 – mf#LL-10024 – us L of C Photodup [340]

Essai sur les sanctuaires primitifs et sur le fetichisme en europe / Toubin, Charles – Besancon: Dodivers, 1864 – 1mf – 9 – 0-8370-9990-0 – (incl bibl ref) – mf#1986-3990 – us ATLA [306]

Essai sur les sources de l'ordo-missae premontre / Luykx, B – Postel, 1947 – 1mf – 8 – €3.00 – ne Slangenburg [241]

Essai sur l'evolution historique / Pattee, Richard – Port-Au-Prince, Haiti. 1944 – 1r – 1 – us UF Libraries [972]

Essai sur l'evolution historique et philosophique des idees morales dans l'egypte ancienne / Amelineau, Emile – Paris: Ernest Leroux, 1895 [mf ed 1992] – (= ser Bibliotheque de l'ecole des hautes etudes. sciences religieuses 6) – 2mf – 9 – 0-524-03474-5 – (in french) – mf#1990-3216 – us ATLA [170]

Essai sur l'habitat sedentaire traditionnel au sahara algerien / Eschallier, J C – Paris: universite de paris, institut d'urbanisme, 1968 – us CRL [960]
Essai sur l'histoire antique d'abyssienie / Kammerer, A – Np, 1926 – 6mf – 8 – mf#A-322 – ne IDC [956]
Essai sur l'histoire de la bible dans la france chretienne au moyen age / Trochon, Charles – Paris: Alphonse Derenne, 1878 [mf ed 1985] – 1mf – 9 – 0-8370-2180-4 – (incl bibl ref) – mf#1985-0180 – us ATLA [220]
Essai sur l'histoire de la formation et des progres du tiers etat : suivi de deux fragments du recueil des monuments inedits de cette histoire / Thierry, Augustin – Bruxelles [u.a. 1853 [mf ed Hildesheim 1995-98] – 1v on 3mf – 9 – €90.00 – ISBN-10: 3-487-25895-1 – ISBN-13: 978-3-487-25895-9 – gw Olms [944]
Essai sur l'histoire des eglises reformees de bretagne, 1535-1808 / Vaurigaud, Benjamin – Paris: Joel Cherbuliez, 1870 – 14mf – 9 – 0-524-08823-3 – mf#1993-3315 – us ATLA [240]
Essai sur l'histoire du culte reforme principalement au 16e et au 19e siecle / Doumergue, Emile – Paris: Fischbacher, 1890 – 1mf – 9 – 0-7905-6225-1 – mf#1988-2225 – us ATLA [240]
Essai sur l'histoire du protestantisme au havre et dans ses environs / Amphoux, Henri – Havre: L. Dombre, 1894 – 2mf – 9 – 0-7905-5445-3 – (incl bibl ref) – mf#1988-1445 – us ATLA [242]
Essai sur l'histoire et la geographie de la palestine : d'apres les thalmuds et les autres sources rabbiniques / Derenbourg, Joseph – Paris: Imprimerie Imperiale, 1867 – 2mf – 9 – 0-7905-3247-6 – (incl bibl ref) – mf#1987-3247 – us ATLA [939]
Essai sur l'histoire et l'etablissement / Dorsainvil, J B – Port-Au-Prince, Haiti. 1892-1893 – 1r – 1 – us UF Libraries [972]
Essai sur l'histoire naturelle du chili / Molina, G J – London. 1844+ (1) 1998+ (5) 1998+ (9) – 6mf – 9 – mf#5451 – ne IDC [918]
Essai sur l'immortalite au point de vue du naturalisme evolutioniste : conferences / Sabatier, Armand – 2. ed. Paris: Librairie Fischbacher, 1895 – 1mf – 9 – 0-524-00093-X – (incl bibl ref) – mf#1989-2793 – us ATLA [240]
Essai sur l'inegalite des races humaines / Gobineau, Arthur – Paris, France. v1-2. 1930 – 1r – 1 – us UF Libraries [301]
Essai sur l'organisation des etudes dans l'ordre des freres precheurs : au treizieme et au quatorzieme siecle (1216-1342): premiere province de provence, province de toulouse / Douais, Celestin – Paris: Picard, 1884 [mf ed 1992] – 1mf – 9 – 0-524-03279-3 – (in french. incl bibl ref) – mf#1990-0890 – us ATLA [241]
Essai sur l'origine et la decadence de la religion chretienne dans l'inde *see*
– Hinduism
– New india
Essai theorique et historique sur la generation des connaissances humaines dans ses rapports avec la morale, la politique et la religion : developpement du memoire couronne par le jury du concours universitaire institue par le gouvernement / Tiberghien, Guillaume – Bruxelles: Th Lesigne, 1844 – 2mf – 9 – 0-524-08654-0 – (incl bibl ref) – mf#1993-2114 – us ATLA [120]
Essaies or rather imperfect offers / Johnson, Robert – 1607 – 9 – us Scholars Facs [840]
Essais / Ernst, Fritz – Zuerich: Fretz & Wasmuth. 3v. c1946 – 1r – 1 – us Wisconsin U Libr [840]
Essais / Hostos, Eugenio Maria De – Paris, France. 1936 – 1r – us UF Libraries [972]
Les essais – n1-9. Paris. avr 1904-avr 1906 [mnthly] – 1 – fr ACRPP [073]
Essais bibliques / Vernes, Maurice – Paris: Ernest Leroux, 1891 – 1mf – 9 – 0-8370-5672-1 – (Incl bibl ref) – mf#1985-3672 – us ATLA [220]
Essais d'art libre – n1-33. Paris. fevr 1892-oct 1894 – 1 – fr ACRPP [073]
Essais de critique religieuse / Reville, Albert – Paris: Joel Cherbuliez, 1860 – 2mf – 9 – 0-524-03420-6 – mf#1990-0974 – us ATLA [240]
Essais de paleoconchologie / Cosmann, M – Paris. 13v. 1895 – 44mf – 9 – mf#Z-2240 – ne IDC [590]
Essais de paleoconchologie comparee / Cossman, Alexandre E – Paris. v1-13. 1895-1925 – 1 – $108.00 – mf#0167 – us Brook [560]
Essais de philosophie religieuse / Laberthonniere, Lucien – Paris: P Lethielleux, c1903 – 1mf – 9 – 0-8370-4027-2 – (incl app) – mf#1985-2027 – us ATLA [200]
Essais de sociologie et psychologie haitienne / Victor, Rene – Port-Au-Prince, Haiti. 1937 – 1r – 1 – us UF Libraries [301]
Essais de vulgarisation scientifique et questions / Dorsainvil, J C – Port-Au-Prince, Haiti. 1952 – 1r – 1 – us UF Libraries [972]
Essais et combats *see* L'etudiant socialiste
Essais historiques et critiques sur la franche-maconerie : ou, Recherches sur son origine, sur son systeme et sur son but... / Laurens, J L – Paris: Chomel 1805 – 3mf – 9 – mf#vrl-153 – ne IDC [366]
Essais historiques sur l'inde : precedes d'un journal de voyages et d'une description geographique de la cote de coromandel – Paris 1769 [mf ed Hildesheim 1995-98] – 1v on 4mf – 9 – €120.00 – 3-487-27405-1 – gw Olms [915]
Essais historiques sur orleans : ou description topographique et critique de cette capitale, et de ses environs / Polluche, Daniel – Orleans 1778 [mf ed Hildesheim 1995-98] – 2mf – 9 – €60.00 – 3-487-29727-2 – gw Olms [914]
Essais historiques sur paris de m de saintfoix / Saint-Foix, Germain F de – Londres [u.a. 1763-1762 [mf ed Hildesheim 1995-98] – 4v on 16mf – 9 – €160.00 – ISBN-10: 3-487-25934-6 – ISBN-13: 978-3-487-25934-5 – gw Olms [944]
Essais poetiques / Lemay, Pamphile – Quebec: G E Desbarats, 1865 [mf ed 1984] – 4mf – 9 – 0-665-45240-3 – mf#45240 – cn Canadiana [810]
Essais sur descartes / Gouhier, Henri Gaston – Paris, France. 1949 – 1r – 1 – us UF Libraries [920]
Essais sur la construction des peuples extra-europeens : ou collection des navires et pirogues construits par les habitant...du grand ocean et de l'amerique... / Paris, Francois Edmond – 1843 – 1r – 1 – mf#pmb doc396 – at Pacific Mss [301]
Essais sur la culture / Blanchet, Jules – Port-Au-Prince, Haiti. 194-? – 1r – 1 – us UF Libraries [972]
Essais sur la politique, l'histoire et les arts – Montreal: Librairie Beauchemin, 1920 [mf ed 1996] – 4mf – 9 – 0-665-79370-7 – mf#79370 – cn Canadiana [320]
Essais sur la valachie et la moldavie, theatre de l'insurrection dite ypsilanti / Salaberry, Charles M de – Paris 1821 [mf ed Hildesheim 1995-98] – 1mf – 9 – €40.00 – 3-487-29090-1 – gw Olms [949]
Essais sur le quebec contemporain = Essays on contemporary quebec / Symposium sur les repercussions sociales de l'industrialisation dans la province de Quebec (1952); ed by Falardeau, Jean-Charles – Quebec: Presses universitaires Laval, 1953 [mf ed 1974] – 1r – 5 – mf#SEM16P209 – cn Bibl Nat [971]
Essais sur les principes de l'harmonie : ou l'on traite de la theorie de l'harmonie en general, des droits respectifs de l'harmonie & de la melodie... / Serre, Jean Adam – Paris: Prault fils 1753 [mf ed 19–] – 3mf – 9 – mf#fiche 940 – us Sibley [780]
Essais sur l'histoire generale et comparee des theologies et des philosophies medievales / Picavet, Francois – Paris: F Alcan, 1913 [mf ed 1991] – 1mf – 9 – 0-7905-9062-X – (1st printed title: esquisse d'une histoire generale et compare des philosophies medievales. incl bibl ref) – mf#1989-2287 – us ATLA [931]
Essay d'un dictionnaire contenant la connaissance du monde, des sciences universelles... / [Feuille, D de la] – Amsterdam: De la Feuille, 1700 – 5mf – 9 – mf#O-659 – ne IDC [700]
Essay d'un dictionnaire contenant la connoissance du monde... / [Feuille, D de la] – Wesel: Chez Jacobus van Wesel, 1700 – 2mf – 9 – mf#O-21 – ne IDC [090]
An essay in aid of a grammar of assent / Newman, John Henry – 5th ed. London: Burns & Oates 1881 [mf ed 1991] – 2mf – 9 – 0-7905-7988-X – mf#1989-1273 – us ATLA [210]
An essay in aid of the better appreciation of catholic mysticism : illustrated from the writings of blessed angela of foligno / Thorold, Algar Labouchere – London: Kegan Paul, Trench, Truebner, 1900 [mf ed 1986] – 1mf – 9 – 0-8370-7270-0 – (incl bibl ref) – mf#1986-1270 – us ATLA [241]
An essay in refutation of atheism / Brownson, Orestes Augustus; ed by Brownson, Henry Francis – Detroit: T Nourse 1882 [mf ed 1987] – 1r – 1 – (filmed with: hildebrand and his times / stephens, w r w) – mf#1932 – us Wisconsin U Libr [210]
Essay of faith : and its connection with good works / Rotheram, John – London, England. 1801 – 1r – 1 – us UF Libraries [240]
An essay on assyriology / Evans, George – London: Williams & Norgate, 1883 [mf ed 1986] – 1mf – 9 – 0-8370-8421-0 – (in english & akkadian) – mf#1986-2421 – us ATLA [470]
Essay on brotherly love / Gilfillan, Samuel – Edinburgh, Scotland. 1807 – 1r – 1 – us UF Libraries [240]
Essay on catholic home missions / Faber, Frederick William – London, England. 1853 – 1r – 1 – us UF Libraries [241]
Essay on catholicism, liberalism and socialism : considered in their fundamental principles = Essayo sobre el catolicismo, el liberalismo y el socialismo / Donoso Cortes, Juan Francisco – Philadelphia: J B Lippincott, 1862 – 1mf – 9 – 0-8370-9228-0 – mf#1986-3228 – us ATLA [241]
An essay on church furniture and decoration / Cutts, Edward Lewes – London 1854 – (= ser 19th c art & architecture) – 2mf – 9 – mf#4.2.1430 – uk Chadwyck [740]
An essay on church polity : comprehending an outline of the controversy on ecclesiastical government, and a vindication of the ecclesiastical system of the methodist episcopal church / Stevens, Abel – New York: Carlton & Porter, c1847 [mf ed 1990] – 1mf – 9 – 0-7905-6452-1 – mf#1988-2452 – us ATLA [242]
An essay on classification / Agassiz, Louis – London, 1859 – (= ser 19th c evolution & creation) – 4mf – 9 – mf#1.1.10879 – uk Chadwyck [575]
An essay on colonization : particularly applied to the western coast of africa / Wadstrom, Carl Bernhard – London. 1794-95 – 1 – us CRL [960]
Essay on colonization : from the royal commonwealth society library / Wadstrom, Carl Bernhard – 1794 – 15mf – 7 – mf#2974 – uk Microform Academic [960]
Essay on comparative agriculture : or, A brief examination into the state of agriculture as it now exists in great britain and canada / Burton, J E – [Montreal?: s.n.] 1828 [mf ed 1983] – 2mf – 9 – 0-665-43093-0 – mf#43093 – cn Canadiana [630]
An essay on dogmatic preaching *see* Some account of the church in the apostolic age / an essay on dogmatic preaching
An essay on gandhian economics / Anjaria, Jashwantrai Jayantilal – Bombay: Vora & Co Publishers, [1944] – (= ser Samp: indian books) – us CRL [330]
Essay on god the holy spirit / Serle, Ambrose – London, England. 1824 – 1r – 1 – us UF Libraries [240]
An essay on hinduism : its formation and future / Ketkar, Shridhar Venkatesh – London: Luzac, 1911 [mf ed 1991] – (= ser History of caste in india 2) – 1mf – 9 – 0-524-00914-7 – mf#1990-2137 – us ATLA [280]
Essay on indifference in matters of religion = Essai sur l'indifference en mati ere de religion / Lamennais, Felicite Robert de – London: John Macqueen, 1895 – 1mf – 9 – 0-8370-7401-0 – (in english. incl bibl ref) – mf#1986-1401 – us ATLA [230]
An essay on laughter : its forms, its causes, its development and its value / Sully, James – London, New York, Bombay: Longmans, Green & Co 1902 [mf ed 1986] – 1r – 1 – (filmed with: religion and culture / schleiter, f) – mf#6688 – us Wisconsin U Libr [150]
An essay on liberty and slavery / Bledsoe, Albert Taylor – Philadelphia: JB Lippincott, 1856 [mf ed 1990] – 1mf – 9 – 0-7905-3757-5 – mf#1989-0250 – us ATLA [976]
Essay on marriage / Jay, William – Bath, England. 1807 – 1r – 1 – us UF Libraries [230]
An essay on military law and the practice of courts-martial / Tytler, Alexander F – 3rd ed. London: T Egerton, 1814 – 5mf – 9 – $7.50 – mf#LLMC 88-118 – us LLMC [355]
Essay on miracles / Hume, David – London, England. 1854 – 1r – 1 – us UF Libraries [240]
Essay on mr w h lynch's pamphlet entitled "scientific butter making" / Barre, Stanislas Morrier – Montreal: s.l, 1884 – 1mf – 9 – (incl some text in french) – mf#02433 – cn Canadiana [630]
An essay on obligations : for lawyers, students and laymen / Foran, Joseph Kearney – Toronto: Carswell, 1886 [mf ed 1980] – 3mf – 9 – 0-665-03145-9 – mf#03145 – cn Canadiana [340]
An essay on original genius and its various modes of exertion in philosophy and the fine arts, particularly in poetry / Duff, William – 1767 – 9 – $20.00 – us Scholars Facs [190]
Essay on ornamental art as applicable to trade and manufactures / Ballantine, James – London 1847 – (= ser 19th c art & architecture) – 3mf – 9 – mf#4.2.1427 – uk Chadwyck [740]
An essay on painting... / Algarotti, F – London, 1764 – 2mf – 8 – mf#O-1188 – ne IDC [750]
An essay on pantheism / Hunt, John – rev ed. London: Gibbings, 1893 [mf ed 1985] – 1mf – 9 – 0-8370-3693-3 – mf#1985-1693 – us ATLA [210]
An essay on personality as a philosophical principle / Richmond, Wilfrid – London: E Arnold, 1900 [mf ed 1991] – 1mf – 9 – 0-7905-9083-2 – mf#1989-2308 – us ATLA [150]
An essay on production, money and government : in which the principle of a natural law is advanced and explained... / Thomson, William Alexander – Buffalo: Wheeler, Matthews & Warren, 1863 – 1mf – 9 – mf#49777 – cn Canadiana [330]
An essay on rural architecture / Elsam, Richard – London 1803 – (= ser 19th c art & architecture) – 2mf – 9 – mf#4.2.1321 – uk Chadwyck [720]
Essay on sisterhoods in the english church / Sellon, W E – London, England. 1849 – 1r – 1 – us UF Libraries [240]
An essay on taste 1759... / Gerard, Alexander – Observations concerning the imitative nature of poetry. 1780 – 9 – us Scholars Facs [700]
An essay on temptation / Wines, Enoch Cobb – Philadelphia: Presbyterian Board of Publ, c1865 [mf ed 1989] – 1mf – 9 – 0-7905-0463-4 – (incl ind) – mf#1987-0463 – us ATLA [150]
An essay on the age and antiquity of the book of nabathaean agriculture : to which is added an inaugural lecture on the position of the shemitic nations in the history of civilization / Renan, Ernest – London: Truebner, 1862 [mf ed 1986] – 1mf – 9 – 0-8370-8855-0 – (incl bibl ref) – mf#1986-2855 – us ATLA [630]
An essay on the ancient topography of jerusalem : with restored plans of the temple etc... / Fergusson, James – London: J Weale, 1847 [mf ed 1992] – 1mf – 9 – 0-524-05034-1 – mf#1992-0287 – us ATLA [930]
Essay on the architecture of the hindus / Ram Raz – 1834 – (= ser Royal asiatic society, oriental translation fund. old series) – 1r – 1 – mf#418 – uk Microform Academic [720]
An essay on the causes of the variety of complexion and figure in the human species : to which are added animadversions on certain remarks made on the 1st ed of this essay by mr charles white [...]: also, strictures on lord kaims' discourse on the original diversity of mankind / Smith, Samuel Stanhope – 2nd enl ed. New Brunswick, [NJ]: J Simpson, 1810 – (incl app. filmed with: la raza negra es la mas antigua de las razas humanas/g fournier) – us CRL [573]
Essay on the character of jesus christ : considered as an evidence of the truth of the christian religion / Carmichael, James – Toronto: Hunter, Rose, 1882 – 1mf – 9 – mf#05850 – cn Canadiana [240]
Essay on the church / Jones, William – London, England. 1800 – 1r – 1 – us UF Libraries [240]
Essay on the church plain chant – London: J P Coghlan 1782 [mf ed 19–] – 3v on 2mf – 9 – (pt1: instructions for learning the church plain song; pt2: several anthems, litanies, proses & hymns; pt3: a supplement of several anthems, litanies, proses & hymns) – mf#fiche 452 – us Sibley [780]
Essay on the common features which appear in all forms of religious belief / Cust, Robert Needham – London: Luzac, 1895 – 1mf – 9 – 0-7905-4222-6 – (Incl bibl ref) – mf#1988-0222 – us ATLA [200]
Essay on the contracted liquid vein affecting the present theory of the science of hydraulics / Steckel, R – Ottawa?: s.n, 1884 (Ottawa: Maclean, Roger) – 2mf – 9 – mf#13923 – cn Canadiana [530]
Essay on the creative imagination = Essai sur l'imagination creatrice / Ribot, Theodule – Chicago: Open Court, 1906 – 1mf – 9 – 0-7905-9082-4 – (In English) – mf#1989-2307 – us ATLA [100]
An essay on the development of christian doctrine / Newman, John Henry – 6th ed. London, New York: Longmans, Green, 1888 [mf ed 1991] – 2mf – 9 – 0-7905-8535-9 – (incl bibl ref) – mf#1989-1760 – us ATLA [241]
An essay on the doric order of architecture / Aikin, Edmund – London 1810 – (= ser 19th c art & architecture) – 3mf – 9 – mf#4.2.1615 – uk Chadwyck [720]
Essay on the economics of detribalization in northern rhodesia / Wilson, G – 2pts – (= ser Institute for social research, university of zambia. papers 5,6) – 4mf – 7 – mf#363/5 – uk Microform Academic [960]
An essay on the education of the eye : with reference to painting / Burnet, John – London 1837 – (= ser 19th c art & architecture) – 2mf – 9 – mf#4.2.21 – uk Chadwyck [750]
Essay on the evils of popular ignorance : and a discourse on the communication of christianity to the people of hindoostan / Foster, John – London: Bell, 1876 – 1r – 1 – 0-8370-0361-X – mf#1984-B434 – us ATLA [240]
Essay on the extent of the death of christ / Polhill, Edward – Berwick, England. 1842 – 1r – 1 – us UF Libraries [240]

An essay on the geography of north-western africa / Bowdich, Thomas E – Paris 1821 [mf ed Hildesheim 1995-98] – 1v on 1mf – 9 – €40.00 – 3-487-27295-4 – gw Olms [916]

An essay on the growth and management of flax in ireland / Sproule, John – Dublin, 1844 – (= ser 19th c ireland) – 1mf – 9 – mf#1.1.5806 – uk Chadwyck [630]

Essay on the headship of the lord jesus christ – Edinburgh, Scotland. 1842 – 1r – 1 – us UF Libraries [240]

Essay on the hessian fly, wheat midge : and other insects injurious to the wheat crops / Hill, George S J – Toronto: s.n, 1858 – 1mf – 9 – (incl bibl ref) – mf#46687 – cn Canadiana [630]

An essay on the history and nature of original titles to land in the province and state of pennsylvania / Huston, Charles – Philadelphia: Johnson, 1849. 484p. LL-650 – 1 – us L of C Photodup [340]

Essay on the history of article 29 and of the 13th elizabeth, cap... / Swainson, Charles Anthony – Cambridge, England. 1856 – 1r – 1 – us UF Libraries [240]

An essay on the history of english church architecture / Scott, George Gilbert – London 1881 – (= ser 19th c art & architecture) – 5mf – 9 – mf#4.2.730 – uk Chadwyck [720]

Essay on the holy sacrament of the lord's supper / Waldo, Peter – London, England. 1803 – 1r – 1 – us UF Libraries [240]

An essay on the improvement of time : with notes of sermons, and other pieces / Foster, John – London: George Bell, 1886. Chicago: Dep of Photodup, U of Chicago Lib, 1973 (1r); Evanston: American Theol Lib Assoc, 1984 (1r) – 1 – 0-8370-0364-4 – mf#1984-B353 – us ATLA [240]

An essay on the improvement to be made in the cultivation of small farms by the introduction of green crops... / Blacker, William – Dublin, 1845 – (= ser 19th c ireland) – 2mf – 9 – mf#1.1.3280 – uk Chadwyck [630]

Essay on the interest and characteristics of the lives of the saint / Faber, Frederick William – London, England. 1853 – 1r – 1 – us UF Libraries [240]

An essay on the juridical history of france : so far as it relates to the law of the province of lower-canada... / Sewell, Jonathan – Quebec: Thomas Cary & Co, 1824 [mf ed 1976] – 1r – 5 – mf#SEM16P255 – cn Bibl Nat [340]

An essay on the law of art / Warden, Robert Bruce – Washington, D.C., The Ernest Institute, 1878. 300 p. LL-355 – 1 – us L of C Photodup [340]

An essay on the law of patents for new inventions / Fessenden, Thomas Green – Boston: Mallory etc. 1810. 229 1p. LL-1064 – 1 – us L of C Photodup [346]

An essay on the learning of partial, and of future interests in chattels personal / Keyes, Wade – Montgomery, Ala., Martin, 1853. 412 p. LL-297 – 1 – us L of C Photodup [340]

An essay on the military architecture of the middle ages / Viollet-Le-Duc, Eugene Emmanuel – Oxford 1860 – (= ser 19th c art & architecture) – 4mf – 9 – mf#4.2.1672 – uk Chadwyck [720]

Essay on the ministerial office : an exposition of the scriptural doctrine as taught in the ev lutheran church / Loy, Matthias – Columbus, O[hio]: Schulze & Gassmann, 1870 – 1mf – 9 – 0-7905-7909-X – mf#1989-1134 – us ATLA [242]

An essay on the nature of credit : as it is connected with the bankrupt law – London: printed by the Philanthropic Soc, 1814 – (= ser 19th c economics) – 1mf – 9 – mf#1.1.481 – uk Chadwyck [332]

An essay on the new analytic of logical forms : being that which gained the prize proposed by sir william hamilton, in the year 1846... / Baynes, Thomas Spencer – Edinburgh: Sutherland & Knox; London: Simpkin, Marshall, 1850 [mf ed 1991] – 1mf – 9 – 0-7905-9133-2 – mf#1989-2358 – us ATLA [160]

Essay on the omission of creeds, liturgies and codes of ecclesiasti... / Whately, Richard – London, England. 1831 – 1r – 1 – us UF Libraries [240]

An essay on the origin and development of window tracery / Freeman, Edward Augustus – Oxford 1851 – (= ser 19th c art & architecture) – 5mf – 9 – mf#4.2.407 – uk Chadwyck [740]

An essay on the origin and formation of the romance languages : containing an examination of m raynouard's theory on the relation of the italian, spanish, provencal and french to the latin / Lewis, George Cornewall – 2nd ed. London: Parkerson & Bourn 1862 [mf ed 1988] – 1r – 1 – (filmed with: lord nelson / forester, c s) – mf#2083 – us Wisconsin U Libr [440]

An essay on the origin and structure of the hindoostanee tongue, or general language of british india : with an account of the principal elementary works on the subject... / Arnot, Sandford & Forbes, Duncan – London: London Oriental Institution, 1828 – (= ser 19th c books on linguistics) – 1mf – 9 – mf#2.1.23 – uk Chadwyck [490]

Essay on the origin, history, and principles, of gothic architecture / Hall, James. baronet – 2nd ed. London 1813 – (= ser 19th c art & architecture) – 5mf – 9 – mf#4.2.1491 – uk Chadwyck [720]

An essay on the origin of the south indian temple / Venkata Ramanayya, N – Madras: Methodist Pub House, 1930 – (= ser Samp: indian books) – us CRL [720]

An essay on the pastoral office : as exemplified in the economy of the methodist episcopal church / Wythe, Joseph Henry – New York: Carlton & Phillips, 1853 [mf ed 1993] – (= ser Methodist coll) – 2mf – 9 – 0-524-07782-7 – mf#1991-3350 – us ATLA [242]

Essay on the person of christ / Kozaki, Hiromichi – Tokyo: [s.n.] 1893 [mf ed 1995] – (= ser Yale coll) – 115p/25p – 1 – 0-524-09989-8 – (in japanese) – mf#1995-0989 – us ATLA [240]

An essay on the place of ecclesiasticus in semitic literature : being the inaugural lecture / Margoliouth, David Samuel – Oxford: Clarendon Press, 1890 [mf ed 1985] – 1mf – 9 – 0-8370-4275-5 – (incl bibl ref & ind) – mf#1985-2275 – us ATLA [221]

Essay on the prevailing methods of the evangelization of the non-christian world / Cust, Robert Needham – London: Luzac, 1894 – 1mf – 9 – 0-8370-6100-8 – (incl ind) – mf#1986-0100 – us ATLA [240]

An essay on the principle of population : or, a view of its past and present effects on human happiness... / Malthus, Thomas Robert – 6th ed. London: J. Murray, 1826 [mf ed 1958] – 2v on 1r – 1 – (filmed with: the business of travel / rae, w f) – mf#1958 – us Wisconsin U Libr [304]

An essay on the principles of circumstantial evidence, illustrated by numerous cases... / Wills, William – Boston, Mass., Boston Book Co., 1905. 448 p. LL-1614 – 1 – (5th english ed. (1902) with american notes by george e beers...and arthur l corbin) – us L of C Photodup [346]

Essay on the principles of sanskrit grammar : pt 1 / Forster, Henry Pitt – Calcutta: Ferris 1810 – (= ser Whsb) – 8mf – 9 – €80.00 – mf#Hu 238 – gw Fischer [490]

Essay on the prize-question : whether the use of distilled liquors, or traffic in them, is compatible, at the present time, with making a profession of christianity? / Stuart, Moses – New York: John P Haven 1830 [mf ed 1989] – 1mf – 9 – 0-7905-2434-1 – mf#1987-2434 – us ATLA [230]

Essay on the probabilities of the duration of human life / Deparcieux, M – Paris, 1746 – 1r – 1 – mf#95692 – uk Microform Academic [120]

Essay on the productive resources of india / Royle, John Forbes – London 1840 – (= ser 19th c british colonization) – 5mf – 9 – mf#1.1.9989 – uk Chadwyck [630]

An essay on the profession of personal religious conviction and upon the separation of church and state : considered with reference to the fulfilment of that duty = Essai sur la manifestation des convictions religieuses et sur la separation de l'eglise et de l'etat / Vinet, Alexandre Rodolphe – London: Jackson & Walford, 1843 [mf ed 1990] – 2mf – 9 – 0-7905-7484-5 – (english trans by charles theodore jones) – mf#1989-0709 – us ATLA [230]

Essay on the promotion of domestic reform among the natives of india / Ganapati Lakshmana – Bombay: Printed at the American Mission Press, 1843 [mf ed 1995] – (= ser Yale coll) – 68p – 1 – 0-524-09882-4 – mf#1995-0882 – us ATLA [306]

Essay on the question whether islam has been beneficial or injurious to human society in general, and to the mosaic and christian dispensations – Lahore: Mohammadan Tract & Book Depot, Punjab, 1891 – us CRL [230]

Essay on the reasons of secession from the national church of scotland / Jaffray, Robert – Kilmarnock, Scotland. 1805 – 1r – 1 – us UF Libraries [242]

An essay on the registry laws of lower canada / Bonner, John – Quebec: printed by John Lovell, 1852 [mf ed 1990] – 2mf – 9 – mf#SEM105P1285 – cn Bibl Nat [340]

Essay on the repeal of the malt-tax : for which a prize of twenty pounds was awarded by the association... / Total Repeal Malt Tax Association – London: printed by Joseph Rogerson, 1846 – (= ser 19th c economics) – 1mf – 9 – mf#1.1.221 – uk Chadwyck [336]

Essay on the sanctification of the lord's day / Gilfillan, Samuel – Edinburgh, Scotland. 1806 – 1r – 1 – us UF Libraries [240]

Essay on the supposed existence of a quadripartite and tripartite d... / Hale, William Hale – London, England. 1832 – 1r – 1 – us UF Libraries [240]

Essay on the theory of the earth : with geological illustrations, by professor jameson / Cuvier, Georges Leopold Chretien Frederic Dagobert de, Baron – [5th ed] [Edinburgh] 1827 – (= ser 19th c evolution & creation) – 7mf – 9 – mf#1.1.10882 – uk Chadwyck [550]

Essay on the times : canada in the 9th decade of the 19th century / Armstrong, William Reginald – S.l: s.n, 1887? – 1mf – 9 – mf#52336 – cn Canadiana [240]

An essay on the utility of collecting the best works...engravers / Cumberland, George – London 1827 – (= ser 19th c art & architecture) – 6mf – 9 – mf#4.2.437 – uk Chadwyck [760]

Essay on the various fears to which god's people are liable / Toplady, Augustus Montague – London, England. 1826 – 1r – 1 – us UF Libraries [240]

An essay on transcendentalism / Ellis, Charles Mayo – 1842 – 9 – $10.00 – us Scholars Facs [190]

An essay on transcendentalism / [Ellis, Charles Mayo] – Boston: Crocker & Ruggles, 1842 [mf ed 1991] – 1mf – 9 – 0-7905-9192-8 – mf#1989-2417 – us ATLA [100]

An essay on uses and trusts, and on the nature and operation conveyances at common law / Sanders, Francis Williams – 2d American ed., from the last London ed. Philadelphia: Small, 1855. 2v in 1. LL-1021 – 1 – us L of C Photodup [346]

Essay on yeoman and peasant proprietorships / Robertson, Thomas – [Athy], 1874 – (= ser 19th c ireland) – 1mf – 9 – mf#1.1.8632 – uk Chadwyck [340]

An essay shewing the expediency of emigration, under certain circumstances : together with a comparative view of every new settlement, a sketch of each... / Elliott, John Frederick – London, 1822 – (= ser 19th c books on british colonization) – 1mf – 9 – mf#1.1.8802 – uk Chadwyck [304]

Essay towards a proposal for catholic communion / Basset, Joshua – London, England. 1801 – 1r – us UF Libraries [241]

An essay towards an indian bibliography : being a catalogue of books relating to the history, antiquities, languages, customs, religion, wars, literature, and origin of the american indians, in the library of thomas w field / Field, Thomas Warren – New York: Scribner, Armstrong, 1873 [mf ed 1980] – 5mf – 9 – 0-665-03100-9 – mf#03100 – cn Canadiana [019]

An essay upon prints : containing remarks upon the principles of picturesque beauty, the different kinds of prints and the characters of the most noted masters / Gilpin, W – Ed 2. London, 1768 – 3mf – 9 – mf#H-10033 – ne IDC [760]

Essay upon the influence of the imagination on the nervous system contributing to a false hope in religion / Powers, Grant – Andover: Flagg and Gould, 1828, c1827 – 1mf – 9 – 0-7905-5793-2 – mf#1988-1793 – us ATLA [150]

Essay upon the sacred use of organs in christian assemblies – Glasgow, Scotland. 1865 – 1r – us UF Libraries [240]

Das essayistische werk zur deutschen literatur in 4 baenden : saemtliche nachtprogramme und aufsaetze / Schmidt, Arno – Zuerich: Arno-Schmidt-Stiftung im Haffmanns Verlag 1988 [mf ed 1992] – 4v on 1r – 1 – mf#3178p – us Wisconsin U Libr [430]

Essayons / United States Army Training Center – Fort Leonard Wood MI. 1988 jun 2-dec 15, 1989 jan 5-jun 24, 1989 jul-dec, 1990 jan-jun – 4r – 1 – (cont: fort leonard wood guidon [fort leonard wood, mo: 1987]; cont by: guidon [fort leonard wood [mo]: 1999]) – mf#1574760 – us WHS [355]

Essays / Bahr, Hermann – Leipzig: Insel-Verlag, 1912 [mf ed 1998] – 255p – 1 – mf#9961 – us Wisconsin U Libr [840]

Essays : literary, critical and historical / O'Hagan, Thomas – Toronto: W Briggs, 1909 – 2mf – 9 – 0-665-75358-6 – mf#75358 – cn Canadiana [840]

Essays : never before published / Godwin, William – London: Henry S King, 1873 – 1mf – 9 – 0-7905-8798-X – mf#1989-2023 – us ATLA [840]

Essays : occasional essays / Chatard, Francis Silas – New York: Catholic Publ Society, 1894 – 1mf – 9 – 0-8370-7926-8 – mf#1986-1926 – us ATLA [840]

Essays : on poetry and music, as they affect the mind; on laughter, and ludicrous composition; on the usefulness of classical learning / Beattie, James – 3rd corr ed, London: printed for E & C Dilly; and W Creech, Edinburgh 1779 [mf ed 19–] – 8mf – 9 – mf#fiche 344 – us Sibley [840]

Essays / Romanes, George John; ed by Morgan, Conwy Lloyd – London, New York: Longmans, Green, 1897 – 1mf – 9 – mf#16906 – cn Canadiana [840]

Essays / Ryder, Henry Ignatius Dudley; ed by Bacchus, Francis – London; New York: Longmans, Green, 1911 – 1mf – 9 – 0-7905-6557-9 – mf#1988-2557 – us ATLA [840]

Essays : theological and literary / Everett, Charles Carroll – Boston:Houghton, Mifflin, 1901 – 1mf – 9 – 0-8370-3717-4 – mf#1985-1717 – us ATLA [840]

Essays / Wilberforce, Samuel – London: J Murray, 1874 – 2mf – 9 – 0-7905-6969-8 – (incl bibl ref) – mf#1988-2969 – us ATLA [840]

Essays aesthetical and philosophical : including the dissertation on the connexion between the animal and spiritual in man / Schiller, Friedrich von – London: G. Bell and Sons, 1875 – 435p – 1 – us Wisconsin U Libr [840]

Essays and addresses : an attempt to treat some religious questions in a scientific spirit / Wilson, James M – London, New York: Macmillan, 1894 [mf ed 1985] – 1mf – 9 – 0-8370-2993-7 – (incl bibl ref) – mf#1985-0993 – us ATLA [210]

Essays and addresses / Dale, Robert William – 2nd ed. New York: AC Armstrong, 1899 [mf ed 1985] – 1mf – 9 – 0-8370-3418-3 – (incl bibl ref) – mf#1985-1418 – us ATLA [242]

Essays and addresses : religious, literary and social / Brooks, Phillips; ed by Brooks, John Cotton – New York: E P Dutton, 1894 [mf ed 1991] – 2mf – 9 – 0-7905-9907-4 – mf#1989-1632 – us ATLA [840]

Essays and addresses chiefly on church subjects / Alford, Henry – London: Strahan, 1869 [mf ed 1985] – 1mf – 9 – 0-8370-4725-0 – mf#1985-2725 – us ATLA [242]

Essays and discourses : with a biographical sketch and a portrait / Ray, Praphulla Candra – Madras: GA Natesan & Co, 1918 – (= ser Samp: indian books) – us CRL [840]

Essays and discourses, practical and historical / Van Rensselaer, Cortlandt – Philadelphia: Presbyterian Board of Publication, c1861 – 1mf – 9 – 0-524-01758-1 – mf#1990-4150 – us ATLA [240]

Essays and dissertations in biblical literature : vol 1: containing chiefly translations of the works of german critics – New-York: G & C & H Carvill, 1829 – 2mf – 9 – 0-7905-0426-X – (no more publ. incl ind) – mf#1987-0426 – us ATLA [220]

Essays and hymns of synesius of cyrene – London, England. v1-2. 1930 – 1r – us UF Libraries [780]

Essays and lectures / O'Malley, Andrew – Barrie [Ont]: Gazette, [1916?] – 3mf – 9 – 0-665-75347-0 – mf#75347 – cn Canadiana [080]

Essays and monographs / Allen, William Francis – memorial ed. Boston: G H Ellis, 1890 [mf ed 1991] – vi/392p – 1 – mf#6854 – us Wisconsin U Libr [080]

Essays and observations on natural history, anatomy, physiology, psychology, and geology... : being his posthumous papers on these subjects, arranged and revised, with notes / Hunter, John – London, 1861 – (= ser 19th c evolution & creation) – 11mf – 9 – mf#1.1.1521 – uk Chadwyck [500]

Essays and other prose fragments / Paratiyar – Madras: Bharati Prachur Alayam, 1937 – (= ser Samp: indian books) – us CRL [840]

Essays and remains of the rev robert alfred vaughan / ed by Vaughan, Robert – London: John W Parker, 1858 – 2mf – 9 – 0-7905-8187-6 – mf#1988-8070 – us ATLA [840]

Essays and review / ed by Parker, J – London, 1860 – (= ser 19th c evolution & creation) – 5mf – 9 – mf#1.1.10914 – uk Chadwyck [100]

Essays and reviews : chiefly on theology, politics, and socialism / Brownson, Orestes Augustus – New York: D & J Sadlier, 1852 – 2mf – 9 – 0-8370-8246-3 – mf#1986-2246 – us ATLA [840]

Essays and reviews / Church, Richard William – London: J. and C. Mozley, 1854 – 2mf – 9 – 0-7905-4849-6 – mf#1988-0849 – us ATLA [240]

Essays and reviews / Hodge, Charles – New York: R Carter, 1857 – 6mf – 9 – 0-7905-9282-7 – mf#1989-2507 – us ATLA [240]

Essays and reviews / Temple, Frederick et al – 9th ed. London: Longman, Green, Longman & Roberts, 1861 [mf ed 1985] – 2mf – 9 – 0-8370-3718-2 – (incl bibl ref) – mf#1985-1718 – us ATLA [240]

"Essays and reviews" and the people of england : a popular refutation of the principal propositions of the essayists – London, England: Houlston & Wright 1861 – 1r – 1 – us UF Libraries [242]

"Essays and reviews" considered : in relation to the current principles and fallacies of the day / Woodgate, Henry Arthur – London: Saunders, Otley, 1861 [mf ed 1985] – xx/155p on 1mf – 9 – 0-8370-5748-5 – (incl bibl ref) – mf#1985-3748 – us ATLA [242]

"Essays and reviews" considered – Toronto?: s.n, 1862 – 1mf – 9 – mf#62247 – cn Canadiana [240]

Essays and soliloquies / Unamuno, Miguel De – New York, NY. 1925 – 1r – us UF Libraries [080]

Essays and speeches / Dawe, Charles G – Boston/NY: Houghton Mifflin, 1915 – 5mf – 9 – $7.50 – (covers anti-trust law and policy, the federal reserve system and banking reform at the turn of the century) – mf#LLMC 96-055 – us LLMC [332]

Essays and speeches / Lilly, William Samuel – London: Chapman & Hall, 1897 – 1mf – 9 – 0-8370-6996-3 – (incl ind) – mf#1986-0996 – us ATLA [080]

Essays and studies / Sinker, Robert – Cambridge: Deighton, Bell; London: George Bell, 1900 – 1mf – 9 – 0-8370-9985-4 – mf#1986-3985 – us ATLA [220]

Essays by german officers and officials – 1991 – 7r – 1 – $910.00 – (incl printed guide) – mf#S3212 – U.S. Naval Historical Center – us Scholarly Res [355]

Essays chiefly on questions of church and state from 1850 to 1870 / Stanley, Arthur Penrhyn – new ed. London: J Murray, 1884 – 1mf – 9 – 0-7905-8910-9 – (incl bibl ref) – mf#1989-2135 – us ATLA [240]

Essays chiefly on the original texts of the old and new testaments / Abbott, Thomas Kingsmill – London: Longmans, Green, 1891 – 1mf – 9 – 0-8370-2031-X – (incl ind) – mf#1985-0031 – us ATLA [220]

Essays chiefly on the science of language / Mueller, Friedrich Max – London: Longmans, Green, 1875 [mf ed 1994] – (= ser Chips from a german workshop 4) – 2mf – 9 – 0-524-08896-9 – mf#1993-4031 – us ATLA [410]

Essays commercial and political : on the real and relative interests of imperial dependent states, particularly those of great britain and her dependencies – Newcastle England: Printed by T Saint...sold by J Johnson...London, 1777 – 2mf – 9 – mf#20540 – cn Canadiana [080]

Essays contributed to the 'quarterly review' / Wilberforce, Samuel – London, 1874 – (= ser 19th c evolution & creation) – 2v on 9mf – 9 – mf#1.1.10896 – uk Chadwyck [200]

The essays, debates, and proceedings / Free Lutheran Diet (2nd: 1878: Philadelphia, Pa); ed by Baum, William M & Kunkelman, J A – Philadelphia: Lutheran Bookstore, 1879 – 1mf – 9 – 0-524-02466-9 – mf#1990-4325 – us ATLA [240]

Essays, descriptive and moral : on scenes in italy, switzerland, and france / Bruen, Matthias – Edinburgh 1823 [mf ed Hildesheim 1995-98] – 1v on 2mf – 9 – €60.00 – 3-487-27755-7 – gw Olms [840]

Essays ecclesiastical and social / Conybeare, William John – London: Longman, Brown, Green, and Longmans, 1855 – 2mf – 9 – 0-524-06467-9 – mf#1990-5241 – us ATLA [240]

Essays ethnological and linguistic / ed by Kennedy, James – London: Williams & Norgate, 1861 [mf ed 1986] – 1mf – 9 – 0-8370-8263-3 – mf#1986-2263 – us ATLA [400]

Essays for sunday reading / Caird, John – London: Pitman, 1906 – 1mf – 9 – 0-7905-3810-5 – mf#1989-0303 – us ATLA [240]

Essays for the day / Munger, Theodore Thornton – Boston: Houghton, Mifflin, 1904 – 1mf – 9 – 0-8370-4013-2 – mf#1985-2013 – us ATLA [840]

Essays for the times : studies of eminent men and important living questions / Dewart, Edward Hartley – S.l: s.n, 1898? – 1mf – 9 – mf#56749 – cn Canadiana [070]

Essays from reviews / Stewart, George – Quebec: Dawson, 1892 – 2mf – 9 – mf#35714 – cn Canadiana [420]

Essays from the edinburgh and quarterly reviews : with addresses and other pieces / Herschel, John Frederick William – London: Longman, Brown, Green, Longmans, & Roberts 1857 [mf ed 1998] – 1r – 1 – mf#film mas 28218 – us Harvard [520]

Essays historical and theological / Mozley, James Bowling – London: Rivingtons, 1878 – 3mf – 9 – 0-7905-5725-8 – mf#1988-1725 – us ATLA [240]

Essays in application / Dyke, Henry van – Toronto: Copp, Clark, 1905 – 4mf – 9 – 0-665-86592-9 – mf#86592 – cn Canadiana [840]

Essays in biblical greek / Hatch, Edwin – Oxford: Clarendon Press, 1889 – 1mf – 9 – 0-8370-9954-4 – (incl ind) – mf#1986-3954 – us ATLA [220]

Essays in criticism – Oxford. 1951+ (1) 1973+ (5) 1976+ (9) – ISSN: 0014-0856 – mf#2505 – us UMI ProQuest [420]

Essays in fallacy / Macphail, Andrew – New York: Longmans, Green, 1910 – 5mf – 9 – 0-665-75883-9 – mf#75883 – cn Canadiana [305]

Essays in history – Charlottesville. 1972-1980 (1) 1977-1980 (5) 1977-1980 (9) – ISSN: 0071-1411 – mf#7946 – us UMI ProQuest [900]

Essays in international economics – Princeton. 2000-2001 (1,5,9) – mf#8016,01 – us UMI ProQuest [332]

Essays in international finance – Princeton. 1943+ (1) 1975+ (5) 1975+ (9) – ISSN: 0071-142X – mf#8016 – us UMI ProQuest [332]

Essays in jurisprudence and ethics / Pollock, Frederick – London: Maxmillan & Co, 1882 – 5mf – 9 – $7.50 – mf#LLMC 95-183 – us LLMC [340]

Essays in jurisprudence and legal history / Salmond, John W – London: Stevens & Haynes, 1891 – 3mf – 9 – $4.50 – mf#LLMC 95-161 – us LLMC [340]

Essays in legal history : read before the international congress of historical studies held in london in 1913 / Vinogradoff, Paul – London: OUP, 1913 – 5mf – 9 – $7.50 – mf#LLMC 95-146 – us LLMC [340]

Essays in literary interpretation / Mabie, Hamilton Wright – Toronto: Morang, 1905 – 3mf – 9 – 0-665-74997-X – mf#74997 – cn Canadiana [410]

Essays in literature – Denver. 1973-1974 (1) – mf#7831 – us UMI ProQuest [400]

Essays in literature – Macomb. 1989-1996 (1,5,9) – ISSN: 0094-5404 – mf#18004 – us UMI ProQuest [400]

Essays in mexican history / Texas University Institute Of Latin American Studies – Austin, TX. 1958 – 1r – us UF Libraries [972]

Essays in modern theology and related subjects : gathered and published as a testimonial to charles augustus briggs, d.d., d. litt., graduate professor of theological encyclopaedia and symbolics in the union theological seminary in the city of new york – New York: Charles Scribner, 1911 – 1mf – 9 – 0-7905-0067-1 – mf#1987-0067 – us ATLA [240]

Essays in national idealism / Coomaraswamy, Ananda Kentish – Madras: GA Natesan and Co, [1911] – (= ser Samp: indian books) – us CRL [954]

Essays in occultism, spiritism, and demonology / Harris, William Richard – Toronto: McClelland, Goodchild & Stewart, 1919 – 3mf – 9 – 0-665-74445-5 – mf#74445 – cn Canadiana [130]

Essays in orthodoxy / Quick, Oliver Chase – London: Macmillan, 1916 – 1mf – 9 – 0-7905-9594-X – mf#1989-1319 – us ATLA [240]

Essays in pentateuchal criticism / Wiener, Harold Marcus – Oberlin, OH: Bibliotheca Sacra, 1909 – 1mf – 9 – 0-8370-5838-4 – (incl indes) – mf#1985-3838 – us ATLA [221]

Essays in philosophy : old and new / Knight, William Angus – Boston: Houghton, Mifflin, 1890 – 1mf – 9 – 0-7905-7880-8 – mf#1989-1105 – us ATLA [100]

Essays in political arithmetic / Kerssebomn, W – The Hague, 1748 – 1r – 1 – mf#95693 – uk Microform Academic [900]

Essays in politics / Macphail, Andrew – London: Longmans, Green, 1909 – 4mf – 9 – 0-665-75904-5 – mf#75904 – cn Canadiana [320]

Essays in puritanism / Macphail, Andrew – Boston: Houghton, Mifflin, 1905 – 4mf – 9 – 0-665-77823-6 – mf#77823 – cn Canadiana [242]

Essays in radical empiricism / James, William – New York: Longmans, Green, 1912 – 1mf – 9 – 0-7905-7590-6 – mf#1989-0815 – us ATLA [100]

Essays in taxation / Seligman, Edwin Robert Anderson – London, Toronto: Macmillan, 1915 – 8mf – 9 – 0-665-65558-4 – (inlc publ list) – mf#65558 – cn Canadiana [336]

Essays in the constitutional history of the united states in the formative period, 1775-1789 : by graduates and former members of the johns hopkins university / ed by Jameson, J Franklin – Boston/New York: Houghton, Mifflin & Co, 1889 – 4mf – 9 – $6.00 – mf#LLMC 95-071 – us LLMC [323]

Essays in the financial history of canada / McLean, James Alexander – New York: Columbia College, 1894 – 1mf – 9 – (p1-20 in mss form and are not incl in vol) – mf#26396 – cn Canadiana [332]

Essays in the history of religious thought in the west / Westcott, Brooke Foss – London: Macmillan, 1891 [mf ed 1984] – (= ser Biblical crit – us & gb 99) – 5mf – 9 – 0-8370-0674-0 – (incl bibl ref) – mf#1984-1099 – us ATLA [200]

Essays in the study of folk-songs / Martinengo-Cesaresco, Evelyn Lilian Hazeldine – London: G. Redway, 1886.395p – 1 – us Wisconsin U Libr [390]

Essays in the study of folk-songs *see* The soviets at work – the international position of the russian soviet republic and the fundamental problems of the socialist revolution

Essays indian and islamic / Khuda Bukhsh, Salahuddin – London: Probsthain, 1912 – (= ser Probsthain's oriental series) – 3mf – 9 – 0-524-00915-5 – (incl bibl ref) – mf#1990-2138 – us ATLA [260]

Essays, lectures, etc : upon select topics in revealed theology / Taylor, Nathaniel William – New York: Clark, Austin & Smith 1859 [mf ed 1990] – 2mf – 9 – 0-7905-9547-8 – mf#1989-1252 – us ATLA [240]

Essays medical and philosophical / Martine, George – London, 1740 – 1 – us Wisconsin U Libr [840]

Essays, moral and religious / Thomson, Edward; ed by Clark, Davis Wasgatt – Cincinnati: Hitchcock & Walden, 1868, c1856 – 1mf – 9 – 0-524-00173-1 – mf#1989-2873 – us ATLA [240]

Essays moral, political, and literary / Hume, David – London, England. v1-2. 1875 – 1r – us UF Libraries [080]

Essays moral, political, and literary / Hume, David – London, England. v2. 1875 – 1r – us UF Libraries [080]

Essays of an americanist / Brinton, Daniel Garrison – Philadelphia: Porter & Coates, 1890 – 6mf – 9 – (incl ind) – mf#27804 – cn Canadiana [305]

Essays of arthur schopenhauer – New York, NY. 192-? – 1r – us UF Libraries [100]

The essays of the literary society : read in montreal during the winter of 1880-81, at the houses of some of its members – Montreal?: s.n, 1881 – 1mf – 9 – mf#55189 – cn Canadiana [420]

Essays of the london architectural society / Architectural Society, London – London 1808 – (= ser 19th c art & architecture) – 2mf – 9 – mf#4.2.1405 – uk Chadwyck [720]

Essays of the times : canada in the 9th decade of the 19th century by vilccxxviii / Armstrong, William Reginald – S.l: s.n, 1887? – 1mf – 9 – mf#52336 – cn Canadiana [240]

Essays on archaeological subjects : and on various questions connected with the history of art, science and literature in the middle ages / Wright, Thomas – London: J R Smith, 1861 – 2v – 1 – us Wisconsin U Libr [930]

Essays on art / Carr, Joseph William Comyns – London 1879 – (= ser 19th c art & architecture) – 3mf – 9 – mf#4.2.1367 – uk Chadwyck [700]

Essays on canadian writing – Toronto. 1990+ (1,5,9) – ISSN: 0316-0300 – mf#17534 – us UMI ProQuest [410]

Essays on catholic life / O'Hagan, Thomas – Baltimore: J Murphy, c1916 – 1mf – 9 – 0-665-77720-5 – mf#77720 – cn Canadiana [241]

Essays on ceremonial – London: De la More Press 1904 [mf ed 1985] – (= ser The library of liturgiology and ecclesiology for english readers 4) – 1mf – 9 – 0-8370-5843-0 – (incl bibl ref & ind) – mf#1985-3843 – us ATLA [242]

Essays on darwinism / Stebbing, Thomas Roscoe Rede – London, 1871 – (= ser 19th c evolution & creation) – 3mf – 9 – mf#1.1.1513 – uk Chadwyck [575]

Essays on education together with the town reports for 1859-1861 : and the school reports for 1861-1862 of concord, massachusetts / Alcott, Amos Bronson – 1830-1862 – 9 – us Scholars Facs [370]

Essays on educational reformers / Quick, Robert Hebert – New York: D Appleton, 1897, c1890 – 2mf – 9 – 0-8370-7576-9 – (incl bibl ref and index) – mf#1986-1576 – us ATLA [370]

Essays on educational subjects / May, John – Ottawa?: s.n, 1880 – 1mf – 9 – mf#09851 – cn Canadiana [370]

Essays on faith and immortality / Tyrrell, George – New York: Longmans, Green, 1914 [mf ed 1990] – xv/277p on 1mf – 9 – 0-7905-7481-0 – mf#1989-0706 – us ATLA [240]

Essays on freethinking and plainspeaking / Stephen, Leslie – New York: G P Putnam, 1905 [mf ed 1985] – 2mf – 9 – 0-8370-5399-4 – (with int essays by james bryce & herbert paul) – mf#1985-3399 – us ATLA [210]

Essays on german literature / Boyesen, Hjalmar Hjorth – New York: C Scribner's Sons 1892 [mf ed 1992] – 1r – 1 – (filmed with: historical chart of english literature for use in schools and colleges) – mf#3160p – us Wisconsin U Libr [430]

Essays on gothic architecture / Warton, Thomas et al – [3rd ed]. London 1808 – (= ser 19th c art & architecture) – 3mf – 9 – mf#4.2.729 – uk Chadwyck [720]

Essays on grace, faith and experience : wherein several gospel truths are stated and illustrated and their opposite errors pointed out – Pictou NS: Printed for J & A Milne, 1832 – 4mf – 9 – mf#64141 – cn Canadiana [210]

Essays on history, biography, geography, engineering, etc / Ellesmere, Francis Egerton, Earl of – London: J Murray, 1858 – 6mf – 9 – mf#34207 – cn Canadiana [080]

Essays on history, philosophy, and theology / Vaughan, Robert – London: Jackson and Walford, 1849 – 2mf – 9 – 0-7905-6141-7 – mf#1988-2141 – us ATLA [240]

Essays on india : written in the intervals of travel and delivered as addresses on various occasions throughout canada / Sing, Saint N – [London, Ont?: s.n.], 1907 – 1mf – 9 – 0-665-88113-4 – mf#88113 – cn Canadiana [840]

Essays on indian art, industry and education / Havell, Ernest Binfield – Madras: GA Natesan & Co, [1915] – (= ser Samp: indian books) – us CRL [840]

Essays on Islam / Sell, Edward – Madras: SPCK Depaot, 1901 – 1mf – 9 – 0-524-01628-3 – mf#1990-2567 – us ATLA [260]

Essays on labour law in the province of quebec / Spector, John Jacob – Montreal? 1952? – 54p – 1 – mf#LL-2223 – us L of C Photodup [344]

Essays on literature, biography, and antiquities / Mueller, Friedrich Max – London: Longmans, Green, 1870 [mf ed 1994] – (= ser Chips from a german workshop 3) – 2mf – 9 – 0-524-08897-7 – mf#1993-4032 – us ATLA [410]

Essays on liturgiology and church history / Neale, John Mason – London: Saunders, Otley, 1863 – 2mf – 9 – 0-524-06474-1 – mf#1990-5248 – us ATLA [240]

Essays on lozi land and royal property / Gluckman, M – (= ser Institute for social research, university of zambia. papers 10) – 2mf – 7 – mf#363/8 – uk Microform Academic [960]

Essays on milton / Thompson, Elbert Nevius Sebring – New Haven: Yale University Press, 1914 – 1r – 1 – (theme of paradise lost is repr fr publ of the modern language association of america) – us Wisconsin U Libr [840]

Essays on mugul art / Solomon, William Ewart Gladstone – London; New York: Humphrey Milford: Oxford University Press, 1932 – (= ser Samp: indian books) – us CRL [700]

Essays on museums : and other subjects connected with natural history / Flower, William Henry – London, 1898 – (= ser 19th c evolution & creation) – 5mf – 9 – mf#1.1.6678 – uk Chadwyck [500]

Essays on mythology, traditions, and customs / Mueller, Friedrich Max – 2nd ed. London: Longmans, Green, 1868 [mf ed 1994] – (= ser Chips from a german workshop 2) – 1mf – 9 – 0-524-08898-5 – mf#1993-4033 – us ATLA [390]

Essays on pentateuchal criticism by various writers : no 1: introductory / Chambers, Talbot Wilson – New York: Funk & Wagnalls, 1887 – 1mf – 9 – 0-7905-0424-3 – mf#1987-0424 – us ATLA [221]

Essays on practical husbandry : addressed to the canadian farmers: shewing the method to cultivate and improve the soil, the advantages of rotation crops... / Grece, Charles Frederick – Montreal: printed by William Gray, 1817 [mf ed 1985] – 2mf – 9 – 0-665-10388-3 – mf#10388 – cn Canadiana [630]

Essays on questions of the day : political and social / Smith, Goldwin – New York, London: MacMillan, 1894 – 5mf – 9 – mf#52698 – cn Canadiana [300]

Essays on questions of the day, political and social / Smith, Goldwin – New York: Macmillan; Toronto: Copp, Clark, 1893 – 4mf – 9 – mf#13734 – cn Canadiana [300]

Essays on religion and literature – London: Longman, Green, Longman, Roberts, & Green, 1865 [mf ed 1986] – 1mf – 9 – 0-8370-8038-X – mf#1986-2038 – us ATLA [230]

Essays on solomon islands life : and missionary review extracts / Metcalfe, John R – c1902-1964 – 1r – 1 – mf#pmb413 – at Pacific Mss [242]

Essays on some biblical questions of the day / ed by Swete, Henry Barclay – London, New York: Macmillan, 1909 – 2mf – 9 – 0-8370-5473-7 – mf#1985-3473 – us ATLA [220]

Essays on some of the modern guides of english thought in matters of faith / Hutton, Richard Holt – New ed. London; New York: Macmillan, 1891 – 1mf – 9 – 0-524-00376-9 – mf#1989-3076 – us ATLA [840]

Essays on some of the prophecies in holy scripture which remain to... / Marsh, Edward Garrard – London, England. 1844 – 1r – us UF Libraries [220]

Essays on some of the testimonies of truth as held by the society of friends / Johnson, Jane – 4th ed. Philadelphia: Friends' Book Assoc, 1882 – 1mf – 9 – 0-524-06630-2 – mf#1991-2685 – us ATLA [840]

Essays on some theological questions of the day / Cunningham, William et al; ed by Swete, Henry Barclay – London; New York: Macmillan, 1905 – 2mf – 9 – 0-7905-9698-9 – mf#1989-1423 – us ATLA [240]

Essays on subjects connected with the reformation in england / Maitland, Samuel Roffey – London: F & J Rivington, 1849 – 2mf – 9 – 0-7905-5192-6 – mf#1988-1192 – us ATLA [242]

Essays on the anatomy of expression in painting / Bell, Charles – London 1806 – (= ser 19th c art & architecture) – 3mf – 9 – mf#4.2.52 – uk Chadwyck [750]

Essays on the bible by the author of "essays on the church" etc / Seeley, Robert Benton – London: Seeley, Jackson, & Halliday, 1869 – 1mf – 9 – 0-8370-5213-0 – (incl bibl ref) – mf#1985-3213 – us ATLA [220]

Essays on the chinese language / Watters, Thomas – Shanghai: Presbyterian Mission Press, 1889 [mf ed 1995] – (= ser Yale coll) – vi/496p – 1 – 0-524-09720-8 – mf#1995-0720 – us ATLA [480]

Essays on the church in canada : the church catholic – national churches – anglican and gallican – the church in canada under french rule... / O'Sullivan, Dennis Ambrose – Toronto: Catholic Truth Society, 1890 [mf ed 1981] – 2mf – 9 – 0-665-11553-9 – (repr fr the American Catholic Quarterly Review) – mf#11553 – cn Canadiana [240]

Essays on the constitution of the united states / Ford, Paul L – Brooklyn: Historical Printing Club, 1892 – 5mf – 9 – $7.50 – (reprints of contemporary newspaper essays by 17 influential citizens) – mf#LLMC 84-809 – us LLMC [342]

Essays on the devolution of land upon the personal representative : and statutory powers relating thereto, with an appendix of statutes / Armour, Edward Douglas – Toronto: Canada Law Book Co, 1903 – 5mf – 9 – 0-665-73558-8 – mf#73558 – cn Canadiana [340]

Essays on the distinguishing traits of christian character / Spring, Gardiner – Boston: Doctrinal Tract and Book Society, 1853 – 1mf – 9 – 0-524-07593-X – mf#1991-3213 – us ATLA [240]

Essays on the english state church in ireland / Brady, William Maziere – London: Strahan, 1869 – 1mf – 9 – 0-7905-4434-2 – mf#1988-0434 – us ATLA [242]

Essays on the future destiny of nova scotia : improvement of female education in nova scotia and on peace – Halifax, NS?: s.n, 1846 – 1mf – 9 – mf#61689 – cn Canadiana [305]

Essays on the greater german poets and writers / Carlyle, Thomas – London: Walter Scott Ltd [19–?] [mf ed 1992] – 1r – 1 – (int by ernest rhys; filmed with: im urteil der dichter / arno mulot [ed]) – mf#3320p – us Wisconsin U Libr [840]

Essays on the history of missions in thailand : by educational missionaries – 294p – 1 – us Southern Baptist [240]

Essays on the intellectual powers of man / Reid, Thomas; ed by Walker, James – 10th ed. Philadelphia: EH Butler, 1864, c1850 – 2mf – 9 – 0-524-08557-9 – mf#1993-2082 – us ATLA [120]

Essays on the Irish church / Byrne, James et al – Oxford: J Parker, 1866 – 1mf – 9 – 0-524-03457-5 – mf#1990-1000 – us ATLA [240]

Essays on the languages, literature, and religion of nepal and tibet : together with further papers on the geography, ethnology, and commerce of those countries / Hodgson, Brian Houghton – London: Truebner, 1874 – 1mf – 9 – 0-524-02207-0 – (incl bibl ref) – mf#1990-2881 – us ATLA [840]

Essays on the languages of the bible and bible-translations / Cust, Robert Needham – London: Elliot Stock, 1890 – 1mf – 9 – 0-8370-7855-5 – (incl bibl ref) – mf#1986-1855 – us ATLA [220]

Essays on the passover / Frey, Joseph Samuel Christian Frederick – New-York: Moore & Payne, 1834 – 1mf – 9 – 0-524-08310-X – mf#1993-0015 – us ATLA [221]

Essays on the philosophy of theism / Ward, William George; ed by Ward, Wilfrid Philip – London: Kegan Paul, Trench, 1884 – 2mf – 9 – 0-7905-7549-3 – mf#1989-0774 – us ATLA [210]

Essays on the preaching required by the times : and the best methods of obtaining it / Stevens, Abel – New-York: Carlton & Phillips, 1856, c1855 – 1mf – 9 – 0-7905-6021-6 – mf#1988-2021 – us ATLA [240]

Essays on the present crisis in the condition of the american indians : first published in the national intelligencer / Evarts, Jeremiah [pseud: William Penn] – Philadelphia: T Kite, 1830 – 2mf – 9 – mf#55682 – cn Canadiana [305]

Essays on the primitive church offices / Alexander, Joseph Addison – New-York: Charles Scribner, 1851 – 1mf – 9 – 0-7905-0848-6 – mf#1987-0848 – us ATLA [240]

Essays on the principles of morality : and on the private and political rights and obligations of mankind / Dymond, Jonathan – abr ed. Philadelphia: Book Cttee...1896 [mf ed 1986] – 2mf – 9 – 0-8370-6111-3 – (incl ind) – mf#1986-0111 – us ATLA [170]

Essays on the punishment of death / Spear, Charles – 13th ed. Boston: self published, 1851 – 3mf – 9 – $4.50 – mf#LLMC 91-077 – us LLMC [840]

Essays on the pursuits of women : reprinted from fraser's and macmillan's magazines / Cobbe, Frances Power – London: Emily Faithfull, 1863 – 1mf – 9 – 0-7905-7709-7 – mf#1989-0934 – us ATLA [840]

Essays on the religion and philosophy of the hindus / Colebrooke, Henry Thomas – new ed. London: Williams & Norgate, 1858 [mf ed 1995] – (= ser Yale coll) – 325p – 1 – 0-524-10148-5 – mf#1995-1148 – us ATLA [280]

Essays on the re-union of christendom / Humble, Henry et al; ed by Lee, Frederick George – London: J T Hayes, 1867 – 1mf – 9 – 0-7905-8825-0 – mf#1989-2050 – us ATLA [240]

Essays on the rise and progress of the christian religion in the west of europe : from the reign of tiberius to the end of the council of trent / Russell, John Russell, Earl – London: Longmans, Green, 1873 – 1mf – 9 – 0-524-03914-3 – (incl bibl ref) – mf#1990-1173 – us ATLA [240]

Essays on the science of religion / Mueller, Friedrich Max – 2nd ed. London: Longmans, Green, 1868 [mf ed 1994] – (= ser Chips from a german workshop 1) – 1mf – 9 – 0-524-08899-3 – mf#1993-4034 – us ATLA [200]

Essays on the social gospel / Harnack, Adolf von; ed by Canney, Maurice Arthur – London: Williams & Norgate; New York: G P Putnam, 1907 – 1mf – 9 – 0-7905-1328-5 – mf#1987-1328 – us ATLA [230]

Essays on the spirit of the inductive philosophy, the unity of worlds, and the philosophy of creation / Powell, Baden – London, 1855 – (= ser 19th c evolution & creation) – 6mf – 9 – mf#1.1.11627 – uk Chadwyck [110]

Essays on the supernatural origin of christianity : with special reference to the theories of renan, strauss, and the tuebingen school / Fisher, George Park – new ed. New York: Charles Scribner's Sons, 1890 [mf ed 1984] – 8mf – 9 – 0-8370-0135-8 – (incl bibl ref) – mf#1984-0021 – us ATLA [240]

Essays on the teaching of history / Maitland, Frederic William et al; ed by Archbold, William Arthur Jobson – Cambridge: University Press, 1901 – 1mf – 9 – 0-524-03766-3 – mf#1990-1113 – us ATLA [900]

Essays on truth and reality / Bradley, Francis Herbert – Oxford: Clarendon Press, 1914 – 2mf – 9 – 0-7905-3608-0 – (incl bibl ref) – mf#1989-0101 – us ATLA [100]

Essays on various subjects / Wiseman, Nicholas Patrick – London: Charles Dolman. 3v. 1853 – 6mf – 9 – 0-8370-7354-5 – mf#1986-1354 – us ATLA [840]

Essays on work and culture / Mabie, Hamilton Wright – Toronto: G N Morang, 1898 – 3mf – 9 – 0-665-92122-5 – mf#92122 – cn Canadiana [306]

The essays or counsels civil and moral of francis bacon / ed by Clarke, George Herbert – New York: Macmillan, 1915 – (= ser Macmillan's pocket series of english classics) – 5mf – 9 – 0-665-88457-5 – (int and notes by ed. incl publ list) – mf#88457 – cn Canadiana [170]

Essays, philosophical and psychological : in honor of william james, professor in harvard university / Fullerton, George Stuart et al – New York: Longmans, Green, 1908 – 2mf – 9 – 0-7905-8792-0 – mf#1989-2017 – us ATLA [840]

Essays practical and speculative / McConnell, Samuel David – New York: T Whittaker, 1900 – 1mf – 9 – 0-7905-9506-0 – mf#1989-1211 – us ATLA [240]

Essays received in response to an appeal by the canadian institute on the rectification of parliament : together with the conditions on which the council of the institute offers to award one thousand dollars for prize essays – Toronto: Copp, Clark, 1893 [mf ed 1980] – 3mf – 9 – (incl bibl ref) – mf#01093 – cn Canadiana [325]

Essays relative to the habits, character, and moral improvement of the hindoos / Bentley, John – London 1823 [mf ed Hildesheim 1995-98] – 1v on 3mf – 9 – €90.00 – 3-487-27424-8 – gw Olms [306]

Essays, reviews, and addresses / Martineau, James – London, New York: Longmans, Green, 1890-91 [mf ed 1990] – 4v on 6mf – 9 – 0-7905-7983-9 – mf#1989-1268 – us ATLA [240]

Essays, reviews, and discourses / Whedon, Daniel Denison – New York: Phillips & Hunt; Cincinnati: Cranston & Stowe, 1887 – 1mf – 9 – 0-524-00208-8 – mf#1989-2908 – us ATLA [240]

Essays, scientific and philosophical : with memoirs of the author / Moore, Aubrey Lackington – London: K Paul, Trench, Trubner, 1890 – 4mf – 9 – mf#16829 – cn Canadiana [210]

Essays, short stories and poems : including a sketch of the author's life / Snell, M S – Chatham, Ont?: s.n, 1881 – 2mf – 9 – mf#17022 – cn Canadiana [800]

Essays submitted in writing contest / Native Sons and Daughters of Kansas – 1955 – 1 – us Kansas [840]

Essays, theological and miscellaneous, second series / Dod, Albert Baldwin – New York: Wiley and Putnam, 1847 – 2mf – 9 – 0-524-00021-2 – mf#1989-2721 – us ATLA [840]

Essays towards a history of painting / Callcott, Maria (Dundas) Graham, lady – London 1836 – (= ser 19th c art & architecture) – 3mf – 9 – mf#4.2.321 – uk Chadwyck [750]

Essays upon heredity and kindred biological problems / Weismann, August (Friedrich Leopold) – aut trans. Oxford, 1889 – (= ser 19th c evolution & creation) – 5mf – 9 – mf#1.1.10888 – uk Chadwyck [574]

Essays upon some controverted questions / Huxley, Thomas Henry – London, 1892 – (= ser 19th c evolution & creation) – 7mf – 9 – mf#1.1.9541 – uk Chadwyck [500]

Essays zur allgemeinen religionswissenschaft / Strauss und Torney, Victor von – Heidelberg: C Winter, 1879 – 1mf – 9 – 0-524-01930-4 – mf#1990-2743 – us ATLA [200]

Essays zur vergleichenden literaturgeschichte / Federn, Karl – Muenchen: G Mueller, 1904 – 1 – us Wisconsin U Libr [410]

Esse continente chamado brasil / Tourinho, Eduardo – Rio de Janeiro, Brazil. 1964 – 1r – us UF Libraries [972]

Essen, L van der *see* Etude critique et litteraire

Essence : issues in the study of ageing, dying, and death – Downsview. 1976-1982 (1,5,9) – ISSN: 0384-8833 – mf#11365 – us UMI ProQuest [618]

Essence – New York. 1970+ (1) 1972+ (5) 1975+ (9) – ISSN: 0014-0880 – mf#6835 – us UMI ProQuest [305]

Essence by mail : [catalog] – 1991 fall – 1r – 1 – mf#4888575 – us WHS [380]

The essence of buddhism / Lakshmi Narasu, Pokala – Bombay: Thacker & Co, 1948 – (= ser Samp: indian books) – (pref by b r ambedkar) – us CRL [280]

The essence of buddhism : with illustrations of buddhist art / Lakshmi Narasu, Pokala – 2nd rev enl ed. Madras: S Varadachari 1912 [mf ed 1991] – 2mf [ill] – 9 – 0-524-01777-8 – (first printed 1907) – mf#1990-2625 – us ATLA [280]

The essence of christianity : a study in the history of definition / Brown, William Adams – New York: Scribner, 1902 – 1mf – 9 – 0-7905-3557-2 – mf#1989-0050 – us ATLA [240]

The essence of christianity : a study in the history of definition / Brown, William Adams – New York: Scribner, 1902 – 1mf – us ATLA [240]

The essence of christianity = Wesen des christenthums / Feuerbach, Ludwig – London: J Chapman, 1854 – (= ser Chapman's Quarterly Series) – 1mf – 9 – 0-7905-3671-4 – (in english) – mf#1989-0164 – us ATLA [240]

Essence of hinduism / Nikhilananda, Swami – New York: Ramakrishna-Vivekananda Center, 1946 – (= ser Samp: indian books) – us CRL [280]

The essence of japanese buddhism / Tsunoda, Ryusaku – Honolulu HI: Advertiser Press 1914 [mf ed 1992] – 1mf – 9 – 0-524-03254-8 – mf#1990-3184 – us ATLA [280]

Essence of religion / Beecher Henry Ward – London, England. 1886? – 1r – us UF Libraries [240]

The essence of religion / Bowne, Borden Parker – Boston: Houghton Mifflin, 1910 – 1mf – 9 – 0-7905-3602-1 – mf#1989-0095 – us ATLA [240]

Essence of the bible – New York, NY. 1930 – 1r – us UF Libraries [220]

Essener allgemeine zeitung *see* Essener general-anzeiger

Essener arbeiter zeitung *see* Die arbeiter-zeitung

Essener arbeiter-zeitung – Essen, Germany. apr 1909-apr 1926 – 24r – 1 – (cont: arbeiter-zeitung) – us L of C Photodup [074]

Essener arbeiter-zeitung *see* Arbeiter-zeitung

Essener general-anzeiger – Essen DE, 1949 26 nov-1954 28 jun – 16r – 1 – (title varies: 1 jan 1918: essener allgemeine zeitung. with suppl) – gw Mikrofilm [074]

Essener kurier *see* Westdeutsche nachrichten

Essener nachrichten *see* National-zeitung / a

Essener stadtanzeiger – Essen DE, 1949 28 oct-1977 1 jul – 20r – 1 – gw Mikrofilm [074]

Essener tageblatt *see* Rhein-ruhr-zeitung

Essener volks-halle – Essen DE, 1849 15 apr-1850 30 jun – 1 – gw Misc Inst [074]

Essener volks-zeitung – Essen DE, 1877, 1908 sep-okt, 1909 apr-mai – 1 – (with suppl: ruhrland 1935 15 aug-1936 1 sep [1r]. cont as suppl: scholle und schacht with: essener allgemeine zeitung) – gw Misc Inst [074]

Essener zeitung *see* Allgemeine politische nachrichten

The essenes : their history and doctrines / Ginsburg, Christian David – London: Longman, Green, Longman, Roberts, and Green, 1864 – 1mf – 9 – 0-8370-9386-4 – (incl bibl ref) – mf#1986-3386 – us ATLA [240]

The essential aldred *see* The works of guy aldred

Essential christianity : a series of explanatory sermons / Hughes, Hugh Price – New York: Fleming H Revell, [1894?] [mf ed 1986] – 1mf – 9 – 0-8370-7390-1 – mf#1986-1390 – us ATLA [240]

The essential nature of law or the ethical basis of jurisprudence / Pattee, William Sullivan – Chicago: Callaghan, 1909 – 3mf – 9 – $4.50 – mf#LLMC 95-197 – us LLMC [340]

The essential of logic : being ten lectures on judgment and inference / Bosanquet, Bernard – London, New York: Macmillan, 1895 – 1mf – 9 – 0-7905-3543-2 – mf#1989-0036 – us ATLA [160]

The essential unity of all religions / Das, Bhagavan [comp] – Benares: Kashi Vidyapitha: Sole agent, Indian Book Shop, Theosophical Society, 1939 – (= ser Samp: indian books) – us CRL [230]

Essentials and non-essentials in religion : six lectures / Clarke, James Freeman – 10th ed. Boston: American Unitarian Association, 1894, c1877 – 1mf – 9 – 0-524-00710-1 – mf#1990-2038 – us ATLA [230]

Essentials in church history / Smith, Joseph Fielding – Salt Lake City, UT. 1950 – 1r – us UF Libraries [240]

The essentials of american constitutional law / Thorpe, Francis Newton – New York, Putnam, 1917. 279 p. LL-1243 – 1 – us L of C Photodup [342]

The essentials of christian belief / Fyffe, David – New York: Hodder and Stoughton, [1912?] – 1mf – 9 – 0-7905-3842-3 – mf#1989-0335 – us ATLA [240]

The essentials of federal finance : a contribution to the problem of financial re-adjustment in india / Chand, Gyan – London: Oxford University Press, 1930 – (= ser Samp: indian books) – us CRL [336]

Essentials of french pronunciation : for use as a supplementary reader in french classes – Toronto: Copp, Clark, 1908 [mf ed 1995] – 1mf – 9 – 0-665-76866-4 – mf#76866 – cn Canadiana [440]

Essentials of french pronunciation and introduction to easy reading – Toronto: Copp, Clark, c1916 [mf ed 1995] – 1mf – 9 – 0-665-76865-6 – mf#76865 – cn Canadiana [440]

Essentials of hinduism : compiled from the speeches and writings of Swami Vivekananda / Vivekananda, Swami – Almora: Advaita Ashrama, 1944 – (= ser Samp: indian books) – us CRL [280]

The essentials of indian philosophy / Hiriyanna, Mysore – London: George Allen & Unwin Ltd, 1949 – (= ser Samp: indian books) – us CRL [180]

Essentials of new testament greek / Huddilston, John Homer – New York: Macmillan, 1896, c1895 – 1mf – 9 – 0-8370-9158-6 – mf#1986-3158 – us ATLA [450]

Essentials of phonography : to accompany the isaac pitman text-book / Kennedy, Alexander Macpherson – Toronto: Central Business College [1904?] [mf ed 1994] – 1mf – 9 – 0-665-71286-3 – mf#71286 – cn Canadiana [650]

Essentials of polish / Fox, Paul – Chicago, IL. 1937 – 1r – us UF Libraries [460]

Essentials of religion briefly considered – London, England. 1824 – 1r – us UF Libraries [240]

The essentials of sanscrit grammar : with examples of parsing / Brown, Thomas Richard – [Southwick], 1841 – (= ser 19th c books on linguistics) – 2mf – 9 – mf#2.1.43 – uk Chadwyck [490]

Essentials of ymca *see* Chi-tu chiao ch'ing nien hui shih yao [ccm93]

Essentials to the principal actions in tort at common law / Schermerhorn, Holden Bovee – Philadelphia: Rees, Welsh, 1913 – 281p – 1 – mf#LL-1127 – us L of C Photodup [346]

Essequie del serenissimo don francesco medici gran duca di toscana 2 / Strozzi, G B – Fiorenza, 1587 – 1mf – 9 – mf#O-1110 – ne IDC [090]

Essequie della sacra cattolica e real maest... / Altoviti, Giovanni – Firenze: Nella stamperia di Bartolommeo Sermartelli e fratelli, 1612 – 1mf – 9 – mf#O-1517 – ne IDC [090]

Essequie della sacra cattolica real maesta del re di spagna don filippo 2. d'austria : celebrate in firenze dalla nobilissima nazione spagnuola / Biondi, A – Fiorenza: Filippo Giunti, 1599 – 1mf – 9 – mf#O-88 – ne IDC [090]

Esser, Ruth Christa *see* Hiv-infizierte monozyten/makrophagen

Esser, Thomas *see* Die lehre des hl. thomas von aquino ueber die moeglichkeit einer anfanglosen schoepfung

Essex 1708-1849 – Oxford, MA (mf ed 1996) – (= ser Massachusetts vital record transcripts to 1850) – 4mf – 9 – 0-87623-248-9 – (Mf 1T: Deaths 1819-25; Births 1799-1825. Mf 1T-2T: Intents & Marriages 1819-49. Mf 2T-3T: Births 1819-43; Deaths 1829-43. Mf 3T: Out-of-Town Marriages 1708-99; Births 1843-49. Mf 4T: Marriages & Deaths 1843-49) – us Archive [978]

Essex 1819-1892 – Oxford, MA (mf ed 1990) – (= ser Massachusetts vital records) – 23mf – 9 – 0-87623-106-7 – (Mf 1-4: Record Book 1819-43. Mf 5-9: Town Records 1819-35. Mf 10-15: Town Records 1835-52. Mf 16: Births 1843-60. Mf 17: Births 1860-61; Marriages 1843-54; Deaths 1843-55. Mf 18-19: Deaths 1856-92. Mf 20-21: Marriages 1854-92. Mf 22-23: Births 1862-92) – us Archive [978]

Essex and herts mercury *see* Kent and essex mercury

Essex and middlesex guardian – London, UK. 9 jan-25 dec 1897 – 1r – 1 – (aka: essex guardian) – uk British Libr Newspaper [072]

Essex antiquarian – v1-7 [1897-1903], v8-13 [1904-09] – 2r – 1 – mf#1668376 – us WHS [071]

Essex Bar Association. Salem, Massachusetts *see* Misuse and abuse of the right of petition for the removal of judicial officers

Essex county chronicle – Chelmsford, England. jun-dec 1832; 1847; 1873; 1877; 1888-89; 1897-99; 1901-16; 1950; 1980- – 98+ r – 1 – uk British Libr Newspaper [072]

Essex gazette – Salem MA. 1768 aug 2-1775 may 2 – 1r – 1 – (continued by: new-england chronicle, or, the essex gazette) – mf#780680 – us WHS [071]

Essex genealogist / Essex Society of Genealogists – v1 n1-v5 n4 [1981 feb-1985 nov] – 1r – 1 – mf#974284 – us WHS [929]

Essex guardian *see* Essex and middlesex guardian

Essex herts and kent mercury *see* Kent and essex mercury

Essex Institute *see* Historical collections of the essex institute

Essex institute historical collections – Salem. 1859-1993 (1) 1971-1993 (5) 1977-1993 (9) – (cont by: peabody essex museum collections) – ISSN: 0014-0953 – mf#3464 – us UMI ProQuest [900]

Essex institute historical collections – v50 [1914] – 1r – 1 – (cont: historical collections of the essex institute; cont by: peabody essex museum collections) – mf#2520061 – us WHS [978]

Essex Society of Genealogists *see* Essex genealogist

Essex times – Havering, UK. 8, 12 dec 1866; 4 may-28 dec 1867; 1869-1871; 1873; 1875-1896; 1898-1911; 31 jan 1912-30 oct 1937 – 103 1/2r – 1 – (incorp with: romford times after 30 oct 1937) – uk British Libr Newspaper [072]

Essex-county historical and genealogical register – v1, v2 n1-7 [1894 jan-1895 jul] – 1r – 1 – mf#1056037 – us WHS [978]

Essig, Hermann *see*
– Des kaisers soldaten
– Der held vom wald

Essig, Montgomery Ford *see* The churchmember's guide and complete church manual

[Esslaoui, Ahmed Ennasiri] *see* Chronique de la dynastie alaouie du maroc

Esslinger allgemeine – Esslingen a. Neckar DE, 1949 16 jul-1955 30 apr – 12r – 1 – gw Misc Inst [074]

Esslinger schnellpost *see* Esslinger schnellpost 1843

Esslinger schnellpost 1843 – Esslingen a. Neckar DE, 1847 3 jul-1857 – 4r – 1 – (title varies: 1850: esslinger tagblatt; 1851: esslinger schnellpost; 19 aug 1854: esslinger wochenblatt) – gw Misc Inst [074]

Esslinger tagblatt *see* Esslinger schnellpost 1843

Esslinger wochenblatt *see* Esslinger schnellpost 1843

Esslinger zeitung – Esslingen a. Neckar DE, 1976- – ca 8r/yr – 1 – gw Misc Inst [074]

Esson, Henry *see* A sermon preached in the presbyterian church, st gabriel street, montreal

Essor / Jeunesses national-populaires – n27-37. Paris. mars-aout 1944 – 1 – (lacking: n35) – fr ACRPP [073]

L'essor de la litterature latine au 12th siecle / Ghellinck, J de – Bruxelles. v1-2. 1946 – v1 5mf v2 8mf – 8 – €25.00 – ne Slangenburg [450]

L'essor du congo – Elisabethville: [s.n.], feb 1960-may 1960 – us CRL [079]

L'essor litteraire and scientifique – Port-au-Prince: Impr. Verrolot, [1912-]. [1ere annee, n1-3eme annee, n27; 3eme annee, 2s, n1-n8 (15 avr 1912-15 juin 1914; mars-oct 1916] – 17r – 1 – us CRL [079]

Az est – Budapest, Hungary 24 jun 1914-14 may 1919; 28 sep 1919-29 jun 1924 – 30r – 1 – uk British Libr Newspaper [077]

L'est central – [Montreal]: Publications associees. v1 n1 30 oct 1947-v23 n43 30 nov 1971 (wkly) [mf ed 2000] – 1 – cn Bibl Nat [071]

L'est ouvrier et paysan / Parti Communiste. Nancy – Nancy. mai 1933-sept 1935 – 1 – fr ACRPP [335]

Esta es guatemala / Juarez Y Aragon, J Fernando – Guatemala, 1950 – 1r – 1 – us UF Libraries [972]

Esta es mi historia / Nunez Portuondo, Ricardo – Habana, Cuba. 1953 – 1r – us UF Libraries [972]

Esta es mi tierra / Flores, Saul – San Salvador, El Salvador. 1948 – 1r – us UF Libraries [972]

Esta es plasencia. 2nd congreso de estudios extremenos. jaime j jimenez garcia – Plasencia: Imp. La Victoria, 1970 – 1 – sp Bibl Santa Ana [946]

Esta noche juega el joker / Sierra Berdecia, Fernando – San Juan, Puerto Rico. 1956 – 1r – us UF Libraries [972]

Esta tierra de gracia / Pardo, Isaac J – Caracas, Venezuela. 1955 – 1r – us UF Libraries [972]

O estabanado : jornal litterario satyrico e illustrado – Recife, PE: Typ Americana, 26 dez 1875 – (= ser Ps 19) – mf#P16,01,69 – bl Biblioteca [870]

Establecimiento de banos mineromedicinales de alange / Gaztelu, Teodoro – Madrid: imp la editora, 1912 – 1 – sp Bibl Santa Ana [946]

Establecimiento...nueva espana / Riva Palacio, Vicente – 1892 – 9 – sp Bibl Santa Ana [972]

Established church : the best means of providing for the pastoral car... / Calvert, Thomas – London, England. 1834 – 1r – us UF Libraries [240]

Established church / Selborne, Roundell Palmer – London, England. 1871 – 1r – us UF Libraries [240]

Established church in england and wales – London, England. 187-? – 1r – 1 – us UF Libraries [240]

Established church versus the "liberation society" / Nevile, Christopher – London, England. 1863 – 1r – us UF Libraries [240]

The establishment and consolidation of imperial government in southern nigeria, 1891-1904 : theory and practice in a colonial protectorate / Anene, J C O – 1952 – us CRL [960]

The establishment and development of youth organizations in the evangelical church 1880-1922 / Plymire, Larry – 1r – 1 – $35.00 – mf#um-316 – us Commission [242]

Establishment of a diocesan clergy retiring pension fund / Fitzgerald, Augustus O – Bath, England. 1874 – 1r – us UF Libraries [240]

The establishment of a great imperial intelligence union as a means of promoting the consolidation of the empire : an address delivered...july 20, 1906 / Fleming, Sandford – [Edinburgh?: s.n, 1906?] [mf ed 1996] – 1mf – 9 – 0-665-78108-3 – mf#78108 – cn Canadiana [327]

Establishment of a service centre by the rural aged / Ramokgopa, Mapula Daphne – Uni of South Africa 2000 [mf ed Johannesburg 2000] – 2mf [ill] – 9 – mf#mfm14928 – sa Unisa [360]

The establishment of blood pressure norms for the indiana university adult fitness program / Southwick, Nancy L – 1996 – 1mf – 9 – $4.00 – mf#PH 1562 – us Kinesology [612]

The establishment of christianity and the proscription of paganism / Huttmann, Maude Aline – New York: Columbia University: Longmans, Green [distributor], 1914 – (= ser Studies in History, Economics and Public Law) – 1mf – 9 – 0-7905-4232-3 – (incl bibl ref) – mf#1988-0232 – us ATLA [240]

Establishment of everglades national park, florida / United States Congress Senate Committee On Publ... – Washington, DC. 1932 – 1r – 1 – us UF Libraries [639]

Establishment of everglades national park hearing / United States Congress. House Committee On Publi... – Washington, DC. 1931 – 1r – us UF Libraries [639]

Establishment of the church in england / Hicks, J W – Cambridge, England. 1885? – 1r – 1 – us UF Libraries [242]

Establishment principle as now interpreted / Wylie, J A – Edinburgh, Scotland. 1870 – 1r – us UF Libraries [240]

Establishment shewn to be "laid prostrate at the feet of the civil... / M'culloch, James Melville – Perth, Australia. 1846 – 1r – us UF Libraries [240]

Establishment with grace – London, England. 1825 – 1r – us UF Libraries [240]

L'establissement d'issiny, 1682-1702 : voyages de ducasse, tibierge et d'amon a la cote de guinee – Paris: E Larose, 1935 – 1 – us CRL [916]

Establissements francais de l'oceanie : arretes du gouverneur, commissaire du roi, 1845-1855 – 1r – 1 – mf#PMB Doc412 – at Pacific Mss [324]

Estacada news – Estacada OR: H A Williams, 1904-08 [wkly] – 1 – (cont by: estacada progress (1908-16)) – us Oregon Lib [071]

Estacada press – Estacada OR: Cascade Pub, 1955- [wkly] – 1 – us Oregon Lib [071]

Estacada progress – Estacada OR: Estacada Progress Inc, 1908-16 [wkly] – 1 – (cont: estacada news; cont by: eastern clackamas news. 1916 incl newspaper pub by estacada high school students) – us Oregon Lib [071]

Estacada's clackamas county news – Estacada OR: James Pub Co, 1991- [wkly] – 1 – (cont: clackamas county news (estacada, or: 1976)) – us Oregon Lib [071]

Estacada's clackamas county news (estacada, or) – Estacada OR: [s.n.] -1976 [wkly] – 1 – (began in 1957. cont by: clackamas county news (estacada, or); cont by: clackamas county news (estacada, or: 1976)) – us Oregon Lib [071]

Estacao sportiva – Rio de Janeiro, RJ. 01 abr-01 jul 1911 – (= ser Ps 19) – mf#DIPER – bl Biblioteca [790]

Estacio na guanabara / Orciuoli, Henrique – Rio de Janeiro, Brazil. 1964 – 1r – us UF Libraries [972]

Estadista da republica / Arinos De Melo Franco, Afonso – Rio de Janeiro, Brazil. v1-3. 1955 – 1r – us UF Libraries [972]

Estadista do imperio / Nabuco, Joaquim – Sao Paulo, Brazil. v1-4. 1949 – 1r – us UF Libraries [972]

Estadistica – Washington. 1943-1989 (1) 1969-1989 (5) 1971-1989 (9) – ISSN: 0014-1135 – mf#1002 – us UMI ProQuest [310]

Estadistica anual 1908-1911 / Panama. Direccion General de Estadistica – (= ser Latin american & caribbean...1821-1982) – 5mf – 9 – (1911 not available) – uk Chadwyck [318]

Estadistica comercial *see* Anuario estadistico 1911-1965

Estadistica comercial de la republica de chile / Chile. Oficina Central de Estadistica – 1844-1915 – 1 – (lacks: 1890, 1907, 1909-10) – us L of C Photodup [318]

Estadistica de la emigracion e inmigracion de espana en 1882-1895 / Instituto Geografico – Madrid, 1882-1895 – 22mf – 9 – sp Cultura [946]

Estadistica de la Navegacion *see* Estadistica de la navegacion exterior de espana en 1924-1926

Estadistica de la navegacion exterior de espana en 1924-1926 / Estadistica de la Navegacion – Madrid, 1927 – 22mf – 9 – sp Cultura [380]

Estadistica de los alcantaristas y pascualistas hacia 1567-1570. memoria de las casas de la provincia de sant josseph / Meseguer Fernandez, Juan – Madrid: Graf. Calleja, 1970 – 1 – sp Bibl Santa Ana [240]

Estadistica de los ferrocarriles en exploitacion / Argentine Republic. Direccion general de ferrocarriles – v.1-15. 1892-1906 – 1 – (lacking: v6) – us L of C Photodup [324]

Estadistica de sangre y de gloria / Bayle, Constantino – Madrid: Razon y Fe, 1939 – 1 – sp Bibl Santa Ana [946]

Estadistica del comercio y de la navegacion de la republica argentina...ano 1885-1892 – Buenos Aires, 1886-1893 – 50mf – 9 – sp Cultura [380]

Estadistica general del comercio de cabotaje entre los puertos de la peninsula e i baleares – 1857-1920 – 594mf – 9 – sp Cultura [380]

Estadistica general del comercio exterior de espana : prov de barcelona...en 1922-1925 – Madrid, 1925 – 13mf – 9 – sp Cultura [380]

Estadistica general del comercio exterior de espana con sus posesiones...en 1857-1926 – Madrid, 1858-1927 – 939mf – 9 – sp Cultura [380]

Estadistica judicial de la isla de puerto-rico, 1877, 1879, 1880 – Puerto-Rico. 3v. 1878-81 – 3mf – 9 – $4.50 – mf#LLMC 92-328 – us LLMC [340]

Estadistica mortuoria de la ciudad de buenos aires / Instituto Geografico – Buenos Aires, 1869-1877 – 2mf – 9 – sp Cultura [304]

Estadistica municipal / Barcelona. (City). Ayuntamiento – 1958-1968 – 1 – us NY Public [350]

Estadisticas del producto e ingreso nacional, 1925 / Tosco, Manuel – Tegucigalpa, Mexico. 1954 – 1r – us UF Libraries [972]

Estadisticas financieras internacionales – Washington. 1976-1992 (1) 1976-1980 (5) 1976-1980 (9) – ISSN: 0252-3078 – mf#6539 – us UMI ProQuest [310]

Estadisticas sangrientas : las victimas del clero secular / Bayle, Constantino – Madrid: Razon y Fe, 1940 – sp Bibl Santa Ana [240]

El estado – Tegucigalpa, Honduras. 1 jul 1904-21 mar 1907 – 3r – 1 – uk British Libr Newspaper [079]

O estado : orgam do partido republicano dederalista – Desterro, SC. 1892-1893 – (= ser Ps 19) – mf#UFSC/BPESC – bl Biblioteca [325]

O estado : orgao dos interesses do departamento – Tarauaca, AC. 29 jan-20 set 1914 – (= ser Ps 19) – mf#P11A,07,12 – bl Biblioteca [079]

O estado : orgao republicano – Maceio, AL: [s.n.] 15 nov 1891; jan-04 fev 1892 – (= ser Ps 19) – mf#P11,01,04 – bl Biblioteca [325]

Estado actual 1939 / Academia De Ciencias Medicas, Fisicas Y Naturales – Habana, Cuba. 1939 – 1r – us UF Libraries [972]

Estado autoritario e a realidade nacional / Amaral, Azevedo – Rio de Janeiro, Brazil. 1938 – 1r – us UF Libraries [972]

Estado cristiano y boliveriano del 13 de junio / Canal Ramirez, Gonzalo – Bogota, Colombia. 1955 – 1r – us UF Libraries [972]

El estado de colima : periodico oficial del gobierno constitucional / Colima. Mexico. (State) – Colima, 1871-1949 – 1 – mf#LL-02019 – us L of C Photodup [342]

Estado de extremadura...isabel la catolica / Barrantes Moreno, Vicente – 1872 – 9 – sp Bibl Santa Ana [946]

Estado de goyaz : orgam do partido republicano federal – Goias, 06 jun 1891-dez 1893; jan-mar, maio-jun, ago-dez 1894; jan 1895-04 jun 1896 – (= ser Ps 19) – mf#P11B,06,04 – bl Biblioteca [325]

O estado de goyaz : orgam do partido republicano federal – Goias, 06 jun 1891-dez 1893; jan-mar, maio-jun, ago-dez 1894; jan 1895-04 jun 1896 – (= ser Ps 19) – bl Biblioteca [079]

El estado de jalisco / Jalisco. Mexico – Guadalajara. aug 1946-nov 1956- – us NY Public [342]

Estado de las cuentas de participacion en los fondos sociales de los senores mutualistas con polizas en vigor en 31 de diciembre de 1962 / Mutua Aseguradora de Transportistas – Caceres: Imp. Moderna, 1963 – 1 – sp Bibl Santa Ana [946]

O estado de sao paulo – Sao Paulo. Brazil. 4 Jan 1875-Aug 1956 – 353r – 1 – us L of C Photodup [079]

O estado de sao paulo – Sao Paulo Brazil, Feb 1940; 9 jun, 31 jul, aug 1954; 11 sep-21 nov 1954; 4 dec-25 dec 1954; 1955-mar 1958; may 1958-feb 1960; feb-dec 1965 – 167r – 1 – uk British Libr Newspaper [079]

O estado de sao paulo – [Sao Paulo: s.n, sep 1956-] – 1 – us CRL [079]

O estado de sao paulo.. / Amaral, Tancredo do – Rio de Janeiro: Alves, 1896. 189p. illus – 1 – us Wisconsin U Libr [972]

O estado de sergipe : jornal official, politico e noticioso – Aracaju, SE. 02-03 ago 1898; mar, ago-set 1899; set 1900; jul-ago 1901; 11 set 1906 – (= ser Ps 19) – mf#P11A,03,09 – bl Biblioteca [320]

El estado de sinaloa / Sinaloa. Mexico – Culiacan, 1902-22, 1925-60, 1963-69 – 1 – (title varies) – mf#LL-02034 – us L of C Photodup [342]

Estado del convento de goatemala del orden de nuestra senora de la merced : redemcion de captivos, y relacion verdadera de los aumentos, que en lo temporal, y espiritual ha tenido... / Mexico [City]. Congregacion de nuestra senora – en Guatemala: Por loseph de Pimeda Ybarra, ano de 1667 – (= ser Books on religion...1543/44-c1800: mercedarios) – 1mf – 9 – mf#crl-239 – ne IDC [241]

El estado del jalisco : periodico oficial del gobierno / Jalisco. Mexico – Guadalajara, 1884-1924, 1926-69 – 1 – mf#LL-02024 – us L of C Photodup [342]

Estado do amazonas : jornal politico, commercial, noticioso e litterario – Manaus, AM. 06 jan-fev, 25 ago 1892 – (= ser Ps 19) – mf#P11B,06,30 – bl Biblioteca [079]

O estado do espirito santo : orgao do partido republicano constructor – Vitoria, ES. 01 jan 1890-dez 1891; mar-dez 1892; jan 1893-dez 1904; jul-ago 1907; ago 1908-dez 1910; jan-abr, jun-25 jul 1911 – (= ser Ps 19) – mf#P11B,05,11 – bl Biblioteca [320]

Estado do para / Lecointe, Paul – Sao Paulo, Brazil. 1945 – 1r – us UF Libraries [972]

O estado do para – Para: Typ d' O Commercio do Para, 17 fev 1890 – (= ser Ps 19) – bl Biblioteca [079]

Estado e capitalismo / Ianni, Octavio – Rio de Janeiro, Brazil. 1965 – 1r – us UF Libraries [972]

Estado e o direito n'os lusiadas / Calmon, Pedro – Rio de Janeiro, Brazil. 1945 – 1r – us UF Libraries [972]

Estado fuerte o caudillo / Laserna, Mario – Bogota, Colombia. 1961 – 1r – us UF Libraries [972]

Estado general de la provincia de san salvador / Gutierrez Y Ulloa, Antonio – San Salvador, El Salvador. 1926 – 1r – 1 – us UF Libraries [972]

Estado general de las fundaciones hechas por don jose escandon...tomos 1, 2 y 3 / Bayle, Constantino – Mexico, 1929-30; Madrid: Razon y Fe, 1932 – 1 – sp Bibl Santa Ana [972]

Estado nacional / Campos, Francisco – Rio de Janeiro, Brazil. 1940 – 1r – us UF Libraries [972]

Estados unidos, cuba y el canal de panama / Rodriguez Lendian, Evelio – Habana, Cuba. 1909 – 1r – 1 – us UF Libraries [972]

Estados unidos mexicanos... / ed by Bayle, Constantino – Madrid: Razon y Fe, 1927 – 1 – sp Bibl Santa Ana [972]
Estados unidos y las antillas / Cestero, Tulio Manuel – Madrid, Spain. 1877-1955 – 1r – us UF Libraries [972]
Estafeta : stikhi / Aseev, Nikolai Nikolaevich – Moskva: Gos izd-vo khudozh lit-ry, 1931 [mf ed 2002] – (= ser Massovaia biblioteka) – 1r – 1 – (filmed with: rafael' / boris zaitsev (1924)) – mf#5238 – us Wisconsin U Libr [810]
L'estafette – Paris: Imp speciale de l'Estafette, apr 23, 26-27, may 2, 4, 6, 8-12, 15-23 1871. Ed du matin – (Filmed as pt of: Commune de Paris newspapers) – us CRL [074]
L'estafette : journal des journaux – ed du matin. 2 fevr-dec 1833-avr 1858 – 1 – fr ACRPP [074]
L'estafette – Paris, mai 1876-21 fevr 1883; 15 fevr 1884; 4 juin 1886-juill 1914 – 1 – fr ACRPP [074]
Estafette du sud – New Orleans LA. 1861 dec 22, 1862 may 12,25, jun 7,18,20, jul 3,11, aug 2,5,11,12,19,20, sep 6,17,20-21,24,26,27,29, dec 6,9,18,23,28 – 1r – 1 – mf#861620 – us WHS [071]
Estaing, Charles Henri *see* Declaration adressee au nom du roi a tous les anciens francois de l'amerique septentrionale
Estamento espiritual del devoto josefino – Trujillo: tip sobrina de b pena, s a – sp Bibl Santa Ana [240]
Estampas criollas / Murillo Gutierrez, Jesus – San Jose, Costa Rica. 1963 – 1r – us UF Libraries [972]
Estampas de espana / Ehrenburg, Ilia Grigorevich – n.p. 1937? – (= ser Blodgett coll) – 9 – mf#fiche w853 – us Harvard [946]
Estampas de honduras / Stone, Doris – Mexico City? Mexico. 1954 – 1r – 1 – us UF Libraries [972]
Estampas de la costa grande / Samayoa Chinchilla, Carlos – Guatemala, 1957 – 1r – us UF Libraries [972]
Estampas de la epoca / Secades, Eladio – Habana, Cuba. 1958 – 1r – us UF Libraries [972]
Estampas del pasado / Perez Valenzuela, Pedro – Guatemala, 1937 – 1r – us UF Libraries [972]
Estampas del vivac / Enamorado-Cuesta, Jose – San Juan, Puerto Rico. 1962 – 1r – us UF Libraries [972]
Estampas guatemaltecas / Rey Soto, Antonio – Ciudad de Guatemala, 1929 – 1r – us UF Libraries [972]
Estampas locales / Lainez, Daniel – Tegucigalpa, Mexico. 1947 – 1r – us UF Libraries [972]
Estampas martianas / Clavijo Tisseur, Arturo – Santiago, Cuba. 1953 – 1r – us UF Libraries [972]
Estampas santaferenas / Hernandez De Alba, Guillermo – Bogota, Colombia. 1938 – 1r – us UF Libraries [972]
Estampas venezolanas / Garcia Hernandez, Manuel – Caracas, Venezuela. 1955 – 1r – us UF Libraries [972]
Estampes, Louis d' *see* La franc-maconnerie et la revolution
Estancelin, L *see* Recherches sur les voyages et decouvertes des navigateurs normands en afrique, dans les indes orientales et en amerique...
Estancelin, Louis *see* Histoire des comtes d'eu
Estancia de isabel la catolica en trujillo / Rubio Cercas, Manuel – n77. oct-dic 1951 – 1 – sp Bibl Santa Ana [946]
O estandarte – Maceio, AL: Typ de Mello Rocha, 22 jun, 28 set, 03 nov 1883; 24 jul, 01 dez 1884; 22 fev 1885 – (= ser Ps 19) – mf#P18B,01,21 – bl Biblioteca [079]
O estandarte – Rio de Janeiro, RJ: Typ Central de Evaristo R da Costa, 10 jan-14 ago 1880 – (= ser Ps 19) – mf#P18A,01,18 – bl Biblioteca [079]
O estandarte – Rio de Janeiro, RJ: Typ de J D da Cruz, 11 maio-06 jun 1851 – (= ser Ps 19) – mf#P17,01,145 – bl Biblioteca [790]
Estandarte obrero / Marxist-Leninist Party, USA – 1979 jan-1986 aug 13 – 1r – 1 – mf#1100755 – us WHS [335]
Estat des reformez en france / [Brousson, C] – Cologne, 1684 – 14mf – 9 – mf#PRS-117 – ne IDC [240]
Estat present de l'eglise et de la colonie francoise dans la nouvelle france / Saint-Vallier, Jean-Baptiste de La Croix de Chevrieres de – Paris: R Pepie, 1688 [mf ed 1976] – 1r – 5 – mf#SEM16P272 – cn Bibl Nat [241]
Estate planning – New York. 1978+ (1,5,9) – ISSN: 0094-1794 – mf#11800 – us UMI ProQuest [332]
Les estats, empires, et principautez du monde : represente's par la description des pays, moeurs des habitans, richesses des provinces, les forces... – Rouen: Chez Clement Malassis...1664 [mf ed 1982] – 16mf – 9 – 0-665-26711-8 – mf#26711 – cn Canadiana [900]
Estaturos de la ve...ano 1977 / Manconumidad Internacional de la Vera – Caceres: Imp. de la Diputacion Provincial, 1977 – sp Bibl Santa Ana [060]
Estatuto del sindicato de obras agricolas y oficios varios / Sindicato Amanecer, Hoyos – Caceres: tip de minerva, s a, 1936 – 1 – sp Bibl Santa Ana [972]
Estatuto dos funcionarios publicos civis da uniao / Brazil – Rio de Janeiro, Brazil. 1950 – 1r – us UF Libraries [972]
Estatutos / APA del Colegio "Santisima Trinidad" de Plasencia – Plasencia: Graf. Sandoval, 1976 – 1 – sp Bibl Santa Ana [060]
Estatutos / Asociacion de Padres de Alumnos y Amigos de la Esuela de EGB – Aldea Moret: Linez XXI, 1979 – 1 – sp Bibl Santa Ana [060]
Estatutos / Asociacion Empresarial de panaderos de la Provincia de Caceres – Caceres: Tip. Extremadura, 1978 – sp Bibl Santa Ana [340]
Estatutos / Asociacion familiar "Los Alamos" Casas de Don Antonio – Caceres: Tip. Extremadura, 1978 – sp Bibl Santa Ana [060]
Estatutos / Asociacion Morala de Padres de Alumnos de Educ. General Basica. Navlamoral de la Mata – Plasencia: Graficas Sandoval, 1977 – sp Bibl Santa Ana [370]
Estatutos / Banca Sanchez SA – Caceres: Tip. El Noticiero S.L., 1958 – 1 – sp Bibl Santa Ana [060]
Estatutos / Caceres. Asociacion Cultural – Caceres: Imprenta Moderna, 1969 – 1 – sp Bibl Santa Ana [060]
Estatutos / Caceres. Mutua Cerealistica – Caceres: Imp. La Minerva, 1970 – 1 – sp Bibl Santa Ana [060]
Estatutos / Caja Rural de Ahorros y Prestamos de los Santos de Maimona – Badajoz: Tip. Lit. y Encuad. de Uceda Hermanos, 1911 – sp Bibl Santa Ana [370]
Estatutos / Central Obrera Nacional Sindicalista – Caceres: Imp. y Enc. Garcia Floriano, 1936 – 1 – sp Bibl Santa Ana [060]
Estatutos / Central Obrera Nacional Sindicalista – Caceres: Imprenta Moderna, 1936 – sp Bibl Santa Ana [340]
Estatutos / Club Alfaya – Plasencia: Imp. Sanchez Rodrigo, S.A. 1976 – 1 – sp Bibl Santa Ana [060]
Estatutos / Cooperativa Local del Campo y Ganaderos "San Isidro". Miajadas – Caceres: Imprenta Moderna, 1971 – 1 – sp Bibl Santa Ana [060]
Estatutos / Plasencia. Aula Medica – Plasencia: Graficas Sandoval, 1970 – 1 – sp Bibl Santa Ana [060]
Estatutos / Sindicato Independiente de trabajadores del Credito (S.I. T.C.). Caceres – Caceres: Imp. La Minerva, 1977 – sp Bibl Santa Ana [331]
Estatutos / Sociedad Economica Amigos del Pais – 1847 – 9 – sp Bibl Santa Ana [060]
Estatutos... / Caceres. Pena. Amigos del Flamenco de Extremadura – Caceres: Editorial Extremadura, 1981 – 1 – sp Bibl Santa Ana [946]
Estatutos... / Cooperativa del Campo – Plasencia: Imprenta La Victoria, 1961 – 1 – sp Bibl Santa Ana [060]
Estatutos aprobados por orden ministerial hacienda de 1960 / Caceres. Mutua Aseguradora de Transportistas de la Provincia de Caceres – Caceres: imp la minerva, 1960 – sp Bibl Santa Ana [240]
Estatutos de... / Terpresa – Caceres: imp la minerva, 1969 – 1 – sp Bibl Santa Ana [350]
Estatutos de la... / Asociacion Empresarial Harino-Panadera de la provincia de Caceres – Caceres: tip extremadura, 1977 – 1 – sp Bibl Santa Ana [350]
Estatutos de la... / Asociacion Provincial de amas de casa – Caceres: imp extremadura, 1971 – 1 – sp Bibl Santa Ana [060]
Estatutos de la... / Cooperativa de Suministros y Consumo de Nuestra Senora de Guadalupe de Caceres – Caceres: tip el noticiero, s.a., 1945? – 1 – sp Bibl Santa Ana [060]
Estatutos de la... / Cooperativa del Campo Union de Cultivadores de tabaco de Jaraiz de la Vera – Caceres: tip el noticiero, s.a. (1948) – sp Bibl Santa Ana [366]
Estatutos de la... / Cooperativa Farmaceutica Extremena "Cofex" – Caceres: tip el noticiero, s.l. 1958 – 1 – sp Bibl Santa Ana [610]
Estatutos de la agencia cooperativa de exportacion / Agencia Cooperativa De Exportacion De Azucar (Havana) – Habana, Cuba. 1930 – 1r – us UF Libraries [380]
Estatutos de la asociacion de padres de familia y amigos de la escuela / Escuela Graduada Jose Luis Cotallo – Caceres: Edit. Extremadura, 1972 – 1 – sp Bibl Santa Ana [060]
Estatutos de la cofradia de la virgen de la montana – 1899 – 9 – sp Bibl Santa Ana [240]
Estatutos de la cofradia de los ramos, cristo de la buena muerte y virgen de la esperanza / Cofradia de los Ramos, Cristo de la Buena Muerte y Virgen de la Esperanza – Caceres: Imp. Sanguino, 1962 – 1 – sp Bibl Santa Ana [360]
Estatutos de la cofradia del santisimo sacramento / Almendralejo. Spain – 1885 – 9 – sp Bibl Santa Ana [946]
Estatutos de la cooperativa agricola de los colonos de valdelacalzada – Badajoz: Imp. Arqueros, 1967 – 1 – sp Bibl Santa Ana [630]
Estatutos de la cooperativa agricola olivarera nuestra senora de santa marta de salvaleon (badajoz) – Badajoz: tip graf extremena, 1961 – sp Bibl Santa Ana [946]
Estatutos de la cooperativa agropecuaria de granja de torrehermosa (badajoz) – Baajoz: imprenta inca, 1965 – sp Bibl Santa Ana [946]
Estatutos de la cooperativa de casas baratas nuestra senora de la asuncion de caceres / Cooperativa se Casas Baratas – Caceres: editor extremadura, 1974 – 1 – sp Bibl Santa Ana [060]
Estatutos de la cooperativa de viviendas de proteccion oficial "san carlos barromero" del sindicato provincial de banca, bolsa y ahorro de caceres / Delegacion Provincial de Sindicatos – Caceres: el noticiero s.l., 1966 – 1 – sp Bibl Santa Ana [350]
Estatutos de la cooperativa del campo "arrago" / Cooperativa del campo "Arrago" – Caceres: imp moderna, 1976 – 1 – sp Bibl Santa Ana [060]
Estatutos de la cooperativa del campo "la benfica" de oliva de la frontera (badajoz) – Jerez de los Caballeros: tip. horizonte, 1960 – sp Bibl Santa Ana [946]
Estatutos de la cooperativa del campo nuestra senora de la estrella de agrado (ciudad real) – Villanueva de la Serena: imp parejo, 1970 – sp Bibl Santa Ana [946]
Estatutos de la cooperativa droguera extremena. codex, detallista de droguerias – Merida: imp rodriguez, 1965 – sp Bibl Santa Ana [946]
Estatutos de la cooperativa ganadera : "alta extremadura" – Caceres: Caceres Tip. El Noticiero, s.a., 1947 – 1 – sp Bibl Santa Ana [334]
Estatutos de la cooperativa provincial de automobiles de alquiler de badajoz – Badajoz: graf jimenez, 1960 – sp Bibl Santa Ana [946]
Estatutos de la cooperativa y caja rural de... – Caceres: tip el noticiero – sp Bibl Santa Ana [340]
Estatutos de la federacion de sindicatos de propietarios de fincas rusticas de la provincia de badajoz – Badajoz: tip de arqueros, 1931 – 1 – sp Bibl Santa Ana [340]
Estatutos de la real...de caceres / Cofradia de Nuestra Senora de la Soledad y del Santo Entierro – Caceres: imp y enc de vda de garcia floriano, 1953 – 1 – sp Bibl Santa Ana [060]
Estatutos de la sociedad de prevision social de merida – Merida: A Rodriguez, 1932 – 1 – sp Bibl Santa Ana [060]
Estatutos de la sociedad obrera denominada la union de fuente de cantos (provincia de badajoz) – Badajoz: tip y enc la minerva extremena, 1906 – 1 – sp Bibl Santa Ana [946]
Estatutos de la...caceres / Caceres. Asociacion de Padres de Alumnos del Colegio San Antonio de Padua – edit extremadura, 1977 – 1 – sp Bibl Santa Ana [366]
Estatutos de la...denominada de santiago y santa margarita / Cooperativa Local de Consumo – Zorita: imp carrasco, 1975 – 1 – sp Bibl Santa Ana [350]
Estatutos de la...en la iglesia parroquial de almendralejo / Cofradia Santisimo Sacramento – 1885 – 9 – sp Bibl Santa Ana [240]
Estatutos de los capellanes (anno 1607) – Avila – 1r – 5,6 – sp Cultura [240]
Estatutos de sociedad de socorros mutuos y caja de ahorros : constituida en villafranca de los barros en 7 de mayo de 1905, s l, s i, s a / Villafranca de los BarrosEl Credito Extremeno – 1 – sp Bibl Santa Ana [332]
Estatutos de...caceres / Marco, S A – tip la minerva, 1962 – 1 – sp Bibl Santa Ana [060]
Estatutos del centro juvenil nuestra senora del rosario de huerta de animas / Huerta de Animas – Trujillo: imp gexme, 1972 – 1 – sp Bibl Santa Ana [060]
Estatutos del hogar extremeno de madrid aprobados en la junta general celebrada al efecto el dia 1st de julio de 1934 – Madrid: imprenta cava baja 17, s.a. – 1 – sp Bibl Santa Ana [340]
Estatutos del patronato de viviendas sociales de alcantara / Alcantara. Ayuntamiento – Caceres: imp la minerva, 1971 – 1 – sp Bibl Santa Ana [060]
Estatutos del sindicato agricola de zarza-capilla (badajoz) – Toledo: imp de viuda e hijos de j pelaez, 1905 – sp Bibl Santa Ana [630]
Estatutos del...y el general de procuradores de los tribunales de espana / Colegio Provincial de Procuradores de Badajoz – Badajoz: Tip. Clasica, 1960 – sp Bibl Santa Ana [340]
Estatutos generales / Caja Ahorros y Monte de Piedad de Caceres. – Caceres: tip extremadura, 1978 – sp Bibl Santa Ana [350]
Estatutos generales / Caja de Ahorros y Monte de Piedad. Caceres – Caceres: edit extremadura, 1976 – 1 – sp Bibl Santa Ana [350]
Estatutos generales de barcelona : para la familia cismontana de la orden de nuestro seraphico padre s francisco... – en Mexice: En casa de Pedro Ocharte, ano 1585 – (= ser Books on religion...1543/44-c1800: franciscanos) – 3mf – 9 – mf#crl-198 – ne IDC [241]
Estatutos o constituciones...de la piedad... almendralejo – 1879 – 9 – sp Bibl Santa Ana [240]
Estatutos para el regimen y administracion de la... / Cofradia de la Santisima Virgen del Pilar de Casas de Don Antonio – Caceres: Tip. El Noticiero – 1 – sp Bibl Santa Ana [060]
Estatutos para el regimen y administracion de la hermandad de caballeros de la santisima virgen de la victoria – Trujillo: Tip. Sobrino de B. Pena, 1944 – sp Bibl Santa Ana [340]
Estatutos por los que ha de regirsela... / Cooperativa Industrial Cacerena – Caceres: Imp. Moderna, 1971 – 1 – sp Bibl Santa Ana [350]
Estatutos por los que se rige la pena "los camborios" / Los Camborios – Plasencia: Imp. Gercilasso, 1974 – 1 – sp Bibl Santa Ana [350]
Estatutos provisionales. octubre, 1946 / Asociacion de Amigos de Guadalupe – Caceres: imp. garcia floriano, s.a. – sp Bibl Santa Ana [060]
Estatutos que han de regular el funcionamiento de la asociacion cultural centro de estudios historicos de caceres – Trujillo: sobrino de b pena, 1960 – sp Bibl Santa Ana [946]
Estatutos sociales del... / Caceres. Club de tenis "Cabeza Rubia" – Caceres: imp. sergio dorado, 1971 – 1 – sp Bibl Santa Ana [366]
Estatutos. sociedad la fortuna emeritense – Merida, 1924 – 1 – sp Bibl Santa Ana [946]
Estatutos y ordenaciones desta provincia de san gabriel...brozas...1615... – S.l, s.i, s.a. – 1 – sp Bibl Santa Ana [946]
Estatutos y ordenaciones...san gabriel – 1602 – 9 – sp Bibl Santa Ana [946]
Estatutos y ordenaciones...san jose – 1802 – 9 – sp Bibl Santa Ana [946]
Estatutos y reglamento / Asociacion de Medicina Extremena – Badajoz: tip clasica, 1965 – sp Bibl Santa Ana [610]
Estatutos y reglamento de la... / Cooperativa de Viviendas "San Antonio" – Caceres: imp rodriguez, 1979 – 1 – sp Bibl Santa Ana [060]
Estatutos y reglamento del casino de almendralejo – Almendralejo: imp y enc de juan bote gonzalez, 1910 – 1 – sp Bibl Santa Ana [340]
Estatutos y reglamento del centro agricola-mercantil de caceres – Caceres: imp la minerva de serafin rodas, 1903 – 1 – sp Bibl Santa Ana [630]
Estatutos y reglamentos / Caja Rural de Ahorros y Prestamos de Almendralejo – Almendralejo: Imprenta Macarro, 1950 – sp Bibl Santa Ana [340]
Estatutos y reglamentos / Federacion Provincial de Empresarios de la Construccion – Caceres: Ed. Extremadura, 1979 – 1 – sp Bibl Santa Ana [350]
Estatutos y reglamentos de la... / Mutua Extremena de Vehiculos – Caceres: imp rodriguez, 1971 – 1 – sp Bibl Santa Ana [350]
Estatutos y reglamentos de la caja rural de ahorros y prestamos de almendralejo. ano de 1906 – Sevilla: imp l vilches, 1915 – 1 – sp Bibl Santa Ana [340]
Estatutos y reglamentos de la caja rural de ahorros y prestamos de monesterio. ano de 1906 – Badajoz: Tip. Antonio Arqueros, 1906 – sp Bibl Santa Ana [240]
Estatutos...catedral de plasencio / Norona, Andres – 1704 – 9 – sp Bibl Santa Ana [946]
Estautos de la... / Obra de Ayuda Nacional-Sindicalista – Caceres: Imp. Garcia Floriano, 1944 – 1 – sp Bibl Santa Ana [060]
Estave Barba, Francisco *see* Descubrimiento y conquista de chile. barcelona, 1946
Estborn, Sigfrid *see* The religion of tagore in the light of the gospel
Est-ce st paul a athenes?...au milieu de l'areopage? : non, c'est le pere giraud dans la province de quebec devant la cour criminelle – Quebec?: s.n, 1900? – 1mf – 9 – mf#04320 – cn Canadiana [920]

Estcourt, Edgar Edmund *see* The english catholic nonjurors of 1715
Este e o livro das chymeras / Pimenta, Alfredo – Lisboa: Portugalia, 1922 – 1 – us Wisconsin U Libr [240]
Este es el cortejo...salamanca 1938 / Castro Albarran, A de – Burgos: Razon y Fe, 1938 – 1 – sp Bibl Santa Ana [946]
Este es un co[m]pe[n]dio breve que tracta d[e]la manera de como se ha[n] de hazer las percessiones / Leuwis, Dionysius de, de Rickel – Se imprimio en...Tenuchtitlan Mexico...:...fray Jua[n] Cumarraga...ano de 1544 – (= ser Books on religion...1543/44-c1800: manuales de rito) – 1mf – 9 – mf#crl-63 – ne IDC [241]
Este es vn co[m]pe[n]dio breue que tracta d[e]la manera de como se ha[n] de hazer las p[ro]cessiones / Leuwis, Dionysius de, de Rickel – Impr[e]ssa e[n] Mexico: Mandado d[e]l senor ob[is]po do[]n fray Cumarraga...[1545?] – (= ser Books on religion...1543/44-c1800: manuales de rito) – 1mf – 9 – mf#crl-459 – ne IDC [241]
Este otro ruben dario / Oliver Belmas, Antonio – Barcelona, Spain. 1960 – 1r – us UF Libraries [440]
Esteban pichardo, 1799-1879 / Massip, Salvador – Habana, Cuba. 1941 – 1r – us UF Libraries [972]
L'est-eclair : quotidien republicain d'information – Troyes, janv-oct 1966 – 1 – fr ACRPP [074]
Esteco y concepcion del bermejo. dos ciudades desaparecidas : buenos aires, 1943 / Torre Revello, Jose – Madrid: Razon y Fe, 1946 – 1 – sp Bibl Santa Ana [946]
La estela de un campesino / ed by Bayle, Constantino – Madrid: Razon y Fe, 1926 – 1 – sp Bibl Santa Ana [920]
La estela de un campesino. muestrario de accion, ideas y sentimientos / Masides Rosado, Severiano – Plasencia: Tip. de Mariano San Jose, 1926 – 1 – sp Bibl Santa Ana [946]
Estella *see* Tra gli eroi i martiri della liberta
Estella, Jose Ramon *see* Historia grafica de la republica dominicana
Estelrique, I *see* Tratado breve y parecer acerca del methodo de curar con sangrias...
Estenger, Rafael *see*
- Caracteres constantes en las letras cubanas
- Cien de las mejores poesias cubanas
- Don pepe
- Hacia un heredia genuino

Estep family newsletter – v1-4 n15 [1973 jul-1977 [sep ?]] – 1r – 1 – mf#379643 – us WHS [929]
Estepas / Laureano, Angel Luis – Santurce, Puerto Rico. 1962 – 1r – us UF Libraries [972]
Ester : la cortesana / Bermudez M, Antonio – Tegucigalpa, Honduras. 1939 – 1r – us UF Libraries [972]
Ester, Carl d' *see* Das zeitungswesen in westfalen von den ersten anfaengen bis zum jahre 1813
Estere-Forriol, Jose *see* Die trauer- und trostgedichte in der romischen literatur
Esterhuizen, Maria *see* Metakognisie, teks en leser
Esterhuyse, J H *see* South west africa, 1880-1894
Esterhuysen, Matthias Wilhelm *see* Synthesis and characterization of palladium complexes exhibiting hemilebile properties
Estermann, Alfred *see*
- Daheim
- Die gartenlaube
- Illustrirte zeitung
- Ueber land und meer
- Wiener theaterzeitung

Estero : lee county / Lamme, Corinne W – s.l, s.l? 1936 – 1r – us UF Libraries [978]
Estes, Charles Sumner *see* Christian missions in china
Estes, David Foster *see* An outline of new testament theology
Estes, Henry B 2 *see* God is in heaven
Estes, Hiram Cushman *see* The christian doctrine of the soul
Estestvenno-istoricheskaia chast' / Materialy dlia otsenki zemel' Samarskoi gubernii – Samara, Spb, 1909-1911. v1-5 – 33mf – 8 – (missing: v1) – mf#RZ-81 – ne IDC [314]
Estestvenno-istoricheskaia chast' / Materialy k otsenke zemel' Ekaterinoslavskoi gubernii – Ekaterinoslav, 1904-1914 v1-7 – 17 – 8 – (missing: v5) – mf#RZ-55 – ne IDC [314]
Estestvenno-istoricheskaia chast' / Materialy k otsenke zemel' Nizhegorodskoi gubernii – Spb, 1884-1886. v1-14 – 56mf – 9 – mf#RZ-68 – ne IDC [314]
Estetica come scienza dell'espressione e linguistica generale *see* Aesthetic as science of expression and general linguistic
Estetica siemprevivas y ensayos / Gomez-Bravo, Vicente – Madrid: La difusora del libro, 1959 – sp Bibl Santa Ana [946]
Estetica y erotismo de la pena de muerte / Cansinos Assens, Rafael – Madrid, Spain. 1916 – 1r – us UF Libraries [025]
Esteve, Claude Louis *see* Etudes philosophiques sur l'expression litteraire
Esteve, Edmond *see* Byron et le romantisme francais
Esteve, Pierre *see* Nouvelle decouverte du principe de l'harmonie
Esteves, Luis Raul *see*
- Barrabases (cosas de mi pueblo)
- Que cosas

Esteves Pereira, F M *see* Historia de minas, ademas sagad, rei de ethiopia
Estevez, Andres Maria *see* Del rosal del arte
Estevez Verdejo, Ramiro *see* Monografia de s. vicente de alcantara
Esther / Adams, Henry – A novel. 1884 – 9 – us Scholars Facs [830]
Esther : a drama of jewish history: being the story of the book of esther / Gill, William Hugh – Philadelphia: George W Jacobs c1899 [mf ed 1985] – 1mf – 9 – 0-8370-3284-9 – mf#1985-1284 – us ATLA [221]
Esther : dramatisches fragment in zwei aufzuegen / Grillparzer, Franz – Wien: M Waizner, 1908 [mf ed 1996] – 20p – 1 – mf#9660 – us Wisconsin U Libr [820]
Esther / Gunkel, Hermann – Tuebingen: J C B Mohr 1916 [mf ed 1992] – (= ser Religionsgeschichtliche volksbuecher fuer die deutsche christliche gegenwart 2/19-20) – 1mf – 9 – 0-524-04399-X – mf#1992-0092 – us ATLA [221]
L'esthetique de richard wagner / Freson, Jules G – Paris: Fischbacher 1893 – (= ser Essais de philosophie de l'art) – 2v on 7mf – 9 – mf#wa-29 – ne IDC [780]
Esti baptisti...25 juubeli aasta malestusets = Historical album of the estonian baptist churches: 25th jubilee year of reminiscences / Feisberg, J & Tetermann, A – Tallinn, 1911 – 178p – 1 – us Southern Baptist [242]
Esti budapest – Hungary, apr 1952-dec 1954; feb 1955-oct 1956 – 15r – 1 – uk British Libr Newspaper [079]
Esti hirlap – Budapest, Hungary. 1962-1990 – 53r – 1 – us L of C Photodup [077]
Esti magyarorszag – Budapest, Hungary. Jun 1942; Sept 1943 (scattered issues) – 2r – 1 – us L of C Photodup [079]
Estienne, H *see*
- The art of making devises
- ...Orationes 2

Estienne, Henri *see*
- De criticis vet. gr. et latinis
- Dictionarium medicum...

Estilistica brasileira / Bueno, Francisco Da Silveira – Sao Paulo, Brazil. 1964 – 1r – us UF Libraries [972]
Estilo y densidad en la poesia de ricardo j bermu... / Alvarado De Ricord, Elsie – Panama, 1960 – 1r – us UF Libraries [972]
Estimate of certain expenses of the civil government of the province of canada : for the year 1857, for which a supply is required – Toronto: Rollo Campbell, [1857] (mf ed 1992) – 1mf – 9 – mf#SEM105P1757 – cn Bibl Nat [336]
Estimates of body composition using a four-component model in individuals with musculoskeletal development / Modlesky, Christopher M – 1995 – 2mf – 9 – $8.00 – mf#PE 3765 – us Kinesology [612]
Estimating body fat percentage using circumference measurements and lifestyle questionnaire data : a multivariate study of 184 college age females / Slack, Jason V – 1997 – 1mf – 9 – $4.00 – mf#PE 3794 – us Kinesology [613]
Estimating body fat percentage using simple measures : A Multivariate Study Of 150 Men / Greenwell, Scott D – 1998 – 1mf – 9 – $4.00 – mf#PE 4029 – us Kinesology [612]
Estimating lumbar spinal loads during a golf swing using an emg-assisted optimization model approach / Lim, Young-Tae – 2000 – 3mf – 9 – $12.00 – mf#PE 4082 – us Kinesology [617]
Estimating soviet and east european hard currency debt : a research paper – Washington, DC: National Foreign Assessment Center, 1980 – 1mf – 9 – (incl bibl ref) – us Wisconsin U Libr [330]
Estimation of vo(2max) from a submaximal 1-mile track jog for relatively fit teenage individuals / Hunt, Brian R – 1993 – 1mf – $4.00 – us Kinesology [612]
Estimauville, Robert Anne d', chevalier de Beaumochel *see* Adresse particulierement aux membres canadiens elus pour le prochain parlement provincial
O estimulo : semanario de revista, propaganda democratica, litt e critica seria – Belem, PA. 10 jun 1877 – (= ser Ps 19) – bl Biblioteca [079]
Estius, G *see*
- In 4 libros sententiarum commentaria
- In omnes beati pauli et septem catholicas apostolorum epistolas commentaria

Estius, Guilielmus *see* In omnes d pauli epistolas
Estlaendische zeitung – Reval (Tallinn EW), 1934 27 aug-1935 29 mar – 1 – gw Misc Inst [077]
Estlin papers, 1840-1844 – London: Dr William's Library – (= ser BRRAM series) – 6r – 1 – £402 / $804 – (with p/g; int by clare taylor) – mf#r02206 – uk Microform Academic [306]
Estoile, Pierre de l' *see*
- Journal de henri 3, roy de france & de pologne
- Journal du regne de henry 4, roi de france et de navarre

Estonia. Riigikogu *see* Protokollid
Estonian Baptist Union *see* Aruanne
Estonian newspapers 1918-1940 – Helsinki: Helsinki University Library, 1994 – ca 525r – 1 – fi Helsinki [077]
Estorino, Abelardo *see* Robo del cochino
Estournelles de Constant, Paul Henri Benjamin, baron d' *see* America and her problems
Estrada, Antonio *see* Vida ejemplar fr. pedro de la purificacion
Estrada de ferro madeira-mamore / Craig, Neville B – Sao Paulo, Brazil. 1947 – 1r – us UF Libraries [972]
Estrada De La Hoz, Julio *see* Belice
Estrada, Enrique *see* Dominicanizacion de la frontera en la era gloriosa
Estrada, Genaro *see* Don juan prim y su labor diploamtica en mexico
Estrada, Juan de *see* Exaltacion gloriosa del mas pequeno sermon panegyrico de n p san francisco
Estrada Molina, Ligia Maria *see* Costa rica de don tomas de acosta
Estrada Monsalve, Joaquin *see*
- Asi fue la revolucion
- Hombres

Estrada, Ricardo *see* Flavio herrera
Estrada, Rodrigo Duque *see* Petroleo no brasil, holding de estado
Estrada S, Julio *see* Inmigracion y extranjeria
Estrada y Escovedo, Pedro de *see* Relacion sumaria del auto particular de fee
Estranger a londres (the foreigner) – London, UK. 1, 15-29 Oct 1904 – 1 – uk British Libr Newspaper [072]
Estrangulados / Robleto, Hernan – Madrid, Spain. 1933 – 1r – us UF Libraries [972]
O estravagante – Rio de Janeiro, RJ. 09 out 1881 – (= ser Ps 19) – mf#P17,01,137 – bl Biblioteca [079]
A estrea – Maceio, AL: Typ do Partido Liberal, 5 ago 1878-5, 13 fev, 5-24 mar 1879 – (= ser Ps 19) – 1,5,6 – mf#P18B,01,22 – bl Biblioteca [079]
El estrecho de magallanes : lo que era y lo que es / ed by Bayle, Constantino – Madrid: Razon y Fe, 1921 – 1 – sp Bibl Santa Ana [946]
A estrella : orgao imparcial – Baependi, MG: Typ do Baependiano, 27 maio-23 jun 1881 – (= ser Ps 19) – mf#P17,02,109 – bl Biblioteca [079]
La estrella – 1961. 842p – 1 – us Southern Baptist [242]
Estrella De Centroamerica *see* El salvador
Estrella de panama – Panama, apr 1904-dec 1914; sep 1951-may 1955; jul 1978; sep 1978-1985 – 161r – 1 – us L of C Photodup [079]
La estrella de panama – Panama City, Panama. 3 Jan-8 Mar 1920 – 6r – 1 – us L of C Photodup [079]
La estrella de panama – Panama: The Star & Herald Co, 1956-oct 1970; han-feb 1971; [oct 1971-jun 1972] – 1 – us CRL [079]
La estrella de panama : steamer edition – Jan 9 1864-Oct 10 1864 – 1r – 1 – us L of C Photodup [079]
A estrella do brazil : folha periodica e liberal – Rio de Janeiro, RJ: Typ Popular, 07 abr-01 jun 1861 – (= ser Ps 19) – mf#P25,03,08 n.01 – bl Biblioteca [079]
A estrella do norte : periodico politico jocoserio – Recife, PE: Typ do Dr Joao de Barros Falcao de Albuquerque Maranhao, out-19 dez 1863 – (= ser Ps 19) – mf#P16,01,28 – bl Biblioteca [079]
A estrella do sul : periodico consagrado aos interesses da religiao – Porto Alegre, RS: Typ do Jornal a Ordem, 15 out-dez 1862; jan-set, nov 1863; jul 1864; jan-maio, out 1866; jan, maio-jun 1867; jun, nov 1868; abr-maio, jul-ago 1869 – (= ser Ps 19) – bl Biblioteca [079]
Estrella Estralla, Jose Emilio *see*
- Dialogos para el futuro
- Juan de borbony battemberg
- La libertad
- Verdad de politica

Estrella Gutierrez, Fermin *see*
- Idolo

Estrellas / Munoz Meany, Enrique – Guatemala, 1960 – 1r – us UF Libraries [972]
Estridge, H W *see* Six years in seychelles
Estrindencias poesias / Ducasse, Angel Braulio – Badajoz: tip vda de a arqueros, 1936 – 1 – sp Bibl Santa Ana [946]
L'estro armonico : concerti...opera terza. libro primo-secondo / Vivaldi, Antonio – Amsterdam: E Roger & M C Le Cene [1720?] [mf ed 1992] – 1r – 1 – mf#pres. film 117 – us Sibley [780]
Estructura economica y banca central / Hidalgo, Carlos F – Madrid, Spain. 1963 – 1r – us UF Libraries [332]
Estructuras demograficas y sociales de colombia / Lannoy, Juan Luis De – Bogota, Colombia. 1961 – 1r – 1 – us UF Libraries [304]
Estructuras sindicales / Di Tella, Torcuato S – Buenos Aires, Argentina. 1969 – 1r – us UF Libraries [972]
Estuardo, Nunez *see* Autores germanos en el peru
Estuaries – Stony Brook. 1978+ (1,5,9) – ISSN: 0160-8347 – mf#11458 – us UMI ProQuest [550]
O estudante – Diamantina, MG: Typ de Luiz Antonio dos Reis, 21 ago-24 out 1873 – (= ser Ps 19) – mf#P17,02,108 – bl Biblioteca [079]
O estudante – Maceio, AL: Typ de Menezes, 10 ago, set, 10 nov 1888 – (= ser Ps 19) – mf#P18B,01,08 – bl Biblioteca [079]
O estudante : semanario critico e litterario – Rio de Janeiro, RJ. 22 out 1881 – (= ser Ps 19) – mf#P17,01,133 – bl Biblioteca [410]
O estudante catholico : religiao e litteratura – Recife, PE: Typ Industrial, 01-31 ago; 03 out 1875 – (= ser Ps 19) – bl Biblioteca [241]
L'estudiant : organe du seminaire de joliette – [Joliette]: [le Seminaire] v1 n1 nov 1936-v37 n8 23 mars 1973 [mf ed 1990] – 2r – 1 – (cont by: cellule) – mf#SEM35P341 – cn Bibl Nat [378]
El estudiante de salamanca / Espronceda, Jose de – Madrid: Est. Tipografico A Marzo, 1903 – sp Bibl Santa Ana [946]
Estudiante poeta / Suarez, Romualdo – Habana, Cuba. 1957 – 1r – us UF Libraries [972]
Estudiantes y politica en america latina / Solari, Aldo E – Caracas, Venezuela. 1968 – 1r – us UF Libraries [320]
Estudio acerca de la guerra de guerrillas en cuba / Consuegra, Walfredo I – Santiago, Cuba. 1914 – 1r – 1 – us UF Libraries [972]
Estudio acerca de una antologia de prosistas guate / Echeverria Barrera, Romeo Amilcar – Guatemala City, 1955 – 1r – us UF Libraries [972]
Estudio bibliografico de don manuel eduardo de gorostiza : mexico, 1932 / Agular, M Maria Esperanza – Madrid: Razon y Fe, 1935 – 1 – sp Bibl Santa Ana [946]
Estudio biografico hernando de soto / Villanueva y Canedo, Luis – Badajoz: tip lit y enc la industria, de uceda hermanos, 1892 – 1 – sp Bibl Santa Ana [910]
Estudio comparativo de la hipoteca minera y la hipoteca comun / Barrientos Lavin, Oscar – Santiago: Talleres graficos "Simiente." 1944 – 45p – 1 – mf#LL-8002 – us L of C Photodup [340]
Estudio critico biografico de juan clemente zenea / Gomez Carbonell, Maria – Habana, Cuba. 1926 – 1r – us UF Libraries [972]
Estudio de la comunidad / Ware, Caroline Farrar – Washington, DC. 1952 – 1r – us UF Libraries [972]
Estudio de la region de upala / Instituto De Tierras Y Colonizacion – San Jose, Costa Rica. 1964 – 1r – us UF Libraries [972]
Estudio de las tierras de los nuevos regadios de la provincia de badajoz / Remon Camacho, Juan – Madrid, 1955 – 1 – sp Bibl Santa Ana [550]
Estudio de un metodo analitico para valoracion cuantitativa conjunta de los acidos organicos en vinos de tierra de barros / Henao Davila, Fernando – Badajoz: Universidad de Extremadura, 1980 – 1 – sp Bibl Santa Ana [946]
Estudio demografico comparativo de espana y la provincia de caceres (decenio 1921-1930) / Campo Cardona, Antonio del – Caceres: imprenta la minerva, 1939 – sp Bibl Santa Ana [304]
Estudio elemental de gramatica historica de la lengua castellana / Alemany Bolufer, Jose – 3d ed. Madrid: Imp de la Rev de Arch, Bibl y Museos, 1911 – 367p – 1 – (incl bibl ref) – mf#2182 – us Wisconsin U Libr [440]
Estudio estadistico de algunos aspectos / Dominican Republic Direccion General De Estadisti – Ciudad Trujillo, Dominican Republic. 1941 – 1r – us UF Libraries [972]
Estudio filosofico...a lopez de ayala – 1882 – 9 – sp Bibl Santa Ana [190]
Estudio geoagronomico de la region oriental de la... / Dondoli B, Cesar – San Jose, Costa Rica. 1954 – 1r – us UF Libraries [630]
Estudio geografico de la isla de cuba / Luzon, A – Toledo, Spain. 1897 – 1r – 1 – us UF Libraries [918]
Estudio geologico de la region de guanacaste, cost... / Dengo, Gabriel – San Jose, Costa Rica. 1962 – 1r – us UF Libraries [550]
Estudio historico sobre el descubrimiento y conquista de la patagonia y de la tierra de fuego / Morla Vicuna, Carlos – Leipzig: F A Brockans, 1903 – 1 – sp Bibl Santa Ana [946]

Estudio petrografico del meteorito de guarena / Calderon, S & Quiroga, F – 1892 – 9 – sp Bibl Santa Ana [550]
Estudio sintetico sobre la iglesia en la constitucion / Tejero Garcia, Angel – Caceres: tip extremadura, 1972 – 1 – sp Bibl Santa Ana [240]
Estudio sobre el fuero de baylio / Mahillo Santos, Juan – Badajoz: imp dip provincial, 1958 – sp Bibl Santa Ana [946]
Estudio sobre el movimiento cientifico y literario / Mitjans, Aurelio – La Habana, Cuba. 1963 – 1r – us UF Libraries [972]
Estudio sobre las condiciones de vida de 179 famil... / Guatemala Direccion General De Estadistica – Guatemala, 1948 – 1r – us UF Libraries [972]
Estudio sobre las condiciones del desarrollo de co... / Mision 'Economia Y Humanismo' – Bogota, Colombia. v1-2. 1958 – 1r – us UF Libraries [972]
Estudio sobre las ideas politicas de jose antonio / Perez, Luis Marino – Habana, Cuba. 1908 – 1r – us UF Libraries [972]
Estudio sobre pesas y medidas en los paises centro / Zertuche C, Albino – s.l, s.l? 1958 – 1r – us UF Libraries [972]
Estudio socio-economico de un municipio rural : malpartida de caceres / Lancho Moreno, Juan Jose – Caceres: tip el noticiero, 1969 – 1 – sp Bibl Santa Ana [300]
Estudio y presentacion de los cuentos de ricardo m... / Rodriguez, Mario Augusto – Panama, 1956 – 1r – us UF Libraries [972]
Estudios / Recinos, Adrian et al – Guatemala: Edit. Universitaria, 1958 – sp Bibl Santa Ana [946]
Estudios / Vitier, Medardo – Habana, Cuba. 1944 – 1r – us UF Libraries [972]
Estudios arqueologicos y etnograficos / Cuervo Marquez, Carlos – Madrid, Spain. v1-2. 1920 – 1r – us UF Libraries [930]
Estudios bibliograficos sobre rafael landivar : guatemala, 1931 / Villacorta, J Antonio – Madrid: Razon y Fe, 1934 – 1 – sp Bibl Santa Ana [946]
Estudios clasicos. morfologia griega. sintaxis griega del p. santiago morillos j / Errandonea, Ignacio – Madrid: Razon y Fe, 1943 – 1 – sp Bibl Santa Ana [574]
Estudios constitucionales / Caro, Miguel Antonio – Bogota, Colombia. 1951 – 1r – us UF Libraries [972]
Estudios constitucionales / Colombia Comision De Estduios Constitucionales – Bogota, Colombia. v1-2. 1953 – 1r – us UF Libraries [323]
Estudios criticos / Gonzalez Serrano, Urbano – 1892 – 9 – sp Bibl Santa Ana [300]
Estudios criticos / Merchan, Rafael Maria – Bogota, Colombia. 1886 – 1r – us UF Libraries [972]
Estudios criticos / Vitier, Cintio – La Habana, Cuba. 1964 – 1r – us UF Libraries [972]
Estudios criticos acerca de la dominacion espanola en america / Cappa, Ricardo S J – 1887 – 9 – (1889 ed. 1890 ed) – sp Bibl Santa Ana [946]
Estudios de etnologia antigua de venezuela / Acosta Saignes, Miguel – Caracas, Venezuela. 1954 – 1r – 1 – us UF Libraries [972]
Estudios de historia del derecho espanol en las in... / Ots Y Capdequi, Jose Maria – Bogota, Colombia. 1940 – 1r – us UF Libraries [972]
Estudios de historiografia de la nueva espana – Mexico City? Mexico. 1945 – 1r – us UF Libraries [972]
Estudios de literatura dominicana / Matos, Esthervina – Ciudad Trujillo, Dominican Republic. 1955 – 1r – us UF Libraries [972]
Estudios de literatura hispanoamericana / Arrom, Jose Juan – Habana, Cuba. 1950 – 1r – us UF Libraries [972]
Estudios de metafisica biblica / Tresmontant, Claude – Madrid, Spain. 1961 – 1r – us UF Libraries [025]
Estudios ecumenicos – Mexico: Centro de Estudios Ecumenicos. [n2 apr/may 1969 [2nd ed only]; n8-41 sep 1970-1980; ns n1-32 dec 1984-1992] – 4r – 1 – us CRL [079]
Los estudios en la orden capuchina en el primer siglo de su existencia / Felder, Hilario – Madrid: Archivo Ibero Americano, 1960 – 1 – sp Bibl Santa Ana [240]
Estudios geograficos, instituto juan sebastian el cano (csic) – Madrid, Ano 1940-1961 – 325mf – 9 – sp Cultura [910]
Estudios gramaticales / Suarez, Marco Fidel – Bogota, Colombia. 1957 – 1r – us UF Libraries [440]
Estudios hispanoamericanos : homenaje a hernan cortes / Bayle, Constantino – Badajoz: imp dip provincial, 1948 – 1 – sp Bibl Santa Ana [972]
Estudios historicos / Bruni Celli, Blas – Caracas, Venezuela. 1964 – 1r – us UF Libraries [972]
Estudios historicos : caracas, 1927 / Rojas, Aristides – Madrid: Razon y Fe, 1930 – 1 – sp Bibl Santa Ana [946]
Estudios historicos / Giraldo Jaramillo, Gabriel – Bogota, Colombia. 1954 – 1r – us UF Libraries [972]
Estudios historicos / Guerra, Jose Joaquin – Bogota, Colombia. v1-4. 1952 – 1r – us UF Libraries [972]
Estudios historicos y literarios / Batres Jauregui, Antonio – Madrid, Spain. 1887 – 1r – us UF Libraries [972]
Estudios juridicos sobre cuestiones practicas de d... / Llano Y Raymat, Gregorio De – Habana, Cuba. 1928 – 1r – us UF Libraries [972]
Estudios literarios / Landarech, Alfonso Maria – San Salvador, El Salvador. 1959 – 1r – us UF Libraries [972]
Estudios literarios y filosoficos / Varona, Enrique Jose – Habana, Cuba. 1883 – 1r – us UF Libraries [100]
Estudios para la historia del arte colonial, vol 1 : arquitectura virreinal seguida de una adicion documental por jose torre...buenos aires, 1934 / Noel, Martin – Madrid: Razon y Fe, 1936 – 1 – sp Bibl Santa Ana [720]
Estudios politicos / Zecena, Mariano – Guatemala, 1957 – 1r – us UF Libraries [320]
Estudios preliminares del plan regulador / Robles Flores, Jose Luis – Guatemala, 1961 – 1r – us UF Libraries [972]
Estudios preliminares parala hora de la provincia de caceres / Rivas Mateos, Marcelo – Caceres: S.L., s.i., 1898 – 1 – sp Bibl Santa Ana [946]
Estudios sobre el simbolismo de la naturaleza = Etudes sur le symbolisme de la nature / La Bouillerie, Francois Alexandre Roullet de – Mexico: Imprenta del comercio de Dublan y Chavez 1877 [mf ed Bloomington IN: Indiana Uni Lib, Preservation Dept 1984] – 2v in 1 on 1r – 1 – us Indiana Preservation [390]
Estudios sobre el vocabulario politico espanol (1931-1971) : tesis doctoral / Rebollo Torio, Miguel Angel – Caceres: La Minerva Cacerena, 1976 – 1 – sp Bibl Santa Ana [321]
Estudios sobre escritores de america / Anderson Imbert, Enrique – Buenos Aires, Argentina. 1954 – 1r – 1 – us UF Libraries [972]
Estudios sobre jesus y su influencia / Nin Frias, Alberto – [Montevideo: s.n., ca 1906] – 1mf – 9 – 0-8370-4592-4 – (incl correspondence between miguel de unamuno and aut) – mf#1985-2592 – us ATLA [240]
Estudios sobre la filosofia de santo tomas / Gonzalez y Diaz Tunon, Ceferino – 2. ed. Madrid: nueva imprenta y libreria de san jose, 1886-1887 – 16mf – 9 – 0-524-05010-4 – mf#1991-2180 – us ATLA [240]
Estudios sobre linguistica aborigen de colombia / Ortiz, Sergio Elias – Bogota, Colombia. 1954 – 1r – 1 – us UF Libraries [490]
Estudios sobre literatura hispanoamericana y espan... / Monguio, Luis – Mexico City? Mexico. 1958 – 1r – us UF Libraries [440]
Estudios sobre literaturas hispano-americanas / Gonzalez, Manuel Pedro – Mexico City? Mexico. 1951 – 1r – us UF Libraries [440]
Estudios sobre los restos de ceramica romana / Barrantes Moreno, Vicente – Madrid, 1877 – 1 – sp Bibl Santa Ana [730]
Estudios sobre personajes y hechos de la historia / Arcaya, Pedro Manuel – Caracas, Venezuela. 1911 – 1r – us UF Libraries [972]
Estudios sobre san pedro de alcantara / Pedro de Alcantara, Saint – Madrid: Archivo Ibero Americano, 1924 – 1 – sp Bibl Santa Ana [240]
Estudios sociales / Moya, Francisco J – 1855 – 9 – sp Bibl Santa Ana [300]
Estudios teologicos : revista semestral de investigacion e informacion religiosa – Guatemala: Instituto Teologico Salesiano [v1 n1-v16 n31/32 (enero jun 1974-enero/dic 1989)] (semiannual) – 4r – 1 – us CRL [240]
Estudios y conferencias / Varona, Enrique Jose – Habana, Cuba. 1936 – 1r – us UF Libraries [972]
Estudio...virgen maria...almoharin (caceres) / Gonzalez y Grez, Juan J – 1898 – 9 – sp Bibl Santa Ana [240]
O estudo : orgao evolucionista – Rio de Janeiro, RJ. 26 jan 1888 – (= ser Ps 19) – bl Biblioteca [073]
O estudo : periodico scientifico e litterario – Recife, PE: Typ da Provincia, 08 maio 1875 – (= ser Ps 19) – bl Biblioteca [079]
Estudo critico dos trabalhos de marcgrave e piso s... / Lichtenstein, Hinrich – Sao Paulo, Brazil. 1961 – 1r – us UF Libraries [972]
Estudos : 2a serie / Lima, Alceu Amoroso – Rio de Janeiro, Brazil. 1934 – 1r – 1 – us UF Libraries [972]
Estudos brasileiros de populacao / Barretto, Castro – Rio de Janeiro, Brazil. 1947 – 1r – us UF Libraries [304]
Estudos criticos / Taunay, Alfredo D'escragnolle Taunay – Rio de Janeiro, Brazil. v1-2. 1881-1883 – 1r – us UF Libraries [972]
Estudos da lingua nacional / Neiva, Artur – Sao Paulo, Brazil. 1940 – 1r – us UF Libraries [440]
Estudos de geografia de bahia / Tricart, Jean – Salvador, Brazil. 1958 – 1r – us UF Libraries [918]
Estudos de historia colonial / Vianna, Helio – Sao Paulo, Brazil. 1948 – 1r – us UF Libraries [972]
Estudos de historia do brasil / Magalhaes, Basilio De – Sao Paulo, Brazil. 1940 – 1r – us UF Libraries [972]
Estudos de historia imperial / Vianna, Helio – Sao Paulo, Brazil. 1950 – 1r – us UF Libraries [972]
Estudos de historia paulista / Taunay, Afonso De E – Sao Paulo, Brazil. 1927 – 1r – us UF Libraries [972]
Estudos de religiao – Sao Bernardo do Campo, Brasil: Impr Metodista. [v1-3 n5 mar 1985-jun 1988] – 1r – 1 – us CRL [079]
Estudos e notas criticas / Tati, Miecio – Rio de Janeiro, Brazil. 1958 – 1r – us UF Libraries [972]
Estudos historicos e politicos / Calogeras, Joao Pandia – Sao Paulo, Brazil. 1936 – 1r – us UF Libraries [972]
Estudos integralistas – Sao Paulo, Brazil. 1933 – 1r – us UF Libraries [972]
Estudos piauienses / Miranda, Agenor Augusto De – Sao Paulo, Brazil. 1938 – 1r – us UF Libraries [972]
Estudos regionaes / Arede, Joao Domingues – Couto de Cucujaes, Portugal. 1925 – 1r – us UF Libraries [960]
Estudos sobre a nova capital do brasil / Demostenes, Manuel – Rio de Janeiro, Brazil. 1947 – 1r – us UF Libraries [972]
Estudos sobre historia do brasil / Ennes, Ernesto – Sao Paulo, Brazil. 1947 – 1r – us UF Libraries [972]
Estudos sobre o negro / Mello, A Da Silva – Rio de Janeiro, Brazil. 1958 – 1r – us UF Libraries [972]
Estudos sociais da guanabara / Souza, Geraldo Sampaio De – Rio de Janeiro, Brazil. 1970 – 1r – us UF Libraries [972]
Estudos teologicos : orgao da faculdade de teologia – Sao Leopoldo, RGS [Rio Grande do Sul]: Federacao Sinodal, Igreja Evangelica de Confissao Luterana no Brasil, [v1-31 (1961-1991] – 4r – 1 – us CRL [079]
Eswau huppeday / Broad River Genealogical Society – 1983 feb-1987 nov – 1r – 1 – mf#1321043 – us WHS [929]
Eszakmagyarorszag – Miskolc, Hungary. 1962-1990 – 54r – 1 – us L of C Photodup [079]
Et cetera – Concord. 1977+ (1) 1977+ (5) 1977+ (9) – (cont: etc: a review of general semantics) – ISSN: 0014-164X – mf#383,01 – us UMI ProQuest [400]
'Et la-ledet ve-kol sason – Tunis, Tunisia. 1910 – 1r – 1 – us UF Libraries [939]
Et omstridt land : studiebog over missionen i manchuriet / Nyholm, J – Kobenhavn: Danske missionsselskab, 1913 [mf ed 1995] – (= ser Yale coll) – 290p (ill) – 1 – 0-524-09807-7 – (in danish) – mf#1995-0807 – us ATLA [951]
Et quand on ne sait pas (encore) lire ? : cycle 2 / Stantina, Andre – 193mf – 9 – €42.69 – mf#250B0144 – fr CRDP [370]
'Et sofer hadash / Landau, Zemah – Vilna, Lithuania. 1844 – 1r – 1 – us UF Libraries [939]
L'etablissement des recollets a l'isle percee, 1673-1690 / Hugolin, pere – Quebec: [s.n.] 1912 [mf ed 1995] – 1mf – 9 – 0-665-75751-4 – mf#75751 – cn Canadiana [241]
L'etablissement des recollets a montreal, 1692 / Hugolin, pere – Montreal: [s.n.] 1911 [mf ed 1995] – 1mf – 9 – 0-665-74640-7 – mf#74640 – cn Canadiana [241]
Etablissement des soeurs de charite a la riviere rouge / Dugas, Georges – S.l: s.n, 18–? – 1mf – 9 – mf#05497 – cn Canadiana [360]
Etablissements ballande : agendas (business diaries) 1933-42, 1950-54 – Port Vila, Santo, New Hebrides – 2r – 1 – (restricted access) – mf#PMB1130 – at Pacific Mss [380]
Etah and beyond : or Life within twelve degrees of the Pole / MacMillan, D B – London, 1928 – 7mf – 9 – mf#N-307 – ne IDC [919]
Il etait une bergere / Rivoire, Andre – Paris, France. 1905 – 1r – us UF Libraries [440]
Il etait une bergere / Rivoire, Andre – Paris, France. 1905 – 1r – us UF Libraries [440]
Il etait une fois– / Palassie, Georges – Bordeux, France. 1930 – 1r – us UF Libraries [440]
Il etait une fois– / Palassie, Georges – Bordeux, France. 1930 – 1r – us UF Libraries [440]
Etapa extremena en la biografia del ministro fernandez negrete / Mota Arevalo, Horacio – Badajoz: (imprenta de la diputacion provincial), 1964 – 1 – sp Bibl Santa Ana [350]
Etapas / Fortun Y Fortun, Joaquin – Habana, Cuba. 1946 – 1r – us UF Libraries [972]
Etapas de la vida colombiana / Santos, Eduardo – Bogota, Colombia. 1946 – 1r – 1 – us UF Libraries [972]
Etape de l'evolution haitienne / Mars, Jean Price – Port-Au-Prince, Haiti. 1929? – 1r – us UF Libraries [972]
Etapes de la guadeloupe religieuse / Guilbaud – Basse-Terre, Guadeloupe. 1936 – 1r – us UF Libraries [972]
Les etapes du rationalisme dans ses attaques contre les evangiles et la vie de jesus-christ : exposition historique et critique / Fillion, Louis-Claude – Paris: P Lethielleux, [1911?] – 1mf – 9 – 0-524-05609-9 – (incl bibl ref) – mf#1992-0464 – us ATLA [220]
Etapes d'un relevement / Haiti (Republic) Service D'information, De Presse – Port-Au-Prince, Haiti. 1956 – 1r – us UF Libraries [972]
Les etapes d'une classe au petit seminaire de quebec, 1859-1868 / Gosselin, David – [Quebec?: H Chasse], 1908 – 4mf – 9 – 0-665-74312-2 – (incl app) – mf#74312 – cn Canadiana [241]
Etapes et perspectives de l'union francaise / Gueye, Lamine – Paris: editions de l'union francaise, [1955] – 1 – us CRL [960]
Etapy zhiznennogo tsikla i byt rabotaiushchei zhenshchiny / Gordon, L et al – Moskva: In-t konkretnykh sotsialnykh issledovanii AN SSSR, 1972 – (Filmed with: Dinamika izmeneniia polozheniia dagestanskoi zhenshchiny i semia/S Gadzhieva) – us CRL [077]
L' etat *see* La cloche
Etat actuel de la corse : caractere et moeurs de ses habitans / Sebastiani, Francois – Paris 1821 [mf ed Hildesheim 1995-98] – 2mf – 9 – €60.00 – 3-487-29168-1 – gw Olms [914]
Etat actuel de l'empire ottoman : contenant des details plus exacts que tous ceux qui ont parus jusqu'a present sur la religion, le gouvernement, la milice, les moeurs et les amusemens des turcs, ... / Habesci, Elias – Paris 1792 [mf ed Hildesheim 1995-98] – 2v on 4mf – 9 – €120.00 – 3-487-29135-5 – gw Olms [932]
Etat actuel des missions protestantes en haiti / Pressoir, Catts – Petion-Ville, Haiti. 194- – 1r – 1 – us UF Libraries [242]
L'etat chretien calviniste a geneve : au temps de theodore de beze / Choisy, Eugene – Geneve: Ch Eggimann; Paris: Fischbacher [1902?] [mf ed 1989] – 2mf – 9 – 0-7905-4204-8 – (in french. incl bibl ref) – mf#1988-0204 – us ATLA [230]
Etat critique du texte d'agee : quatre tableaux comparatifs / Andre, Louis Edward Tony – Paris: Fischbacher, 1895 – 1mf – 9 – 0-8370-2099-9 – mf#1985-0099 – us ATLA [220]
Etat de la cochinchine francaise – Saigon 1879-83, 1885-92, 1899, 1908 – 3r – 1 – us CRL [079]
Etat de la corse : suivi d'un journal d'un voyage dans l'isle et des memoires de pascal paoli; avec une preface du traducteur / Boswell, James [the Elder] – Londres [i e Lausanne] 1769 [mf ed Hildesheim 1995-98] – 2v on 5mf – 9 – €100.00 – 3-487-29170-3 – gw Olms [910]
Etat de l'instruction primaire en 1864, d'apres les rapports officiels des inspecteurs d'academie / France. Ministere de l'Instruction Publique – Paris. 1866 – 1 – fr ACRPP [324]
Etat des colonies et du commerce des europeens dans les deux indes, depuis 1783 jusqu'en 1821 : pour faire suite a l'histoire philosophique et politique des etablissements et du commerce des europeens dans les deux indes, de raynal / Peuchet, Jacques – Paris: Amable Costes & cie...1821 [mf ed 1984] – 9mf – 9 – 0-665-18848-X – mf#18848 – cn Canadiana [380]
Etat des sommes depensees a meme l'octroi de £30, 000 vote dans le but d'aider a l'etablissement des terres vacantes de la couronne dans le bas-canada / Canada (Province). Departement des terres de la couronne – [s.l.]: [s.n.] 1855 [mf ed 1983] – 1mf – 9 – (with ind) – mf#SEM105P239 – cn Bibl Nat [336]
Etat des sommes payees par le gouvernement : et correspondance echangee entre les ingenieurs et autres officiers, relativement a certains chemins de fer – Quebec: impr John Lovell, 1853 [mf ed 1992] – 1mf – 9 – mf#SEM105P1461 – cn Bibl Nat [336]
Etat des travaux legislatifs... / France. Assemblee nationale. Chambre des Deputes – Paris. 1886-Sept 1939. 12 reels – 1 – us L of C Photodup [944]
Etat detaille des depenses faites pendant le voyage en europe des honorables mm mercier : premier ministre, et shehyn, tresorier provincial, et de mm n bernatchez, r ness et alex clement, secretaire, in re l'emprunt provincial et l'etude de la question betteraviere – [S.l.]: [s.n.], [1891] (mf ed 1987) – 1mf – 9 – mf#SEM105P613 – cn Bibl Nat [914]

Etat et avenir du canada en 1854 : tel que retrace dans les depeches du tres-honorable comte d'elgin et kincardine, gouverneur-general du canada, au principal secretaire d'etat de sa majeste pour les colonies / Canada (Province) – Quebec: impr par S Derbishire & G Desbarats, 1855 [mf ed 1993] – 1mf – 9 – mf#SEM105P1990 – cn Bibl Nat [336]

L'etat mahdiste du soudan / Dujarric, Gaston – Paris: J Maisonneuve, 1901 – 1 – us CRL [960]

L'etat pontifical apres le grand schisme : etude de geographie politique / Guiraud, Jean – Paris: A Fontemoing, 1896 [mf ed 1990] – (= ser Bibliotheque des ecoles francaises d'athenes et de rome) – 1mf – 9 – 0-7905-6067-4 – (in french. incl bibl ref) – mf#1988-2067 – us ATLA [930]

Etat present de la noblesse francaise : contenant le dictionaire de la noblesse contemporaine... – 2e ed. Paris: Librairie Bachelin-Deflorenne, 1868 [mf ed 1982] – 10mf – 9 – mf#SEM105P84 – cn Bibl Nat [929]

Etat present des nations et eglises grecque, armenienne, et maronite en turquie / Croix, de la – Paris, 1715 – 4mf – 9 – mf#AR-1683 – ne IDC [956]

L'etat primitif de l'homme dans la tradition de l'eglise avant saint augustin / Slomkowski, A – Paris, 1928 – 3mf – 9 – €7.00 – ne Slangenburg [241]

L'etat religieux et politique de la france contemporaine / Alexis, R P – Quebec: Impr de l'Evenement, 1912 [mf ed 1986] – 1mf – 9 – mf#SEM105P686 – cn Bibl Nat [241]

Etat restitue : ou, le comte de bourgogne / Kotzebue, August Von – Paris, France. 1814 – 1r – us UF Libraries [440]

L'etat sans dieu : mal social de la france / Nicolas, Auguste – 3e ed. Paris: E Vaton, [1873?] [mf ed 1991] – 1mf – 9 – 0-7905-9425-0 – (in french) – mf#1989-2650 – us ATLA [944]

Etats africains d'expression francaise et republique malgache – Paris, France. 1964 – 1r – us UF Libraries [960]

Etats financiers presentes a l'assemblee legislative le 17 octobre 1843 : par ordre de son excellence le gouverneur general, conformement a une resolution de la chambre du 8 septembre 1841 – [S.l: s.n, 1843 ?] (mf ed 1992) – 1mf – 9 – mf#SEM105P1758 – cn Bibl Nat [350]

Etats indiquant le mouvement du commerce et de la marine : presentes au parlement par ordre de son excellence, 1850 – Toronto: impr par Stewart Derbishire & George Desbarats, 1850 [mf ed 1993] – 1mf – 9 – (with ind) – mf#SEM105P1736 – cn Bibl Nat [380]

Les etats-unis : origine, institutions, developpement / DeCelles, Alfred Duclos – Montreal: Librairie Beauchemin, 1913 – 4mf – 9 – 0-665-73904-4 – mf#73904 – cn Canadiana [975]

Les etats-unis : origine, institutions, developpement / DeCelles, Alfred Duclos – 2e ed. Montreal: Librairie Beauchemin ltee, 1913 [i.e. 1924] (mf ed 1985) – (= ser Bibliotheque canadienne. coll jacques cartier) – 4mf – 9 – (with ind and bibl) – mf#SEM105P510 – cn Bibl Nat [370]

Les etats-unis : origine, institutions, developpement / DeCelles, Alfred Duclos – 2e ed. Montreal: Librairie Beauchemin ltee, 1913 [mf ed 1985] – (= ser Bibliotheque canadienne. coll jacques cartier) – 4mf – 9 – (with ind and bibl) – mf#SEM105P511 – cn Bibl Nat [975]

Les etats-unis : origine, institutions, developpement / DeCelles, Alfred Duclos – 3e ed. Montreal: Librairie Beauchemin, ltee, 1925 [mf ed 1985] – (= ser Bibliotheque canadienne. coll jacques cartier) – 4mf – 9 – (with ind and bibl) – mf#SEM105P511 – cn Bibl Nat [975]

Les etats-unis d'amerique *see* America and her problems

Etats-unis d'amerique et la banqueroute d'haiti / Sejourne, Georges – Port-Au-Prince, Haiti. 1932 – 1r – 1 – us UF Libraries [972]

Les etats-unis d'amerique et l'angleterre : annexion du texas; l'oregon / Jollivet, Adolphe – [Paris?: s.n.], 1845 [mf ed 1982] – (= ser Documents americians. Troisieme serie) – 1mf – 9 – mf#18335 – cn Canadiana [975]

Etats-unis de colombie / Pereira, Ricardo S – Paris, France. 1883 – 1r – 1 – us UF Libraries [972]

Les etats-unis d'europe ont commence; la communaute europeene du charbon et de l'acier, discours et allocutions, 1952-54 / Monnet, Jean – Paris: R. Laffont, 1955. 171p – 1 – us Wisconsin U Libr [940]

Etats-unis et le marche haitien / Turnier, Alain – Washington, DC. 1955 – 1r – us UF Libraries [972]

Etats-unis, manitoba, et nord-ouest : notes de voyage / Dionne, Narcisse-Eutrope – Quebec?: L Brousseau, 1882 – 2mf – 9 – mf#64254 – cn Canadiana [917]

Etc : a review of general semantics – San Francisco. 1943-1976 [1]; 1970-1976 [5]; 1976-1976 [9] – (Cont by: et cetera) – ISSN: 0014-164X – mf#383 – us UMI ProQuest [400]

Etched beads in india : decorative patterns and the geographical factors in their distribution / Dikshit, Moreshwar Gangadhar – Poona: Deccan College Postgraduate and Research Institute, 1949 – (= ser Samp: indian books) – us CRL [740]

The etched work of rembrandt...the unauthentic character of certain of those etchings / Haden, Francis Seymour – London 1879 – (= ser 19th c art & architecture) – 1mf – 9 – mf#4.2.1351 – uk Chadwyck [760]

Etcher – London. 1879-1883 (1) – mf#5319 – us UMI ProQuest [700]

The etcher's handbook / Hamerton, Philip Gilbert – London 1871 – (= ser 19th c art & architecture) – 2mf – 9 – mf#4.2.1774 – uk Chadwyck [760]

Etchie, Michael P *see* A submaximal one-mile track jog to estimate vo2max in fit men and women, ages 30-39 years

Etching and etchers / Hamerton, Philip Gilbert – 3rd ed. London 1880 – (= ser 19th c art & architecture) – 6mf – 9 – mf#4.2.133 – uk Chadwyck [760]

Etching and mezzotint engraving / Herkomer, Hubert von – London 1892 – (= ser 19th c art & architecture) – 2mf – 9 – mf#4.2.136 – uk Chadwyck [760]

Etching, engraving : and the other methods of printing pictures / Singer, Hans Wolfgang & Strang, William – London 1897 – (= ser 19th c art & architecture) – 4mf – 9 – mf#4.2.64 – uk Chadwyck [760]

Etching in england / Wedmore, Frederick – London 1895 – (= ser 19th c art & architecture) – 3mf – 9 – mf#4.2.199 – uk Chadwyck [760]

Etchings : representing the best examples of ancient ornamental architecture / Tatham, Charles Heathcote – [2nd ed]. London 1803 – (= ser 19th c art & architecture) – 6mf – 9 – mf#4.2.1682 – uk Chadwyck [760]

Etchings of hollar in the royal library / Windsor Castle. Royal Library – 71mf – 9 – $570.00 – 0-907006-23-X – (most complete coll of hollar's etchings; filming follows sequence of the 1853 parthey catalogue; descriptive catalogue by richard pennington (printed) available fr cambridge up (1982)) – uk Mindata [760]

L'ete du nord, ou voyage autour de la baltique : par le danemarck, la suede, la russie et partie de l'allemagne, dans l'annee 1804 / Carr, John – Paris 1808 [mf ed Hildesheim 1995-98] – 2v on 4mf – 9 – €120.00 – 3-487-27838-3 – gw Olms [914]

Un ete en amerique de l'atlantique aux montagnes rocheuses / Leclercq, Jules – 2e ed. Paris: E Plon, Nourrit, 1886 [mf ed 1984] – 5mf – 9 – 0-665-04565-4 – mf#04565 – cn Canadiana [917]

Eteenpain – New York NY, 1922-aug 1950 – 36r – 1 – (finnish newspaper) – us IHRC [071]

L'etendard – Paris. 27 juin 1866-25 avr 1869 – 1 – fr ACRPP [073]

L'etendard revolutionnaire : organe anarchiste hebdomadaire. – no. 1-12. juil-oct 1882. Remplace: Le Droit social – 1 – (remplace: le droit social) – fr ACRPP [335]

Eternal hope : five sermons preached in westminster abbey nov and dec, 1877 / Farrar, Frederic William – London, New York: Macmillan, 1892 [mf ed 1989] – 1mf – 9 – 0-7905-3017-1 – (incl bibl ref) – mf#1987-3017 – us ATLA [210]

The eternal in man / Vance, James Isaac – New York: Fleming H Revell, c1907 – 1mf – 9 – 0-8370-5647-0 – (incl bibl ref) – mf#1985-3647 – us ATLA [240]

Eternal life : a study of its implications and applications / Huegel, Friedrich, Freiherr von – Edinburgh: T & T Clark, 1912 – 2mf – 9 – 0-7905-7305-9 – (incl bibl ref) – mf#1989-0530 – us ATLA [240]

The eternal life / Muensterberg, Hugo – Boston: Houghton, Mifflin, 1905 – 1mf – 9 – 0-7905-8529-4 – mf#1989-1754 – us ATLA [240]

Eternal life in jesus christ / Sumner, John Bird – London, England. 1840 – 1r – 1 – us UF Libraries [240]

The eternal lotus : a novel / Bandyopadhyaya, Tarasankara – Calcutta: Purvasa Ltd, 1945 – (= ser Samp: indian books) – us CRL [830]

Eternal perfection of the elect in christ / Spurgeon, James A – London, England. 1856 – 1r – 1 – us UF Libraries [240]

The eternal priesthood / Manning, Henry Edward – New York: Catholic Publ Society, [between 1875-1892] – 1mf – 9 – 0-8370-7407-X – mf#1986-1407 – us ATLA [240]

Eternal punishment / the coming one : being the 7th annual lecture and sermon...1884 / Shaw, William Isaac & Parker, William R – Toronto: W Briggs; Montreal: C W Coates, 1884 – 1mf – 9 – mf#13541 – cn Canadiana [210]

The eternal purpose of god in christ jesus our lord : being the fourth series of lectures / Kelly, James – 4th rev enl ed. London: James Nisbet, 1860 – 1mf – 9 – 0-8370-4361-1 – mf#1985-2361 – us ATLA [240]

Eternal realities considered as forming the principle of missionary / Elliott, Richard – London, England. 1820 – 1r – us UF Libraries [240]

The eternal saviour-judge / Clarke, James Langton – 2nd ed. London: J Murray, 1905 – 1mf – 9 – 0-524-05398-7 – (incl bibl ref) – mf#1992-0408 – us ATLA [220]

The eternal values = Philosophie der werte / Muensterberg, Hugo – Boston: Houghton Mifflin, 1909 – 2mf – 9 – 0-7905-9529-X – (in english) – mf#1989-1234 – us ATLA [100]

The eternal verities : a series of plain arguments showing the abundant evidences of the truth of the holy scriptures / Miller, Daniel Long – 1st ed. Elgin, IL: Brethren Pub House, 1902 – 1mf – 9 – 0-524-03382-X – (incl bibl ref) – mf#1990-4694 – us ATLA [220]

Eternity – Philadelphia. 1950-1989 [1,5,9] – ISSN: 0014-1682 – mf#1906 – us UMI ProQuest [240]

Eternity! – London, England. 18– – 1r – us UF Libraries [240]

Eternity of future punishment / Salmon, George – Dublin, Ireland. 1864 – 1r – us UF Libraries [240]

The eternity of future punishment and the place which this doctrine ought to hold in christian preaching : two sermons / Salmon, George – Dublin: Hodges, Smith, 1864 – 1mf – 9 – 0-524-00316-5 – mf#1989-3016 – us ATLA [240]

Eternity of heaven and hell : or, a renunciation of the error of universalism / Fernald, Woodbury Melcher – [S.l.: s.n., 1855?] – 1mf – 9 – 0-524-05077-5 – mf#1991-2201 – us ATLA [240]

Eternity of the universe / Toulmin, G H – London, England. 1825 – 1r – us UF Libraries [240]

Etheart, Liautaud *see* Gouvernement du general boisrond-canal

Etheridge, John Wesley *see*
– Horae aramaicae
– Jerusalem and tiberias
– The targums of onkelos and jonathan ben uzziel on the pentateuch

The ethic of freethought : and other addresses and essays / Pearson, Karl – 2nd rev ed. London: Adam & Charles Black, 1901 [mf ed 1985] – 2mf – 9 – 0-8370-4689-0 – (incl bibl ref) – mf#1985-2689 – us ATLA [170]

The ethic of jesus according to the synoptic gospels / Stalker, James – London: Hodder & Stoughton, [1909?] – 1mf – 9 – 0-7905-3289-1 – mf#1987-3289 – us ATLA [226]

Ethica aristotelica, ad sacrarum literarum normam emendata : ejusdem ethica christiana, seu explicatio virtutum et vitiorum, quorum in sacris literis sit mentio. huic editioni praeter praefixam auctoris vitam, accedit catechesis ecclesiarum polonicarum / Crell, Johann – Cosmopoli: Per Eugenium Philalethem 1681 – (= ser Ethics in the early modern period) – 15mf – 9 – mf#pl-475 – ne IDC [170]

Ethica generalis / Signoriello, Nuntio – 4 ed. Neapoli: bibliothecae catholicae scriptorum, 1889 – 1mf – 9 – 0-8370-7428-2 – (incl bibl ref) – mf#1986-1428 – us ATLA [230]

Ethica naturalis seu documenta moralia e variis rerum naturalium proprietatibus virtutum vitiorumque... / Weigel, Johann Christoph – Norimbergae, [1700] – 3mf – 9 – mf#O-826 – ne IDC [090]

Ethica, politica, oeconomica / Aristoteles [Aristotle] – 15th c – (= ser Holkham library manuscript books 444) – 1r – 1 – (ethica is trans by argyopulus. politica and oeconomica by aretinus) – mf#96532 – uk Microform Academic [170]

Ethica Specialis / Signoriello, Nuntio – 4 ed. Neapoli: Bibliothecae Catholicae Scriptorum, 1889 – 1mf – 9 – 0-8370-7429-0 – (incl bibl ref) – mf#1986-1429 – us ATLA [230]

Ethica symbolica e fabularum umbris in veritatis lucem varia eruditione... / Pexenfelder, M – Monachii: Sumptibus Ioannis Wagneri, & Ioannis Hermanni...Gelder, 1675 – 11mf – 9 – mf#O-398 – ne IDC [090]

Ethicae, seu, de moribus philosophiae brevis et perspicua descriptio : diligenter & ordine perfacili explicata / Valerius, Cornelius – Antverpiae: Ex officina Christophori Plantini 1574 – (= ser Ethics in the early modern period) – 2mf – 9 – mf#pl-463 – ne IDC [170]

Ethical and moral instruction in schools / Palmer, George Herbert – Boston: Houghton, Mifflin, c1909 – 1mf – 9 – 0-8370-8541-1 – mf#1986-2541 – us ATLA [230]

Ethical and religious problems of the war : fourteen addresses / Murray, Gilbert et al; ed by Carpenter, Joseph Estlin – London: Lindsey Press, 1916 – 1mf – 9 – 0-7905-9530-3 – mf#1989-1235 – us ATLA [230]

The ethical approach to theism / Barbour, George Freeland – Edinburgh: W Blackwood, 1913 – 1mf – 9 – 0-7905-3537-8 – mf#1989-0030 – us ATLA [210]

Ethical christianity : a series of sermons / Hughes, Hugh Price – New York: E P Dutton 1892 [mf ed 1990] – (= ser Preachers of the age) – 1mf – 9 – 0-7905-7770-4 – (incl bibl ref) – mf#1989-0995 – us ATLA [242]

Ethical decision making by registered nurses in a bureaucratic context / Nevhutanda, Tshilidzi Rachel – Uni of South Africa 2000 [mf ed Johannesburg 2000] – 4mf – 9 – (incl bibl ref) – mf#mfm837 – sa Unisa [610]

Ethical ideals in India today : delivered at Conway hall, Red Lion Square, WCI on March 22, 1942 / Thompson, Edward John – London: Watts & Co, 1942 – (= ser Samp: indian books) – us CRL [170]

The ethical import of darwinism / Schurman, Jacob Gould – [London] 1888[1887] – (= ser 19th c evolution & creation) – 3mf – 9 – mf#1.1.1582 – uk Chadwyck [170]

The ethical import of darwinism / Schurman, Jacob Gould – New York: Scribner, 1887 – 1mf – 9 – 0-524-07592-1 – mf#1991-3212 – us ATLA [170]

An ethical movement : a volume of lectures / Sheldon, Walter Lorenzo – New York: Macmillan, 1896 [mf ed 1990] – 1mf – 9 – 0-7905-6440-8 – (incl bibl ref) – mf#1988-2440 – us ATLA [170]

The ethical outlook of the current drama / Speer, James Charles – Toronto: W Briggs, 1902 – 1mf – 9 – 0-665-77683-7 – mf#77683 – cn Canadiana [790]

An ethical philosophy of life : presented in its main outlines / Adler, Felix – New York: D Appleton, 1925, c1918 [mf ed 1994] – 1mf – 9 – 0-524-08845-4 – mf#1993-2130 – us ATLA [170]

Ethical principles underlying education / Dewey, John – Chicago: The University of Chicago Press, 1916 [mf ed 1970] – (= ser Library of american civilization 40123) – 34p on 1mf – 9 – (repr fr the third yearbook of the national herbart society) – us Chicago U Pr [170]

Ethical record – London. 1972-1981 (1) 1972-1981 (5) 1976-1981 (9) – ISSN: 0014-1690 – mf#6930 – us UMI ProQuest [170]

Ethical record – Philadelphia. 1888-1890 (1) – mf#2886 – us UMI ProQuest [170]

Ethical religion = Naithi dharma / Gandhi, Mahatma – Madras: S Ganesan, 1922 – (= ser Samp: indian books) – (trans fr Hindi by A Rama Iyer; appreciation of the author by JH Holmes) – us CRL [170]

Ethical religion / Salter, William Mackintire – Boston: Roberts, 1889 – 1mf – 9 – 0-7905-8576-6 – mf#1989-1801 – us ATLA [170]

Ethical review *see* Ethical world series, 1898-1916

Ethical studies / Bradley, Francis Herbert – London: HS King, 1876 – 1mf – 9 – 0-7905-3609-9 – mf#1989-0102 – us ATLA [170]

An ethical sunday school : a scheme for the moral instruction of the young / Sheldon, Walter Lorenzo – London: S Sonnenschein, 1900 [mf ed 1992] – (= ser The ethical library) – 1mf – 9 – 0-524-02931-8 – mf#1990-0747 – us ATLA [170]

The ethical teaching of jesus / Briggs, Charles Augustus – New York: Scribner's, 1904 – 1mf – 9 – 0-8370-2450-1 – (incl indof subjects and biblical passages cited) – mf#1985-0450 – us ATLA [240]

Ethical teachings for the young *see* Chu-tzu hsiao hsueh [ccm111]

The ethical thought of carlyle marney / Blackwell, Michael C – 1982 – 1 – $5.00 – us Southern Baptist [170]

Ethical world (new series) *see* Ethical world series, 1898-1916

Ethical world series, 1898-1916 / ed by Coit, Stanton et al – Charles Albert Watts – (= ser Periodicals connected with owenite socialism and its successors in secularist, freethought and allied movements, 1834-1916) – 7r – 1 – (incl: democracy 1901; ethics 1901-06; ethical review 1906; ethical world (new series) 1907-16) – mf#97146 – uk Microform Academic [073]

The ethical world-conception of the norse people / Fors, Andrew Peter – Chicago: University of Chicago Press, 1904 – 1mf – 9 – 0-524-01058-7 – mf#1990-2206 – us ATLA [390]

Ethices christianae : libri tres in quibus de veris humanarum actionum principiis agitur: atque etiam legis diuinae, siue decalogi, explicatio, illiusque cum scriptis scholasticorum, iure naturali siue philosophico, ciuili romano re... / Daneau, Lambert – editio tertia ab ipso authore recognita. Genevae: Apud Eustath Vignon 1588 – (= ser Ethics in the early modern period) – [11]lea on 8mf – 9 – (with ind) – mf#pl-465 – ne IDC [230]

Ethices historicae specimen : sive, ad genuinam morum humanorum ex historiis cognitionem manuductio, & eorundem characteres ac notas indagandi via: cum singulorum capitum argumentis, & rerum ac verborum praecipuorum indice / Brochmand, Erasmus Johannes – Lugduni Batavorum: Ex officina Adriani Wyngaerden, anno 1653 – (= ser Ethics in the early modern period) – 7mf – 9 – mf#pl-472 – ne IDC [170]

The ethichs of the christian life : or, the science of right living / Robins, Henry Ephraim – Philadelphia: Griffith & Rowland Press, 1904 – 2mf – 9 – 0-7905-9852-3 – mf#1989-1577 – us ATLA [170]

Ethics / Barrett, Clifford – New York, NY. 1933 – 1r – us UF Libraries [170]

Ethics – Chicago. 1890+ (1) 1967+ (5) 1977+ (9) – ISSN: 0014-1704 – mf#482 – us UMI ProQuest [170]

Ethics : descriptive and explanatory / Mezes, Sidney Edward – New York: Macmillan, 1910, c1900 – 2mf – 9 – 0-8370-6283-7 – (incl bibl ref and ind) – mf#1986-0283 – us ATLA [170]

Ethics / Dewey, John – New York: Henry Holt, 1909 [mf ed 1993] – (= ser American science series) – 2mf – 9 – 0-524-08233-2 – mf#1993-2008 – us ATLA [170]

Ethics / Dewey John & Tufts James H – New York/London: Henry Holt & Co; George Bell & Sons, 1909 – 7mf – 9 – $10.50 – mf#LLMC 92-239 – us LLMC [170]

Ethics : an international journal of social, political and legal philosophy – v1-59. oct 1890-jul 1949 [complete] – 13r – 1 – ISSN: 0014-114704 – mf#ATLA S0009 – us ATLA [170]

Ethics / Moore, George Edward – New York: H Holt, [1912?] – (= ser Home university library of modern knowledge) – 1mf – 9 – 0-7905-8856-0 – mf#1989-2081 – us ATLA [170]

Ethics : or, science of duty / Bascom, John – New York: G P Putnam, 1879 – 1mf – 9 – 0-8370-6010-9 – (incl ind) – mf#1986-0010 – us ATLA [170]

Ethics / Rashdall, Hastings – London: TC & EC Jack; New York: Dodge, [1913?] – (= ser The People's Books) – 1mf – 9 – 0-7905-9598-2 – (Incl bibl ref) – mf#1989-1323 – us ATLA [170]

Ethics *see* Ethical world series, 1898-1916

Ethics and atonement / Lofthouse, William Frederick – London: Methuen 1906 [mf ed 1985] – 1mf – 9 – 0-8370-4288-7 – (incl bibl ref & ind) – mf#1985-2288 – us ATLA [230]

Ethics and moral science = Morale et la science des moeurs / Levy-Bruhl, Lucien – London: A Constable, 1905 – 1mf – 9 – 0-7905-9305-X – (In English) – mf#1989-2530 – us ATLA [170]

Ethics and revelation / Nash, Henry Sylvester – New York: Macmillan, 1899 [mf ed 1986] – (= ser The bohlen lectures 1899?) – 1mf – 9 – 0-8370-6508-9 – mf#1986-0508 – us ATLA [230]

Ethics and the environment – Bloomington, 1999+ [1,5,9] – ISSN: 1085-6633 – mf#26020 – us UMI ProQuest [300]

Ethics and the family / Lofthouse, William Frederick – London; New York: Hodder and Stoughton, [1912?] – 1mf – 9 – 0-7905-7904-9 – mf#1989-1129 – us ATLA [170]

Ethics and the materialist conception of history *see* Ethik und materialistische geschichtsauffassung

Ethics and the "new education" / Bryant, Wm M – Chicago: S C Griggs, 1894 [mf ed 1986] – 1mf – 9 – 0-8370-7448-7 – mf#1986-1448 – us ATLA [230]

Ethics Commission of the American Association of Law Libraries *see* Papers on the aall code of ethics

Ethics for children : a guide for teachers and parents / Cabot, Ella Lyman – Boston: Houghton Mifflin, c1910 – 1mf – 9 – 0-8370-7850-4 – (Incl ind) – mf#1986-1850 – us ATLA [370]

Ethics for young people / Everett, Charles Carroll – Boston: Ginn, 1892, c1891 – 1mf – 9 – 0-8370-8736-8 – mf#1986-2736 – us ATLA [170]

Ethics in science and medicine – Oxford. 1973-1980 (1,5,9) – (Cont by: Social science and medicine Pt F: Medical and social ethics) – ISSN: 0306-4581 – mf#49070 – us UMI ProQuest [170]

Ethics in the early modern period / Freedman, Joseph S [comp] – 40 titles on 139mf – 9 – (with finding aids; titles also listed individually) – ne IDC [170]

Ethics management : a challenge to public service in south africa / Nethonzhe, Thonzhe Alpheus – Pretoria: Vista University 2002 [mf ed 2002] – 3mf – 9 – (incl bibl ref) – mf#mmf15215 – sa Unisa [170]

The ethics of aristotle = Nicomachean ethics / Aristotle; ed by Burnet, John – London: Methuen, 1900 – 2mf – 9 – 0-7905-3750-8 – mf#1989-0243 – us ATLA [180]

The ethics of conformity and subscription / Sidgwick, Henry – London: Williams and Norgate, 1870 – 1mf – 9 – 0-524-00345-9 – mf#1989-3045 – us ATLA [240]

The ethics of confucius : the sayings of the master and his disciples upon the conduct of the superior man / Confucius – New York: GP Putnam, c1915 – 1mf – 9 – 0-524-07996-X – mf#1991-0218 – us ATLA [170]

Ethics of fasting / Gandhi, Mahatma; ed by Chander, Jag Parvesh – Lahore: Indian Print Works, [between 1945 and 1955] – (= ser Samp: indian books) – us CRL [230]

The ethics of felix adler / Barth, Joseph – [Chicago?], 1935. Chicago: Dep of Photodup, U of Chicago Lib, 1971 (1r); Evanston: American Theol Lib Assoc, 1984 (1r) – 1 – 0-8370-0275-3 – mf#1984-B164 – us ATLA [170]

The ethics of gambling / Mackenzie, William Douglas – New and enl ed. London: A Melrose, 1911 – 1mf – 9 – 0-7905-7962-6 – mf#1989-1187 – us ATLA [170]

The ethics of jesus / King, Henry Churchill – New York: Macmillan, 1910 – (= ser New Testament Handbooks; William Belden Noble Lectures) – 4mf – 9 – 0-8370-9960-9 – (incl ind) – mf#1986-3960 – us ATLA [220]

The ethics of john stuart mill / Mill, John Stuart; ed by Douglas, Charles – Edinburgh: Blackwood, 1897 – 1mf – 9 – 0-7905-9518-4 – mf#1989-1223 – us ATLA [170]

The ethics of journalism / Crawford, Nelson A – New York: Alfred A Knoff 1924 – 3mf – 9 – $4.50 – mf#llmc92-219 – us LLMC [170]

The ethics of literary art : the carew lectures for 1893, hartford theological seminary / Thompson, Maurice – Hartford, CT: Hartford Seminary Press, 1893 – 1mf – 9 – 0-8370-9190-X – mf#1986-3190 – us ATLA [170]

The ethics of naturalism : a criticism / Sorley, William Ritchie – 2nd ed, rev. Edinburgh: William Blackwood, 1904 – 1mf – 9 – 0-7905-9671-7 – mf#1989-1396 – us ATLA [170]

The ethics of st. paul / Alexander, Archibald Browning Drysdale – Glasgow: James Maclehose, 1910 – 1mf – 9 – 0-7905-0302-6 – (incl bibl ref and index) – mf#1987-0302 – us ATLA [230]

Ethics of the body / Boardman, George Dana – Philadelphia: JB Lippincott, 1903 – 1mf – 9 – 0-7905-9904-X – mf#1989-1629 – us ATLA [170]

The ethics of the christian life = Christliche leben (ethik) / Haering, Theodor – New York: GP Putnam, 1909 – (= ser Theological Translation Library) – 2mf – 9 – 0-7905-3885-7 – (in english) – mf#1989-0378 – us ATLA [170]

Ethics of the dust / Ruskin, John – New York, NY. 1879 – 1r – us UF Libraries [170]

Ethics of the great religions / Gorham, Charles Turner – London: Watts, 1898 – 1mf – 9 – 0-8370-6055-9 – (incl bibl ref) – mf#1986-0055 – us ATLA [170]

The ethics of the hindus / Maitra, Susil Kumar – Calcutta: Calcutta University Press, 1925 – (= ser Samp: indian books) – us CRL [280]

The ethics of the old testament / Bruce, William Straton – 2nd enl ed. Edinburgh: T & T Clark, 1909 – 1mf – 9 – 0-8370-2489-7 – (incl ind) – mf#1985-0489 – us ATLA [221]

De ethiek van ulrich zwingli / Bavinck, Herman – Kampen: G Ph Zalsman, 1880 [mf ed 1990] – 179p on 1mf – 9 – 0-7905-5502-6 – (incl bibl ref) – mf#1988-1502 – us ATLA [230]

Ethier, Joseph Arthur Calixte *see*
– Conference sur chenier

Ethik / Herrmann, Wilhelm – 5. Aufl. Tuebingen: JCB Mohr, 1913 – (= ser Grundriss der theologischen wissenschaften) – 1mf – 9 – 0-7905-3904-7 – mf#1989-0397 – us ATLA [170]

Ethik *see* System of ethics

Die ethik calvins in ihren grundzuegen entworfen : ein beitrag zur geschichte der christlichen ethik / Lobstein, Paul – Strassburg: C. F. Schmidt, 1877. Chicago: Dep of Photodup, U of Chicago Lib, 1975 (1r); Evanston: American Theol Lib Assoc, 1984 (1r) – 1 – 0-8370-0564-7 – (incl bibl ref) – mf#1984-6069 – us ATLA [170]

Die ethik der alten griechen / Schmidt, Leopold – Berlin: W Hertz, 1882 – 3mf – 9 – 0-524-01920-7 – (incl bibl ref) – mf#1990-2733 – us ATLA [170]

Die ethik des apostels paulus / Benz, Karl – Freiburg i B, St Louis MO: Herder, 1912 – (= ser Biblische studien) – 1mf – 9 – 0-7905-2401-5 – mf#1987-2401 – us ATLA [225]

Die ethik des clemens von alexandrien / Winter, Friedrich Julius – Leipzig: Doerffling und Franke, 1882 – 1mf – 9 – 0-7905-6915-9 – (incl bibl ref) – mf#1988-2915 – us ATLA [170]

Die ethik des heiligen augustinus / Mausbach, Joseph – Freiburg im Breisgau; St Louis, Mo: Herder, 1909 – 2mf – 9 – 0-7905-9500-1 – (incl bibl ref) – mf#1989-1205 – us ATLA [170]

Die ethik des judentums / Lazarus, Moritz – Frankfurt a.M. 1904 (mf ed 1996) – 6mf – 9 – €59.00 – 3-8267-3185-9 – mf#DHS 80000 – gw Frankfurter [270]

Ethik des maimonides / Rosin, David – Breslau, Germany. 1876 – 1r – 1 – us UF Libraries [170]

Die ethik huldreich zwinglis / Kuegelgen, C von – Leipzig, 1902 – 2mf – 9 – mf#ZWI-67 – ne IDC [242]

Die ethik huldreich zwinglis / Kuegelgen, Constantin von – Leipzig: R. Woepke, 1902 – 1mf – 9 – 0-7905-6411-4 – (incl bibl ref) – mf#1988-2411 – us ATLA [242]

Ethik in ihren grundzuegen entworfen *see* Collected works

Die ethik jesu / Grimm, Eduard – Hamburg: Grefe & Tiedemann, 1903 – 1mf – 9 – 0-7905-0948-2 – mf#1987-0948 – us ATLA [240]

Die ethik jesu : ihr ursprung und inhre bedeutung vom standpunkte des menschentums / Rau, Albrecht – Giessen: Emil Roth, 1899 [mf ed 1989] – 1mf – 9 – 0-7905-2862-2 – (incl bibl ref) – mf#1987-2862 – us ATLA [230]

Die ethik pascals / Bornhausen, Karl – Giessen: Alfred Toepelmann, 1907 – (= ser Studien Zur Geschichte Des Neueren Protestantismus) – 1mf – 9 – 0-7905-4095-9 – (incl bibl ref) – mf#1988-0095 – us ATLA [170]

Die ethik soeren kierkegaards / Bauer, Wilhelm – Kahla (S-A): J Beck, [1913?] – 1mf – 9 – 0-524-00246-0 – (incl bibl ref) – mf#1989-2946 – us ATLA [170]

Ethik und hyperethik / Coudenhove-Kalergi, Richard Nicolaus – Wien, Austria. 1923 – 1r – us UF Libraries [170]

Ethik und materialistische geschichtsauffassung = Ethics and the materialist conception of history / Kautsky, Karl – Chicago: Charles H Kerr, 1907 – 1mf – 9 – 0-8370-7161-5 – (in english) – mf#1986-1161 – us ATLA [170]

Ethik und mystik in hebbels weltanschauung / Lahnstein, Ernst – Berlin: B Behr, 1913 – 1 – (incl bibl ref) – us Wisconsin U Libr [430]

De ethik van ulrich zwingli / Bavinck, H – Kampen, 1880 – 5mf – 9 – mf#ZWI-17 – ne IDC [242]

Ethiopia. Central Statistical Office *see* Statistical abstract 1963-1976

Ethiopia jamaica society newsletter – 1992 feb/mar – 1r – 1 – mf#4872798 – us WHS [305]

The ethiopian church : historical notes on the church of abyssinia / O'Leary, De Lacy – London: Society for Promoting Christian Knowledge, 1936. Chicago: Dep of Photodup, U of Chicago Lib, 1974 (1r); Evanston: American Theol Lib Assoc, 1984 (1r) – 1 – 0-8370-0063-7 – mf#1984-6021 – us ATLA [240]

Ethiopian economic review – Addis Ababa. [n1-10. dec 1959-jun 1968] – us CRL [079]

Ethiopian herald – Addis Ababa, Ethiopia. 5 sep 1953-26 dec 1954; 1955-aug 1959; 16 jun 1960 – 3 3/4r – 1 – uk British Libr Newspaper [079]

Ethiopian herald – Addis Ababa, Ethiopia. apr 21 1956-may 3 1958 – 1 – us NY Public [079]

Ethiopian herald – Addis Ababa: Ethiopian Press & Information Off, jun 23 1956- – 1 – us CRL [079]

Ethiopian register – 1994 feb-dec, 1995, 1996 – 3r – 1 – mf#3399926 – us WHS [071]

Ethiopian treasurer / Norton, William – London, England. 1862? – 1r – us UF Libraries [240]

Ethiopian tribune – North Hollywood, Los Angeles, CA, 1994 jul 1-1995 dec 1 – 1r – 1 – mf#3399934 – us WHS [071]

Ethiopian Zion Coptic Church *see* Coptic time

Ethiopic book of Enoch *see*
– Das buch henoch
– Liber henoch, aethiopice

The ethiopic liturgy / Mercer, Samuel A B – London, 1915 – 9mf – 8 – €18.00 – ne Slangenburg [243]

The ethiopic version of the book of enoch / ed by Charles, Robert Henry – Oxford. v1-pt11. 1906 – (= ser Anecdota oxoniensia. semitic series) – 9mf – 8 – €18.00 – ne Slangenburg [221]

The ethiopic version of the hebrew book of jubilees / ed by Charles, Robert Henry – Oxford. v1-pt 8. 1895 – (= ser Anecdota oxoniensia. semitic series) – 7mf – 8 – €15.00 – ne Slangenburg [270]

[Ethiopica] – 183 titles on 1440mf – 9 – (coll also mss catalogues & bibl) – ne IDC [960]

Ethiopie d'aujourd'hui – Addis Ababa, Ethiopia. 24 oct 1959; 11 mar-30 dec 1961 – 1/2r – 1 – uk British Libr Newspaper [079]

Ethiopie meridionale : Journal de mon voyage aux pays Amhara, Oromo et Sidama, Septembre 1885...Novembre 1888 / Borelli, J – Paris, 1890 – 10mf – 9 – mf#NE-20313 – ne IDC [916]

Die ethische erscheinung des christlichen lebens / Beck, Johann Tobias; ed by Lindenmeyer, Julius – Guetersloh: Bertelsmann, 1883 – (= ser Vorlesungen ueber christliche Ethik) – 1mf – 9 – 0-7905-3578-5 – (incl bibl ref) – mf#1989-0071 – us ATLA [170]

Der ethische gehalt in grillparzers werken / Freybe, Albert – Guetersloh: C Bertelsmann 1893 [mf ed 1990] – 1r – 1 – (filmed with: grillparzers verhaeltnis zur politischen tendenzliteratur seiner zeit / konrad beste) – mf#2690p – us Wisconsin U Libr [170]

Die ethischen grundfragen : zehn vortraege / Lipps, Theodor – 2. teilweise umgearb. Aufl. Hamburg: Leopold Voss, 1905 – 1mf – 9 – 0-8370-6142-3 – (incl bibl ref) – mf#1986-0142 – us ATLA [170]

Die ethisch-religioese bedeutung der alttestamentlichen namen : nach talmud, targum, und midras / Sarsowsky, Abraham – Kirchhain, N-L: Max Schmersow, 1904 [mf ed 1985] – 1mf – 9 – 0-8370-5051-0 – (incl bibl ref) – mf#1985-3051 – us ATLA [221]

Ethnic affairs / University of Texas at Austin – n1-2 [1987 fall-1988 spr] – 1r – 1 – mf#1551744 – us WHS [305]

Ethnic american news – v3 n1/5-v5 n8/10 [1975 may-1977 fall] – 1r – 1 – (continued by: american ethnic) – mf#635807 – us WHS [305]

Ethnic attitudes of johannesburg youth / Lever, Henry – Johannesburg, South Africa. 1968 – 1r – 1 – us UF Libraries [306]

Ethnic Heritage Council *see* Northwest ethnic news

Ethnic Heritage Training Program (WI) *see* Exploring your ethnic heritage

Ethnic Millions Political Action Committee *see*
– Empac!
– New america

The ethnic trinities and their relations to the christian trinity : a chapter in the comparative history of religions / Paine, Levi Leonard – Boston: Houghton, Mifflin, 1901 – 1mf – 9 – 0-7905-8869-2 – mf#1989-2094 – us ATLA [230]

Ethnicity : multiethnic ministry from the black southern baptist perspective / Southern Baptist Convention – 1987 sep/oct – 1r – 1 – (cont: ebonicity; cont by: black church development) – mf#4878378 – us WHS [242]

Ethnicity in our america – 1st year n1-2nd year n6 [1979 may-1980 apr/jun] – 1r – 1 – mf#633343 – us WHS [305]

Ethnie haitienne / Jacob, Kleber Georges – Port-Au-Prince, Haiti. 1941 – 1r – us UF Libraries [972]

Ethnikon Asteroskopeion Athenon *see* Annales de l'observatoire national d'athenes

Ethnikos kerux = The national herald – New York: Enossis Publ Co, [1915-; jul 1922-29; may 1938-apr 1940; 1949-mar 1967; jul 1967-72 (daily ed only). jan 2 1949-dec 25 1955; jan 6-dec 29 1957; jul 7-dec 28 1958 (Sunday issues only) – 1 – us CRL [071]

Ethnikos kerux – New york, 12 sep 1929-28 jun 1939; 12 feb 1940-18 may 1946 [wkly] – 140r – 1 – (the national herald in greek. imperfect) – uk British Libr Newspaper [071]

Ethnikos keryx = The national herald – New York: Enossis Publ Co, [1915-]. [jul 1922-1929; may 1938-apr 1940; 1949-mar 1967; jul 1967-1972; Daily ed only; jan 2, 1949-dec 25, 1955; jan 6-dec 29 1957; jul 7-dec 28 1958; Sunday issues only] – us CRL [071]

The ethno-geography of the pomo and neighboring indians / Barrett, S A – UCB: 1908 – 1r – 1 – $50.00 – mf#B63015 – us Library Micro [305]

Ethnographic notes in southern india / Thurston, Edgar – Madras: Govt Press, 1906 – (= ser Samp: indian books) – us CRL [390]

Ethnographic notes on south pacific islands / Townsend, Charles H et al – 1899-1900 – 1r – 1 – mf#pmb121 – at Pacific Mss [980]

Ethnographic studies of new ireland (png) / Groves, W C – 1932-66 – 1r – 1 – (available for ref) – mf#pmb1188 – at Pacific Mss [305]

Ethnographie de madagascar / ed by Grandidier, A & Grandidier, G – Paris, 1908-1928 – 4v on 78mf – 8 – mf#A-364 – ne IDC [960]

Ethnographische beobachtungen ueber die voelker des beringsmeeres, 1789-1791 / Merck, C H; ed by Jacobi, A – Berlin, 1936. v19 – 1mf – 9 – mf#N-317 – ne IDC [910]

Ethnography : no 270 flagler county / Goebel, Rubye K – s.l, s.l? 1936 – 1r – us UF Libraries [978]

Ethnography : tampa, florida / Muse, Viola B – s.l, s.l? 193-? – 1r – 1 – us UF Libraries [978]

Ethnography : unusual settlements in Florida / Richardson, Martin D – s.l, s.l? 1937? – 1r – us UF Libraries [307]

Ethnography of the northern territories of the gold coast / Goody, Jack – London, England. 1964 – 1r – 1 – us UF Libraries [960]

Ethnohistory – Durham. 1985+ (1,5,9) – ISSN: 0014-1801 – mf#15459 – us UMI ProQuest [305]

Ethnologia sul-americana / Schmidt, Wilhelm – Sao Paulo, Brazil. 1942 – 1r – us UF Libraries [972]

Ethnological results of the point barrow expedition / Murdoch, J – Washington, 1892 – 12mf – 9 – mf#N-323 – ne IDC [910]

Ethnological Society *see* Bulletin

Ethnological society bulletin / Haile Sellassie University. Institute of Ethiopian Studies – 1 – us AMS Press [306]

Ethnologie und geographie des alten orients / Hommel, F – Muenchen, 1926 – (= ser Handbuch der Altertumswissenschaft) – 12mf – 9 – (Handbuch der Altertumswissenschaft, abt 3 v1 pt1) – mf#NE-417 – ne IDC [956]

Die ethnologischen und anthropogeographischen anschaungen bei j kant und j reinhard forster / Unold, Joh. – Leipzig: [s.n.], 1886 – us CRL [190]

Ethnology – Pittsburgh. 1962+ (1) 1971+ (5) 1976+ (9) – ISSN: 0014-1828 – mf#2400 – us UMI ProQuest [301]

Ethnology of the north california coast indian tribes hupa wyot pomo miwak / Goddard, Pliny E – 3v. 1903-1923 – 3r – 1 – $150.00 – mf#B63016 – us Library Micro [305]

Ethnomusicology – Champaign. 1974+ (1) 1975+ (5) 1976+ (9) – ISSN: 0014-1836 – mf#9325 – us UMI ProQuest [780]

Ethnos – Athens, Greece. -d. 13 June 1917-31 Aug 1918. Imperfect. 2 reels – 1 – uk British Libr Newspaper [949]

Ethnos – Larnaca, Cyprus. 16 apr-dec 1892; 5 jan-12 jan 1893 – 1/4r – 1 – uk British Libr Newspaper [072]

Ethology and sociobiology – New York. 1980-1996 (1) 1980-1996 (5) 1987-1996 (9) – (Cont by: Evolution and human behavior) – ISSN: 0162-3095 – mf#42258 – us UMI ProQuest [300]

Ethos – Arlington. 1979+ (1,5,9) – ISSN: 0091-2131 – mf#12089 – us UMI ProQuest [301]

Ethridge, M Kriss *see* The effectiveness of individualized mental training program on attentional styles, competitive trait anxiety and performance of female softball players

Etia, Abel Moume *see* Le foulbe du nord-cameroun

Etica / Marquez, Gabino – Madrid: Editorial Razon y Fe, 1928 – sp Bibl Santa Ana [946]

Etica elemental / Frutos Cortes, Eugenio – Caceres: imp moderna, 1935 – 1 – sp Bibl Santa Ana [540]

Etica elemental / Nunez Gonzalez, Salvador – Badajoz: libreria la alianza, 1931 – 1 – sp Bibl Santa Ana [170]

Etica general / Conde, Prudencio – Barcelona: luis gili. v1. 1917 – 1 – sp Bibl Santa Ana [170]

La etica o estetica politica de maquiavelo / Santos Neila, Francisco – Badajoz: Imp. Diput. Provincial, 1974 – sp Bibl Santa Ana [320]

Etica y estetica en la danza / Arce, David N – Mexico: impresores unidos, 1949 – 1 – mf#*ZBD-*MGO pv13 – Located: NYPL – us Misc Inst [790]

Etienne, Charles Guillaume *see*
- Arwed
- Deux gendres
- Jeune femme colere
- Plaideurs sans proces

Etienne, Henri *see* Resume de l'histoire de lorraine

Etienne und luise : novelle / Penzoldt, Ernst – Leipzig: P Reclam c1929 [mf ed 1992] – 1r – 1 – (with autobiogr aft. filmed with: der mensch an der wege / rudolf paulsen) – mf#2859p – us Wisconsin U Libr [830]

Etienne vacherot, 1809-1897 / Olle-Laprune, Leon – Paris: Perrin, 1898 – 1mf – 9 – 0-524-01453-1 – mf#1990-2448 – us ATLA [920]

Etika / Lembaga Pendidikan Orang Dewasa – Djakarta, 1954-1955 – 3mf – 9 – (missing: 1954, v1(1-6, 8-9, 11-12); 1955, v2(1)) – mf#SE-751 – ne IDC [959]

L'etincelle : comédie en un acte / Pailleron, Edouard – 8e ed. Paris: Calmann-Levy 1879? [mf ed 1994] – 1r – 1 – us UF Libraries [820]

L'etincelle – Paris, 26 avr 1947-24 30 juin 1954 – 1 – (devenu: le rassemblement) – fr ACRPP [074]

L'etincelle : journal hebdomadaire, politique, litteraire, artistique – Montreal: impr Royale. n1 6 dec 1902-n18 2/9 mai 1903 8mnthly) [mf ed 1984] – 1r – 5 – (journal politique, litteraire, artistique) – mf#SEM16P269 – cn Bibl Nat [073]

L'etincelle – Point-a-Pitre, Guadeloupe. 1944-1962 (1) – mf#67941 – us UMI ProQuest [079]

L'etincelle – Pau. n1-1538. 1933-janv 1952 – 1 – (subtitle varies) – fr ACRPP [073]

L'etincelle socialiste – Paris. 4 sept 1925-19 mars 1927, 1er mai 1932 – 1 – fr ACRPP [320]

Etincelles / Depestre, Rene – Port-Au-Prince, Haiti. 1945 – 1r – us UF Libraries [972]

Etiology of fowl paralysis, leukemia and allied conditions in animals / Emmel, M W – Gainesville, FL. 1935 – 1r – us UF Libraries [636]

Etiology of fowl paralysis, leukemia and allied conditions in animals / Emmel, M W – Gainesville, FL. 1936 – 1r – us UF Libraries [636]

Etiology of fowl paralysis, leukemia and allied conditions in animals / Emmel, M W – Gainesville, FL. 1937 – 1r – us UF Libraries [636]

Etiology of fowl paralysis, leukemia and allied conditions in animals / Emmel, M W – Gainesville, FL. 1946 – 1r – us UF Libraries [636]

Etiopi in palestina : Storia della comunita etiopica di gerusalemme – Roma: Libreria dello Stato, 1943-47 – 1 – us CRL [956]

Etiopia occidentale (dallo scioa alla frontiera del sudan) : note del viaggio, 1927-1928 – Roma: sindicato italiano arti grafiche, [1930] – 1 – us CRL [960]

Etiopia occidentale dallo scioa alla frontiera del sudan / Cerulli, E – Roma, 1933 – 2v on 9mf – 9 – mf#NE-20218 – ne IDC [960]

Etiudy o zapadnoi literature / Lavrov, Petr Lavrovich; ed by Gizetti, Aleksandr & Vitiazev, Petr – Petrograd: "Kolos", 1923 [mf ed 2002] – 1r – 1 – (filmed with: k biografii adama mitskevicha v 1821-1829 godakh / fedor verzhbovskii [teodor wierzbowski], (1898). incl bibl ref) – mf#5239 – us Wisconsin U Libr [410]

Etlich cristlich lider lobgesang, un[d] psalm : dem rainen wort gottes gemess, auss der heylige[n]schrifft, durch mancherley hochgelerter gemacht in der kirchen zu singen, wie es dann zum tayl berayt zu wittenberg in uebung ist / Luther, Martin – [Nuernberg]: [Gutknecht] 1524 – (= ser Hqab. literatur des 16. jahrh.) – 1mf – 9 – €20.00 – mf#1524b – gw Fischer [780]

Etlich cristliche lyeder lobgesang, vnd psalm, dem rainen wort gotes gemess : auss der hailigen gschrifft, durch mancherlay hochgelerter gemacht, in der kirchen zusingen wie es dan[n] zum tail berayt zu wittenberg in yebung ist – [Augsburg]: Ramminger 1524 – (= ser Hqab. literatur des 16. jahrh.) – 1mf – 9 – €20.00 – (incl: luther, martin: ein christenlichs lyed) – mf#1524a – gw Fischer [780]

Etliche fragstueck von der beicht, absolution, vnd vom hochwirdigen sacraments des altars, fuer die christliche jugent / Mathesius, J – Nuernberg, 1568 – 1mf – 9 – mf#TH-1 mf 998 – ne IDC [242]

Etliche gewisse gruende... / Pezelius, C – [Bremen], 1588 – 1mf – 9 – mf#PBA-287 – ne IDC [240]

Etliche schrifften vnd handlungen der wirtenbergischen theologen vnnd victorini strigelij, anno 1563 gesehen : daraus zusehen, was sie von seiner pelagianischen synergia halten / Strigel, V – np, 1564 – 1mf – 9 – mf#TH-1 mf 1450 – ne IDC [242]

Etliche underricht... / Duerer, A – Nuerenberg, 1527 – 2mf – 9 – mf#O-999 – ne IDC [700]

Etlicher gutter teutscher und polnischer tentz... – 1555 – (= ser Mssa) – 6mf – 9 – €80.00 – mfchl 462 – gw Fischer [780]

O etna : hebdomadario illustrado e satyrico – Recife, PE: Typ do Etna, 08 out-nov 1881; 23 jul 1882 – (= ser Ps 19) – mf#P16,01,70 – bl Biblioteca [870]

[Etna mills-] etna standard – CA. Mar-Aug 1898 [wkly] – 1r – 1 – $60.00 – mf#B02205 – us Library Micro [071]

[Etna mills-] scott valley advance – CA. 1897-1912; 1913-17 [wkly] – 10r – 1 – $600.00 – mf#B02206 – us Library Micro [071]

[Etna mills-] western sentinel – CA. 1918; 1920-52 – 13r – 1 – $780.00 – mf#B02207 – us Library Micro [071]

Etnias sergipanas / Bezerra, Felte – Aracaju, Brazil. 1950 – 1r – us UF Libraries [972]

Etnicheskie protsessy i semia / Terenteva, L – Moskva: In-t konkretnykh sotsialnykh issledovanii AN SSSR, 1972 – us CRL [077]

Etnografia angolana / Lima, Augusto Guilherme Mesquitela – Luanda, Angola. 1964 – 1r – us UF Libraries [960]

Etnografia de guatemala / Stoll, Otto – Guatemala, 1958 – 1r – 1 – us UF Libraries [972]

Etnograficheskie dannye / Ocherki prostonarodnogo zhit'ia-byt'ia v Vitebskoi Belorussii i opisanie predmetov obikhodnosti – Vitebsk, 1895 – 12mf – 8 – mf#RZ-176 – ne IDC [314]

Etnograficheskii sbornik, izdavaemyi Imperatorskim russkim geograficheskim obshchestvom – Richmond. 1945+ (1) 1970+ (5) 1976+ (9) – 52mf – 9 – mf#1717 – ne IDC [077]

Etnograficheskoe obozrenie – M., 1889-1916. v1-28(1-112) – 374mf – 9 – mf#1429 – ne IDC [077]

Etnografichnyi visnyk – Kiev. v1-9. 1925-30 – 1 – us NY Public [305]

Etnografiske of antropo-geografiske rejsestudier i nord-gronland / Steensby, H P – (MoG Kobenhavn, 1912 – v50 on 2mf – 9 – mf#N-401 – ne IDC [919]

Etnologia brasileira / Pinto, Estevao – Sao Paulo, Brazil. 1956 – 1r – us UF Libraries [972]

Etnologia caucana / Otero, Jesus M – Popayan, Colombia. 1952 – 1r – us UF Libraries [972]

Etnologia centro-americana / Peralta, Manuel Maria De – Madrid, Spain. 1893 – 1r – us UF Libraries [972]

Etnologia e historia de tierra-firme / Salas, Julio Cesar – Madrid, Spain. 191- – 1r – us UF Libraries [972]

Etnologia y conquistas del tolima y la hoya del qu... / Bedoya, Victor A – Tolima, Colombia. 1952 – 1r – us UF Libraries [972]

Etnologia y etnografia de guatemala / Termer, Franz – Guatemala, 1957 – 1r – us UF Libraries [972]

L'etoile – Paris: Imp Serriere ct co, may 5-10 1871 – (Filmed as pt of: Commune de Paris newspapers) – us CRL [074]

L'etoile : journal du soir – Paris, 1er nov 1820-1er juil 1827 – 1 – fr ACRPP [074]

L'etoile = Nyota – Elisabethville: Service de l'information de la Province, apr 17-may 1 1958; feb 19-dec 31 1959 – (issues filmed as pt of: Herbert J Weiss collection on the Belgian Congo) – us CRL [079]

L'etoile algerienne : mouvement pour le triomphe des libertes democratiques – n.s., n1-3. Alger. juil-aout 1952 – 1 – fr ACRPP [325]

Etoile belge – Brussels Belgium, 1 feb-12 aug 1919; 26, 27 nov 1939 – 1 1/2r – 1 – uk British Libr Newspaper [074]

Etoile d'acadie / Acadian Genealogical and Historical Association – v1 n1-v4 n2 [1981-1984 apr] – 1r – 1 – mf#1277624 – us WHS [929]

L' etoile de france *see* L'etoile francaise

L'etoile de la france – Paris: Impr de Sapia, sep 1848 – us CRL [074]

L'etoile du matin ou les petits mots de madame de verte-allure, ex-religieuse *see* L'observateur feminin

L'etoile du roussillon – Perpignan. n1-213. 7 sept 1849-14 dec 1851 – 1 – fr ACRPP [073]

L' etoile du sud – Rio de Janeiro, RJ: Typ Montenegro, 05 ago-out 1885; dez 1886-abr 1892; jan-dez 1895; jan-dez 1901; fev 1902; jan 1903-dez 1912 – (= ser Ps 19) – mf#P19A,04,85 – bl Biblioteca [380]

L'etoile francaise – Paris. 14 dec 1880-9 dec 1899, 14 fevr-21 nov 1903 – 1 – (puis l' etoile de france.) – fr ACRPP [073]

Eton, William *see* A survey of the turkish empire

Etourderie : ou comment sortira-t-il de la? / Radet, Jean Baptiste – Paris, France. 1808 – 1r – us UF Libraries [440]

Etourneau / Bayard, Jean-Francois-Alfred – Paris, France. 1844? – 1r – us UF Libraries [440]

Etowah baptist church – New York. 1936+ (1) 1970+ (5) 1975+ (9) – 1 – $132.12 – (church minutes 1984-94 157p) – mf#1069 – us Southern Baptist [242]

Etowah baptist church. hendersonville, north carolina : church records – 9 feb 1917-mar 1984 – 1 – us Southern Baptist [242]

Etrange intermede = Strange interlude / O'Neill, Eugene – Paris, France. 1938 – 1r – us UF Libraries [420]

Etranger au theatre / Roussin, Andre – Paris, France. 1950 – 1r – us UF Libraries [790]

Etrangers et le droit de propriete immobiliere / Kernisan, Clovis – Paris, France. 1922 – 1r – us UF Libraries [025]

Etrennes de guitare. 4e annee 1787, ouevre 6 : entierement composees d'airs nouveaux, chansons, romances & c & c avec accompagnement, suivies de douze menuets et autres pieces de j haydn... / Porro, Pierre Jean – Paris: l'auteur et Me Baillon; Versailles: Blaisot 1787 [mf ed 1989] – 1r – 1 – mf#pres. film 50 – us Sibley [780]

Les etrivieres – Londres. n1-2. janv-fevr 1872 – 1 – fr ACRPP [073]

Die etruskische leinwandrolle des agramer national-museums / Herbig, K – Muenchen, 1911 – €5.00 – ne Slangenburg [060]

'Ets ha-da'at / Wiszanski, Aaron Moses – Warsaw, Poland. 1893 – 1r – 1 – us UF Libraries [939]

'Ets ha-da'at be-gan ha-'eden mi-kedem / Rubin, Salomon – Vienna, Austria. 1891 – 1r – 1 – us UF Libraries [939]

'Ets ha-da'ath – New york. v1 n1-2. feb-mar 1896 – 1 – us NY Public [071]

Etscheid, Lisel *see* Das gotterlebnis des germanischen menschen

Das etschoniadzin-evangeliar : beitraege zur geschichte der armenischen, ravennatischen und syro-aegyptischen kunst / Strzygowski, J – Wien, 1891 – 2mf – 9 – mf#AR-1380 – ne IDC [243]

Ette, A van *see* Les chanoines reguliers de saint augustin

Ettelmidh : bulletin mensuel des etudiants musulmans algeriens – Alger. nov 1931-avr 1933 – 1 – fr ACRPP [370]

Ettgroen : verteiln / Fehrs, Johann Hinrich – Braunschweig: G Westermann, [1901] – 1r – 1 – us Wisconsin U Libr [880]

Ettighoffer, Paul Coelestin *see* Feldgrau schafft dividende

Ettinger, Solomon *see* Ale ksovim

Ettinger-Hengstebeck, Irmlind *see*
- Die thematik "illusion und wirklichkeit" in tennessee williams' dramen "the glass menagerie" und "a streetcar named desire"
- Vom ueben und spielen zum gestalten mit dem ball

Ettlinger, Josef *see*
- Aus dem nachlass
- Christian hofmann von hofmannswaldau

Ettmueller, Ludwig *see*
- Des fuersten von ruegen wizlaw's den vierten sprueche und lieder in niederdeutscher sprache
- Engla and seaxna scopas and boceras
- Heinrichs von meissen des frauenlobes leiche, sprueche, streitgedichte und lieder
- Dat spil van der upstandinge
- Theophilus
- Vorda vealhstod engla and seaxna

Ettmueller, Michael *see*
- Methode de consulter et de prescrire les formules de medicine
- Nouvelle chymie raisonnee

Ettrick advance – Blair, Ettrick WI. 1919 oct 3/1921 nov 25-1952/1956 nov 9 [gaps] – 16r – 1 – mf#998624 – us WHS [071]

Ettwein, John *see* Papers of john ettwein

Etty dan erry / Jo, Boen Ek & Khouw, Eng Tie – Batavia: Goedang Tjerita, 1948 [mf ed 1998] – (= ser Goedang tjerita 5; Ngo bie wie kiam kek 2) – 1r – 1 – (coll as pt of the colloquial malay collection. indonesian trans of chinese novel possibly entitled emei wei jianke, or the fierce sword-fighters from emei shan mountain [salmon, claudine. literature in malay by the chinese of indonesia. paris: editions de la maison des sciences de l'homme, c1981]. filmed with: lajangan biroe / im yang tjoe) – mf#10005 – us Wisconsin U Libr [830]

Etty dan erry *see* Etty dan erry

Etude : the music magazine – v1-75. 1883-1957 – 1 – us AMS Press [780]

Etude ayant trait a la solution du probleme de determiner la hauteur atteinte par un projectile qui en retombant au niveau dont il a ete lance, a produit un effet connu : lue par mr baillairge devant la section 3, de la societe royale du canada a sa seance du 27 mai 1891 a montreal – S.l: s.n, 1891? – 1mf – 9 – mf#59380 – cn Canadiana [621]

Etude biographique, m jean raimbault : archipretre, cure de nicolet, etc / Bois, Louis-Edouard – Quebec: A Cote, 1872 – 2mf – 9 – mf#26302 – cn Canadiana [241]

Etude comparative de deux index de periodiques : radar et periodex / Daoust, Daniele et al – [Montreal]: Universite de Montreal, Faculte des arts et des sciences...1982 [mf ed 1993] – mf#SEM105P1914 – cn Bibl Nat [020]

Etude critique du regime special de la zone de tanger (maroc) / Menard, A – Tanger, 1932-1933. 2pts – 9mf – 9 – mf#ILM-2308 – ne IDC [956]

Etude critique et litteraire : sur les vitae des saints merovingiens de l'ancienne belgique / Essen, L van der – Louvain, 1907 – €17.00 – ne Slangenburg [241]

Etude de la qualite des eaux de la riviere chateauguay – [Sainte-Foy]: Services de protection de l'environnement...[1979] (mf ed 1996) – 4mf – 9 – mf#SEM105P2721 – cn Bibl Nat [917]

L'etude de la somme theologique de saint thomas d'aquin / Berthier, Joachim Joseph – nouv ed. Paris: P Lethielleux [1905?] [mf ed 1991] – 2mf – 9 – 0-7905-9237-1 – (in french. 1st ed publ in 1893) – mf#1989-2462 – us ATLA [241]

Etude demonstrative de la langue phenicienne et de la langue libyque / Judas, A-C – Paris: Friedrich Klincksieck, 1847 [mf ed 1986] – 1mf – 9 – 0-8370-7710-9 – (incl ind) – mf#1986-1710 – us ATLA [470]

Etude dogmatique sur la predestination dans calvin / Dadre, Emile – Montauban: typographie de macabiau-vidallet, 1879 – 1mf – 9 – 0-524-07234-5 – mf#1991-2975 – us ATLA [242]

Etude du spectre solaire / Fievez, Charles – Bruxelles: F Hayez 1882 [mf ed 1998] – 1r [pl/ill] – 1 – (extrait des annales de l'observatoire royal de bruxelles, v4 new ser; incl bibl ref) – mf#film mas 28409 – us Harvard [520]

Etude du tshiluba / Gabriel – Bruxelles, Belgium. 1921 – 1r – us UF Libraries [960]

Etude geographique des llanos du venezuela coccide... / Crist, Raymond E – Grenoble, Switzerland. 1937 – 1r – 1 – us UF Libraries [972]

ETUDES

Etude historique et bibliographique sur la discipline ecclesiastique des eglises reformees de France / Frossard, Charles Louis – Paris: Grassart, 1887 – 1mf – 9 – 0-7905-5877-7 – (Incl bibl ref) – mf#1988-1877 – us ATLA [240]
Etude juridique et critique sur les constitutions / Terlonge, Henri – Port-Au-Prince, Haiti. 1933 – 1r – us UF Libraries [972]
Etude lexicographique et grammaticale de la latinite de saint jeraome / Goelzer, Henri – Paris: Hachette, 1884 – 2mf – 9 – 0-7905-8034-9 – (incl bibl ref) – mf#1988-6015 – us ATLA [450]
Etude monographique des plus grandes formations lunaires / Chemla-Lamech, Felix – Toulouse: V Cazellas 1934-35 [mf ed 1998] – 2v on 1r [ill] – 1 – (incl bibl ref) – mf#film mas 28405 – us Harvard [520]
Etude pratique de la legislation civile annamite / Denjoy, Paul – Paris: Challamel 1894 – 271p – 1 – mf#LL-10002 – us L of C Photodup [340]
Etude socio-economique du centre extra-coutumier d'usumbura / Baeck, Louis – Bruxelles, Belgium. 1957 – 1r – us UF Libraries [300]
Etude spectrale des cometes et de leur queue : radiations nouvelles intenses dans la queue / Deslandres, Henri Alexandre – Paris: Imp Merckel 1903-09 [mf ed 1999] – 1r – 1 – (incl 7 papers repr fr: comptes rendus des seances de l'academie des sciences 1903-09) – mf#film mas 28421 – us Harvard [520]
Etude sur alexandre vinet : critique de pascal / Nazelle, L J – Alendcon: impr typ veuve f guy, 1901 – 1mf – 9 – 0-7905-6417-3 – (incl bibl ref) – mf#1988-2417 – us ATLA [190]
Etude sur alexandre vinet : critique litteraire / Molines, Louis – Paris: Fischbacher, 1890 [mf ed 1991] – 2mf – 9 – 0-7905-9352-1 – mf#1989-2577 – us ATLA [242]
Etude sur catulle / Couat, Auguste Henri – Paris: E Thorin, 1875 – xviii/19-295p – 1 – us Wisconsin U Libr [450]
Etude sur "jean rivard" / Roy, Camille – Ottawa: impr pour la Societe Royale du Canada, 1910 – 1mf – 9 – 0-665-77942-9 – mf#77942 – cn Canadiana [440]
Etude sur la baronnie et l'abbaye d'aunay-sur-odon / Hardy, M G Le – Caen, 1897 – 8mf – 8 – €17.00 – ne Slangenburg [241]
Etude sur la condition juridique de letranger en h... / Corvington, Hermann – Port-Au-Prince, Haiti. 1934 – 1r – us UF Libraries [972]
Etude sur la franc-maconnerie / Dupanloup, Felix – 3e ed. Paris: Charles Douniol 1875 – 1mf – 9 – mf#vrl-141 – ne IDC [366]
Etude sur la jeunesse et la conversion de calvin / Boegner, Alfred – Montauban: Impr cooperative, 1873 – 1mf – 9 – 0-524-08735-0 – mf#1993-3240 – us ATLA [242]
Etude sur la loi criminelle du canada en rapport avec les lois penales de la province de quebec... / Chagnon, Joseph Antoine – St-Hyacinthe: Des presses a vapeur de l'Union, 1891 [mf ed 1980] – 1mf – 9 – mf#SEM105P40 – cn Bibl Nat [345]
Etude sur la masculinite / Maurel, E – Paris, 1903 – (= ser Les femmes [coll]) – 1mf – 9 – mf#6396 – fr Bibl Nationale [150]
Etude sur la nationalite de la femme etrangere qui epouse un sujet ottoman / Tawil, A – Le Caire, 1912 – 1mf – 9 – mf#ILM-3433 – ne IDC [956]
Etude sur la reserve des enfants legitimes / Latour, Rene – Toulouse, imprimerie lagarde et sebille, 1899 – 132p – 1 – mf#LL-4039 – us L of C Photodup [340]
Etude sur la theorie du droit musulman / Savvas, Pasha – Paris: Marchal et Billard, 1892-98. 2v. LL-12031 – 1 – us L of C Photodup [340]
Etude sur la veine liquide contractee tendant a modifier la theorie actuelle de l'hydraulique / Steckel, R – Ottawa?: Maclean, Roger, 1885 – 2mf – 9 – mf#13924 – cn Canadiana [530]
Etude sur la vie et les oeuvres de jean paul frederic richter / Firmery, Joseph Leon – Rennes: Typographie Oberthur, 1886 – 1r – 1 – (incl bibl ref) – us Wisconsin U Libr [430]
Etude sur l'adultere au point de vue penal en droit roman et en droit francais / Loustaunau, Joseph – Aire-sur-l'Adour: Dehez, 1889 – (= ser Les femmes [coll]) – 3mf – 9 – mf#11093 – fr Bibl Nationale [306]
Etude sur l'architecture lombarde et sur les origines de l'architecture roma-byzantine / Dartein, F de – Paris, 1865-1882. – 13mf – 9 – mf#OA-130 – ne IDC [720]
Etude sur l'art de parler en public *see* The art of extempore speaking
Etude sur le grec du nouveau testament : le verbe, syntaxe des propositions / Viteau, Joseph – Paris: Emile Bouillon, 1893 – 1mf – 9 – 0-8370-9195-0 – mf#1986-3195 – us ATLA [225]
Etude sur le grec du nouveau testament compare avec celui des septante : sujet, complement et attribut / Viteau, Joseph – Paris: Emile Bouillon, 1896 – 1mf – 9 – 0-8370-9196-9 – (incl bibl ref and ind) – mf#1986-3196 – us ATLA [450]
Etude sur le liber pontificalis – recherches sur les manuscrits archeologiques de jacques grimaldi: archiviste de la basilique de la vaticane au seizieme siecle – etude sur le mystere de sainte agnes / Duchesne, Louis et al – Paris: Ernest Thorin, 1877 – (= ser Bibliotheque des ecoles francaises d'athenes et de rome) – 1mf – 9 – 0-7905-6992-2 – (incl bibl ref) – mf#1988-2992 – us ATLA [240]
Etude sur le principe du protestantisme d'apres la theologie allemande contemporaine / Lichtenberger, Frederic – Strassbourg: Treuttel et Wuertz, 1857 – 1mf – 9 – 0-7905-7114-5 – mf#1988-3114 – us ATLA [242]
Etude sur le senegal : productions agriculture, commerce, geologie, ethnographie...evenements depuis 1884 / Courtet, M – Paris: A Challamel, 1903 – 1 – us CRL [960]
Etude sur le temperament haitien / Magloire, Auguste – Port-Au-Prince, Haiti. 1908 – 1r – us UF Libraries [972]
Etude sur l'education agricole : lue devant le conseil d'agriculture de la province de quebec, le 8 mars 1877 / Beaubien, Louis – Montreal: Cie d'impr canadienne, 1877 – 1mf – 9 – mf#24211 – cn Canadiana [630]
Etude sur les chroniques des comtes d'anjou et des seigneurs d'ambois / Halphen, Louis – Paris. 1906 – 1 – us CRL [944]
Etude sur les colonies de la couronne britannique / Sice, Eugene – Paris, France. 1913 – 1r – us UF Libraries [972]
Etude sur les coutumes des cabrais (togo) / Puig, Francois – Toulouse: imp. toulousaine – lion et fils, 1934 – 204p – 1 – mf#LL-12070 – us L of C Photodup [340]
Etude sur les institutions haitiennes / Justin, Joseph – Paris, France. 2v in 1. 1894-95 – 1r – us UF Libraries [972]
Etude sur les missions nestoriennes en chine au 7 et au 8 siecles d'apres l'inscription syro-chinoise de si-ngan-fou / Cleisz, Augustin – Paris: Alphonse Derenne, 1880 [mf ed 1995] – (= ser Yale coll) – 92p – 1 – 0-524-09194-3 – (in french) – mf#1995-0194 – us ATLA [242]
Etude sur les origines de la penitence chretienne / Ales, A D; ed by Calliste – Paris, 1914 – €21.00 – ne Slangenburg [240]
Etude sur les origines des eglises de l'age apostolique / Faye, Eugene de – Paris: Ernest Leroux, 1909 – 1mf – 9 – 0-7905-0015-9 – (incl bibl ref) – mf#1987-0015 – us ATLA [240]
Etude sur les origines du rosarie : reponse aux articles du p thurston, s j, parus dans le month, 1900 et 1901 / Mezard, Denys – Caluire (Rhone): couvent de la visitation, [1911?] – 2mf – 9 – 0-524-03798-1 – (incl bibl ref) – mf#1990-4870 – us ATLA [241]
Etude sur les poesies de francois-xavier garneau : et sur les commencements de la poesie francaise au canada / Chauveau, Pierre J O – Montreal?: Dawson, 1884 – 1mf – 9 – mf#00591 – cn Canadiana [440]
Etude sur les poesies lyriques de goethe / Lichtenberger, Ernest – 2. rev corr ed. Paris: Librairie Hachette, 1882 [mf ed 2000] – 394p – 1 – mf#10456 – us Wisconsin U Libr [430]
Etude sur les privileges d'exemption et de jurisdiction ecclesiastiques des abbayes normandes (afm44) / Lemarignier, J-F – 1937 – (= ser Archives de la france monastique (afm)) – €18.00 – ne Slangenburg [241]
Etude sur les rapports de l'amerique et de l'ancien continent avant christophe colomb / Gaffarel, Paul – Paris: E Thorin, 1869 – 4mf – 9 – mf#03274 – cn Canadiana [910]
Etude sur les statistiques / Khouri-Saint-Pierre, Anastassia – [mf ed 1976] – (= ser Document de travail) – 1r – 5 – mf#SEM16P262 – cn Bibl Nat [317]
Etude sur l'immacule conception / Perreyve, Henri – Paris: Jules Gervais, 1881 [mf ed 1986] – 1mf – 9 – 0-8370-8055-X – (incl bibl ref) – mf#1986-2055 – us ATLA [240]
Etude sur l'insurrection du dhara (1845-1846) / Richard, Charles Louis Florentin – Alger: A Besancenes, 1846 – 1 – us CRL [960]
Etudes – v2-289. july 1878-june 1956 – 1 – us L of C Photodup [073]
Etudes bakongo : histoire et sociologie / Wing, R P van – Bruxelles: Goemaere, [1921] – 1 – us CRL [960]
Etudes bakongo : notes de sociologie coloniale / Calonne-Beaufaict, Adolphe de – Liege: M Thone, 1912 – 1 – us CRL [300]
Etudes balkaniques – Sofia. 1977-1993 (1) 1977-1980 (5) 1977-1980 (9) – ISSN: 0014-1976 – mf#7613 – us UMI ProQuest [390]
Etudes bibliques / Loisy, Alfred Firmin – 3e rev et augm. ed. Paris: Alphonse Picard, 1903 – 1mf – 9 – 0-8370-9401-1 – (Incl bibl ref) – mf#1986-3401 – us ATLA [220]
Etudes bibliques : Premiere Serie: Ancien Testament = Studies on the old testament / Godet, Frederic Louis – 6th ed. New York: E P Dutton, 1897 – 1mf – 9 – 0-8370-3319-5 – (In English. Incl bibl ref) – mf#1985-1319 – us ATLA [220]
Etudes bibliques. deuxieme serie = Studies on the new testament / Godet, Frederic Louis – 8th ed. New York: E P Dutton, [ca 1873] – 1mf – 9 – 0-8370-3323-3 – (In English) – mf#1985-1323 – us ATLA [220]
Etudes byzantines – 1(1943)-4(1946) – 26mf – 9 – €50.00 – ne Slangenburg [243]
Etudes byzantines – Bucharest: Institut francais d'etudes byzantines, 1943-45 [mf ed 2001] – (= ser Christianity's encounter with world religions, 1850-1950) – 1r – 1 – mf#2001-s104 – us ATLA [931]
Etudes camerounaises – Duala: Institut francais d'Afrique noire. n21/22-52, 56 jun/sep 1948-1958 – us CRL [079]
Etudes christologiques : la doctrine des fonctions diatrices du sauveur / Lobstein, Paul – Paris: Fischbacher 1891 [mf ed 1985] – 1mf – 9 – 0-8370-4337-9 – (incl bibl ref) – mf#1985-2337 – us ATLA [240]
Etudes critiques sur la musique haitienne / Lassegue, Franck – Port-Au-Prince, Haiti. 1919 – 1r – us UF Libraries [780]
Etudes cuneiformes / Lenormant, Frandcois – Paris: Imprimerie nationale. 5v. 1878-80 – 5mf – 9 – 0-8370-8587-X – (comm in french; some texts in sumerian, akkadian and french) – mf#1986-2587 – us ATLA [470]
Etudes de la philosophie de malebranche / Delbos, V – Paris, 1924 – €15.00 – ne Slangenburg [140]
Etudes de lepidopterologie eompaere / ed by Oberthuer, C – Rennes 1904-25 – v1-23+ind on 226mf – 9 – mf#2480/2 – ne IDC [590]
Etudes de mythologie et de folklore germanique / Krappe, Alexander Haggerty – Paris: E Leroux 1928 [mf ed Bloomington IN: Indiana Uni Lib, Preservation Dept 1984] – viii 189p on 1r – 1 – us Indiana Preservation [390]
Etudes de mythologie et d'histoire des religions antiques / Toutain, Jules – Paris: Hachette, 1909 – 1mf – 9 – 0-524-01877-4 – (Incl bibl ref) – mf#1990-2712 – us ATLA [250]
Etudes de theologie et d'histoire / ed by Sabatier, Auguste et al – Paris: Librairie Fischbacher, 1901 – 1mf – 9 – 0-524-02710-2 – mf#1990-0691 – us ATLA [240]
Etudes de theologie moderne / Frommel, Gaston – Saint-Blaise: Foyer Solidariste, 1909 – 1mf – 9 – 0-8370-4768-4 – (incl bibl ref) – mf#1985-2768 – us ATLA [240]
Etudes d'histoire et de psychologie du mysticisme : Les Grands Mystiques Chretiens / Delacroix, Henri – Paris: Felix Alcan, 1908 – 2mf – 9 – 0-524-00020-4 – mf#1989-2720 – us ATLA [240]
Etudes eburneennes – Macon, France: Institut francais d'Afrique noire, Centre de Cote d'Ivoire, 1950-1968. [n1-8 1950-1960] – us CRL [074]
Etudes economiques sur haiti / Roche-Grellier – Paris, France. 1891 – 1r – 1 – us UF Libraries [330]
Etudes entomologiques / ed by Moschulsky, V de – Helsingfors, Dresden [1852]1853-1862 – v1-11 on 18mf – 9 – mf#z-954/2 – ne IDC [590]
Etudes et documents / France. Conseil d'Etat – Paris. 1947-51 – 1 – fr ACRPP [073]
Etudes et documents pour l'histoire missionaire de l'espagne et du portugal / Robert, Ricard – Louvain, 1931; Madrid: Razon y Fe, 1932 – 1 – sp Bibl Santa Ana [240]
Etudes et recherches / Bersaucourt, Albert De – Paris, France. 1913 – 1r – us UF Libraries [025]
Etudes et recherches biographiques sur le chevalier noel brulart de sillery : pretre, commandeur etc de l'ordre de saint-jean de jerusalem, fondateur de la mission de saint-joseph a sillery, pres quebec, etc etc / Bois, Louis-Edouard – Quebec: A Cote, 1855 – 1mf – 9 – (incl bibl ref) – mf#22489 – cn Canadiana [241]
Etudes evangeliques / Loisy, Alfred Firmin – Paris: Alphonse Picard, 1902 – 1mf – 9 – 0-8370-9557-3 – (Incl bibl ref) – mf#1986-3557 – us ATLA [226]
Etudes evangeliques *see* Gospel studies
Etudes folkloriques : recherches sur les migrations des contes populaires et leur point de depart / Cosquin, Emmanuel – Paris: Libraires ancienne H Champion 1922 [mf ed Bloomington IN: Indiana Uni Lib, Preservation Dept 1984] – 634p on 1r – 1 – us Indiana Preservation [390]
Etudes francaises – Montreal. 1971+ (1) 1971+ (5) 1975+ (9) – ISSN: 0014-2085 – mf#5962 – us UMI ProQuest [440]
Etudes historiques et philosophiques sur la franc-maconnerie ancienne et moderne : sur les hauts grades et sur les loges d'adoption / Boubee, Jean-Simon Pierre – Paris: Dutertre 1854 – 4mf – 9 – mf#vrl-17 – ne IDC [366]
Etudes historiques sur la presidence de faustin so... / Bouzon, Justin – Port-Au-Prince, Haiti. 1894 – 1r – us UF Libraries [972]
Etudes historiques sur l'influence de la charite durant les premieres siecles chretiens : et considerations sur son raole dans les societes modernes / Chastel, Etienne – Paris: Capelle, 1853 – 1mf – 9 – 0-7905-4200-5 – (Incl bibl ref) – mf#1988-0200 – us ATLA [240]
Etudes litteraires – Quebec. 1976+ (1,5,9) – ISSN: 0014-214X – mf#11251 – us UMI ProQuest [440]
Etudes litteraires – Quebec: Presses de l'Universite Laval. v1 n1 avril 1968- [mf ed 1978-1991] – 5r – 5 – mf#SEM16P309 – cn Bibl Nat [410]
Etudes litteraires – Quebec: Presses de l'Universite Laval. v1 n1 avril 1968- [mf ed 1992] – 9 – mf#SEM105P1709 – cn Bibl Nat [410]
Etudes litteraires et historiques / Barante, Amable-Guillaume-Prosper Brugiere – Paris, France. v1-2. 1859 – 1r – us UF Libraries [025]
Etudes merovingiennes : actes des journees de poitiers – Paris, 1953 – €19.00 – ne Slangenburg [241]
Etudes morales sur le temps present / Caro, Elme – Paris: L Hachette, 1855 – 1mf – 9 – 0-8370-7205-0 – mf#1986-1205 – us ATLA [170]
Etudes orientales et religieuses : melanges publies a l'occasion de sa 30me annee de professorat / Montet, Edouard Louis – Geneve: Georg, 1917 – 1mf – 9 – 0-524-01972-X – mf#1990-2763 – us ATLA [270]
Etudes philosophiques – Paris. 1974-1994 (1) 1976-1994 (5) 1976-1994 (9) – ISSN: 0014-2166 – mf#8878 – us UMI ProQuest [100]
Etudes philosophiques sur l'expression litteraire / Esteve, Claude Louis – Paris, France. 1938 – 1r – us UF Libraries [025]
Etudes politiques / Salomon, Rene – Port-Au-Prince, Haiti. 1949 – 1r – us UF Libraries [330]
Etudes religieuses et sociales / Frommel, Gaston – Saint-Blaise: Solidariste, 1907 – 1mf – 9 – 0-8370-4952-0 – (incl bibl ref) – mf#1985-2952 – us ATLA [230]
Etudes routieres – Geneve. 1962-69 – 5 – fr ACRPP [073]
Etudes senegalaises, 1785-1826 – Paris: societe de l'histoire des colonies francaises, editions leroux, [19–?] – 1 – us CRL [960]
Les etudes sociales – Paris. v. 1-60 ser 1-10. Jan 1881-Feb 1940 – 1 – us NY Public [300]
Etudes sociales et economiques sur le canada / Bouchette, Errol – Montreal: La Compagnie de Publication de la Revue canadienne, 1905 [mf ed 1985] – 3mf – 9 – mf#SEM105P519 – cn Bibl Nat [300]
Etudes socialistes – Paris. n1-6. 1903 – 1 – fr ACRPP [335]
Etudes sur blaise pascal *see* Studies on pascal
Etudes sur daniel et l'apocalypse / Bruston, Charles – Paris: Librairie Fischbacher, 1896 [mf ed 1985] – 1mf – 9 – 0-8370-2498-6 – (in french) – mf#1985-0498 – us ATLA [220]
Etudes sur la constitution de 1867 / Laporte, Gaston – Aux Cayes? Haiti. 1887? – 1r – us UF Libraries [323]
Etudes sur la methode de la dogmatique protestante / Lobstein, Paul – Lausanne: Georges Bridel, [1885] – 1r – 1 – 0-8370-0566-3 – mf#1984-6064 – us ATLA [242]
Etudes sur la methode de la dogmatique protestante *see* Collected works
Etudes sur la reforme francaise / Hauser, Henri – Paris: Alphonse Picard, 1909 – (= ser Bibliotheque d'histoire religieuse) – 1mf – 9 – 0-7905-4230-7 – (incl bibl ref) – mf#1988-0230 – us ATLA [240]
Etudes sur la religion / Tiberghien, Guillaume – Bruxelles: E Guyot, 1857 – 1mf – 9 – 0-524-00175-8 – (Incl bibl ref) – mf#1989-2875 – us ATLA [200]
Etudes sur la religion des soubbas ou sabeens : Leurs dogmes, leurs moeurs / Siouffi, Nicolas – Paris: Imprimerie nationale, 1880 – 1mf – 9 – 0-524-01923-1 – mf#1990-2736 – us ATLA [290]
Etudes sur la religion romaine lt ee [i.e. et le] moyen age oriental / Sayous, Edouard – Paris: Ernest Leroux 1889 [mf ed 1992] – (= ser Annales du musee guimet) – 1mf – 9 – 0-524-02363-8 – (incl bibl ref) – mf#1990-2974 – us ATLA [230]
Etudes sur la revocation de l'edit de nantes / Puaux, Frank & Sabatier, Auguste – Paris: Grassart, 1886 [mf ed 1990] – 1mf – 9 – 0-7905-5738-X – (incl bibl ref) – mf#1988-1738 – us ATLA [944]
Etudes sur la signification des choses liturgiques / Desloge, T – Paris, 1906 – 10mf – 8 – €19.00 – ne Slangenburg [240]

Etudes sur l'administration de rome au moyen age (751-1252) / Halphen, Louis – Paris: H. Champion, 1907 – (= ser Bibliotheque de l'ecole des hautes etudes) – 1mf – 9 – 0-7905-6472-6 – (incl bibl ref) – mf#1988-2472 – us ATLA [931]
Etudes sur l'americanisme *see* Studies in americanism
Etudes sur l'ancien poeme francais du voyage de charlemagne en orient / Coulet, Jules – Montpellier: Coulet, 1907 – 466p – 1 – us Wisconsin U Libr [440]
Etudes sur le monachisme en espagne...paris, 1966 / Cocheril, Maurice – Madrid: Graf. Calleja, 1966 – 1 – sp Bibl Santa Ana [240]
Etudes sur le theatre contemporain en allemagne : gerhart hauptmann / Besson, Paul – Paris: A Laisney, 1900 [mf ed 1990] – 73p – 1 – mf#7445 – us Wisconsin U Libr [790]
Etudes sur les developpements de la colonisation du bas-canada depuis dix ans, (1851 a 1861) : constatant les progres des defrichements, de l'ouverture des chemins de colonisation et du developpement de la population canadienne francaise / Drapeau, Stanislas – [Quebec?: s.n.], 1863 [mf ed 1982] – 7mf – 9 – mf#34775 – cn Canadiana [304]
Etudes sur les manuscrits des quodlibets *see* Le quodlibet 15 et trois questions ordinaires de godefroid de fontaines
Etudes sur les moines d'egypte / Cauwenbergh, P van – Paris, 1914 – 4mf – 8 – €11.00 – ne Slangenburg [240]
Etudes sur les principaux philosophes : redigees conformement aux programmes officiels du 31 mai 1902 / Adam, Charles – nouv ed. Paris: Hachette, 1903 [mf ed 1990] – 2mf – 9 – 0-7905-7315-6 – (incl bibl ref) – mf#1989-0540 – us ATLA [100]
Etudes sur les religions semitiques / Lagrange, Marie Joseph – Paris, 1903 – 8mf – 8 – €17.00 – ne Slangenburg [270]
Etudes sur les religions semitiques / Lagrange, Marie-Joseph – 2e rev et augm ed. Paris: Victor Lecoffre 1905 [mf ed 1989] – (= ser Etudes bibliques) – 2mf – 9 – 0-7905-1075-8 – (in french & hebrew; incl bibl ref & ind) – mf#1987-1075 – us ATLA [270]
Etudes sur l'espagne / Morel-Fatio, Alfred – Paris, France. 1895 – 1r – us UF Libraries [025]
Etudes sur l'etat interieur des abbayes cisterciennes : et principalement de clairvaux au 12e et au 13e siecle / Arbois de Jubainville, M H d' – Paris, 1858 – 7mf – 8 – €15.00 – ne Slangenburg [241]
Etudes sur l'industrie et la classe industrielle a paris au 13e et au 14e siecle / Fagniez, Gustave Charles – Paris: F Vieweg, 1877 – 1 – us Wisconsin U Libr [380]
Etudes sur l'islam au dahomey : le bas dahomey, le haut dahomey – Paris: E Leroux, 1926 – 1 – us CRL [260]
Etudes sur l'islam au senegal – Paris: E Leroux, 1917 – 1 – us CRL [260]
Etudes sur l'islam en cote d'ivoire – Paris: E Leroux, 1922 – 1 – us CRL [260]
Etudes sur l'islam et les tribus du soudan... – Paris: E Leroux, 1920-21 [v3] – 1 – us CRL [260]
Etudes sur l'islam et les tribus maures : les brakna – Paris: E Leroux, 1921 – 1 – us CRL [260]
Etudes sur l'islam maure : cheikh sidia, les fadelia, les ida ou ali / Marty, Paul – Paris: E Leroux, 1916 [mf ed 1991] – 1mf – 9 – 0-524-01793-X – (in french) – mf#1990-2641 – us ATLA [260]
Etudes sur l'origine des meteores cosmiques et la formation de leurs courants / Bredikhin, Fedor Aleksandrovich – St-Petersbourg: Tipografia Imperatorskoi akademii nauk; Leipzig: Commission Voss' Sortiment (G Haessel) 1903 [mf ed 1998] – 1r [pl/ill] – 1 – (coll of articles, corr & updated, previously publ in various journals; incl bibl ref) – mf#film mas 28405 – us Harvard [520]
Etudes sur schiller – Paris: F Alcan, 1905 – 1 – (incl bibl ref) – us Wisconsin U Libr [430]
Etudes theologiques et religieuses – Montpellier. 1973+ (1) 1974+ (5) 1974+ (9) – ISSN: 0014-2239 – mf#8465 – us UMI ProQuest [240]
Etudes voltaiques : memoires – Ouagadougou, Centre IFAN. ns: n1-5. 1960-1965 – us CRL [079]
L'etudiant – Montreal: Matte & McCoffrey. v1 n1 6 nov 1897-; v1 n1 21 dec 1911-v4 n7 29 janv 1915 (wkly) – (cont: journal des etudiants; cont by: escholier) – mf#SEM35P97 – cn Bibl Nat [378]
L'etudiant noir : Journal de l'association des etudiants martiniquais en France – Paris: Sauphanor, [v1 n1 mar 1935] – (Filmed with les continents and 11 other titles L'Europe democratique. Paris: Boule, Dec 1849) – us CRL [074]
L'etudiant socialiste / L' Association Generale des Etudiants Socialistes. Flenu, Jupille, Liege, etc – 1926-dec 1936 janv 1937 – 1 – (devenu: Essais et combats. fevr 1937; Paris. fevr 1928-janv 1939) – fr ACRPP [335]
Les etudiants tels qu'ils sont / ed by Billy, Valmore-Armand de & Pouliot, Henri – [Quebec?: s.n, 1911?] [mf ed 1997] – 1mf – 9 – 0-665-82779-2 – mf#82779 – cn Canadiana [840]
Etumba – Brazzaville: [Parti congolais du travail, dec 27 1969-1972 – 3r – 1 – us CRL [079]
Etwas fuer alle : dasz ist, eine kurtze beschreibung allerley stands- ambts- und gewerbs-personen mit beygeruckter sittlichen lehre und biblischen concepten... / Abraham...Sancta Clara – Wuertzburg: gedruckt bey hiob hertzen, 1699 – 9mf – 9 – mf#O-1504 – ne IDC [090]
Etwas fuer alle : dasz ist, eine kurze beschreibung allerley stands- ambts- und gewerbs-personen... / Abraham...Sancta Clara – Wuertzburg: gedruckt bey hiob hertzen, 1711-33 – 33mf – 9 – mf#O-1505 – ne IDC [090]
Etwas fuer alle : das ist, eine kurze beschreibung allerley stand-, amts- und gewerkpersonen... / Abraham a Sancta Clara – Halle: Hendel, [18–?] [mf ed 1988] – (= ser Bibliothek der gesamt-literatur des in- und auslandes 376-77) – xv/194p – 1 – mf#6934 n8 – us Wisconsin U Libr [880]
Etwas fuer allerley leser – Flensburg DE, 1768 – 1r – 1 – gw Misc Inst [074]
Etwas fur herz! = Something for the spirit! – Zoar, Tuscarawas, OH – 1r – 1 – (sermons written by the leader of the separatist colony at zoar, in german) – us Western Res [240]
Etwas ueber die natur wunder in nord amerika / Cramer, Charles – St Petersburg: Russisch Kaiserlichen Mineralogischen Gesellschaft. 1837-40 [mf ed 1986] – 2v on 1mf – 9 – 0-665-54942-3 – mf#54942 – cn Canadiana [550]
Etwas von und ueber musik fuers jahr 1777 / Kraus, Joseph Martin – Frankfurt am Mayn: bey den eichenbergschen erben 1778 [mf ed 19–] – 1r – 1 – mf#film 1373 – us Sibley [780]
Etwas zum thee und kaffee fuer teutschlands maedgen und juenglinge – Hamburg DE, 1784 [gaps] – 1r – 1 – gw Misc Inst [640]
Etwelche meistens bayerische denck- und lesswuerdigkeiten *see* Parnassus boicus
Etymachie-traktat (cima36) : ein todsuendentraktat in der katechetisch-erbaulichen sammelhandschrift augsburg, staats- und stadtbibliothek, 2° cod 160 – (mf ed 1995) – (= ser Codices illuminati medii aevi (cima) 36) – 50p on 3 color mf – 15 – €240.00 – 3-89219-036-4 – (int by nigel harris. int to catechical text & description by werner williams-krapp) – gw Lengenfelder [090]
Etymological dictionary of the scottish language : illustrating the words in their different significations, by examples from ancient and modern writers...; in 2 vols / Jamieson, John – Edinburgh: University Press 1808 – (= ser Whsb) – 8mf – 9 – €80.00 – (missing: v2) – mf#Hu 093 – gw Fischer [410]
An etymological gujarati-english dictionary / Belsare, Malhar Bhikaji – Ahmedabad: H K Pathak, 1904 – (= ser Samp: indian books) – us CRL [040]
Etymologicon magnum : or universal etymological dictionary on a new plan: with illustratios drawn from various languages...; the dialects of the slavonic; and the eastern languages, hebrew, arabic, persian...part the first / Whiter, Walter – Cambridge: Hodson 1800 – (= ser Whsb) – 7mf – 9 – €75.00 – (pt1 has the handwritten dedication "von herrn banks an herrn von humbold mit frisingers herzl. grusse aus london") – mf#Hu 006 – gw Fischer [410]
Etymologicum teutonicae linguae sive dictionarium teutonico-latinum / Kilianus, C – Traecti Batavorum. v1-2. 1777 – 8 – €103.00 – (v1 22mf. v2 32mf) – ne Slangenburg [430]
Etymologische studien zum semitischen insbesondere zum hebraeischen lexicon / Barth, J – Leipzig, 1893 – 1mf – 9 – mf#NE-464 – ne IDC [470]
Etymologisches woerterbuch der griechischen sprache : mit besonderer beruecksichtigung des neuhochdeutschen und einem deutschen woerterverzeichnis / Prellwitz, Walther – Goettingen: Vandenhoeck und Ruprecht, 1892 – 1mf – 9 – 0-8370-9264-7 – (incl ind) – mf#1986-3264 – us ATLA [450]
Etymologisches woerterbuch der turko-tatarischen sprachen / Vambery, H – Leipzig, 1878 – 5mf – 8 – mf#U-373 – ne IDC [470]
Etz hadath – New York. Tree of knowledge. 1896 – 1 – us AJPC [939]
Etzbach, Mark E *see* Physical activity motivation of adolescents
Etzel, Anton von *see* Erdumsegelung der koenigl schwedischen fregatte eugenie
Etzenbach, Ulrich von *see*
– Alexander
– Wilhelm von wenden
Etzliche schrifften und acten : daraus man sehen kan, wie der achtbar vnd wolgelarte herr magister victorinvs strigelivs wider in seine profession vnd ampt ist restituirt worden / Strigel, V – [Wittenberg], 1562 – 1mf – 9 – (trans fr latin) – mf#TH-1 mf 1457 – ne IDC [242]
Eu : reu sem crime / Doria, Joao De Seixes – Rio de Janeiro, Brazil. 1965 – 1r – us UF Libraries [972]
Eu, Gastao De Orleans *see* Viagem militar ao rio grande do sul
Euaggras – Nicosia, Cyprus. 2 mar-14 jun 1890; 4 may-dec 1892 – 1/4r – 1 – uk British Libr Newspaper [072]
Euangelium gatianum : quattor euangelia latine translata ex codice monasterii s gatiani turonensis / ed by Heer, Joseph Michael – Friburgi Brisgoviae: Herder, 1910 – 1mf – 9 – 0-524-07649-9 – mf#1992-1090 – us ATLA [220]
Euangelium secundum iohannem : cum variae lectionis delectu / ed by Blass, Friedrich – Lipsiae: B G Teubneri, 1902 [mf ed 1989] – 1mf – 9 – 0-7905-0808-7 – mf#1987-0808 – us ATLA [226]
Eubel, C *see* Hierarchia catholica medii aevi
Eubel, Konrad *see*
– Geschichte der koelnischen minoriten-ordensprovinz
– Hierarchia catholica medii aevi, sive, summorum pontificum, s.r.e. cardinalium, ecclesiarum antistitum series
Eubulus ofte tractaet : vervatende verscheyden aenmerkinghen over de teghenwoordighe staet onser christelicker ghemeynte / Teelinck, W – Middelburch, 1617 – 5mf – 9 – mf#PBA-316 – ne IDC [240]
Eucalypts in florida / Zon, Raphael – Washington, DC. 1911 – 1r – 1 – us UF Libraries [580]
La eucaristia y la vida cristiana. 2nd ed. 2 tomos. barcelona, 1934 / Goma, Isidoro – Madrid: Razon y Fe, 1934 – 1 – sp Bibl Santa Ana [240]
Eucaristicas : antologia / Corredor Garcia, Antonio – Caceres: tip vda floriano, 1955 – 1 – sp Bibl Santa Ana [240]
Eucharis : ballet pantomime en deux actes / Coralli, Eugene – Paris: V Jonas, editeur, libraire de l'Opera, 1844 – 1 – mf#*ZBD-*MGTZ pv2-Res – Located: NYPL – us Misc Inst [790]
The eucharist : five sermons / Maurice, Frederick Denison – London: John E Taylor, 1857 – 1mf – 9 – 0-7905-7984-7 – mf#1989-1269 – us ATLA [240]
Eucharist and penance in the first six centuries of the church = Eucharistie und busssakrament in den ersten sechs jahrhunderten der kirche / Rauschen, Gerhard – St Louis, MO: B Herder, 1913 – 1mf – 9 – 0-7905-5916-1 – (incl bibl ref. in english) – mf#1988-1916 – us ATLA [240]
The eucharistic doctrine and liturgy of the mystagogical catechese of theodore of mopsuestia ((sca2) / Reine, J – Washington DC, 1942 – (= ser Studies in christian antiquity (sca)) – 4mf – 9 – €11.00 – ne Slangenburg [240]
The eucharistic life of jesus christ : preached during the octave of the holy sacrament, in the church of s. andre des arcs, in the year 1657 = Vie de jesus christ dans le st. Sacrement de l'abtel / Biroat, Jacques – London: Swan Sonnenschein, Lowrey, [1886?] – 1mf – 9 – 0-8370-9683-9 – mf#1986-3683 – us ATLA [240]
The eucharistic manuals of john and charles wesley : reprinted from the original editions of 1748-57-94 / Wesley, John; ed by Dutton, William Elliot – London: Bull, Simmons, 1871 – 1mf – 9 – 0-524-00739-X – mf#1990-4008 – us ATLA [242]
The eucharistic offering : spiritual instructions upon the office of holy communion, together with helps for the carrying out of the same... / Walpole, George Henry Somerset – New York: RW Crothers, 1906 – 1mf – 9 – 0-524-03086-3 – mf#1990-4575 – us ATLA [240]
Eucharistic presence, eucharistic sacrifice, and eucharistic adoration : being an examination of "a theological defence for the rev. James De Koven, d.d., warden of Racine College, february 12, 1874" / Buel, Samuel – New-York: T Whittaker, c1874 – 1mf – 9 – 0-7905-3764-8 – mf#1989-0257 – us ATLA [240]
Eucharistic truth and ritual / Dykes, John Bacchus – London, England. 1874 – 1r – us UF Libraries [240]
L'eucharistie : des origines a justin martyr / Goguel, Maurice – Paris: Fischbacher, 1910 [mf ed 1990] – 1mf – 9 – 0-7905-5830-0 – (in french. incl bibl ref) – mf#1988-1830 – us ATLA [240]
Eucharistie und busssakrament in den ersten sechs jahrhunderten der kirche *see* Eucharist and penance in the first six centuries of the church
Die euchiten im 11. jahrhundert : eine dogmengeschichtliche skizze / Schnitzer, Professor – [s.l: Chr Belser? 1838?] [mf ed 1986] – 1mf – 9 – mf#1986-3029 – us ATLA [240]
Euchologion (january) : bessarabskaia kollektsia [bessarabian collection] – late 1400s-early 1500s, and late 1500s – 14mf – 9 – (russian version) – us UMI ProQuest [090]
Euchologion [november] : verkhokamskoe sobranie [upper kamah collection] – late 1400s-early 1500s – 7mf – 9 – (russian version (accented)) – us UMI ProQuest [090]
Euchologion (september) : verkhokamskoe sobranie [upper kamah collection] – late 1400s-early 1500s [1500-25] – 9mf – 9 – (russian version) – us UMI ProQuest [090]
Euchologion to mega – Roma, 1873 – 15mf – 8 – €29.00 – ne Slangenburg [240]
Eucken and bergson : their significance for christian thought / Herman, Emily – 3rd ed. Boston: Pilgrim Press 1912 [i.e. 1913] [mf ed 1991] – 1mf – 9 – 0-524-00270-3 – mf#1989-2970 – us ATLA [190]
Eucken, Rudolf *see*
– Can we still be christians?
– Christianity and the new idealism
– Collected essays of rudolf eucken
– Die geistesgeschichtliche bedeutung der bibel
– Geschichte der philosophischen terminologie im umriss
– Der kampf um einen geistigen lebensinhalt
– Knowledge and life
– Die lebensanschauungen der grossen denker
– The life of the spirit
– Life's basis and life's ideal
– Main currents of modern thought
– The meaning and value of life
– Naturalism or idealism?
– Present-day ethics in their relations to the spiritual life
– Religion and life
– The truth of religion
The euclid era – Toronto: [s.n, 1897?-189- or 19–] – 9 – mf#P05966 – cn Canadiana [242]
Euclid township board of trustees minutes – Euclid Township, Cuyahoga, OH. 29 aug 1908-22 sep 19? – 1r – 1 – us Western Res [978]
Euclides *see*
– Evclide megarense philosopho
– La perspectiva y especularia de euclides
Euclides da cunha / Rabello, Sylvio – Rio de Janeiro, Brazil. 1948 – 1r – us UF Libraries [972]
Euclides da cunha e o socialismo / Jose Aleixo – Sao Jose do Rio Pardo, Brazil. 1960 – 1r – us UF Libraries [972]
Eucoirean eirinn / Smith, Goldwin – [s.l: s.n, 1886?] [mf ed 1983] – 1mf – 9 – 0-665-35571-8 – (trans of original scottish gaelic title: greavances of ireland) – mf#35571 – cn Canadiana [941]
Eudaemonia : oder deutsches volksglueck, ein journal fuer freunde von wahrheit und recht / ed by Goechhausen, E A A von – Leipzig 1795-98 [mf ed Hildesheim 1992-98] – (= ser Dz. historisch-politische abt) – 6v on 24mf – 9 – €240.00 – mf#k/n1740 – gw Olms [320]
Eudemische ethik und metaphysik / Arnim, H von – Wien, 1928 – 1mf – 8 – €3.00 – ne Slangenburg [110]
Eudemon, spirtual and rational : the apology of a preacher for preaching / Newport, David – Philadelphia: J B Lippincott, 1901 [mf ed 1984] – 6mf – 9 – 0-8370-1041-1 – mf#1984-4376 – us ATLA [243]
Eudore et cymodocee / Gary – Paris, France. 1824 – 1r – us UF Libraries [440]
Eugene all-american – 1969 may/jun – 1r – 1 – mf#1583233 – us WHS [071]
Eugene aram / Lytton, Edward Bulwer Lytton, Baron – Boston, MA. 189- – 1r – us UF Libraries [830]
Eugene augur – v3 n17-v5 n10 [1972 aug-1974 may 3] – 1r – 1 – (cont: augur) – mf#701371 – us WHS [071]
Eugene augur – Eugene OR: Eugene Augur Pub Co, 1970 [semimthly] – 1 – (cont: augur; cont by: augur from eugene) – us Oregon Lib [071]
Eugene augur – v1 n8 [1970 feb 3/17] – 1r – 1 – (continued by: augur) – mf#701368 – us WHS [071]
Eugene city guard – Eugene City OR: Buys & Eltzroth, 1870-99 [wkly] – 1 – (related to daily ed: daily eugene guard, 1891-1899. cont: guard (1867-70); cont by: eugene weekly guard (1899-1904)) – us Oregon Lib [071]
Eugene city guard *see* Daily eugene guard
Eugene city register – Eugene OR: Hodson & Yoran, -1889 [wkly] – 1 – (cont by: eugene register (1889-99)) – us Oregon Lib [071]

Eugene city review – Eugene City Or: A Noltner & Co, 1862- [wkly] – 1 – (ceased in 1865. cont: democratic register (eugene, or); cont by: weekly democratic review) – us Oregon Lib [071]
Eugene daily guard – Eugene OR: Campbell Bros, -1924 [daily ex sun] – 1 – (related to wkly ed: eugene weekly guard 1904, and: eugene weekly guard (eugene, or), 1905-10; semiwkly ed: eugene semi-weekly guard 1904, and: twice-a-week guard, 1910-1914. cont: daily eugene guard; cont by: eugene guard) – us Oregon Lib [071]
Eugene daily guard *see*
- Eugene semi-weekly guard
- Eugene weekly guard (eugene, or)
Eugene daily register – Eugene OR: Condon & Edwards [daily ex sun] – 1 – (related to wkly ed: eugene register (1889-99). cont: eugene register (-1898); cont by: eugene morning register (1899-1905)) – us Oregon Lib [071]
Eugene guard – Eugene OR: Guard Print Co, 1924-30 [daily ex sun] – 1 – (cont: eugene daily guard (1904-24). merged with: eugene register (1929-30) to form: eugene register-guard (1930-83)) – us Oregon Lib [071]
[Eugene-] insurgent socialist – OR. 1972-1985 – 4r – 1 – $240.00 – mf#R04990 – us Library Micro [335]
Eugene maximilien haitian collection, 1847-1933 : from the holdings of the schomburg center for research in black culture, manuscripts, archives and rare books division: the new york public library, astor, lenox and tilden foundations – 1995 – 5r – 1 – $425.00 – (guide which covers all coll under "international affairs" sold separately d3305.g5) – mf#D3305P17 – Dist. us Scholarly Res – us L of C Photodup [972]
Eugene morning register – Eugene OR: Gilstrap Bros, 1899-1905 [daily ex mon] – 1 – (related to wkly ed: eugene register (1889-99) and: eugene weekly register (1899-1904) and semiwkly ed: eugene twice a week register (1904-19-?). cont: eugene daily register; cont by: morning register (1905-29).) – us Oregon Lib [071]
Eugene morning register *see*
- Eugene twice a week register
- Eugene weekly register
Eugene news-tribune – Eugene OR: P Huysing, 1978-79 [wkly] – 1 – (cont: valley news-tribune) – us Oregon Lib [071]
Eugene register *see* Eugene daily register
Eugene register (eugene, or: 1889) – Eugene OR: S M Yoran & Son, 1889-99 [wkly] – 1 – (ceased in 1898. related to daily ed: morning register (1895-96) and: eugene register (1896) and: register (eugene, or:1897) and eugene register (1898), 1898) and: eugene daily register, 1898-99 and: eugene morning register, 1899. cont: eugene city register; cont by: eugene weekly register) – us Oregon Lib [071]
Eugene register-guard – Eugene OR: Alton F Baker, 1930-83 [daily] – 1 – (merger of: eugene guard (1924-30) and: eugene register (1929); cont by: register-guard (1983-)) – us Oregon Lib [071]
Eugene scribe and the french theatre, 1815-1860 / Arvin, Neil Cole – Cambridge, MA. 1924 – 1r – us UF Libraries [790]
Eugene semi-weekly guard – Eugene OR: [s.n.] 1904- [semiwkly] – 1 – (related to: eugene daily guard. cont: eugene weekly guard; cont by: eugene weekly guard (eugene, or)) – us Oregon Lib [071]
Eugene twice a week register – Eugene OR: Gilstrap Bros, 1904- [semiwkly] – 1 – (related to: eugene morning register. cont: eugene weekly register) – us Oregon Lib [071]
Eugene weekly – Eugene OR: What's Happening Inc, 1993- [wkly] – 1 – (cont: what's happening (-1993)) – us Oregon Lib [071]
Eugene weekly guard – Eugene OR: I L Campbell, 1899-1904 [wkly] – 1 – (related to daily ed: daily eugene guard, 1899-1903, and: eugene daily guard, 1904-24. cont: eugene city guard; cont by: eugene semi-weekly guard (1904-190?)) – us Oregon Lib [071]
Eugene weekly guard *see* Eugene daily guard
Eugene weekly guard (eugene, or) – Eugene OR: Campbell Bros, -1910 [wkly] – 1 – (related to: eugene daily guard. cont: eugene semi-weekly guard; cont by: twice-a-week guard) – us Oregon Lib [071]
Eugene weekly register – Eugene OR: Gilstrap Bros, 1899-1904 [wkly] – 1 – (related to: eugene morning register. cont: eugene register (eugene, or: 1889); cont by: eugene twice a week register) – us Oregon Lib [071]
Eugenias briefe : nebst einigen episoden und beylagen. theil 1 u.d.t.: hirzel, heinrich: eugenias briefe an ihre mutter. bd 2 und bd 3 / Hirzel, Heinrich – Zuerich 1811, 1820 [mf ed Hildesheim 1995-98] – 5mf – 9 – €100.00 – 3-487-29368-4 – gw Olms [860]
Eugenias briefe an ihre mutter : geschrieben auf einer reise nach den baedern von leuk im sommer 1806 / Hirzel, Heinrich – Zuerich 1809 [mf ed Hildesheim 1995-98] – v1 on 2mf – 9 – €60.00 – 3-487-29369-2 – gw Olms [860]
Eugenio florit – New York, NY. 1943 – 1r – us UF Libraries [972]
Eugenio noel : novela de la vida de un hombre intenso / Caba, Pedro – Valencia: Editorial America, s.a. – sp Bibl Santa Ana [830]
Eugenio sarralbo aquareles, antonio correa y arturo alvarez. (ofm). inventario... / Borges, Pedro – Madrid: Archivo Ibero-Americano, 1959 – 1 – sp Bibl Santa Ana [020]
Eugippii vita sancti severini (mgh1:1/2) / ed by Sauppe, H – 1877 – (= ser Monumenta germaniae historica 1: scriptores – auctores antiquissimi) – €14.00 – ne Slangenburg [240]
Eugippii vita severini (mgh7:26.bd) – 1898 – (= ser Monumenta germaniae historica 7: scriptores rerum germanicarum in usum scholarum (mgh7)) – €5.00 – ne Slangenburg [240]
Eukleria : seu, melioris partis electio: tractatus brevem vitae ejus delineationem exhibens: luc. 10: 41,42: unum necessarium: maria optimam partem elegit / Schurman, Anna Maria van – Altonae: C. van der Meulen, 1673-1685. Chicago: Dep of Photodup, U of Chicago Lib, 1978 (1r); Evanston: American Theol Lib Assoc, 1984 (1r) – 1 – 0-8370-1133-7 – mf#1984-T116 – us ATLA [240]
Eukleria seu meliora partis electio / Schuurman, A M van – Amstelodami, Altonae, 1673-85 – 5mf – 9 – mf#PBA-307 – ne IDC [240]
Eulenberg, A *see* Zeitschrift fuer sexualwissenschaft
Eulenberg, Herbert *see* Belinde
Eulenburg, Albert *see* Real-encyclopaedie der gesammten heilkunde (ael3/8)
Das eulenhaus : roman / Marlitt, Eugenie – Stuttgart-Berlin-Leipzig: Union Deutsche Verlagsgesellschaft [1919] [mf ed 1978] – (= ser Romane und novellen 9) – 1 – mf#film mas C 376 – us Harvard [830]
Das eulenhaus : roman / Marlitt, Eugenie [pseud] – Stuttgart-Berlin-Leipzig: Union Deutsche Verlagsgesellschaft [1919] [mf ed 1978] – (= ser Romane und novellen 9) – 1r – 1 – mf#film mas c376 – us Harvard [830]
Der eulenspiegel *see* Roter pfeffer
Eulenspiegel – Stuttgart 1848-53 [mf ed 1998] – (= ser Satirische zeitschriften (sz) 4) – 18mf – 9 – diazo €140 silver €190 – 3-89131-281-4 – (filmed with: der wieder auferstandene eulenspiegel / eulenspiegel [1862-63]; with suppls: stuttgarter wochenblatt 1863; stuttgarter literarisches wochenblatt 1863) – gw Fischer [870]
Eulenspiegel *see* Ulenspegel
Eulenspiegel, oder, schabernack ueber schabernack : posse in vier aufzuegen / Nestroy, Johann – Wien: Theodor Daberkow, [1895?] – 1r – 1 – us Wisconsin U Libr [790]
Euler, Leonhard *see*
- Tentamen novae theoriae musicae ex certissimis harmoniae principiis dilucide expositae
- Theoria motus lunae exhibens omnes eius inaequalitates
Euling, Karl *see*
- Chronik des johan oldecop
- Heinrich kaufringers gedichte
Eulogium (historiarum sive temporis) (rs9) : chronicon ab orbe condito usque ad annum domini 1366 a monacho quodam malmesbueriensi exaratum / ed by Haydon, F S – (= ser The rolls series (rs)) – (accedunt continuationes duae, quarum una ad annum 1313, altera ad annum 1390 perducta est; v1 1858 €18. v2 1860 €18. v3 1863 €23) – ne Slangenburg [931]
Eulogius und alvar : ein abschnitt spanischer kirchengeschichte aus der zeit der maurenherrschaft / Baudissin, Wolf Wilhelm, Graf von – Leipzig: Fr Wilh Grunow, 1872 – 1mf – 9 – 0-8370-9843-2 – (incl bibl ref) – mf#1986-3843 – us ATLA [240]
Eulogy on chief-justice chase : delivered by william m evarts, before the alumni of dartmouth college, at hanover, june 24 1874 – New York: Appleton, 1874 – 30p – 1 – mf#LL-167 – us L of C Photodup [340]
Eulogy on william henry bartlett : late associate justice of the supreme court of new hampshire: before the alumni of dartmouth college...june 23, 1880 / Smith, Isaac William – Concord, NH: Republican, 1881 – 16p – 1 – mf#LL-1078 – us L of C Photodup [347]
Eulogy upon life and character of george eustis : formerly chief justice of the supreme court...on the 31st mar, 1859, by hon pierre a rost... – New Orleans: Daily Delta, 1859 – 13p – 1 – mf#LL-468 – us L of C Photodup [347]
Eunomia : with brief hints to country gentlemen, and others of tender capacity, on the principles of the new sect of political economical philosophers, termed eunomians – London: Effingham Wilson, 1826 – (= ser 19th c economics) – 1mf – 9 – mf#1.1.291 – uk Chadwyck [332]
Eunomia : eine zeitschrift des neunzehnten jahrhunderts / ed by Fessler, Ignatz Aurelius et al – Berlin 1801-05 – (= ser Dz. abt literatur) – 5jge[zu je 2bdn] on 37mf – 9 – €370.00 – mf#k/n4636 – gw Olms [410]
Der eunuchus des terenz / ed by Fischer, Hermann – Stuttgart: Litterarischer Verein, 1915 (Tuebingen: H Laupp, Jr) [mf ed 1993] – (= ser Blvs 265) – xii/224p – 1 – (incl bibl ref and ind. german trans of a latin text by hans neidhart. int in german) – mf#8470 reel 54 – us Wisconsin U Libr [820]
Der eunuchus des terenz / ed by Fischer, Hermann – Stuttgart: Litterarischer Verein 1915 (Tuebingen: H Laupp, Jr) [mf ed 1993] – (= ser Blvs 265) – 58r – 1 – (incl bibl ref & ind; german trans of latin text by hans neidhart; int in german. filmed with: hermann playders ausgewaehlte werke / gustav bebermeyer [ed] & other titles) – mf#3420p – us Wisconsin U Libr [830]
Eupener nachrichten – Eupen (B), 1927 1 apr-1937 sep, 1938-1940 14 jun – 1 – gw Misc Inst [074]
Eupener zeitung – Eupen (B), 1899 4 jan-30 dec, 1921 1 sep-17 dec, 1922-1941 21 jun, 1942 23 jan-1944 31 mar [gaps] – 1 – gw Misc Inst [074]
Eupener zeitung – Eupen Belgium, 26 mar, 4 jun 1941; 28 jan, 20 feb 1942; sep-dec 1943; 3 jan, 2 aug 1944 – 3r – 1 – uk British Libr Newspaper [074]
Euphemistic liturgical appendixes in the old testament / Grimm, Karl Josef – Baltimore:[s.n.], 1901 (Leipzig: August Pries) – 1mf – 9 – 0-8370-3399-3 – mf#1985-1399 – us ATLA [221]
Euphorion / Lee, Vernon – London, England. v1-2. 1884 – 1r – us UF Libraries [025]
Euphorion : zeitschrift fuer literaturgeschichte – Hamburg, etc. v1-27. 1894-1926 – 417mf – 8 – mf#H-408c – ne IDC [410]
Die euphratlaender und das mittelmeer / Winckler, Hugo – Leipzig: JC Hinrichs, 1905 [mf ed 1989] – (= ser Der alte orient 7/2) – 1mf – 9 – 0-7905-2097-4 – mf#1987-2097 – us ATLA [930]
Euphytica – Dordrecht. 1988-1996 (1,5,9) – ISSN: 0014-2336 – mf#16783 – us UMI ProQuest [630]
Eura spectra – Brussels. 1962-1974 [1]; 1970-1972 [5,9] – ISSN: 0014-2360 – mf#1734 – us UMI ProQuest [540]
Eurarmy – 1976 jan-1978 jun, 1978 jul-1980 dec, 1984 jan-1986 jun, 1986 jul-1988 nov – 4r – 1 – (cont: army in europe) – mf#1507142 – us WHS [355]
Eurasian soil science – Silver Spring. 1992-1996 (1,5,9) – (Cont: Soviet soil science) – ISSN: 1064-2293 – mf#14361,01 – us UMI ProQuest [630]
Eureka baptist church. anderson county. south carolina : church records – 1889-1972 – 1 – us Southern Baptist [242]
[Eureka-] blue lake advocate – CA. 1913-1914; 1947; 1959-1969 – 6r – 1 – $360.00 – mf#B02208 – us Library Micro [071]
Eureka central baptist church. eureka, missouri : church records – oct 1931-jun 1942; aug-sep 1954 – 1 – (formerly: allenton baptist church) – us Southern Baptist [242]
The eureka central draft burner is acknowledged to be a superior oil burner : giving more light than any other burner used in railway cars...williams, page and co...boston, mass – S.l: s.n, 18–? – 1mf – 9 – mf#60750 – cn Canadiana [621]
[Eureka-] daily humboldt times – CA. Jan 1874-Dec 1951 – 223r – 1 – $13,380.00 – mf#BC02213 – us Library Micro [071]
[Eureka-] daily leader – NV. 1878-85 [daily] – 8r – 1 – $480.00 – mf#U04509 – us Library Micro [071]
[Eureka-] daily republican – NV. 1877-78; feb-apr 1878 (scats) [daily] – 3r – 1 – $180.00 – mf#U04510 – us Library Micro [071]
[Eureka-] democratic standard – CA. 1877-83 [wkly] – 1r – 1 – $60.00 – mf#B02215 – us Library Micro [071]
[Eureka-] eureka sentinel – NV. 31 oct 1872; 1875- [daily; wkly; biwkly] – 59r – 1 – $3540.00 (subs $50y) – mf#UN04516 – us Library Micro [071]
[Eureka-] evening leader – NV. 1878 (scattered issues) – 1r – 1 – $60.00 – mf#U04511 – us Library Micro [071]
[Eureka-] evening star – CA. 1876-78 – 1r – 1 – $60.00 – mf#B02210 – us Library Micro [071]
[Eureka-] herald – CA. Jan-May 1905; 1907-13. 3 rolls [daily] – 27r – 1 – $1620.00 – mf#B02216 – us Library Micro [071]
[Eureka-] high school enterprise – NV. 21 nov 1896 – 1r – 1 – $60.00 – mf#U04512 – us Library Micro [071]
[Eureka-] humboldt independent (rio dell) – CA. May 1967-1970 – 2r – 1 – $120.00 – mf#B02212 – us Library Micro [071]
[Eureka-] humboldt life and times – CA. Oct 1976-1980 – 3r – 1 – $180.00 – mf#B02218 – us Library Micro [071]
[Eureka-] humboldt semi-weekly standard – CA. 1899-1905 – 3r – 1 – $180.00 – mf#B02220 – us Library Micro [071]
[Eureka-] humboldt standard – CA. 1884-1939; 1948-51 – 144r – 1 – $8640.00 – (aka: humboldt daily standard) – mf#BC02219 – us Library Micro [071]
[Eureka-] humboldt weekly standard – CA. 1888-98 – 2r – 1 – $120.00 – mf#B02222 – us Library Micro [071]
Eureka journal – Eureka WI. 1867 may 10-1868 may 6 – 1r – 1 – mf#967953 – us WHS [071]
[Eureka-] miner – NV. 1971-1973 – 1r – 1 – $60.00 – mf#N04513 – us Library Micro [622]
[Eureka-] north coast ripsaw; rank and file reporter; humboldt independent news – CA. 1969-70; 1970-80; 1973-74 – 1 – $60.00 – mf#B03215 – us Library Micro [071]
[Eureka-] north county constitution – CA. 1975-76, 1982-89 – 4r – 1 – $240.00 – mf#B03216 – us Library Micro [071]
[Eureka-] northcoast outdoors – CA. 1966-69 – 2r – 1 – $120.00 – mf#B02223 – us Library Micro [071]
[Eureka-] northcoast sporting news – CA. 1973-1976 – 2r – 1 – $120.00 – mf#B05031 – us Library Micro [071]
[Eureka-] northern weekly independent – CA. 1870-1872 – 1r – 1 – $60.00 – mf#B03593 – us Library Micro [071]
[Eureka-] poetry now – CA. v2. 1975 – 1r – 1 – $60.00 – mf#R02224 – us Library Micro [420]
[Eureka-] republican press – NV. 1884-85 [wkly] – 1r – 1 – $60.00 – mf#U04514 – us Library Micro [071]
[Eureka-] ruby hill mining report – NV. 14 aug 1879 – 1r – 1 – $60.00 – mf#U04515 – us Library Micro [071]
Eureka rundschau – Eureka, MN (USA), 1921 17 feb-1932 4 nov [gaps] – 3r – 1 – (title varies: 8 jan 1927: eureka-rundschau und das nordlicht; 23 nov 1928: dakota-rundschau. also publ in winona, mn) – gw Misc Inst [071]
[Eureka-] silver plume – MI. 1877 (scats) – 1r – 1 – $110.00 – mf#U04517 – us Library Micro [071]
[Eureka-] the california – CA. 1898-1909 [wkly] – 5r – 1 – $300.00 – mf#B02209 – us Library Micro [071]
[Eureka-] the daily standard – CA. apr-dec 1876 – 1r – 1 – $60.00 – mf#B02214 – us Library Micro [071]
[Eureka-] the eureka independent – CA. 1952-1958 – 3r – 1 – $180.00 – mf#B02217 – us Library Micro [071]
[Eureka-] the humboldt times – CA. 1854-1908 – 12r – 1 – $720.00 – mf#B02221 – us Library Micro [071]
[Eureka-] the sandpiper – CA. 1969-1990 – 2r – 1 – $120.00 – mf#B05032 – us Library Micro [071]
[Eureka-] the senior news – CA. 1981-1990 – 2r – 1 – $120.00 – mf#B05033 – us Library Micro [071]
[Eureka-] this week news and review – CA. 1988 – 1r – 1 – $60.00 – mf#B05034 – us Library Micro [071]
[Eureka-] tri-weekly standard – NV. 1885-86 – 2r – 1 – $120.00 – mf#U04518 – us Library Micro [071]
[Eureka-] west coast signal – CA. 1871-79 – 2r – 1 – $120.00 – mf#B02225 – us Library Micro [071]
[Eureka-] western watchman – CA. 1886-98 – 4r – 1 – $240.00 – mf#C02226 – us Library Micro [071]
Eureka-census roll of the indians of california – Humboldt Co, CA. 1936-38; 1941-42 – 3r – 1 – $150.00 – mf#B06095 – us Library Micro [317]
Eureka-rundschau und das nordlicht *see* Eureka rundschau
Eures, Robert *see* Lolly bleu
Euringer, Richard *see*
- Aphorismen
- Die arbeitslosen
- Chronik einer deutschen wandlung
- Deutsche passion 1933
- Fliegerschule 4
- Die fuersten fallen
- Reise zu den demokraten
- Der serasker
- Vortrupp "pascha"
- Der zug durch die wueste
Euringer, Sebastian *see*
- Die auffassung des hohenliedes bei den abessiniern
- Die chronologie der biblischen urgeschichte (gen 5 und 11)
- Die kunstform der althebraeischen poesie
- Der masorahtext des koheleth
- Der streit um das deuteronomium
- Die ueberlieferung der arabischen uebersetzung des diatessarons

Euripides *see*
- Bacchae
- The bacchae.
- Euripides medea
- Supplementum euripideum
- Die troerinnen

Euripides medea : mit scholien = Medea / Euripides; ed by Diehl, Ernst – Bonn: A Marcus und E Weber, 1911 – (= ser Kleine texte fuer vorlesungen und uebungen) – 2mf – 9 – 0-524-07217-5 – mf#1991-0079 – us ATLA [450]

Euripides medea und das goldene vliess von grillparzer – Blankenburg, Germany: O Kircher, 1896 – 1r – 1 – us Wisconsin U Libr [820]

Euripidou bakchai – London, England. 1871 – 1r – us UF Libraries [450]

Euro abstracts : scientific and technical publications and patents – Duesseldorf. 1963-1978 [1]; 1971-1978 [5]; 1975-1978 [9] – ISSN: 0014-2352 – mf#1721 – us UMI ProQuest [600]

Euro-American Alliance *see* Talon

Euro-american quarterly – 1980 spr-1983 win – 1r – 1 – mf#679059 – us WHS [327]

Euro-asia business review – Chichester. 1986-1987 (1,5,9) – ISSN: 0264-0155 – mf#16098 – us UMI ProQuest [338]

Euroinvest – London. 2001+ (1,5,9) – ISSN: 1465-4911 – mf#32285 – us UMI ProQuest [332]

Euromoney – London. 1969+ (1) 1974+ (5) 1974+ (9) – ISSN: 0014-2433 – mf#9935 – us UMI ProQuest [332]

Europa : wochenschrift fuer kultur und politik – Berlin DE, 1905 jan-jun – 1 – gw Misc Inst [074]

Europa : wochenzeitung fuer tat und freiheit – Paris (F), 1935 dec-1936 may [gaps] – 1 – (filmed by misc inst: 1935 21 nov-1936 9 may [1r]) – fr ACRPP; gw Misc Inst [327]

Europa : eine zeitschrift / ed by Schlegel, Friedrich – Frankfurt am Main 1803-05 – (= ser Dz. abt literatur) – 2v on 6mf – 9 – €120.00 – mf#k/n4656 – gw Olms [940]

Europa barbara / Navarro, Pedro Juan – Bogota, Colombia. 1942 – 1r – us UF Libraries [025]

Europa en 1949 : comentario a dos discursos de donoso cortes / Calvo Serer, Rafael – Madrid: Arbor, 1949 – 1 – sp Bibl Santa Ana [321]

Europa libre = Freies europa – Bogota (CO), sep 1942 – 1r – 1 – gw Misc Inst [079]

Europa medicophysica – Turin. 1980-1980 (1) 1980-1980 (5) 1980-1980 (9) – ISSN: 0014-2573 – mf#10036 – us UMI ProQuest [610]

Europa oder europa ag? / Schumacher, Kurt – Hannover 1949 – 1 – gw Mikropress [940]

Europa und amerika : oder die kuenftigen verhaeltnisse der civilisirten welt / Schmidt-Phiseldek, Conrad F von – Kopenhagen 1820 [mf ed Hildesheim 1995-98] – 2mf – 9 – €60.00 – 3-487-29361-7 – gw Olms [910]

Europa und die revolution / Goerres, Joseph von – Stuttgart, 1821 (mf ed 1993) – 3mf – 9 – €24.00 – 3-89349-245-3 – mf#DHS-AR 102 – gw Frankfurter [940]

Europa union – Europa Union Verlag. v1-28. 1950-may 1977 [mthly] – 1 – (cont by: europaeische zeitung) – us Wisconsin U Libr [321]

Europa union : europaeische zeitung fuer politik, wirtschaft und kultur – Bonn DE, 1963-77 – 2r – 1 – mf#7143 – gw Mikropress [341]

Europa y america : biografias y semblanzas universales por varios distinguidos escritores – Madrid: tip julian frances, s.a. – 1 – sp Bibl Santa Ana [920]

Europaeische annalen – Tuebingen DE, 1795-97 – 4r – 1 – gw Misc Inst [940]

Die europaeische fama – Hamburg DE, 1685 aug-dec [gaps], 1687 jan-jul [gaps], 1688 [single iss], 1689 jan-nov [gaps], 1695 [single iss] – 1 – gw Misc Inst [940]

Die europaeische fama – Leipzig DE, 1702-21 – 1 – gw Misc Inst [940]

Europaeische hefte : wochenschrift fuer politik, kultur, wirtschaft – Prag (CZ)/Bern (CH)/Paris (F), 1934 19 apr-1935 30 nov – 2r – 1 – (merged with: aufruf [n25 1934]) – gw Misc Inst [940]

Europaeische parlaments-chronik – Leipzig DE, 1848 8 apr-30 jun – 1r – 1 – gw Misc Inst [323]

Die europaeische relation – Hamburg DE, 1676-77 [single iss], 1688, 1698 [single iss], 1701-02, 1703 [single iss] – 11r – 1 – gw Misc Inst [940]

Europaeische revue (klp12) / ed by Rohan, Karl Anton & Moras, Joachim – Leipzig/Berlin/Stuttgart: im Verlag der Neue-Geist 1925/26-1944 [mf ed 2003] – (= ser Marbacher mikrofiche-editionen (mme) 12; Kultur – literatur – politik: deutsche zeitschriften des 19./20. jahrhunderts (klp)) – 20v on 195mf – 9 – €790.00 – 3-89131-370-5 – gw Fischer [320]

Europaeische studiengaenge in der bundesrepublik deutschland : ein modell der europaeisierung der hochschulbildung? / Danthony, Marie-Josephe – (mf ed 1998) – 3mf – 9 – €49.00 – 3-8267-2590-5 – mf#DHS 2590 – gw Frankfurter [378]

Europaeische zeitung – Europa Union Verlag. v28- . jun 1977- [mthly] – 1 – (cont: europa union) – us Wisconsin U Libr [074]

Europaeische zeitung – Hanau DE, 1848-49 – 4r – 1 – (title varies: 26 sep 1774: neue europaeische zeitung; 1784: hanauer neue europaeische zeitung; 1814: hanauer neue zeitung; 1 jan 1826: hanauer zeitung. filmed by other misc inst: 1687, 1690, 1701, 1703-09, 1711-14, 1717-19, 1721-65, 1767-95, 1797-1810 [83r]; 1798-99, 1814-25, 1830-33, 1835-47, 1850-1914 3 aug [148r]. incl suppls) – gw Misc Inst [074]

Europaeischer mercurius – Koenigsberg (Kaliningrad RUS), 1816 jul-1933 mar/apr [gaps] – 264r – 1 – (numerous title changes; 1848: koeniglichen preussischen staats- kriegs- und friedens-zeitung; 1850: koenigsberger hartung'sche zeitung) – gw Misc Inst [074]

Den europaeiske unions tidende *see* The official journal of the european union

Europafaehigkeit der schweizerischen alters- und hinterlassenenversicherung (ahv) : mit blick auf einen beitritt zur europaeischen union / Grieshaber, Christoph – (mf ed 1995) – 2mf – 9 – €40.00 – 3-8267-2083-0 – mf#DHS 2083 – gw Frankfurter [368]

Europa-stunde – Berlin DE, 1929 29 sep-1937 23 oct – 1r – 1 – gw Mikrofilm [074]

Europa-union deutschland – Bonn DE, 2000-mar 2003 – 1r – 1 – (filmed by misc inst: 1987-96, by mikropress: 1963-77 [2r]; title varies: 1950: europa-union; juni 1977: europaeische zeitung) – gw Mikrofilm; gw Misc Inst; gw Mikropress [341]

Europe – Washington. 1979+ (1) 1979+ (5) 1979+ (9) – (Cont: European community) – ISSN: 0191-4545 – mf#6845,01 – us UMI ProQuest [337]

Europe, 1946-1976 / U.S. Central Intelligence Agency – (= ser Cia research reports) – 4r – 1 – $605.00 – 0-89093-452-5 – (with p/g) – us UPA [327]

Europe, 1950-1961 : supplement / U.S. Office of Strategic Services & U.S. State Dept – (= ser Oss/state department intelligence and research reports 10) – 11r – 1 – $1690.00 – 0-89093-294-8 – (with p/g) – us UPA [940]

Europe, Agence Internationale d'Information pour la Presse *see* Europe documents luxembourg

Europe and america : reports of proceedings at an inauguration banquet / Field, Cyrus West – [London?], [1868] – (= ser 19th c ireland) – 1mf – 9 – mf#1.1.7852 – uk Chadwyck [337]

Europe and nato : special studies, 1998-2002: supplement / ed by Lester, Robert E – Bethesda MD: UPA c2003 [mf ed 2003] – (= ser Special studies series) – 14r – 1 – 1-55655-964-X – (with p/g entitled: a guide to the microfilm edition of europe and nato) – us UPA [341]

Europe asks : who is shree krishna: letters written to a christian friend / Pal, Bipin Chandra – Calcutta: New India Print & Pub Co, 1938 – (= ser Samp: indian books) – us CRL [230]

Europe brief notes – 1988- – 1 – us Wisconsin U Libr [073]

L'europe depuis l'avenement du roi louis-philippe : pour faire suite a l'histoire de la restauration du meme auteur / Capefigue, Jean Baptiste Honore Raymond – Paris 1845-1846 [mf ed Hildesheim 1995-98] – 10v on 30mf – 9 – €300 – ISBN-10: 3-487-26079-4 – ISBN-13: 978-3-487-26079-2 – gw Olms [944]

L'europe diplomatique : gazette internationale – Paris, feb 1876-1887 – 1 – (cont as: la gazette diplomatique) – fr ACRPP [327]

Europe documents luxembourg : europe / Europe, Agence Internationale d'Information pour la Presse – 1988- [wkly, irreg] – 1 – (bulletins quotidiens & suppl) – us Wisconsin U Libr [940]

Europe et jupiter, concert francois a deux voix : avec symphonie et la basse continue / Alexandre, Pierre – Paris: aut 1715 [mf ed 19–] – 1r – 1 – mf#pres. film 34 – us Sibley [780]

L'europe et la revolution francaise / Sorel, Albert – v1-8. 1885-1904 – 1 – $120.00 – mf#0562 – us Brook [933]

L'europe et l'amerique comparees / Drouin de Bercy – Paris 1818 [mf ed Hildesheim 1995-98] – 2v on 6mf – 9 – €120.00 – 3-487-26749-7 – gw Olms [910]

L'europe et l'amerique comparees / Drouin de Bercy – Paris: Chez Rosa...2v. 1818 [mf ed 1985] – 2v on 1mf – 9 – mf#39286 – cn Canadiana [910]

Europe financiere – London, UK. 26 nov-10 dec 1870 – 1 – uk British Libr Newspaper [072]

Europe france outremer *see* France-outre-mer

Europe in the nineteenth century (1789-1914) / Grant, A J – New York, NY. 1928 – 1r – 1 – us UF Libraries [940]

Europe in the seventeenth century / Ogg, David – London: A & C Black, 1925 – xi/579p – 1 – us Wisconsin U Libr [940]

L'europe litteraire : journal de la litterature nationale et etrangere – Paris, mars 1833-janv 1834 – 1 – fr ACRPP [410]

Europe litteraire (1833-1834) / Palfrey, Thomas Rossman – Paris, France. 1927 – 1r – us UF Libraries [025]

L'europe nouvelle : revue hebdomadaire des questions exterieures, economiques et litteraires – Paris, 1918-juin 1940 – 1 – fr ACRPP [073]

Europe speaks – London (GB), 1942 mar-1945 10 nov, 1946 20 jun – 1r – 1 – gw Misc Inst [940]

La europeada / Serrano Serrano, Ildefonso – Fuente de Cantos: Imp. Libr. San Jose, 1915 – 1 – sp Bibl Santa Ana [946]

European – Oxford. 1987-1987 (1,5,9) – ISSN: 0892-6824 – mf#49500 – us UMI ProQuest [338]

European and indo-european poets of urdu and persian / Saksena, Ram Babu – Lucknow: Newul Kishore Press, 1941 – (= ser Samp: indian books) – us CRL [490]

European and Mediterranean Plant Protection Organisation *see* Bulletin oepp eppo bulletin

European and north american railway terminus : sydney, cape-breton, the nearest port in british north america to europe – [Sidney, NS?: s.n.], 1851 [mf ed 1986] – 1mf – 9 – 0-665-63539-7 – mf#63539 – cn Canadiana [380]

European archives of psychiatry and clinical neuroscience – Berlin. 1990-1992 (1) – (Cont: European archives of psychiatry and neurological sciences) – ISSN: 0940-1334 – mf#13132,02 – us UMI ProQuest [616]

European archives of psychiatry and neurological sciences – Berlin. 1989 (1) – (cont: archiv fuer psychiatrie und nervenkrankheiten; cont by: european archives of psychiatry and clinical neuroscience) – ISSN: 0175-758X – mf#13132,01 – us UMI ProQuest [616]

European automotive design – Horton Kirby. 1997+ (1) – ISSN: 1368-552X – mf#28191 – us UMI ProQuest [629]

European baptist press service – Boston. 1791-1907 (1) – 1 – mf#5150 – us Southern Baptist [242]

European baptist press service – Rueschlikon. European Baptist Federation Press releases. 1961-81. Single reels available – 1 – us ABHS [242]

European beginnings in west africa, 1454-1578 : aa survey of the first century of white enterprise in west africa / Blake, John William – London, New York [etc]: Pub for the Royal empire society by Longmans, Green and co [1937] [mf ed 1986] – viii/[2]p/212p – 1 – mf#8625 – us Wisconsin U Libr [960]

European business law review – v1-10. 1990-99 – 5,6,9 – $455.00 set – ISSN: 0959-6941 – mf#112241 – us Hein [346]

European cancer news – Dordrecht. 1991-1993 (1,5,9) – ISSN: 0921-3732 – mf#16784 – us UMI ProQuest [616]

European chromatography news – Chichester. 1987-1988 (1) 1987-1988 (5) 1987-1988 (9) – ISSN: 0891-4303 – mf#16167 – us UMI ProQuest [540]

European civilization : protestantism and catholicity compared in their effects on the civilization of europe = Protestantismo comparado con el catolicismo / Balmes, Jaime Luciano – 10th ed. Baltimore: John Murphy, 1868, c1850 – 2mf – 9 – 0-8370-7363-4 – (in english. incl ind) – mf#1986-1363 – us ATLA [240]

European Coal and Steel Community. Common Assembly *see*
- Debats
- Document

European colonies, in various parts of the world : viewed in their social, moral, and physical condition / Howison, John – London, 1834 – (= ser 19th c british colonization) – 2v on 10mf – 9 – mf#1.1.7497 – uk Chadwyck [900]

European community – Washington. 1954-1978 (1) 1972-1978 (5) 1975-1978 (9) – (Cont by: Europe) – ISSN: 0014-2891 – mf#6845 – us UMI ProQuest [337]

European community – London. 1979-1980 (1) 1979-1980 (5) 1979-1980 (9) – mf#9800 – us UMI ProQuest [337]

European eating disorders review – Chichester, 1998+ [1,5,9] – ISSN: 1072-4133 – mf#21190 – us UMI ProQuest [612]

European economic review – Amsterdam. 1970+ (1) 1970+ (5) 1987+ (9) – ISSN: 0014-2921 – mf#42199 – us UMI ProQuest [330]

European education – Armonk. 1991+ – 1,5,9 – (Cont: Western European education) – ISSN: 1056-4934 – mf#13347,01 – us UMI ProQuest [370]

European express and belgian times and news – Brussels, Belgium 9 nov 1901-5 jan 1907 [wkly] – 1 – (cont by: belgian times & news and european express [12 jan 1907-26 sep 1908]) – uk British Libr Newspaper [074]

European herald – London, England. -w. 2 Nov 1933-21 Nov 1936. 2 reels – 1 – uk British Libr Newspaper [072]

European immigration into natal, 1824-1910 / Simmonds, Heather A – Cape Town, South Africa. 1964 – 1r – 1 – us UF Libraries [960]

The european in india : or, anglo-indian's vade-mecum / Hull, Edmund C P – London, 1871 – (= ser 19th c british colonization) – 4mf – 9 – mf#1.1.5687 – uk Chadwyck [954]

European investigations, 1898-1936 – (= ser Records of the immigration and naturalization service, series a: subject correspondence files 4; Research colls in american immigration) – 10r – 1 – $1935.00 – 1-55655-587-3 – (with p/g) – us UPA [324]

European journal of anaesthesiology – Oxford. 1984-1996 (1,5,9) – ISSN: 0265-0215 – mf#15524 – us UMI ProQuest [617]

European journal of applied microbiology – Berlin. (1) 1975-1977 (5) 1975-1977 (9) – (Cont by: European journal of applied microbiology and biotechnology) – ISSN: 0340-2118 – mf#13165 – us UMI ProQuest [576]

European journal of applied microbiology and biotechnology – Berlin. 1978-1983 (1,5,9) – (Cont: European journal of applied microbiology. Cont by: Applied microbiology and biotechnology) – ISSN: 0171-1741 – mf#13165,01 – us UMI ProQuest [576]

European journal of applied physiology – Heidelberg, 2000+ [1,5,9] – (cont: european journal of applied physiology and occupational physiology) – ISSN: 1439-6319 – mf#13166,03 – us UMI ProQuest [612]

European journal of applied physiology and occupational physiology – Heidelberg. 1973-1999 (1) 1973-1999 (5) 1973-1999 (9) – ISSN: 0301-5548 – mf#13166,02 – us UMI ProQuest [612]

European journal of biochemistry – Heidelberg. 1967+ (1) 1967+ (5) 1967+ (9) – ISSN: 0014-2956 – mf#13111 – us UMI ProQuest [574]

European journal of cancer – Oxford. 1990+ (1,5,9) – (Cont: European journal of cancer and clinical oncology) – ISSN: 0959-8049 – mf#49068,01 – us UMI ProQuest [616]

European journal of cancer and clinical oncology – Oxford. 1965-1989 (1,5,9) – (Cont by: European journal of cancer) – ISSN: 0277-5379 – mf#49068 – us UMI ProQuest [616]

European journal of cancer care : english language edition – Oxford. 1993-1995 (1,5,9) – ISSN: 0961-5423 – mf#18770 – us UMI ProQuest [616]

European journal of cancer, pt b : oral oncology – Oxford. 1993-1994 (1,5,9) – ISSN: 0964-1955 – mf#49624 – us UMI ProQuest [616]

European journal of cardiology – Amsterdam. 1979-1979 (1) 1979-1979 (5) (9) – (Cont by: International journal of cardiology) – ISSN: 0301-4711 – mf#42219 – us UMI ProQuest [616]

European journal of cardiovascular nursing – Amsterdam. 2002+ (1,5,9) – ISSN: 1474-5151 – mf#42886 – us UMI ProQuest [610]

European journal of clinical investigation – Oxford. 1980-1996 (1,5,9) – ISSN: 0014-2972 – mf#15526 – us UMI ProQuest [610]

European journal of clinical nutrition – Houndsmill. 1988+ (1,5,9) – ISSN: 0954-3007 – mf#16868 – us UMI ProQuest [613]

European journal of clinical pharmacology – Heidelberg. 1970-1995 (1) 1978-1995 (5) 1978-1995 (9) – (Cont: Pharmacologia clinica: Zeitschrift fuer klinische Pharmakologie und Pharmakotherapie) – ISSN: 0031-6970 – mf#13112,01 – us UMI ProQuest [615]

European journal of disorders of communication – London. 1992-1996 (1) – (cont: british journal of disorders of communication) – ISSN: 0963-7273 – mf#14161,01 – us UMI ProQuest [616]

European journal of heart failure – Amsterdam. 1999+ (1) – ISSN: 1388-9842 – mf#42818 – us UMI ProQuest [616]

European journal of immunogenetics – Oxford. 1991-1995 (1,5,9) – (Cont: Journal of immunogenetics) – ISSN: 0960-7420 – mf#15563,01 – us UMI ProQuest [575]

European journal of information systems – Houndsmill. 1991-1996 (1,5,9) – ISSN: 0960-085X – mf#18479 – us UMI ProQuest [000]

European journal of innovation management – Bradford. 2001+ (1,5,9) – ISSN: 1460-1060 – mf#31601 – us UMI ProQuest [650]

European journal of intensive care medicine – Heidelberg. (1) 1975-1976 (5) – (Cont by: Intensive care medicine) – ISSN: 0340-0964 – mf#13182 – us UMI ProQuest [610]

European journal of international law = Journal europeen de droit international – Oxford, 1998+ [1,5,9] – ISSN: 0938-5428 – mf#27030 – us UMI ProQuest [341]
European journal of marketing – Bradford. 1992-1994 (1) 1992-1994 (5) 1992-1994 (9) – ISSN: 0309-0566 – mf#15748,01 – us UMI ProQuest [650]
European journal of nuclear medicine – Heidelberg. 1981-1993 (1) 1981-1993 (5) 1981-1993 (9) – ISSN: 0340-6997 – mf#13167 – us UMI ProQuest [616]
European journal of obstetrics, gynecology and reproductive biology – Amsterdam. 1971+ (1) 1971+ (5) 1988+ (9) – ISSN: 0301-2115 – mf#42185 – us UMI ProQuest [618]
European journal of operational research – Amsterdam. 1977+ (1) 1977+ (5) 1987+ (9) – ISSN: 0377-2217 – mf#42124 – us UMI ProQuest [650]
European journal of orthodontics – Oxford. 1991-1996 (1) – ISSN: 0141-5387 – mf#13426 – us UMI ProQuest [617]
European journal of pediatrics – Heidelberg. 1979-1996 (1,5,9) – (Cont: Zeitschrift fuer Kinderheilkunde) – ISSN: 0340-6199 – mf#13113,01 – us UMI ProQuest [618]
European journal of personality – Chichester. 1987+ (1,5,9) – ISSN: 0890-2070 – mf#16099 – us UMI ProQuest [150]
European journal of pharmacology – Amsterdam. 1967+ (1) 1967+ (5) 1987+ (9) – ISSN: 0014-2999 – mf#42259 – us UMI ProQuest [615]
European journal of pharmacology : environmental toxicology and pharmacology section – Amsterdam. 1992-1994 (1,5,9) – ISSN: 0926-6917 – mf#42719 – us UMI ProQuest [615]
European journal of pharmacology : molecular pharmacology section – Amsterdam. 1989-1994 (1,5,9) – ISSN: 0922-4106 – mf#42446 – us UMI ProQuest [615]
European journal of physical medicine and rehabilitation – Wien. 1991-1993 (1,5,9) – ISSN: 1017-6721 – mf#19295 – us UMI ProQuest [617]
European journal of political economy – Amsterdam. 1989+ (1,5,9) – ISSN: 0176-2680 – mf#42465 – us UMI ProQuest [330]
European journal of political research – Amsterdam. 1986+ (1,5,9) – ISSN: 0304-4130 – mf#16039 – us UMI ProQuest [320]
European journal of population = Revue europeenne de demographie – Amsterdam. 1990-1992 (1,5,9) – ISSN: 0168-6577 – mf#42537 – us UMI ProQuest [304]
European journal of social psychology – New York. 1971+ (1,5,9) – ISSN: 0046-2772 – mf#11777 – us UMI ProQuest [302]
European journal of soil science – Oxford. 1994+ (1,5,9) – ISSN: 1351-0754 – mf#20777 – us UMI ProQuest [630]
European journal of special needs education – Chichester. 1986-1990 (1) 1986-1988 (5) 1986-1988 (9) – ISSN: 0885-6257 – mf#16100 – us UMI ProQuest [370]
European journal of surgery = Acta chirurgica – Oslo. 1998+ (1,5,9) – ISSN: 1102-4151 – mf#22159,03 – us UMI ProQuest [617]
European legislation on declarations of death : (survey concluded on january 1, 1949) / Office of General Counsel. European Headquarters. American Joint Distribution Committee – Paris: The Committee, [1949?] (mf ed 1995) – 1r – 1 – mf#*ZP-1485 – us NY Public [940]
European magazine and london review – London. 1782-1826 (1) – mf#4248 – us UMI ProQuest [073]
European management journal – London. 1988-1992 (1,5,9) – (Cont by: European management journal) – ISSN: 0263-2373 – mf#17390 – us UMI ProQuest [650]
European management journal – London. 1992+ (1,5,9) – (Cont: European management journal) – ISSN: 0263-2373 – mf#49629 – us UMI ProQuest [650]
European music manuscripts, series 1 : from the british library, london / ed by Bray, Roger – 64r in 4 units (ongoing) – 1 – (unit 1: mss fr egerton, king's, sloane, stowe and add mss 18r C14R-12001. unit 2: add mss 15r C14R-12002. unit 3: add mss 18r C14R-12003. unit 4: add mss, printed books, royal mss and zweig mss 13r C14R-12004. with printed guide based on augustus hughes' catalogue of manuscript music in the british museum) – mf#C14R-12000 – us Primary [780]
European music manuscripts, series 2 : from the biblioteca da ajuda, lisbon / ed by Brito, Manuel Carlos de – 395r in 3 sects (ongoing) – 1 – (sect a: music before 1740 and sacred music 50r in 2 units (unit 1 21r unit 2 29r). sect b: music 1740-1770 ca 196r in 6 units (unit 3 33r unit 4 32r unit 5 32r unit 6 34r unit 7 33r unit 8 32r). sect c: music 1770-1820 149r in 5 units (unit 9 29r unit 10 30r unit 11 27r unit 12 32r unit 13 31r). selection full order of the publ library catalog: biblioteca da ajuda: catalogo de musica manuscrita: mariana amelia machado santos, lisboa 1958-68) – mf#C14R-12100 – us Primary [780]
European music manuscripts, series 3 : from the paco ducal de vila vicosa, portugal / ed by Cranmer, David – [mf ed 2003] – ca 50r – 1 – us Primary [780]
European neurology – Basel. 1968-1974 (1) 1968-1973 (5) 1970-1973 (9) – ISSN: 0014-3022 – mf#2694 – us UMI ProQuest [616]
The european news – Paris. n86-93, 96-117. juin 1870-71 – 1 – fr ACRPP [073]
European official statistical serials, 1841-1984 : detailed information on european political and economic affairs – [mf ed Chadwyck-Healey] – 7214mf – 9 – (in english, french, german, portuguese, russian, spanish. individual titles listed and may be purchased separately) – uk Chadwyck [314]
European Parliament. Secretariat. Luxembourg *see* Ep news
European Parliamentary Assembly *see* L'activitea1 de l'assemblea1e parlementaire europea1ene
European politics in southern rhodesia / Leys, Colin – Oxford, England. 1959 – 1r – 1 – us UF Libraries [960]
European polymer journal – Oxford. 1965+ (1,5,9) – ISSN: 0014-3057 – mf#49069 – us UMI ProQuest [540]
European press – Bremen DE, 1921 – 1r – 1 – gw Misc Inst [074]
European research – Deventer. 1981-1988 (1,5,9) – (Cont by: Marketing and research today) – ISSN: 0304-4297 – mf#42600 – us UMI ProQuest [650]
European review of social psychology – Chichester. 1990-1996 (1,5,9) – ISSN: 1046-3283 – mf#18144 – us UMI ProQuest [301]
European romantic tradition : the sir walter scott manuscripts – 52r – 1 – (pt1: literary and historical mss from the national library of scotland. pt2: the scott correspondence, sect a. pt3: the scott correspondence, sect b. printed guide available) – mf#C35-14700 – us Primary [420]
European rubber journal – London. 1982+ (1) 1982+ (5) 1982+ (9) – (cont: European rubber journal + urethanes today) – ISSN: 0266-4151 – mf#1212,03 – us UMI ProQuest [660]
European rubber journal – London. 1973-1980 (1) 1973-1980 (5) 1976-1980 (9) – (cont: Rubber journal. cont by: European rubber journal + urethanes today) – ISSN: 0305-2222 – mf#1212,01 – us UMI ProQuest [670]
European rubber journal + urethanes today – Croydon. 1980-1981 (1) 1980-1981 (5) 1980-1981 (9) – (cont: European rubber journal. cont by: European rubber journal) – ISSN: 0260-5317 – mf#1212,02 – us UMI ProQuest [660]
European settlements in the far east, china, japan, corea, indo-china, straits settlements, malay states, siam... / Smith, D Warres – London 1900 – (= ser 19th c british colonization) – 5mf – 9 – mf#1.1.7941 – uk Chadwyck [307]
European Society of Parenteral and Enteral Nutrition *see* Clinical nutrition
European spectroscopy news – Chichester. 1982-1988 (1) 1982-1988 (5) 1982-1988 (9) – ISSN: 0307-0026 – mf#13305 – us UMI ProQuest [540]
European surgical research = Recherches chirurgicales europeenes – Basel. 1969-1973 (1) 1969-1972 (5) 1970-1972 (9) – ISSN: 0014-312X – mf#5191 – us UMI ProQuest [617]
The european tour / Allen, Grant – New York: Dodd, Mead, 1899 – 4mf – 9 – mf#05031 – cn Canadiana [914]
European travel and life – New York. 1987-1992 (1,5,9) – ISSN: 0882-7737 – mf#17978 – us UMI ProQuest [914]
European travellers in india : during the fifteenth, sixteenth, and seventeenth centuries, the evidence afforded by them with respect to indian social institutions and the nature and influence of indian government / Oaten, Edward Farley – London: Kegan Paul, Trench, Trubner and Co, 1909 – (= ser Samp: indian books) – us CRL [915]
European twentieth century art : subject collections – (= ser Art exhibition catalogues on microfiche) – 187 catalogues on 206mf – 9 – £1300.00 – (individual titles not listed separately) – uk Chadwyck [700]
European women's periodicals – 284r – 1 – (austrian and belgian periodicals 29r c36-28201; french women's periodicals 24r c36-28202; german women's periodicals 61r c36-28203; dutch women's periodicals 170r c36-28204) – mf#C36-28200 – us Primary [305]
Europeans' guide and medical companion in India / Gangadin – [London], [1895] – (= ser 19th c books on british colonization) – 2mf – 9 – mf#1.1.696 – uk Chadwyck [610]
Europe-asia studies – Abingdon. 1993+ (1) 1999+ (5) 1999+ (9) – (cont: soviet studies) – ISSN: 0966-8136 – mf#2480,01 – us UMI ProQuest [338]
L'europeen : journal des sciences morales et economiques – 1 – (suite de: Journal des sciences morales et politiques. n1-4. 3-24 dec 1831. Paris. dec 1831-oct 1832, oct 1835-oct 1838. devenu: Revue nationale) – fr ACRPP [073]
L'europeen – Paris. n1-239. avr 1929-33 [wkly] – 1 – (lacking no. 104-132, 134-137, 139, 141-158...hebdomadaire economique, artistique et litteraire) – fr ACRPP [073]
Europeesche kolonisatie in suriname / Pijttersen, H – Stockum, Netherlands. 1896 – 1r – 1 – us UF Libraries [972]
Europeo – Milan. 1970-1985 [1]; 1974-1985 [5]; 1979-1985 [9] – ISSN: 0014-3189 – mf#6267 – us UMI ProQuest [073]
Europeo en el caribe / Alonso Quintero, Elfidio – Ciudad Trujillo, Dominican Republic. 1943 – 1r – us UF Libraries [972]
Europeo en el tropico / Aldef – San Salvador, El Salvador. 1956 – 1r – us UF Libraries [972]
Europe's war, america's warning / Macfarland, Charles Stedman – New York: Church Peace Union, [1916?] – (= ser The church and international peace) – 1mf – 9 – 0-7905-9323-8 – mf#1989-2548 – us ATLA [230]
Eurowired – London. 2000+ (1,5,9) – mf#32284 – us UMI ProQuest [380]
Eusebia cosme papers : from the holdings of the schomburg center for research in black culture, manuscripts, archives and rare books division: the new york public library, astor, lenox and tilden foundations – 5r – 1 – $425.00 – (guide which covers all coll under "literature and the arts" sold separately for $20 d3305.g6) – mf#D3305P21 – Dist. us Scholarly Res – us L of C Photodup [790]
Eusebii pamphli evangelicae praeparationis libri 15 / Eusebius of Caesarea, Bishop of Caesarea – Oxonii: E. Typographeo Academico, 1903 – 2r – 1 – 0-8370-0796-8 – mf#1984-B503 – us ATLA [240]
Eusebiou tou pamphilou euaggelikes apodeixeos, logoi deka : cum versione latina donati veronensis = Eusebii pamphili evangelicae demonstrationis, libri decem / Eusebius of Caesarea, Bishop – Oxonii: E Typographeo Academico, 1852 [mf ed 1993] – 2v on 10mf – 9 – 0-524-08322-3 – (in latin & greek) – mf#1993-1017 – us ATLA [240]
Eusebius *see*
– Eusebius kirchengeschichte, buch 6 und 7
– The history of the church from our lord's incarnation to the year of christ
– Die kirchengeschichte des eusebius
– Die palaestinischen maertyrer des eusebius von caesarea
Eusebius als historiker seiner zeit (akg11) / Laquer, R – Berlin-Leipzig, 1929 – (= ser Arbeiten zur kirchengeschichte (akg)) – €11.00 – ne Slangenburg [930]
Eusebius, bishop of caesarea, on the theophania or divine manifestation of our lord and saviour jesus christ / Eusebius of Caesarea, Bishop of Caesarea – Cambridge: University Press, 1843 – 2mf – 9 – 0-524-05143-7 – (in english) – mf#1990-1399 – us ATLA [241]
Eusebius kirchengeschichte : buch 6 und 7 aus dem armenischen uebersetzt / Preuschen, Erwin – Leipzig, 1902 – (= ser Tugal 2-22/3) – 2mf – 9 – €5.00 – ne Slangenburg [240]
Eusebius kirchengeschichte, buch 6 und 7 = Ecclesiastical history, bks 6-7 / Eusebius – Leipzig: J C Hinrichs, 1902 – (= ser Tugal) – 1mf – 9 – 0-7905-1697-7 – (incl bibl ref. in german) – mf#1987-1697 – us ATLA [240]
Eusebius of Caesarea, Bishop *see* Eusebiou tou pamphilou euaggelikes apodeixeos, logoi deka
Eusebius of Caesarea, Bishop of Caesarea *see*
– Eusebii pamphli evangelicae praeparationis libri 15
– Eusebius, bishop of caesarea, on the theophania or divine manifestation of our lord and saviour jesus christ
Eusebius schrift peri toon topikoon onomatoon / Klostermann, Erich – Leipzig, 1902 – (= ser Tugal 2-23/2b) – 1mf – 9 – €3.00 – ne Slangenburg [240]
Eusebius von Caesarea *see*
– Ausgewaehlte schriften, 2.bd (bdk1 2.reihe)
– Leben des kaisers konstantin und des kaisers konstantin rede an die versammlung der heiligen / die martyrer in palestina (bdk9 1.reihe)
Eusebius von nikomedien : versuch einer darstellung seiner persoenlichkeit und seines lebens unter besonderer beruecksichtigung seiner fuehrerschaft im arianischen streit / Lichtenstein, Adolf – Halle a. S: Max Niemeyer, 1903. Chicago: Dep of Photodup, U of Chicago Lib, 1978 (1r); Evanston: American Theol Lib Assoc, 1984 (1r) – 1 – 0-8370-0702-X – mf#1984-T096 – us ATLA [240]
Eusebius werke (gcsej4) – (= ser Griechische christlichen schriftsteller der ersten jahrhunderte (gcsej)) – (bd1: ed by i a heikel 1902 €19. bd2/1 ed by e schwartz 1903 €21. bd2/2 ed by e schwartz 1908 €21. bd2/3 ed by e schwartz 1909 €19. bd3 ed by e klostermann 1904 €23. bd4 ed by e klostermann 1906 €14. bd5 ed by j karst 1911 €17. bd6 ed by i a heikel 1913 €25. bd7/1 ed by r helm 1913 €17. bd7/2 ed by r helm 1926 €32. bd8/1 ed by k mras 1954 €27. bd8/2 ed by k mras 1956 €23. bd9 ed by i ziegler 1975 €18) – ne Slangenburg [240]
Euskadi rojay eri *see* Spanish-basque political periodicals
Eustace, John C *see* A classical tour through italy anno 1802 [eighteen hundred and two]
Eustace, John Chetwode *see* Answer to the charge delivered by the lord bishop of lincoln
Eustache / Duchatelard, Auguste – Paris, France. 1839 – 1r – us UF Libraries [440]
Eustathius : ancienne version latine des neuf homelies sur l'hexameron de basile de cesaree / Mendieta, E A & Rudberg, S Y – Berlin, 1958 – (= ser Tugal 5-66) – 4mf – 9 – €11.00 – ne Slangenburg [240]
Eustathius von sebaste und die chronologie der basilius-briefe : eine patristische studie / Loofs, Friedrich – Halle a S: M Niemeyer, 1898 – 1mf – 9 – 0-7905-4945-X – (incl bibl ref) – mf#1988-0945 – us ATLA [240]
Eustis, florida – Eustis, FL. 1926? – 1r – 1 – us UF Libraries [978]
The eustis news – Eustis, NE: Eustis Pub Co, 1904 (wkly) – 1 – (publ as: the news mar 28-sep 19 1918, mar 25-apr 22 1920 and oct 16 1924-jan 26 1928. publ in curtis ne, jan 5 1967- . issues for 1978-90 include valley voice, the monthly newsletter of medicine valley high school (curtis, ne). issues for v71 n16-v21 n30 also called whole n328-1077 mar 31 1905. july 24 1986 issue called 96th yr n30 but constitutes 84th yr n30. issues for dec 18-25 1986 called v84 n50-51 but constitute v84 n51-52) – us NE Hist [071]
Eutanville baptist church. orangeburg county. south carolina : church records – 1859-1972 [incomplete] – 1 – us Southern Baptist [242]
Eutaxia : or, the presbyterian liturgies. historical sketches / Baird, Charles Washington – New York: MW Dodd, 1855 – 1mf – 9 – 0-524-02442-1 – mf#1990-4301 – us ATLA [242]
Euterpe : eine musik-zeitschrift fur lehrer, kantoren, organisten und freunde der tonkunst uberhaupt – v1-43. 1841-84 – 1 – us L of C Photodup [780]
Euterpeiad : an album of music, poetry and prose – New York. 1830-1831 (1) – mf#3729 – us UMI ProQuest [780]
Euterpeiad : or, musical intelligencer devoted to the diffusion of musical information and belles lettres – Boston. 1820-1823 (1) – mf#3740 – us UMI ProQuest [780]
Euthaliana : studies of euthalius, codex h of the pauline epistles, and the armenian version / Robinson, Joseph Armitage – Cambridge: University Press 1895 [mf ed 1989] – (= ser Texts and studies (cambridge, england) 3/3) – 1mf – 9 – 0-7905-3276-X – (in english & greek) – mf#1987-3276 – us ATLA [225]
Euthanalia : studies of euthalius, codex h of the pauline epistles and the armenian version / Robinson, J A – 1895 – (= ser Texts and studies (ts)) – 3mf – 9 – ne Slangenburg [220]
Euthanasia news – New York. 1972-1978 (1) 1972-1978 (5) 1975-1978 (9) – (Cont by: Concern for dying) – ISSN: 0164-1581 – mf#6771 – us UMI ProQuest [170]
Euthanasia review – New York. 1986-1988 (1,5,9) – mf#14976 – us UMI ProQuest [170]
Euthanasy : or, happy talk towards the end of life / Mountford, William – 2nd ed. London: Edward T Whitfield, 1850 [mf ed 1985] – 2mf – 9 – 0-8370-5979-8 – mf#1985-3979 – us ATLA [230]
Eutiner anzeiger, amtliches verkuendigungsblatt *see* Eutinische woechentliche anzeigen
Eutiner kreis-anzeiger *see* Eutinische woechentliche anzeigen
Euting, Julius *see*
– Nabataeische inschriften aus arabien
– Sammlung der carthagischen inschriften. band 1, tafeln 1-202 und anhang, tafel 1-6
Eutinische woechentliche anzeigen – Eutin DE, 1870 – 1r – 1 – (title varies: 1815: woechentliche anzeigen fuer das fuerstenthum luebeck; 1868: anzeigen fuer das fuerstenthum luebeck; 1879: anzeiger fuer das fuerstenthum luebeck; 1938: anzeiger fuer den landkreis eutin; 16 nov 1949: eutiner kreis-anzeiger; 1 oct 1955: ostholsteiner anzeiger; other earlier titles: eutiner anzeiger, amtliches verkuendigungsblatt. filmed by misc inst: 1802 2 oct-1945 20 jun [gaps], 1949 15 nov-1950 [projected 1951-80]; 1981- [6r/yr]) – gw Misc Inst [074]
Eutrophication-tropic state / Shannon, Earl – s.l, s.l? 1970 – 1r – us UF Libraries [500]

Eutropi breviarium ab urbe condita (mgh1:2.bd) : cum versionibus graecis et pauli landolfique additamentis / ed by Droysen, H – 1879 – (= ser Monumenta germaniae historica 1: scriptores – auctores antiquissimi) – €25.00 – ne Slangenburg [240]

Eutropia : Or, how to find a way out of darkness and doubt into light and certainty / Devine, Pius – London: Burns and Oates, 1880 – 2mf – 9 – 0-8370-7056-2 – mf#1986-1056 – us ATLA [230]

Euvres de jean rotrou – Paris, France. v1-5. 1820 – 1r – us UF Libraries [440]

Euzkadi socialista *see* Spanish-basque political periodicals

Euzko deya *see* Spanish-basque political periodicals

Euzko deya and supplement – Paris, France. 11 jul 1937-10 sep 1939; 7 may-16 dec 1947; 7 sep 1948-22 dec 1950; 1951-56; 1958-31 aug 1962 – 14 1/2r – 1 – uk British Libr Newspaper [074]

Eva von buttler, die messaline und muckerin, als prototype der "seelenbraeute" : ein beitrag zur kenntniss der mysterien des pietismus / Christiany, Ludwig – Stuttgart: J Scheible, 1870 – 1mf – 9 – 0-524-03698-5 – mf#1990-4803 – us ATLA [240]

Les evaeques de quebec : notices biographiques / Taetu, Henri – Quebec: Narcisse-S Hardy, 1889 – 1mf – 9 – 0-524-04425-2 – mf#1992-2030 – us ATLA [240]

Evaeques et dioceses : deuxi eme serie / Houtin, Albert – Paris: A Houtin, 1909 – 1mf – 9 – 0-8370-8829-1 – (incl bibl ref and ind) – mf#1986-2829 – us ATLA [240]

Evagatorium in terrae sanctae, arabiae et egypti peregrinationem / Fabri, Fratris Felicis; ed by Hassler, C D – Stuttgardiae. v1-3. 1843-49 – 3v on 29mf – 8 – €56.00 – ne Slangenburg [243]

evagatorium in terrae sanctae, arabiae et egypti peregrinationen *see* Fratris felicis fabri evagatorium in terrae sanctae, arabiae et egypti peregrinationem

Evagrius *see* A history of the church from a d 322 to the death of theodore of mopsuestia, a d 427. and, from a d 431 to a d 594

Evagrius ponticus / ed by Frankenberg, W – Berlin 1912 – 13mf – 8 – €25.00 – ne Slangenburg [240]

Evaluating and implementing risk management stretegies for the university of south carolina athletics department / Matheny, Tami – 1999 – 1mf – 9 – $4.00 – mf#PE 4047 – us Kinesology [790]

Evaluating and reshaping a model of church renewal at the first baptist church of longwood, florida / Hammock, James W – 1981 – 1 – us Southern Baptist [242]

Evaluating rural housing / Mosier, Charles I – Gainesville, FL. 1942 – 1r – us UF Libraries [360]

Evaluating the boater experience : the interrelationship of recreational use, user contacts, experiential impacts, satisfaction and displacement / Drogin, E B – 1991 – 2mf – 9 – $8.00 – us Kinesology [790]

Evaluating the effectiveness of leadership in schools west of johannesburg / Mathebula, Freddy Masingita – Pretoria: Vista University 2002 [mf ed 2002] – 2mf – 9 – (incl bibl) – mf#mfm15164 – sa Unisa [370]

Evaluation and program planning – New York. 1978+ (1,5,9) – ISSN: 0149-7189 – mf#49300 – us UMI ProQuest [300]

Evaluation and the health professions – Beverly Hills. 1983+ (1,5,9) – ISSN: 0163-2787 – mf#14006 – us UMI ProQuest [610]

Evaluation des asthma-verhaltenstrainings (avt) : auswertung der verlaufsdokumentation / Beys, Martina – (mf ed 1997) – 3mf – 9 – €49.00 – 3-8267-2455-0 – mf#DHS 2455 – gw Frankfurter [150]

An evaluation of a home-based exercise program involving non-exertional hypoxemic and exertional hypoxemic chronic obstructive pulmonary diseased patients / Kotarski, Mark & Berger, Richard A – 1991 – 1mf – 9 – $4.00 – us Kinesology [612]

Evaluation of a sex education programme for indian adolescents / Naran, Shiela – Uni of South Africa 2001 [mf ed Johannesburg 2001] – 4mf – 9 – (incl bibl ref) – mf#mfm14668 – sa Unisa [306]

Evaluation of a static technique for estimating atmospheric... / Harding, Charles Irvin – s.l, s.l? 1959 – 1r – us UF Libraries [025]

An evaluation of an instructor-led and self-managed computer software training course / Falkenberg, Ryan James – Uni of South Africa 2000 [mf ed Pretoria: UNISA 2000] – 4mf – 9 – (incl bibl ref) – mf#mfm14722 – sa Unisa [000]

An evaluation of athletic training support in nata district seven high schools / Liljenquist, Paige – 1996 – 1mf – 9 – $4.00 – mf#PE 3817 – us Kinesology [617]

An evaluation of carolina athletes coming together (act) : a program using student-athletes as educators and mentors / Holliday, Corey L – 1997 – 1mf – 9 – $4.00 – mf#PSY 1988 – us Kinesology [370]

Evaluation of collegiate coaches from the perspective of the student-athlete / DiPuma, Joseph J – 1999 – 2mf – 9 – $8.00 – mf#PE 3959 – us Kinesology [790]

Evaluation of court-annexed arbitration in three federal district courts / Lind, E Allan & Shapard, John E – Washington: FJC, Mar 1981 – 2mf – 9 – $3.00 – mf#LLMC 95-800 – us LLMC [347]

Evaluation of court-annexed arbitration in three federal district courts : september 1983 revision / Lind, E Allan & Shapard, John E – Washington: FJC, 1983 – 2mf – 9 – $3.00 – mf#LLMC 95-313 – us LLMC [347]

An evaluation of customer service : with regard to service quality amongst merchandise retailers / Dhurup, Manilall – Pretoria: Vista University 2001 [mf ed 2001] – 4mf – 9 – (incl bibl ref) – mf#mfm15229 – sa Unisa [650]

Evaluation of exercise tolerance in women : receiving surgery and chemotherapy as treatment for stage 2 breast cancer / Wiley, Lisa D – 1998 – 78p on1mf – 9 – $5.00 – mf#HE 667 – us Kinesology [617]

Evaluation of exercise videotapes performed by fitness experts and celebrities / Robinson, Adrienne – 1998 – 1mf – 9 – $4.00 – mf#HE 649 – us Kinesology [790]

Evaluation of losses in lv feeders using the beta load model / Fourie, Roelof Jacobus – Stellenbosch: U of Stellenbosch 1998 [mf ed 1998] – 4mf – 9 – mf#mf.1360 – sa Stellenbosch [620]

Evaluation of metabotrim(tm) supplementation of body composition, strength and vo2max in training female athletes / Murray, Teena – University of North Carolina at Greensboro, 1996 – 1mf – 9 – mf#PH 1502 – us Kinesology [612]

An evaluation of nature center managers'perceptions of their job responsibilities with regard to educational background, services/resources offered, number of years as administrator, and budget of the nature center / Wilson, D B – 1991 – 2mf – 9 – $8.00 – us Kinesology [150]

Evaluation of oxygen uptake, minute ventilation, and perceived exertion at varying cadences between step and slide aerobics / Santom, Michelle H – 1997 – 1mf – 9 – $4.00 – mf#PH 1637 – us Kinesology [612]

An evaluation of school readiness in an informal settlement / Naidoo, Balendran – Uni of South Africa 2000 [mf ed Johannesburg 2000] – 5mf – 9 – mf#mfm14917 – sa Unisa [370]

Evaluation of stereodirecting potential of camphor-derived carboxylic ester in a-benzylation / Vela, Nomandla M – Pretoria: Vista University 2002 [mf ed 2002] – 3mf – 9 – (incl bibl ref) – mf#mfm15219 – sa Unisa [540]

Evaluation of stretch load capacity and utilization of stored elastic energy in leg extensor muscles during vertical jumps / He, Qin – 1989 – 69p on 1mf – 9 – $4.00 – us Kinesology [612]

An evaluation of student attitudes, intentions, and personal health behaviors as a result of having completed health education 214 at the ohio state university during the spring quarter of 1990 / Michael, J F – 1991 – 2mf – 9 – $8.00 – us Kinesology [613]

An evaluation of the characteristics of successful students : at the brinkman-froemming umpire school / Robertson, Stuart A – 1993 – 1mf – 9 – $4.00 – us Kinesology [370]

An evaluation of the effectiveness of the 1981 health workshops for lane county ceta employees / Paddon, Kathleen S – 1981 – 2mf – 9 – $8.00 – us Kinesology [610]

An evaluation of the effectiveness of training in syndromic management of sexual transmitted diseases / Ngesi, Lechina Buyisile – Uni of South Africa 2000 [mf ed Johannesburg 2000] – 3mf – 9 – (incl bibl ref) – mf#mfm14767 – sa Unisa [610]

An evaluation of the effects of a smoking prevention program on middle school students' knowledge and attitudes concerning cigarette smoking / Tennent, Sylvia R & Baker, Judith A – 1991 – 1mf – 9 – $4.00 – us Kinesology [613]

Evaluation of the first year of the training of english-speaking primary school mathematics teachers in the transvaal / Bezuidenhout, Cynthia Anne – Unisa 1980 [mf ed Pretoria: U of South Africa 1980] – 4mf – 9 – (incl bibl ref) – sa Unisa [370]

Evaluation of the implementation of the financial policy on budgeting in public schools in the free state / Ntseto, Vangeli Emmanuel – Pretoria: Vista University 2001 [mf ed 2001] – 2mf – 9 – (incl bibl) – mf#mfm15190 – sa Unisa [370]

An evaluation of the importance of moderate exercise, t'ai chi, and problem solving in relation to psychological distress / Bond, Dale – 1998 – 2mf – 9 – $8.00 – mf#PSY 2014 – us Kinesology [790]

An evaluation of the physical fitness effects of a high school aerobic dance curriculum / Baldwin, Susan & Pechar, Gary S – 1991 – 1mf – 9 – $4.00 – us Kinesology [612]

An evaluation of the probable impact of selected proposals for imposing mandatory minimum sentences in the federal courts / Eaglin, James B – Washington: FJC, July 1977 – 1mf – 9 – $1.50 – mf#LLMC 95-814 – us LLMC [347]

An evaluation of the progress of rural land reform in south africa / Chawane, Winston Nelson – Pretoria: Vista University 2003 [mf ed 2003] – 3mf – 9 – (incl bibl ref) – mf#mfm15222 – sa Unisa [333]

An evaluation of the relationship between fear of failure, sport confidence, precompetitive affect and batter's run average in baseball / Walker, Brent W – 1997 – 1mf – 9 – $4.00 – mf#PSY 1975 – us Kinesology [150]

Evaluation of the set-point and proportional control models of human thermoregulation during exercise / Ward, Jeffrey J & Quigley, Brian – 1990 – 3mf – 9 – $12.00 – us Kinesology [612]

An evaluation of the square and staggered stance : utilized in amateur wrestling / Goodwin, Ernst C – 2000 – 80p on 1mf – 9 – $5.00 – mf#PE 4159 – us Kinesology [790]

An evaluation of the university of north carolina intramural-recreational sports program / Lands, Craig – 1998 – 1mf – 9 – $4.00 – mf#PE 3955 – us Kinesology [790]

Evaluation of the washington state university intramural sports program / Rinaldi, Nancy E – 1989 – 97p on 1mf – 9 – $4.00 – us Kinesology [378]

An evaluation of water and soil resources for irrigation in the wasbank drainage basin / Gordon, Douglas Hamilton – Uni of South Africa 2001 [mf ed Pretoria: UNISA 2000] – 4mf – 9 – (incl bibl ref) – mf#mfm14713 – sa Unisa [627]

Evaluation practice – Beverly Hills. 1993-1995 (1,5,9) – (cont by: american journal of evaluation) – ISSN: 0886-1633 – mf#17052,01 – us UMI ProQuest [300]

Evaluation review – Beverly Hills. 1983+ (1,5,9) – ISSN: 0193-841X – mf#13550,01 – us UMI ProQuest [300]

Evaluation von assessment-centern (ac) : untersuchung von entscheidungen bei der gestaltung von personalauswahl- und -entwicklungsmassnahmen / Huehnerbein-Sollmann, Christoph Matthias – (mf ed 1998) – 4mf – 9 – €56.00 – 3-8267-2527-1 – mf#DHS 2527 – gw Frankfurter [150]

Evaluation von scoresystemen in der intensivmedizin und deren zusammenhang mit dem langzeitueberleben / Deutschinoff, Gerd – (mf ed 1999) – 2mf – 9 – €40.00 – 3-8267-2622-7 – mf#DHS 2622 – gw Frankfurter [612]

Die evalvation der motivik im musical the phantom of the opera : eine strukturanalytische untersuchung / Petri, Hasso Gottfried – (mf ed 1999) – 2mf – 9 – €40.00 – 3-8267-2664-2 – mf#DHS 2664 – gw Frankfurter [790]

Evangeeliumi kristlaste vabakoguduse ajalooline ulevaade, 1905-1930 = Historical survey of the evangelical christian free church / Laks, Johannes – Tallinn: Evangelical Christian Free Church Publ. House, 1930. Publ. No. 6295 a. One of four items on a reel. 162p – 1 – us Southern Baptist [242]

Evangel – Edinburgh. 1989-1991 (1) 1989-1991 (5) 1989-1991 (9) – mf#16072 – us UMI ProQuest [240]

Evangel – San Francisco. Calif. 1867-Jun 1869 – 1 – us Southern Baptist [242]

The evangel *see* Miscellaneous newspapers of teller county

The evangel of the risen christ : his resurrection triumphs / Varley, Henry – New York: F H Revell, [18–?] – 1mf – 9 – 0-524-00189-8 – mf#1989-2889 – us ATLA [220]

Evangelia apocrypha : adhibitis plurimis codicibus graecis et latinis maximam partem... / ed by Tischendorf, Constantin von – Lipsiae: Avenarius et Mendelssohn, 1853 – 2mf – 9 – 0-8370-9658-8 – mf#1986-3658 – us ATLA [221]

Evangelia apocrypha – Lipsiae: Herm. Mendelssohn, 1876 – 1r – 1 – 0-8370-1111-6 – mf#1984-B527 – us ATLA [240]

Evangelia de communi sanctorum : explicationibus ad mentem sanctorum patrum aliorumque interpretum dilucidata... / Schouppe, Francois Xavier – Bruxellis: H Goemaere, 1869 – 2mf – 9 – 0-8370-7505-X – mf#1986-1505 – us ATLA [240]

Evangelia (siecle 14) – Avila – 1r – 5,6 – sp Cultura [242]

Evangeliar aus weltenburg (cima5) : farbmikrofiche-edition der handschrift wien, oesterreichische nationalbibliothek, cod.1234 – (mf ed 1987) – (= ser Codices illuminati medii aevi (cima) 5) – 24p on 5 color mf – 15 – €370.00 – 3-89219-005-4 – (int by otto mazal) – gw Lengenfelder [090]

Evangeliarium epternacense / evangelistarium (cima9) : farbmikrofiche-edition der handschriften augsburg, universitaetsbibliothek, cod.I.2.4°2. / st peter im schwarzwald, erzbischöfliches priesterseminar, cod.ms.25 – (mf ed 1988) – (= ser Codices illuminati medii aevi (cima) 9) – 45p on 5 color mf – 15 – €335.00 – 3-89219-009-7 – (int & description by daibhi o croinin) – gw Lengenfelder [090]

The evangelic succession : or, the spiritual lineage of the christian church and ministry: being the 29th fernley lecture...london, july 1899 / Lockyer, Thomas Frederick – London: Charles H Kelley 1899 [mf ed 1991] – 1mf – 9 – 0-7905-8832-3 – mf#1989-2057 – us ATLA [240]

Evangelical advocate / Evangelical Church of North America – v9 n1-v10 n12 [1976 jul/aug-1980 jan/mar] – 1r – 1 – (cont: holiness methodist advocate) – mf#668286 – us WHS [242]

Evangelical Alliance *see* Proceedings of the geneva conference..

Evangelical and Reformed Church, Hoisington, KS *see* Church book

Evangelical and sacramental sections / Moule, Handley Carr Glyn – Dorchester? England. 1874? – 1r – us UF Libraries [240]

Evangelical association atlantic – 1876-1907 – 1r – 1 – $35.00 – (in german & english) – mf^{s}-80 – us Commission [242]

Evangelical association black creek circuit, canada 1839-1885 – 1r – 1 – $35.00 – mf^{s}-112 – us Commission [242]

Evangelical association board of publications : minutes, 1860-1876 [german], 1876-1922 [english] – 1r – 1 – $35.00 – mf^{s}-61 – us Commission [242]

Evangelical association board of publications : published reports, 1892-1906 – 1r – 1 – $35.00 – mf^{s}-59 – us Commission [242]

Evangelical association burgh church : trustees minutes, new york conference 1835-1883 – 1r – 1 – $35.00 – mf^{s}-138 – us Commission [242]

Evangelical association / church california – 1r – 1 – $35.00 – (1884-1916 german mss; 1917-21 english mss; 1922-26 printed english) – mf^{s}-125 – us Commission [242]

Evangelical association / church canada – 3r – 1 – $105.00 – (extracts from new york conference concerning canada, 1849-1864; canada minutes, translated typescript, 1865-85; german printed, 1881-87, 1889-92; german & english printed, 1894-1910; printed english, 1911-26) – mf^{s}-114 – us Commission [242]

Evangelical association / church illinois 1885-1892, 1894-1927 – 4r – 1 – $140.00 – mf#um-248 – us Commission [242]

Evangelical association / church kansas 1889-1891, 1893-1946 – 1r – 1 – $35.00 – (in english) – mf^{s}-19 – us Commission [242]

Evangelical association / church south dakota 1921-1922 : as well as south dakota conference deaconess society minutes 1921-40, and historical directory of groton mission – 2r – 1 – $70.00 – mf^{s}-122 – us Commission [242]

Evangelical association / church / united brethren texas 1887-1922, mss – 3r – 1 – $105.00 – (printed, 1923-46, 1946-56) – mf^{s}-69 – us Commission [242]

Evangelical association / church washington 1898-1928 – 1r – 1 – $35.00 – (english mss) – mf^{s}-103 – us Commission [242]

Evangelical association dakota 1884-1920 – 2r – 1 – $70.00 – mf^{s}-122 – us Commission [242]

Evangelical association deaconess society in america : office records, 1904-1920, in german 1904-1909 annual meetings, 1926-1942 executive committee, 1923-1937 – 1r – 1 – $35.00 – mf^{s}-91 – us Commission [242]

Evangelical association des moines 1876-1892, 1894 – 1r – 1 – $35.00 – (no journal publ 1893) – mf^{s}-88 – us Commission [242]

Evangelical association east pennsylvania 1840-1869 – 1r – 1 – $35.00 – (german mss) – mf^{s}-79 – us Commission [242]

Evangelical association eastern 1800-1843 – 1r – 1 – $35.00 – (with ind; in german) – mf^{s}-64 – us Commission [242]

Evangelical association erie conference 1882-1923 – 2r – 1 – $70.00 – (printed german) – mf^{s}-128 – us Commission [242]

Evangelical association evangelical publishing house : board of director's minutes, 1887-1895 – 1r – 1 – $35.00 – mf^{s}-56 – us Commission [242]

Evangelical association general conference minutes 1816-1859 – 1r – 1 – $35.00 – (in german; with ind) – mf^{s}-62 – us Commission [242]

Evangelical association illinois 1845-52 – 2r – 1 – $70.00 – mfs-52 – us Commission [242]

Evangelical association indiana 1852-1892 – 2r – 1 – $70.00 – (in german & english) – mfs-52 – us Commission [242]

Evangelical association iowa 1861-1912 – 1r – 1 – $35.00 – (german mss; also english mss 1869-75) – mfs-99 – us Commission [242]

Evangelical association kansas 1865-1891 – 1r – 1 – $35.00 – (german mss) – mfs-142 – us Commission [242]

Evangelical association minnesota 1868-1887 – 1r – 1 – $35.00 – (trans by h w riegel) – mfs-123 – us Commission [242]

Evangelical association minnesota 1868-1898 – 1r – 1 – $35.00 – (in german) – mfs-71 – us Commission [242]

Evangelical association missionary society : minutes, 1839-1854 [german] – 1r – 1 – $35.00 – mfs-124 – us Commission [242]

Evangelical association nebraska 1879-1915 – 1r – 1 – $35.00 – (german mss) – mfs-102 – us Commission [242]

Evangelical association nebraska 1879-1922 – 1r – 1 – $35.00 – (english mss) – mfs-95 – us Commission [242]

Evangelical association of north america. board of missions. proceedings – Cleveland, 1907-22 [mf ed 2001] – (= ser Christianity's encounter with world religions, 1850-1950) – 1r – 1 – mf#2001-s071 – us ATLA [240]

Evangelical association ohio 1840-1844 – 2r – 1 – $70.00 – mfs-52 – us Commission [242]

Evangelical association oregon 1988-1914 – 1r – 1 – $35.00 – (also abstsracts of pacific 1876-83) – mfs-107 – us Commission [242]

Evangelical association pittsburgh 1852-1882 – 1r – 1 – $35.00 – (english mss) – mfs-17 – us Commission [242]

Evangelical association pittsburgh 1883-1886 – 1r – 1 – $35.00 – (printed english) – mfs-123a – us Commission [242]

Evangelical association pittsburgh 1887-1922 – 3r – 1 – $105.00 – (printed english) – mfs-129 – us Commission [242]

Evangelical association pittsburgh 1892-1922 – 1r – 1 – $35.00 – (printed english) – mf#um-317 – us Commission [242]

Evangelical association platte river 1881-1889 : includes history; 1890-1899 mss – 1r – 1 – $35.00 – (resume of work as related to des moines, 1877-81; printed minutes 1891-1912) – mfs-100 – us Commission [242]

Evangelical association southern indiana 1867-1892 – 2r – 1 – $70.00 – mfs-52 – us Commission [242]

Evangelical association waterloo circuit, canada 1841-1844 – 1r – 1 – $35.00 – mfs-112 – us Commission [242]

Evangelical association west pennsylvania 1840-1860 – 1r – 1 – $35.00 – (german mss) – mfs-55 – us Commission [242]

Evangelical association western 1827-1835 – 1r – 1 – $35.00 – (with ind) – mfs-63 – us Commission [242]

Evangelical beacon / Evangelical Free Church of America – 1981-1983 jun, 1983 jul-1984, 1985-86, 1987-1988 jun – 4r – 1 – (cont: evangelical beacon and evangelist; cont by: beacon [evangelical free church of america]) – mf#1051442 – us WHS [242]

Evangelical beginnings in the arizona territory / Bell, Earl S – 70p – 1 – us Southern Baptist [242]

Evangelical catholic papers. first series : a collection of essays, letters, and tractates from writings of rev william augustus muhlenberg, d.d., during the last forty years – Suffolk County, NY: St Johnland, 1875 – 2mf – 9 – 0-7905-9416-1 – mf#1989-2641 – us ATLA [241]

Evangelical catholic papers. second series : comprising addresses, lectures, and sermons from writings of rev w a muhlenberg, during the last fifty years / Muhlenberg, William Augustus – Suffolk County, NY: St Johnland, 1877 – 2mf – 9 – 0-7905-9417-X – mf#1989-2642 – us ATLA [241]

The evangelical church : or, true grounds for the union of the saints / Ranney, Darwin Harlow – Woodstock, VT: Mercury, 1840 – 1mf – 9 – 0-524-00385-8 – mf#1989-3085 – us ATLA [240]

Evangelical church board of publications : executive committee minutes, 1922-1946 – 2r – 1 – $35.00ea – mfs-58; hs-54 – us Commission [242]

Evangelical church des moines 1901-1926 – 1r – 1 – $35.00 – (in english) – mfs-89 – us Commission [242]

Evangelical Church. Missionary Society *see* Missions of the evangelical church

Evangelical church northwest canada 1928-1946 – 1r – 1 – $35.00 – (in english) – mfs-113 – us Commission [242]

Evangelical church northwestern 1899-1923 – 1r – 1 – $35.00 – (in english) – mfs-73 – us Commission [242]

Evangelical Church of North America *see* Evangelical advocate

Evangelical Free Church of America *see* Evangelical beacon

Evangelical Free Church of Amnerica *see* Chicago-bladet

Evangelical friend – v16-26. 1982-93 – Inquire – 1 – mf#ATLA S0877 – us ATLA [242]

The evangelical guardian – [Rossville OH: J M Christy. v1-4 n7 1843-dec 1846 [mf ed 2004] – 1r – 1 – (lacks: v1 n3-5 p97-240,385-396; cont: christian intelligencer, and evangelical guardian; cont by: united presbyterian and evangelical guardian) – mf#2004-S084 – us ATLA [242]

Evangelical guardian and review – New York. 1817-1819 (1) – mf#3742 – us UMI ProQuest [242]

Evangelical herald *see* The messenger of the evangelical and reformed church

Evangelical inquirer – Virginia. oct 1826-sep 1827 – 1 – us Southern Baptist [242]

Evangelical intelligencer – Philadelphia. 1805-1809 (1) – mf#3573 – us UMI ProQuest [242]

Evangelical Intelligencer, (1805-1809) – Philadelphia, Pa. – 1 reel – 1 – $50.00 – us Presbyterian [242]

The evangelical invasion of brazil : or, a half century of evangelical missions in the land of the southern cross / Gammon, Samuel R – Richmond, VA: Presbyterian Committee of Publ, c1910 [mf ed 1986] – 1mf – 9 – 0-8370-6050-8 – (incl app) – mf#1986-0050 – us ATLA [242]

Evangelical luminary – New York. 1824-1824 (1) – mf#3743 – us UMI ProQuest [190]

Evangelical Lutheran Augustana Synod of North America *see* Luther-baneret

Evangelical Lutheran Augustana Synod of North America. Board of Foreign Missions *see* Board of foreign missions records, 1908-1923

Evangelical Lutheran Augustana Synod of North America. China Mission Society *see* China mission society records, 1901-1919

Evangelical lutheran catechism : or, class-book of religious instruction / Schmucker, Samuel Simon – 10th ed. Baltimore: T Newton Kurtz, 1871 – 1mf – 9 – 0-524-04777-4 – mf#1991-2163 – us ATLA [242]

Evangelical Lutheran Church in America *see* Lutheran

Evangelical lutheran church in america : yearbook – 1988-91 (complete) – 1 – mf#ATLA S0880 – us ATLA [242]

Evangelical lutheran church in america. northeastern pennsylvania synod : minutes – 1988-92 [complete] – 1 – mf#ATLA S0885 – us ATLA [242]

Evangelical lutheran church of finland : news – 1968-91 [complete] – 1 – mf#ATLA S0636 – us ATLA [242]

Evangelical lutheran intelligencer – v1-5. 1826-31 [complete] – (= ser Microfilm corpus of american lutheranism) – 1r – 1 – mf#ATLA S0524A – us ATLA [242]

Evangelical Lutheran Joint Synod of Ohio and Other States *see*
- Reports, 1876-1904
- Reports, 1906-1930

Evangelical Lutheran Joint Synod of Ohio and Other States. Board of Foreign Missions *see*
- Correspondence 1908-1930
- Financial records 1914-1915, 1921-1930
- Minutes 1910-1929

Evangelical Lutheran Joint Synod of Ohio and Other States. Women's Missionary Conference *see*
- Historical records 1927, [1929]
- Minutes and reports 1913-1931

Evangelical Lutheran Joint Synod of Wisconsin and Other States *see*
- Minnesota district bulletin
- Verhandlungen der...versammlung der deutschen evangelischen-lutherischen synode von wisconsin und anderen staaten

Evangelical Lutheran Ministerium of Pennsylvania and Adjacent States. Norristown Conference (Pennsylvania) *see* Jubilee volume, 1517-1917

Evangelical Lutheran Synod *see* Clergy bulletin

Evangelical Lutheran Synod in the Central States *see* Evangelical lutheran synod in the central states

Evangelical lutheran synod in the central states : minutes of the annual convention / Evangelical Lutheran Synod in the Central States – v87-95. 1955-61 [complete] – 1r – 1 – mf#ATLA S0077 – us ATLA [242]

Evangelical lutheran synod of iowa and other states : board of foreign missions. publications 1891-1936 – 2r – 1 – (forms pt of: record group iowa 26 board of foreign missions, new guinea; records consist of pamphlets, bklets, & newsletters relating to new guinea publ prior to the formation of the american lutheran church; mostly in german; with finding aid) – mf#xa0112r – us ATLA [242]

Evangelical Lutheran Synod of Missouri, Ohio and other States *see* Proceedings

Evangelical lutheran synod of missouri, ohio, and other states : proceedings – 1847-1975 [complete] – Inquire – 1 – mf#ATLA S0937 – us ATLA [242]

Evangelical Lutheran Synodical Conference of North America. Missionary Board *see* [Reports of the missionary board]

Evangelical Lutheran Theological Seminary *see* Lutheran church review

Evangelical magazine and gospel advocate – Utica. 1827-1848 (1) – mf#3765 – us UMI ProQuest [242]

Evangelical magazine and missionary chronicle – London. v27. 1819 – 1 – us Southern Baptist [240]

Evangelical meditations = Meditations evangeliques / Vinet, Alexandre Rodolphe – Edinburgh: T & T Clark, 1858 [mf ed 1991] – 1mf – 9 – 0-7905-8617-7 – (english trans by edward masson) – mf#1989-1842 – us ATLA [242]

Evangelical Mennonite Brethren Church et al *see* Gospel tidings

Evangelical mennonite build / Evangelical Mennonite Church – 1977 sum, 1979 spr-1984 sum – 1r – 1 – (cont: evangelical mennonite; cont by: emc today) – mf#863375 – us WHS [243]

Evangelical Mennonite Church *see* Evangelical mennonite build

Evangelical messenger *see* The telescope-messenger

Evangelical Methodist Church *see* Voice of evangelical methodism

Evangelical Mission Covenant Association of California *see*
- California
- Missionstidningen california

Evangelical Mission Covenant Church of America *see*
- Covenant weekly
- Forbundets veckotidning

The evangelical missionaries and the basotho, 1833-1933 / Clark, Sybil G de – 6mf – 9 – (incl bibl ref) – mf#mfm14941 – sa Unisa [240]

Evangelical missions quarterly – South Pasadena. 1985+ (1,5,9) – ISSN: 0014-3359 – mf#15211 – us UMI ProQuest [240]

Evangelical monitor – Woodstock. 1821-1824 (1) – mf#3978 – us UMI ProQuest [242]

Evangelical news and christian leader – [Scotland] Glasgow: Evangelical news and Christian leader 1 nov 1934 (wkly) [mf ed 2003] – 1r – 1 – uk Newsplan [240]

Evangelical nonconformists and higher criticism in the nineteenth century / Glover, W B – London, 1954 – 6mf – 8 – €14.00 – ne Slangenburg [242]

The evangelical pastor / Horn, Edward Traill – Philadelphia: GW Frederick, 1887 – 1mf – 9 – 0-524-03952-6 – (incl bibl ref) – mf#1991-2006 – us ATLA [242]

Evangelical protestantism progressing – London, England. 1850 – 1r – us UF Libraries [242]

Evangelical quarterly – Carlisle. 1929+ (1) 1971+ (5) 1975+ (9) – ISSN: 0014-3367 – mf#5806 – us UMI ProQuest [242]

Evangelical record : and western review – Lexington. 1812-1813 (1) – mf#3779 – us UMI ProQuest [242]

Evangelical recorder – Auburn. 1818-1821 (1) – mf#3744 – us UMI ProQuest [242]

Evangelical Reformirten Schonfeld Gemeinde Kirchenbuch, Barton County, KS *see* Records

Evangelical repertory – Boston. 1823-1824 (1) – mf#3745 – us UMI ProQuest [242]

Evangelical repository – Philadelphia. 1816-1816 (1) – mf#4446 – us UMI ProQuest [920]

Evangelical review of theology – Carlisle. 1980+ (1,5,9) – ISSN: 0144-8153 – mf#12758 – us UMI ProQuest [242]

The evangelical revival in the eighteenth century / Overton, John Henry – New York: ADF Randolph, [1886?] – (= ser Epochs of Church History) – 1mf – 9 – 0-7905-6937-X – mf#1988-2937 – us ATLA [242]

Evangelical sisterhoods : In two letters to a friend / Ayres, Anne; ed by Muhlenberg, William Augustus – New York: T Whittaker, 1867 – 1mf – 9 – 0-524-04668-9 – mf#1990-1295 – us ATLA [242]

Evangelical Theological Society *see*
- Bulletin of the evangelical theological society
- Journal of the evangelical theological society

Evangelical theological society bulletin/journal – v1-11. 1958-81 [complete] – 5r – 1 – ISSN: 0360-8808 – mf#ATLA S0568 – us ATLA [242]

The evangelical type of christianity / Garvie, Alfred Ernest – London: Charles H Kelly, 1916 – (= ser Manuals for Christian Thinkers) – 1mf – 9 – 0-7905-7935-9 – mf#1989-1160 – us ATLA [242]

Evangelical united brethren board of publications : executive committee minutes, 1945-1956 – 1r – 1 – $35.00 – mfs-58 – us Commission [242]

Evangelical united brethren board of publications : minutes, 1946-1966 – 1r – 1 – $35.00 – mfs-53 – us Commission [242]

Evangelical united brethren colorado-new mexico 1947-1951 – 1r – 1 – $35.00 – mf#um-285 – us Commission [242]

Evangelical united brethren missouri 1891-1967 – 3r – 1 – $105.00 – mfs-127 – us Commission [242]

Evangelical united brethren oklahoma 1898-1955 – 2r – 1 – $70.00 – mfs-68 – us Commission [242]

Evangelical united brethren rocky mountain 1951-1967 – 1r – 1 – $35.00 – mf#um-286 – us Commission [242]

Evangelical visitor – v22-65. 1908-52 [gaps] – (= ser Mennonite serials coll) – Inquire – 1 – mf#ATLA 1994-S022 – us ATLA [242]

Evangelical witness – Newburgh. 1822-1826 (1) – mf#4370 – us UMI ProQuest [242]

Evangelical witness and presbyterian review – Dublin, Ireland. jan 1862-oct 1866 – 1r – 1 – uk British Libr Newspaper [072]

The evangelican quarterly – 1(1929)-44(1972) – 264mf – 9 – €503.00 – ne Slangenburg [242]

Evangelie – [M: Anonimnaia tipografiia, 1555] – 15mf – 9 – mf#RHB-19 – ne IDC [460]

Evangelie – [M: Anonimnaia tipografiia, 1560] – 12mf – 9 – mf#RHB-20 – ne IDC [460]

Evangelie – Vil'no: Mamonich Printing House, 1575 – 15mf – 9 – mf#RHB-11 – ne IDC [460]

Evangelie – Vil'no: Mamonich Printing House, 1600 – 14mf – 9 – mf#RHB-14 – ne IDC [460]

Het evangelie in china : drie voorlezingen, gehouden te geneve in de vergadering van het casino = L'evangile et la chine / Watteville, B de – Amsterdam: H Hoeveker, 1844 [mf ed 1995] – (= ser Yale coll) – 127p – 1 – 0-524-10151-5 – (trans fr french into dutch) – mf#1995-1151 – us ATLA [240]

Het evangelie naar johannes / Scholten, Johannes Henricus – Leiden: P Engels, 1864 – 2mf – 9 – 0-8370-1733-5 – mf#1987-6129 – us ATLA [226]

Evangelie uchitel'noe – Vil'no: Mamonich Printing House, 1595 – 15mf – 9 – mf#RHB-34 – ne IDC [460]

Evangelie uchitel'noe – [Vil'no: Vasilii Mikhailovich Garaburda, 1580] – 15mf – 9 – mf#RHB-26 – ne IDC [460]

Evangelie uchitel'noe – Zabludov: Ivan Fedorov and Petr Timofeev Mstislavets, 1569 – 15mf – 9 – mf#RHB-23 – ne IDC [460]

Het evangelie van paulus / Loenen, Jacobus van – Groningen: F Wilkens, 1863 – 1mf – 9 – 0-524-06844-5 – mf#1992-0986 – us ATLA [240]

Die evangelien : nach ihrer entstehung und geschichtlichen bedeutung / Hilgenfeld, Adolf – Leipzig: S Hirzel, 1854 – 1mf – 9 – 0-8370-9547-6 – (incl bibl ref and index) – mf#1986-3547 – us ATLA [225]

Evangelien / ed by Klostermann, Erich – 2. aufl. Bonn: A Marcus & E Weber 1910 [mf ed 1992] – (= ser Kleine texte fuer theologische und philologische vorlesungen und uebungen 8/2) – 1mf – 9 – 0-524-04755-3 – mf#1992-0197 – us ATLA [225]

Die evangelien des markus und lukas : nach der syrischen im sinaikloster gefundenen palimpsesthandschrift / Merx, Adalbert – Berlin: Georg Reimer, 1905 – 2mf – 9 – 0-7905-0104-X – (incl bibl ref) – mf#1987-0104 – us ATLA [225]

Die evangelien des markus und lukas / Weiss, Bernhard & Weiss, Johannes – 8. aufl. Goettingen: Vandenhoeck und Ruprecht, 1892 – (= ser Kritisch Exegetischer Kommentar Ueber Das Neue Testament) – 2mf – 9 – 0-7905-3495-9 – (incl bibl ref) – mf#1987-3495 – us ATLA [225]

Die evangelien des matthaeus und des marcus / Gebhardt, Oscar von – Leipzig, 1883 – (= ser Tugal 1-1/4a) – 3mf – 9 – €7.00 – ne Slangenburg [240]

Die evangelien des matthaeus und des marcus aus dem codex purpureus rossanensis – Leipzig: J C Hinrichs 1883 [mf ed 1989] – (= ser Tugal 1/4) – 1mf – 9 – 0-7905-1701-9 – (in german, greek & latin; incl bibl ref) – mf#1987-1701 – us ATLA [226]

Die evangelien eines alten unzialcodex (b[aleph]-text) : nach einer abschrift des dreizehnten jahrhunderts – Leipzig: J C Hinrichs, 1903 – 1mf – 9 – 0-7905-1847-3 – mf#1987-1847 – us ATLA [221]

Die evangelien und die apostelgeschichte / Schlatter, Adolf von – Calw: Vereinsbuchh, 1908 – 3mf – 9 – 0-7905-0156-2 – mf#1987-0156 – us ATLA [220]

Evangelien von matthaeus und markus *see* Commentary on the gospels of matthew and mark

Die evangeliencitate justin des maertyrers in ihrem wert fuer die evangelienkritik von neuem untersucht / Bousset, Wilhelm – Goettingen:Vandenhoeck und Ruprecht, 1891 – 1mf – 9 – 0-8370-2419-6 – mf#1985-0419 – us ATLA [220]

Die evangelienfrage in ihrem gegenwaertigen stadium / Weisse, Christian Hermann – Leipzig: Breitkopf und Haertel, 1856 – 1mf – 9 – 0-7905-0416-2 – (incl bibl ref) – mf#1987-0416 – us ATLA [220]

Das evangelienfragment von fajjum / Harnack, Adolf von – 1899 – (= ser Tugal 1-5/4b) – 1mf – 9 – €3.00 – ne Slangenburg [240]

Evangelienfragmente : der griechische text des cureton'schen syrers / Baethgen, Friedrich – Leipzig: J C Hinrichs, 1885 – 1mf – 9 – 0-8370-1950-8 – mf#1987-6337 – us ATLA [220]

Evangelienharmonie : die heiligen vier evangelien / Heusser, Theodor – Guetersloh: C Bertelsmann, 1909 – 2mf – 9 – 0-524-04905-X – mf#1992-0248 – us ATLA [225]

Die evangelienzitate des origenes / Hautsch, Ernestus – Leipzig: J C Hinrichs, 1909 – (= ser Tugal) – 1mf – 9 – 0-7905-1716-7 – (incl bibl ref) – mf#1987-1716 – us ATLA [240]

Evangelietroen og den moderne bevidsthed : forelaesninger over jesu liv / Nielsen, Rasmus – Kobenhavn: C A Reitzel, 1849 – 2mf – 9 – 0-524-00454-4 – mf#1989-3154 – us ATLA [220]

Evangelietroen og theologien : tolv forelaesninger / Nielsen, Rasmus – Kobenhavn: CA Reitzel, 1850 – 1mf – 9 – 0-524-00455-2 – mf#1989-3155 – us ATLA [240]

Evangeliets seier : festskrift for hauge synode kinamissions 25 aars jubilaeum 1891-1916 / Oppegaard, A O et al – [S.l.: s.n.], c1916 (Minneapolis, Minn: KC Holter) – 1mf – 9 – 0-524-05252-2 – mf#1991-2244 – us ATLA [240]

Evangelii secundum petrum et petri apocalypseos quae supersunt : ad fidem codicis in aegypto nuper inventi / ed by Lods, Adolphe – Parisiis: E Leroux, 1892 [mf ed 1990] – 1mf – 9 – 0-8370-1881-1 – (in latin & greek) – mf#1987-6268 – us ATLA [226]

Evangelikus elet; orszagos evangelikus hetilap – v32-54. 1967-89 – Inquire – 1 – (lacks v41 1976) – mf#ATLA S0165 – us ATLA [242]

L'evangeline – Moncton, NB. 1887-1910 – 8r – 1 – cn Library Assoc [071]

Evangeline / Longfellow, Henry Wadsworth – Limoges: Eugene Ardant [1894] (mf ed 1985) – 2mf – 9 – mf#SEM105P531 – cn Bibl Nat [810]

Evangeline / Longfellow, Henry Wadsworth – Montreal: editions A Levesque, 1936 [mf ed 1995] – 2mf – 9 – mf#SEM105P2366 – cn Bibl Nat [810]

Evangeline / Longfellow, Henry Wadsworth – [Montreal]: editions Albert Levesque, [1935?] (mf ed 1992) – 2mf – 9 – mf#SEM105P1539 – cn Bibl Nat [810]

"Evangeline" and "the archives of nova scotia" : or, the poetry and prose of history / Anderson, William James – Quebec?: s.n, 1870? – 1mf – 9 – mf#28670 – cn Canadiana [971]

El evangelio comentado conferencias por radio / Bayle, Constantino & Peiro, Francisco – Madrid: Razon y Fe, 1943 – 1 – sp Bibl Santa Ana [240]

Evangelio del amor / Gomez Carrillo, Enrique – Madrid, Spain. 1922 – 1r – us UF Libraries [972]

El evangelio explicado, vol 1 / Goma y Tomas, Isidro – Madrid: Razon y Fe, 1930 – 1 – sp Bibl Santa Ana [240]

Evangelion da-mepharreshe : the curetonian version of the four gospels, with the readings of the sinae palimpsest and the early syriac patristic evidence / ed by Burkitt, F Crawford – Cambridge. v1-2. 1904 – €31.00 – ne Slangenburg [220]

Evangelion da-mepharreshe : the curetonian version of the four gospels, with the readings of the sinai palimpsest and the early syriac patristic evidence / ed by Burkitt, F Crawford – Cambridge [England]: University Press, 1904 – 1r – 1 – 0-8370-0433-0 – mf#1984-B429 – us ATLA [220]

Evangelion da-mepharreshe : the curetonian version of the four gospels, with the readings of the sinai palimpsest and the early syriac patristic evidence / ed by Burkitt, Francis Crawford – Cambridge [England]: University Press, 1904 – 3mf – 9 – 0-8370-1167-1 – mf#1987-6003 – us ATLA [220]

L'evangelisation en pays de langue francaise au 19me siecle : et notamment pendant les 30 dernieres annees / Olivet, Albert – Geneve: P Richter, 1902 [mf ed 1990] – 1mf – 9 – 0-7905-5853-X – (in french) – mf#1988-1853 – us ATLA [240]

The evangelisation of china : addresses delivered at 5 conferences of christian workers, held during august, sep and oct 1896, at chefoo, peking, shanghai, foochow and hankow / ed by Lyon, David Willard – Tientsin: [National Committee of the College of YMCA of China, 1897] – (= ser Yale coll) – vii/141p – 1 – 0-524-09960-X – mf#1995-0960 – us ATLA [240]

Evangelisation of pagan africa : a history of christian missions to the pagan tribes of central africa / Du Plessis, Johannes Christiaan – Cape Town: J C Juta [1930] – 1 – us CRL [240]

The evangelisation of the world : a missionary band: a record of consecration, and an appeal / ed by Broomhall, B – [3rd ed]. London: Morgan & Scott, [1889?] – 1mf – 9 – 0-8370-7129-1 – (incl ind) – mf#1986-1129 – us ATLA [240]

The evangelisation of the world : a missionary band, a record of consecration, and an appeal / Broomhall, Benjamin – 2nd ed. London: Morgan & Scott, [1887] [mf ed 1995] – (= ser Yale coll) – xix, 242p (ill) – 1 – 0-524-10086-1 – (articles by various aut. 1st publ in 1886 under title: a missionary band: a record and an appeal) – mf#1995-1086 – us ATLA [240]

Der evangelisations-bote – v1-29. 1918-39 [gaps] – (= ser Mennonite serials coll) – Inquire – 1 – mf#ATLA 1994-S034 – us ATLA [242]

Evangelisationsbote and gospel tidings – v29-31. 1939-41 [gaps] – (= ser Mennonite serials coll) – Inquire – 1 – mf#ALTA 1994-S035 – us ATLA [242]

Evangelisch commentaar – [Kampen: Kok] v1-9. 1982-91] [semimthly] [mf ed 1987-92] – 9v on 4r – 1 – mf0849 – us ATLA [240]

Das evangelische deutschland : kirchliche rundschau fuer das besamtgebiet der deutschen evangelischen kirche – Berlin, Germany 8 apr 1934, 22 sep 1935-20 aug 1939 – 1 – uk British Libr Newspaper [242]

Der evangelische geistliche : dem nun folgenden geschlechte evangelischer geistlichen dargebracht / Loehe, Wilhelm – Stuttgart: SG Liesching, 1852 – 1mf – 9 – 0-524-07104-7 – mf#1991-2927 – us ATLA [242]

Die evangelische gemeinde in locarno... / Meyer, F – Zuerich, S Haehr, 1836. 2 v – 11mf – 9 – mf#PBU-426 – ne IDC [242]

Die evangelische gemeinde miltenberg und ihr erster prediger : ein zeitbild aus dem 16. jahrhundert / Albrecht, Otto – Halle a S: Verein fuer Reformationsgeschichte, 1896 [mf ed 1992] – (= ser [Schriften fuer das deutsche volk] 28) – 1mf – 9 – 0-524-01941-X – mf#1990-0530 – us ATLA [242]

Die evangelische geschichte und der ursprung des christenthums : auf grund einer kritik der berichte ueber das leiden und die auferstehung jesu / Brandt, Wilhelm – Leipzig: O R Reisland, 1893 – 2mf – 9 – 0-7905-0310-7 – (incl bibl ref) – mf#1987-0310 – us ATLA [220]

Evangelische gesundheitsfuersorge *see* Mitteilungen des deutschen evangelischen krankenhausverbandes (fw3)

Der evangelische glaube nach den hauptschriften der reformatoren... / Wernle, P – Tuebingen, 1919 – 4mf – 9 – mf#ZWI-75 – ne IDC [242]

Der evangelische glaube und die theologie albrecht ritschl's *see* Faith and morals

Der evangelische glaube und die theologie albrecht ritschls : rektoratsrede / Herrmann, Wilhelm – Marburg: NG Elwert, 1890 [mf ed 1991] – 1mf – 9 – 0-7905-9960-0 – mf#1989-1685 – us ATLA [242]

Evangelische glaubenslehre nach schrift und erfahrung / Plitt, Hermann – Gotha: F A Perthes, 1863-1864 – 3mf – 9 – 0-524-08641-9 – mf#1993-2101 – us ATLA [242]

Evangelische jahresbriefe – 6(1937)-17(1953) – 46mf – 9 – €88.00 – ne Slangenburg [242]

Der evangelische kinderfreund – Stuttgart DE, 1895 jan-nov, 1896-1907 – 1 – gw Misc Inst [242]

Die evangelische kirche : ihre organisation und ihre arbeit in der grossstadt / Gruenberg, Paul – Goettingen: Vandenhoeck & Ruprecht, 1910 – (= ser Praktisch-Theologische Handbibliothek) – 1mf – 9 – 0-7905-5891-2 – (incl bibl ref) – mf#1988-1891 – us ATLA [242]

Evangelische kirche der pfalz : amtsblatt – v1. 1921; v7-65. 1927-85 [complete] – 10r – 1 – (title varies) – mf#ATLA S0799 – us ATLA [242]

Evangelische kirche im rheinland : kirchliches amtsblatt – v112-130. 1971-89 – 7r – 1 – (lacking some pp) – mf#atla s0363 – us ATLA [242]

Evangelische kirche in deutschland : amtsblatt – v23-43. 1969-91 [complete] – Inquire – 1 – (Incl ind 1971-80) – ISSN: 0014-343X – mf#ATLA S0371 – us ATLA [242]

Evangelische kirche in deutschland. kirchenkanzlei : berichte – 1948-91 – Inquire – 1 – (lacking: 1953, 1959, 1962, 1964, 1969) – mf#ATLA S0574 – us ATLA [242]

Evangelische kirche in hessen und nassau : amtsblatt – 1968-92 [complete] – Inquire – 1 – mf#ATLA S0777 – us ATLA [242]

Die evangelische kirche in russland / Dalton, Hermann – Leipzig: Duncker & Humblot, 1890 – 1mf – 9 – 0-7905-6103-4 – mf#1988-2103 – us ATLA [242]

Die evangelische kirche und die separatisten und sektierer der gegenwart / Juengst, Johannes – Gotha: FA Perthes, 1881 – 1mf – 9 – 0-524-03286-6 – mf#1990-0897 – us ATLA [242]

Evangelische kirche von westfalen : kirchliches amtsblatt – 1977-89 [complete] – 2r – 1 – mf#ATLA S0801 – us ATLA [242]

Evangelische kirche von westfalen : verhandlungen – 1946-78 – 8r – 1 – (lacks some pp) – mf#atla s0402 – us ATLA [242]

Evangelische kirchengeschichte der elsaessischen territorien bis zur franzoesischen revolution / Adam, J – Strassburg, 1928 – 7mf – 9 – mf#PPE-129 – ne IDC [242]

Evangelische kirchengeschichte der stadt strassburg bis zur franzoesischen revolution / Adam, J – Strassburg, 1922 – 6mf – 9 – mf#PPE-128 – ne IDC [242]

Evangelische kirchenzeitung – 1(1827)-63(1858) – 381mf – 9 – €728.00 – ne Slangenburg [242]

Evangelische landeskirche in greifswald : amtsblatt – 1970-89 [complete] – 3r – 1 – mf#ATLA S0680 – us ATLA [242]

Evangelische landeskirche von wurtemburg : amtsblatt – 1968-91 [complete] – Inquire – 1 – mf#ATLA S0795 – us ATLA [242]

Evangelische landeskirche von wurtemburg : beiblatt zum amtsblatt – n46-50 – 1r – 1 – (lacking: n48) – mf#ATLA S0796 – us ATLA [242]

Die evangelische lehre : auf dem grunde der heiligen schrift und nach ihrem innern zusammenhange fuer freunde des goettlichen wortes / Kritz, Wilhelm – 2. Aufl. Leipzig: J C Hinrichs, 1858 – 1mf – 9 – 0-8370-4418-9 – mf#1985-2418 – us ATLA [242]

Evangelische lutheranische kirk *see* Lutheraneren

Evangelische maenner-choere : fuer gottesdienstliche zwecke – Cleveland, O[hio]: J H Lamb, 1909, c1896 – 2mf – 9 – 0-524-08754-7 – mf#1993-3259 – us ATLA [780]

Das evangelische magazin – 1811-17 [complete] – (= ser Microfilm corpus of american lutheranism 28/1) – 1r – 1 – mf#ATLA B0524F – us ATLA [242]

Das evangelische magazin – 1829-33 [complete] – (= ser Microfilm corpus of american lutheranism 28/2) – 1r – 1 – mf#ATLA S0524G – us ATLA [242]

Die evangelische messe : bis zu luthers deutscher messe / Smend, J – Goettingen, 1896 – €12.00 – ne Slangenburg [242]

Die evangelische mission : geschichte, arbeitsweise, heutiger stand / Baudert, Samuel – Leipzig: BG Teubner, 1913 – (= ser Aus natur und geisteswelt) – 1mf – 9 – 0-524-02970-9 – mf#1990-0757 – us ATLA [242]

Die evangelische mission : ihre laender, voelker und arbeiten / Gundert, Hermann – 4. durchaus verm. Aufl. Calw: Verlag der Vereinsbuchh, 1903 – 2mf – 9 – 0-8370-6259-4 – (incl bibl ref and index) – mf#1986-0259 – us ATLA [242]

Evangelische missions-zeitschrift – Stuttgart: Deutsche Gesellschaft fuer Missionswissenschaft, 1940-74 [mf ed 2001] – (= ser Christianity's encounter with world religions, 1850-1950) – 4r – 1 – (in german) – mf#2001-s161 – us ATLA [242]

Der evangelische ober-kirchenrath in berlin und das concil – Freiburg i.B.: Herder, 1869 – 1mf – 9 – 0-8370-8890-9 – (incl bibl ref) – mf#1986-2890 – us ATLA [242]

Evangelische polemik gegen die roemische kirche / Tschackert, Paul – 2. verb aufl. Gotha: Friedrich Andreas Perthes, 1888 – 2mf – 9 – 0-8370-8625-6 – (incl ind) – mf#1986-2625 – us ATLA [242]

Der evangelische religionsunterricht im zeitalter der reformation / Neumann, Robert – Berlin: R Gaertner, 1899 – 1mf – 9 – 0-8370-7653-6 – (incl bibl ref) – mf#1986-1653 – us ATLA [240]

Evangelische schulordnungen / ed by Vormbaum, R – (bd1: die ev schulordnungen des 16en jhts, guetersloh 1860 €25. bd2: die ev schulordnungen des 17en jhts, guetersloh 1863 €27. bd3: die ev schulordnungen des 18en jhts, guetersloh 1864 €23) – ne Slangenburg [242]

Der evangelische sinn unserer kirchenverfassung / Foerster, Erich – Tuebingen: JCB Mohr, 1904 – (= ser Hefte zur "christlichen welt") – 1mf – 9 – 0-524-03092-8 – mf#1990-0817 – us ATLA [242]

Evangelische stroemungen in der russischen kirche der gegenwart / Dalton, Hermann – Heilbronn: Henninger, 1881 – (= ser Zeitfragen des christlichen volkslebens) – 1mf – 9 – 0-7905-6161-1 – mf#1988-2161 – us ATLA [242]

Evangelische theologie – 25(1965)-34(1974) – 118mf – 9 – €225.00 – ne Slangenburg [242]

Evangelische theologie – Munich. 1973+ (1) 1975+ (5) 1976+ (9) – ISSN: 0014-3502 – mf#8896 – us UMI ProQuest [242]

Die evangelischen erzaehlungen von der geburt und kindheit jesu / Voelter, Daniel – Strassburg: J H Ed Heitz, 1911 – 1mf – 9 – 0-7905-2200-4 – (incl bibl ref) – mf#1987-2200 – us ATLA [220]

Die evangelischen kirchenordnungen des 16. jahrhunderts / ed by Richter, A L – Weimar. bd1-2. 1846 – v1 22mf; v2 30mf – 8 – €99.00 – ne Slangenburg [242]

Die evangelischen kirchenordnungen des 16. jahrhunderts / ed by Sehling, Emil – Thuebingen, 1902-13 – 8 – (v1: sachsen und thueringen 1, 1902 25mf€48. v2: sachsen und thueringen 2, 1904 20mf €38. v3: die mark brandenburg, 1909 17mf €32. v4: herzogthum preussen, 1911 19mf €37. v5: livland-estland, 1913 19mf €37) – ne Slangenburg [242]

Die evangelischen kirchenordnungen des 16. jahrhunderts. erste abtheilung-fuenfter band / ed by Sehling, Emil – Leipzig: OR Reisland, 1902-1913 – 1mf – 9 – 0-524-00602-4 – mf#1990-0102 – us ATLA [242]

Die evangelischen kirchenordnungen des sechszehnten jahrhunderts : urkunden und regesten zur geschichte des rechts und der verfassung der evangelischen kirche in deutschland / ed by Richter, Aemilius Ludwig – Weimar: Landes-Industriecomptoirs, 1846 – 3mf – 9 – 0-524-04192-X – (incl bibl ref) – mf#1990-1231 – us ATLA [242]

Die evangelischen landeskirchen deutschlands im neunzehnten jahrhundert : blicke in ihr inneres leben / Ecke, Gustav – Berlin: Reuther & Reichard, 1904 [mf ed 1989] – (= ser Die theologische schule albrecht ritschls und die evangelische kirche der gegenwart 2) – 2mf – 9 – 0-7905-4355-9 – (incl bibl ref & ind) – mf#1988-0355 – us ATLA [242]

Evangelischer arbeiterbote – Hattingen DE, 1896 1 feb, 1897-1899 1 jan [gaps], 1906 4 jan-1922 21 dec – 4r – 1 – (title varies: 2 apr-6 aug 1908: rheinisch-westfaelischer arbeiterbote [publ in duisburg]; 4 apr-29 dec 1918: rheinisch-westfaelischer volksbote. filmed by misc inst: 1917-20 [1r]; 1887-90 [4r]) – mf#2147 – gw Mikropress; gw Misc Inst [331]

Evangelischer pressedienst fuer oesterreich : epd – 10 oct 1958-1991 – 23r – 1 – (lacks some pp) – ISSN: 0036-6943 – mf#atla s0398 – us ATLA [242]

Evangelische-reformierte kirche in nordwest deutschland : gesetz- und verordnungsblatt – v13-16. 1971-92 [complete] – Inquire – 1 – (Incl ind 1961-71) – mf#ATLA S0836 – us ATLA [242]

Evangelische-reformierte landeskirche des kantons zuerich : jahresbericht – 1974-91 [complete] – Inquire – 1 – (earlier iss filmed as s0439) – mf#ATLA S0502 – us ATLA [242]

Evangelisches christentum in der gegenwart : drei vortraege / Wernle, Paul – Tuebingen: J C B Mohr, 1914 – 1mf – 9 – 0-7905-6399-1 – mf#1988-2399 – us ATLA [242]

Evangelisches concordienbuch : oder, saemmtliche in dem concordienbuche enthaltene symbolische glaubensschriften der evangelisch-lutherischen kirche / ed by Detzer, Johann Andreas – 4. ausg. Nuernberg: Joh Phil Raw 1868 [mf ed 1993] – 2mf – 9 – 0-524-07693-6 – mf#1991-3278 – us ATLA [242]

Evangelisches gemeindeblatt aus und fuer rheinland und westphalen – [Wuppertal-] Elberfeld DE, 1865-67 [gaps] – 1r – 1 – gw Misc Inst [242]

Evangelisches magazin : unter der aufsicht der deutschen evangelisch lutherischen synode – Philadelphia. 1811-1817 (1) – mf#3801 – us UMI ProQuest [242]

Evangelisches monatsblatt – v30-36. 1977-86 [complete] – 2r – 1 – (cont: kirche und mann) – mf#ATLA S0202A – us ATLA [242]

Evangelisches pfarrerblatt / ed by Bund Evangelischer Pfarrer in der Deutschen Demokratischen Republik – Leipzig: Der Bund -[1972] (1960-72) [mf ed 1987] – 1 – (some iss accompanied by suppls; with ind [1959-65] 1v; merged with: glaube + gewissen to form: standpunkt) – ISSN: 0423-8494 – mf0640a – us ATLA [242]

Evangelisches und katholisches schriftprinzip / Kunze, Johannes – Leipzig: Doerffling & Franke, 1899 – 1mf – 9 – 0-524-06213-7 – mf#1992-0851 – us ATLA [220]

Evangelisch-luterisches volksblatt – Dresden DE, 1920-22 – 1r – 1 – gw Misc Inst [242]

Evangelisch-lutherische homiletik : nach der erlaeuterung ueber die praecepta homiletica von dr j j rambach / Pieper, Reinhold – St Louis: Concordia, 1901 – 2mf – 9 – 0-524-05264-6 – (incl ind) – mf#1991-2256 – us ATLA [242]

Die evangelisch-lutherische kirche : die wahre sichtbare kirche gottes auf erden. ein referat fuer die verhandlungen der allgemeinen evangelisch-lutherischen synode von missouri, ohio u. a. staaten bei gelegenheit der sitzungen derselben zu st. louis, mo... / Walther, Carl Ferdinand Wilhelm – St Louis, MO: Lutherischer Concordia-Verlag, 1891 – 1mf – 9 – 0-524-04561-5 – mf#1991-2125 – us ATLA [242]

Evangelisch-lutherische kirche in bayern : amtsblatt – 1969-91 [complete] – Inquire – 1 – ISSN: 0014-3391 – mf#ATLA S0414 – us ATLA [242]

Evangelisch-lutherische kirche in bayern : nachrichten – v16-48. 1961-93 [complete] – Inquire – 1 – mf#ATLA S0338 – us ATLA [242]

Evangelisch-lutherische kirche in bayern. landessynode : verhandlungen der landessynode – 1924-93 [complete] – Inquire – 1 – mf#ATLA S0919 – us ATLA [242]

Evangelisch-lutherische kirche in bayern. landessynode : verhandlungen der ordent... – 1922 [complete] – Inquire – 1 – mf#ATLA S0918 – us ATLA [242]

Evangelisch-lutherische kirche in thueringen : amtsblatt – 1963-88 [complete] – 6r – 1 – ISSN: 0014-326X – mf#ATLA S0733 – us ATLA [242]

Die evangelisch-lutherische kirche ungarns in ihrer geschichtlichen entwicklung : nebst einem anhange ueber die geschichte der protestant. kirchen in den deutsch-slavischen laendern und in siebenbuergen / Borbis, Johannes R – Noerdlingen: CH Beck, 1861 – 2mf – 9 – 0-524-00742-X – (incl bibl ref) – mf#1990-0174 – us ATLA [242]

Evangelisch-lutherische landeskirche mecklenburgs : kirchliches amtsblatt – 1969-77 [complete] – Inquire – 1 – mf#ATLA S0881 – us ATLA [242]

Die evangelisch-lutherische landeskirche schaumburg-lippe : kirchliche amtsblatt – 1970-85 (complete) – 1r – 1 – mf#ATLA S0838 – us ATLA [242]

Evangelisch-lutherische landeskirche schleswig-holsteins : landessynode berichte – 1947-76 [complete] – 4r – 1 – mf#ATLA S0512 – us ATLA [242]

Die evangelisch-lutherische tamulen-mission in der zeit ihrer neubegruendung : ein beitrag zur geschichte der evangelischen mission im 19. jahrhundert / Handmann, Richard – Leipzig: J C Hinrichs'sche Buchhandlung, 1903 [mf ed 1995] – (= ser Yale coll) – x/477p (ill) – 1 – 0-524-09394-6 – (in german) – mf#1995-0394 – us ATLA [242]

Evangelisch-reformierte landeskirche beider appenzell : amtsbericht – v102-117. 1979-92 [complete] – Inquire – 1 – mf#ATLA S0907 – us ATLA [242]

Evangelisch-reformierte landeskirche des kantons argau : jahresbericht – 1958-64 [complete] – 2r – 1 – mf#ATLA S0683 – us ATLA [242]

Evangelisch-reformierte landeskirche des kantons zuerich : jahresbericht – 1929-73 – 3r – 1 – (lacks some pp) – mf#atla s0439 – us ATLA [242]

Evangelisch-reformierte landeskirche des kantons zuerich : kirchliches amtsblatt – v1-27. 1956-85 – 1r – 1 – mf#ATLA S0804 – us ATLA [242]

Evangelisch-reformierte landeskirche von appenzell am rhein : amtsbericht – 1968-79 [complete] – 1r – 1 – mf#ATLA S0401 – us ATLA [242]

Evangelisk kyrklighet / Aulen, Gustaf – Uppsala: Sveriges Kristliga Studentroerelses Foerlag [1916] [mf ed 1991] – (= ser Sveriges kristliga studentroerelses skriftserie 50) – 1mf – 9 – 0-7905-7680-5 – mf#1989-0905 – us ATLA [242]

Evangelisk luthersk dogmatik / Scharling, Carl Henrik – Kobenhavn: GEC Gad, 1913 – 2mf – 9 – 0-7905-8727-0 – (incl bibl ref) – mf#1989-1952 – us ATLA [242]

Evangelisk lutherska augustana-synoden i nord-amerika och dess mission / Norelius, Eric – Lund: Berling 1870 [mf ed 1991] – 1mf – 9 – 0-524-01333-0 – (in swedish) – mf#1990-4082 – us ATLA [242]

Evangeliska luterska kirkjufelag islendinga i vesturheimi *see*
– Arsfundr hins...
– Arsping hins...
– Gjorabok...arsping hins...

Den evangelisk-lutherske kirkes historie i amerika : fra dens begyndelse til nutiden / Andersen, Rasmus – Brooklyn, New York: Fortfatterens Forlag 1889 [mf ed 1992] – 2mf – 9 – 0-524-02441-3 – (incl bibl ref) – mf#1990-4300 – us ATLA [242]

Evangelism old and new : god's search for man in all ages / Dixon, Amzi Clarence – New York: American Tract Society [c1905] [mf ed 1984] – (= ser Revivalism and revival preachers in america 7) – 3mf – 9 – 0-8370-1236-8 – mf#1984-3007 – us ATLA [242]

Evangelist – Hartford. 1824-1825 (1) – mf#4808 – us UMI ProQuest [242]

Evangelist / Jimmy Swaggart Ministries – v16 n1-v17 n9 [1984 feb-1985 sep], v17 n10-v18 n12 [1985 oct-1986 dec] – 2r – 1 – mf#1111444 – us WHS [243]

The evangelist : or, life and labors of rev. jabez s. swan / Swan, Jabez Smith; ed by Denison, Frederic – 2nd ed. Waterford, Conn: WL Peckham, c1873 – 2mf – 9 – 0-524-01671-2 – mf#1990-0492 – us ATLA [240]

Evangelist and religious review – New York. 1830-1902 (1) – mf#5320 – us UMI ProQuest [240]

Evangelist speaks / Church of God in Christ – 1965 mar, 1969 feb,jun, 1970 aug – 1r – 1 – mf#4114503 – us WHS [243]

El evangelista cubano : vinculo de union entre cristianos evangelicos cubanos dispersos por el mundo – Buenos Aires, Argentina: [s.n.] 1971-77 [mf ed 2007] – (= ser Religious periodical literature of the hispanic and indigenous peoples of the americas, 1850-1985) – 7v on 1r – 1 – mf#2007h-s003 – us ATLA [240]

Evangelista, Julio *see* Portugal vis-a-vis the united nations

El evangelista mexicano ilustrado – San Luis Potosi, Mexico: Iglesia Metodista Episcopal del Sur en Mexico -1896 (18893-96) [mf ed 2007] – (= ser Religious periodical literature of the hispanic and indigenous peoples of the americas, 1850-1985) – 1r – 1 – (publ in: san luis potosi, mexico [jan 1 1893]; nashville, tn [jan 15 1893-96]) – mf#2007h-s005 – us ATLA [242]

Evangelistarium *see* Evangeliarium epternacense / evangelistarium (cima9)

Evangelisten – v2-67. 1891-1955 – 18r – 1 – (lacks some pp & iss) – mf#atla s0879 – us ATLA [240]

Evangelistic sermons : With an Essay on the Scriptural and Catholic Creed of Baptism / Mathews, Robert Trott – Cincinnati: Standard Pub Co, 1891 – 1mf – 9 – 0-524-07575-1 – mf#1991-3195 – us ATLA [242]

The evangelists : papers on the four gospels / Bellett, John Gifford – New York: Loizeaux Bros, [18–] – 2mf – 9 – 0-8370-6244-6 – mf#1986-0244 – us ATLA [225]

Evangelists and lay-exhorters / Otts, John Martin Philip – [New York: J M Sherwood, 1877] [mf ed 1984] – (= ser Revivalism and revival preachers in america 15) – 1mf – 9 – 0-8370-1590-1 – (incl bibl ref) – mf#1984-3015 – us ATLA [242]

The evangelists and the mishna : or, illustrations of the four gospels, drawn from jewish traditions / Robinson, Thomas – London: James Nisbet, 1859 – 1mf – 9 – 0-7905-0265-8 – (in english and hebrew) – mf#1987-0265 – us ATLA [225]

Evangelists in the church : From Philip, A.D. 35, To Moody And Sankey, A.D. 1875 / Headley, Phineas Camp – Boston: H. Hoyt, 1875 – 2mf – 9 – 0-7905-6107-7 – mf#1988-2107 – us ATLA [242]

An evangelist's tour round india / Tinling, James Forbes Bisset – London: William Macintosh, 1868 [mf ed 1995] – (= ser Yale coll) – 40p/48p/46p/44p – 1 – 0-524-10125-6 – mf#1995-1125 – us ATLA [240]

Evangelium, briefe und offenbarung des johannes : nach ihrer entstehung und bedeutung / Schmiedel, Paul Wilhelm – Halle (Saale): Gebauer-Schwetschke 1906 [mf ed 1986] – (= ser Religionsgeschichtliche volksbuecher fuer die deutsche christliche gegenwart 1/12) – 1mf – 9 – 0-8370-9901-3 – (incl ind) – mf#1986-3901 – us ATLA [225]

Das evangelium des heiligen johannes *see* Commentary on the gospel of st john

Das evangelium des johannes : nach der syrischen im sinaikloster gefundenen palimpsesthandschrift / Merx, Adalbert; ed by Ruska, Julius – Berlin: Georg Reimer, 1911 – 2mf – 9 – 0-7905-0105-8 – (incl bibl ref and ind) – mf#1987-0105 – us ATLA [226]

Das evangelium des markus : der hausgemeinde ausgelegt / Wenger, Rudolf – 3. aufl. Calw: Verlag der Vereinsbuchh, 1886 – 1mf – 9 – 0-524-05241-7 – mf#1992-0374 – us ATLA [226]

Das evangelium des markus / Wohlenberg, Gustav – 1. und 2. aufl. Leipzig: A Deichert, 1910 – (= ser Kommentar zum Neuen Testament) – 4mf – 9 – 0-7905-2577-1 – mf#1987-2577 – us ATLA [226]

Das evangelium des paulus / Holsten, Carl – Berlin: G Reimer, 1880-1898 – 2mf – 9 – 0-7905-1999-2 – mf#1987-1999 – us ATLA [226]

Das evangelium des petrus : das kuerzlich gefundene fragment seines textes / ed by Zahn, Theodor – Erlangen: A Deichert, 1893 – 1mf – 9 – 0-8370-9598-0 – (incl bibl ref) – mf#1986-3598 – us ATLA [226]

Evangelium eines armen seunders / Weitling, Wilhelm Christian – Muenchen, Germany. 1897 – 1r – us UF Libraries [025]

Das evangelium in den roemischen landen / Fliedner, Fritz – St Louis, MO: A Wiebusch, 1893 – 1mf – 9 – 0-524-02978-4 – mf#1990-0765 – us ATLA [242]

Das evangelium in der apostelgeschichte / Hadorn, Wilhelm – Berlin: Edwin Runge, 1907 – 1mf – 9 – 0-8370-9544-1 – (incl bibl ref) – mf#1986-3544 – us ATLA [225]

Das evangelium in santalistan – Basel: Verlag der Missionsbuchhandlung, 1878 [mf ed 1995] – (= ser Yale coll) – 46p – 1 – 0-524-10012-8 – (in german) – mf#1995-1012 – us ATLA [240]

Das evangelium jesu und das evangelium von jesus (nach den synoptikern) : ein beitrag zur loesung der frage in drei vorlesungen / Schaeder, Erich – Guetersloh: C Bertelsmann, 1906 [mf ed 1992] – (= ser Beitraege zur foerderung christlicher theologie 10/6) – 1mf – 9 – 0-524-05418-5 – mf#1992-0428 – us ATLA [225]

Das evangelium johannis / Wellhausen, Julius – Berlin: Georg Reimer, 1908 – 1mf – 9 – 0-8370-5771-X – (incl bibl ref) – mf#1985-3771 – us ATLA [226]

Das evangelium marci / Wellhausen, Julius – 2. ausg. Berlin: Georg Reimer, 1909 – 1mf – 9 – 0-8370-5773-6 – mf#1985-3773 – us ATLA [226]

Das evangelium marcions in seiner urspruenglichen gestalt : nebst dem vollstaendigsten beweise dargestellt... / Hahn, August – Koenigsberg: Universitaets-Buchh, 1823 – 1mf – 9 – 0-7905-7234-6 – (incl bibl ref) – mf#1988-3234 – us ATLA [226]

Das evangelium matthaei / Wellhausen, Julius – Berlin: Georg Reimer, 1904 – 1mf – 9 – 0-8370-5774-4 – mf#1985-3774 – us ATLA [226]

Das evangelium matthaei vor dem forum der bibel und des talmud / Lippe, Karpel – Jassy: Jsidor Schorr 1889 [mf ed 1985] – 1mf – 9 – 0-8370-4139-2 – mf#1985-2139 – us ATLA [225]

Das evangelium matthaeus : nach der syrischen im sinaikloster gefundenen palimpsesthandschrift / Merx, Adalbert – Berlin: Georg Reimer, 1902 – 2mf – 9 – 0-7905-0106-6 – (incl bibl ref) – mf#1987-0106 – us ATLA [226]

Das evangelium nach johannes *see* The gospel according to john

Das evangelium nach lukas : nebst einleitenden bemerkungen zur evangelischen geschichte / Nast, Wilhelm – Cincinnati: Cranston & Stowe; Bremen: Traktat-Hauses, c1871 – 3mf – 9 – 0-7905-2299-3 – mf#1987-2299 – us ATLA [226]

Das evangelium nach lukas *see* The gospel according to luke

Das evangelium nach matthaeus *see* The gospel according to matthew

Das evangelium nach thomas / Leipoldt, Johannes – Berlin, 1967 – (= ser Tugal 5-101) – 2mf – 9 – €5.00 – ne Slangenburg [240]

Das evangelium nicodemi / Hesler, Heinrich von; ed by Helm, Karl – Stuttgart: Litterarischer Verein, 1902 (Tuebingen: H Laupp, Jr) [mf ed 1993] – (= ser Blvs 224) – c/284p – 1 – mf#8470 reel 46 – us Wisconsin U Libr [221]

Das evangelium nicodemi / Hesler, Heinrich von; ed by Helm, Karl – Stuttgart: Litterarischer Verein 1902 (Tuebingen: H Laupp, Jr) [mf ed 1993] – (= ser Blvs 224) – 58r – 1 – mf#3420p – us Wisconsin U Libr [221]

Evangelium palatinum ineditum : sive, reliquiae textus evangeliorum latini ante hieronymum versi / ed by Tischendorf, Constantin von – Lipsiae: F A Brockhaus, 1847 [mf ed 1990] – 2mf – 9 – 0-8370-1754-8 – mf#1987-6150 – us ATLA [226]

Evangelium secundum matthaeum : cum variae lectionis delectu / ed by Blass, Friedrich – Lipsiae [Leipzig]: B G Teubneri, 1901 [mf ed 1989] – 1mf – 9 – 0-7905-0971-7 – (text in greek; pref & notes in latin) – mf#1987-0971 – us ATLA [226]

Das evangelium und die apokalypse des petrus : die neuentdeckten bruchstuecke / ed by Gebhardt, Oscar von – Leipzig: JC Hinrichs, 1893 [mf ed 1991] – 1mf – 9 – 0-7905-8304-6 – mf#1987-6409 – us ATLA [226]

Das evangelium und die primitiven rassen / Meinhof, Carl – Berlin-Lichterfelde: Edwin Runge 1913 [mf ed 1993] – (= ser Biblische zeit- und streitfragen 7/9) – 1mf – 9 – 0-524-06147-5 – mf#1992-0814 – us ATLA [240]

Das evangelium von jesu in seinen verhaeltnissen zu buddha-sage und buddha-lehre / Seydel, Rudolf – Leipzig: Breitkopf und Haertel, 1882 – 1mf – 9 – 0-7905-2071-0 – (incl bibl ref and indexes) – mf#1987-2071 – us ATLA [240]

Das evangelium von jesus christus / Ihmels, Ludwig – Berlin: Edwin Runge, 1911 – 1mf – 9 – 0-7905-0500-2 – mf#1987-0500 – us ATLA [240]

Evangeliums panier – 1879 [complete] – (= ser Mennonite serials coll) – 1r – 1 – mf#ATLA 1994-S005 – us ATLA [242]

Der evangeliums-bote – Berlin [Kitchener]; Ont: [s.n., 1888-1917?] – 9 – (cont by: the canadian evangel) – mf#P06067 – cn Canadiana [242]

The evangelization of a great city : or, the churches' answer to the bitter cry of outcast london / Smiley, Francis Edward – Philadelphia: Sunshine Pub, 1890 [mf ed 1986] – 1mf – 9 – 0-8370-6380-9 – mf#1986-0380 – us ATLA [362]

The evangelization of the world in this generation / Mott, John Raleigh – New York: Student Volunteer Movement for Foreign Missions, 1905, c1900 – 1mf – 9 – 0-8370-6588-7 – (incl ind) – mf#1986-0588 – us ATLA [242]

Evangel'skii Vestnik Vols 1-199 (Incomplete) = Gospel messenger – Chicago, Wheaton, Illinois: Slavic Gospel Association (Formerly: Russian Gospel Association), 1936-1984 – 2r. 4,452p – 1 – $178.08 – (Lacking vols 2,3,6,52,60,71,167) – us Southern Baptist [242]

Evangelyo – Chikuni, Zambia. 1931 – 1r – us UF Libraries [960]

Evangeri yakanyorwa namateo – London, England. 1959 – 1r – us UF Libraries [960]

Evangerio sante ya yesu kriste yakanyorgwa na luka musante – Salisbury, Zimbabwe. 1937 – 1r – us UF Libraries [960]

Evangerio sante ya yesu kriste yakanyorwa na marko musante – Chishawasha, Zimbabwe. 1947 – 1r – us UF Libraries [960]

L'evangile armenien / Macler, F – Paris, 1920 – 11mf – 9 – mf#AR-419 – ne IDC [243]

L'evangile de jesus-christ (etb) / Lagrange, Marie Joseph – Paris, 1932 – 12mf – 8 – €23.00 – ne Slangenburg [240]

L'evangile de marc et ses rapports avec ceux de mathieu et de luc : essai d'une, introduction critique a l'etude du second evangile / Goguel, Maurice – Paris: Ernest Leroux, 1909 [mf ed 1985] – 1mf – 9 – 0-8370-3328-4 – (in french. incl bibl ref) – mf#1985-1328 – us ATLA [225]

L'evangile de paris – Paris: [s.n.], dec 1848 – us CRL [074]

L'evangile de pierre (etb) / Vaganay, L – Paris, 1900 – 7mf – 8 – €15.00 – ne Slangenburg [226]

Evangile de saint jean – Paris: P Geuthner, 1908 – (= ser Fragments Sahidiques du Nouveau Testament) – 1mf – 9 – 0-8370-1798-X – mf#1987-6186 – us ATLA [220]

L'evangile du jour – Londres: [s.n.] [mf ed 1986-] – 1 – mf#1773 – us Wisconsin U Libr [944]

L'evangile et la chine *see* Het evangelie in china

L'evangile et l'eglise / Loisy, Alfred Firmin – 3e ed. Bellevue: Chez l'auteur, 1904 [mf ed 1990] – 1mf – 9 – 0-7905-3456-8 – (in french) – mf#1987-3456 – us ATLA [241]

Evangile et l'eglise *see* The gospel and the church

L'evangile selon marc / Loisy, Alfred Firmin – Paris: Emile Nourry, 1912 [mf ed 1986] – 2mf – 9 – 0-8370-9558-1 – (in french) – mf#1986-3558 – us ATLA [226]

Evangile selon s luc : traduction et commentaire / Rose, Vincent – Paris: Librairie Bloud, 1904 – (= ser La pensee chretienne) – 1mf – 9 – 0-524-05658-7 – mf#1992-0508 – us ATLA [226]

Evangile selon s marc (etb) / Lagrange, Marie Joseph – Paris, 1942 – 12mf – 8 – €23.00 – ne Slangenburg [226]

Evangile selon s matthieu : traduction et commentaire / Rose, Vincent – Paris: Librairie Bloud, 1905 – (= ser La pensee chretienne) – 1mf – 9 – 0-524-05902-0 – mf#1992-0659 – us ATLA [226]

Evangile selon s matthieu (etb) / Lagrange, Marie Joseph – Paris, 1941 – 13mf – 8 – €25.00 – ne Slangenburg [226]

L'evangile selon saint jean : traduction critique, introduction et commentaire / Calmes, Th – Paris: Victor Lecoffre; Rome: Typographie Polyglotte...1904 [mf ed 1989] – (= ser Etudes bibliques) – 2mf – 9 – 0-7905-0870-2 – (incl ind) – mf#1987-0870 – us ATLA [226]

Evangile selon saint jean / Lavigne, A – Paris: Henri Plon, 1867 – 2mf – 9 – 0-8370-6994-7 – mf#1986-0994 – us ATLA [226]

Evangile selon saint marc / Lagrange, Marie-Joseph – Paris: Victor Lecoffre 1911 [mf ed 1989] – (= ser Etudes bibliques) – 2mf – 9 – 0-7905-1963-1 – (in french & greek; incl ind) – mf#1987-1963 – us ATLA [225]

Les evangiles et la critique au 19e siecle / Meignan, Guillaume Rene – nouv corr augm ed. Paris: Librairie de Victor Palme, 1870 – 2mf – 9 – 0-7905-1437-0 – (incl bibl ref) – mf#1987-1437 – us ATLA [220]

Les evangiles synoptiques : conferences apologetiques. faites a l'institut catholique de paris / Mangenot, Eugene – Paris: Letouzey et Ane, 1911 – 2mf – 9 – 0-524-05925-X – (incl bibl ref) – mf#1992-0682 – us ATLA [220]

Les evangiles synoptiques / Loisy, Alfred Firmin – Ceffonds: A Loisy, 1907-1908 – 17mf – 9 – 0-8370-1907-9 – (incl bibl ref and ind) – mf#1987-6294 – us ATLA [220]
Die evanglienzitate des origenes / Hautsch, E – Leipzig, 1909 – (= ser Tugal 3-34/2a) – 3mf – 9 – €7.00 – ne Slangenburg [240]
Evanglische landeskirche in baden : gesetzes- und verordnungsblatt der evangelischen landeskirche in baden – 1969-91 (incomplete) – 1 – mf#ATLA S0389A – us ATLA [242]
Evans, A Kelly *see* Address given by a kelly evans
Evans and ruffy's farmer's journal – London, dec 1809-jul 1832 [wkly] – 11r – 1 – uk British Libr Newspaper [630]
Evans, Arthur *see*
– Anthropology and the classics
– The mycenaean tree and pillar cult and its mediterranean relations
Evans, Augusta Jane *see* Beulah
Evans avenue baptist church. fort worth, texas : church records – 1904-mar 1963 – 1 – us Southern Baptist [242]
Evans, B *see* On confirmation
Evans, Benjamin *see* Character and reward of a faithful servant of christ
Evans, C *see* Advice to students having in view the christian ministry addressed to them at the academy at britsol
Evans, Charles *see* Friends in the seventeenth century
Evans, Christmas *see*
– Decision of a general congress convened to agree on terms of commun...
– Sermons of christmas evans
Evans, D D *see* The faithful minister of god a burning and a shining light
Evans, Daniel Silvan *see* Cambrian bibliography
Evans, David *see* Brevissima rhetorices institutio
Evans, E P *see* Animal symbolism in ecclesiastical architecture
Evans, Edward Payson *see* Animal symbolism in ecclesiastical architecture
Evans, Elaine Shemoney *see* The margaret cross norton working papers, 1924-1958
Evans, Eliza (Pruitt) *see* Diary
Evans, Elizabeth Edison *see* The christ myth
Evans, Elwood *see* The re-annexation of british columbia to the united states
Evans, F B *see* Madras district gazetteers [madras manuals]
Evans, Frederick William *see*
– Autobiography of a shaker
– Shakers
– Spiritualism on trial
Evans, Gail G *see* The kinematic variables related to the efficiency of throwing
Evans, George *see* An essay on assyriology
Evans, George W *see* A geographical, historical, and topographical description of van diemen's land
Evans, Harold *see* Men in the tropics
Evans, Henry Bentall *see* Our west indian colonies
Evans, Howard Heber *see* St john
Evans, Hugh *see*
– The able minister
– Maryland practice
Evans, Ifor Leslie *see* Native policy in southern africa
Evans, J *see* Monastic life at cluny, 910-1157
Evans, J H *see* Eight sermons on christian union
Evans, James *see*
– The kingdom of god
– Nu-gu-mo-nun o-je-boa an-oad ge-e-se-ueu-ne-gu-noo-du-be-ueng uoo muun-gou-duuz [george henry] gu-ea moo-ge-gee-seg [james evans] ge-ge-noo-ue- muu-ga-oe-ne-ne-oug
Evans, James Cook *see* Letter to the right hon lord lyndhurst
Evans, James Gwallia *see* Complete course in massage, swedish movement and mechanical therapeutics
Evans, John *see*
– Letters written during a tour through south wales, in the year 1803, and at other times
– Recreation for the young and the old
– Reflections on mortality
– Sermon occasioned by the death of the reverend d turner
– A tour through part of north wales
– The welsh nonconformists' memorial
Evans, John Swanton *see*
– Baptizing and teaching
– Christian predestination
– Christian rewards
Evans journal *see* Miscellaneous newspapers of weld county
Evans, Lewis *see* Geographical, historical, political, philosophical and mechanical essays, no 2
Evans, Llewelyn Joan *see*
– Biblical scholarship and inspiration
– Poems, addresses and essays
Evans, Llewelyn Joan et al *see* How shall we revise the westminster confession of faith?
Evans, Luther Harris *see* Virgin islands
Evans, Marshall Blakemore *see* Agnes bernauer
Evans, Maurice Smethurst *see*
– Black and white in southeast africa
– Black and white in the southern states
Evans, Milton G *see* New testament theology, pt 1
Evans, Morgan O *see* The ories and criticisms of sir henry maine
Evans, Nevil Norton *see*
– Elementary chemistry for high schools
– Laboratory manual to accompany "elementary chemistry for high school"
Evans, Philip Saffrey *see* History of connecticut baptist state convention, 1823-1907
Evans, Rachel *see* Effects of warm-up prior to eccentric exercise
Evans, Richard C *see*
– Forty years in the mormon church
– The songs, poems, notes and correspondence of bishop r c evans
Evans, Richardson *see* The age of disfigurement
Evans, Sebastian *see* Church windows
Evans, Silas J *see* A history of persecution for the truth's sake in louisville, ky
Evans, Thomas *see*
– A concise account of the religious society of friends, commonly called quakers
– The friends' library
Evans, W F *see* Mental medicine
Evans, Walter Norton *see*
– Canadian christmas song
– Cartier and hochelaga
– Mount royal
Evans, William *see*
– Agricultural improvement by the education of those who are engaged in it as a profession
– The book of books
– The book-method of bible study
– The friends' library
– The great doctrines of the bible
– How to memorize
– Journal of the life and religious services of william evans
– Supplementary volume to a treatise on the theory and practice of agriculture
Evans, William David *see* A letter to sir samuel romilly,knt. on the revision of the bankrupt law
Evans, William Sanford *see*
– The canadian contingents and canadian imperialism
– Winnipeg welcomes the manufacturer
Evans-Pritchard, Edward Evan *see*
– Nuer religion
– Some features of the nuer religion
Evanston speaks *see* Ai-fan-ssu-tun hu sheng (ccm90)
The evansville argus – Evansville. Ind. Sept. 13, 1940; Dec. 27, 1941 – 1 – us NY Public [071]
Evansville citizen – Evansville WI. 1866 jan 3-1868 dec 30 – 1r – 1 – mf#961939 – us WHS [071]
Evansville post – Evansville WI. 1977 dec 8-1978 apr 30, 1978 may 1-dec 30, 1979-80, 1981-1982 jan 28 – 4r – 1 – (cont: evansille post, inc) – mf#961959 – us WHS [071]
Evansville post – Evansville WI. 1973 mar 29-may 17 – 1r – 1 – (continued by: evansville post, inc) – mf#961955 – us WHS [071]
Evansville post, inc – Evansville WI. 1973 may 24-1974 aug 31, 1974 sep 1-1976 feb 28, 1976 mar 1-1977 mar 31, 1977 apr 1-dec 1 – 4r – 1 – (cont: evansville post [evansville, wi: 1973]; cont by: evansville post [evansville, wi: 1977]) – mf#961957 – us WHS [071]
Evansville review – Evansville WI. 1893 jul 11/1896 jan 30-1961 jan/may 11 [gaps] – 39r – 1 – (cont: brooklyn teller; cont by: evansville weekly review [evansville, wi: 1892]; cont by: evansville review and the brooklyn teller) – mf#1007111 – us WHS [071]
Evansville review – Evansville WI. 1870 mar 15-1871 sep 13, 1871 oct 4-1874 mar 25, 1874 apr 8-1875 sep 29 – 3r – 1 – (cont: evansville citizen; cont by: evansville weekly review [evansville, wi: 1875]) – mf#986196 – us WHS [071]
Evansville review – Evansville WI. 1984 apr 4-dec, 1985-99, 2000 jan-2002 jul-dec – 20r – 1 – (cont: evansville review and the brooklyn teller) – mf#1007114 – us WHS [071]
Evansville review – 1883 feb 17-dec 12, 1884 jan 4-mar 14 – 2r – 1 – (cont: evansville weekly review [evansville, wi: 1875]; cont by: evansville weekly review [evansville, wi: 1884]) – mf#1007100 – us WHS [071]
Evansville review – Evansville WI. 1886 may 21 – 1r – 1 – (cont: evansville weekly review [evansville, wi: 1884]; cont by: weekly review [evansville, wi]) – mf#1007104 – us WHS [071]
Evansville review and the brooklyn teller – Brooklyn, Evansville WI. 1961 may 16/1962-1984 jan/mar 28 [gaps] – 17r – 1 – (cont: evansville review [evansville, wi: 1893]; cont by: evansville review [evansville, wi: 1984]) – mf#1007113 – us WHS [071]
Evansville Seminary [Evansville WI] *see* Bulletin of the evansville...
Evansville weekly enquirer – Evansville IN. 1859 apr 27 – 1r – 1 – (cont: independent pocket; cont by: daily evening enquirer) – mf#856270 – us WHS [071]
Evansville weekly review – Evansville WI. 1875 oct 6-1877 sep 19, 1877 sep 26-1880 nov 17, 1880 nov 24-1883 feb 10 – 3r – 1 – (cont: evansville review [evansville, wi: 1870]; cont by: evansville review [evansville, wi: 1883]) – mf#1007093 – us WHS [071]
Evansville weekly review – Evansville WI. 1884 mar 21-1886 may 14 – 1r – 1 – (cont: evansville review [evansville, wi: 1883]; cont by: evansville review [evansville, wi: 1886]) – mf#1007103 – us WHS [071]
Evansville weekly review – Evansville WI. 1892 nov22-1893 apr 11, 1893 apr 18-jul 4 – 2r – 1 – (cont: weekly review [evansville, wi]; cont by: evansville review [evansville, wi]) – mf#1007108 – us WHS [071]
Evanturel, Eudore *see* Premieres poesies, 1876-1878
O evaristo – Rio de Janeiro, RJ: Typ Fluminense de Brito, 26 set-15 nov 1833 – (= ser Ps 19) – mf#P02,01,19 – bl Biblioteca [320]
Evaristo da veiga / Sousa, Octavio Tarquinio De – Sao Paulo, Brazil. 1939 – 1r – us UF Libraries [972]
Evart review – Big Rapids, MI. 1987-2000 (1) – mf#68339 – us UMI ProQuest [071]
Evarts, Jeremiah [pseud: William Penn] *see* Essays on the present crisis in the condition of the american indians
Evarts, William Maxwell *see* Eulogy on chief-justice chase
Las evas del paraiso / Trigo, Felipe – Madrid: Renacimiento, 5th ed 1923 – sp Bibl Santa Ana [946]
Evatt, Herbert Vere *see* Injustice within the law; a study of the case of the dorsetshire labourers
Evclide megarense philosopho : solo introdvttore delle scientie mathematice / Euclides – Vinegia: Venturino Hoffinelli, 1543 – 1 – us Wisconsin U Libr [510]
Evdokimov, A A *see*
– K teorii kooperatizma
– Kooperativnyi sbyt produktov selskogo khoziaistva v rossii
– Krestianskaia kooperatsiia v svobodnoi rossii
– Narodnyi teatr i kooperatsiia
– Selo i gorod v rossiiskoi kooperatsii
– Selskokhoziastvennye tovarishchestva, kak ikh ustraivat i vesti
– Vozrozhdenie sela i kooperatsiia
Eve *see* Women's journals, 1919-1968
Eve and her daughters : or, heroines of home / McConnell, Thomas Maxwell – Philadelphia: Westminster Press, 1900 [mf ed 1984] – (= ser Women & the church in america) – 4mf – 9 – 0-8370-0683-X – mf#1984-2026 – us ATLA [305]
Eve dans l'humanite / Deraismes, Maria – Paris: L Sauvaitre, 1891 – (= ser Les femmes [coll]) – 3mf – 9 – mf#8877 – fr Bibl Nationale [300]
The eve of catholic emancipation : being the history of the english catholics during the first thirty years of the nineteenth century / Ward, Bernard – London; New York: Longmans, Green, 1911-1912 – 3mf – 9 – 0-7905-7089-0 – mf#1988-3089 – us ATLA [241]
The eve of the reformation : studies in the religious life and thought of the english people in the period preceding the rejection of the roman jurisdiction by henry 8 / Gasquet, Francis Aidan – London: John C Nimmo, 1900 – 2mf – 9 – 0-524-04960-2 – mf#1990-1363 – us ATLA [242]
Evefiala : or, ewe-english dictionary / Westermann, Diedrich – Berlin: D Reimer, [1928] – 1 – us CRL [490]
L'eveil – Sorel: La Cie generale d'impr de Sorel, [ca 1911]-1917?// – mf#SEM35P333 – cn Bibl Nat [073]
L'eveil du peuple / Mouvement Republicain Populaire – Lille. 3no. speciaux. mai 1947-51, 1953 – 1 – fr ACRPP [325]
Eveil politique africain / Deschamps, Hubert Jules – Paris, France. 1952 – 1r – us UF Libraries [321]
Eveing herald *see* [Los angeles-] los angeles herald
Evelyn, J *see*
– Kalendarium hortense
– Silva
– Terra
Evelyn, John *see*
– Kalendarium hortense
– The life of mrs godolphin
– The life of mrs godolphin by john evelyn of wootton. esq
– The miscellaneous writings of john evelyn...
– Silva
Evelyn lascelles : an autobiography / ed by Addison, Julia – London: Thomas Cautley Newby Publ. 3v. 1855 – (= ser 19th c women writers) – 12mf – 9 – mf#5.1.34 – uk Chadwyck [920]
Even me'ir / Gordon, Aharon Ben Me'ir – Piotrkow Trybunalski, Poland. 1909 – 1r – us UF Libraries [939]
Evenbeck, Elizabeth J *see* Rating standards and related factors in high level amateur sports officiating
Evenemens interessans – Bayreuth DE, 1757, 1762-66, 1768-70 – 1 – (with gaps) – gw Misc Inst [074]
L'evenement : journal litteraire quotidien – Paris. n1-284. 11 avr 1868-29 janv 1869 – 1 – fr ACRPP [400]
L'evenement – Paris. Fond., Victor Hugo. Red. en chef, Edmond Magnier. Quot., mens. en 1915. 7 avr 1872-6 juin 1940, 2 fevr 1946-1949, 1951- – 1 – fr ACRPP [800]
Evenement – Paris, France. 10 oct 1916-17 dec 1918; 5 jan-1 apr 1919 – 5r – 1 – uk British Libr Newspaper [072]
Evenement – Quebec, Canada. 15 sep 1873-nov 1875; 1876-1888 – 39r – 1 – uk British Libr Newspaper [071]
L'evenement *see* L'avenement du peuple
Evenements de 1902 / Chancy, Emmanuel – Port-Au-Prince, Haiti. 1906 – 1r – us UF Libraries [972]
Evenements de fevrier, mai, juillet 1911 / Benony, D S – Port-Au-Prince, Haiti. 1911 – 1r – us UF Libraries [972]
Les evenements de mai-juin 1968 / France – (les quotidiens du 2 mai-3 juillet: sept journaux de paris et vingt journaux de province classes par ordre chronologique 54r. les hebdomadaires: aspects de la france; l'express; le figaro-selection hebdomadaire; l'humanite-dimanche; minute; le monde-selection hebdomadaire; le nouvel observateur; reforme; temoignage chretien; la tribune socialiste 4r) – fr ACRPP [074]
Les evenements de paris – Paris. 28 mai-30 sept 1867 – 1 – (devenu: la chronique de paris) – fr ACRPP [073]
Evenepoel, Edmond *see* Le wagnerisme hors d'allemagne bruxelles et la belgique
Evening advertiser – Swindon, England. 1899-1909; sep-dec 1910; 1913-25; 1950- – 396+ r – 1 – uk British Libr Newspaper [072]
Evening advocate – Green Bay WI. 1898 jan 15-may 7, 1898 may 9-sep 27 – 2r – 1 – (cont: green bay daily advocate; cont by: daily advocate [green bay, wi]) – mf#947503 – us WHS [071]
Evening amusement : containing airs, songs, dances, hornpipes, cotillions, reels, waltzs & marches for the german flute or violin – Baltimore: printed & sold at J Carrs Music Store [c1797] [mf ed 19–] – 1mf – 9 – mf#fiche 40 – us Sibley [780]
Evening and morning star – Kirtland, OH. v1-2 n24. jun 1832-sep 1834 – 1 – us NY Public [071]
Evening and sunday journal – East St Louis, IL. 1889-1964 (1) – mf#62613 – us UMI ProQuest [071]
Evening argus – Crawfordsville, IN. 1882-1885 (1) – mf#62752 – us UMI ProQuest [071]
Evening argus [bath] : published as a daily edition of the bath argus (saturday) weekly paper – Bath, England 20 jan 1876-30 oct 1877 – 1 – (cont: bath argus evening telegram [17 may 1875-19 jan 1876]; cont by: bath argus & west of england advertising register (daily ed) 31 oct 1877-29 apr 1892) – uk British Libr Newspaper [072]
Evening argus [brighton] – Brighton, England 25 aug 1896-24 jan 2004 [mf jul-dec 1897, 1899, jul-dec 1910, mar-may 1911, 1987-] – 1 – (wanting: 1898, jan, feb, jun-dec 1911; cont: argus [30 mar 1880-24 aug 1896]; cont by: argus [26 jan 2004-]) – uk British Libr Newspaper [072]
Evening argus [montpelier vt] – Montpelier VT 12 oct 1899 – 1 – uk British Libr Newspaper [071]
Evening astorian budget – Astoria OR: Astorian-Budget Pub Co, 1930-60 [daily ex sun] – 1 – (merger of: morning astorian (1899-1930); astoria evening budget (1914-30); cont by: astorian budget (1960). related to: weekly astorian (astoria, or)) – us Oregon Lib [071]
Evening astorian-budget *see* Weekly astorian (astoria, or)
Evening at home – v1-2. 1874-1975 [complete] – (= ser Mennonite serials coll) – 1r – 1 – mf#ATLA 1993-S019 – us ATLA [242]
Evening at home – v2-5. 1875-78 [complete] – (= ser Mennonite serials coll) – 1r – 1 – mf#ATLA 1993-S021 – us ATLA [073]
Evening auburnian – Auburn, NY. 1878-1885 (1) – mf#68497 – us UMI ProQuest [071]
Evening baker herald – Baker OR: Baker Herald Co, 1928-29 [daily ex sun] – 1 – (cont: baker herald (1911-28). merged with: morning democrat (-1929); baker democrat-herald (1929-63)) – us Oregon Lib [071]
Evening banner – Bluffton, IN. 1908-1929 (1) – mf#62730 – us UMI ProQuest [071]
Evening banner – Greenville, TX. 1915-1954 (1) – mf#66615 – us UMI ProQuest [071]
Evening bee – Toledo, OH. 1884-1899 (1) – mf#65679 – us UMI ProQuest [071]
The evening book: or, fireside talk on morals and manners, with sketches of western life / Kirkland, Caroline Matilda Stansbury – New York: C. Scribner, 1852 – 1 – us Wisconsin U Libr [390]

Evening bulletin – Decatur, IL. 1895-1896 (1) – mf#62594 – us UMI ProQuest [071]

Evening bulletin – Edmonton, Canada. 7 nov 1915-24 nov 1916; 30 jul-6 dec 1917; 4 jun, 29 jun-5 jul 1918 (imperfect) – 6 1/2 r – 1 – uk British Libr Newspaper [071]

Evening bulletin – Providence, RI. 1863-1995 (1) – mf#60574 – us UMI ProQuest [071]

Evening bulletin – Richmond VA. 1854 jul 26-dec 1 – 1r – 1 – mf#886407 – us WHS [071]

Evening call – Lafayette, IN. 1896-1904 (1) – mf#62865 – us UMI ProQuest [071]

The evening call – Topeka, KS. -v3 n36. jul 8 1893 (daily ex sun) [mf ed 1947] – (= ser Negro Newspapers on Microfilm) – 1r – 1 – us L of C Photodup [071]

Evening capital journal – Salem OR: Capital Journal Pub Co, 1888-93 [daily ex sun] – 1 – (ceased with feb 9 or feb 10 1893; cont by: capital journal (1893-95)) – us Oregon Lib [071]

Evening chronicle – Dublin, Ireland. 31 mar 1784 – 1/4r – 1 – uk British Libr Newspaper [072]

Evening chronicle – London, 31 Jan-30 Dec 1835 – 1r – 1 – uk British Libr Newspaper [072]

Evening chronicle / Montgomery Co. Dayton – sep 22-nov 6 1856 [daily] – 1r – 1 – mf#B5000 – us Ohio Hist [321]

Evening chronicle – New Orleans, LA. 1884-1886 (1) – mf#63536 – us UMI ProQuest [071]

Evening chronicle – Providence, RI. 1842-1843 (1) – mf#66293 – us UMI ProQuest [071]

Evening chronicle – [NW England] Oldham Lib 17 mar 1880-17 mar 1882* – 1 – (title change: oldham evening chronicle [20 mar 1882-apr 1974]; oldham chronicle [1974-8 apr 1982]; oldham evening chronicle [15 apr 1982]) – uk Newsplan; uk MLA [072]

Evening chronicle / Tuscarawas Co. Uhrichsville – apr 1940-dec 1949 [daily] – 21r – 1 – mf#B30240-30260 – us Ohio Hist [071]

Evening chronicle / Tuscarawas Co. Uhrichsville – jan 1938-sep 1939 [daily] – 3r – 1 – mf#B30237-30239 – us Ohio Hist [071]

Evening chronicle / Tuscarawas Co. Uhrichsville – jan 1950-dec 1954 [daily] – 16r – 1 – mf#B30261-30276 – us Ohio Hist [071]

Evening chronicle / Tuscarawas Co. Uhrichsville – jan 1955-dec 1959 [daily] – 17r – 1 – mf#B30277-30293 – us Ohio Hist [071]

Evening chronicle / Tuscarawas Co. Uhrichsville – jan 1960-dec 1964 [daily] – 18r – 1 – mf#B30294-30311 – us Ohio Hist [071]

Evening chronicle / Tuscarawas Co. Uhrichsville – jan 1965-dec 1969 [daily] – 18r – 1 – mf#B30312-30329 – us Ohio Hist [071]

Evening chronicle / Tuscarawas Co. Uhrichsville – jan 1970-dec 1974 [daily] – 20r – 1 – mf#B30401-30420 – us Ohio Hist [071]

Evening chronicle / Tuscarawas Co. Uhrichsville – jan 1975-dec 1980 [daily] – 24r – 1 – mf#B30421-30444 – us Ohio Hist [071]

Evening chronicle – Uhrichsville, OH. 1958-1959 (1) – mf#65692 – us UMI ProQuest [071]

Evening chronicle *see* Miscellaneous newspapers of lake county

Evening citizen – Ottawa, Canada. 10 sep-20 sep 1907; 10 jun-30 jun 1908; 9 aug 1909-1921 – 129 1/4r – 1 – (aka: ottawa evening citizen) – uk British Libr Newspaper [071]

Evening citizen – Cairo, IL. 1899-1985 (1) – mf#61312 – us UMI ProQuest [071]

Evening citizen – Glasgow, Scotland, UK. 1882-84; 1887-88; 1890-91. -d. 14 reels – 1 – uk British Libr Newspaper [072]

Evening citizen [glasgow, scotland : 1864] – [Scotland] Glasgow: J Hedderwick & Son 8 aug 1864-dec 1865, jan 1895-dec 1914 (daily) [mf ed 2004] – 85r – 1 – (missing: 1903-06 o/s; cont by: glasgow citizen [jul-dec 1914]) – uk Newsplan [072]

Evening communions : a divine institution and not a "modern invention" / Beddow, J J – London, England. 1886 – 1r – us UF Libraries [240]

Evening communions contrary to the church's mind, and why : being three articles reprinted from the literary churchman, with a letter to the editor / Bright, William – London: William Skeffington, 1870 – 1mf – 9 – 0-524-05873-3 – mf#1990-5167 – us ATLA [240]

Evening communions contrary to the church's mind, and why / Bright, William – London, England. 1870 – 1r – 1 – us UF Libraries [240]

Evening courier – Camden, NJ. 1947-1949 (1) – mf#64804 – us UMI ProQuest [071]

Evening courier – Milwaukee WI. 1847 feb 22-may 18 – 1r – 1 – (continued by: daily wisconsin) – mf#1160448 – us WHS [071]

Evening courier – Lansford, PA. 1932-1933 (1) – mf#65976 – us UMI ProQuest [071]

Evening courier – Tamaqua, PA. 1874-1971 (1) – mf#66093 – us UMI ProQuest [071]

Evening critic – Washington DC. 1883 oct 23 – 1r – 1 – (cont: daily critic [washington, dc: 1872]; cont by: washington critic [washington, dc: 1885]) – mf#846222 – us WHS [071]

The evening daily press – Pawnee City, NE: Press Pub Co. -v5 n132. feb 29 1896 (daily ex sun) [mf ed 1895-96 (gaps) filmed 1975] – 1r – 1 – (absorbed by: pawnee press) – us NE Hist [071]

Evening delta – New Orleans LA. 1862 feb 4,17, mar 20, apr 28, jun 7,18,20,25, jul 24,25,28,31, aug 1,6, sep 27 – 1r – 1 – mf#862506 – us WHS [071]

Evening democrat – Hamilton, OH. 1897-1907 (1) – mf#65519 – us UMI ProQuest [071]

Evening democrat – Missoula, MT. 1896-1897 (1) – mf#64566 – us UMI ProQuest [071]

Evening democrat – Moberly, MO. 1896-1925 (1) – mf#64187 – us UMI ProQuest [071]

Evening democrat – Warren, PA. 1893-1900 (1) – mf#66116 – us UMI ProQuest [071]

Evening despatch – Birmingham, England 14 may 1907-8 apr 1963 [sep,oct 1912] – 1 – (amalg with: birmingham mail & subsequently publ as: birmingham evening mail & despatch; wanting nov,dec 1912; cont: birmingham evening despatch [1 feb 1902-13 may 1907]) – uk British Libr Newspaper [072]

Evening despatch sporting buff – Birmingham, England 26 feb 1909-11 may 1940 [mf jul 1910-dec 1911, jul- dec 1912] – 1 – (not publ between 11 may 1940 & 31 aug 1946; cont: birmingham evening despatch sporting buff [29 mar 1904-25 feb 1909]; cont by: sporting buff [31 aug 1946-16 apr 1960]) – uk British Libr Newspaper [790]

Evening digest – [Scotland] Glasgow: A Guthrie 21 apr 1855 (daily) [mf ed 2004] – 1v on 1r – 1 – (specimen iss for the morning bulletin and evening digest wh began the following mth) – uk Newsplan [072]

Evening dispatch – Auburn, IN. 1898-1899 (1) – mf#62722 – us UMI ProQuest [071]

Evening dispatch – Michigan City, IN. 1881-1942 (1) – mf#62901 – us UMI ProQuest [071]

Evening dispatch – New Martinsville, WV. 1901-1906 (1) – mf#67398 – us UMI ProQuest [071]

Evening dispatch – Providence, RI. 1886-1887 (1) – mf#66294 – us UMI ProQuest [071]

Evening dispatch – Richmond, VA. 1920-1923 (1) – mf#66819 – us UMI ProQuest [071]

Evening dispatch – Seattle, WA. 1872-1878 (1) – mf#67103 – us UMI ProQuest [071]

Evening dispatch – White Plains, NY. 1939-1941 (1) – mf#65286 – us UMI ProQuest [071]

Evening drum *see* Mu ku ku shih (ccm144)

Evening echo – Bournemouth, England 1 jul 1958-4 apr 1997 (daily) [mf 1986-] – 1 – (cont: bournemouth daily echo [20 aug 1900-30 jun 1958]; cont by: daily echo [5 apr 1997-]) – uk British Libr Newspaper [072]

Evening echo – Dublin, Ireland. 17 feb 1894-11 may 1895 – 3r – 1 – uk British Libr Newspaper [072]

Evening echo (city final) – Cork, Ireland. may-dec 1896; jul-dec 1899; jan-20 nov 1926; 1986-1996 – 145r – 1 – uk British Libr Newspaper [072]

Evening empire / Montgomery Co. Dayton – jan-nov 1850 [daily] – 1r – 1 – mf#B5137 – us Ohio Hist [071]

Evening enterprise – Union City, PA. 1906-1911 (1) – mf#66096 – us UMI ProQuest [071]

Evening examiner – [NW England] Manchester ALS 21 nov 1893-5 jan 1894 – 1 – uk MLA; uk Newsplan [072]

Evening expositor – Adrian, MI. 1858-1859 (1) – mf#63673 – us UMI ProQuest [071]

Evening express – Halifax, NS. 1858-74 – 17r – 1 – cn Library Assoc [071]

Evening express – Lancaster, PA. 1856-1876 (1) – mf#65956 – us UMI ProQuest [071]

Evening express – [Scotland] Edinburgh: G Gillies 6 mar 1880-30 jun 1885 (daily ex sun) [mf ed 2003] – 15r – 1 – (missing: jan-jun 1881) – uk Newsplan [072]

Evening express – [NW England] Liverpool jul 1903-jun 1904, jan-jun 1906, jul-sep 1910, jul-dec 1916, jul 1920-jun 1921, jan-jun 1922, 1924, jul 1927-sep 1937, 1939-11 oct 1958 – 1 – uk Newsplan; uk MLA [072]

Evening express – Portland, ME. 1886-1990 (1) – mf#60488 – us UMI ProQuest [071]

Evening express – Rochester, NY. 1859-1882 (1) – mf#65189 – us UMI ProQuest [071]

Evening express – Washington, DC. 1868 feb 11 – 1r – 1 – mf#846226 – us WHS [071]

Evening express *see* Het volksblad

Evening express and star *see* Midland counties evening express

Evening fireside : or literary miscellany – Philadelphia. 1804-1806 (1) – mf#3574 – us UMI ProQuest [420]

Evening free lance *see* [Hollister-] free lance

Evening free press – Easton, PA. 1869-1872 (1) – mf#65884 – us UMI ProQuest [071]

Evening free press – Eau Claire WI. 1890 jul/dec-1900 jul/dec [gaps] – 19r – 1 – mf#964291 – us WHS [071]

Evening free press – Eau Claire, WI. 1898-1900 (1) – mf#67549 – us UMI ProQuest [071]

Evening freeman – Dublin, Ireland. 1837; 1839; 1847; 1850-30 jun 1871 – 62 1/2r – 1 – (incorp with: evening telegraph, sep 1871-96) – uk British Libr Newspaper [072]

Evening gazette – Aberdeen, Scotland. 1890 [daily] – 2r – 1 – uk British Libr Newspaper [072]

Evening gazette / Huron Co. Bellevue – oct 1903-dec 1905 (fire damaged) [daily] – 3r – 1 – mf#B934-936 – us Ohio Hist [071]

Evening gazette – England, 28 aug-dec 1935; 1936-39 – 24r – 1 – (reading gazette: jan-18 oct 1939) – uk British Libr Newspaper [072]

Evening gazette – Shanghai, China. 15 jan-31 dec 1874 [daily] – 2r – 1 – uk British Libr Newspaper [079]

Evening herald – Dublin. Ireland. 17 may 1786-1789 [semiwkly] – 2r – 1 – (aka: morning herald or general advertiser) – uk British Libr Newspaper [072]

Evening herald – Bellingham, WA. 1903-1904 (1) – mf#66940 – us UMI ProQuest [071]

Evening herald – Dayton, OH. 1869-1874 (1) – mf#65460 – us UMI ProQuest [071]

Evening herald – Decatur, IL. 1927-1931 (1) – mf#62595 – us UMI ProQuest [071]

Evening herald – Dublin 1806 – 1r – 1 – ie National [072]

Evening herald – Dublin, Ireland. 19 dec 1891-1899; 24 mar 1900-1907; 1925; 1926; jan-23 dec 1930; 1951; 1952; 1986-30 may 1991; sep 1991-mar 1992; may 1992-1995; 1996; jan-sep 1997 (wanting 1 jan-23 mar 1900) – 268 1/2r – 1 – uk British Libr Newspaper [072]

Evening herald – Dublin, Ireland. 26 jan; 6, 11 apr; 5 aug; 16 sep; 7 dec 1807 – 1/2r – 1 – uk British Libr Newspaper [072]

Evening herald – Duluth, MN. 1892-1947 (1) – mf#63913 – us UMI ProQuest [071]

Evening herald – Huntington, IN. 1903-1911 (1) – mf#62825 – us UMI ProQuest [071]

Evening herald – Joliet, IL. 1906-1915 (1) – mf#62636 – us UMI ProQuest [071]

Evening herald / Montgomery Co. Dayton – jan 1870-mar 1874 – 4r – 1 – mf#B5151-5154 – us Ohio Hist [071]

Evening herald – Shenandoah, PA. 1969-80 – 13 – $25.00r – us IMR [071]

Evening herald – St. John's, Canada. -d. 14 may 1890-13 feb 1892; 18 jul 1898-22 feb 1900; 5 apr 1900-feb 1913; 28 apr 1913-6 dec 1918; 18 mar 1919-27 dec 1920 – 88 1/4r – 1 – uk British Libr Newspaper [072]

Evening herald and ashland daily – Shenandoah, PA., 1967-1969 – 13 – $25.00r – us IMR [071]

Evening herald and commercial intelligencer – Providence, RI. 1840-1840 (1) – mf#66295 – us UMI ProQuest [071]

Evening herald (baker city, or) – Baker City OR: Herald Pub & Engraving Co, 1904- [daily ex sun] – 1 – (cont: Herald (baker city, or); cont by: baker city herald (-1911)) – us Oregon Lib [071]

Evening herald (klamath falls, or) – Klamath Falls OR: Herald Pub Co [daily ex sun] – 1 – (began in jul 1906. absorbed by: klamath news (klamath falls, or); herald and news (klamath falls, or)) – us Oregon Lib [071]

Evening herald (klamath falls, or) *see*
– Herald and news
– Klamath news

Evening herald (new york, ny: 1837) *see* Morning herald (new york, ny: 1837)

Evening herald news – Joliet, IL. 1915-1936 (1) – mf#62637 – us UMI ProQuest [071]

Evening independent – Chippewa Falls WI. 1916 mar 11-apr 16, 1916 apr 17-oct 13, 1916 oct 15-1917 apr 20, 1917 apr 23-oct 31, 1917 nov 1-1918 may 17, 1918 dec 13-1919 jul 13, 1918 may 20-dec 12, 1919 jul 14-dec 11 – 8r – 1 – (cont: daily independent [chippewa falls, wi]; cont by: chippewa daily press) – mf#923784 – us WHS [071]

Evening independent – St Petersburg, FL. 1907-1986 (1) – mf#60440 – us UMI ProQuest [071]

Evening irish times – Dublin, Ireland. 22 oct 1880-sep 1896; 1 jan-31 mar 1915; jan-mar 1920. -d. 66r – 1 – uk British Libr Newspaper [072]

Evening item / Montgomery Co. Dayton – v1 n1. may-jul 1890 [wkly] – 1r – 1 – mf#B4999 – us Ohio Hist [071]

Evening item – Providence, RI. 1886-1886 (1) – mf#66296 – us UMI ProQuest [071]

Evening item – Richmond, IN. 1881-1916 (1) – mf#62957 – us UMI ProQuest [071]

Evening journal – Albany, NY. 1830-1873 (1) – mf#64878 – us UMI ProQuest [071]

Evening journal – Billings, MT. 1909-1918 (1) – mf#64253 – us UMI ProQuest [071]

Evening journal – Portland OR: Journal Print Co, 1902 [daily ex sun] – 1 – (cont by: portland evening journal) – us Oregon Lib [071]

Evening journal – Berlin WI. 1881 may 9-jul 11, 1881 jul 4, 1881 jul 12-oct 6 – 3r – 1 – (cont: daily evening journal; cont by: berlin evening journal) – mf#961579 – us WHS [071]

Evening journal – Plattsmouth, NE: R A & T B Bates, 1902 (daily ex sun) [mf ed 1902, 1905-13 (gaps)] – 5r – 1 – (cont: plattsmouth daily journal (1902). cont by: plattsmouth evening journal. suspended for nearly 2yrs; resumed with v1 n1 jun 19 1905) – us NE Hist [071]

Evening journal – Gadsden, AL. 1908-1908 (1) – mf#62019 – us UMI ProQuest [071]

Evening journal – Hamilton, OH. 1908-1937 (1) – mf#65520 – us UMI ProQuest [071]

Evening journal – Huntingdon, PA. -d 1917-1918 – 13 – $25.00r – us IMR [071]

Evening journal – Lewiston, ME. 1861-1989 (1) – mf#63562 – us UMI ProQuest [071]

Evening journal – Nevada, IA. 1941-1962 (1) – mf#63331 – us UMI ProQuest [071]

Evening journal – New York, NY. 1903-1937 (1) – mf#65071 – us UMI ProQuest [071]

Evening journal – Richmond, VA. 1905-1920 (1) – mf#66820 – us UMI ProQuest [071]

Evening journal – Saratoga Springs, NY. 1883-1886 (1) – mf#65223 – us UMI ProQuest [071]

Evening journal – Shreveport, LA. 1897-1901 (1) – mf#63542 – us UMI ProQuest [071]

Evening journal – St. Catherines, Canada. -d. 13 mar-28 jun 1919; 16 dec 1919-1 may 1920 – 3r – 1 – uk British Libr Newspaper [071]

Evening journal – Vineland, NJ. 1876-1941 (1) – mf#64854 – us UMI ProQuest [071]

Evening journal – Wilmington, Delaware. 1888-1910 (1) – mf#68686 – us UMI ProQuest [071]

Evening journal and post express – Rochester, NY. 1923-1937 (1) – mf#65190 – us UMI ProQuest [071]

Evening journal every evening – Wilmington, DE. 1933-1934 (1) – mf#62381 – us UMI ProQuest [071]

Evening leader / Auglize Co. Saint Marys – oct 1914-feb 1916, jun 1947-82 [daily] – 117r – 1 – mf#B12112-12228 – us Ohio Hist [071]

Evening leader – Tarpon Springs, FL. 1914-1918 – 2r – (gaps) – us UF Libraries [071]

Evening leader – Richmond, VA. 1896-1903 (1) – mf#66821 – us UMI ProQuest [071]

Evening leader – Staunton, VA. 1917-1957 (1) – mf#66874 – us UMI ProQuest [071]

Evening ledger – Gainesville, FL. 1895 feb 22 – 1r – us UF Libraries [071]

Evening ledger – Gainesville, FL. 1895 feb 26 – 1r – us UF Libraries [071]

Evening mail – Charleston, WV. 1893-1896 (1) – mf#67236 – us UMI ProQuest [071]

Evening mail – Oamaru, NZ: apr 1876-apr 1879 – 9r – 1 – (commenced publ apr 1876. title changes to: oamaru mail. aka: oamaru times) – mf#82.6 – nz Nat Libr [079]

Evening mail : a daily paper – [East Midlands] Northampton, Northamptonshire n1 [26 jan 1880]-n245 [5 feb 1881] [mf ed 2002] – 3r – 1 – (discontinued) – uk Newsplan [072]

Evening mail – Halifax, Canada. -d. 29 may 1911-4 nov 1913; 2 mar-17 jun 1914; 14, 17 nov 1914; 21 dec 1917. 17 reels – 17r – 1 – uk British Libr Newspaper [071]

Evening mail – Providence, RI. 1884-1885 (1) – mf#66297 – us UMI ProQuest [071]

Evening mail *see* Dublin evening mail

Evening mail (cardiff) – [Wales] LLGC 13 oct 1884-mar 1885 [mf ed 2004] – 1r – 1 – uk Newsplan [072]

Evening mercantile journal – Boston MA. 1835 jan 1-1839 nov 5, 1843 oct 14-1844 dec 31 – 2r – 1 – (cont: boston mercantile journal [daily]; cont by: boston daily journal) – mf#872551 – us WHS [071]

Evening mercury – St. John's, Canada. -d. 28 dec-31 dec 1883; 2 jan, 15 jul-3 dec 1884; 1885-8 nov 1887 – 6r – 1 – uk British Libr Newspaper [071]

Evening messenger – Valparaiso, IN. 1907-1916 (1) – mf#62985 – us UMI ProQuest [071]

Evening mirror – New york, oct 7 1844-dec 1847 – 1 – (contains several poe contributions) – us NY Public [071]

Evening / morning journal / Columbiana Co. Lisbon – (1928-9/29,6/31-6/33,6/46-1983) [daily] – (= ser Morning Journal) – 93r – 1 – (title changes) – mf#B13262-13354 – us Ohio Hist [071]

Evening news – Accra: Star Pub Co, [-1968]. [aug 21 1958-jul 9 1964] – us CRL [079]

Evening news – Albany, NY. 1928-1937 (1) – mf#64879 – us UMI ProQuest [071]

Evening news – Battle Creek, MI. 1911-1918 (1) – mf#63686 – us UMI ProQuest [071]

Evening news – Beacon, NY. 1961-1975 (1) – mf#64904 – us UMI ProQuest [071]

Evening news – Birmingham, AL. 1888-1890 (1) – mf#61986 – us UMI ProQuest [071]

Evening news – Bluffton, IN. 1893-1929 (1) – mf#62731 – us UMI ProQuest [071]

Evening news – Bridgeton, NJ. 1900-1996 (1) – mf#61597 – us UMI ProQuest [071]

Evening news – Omaha, NE: Fred Nye. v1 n1. may 29 [1878]-v2 n84. sep 1 1879 (daily ex sun) – 2r – 1 – (cont by: omaha news) – us NE Hist [071]

Evening news – Lincoln, NE: News Pub Co. 7v. v11 n139. mar 7 1892-mar 23 1898 (daily ex sun) [mf ed with gaps] – 11r – 1 – (cont: lincoln evening news. cont by: lincoln evening news (1898). numbering ceased with v16 n275 aug 14 1897. semiwkly ed: weekly news (1895-97)) – us NE Hist [071]

Evening news – 1893 mar 4 – 1r – 1 – (cont: post-record) – mf#845714 – us WHS [071]

Evening news – Baraboo WI.1894 jun 4/1894 dec 20-1911 dec 23/1911 dec 30 [gaps] – 48r – 1 – (continued by: baraboo daily news) – mf#1138945 – us WHS [071]

Evening news – Kenosha WI. 1895 jan 11/1895 jun 18-1899 may 27-1899 aug 16 [gaps] – 13r – 1 – (continued by: kenosha evening news) – mf#1145412 – us WHS [071]

Evening news – 1887 jul 16-dec 31, 1888 jan 3-sep 15 – 2r – 1 – (continued by: madison evening news) – mf#922914 – us WHS [071]

Evening news – Washington, DC. 1938 mar 4 – 1r – 1 – (continued by: washington news) – mf#923001 – us WHS [071]

Evening news – Cleveland, OH. v1 n278. mar 1 1869-dec 30 1880 – 9r – 1 – (daily general newspaper. aka: cleveland evening news. suppls accompany some numbers. merged with: evening herald (cleveland, ohio:1883) to form: news and herald. between 1868 and 1869 the evening news absorbed the evening leader. the evening news then continued as the evening edition of the leader) – us Western Res [071]

Evening news – Dannevirke, NZ. 16 oct-31 dec 1909; jul 1910-dec 1919; may-dec 1920; 11 aug-31 dec 1969; sep 1975-mar 2002 – 1 – mf#35.4 – nz Nat Libr [079]

Evening news – Dayton, OH. 1890-1898 (1) – mf#65461 – us UMI ProQuest [071]

Evening news – Daytona Beach, FL. 1937-1986 (1) – mf#61273 – us UMI ProQuest [071]

Evening news – Dublin, Ireland. 2 jan-22 feb 1888 – 1r – 1 – uk British Libr Newspaper [072]

Evening news – Edinburgh: Scotsman Publ Ltd, 1978-81 – 48r – 1 – us CRL [072]

Evening news – Fairbanks, AK. 1906-1906 (1) – mf#62059 – us UMI ProQuest [071]

Evening news / Hamilton Co. Cincinnati – oct 13-nov 7 1887 – 1r – 1 – mf#B37533 – us Ohio Hist [071]

Evening news – Harrisburg, PA. 1917-49. 220 rolls – 13 – $25.00r – us IMR [071]

Evening news – Harrisburg, PA. 1949-1996 (1) – ISSN: 0887-7939 – mf#60106 – us UMI ProQuest [071]

Evening news – Harrisonburg, VA. 1901-1904 (1) – mf#66731 – us UMI ProQuest [071]

Evening news : (Home Edition) – Perth Amboy, NJ. 1943-1946 (1) – mf#64839 – us UMI ProQuest [071]

Evening news – Jeffersonville, IN. 1989-2000 (1) – mf#61387 – us UMI ProQuest [071]

Evening news – Kenosha, WI. 1900-1940 (1) – mf#67566 – us UMI ProQuest [071]

Evening news – London, England. -d. 1881-1945. 299 reels – 1 – uk British Libr Newspaper [072]

Evening news – Monroe, MI. 1915-2000 (1) – mf#61530 – us UMI ProQuest [071]

Evening news / Montgomery Co. Dayton – 1890-jun 1896, jul-dec 1897 [daily] – 15r – 1 – mf#B5202-5216 – us Ohio Hist [071]

Evening news – Newark, NJ. 1883-1969 (1) – mf#60157 – us UMI ProQuest [071]

Evening news – Newburgh, NY. 1962-1990 (1) – mf#69301 – us UMI ProQuest [071]

Evening news – Omaha, NE. 1878-1880 (1) – mf#64718 – us UMI ProQuest [071]

Evening news – Oneonta, NY. 1887-1891 (1) – mf#69303 – us UMI ProQuest [071]

Evening news – Peekskill, NY. 1920-1920 (1) – mf#69307 – us UMI ProQuest [071]

Evening news – Perth Amboy, NJ. 1925-1944 (1) – mf#64838 – us UMI ProQuest [071]

Evening news – Petoskey, MI. 1941-1953 (1) – mf#63836 – us UMI ProQuest [071]

Evening news – Plainfield, NJ. 1884-1893 (1) – mf#64845 – us UMI ProQuest [071]

Evening news – Providence, RI. 1909-1918 (1) – mf#66298 – us UMI ProQuest [071]

Evening news – Dublin, Ireland. 27 may-6 sep 1996 – 3 3/4r – 1 – (publ 27 may 1996-6 sep 1996) – uk British Libr Newspaper [072]

Evening news – Dublin, Ireland.1859-1864. -d 18r – 1 – (publ between 1859 jan 18 and 1864 dec 31) – uk British Libr Newspaper [072]

Evening news – Roseburg OR: B W Bates, 1909-20 [daily ex sun] – 1 – (related to semiwkly ed: umpqua valley news. merged with: roseburg review (roseburg, or: daily) to form: roseburg news-review (roseburg, or: daily)) – us Oregon Lib [071]

Evening news / Richland Co. Mansfield – jan-dec 1890, jul-dec 1893 [daily] – 4r – 1 – mf#B34414-34417 – us Ohio Hist [071]

Evening news – Roanoke, VA. 1903-1913 (1) – mf#66854 – us UMI ProQuest [071]

Evening news – St Joseph, MO. 1879-1885 (1) – mf#64206 – us UMI ProQuest [071]

Evening news : (Sunday Edition) – Buffalo, NY. 1874-1912 (1) – mf#64916 – us UMI ProQuest [071]

Evening news – Sydney, Australia. Feb 1875-Dec 1888.-d. 42 reels – 1 – uk British Libr Newspaper [072]

Evening news – Sydney, nov 1869-dec 1910 – 121r – A$6262.06 vesicular A$6927.56 silver – at Pascoe [079]

Evening news – Warren, PA. 1895-1896 (1) – mf#66117 – us UMI ProQuest [071]

Evening news – Waterford, jan-may 1899 – 1r – 1 – ie National [072]

Evening news – Wilkes-Barre, PA. 1909-1939 (1) – mf#66148 – us UMI ProQuest [071]

Evening news – Youngstown, OH. 1877-1879 (1) – mf#65739 – us UMI ProQuest [071]

Evening news *see* Roseburg review (roseburg, or: daily)

The evening news – Sayre, PA., 1899-1986 – 13 – $25.00r – us IMR [071]

The evening news – Plattsmouth, NE: News Pub Co. v1 n1. nov 9 1891- (daily ex sun) [mf ed -1908 (gaps) filmed 1976] – 17r – 1 – (v1 n1 preceeded by an issue dated nov 5 1891 called sample copy. other ed: semi-weekly news jun 2-dec 1894 and: semi-weekly news-herald 1895-nov 30 1908) – us NE Hist [071]

The evening news – Williamsport, PA., 1895-1907 – 13 – $25.00r – us IMR [071]

Evening news and review – Bayonne, NJ. 1913-1930 (1) – mf#64796 – us UMI ProQuest [071]

Evening news [dundee, scotland] – [Scotland] Dundee: Peddie, Hutcheson & Co 28 mar 1876-12 mar 1879 (daily ex sun) [mf ed 2003] – 9r – 1 – uk Newsplan [072]

Evening news [greenock, scotland] – [Scotland] Inverclyde, Greenock: D Blair 17 jul 1866-dec 1868, jan 1869 & 25 jun 1870 (daily) [mf ed 2004] – 2r – 1 – (cont by: evening news and family journal [14 feb 1867-11 jan 1868]; greenock news and weekly press [18 jan 1868-31 jul 1869]; merged with: public guide to form: greenock news and public guide) – uk Newsplan [072]

Evening news-call – Lincoln, NE: [News-Call] 1v. jul 20 1898-aug 11 1898 (daily ex sun) – 1r – 1 – (formed by the union of: lincoln evening news (1898) and: lincoln evening call. cont by: lincoln evening news and daily call) – us NE Hist [071]

Evening news-dispatch *see* Miscellaneous newspapers of lake county

Evening packet *see* Dublin correspondent / evening packet

Evening packet and correspondent – Dublin, Ireland. 29 jan-21 apr 1829; 26 sep 1829-15 mar 1862 [wkly] – 38r – 1 – uk British Libr Newspaper [072]

Evening penny press – Minneapolis MN. 1896 dec 28-31 – 1r – 1 – (cont: penny press [minneapolis, mn: 1893]; cont by: evening press [minneapolis, mn: 1897]) – mf#851856 – us WHS [071]

Evening penny press – Pittsburgh, PA. 1885-1887 (1) – mf#66035 – us UMI ProQuest [071]

Evening plain dealer / Cuyahoga Co. Cleveland – jan 1886-feb 1893 [daily] – 18r – 1 – mf#B33563-33580 – us Ohio Hist [071]

Evening post – 14 sep-dec 1965; 1966-75; 1977-30 jun 1997 – 378 1/2r – 1 – (aka: reading post) – uk British Libr Newspaper [072]

Evening post – Baltimore, MD. 1805-1811 (1) – mf#63585 – us UMI ProQuest [071]

Evening post – Bridgeport, CT. 1906-1931 (1) – mf#61028 – us UMI ProQuest [071]

Evening post – Charleston, SC. 1894-1991 (1) – mf#60576 – us UMI ProQuest [071]

Evening post – Columbia City, IN. 1917-1927 (1) – mf#62743 – us UMI ProQuest [071]

Evening post – Lincoln, NE: Evening Post Pub Co. v1 n1. aug 25 1896-98// (daily ex sun) [mf ed with gaps] filmed -1977] – 3r – 1 – (cont by: nebraska post. suspended with aug 29 1898; resumed oct 1898. numbering ceased with v3 n4 aug 29 1898) – us NE Hist [071]

Evening post – Gary, IN. 1909-1921 (1) – mf#62791 – us UMI ProQuest [071]

Evening post – Milwaukee, WI. 1939-1942 (1) – mf#67587 – us UMI ProQuest [071]

Evening post – New York. oct 28 1852-dec 28 1854. (incomplete) (not collated) [weekly] – 1 – us NY Public [073]

Evening post – Pawtucket, RI. 1893-1897 (1) – mf#66245 – us UMI ProQuest [071]

Evening post – Providence, RI. 1895-1896 (1) – mf#66299 – us UMI ProQuest [071]

Evening post – Reading, England. 14 sep 1965-1970 [daily] – 63r – 1 – uk British Libr Newspaper [072]

Evening post – Vicksburg, MS. 1990-1994 (1) – mf#61553 – us UMI ProQuest [071]

Evening post – Wellington, 1991- – 12r per y – 1 – us UMI ProQuest [072]

Evening post – Wellington, NZ. 8 feb 1865-feb 1932, apr 1932-apr 1971; 16 jun-30 sep 1977, 16 oct 1977-15 sep 1979, 1 oct 1979-feb1991 – 1 – mf#41.1 – nz Nat Libr [079]

The evening post – Wellington, NZ: Blundell Bros Ltd, jul 1938-1981 – 1 – us CRL [071]

evening post *see* The port elizabeth advertiser

Evening post and chronicle – [NW England] Wigan aug 1963, 28 oct-31 dec 1963, 1 jul-31 dec 1965, 1 jan 1971-dec 1984 – 1 – uk MLA; uk Newsplan [072]

Evening post (dundee) – [Scotland] Dundee: W & D C Thomson 22 jan 1900-16 may 1905 (daily) [mf ed 2003] – 22r – 1 – (merged with: evening telegraph (dundee, scotland) to form: evening telegraph and post (dundee, scotland)) – uk Newsplan [072]

Evening post echo – Watford, England. jan 1980-dec 1981 [daily] – 23 1/2r – 1 – (post echo – watford ed. hemel hempstead) – uk British Libr Newspaper [072]

Evening post essays in review of "the bible for learners" / Schaff, Philip et al – New York: Evening Post, 1880 [mf ed 1985] – 1mf – 9 – 0-8370-3082-X – mf#1985-1082 – us ATLA [225]

Evening post sports post – 21 mar 1936-apr 1975 – 60r – 1 – (aka: sports post) – mf#41.34 – nz Nat Libr [079]

Evening press – Guernsey, Channel Islands. oct 1937-46; 1986; 1993 – 33r – 1 – (aka: guernsey evening press) – uk British Libr Newspaper [072]

Evening press – Belfast, Ireland. 15 May 1873-21 May 1874. -d.1 1/2 reels – 1 – uk British Libr Newspaper [072]

Evening press – Brenham, TX. 1908-1910 (1) – mf#66582 – us UMI ProQuest [071]

Evening press – Greensburg, PA. 1881-1891 (1) – mf#68684 – us UMI ProQuest [071]

Evening press – Hornellsville, NY. 1889-1892 (1) – mf#65001 – us UMI ProQuest [071]

Evening press / Montgomery Co. Dayton – oct 1892-1901, jul-dec 1904 [daily] – 18r – 1 – mf#B5254-5271 – us Ohio Hist [071]

Evening press – Muncie, IN. 1995-1995 (1) – mf#61396 – us UMI ProQuest [071]

Evening press – Newburgh, NY. 1889-1894 (1) – mf#65113 – us UMI ProQuest [071]

Evening press – Princeton, WV. 1917-1923 (1) – mf#67444 – us UMI ProQuest [071]

Evening press – Providence, RI. 1859-1884 (1) – mf#66300 – us UMI ProQuest [071]

Evening press – Dublin, Ireland. 1986-25 may 1995 – 153 1/2r – 1 – (publ 1 sep 1954-25 may 1995) – uk British Libr Newspaper [072]

Evening press – Savannah, GA. 1891-1996 (1) – mf#60449 – us UMI ProQuest [071]

Evening press – Warren, PA. 1901-1901 (1) – mf#66118 – us UMI ProQuest [071]

The evening public – Omaha, NE: [s.n.] v1 n1. jul 4 1892- (daily ex sun) [mf ed sep 8-nov 12 1892 (gaps) filmed 1980] – 1r – 1 – us NE Hist [071]

Evening record – Windsor, Canada. 27 feb-4 aug 1914; 1 oct 1914-aug 1918 (wanting sep 1914) (imperfect) – 22 1/2r – 1 – (aka: windsor record) – uk British Libr Newspaper [071]

Evening record – Allegheny, PA. 1896-1899 (1) – mf#65828 – us UMI ProQuest [071]

Evening record / Columbiana Co. Wellsville – jan-nov 1903 (fire damaged papers) – 2r – 1 – mf#B29327-29328 – us Ohio Hist [071]

Evening record – Marshfield OR: [s.n.] [daily ex sun & hols] – 1 – (cont by: southwestern oregon daily news) – us Oregon Lib [071]

Evening record – Lansford, PA. 1920-1967 (1) – mf#65977 – us UMI ProQuest [071]

Evening record / Portage Co. Ravenna – mar 1928-dec 1955 [daily] – 88r – 1 – mf#B4222-4309 – us Ohio Hist [071]

Evening record – St. Augustine, FL. 1899-1935 (1) – mf#62449 – us UMI ProQuest [071]

Evening recorder – Albion, MI. 1904-1988 (1) – mf#61498 – us UMI ProQuest [071]

Evening recorder – Tamaqua, PA. 1896-1899 (1) – mf#66094 – us UMI ProQuest [071]

Evening recorder *see* [Porterville-] porterville papers

Evening register / Trumbull Co. Niles – may 1923-mar 1924 (damaged) [daily] – 3r – 1 – mf#B31904-31906 – us Ohio Hist [071]

Evening report – Lebanon, PA. 1898-1937 [daily] – 13 – $25.00r – (missing dates) – us IMR [071]

Evening reporter – Ashton-under-Lyne, England 18 dec 1876-13 aug 1914 – 1 – uk British Libr Newspaper [072]

Evening reporter – [NW England] Tameside, Ashton 18 dec 1876-13 aug 1914 [mf ed 2003] – 71r – 1 – (missing: 1880) – uk Newsplan [072]

Evening reporter – Woonsocket, RI. 1873-1908 (1) – mf#66441 – us UMI ProQuest [071]

Evening reporter star – Orlando, FL. 1925-1937 (1) – mf#62438 – us UMI ProQuest [071]

Evening republican – Columbus, IN. 1890-1965 (1) – (filmed with: herald) – mf#61149 – us UMI ProQuest [071]

Evening republican – Janesville WI. [1894 sep 11-1898 jun 3], 1898 jun 15 – 2r – 1 – mf#876320 – us WHS [071]

Evening republican – Meadville, PA. 1951-1955 (1) – mf#65991 – us UMI ProQuest [071]

Evening review – Elkhart, IN. 1872-1879 (1) – mf#62770 – us UMI ProQuest [071]

Evening roseburg review – Roseburg OR: Review Pub Co [daily ex sun] – 1 – (cont: daily roseburg review; cont by: roseburg review (roseburg, or: daily). related to semiwkly ed: roseburg review (roseburg, or)) – us Oregon Lib [071]

Evening roseburg review *see* Roseburg review (roseburg, or)

Evening sentinel – Ansonia, CT. 1896-1992 (1) – mf#61243 – us UMI ProQuest [071]

Evening signal – Milwaukee WI. 1879 aug 4-1880 mar 27, 1880 mar 29-may 29 – 2r – 1 – (continued by: milwaukee evening chronicle) – mf#1126485 – us WHS [071]

Evening song to the virgin [at sea] : a duett / Browne [Miss] – [184-?] [mf ed 19–] – 1r – 1 – (words by felicia dorothea browne hemans) – mf#pres. film 33 – us Sibley [780]

Evening standard – London, 1860-1968 – 514r – 1 – uk British Libr Newspaper [072]

Evening standard – London, England. 1962-1981 – 112r – 1 – us L of C Photodup [072]

Evening standard – Wheeling, WV. 1877-1878 (1) – mf#67516 – us UMI ProQuest [071]

Evening standard and st james's gazette – London: W E Hobbs, [1905-]. [mar 14 1905-sep 1909] – us CRL [072]

Evening star – Ashton-under-Lyne, England 20 apr 1877-10 may, 26 jun 1878 – 1 – uk British Libr Newspaper [072]

Evening star – Auburn, IN. 1990-2000 (1) – mf#61370 – us UMI ProQuest [071]

Evening star – Dunedin, NZ. jan-feb 1974; may-sep 1974; jan-feb 1975; nov-dec 1976; feb 1977; feb-apr 1979; jun-3 nov 1979 – 1 – (ceased publ 3 nov 1979) – mf#81.4 – nz Nat Libr [079]

Evening star – Washington, DC. 1917 apr 2, 6; 1939 dec 31, 1941 dec 8; 1942 jan 23, 1945 aug 15, sep 2 – 1r – 1 – (cont: daily evening star (washington, dc); cont by: washington daily news (washington, dc); evening star and the washington daily news) – us WHS [071]

Evening star – La Crosse WI. 1885 nov 23-24, 28, 1886 apr 7 – 1r – 1 – mf#932229 – us WHS [071]

Evening star – Ipswich, Suffolk, 17 feb 1885-to date – 1 – (lacking: jan 1913; to 30 may 1893 known as: star of the east) – uk Newsplan [072]

Evening star – London, 17 Mar 1856-Dec 1859; Jan 1861-Dec 1862 – 17r – 1 – uk British Libr Newspaper [072]

Evening star – London, 25 jul 1842-28 feb 1843 – 2r – 1 – uk British Libr Newspaper [072]

Evening star – [NW England] Ashton-under-Lyne, Stalybridge Lib 20 apr 1877-26 jun 1878 [incomplete] – 1 – uk MLA; uk Newsplan [072]

Evening star – Pasadena, CA. 1889-1910 (1) – mf#62215 – us UMI ProQuest [071]

Evening star – Thames, NZ. apr 1874-dec 1875; jan-jun 1877; jan 1878-dec 1881; jul 1882-mar 1890; may 1890-apr 1893; may 1893-dec 1902; jul-dec 1903; jul 1904-sep 1906; jan 1907-dec 1909; jul 1910-apr 1919; sep 1919-dec 1938; jan-dec 1943; jan-dec 1945; 12 jan-24 dec 1953; 12 jan-5 may 1969; jan 1976-jun 1986; jan-dec 1987 – 274r – 1 – (title changes to: thames star fr may 1893) – mf#16.1 – nz Nat Libr [079]

The evening star – Cincinnati. Jan. 4-Jun. 29, 1872 – 1 – us NY Public [071]

Evening star 7 o'clock (dunedin) – Dunedin, NZ. sep 1975-mar 1979 – 8r – 1 – mf#81.2 – nz Nat Libr [079]

Evening star (hokitika) – jan 1867-jul 1868 – 3r – 1 – mf#60.10 – nz Nat Libr [079]

Evening star of gwent and south wales times, monmouthshire and border counties advertiser – [Wales] LLGC 10 nov 1877-10 jul 1903 [mf ed 2004] – 87r – 1 – (missing: 1898; cont by: south wales daily star [jul 1892-mar 1901]; south wales daily telegraph [apr 1901-mar 1903]; newport & monmouthshire evening telegraph [apr-jul 1903]) – uk Newsplan [072]

Evening star-telegram – Lakeland, FL. v1 n130-180. 1921 jan-feb – 1r – us UF Libraries [071]

Evening state journal – Lincoln, NE: J C Seacrest. 14v. jun 6 1929-dec 9 1942 (daily) [mf ed llacks sun iss] – 56r – 1 – (cont: evening state journal (and lincoln daily news). cont by: lincoln evening state journal. on sun publ as: sunday state journal 1929-oct 25 1931, sunday state journal and lincoln sunday star nov 1 1931-jan 3 1932, sunday journal and star jan 10 1932-42) – us NE Hist [071]

Evening state journal (and lincoln daily news) – Lincoln, NE: C D Traphagen, J C Seacrest. 10v. mar 16 1920-jun 5 1929 (daily) [mf ed lacks jul 19-20 1922 and sun iss] – 42r – 1 – (cont: evening state journal. cont by: evening state journal (1929)) – us NE Hist [071]

[The evening state journal] lincoln daily news – Lincoln, NE: C D Traphagen, J C Seacrest, Estate of A H Mendenhall, Estate of C H Gere. 4v. feb 21 1916-oct 10 1919 (daily) [mf ed lacks sun iss] – 13r – 1 – (cont: lincoln daily news (the evening state journal); cont by: evening state journal; publ as: evening state journal and lincoln daily news oct 1-dec 7 1917 and: evening state journal (lincoln daily news) dec 8 1917-oct 10 1919) – us NE Hist [071]

Evening statesman – Walla Walla, WA. 1881-1910 (1) – mf#67168 – us UMI ProQuest [071]

Evening sun – Baltimore, MD. 1910-1995 (1) – mf#60490 – us UMI ProQuest [071]

Evening sun – Hamilton, OH. 1902-1906 (1) – mf#65521 – us UMI ProQuest [071]

Evening sun – Woonsocket, RI. 1899-1899 (1) – mf#66442 – us UMI ProQuest [071]

Evening tablet – Dublin, Ireland. 1 jul-12 jul 1850 – 1/2r – 1 – uk British Libr Newspaper [072]

Evening telegram – Portland, OR: [s.n.]. v19 n76-42nd yr n197. jul 15 1886-nov 30 1918; feb 2 1878-jan 31 1881 – (= ser Newspapers in microform, 1948-1983) – 1 – (ceased with nov 30 1918 iss; cont by: portland telegram) – us Oregon Hist [071]

Evening telegram – Portland OR: Evening Telegram Pub Co, 1877- [wkly] – 1 – (cont by: daily evening telegram (1878-). ceased in 1878?) – us Oregon Lib [071]

Evening telegram – Portland OR: H M CLinton, -1918 [daily ex sun] – 1 – (cont: daily evening telegram (portland, or); cont by: portland telegram) – us Oregon Lib [071]

Evening telegram – Sheboygan WI. 1887 sep 6/1888 feb 12-1896 sep 15/sep 19 – 21r – 1 – (cont: morning telegram [sheboygan, wi]; cont by: sheboygan telegram) – mf#953393 – us WHS [071]

Evening telegram – Superior WI. 1929 aug 30, 1954 oct 15, 1976 nov 24, 1977 may 31, 1984 jan 7, 1922 nov 21/1923 jan 3-1993 sep [gaps] – 546r – 1 – (cont: superior telegram) – mf#1153349 – us WHS [071]

Evening telegram – Superior WI. 1890 may 7/dec 31-1904 dec 14/1905 jan 28 [gaps] – 60r – 1 – (continued by: superior telegram) – mf#1150883 – us WHS [071]

Evening telegram – Elyria, OH. 1907-1919 (1) – mf#65480 – us UMI ProQuest [071]

Evening telegram – Green Bay WI. 1908 oct 27-dec 4 – 21r – 1 – mf#920583 – us WHS [071]

Evening telegram : local, general, shipping, commercial & telegraphic news – [Wales] LLGC 1 aug 1870-27 nov 1891 [mf ed 2003] – 59r – 1 – (missing: 1875, jul-dec 1888; cont as: south wales evening telegram [sep 1872-aug 1876]; south wales daily telegram [sep 1876-dec 1885]; south wales (daily) telegram [jan 1886-mar 1891]; south wales evening telegraph [jul 1891-nov 1891]) – uk Newsplan [072]

Evening telegram / Lorain Co. Elyria – jan 1916-jun 1919// [daily] – 9r – 1 – mf#B33030-33038 – us Ohio Hist [071]

Evening telegram – Providence, RI. 1880-1906 (1) – mf#66301 – us UMI ProQuest [071]

Evening telegram – St John, NF: W J Herder, 1879-1909 – 58r – 1 – ISSN: 0839-4199 – cn Library Assoc [071]

Evening telegram – St Johns, Canada. apr 1910-jun 1922 – 71 1/2r – 1 – uk British Libr Newspaper [071]

Evening telegram – Victoria, British Columbia, CN. jul-nov 1866 – 1r – 1 – cn Commonwealth Imaging [071]

The evening telegram – Kearney, NE: Telegram Publishing, 1892 (daily) [mf ed 1st yr n59. dec 2 1892 filmed [1973]] – 1r – 1 – us NE Hist [071]

Evening telegram series / Hamilton Co. Cincinnati – oct 19 1885-feb 7 1889 – 2r – 1 – mf#B37402-37403 – us Ohio Hist [071]

Evening telegraph – 1986-aug 1987; 1988-aug 15 1992; aug 17 1992-jun 30 1997 – 163 1/2r – 1 – (aka: northamptonshire evening telegraph; kettering evening telegraph) – uk British Libr Newspaper [072]

Evening telegraph – North Platte, NE: [Kelly & White] jun 1900-v57 n9. jan 11 1936 (daily ex sun) – 34r – 1 – (cont: daily telegraph (north platte ne); cont by: north platte daily telegraph (1936); iss for may 5 1908-mar 11 1925 called v1 n2-v17 n59; iss for mar 12 1925-jan 11 1936 called v47 n60-v57 n9 to coincide with vol number of north platte telegraph (1904)) – us Bell [071]

Evening telegraph / Crawford Co. Bucyrus – v1 n1. (oct 1887-may 1923) [daily] – 60r – 1 – mf#B51-110 – us Ohio Hist [071]

Evening telegraph – Dublin, 4 oct-31 dec 1884 – 1r – 1 – ie National [072]

Evening telegraph – Dublin, Ireland. 1 jul 1871-5 nov 1873; 30 aug 1875-1896; 1901; jan-jun 1923 – 76 3/4r – 1 – uk British Libr Newspaper [072]

Evening telegraph – Dundee, Scotland. 1890 – 2r – 1 – uk British Libr Newspaper [072]

Evening telegraph – [East Midlands] Northamptonshire jul 1898-dec 1950 [mf ed 2004] – 191r – 1 – (missing: 1900, 1911; cont by: northamptonshire evening telegraph [may 1904-dec 1950]) – uk Newsplan [072]

Evening telegraph – Pittsburgh, PA. 1847-1880 (1) – mf#66036 – us UMI ProQuest [071]

Evening telegraph – Pittsburgh, PA. 1873-1883 (1) – mf#66037 – us UMI ProQuest [071]

Evening telegraph *see*
- Evening freeman
- [Grass valley-] foothill weekly tidings

The evening telegraph – North Platte, NE: [Kelly & White] jun 1900-v57 n9. jan 11 1936 (daily ex sun) [mf ed 1900-06 (gaps)] – 3r – 1 – (cont: daily telegraph. cont by: north platte daily telegraph (1936). issues for may 5 1908-mar 11 1925 called v1 n2-v17 n59. iss for mar 12 1925-jan 11 1936 called v47 n60-v57 n9 to coincide with vol number of north platte telegraph (1904)) – us NE Hist [071]

Evening telegraph and post (dundee, scotland) – Dundee: D C Thomson & Co Ltd 1905- (daily ex sun) [mf ed 2 jan 1997-] – 1 – (formed by union of: evening telegraph (dundee, scotland) and: evening post (dundee, scotland)) – ISSN: 0262-3048 – uk Scotland NatLib [072]

Evening telegraph and star (yorkshire telegraph and star) – Sheffield, England. 1890-1920 [daily] – 76mqn r – 1 – (missing jan-apr 1898, 1911) – uk British Libr Newspaper [072]

Evening times – Akron, OH. 1921-1925 (1) – mf#65362 – us UMI ProQuest [071]

Evening times – Appleton WI. 1874 aug 13, dec 30-1875 mar 5 – 1r – 1 – mf#958907 – us WHS [071]

Evening times / Belmont Co. Martins Ferry – aug 1891-dec 1892 [daily] – 2r – 1 – mf#B2656-2657 – us Ohio Hist [071]

Evening times – Manitowoc WI. 1931 apr-1931 may, 1931 jun-1931 jul, 1931 aug-1931 sep, 1931 oct-1931 nov, 1931 dec-1932 jan, 1932 feb-1932 apr – 6r – 1 – (cont: manitowoc times; cont by: manitowoc herald-news; manitowoc herald-times) – mf#1107472 – us WHS [071]

Evening times – Washington, DC. 1896 apr 1-dec 31, 1898 oct 5 – 1r – 1 – (continued by: washington times) – mf#884634 – us WHS [071]

Evening times / Crawford Co. Bucyrus – v1 n1. may 10-nov 1 1844// [daily] – 1r – 1 – mf#B29842 – us Ohio Hist [071]

Evening times : (final edition) – Trenton, NJ. 1883-1959 (1) – mf#60520 – us UMI ProQuest [071]

Evening times – Glasgow, Scotland, UK. 1882. -d. 2 reels – 1 – uk British Libr Newspaper [072]

Evening times – Hamilton, Canada. 7 sep 1873-1886; 17 sep 1890 – 40r – 1 – uk British Libr Newspaper [071]

Evening times – Kansas City, MO. 1890-1891 (1) – mf#68552 – us UMI ProQuest [071]

Evening times – Grand Island, NE: W H Weckers. ns: v1 n240. dec 8 1883 (daily ex sun) [mf ed jun 11-dec 8 1883 filmed 1996] – 1r – 1 – (lacks: nov 10. cont: daily democrat. weekly ed: weekly news) – us NE Hist [071]

Evening times – Newport News, VA. 1900-1901 (1) – mf#66769 – us UMI ProQuest [071]

Evening times – [Scotland] Glasgow: G Outram 5 jun 1876-78, 1895-dec 1950 (daily ex sun) [mf ed 2003] – 313r – 1 – (not publ: nov 2 1975; incl suppls; cont as: evening times (glasgow) [may 1908-dec 1928]; evening times [jan 1929-dec 1950]; other title: evening times. weekend times mar 31 2001-) – uk Newsplan [072]

Evening times – Pawtucket, RI. 1886-1919 (1) – mf#66246 – us UMI ProQuest [071]

Evening times – Rochester, NY. 1888-1917 (1) – mf#65191 – us UMI ProQuest [071]

Evening times – Sayre, PA. 1993-2000 (1) – mf#66071 – us UMI ProQuest [071]

Evening times / Summit Co. Akron – (nov 1916-apr 1917), jun 1918-20 [daily] – 20r – 1 – mf#B351-370 – us Ohio Hist [071]

Evening times – Warren, PA. 1901-1928 (1) – mf#66119 – us UMI ProQuest [071]

Evening times – West Memphis, AR. 1957-2000 (1) – mf#61216 – us UMI ProQuest [071]

Evening times – West Palm Beach, FL. 1929-1987 (1) – mf#61292 – us UMI ProQuest [071]

The evening times – New York. N.Y. Die abend zeitung. 1906 – 1 – us AJPC [071]

The evening times – Sayre, PA., 1891-1985 – 13 – $25.00r – us IMR [071]

Evening times star *see* [Alameda-] alameda times star

The evening times-globe – St John, New Brunswick, CN. 1970- – 12r per yr – 1 – cn Commonwealth Imaging [071]

Evening times=record – Valley City, ND: F E Packard, jul 2 1906; -v8 n79 feb 27 1915 (daily ex sun – 1 – (official paper of valley city and barnes county 1910-1914; official paper of barnes county 1914; other ed available: valley city times=record (valley city, nd: 1902); valley city times=record and valley city alliance, and: valley city times=record (valley city, nd: 1908); cont by: daily times-record (valley city, nd); missing: 1909 dec 14-28; 1910 jan-feb 10) – mf#11453; 04824; 03521-03526 – us North Dakota [071]

Evening transcript – Boston, MA. 1848-1915 (1) – mf#61196 – us UMI ProQuest [071]

Evening tribune – Beaver Falls, PA. 1915-1928 (1) – mf#65839 – us UMI ProQuest [071]

Evening tribune – Chattanooga, TN. 1940-1940 (1) – mf#66529 – us UMI ProQuest [071]

Evening tribune – Hornell, NY. 1906-1960 (1) – mf#61631 – us UMI ProQuest [071]

Evening tribune – Hornellsville, NY. 1892-1899 (1) – mf#65002 – us UMI ProQuest [071]

Evening tribune / Lawrence Co. Ironton – jan 1926-jun 1960 [daily] – 139r – 1 – mf#B2301-2439 – us Ohio Hist [071]

Evening tribune / Lawrence Co. Ironton – oct-nov 1949 (gap fillers) [daily] – 1r – 1 – mf#B25678 – us Ohio Hist [071]

Evening tribune – Pawtucket, RI. 1888-1900 (1) – mf#66247 – us UMI ProQuest [071]

Evening tribune – Providence, RI. 1906-1929 (1) – mf#66302 – us UMI ProQuest [071]

Evening tribune / Union Co. Marysville – jan 1946-feb 1951 [daily] – 10r – 1 – mf#B12795-12804 – us Ohio Hist [071]

Evening tribune / Union Co. Marysville – jul 1919-apr 1920, mar 1936-45 [daily] – 16r – 1 – mf#B10022-10037 – us Ohio Hist [071]

Evening tribune / Union Co. Marysville – (oct 1898-jun 1936) [daily] – 46r – 1 – mf#B34677-34722 – us Ohio Hist [071]

Evening tribune nuneaton *see* Midland daily tribune

Evening true delta – New Orleans LA. 1862 may 27, jun 5, jul 30 – 1r – 1 – mf#780685 – us WHS [071]

Evening union – Atlantic City, NJ. 1905-1953 (1) – mf#64790 – us UMI ProQuest [071]

Evening union – Newburgh, NY. 1908-1912 (1) – mf#65114 – us UMI ProQuest [071]

Evening virginia sentinel – Alexandria VA. 1860 mar 28, apr 5, aug 11, 1861 mar 1-7, 12-15, apr 2-3, 11-17, may 17-22 – 1r – 1 – mf#881680 – us WHS [071]

Evening volunteer – Carlisle, PA., 1901-1903 – 13 – $25.00r – us IMR [071]

Evening whirl – Saint Louis MO. 1993 jan 12-dec 28, 1994 jan 4-dec 27, 1995 jan 3-nov 28, 1996 jan-dec 31, 1997 jan 7-dec 30/jan 6 – 5r – 1 – (cont: st louis whirl-examiner) – mf#2652378 – us WHS [071]

Evening world – Roanoke, VA. 1911-1913 (1) – mf#66855 – us UMI ProQuest [071]

Evening world-herald – Omaha, NE: [World Pub Co] apr 15 1906-28// (daily ex sun) [mf ed 1918] – 1r – 1 – (cont: omaha world-herald (1890 evening ed). cont by: omaha world-herald (1928 evening ed). morning ed: morning world-herald (1890); weekly ed: weekly world-herald -1899; sunday ed: sunday world-herald (1906)) – us NE Hist [071]

Evenings in the library : bits of gossip about books and those who write them / Stewart, George – St John, NB: R A H Morrow, 1878 – 3mf – 9 – mf#13973 – cn Canadiana [410]

Evenings on a farm near dikanka / Gogol, Nikolai Vasiltevich – New York, NY. 1926 – 1r – us UF Libraries [025]

Evenings with the bible / Errett, Isaac – Cincinnati: Standard Pub Co, 1884-1889 – 3mf – 9 – 0-524-06284-6 – mf#1992-0865 – us ATLA [220]

Evenings with the romanists : with an introductory chapter on the moral results of the romish system / Seymour, Michael Hobart – New York: Robert Carter, 1856 – 2mf – 9 – 0-8370-8944-1 – mf#1986-2944 – us ATLA [240]

Evenings with the skeptics : or, free discussion on free thinkers / Owen, John – New York: J W Bouton, 1881 – 3mf – 9 – 0-7905-7127-7 – (incl bibl ref) – mf#1988-3127 – us ATLA [140]

Evenkijskaia novaia zhizn' – Krasnoyarsk, 1973 – 4r – 1 – us UMI ProQuest [077]

L'evenment : journal quotidien – Paris. 1er nov 1865-15 nov 1866 – 1 – (avec un no spec du 2 oct 1865) – fr ACRPP [074]

L'evenment – Paris. Quot. 1re ed. 2 janv-30 juin 1851 – 1 – fr ACRPP [944]

L'evenment – Paris. Quot. Ed. du soir. 1er sept-30 nov 1850 – 1 – fr ACRPP [074]

L'evenment – Paris, 30 juil 1848-18 sep 1851 – 1 – (suivi de: l' avenement du peuple voir a ce titre) – fr ACRPP [074]

L'eventail – Revue de litterature et d'art. Dir. Francois Laya. Geneve. 1918-oct 1919 – 1 – fr ACRPP [800]

The eventful history of the mutiny and piratical seizure of h m s bounty : its cause and consequences / Barrow, John – London 1831 [mf ed Hildesheim 1995-98] – 1v on 3mf [ill] – 9 – €90.00 – 3-487-26775-6 – gw Olms [355]

An eventful year in north china : a survey of the work of the north china mission of the american board, edited from the reports...for the eight months ending december 31st, 1914 / American Board of Commissioners for Foreign Missions. North China Mission – [s.l: s.n, s.n, 1915?] [mf ed 1995] – (= ser Yale coll) – 57p/4p (ill) – 1 – 0-524-09825-5 – (with foreword) – mf#1995-0825 – us ATLA [951]

Events – [NW England] Liverpool 19 may-4 jun 1855 – 1 – uk MLA; uk Newsplan [072]

Events – Ottawa: "Events" Pub. Co, [1898-19–] – 9 – mf#P04280 – cn Canadiana [071]

Events and epochs in religious history : being the substance of a course of twelve lectures delivered in the lowell institute, boston, in 1880 / Clarke, James Freeman – Boston: J R Osgood, 1881 – 1mf – 9 – 0-7905-4388-5 – mf#1988-0388 – us ATLA [240]

Events and epochs in religious history : being the substance of a course of twelve lectures...lowell institute, boston, in 1880 / Clarke, James Freeman – Boston: J R Osgood 1881 [mf ed 1990] – 1r – 1 – (filmed with: economic survey of argentina / general motors corporation) – mf#7438 – us Wisconsin U Libr [240]

Events at the court of ranjit singh, 1810-1817 : translated from the papers in the alienation office, poona / Garrett, Herbert Leonard Offley – Punjab Govt Record Office Publ, 1935 – 1 – us Wisconsin U Libr [954]

Events in african history / Smith, Edwin William – New York, NY. 1942 – 1r – us UF Libraries [960]

Events in indian history : beginning with an account of the origin of the american indians and early settlements in north america and embracing concise biographies of the principal chiefs and head-sachems of the different indian tribes – Lancaster PA: G Hills, 1841 – 7mf – 9 – mf#54625 – cn Canadiana [305]

Events in the life of s c – London, England. 18– – 1r – us UF Libraries [240]

Ever Hadani *see* Tserif ha-'ets

'Ever ha-yarden ha-yehudi / Klein, Samuel – Vienna, Austria. 1924/25 – 1r – 1 – us UF Libraries [939]

Ever increasing faith messenger / Crenshaw Christian Center – v10 n1-v16 n1 [1989 win-1995 win] – 1r – 1 – mf#2679751 – us WHS [240]

Everest : the challenge / Younghusband, Francis Edward – London; New York: Thomas Nelson & Sons, 1941 – (= ser Samp: indian books) – us CRL [790]

Everest : the unfinished adventure / Ruttledge, Hugh – [London]: Hodder & Stoughton, 1937 – (= ser Samp: indian books) – us CRL [900]

Everest, Harvey William *see* The divine demonstration

Everest, Robert *see*
- A comparison between the rate of wages in some of the british colonies and in the united states; with observations thereupon
- A journey through norway, lapland, and part of sweden

Everett, Alexander H *see* America

Everett, Charles Carroll *see*
- Essays
- Ethics for young people
- Fichte's science of knowledge
- The gospel of paul
- Immortality, and other essays
- Poetry, comedy and duty
- The psychological elements of religious faith
- The science of thought
- Theism and the christian faith

Everett, Craig A *see* Journal of divorce and remarriage

Everett, Edward *see*
- A defence of christianity
- The edward everett papers, 1675-1930

Everett press – Everett, PA., 1881-1983 – 13 – $25.00r – us IMR [071]

Everett press and ledger – Everett, PA., 1892-1893 – 13 – $25.00r – us IMR [071]

Everett press/leader – Pennsylvania. -w 1892-1893 – 13 – $25.00r – us IMR [071]

Everett Public Library [WA] *see* Journal of everett and snohomish county history

Everglades – Clewiston, FL. 1944 – 1r – us UF Libraries [630]

Everglades / United States Sugar Corporation – Clewiston, FL. 1944 – 1r – us UF Libraries [630]

Everglades Engineering Board Of Review *see* Report of everglades engineering board of review

Everglades flood control / Hilsheimer, Ruth – s.l, s.l? 1937 – 1r – us UF Libraries [627]

EVERGLADES

Everglades National Park Commission Committee On... *see* Report to everglades national park commission rela...
Everglades news – Canal Point, FL. 1924 dec 19-1967 oct 5 – 33r – (gaps) – us UF Libraries [071]
Everglades observer – Pahokee, FL. 1968-1978 – 10r – us UF Libraries [071]
Everglades of florida : class 100 : file n120 / Lyons, Isabel J – s.l, s.l? 1936 – 1r – 1 – us UF Libraries [639]
Everglades of florida : class 100 : file n160 / Lyons, Isabel J – s.l, s.l? 1936 – 1r – 1 – us UF Libraries [639]
Everglades of florida : class 100 : file n165 / Lyons, Isabel J – s.l, s.l? 1936 – 1r – 1 – us UF Libraries [639]
Everglades of florida – Washington, DC. 1911 – 1r – 1 – us UF Libraries [639]
Everglades of florida / Wright, Jo – Tallahassee, FL. 1912 – 1r – 1 – us UF Libraries [639]
Everglades sugar institute, 1940 – Clewiston, FL. 1940 – 1r – us UF Libraries [630]
Evergreen – London. 1895-1897 [1] – mf#5181 – us UMI ProQuest [420]
Evergreen : a monthly magazine of new and popular tales and poetry – New York. 1840-1841 (1) – mf#4448 – us UMI ProQuest [800]
Evergreen baptist church. florence, south carolina : church records – 1891-1912, 1941-72 – 1 – us Southern Baptist [242]
Evergreen baptist church. frankfort, kentucky : church records – 1884-1977 – 1 – us Southern Baptist [242]
Evergreen review – New York. 1957-1984 (1) 1971-1984 (5) 1984-1984 (9) – ISSN: 0014-3758 – mf#1806 – us UMI ProQuest [073]
EvergreenCity times – Sheboygan WI. 1854 mar 24-1857 aug 31, 1857 sep-1860 sep 15, 1860 sep 22-1863 dec 31, 1864 jan-1866 feb 3, 1866 feb 10-1869 mar 6, 1869 mar 13-1870 apr 30 – 6r – 1 – (continued by: sheboygan times) – mf#930347 – us WHS [071]
Everhart and Co *see* History of muskingum county
Everitt, Graham *see* English caricaturists and graphic humourists of the nineteenth century
The everlasting arms / Clark, Francis Edward – New York, Boston: T Y Crowell, c1898 – 1mf – 9 – mf#27442 – cn Canadiana [210]
Everlasting gospel / Campbell, John Mcleod – Greenock, Scotland. 1830 – 1r – us UF Libraries [220]
Everlasting punishment / Foreign Chaplain – London, England. 1873 – 1r – us UF Libraries [240]
Everlasting punishment : lectures. delivered at st james's church, piccadilly... / Goulburn, Edward Meyrick – 2nd rev enl ed. London: Rivingtons, 1881 – 1mf – 9 – 0-524-06182-3 – (incl bibl ref) – mf#1991-2438 – us ATLA [240]
Everlasting punishment / Paget, Francis – London, England. 1886 – 1r – us UF Libraries [240]
Everlasting punishment and modern speculation / Reid, William – Edinburgh: W Oliphant, 1874 – 1mf – 9 – 0-7905-9080-8 – (Incl bibl ref) – mf#1989-2305 – us ATLA [240]
Everling, Otto *see* Die paulinische angelologie und daemonologie
Evers, Ernst *see* Christian jensen
Evers, Ernst August *see* Ueber die schulbildung zur bestialitaet
Eversden, great 1538-1992 – (= ser Cambridgeshire parish register transcript) – 4mf – 9 – £5.00 – uk CambsFHS [929]
Eversden, little 1599-1950 – (= ser Cambridgeshire parish register transcript) – 3mf – 9 – £3.75 – uk CambsFHS [929]
Eversholt – (= ser Bedfordshire parish register series) – 2mf – 9 – £5.00 – uk BedsFHS [929]
Eversley, George John Shaw-Lefevre, Baron *see* Freedom of land
Eversmann, Eduard *see* Reise von orenburg nach buchara
Everton – (= ser Bedfordshire parish register series) – 2mf – 9 – £5.00 – uk BedsFHS [929]
Everton's family history magazine – Logan. 2002+ (1,5,9) – ISSN: 1539-1531 – mf#6817,02 – us UMI ProQuest [929]
Everton's genealogical helper – Logan. 1992+ (1) 1992+ (5) 1992+ (9) – (Cont: Genealogical helper) – mf#6817,01 – us UMI ProQuest [929]
Everts and Co *see* Muskingum county atlas-map
Everts, William Wallace *see*
– Baptist layman's book
– The christian apostolate
– William colgate, the christian layman
The every day of life / Miller, James Russell – New York: Thomas Y Crowell, c1892 – 1mf – 9 – 0-8370-7180-1 – mf#1986-1180 – us ATLA [240]
Every day with beginners – 1956-61 – 1 – us Southern Baptist [242]
Every day with primaries – 1956-61 – 1 – us Southern Baptist [242]
Every, E F *see* The anglican church in south america
Every evening – Wilmington, DE. 1871-1932 (1) – mf#62382 – us UMI ProQuest [071]
Every friday – Cincinnati, OH. 1958-65 – 1 – us AJPC [071]
Every gi is a p.o.w. / Fort Ord (CA) – n1-5 [1971 mar-aug], 1972 may – 1r – 1 – mf#917822 – us WHS [355]
Every inch a king / Costa, Sergio Correa Da – New York, NY. 1950 – 1r – us UF Libraries [972]
Every landlord or tenant his own lawyer / Paul, John – London: Strahan and Woodfall, 1775. 144p. LL-717 – 1 – us L of C Photodup [340]
Every man his own art critic at the manchester exhibition, 1887 / Geddes, Patrick – Manchester 1887 – (= ser 19th c art & architecture) – 1mf – 9 – mf#4.2.1385 – uk Chadwyck [700]
Every man's duty to be a teetotaller proved / Dean, John – SANDBACH, England. 1841? – 1r – us UF Libraries [240]
Every other sunday – Boston: Unitarian Sunday-School Society, 1885-1910 [mf ed 2001] – (= ser Christianity's encounter with world religions, 1850-1950) – 2r – 1 – (unitarian paper for the sunday school and home) – mf#2001-s146 – us ATLA [243]
Every saturday : a journal of choice reading – Boston. 1866-1874 (1) – mf#4591 – us UMI ProQuest [321]
Every saturday – Monroe WI. 1879 oct 4 – 1r – 1 – mf#1098902 – us WHS [071]
Every saturday – Ottawa: [Mason and Reynolds, 1886] – 9 – mf#P04257 – cn Canadiana [073]
Every week – Angelica, NY. 1882-1884 (1) – mf#64888 – us UMI ProQuest [071]
Every youth's gazette : a semi-monthly journal devoted to the amusement, instruction, and moral culture of the young – New York. 1842-1842 (1) – mf#4371 – us UMI ProQuest [305]
Everybody – 1974 mar, 1976 jan, sep, nov-1977 jan, 1979 may-1980 mar – 2r – 1 – mf#367505 – us WHS [071]
Everybody's album : a humorous collection of tales, quips, quirks, anecdotes, and facetie – Philadelphia. 1836-1837 (1) – mf#5322 – us UMI ProQuest [880]
Everybody's law book / Shirley, John L – New York?: Lupton? 1886 – iv/124p – 1 – mf#LL-214 – us L of C Photodup [340]
Everybody's law book; legal rights and legal remedies. / Koones, John Alexander – New York: Hitchcock, 1893 – 768p – 1 – mf#LL-358 – us L of C Photodup [340]
Everybody's lawyer and counselor in business... / Crosby, Frank – Philadelphia: John E Potter, 1860 – 4mf – 9 – $6.00 – mf#LLMC 91-083 – us LLMC [346]
Everybody's magazine – New York. 1899-1929 – 1 – mf#2887 – us UMI ProQuest [073]
Everybody's money / Credit Union National Association – 1976 spr-1985 win – 1r – 1 – mf#969336 – us WHS [332]
Everybody's money : us edition – Madison. 1961-1996 (1) 1972-1996 (5) 1976-1996 (9) – ISSN: 0423-8710 – mf#6772 – us UMI ProQuest [332]
Everybody's money canadian edition – Madison. 1966-1980 (1) 1973-1980 (5) 1973-1980 (9) – ISSN: 0046-287X – mf#6965 – us UMI ProQuest [332]
Everybody's monthly *see* Irish temperance league journal
Everybody's schools / Wisconsin Teachers' Association – v1 n2-v2 n37 [1921 mar 5-1922 sep 10] – 1r – 1 – mf#1056088 – us WHS [370]
Every-day / Clarke, James Freeman – Boston: Houghton Mifflin, 1900 – 2mf – 9 – 0-524-08269-3 – mf#1993-3024 – us ATLA [240]
Everyday – Stockholm, Sweden. 2000-01 – 6r – 1 – sw Kungliga [078]
Every-day evangelism : personal, trained, co-operative / Leete, Frederick DeLand – Cincinnati: Jennings and Graham; New York: Eaton and Mains, c1909 – 1mf – 9 – 0-7905-5370-8 – mf#1988-1370 – us ATLA [242]
Everyday life / Western Magazine Co – v11 n2 [1916 nov] – 1r – 1 – mf#940198 – us WHS [640]
Everyday life in ancient india / Sengupta, Padmini Sathianadhan – London; New York: Oxford University Press, 1950 – (= ser Samp: indian books) – us CRL [930]
Everyday life in bengal : and other indian sketches / Hart, William Henry – London: Charles H Kelly [1906] [mf ed 1995] – (= ser Yale coll) – xvi/343p (ill) – 1 – 0-524-09761-5 – mf#1995-0761 – us ATLA [306]
Every-day life in south africa / Lowndes, E E K – London: S W Patridge, 1900 – 1 – us CRL [300]
Everyday life in the old stone age / Quennell, Marjorie (Courtney) – New York, London: G P Putnam's Sons, 1922 – (= ser The everyday life series) – xxii/201p/pl – 1 – us Wisconsin U Libr [573]
Everyday life of the aztecs / Bray, Warwick – London, England. 1968 – 1r – us UF Libraries [972]
Every-day religion : or, the common sense teaching of the bible / Smith, Hannah Whitall – New York: Fleming H Revell, c1893 – (= ser Women & the church in america) – 1mf – 9 – 0-8370-1431-X – mf#1984-2199 – us ATLA [220]
Every-day religion : sermons delivered in the brooklyn tabernacle / Talmage, Thomas De Witt – New York: Funk & Wagnalls, 1886 [mf ed 1993] – 1mf – 9 – 0-524-08621-4 – mf#1993-1071 – us ATLA [242]
Everyday sesotho grammar / Sharpe, M R L – Morija, Zimbabwe. 1952 – 1r – us UF Libraries [470]
Everyday sesotho reader / Sharpe, M R L – Morija, Zimbabwe. 1952 – 1r – us UF Libraries [470]
Everyday soldier life : or a history of the 113th ohio volunteer infantry / McAdams, F M – Columbus, OH: Chas M Cott & Co, 1884 – 1r – 1 – us Western Res [976]
Everyday tsonga / Ouwehand, Mariette – Johannesburg, South Africa. 1965 – 1r – us UF Libraries [470]
Everyman's history of the english church / Dearmer, Percy – London: A R Mowbray, 1909 – 1mf – 9 – 0-524-03143-6 – (incl bibl ref) – mf#1990-4592 – us ATLA [240]
Everyman's history of the prayer book / Dearmer, Percy – London: A R Mowbray, 1912 – 1mf – 9 – 0-524-02817-6 – mf#1990-4438 – us ATLA [240]
Everyman's lawyer... : being a new edition...of the home library of law / Bolles, Albert Sidney – New York: Doubleday, Page, 1908 – 3v – 1 – mf#LL-1069 – us L of C Photodup [340]
Everyone in this house makes babies / Klass, Sheila Solomon – Garden City, NY. 1964 – 1r – us UF Libraries [972]
Everyweek – Middletown. 1965-1969 (1) – mf#1856 – us UMI ProQuest [400]
Everywoman – Los Angeles. 1970-1972 (1) – ISSN: 0014-3766 – mf#7996 – us UMI ProQuest [320]
Everywoman's world : canada's greatest magazine – Toronto: Continental Pub Co, 1914-15, 1919-21 – 2r – 1 – Can$162.00 – cn McLaren [640]
Evesham friends in the olden time : a history of evesham monthly meeting of the society of friends, with notes on worcestershire quarterly meeting, and the circular yearly meetings for the seven western counties / Brown, Alfred W – London: West, Newman, 1885 – 1mf – 9 – 0-524-07557-3 – mf#1991-3177 – us ATLA [240]
Evesham journal, and general advertiser for pershore, alcester – Redditch, England 21 jul 1860-27 jun 1885 [mf 1860-96, 1898-1910, 1912-50, 1986- – 1 – (cont by: evesham journal & four shires advertiser [4 jul 1885-31 dec 1910, 3 feb 1912-31 may 1984]) – uk British Libr Newspaper [072]
Evesham journal & four shires advertiser – [West Midlands] Evesham, England 1860-96, 1898-1910, 1912-50, 1986- [mf ed 2004] – 1 – (cont: evesham journal, & general advertiser for pershore, alcester, etc [21 jul 1860-27 jun 1885]; cont by: evesham journal [7 jun 1984-]) – uk Newsplan; uk British Libr Newspaper [072]
Evetts, B T A *see*
– The churches and monasteries of egypt and some neighbouring countries
Evetts, Basil Thomas Alfred *see* New light on the bible and the holy land
Evgenev-Maksimov, V *see* Iz proshlogo russkoi zhurnalistiki
Evgenii, Mitropolit *see*
– Slovar istoricheskii o byvshikh v rossii pisateliakh dukhovnogo china greko-rossiiskoi tserkvi
– Slovar istoricheskii o byvshikh v rossii pisateliakh dukhovnogo china,greko-rossiiskoi tserkvi
– Slovar russkikh svetskikh pisatelei, sootechestvennikov i chuzhestrantsev, pisavshikh v rossii
Evian conference : statistical tables on the distribution, migration and natural increase of the jews in the world, with special reference to jewish activities in palestine – Jerusalem, 1938 – 1mf – 9 – mf#J-28-138 – ne IDC [956]
Evidence / Natal Colony. Native Affairs Commission – Pietermaritzburg: P Davis & Sons, Govt Printers, 1907 – 1 – us CRL [960]
Evidence : or, religious and moral gazette – Catskill. 1807-1808 (1) – mf#3575 – us UMI ProQuest [240]
Evidence and cross examination of william d haywood in the case of the usa vs wm d haywood, et al – Chicago: General Defense Committee, 1918? – 312p – 1 – mf#LL-308 – us L of C Photodup [347]
Evidence and practice at trials in civil cases / Kingsford, Rupert Etherege – Toronto: Carswell Co, 1911 [mf ed 1995] – 8mf – 9 – 0-665-74761-6 – (incl: references to cases reported since 3rd ed [1908]) – mf#74761 – cn Canadiana [347]
Evidence and report / East African Protectorate. Native Labour Commission, 1912-13 – Nairobi: [s.n.], 1913 – 1 – us CRL [960]
Evidence as to man's place in nature / Huxley, Thomas Henry – [London], 1863 – (= ser 19th c evolution & creation) – 2mf – 9 – mf#1.1.4511 – uk Chadwyck [574]
Evidence, experience, influence / Doane, William Croswell – New York: ES Gorham, 1904 – (= ser The bedell lectures) – 1mf – 9 – 0-7905-7722-4 – mf#1989-0947 – us ATLA [210]
Evidence for a future life = L'ame est immortelle / Delanne, Gabriel; ed by Dallas, Helen Alexandrina – New York: Putnam, 1904 [mf ed 1991] – xvi/264p on 1mf – 9 – 0-524-00829-9 – (trans fr french into english by ed) – mf#1990-2075 – us ATLA [130]
Evidence given – London, England. 1832 – 1r – us UF Libraries [240]
Evidence given – London, England. 1832 – 1r – us UF Libraries [240]
Evidence given – London, England. 1932 – 1r – us UF Libraries [240]
Evidence given before the select committee of the house of commons / Chalmers, Thomas – Edinburgh, Scotland. 1847? – 1r – us UF Libraries [240]
The evidence in the case : an analysis of the diplomatic records submitted by england, germany, russia and belgium in the supreme court of civilization, and the conclusions deducible as to the moral responsibility for the war / Beck, James M – N.Y. & London: G P Putnam's Sons, 1914 – 3mf – 9 – $4.50 – mf#LLMC 92-118 – us LLMC [347]
Evidence of biomechanical functional symmetry : in the presence of lower extremity structural asymmetry during running / McBride, Margaret E & Sanderson, David – 1989 – 2mf – 9 – $8.00 – us Kinesology [612]
The evidence of christian experience / Stearns, Lewis French – New York: Scribner, 1890 – (= ser Ely Lectures) – 2mf – 9 – 0-7905-9679-2 – mf#1989-1404 – us ATLA [240]
Evidence of george washington's religion / Gano, John – 1874-91 – 398p – 1 – us Southern Baptist [242]
The evidence of salvation : or, the direct witness of the spirit / Stackpole, Everett Schermerhorn – New York: Thomas Y Crowell, c1894 – 1mf – 9 – 0-8370-5512-1 – (incl bibl ref) – mf#1985-3512 – us ATLA [210]
Evidence of the christian religion briefly stated – Edinburgh, Scotland. 1796 – 1r – us UF Libraries [240]
Evidence of the laws of manu on the social conditions in india during the third century a d / Ketkar, Shridhar Venkatesh – Ithaca, NY: Taylor and Carpenter, 1909 – (= ser History of caste in india) – 1mf – 9 – 0-524-01185-0 – mf#1990-2261 – us ATLA [954]
The evidence of the motives and objects of the bushman wars, 1769-77. / Moodie, Donald – Cape Town. 1841 – 1 – us CRL [960]
Evidence of the primitive church as to the admission of the laity i... / Lea, John Walter – Aberdeen, Scotland. 1873 – 1r – us UF Libraries [240]
Evidence of the rev john lee / Lee, John – Edinburgh, Scotland. 1837 – 1r – us UF Libraries [240]
Evidence of the truth of the christian religion : derived from the literal fulfilment of prophecy / Keith, Alexander – 37th en ed. London, New York: T Nelson, 1859 [mf ed 1989] – 2mf – 9 – 0-7905-2783-9 – (incl bibl ref) – mf#1987-2783 – us ATLA [221]
Evidence of the truth of the christian religion derived from the li... / Keith, Alexander – Edinburgh, Scotland. 1838 – 1r – us UF Libraries [240]
Evidence on church patronage / Cook, John – Edinburgh, Scotland. 1838 – 1r – us UF Libraries [240]
Evidence on maps for the sources of rhodesian cultures and ndau – s.l, s.l? 19–? – 1r – us UF Libraries [960]
The evidence submitted to the presbytery of new york / Briggs, Charles Augustus – [S.l.: s.n., 1892?] – 1mf – 9 – (incl bibl ref) – mf#1990-4307 – us ATLA [242]
Evidence taken by the commission... / Rhodesia. Northern. Commission Appointed to Enquire into the Disturbances in the Copperbelt – Lusaka: Govt Printer, 1935 – 1 – us CRL [960]

Evidence taken by the commission appointed to enquire into the disturbances in the copperbelt, northern rhodesia, jul-sep 1935 – [Salisbury?: s.n., 193-?]. v1 – us CRL [960]

Evidence-based nursing – London. 1998+ (1,5,9) – ISSN: 1367-6539 – mf#31410 – us UMI ProQuest [610]

The evidences and doctrines of the catholic church : showing that the former are no less convincing than the latter are propitious to the happiness of society / MacHale, John – 3rd ed. Dublin: M H Gill, 1885 – 2mf – 9 – 0-8370-6999-8 – (incl bibl ref) – mf#1986-0999 – us ATLA [230]

Evidences of ancient civilization in america : constituting a lecture delivered on behalf of the mechanics' institute on tuesday, 1st of march, 1870 / Bessey, William E – Guelph: Mechanics' Institute of Guelph, 1870 [mf ed 1979] – 1mf – 9 – mf#00093 – cn Canadiana [930]

Evidences of christianity / Bergen, John Tallmadge – Holland, MI. Wm H Bingham, 1902 [mf ed 1985] – 1mf – 9 – 0-8370-2285-1 – (incl ind) – mf#1985-0285 – us ATLA [240]

Evidences of christianity / Hopkins, Mark – Boston, MA. 1890, c1880 – 1r – us UF Libraries [240]

Evidences of christianity : lectures before the lowell institute, jan 1844... / Hopkins, Mark – 15th ed. Boston: T R Marvin, 1881 [mf ed 1984] – (= ser Lowell institute lectures 1844) – 4mf – 9 – 0-8370-1051-9 – mf#1984-4422 – us ATLA [240]

Evidences of christianity / McGarvey, John William – Cincinnati: Guide Print & Pub Co, 1886-1891 – 1mf – 9 – 0-524-06489-X – mf#1991-2589 – us ATLA [220]

Evidences of christianity / Ragg, Lonsdale – 2nd ed. New York: Edwin S Gorham, 1909 [mf ed 1985] – (= ser Oxford church text books) – 1mf – 9 – 0-8370-4825-7 – (incl ind) – mf#1985-2825 – us ATLA [240]

Evidences of christianity : the supernatural book / Foster, Randolph Sinks – New York: Hunt & Eaton; Cincinnati: Cranston & Stowe 1889 [mf ed 1991] – (= ser Studies in theology) – 1mf – 9 – 0-7905-9273-8 – mf#1989-2498 – us ATLA [220]

Evidences of christianity *see* Die wahrheit des christenthums

The evidences of christianity : in their external divison / McIlvaine, Charles Pettit; ed by Gregory, Olinthus – London: James Blackwood, 1871 – 2mf – 9 – 0-8370-4355-7 – (incl bibl ref) – mf#1985-2355 – us ATLA [230]

The evidences of christianity : with an introduction on the existence of god and the immortality of the soul / Dodge, Ebenezer – Boston: Gould and Lincoln, 1869 – 1mf – 9 – 0-8370-2935-X – mf#1985-0935 – us ATLA [240]

Evidences of christianity briefly stated / Doddridge, Philip – London, England. 1792 – 1r – us UF Libraries [240]

The evidences of natural religion : and the truths established thereby / McArthur, Charles – London: Hodder and Stoughton, 1882 – 1mf – 9 – 0-7905-9501-X – (incl bibl ref) – mf#1989-1206 – us ATLA [240]

Evidences of revealed religion / Thomson, Edward – Cincinnati: Hitchcock & Walden; New York: Nelson & Phillips, 1872 [mf ed 1985] – 1mf – 9 – 0-8370-5523-7 – mf#1985-3523 – us ATLA [230]

Evidences of the authenticity, inspiration, and canonical authority of the holy scriptures / Alexander, Archibald Browning Drysdale – 5th ed. Philadelphia: Presbyterian Bd of Publ, c1836 [mf ed 1985] – 1mf – 9 – 0-8370-2060-3 – (expanded version of: a brief outline of the evidences of the christian religion) – mf#1985-0060 – us ATLA [220]

The evidences of the genuineness of the gospels / Norton, Andrews – 2d ed. Cambridge: John Owen, 1846-1848. Beltsville, Md: NCR Corp, 1978 (17mf); Evanston: American Theol Lib Assoc, 1984 (17mf) – (= ser Biblical crit – us & gb) – 9 – 0-8370-0710-0 – (incl bibl ref) – mf#1984-1030 – us ATLA [226]

Evidences of the human spirit / Vidyarthi, Gurudatta – Chicago ed. Lahore: Mufid-I-Am Press, 1893 – 1mf – 9 – 0-524-08903-5 – mf#1993-4038 – us ATLA [180]

Evidences of the work of the holy spirit / Salmon, George – Dublin, Ireland. 1859 – 1r – us UF Libraries [240]

The evidential value of the acts of the apostles / Howson, John Saul – New York: E P Dutton, 1880 – 1mf – 9 – 0-8370-3681-X – mf#1985-1681 – us ATLA [226]

Evidential value of the early epistles of st paul viewed as history / Lorimer, Peter – London, England. 1874 – 1r – us UF Libraries [227]

Evil and evolution : an attempt to turn the light of modern science on to the ancient mystery of evil / Millin, George Francis – 3rd ed. London, New York: Macmillan, 1899 – 1mf – 9 – 0-8370-6149-0 – mf#1986-0149 – us ATLA [230]

Evil and evolution : an attempt to turn the light of modern science on to the ancient mystery of evil / Millin, George Francis – London, 1896 – (= ser 19th c evolution & creation) – 3mf – 9 – mf#1.1.9365 – uk Chadwyck [210]

Evil consequences of substituting infant-sprinkling for believers... – London, England. 18– – 1r – us UF Libraries [240]

The evil eye : an account of this ancient and widespread superstition / Elworthy, Frederic Thomas – London: J Murray, 1895 – 2mf – 9 – 0-524-06687-6 – (incl bibl ref) – mf#1990-3548 – us ATLA [130]

Evil genius : the story of joseph goebbels / Ebermayer, Erich – London: A Wingate, (1953) – 245p (ill) – 1 – (trans by louis hagen) – us Wisconsin U Libr [934]

The evil of consenting to popery : a sermon / Jones, Hugh – Holywell: W Morris [1849?] [mf ed 1986] – 1mf – 9 – 0-8370-8118-1 – mf#1986-2118 – us ATLA [241]

Evils, constitutional and practical, of the prelatic establishment / Neilson, Thomas – Glasgow, Scotland. 1841 – 1r – us UF Libraries [240]

Evils of disestablishment / Pennant, P P – London, England. 1885 – 1r – us UF Libraries [240]

The evils of infant baptism / Howell, Robert Boyte Crawford – [2nd ed] Charleston, SC: Southern Baptist Publ Society, 1852, c1851 – 1mf – 9 – 0-7905-8662-2 – mf#1989-1887 – us ATLA [242]

Evinrude dock lines / Evinrude Motors – v27 n1-v32 n5 [1976 jan 15-1982 may] – 1r – 1 – mf#648862 – us WHS [380]

Evinrude Motors *see* Evinrude dock lines

Evjen, John Oluf *see*
– Lutheran germany and the book of concord
– Veiledning i den lutherske frikirkes principer

Evliya, Chelebi *see* Travels in europe, asia and africa

Evocacion de jose antonio ramos / Henriquez Urena, Max – Habana, Cuba. 1947 – 1r – 1 – us UF Libraries [972]

Evocacion de pindaro / Selva, Salomon De La – San Salvador, El Salvador. 1957 – 1r – us UF Libraries [972]

Evocacion de rivas / Barrios, Gilberto – Managua, Nicaragua. 1965 – 1r – us UF Libraries [972]

Evocacion de zande / Centeno Guell, Fernando – San Jose, Costa Rica. 1950 – 1r – us UF Libraries [972]

Evocaciones y reflexiones universitarias / Boza Masvidal, Aurelio A – Habana, Cuba. 1946 – 1r – us UF Libraries [378]

Une evocation ; conference faite a la salle de la patrie, jeudi, le 6 decembre 1883 / Buies, Arthur – [Quebec (Province)]: [s.n.], [1883?] [mf ed 1980] – 1mf – 9 – 0-665-03823-2 – mf#03823 – cn Canadiana [060]

Evoking leadership motivation among session members in a presbyterian congregation : the pastor's role / Martin, Charles Copeland – Princeton, New Jersey, 1976. Chicago: Dep of Photodup, U of Chicago Lib, 1976 (1r); Evanston: American Theol Lib Assoc, 1984 (1r) – 1 – 0-8370-1282-1 – mf#1984-T016 – us ATLA [242]

Evolucao : orgao republicano – Manaus, AM. 15 abr-28 jun 1888 – (= ser Ps 19) – mf#P25,02,23 – bl Biblioteca [320]

Evolucao : revista litteraria, scientifica e critica – Fortaleza, CE: Typ Universal, 25 ago 1893 – (= ser Ps 19) – mf#P17,1,46 – bl Biblioteca [073]

Evolucao da prosa brasileira / Grieco, Agrippino – Rio de Janeiro, Brazil. 1933 – 1r – us UF Libraries [972]

Evolucao do estado brasileiro / Cavalacanti De Carvalho, M – Rio de Janeiro, Brazil. 1941 – 1r – us UF Libraries [972]

Evolucao do povo brasileiro / Oliveira Vianna, Francisco Jose De – Sao Paulo, Brazil. 1923 – 1r – us UF Libraries [972]

Evolucao politica do brasil / Prado Junior, Caio – Sao Paulo, Brazil. 1933 – 1r – us UF Libraries [972]

Evolucao politica do brasil / Prado Junior, Caio – Sao Paulo, Brazil. 1963 – 1r – us UF Libraries [972]

Evolucion de la cultura cubana (1608-1927) / Carbonell, Jose Manuel – Habana, Cuba. v1-18. 1928 – 5r – 1 – us UF Libraries [972]

Evolucion de la energia electrica en la provincia de badajoz / Juarez Sanchez-Rubio, Cipriano – Badajoz: dip provincial, 1974 – 1 – sp Bibl Santa Ana [333]

Evolucion de las ideas / Rodriguez Beteta, Virgilio – Paris, France. 1929 – 1r – us UF Libraries [972]

Evolucion del pueblo brasilenoh (2) / Viana, Oliveira – Buenos Aires: [Imprenta Mercatali] 1937 (mf ed 2000) – (= ser Biblioteca de autores brasilenos traducidos al castellano) – 1r – 1 – mf#*Z-9287 – us NY Public [972]

Evolucion e importancia de la cedula / Dominican Republic. Direccion General De La Cedula – Ciudad Trujillo, Dominican Republic. 1948? – 1r – us UF Libraries [972]

La evolucion economica 2 : la produccion / Munoz Casillas, Juan – Madrid: Ed. Iberoam, 1952 – 1 – sp Bibl Santa Ana [330]

La evolucion economica 3 : la empresa / Munoz Casillas, Juan – Barcelona: Tip.Vda. de Golo Saenz, 1952 – 1 – sp Bibl Santa Ana [330]

La evolucion economica 4 : los medios de accion del estado y la colaboracion de la sociedad / Munoz Casillas, Juan – Madrid: Tip.Vda.Galo Saenz, 1952 – 1 – sp Bibl Santa Ana [330]

Evolucion economica de guatemala / Solorzano Fernandez, Valentin – Guatemala, 1963 – 1r – 1 – us UF Libraries [330]

Evolucion economica en el brasil, 1949-1950 / Pan American Union. Division Of Economic Research – Washington, DC. 1950 – 1r – us UF Libraries [972]

Evolucion poetica dominicana / Perez, Carlos Federico – Buenos Aires, Argentina. 1956 – 1r – us UF Libraries [972]

Evolucion politica de ibero-america / Carranca Y Trujillo, Raul – Madrid, Spain. 1925 – 1r – us UF Libraries [972]

Evolucion politica del brasil y otros estudios / Prado Junior, Caio – Buenos Aires, Argentina. 1964 – 1r – us UF Libraries [972]

Evolucion solaire et series astro-chimiques par... / Roso de Luna, Mario – Paris: imprimerie bussiere, 1909 – 1 – sp Bibl Santa Ana [440]

Die evolusie van 'n volksteologie : 'n historiese en dogmatiese ondersoek na die samehang van kerk en afrikanervolk in die teologie van die n g kerk met besondere verwysing na die apartheidsdenke wat daaruit ontwikkel het / Botha, Andries Johannes – U of the Western Cape [mf ed S.l: s.n. 1984] – 7mf – 9 – (abstract in afrikaans & english; incl bibl) – sa Misc Inst [230]

De evolutieleer en het godsdienstig geloof / Groenewegen, H Ij – Baarn: Hollandia-Drukkerij, 1909 [mf ed 1985] – (= ser Redelijke godsdienst 1/4) – 54p on 1mf – 9 – 0-8370-3402-7 – mf#1985-1402 – us ATLA [210]

L'evolution – Port-au-Prince, Haiti: Imp. de l'Abeille, [v1 n2-v2 n90] (1916-1917) – 5 sheets – 9 – us CRL [079]

Evolution / Jevons, Frank Byron – London: Methuen, 1900 – (= ser The Churchman's Library) – 1mf – 9 – 0-7905-7791-7 – mf#1989-1016 – us ATLA [575]

Evolution – Lawrence. 1947+ (1) 1965+ (5) 1976+ (9) – ISSN: 0014-3820 – mf#1460 – us UMI ProQuest [575]

Evolution : or, the divine method of creating and preserving the universe / Frost, A F – Philadelphia: Wm H Alden, 1892 – 1mf – 9 – 0-8370-3211-3 – mf#1985-1211 – us ATLA [210]

Evolution : popular lectures and discussions before the brooklyn ethical association / Thompson, Daniel Greenleaf et al – Boston: James H West, 1889 – 1mf – 9 – 0-7905-8937-0 – (incl bibl ref) – mf#1989-2162 – us ATLA [575]

Evolution and christian faith / Lane, Henry Higgins – Princeton: Princeton University Press, 1923 – xi/214p – 1 – us Wisconsin U Libr [210]

Evolution and christianity / Iverach, James – London, 1894 – (= ser 19th c evolution & creation) – 3mf – 9 – mf#1.1.11588 – uk Chadwyck [140]

Evolution and christianity / Iverach, James – London: Hodder & Stoughton, 1894 – (= ser The Theological Educator) – 1mf – 9 – 0-8370-3735-2 – mf#1985-1735 – us ATLA [240]

Evolution and christianity / Moore, Aubrey Lackington – [London] 1889 – (= ser 19th c evolution & creation) – 1mf – 9 – mf#1.1.11592 – uk Chadwyck [110]

Evolution and christianity : or, an answer to the development infidelity of modern times / Tefft, Benjamin Franklin – Boston: Lee & Shepard; New York: C T Dillingham, 1885 [mf ed 1990] – 2mf – 9 – 0-7905-9705-5 – mf#1989-1430 – us ATLA [210]

Evolution and christianity : a study / Grumbine, Jesse Charles Fremont – Chicago: Charles H Kerr, 1887 – 1mf – 9 – 0-8370-4789-7 – mf#1985-2789 – us ATLA [210]

Evolution and creation / Hardwicke, Herbert Junius – [London?] 1887 – (= ser 19th c evolution & creation) – 3mf – 9 – mf#1.1.9152 – uk Chadwyck [210]

Evolution and dogma / Zahm, John Augustine – Chicago: D H McBride, 1896 – 2mf – 9 – 0-8370-9919-6 – (incl ind) – mf#1986-3919 – us ATLA [230]

Evolution and ethics : and other essays / Huxley, Thomas Henry – London, New York: Macmillan, 1894 – (= ser Romanes lecture; Eversley series) – 1mf – 9 – 0-7905-3921-7 – mf#1989-0414 – us ATLA [230]

Evolution and ethics... : delivered in the sheldonian theatre, may 18 1893 / Huxley, Thomas Henry – London, 1893 – (= ser 19th c evolution & creation) – 1mf – 9 – mf#1.1.10891 – uk Chadwyck [170]

Evolution and human behavior – New York. 1997+ (1,5,9) – (Cont: Ethology and sociobiology) – ISSN: 1090-5138 – mf#42258,01 – us UMI ProQuest [300]

Evolution and its relation to religious thought / Conte, Joseph le – London, 1888 – (= ser 19th c evolution & creation) – 4mf – 9 – mf#1.1.1508 – uk Chadwyck [210]

Evolution and man's place in nature / Calderwood, Henry – London, 1893 – (= ser 19th c evolution & creation) – 4mf – 9 – mf#1.1.11610 – uk Chadwyck [210]

Evolution and natural selection in the light of the new-church / Swift, Edmund, jr – [London], 1879 – (= ser 19th c evolution & creation) – 1mf – 9 – (with app containing an outline of swedenborg's philosophy of creation) – mf#1.1.9236 – uk Chadwyck [210]

Evolution and natural theology / Kirby, William Forsell – London, 1883 – (= ser 19th c evolution & creation) – 3mf – 9 – mf#1.1.1588 – uk Chadwyck [210]

Evolution and progress : an exposition and defence / Gill, William Icrin – New York: Author's Pub Co 1875, c1874 [mf ed 1991] – 1mf – 9 – 0-7905-8796-3 – mf#1989-2021 – us ATLA [575]

Evolution and religion / Beecher, Henry Ward – London, 1885 – (= ser 19th c evolution & creation) – 6mf – 9 – mf#1.1.10915 – uk Chadwyck [210]

Evolution and religion / Beecher, Henry Ward – New York: Fords, Howard, & Hulbert, 1885 – 1mf – 9 – 0-8370-6161-X – mf#1986-0161 – us ATLA [210]

Evolution and religion : from the standpoint of one who believes in both. a lecture. delivered in the philadelphia academy of music... / Savage, Minot Judson – Philadelphia: George H Buchanan, 1886 – 1mf – 9 – 0-8370-5056-1 – mf#1985-3056 – us ATLA [230]

Evolution and religion : or, faith as a part of a complete cosmic system / Bascom, John – New York: G P Putnam, 1897 [mf ed 1985] – 1mf – 9 – 0-8370-2200-2 – mf#1985-0200 – us ATLA [210]

Evolution and religion / Osborn, Henry Fairfield – New York: C Scribner's sons, 1923 – vii/21p – 1 – us Wisconsin U Libr [230]

Evolution and religion : a parent's talks with his children concerning the moral side of evolution / Trumbull, William – New York: Grafton Press, 1907 – 1mf – 9 – 0-8370-5582-2 – (incl bibl ref) – mf#1985-3582 – us ATLA [230]

Evolution and scripture : or the relation between the teaching of scripture and the conclusions of astronomy, geology, and biology with an inquiry into the nature of the scriptures and inspiration / Holborow, Arthur – London, 1892 – (= ser 19th c evolution & creation) – 4mf – 9 – mf#1.1.10779 – uk Chadwyck [210]

Evolution and the fall / Hall, Francis Joseph – New York: Longmans, Green 1910 [mf ed 1989] – (= ser The bishop paddock lectures 1909-10) – 1mf – 9 – 0-7905-0496-0 – (incl bibl ref) – mf#1987-0496 – us ATLA [240]

Evolution and the immanent god : an essay on the natural theology of evolution / English, William F – Boston: Arena, 1894 – 1mf – 9 – 0-7905-7728-3 – mf#1989-0953 – us ATLA [230]

Evolution and the religion of the future / Swanwick, Anna – London, 1894 – (= ser 19th c evolution & creation) – 1mf – 9 – mf#1.1.6613 – uk Chadwyck [210]

Evolution and theology : inaugural address delivered in the new college, edinburgh, at the opening of the session 1874-75 / Rainy, Robert – Edinburgh, 1874 – (= ser 19th c evolution & creation) – 1mf – 9 – mf#1.1.11585 – uk Chadwyck [140]

Evolution as a process / Huxley, Julian – London, England. 1954 – 1r – us UF Libraries [500]

Evolution (as taught) : a myth illusive and degrading – London, 1883 – (= ser 19th c evolution & creation) – 3mf – 9 – mf#1.1.1509 – uk Chadwyck [575]

Evolution, as taught in the bible : a pamphlet for the times / Hasskarl, Gottlieb Christopher Henry – Philadelphia: Lutheran Publication Society, c1887 – 1mf – 9 – 0-8370-3517-1 – mf#1985-1517 – us ATLA [220]

Evolution, creation, and the fall / Wilson, James Maurice – London, England. 1893 – 1r – us UF Libraries [240]

L'evolution creatrice / Bergson, Henri – Paris: F Alcan, 1907 [mf ed 1990] – (= ser [Bibliotheque de philosophie contemporaine]) – 1mf – 9 – 0-7905-7324-5 – (in french) – mf#1989-0549 – us ATLA [110]

Evolution de la doctrine du purgatoire chez saint augustin / Ntedika, Konde – Paris, France. 1966 – 1r – us UF Libraries [960]

L'evolution de la foi catholique / Hebert, Marcel – Paris: Felix Alcan, 1905 [mf ed 1985] – 1mf – 9 – 0-8370-4488-X – (in french. incl bibl ref) – mf#1985-2488 – us ATLA [241]

Evolution de la poesie lyrique en france au dix-neuvieme siecle / Brunetere, Ferdinand – Paris, France. v1-2. 1895 – 1r – us UF Libraries [440]

L'evolution des dogmes / Guignebert, Charles – Paris: Ernest Flammarion, 1910 [mf ed 1986] – (= ser Bibliotheque de philosophie scientifique) – 1mf – 9 – 0-8370-8821-6 – (in french. incl bibl ref) – mf#1986-2821 – us ATLA [240]

L'evolution du marche foncier en peripherie du centre-ville de montreal au cours des annees soixante / Collin, Jean-Pierre – Montreal: Institut national de la recherche scientifique, INRS-Urbanisation, 1977 [mf ed 1998] – (= ser Etudes et documents (inrs-urbanisation)) – 2mf – 9 – mf#SEM105P2968 – cn Bibl Nat [307]

L'evolution du royaume rwanda des origines a 1900 / Vansina, Jan – Bruxelles, 1962 – us CRL [960]

L'evolution feminine : la femme au foyer et dans la cite / Goyau, Lucie Faure – Paris: Perrin, 1917 – (= ser Les femmes [coll]) – 4mf – 9 – mf#6681 – fr Bibl Nationale [305]

The evolution hypothesis : a criticism of the new cosmic philosophy / Martin, William Todd – Edinburgh, 1887 – (= ser 19th c evolution & creation) – 4mf – 9 – mf#1.1.1492 – uk Chadwyck [110]

Evolution in art : as illustrated by the life-histories of designs / Haddon, Alfred Cort – London 1895 – (= ser 19th c art & architecture) – 5mf – 9 – mf#4.2.53 – uk Chadwyck [740]

Evolution in history, language, and science : four addresses delivered...25th session, 1884-85, of the crystal palace company's school of art, and literature / Crystal Palace Co – London, 1884 – (= ser 19th c evolution & creation) – 2mf – 9 – mf#1.1.1576 – uk Chadwyck [575]

Evolution in my mission views : or growth of gospel mission principles in my own mind / Crawford, Tarleton Perry – 1852-1902, 1903 – 1 – us Southern Baptist [240]

Evolution in religion / McLane, William Ward – Boston: Congregational Sunday-School and Pub Society, c1892 – 1mf – 9 – 0-524-08482-3 – mf#1993-3127 – us ATLA [210]

L'evolution intellectuelle de saint augustin / Alfaric, Prosper – Paris: Emile Nourry, 1918 [mf ed 1993] – 2mf – 9 – 0-524-07989-7 – (in french. incl bibl ref) – mf#1991-0211 – us ATLA [240]

Evolution necessaire / Marcelin, Frederic – Paris, France. 1898 – 1r – us UF Libraries [972]

The evolution of ancient hinduism / Floyer, A M – London: Chapman and Hall, 1888 – 1mf – 9 – 0-524-01698-4 – mf#1990-2600 – us ATLA [280]

Evolution of ancient indian law / Sen Gupta, Nares Chandra – London: Arthur Probsthain, 1953 – (= ser Samp: indian books) – us CRL [340]

Evolution of awadhi : a branch of hindi / Saksena, Baburama – Allahabad: Indian Press, 1937 – (= ser Samp: indian books) – us CRL [490]

Evolution of brazil compared with that of spanish / Lima, Oliveira – New York, NY. 1966 – 1r – us UF Libraries [972]

The evolution of christianity / Abbott, Lyman – Boston: Houghton, Mifflin; Cambridge: Riverside Press, 1892 – (= ser Lowell institute lectures) – 1mf – 9 – 0-8370-2209-6 – mf#1985-0209 – us ATLA [240]

The evolution of christianity / Savage, Minot Judson – Boston: Geo H Ellis, 1892 – 1mf – 9 – 0-8370-5129-0 – mf#1985-3129 – us ATLA [240]

The evolution of congregationalism / Edmands, John – Burlington: Free Press Printing, 1916 – 1mf – 9 – 0-524-03146-0 – (incl bibl ref) – mf#1990-4595 – us ATLA [242]

The evolution of decorative art : an essay upon its origin and development as illustrated by the art of modern races of mankind / Balfour, Henry – London: Rivington, Percival & Co, 1893 – (= ser 19th c art & architecture) – 2mf – 9 – mf#4.1.73 – uk Chadwyck [740]

Evolution of fascism / Shah, Khushal Talaksi – Bombay: Popular Book Depot, 1935 – (= ser Samp: indian books) – us CRL [320]

Evolution of harmonic consciousness : a study of pre-eighteenth century technics / Hannas, Ruth – U of Rochester 1934 [mf ed 19–] – 1r – 1 – mf#film 159 – us Sibley [780]

Evolution of hindu moral ideals / Sivaswami Aiyar, Pazhamaneri Sundaram – Calcutta, India. 1935 – 1r – us UF Libraries [280]

Evolution of hindu moral ideals / Sivaswamy Aiyer, P S – Calcutta: Calcutta University, 1935 – (= ser Samp: indian books) – us CRL [280]

The evolution of immortality / McConnell, Samuel David – New York: Macmillan, 1901 – 1mf – 9 – 0-7905-9507-9 – mf#1989-1212 – us ATLA [210]

The evolution of immortality : or, suggestions of an individual immortality based upon our organic and life history / Stockwell, Chester Twitchell – 3rd ed. Chicago: Charles H Kerr, 1890 – 1mf – 9 – 0-524-06232-3 – mf#1991-0025 – us ATLA [210]

The evolution of indian mysticism / Ramaswami Sastri, K S – Bombay: International Book House, [between 1900 and 1948] – (= ser Samp: indian books) – us CRL [280]

Evolution of indian polity / Shama Sastri, Rudrapatna – [Calcutta]: University of Calcutta, 1920 – (= ser Samp: indian books) – us CRL [323]

The evolution of infant baptism and related ideas / Tymms, Thomas Vincent – London: Kingsgate Press, c1913 – 2mf – 9 – 0-7905-9721-7 – mf#1989-1446 – us ATLA [242]

Evolution of italian sculpture / Crawford, David Lindsay – London, England. 1909 – 1r – us UF Libraries [730]

The evolution of law : a historical review / Scott, Henry W – New York: Borden Press, 1908 – 2mf – 9 – $3.00 – mf#LLMC 95-156 – us LLMC [340]

Evolution of malayalam / Sekhar, Anantaramayyar Chandra – Poona: Deccan College Post-graduate and Research Institute, 1953 – (= ser Samp: indian books) – us CRL [490]

The evolution of man : a popular exposition of the principal points of human ontogeny and phylogeny / Haeckel, Ernst Heinrich Philipp August – London, 1879 – (= ser 19th c evolution & creation) – 2v on 12mf – 9 – (fr german of ernst haeckel) – mf#1.1.5410 – uk Chadwyck [573]

The evolution of man and christianity / MacQueary, Howard – rev enl ed. New York: D Appleton, 1891 – 1mf – 9 – 0-8370-4251-8 – (incl bibl ref and index) – mf#1985-2251 – us ATLA [210]

Evolution of mind : speech...mainland teachers' institute, held at vancouver, january 6th, 1896 / Baker, James – S.l: s.n, 1896? – 1mf – 9 – mf#00841 – cn Canadiana [150]

The evolution of morality : being a history of the development of moral culture / Wake, Charles Staniland – London: Truebner, 1878 – 3mf – 9 – 0-524-08166-2 – (incl bibl ref) – mf#1991-0296 – us ATLA [170]

The evolution of new china / Brewster, William Nesbitt – Cincinnati: Jennings & Graham; New York: Eaton & Mains, c1907 [mf ed 1986] – 1mf – 9 – 0-8370-6026-5 – (incl bibl ref) – mf#1986-0026 – us ATLA [951]

The evolution of north-west frontier province : being a survey of the history and constitutional development of n-wf province, in india / Obhrai, Diwan Chand – Peshawar: London Book Co, 1938 – (= ser Samp: indian books) – us CRL [323]

The evolution of old testament religion / Orchard, William Edwin – London: James Clarke, 1908 – 1mf – 9 – 0-8370-4629-7 – (incl ind) – mf#1985-2629 – us ATLA [221]

The evolution of religion : an anthropological study / Farnell, Lewis Richard – New York: Putnam, 1905 – (= ser Crown Theological Library) – 1mf – 9 – 0-7905-3737-0 – mf#1989-0230 – us ATLA [200]

The evolution of religions / Bierer, Everard – New York: Putnam, 1906 – 1mf – 9 – 0-7905-7685-6 – mf#1989-0910 – us ATLA [200]

The evolution of self-government in the colonies : their rights and responsibilities in the empire / Mills, David – [S.l: s.n, 1891?] [mf ed 1981] – 1mf – 9 – (fr: the canadian magazine) – mf#11109 – cn Canadiana [941]

Evolution of superimposed chords and tonalities / Shatzkin, Merton – U of Rochester 1961 [mf ed 19–] – 1r / 7mf – 1,9 – (with bibl) – mf#film 750 / fiche 244, 304 – us Sibley [780]

Evolution of the american flag / Corse, Carita Doggett – s.l, s.l? 193-? – 1r – us UF Libraries [975]

The evolution of the atmosphere as a proof of design and purpose in the creation and of the existence of a personal god : a simple and rigorously scientific reply to modern materialistic atheism / Phin, John – New York: Industrial Publ Co, 1908 [mf ed 1985] – 1mf – 9 – 0-8370-4736-6 – (incl ind) – mf#1985-2736 – us ATLA [210]

The evolution of the english bible : a historical sketch of the successive versions from 1382 to 1885 / Hamilton-Hoare, Henry William – 2nd ed rev corr ed. New York: E P Dutton; London: John Murray, 1902 – 2mf – 9 – 0-8370-9952-8 – (incl bibliography) – mf#1986-3952 – us ATLA [220]

Evolution of the god and christ ideas / Tuttle, Hudson – Berlin Heights, Ohio: Tuttle Pub Co; Chicago: JR Francis, c1906 – 1mf – 9 – 0-524-01319-5 – mf#1990-2355 – us ATLA [210]

The evolution of the hebrew people : and their influence on civilization / Wild, Laura Hulda – New York: Scribner, 1917 – 1mf – 9 – 0-524-04481-3 – (incl bibl ref) – mf#1992-0150 – us ATLA [939]

The evolution of the idea of god : an inquiry into the origins of religion / Allen, Grant – New York: H Holt, 1897 – 5mf – 9 – (incl ind and publ list) – mf#27424 – cn Canadiana [210]

Evolution of the japanese : a study of their characteristics in relation to the principles of social and psychic development / Gulick, Sydney Lewis – 4th rev ed. New York, Chicago: Fleming H Revell [1905] [mf ed 1995] – (= ser Yale coll) – xx/463p – 1 – 0-524-09064-5 – mf#1995-0064 – us ATLA [306]

Evolution of the khalsa / Banerjee, Indubhusan – Calcutta: University of Calcutta, 1936 – (= ser Samp: indian books) – us CRL [954]

The evolution of the monastic ideal : from the earliest times down to the coming of the friars / Workman, Herbert Brook – London: C.H. Kelly, 1913 – 1mf – 9 – 0-7905-6277-4 – (incl bibl ref) – mf#1988-2277 – us ATLA [240]

The evolution of the soul : and other essays / Hudson, Thomson Jay – Chicago: AC McClurg, 1904 – 1mf – 9 – 0-524-04984-X – mf#1990-3442 – us ATLA [130]

The evolution of the sunday school / Cope, Henry Frederick – Boston: Pilgrim Press, c1911 – 1mf – 9 – 0-7905-4262-5 – (incl bibl ref) – mf#1988-0262 – us ATLA [240]

The evolution of theology in the greek philosophers / Caird, Edward – Glasgow: J MacLehose, 1904 – (= ser Gifford lectures) – 2mf – 9 – 0-7905-7618-X – mf#1989-0843 – us ATLA [180]

Evolution, old and new : or, the theories of buffon, dr erasmus darwin, and lamarck, as compared with that of mr charles darwin / Butler, Samuel – London, 1879 – (= ser 19th c evolution & creation) – 5mf – 9 – mf#1.1.4273 – uk Chadwyck [575]

L'evolution religieuse contemporaine chez les anglais, les americains et les hindous *see* The contemporary evolution of religious thought in england, america and india

L'evolution religieuse dans les diverses races humaines / Letourneau, Charles – 2. ed. Paris: Vigot Freres, 1898 [mf ed 1992] – (= ser Bibliotheque anthropologique 20) – 2mf – 9 – 0-524-02025-6 – (in french. incl bibl ref) – mf#1990-2800 – us ATLA [230]

Evolution social and organic / Lewis, Arthur Morrow – Chicago, IL. 1908 – 1r – us UF Libraries [575]

Evolution, the stone book : and the mosaic record of creation / Cooper, Thomas – Cincinnati: Cranston and Curts; New York: Hunt and Eaton, 1893 – 1mf – 9 – 0-7905-0930-X – mf#1987-0930 – us ATLA [220]

Evolution, the stone book : and the mosaic record of creation / Cooper, Thomas – London, 1878 – (= ser 19th c evolution & creation) – 3mf – 9 – mf#1.1.9053 – uk Chadwyck [577]

Evolution von samenglobulin-genen / Braun, Holger – (mf ed 1997) – 2mf – 9 – €40.00 – 3-8267-2464-X – mf#DHS 2464 – gw Frankfurter [574]

Evolution without natural selection : or, the segregation of species without the aid of the Darwinian hypothesis / Dixon, Charles – London, 1885 – (= ser 19th c evolution & creation) – 2mf – 9 – mf#1.1.1526 – uk Chadwyck [575]

The evolutionist at large / Allen, Grant – London, 1881 – (= ser 19th c evolution & creation) – 3mf – 9 – mf#1.1.6829 – uk Chadwyck [575]

The evolutionist at large / Allen, Grant – London: Chatto & Windus, 1884 – 3mf – 9 – (originally appeared in st james's gazette. incl publ list) – mf#26900 – cn Canadiana [575]

Evraziya eurasie hebdomadaire russe – Paris, France. 24 nov 1928-7 sep 1929 – 1/4r – 1 – uk British Libr Newspaper [074]

Evrei = Yehudi / Goldschmidt, Meir; ed by Blagoveshchenskaia, M P – Petrograd: Gos izd-vo, 1919 [mf ed 2004] – 1r – 1 – (filmed with: bog i den'gi / vl krymov (v1-2 1926)) – us Wisconsin U Libr [830]

Evrei i talmud / Brenier, Flavien – Parizh, France. 1928 – 1r – us UF Libraries [270]

Evrei v novorossiiskom kraie / Lerner, Joseph Judah – Odessa, Ukraine. 1901 – 1r – us UF Libraries [939]

Evrei v smolenskie / Ryvkin, Kh D – St Petersburg, Russia. 1910 – 1r – us UF Libraries [939]

Evreinov, G A *see* Reforma denezhnogo obrashcheniia s prilozheniem spravki o nashei bednosti

Evreiskaia biblioteka – St Petersburg. v. 1-10. 1871-1903 – 1 – us NY Public [460]

Evreiskaia mysl' : Samar obl organ sionist org – Samara: Samar sionist kom 1918 [1918 27 [14] marta-[29 sent] – (= ser Asn 1-3) – n1-11 [1918] [gaps] item 132, on reel n28 – 1 – mf#asn-1 132 – ne IDC [077]

Evreiskaia nedelia – Spb., 1910. v1-11 – 9mf – 9 – mf#R-4164 – ne IDC [077]

Evreiskaia nishcheta v odessie / Brodovskii, I – Odessa, Ukraine. 1902 – 1r – us UF Libraries [939]

Evreiskaia starina – Spb., 1909-1916 – 93mf – 9 – mf#R-3408 – ne IDC [077]

Evreiski Vesti – Sofia, Bulgaria. Apr 1945-1955 – 1r – 1 – us L of C Photodup [949]

Evreiskie avtonomisty / Gepshtein, D – n.d. – 1mf – 9 – mf#RPP-96 – ne IDC [325]

Evreiskii al'manakh – Petrograd, Russia. 1923 – 1r – us UF Libraries [939]

Evreiskii rabochii : ezhened organ petrogr kom bunda / ed by Kagan, B I – Pg[Petrograd: s n] 1918 [1918-] – (= ser Asn 1-3) – n3 [1918] item 10, on reel n1 – 1 – mf#asn-2 010 – ne IDC [077]

Evreiskii student – Berlin, 1913-1914. nos 1-16 – 5mf – 9 – (Missing: 1913, no 3; 1914, nos 12, 15) – mf#R-18041 – ne IDC [077]

Evreiskoe Kolonizatsionnoe Obshchestvo *see* Melkii kredit

Evreiskoe obozrienie – St Petersburg. jan-july 1884 – 1 – us NY Public [073]

Evrejskoe slovo : organ sionistkoj mysli. gruppa zhurnalistov – Moscow, Russia, 1918 – 1r – 1 – us UMI ProQuest [077]

L'evridice d'ottavio rinuccini : rappresentata nello sponsalitio della christianiss / Rinuccini, Ottavio – Fiorenza: C Giunti 1600 [mf ed 19–] – 1mf – 9 – mf#fiche 656, 838 – us Sibley [780]

Evsebiana : essays on the ecclesiastical history of eusebius, bishop of caesarea / Lawlor, Hugh Jackson – Oxford: Clarendon Press 1912 [mf ed 1990] – 1mf – 9 – 0-7905-5249-3 – (incl bibl ref) – mf#1988-1249 – us ATLA [240]

Evseev, I *see*
- Kniga proroka daniila v drevneslavjanskom perevode
- Kniga proroka isaii v drevneslavianskom perevode

Evzerov, A et al *see* Izvestiia organizatsionnogo komiteta po sozyvu tret'ego vsesibirskogo sionistskogo sezda

Evzlin, Z *see* Banki i bankirskie kontory v rossii

Evzlin, Z P *see*
- Organizatsiia i tekhnika kreditnykh kooperativov
- Teoriia i praktika kommercheskogo banka

Ew : every wednesday – 1998 mar 25, 1999 feb 3 – 1r – 1 – mf#4322417 – us WHS [071]

Ewald, Alexander Charles *see*
- The life and times of the hon. algernon sydney, 1622-1683
- Our public records: a brief handbook to the national archives

Ewald, Franz [comp] *see* Handbuch der deutschen freimaurerei

Ewald, Georg Heinrich August *see* Abhandlungen zur orientalischen und biblischen literatur

Ewald, Georg Heinrich August von *see* A grammar of the hebrew language of the old testament

Ewald, Heinrich *see*
- Abhandlung ueber den bau der thatwoerter im koptischen
- Abhandlung ueber des aethiopischen buches henokh
- Allgemeines ueber die hebraeische dichtung und ueber das psalmenbuch
- The antiquities of israel
- Commentary on the book of job
- Die dichter des alten bundes
- Die drei ersten evangelien und die apostelgeschichte
- Ewald's introductory hebrew grammar
- Geschichte des apostolischen zeitalters zur zerstoerung jerusalems
- Geschichte des volkes israel
- Die glaubenslehre
- Grammatica critica linguae arabicae
- The history of israel
- Jesaja mit den uebrigen aelteren propheten
- Die komposition der genesis
- Lehrbuch der hebraischen sprache des alten bundes
- Die lehre vom worte gottes
- Old and new testament theology
- Die propheten des alten bundes
- Psalmen und die klagelieder
- Revelation
- Sieben sendschreiben des neuen bundes
- Syntax of the hebrew language of the old testament
- Ueber das leben des menschen und das reich gottes

Ewald, Paul *see*
- Der brief des paulus an die philipper
- Das hauptproblem der evangelienfrage und der weg zu seiner loesung
- Der kanon des neuen testaments

Ewald's introductory hebrew grammar = Hebraeische sprachlehre fuer anfaenger / Ewald, Heinrich – London: Asher, 1870 – 1mf – 9 – 0-8370-9230-2 – (incl ind) – mf#1986-3230 – us ATLA [470]
Ewangeeliumi krislane = The evangelical christian – 1930-1933. Estonian. 776p – 1 – $31.04 – us Southern Baptist [242]
Ewangelists = The evangelist – Riga, 1883, Nos. 1-24, 27. 208p – 1 – $8.32 – us Southern Baptist [242]
Ewart, Felicie *see* Goethes vater
Ewart, Frank Carman *see* Cuba y las costumbres cubanas
Ewart, Henry C *see*
– Toilers in art
Ewart, John Skirving *see*
– The disruption of canada
– Ewart's index of the statutes
– The future of canada / a perplexed imperialist / the canadian flag etc
– An imperial court of appeal
– The kingdom of canada
– A manual of costs in the supreme court of canada, high court of justice, court of appeal, county courts, etc
– The world famine and the duty of canada
Ewart's index of the statutes : being an alphabetical index of all the public statutes passed by the legislatures of the late province of canada – 2nd ed. Toronto: Carswell, 1874 – 191p – 1 – mf#LL-2329 – us L of C Photodup [348]
Ewbank, Thomas *see* Life in brazil
The ewe people and the coming of european rule, 1850-1914 / Amenumey, Divine Edem Kobla – London, 1964 – us CRL [960]
eWeek – New York. 2000+ (1,5,9) – mf#13449,01 – us UMI ProQuest [621]
Ewell, John D *see* Life of rev. william keele
Ewelow, H G *see* Selected works of hyman g enelow
Ewen, Robert *see* Through canada in 1878
Ewenny, glamorgan, parish church of st michael : baptisms 1725-1925, burials 1714-1925, marriages 1755-1837 – [Glamorgan]: GFHS [mf ed [2001?] – 1mf – 9 – £1.25 – uk Glamorgan FHS [929]
Ewenny, st michael, monumental inscriptions – 1mf – 9 – £1.25 – uk Glamorgan FHS [929]
Ewer, Ferdinand Cartwright *see*
– Catholicity in its relationship to protestantism and romanism
– Four conferences touching the operation of the holy spirit
– Sermons on the failure of protestantism, and on catholicity
Ewers, Hanns Heinz *see*
– Alraune
– Reiter in deutscher nacht
– Das wundermaedchen von berlin
– Der zauberlehrling
Ewers, Johann Philipp Gustav *see* Dr friedrich muenter's, professor der theologie an der universitaet zu kopenhagen, handbuch der aeltesten christlichen dogmen-geschichte
Ewerts, Max *see* Erinnerungen
The ewe-speaking peoples of the slave coast of west africa : their religion, manners, customs, laws, languages etc / Ellis, Alfred Burdon – London: Chapman and Hall, 1890 – 1mf – 9 – 0-524-03367-6 – mf#1990-3201 – us ATLA [390]
Ewh, Paul *see* Die begriffe pflicht und tugend in der sittenlehre kant's und schleiermacher's
Der ewig kommende gott / Jatho, Carl – Jena: E Diederichs, 1913 – 1mf – 9 – 0-524-00044-1 – mf#1989-2744 – us ATLA [210]
Ewig wiederkehrt die freude : gedichte / Goltz, Joachim, Freiherr von der – Muenchen: A Langen, G Mueller, 1944, c1942 – 1 – us Wisconsin U Libr [810]
Die ewige gottheit jesu christi / Kunze, Johannes – Leipzig: Doerffling & Franke, 1904 – 1mf – 9 – 0-8370-4404-9 – (incl bibl ref) – mf#1985-2404 – us ATLA [240]
Der ewige jan : roman / Uphoff, Carl Emil – Braunschweig: G Westermann c1937 [mf ed 1991] – 1r – 1 – (filmed with: ararat / arnold ulitz) – mf#2932p – us Wisconsin U Libr [830]
Die ewige ordnung : germanenleben in der bronzezeit / Auerswald, Annmarie von – Berlin: Junge Generation Verlag, [1941?] [mf ed 1988] – 191p – 1 – mf#6969 – us Wisconsin U Libr [890]
Der ewige tag : [poems] / Heym, Georg – Leipzig: E Rowohlt 1911 [mf ed 1990] – 1r – 1 – (filmed with: der wollmarkt / h clauren) – mf#2725p – us Wisconsin U Libr [810]
Die ewige vnnd einige grundfeste auff welchem der seligmachende glaube stehen vnd verharren muss / Huber, S – Vrsel, 1599 – 1mf – 9 – mf#TH-1 mf 723 – ne IDC [242]
Die ewigen gefuehle : roman / Brentano, Bernard von – Wiesbaden: Limes-Verlag, 1947 [mf ed 1989] – 264p – 1 – mf#7082 – us Wisconsin U Libr [830]
Ewiges arkadien! : roman / Bartsch, Rudolf Hans – Leipzig: L Staackmann, 1920 [mf ed 1995] – 275p – 1 – mf#8971 – us Wisconsin U Libr [830]
Ewiges deutschland : [poems] / Brockmeier, Wolfram – 4. aufl. Leipzig: Groten-Verlag, 1934 [mf ed 1989] – 63p – 1 – mf#7089 – us Wisconsin U Libr [810]
Ewiges leben / Seeberg, Reinhold – 2. verb aufl. Leipzig: A Deichert, 1915 – 1mf – 9 – 0-524-00123-5 – mf#1989-2823 – us ATLA [240]
Ewing advocate – Ewing, NE: R B Crellin. 28v. v36 n1-2. mar 4-11 1927; 36th yr n3. mar 18 1927-63rd yr n49. sep 2 1954 (wkly) [mf ed jul 31 1947] – 1r – 1 – (cont: people's advocate) – us NE Hist [071]
Ewing advocate – Ewing, NE: R B Crellin. 28v. v36 n1. mar 4 1927-v36 n2. mar 11 1927; 36th yr n3. mar 18 1927-63rd yr n49. sep 2 1954 – 9r – 1 – (cont: people's advocate (ewing ne)) – us Bell [071]
Ewing, Alexander *see*
– Charge addressed to the clergy of the diocese of argyll and the isl...
– Sermon for christmas-time
Ewing, Alfred Cyril *see* Morality of punishment
Ewing, Arthur Henry *see*
– The hindu conception of the functions of breath
– The mission study movement in india
– Theosophy examined
Ewing brothers' illustrated catalogue of choice vegetable and flower seeds : also garden and farm implements, etc – [Montreal?: s.n.], 1875 [mf ed 1987] – 2mf – 9 – 0-665-67888-6 – mf#67888 – cn Canadiana [635]
Ewing, Clymer and Co. Westport, Mo *see* Account book
The ewing democrat – Ewing, NE: H H Claiborne, oct 15-nov 11 1886 (gaps)] – 1r – 1 – us NE Hist [071]
The ewing democrat – Ewing, NE: O C Bates. v1 n1. jun 6 1888- (wkly) – 1r – 1 – us Bell [071]
Ewing item – Ewing, NE: C Selah. v1 n1. jan 10 1884- (wkly) – 1r – 1 – (cont: item (o'neill ne)) – us Bell [071]
Ewing, J A *see* Papers literary, scientific etc
Ewing, J C *see* On church reforms
Ewing, John *see* Sermons, 1755-1801
Ewing, John Cook *see* The brethren hymnody, with tunes
Ewing, John L, Jr *see* Effects of varying levels of fatigue on the rate of force development in females
Ewing, Joseph L *see* Sketches of the families of thomas ewing
Ewing, Juliana H *see* Blue bells on the lea
Ewing, Juliana Horatia *see* Flat iron for a farthing
Ewing, Martha *see* The relationship of male identity, the mesomorphic image, and anabolic steroid use in bodybuilding
Ewing, Martha E *see* The effects of an evaluative audience upon college males' self-efficacy, perceived ability, anxiety, and learning of a novel motor task
Ewing news – Clearwater, NE: Clearwater Pub Co. 9v. v1 n1. sep 4 1958-v9 n44. jun 22 1967 (wkly) [mf ed 1958-67 (lacks jan 10 1963)] – 5r – 1 – (split from: clearwater record. merged with: clearwater record to form: clearwater record-ewing news) – us NE Hist [071]
Ewing, Thomas *see* The canadian school geography
Ewing, Thomas, Jr *see* Thomas ewing jr. papers
Ewing, Thomas, Sr *see* Papers
Ewing, Tyrone J *see* An historical review of the experiences of eastern washington university african-american male athletes from the 1960s to the 1970s
Ewing, William *see* The sunday-school century
Ewtonville baptist church. dunlap, tennessee : church records – 1836-1952 – 1 – us Southern Baptist [242]
Ex aequo – v1 n1-v7 n1 [1979 oct-1985 sum] – 1r – 1 – mf#1127334 – us WHS [071]
Ex bellarmino epitome controversiarum omnium huius aevi luthero-calvinisticarum / Coppenstein, I A – Moguntiae. v1-3. 1624-1626 – 37mf – 8 – €71.00 – ne Slangenburg [242]
El ex convento e san benito de alcantara en la provincia de caceres / Melida, Jose Ramon – Madrid: tip fortanet – sp Bibl Santa Ana [240]
Ex hippocratis et galeni monumentis isagoge... / Collado, L – Valencia, 1561 – 6mf – 9 – sp Cultura [610]
Ex oriente lux – Leipzig, Germany: E Pfeiffer, 1905-32 [mf ed 2001] – (= ser Christianity's encounter with world religions, 1850-1950) – 1r – 1 – (in german) – mf#2000-s003 – us ATLA [956]
Ex oriente lux / ed by Winckler, Hugo – Leipzig: Eduard Pfeiffer, 1905-06 [mf ed 1989] – 2v on 2mf – 9 – 0-7905-2699-9 – (incl bibl ref) – mf#1987-2699 – us ATLA [930]
Ex rerum anglicarum scriptoribus saec 12 et 13 (mgh5:27.bd) – 1885 – (= ser Monumenta germaniae historica 5: scriptores in folio (mgh5)) – €39.00 – ne Slangenburg [240]
Ex rerum anglicarum scriptoribus saec 13 (mgh5:28.bd) – 1888 – (= ser Monumenta germaniae historica 5: scriptores in folio (mgh5)) – €35.00 – ne Slangenburg [242]
Ex rerum danicarum scriptoribus saec 12 et 13 (mgh5:29.bd) : ex historiis islandicis – 1892 – (= ser Monumenta germaniae historica 5: scriptores in folio (mgh5)) – €32.00 – ne Slangenburg [240]
Ex rerum danicarum scriptoribus saec 12 et 13 (mgh5:29.bd) : ex historiis islandicis – 1892 – (= ser Monumenta germaniae historica 5: scriptores in folio (mgh5)) – €32.00 – ne Slangenburg [240]
Ex rerum francogallicarum scriptoribus (mgh5:26.bd) : ex historiis auctorum flandrensium francogallica lingua scriptis (suppl tomi 24) – 1882 – (= ser Monumenta germaniae historica 5: scriptores in folio (mgh5)) – €44.00 – ne Slangenburg [220]
Ex summa philippi cancellarii questiones de anima / Keeler, L W – Muenster, 1937 – 3mf – 8 – €7.00 – (opusc et textus ser schol fasc 22) – ne Slangenburg [241]
Exaltacion de la divina misericordia en la milagrosa renovacion de la soberana imagen de christo senor nuestro crucificado : que se venera en la iglesia del convento de senor san joseph de religiosas carmelitas descalzas de la antigua fundacion de esta ciudad de mexico... / Velasco, Alfonso Alberto de – en Mexico:...D Joseph de Jauregui, ano de 1776 – (= ser Books on religion...1543/44-c1800: otros milagros) – 2mf – 9 – mf#crl-131 – ne IDC [241]
Exaltacion gloriosa del mas pequeno sermon panegyrico de n p san francisco / Estrada, Juan de – en Mexico: Por Joseph Bernardo de Hogal, ano de 1730 – (= ser Books on religion...1543/44-c1800: sermones en castellano) – 1mf – 9 – mf#crl-298 – ne IDC [241]
O exaltado : jornal litterario, politico e moral – Rio de Janeiro, RJ: Typ de Gueffier & C, 04 ago 1831-jul 1833, 15 abr 1835 – (= ser Ps 19) – mf#P02,04,22 – bl Biblioteca [073]
The exalted life / Castle, Nicholas – Dayton, Ohio: Otterbein Press, 1913 – 1mf – 9 – 0-7905-7380-6 – mf#1989-0605 – us ATLA [240]
Examen chimico-medico...de las aguas termales y bano de fitero / Ramirez, A – Pamplona, SA – 3mf – 9 – sp Cultura [610]
Examen critico-apologeticum super constitutionem dogmaticam de fide catholica editam in sessione tertia SS. oecumenici Concilii Vaticani / Ciasca, Augustino – Romae: S Congreg de Propaganda Fide, 1872 – 1mf – 9 – 0-8370-8411-3 – (Incl bibl ref) – mf#1986-2411 – us ATLA [241]
Examen critique de la soi-disant refutation de la grande guerre ecclesiastique de l'honorable l a dessaulles, sans rehabilitation de celui-ci – Montreal: Societe des ecrivains de bon sens, 1873 – 1mf – 9 – mf#23884 – cn Canadiana [241]
Examen critique de la vie de jesus de m renan / Freppel, Charles – 12e ed. Paris: A Bray, 1864 – 1mf – 9 – 0-8370-3193-1 – mf#1985-1193 – us ATLA [240]
Examen critique de l'histoire du sanctuaire de l'arche / Poels, Henricus Andreas – Louvain: J van Linthout, 1897 [mf ed 1985] – 1mf – 9 – 0-8370-4773-0 – (incl ind) – mf#1985-2773 – us ATLA [270]
Examen de boticarios / Villa, E de – Burgos, 1632 – 9mf – 9 – sp Cultura [615]
Examen de conciencia / Pogolotti, Graziella – Habana, Cuba. 1965 – 1r – us UF Libraries [972]
Examen de ingenios = The examination of mens wits / Huarte de San Juan, Juan – 1594 – 9 – us Scholars Facs [150]
Examen de la situacion economica de mexico / Banco Nacional de Mexico – n1-529. 1925-69 – 1 – us L of C Photodup [339]
Examen de plusieurs prejuges et usages abusifs : concernant les femmes enceintes / Saucerotte, Louis Sebastien – Strasbourg, 1777 – (= ser Les femmes [coll]) – 2mf – 9 – mf#10923 – fr Bibl Nationale [305]
Examen des principales questions critiques soulevees de nos jours au sujet du quatri eme evangile = Pruefung der wichtigsten kritischen streitfragen unserer tage ueber das vierte evangelium / Godet, Frederic Louis – Zuerich: Carl Meyer, 1866 – 1mf – 9 – 0-8370-9543-3 – (In German. Incl bibl ref) – mf#1986-3543 – us ATLA [220]
Examen des principes et recherche dans la conduite des deux f ...s dans une lettre a un membre du parlement : contenant ce qui s'est passe depuis le commencement de la derniere guerre jusqu'a la signature des preliminaires de paix a aix la chapelle – Francfort sur le Mein 1749 [mf ed Hildesheim 1995-98] – 2v on 5mf – 9 – €100.00 – 3-487-26198-7 – gw Olms [944]
Examen doctrinae macarii bulgakow, episcopi russi schismatici, et iosephi langen, neoprotestantis bonnensis, de processione spiritus sancti : paralipomenon tractatus de ss trinitate / Franzelin, Johannes Baptist – Romae: SC de Propaganda Fide, 1876 – 1mf – 9 – 0-7905-9929-5 – mf#1989-1654 – us ATLA [243]
Examen du livre de la reunion du christienisme / [Jurieu, P] – [Orleans], 1671 – 5mf – 9 – mf#PRS-148 – ne IDC [240]
Examen du Livre qui porte pour titre : prejuges legitimes contre les calvinistes / Pajon, C – La Haye, 1683 – 3pts on 8mf – 9 – mf#PRS-175 – ne IDC [242]
Examen du livre qui porte pour titre prejuges legitimes / Pajon, C – Bionne, 1673 – 5mf – 9 – mf#PRS-163 – ne IDC [240]
Examen ethicae clar : geulingii, sive, dissertationum philosophicarum: in quibus praemissa introductione sententiae quaedam paradoxae ex ethica clar. geulingii examinantur, pentas / Andala, Ruardus – Franequerae: Apud Wibium Bleck, Bibliopolam 1656 – (= ser Ethics in the early modern period) – 2v on 2mf – 9 – mf#pl-474 – ne IDC [170]
Examen ofte ondersoeck van de artijckelen : besloten in de synodus nationael... / Wtenbogaert, J – Rotterdam, 1648 – 3mf – 9 – mf#PBA-375 – ne IDC [240]
Examen philosophico-theologicum de ontologismo / Lepidi, Alberto – Lovanii [Louvain]: C-J Fonteyn, 1874 – 1mf – 9 – 0-7905-8503-0 – mf#1989-1728 – us ATLA [110]
Examen rationum quibus rob bellarminus pontificatun romanum adstruere nititur / Marnix van S Aldegonde, P van – n.p, 1602 – 2mf – 9 – mf#PBA-259 – ne IDC [240]
Examen theologicvm, complectens praecipva capita doctrinae christianae, de qvibvs interrogati / Hesshusen, T – [Ienae], 1571 – 4mf – 9 – mf#TH-1 mf 617-620 – ne IDC [242]
Examen variantium lectionum johannes millii...n t / Whitby, Daniel – London: Guil. Bowyer, 1710 – 1r – 1 – 0-8370-0983-9 – mf#1984-B513 – us ATLA [225]
Examen...de la alegacion apologetica medico-physeca / Melero Ximenez, M – Cordoba, SA: 1699 – 4mf – 9 – sp Cultura [612]
The examination and tryal of old father christmas / King, Josiah – London: Charles Brome, 1686 – 1mf – 9 – $1.50 – mf#LLMC 91-094 – us LLMC [870]
Examination before admission to a benefice by the bishop of exeter : followed by a refusal to institute, on the allegation of unsound doctrine respecting the efficacy of baptism / ed by Gorham, George Cornelius – London: Hatchard, 1848 – 1mf – 9 – 0-7905-5223-X – mf#1988-1223 – us ATLA [240]
An examination into the doctrine and practice of confession / Jelf, William Edward – London: Longmans, Green, 1875 [mf ed 1990] – 1mf – 9 – 0-7905-7789-5 – mf#1989-1014 – us ATLA [240]
An examination of a model of burnout in dual-role teacher-coaches / Kelley, B C – 1990 – 3mf – 9 – $12.00 – us Kinesology [150]
Examination of a pamphlet by w f wilkinson, ma, vicar of st wer... / Oates, William – London, England. 18– – 1r – 1 – us UF Libraries [240]
Examination of a pamphlet entitled "considerations on the expedienc... / Henderson, William – Aberdeen, Scotland. 1831 – 1r – us UF Libraries [240]
Examination of "a protestant's" defence of the rev mr Fraser – Aberdeen, Scotland. 1831 – 1r – us UF Libraries [242]
An examination of adolescents' sources of subjective task value in sport / Stuart, Moira E – 1997 – 2mf – 9 – $8.00 – mf#PSY 1961 – us Kinesology [790]
An examination of, and reply to, "a brief statement of facts : for the consideration of the methodist people and the public in general, particularly of eastern canada by r hutchinson...late wesleyan missionary" / Borland, John – Stanstead Quebec: L R Robinson, 1850 – 1mf – 9 – mf#50583 – cn Canadiana [242]
Examination of articles contributed by professor w robertson smith : to the encyclopaedia britannica, the expositor, and the british quarterly review, in relation to the truth, inspiration, and authority of the holy scriptures / Montgomery, John – Edinburgh: James Gemmell, 1877. Princeton: Speer Lib, and Dep of Photodup, U of Chicago Lib, 1978 (1r); Evanston: American Theol Lib Assoc, 1984 (1r) – (= ser Case of william robertson smith in the free church of scotland) – 1 – 0-8370-0618-X – (incl bibl ref) – mf#1984-6290 – us ATLA [220]
An examination of canon liddon's bampton lectures : on the divinity of our lord and saviour jesus christ / [Voysey, Charles] – Boston: Little, Brown, 1872 [mf ed 1993] – 1mf – 9 – 0-524-05756-7 – mf#1992-0599 – us ATLA [240]

Examination of canon liddon's bampton lectures of the divinity of o... / Voysey, Charles – Ramsgate, England. 1870 – 1r – us UF Libraries [240]

Examination of certain opinions – Manchester, England. 1813 – 1r – us UF Libraries [240]

An examination of certain passages in our lord's conversation with nicodemus : eight discourses preached before the university of cambridge in the year 1843 / Marsden, John Howard – London: William Pickering, 1844 [mf ed 1989] – 1mf – 9 – 0-7905-2174-1 – mf#1987-2174 – us ATLA [225]

An examination of certain proceedings and principles of the society of friends, called quakers / Bates, Elisha – St Clairsville: printed...by Horton J Howard, 1837 [mf ed 1993] – (= ser Society of friends (quakers) coll) – 4mf – 9 – 0-524-07390-2 – mf#1991-3050 – us ATLA [243]

An examination of coaching standards in the state of washington / Tucci, Derek C – 1998 – 2mf – 9 – $8.00 – mf#PE 3872 – us Kinesology [790]

An examination of environmental attitudes among college students / McGuire, John R & Graefe, Alan R – 1992 – 2mf – 9 – $8.00 – us Kinesology [790]

An examination of epinephrine and athletic performance : psychophysiological, cognitive, and behavioral indices of arousal / Wilkinson, Michael O – 1982 – 2mf – 9 – $8.00 – us Kinesology [790]

Examination of gillespie : being an analytical criticism of the argument a priori for the existence of a great first cause... / Barrett, Thomas Squire – 2nd ed. London: Provost, 1871 [mf ed 1985] – 1mf – 9 – 0-8370-2186-3 – (incl app) – mf#1985-0186 – us ATLA [210]

An examination of ground reaction forces : in runners with various degrees of pronation / Morely, Joanna B – 2000 – 109p on 2mf – 9 – $10.00 – mf#PE 4118 – us Kinesology [612]

An examination of harnack's 'what is christianity?' : a paper read...oct 24 1901 / Sanday, William – London, New York: Longmans, Green, 1901 [mf ed 1989] – 1mf – 9 – 0-7905-2372-8 – mf#1987-2372 – us ATLA [240]

An examination of leisure in the lives of old lesbians from an ecological perspective / Jacobson, Sharon A – 1996 – 4mf – 9 – $16.00 – mf#RC 503 – us Kinesology [305]

The examination of mens wits *see* Examen de ingenios

Examination of mr isaac marlow's two papers : one called a discourse against singing etc, the other, an appendix / Keach, Benjamin – London. 1691 – 1 – us Southern Baptist [240]

Examination of mr maurice's theological essays / Candlish, Robert Smith – London: J Nisbet, 1854 – 2mf – 9 – 0-7905-0975-X – mf#1987-0975 – us ATLA [240]

Examination of objections made to unitarianism by the rev j c miller, m a : in his sermon preached in st martin's church, birmingham, on sunday, april 2, 1854... / Bache, Samuel & Clarke, Charles – London: Whitfield, Strand, [1854?] – 1mf – 9 – 0-524-07808-4 – mf#1991-3355 – us ATLA [243]

An examination of professor bergson's philosophy / Balsillie, David – London: Williams & Norgate, 1912 [mf ed 1990] – 1mf – 9 – 0-7905-3534-3 – mf#1989-0027 – us ATLA [140]

An examination of professor ferrier's theory of knowing and being / Cairns, John – Edinburgh: T Constable, 1856 [mf ed 1991] – 1mf – 9 – 0-7905-9163-4 – mf#1989-2388 – us ATLA [110]

An examination of rider arousal in the three phases of an equestrian combined training event / Berno, K A – 1990 – 1mf – 9 – $4.00 – us Kinesology [150]

An examination of sir william hamilton's philosophy : and of the principal philosophical questions discussed in his writings / Mill, John Stuart – London: Longman, Green, Longman, Roberts & Green, 1865 [mf ed 1991] – 2mf – 9 – 0-7905-9350-5 – mf#1989-2575 – us ATLA [190]

Examination of some portions of the rev w goode's "letter to the... / Arnold, Thomas Kerchever – London, England. 1850 – 1r – 1 – us UF Libraries [240]

An examination of the academic performance of student-athlete admission exceptions at a divison 1-a institution / Gilmore, Carole A – University of North Carolina at Chapel Hill, 1995 – 1mf – 9 – $4.00 – mf#PE3591 – us Kinesology [370]

An examination of the alleged discrepancies of the bible / Haley, John W – Andover: Warren F Draper; Boston: Estes & Lauriat, 1874 [mf ed 1989] – 2mf – 9 – 0-7905-0713-7 – (int by alvah hovey. incl ind) – mf#1987-0713 – us ATLA [220]

An examination of the ancient orthography of the jews, and of the original state of the text of the hebrew bible : part the first, containing an inquiry into the origin of alphabetic writing; with which is incorporated an essay on the egyptian hieroglyphs / Wall, Charles William – London: Whittaker & Co; Dublin: Milliken & Son, 1835-56 – (= ser 19th c books on linguistics) – 19mf – 9 – (the imprints vary. pt2.1 dated 1840; pt2.2 dated 1841; pt3 dated 1856) – mf#2.1.19 – uk Chadwyck [470]

An examination of the bone mineral density status of women with an intellectual disability and risk factors associated with the acquisition of osteoporosis / Foster, Bernadette L – 1997 – 2mf – 9 – $8.00 – mf#PE 3978 – us Kinesology [617]

An examination of the causes which led to the separation of the religious society of friends in america, in 1827-28 / Janney, Samuel Macpherson – Philadelphia: T Ellwood Zell, 1868 [mf ed 1991] – (= ser Society of friends (quakers) coll) – 1mf – 9 – 0-524-00971-6 – mf#1990-4029 – us ATLA [243]

Examination of the claims of ishmael : as viewed by muhammadans / Bate, John Drew – Banaaras: EJ Lazarus, 1884 [mf ed 1991] – (= ser Missionary's vade-mecum 1 ser; Studies in islam 1/1) – 1mf – 9 – 0-524-01162-1 – mf#1990-2238 – us ATLA [260]

An examination of the comparative statistical results of the labors of elder jacob knapp : in the state of massachusetts / Wilbur, Asa – Boston: Heath & Graves, 1855 [mf ed 1992] – 1mf – 9 – 0-524-03667-5 – mf#1990-1095 – us ATLA [242]

An examination of the corn returns for the years, 1826, 1827, 1828, and 1829 : showing that the defective principle upon which they have been obtained has produced fallacious averages... / Forwood, George – London: Baldwin & Cradock, 1830 – (= ser 19th c economics) – 1mf – 9 – mf#1.1.230 – uk Chadwyck [338]

An examination of the determinants to the overall recreational sports participation among college students / Kiger, John R – 1996 – 2mf – 9 – $8.00 – mf#RC 504 – us Kinesology [790]

An examination of the doctrine of predestination : as contained in a sermon...by daniel haskel, minister of the congregation / Bangs, Nathan – New York: printed...by J C Totten, 1817 [mf ed 1984] – 2mf – 9 – 0-8370-0890-5 – mf#1984-4281 – us ATLA [230]

An examination of the effect of gender, self-confidence and achievement orientation on sport-related attributional styles / Zizzi, Samuel J – 1997 – 2mf – 9 – $8.00 – mf#PSY 1972 – us Kinesology [150]

Examination of the evidence from prophecy in behalf of the christian / Ogilvie, John – Aberdeen, Scotland. 1803 – 1r – us UF Libraries [240]

An examination of the 'friends of carolina' fundraising organizations at the university of north carolina at chapel hill / Torns, Jennifer – 1997 – 1mf – 9 – $4.00 – mf#PE 3839 – us Kinesology [650]

An examination of the motivational differences between adults in structured and unstructured exercise programs / Piepkorn, M B – 1990 – 2mf – 9 – $8.00 – us Kinesology [150]

Examination of the passages contained in the gospels and other books / Smith, J – London, England. 1811 – 1r – us UF Libraries [226]

Examination of the principles of biblical interpretation of ernesti, ammon, stuart : and other philologists / Carson, Alexander – New York: Edward H Fletcher, 1855 [mf ed 1989] – 2mf – 9 – 0-7905-1629-2 – (filmed with: a treatise on the figures of speech and: a treatise on the right and duty of all men to read the scriptures) – mf#1987-1629 – us ATLA [220]

An examination of the relationship among target structures, team motivational climate, and achievement goal orientation / Becker, Susan L – Oregon State University, 1995 – 2mf – 9 – $8.00 – mf#PSY 1880 – us Kinesology [150]

An examination of the relationship between angler specialization and constraints to trout fishing / Lloyd, Gregory S – 1993 – 2mf – $8.00 – us Kinesology [790]

An examination of the relationship between athletic participation, multidimensional self-concept, self-esteem, gender, and collegiate academic performance / Partridge, Julie A – 1998 – 2mf – 9 – $8.00 – mf#PSY 2039 – us Kinesology [150]

An examination of the relationship between teacher enthusiasm and alt-pe / Griffin, Lisa M – West Virginia University, 1995 – 2mf – 9 – $8.00 – mf#PE3594 – us Kinesology [370]

An examination of the relationships between coaching behaviors, sport confidence, and motivational orientation / Diatelevi, Michael P – 1998 – 2mf – 9 – $8.00 – mf#PSY 2022 – us Kinesology [150]

Examination of the scheme of church-power laid down / Foster, Michael – London, England. 1840 – 1r – us UF Libraries [240]

An examination of the sources and levels of perceived competence in male and female interscholastic coaches / Barber, Heather & Weiss, Maureen R – 1992 – 2mf – 9 – $8.00 – us Kinesology [150]

An examination of the test characteristics of the 12 minute aerobic swim test / Fried, Constance R – 1983 – 2mf – 9 – $8.00 – us Kinesology [790]

An examination of the testimony of the four evangelists : by the rules of evidence administered in courts of justice: with an account of the trial of jesus / Greenleaf, Simon – 2nd ed. London: A Maxwell, 1847 [mf ed 1990] – 2mf – 9 – 0-8370-1755-6 – mf#1987-6151 – us ATLA [226]

Examination of the theories of absolution and confession lately pro... / Laurence, Robert French – Oxford, England. 1847 – 1r – us UF Libraries [240]

Examination of the theory and practice of church music in unitarian societies / Angell, Douglas – Chicago, 1946. Chicago: Dep of Photodup, U of Chicago Lib, 1971 (1r); Evanston: American Theol Lib Assoc, 1984 (1r) – 1 – 0-8370-0334-2 – (incl bibl ref) – mf#1984-B193 – us ATLA [780]

An examination of the theory of evolution : and some of its implications / Gresswell, George – [London] 1888 – (= ser 19th c evolution & creation) – 2mf – 9 – mf#1.1.1524 – uk Chadwyck [575]

An examination of the understanding of the human person in the philosophy of john macmurray / Reinecker, Virginia M – 1982 – 1 – $5.00 – us Southern Baptist [190]

An examination of the utilitarian philosophy / Grote, John; ed by Mayor, Joseph Bickersteth – Cambridge: Deighton, Bell, 1870 [mf ed 1990] – 1mf – 9 – 0-7905-7637-6 – (incl bibl ref) – mf#1989-0862 – us ATLA [140]

An examination of the utilitarian theory of morals / Beattie, Francis Robert – Brantford: J & J Sutherland, 1885 [mf ed 1990] – 1mf – 9 – 0-7905-3996-9 – mf#1989-0489 – us ATLA [170]

An examination of the validity of the profile of mood states (poms) in the assessment of mental health in athletes / Sullivan, John P, Jr – Springfield College, 1995 – 1mf – 9 – $4.00 – mf#PSY1867 – us Kinesology [150]

The examination of the validity of the tarskij equation for predicting vo2 max in an active female population / Ellingson, Susan M – 1998 – 1mf – 9 – $4.00 – mf#PH 1638 – us Kinesology [612]

Examination of the validity of the tarskij vo_2 max prediction equation in an active male population / Westby, Christian M – 1998 – 1mf – 9 – $4.00 – mf#PH 1645 – us Kinesology [612]

An examination of two theoretical distributions using three methods of scoring criterion-referenced measures of motor performance / Douglass, Jacqueline A – 1981 – 2mf – 9 – $8.00 – us Kinesology [790]

An examination of weismannism / Romanes, George John – London, 1893 – (= ser 19th c evolution & creation) – 3mf – 9 – mf#1.1.2881 – uk Chadwyck [575]

The examination of witnesses in court : incuding examination in chief, cross-examination, and re-examination / Wrottesley, Frederic John – London: Sweet & Maxwell; Toronto: Carswell, 1919 – 3mf – 9 – 0-665-97119-2 – (incl publ's list and app) – mf#97119 – cn Canadiana [347]

Examination papers in arithmetic in three parts : designed for use of second, third and fourth classes in the public schools / McNaughton, John Alex – St Marys [Ont]: H Fred Sharp, [1883?] [mf ed 1994] – 2mf – 9 – 0-665-94729-1 – mf#94729 – cn Canadiana [510]

Examination questions given in the law school of columbian university / George Washington University. Dept of Law – Washington, D.C., 1898. 27p. LL-414 – 1 – us L of C Photodup [340]

Examination questions in latin and greek / College Entrance Examination Board – Boston, MA. 1910 – 1r – us UF Libraries [450]

Examinations in secondary schools – circulars 113/46 for 1946, and 168/48 for 1948 – 1mf – 9 – mf#86960 – uk Microform Academic [324]

Examinations of secondary schools – circular 996, 1917 – 1mf – 9 – mf#86944 – uk Microform Academic [324]

Examiner / Champaign Co. Saint Paris – jan 1-may 13 1880 [wkly] – 1r – 1 – mf#B11640 – us Ohio Hist [071]

Examiner – Charlottetown, Canada. 16 mar 1863-25 dec 1865; 1866-25 sep 1871 (imperfect) – 3r – 1 – uk British Libr Newspaper [071]

Examiner – Charlottetown, PEI. 1847-1900 – 58r – 1 – ISSN: 1181-263X – cn Library Assoc [079]

Examiner – Colville, WA. 1907-1948 [1] – mf#69163 – us UMI ProQuest [071]

Examiner : containing political essays on the most important events of the time; public laws and official documents – New York. 1813-1816 (1) – mf#4449 – us UMI ProQuest [320]

Examiner – Covington GA. v1 n22 [1866 may 2] – 1r – 1 – mf#859885 – us WHS [071]

Examiner – Dillon, MT. 1893-1957 (1) – mf#64357 – us UMI ProQuest [071]

Examiner / Dwight David Eisenhower Army Medical Center – Fort Gordon GA. v5 n23 [1980 nov18]-1985 dec 13 – 1r – 1 – mf#1099027 – us WHS [355]

Examiner : (first edition) – San Francisco, CA. 1966-1999 (1) – mf#60666 – us UMI ProQuest [071]

Examiner – Franklin, IN. 1850-1852 (1) – mf#62782 – us UMI ProQuest [071]

Examiner – Frederick, MD. 1849-1913 (1) – mf#63609 – us UMI ProQuest [071]

Examiner – Lancaster, PA. 1892-1920 (1) – mf#65957 – us UMI ProQuest [071]

Examiner – London. 1710-1714 (1) – mf#4249 – us UMI ProQuest [900]

Examiner – Moorefield, WV. 1902-1957 (1) – mf#67372 – us UMI ProQuest [071]

Examiner : (News Edition) – San Francisco, CA. 1969-1969 (1) – mf#62270 – us UMI ProQuest [071]

Examiner – [Northern Ireland] Belfast sep 1935-dec 1950 [mf ed 2002] – 12r – 1 – uk Newsplan [072]

Examiner – Omaha, NE. 1900-1924 (1) – mf#64719 – us UMI ProQuest [071]

Examiner – Owensboro, KY. 1875-1878 (1) – mf#63479 – us UMI ProQuest [071]

Examiner – Richmond, VA. 1801-1804 (1) – mf#66822 – us UMI ProQuest [071]

Examiner – San Francisco, CA. 1888-1900 (1) – mf#62269 – us UMI ProQuest [071]

Examiner – Toronto, ON. 1840-55 – 3r – 1 – ISSN: 1181-18032 – cn Library Assoc [071]

Examiner : a weekly paper on politics, literature, music, and the fine arts – London. 1808-1881 (1) – mf#4250 – us UMI ProQuest [320]

Examiner – Yonkers, NY. 1857-1863 (1) – mf#65289 – us UMI ProQuest [071]

Examiner *see*
– Cork examiner
– Launceston examiner

The examiner – Auckland, NZ. oct 1990-aug 1991 – 2r – 1 – (ceased publ aug 1991) – mf#11.68 – nz Nat Libr [079]

The examiner – London. 1852-54.-w. 3 reels – 1 – uk British Libr Newspaper [072]

The examiner : a monthly review of legislation and jurisprudence = L'observateur – Quebec: Printed for the proprietors by J Lovell, [1861] – 9 – (text in english and french) – mf#P05066 – cn Canadiana [340]

The examiner – Omaha, NE: Alfred Sorenson, sep 2 1901- (wkly) [mf ed 1901-24 (gaps) filmed 1980] – 12r – 1 – us NE Hist [071]

The examiner *see* Launceston examiner

Examiner [an sgruiduitheoir] – Dundalk, sep 1930-35 – 6r – 1 – ie National [072]

Examiner and chronicle – New York. 1876 – 1 – us ABHS [071]

Examiner and chronicle – New York. Feb 1865-74. (Complete); The Examiner. 1912 – 1 – us Southern Baptist [242]

Examiner and hardy county news – Moorefield, WV. 1957-1987 (1) – mf#67373 – us UMI ProQuest [071]

Examiner and herald – Lancaster, PA. 1834-1876 (1) – mf#65958 – us UMI ProQuest [071]

Examiner and hesperian – Pittsburgh. 1839-1840 (1) – mf#3979 – us UMI ProQuest [420]

Examiner and melbourne weekly news – Melbourne, Australia. 11 jul 1857-17 oct 1863 – 1 – (very imperfect) – uk British Libr Newspaper [079]

Examiner enterprise – Bartlesville, OK. 1990-2000 (1) – mf#61757 – us UMI ProQuest [071]

Examiner new era – Lancaster, PA. 1920-1923 (1) – mf#65959 – us UMI ProQuest [071]

Examining an educational program in gender equity / Pusateri-Lane, Lori J – 1999 – 1mf – 9 – $4.00 – mf#PE 3969 – us Kinesology [370]

Examining the effects of drug testing in drug use at the secondary education level / Walter, Sandra M – 1997 – 2mf – 9 – $8.00 – mf#HE 600 – us Kinesology [362]

Examining the relationship among measures of anxiety, self-confidence, arousal, and performance of elite field hockey players / Borrelli, Dina M – 1997 – 1mf – 9 – $4.00 – mf#PSY 1938 – us Kinesology [150]

Example *see* Pang yang (ccm287)

EXCISION

Example for young men from real life / Yates, W – London, England. 18– – 1r – us UF Libraries [240]
Example of how an enterprise of citrus is analyzed for use in teaching / Dansby, George William – s.l, s.l? 1930 – 1r – us UF Libraries [634]
Example of ministerial greatness / Law, Joseph – Sunderland, England. 1838 – 1r – us UF Libraries [240]
The example of our lord : especially for his ministers / Hall, Arthur Crawshay Alliston – New York: Longmans, Green, 1906 – 2mf – 9 – mf#1992-1078 – us ATLA [240]
Examples and counsels for the moral guidance of youth / Belfrage, Henry – Edinburgh, Scotland. 1827 – 1r – us UF Libraries [240]
Examples and designs of verandahs / Arundale, Francis – London 1851 – (= ser 19th c art & architecture) – 1mf – 9 – mf#4.1.384 – uk Chadwyck [720]
Examples in historical and geographical antonomasia : for the use of students in history and geography / Borthwick, John Douglas – Montreal?: Owler & Stevenson, 1858 – 1mf – 9 – mf#50554 – cn Canadiana [059]
Examples of ancient and modern furniture, metal work, tapestries, decorations etc / Talbert, Bruce J – London 1876 – (= ser 19th c art & architecture) – 1mf – 9 – mf#4.2.1051 – uk Chadwyck [740]
Examples of antient pulpits existing in england / Dollman, Francis Thomas – London 1849 – (= ser 19th c art & architecture) – 2mf – 9 – mf#4.2.1598 – uk Chadwyck [720]
Examples of architectural art in italy and spain, chiefly of the 13th and 16th centuries / Waring, John Burley & Macquoid, Thomas Robert – London 1850 – (= ser 19th c art & architecture) – 4mf – 9 – mf#4.2.1361 – uk Chadwyck [720]
Examples of chinese ornament / Jones, Owen – London 1867 – (= ser 19th c art & architecture) – 2mf – 9 – mf#4.2.852 – uk Chadwyck [740]
Examples of contemporary art / Carr, Joseph William Comyns – London 1878 – (= ser 19th c art & architecture) – 3mf – 9 – mf#4.2.1210 – uk Chadwyck [700]
Examples of decorative wrought ironwork of the 17th and 18th centuries / Eastlake, Elizabeth (Rigby) et al – London 1879 – (= ser 19th c art & architecture) – 1mf – 9 – mf#4.2.1797 – uk Chadwyck [730]
Examples of french art / Temple, Alfred George – London 1898 – (= ser 19th c art & architecture) – 3mf – 9 – mf#4.1.434 – uk Chadwyck [700]
Examples of indian art at the british empire exhibition, 1924 – London: India Society, 1925 – (= ser Samp: indian books) – (introductory and critical note by Lionel Heath; foreword by the Earl of Ronaldshay) – us CRL [700]
Examples of labourers' cottages : with plans for improving the dwellings of the poor / Birch, John – London 1871 – (= ser 19th c art & architecture) – 1mf – 9 – mf#4.2.1019 – uk Chadwyck [720]
Examples of london and provincial architecture of the victorian age – London 1862 – (= ser 19th c art & architecture) – 2mf – 9 – mf#4.1.435 – uk Chadwyck [720]
Examples of modern etching / Hamerton, Philip Gilbert – London 1875 – (= ser 19th c art & architecture) – 2mf – 9 – mf#4.1.240 – uk Chadwyck [760]
Examples of old english houses and furnitur / Adams, Maurice Bingham – London 1888 – (= ser 19th c art & architecture) – 3mf – 9 – mf#4.2.424 – uk Chadwyck [720]
Examples of ornamental metal work / Shaw, Henry – London 1836 – (= ser 19th c art & architecture) – 2mf – 9 – mf#4.2.1297 – uk Chadwyck [730]
Examples of ornament...from works of art in the british museum / ed by Cundall, Joseph – London 1855 – (= ser 19th c art & architecture) – 2mf – 9 – mf#4.2.1471 – uk Chadwyck [740]
Examples of records in the national archives frequently used in genealogical research / U.S. National Archives and Records Service – 1r – 1 – mf#T325 – us Nat Archives [929]
Examples of the application of trigonometry to crystallographic calculations : drawn up for the use of students in the use of students in the university of toronto / Chapman, Edward John – Toronto?: Lovell and Gibson, 1860 – 1mf – 9 – mf#48456 – cn Canadiana [510]
Excalibur – Cape Town SA, 21 may 1886-22 aug 1890 – 2r – 1 – sa National [960]
Excalibur / Moody Air Force Base (GA). United States – 1982 feb 4-1983 aug 26, 1985 jun-1986 may, 1986 jun 27-1987 may 29, 1987 jun 5-1988 jun 24, 1988 jul-1989 apr, 1989 may-1990 feb, 1990 mar-jun, 1990 jul-dec, 1991 jan-jun, 1991 jul-dec – 10r – 1 – mf#1002312 – us WHS [355]
Excavaciones arqueologicas en la ciudad de merida / Melida, Jose Ramon – Madrid: tip fortanet, 1911 – sp Bibl Santa Ana [930]
Excavaciones arqueologicas en la zona de merida : la casa del anfiteatro / Garcia Sandoval, Eugenio – Sevilla-Malaga, 1963 – sp Bibl Santa Ana [930]
Excavaciones de america / Carmona, Miguel – Badajoz: dip provincial, 1963 – sp Bibl Santa Ana [930]
Las excavaciones de merida / Melida, Jose Ramon – Madrid: Brah, abril 1911 – 1 – sp Bibl Santa Ana [930]
Excavaciones de merida. las. ultimos hallazgos / Melida, Jose Ramon – Madrid: Fortanet, 1913. B.R.A.H. 62. pp. 158-163 – 1 – sp Bibl Santa Ana [930]
Excavaciones de merida. memoria de los trabajos practicados en 1926 y 1927 / Melida, Jose Ramon & Macias, Maximiliano – Madrid: rev arch bibl mus, 1929 – 1 – sp Bibl Santa Ana [930]
Excavaciones de ruinas de epoca visigoda en la aldea de san pedro de merida / Almagro Basch, Martin & Marco Pons, Alejandro – Badajoz: Dip. Provincial, 1958. Sep. REE – 1 – sp Bibl Santa Ana [930]
Excavaciones en la antigua cappara (caparra.caceres) / Floriano Cumbreno, Antonio C – sp Bibl Santa Ana [930]
Excavaciones en la sierra de santa cruz / Roso de Luna, Mario – Caceres: tip enc y lib jimenez, 1902 – 1 – sp Bibl Santa Ana [930]
Excavaciones en merida / Floriano Cumbreno, Antonio C – Madrid, 1944 – 1 – sp Bibl Santa Ana [930]
Excavaciones en merida : memoria / Melida, Jose Ramon – Madrid: tip archivos, 1916 – 1 – sp Bibl Santa Ana [930]
Excavating contractor – Southfield. 1912-1990 (1) 1971-1990 (5) 1976-1990 (9) – ISSN: 0014-3995 – mf#508 – us UMI ProQuest [690]
Excavation at agroha, punjab / Srivastava, H L – Delhi: Manager of Publications, 1952 – (= ser Samp: indian books) – us CRL [930]
The excavation of armageddon / Fisher, Clarence S – 1929 – 9 – $10.00 – us IRC [930]
The excavation of gezer / Macalister, R A S – 3v. 1912 – 9 – $42.00 – us IRC [930]
Excavations : biban el mol-k / Davis, T N et al – London, 1912 – 6mf – 9 – mf#NE-20413 – ne IDC [930]
Excavations at ain shems / Mackenzie, Duncan – 2v. 1911-1913 – 9 – $15.00 – us IRC [930]
The excavations at babylon / Koldewey, Robert – Macmillan. 1914 – 9 – $12.00 – us IRC [930]
Excavations at jerusalem 1894-97 / Bliss, Frederick Jones – London: Cttee of the Palestine Exploration Fund, 1898 [mf ed 1988] – 2mf – 9 – 0-7905-0364-6 – (incl ind) – mf#1987-0364 – us ATLA [933]
Excavations at taxila : the stupas and monasteries at jaulian / Marshall, John Hubert – Calcutta: Supt, Govt Print, 1921 – (= ser Samp: indian books) – us CRL [930]
Excavations at the kesslerloch near thayngen, switzerland : a cave of the reindeer period / Merk, Conrad – London: Longmans, Green, 1876 – viii/68p (ill) – 1 – (trans by john edward lee) – us Wisconsin U Libr [930]
Excavations at ur : a record of twelve years work / Woolley, Leonard – Benn Ltd., 1954 – 9 – $12.00 – us IRC [930]
Excavations in baluchistan 1925, sampur mound, mastung, and sohr damb, nal / Hargreaves, Harold – Calcutta: Govt of India, Central Publication Branch, 1929 – (= ser Samp: indian books) – (app by r b seymour sewell) – us CRL [934]
Excavations in palestine during the years 1898-1900 / Bliss, Frederick Jones – London: Cttee of the Palestine Exploration Fund, 1902 [mf ed 1989] – 2mf – 9 – 0-7905-0671-8 – (incl bibl ref & ind) – mf#1987-0671 – us ATLA [933]
Excavations in swat and explorations in the oxus territories of afghanistan : a detailed report of the 1938 expedition / Barger, Evert & Wright, Philip – Delhi: Manager of Publications, 1941 – (= ser Samp: indian books) – us CRL [934]
Excavations in the cuenca region, ecuador / Bennett, Wendell Clark – New Haven, CT. 1946 – 1r – 1 – us UF Libraries [930]
Excavations in the ft liberte region, haiti / Rainey, Froelich Gladstone – New Haven, CT. 1941 – 1r – 1 – us UF Libraries [930]
Excavations of the hill of ophel / Macalister, R A S – 1926 – 9 – $12.00 – us IRC [930]
Ex-cbi roundup / China-Burma-India Veterans Association – v42 n9-v44 n10 [1987 nov-1989 dec] – 1r – 1 – mf#1653156 – us WHS [360]
Excel / Schenley Imports Co – v3 [1985] – 1r – 1 – (cont: little black book) – mf#4775438 – us WHS [071]
Las excelencias...y...propiedades del tabaco... / Ayo, C – Salamanca, 1645 – 1mf – 9 – sp Cultura [630]
Excell, Edwin O *see* Triumphant songs
Excellence – v3 n8 [1979/80 win], v4 n9-12 [1980 spr-win], v5 n13-14 [1981 spr/sum-fall/win], v6 n15 [1982 spr], v7 n16, 17 [1982 sum/fall, 1983 win/spr], v8 n18, 19-20 [1983 sum/fall, 1984 win/spr-fall/win], v10 n25 [1988 spr/sum], v11 n26 [1989 spr/sum]; v8 n21 [1985 spr/sum], v8 n22-23, [1986 spr/sum-fall/win], v9 n24 [1987 spr/sum], v10 n25 [1987 fall-win/1988 win], v12 n27 [1990 spr/sum], v14 n29 [1991/92] – 2r – 1 – (cont: black excellence) – mf#1699056 – us WHS [071]
Excellence of the liturgy / Dealtry, William – London, England. 1829 – 1r – us UF Libraries [240]
Excellence of the liturgy / Woodd, Basil – London, England. 1810? – 1r – us UF Libraries [240]
Excellency of the liturgy / Simeon, Charles – London, England. 1816 – 1r – us UF Libraries [240]
The excellent and pleasant work collectanea rerum memorabilium / Solinus, Caius Julius – 9 – us Scholars Facs [450]
Excellent encouragements against afflictions : containing david's triumph over distress... / Pierson, Thomas – Edinburgh: James Nichol, 1868 [mf ed 1985] – 1mf – 9 – 0-8370-2159-6 – (incl biogr info) – mf#1985-0159 – us ATLA [221]
L'excelsior : journal illustre quotidien – Paris, 16 nov 1910-avr 1935, 11 juin 1940, 16 juin 1942, 16 juin 1943 – 1 – fr ACRPP [074]
Excelsior – Bucharest, Romania. -w. 2 March 1935-8 June 1940. Imperfect. 4 reels – 1 – uk British Libr Newspaper [949]
Excelsior – Omaha, NE: C Chase. 7v. v20 n1. sep 3 1898-v26 n2. jan 14 1905 (wkly) [mf ed with gaps filmed in 1977] – 4r – 1 – (cont and cont by: omaha excelsior) – us NE Hist [071]
Excelsior – Omaha: Chase Pub Co. 5v. v31 n3. jan 17 1914-v36 n53. dec 27 1919 (wkly) [mf ed with gaps filmed 1977] – 2r – 1 – (cont: omaha excelsior (1905). cont by: omaha excelsior (1920)) – us NE Hist [071]
Excelsior – Mexico City, Mexico. 1917-1996 (1) – mf#60212 – us UMI ProQuest [079]
Excelsior – Mexico: DF: Excelsior, mar 18 1917-1918 – 1 – us CRL [079]
Excelsior – Omaha, NE. 1884-1921 (1) – mf#64720 – us UMI ProQuest [071]
Excelsior! : fuer den katholischen familien kreis – Milwaukee, St Paul WI. 1883 sep 8/1884 dec 25, 1885 jan 1/1886 may 20, 1886 may 27/1887 dec 1, 1887 dec 8/1889 aug 8, 1889 aug 15/1891 apr 23, 1891 apr 30/1892 dec 29, 1893 jan 5/1894 aug 23, 1894 aug 30/1895 sep 26, 1895 oct 3/1898 may 19, 1898 may 26/1900 dec 27, 1901-1946 jan 3-sep – 27r – 1 – mf#852196 – us WHS [071]
Excelsior empire savings bank news – New York NY. 1968 jan – 1r – 1 – mf#4865211 – us WHS [332]
Excelsior journal – Port-au-Prince: Imp Les Presses Libres. [v1 n1-13 feb-may 4 1952] – 2 sheets – 9 – us CRL [079]
Excelsitud del verbo / Rymer K, Roberto – Ciudad Trujillo, Dominican Republic. 1958 – 1r – us UF Libraries [972]
Excentricites du langage / Larchey, Loredan – Paris, France. 1865 – 1r – us UF Libraries [470]
Exceptional children – Reston. 1951+ (1) 1951+ (5) 1951+ (9) – (Cont: Journal of exceptional children) – ISSN: 0014-4029 – mf#12546,02 – us UMI ProQuest [640]
Exceptional parent – Brookline. 1971+ (1) 1971+ (5) 1975+ (9) – ISSN: 0046-9157 – mf#6820 – us UMI ProQuest [640]
Exceptionality – New York. 1990+ – 1,5,9 – ISSN: 0936-2835 – mf#17618 – us UMI ProQuest [370]
Exceptorum ex literis...r p johanne francisco gerbillonio 2. & 3. septemb ex urbe nibchou tartariae orientalis ditionis moschicoe propr sinensis imperii fines – 1mf – 9 – mf#HT-570 – ne IDC [915]
Excerpta cypria : materials for a history of cyprus / Cobham, Claude Delaval – Cambridge: University Press; New York: Putnam [distributor], 1908 – 2mf – 9 – 0-7905-5591-3 – mf#1988-1591 – us ATLA [949]
Excerpta de antiquitatibus constantinopolitani (cshb46) / Georgii Codini; ed by Bekkeri, Imm – Bonnae, 1843 – (= ser Corpus scriptorum historiae byzantinae (cshb)) – €14.00 – ne Slangenburg [243]
Excerpta de antiquitatibus constantinopolitanis (cbh12,2) / Georgii Codini; ed by Medonius, B – Parisiis, 1655 – (= ser Corpus byzantinae historiae (cbh)) – €27.00 – ne Slangenburg [243]
Excerpta de legationibus (cbh1,2) / ed by Cantoclarus, Car & Valesii, Notae H – Parisiis, 1648 – (= ser Corpus byzantinae historiae (CBH)) – €21.00 – ne Slangenburg [243]
Excerpta indonesica – Leiden, 1970-1985 – n1-32. 67mf – 9 – (incl special iss: current studies, leiden 1978-1979) – mf#SE-600 – ne IDC [959]
Excerpta isagogarum et categoriarum : formae tplila 92 – [mf ed 1997] – (= ser ILL – ser a; Cccm 120) – 3mf+27p – 9 – €30.00 – 2-503-64202-0 – be Brepols [400]
Excerpts from her diaries relating to her service with the methodist overseas mission in the solomon islands / Harkness, Effie – 15 jul 1941-7 may 1957 – 3r – 1 – mf#PMB1096 – at Pacific Mss [920]
Excerpts from manifests, speeches and interviews / Vargas, Getulio – Rio de Janeiro, Brazil. 1942 – 1r – us UF Libraries [972]
Excerpts from various colorado newspapers for western americana research / Lewis, M – Denver, CO. 1951 – 1r – 1 – mf#MF C714a – us Colorado Hist [978]
Excess postexercise oxygen consumption and energy expenditure of endurance trained and untrained women / Marshall, Kristin R & Maniscalco, Ignatius A – 1993 – 2mf – $8.00 – us Kinesology [612]
Exchange – 1983 jan/feb-1986 aut – 1r – 1 – mf#510506 – us WHS [071]
Exchange – Livingston, 1993-1993 [1,5,9] – (cont: bellcore exchange) – mf#16438,02 – us UMI ProQuest [380]
Exchange / Native American-Philanthropic News Service – v1 n1-v4 n3 [1976 may-1982/83 win] – 1r – 1 – mf#1043711 – us WHS [071]
Exchange – New York. 1939-1975 (1) 1972-1975 (5) (9) – ISSN: 0014-4436 – mf#8099 – us UMI ProQuest [332]
Exchange – Sacramento. 1975-1975 [1,5,9] – mf#9881 – us UMI ProQuest [360]
Exchange and commissary news – Westbury. 1972-1973 (1) – ISSN: 0014-4452 – mf#7314 – us UMI ProQuest [355]
The exchange and mart – Toronto: [s.n, 1884-18– or 19–] – 9 – mf#P04445 – cn Canadiana [071]
Exchange bibliographies / Council of Planning Librarians – n1-400 – 9 – us Brook [500]
Exchange bibliographies : numerical list / Committee of University Industrial Relations Librarians – n1-1266 – 1 – us NY Public [330]
Exchange bibliographies : subject index / Committee of University Industrial Relations Librarians – n1-1300 – 1 – us NY Public [330]
Exchange control as an instrument to regulate south africa's external equilibrium / Bone, Renee Yolande – Rand Afrikaans University 1979 [mf ed 1980] – 6mf [ill] – 9 – (incl bibl ref; incl summary in afrikaans & english) – sa U of Johannesburg [332]
Exchange herald – [NW England] Manchester ALS 30 sep 1809-23 aug 1825 – 1 – (title change: herald. aston's manchester commercial advertiser [sep 1825-sep 1826]) – uk MLA; uk Newsplan [072]
Exchange National Bank. Atchison, Kansas *see* Records
Exchange news – Montreal: M M Sabiston, [1898-1899] – 9 – (cont by: the exchange news and commerical advertiser) – mf#P04805 – cn Canadiana [332]
Exchange news and commerical advertiser – Montreal: M M Sabiston, [1899-1900] – 9 – (cont: the exchange news) – mf#P04806 – cn Canadiana [332]
Exchange of patent rights and technical information under mutual aid programs / Cardozo, Michael H – Washington: GPO, 1958 – 51p – 1 – mf#LL-2303 – us L of C Photodup [346]
Exchange tables advancing by quarter cents from 4.50 to 4.99 3/4 and ranging from one cent to $100,000 : currency into sterling / Rorie, George [comp] – Vancouver: Grant & Sons, 1918 [mf ed 1996] – 2mf – 9 – 0-665-79567-X – mf#79567 – cn Canadiana [332]
Exchangite / National Exchange Club – 1978 feb-1981 dec/1982 jan, 1982 feb-1985 dec – 2r – 1 – (continued by: exchange today) – mf#592286 – us WHS [071]
Exchequer reports of canada / Canada. Exchequer Court – v1-21. 1877-1922 (all publ) – 123mf – 9 – $184.00 – (cont by: canada law reports, exchequer court. not offered by llmc) – mf#LLMC 81-002 – us LLMC [324]
The excise and hotel laws of the state of new york / Becker, Frank Silvester – Rochester, Williamson, 1895. 159 p. LL-672 – 1 – us L of C Photodup [340]
Excision of half the tongue / Armstrong, George E – S.l: s.n, 1898? – 1mf – 9 – mf#38439 – cn Canadiana [617]
Excision of the knee joint : with report of twenty-eight cases / Fenwick, George Edgeworth – Montreal: Dawson, 1884 – 1mf – 9 – mf#44723 – cn Canadiana [617]

La exclusiva dada por espana contra el cardenal ciustiniani en el conclave de 1830-1831, segun los despachos diplomaticos / March, Jose – Madrid: Razon y Fe, 1932 – 1 – sp Bibl Santa Ana [946]

Exclusive – v18 n17-v21 n38 [1972 apr 21-1975 nov3] – 1r – 1 – mf#367508 – us WHS [071]

Exclusive magazine – 1989 apr – 1r – 1 – mf#4851948 – us WHS [071]

El excmo sr d xavier maria de munibe, conde de penaflorida / Altube, Gregorio de – San Sebastian, 1932; Madrid: Razon y Fe, 1933 – 1 – sp Bibl Santa Ana [946]

Excommunication / Lord, James – London, England. 1846 – 1r – us UF Libraries [240]

Les excommunies : a m auguste vermond, depute de seine-et-oise (de passage a montreal) – S.l: s.n, 189-? – 1mf – 9 – mf#53480 – cn Canadiana [810]

Excurse ueber oesterreichisches buergerliches recht : beilagen zum commentar / Pfaff, Leopold & Hofmann, Franz – Wien, Manz. 2v. 1877-78 – (= ser Civil law 3 coll) – 9mf – 9 – (iss in pts) – mf#LLMC 96-615 – us LLMC [346]

Excursion a fregenal de la sierra y los jarales... / Barras de Aragon, Francisco de las – Caceres: tip enc y lib jimenez, 1905 – 1 – sp Bibl Santa Ana [946]

Excursion a la cote du nord, au dessous de quebec : nouvel etablissement aux escoumins / Faribault, Georges-Barthelemi & Ferland, Jean-Baptiste-Antoine – Quebec: Atelier typographique du "Canadien"...1849 [mf ed 1984] – 1mf – 9 – mf#SEM105P413 – cn Bibl Nat [917]

Une excursion a l'ile aux coudres / Casgrain, Henri-Raymond – Montreal: Librairie Beauchemin, lte, 1925 [mf ed 1992] – (= ser Bibliotheque canadienne. coll dollard) – 2mf – 9 – mf#SEM105P1596 – cn Bibl Nat [390]

Excursion a vueltabajo / Vilaverde, Cirilo – Habana, Cuba. 1961 – 1r – us UF Libraries [972]

Una excursion al real monasterio de guadalupe. madrid, 1927 / Zurbitu, D – Madrid: Razon y Fe, 1928 – 9 – sp Bibl Santa Ana [240]

Excursion al territorio de san martin / Restrepo Echavarria, Emiliano – Bogota, Colombia. 1957 – 1r – us UF Libraries [972]

Excursion al territorio de san martin en diciembre / Restrepo Echavarria, Emiliano – Bogota, Colombia. 1955 – 1r – us UF Libraries [972]

Excursion dans l'amerique du sud : esquisses et souvenirs / Duchatellier, Armand – Paris 1828 [mf ed Hildesheim 1995-98] – 1v on 2mf – 9 – €60.00 – ISBN-10: 3-487-26922-8 – ISBN-13: 978-3-487-26922-1 – gw Olms [918]

Excursion epigrafica por villar del rey, alhambra, venta de los santos, cartagena, logrono y orense / Fita, Fidel – Madrid: Fortanet, 1903 – 1 – sp Bibl Santa Ana [946]

Excursion por el campo teosofico / Bayle, Constantino – Razon y Fe, 1929 – 1 – sp Bibl Santa Ana [240]

Excursion por el campo teosofico 2 / Bayle, Constantino – Madrid: Razon y Fe, 1929 – 1 – sp Bibl Santa Ana [290]

An excursion through the united states and canada during the years 1822-23 / Blane, William N – London 1824 [mf ed Hildesheim 1995-98] – 1v on 4mf – 9 – €120.00 – 3-487-27056-0 – gw Olms [917]

An excursion to the highlands of scotland and the english lakes : with recollections, descriptions, and references to historical facts / Mawman, Joseph – London 1805 [mf ed Hildesheim 1995-98] – 1v on 2mf – 9 – €60.00 – 3-487-27907-X – gw Olms [914]

Excursiones briologicas por, la provincia de badajoz / Fructucso y Tristancho, Gonzalo – Madrid: Imprenta de Fortanet, 1914 – 1 – sp Bibl Santa Ana [946]

Excursiones epigraficas. de monesterio a merida / Fita, Fidel – Madrid: Tip. Fortanet, 1894 – 1 – sp Bibl Santa Ana [240]

Excursions! : a guide to multicultural travel – 1996 spring – 1r – 1 – mf#3661061 – us WHS [910]

Excursions in africa : from the royal commonwealth society library, london / Bowdich, Thomas Edward – 1825 – 8mf – 7 – (incl notes on madeira and porto santo (river gambia), zoological and botanical data) – mf#2978 – uk Microform Academic [916]

Excursions in india : including a walk over the himalaya mountains, to the sources of the jumna and the ganges / Skinner, Thomas – London 1833 [mf ed Hildesheim 1995-98] – 2v on 4mf – 9 – €120.00 – 3-487-27461-2 – gw Olms [915]

Excursions in madeira and porto santo... / Bowdich, S – Ann Arbor. 1975-1975 (1) – 4mf – 9 – mf#8518 – ne IDC [590]

Excursions in new south wales, western australia, and van dieman's land : during the years 1830, 1831, 1832, and 1833 – London 1833 [mf ed Hildesheim 1995-98] – 1v on 3mf – 9 – €90.00 – ISBN-10: 3-487-26803-5 – ISBN-13: 978-3-487-26803-3 – gw Olms [919]

Excursions in north america : described in letters from a gentleman and his young companion, to their friends in england / Wakefield, Priscilla – London: printed & sold by Darton and Harvey...1806 [mf ed 1983] – 5mf – 9 – 0-665-41763-2 – mf#41763 – cn Canadiana [860]

Excursions in southern africa : including a history of the cape colony, an account of the native tribes, etc / Napier, Edward Delaval Hungerford Elers – London 1849 – (= ser 19th c british colonization) – 2v on 10mf – 9 – mf#1.1.2932 – uk Chadwyck [960]

Excursions in southern africa : including a history of the cape colony, an account of the native tribes, etc / Napier, Edward Delaval Hungerford Elers – London: William Shoberl, Publ 1850 – (= ser [Travel descriptions from south africa, 1711-1938]) – 2v on 10mf – 9 – mf#zah-39 – ne IDC [916]

Excursions in the county of cornwall : comprising a concise historical and topographical delineation of the principal towns and villages...; forming a complete guide for the traveller and tourist / Stockdale, Frederick – London 1824 [mf ed Hildesheim 1995-98] – 1v on 3mf – 9 – €90.00 – 3-487-27919-3 – gw Olms [914]

Excursions in the county of suffolk : comprising a brief historical and topographical delineation of every town and village...; forming a complete guide for the traveller and tourist / Cromwell, Thomas – London 1818-19 [mf ed Hildesheim 1995-98] – 2v on 4mf – 9 – €120.00 – 3-487-27921-5 – gw Olms [914]

Excursions in the holy land, egypt, nubia, syria, &c : including a visit to the unfrequented district of the haouran / Madox, John – London 1834 [mf ed Hildesheim 1995-98] – 2v on 6mf – 9 – €120.00 – ISBN-10: 3-487-26695-4 – ISBN-13: 978-3-487-26695-4 – gw Olms [915]

Excursions in the north of europe : through parts of russia, finland, sweden, denmark, and norway, in the years 1830 and 1833 / Barrow, John – London 1834 [mf ed Hildesheim 1995-98] – (= ser Fbc) – 3mf [ill] – 9 – €90.00 – 3-487-28938-5 – gw Olms [914]

Excursions of an evolutionist / Fiske, John – London: MacMillan, 1884 – 1mf – 9 – 0-7905-9272-X – mf#1989-2497 – us ATLA [575]

Excuse de jehan calvin a messieurs les nicodemites : sur la complaincte qu'ilz font de sa trop grand' rigueur / Calvin, J – [Geneva: Jean Girard], 1544 – 1mf – 9 – mf#CL-55 – ne IDC [242]

Excuse de noble seigneur, jaques de bourgoigne, s de fallez et bredam : pour se purger vers la M Imperiale, des calomnies a luy imposees, en matiere de sa foy, dont il rend confession / [Calvin, J] – [Geneva: Jean Girard, 1547-1548] – 1mf – 9 – mf#CL-28 – ne IDC [240]

Excuse me / Hughes, Rupert – New York, NY. 1934 – 1r – us UF Libraries [960]

The execution of justice in england / Burghley, William Cecil – 1583 – 9 – 5.00 – us Scholars Facs [941]

Executive – Boston. 1956-1964 (1) – ISSN: 0531-5190 – mf#5096 – us UMI ProQuest [650]

Executive – Ada. 1990-1993 (1,5,9) – (cont: academy of management executive; cont by: academy of management executive) – mf#16369,01 – us UMI ProQuest [650]

Executive – Ithaca. 1979-1981 (1) 1979-1981 (5) 1979-1981 (9) – (Cont by: Cornell executive) – ISSN: 0145-3963 – mf#12202 – us UMI ProQuest [320]

Executive / Wisconsin Savings and Loan League – 1974 mar-1976 dec – 1r – 1 – (cont: bulletin [wisconsin savings and loan league]; cont by: savings institutions executive) – mf#365076 – us WHS [332]

Executive agreement series / U.S. Dept of State – n1-506. 1929-45 [all publ] – 74mf – 9 – $111.00 – (cont: the treaty series; cont by: treaties and other international acts of the united states) – mf#llmc 80-906 – us LLMC [341]

Executive board minutes / Indiana. State Convention of Baptists – 16 jan 1975-10 jan 1980 – 184p – 1 – us Southern Baptist [242]

Executive board minutes, 1883-86 / South Carolina. Orangeburg Association. First Union Division – 92p – 1 – (treasurer's report, 1883-89; financial report, 1889-1918) – us Southern Baptist [360]

Executive briefing – New York. 1986-1993 (1) 1986-1993 (5) 1986-1993 (9) – ISSN: 0898-7912 – mf#6423,03 – us UMI ProQuest [338]

Executive committee minutes / Typographical Union No 6. New York City – New York. 1870-1884; apr 1893-nov 1899 – 1 – us NY Public [070]

Executive committee minutes, 21 apr 1921-5 mar 1925, and finance board minutes, 25 mar 1925-31 jul 1934 / Melanesian Mission – 2r – 1 – mf#PMB1092 – at Pacific Mss [350]

Executive control of rulemaking : the office of administrative law in california / Price, Monroe E – Washington: GPO, Apr 1981 (all publ) – 2mf – 9 – $3.00 – mf#LLMC 94-347 – us LLMC [340]

Executive educator – Alexandria. 1979-1996 – 1,5,9 – ISSN: 0161-9500 – mf#12394 – us UMI ProQuest [370]

Executive excellence – Provo. 1988+ (1,5,9) – ISSN: 8756-2308 – mf#16428 – us UMI ProQuest [650]

Executive health report – Rancho Santa Fe. 1990-1990 (1) 1990-1990 (5) 1990-1990 (9) – ISSN: 0882-2131 – mf#16728,01 – us UMI ProQuest [360]

Executive housekeeper – Philadelphia. 1968-1979 (1) 1971-1979 (5) 1976-1979 (9) – ISSN: 0014-455X – mf#3377 – us UMI ProQuest [360]

Executive intelligence review / New Solidarity International Press Service – 1977 jan/apr 5-1995 sep 1/dec 15 [gaps] – 53r – 1 – (cont: new solidarity international press service weekly news) – mf#1265489 – us WHS [070]

Executive letter / Wisconsin Association of Manufacturers and Commerce – 1979 jan 5-1984 feb 24 – 1r – 1 – (continued by: legislative report, shaffer, fred; executive/legislative insight) – mf#710888 – us WHS [338]

Executive order – n1-101 [1979-1982 dreyfus], n1-119 [1983-1987 earl] – 2r – 1 – mf#423942 – us WHS [071]

Executive order / U.S. President – Washington, DC. 1845-1936 – n1-7403 – 1 – mf#LL-033 – us L of C Photodup [340]

Executive orders, 1-7521, 1862-1936 / U.S. President – (= ser General records of the United States government) – 17r – 1 – (With printed guide) – mf#M1118 – us Nat Archives [975]

Executive orders in times of war and national emergency : report of the special committee on national emergencies and delegated emergency powers, united states senate / U.S. Congress. Senate. Special Committee on National Emergencies and Delegated Emergency Powers – Washington: govt print off, 1974 – 283p – 1 – mf#LL-2364 – us L of C Photodup [323]

Executive sessions / U.S. Congress. Senate. Foreign Relations Committee – v1-11. 1947-59 – 9 – $275.00 set – 0-89941-270-X – mf#400390 – us Hein [324]

Executive speeches – Dayton. 1986+ (1,5,9) – ISSN: 0888-4110 – mf#16321 – us UMI ProQuest [650]

Executive/legislative insight – v6 n5-43 [1984 mar 9-dec 14], v7 n1-v11 n40, [1985 jan 2-1989 dec 19], v12 n1-10 [1990 jan 2-mar 6] – 1r – 1 – mf#1111449 – us WHS [071]

Die exegese bei den franzoesischen israeliten vom 10. bis 14. jahrhundert / Levy, Antoine – Leipzig: Oskar Leiner, 1873 – 1mf – 9 – 0-8370-4093-0 – (in german and hebrew) – mf#1985-2093 – us ATLA [221]

Die exegese der siebzig wochen daniels in der alten und mittleren zeit / Fraidl, Franz – Graz [Austria]: Leuschner & Lubensky, [ca 1883] – 2mf – 9 – 0-8370-3175-3 – mf#1985-1175 – us ATLA [221]

Exegesen der funf bucher moses / Pollatschek, Isak – s.l, s.l? no date – 1r – 1 – us UF Libraries [221]

Exegesis / Church of Nature – 1978 dec-1986 dec 28 – 1r – 1 – mf#1152846 – us WHS [071]

Exegesis of 1 corinthians 14., 34,35 : and 1 timothy 2., 11,12 / Blackwell, Antoinette Louisa Brown – [Oberlin: Fitch, 1849] [mf ed 1984] – (= ser Women & the church in america 147) – 1mf – 9 – 0-8370-1147-7 – mf#1984-2147 – us ATLA [227]

An exegesis on marriage and divorce : an appeal for reform / Rosenberger, Isaac J – Covington OH: IJ Rosenberger 1899 [mf ed 1992] – 1mf – 9 – 0-524-04234-9 – mf#1990-5025 – us ATLA [241]

Exegesis perspicva et ferme integra contraversiae de sacra coena / Curaeus, J – Heidelbergae, 1575 – 5mf – 9 – mf#TH-1 mf 365-369 – ne IDC [242]

An exegetical commentary on the gospel according to s matthew / Plummer, Alfred – New York: Charles Scribner; London: Elliot Stock, 1909 [mf ed 1986] – 2mf – 9 – 0-8370-9573-5 – (incl ind) – mf#1986-3573 – us ATLA [226]

Exegetical essays on several words relating to future punishment / Stuart, Moses – Philadelphia: Presbyterian Board of Publ c1867 [mf ed 1992] – 1mf – 9 – 0-524-05091-0 – mf#1991-2215 – us ATLA [221]

The exegetical method of the author of hebrews / Martin, Raymond A – Princeton, NJ: Cornell University, 1952. Chicago: U of Chicago Lib, 1976 (1r); Evanston: American Theol Lib Assoc, 1984 (1r) – 1 – 0-8370-1555-3 – mf#1984-T007 – us ATLA [221]

Exegetical study of the original scriptures considered / Black, Alexander – Edinburgh, Scotland. 1856 – 1r – us UF Libraries [220]

Exegetische probleme des hebraeer- und galaterbriefs / Zimmer, Friedrich – Hildburghausen: F W Gadow, 1882 – 1mf – 9 – 0-8370-9599-9 – (Incl indes) – mf#1986-3599 – us ATLA [227]

Die exegetische terminologie der juedischen traditionsliteratur / Bacher, Wilhelm – Leipzig: J.C. Hinrichs, 1905 – 2mf – 9 – 0-8370-1683-5 – (incl bibl ref and index) – mf#1987-6110 – us ATLA [470]

Die exegetische terminologie der juedischen traditionsliteratur : erster teil: die bibelexegetische terminologie der tannaiten – zweiter teil: die bibel-und traditionsexegetische terminologie der amoraeer / Bacher, Wilhelm – Leipzig, 1899 – €12.00 – ne Slangenburg [221]

Exegetische und homiletische schriften / Hippolytus, Antipope; ed by Bonwetsch, Gottlieb Nathanael & Achelis, Hans – Leipzig: JC Hinrichs, 1897 [mf ed 1993] – (= ser Hippolytus werke 1; Griechische christliche schriftsteller der ersten drei jahrhunderte (leipzig, germany)) – 2v on 2mf – 9 – 0-524-07654-5 – mf#1992-1095 – us ATLA [221]

Exegetisches zur irrtumslosigkeit und eschatologie jesu christi / Weiss, Karl – Muenster i W: Aschendorff, 1916 [mf ed 1993] – (= ser Neutestamentliche abhandlungen 5/4-5) – 1mf – 9 – 0-524-06222-6 – (incl bibl ref) – mf#1992-0860 – us ATLA [225]

Exegetisch-kritische aehrenlese zum alten testament / Boettcher, Friedrich – Leipzig: F C W Vogel, 1849 – 1mf – 9 – 0-7905-1026-X – (In German and Hebrew. Incl ind) – mf#1987-1026 – us ATLA [221]

Exegetisch-kritische verhandeling over den brief van paulus aan de galatiers / Baljon, Johannes Marinus Simon – Leiden: E J Brill, 1889 – 1mf – 9 – 0-524-05654-4 – mf#1992-0504 – us ATLA [227]

Exegi monumentum and lyrics / Datta, D C – [Calcutta: Stephen Allen], 1941 – (= ser Samp: indian books) – us CRL [780]

Exeland enterprise – Exeland WI. 1918 may 30 – 1r – 1 – mf#961937 – us WHS [071]

Exelsior – Paris, France. 1 jul 1917-23 dec 1918; 8 feb-11 aug 1919; 26 mar 1929; 30 aug 1939-29 may 1940 – 7 1/2r – 1 – uk British Libr Newspaper [072]

Die exempla des jacob von vitry : ein beitrag zur geschichte der erzaehlungsliteratur des mittelalters. teil 1 / Frenken, Goswin – Muenchen, 1914 (mf ed 1994) – 1mf – 9 – €24.00 – 3-8267-3083-6 – (only pt 1 is publ) – mf#DHS-AR 3083 – gw Frankfurter [430]

Exemplar of divine worship / Nickolls, R B – London, England. 1805 – 1r – us UF Libraries [240]

Exemplar...antonio de la visitacion / Santa Ana, Juan de – 1758 – 9 – sp Bibl Santa Ana [240]

Exemplarische organisten-probe im artikel vom general-bass : welche mittelst 24. leichter und eben so viel etwas schwerer exempel aus allen tonen des endes anzustellen ist... / Mattheson, Johann – Hamburg: Im Schiller- und Kissnerischen Buch-Laden 1719 [mf ed 19–] – 7mf – 9 – mf#fiche 36 – us Sibley [780]

Exemplars of tudor architecture : adapted to modern habitations / Hunt, Thomas Frederick – [London] 1836 – (= ser 19th c art & architecture) – 3mf – 9 – mf#4.2.1125 – uk Chadwyck [720]

O exemplo – Rio Bonito, RJ. 05 ago 1909 – (= ser Ps 19) – bl Biblioteca [079]

Exemptions, special committee of the council of the corporation of toronto, john hallam, chairman, 1876 : letter addressed to the hon oliver mowat, qc, attorney-general and premier, etc etc etc, province of ontario – [Toronto?: s.n, 1876?] – 1mf – 9 – 0-665-92241-8 – mf#92241 – cn Canadiana [336]

Exequiae in templo s nazarii manfredo septalio patritio mediolanensi : eiusdem basilicae canonico celebratae – Mediolani: Apud impressores archiepiscopales, 1680 – 2mf – 9 – mf#O-1967 – ne IDC [090]

Exercice des commercans : contenant des assertions consulaires sur l'edit du mois de nov 1563, le titre 16 de l'ordonnance du mois d'avril 1667... – Paris: Chez Valade...1776 [mf ed 1984] – 8mf – 9 – 0-665-18505-7 – mf#18505 – cn Canadiana [346]

Exercice des commercans : contenant des assertions consulaires sur l'edit du mois de novembre 1563, le titre 16 de l'ordonnance du mois d'avril 1667 – A Paris: Chez Valade, libraire...1776 [mf ed 1984] – 8mf – 9 – 0-665-18505-7 – (incl ind) – mf#18505 – cn Canadiana [380]

Exercice du pouvoir actuel et les desiderata du pe... / Bernard, Dominique F – Port-Au-Prince, Haiti. 1925 – 1r – us UF Libraries [972]
Exercice tres devot envers s antoine de padoue le thaumaturge : de l'ordre seraphique de s francois, avec un petit recueil de quelques principaux miracles / Monceaux, Alexis du & Colnago, Bernard – Montreal: James Brown, 1813 – 1r – 5 – mf#SEM16P26 – cn Bibl Nat [241]
Exercice tres devot envers s antoine de padoue le thaumaturge : de l'ordre seraphique de s francois, avec un petit recueil de quelques principaux miracles / Monceaux, Alexis du & Colnago, Bernard – Quebec: Nouvelle Impr, 1804 – 1r – 5 – mf#SEM16P25 – cn Bibl Nat [241]
Exercice tres devot envers st antoine de padoue, le thaumaturge, de l'ordre seraphique de st francois : avec un petit recueil de quelques principaux miracles / Monceaux, Alexis du – [Montreal?: s.n.] 1843 [mf ed 1985] – 2mf – 9 – 0-665-17055-6 – mf#17055 – cn Canadiana [241]
Exercices et evolutions d'infanterie tels que revises par ordre de sa majeste, 1862 / Suzor, Louis-Timothee [comp] – Quebec: impr par Geo Desbarats, 1863 [mf ed 1991] – 4mf – 9 – mf#SEM105P1286 – cn Bibl Nat [355]
Exercices francais : calques sur les principes de la grammaire selon l'academie / Bonneau & Lucan – nouv ed. Quebec: I P Dery, 1873 [mf ed 1984] – 3mf – 9 – 0-665-38136-0 – mf#38136 – cn Canadiana [440]
Exercices orthographiques : cours de premiere annee mis en rapport avec l'extrait de la grammaire des freres des ecoles chretiennes: livre de l'eleve / FPB – Montreal: freres des ecoles chretiennes, 1875 [mf ed 1992] – 4mf – 9 – mf#SEM105P1655 – cn Bibl Nat [440]
Exercices orthographiques / Fabre, Abel – [S.l.]: [s.n.], 1886 [mf ed 1974] – 1r – 5 – mf#SEM16P80 – cn Bibl Nat [440]
Exercices orthographiques mis en rapport : avec la grammaire francaise a l'usage des ecoles chretiennes – 1st ed. [Montreal?: s.n.] 1846 [mf ed 1984] – 3mf – 9 – 0-665-43085-X – mf#43085 – cn Canadiana [440]
Exercices spirituels d'apres saint ignace = Exercitia spiritualia / Ignatius of Loyola, Saint – Paris: Rene Haton. 3v. 1890 – 6mf – 9 – 0-8370-7069-4 – mf#1986-1069 – us ATLA [240]
Exercicio de las velas...divino maestro – 1815 – 9 – sp Bibl Santa Ana [240]
Exercicios de desagravios de christo senor nuestro en la cruz : en los tres dias de carnestolendas, por las grandissimas injurias, y agravios conque ofenden a su bondad estos tres dias / Espíndola, Nicolas de – en Mexico: impr Biblioteca, enfrente de S Augustin, ano de 1759 – (= ser Books on religion...1543/44-c1800: doctrina cristiana, obras de devocion) – 16lea on 1mf – 9 – mf#crl-40 – ne IDC [241]
Exercicios litterarios do club scientifico – Sao Paulo, SP: Typ Dous de Dezembro de Antonio Louzada Antunes, ago-out 1859 – (= ser Ps 19) – mf#P17,02,219 – bl Biblioteca [500]
Exercise adherence : effectiveness of a broad-based adult fitness program / Culligan, C T – 1991 – 1mf – 9 – $4.00 – us Kinesology [150]
Exercise adoption and adherence of college women / Merrill, Joanne & Mann, Betty J – 1992 – 2mf – $8.00 – us Kinesology [150]
Exercise exchange – Clarion. 1974+ (1) 1974+ (5) 1974+ (9) – ISSN: 0531-531X – mf#10312 – us UMI ProQuest [370]
Exercise induced hypoxemia as a determinant of maximal aerobic capacity / Emery, Michael S – 1997 – 1mf – 9 – $4.00 – mf#PH 1619 – us Kinesology [612]
Exercise mode comparisons of acute energy expenditure during moderate intensity exercise in obese adults / Kim, Jong-Kyung – 1999 – 1mf – 9 – $4.00 – mf#PH 1634 – us Kinesology [612]
Exercise oxygen uptake in 3- through 6-year-old children / Shuleva, K M – 1989 – 1mf – 9 – $4.00 – us Kinesology [612]
Exercise participation, self efficacy, and fear of falling in older adults / Nunley, Danya C – 1999 – 2mf – 9 – $8.00 – mf#PSY 2108 – us Kinesology [612]
Exercise vs imipramine in the treatment of clomipramine-induced depression in male rats / Yoo, Ho S – 1995 – 2mf – 9 – $8.00 – mf#PSY 1970 – us Kinesology [150]
Exercise-induced muscle damage : role of the calpain-calpastatin system in skeletal muscle myofibrillar protein composition / Ball, Chad G – 1988 – 2mf – 9 – $8.00 – mf#PE 3877 – us Kinesology [617]
Exercises d'imagination de differens caracteres et formes humaines... / Goez, J F de – Augsburg, 1783-1784 – 6mf – 9 – mf#O-1152 – ne IDC [700]
Exercises in arithmetic for use in the junior classes of public schools, pt 1 : a collection of problems suitable for first, second and third book classes... / Cuthbert, W Nelson – Toronto: Copp, Clark, 1896 – 2mf – 9 – mf#26056 – cn Canadiana [510]
Exercises in composition for fourth and fifth classes / Henderson, George E et al – Toronto: Educational Pub Co, 1898 – (= ser School helps series) – 2mf – 9 – mf#06750 – cn Canadiana [420]
Exercises in grammar / Henderson, George E et al – Toronto: Educational Pub Co, 1897 – 2mf – 9 – mf#34256 – cn Canadiana [420]
Exercising authority : a critical history of exercise messages in popular magazines, 1925-1968 / Shaulis, Dahn E – 2001 – 286p on 3mf – 9 – $15.00 – mf#PE 4199 – us Kinesology [302]
Exercitation 36...el letargo... / Luna Vega, J – Sevilla, 1617 – 1mf – 9 – sp Cultura [610]
Exercitation 37 : censura al discurso... / Luna Vega, J – s.l, 1617 – 1mf – 9 – sp Cultura [610]
Exercitation medica phylosophica sobre la essencia del morbo gallico / Bonilla Samaniego, A – Cordoba, 1664 – 1mf – 9 – sp Cultura [610]
Exercitationes de gratia universali / Spanheim, F – Leyde, Maire, 1646. 3 v – 29mf – 9 – mf#PFA-175 – ne IDC [240]
Exercitationes in orationem dominicam / Witsius, H – Franequerae, 1689 – 3mf – 9 – mf#PBA-415 – ne IDC [240]
Exercitationes musicae theoreticopracticae curiosae de concordantiis singulis : das ist musicalische wissenschafft und kunst-ubungen von jedweden concordantien, in welchen jeglicher concordantz natur und wesen / Printz, Wolfgang Caspar – Dresden: In Verlegung J C Miethens 1689 [mf ed 19–] – 9v on 6mf – 9 – mf#fiche 648, 832 – us Sibley [780]
Exercitationes sacrae in symbolum quod apostolorum dicitur / Witsius, H – [Franequerae], 1681 – 6mf – 9 – mf#PBA-402 – ne IDC [240]
Exercitationes selectae historico-theologicae / Leydecker, M – Amstelodami, 1712. 2v – 13mf – 9 – mf#PBA-226 – ne IDC [240]
Exercitationes sex varii argumenti / Wagenseil, J C – Altdorfi Noricorum: Joh. Henricus Sch"nnerstaed excudit, 1687 – 3mf – 9 – mf#O-1971 – ne IDC [090]
Exercitationes theologicae / Wittichius, C – Lugduni Batavorum, 1682 – 5mf – 9 – mf#PBA-404 – ne IDC [240]
Exercitia et bibliotheca, studiosi theologiae / Voetius, G – Rheno Trajecti, 1644 – 10mf – 9 – mf#PBA-392 – ne IDC [240]
Exercitia spirituallia : gli esercizii spirituali di sant'ignazio / Ignatius of Loyola, Saint – Torino: S Giuseppe. 2v. 1892 – 4mf – 9 – 0-8370-6907-6 – (in italian) – mf#1986-0907 – us ATLA [240]
Exercitia spiritualia : Meditationibus illustrata ad usum pp. ac ff. societatis jesu / Ignatius of Loyola, Saint – Cincinnati: Typis Hermanni Lehmann, 1849 – 1mf – 9 – 0-8370-9250-7 – mf#1986-3250 – us ATLA [240]
Exercutatio 31...versatur iusta...galeni doctrinam... / Luna Vega, J – s.l, 1613 – 1mf – 9 – sp Cultura [610]
Die exerzitien des ignatius von loyola in den dramen jakob bidermanns sj / Nachtwey, Hermann Joseph – [S.l: s.n.], 1937 (Bochum-Langendreer: Druck H Poeppinghaus) [mf ed 1989] – vii/92p – 1 – (incl bibl) – mf#7019 – us Wisconsin U Libr [430]
Exeter Diocesan Board Of Education *see* Annual report of the exeter diocesan board of education
Exeter enterprise – Exeter, NE: Wm A Connell. 31v. v1 n1. sep 29 1877-v13 n42. may 24 1890; 13th yr n42. jun 1 1890-28th yr n31. may 4 1906; v28 n32. may 11 1906-v31 n36. jun 4 1909 (wkly) [mf ed 1877-82,1884-87,1889-1902,1905-09 (gaps) filmed 1971] – 6r – 1 – (suspended dec 1 1877; resumed jan 12 1878. merged with: nebraska signal (geneva ne) to form: nebraska signal and the exeter enterprise) – us NE Hist [071]
[Exeter-] exeter sun – CA. 1982 – 13+ r – 1 – $780.00 (subs $90/y) – mf#B03217 – us Library Micro [071]
Exeter hall lectures to young men – London. 1845-1865 (1) – mf#2794 – us UMI ProQuest [080]
Exeter mercury or weekly intelligence of news – England.24 Sept 1714-30 Sept 1715; 11 Oct 1715. 1/2 reel – 1 – uk British Libr Newspaper [072]
Exeter, New Hampshire. Exeter Baptist Church *see* Records
Exeter times advocate – Exeter, Ontario, Canada.1980-1979 sic – 1 – cn Commonwealth Imaging [071]
Exhalaciones del alma / Moreno Torrado, Luis – Badajoz: Tip La Minerva Extremena, 1885 – 1 – sp Bibl Santa Ana [440]
Exhaust / International Union, United Automobile, Aircraft and Agricultural Implement Workers of America – v10 n10-v24 n9 [1964 jan-1978 oct] – 1r – 1 – mf#500521 – us WHS [629]
An exhibit of german military documents from the heeresarchiv potsdam, 1679-1935 – (= ser National archives coll of foreign records seized, 1941-) – 2r – 1 – (captured german records) – mf#M129 – us Nat Archives [355]
Exhibition by the italian futurist painters / Sackville Gallery. London – 1912 – 9 – $4.90 – uk Chadwyck [750]
Exhibition catalogs of the hermitage : from the library of the state hermitage museum, st petersberg – 1917-72 [mf ed 2000 Norman Ross Publ] – 379mf – 9 – (in russian. incl printed guide) – us UMI ProQuest [060]
[Exhibition catalogue 1843] / Great Britain. Royal Commission of Fine Arts, Westminster Hall – London 1843 – (= ser 19th c art & architecture) – 1mf – 9 – mf#4.2.671 – uk Chadwyck [700]
[Exhibition catalogue 1844] / Great Britain. Royal Commission of Fine Arts, Westminster Hall – London 1844 – (= ser 19th c art & architecture) – 1mf – 9 – mf#4.2.672 – uk Chadwyck [700]
[Exhibition catalogue 1845] / Great Britain. Royal Commission of Fine Arts, Westminster Hall – London 1845 – (= ser 19th c art & architecture) – 1mf – 9 – mf#4.2.673 – uk Chadwyck [700]
[Exhibition catalogue 1847] / Great Britain. Royal Commission of Fine Arts, Westminster Hall – London 1847 – (= ser 19th c art & architecture) – 1mf – 9 – mf#4.2.674 – uk Chadwyck [700]
[Exhibition catalogue. 1888] : catalogue of the [1st] exhibition / Arts and Crafts Exhibition Society, London – [London] 1888 – (= ser 19th c art & architecture) – 3mf – 9 – mf#4.2.621 – uk Chadwyck [700]
[Exhibition catalogue. 1889] : catalogue of the [2nd] exhibition / Arts and Crafts Exhibition Society, London – [London] 1889 – (= ser 19th c art & architecture) – 3mf – 9 – mf#4.2.622 – uk Chadwyck [700]
[Exhibition catalogue. 1890] : catalogue of the [3rd] exhibition / Arts and Crafts Exhibition Society, London – [London] 1890 – (= ser 19th c art & architecture) – 3mf – 9 – mf#4.2.623 – uk Chadwyck [700]
[Exhibition catalogue. 1893] : catalogue of the [4th] exhibition / Arts and Crafts Exhibition Society, London – [London] 1893 – (= ser 19th c art & architecture) – 2mf – 9 – mf#4.2.624 – uk Chadwyck [700]
[Exhibition catalogue. 1896] : catalogue of the [5th] exhibition / Arts and Crafts Exhibition Society, London – [London] 1896 – (= ser 19th c art & architecture) – 2mf – 9 – mf#4.2.625 – uk Chadwyck [700]
[Exhibition catalogue. 1899] : catalogue of the [6th] exhibition / Arts and Crafts Exhibition Society, London – [London] 1899 – (= ser 19th c art & architecture) – 2mf – 9 – mf#4.2.626 – uk Chadwyck [700]
Exhibition catalogues, 1919-1959 = Catalogues d'exposition, 1919-1959 / National Gallery of Canada & Galerie Nationale du Canada – 2nd printing 1991 – 167mf – 9 – Can$675.00 – (with p/g) – cn McLaren [700]
Exhibition critic illustrated – Montreal: J L Wiseman, [1880] – 9 – mf#P04704 – cn Canadiana [060]
Exhibition exposite and advertiser – Dublin, Ireland. May-oct 1853. -w. 1/2r – 1 – uk British Libr Newspaper [072]
Exhibition herald *see* Courrier des expositions
Exhibition illustrative of the french revival of etching / Burlington Fine Arts Club, London – London 1891 – (= ser 19th c art & architecture) – 1mf – 9 – mf#4.1.304 – uk Chadwyck [760]
Exhibition news – [NW England] Manchester, Stalybridge Lib 8 sep 1880 – 1 – uk MLA; uk Newsplan [072]
Exhibition of drawings and studies by sir edward burne-jones / Burlington Fine Arts Club, London – London 1899 – (= ser 19th c art & architecture) – 1mf – 9 – mf#4.1.303 – uk Chadwyck [740]
Exhibition of fine arts, manufactures, machines...etc : opened on wednesday, 24th may 1865 / Hamilton and gore mechanics' institute – [Hamilton, Ont?: s.n.] 1865 [mf ed 1994] – 1mf – 9 – 0-665-94622-8 – mf#94622 – cn Canadiana [700]
The exhibition of the liverpool academy 1813 – Liverpool [1813] – (= ser 19th c art & architecture) – 1mf – 9 – mf#4.2.1694 – uk Chadwyck [700]
Exhibition of the works of frederick j shields – Manchester 1875 – (= ser 19th c art & architecture) – 1mf – 9 – mf#4.1.285 – uk Chadwyck [700]
Exhibition of the works of industry of all nations, 1851 / Dilke, Charles Wentworth – [London] 1855 – (= ser 19th c art & architecture) – 3mf – 9 – mf#4.2.1792 – uk Chadwyck [700]
Exhibition of the works of sir john e millais / Grosvenor Gallery, London – London 1886 – (= ser 19th c art & architecture) – 2mf – 9 – mf#4.1.290 – uk Chadwyck [700]
Exhibitor's trade review; of, for, and by the motion picture exhibitor – New york. 9 dec 1916-27 feb 1926 – 1 – us L of C Photodup [770]
Exhibits / International Military Tribunal for the Far East – Tokyo. In Japanese. On film: Documents 238-277, 286-473, 476-838, 845-1045, 1107-1225, 1227-1245, 1249-1764, 1766-1820; 1946-48. LL-026 – 1 – us L of C Photodup [340]
Exhibits of the prosecution and of the defense introduced as evidence before the international military tribunal for the far east, 1945-1947 / World War 2. International Prosecution and Defense Section – (= ser Records of allied operational and occupation headquarters, world war 2) – 17r – 1 – mf#M1686 – us Nat Archives [355]
Exhortacion pastoral...con motivo de la epidemia reinante / Perez Munoz, Adolfo – Badajoz: tip uceda hermanos, 1918 – 1 – sp Bibl Santa Ana [240]
Exhortation de notre tres saint-pere pie 10, pape par la divine providence : au clerge catholique a l'occasion du cinquantieme anniversaire de son ordination – Montreal: Arbour & Dupont, [1908?] – 1mf – 9 – 0-665-97262-8 – mf#97262 – cn Canadiana [241]
Exhortation de s s pie 10, pape par la divine providence : au clerge catholique a l'occasion du cinquantieme anniversaire de son sacerdoce – St-Boniface, Man: [s.n.] 1911 – 1mf – 9 – 0-665-73494-8 – (in french and latin) – mf#73494 – cn Canadiana [241]
Exhortation to chastity – London, England. 1798 – 1r – us UF Libraries [240]
Exhortation to the duty of catechising / Pearson, Edward – Nottingham, England. 18– – 1r – us UF Libraries [240]
An exhortation to the ministers of gods woord / Bullinger, Heinrich – London, John Allde, [1575] – 2mf – 9 – mf#PBU-247 – ne IDC [240]
Exhortationes ad monachos / Trithemius, Ioan – Parisiis, 1549 – 7mf – 8 – €28.00 – ne Slangenburg [241]
Exil und literatur : deutsche schriftsteller im ausland 1933-1945 / Wegner, Matthias – Frankfurt/Main, Bonn: Athenaeum Verlag, c1967 [mf ed 1993] – (= ser Athenaeum buecher zur dichtkunst) – 1 – (incl bibl ref & ind) – mf#8292 – us Wisconsin U Libr [430]
The exile – Chicago. n1-3. spring 1927-autumn 1928 – 1 – us NY Public [073]
Exile report – v1 n1 [1971 mar] – 1r – 1 – mf#721944 – us WHS [071]
Exiled jesuits on trespass in england / Bulgin, Robert – London, England. 18– – 1r – 1 – us UF Libraries [241]
L'exilee d'holy-rood / Lamothe-Langon, Etienne L de – Paris 1831 [mf ed Hildesheim 1995-98] – 1v on 3mf – 9 – €90.00 – ISBN-10: 3-487-26076-X – ISBN-13: 978-3-487-26076-1 – gw Olms [941]
The exiles' book of consolation contained in isaiah 40-66 : a critical and exegetical study / Koenig, Eduard – Edinburgh: T & T Clark, 1899. Chicago: Dep of Photodup, U of Chicago Lib, 1975 (1r); Evanston: American Theol Lib Assoc, 1984 (1r) – 1 – 0-8370-0355-5 – (incl bibl footnotes) – mf#1984-B416 – us ATLA [221]
The exiles' book of consolation contained in isaiah 40-66 : a critical and exegetical study / Koenig, Eduard – Edinburgh: T & T Clark, 1899 – 1mf – 9 – 0-8370-3966-5 – (incl bibl ref and ind) – mf#1985-1966 – us ATLA [221]
Eximeno y Pujades, Antonio *see*
– Dell'orgine e delle regole della musica colla storia del suo progresso, decadenza, e rinnovazione
– Dubbio di don antonio eximeno sopra il saggio fondamentale pratico di contrappunto del reverendissimo padre maestro giambattista martini
L'existence des loges de femmes / Fava, Armand-Joseph – Paris: Tequi [1891?] – 2mf – 9 – mf#vrl-177 – ne IDC [366]
The existence of god / Moyes, James – London: Sands; St Louis, MO: Herder, 1906 – 1mf – 9 – 0-8370-7251-4 – mf#1986-1251 – us ATLA [210]
Existencia y vicisitudes del colegio gorjon / Nolasco, Flerida De – Ciudad Trujillo, Dominican Republic. 1947 – 1r – us UF Libraries [972]
Existential psychology / May, Rollo, Ed – New York, NY. 1961 – 1r – us UF Libraries [150]
Existing conditions of choral music in the american high school / Kjerstad, Muriel Adele – U of Rochester 1940 [mf ed 19–] – 1r – 1 – mf#film 2552 – us Sibley [780]
Existing state of theology / Martineau, James – London, England. 1834 – 1r – us UF Libraries [240]
Exito – Fort Lauderdale, FL. 1991-1995 (1) – mf#68904 – us UMI ProQuest [071]

Exlex – Kristiania, Norway; Kobenhavn, Denmark. 1919-20 – 1 – sw Kungliga [073]
Exmouth journal *see* Freeman's exmouth journal
L'exode rural au maroc : etude sociologique de l'exode du tafilalet vers la ville de fes / El Maliki, Abderrahmane – 2mf – 9 – mf#10089 – fr Atelier National [307]
Exodus : an autobiography of moses / Denniston, J M – 2nd ed. London: Morgan & Scott, 1895 [mf ed 1989] – 1mf – 9 – 0-7905-1037-5 – mf#1987-1037 – us ATLA [221]
Exodus / Betteridge, Walter Robert – Philadelphia PA: American Baptist Pub Soc 1914 [mf ed 1993] – (= ser American commentary on the old testament) – 1mf – 9 – 0-524-06456-3 – mf#1992-0884 – us ATLA [221]
Exodus : or, the second book of moses / Lange, Johann Peter – New York: Charles Scribner, c1876 [mf ed 1985] – (= ser A commentary on the holy scriptures. old testament 2/2) – 1mf – 9 – 0-8370-5164-9 – (english trans by charles m mead. incl app) – mf#1985-3164 – us ATLA [221]
Exodus – San Francisco, CA. 1970-78 – 1 – us AJPC [071]
Exodus / Soviet Jewry Action Group – v3 n8-v8 n5 [1973 jul/aug-1978 sep/oct] – 1r – 1 – mf#628630 – us WHS [939]
Exodus – Washington DC. 1880 may 1-oct 16 – 1r – 1 – mf#881899 – us WHS [071]
Exodus / Weidner, Revere Franklin – Chicago: Fleming H Revell, c1903 – 1mf – 9 – 0-524-04117-2 – (Incl ind) – mf#1992-0075 – us ATLA [221]
Exodus 16 : the manna – Ottawa: printed for F Brodie by J Loveday, 1876 – 1mf – 9 – 0-665-89610-7 – mf#89610 – cn Canadiana [221]
Exodus erklaert / Holzinger, Heinrich – Tuebingen: J C B Mohr, 1900 – (= ser Kurzer hand-commentar zum alten testament) – 1mf – 9 – 0-8370-3646-1 – (incl bibl ref and ind) – mf#1985-1646 – us ATLA [221]
Exodus, moses and the decalogue legislation : The Central Doctrine and Regulative Organum of Mosaism / Fluegel, Maurice – Baltimore, MD: M Fluegel, c1910 – 1mf – 9 – 0-524-04455-4 – mf#1992-0124 – us ATLA [221]
Exodus-leviticus-numeri / Baentsch, Bruno – Goettingen: Vandenhoeck und Ruprecht, 1903 – 2mf – 9 – 0-8370-9441-0 – (Incl ind) – mf#1986-3441 – us ATLA [221]
Exodvs; in exodvm vel secundum librum mosis...commentarij / Simler, J – Tigvri, Christoph Froschover, 1584 – 7mf – 9 – mf#PBU-628 – ne IDC [240]
O exorcista – Rio de Janeiro, RJ: Typ Franceza, 02-30 jan 1841 – (= ser Ps 19) – mf#P02,04,32 – bl Biblioteca [320]
Exortacion pastoral / Varela, Cipriano – 1827 – 9 – sp Bibl Santa Ana [240]
Der exote : roman / Wiechert, Ernst Emil – Muenchen: K Desch 1951, c1945 [mf ed 1991] – 1r – 1 – (filmed with: das einfache leben / ernst wiechert) – mf#3043p – us Wisconsin U Libr [830]
L'exotisme : la litterature coloniale / Cario, Louis & Regismanset, Charles – Paris: Mercure de France, 1911 – 1 – us CRL [944]
Expand or explode / Horwitz, Ralph – Cape Town, South Africa. 1957? – 1r – us UF Libraries [960]
Expanded metal : fencing, window guards, lattice, borders, tree guards / Expanded Metal & Fireproofing Co – [Toronto?: s.n, 1900?] [mf ed 1991] – 1mf – 9 – 0-665-99531-8 – mf#99531 – cn Canadiana [670]
Expanded Metal & Fireproofing Co *see* Expanded metal
Expanding access to quality health care : solutions for uninsured americans: hearing... house of representatives, 107th congress, 2nd session...washington, dc, july 9 2002 / United States. Congress. House. Committee on Education and the Workforce. Subcommittee on Employer-Employee Relations – Washington: US GPO 2002 [mf ed 2002] – 2mf – 9 – 0-16-068957-0 – us GPO [362]
Expanding olympic horizons – Olympic Training Center at Colorado Springs, 1981 – (= ser United states olympic academy 5) – 4mf – 9 – $16.00 – us Kinesology [790]
Expansao geographica do brasil colonial / Magalhaes, Basilio De – Sao Paulo, Brazil. 1935 – 1r – us UF Libraries [972]
Expansao para o norte / Miranda, Salm De – Rio de Janeiro, Brazil. 1946 – 1r – us UF Libraries [972]
Expansao portuguesa em mocambique de 1498 a 1530 / Lobato, Alexandre – Lisboa, Portugal. v1. 1954-1960 – 1r – us UF Libraries [960]
Expansion – Madrid, Spain, 1985-present – 12r per yr – 1 – $900.00y – (A daily paper for business and financial news. Backfiles available) – us UMI ProQuest [332]
Expansion – Paris. 1975-1996 (1) 1975-1996 (5) 1975-1996 (9) – ISSN: 0014-4703 – mf#10187 – us UMI ProQuest [338]
La expansion cultural de espana en el extranjero... / Sangroniz, Jose Antonio de – Madrid: Razon y Fe, 1927 – 1 – sp Bibl Santa Ana [946]
L'expansion francaise en afrique occidentale – Paris: Societe de l'histoire des colonies francaises [194-?] – us CRL [960]
The expansion of new england : the spread of new england settlement and institutions to the mississippi river, 1620-1865 / Rosenberry, Lois Kimball Mathews – Boston: Houghton Mifflin, 1909 – 1mf – 9 – 0-524-04498-8 – (incl bibl ref) – mf#1990-1260 – us ATLA [975]
The expansion of religion : six lectures / Donald, Elijah Winchester – Boston: Houghton, Mifflin, 1896 – (= ser Lowell institute lectures) – 1mf – 9 – 0-8370-3572-4 – mf#1985-1572 – us ATLA [240]
The expansion of the christian life / Lang, John Marshall – Edinburgh: William Blackwood, 1897 – 1mf – 9 – 0-8370-6202-0 – (incl bibl ref & index) – mf#1986-0202 – us ATLA [240]
Expansions / Collective Black Artists – 1971 mar, jun, 1973 apr, 1975 win+souvenir iss 74/75 – 1r – 1 – mf#4983049 – us WHS [700]
The expectation of the christ : being a series of lectures on the messianic prophecies. delivered in st. paul's, melbourne / Moorhouse, James – Melbourne: Mason, Firth & M'Cutcheon, [1878?] – 1mf – 9 – 0-524-06049-5 – mf#1992-0762 – us ATLA [240]
Expectations of lutheran military personnel as factors in shaping chaplain ministry / Baldwin, Charles Stealey – 1982 – 1 – us Southern Baptist [242]
The expectations of mothers regarding community participation in antenatal care : at the chinamhora clinic in goromonzi district, zimbabwe / Chitambo, Beritha Ruth – Uni of South Africa 2001 [mf ed Johannesburg 2001] – 3mf – 9 – (incl bibl ref) – mf#mfm14821 – sa Unisa [618]
Expedicao as regioes centrais da america do sul. t. 1 / Castelnau – Sao Paulo, 1949 – 7mf – 8 – ne Slangenburg [918]
Expedicao as regioes centrais da america do sul. t. 2 / Castelnau – Sao Paulo, 1949 – 7mf – 8 – ne Slangenburg [918]
Expedicion cortesiana a las molucas 1527 / Romero, Luis – Mexico: Editorial Jus, 1950 – 1 – sp Bibl Santa Ana [910]
La expedicion de hernando de soto a la florida / Serrano Sanz, Manuel – Madrid: Tip. de Archivos, 1933. B.R.A.Historia – 1 – sp Bibl Santa Ana [917]
Expedicion del adelantado hernando de soto a la florida : notas y documentos... / Hernandez Diaz, Jose – Burgos: Razon y Fe, 1939 – 1 – sp Bibl Santa Ana [917]
Expedicion del maestre de campo bernardo de aldana a hungria / Villela de Aldana, Fr. Juan – 1879 – 9 – sp Bibl Santa Ana [914]
Expedicion geografica a oriente / Nunez Jimenez, Antonio – Habana, Cuba. 1948 – 1r – us UF Libraries [972]
Expediency of christ's departure / Boulding, J W – London, England. 1868 – 1r – 1 – us UF Libraries [240]
Expediente sobre la rebelion de lares, 1868-1869 = Case file on the rebellion of lares, 1868-1869 / Puerto Rico. Spanish Governors – (= ser Records of spanish governors of puerto rico) – 6r – 1 – mf#T1120 – us Nat Archives [972]
Expedionis aethiopicae / Mendez, A; ed by Beccari C – Rome, 1908-1909. 2v – 11mf – 9 – mf#SEP-38 – ne IDC [916]
Expediting patent office procedure; legislative history / U.S. Library of Congress. Legislative Reference Service – Washington, Govt. Print. Off., 1960. 105 p. LL-2310 – 1 – us L of C Photodup [346]
Expediting settlement of employee grievances in the federal sector : an evaluation of the mspb's appeals process / Adams, Arvil V & Figueroa, Jose R – Washington: GPO, 1985 (all publ) – 2mf – 9 – $3.00 – mf#LLMC 94-349 – us LLMC [344]
Expeditio persica, bellum avaricum, heraclias *see* Descriptio templi sanctae sophiae [cshb32]
Expedition – Philadelphia. 1958+ (1) 1972+ (5) 1976+ (9) – ISSN: 0014-4738 – mf#8367 – us UMI ProQuest [301]
Expedition antarctique belge : resultats du voyage du s y belgica en 1897-1898-1899 sous le commandement de a de gerlache de gomery – Anvers, 1902-1912 – 13mf – 9 – mf#2834 – ne IDC [919]
"The expedition at beisan" / Rowe, Alan – University of Pennsylvania Museum Journal, dec 1927 – 9 – $10.00 – us IRC [932]
Expedition de Cochinchine. Puis, de la Cochinchine francaise *see* Bulletin officiel
L'expedition de madagascar : rapport d'ensemble fait au ministre de la guerre le 25 avril 1896 / Duchesne, J – Paris: H C Lavauzelle, 1896 – 1 – us CRL [916]
Expedition de madagascar en 1895 / Anthouard, Albert Francois Ildefonse D' – Paris, France. 1930 – 1r – 1 – us UF Libraries [916]
Expedition du louxor : ou relation de la campagne faite dans la thebaide pour en rapporter l'obelisque occidental de thebes / Angelin, Justin P – Paris 1833 [mf ed Hildesheim 1995-98] – 1v on 1mf [ill] – 9 – €40.00 – 3-487-27370-5 – gw Olms [916]
Expedition et naufrage de la perouse : recueil historique de faits, evenemens [sic], decouvertes, etc, appuyes de documens [sic] officiels... / Hapde, Jean Baptiste Augustin – Paris: Delaunay, 1829 [mf ed 1984] – 1mf – 9 – 0-665-18821-8 – mf#18821 – cn Canadiana [910]
The expedition for the survey of the rivers euphrates and tigris, carried on by order of the british government, in the years 1835, 1836 and 1837 : preceded by geographical and historical notices of the regions situated between the rivers nile and indus / Chesney, F R – Terre Haute. 1963-1980 (1) 1977-1980 (5) 1977-1980 (9) – 22mf – 9 – mf#8462 – ne IDC [915]
Die expedition in die seen von china, japan und ochotsk unter commando von commodore colin ringgold und commodore john rodgers : im auftrage der regierung der vereinigten staaten unternommen in den jahren 1853 bis 1856... / Heine, Wilhelm – Leipzig 1858-59 [mf ed Hildesheim 1995-98] – 5v on 14mf – 9 – €140.00 – 3-487-27559-7 – gw Olms [915]
Expedition in north russia, 1918-1919 / U.S. Army – 1918-19? – 1 – us L of C Photodup [947]
Die expedition nach der halbinsel kola / Kihlman, A O – Sacramento. 1975-1975 (1) 1975-1975 (5) 1975-1975 (9) – 1mf – 9 – mf#9881 – ne IDC [914]
The expedition of the dutch for recovering formosa : in conjunction with the tartars / Montanus, A – London, 1745-1747. v3 – 1mf – 9 – mf#A-271 – ne IDC [915]
Expedition reports of the office of foreign seed and plant introduction of the department of agriculture, 1900-1938 / U.S. Dept of Agriculture – (= ser Records of the bureau of plant industry, soils, and agricultural engineering) – 38r – 1 – (With printed guide) – mf#M840 – us Nat Archives [630]
Expedition scientifique de Moree : Section des sciences physiques / Bory de Saint-Vincent, J B G M – New York. 1967-1984 (1) 1970-1984 (5) 1975-1984 (9) – 17mf – 9 – mf#5860 – ne IDC [910]
Expedition scientifique de moree / Bory de Saint-Vincent, Jean B de – Paris [u a] 1832-36 [mf ed Hildesheim 1995-98] – 6v on 24mf – 9 – €240.00 – 3-487-27946-0 – gw Olms [914]
Expedition scientifique en mesopotamie : executee par ordre du gouvernement de 1851 a 1854 par mm. fulgence fresnel, felix thomas, et jules oppert / Oppert, Jules – Paris: Imprimerie imperiale, 1859-1863 – 7mf – 9 – 0-7905-8312-7 – mf#1987-6417 – us ATLA [915]
Expedition scientifique en mesopotamie. [tome 3, planches] : executee par ordre du gouvernement de 1851 a 1854 par mm. fulgence fresnel, felix thomas, et jules oppert / ed by Oppert, Jules – [S.l.: s.n., 1859?] (Paris: J Claye) – 1r – 1 – 0-7905-8328-3 – mf#1987-B006 – us ATLA [915]
An expedition through the barren lands of northern canada / Tyrrell, J B – London, 1894. v4 – 1mf – 9 – mf#N-429 – ne IDC [917]
Expedition to discover the sources of the white nile, in the years 1840, 1841 / Werne, Ferdinand – London: R Bentley, 1849 – 2v (ill) – 1 – (trans fr german by charles william o'reilly) – us Wisconsin U Libr [916]
The expedition to the river zaire : from the royal commonwealth society library / Tuckey, James Kingston – 1818 – 14mf – 7 – mf#2983 – uk Microform Academic [916]
Expedition tumuc-humac / Maziere, Francis – Garden City, NY. 1955 – 1r – us UF Libraries [910]
Expeditionen : Deutsche Lyrik seit 1945 / ed by Weyrauch, Wolfgang – Muenchen: Paul List, 1959 – 1r – 1 – (Incl bibl ref) – us Wisconsin U Libr [430]
Expeditions of h a lorentz to new guinea, 1903-1914 – [mf ed 2007] – (= ser Science in a colonial context 2) – 120mf – 9 – €1195.00 – (with p/g & concordance; in dutch, also english, french, german, some indonesian) – mf#mmp130 – ne Moran [919]
Expeditions scientifiques du travailleur et du talisman pendant les annees 1880, 1881, 1882, 1883 / Milne-Edwards, A et al – Paris, 1888-1927 – 73mf – 9 – mf#Z-2250 – ne IDC [500]
Expeditionum spiritualium societatis iesu, libri quinque / San Roman, M de – Lugduni, 1644 – 8mf – 9 – mf#CA-24 – ne IDC [241]
Expenses under canadian income tax act / Commerce Clearing House Canadian Limited – 6th ed. Don Mills, Ont. 1968 156 p. LL-2380 – 1 – us L of C Photodup [343]
The experience and spiritual letters of mrs hester ann rogers : with a sermon preached on the occasion of her death by the rev thomas coke... – Toronto: G R Sanderson, 1857 [mf ed 1994] – 3mf – 9 – 0-665-89274-8 – mf#89274 – cn Canadiana [240]
An experience in administration from 1893 to 1899 : the old and new premises of the boards of home and foreign missions of the presbyterian church in the usa / McDougall, Thomas – Cincinnati: Armstrong & Fillmore, 1899 [mf ed 1993] – (= ser Presbyterian coll) – 1mf – 9 – 0-524-06640-X – mf#1991-2695 – us ATLA [242]
Experience in the supreme court of the united states, with some reflections and suggestions as to that tribunal / Garland, Augustus Hill – Washington, D.C.: Byrne, 1898. 100p. LL-491 – 1 – us L of C Photodup [347]
The experience of a planter in the jungles of mysore / Elliot, Robert Henry – London: Chapman and Hall, 1871. 2v. illus – 1 – us Wisconsin U Libr [630]
Experience of german methodist preachers / Miller, Adam; ed by Clark, Davis Wasgatt – Cincinnati: printed...for aut, 1859 [mf ed 1984] – 5mf – 9 – 0-8370-1088-8 – mf#1984-4447 – us ATLA [242]
Experience of the late rev joseph hart, minister of the gospel... – London, England. 1827 – 1r – 1 – us UF Libraries [240]
Experience, the crowning evidence of the christian religion / Granbery, John Cowper – Nashville, TN: Publishing House of the ME Church, South, 1901 [mf ed 1990] – (= ser Cole lectures 1900) – 1mf – 9 – 0-7905-3848-2 – mf#1989-0341 – us ATLA [240]
Experienced christian's magazine – New York. 1796-1806 (1) – mf#3522 – us UMI ProQuest [240]
Experiences and relations in the work of women teacher/coaches : a critical inquiry / Jeffreys, Arcelia T – 1989 – 216p 3mf – 9 – $12.00 – us Kinesology [305]
Experiences d'un israelite : dediees a ses coreligionaires – nouv rev corr ed. Geneve: Librairie Robert; Lyon: Librairie Evangelique, 1912 (mf ed 1995) – 1r – 1 – mf#*ZP-1487 – us NY Public [270]
The experiences of a barrister / Warner, Warren – New York: Cornish, Lamport & Co, 1852 – 3mf – 9 – $4.50 – mf#LLMC 92-144 – us LLMC [340]
The experiences of a barrister and confessions of an attorney / Warren, Samuel – Boston: Wentworth & Co. 2v in one. 1857 – 4mf – 9 – $6.00 – mf#LLMC 92-146 – us LLMC [340]
Experiences of a demarara magistrate / Des Voeux, George William – Georgetown, Guyana. 1948 – 1r – us UF Libraries [340]
Experiences of a diplomatist : being recollections of germany, founded on diaries kept during the years 1840-1870 / Ward, John – London: Macmillan, 1872 – viii/279p – 1 – us Wisconsin U Libr [941]
The experiences of five christian indians of the pequod tribe / Apes, William – [Boston?]: W Apes, 1833 [mf ed 1984] – 1mf – 9 – 0-665-45543-7 – mf#45543 – cn Canadiana [240]
Les experiences religieuses et morales du prophete amos / Aubert, Alexandre – Geneve: Librairie Kuendig, 1911 [mf ed 1988] – 1mf – 9 – 0-7905-0243-7 – mf#1987-0243 – us ATLA [221]
Experiencia pioneira / Rosa, Alberto Machado Da – Salvador, Brazil. 1960 – 1r – us UF Libraries [972]
Experiencia pioneira de intercambio cultural / Rio Grande Do Sul, Brazil (State) Universidade Fe... – Porto Alegre, Brazil. 1963 – 1r – 1 – us UF Libraries [972]
Experiencia pioneira de intercambio cultural / Wisconsin University Luso-Brazilian Center – Porto Alegre, Brazil. 1963 – 1r – us UF Libraries [972]
Experiencias de cuba / Calderio, Francisco – Mexico City? Mexico. 1939 – 1r – 1 – us UF Libraries [972]
Experiencias sobre el poder atrayente, para la mosca del olivo, del fosfato amonico a diversas concentraciones / Moreno Marquez, Victor – Madrid: Direccion Gral. de Agricultura. Seccion de plagas del campo y fitopatologia. Servicio de defensa sanitaria del olivo, 1941. Sep. del Boletin de Patologia Vegetal – 1 – sp Bibl Santa Ana [630]
Experientia – Basel. 1966-1996 (1) 1967-1996 (5) 1977-1996 (9) – (Cont by: Cellular and molecular life sciences) – ISSN: 0014-4754 – mf#1377 – us UMI ProQuest [500]
Experiment and experiment news / Huron Co. Norwalk – mar 1937-feb 1952, may 1952 [wkly] – 7r – 1 – mf#B29911-29917 – us Ohio Hist [071]

Experiment in swaziland / University of Natal Institute for Social Research – Cape Town, South Africa. 1964 – 1r – 1 – us UF Libraries [960]
An experiment to create a soviet jewish homeland : [birobidzhan collection at yivo institute for jewish research] – 942mf 31r – 9,1 – (coll may be divided into foll categories: periodicals, bks & pamphlets fr the soviet union; periodicals, bks & pamphlets fr abroad; archival materials of the pro-soviet organizations "icor" & "ambijan"* [usa]; archival materials fr birobidzhan & other pts of the former soviet union; art albums, posters, slides, videos; titles are also listed individually) – ne IDC [939]
Experimental aging research – Mount Desert. 1975+(1,5,9) – ISSN: 0361-073X – mf#12769 – us UMI ProQuest [618]
Experimental agriculture – London. 1965+ (1) 1972+ (5) 1977+ (9) – ISSN: 0014-4797 – mf#5340 – us UMI ProQuest [630]
Experimental aids training package for nursing personnel – Pretoria: Dept of National Health & Population Development 1993 [mf ed Pretoria, RSA: State Library [199-]] – 250p on 1r with other items – 5 – (incl bibl ref) – mf#op 11326 r26 – us CRL [610]
Experimental Aircraft Association *see* Stool pigeon
Experimental and analytical development of a poroelastic finite element model for tendon / Atkinson, Theresa S – 1998 – 203p on 3mf – 9 – $15.00 – mf#PE 4181 – us Kinesology [617]
Experimental and applied acarology – Amsterdam. 1989-1992 (1,5,9) – ISSN: 0168-8162 – mf#42485 – us UMI ProQuest [574]
Experimental and clinical gastroenterology – Zagreb. 1991-1993 (1) 1991-1993 (5) 1991-1994 (9) – ISSN: 0353-9245 – mf#49612 – us UMI ProQuest [616]
Experimental astronomy – Dordrecht. 1991-1992 (1,5,9) – ISSN: 0922-6435 – mf#16785 – us UMI ProQuest [520]
Experimental brain research – Heidelberg. 1965-1996 (1,5,9) – ISSN: 0014-4819 – mf#13114 – us UMI ProQuest [612]
Experimental chemistry for junior students / Reynolds, James Emerson – London, England. v1-4. 1883-1904 – 1r – us UF Libraries [500]
Experimental evolution : lectures delivered... edinburgh (aug 1891) / Varigny, Henry de – London, 1892 – (= ser 19th c evolution & creation) – 3mf – 9 – mf#1.1.9046 – uk Chadwyck [575]
Experimental gerontology – Oxford. 1964+ (1,5,9) – ISSN: 0531-5565 – mf#49071 – us UMI ProQuest [618]
Experimental heat transfer – Washington. 1987-1993 (1,5,9) – ISSN: 0891-6152 – mf#16653 – us UMI ProQuest [530]
Experimental hematology – Amsterdam. 1999+ (1) – ISSN: 0301-472X – mf#16151 – us UMI ProQuest [616]
An experimental investigation into the flow of marble / Adams, Frank Dawson & Nicolson, John Thomas – [S.l: s.n, 1900?] – 1mf – 9 – 0-665-94281-8 – (incl bibl ref) – mf#94281 – cn Canadiana [550]
Experimental investigation of the spirit manifestations : demonstrating the existence of spirits and their communion with mortals, doctrine of the spirit world respecting heaven, hell, mortality, and god / Hare, Robert – 4th ed. New York: Partridge & Brittan, 1856 – 2mf – 9 – 0-524-08409-2 – mf#1993-0024 – us ATLA [130]
Experimental lung research – Washington. 1986-1995 (1) 1986-1995 (5) 1986-1995 (9) – ISSN: 0190-2148 – mf#14333 – us UMI ProQuest [616]
Experimental mechanics – Bethel. 1961+ (1) 1971+ (5) 1976+ (9) – ISSN: 0014-4851 – mf#5060 – us UMI ProQuest [621]
Experimental physiology – Cambridge. 1992-1996 (1) – ISSN: 0958-0670 – mf#16546,03 – us UMI ProQuest [612]
Experimental radio / Ramsey, Rolla Roy – Bloomington, IL. 1928 – 1r – us UF Libraries [621]
Experimental religion exemplified / Archibald, Alexander – Edinburgh, Scotland. 1819 – 1r – us UF Libraries [240]
Experimental researches in electricity / Faraday, Michael – London, England. 1912 – 1r – us UF Libraries [530]
An experimental rural church *see* I ko shih yen te hsiang tsun chiao hui (ccm109)
Experimental techniques – Westport. 1983+ (1,5,9) – ISSN: 0732-8818 – mf#13549 – us UMI ProQuest [620]
Experimentation and budding industry : orlando, florida – s.l, s.l? 1936 – 1r – 1 – us UF Libraries [978]
Experimentation in the law : report of the fjc advisory committee on experimentation in the law – Washington: GPO, 1981 – 2mf – 9 – $3.00 – mf#LLMC 95-308 – us LLMC [340]
Experimentelle beitrage zur zulu phonetik / Selmer, Ernst Westerlund – Oslo, Norway. 1933 – 1r – us UF Libraries [470]
Experimentelle studien und phantomuntersuchungen zur computertomographie des thorax / Doll, Michael – (mf ed 1997) – 2mf – 9 – €40.00 – 3-8267-2450-X – mf#DHS 2450 – gw Frankfurter [616]
Experimentelle und klinische untersuchungen der kontinenzfaktoren intestinaler harnreservoire / Lampel, H Alexander – (mf ed 1996) – 2mf – 9 – €40.00 – 3-8267-2342-2 – mf#DHS 2342 – gw Frankfurter [616]
Experimentelle und klinische untersuchungen zur therapie von neurogenen blasenfunktionsstoerungen / Hohenfellner, R Markus – (mf ed 1995) – 2mf – 9 – €40.00 – 3-8267-2128-4 – mf#DHS 2128 – gw Frankfurter [616]
Experimentelle untersuchung der antibiotikaproduktion sowie des wachstums von pilzen bei fermentationen im hubstrahl-bioreaktor / Wudtke, Alexander – (mf ed 1994) – 2mf – 9 – €40.00 – 3-8267-2073-3 – mf#DHS 2073 – gw Frankfurter [574]
Experimentelle untersuchung zum problem des zusammenhangs zwischen erfragtem und beobachtetem verhalten / Hueneke, Heinrich – Heidelberg, 1972 – 2mf – 9 – 3-89349-797-8 – gw Frankfurter [150]
Experimentelle untersuchungen : mit einer neuen methode zur pruefung der uv-bestaendigkeit von lichtschutzmitteln und zur bestimmung ihrer schutzwirkung / Lange, Michael – 2000 – 2mf – 9 – 3-8267-2695-2 – mf#DHS 2695 – gw Frankfurter [616]
Experiments and demonstrations in psychology : students' manual / Shaffer, Laurence Frederic et al – New York, London: Harper & Bros, 1942 – xi/230p/pl (ill) – 1 – us Wisconsin U Libr [150]
Experiments for the control of phoma rot of tomatoes / Tisdale, W B – Gainesville, FL. 1937 – 1r – us UF Libraries [630]
Experiments in corn and irish potatoes and analysis of grasses, etc – Lake City, FL. 1890 – 1r – us UF Libraries [630]
Experiments in government and the essentials of the constitution / Root, Elihu – Princeton/ London/Oxford: Princeton UP/Henry Frowde/Oxford UP, 1913 – 1mf – 9 – $1.50 – mf#LLMC 92-202 – us LLMC [323]
Experiments in personal religion *see* Shih yen tsung chiao hsueh chiao cheng (ccm297)
Experiments to determine the density of the earth / Cavendish, Henry – [London: Royal Society of London 1798?] [mf ed 1998] – 1r [pl/ill] – 1 – (repr fr: philosophical transactions of the royal society of london, june 21 1798) – mf#film mas 28292 – us Harvard [520]
Experiments with regard to the pollination of the florida velvet bean / Winters, Rhett Y – s.l, s.l? 1909 – 1r – 1 – us UF Libraries [635]
Expert, Henry *see* Les maitres musiciens de la renaissance francaise
Expert systems – Oxford. 1989+ (1,5,9) – ISSN: 0266-4720 – mf#17524 – us UMI ProQuest [000]
Expert systems with applications – New York. 1990+ (1,5,9) – ISSN: 0957-4174 – mf#49598 – us UMI ProQuest [000]
Expilly, Charles *see* Mulheres e costumes do brasil
Expiratory flow limitation and ventilatory responsiveness interact to determine exercise ventilation / Derchak, P A – 200 – 183p on 2mf – 9 – $10.00 – mf#PH 1698 – us Kinesology [612]
Explanatio psalmorum : qui juxta breviarum romanum in officiis communibus recitantur, ad mentem optimorum interpretum adornata / Schouppe, Frandcois Xavier – Bruxelles: M Closson et Sociorum, 1875 – 2mf – 9 – 0-524-06802-X – (incl bibl ref and ind) – mf#1992-0965 – us ATLA [220]
The explanatio symboli ad initiandos (ts10/1) : a work of st ambrose / Connolly, R H – 1952 – (= ser Texts and studies (ts)) – 1mf – 9 – €3.00 – (a provisionally constructed text, with int, notes and trans) – ne Slangenburg [240]
An explanation of dr watt's hymns for children, in question and answer / Cockle, Mary – 3rd ed. London: printed for C J G & F Rivington, 1829 – (= ser 19th c children's literature) – 2mf – 9 – mf#6.1.29 – uk Chadwyck [240]
Explanation of his design for the proposed new courts of justice / Street, George Edmund – London 1867 – (= ser 19th c art & architecture) – 1mf – 9 – mf#4.2.179 – uk Chadwyck [720]
An explanation of luther's small catechism : a handbook for the catechetical class / Stump, Joseph – Philadelphia: General Council Pub House 1913, c1907 [mf ed 1991] – 1mf – 9 – 0-524-00858-2 – mf#1990-4018 – us ATLA [242]
An explanation of the baltimore catechism of christian doctrine : for the use of sunday-school teachers and advanced classes / Kinkead, Thomas L – 6th ed. New York: Benziger c1891 [mf ed 1993] – 1mf – 9 – 0-524-06310-9 – mf#1991-2483 – us ATLA [241]
An explanation of the common service : with appendices on christian hymnody and liturgical colors, and a glossary of liturgical terms – 4th rev enl ed. Philadelphia: United Lutheran Pub House c1908 [mf ed 1991] – 1mf – 9 – 0-524-00859-0 – mf#1990-4019 – us ATLA [242]
Explanation of the duties of religion / Gilpin, William – London, England. 1800 – 1r – us UF Libraries [240]
An explanation of the eleventh chapter of the book of daniel / Collins, George – [S.l: s.n.], 1870 [mf ed 1980] – 1mf – 9 – 0-665-00713-2 – mf#00713 – cn Canadiana [221]
Explanation of the holy sacraments : a complete exposition of the sacraments and the sacramentals of the church / Rolfus, Hermann – New York: Benziger, c1898 – 1mf – 9 – 0-8370-7187-9 – (also iss under title: illustrated explanation of the holy sacraments) – mf#1986-1187 – us ATLA [240]
Explanation of the picture of chairing the members / Haydon, Benjamin Robert – London 1828 – (= ser 19th c art & architecture) – 1mf – 9 – mf#4.2.749 – uk Chadwyck [750]
Explanation of the rule of st augustine = Expositio in regulam b. augustini episcopi / Hugh of Saint-Victor – London: Sands 1911 [mf ed 1990] – 1mf – 9 – 0-7905-7943-X – (in english) – mf#1989-1168 – us ATLA [240]
Explanation of the system of the catalogue / Blackstone, Frederick Elliot – 2nd ed. [London], 1889 – (= ser 19th c publishing...) – 1mf – 9 – mf#3.1.58 – uk Chadwyck [020]
An explanation of the thirty-nine articles : with an epistle dedicatory to the late rev e b pusey / Forbes, Alexander Penrose – 5th ed. New York: E P Dutton; Oxford: Parker [c1875] [mf ed 1986] – 3mf – 9 – 0-8370-8668-X – (incl bibl ref) – mf#1986-2668 – us ATLA [242]
An explanation of the visions of the four beasts, daniel 7 / Collins, George – [Ottawa?: s.n, 1870?] [mf ed 1980] – 1mf – 9 – 0-665-00714-0 – mf#00714 – cn Canadiana [221]
Explanations : a sequel to "vestiges of the natural history of creation" / Chambers, Robert – 2nd ed. London, 1846 – (= ser 19th c evolution & creation) – 3mf – 9 – mf#1.1.10900 – uk Chadwyck [577]
Explanations relative to the training of educated native ministers in connexion with the general assembly's mission : a letter addressed to the chairman of the corresponding board of the church of scotland's calcutta mission / Ogilvie, James – Calcutta 1867 – (= ser 19th c british colonization) – 2mf – 9 – mf#1.1.505 – uk Chadwyck [242]
An explanatory address, and vindication, to the legislature : of practical means, proposed for remedying our present distressed, and very dangerous situation / Edwards, George – Barnardcastle: printed at the office of J Atkinson, 1820 – (= ser 19th c economics) – 1mf – 9 – mf#1.1.114 – uk Chadwyck [330]
Explanatory analysis of st paul's epistle to the romans / Liddon, Henry Parry – 3rd ed. London; New York: Longmans, Green, 1897 [mf ed 1985] – 1mf – 9 – 0-8370-4116-3 – mf#1985-2116 – us ATLA [227]
Explanatory analysis of st paul's first epistle to timothy / Liddon, Henry Parry – London, New York: Longmans, Green, 1897 [mf ed 1988] – 1mf – 9 – 0-7905-0097-3 – mf#1987-0097 – us ATLA [227]
An explanatory commentary on esther : with 4 appendices, consisting of the second targum...mithra, the winged bulls of persepolis, and zoroaster = Das buch esther / Cassel, Paulus – Edinburgh: T & T Clark, 1888 [mf ed 1985] – (= ser Clark's foreign theological library. new series 34) – 1mf – 9 – 0-8370-2608-3 – (english trans by aaron bernstein) – mf#1985-0608 – us ATLA [221]
Explanatory lecture on visible speech, the science of universal alphabetics : delivered before the college of preceptors, feb 9, 1870 / Bell, Alexander Melville – London: Simpkin, Marshall, 1870 – 1mf – 9 – mf#07075 – cn Canadiana [410]
Explanatory memorandum, basuto national treasury – [S.l: s.n., 1944?] – us CRL [324]
Explanatory notes upon the old testament / Wesley, John – Bristol, Eng.: William Pine, 1765. Chicago: Dep of Photodup, U of Chicago, 1966 (3r); Evanston: American Theol Lib Assoc, 1984 (2r) – 1 – 0-8370-1486-7 – mf#1984-B029 – us ATLA [221]
Explanatory statement addressed to the friends of the India mission / Duff, Alexander – Edinburgh, Scotland. 1844 – 1r – us UF Libraries [240]
An explanatory treatise on the valuation of friendly societies : with a full description of the method employed in the calculations, with examples etc / Watson, Reuben – Brighton: Curtis Bros & Towner, 1878 – (= ser 19th c economics) – 1mf – 9 – mf#1.1.463 – uk Chadwyck [360]
Explicacion de el catechismo en lengua guarani / Yapuguay, Nicolas – En el Pueblo de Santa Maria la Mayor: [s.n.], ano de 1724 – (= ser Books on religion...1543/44-c1800: catecismos) – 5mf – 9 – mf#crl-10 – ne IDC [241]
Explicacion de la sagrada pasion de nuestro senor jesucristo.. – [S l: s n, 196-?] [mf ed Bloomington IN: Indiana Uni Lib, Preservation Dept 1984] – 340p on 1r – 1 – us Indiana Preservation [390]
Explicacion de los estados que forman la balanza del comercio reciproco que hizo espana...en 1795 / Archivo Ministerio de Hacienda MS. – 5mf – 9 – sp Cultura [380]
Explicacion literal del catecismo del padre astete con una exposicion y refutacion de los erores modernos y la explicacion de la bula de la santa cruzada / Marquez, Gabino – Madrid: Razon y Fe 1929 – 1 – sp Bibl Santa Ana [240]
Explicacion...confesores.. / Blazquez del Barco, Juan – 1721 – 9 – sp Bibl Santa Ana [240]
Explicacion...regla de san agustin / Logrosan, Juan de – 1716 – 9 – sp Bibl Santa Ana [240]
Explicatio brevis, simplex et catholica libelli rvth... / Pellican, C – Tigvri, Christoph Froschouer, 1531 – 1mf – 9 – mf#PBU-560 – ne IDC [241]
Explicatio epistolae pavli ad galatas / Hesshusen, T – [Helmstedt], 1579 – 8mf – 9 – mf#TH-1 mf 621-628 – ne IDC [242]
Explicatio epistolae pavli ad romanos / Hesshusen, T – Ienae, 1571 – 10mf – 9 – mf#TH-1 mf 629-638 – ne IDC [242]
Explicatio malachiae prophetae / Chytraeus, D – Rostochii, 1568 – 2mf – 9 – mf#TH-1 mf 274-275 – ne IDC [242]
Explicatio prioris epistolae pavli ad corinthios / Hesshusen, T – Ienae, 1573 – 8mf – 9 – mf#TH-1 mf 639-646 – ne IDC [242]
Explicatio psalmi 110 in qva doctrina de spirituali regno / Hesshusen, T – Helmstadii, 1580 – 2mf – 9 – mf#TH-1 mf 667-668 – ne IDC [242]
Explicatio secvndae epistolae pavli ad corinthios / Hesshusen, T – Helmstadii, 1580 – 7mf – 9 – mf#TH-1 mf 647-653 – ne IDC [242]
Explication de l'appel d'un protestant au pape – Geneve: Vve Auguste Garin, 1869 – 1mf – 9 – 0-8370-6940-8 – (incl bibl ref) – mf#1986-0940 – us ATLA [327]
Explication de l'edict de nantes par les autres edicts de pacification, declarations et arrests de reglement / Bernard, P – Paris, 1666 – 12mf – 9 – mf#CA-109 – ne IDC [241]
L'explication de l'edit de nantes de m bernard, avec de nouvelles observations... / Soulier, [P] – Paris, 1683 – 7mf – 9 – mf#CA-107 – ne IDC [240]
Explication de l'evangile selon saint jean : contenant une preface... / Astie, Jean-Frederic – Geneve: Joel Cherbuliez, 1863 – 2mf – 9 – 0-7905-0662-9 – (incl bibl ref) – mf#1987-0662 – us ATLA [241]
Explication du systeme de l'harmonie : pour abreger l'etude de la composition, & accorder la pratique avec la theorie / Lirou, Jean-Francois Espic de Lirou, Chevalier de – Londres, Paris: Chez Merigot..., Bailly..., Bailleux..., Boyer...1785 [mf ed 19–] – 4mf / 1mf – 9 – mf#fiche 605 / fiche 782 – us Sibley [780]
Explication historique des fables / Banier, Antoine – A Paris: Chez Briasson... 1742 [mf ed Bloomington IN: Indiana Uni Lib, Preservation Dept 1984] – 1r – 1 – us Indiana Preservation [390]
Explicationes brevis, in 16 22 110 118 psalmos davidis / Wigand, J – [Gedani], 1575 – 2mf – 9 – mf#TH-1 mf 1624-1625 – ne IDC [242]
Explicator – Fredericksburg, WV. 1942-1956 (1) – mf#67292 – us UMI ProQuest [071]
Explicator – Washington. 1942+ (1) 1942+ (5) 1942+ (9) – ISSN: 0014-4940 – mf#1073 – us UMI ProQuest [400]
Expliquez par des vers francois – Paris: Pierre Mariette, n.d. – 3mf – 9 – mf#O-1340 – ne IDC [090]
L'exploit de dollard : recit de l'heroique fait d'armes du long-sault, d'apres les relations du temps / Faillon, Etienne Michel – [Montreal: Action francaise, 1920?] [mf ed 1995] – 1mf – 9 – 0-665-74212-6 – mf#74212 – cn Canadiana [971]
L'exploite : organe socialiste revolutionnaire de la region nord. – Douai. aout-nov 1884 – 1 – fr ACRPP [335]
Les exploits de masire isse dieye : un episode de l'epopee du kayor / Dioum, Abdoulaye – 1978 – us CRL [960]

Exploits in the tropics / Moulton, C O – Bridgetown, Barbados. 1907 – 1r – us UF Libraries [972]

Les exploits policiers du domino noir : une autre aventure extraordinaire du domino noir – Montreal: ed Police journal. n1 9 avril 1948-n883 27 janv 1965 [mf ed 1982] – 10r – 5 – mf#SEM16P325 – cn Bibl Nat [073]

Explor – Evanston. 1988-1988 (1,5,9) – ISSN: 0362-0867 – mf#15669 – us UMI ProQuest [240]

Exploracion de guatemala / Alvarado, Huberto – Guatemala, 1961 – 1r – 1 – us UF Libraries [972]

Exploracoes botanicas em timor / Gomes, Ruy Cinatti Vaz Monteiro – Lisbon, 1950 – 1 – us CRL [580]

Exploradores y conquistadores de indias / Dantin Cereceda, Juan – Madrid, Spain. 1922 – 1r – us UF Libraries [972]

Exploradores y conquistadores extremenos en america : estudio biografico: hernando de soto / Villanueva y Canedo, Luis – 2nd ed. Badajoz: tip lib y enc de a arqueros, 1929 – 1 – sp Bibl Santa Ana [972]

L'explorateur : journal geographique et commercial – Paris. v1-4. 1875-aout 1876 – 1 – fr ACRPP [910]

Un explorateur de la louisiane : jean-baptiste benard de la harpe, 1683-1765 / Villiers du Terrage, Marc de, Baron – Rennes: impr Oberthur, 1934 [mf ed 1988] – 1mf – 9 – mf#SEM105P943 – cn Bibl Nat [917]

Explorateurs de l'afrique / Bory, Paul – Tours, France. 1890 – 1r – us UF Libraries [960]

Exploratio philosophica : rough notes on modern intellectual science / Grote, John – Cambridge: Deighton, Bell, 1865-1900 – 2mf – 9 – 0-524-08240-5 – mf#1993-2015 – us ATLA [190]

The exploration and colonization of africa, 1794-1844 : british colonial office files 2 and 392 – 1979 – 14r – 1 – (with printed guide) – mf#D3256 – us Scholarly Res [960]

Exploration and control of leg movements in infants / Angulo-Kinsler, Rosa M – 1997 – 2mf – 9 – $8.00 – mf#PSY 2031 – us Kinesology [612]

Exploration archeologique de la galatie et de la bithynie / Perrot, G, Guillaume E & Delbet, J – Paris, 1872 – 33mf – 9 – mf#NE-105 – ne IDC [915]

Exploration botanique de l'Afrique occidentale francaise... / Chevalier, A – Paris, 1920 – 16mf – 9 – mf#5230 – ne IDC [916]

L'exploration du sahara : etude historique et geographique / Vuillot, Paul – Paris: A Challamel, 1895 – 1 – us CRL [916]

Exploration in orissa / Chanda, Ramaprasad – Calcutta: govt of india, central publication branch, 1930 – (= ser Samp: indian books) – us CRL [930]

Exploration in tibet / Pranavananda, Swami – [Calcutta]: University of Calcutta, 1950 – (= ser Samp: indian books) – (int by syamaprasad mookerjee; foreword by s p chatterjee) – us CRL [915]

Exploration of a munsee cemetery near montague, new jersey / Heye, George Gustav – New York, NY. 1915 – 1r – 1 – us UF Libraries [978]

Exploration of space / Clarke, Arthur Charles – New York, NY. 1951 – 1r – us UF Libraries [520]

The exploration of the caucasus / Freshfield, D W – London, New York, 1896. 2v – 11mf – 9 – mf#AR-1873 – ne IDC [914]

Exploration of the great lakes, 1669-1670 / Dollier de Casson, Francois & Galinee, Rene de Brehant de; ed by Coyne, James H – Toronto: Ontario Historical Society, 1903 – 2mf – 9 – 0-665-74339-4 – (trans by ed) – mf#74339 – cn Canadiana [917]

Exploration of the nile tributaries of abyssinia : the sources, supply, and overflow of the nile... / Baker, Samuel White – Hartford: O D Case & Co, 1868 [mf ed 1985] – xx/[23]-624p/pl – 1 – mf#1541 – us Wisconsin U Libr [916]

An exploration of the opinions of recreation and parks/leisure studies faculty and public sector practitioners concerning the computer competency skills of recreation and parks/leisure studies bacca-laureate students / Case, Alan J & Christiansen, Monty L – 1991 – 3mf – 9 – $12.00 – us Kinesology [790]

Exploration of the red river of louisiana, in the year 1852 / Marcy, R B – Ann Arbor. 1976-1977 (1,5,9) – 5mf – 9 – mf#11153 – ne IDC [590]

Exploration of the valley of the amazon / Herndon, William – Washington, DC 1854 [mf ed Hildesheim 1995-98] – 1v on 5mf – 9 – €100.00 – ISBN-10: 3-487-26763-2 – ISBN-13: 978-3-487-26763-0 – gw Olms [918]

An exploration of various clinical settings for the educational preparation of student nurses / Pilane, Cynthia Nkhumisang – Uni of South Africa 2000 [mf ed Johannesburg 2000] – 5mf [ill] – 9 – (incl bibl ref) – mf#mfm15057 – sa Unisa [610]

Exploration scientifique de l'algerie : Flore d'algerie. botanique 2. phanerogamie / Durieu de Maisonneuve, M C & Cosson, E S C – Chicago. 1970-1972 (1) 1965-1972 (5) (9) – 14mf – 9 – mf#5884 – ne IDC [916]

Exploration scientifique de l'algerie : flore d'algerie. [botanique i]. cryptogamie / Durieu de Maisonneuve, M C – Hyderabad. 1972-1989 (1) 1975-1989 (5) 1975-1989 (9) – 29mf – 9 – mf#7660 – ne IDC [916]

Exploration scientifique du maroc. botanique (1912) / Pitard, C J M – Paris, 1913 – 3mf – 9 – mf#11911 – ne IDC [956]

Explorations : studies in culture and communications – New York. 1953-1959 (1) – ISSN: 0531-5697 – mf#2992 – us UMI ProQuest [301]

Explorations among the watershed rockies of canada / Allen, Samuel E S – S.l: s.n, 1894? – 1mf – 9 – mf#06576 – cn Canadiana [917]

Explorations and adventures in equatorial africa / Chaillu, P B du – London, 1861 – 10mf – 9 – mf#A-158 – ne IDC [916]

Les explorations au senegal et dans les contrees voisines depuis l'antiquite jusqu'a nos jours / Ancelle, J – Paris: Maisonneuve & C Leclerc, 1886 – 1 – (filmed with: les francais au niger par pietri camille) – us CRL [550]

Explorations en guyane / Coudreau, Henri Anatole – Rouen, France. 1892 – 1r – us UF Libraries [972]

Explorations et missions de doudart de lagree : capitaine de fregate premier representant du protectorat francais au cambodge, hef de la mission d'exploration du me-king et du haut song-koi. extraits de ses manuscrits mis en ordre par m.a.b. de villemereuil / Doudart de Lagree, Ernest Marc Louis de Gonzague – Paris: J Tremblay 1883 [mf ed 1989] – 1r with other items – 1 – mf#mf-10289 seam reel 009/05 [§] – us CRL [500]

Explorations in economic research – New York. 1977-1978 (1,5,9) – ISSN: 0094-0852 – mf#11419 – us UMI ProQuest [330]

Explorations in hittite asia minor / Von der Osten, Hans Henning – 1929 – 9 – $10.00 – us IRC [930]

Explorations in jarvis inlet and desolation sound, british columbia / Downie, William – S.l: s.n, 1859? – 1mf – 9 – mf#18050 – cn Canadiana [917]

Explorations in sind : being a report of the exploratory survey carried out during the years 1927-28, 1929-30, and 1930-31 / Majumdar, Nani Gopal – Delhi: Manager of Publications, 1934 – (= ser Samp: indian books) – us CRL [930]

Explorations in south-west africa / Baines, Thomas – Farnborough, England. 1864 – 1r – 1 – us UF Libraries [960]

Explorations in south-west africa : being an account of a journey in the years 1861 and 1862... / Baines, T – London, 1864 – 7mf – 9 – mf#HT-4 – ne IDC [916]

Explorations in the pictou coal field / Haliburton, Robert Grant – Halifax, NS?: T Chamberlain, 1868 – 1mf – 9 – mf#05329 – cn Canadiana [622]

Explorations in the tyropoeon valley / Crowfoot, J W – 1929 – 9 – $10.00 – us IRC [915]

An exploratory factor analysis of collegiate athletes' perceptions of psychological adjustment to sport disengagement / Deaner, Heather R – 2000 – 104p on 2mf – 9 – $10.00 – mf#PSY 2153 – us Kinesology [150]

An exploratory investigation into the effects of tai chi exercise on balance and gait performance for hip replacement patients / Krugger, Tammy Marie – 2001 – 63p on 1mf – 9 – $5.00 – mf#PSY 2160 – us Kinesology [617]

An exploratory study of grasping in preterm, low birthweight infants / Deman, Daniela – 1999 – 1mf – 9 – $4.00 – mf#PSY 2076 – us Kinesology [150]

An exploratory study of quality assurance methodology in therapeutic recreation using the delphi technique / Riley, Robert G – 1989 – 332p 4mf – 9 – $16.00 – us Kinesology [790]

Exploratory survey of 1871 : general instructions to engineers in charge of parties, transit-men and levelers / Fleming, Sandford – [Ottawa?: Canadian Pacific Railway Co, 1871?] – 1mf – 9 – 0-665-94090-4 – mf#94090 – cn Canadiana [380]

Exploratory survey of part of the lewes, tat-on-duc, porcupine, bell, trout, peel and mackenzie rivers / Ogilvie, William – Ottawa: B Chamberlain, 1890 – 2mf – 9 – mf#09227 – cn Canadiana [520]

Explore – St. Louis. 1969-1974 (1) 1972-1973 (5) (9) – ISSN: 0014-4991 – mf#7251 – us UMI ProQuest [400]

The exploring expedition to the rocky mountains, oregon and california / Fremont, John Charles – New York, Auburn NY: Miller, Orton & Mulligan, 1856 – 5mf – 9 – mf#04685 – cn Canadiana [917]

Exploring forgiveness : a narrative approach to stories of hurt / Von Krosigk, Beate – Uni of South Africa 2000 [mf ed Johannesburg 2000] – 5mf – 9 – (incl bibl ref) – mf#mfm14931 – sa Unisa [150]

Exploring ownership of learning in children's choreographic projects / Muehlhauser, Emmely K – 1998 – 2mf – 9 – $8.00 – mf#PE 3884 – us Kinesology [790]

Exploring stories of coping with childhood cancer in a support group for parents / Papaikonomou, Maria – Uni of South Africa 2001 [mf ed Johannesburg 2001] – 5mf – 9 – (incl bibl ref) – mf#mfm14831 – sa Unisa [150]

Exploring the impact of an imagery/relaxation program : on athletes with a knee injury requiring surgery / Schriml, Carla M – 2000 – 107p on 2mf – 9 – $10.00 – mf#PSY 2129 – us Kinesology [617]

Exploring the karate way of life : coping, commitment, and psychological well-being among traditional karate participants / Wingate, Catherine F & Sachs, Michael L – 1993 – 4mf – 9 – $16.00 – us Kinesology [150]

Exploring your ethnic heritage / Ethnic Heritage Training Program (WI) – n1-6 [1981 jan-aug/sep] – 1r – 1 – mf#641413 – us WHS [305]

Explosion de mayo / Berardo Garcia, Jose – Cali, Colombia. 1957 – 1r – us UF Libraries [972]

Explosiones del sentimiento / Moreno Torrado, Luis – 1884 – 9 – sp Bibl Santa Ana [810]

Explosives engineer – Salt Lake City. 1950-1961 (1) – mf#728 – us UMI ProQuest [622]

Explotacion racional del serdo / Moreno de Arteaga, Antonio – Madrid: MAG, 1954 – 1 – sp Bibl Santa Ana [946]

Expomat actualites : magazine technique des travaux publics et de la construction – Paris. n25-42. feb 1971-73 – 5 – fr ACRPP [624]

Exponent – Bozeman, MT. 1902-1974 (1) – mf#64280 – us UMI ProQuest [071]

Exponent – Chagrin Falls, OH. 1967-1967 (1) – mf#65407 – us UMI ProQuest [071]

Exponent – Clarksburg, WV. 1995-2000 (1) – mf#67249 – us UMI ProQuest [071]

Exponent – Culpeper, VA. 1881-1929 (1) – mf#61174 – us UMI ProQuest [071]

Exponent / Wisconsin University – Platteville – 1943 oct 1-1954 may 31, 1978 apr 27-1981, 1982-88, 1989 feb-1991 may 2, 1991 sep 12-1994 may 5 – 6r – 1 – mf#579292 – us WHS [378]

Exponent 2 – 1974 jul-1984 fall – 1r – 1 – mf#966184 – us WHS [071]

Exponent and enterprise – Culpeper, VA. 1907-1908 (1) – mf#66696 – us UMI ProQuest [071]

Exponential outline with definitions of blackstone's commentaries. / Bates, W C – Columbus OH, London, 1893 – 94p – 1 – mf#LL-209 – us L of C Photodup [340]

Export, 1937-70 : the official journal of the institute of export – v1-33 – 10r – 1 – mf#96519 – uk Microform Academic [380]

Export administration act of 2001 – h.r. 2581 : hearing...house of representatives, 107th congress, 2nd session...feb 28 2002 / United States. Congress. House. Committee on Armed Services – Washington: US GPO 2002 [mf ed 2002] – 3mf – 9 – 0-16-069009-9 – (incl bibl ref) – us GPO [337]

Export america – Washington. 1999+ (1) – mf#30059 – us UMI ProQuest [380]

La exportacion del uruguay – Montevideo. 1948-55 – 1 – $46.00 – us L of C Photodup [972]

Exportdienst – Duesseldorf DE, 1946-49 – 2r – 1 – (suppl of handelsblatt duesseldorf) – gw Misc Inst [380]

Exporting to latin america / Filsinger, Ernst B – New York, NY. 1916 – 1r – 1 – us UF Libraries [380]

Exports of merchandise from afghanistan 1959/60-1973/74 / Afghanistan. Ministry of Commerce – Kabul – 39mf – 8 – (cont as: central statistics office. exports of merchandise from afghanistan 1977/78-1978/79. kabul) – mf#AS-15 – ne IDC [380]

Expose budgetaire : chambre des communes, mardi 23 juin 1891 / Foster, George Eulas – Ottawa: B Chamberlin, 1891 – 1mf – 9 – mf#04336 – cn Canadiana [336]

Expose budgetaire : chambre des communes, mardi, 5 mars 1889 / Foster, George Eulas – Ottawa?: A Senecal, 1889 – 1mf – 9 – 0-665-04341-4 – mf#04341 – cn Canadiana [336]

Expose budgetaire par l'hon george e foster...ministere des finances : chambre des communes, mardi 14 fevrier 1893 / Foster, George Eulas – Ottawa: S E Dawson, 1893 – 1mf – 9 – mf#47754 – cn Canadiana [336]

Expose chronologique des relations du cambodge avec le siam, l'annam & la france / Lemire, Charles – Paris: Challamel Aine 1879 [mf ed 1989] – 1r with other items – 1 – mf#mf-10289 seam reel 015/10 [§] – us CRL [327]

Expose de la doctrine catholique / Girodon, P – Paris: Plon-Nourrit, c1898 – 2mf – 9 – 0-8370-8259-5 – (incl bibl ref) – mf#1986-2259 – us ATLA [241]

Expose de la reforme de l'islamisme : commencee au 3eme siecle de l'hegire par abou-al-hasan ali el-ashari et continuee par son ecole: avec des extraits du texte arabe d'ibn asaakir / Mehren, August Ferdinand – [Leide: E J Brill, 1878?] [mf ed 1991] – 1mf – 9 – 0-524-01573-2 – (in french and arabic) – mf#1990-2527 – us ATLA [260]

Expose de la religion des druzes, tire des livres religieux de cette secte / Sacy, S de – Paris, 1838. 2v – 16mf – 9 – mf#NE-304 – ne IDC [956]

Expose de monsieur thiounn prasith, chef de la delegation du funk et du grunc / Thiounn Prasith – [s.l: s.n. 1973?] [mf ed 1989] – 1r with other items – 1 – (at head of title: reunion du comite executif de la conference de stockholm du 2-3 juin 1973) – mf#mf-10289 seam reel 017/34 [§] – us CRL [959]

Expose du marechal lon nol, president de la republique, a la reunion du 13-7-72 au palais d'etat a chamcar mon [texte original] / Lon Nol – [n.p. 1972?] [mf ed 1989] – 1r with other items – 1 – (in khmer & french) – mf#mf-10289 seam reel 028/10 [§] – us CRL [959]

Expose financier de sir francis hincks, mardi, 30 avril 1872 = Sir francis hincks' budget speech, tuesday, april 30, 1872 / Hincks, Francis – Ottawa?: I B Taylor, 1872 – 1mf – 9 – mf#23768 – cn Canadiana [336]

Expose of odd fellowship : containing all the lectures complete, with regulations for opening, conducting and closing a lodge... / Lander, Edwin F – Toronto, Clifton Niagara Falls, Ont: Toronto News Co, between 1878 and 1881 – 1mf – 9 – mf#28405 – cn Canadiana [360]

Expose of the royal academy of arts / Skaife, Thomas – London 1854 – (= ser 19th c art & architecture) – 2mf – 9 – mf#4.2.1365 – uk Chadwyck [700]

Exposicao apresentada ao chefe do governo provisorio da republica dos estados unidos do Brazil / Brazil. Ministerio da Justica – Rio de Janeiro: Impr Nacional [jan 1891] (annual) – 1r – 1 – us CRL [340]

Exposicao do venerando corpo do glorioso apostolo das indias, s francisco xavier, em 1890 : memoria historico-descriptiva / Albuquerque, Viriato Antonio Caetano Bras de – Nova Goa: Imprensa Nacional, 1891 [mf ed 1995] – (= ser Yale coll) – 44p/95p (ill) – 1 – 0-524-09967-7 – (in portuguese) – mf#1995-0967 – us ATLA [241]

Exposicao machado de assis / Brazil. Ministerio da Educacao e Saude Publica – Rio de Janeiro, Brazil. 1939 – 1r – us UF Libraries [972]

Exposicao sobre los livros de beato dionisio areopagita / Pedro Hispano (Pedro Juliao); ed by Alonso, M – Lisboa, 1957 – €25.00 – ne Slangenburg [240]

Exposicion... = Memoria de hacienda 1839 / Colombia. Secretaria de Hacienda – Bogota: Impr de B Espinosa [1833-1860] (annual) – 4r – 1 – us CRL [336]

Exposicion... = Memoria de la secretaria del interior i relaciones esteriores de la nueva granada presentada al congreso de...-1839 / Colombia. Secretaria de lo Interior i Relaciones Exteriores – Bogota: Impr de B Espinosa [1833-1834, 1839] (annual) – 2r – 1 – us CRL [972]

Exposicion... = Memoria del interior y relaciones exteriores / Ecuador. Ministerio de lo Interior y Relaciones Exteriores – Quito: Impr del Gobierno, [1839, 1849, 1853, 1855] (annual) – 1r – 1 – us CRL [972]

Exposicion a la real academia de la historia en favor de la aparicion de la virgen de guadalupe en mexico... / Fabie, Antonio Maria – Madrid: ed Reus, 1922 – 1 – sp Bibl Santa Ana [240]

Exposicion al nuncio / Caceres, Diego de – S.l., s.i., s.a. 1642? – 1 – sp Bibl Santa Ana [946]

Exposicion de la obra grafica original miro en extremadura / Malpartida de Caceres. Ayuntamiento – Caceres: Edit. Extremadura, 1981 – 1 – sp Bibl Santa Ana [700]

Exposicion de libros del siglo 15th al 20th ano 1944 – S e u. tip. el noticiero, s.a. – 1 – sp Bibl Santa Ana [020]

Exposicion de reproducciones en color de la unesco 90 anos de pintura universal : catalogo / Diputacion Provincial – Badajoz: dip provincial, 1954 – sp Bibl Santa Ana [946]

Exposicion del dogma catolico / Garcia Garces, Narciso – Madrid, 1943 – 1 – sp Bibl Santa Ana [241]

Exposicion del ministro de hacienda a las camaras legislativas de... / Ecuador. Ministerio de Hacienda – Quito: Impr Nacional por Mariano Mosquera [1867, 1871, 1873] (annual) – 1r – 1 – us CRL [336]

Exposicion del ministro de hacienda y comercio, mensaje del presidente de la republica al congreso nacional e informes de la comision de hacienda y comercio / Dominican Republic. Secretaria de Estado de Hacienda y Comercio – Santo Domingo: Impr "la Cuna de America" [1902] (annual) – 1r – 1 – us CRL [336]

Exposicion del plan secreto para establecer un soviet en espana – Bilbao, 1939? – (= ser Blodgett coll) – 9 – mf#fiche w865 – us Harvard [946]

Exposicion del...ayuntamiento...badajoz / Gavino Rodriguez, Martin – 1840 – 9 – sp Bibl Santa Ana [946]

Exposicion historia de la feria de mayo en caceres (documentacion del archivo municipal) catalogo 1973 / Caceres. Ayuntamiento – Caceres: tip extremadura, 1973 – 1 – sp Bibl Santa Ana [390]

La exposicion internacional de barcelona / Bayle, Constantino – Madrid: Razon y Fe, 1929 – 1 – sp Bibl Santa Ana [946]

Exposicion juez / Juez, Antonio – Badajoz, 1917 – 1 – sp Bibl Santa Ana [946]

La exposicion misional del vaticano / Bayle, Constantino – Madrid: Razon y Fe, 1925 – 1 – sp Bibl Santa Ana [241]

Exposicion provincial de arte / Obra Sindical Educacion y Descanso. Caceres – Caceres: s.i. 1949 – sp Bibl Santa Ana [700]

Exposicion que a la legislatura nacional presenta el ministro de hacienda... – Caracas: imprenta del teatro de legislacion de pedro p. del castillo e hijos, 1866-67 – us CRL [972]

Exposicion que dirige al congreso de venezuela en [...] el secretario de hacienda sobre los negocios de su cargo – Caracas: imprenta de valentin espinal, 1833-61 – us CRL [972]

Exposicion que dirige al congreso nacional de los estados unidos de venezuela el ministro de hacienda en [...] – Caracas: impr de "la concordia," de evaristo fombona, 1874-77 – us CRL [972]

Exposicion que dirije al congreso de obras nacional de los estados unidos de venezuela el ministro de hacienda en [...] – Caracas: impr bolivar, 1892, 1894-1909 – us CRL [972]

Exposicion que dirije al congreso nacional de los estados unidos de venezuela el ministro de hacienda y credito publico en [...] – Caracas: empresa el cojo, 1910-12 – us CRL [972]

Exposicion que dirije al presidente de los estados unidos de venezuela el ministro de hacienda en [...] – Caracas: impr de "la concordia," de evaristo fombona, 1873 – us CRL [972]

Exposicion que el ministro de obras publicas presenta al jefe del poder ejecutivo nacional de los asuntos de su departamento... – Caracas: imprenta bolivar, 1893 – us CRL [972]

Exposicion que presenta el ministerio de hacienda de los estados unidos de venezuela al congreso nacional en [...] – Caracas: impr de pedro coll otero, [1878-] – us CRL [972]

Exposicion regional extremena – 1892 – 9 – sp Bibl Santa Ana [946]

Exposicion sobre el tratado de limites de 1916 ent... / Munos Vernaza, Alberto – Quito, Ecuador. 1928 – 1r – us UF Libraries [972]

Exposite in terentium... / De Cass, Anthonius de Petrianis – 14th, 15th c – (= ser Holkham library manuscript books 301,426,427,489,496,511) – 1r – 1 – (filmed with: carmen de bello parthico by thomas de chaula; carmina by thomas seneca; oratio by jacobus antiquarius; orationes 4 by johannes lucidus) – mf#2197 – uk Microform Academic [090]

Expositer *see* Alamosa county miscellaneous newspapers

Expositio actuum apostolorum. retractatio in actus apostolorum. nomina regionum atque locorum de actibus apostolorum. in epistulas 7 catholicas : formae tplila 12 / Beda Venerabilis – 1983 – (= ser ILL – ser a; Ccsl 121) – 9mf+92p – 9 – €30.00 – 2-503-61212-1 – be Brepols [400]

Expositio alexandri...ordinis...san jacobi / Ramirez, Juan – 1599 – 9 – sp Bibl Santa Ana [946]

Expositio decalogi, symboli, apostolici, sacramentoru, et dominicae praecationis / Corvinus, A – [Lipsiae], 1540 – 2mf – 9 – mf#TH-1 mf 352-353 – ne IDC [242]

Expositio doctrinae augustini de creatione mundi, peccato, gratia / Ritschl, Albrecht – Halis: formis expressum hendelianis, [1843?] [mf ed 1990] – 1mf – 9 – 0-7905-6722-9 – mf#1988-2722 – us ATLA [240]

Expositio hystorica in librum regum : formae tplila 91 / Andreas a s Victore – [mf ed 1997] – (= ser ILL – ser a; Cccm 53a) – 4mf+54p – 9 – €30.00 – 2-503-63534-2 – be Brepols [450]

Expositio in epistolam ad romanos / Guillelmus a Sancto Theodorico – 1990 – (= ser ILL – ser a; Cccm 86) – 6mf+60p – 9 – €40.00 – 2-503-63862-7 – be Brepols [400]

Expositio in matthaeum : formae tplila 145 / Rabanus Maurus – [mf ed 2003] – (= ser ILL – ser a; Cccm 174-174a) – 19mf+viii/168p – 9 – €99.00 – 2-503-64742-1 – be Brepols [400]

Expositio in octo libros phisicorum aristotelis... / Celaya, J de – Paris, 1517 – 7mf – 9 – sp Cultura [600]

Expositio in psalmum 44 : formae tplila 69 / Radbertus, Pascasius – 1991 – (= ser ILL – ser a; Cccm 94) – 3mf+39p – 9 – €30.00 – 2-503-63942-9 – be Brepols [400]

Expositio super cantica canticorum : formae tplila 89 / Guillelmus a Sancto Theodorico – [mf ed 1998] – (= ser ILL – ser a; Cccm 87) – 10mf+119p – 9 – €70.00 – 2-503-63872-4 – be Brepols [400]

Expositio super danielem : formae tplila 60 / Victore, Andreas de Sancto – 1991 – (= ser ILL – ser a; Cccm 53f) – 3mf+41p – 9 – €30.00 – 2-503-50051-X – be Brepols [400]

Expositio super genesim : formae tplila 115 / Remigius Autissiodorensis – [mf ed 2000] – (= ser ILL – ser a; Cccm 136) – 6mf+70p – 9 – €40.00 – 2-503-64362-0 – be Brepols [400]

Expositio super lamentationes hieremiae : formae tplila 49 / Radbertus, Pascasius – 1989 – (= ser ILL – ser a; Cccm 85) – 8mf+76p – 9 – €40.00 – 2-503-63852-X – be Brepols [400]

Exposition agricole et industrielle de la province de quebec = Agricultural and industrial exhibition of the province of quebec – [Quebec?: s.n, 1877?] – 1mf – 9 – 0-665-92147-0 – mf#92147 – cn Canadiana [630]

An exposition and defence of universalism : in a series of sermons / Williamson, Isaac Dowd – New York: P Price, 1840 [mf ed 1992] – (= ser Unitarian/universalist coll) – 1mf – 9 – 0-524-04284-5 – mf#1991-2068 – us ATLA [243]

An exposition and defense of the scheme of redemption : as it is revealed and taught in the holy scriptures / Milligan, Robert – rev ed. St Louis: Christian Pub Co, 1894 [mf ed 1993] – 2mf – 9 – 0-524-07183-7 – mf#1992-1053 – us ATLA [240]

Exposition and defense of the westminster assembly's confession / Annan, Robert – 1855 – 1 – $50.00 – us Presbyterian [240]

Exposition critique de la methode de wronski : pour la resolution des problemes de mecanique celeste / Lagrange, Charles – Bruxelles: F Hayez 1882 [mf ed 1998] – 1r – 1 – (on tp: annexe aux annales. nouvelle serie, astronomie; no more publ after v1; incl bibl ref) – mf#film mas 28292 – us Harvard [520]

Exposition de la doctrine de l'eglise catholique orthodoxe : accompagnee des differences qui se rencontrent dans les autres eglises chretiennes / Guettee, Wladimir, abbe – 2e ed. Paris: Fischbacher, 1884 [mf ed 1990] – 2mf – 9 – 0-7905-6595-1 – (in french) – mf#1988-2595 – us ATLA [241]

Exposition de la doctrine de l'eglise catholique sur les matieres de controverse / Bossuet, J-B – Paris, 1671 – 3mf – 9 – mf#CA-98 – ne IDC [241]

Exposition de la doctrine de l'eglise catholique sur les matieres de controverse / Bossuet, J-B – Paris, 1679 – 5mf – 9 – mf#CA-117 – ne IDC [241]

Exposition de la theorie et de la pratique de la musique : suivant les nouvelles decouvertes / Bethisy, Jean Laurent de – Paris: M Lambert 1754 [mf ed 19–] – 6mf – 9 – mf#fiche 360 – us Sibley [780]

L'exposition de paris (1889) – Paris: Librairie illustree, 1889. n1-80 oct 1888-feb 1890 – us CRL [900]

L'exposition de paris (1900) – Paris: Montgredien et Cie. v1-120. 1898-1900 – us CRL [900]

Exposition des faits et de la situation actuelle de la societe de colonisation du temiscamingue vis-a-vis des actionnaires francais / Bodard, Auguste – Montreal?: s.n, 1892 – 1mf – 9 – mf#00854 – cn Canadiana [366]

Exposition des produits de la republique / Dumanoir, Philippe – Paris, France. 1849? – 1r – us UF Libraries [440]

Exposition et critique de l'ecclesiologie de calvin / Grosclaude, Charles – Geneve: W Kuendig, 1896 – 2mf – 9 – 0-524-07878-5 – (incl bibl ref) – mf#1991-3423 – us ATLA [242]

The "exposition" expounded, defended and supplemented / Carroll, John – Toronto: Methodist Book and Publishing House, 1881 – 2mf – 9 – mf#02552 – cn Canadiana [242]

Exposition familiere des principaux points du catechisme... / Viret, P – [Geneve], Rivery, 1561 – 5mf – 9 – mf#PFA-201 – ne IDC [240]

Exposition internationale du canada...et seculaire de selkirk, winnipeg, 1912 – [Winnipeg?: s.n, 1909?] [mf ed 1996] – 1mf – 9 – 0-665-79807-5 – (also available in english) – mf#79807 – cn Canadiana [338]

An exposition of facts connected with the late prosecutions in the methodist episcopal church of cincinnati / Fisher, David – Cincinnati, OH: Looker & Reynolds, 1828 – 1r – 1 – mf#F34Y C574Rl E8 – us Western Res [242]

An exposition of fallacies in the hypothesis of mr darwin / Bree, Charles Robert – London, 1872 – (= ser 19th c evolution & creation) – 6mf – 9 – mf#1.1.1578 – uk Chadwyck [575]

Exposition of part of the 24th and 25th chapters of st matthew / Lillingston, I W – Edinburgh, Scotland. 1838 – 1r – us UF Libraries [225]

Exposition of romans, chap 9 ver 6-24 / Payne, George – Edinburgh, Scotland. 1816 – 1r – us UF Libraries [221]

An exposition of some of the transactions, that have taken place at st helena : since the appointment of sir hudson lowe as governor of that island in answer to an anonymous pamphlet, entiteled, "facts illustrative of the treatment of napoleon bonaparte," etc... / O'Meara, Barry E – London 1819 [mf ed Hildesheim 1995-98] – 1v on 2mf – 9 – €60.00 – 3-487-26364-5 – gw Olms [941]

An exposition of st paul's epistle to the romans / Williams, Henry Wilkinson – London: Wesleyan Conference Office, 1869 [mf ed 1989] – 2mf – 9 – 0-7905-2571-2 – mf#1987-2571 – us ATLA [227]

An exposition of the apocalypse of st john the apostle / Putnam, Edward – Boston: Patrick Donahoe, 1858 [mf ed 1986] – 1mf – 9 – 0-8370-7425-8 – mf#1986-1425 – us ATLA [225]

An exposition of the book of ecclesiastes / Bridges, Charles – New York: Robert Carter, 1860 [mf ed 1985] – 1mf – 9 – 0-8370-2445-5 – (incl ind) – mf#1985-0445 – us ATLA [221]

Exposition of the book of proverbs / Lawson, George – Edinburgh: William Oliphant, 1829. Chicago: Dep of Photodup, U of Chicago Lib, 1978 (1r); Evanston: American Theol Lib Assoc, 1984 (1r) 35 mm – 1 – 0-8370-0599-X – mf#1984-T077 – us ATLA [221]

An exposition of the book of revelation / Burgh, William – 4th rev enl ed. London: J Shaw 1845 [mf ed 2005] – 1r – 1 – 0-524-10532-4 – (with app of notes not in former ed; incl bible text; incl bibl ref) – mf#b00742 – us ATLA [225]

An exposition of the book of solomon's song : commonly called canticles: wherein the authority of it is established and vindicated against objections, both ancient and modern / Gill, John – London: WH Collingridge, 1854 [mf ed 1994] – 4mf – 9 – 0-524-08780-6 – mf#1993-0055 – us ATLA [221]

Exposition of the case of lieutenant-colonel bouchette, surveyor-general : before the house of assembly of lower canada / Equitas – [Quebec?: s.n.] 1826 [mf ed 1983] – 1mf – 9 – 0-665-44462-1 – (in english and french) – mf#44462 – cn Canadiana [336]

An exposition of the confession of faith of the westminster assembly of divines / Shaw, Robert – 9th ed. London: Blackie, 1861 [mf ed 1986] – 1mf – 9 – 0-8370-8787-2 – (incl bibl ref & ind) – mf#1986-2787 – us ATLA [242]

Exposition of the doctrines of the catholic church / Boussuet, Jacques Benigne – London, England. 1829 – 1r – us UF Libraries [241]

An exposition of the educational significance of the principles and rules embodied in the indigenous proverbs : with reference to the setswana language / Monyai, Reginald Botshabeng – Pretoria: Vista University 2003 [mf ed 2003] – 4mf – 9 – (incl bibl ref) – mf#mfm15244 – sa Unisa [470]

An exposition of the epistle of james : in a series of discourses / Adam, John – Edinburgh: T & T Clark, 1867 [mf ed 1985] – 2mf – 9 – 0-8370-2039-5 – (incl app) – mf#1985-0039 – us ATLA [227]

An exposition of the epistle of paul to the romans / Williams, William George – Cincinnati: Jennings & Pye; New York: Eaton & Mains, c1902 [mf ed 1989] – 1mf – 9 – 0-7905-2997-1 – mf#1987-2997 – us ATLA [227]

An exposition of the epistle of saint paul to the colossians = Sermons de jean daille sur l'epitre de l'apotre s paul aux colossians / Daille, Jean – Edinburgh: James Nichol, 1863 [mf ed 1985] – 1mf – 9 – 0-8370-5994-1 – (rev & corr by james sherman; incl ind) – mf#1985-3994 – us ATLA [227]

An exposition of the epistle of saint paul to the philippians = Exposition sur la divine eptre de l'aptre s paul aux filippiens / Daill, Jean – Edinburgh: James Nicol, 1865 [mf ed 1985] – 1mf – 9 – 0-8370-5993-3 – (english trans by james sherman; incl ind) – mf#1985-3993 – us ATLA [227]

An exposition of the epistle to the hebrews / Williams, Henry Wilkinson – London: Wesleyan Conference Office, 1871 [mf ed 1989] – 2mf – 9 – 0-7905-2572-0 – (incl ind) – mf#1987-2572 – us ATLA [227]

An exposition of the epistles of st paul = Triplex expositio epistolarum sancti pauli / Bernadine a Piconio; ed by Prichard, A H – 2nd ed. London: John Hodges, 1889-90 [mf ed 1993] – (= ser Catholic standard library) – 3v on 3mf – 9 – 0-524-05969-1 – (english trans fr latin by ed) – mf#1992-0706 – us ATLA [227]

An exposition of the faith of the religious society of friends : commonly called quakers, in the fundamental doctrines of the christian religion – 5th american ed. Philadelphia: For sale at Friends' Book Store, 1878 [mf ed 1986] – 1mf – 9 – 0-8370-8897-6 – (incl ind) – mf#1986-2897 – us ATLA [243]

An exposition of the first epistle to the corinthians / Hodge, Charles – New York: R Carter, 1857 [mf ed 1984] – 5mf – 9 – 0-8370-0895-6 – mf#1984-4276 – us ATLA [227]

An exposition of the gospel of mark / Kelly, William; ed by Whitfield, E E – London: Elliot Stock, 1907 [mf ed 1985] – 1mf – 9 – 0-8370-3875-8 – (incl ind) – mf#1985-1875 – us ATLA [226]

An exposition of the gospel of st john : consisting of an analysis of each chapter, and of a commentary, critical, exegetical, doctrinal, and moral, having the text, english and latin, prefixed in full to each chapter / MacEvilly, John – New York: Benziger Bros, 1889 [mf ed 1993] – 1mf – 9 – 0-524-06520-9 – mf#1992-0904 – us ATLA [226]

Exposition of the gospel of st john / Govett, R – London: Bemrose, [1881?] – 3mf – 9 – 0-7905-1819-8 – (V2 incorrectly numbered as v1) – mf#1987-1819 – us ATLA [226]

An exposition of the gospels : consisting of an analysis of each chapter and of a commentary, critical, exegetical, doctrinal, and moral / MacEvilly, John – 3rd rev corr ed. New York: Benziger Bros, 1888 [mf ed 1993] – 2mf – 9 – 0-524-06922-0 – mf#1992-1015 – us ATLA [226]

An exposition of the gospels of st matthew and st mark : and of some other detached parts of holy scripture / Watson, Richard – New York: Carlton & Phillips, 1855 [mf ed 1993] – 2mf – 9 – 0-524-05757-5 – mf#1992-0600 – us ATLA [226]

An exposition of the gospels of the church year on the basis of nebe / Wolf, Edmund Jacob – Philadelphia: Lutheran Publ Soc, c1900 [mf ed 1989] – 3mf – 9 – 0-7905-0659-9 – (incl ind) – mf#1987-0659 – us ATLA [226]

Exposition of the ninth chapter of the epistle to the romans / Morison, James – new ed. London: Hodder and Stoughton, 1888 – 1mf – 9 – 0-8370-4492-8 – mf#1985-2492 – us ATLA [227]

An exposition of the parables of our lord : showing their connection with his ministry, their prophetic character, and their gradual development of the gospel dispensation / Bailey, B – London: Printed for John Taylor, 1828. Chicago: Dep of Photodup, U of Chicago Lib, 1978 (1r); Evanston: American Theol Lib Assoc, 1984 (1r) – 1 – 0-8370-0701-1 – (incl bibl ref) – mf#1984-T095 – us ATLA [220]

An exposition of the pretensions of baptists to antiquity : as viewed from scripture and history / Clement, James A – Nashville, TN: publ for aut, c1860 [mf ed 1992] – 1mf – 9 – 0-524-03576-8 – mf#1990-1036 – us ATLA [242]

Exposition of the principles of church-government adopted by the me... / Allin, Thomas – Sheffield, England. 1833 – 1r – us UF Libraries [240]

An exposition of the principles of code pleading / Phillips, George Lemon – 2nd ed. Chicago: Callaghan, 1932. 775p. LL-1178 – 1 – us L of C Photodup [348]

An exposition of the principles of partnership / Parsons, James – Boston: Little, Brown, 1889. 709p. LL-1405 – 1 – us L of C Photodup [346]

An exposition of the principles of pleading under the codes of civil procedure / Phillips, George Lemon – Chicago: Callaghan, 1896. 604p. LL-1338 – 1 – us L of C Photodup [348]

An exposition of the relations of the british government with the sultaun and state of palembang : and the designs of the netherlands' government upon that country, with descriptive accounts and maps of palembang and the island of banca / Court, H – London 1821 [mf ed Hildesheim 1995-98] – 1v on 2mf – 9 – €60.00 – 3-487-27443-4 – gw Olms [327]

An exposition of the second epistle to the corinthians / Hodge, Charles – New York: R Carter, 1864 [mf ed 1984] – 4mf – 9 – 0-8370-0894-8 – mf#1984-4277 – us ATLA [227]

An exposition of the shorter catechism : or, a scripture catechism in the method of the assemblies / Henry, Matthew – Edinburgh: J Lowe, 1857 [mf ed 2004] – 1r – 1 – 0-524-10473-5 – mf#b00690 – us ATLA [240]
An exposition of the shorter catechism / ed by Salmond, Stewart Dingwall Fordyce – Edinburgh: T & T Clark, [18–?] [mf ed 1993] – (= ser Bible class primers; Presbyterian coll) – 3v on 3mf – 9 – 0-524-07914-5 – mf#1991-3459 – us ATLA [240]
An exposition of the tabernacle, the priestly garments, and the priesthood / Soltau, Henry W – London: Morgan & Chase, [1865?] [mf ed 1989] – 2mf – 9 – 0-7905-3358-8 – mf#1987-3358 – us ATLA [240]
An exposition of the thirty-nine articles : historical and doctrinal / Browne, Edward Harold – 14th ed. London; New York: Longmans, Green, 1894 [mf ed 1986] – 2mf – 9 – 0-8370-8654-X – (incl bibl ref & ind) – mf#1986-2654 – us ATLA [242]
An exposition of views respecting the principal facts, causes, and peculiarities involved in spirit manifestations : together with interesting phenomenal statements and communications / Ballou, Adin – Boston: Bela Marsh, 1852 [mf ed 1992] – (= ser Unitarian/universalist coll) – 1mf – 9 – 0-524-04421-X – mf#1992-2026 – us ATLA [243]
Exposition provinciale (1887: Quebec, Quebec) *see* Liste des prix de l'exposition
Exposition provinciale (1916: Quebec, Quebec) *see* Le merite agricole a l'exposition provinciale de quebec
Exposition provinciale (1917: Quebec, Quebec) *see* Le merite agricole a l'exposition provinciale de quebec
Exposition provinciale (1918: Quebec, Quebec) *see* Le merite agricole a l'exposition provinciale de quebec
Exposition scolaire de la province de quebec : catalogue / Verreau, Hospice Anthelme – Montreal?: J B Plante, 1880 – 1mf – 9 – (incl english text) – mf#25340 – cn Canadiana [370]
Exposition sur la divine eptre de l'aptre s paul aux filippiens *see* An exposition of the epistle of saint paul to the philippians
Exposition Universelle 1900. Paris *see* Bulletin des lois, dea1crets et documents officiels
Exposition universelle internationale de 1878, a Paris. Commissariat General *see* Catalogue officiel
Exposition Universelle. Paris, 1878 *see* Catalogue officiel
An exposition upon the epistle of jude / Jenkyn, William – Edinburgh: James Nichol, 1865 [mf ed 1985] – 4mf – 9 – 0-8370-3778-6 – (incl ind) – mf#1985-1778 – us ATLA [227]
An exposition with notes : unfolded and applied on john 17 / Newton, George – Edinburgh: James Nichol, 1867 [mf ed 1985] – (= ser Nichol's series of commentaries) – 1mf – 9 – 0-8370-4578-9 – (incl ind) – mf#1985-2578 – us ATLA [225]
Expositiones historicae in libros salomonis : formae tplila 67 / Victore, Andreas de Sancto – 1991 – (= ser ILL – ser a; Cccm 53b) – 5mf+54p – 9 – €30.00 – 2-503-63536-9 – be Brepols [400]
Expositiones pauli epistolarum : formae tplila 90 – [mf ed 1997] – (= ser ILL – ser a; Cccm 151) – 8mf+63p – 9 – €50.00 – 2-503-64512-7 – be Brepols [400]
Expositions of the epistles of paul to the philippians and colossians / Calvin, J – Edinburgh: Thomas Clark, 1842 – 1mf – 9 – 0-7905-0974-1 – (In English and Greek) – mf#1987-0974 – us ATLA [227]
Expositor – 1875-1925 – 1 – us L of C Photodup [240]
Expositor – Sturgeon Bay WI. 1873 oct 24-1875, 1876-78, 1879-1880 jun 25 – 3r – 1 – (continued by: weekly expositor) – mf#934403 – us WHS [071]
Expositor – Geneva, NY. 1806-1809 (1) – mf#64975 – us UMI ProQuest [071]
Expositor : a weekly journal of foreign and domestic intelligence, literature, science and the fine arts – New York. 1838-1839 – 1 – mf#5323 – us UMI ProQuest [073]
O expositor – Desterro, SC: Typ da Sociedade Patriotica, 08 dez 1832; 16 fev 1833 – (= ser Ps 19) – bl Biblioteca [079]
Expositor and universalist review – Boston. 1831-1840 (1) – mf#3980 – us UMI ProQuest [240]
El expositor bautista : paper of the argentine baptist convention – 1909-13 – 960p – 1 – us Southern Baptist [242]
El expositor biblico – 1890, 1916-17, 1929 – 1212ü – 1 – us Southern Baptist [220]
Expositor christao – Rio de Janeiro, RJ. 04 ago 1904 – (= ser Ps 19) – mf#DIPER – bl Biblioteca [079]
The expositor in the pulpit / Vincent, Marvin Richardson – New York: A D F Randolph, c1884 – 1mf – 9 – 0-7905-3106-2 – mf#1987-3106 – us ATLA [220]
The expositor of holiness – Toronto: Publ under the auspices of the Canada Holiness Assoc, [1882-189- or 19–] – 9 – mf#P04390 – cn Canadiana [240]
The expositor's dictionary of texts : containing outlines, expositions, and illustrations of bible texts, with full references to the best homiletic literature / ed by Nicoll, William Robertson, Sir et al – New York: Hodder and Stoughton: George H Doran, c1910 – 5mf – 9 – 0-7905-8311-9 – (incl bibl ref) – mf#1987-6416 – us ATLA [052]
The expositor's greek testament / Bruce, Alexander Balmain et al; ed by Nicoll, William Robertson, Sir – London; New York: Hodder and Stoughton, [1897-1910] – 8mf – 9 – 0-8370-1686-X – mf#1987-6113 – us ATLA [220]
Expository and practical lectures on haggai and zechariah / Eaton, John van; ed by Robinson, W J – Pittsburgh: United Presbyterian Bd of Publ, [1882?] – 1mf – 9 – 0-8370-5617-9 – mf#1985-3617 – us ATLA [221]
Expository discourses on various scripture facts and characters / Jackson, Thomas – London: John Mason, 1839 – 2mf – 9 – 0-524-07783-5 – mf#1992-1107 – us ATLA [220]
Expository lectures on st paul's epistles to the corinthians / Robertson, Frederick William – new ed. London: Kegan Paul, Trench, 1883 [mf ed 1985] – 2mf – 9 – 0-8370-4921-0 – mf#1985-2921 – us ATLA [227]
Expository lectures on the Heidelberg catechism / Bethune, George Washington – New York: Sheldon and Co, 1866 – 3mf – 9 – 0-524-04250-0 – (Incl bibl ref) – mf#1991-2034 – us ATLA [240]
Expository notes on the book of joshua / Crosby, Howard – New York: Robert Carter, 1875 – 1mf – 9 – 0-8370-2781-0 – (Includes appendix) – mf#1985-0781 – us ATLA [221]
Expository thoughts on the gospels : st luke for family and private use / Ryle, John Charles – New York: Robert Carter, 1859-60 [mf ed 2002] – 2v on 1r – 1 – mf#b00643 – us ATLA [226]
Expository thoughts on the gospels : st matthew for family and private use / Ryle, John Charles – New York: Robert Carter, 1857 [mf ed 1986] – 1mf – 9 – 0-8370-9815-7 – mf#1986-3815 – us ATLA [226]
The expository times – 1889-1999+ – 42r – 1 – £1980.00 – (an inter-denominational monthly magazine) – mf#EXT – uk World [073]
The expository times – 1889- – 1 – enquire for prices – (yrly reel count varies) – us UMI ProQuest [072]
Expossitio latinitatis : formae tplila 74 – 1992 – (= ser ILL – ser a; Cccm 133d) – 5mf+56p – 9 – €30.00 – 2-503-50281-4 – be Brepols [400]
Expostulation addressed to the friends of the reformation in the un... / Lang, B – Glasgow, Scotland. 1838 – 1r – us UF Libraries [241]
Ein expostulation oder klag jhesu zu dem menschen, der vss eygnem mutwill verdampt wuert... / Erasmus – [Zuerich, Christoph Froschouer, 1522] – 1mf – 9 – mf#PBU-538 – ne IDC [240]
The exposure – [Lilongwe: s.n.] [mar 8-21 1994] (biwkly) – 1r – 1 – us CRL [079]
Exposure of an attempt recently made by certain west-indian agents to mislead parliament on the subject of colonial slavery – [London, 1831] – (= ser 19th c books on british colonization) – 1mf – 9 – mf#1.1.1322 – uk Chadwyck [972]
Exposure of popery : being a free translation, with some additions, of popiyat ka ahwal, a work written in hindustani = Popiyat ka ahwal / Ullmann, Julius Frederick – Bombay: A W Prautch, [1892?] – 1mf – 9 – 0-8370-8394-X – (in english) – mf#1986-2394 – us ATLA [240]
Exposure of the jesuits – Cheltenham, England. 18– – 1r – us UF Libraries [241]
An exposure of the mischievious perversions of holy scripture in the national temperance society's publications : addressed to men of sense and candour / Carry, John – [Toronto: s.n.], 1885 [mf ed 1980] – 1mf – 9 – 0-665-00493-1 – mf#00493 – cn Canadiana [220]
Ex-pow bulletin / American Ex-Prisoners of War, inc – 1983 feb-1985 dec, 1986-1987 jun, 1987 jul-1988 – 3r – 1 – mf#998795 – us WHS [355]
Expozicao dos direitos que a constituicao – Rio de Janeiro, Brazil. 1833 – 1r – us UF Libraries [972]
Expresion festiva a la funcion que la excellentissima senora maria ventura guirior, virreyna de los reynes del peru : hizo al glorioso apostol de las indias san francisco xavier, su especial protector, el dia 3. de diciembre del ano de 1777 / Guirior, Maria Ventura – Mexico: Impr de Luis Abodiano y Valdes [1777?] – (= ser Books on religion...1543/44-c1800: jesuitas) – 1mf – 9 – mf#crl-238 – ne IDC [241]
Expresion literaria de nuestra vieja raza / Herrera Vega, Adolfo – San Salvador, El Salvador. 1961 – 1r – us UF Libraries [972]
Expreso de miami – Miami, FL. 1976 apr 2-1999 jul 30 – 18r – 1 – (gaps) – us UF Libraries [071]
L'express : journal quotidien republicain independant – Paris, 14 janv 1881-4 fevr 1883 – 1 – fr ACRPP [074]
Express – Big Timber, MT. 1896-1901 (1) – mf#64248 – us UMI ProQuest [071]
Express – Chicago.18 Dec 1936. English & Yiddish. Ceased publ – 1 – us AJPC [071]
Express – Dallas, TX. 1941-1970 (1) – mf#66590 – us UMI ProQuest [071]
Express – Dublin, Ireland. 8 oct 1832-2 feb 1833 – 1r – 1 – uk British Libr Newspaper [072]
Express / Franklin Co. Columbus – 1894-95,97-8/05,11/05-06,5/07-7/17 [wkly] – 9r – 1 – (In German) – mf#B11738-11746 – us Ohio Hist [071]
Express / Franklin Co. Columbus – oct 1891-jul 1903 [daily] – 26r – 1 – (In German) – mf#B1524-1549 – us Ohio Hist [071]
Express – Fremantle, Australia. 5 jan-31 mar 1870; 1 jul-31 dec 1870 (imperfect) – 1 1/2r – 1 – uk British Libr Newspaper [072]
Express – Hurstville – 10r – A$628.32 vesicular A$683.32 silver – at Pascoe [079]
Express / Licking Co. Newark – dec 1894-sep 1917 [wkly] – 8r – 1 – (In German) – mf#B3418-3437 – us Ohio Hist [071]
Express / Lucas Co. Toledo – jul 1903-09, jul-sep 1912 [daily] – 23r – 1 – (In German) – mf#B10460-10482 – us Ohio Hist [071]
Express : muelhauser zeitung – Muelhausen / Elsass (Mulhouse F), 1884-1914 14 sep, 1919-31 – 1 – (filmed by misc inst: 1879-83 (gaps) [6r]) – fr ACRPP; gw Misc Inst [074]
Express – Muenchen DE, 1948 11 jan-25 sep – 1 – uk British Libr Newspaper [074]
Express – Pasco, WA. 1905-1919 (1) – mf#67069 – us UMI ProQuest [071]
Express – Petersburg, VA. 1856-1869 (1) – mf#66795 – us UMI ProQuest [071]
Express – Pony, MT. 1908-1911 (1) – mf#64612 – us UMI ProQuest [071]
Express / Richland Co. Shelby – v1 n3. may 6-oct 15 1862 (short) [wkly] – 1r – 1 – mf#B6576 – us Ohio Hist [071]
Express – Salem, WV. 1913-1916 (1) – mf#67467 – us UMI ProQuest [071]
Express – Sanford, NC. 1887-1937 (1) – mf#65336 – us UMI ProQuest [071]
Express – Spartanburg, SC. 1857-1857 (1) – mf#66521 – us UMI ProQuest [071]
Express – Sydney, Australia. 17 jan 1880-25 dec 1880; 1881-29 dec 1883 – 4r – 1 – uk British Libr Newspaper [072]
Express – Sydney, jan 1880-jun 1887 – 3r – A$258.59 vesicular A$275.09 silver – at Pascoe [079]
Express – Tallmadge, OH. 1996-2000 (1) – mf#65672 – us UMI ProQuest [071]
Express – Thomaston, CT. 1980-2000 (1) – mf#68132 – us UMI ProQuest [071]
Express – Koeln DE, 1964 11 mar-1966 – 1 – (until 1966 n21: koelner stadt-anzeiger / express. filmed by other misc inst: 1976- [ca 6r/yr]) – gw Misc Inst [074]
Express – Watkins, NY. 1872-1976 (1) – mf#65275 – us UMI ProQuest [071]
Express / Wisconsin Center University – Richland – 1978 feb 21-1985 may – 1r – 1 – mf#1565648 – us WHS [378]
Express – Wollongong, jul 1973-jun 1975 – 3r – at Pascoe [079]
Express *see* The uitenhage times
The express – Dar es Salaam: Media Holdings Ltd, [feb 1992-] – us CRL [079]
The express – Dublin. Ireland. -d. 8 Oct 1832-2 Feb 1833. (1 reel) – 1 – uk British Libr Newspaper [072]
The express – Freemantle, Australia. 5 Jan-31 Mar, 1 Jul-31 Dec 1870 (imperfect).-w. 1mqn reels – 1 – uk British Libr Newspaper [079]
The express – Sydney, Australia. -w. 17 Jan 1880-29 Dec 1883. 4 reels – 1 – uk British Libr Newspaper [079]
Express and orange free state advertiser – Bloemfontein, South Africa. Express en Oranjevrijstaatsch advertentieblad. -w. March 1875-99. 1880, 1881 imperfect. 14 1 2 reels – 1 – uk British Libr Newspaper [079]
Express and the superior sun – Superior, NE: Express Print Co. 1v. v1 n6. feb 15 1900-// (wkly) [mf ed -mar 29 1900 (gaps)] – 1r – 1 – (formed by the union of: express (superior ne) and: superior sun. cont by: superior express) – us NE Hist [071]
Express [bromley ed] – Bromley, England 23 sep 2004-19 jan 2006 – 1 – (cont: bromley & beckenham express [10 jan 2001-15 sep 2004]) – uk British Libr Newspaper [072]
L'express de toronto – Ontario, CN. 1985- – 1r/yr – 1 – Can$93.00r – cn Commonwealth Imaging [071]
Express dispatch *see* Mckinleyville express
L'express du midi – Toulouse, 1924, 1929 – 1 – fr ACRPP [074]
Express extra – [NW England] Stockport 1980-jul 1981 – 1 – (title change: express advertiser Plus [aug-dec 1981, jan-oct 1982]) – uk MLA; uk Newsplan [072]
Express ilustrowany – Lodz, Poland. dec 1952-aug 1953; 8 jun 1959 – 1r – 1 – us L of C Photodup [077]
Express magazine – v1 n2 [1988:dec], v1 n3-6 [1989:jan/fev-mai/juin] – 1r – 1 – (continued by: haiti express magazine) – mf#3070979 – us WHS [071]
Express mail – Blantyre: [s.n.] [jan 1994] – 1r – 1 – us CRL [079]
Express [moston middleton blackley crumpsall] – [North West] Manchester ALS jan 1992-dec 1996 – 1 – uk MLA; uk Newsplan [072]
Express [moston, middleton, blackley, crumpsall] – [NW England] Manchester ALS jan 1992-dec 1996 – 1 – uk MLA; uk Newsplan [072]
Express news pm – San Antonio, TX. 1984-1994 (1) – (cont: news) – mf#60130,01 – us UMI ProQuest [071]
De express / orange vrijstaasth blad – Bloemfontein SA, 1875-99 – 21r – 1 – sa National [079]
Express poznanski – Poznan, Poland. Jul 1952-1959 – 11r – 1 – (some iss missing) – us L of C Photodup [077]
Express series / Montgomery Co. Dayton – aug 1964-oct 1971 [wkly] – 4r – 1 – mf#B5449-5452 – us Ohio Hist [976]
Express / standard / Jackson Co. Jackson C.H. – jun 1858-may 1866 [wkly] – (= ser Standard) – 3r – 1 – (title changes) – mf#B2028-2030 – us Ohio Hist [071]
Express times – Easton, PA. 1947-2000 (1) – mf#61779 – us UMI ProQuest [071]
Express Tribe *see* Queen city express
Express und westbote / Franklin Co. Columbus – aug 1903-aug 1918 [daily] – 37r – 1 – (in german) – mf#B5962-5993 – us Ohio Hist [071]
Express views – Blantyre: [s.n.] [jan 27, feb, apr 8/14 1994] – 1r – 1 – us CRL [079]
Express wieczorny – Warsaw, Poland. jul 1946-aug 1948; mar 1950-jun 1970 – 43r – 1 – (some missing iss) – us L of C Photodup [077]
Expressen – Stockholm, Sweden. 1944-78. 473 reels – 1 – (newsbills, 1944-63 34r) – sw Kungliga [078]
Expressen – Stockholm, Sweden. 1979- – 1 – sw Kungliga [078]
Expressing universal themes through storydance choreography : the creation and production of two narratives / Cambridge, Lark A – 1993 – 2mf – $8.00 – us Kinesology [790]
Expression – 1997 apr 10, may 19-jun 4, aug 20 – 1r – 1 – mf#4695593 – us WHS [071]
Expression der angiogenese-modulierenden faktoren b-fgf, vegf, tf und tsp in in-vitro-kultivierten und der nacktmaus transplantierten pankreaskarzinomen / Schult, Ricardo – (mf ed 2000) – 1mf – 9 – €30.00 – 3-8267-2708-8 – mf#DHS 2708 – gw Frankfurter [616]
L'expression du chant gregorien, vol 2 : le temporal de paques a l'avent / Baron, L – Plouharnel, 1948 – €15.00 – ne Slangenburg [241]
Expression du haut degre en francais contemporain / Berthelon, Christiane – Bern, Switzerland. 1955 – 1r – us UF Libraries [960]
The expression of the emotions in man and animals / Darwin, Charles Robert – London, 1872 – (= ser 19th c evolution & creation) – 5mf – 9 – (with ill) – mf#1.1.4598 – uk Chadwyck [574]
Expressionism : subject collections – (= ser Art exhibition catalogues on microfiche) – 80 catalogues on 109mf – 9 – £685.00 – (individual titles not listed separately) – uk Chadwyck [700]
Expressionismus : aufzeichnungen und erinnerungen der zeitgenossen / ed by Raabe, Paul – Olten: Walter-Verlag, c1965 [mf ed 1993] – (= ser Walter texte und dokumente zur literatur des expressionismus) – 422p – 1 – (incl bibl ref and ind. ann by ed) – mf#8280 – us Wisconsin U Libr [430]
Expressionismus : gestalten einer literarischen bewegung / Friedmann, Hermann & Mann, Otto – Heidelberg: W Rothe, 1956 [mf ed 1993] – 375p – 1 – mf#8271 – us Wisconsin U Libr [430]
Expressionismus und religion : gezeigt an der neuesten deutschen expressionistischen lyrik / Knevels, Wilhelm – Tuebingen: Verlag der J C B Mohr (Paul Siebeck), 1927 – 40p – 1 – (incl bibl ref) – us Wisconsin U Libr [430]
L'expressionnisme allemand / Garnier, Ilse – Paris: A Silvaire, c1962 [mf ed 1993] – (= ser Coll "ecoles et mouvements") – 1 – (incl bibl ref) – mf#8271 – us Wisconsin U Libr [430]
Expressions figurees de la langue malgache / Grandidier, Guillaume – Paris: C Lamy, 1902 – 1 – us CRL [490]

Expressions of law and fact construed by the courts of georgia / Dutcher, Salem – Atlanta, Franklin, 1899 – 172p – 1 – mf#LL-605 – us L of C Photodup [347]

L'expressivite chez salvien de marseille, premiere partie : lat christ primaeva 7 / Janssen, O – Noviomagi, 1937 – 5mf – 8 – €12.00 – ne Slangenburg [240]

Expresso – Lisbon. Nos.1-380, 6 Jan 1973-9 Feb 1980. Incomplete. (Portuguese Revolution of 1974. Newspapers from Portugal publ. from 21 Feb 1971 to 15 Feb 1980, collected and filmed by University of Wisconsin-Madison libraries.) – 1 – us Wisconsin U Libr [074]

Expropiacion / Morales Saenz, Julio Cesar – Panama, 1964 – 1r – us UF Libraries [972]

Expulsion de los moriscos de denia (anno 1596-1621) – Barcelona – 1r – 5,6 – sp Cultura [946]

Expunerea situatiunei financiare a tesaurului public / Romania.Ministerul Finantelor – Bucharest. 31v.1880-1911 – 1 – 69.00 – us L of C Photodup [336]

Exquemelin, A O *see* Buccaneers of america

Exquiarum [sic] ordo – Mariannhill, South Africa. 1909 – 1r – us UF Libraries [960]

Exsae international / Starlite, Inc – 1993 spr, 1994 spr-sum, 1996 spr – 1r – 1 – (continued by: exsae) – mf#2681050 – us WHS [071]

Ex-situ bioremediation of hydrocarbon contaminated soil using the biopile technique / Zyl, Helena Dorathea van – Stellenbosch: U of Stellenbosch 1998 [mf ed 1998] – 4mf – 9 – mf#mf.1300 – sa Stellenbosch [660]

Exspectatio gloriae futurae jesu christi... / Marck, J – Lugduni Batavorum, 1730 – 10mf – 9 – mf#PBA-248 – ne IDC [240]

Extases de m hochenez / Marc-Michel, M – Paris, France. 1850 – 1r – us UF Libraries [440]

The extel records : archives of the exchanges telegraph co ltd – 3pts. 1872-1966 – 122r – 1 – £5,650.00 coll – mf#EXL – uk World [380]

Extended Opportunity Program and Services [Gendale CA] et al *see* Otro vacquero

Extension bulletin – New York. v6 n1-5, 7 n1-4, 6; 14-17. jan-may 1903, jan-apr, jun 1904; sep 1910-jun 1914 – 1 – us NY Public [780]

Extension of empire, weakness? : deficits? ruin? / Lloyd, Francis – London, 1880 – (= ser 19th c british colonization) – 2mf – 9 – mf#1.1.8364 – uk Chadwyck [320]

Extension review – Washington. 1978-1989 (1) 1978-1989 (5) 1978-1989 (9) – (cont: extension service review) – ISSN: 0162-9875 – mf#5766,01 – us UMI ProQuest [370]

Extension service annual reports : alabama, 1909-1944 – (= ser The 1910 federal population census (1996)) – 115r – 5 – mf#T845 – us Nat Archives [317]

Extension service field representatives annual reports / U.S. Federal Extension Service – (= ser Records of the federal extension service) – 5 – (alabama 1909-44 115r t845. alaska 1930-44 2r t846. arizona 1915-44 22r t847. arkansas 1909-44 106r t848. california 1913-44 54r t849. colorado 1913-44 77r t850. connecticut 1913-44 30r t851. delaware 1914-44 13r t852. district of columbia 1917-19 1r t853. florida 1909-44 46r t854. georgia 1909-44 141r t855. hawaii 1929-44 7r t856. idaho 1913-44 47r t857. illinois 1914-44 82r t858. indiana 1912-44 80r t859. iowa 1912-44 195r t860. kansas 1913-44 186r t861. kentucky 1912-44 89r t862. louisiana 1909-44 67r t863. maine 1915-44 48r t864. maryland 1912-44 70r t865. massachusetts 1914-44 59r t866. michigan 1913-44 77r t867. minnesota 1914-44 105r t868. mississippi 1909-44 96r t869. missouri 1914-44 90r t870. montana 1914-44 64r t871. nebraska 1913-44 89r t872. nevada 1915-44 19r t873. new hampshire 1914-44 34r t874. new jersey 1913-44 45r t875. new mexico 1914-44 30r t876. new york 1912-44 90r t877. north carolina 1909-44 144r t878. north dakota 1912-44 68r t879. ohio 1915-44 98r t880. oklahoma 1909-44 135r t881. oregon 1914-44 73r t882. pennsylvania 1914-44 43r t883. puerto rico 1930-44 14r t884. rhode island 1914-44 12r t885. southern states: report of progress 1913-14 1r t886. south carolina 1909-44 91r t887. south dakota 1913-44 65r t888. tennessee 1910-44 64r t889. texas 1909-44 182r t890. utah 1914-44 30r t891. vermont 1912-44 31r t892. virginia 1908-44 82r t893. washington 1913-44 45r t894. west virginia 1912-44 47r t895. wisconsin 1913-44 49r t896. wyoming 1914-44 37r t897) – us Nat Archives [630]

Extension service review – Washington. 1930-1978 (1) 1973-1978 (5) 1975-1978 (9) – (Cont by: Extension review) – ISSN: 0014-5408 – mf#5766 – us UMI ProQuest [630]

Extension urbana de la merida romana / Gil Farres, Octavio – 1 – sp Bibl Santa Ana [946]

Extensive and systematic colonisation in connection with the construction of the intercolonial railway through the canada dominion : being a series of letters published in the "glasgow sentinel", and addressed to alexander campbell, esq... / Doull, Alexander – Glasgow?: s.n, 1868 – 1mf – 9 – mf#13025 – cn Canadiana [380]

An extensive system of emigration considered : with a practical mode of raising the necessary funds / Shaw, Charles – 2nd ed. London 1848 – (= ser 19th c british colonization) – 1mf – 9 – mf#1.1.527 – uk Chadwyck [304]

The extent of the atonement : in its relation to god and the universe / Jenkyn, Thomas William – 3rd ed. Boston: Gould & Lincoln 1859 [mf ed 1993] – 1mf – 9 – 0-524-07317-1 – mf#1991-3032 – us ATLA [240]

The extent to which the title of a purchaser to land, bought at a sheriff's sale, is affected by error in the proceedings, in pennsylvania / Lancaster, Joseph Campbell, 1861-1916 – Philadelphia, Welsh, 1884. 92 p. LL-324 – 1 – us L of C Photodup [346]

Exterior ballistics / Hayes, Thomas Jay – New York, NY. 1938 – 1r – us UF Libraries [500]

Exterior, revistas politicas y literarias / Sierra, Justo – Mexico City? Mexico. 1948 – 1r – us UF Libraries [972]

Exterminacion anorada / Rosario Perez, Angel S, Del – Ciudad Trujillo, Dominican Republic. 1957 – 1r – us UF Libraries [972]

The external evidence of the bible / Hughes, Isaac C – Rock Island, IL: R Crampton, 1877 – 1mf – 9 – 0-524-06208-0 – mf#1992-0846 – us ATLA [220]

External religion : its use and abuse / Tyrrell, George – 2nd ed. London: Sands, 1900 [mf ed 1986] – ix/166p on 1mf – 9 – 0-8370-8952-2 – mf#1986-2952 – us ATLA [241]

Extinct and dormant baronetcies of england / Burke, J B & Burke, J B – 1r – 1 – mf#2153 – uk Microform Academic [920]

The extinction of evil : three theological essays = Fin du mal / Petavel, E – Boston: Charles H. Woodman, 1889 – 1mf – 9 – 0-7905-3098-8 – (in english) – mf#1987-3098 – us ATLA [240]

The extinction of the christian churches in north africa / Holme, Leonard Ralph – London: C J Clay 1898 [mf ed 1990] – 1mf [ill] – 9 – 0-7905-5657-X – (incl bibl ref) – mf#1988-1657 – us ATLA [240]

Extirpacion de la idolatria del piru : dirigido al rey n s en su real consejo de indias / Arriaga, Pablo Jose de – en Lima: Por Ceronymo de Contreras...ano 1621 – (= ser Books on religion...1543/44-c1800: evangelizacion) – 2mf – 9 – mf#crl-340 – ne IDC [241]

La extirpacion de la idolatria en el peru del p. pablo joseph de arriaga / Bayle, Constantino – Madrid: Razon y Fe, 1922 – 1 – sp Bibl Santa Ana [240]

Extirpacion total de la laringe por cacinoma / Gisneros y Sevillano, Juan – Madrid: establecimiento tipografico de fortanet, 1961 – 1 – sp Bibl Santa Ana [946]

Extra : african nationalist pioneer movement news / African Nationalist Pioneer Movement – 1966 may, 1967 jan, v1 n9 – 1r – 1 – mf#4848565 – us WHS [320]

Extra – Lancaster, PA. 1982-1988 (1) – mf#68061 – us UMI ProQuest [071]

Extra – Minot, ND. 1986-1989 (1) – mf#68491 – us UMI ProQuest [071]

Extra – Providence RI. v1 n1-26 [1968 oct 1/14-nov 4/18], v2 n1-9 [1969 dec 2-1970 mar 12] – 1r – 1 – mf#154679 – us WHS [071]

Extra (ayr and south ayrshire ed) – Ayr: Archant Regional Ltd 2003- (wkly) – 1 – (cont: leader (ayr & south ed)) – uk Scotland NatLib [072]

Extra [basingstoke & north hampshire] – Basingstoke, England 24 nov 1999-9 feb 2000 – 1 – (cont: basingstoke gazette extra [13 may 1981-17 nov 1999); cont by: gazette extra: for basingstoke & north hampshire [16 feb 2000-5 sep 2001]) – uk British Libr Newspaper [072]

Extra cambodja-nummer : een extra-uitgave met diskussie-stukken voor een seminar over een anti-imperialisme kampagne op het karl marx instituut, 9 en 10 mei 1970 / De Vliegende tering – Amsterdam [Netherlands]: ASVA [mf ed 1989] – 1r with other items – 1 – mf#mf-10289 seam reel 017/01 [§] – us CRL [959]

The extra examined : a reply to mr. a. campbell's m. harbinger / Broaddus, Andrew – 1836 – 1 – $50.00 – us Presbyterian [240]

Extra (glasgow south & eastwood ed) – Glasgow: Archant Regional Ltd 2003- (wkly) – 1 – (cont: glasgow south & eastwood extra) – uk Scotland NatLib [072]

Extra (glasgow south & eastwood ed) – Glasgow: Community Media Ltd 1997-2003 (wkly) – 1 – (cont: glasgow south & eastwood extra; cont by: extra (glasgow south & eastwood ed)) – uk Scotland NatLib [072]

Extra (hamilton ed) – Motherwell: Archant Regional Ltd 2003 (wkly) – 1 – (cont: hamilton extra; cont by: extra (south lanarkshire ed)) – uk Scotland NatLib [072]

Extra oestergoetland – Linkoeping, Sweden. 2004- – 1 – sw Kungliga [078]

Extra (south lanarkshire ed) – Motherwell: Archant Regional Ltd 2003- (wkly) – 1 – (cont: extra (hamilton ed)) – uk Scotland NatLib [072]

Extra-biblical sources for hebrew and jewish history – New York: Longmans, Green, 1913 – 1mf – 9 – 0-7905-1132-0 – (incl bibl ref and ind) – mf#1987-1132 – us ATLA [939]

Extrablatt, illustriertes wiener neustadter – Vienna, mar 1872-dec 1928 – 254r – 1 – (liberal) – us UMI ProQuest [074]

Extrablatt, neues wiener neustadter – Vienna, jan 1928-dec 1933 – 15r – 1 – (liberal leftist) – us UMI ProQuest [074]

Extracellular calcium and the inotropic effect of epinephrine on isolated skeletal muscle / Williams, Jay H – 1988 – 125p 2mf – 9 – $8.00 – us Kinesology [612]

Extract derer eingelauffenen nouvellen – Leipzig DE, 1739-41, 1742-44, 1745, 1746, 1747, 1748, 1749-50 – 3r – 1 – gw Misc Inst [900]

Extract from a report on the drainage of the everglades of florida / Wright, J O – Tallahassee, FL. 1909 – 1r – 1 – us UF Libraries [639]

Extract from a return dated the 2nd of december, 1854 : to addresses presented by the legislative council to his exellency the governor general, on the subject of the seigniorial tenure / Canada. Parlement. Conseil legislatif – Quebec: Rollo Cambell...1855 [mf ed 1983] – 1mf – 9 – mf#SEM105P277 – cn Bibl Nat [324]

Extract from a sermon preached in st andrew's church, toronto, on the 30th april, 1865 : by the rev dr barclay, on the occasion of the sudden death of colonel e w thomson, one of the elders of the congregation / Barclay, John – Montreal?: s.n, 1865 – 1mf – 9 – mf#67291 – cn Canadiana [240]

An extract from ibn kutaiba's adab al-kaatib : or, the writer's guide: with translation and notes = Adab al-katib / Ibn Qutaybah, Abd Allah ibn Muslim – Leipsic: In commission with Th Stauffer, 1877 [mf ed 1986] – 1mf – 9 – 0-8370-7800-8 – (in english & arabic) – mf#1986-1800 – us ATLA [470]

Extract from pres azana's speech at valencia university, july 18, 1937 / Azana, Manuel – Washington, DC. 1937 – (= ser Blodgett coll) – 9 – mf#fiche w739 – us Harvard [946]

Extract from the advice of william penn to his children, pt 1 – London, England. 1819 – 1r – 1 – us UF Libraries [240]

Extract from the journals of the legislative council of the 2d of march, 1814 : extrait des journaux du conseil legislatif du 2e mars, 1814 / Bas-Canada. Parlement. Conseil legislatif – Quebec: John Neilson, [1814] (mf ed 1987) – 1mf – 9 – mf#SEM105P809 – cn Bibl Nat [324]

Extract of a despatch from lord glenelg to the earl of durham : dated downing-street, 20th january 1838 – [London, England: s.n., 1838] (mf ed 1991) – 1mf – 9 – mf#SEM105P1390 – cn Bibl Nat [323]

Extract of the district of columbia code : 1935, rev 1936 / Eby, Herbert Oscar – Washington, DC: Students Law Book Corp, 1936 – 1 – (various pagings) – mf#LL-870 – us L of C Photodup [348]

Extract uit het register der resolutien van de hoog mogende heeren staaten generaal der vereenigde nederlanden, veneris den 21 october 1785 – ['s-Gravenhage: s.n, 1785?) – 1 – us CRL [949]

Extracto de la tesis de sobre la accion cultural de espana en marruecos / Garcia Carrasco, Francisco Andres – Caceres: Tip. Extremadura, 1977 – 1 – sp Bibl Santa Ana [306]

Extracto de las ordenanzas de los distintos arbitrios votados para el presupuesto municipal para 1924-25 – Almendralejo: Imprenta Bote, 1924 – 1 – sp Bibl Santa Ana [350]

Extracto de las siete partidas : formado para facilitar la lectura, inteligencia y memoria de sus disposiciones / Valdelomar, Juan de la Reguera – Madrid: Impr de la viuda e hijo de Marin, 1799 – 1 – us Wisconsin U Libr [342]

Extracto estadistico 1918-1943 / Peru. Direccion de Estadistica – (= ser Latin american & caribbean...1821-1982) – 84mf – 9 – (1918-22, 1925 not available) – uk Chadwyck [318]

Extracto estadistico de la republica argentina 1915 / Argentine Republic. Direccion General de Estadistica – (= ser Latin american & caribbean...1821-1982) – 7mf – 9 – uk Chadwyck [318]

Extracto estadistico de la republica de panama 1941-1943, 1944-1946, 1951-1952 / Panama. Direccion de Estadistica y Censos – (= ser Latin american & caribbean...1821-1982) – 40mf – 9 – uk Chadwyck [318]

Extracto puntual de todas las pragmaticas, cedulas, provisiones publicadas en el reinado de carlos 3 / Sanchez, S – Madrid, 1794 – 2v on 14mf – 9 – sp Cultura [946]

Extractos da historia da conquisto do iaman pelos otomanos / [Nahrawali, Muhamud Ibn Ahmadal] – Lisboa, 1892 – 2mf – 9 – mf#SEP-81 – ne IDC [956]

Extracts form the letters and other writings of the late joseph gurney bevan : preceded by a short memoir of his life / Bevan, Joseph Gurney – London: W Phillips, 1821 – 3mf – 9 – 0-524-07851-3 – mf#1991-3396 – us ATLA [240]

Extracts from a correspondence with the academies of vienna and st petersburg / Hoare, Prince – London 1802 – (= ser 19th c art & architecture) – 1mf – 9 – mf#4.2.1644 – uk Chadwyck [700]

Extracts from a journal : written on the coasts of chili, peru, and mexico, in the years 1820, 1821, 1822 / Hall, Basil – Edinburgh 1824 [mf ed Hildesheim 1995-98] – 2v on 6mf – 9 – €120.00 – ISBN-10: 3-487-26918-X – ISBN-13: 978-3-487-26918-4 – gw Olms [918]

Extracts from a pamphlet on the present state of the irish poor / O'Flynn, James – [London], 1836 – (= ser 19th c ireland) – 1mf – 9 – mf#1.1.1928 – uk Chadwyck [360]

Extracts from a speech delivered by the pres of the spanish republic, january 21, 1937 / Azana, Manuel – Washington, DC. 193? – (= ser Blodgett coll) – 9 – mf#fiche w740 – us Harvard [946]

Extracts from a teacher's observations on school government : with introductory and concluding remarks / Fordyce, Alexander Dingwall – [Elora, Ont?: Observer], 1870 – 1mf – 9 – 0-665-91823-2 – mf#91823 – cn Canadiana [370]

Extracts from a work entitled the spirit of prayer / Law, William – York, England. 1815 – 1r – us UF Libraries [240]

Extracts from china mainland publications / U.S. Consulate General. Hong Kong – 1 Apr 1962-31 Aug 1964 (ceased publication) – (= ser Hong kong press summaries) – 1 – $52.00 – us L of C Photodup [951]

Extracts from correspondence with mission stations at rotuma / Roman Catholic Mission, Fiji – 1846-1889 – 1r – 1 – mf#pmb429 – at Pacific Mss [241]

Extracts from midwestern newpapers concering the armistice – s.l, s.l? no date – 1r – us UF Libraries [025]

Extracts from regole brievi della volgare grammatica / Fortunio, Giovanni Francesco – c1700 – (= ser Holkham library manuscript books 536,617,702) – 1r – 1 – (filmed with: lorenzo priuli: ambassadorial relazione, 1565. luca giordane: descrittione) – mf#97247 – uk Microform Academic [450]

Extracts from s hieronymus and others / Quartigianis, Philippus de Diversis de – Venice, 1455 – (= ser Holkham library manuscript books 131) – 1r – 1 – mf#96547 – uk Microform Academic [090]

Extracts from the anglo-saxon laws / Cook, Albert Stanburrough – New York, Holt, 1880. 19 p. LL-109 – 1 – us L of C Photodup [348]

Extracts from the chief superintendent's report on education in upper canada for the year 1857 / Canada (Province). Surintendant des ecoles du haut-Canada – Toronto: printed by Lovell & Gibson, 1859 [mf ed 1992] – 2mf – 9 – mf#SEM105P1683 – cn Bibl Nat [370]

Extracts from the diary and correspondence...with a brief account of some incidents in his life / Lawrence, Amos; ed by Lawrence, William R – Boston: Gould and Lincoln, 1855 – viii/369p (ill) – 1 – us Wisconsin U Libr [920]

Extracts from the diary of a field officer of the bengal army : during a journey overland from bombay, via aden, suez, alexandria, trieste, vienna, dresden, prague, hanover, brussels and ostend, in march and april, 1853... – London, 1853 – (= ser 19th c books on british colonization) – 1mf – 9 – mf#1.1.8230 – uk Chadwyck [920]

Extracts from the fathers – London, England. 18– – 1r – us UF Libraries [240]

Extracts from the journals of the legislative council of the province of lower-canada from the year 1795 to 1813 inclusive = Extraits des journaux du conseil legislatif de la province du bas-canada depuis l'annee 1795 jusqu'a 1813... / Bas-Canada. Parlement. Conseil legislatif – Quebec: P E Desbarats, 1821 [mf ed 1987] – 1r – 1 – mf#SEM35P280 – cn Bibl Nat [324]

Extracts from the journals of the legislative council of the province of lower-canada from the year 1795 to 1813 inclusive : printed by order of the legislative council of the 12th march, 1821 = Extraits des journaux du conseil legislatif de la province du bas-canada depuis l'annee 1795 jusqu'a 1813 inclusivement, imprime par l'ordre du conseil legislatif en date du 12e mars, 1821 / Bas-Canada. Parlement. Conseil legislatif – Quebec: P E Desbarats, 1821 [mf ed 1987] – 1r – 1 – mf#SEM35P280 – cn Bibl Nat [323]

Extracts from the letters, and journal of daniel wheeler : now engaged in a religious visit to the inhabitants of some of the islands of the pacific ocean, van diemen's land, and new south wales... / Wheeler, D – London, 1839 – 4mf – 9 – mf#HTM-211 – ne IDC [919]

Extracts from the letters and journals of george fletcher moore : now filling a judicial office at the swan river settlement / Moore, George – London 1834 [mf ed Hildesheim 1995-98] – 1v on 2mf – 9 – €60.00 – ISBN-10: 3-487-26811-6 – ISBN-13: 978-3-487-26811-8 – gw Olms [880]

Extracts from the minutes / United States Military Philosophical Society – Washington. 1808-1809 (1) – mf#3669 – us UMI ProQuest [355]

Extracts from the municipal records of the city of york : during the reigns of edward 4, edward 5 and richard 3, with notes / Davies, Robert – London: J B Nichols, 1843 – vii/304p – 1 – (incl app) – us Wisconsin U Libr [941]

Extracts from the religious literature of the hindus / Schanzlin, G [comp] – Calcutta: G Schanzlin, 1916 [mf ed 1995] – (= ser Yale coll) – 52p – 1 – 0-524-10166-3 – mf#1995-1166 – us ATLA [280]

Extracts from the reports on the coals of pictou county, nova scotia / Hartley, Edward – Montreal?: s.n, 1871 (Montreal: Gazette) – 1mf – 9 – mf#42114 – cn Canadiana [550]

Extracts from the votes and proceedings of the american continental congress held at Philadelphia on the fifth day of september, 1774 : containing the bill of rights, a list of grievances, occassional resolves, the association, an address to the people of great-britain... – [Norwich, England]: Philadelphia printed, Norwich repr by Robertson & Trumbull, 1774 [mf ed 1986] – 1mf – 9 – 0-665-56039-7 – mf#56039 – cn Canadiana [975]

Extracts from the votes and proceedings of the american continental congress held at Philadelphia, on the fifth of september, 1774 : containing the bill of rights, a list of grievances, occasional resolves, the association, an address to the people of great-britain... – [London]: Philadelphia printed, London repr for J Almon...1774 [mf ed 1986] – 1mf – 9 – 0-665-54219-4 – (incl publ list) – mf#54219 – cn Canadiana [975]

Extracts from the writings of william penn and richard claridge – London, England. 1823 – 1r – 1 – us UF Libraries [240]

Extracts from woman's service on the lord's day / Bickersteth, Emily – Philadelphia: McCalla & Stavely, 1865 [mf ed 1984] – (= ser Women & the church in america 144) – 1mf – 9 – 0-8370-1413-1 – (pref by lord bishop of rochester) – mf#1984-2144 – us ATLA [305]

Extracts of letters : from poor persons who emigrated last year to canada and the united states. printed for the information of the labouring poor and their friends in this country / Scrope, George Julius Duncombe Poulett – London 1831 – (= ser 19th c british colonization) – 1mf – 9 – mf#1.1.457 – uk Chadwyck [304]

Extracts of letters on the object and connexions of the british and... / Owen, John – London, England. 1819 – 1r – us UF Libraries [240]

Extraict des lettres d'vn gentil homme de la suitte de monsieur de rambouillet, ambassadeur du roy au royaume de pologne, a vn seigneur de la court – Paris, 1574 – 1mf – 9 – mf#H-8198 – ne IDC [956]

Extrait de la grammaire francaise – [Montreal?: s.n.] 1845 [mf ed 1984] – 1mf – 9 – 0-665-45459-7 – mf#45459 – cn Canadiana [440]

Extrait de la relation des aventures et voyage de mathieu sagean – [New York: s.n.] 1863 [mf ed 1984] – 1mf – 9 – 0-665-20050-1 – mf#20050 – cn Canadiana [917]

Extrait des annales intitulees uvres de st augustin et de sainte-monique : offert en souvenir de recompense aux protecteurs de l'orphelinat d'afrique / Charmetant, Felix – Montreal: E Senecal, 1876 – 1mf – 9 – mf#64736 – cn Canadiana [241]

Extrait des instructions royales a son excellence le tres-honorable george, comte de dalhousie... / Bas-Canada. Parlement. Chambre d'assemblee – [Quebec (Province): s.n, 1823?] (mf ed 2000) – 1mf – 9 – mf#SEM105P3238 – cn Bibl Nat [971]

Extrait des oeuvres du pdg – Conakry: le bureau de presse de la presidence de la republique, 1978 – us CRL [324]

Extrait du voyages des hollandais envoyez es annees 1656 & 1657 : en qualite d'ambassadeurs vers l'empereur des tartares, maintenant maistre de la chine, traduit du manuscrit hollandois / Nieuhof, J – Paris, 1696 – v2 on 2mf – 9 – mf#HT-682 – ne IDC [915]

Extrait d'un commentaire et d'une traduction nouvelle du vendidad sade : l'un des livres de zoroastre / Burnouf, Eugene – [Paris] [1829] – (= ser Whsb) – 1mf – 9 – €20.00 – (extract of nouveau journal asiatique) – mf#Hu 223 – gw Fischer [410]

Extrait d'une lettre, crite de pekin : sur le musc. le 2 novembre 1717 / Collas, J P L – 1mf – 9 – (extract fr: martin, l a: lettres, difiantes et curieuses...paris, 1838-43 v3) – mf#HT-572 mf. 22 – ne IDC [590]

Extrait d'une lettre de m schmidt, de st-petersbourg : en reponse a l'examen des extraits d'une histoire des khans mongols / Klaproth, Heinrich Julius von – Paris: [Societe asiatique] 1823 – (= ser Whsb) – 1mf – 9 – €20.00 – (fr: journal asiatique 5 [1823]) – mf#Hu 480 – gw Fischer [951]

Extrait d'une lettre du reverend pere laureati ecrite de fo-kien : le 26 juillet 1714, et traduite de l'italien / Laureati, G – Paris, 1838-1843 – v3 on 1mf – 9 – mf#HT-572 – ne IDC [915]

Extraits : ou precedents des arrests tires des registres du conseil superieur de quebec et dedies a son honneur sir francis nathaniel burton, lieutenant-gouverneur, et aux autres honorables membres de la cour d'appel de la province du bas-canada / Perrault, Joseph-Francois – [Quebec?: s.n.], 1824 [mf ed 1983] – 1mf – 9 – mf#21184 – cn Canadiana [340]

Extraits des annales manuscrites de l'hotel-dieu du precieux-sang, quebec – [Quebec: s.n, 1923] (mf ed 1992) – 1mf – 9 – mf#SEM105P1538 – cn Bibl Nat [241]

Extraits des memoires relatifs a l'histoire de france : depuis l'annee 1757 jusqu'a la revolution / ed by Aignan, Etienne – Paris 1824 [mf ed Hildesheim 1995-98] – 2v on 7mf – 9 – €140.00 – 3-487-25852-8 – gw Olms [944]

Extraits des minutes du comite nomme le 2e mars, 1816 / Compagnie d'assurance du feu de Quebec – Quebec: Imprime par John Neilson...1816 – 1mf – 9 – mf#55042 – cn Canadiana [360]

Extraits des proces-verbaux / Bordeaux. France. Chambre de Commerce – v1-89. 1850-1938 – 1 – us NY Public [073]

Extraits des titres des anciennes concessions de terre en fief et seigneurie... / Vondenvelden, William [comp] – Quebec: P E Desbarats, 1803 [mf ed 1971] – 1r – 5 – mf#SEM16P92 – cn Bibl Nat [971]

Extraits ou precedents, tires des registres de la prevoste de quebec 1726-1756, et dedies aux honorables juges, aux gens du roi, aux avocats, procureurs, et praticiens de la province du bas-canada / Perrault, Joseph Francois – Quebec: Cary, 1824 – 88p – 1 – mf#LL-2221 – us L of C Photodup [340]

Extranjeros (foreigners) in puerto rico, 1872-1880 / Puerto Rico. Spanish Governors – (= ser Records Of Spanish Governors Of Puerto Rico) – 19r – 1 – mf#T1170 – us Nat Archives [972]

Extrano habitante (mexico, 3 am) / Menen Desleal, Alvaro – San Salvador, El Salvador. 1964 – 1r – 1 – us UF Libraries [972]

Die extraordinaire relation *see* Nordischer mercurius

El extraordinario poder de las plantas / Fernandez Casco, Juan – Plasencia: Imp Garcilasso, 1973 – 1 – sp Bibl Santa Ana [946]

Extraordinary cases / Clinton, Henry Lauren – New York: Harper, 1896. 403p. LL-453 – 1 – us L of C Photodup [340]

Extraordinary trial by a sister of mercy – London, England. 1869 – 1r – us UF Libraries [240]

Extrapolation – Kent. 1979+ (1,5,9) – ISSN: 0014-5483 – mf#11919 – us UMI ProQuest [400]

Extrapolation – Wooster OH: MLA Seminar on Science Fiction. v16-19. dec 1974-may 1978 – (= ser Science fiction periodicals, 1926-1978. series 2) – 1r – 1 – $105.00 – us UPA [830]

Extrapolation – Wooster OH. v1-15. dec 1959-may 1974 – (= ser Science fiction periodicals, 1926-1978. series 1) – 2r – 1 – $165.00 – us UPA [830]

Extrapost – Temeschburg (Timisoara RO), 1937 1 oct-1943 30 dec – 6r – 1 – (lacking: 1924 & 1943) – gw Misc Inst [077]

Extraterritorial cases : the us court for china, 1844-1924 / Lobinger, C S – Manila: Bureau of Printing. 2v. 1920-1928 (all publ) – 22mf – 9 – $33.00 – mf#LLMC 81-483 – us LLMC [347]

A extremadura / Pelaez, Florentino – S.l. Imp. de los Menores de Ramos, s.a. – 1 – sp Bibl Santa Ana [946]

Extremadura – Badajoz, 1900 2 numeros – 5 – sp Bibl Santa Ana [073]

Extremadura – Caceres, 1923-1930 y 1934-1936 – 5 – sp Bibl Santa Ana [073]

Extremadura : conferencia dada en el ateneo espanol de mexico, el 2 de mayo de 1950 / Castillo, Manuel – Mexico D.F. s.i, 1950 – 1 – sp Bibl Santa Ana [972]

Extremadura / Diaz Perez, Nicolas – 1887 – 9 – sp Bibl Santa Ana [946]

Extremadura artistica e industrial (1930-1931) – Sevilla: Imp. Bergali, s.a. – sp Bibl Santa Ana [700]

La extremadura del s.15 en tres de sus paladines / Munoz de San Pedro, Miguel – Madrid: Tall. Asilo del Corazon de Jesus, 1964 – 1 – sp Bibl Santa Ana [946]

Extremadura en 1829 (datos de sus partidos y localidades) / Munoz de San Pedro, Miguel – Badajoz: imp. de la diputacion provincial, 1963 – sp Bibl Santa Ana [946]

Extremadura en la guerra de la independencia espanola / Gomez Villafranca, Ramon – Badajoz: tip lit y enc de uceda hnos, 1908 – 1 – sp Bibl Santa Ana [320]

Extremadura en la guerra de la independencia (informe de gomez villafranca) / Blazquez, Antonio – Madrid: Fortanet, 1911 – sp Bibl Santa Ana [946]

Extremadura en las obras de cervantes / Berjano Escobar, Daniel – Caceres: revista de extremadura, 1905 – 1 – sp Bibl Santa Ana [946]

Extremadura en toledo. impresiones de turista / Hurtado de Mendoza, Publio – Caceres: tip enc y lib de luciano jimenez merino, 1920 – sp Bibl Santa Ana [946]

Extremadura, fotografias de josip ciganovic, ensayo preliminar de pedro de lorenzo : textos de joaquin fernandez, direccion de fermin h garbayo – Madrid: Hauser y Menet, 1968 – 1 – sp Bibl Santa Ana [770]

Extremadura, la fantasia heroica / Lorenzo, Pedro de – Madrid: Edit. Nacional, 1961 – 1 – sp Bibl Santa Ana [946]

Extremadura. Spain (Province.) *see*
– Calendario de extremadura para el ano 1863...y aumentado con el calendario portugues de barda d'agua
– Calendario para la provincia de extremadura..

Extremadura y america : sevilla, 1929 / Rubio y munoz-Bocanegra, A – Madrid: Razon y Fe, 1930 – 1 – sp Bibl Santa Ana [972]

Extremadura y el mar / Asociacion Amigos de Guadalupe – Caceres: imp sanguino, 1952 – sp Bibl Santa Ana [946]

Extremadura y espana. conferencias familiares sobre la raza de los conquistadores / Lopez Prudencio, Jose – Badajoz: ed arqueros, 1929 – 1 – sp Bibl Santa Ana [946]

Extremadura y los extremenos / Hernandez Pacheco, Eduardo – Madrid, 1931 – 1 – sp Bibl Santa Ana [946]

Extremadura y sus hombres : las escuelas parroquiales de los santos / Suarez Murillo, Marcos – Los Santos: Tip.Sanchez Hnos., 1914 – 1 – sp Bibl Santa Ana [377]

Extremadura...inscripciones y documentos / Viu, Jose de – v1. 1852 – 9 – sp Bibl Santa Ana [946]

Los extremanos en las cortes de cadiz / Gomez Villafranca, Ramon – Badajoz: Top. y Libreria de A. Arqueros, 1912 – 1 – sp Bibl Santa Ana [946]

L'extreme orient – Hanoi: Extreme Orient 1894 aug 9-dec 30, 1895 jan 6, jan 13-jul 21, jul 28-dec 29, 1896 jan 5-jun 7, jun 18-dec 31, 1897 sep 2-23, dec 9, 16-23, 1898 jan 6-may 12, may 18-jun 30, jul 7-aug 7, aug 14-21, sep 4,11,18,25, oct 6-23,30, nov 3,10,17-27, dec 4-29, 1899 sep 24-oct 12 – 2r – 1 – mf#mf-11751 seam – us CRL [079]

Extreme-asie : revue indochinoise – Saigon. juil 1926-mars 1935 – 1 – fr ACRPP [959]

Extreme-asie *see* La revue indochinoise

Extremena de Seguros *see* Memoria ejercicio 1978

El extremeno : almanaque satirico-literario – 1868 – 9 – sp Bibl Santa Ana [870]

El extremeno – Plasencia, 1879-1884 – 5 – (es continuacion del siguiente(72)en su primera epoca que empezo en 1869) – sp Bibl Santa Ana [073]

Un extremeno en la corte de los austrias / Munoz de San Pedro, Miguel – Badajoz: Dip. Provincial, 1947 – 1 – sp Bibl Santa Ana [946]

Extremetech – New York. 2001+ (1,5,9) – mf#32113 – us UMI ProQuest [000]

Exvacations in the tyropoeon valley, jerusalem 1927 / Crowfoot, J W & Fitzgerald, G M – London, 1929 – 2mf – 9 – mf#H-3076 – ne IDC [956]

Exvacations on the hill of ophel, jerusalem, 1923-1925... / Macalister, R A S & Duncan, J G – London, 1926 – 4mf – 9 – mf#H-2936 – ne IDC [956]

Exxon usa – Houston. 1975-1986 (1) 1976-1986 (5) 1976-1986 (9) – (cont: humble way) – mf#10110,01 – us UMI ProQuest [550]

Exzitonentransfer in semimagnetischen cdte/cdmnte-doppel-quantengrabenstrukturen / Hiecke, Katharina – (mf ed 1995) – 2mf – 9 – €40.00 – 3-8267-2260-4 – mf#DHS 2260 – gw Frankfurter [530]

Ey, Adolf *see* Schillers balladen

E'yanpaha reservation news / Devils Lake Sioux Tribe – 1977 jul-1979 jun, 1979 jun-1980 dec 15 – 2r – 1 – mf#604893 – us WHS [305]

Eybers, George Von Welfling *see* Select constitutional documents illustrating south african history

Eybike tfise / Granitstein, Moses – Buenos Aires, Argentina. 1951 – 1r – us UF Libraries [939]

Der eyd / Lavater, L – Zuerych, Johan Wolff, 1592 – 3mf – 9 – mf#PBU-327 – ne IDC [240]

Eydoux, J F T *see* Voyage autour du monde par les mers de l'inde et de chine execute sur la corvette de l'etat la favorite pendant les annees 1830, 1831, 1832

Eye : by obadiah optic – Philadelphia. 1808-1808 (1) – mf#3576 – us UMI ProQuest [420]

Eye – Burwell, NE: Radle L Miller. 4v. v6 n1. jan 3 1895-v9 n1. may 12 1898 (wkly) [mf ed with gaps)] – 1r – 1 – (cont: loup valley alliance. absorbed by: burwell mascot) – us NE Hist [071]

Eye – n1-13 [1970 sep-1971 may] – 1r – 1 – mf#1051817 – us WHS [071]

Eye – Needles, CA. 1891-1894 (1) – mf#62196 – us UMI ProQuest [071]

Eye – Snohomish, WA. 1882-1897 (1) – mf#67130 – us UMI ProQuest [071]

Eye – Cleveland, OH, feb 3 1927-may 28 1992 – 4r – 1 – (student newspaper of st. ignatius high school) – us Western Res [373]

The eye – Monrovia, Liberia: Visual Professional Associates [apr 30 1991-jul 31 1995] (3 times/wk) – 4r – 1 – us CRL [079]

Eye and star – Dodgeville WI. 1889 sep 5-1890 oct 16, 1890 oct 23-1893 dec 15, 1893 dec 22-1895 dec 6 – 3r – 1 – (cont: rural eye; dodgeville star; cont by: new star [dodgeville, wi]) – mf#962738 – us WHS [071]

Eye doctor – v2 n3 [[198-?]] – 1r – 1 – mf#4877114 – us WHS [071]

Eye, ear, nose and throat monthly – New York. 1976-1976 (1,5,9) – (Cont by: Ear, nose and throat journal) – ISSN: 0014-5491 – mf#11129 – us UMI ProQuest [617]

An eye for an eye / Darrow, Clarence S – Girard, Kansas: Haldeman-Julius Co, 1905? – 1mf – 9 – $1.50 – mf#LLMC 91-061 – us LLMC [340]

An eye for an eye / West, John B – [New York]: New American Library, [c1959] (mf ed 1978) – 2mf – 9 – mf#Sc Micro F-1156 – us NY Public [830]

The eye for spiritual things : and other sermons / Gwatkin, Henry Melvill – Edinburgh: T & T Clark, 1906 – 1mf – 9 – 0-7905-7641-4 – mf#1989-0866 – us ATLA [240]

The eye glass – Stratford, Ont: H C Brown, [19–] – 9 – mf#P05003 – cn Canadiana [680]

Eye movement desensitisation and reprocessing (emdr) facilitating rational emtive behaviour therapy (rebt) in the treatment of text anxiety / Ten Cate, Hein – Stellenbosch: U of Stellenbosch 1998 – 2mf – 9 – mf#mf.1264 – sa Stellenbosch [616]

Eye of light – v1 n4 [1967 may] – 1r – 1 – mf#1583242 – us WHS [071]

Eye of the beast / Church of the Apocalypse – 1970 sep 4, 23-oct 26, nov 17, 30, 1971 jan 4 – 1r – 1 – mf#1107175 – us WHS [243]

Eye of the storm – iss n1-3 [1995 win, 1996 spr, 1997 rebirth] – 1r – 1 – mf#3239587 – us WHS [071]

Eye opener – Butte, MT. 1934-1941 (1) – mf#64295 – us UMI ProQuest [071]

Eye opener / International Union, United Automobile, Aerospace, and Agricultural Implement Workers of America – v32 n9-v39 n2 [1975 sep-1982 feb/mar] – 1r – 1 – mf#615770 – us WHS [331]

Eye opener / Socialist Party (US) – v9 n5-v12 n11 [1917 aug 25-1920 jun 1] – 1r – 1 – mf#1111458 – us WHS [071]

Eye opener / United Electrical, Radio and Machine Workers of America – n4-27 [1952 nov 20-1953 may 13] – 1r – 1 – mf#367509 – us WHS [331]

An eye to the ermine : a dream / Albyn [i.e. Andrew Shiels] – Halifax, NS?: s.n, 1871 – 1mf – 9 – mf#06128 – cn Canadiana [880]

Eyebers, Cornelia *see* Die komplementariteit tussen intimiteit en afstand in die terapeutiese verhouding

Eye-gate : or, the value of native art in the mission field, with special reference to the evangelization of china / Wilson, William – 2nd ed. London: S W Partridge, [1897] [mf ed 1995] – (= ser Yale coll) – 20p (ill) – 1 – 0-524-10085-3 – (ill by 30 facs col-repr of chinese paintings) – mf#1995-1085 – us ATLA [951]

Eyes / Blackside, Inc – 1991 win-1992 win – 1r – 1 – (continued by: inside blackside) – mf#3167185 – us WHS [071]

"Eyes alone" correspondence of general joseph w stilwell : january 1942-october 1944 – (= ser Records of united) states) theaters of war, world war 2) – 5r – 1 – (with p/g) – mf#m1419 – us Nat Archives [934]
Eyes left – n1-7 [1969 may-dec] – 1r – 1 – mf#679325 – us WHS [071]
Eyes of the eagle / 552nd airborne warning and control / Tinker Air Force Base (OK). United States – 1981 may-1988 – 1r – 1 – mf#1702835 – us WHS [355]
Eyes right – v1 n1 [1969 jul 4] – 1r – 1 – mf#721943 – us WHS [071]
Eyestone, Edward D *see* Effect of water running and cycling on vo2max and 2-mile performance
Eyewitness – v2 n3-v5 n2 [1970 mar-[1975. oct ?] – 1r – 1 – mf#367507 – us WHS [071]
Eyeworth – (= ser Bedfordshire parish register series) – 1mf – 9 – £3.00 – uk BedsFHS [929]
Eyles, Fred *see* Zulu self-taught
Eyn christlich vnterricht eynes gottseligen lebens / Bugenhagen, J – [Altenburg, 1526] – 1mf – 9 – mf#TH-1 mf 157 – ne IDC [242]
Eyn sendebrieff / Bugenhagen, J – [Wittenberg, 1525] – 1mf – 9 – mf#TH-1 mf 143 – ne IDC [242]
Eyn sermon von der eygenschafft vnd weyse des sacraments der tauff / Bugenhagen, J – [Hagenau], 1529 – 1mf – 9 – mf#TH-1 mf 185 – ne IDC [242]
Eynard, Samuel *see* Madagascar illustre
Eynatten, Carola, Freiin von *see* Brandenburger sagen
Eynde, Karel Van Den *see* Fonologie en morfologie van het cokwe
Eyne schoene artzney : dadurch der leidenden christen sorge und trubnus gelindert werden / Magdeburg, J – [Luebeck, 1555] – 4mf – 9 – mf#TH-1 mf 903-906 – ne IDC [242]
Eynigkeyt : a jewish labor weekly = Einigkeit – New york [NY]: unity comm of the joint board cloakmaker's union and joint board furriers' union. v1 n1. mar 25 1927- (wkly) [mf ed 197-?] – (in yiddish and english) – mf#*ZAN-*P927 – us NY Public [071]
Eyquem, Marie-Therese *see* France
Eyre, Archibald *see* The custodian
Eyre, Francis *see* Letter to the rev mr ralph churton
Eyre's weekly journal : or, the warrington advertiser – [NW England] Warrington Lib 4,11,18 may, 29 jun, 6,13,27 jul 1756 – 1 – uk MLA; uk Newsplan [072]
Eyring, Jeremias Nikolaus *see* Paedagogisches jahrbuch
Eyser, J *see* Farrago
Eyth, Max *see*
– Blut und eisen
– Feierstunden
– Geld und erfahrung
– Hinter pflug und schraubstock
– Der kampf um die cheopspyramide
– Moench und landsknecht
– Der schneider von ulm
Eyton, John *see* Sermon on the mount
Eyzaguirre, Raphaele *see* Apocalipseos interpretatio litteralis
Ez flash : late breaking news from the upper manhattan empowerment zone / Upper Manhattan Empowerment Zone – 1997 mar 14 – 1r – 1 – mf#5265275 – us WHS [338]
Ez works / Upper Manhattan Empowerment Zone Development Corporation – 1996 jan-1998 aug – 1r – 1 – mf#3995658 – us WHS [338]
Ezasekhaya / Malcolm, D Mck – London, England. 1942 – 1r – us UF Libraries [960]
Der ezechielische tempel : eine exegetische studie ueber ezechlel 40 ff / Richter, Georg – Guetersloh: C Bertelsmann, 1912 – (= ser Beitraege zur foerderung christlicher theologie) – 2mf – 9 – 0-524-06681-7 – mf#1992-0934 – us ATLA [220]
Ezechielstudien / Herrmann, Johannes – Leipzig: J C Hinrichs, 1908 – (= ser Beitraege zur wissenschaft vom alten testament) – 1mf – 9 – 0-7905-1893-7 – mf#1987-1893 – us ATLA [221]
Ezekiel : introduction, revised version with notes and index / Lofthouse, William Frederick – New York: Henry Frowde; Edinburgh: T C & E C Jack, [1907] – 1mf – 9 – 0-8370-4168-6 – (incl bibl ref and ind) – mf#1985-2168 – us ATLA [221]
Ezekiel and daniel / Cobern, Camden McCormack – New York: Eaton & Mains; Cincinnati: Jennings & Pye c1901 [mf ed 1989] – (= ser Commentary on the old testament 8) – 1mf – 9 – 0-7905-0977-6 – mf#1987-0977 – us ATLA [221]
Ezekiel and the book of his prophecy : an exposition / Fairbairn, Patrick – 4th ed. Edinburgh: T & T Clark, 1876 – 2mf – 9 – 0-7905-0078-7 – (incl ind) – mf#1987-0078 – us ATLA [221]
Ezekiel gilman robinson : an autobiography / ed by Johnson, Elias Henry – New York: Silver, Burdett, 1896 [mf ed 1991] – 1mf – 9 – 0-524-00087-5 – mf#1989-2787 – us ATLA [920]
The ezekiel price papers, 1754-1785 – [mf ed 1968] – 1r – 1 – us MA Hist [978]
Ezekiels vision und die salomonischen wasserbecken / Venetianer, Ludwig – Budapest: Friedrich Kilian Nachfolger, 1906 – 1mf – 9 – 0-8370-9326-0 – (Incl bibl ref) – mf#1986-3326 – us ATLA [221]
Ezhednevnaia bespartiinaia gazeta / ed by Avksentev et al – Paris, 1914-1915. nos 1-248 – 3mf – 9 – mf#R-18119 – ne IDC [077]
Ezhednevnaia obshchestvennaia i politicheskaia gazeta / ed by Mesheriakov, M – Paris, 1916-1917. nos 1-147 – 1mf – 9 – (Missing: 1916. nos 51, 63) – mf#R-18113 – ne IDC [077]
Ezhednevnaia obshchestvennaia i politicheskaia gazeta – Paris, 1915-1916 – 12mf – 9 – mf#R-18115 – ne IDC [077]
Ezhednevnaia politicheskaia gazeta – Paris, 1914-15. nos 1-108 – 6mf – 9 – (Missing: 1914 nos 15-16) – mf#R-18034 – ne IDC [077]
Ezhednevnyi vechernii biulleten' : ofits soobshch shtaba dobrovol armii: dlia khar'k otd-niia otd propagandy, izgotovlen gaz "narodnoe slovo" – Khar'kov: [s n] 1919 [1919 n84 [27 avg], 116 [4 okt] – (= ser Asn 1-3) – n84, 116 [1919] item 136, on reel n29 – 1 – mf#asn-1 136 – ne IDC [077]
Ezhegodnik eksperimentalnoi pedagogiki – Spb., 1908-1914. nos 1-8 – 27mf – 9 – mf#R-3797 – ne IDC [077]
Ezhegodnik glukhovskogo uchitelskogo instituta – Kiev, 1912. v1-2 – 8mf – 9 – mf#R-3413 – ne IDC [077]
Ezhegodnik Imperatorskogo russkogo geograficheskogo obshchestva – Chicago. 1966-1974 (1) 1970-1974 (5) – 23mf – 9 – mf#1719 – ne IDC [077]
Ezhegodnik ministerstva finansov vypusk : 1916 goda – Pg, 1917 – 9mf – 9 – mf#REF-189 – ne IDC [332]
Ezhegodnik ministerstva inostrannykh del – Spb., 1861-1916 – 43v on 253mf – 9 – (missing: 1867-68 v7-8) – mf#R-9312 – ne IDC [077]
Ezhegodnik narodnoi shkoly – Moscow 1908 – 9mf – 9 – mf#r-3415/2 – ne IDC [590]
Ezhegodnik proizvodstvennoi kooperatsii 1925 g – 1925 – 7mf – 9 – mf#COR-585 – ne IDC [335]
Ezhegodnik rossii 1904-1911 = Annuaire de la russe 1904-1911 / Russia. Tsentral'nyi Statisticheski Komitet Ministerstva Vnutrennykh Del – (= ser European official statistical serials, 1841-1984) – 59mf – 9 – uk Chadwyck [314]
Ezhegodnik russkikh kreditnykh uchrezhdenii / ed by Ivashchenko, I S – Spb, 1880-1886. 4v – 52mf – 9 – mf#REF-142 – ne IDC [332]
Ezhegodnik sovetskogo stroitel' stva i prava – Moskva: gosudarstvennoe sotsial' neokonomicheskoe izdatel'stvo, 1931 – 1v – 1 – mf#LL-4212 – us L of C Photodup [340]
Ezhegodnik tobolskogo gubernskogo muzeia, sostoiashchego pod avgusteishim ego imperatorskogo velichestva pokrovitelstvom – Tobolsk, 1893-1918 – pt1-29 on 92mf – 9 – (missing: 1897 pt6) – mf#RET-1 – ne IDC [077]
Ezhegodnik vladimirskogo gubernskogo statisticheskogo komiteta – Vladimir, [1875-1885]. v1-5 – 21mf – 9 – mf#RET-2 – ne IDC [077]
Ezhegodnoe iumoristicheski-satiricheski-skandalnoe obozrenie – Karlsruhe – 3mf – 9 – mf#R-18173 – ne IDC [077]
Ezhemesiachnaia sotsial-demokraticheskala rabochaia gazeta – Paris, 1911. nos 1-3. – 1mf – 9 – mf#R-18149 – ne IDC [077]
Ezhemesiachnik dlia liubitelei iskusstva i stariny – Spb., 1907-1916 – 229mf – 9 – mf#1910 – ne IDC [077]
Ezhemesiachnik iskusstva i teatra – M., 1912-1914 – 35mf – 9 – mf#R-3360 – ne IDC [077]
Ezhemesiachnik stikhov i kritiki – Spb., Noiabre, 1912-Oktiabre, 1913(8) – 4mf – 9 – (Missing: 1912(1, 3); 1913(5, 7)) – mf#R-1850 – ne IDC [077]
Ezhemesiachnoe illiustrirovannoe izdanie : bibliograficheskie zapiski – M., 1892. v1-12 – 34mf – 9 – mf#R-4301 – ne IDC [077]
Ezhemesiachnoe illiustrirovannoe izdanie imperatorskogo obshchestva pooshchreniia khudozhestv v peterburge – Spb., 1898-1902. v1-38 – 120mf – 9 – mf#R-3217 – ne IDC [077]
Ezhemesiachnoe istoricheskoe izdanie – London. 1949-1994 (1) 1972-1994 (5) 1975-1994 (9) – 2399mf – 9 – (Missing: 1918. v174-176) – mf#1283 – ne IDC [077]
Ezhemesiachnoe istoriko-literaturnoe izdanie – Riga, 1881. v1-12 – 14mf – 9 – mf#R-3410 – ne IDC [077]
Ezhemesiachnoe izdanie / ed by Averkiev, D V – New York. 1973-1979 (1) 1979-1979 (5) 1979-1979 (9) – 15mf – 9 – mf#1706 – ne IDC [077]
Ezhemesiachnoe izdanie – Chicago. 1937+ (1) 1971+ (5) 1975+ (9) – 4mf – 9 – mf#2032 – ne IDC [077]
Ezhemesiachnoe izdanie – M., 1897-1914. v1-36 – 594mf – 9 – mf#R-8129 – ne IDC [077]
Ezhemesiachnoe izdanie – Spb., 1777-1780. v1-9 – 42mf – 9 – (Missing: 1778, v4(p 328-end); 1779, v5(p 161-261); v6(p 1-83); v8(p 1-99; 195-291)) – mf#R-4018 – ne IDC [077]
Ezhemesiachnoe izdanie / Nevskii zritele – New York. 1903-1982 (1) 1971-1982 (5) 1975-1982 (9) – 19mf – 9 – (Missing: 1820(6, 9, 11); 1821(7-12)) – mf#1805 – ne IDC [077]
Ezhemesiachnoe izdanie – New York. 1955+ (1) 1960+ (5) 1955+ (9) – 13mf – 9 – mf#1452 – ne IDC [077]
Ezhemesiachnoe izdanie... / Beseduiushchii grazhdanin – Spb., 1789. 3 pts – 24mf – 9 – mf#R-18543 – ne IDC [077]
Ezhemesiachnoe izdanie dlia vsekh, ishchushchikh istinu i liubiashchikh gospoda – Spb., 1906-1912, 1914 – 128mf – 9 – (missing: 1906(1); 1907(4); 1914) – mf#R-1565 – ne IDC [077]
Ezhemesiachnoe izdanie dukhovnogo soderzhaniia / Dushepoleznoe chtenie – M., 1860-1917 – 476mf – 9 – (Missing: 1860(4)-1891; 1892(3); 1895-1900; 1901(2); 1902(1-2); 1904-1905; 1909; 1916-1917) – mf#R-2288 – ne IDC [077]
Ezhemesiachnoe izdanie soiuza sotsialistov-revoliutsionerov / ed by Teplov, A – Paris, Geneva, 1894-99 – 7mf – 9 – (missing: 1895 n4) – mf#R-18164 – ne IDC [077]
Ezhemesiachnoe literaturno-istoricheskoe izdanie – Tashkent, 1910-1911 – 59mf – 9 – (Missing: 1910(3, 8)) – mf#R-1622 – ne IDC [077]
Ezhemesiachnoe literaturno-politicheskoe izdanie – Stamford. 1951-1982 (1) 1970-1982 (5) 1977-1982 (9) – 90mf – 9 – mf#1854 – ne IDC [077]
Ezhemesiachnoe obozrenie / ed by Biriukov, P – Geneva, Paris, 1899, v1, nos 1-5; 1900-1901, v2-3, nos 1-16 – 8mf – 9 – mf#R-18169 – ne IDC [077]
Ezhemesiachnoe obozrenie periodicheskoi literatury, uchebnykh posobii i knig po pedagogike i uchilishchevedeniiu – Spb., 1876-1880 – 80mf – 9 – mf#R-4157 – ne IDC [077]
Ezhemesiachnoe obshchedostupnoe izdanie / ed by Kruglov, A V – M., 1907-1909 – 72mf – 9 – (Missing: 1909, no 4) – mf#1707 – ne IDC [077]
Ezhemesiachnoe politicheskoe izdanie, posviashchennoe tekushchim russkim delam / ed by Dementev, P A – London, 1897. nos 1-3 – 2mf – 9 – mf#R-18174 – ne IDC [077]
Ezhemesiachnoe politicheskoe, literaturnoe i khudozhestvennoe izdanie – M., 1908. nos 1-10 – 3mf – 9 – mf#R-3354 – ne IDC [077]
Ezhemesiachnoe prilozhenie k zhurnalu "seleskii vestnik" : khronika uchrezhdenii melkogo kredita – Kiev, 1907-1912 – 54mf – 9 – (missing: 1907(2, 4, 7-8, 10-12); 1910(3, 5); 1912(1-7, 10)) – mf#R-1572 – ne IDC [077]
Ezhemesiachnoe sochinenie – Iaroslavl, 1786. 2 pts – 22mf – 9 – mf#R-18824 – ne IDC [077]
Ezhemesiachnoe sochinenie – Spb., 1785-1787 – 27mf – 9 – (Missing: 1785(1-3); 1787(4-12)) – mf#R-3794 – ne IDC [077]
Ezhemesiachnoe sochinenie – Spb., 1775-1776 – 20mf – 9 – mf#R-18805 – ne IDC [077]
Ezhemesiachnoe sochinenie, izdavaemoe ot tobolskogo glavnogo narodnogo uchilishchia – Tobolsk, 1789-91. n1-28 – 29mf – 9 – (missing: 1789, jan-aug; 1790, jan-apr; 1790, jun-dec) – mf#R-18553 – ne IDC [077]
Ezhemesiachnye oboroty po glavneishim aktivnym i passivnym schetam po operatsiiam pravleniia Moskovskikh gorodskikh i inogorodnikh otdelenii i svodnye po banku v tselom / Torgovo-Promyshlennyi Bank SSSR (Prombank) – M, 1926 – 1mf – 9 – mf#REF-85 – ne IDC [332]
Ezhemesiachnye sochineniia k poleze i uveseleniiu sluzhashchie – Spb., 1755-1764 – 173mf – 9 – mf#1412 – ne IDC [077]
Ezhemesiachnyi belletristicheskii i publitsisticheskii zvon : krasnyi zvon – Spb. n1. 1908-10 – 151mf – 9 – mf#R-3195 – ne IDC [077]
Ezhemesiachnyi bibliograficheskii zhurnal / Kavkazskii knizhnyi vestnik – Tiflis, 1900-1901 – 4mf – 9 – mf#R-4315 – ne IDC [077]
Ezhemesiachnyi bibliograficheskii zhurnal – Springfield. 1965+ (1) 1979+ (5) 1979+ (9) – 56mf – 9 – (Missing: 1897(1-2)) – mf#1760 – ne IDC [077]
Ezhemesiachnyi bibliograficheskii zhurnal : novosti kommercheskoi literatury – Spb., 1913(1-4) – 4mf – 9 – (missing: 1913(3)) – mf#R-4333 – ne IDC [077]
Ezhemesiachnyi dukhovnyi zhurnal – M., 1913-1914. v1-12 – 18mf – 9 – mf#R-3997 – ne IDC [077]
Ezhemesiachnyi ekonomicheskii zhurnal : russkoe ekonomicheskoe obozrenie – Spb., 1897-1905 – 401mf – 9 – mf#R-3770 – ne IDC [330]
Ezhemesiachnyi filosofskii i obshchestvenno-ekonomicheskii zhurnal – Washington. 1955-1995 (1) 1971-1995 (5) 1976-1995 (9) – 9mf – 9 – mf#1795 – ne IDC [077]
Ezhemesiachnyi illiustrirovannyi zhurnal / Vseobshchii zhurnal literatury, iskusstva, nauki i obshchestvennoi zhizni – Wellington. 1937+ (1) 1971+ (5) 1977+ (9) – 28mf – 9 – (Missing: 1910(1-12); 1911(1, 3, 5-6); 1912(3-12)) – mf#1986 – ne IDC [077]
Ezhemesiachnyi illiustrirovannyi istoriko-literaturnyi zhurnal – Washington. 1954+ (1) 1970+ (5) 1975+ (9) – 10mf – 9 – (Missing: 1888(1-12)) – mf#1710 – ne IDC [077]
Ezhemesiachnyi illiustrirovannyi literarurno-nauchnyi zhurnal dlia vsekh – New York. 1931-1980 (1) 1970-1980 (5) 1977-1980 (9) – 143mf – 9 – mf#1918 – ne IDC [077]
Ezhemesiachnyi illiustrirovannyi literaturnyi, obshchestvenno-ekonomicheskii i nauchno-populiarnyi zhurnal – Columbus. 1953+ (1) 1968+ (5) 1970+ (9) – 60mf – 9 – (Missing:1910(9-12); 1912(1, 10)) – mf#1799 – ne IDC [077]
Ezhemesiachnyi illiustrirovannyi politicheskii, nauchnyi i literaturnyi zhurnal – Ithaca. 1964-1981 (1) 1971-1981 (5) 1975-1981 (9) – 75mf – 9 – (Missing: 1900(6); 1902(5-12); 1903(4-12)) – mf#1814 – ne IDC [077]
Ezhemesiachnyi illiustrirovannyi sbornik – Arlington. 1888+ (1) 1907+ (5) 1888+ (9) – 41mf – 9 – mf#1829 – ne IDC [077]
Ezhemesiachnyi illiustrirovannyi voenno-obshchestvennyi zhurnal / Voina i mir – M., 1906-1907 – 75mf – 9 – (Missing: 1907(10-12)) – mf#R-4175 – ne IDC [077]
Ezhemesiachnyi illiustrirovannyi zhurnal izdateleskoi deiatelenosti i graficheskikh iskusstv : posrednik pechatnogo dela – M., 1892 – 4mf – 9 – mf#R-4331 – ne IDC [077]
Ezhemesiachnyi istoricheskii zhurnal – London. 1936-1968 (1) – 1303mf – 9 – (cont as: ukraina [kiev 1907] 1v in 4 iss; ind 1882-1906 [poltava 1911]) – mf#1239 – ne IDC [077]
Ezhemesiachnyi istoriko-literaturnyi i politicheskii zhurnal : galitsko-russkii vestnik – Spb., 1894. v1-2 – 4mf – 9 – mf#R-3980 – ne IDC [077]
Ezhemesiachnyi istoriko-literaturnyi zhurnal : panteon literatury – Chicago. 1952-1974 (1) 1973-1974 (5) 1974-1974 (9) – 250mf – 9 – (missing: 1892(3); 1895(2-4)) – mf#1827 – ne IDC [077]
Ezhemesiachnyi kooperativnyi i obshchestvenno-ekonomicheskii zhurnal – Nikolaev, 1923(1/2) – 1mf – 9 – mf#COR-605 – ne IDC [335]
Ezhemesiachnyi kritiko-bibliograficheskii zhurnal – Spb., 1909 – 3mf – 9 – mf#R-4336 – ne IDC [077]
Ezhemesiachnyi literaturniy i nauchnopopuliarnyi zhurnal – Spb., 1892-1906 – 1265mf – 9 – mf#R-3356 – ne IDC [077]
Ezhemesiachnyi literaturno-istoricheskii illiustrirovannyi sbornik : besplatnoe prilozhenie k zhurnalu "russkii palomnik" – Philadelphia. 1965-1967 (1) – 41mf – 9 – (missing: 1910(1-12); 1911(10)) – mf#1917 – ne IDC [077]
Ezhemesiachnyi literaturno-istoricheskii zhurnal – Decatur. 1965+ (1) 1979+ (5) 1979+ (9) – 39mf – 9 – mf#1835 – ne IDC [077]
Ezhemesiachnyi literaturno-istoricheskii zhurnal / Vestnik inostrannoi literatury – New York. 1898+ (1) 1965+ (5) 1970+ (9) – 1832mf – 9 – (Missing: 1908, v18(8-12); 1909, v19(1-12); 1916, v26(9-12)) – mf#1965 – ne IDC [077]
Ezhemesiachnyi literaturno-khudozhestvennyi, nauchno-populiarnyi zhurnal – Washington. 1926+ (1) 1971+ (5) 1975+ (9) – 12mf – 9 – (Missing: 1907(3, 5, 7-10, 12); 1908(1-12)) – mf#2026 – ne IDC [077]
Ezhemesiachnyi literaturno-nauchnyi zhurnal – Spb., 1880-1882 – 84mf – 9 – mf#1794 – ne IDC [077]
Ezhemesiachnyi literaturno-patrioticheskii zhurnal – Tambov, 1907. v1 – 1mf – 9 – mf#R-3785 – ne IDC [077]
Ezhemesiachnyi literaturno-politicheskii zhurnal – Nashville. 1948+ (1) 1970+ (5) 1977+ (9) – 74mf – 9 – mf#1954 – ne IDC [077]
Ezhemesiachnyi literaturnyi i nauchnyi zhurnal – Spb., 1906 no 1 – 8mf – 9 – mf#R-3767 – ne IDC [077]
Ezhemesiachnyi literaturnyi i nauchnyi zhurnal – Spb., 1909. v1 – 3mf – 9 – mf#R-4004 – ne IDC [077]

Ezhemesiachnyi literaturnyi i obshchestvenno-ekonomicheskii zhurnal – Spb., 1909. v1-3 – 16mf – 9 – mf#R-3784 – ne IDC [077]

Ezhemesiachnyi literaturnyi i obshchestvennyi zhurnal – Ventnor. 1963-1970 (1) – 39mf – 9 – mf#1841 – ne IDC [077]

Ezhemesiachnyi literaturnyi i politicheskii zhurnal – Concord. 1957-1971 (1) – 185mf – 9 – (Missing: 1883(11-12)) – mf#1958 – ne IDC [077]

Ezhemesiachnyi literaturnyi, nachnyi i politicheskii zhurnul – Pg., 1914-1917 – 211mf – 9 – mf#R-3775 – ne IDC [077]

Ezhemesiachnyi literaturnyi, nauchnyi i politicheskii zhurnal – Oxford. 1925+ (1) 1971+ (5) 1975+ (9) – 820mf – 9 – (Missing: 1915(7); 1917(2-12)) – mf#1200 – ne IDC [077]

Ezhemesiachnyi literaturnyi, nauchnyi i politicheskii zhurnal / Raduga – Geneva, 1907-1908. nos 1-4 – 7mf – 9 – mf#R-18156 – ne IDC [077]

Ezhemesiachnyi, literaturnyi, nauchnyi i politicheskii zhurnal – Sausalito. 1959-1965 (1) – 118mf – 9 – mf#1194 – ne IDC [077]

Ezhemesiachnyi literaturnyi zhurnal – Chicago. 1940-1986 (1) 1971-1986 (5) 1976-1986 (9) – 70mf – 9 – mf#1810 – ne IDC [077]

Ezhemesiachnyi literaturnyi zhurnal – Denton. 1951-1972 (1) – 45mf – 9 – mf#1716 – ne IDC [077]

Ezhemesiachnyi mezhdunarodnyi zhurnal : russkii otdel – Spb., 1897-1898 – 40mf – 9 – mf#R-3196 – ne IDC [077]

Ezhemesiachnyi nauchno-literaturnyi i obshchestvenno politicheskii zhurnal / Ukrainskaia zhizne – Westbrook. 1949+ (1) 1970+ (5) 1976+ (9) – 110mf – 9 – (Missing: 1917(7-12)) – mf#1404 – ne IDC [077]

Ezhemesiachnyi nauchno-literaturnyi i politicheskii zhurnal – Davos, 1909-1910. nos 1-3 – 6mf – 9 – mf#R-18047 – ne IDC [077]

Ezhemesiachnyi nauchno-populiarnyi, khudozhestvennyi i literaturnyi zhurnal – Washington. 1964+ (1) 1971+ (5) 1972+ (9) – 109mf – 9 – (1905(10-12); 1906(1, 4-12) are missing.) – mf#1801 – ne IDC [077]

Ezhemesiachnyi nauchnyi i kritiko-bibliograficheskii zhurnal – Elmhurst. 1963-1968 (1) – 370mf – 9 – (Missing: 1870(1-9); 1875(7-12); 1877(5-12)) – mf#2030 – ne IDC [077]

Ezhemesiachnyi obshchestvenno-literaturnyi illiustrirovannyi zhurnal – London. 1872-1989 (1) 1958-1989 (5) 1958-1989 (9) – 51mf – 9 – (Missing: 1906(45)) – mf#1225 – ne IDC [077]

Ezhemesiachnyi obshchestvenno-nauchno-literaturnyi zhurnal, posviashchennyi zhenskomu voprosu / Zhenskii vestnik – Cincinnati. 1897-1986 (1) 1979-1986 (5) 1979-1986 (9) – 111mf – 9 – (Misisng: 1904(4-12); 1905(4-12); 1907(7-8); 1913(10); 1915(2, 5-6, 11-12); 1916(7-11); 1917(3-12)) – mf#2017 – ne IDC [077]

Ezhemesiachnyi politicheskii organ / Otechestvennaia oborona; ed by Kochubei, M – Paris, 1906-1907. nos 1-7 – 9mf – 9 – (Missing: 1907. no 6) – mf#R-18129 – ne IDC [077]

Ezhemesiachnyi professionalenyi organ uchitelestva nachalenoi i srednei shkoly – Spb., 1912-1913 – 17mf – 9 – mf#R-3779 – ne IDC [077]

Ezhemesiachnyi zhurnal – Aberdeen. 1903-1991 (1) 1971-1991 (5) 1977-1991 (9) – 345mf – 9 – mf#1226 – ne IDC [077]

Ezhemesiachnyi zhurnal – Dallas. 1859+ (1) 1974+ (5) 1976+ (9) – 45mf – 9 – mf#1840 – ne IDC [077]

Ezhemesiachnyi zhurnal / Ekonomicheskoe vozrozhdenie – 1922. v1-2 – 4mf – 8 – mf#R-3401 – ne IDC [335]

Ezhemesiachnyi zhurnal – M., 1911 – 4mf – 9 – mf#R-3982 – ne IDC [077]

Ezhemesiachnyi zhurnal – Washington. 1964+ (1) 1971+ (5) 1971+ (9) – 64mf – 9 – (Missing: 1875; 1876(1-2, 7-8, 11-12); 1877-1879; 1880(1-2, 5-6, 9-10)) – mf#1691 – ne IDC [077]

Ezhemesiachnyi zhurnal – Boston. 1961+ (1) 1967+ (5) 1967+ (9) – 2700mf – 9 – (Missing: 1879(2-12); 1882(9-12)) – mf#1407 – ne IDC [077]

Ezhemesiachnyi zhurnal – Spb., 1890-1902 – 63mf – 9 – (Missing: 1891(16)) – mf#R-3203 – ne IDC [077]

Ezhemesiachnyi zhurnal – Nizhnii-Novgorod, 1906-1907 – 53mf – 8 – (missing: 1907(9)) – mf#1443 – ne IDC [243]

Ezhemesiachnyi zhurnal : organ soveta truda i oborony – 1923-1930 – 299mf – 9 – mf#R-2441 – ne IDC [335]

Ezhemesiachnyi zhurnal : organ tserkovno-obshchestvennoi zhizni staroobriadchestva – Gottingen. 1959-1963 (1) – 30mf – 9 – mf#1909 – ne IDC [077]

Ezhemesiachnyi zhurnal – Pg., 1915. v1-6 – 9mf – 9 – mf#R-2332 – ne IDC [077]

Ezhemesiachnyi zhurnal : russkii nachalenyi uchitel – Spb., 1880-1911 – 539mf – 9 – (missing: 1881(12); 1883(2); 1892(10); 1893(10); 1898(1-12)) – mf#R-3745 – ne IDC [077]

Ezhemesiachnyi zhurnal : russkoe sudokhodstvo, torgovoe i promyslovoe, na rekakh, ozerakh i moriakh – Spb., 1886-1901. n1-237 – 428mf – 9 – (missing: 1897(178, 180, 185, 189)) – mf#R-3744 – ne IDC [077]

Ezhemesiachnyi zhurnal – Spb., 1906. no 1 – 6mf – 9 – mf#R-4183 – ne IDC [077]

Ezhemesiachnyi zhurnal / Voprosy narodnogo obrazovaniia – Spb., 1912-1913. v1-15 – 14mf – 9 – (Missing: 1913. v6-7) – mf#R-8036 – ne IDC [077]

Ezhemesiachnyi zhurnal – Washington. 1948+ (1) 1971+ (5) 1976+ (9) – 4mf – 9 – mf#1540 – ne IDC [077]

Ezhemesiachnyi zhurnal iskusstva, literatury, obshchestvennoi zhizni – M., 1904-1906 – 274mf – 9 – (Missing: 1906(6-10)) – mf#1838 – ne IDC [077]

Ezhemesiachnyi zhurnal istoricheskoi literatury i nauki / Vestnik vsemirnoi istorii – New York. 1958+ (1) 1965+ (5) 1970+ (9) – 154mf – 9 – mf#1973 – ne IDC [077]

Ezhemesiachnyi zhurnal literaturno-nauchnyi – Philadelphia. 1887-1993 (1) 1970-1993 (5) 1972-1993 (9) – 217mf – 9 – (Missing: 1908(3-12); 1913(4-12)) – mf#1915 – ne IDC [077]

Ezhemesiachnyi zhurnal literatury, kritiki, bibliografii, pechatnogo i izdateleskogo dela – M., 1895-1896 – 20mf – 9 – (1895(12)) – mf#R-4326 – ne IDC [077]

Ezhemesiachnyi zhurnal literatury, nauki i obshchestvennoi zhizni – Spb., 1914-1917 – 175mf – 9 – mf#1723 – ne IDC [077]

Ezhemesiachnyi zhurnal literatury, politiki i nauki – Pg., 1915. v1-5 – 3mf – 9 – mf#R-3757 – ne IDC [077]

Ezhemesiachnyi zhurnal literatury, politiki, istorii, iskusstva i obshchestvennoi zhizni – Spb., 1911-1915 – 350mf – 9 – (Mising: 1915. v6-9, 11-12) – mf#1199 – ne IDC [077]

Ezhemesiachnyi zhurnal po natsionalenomu i oblastnomu voprosam – M., 1914. v1-8 – 6mf – 9 – mf#R-3986 – ne IDC [077]

Ezhemesiachnyi zhurnal po vsem otrasliam obshchestvennykh znanii / Volshebnyi fonar – Spb., 1878. v1-2 – 6mf – 9 – mf#R-4193 – ne IDC [077]

Ezhemesiachnyi zhurnal, posviashchaemyi izucheniiu tikhvina i nagornogo obonezheia – Spb., 1914. v1-2 – 2mf – 9 – mf#R-4005 – ne IDC [077]

Ezhemesiachnyi zhurnal, posviashchennyi tserkovno-obshchestvennoi zhizni staroobriadchestva – Washington. 1954-1971 (1) 1968-1971 (5) – 151mf – 9 – (Missing: 1912(9)) – mf#1434 – ne IDC [077]

Ezhemesiachnyi zhurnal, posviashchennyi voprosam emigratsii i kolonizatsii – Spb., 1913. v1-6 – 7mf – 9 – mf#R-4165 – ne IDC [077]

Ezhemesiachnyi zhurnal, posviashchennyi vospitaniiu i obrazovaniiu evreev – Spb., 1904-1905 – 25mf – 9 – (Missing: 1904(3, 6-7); 1905(6-7, 9-10)) – mf#R-3407 – ne IDC [077]

Ezhemesiachyi zhurnal – Lucknow. 1962-1995 (1) 1970-1995 (5) 1977-1995 (9) – 157mf – 9 – (Missing: 1891. no 1-4) – mf#1928 – ne IDC [077]

Ezhemiesiachnyi biulleten povgodu moskvie / Moscow. Gorodskaia Uprava. Statisticheskii Otdiel – 1888-1915, 1925-29 – 1 – us L of C Photodup [315]

Ezhemiesiachnyi literaturnyi zhurnal – 1885-1902. – 1 – (1892-95, 1900-01, and 3 scattered issues wanting) – us L of C Photodup [460]

Ezhenedelenaia politicheskaia gazeta / Russkoe tsarstvo – M., 1907. v1-8 – 4mf – 9 – (Missing: 1907(2, 4, 6, 7)) – mf#R-2250 – ne IDC [077]

Ezhenedelenik, posviashchennyi evreiskim interesam – Spb., 1910-1915. v1-6 – 204mf – 9 – mf#R-4150 – ne IDC [077]

Ezhenedelenik s karikaturami : listok dlia svetskikh liudei – Spb. 1843 n1-48; 1844 n1-48 – 14mf – 9 – mf#R-3365 – ne IDC [077]

Ezhenedelenoe illiustrirovannoe izdanie – Spb., 1858-1863. v1-12 – 305mf – 9 – mf#R-3382 – ne IDC [077]

Ezhenedelenoe obshchestvenno-politicheskoe i kuleturno-filosofskoe izdanie – Spb., 1905-1906. nos 1-14 – 28mf – 9 – (cont as: svoboda i kuletura. ezhenedelenyi zhurnal [spb. 1906] n1-8) – mf#1442, R-3733 – ne IDC [077]

Ezhenedelenoe prilozhenie k zhurnalu "Russkii invalid" – Spb., 1874-1965 – 52mf – 9 – mf#R-3778 – ne IDC [077]

Ezhenedelenyi illiustrirovannyi zhurnal – M., 1915. nos 1-31 – 31mf – 9 – mf#R-3740 – ne IDC [077]

Ezhenedelenyi illiustrirovannyi zhurnal / Vsemirnaia nov – Pittsburgh. 1945+ (1) 1971+ (5) 1977+ (9) – 111mf – 9 – (Missing: 1910(1-52); 1911(1-52); 1912(1-21, 23, 27-28, 30, 32, 35-36, 38-52); 1914(2, 4-5, 7-26, 44, 48, 50, 52); 1915(1, 5-9, 15-17, 19-22, 32-52); 1916(1-46, 48-52)) – mf#1981 – ne IDC [077]

Ezhenedelenyi khudozhestvenno literaturnyi i iumoristicheskii zhurnal – Spb., 1912-1917 – 299mf – 9 – mf#R-10494 – ne IDC [077]

Ezhenedelenyi kritiko-bibliograficheskii zhurnal – Springfield. 1940+ (1) 1978+ (5) 1978+ (9) – 18mf – 9 – (Missing: 1907, v22-23) – mf#1759 – ne IDC [077]

Ezhenedelenyi literaturno-khudozhestvennyi zhurnal – St. Louis. 1962-1976 (1) 1971-1976 (5) – 200mf – 9 – (Missing: 1888, v1; 1889, v2(1-23); 1890, v3(1); 1894, v7(1)) – mf#1883 – ne IDC [077]

Ezhenedelenyi literaturno-kriticheskii zhurnal : sibirskaia nove – Tomsk, 1910. v1-7 – 4mf – 9 – mf#R-3760 – ne IDC [077]

Ezhenedelenyi, nauchnyi i politicheskii zhurnal, posviashchennyi ekonomicheskoi i finansovoi zhizni, russkoi i zagranichnoi / ed by Migulin, P P – Spb., 1913-1917. v1-5 – 122mf – 9 – (Missing:1917, no 8-end) – mf#R-2320 – ne IDC [077]

Ezhenedelenyi nauchnyi zhurnal – Spb., 1894-1903. nos 1-5 – 488mf – 9 – (Missing: 1895, nos 1-12; 1896, nos 1-12) – mf#1800 – ne IDC [077]

Ezhenedelenyi obshestvenno-politicheskii, literaturno-khudozhestvennyi i ekonomicheskii zhurnal – Pg., 1915. v1-4 – 4mf – 9 – mf#R-3988 – ne IDC [077]

Ezhenedelenyi, populiarnyi, literaturnyi i nauchnyi zhurnal – M., 1907. nos 1-16 – 3mf – 9 – mf#R-8135 – ne IDC [077]

Ezhenedelenyi vestnik russkoi pechati – Trier. 1965-1996 (1) 1966-1996 (5) 1966-1996 (9) – 67mf – 9 – mf#1848 – ne IDC [077]

Ezhenedelenyi zhurnal literatury, iskusstv i sovremennykh novostei – Spb., 1860. v1-52 – 48mf – 9 – mf#1882 – ne IDC [077]

Ezhenedelenyi zhurnal politicheskii, literaturnyi, teatralenyi i khudozhestvennyi – M., 1876-1878 – 50mf – 9 – (Missing: 1876(1-52); 1877(1-8, 49-52); 1878(2-5, 7-48)) – mf#1790 – ne IDC [077]

Ezhenedelnaia gazeta / ed by Burtsev, V – Paris, 1911-14. n1-48 – 8mf – 9 – (missing: 1911 n1) – mf#R-18005 – ne IDC [074]

Ezhenedelnaia iuridicheskaia gazeta – Spb., 1898-1917. Systematic ind 1898-1909 – 907mf – 9 – mf#R-4160 – ne IDC [077]

Ezhenedel'naia klinicheskaia gazeta – St Petersburg, 1881-89 – 1 – us UMI ProQuest [077]

Ezhenedelnaia obshchestvenno-politicheskaia gazeta – M., 1905. nos 1-3 – 3mf – 9 – mf#R-4133 – ne IDC [077]

Ezhenedelnaia obshchestvenno-politicheskaia gazeta – M., 1906-10 – 146mf – 9 – (Missing: 1906(9-10); 1907(1, 20-21, 29); 1908(3, 43-50)) – mf#R-18349 – ne IDC [077]

Ezhenedelnaia politicheskaia i literaturnaia gazeta – Geneva, 1882-1883. nos 1-20 – 3mf – 9 – mf#R-18140 – ne IDC [077]

Ezhenedelnaia politicheskaia i obshchestvennaia gazeta – M., 1881. v1-57 – 45mf – 9 – mf#R-3530 – ne IDC [077]

Ezhenedelnik : illiustrirovannaia gazeta – Spb., 1864-1867 – 199mf – 9 – (missing: 1864(34-37)) – mf#R-3381 – ne IDC [077]

Ezhenedel'nik petrogradskikh gosudarstvennykh akademicheskikh teatrov – Petrograd, 1922-23 – 31mf – 9 – us UMI ProQuest [780]

Ezhenedelnik teatrov – Moscow, sep 1923-jun 1924 – 22mf – 9 – us UMI ProQuest [790]

Ezhenedelnoe illiustrirovannoe izdanie / Zhenskoe delo – M., 1910-1918 – 234mf – 9 – (Missing: 1910(6, 11-12, 29-30); 1911(12-13, 16-17, 21); 1913(8-9, 12-13, 18-19, 21-23); 1914(1, 5-6, 12, 14); 1915(1-5, 8-9, 12); 1916(2, 7, 12, 16); 1917(5-8, 10-24); 1918(1-11)) – mf#R-8199 – ne IDC [077]

Ezhenedelnoe illustrirovannoe izdanie – Kiev, 1906. v1-4 – 2mf – 9 – mf#R-3976 – ne IDC [077]

Ezhenedelnoe izdanie – Spb., 1906-1907. nos 1-14 – 8mf – 9 – (Missing: 1907, no 14) – mf#R-1536 – ne IDC [077]

Ezhenedelnoe izdanie – Spb., 1772-1773. 2 pts – 8mf – 9 – mf#R-18830 – ne IDC [077]

Ezhenedelnoe izdanie – Spb., 1788-1789. 4 pts – 16mf – 9 – mf#R-18826 – ne IDC [077]

Ezhenedelnoe izdanie / Vestnik knigoprodavtsev – M., 1900-1901 – 46mf – 9 – mf#R-4306 – ne IDC [077]

Ezhenedelnoe izdanie, posviashchennoe interesam mestnogo selskogo khoziaistva, promyshlennosti i torgovli – Poltava, 1896-1917 – 294mf – 9 – (Missing: several yrs) – mf#R-1575 – ne IDC [077]

Ezhenedelnyi illiustrirovannyi zhurnal – Spb., 1897-1918 – 615mf – 9 – (Missing: 1903(1-2, 12-13, 17, 19, 27, 30-31, 34, 37, 43-44, 51-52); 1904(2, 5, 10, 12, 15, 20, 22, 43-44, 50-52); 1911 (5, 36); 1914 (8, 14, 17, 23-25, 30-41, 44); 1915(6, 18, 34, 42); 1916(44, 52); 1917(4-52)) – mf#1403 – ne IDC [077]

Ezhenedelnyi illiustrirovannyi zhurnal / Zhivopisnoe obozrenie – Spb., 1872-1902, 1904-1905. Zhivopisnoe obozrenie. Ezhemesiachnyi literaturnyi i politicheskii zhurnal. Prilozhenie k zhurnalu "Zhivopisnoe obozrenie" Spb., 1882-1902 (incomplete) – 1514mf – 9 – mf#2024, 2025 – ne IDC [077]

Ezhenedelnyi illiustrirovannyi zhurnal, posviashchennyi knizhnomu, zhurnalenomu i pechatnomy dely – M., 1903 – 2mf – 9 – mf#R-4329 – ne IDC [077]

Ezhenedelnyi khudozhestvenno-illiustrirovannyi zhurnal – Spb., 1909. v1-3 – 2mf – 9 – mf#R-1567 – ne IDC [077]

Ezhenedelnyi khudozhestvenno-literaturnyi i nauchnyi zhurnal / Zhivaia mysl – Montvale. 1937+ (1) 1971+ (5) 1971+ (9) – 17mf – 9 – (Missing: v19-21, 24) – mf#2018 – ne IDC [077]

Ezhenedelnyi khudozhestvenno-literaturnyi zhurnal – New York. 1932-1985 (1) 1971-1985 (5) 1976-1985 (9) – 470mf – 9 – (Missing: 1900-1907; 1908(15, 51); 1909(1, 3-5, 11, 14, 52); 1915(22, 43); 1916(3); 1917(25, 27-29, 39-40, 42-52)) – mf#1818 – ne IDC [077]

Ezhenedelnyi khudozhestvenno-literaturnyi zhurnal – Weaverville. 1942-1987 (1) 1964-1987 (5) 1964-1987 (9) – 14mf – 9 – mf#1715 – ne IDC [077]

Ezhenedelnyi khudozhestvennyi i politiko-satiricheskii zhurnal – Spb., 1906. v1-5 – 5mf – 9 – mf#R-3477 – ne IDC [077]

Ezhenedelnyi nauchnyi, literaturnyi i politicheskii zhurnal / Vestnik zhizni – Spb., 1906-1907 – 37mf – 9 – mf#R-4865 – ne IDC [077]

Ezhenedelnyi obshchedostupnyi nauchno-literaturnyi illiustrirovannyi zhurnal – Nikolsk-Ussuriiskii, 1910. nos 1-4 – 4mf – 9 – mf#R-8126 – ne IDC [077]

Ezhenedelnyi obshchestvenno-politicheskii i literaturnyi zhurnal – Spb., 1906. v1-12 – 16mf – 9 – mf#1962 – ne IDC [077]

Ezhenedelnyi selskokhoziaistvennyi i ekonomicheskii zhurnal – Kiev, 1906-1916 – 231mf – 9 – (Missing: 1906-1908; 1909(22, 43); 1911(36); 1913(4-10); 1914(28-50); 1915(1-50); 1916(7-8, 11-14, 21-28, 31-50)) – mf#R-2300 – ne IDC [077]

Ezhenedelnyi zhurnal – M., 1905-1906 – 50mf – 9 – mf#R-4166 – ne IDC [077]

Ezhenedelnyi zhurnal / Vestnik popechitelestv o narodnoi trezvosti – Spb., 1903-1905 – 54mf – 9 – (Missing: 1903(9); 1904(13-17, 26, 30, 33, 36, 42-end); 1905 (28, 49, 50)) – mf#R-9273 – ne IDC [077]

Ezhenedelnyi zhurnal / Za svobodu – Odessa, 1905. nos 1-2 – 1mf – 9 – (Missing: 1905 (no 2)) – mf#R-8191 – ne IDC [077]

Ezhov, V *see*
– Puti rabochei kooperatsii na zapade i u nas
– Rabochaia kooperatsiia i ee zadachi

Ezop – Istanbul, 1908. Sahib-i Imtiyaz ve Mueduer-i Mes'ul: Tuerkiye Kuetuephanesi Sahibi Mihran; Muharrir ve Musavveri: Mehmed Sedad, Agah. n1-2. 6 se'ban 1326-9 se'ban 1326 [20-23 agustos 1324] [2-5 sep 1908S] [all publ] – (= ser O & t journals) – 3mf – 9 – $55.00 – (cont by: karakus ezop. only n1 & 2 publ under this title. only n3 publ under the title karakus ezop) – us MEDOC [956]

Ezov, G A *see* Snosheniia petra velikogo s armianskim narodom

'Ezra – Providans, R Ay: Aroysgegeben fun dem Order 'Ezra. v1 n1 1911 (mthly) [mf ed 197-?] – 1 – (No more publ?) – mf#*ZAN-*P492 – us NY Public [270]

Ezra abbot – Cambridge: publ for the alumni of the harvard divinity school, 1884 – 1mf – 9 – 0-7905-1698-5 – mf#1987-1698 – us ATLA [240]

Ezra and nehemiah : their lives and times / Rawlinson, George – New York: Anson D F Randolph, [1890?] – 1mf – 9 – 0-8370-4846-X – (incl bibl ref) – mf#1985-2846 – us ATLA [221]

Ezra Family International *see* Your koopischik's nooze

Ezra, nehemiah and esther : introduction, revised version with notes, maps and index / ed by Davies, Thomas Witton – New York: Henry Frowde, [ca 1908] – 1mf – 9 – 0-8370-2846-9 – (incl ind) – mf#1985-0846 – us ATLA [221]

Ezra, nehemiah, and esther / Adeney, Walter Frederic – New York: A C Armstrong, 1893 – (= ser The expositor's bible) – 1mf – 9 – 0-8370-2050-6 – mf#1985-0050 – us ATLA [221]

Ezra stiles gannett, unitarian minister in boston, 1824-1871 : a memoir / Gannett, William Channing – Boston: Roberts Bros, 1875 – 2mf – 9 – 0-524-04296-9 – (incl bibl ref) – mf#1992-2016 – us ATLA [242]

Ezra studies / Torrey, Charles Cutler – Chicago: U of Chicago Press, 1910 – 1mf – 9 – 0-7905-0403-0 – (Incl bibl ref and indexes) – mf#1987-0403 – us ATLA [221]

The ezra-apocalypse : being chapters 3-14 of the book commonly known as 4 ezra (or ii esdras) – London: Sir Isaac Pitman, 1912 – 2mf – 9 – 0-7905-0814-1 – (text in english and latin. incl indes) – mf#1987-0814 – us ATLA [221]

Ezras : liber primvs ezrae, homiliis 38...expositus / Lavater, L – Tiguri, Officina Froschoviana, 1586 – 2mf – 9 – mf#PBU-325 – ne IDC [221]

Ezrath Torah *see* Lua ha-yovel

f *see* Alcide al brivo

F a michaux's, m d mitglieds der naturhistorischen gesellschaft zu paris, korrespondenten der ackerbau-gesellschaft des dept der seine und oise usw... – Weimar 1805 [mf ed Hildesheim 1995-98] – XIV/250 S. 2mf – 9 – €60.00 – ISBN-10: 3-487-26582-6 – ISBN-13: 978-3-487-26582-7 – (trans fr french) – gw Olms [500]

F and b topics / AFL-CIO – v1 n1-v3 n4 [1977 may-1979 winter] – 1r – 1 – mf#666004 – us WHS [331]

F and i war / 18th Century Society [New Alexandria PA]. French and Indian War Associators for Re-Enactments – 1982 winter-1986 jul – 1r – 1 – mf#1336449 – us WHS [366]

F depons vormaligen agenten...historisch-geographisch-statistische nachrichten von der general-hauptmannschaft caracas : oder dem oestlichen theile der spanischen landschaft terraf, gesammelt auf einer reise und waehrend des aufenthalts in diesem lande in den jahren 1801 bis 1804 / Pons, Francois R de – Weimar 1807 [mf ed Hildesheim 1995-98] – 1v on 4mf – 9 – €120.00 – ISBN-10: 3-487-26554-0 – ISBN-13: 978-3-487-26554-4 – (trans fr french) – gw Olms [918]

A f e langbein's saemmtliche gedichte / Langbein, August Friedrich Ernst – new ed. Stuttgart: Rieger 1855 [mf ed 1990] – 1r – 1 – (filmed with: kristin / mariluise lange & other titles) – mf#2815p – us Wisconsin U Libr [810]

F e w's news and views from federally employed women, inc / Federally Employed Women, Inc – 1982 jan/feb-1983 sep/oct – 1r – 1 – (cont: news and views from federally employed women, inc; cont by: news and views [federally employed women [association]]) – mf#630002 – us WHS [331]

F geminiani's violin methods / Gutheil, Crystal H – U of Rochester 1943 [mf ed 19–] – 3mf – 9 – mf#fiche 184 – us Sibley [780]

F l w meyer : sein leben und seine schriftstellerische wirksamkeit; ein beitrag zur litteraturgeschichte des 18. und 19. jahrhunderts / Zimmermann, Curt – [S.l: s.n.] (Halle a.S: Druck von E Karras), 1890 – 1 – (incl bibl ref) – us Wisconsin U Libr [430]

F marini mersenni minimi cogitata physico-mathematica : in quibus tam naturae qua'm artis effectus admirandi certissimis demonstrationibus explicantur / Mersenne, Marin – Parisiis: Sumptibus Antonii Bertier, vialacobaea 1644 [mf ed 19–] – 6v in 1 on 1r – 1 – mf#film 88 – us Sibley [510]

F nordy hoffmann, senate service 1975-1981 : sergeant-at-arms for the senate – (= ser Us senate historical office oral history coll) – 3mf – 9 – $15.00 – us Scholarly Res [323]

A f skjoeldebrand's koenigl schwed obersten und ritters des schwerdordens beschreibung der wasserfaelle und des kanals von trollhaetta in schweden / Skjoeldebrand, Anders F – Weimar 1805 [mf ed Hildesheim 1995-98] – 1v on 2mf – 9 – €60.00 – 3-487-26566-4 – gw Olms [914]

F w dodge northwest construction data and news : washington-alaska ed – Portland, 1998+ [1,5,9] – mf#25667,02 – us UMI ProQuest [690]

F w dodge southeast construction – New York. 2000+ (1,5,9) – ISSN: 1538-6570 – mf#33169 – us UMI ProQuest [690]

F w t a o newsletter / Federation of Women Teachers' Associations of Ontario – 1983-84, 1985-86 – 2r – 1 – (cont by: fwtao/faeo newsletter) – mf#1111462 – us WHS [370]

F x garneau et francis parkman / Casgrain, Henri Raymond – Montreal: Librairie Beauchemin, 1912 – 2mf – 9 – 0-665-76350-6 – mf#76350 – cn Canadiana [971]

F y i : information from the american federation of teachers, public relations department / American Federation of Teachers – 1985 sep 5-1986 feb 19 – 1r – 1 – (cont: aft news clips) – mf#1130994 – us WHS [370]

Fa chueh / Liu, Chung-an – Shang-hai: T'ien ma shu tien, Min kuo 23 [1934] – (= ser P-k&k period) – us CRL [480]

Fa, hsueh *see* Hsin pan chih chieh shui fa kuei hui pien

Fa lu wai ti hang hsien / Sha, T'ing – Shang-hai: Hsin k'en shu tien, 1932 – (= ser P-k&k period) – us CRL [480]

Fa ti ku shih / Pa, Chin – Shang-hai: Wen hua sheng huo ch'u pan she, Min kuo 25 [1936] – (= ser P-k&k period) – us CRL [480]

Faa aviation news – Washington. 1962-1975 (1) 1971-1975 (5) 1975-1975 (9) – (cont by: faa general aviation news: a dot/faa flight standards safety publication) – ISSN: 0014-553X – mf#1572 – us UMI ProQuest [629]

Faa aviation news – Washington. 1987+ (1,5,9) – (cont: faa general aviation news: a dot/faa flight standards safety publication) – ISSN: 1057-9648 – mf#1572,02 – us UMI ProQuest [629]

Faa general aviation news : a dot/faa flight standards safety publication – Washington. 1975-1987 (1,5,9) – (cont: faa aviation news; cont by: faa aviation news) – ISSN: 0362-7942 – mf#1572,01 – us UMI ProQuest [629]

Faar and ncn : a newsletter of the feminist alliance against rape and the national communications network – 1978 jul/aug – 1r – 1 – mf#1330332 – us WHS [305]

Faar news / Feminist Alliance Against Rape [US] – 1977 jul/aug-1978 jan/feb – 1r – 1 – (cont: newsletter [feminist alliance against rape [us]]) – mf#609201 – us WHS [305]

Faasen, Nicol *see* Die rol van onderwysersentrums in skoolontwikkeling

Fabbiani Ruiz, Jose *see* A orillas del sueno

Fabbrizi, Pietro *see* Regole generali di canto fermo

Fabel, Karel *see* Umeni a remesla

Fabel vom kranken Loewen *see* Die zehn gebote (mxt3)

Fabel vom kranken loewen *see* Apokalypse / ars moriendi / biblia pauperum / antichrist / fabel vom kranken loewen / kalendarium und planetenbuecher / historia david (mxt2)

Fabela, Isidro *see*
- Belice
- International controversy

Fabeln : der helvetischen gesellschaft gewidmet / Pfeffel, Gottlieb Konrad – Basel: bey J Jacob Thurneysen, dem Juengeren, 1783 [mf ed 1993] – 208p – 1 – mf#8361 – us Wisconsin U Libr [390]

Fabeln / Pestalozzi, Johann Heinrich – Bern: H Feuz, [1940] – 1r – 1 – us Wisconsin U Libr [390]

Die fabeln des erasmus alberus / ed by Braune, Wilhelm – Halle: Max Niemeyer, 1892 – (incl bibl ref) – us Wisconsin U Libr [430]

Fabens, Joseph Warren *see* Resources of santo domingo

Faber – Stockville, NE: A.G. Williams. 16v. v29 n30. may 1 1913-v44 n21. jan 26 1928 (wkly) [mf ed with gaps filmed 1970] – 8r – 1 – (cont: republican-faber. absorbed by: curtis enterprise) – us NE Hist [071]

The faber – Stockville, NE: Chadderton & Reed, 1895-v21 n28. nov 24 1904 (wkly) [mf ed with gaps filmed 1970] – 6r – 1 – (cont: frontier county faber. merged with: republican to form: republican-faber. issues for 1895-99 also carry whole numbering. occasional text in german) – us NE Hist [071]

Faber, Basilius *see* Thesaurus eruditionis scholasticae

Faber, Benedikt *see*
- Canticum gratulatorium...
- Colloquium metricum...

Faber, Ernest *see* Introduction to the science of chinese religion

Faber, Ernst *see*
- Bilder aus china
- China in historischer beleuchtung
- Introduction to the science of chinese religion
- The mind of mencius
- The principal thoughts of the ancient chinese socialism
- Problems of practical christianity in china
- Systematical digest of the doctrines of confucius

Faber, F *see* Prodromus der islaendischen ornithologie

Faber, Frederick William *see*
- Bethlehem
- The blessed sacrament
- Essay on catholic home missions
- Essay on the interest and characteristics of the lives of the saint
- The foot of the cross
- Grounds for remaining in the anglican communion
- Letter to the members of the confraternity of the most precious blo...
- The spirit and genius of st philip neri, founder of the oratory

Faber, G H von *see*
- Er werd een stad geboren
- Oud soerabaia

Faber, Georg *see* Buddhistische und neutestamentliche erzaehlungen

Faber, George Stanley *see*
- Facts and assertions
- An inquiry into the history and theology of the ancient vallenses and albigenses
- Rome and the bible

Faber, Heinrich *see*
- Musica
- Musicae compendium

Faber, J E E *see* Magazin der geographie, staatenkunde und geschichte

Faber, Joh Heinr *see* Der sammler

Faber, Michael L O *see* Zambia

Faber, oder, die verlorenen jahre : roman / Wassermann, Jakob – Berlin: S Fischer, 1925, c1924 – 1r – 1 – us Wisconsin U Libr [830]

Fabian : die geschichte eines moralisten / Kastner, Erich – Frankfurt am Main, Germany. 1958 – 1r – us UF Libraries [170]

Fabian, Bernhard *see* Kataloge der frankfurter und leipziger buchmessen 1594-1860

Fabian, Bernhard [comp] *see* Deutsches biographisches archiv (dba1)

Fabian economic and social thought, series 1 : the papers of edward carpenter, 1844-1929, from sheffield archives – [mf ed Marlborough 1993] – 2pt – 1 – (pt1: correspondence & mss [22r] £2050; pt2: mss, cuttings, pamphlets & selected publ [26r] £2450; with d/g) – uk Matthew [300]

Fabian economic and social thought, series 2 : the papers of hugh dalton, 1887-1962 from the british library of political and economic science – 1pt – 1 – (pt1: complete diaries 1916-60 [7r] £650; with d/g) – uk Matthew [300]

Fabian, Juergen *see* Sozialpsychologische kleingruppenforschung im kontext von ansaetzen einer theorie selbstreferentieller systeme

Fabian news – v. 1-67. Mar 1891-Dec 1956. Apr 1938-Aug 1941 and scattered issues wanting – 1 – 52.00 – us L of C Photodup [335]

Fabian Society *see* Young fabian pamphlets, 1961-82

Fabian Society (Great Britain) *see* Unprotected protectorates, basutoland, bechuanaland

Fabian Society (Great Britain). Colonial Bureau *see* Empire

Fabian Society, London *see*
- Fabianism and the empire
- Local government in ireland
- State railways for ireland

Fabian society research pamphlets, 1931-81 – n1-347 – 9r 17mf – 1,9 – mf#8/96302 – uk Microform Academic [335]

Fabian society tracts, 1884-1980 – tract n428-472 [1974-80] – 21mf – 9 – £126 / $252 – (also available on 9r [r96301]: tract n1-427 [1884-1973]) – mf#f86301 – uk Microform Academic [335]

Fabian y Fuero, Francisco *see*
- Coleccion de providencias dadas a fin de establecer la santa vida comun
- Coleccion de providencias diocesanas del obispado de la puebla de los angeles
- Nos don francisco fabian y fuero por la divina gracia y de la santa sede apostolica de la puebla de los angeles

Fabianelo / Diaz Maciaz, Jose – 1896 – 9 – sp Bibl Santa Ana [830]

Fabianism and the empire : a manifesto by the fabian society / Fabian Society, London; ed by Shaw, Bernard – [London], 1900 – (= ser 19th c books on british colonization) – 2mf – 9 – mf#1.1.3505 – uk Chadwyck [941]

Fabie, Antonio Maria *see*
- Exposicion a la real academia de la historia en favor de la aparicion de la virgen de guadalupe en mexico...
- Mi gestion ministerial respecto a la isla de cuba

Fabius *see* 52 questions on the nationalization of canadian railways

Une fable de florian / Ristelhuber, Paul – Paris: J Baur 1881 [mf ed Bloomington IN: Indiana Uni Lib, Preservation Dept 1984] – 40p on 1r – 1 – us Indiana Preservation [390]

Fabled tribe / Cowley, Clive – New York, NY. 1968 – 1r – us UF Libraries [420]

Fables / Lemay, Pamphile – Montreal: Librairie Granger, 1903 [mf ed 1997] – 2mf – 9 – 0-665-85792-6 – mf#85792 – cn Canadiana [810]

Les fables de faerne / Faerno, G – Amsterdam: Chez Gerard Onder de Linden, 1718 – 3mf – 9 – mf#O-1934 – ne IDC [090]

Les fables de la fontaine / Bourassa, Gustave – Montreal: C O Beauchemin, 1899 – 1mf – 9 – mf#03886 – cn Canadiana [390]

Fables in slang / Ade, George – Chicago, IL. 1899 – 1r – us UF Libraries [390]

Fables of infidelity and facts of faith : being an examination of the evidences of infidelity / Patterson, Robert – rev enl. Cincinnati: Western Tract Society, 1875 [mf ed 1991] – 2mf – 9 – 0-7905-8872-2 – (1st printed 1859) – mf#1989-2097 – us ATLA [230]

Fables of the veld / Posselt, Friedrich Wilhelm Traugott – Oxford, England. 1929 – 1r – us UF Libraries [390]

Fabliaux or tales / Le Grand d'Aussy, Pierre Jean Baptiste – London: Printed for J Rodwell 1815 [mf ed Bloomington IN: Indiana Uni Lib, Preservation Dept 1984] – 3v on 1r [ill] – 1 – us Indiana Preservation [390]

Fabo de Maria *see* San agustin de joven

Fabo, Pedro *see* Liberalades de una revolucion

Fabr, Johann Ernst *see* Neue geographisches magazin

Fabra, Pompeu *see*
- Gramatica de la lengua catalana
- Grammaire catalane

Fabre, Abel *see* Exercices orthographiques

Fabre, Auguste *see* La revolution de 1830

Fabre de Parrel, R *see* Observations sur les lois de naturalisation des etrangers en algerie

Fabre D'eglantine, P-F-N (Philippe Francois Naz *see* Philinte de moliere

Fabre D'eglantine, P-F-N (Philippe Francois Naz... *see*
- Precepteurs

Fabre D'eglantine, P-F-N (Philippe-Francois-Naz *see* Philinte de moliere

Fabre, Edouard Charles *see* Nous aimerions a voir le clerge, les communautes religieuses

Fabre, F *see* Le college anglais de douai

Fabre, Hector *see*
- Chroniques
- Confederation, independence, annexion

Fabre, Jean *see* Histoire secrete du directoire

Fabre, Jean Henri *see*
- The glow-worm
- The life of the fly
- The mason-wasps

Fabre, Jean-Henri *see* Ha-instinkt mahu?

Fabrega, Demetrio *see*
- Cuerpo amoroso
- Libro de la mal sentada

Fabrega P, Jorge *see* Enriquecimiento sin causa

Fabregas, Juan P *see* Los factores economicos de la revolucion espanola

Fabres y Fernandez, Jose Clement *see* Obras completas de don jose clemente fabres

Fabretti, A, Rossi, F and Lanzone, R *see* Catalogo generale dei musei di antichita...regio museo di torino

Fabri, Felix *see*
- Fratris felicis fabri evagatorium in terrae sanctae, arabiae et egypti peregrinationem
- Fratris felicis fabri tractatus de civitate ulmensi, de eius origine, ordine, regimine, de civibus eius et statu

Fabri, Fratris Felicis *see* Evagatorium in terrae sanctae, arabiae et egypti peregrinationem

Fabri, Friedrich *see* Ein dunkler punkt

Fabri, Joh Ernst *see* Beitraege zur geographie, geschichte und statenkunde

Fabri, Joh Ernst et al *see* Historisch-geographisches journal

Fabri, Johann Ernst *see*
- Geographisches lesebuch zum nutzen und vergnuegen
- Geographisches magazin
- Neues geographisches magazin

Fabri, Joseph *see* Les belges au guatemala, 1840-1845

Fabric of terror / Teixeira, Bernardo – New York, NY. 1965 – 1r – us UF Libraries [400]

Fabrica de corporis humani, libri septem / Vesalius, A – Brasilea, 1555 – 15mf – 9 – sp Cultura [611]

Fabricii, P *see* Delle allusioni, imprese, et emblemi del sig. principio fabricii...

Fabricio Diaz, Francisco *see* Ecos de amor y dolor

Fabricius, Georg *see* Syn theo kai no m georg andreae fabrici poetae l caes gymnasiae mulhusini gymnasium ethicum, viro bono cuilibet vitae praeparando quindecim excercitationibus instructum et nobilium atque ingenuorum juvenum

Fabricius, J A *see*
- Opera graece et latine
- Sexti empirici
- Sexti empirici opera graece et latine...

Fabricius, Johann *see* Voyage en norwege [norvege]

Fabricius, Just Friedrich Erdmann *see* Vermischte gedichte

Fabricius, O *see* Udforlig beskrivelse over de gronlandske saele 1/2

Fabricius, Otho *see* Forsog til en forbedret gronlandsk grammatica

Fabricius, Wilhelm *see* Joh georg schoch's comoedia vom studentenleben

Fabricus, Cajus *see* Concordance to gregory of nyssa

Fabrig, Peter *see* Festschrift anlaesslich der emeritierung von prof. dr.-ing. walter raab

Die fabrik zu niederbronn : schauspiel in 5 aufzuegen / Wichert, Ernst – Leipzig: P Reclam, [1874?] – 1r – 1 – us Wisconsin U Libr [820]

Fabrique, Andrew Hinsdale *see* Medical records

Fabrique de Notre Dame (Montreal, Quebec) *see* Replique des marguilliers de notre-dame de montreal

Fabrotus, C *see*
- Breviarium historicum
- Historia ecclesiastica sive chronographia tripertita

Fabrotus, C A *see*
- De vita et honestate clericorum
- Historia
- Historia de vitis romanorum pontificum

Fabry, Jean *see*
– Itineraire de buonaparte, de l'ile d'elbe a l'ile sainte-helene
– Les missionnaires de 93

Fabula del tiburon y las sardinas / Arevalo, Juan Jose – Habana, Cuba. 1960 – 1r – us UF Libraries [972]

Fabulae centum ex antiquis auctoribus delectae, carminibusque explicatae / Faerno, G – Patavii: Excudebat Josephus Cominus, 1718 – 2mf – 9 – mf#O-1855 – ne IDC [090]

Fabulas / Zuniga, Luis Andres – Tegucigalpa, Mexico. 1931 – 1r – us UF Libraries [972]

Fabulas de salon y poesias / Sanchez Arjona y Sanchez Arjona, Francisco – 1880 – 9 – sp Bibl Santa Ana [830]

Fabulas dominicanas / Rodriguez Demorizi, Emilio – Ciudad Trujillo, Dominican Republic. 1946 – 1r – us UF Libraries [972]

Fabulas e alegorias / Cearense, Catullo Da Paixao – Rio de Janeiro, Brazil. 1946 – 1r – us UF Libraries [390]

Fabulas morales satiricas y... / Doncel y Ordaz, Jose – 1895 – 9 – sp Bibl Santa Ana [830]

Fabulista / Llopis, Regelio – Habana, Cuba. 1963 – 1r – us UF Libraries [972]

The fabulous gods denounced in the bible = de diis syris syntagmata 2 / Selden, John – Philadelphia: J B Lippincott, 1880 – 1mf – 9 – 0-7905-0959-8 – mf#1987-0959 – us ATLA [220]

Fabulous monster / Chamier, Jacques Daniel – New York: Longmans, Green & Co 1934 [mf ed 1987] – 1r – 1 – mf#9470 – us Wisconsin U Libr [943]

Facal, Angel *see* Villa de la union

Face a face / Lumanyisha, Dikonda Wa – Bruxelles, Belgium. 1964 – 1r – us UF Libraries [960]

Face a face : ou, luttes mentales d'un catholique roman / Beaudry, Louis Napoleon – Montreal: L E Rivard, 1882 – 3mf – 9 – mf#03521 – cn Canadiana [241]

Face a la delinquance juvenile latente / Bistoury, Andre F – Port-Au-Prince, Haiti. 1960 – 1r – us UF Libraries [972]

Face a main – Brussels Belgium, 16 sep 1944-9 jun 1945 – 1/2r – 1 – uk British Libr Newspaper [074]

The face and the mask / Barr, Robert – New York: F Stokes, c1895 – 3mf – 9 – (ill by a hencke) – mf#03345 – cn Canadiana [830]

Face au peuple et a l'histoire / Duvalier, Francois – Port-Au-Prince, Haiti. 1961 – 1r – us UF Libraries [972]

Face aux realites, la direction des finances francaises sous l'occupation / Cathala, Pierre Adolphe Juste – Paris: Editions du Triolet 1948 [mf ed 1978] – (= ser Pensee libre) – 1r – 1 – mf#20 – us Wisconsin U Libr [332]

La face de l'eglise primitive / Martin, G – Paris, 1656 – 5mf – 9 – mf#CA-139 – ne IDC [240]

A face illumined / Roe, Edward Payson – Toronto: J Campbell & Son, [1878?] [mf ed 1994] – 8mf – 9 – 0-665-94711-9 – mf#94711 – cn Canadiana [830]

The face of china : travels in east, north, central and western china / Kemp, Emily Georgiana – London: Chatto & Windus, 1909 [mf ed 1995] – (= ser Yale coll) – xv/275p (ill) – 1 – 0-524-09272-9 – mf#1995-0272 – us ATLA [915]

The face of china : travels in east, north, central and western china... / Kemp, Emily Georgiana – Toronto: Musson, 1909 [mf ed 1996] – 5mf – 9 – 0-665-80954-9 – mf#80954 – cn Canadiana [915]

The face of mother india / Mayo, Katherine – London: Hamish Hamilton Ltd, [193-] – (= ser Samp: indian books) – us CRL [915]

El facedor de un entuerto...agravios / Hurtado, Antonio – 1869 – 9 – sp Bibl Santa Ana [830]

Facetas de marti / Quesada Y Miranda, Gonzalo De – Habana, Cuba. 1939 – 1r – us UF Libraries [972]

Facey, James William *see* Elementary decoration...of dwelling-houses

Fachbereich Politische Wissenschaft der Freien Universitaet Berlin, Pressearchiv des Bibliotheks- und Informationssystems *see* Pressearchiv zur geschichte deutschlands sowie zur internationalen politik von 1949-60

Fachenetti, Cesar *see*
– Auto del nuncio contra una asserta sentencia impressa...
– Puntos de la sentencia que dio el senor nuncio de su santidad contra los tres diputados del capitulo privado de la orden de san jeronimo en tres de abril de 1642
– Le saint du peuple antonie de padoue
– Sentencia dada por...nuncio...en el pleyto que ha tratado el...padre general de la orden de san geronimo y otros religiosos del monasterio de yuste con el prior de...
– Sentencia traducido del latin en romance que dio el nuncio en las controversias que han pasado entre el capitulo difinitorio de la orden de san geronimo y el padre general caceres della
– Sententia...nuntii apstolici...
– Tres diputados del capitulo privado de la san geronimo

Facheuse aventure / Temiriazev, B – Paris, France. 1946 – 1r – us UF Libraries [440]

Fachfunk *see* Das deutsche funkprogramm

Fachgruppe deutsch-geschichte. interpretationen moderner lyrik : anlaesslich der germanistenverbandstagung in nuernberg / ed by Bayrischer Philologenverband – Frankfurt am Main: M Diesterweg, 1956 – 112p – 1 – (incl bibl ref) – us Wisconsin U Libr [430]

Fachinger, Josef *see* Das staatliche aufsichtsrecht ueber gemeinden und gemeindeverbaende in rechtsvergleichender darstellung

Fach-zeitung des bundes deutscher koche in grossbrittannien und irland – 1 Sept 1902-1 Oct 1903 – 1 – uk British Libr Newspaper [072]

Fachzeitung fuer den colportage-buchhandel – Berlin DE, 1890 [mpf], 1891-1910 – 2r – 1 – gw Misc Inst [070]

Fachzeitung fuer schneider – Verband der Schneider.... v11, no.52-53; v20, no.24. 1898-1907. (Serial publications of German trade unions in the Memorial Library, University of Wisconsin-Madison.) – 1 – us Wisconsin U Libr [330]

Faci, Roque Alberto *see* Novena de santa lucia la siracusana, virgen y martir soberana, que para aumento de su devocion y lograr su patrocinio en los ojos del alma y cuerpo

Facial paintings of the indians of northern british columbia / Boas, Franz – S.l: s.n, 1898? – 1mf – 9 – mf#02417 – cn Canadiana [390]

Facilitators and learners : co-creating a better understanding of one another / Jager, Esme de – Uni of South Africa 2001 [mf ed Johannesburg 2001] – 3mf – 9 – (incl bibl ref) – mf#mfm14683 – sa Unisa [370]

Facilities design and management – New York. 1984+ (1,5,9) – ISSN: 0279-4438 – mf#14268 – us UMI ProQuest [650]

Facing forward : organ of the martin luther king, jr alumni assn / Martin Luther King, Jr Alumni Association – 1978 spring – 1r – 1 – mf#4876466 – us WHS [320]

Facing Reality Publishing Committee *see* Speak out

Facing the crisis / Eddy, Sherwood – New York, NY. 1922 – 1r – us UF Libraries [025]

Facing the situation : addresses. delivered at the fourth general convention of the laymen's missionary movement, presbyterian church in the u.s... – Athens, Ga.: Laymen's Missionary Movement, Presbyterian Church in the United States, [1915?] – 4mf – 9 – 0-524-07439-9 – mf#1991-3099 – us ATLA [242]

Facio, Justo A *see* Cultura literaria

Facio, Rodrigo *see*
– Moneda y la banca central en costa rica
– Planification economica en regimen dem...

Facius, Friedrich *see* Deutscher industrie- und handelstag / reichswirtschaftskammer [bestand r 11] bd 12

Die fackel – Chicago, IL: Socialist Pub Society, apr 8 1883-jan 11 1891 – 1 – us CRL [071]

Die fackel – Leipzig DE, 1877 22 jun-1878 29 sep – 1r – 1 – gw Misc Inst [074]

Die fackel – Muelhausen / Elsass (Mulhouse F), 1922-1923 apr [gaps] – 1 – fr ACRPP [074]

Die fackel – St Petersburg, Russia 1917 [mf ed Norman Ross] – 1r – 1 – mf#nrp-1696 – us UMI ProQuest [077]

Die fackel – Temeschburg (Timisoara RO), 1924-25, 1929 – 1 – gw Misc Inst [077]

Die fackel – Temeschburg (Timisoara RO), 1924-25, 1929 – 1 – gw Misc Inst [077]

Die fackel *see* Reichs-geldmonopol

Fackel – Chicago IL. 1903 jan 4-1904 jan 17-1919 aug 12-oct 12 – 20r – 1 – mf#855347 – us WHS [071]

Fackert, Juergen *see* Hugo von hofmannsthals nachgelassenes lustspielfragment "die rhetorenschule" oder "timon der redner"

Facklan – Chicago: Scandinavian Socialist Federation, 1921-22. jul 22 1921 – us CRL [071]

Facklan – Chicago IL. 1921 jul 29-1922 feb 24 – 1r – 1 – (cont: svenska socialisten; cont by: folket; ny tid) – mf#846338 – us WHS [071]

Facktz, P N *see* Canada and the united states compared

Faclia – Cluj, Romania. 1962-70; Jan 1971-75; 1976-89 – 34r – 1 – us L of C Photodup [949]

Faco, Rui *see*
– Brasil siglo 20
– Cangaceiros e fanaticos

Facsimile del acta de independencia de centro amer... – Guatemala, 1948 – 1r – us UF Libraries [972]

Facsimile of manuscripts in european archives relating to america, 1773-1783 / Stevens, B F – v. 1-25. 1889-98 – 1 – us AMS Press [975]

Fac-similes of certain portions of the gospel of st. matthew and of the epistles of ss. james and jude : written on papyrus in the first century, and preserved in the egyptian museum of joseph mayer, esq., liverpool... / ed by Simonides, Konstantinos – London: Truebner, 1861 – 1r – 1 – 0-524-02757-9 – mf#1987-B007 – us ATLA [090]

Facsimiles of egyptian hieratic papyri in the british museum / Budge, Ernest Alfred Wallis – London, 1910 – 6mf – 9 – mf#NE-458 – ne IDC [930]

Facsimiles of horae de b m v 11th century (hbs21) / Dewick, E S – 1902 – (= ser Henry bradshaw society (hbs)) – 4mf – 8 – €11.00 – ne Slangenburg [241]

Facsimiles of the athos fragments of codex h of the pauline epistles / Lake, Kirsopp – Oxford: Clarendon Press 1905 [mf ed 1986] – 1mf – 9 – 0-8370-9485-2 – mf#1986-3485 – us ATLA [090]

Facsimiles of the creeds from early manuscripts (hbs36) / Burn, A U – 1909 – (= ser Henry bradshaw society (hbs)) – 4mf – 8 – €11.00 – ne Slangenburg [220]

Fac-similes of the miniatures and ornaments of anglo-saxon and irish manuscripts / Westwood, John Obadiah – London 1868 – (= ser 19th c art & architecture) – 17mf – 9 – mf#4.2.1730 – uk Chadwyck [740]

Fact / Democratic Party [US] – [v1 n3?]-v3 n13 [1970 jul 2-1972 jun 23] – 1r – 1 – mf#1111465 – us WHS [325]

Fact – New York. 1964-1967 [1,5,9] – 1,5,9 – ISSN: 0429-9825 – mf#1808 – us UMI ProQuest [073]

Fact and fiction / Posselt, Friedrich Wilhelm Traugott – Bulawayo, Zimbabwe. 1935 – 1r – us UF Libraries [960]

The fact divine : an historical study of the christian revelation and of the catholic church / Broeckaert, Joseph – Portland, ME: McGowan & Young, 1885 – 1mf – 9 – 0-8370-7203-4 – (incl bibl ref) – mf#1986-1203 – us ATLA [230]

The fact of christ : a series of lectures / Simpson, Patrick Carnegie – [2nd ed]. NY: Fleming H Revell [1901?] [mf ed 1985] – 1mf – 9 – 0-8370-5386-2 – (incl bibl ref) – mf#1985-3386 – us ATLA [240]

The fact of conversion / Jackson, George – New York: FH Revell, c1908 – (= ser Cole lectures) – 1mf – 9 – 0-7905-7518-3 – (incl bibl ref) – mf#1989-0743 – us ATLA [200]

Le facteur – Port-au-Prince: [s.n.], 1902. [1ere annee, n1-n43. 6 aout-18 nov 1902] – 3 sheets – 9 – us CRL [079]

Faction defeated : or, the political crisis in victoria / Democritus [pseud] – Melbourne, 1861 – (= ser 19th c books on british colonization) – 1mf – 9 – mf#1.1.7007 – uk Chadwyck [971]

A factor analysis of selected badminton skills tests for college students / Louie, L H – 1990 – 2mf – 9 – $8.00 – us Kinesology [790]

Los factores economicos de la revolucion espanola / Fabregas, Juan P – Barcelona, 1937. Fiche W 866. (Blodgett Collection of Spanish Civil War Pamphlets) – 9 – us Harvard [946]

Factores humanos de la cubanidad / Ortiz, Fernando – Habana, Cuba. 1940 – 1r – us UF Libraries [972]

Factorial validity of a teacher effectiveness scale for the teacher preparation program in hong kong / Ho, Winnie WY – Springfield College, 1995 – 2mf – 9 – $8.00 – mf#PE3597 – us Kinesology [370]

Factors affecting attendance in the national hockey league : a multiple regression model / Wiedeke, Jennifer – 1999 – 1mf – 9 – $4.00 – mf#PE 3947 – us Kinesology [650]

Factors affecting composition of everglades grasses and legumes : with special reference to proteins / Neller, J R – Gainesville, FL. 1944 – 1r – us UF Libraries [630]

Factors affecting cucumber yields, costs, and profits / Rochester, Morgan Columbus – s.l, s.l? 1933 – 1r – us UF Libraries [630]

Factors affecting easter lily flower production in florida / Shippy, William B – Gainesville, FL. 1937 – 1r – us UF Libraries [630]

Factors affecting farm profits in the williston area / Turlington, J E – Gainesville, FL. 1925 – 1r – us UF Libraries [630]

Factors affecting farming returns in jackson county, florida / Brunk, Max E – Gainesville, FL. 1942 – 1r – us UF Libraries [630]

Factors affecting risk perception about drinking water and response to public notification / Anadu, Edith C – 1997 – 2mf – 9 – $8.00 – mf#HE 586 – us Kinesology [614]

Factors affecting the performance of intramural officials in competitive situations / Blumenthal, Harvey – 1989 – 105p 2mf – 9 – $8.00 – us Kinesology [790]

Factors associated with the recall of physical activity / Cunningham, Lynda F & Ainsworth, Barbara E – 1992 – 2mf – $8.00 – us Kinesology [150]

Factors contributing to withdrawal behaviour in early adolescents / Singh, Nandkissor – Uni of South Africa 2000 [mf ed Johannesburg 2000] – 4mf – 9 – mf#mfm14925 – sa Unisa [150]

Factors indiana high school athletic directors consider when hiring a boys' varsity basketball coach / Mehaffey, Chip A – 1999 – 1mf – 9 – $4.00 – mf#PE 4065 – us Kinesology [790]

Factors influencing african-american students at historically black colleges and universities to enter the adapted physical education profession / Webb, Daniel – University of Wisconsin-La Crosse, 1995 – 1mf – 9 – $4.00 – mf#PSY1870 – us Kinesology [150]

Factors influencing family use of health care services in tamil nadu (india) villages / Harding, AKW – 1990 – 3mf – 9 – $12.00 – us Kinesology [613]

Factors influencing gait transition in adolescents / Tseh, Wayland – 2000 – 1mf – 9 – $4.00 – mf#PSY 2120 – us Kinesology [612]

Factors of faith in immortality / Denney, James – London: Hodder and Stoughton, [1910?] – (= ser Little books on religion) – 1mf – 9 – 0-7905-0756-0 – mf#1987-0756 – us ATLA [240]

The factors of organic evolution / Spencer, Herbert – [London] 1887 – (= ser 19th c evolution & creation) – 2mf – 9 – mf#1.1.11608 – uk Chadwyck [574]

Factors related to fasting behavior among american adults / Cho, Ho S – Temple University, 1995 – 1mf – 9 – $4.00 – mf#PSY1840 – us Kinesology [150]

Factors related to hunting and fishing participation in the united states / Duda, Mark Damian – Harrisonburg VA: Responsive Management, Western Assoc of Fish & Wildlife Agencies, 1993-95 [i.e. 1996] [mf ed 1994-97] – 5v on 9mf – 9 – (incl bibl ref) – us GPO [639]

Factors related to obesity among female african american university students / Wagner, Sally B – 1997 – 2mf – 9 – $8.00 – mf#PH 1584 – us Kinesology [612]

Factors related to stability in the education profession : with specific reference to mooifontein circuit in north west province / Matseke, Livian Molefe – Pretoria: Vista University 2000 [mf ed 2000] – 3mf – 9 – (incl bibl) – mf#mfm15166 – sa Unisa [370]

The factors that division 1a football players at ball state university considered most important when deciding which university to attend during the recruiting process / Baldwin, Brent T – 1999 – 1mf – 9 – $4.00 – mf#PE 3916 – us Kinesology [790]

The factors that head football coaches at ncaa division 1a universities use to evaluate a potential athlete during the recruiting process / Baldwin, Brent T – 1999 – 1mf – 9 – $4.00 – mf#PE 3896 – us Kinesology [790]

Factors that influence division 2 recruited female intercollegiate soccer student-athletes in selecting their university of their choice / Baumgartner, Amy – 1999 – 70p on 1mf – 9 – $5.00 – mf#PE 4095 – us Kinesology [150]

Factors which influence patient satisfaction with sexuality education / Young, Elaine W – 1982 – 3mf – 9 – $12.00 – us Kinesology [613]

Factory – New York. 1968-1976 (1) 1970-1976 (5) (9) – (cont by: factory management) – mf#5330 – us UMI ProQuest [338]

Factory labour in india / Mukhtar, Ahmad – Madras: Annamalai University: Methodist Pub House, 1930 – (= ser Samp: indian books) – (int note by se runganadhan) – us CRL [331]

Factory management – New York. 1977-1977 (1) 1977-1977 (5) 1977-1977 (9) – (cont: factory) – ISSN: 0146-3314 – mf#5330,01 – us UMI ProQuest [338]

Facts – Redlands, CA. 1891-1893 (1) – mf#62251 – us UMI ProQuest [071]

Facts / Republican Party [WI] – 1959 jan 16-1960 feb 13 – 1r – 1 – (cont by: facts about your government) – mf#3564693 – us WHS [325]

Facts – Seattle, WA. 1965-1978 (1) – mf#67104 – us UMI ProQuest [071]

Facts – Seattle WA. 1993 may 5-dec 29, 1994 jan 5-1994 dec 7, 1995 apr 5, 19, jun 14, 1997 jan 15, jun 25, dec 3 – 3r – 1 – mf#851063 – us WHS [071]

Facts! : respecting the scheme for the annexation of london south to the city of london – [S.l: s.n, 1890?] [mf ed 1986] – 1mf – 9 – 0-665-55777-9 – mf#55777 – cn Canadiana [350]

Facts about bankruptcy you ought to know / Isaac, Max – New York, American Bankruptcy Review, 1927. 347 p. LL-538 – 1 – us L of C Photodup [346]

Facts about cuba / Aldama, Miguel De – New York, NY. 1875 – 1r – us UF Libraries [972]

Facts about cuba / Junta Cubana De Nueva York – New York, NY. 1870 – 1r – us UF Libraries [972]
Facts about ireland : a curve-history of recent years / MacDowell, Alexander B – London: E. Stanford, 1888. 32p.illus – 1 – us Wisconsin U Libr [941]
The facts about sexuality = Die feite van seksualiteit / South Africa. Department of National Health and Population Development [Departement van Nasionale Gesondheid en Bevolkingsontwikkeling [Departement van Nasionale Gesondheid en Bevolkingsontwikkeling] – Pretoria: The Dept: Die Dept 1987 [mf ed Pretoria, RSA: State Library [199-]] – 34p [ill] on 1r with other items – 5 – (lacks: p25-27 [english sect]; in english & afrikaans; incl bibl ref; filmed in cooperation with the advisory committee on health education) – mf#op 08462 r24 – us CRL [613]
Facts about spain to be used by speakers / North American Committee to Aid Spanish Democracy – New York City 193-?] [mf ed 1977] – (= ser Blodgett coll) – 1mf – 9 – mf#w1082 – us Harvard [946]
Facts about the iron ore deposits of british columbia : including vancouver island – [Victoria BC?: s.n, 1918?] [mf ed 1995] – 1mf – 9 – 0-665-76873-7 – mf#76873 – cn Canadiana [550]
Facts about your government / Republican Party [WI] – 1960 feb 20-1967 dec 16 – 1r – 1 – (cont: facts [madison, wi]) – mf#1056122 – us WHS [325]
Facts and arguments for darwin... / Mueller, Johann Friedrich Theodore – London, 1869 – (= ser 19th c evolution & creation) – 2mf – 9 – (with additions by aut. trans fr german by w s dallas) – mf#1.1.1519 – uk Chadwyck [575]
Facts and assertions / Faber, George Stanley – London, England. 1835 – 1r – us UF Libraries [240]
Facts and comments / Grain Services Union [CLC] – 1983 jun-1984 apr – 1r – 1 – (cont by: gsu news) – mf#927428 – us WHS [331]
Facts and fancies : being studies in popular problems / Gour, Hari Singh – Saugor: Saugor Book Depot, 1948 – (= ser Samp: indian books) – us CRL [301]
Facts and fancies about java / Wit, Augusta de – 2nd rev enl ed. The Hague: W P van Stockum 1900 [mf ed 1987] – 1r [ill] – 1 – (with: dara shukoh / qanungo, k) – mf#1823 – us Wisconsin U Libr [915]
Facts and fancies in modern science : studies of the relations of science to prevalent speculations and religious belief being the lectures... / Dawson, John William – Philadelphia:American Baptist Pub Soc, c1882 [mf ed 1985] – 1mf – 9 – 0-8370-2851-5 – mf#1985-0851 – us ATLA [210]
Facts and figures / Kennerly, Clarence Hickman – St Augustine, FL. 1911 – 1r – us UF Libraries [500]
Facts and findings / Frankfort Area Genealogy Society – v1 n4 [1976 winter], v3 n1-? [1978 jan/mar-1983] – 1r – 1 – mf#966270 – us WHS [929]
Facts and fossils adduced to prove the deluge of noah : and modify the transmutation system of darwin, with some notices regarding indus flint cores / Twemlow, George – London, [1868] – (= ser 19th c evolution & creation) – 3mf – 9 – mf#1.1.4280 – uk Chadwyck [560]
Facts and information about brazil / Brazil. Departamento de Imprensa e Propaganda – Rio de Janeiro, Brazil. 1942 – 1r – us UF Libraries [972]
Facts and observations on the culture of vines, olives, capers, alm / Chazotte, Peter Stephen – Philadelphia, PA. 1821 – 1r – us UF Libraries [634]
Facts and observations on the irish land question / Sproule, John – Dublin, 1870 – (= ser 19th c ireland) – 1mf – 9 – mf#1.1.1931 – uk Chadwyck [333]
Facts and observations respecting canada, and the united states of america : affording a comparative view of the inducements to emigration presented in those countries. to which is added an appendix of practical instructions... / Grece, Charles Frederick – London, 1819 – (= ser 19th c books on british colonization) – 2mf – 9 – mf#1.1.1048 – uk Chadwyck [970]
Facts and observations respecting canada, and the united states of america : affording a comparative view of the inducements to emigration presented in those countries; to which is added an appendix of practical instructions to emigrant settlers in the british colonies / Grece, Charles – London 1819 [mf ed Hildesheim 1995-98] – 1v on 2mf – 9 – €60.00 – 3-487-27103-6 – gw Olms [917]
Facts and reflections bearing on annexation, independence and imperial federation / Douglas, James – Ottawa: Mortimer, [18–] [mf ed 1980] – 1mf – 9 – 0-665-02745-1 – mf#02745 – cn Canadiana [320]
Facts and reports – Amsterdam: Angola Comite, [1979-]. [v1-5 n25 nov 1970-dec 13 1975] – us CRL [074]
The facts and the faith : a study in the rationalism of the apostles' creed / Warner, Beverley Ellison – New York: Thomas Whittaker, c1897 – 1mf – 9 – 0-8370-5697-7 – (incl bibl ref and index) – mf#1985-3697 – us ATLA [210]
Facts and trends – 1955-72. (Education Division Newsletter. 1955-56; SSB Newsletter. 1957-65; Facts and Trends. 1966-71) – 1 – 55.09 – us Southern Baptist [242]
Facts are : a guide to falsehood and propaganda / Seldes, George – New York, NY. 1942 – 1r – us UF Libraries [025]
Facts for baptist churches / Foss, A T & Mathews, E – 1850 – 1 – us Southern Baptist [242]
Facts for behaists / ed by Kheiralla, Ibrahim George – Chicago: IG Kheiralla, 1901 [mf ed 1991] – 1mf – 9 – 0-524-01567-8 – (trans by ed) – mf#1990-2521 – us ATLA [290]
Facts for businessmen interested in establishing... / Puerto Rico Development Company, San Juan – San Juan, Puerto Rico. 1948 – 1r – us UF Libraries [650]
Facts for faith – Glendora, 2000+ [1,5,9] – ISSN: 1534-7176 – mf#31907 – us UMI ProQuest [210]
Facts for the electors : the provincial finances as administered by the joly and chapleau governments – [s.l: s.n, 1881?] [mf ed 1983] – 1mf – 9 – 0-665-44224-6 – mf#44224 – cn Canadiana [336]
Facts for the farmers – Toronto?: Industrial League, 1887? – 1mf – 9 – mf#06742 – cn Canadiana [380]
Facts, illustrative of the treatment of napoleon buonaparte in saint helena : being the result of minute inquiries and personal research in that island / Hook, Theodore Edward – London 1819 [mf ed Hildesheim 1995-98] – 1v on 1mf – 9 – €40.00 – ISBN-10: 3-487-26368-8 – ISBN-13: 978-3-487-26368-7 – gw Olms [920]
Facts, not falsehoods / Lockhart, Lawrence – Edinburgh, Scotland. 1845 – 1r – us UF Libraries [240]
Facts of economic life / West Bend Co – n1-9 [i.e. 10] [1977 spring-1979 summer], n1010 [i.e. 11] – 1r – 1 – mf#672542 – us WHS [330]
The facts of life in relation to faith / Simpson, Patrick Carnegie – London; New York: Hodder and Stoughton, 1913 – 1mf – 9 – 0-7905-9660-1 – (incl bibl ref) – mf#1989-1385 – us ATLA [240]
Facts on file 5-year indexes – 1946-90 – 9 – $230.00 – (1986-90. $90.00) – us Facts [020]
Facts on file yearbooks – 1941-91. Annual – 9 – $1,595.00 – (1941-49. $338.00; 1950-59. $375.00; 1960-69. $375.00; 1970-79. $375.00; 1980-89. $375.00; 1990-91. $90.00; 1992-93. $90.00) – us Facts [030]
Facts, statements, and explanations : connected with the publication of the second volume of the tenth edition of horne's introduction to the study of the holy scriptures, entitled "the text of the old testament considered", etc etc / Davidson, Samuel – London: Longman, Brown, Green, Longmans, & Roberts, 1857 – 1mf – 9 – 0-7905-3369-3 – mf#1987-3369 – us ATLA [220]
Facts to know florida / Mullen, John M – Jacksonville, FL. 1938 – 1r – us UF Libraries [978]
Factsheet five – 1982 may-1987 nov – 1r – 1 – mf#1231392 – us WHS [071]
Factum : pour les directeurs et associez de la compagnie de la nouvelle france, demandeurs et complaignants... / Compagnie des Cent-Associes – [s.l: s.n, 1647?] [mf ed 1993] – 1mf – 9 – 0-665-92966-8 – mf#92966 – cn Canadiana [338]
Factum du proces entre jean de biencourt, sr de poutrincourt et les peres biard et masse, jesuites / Biencourt de Poutrincourt et de Saint-Just, Jean de, Baron – Paris: Maisonneuve et C Leclerc, 1887 – 2mf – 9 – (int by gabriel marcel) – mf#03585 – cn Canadiana [971]
Factum pour mademoiselle petit : danseuse de l'opera, revoquee complaignante au public / Petit, Marie-Antoinette – [Paris? 1740?] – 1 – mf#*ZBD-*MGO pv30 – Located: NYPL – us Misc Inst [790]
Facula veritatis / Pennotto, Gabr – Colonia Agrippina, 1644 – 32mf – 8 – €61.00 – ne Slangenburg [240]
Faculdade do recife : jornal academico – Recife, PE: Typ de Freitas Irmaos, 15 maio-30 ago 1863 – (= ser Ps 19) – mf#P16,01,15 – bl Biblioteca [321]
Facultad de filosofia y letras. universidad de buenos aires. publicaciones historicas / Bayle, Constantino – Madrid: Razon y Fe, 1925 – 1 – sp Bibl Santa Ana [100]
Faculte libre de theologie protestante de Montauban *see* Seance publique de rentree
Faculties and difficulties for belief and disbelief / Paget, Francis – London: Rivingtons, 1887 – 1mf – 9 – 0-7905-9047-6 – mf#1989-2272 – us ATLA [240]
The faculty of laws, and the idea of law / Hopkinson, Alfred – Manchester: Cornish, 1875. 20p. LL-374 – 1 – us L of C Photodup [340]
Faculty of medicine, 1871 : degree of m b / University of Toronto. Faculty of Medicine – [Toronto?: s.n, 1871?] [mf ed 1984] – 1mf – 9 – 0-665-01587-9 – mf#01587 – cn Canadiana [378]
Faculty-focus / Smithtown Teachers Association – v5 n4-17 [1979 oct 17-1980 jun 11] – 1r – 1 – (cont: scta faculty focus) – mf#647939 – us WHS [370]
Fadali, Muhammad ibn al-Shafii *see* Muhammedanische glaubenslehre
Faddegon, Barend *see*
– Studies on the samaveda
– The vaidcesika-system
Faded myths / Peake, Arthur Samuel – London: Hodder & Stoughton, 1908 [mf ed 1989] – (= ser Aids to the devotional study of scripture 3) – 1mf – 9 – 0-7905-1777-9 – mf#1987-1777 – us ATLA [221]
Faden, William *see* A short topographical description of his majesty's province of upper canada, in north america
Faderneslandet – Stockholm, Sweden. 1830-1954 – 89r – 1 – sw Kungliga [078]
Fadimata, la princesse du desert : suivi du drame de deguembere / Ouane, Ibrahim Manadou – [Avignon]: Presses universelles, [1955] – 1 – us CRL [820]
Fadjar – Jakarta, Indonesia. 1966-1969 (1) – mf#68468 – us UMI ProQuest [079]
Fadjar *see* Pp muhammadijah madjlis taman pustaka
Fadon Sanchez, Antonio *see*
– Ligeras e insignificantes observaciones sobre algunos puntos cuestionables y dudosos en la historia de merida
– Sucinta...manicomio del carmen de merida
Faechererweiterungen symmetrischer moduale und homotopiemengen von produktabbildungen auf sphaeren / Endres, Norbert – (mf ed 2001) – 158p 2mf – 9 – €40.00 – 3-8267-2770-3 – mf#DHS 2770 – gw Frankfurter [510]
Faederneslandet – Stockholm, 1849 – 1r – 1 – sw Kungliga [078]
Faedrelandet – Copenhagen, Denmark. 25 sep 1914-1915; 24 jan, feb-5 mar 1916; jul-dec 1917; 1918-3 aug 1919 – 10 1/4r – 1 – uk British Libr Newspaper [074]
Faedrelandet – Copenhagen, Denmark. -d. 1875-81; 18 jun-14 oct 1944; 3 nov 1944-4 may 1945 – 19r – 1 – uk British Libr Newspaper [072]
Das faehnlein der sieben aufrechten : erzaehlung / Keller, Gottfried – Wiesbaden: Verlag des Volksbildungsvereins zu Wiesbaden 1913 [mf ed 1990] – (= ser Wiesbadener volksbuecher 16) – 1r – 1 – (int by m cornicelius. filmed with: on the eve / leopold kampf) – mf#2752p – us Wisconsin U Libr [830]
Faehnrich charlotte : geschichte einer liebe / Gerstner, Hermann – 9. aufl. Muenchen: Zentralverlag der NSDAP, F Eher 1944 [mf ed 1990] – 1r – 1 – (filmed with: die regulatoren in arkansas / friedrich gerstacker) – mf#2609p – us Wisconsin U Libr [830]
Der faehrmann an der weichsel : zwei erzaehlungen / Planner-Petelin, Rose – Berlin: Furche-Verlag 1941 [mf ed 1992] – 1r – 1 – (filmed with: das heilige band & other titles) – mf#3072p – us Wisconsin U Libr [830]
Die faelschungen erzbischof lanfranks von canterbury / Boehmer, Heinrich – Leipzig: Dieterich, 1902 – (= ser Studien Zur Geschichte Der Theologie Und Der Kirche) – 1mf – 9 – 0-7905-6282-0 – (incl bibl ref) – mf#1988-2282 – us ATLA [241]
Faems weer-galm der neder-duytsche poesie van cornelio de bie tot lyer... / Bie, C de – Mechelen: Jan Jaye, 1670 – 5mf – 9 – mf#O-147 – ne IDC [090]
Faena – Chimbote: Instituto de Promoción y Educacion Popular de Chimbote, 1981- (mf ed 1986) – 1r – 1 – mf#*ZAN-6163 1 (dic 1980) – us NY Public [073]
Faena intima / Joglar Cacho, Manuel – San Juan, Puerto Rico. 1955 – 1r – us UF Libraries [972]
Faerber, Rubin *see* Koenig salomon in der tradition
Faerber, Wilhelm *see* Katechismus fuer die katholischen pfarrschulen der vereinigten staaten
Faerber-zeitung – Berlin DE, 1902-09 – 4r – 1 – uk British Libr Newspaper [074]
Faerno, G *see*
– Cent fables
– Centum fabulae ex antiquis...
– Les fables de faerne
– Fabulae centum ex antiquis auctoribus delectae, carminibusque explicatae
Faesi, Robert *see*
– Gerhart hauptmanns "emanuel quint"
– Paul ernst und die neuklassischen bestrebungen im drama
– Das poetische zuerich
Faesser, Johann Chr *see* Geschichte der wiedertaeufer zu muenster
Faeulhammer, Adalbert *see* Franz grillparzer
Fafard, Francois-Xavier [comp] *see* Les cantons de la province de quebec
Fag rag – Boston, MA. 1983. OCLC 9353865 – 1 – us Wisconsin U Libr [360]
Fagan *see* Originaux, comedie en un acte et en prose
Fagan, Brian M *see* Short history of zambia
Fagan, G H *see* School rates in their religious and financial aspect
Fagan, George Hickson *see* Church endowments
Fagan, Henry Allan *see* Our responsibility
Fagan, Louis Alexander *see*
– A catalogue raisonne of the engraved works of william woollett
– Collectors' marks
– An easy walk through the british museum
– Handbook to the department of prints and drawings in the british museum
Fagan, Myron C (Myron Coureval) *see* Red treason on broadway
Fager, Karl E *see* A survey of division 1 athletic administrator's competency dimensions for administration and mentoring
Fagerstaposten – Hedemora, Sweden. 1963-78 – 57r – 1 – sw Kungliga [078]
Fagerstaposten – Sala, Sweden. 1943-62, 1979- – 39r – 1 – sw Kungliga [078]
Fagg, John Gerardus *see* Forty years in south china, the life of rev. john van nest talmage, d.d
Fagnani, Charles Prospero *see* A primer of hebrew
Fagniez, Gustave Charles *see* Etudes sur l'industrie et la classe industrielle a paris au 13e et au 14e siecle
Fahie, John Joseph *see* Memorials of galileo galilei, 1564-1642
Fahlu tidning – Falun, Sweden. 1822-28 – 3r – 1 – sw Kungliga [078]
Fahlu weckoblad – Falun, Sweden. 1786-1821 – 16r – 1 – sw Kungliga [078]
Fahndungsverzeichnis zu dem zentralpolizeiblatte, dem wiener, grazer und innsbrucker taeglichen fahndungsblatte / Austria. Bundespolizeidirektion, Vienna – 28 Feb 1947-15 Nov 1953. Scattered issues wanting – 1 – us L of C Photodup [360]
Die fahne : kampfblatt deutscher jugend in polen – Posen (Poznan PL), 1936 5 jan-20 sep – 1r – 1 – gw Misc Inst [934]
Die fahne der solidaritaet : deutsche schriftsteller in der spanischen freiheitsarmee, 1936-1939 / Weinert, Erich [comp] – Berlin: Aufbau-Verlag, 1953 – 1 – us Wisconsin U Libr [430]
Die fahne des kommunismus – Berlin. v. 1-4. 1927-30 – 1 – us NY Public [335]
Der fahrende schueler : eine dichtung / Wolff, Julius – Berlin: G Grote 1900 [mf ed 1995] – (= ser Grote'sche sammlung von werken zeitgenoessischer schriftsteller 68) – 1r – 1 – (filmed with: die papenheimer & other titles) – mf#3765p – us Wisconsin U Libr [810]
Fahrner, Rudolf *see*
– Hoelderlins begegnung mit goethe und schiller
– K ph. moritz' goetterlehre
– Wortsinn und wortschoepfung bei meister eckehart
Fahrt frei – Berlin DE, 1953 7 jul-1990 sep [gaps] – 16r – 1 – (deutsche reichsbahn, ministerium fuer verkehrswesen) – gw Misc Inst [380]
Fahrt frei : die wochenzeitung der deutschen eisenbahner – Berlin DE, 1951-1989 3 dec – 22r – 1 – gw Misc Inst [380]
Die fahrt nach der ahnfrau / Fechter, Paul – 1. Aufl d Feldausg. Guetersloh: C Bertelsmann, 1944 – 1r – 1 – us Wisconsin U Libr [880]
Die fahrt nach letztesand / Luserke, Martin – Wien: W Frick, 1943 – 1r – 1 – us Wisconsin U Libr [830]
Fahrtgenoss : monatsschrift fuer proletarisches wandern – Berlin DE, 1922 n1-1929 n6 – 1r – 1 – gw Misc Inst [790]
Fahs, Charles Harvey et al *see* The open door
Fa-hsien *see*
– Record of buddhistic kingdoms
– Record of the buddhistic kingdoms
– Travels of fah-hian and sung-yun, buddhist pilgrims, from china to india (400 a.d. and 518 a.d.)
Fa-hsi-ssu chu i chih ching chi chi ch'u / Einzig, Paul – Shang-hai: Li ming shu chu, Min kuo 25 [1936] – (= ser P-k&k period) – us CRL [945]
Fa-hsi-ssu chu i chih li lun t'i hsi / Ts'ai, Chih-hua – Shang-hai: Shang wu yin shu kuan, Min kuo 24 [1935] – (= ser P-k&k period) – us CRL [951]
Fa-hsi-ssu yun tung wen t'i / Wu, Yu-san – Shang-hai: Shang wu yin shu kuan, 1937 – (= ser P-k&k period) – us CRL [951]

Fahzi ribao = The legal daily – 1981- – 1 – (yrly reel count varies) – us UMI ProQuest [079]

Faidherbe, Louis Leon Cesar *see* Notice sur la colonie du senegal et sur les pays qui sont en relations avec elle

Faiershtein, M *see* Sovetskaia kooperatsiia

Faik, Mehmet *see* Suevarilere mahsus malumat ve terbiye-i askeriye

Faillite d'une democratie / Depestre, Edourard – Port-Au-Prince, Haiti. 1916 – 1r – us UF Libraries [972]

Faillon, Etienne Michel *see*
- L'exploit de dollard
- Memoires particuliers pour servir a l'histoire de l'eglise de l'amerique du nord
- Vie de m olier, fondateur du seminaire de s-sulpice

The failure experienced by the united states in their dealings with the "third world" : viewed in the light of cambodia's own experience, by norodom sihanouk; editorial / Norodom Sihanouk, Prince – [Phnom Penh? 1968?] [mf ed 1989] – 1r with other items – 1 – mf#mf-10289 seam reel 015/23 [§] – us CRL [327]

The failure of liberal christianity : and, some thoughts on the athanasian creed. two addresses / Burkitt, Francis Crawford – Cambridge: Bowes & Bowes, 1910 – 1mf – 9 – 0-7905-4252-8 – mf#1988-0252 – us ATLA [240]

The failure of the churches – London: Eveleigh Nash 1903 [mf ed 1985] – 1mf – 9 – 0-8370-2002-6 – mf#1985-0002 – us ATLA [240]

The failure of the "higher criticism" of the bible / Reich, Emil – Cincinnati: Jennings & Graham; New York: Eaton & Mains, c1905 – 1mf – 9 – 0-8370-5442-7 – (incl bibl ref) – mf#1985-3442 – us ATLA [220]

The failure of the rebel policy – Valencia, 193? Fiche W 868. (Blodgett Collection of Spanish Civil War Pamphlets) – 9 – us Harvard [946]

Fain, Agathon *see*
- Manuscrit de l'an trois
- Manuscrit de mil huit cent douze
- Manuscrit de mil huit cent quatorze
- Manuscrit de mil huit cent treize

Fainburg, Z *see* Vliianie emotsialnykh otnoshenii v seme na ee stabilizatsiiu

Faine, Jules *see* Philologie creole

Fair employment newsletter – CA. n13-14 [1963 may/jun-1963 jul/aug] – 1r – 1 – (cont by: fair practices news) – mf#625743 – us WHS [331]

The fair grit : or, the advantages of coalition: a farce / Davin, Nicholas Flood – Toronto: Belford, 1876 – 1mf – 9 – mf#24107 – cn Canadiana [790]

Fair immigration report / Federation for American Immigration Reform – v4 n11-v10 n4 [1984 aug-1990 apr] – 1r – 1 – (cont by: immigration report) – mf#1746073 – us WHS [323]

The fair journal – Cincinnati. Ohio. 1881 – 1 – us AJPC [071]

The fair maid of taunton : a tale of the siege / Alford, Elizabeth Mary – London: Samuel Tinsley & Co, 1878 – (= ser 19th c women writers) – 3mf – 9 – mf#5.1.114 – uk Chadwyck [830]

Fair measure / Southerners for Economic Justice – v1 n1, v2 n2, v4 n1-v8 n1 [1978 jul, 1979 jul, 1981 mar-1985 winter] – 1r – 1 – mf#1399720 – us WHS [323]

[Fair oaks-] fair oaks progress – CA. 1918-1931 – 2r – 1 – $120.00 – mf#C03218 – us Library Micro [071]

[Fair oaks-] san juan record – CA. 1933-35 (broken series); 1935-65 – 23r – 1 – $1380.00 – mf#BC02232 – us Library Micro [071]

[Fair oaks-] the fair oaks post – CA. 1984-1989 – 2r – 1 – $120.00 – mf#B05035 – us Library Micro [071]

[Fair oaks-] the gazette – CA. 1946-48 – 1r – 1 – $60.00 – mf#C02233 – us Library Micro [071]

Fair oaks/folsom – 1992- – (= ser California telephone directory coll) – 5r – 1 – $250.00 – mf#P00026 – us Library Micro [917]

Fair play – v1 n1-v2 n23/24=75/76 [1888 may 19-1890 aug 2/16] – 1r – 1 – (cont by: fair play [sioux city, ia]) – mf#1096552 – us WHS [071]

Fair play – v2 n25 [i.e. v3 n1]=77-v3 n3=79 [1891 jan-mar] – 1r – 1 – (cont: fair play [valley city, ks]) – mf#1096553 – us WHS [071]

Fair play – nov 1893-nov 1894 – 1r – 1 – mf#ZB 15 – nz Nat Libr [079]

"Fair play everyday" : a sportsmanship training program for high school coaches / Hansen, David E – 1999 – 2mf – 9 – $12.00 – mf#PE 4042 – us Kinesology [790]

Fair practices news – CA. n15-37 [1963 nov/dec-1969 jan/feb] – 1r – 1 – (cont: fair employment newsletter; cont by: fepc news) – mf#625737 – us WHS [331]

Fair women : reproductions...of some of the principal works exhibited / Grafton Galleries, London – London 1894 – (= ser 19th c art & architecture) – 2mf – 9 – mf#4.2.1266 – uk Chadwyck [700]

Fairbairn, A M *see* Religion in history and in modern life

Fairbairn, Andrew Martin *see*
- Catholicism, roman and anglican
- Christianity in the first century
- The influences of greek ideas and useages upon the christian church
- The philosophy of the christian religion
- The place of christ in modern theology
- Studies in the life of christ
- Studies in the philosophy of religion and history
- The united free church of scotland

Fairbairn, Andrew Martin et al *see* Jubilee lectures

Fairbairn, J *see* Book of crests

Fairbairn, John *see* Farewell sermon

Fairbairn, Patrick *see*
- The book of the prophet ezekiel
- Ezekiel and the book of his prophecy
- Hermeneutical manual
- Jonah
- The pastoral epistles
- Prophecy viewed in respect to its distinctive nature, its special function, and proper interpretation
- Real opinions of the most eminent reformers
- The typology of scripture
- The typology of scripture viewed in connection with the whole series of the divine dispensations

Fairbairn, Patrick et al *see* Divine revelation explained and vindicated

Fairbairn, Robert Brinckerhoff *see*
- Of the doctrine of morality in its relation to the grace of redemption
- The unity of the faith

Fairbank view – Fairbank IA. 1888 oct 4, 1889 jan 24 – 1r – 1 – (cont by: four county advocate) – mf#851153 – us WHS [071]

Fairbanks, Arthur *see* A handbook of greek religion

Fairbanks daily news-miner – Fairbanks AK. [1917 nov 29/1928 mar 16]-[1955 feb 25-28; 1956 feb 6; 1957 sep 18] – 5r – 1 – (cont: fairbanks daily news; tanana miner; tanana tribune) – mf#761906 – us WHS [071]

Fairbanks daily news-miner – Fairbanks, Alaska. 1959 nov – 1r – us UF Libraries [071]

Fairbanks Native Association *see* New river times

Fairbrother, William Henry *see*
- The philosophy of thomas hill green

Fairbury daily news – Fairbury, NE: Fairbury Daily News. v76 n23. apr 1 1946-95th yr n250. aug 27 1965 (daily ex sat & sun) – 6r – 1 – (cont: fairbury news and the fairbury gazette; absorbed: western wave and: plymouth news; merged with: fairbury journal to form: fairbury journal-news. issue for jan 4 1965 called 95th yr n87 but constitutes 95th yr n82) – us Bell [071]

Fairbury daily news – Fairbury, NE: Fairbury Daily News. v76 n23. apr 1 1946-95th yr n250. aug 27 1965 (daily ex sat & sun) [mf ed with gaps filmed -1975] – 30r – 1 – (cont: fairbury news and the fairbury gazette. absorbed: western wave and: plymouth news. merged with: fairbury journal to form: fairbury journal-news. v77 n1-101 not publ. 93rd yr n1-31 not publ. issue for jan 4 1965 called 95th yr n87 but constitutes 95th yr n82) – us NE Hist [071]

The fairbury enterprise – Fairbury, NE: Cash M Taylor, 1888-v15 n7. aug 30 1902 (wkly) [mf ed 1895-1902 (gaps) filmed 1979] – 3r – 1 – (cont: jefferson county democrat. absorbed by: fairbury gazette) – us NE Hist [071]

Fairbury gazette – Fairbury, NE: Geo Cross. v1 n[1] sep 3 1870-sep 15 1911// (wkly) [mf ed with gaps] – 13r – 1 – (absorbed: fairbury enterprise. merged with: fairbury news to form: fairbury news and the fairbury gazette. issue for sep 10 1870 not publ. some irregularities in numbering. issues for first and last week of year not publ) – us NE Hist [071]

Fairbury journal – Fairbury, NE: W F Cramb. 11th yr n10. may 31 1902-aug 27 1965// (3 times/wk) [mf ed -1964 (gaps) filmed - 1975] – 46r – 1 – (cont: jefferson county journal. absorbed: jansen news 1925, diller record 1954. merged with: fairbury daily news to form: fairbury journal-news. issues for dec 25 1903 and dec 27 1907 not publ. vol numbering irregular: v65, 66, and 68 repeated) – us NE Hist [071]

The fairbury journal-news – Fairbury, NE: Fairbury Journal Inc. 71st yr n55. aug 31 1965- (semiwkly) [mf ed 1968- filmed 1975-] – 1 – (formed by the union of: fairbury journal and: fairbury daily news. some irregularities in numbering) – us NE Hist [071]

The fairbury news and the fairbury gazette – Fairbury, NE: Shelley & Hinshaw. 36v. v41 n4. sep 22 1911-v76 n22. mar 28 1946 (wkly) [mf ed with gaps] – 30r – 1 – (formed by the union of: fairbury news and: fairbury gazette. cont by: fairbury daily news. v41 n4-v60 n18 called also v15 n42-v33 n35 cont the numbering of: fairbury news. v60 n20-36 not publ) – us NE Hist [071]

Fairbury republican – Fairbury, NE: E T & C C Bartruff, 1885 (wkly) [mf ed 1885-jul 22 1887 (gaps)] – 1r – 1 – (cont: southern nebraskan) – us NE Hist [071]

Fairchild, Ashbel Green *see* The unpopular doctrines of the bible

Fairchild crown – Fairchild AFB WA. 1986 jun 6-1987 aug 28, 1987 sep 4-1989 feb 24 – 2r – 1 – (cont: inland empire crown; cont by: strikehawk) – mf#1498071 – us WHS [355]

Fairchild, Edwin Milton *see* The function of the church

Fairchild, George Moore *see*
- Canadian leaves
- Gleanings from quebec
- Quebec

Fairchild graphic – Fairchild WI. 1895 sep 6 – 1r – 1 – (cont: friday evening post [black river falls, wi]) – mf#1313027 – us WHS [071]

Fairchild, James H *see*
- Oberlin
- Woman's rights and duties

Fairchild, James Harris *see*
- Elements of theology natural and revealed
- Lectures on systematic theology
- Moral science

Fairchild observer – Fairchild WI. 1905 aug 3-1907, 1908-10, 1911-13, 1914-1918 oct 10 – 4r – 1 – (cont by: fall creek journal [fall creek, wi: 1916]; augusta eagle [august, wi: 1915]; cooperative news-budget) – mf#1044338 – us WHS [071]

Fairchild observer – 1927 jan 13, feb 10-mar 3 – 1r – 1 – (cont: eau claire county union; cont by: augusta eagle times; fall creek journal [fall creek, wi: 1927]; union [augusta, wi]) – mf#1044310 – us WHS [071]

Fairchild times – Fairchild Air Force Base WA. 1981 jun 12-1983 apr, 1983 may-1984, 1985-1986 apr 11 – 3r – 1 – (cont by: inland empire crown) – mf#661589 – us WHS [355]

Fairfax chronicles – Fairfax County [VA] 1977 winter-1983 jan – 1r – 1 – mf#639080 – us WHS [071]

Fairfax county journal – Alexandria, VA. 1969-1969 (1) – mf#66662 – us UMI ProQuest [071]

Fairfax, Edward *see* Daemonologia

[Fairfax-] fairfax gazette – CA. 1927-47 – 7r – 1 – $420.00 – mf#B02227 – us Library Micro [071]

Fairfax first baptist church (formerly bethlehem). fairfax, south carolina : church records – 1917-75 – 1 – us Southern Baptist [242]

Fairfield advance – Fairfield – 11r – A$705.14 vesicular A$765.64 silver – at Pascoe [079]

Fairfield advance – Fairfield, apr 1967-dec 1969, jan 1980-jun 1997 – at Pascoe [079]

The fairfield auxiliary – Fairfield, NE: F M & H W Coleman, 1911-oct 7 1965// (wkly) [mf ed 1914-23,1925-65 (gaps) filmed -1970] – 17r – 1 – (absorbed by: clay county sun. v16 n19-v17 n18 not publ. vol numbering irregular: v21 and 32 repeated. v26 n6-v27 n5 not publ) – us NE Hist [071]

Fairfield baptist church – Winona. 1972-1977 (1) 1957-1977 (5) 1975-1977 (9) – 1r – 1 – $10.00 – mf#6479 – us Southern Baptist [242]

Fairfield cabramatta chronicle – Fairfield. oct 1966-oct 1972, mar 1973-oct 1975 – 3r – at Pascoe [079]

Fairfield cabramatta guardian – Fairfield, jan 1981-feb 1983 – 1r – at Pascoe [079]

Fairfield city champion – Fairfield, jan 1980-jun 1997 – at Pascoe [079]

Fairfield Co. Baltimore *see*
- Fairfield county news
- Fairfield leader
- Fairfield recorder series
- Twin city news series

Fairfield Co. Bremen *see*
- Derrick

Fairfield Co. Lancaster *see*
- Daily eagle
- Daily gazette series
- Eagle=gazette
- Fairfield county democrat
- Fairfield times
- Gazette
- Ohio eagle
- Times tribune

Fairfield Co. Pleasantville *see* News

Fairfield county ad – Westport, CT. 1989-1998 (1) – mf#61265 – us UMI ProQuest [071]

Fairfield county atlas, 1866 – 1r – 1 – mf#B7070 – us Ohio Hist [978]

Fairfield county business journal – Stamford, 1998+ [1,5,9] – ISSN: 0898-9818 – mf#17865,04 – us UMI ProQuest [338]

Fairfield county democrat / Fairfield Co. Lancaster – apr 1908-dec 1909 [semiwkly] – 2r – 1 – mf#B12383-12384 – us Ohio Hist [071]

Fairfield county news / Fairfield Co. Baltimore – v1 n1. oct 1889-dec 1896// [wkly] – 3r – 1 – mf#B32331-32333 – us Ohio Hist [071]

Fairfield County, OH *see* Atlas, 1875

Fairfield, Edmund Burke *see* Letters on baptism

[Fairfield-] enterprise – CA. 1915-23 – 2r – 1 – $120.00 – (missing issues: 25 jan, 27 sep 1919; 28 dec 1921) – mf#B02229 – us Library Micro [071]

Fairfield first baptist church. freestone county. fairfield, texas : church records – Oct 1871-Feb 1961 – 1 – us Southern Baptist [242]

Fairfield herald – Fairfield, NE: Wm H Wheeler, 1902 (wkly) [mf ed -dec 19 1907 (gaps)] – 2r – 1 – (cont: fairfield news-herald. issues for may 15-dec 4 1903 called v26 n20-49 but constitute v12 n20-49. v13 n3-v27 n2 not publ) – us NE Hist [071]

Fairfield independent – Winnsboro, SC. 1979-1993 (1) – mf#68774 – us UMI ProQuest [071]

The fairfield independent – Fairfield, NE: I W Evans, oct 1902 (wkly) [mf ed 1908-17 (gaps)] – 3r – 1 – (issue for jan 7 1910 not publ. issues for jan 14-feb 25 1910 misdated 1909. some irregularities in numbering) – us NE Hist [071]

Fairfield leader / Fairfield Co. Baltimore – nov 1977-dec 1986 [wkly] – 6r – 1 – mf#B29250-29255 – us Ohio Hist [071]

Fairfield messenger – Fairfield, NE: Coleman & Corey. v8 n22. feb 16 1900- (wkly) [mf ed -dec 6 1901 (gaps)] – 1r – 1 – (formed by the union of: fairfield tribune and: true light) – us NE Hist [071]

Fairfield news-herald – Fairfield, NE: A J Mercer & Son, 1892-1902// (wkly) [mf ed 1895-may 16 1902 (gaps)] – 2r – 1 – (formed by the union of: fairfield news and: fairfield herald. cont by: fairfield herald (1902). issue for jul 5 1901 not publ. some irregularities in numbering) – us NE Hist [071]

Fairfield, Oliver Jay *see* Stories from the new testament

Fairfield recorder series / Fairfield Co. Baltimore – jan 1971-aug 1974 [wkly] – 2r – 1 – mf#B29256-29257 – us Ohio Hist [071]

[Fairfield-] the daily republic – CA. 1962- – 357r – 1 – $21,420.00 (subs $570/y) – mf#B02228 – us Library Micro [071]

Fairfield times / Fairfield Co. Lancaster – jan-jul 1943 [wkly] – 1r – 1 – mf#B29537 – us Ohio Hist [071]

Fairfield tribune – Fairfield, NE: F M Coleman, 1892-feb 1900// (wkly) [mf ed 1895-feb 3 1899 (gaps)] – 1r – 1 – (absorbed: glenville surprise. merged with: true light to form: fairfield messenger) – us NE Hist [071]

[Fairfield-] weekly solano herald – CA. 1863-69 – 2r – 1 – $120.00 – mf#C02230 – us Library Micro [071]

[Fairfield-] weekly solano republican – CA. 1866-71; 1878-81; 1891-1905; 1916-61 – 63r – 1 – $3780.00 – mf#BC02231 – us Library Micro [071]

Fairfield/vacaville – 1992- – (= ser California telephone directory coll) – 3r – 1 – $150.00 – mf#P00027 – us Library Micro [917]

Fairford, Ford *see* Cuba

Fairford graves : a record of researches in an anglo-saxon burial-place in gloucestershire / Wylie, W M – Oxford, 1852 – 3mf – 8 – mf#H-1142 – ne IDC [700]

Fairforest baptist church – Spartanburg Co, SC. 701p. 1885-1980 – 1r – 1 – $31.55 – mf#6502 – us Southern Baptist [242]

Fairhaven 1733-1910 – Provo UT (mf ed 2004) – (= ser Massachusetts vital records) – 19v on 59mf – 9 – 0-87623-433-3 – (mf1-4: town records 1815-32. mf5-7: town & vitals 1777-1846. mf8: births & deaths 1733-1888. mf9-12: marriages 1834-45. mf13-17: intentions 1835-79. mf17: marriages 1787-1850. mf18-22: intentions 1879-1918. mf23-24: birth index 1843-92. mf25-26: marriage index 1844-92. mf27-28: death index 1844-92. mf29-32: births 1843-61. mf32-33: marriages 1844-55. mf33-34: deaths 1844-54. mf35-36: births 1861-92. mf37-39: marriages 1855-92. mf40-43: deaths 1855-92. mf44-46: birth index 1893-1949. mf47-49: marriage index 1893-1941. mf50-52: death index 1893-1943. mf53-54: births 1893-1912. mf55-56: marriages 1893-1911. mf57-59: deaths 1893-1912) – us Archive [978]

Fairholt, F W *see* A dictionary of terms in art...with 500 engravings on wood

Fairholt, Frederick William *see*
- A dictionary of terms in art
- Homes, haunts, and works of rubens, vandyke, rembrandt, and cuyp...michael angelo and raffaelle
- Homes, works, and shrines of english artists with specimens of their styles

Fairhope courier / Fairhope Industrial Association – Battles Wharf AL, Des Moines IA, Fairhope AL. 1896 jul 20, 1900 nov 1, 1904 jun 15-jul 1, aug 15, sep 15, 1905 jan 13-may 5, 1908 feb 14-1910 oct 7, 1910 oct 14-1912 mar 8, 1912 mar 15-1913 aug 15, 1913 aug 27-1914 dec 25, 1931 may 28, dec 10 – 4r – 1 – mf#870881 – us WHS [338]

Fairhope Industrial Association *see* Fairhope courier

Fairleigh dickinson university business review – Madison. 1961-1977 (1) 1974-1977 (5) 1975-1977 (9) – ISSN: 0427-931X – mf#9624 – us UMI ProQuest [650]

Fairless union news : the voice of local 4779, uswa, afl-cio / United Steelworkers of America – v18 n1-v20 n3 [1972:1st qtr-1974:3rd qtr], v21 n1-3, [1975:1st-4th qtr], v22 n1-6 [1976], v23 n1-v24 n1 [1977:1st qtr-1978:2nd qtr], v24 n3-v25 n1 [1978:4th qtr-1979:1st qtr], v25 n3-v27 n1 [1979:1st qtr-1981:1st qtr]; v27:n3 [1981:4th qtr], v28:n2-v32:n1 [1982:2nd qtr-1986:1st qtr] – 1r – 1 – mf#1565737 – us WHS [331]

Fairley, Barker *see*
- Goethe as revealed in his poetry
- Goethe's faust

Fairley, Margaret *see* New frontiers

Fairlie, Margaret Carrick *see* History of florida

Fairman, Charles E *see* Art and artists of the capitol of the united states of america

Fairmont bulletin – Fairmont, NE: W T Strother, 1875?-v14 n33. dec 16 1885 (wkly) [mf ed 1876-85 (gaps) filmed -1986] – 3r – 1 – (cont: fillmore county bulletin. cont by: fillmore weekly chronicle) – us NE Hist [071]

The fairmont news – Fairmont, NE: R G Strother (wkly) [mf ed v1 n19. dec 11 1886] – 1r – 1 – us NE Hist [071]

Fairmont. North Carolina. First Baptist Church *see* The baptist informer, bulletin

The fairmont tribune – Fairmont, NE: Risler & Jackson. v1 n1. jan 15 1897-99// (wkly) [mf ed with gaps] – 1r – 1 – us NE Hist [071]

Fairplay – Alzada, MT. 1923-1934 (1) – mf#64220 – us UMI ProQuest [071]

Fairplay : a journal for the consideration of financial, shipping and commercial subjects – London, England. 18 may 1883-dec 1918 – 96r – 1 – uk British Libr Newspaper [073]

Fairplay sentinel *see* Miscellaneous newspapers of park county

Fairpress – Norwalk, CT. 1990-1992 (1) – mf#68613 – us UMI ProQuest [071]

Fair-trade – A weekly journal devoted to industry and commerce. London. -w. 16 Oct 1885-25 Dec 1891. (5 reels) – 1 – uk British Libr Newspaper [380]

Fairview baptist church – Champaign. 1950-1985 (1) 1950-1985 (5) 1950-1985 (9) – 1r – 1 – $56.93 – mf#6495 – us Southern Baptist [242]

Fairview baptist church – Cincinnati. 1967-1988 (1) 1972-1988 (5) 1975-1988 (9) – 1r – 1 – $31.05 – mf#6492 – us Southern Baptist [242]

Fairview baptist church. mohawk, tennessee : church records – 1912-Jul 1924 – 1 – 7.56 – us Southern Baptist [242]

Fairview baptist church. spartanburg county. south carolina : church records – 1852-1982 – 1 – us Southern Baptist [242]

Fairview baptist church. stewart county. fairview, tennessee : church records – 1935-40, 1948, 1965-Jan 1968 – 1 – 5.94 – us Southern Baptist [242]

Fairview herald – Fairview Park, OH: Emil Uschelbec, feb 27 1947-feb 5 1959 – 10r – 1 – (weekly newspaper publ in a western cleveland suburb) – mf#(M) 34 C9.3 192 – us Western Res [071]

[Fairview-] news – NV. 1906-08 [wkly] – 1r – 1 – $60.00 – mf#U04519 – us Library Micro [071]

Fairwater register – Fairwater WI. 1903 apr 24-1904 jul 1 – 1r – 1 – mf#916115 – us WHS [071]

Fairweather, George Edwin *see* Rothesay and other verses

Fairweather, William *see*
- The background of the gospels
- From the exile to the advent
- Jesus and the greeks: or, early christianity in the tideway of hellenism
- The pre-exilic prophets

Fairweather, WilliamBlack, John Sutherland *see* The first book of maccabees

The fairy of the alps : a novel / Werner, E – New York: John W Lovell, [18–?] [mf ed 1989] – 285p – 1 – mf#7096 – us Wisconsin U Libr [830]

The fairy of the woodland glades : from the sleeping beauty (prologue) / Petipa, Marius – [n.p., n.d.] – 1 – mf#*ZBD-*MGO pv25 – Located: NYPL – us Misc Inst [790]

Fairy tales / Allyn, Rose – Chicago, IL. 1918 – 1r – us UF Libraries [390]

Fairy tales : their origin and meaning / Bunce, John Thackray – London, England. 1878 – 1r – us UF Libraries [390]

Fairy tales and stories / Andersen, Hans Christian – Boston, MA. 1887 – 1r – us UF Libraries [390]

Fais ce que dois / Coppee, Francois – Paris, France. 1871 – 1r – us UF Libraries [440]

Le fait colonial et l'imperialisme en afrique : parti democratique de guinee – Conakry: l'Imprimerie Nationale "Patrice Lumumba", [1968?] – us CRL [960]

Faith : treated in a series of discourses / Alexander, James Waddel – New York: Scribner, 1862 – 1mf – 9 – 0-7905-8757-2 – mf#1989-1982 – us ATLA [240]

Faith and character / Vincent, Marvin Richardson – New York: Charles Scribner, 1880 [mf ed 1989] – 1mf – 9 – 0-7905-0445-6 – (incl bibl ref) – mf#1987-0445 – us ATLA [243]

Faith and criticism : essays by congregationalists / Bennett, William Henry et al – London: Sampson Low Marston, 1893 – 2mf – 9 – 0-8370-9924-2 – mf#1986-3924 – us ATLA [242]

Faith and doubt in the century's poets / Armstrong, Richard Acland – New York: Thos Whittaker, 1898 – 1mf – 9 – 0-524-00242-8 – mf#1989-2942 – us ATLA [420]

Faith and fact : a study of ritschlianism / Edghill, Ernest Arthur – London: Macmillan and co., 1910 – 1r – 1 – 0-8370-0351-2 – mf#1984-B103 – us ATLA [240]

Faith and fellowship / Church of the Lutheran Brethren et al – 1977-79, 1980-82, 1983-86 – 3r – 1 – mf#555784 – us WHS [242]

Faith and folly / Vaughan, John Stephen – 2nd ed. London: Burns & Oates; New York: Benziger, 1905 [mf ed 1986] – 2mf – 9 – 0-8370-7032-5 – (incl bibl ref & ind) – mf#1986-1032 – us ATLA [241]

Faith and form : an attempt at a plain re-statement of christian belief in the light of to-day / Varley, Henry – London: James Clarke, 1908 – 1mf – 9 – 0-8370-5667-5 – mf#1985-3667 – us ATLA [240]

Faith and free press – High Point. N.C. v. 13-16. May 1957-Aug 1960 – 1 – 7.49 – us Southern Baptist [242]

Faith and freedom / Central Florida Christian School et al – v3 n10-v16 n11 [1967 dec-1980 dec] – 1r – 1 – mf#664936 – us WHS [242]

Faith and freedom – Oxford. 1989+ (1,5,9) – ISSN: 0014-701X – mf#15264 – us UMI ProQuest [210]

Faith and friends / Harris, Carrie Jenkins – Windsor, NS?: s.n, 1895 – 1mf – 9 – mf#08112 – cn Canadiana [890]

Faith and health / Brown, Charles Reynolds – New York: TY Crowell, 1910 – 1mf – 9 – 0-7905-7694-5 – mf#1989-0919 – us ATLA [130]

Faith and its effects : or, fragments from my portfolio / Palmer, Phoebe – New York: Palmer & Hughes [1867] [mf ed 1984] – (= ser Women & the church in america 174) – 1mf – 9 – 0-8370-1462-X – mf#1984-2174 – us ATLA [240]

Faith and its psychology / Inge, William Ralph – New York: Scribner 1910 [mf ed 1992] – (= ser Studies in theology) – 1mf – 9 – 0-524-05147-X – (incl bibl ref) – mf#1990-1403 – us ATLA [210]

Faith and knowledge / Inge, William Ralph – 2nd ed. Edinburgh: T & T Clark, 1905 – 1mf – 9 – 0-7905-7648-1 – mf#1989-0873 – us ATLA [240]

Faith and life : conferences in the oratory of princeton seminary / Warfield, Benjamin Breckinridge – New York: Longmans, Green, 1916 [mf ed 1991] – 2mf – 9 – 0-7905-9741-1 – mf#1989-1466 – us ATLA [242]

The faith and life of the early church : an introduction to church history / Slater, William Fletcher – London: Hodder and Stoughton, 1892 – 1mf – 9 – 0-7905-8899-4 – mf#1989-2124 – us ATLA [240]

Faith and mission – Wake Forest. 1985+ (1,5,9) – ISSN: 0740-0659 – mf#15288 – us UMI ProQuest [240]

Faith and modern thought / Welch, Ransom Bethune – New York: G P Putnam, 1876 – 1mf – 9 – 0-8370-5775-2 – (incl bibl ref) – mf#1985-3775 – us ATLA [210]

The faith and modern thought : six lectures / Temple, William – Shilling ed. London: Macmillan, 1913, c1910 – 1mf – 9 – 0-7905-7479-9 – mf#1989-0704 – us ATLA [240]

Faith and morals : 1. faith as ritschl defined it 2. the moral law as understood in romanism and protestantism / Hermann, Wilhelm – New York: G P Putnam; London: Williams & Norgate, 1904 [mf ed 1985] – (= ser Crown theological library 6) – 1mf – 9 – 0-8370-3811-1 – (english trans fr german by donald matheson and robert w stewart. incl ind and app) – mf#1985-1811 – us ATLA [241]

Faith and Order Commission *see* Minutes

Faith and order : meetings amsterdam-baarn – 1948 – 2mf – 8 – €5.00 – ne Slangenburg [240]

Faith and order : meetings bievres – 1950 – 1mf – 8 – €3.00 – ne Slangenburg [240]

Faith and order : meetings chichester – 1949 – 1mf – 8 – €3.00 – ne Slangenburg [240]

Faith and order : meetings clarence – 1947 – 2mf – 8 – €5.00 – ne Slangenburg [240]

Faith and order : meetings clarence – 1951 – 2mf – 8 – €5.00 – ne Slangenburg [240]

Faith and order : meetings geneve – 1946 – 1mf – 8 – €3.00 – ne Slangenburg [240]

Faith and order : meetings london – 1952 – 1mf – 8 – €3.00 – ne Slangenburg [240]

Faith and order : reports 1951 : (paper 10) churchdivision – 1mf – 8 – €3.00 – ne Slangenburg [240]

Faith and order : reports 1951 : (paper 10) division des eglises – 1mf – 8 – €3.00 – ne Slangenburg [240]

Faith and order : reports 1951 : (paper 10) kirchenspaltung – 1mf – 8 – €3.00 – ne Slangenburg [240]

Faith and order : reports lund : (paper 5) intercommunion – 1952 – 1mf – 8 – €3.00 – ne Slangenburg [240]

Faith and order : reports lund : (paper 6) formen des gottesdienstes – 1952 – 1mf – 8 – €3.00 – ne Slangenburg [240]

Faith and order : reports lund : (paper 6) ways of worship – 1952 – 1mf – 8 – €3.00 – ne Slangenburg [240]

Faith and order : reports lund : (paper 7) die kirche – 1952 – 2mf – 8 – €5.00 – ne Slangenburg [240]

Faith and order : reports lund : (paper 7) the church – 1952 – 2mf – 8 – €5.00 – ne Slangenburg [240]

Faith and philosophy : discourses and essays / Smith, Henry Boynton; ed by Prentiss, George Lewis – New York: Scribner, Armstrong, 1877 – 2mf – 9 – 0-7905-8734-3 – mf#1989-1959 – us ATLA [240]

Faith and philosophy : journal of the society of christian philosophers / Society of Christian Philosophers – Wilmore. 1984+ (1,5,9) – ISSN: 0739-7046 – mf#15246 – us UMI ProQuest [110]

The faith and progress of the brahmo somaj / Mozoomdar, Protap Chunder – Calcutta: Calcutta Central Press, 1882 – 1mf – 9 – 0-524-01626-7 – mf#1990-2565 – us ATLA [280]

Faith and rationalism : with short supplementary essays on related topics / Fisher, George Park – new enl ed. New York: Charles Scribner's Sons, 1885 [mf ed 1984] – 3mf – 9 – 0-8370-0136-6 – (incl bibl ref) – mf#1984-0022 – us ATLA [210]

Faith and science, or, how revelation agrees with reason, and assists it / Brownson, Henry Francis – Detroit: HF Brownson, 1895 – 1mf – 9 – 0-524-00249-5 – mf#1989-2949 – us ATLA [210]

Faith and southern baptists – Mayfield. Ky. v. 1-13. Apr 1945-Mar 1957 – 1 – 57.12 – us Southern Baptist [242]

The faith and the war : a series of essays / Gardner, Percy et al; ed by Foakes-Jackson, Frederick John – London, New York: Macmillan, 1916, c1915 – 1mf – 9 – 0-7905-4638-8 – mf#1988-0638 – us ATLA [240]

Faith and verification, with other studies in christian thought and life / Griffith-Jones, Ebenezer – London: J Clarke, 1907 – 1mf – 9 – 0-7905-3852-0 – mf#1989-0345 – us ATLA [240]

Faith and works – Providence, RI. 1896-1899 (1) – mf#66303 – us UMI ProQuest [071]

Faith and works / Willis, J T – London, England. 1874? – 1r – us UF Libraries [240]

The faith as unfolded by many prophets : an essay / Martineau, Harriet – Boston: Leonard C Bowles, 1833 – 1mf – 9 – 0-524-01846-4 – mf#1990-2681 – us ATLA [230]

Faith baptist church. kings mountain, north carolina : church records – 1953-63 – 1 – 6.12 – us Southern Baptist [242]

Faith baptist church. saskatoon, saskatchewan. canada : church records – Nov 1955-4 Apr 1973 – 1 – us Southern Baptist [242]

Faith building / Merrill, William P – Philadelphia: Presbyterian Board of Publication and Sabbath-School Work, 1896 – 1mf – 9 – 0-8370-4119-8 – mf#1985-2119 – us ATLA [240]

A faith for a new age *see* Hsin shih-tai te shin yang (ccm170)

A faith for to-day : suggestions towards a system of christian belief / Campbell, Reginald John – London: James Clarke, 1900 [mf ed 1985] – 1mf – 9 – 0-8370-3154-0 – mf#1985-1154 – us ATLA [210]

Faith, freedom, and the future / Forsyth, Peter Taylor – New York: Hodder and Stoughton, [1912?] – 1mf – 9 – 0-7905-3678-1 – mf#1989-0171 – us ATLA [240]

Faith, hope, love, and duty / Wise, Daniel – New York: Hunt & Eaton, 1891 – 1mf – 9 – 0-524-07775-4 – mf#1991-3343 – us ATLA [240]

Faith in god and modern atheism compared : in their essential nature, theoretic grounds, and practical influence / Buchanan, James – Edinburgh: J Buchanan, Jr; London: Groombridge, 1855 [mf ed 1990] – 2v on 3mf – 9 – 0-7905-7804-2 – mf#1989-1029 – us ATLA [210]

Faith in its relation to creed, thought and life : three short addresses / Swete, Henry Barclay – London: SPCK, 1895 – 1mf – 9 – 0-8370-5561-X – mf#1985-3561 – us ATLA [240]

Faith in practice *see* Shih chien ti hsin yang (ccm155)

Faith in unity / Gowon, Yakubu – Lagos, Printed by Academy Press, [n.d.] – us CRL [320]

Faith justified by progress / Wright, Henry Wilkes – New York: Scribners, 1916 – (= ser Bross lectures) – 1mf – 9 – 0-7905-9772-1 – mf#1989-1497 – us ATLA [240]

Faith made easy : or, what to believe and why: a popular statement of the doctrines and evidences of christianity in the light of modern research and sound biblical interpretation / Potts, James Henry – Toronto: W Briggs, 1889 [mf ed 1994] – 1mf – 9 – 0-665-94620-1 – mf#94620 – cn Canadiana [210]

Faith missionary – Oberlin, OH. 1882-88 – 1r – 1 – us Western Res [071]

The faith of a modern christian / Orr, James – New York: Hodder & Stoughton [1910?] [mf ed 1989] – 1mf – 9 – 0-7905-1618-7 – (incl ind) – mf#1987-1618 – us ATLA [240]

The faith of a modern protestant = Unser gottesglaube / Bousset, Wilhelm – New York: Scribner's, 1909 [mf ed 1985] – 1mf – 9 – 0-8370-2433-1 – (english trans fr german by florence b low) – mf#1985-0433 – us ATLA [242]

Faith of abraham lincoln / Taggart, David Raymond – Topeka, KS. 1943 – 1r – us UF Libraries [976]

The faith of catholics : confirmed by scripture and attested by the fathers of the first five centuries of the church – New York: Fr Pustet, 1885, c1884 [mf ed 1986] – 6mf – 9 – 0-8370-8887-9 – (incl bibl ref and ind) – mf#1986-2887 – us ATLA [241]

The faith of centuries : addresses and essays on subjects connected with the christian religion – 2nd ed. New York: T Whittaker, [189-?] – (= ser Popular Biblical Library) – 1mf – 9 – 0-7905-7704-6 – mf#1989-0929 – us ATLA [240]

The faith of france : studies in spiritual differences and unity / Barres, Maurice – Boston & New York: Houghton Mifflin Co, [1918] [mf ed 1985] – xxiv/294p – 1 – (trans by elisabeth marbury. foreword by henry van dyke) – mf#1309 – us Wisconsin U Libr [240]

The faith of islam / Sell, Edward – 3rd ed. rev. and enl. London: Society for Promoting Christian Knowledge; New York: E.S. Gorham, 1907. xvi,427p – 1 – us Wisconsin U Libr [260]

The faith of islam / Sell, Edward – London: Truebner, 1880 – 1mf – 9 – 0-524-02367-0 – mf#1990-2978 – us ATLA [260]

The faith of japan / Harada, Tasuku – New York: Macmillan, 1914 – (= ser Hartford-Lamson Lectures on the Religions of the World) – 1mf – 9 – 0-524-00883-3 – mf#1990-2106 – us ATLA [290]

Faith of our fathers : a historical drama presented by members of the schwenckfelder churches in pennsylvania, august 26 1934 – Norristown PA: Board of Pub of the Schwenckfelder Church, 1955 [mf ed 2003] – (= ser Schwenckfeldiana 2/5) – 1r – 1 – (incl music) – mf#2003-s008k – us ATLA [242]

The faith of our fathers : being an exposition and vindication of the church founded by our lord jesus christ / Gibbons, James – 76th rev ed. Baltimore, MD: John Murphy, c1904 – 1mf – 9 – 0-8370-8344-3 – (incl bibl ref and index) – mf#1986-2344 – us ATLA [230]

The faith of our forefathers : an examination of archbishop gibbons's faith of our fathers / Stearns, Edward Josiah – 2nd rev ed. New York: T Whittaker, 1879 – 1mf – 9 – 0-524-03661-6 – mf#1990-1089 – us ATLA [241]

The faith of reason : a series of discourses on the leading topics of religion / Chadwick, John White – Boston: Roberts Bros, 1879 – 1mf – 9 – 0-8370-2621-0 – mf#1985-0621 – us ATLA [200]

The faith of the artist : essays / Cousins, James Henry – Madras: Kalakshetra, 1941 – (= ser Samp: indian books) – us CRL [700]

Faith of the christian in its relations to the two advents / Nicolson, William Millar – Edinburgh, Scotland. 1870 – 1r – us UF Libraries [240]

The faith of the crescent / Takle, John – Calcutta: Association Press, 1913 – 1mf – 9 – 0-524-02723-4 – (incl bibl ref) – mf#1990-3126 – us ATLA [260]

The faith of the cross / Rhinelander, Philip Mercer – New York: Longmans, Green, 1916 – (= ser The bishop paddock lectures) – 1mf – 9 – 0-524-05057-0 – mf#1992-0310 – us ATLA [240]
The faith of the eastern church : a catechism / Platon, Metropolitan of Moscow – [New York]: G P Putnam, 1867 [mf ed 1986] – 1mf – 9 – 0-8370-7497-5 – mf#1986-1497 – us ATLA [243]
The faith of the gospel : a manual of christian doctrine / Mason, Arthur James – [3rd ed] New York: EP Dutton, 1903 – 2mf – 9 – 0-7905-8846-3 – mf#1989-2071 – us ATLA [240]
The faith of the millions. first series : a selection of past essays / Tyrrell, George – London, New York: Longmans, Green, 1901 [mf ed 1990] – xxv/344p on 1mf – 9 – 0-7905-7482-9 – mf#1989-0707 – us ATLA [241]
The faith of the millions. second series : a selection of past essays / Tyrrell, George – London, New York: Longmans, Green, 1901 [mf ed 1990] – 369p on 1mf – 9 – 0-7905-7483-7 – mf#1989-0708 – us ATLA [241]
The faith of the old testament / Nairne, Alexander – London; New York: Longmans, Green, 1914 – (= ser The Layman's Library) – 1mf – 9 – 0-7905-1479-6 – (incl ind) – mf#1987-1479 – us ATLA [221]
The faith of the unitarian christian explained, justified and distinguished : a discourse delivered at the dedication of the unitarian church, montreal, on sunday, may 11 1845 / Gannett, Ezra Stiles – Boston: W Crosby & H P Nichols, 1845 [mf ed 1983] – 1mf – 9 – 0-665-44521-0 – mf#44521 – cn Canadiana [243]
The faith once delivered to the saints : a sermon delivered at worcester, ma oct 15 1823, at the ordination of the rev loammi ives hoadly... / Beecher, Lyman – 2d ed. Boston: Printed by Crocker and Brewster, 1824. Beltsville, Md: NCR Corp, 1978 (1mf); Evanston: American Theol Lib Assoc, 1984 (1mf) – (= ser Revivalism and revival preachers in america) – 9 – 0-8370-0692-9 – (incl bibl ref) – mf#1984-3002 – us ATLA [240]
Faith or fact / Taber, Henry Moorehouse – New York: Peter Eckler, 1897 [mf ed 1985] – 1mf – 9 – 0-8370-5475-3 – mf#1985-3475 – us ATLA [210]
Faith papers : a treatise on experimental aspects of faith / Keen, Samuel Ashton – Cincinnati: Cranston & Curts, 1894, c1888 – 1mf – 9 – 0-8370-4207-0 – mf#1985-2207 – us ATLA [240]
Faith the greatest power in the world / McComb, Samuel – New York: Harper, 1915 – 1mf – 9 – 0-7905-9502-8 – mf#1989-1207 – us ATLA [210]
Faith, war, and policy : addresses and essays on the european war / Murray, Gilbert – Boston: Houghton Mifflin, 1917 – 1mf – 9 – 0-524-01950-9 – mf#1990-0539 – us ATLA [940]
Faith work under dr. cullis in boston / Boardman, William Edwin – Boston: Willard Tract Repository, 1874, c1873 – 1mf – 9 – 0-7905-7272-9 – mf#1989-0497 – us ATLA [240]
Faithful letter – London, England. 18-- – 1r – us UF Libraries [240]
The faithful love *see* The master of the isles / an afterword / a robin song / the tragedy of willow / the faithless lover / the faithful love
Faithful men / Swaine, S A – 402p – 1 – us Southern Baptist [242]
Faithful minister / Beddy, Joseph Fawcett – London, England. 1846 – 1r – us UF Libraries [240]
The faithful minister of god a burning and a shining light : a sermon, preached to the congregation of st andrew's church, st john's newfoundland, on the evening of february 16th, 1845... / Evans, D D – [St John's, Nfld?: s.n.], 1845 – 1mf – 9 – 0-665-91082-7 – mf#91082 – cn Canadiana [240]
Faithful minister's character and reward / Campbell, John – Edinburgh, Scotland. 1818 – 1r – us UF Libraries [240]
Faithful nurse – London, England. 18-- – 1r – us UF Libraries [240]
Faithful pastor / Jackson, Thomas – London, England. 1829 – 1r – us UF Libraries [240]
Faithful religious teacher / Gaskell, William – London, England. 1858 – 1r – us UF Libraries [240]
Faithful saying – London, England. 18-- – 1r – us UF Libraries [240]
Faithful steward / Moore, Robert – Blandford, England. 1824 – 1r – us UF Libraries [240]
Faithful unto death : an account of the sufferings of the english franciscans during the 16th and 17th centuries / Stone, Jean Mary – London: Kegan Paul, Trench, Truebner, 1892 – 1mf – 9 – 0-7905-6954-X – (incl bibl ref) – mf#1988-2954 – us ATLA [240]
Faithful unto death – London, England. 18-- – 1r – us UF Libraries [240]
Faithfulness to grace / Seager, Charles – London, England. 1850 – 1r – us UF Libraries [240]
Faith-healing, christian science and kindred phenomena / Buckley, James Monroe – New York: Century, 1892 – 1mf – 9 – 0-7905-4724-4 – mf#1988-0724 – us ATLA [130]
The faithless lover *see* The master of the isles / an afterword / a robin song / the tragedy of willow / the faithless lover / the faithful love
Faiths, fairs, and festivals of india / Buck, Cecil Henry – Calcutta: Thacker, Spink, 1917 – 1mf – 9 – 0-524-02074-4 – (incl bibl ref) – mf#1990-2838 – us ATLA [280]
Faiths of famous men in their own words : comprising religious views of the most distinguished scientists, statesmen,... / Kilbourn, John Kenyon – Philadelphia: Henry T Coates, 1900 – 1mf – 9 – 0-8370-3896-0 – (incl ind) – mf#1985-1896 – us ATLA [210]
Faiths of man : a cyclopaedia of religions / Forlong, James George Roche – London: B Quaritch, 1906 – 4mf – 9 – 0-524-04335-3 – mf#1990-3319 – us ATLA [052]
The faiths of the world : an account of all religions and religious sects, their doctrines, rites, ceremonies, and customs / Gardner, James – Edinburgh: A Fullarton, [1858-1860?] – 5mf – 9 – 0-524-06379-6 – mf#1990-3546 – us ATLA [052]
The faiths of the world : a concise history of the great religious systems of the world / Caird, John et al – Edinburgh: William Blackwood, 1882 – (= ser St. Giles Lectures) – 1mf – 9 – 0-524-01048-X – mf#1990-2196 – us ATLA [200]
Faitlovitch, J *see* Proverbes abyssins, traduits, expliques et annotes
Faits contemporaines / Chancy, Emmanuel – Port-Au-Prince, Haiti. 1905 – 1r – us UF Libraries [972]
Faits et chiffres – [Canada: s.n.] 1908 [mf ed 1995] – 1mf – 9 – 0-665-76878-8 – mf#76878 – cn Canadiana [630]
Faivelson, Israel Benjamin *see* Li-vene yisra'el
Fajardo, Eliecer L *see* La rosa del guayas
Fajardo, Heraclio C *see* Arenas del uruguay
Fajardo, Raoul J *see* Marti en dos rios
Fajgenbaum, Moses Joseph *see* Podliashe in umkum
Fak-Ao Tinapua Mata, Dick *see* Two texts in the language of tongoa island in the shepherd islands group, vanuatu
Fakel – St Petersburg, Russia, 1917 – 2r – 1 – us UMI ProQuest [077]
Fakely – Spb., 1906-1908. v1-3 – 12mf – 9 – mf#R-1846 – ne IDC [077]
Fa-kheu-pi-u *see* Texts from the buddhist canon commonly known as dhammapada
Fakhretdinov, A *see* Vestnik pravitel'stva bashkirii
Fakire und fakirtum im alten und modernen indien : yoga-lehre und yoga-praxis / Schmidt, Richard – Berlin: H Barsdorf, 1908 – 1mf – 9 – 0-524-03682-9 – mf#1990-3260 – us ATLA [280]
Faklya – Oradea, Romania. 1967; 1975-80 – 6r – 1 – us L of C Photodup [949]
Faks – 1976- – 1r/y – 1 – (cont: mladez and narodna mladez) – us UMI ProQuest [070]
Fakultas hukum dan pengetahuan masjarakat, universitas negeri padjadjaran – Bandung, 1958-1971. v1-3(4) – 14mf – 9 – (missing: 1959 v2(2-4)) – mf#SE-500 – ne IDC [959]
Fakultas ilmu pendidikan, universitas gadjah mada – Jogjakarta, 1951-1971 – 79mf – 9 – (missing: several iss) – mf#SE-469 – ne IDC [378]
Fakultas ilmu rumah tengga buku pedoman / Universitas Wanita Kartini – Bandung, 1968 – 1mf – 9 – mf#SE-1980 – ne IDC [959]
Fakultas kedokteran madjallah journal lembaga penelitian kedokteran, fakultas kedokteran unsrat / Universitas Sam Ratulangi – Manado, 1968-1970. v1-2(2) – 5mf – 9 – (missing: 1969 v1(3-4)) – mf#SE-1979 – ne IDC [950]
Fakultas Kedokteran Universitas Negeri Seriwidjaja *see* Madjalah kedokteran seriwidjaja
Fakultas mekanisasi dan teknologi hasil pertanian katalog / Institut Pertanian Bogor – Bogor, 1969 – 2mf – 9 – mf#SE-1710 – ne IDC [378]
Fakultas pertanian dan kehutanan, universitas gadjah mada : kehutanan – Jogjakarta. v1(1). 1961 – 1mf – 9 – mf#SE-1739 – ne IDC [378]
Fakultet Ekonomi Pedoman *see* Universitet krisnadwipajana
Falaise, Rayliane de la *see* Caraja...kou trois ans chez les indiens du brasil
La falange : revista de cultura latina – Mexico. dec 1922-feb 1923, july-oct 1923 – 1 – us NY Public [073]
Falange desde febrero de 1936 al gobierno nacional / Garceran, Rafael – n.p. 1936? Fiche W 906. (Blodgett Collection of Spanish Civil War Pamphlets) – 9 – us Harvard [946]
Falange Espanola de las Jons *see* Reglamento de primera linea
Falange Espanola Tradicionalista y de la Jons *see*
– Ordenanza
– Que es el servicio de administracion local (la falange y el ayuntamiento rural)
Falange Espanola Tradicionalista y de las Juntas Ofensivas Nacional-Sindicalistas *see*
– La falange y el combatiente
– Nacionalsindicalismo
– Primer discurso de la falange a cataluna
Falange Espanola Tradicionalista y de las Juntas Ofensivas nacional-Sindicalistas *see* El futuro de la agricultura nacional-sindicalista
Falange exterior : boletín decenal informativo de la delegacion nacional del servicio exterior de falange espanola tradicionalista y de las j o n s / Departamento de Intercambio y Propaganda Exterior – San Sebastian: El Departamento 1938 [mf ed 1977] – (= ser Blodgett coll) – 1mf – 9 – mf#w873 – us Harvard [946]
La falange y el combatiente / Falange Espanola Tradicionalista y de las Juntas Ofensivas Nacional-Sindicalistas – Bilbao, 1938. Fiche W 874. (Blodgett Collection of Spanish Civil War Pamphlets) – 9 – us Harvard [946]
Falaquera, Shem Tov Ben Joseph *see* Mevakesh
Falardeau, Edith *see* L'artisanat au canada francais (1900-1950)
Falardeau, Emile *see* Artistes et artisans du canada
Falardeau, Jean-Charles *see* Essais sur le quebec contemporain
Falcao, Ancino Pinto *see* Novas instituicoes do direito politico brasileiro
Falcao, Annibal *see* Formula da civilisacao brasileira
Falcao Espalter, Mario *see* Entre dos siglos
Falcato, Joao *see*
– Angola do eu coracao
– Raizes de angola
Fal'chenko, G I *see* Iug rossii
Falcioni, Zeffirino *see* Coup-d'oeil sur le christianisme
Falck, Anton Reinhard *see* Ambts-brieven
Falck, P T *see* Der dichter j m r lenz in livland
Falckenberg, Richard *see* History of modern philosophy
Falck-Ytter, Yngve *see* Die klinische-praktische evaluation aerztlicher kompetenz im medizinstudium
Falco, Francesco Federico *see* Inmigracion italiana y la colonizacion en cuba
Falcon – Mansfield. 1970-1980 (1) 1970-1980 (5) 1974-1980 (9) – ISSN: 0014-7079 – mf#6395 – us UMI ProQuest [400]
Falcon, Cesar *see* Similar fates: towns of england, towns of france you will also be raided
Falcon facts – 1981 apr 30-1985 dec 13 – 1r – 1 – mf#1099032 – us WHS [071]
Falcon flyer / United States Air Force Academy – 1981 may-1982 jun, 1982 jul 9-1983 aug, 1983 sep-1985 apr, 1985 may 10-1986 may, 1986 jun 27-1987 jul 24, 1987 jul 31-1988 may 27, 1988 jun-1989 may – 7r – 1 – mf#619792 – us WHS [355]
Falcon herald *see* El paso county miscellaneous newspapers, reel 2
Falconbridge, John Delatre *see* The canadian law of banks and banking
Falconer, Hugh *see*
– Palaeontological memoirs and notes of the late hugh falconer
– The unfinished symphony
Falconer, Robert *see* Idealism in national character
Falconer, Thomas *see* 1. on the nomination of agents formerly appointed to act in england for the colonies in north america. 2. a brief statement of the dispute between sir c metcalfe and the house of assembly of the province of canada
Falda, G B *see*
– Le fontane di roma nelle piazze...
– Li giardini di roma con le loro piante alzate e vedute in prospettiva
Falise, Jean Baptiste *see* Liturgiae practicae compendium
Falk, Brigitte *see* Kreative prozesse im unterricht
Falk, Genevieve *see* Harmonic equipment of rameau
Falk, Maryla *see* Nama-rupa and dharma-rupa
Falk und goethe : ihre beziehungen zu einander nach neuen handschriftlichen quellen / Schultze-Gallera, Siegmar Baron von – Halle a.S.: C A Kaemmerer, 1900 – 1r – 1 – us Wisconsin U Libr [920]
Falk, Victor von *see* Die todtenfelder von sibirien
Falke, Gustav *see*
– Ausgewaehlte gedichte
– Timm kroeger
– Vaterland, heilig land
Falke, Robert *see* Christentum und buddhismus
Der falke vom mons regius : geschichte einer jagd- und liebesleidenschaft / Bartsch, Rudolf Hans – Berlin: Deutsche Buch-Gemeinschaft, c1930 [mf ed 1995] – 252p – 1 – mf#8971 – us Wisconsin U Libr [830]
Falkenberg, Heinrich *see* Wir katholiken und die deutsche literatur
Falkenberg, Ryan James *see* An evaluation of an instructor-led and self-managed computer software training course
Falkenberger kreisblatt – Falkenberg (Niemodlin PL), 1919, 1921, 1922 – 1 – gw Misc Inst [074]
Falkenbergs tidning – Falkenberg, Sweden. 1876-1954 – 156r – 1 – sw Kungliga [078]
Falkenbergsposten – Falkenberg, Sweden. 1905-07; 1909-19 – 13r – 1 – sw Kungliga [078]
Falkener, Edward *see*
– Daedalus
– Ephesus and the temple of diana
– The museum of classical antiquities
Falkenstein, A *see*
– Archaische texte aus uruk
– Literarische keilschrifttexte aus uruk
Falkenstein, George N *see* History of the german baptist brethren church
Falkenstern, Anna Maria *see* Zwischen den maechten
Falkirk advertiser – Grangemouth: F Johnston & Co (wkly) [mf ed 7 jan 1998-] – 1 – uk Scotland NatLib [072]
Falkirk herald and scottish midlands journal – Falkirk: F Johnston & Co 1913- (semiwkly, wkly) [mf ed 1 jan 1998-] – 1 – (not publ: 29 dec 1973, 10 jul-7 aug 1981; cont: falkirk herald & midland counties journal; suppl: falkirk today) – uk Scotland NatLib [072]
Falkirk journal – [Scotland] Falkirk: J Duncan 2 aug 1862-25 nov 1865 (wkly) [mf ed 2003] – 2r – 1 – (cont by: falkirk journal and linlithgow or westlothian advertiser) – uk Newsplan [072]
Falkirk journal – [Scotland] Falkirk: J Courbough 14 nov-26 dec 1857 (wkly) [mf ed 2003] – 7v on 1r – 1 – (incl "falkirk burghs election – supplemental edition to the "falkirk herald") – uk Newsplan [072]
Falkirk mail and stirlingshire liberal – [Scotland] Falkirk: J Macgregor jan 1893-dec 1950 (wkly) [mf ed 2003] – 52r – 1 – (missing: 1894, 1896, 1919; cont: falkirk mail, bo'ness advertiser, and stirlingshire liberal; cont as: falkirk mail [jan 1906-dec 1950]) – uk Newsplan [072]
Falkland Islands *see* Falkland islands gazette
Falkland islands gazette / Falkland Islands – Stanley. 1958-Dec. 7, 1967 – 1 – us NY Public [972]
Falkland / malvinas : der umstrittene archipel im suedatlantik / Mann, Gerald H – (mf ed 1995) – 1mf – 9 – €30.00 – 3-8267-2158-6 – mf#DHS 2158 – gw Frankfurter [321]
Falkopings nyheter – Goteborg, Sweden. 1912-14 – sw Kungliga [078]
Falkopings tidning – Falkoping, Sweden. 1879-1978 – 176r – 1 – (aka: tidning fran falkopings stad och falbygden, 1857-79) – sw Kungliga [078]
Falkopings tidning – Falkoping, Sweden. 1857- – 1 – sw Kungliga [078]
Falkopings tidning *see* Vastgotabladet
Falkopingsposten – Falkoeping, 1899-1916 – 9r – 1 – sw Kungliga [078]
Falksadvokat = The volks advocate – New York, NY. 1888-89 – 1 – us AJPC [071]
Der fall : erzaehlung / Mueller, Wolfgang – Stuttgart: E Klett 1948 [mf ed 1990] – 1r – 1 – (filmed with: der schopfer / hans muller) – mf#2840p – us Wisconsin U Libr [880]
Fall, and exile and the kingdom / Camus, Albert – New York, NY. 1964 – 1r – us UF Libraries [025]
Fall army-worm, southern grass worm (laphygma frugiperda, smith and abbott) / Quaintance, A L – Lake City, FL. 1897 – 1r – us UF Libraries [630]
Fall, Bernard B *see* Le probleme de l'administration des minorites ethniques au cambodge, au laos et dans les deux zones du viet-nam
Fall campaign : western advertiser! london ont, from now until new year's gratis – [s.l: s.n. 1873] [mf ed 1987] – 1mf – 9 – 0-665-37371-6 – mf#37371 – cn Canadiana [071]
Fall creek cultivator – Fall Creek WI. 1910 dec 21-1913 feb 20, 1913 mar 13-1914 aug 27 – 2r – 1 – (cont by: eau claire county journal) – mf#1044385 – us WHS [071]
Fall Creek journal – Fall Creek WI. 1916 dec 16-1918 feb 22 – 1r – 1 – (cont: eau claire county journal; cont by: augusta eagle [augusta wi: 1915]; fairchild observer [fairchild wi: 1897]; cooperative news-budget) – mf#1044377 – us WHS [071]
Fall creek journal – Fall Creek WI. 1927 jan 13, feb 10-mar 3 – 1r – 1 – (cont: eau claire county union; cont by: augusta eagle times; fairchild observer [fairchild wi: 1927]; union [augusta wi]) – mf#1044314 – us WHS [071]
Fall creek times – Fall Creek WI. 1931 dec 3-1933 apr 6 – 1r – 1 – mf#1044339 – us WHS [071]
Fall creek tribune – Fall Creek WI. 1953 sep 24-1957 mar 28 – 1r – 1 – (cont by: eau claire county herald) – mf#964800 – us WHS [071]

Fall, George Howard *see* The law of the apothecary: a compendium of both the common and statutory law governing druggists and chemists in massachusetts, maine, new hampshire, vermont, rhode island and connecticut
Der fall hebbel : ein kuenstler-problem / Friedrich, Paul – Leipzig: Xenien-Verlag 1908 [mf ed 1990] – 1r – 1 – (filmed with: hebbels dithmarschenfragment / heinrich bender) – mf#2704p – us Wisconsin U Libr [430]
Fall in, at ease / Pacific Counseling Service [Tokyo, Japan] – 1970 autumn-1972 jan – 1r – 1 – (cont: we got the brass [asian ed]) – mf#721941 – us WHS [071]
Fall, Lynn A *see* A critical review of validity in alignment and dance performance studies using imagery training
Der fall maurizius : roman / Wassermann, Jakob – Berlin: S Fischer 1928 [mf ed 1991] – 1r – 1 – (filmed with: die kunst der erzaehlung) – mf#3036p – us Wisconsin U Libr [830]
Fall of babylon the great / Mason, Archibald – Glasgow, Scotland. 1821 – 1r – us UF Libraries [240]
The fall of constantinople : being the story of the fourth crusade / Pears, Edwin, Sir – London: Longmans, Green, 1885 – 1mf – 9 – 0-524-00644-X – mf#1990-0144 – us ATLA [940]
The fall of enron : how could it have happened?: hearing...u.s. senate, 107th congress, 2nd session, jan 24 2002 / United States. Congress. Senate. Committee on Governmental Affairs – Washington: US GPO 2002 [mf ed 2002] – 2mf – 9 – us GPO [364]
Fall of kruger's republic / Marais, Johannes Stephanus – Oxford, England. 1961 – 1r – us UF Libraries [960]
The fall of man and other sermons / Farrar, Frederic William – 3rd ed. London; New York: Macmillan, 1876 – 1mf – 9 – 0-8370-9777-0 – mf#1986-3777 – us ATLA [240]
Fall of mevar : a drama in five acts / Roy, Dwijendra Lal – Bombay: Nalanda Publications, 1946 – (= ser Samp: indian books) – (trans by harindranath chattopadhyaya and dilip kumar roy; int by bryan rhys) – us CRL [820]
The fall of nineveh – No. 21,901 in the British Museum. Edited, with transliteration, translation, notes, etc. by C.J. Gadd.1923 – 1 – us Wisconsin U Libr [900]
Fall of robespierre / Mathiez, Albert – New York, NY. 1927 – 1r – us UF Libraries [944]
Fall of rome / Christie, Thomas William – London, England. 1872 – 1r – us UF Libraries [240]
Fall of the mughal empire / Sarkar, Jadunath – Calcutta: MC Sarkar & Sons, 1932- – (= ser Samp: indian books) – us CRL [954]
Fall of the planter class in the british caribbean / Ragatz, Lowell Jospeh – New York, NY. 1928 – 1r – us UF Libraries [305]
The fall of turkey – London: Wyman & Sons, 1875 – (= ser 19th c economics) – 1mf – 9 – mf#1.1.119 – uk Chadwyck [956]
Fall river 1803-1889 – Oxford, MA (mf ed 1985) – (= ser Massachusetts vital records) – 318mf – 9 – 0-931248-71-X – (mf 1-18: vital records 1724-1856. mf 19-26: vitals index 1724-1856. mf 27-33: vital records 1843-49. mf 34-36: vitals index 1843-49. mf 37-38: vital records 1857-62. mf 39-40: vitals index 1857-62. mf 41-57: births 1843-64. mf 58-87: births 1865-78. mf 88-113: births 1879-86. mf 114-128: births 1887-90. mf 129-200: intentions 1843-90. mf 201-213: ints index 1843-92. mf 214-220: marriages 1850-64. mf 221-226: marriages 1865-70. mf 227-233: marriages 1871-74. mf 234-239: marriages 1875-78. mf 240-246: marriages 1879-82. mf 247-255: marriages 1883-86. mf 256-262: marriages 1887-89. mf 263-271: deaths 1850-64. mf 272-278: deaths 1865-70. mf 279-287: deaths 1871-74. mf 288-296: deaths 1875-78. mf 297-304: deaths 1879-82. mf 305-312: deaths 1883-86. mf 313-318: deaths 1887-89) – us Archive [978]
Fall river herald – Fall River WI. 1916 oct 30 – 1r – 1 – mf#5258660 – us WHS [071]
Fall river new era – Fall River WI. 1911 sep 29-1912 mar 1 – 1r – 1 – (cont: new era; cont by: columbus republican [columbus wi]) – mf#916192 – us WHS [071]
Fallacies and fictions relating to the irish church establishment... / Gayer, Arthur Edward – Dublin, Ireland. 1868 – 1r – us UF Libraries [240]
Fallacies and vagaries of misinterpretation / Ray, Charles Walker – Philadelphia: American Baptist Publ Soc, 1914 – 1mf – 9 – 0-524-07709-6 – mf#1991-3294 – us ATLA [220]
Fallacies of race theories as applied to national characteristics / Babington, William Dalton; ed by MacDonnell, Hercules Henry Graves – London, New York: Longmans, Green & Co, 1895 [mf ed 1987] – 277p – 1 – mf#1823 – us Wisconsin U Libr [572]
The fallacies of the alleged antiquity of man proved : and the theory shown to be a mere speculation / Cooke, William – London, 1872 – (= ser 19th c evolution & creation) – 2mf – 9 – mf#1.1.11622 – uk Chadwyck [573]
The fallacy of atheism : a lecture delivered by professor wm seymour, phrenologist and psychologist, in shaftesbury hall, toronto, ont, november 2nd, 1888 / Seymour, William – [Toronto?: s.n, 1888?] [mf ed 1993] – 1mf – 9 – 0-665-91765-1 – mf#91765 – cn Canadiana [210]
The fallacy of free food : the tariff is not a contributing cause to the high cost of living – [Canada: s.n, 1914?] [mf ed 1995] – 1mf – 9 – 0-665-76104-X – mf#76104 – cn Canadiana [380]
Fallacy of free trade in farm products : high prices at home would be exchanged for low prices in united states... – [Canada: s.n, 1911?] [mf ed 1997] – 1mf – 9 – 0-665-85224-X – mf#85224 – cn Canadiana [380]
Fallada, Hans *see*
- Altes herz geht auf die reise
- Bauern, bonzen und bomben
- Heute bei uns zu haus
- Little man, what now?
- Wolf among wolves
- The world outside

Fallas, Carlos Luis *see*
- Mamita yunai
- Marcos ramirez
- Tres cuentos

Fallbericht eines 16 wochen alten menschlichen anenzephalus mit rhachischisis / Lang, Marek – (mf ed 2000) – 1mf – 9 – €30.00 – 3-8267-2718-5 – mf#DHS 2718 – gw Frankfurter [617]
[Fallbrook-] fallbrook enterprise – CA. 1911-1959; 1978 – 48r – 1 – $2880.00 (subs $50y) – mf#H03219 – us Library Micro [071]
Fallecimiento / Castalleda, Vicente & Perez Jimenez, Nicolas – Madrid: Rev. Arch. Bibl. y mus, 1927. B.R.A.H. 90, p. 242 – sp Bibl Santa Ana [946]
Fallecimiento / Perez de Guzman, Juan & Sanguino y Michel, Juan – Madrid: Editorial Reus, 1921. B.R.A.H 78, p. 786 – sp Bibl Santa Ana [946]
Fallecimiento de d. francisco jarrin, obispo de plasencia / Fita, Fidel – Madrid: Fortanet, 1912. B.R.A.H. 61. p. 533 – 1 – sp Bibl Santa Ana [240]
Fallecimiento de don vicente paredes guillen / Perez de Guzman, Juan – Madrid: Fortanet, 1916. B.R.A.H. 68, pp. 327 – sp Bibl Santa Ana [920]
Fallecimiento del marques de monsalud / Fita, Fidel – Madrid: Fortanet, 1910. B.R.A.H. 56, 1910, p. 160 – sp Bibl Santa Ana [946]
Fallecimiento en caceres / Acedo, Federico – Madrid: Ed. Reus, 1922. B.R.A.H. 80. p. 494 – 1 – sp Bibl Santa Ana [946]
Fallecimiento en plasencia de don jose benavides checa / Fita, Fidel – B.R.A.H. 61. pp. 458-459. 1912 – 1 – sp Bibl Santa Ana [946]
Fallen angels / Heath, D I – Ryde, England. 1857 – 1r – us UF Libraries [240]
Fallen leaves from the note book of henry moorhouse – London, England. 18- – 1r – us UF Libraries [240]
Faller, O *see* S ambrosii de virginibus (fp31)
Fallersleben, Hoffmann von *see*
- Geschichte des deutschen kirchenliedes bis auf luthers zeit
- In dulci iubilo, nun singet und seid froh

Falling creek baptist church : minutes – Elbert Co, GA. 1835, 1837, 1839, 1841-45 – 1 – $10.00 – mf#6878 – us Southern Baptist [242]
Falling in love : with other essays on more exact branches of science / Allen, Grant – London: Smith, Elder, 1891 (London: Spottiswoode) – 4mf – 9 – mf#22827 – cn Canadiana [500]
Fallmerayer, Jakob Philipp *see* Der heilige berg athos
[Fallon-] churchill county courier – NV. 1961-62 [wkly] – 1r – 1 – $60.00 – mf#U04522 – us Library Micro [071]
[Fallon-] churchill county eagle – NV. 1906-49 (incomplete) [wkly] – 17r – 1 – $1020.00 – (aka: fallon eagle) – mf#U04523 – us Library Micro [071]
[Fallon-] churchill standard – NV. 1904-50 [wkly] – 18r – 1 – $1080.00 – (aka: churchill county standard) – mf#U04524 – us Library Micro [071]
[Fallon-] citizen – NV. 1964-1967 – 3r – 1 – $180.00 – mf#N04521 – us Library Micro [071]
Fallon county times – Baker, MT. 1916-1949 (1) – mf#64232 – us UMI ProQuest [071]
Fallon eagle *see* [Fallon-] churchill county eagle
[Fallon-] eagle standard – CA. 1949-1985 – 53r – 1 – $3180.00 – mf#N04526 – us Library Micro [071]
[Fallon-] in focus and muse news – NV. 1987- – 2r – 1 – $120.00 (subs $50y) – mf#U04840 – us Library Micro [071]
[Fallon-] lahontan daily news – NV. 1974-1979 – 7r – 1 – $420.00 – (cont by: lahontan valley news) – mf#N03705 – us Library Micro [071]
[Fallon-] lahontan valley news and fallon eagle standard – NV. 1972; 1980-1982 – 37r – 1 – $2220.00 (subs $240y) – mf#N04529 – us Library Micro [071]
Fallon, Michael Francis *see*
- The declaration against catholic doctrines which accompanies the coronation oath of the british sovereign

Fallon, P J *see* Voyage a holyrood pendant l'automne de 1831
[Fallon-] standard – NV. 1920-1947; 1949-1958 – 10r – 1 – $600.00 – mf#N04527 – us Library Micro [071]
[Fallon-] the ballot box – NV. 1911-13 [wkly] – 1r – 1 – $60.00 – mf#U04520 – us Library Micro [071]
[Fallon-] the co-operative colonist – NV. 1916-18 [wkly] – 1r – 1 – $60.00 – mf#U04525 – us Library Micro [071]
[Fallon-] the high view – NV. 1926-1927 – 1r – 1 – $60.00 – (aka: low down) – mf#U04528 – us Library Micro [071]
Fallos...: con la relacion de sus respectivas causas / Argentine Republic. Corte Suprema de Justicia de la Nacion – 1864-. Annual – 1 – us Wisconsin U Libr [324]
Fallouard, Pierre Jean Michel *see* Les musiciens normands, esquisse biographiques
Fallow, Thomas McCall *see* The cathedral churches of ireland
Fallows, Samuel *see* The home beyond
Falls Church. Virginia. Columbia Baptist Church *see* Index of materials at beginning, associational church letters at end
Falls city daily news – Falls City, NE: Davis & Davis. 16v. 45th yr n169. jul 1 1919-60th yr n84. apr 11 1934 (daily ex mon) [mf ed with gaps)] – 17r – 1 – (cont: falls city news. absorbed: falls city times (1928). absorbed by: falls city journal) – us NE Hist [071]
Falls city enterprise – Falls City OR: Falls City Chamber of Commerce [mf 1909 apr 8-1918 jul 27] – 1 – us Oregon Lib [071]
Falls city journal – Falls City, NE: May & Pepoon. v15 n734. feb 11 1882-41st yr whole n1233 [ie 2133] dec 4 1908; v1 n1 dec 8 1908-v1 n56 feb 10 1909; v41 n57 feb 11 1909- (semiwkly ex hols) – 1 – (cont: globe-journal; absorbed: falls city daily news 1934 and: shubert citizen 1942; issues for oct 2-dec 4 1908 called 41st yr whole n1224-1233 but constitute 41st yr whole n2124-2133; numbering very irreg) – us Bell [071]
Falls city journal – Falls City, NE: May & Pepoon. v15 n734. feb 11 1882-41st yr whole n1233 [ie 2133] dec 4 1908]; v1 n1. dec 8 1908-v1 n56. feb 10 1909; v41 n57. feb 11 1909- (semiwkly) [mf ed 1882-83,1885-1949 (gaps) filmed -1977 – 95r – 1 – (numbering very irregular. cont: globe-journal. absorbed: falls city daily news 1934, and: shubert citizen 1942) – us NE Hist [071]
Falls city news – Falls City OR: French Bros Pubs [wkly] – 1 – us Oregon Lib [071]
The falls city news – Falls City, NE: News Print Co. v1 n1. nov 20 1879-45th yr n56. jun 27 1919=whole n1-250-2323 (semiwkly) – us NE Hist [071]
Falls city press – Falls City, NE: Spurlock & Martin. v1 n1. feb 10 1875-v5 n41. nov 13 1879 (wkly) – 2r – 1 – (cont by: falls city news. issues for jul 14 1875-nov 13 1879 also called whole n22 [ie 23]-249) – us NE Hist [071]
Falls city register – Paterson, NJ. 1855-1928 (1) – mf#60224 – us UMI ProQuest [071]
Falls city times – Falls City, NE: A M Baughman. v1 n1. aug 2 1928-v8 n47. nov 22 1929 (wkly) [mf ed 1970] – 1r – 1 – (absorbed: verdon visitor and salem standard. absorbed by: falls city daily news. v1 n11-v7 n41 not publ) – us NE Hist [071]
Falls city times – Falls City, NE: Bean & Hill. v5 n33. jun 17 1898-// (wkly) [mf ed jun 17-jul 8 1898 (lacks jun 24 1898)] – 1r – 1 – (cont: populist. cont by: semi-weekly times (falls city, ne)) – us NE Hist [071]
Falls city tribune – Falls City, NE: Ross & Ray. v1 n1. jan 8 1903 [ie 1904]- (wkly) [mf ed 1904-11 (gaps) filmed 1958-[1965?]] – 3r – 1 – (absorbed: humboldt enterprise 1905, rulo record, crocker's educational journal, dawson outlook 1910. issues for jan 6 1905-may 12 1905 also called whole n53-71. issues for may 19 1905-jan 17 1908 called v2 whole n72-v4 whole n208) – us NE Hist [071]
Falls creek baptist assembly. oklahoma : church records – 1917. 58p – 1 – us Southern Baptist [242]
The falls of niagara : being a complete guide to all the points of interest around and in the immediate neighbourhood of the great cataract, with views taken from sketches – New York: T Nelson, [187-?] [mf ed 1984] – 1mf – 9 – 0-665-12840-1 – mf#12840 – cn Canadiana [917]
Fallston baptist church. kings mountain association. north carolina : church records – 1902-56 – 1 – us Southern Baptist [242]
Falmouth 1688-1892 – Oxford, MA (mf ed 1987) – (= ser Massachusetts vital records) – 46mf – 9 – 0-931248-92-2 – (mf 1-2: b,m,d 1668-1753. mf 3-7: vital records 1681-1757. mf 8-13: vital records 1750-1831. mf 14-20: vital records index 1756-1831. mf 21-26: b,m,d 1780-1858. mf 27-29: index to records 1780-1858. mf 30-32: b,m,d 1855-92. mf 33-38: b,m,d 1855-1892. mf 39-41: vital records index 1843-92. mf 42-46: congregational church records 1731) – us Archive [978]
Falmouth and penryn weekly times – Falmouth, England. jun 1861-oct 1952 – 73r – 1 – (lacking: 1897) – uk British Libr Newspaper [072]
Falmouth gazette and jamaica general advertiser – Jamaica, 3 Jan 1879-29 Dec 1888 – 5r – 1 – uk British Libr Newspaper [072]
Falmouth packet – Falmouth, England. apr 1829-48; 1858-1971; 1988- – 113+ r – 1 – uk British Libr Newspaper [072]
Falmouth penryn leader – Jun 4-Dec 24 1988; 1989-Jun 1990; Jul 7-Dec 22 1990; 1991-92; Jan 9-Jun 26 1993; Jul-Dec 1993; Jan 8-Jun 25 1994; Jul 2-Dec 24 1994; Jan-Dec 16 1995 – 13 1/2r – 1 – (discontinued) – uk British Libr Newspaper [072]
Falques, Marianne-Agnes Pillement *see* La derniere guerre des betes
Der falsche demetrius in der dichtung / Popek, Anton – Linz: Verlag des k.k. Staats-Gymnasiums 1893, 1895 [mf ed 1991] – 3v in 1 on 1r – 1 – (filmed with: schillers demetrius / martin greif) – mf#2872p – us Wisconsin U Libr [820]
Falsche extreme in der neueren kritik des alten testaments / Koenig, Eduard – Leipzig: Alexander Edelmann, 1885 – 1mf – 9 – 0-8370-7163-1 – (incl bibl ref) – mf#1986-1163 – us ATLA [225]
Falscher zuschauer *see* Deutscher zuschauer
Die falschmuenzerische theologie albrecht ritschls und die christliche wahrheit / Claassen, Johannes – Guetersloh: C Bertelsmann, 1891 [mf ed 1990] – 1mf – 9 – 0-7905-7277-X – mf#1989-0502 – us ATLA [242]
False christs and the true / Cairns, John – Edinburgh, Scotland. 1864 – 1r – us UF Libraries [240]
Le false consonanse della musica : per toccar la chittara sopra alla parte in breue / Matteis, Nicola – [168-?] [mf ed 19-] – 3mf / 1r – 9,1 – mf#fiche 1098 / pres. film 62 – us Sibley [780]
The false decretals / Davenport, Ernest Harold – Oxford: BH Blackwell, 1916 – 1mf – 9 – 0-7905-7714-3 – mf#1989-0939 – us ATLA [240]
False gods / Dearmer, Percy – London: AR Mowbray, 1914 [mf ed 1991] – 1mf – 9 – 0-7905-7720-8 – mf#1989-0945 – us ATLA [230]
False insurance methods / Ferguson, John – [s.l: s.n. 189-?] [mf ed 1987] – 1mf – 9 – 0-665-35437-1 – mf#35437 – cn Canadiana [368]
False liberality, and the power of the keys / Copleston, Edward – London, England. 1841 – 1r – us UF Libraries [240]
The false nation and its "bases" : or, why the south can't stand / Partridge, J Arthur – London: Edward Stanford, 1864 (mf ed: Louisville, KY: Lost Cause Press, 1984) – 2mf – 9 – mf#Sc Micro F-13740 – Located: NYPL – us Misc Inst [976]
False pleas and deceptive pretences / Coxe, R C – London, England. 1855 – 1r – us UF Libraries [240]
False witnesses answered / Clarke, James Freeman – Boston: Amer. Unitarian Assoc., 1835 – 1mf – 9 – 0-8370-1648-7 – mf#1984-6248 – us ATLA [240]
False worship : an essay / Maitland, Samuel Roffey – London: Rivingtons, 1856 – 1mf – 9 – 0-7905-0141-4 – (incl bibl ref and index) – mf#1987-0141 – us ATLA [240]
Falsehood of protestantism demonstrated / Malou, Jean Baptiste – London, England. 1858 – 1r – us UF Libraries [242]
Falsher hertsog / Shaikewitz Nahum Meir – Odessa, Ukraine. 1902 – 1r – us UF Libraries [939]
Falsified departmental reports : a letter to his excellency the marquis of lorne, governor general of canada / Hind, Henry Youle – Windsor, NS: C W Knowles, 1880 – 1mf – 9 – mf#08852 – cn Canadiana [317]
Falso cubanidad de saco, luz y del monte / Soto Paz, Rafael – Habana, Cuba. 1941 – 1r – us UF Libraries [972]
Falsos precursores de alvares cabral / Leite, Duarte – Lisboa, Portugal. 1946 – 1r – us UF Libraries [972]
Falukuriren – Falun, Sweden. 1894-1978 – 393r – 1 – sw Kungliga [078]

Falukuriren – Falun, Sweden. 1979- – 1 – sw Kungliga [078]
Faluposten – Falun, Sweden. 1869-90 – 10r – 1 – sw Kungliga [078]
Falvo, Lisa A *see* Mechanical and physiological differences between running and walking at various velocities
Fama, eclipse y resurreccion de donoso / Armas, Gabriel – Madrid: Imp. Aguirre, 1969. En Verbo, ano VIII, no 74, pp. 321-333 – sp Bibl Santa Ana [946]
Fama prognostica ad cunas serenissimi principis Maximiliani *see* Emmanuelis...in communi patriae plausu celebrata et demississime dicata...
Fambach, Oscar *see*
- Der aufstieg zur klassik in der kritik der zeit
- Das grosse jahrzehnt in der kritik seiner zeit
- Ein jahrhundert deutscher literaturkritik
- Der romantische rueckfall in der kritik der zeit

Fambai – Cape Town, South Africa. 1956 – 1r – us UF Libraries [960]
Fambai – Salisbury, Zimbabwe. 1966 – 1r – us UF Libraries [960]
La familia cristiana / Perez Munoz, Adolfo – Cordoba: Imprenta El defensor, 1921 – 1 – sp Bibl Santa Ana [240]
La familia de don pedro de valdivia / Roa y Ursua, Luis – Sevilla: Imprenta de la Gavidra, 1935 – 1 – sp Bibl Santa Ana [920]
Una familia de ingenios, los ramirez de prado / Hornedo, A M & Entrambasaguas, Joaquin de – Madrid: Razon y Fe, 1944 – 1 – sp Bibl Santa Ana [920]
La familia de miguel servet / Pano, Mariano de – [S.l.: s.n.], 1901 – 1mf – 9 – 0-524-02648-3 – mf#1990-0672 – us ATLA [240]
La familia en directo / Aradillas Agudo, Antonio & Espias-Sanchez, Manuel – Madrid: Doncel, 1973 – 1 – sp Bibl Santa Ana [306]
Familia gutierrez / Magarinos Borja, Mateo A – Montevideo, Uruguay. 1918 – 1r – us UF Libraries [972]
A familia maconica (1872-1873) : jornal dedicado aos interesses da maconaria da civilizacao e da hum.. – Rio de Janeiro, RJ : Typ da Família Maconica, 01 jun 1874-jan 1876; jan-set,nov-dez 1880; jan,mar-jul,out-dez 1881; jan-jul 1882; out-20 dez 1883 – (= ser Ps 19) – mf#P18A,1,13 – bl Biblioteca [079]
La familia preocupacion fundamental del estado espanol / Leal Ramos, Leon – Caceres: Tip. El noticiero, s.a. 1950 – sp Bibl Santa Ana [946]
La familia rural hacia la conquista de un mejor nivel de cultura. encuesta campana experimental 1963-1964 / Comision Diocesana de Apostolada Rural – Plasencia: Imp. Padilla, 1963 – 1 – sp Bibl Santa Ana [240]
La familia, segun el derecho natural y cristiano... / Goma, Isidoro – Madrid: Razon y Fe, 1927 – 1 – sp Bibl Santa Ana [360]
La familia sobre todo / Mendoza, Luis de – 1879 – 9 – sp Bibl Santa Ana [830]
A familia universal : orgao da sociedade universal dos macons – Recife, PE: Typ Mercantil, 01-22 jun 1872 – (= ser Ps 19) – bl Biblioteca [079]
Familia y fecundidad en puerto rico / Stycos, J Mayone – Mexico City? Mexico. 1958 – 1r – us UF Libraries [972]
Familial patterns of vo(2max) and physical activity levels / Guion, Willie K – 1994 – 2mf – $8.00 – us Kinesology [612]
A familial study of growth and health-related fitness among canadians of aboriginal and european ancestry / Katzmarzyk, Peter T – 1997 – 4mf – 9 – $16.00 – mf#HE 635 – us Kinesology [614]
O familiar – Bagagem, MG. 17 set 1891 – (= ser Ps 19) – mf#P17,02,82 – bl Biblioteca [079]
Familiar architecture : consisting of original designs of houses for gentlemen and tradesmen, parsonages and summer retreats, with back-fronts, sections, etc... / Rawlins, Thomas – [S.l.]: printed for the author, 1768 [mf ed 1975] – 1r – 1 – mf#SEM35P78 – cn Bibl Nat [720]
Familiar lectures on scientific subjects / Herschel, John Frederick William – London: A Strahan 1867 [mf ed 1998] – 1r [ill] – 1 – mf#film mas 28211 – us Harvard [520]
Familiar lectures on the pentateuch : delivered before the morning class of bethany college, during the session of 1859-60 / Campbell, Alexander; ed by Moore, William Thomas – St Louis: Christian Pub Society, c1867 – 1mf – 9 – 0-524-08353-3 – mf#1993-3053 – us ATLA [221]
Familiar letters : containing an account of his travels as one of the deputation sent out by the church of scotland on a mission of inquiry to the jews in 1839 / M'Cheyne, R M – London, 1848 – 2mf – 9 – mf#HTM-125 – ne IDC [915]
Familiar letters on population, emigration, home colonization etc / Burn, John Ilderton – London, 1832 – (= ser 19th c books on british colonization) – 3mf – 9 – mf#1.1.7135 – uk Chadwyck [941]
Familiar prayers : their origin and history / Thurston, H – London, 1953 – €11.00 – ne Slangenburg [240]
Familie dawidsohn / Frenk, Azriel Nathan – Warszawa, Poland. 1924 – 1r – us UF Libraries [939]
Die familie ghonorez / Kleist, Heinrich von – Berlin: Weidmann, 1927 [mf ed 1996] – (= ser Schriften der kleist-gesellschaft) – 164pl – 1 – mf#9701 – us Wisconsin U Libr [820]
Familie mendelssohn / Hensel, Sebastian – Berlin, Germany. v1-2. 1911 – 1r – us UF Libraries [939]
Die familie mendelssohn 1729-1847 / Hensel, Sebastian – Berlin. 1879. 3 v – 1 – us L of C Photodup [780]
Familie Wawroch : ein oesterreichisches drama in vier akten / Adamus, Franz – 2. aufl. Muenchen: A Langen, 1900 [mf ed 1989] – 177p – 1 – mf#7090 – us Wisconsin U Libr [820]
Familienbriefe jeremias gotthelfs / ed by Waeber, Hedwig – Frauenfeld und Leipzig: Huber, 1929 [mf ed 1989] – 121p (ill) – 1 – mf#7028 – us Wisconsin U Libr [860]
Der familienfreund – Berlin DE, 1872 & 1875 – 1 – gw Misc Inst [640]
Das familiengueterrecht in dem entwurfe eines buergerlichen gesetzbuches fuer das deutsche reich / Schroeder, Richard – Berlin: J Guttentag, 1889 – (= ser Civil law 3 coll; Beitraege zur erlaeuterung und beurtheilung des entwurfes eines buergerlichen gesetzbuches fuer das deutsche reich) – 1mf – 9 – (incl bibl ref) – mf#LLMC 96-605 – us LLMC [346]
Familien-journal – New York. N.Y. 1911-12 – 1 – us AJPC [306]
Familien-kalender – 1871-1940 [complete] – (= ser Mennonite serials coll) – 4r – 1 – mf#ATLA 1994-S004 – us ATLA [242]
Das familienrecht des buergerlichen gesetzbuchs : erster und zweiter abschnitt; buergerliche ehe; verwandtschaft / Opet, Otto & Blume, Wilhelm von – Berlin: Carl Heymann, 1906 – (= ser Civil law 3 coll; Kommentar zum buergerlichen gesetzbuche und seinen nebengesetzen) – 8mf – 9 – (incl bibl ref and index) – mf#LLMC 96-558A – us LLMC [348]
Das familienrecht des buergerlichen gesetzbuchs; dritter abschnitt; vormundschaftsrecht / Opet, Otto & Blume, Wilhelm von – Berlin: C Heymann, 1904 – (= ser Civil law 3 coll; Kommentar zum buergerlichen gesetzbuche und seinen nebengesetzen) – 3mf – 9 – (incl bibl ref and index) – mf#LLMC 96-558B – us LLMC [348]
Familienvaeter / Eckart, Dietrich – 3. Aufl. Muenchen: Hoheneichen-Verlag, [1920] (mf ed 1990) – 1r – 1 – (filmed with: ebner-eschenbach) – us Wisconsin U Libr [820]
Familiere...instruction...touchant la divine providence et predestination / Viret, P – [Geneve], Rivery, 1559 – 11mf – 9 – mf#PFA-200 – ne IDC [240]
Families in society – Milwaukee. 1990+ (1,5,9) – (cont: social casework) – ISSN: 1044-3894 – mf#984,01 – us UMI ProQuest [306]
Families systems and health – Rochester. 1996+ (1,5,9) – (cont: family systems medicine) – mf#13361,01 – us UMI ProQuest [360]
Familjetidningen smalaenningen – Stockholm, Sweden. 1924-62 – 39r – 1 – sw Kungliga [078]
Familjetidningen smalaenningen – Stockholm, Sweden. 1925 – 1 – sw Kungliga [078]
La famille – Paris: N Chaix, may 6-jun 17 1848 – us CRL [074]
Famille alexis reau : petites notes biographiques et genealogiques / Marguerite-Marie, soeur – Les Trois-Rivieres: impr le Bien public, 1923 [mf ed 1993] – 1mf – 9 – mf#SEM105P1830 – cn Bibl Nat [920]
Famille benoiton / Sardou, Victorien – Paris, France. 1866 – 1r – us UF Libraries [440]
La famille bonaparte devant le tribunal du peuple – Paris, 1848 – us CRL [944]
Famille charles-edouard gagnon : petites notices biographiques et genealogiques / Gagnon, Ernest – Quebec?: C Darveau, 1898 – 1mf – 9 – mf#05755 – cn Canadiana [920]
Famille charles-edouard gagnon : petites notices biographiques et genealogiques – Quebec?: C Darveau, 1898 – 1mf – 9 – mf#07899 – cn Canadiana [920]
La famille chretienne – Masson [Quebec]: Jeanne d'Arc, [1898-1902] – 9 – mf#P04237 – cn Canadiana [241]
La famille chretienne / Pressense, Edmond de – 2e ed. Paris: C Meyrueis, 1857 – 1mf – 9 – 0-524-02865-6 – mf#1990-0722 – us ATLA [240]
La famille cousineau / Girouard, Desire – S.l: s.n, 1884? – 1mf – 9 – mf#03438 – cn Canadiana [920]
La famille de nicolas gendron : supplement au dictionnaire genealogique de 1929 / Gendron, Pierre-Saul – Saint-Hyacinthe: Seminaire de Saint-Hyacinthe, 1930 [mf ed 1995] – 2mf – 9 – mf#SEM105P2403 – cn Bibl Nat [929]
La famille de ramezay / Roy, Pierre-Georges – Levis: [s.n.] 1910 [mf ed 1986] – 1mf – 9 – mf#SEM105P717 – cn Bibl Nat [920]
La famille de salaberry / Daniel, Francois – [S.l: s.n, 18–?] – 1mf – 9 – 0-665-94556-6 – mf#94556 – cn Canadiana [929]
La famille de salaberry – [S.l.]: [s.n.], [18–] (mf ed 1987) – 1mf – 9 – mf#SEM105P654 – cn Bibl Nat [920]
La famille demers d'etchemin, p q / Demers, Benjamin – [Levis, Quebec?: s.n.], 1905 – 2mf – 9 – 0-665-73676-2 – mf#73676 – cn Canadiana [929]
Famille des innocens : ou, comme l'amour vient / Sewrin, M – Paris, France. 1807 – 1r – us UF Libraries [440]
La famille d'irumberry de salaberry / Roy, Pierre-Georges – Levis: [s.n.] 1905 [i.e. 1906] (mf ed 1986) – 3mf – 9 – (with ind and bibl) – mf#SEM105P556 – cn Bibl Nat [920]
La famille d'orleans : depuis son origine jusqu'a nos jours / Marchal, Charles – Bruxelles 1846 [mf ed Hildesheim 1995-98] – 1v on 3mf – 9 – €90.00 – ISBN-10: 3-487-26069-7 – ISBN-13: 978-3-487-26069-3 – gw Olms [929]
Famille du fumiste / Varner, Antoine-Francois – Paris, France. 1840 – 1r – us UF Libraries [440]
Famille du porteur d'eau / Francis, M – Paris, France. 1824 – 1r – us UF Libraries [440]
La famille girouard / Girouard, Desire – S.l: s.n, 1884? – 1mf – 9 – mf#03439 – cn Canadiana [920]
La famille girouard en france / Girouard, Desire – Levis [Quebec]: Bulletin des recherches historiques – 1mf – 9 – 0-665-72332-6 – (incl bibl ref) – mf#72332 – cn Canadiana [929]
Famille glinet : ou, les premiers temps de la ligue / Merville, M – Paris, France. 1818 – 1r – us UF Libraries [440]
Famille glinet : ou, les premiers temps de la ligue / Merville, M – Paris, France. 1818 – 1r – us UF Libraries [440]
La famille janelle : histoire et genealogie / Janelle, Joseph-Emile – Drummondville: "La Parole", ltee, 1928 [mf ed 1993] – 5mf – 9 – (pref by elphege-j-b janelle and j-a janelle) – mf#SEM105P1838 – cn Bibl Nat [920]
La famille kerdalec au soudan : essai de vulgarisation coloniale / Decourt, Fernand – Paris: Vuibert, [1910] – 1 – us CRL [306]
La famille martiniquaise : analyse et dynamique / Dubreuil, Guy – Fonds St-Jacques: Centre de recherches caraibes, Universite de Montreal, [1976?] (mf ed: Bethlehem, PA: Mid-Atlantic Preservation Service, 1989) – 1mf – 9 – (incl bibl ref) – mf#Sc Micro F-11761 – Located: NYPL – us Misc Inst [306]
Famille renneville / Leonce – Paris, France. 1843 – 1r – us UF Libraries [440]
Les familles au sacre-coeur / Archambault, Joseph-Papin – Quebec: Secretariat des oeuvres de l'A S C, 1916 – 1mf – 9 – 0-665-71091-7 – mf#71091 – cn Canadiana [241]
Familles des plantes / Adanson, Michel – 2v. 1763 – 2r – 5 – mf#9/10 [sic] – uk Microform Academic [580]
Les familles royales actuelles en europe : complement au programme d'histoire du cours secondaire / St-Nom-de-Jesus, soeur – 1962 [mf ed 1978] – (= ser Bibliographies du cours...1947-66) – 1mf – 9 – (with ind) – mf#SEM105P4 – cn Bibl Nat [929]
Family : the nursery of the church / Chrichton, David – Edinburgh, Scotland. 1865 – 1r – us UF Libraries [240]
Family advocate – Chicago. 1990+ (1,5,9) – ISSN: 0163-710X – mf#17376 – us UMI ProQuest [346]
Family advocate – v1-23. 1978-2001 – 9 – $397.00 set – (cont: family law newsletter) – ISSN: 0163-710X – mf#102651 – us Hein [640]
Family almanac – 1871-1955 [complete] – (= ser Mennonite serials coll) – 3r – 1 – mf#ATLA 1993-S002 – us ATLA [242]
The family altar – Richmond, VA: [s.n., 1905?] – 1mf – 9 – 0-524-03735-3 – mf#1990-4840 – us ATLA [240]
The family and alzheimer's disease : a look into leisure experiences and adjustments / Hubley, Melissa – 1996 – 3mf – 9 – $12.00 – mf#RC 523 – us Kinesology [616]
Family and child mental health journal : journal of the jewish board of family and children's services – New York. 1980-1981 (1,5,9) – (cont: issues in child mental health) – ISSN: 0190-230X – mf#11188,02 – us UMI ProQuest [150]
Family and community health – Gaithersburg. 1978+ (1,5,9) – ISSN: 0160-6379 – mf#12730 – us UMI ProQuest [360]
Family and consumer sciences research journal – Thousand Oaks. 1994+ (1,5,9) – (cont: home economics research journal) – ISSN: 1077-727X – mf#10308,01 – us UMI ProQuest [640]
The family and heirs of sir francis drake / Fuller-Eliott-Drake, Elizabeth Douglas – London: Smith, Elder 1911. 2v. Illus. Plates, maps, geneal. table. With: Antiquities from San Tome and Mylapore by H. Hosten. 1 reel. 1263 – 1 – us Wisconsin U Libr [920]
Family and home office computing – New York. 1987-1988 (1,5,9) – (cont: family computing; cont by: home office computing) – ISSN: 0896-6028 – mf#13402,01 – us UMI ProQuest [000]
Family and population control / Hill, Reuben – New Haven, CT. 1965 – 1r – us UF Libraries [304]
The family and the state : select documents / Breckinridge, Sophonisba Preston – Chicago, IL: University of Chicago Press, 1934 [mf ed 1970] – (= ser The university of chicago social service series; Library of american civilization 14951) – xiv/565p on 1mf – 9 – (with bibl) – us Chicago U Pr [346]
Family association newsletter, droddy, drody, drawdy and variants – v1-v3 n2 [1981 sep-1982 sep], 1983 winter-fall, 1984 winter – 1r – 1 – (cont by: drady, drawdy, droddy, drody, drude & variants [o'grady, a variant of draddy]) – mf#1832782 – us WHS [366]
Family backtracking / Olympic Genealogical Society et al – v1 n1,4 [1976 feb,nov], v2 n1-v10 n4 [1977 feb-1985 nov], v11 n1-4 [1986 feb-aug], v12 n1-v14 n4 [1987:feb-1989 nov] – 1r – 1 – (cont by: backtracker) – mf#1685737 – us WHS [929]
The family bible : containing the old and new testaments – New York: American Tract Society, c1851-1856 – 15mf – 9 – 0-524-08305-3 – mf#1993-0010 – us ATLA [220]
Family business review – San Francisco. 1988+ (1,5,9) – ISSN: 0894-4865 – mf#17615 – us UMI ProQuest [650]
Family circle – Bloomington, IL. 1929-1986 (1) – mf#62518 – us UMI ProQuest [071]
Family circle – New York. 1974+ (1,5,9) – ISSN: 0014-7206 – mf#60014 – us UMI ProQuest [640]
The family circle : a journal of health, instruction, amusement and choice literature – London, Ont: J T Latimer, [187-?-188- or 19–] – 9 – ISSN: 1190-7215 – mf#P04274 – cn Canadiana [640]
Family computing – New York. 1983-1987 (1,5,9) – (cont by: family and home office computing) – ISSN: 0738-6079 – mf#13402 – us UMI ProQuest [000]
Family cook and home journal – London, UK. Dec 1906 – 9ft – 1 – uk British Libr Newspaper [640]
Family coordinator – Minneapolis. 1952-1979 (1) 1968-1979 (5) 1976-1979 (9) – (cont by: family relations) – ISSN: 0014-7214 – mf#3492 – us UMI ProQuest [306]
Family correspondence / Ward, Allen T – 1828-1964, Letters of a Methodist Indian missionary in Kansas, 1828-1928, 1928-1964 (2 copies) – 1 – us Kansas [920]
Family court review – Thousand Oaks, 2001+ [1,5,9] – ISSN: 1531-2445 – mf#21524,03 – us UMI ProQuest [347]
Family dialogues / Hughes, Mary – London, England. pt1. 1823- – 1r – us UF Libraries [640]
Family digest – Minneapolis. 1945-1975 (1) 1970-1973 (5) 1970-1971 (9) – ISSN: 0040-8506 – mf#2446 – us UMI ProQuest [240]
Family economics and nutrition review – Washington. 1995+ (1,5,9) – (cont: family economics review) – ISSN: 1085-9985 – mf#6947,01 – us UMI ProQuest [640]
Family economics review – Washington. 1957-1995 (1) 1976-1995 (5,9) – (cont by: family economics and nutrition review) – ISSN: 0425-676X – mf#6947 – us UMI ProQuest [640]
Family Enhancement Program [Madison WI] *see*
- Parent center news
- Parents' place
- Parents' place south

Family expenditure survey 1957/59-1977 – [mf ed Chadwyck-Healey] – (= ser British government publications...1801-1977) – 31mf – 9 – uk Chadwyck [339]
Family favorite and temperance journal – Adrian. 1849-1850 – 1 – mf#3981 – us UMI ProQuest [073]
The family favorite and temperance journal – v1 n1 [dec 1849] through v1 n12 [dec 1850] – 1r – 1 – $35.00 – mf#um-40 – us Commission [242]
A family flight over egypt and syria / Hale, Edward Everett & Hale, Susan – Boston: D. Lothrop & Co, c1882 [mf ed 1986] – 387p/pl – 1 – mf#1797 – us Wisconsin U Libr [916]
Family friend – Monticello, FL. 1860-1861; 1959 feb 22-dec 24 – 1r – us UF Libraries [071]
Family genealogy, 1875-1885 / Thomas, N M – 1r – 1 – mf#B26341 – us Ohio Hist [978]
Family grocer and wine merchant *see* Wine merchant and grocers review
Family handyman – Minneapolis. 1976-1997 (1) 1976-1997 (5) 1976-1997 (9) – ISSN: 0014-7230 – mf#10998 – us UMI ProQuest [690]

Family health – New York. 1969-1981 (1) 1975-1981 (5) 1971-1981 (9) – (cont by: health) – ISSN: 0014-7249 – mf#10761 – us UMI ProQuest [640]
Family herald – London. v1-67. May 1843-Oct 1891. Incomplete – 1 – us NY Public [640]
The family herald – New York. Weekly. Jan 6 1858-Nov 20 1862. Incomplete. Not collated – 1 – us NY Public [640]
Family herald and weekly star – Montreal, Canada. -w. 18 nov 1903; 6,20 nov, 11, 25 dec 1912; 1 jan-19 mar 1913; apr 1913-31 mar 1915 – 7 1/2r – 1 – uk British Libr Newspaper [071]
Family herald veterinary adviser : answers to veterinary questions reprinted from the family herald and weekly star – Montreal: [s.n.], 1900 [mf ed 1986] – 1mf – 9 – 0-665-63086-7 – (incl ind) – mf#63086 – cn Canadiana [636]
Family historian / Madison County Genealogical Society [TX] – v2 n1-4 [1982 aug /oct-1983 may/jul] – 1r – 1 – (cont by: past & present [madisonville tx]) – mf#687230 – us WHS [929]
Family historian quarterly / Madison County Genealogical Society [Madison County, TX] – 1984 oct-1989 jul – 1r – 1 – (cont: past & present [madisonville tx]) – mf#1685330 – us WHS [929]
Family history – Canterbury. 1967-1996 (1) 1973-1996 (5) 1973-1996 (9) – ISSN: 0014-7265 – mf#3378 – us UMI ProQuest [929]
Family history capers / Genealogical Society of Washtenaw Co. Michigan – v2 n3-v12 n4 [1979 apr-1989 spring] – 1r – 1 – mf#1056146 – us WHS [929]
Family history, stairs, morrow : including letters, diaries, essays, poems, etc / Stairs, William James – Halifax, NS: McAlpine Pub Co, 1906 – 3mf – 9 – 0-665-73235-X – mf#73235 – cn Canadiana [929]
The family in its civil and churchly aspects : an essay in two parts / Palmer, Benjamin Morgan – Richmond: Presbyterian Committee of Publication, c1876 – 1mf – 9 – 0-524-07906-4 – mf#1991-3451 – us ATLA [240]
Family journal – Thousand Oaks. 1994+ (1,5,9) – ISSN: 1066-4807 – mf#19611 – us UMI ProQuest [640]
Family journal and northern new-york organ – Troy NY. 1855 jan 6-dec 29, 1856 jan 6-1857 dec 26, 1858 jan 2-1859 dec 31, 1860 jan 7-1861 may 4 – 4r – 1 – (cont: family journal [troy, ny]; northern new-york organ) – mf#861732 – us WHS [929]
Family law newsletter (aba) – v1-18. 1960-78 (all publ) – 9 – $45.00 – (title varies: v1-10 1960-69 as family lawyer; cont by: family advocate) – mf#102661 – us Hein [346]
Family law quarterly (aba) – v1-34. 1967-2001 – 5,6,9 – $784.00 set – (v1-18 1967-85 on reel $251. v19-34 1985-2001 on mf $533. cont: american bar association. section of family law proceedings) – ISSN: 0014-729X – mf#102671 – us Hein [346]
Family leader – Lurgan, Ireland. Oct 1986 – 1/4r – 1 – uk British Libr Newspaper [072]
Family life – Aylmer. 1968+ (1) 1976+ (5) 1976+ (9) – ISSN: 0014-7303 – mf#10108 – us UMI ProQuest [240]
The family life of heinrich heine : illustrated by one hundred and twenty-two hitherto unpublished letters addressed by him to different members of his family / ed by Embden, Ludwig von – London: W Heinemann, 1896 – 1 – us Wisconsin U Libr [920]
Family Limbs and Branches *see* Tree of lives
Family lyceum : designed for instruction and entertainment, and adapted to families, schools, and lyceums – Boston, 1832-1833 [1,5,9] – mf#3730 – us UMI ProQuest [640]
Family magazine : or monthly abstract of general knowledge – New York. 1833-1841 – 1 – mf#3982 – us UMI ProQuest [073]
The family magazine (london) – 1830 – (= ser 19th c british periodicals) – r34 – 1 – us Primary [073]
The family magazine (london) – aug 1834-dec 1837 – (= ser 19th c british periodicals) – r35 – 1 – us Primary [073]
Family members' experiences of saturation, bonding, and leisure : a feminist perspective / Zangari, Mary-Eve C – 1997 – 199p on 3mf – 9 – $15.00 – mf#RC 546 – us Kinesology [150]
Family minstrel : a musical and literary journal – New York. 1835-1836 (1) – mf#5343 – us UMI ProQuest [780]
Family minstrel – New York: s.n. [mf ed 19–] – 1mf – 9 – mf#fiche 768 – us Sibley [780]
Family mirror – Dar es Salaam: General Publications Ltd, feb 1 1991; feb 15-sep 1 1992; nov-dec 1992; jan-oct 1 1993; nov-dec 1993; jan-dec 1994; jan-nov 21/30 1995; dec 21/31 1995 – 2r – us CRL [079]
Family news / Arctic Missions, Inc – Portland OR: Arctic Missions, Inc 1959- [irreg] [mf ed 2006] – v1-2 n7 (1959-spring/summer 1963) [complete] on 1r – 1 – (cont by: arctic news (portland, or) [2006i-s003]) – mf#2006i-s002 – us ATLA [240]
Family of Captain Daniel Little, Esquire *see* Little bit
Family papers – 1906-1962 – 1 – us CRL [920]
Family papers / Banvard, John – 1752-1985. 2 rolls including filmed inventory – 1 – $30.00r – us Minn Hist [920]
Family papers / Bartlett, Josiah – 1713-1931 – 1 – us L of C Photodup [920]
Family papers / Reid, Whitelaw – (in part). 1829-1912 – 1 – us L of C Photodup [920]
Family papers / Rivers, Elias Lynch – c1750-1913 [mf ed Spartanburg SC: Reprint Co, 1981] – 3mf – 9 – mf#51-542 – us South Carolina Historical [976]
Family papers ms 3231 / Harper, Alexander – Ashtabula, 1755-1935 – 24r – 1 – (letters, financial accounts, business files, legal docs, military records, & other documents related to the business and personal interest of the harper family, early settlers of ashtabula county, ohio) – us Western Res [929]
Family papers, ms 3893 / Hudson, David – Hudson, Summit, OH. 1799-1836 – 1r – 1 – (personal journals of david hudson, 1799-1801, and david hudson, jr, 1820-36. david hudson, sr, was the founder of the town of hudson, summit co, ohio) – us Western Res [920]
Family papers, ms 4675 / Breck, John – Brecksville, Cuyahoga, OH. 1782-1993 – 2r – 1 – (correspondence, writings, financial, legal, and other documents pertaining to john breck, founder of brecksville, and family) – us Western Res [929]
The family papers of james g. blaine – 21r – 1 – $735.00 – Dist. us Scholarly Res – us L of C Photodup [975]
Family papers of james parker, 1760-1795 – Liverpool Central Library – (= ser BRRAM series) – 4r – 1 – £268 / $536 – (int by w e minchinton) – mf#r95803 – uk Microform Academic [976]
The family physician and the farmer's companion – S.l: s.n, 18–? – 1mf – 9 – mf#55829 – cn Canadiana [610]
Family planning concerns you = Gesinsbeplanning is ook u verantwoordelikheid / South Africa. Department of Health [Departement van Gesondheid] – Pretoria: Dept of Health [1979?] [mf ed Pretoria, RSA: State Library [199-]] – 6p on 1r with other items – 5 – mf#op 06847 r24 – us CRL [360]
Family planning digest – Washington. 1972-1975 (1) 1972-1975 (5) (9) – mf#6828 – us UMI ProQuest [304]
Family planning perspectives – New York. 1969+ (1) 1972+ (5) 1975+ (9) – ISSN: 0014-7354 – mf#6785 – us UMI ProQuest [304]
The family planning programme of the department of health = Gesinsbeplanningsprogram van die Departement van Gesondheid / South Africa. Department of Health [Departement van Gesondheid] – [Pretoria: Dept of Health 1977?] [mf ed Pretoria, RSA: State Library [199-]] – 4p [ill] on 1r with other items – 5 – mf#op 06707 r24 – us CRL [360]
Family planning, your employee and you = Gesinsbeplanning, u en u werknemer / South Africa. Department of National Health and Population Development [Departement van Nasionale Gesondheid en Bevolkingsontwikkeling – [Pretoria: Dept of National Health & Population Development 1988?] [mf ed Pretoria, RSA: State Library [199-]] – 3p on 1r with other items – 5 – (in english & afrikaans) – mf#op 08727 r24 – us CRL [360]
Family practice – Oxford. 1989+ (1,5,9) – ISSN: 0263-2136 – mf#14050 – us UMI ProQuest [610]
Family practice news – New York. 1971-1995 (1) 1975-1981 (5) 1975-1981 (9) – ISSN: 0300-7073 – mf#6868 – us UMI ProQuest [610]
The family practice of physic / Hill, John – London, 1769 – 1 – us Wisconsin U Libr [610]
Family practice research journal – New York. 1983-1994 (1,5,9) – ISSN: 0270-2304 – mf#14130 – us UMI ProQuest [610]
Family prayer – Dublin, Ireland. 1819 – 1r – us UF Libraries [240]
Family process – Rochester. 1962+ (1) 1973+ (5) 1973+ (9) – ISSN: 0014-7370 – mf#9860 – us UMI ProQuest [150]
Family record : devoted for 1897 to the sackett, the weygant and the mapes families, and to ancestors of their intersecting lines – 1897 jan-dec – 1r – 1 – mf#2800195 – us WHS [929]
Family record of the name of dingwall fordyce in aberdeenshire, vol 1 : showing descent from the first known progenitor of either name – both direct and collateral / Fordyce, Alexander Dingwall – Fergus, Ont?: C B Robinson, 1885 – v1 on 5mf – 9 – mf#05793 – cn Canadiana [929]
Family record of the name of dingwall fordyce in aberdeenshire, vol 2 : including relatives of both names separately and connections / Fordyce, Alexander Dingwall [comp] – Fergus, Ont?: C B Robinson, 1888 – v2 on 3mf – 9 – mf#05794 – cn Canadiana [929]
Family record of the name of dingwall fordyce in aberdeenshire, vols 1 and 2 / Fordyce, Alexander Dingwall [comp] – Fergus, Ont?: C B Robinson, 1885 – 2v on 1mf – 9 – mf#05792 – cn Canadiana [929]
Family records : containing memoirs of major general sir issac brock...lieutenant e w tupper... and colonel william de vic tupper... / Tupper, Ferdinand Brock – Guernsey, Great Britain: S Barbet, 1835 [mf ed 1983] – 3mf – 9 – 0-665-41422-6 – mf#41422 – cn Canadiana [355]
Family records: early lyon county settlers / Kansas State Historical Society – ca1911 – 1 – us Kansas [978]
Family records today / American Family Records Association – v1 n1-v7 n5 [1980 apr-1986 nov] – 1r – 1 – mf#1573054 – us WHS [929]
Family relations – Minneapolis. 1980+ (1,5,9) – (cont: family coordinator) – ISSN: 0197-6664 – mf#3492,01 – us UMI ProQuest [306]
Family safety – Chicago. 1942-1984 (1) 1975-1984 (5) 1976-1984 (9) – (cont by: family safety and health) – ISSN: 0014-7397 – mf#10610 – us UMI ProQuest [360]
Family safety and health – Chicago. 1984+ (1,5,9) – (cont: family safety) – mf#10610,01 – us UMI ProQuest [360]
Family Service of Milwaukee *see* Annual report
Family services newsletter / Toiyabe Indian Health Project – v3 n1-v8 n1 [1981 oct-1987 jan] – 1r – 1 – (cont: indian child welfare study, 1980-1981; cont by: health newsletter [bishop, ca]) – mf#1278342 – us WHS [362]
Family structure in 17th-century windsor, connecticut / Holbrook, Jay Mack – Oxford, MA (mf ed 1990) – 1mf – 9 – 0-87623-121-0 – (an argument illustrating the possibility of a modified-extended family structure in a pre-industrial community through an analysis of mortality, migration and fertility data) – us Archive [978]
A family systems analysis of anxiety, depression, and somatization in graduate nursing students / Kleeman, Karen M – 1983 – 3mf – 9 – $12.00 – us Kinesology [150]
Family systems medicine – New York. 1983-1995 (1) 1983-1995 (5) 1983-1995 (9) – (cont by: families systems and health) – ISSN: 0736-1718 – mf#13361 – us UMI ProQuest [610]
Family therapy – San Diego. 1972+ (1,5,9) – ISSN: 0091-6544 – mf#10972 – us UMI ProQuest [360]
A family tour through south holland : up the rhine; and across the netherlands to ostend / Barrow, John – London 1831 [mf ed Hildesheim 1995-98] – (= ser Fbc) – 4mf – 9 – €120.00 – 3-487-27512-0 – gw Olms [914]
A family tour through the british empire : containing some account of its manufactures, natural and artificial curiosities, history and antiquities / Wakefield, Priscilla – London 1804 [mf ed Hildesheim 1995-98] – (= ser Fbc) – 3mf – 9 – €90.00 – 3-487-28893-1 – gw Olms [910]
A family tour through the british empire : containing some account of its manufactures, natural and artificial curiosities, history and antiquities / Wakefield, Priscilla (Bell) – corr enl 15th ed. London: Harvey & Darton, 1840 – (= ser 19th c children's literature) – 6mf – 9 – mf#6.1.52 – uk Chadwyck [910]
A family tour through the british empire : containing some account of its manufactures, natural and artificial curiosities, history and antiquities / Wakefield, Priscilla (Bell) – London: printed and sold by darton & harvey, 1804 – (= ser 19th c children's literature) – 5mf – 9 – (pt of map missing) – mf#6.1.25 – uk Chadwyck [910]
Family trails / Historical and Genealogical Association of Mississi – 1981 aug-1984 may, 1984 aug-1988 may – 2r – 1 – mf#629806 – us WHS [929]
Family tree / Howard County Genealogical Society – iss n3-126 [1977 mar 30-1989 dec] – 1r – 1 – (cont: hcgs newsletter) – mf#1685335 – us WHS [071]
Family tree / Ohio Genealogical Society – v2 n1-v9 n12 [1982 jan-1989 dec] – 1r – 1 – (cont: montgomery county chapter-ohio genealogical society) – mf#1111475 – us WHS [929]
Family treebune / Thomas Nash Descendants Association – v1:[n1]-v2 n8 [1977 fall-1979 summer] – 1r – 1 – mf#639189 – us WHS [929]
Family visitor – Cleveland, OH. 1850-1851 (1) – mf#65423 – us UMI ProQuest [071]
Family visitor – Hudson, OH. 1850-1853 (1) – mf#65530 – us UMI ProQuest [071]
Family voice / Environmental Peace Action Workshop Coalition – n1,4-6 [1972 apr/may, oct 15- dec 1/1973 jan 1] – 1r – 1 – mf#1583265 – us WHS [333]
Family weekly – New York. 1953-1980 (1) 1979-1980 (5) 1979-1980 (9) – (cont by: usa weekend) – ISSN: 0014-7427 – mf#6494 – us UMI ProQuest [073]
Family Welfare Association (Milwaukee WI) *see* Annual report
Family welfare work in the metropolitan community : selected case records / Breckinridge, Sophonisba Preston – Chicago, IL: The University of Chicago Press, 1924 [mf ed 1970] – (= ser University of chicago social service series; Library of american civilization 15814) – 1mf – 9 – us Chicago U Pr [360]
The famine campaign in southern india (madras and bombay presidencies and province of mysore) 1876-1878 / Digby, William – London. 2v. 1878 – (= ser 19th c books on british colonization) – 12mf – 9 – mf#1.1.4864 – uk Chadwyck [630]
The famine in india / Forrest, George William – London, 1897 – (= ser 19th c books on british colonization) – 1mf – 9 – mf#1.1.6633 – uk Chadwyck [630]
Famines in bengal, 1770-1943 / Ghosh, Kali Charan – Calcutta: Indian Associated Pub Co, 1944 – (= ser Samp: indian books) – us CRL [954]
Famines in india : their causes and possible prevention. being the cambridge university le bas prize essay, 1875 / Williams, Arthur Lukyn – [London] 1876 – (= ser 19th c british colonization) – 3mf – 9 – mf#1.1.4006 – uk Chadwyck [630]
O famoso botao de ancora (1600-1895) / Marques Esparteiro, Antonio – Madrid: Archivo Ibero Americano, 1960 – 1 – sp Bibl Santa Ana [946]
Famossisimos romances / Cepeda, Joaquin – S.L. s.i.,s.a. – 1 – sp Bibl Santa Ana [946]
Famous algonquins : algic legends / Hamilton, James Cleland – Toronto?: Murray Print Co, 1899 – 1mf – 9 – mf#07388 – cn Canadiana [305]
Famous cities of ireland / Gwynn, Stephen Lucius – Dublin, Ireland. 1915 – 1r – us UF Libraries [710]
Famous dutch writer denounces rebel atrocities / Brouwer, Johannes – [s.l: s.n. 1936?] – 9 – mf#w766 – us Harvard [946]
Famous edison fish story / Frost, Jules A – s.l, s.l? 1936 – 1r – us UF Libraries [978]
Famous firesides of french canada / Alloway, Mary Wilson – Montreal?: J Lovell, 1899 – 3mf – 9 – mf#00040 – cn Canadiana [720]
Famous irish preachers / Irwin, Clarke Huston – Dublin, Ireland. 1889 – 1r – us UF Libraries [240]
Famous missionaries of the reformed church / Good, James Isaac – 1st ed. [S.l.]: Sunday-School Board of the Reformed Church in the United States, 1903 – 1mf – 9 – 0-7905-5218-3 – mf#1988-1218 – us ATLA [242]
Famous modern battles / Atteridge, Andrew Hilliard – Boston: Small, Maynard & Co, c1913 [mf ed 1986] – viii/401p – 1 – mf#6851 – us Wisconsin U Libr [355]
Famous modern negro musicians / Lovingood, Penman – New York: Da Capo Press, 1978 [c1921] (mf ed 1969) – 1r – 1 – (repr of the ed publ by press forum co, brooklyn) – mf#Sc Micro R-1317 – us NY Public [780]
The famous negro robber, and terror of jamaica : or, the history and adventures of jack mansong – Glasgow: Printed for the booksellers, [18–?] (mf ed 1969) – 1r – 1 – mf#Sc Micro R-1318 – us NY Public [972]
Famous places of the reformed churches : a religious guidebook to europe / Good, James Isaac – Philadelphia: Heidelberg Press 1910 [mf ed 1986] – 2mf – 9 – 0-8370-8673-6 – mf#1986-2673 – us ATLA [914]
Famous reformers of the reformed and presbyterian churches : a mission study manual on the reformation / Good, James Isaac – Philadelphia PA: Home & Foreign Mission Boards...1916 [mf ed 1993] – 1mf – 9 – 0-524-06412-1 – mf#1991-2534 – us ATLA [242]
Famous trials of history / Birkenhead, Earl of – Garden City NY: Garden City Publ Co 1926 – 4mf – 9 – $6.00 – mf#llmc92-218 – us LLMC [347]
Famous urdu poets and writers / Qadir, Abdul – Lahore: New Book Society, 1947 – (= ser Samp: indian books) – (foreword by sachchidananda sinha) – us CRL [490]
Famous women of history : containing nearly three thousand brief biographies and over one thousand female pseudonyms / Browne, William Hardcastle – Philadephia: Arnold, 1895 [mf ed 1984] – (= ser Women & the church in america 113) – 1mf – 9 – 0-8370-1380-1 – mf#1984-2113 – us ATLA [920]
Famu cluster : florida a and m university/ industry cluster – 1993 spring – 1r – 1 – mf#5319856 – us WHS [071]

Famuan : the voice of the students of florida a and m / Florida Agricultural and Mechanical University – 1993 sep 9-dec 2, 1994 jan 13-apr 14, sep 22-dec 8, 1995 jan 12-apr 6, sep 28-dec 7, 1996 jan 11,25, mar 28, apr 4,11 – 1r – 1 – mf#2836421 – us WHS [378]

The fan = l'eventael / Uzanne, Louis Octave – London: .C Nimmo & Bain, 1883 – (= ser 19th c art & architecture) – 2mf – 9 – (trans fr french; ill by paul avril) – mf#4.1.64 – uk Chadwyck [740]

Fan chan fan fa-hsi-ssu tou cheng ti tang ch'ien wen t'i – [China]: Chung-kuo ch'u pan she, Min kuo 27 [1938] – (= ser P-k&k period) – us CRL [951]

Fan, Ch'ang-chiang *see*
- Hsi hsien feng yun
- Hsi hsien ti hsueh chan
- Hsi pei hsien
- Hua pei liu sheng k'ang jih hsueh chan shih shang chi
- Kan k'ai kuo chin-ling

Fan, Ch'ang-chiang, 1907- *see* Hsi pei chin ying

Fan cheng ch'ien hou / Kuo, Mo-jo – Ch'ung-ch'ing: Tso chia shu wu, 1943 – (= ser P-k&k period) – us CRL [951]

Fan, Ch'uan, 1918- *see* Chan cheng yu wen hsueh

Fan chu-hsien : san mu chu / Yao, Ya-ying – Ch'ung-ch'ing: Ya-chou shu chu, Min kuo 32 [1943] – (= ser P-k&k period) – us CRL [820]

Fan, Chung-yun *see*
- I chiu san erh nien chih kuo chi cheng chih ching chi
- K'ang chan yu kuo chi hsing shih

Fan fa-hsi-ssu / Ai, Ch'ing – Ch'ung-ch'ing: Tu shu ch'u pan she, Min kuo 36 [1947] – (= ser P-k&k period) – us CRL [810]

Fan hou t'an hua / Yu, Ch'ieh – [China]: Liang yu t'u shu yin shua kung ssu, 1933 – (= ser P-k&k period) – us CRL [840]

Fan, Hung *see*
- Kung tzu li lun chih fa chan
- Lao tung li fa yuan li

Fan, I *see* T'a shan shih yu (ccm123)

Fan i lei biyan bithe *see* Fan yi lei bian

Fan i lun chi / Huang, Chia-te – Shang-hai, Hsi feng she, Min kuo 29 [1940] – (= ser P-k&k period) – us CRL [400]

Fan kung / Chang, T'ien-i – Shang-hai: Sheng huo shu tien, Min kuo 23 [1934] – (= ser P-k&k period) – us CRL [830]

Fan lao huan t'ung : san mu hsi chu / Lu, Ssu-an – Lung-ch'uan: Ch'ing nien shu tien, Min kuo 33 [1944] – (= ser P-k&k period) – us CRL [820]

Fan liao chai ch'u chi / Wu, Ch'i-yuan – Shang-hai: Ta chung shu chu, Min kuo 23 [1934] – (= ser P-k&k period) – us CRL [480]

Fan loyalty : the structure and stability of an individual's loyalty toward an athletic team / Funk, Daniel C – 1998 – 4mf – 9 – $16.00 – mf#PSY 2078 – us Kinesology [150]

Fan men / Han, Chen-yeh pien chi – Shang-hai: T'ien ma shu tien, 1934 – (= ser P-k&k period) – us CRL [830]

Fan, Shou-k'ang *see*
- Chiao yue che hsueh ta kang
- Ko hsing chiao yu

Fan, Tien-hsiang *see* Ming chung sheng ko chi (ccm75)

Fan tsui she hui hsueh / Li, Chien-hua – Shang-hai: Hui wen t'ang hsin chi shu chu, Min kuo 24 [1935] – (= ser P-k&k period) – us CRL [360]

Fan tui t'o-lo-ssu-chi ti huang miou / P'an, Wen-yu – Han-k'ou: K'ang chan ch'u pan she, 1937 – (= ser P-k&k period) – us CRL [335]

Fan tz'u chi / Jung, Lu – Shang-hai: Shih chieh shu chu, 1939 – (= ser P-k&k period) – us CRL [480]

Fan, Yen-ch'iao *see*
- Ch'a yen hsieh
- Ch'in huai shih chia

Fan yi lei bian = Fan i lei biyan bithe / Guan, Juchuang – [China: s.n.], Qianlong 14 nian [1749] [mf ed 1966] – (= ser Tenri coll of manchu-books in manchu-characters. series 1, linguistics 48; Mango bunkenshu. 1, gogaku hen) – 4v on 1r – 1 – (cover title also in manchu : fan i lei biyan bithe) – ja Yushodo [480]

Fan yu chi / Wu, Tsu-hsiang – Shang-hai: Wen hua sheng huo ch'u pan she, Min kuo 25 [1935] – (= ser P-k&k period) – us CRL [840]

Fanal – Berlin DE, 1926 oct-1931 jul – 1r – 1 – gw Mikrofilm [074]

Fanal – Berlin. v. 1-5. 1926-1931 – 1 – us NY Public [073]

Fanaticism / Taylor, Isaac – New-York: J Leavitt, 1834 – 1mf – 9 – 0-524-05162-3 – mf#1990-1418 – us ATLA [150]

Fanaticism. by the author of natural history of enthusiasm / Taylor, Isaac – Boston: Crocker & Brewster, 1834 – 1 – us Wisconsin U Libr [240]

Fanchon la vielleuse / Bouilly, Jean Nicolas – Paris, France. 1803 – 1r – us UF Libraries [440]

Fanchon la vielleuse / Bouilly, Jean Nicolas – Paris, France. 1809 – 1r – us UF Libraries [440]

Fancourt, Charles Saint John *see* The history of yucatan

Fancy dress described : or, what to wear at fancy balls / Holt, Ardern – London [1879] – (= ser 19th c art & architecture) – 2mf – 9 – mf#4.2.1015 – uk Chadwyck [740]

Fancy fairs / Nevin, John Williamson – [S.l.: s.n., 1843?] – 1mf – 9 – 0-524-08767-9 – mf#1993-3272 – us ATLA [390]

Fane, Cecil *see* Short paper on the productive capabilities of newfoundland

Fanelli, F *see* Fontaines et iets d'eau

Fanfan et colas / Beaunoir, M De – Paris, France. 1806 – 1r – us UF Libraries [440]

Fanfant, J E *see* De la recherche de la paternite naturelle

Die fanfare – Berlin DE, 1932-1933 jan – 1r – 1 – (filmed by other misc inst: 1924-1925 n42 [1r]) – gw Misc Inst [074]

Die fanfare : gedichte der deutschen erhebung / Anacker, Heinrich – 5. aufl. Muenchen: Zentralverlag der NSDAP, F Eher, 1936 [mf ed 1988] – 116p – 1 – mf#6939 n12 – us Wisconsin U Libr [810]

Die fanfare *see* Land und stadt

Fanfare – Goettingen DE, 1931 1 nov-1933 23 mar [gaps] – 1r – 1 – (title varies: 1932?: goettinger fanfare) – gw Mikrofilm [074]

Fanfare : waschmittelwerk – Genthin DE, 1971 14 jan-1990 2 oct [gaps] – 4r – 1 – gw Misc Inst [074]

Die fanfare im pariser einzugsmarsch : eine preussische novelle / Welk, Ehm – Berlin: Im Deutschen Verlag c1942 [mf ed 1991] – 1r – 1 – (ill by fritz busse. filmed with: vereinsamtes herz / josef weinheber) – mf#2982p – us Wisconsin U Libr [830]

Fang, Chao *see* Ju tz'u huang chun

Fang, Ching *see* Yu ching

Fang, Chi-sheng *see* Tsung li i chiao yu chang hsueh-liang

Fang, Ch'iu-wei *see* Tsui chin ti jen ch'in hua chun shih hsing shih

Fang, Chun-i *see*
- Man t'ing fang
- Ssu chieh mei
- Yin hsing meng

Fang, Hsi *see* Chen shang chi

Fang huo kai lun / Huang, Chin-fu – Shang-hai: Shang wu yin shu kuan, Min kuo 36 [1947] – (= ser P-k&k period) – us CRL [360]

Fang, I *see* Kuo yin hsueh sheng tzu hui

Fang, J *see* Prediction of the world men's best performances in high jump and long jumb by the top average performance

Fang kung chou hsin lun / Ch'en, Shu-shih – [Ch'ung-ch'ing]: Cheng chung shu chu, Min kuo 28 [1939] – (= ser P-k&k period) – us CRL [327]

Fang k'ung yu kuo fang / Shen, Kuo-chun – Shang-hai: Chun shih pien i she, Min kuo 22 [1933] – (= ser P-k&k period) – us CRL [355]

Fang, Shou-ch'u *see* Mo hsueh yuan liu

Fang, Ta-tsai *see* Hsiang ts'un hsiao hsueh lao tso chiao yu

Fang, Tung-mei *see* K'o hsueh che hsueh yu jen sheng

Fang yen chu shang / Wu, Yu-t'ien – Shang-hai: Shang wu yin shu kuan, Min kuo 22 [1933] – (= ser P-k&k period) – us CRL [480]

Fang, Yin *see* P'ing fan ti yeh hua

Fang ying chien pi / Hang, Li-wu – Ch'ung-ch'ing: Chung-hua shu chu, 1944 – (= ser P-k&k period) – us CRL [915]

Fang Yu-yen *see* Hsiang ts'un chiao yu ts'ung chi

Fang, Yu-yen *see* Hsiao-chuang chih i yeh

Fangst, jakt och fiske – Stockholm: A Bonnier 1955 [mf ed Bloomington IN: Indiana Uni Lib, Preservation Dept 1984] – 1r – 1 – us Indiana Preservation [390]

Fan-hy-cheu : a tale, in chinese and english / Weston, Stephen – London: Robert Baldwin, 1814 – (= ser 19th c books on linguistics) – 1mf – 9 – (with notes, and a short grammar of the chinese language) – mf#2.1.21 – uk Chadwyck [480]

[Fani (Muhsin, Muhammad)] *see* The dabistan

Fann al-sinima – Cairo: Hasan 'Abd al-Wahhab (Jama'at al-Naqqad al-Sinima'iyin) 1933-34. yr 1 n1-8. 15 oct 1933-17 nov 1934 – (= ser Arabic journals and popular press) – 1r – 1 – $375.00 – us MEDOC [956]

Fann street foundry, london : specimen of printing types / W Thorowgood and Co, London – London: W Thorowgood & Co, 1839 – (= ser 19th c publishing...) – 4mf – 9 – mf#3.1.111 – uk Chadwyck [680]

The fannie farmer junior cook book / Perkins, Wilma Lord – 1st ed. Boston: Little, Brown and Co, 1942 (mf ed 1993) – 1r – 1 – mf#*Z-6746 n2 – us NY Public [640]

Fannie fox's cook book / Fox, Fannie Ferber – Boston: Little, Brown, and Co, 1923 (mf ed 1993) – 1r – 1 – mf#*Z-6975 – us NY Public [640]

Fanning, David *see* Col david fanning's narrative of his exploits and adventures as a loyalist of north carolina in the american revolution

Fanning, Edmund *see* Voyages round the world

Fanny kelley v. sarah l. larimer, et al / Kelley, Fanny – 1869-77 – 1 – us Kansas [978]

Fanon Research and Development Center *see* Research bulletin

The fan-qui in china : in 1836-1837 / Downing, C T – London: Henry Colburn, 1838. 3v – 4mf – 9 – (missing: v1, 3) – mf#HT-654 – ne IDC [915]

Fans and fan leaves english / Schreiber, Charlotte Elizabeth (Beatie) Guest, lady – London 1888-90 – (= ser 19th c art & architecture) – 18mf – 9 – mf#4.2.1225 – uk Chadwyck [740]

Fans of japan / Salwey, Charlotte Maria (Birch) – London: Kegan Paul, Trench, Truebner, & Co Ltd, 1894 – (= ser 19th c art & architecture) – 3mf – 9 – mf#4.1.159;c.4.1.234 – uk Chadwyck [740]

Fan-shan kung tu – Shang-hai: Ta ta t'u shu kung ying she, Min kuo 22 [1933] – (= ser P-k&k period) – us CRL [340]

Fanshawe, Anne Harrison *see* Memoirs

Fantaisie pour flute avec accompagnement de piano...op 29 / Tulou, Jean Louis – Paris: I Pleyel & Fils aine [182-] [mf ed 1992] – 1r – 1 – mf#pres. film 112 – us Sibley [780]

Fantaisie pour flute et piano sur l'air : que ne suis-je la fougere!...opera 6 / Farrenc, Aristide – Paris: Frey [182-?] [mf ed 1992] – 1r – 1 – mf#pres. film 112 – us Sibley [780]

Fantaisie pour piano sur un vieil air de ronde francais[e], [op 99] / Indy, Vincent d' – [1930] [mf ed 1988] – 1r – 1 – mf#pres. film 7 – us Sibley [780]

Fantasia boricua / Babin, Maria Teresa – New York, NY. 1956 – 1r – us UF Libraries [972]

Fantasia del dibujo popular – Santa Clara, Cuba. 1960 – 1r – us UF Libraries [972]

La fantasia en !sonetos! / Sanchez-Arjona, Vicente – Sevilla: Imp. Alvarez, Tomo 1. 1955 – 1 – sp Bibl Santa Ana [810]

La fantasia en...!sonetos! / Sanchez-Arjona, Vicente – Sevilla: Imp. Alvarez, Tomo 2. 1955 – 1 – sp Bibl Santa Ana [810]

La fantasia en...!sonetos! / Sanchez-Arjona, Vicente – Sevilla: Imp. Alvarez, Tomo 3. 1955 – 1 – sp Bibl Santa Ana [810]

La fantasia en...!sonetos! / Sanchez-Arjona, Vicente – Sevilla: Imp. Alvarez, Tomo 4. 1955 – 1 – sp Bibl Santa Ana [810]

Fantasie en verbeelding as moontlikheidsvoorwaardes vir kreatiewe denke gedurende aanvangsonderwys / Antwerp, Gertruida Cornelia van – Uni of South Africa 2000 [mf ed Johannesburg 2000] – 5mf [ill] – 9 – (text in afrikaans; abstract in english and afrikaans; incl bibl ref) – mf#mfm14832 – sa Unisa [370]

Fantasien : a series of subjects in outline = Fancies / Retzsch, Friedrich August Moritz – London: Saunders & Otley, A Richter & Co; Strasburgh, Paris...Leipzig...1834 – (= ser 19th c art & architecture) – 1mf – 9 – (parallel english, french & german descriptions) – mf#4.1.21 – uk Chadwyck [760]

Fantasies off 3 partes / jo wythie his booke – 17th c [mf ed 19–] – 1r / 2mf – 1,9 – (incl notes on mensural notation; contains works of orlando gibbons & john jenkins; also unverified attributions to john wythie & thomas tomkins) – mf#film 300 / fiche 1064 – us Sibley [780]

O fantasma – Rio de Janeiro, RJ: Typ de J R Alves e Companhia, 04-22 jun 1850 – (= ser Ps 19) – mf#P15,01,64 – bl Biblioteca [870]

Fantasmas da sao paulo antiga / Milano, Miguel – Sao Paulo, Brazil. 1949 – 1r – us UF Libraries [972]

Le fantasque – Quebec: O Cote, Proulx, 1857-1858 – 9 – mf#P04213 – cn Canadiana [073]

Fantastic adventures – New York. v1-7. may 1939-oct 1945 – (= ser Science fiction periodicals, 1926-1978. series 1) – 7r – 1 – $875.00 – us UPA [830]

Fantastic stories – New York. ser 1: v8-15. feb 1946-mar 1953. ser 2: v2 n3-v24. may-jun 1953-oct 1975 – (= ser Science fiction periodicals, 1926-1978. series 2) – 25r – 1 – $3745.00 – us UPA [830]

Fantasy and science fiction – Hoboken. 1987+ (1) 1987+ (5) 1987+ (9) – (cont: magazine of fantasy and science fiction) – ISSN: 1095-8258 – mf#6119,01 – us UMI ProQuest [400]

Fanti confederation : a reconsideration / Brown, James W – Brighton, 1967 – us CRL [320]

Fanti law report of decided cases on fanti customary laws : second selection / Sarbah, John Mensah – London: W Clowes, 1904 – us CRL [340]

Fantin Desodoards, Antoine *see* Histoire de france

Fanua, Tupou Posesi *see* Papers

Fao plant protection bulletin / Food and Agricultural Organization of the United Nations. World Reporting Service on Plant Diseases and Pests – Rome. 1954-1967 (1) – ISSN: 0014-5637 – mf#8758 – us UMI ProQuest [580]

FAO/SIDA Workshop for Intermediate Level Instructors in Home Economics and Rural Family-Oriented Programmes in East and Southern Africa, (1974: Njoro, Kenya) *see* The changing roles of women in east africa

Far away and long ago / Hudson, William Henry – New York, NY. 1918 – 1r – us UF Libraries [025]

Far east : a periodical devoted to the conversion of china – Mentone VIC: Far East Office, v2-32 n11(1921-nov 1950) [gaps] [mf ed 2005] – 4r – 1 – (publ by: irish mission to china, st columban's mission house, oct 1921-1928; st columban's mission society, 1929-nov 1950; lacking several iss; some pgs damaged) – mf#2005C-s013 – us ATLA [241]

Far east – amer ed. Omaha NE: St Columban's Foreign Mission Society, v1 n3-v34 (jun 1918-51) [mf ed 2006] – 6r – 1 – (publ by: irish mission to china, jun-oct 1918; chinese mission society of st columban, nov 1918-oct 1935; st columban's foreign mission society, nov 1935-1951; cont by: columban fathers missions; lacking several iss; some pgs damaged) – mf#2005C-s092 – us ATLA [241]

The far east : internal affairs and foreign affairs, 1945-jan 1963 / U.S. State Dept – (= ser Confidential u s state department central files) – 1 – $15,630.00 coll – (1945-49 21r $4070 isbn 1-55655-314-5. 1950-54 26r $5040 isbn 1-55655-315-3. 1955-59 30r $5810 isbn 1-55655-316-1. 1960-jan 1963 8r* isbn 1-55655-973-9 $1550. with p/g) – us UPA [950]

The far east : a magazine devoted to the organization of an irish national mission to china – Galway: Maynooth Mission to China 1918- [mthly] [v1-26 (1918-43)] [mf ed 2005] – 1r – 1 – (printed at dublin, 1918; st. columban's office at: shrule, galway 1918-jun 1927; navan, jul 1927-dec 1943; lacks: v1 n6,9,10 (jun, sep, oct 1918), v3 (1920), v7 (1924); some pgs damaged; publ by: maynooth mission to china) – mf#2005c-s035 – us ATLA [241]

Far east reporter – Houston. 1977-1979 (1) 1977-1979 (5) 1977-1979 (9) – ISSN: 0014-7575 – mf#9294 – us UMI ProQuest [950]

Far east reporter / Release the World for Christ – v22 n120-v31 n128 [1976 aug-1977 jun/jul], 1978 mar-1982 jul/aug/sep – 1r – 1 – mf#656914 – us WHS [242]

Far east views / U.S. Library of Congress. Prints and Photographs Division – six albums of travel views by major commercial studios active in the Orient in the 1870's and 1880's. 1 reel. P&P6615-6620 – 1 – us L of C Photodup [080]

Far eastern economic review – Hong Kong. 1971+ (1) 1961+ (5) 1977+ (9) – ISSN: 0014-7591 – mf#6396 – us UMI ProQuest [330]

Far eastern economic review – Hong Kong. 1946- updates twice per yr – 1,5 – (index available) – us Primary [330]

Far eastern economic review, 1947-1956 – 12r – 1 – $420.00 in US $40.00r outside – mf#L9000001 – Dist. us Scholarly Res – us L of C Photodup [321]

Far eastern law review – Manila. 1977-1980 (1) 1977-1980 (5) 1977-1980 (9) – ISSN: 0046-3272 – mf#7993 – us UMI ProQuest [950]

Far eastern review : engineering, commerce, finance – Manila: GB Rea, 1904-41. [v6, n9 feb 1910] – us CRL [073]

Far eastern review – Shanghai, China. -m. Jun 1907; 1919-oct 1941 – 35 1/4r – 1 – uk British Libr Newspaper [079]

The far eastern review, 1904-1933 – 19r – 1 – $665.00 in US $40.00r outside – (vol 23 no 7 is missing from this collection) – mf#L9000002 – Dist. us Scholarly Res – us L of C Photodup [330]

Far eastern survey – New York. 1950-1961 (1) – mf#300 – us UMI ProQuest [327]

Far north in india : a survey of the mission field and work of the united presbyterian church in the punjab / Anderson, William B & Watson, Charles Roger – Philadelphia, PA: Board of Foreign Missions of the United Presbyterian Church of North America, c1909 – 1mf – 9 – 0-8370-6002-8 – (incl ind) – mf#1986-0002 – us ATLA [242]

Far out : rovings retold / Butler, William Francis – London: W Isbister, 1881 – 4mf – 9 – mf#26663 – cn Canadiana [910]

Far – the french-american review – Fort Worth. 1980-1981 (1,5,9) – ISSN: 0160-0419 – mf#12006 – us UMI ProQuest [400]

Far undzer shul / Bastomski, Solomon – Wilno, Lithuania. 1933 – 1r – us UF Libraries [939]

Far unzer shul *see* Literarisze tribune

Faraday discussions – London. 1991+ (1) 1991+ (5) 1991+ (9) – (cont: faraday discussions of the chemical society) – mf#7188,01 – us UMI ProQuest [540]
Faraday discussions of the chemical society / Chemical Society (Great Britain) – London. 1972-1991 (1) 1972-1991 (5) 1976-1991 (9) – (cont by: faraday discussions) – ISSN: 0301-7249 – mf#7188 – us UMI ProQuest [540]
Faraday, Michael *see*
– Experimental researches in electricity
– Faraday's diary
– Lectures on the forces of matter and on the chemical history of a candle
– The manuscripts of michael faraday
Faraday's diary / Faraday, Michael – London. v1-7. 1932-36 – 9 – $267.00 – mf#0199 – us Brook [500]
Farago, L *see* Abyssinia on the eve
Farago, Lydia *see* "Ein jeder wird nach seinem mass gerichtet..."
Faragual / Chang Marin, Carlos Francisco – Panama, 1960 – 1r – us UF Libraries [972]
Farbe und lack – Hannover. 1970-1996 (1) 1972-1996 (5) 1974-1996 (9) – ISSN: 0014-7699 – mf#5813 – us UMI ProQuest [660]
Farce du chaudronnier / Chancerel, Leon – Paris, France. 1949? – 1r – us UF Libraries [440]
Fardon, Ian *see* Papers on the methodist church in rabaul
Fare, Charles Auguste de la *see* Memoires et reflexions sur les principaux evenemens du regne de louis 14
O fareco militar – Ouro Preto, MG: Typ de Leyraud, 27 jul,18 out 1833 – (= ser Ps 19) – mf#P17,02,100 – bl Biblioteca [079]
Fareham standard – Gosport, England. 28 Sept 1978-20 Dec 1979; Jan-24 Dec 1980; 8 Jan 1981-22 Jul 1982. -w.3 reels – 1 – uk British Libr Newspaper [072]
Farel, Guillaume *see*
– De la saincte cene de nostre seigneur jesus
– Du vray usage de la croix...
– Du vray usage de la croix de jesus christ
– Epistre envoye aux reliques de la dissipation horrible de l'antechrist
– Epistre envoyée au duc de lorraine
– Epistre exhortatoire...tous ceux qui ont congnoissance de l'evangile
– Forme d'oraison pour demander...dieu la saincte predication de l'evangile
– Le glaive de la parolle veritable
– Letres certaines d'aucuns grandz toubles... advenus...geneve l'an 1534
– La maniere et fasson qu'on tient en baillant le sainct baptesme...
– Oraison tres devote
– Sommaire
– Le sommaire de g. farel . . . reimprime d'apres . . . 1534
– Sommaire et briefve declaration
– Sommaire et briefve declaration
– Sommaire, une briefve declaration...
Farese, Giuseppe *see* Poesia e rivoluzione in germania, 1830-1850
Farewel sermon / Dalton, William – Liverpool, England. 1834? – 1r – us UF Libraries [240]
Farewell address / Duff, Alexander – Edinburgh, Scotland. 1839 – 1r – us UF Libraries [240]
Farewell address from a pastor to his flock / Langdale, G A – Wendover? England. 1843? – 1r – us UF Libraries [240]
Farewell discourse to the congregation and parish of st john's, glasgow / Irving, Edward – Glasgow, Scotland. 1822 – 1r – us UF Libraries [240]
Farewell, J E *see* County of ontario
Farewell letters to a few friends in britain and america, on returning to bengal in 1821 / Ward, William – London 1821 [mf ed Hildesheim 1995-98] – 1v on 2mf – 9 – €60.00 – 3-487-27433-7 – gw Olms [860]
Farewell sermon / Donaldson, John William – Bury St Edmunds, England. 1855 – 1r – us UF Libraries [240]
Farewell sermon / Dunbar, W – London, England. 1855 – 1r – us UF Libraries [240]
Farewell sermon / Fairbairn, John – Glasgow, Scotland. 1842 – 1r – us UF Libraries [240]
Farewell sermon / Hills, George – London, England. 1848? – 1r – us UF Libraries [240]
Farewell sermon / Hook, Walter Farquhar – London, England. 1837 – 1r – us UF Libraries [240]
Farewell sermon / Smith, John – Berwick-on-Tweed, England. 1885? – 1r – us UF Libraries [240]
farewell sermon, preached in the episcopal churches, st john, n b : on sunday, 7th sep 1840 / Carey, John – s.l. s.n, 1840 [mf ed 1983] – 1mf – 9 – 0-665-44015-4 – mf#44015 – cn Canadiana [240]
Farewell sermon preached in the parish church of hodnet / Heber, Reginald – Shrewsbury, England. 1826 – 1r – us UF Libraries [240]
Farewell sunday – London, England. 18– – 1r – us UF Libraries [240]
Farewell the little people / Pohl, Victor – London, England. 1968 – 1r – us UF Libraries [890]
A farewell to india / Thompson, Edward John – London: Ernest Benn, 1931 – (= ser Samp: indian books) – us CRL [954]
Farewell to paradise / Thiess, Frank – New York: A A Knopf, 1929 – 1r – 1 – us Wisconsin U Libr [430]
Farewell words to the first german reformed church, race street, philadelphia : delivered march 14, 1852 / Berg, Joseph Frederick – Philadelphia: Lippincott, Grambo, 1852 – 1mf – 9 – 0-524-08710-5 – mf#1993-1080 – us ATLA [242]
Farewells to the pope! / Maurette, Jean Jacques – London, England. 1846 – 1r – us UF Libraries [240]
Farfalla : pour violin et piano [par] mile auret / Sauret, Emile – [Berlin 1907] [mf ed 19–] – 1r – 1 – mf#pres. film 8 – us Sibley [780]
Farfan de los Godos, Antonio *see* Discursos en defensa de la religion catholica
Farges-Mericourt, P J *see* Description de la ville de strasbourg
Fargo forum – Fargo ND. 1922 dec 11-30, 1923 jan 2-feb 28, 1923 jul 2-aug 31, 1923 mar 1-apr 30, 1923 may-jun, 1923 nov-dec, 1923 sep-oct, 1924 jan-feb, 1924 mar – 9r – 1 – (cont: fargo forum and daily republican; cont by: forum [fargo, nd]) – mf#854828 – us WHS [071]
Farhan, Yahya Isa *see* Al-tatbiq al-handasi lil-kharait al-jiyumurfulujiyah
Farhang – Rasht. sal-i 1, shumarah-i 1-7 burj-i jadi 1298-burj-i saratan 1299 [nov 1919-may 1920]; sal-i 2, shumarah-i 1-11 burj-i hamal 1304-bahman 1304 [mar 1925-jan 1926]; sal-i 3, shumarah-i 1-12 farvardin-isfand 1305 [mar 1926-feb 1927]; sal-i 4, shumarah-i 1-7 farvardin-shahrivar va mihr 1307 [mar-aug, sep 1928] – 1r – 1 – $350.00 – us MEDOC [079]
Farhang-i khurasan – Khurasan. sal-i 1, shumarah-i 1-6 farvardin-shahrivar 1336 [mar-aug 1957]; sal-i 2, shumarah-i 1-12 farvardin 1337-bahman 1338 [mar 1958-jan 1960] – 1r – 1 – $225.00 – us MEDOC [079]
Farhangistan-i iran : (lughat'ha-yi naw) – Tehran. shumarah-i 4-7. farvardin 1317-isfand 1319 [mar 1937-23 mar 1941] – 1r – 1 – $175.00 – us MEDOC [079]
Faria, Alberto De *see* Maua
Faria, Francisco Xavier de *see* Vida y heroycas virtudes del v[enerabl]e padro pedro de velasco, provincial
Faria, Julio Cezar De *see* Jose bonifacio o moco
Faria, Octavio De *see* Luocos
Farias brito / Serrano, Jonathas – Sao Paulo, Brazil. 1939 – 1r – us UF Libraries [972]
Farias Galindo, Jose *see* El simbolismo medico indigena
Faribault, Eugene Rodolphe *see* The gold measures of nova scotia and deep mining
Faribault, Georges-Barthelemi *see*
– Excursion a la cote du nord, au dessous de quebec
– [Lettre]
– [Lettre]: a m la [sic] redacteur de la gazette de quebec
– Notice sur la destruction des archives et bibliotheques des deux chambres legislatives du canada
Faridi, Abid Hasan *see* An outline history of persian literature, a d 822-1926
Farina, P Jose Agustin *see* Vida de la sierva de dios sor asuncion galan de san cayetano
Farinelli : ou, la piece de circonstance / Dupin, Henri – Paris, France. 1816 – 1r – us UF Libraries [440]
Farinelli, Arturo *see*
– Goethe
– Paul heyse
Farington diary / Farington, Joseph – London. v1-8. 1922-28 – 9 – $60.00 – mf#0200 – us Brook [700]
Farington, Joseph *see*
– The diary of joseph farington, 1788-1821
– Farington diary
– Memoirs of the life of sir joshua reynolds
Farini, Gilarmi Antonio *see* Through the kalahari desert
Faris, Lillie Anne *see* The sand-table
Faris, Nabih Amin *see* The mysteries of almsgiving
Farjenel, Fernand *see* La morale chinoise
Farland, Merle *see* Solomon islands diary and index
Farley, Frederick Augustus *see*
– Unitarianism defined
– Unitarianism exhibited in its actual condition
Farley, James Lewis *see* Egypt, cyprus and asiatic-turkey
Farley, R *see* Kalendarium humanae vitae
Farley's bath journal – Bath, England 11,18 oct 1756 – 1 – uk British Libr Newspaper [072]
Farm *see* Irish farm forest and garden
Farm and country – 1981 mar-1984 nov – 1r – 1 – (cont: polk county farmer) – mf#998884 – us WHS [630]
Farm and dairy and rural home – Peterboro [Peterborough], Ont: Rural Pub Co, [1909-1918] – 9 – (cont: the canadian dairyman and farming world. cont by: farm and dairy) – mf#P05026 – cn Canadiana [630]
Farm and fireside / Clark Co. Springfield – (oct 1882-dec 1921) [semimthly, mthly] – 21r – 1 – mf#B34567-34587 – us Ohio Hist [071]
The farm and fireside – Toronto: [s.n., 18–?] – 9 – mf#P04435 – cn Canadiana [630]
Farm and fireside (eastern) / Clark Co. Springfield – (oct 1899-sep 1916) [semimthly] – 5r – 1 – mf#B34620-34624 – us Ohio Hist [071]
Farm and home – Montreal: Phelps Pub. Co, [1880?-1925] – 9 – mf#P04698 – cn Canadiana [630]
Farm and market journal – Spokane, WA. 1924-1931 (1) – mf#69259 – us UMI ProQuest [071]
Farm and ranch – Nashville. 1951-1963 (1) – mf#334 – us UMI ProQuest [630]
Farm ballads / Carleton, Will – New York: Harper & Bros, 1875 – 2mf – 9 – $3.00 – mf#LLMC 91-008 – us LLMC [810]
Farm chemicals – Willoughby. 1973-2000 (1) 1973-2000 (5) 1976-2000 (9) – ISSN: 0092-0053 – mf#1414,01 – us UMI ProQuest [630]
Farm chemicals and crop life – Willoughby. 1940-1973 [1]; 1971-1973 [5] – ISSN: 0014-7885 – mf#1414 – us UMI ProQuest [630]
Farm Equipment and Metal Workers of America *see* Cio news
Farm, field and fireside – 1894 feb 3-oct 27, 1894 nov 3-1895 jul 13, 1895 jul 20-1896 apr 4, 1896 apr 11-dec 26 – 4r – 1 – (cont: farm, field and stockman) – mf#573451 – us WHS [630]
Farm, field and fireside monthly – v1 n8-v4 n2 [1901 jul-1903 feb] – 1r – 1 – (cont: american fruit and vegetable journal) – mf#571859 – us WHS [630]
Farm, field and stockman – 1890 jan 4-dec 31, 1891 jan 3-dec 19 – 2r – 1 – (cont by: farm, field and fireside) – mf#604890 – us WHS [630]
Farm holiday news – St Paul etc. v. 1-3, no. 15. Feb 20 1933-Aug 14 1936 – 1 – us NY Public [630]
Farm income situation – Washington. 1975-1975 (1) 1975-1975 (5) 1975-1975 (9) – ISSN: 0014-7974 – mf#9151 – us UMI ProQuest [630]
Farm index – Washington. 1974-1979 (1) 1975-1979 (5) 1975-1979 (9) – ISSN: 0014-7982 – mf#7352 – us UMI ProQuest [630]
Farm industry news – Minneapolis. 1986-1996 (1) 1986-1996 (5) 1986-1996 (9) – ISSN: 0892-8312 – mf#15038,01 – us UMI ProQuest [630]
Farm journal : [midwest/central edition] – Philadelphia. 1877+ (1) 1968+ (5) 1960+ (9) – ISSN: 0014-8008 – mf#895 – us UMI ProQuest [630]
Farm journals / Burroughs, Joseph Washington – undated, Business dealings, life, and activities of a Cloud county, KS, family – 1 – us Kansas [920]
Farm labor camp design in rural marion county / Gossman, Stephen J – 1989 – 56p 1mf – 9 – $4.00 – us Kinesology [790]
Farm labor contracting in the united states, 1981 / Pollack, Susan L – Washington DC: US Dept of Agriculture, Economic Research Service...[mf ed 1984] – (= ser Agricultural economic report 542) – 9 – (with bibl) – us GPO [331]
Farm labor news *see* Sharecropper's voice, 1935-1937 / southern farm leader, 1936 / stfu news, 1938-1939 / tenant farmer, 1941-1942 / farm worker, 1943-1944 / farm labor news, 1946-1951 / the union farmer, 1952-1953 / agricultural unionist, 1952-1954
Farm Labor Organizing Committee *see* Update
Farm Labor Organizing Committee [OH] *see* Nuestra lucha
Farm Labor Organizing Committee [Ohio] *see* Boycott update
Farm lands in florida – Chicago, IL. 1910? – 1r – us UF Libraries [630]
Farm management studies of truck and citrus farms in florida / Hamilton, H G – s.l, s.l? 1923 – 1r – us UF Libraries [630]
Farm market guide / National Producers Alliance – [some issues mutilated] v1 n1-v3 n2 [1924 jul-1927 feb] – 1r – 1 – (cont by: cooperators herald; farmers union herald) – mf#1494411 – us WHS [630]
Farm mortgage loan experience in four florida counties / Miley, D Gray – s.l, s.l? 1939 – 1r – us UF Libraries [630]
Farm news – Bakersfield, CA. 1954-1969 (1) – mf#62091 – us UMI ProQuest [071]
Farm news / Dairyland Fertilizers, Inc – 1964 jan-1970 spring – 1r – 1 – mf#1056166 – us WHS [630]
The farm of the dagger : [novel] / Phillpotts, Eden – Toronto: Musson, c1904 – 4mf – 9 – 0-659-90445-4 – (ill by f m relyea) – mf#9-90445 – cn Canadiana [830]
The farm pasteuriser : or, the improvement of gathered cream butter and the sanitation of milk, cream and their products / Barre, Stanislas Morrier – Winnipeg: [s.n, 1903] – 1mf – 9 – 0-659-90227-3 – mf#9-90227 – cn Canadiana [630]
Farm poultry / Elford, Frederic C – [Ste Anne de Bellevue, Quebec?: s.n.] 1912 [mf ed 1997] – 1mf – 9 – 0-665-85281-9 – mf#85281 – cn Canadiana [636]
Farm quarterly – Cincinnati. 1946-1972 (1) 1971-1972 (5) – ISSN: 0014-8091 – mf#3330 – us UMI ProQuest [630]
Farm relief news / Wisconsin Farm Service Bureau – v1 n1-v7 n2 [1930 jan-1936 feb] – 1r – 1 – (cont by: midwest mutual news) – mf#3557725 – us WHS [630]
Farm safety review – Itasca. 1975-1980 (1) 1976-1980 (5) 1976-1980 (9) – ISSN: 0014-8105 – mf#10448 – us UMI ProQuest [630]
Farm security administration-office of war information collection / U.S. Library of Congress. Prints and Photographs Division – The most famous pictorial record of American life in the 1930's and early 1940's. 109 reels. P&P3 – 1 – us L of C Photodup [080]
Farm store – Minnetonka. 1989-1992 (1) 1989-1992 (5) 1989-1992 (9) – (cont: farm store merchandising) – ISSN: 1057-3542 – mf#1665,01 – us UMI ProQuest [630]
Farm store merchandising – Minnetonka. 1964-1989 (1) 1971-1989 (5) 1976-1989 (9) – (cont by: farm store) – ISSN: 0014-8121 – mf#1665 – us UMI ProQuest [630]
Farm technology and agri-fieldman – Willoughby. 1961-1973 [1]; 1972-1973 [5] – (cont by: agri-fieldman) – ISSN: 0191-0205 – mf#1415 – us UMI ProQuest [630]
Farm tenancy in jackson county, florida / Brooker, Marvin A – s.l, s.l? 1927 – 1r – us UF Libraries [630]
Farm truckers news of wisconsin – v3-v8 n47 [1948 jul 1-1954 may 27] – 1r – 1 – mf#1056167 – us WHS [630]
Farm week – Lurgan, Ireland. jan-19 dec 1986 – 2r – 1 – (aka: northern irelands farmweek) – uk British Libr Newspaper [072]
Farm worker *see* Sharecropper's voice, 1935-1937 / southern farm leader, 1936 / stfu news, 1938-1939 / tenant farmer, 1941-1942 / farm worker, 1943-1944 / farm labor news, 1946-1951 / the union farmer, 1952-1953 / agricultural unionist, 1952-1954
Farmaco : edizione pratica – Pavia, 2001+ [1,5,9] – ISSN: 0014-827X – mf#13602,01 – us UMI ProQuest [615]
Farmaco : edizione scientifica – Pavia, 2001+ [1,5,9] – ISSN: 0014-827X – mf#13603 – us UMI ProQuest [615]
Farmacopea tradicional indigena y practicas rituales – Lima. 1946 – 1 – us CRL [615]
Farmer – Minneapolis. 1986-1989 (1,5,9) – mf#15037 – us UMI ProQuest [630]
Farmer and Brindley *see* Specimens of marble pavements
The farmer and mechanic – Toronto, CW [Ont]: Eastwood, [1848-18–] – 9 – mf#P04276 – cn Canadiana [630]
Farmer and settler – Sydney, Australia. -w. 3 May 1912-7 Aug 1914. Imperfect. 3 reels – 1 – uk British Libr Newspaper [079]
Farmer b... – London, England. 18– – 1r – us UF Libraries [240]
Farmer, Frances *see* Will there really be a morning?
Farmer, James Eugene *see* Versailles and the court under louis 14
Farmer, John Stephen *see* Americanisms, old and new
Farmer, Larry *see* The impact of word processing and electronic mail on us courts of appeals
Farmer, Larry C *see*
– Appeals expediting systems
– Observation and study
The farmer patriot – Beaver Crossing, NE: Chas E Miller. ns: v1 n1. sep 5 1891- =old ser. v3 n7- (wkly) – 1 – (successor to the journal and bugle) – us NE Hist [071]
Farmer review – Farmer, NY. 1887-1904 (1) – mf#64963 – us UMI ProQuest [071]
Farmer stockman – Cozad, NE: David F Stevens, Jr. 1967// (mthly) [mf ed v6 n3. jan 1967 filmed 1973] – 1r – 1 – (cont by: platte valley farmer-stockman) – us NE Hist [630]
Farmer tomkins and his bibles / Beecher, Willis Judson – Philadelphia: Presbyterian Bd of Publ, 1874 – 1mf – 9 – 0-8370-2240-1 – mf#1985-0240 – us ATLA [830]
Farmer, Wilmoth Alexander *see* Ada beeson farmer
Farmer-herald – Oconto Falls WI. 1916 may 5-1917 jun 29, 1917 jul 6-1919 jan 17, 1919 jan 24-1920 may 21 – 3r – 1 – (cont: union farmer-herald; cont by: oconto falls herald [oconto falls, wi: 1920]) – mf#1003885 – us WHS [630]
Farmer-Labor Association *see* Newsletter
Farmer-labor herald – North Platte, NE: Farmer-Labor Pub Co, -mar 1925// (wkly) [mf ed 1924-25 (gaps)] – 2r – 1 – (cont by: north platte herald) – us NE Hist [071]

Farmer-labor monitor and farmers leader : [the official newspaper of bowman county] – Bowman, ND: H B French. v6 n36 aug 2 1923-v8 n34 jun 18 1925 (wkly) – 1 – (publ as: farmer-labor monitor aug 2 1923; also bears whole numbering: n308-n398; cont: farmers leader; cont by: bowman county leader and farmer-labor monitor) – mf#10375-10376 – us North Dakota [071]

Farmer-labor press / Council Bluffs Central Labor Union – 1962 oct 25-1966 mar 31, 1966 apr 7-1969 aug 28, 1969 sep 4-1972 dec 28, 1973 jan 4-1975 jan 2, 1974 mar 24-1976 dec 30, 1977 jan-1982 dec – 6r – 1 – mf#696363 – us WHS [331]

Farmer's advocate – Independence IA. 1897 jan 30, jul 3, 1901 oct 26 – 1r – 1 – (cont: farmer's alliance) – mf#851172 – us WHS [630]

Farmer's advocate : eastern edition – 1867-1920 – 41r – 1 – cn Library Assoc [630]

Farmer's advocate : western edition – 1890-1910 – 2r – 1 – cn Library Assoc [630]

Farmer's advocate : western edition – 1901-10 – 2r – 1 – (2 trailer reels (issues that were missing for 1901-10)) – cn Library Assoc [630]

Farmers advocate – Charles Town, WV. 1897-1947 (1) – mf#67232 – us UMI ProQuest [071]

Farmers advocate – Topeka KS. 1900 jan 31-1908 feb 27, 1908 mar 5-dec 5 – 2r – 1 – (cont: advocate and news; jerry simpson's bayonet; kansas state register [topeka, ks]; central farmer; western life; western breeders journal; cont by: kansas farmer) – mf#1037495 – us WHS [630]

Farmers advocate – Bottineau, ND: Northwestern Farmers Pub Co. v19 n47 apr 5 1918-v28 n44 mar 3 1927 (wkly) – 1 – (cont: bottineau county news and omemee herald. absorbed: souris messenger. cont by: bottineau county herald. missing: 1919 dec 26; 1920-22) – mf#06992-06994; 01628-01631 – us North Dakota [071]

Farmers advocate *see* Hawkesbury chronicle / farmers advocate

Farmers' advocate – Humboldt, NE: Humboldt Print Co, jul 9 1881-v2 n43. apr 28 1883 (wkly) [mf ed with gaps filmed 1990] – 1r – 1 – us NE Hist [071]

Farmers advocater and advertiser – Ithaca, NY. 1823-1835 (1) – mf#65008 – us UMI ProQuest [071]

Farmer's alliance / National Farmers' Alliance – Independence IA. 1891 dec 17 – 1r – 1 – (cont: national advocate [independence, ia: 1878]; cont by: farmer's advocate) – mf#851156 – us WHS [630]

The farmers' alliance – Lincoln, NE: Alliance Pub Co. 3v. v1 n26. dec 14 1889-v3 n42. mar 31 1892 (wkly) [mf ed 1962?] – 2r – 1 – (cont: alliance. merged with: nebraska independent to form: farmers' alliance and nebraska independent) – us NE Hist [630]

Farmers' alliance and nebraska independent – Lincoln, NE: Alliance Pub Co. 2v. v3 n43. apr 7 1892-v4 n2. jun 23 1892 (wkly) [mf ed 1962?] – 1r – 1 – (formed by the union of: farmers' alliance and: nebraska independent. cont by: alliance-independent) – us NE Hist [071]

Farmers' alliance in florida / Knauss, James Owen – Durham, North Carolina, 1926 – 1r – us UF Libraries [630]

Farmers' and Laborers' Union of America *see* Proceedings of the annual session of the farmers and laborers union of america

Farmers' and Mechanics' Institute of Streetsville (Ont) *see* Constitution and general laws of the farmers' and mechanics' institute of streetsville, in the county of peel

Farmers and miners journal – Lykens, PA., 1856-1857 – 13 – $25.00r – us IMR [071]

Farmers' and planters' friend – Philadelphia. 1821-1821 (1) – mf#3802 – us UMI ProQuest [630]

Farmers and the clergy / Rose, Hugh James – London, England. 1831 – 1r – us UF Libraries [240]

Farmer's bulletin – Washington. 1889-1930 (1) – mf#5657 – us UMI ProQuest [630]

Farmers' cabinet and american herd-book : devoted to agriculture, horticulture, and rural and domestic affairs – Philadelphia. 1836-1848 (1) – mf#3983 – us UMI ProQuest [630]

Farmer's cooperative associations in florida / Hamilton, H G – Gainesville, FL. 1939 – 1r – us UF Libraries [334]

Farmers' cooperative associations in florida / Brooker, Marvin A – Gainesville, FL. 1932 – 1r – us UF Libraries [334]

Farmers' cooperative associations in florida / Brooker, Marvin A – Gainesville, FL. 1933 – 1r – us UF Libraries [334]

Farmers' cooperative associations in florida / Hamilton, H G – Gainesville, FL. 1935 – 1r – us UF Libraries [334]

Farmers' cooperative associations in florida citrus cooperative / Hamilton, H G – Gainesville, FL. 1943 – 1r – us UF Libraries [334]

Farmers' Co-operative Packing Company [Madison WI] *see* Co-operative educator

Farmers courant *see* The natal standard / farmers courant

Farmer's daughter in the west of england / Teall, John – London, England. 18– – 1r – us UF Libraries [240]

Farmer's digest – Brookfield. 1937+ [1]; 1971+ [5]; 1977+ [9] – ISSN: 0046-3337 – mf#1809 – us UMI ProQuest [630]

Farmers Educational and Cooperative Equity Union *see* Farmers equity union news

Farmers equity union news / Farmers Educational and Cooperative Equity Union – 1935 jan-1942 oct – 1r – 1 – (cont: equity union farmer; cont by: wisconsin farmers union news) – mf#1427833 – us WHS [331]

Farmer's exchange – v1 n1-v3 n14 [1979 nov 15-1981 jul 16] – 1r – 1 – (cont by: farmer's exchange [verona, wi: central ed]) – mf#607499 – us WHS [630]

Farmer's exchange – 1934 dec 14-1936 apr 24, 1936 may-1937, 1938 jan 7-dec 30 – 3r – 1 – (cont: farmer's exchange and the noble farmer) – mf#955627 – us WHS [630]

Farmer's exchange – 1981 jul 27-1982 jun, 1982 jul-1983 jun, 1983 jul 14-1984 may 3 – 3r – 1 – (cont: farmer's exchange [verona, wi: wisconsin southwest ed]; cont by: dairy farmer's exchange) – mf#719449 – us WHS [630]

The farmers' exchange – Bayard, NE: C W Clifton. v5 n39. may 25 1922- (wkly) – 2r – 1 – us Bell [071]

Farmers exchange and the morrill county news, combined – Bayard, NE: C W Clifton. 4v. sep 1918-v5 n38. may 18 1922 (wkly) [mf ed v2 n4. sep 25 1918-may 22 1919 (gaps) filmed 1999] – 1r – 1 – (cont: morrill county news. cont by: farmers exchange) – us NE Hist [071]

Farmer's exchange bulletin *see* Miscellaneous newspapers of saguache county

Farmers' federation news – Asheville. 1949-1958 (1) – mf#423 – us UMI ProQuest [630]

Farmers' free press – Fairmont, WV. 1892-1925 (1) – mf#67275 – us UMI ProQuest [071]

Farmer's friend – Green Bay WI. 1942 dec 17, 1943 oct 22-1946, 1947-49, 1950-52, 1953-55, 1956-58, 1959-61, 1962 jan-jul 13 – 7r – 1 – (cont by: farmer's friend and rural reporter) – mf#1001469 – us WHS [630]

Farmer's friend – Mechanicsburg, PA. -w 1891-1912. 8 rolls – 13 – $25.00r – us IMR [071]

Farmer's friend and rural reporter – Denmark, WI, Green Bay WI. 1962 jul 20-1963, 1964, 1965-1967 mar 31, 1967 apr 7-1969 mar 13, 1969 mar 20-1971 apr 1, 1971 apr 8-1972 dec 28, 1973 jan 1-1974 sep 30, 1974 oct 3-1976 sep 30, 1976 oct 7-28 – 9r – 1 – (cont: farmer's friend; cont by: new farmer's friend and rural reporter) – mf#1001463 – us WHS [630]

Farmers gazette *see* Farmers gazette and journal of practical horticulture

Farmers' gazette – Dublin, nov 1842-1849, 1863 – 8r – 1 – ie National [072]

Farmers gazette and journal of practical horticulture – Dublin, Ireland. 1 nov 1845-29 nov 1846; 1850-1896 – 50r – 1 – (aka: irish farmers gazette and journal of practical horticulture; farmers gazette) – uk British Libr Newspaper [072]

Farmers gazette and journal of practical horticulture *see* Irish farmers gazette and journal of practical horticulture

Farmers' gazette and midland counties advertiser – Oxford, England. -w. 7 Nov 1843-2 July 1844. 20 ft – 1 – uk British Libr Newspaper [072]

Farmers' guide (perth) – [Scotland] Perth: D Robertson 17 sep 1897 (wkly) [mf ed 2004] – 1v on 1r – 1 – uk Newsplan [630]

Farmers' Institute and Subscription Library Society *see* Catalogue of books and an abridgement of the constitution and rules...

Farmers' Joint Stock Banking Co *see* Deed of settlement...

Farmers' journal *see* Niagara peninsula newspapers, pt 2

Farmer's journal for orkney and shetland and the north of scotland – [Scotland] Orkney, Kirkwall: J Calder nov 1876-dec 1879 (mthly) [mf ed 2004] – 2r – 1 – (cont: farmer's monthly journal and agricultural and general magazine for orkney and shetland; cont by: orkney & shetland telegraph and farmer's journal) – uk Newsplan [630]

Farmers leader – Bowman, ND: Farmers Pub Co. v1 n21 mar 14 1918-v6 n35 jul 26 1923 (wkly) – 1 – (exclusive official newspaper of bowman county & 6th judicial district of north dakota (later just official newspaper of bowman county, 1920-23; official paper of the nonpartisan league for bowman county, 1917-19; also bears whole numbering: n21-n307; cont: farmers leader and gascoyne gazette; cont by: farmer-labor monitor and farmers leader; missing: 1918 apr 18; 1920 mar 25; 1922 may 11) – mf#10373-10375 – us North Dakota [630]

Farmers leader and gascoyne gazette : [official paper of the nonpartisan league for bowman county] – Bowman, ND: Farmers Pub Co. v1 n4 nov 15 1917-v1 n20 mar 7 1918 (wkly) – 1 – (also bears the numbering of the gascoyne gazette: [v3 n2]-v3 n47; and whole numbering: n4-n20; formed by the union of: gascoyne gazette and farmers leader; cont by: farmers leader; missing: 1918 jan 10-31, feb 7) – mf#10373 – us North Dakota [071]

Farmer's magazine – Edinburgh. 1800-1825 (1) – mf#5344 – us UMI ProQuest [630]

The farmer's manual – Fredericton [NB]: Print and pub by P A Phillips, [1844-1845?] – 9 – ISSN: 1190-6634 – mf#P04533 – cn Canadiana [630]

The farmer's manual – Kentville, NS: G W Woodworth, [1880-1881?] – 9 – mf#P04624 – cn Canadiana [630]

The farmer's manual and veterinary guide – Montreal: Family Herald and Weekly Star, [191-?] – 3mf – 9 – 0-665-74253-3 – mf#74253 – cn Canadiana [630]

Farmers National Committee for Action *see* Farmers' national weekly

Farmers National Educational Association *see* – Farmers national weekly

Farmers national weekly / Farmers National Educational Association – Chicago IL, Minneapolis MN. 1934 jan 15 – 1r – 1 – (cont: producers news [plentywood, mt: national ed]; farmers national weekly [washington, dc: 1933]; workers and farmers cooperative bulletin; cont by: farm holiday news; national farm holiday news) – mf#3910478 – us WHS [630]

Farmers national weekly – Minneapolis. n.s. v. 1-3 no. 25. Jan 15 1934-Aug 21 1936 – 1 – us NY Public [630]

Farmers national weekly : voice of the toiling farmers / Farmers National Educational Association – 1936 jan 3 – 1r – 1 – (cont: workers and farmers cooperative bulletin; producers news; farmers national weekly; cont by: farm holiday news; national farm holiday news) – mf#3925816 – us WHS [630]

Farmers' national weekly / Farmers National Committee for Action – Washington DC. 1933 aug 12 – 1r – 1 – (cont: farm news letter; organized farmer [dublin, pa: 1932]; cont by: producers news [plentywood, mt: national ed]; workers and farmers cooperative bulletin; farmers national weekly [chicago: 1934]) – mf#3910465 – us WHS [630]

Farmer's odd man – London, England. 1864? – 1r – us UF Libraries [240]

Farmer's register – Greensburg, PA. 1799-1840 (1) – mf#65912 – us UMI ProQuest [071]

Farmer's register : a monthly publication – Shelbanks. 1833-1843 (1) – mf#4372 – us UMI ProQuest [630]

Farmers register – Greensburg, PA., 1808-1811 – 13 – $25.00r – us IMR [071]

Farmers review : [official paper of bowman county] – Rhame, Bowman Co, ND: A D Fuller, jun 1918? -v12 n38 jan 1 1920 (wkly) – 1 – (cont: rhame review (1908); cont by: rhame review (1920)) – mf#10358-10359 – us North Dakota [071]

Farmers' review – v21 n1-v22 n51 [1890 jan 1-1891 dec 23] – 1r – 1 – (cont by: orange judd farmer) – mf#761698 – us WHS [630]

Farmers' sentinel – Milwaukee WI. 1906 aug 2, 16-sep 27, oct 11, 25 – 1r – 1 – mf#1167080 – us WHS [630]

The farmers' stock book : a manual on the breeding, feeding, management, and care of live stock, and common sense treatment and prevention of diseases of farm animals / Periam, Jonathan – Toronto, Chicago: International Pub Co, [1887?] – 5mf – 9 – 0-665-91787-2 – (incl app) – mf#91787 – cn Canadiana [636]

Farmer's telegram – Superior WI. 1923 jan 6-apr 28, sep 6 – 1r – 1 – mf#927725 – us WHS [630]

The farmer's tour through the east of england... / Young, A – London, 1771. 4v – 25mf – 9 – mf#HT-190 – ne IDC [914]

Farmers tribune – Des Moines IA. n1431-1501 [1894 may 30-1895 sep 1] – 1r – 1 – mf#948830 – us WHS [630]

Farmers Union Central Exchange [Saint Paul MN] *see* Co-op country news

Farmers union herald – 1927-1972/1974 sep 16 – 12r – 1 – (cont: farm market guide; cooperators herald; cont by: co-op country news) – mf#1494937 – us WHS [331]

Farmers union in washington / National Farmers' Union [US] – 1948 may 14-1952 aug 29, 1952 sep 5-1954 jul 2 – 2r – 1 – mf#1056176 – us WHS [331]

Farmers union news – Helena, MT. 1935-1969 (1) – mf#64455 – us UMI ProQuest [071]

Farmers weekly – Sutton. 1978-1980 (1,5,9) – ISSN: 0014-8466 – mf#11286 – us UMI ProQuest [630]

The farmer's wife – St Paul: Webb Publ Co [-1935]. [v15-16, n8 may 1910-11]; v14, n9-11 jan-mar 1912; v15-18 may 1912-may 1916; v19, n8-v21, n7 1917-1918; v26, n8-v38 1924-nov 1935] – us CRL [071]

The farmer's wife magazine – St Paul: Webb Pub Co. [v38 n12 dec 1935; v39-42 n4 1939-apr 1939] – us CRL [071]

Farmers'journal – Antigo WI. 1911 oct 3-1913 jan 28, 1913 feb 4-1914 oct 20, 1914 oct 27-1916 oct 3, 1916 oct 10-1919 jul 8, 1919 jul 15-1921 dec 27, 1922 jan 3-1923 dec 25 – 6r – 1 – mf#958601 – us WHS [630]

Farmer-stockman of nebraska – Cozad, NE: David F Stevens, Jr. 6v. v9 n3. oct 1968-v14 n37. sep 15 1975 (wkly) [mf ed with gaps filmed 1973-87] – 5r – 1 – (cont: platte valley farmer-stockman. cont by: farmer-stockman of the midwest. publ in cozad ne oct 1968-may 1972; in superior ne, jan 22 1973-sep 15 1975) – us NE Hist [630]

Farmer-stockman of the midwest – Superior, NE: Nebraska Farmer-Stockman. v14 n38. sep 22 1975- (wkly) [mf ed with gaps filmed 1987-] – 1 – (cont: farmer-stockman of nebraska. publ in superior ne, sep 22 1975-may 26 1986; in belleville ks, jun 2 1986-) – us NE Hist [630]

Farming – Tampa, FL. 1926? – 1r – us UF Libraries [630]

Farming – Toronto: Bryant Press, [1895?-1900?] – 9 – mf#P04047 – cn Canadiana [630]

The farming and account books of 1641 : from the library of the surtees society / Best, Henry; ed by Robinson, Charles Best – 1857 – 3mf – 7 – mf#275 – uk Microform Academic [650]

Farming for profit : a hand-book for the american farmer / Read, John Elliot – Brantford, Ont: Bradley, Garretson, 1880 – 10mf – 9 – 0-665-91860-7 – mf#91860 – cn Canadiana [630]

Farming world – Toronto: D T McAinsh, [1900-1902] – 9 – (cont: farming; merged with: canadian farm and home, to become: the farming world and canadian farm and home) – mf#P04236 – cn Canadiana [630]

Farming world – Toronto: Farming World, [1907-1908?] – 9 – (cont: the farming world and canadian farm and home; merged with: the canadian dairyman, to become: canadian dairyman and farming world) – mf#P05124 – cn Canadiana [630]

Farming world – Toronto, Canada. -w. 2 nov 1903-15 dec 1904; 1905-1 feb 1908 – 4r – 1 – uk British Libr Newspaper [630]

Farming world and canadian farm and home – Toronto: Dominion Phelps, [1903-1907] – 9 – (merger of: the farming world; canadian farm and home; cont by: the farming world) – mf#P05015 – cn Canadiana [630]

Farmingdale, South Dakota. Farmingdale Baptist Church *see* Records

Farmington. Ohio. United Presbyterian and Congregational Church *see* Church records, ms 2125

Farm-labor times – 1934 aug 9 – 1r – 1 – mf#3925796 – us WHS [331]

Farmland news / Fulton Co. Archbold – jul 1972-dec 1982 [wkly] – 17r – 1 – mf#B13121-13137 – us Ohio Hist [071]

Farmline – Washington. 1980-1992 (1) 1980-1992 (5) 1980-1992 (9) – ISSN: 0270-5672 – mf#12429 – us UMI ProQuest [630]

Farms, how and where to obtain them : manitoba, assiniboia, alberta, saskatchewan: the four great fertile provinces of the canadian northwest described and illustrated – s.l: s.n, 1891 – 1mf – 9 – mf#05649 – cn Canadiana [630]

Far'n folk – London, UK. 7 Mar-2 May 1916 – 1 – uk British Libr Newspaper [072]

The farnam echo – Farnam, NE: [s.n.] dec 1903-35th yr n22. feb 1 1940 (wkly) [mf ed with gaps] – 7r – 1 – (cont: public press. merged with: gothenburg times and gothenburg independent to form: gothenburg times, gothenburg independent and farnam echo. issues for jan 7-28 1915 incorrectly dated jan 7-28 1914. numbering very irregular) – us NE Hist [071]

Farnam, Jonathan Everett *see* Open communion shown to be unscriptural and deleterious. a history of infant baptism

The farnam press – Farnam, NE: L H Whitman. 25v. v1 n1. sep 26 1940-v25 n34. jun 24 1965 (wkly) [mf ed lacks jun 30 1960, aug 22 1963 filmed 1970-] – 7r – 1 – (absorbed by: hi-line enterprise. iss for jul 18-aug 15 1963 called v23 n38-42 but constitute v23 n37-41) – us NE Hist [071]

Farndish – (= ser Bedfordshire parish register series) – 1mf – 9 – £3.00 – uk BedsFHS [929]

Farndish, st michael monumental inscriptions – L H Chambers 1928 – (= ser Bedfordshire parish register series) – 1mf – 9 – £1.25 – uk BedsFHS [929]

Farnell, Lewis Richard *see*
- The cults of the greek states
- The evolution of religion
- Greece and babylon
- The higher aspects of greek religion

Farney, Roger *see* La religion de l'empereur julien et le mysticisme de son temps

Farnham, Charles Haight *see* A life of francis parkman
Farnol, Jeffery *see*
- The amateur gentleman
- Beltane the smith
- The broad highway
Farnsworth, E M *see* Kamba grammar
Farnworth chronicle – [NW England] Farnworth 1906-21 Dec 1917 [wkly] – 1 – uk MLA; uk Newsplan [072]
Farnworth express and district effective advertiser – [NW England] Farnworth 10 apr 1890-4 jun 1891* – 1 – uk MLA; uk Newsplan [072]
Farnworth messenger – Farnworth, England. -m. July 1874-Aug 1876. 19 ft – 1 – uk British Libr Newspaper [072]
Farnworth observer – [NW England] Farnworth 7 nov 1868-apr 1873* – 1 – uk MLA; uk Newsplan [072]
Farnworth weekly – [NW England] Farnworth dec 1952-, 12 jan-11 dec 1953, 11 mar-12 aug 1954 – 1 – uk MLA; uk Newsplan [072]
Farnworth weekly journal – [NW England] Farnworth 30 aug 1873-7 aug 1875 – 1 – (title change: farnworth journal & observer [14 aug 1875-16 jun 1883]; farnworth weekly journal & observer [23 jun 1883-3 feb 1950]; farnworth & worsley journal [10 feb 1950-61, 1963-75]) – uk MLA; uk Newsplan [072]
Faro – Miami, FL. 1991 may 01-1999 jul 15 – 1r – us UF Libraries [071]
El faro dominical – 1927 – 260p – 1 – us Southern Baptist [242]
Faro nell and her friends : wolfville stories / Lewis, Alfred Henry – Toronto: Bell & Cockburn, c1913 [mf ed 1995] – 5mf – 9 – 0-665-74857-4 – (ill by w herbert dunton & j n marchand) – mf#74857 – cn Canadiana [830]
Farol / Soldevilla, Dolores – Habana, Cuba. 1964 – 1r – us UF Libraries [972]
O farol do norte – Recife, PE. 01-04 mar 1876 – (= ser Ps 19) – bl Biblioteca [079]
Farol indiano, y guia de curas de indios : summa de los cinco sacramentos que administran los ministros evangelicos en esta america. con todos los casos morales que suceden entre indios / Perez, Manuel – en Mexico: Por Francisco de Rivera Calderon, ano de 1713 – (= ser Books on religion...1543/44-c1800: confesionarios) – 3mf – 9 – mf#crl-24 – ne IDC [241]
El farouk : revue musulmane litteraire, economique et d'education sociale – Alger. 1913-15, 1920-21. (v1-3, v8-9) – 1 – fr ACRPP [073]
Farquhar, J N *see* Modern religious movements in india
Farquhar, John Nicol *see*
- The approach of christ to modern india
- The crown of hinduism
- Gita and gospel
- Modern religious movements in india
- An outline of the religious literature of india
- Permanent lessons of the gita
- A primer of hinduism
- Primer of hinduism
Farquhar, William *see*
- Arguments in favour of lay representation in ecclesiastical synods
- Few thoughts on the eucharistic question
Farr, Frederic William *see* A manual of christian doctrine
Farrago / Eyser, J – St Louis: C Witter, 1876 – 1r – 1 – us Wisconsin U Libr [880]
Farrago confvsanearvm et inter se dissidentivm opinionum de coena domini / Westphal, J aus Hamburg – [Magdeburg, 1552] – 1mf – 9 – mf#TH-1 mf 1482 – ne IDC [242]
Farrago sententiarvm consentientivm in vera et catholica doctrina, de coena domini, quam firma assensione / Timan, J – Francoforti, 1555 – 7mf – 9 – mf#TH-1 mf 1458-1464 – ne IDC [242]
Farrand, Max *see*
- The framing of the constitution of the united states
- Record of the federal convention of 1789
- The records of the federal convention of 1787
Farrar, Adam Storey *see*
- A critical history of free thought in reference to the christian religion
- Science in theology
Farrar, Ephraim H *see* The greatest of the world's forces applied through a half-day perpetual, industrial, and universal school
Farrar, Frederic William *see*
- The art annual of 1893 william holman hunt
- A brief greek syntax and hints on greek accidence
- Companions for the devout life
- The early days of christianity
- Ephphatha
- Eternal hope
- The fall of man and other sermons
- The first book of kings
- The gospel according to st luke
- The herods
- History of interpretation
- The influence of the revival of classical studies on english literature during the reigns of elizabeth and james 1
- The life and work of st paul
- The life of christ
- The life of christ as represented in art
- Lives of the fathers
- The lord's prayer
- Men i have known
- The messages of the books
- The minor prophets
- The second book of kings
- The silence and the voices of god
- Sin and its conquerors
- Solomon
- True religion
- The voice from sinai
- The witness of history to christ
Farrar, Fredric William *see* Seekers after god
Farrar, John *see* An ecclesiastical dictionary
Farrar, John (Mrs) *see* Children's robinson crusoe
Farrar, Reginald *see* The life of frederic william farrar
Farrell, Alfred *see*
- Negro churches
- Negro education
- Negro ethnography
- Negro history
- Negro religion
Farrell, Gerard J *see* Accompaniment to gregorian chant in medieval harmony
Farrell, Hugh *see* The sherman law, an anchor, to yesterday
Farrell, John P *see* An outline of the laws of the ttpi relating to real property
Farrell, Kevin P *see* Effects of a proximal provocation on carpal tunnel syndrome
Farrell, Warren *see* Liberated man
Farrenc, Aristide *see* Fantaisie pour flute et piano sur l'air
Farrer, James Anson *see*
- Paganism and christianity
- Zululand and the zulus
Farrer, Reginald John *see* In old ceylon
Farrer, Thomas Henry *see*
- The neo-protection scheme of the right hon joseph chamberlain
- The sugar convention
Farrier's magazine – Philadelphia, 1818-1818 [1,5,9] – mf#3803 – us UMI ProQuest [636]
Farries, Francis Wallace *see* Anniversary sermon preached in knox church, november 30th, 1890
Farrington, Joseph R *see* The american samoan commission's visit to samoa, september-october 1931
Farrington, S *see*
- Ideal of religion
Farrow, Edward Samuel *see* Gas warfare
Farsat, Henri *see* L'eglise d'apres calvin
Farsi, S S *see* Swahili sayings from zanzibar
Farsy, Muhammad Saleh *see* Kurwa na doto
Farther defence of the methodists / Benson, Joseph – London, England. 1794 – 1r – us UF Libraries [242]
Farthest-north collegian – College AK. 1935 aug 1, 1937 jun 1, aug 1-sep 1, dec 1, 1938 sep 1, 1941 aug 1 – 1r – 1 – mf#853357 – us WHS [071]
Faruk – Athens. Idare Mueduerue: Ivanaki Pasfidi; Muharriri: Mevlanazade Rifat. n1. 25 tesrinisani 1327 [1911] – (= ser O & t journals) – 1mf – 9 – $25.00 – us MEDOC [956]
Faruki, Zahiruddin *see* Aurangzeb and his times
Farun, Fred Nagib *see* Economic study of the lake hamilton citrus growers' association
Faruqi, Burhan Ahmad *see* Imam-i rabbani mujaddid-i-alf-i thani shaikh ahmad sirhindi's conception of tawhid
Farvos kemfn mire kegn religye / Sudarskivi, I – Kharkov, Ukraine. 1931 – 1r – us UF Libraries [939]
Farwell, Arthur *see* Works
Farwell, Brice *see* Works
Farwell, Mary E *see* The life of william carey
Farwell, William Washington *see*
- Questions for law students on cooley's constitutional limitations
- Questions for law students on story's equity pleadings
Faryad-i gawd'nishinan – Tehran: [Sazman-i Mujahidin-i Khalq-i Iran], 1979. shumarah-'i 1-66. 30 tir 1358-7 aban 1359 [21 jul-29 oct 1980?] – 1r – 1 – $53.00 – (missing: n51-54, 56-57) – us MEDOC [079]
Fas reports / University of Pennsylvania – 1981 nov – 1r – 1 – mf#5307313 – us WHS [071]
Faschismus – Amsterdam NL, 1939*; 1941-45* – 1 – Dist. gw Mikrofilm – gw Misc Inst [074]
Fasciculi aliquot sacrarum cantionum cum 4, 5, 6 & 8 vocibus / Lasso, Orlando di – 1589 – (= ser Mssa) – 10mf – 9 – €110.00 – mfchl 307 – gw Fischer [780]
Fasciculi zizaniorum magistri johannis wyclif : cum tritico / Netter, Thomas; ed by Shirley, Walter Waddington – London: Longman, Brown, Green, Longmans & Roberts, 1858 [mf ed 1992] – (= ser Rerum brittannicarum medii aevi scriptores 5) – 2mf – 9 – 0-524-03355-2 – (text in latin, int in english. incl bibl ref) – mf#1990-0936 – us ATLA [931]
Fasciculi zizaniorum magistri wyclif cum tritico (rs5) / Netter, Thomas; ed by Schirley, W W – 1858 – (= ser The rolls series (rs)) – €21.00 – ne Slangenburg [242]
Fasciculus aliquot cantionum sacrarum quinque vocum / Lasso, Orlando di – 1572 – (= ser Mssa) – 3mf – 9 – €50.00 – mfchl 290 – gw Fischer [780]
Fasciculus disputationum theologicarum de socianismo / Heidanus, A – Lugduni Batavorum, 1659 – 1mf – 9 – mf#PBA-187 – ne IDC [240]
Fasciculus sanctorum ordinis cisterciensis / Henriquez, Chrysostomus – Bruxellae, 1623 – 19mf – 8 – €37.00 – ne Slangenburg [241]
The fascination of the book / Work, Edgar Whitaker – New York: Fleming H Revell, c1906 – 1mf – 9 – 0-524-06005-3 – mf#1992-0742 – us ATLA [220]
Fascism, 1933-1945 : from the international transport workers' federation archives held at the modern records centre, university of warwick library – 22mf – 9 – mf#87545 – uk Microform Academic [320]
El fascismo al desnudo – Madrid, 1937 – (= ser Blodgett coll) – 9 – mf#fiche w876 – us Harvard [946]
El fascismo intenta destruir el museo del prado – Madrid, 1936? – (= ser Blodgett coll) – 9 – mf#fiche w877 – us Harvard [946]
El fascismo internacional y la guerra antifascista espanola / Garcia Oliver, Juan – Barcelona? 1937? – (= ser Blodgett coll) – 9 – mf#fiche w908 – us Harvard [946]
El fascismo pretende encarcelar espana / Catalonia. Comissariat de Propaganda – n.p. 193? – (= ser Blodgett coll) – 9 – mf#fiche w878 – us Harvard [946]
El fascismo y las armas y las letras espanolas / Albornoz, Alvaro de – Madrid, 1938 – (= ser Blodgett coll) – 9 – mf#fiche w707 – us Harvard [946]
The fascist *see* Fascist and anti-fascist newspapers
Fascist and anti-fascist archives from the hackney archives, london – (= ser Fascism and reactions to fascism in britain (1918-89)) – 5r – 1 – (with guide. incl: the jewish workers circle minutes (1935-1952); surveys of fascism in britain) – mf#97575 – uk Microform Academic [320]
Fascist and anti-fascist archives from the imperial war museum, london – (= ser Fascism and reactions to fascism in britain (1918-89)) – 3r – 1 – (with guide. incl: capt luttman johnson's archive and correspondence. the personal papers of j macnab, r ling, memoirs and accounts detailing personal responses to fascism) – mf#97574 – uk Microform Academic [320]
Fascist and anti-fascist newspapers – (= ser Fascism and reactions to fascism in britain series (1918-1989)) – 1 – (incl: action, oct-dec 1931 1r 97591; challenge, 1935-39 2r 97593; the fascist, mar 1929-sep 1939 1r 97595; fascist bulletin, jun 1925-jun 1934 1r 97594; fascist week, nov 1933-may 1934 1r 97596) – uk Microform Academic [320]
Fascist bulletin *see* Fascist and anti-fascist newspapers
Fascist spain, menace to world peace / Wolff, Milton – New York: New Century, 1947. 16p – 1 – us Wisconsin U Libr [946]
Fascist spain, menace to world peace / Wolff, Milton – New York: New Century Publ 1947 – 9 – mf#w1255 – us Harvard [946]
The fascist threat to culture; a speech delivered on march 8, 1937, in the new lecture hall, harvard university / Malraux, Andre – Under the auspices of the Cambridge Union of University Teachers and the Harvard Student Union. Cambridge, 1937. Fiche W1017. (Blodgett Collection of Spanish Civil War Pamphlets) – 9 – us Harvard [946]
Fascist week – nov 1933-may 1934 – 1r – 1 – (incorp in: blackshirt in 1934 (not pt of the fascism and reactions to fascism series.)) – mf#97596 – uk Microform Academic [320]
Fascist week *see* Fascist and anti-fascist newspapers
Faseb journal – Bethesda. 1987+ (1,5,9) – (cont: federation proceedings / federation of american societies for experimental biology) – ISSN: 0892-6638 – mf#16278 – us UMI ProQuest [574]
Die faserpflanze flachs/lein : materialien zur kunst des anbaus, der verarbeitung, veredelung und handels / Heubach, Helga – 2000 – 2mf – 9 – 3-8267-2597-2 – mf#DHS 2597 – gw Frankfurter [300]
Die faserpflanze hanf : materialien zur geschichte ihres anbaus, ihrer verarbeitung und handels / Heubach, Helga – (mf ed 1995) – 2mf – 9 – €40.00 – 3-8267-2104-7 – mf#DHS 2104 – gw Frankfurter [631]
Fashion advertising collection, 1942-1982 – 7 sections. 1r per y – 1 – $8850.00 diazo; $9990.00 silver – us Alper [740]
Fashion in deformity as illustrated in the customs of barbarous and civilised races / Flower, Sir, William Henry – London: Macmillan & Co, 1881 – (= ser 19th c art & architecture) – 2mf – 9 – mf#4.1.47 – uk Chadwyck [740]
Fashionable lady : or, harlequin's opera. in the manner of a rehearsal. as it is perform'd at the theatre in goodman's-fields / Ralph, James – London: J Watts 1730 [mf ed 19–] – 2mf – 9 – mf#fiche 9 – us Sibley [790]
Fassbinder, Franz *see*
- Eichendorffs lyrik
- Friedrich hebbel
Fassel, Hort *see* Beitraege zur literaturgeschichte und – methodologie
The fast and thanksgiving days of new england / Love, William De Loss – Boston: Houghton, Mifflin, 1895 – 2mf – 9 – 0-524-06958-1 – (incl bibl ref) – mf#1990-5322 – us ATLA [390]
Fast before communion / Poyntz, Newdigate – London, England. 1872 – 1r – us UF Libraries [240]
Fast ferry international – Kingston-Upon-Thames. 1989-1996 (1) 1989-1996 (5) 1989-1996 (9) – (cont: high-speed surface craft) – ISSN: 0954-3988 – mf#2997,02 – us UMI ProQuest [629]
Fast food – New York. 1974-1974 (1) – (cont by: restaurant business) – ISSN: 0014-8725 – mf#2967 – us UMI ProQuest [640]
Fast forward for atlantans onthe move – 1989 feb-mar, 1990 feb-1990 dec – 1r – 1 – mf#3072052 – us WHS [071]
Fast, Howard *see* Spartacus
Ein fast nutzlich vslegung des ersten psalmen... / Erasmus – [Basel, Adam Petri, 1520] – 1mf – 9 – mf#PBU-539 – ne IDC [240]
Fast sermon : in which the real cause of all wars and public calamities / Proud, J – Birmingham, England. 1796 – 1r – us UF Libraries [240]
Fast sermons to parliament: reproductions in facsimile with notes / ed by Jeffs, Robin – London: Cornmarket Press, 1970-1971 – 1 – us Wisconsin U Libr [240]
Les fastes de Louis 15, de ses ministres, maitresses, gene'raux, et autres notables personnages de son regne / Bouffonidor – Ville-Franche 1782 [mf ed Hildesheim 1995-98] – 3v on 10mf – 9 – €100.00 – ISBN-10: 3-487-26209-6 – ISBN-13: 978-3-487-26209-3 – gw Olms [944]
Fastes de napoloeon : ou traits les plus remarquables de son regne – Bruxelles 1818 [mf ed Hildesheim 1995-98] – 1v on 2mf [ill] – 9 – €60.00 – 3-487-26361-0 – gw Olms [944]
Fastes episcopaux de l'ancienne gaule, tome 1 : provinces du sud-est / Duchesne, Louis – Paris, 1894 – €15.00 – ne Slangenburg [241]
Fastes episcopaux de l'ancienne gaule, tome 2 : l'aquitaine et les lyonnaises / Duchesne, Louis – Paris, 1900 – €18.00 – ne Slangenburg [241]
Fastes episcopaux de l'ancienne gaule, tome 3 : les provinces du nord et de l'est / Duchesne, Louis – Paris, 1915 – €12.00 – ne Slangenburg [241]
Les fastes historiques du vieux montreal = The historical records of old montreal / Morin, Victor – Montreal: editions des Dix, 1944 [mf ed 1987] – 2mf – 9 – mf#SEM105P745 – cn Bibl Nat [971]
Fasti – Toronto: University College, [1884-18– or 19–] – 9 – mf#P04405 – cn Canadiana [378]
Fasti apostolici : a chronology of the years between the ascension of our lord and the martyrdom of ss. peter and paul / Anderdon, William Henry – London: Kegan Paul, Trench, 1882 – 1mf – 9 – 0-8370-2090-5 – mf#1985-0090 – us ATLA [220]
Fasti consulares imperii romani von 30 v. chr. bis 565 n. chr : mit kaiserliste und anhang / Liebenam, Wilhelm – Bonn: A Marcus & E Weber 1909 [mf ed 1992] – (= ser Kleine texte fuer theologische und philologische vorlesungen und uebungen 41-43) – 1mf – 9 – 0-524-04619-0 – (incl bibl ref) – mf#1990-1279 – us ATLA [930]
Fasti monastici aevi saxonici : or, an alphabetical list of the heads of religious houses in england previous to the norman conquest / Birch, Walter de Gray – London: Truebner, 1873 [mf ed 1990] – 1mf – 9 – 0-7905-5568-9 – mf#1988-1568 – us ATLA [241]
Fasti of the irish presbyterian church, 1613-1840 / McConnell, James – Pts. 5-12 – 1 – $50.00 – us Presbyterian [242]
Fasti sacri : or, a key to the chronology of the new testament / Lewin, Thomas – London: Longmans, Green, 1865 [mf ed 1985] – 2mf – 9 – 0-8370-4098-1 – (incl bibl ref & ind) – mf#1985-2098 – us ATLA [225]
Fasting / Frere, John – London, England. 1840 – 1r – us UF Libraries [240]

Fasting-day / M'dowall, P – Edinburgh, Scotland. 1855 – 1r – us UF Libraries [240]
Fastnachtspiele aus dem fuenfzehnten jahrhundert – Stuttgart: Litterarischer Verein, 1853-1858 [mf ed 1993] – (= ser Blvs 28-30, 46) – 4v – 1 – (incl bibl ref and ind) – mf#8470 reel 10 – us Wisconsin U Libr [820]
Fastre, Paul, Father *see* Notes sur les moeurs et coutumes des fujuges
Faswoc newsletter / Food and Service Workers of Canada – 1985 may 1-1986 dec – 1r – 1 – (cont: faswoc union news) – mf#1477252 – us WHS [660]
Faswoc union news / Food and Service Workers of Canada – v1 n1-2 [1982 jul-oct], v2 n1-2 [1983 jan-may], v3 n1, [1984 may], 1985 feb 23 – 1r – 1 – (cont by: faswoc newsletter) – mf#1477256 – us WHS [331]
Fat intake of university students / Brown, Stefani – 1998 – 1mf – 9 – $4.00 – mf#HE 647 – us Kinesology [613]
Fatal consequences of gambling / Scott, John – Hull, England. 1856? – 1r – us UF Libraries [360]
The fatal opulence of bishops : an essay on a neglected ingredient of church reform / Handley, Hubert – London: Adam and Charles Black, 1901 – 1mf – 9 – 0-524-03342-0 – mf#1990-0923 – us ATLA [240]
Fatal use of the sword / Madan, Spencer – Birmingham, England. 1805 – 1r – us UF Libraries [360]
Fatalism or freedom / Herrick, C Judson – New York, NY. 1926 – 1r – us UF Libraries [025]
Fatalisme et liberte dans l'antiquite grecque / Amand, D – Louvain, 1945 – 11mf – 8 – €21.00 – ne Slangenburg [180]
Fatat boston – Boston: Publ by W E Shakir, [dec 28 1917-may 8 1919] – us CRL [071]
Fatat misr al-fatah – Cairo: Amali 'Abd al-Masih (Jam'iyat Fatat Misr al-Fatah) 1921-? v1 n1-v3 n10. apr 1921-jan 1923 – (= ser Arabic journals and popular press) – 1r – 1 – $75.00 – us MEDOC [956]
Fate – St. Paul. 1970+ (1) 1948+ (5) 1974+ (9) – ISSN: 0014-8776 – mf#5876 – us UMI ProQuest [130]
Le fate *see* Due novelline toscane
Fate rides a tortoise; a biography of ellen spencer mussey / Hathaway, Grace – Philadelphia: Winston Co., 1937. 204p. LL-206 – 1 – us L of C Photodup [340]
Fateful year : being the speeches and writings during the year of presidentship of congress / Kripalani, Jiwatram Bhagwandas – Bombay: Vora & Co, 1948 – (= ser Samp: indian books) – us CRL [954]
Fath : newsletter – [S.l: Fath, jun 1969-mar 1970] – 1r – us CRL [071]
Fath (Organization) *see* Balagh 'askari raqm
Father beschi of the society of jesus : his times and his writings / Besse, Leon – Trichinopoly: St Joseph's Industrial School Press, 1918 [mf ed 1995] – (= ser Yale coll) – iv/246p (ill) – 1 – 0-524-09907-3 – mf#1995-0907 – us ATLA [241]
Father clark, or, the pioneer preacher : sketches and incidents of rev. john clark / Peck, John Mason – New York: Sheldon, Lamport & Blakeman, 1855 – 1mf – 9 – 0-7905-5543-3 – mf#1988-1543 – us ATLA [240]
"Father corson" : or, the old style canadian itinerant: embracing the life and gospel labours of the rev robert corson, fifty-six years a minister in connection with the central methodism of upper canada / ed by Carroll, John – Toronto: publ by Samuel Rose at the Methodist Book Room, 1879 [mf ed 1980] – 4mf – 9 – 0-665-00483-4 – mf#00483 – cn Canadiana [242]
Father damen's lecture : the answer to popular objections against the catholic religion: a verbatim report / Waller, William Henry – Ottawa?: I B Taylor, 1871 [mf ed 1984] – 1mf – 9 – 0-665-04199-3 – mf#04199 – cn Canadiana [240]
Father damen's lecture, thursday evening, 14th december : "the catholic church the only true church of god"; the fallacy of private interpretation clearly proved: the orthodoxy of the catholic religion established / Waller, William Henry – Ottawa?: I B Taylor, 1871 – 1mf – 9 – mf#23716 – cn Canadiana [241]
Father damen's lectures : 1. the private interpretation of the bible; 2. the catholic church the only true church of god; 3. confession; 4. the real presence; 5. answers to popular objections against the catholic church – London, Ont: Catholic Record Pub House, 1900 – 2mf – 9 – mf#56016 – cn Canadiana [241]
Father eells : or, the results of fifty-five years of missionary labors in washington and oregon. a biography of rev cushing eells, dd / Eells, Myron – Boston: Congregational Sunday-School & Publ Society, c1894 – 1mf – 9 – 0-7905-4463-6 – mf#1988-0463 – us ATLA [240]
Father flynn / Needham, Geo. C – New York: James A. O'Connor, c1890 – 1mf – 9 – 0-8370-8283-8 – mf#1986-2283 – us ATLA [240]
Father hecker / Sedgwick, Henry Dwight – Boston: Small, Maynard, 1906 [mf ed 1992] – (= ser Beacon biographies of eminent americans; Roman catholic coll) – 1mf – 9 – 0-524-04484-8 – (incl bibl ref) – mf#1992-2033 – us ATLA [241]
Father joques at the lake of the holy sacrament / Costa, Benjamin Franklin de – S.l: s.n, 1900 – 1mf – 9 – mf#27820 – cn Canadiana [978]
Father marquette / Thwaites, Reuben Gold – New York: D. Appleton, 1902 – (= ser Appleton's life histories) – 1mf – 9 – 0-7905-6327-4 – mf#1988-2327 – us ATLA [240]
Father murphy's reply to the witness correspondence evoked by his defence of papal infallibility : a lecture delivered in the mechanics' hall, monday, october 18th, 1875 / Murphy, James – Montreal: D & J Sadlier, 1875? – 1mf – 9 – mf#11175 – cn Canadiana [241]
Father taylor / Collyer, Robert – Boston: American Unitarian Association, 1906 – 1mf – 9 – 0-524-04292-6 – mf#1992-2012 – us ATLA [240]
Father-daughter relationships during six developmental periods and sexual victimization : comparing women from intact and divorced families / Nicharat, Wasaporn – 1997 – 1mf – 9 – $4.00 – mf#PSY 1953 – us Kinesology [640]
The fatherhood of god : and its relation to the person and work of christ and the operations of the holy spirit / Wright, Charles Henry Hamilton – Edinburgh: T & T Clark, 1867 – 1mf – 9 – 0-8370-6551-8 – (incl bibl ref and indexes) – mf#1986-0551 – us ATLA [210]
The fatherhood of god : being the first course of the cunningham lectures / Candlish, Robert Smith – 5th ed. Edinburgh: Adam and Charles Black. 2v. 1870 – 2mf – 9 – 0-7905-0871-0 – mf#1987-0871 – us ATLA [210]
The fatherhood of god : considered in its general and special aspects and particularly in relation to the atonement... / Crawford, Thomas Jackson – 3rd ed. Edinburgh, London: William Blackwood 1868 [mf ed 1989] – 1mf – 9 – 0-7905-0980-6 – (incl bibl ref) – mf#1987-0980 – us ATLA [240]
Fatherland – New York NY (USA), 1914 24 aug-1917 24 jan – 2r – 1 – gw Misc Inst [071]
The fatherland. (viereck's the american weekly.-vireck's the american monthly.-american monthly) – New York. 24 Aug 1914-Jan 1933.-w,m. 10 reels – 1 – uk British Libr Newspaper [071]
Fathers and sons / Turgenev, Ivan Sergeevich – New York, NY. no date – 1r – us UF Libraries [460]
Father's barmitzvah exhortation / Adler, Hermann – London, England. 1889 – 1r – us UF Libraries [939]
The fathers of greek philosophy / Hampden, Renn Dickson – Edinburgh: A and C Black, 1862 – 1mf – 9 – 0-7905-7401-2 – (incl bibl ref) – mf#1989-0626 – us ATLA [180]
The fathers of jesus : a study of the lineage of the christian doctrine and traditions / Cook, Keningale – London: K Paul, Trench, 1886 – 2mf – 9 – 0-524-04326-4 – (incl bibl ref) – mf#1990-3310 – us ATLA [240]
Fathers of the catholic church : a brief examination of the "falling away" of the church in the first three centuries / Waggoner, Ellet Joseph – Oakland, Cal[if]: Pacific Press, 1888 – 1mf – 9 – 0-524-04855-X – mf#1990-1347 – us ATLA [241]
The fathers of the desert; or, an account of the origin and practice of monkery among heathen nations; its passage into the church...and stories. / Ruffner, Henry – New York: Baker and Scribner, 1850. 2v – 1 – us Wisconsin U Libr [240]
The fathers of the german reformed church in europe and america / Harbaugh, Henry – Lancaster: Sprenger & Westhaeffer 1857-[1956] [mf ed 1984] – 12v on 4r – 1 – $85.00 set – 0-8370-0404-7 – mf#1984-b146 – us ATLA [242]
The fathers of the third century / Jackson, George Anson – New York: D Appleton, 1881 – (= ser Early Christian Literature Primers) – 1mf – 9 – 0-524-02702-1 – mf#1990-0683 – us ATLA [240]
Father's treatment of the lost son on his return / Wright, R – London, England. 1824 – 1r – us UF Libraries [240]
Fathom – Washington. 1973-1998 (1) 1974-1998 (5) 1975-1998 (9) – ISSN: 0014-8822 – mf#7353 – us UMI ProQuest [380]
Fatiche spirituali di simone molinaro, a sei voci – 1610 – (= ser Mssa) – 2mf – 9 – €35.00 – mfchl 155 – gw Fischer [780]
Fatigue / Mosso, Angelo – 1972 – 5mf – 9 – $15.00 – us Kinesology [612]
Fatigue press – iss n8 n10-31 [1968 sep-1971 dec] – 1r – 1 – mf#721621 – us WHS [071]
Fatima et les filles de mahomet : notes critiques pour l'etude de la sira / Lammens, Henri – Romae: Sumptibus Pontificii Instituti Biblici, 1912 – (= ser Scripta Pontificii Instituti Biblici) – 1mf – 9 – 0-524-01778-6 – (incl bibl ref) – mf#1990-2626 – us ATLA [260]
Fatin *see* The divan project
Fato urbano na bacia do rio paraiba, estado de sao / Muller, Nice Lecocq – Rio de Janeiro, Brazil. 1969 – 1r – us UF Libraries [972]
Fatores adversos na formacao brasileira / Berlinck, Eodoro Lincoln – Sao Paulo, Brazil. 1954 – 1r – us UF Libraries [972]
Fattening market hogs in dry lot : using dried grapefruit pulp, blackstrap molasses and alfalfa / Kirk, W Gordon – Gainesville, FL. 1947 – 1r – us UF Libraries [636]
Fattening steers on winter pasture with ground snapped corn, ground shallu heads, molasses / Kidder, Ralph W – Gainesville, FL. 1943 – 1r – us UF Libraries [636]
Fattorini, Gabriele *see* I sacri concerti a due voci...
Faublee, Jacques *see*
– Esprits de la vie a madagascar
– Introduction au malgache
Le faubourg – Paris: Assoc generale typogr, mar 26 1871 – (filmed as pt of: commune de paris newspapers) – us CRL [074]
Faucett baptist church. faucett, missouri : church records – 1895-1972 – 1 – us Southern Baptist [242]
Faucher de Saint-Maurice *see*
– A la brunante
– A la veillee
– De quebec a mexico, souvenirs de voyage, de garnison, de combat et de bivouac
– De tribord a babord
– La gaspesie
– La gaspesie, promenades dans le golfe saint-laurent
– Les larmes du christ
– Loin du pays, souvenirs d'europe, d'afrique, et d'amerique
Fauchois, Rene *see*
– Augusta
– Beethoven
– Masques et bergamasques
Faucon, Maurice *see* La librairie des papes d'avignon
Fauconberg, Thomas *see* Terrar of the estate of lord thomas fauconberg
The fauconberge memorial: : an account of henry fauconberge, ll.d. of beccles and of the endowment...to encourage learning and the instruction of youth / Rix, Samuel Wilton – Ipswich: J.M. Burton, 1849 – iv/84p – 1 – us Wisconsin U Libr [370]
Fauconnet, Andre *see* Liebe und hass
Faughnan, Thomas *see*
– Stirring incidents in the life of a british soldier
– The young hussar
Faukelius, H *see* Babel
Faulhaber, Johann *see* Newe geometrische und perspectiuische inuentiones etlicher sonderbahrer instrument
Faulhaber, Johannes *see* Arithmetischer tausendkuenstler
Faulhaber, Michael Von *see*
– Judentum, christentum, germanentum
Faulkner, Arthur *see*
– Rambling notes and reflections
– Visit to germany and the low countries, in the years 1829, 30 and 31
Faulkner, John Alfred *see*
– Cyprian the churchman
– Methodists
Faulkner journal – Orlando. 1985+ (1,5,9) – ISSN: 0884-2949 – mf#17768 – us UMI ProQuest [420]
Faulkner, William *see* Light in august
Faulkners dublin postboy – Dublin, Ireland. 20 dec 1725-12 jan 1726 – 1/4r – 1 – uk British Libr Newspaper [072]
Faull, Joseph Horace *see* The anatomy of the osmundaceae
The fault of one / Albanesi, Effie Adelaide Maria – London: Kegan Paul, Trench, Truebner & Co, 1897 – (= ser 19th c women writers) – 4mf – 9 – mf#5.1.31 – uk Chadwyck [830]
Faultes faults, and nothing else but faultes / Rich, Barnaby – 1606 – 9 – us Scholars Facs [840]
The faults of speech : a self-corrector and teachers' manual / Bell, Alexander Melville – Boston?: J P Burbank?, c1880 – 1mf – 9 – mf#28683 – cn Canadiana [616]
Faulwasser, Julius *see* Grosse brand und der wiederaufbau von hamburg
Faun – Berlin DE, 1918 23 dec-1919 9 jun – 1r – 1 – gw Misc Inst [074]
Fauna / Harris, Jack D – s.l, s.l? 1936 – 1r – us UF Libraries [590]
Fauna : palm beach county, florida – s.l, s.l? 193-? – 1r – us UF Libraries [590]
Fauna : tropical fish / Sweett, Zelia Wilson – s.l, s.l? 1936 – 1r – us UF Libraries [590]
The fauna and geography of the maldive and laccadive archipelagoes : being the account of the work carried on and of the collections made by an expedition during the years 1899 and 1900 / Gardiner, J S – Cambridge, 1903-1906. 2v – 24mf – 9 – mf#Z-2248 – ne IDC [590]
Fauna japonica : sive descriptio animalium, quae in itinere per japoniam... / Siebold, P F [B] von – Basel. 1966-1974 (1) 1971-1974 (5) 1974-1974 (9) – 64mf – 9 – mf#2054 – ne IDC [910]
Fauna japonica / Thunberg, C P – Bonn. 1972-1980 (1) 1972-1980 (5) 1975-1980 (9) – 1mf – 9 – mf#7393 – ne IDC [590]
Fauna, labelle, florida, hendry county / Huss, Veronica E – s.l, s.l? 193-? – 1r – us UF Libraries [978]
Faunce, Daniel Worcester *see*
– Advent and ascension
– The christian experience
– The christian in the world
– Hours with a sceptic
– The mature man's difficulties with his bible
– Prayer as a theory and a fact
– Shall we believe in a divine providence?
– A young man's difficulties with his bible
Faunce, William Herbert Perry *see*
– The educational ideal in the ministry
– New horizon of state and church
– The social aspects of foreign missions
– What does christianity mean?
Faunce, William Herbert Perry et al *see* A guide to the study of the christian religion
Fauquier citizen – Warrenton, VA. 1989-2000 (1) – mf#68593 – us UMI ProQuest [071]
Faur, Louis Francois *see* Confident par hasard
Faure, D P *see* My life and times
Faure, Gabriel *see* La derniere journee de sappho
Faure, Raymond *see* Souvenirs du midi
Faure requiem / Borgman, Jean Pawley – U of Rochester 1948 [mf ed 19–] – 3mf – 9 – mf#fiche375 – us Sibley [780]
Faurie, Annari *see* The admissibility and evaluation of scientific evidence in court
Fausboell, Viggo *see* Indian mythology according to the mahabharata
Fausse agnes : ou, le poete campagnard / Destouches, Nericault – Paris, France. 1802 – 1r – us UF Libraries [440]
Fausse agnes : ou, le poete campagnard / Destouches, Nericault – Paris, France. 1823 – 1r – us UF Libraries [440]
Fausse-Lendry *see* Memoires sur charles 10
Fausset, Hugh l'anson *see* Proving of psyche
Faussett, Godfrey *see*
– Revival of popery
– Sermon on the necessity of educating the poor
Faust : adaptation en douze tableaux des deux faust de goethe / Vedel, Emile – Paris: Librairie theatrale, artistique et litteraire, 1913 (mf ed 1990) – 1 – (filmed with: la tragedie du docteur faust de goethe en vers francais) – us Wisconsin U Libr [820]
Faust : chast pervaia / Goethe, Johann Wolfgang von; ed by Kholodkovskogo, N A – Moskva: Gos izdatel stvo 'Khudozhestvennaia literatura', 1936 [mf ed 1993] – 267p – 1 – (in russian) – mf#8660 – us Wisconsin U Libr [430]
Faust : eine dichtung / Avenarius, Ferdinand – Muenchen: G D W Callwey, [1919] [mf ed 1988] – 133p – 1 – mf#6970 – us Wisconsin U Libr [810]
Faust : a dramatic poem / Goethe, Johann Wolfgang von – 7th ed. Edinburgh, London: W Blackwood, 1879 – 1r – 1 – us Wisconsin U Libr [820]
Faust : a dramatic poem / Goethe, Johann Wolfgang von – Edinburgh: W Blackwood, 1887 – 1r – 1 – us Wisconsin U Libr [820]
Faust : a dramatic poem / Goethe, Johann Wolfgang von – new ed. Boston: Tricknor, Reed, and Fields, 1853 [mf ed 1993] – 322p – 1 – (incl bibl ref. trans into english with notes by abraham hayward) – mf#8611 – us Wisconsin U Libr [810]
Faust : ein dramatisches gedicht in drei abschnitten / Wolfram, Ludwig Hermann; ed by Neurath, Otto – Berlin: Frensdorff, [1906] – 1 – (incl bibl ref and indexes) – us Wisconsin U Libr [810]
Faust : erster teil, zweiter teil, urfaust / Goethe, Johann Wolfgang von; ed by Heimann, Moritz – Leipzig: Tempel-Verlag, [1921?] – 1r – 1 – us Wisconsin U Libr [430]
Faust : erster theil / Goethe, Johann Wolfgang von; ed by Hart, James Morgan – New York: G P Putnam, 1893 (mf ed 1990) – 1r – 1 – (filmed with: ueber goethe's egmont. incl bibl ref) – us Wisconsin U Libr [820]
Faust : an exposition of "goethe's faust" / Reichlin-Meldegg, Karl Alexander, Freiherr von – New York: J Miller, [1864?] – 1r – 1 – (incl bibl ref) – us Wisconsin U Libr [430]
Faust : ein fragment / Goethe, Johann Wolfgang von; ed by Seuffert, Bernhard – Heilbronn: Henninger, 1882 [mf ed 1993] – (= ser Deutsche litteraturdenkmale des 18. und 19. jahrhunderts 5) – xv/89p – 1 – (repr fr: leipzig, 1790. incl bibl ref and int) – mf#8676 reel 1 – us Wisconsin U Libr [430]

Faust : fragment / Lenau, Nicolaus – Stuttgart: F Brodhag, 1835 [mf ed 1996] – 134p – 1 – mf#9706 – us Wisconsin U Libr [820]
Faust / Goethe, Johann Wolfgang von – Paris: A Michel, 1947 (mf ed 1990) – 1r – 1 – (filmed with: faust) – us Wisconsin U Libr [820]
Faust / Goethe, Johann Wolfgang von – Paris: Delarue, [19–?] (mf ed 1990) – 1r – 1 – (filmed with: faust) – us Wisconsin U Libr [820]
Faust / Goethe, Johann Wolfgang von – Dublin: Hodges, Figgis, 1880 (mf ed 1990) – 1r – 1 – (filmed with: faust und urfaust. incl bibl ref) – us Wisconsin U Libr [820]
Faust / Goethe, Johann Wolfgang von – Firenze: G C Sansoni, 1927 (mf ed 1990) – 1r – 1 – (filmed with: goethes faust. incl bibl ref) – us Wisconsin U Libr [820]
Faust / Goethe, Johann Wolfgang von – Paris: A Lemerre. 2v in 1. [1908] (mf ed 1990) – 1r – 1 – (filmed with: goethe's faust: pt i) – us Wisconsin U Libr [820]
Faust / Goethe, Johann Wolfgang von – Paris: A Lemerre. 2v. 1891 (mf ed 1990) – 1 – (filmed with: le faust de goethe) – us Wisconsin U Libr [820]
Faust / Goethe, Johann Wolfgang von – 4., durchaus rev Aufl. Leipzig: O R Reisland. 2v. 1898 (mf ed 1990) – 1r – 1 – (filmed with: studien zu goethes egmont. incl bibl ref) – us Wisconsin U Libr [820]
Faust / Goethe, Johann Wolfgang von – Berlin: Askanischer Verlag, C A Kindle, 1924 – 1r – 1 – (incl bibl ref) – us Wisconsin U Libr [430]
Faust / Goethe, Johann Wolfgang von; ed by Schroeer, Karl Julius – Leipzig: O R Reisland, 1896-1898 [mf ed 1990] – 2v – 1 – (incl bibl ref and ind) – mf#7323 – us Wisconsin U Libr [820]
Faust / Gounod, Charles – Paris, France. 187- – 1r – us UF Libraries [780]
Faust : part one / Goethe, Johann Wolfgang von – London: G Routledge; New York: E P Dutton, [1927] (mf ed 1990) – 1r – 1 – (filmed with: ueber goethe's egmont) – us Wisconsin U Libr [820]
Faust : das persoenlich gepraegte abbild des deutschen geistes in seiner art und entartung... / Freybe, A – Halle (Saale): R Muehlmann, 1911 – 1r – 1 – us Wisconsin U Libr [430]
Faust : tragedia / Goethe, Johann Wolfgang von – Firenze: G C Sansoni, 1900 (mf ed 1990) – 1r – 1 – (filmed with: la tragedie du docteur faust de goethe en vers francais) – us Wisconsin U Libr [820]
Faust : eene tragedie / Goethe, Johann Wolfgang von – Brussel: C Muquardt, 1842 (mf ed 1990) – 1r – 1 – (filmed with: faust und urfaust) – us Wisconsin U Libr [820]
Faust : tragedie / Goethe, Johann Wolfgang von – Paris: Librairie de la Bibliotheque Nationale, [1925?] (mf ed 1990) – 1r – 1 – (filmed with: faust und urfaust) – us Wisconsin U Libr [820]
Faust : tragedie / Goethe, Johann Wolfgang von – Paris: C Delgrave, 1884 (mf ed 1990) – 1r – 1 – (filmed with: goethes faust in urspruenglicher gestalt. german text of pt one of faust with french notes) – us Wisconsin U Libr [820]
Faust : tragedie / Goethe, Johann Wolfgang von – Paris: Perrin, 1905 (mf ed 1990) – 1r – 1 – (filmed with: la tragedie du docteur faust de goethe en vers francais) – us Wisconsin U Libr [820]
Faust : une tragedie / Goethe, Johann Wolfgang von – Paris: La Renaissance du Livre [1920] [mf ed 1990] – 2v in 1/2pl – 1 – (incl bibl ref) – mf#7078 – us Wisconsin U Libr [820]
Faust : a tragedy / Goethe, Johann Wolfgang von – Dublin: W Robertson, 1860 (mf ed 1990) – 1r – 1 – (filmed with: faust und urfaust. incl bibl ref) – us Wisconsin U Libr [820]
Faust : a tragedy / Goethe, Johann Wolfgang von – New York: J Cape & H Smith, 1930 [mf ed 1993] – xxi/262p/6pl (ill) – 1 – (trans by alice raphael. int by mark van doren. woodcuts by lynd ward) – mf#8611 – us Wisconsin U Libr [820]
Faust : a tragedy: the second pt: vol ii / Goethe, Johann Wolfgang von – London: Longmans, Green, 1889 (mf ed 1990) – 1r – 1 – (filmed with: ueber goethe's egmont) – us Wisconsin U Libr [820]
Faust : tragoedie / Goethe, Johann Wolfgang von – 2. aufl. Leipzig: E Avenarius 1900 [mf ed 1990] – 1r – 1 – (filmed with: goethes faust) – mf#7327 – us Wisconsin U Libr [820]
Faust : eine tragoedie / Goethe, Johann Wolfgang von – 2. Bearb. Berlin: G Hempel. 2v in 1. 1879 (mf ed 1990) – 1r – 1 – (filmed with: ueber goethe's egmont. incl bibl ref) – us Wisconsin U Libr [820]
Faust : eine tragoedie / Goethe, Johann Wolfgang von; ed by Loeper, G von – Leipzig: Hempel. 2v in 1. [19–?] (mf ed 1990) – 1r – 1 – (filmed with: goethes faust in urspruenglicher gestalt) – us Wisconsin U Libr [820]
Faust : eine tragoedie / Goethe, Johann Wolfgang von – Muenchen: T Stroefer, [1887?] – 1r – 1 – us Wisconsin U Libr [820]
Faust : der tragoedie 1. und 2. teil / Goethe, Johann Wolfgang von – Leipzig: W Borngraeber, [1923] (mf ed 1990) – 1r – 1 – (filmed with: goethes faust) – us Wisconsin U Libr [820]
Faust : der tragoedie erster teil, synoptisch / Goethe, Johann Wolfgang von; ed by Lebede, Hans – Berlin: W Borngraeber, Verlag Neues Leben, [1912] [mf ed 1993] – 240p – 1 – (int by ed) – mf#8603 – us Wisconsin U Libr [820]
Faust : der tragoedie letzter akt / Goethe, Johann Wolfgang von; ed by Wahl, Hans – Weimar: Goethe-Gesellschaft, 1929 [mf ed 1993] – (= ser Schriften der goethe-gesellschaft 42) – 31p – 1 – mf#8657 reel 10 – us Wisconsin U Libr [430]
Faust : das volksbuch und das puppenspiel: nebst einer einleitung ueber den ursprung der faustsage / ed by Simrock, Karl – 3. Aufl. Basel: B Schwabe. 2v in1. 1903 – 1 – us Wisconsin U Libr [820]
Faust : eine weltdichtung / Koester, Albert – Muenchen: Verlag fuer Kulturpolitik, 1924 – 1r – 1 – us Wisconsin U Libr [430]
Il faust : verzione integra dell' edizione critice di weimar con introduzione e commento e cura di guido manacorda / Goethe, Johann Wolfgang von – Milano: A Mondadori, 1932 (mf ed 1990) – 1r – 1 – (filmed with: fausto. incl bibliographies) – us Wisconsin U Libr [820]
Le faust / Goethe, Johann Wolfgang von – Paris: P Ollendorff, 1895 (mf ed 1990) – 1r – 1 – (filmed with: goethes faust: in german and french on facing pages) – us Wisconsin U Libr [430]
Faust 2. teil : in der sprachform gedeutet / May, Kurt – Berlin: Junker und Duennhaupt, 1936 – 1r – 1 – (incl bibl ref) – us Wisconsin U Libr [430]
Faust 2. teil als politische dichtung / Heilbrunn, Ludwig – Frankfurt (Main): Neuer Frankfurter Verlag, 1925 – 1r – 1 – us Wisconsin U Libr [430]
Faust, a tragedy / Goethe, Johann Wolfgang von – Transl. into English verse, with notes and preliminary remarks by John Stuart Blackie. 2nd ed., carefully rev. and largely rewritten. London: Macmillan, 1880. ixxvii, 296p – 1 – us Wisconsin U Libr [820]
Faust, Albert Bernhardt *see*
– Charles sealsfield, (carl postl) der dichter beider hemisphaeren
– The german element in the united states
Faust als fuehrer : [ein buch fuer nachdenkliche menschen: der tragoedie erster und zweiter teil] – Muenchen: Faust-Verlag, 1919 [mf ed 1993] – 111p (ill) – 1 – mf#8604 – us Wisconsin U Libr [430]
Faust als tragoedie / Wiese, Benno von – Stuttgart: W Kohlhammer, [1946] – 1r – 1 – (incl bibl ref) – us Wisconsin U Libr [430]
Faust, Bernhard *see* Reiterliebe
Faust de goethe : essai d'adaption scenique integrale: precede d'une etude critique et d'une bibliographie dramatique / Goethe, Johann Wolfgang von – Paris: R Chiberre, c1922 (mf ed 1990) – 1r – 1 – (filmed with: goethe's faust: pt i) – us Wisconsin U Libr [820]
Le faust de goethe : 1re et 2e parties: en 7 tableaux et 1 prologue (premiere adaptation francaise) / Goethe, Johann Wolfgang von – Paris: Societe generale d'editions, 1908 (mf ed 1990) – 1r – 1 – (filmed with: faust) – us Wisconsin U Libr [820]
Le faust de goethe / Blaze, M Henri – Paris: Dutertre; M Levy freres, 1847 [mf ed 1993] – 373p/10pl (ill) – 1 – (essay on goethe by henri blaze. ill by tony johannot) – mf#8631 – us Wisconsin U Libr [430]
Le faust de goethe : essai de critique impersonelle / Lichtenberger, Ernest – Paris: F Alcan, 1911 [mf ed 1993] – 223/[1]p – 1 – (incl bibl ref) – mf#8604 – us Wisconsin U Libr [430]
Le faust de goethe / Goethe, Johann Wolfgang von – 2e ed rev et augm d'une preface et d'une appendice. Paris: Sandoz et Fischbacher, 1883 (mf ed 1990) – 1r – 1 – (filmed with: faust) – us Wisconsin U Libr [820]
Le faust de goethe / Goethe, Johann Wolfgang von – Paris: F Plon, 1881 (mf ed 1990) – 1r – 1 – (filmed with: faust) – us Wisconsin U Libr [820]
Le faust de goethe / Goethe, Johann Wolfgang von – Paris: Librairie nouvelle, A Bourdilliat, 1859 (mf ed 1990) – 1r – 1 – (filmed with: goethe's faust: pt 1) – us Wisconsin U Libr [820]
Le faust de goethe / Goethe, Johann Wolfgang von – Berlin: Wedekind, 1911 (mf ed 1990) – 1r – 1 – (filmed with: goethe's faust: pt i) – us Wisconsin U Libr [820]
Le faust de goethe / Goethe, Johann Wolfgang von – Paris: C Delagrave, 1893 (mf ed 1990) – 1r – 1 – (filmed with: la tragedie du docteur faust de goethe en vers francais. in german and french) – us Wisconsin U Libr [820]
Faust, der nichtfaustische / Boehm, Wilhelm – Halle a/S: M Niemeyer, 1933 [mf ed 1990] – 135/[1]p – 1 – (incl bibl ref) – mf#7341 – us Wisconsin U Libr [430]
Faust, ein menschenleben : versuch einer harmonistischen analyse des goetheschen faust / Schmidt, C – [S.l.: s.n., 1895?]; Berlin: Druck von Rosenbaum & Hart – 1r – 1 – us Wisconsin U Libr [430]
Faust, first part / Goethe, Johann Wolfgang von; ed by Salm, Peter – New York: Bantam Books, 1967 – 1 – (german and english parallel texts on facing pages with an introduction in english. incl bibl ref) – us Wisconsin U Libr [820]
Der faust goethes : einfuehrung und erklaerung / Witkowski, Georg – Leipzig: Duerr & Weber 1923 [mf ed 1990] – 1r – 1 – (filmed with: goethes faust / kurt wagner) – mf#7362 – us Wisconsin U Libr [430]
Faust ii : la folle nuit de walpurgis / Masclaux, Pierre – Paris: Presses universitaires de France, 1923 (mf ed 1990) – 1r – 1 – (filmed with: goethe's faust: pt i) – us Wisconsin U Libr [820]
Faust im zeichen des kreuzes : eine neue deutung der faustgestalt als einfuehrung in die lebensphilosophie / Kochheim, Gustav – Hamburg: Agentur des Rauhen Hauses, 1930 – 1 – us Wisconsin U Libr [430]
Faust in monbijou : roman / Bloem, Walter – Leipzig: K F Koehler, c1931 [mf ed 1989] – 226/[1]p – 1 – (incl bibl) – mf#7032 – us Wisconsin U Libr [830]
Faust, Jan *see* Journal of trauma practice
Faust, Jean Jacques *see* Bresil
Faust, opera [in three acts by j c bernard music by] spohr : for voice and pianoforte, with english and german words / Spohr, Louis – London: Boosey & Sons [187-?] [mf ed 19–] – 6mf – 9 – mf#fiche 211, 672 – us Sibley [780]
Faust papers : containing critical and historical remarks on faust and its translations, with some observations upon goethe / Koller, W H – London: Printed for Black, Young, and Young, 1835 – 1r – 1 – us Wisconsin U Libr [430]
Faust, Philipp *see* Das haus
Faust, R *see* Princeps christiano-politicus
Faust und der weg zum leben : fausts heimkehr / Wizemann, Karl – 9. Aufl. Stuttgart: Wege-Verlag Wilhelm Kaz, 1932 – 1r – 1 – us Wisconsin U Libr [410]
Faust und luther : ein beitrag zur entstehung der faust-dichtung / Wolff, Eugen – Halle (Saale): M Niemeyer, 1912 – 1r – 1 – (incl bibl ref) – us Wisconsin U Libr [430]
Faust und urfaust / Goethe, Johann Wolfgang von – Leipzig: Dieterich, [1939] (mf ed 1990) – 1 – (filmed with: goethes faust. incl bibl ref) – us Wisconsin U Libr [820]
Das faustbuch des christlich meynenden : nach dem druck von 1725 / ed by Szamatolski, Siegfried – Stuttgart: G J Goeschen 1891 [mf ed 1993] – (= ser Deutsche litteraturdenkmale des 18. und 19. jahrhunderts 39) – 1r [ill] – 1 – (incl bibl ref. filmed with: die ammen-uhr / dresdener kuenstlern) – mf#8359 – us Wisconsin U Libr [430]
Das faustbuch des christlich meynenden / ed by Szamatolski, Siegfried – Stuttgart: G J Goeschen, 1891 [mf ed 1993] – (= ser Deutsche litteraturdenkmale des 18. und 19. jahrhunderts 39) – xxvi/30p/[3]pl – 1 – (incl bibl ref) – mf#8359 – us Wisconsin U Libr [390]
Fauste socin : biographie et critique / Lecler, Paul – Geneve: Charles Schuchardt, 1885 – 1mf – 9 – 0-524-07755-X – mf#1991-3323 – us ATLA [240]
Der faustische mensch : vierzehn betrachtungen zum zweiten teil von goethes faust / Obenauer, Karl Justus – Jena: E Diederichs 1922 [mf ed 1990] – 1r – 1 – (filmed with: goethes faust: eine evangelische auslegung / friso melzer) – mf#7356 – us Wisconsin U Libr [430]
Faustischer glaube : versuch ueber das problem humaner lebenshaltung / Korff, Hermann August – Leipzig: J J Weber, 1938 [mf ed 1990] – 167p – 1 – mf#7353 – us Wisconsin U Libr [430]
Faustisches christentum / Bornhausen, Karl – Gotha: L Klotz 1925 [mf ed 1990] – (= ser Buecherei der christlichen welt) – 1r – 1 – (filmed with: "old-iniquity": der schluessel zu goethes 'faust' / ottomar beta) – mf#7341 – us Wisconsin U Libr [430]
Faust-mephisto, der deutsche mensch : mit erlaeuternder darlegung des romantischen und des realinhalts von goethes "faust" / Gabler, Karl – Berlin: T Fritsch, [1938] – 1r – 1 – (incl bibl ref) – us Wisconsin U Libr [430]
Fausto / Goethe, Johann Wolfgang von – [Mexico]: Universidad nacional de Mexico, 1924 (mf ed 1990) – 1r – 1 – (filmed with: fausto) – us Wisconsin U Libr [820]
Fausto : tragedia / Goethe, Johann Wolfgang von – Madrid: Hernando, 1925 (mf ed 1990) – 1 – (filmed with: doctor johannes faust: puppenspiel in vier aufzuegen) – us Wisconsin U Libr [820]
Fausto : tragedia / Goethe, Johann Wolfgang von – Firenze: Le Monnier, 1866 (mf ed 1990) – 1r – 1 – (filmed with: il faust) – us Wisconsin U Libr [820]
Fausto : tragedia / Goethe, Johann Wolfgang von – Firenze: A Salani, 1895 (mf ed 1990) – 1r – 1 – (filmed with: la tragedie du docteur faust de goethe en vers francais) – us Wisconsin U Libr [820]
Fausto : tragedia / Goethe, Johann Wolfgang von – Firenze: A Salani, 1922 (mf ed 1990) – 1r – 1 – (filmed with: la tragedie du docteur faust de goethe en vers francais) – us Wisconsin U Libr [820]
Fausto : y el segundo fausto / Goethe, Johann Wolfgang von – Paris: H Garnier, [19–?] (mf ed 1990) – 1r – 1 – (filmed with: fausto) – us Wisconsin U Libr [820]
Fausto sozzini da siena e il razionalismo umanistico nella reforma religiosa del secolo 16 / Mazzei, Antonio – Firenze: A Meozzi, 1910 – 1mf – 9 – 0-524-08481-5 – mf#1993-3126 – us ATLA [012]
Fausts leben / Mueller, Friedrich; ed by Seuffert, Bernhard – Heilbronn: Henninger, 1881 [mf ed 1993] – (= ser Deutsche litteraturdenkmale des 18. und 19. jahrhunderts 3) – xxvi/116p – 1 – (original ed, mannheim 1778 and 1776) – mf#8676 reel 1 – us Wisconsin U Libr [430]
Fausts leben / Widmann, Georg Rudolf; ed by Keller, Adelbert von – Stuttgart: Litterarischer Verein, 1880 (Tuebingen: H Laupp) – us Wisconsin U Libr [430]
Fausts leben / Widmann, Georg Rudolf; ed by Keller, Adelbert von – Stuttgart: Litterarischer Verein, 1880 (Tuebingen: H Laupp) [mf ed 1993] – (= ser Blvs 146) – 737p – 1 – mf#8470 reel 31 – us Wisconsin U Libr [390]
Fausts leben, taten und hoellenfahrt / Klinger, Friedrich Maximilian – Berlin: Aufbau-Verlag, 1958 – 1 – (incl bibl ref) – us Wisconsin U Libr [830]
Fausts rettung / Schneider, Reinhold – Berlin: Suhrkamp, 1946 – 1r – 1 – us Wisconsin U Libr [430]
Die faustsage und der goethe'sche faust / Kuechler, Carl – Leipzig: G Fock, 1893 – 1r – 1 – us Wisconsin U Libr [430]
Ein faustschlag : schauspiel in drei akten / Anzengruber, Ludwig – Wien: L Rosner, 1878 [mf ed 1988] – 70p – 1 – mf#6947 – us Wisconsin U Libr [820]
Faust-studien : ein beitrag zum verstaendnis goethes in seiner dichtung / Wood, Henry – Berlin: G Reimer, 1912 [mf ed 1998] – 1r – 1 – (filmed with: goethes faust / von ernst ziegeler & other titles. incl bibl ref & ind) – mf#9981 – us Wisconsin U Libr [430]
Fauststudien / Buechner, Wilhelm – Weimar: H Boehlau 1908 [mf ed 1990] – 1r – 1 – (filmed with: a passage in the night / sholem asch) – mf#7342 – us Wisconsin U Libr [430]
Eine faust-trilogie : dramaturgische studie / Dingelstedt, Franz, Freiherr von – Berlin: Paetel, 1876 (mf ed 1990) – 1r – 1 – (filmed with: goethes faust in seiner haltesten gestalt) – us Wisconsin U Libr [890]
Faustus : ein gedicht / Bechstein, Ludwig – Leipzig: F A Leo, 1833 [mf ed 1989] – (= ser Bibliothek der deutschen literatur 18837) – iv/195p/[8]pl – 1 – mf#7001 – us Wisconsin U Libr [810]
Faustus *see* An epoch in printing
Faustus, his life, death, and doom : a romance in prose / Klinger, Friedrich Maximilian – London: W Kent, 1864 – 1 – us Wisconsin U Libr [830]
Il faut qu'une porte soit ouverte ou fermee / Musset, Alfred De – Paris, France. 1848? – 1r – us UF Libraries [440]
Il faut qu'une porte soit ouverte ou fermee / Musset, Alfred De – Paris, France. 1848? – 1r – us UF Libraries [440]
Faut, S *see* Die christologie seit schleiermacher
Fautes a corriger une chaque jour / Lusignan, Alphonse – Quebec?: s.n, 1890 – 3mf – 9 – (incl ind) – mf#07727 – cn Canadiana [440]
Fauteuil 47 / Verneuil, Louis – Paris, France. 1924 – 1r – us UF Libraries [025]
Fauteux, Aegidius *see*
– Les carnets d'un curieux
– L'introduction de l'imprimerie au canada
Fauteux, Albina *see* La venerable mere d'youville
Fauth, Gertrud *see* Joerg wickrams romantechnik
Fauvel, Andre-Joseph *see* Trois quatuor pour deux violons, alto et basse, oeuvre 6, 2e livre de quatuor
Fauvette / Chauvigne, Auguste – Tours, France. 188-? – 1r – us UF Libraries [440]
Faux billet / Delapierre, Andre – Paris, France. 19– – 1r – us UF Libraries [440]
Faux bonshommes / Barriere, Theodore – Paris, France. 1857 – 1r – us UF Libraries [440]

Les faux brillants : comedie en cinq actes et en vers / Marchand, Felix-Gabriel – Montreal: Prendergast, 1885 [mf ed 1976] – 1r – 5 – mf#SEM16P265 – cn Bibl Nat [820]
Les faux liberaux de l'eglise romaine : reponse au p. perraud et a ses adherents / Michaud, Eugene – Paris: Sandoz et Fischbacher, 1872 – 1mf – 9 – 0-8370-8534-9 – (incl bibl ref) – mf#1986-2534 – us ATLA [240]
Faux, William *see* Memorable days in america
Fava, Armand-Joseph *see*
– L'existence des loges de femmes
– Lettre pastorale de monseigneur l'eveque de grenoble
Favart, Antoine Pierre Charles *see* Memoires et correspondance litteraires, dramatiques et anecdotiques...
Favart, Charles-Simon *see* Memoires et correspondance litteraires, dramatiques et anecdotiques...
Favelas do rio de janeiro / Parisse, Luciano – Rio de Janeiro, Brazil. 1969 – 1r – us UF Libraries [972]
Favero, Terence G *see* The ability of sarcoplasmic reticulum to regulate intracellular calcium following a fatiguing bout of exercise
[Faverot, I] *see* Reveille-matin a double montre
Faversham labour party records, 1918-1994 – (= ser Labour party in britain, origins and development at local level. series 2) – 18r – 1 – (with p/g. int by lawrence black) – mf#97563 – uk Microform Academic [325]
Faversham mercury – [London & SE] BLNL 14 apr 1860-31 dec 1938 [mf 1897, 1900] – 1 – (lacking: 1896; cont as: faversham times [1 jan 1939-]) – uk Newsplan [072]
Faversham news and east kent journal – Faversham, England. 24 feb-dec 1883; 1884-96; 1898-1906; 1908-11; 1913-35 [wkly] – 48r – 1 – (aka: north east kent news, 1894-96) – uk British Libr Newspaper [072]
Favieres, Edme Guillaume Francois De *see* Herman et verner, ou, les militaires
Favieres, M de Edme-Guillaume-Francois *see* Le concert interrompu
Favole heroiche contenenti le vere massime della politica, et della morale...parte prima / Audin, M – Venetia: Presso Gio. Giacomo Hertz, 1667 – 4mf – 9 – mf#O-2042 – ne IDC [090]
The favor of your attendance at the funeral of the late mr russell : ...york, 3d october, 1808... – [Toronto?: s.n. 1808?] [mf ed 1983] – 1mf – 9 – 0-665-38690-7 – mf#38690 – cn Canadiana [090]
Favorite / Royer, Alphonse – Paris, France. 184-? – 1r – us UF Libraries [440]
Favorite / Royer, Alphonse – Paris, France. 1892 – 1r – us UF Libraries [440]
The favorite – Montreal: G E Desbarats, [1873-1874] – 9 – mf#P04749 – cn Canadiana [420]
Favorite air introduced and sung...in the comic opera l'albero di diana [by martin y soler] / Mazzinghi, Joseph – London: printed for G Goulding...[179-?] [mf ed 1991] – 1r – 1 – (additional aria by mazzinghi, inserted in martin y soler's l'arbore di diana; italian words) – mf#pres. film 108 – us Sibley [780]
Favorite ballet music in the entertainment of raymond and agnes : performed at the theater royal covent garden / Reeve, William – London: printed & sold by Preston & Son... [1797?] [mf ed 1991] – 1r – 1 – (arr for pianoforte) – mf#pres. film 108 – us Sibley [780]
Favorite ballet of la fille sauvage : ou le pouvoir de la musique, performed at the king's theatre haymarket / Mortellari, Michele – London: Rt Birchall [1805?] [mf ed 1989] – 1r – 1 – mf#pres. film 49 – us Sibley [790]
Favorite divertissement of le marchand de smyrne : by mr barree as performed at the king's theater, composed and arrang'd for the piano forte, by sigr cesare bossi / Bossi, Cesare – London: Goulding, Phipps & D'Almaine [1799] [mf ed 1988] – 1r – 1 – mf#pres. film 44, 45 – us Sibley [780]
Favorite march as performed at the king's theater, haymarket : in the serious opera of evelina / Sacchini, Antonio – London: R Birchall [179-?] [mf ed 1991] – 1r – 1 – mf#pres. film 108 – us Sibley [780]
Favorite quinetetto for two violins, two tenors, and a bass / Haydn, Joseph – London: G Gardom [179-?] [mf ed 19–] – 6pt on 1r – 1 – mf#film 2543 – us Sibley [780]
Favorite recitative and air tu gran dio... : in the serious opera of merope / Bianchi, Francesco – London: L Lavenu [c1798] [mf ed 1992] – 1r – 1 – (italian words) – mf#pres. film 113 – us Sibley [780]
Favorite rondo : sung by sigra sestini, in the new comic-opera il marchese tulipano [by paisiello] / Toniolo, A – London: John Preston [1786?] [mf ed 19–] – 1r – 1 – mf#pres. film 73 – us Sibley [780]
A favorite sonata for the piano forte with an accompaniment for the violin / Willson, Joseph – New Brunswick: Printed for the Author 1801-1804. MUSIC 123, Item 7 – 1 – us L of C Photodup [780]
Favorite songs from the opera il marchese tulipano : [composed by paisiello. given in london, 1786, with additional music by cherubini. text by p chiari] / Paisiello, Giovanni – London: Longman & Broderip [1786?] [mf ed 19–] – 1r – 1 – (italian words) – mf#pres. film 73 – us Sibley [780]
Favour and fortune : a novel – Toronto: W Bryce, 1889 – 3mf – 9 – mf#03079 – cn Canadiana [830]
Favourite solo for the violin and harpsicord : bk 1-2 / Angelini, Carlo Antonio – London: printed for C & S Thompson [1770?] [mf ed 19–] – 1r – 1 – mf#pres. film 70 – us Sibley [780]
Favourite solo for the violin & harpsichord by e bach / Bach, Carl Philipp Emanuel – London: C & S Thompson [c1775] [mf ed 1992] – 1r – 1 – mf#pres. film 114 – us Sibley [780]
Favre, Francois [comp] *see* Documents maconniques
Favre, Norette *see* L'abc du hatha-yoga pour enfants de 6 a 12 ans
Fawcett and lister papers, 1733-1775 – Halifax Central Library: Archives Dept – (= ser BRRAM series) – 2r – 1 – £134 / $268 – (int by d w ockleton) – mf#r16595 – uk Microform Academic [975]
Fawcett, Charles *see*
– The english factories in india
– The first century of british justice in india
Fawcett, Edgar *see* A gentleman of leisure
Fawcett, H S *see*
– Citrus scab
– Fungi parasitic upon aleyrodes citri...
– Scaly bark of citrus
– Scaly bark or nail-head rust of citrus
– Stem-end rot of citrus fruits
Fawcett, Henry *see*
– Art in everything
– Speech on indian finance...
Fawcett, J *see* Correct system of chanting, made easy
Fawcett, Millicent Garrett *see* Women's suffrage collection from manchester central library
Fawcett, Trevor *see*
– Nineteenth century books on art and architecture collection
– Visual arts and architecture collection
Fawcett, William *see* Banana
Fawkes, Alfred *see* Studies in modernism
Fawq al-'adah-'i khabari-i kar – Sazman-i Chirik'ha-yi fada-i-i khalq, 1981. sal-i 3, shumarah-'i 1-4. 28 khurdad-8 tir 1360 [18 jun-29 jun 1981] – 1r – 1 – $53.00 – us MEDOC [079]
Faxon's illustrated handbook of travel to saratoga, lakes george and champlain, the adirondacks, niagara falls, montreal... – Boston: C A Faxon, 1874 – 3mf – 9 – (incl ind) – mf#32833 – cn Canadiana [917]
Fay, F R *see* The book of joshua
Fay, Robert Vernon *see*
– Study of the chansons of gilles binchois
– Vocal style of michael praetorius
Faye, Amad *see* La poesie funebre en pays seereer du sine
Faye, E de *see*
– Clement d'alexandrie
– Origene
Faye, Eugene de *see*
– Les apocalypses juives
– Clement d'alexandrie
– Etude sur les origines des eglises de l'age apostolique
– Gnostiques et gnosticisme
Fayette Co. Washington Court House *see*
– Cyclone and fayette republican
– Cyclone and fayette republican series
– Fayette county herald
– Fayette county record
– Fayette republican
– Fayette times
– Ohio state register
– Ohio state register series
– Record republican
– Register
– Register and peoples advocate
Fayette County Genealogical Society [IL] *see* Fayette facts
Fayette county herald / Fayette Co. Washington Court House – (mar 1867-mar 1881) [wkly] – 4r – 1 – mf#B6612-6615 – us Ohio Hist [071]
Fayette county herald – Washington, OH. 1862-1864 (1) – mf#65711 – us UMI ProQuest [071]
Fayette County, OH *see* Atlas, 1875
Fayette county record / Fayette Co. Washington Court House – aug 1901-nov 1905 [wkly, twice wkly] – 2r – 1 – mf#B11185-11186 – us Ohio Hist [071]
Fayette county union – West Union IA. 1886 dec 7-21 – 1r – 1 – (cont: fayette county pioneer; cont by: west union argo-gazette; fayette county union and west union argo-gazette) – mf#851198 – us WHS [071]
Fayette democrat – Fayetteville, WV. 1916-1941 (1) – mf#67282 – us UMI ProQuest [071]
Fayette facts / Fayette County Genealogical Society [IL] – 1978 mar-1981 dec, 1982 mar-1984 dec, 1985 mar-1987 dec – 3r – 1 – mf#579496 – us WHS [929]
Fayette republican / Fayette Co. Washington Court House – v1 n1. sep 1879-apr 1888 [wkly] – 4r – 1 – mf#B10936-10939 – us Ohio Hist [071]
Fayette times / Fayette Co. Washington Court House – jan 1939-may 1949 [wkly] – 3r – 1 – mf#B11219-11221 – us Ohio Hist [071]
Fayette tribune – Oak Hill, WV. 1908-1988 (1) – mf#67403 – us UMI ProQuest [071]
Fayle, C Ernest *see* A history of lloyd's from the founding of lloyd's coffee house to the present day
Fayrer, Joseph *see* Inspector-general sir james ranald martin
Fayssoux, Callender I *see* [Callender 1] fayssoux collection of william walker papers, 1856-1860
Fazal, Cyril P K *see* A guide to punjab government reports and statistics
Fazies, diagenese und geochemie des unteren muschelkalks am suedwestrand der querfurter mulde (sachsen-anhalt) / Kleinschnitz, Markus – (mf ed 1996) – 4mf – 9 – €56.00 – 3-8267-2358-9 – mf#DHS 2358 – gw Frankfurter [550]
Fazil *see* The divan project
Fazl-i-Hussain, Khan Bahadur Mian *see* Presidential address
Faz-magazin – 1980-1994 – 47r – 1 – gw Mikropress [943]
Fbi american legion contact program / U.S. Federal Bureau of Investigation; ed by Theoharis, Athan – 1985 – 1r – 1 – $130.00 – (documents from 1940s-1960s) – mf#S1751 – us Scholarly Res [360]
Fbi file : robert f. kennedy – 1991 – 1r – 1 – $130.00 – (with printed guide) – mf#S3242 – us Scholarly Res [360]
Fbi file: miburn (mississippi burning) : the investigation of the murders of henry schwerner, andrew goodman, and james earl chaney, june 21 1964 / U.S. Federal Bureau of Investigation – 1990 – 1r – 1 – $130.00 – (with printed guide) – mf#S3244 – us Scholarly Res [360]
Fbi file on a. philip randolph / U.S. Federal Bureau of Investigation – 1990 – 1r – 1 – $130.00 – (with printed guide) – mf#S3202 – us Scholarly Res [331]
Fbi file on abbie hoffman – 8r – 1 – $130.00r – mf#S3432 – us Scholarly Res [360]
Fbi file on albert einstein / U.S. Federal Bureau of Investigation – 1986 – 1r – 1 – $130.00 – mf#S1764 – us Scholarly Res [323]
Fbi file on cesar chavez and united farm workers – 1996 – 2r – 1 – $260.00 – (guide can also be purchased separately s3354.g $10) – mf#S3354 – us Scholarly Res [331]
Fbi file on charles lindbergh – 1r – 1 – $130.00 – (guide also sold separately $15 s3515.g) – mf#S3515 – us Scholarly Res [360]
Fbi file on eleanor roosevelt – 3r – 1 – $390.00 – (guide also sold separately $10) – mf#S3355 – us Scholarly Res [360]
Fbi file on elijah muhammed – 3r – $390.00 – (with guide which is also sold separately s3342.g $10) – mf#S3342 – us Scholarly Res [360]
Fbi file on howard hughes – 2r – 1 – $260.00 – (guide also sold separately $15 s3514.g) – mf#S3514 – us Scholarly Res [360]
Fbi file on huey long – 2r – 1 – $260.00 – (guide also sold separately $15 s3516.g) – mf#S3516 – us Scholarly Res [360]
Fbi file on joseph mccarthy – 1996 – 4r – 1 – $520.00 – (guide also sold separately $10 s3353.g) – mf#S3353 – us Scholarly Res [360]
Fbi file on malcolm x / U.S. Federal Bureau of Investigation – 1996 – (= ser Malcolm X: FBI surveillance file) – 10r – 1 – $1300.00 – (supersedes: malcolm x – fbi surveillance file) – mf#S3341 – us Scholarly Res [320]
Fbi file on muslim mosque, inc. – 3r – $390.00 – (with guide which is also sold separately s3343.g $10) – mf#S3343 – us Scholarly Res [360]
Fbi file on osage indian murders / U.S. Federal Bureau of Investigation – 1986 – 3r – 1 – $390.00 – (with guide) – mf#S3022 – us Scholarly Res [360]
Fbi file on paul robeson / U.S. Federal Bureau of Investigation – 1986 – 2r – 1 – $260.00 – (printed guide sold separately $10 s3040.g) – mf#S3040 – us Scholarly Res [780]
Fbi file on roy wilkins / U.S. Federal Bureau of Investigation – 1990 – 1r – 1 – $130.00 – (with printed guide) – mf#S3201 – us Scholarly Res [320]
Fbi file on the american churchwomen killed in el salvador, december 1980 / U.S. Federal Bureau of Investigation – 1990 – 2r – 1 – $260.00 – (with printed guide) – mf#S3204 – us Scholarly Res [360]
Fbi file on the atlanta child murders (atkid) / U.S. Federal Bureau of Investigation – 1990 – 3r – 1 – $390.00 – (with printed guide) – mf#S3200 – us Scholarly Res [360]
Fbi file on the black panther party, north carolina / U.S. Federal Bureau of Investigation – 1986 – 2r – 1 – $260.00 – (with printed guide) – mf#S3039 – us Scholarly Res [320]
Fbi file on the committee for public justice / U.S. Federal Bureau of Investigation; ed by Theoharis, Athan – 1r – 1 – $130.00 – mf#S1752 – us Scholarly Res [360]
Fbi file on the fire bombing and shooting at kent state university / U.S. Federal Bureau of Investigation – 1986 – 7r – 1 – $910.00 – mf#S1763 – us Scholarly Res [360]
Fbi file on the highlander folk school / U.S. Federal Bureau of Investigation – 1990 – 1r – 1 – $130.00 – (with guide) – mf#S3241 – us Scholarly Res [331]
Fbi file on the house committee on un-american activities (huac) / U.S. Federal Bureau of Investigation – 1986 – 9r – 1 – $1170.00 – (with printed guide) – mf#S1765 – us Scholarly Res [360]
Fbi file on the ku klux klan murder of viola liuzzo / U.S. Federal Bureau of Investigation – 1990 – 1r – 1 – $130.00 – (with printed guide) – mf#S3243 – us Scholarly Res [360]
Fbi file on the moorish science temple of america (noble drew ali) – 1990 – 3r – $255.00 – (with guide which is also sold separately s3352.g $10.00) – mf#S3245 – us Scholarly Res [360]
Fbi file on the national association for the advancement of colored people (naacp) / U.S. Federal Bureau of Investigation – 1990 – 4r – 1 – $520.00 – (with printed guide) – mf#S3203 – us Scholarly Res [320]
Fbi file on the national negro congress / U.S. Federal Bureau of Investigation – 1986 – 2r – 1 – $260.00 – (with printed guide) – mf#S3045 – us Scholarly Res [320]
Fbi file on the organization of afro-american unity (oaau) – 1r – $130.00 – (with guide which is also sold separately s3344.g $10) – mf#S3344 – us Scholarly Res [360]
Fbi file on the student nonviolent coordinating committee (sncc) / U.S. Federal Bureau of Investigation – 1990 – 2r – 1 – $260.00 – (with guide) – mf#S3245 – us Scholarly Res [320]
Fbi file on the students for a democratic society and the weatherman underground organization / U.S. Federal Bureau of Investigation – 1990 – 8r – 1 – $1040.00 – (with printed guide) – mf#S3246 – us Scholarly Res [320]
Fbi file on w.e.b. du bois – 1r – $130.00 – (with guide which is also sold separately s3345.g $10) – mf#S3345 – us Scholarly Res [360]
The fbi files on the american indian movement and wounded knee / ed by Dewing, Rolland – 26r – 1 – $4080.00 – 0-89093-989-6 – (with p/g) – us UPA [322]
Fbi files on the reverend jesse jackson / U.S. Federal Bureau of Investigation – 1988 – 1r – 1 – $130.00 – (with printed guide) – mf#S3158 – us Scholarly Res [320]
Fbi filing and records procedures / U.S. Federal Bureau of Investigation; ed by Theoharis, Athan & O'Reilly, Kenneth – 1984 – 1r – 1 – $130.00 – mf#S1755 – us Scholarly Res [360]
Fbi law enforcement bulletin / United States. Federal Bureau of Investigation – Washington. 1978+ (1,5,9) – ISSN: 0014-5688 – mf#11747 – us UMI ProQuest [360]
Fbi manuals of instruction, investigative procedures, and guidelines, 1927-1978 / U.S. Federal Bureau of Investigation; ed by Theoharis, Athan & O'Reilly, Kenneth – 1984 – 2r – 1 – $260.00 – (four manuals dated 1927, 1936, 1941 and 1978. printed guide) – mf#S1759 – us Scholarly Res [360]
Fbi reports of the fdr white house – 24r – 1 – $4640.00 – 1-55655-951-8 – (with p/g) – us UPA [360]
Fbi wiretaps, bugs, and break-ins : the national security electronic surveillance card file and the surreptitious entries file / ed by Theoharis, Athan – (= ser Federal bureau of investigation confidential files) – 4r – 1 – $855.00 – 1-55655-088-X – (with p/g) – us UPA [322]
Fcc record : a comprehensive compilation of documents, reports, public notices, and other documents of the fcc / U.S. Federal Communications Commission – Washington: GPO. v1-16 no 16. 1986-94 – 2206mf – 9 – $3309.00 – (with a cumulative index for v1-12. updates planned) – mf#LLMC 88-006 – us LLMC [324]

Fcc record / United States Federal Communications Commission – Washington. 1986-1994 (1) 1986-1994 (5) 1986-1994 (9) – (cont: federal communications commission reports) – mf#16697 – us UMI ProQuest [380]
Fcc telephone equipment registration list / U.S. National Technical Information Service – Quarterly. Listed in order of equipment type – 9 – us NTIS [000]
Fcnl action / Friends Committee on National Legislation [US] – 1956 feb 28-1972 apr 13 – 1r – 1 – mf#637107 – us WHS [323]
Fcnl action / Friends Committee on National Legislation [US] – 1956-1961 – 1r – 1 – mf#644044 – us WHS [323]
Fcnl memo / Friends Committee on National Legislation [US] – n1-195 [1954 jun 25-1970 jan 20] – 1r – 1 – mf#1111460 – us WHS [323]
Fcnl washington newsletter / Friends Committee on National Legislation [US] – 1961 feb-1981 – 1r – 1 – (cont: washington newsletter of the friends committee on national legislation) – mf#612382 – us WHS [323]
Fcnl washington newsletter / Friends Committee on National Legislation, Washington, DC – Washington. 1943-1996 (1) 1970-1981 (5) 1976-1981 (9) – ISSN: 0014-5734 – mf#2491 – us UMI ProQuest [320]
Fda consumer / United States Food and Drug Administration – Rockville. 1967+ (1) 1970+ (5) 1975+ (9) – ISSN: 0362-1332 – mf#2979 – us UMI ProQuest [380]
Fda handbook of total drug quality / U.S. Food and Drug Administration – Washington, 1970 92 p. LL-2238 – 1 – us L of C Photodup [344]
FDA-HEW *see* Radiation control for health and safety act, 1968
Fdcac newsletter / Frederick Douglass Creative Arts Center – 1982 sep, 1983 sep – 1r – 1 – mf#5294194 – us WHS [700]
Fdcc quarterly – Tampa. 2001+ (1,5,9) – mf#6495,03 – us UMI ProQuest [347]
Fdm – Chicago. 1957+ (1) 1971+ (5) 1975+ (9) – ISSN: 1098-6812 – mf#1798 – us UMI ProQuest [740]
Fe – Morristown. 1985-1986 (1) 1985-1986 (5) 1985-1986 (9) – (cont: financial executive. cont by: financial executive) – ISSN: 0883-7481 – mf#14465 – us UMI ProQuest [650]
Fe – Miami, FL. 1972 oct 01-1973 sep 01 – 1r – us UF Libraries [071]
A fe christa : hebdomadario dedicado aos interesses da religiao catholica – Penedp, AL: Typ do Trabalho, 11 jan 1902-13 jul 1907 – (= ser Ps 19) – 1,5,6 – bl Biblioteca [079]
Fe de erratas de la antologia / Soto Ramos, Julio – Roosevelt, Puerto Rico. 1953 – 1r – us UF Libraries [972]
Fe news / United Farm Equipment and Metal Workers of America – 1943-49 – (= ser Labor union periodicals, pt 1: the metal trades) – 1r – 1 – $210.00 – 1-55655-236-X – us UPA [331]
Fe news / United Farm Equipment and Metal Workers of America – v1-6 n115 [1943 mar 15-1949 sep] – 1r – 1 – mf#1056111 – us WHS [331]
La fe triunfante del amor / Garcia de la Huerta, Vicente – Madrid: Pantaleon Aznar, 1748 – 1 – sp Bibl Santa Ana [946]
Fe y compromiso humano / Perez Lozano, Jose Maria & Ruiz Ginenez, Joaquin y Jose Maria Perez Lozano – Salamanca: Ediciones Secratariado Trinitario, 1969 – 1 – sp Bibl Santa Ana [946]
Fe y solidaridad – Santiago, Chile: Educacion y Comunicaciones [n44-70 (abr 1983-sep 1991)] (irreg) – 1r – 1 – us CRL [230]
Fear / Mosso, Angelo – Trans. from the 5th ed. of the Italian by E. Lough and F. Kiesow. London, New York: Longmans, Green and Co., 1896. 278p. illus., plates – 1 – us Wisconsin U Libr [150]
Fear of falling among community elders / Ferrari, Anne – Temple University, 1996 – 2mf – 9 – $8.00 – mf#PSY 1885 – us Kinesology [150]
Fear or freedom – Johannesburg: SPRO-CAS 2, [n.d.] – us CRL [079]
Fearless bible reading : a voice from the pews / Hawley, John S – New York: Abbey Press, c1901 – 1mf – 9 – 0-8370-4527-4 – mf#1985-2527 – us ATLA [220]
Fearnley, Thomas *see* The harmony of scripture
Fearon, Henry *see* Sketches of america
Fearon, Henry Bradshaw *see* Sketches of america
Feasey, Henry John *see* Westminster abbey historically described
The feasibility of a commercial union between the united states and canada : interview with erastus wiman in the "chicago tribune", october 5, 1889 – New York: s.n, 1889 – 1mf – 9 – mf#25967 – cn Canadiana [337]
The feasibility of punishing negligent assault / Plessis, Anton du – Uni of South Africa 2000 [mf ed Pretoria: UNISA 2000] – 1mf – 9 – (incl bibl ref) – mf#mfm14728 – sa Unisa [345]
Feast of fools / Voice of the Turtle Publications – v1 n1-5 [1971 dec-1972 aug] – 1r – 1 – mf#714269 – us WHS [071]
The feast of saint anne and other poems / Hamilton, Pierce Stevens – Montreal?: J Lovell, 1890 – 2mf – 9 – mf#29238 – cn Canadiana [810]
The feast of youth : poems / Chattopadhyaya, Harindranath – Madras, India: Theosophical Pub House, 1918 – (= ser Samp: indian books) – us CRL [810]
Feather, A G *see* Thrilling tales of the frozen north
Feather river bulletin – Quincy, CA. 1931-1976 (1) – mf#62237 – us UMI ProQuest [071]
Feathers and stones : "my study windows" / Pattabhi Sitaramayya, Bhogaraju – Bombay: Padma Publications, 1946 – (= ser Samp: indian books) – us CRL [870]
Featherston, Howell Colston *see* Featherston's index to the virginia corporation law, acts 1902-3, page 437 et seq. supplement to the virginia law register, september 1903
Featherstonhaugh, George W *see* Papers
Featherstonhaugh, George William *see*
– Observations on the application of human labour under different circumstances
Featherston's index to the virginia corporation law, acts 1902-3, page 437 et seq. supplement to the virginia law register, september 1903 / Featherston, Howell Colston – Lynchburg, Bell, 1903. 16 p. LL-865 – 1 – us L of C Photodup [348]
Feature : arts and crafts at new smyrna and environs / Sweett, Zelia Wilson – s.l, s.l? 1936 – 1r – us UF Libraries [700]
Feature – New York. 1979-1979 (1,5,9) – (cont: crawdaddy) – ISSN: 0163-9404 – mf#10595,01 – us UMI ProQuest [780]
Febles, Horacio A A *see* Cronicas del centenario
Febrer, Jaume *see* Trobes de mosen jaume febrer, caballer
Febres, Andres *see*
– Arte de la lengua general del reyno de chile, con un dialogo chileno-hispano muy curioso
Febres Cordero, Focion *see* Autonomia universitaria
Febres Cordero, Julio *see*
– Archivo de historia y variedades
– Coleccion de cuentos
– Don quijote en america
Febres Cordero, Luis *see*
– Del antiguo cucuta
– Terremoto de cucuta, 1875-1925
Febres Cordero, Tulio *see* Procedencia y lengua de los aborigenes
Febriologiae lectiones pincianae, aprendix ad febrilogiam, doloris diagnosim... / Gutierrez, J L – Lyon, 1668 – 7mf – 9 – sp Cultura [610]
Febs letters / Federation of European Biochemical Societies – Amsterdam. 1968+ (1) 1968+ (5) 1987+ (9) – ISSN: 0014-5793 – mf#42260 – us UMI ProQuest [574]
Febus, Sixto *see* Diez de mis cuentos
Febve De Vivy, Leon *see* Verlaine
La fecha en la conquista de caceres ante los documentos (la carta populationis) / Floriano Cumbreno, Antonio C – Gran Canaria: Caja Insular de Ahorros, 1975 – 1 – sp Bibl Santa Ana [946]
Fechamento do partido comunista do brasil / Barbedo, Alceu – Rio de Janeiro, Brazil. 1947 – 1r – us UF Libraries [972]
Fechas de la historia de honduras / Caceres Lara, Victor – Tegucigalpa, Mexico. 1964 – 1r – us UF Libraries [972]
Fechner, Ellen *see* Meine frau theresa
Fechner, Gustav Theodor *see*
– Die drei motive und gruende des glaubens
– The little book of life after death
– Revision der hauptpuncte der psychophysik
– Ueber die seelenfrage
Fechner, Helmuth *see* Deutschland und polen, 1772-1945
Fechos e subcesos de la mia cibdad / Matos-Hurtado, Belisario – Bogota, Colombia. 1948 – 1r – us UF Libraries [972]
Fecht, Friedrich *see* Lessing-galerie
Fecht- und ringbuch / vermischtes kampfbuch (cf-lp2) : farbmikrofiche-edition der handschrift augsburg, universitaetsbibliothek, cod.1.6.4°2 / ed by Hils, Hans-Peter – (mf ed 1991) – (= ser Codices figurati – libri picturati (cf-lp) 2) – 30p on 3 color mf – 15 – €280.00 – 3-89219-301-0 – (int & description by hans-peter hils) – gw Lengenfelder [090]
Fechter, Paul *see*
– Die fahrt nach der ahnfrau
– Geschichte der deutschen literatur
Der fechter von ravenna : trauerspiel in fuenf akten / Halm, Friedrich – 4. aufl. Wien: C Gerold 1894 [mf ed 1999] – 1r – 1 – (filmed with: held und kaiser / gregor samarow) – mf#10146 – us Wisconsin U Libr [820]
Fechter, Werner *see* Das publikum der mittelhochdeutschen dichtung
Fechter-zeitung – Hanau a.M. 1931-1938 – 1 – us NY Public [790]
Fed up! / American Servicemen's Union [Fort Lewis, WA] – v1 n1-v2 n4 [1969 oct 13-1971 jun 18], 1972 jan 1, 1973 mar-dec – 1r – 1 – mf#917634 – us WHS [355]
Fedan / Mendelssohn, Moses – Elk, Poland. 1862 – 1r – us UF Libraries [939]
Feddy, Beatrice A *see* Perceptions of competence, affect, and persistence of ghanaian elementary school students
Feder, Alfred Leonhard *see* Justins des maertyrers lehre von jesus christus
Feder, J M *see* Magazin zur befoerderung des schulwesens im katholischen deutschland
Feder, Johann Georg Heinrich et al *see* Philosophische bibliothek
A federacao : orgam do partido republicano federal – Manaus, AM, 11-29 dez 1895; jan-maio, set 1896; out 1898-dez 1899; maio, jul-30 dez 1900 – (= ser Ps 19) – 1,5,6 – mf#P11,01,46 – bl Biblioteca [079]
A federacao : orgao do partido republicano – Porto Alegre, RS: [s.n.] 28 fev 1884-jun 1887; jan-jun 1888; jun 1889-jun 1890; jan 1891-jun 1893; jan-jun 1894; jan-dez 1895; jul 1899-jun 1901; jul 1902-dez 1929; jan 1931-16 nov 1937 – (= ser Ps 19) – 1 – mf#P11A,05,55 – bl Biblioteca [079]
La federacion / Castillo, Marciano – San Salvador 1906. 91p. LL-8004 – 1 – us L of C Photodup [340]
Federacion Cuban del Medio Oeste *see* Boletin de la federacion
Federacion del Trabajo de Filipinas *see* Trabajo
Federacion Empresarial Cacerena *see* Presentacion de la federacion empresarial cacerena
Federacion en colombia 1810-1912 / Vega, Jose De La – Bogota, Colombia. 1952 – 1r – us UF Libraries [972]
Federacion Espanola de Trabajadores de la Ensenanza *see* Les professionnels de l'enseignement luttent pour la liberation du peuple espagnole
Federacion Extremena de Futbol *see* Calendario del campeonato 1972-73
La federacion interamericana de abogados; memoria de prueba para optar al grado de licenciado en la facultad de ciencias juridicas y sociales de la universidad de chile / Gutierrez Carrasco, Octavio – Santiago? Imprenta Sanchez 1946. 66 2 p. LL-8010 – 1 – us L of C Photodup [340]
Federacion Nacional De Cafeteros De Colombia *see* Manual del cafetero colombiano
Federacion Nacional De Comerciantes (Colombia) *see* Comercio colombiano y la economia nacional
Federacion obrera de la industria tabaquera y otras trabajadores de filipinas (foitaf). convention. program – Manila?: FOITAF, 1969? [mf ed 1985] – (= ser Philippine labor publications 1/4) – 1v (ill) – (chiefly in tagalog. also in spanish and english) – mf#6580 reel 1 n4 – us Wisconsin U Libr [331]
Federacion Ornitologica Espanola *see* 15th campeonato nacional federal de canaricultura y pajaros exoticos e indigenas y 7th concurso exposicion de la u c e...
Federacion Provincial de Empresarios de la Construccion *see* Estatutos y reglamentos
Federacion Provincial de Escritores de la Habana *see* 6 poesias y 5 cuentos premiados
Federal – Townsville. sep 1913-nov 1919 (misc iss) – 1r – A$30.32 vesicular A$35.82 silver – at Pascoe [079]
Federal accountant – Washington. 1956-1975 (1) 1975-1975 (5) (9) – (cont by: government accountants journal) – ISSN: 0014-9004 – mf#10381 – us UMI ProQuest [336]
Federal administrative law judge hearings, statistical reports : reports for 1975 and 1976-1978 / Administrative Conference of the US (ACUS) – Washington: GPO, 1977 and 1980 (all publ) – 7mf – 9 – $10.50 – mf#LLMC 94-339 – us LLMC [340]
Federal administrative regulatory agencies and the doctrine of the separation of powers / Lattin, Ward Elgin – Washington, D.C., 1938. 95 p. LL-283 – 1 – us L of C Photodup [340]
Federal agency protection of privacy act : hearing...house of representatives, 107th congress, 2nd session on h.r. 4561, may 1 2002 / United States. Congress. House. Committee on the Judiciary. Subcommittee on Commercial and Administrative Law – Washington: US GPO 2002 [mf ed 2002] – 1mf – 9 – 0-16-068756-X – (incl bibl ref) – us GPO [342]
Federal and state constitutions : colonial charters and other organic acts of the states, territories and colonies of the united states / Thorpe, Francis N – Washington: GPO. v1-7. 1909 – 49mf – 9 – $73.00 – mf#LLMC 82-713 – us LLMC [323]
The federal and state constitutions, colonial charters, and other organic laws of the state, territories, and colonies. / Thorpe, Francis Newton – Washington, Govt. Print. Off., 1909. 7 v. LL-466 – 1 – us L of C Photodup [342]
Federal and state criminal reporter / ed by Silvernail, WM H – Albany: W C Little & Co. v1-3. 1896-97 (all publ) – 21mf – 9 – $31.50 – mf#LLMC 84-264 – us LLMC [360]
Federal antitrust decisions, 1890-1931 / U.S. Courts – Washington: GPO. v1-12 + add vol. 1890-1931 – (= ser Decrees And Judgements In Antitrust Cases) – 141mf – 9 – $211.00 – (add vol entitled: decrees and judgements in antitrust cases) – mf#LLMC 79-430A/B – us LLMC [340]
The federal appellate judiciary in the 21st century / ed by Harrison, Cynthia & Wheeler, Russel R – Washington: FJC, 1989 – 3mf – 9 – $4.50 – mf#LLMC 95-367 – us LLMC [340]
Federal bar association journal *see*
– Federal bar journal
– Federal lawyer
Federal bar journal – v1-39. 1931-80 – 9 – $523.00set – (title varies: v1-5 1931-58 as federal bar association journal. merged with: federal bar news and became federal bar news and journal) – mf#102691 – us Hein [340]
Federal bar news *see*
– Federal bar journal
– Federal lawyer
Federal bar news and journal *see* Federal lawyer
federal bar news and journal *see* Federal bar journal
Federal bill of lading act, to take effect january 1, 1917. / U.S. Laws, Statutes, etc – New York National City Bank of New York 1916 32 p. LL-388 – 1 – us L of C Photodup [348]
Federal cases : comprising cases argued and determined in the circuit and district courts of the u.s. from the earliest times to the beginning of the federal reporter / U.S. – St Paul: West Pub Co. v1-30. 1894-98 (all publ) – (= ser U s circuit & district court reports) – 438mf – 9 – $657.00 – (west's major retrospective repr of all lower federal case reports prior to coverage provided by "federal reporter") – mf#LLMC 78-054 – us LLMC [347]
Federal cases – Law Library Microform Consortium. v1-30. 1788-1879 – 9 – $657.00 set – (with digest) – mf#402070 – us Hein [340]
Federal cases 1789-1879 / U.S. Federal Court – 16r – 1 – $550.00 – us Trans-Media [340]
The federal cases: comprising cases argued and determined in the circuit and district courts of the united states from the earliest times to the beginning of the federal reporter, 1789-1880. / U.S. Circuit and District Courts – St. Paul, West, 1894-97. 30 v. LL-1712 – 1 – (table. st. paul, 1894. 365 p. suppl. digest. st. paul, 1898. 83, 259 p. suppl. 2) – us L of C Photodup [347]
Federal circuit bar journal – v1-11. 1991-2002 – $180.00 set – (filming in process) – ISSN: 1055-8195 – mf#113801 – us Hein [347]
Federal communications bar journal – Washington. 1937-1976 (1) 1970-1976 (5) 1976-1976 (9) – (cont by: federal communications law journal) – ISSN: 0014-9055 – mf#3441 – us UMI ProQuest [340]
Federal communications bar journal *see* Federal communications law journal
Federal communications commission reports / United States Federal Communications Commission – Washington. 1934-1957 (9) – (cont by: fcc record) – mf#6228 – us UMI ProQuest [380]
Federal communications commission reports / U.S. Federal Communications Commission – 1st series: v1-45. 1934-65. 2nd series: v1-104 no 4 + index/digests 1965-86 (all publ) – 2054mf – 9 – $3082.00 – mf#LLMC 78-216 – us LLMC [324]
Federal communications law journal – Bloomington. 1977+ (1) 1977+ (5) 1977+ (9) – (cont: federal communications bar journal) – ISSN: 0163-7606 – mf#3441,01 – us UMI ProQuest [340]
Federal communications law journal – University of California at Los Angeles. v1-53. 1937-2001 – 5,6,9 – $782.00 set – (v1-36 1937-84 on reel $363. v37-53 1985-2001 on mf $419. title varies: v1-29 1937-76 as federal communications bar journal) – ISSN: 0163-7606 – mf#102711 – us Hein [340]
Federal Council of the Churches of Christ in America *see*
– Handbook of the churches
– The korean situation; the korean situation no. 2
– Selected quotations on peace and war
Federal Council of the Churches of Christ in America. Committee on Financial and Fiduciary Matters *see* Wills, why make them, how make them; an effort to be of service.
The federal courts / Simonton, Charles Henry – Richmond, Va.: Johnson, 1896. 120p. LL-1357 – 1 – (the federal courts. 2d ed. richmond, va.: johnson, 1898. 248 (i.e.249), xxixp. ll-1689) – us L of C Photodup [347]

Federal courts and practice: all sherman law trust prosecutions and syllabus of equity, jurisdiction, pleading and practice / Shields, John A – New York: Banks, 1912. 874p. LL-1139 – 1 – us L of C Photodup [347]

The federal courts and the orders of the interstate commerce commission / Newcomb, Harry Turner – Washington, D.C.: Gibson, 1905. 206p. LL-1045 – 1 – us L of C Photodup [347]

Federal courts and what they do – Washington: FJC, n.d. (1987?) – 1mf – 9 – $1.50 – mf#LLMC 95-840 – us LLMC [347]

Federal criminal procedure, with forms for the defense. / Byrne, John Elliott – Chicago: Callaghan, 1916. 446p. LL-1274 – 1 – us L of C Photodup [345]

Federal decisions : cases in the supreme, circuit and district courts of the united states / U.S. Supreme Court; ed by Myer, William G – St Louis: Gilbert Book Co. v1-30. 1790-1884 (all publ) – (= ser U s circuit & district court reports) – 306mf – 9 – $459.00 – mf#LLMC 81-426 – us LLMC [347]

Federal design matters / National Endowment for the Arts – iss n1-26 [1974 jan-1981 winter] – 1r – 1 – mf#1940321 – us WHS [740]

Federal election campaign laws – June 1986 ed. Washington: GPO, 1986 – 2mf – 9 – $3.00 – (oct 1990 ed. washington: gpo, 1990, 2mf llmc 95-024 $3.00) – mf#LLMC 95-024 – us LLMC [340]

Federal Election Commission *see* Campaign finance law

Federal election commission record – v1-20 – 9 – mf#LLMC 90-008 – us LLMC [340]

Federal employee / National Federation of Federal Employees – 1965 may-1971 jun, 1971 jul-1975 jul 10, 1975 aug 10-1982 jun, 1982 feb 15-1994 – 4r – 1 – mf#610207 – us WHS [331]

[Federal environmental impact statement : wisconsin highways and bridges] – 1971 n1-41; 1972 8d-10f; 1972 n11d-16f; 1972 n1d-2d; 1972 n2f-7f; 1973 n12d-13f; 1973 n14d-17f; 1973 n18d-20f; 1973 n1d-7d; 1973 n7f-11f; 1974 n10d-14f; 1974 n1d-4d; 1974 n4d-4f; 1974 n5d-9d; 1975 n11d-13f; 1975 n14d-16f; 1975 n1d-5d; 1975 n5f-8d; 1975 n8f-10f; 1975 n11d-13f; 1975 n14d-16f – 1 – mf#555556 – us WHS [071]

Federal environmental pesticides control act: hearing...may 1, 1975 / U.S. Congress. Senate. Committee on Commerce. Subcommittee on the Environment – Washington, Govt. Print. Off., 1975. 41 p. LL-2391 – 1 – us L of C Photodup [344]

Federal facilities environmental journal – v1-4. 1990-94 – 9 – $105.00 set – ISSN: 1048-4078 – mf#113311 – us Hein [333]

Federal Financial Institutions Examination Council (US) *see* Hmda. msa 1160, bridgeport-milford, ct

Federal food and drug act decisions : decisions of the courts in cases under the federal food and drug acts / Gates, Otis H – Washington, GPO, 1934 (all publ) – 16mf – 9 – $24.00 – mf#LLMC 84-111 – us LLMC [347]

Federal gazette, 1788-1802 / Philadelphia, Pennsylvania – 1972 – 28r – 1 – $3640.00 – mf#S1711 – us Scholarly Res [071]

Federal gazette and philadelphia daily advertiser – Philadelphia PA. 1790 apr-sep 8, 1790 sep 9-1791 mar 18 – 2r – 1 – (cont: federal gazette, and philadelphia evening post; cont by: philadelphia gazette) – mf#780688 – us WHS [071]

Federal gazette, and philadelphia evening post – Philadelphia PA. 1788 oct-1789 may 30, 1789 jun-1790 jan, 1790 feb-mar 31 – 3r – 1 – (cont by: federal gazette, and philadelphia daily advertiser) – mf#859720 – us WHS [071]

Federal glass company catalogs and price lists, 1910-1979 – 4r – 1 – mf#B27445-27448 – us Ohio Hist [338]

Federal government gazette / Federation of Rhodesia and Nyasaland – Salisbury. 1956-1963 – 1 – us NY Public [960]

Federal government in canada / Bourinot, John George – Toronto: Carswell, 1889 – 1mf – 9 – mf#06474 – cn Canadiana [323]

Federal guardian and commercial advertiser – Kuala Lumpur. Malaysia. -w. 4 Sep 1915-28 Oct 1916. (40 ft) – 1 – uk British Libr Newspaper [079]

Federal habeas corpus review of state judgements, 27 may 1988 – Washington: GPO, 1988 – (= ser Office Of Legal Policy, Reports To The Attorney General) – 2mf – 9 – $3.00 – mf#LLMC 94-365 – us LLMC [340]

Federal Home Loan Bank *see* Savings and home financing sourcebook

Federal home loan bank board annual reports – 1960-75 – 26mf – 9 – $39.00 – (includes 1989 financial report. lacking: 1972) – mf#LLMC 90-378 – us LLMC [336]

Federal home loan bank board journal / United States Federal Home Loan Bank Board – Washington. 1968-1984 (1) 1972-1984 (5) 1975-1984 (9) – ISSN: 0737-0725 – mf#6832 – us UMI ProQuest [346]

Federal home loan bank board journal – v1-17 no3. 1968-Apr 1984 – 184mf – 9 – $276.00 – mf#LLMC 90-379 – us LLMC [336]

Federal home loan bank review – v1-13. 1934-47 (all publ) – 71mf – 9 – $106.00 – mf#LLMC 84-461 – us LLMC [332]

Federal housing and home finance agency annual reports – 1947-64 – 9 – (lacking: 1947-52. 1956. 1960. this agency, founded 27 jul 1947, became part of hud on 9 nov 1965) – mf#llmc 90-376 – us LLMC [360]

Federal income tax record for individuals / D B Lewis & Co – Boston: Thomas Groom & Co, 1914? – 1mf – 9 – $1.50 – (an early example of a compliance manual for the newly enacted income tax) – mf#LLMC 94-269 – us LLMC [336]

Federal india / Haksar, Kailas Narayan & Panikkar, K M – London: Martin Hopkinson Ltd, 1930 – (= ser Samp: indian books) – us CRL [954]

The federal investigations *see* Oil

Federal judicial workload statistics – Admin Office of the US Courts, 1978-85 (all publ) – 21mf – 9 – $31.50 – mf#LLMC 95-005 – us LLMC [340]

The federal judiciary acts of 1875 and 1887 / Foster, Roger – New York: Strouse, 1887. 109p. LL-1290 – 1 – us L of C Photodup [348]

Federal juror – Grand Jury Association, Southern District of New York. v1-31. 1929-61 – 11mf – 9 – $16.50 – (title may have add vols) – mf#LLMC 84-462 – us LLMC [340]

Federal labor relations authority decisions and orders – v1-17. 1979-85 – 910mf – 9 – $1365.00 – (cont: federal labor relations council and interpretations) – mf#LLMC 82-603 – us LLMC [344]

Federal labor relations authority / federal service impasse panel annual reports – 1st-6th. 1979-84 – 8mf – 9 – $12.00 – mf#LLMC 95-003 – us LLMC [344]

Federal labor relations council decisions and interpretations – v1-6. 1970-78 (all publ) – 57mf – 9 – $85.00 – (cont by: federal labor relations authority decisions and orders: v1-17 1979-1985. cite als flra. 185mf llmc 82-603 $277.00) – mf#LLMC 80-507 – us LLMC [344]

Federal land records for idaho, 1860-1934 / U.S. Bureau of Land Management – (= ser Records of the bureau of land management) – 23r – 1 – mf#M1620 – us Nat Archives [333]

Federal land records for idaho, 1860-1934 / U.S. Bureau of Land Management – (= ser Records of the bureau of land management) – 23r – 1 – mf#M1620 – us Nat Archives [333]

Federal land records for oregon / U.S. Bureau of Land Management – (= ser Records of the bureau of land management) – 93r – 1 – mf#M1621 – us Nat Archives [333]

Federal land records for washington, 1860-1910 / U.S. Bureau of Land Management – (= ser Records of the bureau of land management) – 72r – 1 – mf#M1622 – us Nat Archives [333]

Federal law of science and technology / ed by Reams, Bernard D Jr – 1945-1984 – 9 – $2795.00 set – 0-89941-702-7 – mf#4014401 – us Hein [346]

Federal laws governing licensed dealers / Capers, John G – Chicago: Criterion, 1910. 187p. LL-920 – 1 – us L of C Photodup [340]

Federal lawyer – v1-48. 1953-2001 – 9 – $863.00 set – (title varies: v1-26 1953-81 as federal bar news. v27 1981-94 as federal bar news and journal) – ISSN: 0279-4691 – mf#102701 – us Hein [340]

Federal Life Assurance Co *see*
- Application for assurance in the federal life assurance company of ontario
- Guaranteed 4% insurance bonds

Federal maritime administration and maritime subsidy board annual reports / U.S. Dept of Commerce – 1962-1979 – (= ser U.S. Maritime Commission, Maritime Administration, Federal Maritime Board, Federal Maritime Commission Decisions) – 28mf – 9 – $42.00 – mf#LLMC 81-227B – us LLMC [380]

Federal maritime board and maritime administration annual reports / U.S. Dept of Commerce – 1950-61 (all publ) – (= ser U.S. Maritime Commission, Maritime Administration, Federal Maritime Board, Federal Maritime Commission Decisions) – 13mf – 9 – $19.50 – mf#LLMC 81-227A – us LLMC [324]

Federal maritime commission annual reports – 1st 1962; 15th 1976; 17th 1978 – (= ser Federal Maritime Commission Decisions) – 18mf – 9 – $27.00 – mf#LLMC 94-355 – us LLMC [341]

Federal maritime commission decisions – v1-27. 1919-85 – 224mf – 9 – $336.00 – (lacking: v21-22) – mf#LLMC 78-024 – us LLMC [341]

Federal mediation and conciliation service annual reports – 1st-32nd. 1948-79 – 37mf – 9 – $55.00 – mf#LLMC 81-223 – us LLMC [324]

Federal mine safety and health review commission decisions and orders / U.S. Federal Mine Safety and Health Review Commission – v1-16 no 10. 1979-94 – 512mf – 9 – $768.00 – (incl indfor 1978-92. updates planned) – mf#LLMC 82-601 – us LLMC [344]

Federal mortality census schedules, 1850-1880 (formerly in the custody of the daughters of the american revolution) and related indexes / U.S. Bureau of the Census – (= ser Federal nonpopulation census schedules) – 30r – 1 – mf#T655 – us Nat Archives [317]

Federal news / National Federation of Federal Employees – v1-2 n52 [1931 jun 20-1933 jun 10] – 1r – 1 – mf#1111494 – us WHS [331]

Federal news clip sheet – n72-193 [1968 sep-1981 mar] – 1r – 1 – (cont: federal news clip sheet; united states. civil service commission. public information office; cont by: federal news clip sheet, united states. office of personnel management) – mf#166936 – us WHS [350]

Federal offenders in the u.s. courts – Admin Office of the US Courts, 1968-74, 1979-90 – (= ser Federal Offenders In The United States District Courts) – 55mf – 9 – $82.00 – (1980-1983 are titled "federal offenders in the united states district courts". coverage for the years 1986-1990 is combined into one report) – mf#LLMC 95-004 – us LLMC [347]

Federal one / George Mason University – v1:no-v3 n2 [1975 nov-1978 sep] – 1r – 1 – (cont by: new federal one) – mf#666494 – us WHS [071]

Federal organizer / United Federal Workers of America – v1 n1-v2 n4 [1937 sep 3-1938 sep 3] – 1r – 1 – (cont by: federal record) – mf#930558 – us WHS [331]

Federal power commission annual reports / U.S. Federal Power Commission – 1921-76 – 106mf – 9 – $159.00 – (no report publ for 1940-45, 1977. lacking: 1928-34) – mf#LLMC 81-224 – us LLMC [340]

Federal power commission opinions and decisions / U.S. Federal Power Commission – v1-58. 1931-77 (all publ) – 981mf – 9 – $1471.00 – (incl ind/digest) – mf#LLMC 78-060 – us LLMC [324]

Federal power commission reports / United States Federal Power Commission – Washington. 1931-1967 (1) – ISSN: 0196-1667 – mf#6297 – us UMI ProQuest [350]

The federal power over commerce and its effect on state action / Lewis, William Draper – Philadelphia, University of Pennsylvania, 1892. 145 p. LL-185 – 1 – us L of C Photodup [346]

Federal practitioner – Chatham. 1994+ (1,5,9) – (cont: va practitioner) – ISSN: 1078-4497 – mf#16065,01 – us UMI ProQuest [615]

Federal Prison Camp [Mill Point W VA] *see* Pioneer

Federal probation / U.S. Dept of Justice – v1-63. apr 1937-99 – (= ser Federal Probation Newsletter) – 280mf – 9 – $420.00 – (v1-4 entitled: federal probation newsletter. v5-13: federal probation – a quarterly journal of correctional philosophy and practice. v14-53: federal probation – a journal of correctional..... updates planned) – mf#LLMC 80-504 – us LLMC [340]

Federal probation – Washington. 1937+ (1) 1971+ (5) 1975+ (9) – ISSN: 0014-9128 – mf#2225 – us UMI ProQuest [360]

Federal procedure at law. / Bates, Chrisenberry Lee – Chicago, Flood, 1908. 2 v. LL-536 – 1 – us L of C Photodup [340]

Federal record / United Federal Workers of America – v2 n5-v7 n11 [1938 sep 17-1946 may] – 1r – 1 – (cont: federal organizer; cont by: news of state, county and municipal workers; public record [new york, ny]) – mf#930557 – us WHS [331]

Federal register – Washington, DC: US GPO, backfile 1936-2001 – 9 – $19,210.00 set (2002 subs $640 ea) – 0-89941-214-9 – (incl presidential docs and annual cumulation ind. updated as released by gpo) – ISSN: 0042-1219 – mf#400000 – us Hein [324]

Federal register / National Archives and Records Administration, Office of the Federal Register – 9 – $433.00y in US $541.25 outside – 0-16-012696-7 – (issued daily) – mf#769-003-00000-2 – us GPO [324]

Federal register – Washington. 1936+ (1) 1969+ (5) 1964+ (9) – ISSN: 0097-6326 – mf#2575 – us UMI ProQuest [347]

The federal register, 1936-1983 / U.S. National Archives and Records Service – (= ser Records of the national archives and records administration) – 432r – 1 – (with printed guide) – mf#M190 – us Nat Archives [324]

Federal register on microfiche / U.S. – v1-1936- – 9 – apply for price – (printed index available separately) – us CIS [324]

Federal regulatory libraries – 9 – $785.00. Annual revision ca $470.00 – (all required faa publications including all far's (except airspace), ad's (v1-2), type certificate data sheets. summary of supplemental type certificates, advisory circulars, faa handbooks, manufacturer service bulletins and appliances for aircraft, engines, propellers. canadian and australian regulatory libraries also available. updated biweekly) – us Aircraft Tech [629]

The federal reporter – 1st series: v1-258. 1880-1919 – (= ser National reporter system, 1879 thru 1919) – 2782mf – 9 – $4173.00 – (title pt of: national reporter system publ by west pub co. updates planned) – mf#LLMC 79-404B – us LLMC [340]

Federal republican – Georgetown DC. 1814 jun 10-1816 apr 2 – 1r – 1 – (cont: federal republican and commercial gazette [baltimore, md]; cont by: baltimore telegraph and mercantile advertiser; federal republican and baltimore telegraph) – mf#851694 – us WHS [071]

Federal Republican and commercial gazette – Baltimore MD; Washington DC. 1808 jul 20-1813 may 31 – 1r – 1 – (cont: north american and mercantile daily advertiser [baltimore md: daily]; cont by: federal republican [georgetown, washington, dc: daily]) – mf#851103 – us WHS [071]

Federal Reserve Bank of Atlanta *see*
- Economic review
- Monthly review federal reserve bank of atlanta

Federal Reserve Bank of Boston *see* Landmark '76

Federal reserve bank of boston conference series – Boston. 1969+ (1,5,9) – ISSN: 0361-8714 – mf#12275 – us UMI ProQuest [332]

Federal Reserve Bank of Chicago *see* Frb chicago economic perspectives

Federal reserve bank of cleveland economic commentary – Cleveland. 1985-1996 (1,5,9) – ISSN: 0428-1276 – mf#15739,01 – us UMI ProQuest [332]

Federal Reserve Bank of Dallas *see*
- Economic and financial review
- Economic review

Federal Reserve Bank of Kansas City *see* Economic review federal reserve bank of kansas city

Federal reserve bank of kansas city monthly review – Kansas City. 1916-1977 (1) 1974-1977 (5) 1974-1977 (9) – (cont by: economic review federal reserve bank of kansas city) – ISSN: 0014-9152 – mf#8798 – us UMI ProQuest [332]

Federal Reserve Bank of Minneapolis *see*
- Monthly review
- Region

Federal reserve bank of minneapolis ninth district conditions – Minneapolis. 1972-1973 (1) 1972-1973 (5) 1972-1972 (9) – ISSN: 0029-0580 – mf#7932 – us UMI ProQuest [332]

Federal reserve bank of minneapolis ninth district quarterly – Minneapolis. 1976-1977 (1,5,9) – ISSN: 0364-4529 – mf#11307 – us UMI ProQuest [332]

Federal reserve bank of minneapolis quarterly review – Minneapolis. 1979+ (1,5,9) – ISSN: 0271-5287 – mf#11983 – us UMI ProQuest [332]

Federal Reserve Bank of New York *see* Economic policy review

Federal reserve bank of new york annual report – New York. 1914+ (1) 1971+ (5) 1976+ (9) – ISSN: 0361-7998 – mf#5165 – us UMI ProQuest [332]

Federal reserve bank of new york monthly review – New York. 1919-1976 (1) 1971-1976 (5) 1975-1976 (9) – (cont by: federal reserve bank of new york quarterly review) – ISSN: 0014-9160 – mf#2268 – us UMI ProQuest [332]

Federal reserve bank of new york quarterly review – New York. 1976-1994 [1,5,9] – (cont by: economic policy review) – ISSN: 0147-6580 – mf#11171 – us UMI ProQuest [332]

Federal reserve bank of new york quarterly review – New York. 1976-1994 (1) 1976-1994 (5) 1976-1994 (9) – (cont: federal reserve bank of new york monthly review) – ISSN: 0147-6580 – mf#11171 – us UMI ProQuest [332]

Federal Reserve Bank of Philadelphia *see* Business review [1950]

Federal Reserve Bank of Richmond *see*
- Economic quarterly federal reserve bank of richmond
- Economic review federal reserve bank of richmond
- Monthly review federal reserve bank of richmond

Federal Reserve Bank of San Francisco *see*
- Business review [1973]
- Economic review federal reserve bank of san francisco
- Monthly review federal reserve bank of san francisco

Federal Reserve Bank of St Louis *see* Review federal reserve bank of st louis

FEDERATION

Federal reserve bank of st louis national economic trends – St. Louis. 1986+ (1,5,9) – ISSN: 0430-1986 – mf#15083 – us UMI ProQuest [332]
Federal reserve bulletin / U.S. Federal Reserve Board – v1-69. 1915-83 – 1146mf – 9 – $1719.00 – mf#LLMC 81-225 – us LLMC [332]
Federal reserve bulletin – Washington. 1915+ (1) 1968+ (5) 1975+ (9) – ISSN: 0014-9209 – mf#1515 – us UMI ProQuest [332]
Federal reserve monthly chart book – 1947-76 – 1 – $810.00 – mf#0201 – us Brook [332]
Federal reserve update / Honest Money for America – 1983 aug-nov, v2 n1-v5 n1, v6 n1 [1984 jan-1985 aug, 1986 feb-1988 may, 1989 sep] – 1r – 1 – mf#1056190 – us WHS [332]
Federal reserve's first monetary policy report for 2002 : hearing...united states senate, 107th congress, 2nd session on oversight on the monetary policy report to congress pursuant to the full employment and balanced growth act of 1978, march 7 2002 / United States. Congress. Senate. Committee on Banking, Housing, and Urban Affairs – Washington: US GPO 2002 [mf ed 2002] – 1mf – 9 – us GPO [332]
Federal rulemaking : problems and possibilities / Brown, Winifred R – Washington: FJC, June 1981 – 2mf – 9 – $3.00 – mf#LLMC 95-309 – us LLMC [340]
Federal rules of evidence : legislative histories and related documents / ed by Bailey, James F & Trelles, Oscar M – 1980 – 4v – 9 – $120.00 set – 0-89941-235-1 – (covers the major publications considered by congress prior to the effective date, july 1, 1975) – mf#400100 – us Hein [323]
Federal sentencing reporter: fsr – New York. 1992-1992 (1,5,9) – ISSN: 1053-9867 – mf#18785 – us UMI ProQuest [360]
Federal services impasses panel releases – n1-299. 1970-90 – 8mf (1:42) 10mf (1:24) – 9 – $51.00 – (lacking nos 61-73, 111-132, 175-189) – mf#LLMC 90-379 – us LLMC [340]
A federal south africa : a comparison of the critical period of american history with the present position of the colonies and states of south africa, and a consideration of the advantages of a federal union...with maps / Molteno, Percy Alport – London 1896 – (= ser 19th c british colonization) – 4mf – 9 – mf#1.1.9756 – uk Chadwyck [960]
The federal statutes annotated...1789-1903 / McKinney, William M & Moore, Charles C – New York: Thompson. 10v + suppls. 1903-25 (all publ) – 252mf – 9 – $378.00 – mf#llmc 80-032 – us LLMC [348]
Federal supplement – 1932- – (= ser National reporter system, 1879 thru 1919) – 9 – (for copyright reasons, filming of vols not expected before year 2007) – us LLMC [340]
Federal surveillance of afro-americans (1917-25) : the first world war, the red scare, and the garvey movement / U.S. Federal Bureau of Investigation; ed by Kornweibel, Theodore – 25r – 1 – $4465.00 – 0-89093-741-9 – (with p/g) – us UPA [360]
Federal tax policy for economic growth and stability / U.S. Congress. Joint Committee on the Economic Report – Washington: GPO. 1v. 1955 – 10mf – 9 – $15.00 – mf#LLMC 82-702 – us LLMC [336]
Federal tax regulations – 1954-v3 1979 – 9 – $695.00 set – mf#402250 – us Hein [336]
The federal tax system : facts and problems / U.S. Congress. Joint Economic Committee – Washington: GPO. 1v. 1964 – 4mf – 9 – $6.00 – mf#LLMC 82-704 – us LLMC [336]
Federal times – Washington. 1965+ (1) 1979+ (5) 1979+ (9) – ISSN: 0014-9233 – mf#5854 – us UMI ProQuest [350]
Federal trade commission : advisory opinion digests, no 1-313. 1 jun 1962-31 dec 1968 / ed by McMahill, Richard B – Washington: GPO, 1969? (all publ) – 4mf – 9 – $6.00 – mf#LLMC 94-320 – us LLMC [343]
Federal trade commission annual reports / U.S. Federal Trade Commission – 1915-79 – 125mf – 9 – $187.00 – (lacking: 1972. updates planned) – mf#LLMC 80-509 – us LLMC [380]
Federal trade commission decisions, findings, orders and stipulations / U.S. Federal Trade Commission – v1-114. 1915-91 – 1502mf – 9 – $2,253.00 – (updates planned) – mf#LLMC 78-028 – us LLMC [324]
Federal trade commission, statutes and court decisions / U.S. Federal Trade Commission – v1-16. 1914-82 (all publ?) – 203mf – 9 – $304.00 – (add vols will be added if any) – mf#LLMC 81-226 – us LLMC [324]
Federal user fees : proceedings of a symposium washington, dc 1988 / Administrative Conference of the US (ACUS); ed by Hopkins, Thomas D – Acus: n.p., 1988? (all publ) – 2mf – 9 – $3.00 – mf#LLMC 94-350 – us LLMC [340]
Federal Writer's Project *see* Intracoastal waterway, norfolk to key west
Federal Writers' Project (FL) *see*
– Court records, 1811-1834
– Jefferson county 1939
Federal Writers' Project (Fla) *see* Negro in florida, 1528-1940
Federalist – Dublin, Ireland. jan-6 may 1871 – 1/2r – 1 – uk British Libr Newspaper [072]
The federalist : a commentary on the constituion of the united states / ed by Lodge, Henry Cabot – New York/London: G P Putnam's Sons, 1888 – 7mf – 9 – $10.50 – (with index) – mf#LLMC 90-362 – us LLMC [323]
The federalist : a commentary on the constitution of the united states, a collection of essays / Hamilton, Alexander et al – Philadelphia: J B Lippincott, 1864 – 9mf – 9 – $13.50 – mf#LLMC 95-062 – us LLMC [323]
The federalist – St. George's, Grenada. -w. 11 March 1896-23 Jan 1901; 2 March 1901-17 April 1907; 30 Oct 1907-25 June 1908; 29 Dec 1909-25 Dec 1920. Very imperfect. 6 reels – 1 – uk British Libr Newspaper [072]
The federalist and other constitutional papers / Hamilton, Alexander; ed by Scott, E H – Chicago: Scott, Foresman & Co, 1902 – 10mf – 9 – $15.00 – mf#LLMC 90-363 – us LLMC [323]
Federalist letter – Washington. 1951-1973 (1) 1971-1973 (5) – ISSN: 0014-9241 – mf#2241 – us UMI ProQuest [327]
O federalista : periodico republicano – Sao Paulo, SP: Typ do Farol Paulistano, 03-10 maio 1832 – (= ser Ps 19) – mf#P17,02,239 – bl Biblioteca [320]
O federalista alagoense : jornal politico, litterario e moral – Maceio, AL: Typ Federal, 03 out 1832 – (= ser Ps 19) – mf#P18B,01,23 – bl Biblioteca [321]
Le federaliste – Paris: Imp Schiller, may 1871 – (filmed as pt of: commune de paris newspapers) – us CRL [074]
Federally Employed Women, Inc *see*
– F e w's news and views from federally employed women, inc
– Few's news and views
– News and views
– News and views from federally employed women, inc
Federated Association of Letter Carriers (Canada) *see* Convention souvenir
Federated Canadian Mining Institute *see* The journal of the federated canadian mining institute
Federated circles of the garden club of jacksonville / Shepherd, Rose – s.l, s.l? 1937 – 1r – us UF Libraries [630]
Federated press records : american labor journalism in the mid-twentieth century – [mf ed 2003] – ca 140r – 1 – us Primary [331]
Federated railwayman / American Federation of Labor – 1943 mar-1951, 1952-55, 1956-64, 1965-1969 mar – 2r – 1 – mf#1056191 – us WHS [331]
Federated States of Micronesia *see*
– Chuuk – truk district charter, 1977
– Compact of free association and related agreements between the federated states of micronesia and the united states, 1 october 1982
– Constitution of the state of yap, 1982
– The federated states of micronesia
– Final report of the plebiscite commission on the public information program and plebiscite on the future political status of the federated states of micronesia
– Foreign investment law 2-5 and regulations, 1981-1986
– Kosrae district charter, 1978
– Laws and resolutions of the federated states of micronesia, 1979-1986
– Laws and resolutions of the state of yap, 1983-1984
– The pohnpei constitution and the legislature rules of order, 1984
– Pohnpei – ponape district code, 1978
– Report of the commission on future political status and transition
– Yap district charter, 1978
– Yap state code
The federated states of micronesia : report issued by the fsm representative office / Federated States of Micronesia – Washington, 1 dec 1983 – 1mf – 9 – $1.50 – mf#LLMC 82-100H Title 9 – us LLMC [324]
Federated Trades and Building Council [Fresno, CA] *see* Fresn labor citizen
Federated Trades and Labor Council *see* Tri-county labor news
Federated Trades Council *see* Union sentinel
Federated Union of Black Arts [South Africa] *see*
– Forum
– Fuba forum
Federation – London, UK. 24 Aug-28 Sept 1872; 15 Mar 1875 – 1 – uk British Libr Newspaper [072]
Federation – Miami, FL. Mar 1973-Jun 1979. Some issues missing. Continued by: Federation (1979) – 1 – us AJPC [071]
Federation – Miami, FL. Oct 1979-1986 – 1 – us AJPC [071]
Federation / New York State AFL-CIO – v17 n2-v24 n4 [1976 feb-1983] – 1r – 1 – (cont: new york state afl-cio news) – mf#365179 – us WHS [331]
Federation : or, a machiavelian solution of the australian labour problem. an address... sydney on 28th may, 1891 / Haynes, H Valentine – Sydney, 1891 – (= ser 19th c british colonization) – 1mf – 9 – mf#1.1.4933 – uk Chadwyck [331]
La federation – n1-7. Londres. aout 1872-janv 1873 – 1 – fr ACRPP [073]
La federation : Revue de l'ordre vivant – Paris. n48-141.1949-oct 1956 – 1 – fr ACRPP [073]
Federation balkanique – Vienna, Austria. 15 may 1928-1 dec 1929 – 1r – 1 – uk British Libr Newspaper [072]
La federation balkanique – I-VI, no. 1-123. Wien. juil 1924-nov 1929. Texte multilingue suivi de: La Federation balkanique. Organe des peuples opprimes et minorites nationales des Balkans. Edition francaise-allemande puis europeenne. Wien puis Frankfurt am Main. VI-VIII, no. 124-147. dec 1929-fevr 1932 – 1 – fr ACRPP [949]
La Federation Communiste des Soviets *see* Le soviet
La federation congolaise – Leopoldville: E Nzeza-Nlandu, jun 11, jul 9, sep 5 1961 – (issues filmed as pt of: herbert j weiss collection on the belgian congo) – us CRL [960]
La federation congolaise – [Leopoldville]: E Nzeza-Nlandu [jun 11, jul 9, sep 5 1961] (wkly) – (= ser Herbert J [i.e. F] Weiss coll on the Belgian Congo) – 1r – 1 – us CRL [079]
La Federation de Syndicats Chretiens de Mineurs *see* Le mineur
Federation d'education physique et de recreation du Quebec *see* Mouvement
La Federation des Comites d'Alliance Ouvriere *see* Informations ouvrieres
Federation des medecins specialistes du Quebec *see* Memoire aux membres de l'assemblee nationale
Federation des Mineurs du Nord et du Pas-de-Calais *see* La voix du mineur 1907-1914
La Federation des Ouvriers des Metaux et Similaires de France *see* L'union des metaux
Federation des Ouvriers des Metaux et Similaires de France. 4e-6e Congres National *see* Documents
Federation des scouts catholiques, Canada *see* Cibles
Federation des societes de gynecologie et d'obstetrique de langue francaise *see* Bulletin de la federation des societes de gynecologie...
La Federation des Syndicats de Cultivateurs de la Region de Moulins *see* Le travailleur rural
La Federation des travailleurs de la metallurgie *see* L'union des metallurgistes
Federation des travailleurs du Quebec [FTQ] *see* Monde ouvrier
La Federation des Travailleurs Socialistes de France *see* La france socialiste
La Federation des Travailleurs Socialistes des Ardennes *see* Le socialiste ardennais
Federation du Nord du Parti Socialiste *see* La bataille
La Federation du Nord. (SFIO) *see* Le travailleur
Federation for American Immigration Reform *see* Fair immigration report
Federation forum : official publication, michigan federation of teachers, aft, afl-cio / Michigan Federation of Teachers – 1988 feb-1992 jul – 1r – 1 – (cont: forum [detroit, mi]; cont by: forum [southfield, mi]) – mf#2899823 – us WHS [370]
Federation forum : official publication, michigan federation of teachers, aft, afl-cio / Michigan Federation of Teachers – 1982 mar-1986 may/jun – 1r – 1 – (cont: michigan teacher; cont by: forum [detroit, mi]) – mf#1140605 – us WHS [370]
La Federation Generale des Functionnaires *see* La tribune des fonctionnaires et des retraites
Federation guide and market news / Wisconsin Cheese Producers Federation – 1920 jul-1934 feb – 1r – 1 – mf#1056196 – us WHS [660]
Federation highlights / water environment federation – Alexandria. 1988-1995 (1) 1988-1995 (5) 1988-1995 (9) – (cont: water pollution control federation highlights. cont by: wef highlights) – ISSN: 1048-3063 – mf#8361,01 – us UMI ProQuest [333]
Federation letter / Federation of Historical Bottle Clubs – v7-8 n4 [1977 jan-1978 apr] – 1r – 1 – mf#367511 – us WHS [366]
Federation liberale nationale du Canada *see* Ce que le gouvernement a fait pour quebec
La federation nationale des canadiens-francais / Derouet, Camille – S.l: s.n, 1897? – 1mf – 9 – mf#56246 – cn Canadiana [305]
Federation nationale des enseignants et des enseignantes du Quebec *see*
– Info-fneeq
Federation Nationale des Ouvriers Metallurgistes de France *see* Bulletin officiel
Federation Nationale des Syndicats du Cuivre et Similaires *see* Le cuivre
La Federation Nationale des Travailleurs des Chemins de Fer. CGT *see* La tribune des cheminots
Federation nationale des Travailleurs du Sous-Sol *see* Documents du congres
La Federation Nationale des Travailleurs du Sous-sol et Parties Similaires (Mineurs, Miniers et Ardoisiens) *see* Le travailleur du sous-sol
Federation nationale Saint-Jean-Baptiste. Congres (2e: 1909: Montreal, Quebec) *see* Deuxieme congres de la federation nationale saint-jean-baptiste
Federation news / Chicago Building Trades Council et al – 1924 aug 16/1925-1992/94 – 26r – 1 – (with small gaps) – mf#1111504 – us WHS [331]
Federation news : official organ of the scottish building trades federation – [Scotland] Glasgow: Carter & Pratt for the Federation 20 jun 1899 (mthly) [mf ed 2003] – 1r – 1 – uk Newsplan [690]
Federation news / Wisconsin Federation of Teachers, Local 3605 [Madison, WI] – 1987 jan-1992 may – 1r – 1 – (cont: we're for teachers) – mf#2534992 – us WHS [370]
Federation of American Health Systems *see* Review – federation of american health systems
Federation of american hospitals review – Little Rock. 1979-1986(1,5,9) – (cont by: review – federation of american health systems) – ISSN: 0148-9496 – mf#12216,02 – us UMI ProQuest [360]
Federation of american societies for experimental biology federation proceedings – Bethesda. 1942-1987 [1]; 1965-1987 [5]; 1970-1987 [9] – (cont by: faseb journal) – ISSN: 0014-9446 – mf#1893 – us UMI ProQuest [574]
Federation of Architects, Engineers, Chemists, and Technicians *see*
– Officers' report...national convention
– Proceedings...national convention of the federation of architects, engineers, chemists and technicians
Federation of Black Community Partisans *see* Black autonomy
Federation of Boards of Trade and Municipalities *see* Canada's canal problem and its solution
Federation of Catholic Teachers, Local 2092 AFT *see* Union catholic teacher
Federation of defense and corporate counsel quarterly *see* Fdcc quarterly
Federation of European Biochemical Societies *see* Febs letters
Federation of European Microbiological Societies *see* Fems microbiology
Federation of European National Societies of the Theosophical Society. Congress *see*
– Congreso de barcelona
– Emlekkoenyve
– Transactions of the...congress of the federation of european national societies of the theosophical society
Federation of European Sections of the Theosophical Society. Congress *see* Transactions of the...annual congress of the federation of european sections of the theosophical society
Federation of Flat Glass Workers of America *see* C i o news
Federation of Genealogical Societies [US] *see* Newsletter
Federation of Glass, Ceramic and Silica Sand Workers of America *see* Cio news
Federation of Historical Bottle Clubs *see* Federation letter
Federation of insurance and corporate counsel quarterly – Iowa City. 1985+ (1) 1985+ (5) 1985+ (9) – (cont: federation of insurance counsel quarterly) – ISSN: 0887-0942 – mf#6495,01 – us UMI ProQuest [346]
Federation of insurance counsel quarterly – Champaign. 1950-1985 [1,5,8] – (cont by: federation of insurance and corporate counsel quarterly) – ISSN: 0430-2583 – mf#6495 – us UMI ProQuest [346]
Federation of Organized Trades and Labor Unions of the State of New Jersey *see* Report of the...annual session of the federation of organized trades and labor unions of the state of new jersey
Federation of Organized Trades and Labor Unions of the United States and Canada *see* Report of the...annual session of the federation of organized trades and labor unions of the united states and canada
The federation of religions / Vrooman, Hiram – Philadelphia: Nunc Licet Press, 1903 – 1mf – 9 – 0-524-02619-X – mf#1990-3069 – us ATLA [200]
Federation of Rhodesia and Nyasaland *see* Federal government gazette
Federation of Saskatchewan Indians *see* Saskatchewan indian

FEDERATION

Federation of Westinghouse Independent Salaried Unions *see* Regulator
Federation of Women Teachers' Associations of Ontario *see* F w t a o newsletter
Federation of Woodworkers *see* Timber worker
Federation ouvriere de l'industrie textile de France *see* L'ouvrier textile
Federation provinciale du travail du Quebec *see*
– Memoire de la federation du travail du quebec presente a l'honorable juge thomas tremblay, president
– Memoire...presente a l'honorable juge thomas tremblay, president
Federation radicale et radicale-socialiste de Guyane. Comite Executif *see* Bulletin
La Federation Republicaine de France *see* La nation
La Federation Revolutionnaire de la Region du Nord *see* Le vengeur
Federation teacher / Cincinnati Federation of Teachers – v10 n7 [1988 may], v11 n1-v14 n3, 5 [1988 oct-1992 feb,jun], v15 n2-3,4 [1992 nov, 1993 feb, oct]; v3 n1 [1979 oct], v3 n2-3 [1980 mar-apr], v4 n2-v6 n3 [1980 sep-1982 dec], v6 n6 [1983 apr], v10 n1 [1986 oct] – 2r – 1 – mf#1565736 – us WHS [370]
La Federation unitaire des Travailleurs du Sous-Sol et similaires *see* Le mineur unitaire
Federation voice – Providence, RI. 1988-1992 (1) – mf#68325 – us UMI ProQuest [071]
Federation world – Indianapolis, Indiana – 1 – (vol. 23, no. 7 (july 1983)-v. 26, no. 1 (jan. 1986); continues jwf report) – us AJPC [978]
Federatsiia – Russia, 1999- – 2r per y – 1 – $160.00 standing order – (backfile through 1998 $85/r) – us UMI ProQuest [320]
Federbusch, Simon *see* 'Iyunim
Federbush-Resheff, H Zvi *see* Requirements for the education and vocational training of the blind in liberia
Federe – Paris, France. 1816 – 1r – us UF Libraries [440]
Federer, Heinrich *see*
– Am fenster
– Berge und menschen
Federici, Francesco *see* Pigmalione
Federici, Vincenzo *see* Affani crudeli
Federmann, Arnold *see* Goethe als bildender kuenstler
Federmann, Nikolaus *see*
– Historia indiana
– N federmanns und h stades reisen in suedamerica
– Viaje a las indias del mar oceano
Federn, Etta [Etta Federn-Kohlhaas] *see*
– Friedrich hebbel
– Goethe
– Goethes faust
Federn, Karl *see*
– Essays zur vergleichenden literaturgeschichte
– Neun essays
Fedgazette – Minneapolis. 1992-1996 (1,5,9) – ISSN: 1045-3334 – mf#18392 – us UMI ProQuest [338]
Fedha, Nathan W *see* A catalogue of the kenya national archive collection on microfilm at syracuse university
Fedler, Joan M *see* The effect of a lifetime of physical activity on the quantity of bone in the canine
Fednews / National Association of Government Employees [US] – 1976 feb-1984 – 1r – 1 – mf#599401 – us WHS [331]
Fedorov, A I *see* Nash put' [chita: 1918-1919]
Fedorov, Ia *see* Starye i novye den'gi denezhnaia reforma 1924 goda
Fedorovich, A I *see* Vedomosti turgaiskogo oblastnogo komissara
Fedorovskie chteniia – Moskva: [Gos. biblioteka SSSR im V I Lenina], 1976- (publ 1978) – 1 – us CRL [460]
Fedrici, C de *see* Viaggio di m cesare de i fredrici nell' india orientale et, oltra l'india...
Fee, John Gregg *see* Christian baptism
Feed industry – Eden Prairie. 1980-1982 (1) 1980-1982 (5) 1980-1982 (9) – (cont: feed industry review) – mf#2464,01 – us UMI ProQuest [630]
Feed industry review – Eden Prairie. 1925-1979 (1) 1971-1979 (5) 1972-1979 (9) – (cont by: feed industry) – ISSN: 0191-9334 – mf#2464 – us UMI ProQuest [630]
Feed management – Mount Morris. 1974-1996 (1) 1975-1996 (5) 1977-1996 (9) – ISSN: 0014-956X – mf#9650 – us UMI ProQuest [630]
Feed outlook and situation – Washington. 1981-1982 (1) 1981-1982 (5) 1981-1982 (9) – (cont: feed situation. cont by: feed outlook and situation report) – ISSN: 0278-0127 – mf#9157,01 – us UMI ProQuest [630]
Feed outlook and situation report – Washington. 1983-1986 (1) 1983-1986 (5) 1983-1986 (9) – (cont: feed outlook and situation. cont by: situation and outlook report feed) – ISSN: 8755-853X – mf#9157,02 – us UMI ProQuest [630]
Feed situation – Washington. 1975-1980 (1) 1975-1980 (5) 1975-1980 (9) – (cont by: feed outlook and situation) – ISSN: 0014-9578 – mf#9157 – us UMI ProQuest [630]
Feed/back – San Francisco. 1975-1981 (1,5,9) – ISSN: 0145-6261 – mf#10737 – us UMI ProQuest [070]
Feeding for mild production / Scott, John M – Gainesville, FL. 1918 – 1r – us UF Libraries [630]
Feeding horses and mules on home-grown feed-stuffs / Conner, Charles M – Lake City, FL. 1904 – 1r – us UF Libraries [636]
Feeding value and nutritive properties of citrus by-products 2 / Arnold, P T – Gainesville, FL. 1941 – 1r – us UF Libraries [634]
Feeding value and nutritive properties of citrus by-products the digestible nutrients / Neal, W M – Gainesville, FL. 1935 – 1r – us UF Libraries [634]
Feeding with florida feed stuffs / Stockbridge, Horace E – Lake City, FL. 1900 – 1r – us UF Libraries [630]
Feedlot management – Minneapolis. 1964-1987 (1) 1971-1987 (5) 1977-1987 (9) – ISSN: 0014-9616 – mf#1664 – us UMI ProQuest [636]
Feedstuffs – Minneapolis. 1950+ (1) 1966+ (5) 1980+ (9) – ISSN: 0014-9624 – mf#377 – us UMI ProQuest [630]
Feeling after him : sermons / Wilberforce, Basil – London: Elliot Stock, 1902 [mf ed 1991] – 1mf – 9 – 0-7905-8976-1 – mf#1989-2201 – us ATLA [242]
Feelosofia prava / Chicherin, Boris N – Moskva: Kushnerev, 1900. 336p. LL-4054 – 1 – us L of C Photodup [340]
Feeman, Harlan Luther *see* The kingdom and the farm
Feeney, Bernard *see* How to get on
Feeney, Tara B *see* Perceptions of high school student-athletes of coaching competence
Fees and taxes charged insurance companies under the laws of new york : together with abstracts of fees, taxes and other requirements of other states / New York. Insurance Dept – 1909-78. 85 fiches. (Harvard Law School Library Collection.) – 9 – us Harvard Law [336]
Feetham, Richard *see* Fei-t'ang fa kuan yen chiu shang-hai kung kung tsu chieh ch'ing hsing pao kao shu 1-3 chuan
Fehde um brandenburg : geschichte eines rebellen / Helke, Fritz – Stuttgart: Union Deutsche Verlagsgesellschaft, 1943 – 1r – 1 – us Wisconsin U Libr [830]
Fehim, Sueleyman *see* The divan project
Fehler bei der messung und auswertung von festkoerper-mas-nmr-spektren / Jeschke, Gunnar – (mf ed 1992) – 1mf – 9 – €37.50 – 3-89349-582-7 – mf#DHS 582 – gw Frankfurter [540]
Fehlhaltung beim instrumentalspiel und ihre vermeidung / Wunsch, Hildrun – (mf ed 1994) – 2mf – 9 – €40.00 – 3-8267-2074-1 – mf#DHS 2074 – gw Frankfurter [780]
Fehling, Ferdinand *see* Urkunden und aktenstuecke zur geschichte des kurfuersten friedrich wilhelm von brandenberg
Fehmarnsches tageblatt – Burg auf Fehmarn DE, 1983 1 jun- – 4r/yr – 1 – gw Misc Inst [074]
Fehme, Hermann *see* Ueber das verhaeltnis heinrich von kleists zu c.m. wieland
Fehn, Andreas *see* Die geschichtsphilosophie in den historischen dramen julius mosens
Fehrle, Eugen *see* Deutsche feste und volksbraeuche
Fehrs, Johann Hinrich *see*
– Allerhand slag lued
– Ettgroen
– Kattengold
– Luettj hinnerk
– Maren
– Ut ilenbeck
Fehse, Wilhelm *see* Raabe und jensen
Fei chan kung yueh yu shih chieh ho p'ing / Hsu, Ching-wei – Nan-ching: Wai chiao p'ing lun she, Min kuo 21 [1932] – (= ser P-k&k period) – us CRL [327]
Fei ch'ang shih ch'i chih ching ch'a / Hsu, Tseng-ming & Lei, Chen teng – Shang-hai: Chung-hua shu chu, Min kuo 26 [1937] – (= ser P-k&k period) – us CRL [360]
Fei ch'ang shih ch'i chih ching chi cheng ts'e / Lo, Tun-wie et al – Shang-hai: Chung-hua shu chu, min kuo 26 [1937] – (= ser P-k&k period) – us CRL [330]
Fei ch'ang shih ch'i chih chun shih chih shih / Ch'en, Mu et al – Shang-hai: Chung hua shu chu, min kuo 26 [1937] – (= ser P-k&k period) – us CRL [355]
Fei ch'ang shih ch'i chih hsien cheng / Hu, Ming-lung – Shang-hai: Chung-hua shu chu, Min kuo 26 [1937] – (= ser P-k&k period) – us CRL [350]
Fei ch'ang shih ch'i chih kuo fang chien she / Ch'eng, Ch'ing-fang & Lei, Chen teng – Shang-hai: Chung-hua shu chu, Min kuo 26 [1937] – (= ser P-k&k period) – us CRL [355]
Fei ch'ang shih ch'i chih pao chih / Wu, Ch'eng & Lei Chen teng – Shang-hai: Chung-hua shu chu, Min kuo 26 [1937] – (= ser P-k&k period) – us CRL [070]
Fei ch'ang shih ch'i chih she hui cheng ts'e / Li, Chien-hua – Shang-hai: Chung-hua shu chu, Min kuo 26 [1937] – (= ser P-k&k period) – us CRL [350]
Fei chiang chun : [tu mu chu] / Hung, Shen – Han-k'ou: Shang-hai tsa chih kung ssu, Min kuo 27 [1938] – (= ser P-k&k period) – us CRL [951]
Fei, Chien-chao *see* Lang man yun tung
Fei ching yueh pao – Taipei. 1979-1981 (1) 1979-1981 (5) 1979-1981 (9) – ISSN: 0014-9675 – mf#9063 – us UMI ProQuest [320]
Fei ch'u pu p'ing teng t'iao yueh / Yeh, Tsu-hao – Ch'ung-ch'ing: Tu li chu pan she, Min kuo 33 [1944] – (= ser P-k&k period) – us CRL [327]
Fei hsu chi / Miao, Ch'ung-ch'un – Kuei-lin: Wen hua sheng huo ch'u pan she, Min kuo 31 [1942] – (= ser P-k&k period) – us CRL [840]
Fei hua ch'u / Hsien, Ch'un – Ch'ung-ch'ing: Kuo hsun shu tien, 1943 – (= ser P-k&k period) – us CRL [820]
Fei tao tsa shih / Su, Su – [China: sn, 1940] – (= ser P-k&k period) – us CRL [810]
Fei tsung-chiao lung (ccm124) – Pei-ching. 1v. 1922 [mf ed 198?] – (= ser Ccm 124) – 1 – mf#1984-B500 – us ATLA [210]
Fei wo ti t'u ti / Pi-yeh – Kuei-lin: San hu t'u shu she ching shou, 1944 – (= ser P-k&k period) – us CRL [830]
Fei yueh yun tung shih mo / Sung, Chia-hsiu & Cheng, Jui-mei – [Yung-an]: Yung-an ko chieh ch'ing chu Chung Mei Chung Ying ting li p'ing teng hsin yueh ta hui, Min kuo 32 [1943] – (= ser P-k&k period) – us CRL [327]
Feicht,Thomas *see* Die pseudocyprianische schrift "de rebaptismate"
Feiczewicz, Louis [comps] *see* The quebec tercentenary commemorative history
Feier der einweihung des israelitischen gotteshauses zu kopenhagen / Wolff, A A – Kopenhagen, Denmark. 1833? – 1r – us UF Libraries [939]
Feierfell, Georg *see* Otto ludwigs lehre von der tragischen schuld
Das feierliche geluebde als ehehindernis : in seiner geschichtlichen entwicklung / Scharnagl, Anton – Freiburg im Breisgau; St Louis, MO: Herder, 1908 – (= ser Strassburger theologische Studien) – 1mf – 9 – 0-7905-6827-6 – (incl bibl ref) – mf#1988-2827 – us ATLA [240]
Die feierstunde – Freiburg Br DE, 1921-1933 11 mar – 1 – gw Misc Inst [074]
Feierstunde – Wertheim DE, 1859 1 jan-1859 14 nov – 1 – (suppl to: wertheimer woechentliche anzeigen und nachrichten zum nutzen und vergnuegen) – gw Misc Inst [074]
Feierstunden : [collected short stories, plays, poems] / Eyth, Max – 5. aufl. Stuttgart: Deutsche Verlags-Anstalt [1904?] [mf ed 1989] – 1r – 1 – (filmed with: blut und eisen) – mf#7228 – us Wisconsin U Libr [880]
Feierstunden *see* Strassburger buergerzeitung
Fei-fu-na *see* Kuan min lien hsi
Feigenbaum, Benjamin *see* Vi azoy vert men poter fun der hefker velt?
Feigenberg, Rachel *see* Bay di bregen fun dnyester
Feigenberg-Eamri, Rachel *see*
– Kinder-yohren
– Susato shel mendeli ve-shot ha-yidisha'im
Feigensohn, Samuel Shraga *see* 'Elbonah shel torah
Feigin, V *see* Kustarno-remeslennaia promyshlennost sssr
Feigina, S A *see* Alandskii kongress
Der feigling, die belagerung von neuss / Beumelburg, Werner – Leipzig: Quelle & Meyer, [1934?] [mf ed 1989] – (= ser Deutsche novellen des 19. und 20. jahrhunderts 52) – 62p – 1 – mf#7017 – us Wisconsin U Libr [830]
Feijo / Azevedo, Vitor De – Sao Paulo, Brazil. 1942 – 1r – us UF Libraries [972]
Feijo Bittencourt *see* Instituto historico
Feijo e a primeira metade do seculo 19 / Ellis Junior, Alfredo – Sao Paulo, Brazil. 1940 – 1r – us UF Libraries [972]
Feijoo, Samuel *see*
– Alcancia del artesano
– Azar de lecturas
– Cantos a la naturaleza cubana del siglo 19
– Carta en otono
– Cuentos pouplares cubanos
– Cuerda menor
– Decima culta en cuba
– Decima popular
– Diario abierto
– Juan quinquin en pueblo mocho
– Libreta de pasajero
– Movimiento de los romances cubanos del siglo 19
– Poemas del bosquezuelo, 1954
– Ser fiel, 1948-62
– Sonetos en cuba
– Tumbaga
Feilchenfeld, Alfred *see* Zur geschichte der israelitischen realschule
Feilchenfeld, W *see* Das hohelied
Feilding express – 1954-55 – 2r – 1 – mf#45.1 – nz Nat Libr [079]
Feilding guardian – 21 may 1879-13 oct 1880 – 2r – 1 – mf#46.09 – nz Nat Libr [079]
Feilding herald – jan 1974-dec 1987; jan-dec 1989 – 31r – 1 – mf#45.1 – nz Nat Libr [079]
Feilding star – jun 1882-apr 1934 – 196r – 1 – mf#45.13 – nz Nat Libr [079]
Feilner, Franz *see* General-geschichte der fuerchterlichsten giftmischerinn gesche margarethe gottfried, gebornen timm
Fei-ming *see* Shui pien
Fein, Yosef *see* Yosef fain
Feinberg, N *see* Some problems of the palestine mandate
Der feind im haus : lebensbild mit gesang in drei aufzuegen – Wien: L Rosner 1878 [mf ed 1996] – (= ser Neues wiener theater 79) – 1r – 1 – (filmed with: donauland-almanach 1918 / alois veltze [ed]) – mf#4095p – us Wisconsin U Libr [820]
Feinde des volkes : eine erzaehlung im rahmen der geschichtlichen ereignisse im erzstift bremen fruehjahr 1557 / Holscher, Kurt Heimart – Berlin: Nordland Verlag, c1939 – 1r – 1 – us Wisconsin U Libr [830]
Feine leute, oder, die grossen dieser erde : roman / Edschmid, Kasimir – Berlin: P Zsolnay, 1931 – 1r – 1 – us Wisconsin U Libr [830]
Feine, Paul *see*
– Die abfassung des philipperbriefes in ephesus
– Die erneuerung des paulinischen christentums durch luther dekanatsrede gehalten am 31. oktober 1902 in wien
– Paulus als theologe
– Eine vorkanonische ueberlieferung des lukas in evangelium und apostelgeschichte
Feiner, Ruth *see*
– Sunset at noon
– Yesterday's dreams
Feinstein, Aryeh Loeb *see* Ir tehilah
Feinstone, Sol *see* The sol feinstone collection of the american revolution
The feisal-weizmann agreement – [London, January 3, 1919] – 1mf – 9 – mf#J-28-165 – ne IDC [956]
Feisberg, J *see* Esti baptisti...25 juubeli aasta malestusets
Feise, Ernst *see*
– Der knittelvers des jungen goethe
– Die leiden des jungen werthers
Feit, Edward *see* African opposition in south africa
Feit, Marvin D *see*
– Journal of health and social policy
– Journal of human behavior in the social environment
Feit, Marvin D et al *see* Journal of evidence-based social work
Fei-t'ang fa kuan yen chiu shang-hai kung kung tsu chieh ch'ing hsing pao kao shu 1-3 chuan / Feetham, Richard – [Shanghai]: Kung pu chu], 1931-1932 – (= ser Samp: indian books) – us CRL [951]
Feiticeiro / Marques, Xavier – Rio de Janeiro, Brazil. 1922 – 1r – us UF Libraries [972]
Fejer megyei hirlap – Szekesfehervar, Hungary. 1962-79 – 34r – 1 – us L of C Photodup [079]
Fekar, Ben Ali *see* Lecons d'arabe dialectal marocain, algerien
Feland, Jeffrey B *see* A comparison of different durations of static stretch of the hamstring muscle group in an elderly population
Felbermann, Lajos *see* Hungary and its people.
Das feld unserer ehre : roman aus dem hunsrueck / Bauer, Albert – Leipzig: P List, c1933 [mf ed 1989] – 266/[1]p – 1 – mf#6982 – us Wisconsin U Libr [830]
Felda, hendry county, florida / Huss, Veronica E – s.l, s.l? no date – 1r – us UF Libraries [978]
Feldblatt posen : nachrichtenblatt des wehrkreises 21 – Posen (Poznan PL), 1939 23 sep-1941 26 dec – 1r – 1 – gw Misc Inst [355]
Feldborg, Andreas *see* Denmark delineated
Feldborg, Andreas A *see* A andersen's (eines gebornen daenen) kleine fuss-reise durch einen theil von seeland
Felden, Emil *see* Menschen von morgen
Felder, Franz Michael *see* Aus meinem leben
Felder grinen / Rajzman, E – Warszawa, Poland. 1950 – 1r – us UF Libraries [939]
Felder, Hilario *see* Los estudios en la orden capuchina en el primer siglo de su existencia
Felder, Vada P *see* A mighty army
Feldgeistlicher bei legion condor : spanisches kriegstagebuch eines evangelischen legionspfarrers / Keding, Karl – Berlin: Ostwerk [1939?] – 9 – mf#w978 – us Harvard [946]

Feldgrau schafft dividende / Ettighoffer, Paul Coelestin – Koeln: Gilde-Verlag, 1932 (mf ed 1990) – 1r – 1 – (filmed with: manfred und beatrice) – us Wisconsin U Libr [430]

Feldkeller, Paul *see* Untersuchungen ueber normatives und nicht-normatives denken

Feldman, Jozef *see* Polska w dobie wielkiej wojny polnocnej, 1704-1709

Fel'dman, M S *see* Golos naroda

Feldman, Simhah Bunam *see* Yesodot ha-mediniyut ha-le-umit

Feldmann, Franz *see*
– Der knecht gottes in isaias kap 40-55
– Die weissagungen ueber den gottesknecht im buche jesaias

Feldmann, Fritz *see* Der codex mf 2016 des musikalischen instituts bei der universitaet breslau

Feldmann, Joseph *see* Paradies und suendenfall

Feldmuenster : roman aus einem jesuiteninternat / Zedtwitz, Franz Xaver, Graf – Berlin: Nordland-Verlag, 1943 – 1 – us Wisconsin U Libr [830]

Feldmuenster : roman aus einem jesuiteninternat / Zedtwitz, Franz Xaver, Graf – Berlin: Nordland-Verlag, 1943 – 1r – 1 – us Wisconsin U Libr [830]

Feldner, Gundisalv *see*
– Die lehre des heiligen thomas von aquin ueber die willensfreiheit der vernuenftigen wesen
– Die lehre des hl thomas ueber den einfluss gottes auf die handlungen der vernuenftigen geschoepfe

Feldner, Wilhelm C von *see* Wilh christ gotthelf v feldner's koenigl portugiesisch-brasilischen obristlieutenants vom generalstabe der artillerie, des koenigl portugiesischen christordens und des kaiserl o

Feldzug von sennaar nach taka, basa und beni-amer, mit besonderem hinblick auf die voelker von bellad-sudan / Werne, Ferdinand – Stuttgart 1851 [mf ed Hildesheim 1995-98] – 1v on 2mf – 9 – €60.00 – 3-487-27346-2 – gw Olms [916]

Felgas, Helio A Esteves *see* Populacoes nativas do norte de angola

Felgate archer / Abbott, W – London, England. 18– – 1r – us UF Libraries [240]

Felgner, Harald *see* Konsens und dissens im bildungssystem von baden-wuerttemberg

Felibien, A *see* Recueil de descriptions de peintures et de'autres ouvrages faits pour le roy

[Felibien, A] *see*
– Tapisseries du roi
– Tapisseries du roy

Felibien, A Sieur des Avaux et de Javercy *see* Des principes de l'architecture, de la sculpture, de la peinture, et des autres arts qui en dependent...

Felibien, M *see* Histoire de l'abbaye royale de saint-denys en france

Felice, Algernon A D *see* Perceived exertion of paraplegics during submaximal arm crank ergometry

Felice Cardot, Carlos *see* Decadas de una cultura

Felice, G de *see* History of the protestants of france

Felice, Guillaume de *see*
– Histoire des synodes nationaux des eglises reformees de france
– History of the protestants of france

Felice, Paul de *see*
– Jean calvin
– Lambert daneau (de baugency-sur-loire

Felices, Jorge *see* Enrique abril, heroe

Felici, Osea *see* Il brasile com'e

Feliciano Mendoza, Ester *see*
– Literatura infantil puertorriquena
– Nanas de la navidad
– Voz de mi tierra

Felicidad de mexico en la admirable aparicion de la virgen maria nra. senora de guadalupe, y origen de su milagrosa imagen / Becerra Tanco, Luis – en Mexico: Por D Felipe de Zuniga y Ontiveros, Calle de la Palma, ano de 1780 – (= ser Books on religion...1543/44-c1800: milagros y culto de la virgen) – 2mf – 9 – mf#crl-101 – ne IDC [241]

Felicidad de mexico...guadalupe extremuros / Bezerra Tanco, Luis – 1685 – 9 – sp Bibl Santa Ana [946]

Felicitas : historischer roman aus der voelkerwanderung (a 476 n chr) / Dahn, Felix – 10. aufl. Leipzig: Breitkopf und Haertel, 1886 [mf ed 1989] – (= ser Kleine romane aus der voelkerwanderung 1) – 275p – 1 – mf#7161 – us Wisconsin U Libr [830]

Felicitas : mga sugilanong binisaya / Alviola, Uldarico – Sugbo: Falek 1912 [mf ed Bloomington IN: Indiana Uni Lib, Preservation Dept 1984] – (= ser Coll...in the bisaya language 1) – 1r – 1 – us Indiana Preservation [490]

Felicitas : a romance / Dahn, Felix – Chicago: A C McClurg, 1903 [mf ed 1989] – xxiv/341p – 1 – (trans fr german by mary j safford) – mf#7183 – us Wisconsin U Libr [830]

Feliciter : linking canada's information professionals – Ottawa, ON: Canadian Library Association, 1956- – 1 – cn Library Assoc [020]

Feline, Marie Charles *see* L'artillerie au maroc

Feline practice – Santa Barbara. 1971-1987 (1) 1974-1987 (5) 1976-1987 (9) – ISSN: 0046-3639 – mf#9793 – us UMI ProQuest [636]

Felipe 2nd y la evangelizacion de america / Bayle, Constantino – Madrid: Imprenta del Ministerio de Asuntos Exteriores, 1947 – 1 – sp Bibl Santa Ana [240]

Felipe 3 en merida de paso para portugal / Lopez Martinez, Antonio – Badajoz: Dip. Provincial, 1970. Sep. REE – 1 – sp Bibl Santa Ana [946]

Felipe 5 en moraleja, ano de 1704 / Doncel, Fernando – Madrid: Tip. de Fortanet, 1895 – 1 – sp Bibl Santa Ana [946]

Felipe 5 y portugal. matrimonios reales en caya (1729) / Rodriguez Amaya, Esteban – Badajoz: Diputacion Provincial, 1945 – sp Bibl Santa Ana [946]

Felipe, Dionisio de *see* El padre cristobal, otro gigante

Felipe, Israel *see* Historia do cabo

Felipe trigo. exposicion y glosa de su vida, su filosofia, su moral, su arte, su estilo / Abril, Manuel – Madrid: Renacimiento, 1917 – sp Bibl Santa Ana [170]

Felipe y su piel / Iznaga, Alcides – Habana, Cuba. 1954 – 1r – us UF Libraries [972]

Felis, Stefano *see* Liber secundus motectorum quinis senis octonisque vocibus

Felix dahn's saemtliche werke poetischen inhalts / Dahn, Felix – Leipzig: Breitkopf & Haertel. 21v. 1898-1911 – us Wisconsin U Libr [810]

Felix dahn's saemtliche werke poetischen inhalts / Dahn, Felix – Leipzig: Breitkopf & Haertel. 21v. 1898-1911 – 1 – us Wisconsin U Libr [800]

Felix farley's bristol journal – [SW England] Bristol jan 1800-26 mar 1853 [mf ed 2003] – 31r – 1 – uk Newsplan; uk Microform Academic [072]

The felix frankfurter papers – 3pt – (= ser American legal manuscripts from the harvard law school library) – 1 – (pt1: supreme court...case files of opinions & memoranda, oct terms, 1938-52 74r isbn 0-89093-809-1 $13,245. pt2: 1953-61 92r isbn 0-89093-810-5 $16,450. pt3: correspondence & related material 43r isbn 0-89093-811-6 $7685. guide only to all 3pt $335) – us UPA [348]

Felix Holt : the radical / Eliot, George – Toronto: G N Morang, 1902 [mf ed 1995] – 6mf – 9 – 0-665-74182-0 – (1st publ edinburgh: w blackwood, 1866) – mf#74182 – cn Canadiana [830]

Felix Maria *see* Como es la guajira

Felix matos bernier / Diaz De Olano, Carmen R – San Juan, Puerto Rico. 1955 – 1r – us UF Libraries [972]

Felix mendelssohn-bartholdy (1809-1847) : collected edition / ed by Rietz, Julius – Leipzig: Breitkopf & Haertel. 19 ser. 1874-77 – 11 – $270.00 set – us Univ Music [780]

Felix, P Joseph *see* Der socialismus und die gesellschaft

Felix panoramas intimos / Valverde Grimaldi, Felix – Merida: Tipografia Rodriguez, 1958 – sp Bibl Santa Ana [946]

Felix poutre : drame historique en 4 actes / Frechette, Louis – Montreal: C O Beauchemin, 1871? – 1mf – 9 – mf#07368 – cn Canadiana [820]

Felix ravenna – Ravenna, 1911-1917; 1919 – 28mf – 9 – mf#O-500 – ne IDC [720]

Felix, Scott D *see* Swimming peformance following different recovery protocols

Felixstowe times – Suffolk, 19 april 1924-1983 (wkly) – 1 – ([may 1920]-n8544 7 oct 1983 (publ apr-oct until 1928; discont as a paid paper) – uk Newsplan [072]

Fell and langley's british columbia speaker's decisions / British Columbia. Canada – v1-3. 1877-1943 (all publ) – 5mf – 9 – $7.50 – mf#LLMC 81-021 – us LLMC [340]

Le fellah souvenirs d'egypte / About, Edmond – Paris 1869 [mf ed Hildesheim 1995-98] – 1v on 3mf – 9 – €90.00 – 3-487-27374-8 – gw Olms [916]

Feller, Dirk *see* Der erfolg der werbung

Felling gazette – England. -f. 20 Feb 1880-22 Nov 1889. (Wanting Jan-Apr 1886). (4 reels) – 1 – uk British Libr Newspaper [072]

Fellner, F *see* Die weingartner liederhandschrift

Fellner, H *see* Nineteenth century books on publishing, the booktrade and the diffusion of knowledge collection

Fellow of the Colonial Society *see* Pauperism and emigration

Fellow travellers : a personnally conducted journey in three continents, with impressions of men, things and events / Clark, Francis Edward – New York, Chicago: F H Revell, 1898 – 4mf – 9 – mf#27443 – cn Canadiana [910]

Fellow worker / Industrial Workers of the World – v1 n5 [1920 mar 1] – 1r – 1 – mf#487248 – us WHS [071]

Fellowes, William D *see* A visit to the monastery of la trappe in 1817

Fellows, Charles *see* Travels and researches in asia minor: more particularly in the province of lycia

Fellows, John *see* Mysteries of freemasonry [the]

Fellowship – Nyack. 1964+ (1) 1971+ (5) 1976+ (9) – ISSN: 0014-9810 – mf#5055 – us UMI ProQuest [327]

Fellowship / Union Congregational Church [Green Bay, WI] – 1904 nov-1905 mar – 1r – 1 – (cont: unionist [green bay, wi]) – mf#2962944 – us WHS [242]

Fellowship – v1 n6-v16 n11. 1935-50 [complete] – Inquire – 1 – mf#ATLA 1993-S514 – us ATLA [073]

Fellowship baptist church. dubberly, louisiana : church records – 1848-64 – 1 – 5.94 – us Southern Baptist [242]

Fellowship baptist church. meridian, mississippi : church records – 1838-98, Aug 1906 – 1 – us Southern Baptist [242]

Fellowship baptist church. mount moriah, alabama : church records – 1828-1909 – 1 – us Southern Baptist [242]

Fellowship forum – Washington DC. 1925 may 2, 23, 1926 jul 10-1927 dec 3, 1927 dec 10-1929 jun 29, 1929 jul 6-1932 apr 15 – 3r – 1 – (cont: protestant; cont by: nation's forum [washington, dc]) – mf#874763 – us WHS [242]

Fellowship in the life eternal : an exposition of the epistles of st. john / Findlay, George Gillanders – London; New York: Hodder and Stoughton, [ca 1900] – 1mf – 9 – 0-8370-3132-X – mf#1985-1132 – us ATLA [220]

Fellowship of Brethren Genealogists *see* Newsletter

The fellowship of silence : being experiences in the common use of prayer without words / Hodgkin, Thomas et al; ed by Hepher, Cyril – London: Macmillan, 1915 – 1mf – 9 – 0-524-02690-4 – mf#1990-4397 – us ATLA [240]

The fellowship of the mystery / Figgis, John Neville – London, New York: Longmans, Green, 1914 – (= ser The bishop paddock lectures) – 1mf – 9 – 0-7905-4518-7 – mf#1988-0518 – us ATLA [240]

Fellowship of the veld / Callaway, Godfrey – London, England. 1926 – 1r – us UF Libraries [960]

The fellowship porters and tacklehouse and ticket porters : minute books, accounts and other records, 1566-1895 – 14r – 1 – £650.00 – mf#TTP – uk World [366]

Fellowship primitive baptist church : church records – Peach Co, GA. 1827-1978 – 1 – $24.30 – mf#6822 – us Southern Baptist [242]

Fellowship with the spirits of just men made perfect / Ferguson, Archibald – Alyth, Scotland. 1886 – 1r – us UF Libraries [240]

The fells of swarthmoor hall, and their friends : with an account of their ancestor, anne askew, the martyr / Webb, Maria – 2nd ed. London: F Bowyer Kitto, 1867 – 1mf – 9 – 0-524-03088-X – mf#1990-4577 – us ATLA [240]

Fellsmere farms of florida – Chattanooga, TN. 1912 – 1r – us UF Libraries [630]

Fell-Smith, Charlotte *see*
– James parnell
– Steven crisp and his correspondents, 1657-1692

Y fellten : newyddiadur wythnosol i cymru, etc – Merthyr Tydfil, Wales oct 1868-sep 1876 [mf 1874] – 1 – (discontinued) – uk British Libr Newspaper [072]

Felmersham – (= ser Bedfordshire parish register series) – 1mf – 9 – £3.00 – uk BedsFHS [929]

Der fels von erz : roman / Brachvogel, Albert Emil – Philadelphia: Hoffman & Morwitz [189-?] [mf ed 1993] – 2v in 1 on 1r – 1 – (filmed with: fragments politiques et litteraires / ludwig boerne) – mf#8524 – us Wisconsin U Libr [830]

Felsefe ve ictimaiyyat mecmuasi – Istanbul: Matbaa-i Ebuezziya, 1927-? Yayimliyan: Tuerk Felsefe Cemiyeti; Mueessisi: Agah Sirri [Levend]; Mueduerue: Mehmed Servet. n3 (temmuz 1927), 7 (tesrinievvel 1928) – (= ser O & t journals) – 3mf – 9 – $75.00 – us MEDOC [956]

Felsenbrunner hof : eine gutsgeschichte / Croissant-Rust, Anna – 3. aufl. Muenchen: G Mueller, 1910 [mf ed 1989] – 388p – 1 – mf#7160 – us Wisconsin U Libr [830]

Die felsengebirge oregon und nordcalifornien / Fremont, John Charles – Stuttgart: Franckh'sche Verlagsbuchhandlung, 1851 – 4mf – 9 – (trans fr english by kottenkamp) – mf#16685 – cn Canadiana [917]

Die felseninschriften von hatnub nach den aufnahmen georg moellers / Anthes, R – Leipzig, 1928 – (= ser Untersuchungen zur geschichte und altertumskunde egyptens v9) – 7mf – 8 – (untersuchungen zur geschichte und altertumskunde egyptens. v9) – mf#H-107 – ne IDC [930]

Felsenthal, Emma *see* Bernhard felsenthal

Felsobirosagaink elvi hatarozatai / Hungary. Kuria – Budapest. On film: v1-20; 1891-1910. LL-0276 – 1 – us L of C Photodup [340]

Felt, Joseph Barlow *see*
– Did the first church of salem originally have a confession of faith distinct from their covenant?
– The ecclesiastical history of new england
– A memoir or defence of hugh peters

Felten, J *see* Die apostelgeschichte uebersetzt und erklaert

Felten, Joseph *see*
– Die apostelgeschichte
– Papst gregor 9

Feltham, John *see* A guide to all the watering and sea-bathing places

Feltoe, Ch Lett *see* Sacramentarium leonianum

Feltoe, Charles Lett *see* The letters and other remains of dionysius of alexandria

Felton, Cornelius Conway *see* Greece, ancient and modern

[Felton-] valley press – CA. 1989-1994 – 5r – 1 – $300.00 – mf#B06025 – us Library Micro [071]

Felton, William Bowman *see* Observations soumises a la consideration des membres de la legislature et du public en general sur un rapport d'un comite de la chambre d'assemblee

Felton, William Bowman [comp] *see* A report from the special committee...to whom the petition from several merchants and ship-owners of the port of quebec, was referred

Feltz, Deborah D *see* Effects of same-sex and coeducational physical education on perceptions of self-confidence and class environment

Female Benevolent Society of Montreal *see* A number of ladies

Female body-building : exploring muscularity, femininity and bodily empowerment / Lang, Margot C – 1998 – 2mf – 9 – $8.00 – mf#PSY 2053 – us Kinesology [790]

Female characters of holy scripture : in a series of sermons / Williams, Isaac – London: Rivingtons, 1873 – 1mf – 9 – 0-8370-5857-0 – mf#1985-3857 – us ATLA [220]

Female charitable association minute book, 1824-1860 – [mf ed 1981] – 1v – 9 – mf#0-260 – us South Carolina Historical [366]

Female chartists' visit to the parish church / Close, Francis – Edinburgh, Scotland. 1840 – 1r – us UF Libraries [240]

Female education : the importance of public institutions for the education of young women: an address before the officers and students of mount holyoke female seminary, july 18, 1867 / Boardman, George Nye – New York: Charles Scribner, 1867 – 1mf – 9 – 0-8370-7846-6 – mf#1986-1846 – us ATLA [376]

Female education from a medical point of view / Clouston, T S – Edinburgh: Macniven & Wallace, 1882 – us CRL [618]

Female influence and obligations – Glasgow, Scotland. 18– – 1r – us UF Libraries [240]

Female missionaries in india : letters from a missionary's wife abroad to a friend in england / Weitbrecht, M – London, 1843 – 2mf – 9 – mf#HTM-204 – ne IDC [915]

Female patient : ob/gyn edition – Chatham. 1980-1994 (1) 1980-1994 (5) 1980-1994 (9) – mf#12560 – us UMI ProQuest [618]

Female scripture biography : including an essay on what christianity has done for women / Cox, Francis Augustus – Boston: Lincoln & Edmands, 1831 [mf ed 1984] – 2v on 9mf – 9 – 0-8370-0263-X – (incl bibl ref) – mf#1984-0062 – us ATLA [220]

Female spectator – London. 1775-1775 (1) – mf#5546 – us UMI ProQuest [640]

Female tatler – London. 1709-1710 (1) – mf#4251 – us UMI ProQuest [640]

The female voyeur and the possibility of a pornography for women : redefining the gaze of desire / Schroeder, Kathleen Mary – Uni of South Africa 2000 [mf ed Johannesburg 2000] – 2mf – 9 – (incl bibl ref) – mf#mfm15069 – sa Unisa [420]

Der femhof : roman / Berens, Josefa – Jena: E Diederichs, 1943, c1935 [mf ed 1989] – 288p – 1 – mf#7006 – us Wisconsin U Libr [830]

The feminin 'monarchi', or the histori of the bee's / Butler, Charles – 3rd ed. Oxford, 1634 – 1r – 1 – mf#97089 – uk Microform Academic [400]

Feminine focus / Intercollegiate Association of Women Studies – v11 n1-v18 n1 [1974 sep-1981 oct] – 1r – 1 – mf#603326 – us WHS [305]

The feminine soul : its nature and attributes / Strutt, Elizabeth – Boston: Henry H. and T. W. Carter, 1870. Beltsville, Md: NCR Corp, 1978 (3mf); Evanston: American Theol Lib Assoc, 1984 (3mf) – (= ser Women & the church in america) – 9 – 0-8370-1242-2 – mf#1984-2097 – us ATLA [240]

Femininity and masculinity : sport and social change / Karwas, Marcia R & DePauw, Karen P – 1993 – 2mf – 9 – $8.00 – us Kinesology [150]

FEMINISM

Feminism lives! : publication of the radical feminist organizing committee / Radical Feminist Organizing Committee – 1980 jun 11-nov 3, 1981 mar 25, jun5-1984 nov – 1r – 1 – mf#941367 – us WHS [305]
Le feminisme / Avril de Saint-Croix, Mme. Ghenia – Paris: Giard et Briere, 1907 – (= ser Les femmes [coll]) – 3mf – 9 – mf#12747 – fr Bibl Nationale [305]
Le feminisme dans le socialisme francais, de 1830 a 1850 / Thibert, Marguerite – Paris: Giard, 1926 – (= ser Les femmes [coll]) – 4mf – 9 – mf#10630 – fr Bibl Nationale [305]
Feminist Alliance Against Rape [US] *see*
– Faar news
– Newsletter
Feminist art journal – Brooklyn. 1972-1977 (1) 1977-1977 (5) 1977-1977 (9) – ISSN: 0300-7014 – mf#7877 – us UMI ProQuest [700]
Feminist bookstore news – v7 n1-v11 n6 [1983 sep-1989 mar/apr] – 1r – 1 – (cont: feminist bookstore newsletter) – mf#1727826 – us WHS [305]
Feminist bulletin / Center for Women's Studies and Services [San Diego, CA] – v7 n6-v12 n10 [1977 nov-[1982] oct] – 1r – 1 – (cont: cwss newsletter) – mf#669927 – us WHS [305]
Feminist collections / University of Wisconsin System – 1980 feb-1985 spring – 1r – 1 – mf#822621 – us WHS [305]
Feminist connection – v1 n1-v5 n3 [1980 sep-1984 dec] – 1r – 1 – mf#711178 – us WHS [305]
Feminist issues – New Brunswick. 1980-1997 (1) 1980-1997 (5) 1980-1997 (9) – (cont by: gender issues) – ISSN: 0270-6679 – mf#12222 – us UMI ProQuest [305]
Feminist Organization for Communication, Action, and Service *see* Majority report
Feminist periodicals : a current listing of contents / University of Wisconsin System – 1981 mar-1984 dec, 1985 mar-1989 winter – 2r – 1 – mf#547787 – us WHS [305]
Feminist Publications of Ottawa *see* Upstream
Feminist studies – College Park. 1972+ (1) 1972+ (5) 1978+ (9) – ISSN: 0046-3663 – mf#8163 – us UMI ProQuest [322]
Feminist Writers' Guild [US] *see* National newsletter
Femme a deux maris / Pixerecourt, Rene-Charles Guilbert De – Paris, France. 1802 – 1r – us UF Libraries [440]
La femme amoureuse dans la vie et dans la litterature / Almeras, Henri d' – Paris: Albin-Michel, n.d. – (= ser Les femmes [coll]) – 6v on 21mf – 9 – mf#5184-89 – fr Bibl Nationale [410]
La femme au dix-huitieme siecle / Goncourt, Edmond & Goncourt, Jules – Paris: Flammarion, 1939 – (= ser Les femmes [coll]) – 2mf – 9 – mf#11918 – fr Bibl Nationale [305]
La femme biblique : sa vie morale et sociale, sa participation au developpement de l'idee religieuse / Bader, Clarisse – Paris: Didier, 1866 – 2mf – 9 – 0-8370-2146-4 – mf#1985-0146 – us ATLA [220]
La femme cambodgienne a l'ere du sangkum / Cambodia. Ministere de l'information – Phnom-Penh [1965] [mf ed 1989] – 1r with other items – 1 – mf#mf-10289 seam reel 014/06 [§] – us CRL [305]
La femme compositeur / Soleniere, Eugene de – Paris: La Critique 1895 [mf ed 19–] – 1mf – 9 – mf#fiche 48 – us Sibley [780]
La femme dans la societe / Toure, Ahmed Sekou – Conakry: Bureau de presse de la Presidence de la Republique populaire revolutionnaire de Guinee, [197-?] – us CRL [960]
La femme dans le droit penal du proche-orient ancien / Demare, Sophie – 3mf – 9 – (10513) – fr Atelier National [345]
La femme dans l'islam moderne / Decroux, Paul – Casablanca: Gazette des Tribunaux du Maroc, 1947 – (= ser Les femmes [coll]) – 1mf – 9 – mf#13117 – fr Bibl Nationale [306]
Femme de quarante ans / Galoppe D'onquaire, Jean Hyacinthe Adonis – Paris, France. 1845 – 1r – us UF Libraries [440]
La femme en culotte / Grand-Carteret, John – Paris: Flammarion, 1899 – (= ser Les femmes [coll]) – 5mf – 9 – mf#7712 – fr Bibl Nationale [305]
La femme en lutte pour ses droits / Pelletier, Madeleine – Paris: Giard et Briere, 1908 – (= ser Les femmes [coll]) – 1mf – 9 – mf#12944 – fr Bibl Nationale [305]
La femme esclave / Chaughi, Rene – Conflans-Ste-Honorine: L'Idee Libre, 1920 – (= ser Les femmes [coll]) – 1mf – 9 – mf#8723 – fr Bibl Nationale [305]
La femme et la democratie de notre temps / Allart de Meritens, Hortense – Paris: Delaunay, 1836 – (= ser Les femmes [coll]) – 2mf – 9 – mf#6911 – fr Bibl Nationale [305]
La femme et la famille – Paris, Gautier, 1834 – (= ser Les femmes [coll]) – 1mf – 9 – mf#7005 – fr Bibl Nationale [305]
La femme et le feminisme : colllection de livres, periodiques etc / Jacobs, Aletta Henriette – Faisant partie de la bibliotheque de M. et Mme. C.V. Gerritsen a Amsterdam. Paris: V. Giard & E. Briere, 1900.xvi,240p. Comp. by H.J. Mehler – 1 – us Wisconsin U Libr [305]
La femme et le socialisme / Bebel, Auguste – Gand: Volksdruckkerij 1911 – (= ser Les femmes [coll]) – 8mf – 9 – mf#8677 – fr Bibl Nationale [335]
La femme et l'enfant dans la francmaconnerie universelle / Clarin de la Rive, Abel – Paris: Delhomme & Briguet 1894 – 8mf – 9 – mf#vrl-135 – ne IDC [366]
La femme future / Desmarest, Henri – Paris: Victor-Harvard, 1890 – (= ser Les femmes [coll]) – 3mf – 9 – mf#7440 – fr Bibl Nationale [305]
Femme juive / Weill, Emmanuel – Paris, France. 1907 – 1r – us UF Libraries [939]
La femme libre puis nouvelle *see* La tribune des femmes
La femme mariee et les charges du menage / Brunel, J – Montpellier: Firmin, 1910 – (= ser Les femmes [coll]) – 3mf – 9 – mf#11471 – fr Bibl Nationale [640]
La femme nouvelle : apostolat des femmes – Paris, 1832-34 – (= ser Les femmes [coll]) – 6mf – 9 – mf#6904- – fr Bibl Nationale [305]
La femme nouvelle et la classe ouvriere / Kollontai, Alexandra – Paris: L'Eglantine, 1932 – (= ser Les femmes [coll]) – 2mf – 9 – mf#11211 – fr Bibl Nationale [305]
Femme nue dans la sculpture / Cinotti, Mia – Paris, France. 1951 – 1r – us UF Libraries [730]
La femme pauvre au 19e siecle : condition economique. condition professionelle / Daube, Julie – Paris: Torin, 1869-70 – (= ser Les femmes [coll]) – 3v on 15mf – 9 – mf#1139-41 – fr Bibl Nationale [305]
La femme sans peche / Lemonnier, Leon – Paris: Flammarion, 1927 – (= ser Les femmes [coll]) – 3mf – 9 – mf#10618 – fr Bibl Nationale [410]
La femme seule / Brieux, Eugene – Paris: Stock, 1913 – (= ser Les femmes [coll]) – 3mf – 9 – mf#12794 – fr Bibl Nationale [305]
La femme socialiste – Organe feministe socialiste. Dir. Louise Saumoneau. Paris. mai 1901-sept 1902, mars 1912- fevr mars 1940, oct 1947-juil sept 1949 – 1 – fr ACRPP [335]
Femmes : ou, le marite des femmes / Antier, Benjamin – Paris, France. 1824 – 1r – us UF Libraries [440]
Femmes / Prevost, Marcel – Paris: Lemerre, 1907 – (= ser Les femmes [coll]) – 4mf – 9 – mf#8829 – fr Bibl Nationale [305]
Les femmes [collection] : rare works from the bibliotheque nationale covering the history of women in france from the 17th to the 20th century – 17th-20th c [mf ed Bibliotheque Nationale] – 127 titles on 478mf – 9 – (individual titles listed separately) – fr Bibl Nationale [305]
Femmes criminelles / Mace, G – Paris: Bibliotheque-Charpentier, 1904 – us CRL [360]
Femmes d'algerie / L'Union des Femmes d'Algerie – Revue mensuelle. Alger. sept 1944-avr 1946 – 1 – fr ACRPP [305]
Femmes francaises – Paris, France. 28 sep 1944-9 nov 1945 – 1r – 1 – uk British Libr Newspaper [072]
Femmes galantes du 17e siecle : mme de villedieu (1632-1692) / Magne, Emile – Paris: Mercure de France, 1907 – (= ser Les femmes [coll]) – 5mf – 9 – mf#6811 – fr Bibl Nationale [920]
Femmes heroiques! : les soeurs grises dans l'extreme-nord / Duchaussois, Pierre – 8emille. Lyon [etc]: Oeuvre apostolique de Marie Immaculee; [Montreal]: [Librairie Beauchemin ltee], [1920 ?] (mf ed 1986) – 3mf – 9 – mf#SEM105P669 – cn Bibl Nat [241]
Les femmes pieuses de la france / Drohojowska, Antoinette – Paris [u.a. 1856 [mf ed Hildesheim 1995-98] – 1v on 6mf – 9 – €120.00 – ISBN-10: 3-487-25857-9 – ISBN-13: 978-3-487-25857-7 – gw Olms [305]
Les femmes poetes dans l'inde / Tassy, M Garcin de – Paris: J Rouvier, 1854 – us CRL [305]
Femmes revees / Ferland, Albert – Montreal: Chez l'auteur, 1899 – 1mf – 9 – (pref by louis frechette. ill by geo delfosse. engravings by a morissette) – mf#57051 – cn Canadiana [305]
Femmes savantes / Moliere – Paris, France. 1817 – 1r – us UF Libraries [440]
Les femmes vengees : opera comique en un acte et en vers par m sedaine, mis en musique par a d philidor / Philidor, Francois Danican – a Paris: ches l'auteur [1775] [mf ed 1980] – 1r – 1 – mf#film 2508 – us Sibley [780]
Fems microbiology / Federation of European Microbiological Societies – Amsterdam. 1977-1991 (1) 1977-1991 (5) 1987-1991 (9) – ISSN: 0378-1097 – mf#42131 – us UMI ProQuest [576]
Fems microbiology ecology – Amsterdam. 1992-1995 (1,5,9) – ISSN: 0168-6496 – mf#42431 – us UMI ProQuest [576]
Fems microbiology immunology – Amsterdam. 1992-1992 (1,5,9) – ISSN: 0920-8534 – mf#42685 – us UMI ProQuest [576]
Fems microbiology letters – Lausanne. 1992-1992 (1,5,9) – ISSN: 0378-1097 – mf#42686 – us UMI ProQuest [576]
Fems microbiology reviews – Amsterdam. 1992-1995 (1,5,9) – ISSN: 0168-6445 – mf#42432 – us UMI ProQuest [576]
Fems yeast research – Amsterdam. 2001+ (1,5,9) – ISSN: 1567-1356 – mf#42881 – us UMI ProQuest [576]
Fen ditton 1538-1950 – (= ser Cambridgeshire parish register transcript) – 9mf – 9 – £11.25 – uk CambsFHS [929]
Fen drayton 1559-1950 – (= ser Cambridgeshire parish register transcript) – 5mf – 9 – £6.25 – uk CambsFHS [929]
Fen ho shang / Wu, Hsi-ju – Shang-hai: Pei yeh shu tien, [1940] – (= ser P-k&k period) – us CRL [830]
Fenaroli, Fedele *see* Partimenti ossia basso numerato opera completa
Fence industry – New York. 1958-1980 (1) 1972-1980 (5) 1974-1980 (9) – ISSN: 0014-9977 – mf#7706 – us UMI ProQuest [690]
Fenchow – Fenchow, Shansi, China: North China Mission, Fenchow Station of the American Board of Commissioners for Foreign Missions. v1-19 n1. aug 1919-dec 1936 (frequency varies) [all publ?] – (= ser Missionary periodicals from the china mainland) – 1r – 1 – $165.00 – us UPA [242]
Fendrich, Anton *see* Land meiner seele
Fendrikov, F N *see* Kubanskii put'
Fendt, Leonhard *see* Die dauer der oeffentlichen wirksamkeit jesu
Fendtius, T *see* Monumenta illustrium virorum et elogia
Fenelon *see* Le gnostique de clement d'alexandrie
Fenelon, archbishop of cambrai : a biographical sketch / Lear, H L Sidney – new ed. London: Rivingtons 1877 [mf ed 1990] – 2mf – 9 – 0-7905-6813-6 – mf#1988-2813 – us ATLA [241]
Fenelon, Francois De Salignac De La Mothe- *see*
– Counsels to those who are living in the world
– Letter on frequent communion
Fenelon, Francois de Salignac de La Mothe *see* Oeuvres choisies
Fenelon, Francois de Salignac de la Mothe- *see* The spiritual letters of archbishop fenelon
Fenelon, his life and works = Fenelon / Janet, Paul; ed by Leuliette, Victor – London: Isaac Pitman, 1914 – 1mf – 9 – 0-7905-9974-0 – (incl bibl ref. in english) – mf#1989-1699 – us ATLA [944]
Fenestra steel window sash and casements – [Toronto?: s.n, 1909?] [mf ed 1991] – 1mf – 9 – 0-665-99518-0 – mf#99518 – cn Canadiana [690]
Fenety, George Edward *see*
– The city hall clock
– The lady and the dress-maker
– Longevity
– Parliamentary reminiscences
– Political notes and observations, vol 1
– Political notes and observations, vols 1 and 2
– Political notes, vol 2
– Water works for fredericton
Feng / Chao, Ch'ing-ko – Ch'ung-ch'ing: Tzu li shu tien, 1944 – (= ser P-k&k period) – us CRL [830]
Feng / Yu, Ch'ieh – Shang-hai: Liang yu t'u shu kung ssu, Min kuo 34 [1945] – (= ser P-k&k period) – us CRL [830]
Feng, Ch'eng-chun *see* Ching chiao pei kao [ccm129]
Feng, Chih *see*
– Die analogie von natur und geist als stilprinzip in novalis' dichtung
– Shan shui
Feng, Ching-yuan *see* Nung ts'un ching chi chi ho tso
Feng feng yu yu / Chou, Leng-ch'ieh – Shang-hai: Wei po ch'u pan she, [Min kuo 25 [1936]] – (= ser P-k&k period) – us CRL [830]
Feng, Hsueeh-feng *see* Chen shih chih ko
Feng, Hsueh-feng *see* Hsiang feng yu shih feng
Feng, Hui-t'ien *see* Min tsu hsin li hsueh
Feng jao ti yuan yeh ti 1 pu, ch'un t'ien / Ai, Wu – Shang-hai: Liang yu t'u shu yin shua kung ssu, 1940 – (= ser P-k&k period) – us CRL [830]
Feng k'uang pa yueh chi / Lo, Feng – Shang-hai: Tsa chih she, 1944 – (= ser P-k&k period) – us CRL [480]
Feng, Mi Tao-Jen *see* Siauw ngo gie
Feng nien / Shan-ting – Pei-ching: Hsin min yin shu kuan, 1944 – (= ser P-k&k period) – us CRL [480]
Feng nu : san mu chu / Chu, T'ung – Ch'ung-ch'ing: Ta shih tai shu chu, Min kuo 34 [1945] – (= ser P-k&k period) – us CRL [820]
Feng, P'in-lan *see* She hui hsueh kang yao
Feng, San-mei *see* Hsiao p'in wen yen chiu
Feng, Shang-li *see* Wen ku chih hsin (ccm125)
Feng shou i hou / Kung-sun, Chia – Ch'ung-ch'ing: Hua yen ch'u pan she, 1943 – (= ser P-k&k period) – us CRL [820]
Feng, Shou-chu *see* Hsin shih ho hsin shih jen
Feng tung shan ch'uan ch'i / Wu, Mei – Shang-hai: Feng yu shu wu, Min kuo 27 [1938] – (= ser P-k&k period) – us CRL [951]
Feng, Tzu-k'ai *see*
– Ch'e hsiang she hui
– I shu ch'u wei
– Kan mei ti hui wei
– Sui pi erh shih p'ien
– Tzu-k'ai sui pi
– Yuan yuan t'ang sui pi
– Yuan yuan t'ang tsai pi
Feng, tzu-kang *see* Nan-yang nung t'un she hui tiao c'a pao kao
Feng, Tzu-yu *see* Hua ch'iao ko ming shih hua shang ts'e
Feng yu chih yeh / Ch'en, Pai-ch'en – Shang-hai: Ta tung shu chu, Min kuo 22 [1933] – (= ser P-k&k period) – us CRL [480]
Feng yu kuei chou : ssu mu chu / T'ien, Han – Kuei-lin: Chi mei shu tien, 1942 – (= ser P-k&k period) – us CRL [840]
Feng, Yuan-chun *see* Yuan-chun sa ch'ien hsuan chi
Feng, Yu-hsiang *see*
– Feng yu-hsiang shih chi
– Jung kuan chi hsing
– K'ang chan che hsueh
– Kang chan shih ko chi, vol 5
– K'ang jih ti mo fan chun jen
– Pu wang kuo ch'ou wen ta
– Wei le hsien chin chiu kuo kei ai kuo peng yu te shih ssu feng hsin
– Wo ti sheng huo
Feng yu-hsiang shih chi / Feng, Yu-hsiang – [China: sn], 1931 – (= ser P-k&k period) – us CRL [810]
Feng, Yu-lan *see* Hsin li hsueh
Fenger, J F *see* History of the tranquebar mission...
Fenger, J Ferd *see* History of the tranquebar mission
Feng-huang shan : li shih ch'i ch'ing ch'ang p'ien shuo pu: erh chuan ch'i shih erh hui / Hu, Hsieh-yin – Shang-hai: Kuang i shu chu, Min kuo 26 [1937] – (= ser P-k&k period) – us CRL [830]
Feng-tu tsung chiao hsi su tiao ch'a / Wei, Hui-lin – Ssu-ch'uan: Hsiang ts'un chien she hsueh yuan yen chiu shih yen pu, Min kuo 24 [1935] – (= ser P-k&k period) – us CRL [390]
The fenian invasions of canada of 1866 and 1870 : and the operations of the montreal militia brigade in connection there with: a lecture delivered before the montreal military institute, april 23rd, 1898 / Campbell, Francis Wayland – [Montreal?: J Lovell], 1904 – 1mf – 9 – 0-665-99757-4 – mf#99757 – cn Canadiana [971]
Fenian nights' entertainments : being a series of ossianic legends / Mccall, P(Atrick) J(Oseph) – Dublin, Ireland. 1897 – 1r – us UF Libraries [490]
The fenian raid on fort erie : with an account of the battle of ridgeway, june 1866 / Denison, George Taylor – Toronto: Rollo & Adam, 1866 – (= ser History of the fenian raid on fort erie) – 2mf – 9 – (also issued under title: history of the fenian raid on fort erie) – mf#23347 – cn Canadiana [971]
The fenian raid!!; the queen's own! : poems on the events of the hour / Breeze, James T – [Napanee, Ont?: s.n.] 1866 [mf ed 1993] – 1mf – 9 – 0-665-93060-7 – mf#93060 – cn Canadiana [810]
Fenland notes and queries – Peterborough. 1889-1909 (1) – mf#3902 – us UMI ProQuest [420]
Fenn, Courtenay Hughes *see*
– Diary of courtenay hughes fenn (1866-1953) for the period 1866-1927
– Over against the treasury
Fenn, Harry *see* The niagara book
Fenn, William Wallace *see*
– The bible in theology
– The teaching of jesus
Fenne, Christina *see* Anselm kiefer
Fennell, D A *see* A provile of ecotourists and the benefits derived from their experience
Fennell, Dorothy Codner *see* Contemporary rhythmic devices
Fennell, Frederick *see* Orchestral development of the kettledrum from purcell through beethoven
Fennelly, John, Bishop *see* Pastoral do illustrissimo doutor fennelly, vigario apostolico em madrasta, datada de 8 de janeiro de 1863
Fenner, Charles Erasmus *see* The civil code of louisiana as a democratic institution
Fenner, D *see* A counter-poyson

Fennimore journal – Fennimore WI. 1947 nov-dec, 1948 jan-1950 mar, 1954 apr 29-1958, 1959-62, 1962-1965 apr 15 – 5r – 1 – (cont: fennimore buyer's guide) – mf#942280 – us WHS [071]
Fennimore times – 1892 oct 6-1893 feb 23, 1889 sep-6-1892 sep 29 – 2r – 1 – (cont by: times review [fennimore, wi]) – mf#942509 – us WHS [071]
Fennimore times – Fennimore WI. 1900 dec 12/1902 apr 2-2001 jan/jun – 88r – 1 – (cont: times review [fennimore, wi]) – mf#904629 – us WHS [071]
Fenny stratford times *see* Fenny stratford weekly times
Fenny stratford weekly times – Fenny, Stratford, England. 21 aug 1879-1918; 1942 [wkly] – 38r – 1 – (aka: fenny stratford times 1882-86. north bucks times etc 1887-) – uk British Libr Newspaper [072]
Fenollosa, Ernest Francisco *see* "Noh"
Fenster, Samuel Benjamin *see* Fenster's georgia law problems and answers.
Fenster's georgia law problems and answers. / Fenster, Samuel Benjamin – Atlanta, 1941. LL-606 – 1 – us L of C Photodup [340]
Fenton family newsletter – v1 iss 1-v2 iss 3 [1982 winter-1989 fall] – 1r – 1 – mf#967750 – us WHS [929]
Fenton, Jennifer M *see* Linking girls' experiences in physical activity to school culture and social and political contexts
Fenton, W J *see* Letter to rev a b simpson
Fentster tsu der velt / Glazman, Ari – Kaunas, Lithuania. 1938 – 1r – us UF Libraries [939]
Fenwick, George E *see* Canada medical journal and monthly record of medical and surgical science
Fenwick, George Edgeworth *see*
– Antiseptic surgery
– Excision of the knee joint
– Scrofulous glands
– Valedictory address to the graduates in medicine and surgery, mcgill university
Fenwick, John *see* Biographical sketches of joshua marshman
Fenwick, Malcolm C *see* The church of christ in corea
Fenzl, E *see*
– Illustrationes et descriptiones plantarum novarum syriae et tauri occidentalis
– Pugillus plantarum novarum syriae et tauri occidentalis primus
Feofan prokopovich i ego vremia / Chistovich, I A – 1868 – 14mf – 8 – mf#R-6011 – ne IDC [947]
Feofan prokopovich kak pisatel / Morozov, P O – 1880 – 7mf – 8 – mf#1545 – ne IDC [947]
Feoktistov, L G et al *see* Novosti elektrokhimii organicheskikh soedinenii, 1973
Fepc news – n38-n53, n54 [1969 jul/aug-1977 feb/mar, 1977 summer] – 1r – 1 – (cont: fair practices news) – mf#625738 – us WHS [380]
Fer, N *see* Les forces de l'europe
Ferber, Reed *see* Effect of proprioceptive neuromuscular facilitation stretch techniques in trained and untrained older adults
Ferber, Ronald R *see* Gait perturbation response in anterior cruciate ligament deficiency and surgery
Ferdi, Katip *see* Mardin mueluek-i artukiye tarihi ve kitabeleri ve sair vesaik-i muehimme
Ferdinand, Archduke of Austria *see* Speculum vitae humanae
Ferdinand christian baur : rede zur akademischen feier seines 100. geburtstages 21. juni 1892 in der aula in tuebingen / Weizsaecker, Carl – Stuttgart: F. Frommanns Verlag (E. Hauff), 1892. Chicago: Dep of Photodup, U of Chicago Lib, 1972 (1r); Evanston: American Theol Lib Assoc, 1984 (1r) – 1 – 0-8370-0086-6 – mf#1984-B321 – us ATLA [920]
Ferdinand christian baur, der begruender der tuebinger schule : als theologe, schriftsteller und charakter / Fraedrich, Gustav – Gotha: FA Perthes, 1909 – 1mf – 9 – 0-7905-7933-2 – mf#1989-1158 – us ATLA [240]
Ferdinand freiligrath : ein biographisches denkmal / Schmidt-Weissenfels, Eduard – Stuttgart: W Mueller, 1876 (mf ed 1990) – 1r – 1 – (filmed with: freiligraths einfluss auf die lyriker der muenchener dichterschule) – us Wisconsin U Libr [430]
Ferdinand freiligrath : ein dichterleben in briefen / Freiligrath, Ferdinand – Lahr: M Schauenburg. 2v. 1882 (mf ed 1990) – 1r – 1 – (filmed with: ferdinand freiligrath's gesammelte dichtungen) – us Wisconsin U Libr [860]
Ferdinand freiligrath's gesammelte dichtungen / Freiligrath, Ferdinand – Stuttgart: G J Goeschen. 6v in 3. 1898 (mf ed 1990) – 1r – 1 – (filmed with: die juden von barnow) – us Wisconsin U Libr [802]
Ferdinand freiligrath's gesammelte dichtungen / Freiligrath, Ferdinand – Stuttgart: G J Goeschen. 6v in 3. 1871 (mf ed 1990) – 1r – 1 – (filmed with: wir sind die kraft) – us Wisconsin U Libr [810]
Ferdinand freiligraths saemtliche werke in zehn bznden = Works / Freiligrath, Ferdinand; ed by Schroeder, Ludwig – Leipzig: M Hesse. 10v in 2. [1907?] – 1 – us Wisconsin U Libr [430]
Ferdinand gregorovius als dichter / Hoenig, Johannes – Stuttgart: Metzler, 1914 – 1r – 1 – (incl bibl ref) – us Wisconsin U Libr [430]
Ferdinand kuernbergers briefe an eine freundin, 1859-1879 / ed by Deutsch, Otto Erich – Wien: Literarischer Verein, 1907 – xxv/453/16p – 1 – us Wisconsin U Libr [860]
Ferdinand raimund : lebenswerk und wirkungsraum eines deutschen volksdramatikers / Kindermann, Heinz – Wien: Adolf Luser, 1940 – 1r – 1 – (incl indes) – us Wisconsin U Libr [430]
Ferdinand raimunds saemtliche werke in drei teilen = Works / Raimund, Ferdinand; ed by Castle, Eduard – Leipzig: Hesse & Becker, [1923?] – 1 – (incl bibl ref) – us Wisconsin U Libr [800]
Ferdinand, Roger *see* 'J 3'
Ferdinand v hayden papers *see* Hayden, ferdinand v, papers, ms 3154
Ferdinand v wrangel und seine reise laengs der nordkueste von sibirien und auf dem eismeere / Engelhardt, Lisa von – Leipzig 1885 [mf ed Hildesheim 1995-98] – 2mf – 9 – €60.00 – 3-487-28961-X – gw Olms [910]
Ferencz, Jozsef *see* Kleiner unitarier-spiegel
Feret, Pierre *see* Notice sur dieppe, arques et quelques monumens circonvoisins
Feret, Pierre J *see* Dieppe en 1826 [dix-huit cent vingt-six]
Fergus county argus – Lewistown, MT. 1891-1946 (1) – mf#64526 – us UMI ProQuest [071]
Fergus county democrat – Lewistown, MT. 1919-1920 (1) – mf#64527 – us UMI ProQuest [071]
Fergus' historical series – v3 n16-19 – 1r – 1 – mf#772558 – us WHS [071]
Ferguson, Archibald *see*
– Afflicted saviour rising from the depths in the garment of praise
– Christ dying for the helpless and ungodly
– Fellowship with the spirits of just men made perfect
– Heavenly bridegroom's desire for his bride
Ferguson, Charles *see* The affirmative intellect
Ferguson, Donald *see* Agricultural education
Ferguson, J A *see* Bibliography of the new hebrides islands, 1610-1942
Ferguson, Jan Helenus *see* Manual of international law
Ferguson, John *see*
– Bibliographical notes on histories of inventions and books of secrets
– False insurance methods
Ferguson, Joyce *see* Salutatory speech
Ferguson, Mary Catharine Guinness *see* The story of the irish before the conque from the mythical period to the invasion under strongbow
Ferguson, Phyllis *see* Catalogue of arabic manuscripts from ghana and adjacent territories from institute of african studies
Ferguson, R S *see* Seal used by the archdeacon of carlisle, with notes of seal of chancellor lowther...
Ferguson, Samuel *see* The cromlech on howth
Ferguson surname quarterly – v1 iss 1-2 [[1978 mar]-[1978 jun]] – 1r – 1 – mf#638831 – us WHS [929]
Ferguson, Thomas James *see* 1734-1884
Ferguson, William *see* The early years of john calvin
Ferguson, William Duncan *see* The legal and governmental terms common to the macedonian greek inscriptions and the new testament
Fergusson, Adam *see*
– Notes made during a visit to the united states and canada in 1831
– On the agricultural state of canada
– Practical notes made during a tour in canada
– Practical notes made during a tour of canada
Fergusson, Archibald *see* Baptismal regeneration opposed to the doctrines and facts of the bi...
Fergusson, Erna *see*
– Cuba
– Guatemala
– Venezuela
Fergusson, James *see*
– The cave temples of India
– An essay on the ancient topography of jerusalem
– A history of architecture in all countries
– History of indian and eastern architecture
– History of the modern styles of architecture
– The holy sepulchre and the temple at jerusalem
– Illustrations of various styles of indian architecture
– The mausoleum at halicarnassus restored in conformity with recently discovered remains
– On the study of indian architecture
– The palaces of nineveh and persepolis restored
– Picturesque illustrations of ancient architecture in hindostan
– Rude stone monuments in all countries
Feria de agosto, 1927 / Almendralejo – Almendralejo: Imp. de J. Bote, 1927 – 1 – sp Bibl Santa Ana [946]
Feria de guaicanama / Enriquez, Carlos – Habana, Cuba. 1960 – 1r – us UF Libraries [972]
Feria de las mercedes en almendralejo / Almendralejo – Badajoz: Imp. V. Rodriguez, 1913 – 1 – sp Bibl Santa Ana [946]
Feria de las mercedes en almendralejo de 1913. concurso de ganaderia / Almendralejo – Badajoz: La Minerva Extremana – 1 – sp Bibl Santa Ana [946]
Feria de san miguel – Caceres: Caceres, Imp. Sanguino, s.a. – 1 – sp Bibl Santa Ana [240]
Feria de san miguel, 1954 / Caceres. Ayuntamiento – Caceres: Tip. El Noticiero – sp Bibl Santa Ana [390]
Feria de santiago de toda clase de ganado. durante...julio, 1961 / Casatejada. Ayuntamiento – Caceres: La Minerva, 1961 – 1 – sp Bibl Santa Ana [390]
Feria de santiago de toda clase de ganados. durante...julio, 1962 / Casatejada. Ayuntamiento – Caceres: Tip. La Minerva, 1962 – 1 – sp Bibl Santa Ana [390]
Feria de santiago. para toda clase de ganados. julio 1973 – Caceres: Imp. La Minerva, 1973 – 1 – sp Bibl Santa Ana [390]
Feria de septiembre, 1947. arroyo de la luz – Imp. Moderna. Caceres. 1947, 1948, 1954, 1963, 1972 – 1 – sp Bibl Santa Ana [946]
Feria. Spain. Ayuntamiento *see* Ordenanzas municipales
Feria y feistas de...agosto 1970 / Olivenza – (Olivenza Tip. Martinez – Reginfo, 1970) – 1 – sp Bibl Santa Ana [946]
Feria y fiestas – Valencia de Alcantara: Tip. Avila, 1959 – 1 – sp Bibl Santa Ana [390]
Feria y fiestas 1945 / San Vicente de Alcantara – Olivenza: Imprenta Marquez, 1945 – sp Bibl Santa Ana [390]
Feria y fiestas, 1945 / Jerez de los Caballeros. Badajoz – Jerez de los Caballeros: Tip. Horizonte, 1945 – sp Bibl Santa Ana [390]
Feria y fiestas, 1947. cabeza del buey / Cabeza del Buey. Ayuntamiento – Imprenta Juan F. Lozano, 1947 – 1 – sp Bibl Santa Ana [390]
Feria y fiestas 1958 – Plasencia: Imp. La Victoria, 1958 – 1 – sp Bibl Santa Ana [390]
Feria y fiestas. 1967 / Don Benito – Don Benito: Tip. Trejo – sp Bibl Santa Ana [390]
Feria y fiestas 1975 – Jaraiz de la Vera: Imp. La Verata, 1975 – 1 – sp Bibl Santa Ana [390]
Feria y fiestas 1977 / Valencia de Alcantara. Ayuntamiento – Valencia de Alcantara: Tip. Avila, 1977 – sp Bibl Santa Ana [390]
Feria y fiestas de 1945 / Zarza de Alange – Olivenza: Tip. M. Reginfo, 1945 – sp Bibl Santa Ana [390]
Feria y fiestas de 1946 / Llerena – Badajoz: Tip. Clasica, 1946 – sp Bibl Santa Ana [390]
Feria y fiestas de jaraiz de la vera 1974 / Jaraiz de la Vera. Ayuntamiento – Jaraiz de la Vera: Imp. La Verata, 1974 – 1 – sp Bibl Santa Ana [390]
Feria y fiestas de junio 1975 – Plasencia: Imp. Padilla, 1975. Tambien Ano 1976 – 1 – sp Bibl Santa Ana [390]
Feria y fiestas de primavera 1974 / Canaveral. Ayuntamiento – Plasencia: Talleres Sanguino, 1973 – 1 – sp Bibl Santa Ana [390]
Feria y fiestas de primavera 1975 / Canaveral. Ayuntamiento – Plasencia: Sandoval, 1975 – 1 – sp Bibl Santa Ana [390]
Feria y fiestas de san bartolome, 1973 – Caceres: Imp. M. Sergio Dorado, 1973 – 1 – sp Bibl Santa Ana [390]
Feria y fiestas en honor de nuestro excelso patron el santisimo cristo de la expiracion... – Jerez de los Caballeros: Imp. Horizonte, 1970 – 1 – sp Bibl Santa Ana [240]
Feria y fiestas en honor de san agustin... 1970 – Fregenal de la Sierra: Imp. Angel Verde, 1970 – 1 – sp Bibl Santa Ana [240]
Feria y fiestas en la ciudad de alfonso 8, 1965 / Plasencia. Ayuntamiento – Caceres, 1965 – 1 – sp Bibl Santa Ana [390]
Feria y fiestas. junio, 1965 / Plasencia. Ayuntamiento – Plasencia: Imprenta Padilla, 1965 – 1 – sp Bibl Santa Ana [390]
Feria y fiestas mayo 1970 / Canaveral. Ayuntamiento – Plasencia: Imp. La Victoria, 1970 – 1 – sp Bibl Santa Ana [390]
Feria y fiestas mayo de 1952. guia comercial / Caceres. Ayuntamiento – Caceres: Imp. Sanguino – sp Bibl Santa Ana [390]
Feria y fiestas mayo-junio, 1953 / Caceres. Ayuntamiento & Trujillo. Ayuntamiento – sp Bibl Santa Ana [390]
Feria y fiestas. merida – Badajoz: Argreus, 1933 – 1 – sp Bibl Santa Ana [946]
Feria y fiestas merida. 1932 – 1 – sp Bibl Santa Ana [946]
Feria y fiestas. programa oficial 1972 / Valencia de Alcantara. Ayuntamiento – Valencia de Alcantara: Tip. Avila, 1972 – 1 – sp Bibl Santa Ana [390]
Feria y fiestas septiembre 1970 / Monterrubio de la Serena, Ayuntamiento – (Castuera: Imprenta de Eladio Fernandez, 1970) – 1 – sp Bibl Santa Ana [946]
Feria y fiestas...1960 / Malagon (Ciudad Real). Ayuntamiento – Plasencia: Tip. La Victoria, 1960 – 1 – sp Bibl Santa Ana [390]
Ferias / Fleiuss, Max – Sao Paulo, Brazil. 1897 – 1r – us UF Libraries [972]
Ferias de ganados durante...septiembre 1962 / Malpartida de Plasencia. Ayuntamiento – Caceres: Tip. La Minerva, 1962 – 1 – sp Bibl Santa Ana [390]
Ferias de san juan. junio 1956 / Coria. Ayuntamiento – Plasencia: Imp. La Victoria, 1956 – 1 – sp Bibl Santa Ana [390]
Ferias de santiago de toda clase de ganados – Caceres: Tip. La Minerva, 1963 – 1 – sp Bibl Santa Ana [390]
Ferias y fiesta en honor de nuestra senora del salor, patrona de...septiembre 1975 – Caceres: La Minerva, 1975 – 1 – sp Bibl Santa Ana [240]
Ferias y fiestas / Oliva de la Frontera – Jerez de los Caballeros. I. Horizontes, 1970 – 1 – sp Bibl Santa Ana [946]
Ferias y fiestas 1928. merida – 1 – sp Bibl Santa Ana [946]
Ferias y fiestas 1944 : nuestra senora de la piedad – 1 – sp Bibl Santa Ana [946]
Ferias y fiestas, 1945 / Don Benito – Badajoz: Arqueros, 1945 – sp Bibl Santa Ana [390]
Ferias y fiestas 1952 / Caceres. Ayuntamiento – Caceres: Imp. Sanguino, y 1953 – sp Bibl Santa Ana [390]
Ferias y fiestas 1954 – Badajoz: Imp. Arqueros, 1954 – 1 – (tambien anos 1955, 1956, 1957, 1958, 1959, 1960, 1961, 1962, 1965, 1968) – sp Bibl Santa Ana [390]
Ferias y fiestas, 1954 / Hervas. Ayuntamiento – Valencia: Viuda de Climent, 1954 – sp Bibl Santa Ana [390]
Ferias y fiestas 1961 / Hervas. Ayuntamiento – Plasencia: La Victoria, 1961 – 1 – sp Bibl Santa Ana [390]
Ferias y fiestas 1964 / Hervas. Ayuntamiento – Plasencia: El autor, 1964 – 1 – sp Bibl Santa Ana [390]
Ferias y fiestas. 1970 – Fregenal de la Sierra: Imp. Angel Verde, 1970 – 1 – sp Bibl Santa Ana [390]
Ferias y fiestas 1971 – Caceres: Imp. La Minerva, 1971 – 1 – sp Bibl Santa Ana [390]
Ferias y fiestas 1971 / Cilleros. Ayuntamiento – Coria: Imp. Fernandez, 1971 – 1 – sp Bibl Santa Ana [390]
Ferias y fiestas 1972 / Sotillo de la Andrada. Ayuntamiento – Jaraiz de la Vera: Imp. La Verata, 1972 – 1 – sp Bibl Santa Ana [390]
Ferias y fiestas 1973 / Asociacion Cultural "Pedro de Trejo" – Plasencia, Caceres: A.C.P. Pedro de Trejo, 1973 – 1 – sp Bibl Santa Ana [390]
Ferias y fiestas 1973 / Malpartida de Caceres. Ayuntamiento – Caceres: Imp. Tomas Rodriguez Santano, 1973 – 1 – sp Bibl Santa Ana [390]
Ferias y fiestas. 1973. guia oficial / Caceres. Ayuntamiento – Caceres: Ed. Extremadura, 1973 – 1 – sp Bibl Santa Ana [390]
Ferias y fiestas 1974 – Caceres: Tip. La Minerva, 1974 – 1 – sp Bibl Santa Ana [390]
Ferias y fiestas 1974 / Navas del Madrono. Ayuntamiento – Caceres: Tip. Extremadura, 1974 – 1 – sp Bibl Santa Ana [390]
Ferias y fiestas 1975 – Caceres: Imp. La Minerva, 1975 – 1 – sp Bibl Santa Ana [390]
Ferias y fiestas 1977 – Caceres: Edit. Extremadura, s.a. 1977 – 1 – sp Bibl Santa Ana [390]
Ferias y fiestas. agosto, 1973 – Caceres: Imp. La Minerva, 1973 – 1 – sp Bibl Santa Ana [390]
Ferias y fiestas agosto 1976 – Caceres: Tip. Extremadura, 1976 – 1 – sp Bibl Santa Ana [390]
Ferias y fiestas. agosto de 1945 / Almendralejo – Almendralejo: Imp. Macarro, 1945 – 1 – sp Bibl Santa Ana [390]
Ferias y fiestas. caceres, mayo 1958 / Odarial Publicidad – Caceres: Odarial Publicidad, 1958 – 1 – sp Bibl Santa Ana [390]
Ferias y fiestas. caceres, trujillo, logrosan / Publicidad Fefa – Caceres: Publicidad "Fefa", 1954 – 1 – sp Bibl Santa Ana [390]
Ferias y fiestas de 1974 / Malpartida de Plasencia. Ayuntamiento – Plasencia: Imp. Padilla, 1974 – 1 – sp Bibl Santa Ana [390]
Ferias y fiestas de agosto. 1975 / Olivenza. Ayuntamiento – Zafra: Ind. Tip. Extremenas, 1975 – 1 – (tambien 1980) – sp Bibl Santa Ana [390]
Ferias y fiestas de caceres, 1947 – Caceres: Tip. Garcia Floriano, 1947 – 1 – sp Bibl Santa Ana [390]
Ferias y fiestas de junio, 1960 / Trujillo. Ayuntamiento – s.l. s.i., 1960 – sp Bibl Santa Ana [390]

Ferias y fiestas de junio. 1963 / Malpartida de Plasencia. Ayuntamiento – Caceres: Tip. La Minerva, 1963 – 1 – sp Bibl Santa Ana [390]
Ferias y fiestas de la velada, 1945 / Zorita. Ayuntamiento – Caceres: Tip. El Noticiero, s.a. 1945 – sp Bibl Santa Ana [390]
Ferias y fiestas de mayo 1954 / Caceres. Ayuntamiento – Caceres: Imp. Sanguino – sp Bibl Santa Ana [390]
Ferias y fiestas de mayo de 1950 guia comercial / Caceres. Ayuntamiento – Caceres: Imp. Moderna – sp Bibl Santa Ana [390]
Ferias y fiestas de miajadas 1972 / Miajadas. Ayuntamiento – Caceres: Tip. Extremadura, 1972 – 1 – sp Bibl Santa Ana [390]
Ferias y fiestas de nuestra senora de la asuncion 1974 – Plasencia: J. Luis Heras, 1974 – 1 – sp Bibl Santa Ana [240]
Ferias y fiestas de nuestra senora de la asuncion, 1975 – Plasencia: Imp. Las Heras, 1975 – 1 – sp Bibl Santa Ana [240]
Ferias y fiestas de nuestra senora de la asuncion 1976 – Plasencia: Imp. Luis Heras, 1976 – 1 – sp Bibl Santa Ana [240]
Ferias y fiestas de nuestra senora de la asuncion durante...agosto 1962 / Galisteo. Ayuntamiento – Plasencia: Imp. Luis Heras, 1962 – 1 – sp Bibl Santa Ana [390]
Ferias y fiestas de nuestra senora de la asuncion. galisteo 1973. agosto – Plasencia: Imp. Luis Heras, 1973 – 1 – sp Bibl Santa Ana [240]
Ferias y fiestas de nuestra senora de la asuncion los dias...agosto 1960 / Galisteo. Ayuntamiento – Plasencia: Imp. Luis Heras, 1960 – 1 – sp Bibl Santa Ana [390]
Ferias y fiestas de nuestra senora de la piedad – Almendralejo, 1947 – 1 – sp Bibl Santa Ana [946]
Ferias y fiestas de plasencia 1973 – Plasencia: Imp. Sanguino, 1973 – 1 – sp Bibl Santa Ana [390]
Ferias y fiestas de plasencia 1975 / Asociacion Cultural Placentina Pedro de Trejo – Plasencia: Imp. Padilla, 1975 – 1 – sp Bibl Santa Ana [390]
Ferias y fiestas de plasencia. junio, 1959 / Plasencia. Ayuntamiento – Plasencia: Imp. Gabriel y Galan, 1959 – 1 – sp Bibl Santa Ana [390]
Ferias y fiestas de primavera 1972 / Aleantara. Ayuntamiento – Caceres: Imp. La Minerva, 1972 – 1 – sp Bibl Santa Ana [390]
Ferias y fiestas de san agustin – Villanueva de la Serena: Hijo de Pedro Parejo, 1971 – 1 – sp Bibl Santa Ana [240]
Ferias y fiestas de san gil abad, septiembre, 1953 / Jerte, Ayuntamiento de – Caceres: Tip. El Noticiero – sp Bibl Santa Ana [390]
Ferias y fiestas de san gil abad, septiembre de 1952 / Jerte, Ayuntamiento de – Caceres: Tip. El Noticiero – sp Bibl Santa Ana [390]
Ferias y fiestas de san juan. 1946 / Badajoz – Badajoz: A. Arqueros, 1946 – sp Bibl Santa Ana [390]
Ferias y fiestas de san juan. guia de espectaculos 1953 / Badajoz – Badajoz: Casa Arqueros, 1953 – sp Bibl Santa Ana [390]
Ferias y fiestas de septiembre 1970 / Barcarrota – Fregenal de la Sierra: Imp. Angel Verde, 1970 – sp Bibl Santa Ana [390]
Ferias y fiestas durante los dias 28, 29 y 30...1960 / Navalmoral de La Mata. Ayuntamiento – Caceres: Imp. Rivero, 1960 – 1 – sp Bibl Santa Ana [390]
Ferias y fiestas durante...8, 9 y 10 de junio, 1960 / Plasencia. Ayuntamiento – Plasencia: Imp. Luis Heras, 1960 – 1 – sp Bibl Santa Ana [390]
Ferias y fiestas en 1974 / Plasencia. Asociacion Cultural Placentina Pedro de Trejo – S.l., s.i, s.a. – 1 – sp Bibl Santa Ana [390]
Ferias y fiestas en barcarrota. septiembre de 1946 / Barcarrota – Jerez de los Caballeros: Tip. Horizonte, 1946 – sp Bibl Santa Ana [390]
Ferias y fiestas en honor a san miguel. 1979 / Cabeza del Buey. Ayuntamiento – Cabeza del Buey: Imp. Lorenzo Gonzalez, 1979 – 1 – (tambien 1980) – sp Bibl Santa Ana [390]
Ferias y fiestas en honor de la santisima virgen de carrion, patrona de alburquerque / Alburquerque. Ayuntamiento – Valencia de Alcantara: Tip. Avila, 1972 – 1 – sp Bibl Santa Ana [390]
Ferias y fiestas en honor de ntra. sra. de gracia. 1974 / Membrio. Ayuntamiento – Valencia de Alcantara: Imp. Avila, 1974 – 1 – sp Bibl Santa Ana [390]
Ferias y fiestas en honor de ntra. sra. de la asuncion. 1971 / Galisteo. Ayuntamiento – Plasencia: Imp. Luis Heras, 1971 – 1 – sp Bibl Santa Ana [390]
Ferias y fiestas en honor de nuestra senora de gracia. agosto 1975 – Caceres: La Minerva, 1975 – 1 – sp Bibl Santa Ana [240]
Ferias y fiestas en honor de san isidro labrador. mayo de 1951 / Aldea de Trujillo. Ayuntamiento – Caceres: Imp. El Noticiero – sp Bibl Santa Ana [390]
Ferias y fiestas en malpartida de caceres. septiembre 1972 – Caceres: Edit. Extremadura, 1972 – 1 – sp Bibl Santa Ana [390]
Ferias y fiestas en medellin (badajoz) durante los dias 20, 21, 22 de febrero – Villanueva de la Serena: Imp. Julian Gil y Cia, 1957 – 1 – sp Bibl Santa Ana [390]
Ferias y fiestas en plasencia, 1958 / Plasencia. Ayuntamiento – Plasencia: Tip. La Victoria – sp Bibl Santa Ana [390]
Ferias y fiestas. hervas, 1962 / Hervas. Ayuntamiento – Plasencia: Imp. Victoria, 1962 – 1 – sp Bibl Santa Ana [390]
Ferias y fiestas hervas 1971 / Hervas. Ayuntamiento – Plasencia: Imp. Garcilasso, 1971 – 1 – sp Bibl Santa Ana [390]
Ferias y fiestas junio, 1961 – Plasencia: Imprenta Gabriel y Galan, 1961 – 1 – sp Bibl Santa Ana [390]
Ferias y fiestas los dias...1960 / Malpartida de Plasencia. Ayuntamiento – Caceres: Imprenta La Minerva, 1960 – sp Bibl Santa Ana [390]
Ferias y fiestas, malagon 1961 / Malagon. Ayuntamiento – Plasencia: La Victoria, 1961 – 1 – sp Bibl Santa Ana [390]
Ferias y fiestas mayo 1947 / Caceres. Ayuntamiento – Caceres: Tip. El Noticiero – sp Bibl Santa Ana [390]
Ferias y fiestas mayo 1963 / Caceres. Ayuntamiento – Caceres: Tip. El Noticiero, 1963 – 1 – sp Bibl Santa Ana [390]
Ferias y fiestas patronales en honor de la virgen del soterrano. 1980 / Barcarrota – Fregenal de la Sierra: Imp. Angel Verde, 1980 – 1 – (tambien ano 1979) – sp Bibl Santa Ana [390]
Ferias y fiestas patronales. san jorge, 1960 / Caceres. Ayuntamiento – Caceres: Imprenta La Minerva, 1960 – 1 – sp Bibl Santa Ana [390]
Ferias y fiestas s gil abad, 1973 – Plasencia: Graf. Sandoval, 1973 – 1 – sp Bibl Santa Ana [390]
Ferias y fiestas, septiembre, 1961 / Torrecillas de la Tiesa. Ayuntamiento – Caceres: Imp. Sanguino, 1961 – 1 – sp Bibl Santa Ana [390]
Ferias y fiestas. septiembre 1973 – Caceres: Imp. La Minerva, 1973 – 1 – sp Bibl Santa Ana [390]
Ferias y fiestas...1972 / Santiago de Alcantara. Ayuntamiento – Caceres: Imp. La Minerva, 1972 – 1 – sp Bibl Santa Ana [390]
Ferias y fiestas...mayo 1959...feria de ganados de todas clases / Navas del Madrono. Ayuntamiento – Caceres: Imprenta La Minerva, 1959 – 1 – sp Bibl Santa Ana [390]
Ferias y fiestas...septiembre, 1960 en honor de la santisima virgen de carrion / Alburquerque. Ayuntamiento – Valencia de Alcantara: Tip. Avila, 1960 – 1 – sp Bibl Santa Ana [390]
Ferias y fietas 1963 / Hervas. Ayuntamiento – Jaraiz de la Vera: Imp. Los Veratos, 1963 – 1 – sp Bibl Santa Ana [390]
Ferienreise eines evangelischen predigers : zeitgeschichtliche studien / Dalton, Hermann – Bremen: C Ed Mueller, 1886 – 1mf – 9 – 0-7905-4284-6 – (incl bibl ref) – mf#1988-0284 – us ATLA [242]
Ferishta's history of dekkan : from the first mahummedan conquests / ed by Scott, Jonathan – Shrewsbury: Printed by J & W Eddowes, for John Stockdale, 1794 [mf ed 1995] – (= ser Yale coll) – 2v – 1 – 0-524-10110-8 – mf#1995-1110 – us ATLA [954]
Ferity / Young Women's Christian Associations – 1980 nov-1981 aug/sep – 1r – 1 – (cont: woman alive; cont by: all she wrote) – mf#622098 – us WHS [366]
Ferland, Albert *see*
– Le canada chante, vol 1
– Le canada chante, vol 2
– Le canada chante, vol 3
– Le canada chante, vol 4
– La consolatrice
– Femmes revees
– Sur une tombe
Ferland, Jean-Baptiste-Antoine *see*
– Cours d'histoire du canada
– Excursion a la cote du nord, au dessous de quebec
– La gaspesie
– Mgr joseph octave plesis
– Observations sur un ouvrage intitule histoire du canada, etc
– Opuscules
Fermanagh herald – 14 mar 1903-12 mar 1904; 30 apr 1904-1950; 1986-89; 13 jan 1990-1998 – 67 1/2r – 1 – uk British Libr Newspaper [072]
Fermanagh mail and enniskillen chronicle *see* Enniskillen chronicle and erne packet
Fermanagh mail etc – Enniskillen, Ireland. 23 Aug 1849-4 Nov 1850; 20 Mar-25 Dec 1851; 1852-89; 2 Mar-Dec 1890. -w. 32 3/4 reels. Missing: 1860, 1883 – 1 – uk British Libr Newspaper [072]
Fermanagh news – Enniskillen, Ireland. 1986-90; 11 jan 1991-92 – 15r – 1 – (aka: fermanagh news and west ulster observer) – uk British Libr Newspaper [072]
Fermanagh news and cavan leitrim monaghan and south tyrone advertiser – Dungannon, Ireland. 30 jul 1896-jun 1907; oct 1907-aug 1912; oct 1912-1913; 10 jan 1914-1 may 1920 – 15r – 1 – (aka: fermanagh news and enniskillen press) – uk British Libr Newspaper [072]
Fermanagh news and enniskillen press *see* Fermanagh news and cavan leitrim monaghan and south tyrone advertiser
Fermanagh news and west ulster observer *see* Fermanagh news
Fermanagh sentinel and north west advertiser – Enniskillen, Ireland. 7 mar-26 dec 1854; 2 jan 1855. -w – 1/2r – 1 – uk British Libr Newspaper [072]
Fermanagh times – Enniskillen, Fermanagh. 1930 – 1r – 1 – ie National [072]
Fermanagh times – Enniskillen, Ireland. 4 mar 1880-1900; 27 feb-25 dec 1930 – 9 3/4r – 1 – uk British Libr Newspaper [072]
Fermanagh times – [Northern Ireland] Belfast jan 1901-21 apr 1949 [mf ed 2002] – 43r – 1 – uk Newsplan [072]
Ferme, Charles *see* Logical analysis of the epistle of paul to the romans
Fermenta cognitionis / Baader, Franz von – Berlin, 1822-24 (mf ed 1992) – 2mf – 9 – €24.00 – 3-89349-075-2 – mf#DHS-AR 49 – gw Frankfurter [100]
Fermin, Philippe *see*
– D philipp fermins ausfuehrliche historisch-physikalische beschreibung der kolonie surinam
– Description generale, historique, geographique et physique de la colonie de surinam
Fermor, Patrick Leigh *see* Traveller's tree
Fermoso Estebanez, Paciano *see* Catolocismo de la juventud colombiana
Fermoso Palmeco, Aristides *see* Normas de trabajo vigentes en las industrias de la construccion yunobras publicas y tablas de liquidacion de salarios
Fernald, Raymond T *see* Florida scrub jay
Fernald, Woodbury Melcher *see*
– Emanuel swedenborg
– Eternity of heaven and hell
– God in his providence
– The true christian life and how to attain it
Fernand braudel center for the study of economies, historical systems, and civilizations review – Binghamton. 1989+ (1,5,9) – ISSN: 0147-9032 – mf#14007 – us UMI ProQuest [900]
Fernandes, Albino Goncalves *see* Sincretismo religioso no brasil
Fernandes, Anibal *see* Nabuco
Fernandes, Benjamin Dias *see* A series of letters on the evidences of christianity
Fernandes, Florestan *see*
– Padrao de trabalho cientifico dos sociologos brasi...
– 'Trocinhas' Do Bom Retiro
Fernandes Gaytan, J *see* Marinos extremenos
Fernandes, Jose Fonseca *see* Caminhos do novo mundo
Fernandes Pinheiro, Jose Feliciano *see* Anais da provincia de s pedro
Fernandex Mato, Romas *see* El generalisimo trujillo
Fernandez Abelehira, Maria Isabel *see* Como es la vida
Fernandez Almagro, Melchor *see* La emancipacion de america y su reflejo en la cultura espanola. madrid, 1944
Fernandez Almuzara, E *see* Cartas literarias... sobre gregorio silvestre
Fernandez, Angel Luis *see* Nueva noche
Fernandez, Antonio J *see* La voz libre
Fernandez Arias, Evaristo *see* El beato sanz y companeros martires del orden de predicadores
Fernandez, Aristides *see* Cuentos
Fernandez Ballesteros, Alberto *see* Toulon
Fernandez Barea, M *see* Juicio practico sobre las virtudes medicinales...
Fernandez, Benjamin Dias *see* Letters of benjamin dias fernandez on the evidences of christianity
Fernandez Bolandi, Tomas *see* Cartilla de correspondencia y legislacion mercanti
Fernandez Cabrera, Manuel *see* Cronicas y devaneos
Fernandez Casco, Juan *see* El extraordinario poder de las plantas
Fernandez Caton, Jose Maria *see* Catalogo de los materiales codigologicos....de auspach 1966...
Fernandez Cortes, Gil *see* Memorial...d. isidro de carvajal
Fernandez D, Cesareo *see*
– Colon estremena
– Colon extremeno? de vicente paredes
– Noticias
Fernandez, D E *see* Nuevo vocabulario, o manual de conversaciones en espanol, tagalo y pampango
Fernandez, David *see*
– Arbol y luego bosque
– Diecisiete anos
Fernandez de Bethencourt, Francisco *see*
– Carlos 4 y maria luisa, de juan perez de guzman y gallo
– El marques de monsalud es sustituido por d adolfo bonilla
Fernandez de Castro, Eduardo Felipe *see*
– Dos palabras sobre el presente numero-homenaje. benito arias montano
– Iconografia de benito arias montano por carlos doetsch
Fernandez De Castro, Jose Antonio *see*
– Barraca de feria
– Esquema historico de las letras en cuba
– Impugnador cubano de ernesto renan
– Tema negro en las litras de cuba
Fernandez De Castro, Rafael *see*
– Meeting de tacon
– Para la historia de cuba
Fernandez de Enciso, M *see* Suma de geographia que trata de todos los partidos e provincias del mundo...
Fernandez de los Rios, Angel *see* Munoz torrero
Fernandez de Molina Menitez Donoso, Antonio *see* Cordobesas
Fernandez de Moratin, Nicolas *see* Las Naves De Cortes Destruidas
Fernandez de Oviedo Valdes, Gonzalo *see*
– De la natural historia de las indias. sumario de historia natural de las indias. con un estudio preliminar y notas por enrique alvarez lopez
– General y natural historia de las indias
– Historia...indias...mar oceano
– Sumario de la natural y general historia de las indias
Fernandez De Oviedo Y Valdes, Gonzalo *see* Historia general y natural de las indias
Fernandez de Quesada, Juan Antonio *see* Memorial ajustado...convento del amparo de almendralejo...dona mariana golfin...duque de villahermosa y d. garcia golfin...vinculo de dona juana...
Fernandez de Ribera, Rodrigo *see* Lagrimas de san pedro
Fernandez de S Corde, J *see* Bullarium ordinis recollectorum sancti augustini
Fernandez de Soria, Rafael *see* Discurso... ciencias morales
Fernandez de Soria y Villanueva, Fernando *see* Arrendamientos rusticos protegidos
Fernandez De Soto, Mario *see* Revolucion en colombia
Fernandez De Tinoco, Maria *see* Zulai
Fernandez del Castillo, Francisco *see* Don pedro de alvarado. obra postuma revisada por antonio fernandez del castillo
Fernandez del Valle, J *see*
– Tratado completo de la flebotomia u operaciones de la sangria
– Tratado teorico y practico de las hernias en general y de las estranguladas
Fernandez Duro, Cesareo *see*
– D francisco fernandez de la cueva
– Juan de la torre (de j.a. lavalle. informe)
– Un soldado de la conquista de chile
Fernandez, F *see*
– De facultatibus naturalibus disputationes medicae et phylosophicae
– Disertacion fisico-legal de los sitios...para sepulturas
– Instrucciones para...la conservacion y aumento de las poblaciones
– El juicio de paris verdadero desengano del agua
Fernandez Fernandez, Juan *see*
– Amor alos enemigos en el antiguo testamento
– La caridad cristiana
Fernandez, Francisco *see* Prontuario de ortografia
Fernandez Garcia, Manuel *see* Entretenimientos poeticos
Fernandez Golfin, Luis *see* Breves...isla de cuba
Fernandez Gomez, Otto *see* Dias repartidos
Fernandez Guardia, Leon *see* Historia de costa rica, adapta al programa oficial
Fernandez Guardia, Ricardo *see*
– Cartilla historica de costa rica
– Cosas y gentes de antano...
– Cosas y gentes de antano
– Costa rica en el siglo 19
– Guerra de la liga y la invasion de quijano
– Historia de costa rica
– Independencia y otro episodios
– Miniatura
– Morazan en costa rica
– Resena historica de talamanca
Fernandez Guerra, Aureliano *see* Lapidas romanas de burguillos
Fernandez, Jesus Maria *see* Obra civilizadora de la iglesia en colombia
Fernandez, Jose Manuel *see*
– Todo angel es terrible
– Tren de las 11:30
Fernandez, Juan *see* Un manuscrito de pedro de valencia que lleva por titulo en su portada
Fernandez, Juan Antonio *see* Madrid. archivo historico nacional. seccion de ordenes militares. inventario del archivo de ucles, tomo 1-3

Fernandez Juncos, Manuel *see*
– Antologia de sus obras
– Cuentos y narraciones
– Galeria puertorriquena
– Ultima hornada
Fernandez Larrain, Sergio *see* Julio cejador y frauca
Fernandez, Leon *see*
– Coleccion de documentos para la historia de costa-...
– Historia de costa rica durante
Fernandez Madrid, Jose *see* Jose fernandez de madrid y su obra en cuba
Fernandez, Manuel *see* Bosquejo fisico, politico e historico
Fernandez Mato, Ramon *see* Trujillo
Fernandez Mejia, Abel *see* Adolescente y nubes. poemas, 1947-1954
Fernandez Mendez, Eugenio *see*
– Historia de la cultura en puerto rico
– Identidad y la cultura
– Salvador brau y su tiempo
Fernandez Molina, Antonio *see* Biografia de roberto g
Fernandez Montufar, Joaquin *see* Vibraciones y recuerdos
Fernandez Mora, Carlos *see* Calderon guardia, lider y caudillo
Fernandez Morejon, Antonio *see* Historia bibliografica de la medicina espanola
Fernandez Moreno, Ramon *see* Restauracion de la patrona de los santos de maimona. nuestra senora de la estrella
Fernandez Navarro, Lucas *see* El meteorito de olivenza (badajoz)
Fernandez, Pablo Armando *see*
– Himnos
– Libro de los heroes
– Toda la poesia
Fernandez Padron, Artemio *see* Sueno y vigilia
Fernandez Pascual, Alfonso *see* Madre, ya tenemos bandera
Fernandez, Pedro Villa *see* Latinoamerica
Fernandez Pesquero, J *see* America
Fernandez, Placido Jose *see*
– El seminarista
– Teatro nacional...
Fernandez Retamar, Roberto *see*
– Alabanzas, conversaciones, 1951-1955
– Historia antigua
– Papeleria
– Vuelta de la antigua esperanza
Fernandez, Salvador Diego *see* Los pactos de bucaseli y el tratado de la mesilla...
Fernandez Sanchez, Teodoro *see*
– Gran matematico y fecundo poeta. arsenio gallego hernandez
– Historia de la imagen de nuestra senora de la fuente santa excelsa patrona de zorita
– Juan macias, inclita gloria de extremadura y de la iglesia
Fernandez Santana, Ezequiel *see*
– Apuntes de pedagogia deportiva
– Catecismo social
– Conferencia
– La cuestion social en extremadura a la luz de las enciclicas rerum novarum y quadregesimo anno
– Homenaje de gratitud a dr. ezequiel fernandez santana
– Organizacion y procedimiento pedagogicos de las esouelas parroquiales de los santos
Fernandez Serrano, Francisco *see* De re bibliographica
Fernandez Spencer, Antonio *see*
– A orillas del filosofar
– Caminando por la literatura hispanica, 1948-1964
– Ensayos literarios
– Nueva poesia dominicana
Fernandez, T *see* Defensa de la china y verdadera respuesta a las falsas razones ove para su reprobacion trae el doctor don jose colmenero
Fernandez Valbuena, Ramiro *see*
– Catolico o krausista?
– De santo tomas a krause?
– Discursos leidos...san benito de villanueva
– El testimonio de las piedras-discurso
Fernandez Vanga, Epifanio *see*
– Idioma de puerto rico y el idioma escolar de puert...
– Idioma de puerto rico y el idioma escolar de puert
Fernandez Vega, Wifredo *see* Alma y tierra, problemas cubanos
Fernandez y Fernandez, Juan *see*
– La caridad misional y la epistola de san pablo a los filipenses. badajoz
– El misterio del cristo mistico
Fernandez y Perez, Gregorio *see* Historias... merida
Fernandez-Davila, Guillermo *see*
– El asesinato de don francisco pizarro
– El asesinato del conquistador del peru
Fernandez-Guerra Y Orbe, Luis *see* D juan ruiz de alarcon y mendoza
Fernandez-Madrid, Pedro *see* Rasgos de la vida publica del jeneral francisco de...
Fernandez-Marina, Ramon *see* Horizons of the mind
Fernandez-Ouesta, Raimundo *see* Discursos
Fernandina – s.l, s.l? 193-? – 1r – us UF Libraries [978]
Fernandina beach news-leader – Fernandina Beach, FL. 1959-1987 – 34r – (gaps) – us UF Libraries [071]
Fernandina express – Fernandina FL: B C Cuvellier 12 jun 1800-1 jul 1882 (wkly) (gaps) [mf ed 1995] – 1r – 1 – (a conservative newspaper of the period) – us UF Libraries [071]
Fernandina history / Johnson – s.l, s.l? 193-? – 1r – us UF Libraries [978]
Fernandina 'in the long ago' / Wolff, George E – s.l, s.l? 193-? – 1r – us UF Libraries [978]
Fernandina news-leader – Fernandina Beach, FL. 1949 feb 4-1958 – 10r – us UF Libraries [071]
Fernandina notes – s.l, s.l? 193-? – 1r – us UF Libraries [978]
Fernandina pirates – s.l, s.l? 193-? – 1r – us UF Libraries [978]
Fernando 1 *see* Commune sigilli secret (anno 1412-1416)
Fernando 2 *see*
– Curiae sigilli secreti (anno 1479-1516)
– Diversorum (anno 1479-1516)
Fernando de gabriel / Blanco Garcia, Francisco – Madrid: Saenz de Jubera, 1909 – sp Bibl Santa Ana [440]
El fernando do sevilla restaurada / Vera y Figueroa, Juan Antonio – 1632 – 9 – sp Bibl Santa Ana [946]
Fernando, Henry *see* Sinhalese diary, 1893
Fernando, J S A *see*
– Centenary souvenir, 1851-1951
– Jubilee memorials, 1860-1910
Fernando, Solomon *see* Lectures on buddhism
Ferndale cemetery register – Ferndale, CA. 1876-1982 – 1r – 1 – $50.00 – mf#B40223 – us Library Micro [920]
[Ferndale-] the ferndale enterprise – CA. 1878-[wkly] – 78r – 1 – $4680.00 (subs $50y) – mf#B02234 – us Library Micro [071]
Der ferne sohn : erzaehlung / Vring, Georg von der – Muenchen: R Piper 1942 [mf ed 1991] – 1r – 1 – (filmed with: die spur im hafen & other titles) – mf#2970p – us Wisconsin U Libr [830]
Ferne stimmen : erzaehlungen / Raabe, Wilhelm Karl – Berlin: Otto Janke, 1865 – 1 – us Wisconsin U Libr [830]
[Fernley-] leader-dayton courier – $110.00 – (see: yerington) – mf#U04841 – us Library Micro [071]
[Fernley-] tri-county express – NV. 1988-1992 – 5r – 1 – $300.00 – mf#U04842 – us Library Micro [071]
[Fernley-] tri-town times – NV. 1957-1958; 1965 – 1r – 1 – $60.00 – mf#N04531 – us Library Micro [071]
Fernos Isern, Antonio *see*
– Necesidades educativas de puerto rico
– Puerto rico libre y federado
Fernos-Isern, A *see* Text of the constitution of the commonwealth of puerto rico
Ferns of florida / Small, John Kunkel – New York, NY. 1931 – 1r – us UF Libraries [580]
Ferns of the southeastern states / Small, John Kunkel – Lancaster, PA. 1938 – 1r – us UF Libraries [580]
Fernsehen und bildung – Munich. 1970-1973 [1]; 1972-1972 [5] – ISSN: 0015-0150 – mf#5864 – us UMI ProQuest [380]
Fernseh-informationen – Muenchen DE, 1950 1 nov- – 23r until 2002 – 1 – (began in gauting) – gw Mikrofilm [790]
Fernsworth, Lawrence A *see* Back of the spanish rebellion
Fernwood baptist church : church bulletins and newsletters – Milan. 1970-1985 (1) 1974-1985 (5) 1979-1985 (9) – 6r – 1 – $321.98 – mf#6267 – us Southern Baptist [242]
Feron, Jean *see*
– L'aveugle de saint-eustache
– La besace d'amour
– Les cachots d'haldimand
– La fin d'un traitre
– La revanche d'une race
Ferotin, M *see*
– Le liber mozarabicus sacramentorum et les manuscrits mozarabes
– Le liber ordinum dans l'eglise wisigoth. et mozar. d'espagne, du 5e-19ths
Ferrall, Simon A *see* A ramble of six thousand miles through the united states of america
Ferran, H R *see* Early florida citrus fruits in northern markets
Ferrand, Gabriel *see* Les comalis
Ferrandez / Linares, Manuel – Barcelona, Spain. 1965 – 1r – us UF Libraries [972]
Ferrandi, Ugo *see* Lugh
Ferrandis Torres, Manuel *see* El mito del oro en la conquista de america
Ferrando, Juan *see* Historia de los pp dominicos en las islas filipinas y en sus misiones del japon, china, tung-kin y formosa
O ferrao : jornalzinho humoristico, critico e noticioso – Laguna, SC. 23 mar 1934 – (= ser Ps 19) – mf#UFSC/BPESC – bl Biblioteca [073]
O ferrao : orgao critico, humoristico e noticioso – Florianopolis, SC. 07 ago 1927 – (= ser Ps 19) – mf#UFSC/BPESC – bl Biblioteca [320]
The ferrar papers, 1590-1790 : from magdalene college, cambridge – 1992 – 14r – 1 – (with guide. int by david ransome) – mf#97513 – uk Microform Academic [941]
Ferrar, W A *see* Narrative of a shipwreck off the coast of north america, in the winter of 1814
Ferrar, William Hugh *see* Baptismal regeneration
Ferrara, Orestes *see* Ensenanzas de una revolucion
Ferrari, Anne *see* Fear of falling among community elders
Ferrari, Giacomo Gotifredo *see* Trois grandes sonates pour harpe avec violon et violoncelle, op 18
Ferrari, Joseph R *see* Journal of prevention and intervention in the community
Ferraro, Joseph A *see* A comparative study of injuries in division i and division iii men's lacrosse
Ferrars limerick chronicle *see* Limerick chronicle
Ferraz, Paulo Malta *see* Apontamentos para a historia da colonizacao de blu...
Ferreira, Athos Damasceno *see* Jornais criticos e humoristicos de porto alegre no...
Ferreira da Silva, Francisco *see* Apontamentos para a historia da administracao da diocese e da organiscao do seminario lyceu
Ferreira de Castro, Jose *see* Obras completas
Ferreira, Ignatius Wilhelm *see* A critical analysis of elements of educational management
Ferreira, Jan Hendrik Strauss *see* Dieback of grapevines in south africa
Ferreira, M *see* Cinvanja hulpboekie
Ferreira, Manoel Rodrigues *see* Ma conaria na independencia brasileira
Ferreira, Maria Celeste *see* Indianismo na literatura romantica brasileira
Ferreira Pinto, Julio *see* Angola
Ferreira, Tito Livio *see*
– Genese social da gente bandeirante
– Historia da civilizacao brasileira
Ferreira, Waldemar Martins *see* Directrizes do direito mercantil brasileiro
Ferrell, M D *see* An analysis of the bernoulli lift effect as a propulsive component of swimming strokes
Ferrer Caja, Emilio *see* Determinants of intrinsic motivation among female and male adolescent students in physical education
Ferrer Canales, Jose *see* Imagen de varona
Ferrer De Couto, Jose *see* Cuba puede ser independiente
Ferrer de Esparza, T *see* Tratado...de la facultad medicamentosa que se halla en el agua de los banos...
Ferrer Deulofeu, Agustina Surama De Las Mercedes *see* Romelia vargas
Ferrer Hernandez, Gabriel *see* Anhelos y esperanzas
Ferrer, Margarita *see* Siete cantos
Ferrer, Rolando *see* Teatro
Ferrer, Vicente *see* Guerra dos mascates
Ferrer, William Hugh *see* Christian sacrifice
Ferrere, F *see* La situation religieuse de l'afrique romaine
Ferreres, Juan Bautista *see*
– Death real and apparent
– The decree on daily communion
Ferrerias baptisms – Minorca, Spain. v1-8. 1570-1816 – 3r – us UF Libraries [324]
Ferrerias deaths – Minorca, Spain. v1-5. 1571-1830 – 2r – us UF Libraries [324]
Ferrerias marriages – Minorca, Spain. v1-5. 1570-1816 – 2r – us UF Libraries [324]
Ferrero DellaMarmora, Alberto *see* Voyage en sardaigne
Ferrero, Guglielmo *see* Ancient rome and modern america
Ferresheim, Fritz *see* Schiller als herausgeber der rheinischen thalia, thalia, und neuen thalia, und sein mitarbeiter
Ferret : or south wales ratepayer – [Wales] LLGC dec 1870-dec 1873, 28 feb 1874-5 apr 1879 [mf ed 2004] – 11r – 1 – uk Newsplan [072]
Ferret, Pierre Victoire *see* Voyage en abyssinie dans les provinces du tigre
Ferretti, Augustus *see* Institutiones philosophiae moralis
Ferretti, Giovanni *see*
– Il primo libro delle canzoni alla napolitana
– Il quarto libro delle napolitane
– Il secondo libro delle canzoni a sei voci
– Il secondo libro delle canzoni alla napolitana
– Il terzo libro delle napolitane
Ferrey, Benjamin *see* Recollections of a n welby pugin and his father, augustus pugin
Ferri *see* Scultori peregrini a emerita
Ferri, Giovanni *see* London und die englaender
Ferri, Silvia *see* La testa di merida cenni sulla critica iconografica
Ferrie, Adam *see* Letter to the right hon earl grey, one of her majesty's most honorable privy council, and secretary of state for colonial affairs
Ferrier de Tourettes, Alexandre *see*
– Guide pittoresque du voyageur en belgique
– Manuel du voyageur sur le chemin de fer belge
Ferrier, Jeanne-Paul *see* Lezard
Ferrier, Walter Frederick *see* Short notes on some canadian minerals
Ferriere, Emile *see*
– Les apaotres
– Paganisme des hebreux jusqu' a la captivite de babylone
Ferrieres, Charles E de *see* Memoires du marquis de ferrieres
Ferriol, A *see* Wahreste und neueste abbildung des tuerckischen hofes
Ferris, Benjamin *see* A history of the original settlements on the delaware
Ferris, David *see*
– Memoirs of the life of david ferris
– Memoirs of the life of david ferris, an approved minister of the society of friends
Ferris, Isaac *see* Jubilee memorial of the american bible society
O ferro : periodico liberal e progressista – Bahia, 07 jul 1858 – (= ser Ps 19) – bl Biblioteca [320]
Ferro, Giovanni *see*
– Ombre apparenti nel teatro d'impresse di giovanni ferro...
– Teatro d'imprese...
Ferrocarriles de colombia en 1925-26 / Escobar, Paulo Emilio – Bogota, Colombia. 1926 – 1r – us UF Libraries [972]
Ferroequinologist / Central Coast Railway Club – n266-281 [1974 apr-1975 jul] – 1r – 1 – mf#367510 – us WHS [380]
Ferroli, Domenico *see* The jesuits in malabar
Ferrus Roig, Francisco *see* General mayor de la universidad de san carlos en g...
Ferry, Christopher *see* Internal and external rotation strength values of female swimmers and water polo players
Fersterra, Holger *see* Kopplung von schadstoffabbau und nutzstoffproduktion mit halomonas elongata
Fertig, James Walter *see* The secession and reconstruction of tennessee
Fertile land, brazil / Greenbie, Sydney – Evanston, IL. 1943 – 1r – us UF Libraries [972]
Fertility program for celery production on everglades organic soils / Beckenbach, J R – Gainesville, FL. 1939 – 1r – us UF Libraries [630]
Fertilizer experiments with pecans / Blackmon, G H – Gainesville, FL. 1934 – 1r – us UF Libraries [634]
Fertilizer experiments with potatoes on the marl soils of dade county / Fifield, W M – Gainesville, FL. 1940 – 1r – us UF Libraries [630]
Fertilizer experiments with truck crops / Skinner, J J – Gainesville, FL. 1930 – 1r – us UF Libraries [630]
Fertilizer research – Dordrecht. 1989-1995 (1,5,9) – ISSN: 0167-1731 – mf#16786 – us UMI ProQuest [630]
Fertilizer suggestions / Flint, E R – Lake City, FL. 1905 – 1r – us UF Libraries [630]
Fertilizers : how to make and how to use them / Persons, A A – Lake City, FL. 1893 – 1r – us UF Libraries [630]
Fertilizers for japanese cane / Scott, John M – Gainesville, FL. 1918 – 1r – us UF Libraries [630]
Fertilizing the irish potato crop / Floyd, B F – Gainesville, FL. 1920 – 1r – us UF Libraries [630]
Fertin, Pierre *see* L'eglise en notre temps
Fertsig yohr in midber / Saphire, Saul – New York, NY. 1934 – 1r – us UF Libraries [939]
Ferus, Ioan *see*
– In sacrosanctum iesu christi...
Ferussac, A de *see* Histoire naturelle generale et particuliere des c,phalopodes ac,tabuliferes vivants et fossiles
Feryad – Nigde. Cumhuriyetci ve Birlikci siyasi haftalik halk gazete. Sahibi: Cemil Sakir. Mueduerue: Riza Ratib. n8. 4 eyluel 1340 [1924] – (= ser O & t journals) – 1mf – 9 – $25.00 – us MEDOC [956]
Fesch, Willem de *see* 10 sonatas for 2 german flutes or 2 violins
Fescourt *see* Histoire de la double conspiration de 1800
Fespaco '97 : festival panafricain du cinema et de la television de ouagadougou – Bethlehem, PA, 1997 – us CRL [770]
Fessenden, Thomas Green *see* An essay on the law of patents for new inventions
Fessenden, William P *see* Papers
Fessenden's silk manual and practical farmer – Boston. 1835-1837 (1) – mf#4152 – us UMI ProQuest [630]
Fessler, Ignatz Aurelius et al *see* Eunomia
Fessler, J *see* Institutiones patrologiae quas denuo recensuit auxit

Fessler, Joseph *see*
- Das letzte und das naechste allgemeine concil
- Sammlung vermischter schriften ueber kirchengeschichte und kirchenrecht
- The true and the false infallibility of the popes

Fest, Joh Samuel *see* Beitraege zur beruhigung und aufklaerung ueber diejenigen dinge, die dem menschen unangenehm sind oder sein koennen

Fest- und gelegenheitspredigten gehalten in den synagogen kolns / Rosenthal, Ludwig – Frankfurt am Main, Germany. 1901 – 1r – us UF Libraries [270]

Festa christianorvm... / Hospinian, R – Tigvri, Ioannes Wolph, 1593 – 3mf – 9 – mf#PBU-474 – ne IDC [240]

Festa fatta in roma... / Mascardi, V – Roma, n.d. – 2mf – 9 – mf#0-39 – ne IDC [090]

The festal letters of athanasius in an ancient syriac version / ed by Cureton, W – London, 1848 – €11.00 – ne Slangenburg [243]

Festas e tradicoes populares do brasil / Morais Filho, Melo – 3a. ed. Rio de Janeiro. 1946 – 1 – us CRL [390]

Festbrevier und kirchenjahr der syrischen jakobiten / Baumstark, Anton – Paderborn, 1910 – 6mf – 8 – €14.00 – ne Slangenburg [243]

Festejos en honor de la santisima virgen de carrion 1975 – Valencia de Alcantara: Tip. Avila, 1975 – 1 – sp Bibl Santa Ana [240]

Festejos en honor de la santisima virgen de carrion, 1977 / Alburquerque. Ayuntamiento – Valencia de Alcantara: Imp. Avila, 1977 – sp Bibl Santa Ana [390]

Festejos en honor de la santisima virgen de carrion. alburquerque, 1980 – Valencia de Alcantara: Tip. Avila, 1980 – 1 – sp Bibl Santa Ana [240]

Festejos en honor de nuestra madre la santisima virgen de carrion. alburquerque, 1979 – Valencia de Alcantara: Tip. Avila, 1979 – 1 – sp Bibl Santa Ana [240]

Festejos en honor del principe de la paz habidos en badajoz en 1807 / Guerra Guerra, Arcadio – Badajoz: Dip. Provincial, 1967 – sp Bibl Santa Ana [946]

Festejos taurinos tradicionales en guadalupe (caceres). 1971 – S.l., s.i., s.a. – 1 – sp Bibl Santa Ana [390]

Festejos y romeria en honor del santo patron de las hermandades. "san isidro labrador" / Villar del Rey – Badajoz: Imp. Inca, 1970 – sp Bibl Santa Ana [390]

Fester, Richard *see*
- Johann daniel schoepflins brieflicher verkehr
- Johann daniel schoepflins brieflicher verkehr mit goennern, freunden und schuelern

Festgabe fuer eduard berend, zum 75. geburtstag, am 5. dezember 1958 / ed by Seiffert, Hans Werner & Zeller, Bernhard – Weimar: H Boehlaus Nachfolger, 1959 – 1 – (incl bibl ref) – us Wisconsin U Libr [430]

Festgabe herrn dr. rudolph von jhering zum doktorjubilaum am 6, august 1892 / Tubingen. Universitat. Juristische Facultat – Tubingen, Laupp, 1892 185 p. LL-4117 – 1 – us L of C Photodup [340]

Festgabe philipp strauch : zum 80. geburtstage am 23. september 1932 dargebracht von fachkollegen und schuelern / ed by Baesecke, Georg & Schneider, Ferdinand Joseph – Halle (Saale): M. Niemeyer, 1932 [mf ed 1993] – (= ser Hermaea 31) – 157p – 1 – (incl bibl ref) – mf#8084 – us Wisconsin U Libr [430]

Festgabe von fachgenossen und freunden a von harnack : zum siebzigsten geburtsdag dargebracht – Tuebingen, 1921 – 8mf – 8 – €17.00 – ne Slangenburg [240]

Festgabe zum neunzigsten geburtstag leopolds von ranke : marburg am 21. december 1885 / Ranke, Ernst Constantin – Marburg: Universitaets-Buchdruckerei, 1885 – (= ser Zur Beurtheilung Wielands) – 1r – 1 – (caption title: zur beurtheilung wielands: ein kritischer versuch) – us Wisconsin U Libr [943]

Der festgeankerte / anakephalaios / gegen die antikomarianiten (bdk38 1.reihe) / Epiphanius von Salamis (Epiphanius of Constantia, Saint) – (= ser Bibliothek der kirchenvaeter. 1. reihe (bdk 1.reihe)) – €12.00 – ne Slangenburg [241]

Festgruss bernhard stade : zur feier seiner 25 jaehrigen wirksamkeit als professor – Giessen: J Ricker (Alfred Toepelmann) 1900 [mf ed 1985] – 1mf – 9 – 0-8370-3120-6 – (incl bibl ref) – mf#1985-1120 – us ATLA [220]

Le festin d'esope – Revue des belles-lettres. Red. en chef Guillaume Apollinaire. no. 1-9. Paris. nov 1903-aout 1904 – 1 – fr ACRPP [800]

Festino nella sera del giovedi grasso... / Banchieri, Adriano – 1608 – (= ser Mssa) – 2mf – 9 – €35.00 – mfchl 178 – gw Fischer [780]

Le festival artistique et culturel et le panafricanisme – Conakry: Impr National "Patrice Lumumba", 1975 – us CRL [700]

Festival de Folklore Hispanoamericano 1, 1958 *see* Espana 1958

Festivales de Espana N. Plasencia, 1972 *see* 4 festivales de espana. plasencia 23-29 junio 1972

The festival-hallof csorkon 2 in the great temple of bubastis / Naville, E – London, 1892 – (= ser Mees 10) – 6mf – 8 – €14.00 – ne Slangenburg [720]

The festivals of the lord : as celebrated by the house of israel in every part of the world – London: Hebrew Review Office, 5599 [1838 or 1839] – 1mf – 9 – 0-524-05466-5 – mf#1990-3492 – us ATLA [270]

Festividad de san fernando 30 de mayo 1959, valencia de alcantara / Valencia de Alcantara. Estacion de Renfe – Avila, 1959 – 1 – sp Bibl Santa Ana [390]

Das festland am suedpol : die expedition zum suedpolarland in den jahren 1898-1900 / Borchgrevink, C – Breslau, 1905 – 7mf – 9 – mf#H-6185 – ne IDC [919]

Festoni, Abbate *see* Sonata or trio for two violins and a bass

Festorvm diervm...sermones / Bullinger, Heinrich – Tigvri, Christoph Froschover, 1558 – 5mf – 9 – mf#PBU-204 – ne IDC [240]

Festschrift anlaesslich der emeritierung von prof. dr.-ing. walter raab / ed by Fabrig, Peter – (mf ed 1994) – 2mf – 9 – €40.00 – 3-8267-2023-7 – mf#DHS 2023 – gw Frankfurter [620]

Festschrift des zionistischen vereines "jeschurun'-troppau... – Troppau, Czechoslovakia . 1911? – 1r – us UF Libraries [939]

Festschrift eduard sachau zum siebzigsten geburtstage / ed by Weil, Gotthold – Berlin: G Reimer, 1915 – 5mf – 9 – 0-524-06934-4 – (incl bibl ref) – mf#1990-3560 – us ATLA [956]

Festschrift for ralph farrell / ed by Stephens, Anthony et al – Bern: P. Lang, 1977 – 1r – 1 – (contributions in english or german. incl bibl ref) – us Wisconsin U Libr [430]

Festschrift fuer berthold litzmann zum 60. geburtstag 18.4.1917 : im auftrage der literarhistorischen gesellschaft bonn / ed by Enders, Carl – Bonn: F Cohen, 1920 – 1 – (incl bibl ref and index) – us Wisconsin U Libr [430]

Festschrift fuer eduard castle zum achtzigsten geburtstag : gewidmet von seinen freunden und schuelern / ed by Gesellschaft fuer Wiener Theaterforschung; Wiener Goethe-Verein – Wien: Notring der Wissenschaftlichen Verbaende OEsterreichs, 1955 – 1 – (incl bibl ref) – us Wisconsin U Libr [790]

Festschrift fuer friedrich keinecker zum 60. geburtstag : gewidmet von seinen kollegen, schuelern und mitarbeitern / ed by Michels, Gerd – Heidelberg: J Groos, c1980 – 1r – 1 – (incl bibl ref) – us Wisconsin U Libr [430]

Festschrift fuer wolfgang stammler zu seinem 65. geburtstag – Berlin: E Schmidt, c1953 – 1r – 1 – (incl bibl ref) – us Wisconsin U Libr [943]

Festschrift meinhof – Gluckstadt, Germany. 1927 – 1r – us UF Libraries [470]

Festschrift meinhof – Hamburg: Kommissionsverlag von L Friederichsen, 1927 – 1 – us CRL [920]

Festschrift meinhof. sprachwissenschaftliche und andere studien – (Gluckstadt und J.J. Augustin, 1927). xii,514p.Illus. (inc. music, map). With: Midrash Hasirot ve-Yiterot. 1 reel. 1251 – 1 – us Wisconsin U Libr [400]

Festschrift zum 50jahrigen jubilaum des israelitischen mannerverein / Osterberg, Max – Stuttgart, Germany. 1925 – 1r – us UF Libraries [939]

Festschrift zum funfzigjahrigen bestehen des vereins, 1874-1924 – Frankfurt am Main, Germany. 1924 – 1r – us UF Libraries [939]

Festschrift zum siebzigsten geburtstage jakob guttmanns / Cohen, Hermann – Leipzig: Gustav Fock, 1915 – (= ser Schriften (gesellschaft zur foerderung der wissenschaft des judentums (germany))) – 1mf – 9 – 0-524-00363-7 – mf#1989-3063 – us ATLA [270]

Festschrift zur einweihung des neuen tempels zu steinamanger / Stier, Josef – Dessau, Germany. 1880 – 1r – us UF Libraries [939]

Festschrift zur erinnerung an die feierliche einweihung – Linz, Austria. 1877 – 1r – us UF Libraries [939]

Festschrift zur feier des hundertjaehrigen bestehens der wetterauischen gesellschaft fuer die gesammte naturkunde – Hanau, 1908 – 2mf – 9 – mf#8618 mf. A3-A4 – ne IDC [590]

Festschrift zur grundungsfeier der augustin keller=loge in zuerich... – Zuerich, Switzerland. 1909? – 1r – us UF Libraries [939]

Festschrift zur jahrhundertfeier des allgemeinen buergerlichen gesetzbuches : 1. juni 1911 – Wien, Manz. 2v. 1911 – (= ser Civil law 3 coll) – 20mf – 9 – (incl bibl ref) – mf#LLMC 96-619 – us LLMC [346]

Festskrift til den norske synodes jubilaeum, 1853-1903 / ed by Halvorsen, Halvor – Decorah, Iowa: Norske Synodes Forlag, 1903 – 2mf – 9 – 0-524-01726-3 – mf#1990-4118 – us ATLA [240]

Festugiere, A *see* L'ideal religieux des grecs et l'evangile

Festugiere, A J *see*
- Antioche paienne et chretienne
- L'enfant d'argigente

Der festungs-bote – Rastatt DE, 1849 7 jul-22 jul – 1 – gw Misc Inst [074]

Der festungsbote : nachrichtenblatt fuer den festungsbereich. – La Rochelle (F), 1944 sep-1945 may – 1 – fr ACRPP [074]

Fest-zeitung : erscheint im interesse der in sheboygan stattzufindenden 4. general-versammlung der katholischen unterstutzungs-vereine wisconsin / Catholic Central Verein of America – n1-5 [den 15 dez 1893-den 12 mai 1894] – 1r – 1 – (cont: fest-zeitung, deutsch-katholische unterstutzungs-verein von wisconsin) – mf#949334 – us WHS [241]

Fest-zeitung : herausgegeben von der festbehorde der 3. staats-versammlung der deutschen katholischen unterstutzungs-vereine von wisconsin / ed by Catholic Central Verein of America – 1892 jun 6/8 – 1r – 1 – (cont by: fest-zeitung, deutsch-katholische unterstutzungs-verein von wisconsin) – mf#1002897 – us WHS [241]

Fest-zeitung / Saengerbund des Nordwestens Saengerfest. 18th. Davenport, Iowa, 1898 – Davenport, Iowa. Aug 1898. nr. (13) – 1 – us NY Public [780]

Fest-zeitung / Saengerbund des Nordwestens Saengerfest. 24th. Omaha. 1910 – 10 Feb-5 Jul 1910. (Nr. 1-6) – 1 – us NY Public [780]

Fete de st gregoire 7 : pape et confesseur, patron de l'union allet, 25 mai 1073-25 mai 1873 – S.l: s.n, 1873? – 1mf – 9 – mf#29754 – cn Canadiana [241]

Fete des fous / Arnould, Auguste Jean Francois – Paris, France. 1843 – 1r – us UF Libraries [440]

Fete d'un bourgeois de paris : ou, le jour et le le... / Dumersan, Theophile Marion – Paris, France. 1816 – 1r – us UF Libraries [440]

Fete nationale des canadiens francais, 23 juin 1908 – [Quebec (Province): s.n, 1908?] (mf ed 1994) – 9 – cn Bibl Nat [971]

Fete nationale des canadiens-francais celebree a windsor, ontario, le 25 juin 1883 / Dionne, Narcisse Eutrope – Quebec: Impr Leger Brousseau, 1883 [mf ed 1979] – 2mf – 9 – mf#SEM105P33 – cn Bibl Nat [971]

Les fetes annuellement celebrees a emoui (amoy) : etude concernant la religion populaire des chinois = Jaarlijksche feesten en gebruiken van de emoy-chineezen / Groot, Jan Jakob Maria de – Paris: E Leroux 1886 [mf ed 1991] – (= ser Annales du musee guimet 11-12) – 2v on 3mf [ill] – 9 – 0-524-01904-5 – (french trans fr dutch by g g chavannes; ill by felix regamey) – mf#1990-2717 – us ATLA [390]

Les fetes colombiennes a quebec : compte-rendu et discours / Institut canadien (Montreal, Quebec) – Quebec?: Leger Brousseau, 1893 – 1mf – 9 – mf#07437 – cn Canadiana [910]

Les fetes de l'amour et de baccus / Lully, Jean Baptiste – [17-?] [mf ed 19–] – 1r – 1 – (text by philippe quinault in collaboration with moliere and isaac de benserade) – mf#film 615 – us Sibley [780]

Fetes d'eleusis : ou, les jeux de la grece / Hapde, Jean-Baptiste-Augustin – Paris, France. 1810 – 1r – us UF Libraries [440]

Les fetes du troisieme centenaire de l'hotel-dieu de quebec, 1639-1939 : bibliographie analytique / Saint-Clement, soeur – 1964 [mf ed 1979] – (= ser Bibliographies du cours...1947-66) – 7mf – 9 – mf#SEM105P4 – cn Bibl Nat [360]

Les fetes eucharistiques de saint-thomas-de-pierreville, les 27, 28, 29, et 30 aout 1916 – [Arthabaska, Quebec: s.n, 1917?] – 1mf – 9 – 0-665-66418-4 – mf#66418 – cn Canadiana [241]

Fetes jubilaires celebrees a ottawa les 25 et 26 octobre 1899 – Ottawa: C Boudreault, 1899 – 3mf – 9 – mf#03097 – cn Canadiana [241]

Fetes jubilaires de la congregation des hommes du tres saint sacrement, 1894-1919 – [Quebec (Province)?: s.n, 1919?] – 1mf – 9 – 0-665-76946-6 – mf#76946 – cn Canadiana [241]

Fetes jubilaires en l'honneur du rev jean antoine boissonnault, cure : noces d'argent 1866-11 novembre 1891, jubile curial, 1874-29 juillet 1899 – St Johnsbury, VT: s.n, 1899? – 1mf – 9 – (incl english and latin text) – mf#58596 – cn Canadiana [241]

Fetes nationale des canadiens-francais, le 24 juin 1890 / Societe Saint-Jean-Baptiste de Quebec – [Quebec?: s.n, 1890?] [mf ed 1981] – 6mf – 9 – 0-665-13824-5 – mf#13824 – cn Canadiana [390]

Les fetes, offices, ceremonies et usages de l'ancienne eglise cathedrale de tournai / Vos, Chan – Tournai, 1894 – €5.00 – ne Slangenburg [241]

Fetes patriotiques celebrees en 1919 : et recits populaires des evenements qui s'y rapportent / Tourigny, Joseph-Donat – [Montreal: impr de La Salle], 1920 [mf ed 1995] – 2mf – 9 – mf#SEM105P2316 – cn Bibl Nat [971]

Fetes patriotiques et recits populaires des evenements qui s'y rapportent / Tourigny, Joseph-Donat – 2e rev augm ed. Montreal: impr de La Salle, 1921 [mf ed 1995] – 3mf – 9 – mf#SEM105P2317 – cn Bibl Nat [971]

Fetes...l'occasion du mariage de s m napoleon...avec marie-louise... / Goulet, M – Paris, 1810 – 2mf – 9 – mf#0-1107 – ne IDC [700]

Il "fetha nagast," o "legislazione dei re" : codice ecclesiastico e civile di abissinia, pubblicato da ignazio guidi a spese del r istituto orientale in napoli – Roma, Tip della Casa editrice italiana, 1897-99 – us CRL [074]

Fetherstonhaugh, Edward J *see* Epitome of the patent laws in canada and united states

Feth-i celil-i konstantiniye / Muhtar, Ahmet – [Istanbul]: Matbaa-yi Tahir Bey, 1320 [1904] – (= ser Ottoman histories and historical sources) – 4mf – 9 – $60.00 – us MEDOC [956]

Fetich in theology, or, doctrinalism twin to ritualism / Miller, John – New York:Dodd & Mead, 1874 – 1mf – 9 – 0-8370-4434-0 – mf#1985-2434 – us ATLA [240]

Fetichism : a contribution to anthropology and the history of religion = Fetischismus / Schultze, Fritz – New-York: Humboldt Pub Co, c1885 – (= ser The Humboldt Library of Science) – 1mf – 9 – 0-524-02366-2 – (incl bibl ref. in english) – mf#1990-2977 – us ATLA [390]

Fetichism and fetich worshipers / Baudin, Noel – New York: Benziger Bros, 1885 – 1mf – 9 – 0-524-06686-8 – mf#1990-3547 – us ATLA [210]

Fetichism and fetich worshipers / Baudin, P – New York, Cincinnati [etc]: Benziger Bros, 1885 – 1 – us CRL [290]

Fetichism in west africa : forty years' observation of native customs and superstitions / Nassau, Robert Hamill – New York: Scribner, 1904 – 1mf – 9 – 0-524-01068-4 – mf#1990-2216 – us ATLA [390]

Fetis, Francois J *see* Biographie universelle des musiciens et bibliographie generale de la musique

Fetis, Francois-Joseph *see* Traite du contrepoint et de la fugue, contenant l'expose analytique des regles...

Fetler, Robert *see* Blagovestnik

Fetler, V A *see*
- Porkorneishee khodataistvo
- The stundist in siberian exile and other poems

Fetler, William *see* Ka es atklahju modernisumu (wiltigu mahzibu) starp amerikanu baptisteem un kapehz es nodibinaju anglu-amerikanu missiones beedribu

The fetters of freedom / Brady, Cyrus Townsend – Toronto: W Briggs, 1913 – 5mf – 9 – 0-665-73689-4 – mf#73689 – cn Canadiana [890]

Fettsaeurebindende proteine humaner keratinozyten / Schuerer, Nanna Y – (mf ed 1996) – 2mf – 9 – €40.00 – 3-8267-2351-1 – mf#DHS 2351 – gw Frankfurter [616]

Fetzer, F *see* Chan shih shih yu cheng ts'e

Fetzer, Johann Jakob *see* Das papstthum im widerspruch mit vernunft, moral und christentum

Feu come seraphin cherrier : conference faite a la salle de "la patrie", vendredi, le 16 octobre 1885 / Mercier, Honore – Montreal: s.n, 1885? – 1mf – 9 – mf#10097 – cn Canadiana [920]

Feu lionel : ou, qui vivra, verra / Scribe, Eugene – Paris, France. 1858 – 1r – us UF Libraries [440]

Feu peterscott / Ennery, Adolphe D' – Paris, France. 1841 – 1r – us UF Libraries [440]

Feuchtersleben, Ernst, Freiherr von *see* Ernst freiherrn von feuchtersleben's saemmtliche werke

Feuchthofen, Joerg E *see* Qualitaetsmanagement und qualitaetssicherung in der weiterbildung

Feuchtwanger, Lion *see*
- Double, double, toil and trouble
- Die haessliche herzogin margarete maultasch
- Josephus
- Jud suess
- Marianne in india
- The oppermanns
- Pep, j.l. wetcheek's american song book
- Power
- The pretender
- Proud destiny
- Stories from far and near
- The ugly duchess

Feudel, Werner *see* Morgenruf

Feuer, Abraham *see* Sefer zikhron avraham

Feuer der nacht : gedichte und briefauszuege eines gefallenen soldaten / Behrmann, Willi; ed by Meichner, Fritz – Heidelberg: Huethig, c1943 [mf ed 1989] – 80p (ill) – 1 – mf#7004 – us Wisconsin U Libr [800]

Feuer im wind : leben und vergehen des dichters johann christian guenther / Wille, Hanns Julius – Berlin: Verlag der Nation, 1955 [mf ed 1993] – 270p – 1 – mf#8668 – us Wisconsin U Libr [830]

Die feuer sind entglommen : roman aus dem grossen kampf gegen die zwingherren der erde, der im jahre 1914 begann / Schworm, Karl – 4. Aufl. Muenchen: F Eher, 1944 – 1r – 1 – us Wisconsin U Libr [830]

Feuer, Willie *see* The relationship between the timing of arm movement and the force of landing

Feuerbach, Henriette *see* Uz und cronegk

Feuerbach, Ludwig *see*
- Die akte ludwig feuerbach
- The essence of christianity

Feuerbach, the roots of the socialist philosophy = Ludwig feuerbach und der ausgang der klassischen deutschen philosophie / Engels, Friedrich – Chicago: CH Kerr, c1903 – 1mf – 9 – 0-7905-9193-6 – (in english) – mf#1989-2418 – us ATLA [190]

Der feuerberg : eine erzaehlung von deutschen siedlern in amerika / Blunck, Hans Friedrich – Jena: E Diederichs, c1934 [mf ed 1989] – (= ser Deutsche reihe 15) – 71p – 1 – mf#7036 – us Wisconsin U Libr [880]

Feuerberg, Mordecai Zeev *see* Kovets sipurav u-ketavav

Feuerbrand in kaernten : der heldenkampf eines volkes / Reinhardstein, Joachim – Berlin: Im Deutschen Verlag, c1937 [mf ed 1989] – 223p (ill) – 1 – mf#6981 – us Wisconsin U Libr [830]

Die feuerkugel : welche am abende des 3. decembers 1861 in deutschland gesehen worden ist / Heis, Eduard – Halle: H W Schmidt 1862 [mf ed 1998] – 1r [pl/ill] – 1 – (fr: wochenschrift fuer astronomie etc [jahrg 1862]) – mf#film mas 28415 – us Harvard [520]

Feuerlein, Emil *see* Die sittenlehre des christenthums in ihren geschichtlichen hauptformen

Die feuermeteore : insbesondere die meteoriten, historisch und naturwissenschaftlich betrachtet / Buchner, Otto – Giessen: J Ricker 1859 [mf ed 1998] – 1r – 1 – (incl ind) – mf#film mas 28417 – us Harvard [520]

Der feuerreiter *see* Duesseldorfer sonntagsblatt

Der feuerspeiende berg : roman / Berglar-Schroeer, Paul – Salzburg: Verlag 'Das Bergland-Buch', c1943 [mf ed 1989] – 379p – 1 – mf#7010 – us Wisconsin U Libr [830]

Der feuerwachtturm – Wernigerode DE, 1955 8 nov-1959 5 mar [gaps] – 1r – 1 – (forstwirtschaft) – gw Misc Inst [634]

Feuerwerk im juli : begegnungen in paris, 1789-1871 / ed by Weber, Rolf – Berlin: Der Morgen, 1978 [mf ed 1993] – 379p – 1 – mf#8156 – us Wisconsin U Libr [430]

Feuerwerkbuch *see* Bellifortis / feuerwerkbuch (cf-lp3)

Das feuerzeichen : roman / Bergengruen, Werner – Muenchen: Nymphenburger Verlagshandlung, c1949 [mf ed 1989] – 259p – 1 – mf#7008 – us Wisconsin U Libr [830]

La feuille – Paris. n1-61. 3 aout 1916-10 janv 1918 – 1 – (socialiste, syndicaliste, revolutionnaire) – fr ACRPP [325]

la feuille *see* l'avant-coureur

Feuille bimensuelle d'informations syndicales / Centre Syndical d'Action Contre la Guerre – n1-24. Paris. 14 juil 1938-28 aout 1939 – 1 – (mq n18-23) – fr ACRPP [325]

[Feuille, D de la] *see*
- Essay d'un dictionnaire contenant la connaissance du monde, des sciences universelles...
- Essay d'un dictionnaire contenant la connoissance du monde...
- Methode nouvelle pour apprendre l'art du blason
- La science des hieroglyphes
- Science hieroglyphique

La feuille d'erable – Montreal: L J Beliveau. v1 n1 10 avril 1896-v1 n6 25 juin 1896 (bimthly) – 1mf – 9 – mf#SEM105P292 – cn Bibl Nat [073]

La feuille d'erable (new-york) – New-York, NY: [s.n.] v1 n1 1er janv 1887- (mthly) [mf ed 1988] – 1r – 1 – (incl english text; ceased 1891?) – mf#SEM35P320 – cn Bibl Nat [071]

Feuille des jeunes naturalistes – Rennes 1870-1914 – n1-528 on 97mf – 9 – mf#8601/1 – ne IDC [590]

Feuille du commerce – Port-au-Prince: Jh Courtois, may 1843-dec 1861 – 46 sheets – 9 – us CRL [079]

La feuille du jour – Paris. dec 1790-aout 1792 – 1 – fr ACRPP [073]

Feuille hebdomadaire patriotique – Strassburg (Strasbourg F), 1789 6 dec-1790 8 may – 1 – (title varies: 20 dec 1789: patriotisches wochenblatt) – fr ACRPP [074]

La feuille maritime de nantes *see* La feuille nantaise

La feuille nantaise – Nantes. nov 1793-1810 – 1 – (suite de: la feuille maritime de nantes) – fr ACRPP [073]

La feuille necessaire – Contenant divers details sur les sciences, les lettres et les arts. no. 1-47. Paris. 1759. devenu: L' Avant-coureur voir a ce titre – 1 – (contenant divers details sur les sciences, les lettres et les arts) – fr ACRPP [073]

La feuille villageoise adresse chaque semaine a tous les villages de la france – Par Cerutti, Rabaud Saint-Etienne, Grouvelle et Guinguene. Paris. avr 1793-aout 1795 – 1 – fr ACRPP [073]

Feuillee, L *see* Journal des observations physiques, mathematiques et botaniques...

Les feuilles libres – Lettres et arts. Revue mensuelle. Dir. M. Raval. no. 1-48. Paris. dec 1918-juin 1928 – 1 – fr ACRPP [800]

Feuilles libres (journal d'alain) – ns 9, n10-12. 1 dec 1935- – 1 – (running title: propos d'alain) – us Wisconsin U Libr [073]

Feuilles volantes : et pages d'histoire / Gagnon, Ernest – [Quebec?: Laflamme & Proulx], 1910 – 5mf – 9 – 0-665-71190-5 – mf#71190 – cn Canadiana [971]

Feuilles volantes / Frechette, Louis – Montreal: Granger, 1891 – 3mf – 9 – mf#28779 – cn Canadiana [810]

Feuillet, Octave *see*
- Montjoye
- Peril en la demeure
- Redemption
- Roman d'un jeune homme pauvre
- Tentation
- Village
- Le village

Le feuilleton : ou, supplement du fantasque – Quebec: N Aubin, [1838] – 9 – mf#P04156 – cn Canadiana [320]

Feuilleton aus der niederschlesischen zeitung – Goerlitz, Okt 1850-Juni 1851 – 1r – 1 – gw Mikropress [943]

Feuilleton illustre – Montreal: Houle, [1880-188-?] – 9 – mf#P04960 – cn Canadiana [440]

Feuilleton-korrespondenz – Berlin DE, 1896-98, 1900-01, 1903-04, 1907-11, 1913 – 4r – 1 – gw Misc Inst [074]

Les feux-follets / Beaugrand, Honore – [S.l: s.n, 189-?] – 1mf – 9 – 0-665-61706-2 – mf#61706 – cn Canadiana [830]

Fevaid – Bursa: Matbaa-i Emri, 1896-? Sahib-i Imtiyaz: Murad Emri; Sermuharriri: Mehmed Rifat. n1. 3 tesrinisani 1312 [1896] – (= ser O & t journals) – 1mf – 9 – $25.00 – us MEDOC [956]

Feval, Paul *see*
- Jesuits!
- The two wives of the king

Fevralskaia revoliutsiia *see* Revoliutsiia i grazhdanskaia voina v opisaniakh belogvardeitsev

Fevre, Justin Louis Pierre *see* Histoire critique du catholicisme liberal en france

Fevzi, Muntahabat-i Divan-i *see* The divan project

A few brief hints on the causes of the present distress / Gore, Montague – London: James Ridgway, 1830 – (= ser 19th c economics) – 1mf – 9 – mf#1.1.431 – uk Chadwyck [339]

A few chapters in work-shop re-construction and citizenship / Ashbee, Charles Robert – [London]: publ by the Guild & School of Handicraft, 1894 – (= ser 19th c art & architecture) – 2mf – 9 – mf#4.1.100 – uk Chadwyck [740]

Few comments on dr pusey's letter to the bishop of london / Dodsworth, William – London, England. 1851 – 1r – us UF Libraries [240]

Few comments on mr gladstone's expostulation / Neville, Henry – London, England. 1875 – 1r – us UF Libraries [240]

A few comments upon mr macaulay's remarks on the internal water communications of the canadas : as published in the quebec gazette of the 8th february / George, James – S.l: s.n, c1837 – 1mf – 9 – (in dble clms) – mf#44518 – cn Canadiana [380]

Few days on the continent / Johnstone, James – Edinburgh, Scotland. 1875 – 1r – us UF Libraries [240]

A few facts respecting the regina district in the great grain growing and stock raising province of assiniboia, north-west territories, canada – Regina: The Board, 1889 [mf ed 1984] – 1mf – 9 – 0-665-30715-2 – mf#30715 – cn Canadiana [630]

Few happy ones / Van Der Veer, Judy – New York, NY. 1943 – 1r – us UF Libraries [025]

A few hints on colour and printing in colours / Watt, P B – London, Manchester, Glasgow, 1872 – (= ser 19th c publishing...) – 1mf – 9 – mf#3.1.21 – uk Chadwyck [680]

A few hints on...study of ecclesiastical architecture / Cambridge Camden Society – [4th ed] Cambridge 1843 – (= ser 19th c art & architecture) – 1mf – 9 – mf#4.2.81 – uk Chadwyck [720]

Few incidents in the life of samuel lines, sen : written by himself at the request of a friend – Birmingham: printed by Josiah Allen, jr. 1862 – (= ser 19th c art & architecture) – 1mf – 9 – mf#4.1.106 – uk Chadwyck [920]

A few notes on the gospels according to st mark and st matthew : based chiefly on modern greek / Palles, Alexandros – Liverpool: Liverpool Booksellers, 1903 [mf ed 1992] – 1mf – 9 – 0-524-05052-X – mf#1992-0305 – us ATLA [226]

Few observations on the union of professing episcopalians in scotland / Ramsay, Edward Bannerman – Edinburgh, Scotland. 1831 – 1r – us UF Libraries [240]

A few plain and candid objections : to the committee appointed by the rev j m hanson, to try the charges exhibited against the reforming local ministers – Baltimore: Woddy 1827 – 1r – 1 – $35.00 – mf#um-15 – us Commission [242]

Few plain answers to the question, why do you receive the testimony / Clowes, J – Birmingham, England. 1807 – 1r – us UF Libraries [240]

Few plain observations on the enactment of the general assembly, 18... / Cook, George – Edinburgh, Scotland. 1834 – 1r – us UF Libraries [240]

Few plain questions addressed to parents whose children are about t... – London, England. 1865? – 1r – us UF Libraries [240]

Few plain reasons why we should believe in christ and adhere to his... / Cumberland, Richard – London, England. 1801 – 1r – us UF Libraries [240]

A few plain words in respect to the american sunday-school union / Westbrook, R B – Philadelphia: Isaac Ashmead, 1855 [mf ed 1993] – 1mf – 9 – 0-524-08702-4 – mf#1993-3227 – us ATLA [240]

A few remarks about the niagara gorge / Buck, Leffert Lefferts – S.l: s.n, 1894? – 1mf – 9 – mf#60252 – cn Canadiana [917]

A few remarks on internal improvements in the canadas / George, James – S.l: s.n, 1835? – 1mf – 9 – mf#21491 – cn Canadiana [380]

A few remarks on religious corporations and american examples of them / Ryerson, Egerton – Toronto?: s.n, 1851 (Toronto: T H Bentley) – 1mf – 9 – mf#47736 – cn Canadiana [338]

A few remarks on the "new library" question / Coddington, Henry – Cambridge 1831 – (= ser 19th c art & architecture) – 1mf – 9 – mf#4.1.372 – uk Chadwyck [700]

A few sheaves of devon bibliography gleaned / Dredge, John Ingle – Plymouth: W Brendon & Son, 1889-96 – (= ser 19th c publishing...) – 3mf – 9 – mf#3.1.72 – uk Chadwyck [019]

A few steps of the road to zion : as understood by travellers in 1854; also, graveyard flowers, or, a collection of elegies etc / Megowan, Agnes – [St John NB: s.n.] 1855 [mf ed 1983] – 1mf – 9 – 0-665-38235-9 – mf#38235 – cn Canadiana [810]

Few strictures on a late publication by the reverend george lawson / Thomson, John – Glasgow, Scotland. 1797 – 1r – us UF Libraries [240]

A few suggestions on the problems of the indian constitution / Beniprasada – Allahabad: Indian Press, 1928 – (= ser Samp: indian books) – us CRL [954]

Few thoughts on the eucharistic question / Farquhar, William – Forfar, Scotland. 1858 – 1r – us UF Libraries [240]

A few thoughts on the state of affairs in south africa / Sartorius, George Rose – London 1879 – (= ser 19th c british colonization) – 1mf – 9 – mf#1.1.4946 – uk Chadwyck [320]

Few thoughts on the supreme authority of the word of god / Butcher, Samuel – Dublin, Ireland. 1864 – 1r – us UF Libraries [240]

Few words about the devil / Bradlaugh, Charles – London, England. 1890 – 1r – us UF Libraries [240]

A few words addressed to the labouring classes = Aux ouvriers: du pain, du travail, et la verite / Schmit, Jean Philippe – London: Effingham Wilson, 1848 – (= ser 19th c economics) – 1mf – 9 – (trans fr french) – mf#1.1.447 – uk Chadwyck [331]

Few words of hope on the present crisis of the english church / Neale, J M – London, England. 1850 – 1r – us UF Libraries [240]

A few words on evolution and creation : a thesis maintaining that the world was not made of matter by the development of one potency, but by that of innumerable specific powers / Boase, Henry Samuel – London, 1882 – (= ser 19th c evolution & creation) – 3mf – 9 – mf#1.1.9154 – uk Chadwyck [210]

Few words on the petition against the catholics – Aberdeen, Scotland. 1832 – 1r – us UF Libraries [240]

A few words on the promoting and encouraging of free emigration to the west india colonies : addressed to the right honourable lord john russell... – Liverpool, 1840 – (= ser 19th c books on british colonization) – 1mf – 9 – mf#1.1.446 – uk Chadwyck [304]

Few words on the spirit in which men are meeting the present crisis / Monro, Edward – Oxford, England. 1850 – 1r – us UF Libraries [240]

A few words to church builders / Cambridge Camden Society – [2nd ed] Cambridge 1842 – (= ser 19th c art & architecture) – 1mf – 9 – mf#4.2.1006 – uk Chadwyck [720]

Few words to churchwardens on churches and church ornaments / Neale, J M – London, England. 1846 – 1r – us UF Libraries [240]

A few words to churchwardens on churches and church ornaments no 1 : suited to country parishes / Neale, John Mason – 4th ed. Cambridge, Oxford, London: Cambridge Camden Society, 1841 – (= ser 19th c art & architecture) – 1mf – 9 – (n2: suited to town and manufacturing parishes – anonymous. by john mason neale – pt1 only is of the 4th ed) – mf#4.1.111 – uk Chadwyck [720]

Few words to some of the women of the church of god in england / Sellon, Priscilla Lydia – London, England. 1850 – 1r – us UF Libraries [240]

Few words to those churchmen : being members of convocation, who pur... / Oakeley, Frederick – London, England. 1845 – 1r – us UF Libraries [240]

A few words with bishop colenso on the subject of the exodus of the israelites and the position of mount sinai / Beke, Charles Tilstone – 2nd ed. London: Williams & Norgate, 1862 [mf ed 1993] – 1mf – 9 – 0-524-05657-9 – mf#1992-0507 – us ATLA [221]

Fewkes, Jesse Walter *see* Antiquities of the mesa verde national park, cliff palace

Few's news and views / Federally Employed Women, Inc – 1969 oct-1978 feb/mar, 1987 july/aug-1994 nov/dec – 2r – 1 – (cont by: news and views from federally employed women, inc) – mf#629987 – us WHS [331]

Fewster, Jonathan B *see* The role of musculoskeletal forces in the human walk-run transition

Feyerabend, M *see* Saemtliche briefe

Feyjoo, B Geronimo *see* Justa repulsa...teatro critico...francisco soto y marne

Feyrabend, S *see* Reyszbuch desz heyligen lands

Feyrol, Jacques *see* Les francais en amerique

Feyzi, Emin *see* Icmal-i netayic yahut mutira-i funun-i idadiye

Feyz-i Huerriyet *see* Uec gazete [metin, feyz-i huerriyet, tasvir-i hayal]

Fezkeke-i tarih / Celebi, Katib – [Istanbul]: Ceride-i Havadis Matbaasi, 1287 [1870] – (= ser Ottoman histories and historical sources) – 11mf – 9 – $200.00 – us MEDOC [956]

Ff funk und fernsehen der ddr *see* Der rundfunk [main edition]

Fff-courier *see* Tv-contact

Fff-press – Hamburg DE, 1952 jul-1966 22 dec – 25r – 1 – (incl suppl) – gw Mikrofilm [074]

Ffoulkes, Charles John *see* Armour and weapons

Ffoulkes, Edmund S *see*
- Christendom's divisions
- Greeks and latins

Ffoulkes, Edmund Salisbury *see*
- The athanasian creed
- Church's creed or the crown's creed?

Ffoulkes, Edmund Salusbury *see*
- Is the western church under anathema?
- The roman index and its proceedings

Ffrench, James Frederick Metge *see* Prehistoric faith and worship

Fi rubu'al-azbakiyah / Kilani, Muhammad Sayyid – Dar al-'Arab, Egypt. 1958-1959 – 1r – us UF Libraries [025]

La fiaccola – New york. sept 5 1912-feb 10 1921. (not collated) (incomplete) [wkly] – 1 – (in italian) – us NY Public [073]

Fiallo, Fabio *see*
- Cancion de una vida
- Crime of wilson in santo domingo
- Cuentos fragiles

Fiallos Gil, Mariano *see* Proceso cultural centroamericano

La fiammetta / Monte, Philipp de – 1599 – (= ser Mssa) – 3mf – 9 – €50.00 – mfchl 352 – gw Fischer [780]

Fiancee / Scribe, Eugene – Paris, France. 1839 – 1r – us UF Libraries [440]

Les fiances de 1812 : essai de litterature canadienne / Doutre, Joseph – Montreal: L Perreault, 1844 [mf ed 1975] – 1r – 5 – mf#SEM16P23 – cn Bibl Nat [420]

Les fiances de 1812 : essais de litterature canadienne / Doutre, Joseph – Montreal?: L Perrault, 1884 – 6mf – 9 – mf#34770 – cn Canadiana [440]

Fianna – Buenos Aires: [s.n.], [1910-]. [v1 n1-v2 n7 mar 17 1910-jul 1913] – 1r – 1 – us CRL [079]
Fians, fairies and picts / MacRitchie, David – London: K Paul, Trench, Truebner, 1893 – 1mf – 9 – 0-524-01198-2 – (incl bibl ref) – mf#1990-2274 – us ATLA [390]
Fiat justitia / Biber, George Edward – London, England. 1850 – 1r – us UF Libraries [240]
Fiat lux : orgao republicano – Teresina, PI: Typ do Fiat Lux, 10 fev 1890 – (= ser Ps 19) – mf#P11,03,06 – bl Biblioteca [321]
Fiat lux / Solar y Taboada, Antonio – Badajoz: La Minerva Extremena, 1913 – 1 – sp Bibl Santa Ana [946]
Die fibel : auswahl erster verse / George, Stefan Anton – Berlin: G Bondi, 1901 (mf ed 1990) – 1r – 1 – (filmed with: gellerts lustspiele) – us Wisconsin U Libr [810]
Fiber and integrated optics – New York. 1977-1996 (1,5,9) – ISSN: 0146-8030 – mf#11362 – us UMI ProQuest [530]
Fiber producer – Atlanta. 1975-1976 (1) 1975-1976 (5) 1975-1976 (9) – ISSN: 0361-4921 – mf#10432 – us UMI ProQuest [670]
Fibonacci quarterly – Santa Clara. 1963+ (1) 1974+ (5) 1976+ (9) – ISSN: 0015-0517 – mf#9488 – us UMI ProQuest [510]
Fibre chemistry – New York. 1969-1994 (1) 1969-1994 (5) 1993-1994 (9) – ISSN: 0015-0541 – mf#10883 – us UMI ProQuest [660]
Ficc quarterly – Walpole. 1999-2000 (1,5,9) – mf#6495,02 – us UMI ProQuest [340]
Fichas del romulato / Pepper B, Jose Vicente – Ciudad Trujillo, Dominican Republic. 1947 – 1r – us UF Libraries [972]
Ficheleff, S *see* Le statut international de la palestine orientale
Fichero bibliografico hispanoamericano – Buenos Aires. 1961-1992 (1) 1971-1992 (5) 1973-1992 (9) – ISSN: 0015-0592 – mf#2093 – us UMI ProQuest [972]
Fichier d'autorite / Bibliotheque Nationale. Quebec – [Montreal: la Bibliotheque] 26 nov 1993-janv 1996 – 9 – (cont: autorites) – cn Bibl Nat [020]
Fichot, Eugene *see* Lecons de mecanique celeste
Fichte / Adamson, Robert – Edinburgh: William Blackwood, 1881 – (= ser Philosophical Classics for English Readers) – 1mf – 9 – 0-7905-9115-4 – mf#1989-2340 – us ATLA [190]
Fichte als religioeser denker / Gogarten, F – Jena, 1914 – 2mf – 8 – €5.00 – ne Slangenburg [190]
Fichte, Johann Gottlieb *see*
– Anthologie aus den werken von johann gottlieb fichte
– Nachgelassene schriften
Fichte, Werner von *see* Spukflieger
Fichtelgebirgs-warte – Marktredwitz DE, 1943 1 sep-1944 – 2r – 1 – gw Misc Inst [074]
Fichtelgebirgs-warte – Selb DE, 1943 1 sep-1944 30 jun – 1r – 1 – gw Misc Inst [074]
Fichte's science of knowledge : a critical exposition / Everett, Charles Carroll – Chicago: S C Griggs, 1884 [mf ed 1984] – (= ser German philosophical classics for english readers and students) – 4mf – 9 – 0-8370-1023-3 – (incl bibl ref) – mf#1984-4395 – us ATLA [120]
Ficino, M *see* Libro...para curar...de pestilencia...
Ficinus, Marsilius *see*
– Platonis opera omnia...
– Plotini enneades
Fick, Hermann *see* Life and deeds of dr. martin luther
Fick, Johann *see* Meine neueste reise zu wasser und land
Fick, Richard *see* The social organisation in north-east india in buddha's time
Ficker, J *see*
– Handschriftenproben des sechzehnten jahrhunderts nach strassburger originalen
– Thesaurus baumianus
Ficker, Johannes *see* Die konfutation des augsburgischen bekenntnisses
Fickler, Liane *see* Die substantia nigra im menschlichen gehirn bei aids-encephalopathie im vergleich mit alterskorrelierten kontrollen und in relation zur ausbildung eines parkinson-syndroms
Fico : minas e os mineiros na independencia / Vasconcellos, Salomao De – Sao Paulo, Brazil. 1937 – 1r – us UF Libraries [972]
Fiction – New York. 1972+ (1) 1978+ (5) 1978+ (9) – ISSN: 0046-3736 – mf#10441 – us UMI ProQuest [400]
Fiction international – San Diego. 1973+ (1) 1974+ (5) 1975+ (9) – ISSN: 0092-1912 – mf#7521 – us UMI ProQuest [400]
Fiction on fiche / British Library. National Bibliographic Service – Adult Fiction. Bibliographic records from 1950 to present, updated annually. First available 1992 – £175.00 – uk British Libr [830]
Fictions and errors in a book on "the origin of the world according to revelation and science by j w dawson...principal of mcgill university, montreal" : exposed and condemned on the authority of divine revelation / Marshall, John George – Halifax, NS?: Methodist Book Room, 1877 – 1mf – 9 – mf#09922 – cn Canadiana [210]
Fictuld, Hermann *see* Des langst gewunschten und versprochenen chymisch-philosophischen probier-steins, erste classe, in welcher der wahren und achten adeptorum und anderer wurdig erfundenen schrifften...
Fid amchitka / Concerned Servicemen's Movement – v1 iss 7 [1971 dec?] – 1r – 1 – mf#721999 – us WHS [355]
Fid news bulletin / International Federation for Documentation General Secretariat – The Hague. 1951+ (1) 1976+ (5) 1976+ (9) – ISSN: 0014-5874 – mf#3205 – us UMI ProQuest [020]
Fidalgo Carasa, Pilar *see* Une jeune mere dans les prisons de franco
The fiddlehead – no. 1-87. 1945-70 – 1 – us AMS Press [800]
Fidel castro / Hernandez Sanchez, Jesus – Rio Piedras, Puerto Rico. 1959 – 1r – us UF Libraries [972]
Fidele blaetter – Berlin DE, 1911-13 – 1r – 1 – gw Misc Inst [074]
Fidelity / Ultramontane Associates, Inc et al – v7 n1-v8 n11 [1987 dec-1989 nov] – 1r – 1 – mf#1700793 – us WHS [071]
Fidelity and filial affection – Belfast, Northern Ireland. 1816 – 1r – us UF Libraries [240]
Fidelity to conscience / Mclaren, Alexander – London, England. 1862? – 1r – us UF Libraries [240]
Fides : [a short novel] / Busse, Hermann Eris – 5. aufl. Guetersloh: C Bertelsmann, 1943 [mf ed 1989] – (= ser Kleine feldpost-reihe) – 64p – 1 – mf#7097 – us Wisconsin U Libr [830]
Fides et historia – Terre Haute. 1984+ (1,5,9) – ISSN: 0884-5379 – mf#15124 – us UMI ProQuest [900]
Fides et ratio collatae : ac suo utraque loco redditae, adversus principia joannis lockii / Poiret, P – Amsterdam, 1707 – 7mf – 9 – mf#PPE-220 – ne IDC [240]
Fides implicita : eine untersuchung ueber koehlerglauben, wissen und glauben, glauben und kirche / Ritschl, Albrecht; ed by Ritschl, Otto – Bonn: Adolph Marcus, 1890 [mf ed 1990] – 1mf – 9 – 0-7905-9615-6 – (incl bibl ref) – mf#1989-1340 – us ATLA [210]
Fides, religio, moresque aethiopum / Goes, D de – Lovanii, 1544 – 4mf – 9 – mf#SEP-37 – ne IDC [960]
Fides. Service de bibliographie et de documentation *see* Lectures
Fidibus-herald – Denver TX (USA), 1920 2 jun-1922 28 jun 28 – 1 – gw Misc Inst [071]
Fidler, Merrie A *see* The development and decline of the all-american girls baseball league, 1943-1954
Fiducia *see* Madjalah gerakan mahasiswa kristen di indonesia
Fiduciary reporter – First National City Bank: New York, 1940-62 – 7mf – 9 – $31.50 – (lacking: 1956) – mf#LLMC 84-463 – us LLMC [332]
Fiebig, Paul *see*
– Altjuedische gleichnisse und die gleichnisse jesu
– Die aufgaben der neutestamentlichen forschung in der gegenwart
– Babel und das neue testament
– Die gleichnisreden jesu im lichte der rabbinischen gleichnisse des neutestamentlichen zeitalters
– Jesu blut, ein geheimnis?
– Das judentum
– Juedische wundergeschichten des neutestamentlichen zeitalters
– Der menschensohn
– Rabbinische wundergeschichten des neutestamentlichen zeitalters
– Talmud und theologie
– Der tosephtatraktat ros hassana
Fiebig, Wilfried *see* Abstrakte arbeit und abstraktwerden der kunst
Fiedler, Hermann Georg *see* A w schlegel's lectures on german literature
Fiedler Leonhard *see* Deutsche literaturgeschichte
Fiedler, Reginald Hobson *see* Wholesale trade in fresh and frozen fishery products
Field / Clark, Thomas G – Edinburgh, Scotland. 1862 – 1r – us UF Libraries [240]
Field – Conway, SC. 1934-1964 (1) – mf#66479 – us UMI ProQuest [071]
The field – London. 1-82. 1853-1893 – 1 – us NY Public [073]
Field and farm – Denver, CO: Field and Farm Pub Co, 1886 – 1r – 1 – mf#MF F453f – us Colorado Hist [630]
Field and herald – Conway, SC. 1965-1989 (1) – mf#61824 – us UMI ProQuest [071]
Field and stream – New York. 1896-1984 (1) 1974-1984 (5) 1975-1984 (9) – ISSN: 0015-0673 – mf#10066 – us UMI ProQuest [790]
Field and stream : south edition – Los Angeles. 1984-1999 (1) 1984-1999 (5) 1984-1999 (9) – ISSN: 8755-8602 – mf#10066,01 – us UMI ProQuest [790]
The field and the men for it : an address to the divinity students of queen's college, kingston, at the close of the session, 1859-60 / George, James – Montreal?: J Lovell, 1860 – 1mf – 9 – mf#43323 – cn Canadiana [378]
Field, Annis S *see* Visiilano
Field artillery – Fort Sill. 1987+ (1) 1987+ (5) 1987+ (9) – (cont: field artillery journal) – ISSN: 0899-2525 – mf#556,01 – us UMI ProQuest [355]
Field artillery journal – 1911-50 – 1 – 579.00 – us L of C Photodup [355]
Field artillery journal – Fort Sill. 1911-1987 (1) 1973-1987 (5) 1973-1987 (9) – (cont by: field artillery) – ISSN: 0191-975X – mf#556 – us UMI ProQuest [355]
Field, Barron *see* Geographical memoirs on new south wales
Field, Benjamin *see* The student's handbook of christian theology
Field characteristics and partial chemical analyses off the humus layer of longleaf pine forest soil / Heyward, Frank – Gainesville, FL. 1936 – 1r – us UF Libraries [630]
Field crops research – Amsterdam. 1978-1995 (1) 1978-1995 (5) 1987-1995 (9) – ISSN: 0378-4290 – mf#42017 – us UMI ProQuest [630]
Field, Cyrus West *see* Europe and america
Field, David Dudley *see* The civil code
Field diary / Andrews, Caesar – 1835 – 3 fiche – 9 – sa National [960]
Field, Dorothy *see* The religion of the sikhs
Field experiments / Kost, John – Lake City, FL. 1888 – 1r – us UF Libraries [630]
Field experiments / Kost, John – Lake City, FL. 1888 – 1r – us UF Libraries [630]
Field experiments in the use of sulfur to control lice, fleas and mites of chickens / Emmel, M W – Gainesville, FL. 1942 – 1r – us UF Libraries [630]
Field, Frederick *see*
– Notes on the translation of the new testament
– Otium norvicense, sive, tentamen de reliquiis aquilae
– Tentamen de quibusdam vocabulis syro-graecis in r. payne smith s.t.p. thesauri syriaci
Field, Frederick William *see* Capital investments in canada
Field, George *see*
– Chromatics
– Chromatography
Field, Grenville O *see* Opened seals, open gates
Field, Harry Hubert *see* After mother india
Field, Henry M *see* Bright skies and dark shadows
Field, Henry Martyn *see*
– History of the atlantic telegraph
– The story of the atlantic telegraph
Field, J *see* An admonition to the parliament
Field, Jasper Newton *see* Isms, fads and fakes
Field, John *see* The life of john howard
Field, John Edward *see*
– The apostolic liturgy and the epistle to the hebrews
– Saint berin, the apostle of wessex: the history, legends and traditions of the beginning of the west-saxon church
Field, M J *see* Ada field notes
Field, Maunsell Bradhurst *see* Memories of many men and of some women: being personal recollections of emperors, kings, queens, princes, presidents, statesmen, authors.
Field museum of natural history. chicago fieldiana, zoology. zoological series – Chicago. v1-73. 1895-1978 – 9 – $1417.00 – (v18-30 1930 (1) [0204]) – mf#0205 – us Brook [590]
Field naturalist : and scientific student – Manchester. 1883-1883 (1) – mf#4771 – us UMI ProQuest [500]
Field naturalist – London. 1833-1834 (1) – mf#2889 – us UMI ProQuest [639]
Field notebooks / Brower, Jacob V – Archaeological notebooks (1897-1905) of maps of ancient midwestern Indian mounds.Sketches and drawings of mounds, sites, Indian implements and other artifacts. 3 reels including filmed inventory – 1 – $90.00; $30.00r – us Minn Hist [930]
Field notes / Shawnee County. Kansas. Surveyor – 1861-64 – 1 – us Kansas [978]
Field notes and correspondence regarding indian and military reservations – 1839-83 – 1 – us Kansas [978]
Field notes from selected general land office township surveys / U.S. General Land Office – (= ser Records of the bureau of land management) – 280r – 1 – mf#T1240 – us Nat Archives [333]
Field notes of the survey of the cherokee lands / McCoy, John Calvin – 1836-37 – 1 – us Kansas [978]
The field of ethics : being the william belden noble lectures for 1899 / Palmer, George Herbert – Boston: Houghton, Mifflin, 1901 – 1mf – 9 – 0-8370-6293-4 – mf#1986-0293 – us ATLA [170]
The field of ethics / Palmer, George Herbert – Boston/New York: Houghton, Mifflin & Co, 1901 – 3mf – 9 – $4.50 – mf#LLMC 92-174 – us LLMC [170]
Field, Richard H *see* The richard h. field papers
Field, Samuel *see* In memoriam...and other genealogical data on the field family
Field songs of chhattisgarh / Dube, Shyama Charan – Lucknow: Universal Publ, 1947 – (= ser Samp: indian books) – us CRL [780]
Field sports of the north of europe : comprised in a personal narrative of a residence in sweden and norway, in the years 1827-28 / Lloyd, Llewellyn – London 1830 [mf ed Hildesheim 1995-98] – 2v on 6mf – 9 – €120.00 – 3-487-28922-9 – gw Olms [790]
A field systems analysis of dual learning environments / Neel, Wallace B – West Virginia University, 1995 – 2mf – 9 – $8.00 – mf#PSY 1896 – us Kinesology [150]
Field tables of lepidoptera / Forbes, William Trowbridge Merrifield – Worcester, MA. 1906 – 1r – us UF Libraries [590]
Field, Thomas Meagher *see* Unpublished letters of charles carroll of carrollton and of his father, charles carroll of doughoregan
Field, Thomas Warren *see* An essay towards an indian bibliography
Field trials / Harold, William G – s.l, s.l? 1936 – 1r – us UF Libraries [978]
Fielde, Adele Marion *see* Alltagsleben in china
Fielding, Henry *see* A dialogue between a gentleman of london...and an honest alderman of the country party
Fielding, Theodore Henry Adolphus *see*
– The art of engraving, with the various modes of operation, under the following different divisions
– On the theory and practice of painting
Fielding, William Stevens *see*
– Canadian politics in war and peace
– Discours de l'hon w s fielding, m p
– Discours de l'honorable w s fielding...
Fielding-Hall, Harold *see* The inward light
Fields, Annie Adams *see* The annie adams field papers, 1852-1912
Fields within fields within fields – New York. 1968-1972 (1) – ISSN: 0015-0770 – mf#7331 – us UMI ProQuest [333]
A fieldwork study of how young children learn fundamental motor skills and how they progress in the development of striking / Garcia, Clersida & Branta, Crystal – 1991 – 3mf – 9 – $12.00 – us Kinesology [150]
Fieles amantes na si floro asin clavela sa isla nin sierra nevada – [Nueva Caceres: Libreria Mariana 190-?] [mf ed Bloomington IN: Indiana Uni Lib, Preservation Dept 1984] – (= ser Coll...in the bikol language) – 1r – 1 – us Indiana Preservation [490]
Fient, G *see* Lustig g'schichtenae
Fier yohr in der velt-milhome, 1914-1918 / Leipuner, I – Varsha, Poland. 1923 – 1r – us UF Libraries [939]
Fiercest heart / Cloete, Stuart – Boston, MA. 1960 – 1r – us UF Libraries [890]
Fierle, Karen M *see* Development and evaluation of a leisure education module for use in a college resource center
Fierte tsienistishe konferents in poylen (22-26 menahem av 679, 18-...) / Zionist Organisation Poland – Varsha, Poland. 1920 – 1r – us UF Libraries [939]
Fiery cross – [Scotland] Edinburgh: J B Fairgrieve jan 1901-jul 1912 [mf ed 2004] – 43v on 1r – 1 – uk Newsplan [320]
Fiery cross / United Klans of America – 1972 may-1974 jan/feb/mar – 1r – 1 – (cont: united klan's of america's fiery cross) – mf#1056235 – us WHS [366]
The fiery cross – Atlanta: Knights of the Ku Klux Klan. [v1-4 n6 jul 1939-sep/oct 1942] – 1r – 1 – us CRL [071]
Fiesel, Eva (Lehmann) *see* Die sprachphilosophie der deutschen romantik
Fiesta de la hispanidad. 1972 / Plasencia. Ayuntamiento – Plasencia: Talleres Graficos Sanguino, 1972 – 1 – sp Bibl Santa Ana [390]
Fiesta de luciernagas / Acuna, Angelina – Guatemala, 1953 – 1r – us UF Libraries [972]
Fiesta de nuestra senora, de la victoria / Trujillo – Trujillo: Imp. Sobrino de B. Pena, 1966 – sp Bibl Santa Ana [240]
Fiesta de san jorge, 1953 / Caceres. Ayuntamiento – Caceres: Tip. Jomarin – sp Bibl Santa Ana [390]
Fiestas 1976 – Caceres: Edit. Extremadura, 1976 – 1 – sp Bibl Santa Ana [390]
Fiestas 1977 / Torrejoncillo. El Pechin – Caceres: Imp. Extremadura, 1977 – 1 – sp Bibl Santa Ana [390]
Fiestas. agosto 1954 – Madrid: Graf. Garcia, 1954? – 1 – sp Bibl Santa Ana [390]
Fiestas barriada maria auxiliadora – Merida: Imp. Vadillo, 1971 – 1 – sp Bibl Santa Ana [390]
Fiestas comarcales a santa rita 1973 – Caceres: Imp. Tomas Rodriguez Santano, 1973 – 1 – sp Bibl Santa Ana [240]
Fiestas comarcales de santa cruz de la sierra...mayo 1981 en honor de santa rita... / Pastor Serrano, Juan Jose – Santa Cruz de la Sierra: Cofradia de Santa Rita, 1981 – 1 – sp Bibl Santa Ana [390]

FIFTEEN

Fiestas de agosto, 1979 en logrosan. nuestra senora del consuelo – Zorita: Imp. Carrasco, 1979 – 1 – sp Bibl Santa Ana [390]
Fiestas de agosto de 1971 en honor de ntra. sra. de la asuncion / Valverde del Fresno – Coria: Imp. Fernandez, 1971 – 1 – sp Bibl Santa Ana [390]
Fiestas de la juventud. 1974 / Madronera. Ayuntamiento – Trujillo: Imp. Gexme, 1974 – 1 – sp Bibl Santa Ana [390]
Fiestas de la juventud. caceres 26 abril 30 mayo 1973 / Delegacion Provincial de la Juventudes – Caceres: N. Sergio Dorado, 1973 – 1 – sp Bibl Santa Ana [390]
Fiestas de la santisima virgen de la vendimia / Trujillo – Trujillo: Tip. Sobrino de B. Pena, 1968 – sp Bibl Santa Ana [240]
Fiestas de la soledad, 1961 / Casatejada. Ayuntamiento – Plasencia: Imprenta La Victoria, 1961 – 1 – sp Bibl Santa Ana [390]
Fiestas de la soledad...1960 / Casatejada. Ayuntamiento – Plasencia: Imp. La Victoria, 1960 – 1 – sp Bibl Santa Ana [390]
Fiestas de la vela en zorita...en honor... la virgen de fuente santa 1979 / Zorita. Ayuntamiento – Imp. Carrasco, 1979 – 1 – sp Bibl Santa Ana [390]
Fiestas de ntra. sra. de los remedios. abril 1964 – Fragenal de la Sierra: Angel Verde, 1964 – 1 – sp Bibl Santa Ana [946]
Fiestas de nuestra senora de la piedad. 1975 – Caceres: Imp. Padilla, 1975 – 1 – sp Bibl Santa Ana [240]
Fiestas de san agustin 1980 – Caceres: Imp. Offset Rodriguez, 1980 – 1 – sp Bibl Santa Ana [390]
Fiestas de san anton. 1977 – Caceres, 1977 – 1 – sp Bibl Santa Ana [390]
Fiestas de san buenaventura. junio 1969 – Coria: Imp. Fernandez, 1969 – 1 – sp Bibl Santa Ana [390]
Fiestas de san cristobal 1974 / Cofradia de San Cristobal – Coria: Imp. Fernandez, 1974 – 1 – sp Bibl Santa Ana [390]
Fiestas de san cristobal 1976 / Cofradia de San Cristobal – Coria: Imp. Fernandez, 1976 – 1 – sp Bibl Santa Ana [390]
Fiestas de san cristobal 1976 / Cofradia de San Cristobal – Coria: Imp. Fernandez, 1976 – 1 – sp Bibl Santa Ana [390]
Fiestas de san cristobal. patrono de los automovilistas – Caceres: Imp. Fernandez, 1973 – 1 – sp Bibl Santa Ana [240]
Fiestas de san jose. 1971 / Badajoz – Badajoz: Graf. Tejado, 1971 – sp Bibl Santa Ana [390]
Fiestas de san juan 1971 / Badajoz – Badajoz: La Minerva Extremena, 1971 – sp Bibl Santa Ana [390]
Fiestas de san juan 1972 / Badajoz – Badajoz: Imp. Manuel Barrena, 1972 – 1 – sp Bibl Santa Ana [390]
Fiestas de san juan 1975 / Badajoz – Zafra: Ind Tip. Extremenas, 1974 – 1 – sp Bibl Santa Ana [390]
Fiestas de san juan 1976 / Pena Viva la Gente – Plasencia: Imp. Sanchez Rodrigo, 1976 – 1 – sp Bibl Santa Ana [390]
Fiestas de san juan 1977 / Pena Viva la Gente – Plasencia: Imp. Sanchez Rodrigo, 1977 – 1 – sp Bibl Santa Ana [390]
Fiestas de san marcos. desde el cerro de la salve / Oliva de la Frontera – (Badajoz): Imprenta Espanola, 1970 – 1 – sp Bibl Santa Ana [946]
Fiestas de san pedro 1973 – Caceres: Imp. La Minerva, 1973 – 1 – sp Bibl Santa Ana [390]
Fiestas de san pedro 1974 – Caceres: Imp. La Minerva, 1974 – 1 – sp Bibl Santa Ana [240]
Fiestas de san pedro 1976 – Caceres: Imp. La Minerva, 1976 – 1 – sp Bibl Santa Ana [390]
Fiestas de san roque en almaraz 1980 / Almaraz. Ayuntamiento – Navalmoral de la Mata: Imp. Rivero, 1980 – 1 – sp Bibl Santa Ana [390]
Fiestas de santo domingo de guzman. 1971 – s.l, s.i, s.a. – 1 – sp Bibl Santa Ana [390]
Fiestas de septiembre de 1970 / Zarza de Alange – Olivenza: Tip. M. Reginfo, 1970 – sp Bibl Santa Ana [390]
Fiestas de su patrona la virgen de la victoria / Trujillo – Plasencia: Imp. La Victoria, 1970 – sp Bibl Santa Ana [240]
Fiestas de su patrona la virgen de la victoria / Trujillo – Plasencia: Imp. La Victoria, 1972 – sp Bibl Santa Ana [240]
Fiestas de su santo patron san buenaventura durante...julio 1958 / Moraleja. Ayuntamiento – Caceres: Tip. El Noticiero, 1958 – 1 – sp Bibl Santa Ana [390]
Fiestas de su santo patron. san buenaventura. durante...julio 1962 / Moraleja. Ayuntamiento – Caceres: Tip. El Noticiero, 1962 – 1 – sp Bibl Santa Ana [390]
Fiestas de su santo patron san buenaventura julio, 1959 – Caceres: Tip. El Noticiero, S.L., 1959 – 1 – sp Bibl Santa Ana [240]
Fiestas de tentudia, 1969 / Calera de Leon – Fregenal de la Sierra: Imprenta Angel Verde, 1969 – sp Bibl Santa Ana [390]
Fiestas de tentudia, 1970 / Calera de Leon – Fregenal de la Sierra: Imp. Angel Verde, 1970 – sp Bibl Santa Ana [390]
Fiestas del carmen / Villafranca de los Barros – Villafranca de los Barros: Imp. Machuca, 1967 – 1 – sp Bibl Santa Ana [390]
Fiestas del carmen / Villafranca de los Barros – Villafranca de los Barros: Imp. Machuca, 1968 – 1 – sp Bibl Santa Ana [390]
Fiestas del carmen, 1964 / Villafranca de los Barros – Villafranca de los B. Graficas Gisver e Imprenta Machuca, 1964 – 1 – sp Bibl Santa Ana [390]
Fiestas del carmen 1976 / Villafranca de los Barros – Villafranca: Imp. Machuca, 1976 – 1 – sp Bibl Santa Ana [390]
Fiestas del carmen, julio 1963 / Vilia Franca de Los Barros. Ayuntamiento – Villafranca: Graf. Grisferv, 1963 – 1 – sp Bibl Santa Ana [390]
Fiestas del cristo de pardaleras – Badajoz: Graficas Tejado, 1970 – 1 – sp Bibl Santa Ana [240]
Fiestas del risco 1979 / Asociacion Cultural Chambra – Caceres: Imp. Marosa, 1979 – 1 – sp Bibl Santa Ana [390]
Fiestas del santisimo cristo del risco – Sierra de Fuentes, Caceres: Imp. Offset T. Rodriguez, 1980 – 1 – sp Bibl Santa Ana [240]
Fiestas del tabaco y del pimiento 1980 / Jaraiz de la Vera. Ayuntamiento – Navalmoral de la Mata: Imp. Rivero, 1980 – 1 – sp Bibl Santa Ana [390]
Fiestas en belen de trujillo 1974 / Parroquia de Nuestra Senora de Belen – Trujillo: Imp. Gexme, 1974 – 1 – sp Bibl Santa Ana [240]
Fiestas en garganta la olla 1974 / Garganta la Olla. Ayuntamiento – Jaraiz de la Vera: Imp. La Verata, 1974 – 1 – (tambien ano 1973) – sp Bibl Santa Ana [390]
Fiestas en garganta la olla del...julio de 1972 – Jaraiz de la Vera: Imp. La Verata, 1972 – 1 – sp Bibl Santa Ana [390]
Fiestas en honor de la santisima virgen del pilar – Coria: Imp. Fernandez, 1972 – 1 – sp Bibl Santa Ana [240]
Fiestas en honor de la santisima virgen del rosario, 1971 / Alcuescar. Ayuntamiento – Caceres: Imp. La Minerva, 1971 – sp Bibl Santa Ana [390]
Fiestas en honor de la virgen de fuensanta. 1975 – Zorita: Imp. Carrasco, 1975 – 1 – sp Bibl Santa Ana [240]
Fiestas en honor de la virgen de la soterrana. 1975 – Trujillo: Imp. Gexme, 1975 – 1 – sp Bibl Santa Ana [240]
Fiestas en honor de la virgen de la soterrana (nuestra senora de las nievas) agosto 1973 – Trujillo: Imp. Gex, 1973 – 1 – sp Bibl Santa Ana [240]
Fiestas en honor de los emigrantes. alcollarin, 7 y 8 agosto 1971 / Alcollarin. Ayuntamiento – Zorita: Imp. Carrasco, 1971 – 1 – sp Bibl Santa Ana [390]
Fiestas en honor de ntra. sra. de soterrana (virgen de las nieves). 1971 / Madronera. Ayuntamiento – Trujillo: Imp. Gexme, 1971 – 1 – sp Bibl Santa Ana [390]
Fiestas en honor de nuestra excelsa patrona la santisima virgen de fuente santa. 1974 / La Velada – Tambien 1976 – 1 – sp Bibl Santa Ana [240]
Fiestas en honor de san isidro labrador – Badajoz: Imp. Vicente Campini, 1970 – 1 – sp Bibl Santa Ana [240]
Fiestas en honor de san jose obrero 1971 / Rincon del Obispo-Coria. Ayuntamiento – Coria: Imp. Fernandez, 1971 – 1 – sp Bibl Santa Ana [390]
Fiestas en honor de san roque 1945 / Badajoz – Badajoz: Graf. Iberia, 1945 – sp Bibl Santa Ana [390]
Fiestas en honor de sus emigrantes. agosto de 1972 – Zorita: Imp. Carrasco, 1972 – 1 – sp Bibl Santa Ana [390]
Fiestas en honor del emigrante. 1971 / Torrejoncillo. Ayuntamiento – Caceres: Tip. La Minerva, 1971 – 1 – sp Bibl Santa Ana [390]
Fiestas en honor del santisimo cristo 1974 – Caceres: Tip. La Minerva, 1974 – 1 – sp Bibl Santa Ana [240]
Fiestas en monroy durante...septiembre 1962 / Monroy. Ayuntamiento – Caceres: Tip. La Minerva, 1962 – 1 – sp Bibl Santa Ana [390]
Fiestas en rincon del obispo en honor de san jose obrero. mayo 1973 / Rincon del Obispo – Caceres: Imp. Fernandez, 1973 – 1 – sp Bibl Santa Ana [240]
Fiestas en sotillo de andrada ano 1973 – Jaraiz de la Vera: Imp. La Verata, 1973 – 1 – sp Bibl Santa Ana [390]
Fiestas jubilares de la adoracion nocturna. trujillo 1971 – Plasencia: Imp. La Victoria, 1971 – 1 – sp Bibl Santa Ana [390]
Fiestas locales y feria deganados...los dias 1,2,3 y 4 de mayo de 1981 / Zarza de Alange – Trujillo: Imprenta Gezme, 1981 – 1 – sp Bibl Santa Ana [390]
Fiestas patronales 1971 / Talavan. Ayuntamiento – Caceres: Imp. La Minerva, 1971 – 1 – sp Bibl Santa Ana [390]
Fiestas patronales agosto 1974 – Caceres: Tip. La Minerva, 1974 – 1 – sp Bibl Santa Ana [240]
Fiestas patronales agosto de 1973 – Caceres: Imp. La Minerva, 1973 – 1 – sp Bibl Santa Ana [390]
Fiestas patronales de ntra. sra. de la estrella. 1975 / Santos de Maimona, Los. Ayuntamiento – Zafra: Ind Tip. Extremenas, 1975 – 1 – sp Bibl Santa Ana [946]
Fiestas patronales de nuestra senora la santisima virgen de la soledad / Aceuchal – Almendralejo: V. Rodriguez, 1970 – 1 – sp Bibl Santa Ana [240]
Fiestas patronales de santa ana y santiago los dias 25-26 de julio 1973 – Zorita: Imp. Carrasco, 1973 – 1 – sp Bibl Santa Ana [240]
Fiestas patronales en honor de la santisima virgen del rosario / Huerta de Animas – Trujillo: Imp. Gexme, 1972 – 1 – sp Bibl Santa Ana [240]
Fiestas patronales en honor de la virgen del rosario. 1973 / Huerta de Animas – Trujillo: Imp. Gexme, 1973. Tambien ano 1974 – 1 – sp Bibl Santa Ana [240]
Fiestas patronales en honor de ntra sra de los milagros, 1979 / Bienvenida. Ayuntamiento – Los Santos de Maimona: Grafisur, 1979 – 1 – sp Bibl Santa Ana [390]
Fiestas patronales en honor de ntra. sra. del consuelo 1971 / Logrosan. Ayuntamiento – Caceres: Imp. La Minerva, 1971 – sp Bibl Santa Ana [390]
Fiestas patronales en honor de nuestra senora de los milagros. septiembre 1970 / Bienvenida. Ayuntamiento – Fregenal de la Sierra: Imp. Angel Verde, 1970 – sp Bibl Santa Ana [240]
Fiestas patronales en honor de san bartolome / Herguijuela. Ayuntamiento – Trujillo: Imp. Gexme, 1976 – 1 – sp Bibl Santa Ana [390]
Fiestas patronales en la barriada del sagrado corazon... / Olivenza, Teleclub no 23 – Olivenza: Tip. Martinez Reginfo, 1970 – 1 – sp Bibl Santa Ana [946]
Fiestas patronales san jorge, 1955 / Caceres. Ayuntamiento – Caceres: Imp. Jomarin – sp Bibl Santa Ana [390]
Fiestas patronales san jorge 1959 / Caceres. Ayuntamiento – Caceres: Comision de Festejos Patronales, 1959 – 1 – sp Bibl Santa Ana [650]
Fiestas patronales. santisimo cristo de la agonia – Caceres: Edit. Extremadura, 1974 – 1 – sp Bibl Santa Ana [240]
Fiestas patronales. septiembre 1975 / Fuente del Maestre – Zafra: Ind. Tip. Extremenas, 1974 – 1 – sp Bibl Santa Ana [390]
Fiestas patronales septiembre-octubre, 1977 / Hogar Extremeno de Zaragoza – sp Bibl Santa Ana [390]
Fiestas populares. agosto 1980 / Orellana la Vieja. Ayuntamiento – Villanueva de la Serena: Graf. Samat, 1980 – 1 – sp Bibl Santa Ana [390]
Fiestas populares en honor de la virgen de argeme, patrona de... / Puebla de Argeme. Ayuntamiento – Plasencia: Imp. La Victoria, 1972 – 1 – sp Bibl Santa Ana [390]
Fiestas y danzas en el cuzco y en los andes / Verger, Pierre – Madrid: Missionalia Hispanica, 1948 – 1 – sp Bibl Santa Ana [390]
Fiestas y feria de la barriada de san roque... 1971 / Badajoz – Badajoz: Graf. Tejado, 1971 – sp Bibl Santa Ana [390]
Fiestas y festejos en honor de su patrona la santisima virgen del rosario 1977 / Valdemorales. Ayuntamiento – Caceres: Tip. Extremadura, 1977 – sp Bibl Santa Ana [390]
Fiestas...septiembre 1960 / Tornavascas. Ayuntamiento – Plasencia: Imp. La Victoria, 1960 – 1 – sp Bibl Santa Ana [350]
Fietkau, Rebecca *see* Comparison of one continous bout versus a split bout of aerobic exercise on 13-hour ambulatory blood pressure in healthy females
Fievee, Joseph *see* Histoire de la session de 1815
Fievez, Charles *see* Etude du spectre solaire
Fievres d'afrique / Carbonneau, Louis – Paris: J Ferenczi, [1926] – 1 – us CRL [960]
Fiey, J M *see* Communautes syriaques en iran et irak des origines a 1552
Fife and kinross extra – Dunfermline: Dunfermline Press 1981- (wkly) [mf ed 7 jan 1995-] – 1 – uk Scotland NatLib [072]
Fife circular, etc – [Scotland] Fife, Kirkcaldy: J Crawford 7 apr 1860-1 mar 1877 (wkly) [mf ed 2004] – 8r – 1 – (missing: 1865 & 1868; cont: fife circular of advertisements) – uk Newsplan [072]
Fife free press – Kirkcaldy: J Strachan & W G Livingston 1892- (wkly) [mf ed 7 jan 1994-] – 1 – (cont: fife free press & kirkcaldy guardian) – ISSN: 1354-6058 – uk Scotland NatLib [072]
Fife free press – Kirkcaldy, Scotland. -w. 1871-99. 24 reels – 1 – uk British Libr Newspaper [072]
Fife herald – Cupar, Scotland, UK. 18 Mar 1824-9 Mar 1826; 11 Mar 1830-27 Dec 1832; 3 Jan-28 Feb 1833; 5 Mar 1835-28 Dec 1837; 1838-43. -w. 5 reels – 1 – uk British Libr Newspaper [072]
Fife herald – Kirkaldy, Scotland. -w. 1880. 1 reel – 1 – uk British Libr Newspaper [072]
Fife herald (1985) : incorporating the fife news and kinross-shire advertiser – Cupar: Strachan & Livingston Ltd 1985- (wkly) [mf ed 7 jul 1995-] – 1 – (cont: fife herald news) – ISSN: 1354-6082 – uk Scotland NatLib [072]
Fife herald & journal – [Scotland] Fife, Cupar: J & G Innes jan 1890-dec 1950 (wkly) [mf ed 2004] – 58r – 1 – (formed by union of: fife herald (1881) and: fifeshire journal; cont by: fife herald (1968); cont the numbering of: fife herald (1881); numbering is irreg) – uk Newsplan [072]
Fife, LK *see* The reliability of the dynavec (tm) lvd
Fife news [cupar, scotland : 1870] – [Scotland] Fife, Cupar: R Tullis 19 mar 1870-dec 1878, jan 1891-dec 1950 (wkly) [mf ed 2004] – 76r – 1 – (merged with: coast chronicle for leven, methil, buckhaven, & wemyss, etc to form: fife news and coast chronicle [jan 1911-dec 1921]; cont as: fife news [jan 1922-dec 1950]) – uk Newsplan [072]
Fife sentinel – Scotland, UK. 12 Jan 1843-30 Jan 1845. -w. 1 reel – 1 – uk British Libr Newspaper [072]
Fifeshire advertiser – Kirkcaldy, Scotland. -w. 1849-99. 31 reels – 1 – uk British Libr Newspaper [072]
Fifeshire catholic herald – [Scotland] Dunfermline: Scottish Catholic Printing Press for the Proprietors 11 oct 1919-dec 1938 [wkly] [mf ed 2004] – 38r – 1 – (ceased with n1073 publ jan 6 1939; absorbed by: scottish catholic herald) – uk Newsplan [241]
Fifeshire express – Cupar, Scotland, UK. 4 Aug-23 Jan 1856.-w. 24 feet – 1 – uk British Libr Newspaper [072]
Fifeshire journal – Cupar, Scotland, UK. 7 Jan-Dec 1836; 1839-43; 1844-49; 26 Nov 1850; 3 Dec 1850-28 Jan 1851; 4 Feb-Dec 1851; 1852-92; Jan-31 Aug 1893. -w. 48 reels – 1 – uk British Libr Newspaper [072]
Fifeshire journal – Kirkaldy, Scotland, UK. 26 Jan 1833-7 Jan 1835.-w. 1 reel – 1 – uk British Libr Newspaper [072]
Fifeshire news *see* The news
Fifield advocate – Fifield WI. 1883 oct 25, 1889 mar 21 – 1r – 1 – (cont by: phillips times [phillips, wi: 1877]; phillips times and fifield advocate) – mf#916190 – us WHS [071]
Fifield, W M *see*
– Fertilizer experiments with potatoes on the marl soils of dade county
– Potato growing in florida
Fifille a sa memere / Coolus, Romain – Paris, France. 1925 – 1r – us UF Libraries [440]
Fifteen drypoints / Dey, Mukul – Calcutta: M Dey, 1939 – (= ser Samp: indian books) – (interpreted in verse by harindranath chattopadhyaya) – us CRL [810]
Fifteen months' pilgrimage through untrodden tracts of khuzistan and persia : in a journey from india to england, through parts of turkish arabia, persia, armenia, russia, and germany; performed in the years 1831 and 1832 / Stocqueller, Joachim – London 1832 [mf ed Hildesheim 1995-98] – 2v on 4mf – 9 – €120.00 – 3-487-27629-1 – gw Olms [910]
Fifteen sermons preached before the university of oxford between ad 1826 and 1843 / Newman, John Henry – 3d ed. London: Rivingtons, 1872. Chicago: Dep of Photodup, U of Chicago Lib, 1972 (1r); Evanston: American Theol Lib Assoc, 1984 (1r) – 1 – 0-8370-0339-3 – (includes biographical references) – mf#1984-B295 – us ATLA [240]
Fifteen solemn facts – London, England. 18– – 1r – us UF Libraries [240]
Fifteen thirty-two's news and views : official publication of united food and commercial workers union, local 1532 / United Food and Commercial Workers International Union – 1988 aug-1991 jun – 1r – 1 – (cont by: news & views [united food and commercial workers international union. local 1532]) – mf#2699828 – us WHS [331]
Fifteen years among the top-knots : or, life in korea / Underwood, Lillias Horton – 2nd rev enl ed. Boston: American Tract Society, c1908 [mf ed 1990] – 1mf – 9 – 0-7905-6899-3 – (1st printed 1904) – mf#1988-2899 – us ATLA [915]
Fifteen years among the top-knots : or, life in korea / Underwood, Lillias Horton – Boston: American Tract Society [1904] [mf ed 1995] – (= ser Yale coll) – xviii/271p (ill) – 1 – 0-524-09831-X – (int by frank f ellinwood) – mf#1995-0831 – us ATLA [915]

Fifteen years in canada : being a series of letters on its early history and settlement...its agricultural progress and wealth compared with the united states / Haw, William – Edinburgh, 1850 – (= ser 19th c british colonization) – 2mf – 9 – mf#1.1.8737 – uk Chadwyck [971]

Fifteen years in india; or, sketches of a soldier's life : being an attempt to describe persons and things in various parts of hindostan / Wallace, Robert – London 1823 [mf ed Hildesheim 1995-98] – 1v on 4mf – 9 – €120.00 – 3-487-27439-6 – gw Olms [880]

Fifteen years in the chapel of yale college : 1871-1886 / Porter, Noah – New York: Charles Scribner's sons, 1888 [mf ed 1984] – 5mf – 9 – 0-8370-1001-2 – mf#1984-4357 – us ATLA [080]

Fifteen years of prayer in the fulton street meeting / Prime, Samuel Irenaeus – New York: Scribner, Armstrong, 1872 – 1mf – 9 – 0-7905-5737-1 – mf#1988-1737 – us ATLA [240]

Fifteen years of the drink question in massachusetts / Stoddard, Cora Frances – Westerville, OH. 1929 – 1r – us UF Libraries [362]

Fifteenth century bibles : a study in bibliography / Prime, Wendell – New York: Anson D F Randolph, 1888 – 1mf – 9 – 0-7905-0146-5 – (incl ind) – mf#1987-0146 – us ATLA [220]

Fifteenth century italian manuscripts – 15th c – (= ser Holkham library manuscript books 317,327,328,331,341,369,388,409,418) – 1r – 1 – (copies of works of caesar, cicero, horatius and others) – mf#96482 – uk Microform Academic [450]

The fifth army history, 1943-1945 / U.S. Army. Fifth Army – 2r – 1 – $260.00 – mf#S1684 – us Scholarly Res [355]

The fifth book of moses called deuteronomy – London: J M Dent; Philadelphia: J B Lippincott 1902 [mf ed 1989] – (= ser The temple bible) – 1mf – 9 – 0-7905-1859-7 – mf#1987-1859 – us ATLA [221]

Fifth census of the united states, 1830 / U.S. Bureau of the Census – (= ser 1790-1840 census schedules) – 201r – 1 – mf#M19 – us Nat Archives [317]

Fifth commission / National Council of the Churches of Christ in the United States of America – v2 n2-v8 n? [1976 feb-1982 spring] – 1r – 1 – mf#657344 – us WHS [242]

Fifth estate – 1966 oct 16-1970 jun 10, 1970 jun 11-1972 jan 12, 1972 jan 13-1973 jan 5, 1973 jan 6-1974 jul 31, 1974 aug 1-1975 sep, 1975 oct 31-1986 summer – 6r – 1 – mf#806307 – us WHS [071]

Fifth estate – Detroit. 1970+ (1) 1985-1986 (5) 1985-1986 (9) – ISSN: 0015-0800 – mf#6015 – us UMI ProQuest [073]

Fifth estate – Detroit, MI. 1966-1971 (1) – mf#63721 – us UMI ProQuest [071]

The fifth gospel : the land where jesus lived / Otts, John Martin Philip – New York: Fleming H Revell, c1892 – 1mf – 9 – 0-8370-4651-3 – mf#1985-2651 – us ATLA [226]

Fifth letter to n wiseman / Palmer, William – Oxford, England. 1841 – 1r – us UF Libraries [240]

Fifth of november – London, England. 1814 – 1r – us UF Libraries [240]

Fifth ohio volunteer infantry : civil war recollections of john m paver / Paver, John M – 1r – 1 – mf#B29618 – us Ohio Hist [355]

The fifth report from the select committee of the house of commons on the affairs of the east india company : dated 28th july 1812 / Great Britain Parliament House of Commons. Select Committee on the East India Company; ed by Firminger, Walter Kelly – Calcutta: R Cambray & Co, 1917 – (= ser Samp: indian books) – (notes and int by ed) – us CRL [380]

Fifth report of the aberdeen auxiliary bible society / Aberdeen Auxiliary Bible Society – Aberdeen, Scotland. 1816 – 1r – us UF Libraries [240]

Fifth report of the proceedings of the church... / Church Society of the Archdeaconry of New Brunswick – [St John, NB?: s.n.] 1841 [mf ed 1983] – 1mf – 9 – 0-665-43837-0 – mf#43837 – cn Canadiana [240]

Fifth report of the standing committee on roads and public improvements = Cinquieme rapport du comite permanent des chemins et des ameliorations publiques / Bas-Canada. Parlement. Chambre d'assemblee – [S.l: s.n, 1833?] (mf ed 1992) – 1mf – 9 – (in english and french) – mf#SEM105P1611 – cn Bibl Nat [380]

Fifth wheel / Nor-Cal Teamsters for a Democratic Union – [v1? n1?]-v8 n2 [1971 mar 1-1978 apr] – 1r – 1 – mf#398293 – us WHS [331]

The fifth world congress of free christians : and other religious liberals at berlin, germany, august 5-11, 1910: a summary and appreciation / Wendte, Charles William – Boston: American Unitarian Assoc 1910 [mf ed 1991] – 1mf – 9 – 0-524-01339-X – mf#1990-4088 – us ATLA [243]

Fiftie godlie and learned sermons / Bullinger, Heinrich – London, Ralphe Newberrie, 1577 – 1mf3mf – 9 – mf#PBU-163 – ne IDC [240]

The fiftieth anniversary of the formation of the carleton baptist church in carleton : saint john, may 16, 1841, sunday may 17, 1891 – St John, NB: G Day, 1891 – 1mf – 9 – mf#05581 – cn Canadiana [242]

Fiftieth report...1885 : pt 1 / Diocesan Church Society of New Brunswick – [St John NB: s.n.] 1885 [mf ed 1983] – 2mf – 9 – 0-665-43882-6 – mf#43882 – cn Canadiana [242]

Fiftieth report...1885 : pt 2: subscription lists / Diocesan Church Society of New Brunswick – [St John NB: s.n.] 1885 [mf ed 1983] – 1mf – 9 – 0-665-43883-4 – mf#43883 – cn Canadiana [242]

Fifty helps : for the beginner in the use of the japanese language being an adaptation of mrs baird's fifty helps / Winn, George H – Seoul: Korean Religious Tract Society, 1914 [mf ed 1995] – (= ser Yale coll) – 91p – 1 – 0-524-09565-5 – mf#1995-0565 – us ATLA [480]

Fifty lessons in training for service / Moninger, Herbert – rev ed. Cincinnati, Ohio: Standard Pub Co, c1908 – 1mf – 9 – 0-524-04078-8 – mf#1991-2023 – us ATLA [240]

Fifty sermons and evangelistic talks / Moody, Dwight Lyman – Cleveland: Union Gospel News, c1899 [mf ed 1991] – 1mf – 9 – 0-7905-9414-5 – mf#1989-2639 – us ATLA [240]

Fifty years ago – London, England. 1848 – 1r – us UF Libraries [240]

Fifty years among the baptists / Benedict, David – New York and Boston. 1860 – 1 – $15.96 – us Southern Baptist [242]

Fifty years among the baptists / Benedict, David – New York: Sheldon, 1860 c1859 [mf ed 1990] – 1mf – 9 – 0-7905-5565-4 – mf#1988-1565 – us ATLA [242]

Fifty years as a presiding elder / Cartwright, Peter; ed by Hooper, William Story – Cincinnati: Hitchcock and Walden; New York: Nelson and Phillips, c1871 – 1mf – 9 – 0-7905-6803-9 – mf#1988-2803 – us ATLA [240]

Fifty years at east brent : the letters of george anthony denison, 1845-1896, archdeacon of taunton = Correspondence / Denison, George Anthony; ed by Denison, Louisa Evelyn – New York: E P Dutton; London: John Murray, 1902 – 1mf – 9 – 0-7905-4397-4 – us ATLA [240]

Fifty years at panama... / Robinson, Tracy – New York, NY. 1907 – 1r – us UF Libraries [972]

Fifty years in amoy : or, a history of the amoy mission, china / Pitcher, Philip Wilson – New York: Board of Publ of the Reformed Church in America, 1893 [mf ed 1995] – (= ser Yale coll) – 207p (ill) – 1 – 0-524-09431-4 – mf#1995-0431 – us ATLA [240]

Fifty years in china : being some account of the history and conditions in china and of the missions of the presbyterian church in the united states there from 1867 to the present day / by samuel Isett Woodbridge / Woodbridge, Samuel Isett – Richmond: Presbyterian Committee of Publ [1919] [mf ed 1995] – (= ser Yale coll) – 231p (ill) – 1 – 0-524-09189-7 – mf#1995-0189 – us ATLA [242]

Fifty years in constantinople : and recollections of robert college / Washburn, G – Boston, New York, 1911 – 4mf – 9 – mf#HT-172 – ne IDC [915]

Fifty years in south africa : being some recollections and reflections of a veteran pioneer / Nicholson, G – London, 1898 – 4mf – 9 – mf#HT-98 – ne IDC [916]

Fifty years in the church of rome / Chiniquy, Charles Paschal Telesphore – Chicago: Craig & Barlow, 1885, c1884 – 2mf – 9 – 0-8370-8971-9 – mf#1986-2971 – us ATLA [240]

Fifty years of british art / Hodgson, John Evan – Manchester 1887 – (= ser 19th c art & architecture) – 2mf – 9 – mf#4.2.24 – uk Chadwyck [700]

Fifty years of concessions to ireland 1831-1881 / O'Brien, Richard Barry – London, [1883-1885] – (= ser 19th c ireland) – 13mf – 9 – mf#1.1.6203 – uk Chadwyck [941]

Fifty years of concessions to ireland, 1831-1881 / O'Brien, Richard Barry – London: S. Low, Marston, Searle & Rivington, 1883-85. 2v. illus – 1 – us Wisconsin U Libr [941]

Fifty years of foreign missions / Smith, George – Edinburgh, Scotland. 1879 – 1r – us UF Libraries [240]

Fifty years of plymouth church, minneapolis, minnesota : full account of the semi-centennial celebration, april 25-28, 1907 and other items of historic interest, with illustrations / Hallock, Leavitt Homan – Minneapolis: Hall, Black, 1907 – 1mf – 9 – 0-524-06753-8 – (incl ind) – mf#1990-5279 – us ATLA [240]

Fifty years of public work of sir henry cole – London 1884 – (= ser 19th c art & architecture) – 10mf – 9 – mf#4.2.971 – uk Chadwyck [740]

Fifty years of the history of the republic of south africa / Voigt, Johan Carel – New York, NY. v1-2. 1969 – 1r – us UF Libraries [960]

Fifty years with the baptist ministers and churches of the maritime provinces of canada / Bill, Ingraham Ebenezer – St John NB: printed by Barnes 1880 [mf ed 1990] – 2mf [ill] – 9 – 0-7905-5630-8 – mf#1988-1630 – us ATLA [242]

Fifty-fifth report...1890 / Diocesan Church Society of New Brunswick – St John NB: Barnes, 1890 [mf ed 1983] – 3mf – 9 – 0-665-43888-5 – (incl ind) – mf#43888 – cn Canadiana [242]

Fifty-first report...1886 / Diocesan Church Society of New Brunswick – [St John NB: s.n.] 1886 [mf ed 1983] – 3mf – 9 – 0-665-43884-2 – mf#43884 – cn Canadiana [242]

Fifty-fourth report...1889 / Diocesan Church Society of New Brunswick – St John NB: Barnes, 1889 [mf ed 1983] – 3mf – 9 – 0-665-43887-7 – (incl ind) – mf#43887 – cn Canadiana [242]

Fifty-nine years of history : an address. delivered at the annual meeting of the ohio christian missionary society, new lisbon... / Errett, Isaac – Cincinnati: Standard Pub, 1886 – 1mf – 9 – 0-524-02251-8 – mf#1990-4258 – us ATLA [240]

Fifty-one photographic illustrations / Cole, Henry Hardy – London 1883 – (= ser 19th c art & architecture) – 2mf – 9 – mf#4.2.519 – uk Chadwyck [770]

Fifty-second report...1887 / Diocesan Church Society of New Brunswick – St John NB: Barnes, 1887 [mf ed 1983] – 3mf – 9 – 0-665-43885-0 – mf#43885 – cn Canadiana [242]

Fifty-third report...1888 / Diocesan Church Society of New Brunswick – St John NB: Barnes, 1883 [mf ed 1983] – 4mf – 9 – 0-665-43886-9 – (incl ind) – mf#43886 – cn Canadiana [242]

Fifty-three years in syria / Jessup, Henry Harris – New York: Fleming H Revell, c1910 – 3mf – 9 – 0-8370-6669-7 – (incl tables and ind) – mf#1986-0669 – us ATLA [240]

Fifty-two letters to dr. john ryland / Carey, William – 1781-1825 – 1 – us Southern Baptist [242]

Fifty-two primary missionary stories : including 52 drawings and verses / Applegarth, Margaret Tyson – New York City: Board of Publ and Bible School Work, c1917 – 1mf – 9 – 0-524-05246-8 – mf#1991-2238 – us ATLA [240]

Fifty-two sermons – Princeton: Princeton Theological Seminary, [18–] – 1r – 1 – 0-8370-1142-6 – mf#1984-B474 – us ATLA [240]

Figaro : [humor magazine] – Vienna, Graz. jan 1857-dec 1919 – 27r – 1 – (missing: 1863-64) – us UMI ProQuest [074]

Figaro – Stockholm, Sweden. 1878-1925 – 20r – 1 – sw Kungliga [790]

Le figaro – Edition hebdomadaire internationale. no. 1-60. Paris. 1974-1er mars 1975 – 1 – fr ACRPP [074]

Le figaro – Journal non politique. Paris. 1 avr 1854-24 nov 1942, 23 aout 1944-1990 – 1 – fr ACRPP [074]

Le figaro – Journal non politique. Dir. Lepoitevin-Saint-Alme et M. Alhoy. Paris. 1826-34 – 1 – (le titre de figaro a ete repris successivement par:. figaro. electeur, jure, contribuable, artiste, financier puis nouvelliste du soir; journal quotidien, politique et litteraire; journalivre. revue quotidienne; journal quotidien, politique et litteraire. 15 fevr-20 aout 1835 no. 1-180, 16 mai-10 juil 1836 n.s., no. 1-56, 1er oct-24 dec 1836 no. 1-86, 26-27 dec 1836, 1er janv-13 juin 1837, 16 juin-15 aout 1837, 15 oct 1837-6 sept 1838 no. 1-322. figaro. 30 dec 1838-28 fevr 1839. le figaro. journal litteraire et d'arts puis de litterature et d'arts. no. 1-188. 3 mars 1839-27 dec 1840. le nouveau figaro. 19 dec 1841-mai 1842. absorbe par: les coulisses. le figaro. journal de l'apres-midi. 25 dec 1847-8 mars 1848. figaro. no. 1, 15. mai 1848. nouveau figaro. programme des theatres, journal quotidien du soir, politique, litteraire et satirique. no. 1-6. 8-15 juin 1848. figaro. journal non politique. no. 1-3. 18-20 aout 1852) – fr ACRPP [074]

Le figaro – Paris: Figaro, 1953-54 – us CRL [074]

Le figaro – Supplement litteraire. Paris. 1876-1895, dec 1905-aout 1914 – 1 – fr ACRPP [074]

Le figaro – 1854 – 1 – (yrly reel count varies) – us UMI ProQuest [074]

Figaro and irish gentlewoman *see* Irish life

Figaro in london – London. 1831-1839 (1) – mf#4252 – us UMI ProQuest [790]

Figaro litteraire *see* Litteraire

Le figaro litteraire – Paris. mars 1946-1978 – 1 – fr ACRPP [410]

Figarola-Caneda, Domingo *see* Gertrudis gomez de avellaneda

Figaro's chronik *see* Landstreicher

Figatner, IU P *see* Sostav sovetskikh i torgovykh sluzhashchikh

Figgis, John Neville *see*
- Antichrist and other sermons
- Christianity and history
- Churches in the modern state
- Civilisation at the cross roads
- The divine right of kings
- The fellowship of the mystery
- The gospel and human needs
- Historical essays and studies
- The history of freedom
- Religion and english society
- Studies of political thought from gerson to grotius, 1414-1625

Figgis, john Neville *see* Lectures on modern history

Figgis, John Neville et al *see* Typical english churchmen

Fight back! – 1969 jun – 1r – 1 – mf#5266111 – us WHS [071]

Fight back! / G I Rights and Information Center – v2 n2, v3 n3 [1974 feb/mar, sep] – 1r – 1 – mf#721615 – us WHS [355]

Fight back! / Hospital Rank and File Action Committee. Service Employee's International Union – n1-4 [197?], – 1r – 1 – mf#721568 – us WHS [071]

Fight back! / Revolutionary Student Brigade – v2 n1 [1972 sep 15] – 1r – 1 – mf#721579 – us WHS [320]

Fight back news / National Fight Back Organization – v1 n1 [1977 dec] – 1r – 1 – mf#367513 – us WHS [366]

Fight for light : and other sermons / Rader, Paul – New York: Book Stall, c1916 [mf ed 1992] – (= ser Christian & missionary alliance coll) – 1mf – 9 – 0-524-02264-X – mf#1990-4271 – us ATLA [240]

Fight for santiago / Bonsal, Stephen – New York, NY. 1899 – 1r – us UF Libraries [972]

The fight in the beechwoods : a study in canadian history / Cruikshank, Ernest Alexander – Welland Ont: Lundy's Lane Historical Society, 1895 – 1mf – 9 – mf#13790 – cn Canadiana [355]

The fight of faith : sermons / Brooke, Stopford Augustus – 2nd ed. London: HS King, 1877 – 1mf – 9 – 0-7905-7500-0 – mf#1989-0725 – us ATLA [240]

A fight with distances : the states, the hawaiian islands, canada, british columbia, cuba, the bahamas / Aubertin, John James – London: Kegan Paul, Trench, 1888 – 5mf – 9 – mf#14068 – cn Canadiana [917]

Fighter : official publication of the revolutionary council of the hungaria freedom fighter movement – Cleveland: The Council, v5-8 n1/2 feb 1979-mar 1982 – 1r – 1 – us CRL [071]

Fighter forum – 1981 may 1-1982, 1983-84, 1985, 1986 jan 10-1987 mar 20, 1987 apr-1988 may, 1988 jun-1989 aug 18 – 6r – 1 – mf#645311 – us WHS [355]

"The fighting custers" / Sibrava, Frank – 1 – us Kansas [978]

Fighting for the vote : the suffragette fellowship – 14r – 1 – (coll compiles the papers of 12 leading suffragists and 2 of the movement's key organizations, the women's social and political union and mrs charlotte despard's women's freedom league. includes complete listing) – mf#C36-28070 – us Primary [322]

Fighting talk, 1954-62, johannesburg – 1r – 1 – (variously publ as: an independent monthly review, a monthly journal for democrats, a monthly journal for democrats of all races) – mf#97295 – uk Microform Academic [960]

Fighting the mill creeks / Anderson, Robert – 1909 – 1r – 1 – $50.00 – mf#B63017 – us Library Micro [978]

Fighting times / Revolutionary Union – v2 n7/8, v3 n3-4 [1973 nov/dec, 1974 jul-aug] – 1r – 1 – mf#367515 – us WHS [320]

Fighting times / United Workers Caucus, Local 72 [Racine, WI] et al – 1979 feb-1984 jan – 1r – 1 – mf#518058 – us WHS [331]

Fighting words : the street journal of anti-racist action / Anti-Racist Action [Organization] – 1995 fall-1997 spring – 1r – 1 – mf#3622075 – us WHS [320]

Fighting worker : organ of the revolutionary workers league of the u s [marxist internationalists] / Revolutionary Workers League of the US – v1 n0-v2 n6 [1935 nov 30-1937 aug 1], v3 n1-v6 n7 [1938 mar 1-jul 1941], v9 n1-12 [1944 jan-dec] – 1r – 1 – mf#1399404 – us WHS [335]

Fighting worker : revolutionary workers league of the us – v1-12 n11. 1936-47 [all publ] – (= ser Radical periodicals in the united states, 1881-1960. series 2) – 1r – 1 – $200.00 – us UPA [335]
Figliucci, Felice *see* De la filosofia morale libri dieci
Figner, V N *see* Zhurnal zagranichnykh organizatsii pomoshchi politicheskim ssylnym i zakliuchennym v rossii
Fignole, Daniel *see*
– Cuba et haiti
– Quelques realisations du second empire
Figo, Azariah *see* Binah la-'itim
Figueira, Luiz – 4. impr. Lisboa: Officina patriarcal 1795 – (= ser Whsb) – 2mf – 9 – €30.00 – mf#Hu 398 – gw Fischer [440]
Figueiredo de estudios de historia americana / Bayle, Constantino – S. Paulo, 1929; Madrid: Razon y Fe, 1931 – 1 – sp Bibl Santa Ana [370]
Figueiredo, Guilherme *see*
– Raposa e as uvas
– Trinta anos sem paisagem
Figueiredo, Jackson De *see* Correspondencia
Figueiredo, Jose De Lima *see* Limites do brasil
Figuelo – Hungary, 1999- – 3r per y standing order – 1 – us UMI ProQuest [079]
Figueredo, Candelaria *see* La abanderada de 1868 candelaria figueredo (hija de perucho)
Figueres Ferrer, Jose *see* Cartas a un ciudadano
Figueroa, Carlos Alberto *see* Carruaje bajo la lluvia
Figueroa De Cifredo, Patria *see* Apuntes biograficos en torno a la vida
Figueroa, Edwin *see* Sobre este suelo
Figueroa, F *see*
– De una especie de garrotillo o esquilencia mortal
– Libro de las calidades y effectos de la aloxa
Figueroa Fernandez, Cotidio *see* Judios en america
Figueroa, Jose R *see* Expediting settlement of employee grievances in the federal sector
Figueroa, Loida *see* Arenales
Figueroa, Marco *see* Por los archivos del tachira
Figueroa Marroquin, Horacio *see*
– Enfermedades de los conquostadores
– Historia de la fisiologia en guatemala
Figueroa, Pedro Pablo *see* Historia de francisco bilbao
Figueroa y Melgar, Alfonso de *see*
– Espanoles fuera de espana
– Literatos cacerenses
– Pregon de las fiestas patronales de trujillo
– Los suarez de figueroa, de feria y zafra
Figueroa-Cruz, Blas E *see* A historical documentation, an instructional manual and an annotated bibliography of selected folk dances of puerto rico
Figuier, Louis *see*
– The day after death
– Reptiles and birds
– The to-morrow of death
Die figur als signifikante spur : zu den gedichten esther und david und jonathan aus dem zyklus hebraeische balladen von else lasker-schueler / Krug, Marina – (mf ed 1999) – 4mf – 9 – €56.00 – 3-8267-2671-5 – mf#DHS 2671 – gw Frankfurter [430]
Figuracion de puerto rico y otros estudios / Melendez, Concha – San Juan, Puerto Rico. 1958 – 1r – us UF Libraries [972]
Figuras contemporaneas – Madrid: Imp. Plaza de los Montenses 7, 1913 – 1 – sp Bibl Santa Ana [946]
Figuras de azulejo, perfis e cenas da historia do... / Calmon, Pedro – Rio de Janeiro, Brazil. 1940? – 1r – us UF Libraries [972]
Figuras do imperio e outros ensaios / Baptista Pereira, Antonio – Sao Paulo, Brazil. 1934 – 1r – us UF Libraries [972]
Figuras ilustres / Gutierrez Macias, Valeriano – Badajos: Imp. Diputacion Provincial, 1965. Sep. REE – sp Bibl Santa Ana [920]
Figuras politicas de colombia / Caballero Calderon, Lucas – Bogota, Colombia. 1945 – 1r – us UF Libraries [972]
Figuras y figurones, biografias de los hombres que mas figuran actualmente en espana. 2 vol / Segovia, A – Madrid, 1877-78 – 52mf – 9 – sp Cultura [946]
Figure emblematique en trois langues : et seulement en une visible de soy / [Claviere, E de] – Paris: Rovert Foet, 1607 – 1mf – 9 – mf#0-1337 – ne IDC [090]
Figure sketching / Oehler, Bernice Olivia – Pelham, NY. 1929 – 1r – us UF Libraries [740]
Figuren- und konfliktdarstellung bei friedrich spielhagen, theodor fontane, ferdinand von saar, eduard von keyserling : eine vergleichende untersuchung der erzaehlungen "zum zeitvertreib", "effie briest", "schloss kostenitz" und "am suedhang" / Manko, Mandane – (mf ed 1995) – 3mf – 9 – €49.00 – 3-8267-2105-5 – mf#DHS 2105 – gw Frankfurter [430]
Figures and descriptions of the palaeozoic fossils of cornwall, devon, and west somerset : observed in the course of the ordnance geological survey of that district / Phillips, John – London, 1841 – (= ser 19th c evolution & creation) – 4mf – 9 – mf#1.1.1208 – uk Chadwyck [560]
Figures d'hier et d'aujourd'hui a travers saint-laurent, i o, vol 1 / Gosselin, David – Quebec: Impr franciscaine missionnaire, 1919 – 9 – 0-665-71521-8 – mf#71521 – cn Canadiana [971]
Figures d'hier et d'aujourd'hui a travers saint-laurent, i o, vol 2 / Gosselin, David – Quebec: Impr franciscaine missionnaire, 1919 – 9 – 0-665-71522-6 – mf#71522 – cn Canadiana [971]
Figures d'hier et d'aujourd'hui a travers saint-laurent, i o, vol 3 / Gosselin, David – Quebec: Impr franciscaine missionnaire, 1919 – 9 – 0-665-71523-4 – mf#71523 – cn Canadiana [971]
Figures et recits de carthage chretienne : etudes sur le christianisme africain aux 2e et 3e siecles / Alcais, Abel – Paris: Fischbacher, 1908 – 1mf – 9 – 0-524-02577-0 – (incl bibl ref) – mf#1990-0629 – us ATLA [240]
Figures of speech : or, figures of thought: collected essays on the traditional, or, "normal" view of art / Coomaraswamy, Ananda Kentish – London: Luzac, 1946 – (= ser Samp: indian books) – us CRL [700]
The figures or types of the old testament *see* The gospel of the old testament
Figures, vases, fountains etc : executed in marble and artificial stone / Wyatt, Parker and Co – London [1841] – (= ser 19th c art & architecture) – 1mf – 9 – mf#4.2.1757 – uk Chadwyck [730]
Figurones y estampas. caracas, 1937 / Zaraza, Lorenzo A – Madrid: Razon y Fe, 1940 – 1 – sp Bibl Santa Ana [700]
Figyelo – Budapest, 1976-1993ff – 22r – 1 – gw Mikropress [949]
Figyelo – 1976- – 1 – (yrly reel count varies) – us UMI ProQuest [070]
Fihrist al-kutub al-makhtutah bi-maktabat al-ahqaf bi tarim lil-mu'allifin al-yamaniyin – Hadramamt: al-Markaz al-Yamani lil-Abhath al-Thaqafiyah wa-al-Athar wa-al-Matahif, sep 1988 – (= ser Arabic research materials) – 3mf – 9 – $55.00 – us MEDOC [956]
Fihrist al-kutub (al-makhtutat) al-'arabiyah al-mahfuzah bi-al-kutubkhanah al-khidiwiyah – Cairo, Egypt: al-Kutubkhanah al-Khidiwiyah, 1308-10 H. (1890-92). 2d impr. v1-7 – (= ser Arabic research materials) – 1r – 1 – $200.00 – us MEDOC [956]
Fihrist maktabat al-ahqaf lil-makhtutat bi-tarim – Hadramamt: al-Markaz al-Yamani lil-Abhath al-Thaqafiyah wa-al-Athar wa-al-Matahif. 3v. oct 1988 – (= ser Arabic research materials) – 6mf – 9 – $90.00 – us MEDOC [956]
Fiji agricultural journal / Fiji Dept of Agriculture [Ministry of Agriculture, Fisheries and Forests] – v1-52. 1928-90 – r1-5 – 1 – (incl: fiji farmer v1 n1-v3 n1 mar 1965-mar 1967. available for ref) – mf#pmb doc457 – at Pacific Mss [630]
Fiji argus – Levuka, Fiji 6 aug 1876-11 aug 1882, 7 dec 1883 – 2 1/2r – 1 – uk British Libr Newspaper [079]
Fiji constitution review commission : copies of written submissions and verbatim notes – jul 1995-jan 1996 – 8r – 1 – mf#pmb1149 – at Pacific Mss [342]
Fiji constitution review commission : copies of written submissions, vols 1-8, and verbatim notes – jul 1995-jan 1996 – 8r – 1 – (available for reference) – mf#pmb1149 – at Pacific Mss [323]
Fiji Court of Appeal *see* Judgements
Fiji court of appeal : judgements 1949-1996, together with privy council judgements relating to fiji cases, 1936-1986 – 12r – 1 – (available for reference) – mf#PMB1137 – at Pacific Mss [347]
Fiji Dept of Agriculture [Ministry of Agriculture, Fisheries and Forests] *see* Fiji agricultural journal
Fiji diary / Parfitt, Peter – n.d. – 1r – 1 – mf#pmb995 – at Pacific Mss [980]
Fiji diary and narratives / Turpin, Edwin James – 1870-1892 – 1r – 1 – mf#pmb1209 – at Pacific Mss [980]
Fiji diary and narratives / Turpin, Edwin James – 1870-92 – 1r – 1 – (available for ref) – mf#pmb1209 – at Pacific Mss [980]
Fiji Executive Council *see* Minutes if sitting for the rehearing of claims to land
Fiji farmer *see* Fiji agricultural journal
Fiji Independent News Service *see* Archives
Fiji Independent News Service, Sydney *see*
– Fiji situation report
– Fiji voice
Fiji journals and letters (wesleyan mission in fiji) / Jaggar, Thomas James – 1838-46 – 1r – 1 – (available for ref) – mf#pmb1185 – at Pacific Mss [240]
Fiji labour sentinel / Fiji Trade Union Congress – n1-47, 49-98. 1978-98 – 1r – 1 – mf#pmb doc433 – at Pacific Mss [331]
Fiji. Methodist Church *see* Circuit reports, 1835-1898, and the swanston collection on the ra and ba military campaigns, 1873
The fiji of to-day / Burton, John Wear – London: Charles H Kelly, [1910] [mf ed 1995] – (= ser Yale coll) – 364p (ill) – 1 – 0-524-09531-0 – (int by a j small) – mf#1995-0531 – us ATLA [919]
Fiji oral history project in association with the fiji museum : pt 1: part-europeans and europeans, transcripts of audio recording series / Mackinnon, Marsali – 1998-1999 – 1r – 1 – mf#pmb1235 – at Pacific Mss [980]
Fiji planters journal / Planters Association of Fiji – 1913-17 – r1-2 – 1 – (available for ref) – mf#pmb doc455 – at Pacific Mss [630]
Fiji situation report / Fiji Independent News Service, Sydney – oct 1987-nov 1990 – 1r – 1 – mf#pmb doc419 – at Pacific Mss [380]
Fiji times – Levuka, Fiji. 4 sep 1869-1876 – 6r – 1 – uk British Libr Newspaper [072]
Fiji times *see* Western pacific herald
Fiji times and herald *see* Western pacific herald
Fiji Trade Union Congress *see* Fiji labour sentinel
Fiji Trades Union Congress *see* Archives
Fiji voice / Fiji Independent News Service, Sydney – n1-24. sep 1987-dec 1992 – 1r – 1 – mf#pmb doc418 – at Pacific Mss [380]
'Fiji-70 years and one month' : the memoirs of william (tui) johnson / Johnson, William (Tui) Grainger – 1900-1970 – 1r – 1 – mf#pmb1017 – at Pacific Mss [980]
Fijian pamphlets collected by sir arthur gordon – v1-3. 1877-83 – r1-2 – 1 – (available for ref) – mf#pmb1213 – at Pacific Mss [980]
Fijian pamphlets collected by sir arthur gordon, vols 1-5 / Gordon, Arthur – 1870-1883 – 2r – 1 – mf#pmb1213 – at Pacific Mss [980]
Fikenscher, Georg Wolfgang Augustin *see* Gelehrtes fuerstenthum baireut
Fikir hareketleri – Istanbul, 1933-? Sahibi: Hueseyin Cahit [Yalcin]. n1-364 (29 Tesrinievvel 1933-12 Tesrinievvel 1940) – (= ser O & t journals) – 159mf – 9 – $2385.00 – us MEDOC [079]
Filalet Khristofor *see* Apokrisis
Filantropia sospechosa / Bayle, Constantino – Madrid: Razon y Fe, 1925 – 1 – sp Bibl Santa Ana [946]
Filaret, Archbishop of Chernigov *see* Geschichte der kirche russlands
Filaret, Arkhiepiskop *see* Obzor russkoi duchovnoi literatury, 862-1720
Filaret, Mitropolit Moskovskii *see*
– Pisma k arkhimandritu antoniiu, 1831-1867
– Pisma k rodnym
– Pisma k vysochaishim osobam i raznym drugim litsam
– Russkie sviatie chtimye vseiu tserkoviiu ili mestno...
– Sviatye iuzhnykh slavian
– Zhitiia sviatykh, chtimykh pravolsavianoiu tserkoviiu...
[Filarete] Oettingen, W von *see* Antonio averli filarete's tractat ueber die baukunst
Filastin – Jaffa: I D Elissa, 1956-mar 21 1967 – 31r – us CRL [079]
Filastin – Jaffa, Jerusalem, 1911-1914, 1921-1967 – 99r – 1 – (missing: 1948(may-dec); 1949(jan, sep-dec); 1951(jul-dec); 1952(sep-dec); 1953-1955(jan)) – mf#J-93-7 – ne IDC [956]
Filatov, V B *see*
– Novyi zakon o promyslovoi kooperatsii 11 maia 1927 g
– Promyslovoe kooperativnoe tovarishchestvo
Filbeck, David Lee *see* T'in culture
File note : manpower assessment / Swaziland Labour Dept – [S.l: s.n., 196-?] – us CRL [331]
File of correspondence relating to emin pasha extracted from the zanzibar archives / ed by Gray, John – [s.l: s.n.], 1971 – 1 – us CRL [960]
Filene, Edward Albert *see* The present status and future prospects of chains of department stores
Filene house news : a biweekly newsletter for filene house employees – n20/74-n19/79 [1974 oct 30-1979 oct 23] – 1r – 1 – (cont by: window [credit union national association]) – mf#700872 – us WHS [366]
Fileppeli, Ronald L *see* The socialist party of the united states
Files / Baptist Faith and Message Committee – 1962-63. 728p – 1 – us Southern Baptist [242]
Files concerning bengal, 1927-1947 / Indian National Congress. All-India Congress Committee – Chicago, IL: Uni of Chicago Photodup Dept, 1973 (mf ed) – 1 – us CRL [954]
Files on the john frum movement / New Hebrides British Service. Southern District Administration – Tanna, 1947-56 – 1r – 1 – (restricted access) – mf#PMB1133 – at Pacific Mss [350]
Filesi, Teobaldo *see* Relazioni tra il regno del congo e la sede apostolica nel 16...
La fileuse : ou legende de mon pays / Prevost, Marcel – [Ste-Hyacinthe, Quebec?: s.n.] 1880 [mf ed 1984] – 1mf – 9 – 0-665-04861-0 – mf#04861 – cn Canadiana [830]
O filho de minas : orgao litterario – Ouro Preto, MG. 01 abr 1900 – (= ser Ps 19) – bl Biblioteca [410]
O filho do brasil – Rio de Janeiro, RJ: Imparcial de Brito, 04 jul-13 out 1840 – (= ser Ps 19) – mf#P14,4,22 – bl Biblioteca [320]
Filial tribute of justice, affection, and gratitude / White, Verner Moore – Liverpool, England. 1862? – 1r – us UF Libraries [240]
Filiarchi, C *see* Trattato della gverra
Filiatreault, Aristide *see* Assurance, banque et stocks
Filices in charles wilkes' u.s. exploring expedition / Brackenridge, W D – Philadelphia. v16. 1854 – 1r – 1 – mf#95677 – uk Microform Academic [580]
Les filigraines : a historical dictionary of watermarks / Briquet, C M – 2nd ed. Leipzig, 1928 – 4r – 1 – mf#305 – uk Microform Academic [010]
Filimonov, D *see* Vestnik kubanskogo kraevogo pravitel'stva
Filimonov, F F *see* Svobodnaia sibir'
Filion, Laetitia *see* A deux
Filipinas ante europa – Madrid: Filipinas ante Europa. ano1 n1 (25 de oct de 1899)- [mf ed 1985] – (= ser Rare philippine newspapers and serials 1) – 1 – (cont: defensor de filipinas) – mf#5771 reel 8 & 10 – us Wisconsin U Libr [959]
Filipino language lexicon / Enriquez, Jose T et al; ed by Balmaceda, Julian C – Manila: Jose C Velo c1958 [mf ed [Moskva]: Gosudarstvennaia biblioteka SSSR im V I Lenina [19–]] – 1r – 1 – us CRL [490]
El filipino libre – Manila: M Xerez y Burgos, feb 13 1900 – us CRL [079]
The filipino national language / Panganiban, Cirio H – Bruxelles: International Institute of Differing Civilizations, 1952 – us CRL [490]
Filipino-english vocabulary, with practical examples of filipino and english grammars / Daluz, Eusebio T – First ed. Manila, 1915 – 1 – us Wisconsin U Libr [490]
Filipov, Iu D *see* Slovar' iuridicheskikh i gosudarstvennykh nauk
Filippenko, A K *see* Utro
Filippi, Filippo *see* Richard wagner
Filippi, Ludovico *see* Lettera pastorale di monsig. luigi filippi
Filippov, A N *see*
– O nakazanii po zakonodatelstvu petra velikogo,v sviazi s reformoiu
– Pravitelstvuiushchii senat pri petre velikom i ego blizhaishikh preemnikakh (1711-1741 gg
Filippov, L P et al *see* Dissipativnye svoistva metallov i metallicheskikh splavov
Filippov, T I *see* Sovremennye tserkovnye voprosy
Filippov, V V *see* Gelmintozy cheloveka, zhivotnykh, rastenii i mery borby s nimi
Filippova, L P *see* Dissipativnye svoistva metallov i metallicheskikh splavov
Filippova, L P et al *see* Primenenie metoda integralnykh kharakteristik kissledovaniiu problemy vosstanovleniia parametrov teplomassoperenosa
Filips, Teri L *see* Meaning and survival
Filipstads stads och bergslags tidning – Filipstad, Sweden. 1850-1958 – 1 – sw Kungliga [078]
Filipstads tidning – Filipstad, Sweden. 1959- – 1 – (varmlandsberg, 1963-74) – sw Kungliga [078]
Filipstads tidning *see* Varmlandsberg
Filipstadsposten – Filipstad, 1991-92 – 9 – sw Kungliga [078]
Fille bien gardee / Labiche, Eugene – Paris, France. 1850 – 1r – us UF Libraries [440]
Fille de roland / Bornier, Henri – Paris, France. 1875 – 1r – us UF Libraries [440]
Fille d'haiti / Chauvet, Marie – Paris, France. 1954 – 1r – us UF Libraries [972]
La fille du peuple / Michel, Louise – Paris: Librairie Nationale, 1883 – 9mf – 9 – mf#10330 – fr Bibl Nationale [830]
Fille sauvage / Curel, Francois De – Paris, France. 1919 – 1r – us UF Libraries [440]
Fillebrown, C B *see* The a b c of taxation
Filleul, P Valpy M *see* Limits of toleration
The filley farmer – Filley, NE: J H Brayton, 1887-89// (wkly) [mf ed v1 n47. jan 27 1888 filmed [1979]] – 1r – 1 – (cont by: filley review) – us NE Hist [071]
The filley republican – Filley, NE: J M Linscott & Son. v1 n1. apr 13 1900- (wkly) [mf ed -may 251900 (lacks may 181900) filmed [1979]] – 1r – 1 – us NE Hist [071]

The filley republican – Filley, NE: Trimmer & Montgomery, 1894-1902// (wkly) [mf ed 1895-99 (gaps) filmed 1979] – 1r – 1 – (some irregularities in numbering) – us NE Hist [071]

Filley review – Filley, NE: Filley Pub Co, sep 1889 (wkly) [mf ed 1890,1892 (gaps) filmed 1980] – 1r – 1 – (cont: filley farmer. v1 n37-38 not publ; v1 n51-52 repeated) – us NE Hist [071]

The filley spotlight – Filley, NE: George T Edson. v1 n1. nov 5 1915- (wkly) [mf ed -1926 (gaps) filmed 1978] – 2r – 1 – (issues for nov 5 1915-oct 26 1917 and aug 15 1919-dec 3 1926 also carry whole numbering) – us NE Hist [071]

Fillieux, Claude *see* Merveilleux cambodge

Filling the "gaps" and "dramatic art" in d t mtywaku's dramatic works / Piko, Phindiwe – Pretoria: Vista University 2001 [mf ed 2001] – 4mf – 9 – (incl bibl ref) – mf#mfm15185 – sa Unisa [470]

Fillion, Louis-Claude *see*
- Atlas archeologique de la bible
- Atlas d'histoire naturelle de la bible
- Les etapes du rationalisme dans ses attaques contre les evangiles et la vie de jesus-christ

Fillmore chronicle – Fairmont, NE: Lou W Frazier. 43rd yr n48. mar 26 1915-85th yr n52. apr 25 1957 (wkly) [mf ed with gaps] – 28r – 1 – (cont: fillmore weekly chronicle (1912). absorbed by: nebraska signal (1913)) – us NE Hist [071]

Fillmore chronicle – Fairmont, NE: Lou W Frazier. 15v. 26th yr n1. may 6 1897-40th yr n50. apr 12 1912 (wkly) [mf ed with gaps] – 11r – 1 – (cont: fillmore weekly chronicle. cont by: fillmore weekly chronicle (1912)) – us NE Hist [071]

Fillmore county bulletin – Fairmont, NE: Will R Gaylord. v1 n26. nov 2 1872-1874?// (wkly) [mf ed -oct 161873 (gaps) filmed 1986] – 1r – 1 – (cont: weekly nebraska bulletin. cont by: fairmont bulletin) – us NE Hist [071]

Fillmore county democrat – Exeter, NE: W H Wallace, 1892-dec 1898// (wkly) [mf ed 1893-may 28 1898 (gaps)] – 2r – 1 – (cont by: fillmore county news. v6 n17-v7 n16 not publ) – us NE Hist [071]

Fillmore county news – Exeter, NE: W H Wallace. v8 n48. jan 7 1899- (wkly) [mf ed with gaps] – 1 – (cont: fillmore county democrat) – us NE Hist [071]

Fillmore county republican – Geneva, NE: [T Wilkins] jun 1889-v19 n18. feb 7 1894 (wkly) [mf ed with gaps] – 3r – 1 – (cont: geneva review. merged with: geneva journal to form: republican-journal) – us NE Hist [071]

Fillmore county review – Geneva, NE: Mark M Neeves. 8v. oct 1875-v8 n10. dec 28 1882 (wkly) [mf ed 1876,1878-80,1882 (gaps)] – 2r – 1 – (cont by: geneva review) – us NE Hist [071]

[Fillmore-] the fillmore sun – CA. 1917-1918 – 1r – 1 – $60.00 – mf#C03595 – us Library Micro [071]

Fillmore weekly chronicle – Fairmont, NE: Frazier & Frazier. 12v. v14 n34. dec 23 1885-v25 n52. apr 29 1897 (wkly) [mf ed with gaps filmed -1980] – 5r – 1 – (cont: fairmont bulletin. cont by: fillmore chronicle) – us NE Hist [071]

Fillmore weekly chronicle – Fairmont, NE: Lou W Frazier. 4v. 40th yr n51. apr 19 1912-43rd yr n47. mar 19 1915 (wkly) – 3r – 1 – (cont: fillmore chronicle. cont by: fillmore chronicle (1915)) – us NE Hist [071]

Der film – Berlin DE, 1927 15 nov-1931 24 dec – 1 – (filmed by mikropress: 1916-1943 23 apr [26r] order#5066) – gw Mikrofilm; gw Mikropress [790]

Film : the journal of the british federation of film societies – 1954-99+ – 25r – 1 – £820.00 – mf#FIL – uk World [790]

Le film – Hebdomadaire illustre. Cinematographe theatre, concert, music-hall. Paris. fevr-juil 1914, avr 1916-18, dec 1919-20 – 1 – fr ACRPP [790]

Film and broadcasting review / United States Catholic Conference – v41-v45 n17 [1976 jan 15-1980 sep 1] – 1r – 1 – (cont: catholic film newsletter) – mf#505533 – us WHS [241]

Film and video news – La Salle. 1984-1984 (1) 1984-1984 (5) 1984-1984 (9) – (cont: film news) – ISSN: 8750-068X – mf#2173,01 – us UMI ProQuest [790]

Film and video review index – Pasadena. 1978-1978 (1,5,9) – mf#11897 – us UMI ProQuest [790]

Film art – London. v. 1 no. 3-v. 3 no. 8. Spring 1934-1936 – 1 – us NY Public [790]

Film comment – New York. 1962+ (1) 1971+ (5) 1976+ (9) – ISSN: 0015-119X – mf#5026 – us UMI ProQuest [790]

Film culture – New York. 1955-1996 [1]; 1972-1996 [5]; 1977-1996 [9] – ISSN: 0015-1211 – mf#6518 – us UMI ProQuest [790]

Film daily and film daily yearbook : the complete collection, 1915-1970 – 126r – 1 – (previous title: major film periodicals for media research. list accompanies coll and supplies reel contents and ind to the feature films reviewed in film daily) – mf#C39-10820 – us Primary [790]

Film daily and predecessors, 1915-1970 – 5pt-coll on 107r – 1 – (previous title: major film periodicals for media research. coll consists of pt1: wid's film and film folk 1915-16; wid's independent review of feature films 1916-18; wid's daily 1918-21; film daily, 1922-27 22r c39-10801. pt2: film daily 1928-37 21r c39-10802. pt3: 1938-48 22r c39-10803. pt4: 1945-59 22r c39-10804; pt5: 1960-70 20r c39-10805. incl printed guide) – mf#C39-10800 – us Primary [790]

Film fuer alle – Berlin DE, 1928-1935 feb [gaps] – 1r – 1 – gw Mikrofilm [790]

Film heritage – Dayton. 1965-1977 (1) 1965-1977 (5) 1976-1977 (9) – ISSN: 0015-1270 – mf#6273 – us UMI ProQuest [790]

Film history – London. 1989-1996 (1,5,9) – ISSN: 0892-2160 – mf#17318 – us UMI ProQuest [790]

Film information – New York. 1970-1978 (1) 1972-1978 (5) 1975-1978 (9) – ISSN: 0015-1297 – mf#6724 – us UMI ProQuest [790]

Film journal – Hollins College. 1971-1975 (1) 1971-1975 (5) (9) – ISSN: 0046-3787 – mf#7546 – us UMI ProQuest [790]

Film journals from great britain and australia – (= ser Cinema history microfilm series) – 6r – 1 – $865.00 set; $165.00 ea – (afterimage, 1970-85. the australian journal of screen theory, 1976-85. framework, 1975-85. historical journal of film, radio and television, 1981-85. primetime, 1981-85. undercut, 1981-85. with p/g) – us UPA [790]

Film journals from the united states and canada – (= ser Cinema history microfilm series) – 6r – 1 – $865.00 set $165.00 ea – (camera obscura: n1-12 fall 1976-summer 1984 1r. cine-tracts: n1-17 spring 1977-fall 1982 1r. dreamworks: v1 n1-v3 n4 spring 1980-84 1r. film criticism: v1 n1-v9 n1 spring 1976-fall 1984 1r. film reader n1-5 1975-82 1r. millennium film journal: n1-13 winter 1977/78-fall/winter 1983/84 1r. with p/g) – us UPA [790]

Film news – New York. 1939-1981 (1) 1971-1981 (5) 1976-1981 (9) – (cont by: film and video news) – ISSN: 0195-1017 – mf#2173 – us UMI ProQuest [790]

Film quarterly – Berkeley. 1945+ (1) 1970+ (5) 1970+ (9) – ISSN: 0015-1386 – mf#306 – us UMI ProQuest [790]

Film review index – Pasadena. 1972-1972 (1) 1972-1972 (5) (9) – (cont by: international index to multi-media information) – ISSN: 0046-3809 – mf#6793 – us UMI ProQuest [790]

Film, television and mass communication 1949-1977 *see* Publications of the venice biennale, 1895-1977 (pvb)

Film und brettl – Berlin DE, 1919 apr-1923 aug – 1 – gw Mikrofilm [790]

Film und kino – Berlin, Muenchen DE, 1918-1919 26 jan – 1 – gw Mikrofilm [790]

Film und presse – Berlin DE, 1920 jul-1921 – 1r – 1 – gw Mikrofilm [790]

Film und wissen – Berlin DE, 1920 1 apr-sep – 1 – (with suppls: der industrielle werbefilm 1920 n4-6; der volks- und jugendfilm 1920 n6) – gw Mikrofilm [790]

Film user – Croydon. 1964-1971 (1) 1971-1971 (5) – ISSN: 0015-1459 – mf#1343 – us UMI ProQuest [790]

Film weekly, 1928-1939 – oct 1928-sep 1939 [mf ed Chadwyck-Healey] – 235mf – 9 – uk Chadwyck [790]

Film world : a current study of international films and filmfolk – v1-6, no. 4. 1964-70 – 1 – us AMS Press [790]

Film world and a-v news magazine – Los Angeles. 1962-1966 (1) – mf#1517 – us UMI ProQuest [790]

Filma – Revue photo-phonocinematographique puis Revue cinematographique. Dir. A. Millo. Paris. juin-dec 1908, fevr 1917-mai 1921, 1924-juil 1928, 1930-sept 1936 – 1 – fr ACRPP [490]

Film-almanach – Berlin DE, 1917, 1920 – 1 – gw Mikrofilm [790]

Filmando janio / Carneiro, Milton – Curitiba, Brazil. 1961 – 1r – us UF Libraries [972]

Film-atelier : die zeitung fuer den produktionsfachmann – Berlin DE, 1929 10 apr-1936 15 mar – 1r – 1 – gw Mikrofilm [790]

Der filmbote – Wien (A), 1918 aug-1919 jun, 1919 aug-1926 – 19r – 1 – gw Mikrofilm [790]

Filmburg – Berlin DE, 1920 jan-2 feb – 1 – gw Mikrofilm [790]

Filmcritica – Rome. 1973-1980 (1) 1976-1980 (5) 1976-1980 (9) – ISSN: 0015-1513 – mf#6940 – us UMI ProQuest [790]

Film-echo – Wiesbaden DE, 1960 6 jan-1999 25 sep [gaps] – 51r – 1 – (title varies: 1962 9/10: film-echo, filmwoche) – gw Mikrofilm [790]

Filmer, Harry J *see* Usutu!

Film-express / ed by Redaktion der "Lichtbild-Buehne" – Berlin DE, 1920 oct-dec – 1 – gw Mikrofilm [790]

Filmforum – Emsdetten 1951-1960 n9 [gaps] [mf ed 2004] – 1r – 1 – gw Mikrofilm [790]

Film-funken – Wolfen DE, 1947 dec-1990 22 jun [gaps] – 14r – 1 – (veb filmfabrik) – gw Misc Inst [790]

Der filmhandel – Berlin DE, 1919 6 jul-1921 21 jan [gaps] – 1 – gw Mikrofilm [790]

Film-hoelle – Berlin DE, 1922, jan, jun, jul, dec, 1923 apr, may – 1 – gw Mikrofilm [790]

Filmjournalen – Stockholm, 1919-53 – 28r – 1 – sw Kungliga [790]

Filmkultura – Budapest. 1972-1973 (1) – ISSN: 0015-1580 – mf#7034 – us UMI ProQuest [790]

Filmkunst – Berlin DE, 1913 30 may-1 aug [gaps], 1919-20 – 2r – 1 – gw Mikrofilm [790]

Film-kurier – Berlin DE, 1919 30 apr-1944 29 sep – 37r – 1 – gw Mikrofilm [790]

Filmmakers – New York. 1977-1982 (1) 1977-1982 (5) 1977-1982 (9) – (cont: filmmakers newsletter) – ISSN: 0194-4339 – mf#7439,01 – us UMI ProQuest [790]

Filmmakers newsletter – New York. 1969-1977 (1) 1972-1977 (5) 1976-1977 (9) – (cont by: filmmakers) – ISSN: 0015-1610 – mf#7439 – us UMI ProQuest [790]

Film-nachrichten – Berlin DE, 1944 7 oct-1945 24 mar – 1r – 1 – gw Mikrofilm [790]

Das filmrecht *see* Erste internationale film-zeitung 1909

Films for television / Standard Rate & Data Service – 1953 jul 9-dec 9, 1954 jan 9-dec 25, 1955 jan 25-mar 25, 1955 apr-1962 sep/dec, 1963 jul-1964 jul – 12r – 1 – (cont: television rates and data) – mf#864227 – us WHS [380]

Films in review – New York. 1950-1997 (1) 1950-1997 (5) 1950-1997 (9) – ISSN: 0015-1688 – mf#12499 – us UMI ProQuest [790]

Filmschau – Berlin DE, 1919 25 oct, 20 dec, 1920 n4 – 1 – gw Mikrofilm [790]

Filmspiegel – 1970-1974, 1976-1983 – 437mf – 1 – gw Mikropress [790]

Filmtechnik – Berlin DE, 1929-44, 1947 oct-1948 jun – 5r – 1 – gw Mikrofilm [790]

Film-tribuene – Berlin DE, 1919 30 jun-1920 23 jan, 1920 14 nov-1921 22 nov – 1 – (with gaps) – gw Mikrofilm [790]

Die filmwelt – Wien (A), 1919 iss1-3, 1921 iss6-25, 1922 n1-1925 n11 – 2r – 1 – gw Mikrofilm [790]

Die filmwoche – Wien (A), 1913 16 mar-1918 21 sep – 7r – 1 – gw Mikrofilm [790]

Filmwoche *see* Film-echo

Filo del ensueno / Bauza, Guillermo – San Juan, Puerto Rico. 1962 – 1r – us UF Libraries [972]

Il filocolo (cima54) : la storia di florio e biancifiore. farbmikrofiche-edition der handschrift kassel, gesamthochschul-bibliothek, landesbibliothek und murhardsche bibliothek, 2° ms poet et roman 3 / Boccaccio, Giovanni – (mf ed 1999) – (= ser Codices illuminati medii aevi (cima) 54) – 49p on 11 color mf – 15 – €370.00 – 3-89219-054-2 – (int by michael dallapiazza) – gw Lengenfelder [090]

Filogeus and various italian poems *see* Sonetti e poesie italiane de vari autori

Filologicheskie zapiski – Washington. 1963-1988 (1) 1971-1988 (5) 1976-1988 (9) – 811mf – 9 – (missing: 1861, v1; 1864, v6; 1865, v4-6; 1866, v6; 1868, v4-6) – mf#1727 – ne IDC [077]

Filologisze szriftn – Vilnius LI, 1926-29 – 1r – 1 – (in yiddish & english. with: togblat [l'viv, ukraine] 1926) – us UMI ProQuest [939]

Filopanti, Quirico [pseud] *see* Bartolini e la cerrito; ossia, dell'onorare e premiare gli artisti

Filosofia cristiana. tomo 2: prolegomenos / Torre Isunza de Hita, Ramon – Madrid: Imp. y Lit. de Felipe Gonzalez Rojas, 1901. Tambien Tomo 3 – 9 – sp Bibl Santa Ana [240]

La filosofia de calderon en sus autos sacramentales / Frutos Cortes, Eugenio – Zaragoza: Institucion Fernando el Catolico de la Dip. Prov., 1952 – sp Bibl Santa Ana [440]

Filosofia de la naturaleza / Nieto y Serrano, M – Madrid, 1884 – 6mf – 9 – sp Cultura [100]

La filosofia de la naturaleza y la psicologia segun ibn hazm / Gomez Nogales, Salvador – Milano: Societa Editrice Vita e Pensiero, 1964 – 9 – sp Bibl Santa Ana [180]

Filosofia de la presencia humana / Caba, Pedro – Mexico: Editorial Herrero, 1961 – sp Bibl Santa Ana [100]

Filosofia de machado de assis / Coutinho, Afranio – Rio de Janeiro, Brazil. 1940 – 1r – us UF Libraries [972]

La filosofia del conocimiento de san agustin / Caba, Pedro – Madrid – sp Bibl Santa Ana [240]

La filosofia del no-ser en el pensamiento griego (anaximandro-platon) / Caba, Pedro – Madrid, 1957. Revista Crisis, Ano 4, no 13. Enero-Marzo 1957 – sp Bibl Santa Ana [100]

Filosofia e teologia della storia / Padovani, Umberto Antonio – Brescia: Morcelliana, 1953 – 1mf – 9 – 0-524-08128-X – mf#1993-9034 – us ATLA [100]

Filosofia en cuba / Vitier, Medardo – Mexico City? Mexico. 1948 – 1r – us UF Libraries [100]

Filosofia en el brasil / Gomez Robledo, Antonio – Mexico City? Mexico. 1946 – 1r – us UF Libraries [100]

La filosofia en la ciencia. ensayo sobre el concepto y condiciones de ambas... / Moreno Izquierdo, Juan – Madrid: Enrique Teodoro, 1882 – 1 – sp Bibl Santa Ana [100]

La filosofia juridica del profesor de asis garrote / Elias de Tejada Spinola, Francisco – Sevilla: Imp. Gonzalez-Cabanas, 1970 – 1 – sp Bibl Santa Ana [190]

La filosofia juridica en la espana actual / Elias de Tejada Spinola, Francisco – Madrid: Instituto Editorial Reus, 1949 – 1 – sp Bibl Santa Ana [190]

Filosofia moral / Aristotle – 1692. Derivada de Aristoteles de Manuel Thesauro, traducela Don Gomez de la Roca y Figueroa – 9 – sp Bibl Santa Ana [170]

Filosofia moral / Aristotle – 1715. Derivada de Aristoteles de Manuel Thesauro, traducela Don Gomez de la Roca y Figueroa – 9 – sp Bibl Santa Ana [170]

Filosofia moral / Marquez, Gabino – Madrid: Prensa Nueva, 4th ed 1927 – 1 – sp Bibl Santa Ana [170]

Filosofia moral / Rocha y Figueroa, Gomez – Lisboa: Antonio de Creesbeck de Mello, 1682 – 1 – sp Bibl Santa Ana [170]

Filosofia moral derivada de aristoteles / Tesauro, Emanuel – 1770, 1750, 1723 – 9 – sp Bibl Santa Ana [170]

Filosofia moral, derivada de la alta fuente del grande aristoteles / Tesauro, Emanuel – Madrid: Juan de Zuniga, 1733 – 1 – sp Bibl Santa Ana [170]

Filosofia moral derivada de la alta...del grande aristoteles / Thesauro, Manuel – Lisboa: Imp. de Antonio Craesbeeck Mello, 1682 – 1 – (traducida en espanol d iuan luis de orleans...) – sp Bibl Santa Ana [170]

Filosofia moral para la juventud espanola / Piquer, Andres – Madrid, 1775 – 12mf – 9 – sp Cultura [170]

Filosofia moral. tomo 2. la moralidad en particular, o sea, el derecho natural / Marquez, Gabino – Madrid: Escelicer, S.A. 5th ed 1942 – 1 – sp Bibl Santa Ana [170]

Filosofia y fisiologia comparadas en su historia con el criterio de la ciencia viviente... 3 vol / Nieto y Serrano, M – Madrid, 1889-1890 – 17mf – 9 – sp Cultura [612]

Filosofische und patriotische traeume eines menschenfreundes / Iselin, Isaak – Freiburg: s.n., 1755. 192p – 1 – us Wisconsin U Libr [320]

Filosofo y la comprension internacional / Agramonte Y Pichardo, Roberto Daniel – Habana, Cuba. 1950 – 1r – us UF Libraries [972]

Filosofskoe uchenie marksa / Plekhanov, Georgii Valentinovich – Moskva: OGIZ Gos sotsial'no-ekonomicheskoe isd-vo, 1933 [mf ed 2002] – 1r – 1 – (filmed with: osnovnye momenty dialekticheskogo protsessa poznaniia / g. obichkin (1933). incl bibl ref) – mf#5224 – us Wisconsin U Libr [120]

Filoz, Auguste Achille Hippolyte *see*
- Cambodge et siam

Fils de famille / Bayard, Jean-Francois-Alfred – Paris, France. 1853 – 1r – us UF Libraries [440]

Fils du bravo / Bouchardy, Joseph – Paris, France. 1843 – 1r – us UF Libraries [440]

Fils naturel / Dumas, Alexandre – Paris, France. 1858 – 1r – us UF Libraries [440]

Fils naturel / Dumas, Alexandre – Paris, France. 1925 – 1r – us UF Libraries [440]

Filsinger, Ernst B *see*
- Commercial travelers' guide to latin america
- Exporting to latin america

Filson club history quarterly – Louisville. 1973-2000 (1) 1973-2000 (5) 1973-2000 (9) – ISSN: 0015-1874 – mf#9324 – us UMI ProQuest [978]

Filson, FV *see* The new testament against its environment

Filson history quarterly – Louisville, 2001+ [1,5,9] – (cont: filson club history quarterly) – mf#9324,01 – us UMI ProQuest [978]

Filson, John *see* Histoire de kentucke, nouvelle colonie a l'ouest de la virginie

Filteau, Louis Honore *see* Genealogy of the family normandeau

Filtration and separation – Croydon. 1990+ (1,5,9) – ISSN: 0015-1882 – mf#42637,01 – us UMI ProQuest [660]

Filtsch, Eugen *see* Goethes stellung zur religion

Filtsch, Johann et al *see* Siebenbuergische quartalschrift

Fin – Miami, FL. 1983 jun – 1r – us UF Libraries [071]

Fin back – Seattle, WA. 1879-1881 (1) – mf#67105 – us UMI ProQuest [071]

Fin de la dominacion de espana en cuba / Torriente Y Peraza, Cosme De La – Habana, Cuba. 1948 – 1r – us UF Libraries [972]

Fin de la souverainete belge au congo / Ganshof Van Der Meersch, W J – Bruxelles, Belgium. 1963 – 1r – us UF Libraries [960]

La fin de l'empire espagnol d'amerique par marins andre / Bayle, Constantino – Madrid: Razon y Fe, 1922 – 1 – sp Bibl Santa Ana [972]

Fin de siecle / Bahr, Hermann – Berlin: A Zoberbier, 1891 [mf ed 1995] – 192p – 1 – mf#8920 – us Wisconsin U Libr [830]

Fin de siecle – Paris. 27 dec 1890-1910 – 1 – (subtitle varies. after 21 nov 1909 published as: le nouveau siecle. ex fin de siecle) – fr ACRPP [073]

La fin du mandat francais en syrie... / Jones, J M – Paris, 1938 – 2mf – 9 – mf#ILM-1924 – ne IDC [956]

La fin du paganisme : etude sur les dernieres luttes religieuses en occident au quatrieme siecle / Boissier, Gaston – Paris: Hachette, 1891 – 3mf – 9 – 0-7905-5514-X – mf#1988-1514 – us ATLA [240]

La fin d'un christianisme : trois conferences / Monod, Wilfred – Paris: Fischbacher, 1903 – 1mf – 9 – 0-8370-3825-1 – mf#1985-1825 – us ATLA [240]

La fin d'un traitre : roman canadien inedit / Feron, Jean – Montreal: publie par "Le Roman canadien" Editions Edouard Garand, 1930 [mf ed 1987] – (= ser Le Roman canadien) – 1mf – 9 – (ill by albert fournier) – mf#SEM105P848 – cn Bibl Nat [830]

Fin, feather and fur on the british columbia coast : containing descriptions of a few select sport districts in this chosen land for rod and gun / Tweedale, Aitken [comp] – Vancouver: Tower Pub Co, [1918 or 19] – 1mf – 9 – 0-665-99312-9 – mf#99312 – cn Canadiana [639]

Fin, fur and feather – Amherst, NS: C de L Black, [1893-1894?] – 9 – mf#P04293 – cn Canadiana [639]

Fin mot / Dandre, Paul [Pseud] – Paris, France. 1843 – 1r – us UF Libraries [440]

Fina Garcia, Francisco *see* Arcoiris

Fina, Jose Augusto *see* Tesoros de nuestra biblioteca nacional

Final act / Inter-American Conference (10th : 1954 : Caracas) – Caracas, Venezuela. 1957 – 1r – us UF Libraries [972]

Final appeal in matters of faith / Wiseman, Nicholas Patrick – London, England. 1850? – 1r – us UF Libraries [240]

Final call / World Community of al-Islam in the West – 1979 may-1987 jun, 1987 jul 15-1989 nov, dec 31, 1990 jul 23-dec 24, 1991 jan-dec, 1992 jan 6-dec 28, 1993 jan 11-dec 22, 1994 jan 5-dec 28, 1995 jan 11-dec 20, 1996 jan 6-dec 31, 1997 jan 7-jun 10, 1997 jul-dec, 1998 jan 6-jul 14, 1998 jul 21-dec 29, 1999 aug 3-dec 28, 1999 jan 5-may 4, 1999 may 11-jul 27, 2000 jan-jun – 17r – 1 – mf#1353088 – us WHS [260]

Final draft/only for life / American Friends Service Committee – n1-5 [1972 jun-nov] – 1r – 1 – (cont by: peacework) – mf#683799 – us WHS [320]

The final faith : a statement of the nature and authority of christianity as the religion of the world / Mackenzie, William Douglas – New York: Macmillan, 1910 – 1mf – 9 – 0-7905-7663-5 – mf#1989-0888 – us ATLA [240]

Final flight – v1 n1-7 [1969 may-dec], 1970 jan/feb-1971 feb – 1r – 1 – mf#679330 – us WHS [071]

Final judgment / Reed, Andrew – London, England. 1829 – 1r – us UF Libraries [240]

Final proof : g a p's newsletter / Group for Advertising Progress – 1972 jul – 1r – 1 – mf#4877604 – us WHS [650]

Final record books of the u.s. circuit court for west tennessee, 1808-1839, and of the u.s. circuit court for the middle district of tennessee, 1839-1865 / U.S. Circuit and District Courts – (= ser Records of district courts of the united states) – 10r – 1 – (with printed guide) – mf#M1212 – us Nat Archives [347]

Final record books of the u.s. district court for west tennessee, 1803-1839, and of the u.s. district court for the middle district of tennessee, 1839-1850; land claims records for west tennessee, 1807-1820 / U.S. District Court – (= ser Records of district courts of the united states) – 1r – 1 – (with printed guide) – mf#M1215 – us Nat Archives [324]

Final report of the commissioners of inquiry into the affairs of king's college university and upper canada college / University of King's College (Toronto, Ont) – Quebec: printed by Rollo Campbell, 1852 [mf ed 1995] – 5mf – 9 – mf#SEM105P1903 – cn Bibl Nat [378]

Final report of the drought investigation commission : october 1923 – Cape Town: Cape Times Govt Printers, 1923 – 1 – (filmed with: transvaal (province) local government commission report) – us CRL [960]

Final report of the plebiscite commission on the public information program and plebiscite on the future political status of the federated states of micronesia / Federated States of Micronesia – Kolonia, Ponape: the Commission, 15 Jul 1983 – 6mf – 9 – $9.00 – mf#LLMC 82-100H Title 8 – us LLMC [324]

Final report on nuernberg war crimes trials : final report to the secretary of the army on the nuernberg war crimes trials under control council law no. 10 / Taylor, Telford – Washington: GPO, 1949 – 4mf – 9 – $6.00 – mf#LLMC 97-007 – us LLMC [327]

Final report to the hon commissioners of public works : on the completion of the improvements in the north-east wing of the common gaol at montreal... / McGinn, Thomas – Montreal?: Salter & Ross, 1857 – 1mf – 9 – mf#36825 – cn Canadiana [365]

Final report, together with...minutes of proceedings / Papua-New Guinea. House of Assembly. Select Committee on Constitutional Development – Port Moresby, 1971. 32p. LL-2338 – 1 – us L of C Photodup [340]

Final reports of the u.s. strategic bombing survey, 1945-1947 / U.S. Strategic Bombing Survey – (= ser Records of the united statgic bombing survey) – 25r – 1 – (with printed guide) – mf#M1013 – us Nat Archives [355]

Final revolutionary war pension payment vouchers : georgia / U.S. Treasury Dept – (= ser Records of the accounting officers of the department of the treasury) – 6r – 1 – (with printed guide) – mf#M1746 – us Nat Archives [360]

Final rolls of citizens and freedmen of the five civilized tribes in indian territory (...on or before march 4 1907, with supplements dated sept 25 1914) / U.S. Dept of the Interior. Office of the Secretary – (= ser Records relating to census rolls and other enrollments) – 3r – 1 – mf#T529 – us Nat Archives [317]

Final statement to the indian penetration commission appointed by the union government of south africa, 1940-1941 / Natal Indian Association – [Durban: The Association, 1941] – 1 – us CRL [960]

Finale verslag van die kommissie van ondersoek na sekere organisasies 109=christelike instituut van suidelike afrika = Final report of the commission of inquiry into certain organisations. christian institute of southern africa – Pretoria: Staatsdrukker, 1975 – us CRL [360]

The finality of the christian religion / Foster, George Burman – Chicago: University of Chicago Press, 1906 – (= ser The Decennial Publications of the University of Chicago) – 2mf – 9 – 0-7905-3743-5 – (incl bibl ref) – mf#1989-0236 – us ATLA [240]

Financas e desenvolvimento – Washington. 1972-1973 [1,5,9] – ISSN: 0255-7622 – mf#6534 – us UMI ProQuest [332]

Finance – (= ser The Goldsmiths'-Kress Library of Economic Literature) – 564r – 1 – us Primary [332]

Finance – New York. 1972-1978 (1) 1972-1978 (5) 1975-1978 (9) – ISSN: 0015-1912 – mf#7527 – us UMI ProQuest [332]

Finance accounts of great britain for the year ended 5th january...1801-1817 – [mf ed Chadwyck-Healey] – (= ser British government publications...1801-1977) – 2r – 1 – uk Chadwyck [336]

Finance accounts of ireland for the year ended 5th january 1801-1817 – [mf ed Chadwyck-Healey] – (= ser British government publications...1801-1977) – 1r – 1 – uk Chadwyck [336]

Finance accounts of the united kingdom of great britain and ireland for the financial year ended 31st march...1818-1966 – [mf ed Chadwyck-Healey] – (= ser British government publications...1801-1977) – 9r – 1 – uk Chadwyck [336]

Finance and commerce – Shanghai, China. 1934-1940; 8 jan-7 may 1941 [wkly] – 11r – 1 – uk British Libr Newspaper [332]

Finance and development – Washington. 1964+ [1]; 1971+ [5,9] – ISSN: 0015-1947 – mf#1691 – us UMI ProQuest [332]

Finance anglaise – London, UK. 11 Jul 1891 – 1 – uk British Libr Newspaper [072]

Finance facts – Washington. 1958-1990 (1) 1975-1990 (5) 1975-1990 (9) – ISSN: 0015-1963 – mf#9608 – us UMI ProQuest [332]

La finance internationale et la guerre d'espagne / Bougouin, E – Paris, 1938. Fiche W 761. (Blodgett Collection of Spanish Civil War Pamphlets) – 9 – us Harvard [946]

Finance of the free church of scotland / Lewish, James – Edinburgh, Scotland. 1843 – 1r – us UF Libraries [242]

Finance sector circular keu / Commercial Advisory Foundation in Indonesia – Djakarta, 1970-1972 Nos 1-51 – 22mf – 9 – (missing: 1972(39, 50)) – mf#SE-1389 – ne IDC [959]

Finance week – Johannesburg. 1998+ (1) – mf#28186 – us UMI ProQuest [332]

Finances and developpement – Washington. 1972-1994 [1]; 1964-1980 [5]; 1972-1980 [9] – ISSN: 0430-473X – mf#6535 – us UMI ProQuest [332]

Finances communales et urbaines au congo belge / Parisis, Albert – Bruxelles, Belgium. 1960 – 1r – us UF Libraries [336]

Financial accountability and management – Oxford. 1985+ (1,5,9) – ISSN: 0267-4424 – mf#14842 – us UMI ProQuest [650]

The financial accounting standards board act : hearing...house of representatives, 107th congress, 2nd session, june 26 2002 / United States. Congress. House. Committee on Energy and Commerce. Subcommittee on Commerce, Trade, and Consumer Protection – Washington: US GPO 2002 [mf ed 2002] – 1mf – 9 – 0-16-068789-6 – (incl bibl ref) – us GPO [343]

Financial analysts journal – New York. 1971+ (1) 1945+ (5) 1975+ (9) – ISSN: 0015-198X – mf#6114 – us UMI ProQuest [332]

Financial and accounting systems – Boston. 1990-1991 (1,5,9) – (cont: journal of accounting and edp) – ISSN: 1053-2579 – mf#14375,01 – us UMI ProQuest [650]

Financial and economic annual of japan, 1901-1929 – 3r – 1 – $105.00 in US $40.00r outside – (in english) – mf#L9400080 – Dist. us Scholarly Res – us L of C Photodup [332]

Financial and economic survey / Brown, G A – Kingston, Jamaica. 1959 – 1r – us UF Libraries [332]

Financial and product sponsorship within athletic training programs / Sanderson, Natalie – Ball State University, 1996 – 1mf – 9 – mf#PE 3669 – us Kinesology [790]

Financial burden of the war on india / Vakil, Chandulal Nagindas – Bombay: CN Vakil, 1943 – (= ser Samp: indian books) – us CRL [330]

The financial collapse of enron : hearing...house of representatives, 107th congress, 2nd session / United States. Congress. House. Committee on Energy and Commerce. Subcommittee on Oversight and Investigations – Washington: US GPO 2002 [mf ed 2002] – 9 – us GPO [345]

Financial control and compliance manual for presidential primary candidates receiving public financing – Washington: The Commission, July 1979 (rev 1983, 1987 and 1992) – 3mf – 9 – $4.50 – mf#LLMC 95-025 – us LLMC [340]

Financial crises : their causes and effects / Carey, Henry Charles – Philadelphia: H C Baird, 1864 – 1mf – 9 – mf#27093 – cn Canadiana [330]

Financial developments in modern india, 1860-1924 / Vakil, Chandulal Nagindas – Bombay: DB Taraporevala Sons & Co, [1924] – (= ser Samp: indian books) – (foreword by sir basil p blackett) – us CRL [332]

Financial executive – New York. 1932-1984 (1) 1971-1984 (5) 1975-1984 (9) – (cont by: fe) – ISSN: 0015-1998 – mf#355 – us UMI ProQuest [650]

Financial executive – Morristown. 1987+ (1,5,9) – (cont: fe) – ISSN: 0895-4186 – mf#14465,01 – us UMI ProQuest [650]

Financial executives institute bulletin – Morristown. 1970-1980 (1) 1976-1980 (5) 1976-1980 (9) – mf#8640 – us UMI ProQuest [332]

The financial expert : a novel / Narayan, R K – London: Methuen & Co, 1952 – (= ser Samp: indian books) – (int by graham greene) – us CRL [830]

Financial express – Bombay, India. 1962-1993 – 125r – 1 – us L of C Photodup [079]

Financial facts concerning alachua county public s... / Alachua County (FL) Board Of Public Instruction – Gainesville, FL. 1933 – 1r – us UF Libraries [336]

Financial history of the united states / Dewey, Davis R – 7th ed. New York: Longmans, Green & Co, 1920 – 7mf – 9 – $10.50 – mf#LLMC 92-148 – us LLMC [346]

Financial management – London, 2000+ [1,5,9] – (cont: management accounting) – ISSN: 1471-9185 – mf#11861,02 – us UMI ProQuest [350]

Financial management – Tampa. 1972+ (1,5,9) – ISSN: 0046-3892 – mf#11664 – us UMI ProQuest [650]

Financial manager – Boston. 1988-1990 (1,5,9) – (cont: corporate accounting. cont by: small business controller) – ISSN: 1040-0842 – mf#16636 – us UMI ProQuest [650]

Financial managers' statement: fms – Chicago. 1986-1992 (1) 1986-1992 (5) 1986-1992 (9) – ISSN: 0887-4808 – mf#15943,01 – us UMI ProQuest [650]

Financial market trends – Paris. 1983+ (1,5,9) – ISSN: 0378-651X – mf#14982 – us UMI ProQuest [332]

Financial measures for india : speech of the right hon james wilson, delivered before the legislative council of calcutta...18th feb 1860 / Wilson, James – London 1860 – (= ser 19th c british colonization) – 1mf – 9 – mf#1.1.3663 – uk Chadwyck [350]

Financial news – London. -d. Jan 1885-Dec 1887. (6 reels) – 1 – uk British Libr Newspaper [330]

Financial news – Providence, RI. 1890-1891 (1) – mf#66304 – us UMI ProQuest [071]

The financial observer : malawi's own business paper – Blantyre, Malawi: West Publ [v5 n12-v7 n5 (aug 1993-may 18 1995)] (semimthly) – 1r – 1 – us CRL [332]

The financial outlook in canada : an address delivered before the canadian club, toronto, december 4th, 1913 / Paish, George – Toronto: Warwick Bro's & Rutter, 1913 – 1mf – 9 – 0-665-97867-7 – mf#97867 – cn Canadiana [336]

Financial post magazine – Toronto. 1990-1993 (1,5,9) – (cont: financial post moneywise) – ISSN: 1182-0713 – mf#16327,05 – us UMI ProQuest [332]

Financial post moneywise – Toronto. 1989-1990 (1,5,9) – (cont by: financial post magazine) – ISSN: 0843-2317 – mf#16327,04 – us UMI ProQuest [332]

Financial post of canada – Woodstock, Canada. -w. 12 jun 1909-6 sep 1913; 24 jan 1914-10 sep 1920; 2 jan-21 jun 1921; 1 jul 1921-29 may 1925 – 19 1/2r – 1 – uk British Libr Newspaper [072]

Financial problems of indian states under federation / Khan, Abdul Wajid – London: Jarrolds Publishers, 1935 – (= ser Samp: indian books) – (pref by hugh dalton) – us CRL [332]

Financial records / Pratt, Hilton – undated. Records of Cottonwood Ranch, Sheridan County, KS – 1 – us Kansas [630]

Financial records *see* Bethel baptist church

Financial records 1914-1915, 1921-1930 / Evangelical Lutheran Joint Synod of Ohio and Other States. Board of Foreign Missions – [mf ed 2004] – 1r – 1 – (comprises primarily business correspondence of treasurers of the board; mostly in english with small amount in german) – mf#xa0090r – us ATLA [242]

Financial records and journals / Ridgecrest. North Carolina. Ridgecrest Baptist Assembly – 1909-12, 1911-12; hotel building fund, 1918-20; blue prints; guide map for Blue Mont (now Ridgecrest). 390p – 1 – us Southern Baptist [242]

Financial reformer – Liverpool, England. -w. 1 July-Dec 1878. 4 reels – 1 – uk British Libr Newspaper [072]

Financial report of the children's home society of wisconsin, 1897-1898 / Children's Home Society of Wisconsin – 1897-1898 – 1r – 1 – mf#2792936 – us WHS [362]

Financial report of the secretary-treasurer / International Printing Pressmen and Assistants' Union of North America – 1916 may 23/1919 jun 30, 1919 sep 1/nov 30, 1928 jul 20/1940 may 31 – 1r – 1 – (cont: annual report of secretary-treasurer, international printing pressmen and assistants' union of north america) – mf#3389956 – us WHS [331]

Financial reports / Massachusetts. Comptroller's Division – 1922-77. 154 fiches. (Harvard Law School Library Collection.) – 9 – us Harvard Law [324]

Financial review : the official publication of the eastern finance association – Tallahassee. 1978-2000 (1,5,9) – ISSN: 0732-8516 – mf#14474,01 – us UMI ProQuest [332]

Financial services advisor – Lexington. 1999+ (1) 1999+ (5) 1999+ (9) – (cont: life and health insurance sales) – mf#7745,03 – us UMI ProQuest [360]

Financial services review – Greenwich. 1998+ (1,5,9) – ISSN: 1057-0810 – mf#19770 – us UMI ProQuest [332]

Financial statistics – Norwich. 1975-1992 (1) 1975-1992 (5) 1975-1992 (9) – ISSN: 0436-3663 – mf#9883 – us UMI ProQuest [332]

The financial system of india / Chand, Gyan – London: Kegan Paul, Trench, Trnbner & Co, 1926 – (= ser Samp: indian books) – (foreword by edward hilton young) – us CRL [332]

Financial times : for the decision-maker – Toronto, Canada. 1981-1986 (1) 1981-1986 (5) 1981-1986 (9) – (cont: financial times of canada) – ISSN: 0711-5938 – mf#3048,01 – us UMI ProQuest [332]

Financial times – Frankfurt/M DE, 1979- – ca 10r/yr – 1 – gw Misc Inst [074]

Financial times – London, England.1888- mthly updates – 1 – (index 1981- available. backfile available. ft not publ in jun/jul 1983) – us Primary [072]

Financial times – Montreal, Canada. 6 dec 1913-27 dec 1913; jan-jun 1914; 1915-1917; 5 jan-27 dec 1918; 24 may-27 dec 1919; 1920-21; 6 jul-28 dec 1951; 4 jan-25 jul 1952 – 11r – 1 – uk British Libr Newspaper [071]

Financial times – Toronto, Canada. 1986-1995 (1) 1986-1995 (5) 1986-1995 (9) – ISSN: 0839-2188 – mf#3048,02 – us UMI ProQuest [332]

The financial times – 1888- – (considered the premier business and financial newspaper of europe, this paper is divided into sections on world trade, company news, share information, marketing, small businesses, insurance and shipping) – us Primary [072]

Financial times / german edition – Hamburg DE, 2000 21 feb – ca 9r/yr – 1 – gw Misc Inst [332]

Financial times of canada – Toronto, Canada. 1912-1981 (1) 1968-1981 (5) 1979-1981 (9) – (cont by: financial times: for the decision-maker) – ISSN: 0015-2056 – mf#3048 – us UMI ProQuest [332]

Financial trend – Dallas. 1970-1985 (1) 1978-1985 (5,9) – (cont by: american law review) – ISSN: 0040-4195 – mf#9313 – us UMI ProQuest [332]

Financial world – New York. 1902-1997 [1]; 1966-1997 [5]; 1975-1997 [9] – ISSN: 0015-2064 – mf#2009 – us UMI ProQuest [332]

Financier franco-anglais – London, UK. 14 Sept 1899-14 Jun 1900 – 1 – uk British Libr Newspaper [072]

El financiero – Mexico City, Mexico. 1990-June 1991 – 18r – 1 – us L of C Photodup [079]

Financing colorado public schools *see* University of denver theses

Financing schools through establishment of sound external relations / Khoetha, Lefau Nathaniel – Pretoria: Vista University 2002 [mf ed 2002] – 2mf – 9 – (abstract in afrikaans & english; incl bibl ref) – mf#mfm15232 – sa Unisa [370]

Finans vision – Stockholm, Sweden. 2002-03 – 8r – 1 – sw Kungliga [078]

Finansi i kredit – Sofia. May 1956-Dec. 1962 – 1 – us NY Public [332]

Die finansiele posisie van gades na egskeiding met spesifieke verwysing na die "clean break"-beginsel / Engelbrecht, Ockert Michiel – Uni of South Africa 2001 [mf ed Pretoria: UNISA 2000] – 2mf – 9 – (incl bibl ref; summary in english) – mf#mfm14723 – sa Unisa [346]

Finansirovanie vneshnei torgovli / Frei, Lazar' Isaevich et al; ed by Stefanov, N V – Moskva: Vneshtorgizdat, 1935 [mf ed 2002] – 1r – 1 – (Filmed with: morozovskaia stachka 1885 / s predisloviem v i nevskogo (1925). incl bibl ref) – mf#5232 – us Wisconsin U Libr [380]

Finansovaia entsiklopediia / ed by Bogolepov, D P et al – Ed 2. M, L, 1927 – 11mf – 9 – mf#REF-150 – ne IDC [332]

Finansovaia Gazeta *see* Denezhnaia reforma, snizhenie tsen, zarabotnaia plata

Finansovaia gazeta – Moscow. no.1-25. Nov. 16 1937-June 13 1941 – 1 – 94.00 – us L of C Photodup [947]

Finansovaia gazeta – Moscow, Russia, 1918-22 – 25r – 1 – us UMI ProQuest [077]

Finansovaia politika sovetskoi vlasti / Potiaev, A – Pg, 1919 – 1mf – 9 – mf#REF-19 – ne IDC [332]

Finansovaia politika v period velikoi oktiabr'skoi sotsialisticheskoi revoliutsii / Rivkin, B – M, 1957 – 4mf – 9 – mf#REF-14 – ne IDC [332]

Finansovaia politika za period s dekabria 1920 g po dekabr' 1921 g : (otchet k 9 vseross sezdu sovetov) – M, 1921 – 2mf – 9 – mf#REF-26 – ne IDC [332]

Finansovaia reforma v rossii / Genzel', P P & Sokolov, A – Pg, 1917 – 3mf – 9 – mf#REF-156 – ne IDC [332]

Finansovaia reforma v rossii : otkuda u nas gosudarstvo beret den'gi i na chto ikh raskhoduet / Ozerov, I Kh – M, 1906 – 3mf – 9 – mf#REF-155 – ne IDC [332]

Finansovoe ozdorovlenie ekonomiki : opyt nepa / Kaz'min, A I – M, 1990 – 3mf – 9 – mf#REF-28 – ne IDC [332]

Finansovo-kreditnye problemy v period natsionalizatsii promyshlennosti v sssr / Mekhanik, S – M, 1957 – 4mf – 9 – mf#REF-33 – ne IDC [332]

Finansovo-kreditnyi slovar' / ed by D'iachenko, VP – M, 1961-1964. 2v – 24mf – 9 – mf#REF-151 – ne IDC [332]

Finansovo-statisticheskii atlas rossii / Antropov, P A – Spb, 1898 – 3mf – 9 – mf#REF-147 – ne IDC [332]

Finansovye problemy planovogo khoziaistva – 1922-31 – 1 – us L of C Photodup [947]

Finansovyi kapital v rossii nakanune mirovoi voiny : opyt istoriko-ekonomicheskogo issledovaniia sistemy finansovogo kapitala v rossii / Vanag, N N – Ed 3. M, 1930 – 5mf – 9 – mf#REF-157 – ne IDC [332]

Finansovyi kontrol' / Pontovich, E E – L, 1928 – 2mf – 9 – mf#REF-47 – ne IDC [332]

Finansovyi kontrol' v dorevoliutsionnoi rossii : ocherki istorii / Koniaev, A – M, 1959 – 3mf – 9 – mf#REF-225 – ne IDC [332]

Finanstidningen – Stockholm, Sweden. 1989- – 1 – sw Kungliga [078]

Finansy i novaia ekonomicheskaia politika : lektsiia, prochitannaia 15 oktiabria 1921 goda / Preobrazhenskii, E – M, 1921 – 1mf – 9 – mf#REF-22 – ne IDC [332]

Finansy posle oktiabria / Sokol'nikov, G – [M], 1923 – 1mf – 9 – mf#REF-15 – ne IDC [332]

Finansy rossii : vedomosti [po otchetam gosudarstvennogo kontrolia] – Spb, 1907 – 3mf – 9 – mf#REF-217 – ne IDC [332]

Finansy rossii 19 stoletiia : istoriia – statistika / Bliokh, I S – Spb, 1882. 4v – 25mf – 9 – mf#REF-152 – ne IDC [332]

La finanza internacional y la guerra de espana / Bougouin, E – Paris, 1938. Fiche W 762. (Blodgett Collection of Spanish Civil War Pamphlets) – 9 – us Harvard [946]

La finanza italiana – Rome, Italy. 22 nov 1924-14 jun 1930 – 1 – (imperfect) – mf#m.f.831 – uk British Libr Newspaper [074]

Finanzas publicas y el desarrollo economico de gua... / Adler, John Hans – Mexico City? Mexico. 1952 – 1r – us UF Libraries [336]

Finanzas y desarrollo – Washington. 1972-1996 (1) 1964-1996 (5) 1972-1996 (9) – ISSN: 0250-7447 – mf#6537 – us UMI ProQuest [332]

Finanz-chronik *see* Reuter's finanz-chronik

Die finanzen frankreichs / Kaufmann, Richard von – Leipzig 1882 [mf ed Hildesheim 1995-98] – 1v on 10mf – 9 – €100.00 – 3-487-25929-X – gw Olms [332]

Der finanzer : erzaehlung vom bodensee / Achleitner, Arthur – Leipzig: M Hesse, [19–] [mf ed 1988] – 92p – 1 – mf#6934 n13 – us Wisconsin U Libr [880]

Finanzierung und entwicklung – Washington. 1972-1973 (1) 1970-1972 (5) 1972-1972 (9) – ISSN: 0250-7439 – mf#6536 – us UMI ProQuest [332]

Die finanzskandale des kaiserreichs / Heinig, Kurt – Berlin: Verlag fuer Sozialwissenschaft, 1925. 80p. bibliog. index – 1 – us Wisconsin U Libr [943]

Finat *see* The divan project

Finati, Giovanni *see* Narrative of the life and adventures of giovanni finati, native of ferrara

Finauer, Peter Paul *see*
- Magazin fuer die neueste litteratur, kenntnis bayerscher schriftsteller, diplomatik, genealogie und heraldik, topographie, dann ueberhaupts fuer die alte und neuere geschichte in baiern
- Miscellanien fuer das schul- und erziehungswesen in baiern

Finazzi, Giovanni *see* Dell'immacolato concepimento di maria e della sua dogmatica definizione

Finch, A Elley *see* Erasmus

Finch, Arthur Elley *see* The victories of science in its warfare with superstition

Finch, G *see*
- The sketch of the romish controversy, pt 1
- The sketch of the romish controversy, pt 2

Finch, George *see* Rome, the babylon of the apocalypse

Finch, John *see*
- The natural boundaries of empires
- To south africa and back
- Travels in the united states of america and canada

Finch, Laura M *see* An assessment of the factor validity of the precompetitive stress inventory

Finch, Merry B *see* The effects of exercise on myocardial capillary bed and connective tissue in senescent rats

Fincham, Jack E *see* Journal of pharmacoepidemiology

Finchley advertiser – Barnet, England. 3 apr 1986-10 mar 1994 [mf 1986-] – 1 – (cont: finchley local advertiser) – mf#[1986-:]sp624 – uk British Libr Newspaper [072]

Finchley free press – London, 9 nov 1895-1960; 1970-31 jul 1986 [wkly] – 94r – 1 – (incorp with: barnet press and publ as: the press) – uk British Libr Newspaper [072]

Finchley guardian – London UK, 1 nov 1902-15 apr 1905 – 2r – 1 – uk British Libr Newspaper [072]

Finchley local advertiser – Barnet, England. 10 jan 1985-27 mar 1986 – n62-125 – 1 – (cont as: finchley advertiser) – mf#[1986-:]SP624 – uk British Libr Newspaper [072]

Finchley press – London, UK. jan-jul 1986 [wkly] – 2 1/2r – 1 – uk British Libr Newspaper [072]

Finchley telegraph & barnet times – Barnet, England 1 jan 1892-2 jun 1896 – 1 – (cont: barnet times [13 dec 1890-25 dec 1891]; cont by: barnet times & finchley telegraph [19 jun 1896-1 nov 1907]) – uk British Libr Newspaper [072]

Finchley times *see* Finchley times and guardian

Finchley times and guardian – London. 20 jan 1961-64; 1982-84 [wkly] – 18r – 1 – (aka: finchley times) – uk British Libr Newspaper [072]

Finck, William J *see* Lutheran landmarks and pioneers in america

Finck, William John *see* Lutheran landmarks and pioneers in america

Finckenstein, Ottfried, Graf *see*
- Daemmerung
- Maenner am brunnen

Finckh, Ludwig *see*
- Ahnenbuechlein
- Der ahnenring
- Der bodenseher
- Das deutsche ahnenbuch
- Der deutsche finckh
- Der goettliche ruf
- Das goldene erbe
- Herzog und vogt
- Hinterm gartenbusch
- Die jakobsleiter
- Die kaiserin, der koenig und ihr offizier
- Rapunzel
- Das vogelnest

Findeisen, Kurt Arnold *see* Ich blas auf gruenen halmen

Findel, J G *see* Die bauhuette

Findel, Joseph Gabriel *see*
- Geest en vorm der vrijmetselarij
- Histoire de la franc-maconnerie

Fin-de-siecle symbolist and avant-garde periodicals – 236r – 1 – (based on "a chronological list of the most important periodicals in the symbolist movement" compiled by kenneth cornell in his book, the symbolist movement. new haven, 1951, this coll provides insight into the context in which symbolism and other literary currents emerged) – mf#C39-28550 – us Primary [410]

Finding aids to national archives photographs relating to the third german reich / U.S. National Archives and Records Service – (= ser Records of the national archives and records administration) – 73mf – 9 – mf#M1137 – us Nat Archives [324]

The finding of the cross = Etudes sur les souvenirs de la passion / Combes, Louis de – New York: Benziger, 1907 – (= ser The International Catholic Library) – 1mf – 9 – 0-8370-7614-5 – (incl bibl ref and index. in english) – mf#1986-1614 – us ATLA [240]

Finding the lost story of the cowgirls / Robison, Kristenne M – 1999 – 1mf – 9 – $4.00 – mf#PE 3939 – us Kinesology [790]

Finding the missing link / Broom, Robert – London, England. 1950 – 1r – us UF Libraries [890]

Finding the way / Miller, James Russell – New York: Thomas Y Crowell, c1904 – 1mf – 9 – 0-8370-7311-1 – mf#1986-1311 – us ATLA [240]

The findings of the continuation committee conferences held in asia, 1912-1913 : arranged by topics – New York: Student Volunteer Movement for Foreign Missions, 1913 – 1mf – 9 – 0-7905-8137-X – mf#1988-6084 – us ATLA [240]

Findings on five year movement *see* Hua tung chiao hui wu nien yun tung chi hua (ccm254)

Findlay, Alexander George *see* A description and list of the lighthouses of the world, 1863

Findlay, George *see* Irish railways and state purchase

Findlay, George G *see* Wesley's world parish

Findlay, George Gillanders *see*
- The apostle paul
- Christian doctrine and morals viewed in their connexion
- The epistle to the galatians
- The epistles of paul the apostle
- Fellowship in the life eternal

Findlay, Mary Grace *see* Wesley's world parish

Findlay, William *see* The book of the prophet ezekiel

Fine art / Victoria and Albert Museum. London – (= ser Fine art and design in the victoria and albert museum) – 152mf – 9 – $1125.00 – 0-907006-35-3 – (over 9000 reproductions; with ind to artists) – uk Mindata [700]

Fine art and design in the victoria and albert museum / Victoria and Albert Museum. London – 489mf – 9 – $3450.00 coll – 0-907006-50-7 – (pictorial record of fine & graphic art & design in 3 sects; titles also listed individually) – uk Mindata [700]

Fine art as a branch of university study : inaugural address / Brown, Gerard Baldwin – Edinburgh: David Douglas, 1881 – (= ser 19th c art & architecture) – 1mf – 9 – mf#4.1.151 – uk Chadwyck [700]

The fine art circular and print collector's manual – London [1857] – (= ser 19th c art & architecture) – 5mf – 9 – mf#4.1.436 – uk Chadwyck [760]

Fine art society catalogue – London, 1878-1976 – (= ser Art exhibition catalogues on microfiche) – 85 catalogues on 91mf – 9 – £670.00 – (individual titles not listed separately) – uk Chadwyck [700]

The fine arts in italy in their religious aspect : letters from rome, naples, pisa, &c = des beaux-arts en italie au point de vue religieux / Coquerel, Athanase – London: Edward T Whitfield, 1859 – 1mf – 9 – 0-7905-4333-8 – (in english) – mf#1988-0333 – us ATLA [240]

The fine arts in italy in their religious aspect : letters from rome, naples, pisa... / Coquerel, Athanase – London: Edward T. Whitfield, 1859 – 1mf – (with an appendix on the iconography of the immaculate conception) – us ATLA [700]

Fine arts journal – London. 1846-1847 (1) – mf#5547 – us UMI ProQuest [700]

The fine arts of the english school / Britton, John – [London] 1812 – (= ser 19th c art & architecture) – 3mf – 9 – mf#4.2.1507 – uk Chadwyck [700]

Fine arts quarterly review – London. 1863-1867 (1) – mf#5548 – us UMI ProQuest [700]

Fine arts quarterly review – London. v. 1-3, n.s. v. 1-2. May 1863-June 1867 – 1 – us NY Public [700]

Fine, Deborah L *see* The influence of fitness-oriented physical activity on the physical self-perception and global self-worth of boys and girls

Fine prints / Wedmore, Frederick – London 1897 – (= ser 19th c art & architecture) – 3mf – 9 – mf#4.2.531 – uk Chadwyck [760]

Fine, Sidney *see* Frank murphy in world war i

Fine tuning : the membership magazine of the channel 10/36 friends / WMVS [Television station : Milwaukee, WI] et al – 1983 feb-1985 jun, 1985 jul-1989 jun – 2r – 1 – mf#966092 – us WHS [380]

Finegan, Mark T *see* The status of continuous ecg monitoring in phase 2 cardiac rehabilitation programs

Fineman, Helene H *see* The papers of albert gallatin

Finer, Herman *see* Hsien tai cheng fu chih li lun yu shih chi

Fines y utilidad de la sindicacion ganadera por... / Villamor Angulo, Hilario – Caceres: Tip. Garcia Floriano, 1938 – 1 – sp Bibl Santa Ana [946]

Le finezze de pennelli italiani, ammirate e studiate da girupe(perugino) sotto la scorta e disciplina del genio di raffaello d'urbino / Scaramuccia, L – Pavia, [1674] – 4mf – 9 – mf#0-1172 – ne IDC [700]

Finffakher mord / Shevits, S E – London, England. 1905 – 1r – us UF Libraries [939]

Fingal herald (1898) – Fingal, ND: Albert O Wold. 1898; -v40 n32 jan 27 1938 (wkly) – 1 – (missing: 1905: aug 17-24; official paper of barnes co 1906, 1908, 1910, 1912, 1914, 1916; official paper village of fingal 1934; iss with: mirror (nome, n.d) 1905; incl sect: lucca ledger 1905-1909; merged with: nome informer to form: fingal herald and the nome informer) – mf#03949-03959 – us North Dakota [071]

Fingal herald (1898) *see* Mirror

Fingal herald (1939) – Fingal, ND: L J Morth. v42 n19 oct 26 1939-v45 n2 jun 25 1942 (wkly) – 1 – (cont: fingal herald and the nome informer) – mf#03960-03961 – us North Dakota [071]

Fingal herald and the nome informer : [official paper for the villages of nome and fingal 1938-1939] – Fingal, ND: L J Morth. v40 n33 feb 3 1938-v42 n18 oct 19 1939 (wkly) – 1 – (formed by the union of: fingal herald (fingal, nd: 1898) and: nome informer; cont by: fingal herald (fingal, nd: 1939)) – mf#03959-03960 – us North Dakota [071]

Finger – v1 n1-v2 n4 [[1968? oct-1970? feb]] – 1r – 1 – mf#766139 – us WHS [380]

Finger, Charles Joseph *see* Historic crimes and criminals

Finger lakes times – Geneva, NY. 1977-2000 (1) – mf#61626 – us UMI ProQuest [071]

Finger, Richard *see* Heinrich von kleists geheimnis

Fingerprint and identification magazine – Chicago. 1975-1977 (1) 1975-1977 (5) 1975-1977 (9) – ISSN: 0015-2323 – mf#10932 – us UMI ProQuest [360]

Fininberg, Ezra *see*
- Geshikhtes
- Lirik, 1920-1940

Finishers' management – Glenview. 1972-1991 (1) 1972-1981 (5) 1974-1981 (9) – ISSN: 0015-2358 – mf#7679 – us UMI ProQuest [660]

Finishing – Watford. 1985-1985 (1,5,9) – ISSN: 0264-2506 – mf#14426,01 – us UMI ProQuest [660]

A finishing stroke to the high claims of ecclesiastical sovereignty : in reply to the address of a meetingof male members of the methodist episcopal church church in baltimore / Shinn, Asa – Baltimore: Toy 1827 – 1r – 1 – $35.00 – mf#um-15 – us Commission [242]

Finite elements in analysis and design – Amsterdam. 1985-1992 (1,5,9) – ISSN: 0168-874X – mf#42549 – us UMI ProQuest [000]

Fink, George *see* Mich hungert

Fink, J *see* Ueber die politischen unterhandlungen des churfuersten johann wilhelm von der pfalz zur befreiung der christenheit in armenien vom joche der unglaeubigen, von 1698 bis 1705

Fink, R A *see* Women in the church

Fink, Reinhard *see* Universitaetsbibliothek leipzig

FIRELIGHT

Fink, Susan E *see* Environmental law in a developing country, botswana
Finke, Edmund *see* Der tod vor dem spiegel
Finke, G *see* Das schreien der steine
Finke, George *see* The verdict of the monuments
Finke, Heinrich *see*
– Aus den tagen bonifaz 8
– Forschungen und quellen zur geschichte des konstanzer konzils
– Konzilienstudien zur geschichte des 13. jahrhunderts
– Der madonnenmaler franz ittenbach
– Papsttum und untergang des templerordens
– Ueber friedrich und dorothea schlegel
Finkel, Robert J *see* A revision of the heart disease locus of control scale
Finkelstein, L *see* Corpus tannaiticum
Finkelstein, Leo *see* Grunt-shtrikhn fun der yidisher filozofye
Finkelstein, Simon Isaac *see* 'En shim'on
Finkenwaerder nachrichten – Hamburg DE, 1879-1934 – 99r – 1 – gw Misc Inst [914]
Finland *see* Suomen virallinen lehti
Finland. Statisticka Centralbyran *see* Bidrag till finlands officiela statistik 1885-1914
Finland. Tilastollinen Paatoimisto *see* Suomen tilastollinen vuosikirja arsbok for finland 1879-1970
Finlason, C E *see* Nobody in mashonaland
Finlason, W F *see* Report of the trial and preliminary proceedings in the case of the...
Finlason, William Francis *see* Report of the trial and preliminary proceedings in the case of the queen on the prosecution of g. achilli v. dr. newman
Finlay, Carlos Juan *see* Trabajos selectos del dr carlos j finlay
Finlay, John *see*
– Scottish historical and romantic ballads, chiefly ancient
Finlayson, G *see* The mission to siam and hue the capital of cochin china
Finlayson, George *see*
– Die gesandtschafts-reise nach siam und hue, der hauptstadt von cochinchina, in den jahren 1821 bis 1822
– The mission to siam, and hue, the capital of cochin china, in the years 1821-2
Finley, Gordon E *see* Adoption quarterly
Finley, James B *see* Sketches of western methodism
Finley, James Bradley *see*
– Autobiography of rev james b finley
– History of the wyandott mission at upper sandusky, ohio
– Memorials of prison life
Finley, Martha *see* Elsie dinsmore
Finleys of virginia – s.l, s.l? 193-? – 1r – us UF Libraries [978]
Finn, D E *see* Poetry in rhodesia
Finn, George *see* Datus
Finn sanomat – Goteborg, Sweden. 1979-85 – 1 – sw Kungliga [078]
Finnan, Alan Pierson *see* Variant uses of aleph in iqisa
Finnegan, Dana *see* Journal of chemical dependency treatment
Finnegan, Gregory Allan [comp] *see* Africana archives
Finnell, Rueben Ashford *see* A study of the book of revelation
Finnelly, W *see*
– Clark and finnelly's reports
– House of lords cases (clark and finnelly)
Finn-Enotaevskii, A E *see* Nashi banki i birzha
Finney, Charles G *see* Memoirs
Finney, Charles Grandison *see*
– The character, claims and practical workings of freemasonry
– Lectures on revivals of religion
– Lectures on systematic theology
– Lectures to professing christians
– Letters on revivals of religion
– Memoirs of rev. charles g. finney
– Sermons on important subjects
– Sermons on the way of salvation
– Sermons on various subjects
– Views of sanctification
Finney, Charles Herbert *see* British theorists of the nineteenth century
Finney, Ross Lee *see* The american public school
Finnische novelle / Kurzbach, Herbert – Muenchen: Deutscher Volks-Verlag, 1943 – 1r – 1 – us Wisconsin U Libr [830]
Finnish Baptist Mission Union of America *see* Missionsposten
Finnish chemical letters – Helsinki. 1980-1980 (1,5,9) – ISSN: 0303-4100 – mf#11282 – us UMI ProQuest [540]
Finnish Federation [US] *see* Tyolaisnainen
Finnish newspapers – ca 30,000r – 1 – (the collection increases annually with ca 900r) – fi Helsinki [079]
Finnish Organization of Canada *see* Finnish-canadian play and operetta manuscript collection
Finnish periodicals on microfilm – Helsinki: Helsinki University Library, 1994 – ca 215r 10,000mf – 1,9 – fi Helsinki [073]
Finnish Socialist Organization of the United States *see* Kuukausi-julkaisu
Finnish-canadian play and operetta manuscript collection / Finnish Organization of Canada – Toronto: McLaren Micropublishing 1974 (mf ed) – 76mf – 9 – Can$575.00 – (51 items in finnish; these works were written by first generation immigrants to canada and the us; coll also of interest to historians of radical political movts; with printed finding aid) – cn McLaren [790]
Finn's leinster journal *see* Kilkenny journal [finn's leinster journal]
Finnveden – Jonkoping, Sweden. 1981-89 – 1 – sw Kungliga [078]
Finnveden fredag – Vaernamo, 1992- – 9 – sw Kungliga [078]
Finnveden nu – Vaernamo, Sweden. 2003- – 1 – sw Kungliga [078]
Finnveden onsdag – Vaernamo, 1992-97 – 6r – 1 – sw Kungliga [078]
Finot, Ed *see* Port-royal et magny
Finotti, Joseph M *see* The mystery of the wizard clip (smithfield, w. va.)
Finotti, Joseph Maria *see* Bibliographia catholica americana, pt 1
Finsbury – London, England. Published by the Finsbury Independent Labour Party. -m. Jan-May 1900 1/4r – 1 – uk British Libr Newspaper [072]
Finsbury and holborn guardian *see* Holborn guardian and bloomsbury chronicle
Finsbury weekly news and chronicle *see* The clerkenwell chronicle, st luke's examiner, holborn reporter and north london observer
Finsbury weekly news and clerkenwell chronicle and st lukes examiner *see* The clerkenwell chronicle, st luke's examiner, holborn reporter and north london observer
Finsch, Otto *see* Samoafahrten
Finska kemistsamfundets meddelanden – Helsinki. 1973-1973 [1] – ISSN: 0015-2498 – mf#8796 – us UMI ProQuest [540]
Finskii vestnik *see* Ucheno-literaturnyi zhurnal
Finsler, G *see*
– Die chronik des bernhard wyss 1519-1530
– Ulrich zwingli
Finsler, Georg *see* Kirchliche statistik der reformirten schweiz
Finsler, Rudolf *see* Darstellung und kritik der ansicht wellhausens von geschichte und religion des alten testaments
Finster, Reinhard et al *see* A dual concordance to leibniz's philosophische schriften, teil 2
Finsterwalder, Florian *see* Untersuchungen zur plasmapolymerisation und zur methanol-diffusion ionenleitender polymerelektrolytmembranen
Finzer, Michael *see* Erlernte hilflosigkeit
Fioletov, N N *see* Tserkov i gosudarstvo po sovetskomu pravu
Fior angelico di musica / Angelo da Picitono – In Vinegia: Per Agostino Bindoni 1557 [mf ed 19–] – 1r – 1 – mf#film 100 – us Sibley [780]
Fiore, Dolores Ackel *see* Ruben dario in search of inspiration
The fiorello la guardia papers – Rare Books and Manuscripts Division: The New York Public Library, Astor, Lenox and Tilden Foundations 1995 – ca 54r – 1 – ca $4,590.00 – (with guide) – mf#D3340 – us NY Public [320]
Fioretti, petites fleurs de s francois d'assise : legendes du moyen age / Francois d'Assise, Saint – [Montreal?: s.n.] 1901 [mf ed 1996] – 5mf – 9 – 0-665-79321-9 – mf#79321 – cn Canadiana [241]
Fiorillo, Federigo *see*
– Five pieces selected from the celebrated opera of la clemenza di tito
– Six quatuors concertants pour deux violons, alto et basse, op 6
– Six quatuors concertants pour flute, violon, alto et basse, oeuvre 4
– Trois duo concertants, pour deux violons...opera 31
Fiorillo, J D *see* Geschichte der zeichnenden kuenste von ihrer wiederauflebung bis auf die neuesten zeiten
Fiorvanti, Leonardo *see* Three exact pieces..
Fipa and related peoples of south-west tanzania / Willis, Roy G – London, England. 1966 – 1r – us UF Libraries [960]
Firbank, Mr *see*
– La cybelline
– Mr caverley's slow minuet
Firdawsi *see* Rostem und suhrab
Firdousi and the shahnama : a study of the great persian epic of the homer of the east / Vachha, Phirozeshah Benjani – Bombay: New Book Co, 1950 – (= ser Samp: indian books) – us CRL [490]
Fire! / Students for a Democratic Society [US] – 1969 nov7-1970 jan 30 – 1r – 1 – (cont: new left notes [chicago, il: 1966]; cont by: new left notes [boston, ma]) – mf#713508 – us WHS [320]
Fire!! : devoted to younger negro artists – New York. v1 n1. 1926 [all publ] – (= ser Black journals, series 1) – 1mf – 9 – $20.00 – us UPA [700]
Fire and frost : the meadow lea tragedy / Dezell, Robert – Toronto: W Briggs, 1907 – 2mf – 9 – 0-665-65484-7 – mf#65484 – cn Canadiana [920]
Fire and materials – Chichester. 1976+ (1,5,9) – ISSN: 0308-0501 – mf#13306 – us UMI ProQuest [540]
Fire and sword in shansi : the story of the martyrdom of foreigners and chinese christians / Edwards, E H – new enl ed. Edinburgh: Oliphant Anderson & Ferrier, 1907 – 4mf – 9 – 0-524-07868-8 – (incl ind) – mf#1991-3413 – us ATLA [951]
Fire and sword in shansi : the story of the martyrdom of foreigners and chinese christians / Edwards, E H – New York: Fleming H Revell [1903] [mf ed 1995] – (= ser Yale coll) – 325p (ill) – 1 – 0-524-09013-0 – (int note by alexander maclaren) – mf#1995-0013 – us ATLA [951]
Fire and sword in the caucasus / Villari, Luigi – London: T.F. Unwin, 1906. 347p.64 pl – 1 – us Wisconsin U Libr [947]
Fire and sword in the sudan : a personal narrative of fighting and serving the dervishes, 1879-1895 = Feuer und schwert im sudan / Slatin, Rudolf Carl, Freiherr von – London: E Arnold, 1896 – 8mf – 9 – 0-524-07954-4 – (in english) – mf#1991-0204 – us ATLA [960]
Fire bringer – v1 n1-6/7 [1971 nov-1973 jul] – 1r – 1 – mf#1056276 – us WHS [071]
Fire chief – Atlanta. 1989+ (1,5,9) – ISSN: 0015-2552 – mf#17088 – us UMI ProQuest [360]
Fire command – Boston. 1933-1981 (1) 1971-1981 (5) 1973-1981 (9) – (cont by: fire service today: a publication of the public fire protection division of the nfpa) – ISSN: 0015-2560 – mf#8037 – us UMI ProQuest [360]
Fire command – Quincy. 1984-1990 (1) 1984-1990 (5) 1984-1990 (9) – (cont: fire service today: a publication of the public fire protection division of the nfpa) – ISSN: 0746-9586 – mf#8037,02 – us UMI ProQuest [360]
Fire control notes – Washington. 1936-1973 (1) 1970-1973 (5) (9) – (cont by: fire management) – ISSN: 0015-2579 – mf#5767 – us UMI ProQuest [634]
Fire department news / International Association of Fire Fighters – iss 229-319 [1978 jul/aug-1986 may] – 1r – 1 – mf#998656 – us WHS [360]
Fire engineering – Tulsa. 1926+ (1) 1971+ (5) 1976+ (9) – ISSN: 0015-2587 – mf#1971 – us UMI ProQuest [628]
De fire evangeliers harmoni : eller, jesu historie efter de fire evangelier i kronologisk sammenstilling / Ylvisaker, Johannes – Decorah, Iowa: Den Norske Synodes Forlag, 1896 [mf ed 1992] – 160p on 1mf – 9 – 0-524-05207-7 – mf#1992-0340 – us ATLA [225]
Fire fighter / International Association of Fire Fighters – v23 n1-v32 n1 [1973 jan/feb-1982 mar] – 1r – 1 – (cont by: capital city fire fighter) – mf#647037 – us WHS [360]
Fire from heaven / Rees, Seth Cook – Cincinnati, OH: God's Revivalist Office, c1899 [mf ed 1993] – (= ser Society of friends (quakers) coll) – 1mf – 9 – 0-524-06440-7 – mf#1991-2562 – us ATLA [240]
Fire from strange altars / Fradenburgh, Jason Nelson – Cincinnati: Cranston and Stowe; New York: Hunt and Eaton, c1891 – 1mf – 9 – 0-524-01481-7 – mf#1990-2457 – us ATLA [200]
Fire in the woods / De Mille, James – Boston: Lee & Shepard, 1871? – 4mf – 9 – mf#05999 – cn Canadiana [830]
Fire insurance maps from the sanborn map company archives : late 19th century to 1900 (p-w) – 603r entire coll (a-w) – 1 – $123,890.00 entire coll $340.00r – (pennsylvania 1935-90 40r $10,275 isbn 1-55655-323-4. rhode island 1941-90 4r $1040 isbn 1-55655-328-5. south carolina 1946-74 3r $760 isbn 1-55655-334-X. south dakota 1r $260 isbn 1-55655-365-X. tennessee 1947-69 7r $1795 isbn 1-55655-342-0. texas 1940-86 18r $4620 isbn 1-55655-335-8. utah 2r $510 isbn 1-55655-366-8. vermont 1946-89 1r $260 isbn 1-55655-327-7. virginia 1946-89 11r $2810 isbn 1-55655-319-6. washington 1912-89 13r $3345 isbn 1-55655-350-1. west virginia 1946-89 5r $1295 isbn 1-55655-320-X. wisconsin 1945-86 9r $2320 isbn 1-55655-338-2. wyoming 1r $260 isbn 1-55655-367-6) – us UPA [978]
Fire insurance maps from the sanborn map company archives : late 19th century to 1990 (a-o) – 603r entire coll (a-w) – 1 – $123,890.00 entire coll $340.00r – (alabama 1945-81 6r $1540 isbn 1-55655-343-9. alaska 1r $260 isbn 1-55655-353-6. arizona 2r $510 isbn 1-55655-354-4. arkansas 1944-70 3r $760 isbn 1-55655-351-X. california 88r $22,620 isbn 1-55655-348-X. colorado 5r $1295 isbn 1-55655-355-2. connecticut 1945-90 7r $1795 isbn 1-55655-337-4. delaware 1951-88 1r $260 isbn 1-55655-321-8. district of columbia 1984-85 2r $510 isbn 1-55655-318-8. florida 1945-87 9r $2320 isbn 1-55655-332-3. georgia 1940-78 6r $1540 isbn 1-55655-333-1. hawaii 9r $2320 isbn 1-55655-356-0. idaho 2r $510 isbn 1-55655-357-9. illinois 1938-89 36r $9245 isbn 1-55655-337-4. indiana 1933-80 13r $3345 isbn 1-55655-340-4. iowa 1946-80 6r $1540 isbn 1-55655-346-3. kansas 5r $1295 isbn 1-55655-358-7. kentcuky 1946-92 7r $1795 isbn 1-55655-341-2. louisiana 1940-94 9r $2320 isbn 1-55655-352-8. maine 1943-88 3r $760 isbn 1-55655-329-3. maryland 1946-82 9r $2320 isbn 1-55655-317-X. massachusetts 1940-90 16r isbn 1-55655-325-0 $4115. michigan 1946-92 24r $6165 isbn 1-55655-339-0. minnesota 1946-69 8r $2065 isbn 1-55655-345-5. mississippi 1945-70 2r $510 isbn 1-55655-344-7. missouri 1933-92 13r $3345 isbn 1-55655-347-1. montana 6r $1540 isbn 1-55655-359-5. nebraska 3r $760 isbn 1-55655-360-9. nevada 1r $260 isbn 1-55655-362-5. new hampshire 1937-89 3r $760 isbn 1-55655-326-9. new jersey 1945-91 24r $6165 isbn 1-55655-331-5. new mexico 2r $510 isbn 1-55655- 361-7. new york 1909-90 32r $8225 isbn 1-55655-324-2. new york city 1931-94 78r $20,125 isbn 1-55655-368-4. north carolina 1945-79 7r $1795 isbn 1-55655-322-6. north dakota 1r $260 isbn 1-55655-363-3. ohio 1946-89 26r $6670 isbn 1-55655-336-6. oklahoma 5r $1295 isbn 1-55655-364-1. oregon 1912-78 8r $2065 isbn 1-55655-349-8) – us UPA [978]
Fire journal – Boston. 1907-1990 [1]; 1972-1990 [5]; 1973-1990 [9] – ISSN: 0015-2617 – mf#8035 – us UMI ProQuest [360]
The fire journal – Toronto: Toronto Pub Co, [1879-18–?] – 9 – mf#P04429 – cn Canadiana [360]
Fire management – Washington. 1973-1976 (1) 1973-1976 (5) 1976-1976 (9) – (cont: fire control notes. cont by: fire management notes) – ISSN: 0095-5450 – mf#5767,01 – us UMI ProQuest [634]
Fire management notes – Washington. 1976-1999 (1) 1976-1999 (5) 1976-1999 (9) – (cont: fire management. cont by: fire management today) – ISSN: 0194-214X – mf#5767,02 – us UMI ProQuest [634]
Fire management today – Washington. 2000+ (1) 2000+ (5) 2000+ (9) – (cont: fire management notes) – mf#5767,03 – us UMI ProQuest [634]
Fire news – Boston. 1985-1988 (1) 1985-1987 (5) 1985-1987 (9) – (cont by: nfpa update) – ISSN: 0015-2625 – mf#14476,01 – us UMI ProQuest [360]
The fire of god's anger : or, light from the old testament upon the new testament teaching concerning future punishment / Baker, Lewis Carter – Philadelphia, PA: Office of "Words of Reconciliation", 1887 – 1mf – 9 – 0-7905-0903-2 – (incl bibl ref and indexes) – mf#1987-0903 – us ATLA [220]
Fire prevention – London. 1973-1976 (1) 1975-1976 (5) 1975-1976 (9) – mf#9723 – us UMI ProQuest [360]
Fire safety journal – Lausanne. 1977+ (1) 1977+ (5) 1987+ (9) – ISSN: 0379-7112 – mf#42018 – us UMI ProQuest [360]
Fire service newsletter / Wisconsin Fire Service Training et al – 1967 apr-1979 jul – 1r – 1 – mf#601709 – us WHS [360]
Fire service today : a publication of the public fire protection division of the nfpa – Quincy. 1981-1983 (1) 1981-1983 (5) 1981-1983 (9) – (cont: fire command. cont by: fire command) – ISSN: 0279-3563 – mf#8037,01 – us UMI ProQuest [360]
Fire, snow and water : or, life in the great lone land: a tale of northern canada / Ellis, Edward Sylvester – Toronto: Musson, c1908 – (= ser Canada series (musson)) – 4mf – 9 – 0-665-74173-1 – (ill by louis r dougherty) – mf#74173 – cn Canadiana [830]
Fire technology – Quincy. 1965+ (1) 1972+ (5) 1973+ (9) – ISSN: 0015-2684 – mf#8036 – us UMI ProQuest [360]
Fire this time – 1992 mar – 1r – 1 – mf#5327595 – us WHS [071]
[Firebaugh-] firebaugh-mendota journal – CA. 1991- – 8r – 1 – $480.00 (subs $50y) – mf#B02235 – us Library Micro [073]
Firebrand : "for the burning away of the cobwebs of ignorance and superstition" – Portland OR: Firebrand Pub Cttee, [wkly] – 1 – (began jan 27 1895. ceased sep 1897; cont by: free society (san francisco, ca: 1897)) – us Oregon Lib [320]
Fireflies / Tagore, Rabindranath – New York: Macmillan Co, 1928 – (= ser Samp: indian books) – (decorations by boris artzybasheff) – us CRL [490]
Firelands farmer / Huron Co. New London – may 1977-dec 1987 [biwkly, wkly] – 12r – 1 – mf#B29750-29761 – us Ohio Hist [071]
Firelight entertainments / Soifer, Margaret K – Brooklyn, NY. 1935 – 1r – us UF Libraries [025]

Firemen and oilers' journal / International Brotherhood of Firemen and Oilers – v19 n11-v30 n1 [1919 oct-1930 jan] – 1r – 1 – (cont: stationary firemen's journal) – mf#3125330 – us WHS [366]

Firemen's magazine / Brotherhood of Locomotive Firemen [US] – v6 n1-v9 n12 [1882 jan-1885 dec] – 1r – 1 – (cont: locomotive firemen's monthly magazine; cont by: locomotive firemen's magazine) – mf#2596282 – us WHS [360]

Firenze citt...nobilissima illustrata / Migliore, F I del. – Firenze, 1684 – 8mf – 9 – mf#O-958 – ne IDC [700]

Fires and fire-proof construction / Baillairge, Charles P Florent – S.l: s.n, 1898? – 1mf – 9 – mf#60882 – cn Canadiana [628]

Fireside book of chess / Chernev, Irving – New York, NY. 1949 – 1r – us UF Libraries [790]

Fireside chats / Franklin D. Roosevelt Philatelic Society – v12-v20 n6, v21-v23 n2 [1977-1988 summer] – 1r – 1 – mf#1056280 – us WHS [730]

Fire-side missionary / Youatt, Elizabeth – London, England. 1851 – 1r – us UF Libraries [240]

Fireside stories of ireland / Kennedy, Patrick – Dublin, Ireland. 1870 – 1r – us UF Libraries [830]

Fireside tales from the north / Savory, Phyllis – Cape Town, South Africa. 1966 – 1r – us UF Libraries [390]

Fireside tales of the hare and his friends / Savory, Phyllis – Cape Town, South Africa. 1965 – 1r – us UF Libraries [390]

Firework / Women in Fire Suppression [Organization] – 1982 nov-1986 jan – 1r – 1 – mf#1043797 – us WHS [366]

Firing line / JOIN Community Union – v1 n2-10 [1967 jun 15-oct 15], 1968 jan 16 – 1r – 1 – mf#629213 – us WHS [071]

Firing line – n12-57 [1985 nov-1989 aug], – 1r – 1 – mf#1704077 – us WHS [071]

Firishta, Abdulmacid ibn *see* Ishq-name

Firishtah, Muhammad Qasim Hindu Shah Astarabadi *see* Ferishta's history of dekkan

The firm foundation of the christian faith : a handbook of christian evidences for sunday school teachers / Beet, Joseph Agar – London: Wesleyan Methodist Sunday School Union, 1891Beltsville, Md: NCR Corp, 1978 (2mf); Evanston: American Theol Lib Assoc, 1984 (1mf) – 9 – 0-8370-0845-X – mf#1984-4223 – us ATLA [240]

Firmas del ciclo heroico / Rosa Y Meano, Andres Eloy De La – Lima, Peru. 1938 – 1r – us UF Libraries [972]

Firme de sangre / Branly, Roberto – Habana, Cuba. 1962 – 1r – us UF Libraries [972]

Firmery, Joseph Leon *see* Etude sur la vie et les oeuvres de jean paul frederic richter

Firmes / Aradillas Agudo, Antonio & Castaneda, P – Madrid: Editorial Atenas, 1964 – 1 – sp Bibl Santa Ana [338]

Firmici materni consultationes zacchaei et apollonii (fp39/1) / ed by Morin, G – 1935 – (= ser Florilegium patristicum (fp)) – €7.00 – ne Slangenburg [240]

Firmin, Antenor *see*
- Defense
- Diplomates et diplomatie
- France et haiti
- Haiti au point de vue politique
- Lettres de saint thomas
- M roosevelt, president des etats-unis et la repub...

Firminger, Walter Kelly *see*
- The alterations in the ordinal of 1662
- The fifth report from the select committee of the house of commons on the affairs of the east india company
- Religion

The first 100 years of the first baptist church, denton, texas / Floyd, LP – 1858-1958. 470p – 1 – us Southern Baptist [242]

First account / First Wisconsin National Bank of Milwaukee – 1977 jan/feb-1983 winter – 1r – 1 – mf#1477279 – us WHS [332]

A first account of labour organisation in south africa / Gitsham, Ernest & Trembath, James F – Durban: Printed by E P & Commercial Printing Co, 1926 – 1 – us CRL [331]

The first african baptist informer – Savannah GA: First African Baptist Church. v1 n9 (jun 1943) [mf ed 2005] – 1r – 1 – (reel also incl 2005-s055: years' report of the colored y w christian association) – mf#2005-s055 – us ATLA [242]

The first age of christianity and the church = Christenthum und kirche in der zeit der grundlegung / Doellinger, Johann Joseph Ignaz von – 4th ed. London: Gibbings, 1906 – 2mf – 9 – 0-7905-5652-9 – (incl bibl ref. in english) – mf#1988-1652 – us ATLA [240]

The first american catholic missionary congress : held under the auspices of the catholic church extension society of the usa – Chicago, IL: J S Hyland [1909?] [mf ed 1986] – 2mf – 9 – 0-8370-6880-0 – mf#1986-0880 – us ATLA [241]

First american edition of the works of the rev d w cahill : the highly distinguished irish priest, patriot and scholar... – New York: A Franchi, 1854 [mf ed 1986] – 1mf – 9 – 0-8370-9688-X – mf#1986-3688 – us ATLA [241]

The first and chief groundes of architecture used in all the auncient and famous monymentes / Shute, J – London, 1563. Facs ed 1912 – 1mf – 9 – mf#OA-50 – ne IDC [720]

First and fastest / Shore Line Interurban Historical Society – v1 n1-v3 n4 [1984 spring-1986/1987 winter] – 1r – 1 – mf#1214471 – us WHS [978]

First and fundamental truths : being a treatise on metaphysics / McCosh, James – New York: Scribner, 1889 – 1mf – 9 – 0-7905-9806-X – mf#1989-1531 – us ATLA [110]

The first and last words of a pastor to his people / Cartwright, Robert David – [Kingston, Ont?: s.n.] 1843 [mf ed 1984] – 1mf – 9 – 0-665-28139-0 – mf#28139 – cn Canadiana [242]

The first and second books of chronicles / ed by Hughes-Games, A – London: J M Dent; Philadelphia: J B Lippincott 1902 [mf ed 1989] – (= ser The temple bible) – 1mf – 9 – 0-7905-1826-0 – mf#1987-1826 – us ATLA [221]

The first and second books of kings / ed by Robertson, James – London: J M Dent; Philadelphia: J B Lippincott 1902 [mf ed 1989] – (= ser The temple bible) – 1mf – 9 – 0-7905-1913-5 – mf#1987-1913 – us ATLA [221]

The first and second books of samuel – London: J M Dent; Philadelphia: J B Lippincott 1902 [mf ed 1989] – (= ser The temple bible) – 1mf – 9 – 0-7905-1849-X – mf#1987-1849 – us ATLA [221]

The first and second epistles to the corinthians : with notes critical and practical / Sadler, Michael Ferrebee – London: G Bell, 1897 – 2mf – 9 – 0-524-04923-8 – mf#1992-0266 – us ATLA [227]

First and second interim reports of the commission of inquiry into certain organisations : south africa commission of inquiry into certain organisations. – [s.l: s.n., 197-] – us CRL [360]

First and second preliminary report of the egyptian expedition / Breasted, James Henry – Chicago, 1906-1907, v23; 1908-1909, v25 – 4mf – 9 – mf#H-108 – ne IDC [916]

First and second reports of governor shoup's committee on child welfare legislation for colorado / Colorado. Committee on Child Welfare Legislation – Denver, Eastwood 1923? 160 p. LL-600 – 1 – us L of C Photodup [305]

First and second reports of the royal commission on technical instruction, 1882 and 1884 : command n3171, 3981 to 3981/4 – 32mf – 9 – mf#87119 – uk Microform Academic [324]

First and second reports of the select committee of the legislative assembly : appointed to inquire into the public income and expenditure of the province: third session, third parliament of canada, 1850 / Canada (Province). Parlement. Assemblee legislative – Toronto: Printed by Lovell & Gibson, 1850 [mf ed 1982] – 3mf – 9 – mf#SEM105P140 – cn Bibl Nat [336]

First annual catalog, 1901-02 / Theodore Harris Institute. Pineville, Kentucky – 24p – 1 – 5.00 – us Southern Baptist [242]

The first annual report of the trinity church district visiting society, 1843 / Trinity Church (Montreal, Quebec). District Visiting Society – [Montreal?: s.n, 1843?] [mf ed 1983] – 1mf – 9 – 0-665-41585-0 – mf#41585 – cn Canadiana [240]

First annual report of the young men's christian association, portsmouth, va : embracing the reports of the president and treasurer... – Philadelphia: JM Wilson, 1857 [mf ed 1994] – 1mf – 9 – 0-524-08843-8 – mf#1993-1102 – us ATLA [360]

First annual review of the copper mining industry of lake superior: containing a carefully and concisely written account of early explorations and discoveries in the lake superior copper region / Russell, James – Marquette, Mich.: Mining Journal Co., 1899 – 1 – us Wisconsin U Libr [622]

The first attempts of the dutch to trade in china, and settlement at tay wan / Rechteren, Z van – London, 1745-1747. v3 – 2mf – 9 – mf#A-271 – ne IDC [915]

First Baptist Church *see*
- Our church home
- Visitor
- Visitor of first baptist church

First baptist church – Columbus, MS – 1 – $516.33 – (armstrong society, wmu record book 1889-1907, minutes of contents of sunday school lessons 1894-1896; church minutes and member rolls 1886-1927, baraca sunday school 1914-29, wmu minutes 1911-14; church bulletins 1925-78) – mf#6977 – us Southern Baptist [242]

First baptist church – Athens, TN. 250p. 1871-1904 – 1 – $11.25 – mf#6517 – us Southern Baptist [242]

First baptist church – Birmingham. 1972-1998 (1) 1974-1998 (5) 1974-1998 (9) – 2r – 1 – $88.79 – mf#6654 – us Southern Baptist [242]

First baptist church – Bombay. 1980-1980 (1) 1980-1980 (5) 1980-1980 (9) – 1r – 1 – $16.83 – mf#6611 – us Southern Baptist [242]

First baptist church : church minutes – Baxley, GA. 1986-93 – 1 – $41.13 – mf#6852 – us Southern Baptist [242]

First baptist church : church minutes – Bunkie, LA. 2921p. 1892-1994 – 1 – $131.45 – mf#6983 – us Southern Baptist [242]

First baptist church – Franklinton, LA – 1 – $257.90 – (church minutes 1923-1981, newsletters, church history, miscellaneous information, deacon's meeting minutes 1941-1994, wmu annuals, and scrapbooks) – mf#7081 – us Southern Baptist [242]

First baptist church : church minutes and membership rolls – Tecumseh, OK. 1914-88 – 1 – $50.04 – mf#6810 – us Southern Baptist [242]

First baptist church – TN. Lenoir City – 1 – $982.40 – (church minutes, deacon minutes, membership records, financial records, church bulletins, newsletter, mid-weekly, uniform associational letters. covers yrs 1905-97) – mf#6712 – us Southern Baptist [242]

First baptist church : church minutes, records, and financial reports – Athens, GA. 1914-97 – 1 – $90.59 – (incl members received, transferred, and erased, and member deaths) – mf#7010 – us Southern Baptist [242]

First baptist church : church records – Dresden, TN. 1910-45;1951-83 – 1 – $67.50 – mf#6726 – us Southern Baptist [242]

First baptist church – Knoxville, TN. 7506p. 1843-1952 – 1 – $532.00 – (church records 1950-1993. 4327p) – mf#0224 – us Southern Baptist [242]

First baptist church – Iva, Anderson Co, SC. 1637p. 1906-16, 1922-49, 1967-72, 1976-90 – (= ser Mizpah) – 1 – $73.67 – (deacons' minutes, 1916-24, 1927, 1958-68, 1971-73, 1975-90. wmu minutes 1926-66, 1973-90. formerly: mizpah 1890-1912) – mf#5003-46a – us Southern Baptist [242]

First baptist church – Birmingham. 1972-1998 (1) 1972-1998 (5) 1972-1998 (9) – 5r – 1 – $305.28 – (deacons' minutes 1946-53. wmu minutes 1927-jan 1965. bulletins, newsletters, other misc items) – mf#6651 – us Southern Baptist [242]

First baptist church – Montmorenci, SC. 1934-70, 1974-89 – 1 – $81.27 – (deacons' minutes 1952-59) – mf#5003-3c – us Southern Baptist [242]

First baptist church – Conway, Horry Co, SC. 836p. 1979-80 – 1 – $37.62 – (deacons' minutes 1983-87) – mf#6498 – us Southern Baptist [242]

First baptist church – Columbia, Richland Co, SC. 1809-40, 1870-1930; 1927-49; 1984-86; 1986-88 – 1 – $140.22 – (deacons' minutes 1985-87 3116p. reel 1 1656p $74.52 reel 2 734p $33.03 reel 3 278p $12.51 reel 4 448p $20.16) – mf#5003-56 – us Southern Baptist [242]

First baptist church – Dillon Co, SC. 1434p. 1891-1902, 1905-16, 1924-41, 1947-72, 1973-89 – 1 – $64.53 – mf#5003-40a&c – us Southern Baptist [242]

First baptist church – Eastman, GA. 770p. 1879-1910, 1912-42, misc papers 1894-1942 (770p); sep 1942-aug 1990 (1619p) – 1 – $107.51 – mf#6706 – us Southern Baptist [242]

First baptist church – Elton, LA. 1910-96 – 1 – $107.87 – mf#2108 – us Southern Baptist [242]

First baptist church – Taylor, Greenville Co, SC. 426p. 1864-1917 – (= ser Chick's Springs Baptist Church) – 1 – $19.17 – (formerly: chick's springs baptist church) – mf#6481 – us Southern Baptist [242]

First baptist church – Laurens, SC. 385p. 1961-86 – 1 – $17.33 – (historical sketches 1834-1934, 1834-1959. dedication service brochure may 18 1958. membership rolls 1898-jan 1987) – mf#0982 – us Southern Baptist [242]

First baptist church : history, church minutes and membership rolls – Wickliffe, KY. 1034p. 1901-91 – 1 – $46.53 – mf#6724 – us Southern Baptist [242]

First baptist church – Nashville, TN. 15,921p. 1820-1966, dec 1985-98 – 1 – $716.45 – (incl business meeting minutes, deacons meeting minutes, the evangel, and financial reports) – mf#0294 – us Southern Baptist [242]

First baptist church – Cross Hill, SC. 1381p. 1879-1945, 1968-78 – 1 – $62.15 – (lacking: 1974) – mf#5618 – us Southern Baptist [242]

First baptist church – Nashville, IL. 4322p. aug 1873-sep 1990 – 1 – $194.49 – (lacking: nov 1900-mar 1903, jan-nov 1909) – mf#5830 – us Southern Baptist [242]

First baptist church : membership, minutes, history, scrapbook – Irvington, AL. 1958-83 – 1 – $17.46 – mf#6897 – us Southern Baptist [242]

First baptist church : minutes – Homer, LA. 1851-1903 – 1 – $30.06 – mf#6848 – us Southern Baptist [242]

First baptist church – Darlington, SC – 1 – $135.32 – (minutes 1856-96, 1923-52, miscellaneous records 1940-46. deacon's meetings 1949-59. wmu minutes 1895-1908, and and misc materials 2735p. deacon's minutes apr 8 1968-89 272p) – mf#1659 – us Southern Baptist [242]

First baptist church – Alma, GA – 1 – $14.76 – (minutes 1895-1911, 1917-1945. church history (formerly patrick's chapel) 1895-1995. minutes 1895-1911 not from original copy, but photocopied) – mf#6876 – us Southern Baptist [242]

First baptist church : newsletter, helping words – Augusta, GA. v1-5. mar 1887-jan 1893 – 1 – $47.39 – mf#6531 – us Southern Baptist [242]

First baptist church – Savannah, GA. Chatham Co. 2619p. 1805-22, 1831-36, 1852-1922, 1925-53, 1966-86 – 1 – $117.86 – mf#6487 – us Southern Baptist [242]

First baptist church – Venice, FL. 5388p. 1934-1995 – 1 – $242.46 – mf#4355 – us Southern Baptist [242]

First baptist church – Washington. 1972+ (1) 1972+ (5) 1975+ (9) – 1r – 1 – $24.93 – mf#6608 – us Southern Baptist [242]

First baptist church – Gaffney, SC. Cherokee Co. 649p. october 1980-aug 1992 – 1 – $29.21 – (wmu minutes 1977, 1981-88. deacons' minutes, nov 1986-sep 1992) – mf#6381 – us Southern Baptist [242]

First baptist church, farmersville, texas centennial story / Rike, Charles Jesse – 1865-1965. 1961 – 1 – 5.00 – us Southern Baptist [242]

First Baptist Church [Madison, WI] *see* First church visitor

First baptist church. minutes – Seymour, TN. jul 1893-oct 1994 – 1 – $81.14 – mf#6890 – us Southern Baptist [242]

First Baptist Church, Ottawa, KS *see* Records

First baptist church pulpit – Yarmouth, NS: [s.n., 1889?-18–?] – 9 – mf#P05057 – cn Canadiana [242]

A first book for children / Murray, Lindley – Quebec: P Sinclair, 1856 [mf ed 1984] – 1mf – 9 – 0-665-38145-X – mf#38145 – cn Canadiana [420]

A first book in organic evolution / Shute, Daniel Kerfoot – London, 1899 – (= ser 19th c evolution & creation) – 4mf – 9 – mf#1.1.2879 – uk Chadwyck [574]

The first book of kings / Farrar, Frederic William – New York: A C Armstrong, 1893 – (= ser The expositor's bible) – 2mf – 9 – 0-8370-2404-8 – mf#1985-0404 – us ATLA [221]

The first book of maccabees : with introduction and notes / Fairweather, WilliamBlack, John Sutherland – Cambridge: University Press; New York: Macmillan [distributor], 1897 – (= ser The Cambridge Bible for Schools and Colleges) – 1mf – 9 – 0-8370-6736-7 – mf#1986-0736 – us ATLA [221]

The first book of marcus paulus venetus : or of master marco polo, a gentleman of venice, his voyages – London, 1625-1626. v3 – 2mf – 9 – mf#HT-679 – ne IDC [910]

The first book of moses called genesis / ed by Sayce, Archibald Henry – London: J M Dent; Philadelphia: J B Lippincott 1901 [mf ed 1989] – (= ser The temple bible) – 1mf – 9 – 0-7905-1845-7 – mf#1987-1845 – us ATLA [221]

The first book of reading lessons – new enl ed. Montreal: publ for the Christian Bros, by D & J Sadlier, 1862 [mf ed 1984] – 1mf – 9 – 0-665-44861-9 – mf#44861 – cn Canadiana [420]

First book of samuel – Cambridge, England. 1894 – 1r – us UF Libraries [939]

The first book of samuel : with map, notes and introduction / Kirkpatrick, Alexander Francis – Cambridge: University Press, 1886 – 1mf – 9 – 0-8370-3906-1 – (includes appendix on special interpretations of 1st samuel. incl ind) – mf#1985-1906 – us ATLA [221]

The first book of samuel in hebrew : with a vocabulary – Morgan Park, IL: American Publ Society of Hebrew, 1884 – 1mf – 9 – 0-524-06114-9 – mf#1992-0781 – us ATLA [221]
First break – Oxford. 1983-1995 (1) 1983-1995 (5) 1983-1995 (9) – ISSN: 0263-5046 – mf#15544 – us UMI ProQuest [550]
First british occupation of the cape of good hope, 1795-1803 / Wagner, Mary St Clair – Cape Town, South Africa. 1946 – 1r – us UF Libraries [960]
First calvary baptist church (formerly: pleasant hill baptist church). lexington county. south carolina : church records – 1886-88, 1922-48, 1959-71 – 1 – us Southern Baptist [242]
The first catechism of christian instruction and doctrine in the cree language – London: SPCK, 1910 – 1mf – 9 – 0-524-06178-5 – mf#1991-2434 – us ATLA [240]
First Catholic Slovak Ladies Association [US] *see* Fraternally yours zenska jednota
First cedar creek baptist church – Cincinnati. 1824-1825 (1) – 1 – $96.48 – mf#4438 – us Southern Baptist [242]
First census of the united states, 1790 / U.S. Bureau of the Census – (= ser 1790-1840 census schedules) – 12r – 1 – (original schedules for 1790. schedules for some counties missing) – mf#M637 – us Nat Archives [317]
First census of the united states, 1790 / U.S. Bureau of the Census – (= ser 1790-1840 census schedules) – 3r – 1 – (printed schedules as publ in 1907-8. end of roll 3 includes the 1840 census of pensioners for revolutionary or military services) – mf#T498 – us Nat Archives [317]
The first century of british justice in india : an account of the court of judicature at bombay, established in 1672, and of other courts of justice in madras, calcutta, and bombay, from 1661 to the later part of the eighteenth century / Fawcett, Charles – Oxford: Clarendon Press, 1934 – (= ser Samp: indian books) – us CRL [340]
The first century of methodism in canada / Sanderson, Joseph Edward – Toronto: W. Briggs, 1908-1910 – 3mf – 9 – 0-7905-8070-5 – mf#1988-6051 – us ATLA [242]
The first century of the colonial episcopate / Torrey, Henry Warren – [Canada?: s.n, 1887?] [mf ed 1994] – 1mf – 9 – 0-665-94610-4 – mf#94610 – cn Canadiana [242]
The first century of the liberal movement in american religion / Wilbur, Earl Morse – Boston: American Unitarian Association, 1918 – 1mf – 9 – 0-524-08704-0 – mf#1993-3229 – us ATLA [240]
The first chapter of genesis as the rock foundation for science and religion / Gridley, Albert Leverett – Boston: Richard G Badger, c1913 – 1mf – 9 – 0-524-05673-0 – mf#1992-0523 – us ATLA [221]
The first christian century : notes on dr moffatt's introduction to the literature of the new testament / Ramsay, William Mitchell – London, New York: Hodder and Stoughton, 1911 – 1mf – 9 – 0-7905-0197-X – (incl bibl ref) – mf#1987-0197 – us ATLA [225]
First christian creed / Laidlaw, John – Aberdeen, Scotland. 1873 – 1r – us UF Libraries [240]
The first christian mission to the great mogul : or, the story of blessed rudolf acquaviva, and of his four companions in martyrdom of the society of jesus / Goldie, Francis – Dublin: MH Gill, 1897 – 1mf – 9 – 0-524-04071-0 – (incl bibl ref) – mf#1991-2016 – us ATLA [240]
The first christians : or, christian life in new testament times / Veitch, Robert – London: J Clarke; Leicester: J J Townsend, 1906 – 1mf – 9 – 0-7905-8954-0 – mf#1989-2179 – us ATLA [240]
First church endeavorer – Hamilton, Ont: Young People's Society of Christian Endeavor, [1890-189- or 19–] – 9 – mf#P04350 – cn Canadiana [242]
First Church of God in Christ *see* Pentecostal perspective newsletter
First Church of God [Inglewood, CA] *see* Focus
First Church of Love, Faith and Deliverance *see* Lifeline
First church visitor / First Baptist Church [Madison, WI] – ns: v2 n1-v8 n11 [1957 sep 14-1965 jun 25] – 1r – 1 – (cont by: visitor [madison, wi]) – mf#1006376 – us WHS [242]
First citizen / First Nations Confederacy – [1982 dec], [1983 mar-aug, oct-dec], [1984 jan-may, sep], [1985 aug-nov], [1986 jan-jul] – 1r – 1 – mf#1303306 – us WHS [071]
First citizen – [v1] n1-19 [1969 nov-1972 [jul?]] – 1r – 1 – mf#1056283 – us WHS [071]
The first code of laws.. / Russia. (1917-R.S.F.S.R.). Laws, Statutes, etc – Petrograd: Peoples Commissariat of Justice, 1919 – 1 – us Wisconsin U Libr [348]
First communion with prayers and devotions for the newly confirmed / Maclear, George Frederick – New York: Macmillan, 1917 – 1mf – 9 – 0-524-04663-8 – mf#1990-5059 – us ATLA [240]
First Community Interfaith Institute of Ohio *see* News update
First comptroller, department of the treasury, decisions : 1880-1894 / U.S. Treasury Dept. Comptroller – v1-7 (all publ) – 50mf – 9 – $75.00 – (none publ 1886-93. the office was abolished in 1894. cont by: comptroller of the treasury decisions) – mf#LLMC 80-505 – us LLMC [336]
First Congregational Church *see* Our paper
First Congregational Church, Emporia KS *see* The early church
First Congregational Church, Emporia, KS *see* Record
First Congregational Church, Los Angeles *see* Freedom club bulletin
A first course in chemistry : for the use of students at high schools and normal schools and for beginners' classes in general / Brittain, John – Toronto: Educational Book Co, c1912 – 1mf – 9 – 0-665-71583-8 – mf#71583 – cn Canadiana [540]
First creek baptist church. anderson county. south carolina : church records – 1824-1911, 1919-82. Deacons' Minutes: 1824-1923, misc. material. 1802p – 1 – 81.09 – us Southern Baptist [242]
First Day Cover Collectors of Wisconsin *see* Wisconsin cover sheet
The first decade of the circuit court executive : an evaluation / Macy, John W Jr. – Washington: FJC, 1985 – 1mf – 9 – $1.50 – mf#LLMC 95-361 – us LLMC [347]
First decade of the woman's foreign missionary society of the methodist episcopal church : with sketches of its missionaries / Wheeler, Mary Sparkes – New York: Phillips & Hunt, 1881 [mf ed 1984] – (= ser Women & the church in america 138) – 1mf – 9 – 0-8370-1410-7 – mf#1984-2138 – us ATLA [242]
The first divorce of henry 8 : as told in the state papers / Hope, Anne Fulton; ed by Gasquet, Francis Aidan, Cardinal – London: K Paul, Trench, Truebner, 1894 – 1mf – 9 – 0-7905-4688-4 – (incl bibl ref) – mf#1988-0688 – us ATLA [941]
The first easter dawn : an inquiry into the evidence for the resurrection of jesus / Gorham, Charles Turner – London: Watts, 1908 – 1mf – 9 – 0-8370-3347-0 – (includes bibliography and index) – mf#1985-1347 – us ATLA [240]
First ecumenical council : that is, the first council of the whole christian world, which was held a d 325 at nicaea in bithynia – Jersey City, NJ: J Chrystal, 1891 – (= ser Authoritative christianity) – 2mf – 9 – 0-7905-5083-0 – (incl bibl ref) – mf#1988-1083 – us ATLA [240]
First edition / First Wisconsin National Bank of Madison – 1976 jun-1980 jun – 1r – 1 – mf#679068 – us WHS [332]
First elements of sacred prophecy : including an examination of several recent expositions and of the year-day theory / Birks, Thomas Rawson – London: W E Painter, 1843 – 2mf – 9 – 0-7905-3310-3 – mf#1987-3310 – us ATLA [220]
The first english conquest of canada : with some account of the earliest settlements in nova scotia and newfoundland / Kirke, Henry – London, 1871 – (= ser 19th c british colonization) – 3mf – 9 – mf#1.1.6850 – uk Chadwyck [971]
The first englishmen in india : letters and narratives of sundry elizabethans written by themselves and edited with an introduction and notes / ed by Locke, J Courtenay – London: George Routledge & Sons, 1930 – (= ser Samp: indian books) – us CRL [915]
First epistle of clemens romanus to the church at corinth / Clement 1, Pope – Dundee, Scotland. 1803 – 1r – us UF Libraries [240]
First epistle of john – Kelso, Scotland. 18– – 1r – us UF Libraries [227]
The first epistle of john / Candlish, Robert Smith – 3rd ed. Edinburgh: Adam and Charles Black, 1877 – 2mf – 9 – 0-7905-2404-X – mf#1987-2404 – us ATLA [227]
The first epistle of paul the apostle to the corinthians : in the revised version: with introduction and notes / Parry, Reginald St John – Cambridge: University Press; New York: G P Putnam, 1916 – 1mf – 9 – 0-8370-6830-4 – (incl ind) – mf#1986-0830 – us ATLA [227]
The first epistle of paul to the corinthians = Der erste brief pauli an die korinther / Kling, Christian Friedrich – 2nd rev ed. New York: Charles Scribner, 1886, c1868 [mf ed 1986] – (= ser A commentary on the holy scriptures. new testament 6/1) – 1mf – 9 – 0-8370-6200-4 – (trans fr 2nd rev ed german ed, with additions by daniel warren poor) – mf#1986-0200 – us ATLA [227]
The first epistle of paul to the corinthians / Moffatt, J – 1938 – 9 – $12.00 – us IRC [240]
The first epistle of peter : revised text, with introduction and commentary / Johnstone, Robert – Edinburgh: T & T Clark, 1888 – 2mf – 9 – 0-8370-3792-1 – mf#1985-1792 – us ATLA [227]
The first epistle of s. peter / Masterman, John Howard Bertram – London; New York: Macmillan, 1900 – 1mf – 9 – 0-7905-1357-9 – (in english and greek. incl indes) – mf#1987-1357 – us ATLA [240]
First epistle of st john / Haupt, Erich – Edinburgh: T.&T. Clark, 1893. 1 fiche – 9 – us ATLA [240]
The first epistle of st john : a contribution to biblical theology = Der erste brief des johannes / Haupt, Erich – Edinburgh: T & T Clark 1893 [mf ed 1985] – (= ser Clark's foreign theological library. new series 64) – 1mf – 9 – 0-8370-3526-0 – (translated fr german, with int by william burt pope) – mf#1985-1526 – us ATLA [227]
The first epistle of st john / Lias, John James – London: J Nisbet 1887 [mf ed 1989] – 9 – 0-7905-1284-X – (with exposition & homiletical treatment by aut) – mf#1987-1284 – us ATLA [225]
The first epistle of st peter / Selwyn, E G – 1946 – 9 – $18.00 – us IRC [240]
The first epistle of st peter 1. 1-2. 17 : the greek text with introductory lecture, commentray, and additional notes / Hort, Fenton John Anthony – London, New York: Macmillan, 1898 – 1mf – 9 – 0-8370-3662-3 – (text in greek. introduction & notes in english. includes bibliographical references & index) – mf#1985-1662 – us ATLA [227]
The first epistle of st peter, 1.1-2.17 / Hort, Fenton John Anthony – 1898 – 9 – $10.00 – us IRC [240]
The first epistle to the corinthians / Dods, Marcus – New York: A C Armstrong, 1890 – (= ser The expositor's bible) – 1mf – 9 – 0-8370-2937-6 – mf#1985-0937 – us ATLA [227]
The first epistle to the corinthians : with introduction and notes / Goudge, Henry Leighton – London: Methuen, 1903 – (= ser Westminster Commentaries) – 1mf – 9 – 0-8370-3350-0 – (incl ind) – mf#1985-1350 – us ATLA [227]
The first epistle to the thessalonians : analysis and notes / Garrod, George Watts – London; New York: Macmillan, 1899 – 1mf – 9 – 0-8370-9946-3 – (incl ind) – mf#1986-3946 – us ATLA [227]
First exercises for children in light, shade, and colour / Cole, Henry Hardy – London 1840 – (= ser 19th c art & architecture) – 2mf – 9 – mf#4.1.364 – uk Chadwyck [700]
First exhibition of works of art in black and white, february, 1881 : the catalogue / Art Association of Montreal – Montreal?: The Association?, 1881? – 1mf – 9 – mf#02465 – cn Canadiana [750]
First expedition of vargas into new mexico, 1692 / Bayle, Constantino & Espinosa, J Manuel – Alburquerque, 1940; Madrid: Missionalia Hispanica, 1945 – 1 – sp Bibl Santa Ana [917]
First five chapters of an excellent essay on the holy sacrament of... / Waldo, Peter – Dublin, Ireland. 1812 – 1r – us UF Libraries [240]
First footsteps in east africa : or, an exploration of harar / Burton, R – London, 1894 – 7mf – 9 – mf#NE-20290 – ne IDC [916]
The first general conference of lutherans in america : held in philadelphia, december 27-29, 1898 – Philadelphia: Council Publication Board: Lutheran Publication Society, 1899 – 1mf – 9 – 0-524-00969-4 – mf#1990-4027 – us ATLA [242]
The first general epistle of st john the apostle / Hardy, Nathaniel – Edinburgh: James Nichol, 1865 [mf ed 1985] – (= ser Nichol's series of commentaries) – 2mf – 9 – 0-8370-3470-1 – mf#1985-1470 – us ATLA [227]
First grammar of the language spoken by the bontoc igorot / Seindenadel, Carl Wilhelm – Chicago, IL. 1909 – 1r – us UF Libraries [490]
First greek colony in america / Anastasion, Georgios – s.l, s.l? 1939? – 1r – us UF Libraries [305]
The first half century of the northumberland baptist association : situated in northumberland, montour, columbia, sullivan, lycoming, clinton, union and snyder counties, pennsylvania / Worden, Oliver N – Philadelphia: JA Wagenseller, 1871 – 1mf – 9 – 0-524-08887-X – mf#1993-3351 – us ATLA [242]
First highway along the southeast coast of florida / Comstock, Bertha A – s.l, s.l? 193-? – 1r – us UF Libraries [380]
First historical transformations of christianity = Des premieres transformations historiques de christianisme / Coquerel, Athanase – Boston: WV Spencer, 1867 – 1mf – 9 – 0-7905-5522-0 – (in english) – mf#1988-1522 – us ATLA [240]
First impressions on a tour upon the continent in the summer of 1818 : through parts of france, italy, switzerland, the borders of germany, and a part of french flanders / Baillie, Marianne – London 1819 [mf ed Hildesheim 1995-98] – (= ser Fbc) – 1v on 3mf [ill] – 9 – €90.00 – 3-487-27780-8 – gw Olms [914]
The first indian member of the imperial parliament : being a collection of the main incidents relating to the election of mr dadabhai naoroji to parliament / Naoroji, Dadabhai – Madras 1892 – (= ser 19th c british colonization) – 2mf – 9 – mf#1.1.7373 – uk Chadwyck [323]
First international convention of reformed presbyterian churches, scotland : june 27-july 3 1896 – Glasgow: Alex Malcolm [1896?] [mf ed 1992] – 2mf [ill] – 9 – 0-524-03162-2 – mf#1990-4611 – us ATLA [242]
The first international railway and the colonization of new england : life and writings of john alfred poor / ed by Poor, Laura Elizabeth – New York, London: G P Putnam's Sons, 1892 – 5mf – 9 – mf#12073 – cn Canadiana [380]
The first interpreters of jesus / Gilbert, George Holley – New York: Macmillan, 1901 – 1mf – 9 – 0-8370-3278-4 – (incl bibl ref and index) – mf#1985-1278 – us ATLA [220]
First issue – Ithaca NY. n9 [1968 dec] – 1r – 1 – mf#5187111 – us WHS [071]
First kafir course / Crawshaw, C J – Cape Town, South Africa. 1903 – 1r – us UF Libraries [470]
First korean congress : held in the little theatre, 17th and delancey streets, april 14, 15, 16 / Korean Congress (1st: 1919: Philadelphia) – Philadelphia: [s.n.] 1919 – (= ser Yale coll) – 82p (ill) – 1 – 0-524-10024-1 – mf#1995-1024 – us ATLA [950]
The first law reporter in upper canada and his reports / Riddell, William Renwick – Toronto: Canadian Bar Association, 1916 – 1mf – 9 – mf#77542 – cn Canadiana [340]
First lessons for the harpsichord or spinnet : wherein the italian manner of fingering is shewn in all the different tones... / Colizzi, Johannes – London: printed & sold by Longman, Lukey & Co [18–?] [mf ed 1989] – 1r – 1 – mf#pres. film 45 – us Sibley [780]
First lessons in christian morals : for canadian families and schools / Ryerson, Egerton – Toronto: Copp, Clark, 1871 – 2mf – 9 – mf#12796 – cn Canadiana [230]
First lessons in english and tamul. : designed to assist tamul youth in the study of the english language – Manepy: Press of the American Mission, 1835-36 – 1 – us CRL [490]
First lessons in urdu / Dann, George James – Calcutta: Baptist Mission Press, 1911 [mf ed 1995] – (= ser Yale coll) – iv/152p – 1 – 0-524-09348-2 – (urdu text, roman transliteration and english trans arr in 3 parallel clms) – mf#1995-0348 – us ATLA [490]
First lessons on agriculture : for canadian farmers and their families / Ryerson, Egerton – Toronto: Copp, Clark, 1871 – 3mf – 9 – mf#12797 – cn Canadiana [630]
First letter on the present position of the high church party in th... / Maskell, William – London, England. 1850 – 1r – us UF Libraries [240]
First letter to the rev. father gratry : priest of the oratoire, member of the academy / Dechamps, Victor Auguste – London: JT Hayes, [1870?] – 1mf – 9 – 0-524-08282-0 – mf#1993-3037 – us ATLA [240]
First letter to the right honourable lord john russell / Bennett, William J E – London, England. 1850 – 1r – us UF Libraries [240]
First letter to the very rev. j. h. newman, d.d : in explanation, chiefly in regard to the reverential love due to the ever-blessed theotokos, and the doctrine of her immaculate conception / Pusey, Edward Bouverie – London: Rivingtons, 1869 – (= ser Eirenicon) – 2mf – 9 – 0-7905-7455-1 – mf#1989-0680 – us ATLA [240]
First line index of manuscript poetry in the folger shakespeare library – [mf ed Marlborough 1996] – 3r – 1 – £295.00 – (with d/g) – uk Matthew [090]
First line index of manuscript poetry in the huntington library – [mf ed Malborough 1991] – 15mf – 9 – £110.00 – (with d/g) – uk Matthew [090]
The first man and his place in creation : considered on the principles of science and common sense from a christian point of view with an appendix on the negro / Moore, George – London, 1866 – (= ser 19th c evolution & creation) – 5mf – 9 – mf#1.1.1532 – uk Chadwyck [573]

First measures in malarial prevention for farmers and settlers = Eerste maatreels teen malaria vir boere en setlaars / Hay, G G – Johannesburg: SA Red Cross Society [1924?] [mf ed Pretoria, RSA: State Library [199-]] – 36p [ill] on 1r with other items – 5 – (rev by g g hay, with chapter on mosquitoproof housing by g a park ross) – mf#op 09455 r23 – us CRL [614]

First metatarsophalangeal joint range of motion as a factor in turf toe injuries / Eggert, KE – 1990 – 1mf – 9 – $4.00 – us Kinesology [790]

First Methodist Episcopal Church *see* Madison methodist

First Missionary Baptist Church *see* Newsletter

First national bank of chicago business and economic review – Chicago. 1957-1972 [5] – ISSN: 0015-2773 – mf#6477 – us UMI ProQuest [332]

First National City Bank *see* Monthly economic letter

First national poll – Monrovia, Liberia: Infotech Consultants, Inc [jun 9/15 1992-dec 31 1994/jan 3 1995] (wkly) – 1r – 1 – us CRL [079]

First Nations Confederacy *see* First citizen

First new jerusalem society [chillicothe, ohio] records, 1838-1879 / Chillicothe. Ohio. First New Jerusalem Society – [mf ed 1974] – 1r – 1 – mf#ms399 – us Western Res [243]

First nzef reserve and military defaulters – ww1 – (mf ed 1999) – 10mf – 9 – NZ$40.00 – 0-908989-46-6 – nz BAB [355]

The first of empires : "babylon of the bible" in the light of latest research / Boscawen, William Saint Chad – London; New York: Harper, 1903 – 1mf – 9 – 0-7905-0908-3 – (incl bibl ref and ind) – mf#1987-0908 – us ATLA [930]

First of may in the midst of tragedy / Prieto, Indalecio – [1936?] [mf ed 1977] – (= ser Blodgett coll) – 1mf – 9 – mf#w1177 – us Harvard [946]

First of may magazine / Workers' International Industrial Union – 1919 may 1 – 1r – 1 – mf#1807987 – us WHS [331]

The first one hundred schwenckfelder memorial days, 1734-1834 : pt 1: 1737-1784 – Norristown PA: Board of Pub of the Schwenckfelder Church, 1952 [mf ed 2003] – (= ser Schwenckfeldiana 2/3,4) – 1r – 1 – (pt2: 1785-1834 (1954) [mf ed 2003]. english trans fr german) – mf#2003-s008i; 2003-s008j – us ATLA [242]

The first page of the bible = Erste blatt der bibel / Bettex, Frederic – Burlington, Iowa: German Literary Board, 1908 – 1mf – 9 – 0-524-04392-2 – (in english) – mf#1992-0085 – us ATLA [220]

First part of the collection of detached enrichments : and various articles of taste and furniture... / Jackson, George & Sons – London, 1836 – (= ser 19th c art & architecture) – 1mf – 9 – mf#4.1.26 – uk Chadwyck [740]

[First part of the] musician's companion / Howe, Elias – Boston: Elias Howe, Jr, 1844 – 1 – (containing 18 setts of cotillions) – us L of C Photodup [780]

First pious youth / Symington, William – Edinburgh, Scotland. 1843 – 1r – us UF Libraries [240]

First prayer book of edward 6 / Great Britain Parliament – London, England. 1894? – 1r – us UF Libraries [240]

First prayer in the family – London, England. 18– – 1r – us UF Libraries [240]

First Presbyterian Church (Pensacola, FL) *see* Year book and directory

First Presbyterian Church, Topeka, KS *see*
– Church records
– Records

First principles / Center for National Security Studies [Washington, DC] et al – v1 n7-8 [1968 mar-apr], v2 n3,6,9 [1976 nov, 1977 feb, may], v3 n5-v10 n2 [1978 jan-1984 dec] – 1r – 1 – mf#1117830 – us WHS [322]

First principles / Davis, Morrison Meade – Cincinnati: Standard Pub Co, c1904 – 1mf – 9 – 0-524-06402-4 – mf#1991-2524 – us ATLA [220]

First principles / Spencer, Herbert – London, 1862 – (= ser 19th c evolution & creation) – 6mf – 9 – mf#1.1.3923 – uk Chadwyck [170]

First principles / Spencer, Herbert – New York: D Appleton, 1888 [mf ed 1993] – (= ser A system of synthetic philosophy 1) – 2mf – 9 – 0-524-08648-6 – mf#1993-2108 – us ATLA [190]

First principles of a new system of philosophy / Spencer, Herbert – New York, NY. 1877 – 1r – us UF Libraries [140]

First principles of the reformation : or the ninety-five theses and the three primary works of dr martin luther translated into english / Luther, Martin; ed by Wace, Henry & Buchheim, Carl Adolf – LOndon: J Murray c1883 – 1r – 1 – 0-524-10476-X – (in english) – mf#b00693 – us ATLA [242]

The first publishers of truth : being early records (now first printed) of the introduction of quakerism into the counties of england and wales / ed by Penney, Norman – London: Headley; Philadelphia: H. Newman, 1907 – 1mf – 9 – 0-7905-6311-8 – mf#1988-2311 – us ATLA [243]

First questions on religion / Blomfield, Charles James – London, England. 18– – 1r – us UF Libraries [200]

A first reader in new testament greek / Moulton, James Hope – London: Charles H Kelly, 1896 [mf ed 1986] – (= ser Books for bible students) – 1mf – 9 – 0-8370-9297-3 – mf#1986-3297 – us ATLA [450]

A first reading book in the micmac language : comprising the micmac numerals, and the names of the different kinds of beasts, birds, fishes, trees, etc of the maritime provinces of canada... / Rand, Silas Tertius – Halifax, NS?: s.n, 1875 – 2mf – 9 – mf#12362 – cn Canadiana [490]

First rejection of christ : a warning to the church / Crosthwaite, John Clarke – London, England. 1837 – 1r – us UF Libraries [240]

First report : the select committee appointed to investigate and report on the outrages alleged to have been committed at the general election in the counties of terrebonne, montreal, vaudreuil, beauharnois, chambly and rouville / Neilson, John – [Kingston?: s.n, 1843?] (mf ed 1997) – 1mf – 9 – mf#SEM105P2805 – cn Bibl Nat [325]

First report from the select committee on emigration, scotland : together with the minutes of evidence and appendix / Grande-Bretagne. Parliament. House of Commons – [s.l.]: House of Commons, 1841 [mf ed 1983] – 3mf – 9 – mf#SEM105P157 – cn Bibl Nat [324]

First report of the commissioners appointed to enquire into the losses occasioned by the troubles during the years 1837 and 1838 : and into the damages arising therefrom / Canada (Province) – Montreal: printed by Lovell & Gibson, [1846] (mf ed 1994) – 2mf – 9 – mf#SEM105P2176 – cn Bibl Nat [971]

First report of the committee on research and development in modern languages, 1968 – 1mf – 9 – mf#87025 – uk Microform Academic [324]

First report of the committee...on that part of the speech of his excellency the governor in chief : which relates to the settlement of the crown lands with the minutes of evidence taken before the committee / Bas-Canada. Parlement. Chambre d'assemblee – Quebec: J Neilson, 1821 [mf ed 1990] – 3mf – 9 – mf#SEM105P1281 – cn Bibl Nat [324]

The first report of the general conference of christians expecting the advent of the lord jesus christ : held in boston, oct. 14, 15, 1840 / Litch, Josiah et al – Boston: [s.n.], 1842 [mf ed 1993] – 1v on 1mf – 9 – 0-524-06352-4 – (incl bibl ref) – mf#1990-1535 – us ATLA [242]

First report of the glasgow association in aid of chinese medical missions – [Hackney], 1846 – 1mf – 9 – mf#HT-1137 – ne IDC [915]

First report of the national advisory council on art education, 1960 – 1mf – 9 – mf#87031 – uk Microform Academic [350]

First report of the public schools commission, 1968 – 2v – 9mf – 9 – (v1 report. v2 app) – mf#87034/5 – uk Microform Academic [324]

First report of the special committee appointed : to inquire into the causes which retard the settlement of the eastern townships of lower canada – [Toronto?: s.n.] 1851 [mf ed 1994] – 2mf – 9 – 0-665-94698-8 – mf#94698 – cn Canadiana [917]

First report of the special committee appointed to inquire into the causes which retard the settlement of the eastern townships of lower canada = Premier rapport...nomme pour s'enquerir des causes qui retardent la colonisation des townships de l'est du bas-canada / Canada (Province). Parlement. Assemblee legislative – Toronto: printed by Lovell & Gibson, 1851 [mf ed 1995] – 1mf – 9 – mf#SEM105P2027 – cn Bibl Nat [971]

First report of the special committee of the house of assembly on the engrossed bill from the legislative council to repeal certain parts of the judicature act : and to make further provision for the more certain and uniform administration of justice within this province / Bas-Canada. Parlement. Chambre d'assemblee – Quebec: printed by P E Desbarats, [1824] (mf ed 1994) – 1mf – 9 – (in french and english) – mf#SEM105P1976 – cn Bibl Nat [323]

First report of the standing committee on roads and public improvements = Premier rapport du comite permanent sur les chemins et les ameliorations publiques / Bas-Canada. Parlement. Chambre d'assemblee – [S.l: s.n, 1832?] (mf ed 1992) – 2mf – 9 – (in english and french) – mf#SEM105P1229 – cn Bibl Nat [625]

First report on conveyance as adopted by the executive committee / British Association for the Advancement of Science. Canada – Montreal?: The Gazette Print Co, 1884 – 1mf – 9 – mf#10170 – cn Canadiana [380]

First resurrection / Tregelles, Samuel Prideaux – London, England. 1876 – 1r – us UF Libraries [240]

First, Ruth *see* South west africa

First, second and third conferences / United Nations. Conferences on the Law of the Sea – 22v. 81-906 – 9 – $40.00 – us LLMC [341]

First six bishops of pennsylvania / Hotchkin, Samuel Fitch – [S.l: s.n., 1911?] – 1mf – 9 – 0-524-03795-7 – mf#1990-4867 – us ATLA [240]

The first social experiments in america. a study in the development of spanish indian policy...cambridge (usa), 1935 / Hanke, Lewis – Madrid: Razon y Fe, 1936 – 1 – sp Bibl Santa Ana [320]

The first soprano / Hitchcock, Mary – 4th ed. New York: Christian Alliance Pub Co, 1913 [mf ed 1992] – (= ser Christian & missionary alliance coll) – 3mf – 9 – 0-524-02257-7 – mf#1990-4264 – us ATLA [240]

First southern baptist church – Buckeye, AZ. 2402p. 1925-95 – 1 – $108.09 – mf#4367 – us Southern Baptist [242]

First southern baptist church. omaha, nebraska : church records – 1956-61 – 1 – 6.75 – us Southern Baptist [242]

First steps in assyrian : a book for beginners: being a series of historical, mythological, religious, magical, epistolary and other texts printed in cuneiform characters with interlinear transliteration and translation: and a sketch of assyrian grammar, sign-list and vocabulary / King, Leonard William – London: Kegan Paul, Trench, Truebner, 1898 – 2mf – 9 – 0-8370-8915-8 – (includes glossary) – mf#1986-2915 – us ATLA [470]

First steps in civilizing rhodesia : being a true account... / Boggie, Jeannie M – Bulawayo, Zimbabwe. 1940 – 1r – us UF Libraries [960]

First steps in new testament greek / Clapperton, John Alexander – London: Charles H Kelly, 1901 – 1mf – 9 – 0-8370-9134-9 – mf#1986-3134 – us ATLA [450]

First steps in zulu / Colenso, John William – Pietermaritzburg, South Africa. 1903 – 1r – us UF Libraries [470]

The first steps to irish liberty / Beggs, Charles – Dublin, 1857 – (= ser 19th c ireland) – 1mf – 9 – mf#1.1.1929 – uk Chadwyck [330]

First steps toward church unity / Parkhurst, Charles Henry – New York: F H Revell, c1891 – 1mf – 9 – 0-7905-6716-4 – mf#1988-2716 – us ATLA [240]

First swahili book / Werner, Alice – London, England. 1930 – 1r – us UF Libraries [470]

First territorial census for oklahoma, 1890 / U.S. Bureau of the Census – (= ser Records of the bureau of the census – state and special census schedules) – 1r – 1 – mf#M1811 – us Nat Archives [317]

First things : a series of lectures on the great facts and moral lessons first revealed to mankind / Spring, Gardiner – New York: MW Dodd, 1851 – 2mf – 9 – 0-7905-8737-8 – mf#1989-1962 – us ATLA [220]

First things first / Spurgeon, C H – London, England. 1885? – 1r – us UF Libraries [240]

First three centuries of appalachian travel – 681mf – 9 – (coll based on the travels of desoto to the area of chattanooga, as recorded by the gentleman of elvas. with printed guide) – mf#C39-27390 – us Primary [917]

The first three christian centuries : a history of the church of christ with a special view to the delineation of christian faith and life from a.d. 1 to a.d. 313 / Burns, Islay – London: T Nelson & Sons, 1884 – 1mf – 9 – 0-524-05838-5 – (incl bibl ref) – mf#1990-3502 – us ATLA [240]

The first traveler : and nine idle essays on the trans-canadian portage / Cooke, Britton Bertrand – [Toronto?: s.n.], c1911 – 1mf – 9 – 0-665-98559-2 – mf#98559 – cn Canadiana [380]

First twelve mile baptist church. campbell county. kentucky : church records – 1818-1957. Lacks 1885-1926 – 640p – 1 – $28.80 – us Southern Baptist [242]

First twenty years of australia : history founded on official documents / Bonwick, James – London, 1882 – (= ser 19th c books on british colonization) – 3mf – 9 – mf#1.1.8132 – uk Chadwyck [980]

The first two nawabs of oudh : critical study based on original sources / Srivastava, Ashirbadi Lal – Lucknow: Upper India Pub House, 1933 – (= ser Samp: indian books) – (foreword by sir jadunath sarkar) – us CRL [954]

First Unitarian Society *see*
– Madison unitarian
– Religion for today

The first unitarian society of chicago : its relation to a changing community / Thompson, Donald Alexander – Chicago, 1933.Chicago: Dep of Photodup, U of Chicago Lib, 1971 (1r); Evanston: American Theol Lib Assoc, 1984 (1r) – 1 – 0-8370-0286-9 – mf#1984-B163 – us ATLA [243]

First United Church of Jesus Christ Apostolic *see* National news-letter

First United Methodist Church, Wellington, KS *see* Membership register

First u.s. army report of operations, 1943-1945 / U.S. Army. First Army – 2r – 1 – $260.00 – mf#S1682 – us Scholarly Res [355]

First victoria directory, fifth issue, and british columbia guide : comprising a general directory of business-men and householders in victoria... / Mallandaine, Edward – Victoria, BC: E Mallandaine, 1874 [mf ed 1980] – 2mf – 9 – (incl ind) – mf#15578 – cn Canadiana [971]

First victoria directory, second issue, and british columbia guide : comprising a general directory of business-men and householders in victoria... / Mallandaine, Edward – Victoria [BC]: E Mallandaine, 1868 [mf ed 1981] – 1mf – 9 – (incl ind) – mf#15575 – cn Canadiana [971]

First victoria directory, third [i e fourth] issue, and british columbia guide : comprising a general directory of business-men and householders in victoria... / Mallandaine, Edward – Victoria [BC]: E Mallandaine, 1871 [mf ed 1980] – 2mf – 9 – (incl ind) – mf#15577 – cn Canadiana [971]

First victoria directory, third issue, and british columbia guide : comprising a general directory of business-men and householders in victoria... / Mallandaine, Edward – Victoria [BC]: E Mallandaine, 1869 [mf ed 1980] – 1mf – (incl ind) – mf#15576 – cn Canadiana [971]

First Wisconsin National Bank of Madison *see* First edition

First Wisconsin National Bank of Milwaukee *see* First account

First women physicians to the orient / Baker, Frances J – Boston: Woman's Foreign Missionary Society Methodist Episcopal Church, [1904?] [mf ed 1984] – (= ser Women & the church in america 111) – 1mf – 9 – 0-8370-1406-9 – mf#1984-2111 – us ATLA [305]

The first words from god : or, truths made known in the first two chapters of his holy word. also, the harmonizing of the records of the resurrection morning / Upham, Francis William – New York: Hunt & Eaton; Cincinnati: Cranston & Curts, 1894 – 1mf – 9 – 0-8370-6434-1 – (incl bibl ref and appendixes on the topics of time and creation) – mf#1986-0434 – us ATLA [220]

First words in australia : sermons / Barry, Alfred – London: Macmillan 1884 [mf ed 1990] – 1mf – 9 – 0-7905-7318-0 – mf#1989-0543 – us ATLA [242]

First world – Atlanta. 1977-1980 (1,5,9) – mf#11441 – us UMI ProQuest [305]

First world news / Lehigh Valley Black Interest Coalition [PA] – Allentown PA. v1 n5 [1993 nov/dec], v3 n2 [1995 apr/may] – 1r – 1 – mf#3433283 – us WHS [321]

The first world war : political, social and military manuscript sources – 10r – 1 – (coll from the national library of scotland, edinburgh contains the papers of field marshall sir douglas haig) – mf#C39-15500 – us Primary [920]

First world war: a documentary record, series 1 : european war 1914-1919, the war reserve collection from cambridge university library – 10pt – 1 – (pt1: card catalogue ind & mss listings [7r] £650; pt2: trench journals, personal narratives & reminiscences [20r] £1850; pt3: allied propaganda of ww1 [20r] £1850; pt4: german propaganda of ww1 [20r] £1850; pt5: royal army medical corps, red cross & other auxiliary services [25r] £2350; pt6: war at sea & the war in the air [22r] £2050 [mf ed 2000]; pt7: economics, finance & socialism [15r] £1400; pt8: russian affairs, bolshevism & the eastern front [10r] £950; pt9: peace, versailles & the league of nations [20r] £1850 [mf ed fall 2004]; pt10: memory of war [20r] £1850; with d/g) – uk Matthew [933]

First world war: the home front : the diaries of andrew clark, rector of great leighs, essex from the bodleian library, oxford – 2pt – 1 – (pt1: diaries, 2 aug 1914-31 dec 1916 [17r] [mf ed 2004]; pt2: diaries, 1 jan 1917-dec 1919 [12r]; with d/g) – uk Matthew [933]

The first years of emerita augusta. (los primeros anos de emerita augusta) / Richmond, I A – London: Imp. Instituto Lancaster House, 1931. Reimp. Archaeological Journal. vol 86. 1930. Original y traduccion – 1 – sp Bibl Santa Ana [946]

First-aid : anywhere, anytime = Noodhulp / Tucker, Cyril – Cape Town: SA Pharmaceutical Journal on behalf of the Pharmaceutical Society of SA, Cape Western Province Branch c1986 [mf ed Pretoria, RSA: State Library [199-]] – 102p [ill] on 1r with other items – 5 – (in english & afrikaans; incl bibl ref) – mf#op 70-0243 r26 – us CRL [362]
Firth, Charles Harding *see*
– Cromwell's army
– The last years of the protectorate, 1656-1658
– Oliver cromwell and the rule of the puritans in england
Firth, Frank Jones *see* The acts of the apostles, the epistles and the revelation of st john the divine
The firth graphic – Firth, NE: Minnie H Damrow (wkly) [mf ed 1908-10 (gaps) filmed 1978] – 2r – 1 – (cont: firth weekly graphic. v13 n37 not publ) – us NE Hist [071]
Firth, John B *see* Constantine the great
Firth, John Benjamin *see* Augustus caesar and the organization of the empire of rome
Firth weekly graphic – Firth, NE: Graphic Pub Co, 1893 (wkly) [mf ed 1896-99 (gaps) filmed 1978] – 1r – 1 – (cont by: firth graphic) – us NE Hist [071]
Fiscal and staff news / American Federation of State, County, and Municipal Employees – 1976 jun-1977 aug – 1r – 1 – mf#367514 – us WHS [350]
Fiscal policy forum – Washington. 1983-1986 (1) 1983-1986 (5) 1983-1986 (9) – ISSN: 0737-3481 – mf#13440 – us UMI ProQuest [336]
Fiscal studies – London. 1982+ (1,5,9) – ISSN: 0143-5671 – mf#13526 – us UMI ProQuest [336]
Fisch, J-C-A *see*
– Initiation a la philosophie de la francmaconnerie
– Instruction maconnique pour le grade d'apprenti
Fisch, Richard *see* Generalmajor v. stille und friedrich der grosse contra lessing
Fischart, Johann *see*
– Aller praktik grossmutter
– Der floehhaz
– Das glueckhafte schiff von zuerich
Fischbach, Friedrich Ludwig Joseph *see* Historische, politisch-geographisch-statistisch und militaerische beytraege die kgl preussischen und benachbarten staaten betreffend
Fischer *see* [Statement of political beliefs(
Fischer, Abraham Eliezer *see* In veg un andere dertseylungen
Fischer, Adolf *see* Bilder aus japan
Fischer, Arwed *see*
– Brun von schonebeck
Fischer, Axel *see*
– Kammermusik und klaviermusik [die sammlung der sing-akademie zu berlin...pt 4]
– Die sammlung der sing-akademie zu berlin
– Sinfonien, konzerte und ouvertueren [die sammlung der sing-akademie zu berlin...pt 3]
Fischer, Axel [comp] *see*
– Die bach-sammlung
– Die georg philipp telemann-sammlung
– Oratorien, messen, geistliche und weltliche kantaten, arien und lieder [die sammlung der sing-akademie zu berlin...pt 1]
Fischer, Bernard *see* Bibel und talmud in ihrer bedeutung fuer philosophie und kultur
Fischer, Bernhard *see*
– Morgenblatt fuer gebildete staende / gebildete leser
– Otto ludwigs trauerspielplan "der sandwirt von passeier"
Fischer, Bernhard [comp] *see* Die augsburger "allgemeine zeitung" 1798-1866
Fischer, Christian *see*
– Description de valence
– Gemaelde von madrid
– Gemaelde von valencia
– Letters written during a journey to montpellier
– Reise nach hyeres
– Reise nach montpellier
– Reise von amsterdam ueber madrid und cadiz nach genua in den jahren 1797 und 1798
– Travels to hyeres, in the south of france
– Voyage en espagne, aux annees 1797 [mil sept cent quatre-vingt-dix-sept] et 1798
Fischer, Christian A *see*
– Allgemeine unterhaltende reise-bibliothek
– Bergreisen
– Beytraege zur genauern kenntniss der spanischen besitzungen in amerika
– Briefe eines suedlaenders
– Kriegs- und reisefahrten
– Neue kriegs- und reisefahrten
– Reise von livorno nach london
– Reiseabentheuer
Fischer, Engelbert Lorenz *see*
– Heidenthum und offenbarung
– Der triumph der christlichen philosophie
Fischer, Erika Sigrid *see* Magnesiummangel an hoeheren pflanzen
Fischer, Ernst *see* Goethe der grosse humanist
Fischer, Ernst Gottfried *see* Lehrbuch der mechanischen naturlehre
Fischer, Eugen *see* Woodrow wilsons entschluss
Fischer, Ferdinand *see* Das studium der technischen chemie an den universitaten und technischen hochschulen deutschlands und das chemiker-examen
Fischer, Friedrich *see* Die basler hexenprozesse in dem 16ten und 17ten jahrhundert
Fischer, Fritz *see* Ludwig nicolovius
Fischer, Gerhard *see* Gesamtverantwortung und spezifik: die cdu-presse in der entwicklung des ddr-journalismus 1957 bis 1961
Fischer, Gottlob Nathanael *see*
– Deutsche monatsschrift
– Fliegende blaetter fuer freunde der toleranz, aufklaerung und menschenbesserung
Fischer, Gottlob Nathanael et al *see* Berlinisches journal fuer aufklaerung
Fischer, Hanns *see*
– Aberglaube oder volksweisheit?
– Eine schweizer kleinepiksammlung des 15. jahrhunderts
– Verserzaehlungen
– Vier erzaehlungen
Fischer, Henri *see* Le role de la femme
Fischer, Hermann *see*
– Briefwechsel zwischen albrecht von haller und eberhard friedrich von gemmingen
– Der eunuchus des terenz
– Gedichte
– Georg rudolf weckherlins gedichte
– Hermann kurz' saemtliche werke
– Die krankheit des apostels paulus
– Die reise der soehne giaffers
Fischer, J *see*
– The discoveries of the norsemen in america
– Die erkenntnislehre anselms von canterbury
Fischer, J G *see*
– Schillers werke
– Schiller's works
Fischer, Johann Georg *see*
– Auf dem heimweg
– Aus frischer luft
– Gedichte
Fischer, John L *see*
– Contemporary ponapean land tenure
– Native land tenure in the truk district
Fischer, Joseph *see*
– Die davidische abkunft der mutter jesu
– Der "deutsche ptolemaeus"
– Ehe und Jungfraeulichkeit im Neuen Testament
Fischer, Joseph C *see* The construct validity of a scale to measure teacher enthusiasm in secondary physical education
Fischer, Julius *see* Reisen durch oesterreich, ungarn, steyermark, venedig, boehmen und maehren, in den jahren 1801 und 1802
Fischer, Karl Philipp *see*
– Die speculative dogmatik von dr. david friedr. strauss
– Zur hundertjaehrigen geburtsfeier franz von baaders
Fischer, Kuno *see*
– Briefwechsel zwischen goethe und k goettling in den jahren 1824-1831
– A critique of kant
– Francis bacon und seine nachfolger
– Goethes faust
– Goethes tasso
– Die hundertjaehrige gedaechtnissfeier der kantischen kritik der reinen vernunft; johann gottlieb fichtes leben und lehre; spinozas leben und charakter
– Kant's leben und die grundlagen seiner lehre
– Ueber david friedrich strauss
Fischer, Kurt *see* Die herberge am tartaro
Fischer, Kurt W *see* Anna und greite
Fischer, L *see* Die kirchlichen quatertember
Fischer, L H *see* Gedichte des koenigsberger dichterkreises aus heinrich alberts arien und musicalischer kuerbshuette, 1638-1650
Fischer, Louis *see*
– Gandhi and stalin
– The life of mahatma gandhi
– The war in spain
– Why spain fights on
Fischer, Max *see*
– Heinrich von kleist
– Die religion und das leben
– Schleiermacher
Fischer, Otto *see* Recht und rechtsschutz
Fischer, Ottokar *see* H w v gerstenbergs rezensionen in der hamburgischen neuen zeitung
Fischer, Paul *see* Gott-natur
Fischer, Philipp *see* Worte gesprochen an der bahre meiner mutter
Fischer, Richard *see* Conrad ferdinand meyer
Fischer, Robert *see*
– Briefe ueber freimaurerei
– Deutsche geistesheroen
Fischer, Thomas *see*
– Auswahl von schraubwerkzeugen
– Ergebnisdarstellung einer bem-berechnung mit cad-system icem ddn
– Systematische untersuchungen an einem offline-fertigungsplanungssystem in einer modellierten experimentierumgebung
Fischer, Tim *see* Abgasuntersuchungen an einem pflanzenoelbetriebenen vorkammer-dieselmotor
Fischer, Uwe *see* Liberalismus und staatsumfang
Fischer von Erlach, J B *see* Entwurf einer historischen architektur, in abbildung unterschiedener beruehmten gebaeude, des alterthums und fremder voelcker...
Fischer, William Joseph *see* Child of destiny
Fischerei-zeitung, oesterreich-ungarische – Vienna. jan 1880-dec 1882 – 1r – 1 – us UMI ProQuest [074]
Fischer-Friesenhausen, Friedrich *see*
– Nicht mutlos werden, beharrlichkeit
– Sieghaftes blut
Fischer-Stockern, Hans *see* Der preis ist – evi!
Fischer-Wildhagen, Rita *see* Adalbert gyrowetz (1763-1850)
Fischle, Ernst *see* Ein volk ohne suende und die kirche christi
Fischlowitz, Estanislau *see* Principais problemas da migracao nordestina
Fiscus *see* The principles of money in relation to a national currency
Fiscus, Douglas L *see* Comparison of the mid-ameriacn conference athletic department regarding compliance with title 9
Fish and game club in connection with the fish and game protection club, province of quebec : constitution and list of members, september, 1887 – Montreal: Becket, 1887 – 1mf – 9 – mf#28365 – cn Canadiana [639]
Fish cheer – v1 n1-3 [1971 feb] – 1r – 1 – (cont by: gulf coast fish cheer) – mf#1583293 – us WHS [071]
Fish, E J *see* Ecclesiology
Fish, Henry Clay *see*
– Handbook of revivals
– History and repository of pulpit eloquence, deceased divines
– Primitive piety revived
– Pulpit eloquence of the nineteenth century
– Romanism and the common schools
– The voice of our brother's blood
Fish, Simon *see* A supplication for the beggars
Fish tales from seafood workers, local p-554 / Seafood Workers, Local P-554 – 1979 feb 13, aug 1 – 1r – 1 – mf#678852 – us WHS [331]
Fishback, James *see* A defence of the elkhorn association in sixteen letters
Fishberg, Maurice *see* Materials for the physical anthropology of eastern european jews
Fishburne, Benjamin Postell *see* The patent application, preparation and prosecution
Fisher, A Garth *see*
– A comparison of body composition changes in moderately obese and extremely obese women who experience the same caloric deficit
– The effect of transverse pedal spacing on cycling efficiency
– Effect of water running and cycling on vo2max and 2-mile performance
– A submaximal one-mile track jog to estimate vo2max in fit men and women, ages 30-39 years
Fisher, Alexander *see* A journal of a voyage of discovery to the arctic regions
Fisher, Arabella Burton (Buckley) *see* Moral teachings of science
Fisher, Arthur A'Court *see* Personal narrative of three years' service in china
Fisher, Carl Melchior *see* The relationship of financial subsidy to the growth and development of the lutheran church in malaysia
Fisher, CG *see* General land office circulars and general regulations
Fisher, Charles A *see* A study of the guidance considered essential for teacher's intelligent self-direction
Fisher, Charles Milton *see* Analysis of the part writing technic in the later works of vaughan williams
Fisher, Clarence S *see*
– "Bethshean"
– The excavation of armageddon
– The throne room of merenptah
Fisher, Clyde et al *see* Astronomy
Fisher, David *see* An exposition of facts connected with the late prosecutions in the methodist episcopal church of cincinnati
Fisher, Dorothy Canfield *see*
– The bent twig
– Home-maker
Fisher, Edward *see* The marrow of modern divinity
Fisher, Ezra *see* Correspondence of the reverend ezra fisher
Fisher, Frederic I *see* Study of humor in keyboard music
Fisher, Frederick Bohn *see* India's silent revolution
Fisher, George Park *see*
– The beginnings of christianity
– The christian religion
– Colonial era
– A discourse
– Discussions in history and theology
– Essays on the supernatural origin of christianity
– Faith and rationalism
– The grounds of theistic and christian belief
– History of christian doctrine
– History of the christian church
– Manual of christian evidences
– The nature and method of revelation
– The reformation
– An unpublished essay of edwards on the trinity
Fisher, Harold *see* Canadian patent law and practice
Fisher, Herbert Albert Laurens *see*
– Frederic william maitland, downing professor of the laws of england; a biographical sketch
– The medieval empire
Fisher, Hugh Dunn *see*
– The gun and the gospel
– Papers of hugh dunn fisher
Fisher Institute *see* Enterprise newsletter
Fisher, Irving *see*
– Comment vivre longtemps
– Wen ting huo pi yun tung shih
Fisher, Jacob Carney *see* A selected list of probate attorneys with general practice, as compiled for fisher's probate law directory... 1935-36
Fisher, James *see*
– The manitoba school question
– The school question
– The school question in manitoba
– Water transportation and freight rates
Fisher, Janet M *see*
– Effects of visual training on visual pursuit, catching, and attentiveness
– Functional motor skills and the developmentally disabled
Fisher, John *see*
– An illustrated record of the retrospective exhibition held at south kensington
– National competitions 1896-1897
– National competitions 1896-97
Fisher, Joseph Robert *see* The end of the irish parliament
Fisher, M K *see*
– Ludna-ndembu handbook
– Lunda for beginners
Fisher, Michele M *see* The effect of submaximal exercise in a neutral or hot-humid environment on recovery hemodynamics in men and women
Fisher, N M *see* Muscle rehabilitation of patients with osteoarthritis of the knees
Fisher, Peter *see* Proportional yields and processing of pork derived from different halothane hhyperthermia pig genotypes
Fisher, R *see* Fisher's prize cases in pennsylvania
Fisher, Robert Howie *see* The four gospels
Fisher, Ronald Aylmer *see* Statistical tables for biological, agricultural
Fisher, S B *see*
– Problems and quiz. bills and notes
– Problems and quiz on common-law pleading
Fisher, S S *see* Fisher's patent reports in the u.s. supreme and circuit courts
Fisher, Samuel *see*
– Christian warfare
– Christian's monitor
– Conjugal and parental duties stated and enforced
Fisher, Sidney G *see* The trial of the constitution
Fisher, Sydney *see* Conscription and true liberalism
Fisher, Sydney George *see* The true william penn
Fisher, William A *see* Notes on music in old boston
Fisher, William Edward Garrett *see* Transvaal and the boers
Fisher, William H *see* Fisher's patent cases in the u.s. circuit courts
Fisher, William Logan *see*
– History of the institution of the sabbath day, its uses and abuses
– An inquiry into the laws of organized societies
Fisher, William Richard *see* The law of mortgage and other securities upon property, vol 1
The fisheries dispute and annexation of canada / De Ricci, James Herman – London, 1888 – (= ser 19th c books on british colonization) – 4mf – 9 – mf#1.1.1176 – uk Chadwyck [343]
The fisheries dispute and annexation of canada / De Ricci, James Herman – London: S Low, Marston, Searle & Rivington, 1888 [mf ed 1981] – 4mf – 9 – mf#06409 – cn Canadiana [343]
Fisheries management – Oxford. 1980-1984 (1,5,9) – (cont by: aquaculture and fisheries management) – ISSN: 0141-9862 – mf#15503 – us UMI ProQuest [639]
Fisheries of key west and the clam industry of sou... / Schroeder, William Charles – Washington, DC. 1924 – 1r – us UF Libraries [639]
Fisheries research – Amsterdam. 1981-1994 (1,5,9) – ISSN: 0165-7836 – mf#42495 – us UMI ProQuest [639]
Fisherman / Pacific Coast Fishermen's Union et al – 1983 jan 14-1987 dec 11, 1988 jan 22-1991 dec 13, 1992 jan 20-1994 dec 13 – 3r – 1 – mf#1320931 – us WHS [331]
Fishermen's net / Grand United Order of Galilean Fishermen – 1906 jun 30 – 1r – 1 – mf#4024193 – us WHS [639]
Fishermen's Union of the Atlantic *see* Official year book of the fishermen's union of the atlantic

Fishers irish railway advertiser and general advertiser – Dublin, Ireland. Sep 1846-jun 1849 – 1/4r – 1 – uk British Libr Newspaper [072]

Fisher's juvenile scrapbook – 1836-50 – (= ser English gift books and literary annuals, 1823-1857) – 30mf – 9 – uk Chadwyck [800]

Fishers of men : or, practical hints to those who would win souls / Roberts, Benjamin Titus – Rochester, NY: G L Roberts, 1878 [mf ed 1984] – 4mf – 9 – 0-8370-1084-5 – mf#1984-4443 – us ATLA [240]

Fishers of men / Robberds, J G – London, England. 1846 – 1r – us UF Libraries [240]

Fisher's patent cases in the u.s. circuit courts / Fisher, William H – Cincinnati: Clarke & Co. v1-6. 1840-73 (all publ) – 48mf – 9 – $72.00 – mf#LLMC 81-449 – us LLMC [346]

Fisher's patent reports in the u.s. supreme and circuit courts / Fisher, S S – Cincinnati: Clarke & Co. 1v. 1821-50 (all publ) – 9mf – 9 – $13.50 – mf#LLMC 84-325 – us LLMC [346]

Fisher's prize cases in pennsylvania / Fisher, R – Philadelphia: Redwood Fisher. 1v. 1813 (all publ) – (= ser Early federal nominative reports) – 1mf – 9 – $1.50 – mf#LLMC 81-450 – us LLMC [343]

Fisher's probate law directory, 1914-1918 – St Louis: The Legal Directory Publ Co, 1918 – 4mf – 9 – $6.00 – (digests the probate laws for the us states and territories of the period) – mf#LLMC 96-053 – us LLMC [343]

Fisher's railway advertiser – Dublin, Ireland.Sept 1846-1849. -w. 1/4 reel – 1 – uk British Libr Newspaper [072]

Fisher-Stitt, Norma S *see* Effect of an interactive multimedia computer tutorial on students' understanding of ballet allegro terminology

Fishery bulletin – Washington. 1972+ (1) 1972+ (5) 1976+ (9) – ISSN: 0090-0656 – mf#7355 – us UMI ProQuest [639]

The fishery question : its imperial importance / Bourinot, John George – Ottawa: J Durie, 1886 – 1mf – 9 – mf#00189 – cn Canadiana [639]

The fishery question : its origin, history and present situation / Isham, Charles – New York, London: G P Putnam's Sons, 1887 – 2mf – 9 – mf#07340 – cn Canadiana [639]

The fishery question : letters from the n y herald's special commissioners – S.l: s.n, 1870? – 1mf – 9 – mf#12598 – cn Canadiana [343]

The fishery question : or, american rights in canadian waters / Kerr, William Hastings – Montreal?: D Rose, 1868 – 1mf – 9 – mf#23510 – cn Canadiana [343]

Fishery statistics of the us and annual reports *see* Us fish and wildlife service. fishery statistics of the us and annual reports

Fishery worker / Cannery Workers' Union of the Pacific, Los Angeles County Harbor District – 1948-54 – (= ser Labor union periodicals, pt 3: food and agricultural industries) – 1r – 1 – $210.00 – 1-55655-629-2 – us UPA [660]

Fishes of silver springs, florida / Allen, Ross – Silver Springs, FL. 1946 – 1r – us UF Libraries [590]

Fishgendler, A M *see*
- Dva puti
- Kooperatsiia v zapadnoi evrope i rossii
- Osnovnye voprosy sovetskoi kooperatsii
- Uspekhi potrebitelskoi kooperatsii

Fishing and fish along florida's coast – s.l, s.l? 193-? – 1r – us UF Libraries [639]

Fishing and shooting along the line of the canadian pacific railway : in the provinces of ontario, quebec, british columbia, the maritime provinces, and the prairies and mountains of western canada – Montreal: CPR, 1893 [mf ed 1982] – 1mf – 9 – mf#26779 – cn Canadiana [639]

Fishing and shooting along the lines of the canadian pacific railway : in the provinces of ontario, quebec, british columbia, the maritime provinces, the prairies and mountains of western canada, and in the state of maine – Montreal: CPR, 1896 [mf ed 1981] – 1mf – 9 – mf#26778 – cn Canadiana [639]

Fishing and shooting along the lines of the canadian pacific railway : in the provinces of ontario, quebec, british columbia, the prairies and mountains of western canada, the maritime provinces, the state of maine, and in newfoundland – 12th ed. Montreal: CPR, 1899 [mf ed 1982] – 1mf – 9 – mf#26780 – cn Canadiana [639]

Fishing and shooting along the lines of the canadian pacific railway : in the provinces of ontario, quebec, british columbia, the prairies and mountains of western canada, the maritime provinces, the state of maine, and in newfoundland – 13th ed. Montreal: CPR, 1900 [mf ed 1981] – 1mf – 9 – mf#26781 – cn Canadiana [639]

Fishing creek baptist church. somerset, kentucky : church records – 1813-1949 – 1 – us Southern Baptist [242]

Fishing hazards / Crowe, F Hilton – s.l, s.l? 1936 – 1r – us UF Libraries [639]

Fishing news international – London. 1968+ (1) 1971-1977 (5) 1976-1977 (9) – ISSN: 0015-3044 – mf#2977 – us UMI ProQuest [639]

Fishing resorts along the canadian pacific railway, eastern division : where to go for trout, bass and maskinonge, and what it costs to get there – Montreal: Canadian Pacific Railway, 1887 – 1mf – 9 – mf#00431 – cn Canadiana [639]

Fishing vessels of the florida west coast / Smith, Gerard – s.l, s.l? 193-? – 1r – us UF Libraries [639]

Fishkill daily herald – Fishkill, NY. 1903-1913 (1) – mf#65115 – us UMI ProQuest [071]

Fishlake, John Roles *see* Lexilogus

Fishman, Joshua A *see* Language problems of developing nations

Fishrapper – Wheeler OR: Paul Rouse, 1980-85 [wkly] – 1 – (cont: nehalem bay fishrapper (1975-80)) – us Oregon Lib [071]

Fisica moderna racional y experimental. 1er t. / Piquer, Andres – Valencia, 1745 – 8mf – 9 – sp Cultura [610]

Fisiografia del guadiana / Hernandez Pacheco, Eduardo – Badajoz: Imprenta del Hospicio Provincial, 1929 – 1 – sp Bibl Santa Ana [550]

Fisk, James L *see* Idaho, her gold fields and the route to them: a handbook for emigrants

Fisk news – v42-[v49? n4?] [1967 fall-1974 summer] – 1r – 1 – mf#367518 – us WHS [378]

Fisk University *see* Challenge

Fiske, Amos Kidder *see*
- The great epic of israel
- The jewish scriptures
- The myths of israel
- West indies

Fiske, Asa Severance *see* Reason and faith

Fiske, Daniel Taggart *see* The creed of andover theological seminary

Fiske, Daniel Taggart et al *see* Jubilee anniversary of the pastorate of rev. d.t. fiske, d.d

Fisk[e] family association newsletter – n1-15 [1977 spring-1982 winter], n16-17/18 [1984 spring/summer-1985] – 1r – 1 – mf#1288267 – us WHS [929]

Fiske, John *see*
- The beginnings of new england
- A century of science
- Civil government in the united states
- Darwinism and other essays
- The destiny of man viewed in the light of his origin
- The discovery of america, vol 1
- The discovery of america, vol 2
- The discovery of america, vols 1-2
- Excursions of an evolutionist
- The idea of god as affected by modern knowledge
- Life everlasting
- Myths and myth-makers
- Outlines of cosmic philosophy
- Through nature to god

Fison, Lorimer *see*
- Articles, letters and miscellaneous papers, 1873-1907
- Correspondence, 16 october 1873-15 october 1878
- Correspondence from lewis henry morgan and some others,1870-81
- Letterbooks
- Linguistic material and correspondence on local customs
- Miscellaneous papers on fiji, letters, notes, book draft
- Various manuscripts and papers, 188?-19-?, and press copy book

Fisonomia moral de a montano / Roldan, Federico – Malaga: Revista Espanola de Estudios Biblicos, 1928 – 1 – sp Bibl Santa Ana [170]

Fiss, Sabine *see*
- Das beurteilen in der fachsprachlichen kommunikation
- Statistische methoden zur untersuchung von assoziationen zwischen antigenen und krankheiten

"Fit for freedom" : the slaves, slavery, and emancipation in the cape colony, south africa, 1806 to 1842 / Mason, John Edwin – Yale University 1992 [mf ed Ann Arbor MI: UMI 1993] – 7mf – 9 – (incl bibl) – us UMI ProQuest [960]

Fita, Fidel *see*
- Alfar moruno de badajoz
- Antiguas epigrafes de tanger, jerez y arcos de la frontera
- Ara romana de barcarrota
- Cartas de barolome jose gallardo. noticia
- El castro romano de caceres el viejo. nuevas inscripciones
- Colon estremena
- Colon extremeno? de vicente paredes
- Coria compostelana y templaria
- Disquisiciones americanas 2. don martin cortes y don diego colon, caballeros de santiago
- Duque de t'serclaes toma posesion academico numero real de la historia. noticias
- Durante el semestre pasado fallecieron...tambien d. pedro maria plano, en merida...
- En la sesion del 23 octubre...vacantes...fueron elegidos...academicos de numero...duque de t'serclares...
- Epigrafes hebreos de bejar y salamanca
- Epigrafia romana de merida
- Epigrafia romana de montanchez, rena, banos de la encina. santisteban del puerto, cartagena y cadiz
- Epigrafia romana y griega de la provincia de caceres. nuevas ilustraciones
- Epigrafia romana y visigotica
- Epigrafia romana y visigotica de garlitos, capilla, belalcazar y el guijo
- Epigrafia romana y visigotica de montemolin
- Epigrafia romana y visigotica. poza de la sal. merida. alburquerque
- Epigrafia visigotica y romana de barcelona, merida, morente y bujalance
- Excursion epigrafica por villar del rey, alhambra, venta de los santos, cartagena, logrono y orense
- Excursiones epigraficas. de monesterio a merida
- Fallecimiento de d. francisco jarrin, obispo de plasencia
- Fallecimiento del marques de monsalud
- Fallecimiento en plasencia de don jose benavides checa
- El guijo, belalcazar y capilla
- Han sido nombrados...en plasencia d. vicente paredes...
- La inquisicion en guadalupe
- Inscripcion hemisferica de santa cruz y lapida de solans de cabanas. notas a una carta de roso de luna
- Inscripcion romana de la parra y de almendralejo
- Inscripcion romana de merida
- Inscripcion romana de riolobos
- Inscripcion romana de valera la vieja, junto a fregenal
- Inscripciones constantinianas de merida
- Inscripciones ineditas de merida, badajoz, alanje, canete de las torres y vilches
- Inscripciones romanas de caceres, ubeda y alcala de henares
- Inscripciones romanas de merida y nava de rico malillo
- Inscripciones romanas ineditas de caceres, brandomil, naranco y lerida
- Inscripciones romanas ineditas de trujillo
- Inscripciones visigoticas
- Lapida romana inedita de almendralejo
- Lapida romana inedita, merida
- Lapidas ineditas de marchamalo, caceres, palencia y lugo
- Lapidas romanas de garlitos, arroyo del puerco y araya, en extremadura
- Lapidas romanas de jerez de los caballeros y moron de la frontera
- Lapidas romanas ineditas
- El marques de monsalud es sustituido por d adolfo bonilla
- El marques de monsalud miembro del instituto arqueologico de Berlin
- Montanchez. nueva inscripcion romana
- Monumentos hebreos (salamanca, bejar, plasencia, bemibre)
- Necrologia del marques de monsalud
- Nertobriga beturica
- Noticia sobre revista de extremadura y lapidas
- Noticias
- Noticias de...
- Noticias sobre el estado de la basilica de santa eulalia y nuevos hallazgos en merida
- Nueva inscripcion romana de santa amalia
- Nueva lapida romana del escurial
- Nueva lapida romana en serradilla
- Nuevas inscripciones de merida y sevilla
- Nuevas inscripciones romanas (coria, gijon)
- Nuevas inscripciones romanas de ibahernando
- Nuevas inscripciones romanas de merida
- Nuevas inscripciones romanas y visigoticas
- Nuevas inscripciones romanas y visigoticas de talavan y merida
- Nuevas lapidas romanas de noya, cando, cerezo y jumilla
- Nuevas lapidas romanas de santisteban del puerto, berlanga (badajoz), avila y retortillo
- Nuevas lapidas romanas de tarragona, palencia, salvarierra de los barros baeza y nava de mena
- Nunez de balboa. congreso de historia y geografia hispanoamericana, celebrado en sevilla
- Pasa a la comision de recompensar de la real academia de la historia
- Presento...nuevas inscripciones romanas que acaban de hacer en plasenica...extension de la peninsula
- Resena epigrafica. san martin e trebejo, mestanza, mazarron, tobarra, oreto, perales de milla, liria rubi, ampurias y olleros de pisuerga
- Un sarcofago romano de merida
- Sello legionario (de azuaga)
- Talavera la vieja. lapida
- Tesera romana de plomo extremena, que posee don antonio vives
- Tres lapidas visigodas de merida

Fitch, Adelaide Paddock *see* East and west

Fitch, Ernest Robert *see* The baptists of canada

Fitch, J W *see* A debate on the beginning of messiah's reign, the abrogation of the mosaic law, and first proclamation of the gospel

Fitch, John *see*
- Papers
- Steamboat invention drawings, ca 1784-1826

Fitch, Joshua Girling *see*
- Art of questioning
- Art of securing attention in a sunday school class
- Thomas and matthew arnold

Fitch, Robert F *see* Ma-li-hsun hsiao chuan

Fitchburg 1751-1895 – Oxford, MA (mf ed 1987) – (= ser Massachusetts vital records) – 112mf – 9 – 0-87623-041-9 – (mf 1-3: index: b,m,d 1751-1872. mf 4: marriages & intentions 1754-96. mf 5-14: town & vital records 1776-1842. mf 15-17: town & vital records 1814-50. mf 18-23: births & deaths by family 1797-1843. mf 24-37: index to births 1844-1915. mf 38-39: b,m,d 1843-59. mf 40-58: births 1860-95. mf 59-63: marriages 1851-72. mf 64-70: index to marriages 1873-1916. mf 71-81: marriages 1873-1900. mf 82-90: index to deaths 1844-1916. mf 91-112: deaths 1856-1907) – us Archive [978]

Fitchburg star – Fitchburg, Verona WI. 1975 jan 30-1976 dec 30, 1977 jan 6-1978 may 25, 1978 jun 1-dec 28, 1979-1998 – 20r – 1 – mf#942021 – us WHS [071]

Fitchett, William Henry *see*
- The beliefs of unbelief
- The unrealized logic of religion

Fite, Warner *see* An introductory study of ethics

Fitnam, John Christopher *see* A practical treatise on the code summons and the mode of serving it as prescribed by the civil codes of colorado, wyoming, kansas, nevada, nebraska and other states.

Fitness for living – Emmaus. 1972-1974 (1) 1967-1974 (5) (9) – ISSN: 0015-315X – mf#6307 – us UMI ProQuest [360]

Fitness levels of children in north carolina / Baines, KC – 1991 – 1mf – 9 – $4.00 – us Kinesology [613]

The fitness of christianity to man / Huntington, Frederic Dan – New York: Thomas Whittaker, 1878. Beltsville, Md: NCR Corp, 1978 (2mf); Evanston: American Theol Lib Assoc, 1984 (2mf) – 9 – 0-8370-0819-0 – mf#1984-4163 – us ATLA [240]

The fitness of christianity to man / Huntington, Frederic Dan – New York: Thomas Whittaker 1878 [mf ed 1985] – (= ser The bohlen lectures 1878) – 1mf – 9 – 0-8370-3698-4 – mf#1985-1698 – us ATLA [240]

Fitoterapia : revista di studi ed applicazioni delle piante medicinali – Milano: Inverni and Della Beffa SpA, [v51-52. 1980-81] – 1r – 1 – us CRL [074]

Fitting human performance data: a comparison of three methods of data smoothing with mlab / York, Sherril L – 1981 – 2mf – 9 – $8.00 – us Kinesology [790]

Fitton, James *see* Sketches of the establishment of the church in new england

Fittz, Hervey *see* The friends of christ keep his commandments or obedience the test of discipleship

Fitz-edward / Colchester, Elizabeth Susan (Law) Abbot, Baroness – [London], 1875 – (= ser 19th c women writers) – 3mf – 9 – mf#5.1.61 – uk Chadwyck [830]

Fitzgerald, Augustus O *see* Establishment of a diocesan clergy retiring pension fund

Fitzgerald, D *see* Ritualistic teaching not the teaching of the church of england

Fitzgerald, Dani J *see* Cardiovascular endurance effects of a required college health, physical education, and recreation class

Fitzgerald, G M *see* Evacations in the tyropoeon valley, jerusalem 1927

Fitzgerald, James Edward *see*
- Irish migration
- Vancouver's island

Fitz-Gerald, John Driscoll *see* Historia de la universidad de arizona

Fitzgerald, Percy Hetherington *see*
- Croker's boswell and boswell. studies in the "life of johnson"
- The great canal at suez
- Life and times of john wilkes

Fitzgerald, Thomas Edward *see* Historical highlights of volusia county

Fitzgerald, W J *see* Report on the local administration of jerusalem

Fitzgerald, William *see*
- Cautions for the times
- Connexion of morality with religion
- Episcopacy, tradition, and the sacraments

Fitzgerald, William Forster Vesey. *see* Egypt, india, and the colonies

Fitzgerald, William G *see*
- The new el-dorado on the klondike
- The romance of seal hunting

Fitzgerald, William Walter Augustine *see* Travels in the coastlands of british east africa and the islands

Fitzgibbon, Gerald *see*
- Ireland in 1868
- The land difficulty of ireland

FitzGibbon, Mary Agnes *see*
- Cot and cradle stories
- A historic banner
- A trip to manitoba
- A veteran of 1812

Fitzgibbon, Maurice *see* Arts under arms
Fitzgibbon, Russell Humke *see*
- Constitutions of the americas
- Cuba and the united states

Fitz-James, James *see* Bahamian folk lore
Fitzmaurice-Kelly, James *see* Miguel de cervntes saavedra
Fitzpatrick, B B *see* Let's learn shona
Fitzpatrick, Charles *see* Les ecoles du manitoba
Fitzpatrick, Florence Baillie *see* A life of christ for children
Fitzpatrick, James Percy *see* The transvaal from within
Fitzpatrick, Mary Ann *see* Restraint reduction among the hospitalized elderly in intensive care units
Fitzpatrick, Percy *see* Transvaal from within
Fitzpatrick, William John *see*
- Correspondence of daniel o'connell, the liberator
- The life of the very rev. thomas n. burke, o.p
- The life, times and correspondence of the right rev. dr. doyle, bishop of kildare and leighlin
- The life, times, and correspondence of the right rev. dr. doyle, bishop of kildare and leighlin

FitzRalph, Richard *see* De dominio divino libri tres – de pauperie salvatoris
Fitzroy, Ed M *see* Papers, 1952-1953
FitzSimons, Mabel Trott *see* Hot words and hairtriggers
Fitzwater, Perry Braxton *see* The church and modern problems in the light of the teachings of paul in first corinthians
Fitzwilliam, Charles William Wentworth Fitzwilliam, earl of *see* A letter to the rev john sergeaunt
Fiumi, Lionello *see*
- Images des antilles

Fivaz, Derek *see*
- Shona language lessons
- Shona morphophonemics and morphosyntax
- Some aspects of shona structure
- Some aspects of the ideophone in zulu

Five african states / Carter, Gwendolen Margaret – Ithaca, NY. 1963 – 1r – us UF Libraries [960]
The five books of moses / Allis, O T – 2nd ed. Presbyterian and Reformed, 1949 – 9 – $12.00 – us IRC [221]
The five books of moses : a lecture. delivered at haarlem in 1870 / Kuenen, Abraham – London: Williams & Norgate, 1877 [mf ed 1992] – 1mf – 9 – 0-524-05226-3 – (english trans fr dutch by john muir. incl bibl ref) – mf#1992-0359 – us ATLA [221]
Five contemporary liberal preachers / Dahir, James Safady – Chicago, 1932. Chicago: Dep of Photodup, U of Chicago Lib, 1971 (1r); Evanston: American Theol Lib Assoc, 1984 (1r) – 1 – 0-8370-0272-9 – mf#1984-B159 – us ATLA [240]
Five discourses preached before the university of cambridge / Brown, John – London, England. 1840 – 1r – us UF Libraries [240]
Five feather news – 1971-n.d. – (= ser American indian periodicals... 1) – 9mf – 9 – $105.00 – us UPA [305]
Five feathers news / Tribe of Five Feathers [US] – v1 n10-11 [1971 oct/nov ?], v2 n2-10 [1972 feb-oct?] v2 n12-v3 n1 [1972 dec-1973 jan ?], v3 n11-v4 n8 – 1r – 1 – (cont by: tribe of five feathers) – mf#516580 – us WHS [305]
The five great duties of the aryans : being 1. meditation of first principles, 2. purification of atmosphere, 3. service of elders, 4. support of fellow-creatures, 5. hospitality = Panchamahayajnavidhi / Dayananda Sarasvati – 2nd ed. Lahore: Harbinger, [1913?] – 2mf – 9 – 0-524-03599-7 – (in english) – mf#1990-3243 – us ATLA [280]
Five great oxford leaders : keble, newman, pusey, liddon and church / Donaldson, Augustus Blair – London: Rivingtons, 1900 – 1mf – 9 – 0-524-00536-2 – mf#1990-0036 – us ATLA [240]
Five great painters / Eastlake, Elizabeth (Rigby) – London 1883 – (= ser 19th c art & architecture) – 6mf – 9 – mf#4.2.359 – uk Chadwyck [750]
The five great philosophies of life / Hyde, William De Witt – [2nd ed]. New York: Macmillan, 1911 – 1mf – 9 – 0-7905-0376-X – (incl ind) – mf#1987-0376 – us ATLA [180]
Five hundred bible readings : or, light from the lamp of truth / Marsh, Frederick Edward – 4th ed. New York: Gospel Pub House, [1904?] [mf ed 1992] – (= ser Christian & missionary alliance coll) – 1mf – 9 – 0-524-02481-2 – mf#1990-4340 – us ATLA [220]
Five hundred questions on subjects requiring investigation in the social condition of natives / Long, James – Calcutta, 1862 – (= ser 19th c british colonization) – 1mf – 9 – mf#1.1.7069 – uk Chadwyck [360]
Five italian madrigal books of the late 16th century : a transcription and study of the first books a cinque by antonio il verso, bartolomeo roy, bernardino scaramella, pietro paolo quartieri, and emilio virgelli / Watanabe, Ruth Taiko – U of Rochester 1951 [mf ed 19–] – 6v on 1r – 1 – (contents: v1: commentary; v2: madrigals of il verso in transcription; v3: madrigals of roy in transcription; v4: madrigals of scaramella in transcription; v5: madrigals of quartieri in transcription; v6: madrigals of virgelli in transcription) – mf#film 187 – us Sibley [780]
Five jatakas – Copenhagen: C A Reitzel 1861 [mf ed Bloomington IN: Indiana Uni Lib, Preservation Dept 1984] – 71p on 1r – 1 – us Indiana Preservation [390]
Five lectures on the character of st paul : with a sermon preached before the university on ascension day, 1863 / Howson, John Saul – London: Longman, Green, Longman, Roberts & Green; Cambridge: Deighton, Bell, 1864 – 1mf – 9 – 0-7905-1112-6 – (incl bibl ref) – mf#1987-1112 – us ATLA [225]
Five lectures on the gospel of st john as bearing testimony to the... / Blomfield, Charles James – London, England. 1823 – 1r – us UF Libraries [226]
Five letters on confirmation / Piers, Octavius – London, England. 1841 – 1r – us UF Libraries [242]
The five ministers : a sermon in west church / Bartol, Cyrus Augustus – Boston: A Williams, 1877 – 1mf – 9 – 0-524-06596-9 – mf#1991-2651 – us ATLA [240]
Five months in labrador and newfoundland : during the summer of 1838 / Tucker, Ephraim W – Concord [NH]: I S Boyd & W White, 1839 [mf ed 1983] – 2mf – 9 – 0-665-41420-X – mf#41420 – cn Canadiana [917]
Five percenter – 1989 jan – 1r – 1 – mf#5305295 – us WHS [071]
Five pieces selected from the celebrated opera of la clemenza di tito : adapted for the piano forte with an accompaniment for the flute by f fiorillo. 1st set / Mozart, Wolfgang Amadeus – London: R Birchall [179-?] [mf ed 1989] – 1r – 1 – mf#pres. film 41 – us Sibley [780]
Five plays / Chattopadhyaya, Harindranath – Madras: Shakti Karyalayam, 1937 – (= ser Samp: indian books) – us CRL [820]
Five points from barclay / ed by Wilbur, Henry Watson – Philadelphia, PA: Friends' General Conference Advancement Cttee, 1912 [mf ed 1993] – (= ser Society of friends (quakers) coll) – 1mf – 9 – 0-524-06674-4 – mf#1991-2729 – us ATLA [243]
The five points of calvinism / Dabney, Robert Lewis – Richmond VA: Presbyterian Cttee of Publ 1895 [mf ed 1993] – 1mf – 9 – 0-524-07233-7 – mf#1991-2974 – us ATLA [242]
The five practical plans : whereby we are able effectually to meet and remedy our present distressed and dangerous situation... / Edwards, George – Barnardcastle, 1820 – (= ser 19th c economics) – 2mf – 9 – mf#1.1.112 – uk Chadwyck [360]
Five problems of state and religion / Wood, William Converse – Boston: H Hoyt, 1877 – 1mf – 9 – 0-7905-3626-9 – mf#1989-0119 – us ATLA [240]
Five republics of central america / Munro, Dana Gardner – New York, NY. 1918 – 1r – us UF Libraries [972]
Five sermons / Douglas, Andrew Halliday – London: Hodder and Stoughton, 1903 – 1mf – 9 – 0-7905-7507-8 – mf#1989-0732 – us ATLA [240]
Five sermons on the principles of faith and church authority / Marriott, Charles – Littlemore, England. 1850 – 1r – us UF Libraries [240]
Five sermons on the temptation of christ our lord in the wilderness : preached before the university of cambridge in lent 1844 / Mill, William Hodge; ed by Webb, Benjamin – 2nd ed. Cambridge: Deighton, Bell; London: Bell and Daldy, 1873 – 1mf – 9 – 0-7905-2296-9 – mf#1987-2296 – us ATLA [240]
Five sermons preached before the university of cambridge / Trench, Richard Chenevix – London, England. 1843 – 1r – us UF Libraries [240]
Five sermons preached in oxford / Tait, Archibald Campbell – London, England. 1843 – 1r – us UF Libraries [240]
The five theological orations of gregory of nazianzus = Five theological orations / Gregory of Nazianzus, Saint; ed by Mason, Arthur James – Cambridge: University Press, 1899 – (= ser Cambridge Patristic Texts) – 1mf – 9 – 0-7905-9944-9 – mf#1989-1669 – us ATLA [240]
Five tomes against nestorius : scholia on the incarnation: christ is one: fragments against diodore of tarsus, theodore of mopsuestia, the synousiasts / Cyril, Saint, Patriarch of Alexandria – Oxford: J Parker and Rivingtons, 1881 – (= ser A library of fathers of the holy catholic church) – 1r – 1 – 0-8370-0268-0 – mf#1984-B113 – us ATLA [240]
Five ventures / Buckley, Christopher – London, England. 1954 – 1r – us UF Libraries [890]
Five years at panama : the trans-isthmian canal / Nelson, Wolfred – Montreal: W Drysdale, 1891 – 4mf – 9 – mf#11341 – cn Canadiana [918]
Five years' church work in the kingdom of hawaii / Staley, T N – London, 1868 – 3mf – 9 – mf#HTM-182 – ne IDC [917]
Five years in a sailor's life / Bech, Birger – Toronto: Queen's City Pub Co, 1886 – 1mf – 9 – mf#03533 – cn Canadiana [920]
Five years in china : from 1842 to 1847 / Forbes, F E – London: Richard Bentley, 1848 – 5mf – 9 – mf#HT-711 – ne IDC [915]
Five years in damascus : including an account of the history, topography, and antiquities of that city / Porter, J L – London: John Murray, 1855 – 2mf – 9 – 0-7905-1839-2 – (incl ind) – mf#1987-1839 – us ATLA [915]
Five years in kaffirland : with sketches of the late war in that country, to the conclusion of peace / Ward, Harriet – London: H Colbourn, 1848 – 1 – us CRL [960]
Five years in ludhiana : or, work amongst our indian sisters / Greenfield, M Rose – London: S W Partridge; Edinburgh: Religious Tract & Book Society, 1886 [mf ed 1995] – (= ser Yale coll) – vi/128p (ill) – 1 – 0-524-09418-7 – mf#1995-0418 – us ATLA [920]
Five years in madagascar... / Maude, F C – London, [1895] – 4mf – 9 – mf#HT-89 – ne IDC [916]
Five years in siam : from 1891 to 1896 / Smyth, Herbert Warington – Maps and illus. by author. London: J. Murray, 1898. illus. 9 maps. 2 v – 1 – us Wisconsin U Libr [959]
Five years in unknown jungles for god and empire : being an account of the founding of the lakher pioneer mission... / Lorrain, Reginald Arthur – London: Lakher Pioneer Mission [1912] [mf ed 1995] – (= ser Yale coll) – xii/274p (ill) – 1 – 0-524-09124-2 – mf#1995-0124 – us ATLA [920]
Five years of prayer : with the answers / Prime, Samuel Irenaeus – New York: Harper, 1864, c1863 [mf ed 1990] – 1mf – 9 – 0-7905-5673-1 – mf#1988-1673 – us ATLA [240]
A five years' residence in buenos ayres : during the years 1820 to 1825 containing remarks on the country and inhabitants; and a visit to colonia del sacramento / Love, George T – London 1827 [mf ed Hildesheim 1995-98] – 1v on 2mf – 9 – €60.00 – 3-487-26858-9 – (incl app) – gw Olms [918]
Five years' residence in the canadas : including a tour through part of the united states of america, in the year 1823 / Talbot, Edward – London 1824 [mf ed Hildesheim 1995-98] – 2v on 6mf – 9 – €120.00 – 3-487-27086-2 – gw Olms [917]
Five years' residence in the west indies / Day, Charles William – London, England. v1-2. 1852 – 1r – us UF Libraries [972]
A five-mile mountain bicycle test to predict vo2max / Veldhuis, Robert J & Butts, Nancy Kay – 1993 – 2mf – $8.00 – us Kinesology [612]
Five-minute sermons to children / Armstrong, William – New York: Methodist Book Concern, c1914 – 1mf – 9 – 0-524-08330-4 – mf#1993-2020 – us ATLA [240]
A five-year plan of ministry for the first baptist church, madison, illinois / Lindsey, Wilford Daniel – 1982 – 1 – $5.04 – us Southern Baptist [242]
Die fixsterne : darstellung der wichtigsten beobachtungs-ergebnisse und erklaerungs-versuche / Plassmann, Joseph – Kempten: J Koesel 1906 [mf ed 1998] – 1r [pl/ill] – 1 – (incl ind) – mf#film mas 28425 – us Harvard [520]
Fjaelloernen – Skelleftea, Sweden. 1848-49 – 1r – 1 – sw Kungliga [078]
Fjallkonan – Reykjavik, Iceland. -w. 29 Feb 1884-7 April 1911. 6 reels – 1 – uk British Libr Newspaper [079]
FJC Prisoner Civil Rights Committee *see* Recommended procedures for handling prisoner civil rights cases in the federal courts
Fjcff newsletter – v4 n5, v5 n1 [1975 jul, 1976 oct] – 1r – 1 – (cont: newsletter [american federation of teachers. local 2397 [jacksonville [fl]; cont by: news [florida junior college faculty federation]) – mf#647953 – us WHS [373]
Fjellstedt, Peter *see* Kwad laerer bibeln om foersoningen?
Fjerde juli og dakota – Fargo, Grand Forks ND. 1898 may 3-may 11 – 1r – 1 – (cont: dakota; fjerde juli; cont by: rodhuggeren; fram) – mf#901912 – us WHS [071]
Fkb mitteilungen – Berlin DE, 1929-32 – 1r – 1 – gw Misc Inst [074]
Fl merobaudis reliquiae (mgh1:14.bd) : blosii aemilii dracontii carmina – eugenii toletani episcipi carmina et epistulae / ed by Vollmer, F – 1905 – (= ser Monumenta germaniae historica 1: scriptores – auctores antiquissimi) – €25.00 – ne Slangenburg [240]
Flacara iasului – Iasi, Romania. 1962-Jun 1980; Apr-Oct 1981; 1982-88 – 26r – 1 – us L of C Photodup [949]
Flacara sibiului – Sibiu, Romania. 4 Jan 1956-1957; 4 Jan 1959-1962 – 3r – 1 – us L of C Photodup [949]
Flaccus, Horatius *see* Life, its true genesis
Flacius Illyricus d A, M *see*
- Antwort matthiae flacii illirici, auff das stenckfeldische buechlein iudicium etc genant
- Apologia matthiae flacij illyrici ad scholam vitebergensem in adiaphororum causa
- Das die buss rewe oder erkentnis des zorns vnd der suenden eigentlich allein aus dem gesetz
- Eine christliche predigt vber der leiche des herrn m: matthiae flacij jllyrici gestellet
- Compendiaria expositio doctrinae de essentia orig
- De essentia originalis ivstitiae et inivstitiae seu imaginis dei et contrariae
- De translatione imperii romani ad germanos
- De vocabvlo fidei et aliis qvibvsdam vocabvlis, explicatio uera et utilis, sumta ex fontibus ebraicis
- Defensio sanae doctrinae de originali ivstitia ac iniustitia, aut peccato
- Demonstrationes evidentissimae doctrinae de essentia imaginis dei et diaboli
- Liber de veris et falsis adiaphoris, in quo integre propemodum adiaphorica controuersia explicatur
- Matthiae flacij illyrici, de voce et re fidei, quodque sola fide iustificemur, contra pharisaicum hipocritarum fermentum
- Omnia latina scripta
- Refvtatio invectivae brvni contra centurias historiae ecclesiasticae

[Flacius Illyricus d A, M] *see*
- Antilogia papae
- Defensio confessionis ministrorvm iesv christi, ecclesiae antuerpiensis, quae augustanae confessioni adsentitur, contra ivdoci tiletani uaria sophismata
- Pia et necessaria admonitio de decretis et canonibvs concilii tridentini, sub pio quarto rom pontifice, anno etc 62 et 63 celebrati

Flacius Illyricus, Matthias *see* Disputatio de originali peccato et libero arbitrio
Flacius, M *see* Catalogus testium veritatis
Flack, A G *see*
- Democracy
- Moral education

Flackton, William *see* Chace
Flad, Johann Martin *see* Zehn Jahre in china
Flad, Joseph *see* Latin de l'eglise
Flaeming-Echo – Belzig DE, 1963 10 may-1965 18 sep – 1r – 1 – gw Misc Inst [074]
Flag Cancel Society *see* Flag cancellations
Flag cancellations : journal of the flag cancel society / Flag Cancel Society – 1983 feb/apr-1986 sep/oct – 1r – 1 – (cont by: machine cancel forum [1987]) – mf#1289931 – us WHS [366]
Flag of '98 – Warrenton VA. 1944 nov30, 1847 feb 6, 20, 1848 jun 17, 1850 jan 5, 1853 jun 18, aug 27-sep 3 – 2r – 1 – (cont by: warrenton flag of '98) – mf#882677 – us WHS [071]
A flag of distress / Campbell, Robert – London, Ont?: s.n, 1878 – 1mf – 9 – mf#26785 – cn Canadiana [242]
Flag of ireland – Dublin, Ireland. 5 sep 1868-24 dec 1869; 1870-feb 1882; 18 apr-8 dec 1882; 1883-24 dec 1886; 1887; 1895-10 sep 1898 – 28 1/2r – 1 – (aka: united ireland) – uk British Libr Newspaper [072]
Flag of ireland – Dublin, 13 dec 1890-24 jan 1891 – 1r – 1 – (cont as: united ireland 13 aug 1881-10 sep 1898; suppressed in dec 1890; 15 dec 1890-24 jan 1891 appeared as insuppressible, then as suppressed united ireland, then as united ireland again) – ie National [072]
Flag of ireland *see* United ireland
Flag of our union – Boston. 1854-1870 – 1 – mf#5549 – us UMI ProQuest [073]
Flag of seventy-six / Perry Co. Somerset – v1 n1. oct 1842-sep 1844 [wkly] – 1r – 1 – mf#B5530 – us Ohio Hist [071]
Flag on devil's island / Lagrange, Francis – Garden City, NY. 1961 – 1r – us UF Libraries [972]
Flag staff banner / Star Spangled Banner Flag House Association – 1972 winter-1975 autumn – 1r – 1 – mf#367517 – us WHS [366]
Flagellation and the flagellants : a history of the rod / Cooper, William M – London: John Camden Hotten, 1869? – 7mf – 9 – $10.50 – (incl ind) – mf#LLMC 91-082 – us LLMC [340]
Flagg, Elisha *see* A paper on the symbols, emblem, color and motto of the metaphysical society
Flagg, William Joseph *see* Yoga
Flagg, Wilson *see* Studies in the field and forest
Flagler beach – s.l, s.l? 193-? – 1r – us UF Libraries [978]
Flagler beach coast guard station / Scoville, Dorothy R – s.l, s.l? 1936 – 1r – us UF Libraries [978]

Flagler tribune – Bunnell, FL. 1918 dec 12-1981 sep – 35r – (gaps) – us UF Libraries [071]
Flagler/palm coast news-tribune – Bunnell, FL. 1982 jul 14-1997 may – 41r – us UF Libraries [071]
Flagler/palm coast news-tribune – Bunnell, FL. 1988 jul-1989 jul – 5r – (missing: 1988 jul 2,20; aug 3, 10,17; sep 3,7,14) – us UF Libraries [071]
Flagstaff – Davison, MI. 1992+ (1) – mf#68842 – us UMI ProQuest [071]
Flaherty, R J *see* The belcher islands of hudson bay
Flaherty, Robert F *see* Running economy and kinematic differences among running with the foot shod, with the foot bare, and with the bare foot equated for weight
Flaire – 1961 jun, nov – 1r – 1 – mf#4717672 – us WHS [071]
Flaischlen, Caesar *see*
– Hauff's werke
– Neuland
Flaka e vellazerimit – Skopje, Yugoslavia. Apr 1945-1955; Apr 1957-1960 – 4r – 1 – (some issues missing) – us L of C Photodup [949]
Flaka e vllaznimit – Skoplje, Yugoslavia. -w. 4 April-7 May 1945; Jan 1960-Dec 1970. 16 reels – 1 – uk British Libr Newspaper [949]
Flake, Otto *see*
– Christa
– Dinge der zeit
– Es ist zeit
– Freitagskind
– Freund aller welt
– Der gute weg
– Horns ring
– Montijo, oder, die suche nach der nation
– Nein und ja
– Schritt fuer schritt
– Die stadt des hirns
– Villa u.s.a.
Flakoll, D J *see* New voices of hispanic america
Flambeau – Clermont Ferrand, France. 6 apr 1941-5 jul 1942 – 1r – 1 – uk British Libr Newspaper [072]
Le flambeau democratique – Port-au-Prince. Haiti. mar. 8, 20, 1948 – 1 – us NY Public [079]
Le flambeau des anciens combattants de l'avant / Mouvement Croix de Feu – Paris. nov 1929-aout 1937, mars 1941-juil 1942 – 1 – fr ACRPP [073]
Le flambeau du macon – nouv ed. Bordeaux: Lawalle 1811 – 3mf – 9 – mf#vrl-116 – ne IDC [366]
Le flambeau; revue belge des questions politiques et literaires – v. 1-23. 1918-40. N.S. v. 31-32. 1948-49. Jan-Apr 1936 wanting – 1 – $277.00 – us L of C Photodup [949]
Flambeau sagueneen : monsieur l'abbe charles-elzear tremblay: (bio-bibliographie) / Tremblay-Boily, Germaine – [1964?] (mf ed 1979) – (= ser Bibliographies du cours...1947-66) – 4mf – 9 – (with ind; pref by victor tremblay) – mf#SEM105P4 – cn Bibl Nat [241]
The flame – Blantyre: Central Publ Ltd, 1993- (Lilongwe: Alpha Printers) [feb24/mar9-mar24/apr6, may 3/9, jun, oct, nov 29/dec 12 1993] – 2r – 1 – us CRL [079]
Flame heart / Claude McKay Secondary School [May Pen, Jamaica] – 1980, 1989 – 1r – 1 – mf#5307133 – us WHS [373]
The flame of hispanicism / Gonzalez, Palencia Angel – New York, 1938. Fiche W928. (Blodgett Collection of Spanish Civil War Pamphlets) – 9 – us Harvard [946]
Flamen, A *see* Devises et emblesmes d'amour moralisez
Flametti : oder, vom dandysmus der armen: roman / Ball, Hugo – Berlin: E Reiss, 1918 [mf ed 1987] – 224p – 1 – mf#7144 – us Wisconsin U Libr [830]
Flamingo feather / Munroe, Kirk – New York, NY. 1923 – 1r – us UF Libraries [978]
Flamion, Joseph *see* Les actes apocryphes de l'apotre andre
Flamman – Stockholm, Sweden. 1998- – 1 – sw Kungliga [078]
Flammarion, Camille *see*
– Contemplations scientifiques
– Gott in der natur
– La planete mars et ses conditions d'habitabilite
Die flamme – Dessau DE, 1964 31 dec-1989 13 oct [gaps] – 4r – 1 – (notes: gasgeraetewerk) – gw Misc Inst [621]
Flammen und winde : neue gedichte und gesaenge / Lissauer, Ernst – Stuttgart: Deutsche Verlags-Anstalt, 1923 – 1r – 1 – us Wisconsin U Libr [810]
Der flammenbaum : balladen / Blunck, Hans Friedrich – Muenchen: A Langen/G Mueller, 1935 [mf ed 1989] – (= ser Die kleine buecherei 46) – 49p – 1 – mf#7036 – us Wisconsin U Libr [780]
Der flammende pfeil : erzaehlung / Ehmer, Wilhelm – Stuttgart: J Engelhorns Nachf Adolf Spemann c1939 [mf ed 1990] – 1r – 1 – (filmed with: die geburt des jahrtausends / kurt eggers) – mf#7205 – us Wisconsin U Libr [830]
Flammenzeichen : ausgewaehlte zeitgedichte / Juengst, Hugo C – 2. Aufl. Dresden-Blasewitz: Verlag der "Deutschen Litteratur- und Kunstzeitung", [194-?] – 1r – 1 – us Wisconsin U Libr [810]
Flammenzeichen – Leonberg, Stuttgart DE, 1927 2 apr-1939 mar – 1 – gw Misc Inst [074]
A flammigera : revista maconica – Belem, PA: Typ do Santo Officio, 16 out 1873 – (= ser Ps 19) – mf#P17,02,136 – bl Biblioteca [240]
Flammulae amoris s p augustini versibus et iconibus exonatae... / Hoyer, M – Antverpiae: Apud Henricum Aertssens, 1629 – 2mf – 9 – mf#O-308 – ne IDC [090]
Flamsteed, John *see*
– The correspondence of john flamsteed
– Historiae coelestis britannicae..
Flamura prahovei – Ploiesti, Romania. 1962-81 – 23r – 1 – us L of C Photodup [949]
Flamuri : organ i ballit kombetar – Rome, Italy. jan-dec 1950; mar 1951-28 nov 1964; 28 nov 1965; 28 nov 1966; 20 may 1967 etc [mf 1950-70] – 1 – (albanian, french, english and italian) – mf#[1950-70:] m.f.877.h – uk British Libr Newspaper [074]
Flanagan, Brigid *see* Henry cloete in natal, 1843-1855
Flanagan, Lance *see* The history of volleyball in the united states
Flanders, George E *see* Civil war letters
Flanders, Steven *see*
– The 1979 federal district court time study
– Case management and court management in u.s. district courts
– Operation of the federal judicial councils
Flandin, Jean *see*
– Jugement sur les ministres actuels
– Revelations sur la fin du ministere de m le cte de villele
Flandrau, Grace *see* Being respectable
Flandre liberale – Ghent Belgium, 18 oct 1944-10 jul 1945 – 1r – 1 – uk British Libr Newspaper [074]
Flandreau Indian School *see* Spirit
Flandreau Indian School [SD] *see* Flandreau spirit
Flandreau spirit / Flandreau Indian School [SD] – v15 n31 [1982 apr 30], v16 n28-29 [1983 apr 22-29], v19 n1-2, 5-8 [1984 sep 20-oct 5, nov 2-dec 14], v19 n9 – 1r – 1 – (cont by: spirit [flandreau, sd: 1985]) – mf#819195 – us WHS [071]
Flandria illustrata / Sanderus, A – 's-Gravenhage. v1-3. 1735 – €201.00 – ne Slangenburg [240]
Le flaneur – Paris: H V de Surcy et Cie, may 1848 – us CRL [074]
Le flaneur des deux rives : bulletin d'etudes apollinariennes – Paris. n1-7 8. mars 1954-sept dec 1955 – 1 – fr ACRPP [440]
Flanigen, J R *see* Methodism old and new
Flannery o'connor bulletin – Milledgeville. 1972-1996 (1) 1974-1996 (5) 1974-1996 (9) – mf#8747 – us UMI ProQuest [920]
Flannery o'connor review – Milledgeville. 2001+ (1,5,9) – mf#33064 – us UMI ProQuest [920]
Flapdoodle : a political encyclopaedia and manual for public men / ed by Fuller, William Henry – Toronto: printed for the publ, 1881 – 1mf – 9 – (ill by bengough) – mf#27641 – cn Canadiana [320]
Die flasche und mit ihr auf reisen / Ringelnatz, Joachim [Hans Boetticher] – Berlin: Rowohlt, 1932 [mf ed 1989] – 178p – 1 – mf#7055 – us Wisconsin U Libr [830]
Flaschner, Gotthelf Benjamin *see* Zwanzig lieder vermischten inhalts
The flash light – Mansfield, PA. 1916-88. 7 rolls – 13 – $25.00r – us IMR [071]
Flash on angola – Lusaka: Dept of Information and Propaganda [sic] of the M P L A, nov 1971; apr-may 1972 – us CRL [960]
Flash report! : official publication of the madison area local postal workers union, afl-cio / Madison Area Local Postal Workers Union [Madison, WI] – 1978 dec-1985 dec, 1986 jan-1994 winter – 2r – 1 – mf#1701202 – us WHS [331]
A flashlight and compass : a collection of tools to promote instructional coherence / Copeland, Glenda [et al] – Austin TX: Southwest Educ Devt Lab; [Washington DC]: US Dept of Education, Office of Educ Research & Improvement...[2000] [mf ed 2001] – 3mf – 9 – us GPO [370]
Flashlights on evangelical history : a volume of entertaining narratives, anecdotes and incidents... / Stapleton, Ammon – 1st ed. York, PA: A Stapleton, 1908 [mf ed 1990] – 1mf – 9 – 0-7905-6627-3 – mf#1988-2627 – us ATLA [242]
Flashlights on nature / Allen, Grant – New York: Doubleday & McClure, 1898 – 4mf – 9 – (ill by frederick enock) – mf#28146 – cn Canadiana [590]
Flashpoint / Anarchist Communist Federation [Regina, SK] – v2 n8-v3 n4 [1978 sep-1980 oct] – 1r – 1 – mf#665059 – us WHS [335]
Flat bark beetles of florida / Thomas, M C – Gainesville, FL. 1993 – 1r – us UF Libraries [590]
Flat creek baptist church. petis county. missouri : church records – 1846-81 – 1 – us Southern Baptist [242]
Flat creek baptist church. weaverville, north carolina : church records – 1833-1931. WMU Minutes. 1915-21 – 1 – us Southern Baptist [242]
Flat gap baptist church. enterprise association. kentucky : church records – Apr 1869-Sep 1914 – 1 – 8.82 – us Southern Baptist [242]
Flat glass worker / United Glass and Ceramic Workers of North America – v1-v2 n8 [1934 sep-1936 apr] – 1r – 1 – mf#1111547 – us WHS [331]
Flat iron for a farthing / Ewing, Juliana Horatia – Leipzig, Germany. 1891 – 1r – us UF Libraries [025]
Flat, Paul *see* Lettres de bayreuth
Flat River Association. North Carolina *see* Manuscript minutes
Flat rock baptist church. anderson county. south carolina : church records – 1871-1911, 1913-54. Deacons' Minutes. 1970-72 – 1 – us Southern Baptist [242]
Flat top flash : published by kaiser co inc for the 36,000 workers in vancouver – Vancouver WA: Kaiser Co Inc [1943-44] [wkly] – 1r – 1 – (absorbed by: the bo's'n's whistle 1944) – us Oregon Lib [623]
Flat top flash *see* Bo's'n's whistle
Flatbush, Adda M *see* Methods and results of rescue work
Flathe, Theodor *see* Das zeitalter der restauration und revolution
Flathead county news – Polson, MT. 1808-1915 (1) – mf#64607 – us UMI ProQuest [071]
Flathead courier – Polson, MT. 1911-1974 (1) – mf#64608 – us UMI ProQuest [071]
Flathead herald journal – Kalispell, MT. 1893-1907 (1) – mf#64502 – us UMI ProQuest [071]
Flathead monitor and times monitor – Kalispell, MT. 1920-1950 (1) – mf#64503 – us UMI ProQuest [071]
Flather, John Joseph *see* Dynamometers and the measurement of power
Flatlands – v1 n3, 5-10,12-14,17,19-22, v2 n5-7,9 [1966 apr 9-1967 nov 1/14], v3 n2 [1968 aug 13] – 1r – 1 – mf#705460 – us WHS [071]
Flatworm as an enemy of florida oysters / Danglade, Ernest – Washington, DC. 1919 – 1r – us UF Libraries [639]
Flaubert et ses projets inedits / Durry, Marie Jeanne – Paris, France. 1950 – 1r – us UF Libraries [440]
Flaubert, Gustave *see*
– Briefe ueber seine werke
– Madame bovary
– Die versuchung des heiligen antonius
Flavelle, Joseph *see*
– Canada and its relations to the empire
– Munitions in canada
– An open letter addressed to the honourable the minister of agriculture for ontario
Flavian, Saint, Patriarch of Constantinople *see* Appellatio flaviani
Flavii iosephi antiquitatum iudaicarum epitome = Antiquitate judaicae. 1896 / Josephus, Flavius; ed by Niese, Benedikt – Berolini: apud Weidmannos, 1896. Chicago: Dep of Photodup, U of Chicago Lib, 1979 (1r); Evanston: American Theol Lib Assoc, 1984 (1r) – 1 – 0-8370-1331-3 – mf#1984-T180 – us ATLA [930]
Flavii iosephi opera / ed by Niese, Benedikt – Berolini [Berlin]: Apud Weidmannos, 1885-95 [mf ed 1992] – 7v on 8mf – 9 – 0-524-02782-X – (text in greek. crit app in latin & greek) – mf#1987-6476 – us ATLA [939]
Flavio herrera / Estrada, Ricardo – Guatemala, 1960 – 1r – us UF Libraries [972]
Flavio Josefo *see* Los siete libros...de la guerra que tuvieron los judios con los romanos...
Flavius *see* Arriani historici et philosophi ponti euxini et maris erythraei periplus...
Flavius, Josephus *see* Vom juedischen kriege
Die flavius josephus beigelegte schrift ueber die herrschaft der vernunft (4 makkabaeerbuch) : eine predigt aus dem ersten nachchristlichen jahrhundert / Freudenthal, Jacob – Breslau [Wroclaw]: Schletter, 1869 – 1mf – 9 – 0-8370-3194-X – (incl bibl ref) – mf#1985-1194 – us ATLA [270]
Flavor – 1992 mar-1995 jan, 1995 feb-1996 – 2r – 1 – mf#2947385 – us WHS [071]
Flavour and fragrance journal – Chichester. 1985-1996 (1,5,9) – ISSN: 0882-5734 – mf#16101 – us UMI ProQuest [640]
Flaxman, John *see*
– Anatomical studies of the bones and muscles, for the use of artists
– Flaxman's classical outlines...with a brief memoir of the artist
– Lectures on sculpture...before...the royal academy
Flaxman's classical outlines...with a brief memoir of the artist / Flaxman, John – London 1879 – (= ser 19th c art & architecture) – 4mf – 9 – mf#4.2.1495 – uk Chadwyck [700]
Flaxmere and western suburbs gazette – Napier, NZ. 1981-84 – 4r – 1 – (aka: leader) – mf#35.6 – nz Nat Libr [079]
Flayder, Friedrich Hermann *see* Hermann flayders ausgewaehlte werke
A flecha – Maranhao: Typ do Frias, 15 mar 1879-09 out 1880 – (= ser Ps 19) – 1,5,6 – mf#P30,04,18 – bl Biblioteca [079]
Flecha de sombra / Guerra Flores, Jose – Habana, Cuba. 1961 – 1r – us UF Libraries [972]
Flechazos / Aguila, Gilberto R – Santa Tecla, El Salvador. 1956 – 1r – us UF Libraries [972]
La fleche : de Paris / Frontisme – Paris. aout 1934-aout 1939 – 1 – fr ACRPP [073]
Flechier, Esprit *see* Voyage de flechier en auvergne
Fleck, Konrad *see* Flore und blanscheflur
Fleckenstein, J *see*
– Die hofkapelle der deutschen koenige (mgh schriften:16.bd 1.teil)
– Die hofkapelle im rahmen der ottonisch-salischen reichskirche (mgh schriften..:16.bd. 2.teil)
Flecker, Eliezer *see* Shemot ha-katvim
Flee fornication – London, England. 1843 – 1r – us UF Libraries [240]
Fleet equipment – Lincolnwood. 1988-1996 (1,5,9) – ISSN: 0747-2544 – mf#16427,01 – us UMI ProQuest [380]
Fleet owner : big fleet edition – Overland Park. 1982-1989 (1) 1982-1989 (5) 1982-1989 (9) – (cont: fleet owner) – ISSN: 0731-9622 – mf#751,01 – us UMI ProQuest [380]
Fleet owner – Overland Park. 1928-1982 (1) 1967-1982 (5) 1976-1982 (9) – (cont by: fleet owner big fleet edition) – ISSN: 0015-3567 – mf#751 – us UMI ProQuest [380]
Fleet owner – Overland Park. 1989+ (1) 1989+ (5) 1989+ (9) – (cont: fleet owner big fleet edition) – ISSN: 1070-194X – mf#751,02 – us UMI ProQuest [380]
Fleet owner : small fleet edition – New York. 1980-1989 (1) 1980-1989 (5) 1980-1989 (9) – ISSN: 0162-1025 – mf#12317 – us UMI ProQuest [380]
Fleet papers : being letters from richard oastler with occasional communications from friends – v1-4. 1841-44 [all publ] – (= ser Radical periodicals of great britain, 1794-1914. period 1) – 20mf – 9 – $125.00 – us UPA [330]
Fleet's in! / Holman, Russell – New York, NY. 1928 – 1r – us UF Libraries [025]
Fleetwood chronicle – [NW England] Fleetwood 12 apr 1845-26 oct 1849, 1851-81, 1889-94, 1925, 1948, 1956-57, 1959, 1984 – 1 – (title change: fleetwood chronicle & general advertiser for blackpool, poulton, kirkham, lytham, ulverston & lonsdale north of the sands) – uk MLA; uk Newsplan [072]
Fleetwood express – [NW England] Fleetwood Lib jan 1896-dec 1920 [wkly] – 35r – 1 – uk MLA; uk Newsplan [072]
Fleetwood, John *see* The life of our blessed lord and saviour jesus christ
Fleetwood weekly news – [NW England] Fleetwood 4 sep 1984- – 1 – uk MLA; uk Newsplan [072]
Flegel, Eduard *see* Vom niger-benue
Flegmatov, Andrei *see* Besedy po russkomu raskolu i sektantstvu
Fleisch, Urban *see* Die erkenntnistheoretischen und metaphysischen grundlagen der dogmatischen systeme von a.e. biedermann und r.a. lipsius
Fleischer, Arthur C *see* Social and administrative problems of labour migration in south africa
Fleischer, George W *see* Civil war letters
Fleischer, Hans-Heinrich *see* General-inspektion des militaer-verkehrswesens (bestand ph 9 5) / inspektion des militaer-luft- und kraftfahrwesens (bestand ph 9 20) bd 25
Fleischer, Hans-Heinrich et al *see* Kaiserliches marinekabinett (bestand rm 2) bd 28
Fleischer, Nat *see* Jack dempsey
Fleischhack, Marianne *see*
– Sein bauernmaedchen
Fleischlin, B *see*
– Studien und beitraege zur schweizerischen kirchengeschichte
Fleischmann, Carl *see* Wegweiser und rathgeber nach und in den vereinigten staaten von nord-amerika
Fleisher, Gerald *see* Concerto for clarinet and string orchestra by aaron copland
Fleiuss, Max *see*
– Ferias
– Historia adminisdtrativa do brasil
– Institut historique et geographique du bresil
Fleming, Alexander *see*
– Historical lecture on teinds or tithes
– Letter to the right hon sir robert peel
Fleming, Ann Cuthbert *see* A year in canada
Fleming, Christopher Alexander *see* How to write a business letter

Fleming, Daniel Johnson *see*
– Church formation in india
– Devolution in mission administration
– The social mission of the church in india

Fleming, David Hay *see*
– Mary, queen of scots
– The reformation in scotland
– The scottish reformation
– Six saints of the covenant
– The story of the scottish covenants in outline

Fleming, Francis P *see* Did the florida legislature of 1891 elect a senato...

Fleming, Henry *see* Papers of henry fleming, 1772-1795

Fleming, J *see*
– Greatness at the feet of jesus
– Little means and large results
– Rich made low
– Salvation through dreams

Fleming, J S *see* What is ku kluxism?

Fleming, James *see*
– Catalogue of garden, agricultural and flower seeds for sale by james fleming, seedsman and florist, yonge street, toronto
– Darkness fleeing before light

Fleming, John Robert *see* The secession of 1733

Fleming, Paul *see*
– Paul flemings deutsche gedichte
– Paul flemings lateinische gedichte

Fleming, Peter *see*
– Brazilian adventure

Fleming, Rachel M *see* Stories from the early world

Fleming, Robert *see*
– Apocalyptical key
– The rise and fall of papacy
– Seculum davidicum redivivum
– Sketch of the life of elder humphrey posey

Fleming, Robert Alexander *see* A short practice of medicine

Fleming, Samuel Todd *see* Agricultural college organization in land-grant institutions

Fleming, Sandford *see*
– Address delivered in convocation hall, queen's college, kingston, april 28th, 1885
– An address on build up canada
– An appeal to the canadian institute on the rectification of parliament
– Canadian pacific railway
– Canadian pacific railway, ottawa, 1st july, 1880
– Cheap telegraph rates
– England and canada
– The establishment of a great imperial intelligence union as a means of promoting the consolidation of the empire
– Exploratory survey of 1871
– Imperial intelligence department
– Letter to his honour the lieut-governor
– Letter to the president of the america sic society for the advancement of science
– Letter to the secretary of state, canada
– Memorandum on the canadian pacific railway
– The new canadian trans-continental railway
– The pacific cable
– Postal and telegraphic communication by the canadian route
– Postal telegraph service by sea and land
– Progress report on the canadian pacific railway exploratory survey
– Rapport sur l'exploration preliminaire du chemin de fer intercolonial
– Report and documents in reference to the canadian pacific railway
– Report in reference to the canadian pacific railway
– Report on surveys and preliminary operations on the canadian pacific railway up to january 1877
– Report on the intercolonial railway exploratory survey
– The story of the steamship
– Views of many eminent canadians on the establishment of an imperial intelligence service on a comprehensive scale

Fleming, Sarah Hollis *see* Comparative study of some aspects of the supernatural

Fleming, Thomas *see* Nature, importance and right exercise of christian zeal

Fleming, Wallace Bruce *see* The history of tyre

Fleming, William Kaye *see* Mysticism in christianity

Flemingston, glamorgan, parish church of st michael : baptisms 1576-1900, burials 1576-1900, marriages 1578-1839 – 1mf – 9 – £1.25 – uk Glamorgan FHS [929]

Flemingston, st michael; and llanmaes, st catwg, monumental inscriptions – 1mf – 9 – £1.25 – uk Glamorgan FHS [929]

Flemming, J *see*
– Das buch henoch
– Die syrische didascalia

Flemming, Johannes *see* Die grosse steinplatteninschrift nebukadnezars 2. in transcribiertem babylonischen grundtext

Flemming, Wilhelm *see* Zur beurteilung des christentums justins des maertyrers

Flemming, Willi *see* Catharina von georgien

Flemming-Benz, Hasso graf von [comp] *see* Der kreis cammin

Flemyng, Francis Patrick *see*
– Kaffraria, and its inhabitants
– Southern africa

Flensborg avis : zweisprachige tageszeitung – Flensburg DE, 1951-1960 30 nov – 19r – 1 – (filmed by misc inst: 1947 2 may-1978 29 apr; 1976 1 may- [ca 6r/yr]) – gw Mikrofilm; gw Misc Inst [074]

Flensburger fackel – Flensburg DE, 1930 oct-1931 feb – 1 – gw Misc Inst [074]

Flensburger nachrichten der militaerregierung – Flensburg DE, 1945 16 aug-1946 28 mar 28 – 1r – 1 – gw Misc Inst [943]

Flensburger ns-zeitung – Flensburg DE, jul 16 1932-jun 24 1933 – 1 – gw Misc Inst [943]

Flensburger stimme – Flensburg DE, 1949 2 jul-1951 4 aug – 1r – 1 – gw Misc Inst [074]

Flensburger tageblatt – Flensburg DE, 1946 6 apr-1956 – 30r – 1 – uk British Libr Newspaper [074]

Flensburger tageblatt – Flensburg DE, 1969- – ca 7r/yr – 1 – gw Misc Inst [074]

Flesch, Fritz, collector *see* Fritz flesch collection on jews in south africa

Fleshman, Arthur Cary *see* Human thinking

Fletcher *see* Reflections on the spirit

Fletcher, Banister *see*
– A history of architecture for the student, craftsman, and amateur
– Model houses for the industrial classes

Fletcher, Banister Flight *see*
– A book about fans
– A history of architecture for the student, craftsman, and amateur
– The influence of material on architecture

Fletcher, Edward Taylor *see*
– The lost island
– The lost island of atlantis
– On languages as evincing special modes of thought
– Our lord at bethany

Fletcher, Edwin W *see* Hellenism in england

Fletcher, Ella Adelia *see* The law of the rhythmic breath

Fletcher forum *see* Fletcher forum of world affairs

Fletcher forum of world affairs – v1-25. 1976-2001 – 9 – $393.00 set – (title varies: v1-12 1976-88 as fletcher forum) – ISSN: 1046-1868 – mf#110141 – us Hein [337]

Fletcher, Henry Charles *see*
– A lecture delivered at the literary and scientific institute, ottawa
– Memorandum on the militia system of canada
– A volunteer force

Fletcher, John *see*
– Difficulties of protestantism
– The portrait of st. paul
– Second letter to the right honourable lord
– Works

Fletcher, John Joseph Kilpin *see* The sign of the cross in madagascar

Fletcher, Joseph *see*
– Devout observation of national calamities enforced
– The funeral discourse, occasioned by the death of the rev robert morrison
– Funeral discourse on the death of the rev william orme
– Holy scriptures the only standard of divine truth
– On the attention due to unfulfilled prophecies
– Reforamation

Fletcher, Joseph Smith *see* A short life of cardinal newman

Fletcher, L J *see* Guide to salvation

Fletcher, M Scott *see* The psychology of the new testament

Fletcher, Mary E *see* Crustula juris

Fletcher, Norman *see* Wutomi gi nene

Fletcher of madeley / Macdonald, Frederic William – New York: AC Armstrong, 1886 – (= ser Heroes of Christian History) – 1mf – 9 – 0-7905-8510-3 – mf#1989-1735 – us ATLA [900]

Fletcher, Reginald James *see* A study of the conversion of st paul

Fletcher, S *see* The late earl stanhope's political opinions

Fletcher, William Meade *see* Cyclopedia of the law of private corporations

Fletcher, William S *see* At sea and in port

Flett, Austin T *see* United states as a satellite nation

Fleuchaus, Andrea *see* Rekombinante modifizierte vaccinia ankara viren zur expression von siv-antigenen

Fleur de lys / Ledoux, Albert H – v1 n1-v2 n4 [1979 mar-1980 winter] – 1r – 1 – mf#671674 – us WHS [071]

Une fleur du carmel : la premiere carmelite canadienne, marie-lucie-hermine fremont: en religion soeur therese de jesus / Braun, Antoine – Quebec?: L Brousseau, 1881 – 7mf – 9 – mf#26688 – cn Canadiana [241]

Fleurette : ou le premier amour de henri 4 / Thierry, Auguste Francois – Paris, France. 1835 – 1r – us UF Libraries [440]

Fleuriau d'Armenonville, T *see* Nouveaux memoires des missions de la compagnie de jesus dans le levant

Fleurieu, C P C de *see* Voyage autour du monde, pendant les annees 1790, 1791 et 1792

Les fleurs boreales / les oiseaux de neige : poesies canadiennes couronnees par l'academie francaise / Frechette, Louis – Montreal: C O Beauchemin, 1886 – 4mf – 9 – 0-665-90972-1 – mf#90972 – cn Canadiana [810]

Les fleurs de givre / Chapman, William – Paris: editions de la Revue des poetes, 1912 – 3mf – 9 – 0-665-75958-4 – mf#75958 – cn Canadiana [810]

Les fleurs de la charite – [S.l: s.n, 1897?-19–] – 9 – (cont: bibliotheque canadienne-francaise. ceased 1950? incl ind) – mf#P04591 – cn Canadiana [440]

Fleurs de la palestine : 54 feuilles en chromolithographie = Wild flowers of the holy land / Zeller, Hannah – Bale: C F Spittler, [1876?] – (= ser Feldblumen Aus Dem Heiligen Land) – 1mf – 9 – 0-8370-5955-0 – (also publ in german under title: feldblumen aus dem heiligen land) – mf#1985-3955 – us ATLA [580]

Fleurs des alpes : episode de la vie du roi louis 2 de bavioere / Baltz, Johanna – Lausanne: Bureau de la Bibliotheque Universelle, 1888 [mf ed 1993] – 79/12p (ill) – 1 – (trans of: alpenrosen und gentianen) – mf#8509 – us Wisconsin U Libr [830]

Les fleurs des histoires de la terre d'orient : divisees en cinq parties – Lion, 1585 – 3mf – 9 – mf#H-8421 – ne IDC [956]

Fleury, Alfred *see*
– Instructions philosophiques sur la francmaconnerie
– Raison et religion

Fleury, Amedee *see* Saint paul et seneque

Fleury, Claude *see* Catechisme historique

Fleury de Chaboulon, Pierre *see*
– Memoires pour servir a l'histoire de la vie privee
– Memoires pour servir a l'histoire de la vie privee, du retour, et du regne de napoleon en 1815

Fleury, Francois *see* Courte reponse aux dernieres attaques contre la brochure calvin a geneve

Fleury, Jules *see* Monsieur de boisdhyver

Fleury mesplets, pionnier de l'imprimerie a montreal : causerie faite au diner annuel des maitres imprimeurs de montreal le 19 avril 1939 / Morin, Victor – Montreal: Compagnie de papier Rolland, 1939 [mf ed 1987] – 1mf – 9 – (mesplet, fleury) – mf#SEM105P763 – cn Bibl Nat [920]

Fleury-Giroux, Marie *see* Les occupations en milieu urbain

Fleuve st-laurent : etudes biologiques: vol 1: bibliographie generale annotee / Laperle, Marcel & Lamoureux, J-P – [Montreal]: Dimension environnement ltee, 1975 [mf ed 1992] – 6mf – 9 – mf#SEM105P1499 – cn Bibl Nat [574]

Flewelling, Ralph Tyler *see*
– Christ and the dramas of doubt
– Personalism and the problems of philosophy

Flex, Konrad *see* Walter flex

Flex, Oscar Theodor *see* Aus dem palmenlande

Flex, Walter *see*
– Briefe
– Ihr lebt!
– Lothar
– Wallensteins antlitz
– Wolf eschenlohr
– Zwei bismarcks unter schwedischen fahnen
– Zwoelf bismarcks

Flexibilisierung der kardiologischen rehabilitation durch teilstationaere behandlungsmodelle / Weinheimer, Heike Birgit Karin – (mf ed 1997) – 2mf – 9 – €40.00 – 3-8267-2417-8 – mf#DHS 2417 – gw Frankfurter [617]

Flexible arbeitszeit : bedingungen und ziele der durchfuehrung flexibler arbeitszeiten unter besonderer beruecksichtigung ihrer anwendung bei mechanischer technologie / Utsch, Juergen – (mf ed 1994) – 2mf – 9 – €49.00 – 3-89349-864-8 – mf#DHS 864 – gw Frankfurter [331]

Flibustier / Vilaire, Etzer – Port-Au-Prince, Haiti. 1902 – 1r – us UF Libraries [972]

Fliche, A *see* Histoire de l'eglise (he)

Flick, Alexander Clarence *see* The rise of the mediaeval church

Flickinger, Daniel Kumler *see* History of the origin and development and condition of missions among the sherbro and mendi tribes in western africa

Fliedner, Fritz *see* Das evangelium in den roemischen landen

Fliedner, Wilhelm *see* Goethe und christentum

Fliegen, Ina *see* Berufsbezogene possible selves in der betrieblichen weiterbildung

Fliegende blaetter aus dem rauhen hause zu horn bei hamburg (fw1) – 1844/45-1905 [mf ed 2004] – (= ser Freie wohlfahrtspflege (fw) 1) – 62v on 645mf – 9 – €2900.00 – 3-89131-451-5 – (die innere mission im evangelischen deutschland: ns: v1=63 1906-v15=77 1921; v16-26 1921-31; die innere mission v27-36 1932-41; with suppls: das beiblatt der fliegenden blaetter aus dem rauhen hause...v1-33 1850-82; geschichten und bilder zur foerderung der inneren mission v34-57 1883-1906; geschichten und bilder aus der christlichen liebestätigkeit v58-71 1907-20; die rundschau v1-12 1930-41) – gw Fischer [242]

Fliegende blaetter fuer freunde der toleranz, aufklaerung und menschenbesserung / ed by Fischer, Gottlob Nathanael – Dessau [Leipzig] 1783 – (= ser Dz) – 1jg on 3mf – 9 – €90.00 – mf#k/n5120 – gw Olms [302]

Fliegende blaetter (sz1) – Muenchen. v1 n1-160 n4099 1845-1924, ann v80 n4100-100 n5174 1924-1944; suppl v52-70 n9 1870-1879; suppl v70 n10-158 1879-1923 [mf ed 1998] – (= ser Satirische zeitschriften (sz) 1) – 974mf – 9 – diazo €4860 silver €6560 – 3-89131-278-4 – gw Fischer [870]

Der fliegende hollaender : romantische oper in 3 aufzuegen / Wagner, Richard – Berlin: C F Meser [18–?] [mf ed 1992] – 1r – 1 – (filmed with: lohengrin) – mf#7806 – us Wisconsin U Libr [790]

Die fliegende taube – Aubel (B), 1918 19 nov-1922, 1924-1940 24 apr [gaps] – 1 – gw Misc Inst [074]

Fliegende volksblaetter [...] – Bayreuth DE, 1797 jul-nov, 1798 – 1r – 1 – gw Misc Inst [074]

Flieger, Heinrich *see* Die oeffentliche meinung in der staatsphilosophie von thomas hobbes

Fliegerschule 4 : buch der mannschaft / Euringer, Richard – Hamburg: Hanseatische Verlagsanstalt 1942, c1929 [mf ed 1989] – 1r – 1 – (filmed with: die arbeitslosen) – mf#7226 – us Wisconsin U Libr [830]

Fliegner, Ferdinand *see* Bilder aus constantinopel

Fliegt der blaufuss? : roman aus der flaemischen bewegung unserer tage / Bruees, Otto – Berlin: G Grote, 1935 [mf ed 1989] – (= ser Grote'sche sammlung von werken zeitgenoessischer schriftsteller 216) – 219p – 1 – mf#7092 – us Wisconsin U Libr [830]

Flierl, Johann *see*
– Dreissig jahre missionsarbeit in wuesten und wildnissen
– Evangelical lutheran synod of iowa and other states
– Gedenkblatt der neuendettelsauer heidenmission in queensland und neu-guinea, 1885-1910

Flight – Fort Worth. 1934-1974 (1) 1971-1974 (5) – (cont by: flight operations) – ISSN: 0015-3729 – mf#1539 – us UMI ProQuest [629]

Flight aircraft engineer – 1909-58 – 1 – us L of C Photodup [629]

Flight for life and an inside view of mongolia / Roberts, James Hudson – Boston: Pilgrim Press, [1903] [mf ed 1995] – (= ser Yale coll) – 402p (ill) – 1 – 0-524-09866-2 – mf#1995-0866 – us ATLA [915]

Flight international – London. 1909+ (1) 1974+ (5) 1974+ (9) – ISSN: 0015-3710 – mf#662 – us UMI ProQuest [629]

Flight jacket – Mission Viejo CA. 1987-90 – 4r – 1 – mf#3611060 – us WHS [071]

Flight journal – Ridgefield. 1997+ (1) – ISSN: 1095-1075 – mf#22546,01 – us UMI ProQuest [629]

Flight lines : a publication of the public affairs office, 144th fighter interceptor wing, fresno ang base, – Fresno CA. 1981 jun-jul, sep-dec, 1982 jan-jul, sep-dec, 1983 jan-apr, jun, aug-nov, 1984 sep, nov-dec, 1985 jan-jun, oct-dec, 1986 jan-mar – 1r – 1 – (cont by: freddie phantom's newsletter) – mf#1221025 – us WHS [355]

Flight magazine – 1934-63 – 1 – us L of C Photodup [073]

Flight operations – Dallas. 1975-1979 (1) 1975-1979 (5) 1976-1979 (9) – (cont: flight) – ISSN: 0361-5030 – mf#1539,01 – us UMI ProQuest [629]

Flightline – Antelope Valley, CA. 1966-1967 (1) – mf#62080 – us UMI ProQuest [071]

Flinders, M *see* A voyage to terra australis

Flinders, Matthew *see* Reise nach dem australlande

Flinders Petrie, W M *see*
– Abydos
– Dendereh
– Deshasheh
– Diospolis parva
– Ehnasya
– Naukratis
– The royal tombs of the earliest dynasties
– The royal tombs of the first dynasty
– Tanis

Flinn, John William *see* Complete works of rev thomas smyth

Flint central baptist church. flint, michigan : church records – 1953-Feb 1969 – 1 – us Southern Baptist [242]

Flint chips. a guide to pre-historic archaeology, as illustrated by the collection in the blackmore museum, salisbury / Stevens, Edward Thomas – London: Bell and Daldy, 1870. illus – 1 – us Wisconsin U Libr [930]

Flint, E R *see* Fertilizer suggestions

Flint, Grover *see* Marching with gomez

Flint hill baptist church : minutes and membership rolls – Mcduffie Co, GA. dec 22 1874-feb 5 1909 – 1 – $12.96 – mf#5495 – us Southern Baptist [242]

Flint hill baptist church. shelby, north carolina : church records – 1909-63 – 1 – us Southern Baptist [242]

Flint, James *see* Letters from america

Flint, James Henry *see* The law of trusts and trustees as determined by the decisions of the principal english and american courts

Flint journal – Flint, Mich.. 1898+ (1) – mf#60161 – us UMI ProQuest [071]

Flint, Matthew O *see* The influence of health behavior contracting on internal locus of control

Flint ridge baptist church. lancaster county. south carolina : church records – 1942-45; 1949-60; 1965-83. Deacons' Minutes, 1963-65 – 1 – us Southern Baptist [242]

Flint river baptist church. huntsville, alabama : church records – Oct 1808-Oct 1868 – 1 – 6.93 – us Southern Baptist [242]

Flint, Robert *see*
- Agnosticism
- Anti-theistic theories
- Christ's kingdom upon earth
- Duty of divinity students
- Historical philosophy in france and french belgium and switzerland
- Introductory lecture delivered at the opening of the class of moral...
- On theological, biblical, and other subjects
- Philosophy as scientia scientiarum
- The philosophy of history in france and germany
- Sermons and addresses
- Socialism
- Vico

Flint voice – Burton MI. v6 n4, v6 n1-18 [1982 jan 8/21, 1982 feb 19/mar 4-dec] – 1r – 1 – (cont by: michigan voice) – mf#1043387 – us WHS [071]

Flintknappers' exchange – v1 n1-v4 n3 [1978 jan-1981 dec] – 1r – 1 – mf#626166 – us WHS [071]

Flintshire observer – Mold, Colwyn Bay & Holywell, Wales. Flintshire Observer & News. -w. Jan 1857-Feb 1916; April 1919-Dec 1932. Lacking Jan-Dec 1896; Jan-Feb 1899. 50 reels – 1 – uk British Libr Newspaper [072]

Flintshire observer mining journal and general advertiser for the counties of flint and denbigh – [Wales] Flintshire jan 1857-dec 1950 [mf ed 2003] – 75r – 1 – (cont by: flintshire observer, mining journal and general advertiser for the counties of flint and denbigh [jan 1864-dec 1913]; flintshire observer and news [jan 1914-dec 1915, jan-17 feb 1916 (then suspended) 10 apr-23 dec 1919, jan 1920-dec 1950]) – uk Newsplan [622]

Flippen, W S *see* Flippen's reports of cases in the sixth circuit, 1859-1881

Flippen's reports of cases in the sixth circuit, 1859-1881 / Flippen, W S – Chicago: Callaghan. v1-2. 1881-89 (all publ) – (= ser Early federal nominative reports) – 17mf – 9 – $25.50 – mf#LLMC 81-451 – us LLMC [340]

Flippin, Percy Scott *see* The archives of the u s government

Flitner, J *see*
- Nebulo nebulonum

Flitton with silsoe – (= ser Bedfordshire parish register series) – 2mf – 9 – £5.00 – uk BedsFHS [929]

Flitwick – (= ser Bedfordshire parish register series) – 1mf – 9 – £3.00 – uk BedsFHS [929]

Flm joint board tempo / Joint Board Fur, Leather & Machine Workers Union – 1967 dec, v2 n1-v18 n1 [1968 feb-1984 feb/mar] – 1r – 1 – (cont by: local 1-flm tempo) – mf#367512 – us WHS [331]

Floating island : a tragi-comedy, acted before his majesty at oxford, aug 29, 1636... / Strode, William – London: printed by T C for H Twiford...1655 [mf ed 19–] – 1mf – 9 – mf#fiche 902 – us Sibley [780]

Floeck, Oswald *see*
- Die tagebuecher des dichters zacharias werner
- Die tagebuecher des dichtes zacharias werner

Der floehhaz / Fischart, Johann; ed by Wendeler, Camillus – Halle a/S: Max Niemeyer 1877 [mf ed 1993] – (= ser Neudrucke deutscher literaturwerke des 16. und 17. jahrhunderts 5) – 11r – 1 – (int by ed) – mf#3387p – us Wisconsin U Libr [830]

Floering, Friedrich *see* Das alte testament im evangelischen religionsunterricht

Die floia und andere deutsche maccaronische gedichte / ed by Bluemlein, Carl – Strassburg: J H E Heitz 1900 [mf ed 1993] – (= ser Drucke und holzschnitte des 15. und 16. jahrhunderts in getreuer nachbildung 4) – 1r [ill] – 1 – (filmed with: kleines deutsches sagenbuch / will-erich peuckert [ed]) – mf#3367p – us Wisconsin U Libr [810]

Floigl, Victor *see*
- Die chronologie der bibel des manetho und beros
- Geschichte des semitischen altertums in tabellen

Flood insurance studies – 102r – 1 – (cont by: flood insurance study) – mf#665498 – us WHS [368]

Flood insurance study – 118r – 1 – (cont by: flood insurance study [1979]) – mf#665484 – us WHS [368]

Flood, J M *see* Discussion on the trinity, church constitutions and disciplines, and human depravity

Flood, Johan *see* Forindien

Flood, John Charles Henry *see* A tractate on the rule of practice in english law, embodying what is known as the equitable doctrine of election.

Flood, John L [comp] *see* Incunabula

"The flood-1903" / Imes, Merle Graybill – 1 – us Kansas [978]

A flor : periodico litterario e politico – Rio de Janeiro, RJ: Typ e Lith Esperanca de Santos & Velloso, 05 mar 1871 – (= ser Ps 19) – mf#P17,03,91 – bl Biblioteca [079]

Flor de cinco petalos / Rivera Landron, Francisco – San Juan, Puerto Rico. 1951 – 1r – us UF Libraries [972]

Flor de corralitos de piedra / Lemaitre, Daniel – Cartagena, Colombia. 1961 – 1r – us UF Libraries [972]

Flor de mesoamerica / Valle, Rafael Heliodoro – San Salvador, El Salvador. 1955 – 1r – us UF Libraries [972]

Flor, Elmer Nicodemo *see* "Amen" in old testament liturgical texts

Flor, Karen K *see* The relationship between personality hardiness, stress and burnout in selected collegiate athletes

Flora : deland environs / Goebel, Rubye K – s.l, s.l? 1936 – 1r – us UF Libraries [574]

Flora : lue gim gong / Trainor, A W – s.l, s.l? 1936 – 1r – us UF Libraries [580]

Flora : oder [allgemeine] botanische zeitung / ed by Hoppe, D H – Regensburg, 1818-1926. v1-120 – 1416mf – 8 – (preceded by: botanische zeitung. hannover) – mf#86c – ne IDC [580]

Flora : papaya (carica papaya) / Trainor, A W L – s.l, s.l? 1936 – 1r – us UF Libraries [580]

Flora aegyptiaco-arabica : sive descriptiones plantarum, quas per aegyptum inferiorum et arabium felicem detexit... / Forsskal, P – Havniae, 1775 – 8mf – 8 – mf#5045 – ne IDC [580]

Flora altaica / Ledebour, C F von – Berolini, 1829-1833. 4 v – 27mf – 8 – mf#5296 – ne IDC [580]

Flora and fauna : aloe vera / Harold, William G – s.l, s.l? 1936 – 1r – us UF Libraries [580]

Flora cubana / Sauvalle, Francisco Adolfo – Havana, Cuba. 1873 – 1r – us UF Libraries [580]

Flora da bahia / Menezes, Antonio Inacio De – Sao Paulo, Brazil. 1949 – 1r – us UF Libraries [580]

Die flora der aegyptisch-arabischen wueste auf grundlage anatomisch-physiologischer forschungen dargestellt... / Volkens, G L A – Berlin, 1887 – 3mf – 9 – mf#8710 – ne IDC [580]

Flora des tropischen arabien... / Schwartz, O – Hamburg, 1939 – 5mf – 9 – mf#13058 – ne IDC [956]

Flora in fort myers / Crowe, F Hilton – s.l, s.l? 1936 – 1r – us UF Libraries [580]

Flora journal – Flora OR: Flora Pub Co [wkly] – 1 – us Oregon Lib [071]

Flora libycae specimen sive plantarum enumeratio cyrenaicam, pentapolim, magne syrteos... / Viviani, D – Genuae, 1824 – 3mf – 8 – mf#6482 – ne IDC [580]

Flora mexicana : jardin botanico 4-11 / Sesse y Lacasta, Martin de & Mocino, Jose Mariano – 1,305mf – 9 – sp Cultura [580]

Flora of afghanistan / Kitamura, S – Kyoto, 1960 – 9mf – 8 – (results of the kyoto university scientific expedition to the karakoram and hindukush, 1955 v2) – mf#990 – ne IDC [956]

Flora of bermuda / Britton, Nathaniel Lord – New York, NY. 1965 – 1r – us UF Libraries [580]

The flora of montreal island / Campbell, Robert – S.l: s.n, 1892? – 1mf – 9 – (repr fr the canadian record of science) – mf#07095 – cn Canadiana [580]

Flora of syria, palestine, and sinai / Post, G E – Beirut, [1896] – 10mf – 9 – mf#11452 – ne IDC [956]

The flora of the rocky mountains / Campbell, Robert – S.l: s.n, 1900? – 1mf – 9 – (repr fr the canadian record of science) – mf#02094 – cn Canadiana [580]

Flora of the sand keys of florida / Millspaugh, Charles Frederick – Chicago, IL. 1907 – 1r – us UF Libraries [580]

Flora peruana..., 1778-1783 : jardin botanico 4-1-1 / Ruiz Lopez, Hipolito – 1,304mf – 9 – sp Cultura [580]

Flora sinensis ou trait, des fleurs, des plantes et des animaux particuliers a la chine / Boym, M P – 1mf – 9 – (an extract from: thevenot, m: relations de divers voyages curieux...paris, 1696 v1) – mf#HT-682 mf. 15 – ne IDC [590]

Flora temiscouatensis / Ami, Henry Marc – S.l: s.n, 1888? – 1mf – 9 – mf#02410 – cn Canadiana [580]

Les floraisons matutinales / Beauchemin, Neree – Trois Rivieres: V Ayotte, 1897 – 3mf – 9 – mf#03518 – cn Canadiana [810]

The floral fortune-teller : a game for the season of flowers / Edgarton, S C – Boston: A Tompkins, c1846 – us CRL [790]

Florante at laura sa kahariang albania *see* Pinagdaanang buhay ni florante at ni laura sa kahariang albania

Floras lake banner – Lakeport OR: Lakeport Pub Co [wkly] – 1 – us Oregon Lib [071]

Flore canadienne : ou description de toutes les plantes des forets, champs, jardins et eaux du canada... / Provancher, Leon – Quebec: J Darveau. 2v. 1862 [mf ed 1984] – 2v on 1mf – 9 – mf#47209 – cn Canadiana [580]

Flore de l'algerie : ou catalogue des plantes indigenes de royaume d'alger, accompagne des descriptions de quelques especes nouvelles ou peu connues... / Munby, G – Paris, Alger, 1847 – 3mf – 9 – mf#6197 – ne IDC [956]

Flore francaise / Lamarck, Jean Baptiste – Paris: Desray,. v1-5. 1815 – 1 – $120.00 – mf#0317 – us Brook [580]

Flore und blanscheflur : eine erzaehlung / Fleck, Konrad; ed by Sommer, Emil – Quedlinburg, Leipzig: G Basse, 1846 [mf ed 1993] – (= ser Bibliothek der gesammten deutschen national-literatur von den aeltesten bis auf die neuere zeit sect1/12) – xxxviii/341p – 1 – (incl ind) – mf#8438 reel 4 – us Wisconsin U Libr [810]

Florecillas de san francisco de san francisco contadas a los ninos / Sobral, Maria da Luz – Barcelona, 1931; Madrid: Razon y Fe, 1931 – 1 – sp Bibl Santa Ana [580]

Florence / Allen, Grant – London: E G Richards, 1906 – (= ser Grant allen's historical guides) – 4mf – 9 – 0-665-65592-4 – mf#65592 – cn Canadiana [914]

Florence baptist church. kentucky : church records – Oct 1893-Nov 1894, May 1902-Jun 1962 – 1 – 55.98 – us Southern Baptist [242]

Florence baptist church. lexington county. south carolina : church records – 1938-1973 – 1 – 5.00 – us Southern Baptist [242]

Florence baptist church. texas : church records – 1856-1942 – 1 – us Southern Baptist [242]

Florence black sun / South Carolina Black Media Group – Florence SC. 1988 dec 1/3-29/31, 1989 jan 5/7-aug 17, 1994 apr 21/27-aug 25/31, 1994 sep 1/7-dec 29/jan 4, 1995, 1995 jan 5/11-apr 27/may 3, 1995 may 4/10-aug 31/sep 6, 1995 sep 7/13-dec 28/jan 3, 1996, 1996 jan 4/10-mar 28/apr 3, 1996 apr 4/10-sep 26/oct 2 – 8r – 1 – mf#1663902 – us WHS [321]

Florence courier – Florence, NE: James C Mitchell, dec 1856 (wkly) [mf ed with gaps filmed 1971] – 1r – 1 – (daily ed: daily florence courier) – us NE Hist [071]

Florence first baptist church. florence, alabama : church records – 1907-59 – 1 – 72.24 – us Southern Baptist [242]

Florence first baptist church. florence, south carolina : church records – 1866-1946 – 1 – us Southern Baptist [242]

Florence fontenelle – Florence, NE: J M Myers, jul 9 1915-jun 1928// (wkly) [mf ed 1919-28 (gaps) filmed 1978] – 2r – 1 – (cont by: florence fontenelle and minne lusa review) – us NE Hist [071]

Florence fontenelle and minne lusa review – Florence, NE: Fontenelle Pub Co. v14 n1. jun 15 1928-1933// (wkly) [mf ed with gaps filmed 1978] – 2r – 1 – (cont: florence fontenelle. absorbed by: milligan review. publ in "omaha, florence station" jul 20 1928-33. numbering very irregular) – us NE Hist [071]

Florence gazette – Florence, NE: Geo H Holton, jul 1908 (wkly) [mf ed with gaps filmed 1979] – 1r – 1 – (cont: florence items) – us NE Hist [071]

Florence gazette – Florence, Italy. Italian Gazette. -m. Sept 1890-12 May 1894; 3 Nov 1894-3 Nov 1903; 1 Nov 1904-4 June 1907. 6 reels – 1 – uk British Libr Newspaper [072]

Florence, Hercules *see* Viagem fluvial do tiete ao amazonas de 1825 a 1829

Florence items – Florence, NE: F B Nichols. v1 n1. jun 5 1903-jul 1908// (wkly) [mf ed with gaps filmed 1979-[89]] – 2r – 1 – (cont by: florence gazette. issues for may 25 1908-jul 10 1908 called v5 n1-v5 n7 but constitute v6 n1-v6 n7) – us NE Hist [071]

Florence; its history, the medici, the humanists, letters, arts / Yriarte, Charles – New ed. rev...by Maria Hornor Lansdale. Philadelphia: H.T. Coates, 1897. viii,478p. plates, maps, geneal. tab – 1 – us Wisconsin U Libr [945]

Florence mining news – Florence WI. 1881 jan-nov, 1881 jun 11, 1881 dec 3/1883 sep 22-2004 jul-dec – 79r – 1 – (with gaps) – mf#943171 – us WHS [622]

Florence news *see*
- Siuslaw news
- Siuslaw oar

Florence nightingale's indian letters : a glimpse into the agitation for tenancy reform, bengal, 1878-82 / ed by Sen, Priyaranjan – Calcutta: Mihir Kumar Sen, 1937 – (= ser Samp: indian books) – us CRL [610]

Florence of Worcester *see* Florentii wigorniensis monachi

Florence, Philip Sargant *see* Economics and human behaviour

Florence times – Florence OR: R Moore, 1925-[wkly] – 1 – (ceased in 1926?) – us Oregon Lib [071]

The florence tribune – Florence, NE: Lubold & Platz. v1 n1, jun 4 1909- (wkly) [mf ed -1916 (gaps) filmed 1978 – 3r – 1 – us NE Hist [071]

Florence, William James *see* The gentlemen's handbook on poker

Florence-du-Sacre-Coeur, soeur *see* Bibliographie analytique sur le forum catholique de montreal (catholic inquiry forum) 1952-1962

Florencia, Francisco de *see*
- Descripcion historical y moral del yermo de san migvel, de las cvevas en el reyno de la nueva-espana
- Historia de la provincia de la compania de jesus de nueva espana
- Menologio de los varones mas senalados en perfeccion religiosa de la provincia de la compania de jesus de nueva-espana
- La milagrosa invencion de un thesoro escondidio en un campo
- Origen de los celebres santuarios de la nueva galicia
- Origen del celebre santuario de nuestra senora de san juan
- Vida admirable, y mverte dichosa del religioso p geronimo de figveroa
- Zodiaco mariano

Florentii wigorniensis monachi / Florence of Worcester – English Historical Society, Publications, 1848-49. 1 reel. 1247 – 1 – us Wisconsin U Libr [941]

Florentine history, from the earliest authentic records to the accession of ferdinand the third, grand duke of tuscany / Napier, Henry Edward – London: E. Moxon, 1846-47. 6v. plates, maps – 1 – us Wisconsin U Libr [945]

Florentine painters of the renaissance / Berenson, Bernard – New York, NY. 1909 – 1r – us UF Libraries [750]

Florentinische naechte = Florentine nights / Heine, Heinrich – London: Methuen, 1927 – 1 – (in english) – us Wisconsin U Libr [830]

Florenz, Karl *see* Japanische mythologie

Florer, Warren Washburn *see* German liberty authors

Florero de llorente / Abella Rodriguez, Arturo – Medellin, Colombia. 1964 – 1r – us UF Libraries [972]

Flores, Antonio *see* Sermon...catedral de badajoz

Flores de heroismo. sevilla, 1939 / Garcia Alonso, Francisco – Madrid: Razon y Fe, 1940 – 1 – sp Bibl Santa Ana [946]

Flores de miraflores, hieroglificos sagrados...del mysterio de la concepcion de la virgen, y madre de dios maria senora nuestra / Iglesia, N de la – Burgos: Diego de Nieva y Murillo, 1659 – 5mf – 9 – mf#0-08 – ne IDC [090]

Flores de otono / Cruz Marquez Espinosa, J – Mexico City? Mexico. 1965 – 1r – us UF Libraries [580]

Flores de Ribera, Joseph Antonio *see*
- Meritos del dr d joseph antonio flores de ribera
- Relacion de meritos del dr don j a flores de ribera

Flores de san bernardo. de la lamentacion de la virgen maria / Bravo Riesco, Agustin – S.l., s.i, s.a. – 1 – sp Bibl Santa Ana [240]

Flores del destierro / Marti, Jose – Habana, Cuba. 1933 – 1r – us UF Libraries [580]

Flores del sendero / Vargas, Leon – Alajuela, Costa Rica. 1957 – 1r – us UF Libraries [580]

Flores del valle / Garcia Miranda, Vicenta – 1855 – 9 – sp Bibl Santa Ana [580]

Flores, Elsa Mercedes *see* Danzas clasicas

Las flores en la tradicion extremena / Gil Garcia, Bonifacio – Badajoz: Dip Provincial, 1962 – sp Bibl Santa Ana [946]

Flores epytaphii sanctorum : formae tplila 98 / Thiofridus Epternacensis – [mf ed 2000] – (= ser ILL – ser a; Cccm 133) – 4mf+68p – 9 – €40.00 – 2-503-64332-9 – be Brepols [400]
Flores guadalupanas, o, sonetos alusivos a la celestial imagen de maria santisima neustra senora en su advocacion de guadelupe : especialmente quanto a el vestido y adornos / Plancarte, Joseph Antonio – en Mexico: En la imprenta de D Felipe de Zuniga y Ontiveros, calle Espiritu Santo, ano de 1785 – (= ser Books on religion...1543/44-c1800: milagros y culto de la virgen) – 1mf – 9 – mf#crl-103 – ne IDC [241]
Flores historiarum (rs95) : per matthaeum west-monasteriensem collecti / Matthew of Westminster; ed by Luard, H R – (= ser The rolls series (rs)) – (v1 1890 €21. v2 1890 €18. v3 1890 €23) – ne Slangenburg [931]
Flores Lopez, Santos *see* Ruben dario
Flores marchitas / Rivera Natal, Facundo – San Juan, Puerto Rico. 1958 – 1r – us UF Libraries [580]
Flores Morales, Angel *see* Africa a traves del pensamiento espanol
Flores musice omnis cantus gregoriani – 1488 – (= ser Mssa) – 3mf – 9 – €50.00 – mfchl 458 – gw Fischer [780]
Flores musice omnis cantus gregoriani / Spechtshart of Reutlingen, Hugo; ed by Beck, Carl – Stuttgart: Litterarischer Verein, 1868 [mf ed 1993] – (= ser Blvs 89) – 77p/[3]pl – 1 – (incl bibl ref. german trans by ed) – mf#8470 reel 18 – us Wisconsin U Libr [810]
Flores omnium doctorum illustrium... / Hibernicus, Thomas – Lugduni. v1-2. 1678 – 2v on 14mf – 9 – €27.00 – ne Slangenburg [240]
Flores, S L de *see* Desempeno al metodo racional en la curacion de las calenturas tercianas
Flores, Saul *see*
– Esta es mi tierra
– Lecturas nacionales de el salvador
Flores sin aroma / Sanchez-Arjona, Vicente – Sevilla: Imp. Carlos Acuna, Tomo 1-4. 1956 – 1 – sp Bibl Santa Ana [810]
Flores y abrojos / Vargas, Adolfo de – 1883 – 9 – sp Bibl Santa Ana [810]
Flores y espinas : coleccion de poesias para el ofrecimiento de las flores de mayo / Castro Bajo, Julian – Barcelona: Eugenio Subirana, 1912 – 1 – sp Bibl Santa Ana [580]
Flores y frutos de mi corazon dedicacdos a ti / Cruz Marin, Eugenio de la – Badajoz: Tip. Espanola, 1947 – 1 – sp Bibl Santa Ana [810]
Floresta de disertaciones historico-medicas... / Baguer y Oliver, J – Valencia, 1741 – 22mf – 9 – sp Cultura [610]
Floresta de exemplos / Ribeiro, Joao – Rio de Janeiro, Brazil. 1931 – 1r – us UF Libraries [972]
Florez Alvarez, Leonidas *see* Campana libertadora de 1821
Florez de Ocariz, Juan *see* Genealogias del nuevo reyno de granada. 2 vol. bogota, 1943
Florez, E et al *see* Espana sagrada...tomos 1-51
Florez, Enrique *see* Relacion del viage que ambrosio de morales...hizo por su mandado al ano de 1572 en galicia y asturias
Florez, F *see* Regimiento de sanidad de todas las cosas que se comen y beven, con muchos consejos
Florez, Henrique *see* Espana sagrada
Florez, Jose Segundo *see*
– Espartero
– Primeras nociones de cronologia
Florez, Luis *see*
– Espanol hablado en santander
– Habla y cultura popular en antioquia
– Lengua espanola
– Lexico de la casa popular urbana en bolivar, colom...
– Pronunciacion del espanol en bogota
Florhaug, Jessica A *see* The effect of different interval magnitudes on measures of exercise intensity
Flori, Jakob *see* Cantiones sacrae quinque vocum
Floriad – Schenectady. 1811-1811 (1) – mf#3805 – us UMI ProQuest [420]
Florian geyer : [a novel] / Bauer, Heinrich – Berlin: F Eher, 1936 [mf ed 1989] – (= ser Deutsche kulturbuchreihe) – 312p – 1 – mf#6982 – us Wisconsin U Libr [830]
Floriani, P P *see* Diffesa et offesa delle piazze
Floriano, Antonio *see* Floriano, antonio. la iglesia de santiago de los caballeros en caceres
Floriano, antonio. la iglesia de santiago de los caballeros en caceres / Ortega, Angel & Floriano, Antonio – Madrid: Archivo Ibero-Americano, 1916 – 1 – sp Bibl Santa Ana [240]
Floriano Cumbreno, Antonio C *see*
– Caceres ante la historia. la cuestion critica de la fundacion y el nombre de caceres
– Catalogo del archivo de la diputacion provincial de teruel
– Curso general de didactica
– Curso general de paleografia y diplomatica espanoles. seleccion diplomatica
– Excavaciones en la antigua cappara
– Excavaciones en merida
– La fecha en la conquista de caceres ante los documentos (la carta populationis)
– Las fuentes para la historia de la pedagogia espanola
– Hallazgo de la necropolis judaica de la ciudad de teruel
– La iglesia de santiago de los caballeros de caceres y el escultor alonso berruguete
– La iglesia de santiago de los caballeros. descripcion historico-artistica
– Informe sobre la catalogacion de la coleccion numismatica del museo de caceres
– Origenes historicos de la agricultura y de la ganaderia en caceres
– Pregon de la semana santa cacerena
– Los problemas de su reconquista y de su nombre
– El retablo de santiago de los caballeros de caceres y el escultor alonso berruguete
– Teruel en el siglo 15. la vida economica y la cuestion monetaria
– Transcripcion paleografica y version castellana de la carta de poblacion o fuero latino de caceres
– El tribunal del santo oficio en aragon establecimiento de la inquisicion en teruel
– La villa de caceres y la reina catolica: 2. ordenanzas sobre las labranzas y pastos de zafra y zafrilla, de la "penas de ganados" y administracion de los bienes propios del concejo de caceres...los juramentos de los reyes catolicos
Floriated ornament : a series of thirty-one designs / Pugin, Augustus Welby Northmore – London: Henry G Bohn, 1849 – (= ser 19th c art & architecture) – 1mf – 9 – mf#4.1.37;c.4.1.120 – uk Chadwyck [740]
Florida – (= ser General education board: the early southern program) – 6r – 1 – $780.00 – us Scholarly Res [370]
Florida : an advancing state, 1907-1917-1927 / Florida Dept Of Agriculture – St Petersburg, FL. 1928 – 1r – us UF Libraries [630]
Florida : america's first name – s.l, s.l? 193-? – 1r – us UF Libraries [978]
Florida : beauties of the east coast / Jacksonville, St Augustine And Indian River Railway – St Augustine, FL. 1893 – 1r – us UF Libraries [978]
Florida : 'the dawn of a new day' / Rose, Walter W – Orlando, FL. no date – 1r – us UF Libraries [978]
Florida : the farmer's sportsmen's and tourist's pa... / Florida Farms And Homes Company Orange Springs – s.l, s.l? 1916 – 1r – us UF Libraries [790]
Florida / Florida Bureau Of Immigration – Tallahassee, FL. 1882 – 1r – us UF Libraries [978]
Florida – Gainesville, FL. 1898 – 1r – us UF Libraries [978]
Florida : health, climate, transportation, recreation / Florida State Committee On National Soldiers Home – Tallahassee, FL. 1930? – 1r – us UF Libraries [362]
Florida : the healthiest state in the union / Gano, W – Jacksonville, FL. 1800 – 1r – us UF Libraries [362]
Florida : an ideal cattle state / Florida State Live Stock Association – Jacksonville, FL. 1918 – 1r – us UF Libraries [636]
Florida : information for those who desire to know of the state / Alden, George J – New Smyrna, FL. 1875 – 1r – us UF Libraries [978]
Florida : its climate, soil, and productions / Florida Commissioner Of Lands – New York, NY. 1869 – 1r – us UF Libraries [630]
Florida : its climate, soil, production, and agriculture / United States Dept Of Agriculture – Washington, DC. 1882 – 1r – us UF Libraries [630]
Florida : its soil, climate, health, productions, resources and adva... / Florida Land Agency – Jacksonville, FL. 1875 – 1r – us UF Libraries [630]
Florida – Jacksonville, FL. 1890 – 1r – us UF Libraries [978]
Florida : jefferson county / Bailey, E B – Monticello, FL. 1887 – 1r – us UF Libraries [978]
Florida : land of change / Hanna, Kathryn Abbey – Chapel Hill, North Carolina. 1941 – 1r – us UF Libraries [978]
Florida : a land of homes / Florida Dept of Agriculture – Tallahassee, FL. 1934 – 1r – us UF Libraries [640]
Florida : the land of opportunity – Madison, FL. 192- – 1r – us UF Libraries [630]
Florida / Lanier, Sidney – Philadelphia, PA. 1875 – 1r – us UF Libraries [978]
Florida : the march of progress – Tallahassee, FL. 1939? – 1r – us UF Libraries [630]
Florida : our last frontier – Savannah, GA. 1924 – 1r – us UF Libraries [978]
Florida : session laws of american states and territories – 1822-2001 – 9 – $4194.00 set – mf#402590 – us Hein [348]
Florida : 'the state distinctive' / Gilbert, D H – Tallahassee, FL. 1926 – 1r – us UF Libraries [978]
Florida : sub-tropical exposition, jacksonville, florida – s.l, s.l? 1888 – 1r – us UF Libraries [978]
Florida – Tallahassee, FL. 1941 – 1r – us UF Libraries [978]
Florida : the under-ground wealth and prehistoric wo... / Shrader, Jay – Bartow, FL. 1890 – 1r – us UF Libraries [978]
Florida / United States Railroad Administration – St Augustine, FL. 1919? – 1r – us UF Libraries [380]
Florida : west's florida statutes annotated – St Paul: West Pub Co, 1943-jun 2002 update – 9 – $5404.00 set – mf#401191 – us Hein [348]
Florida : what has been and can be done on florida / Tomlinson, E H – Jacksonville, FL. 1915? – 1r – us UF Libraries [978]
Florida : where industry is rewarded / Seaboard Air Line Railway Company General Develop... – Norfolk, VA. 1920? – 1r – us UF Libraries [338]
Florida : the winter garden of america – Fort Pierce, FL. 1912 – 1r – us UF Libraries [630]
Florida *see*
– Citrus fruit laws
– Digest of the laws of the state of florida
– Reports and opinions
– Reports, post-nrs
– Reports, pre-nrs
Florida, 1513-1913 : past and future / Chapin, George M – Chicago, IL. v1-2. 1914 – 1r – us UF Libraries [978]
Florida 1781-1900 – Oxford, MA (mf ed 1990) – (= ser Massachusetts vital records) – 6mf – 9 – 0-87623-114-8 – (mf 1: births 1781-1859. mf 2: births 1826-63; marriages 1810-60; deaths 1806-60. mf 3: births 1861-96. mf 4: births 1896-1900. mf 5: marriages 1861-1901. mf 6: deaths 1860-1914) – us Archive [978]
Florida Academy Of Sciences *see* Proceedings of the florida academy of sciences
Florida AFL-CIO [American Federation of Labor and Congress of Industrial Organizations] *see* More
Florida Agricultural and Mechanical University *see* Famuan
Florida Agricultural College, Lake City *see* Pinakidia
Florida alligator – Gainesville, FL. 1915 oct-1917 aug 11 – 1r – us UF Libraries [071]
Florida alligator – Gainesville, FL. 1912-1955 – 29r – (gaps) – us UF Libraries [071]
Florida and mexico competition for the winter fresh vegetable market / Buckley, Katharine C – Washington DC: US Dept of Agriculture, Economic Research Service...1986 – (= ser Agricultural economic report 556) – 9 – (with bibl) – us GPO [635]
Florida and miscellaneous prose / Lanier, Sidney – Baltimore, MD. 1945 – 1r – us UF Libraries [978]
Florida and texas – Ocala, FL. 1866 – 1r – us UF Libraries [978]
Florida argus – Pensacola, FL. 1828 jun-nov – 1r – us UF Libraries [071]
Florida armored scale insects / Dekle, George Wallace – Gainesville, FL. 1965 – 1r – us UF Libraries [590]
Florida as a health-resort / Lente, Frederic D – New York, NY. 1876 – 1r – us UF Libraries [338]
Florida as a permanent home / Jacques, Dh – Jacksonville, FL. 1877 – 1r – us UF Libraries [640]
Florida as the nation's editors see it – Tallahassee, FL. 1941 – 1r – us UF Libraries [070]
Florida asian-american – Hollywood, FL. 1985-1987 – 1r – us UF Libraries [071]
Florida at rockefeller center in new york city / Florida National Exhibits – Jacksonville, FL. 1936? – 1r – us UF Libraries [978]
Florida attorney general reports and opinions – 1845-2000 – 6,9 – $737.00 set – (1845-1978 on reel $455. 1979-2000 on mf $282) – mf#408180 – us Hein [340]
Florida. Baptist Associations *see* Annuals
Florida baptist witness : index – 1908-56 – 1,5 – 80.32 – us Southern Baptist [242]
Florida baptist witness – Jacksonville, Florida: Florida Baptist Convention, 1885-1991 – 1 – $2,914.96 – us Southern Baptist [242]
Florida bar journal – Tallahassee. 1927+ (1) 1971+ (5) 1977+ (9) – ISSN: 0015-3915 – mf#2450 – us UMI ProQuest [340]
Florida bar journal – v1-75. 1927-2001 – 5,6,9 – $1187.00 set – (v1-58 1927-84 on reel or mf $688. v59-75 1985-2001 on mf $499. title varies: v1-8 1927-34 as florida state bar association law journal. v8-27 1934-53 as florida law journal) – ISSN: 0015-3915 – mf#112771 – us Hein [340]
Florida bar news – Tallahassee. 1979+ (1) 1979-1983 (5) 1979-1983 (9) – ISSN: 0360-0114 – mf#12203 – us UMI ProQuest [340]
Florida "better farming special" / Yonge, Philip Keyes – s.l, s.l? 1911? – 1r – us UF Libraries [630]
Florida bird life / Howell, Arthur Holmes – New York, NY. 1932 – 1r – us UF Libraries [590]
Florida b'nai b'rith jewish news and views – Miami, FL. v1 n1-v5 n2. 1980 apr-1984 apr – 1r – (missing: 1984 jan) – us UF Libraries [071]
Florida boom / Hunter, C M – s.l, s.l? 1936 – 1r – us UF Libraries [978]
Florida boom / Sweett, Zelia Wilson – s.l, s.l? 1936 – 1r – us UF Libraries [978]
Florida breezes / Long, Ellen Call – Jacksonville, FL. 1883 – 1r – us UF Libraries [978]
Florida buggist – Gainesville, FL. v1-3. 1917/1918-1919/1920 – 1r – us UF Libraries [630]
Florida bulletin – Gainesville, FL. 1904 aug 12 – 1r – us UF Libraries [071]
Florida Bureau Of Immigration *see*
– Florida
– Florida facts for tourists
– Semi-tropical florida
Florida. Bureau Of Immigration *see* All Florida
Florida butterflies / Gerberg, Eugene J – Baltimore, MD. 1989 – 1r – us UF Libraries [590]
Florida catholic – Miami, FL. 1939 dec-1967 – 20r – us UF Libraries [071]
Florida catholic – Orlando, FL. 1939-1997 (1) – mf#62439 – us UMI ProQuest [071]
Florida cattleman – Kissimmee, FL. v1 n1-3. 1936 oct-dec – 1r – us UF Libraries [636]
Florida cattleman and dairy journal – Kissimmee, FL. v1 n4-v8 n6. 1937:jan-1944 mar – 2r – us UF Libraries [636]
Florida cattleman and livestock journal – Kissimmee, FL. v8 n7-v31. 1944 apr-1966 sep – 21r – us UF Libraries [636]
Florida Centennial Celebration *see* Official souvenir program
Florida Citizens Finance And Taxation Committee *see* Preliminary report of perry g wall
Florida Citrus Commission *see* Florida freeze of january 1940
Florida citrus diseases / Stevens, H E – Gainesville, FL. 1918 – 1r – us UF Libraries [634]
Florida citrus prices / Spurlock, A H – Gainesville, FL. 1937 – 1r – us UF Libraries [634]
Florida citrus prices, 2 / Spurlock, A H – Gainesville, FL. 1937 – 1r – us UF Libraries [634]
Florida coastal law journal – v1. 1999-2000 – 9 – $23.00 – mf#118111 – us Hein [343]
Florida Commissioner Of Lands *see* Florida
Florida Confederation of Historical Societies *see* Florida history newsletter
Florida conservation lands, 1998 / Blanchard, Jon David – Tallahassee, FL. 1998 – 1r – us UF Libraries [333]
Florida conservator – Tallahassee, FL. v1. 1934-1935 – 1r – us UF Libraries [639]
Florida. Constitution Revision Commission *see* Commission minutes, committee minutes and proposals, january 1966-june 1966
Florida Constitutional Convention (1885) *see* Journal of the proceedings of the constitutional c...
Florida Cooperative Sugar Association *see* Organization agreement of florida cooperative sugar association
Florida courier – Fort Pierce FL. 1994 nov3/9-dec 29/jan 4, 1995 jan 5/11-dec 21/27, 1995 nov 30/dec 6 [v3 n48], 1996 jan 4/10-dec 26/jan 1, 1997 jan 2/8-dec 25/31, 1998 jan 1/7-dec 31/jan 6 – 5r – 1 – mf#3143907 – us WHS [071]
Florida days / Deland, Margaret Wade Campbell – London, England. 1889 – 1r – us UF Libraries [630]
La florida del inca : historia del adelantado hernando de soto... / Laso de la Vega, Garcia – Mexico: Fondo de Cultura Economica, 1956 – 1 – sp Bibl Santa Ana [972]
La florida del Inca. historia del adelantado hernando de soto. tomo 1 / Laso de la Vega, Garcia – Cuzco (Peru): Imp. H.G. Rozas, 1958 – sp Bibl Santa Ana [350]
Florida democrat – Gainesville, FL. 1896 mar 13,27 – 1r – us UF Libraries [071]
Florida Dept Of Agriculture *see*
– Florida
– Florida resources and sports
– Fourth census of the state of florida taken in the year 1915
– North and northwest florida
– Report of manufacturing in florida, 1937
– Safety on the farm
– Third census of the state of florida taken in the year 1905
– Why i like florida
Florida Dept of Agriculture *see* Florida
Florida. Dept Of Agriculture *see* Agricultural statistics of florida
Florida dispatch – Jacksonville, FL. 1882-1888 (1) – mf#62422 – us UMI ProQuest [071]

Florida early settlers : jackson county – s.l, s.l? 193-? – 1r – us UF Libraries [978]

Florida early settlers : leon county – s.l, s.l? 193-? – 1r – us UF Libraries [978]

Florida early settlers : liberty county – s.l, s.l? 193-? – 1r – us UF Libraries [978]

Florida early settlers – s.l, s.l? 193-? – 1r – us UF Libraries [978]

Florida East Coast Drainage And Sugar Company *see* Prospectus of the florida east coast drainage and sugar company

Florida East Coast Railway *see*
- Brief history of the florida east coast railway...
- Homes on the east coast of florida
- Livestock farming in florida along the lines of...

Florida East Coast Railway. Land Dept *see* Climatic data of the east coast of florida

Florida enchantments / Dimock, A W – New York, NY. 1908 – 1r – us UF Libraries [978]

Florida enchantments / Dimock, Anthony Weston – Peekamose, NY. 1915 – 1r – us UF Libraries [978]

Florida entomologist – Winter Haven, FL. v4-58. 1920/1921-1975 – 8r – us UF Libraries [630]

The florida evangelist – Jacksonville, FL: Rev J Milton Waldron (wkly) [mf ed 1947] – (= ser Negro Newspapers on Microfilm) – 1r – 1 – us L of C Photodup [071]

Florida everglades : class 100 : file no 170 / Lyons, Isabel J – s.l, s.l? 1936 – 1r – us UF Libraries [978]

Florida everglades / Florida Everglades Engineering Commission – Washington, DC. 1914 – 1r – us UF Libraries [627]

Florida everglades / Lyons, Isabel J – s.l, s.l? 1936 – 1r – us UF Libraries [574]

Florida Everglades Engineering Commission *see* Florida everglades

Florida facts for tourists / Florida Bureau Of Immigration – Tallahassee, FL. 1923? – 1r – us UF Libraries [338]

Florida Fair And Gasparilla Association, Inc *see* Official guide and souvenir program

Florida farm / Whitmore, Frederic – Springfield, MA. 1903 – 1r – us UF Libraries [630]

Florida farm prices / Spurlock, A H – Gainesville, FL. 1944 – 1r – us UF Libraries [630]

Florida farmer and fruit grower – Jacksonville, FL. 1888-1898 (1) – mf#62423 – us UMI ProQuest [071]

Florida farmer and fruit grower – Jacksonville, FL. 1888-1899 feb – 11r – (missing: 1888 jan 11, 18; feb 22; sep 19; oct 3; 1889 jan 23, 30) – us UF Libraries [634]

Florida Farms And Homes Company *see* Tour of the lands of the florida farms and homes

Florida Farms And Homes Company Orange Springs *see* Florida

Florida farms at lawtey, florida / Raley-Hamby Company Jacksonville – Jacksonville, FL. 1910 – 1r – us UF Libraries [630]

Florida fishing industry – s.l, s.l? 193-? – 1r – us UF Libraries [639]

Florida flashlights / Reese, Joseph Hugh – Miami, FL. 1917 – 1r – us UF Libraries [978]

Florida folk lore and customs – s.l, s.l? 193-? – 1r – us UF Libraries [390]

Florida for tourists / Barbour, George M – New York, NY. 1882 – 1r – us UF Libraries [338]

Florida forest fire fighters manual – Tallahassee, FL. 1943 – 1r – us UF Libraries [634]

Florida forestry – Jacksonville, FL. 1932 – 1r – us UF Libraries [634]

Florida freeze of january 1940 / Florida Citrus Commission – Lakeland, FL. 1940 – 1r – us UF Libraries [634]

Florida from the air / Aero-Graphic Corporation – Louisville, KY. 1936 – 1r – us UF Libraries [978]

Florida fruit and produce news – Jacksonville, FL. v1-3 n18. 1908 oct-1911 jan – 1r – (gaps) – us UF Libraries [634]

Florida fruits and how to raise them / Warner, Helen Garnie – Louisville, KY. 1886 – 1r – us UF Libraries [634]

Florida Game And Fresh Water Fish Commission *see* Closing the gaps in florida's wildlife habitat conservation

Florida gazette – St Augustine FL. 1821 jul 28 – 1r – 1 – mf#846034 – us WHS [071]

Florida genealogical journal – 1967 aug-1975 – 1r – 1 – (cont: florida genealogist's journal; cont by: journal [florida genealogical society]) – mf#619987 – us WHS [929]

Florida genealogical journal – v14 n1 [1978] – 1r – 1 – (cont: journal [florida genealogical society: 1976]; cont by: journal [florida genealogical society: 1978]) – mf#2593871 – us WHS [929]

Florida Genealogical Society *see*
- Journal

Florida genealogist – v2, n2-v10 n4 [1979 winter-1987 summer] – 1r – 1 – mf#1265535 – us WHS [929]

Florida genealogist's journal – v3 n1 [1967 feb] – 1r – 1 – (cont: florida genealogist's newsletter; cont by: florida genealogical journal) – mf#619974 – us WHS [929]

Florida genealogist's newsletter – v1 n1-v2 n1 [1965 oct-1966 aug] – 1r – 1 – (cont by: florida genealogist's journal) – mf#619545 – us WHS [929]

Florida Geological Survey *see*
- Annual report
- Biennial report
- Biennial report to state board of conservation
- Bulletin
- Geological bulletin
- Information circular
- Leaflet
- Open file report
- Report of investigation
- Report of investigations – division of geology
- Report of investigations – florida geological survey
- Special publication

Florida Governor's Committee On Forest Conservation *see* Report of governor's committee on forest conservation april 25, 193...

Florida Governor's Council on Indian Affairs *see* Other side of the coin

Florida grower – Willoughby. 1998+ (1) – (cont: florida grower and rancher) – mf#294,01 – us UMI ProQuest [630]

Florida grower – Tampa, FL. v3 n19-44. 1911 feb-jul – 29r – us UF Libraries [630]

Florida grower and rancher – Orlando. 1949-1994 (1) 1973-1994 (5) 1977-1994 (9) – (cont by: florida grower) – ISSN: 0015-4091 – mf#294 – us UMI ProQuest [630]

Florida grower and rancher – Tampa, FL. v61 n8-v90. 1953 aug-1997 – 19r – us UF Libraries [636]

Florida herald – St Andrews, FL. 1829-1839 nov 15 – 4r – (gaps) – us UF Libraries [071]

Florida herald and southern democrat – St Andrews, FL. 1839-1849 – 2r – us UF Libraries [071]

Florida historical, 1930-1935 / Shepherd, Rose – s.l, s.l? 1936 – 1r – us UF Libraries [978]

Florida historical chronicle – 1975 jan 5-1976 feb 2 – 1r – 1 – mf#1056336 – us WHS [978]

Florida historical pageant / Jacksonville (Fla) Community Service – Jacksonville, FL. 1922 – 1r – us UF Libraries [978]

Florida Historical Records Survey *see* Translations of unique spanish land grants and deeds

Florida history newsletter / Florida Confederation of Historical Societies – v5 n1-v14 n1 [1979 mar-1988 mar] – 1r – 1 – mf#1698493 – us WHS [978]

Florida hurricane of september 18-20, 1926 / Mitchell, Charles Lyman – Washington, DC. 1926? – 1r – us UF Libraries [550]

Florida immigrant – Tallahassee, FL. 1877 jul-1878 jun – 1r – us UF Libraries [071]

Florida in the spanish-american war / Shepherd, Rose – s.l, s.l? 1936 – 1r – (was filmed as mn02097 (error)) – us UF Libraries [977]

Florida in the spanish-american war, 1898-1899 – s.l, s.l? 193-? – 1r – us UF Libraries [977]

Florida index – Lake City, FL. v1 n1-v12 n2. 1899 jun 16-1909 jul – 3r – us UF Libraries [071]

Florida indian war claim / Wailes, Sidney I – Tallahassee, FL. 1905 – 1r – us UF Libraries [978]

Florida Inland And Coastal Waterways Association *see* Inland and coastal waterways of florida

Florida Inland Navigation District Commissioners *see* Florida intracoastal waterway

Florida intelligencer – Tallahassee, FL. 1826 feb 24-dec 08 – 1r – us UF Libraries [071]

Florida international law journal *see* Florida journal of international law

Florida intracoastal waterway : from the st johns... / Florida Inland Navigation District Commissioners – Jacksonville, FL. 1935? – 1r – us UF Libraries [380]

Florida, its climate, soil, and productions – Jacksonville, FL. 1868 – 1r – us UF Libraries [630]

Florida, its climate, soil, and productions – New York, NY. 1881 – 1r – us UF Libraries [630]

Florida jewish news – Jacksonville, FL. 1936 apr-1938 nov – 1r – (missinng: 1936 jul, may-jun, sep; 1938 mar) – us UF Libraries [071]

Florida journal of anthropology / University of Florida Anthropology Student Association – 1976 winter-1983 v8 n2 pt 2, 1983:v9 n1-1988 winter – 2r – 1 – mf#1494616 – us WHS [301]

Florida journal of commerce – Jacksonville. 1972-1973 (1) 1973-1973 (5) (9) – (cont by: florida journal of commerce, american shipper) – ISSN: 0015-413X – mf#8118 – us UMI ProQuest [380]

Florida journal of commerce, american shipper – Jacksonville. 1975-1976 (1) 1975-1976 (5) 1975-1976 (9) – (cont: florida journal of commerce. cont by: american shipper) – ISSN: 0097-6237 – mf#8118,01 – us UMI ProQuest [380]

Florida journal of international law – v1-13. 1984-2001 – 9 – $244.00 set – (title varies: v1-5 1984-90 as florida international law journal) – ISSN: 0882-6420 – mf#109711 – us Hein [341]

Florida keys / Kennedy, Stetson – s.l, s.l? 1940 – 1r – us UF Libraries [978]

Florida keys / Manucy, Albert C – s.l, s.l? 1936 – 1r – us UF Libraries [978]

Florida keys keynoter – Marathon, FL. 1983-2001 (1) – mf#62429 – us UMI ProQuest [071]

Florida labor advocate – Tampa, FL. 1950 mar 3-dec 29 – 1r – us UF Libraries [071]

Florida labor news – Tallahassee, FL. 1948-1969 – 2r – us UF Libraries [071]

Florida Land Agency *see* Florida

Florida Land And Improvement Company *see* Descriptive list catalogue of the disston lands...

Florida Land And Settlement Company *see* Ten-acre farms for florida settlers

Florida law journal *see* Florida bar journal

Florida law review – v1-53. 1948-2001 – 5,6,9 – $1032.00 set – (v1-36 1948-84 on reel $542. v37-53 1985-2001 on mf $490. title varies: v1-40, 51 n4 1948-88, 1999 as university of florida law review) – ISSN: 1045-4241 – mf#107721 – us Hein [340]

Florida Laws, Statutes *see* County boundaries

Florida legionnaire – Arcadia, FL. 1940 jun-1947 jun – 1r – (1942 apr) – us UF Libraries [071]

Florida Legislative Tax Committee *see* Report of legislative tax committee

Florida Legislature *see* Investigationharrison reed...

Florida libraries – Winter Park. 1949-1983 (1) 1975-1983 (5) 1976-1983 (9) – ISSN: 0046-4147 – mf#9105 – us UMI ProQuest [020]

Florida lighthouses : st augustine / Crowe, F Hilton – s.l, s.l? 1938 – 1r – us UF Libraries [380]

Florida limestone – s.l, s.l? 193-? – 1r – us UF Libraries [550]

Florida loafing / Roberts, Kenneth Lewis – Indianapolis, IN. 1925 – 1r – us UF Libraries [978]

Florida Marketing Bureau *see* Statistics of florida agriculture and related enterprises

Florida material in us congressional documents / Florida State Library – Tallahassee, FL. 193- – 1r – us UF Libraries [324]

Florida Medical Association, Inc *see* Journal of the florida medical association, inc

Florida Memorial College *see* Catalogs

Florida metropolis – Jacksonville, FL. 1901-1922 (1) – mf#62424 – us UMI ProQuest [071]

Florida mirror – Fernandina Beach, FL. 1878 nov 30-1886 jul – 7r – (gaps) – us UF Libraries [071]

Florida muck farms – Miami, FL. 1926 – 1r – us UF Libraries [630]

Florida museum catalog cards / Florida State Museum – Gainesville, FL. cards 1493-76807. 1964 – 1r – us UF Libraries [060]

Florida museum catalog cards / Florida State Museum – Gainesville, FL. cards 76809-98563. 1964 – 1r – us UF Libraries [060]

Florida National Exhibits *see* Florida at rockefeller center in new york city

Florida national forests / United States Forest Service Southern Region – Washington, DC. 1939 – 1r – us UF Libraries [634]

Florida naturalist – Casselberry. 1972+ (1) 1972+ (5) 1974+ (9) – ISSN: 0015-4172 – mf#7229 – us UMI ProQuest [639]

Florida news – Gainesville, FL. no date – 1r – us UF Libraries [071]

Florida news and mirror – Fernandina Beach, FL. 1891 – 1r – us UF Libraries [071]

Florida newspapers, 1885-1898 – Jacksonville, FL. 1937 – 1r – us UF Libraries [071]

Florida newspapers and periodicals – s.l, s.l? 193-? – 1r – us UF Libraries [070]

Florida nurse – Orlando. 1991+ (1,5,9) – ISSN: 0015-4199 – mf#14454,01 – us UMI ProQuest [610]

Florida occidental – Sevilla – 11r – 5,6 – sp Cultura [977]

Florida occidental y luisiana : correspondencia (anno 1775-1814) – Sevilla – 112r – 5,6 – sp Cultura [977]

Florida occidental y luisiana : correspondencia gobernadores (anno 1766-) – Sevilla – 73r – 5,6 – sp Cultura [977]

Florida occidental y luisiana. caja real (anno 1783-1821) – Sevilla – 23r – 5,6 – sp Cultura [977]

Florida occidental y luisiana. correspondencia gobernadores (anno 1764-) – Sevilla – 122r – 5,6 – sp Cultura [977]

Florida of today / Davidson, James Wood – New York, NY. 1889 – 1r – us UF Libraries [978]

Florida orange groves for sale / Robinson, M F – Sanford, FL. 1911 – 1r – us UF Libraries [634]

Florida oriental – Sevilla – 8r – 5,6 – sp Cultura [977]

Florida oriental. caja real (anno 1783-1821) – Sevilla – 93r – 5,6 – sp Cultura [977]

Florida ostrich farm – St Augustine, FL. 1904? – 1r – us UF Libraries [636]

Florida park, parkway and recreational-area study... / Florida State Planning Board – Tallahassee, FL. 1940 – 1r – us UF Libraries [790]

Florida peninsular – Tampa, FL. 1860 jun 23-1871 – 2r – (gaps) – us UF Libraries [071]

Florida peninsular – Tampa, FL. 1855 mar-1860 jun – 1r – us UF Libraries [071]

Florida photo news – 1987 sep 10/16-1988 dec 21/28, 1989-94, 1995 jan 5/12-dec 21/27, 1996 jan 4/10-dec 26/jan 1, 1997, 1997 jan 9/15-dec 18/24, 1998 jan 15/21-nov 19/25, 1999 mar 11/17-dec 30/jan 5, jan, feb,mar 4/10 mutilated – 12r – 1 – mf#1062008 – us WHS [071]

Florida plant immigrants / Bailey, L H – Coconut Grove, FL. 1940 – 1r – us UF Libraries [630]

Florida plantation records from the papers of george noble jones – St Louis, MO. 1927 – 1r – us UF Libraries [630]

Florida plantations – s.l, FL. 1914 – 1r – us UF Libraries [630]

Florida portrayed – London, England. 1880 – 1r – us UF Libraries [630]

Florida post-war agriculture – s.l, s.l? 1944 – 1r – us UF Libraries [630]

Florida preparatory school : daytona beach, florida / Goebel, Rubye K – s.l, s.l? 1936 – 1r – us UF Libraries [370]

Florida reading quarterly – Orlando. 1974+ (1) 1974+ (5) 1975+ (9) – ISSN: 0015-4261 – mf#8741 – us UMI ProQuest [370]

Florida republican – Jacksonville, FL. 1849-1857 apr – 3r – us UF Libraries [071]

Florida resources and sports / Florida Dept Of Agriculture – St Augustine, FL. 193-? – 1r – us UF Libraries [790]

Florida scrub jay / Fernald, Raymond T – Tallahassee, FL. 1991 – 1r – us UF Libraries [574]

Florida sea shells / Aldrich, Bertha – Boston, MA. 1936 – 1r – us UF Libraries [590]

Florida seminole agency special report of the flo... / United States Office Of Indian Affairs – Washington, DC. 1921 – 1r – us UF Libraries [350]

Florida sentinel – Gainesville, FL. 1890 nov 14; 1891 jul 17; 1892 oct 28; 1893 dec 08 – 1r – us UF Libraries [071]

Florida sentinel – Tampa, FL. 1955 dec 17-1957 oct 12 – 2r – (gaps) – us UF Libraries [071]

Florida sentinel – Jacksonville, Pensacola FL. 1947 jan 18 – 1r – 1 – mf#849706 – us WHS [071]

Florida sentinel – Tallahassee, FL. 1854 may 30-1855 – 1r – us UF Libraries [071]

The florida sentinel – Pensacola, FL: [s.n.] [mf ed 1947] – (= ser Negro Newspapers on Microfilm) – 1r – 1 – us L of C Photodup [071]

Florida sketch-book / Torrey, Bradford – Boston, MA. 1924 – 1r – us UF Libraries [978]

Florida, south carolina, and canadian phosphates / Millar, C C Hoyer – London, England. 1892 – 1r – us UF Libraries [630]

Florida squatters / Darsey, Barbara Berry – s.l, s.l? 1939 – 1r – us UF Libraries [307]

Florida standard – Jacksonville, FL. 1859 feb-may – 1r – us UF Libraries [071]

Florida standard guide / Reynolds, Charles B – New York, NY. 1928 – 1r – us UF Libraries [978]

Florida star – Jacksonville, FL. 1956-1998 jun – 43r – (gaps) – us UF Libraries [071]

Florida star – Jacksonville FL. 1970 jan 31-feb 21 – 1r – 1 – mf#780692 – us WHS [071]

Florida star – Titusville, FL. 1880-1914 jan 30 – 21r – us UF Libraries [071]

Florida. State Bar Association *see* Proceedings

Florida state bar association law journal *see* Florida bar journal

Florida state bar association proceedings – v1-20. 1907-27 (all publ) – 22mf – 9 – $33.00 – (lacking: v4-11) – mf#LLMC 84-464 – us LLMC [340]

Florida State Committee On National Soldiers Home *see* Florida

Florida state farm, raiford, florida / Sheffield, L C – s.l, s.l? 193-? – 1r – us UF Libraries [630]

Florida state horticultural society proceedings – Lake Alfred. 1892-1974 (1) – mf#10059 – us UMI ProQuest [630]

Florida State Library *see* Florida material in us congressional documents

Florida State Live Stock Association *see* Florida

Florida State Museum *see*
- Florida museum catalog cards

Florida State Planning Board *see* Florida park, parkway and recreational-area study...

Florida State Road Dept Division Of Statewide Highways *see* Highways of florida
Florida state university law review – v1-27. 1973-2000 – 5,6,9 – $643.00 set – (v1-12 1973-85 on reel $237. v13-27 1985-2000 on mf $406) – ISSN: 0096-3070 – mf#102781 – us Hein [340]
Florida sugar lands / Malabar Sugar Company – New York, NY. 1924 – 1r – us UF Libraries [630]
Florida sun – Jacksonville, FL. 1877 jan 16-feb 25 – 1r – us UF Libraries [071]
Florida sun-land company – Jacksonville, FL. 1935 – 1r – us UF Libraries [630]
Florida. Supreme Court *see* Florida supreme court reports
Florida supreme court reports / Florida. Supreme Court – v1-103. 1846-1931 – 938mf – 9 – $1407.00 – (pre-nrs: v1-22 1846-86 168mf $252.00. updates not poss until year 2006) – mf#LLMC 80-821 – us LLMC [347]
Florida supreme court reports – Tallahassee. 1846-1886 [1,5,9] – mf#1785 – us UMI ProQuest [323]
Florida tattler – Jacksonville FL. 1934 dec 1, 1935 jun 29, jul 13, 20, 1945 sep 29 – 1r – 1 – mf#4039314 – us WHS [071]
The florida tattler – Jacksonville. Fla. oct. 4, 11, 1941 – 1 – us NY Public [071]
Florida tax comparisons with california, illinois... / Prentice-Hal, Inc – s.l, s.l? 1936 – 1r – us UF Libraries [336]
Florida tax review – v1-4. 1992-2001 – $136.00 set – mf#115591 – us Hein [343]
Florida Taxpayers Of The Drainage District *see* Plain answer to governor broward's open letter to...
Florida telegraph – Starke, FL. 1879 dec. 27 – 1r – us UF Libraries [071]
Florida times – Jacksonville, FL. 1865-1866 [scattered] – 1r – us UF Libraries [071]
Florida times union : (star edition) – Jacksonville, FL. 1950-1966 (1) – mf#62425 – us UMI ProQuest [071]
Florida times union (state news) – Jacksonville, FL. 1963-1967 (1) – mf#62426 – us UMI ProQuest [071]
Florida times-union – Jacksonville FL. 1893 may 29-31 – 1r – 1 – (cont: daily florida union; florida daily times) – mf#856495 – us WHS [071]
Florida times-union – Jacksonville, FL. 1883 feb 04-dec 30 – 2r – (gaps) – us UF Libraries [071]
Florida times-union – Jacksonville, FL. 1881+ (1) – mf#60434 – us UMI ProQuest [071]
Florida times-union index – Jacksonville, FL. 1896-1964 – 18r – (gaps) – us UF Libraries [071]
Florida times-union index – Jacksonville, FL. 1961pt 2 S-Z-1963 – 3r – us UF Libraries [071]
Florida today – Cocoa, FL. 1966+ (1) – mf#61271 – us UMI ProQuest [071]
Florida today and tomorrow / Miller, G L – Miami, FL. 1923 – 1r – us UF Libraries [630]
Florida trails as seen from jacksonville to key west / Packard, Winthrop – Boston, MA. 1910 – 1r – us UF Libraries [790]
Florida trees / Snyder, Ethel – Fort Pierce, FL. 1940 – 1r – us UF Libraries [634]
Florida trend – St Petersburg. 1958+ (1) 1970+ (5) 1976+ (9) – ISSN: 0015-4326 – mf#1623 – us UMI ProQuest [332]
Florida truck and garden insects / Watson, J R – Gainesville, FL. 1917 – 1r – us UF Libraries [630]
Florida truck and garden insects / Watson, J R – Gainesville, FL. 1919 – 1r – us UF Libraries [630]
Florida truck and garden insects / Watson, J R – Gainesville, FL. 1931 – 1r – us UF Libraries [630]
Florida truck crop competition / Wann, John L – Gainesville, FL. 1931 – 1r – us UF Libraries [630]
Florida truck crop competition... / Noble, C V – Gainesville, FL. 1931 – 1r – us UF Libraries [630]
Florida truck crop competition, intra-state / Wann, John L – s.l, s.l? 1932 – 1r – us UF Libraries [630]
Florida trucking for beginners / Bateman, Lee La Trobe – Deland, FL. 1913 – 1r – us UF Libraries [380]
Florida union – Jacksonville, FL. 1865 feb 11-1877 – 4r – (gaps) – us UF Libraries [071]
Florida union – Jacksonville, FL. aug 18 1866 [wkly] – (= ser Confederate newspapers) – 1 – us Western Res [071]
Florida vegetables, irish potatoes, melons and cucumbers – Jacksonville, FL. 1919 – 1r – us UF Libraries [630]
Florida vocational journal – Tallahassee. 1986-1991 (1,5,9) – ISSN: 0145-9376 – mf#12843 – us UMI ProQuest [331]
Florida weekly advocate – Starke, FL. 1898 jan . 27-1900 sep 27 – 1r – us UF Libraries [071]
Florida wild life / Simpson, Charles Torrey – New York, NY. 1932 – 1r – us UF Libraries [500]
Florida womans world – 1984 spring, v1 n1-3 [1984 summer-1984/85 winter], v2 n1-2 [1987 summer-fall] – 1r – 1 – (cont by: independent woman [saint petersburg, fl]) – mf#2801800 – us WHS [071]
Floridablanca, Conde de *see*
- Censo espanol
- Censo espanol executado por orden del rey en el ano 1787

Floridagriculture – Gainesville, FL. v46 n1-v50 n12. 1987 jan-1991 dec – 1r – us UF Libraries [071]
Floridale farms and groves / Ringling And White, Inc – New York, NY. 1925 – 1r – us UF Libraries [630]
Florida's $500,000,000 citrus industry – Jacksonville, FL. 1931 – 1r – us UF Libraries [634]
Florida's central lake region – Norfolk, VA. 1915 – 1r – us UF Libraries [630]
Florida's deutsches echo – Miami FL (USA), 1929 8 jun-1933 19 may – 1r – 1 – gw Misc Inst [071]
Florida's first coast carriers / Northeast Florida Letter Carriers Local Union Branch 53 – 1984 sep-1985 may, aug-sep, nov-dec, 1987 may-1988 jul, oct-1989 may, jul-aug – 1r – 1 – (cont: serving america [jacksonville, fl]; cont by: north florida letter carrier) – mf#1614439 – us WHS [380]
Florida's geological history and geological resources – Tallahassee, FL. 1994 – 1r – us UF Libraries [550]
Florida's latest accomplishment insuring a progres... – Ft Lauderdale, FL. 1928 – 1r – us UF Libraries [978]
Florida's opportunity / Corey, Merton L – s.l, s.l? 1932 – 1r – us UF Libraries [630]
Florida-via camera / Shrader, Welman Austin – New Albany, IN. 1939 – 1r – us UF Libraries [978]
Floridian – Tallahassee FL. 1828 nov 18 – 1r – 1 – (cont: floridian and advocate; cont by: southern journal; floridian and journal) – mf#846234 – us WHS [071]
Floridian – St. Petersburg, FL. 1944-1959 (1) – mf#62451 – us UMI ProQuest [071]
Floridian – Tallahassee, FL. 1831 oct-1848 – 4r – us UF Libraries [071]
Floridian and journal – Tallahassee FL. 1849 feb 24 – 1r – 1 – (cont: floridian; southern journal; cont by: weekly floridian) – mf#787005 – us WHS [071]
Floridian and journal : weekly confederate newspaper – Tallahassee, FL. mar 5 1864 – (= ser Confederate newspapers) – 1r – 1 – us Western Res [071]
Floridian journal – Tallahassee, FL. 1858 jun-1860 dec – 1r – us UF Libraries [071]
Floridsdorf : the vienna workers in revolt: a play / Wolf, Friedrich – 1st American ed. New York: Universum Publishers and Distributors, c1935 – 1r – 1 – us Wisconsin U Libr [820]
Florilegia : florilegium frisingense. testimonia divinae scripturae et patrum (ccsl 108d). formae tplila 42 – 1987 – (= ser ILL – ser a; Cccm) – 3mf+39p – 9 – €30.00 – 2-503-61082-X – be Brepols [400]
Florilegii hebraici lexicon : quo illius vocabula latine et germanice versa continentur / ed by Lindemann, Hubert – Friburgi Brisgoviae; S Ludovici Americae: Herder, 1914 – 1mf – 9 – 0-8370-1947-8 – mf#1987-6334 – us ATLA [052]
Florilegio a las madres / Sanz Agramonte, Altagracia – Camaguey, Cuba. 1949 – 1r – us UF Libraries [972]
Florilegio de sonetos / Abad Mendez, Ramon Antonio – Santo Domingo, Dominican Republic. 1935 – 1r – us UF Libraries [972]
Florilegio...iglesia...dividido en discursos / Soto y Marne, Francisco de – 1738 – 9 – sp Bibl Santa Ana [240]
Florilegium historico-criticum librorum rariorum / Gerdes, Dan – ed 3a. Croningae Bremae, 1763 – 9mf – 8 – €18.00 – ne Slangenburg [240]
Florilegium patristicum : tam veteris quam medii aevi auctores complectens – Bonn. v1-45. 1911-1938 – 112mf – 9 – €61.00 – ne Slangenburg [240]
Florimonte, Galeazzo *see* Ragionamenti di mons galeazzo florimonte, vescovo di sessa, sopra l'ethica di aristotile
Florio, John *see* Second frutes
Floris, C *see*
- Veelderleij niewe inventien van antycksche sepultueren...libro secundo
- Veelderley veranderinghe van grotissen ende compertimenten...

Florissant eagle *see* Miscellaneous newspapers of teller county
Florissant valley baptist church. missouri : church records – 1955-60. and Bulletins (Voice). 1956-59 – 1 – 48.60 – us Southern Baptist [242]
Florissant Valley Historical Society *see* Quarterly
Florists' review – Chicago. 1972+ (1) 1974+ (5) 1974+ (9) – ISSN: 0015-4423 – mf#7094 – us UMI ProQuest [630]
Florit, Eugenio *see*
- Asonante final
- Cuatro poemas de eugenio florit
- Doble acento
- Habito de esperanza
- Tropico

Floro bartolomeu / Macedo, Nertan – Rio de Janeiro, Brazil. 1970 – 1r – us UF Libraries [972]
Florovskij, G V *see* Puti russkago bogoslovija
Florus, Publius Annius *see* In orationes quasdam ciceronis...
Flory, J S *see*
- Love's sweet dream fully realized through holy matrimony and a sanctified home
- Mind mysteries

Flory, John Samuel *see* Literary activity of the german baptist brethren in the eighteenth century
Flos florum : emblema pulcherrimum et homine christiano dignissimum / [Ecclesia, S ab] – Mediolani: Marchisini, 1550 – 1mf – 9 – mf#0-1997 – ne IDC [090]
Die flotte – Berlin DE, 1900-06, 1915-17 [single iss] – 1mf=2df – 1 – gw Mikrofilm [074]
Flou, Karel de *see* Die bedudinghe naden sinne van sunte augustijns regule
Flour from canada's far north west : with some account of wheat growing and flour milling ancient and modern – Winnipeg: Free Press Job Print, 1907 [mf ed 1995] – 1mf – 9 – 0-665-76893-1 – mf#76893 – cn Canadiana [660]
Flournoy, Parke Poindexter *see* New light on the new testament
Flournoy, Theodore *see* Metaphysique et psychologie
Flow measurement and instrumentation – Kidlington. 1989-1996 (1,5,9) – ISSN: 0955-5986 – mf#17228 – us UMI ProQuest [620]
Flow, turbulence and combustion – The Hague. 1998+ (1) – (cont: applied scientific research) – ISSN: 1386-6184 – mf#16767,01 – us UMI ProQuest [500]
Flower and garden – Kansas City. 1957+ (1) 1972+ (5) 1973+ (9) – ISSN: 0891-9534 – mf#6692 – us UMI ProQuest [630]
Flower, Benjamin Orange *see* Christian science
Flower bud differentiation and growth studies in the gladiolus / Watkins, John V – s.l, s.l? 1931 – 1r – us UF Libraries [580]
Flower, Elliott, 1863-1920 *see* Policeman flynn
Flower, fruit, and thorn pieces / Jean Paul – London, England. 1888 – 1r – us UF Libraries [890]
The flower garden and window gardening – Windsor, NS: The Pidgeon Fertilizer Co, [18–?] [mf ed 1987] – 1mf – 9 – 0-665-68288-3 – mf#68288 – cn Canadiana [635]
Flower, George Edward *see* Life and writings of george edward flower
The flower queen : or, the coronation of the rose: a juvenile cantata / Root, George Frederick – [Toronto?: s.n.], 1873 [mf ed 1984] – 1mf – 9 – 0-665-28886-7 – mf#28886 – cn Canadiana [780]
Flower, Sir, William Henry *see* Fashion in deformity as illustrated in the customs of barbarous and civilised races
Flower thrips / Watson, J R – Gainesville, FL. 1922 – 1r – us UF Libraries [630]
Flower, William *see* Pedigrees recorded at the visitations of the county palatine of...
Flower, William Henry *see* Essays on museums
Flowering, fruiting, yield and growth habits of tung trees / Dickey, R D – Gainesville, FL. 1940 – 1r – us UF Libraries [634]
Flowering of indian art / Mukerjee, Radhakamal – New York, NY. 1964 – 1r – us UF Libraries [700]
Flowers and fruits in the wilderness / Morrell, Z N – 1886 – 1 – us Southern Baptist [242]
Flowers and serpents – London, England. 18– – 1r – us UF Libraries [240]
Flowers and their pedigrees / Allen, Grant – London: Longmans, Green, 1883 – 3mf – 9 – mf#27754 – cn Canadiana [580]
Flowers by the wayside : a miscellany of prose and verse, including marion somers, the old man's desire, precious memories, siege of lucknow, absent friends etc / Herbert, Mary E – [Halifax, NS?: s.n.] 1865 [mf ed 1994] – 1mf – 9 – 0-665-94733-X – mf#94733 – cn Canadiana [800]
Flowers from a canadian garden / ed by Burpee, Lawrence J – Toronto: Musson, [1909?] [mf ed 1994] – (= ser Canadian treasury series) – 2mf – 9 – 0-665-72221-4 – mf#72221 – cn Canadiana [810]
Flowers of a mystic garden = Selections. 1912 / Ruusbroec, Jan van – London: JM Watkins, 1912 – 1mf – 9 – 0-524-00092-1 – (in english) – mf#1989-2792 – us ATLA [240]
Flowers of literature (london) – 1801-04 – (= ser 19th c british periodicals) – r36 – 1 – us Primary [410]
Flowers of literature (london) : 1805-06 – (= ser 19th c british periodicals) – r37 – 1 – us Primary [410]
Flowers of literature (london) – 1807-09 – (= ser 19th c british periodicals) – r38 – 1 – us Primary [410]
Flowers of modern voyages and travels : comprising those most worthy of record between the years 1806 and 1820 / Adams, William – London 1820 [mf ed Hildesheim 1995-98] – 4v on 12mf – 9 – €120.00 – 3-487-29934-8 – gw Olms [910]
Flowing road / Whitney, Casper – Philadelphia, PA. 1912 – 1r – us UF Libraries [972]
Floy, James *see* Old testament characters
Floyd, B F *see*
- Dieback or exanthema of citrus trees
- Fertilizing the irish potato crop
- Melanose and stem-end rot
- Some cases of injury to citrus tres apparently induced by ground limestone

Floyd county press – Rome, GA. 1984-1985 (1) – mf#62467 – us UMI ProQuest [071]
Floyd, LP *see* The first 100 years of the first baptist church, denton, texas
Floyd m riddick, senate service 1947-1974 : senate parliamentarian – (= ser Us senate historical office oral history coll) – 6mf – 9 – $30.00 – us Scholarly Res [323]
Floyd, Thomas *see* A new collection of instrumental music, in three parts, arranged for the violin, clarionett, bass-viol, etc
Floyd, William *see*
- A comparison of body density and percent body fat using functional residual capacity and residual volume and development of immersed functional residual capacity and residual volume prediction formulas
- A comparison of selected coronary heart disease risk factors in weight trained males

Floyer, A M *see* The evolution of ancient hinduism
Flq – film library quarterly – New York. 1967-1984 [1]; 1970-1984 [5]; 1975-1984 [9] – ISSN: 0160-7316 – mf#5860 – us UMI ProQuest [790]
Der fluch der schoenheit : novelle / Riehl, Wilhelm Heinrich; ed by Leonard, Arthur N – Boston: Ginn c1908 [mf ed 1995] – (= ser International modern language series) – 1r – 1 – (german text, int etc in english. filmed with: gesammelte geschichten und novellen & other titles) – mf#3716p – us Wisconsin U Libr [830]
Die flucht : novellen / Alverdes, Paul – Potsdam: L Voggenreiter, c1935 [mf ed 1988] – 125p – 1 – mf#6939 n5 – us Wisconsin U Libr [830]
Die flucht : roman / Wiechert, Ernst Emil – Berlin: G Grote, 1936 – 1r – 1 – us Wisconsin U Libr [830]
Flucht aus dem paradies : arno schmidts erzaehlung 'caliban ueber setebos' / Neuner, Michael – (mf ed 1996) – 2mf – 9 – €31.00 – mf#DHS-AR 3208 – gw Frankfurter [430]
"Flucht aus der zeit?" : anarchismus, kulturkritik und christliche mystik – hugo balls konversionen / Steinbrenner, Manfred – Frankfurt a.M., 1983 (mf ed 1994) – 2mf – 9 – €31.00 – 3-89349-879-6 – mf#DHS-AR 879 – gw Frankfurter [430]
Flucht- und werbungssagen in der legende / Schmeing, Karl – Muenster: Aschendorff, 1911. 50p – 1 – us Wisconsin U Libr [390]
Flucht zum fakir von ipi : roman / Gaebert, Hans Walter – Leipzig: Wehnert, 1943 (mf ed 1990) – 1r – 1 – (filmed with: gustav freytag, ein publizist) – us Wisconsin U Libr [830]
Fluchtlingspolitik der schweiz seit 1933 bis zur gegenwart – Bern, Switzerland. 1957 – 1r – us UF Libraries [943]
Fluctuat nec mergitur. la prevote des marchands et l'urbanisme parisien au 15 siecle d'apres la jurisprudence du parlement (1380-1500) / Auzary, Bernadette – 2mf – 9 – (10299) – fr Atelier National [944]
Fludd, Robert *see* Tractatvs apologeticvs integritatem societatis de rosea crvce defendens
Fluecht, Liselott *see*
- Im foehn

Fluechtige bemerkungen auf einer reise von nuernberg ueber wuerzburg, frankfurt, mainz und koblenz in die baeder am taunus im jahr 1825 / Schaller, Christian – Nuernberg 1826 [mf ed Hildesheim 1995-98] – 2mf – 9 – €60.00 – 3-487-29419-2 – gw Olms [914]
Fluechtige bemerkungen auf einer reise von st petersburg ueber moskwa, grodno, warschau, breslau nach deutschland im jahre 1805 : in briefen / Reinbeck, Georg – Leipzig 1806 [mf ed Hildesheim 1995-98] – 2v on 5mf – 9 – €100.00 – 3-487-27802-2 – gw Olms [880]
Fluechtlings-kurier – Bremen DE, 1948 – 1 – gw Misc Inst [074]
"Fluegel auf!" : novellen / Frapan, Ilse – Berlin: Gebrueder Paetel, 1895 [mf ed 1995] – 376p – 1 – mf#8918 – us Wisconsin U Libr [830]

Fluegel, G *see*
– Kitab-al-fihrist mit anmerkungen
– Mani
Fluegel, Gustav *see*
– Al-kindi
– Concordantiae corani arabicae
– Die grammatischen schulen der araber
– Mani, seine lehre und seine schriften
Fluegel, Maurice *see*
– Exodus, moses and the decalogue legislation
– The humanity, benevolence and charity legislation of the pentateuch and the talmud
– Israel, the biblical people
– The messiah-ideal
– Der parsismus und die biblischen religionen
– The zend-avesta and eastern religions
Fluegel, Otto *see*
– A ritschl's philosophische und theologische ansichten
– Das wunder und die erkennbarkeit gottes
Fluegelrad *see* Das laeutewerk
Flug von der nordsee zu montblank, durch westphalen, niederrhein, schwaben, die schweiz, ueber baiern, franken, niedersachsen zurueck : skizze zum gemaelde unserer zeit / Mueller, Wilhelm – Altona 1821 [mf ed Hildesheim 1995-98] – 2v on 5mf – 9 – €100.00 – 3-487-27813-8 – gw Olms [914]
Flugmeldehelferin inge berger : roman / Gaebert, Hans Walter – Leipzig: Wehnert, 1943 (mf ed 1990) – 1r – 1 – (filmed with: gustav freytag, ein publizist) – us Wisconsin U Libr [830]
Flugschriftensammlung gustav freytag : vollstaendige wiedergabe der 6265 flugschriften aus dem 15.-17. jahrhundert sowie des katalogs von paul hohenemser auf microfiche = Pamphlet collection of gustav freytag / Frankfurt. Stadt- und Universitaetsbibliothek. Flugschriftensammlung Gustav Freytag – [mf ed 1980-81] – 746mf (1:42) – 9 – silver €7250.00 – ISBN-10: 3-598-21189-9 – ISBN-13: 978-3-598-21189-8 – gw Saur [430]
Fluharty, Shawn K *see* A model for improving summative ratings of student teachers utilizing generalizability theory
Fluharty, Vernon Lee *see* Dance of the millions
Fluid dynamics – New York. 1967-1974 (1) 1967-1974 (5) – ISSN: 0015-4628 – mf#10906 – us UMI ProQuest [530]
Fluid dynamics research – Amsterdam. 1986-1989 (1,5,9) – ISSN: 0169-5983 – mf#42505 – us UMI ProQuest [530]
Fluid mechanics research : english ed – New York. 1992-1992 (1) 1992-1992 (5) 1992-1992 (9) – (cont: fluid mechanics: soviet research) – ISSN: 1064-2277 – mf#14351,01 – us UMI ProQuest [627]
Fluid mechanics: soviet research – Washington. 1972-1992 (1) 1972-1992 (5) 1972-1992 (9) – (cont by: fluid mechanics research: english ed) – ISSN: 0096-0764 – mf#14351 – us UMI ProQuest [627]
Fluid phase equilibria – Amsterdam. 1977+ (1) 1979+ (5) 1986+ (9) – ISSN: 0378-3812 – mf#42186 – us UMI ProQuest [540]
O fluminense – Rio de Janeiro, RJ: Typ de C Ogier & C, 23 dez 1840 – (= ser Ps 19) – mf#P03A,03,17 n01 – bl Biblioteca [073]
Fluoreszenz des ensembles und des individuellen molekuels / Zander, Christoph – (mf ed 2000) – 3mf – 9 – €49.00 – 3-8267-2679-0 – mf#DHS 2679 – gw Frankfurter [574]
Fluoride – Auckland. 1968-1992 (1) 1972-1980 (5) 1975-1980 (9) – ISSN: 0015-4725 – mf#6397 – us UMI ProQuest [574]
Flurhaim, Christophorum *see* Alle kirchen gesaeng vnd gebeet des gantzen jars
Flurnamenstudien anhand einer sammlung der alten namen von offenbach/m.-bieber / Schwarz, Werner – Frankfurt a.M., 1967 – 2mf – 9 – 3-89349-707-2 – gw Frankfurter [943]
Die flusspiraten des mississippi : aus dem waldleben amerikas / Gerstaecker, Friedrich – Berlin: H Costenoble, 1901 (mf ed 1990) – 1r – 1 – (filmed with: die regulatoren in arkansas) – us Wisconsin U Libr [830]
Die flusspiraten des mississippi : aus dem waldleben amerikas: zweite abtheilung / Gerstaecker, Friedrich – 9. Aufl. Jena: H Costenoble [18–?] (mf ed 1990) – 1r – 1 – (filmed with: die regulatoren in arkansas) – us Wisconsin U Libr [830]
Fluszgebiet der ribeira de iguape im suden des sta... / Stutzer, Gustav – Berlin, Germany. 1912 – 1r – us UF Libraries [972]
Flute book / Beck, Henry – Manuscript book of 280 tunes for treble instrument copied by Beck in 1786. Pages 1-3 missing. Includes: Drink to Me Only; God Save the King; and College Hornpipe. Music-838 – 1 – us L of C Photodup [780]
Flute book, ms. no. 1 – Manuscript book of tunes for a treble instrument, ca. 1791. Solos and duets, one song with melody and text. Composers mentioned are Arne, Frederick Granger, and Reinagle. MUSIC 3079 – 1 – us L of C Photodup [780]
Flute method for children / Hickok, Dorothy Jane – U of Rochester 1939 [mf ed 19–] – 2mf – 9 – mf#fiche 1008 – us Sibley [780]
Flutist – Asheville NC. v1-10 n2 1920-feb 1929 – 1 – $60.00 – mf#0206 – us Brook [780]
The flutist – Asheville. 1-10, n2. January 1920-February 1929 – 1 – us NY Public [780]
Die flutsagen : ethnographisch betrachtet / Andree, Richard – Braunschweig: Friedrich Vieweg, 1891 [mf ed 1985] – 1mf – 9 – 0-8370-2101-4 – mf#1985-0101 – us ATLA [390]
Fly : or juvenile miscellany – Boston. 1805-1806 (1) – mf#4450 – us UMI ProQuest [305]
Fly by night – v1 n3 [1971 mar] – 1r – 1 – mf#1583315 – us WHS [071]
Fly, Elijah M *see* The bible true
Fly fisherman – Harrisburg. 1985-1994 (1) 1985-1986 (5) 1985-1986 (9) – ISSN: 0015-4741 – mf#11920 – us UMI ProQuest [790]
Fly plague : a manual = Vliegoorlas: 'n handleiding / South Africa. Department of National Health and Population Development [Departement van Nasionale Gesondheid en Bevolkingsontwikkeling – 2nd ed. Pretoria: Dept of National Health & Population Development 1988 [mf ed Pretoria, RSA: State Library [199-]] – 11p [ill] on 1r with other items – 5 – (in english & afrikaans) – mf#op 08747 r25 – us CRL [614]
Flyer / Langley Air Force Base [VA] – Virginia. 1981 may-1982 oct 8, 29, nov 5-12, 1983-84, 1985 jan 11-1986 mar, 1986 apr-1987 sep, 1987 oct-1988 oct, 1988 nov-1989 aug – 6r – 1 – mf#646416 – us WHS [355]
Flyer – v4 n18-20 [1945 sep 15-29] – 1r – 1 – mf#2892687 – us WHS [071]
Flyers : six months of toronto junk mail – 2r – 1 – Can$150.00 – (a complete photographic record (ca 2000 exposures) of the flood of advertising flyers and other unaddressed mail delivered to a spruce street house in toronto's cabbagetown neighbourhood, jul 1-dec 31 1996) – cn McLaren [650]
Flygposten – Malmoe, 1876-78 – 9 – sw Kungliga [078]
Flying – 1927-63 – 1 – 840.00 – us L of C Photodup [629]
Flying – New York. 1927+ (1) 1968+ (5) 1963+ (9) – ISSN: 0015-4806 – mf#257 – us UMI ProQuest [629]
Flying models – Canton, OH. v1-62. 1928-55 – 1 – us L of C Photodup [600]
Flying post : or, the postmaster – Dublin, Ireland. 6 jan, 3 jun, 2, 24 aug, 4, 23, 29, 30 sep, 29 nov, 2 dec 1708; 17, 26 mar, 25, 27 apr, 2 may, 29 jun, 4, 14, 28 jul, 8 aug 1709; 10 may, 7 jun, 4, 10 jul, 24 aug 1710.-d – 1/4r – 1 – (publ only 6 jan 1708-24 aug 1710) – uk British Libr Newspaper [072]
Flying post (james esdall) – Dublin, Ireland. 31 mar, 3 may 1744 – 1/4r – 1 – (publ only 31 mar, 3 may 1744) – uk British Libr Newspaper [072]
"The flying priest" : fr glover's account of his flying experiences in new guinea, mainly during the pacific war, including the evacuation to kainantu and his attempted flight to thursday island / Glover, John Corbett – 1936-1942 – 1r – 1 – mf#pmb1233 – at Pacific Mss [934]
Flying quill / Goodspeed's Book Shop – 1939 nov, 1940 jan-1984 winter – 1r – 1 – mf#1671041 – us WHS [070]
Flying safety – Washington. 1981+ (1,5,9) – (cont: aerospace safety) – ISSN: 0279-9308 – mf#6287,01 – us UMI ProQuest [629]
Flying saucers – Amherst. 1973-1976 (1) 1976-1976 (5) 1976-1976 (9) – ISSN: 0015-489X – mf#7396 – us UMI ProQuest [629]
Flying trip to the tropics / Robinson, Wirt – Cambridge, MA. 1895 – 1r – us UF Libraries [972]
Flynn, Edmund James *see*
– Affaire riel
– Chemins de fer dans la province de quebec
– Discours de l'honorable e-james flynn, prononce a l'assemblee legislative aux seances des 8, 13 et 16 mai 1884
– Discours prononce par l'honorable m flynn sur la deuxieme lecture du bill pour diviser les districts electoraux de montreal-est, montreal centre et montreal-ouest, quebec-est, drummond et arthabaska, chicoutimi et saguenay
– Discours prononces par l'hon depute de gaspe
– Le gouvernement provincial devant l'opinion
– Manual for crown land and timber agents
– La mauvaise politique qui depouille nos forets sans profit pour le travail canadien
– Projet de loi concernant les mines
– Speech of hon e j flynn
Flynn, Edmund James [comp] *see*
– 1894 settler's guide, province of quebec
– 1896 guide du colon
Flynn, G *see* Survey of the prospects of agricultural and industrial development in tuvalu
Flynn, Priscilla M *see* Mammography adoption of winona county women using the transtheoretical model
Fly-switch from the sultan / Bates, Darrell – London, England. 1961 – 1r – us UF Libraries [960]
Fm engineering data base in order by channel and location / U.S. National Technical Information Service – Monthly.Sorted by frequency, and secondarily by state.Includes call letters, city, state, frequency, licensee's name, power, height, etc – 9 – us NTIS [000]
Fm engineering data base in order by location / U.S. National Technical Information Service – Monthly.Sorted by state and secondarily by city – 9 – us NTIS [000]
La F-M et l'ouvrier / Hello, Henri – Paris: Renaissance francaise 1909 – 2mf – 9 – mf#vrl-146 – ne IDC [366]
Fo chiao kai lun / Chiang, Wei-ch'iao – Shanghai: Chung-hua shu chu, Min kuo 39 [1940] – (= ser P-k&k period) – us CRL [280]
Fo fa tao lun / Li, Yuan-ching – [China: sn, 1940] – (= ser P-k&k period) – us CRL [280]
Fo hebdo : organe officiel de la cgtfo – Paris: s.n., [1966-]. [n1068-1203 nov 23 1966-69] – us CRL [074]
Fo hua chi-tu chiao (ccm10) / Chang, Ch'un-i – 13th ed. Shanghai, 1930 [mf ed 198?] – (= ser Ccm 10) – 1 – mf#1984-b500 – us ATLA [230]
Fo kuo chi *see* Record of buddhistic kingdoms
Fo magazine – Paris: Imp Chaix-Desfosses, Neogravure. [n1-39. nov 1965-69] – us CRL [074]
Fo veckotidningen – Norrkoping, Sweden. 1980-83 – 1 – sw Kungliga [078]
Foa, Edouard *see*
– Le dahomey
– Mes grandes chasses dans l'afrique centrale
Foakes-Jackson, Frederick John *see*
– The biblical history of the hebrews
– The biblical history of the hebrews to the christian era
– A brief biblical history
– Christian difficulties in the second and twentieth centuries
– The faith and the war
– History of the christian church
– Josephus and the jews
– The parting of the roads
– St luke and a modern writer
Focal spot – Ottawa. 1944-1970 (1) – mf#7249 – us UMI ProQuest [610]
Fochriw pentwyn cemetery burial records – 1mf – 9 – £1.25 – uk Glamorgan FHS [929]
Fochriw, pentwyn cemetery, monumental inscriptions – 2mf – 9 – £2.50 – uk Glamorgan FHS [929]
Fock, Gorch *see* Schullengrieper und tungenknieper
Fock, Otto *see*
– Der socinianismus
Focke, Friedrich *see*
– Die entstehung der weisheit salomos
– Odyssee
Focken, Charles Melbourne *see* Dimensional methods and their applications
Focus : 1947 news / International Association of Machinists and Aerospace Workers – v6 n1-v9 n6 [1983 jan-1986 jul] – 1r – 1 – mf#905968 – us WHS [629]
Focus – New York. 1950-1998 (1) 1968-1998 (5) 1968-1998 (9) – (cont by: focus on geography) – ISSN: 0015-5004 – mf#2501 – us UMI ProQuest [900]
Focus – Albuquerque NM. 1981 may 21-1982 apr, 1982 may 7-1983 feb 18, 1983 feb 25-1985 feb, 1985 mar-1987 mar 17, 1987 apr-1988 may, 1988 jun-1989 may 12 – 6r – 1 – (cont by: kirtland focus) – mf#646902 – us WHS [071]
Focus – Kampala, Uganda. jun 12 1984-dec 27 1987; jan 5 1988-sep 28 1990 – 2r – 1 – (cont: weekly focus) – us L of C Photodup [079]
Focus / First Church of God [Inglewood, CA] – 1995 feb – 1r – 1 – mf#4024277 – us WHS [242]
Focus : technical cooperation = Focus cooperacion tecnica – Washington. 1972-1978 (1) 1975-1978 (5) 1975-1978 (9) – ISSN: 0146-8502 – mf#9626 – us UMI ProQuest [337]
Focus / Franklin Co. Columbus – (apr 1967-may 1980) [mthly] – 2r – 1 – mf#B10143-10144 – us Ohio Hist [331]
Focus : an independent monthly newspaper – v2 n4-v5 n7 [1977 may-1980 aug] – 1r – 1 – mf#646533 – us WHS [071]
Focus / Joint Center for Political Studies [US] et al – 1972 dec, 1973 apr-1985 jan – 1r – 1 – mf#606296 – us WHS [320]
Focus / Livonia Education Association [MI] – v9 n1-v15 n1 [1975 nov-1980 oct/nov] – 1r – 1 – mf#675812 – us WHS [370]
Focus : metropolitan philadelphia's business newsmagazine – Philadelphia. 1986-1990 (1) 1986-1990 (5) 1986-1990 (9) – ISSN: 0193-502X – mf#15029 – us UMI ProQuest [650]
Focus / Ohio AFL-CIO – v1 n1-v14 n5 [1967 apr 21-1980 oct] – 1r – 1 – (cont by: focus quarterly) – mf#898326 – us WHS [331]
The focus – Blantyre: Blantyre Print and Packaging, dec 1994-apr/may 1996 – 1r – 1 – us CRL [079]
Focus [banbury ed] – Banbury, England 26 nov 1971-10 oct 1985 – 1 – (cont: banbury focus [5 nov 1970-19 nov 1971; cont by: banbury focus [17 oct 1985-23 feb 1989]; very incomplete for 1970-71) – uk British Libr Newspaper [072]
Focus (harrow ed) – London, UK. 4 jan-20 dec 1986; 3 jan-12 sep 1987 – 2r – 1 – (aka: focus magazine) – uk British Libr Newspaper [072]
Focus magazine *see* Focus (harrow ed)
Focus mda / Missouri Dental Association – Jefferson City. 1998+ (1) – (cont: missouri dental journal: the journal of the missouri dental association) – mf#2681,04 – us UMI ProQuest [617]
Focus on critical care – St. Louis. 1983-1992 (1) 1983-1992 (5) 1983-1992 (9) – ISSN: 0736-3605 – mf#14236,01 – us UMI ProQuest [610]
Focus on exceptional children – Denver. 1969+ (1) 1975+ (5) 1975+ (9) – ISSN: 0015-511X – mf#10347 – us UMI ProQuest [370]
Focus on film – London. 1970-1981 [1,5]; 1976-1981 [9] – ISSN: 0015-5128 – mf#6561 – us UMI ProQuest [790]
Focus on full employment / National Committee for Full Employment – 1976 jan-1977 jun – 1r – 1 – mf#379641 – us WHS [331]
Focus on geography – New York, 2000+ [1,5,9] – (cont: Focus) – mf#2501,01 – us UMI ProQuest [910]
Focus on guidance – Denver. 1968-1977 (1) 1975-1977 (5) 1975-1977 (9) – (cont by: counseling and human development) – ISSN: 0015-5136 – mf#10348 – us UMI ProQuest [370]
Focus on indiana libraries – Indianapolis. 1947+ [1]; 1970-1995 [5]; 1975-1995 [9] – ISSN: 0015-5152 – mf#5938 – us UMI ProQuest [020]
Focus quarterly / Ohio AFL-CIO – v15 n1-v21 n1 [1981 apr-1987 oct] – 1r – 1 – (cont: focus [columbus, oh]) – mf#1302689 – us WHS [331]
Fodor, Carel Anton *see*
– Ouverture de la dot
– Quatuor pour le piano forte accompagne d'un violon, viola & violoncelle, oeuvre 7, livre 4
Fodor, Istvan *see* Problems in the classification of the african languages
Fodor, Josephus *see*
– Six quatuors concertans pour deux violons, alto et basse, 4e livre de quatuors
– Trois duos pour deux violons, opera 18. 2e partie
Foe koue ki : ou, relation des royaumes bouddiques... / Remusat, J P A – Paris: l'Imprimerie Royale, 1836 – 9mf – 9 – mf#HT-874 – ne IDC [915]
Foebadius Aginnensis et al *see* Contra arianos; de laude sanctorum; libellus emendationis; epistulae; commonitorium. excerptis ex operibus s. augistini; altercatio legis inter simonem iudaeum et theophilum christianum
Foedera, conventiones, literae et cujuscunque generis acta publica, inter reges angliae : et alios quosvis imperatores, reges pontifices, principes, vel communitates / Rymer, Thomas – ed 2a. Londini. v1-20. 1727-35 – 9 – €1424.00 – ne Slangenburg [240]
Foederal american monthly – New York. 1833-1865 (1) – mf#4572 – us UMI ProQuest [870]
Foegl d'engiadina – Zuoz, Switzerland. 23 Dec 1857-28 Dec 1878. 5 reels – 1 – uk British Libr Newspaper [949]
Foehrer lokalanzeiger – Wyk (Foehr) DE, 1909 22 sep-1910 [gaps], 1919 30 jul-1941 30 aug – 1 – (further title: neue foehrer nachrichten) – gw Misc Inst [074]
Foehrer nachrichten – Wyk (Foehr) DE, 1892-1902, 1904-44 – 1 – (further title: foehrer zeitung) – gw Misc Inst [074]
Foehrer zeitung *see* Foehrer nachrichten
Foels, Tracie L *see* The fundraising process for the mccaskill soccer center
Das foerderband – Mumsdorf DE, 1950 30 nov-1968 28 jun [gaps] – 4r – 1 – (title varies: auch: unser foerderband, veb braunkohlenwerk phoenix) – gw Misc Inst [622]
Foeretags ekonomi – Goeteborg. 1977-1980 (1,5,9) – ISSN: 0015-7619 – mf#11152 – us UMI ProQuest [336]
Foersoek till en grundlig och dock laettfattlig foerklaring af pauli bref till eferserna / Hasselquist, Tufve Nilsson – Rock Island, IL: Augustana Book Concern, 1887 [mf ed 1992] – 1mf – 9 – 0-524-05299-9 – (in swedish) – mf#1992-0400 – us ATLA [227]
Foerste mosebok, eller, genesis : normaluppl agans text – Uppsala: L. Norblad, [1911?] – 1mf – 9 – 0-7905-2143-1 – mf#1987-2143 – us ATLA [221]
Foerster, C F L *see* Allgemeine bauzeitung
Foerster, D *see* Luthers wartburgsjahr, 1521-1522
Foerster, Emanuel Aloys *see* Variations to favorite airs for the piano forte, no [21]

Foerster, Erich *see*
- Die christliche religion im urteil ihrer gegner
- Die entstehung der preussischen landeskirche unter der regierung koenig friedrich wilhelms des dritten
- Der evangelische sinn unserer kirchenverfassung
- Weshalb wir in der kirche bleiben!

Foerster, Ernst *see*
- Handbuch fuer reisende in italien
- Manuel du voyageur en italie

Foerster, Friedrich C *see* Briefe eines lebenden
Foerster, Gerhard *see* Das mosaische strafrecht in seiner geschichtlichen entwickelung
Foerster, Matthias *see* Entwurf, aufbau und erprobung eines rastertunnel-messkopfes fuer den einsatz in einem rasterelektronenmikroskop
Foerster, Max [comp] *see* Der vercelli-codex 118
Foerster, Theodor *see* Der altkatholicismus
Foerster, Wilhelm Julius *see* Lebenserinnerungen und lebenshoffnungen
Foertsch, Richard *see* Vergleichende darstellung des code civil und des buergerlichen gesetzbuches fuer das deutsche reich
Fog bells : a sequel to nazareth / Dall, Caroline Wells Healey – Boston:...Little, Brown, 1905 [mf ed 1985] – 1mf – 9 – 0-8370-2807-8 – mf#1985-0807 – us ATLA [240]

Fogazzaro, Antonio *see*
- The saint

Fogazzaro, Antonio, 1842-1911 *see* The patriot
Fogg, F M *see* Banking octopus and the silver question
Foggini, N *see* Il museo capitoli...
Fogginius, P *see* Nova appendix (cbh35)
Foghorn / Spark [Organization : Baltimore, MD] – n11-14,17-19 [1972 aug 30-oct 11, nov22-dec 20], n20-23,25-26 [1973 jan 9-mar 7, 21-apr 4] – 1r – 1 – (cont by: spark foghorn) – mf#957973 – us WHS [071]
Fogliano, Lodovico *see* Mvsica theorica ludouici foliani mutinensis

Foglietta, U *see*
- De causis magnitudinis imperii tvrcici
- De sacro foedere in selimvm
- Historiae genuensium libri 12
- Istoria...della sacra lega contra selim...

Foguet, Juan *see* El taumaturgo catalan beato salvador de horta
O foguete – Nazare, PE: Typ SOS do P L I de A Lima, 29 jun 1844 – (= ser Ps 19) – mf#P19,03,34 n01 – bl Biblioteca [320]
O foguete : periodico critico, literario e noticioso – Paraiba, 07 ago 1862 – (= ser Ps 19) – mf#P17,02,131 – bl Biblioteca [079]
Fohne (the hebrew standard) – London, UK. 5 Sept-5 Nov 1897 – 1 – uk British Libr Newspaper [072]
Fo-hsi hsi chu ti san chi, ti ssu chi / Hsiung, Fo-hsi – Shang-hai: Shang wn yin shu kuan, Min kuo 22 [1933] – (= ser P-k&k period) – us CRL [820]
La foi catholique dans ses relations avec la raison et la volonte : conferences donnees...le 19 janvier et le 2 fevrier 1898 / Auclair, Elie-Joseph – Montreal?: Arbour & Laperle, 1897 i.e. 1898 – 1mf – 9 – mf#11919 – cn Canadiana [230]
Foi digest / University of Missouri – Columbia – 1960 mar-1983 jun – 1r – 1 – mf#971513 – us WHS [071]
Foi digest – Columbia. 1974-1985 (1) 1974-1985 (5) 1974-1985 (9) – ISSN: 0015-5349 – mf#9289 – us UMI ProQuest [320]
Foi et gnose : introduction a l'etude de la connaissance mystique chez clement d'alexandrie / Camelot, P Th – Paris, 1945 – 3mf – 8 – €7.00 – ne Slangenburg [240]
La foi et la devotion a marie toujours immaculee : expliquee et proposee d'apres les sentiments et les paroles des ss. peres = Fede e la devozione a maria sempre immacolata / Parodi, Louis – Paris: H. Casterman, 1858 – 1mf – 9 – 0-8370-8052-5 – (incl bibl ref. in french) – mf#1986-2052 – us ATLA [240]
La foi et l'acte de foi / Bainvel, Jean Vincent – Paris: P Lethielleux, [1898?] – 1mf – 9 – 0-7905-9126-X – mf#1989-2351 – us ATLA [240]
Foi nouvelle : livre des actes – Paris: Alexandre Johanneau, 1833 – (= ser Les femmes [coll]) – 3mf – 9 – mf#6905 – fr Bibl Nationale [335]
Foillet, J *see* New modelbuch...
Foire de seville / Le Roy, Adrian – Paris, France. c1895 – 1r – us UF Libraries [440]
Fok-lor en la musica cubana / Fuentes, Eduardo Sanchez – Habana, Cuba. 1923 – 1r – us UF Libraries [390]
Le fokonolona et le pouvoir : memoire pour le diplome d'etudes superieures de science politique / Lejamble, Georges – Tananarive: Publication du Centre de droit publique et de sciences politique, 1963 – us CRL [079]

Folarin, Abedesin *see*
- The laws and customs of egba-land
- A short historical review of the life of the egbas from 1829 to 1930

Folclore no brasil / Bettencourt, Gastao De – Salvador, Brazil. 1957 – 1r – us UF Libraries [390]
Folded lambs, or, infants in their heavenly home / Nevin, Alfred – Philadelphia: William Syckelmoore, 1885 – 1mf – 9 – 0-8370-3918-5 – mf#1985-1918 – us ATLA [240]
Foleshill & bedworth express – Coventry, England 2 may 1874-8 apr 1876 [mf 1874] – 1 – (discontinued) – uk British Libr Newspaper [072]
Foley, George Cadwalader *see* Anselm's theory of the atonement
Foley, James Gervase *see* [Resume of general elections, 1896-1911]
Foley Lumber Company *see* Foley lumber industries of florida
Foley lumber industries of florida / Foley Lumber Company – Jacksonville, FL. 1942 – 1r – us UF Libraries [634]
Foley, Thomas S *see* The effects of the cross walk#zy's resistive arm poles on the metabolic costs of treadmill walking
Folgen der technik – verantwortung der ingenieure? : zu einem ethischen problem und seiner bedeutung fuer die paedagogik / Schlotter, Herbert – (mf ed 1994) – 3mf – 9 – €49.00 – 3-8267-2007-5 – mf#DHS 2007 – gw Frankfurter [170]
Folha 8 – Luanda, Angola. n170-407. 1997 jan 03-1999 may 08 – 1r – us UF Libraries [079]
Folha academica : orgao dos alumnos do instituto polytechnico de florianopolis – Florianopolis, SC. 01 jun 1923; jun 1929; jan-abr, jun 1930; jun 1931 – (= ser Ps 19) – mf#UFSC/BPESC – bl Biblioteca [370]
A folha da victoria – Vitoria, ES: 08 jul 1883-nov 1884; out 1885-dez 1888; jan-mar, maio-dez 1889; jan-20 jul 1890 – (= ser Ps 19) – mf#P11B,05,09 – bl Biblioteca [079]
Folha de annuncios – S Tome: Ezeguiel Pires dos Santos Ramos, jul 10-20 1911 – us CRL [079]
Folha de inferneiras e diabruras *see* O simplicio endiabrado
Folha de minas : orgao da lavoura, commercio e industria – Cataguazes, MG: Typ da Folha de Minas, 09 nov-dez 1884; jan 1885; 25 mar 1888 – (= ser Ps 19) – bl Biblioteca [079]
Folha de s paulo – Sao Paulo, Brazil. 1994 apr 01-2000 feb – 158r – (gaps) – us UF Libraries [079]
Folha de sao paolo – Brazil, 1971-98 – 1 – (yrly reel count varies) – us UMI ProQuest [079]
Folha de sergipe – Aracaju, SE. 20 nov 1890 – (= ser Ps 19) – bl Biblioteca [079]
Folha de sergipe : orgao do partido republicano – Aracaju, SE: Typ da Gazeta de Aracaju, 15 nov 1890; mar, nov 1895; 24 jan 1896 – (= ser Ps 19) – bl Biblioteca [325]
Folha do acre : orgam do partido constructor acreano – Rio Branco, RJ: Typ da Folha do Acre, 14 ago 1910-jun 1915; out 1917-ago 1918; jan 1920-maio 1923; jan 1926-dez 1927; fev-24 mar 1946 – (= ser Ps 19) – mf#P11A,07,08 – bl Biblioteca [321]
Folha do rio – Rio de Janeiro, RJ. 21 nov 1909 – (= ser Ps 19) – mf#DIPER – bl Biblioteca [079]
Folha esportiva – Florianopolis, SC. 31 maio 1933 – (= ser Ps 19) – mf#UFSC/BPESC – bl Biblioteca [790]
A folha fluminense – Rio de Janeiro, RJ, 08 mar 1889 – (= ser Ps 19) – mf#DIPER – bl Biblioteca [079]
Folha israelita – Rio de Janeiro, Brazil, v1, no. 0 (Dec. 1984); v2, no. 2 (1985); v3, no. 10 (June 1986) – us AJPC [270]
A folha novinha : bucolicosa, critica e carnavalesca – Rio de Janeiro, RJ, 24 fev-03 maio 1884 – (= ser Ps 19) – mf#P19A,04,144 – bl Biblioteca [079]
Folha politica, commercial e noticiosa *see* O mercantil
Folha popular – Sao Jose do Paraizo, MG: Typ da Folha Popular, 14 abr 1912 – (= ser Ps 19) – mf#P11B,3,57 – bl Biblioteca [079]
A folha semanal : jornal independente – Sao Jose, SC, 21 jun 1931 – (= ser Ps 19) – bl Biblioteca [079]
Folia allergologica – v1-16. 1954-69 – 1 – us AMS Press [616]
Folia clinica et biologica – v1-36. 1929-67 – 1 – us AMS Press [610]
Folia endrocrinologica – v1-22. 1948-69 – 1 – us AMS Press [616]
Folia pharmacologica japonica – Kyoto. 1972-1980 (1) 1972-1980 (5) 1974-1980 (9) – ISSN: 0015-5691 – mf#7258 – us UMI ProQuest [615]
Folia phoniatrica – Basel. 1966-1974 (1) 1970-1974 (5) – ISSN: 0015-5705 – mf#2056 – us UMI ProQuest [610]
Folia primatologica – Basel. 1966-1974 (1) 1971-1972 (5) – ISSN: 0015-5713 – mf#2057 – us UMI ProQuest [574]
Foliani, Ludovicus *see* Musica theoretica
Folio – Birmingham. 1965-1973 (1) 1972-1973 (5) (9) – ISSN: 0015-5756 – mf#7756 – us UMI ProQuest [400]
Folio : the magazine for magazine management – Overland Park. 1972+ (1) 1972+ (5) 1974+ (9) – ISSN: 0046-4333 – mf#6818 – us UMI ProQuest [070]
The folio : f.c.c. [forman christian college] magazine – v26-34. 1934-43 [complete] – 1r – 1 – mf#ATLA S0540 – us ATLA [378]
Folk – Leipzig: Verlag von S Hirzel 1937 [mf ed Bloomington IN: Indiana Uni Lib, Preservation Dept 1984] – 1r – 1 – us Indiana Preservation [390]
Folk arts, arts and crafts : architecture, applied arts, studio arts – (= ser Art exhibition catalogues on microfiche) – 43 catalogues on 52mf – 9 – £380.00 – (individual titles not listed separately) – uk Chadwyck [740]
Folk customs and folk lore, class 3 b : crackers / Ramsdell, Nellie B – s.l, s.l? 1936 – 1r – us UF Libraries [390]
Folk customs and lore / Scoville, Dorothy R – s.l, s.l? 1936 – 1r – us UF Libraries [390]
Folk lore : labelle, hendry county, florida / Huss, Veronica E – s.l, s.l? 193-? – 1r – us UF Libraries [390]
Folk music in america / Barry, Phillips – [New York]: Works Progress Administration, Federal Theatre Project, National Service Bureau 1939 [mf ed Bloomington IN: Indiana Uni Lib, Preservation Dept 1984] – 1r – 1 – us Indiana Preservation [390]
Folk music journal – London. 1975-1989(1,5,9) – ISSN: 0531-9684 – mf#10535 – us UMI ProQuest [780]
Folk music periodicals, 1946-1987 – Clearwater Publ Co – 45mf (24:1) – 9 – $350.00 set – (also available for separate sale: broadside alone $250, people's songs $105, new city songster $105. with p/g) – us UPA [780]
Folk og kirke i middelalderen : studier til norges historie / Bull, Edvard – Kristiania: Gyldendal, 1912 [mf ed 1990] – 1mf – 9 – 0-7905-5516-6 – (incl bibl ref) – mf#1988-1516 – us ATLA [241]
Folk og kirke paa madagaskar / Jorgensen, Simon Emanuel – Kristiania: T. Steens Vorlagsekspedition, 1887 – 1r – 1 – 0-8370-1520-0 – mf#1984-B232 – us ATLA [240]
Folk religion of bengal / Das, Sudhir Ranjan – Calcutta: SC Kar, 1953- – (= ser Samp: indian books) – (foreword by nirmal kumar bose) – us CRL [280]
Folk song society journal – London. 1899-1931 (1) – mf#2112 – us UMI ProQuest [780]
Folk songs and stories of the americas / Pan American Union – Washington, DC. 1943 – 1r – us UF Libraries [390]
Folk songs from somerset / Sharp, Cecil James – London: Simpkin 1904-06 [mf ed Bloomington IN: Indiana Uni Lib, Preservation Dept 1984] – xv 76p on 1r – 1 – us Indiana Preservation [390]
Folk tales from iceland and other countries / Scargill, M H – Reykjavik, Iceland . 1943 – 1r – us UF Libraries [390]
Folk tales of sind and guzarat / Kincaid, Charles Augustus – Karachi: Daily Gazette Press, 1925 – (= ser Samp: indian books) – us CRL [390]
Folk un land – Vilna, Lithuania. 1910 – 1r – us UF Libraries [939]
Folk un velt – New York. N.Y. 1952-57 – 1 – us AJPC [071]
Folk university's foolkiller – [v1 n1?]-v4 [1975 apr/may-1979 winter] – 1r – 1 – mf#382321 – us WHS [071]
Folkard, Henry Coleman *see* Folkard's starkie on slander and libel.
Folkard's starkie on slander and libel. / Folkard, Henry Coleman – 4th Eng. ed. New York: Banks, 1877. 992p. LL-591 – 1 – us L of C Photodup [340]
Folkbladet – Norrkoping, Sweden. 1981- – 1 – sw Kungliga [078]
Folkbladet – Stockholm, Sweden. 1894-1907 5r; 1930-35 4r – 1 – sw Kungliga [078]
Folkbladet For Vastmanland *see* Vastmanlands folkblad
Folkbladet for vastmanland – Stockholm, Sweden. 1959-66 – sw Kungliga [078]
Folkbladet joenkoeping – Joenkoeping, 1992- – 9 – sw Kungliga [078]
Folkbladet ostgoten – Norrkoping, Sweden. 1979-81 – 1 – sw Kungliga [078]
Folkbladet ostgoten – Norrkoping, Sweden – 377r – 1 – (ostergotlands folkblad 1905-66. folkbladet ostgoten 1966-78) – sw Kungliga [078]
The folk-dance of india / Banerji, Projesh – Allahabad: Kitabistan, 1944 – (= ser Samp: indian books) – us CRL [390]
Folk-dances of south india / Spreen, Hildegard L – London; New York: Humphrey Milford: Oxford University Press, 1945 – (= ser Samp: indian books) – (foreword by marie buck) – us CRL [790]
Folkebladet *see* Schleswigsche grenzpost
The folk-element in hindu culture : a contribution to socio-religious studies in hindu folk-institutions / Sarkar, Benoy Kumar – London: Longmans, Green, 1917 – 1mf – 9 – 0-524-04990-4 – (incl bibl ref) – mf#1990-3448 – us ATLA [280]
Folkert der schoeffe : roman / Bauer, Albert – Leipzig: P List, c1935 [mf ed 1989] – 350p – 1 – mf#6982 – us Wisconsin U Libr [830]
Folkerth, Jean E *see* Parents' attitudes regarding the leisure behavior of their handicapped child
Folkestone argus – Folkestone, England 30 mar 1889-23 may 1891 – 1 – (discontinued) – uk British Libr Newspaper [072]
Folkestone chronicle – Folkestone, England. -w. 1886-1906. 20 reels – 1 – uk British Libr Newspaper [072]
Folkestone express – Folkestone, England. -w. 14 March 1868-Dec 1902. Lacking 1898, 1901. 36 reels – 1 – uk British Libr Newspaper [072]
Folkestone observer – Folkestone, England. -w. 8 Dec 1860-29 Sept 1870. 5 reels – 1 – uk British Libr Newspaper [072]
Folket – Eskilstuna, Sweden. 1979-82 – 1 – sw Kungliga [078]
Folket – Eskilstuna, Sweden. 1989- – 1 – sw Kungliga [078]
Folket : morgonedition [foer vaestmanland] – Eskilstuna, 1966-70 – 39r – 1 – sw Kungliga [078]
Folket : morgonedition foer vaestmanland – Eskilstuna, Sweden. 1906-82 – 1 – (sormlandskuriren, 1975) – sw Kungliga [078]
Folket *see* Vastmanlands folkblad
Folket i bild – Stockholm, Sweden. 1934-63:4 – 1 – sw Kungliga [078]
Folket nord – Eskilstuna, Sweden. 1982-89 – 1 – sw Kungliga [078]
Folket syd – Eskilstuna, Sweden. 1982-89 – 1 – sw Kungliga [078]
Folketru / ed by Lid, Nils – Stockholm: A Bonnier 1935 [mf ed Bloomington IN: Indiana Uni Lib, Preservation Dept 1984] – 171p on 1r – 1 – us Indiana Preservation [390]
Folkets avis – Hillsboro ND. 1899 jan 7-1901 dec 28 – 1r – 1 – (cont: afholds-basunen; cont by: statstidende) – us WHS [071]
Folkets avis – Racine WI. 1900-03 – 1r – 1 – mf#1097970 – us WHS [071]
Folkets dagblad – Stockholm, Sweden. 1917-40, 1942-45 – 153r – 1 – (politiken, 1916-17. folkets dagblad, 1917-45) – sw Kungliga [078]
Folkets degblad oolitiken – Stockholm, Sweden. -d. 1 July 1918-8 Aug 1919. 5 reels – 1 – uk British Libr Newspaper [079]
Folkets roest – Goteborg, Sweden. 1887-90 – 1 – sw Kungliga [078]
Folkets roest – Stockholm, 1849-61 – 9 – sw Kungliga [078]
Folkets vaen – Gaevle, Sweden. 1880-86 – 2r – 1 – sw Kungliga [078]
Folkets van – Stromsburg, NE: C A Wenngren & Co, dec 17 1885-87// (wkly) [mf ed with gaps filmed 1973] – 1r – 1 – (chiefly in swedish with some english) – us NE Hist [071]
Folke-vennen – Chicago IL. 1899 jan 5/1900 jul 19-1930 sep 1/1933 nov 15 [gaps] – 22r – 1 – mf#866401 – us WHS [071]
Folkevisor – Stockholm: A Bonnier 1931 [mf ed Bloomington IN: Indiana Uni Lib, Preservation Dept 1984] – 1r – 1 – us Indiana Preservation [390]
Folk-literature of the galla of southern abyssinia / Cerulli, E – [Cambridge, 1917] – 4mf – 9 – mf#NE-20239 – ne IDC [956]
Folkliv – [Stockholm]: K Foerlags Thule 1937 [mf ed Bloomington IN: Indiana Uni Lib, Preservation Dept 1984] – 1r – 1 – (cont by: folk-liv) – us Indiana Preservation [390]
Folkliv – [Stockholm]: K Foerlags Thule 1938-70 [mf ed Bloomington IN: Indiana Uni Lib, Preservation Dept 1984] – 3r – 1 – us Indiana Preservation [390]
Fol'klor chkalovskoi oblasti / Bardin, A V – [Chkalov]: Chkalovskoe izd-vo 1940 [mf ed Bloomington IN: Indiana Uni Lib, Preservation Dept 1984] – 1r – 1 – (russian folk literature) – us Indiana Preservation [390]
Folk-lore – Calcutta. 1960-1992 (1) 1971-1992 (5) 1976-1992 (9) – ISSN: 0015-5896 – mf#3080 – us UMI ProQuest [390]
Folklore : city guide / Virgin, Martina – s.l, s.l? 1936 – 1r – us UF Libraries [390]
Folklore – London. 1958+ [1]; 1989+ [5]; 1989+ [9] – ISSN: 0015-587X – mf#1219 – us UMI ProQuest [390]
Le folk-lore : litterature orale et ethnographie traditionnelle / Sebillot, Paul – Paris: O Doin 1913 [mf ed Bloomington IN: Indiana Uni Lib, Preservation Dept 1984] – xxii 393p on 1r – 1 – us Indiana Preservation [390]
Le folklore : croyances et coutumes populaires francaises / Gennep, Arnold van – Paris: Librairie Stock 1924 [mf ed Bloomington IN: Indiana Uni Lib, Preservation Dept 1984] – 125p on 1r [ill] – 1 – us Indiana Preservation [390]
Folklore and folk music archivist – Bloomington. 1958-1968 (1) – mf#1630 – us UMI ProQuest [390]

Le folklore brabancon – 9e annee: n40-50(1929-1930) – 7mf – 9 – €15.00 – (16e annee n91-96(1936-1937) 11mf €21) – ne Slangenburg [390]

Folklore brasileiro / Ribeiro, Joaquim – Rio de Janeiro, Brazil. 1944 – 1r – us UF Libraries [390]

Folk-lore chinois moderne / Wieger, Leon – [Sienhsien]: Impr de la Mission Catholique, 1909 [mf ed 1995] – (= ser Yale coll) – 422p (ill) – 1 – 0-524-10163-9 – (in french) – mf#1995-1163 – us ATLA [390]

Folklore de constantinople / Carnoy, H & Nicolaides, J – Paris, 1894 – 3mf – 9 – mf#AR-1817 – ne IDC [956]

Folklore de la republica dominicana / Andrade, Manuel Jose – Ciudad Trujillo, Dominican Republic. pt1-2. 1948 – 1r – us UF Libraries [390]

El folklore de madrid / Olavarrieta y Huarte, Eugenio – 1884 – 1 – sp Bibl Santa Ana [390]

Folklore extremeno. extremadura y la posible regionalizacion de su musica popular. la tradicion en la cancion extremena y su evolucion / Gil Garcia, Bonifacio – Badajoz: Dip. Provinicial, 1938. Comunic. 3 Congreso Internacional de Musicologia. Barcelona. 1936 – 1 – sp Bibl Santa Ana [780]

Le folklore flammand – Bruxelles: C Rozez 1895 [mf ed Bloomington IN: Indiana Uni Lib, Preservation Dept 1984] – 165p on 1r – 1 – us Indiana Preservation [390]

Folklore in the works of mark twain / West, Victor Royce – Lincoln, NE. 1930 – 1r – us UF Libraries [420]

Folk-lore journal – London. 1883-1889 (1) – mf#2890 – us UMI ProQuest [390]

The folklore of bombay / Enthoven, Reginald Edward – Oxford: Clarendon Press, 1924 – (= ser Samp: indian books) – us CRL [390]

Folklore of the santal parganas – London: D Nutt, 1909 – 2mf – 9 – 0-524-05450-9 – mf#1990-3476 – us ATLA [390]

Folklore of wells : being a study of water-worship in east and west / Masani, Rustom Pestonji – Bombay: DB Taraporevala Sons & Co, 1918 – (= ser Samp: indian books) – us CRL [390]

Folk-lore record – London. 1878-1882 (1) – mf#2891 – us UMI ProQuest [390]

Folk-lore savladoreno / Espinosa, Francisco – San Salvador, El Salvador. 1946 – 1r – us UF Libraries [390]

Folklore women's communication – 1978 winter-1991 spring, 1985 suppl – 1r – 1 – (cont: folklore feminists' communication) – mf#2571781 – us WHS [305]

Folkmimen och folktankar – v1-15. 1914-28 – 1 – us Indiana Preservation [390]

Folkminnen och folktankar – Lund: Folkminnesfoereningen i Lund 1914-44? [mf ed Bloomington IN: Indiana Uni Lib, Preservation Dept 1984] – 1 – (cont by: arv) – us Indiana Preservation [390]

Folkrakningen den 1 nov 1960 / Sweden. Statistika centralbyran – Stockholm, 1961-65. 11v. – 1 – us Wisconsin U Libr [314]

Folks tsaytung – Los Angeles. 1936-37 – 1 – us AJPC [071]

Folkscajtung – Warsaw. Poland. -m. Aug 1946-Jan 1949. (1 reel) – 1 – uk British Libr Newspaper [947]

Folksfraint – Pittsburgh. Jan. 15, 1892-Dec. 13, 1901, Dec. 27, 1901-1923 – 1 – us NY Public [071]

Folksfrajnd – Przemysl PL, 1928-29 – 1r – 1 – (in yiddish. with: tomaszower wochenblat [tomaszow mazowiecki, poland] 1929) – us UMI ProQuest [939]

Folk-songs of america / Gordon, Robert Winslow – 1938 – 1 – us Indiana Preservation [390]

Folk-songs of america / Gordon, Robert Winslow – New York: National Service Bureau 1938 [mf ed Bloomington IN: Indiana Uni Lib, Preservation Dept 1984] – 1r – 1 – (articles originally publ in the new york times sunday magazine) – us Indiana Preservation [780]

Folk-songs of chhattisgarh / Elwin, Verrier – London: Published for Man in India by Oxford University Press, 1946 – (= ser Samp: indian books) – us CRL [780]

Folk-songs of the maikal hills / Elwin, Verrier – [London; New York: Humphrey Milford, 1944] – (= ser Samp: indian books) – us CRL [390]

Folksshtime – Wilna. v1-16. dec. 1, 1906-aug. 17, 1907 – 1 – (includes: supplement Di Shtime. 1907) – us NY Public [073]

Folks-sztvme – Warsaw, Poland. In Yiddish. -d. Feb 1952-Dec 1955. 12 reels – 1 – uk British Libr Newspaper [947]

Folks-sztyme – Warsaw, Poland. 1952-1991 – 48r – 1 – (nasz glos (suppl) 1957-jun 1968 2r) – us L of C Photodup [947]

Folk-tales of angola : fifty tales, with ki-mbundu text, literal english translation, introduction, and notes / ed by Chatelain, Heli – Boston: publ for the American Folk-Lore Society..., 1894 [mf ed 1991] – (= ser Memoirs of the american folk-lore society 1) – 1mf – 9 – 0-524-00709-8 – (text in english & kimbundu; int & notes in english) – mf#1990-2037 – us ATLA [390]

Folk-tales of bengal / Day, Lal Behari – London: Macmillan and Co, 1912 – (= ser Samp: indian books) – us CRL [390]

Folk-tales of hindustan / Vasu, Srisa Chandra – Allahabad: Panini Office, 1913 – (= ser Samp: indian books) – us CRL [390]

Folk-tales of mahakoshal / Elwin, Verrier – London: Published for Man in India by Oxford University Press, 1944 – (= ser Samp: indian books) – us CRL [390]

Folktro och trolldom – Helsingfors: [s.n.] 1919-55 [mf ed Bloomington IN: Indiana Uni Lib, Preservation Dept 1984] – 5v [ill] on 3r – 1 – us Indiana Preservation [390]

Folkviljan – Malmo, Sweden. 1906-20 – 1 – (folkviljan 1906; nya folkviljan 1906-20) – sw Kungliga [078]

Folkviljan – Lulea, Sweden. 1980-89 – 1 – sw Kungliga [078]

Folkviljan – Malmo, Sweden. 1882-85 – 1r – 1 – sw Kungliga [078]

Folkviljan – Stockholm, Sweden. 1891-92 – 1 – sw Kungliga [078]

Folkvisor – Helsingfors: [s.n.] 1934-67 [mf ed Bloomington IN: Indiana Uni Lib, Preservation Dept 1984] – v1,3 on 1r – 1 – us Indiana Preservation [390]

Folkwang – Hagen, Westf DE, 1921-22 – 1 – gw Misc Inst [074]

Follen, August Adolf Ludwig *see*
– Tristans eltern

Follenius, Christopher *see* The effects of enhanced eccentric training on improvement of strength

Follet – London, UK. Oct 1846-Jul 1900 – 1 – uk British Libr Newspaper [072]

Folleto de las conferencias dadas durante la semana agricola de badajoz del 12-18 nov. 1912 / Conseso Provincial de Fomento – Badajoz: C.P.F., 1913 – 1 – sp Bibl Santa Ana [630]

Folleto del combatiente – Salamanca, 1938. Fiche W 887. (Blodgett Collection of Spanish Civil War Pamphlets) – 9 – us Harvard [946]

Folleto informativo sobre el servicio municipal de limpieza / Caceres – Caceres: Imp. T. Rodriguez, s.a. 1965? – sp Bibl Santa Ana [350]

Folleto primero don benito... / Don Benito – Plasencia: Imprenta La Victoria, s.a. – sp Bibl Santa Ana [946]

Folleto refutado de injurias...a paredes – 1888 – 9 – sp Bibl Santa Ana [946]

Follett, M P *see* The speaker of the house of representatives

La Follette, Robert M *see* Papers of robert m la follette, 1876-1924

Follick, Mont *see* Twelve republics

Folliculaire / Ville De Mirmont, Alexandre Jean Joseph De La – Paris, France. 1820 – 1r – us UF Libraries [440]

Follow thou me / Hatton, Eleanor Beard – New York: Christian Alliance, c1916 [mf ed 1991] – (= ser Christian & missionary alliance coll) – 1mf – 9 – 0-524-01728-X – mf#1990-4120 – us ATLA [830]

Follower's of Jesus [Mineola, NY] *see* Happy world

Following on to know the lord / Wilberforce, Basil – [London?]: Hodder and Stoughton; New York: GH Doran, [1903?] – (= ser The expositor's library) – 1mf – 9 – 0-7905-7491-8 – mf#1989-0716 – us ATLA [240]

Following the color line / Baker, Ray Stannard – New York, NY. 1908 – 1r – us UF Libraries [025]

Following the danforth report : defining the next step on the path to peace in sudan: hearing...house of representatives, 107th congress, 2nd session, june 5 2002 / United States. Congress. House. Committee on International Relations – Washington: US GPO 2002 [mf ed 2002] – 1mf – 9 – 0-16-068941-4 – (incl bibl ref) – us GPO [327]

Following the sunrise : a century of baptist missions, 1813-1913 / Montgomery, Helen Barrett – Philadelphia: American Baptist Publication Society, c1913 – 1mf – 9 – 0-7905-5073-3 – (incl bibl ref) – mf#1988-1073 – us ATLA [242]

Following the sunrise : a century of baptist missions, 1813-1913 / Montgomery, Helen Barrett – Philadelphia: American Baptist Publication Society, c1913 – 1mf – us ATLA [240]

Follow-up study of word processing and electronic mail in the 3rd circuit court of appeals / Greenwood, J Michael – Washington: FJC, June 1980 – 1mf – 9 – $1.50 – mf#LLMC 95-816 – us LLMC [347]

Folsom, Justus Watson *see* Entomolgy

[Folsom-] telegraph/orangevale news – CA. 1980- – 28r – 1 – $1680.00 (subs $100y) – mf#H04078 – us Library Micro [071]

[Folsom-] the folsom telegraph – CA. 1873-88; 1889-1982 (irregular) – 53r – 1 – $3180.00 – mf#C02236 – us Library Micro [071]

Folz, August *see* Kaiser friedrich 2. und papst innocenz 4

Folz, Hans *see*
– Hans folz
– Die meisterlieder des hans folz

Fomenko, Kliment *see* Razbor desyati pravil verocheniya shtundistov

El fomento – Caceres, 1901. 1 numero – 5 – sp Bibl Santa Ana [073]

Fomento de Caceres *see*
– Comision organizadora de festejos. memoria de 1920
– Reglamento de la sociedad...1921 y 1923

Fomento Del Trabajo Nacional (Spain) *see* Informe sobre aranceles antillanos

Fomento e mercantilismo / Dias, Manuel Nunes – Belem, Brazil. v1-2. 1970 – 1r – us UF Libraries [972]

Fon of bafut / Ritzenthaler, Pat – New York, NY. 1966 – 1r – us UF Libraries [960]

Fonck, Leopold *see*
– Documenta ad pontificiam commissionem de re biblica spectantia
– The parables of the gospel

La fonction de la femme dans l'evolution sociale / Lind af Hageby, Louise – Conflans-Ste-Honorine: Edition de l'Idee Libre, 1922 – (= ser Les femmes [coll]) – 1mf – 9 – mf#8745 – fr Bibl Nationale [305]

La fonction religieuse et sociale des cloches : sermon prononce a la ceremonie de benediction des cloches du sanctuaire de notre-dame de roc-amadour... / Leve, Martial – St-Francois d'Assise, Quebec: [s.n, 1920?] [mf ed 1996] – 1mf – 9 – 0-665-81352-X – mf#81352 – cn Canadiana [240]

Le fonctionnaire indochinois : l'organe officiel de l'agfali – Saigon: Impr J Viet [1936- [v1 n1-48 [jan 31-dec 1936], v3 [1938] [mf ed Hanoi, Vietnam: National Library of Vietnam 1998] – 1r – 1 – (master neg held by crl) – mf#mf-11780 seam – us CRL [350]

Le fonctionnement du marche du logement et le peuplement residentiel / Vervaeke, Monique – 1mf – 9 – (10361) – fr Atelier National [307]

Fonctionnement du pouvoir revolutionnaire local (prl) – Conakry: Impr nationale "Patrice Lumumba", 1974 – us CRL [079]

Fond du Lac Association of Commerce *see* Fond du lac trade extension

Fond du lac [city directory : listing] – 1884, 1893-94, 1895-96, 1899-1900, 1901 – 5r – 1 – mf#3116674 – us WHS [071]

Fond du lac commentator – Fond Du Lac WI. 1937 apr 15-july 23 – 1r – 1 – mf#918090 – us WHS [071]

Fond du lac commonwealth – Fond Du Lac WI. 1865 aug 16-1866, 1867-70, 1871-72, 1873 jan 4-1876 jul 22, 1876 jul 29-1880 jan 10, 1880 jan 17-1883 mar 31, 1883 apr 7-1885 mar 27, 1885 jan-mar 27 – 8r – 1 – (cont by: commonwealth [fond du lac, wi]) – mf#941398 – us WHS [071]

Fond du lac commonwealth reporter – Fond Du Lac WI. 1926 oct 1/dec 24-1965 may/jun [gaps] – 198r – 1 – (cont: daily commonwealth [fond du lac, wi: 1912]; fond du lac daily reporter; cont by: fond du lac reporter) – mf#941530 – us WHS [071]

Fond du Lac County Genealogical Society *see* Magnifier

Fond du lac county maturity times – 1987 dec-1989 dec, 1990 jan-1991 dec, 1992 jan-1993 dec, 1994 jan-1996 dec [v8 n1-v10 n12] – 4r – 1 – (cont by: maturity times) – mf#1897875 – us WHS [071]

Fond du lac daily – Fond Du Lac WI. 1870 aug 22-29 – 1r – 1 – (cont by: fond du lac daily commonwealth [fond du lac, wi: 1870]) – mf#941484 – us WHS [071]

Fond du lac daily commonwealth – Fond Du Lac WI. 1866 sep 8-30 – 1r – 1 – (cont by: daily commonwealth [fond du lac, wi: 1866]) – mf#941477 – us WHS [071]

Fond du lac daily commonwealth – Fond Du Lac WI. 1875 nov 26/1876 apr 10-1884 aug 13/1885 feb 24 [gaps] – 16r – 1 – (cont: daily commonwealth [fond du lac, wi: 187?]; cont by: daily commonwealth [fond du lac, wi: 1885]) – mf#941488 – us WHS [071]

Fond du lac daily commonwealth – Fond Du Lac WI. 1870 aug 30-dec 21 – 1r – 1 – (cont: fond du lac daily) – mf#941485 – us WHS [071]

Fond du lac daily reporter – Fond Du Lac WI. 1921 jul 26/sep-1926 jul/sep – 21r – 1 – (cont: daily reporter [fond du lac, wi: 1901]; cont by: daily commonwealth [fond du lac, wi: 1912]; fond du lac commonwealth reporter) – mf#941540 – us WHS [071]

Fond du lac journal – Fond Du Lac WI. 1857 feb 21-1858 may 8, 1857 dec 26 – 2r – 1 – (cont by: fond du lac union; democratic press [fond du lac, wi: 1858]) – mf#917536 – us WHS [071]

Fond du lac journal – Fond Du Lac WI. 1846 oct 8-1849, 1846 dec 31, 1850-1853 jun 15 – 3r – 1 – (cont by: national democrat [fond du lac, wi: 1852]; fond du lac union) – mf#917523 – us WHS [071]

Fond du lac journal – Fond Du Lac WI. 1867 may 2-1869, 1870-1872 feb 22, 1873 sep 25-1874, 1875-89, 1890-1892 dec 29 – 8r – 1 – (cont by: saturday reporter) – mf#917592 – us WHS [071]

Fond du lac times – Fond Du Lac WI: 1949 may 26-1951 jun 28, 1951 jul 12-193 dec 24 – 2r – 1 – mf#918291 – us WHS [071]

Fond du lac times – Fond Du Lac WI: 1976 jul 21-1977 jul 27, 1977 aug 3-1978 jun 28 – 2r – 1 – mf#918755 – us WHS [071]

Fond du lac times, rural farmer advertiser – Fond Du Lac WI: 1949 jan 8-may 12 – 1r – 1 – mf#918289 – us WHS [071]

Fond du lac trade extension / Fond du Lac Association of Commerce – 1918 feb 13 – 1r – 1 – mf#917908 – us WHS [380]

Fond du lac union – Fond Du Lac WI. 1853 jun 24-1855 dec 27, 1856 jan 3-1858 may 20 – 2r – 1 – (cont: fond du lac journal [fond du lac, wi: 1846]; national democrat [fond du lac, wi: 1852]; cont by: fond du lac journal [fond du lac, wi: 1857]; democratic press [fond du lac, wi: 1858]) – mf#917532 – us WHS [071]

Fond du lac weekly commonwealth – Fond Du Lac WI. 1856 sep 10-1858, [1958 mar 7-dec 15], 1859-62, 1863-1865 aug 9 – 4r – 1 – (cont: fountain city herald; western freeman; cont by: fond du lac commonwealth) – mf#890519 – us WHS [071]

Fond du lac whig – Fond Du Lac WI. 1846 dec 14-1847 jan 27, feb 18-jul 22, aug 5-sep 30, oct 14, 28-nov 22 – 1r – 1 – (cont by: wisconsin republican [green bay, wi: 1844]; wisconsin republican [fond du lac, wi: 1848]) – mf#916203 – us WHS [071]

Le fondateur des religieuses de l'assomption / Desaulniers, Francois Lesieur – Montreal: Arbour & Dupont, 1911 – 1mf – 9 – 0-665-73937-0 – mf#73937 – cn Canadiana [241]

Fondateur devant l'histoire / Jean-Baptiste, St Victor – Port-Au-Prince, Haiti. 1954 – 1r – us UF Libraries [972]

Les fondateurs de l'union nationale / Liberator – [Quebec (Province): s.n. 1936 ?] (mf ed 1992) – 2mf – 9 – mf#SEM105P1755 – cn Bibl Nat [325]

Les fondateurs d'empire : ceremonies en hommage a la mission saharienne foureau-blide (avril 30) / Lehuraux, Leon Joseph – Alger: Ancienne Maison Bastide Jourdan J Carbonel, 1931 – 1 – us CRL [960]

Fondateurs superieurs, professeurs et eleves du college de l'assomption 1833-1893 – Montreal: C O Beauchemin, 1893 – 1mf – 9 – mf#53561 – cn Canadiana [378]

Fondation de la republique d'haiti / Dalencour, Franciois Stanislas Ranier – Port-Au-Prince, Haiti. 1944 – 1r – us UF Libraries [972]

The fonds curtin, i f a n, universite de dakar, senegal : arabic documents photographed in the dept de bakel, mar and apr 1966 – Bloomington, 1966 – (incl ind) – us CRL [470]

Fonds Du Bienetre Indigene *see* Work of co-operation in development

Fonds FCAC *see* Action concertee cablodistribution

Fonds pamphile-lemay / LeMay, Pamphile – [mf ed 2000] – 7r – 1 – mf#SEM35P483 – cn Bibl Nat [025]

Foner, Meir *see* Yosef delah rainah

Fonetic herald : devoted to orthoepi and orthografi – Port Hope [Ont: s.n. 1885?-1886] – 9 – (cont by: the herald) – mf#P04541 – cn Canadiana [420]

The fonetic primer : offering the universal alfabet and the science of spelling / Story, Charles A – New York City, NY: Isaac H Blanchard, c1907 – 1mf – 9 – 0-8370-8151-3 – mf#1986-2151 – us ATLA [420]

Fon-fon : semanario alegre, politico, critico e efusiante – Rio de Janeiro, RJ: Officina Typographica de J Schmidt, 13 abr 1907-28 dez 1945 – (= ser Ps 19) – bl Biblioteca [079]

Fonfrias, Ernesto Juan *see*
– Conversao en el batey
– Cosecha, ensayos y articulos
– Guasima
– Presencia jibara desde manuel alonso hasta don flo...
– Voz en la montana

Fonk, Friedrich Hermann *see* Das staatliche mischehenrecht in preussen vom allgemeinen landrecht an

Fonkich, B L *see* La russie dans lasie-mineure

'N fonnie bisnis / Nieland, Dirk – Grand Rapids MI: Eerdmans [1929?] – 3mf – 9 – 0-524-07446-1 – mf#1991-3106 – us ATLA [949]

Fono – Tonbridge. 1999+ (1,5,9) – ISSN: 1464-9403 – mf#32328 – us UMI ProQuest [780]

Fonograph (the phonograph) – London, UK. 22 Nov 1907-7 Aug 1914 – 1 – uk British Libr Newspaper [072]

Fonologie en morfologie van het cokwe / Eynde, Karel Van Den – Leuvenia, Belgium. 1960 – 1r – us UF Libraries [960]

Fonologie en morfologie van westelike shona / Wentzel, Petrus Johannes – s.l, s.l? 1961 – 1r – us UF Libraries [960]

Fons vitae / Baeumker, Cl Avencebrolis (Ibn Gebirol) – Muenster, 1892-1895 – (= ser Bgphma 1/2-4) – 10mf – 8 – €19.00 – ne Slangenburg [100]

Fonseca *see* Historia de la milagrosa aparicion de nuestra sra de la caridad

Fonseca, Gondin Da *see*
- Biografia do jornalismo carioca
- Gorilas
- Machado de assis e o hipopotamo

Fonseca, Hermes Da *see* Pinheiro machado

Fonseca, J M *see*
- Annales minorum
- Syllabus universus annalium minorum p

Fonseca, Joao Severiano Da *see* Viagem ao redor do brasil

Fonseca, Jose Nicolau da *see* An historical and archaeological sketch of the city of goa

Fonseca, Luis Gonzaga Da *see* Historia de oliveira

Fonseca, Manuel Da *see* Vida do veneravel padre belchior de pontes

Fonseca, Miguel Angel *see* Compendio de historia de cuba

[Fonseca, P] *see* Institutionum dialecticarum...libro octo

Fonseca, Pedro S *see*
- Geografia ilustrada de el salvador, c a
- Moneda salvadorene
- Prontuario geografico y estadistico

Fonsegrive, George *see* Essai sur le libre arbitre

Fonssagrives, E *see* Marie de bretagne

Font Obrador, Bartolome *see* El padre boscana, historiador de california

Fontaine – Revue mensuelle de la poesie et des lettres francaises. Dir. Max-Pol Fouchet. no. 1-63. Alger puis Paris. nov 1938-1947. Les no. 1-2 ont paru sous le titre de: Mithra – 1 – fr ACRPP [800]

Fontaine, Antoine Nicolas Marie *see* Air allemand varie pour le violon avec accompagnement de piano

Fontaine, Camille *see* Poetes francais du 19e siecle

Fontaine, L Urgele *see* Cent trente-cinq ans apres ou la renaissance acadienne

Fontaine, Nicolas *see* Buez ar saint, gant reflexionon spirituel

Fontaine, P F L *see* Recueil de decorations interieures, comprenant tout ce qui a rapport... l'ameublement

Fontaine, Pamela *see* Wheelchair basketball

Fontaine, Pierre *see* Marche des braves

Fontaine, Raphael Ernest *see*
- Un duel a poudre
- Un parti de tire!

Fontaines et iets d'eau : dessines d'apres les plus beaux lieux d'italie / Fanelli, F – Paris, [1690] – 1mf – 9 – mf#GDI-11 – ne IDC [710]

Fontana, C *see* Utilissimo trattato dell'acque correnti, diviso in tre libri...

Fontana, Ferdinando *see* Edgar

[Fontana-] fontana news herald – CA. 1989-1990 – 5r – 1 – $300.00 – mf#R04024 – us Library Micro [071]

[Fontana-] the herald news – CA. 1958 – 2r – 1 – $120.00 – mf#B03596 – us Library Micro [071]

Fontane / Spiero, Heinrich – Wittenberg (Halle): A Ziemsen, c1928 (mf ed 1990) – 1r – 1 – (filmed with: bozena) – us Wisconsin U Libr [430]

Le fontane di roma nelle piazze... / Falda, G B – Roma, [1675-1691]. 4v – 5mf – 9 – mf#GDI-9 – ne IDC [700]

Fontane, Friedrich *see*
- Bilderbuch aus england
- Heiteres darueberstehen

Fontane, Marius *see* La papaute

Fontane, Theodor *see*
- Allerlei gereimtes
- Aus dem nachlass
- Aus den tagen der occupation
- Aus england
- Berlinerinnen
- Bilderbuch aus england
- Briefe an seine freunde
- Christian friedrich scherenberg
- Das fontane-buch
- Fuenf schloesser
- Gedichte
- Gesammelte werke
- Gesamtausgabe der erzaehlenden schriften
- Grete minde
- Heiteres darueberstehen
- Mathilde moehring
- Quitt
- Theodor fontane's gesammelte romane und novellen
- Vor dem sturm

Das fontane-buch : beitraege zu seiner charakteristik: unveroeffentlichtes aus seinem nachlass... / Fontane, Theodor; ed by Heilborn, Ernst – 1-5.aufl. Berlin: S Fischer, 1919 [mf ed 1989] – 227p/[3pl] – 1 – mf#7074 – us Wisconsin U Libr [920]

Fontanes, Ernest *see* Le christianisme moderne

Fontaney, J de *see* Voyage du pere jean de fontaney, jesuite, de peking... kyang-cheu dans la province de chansi, & de-l...nan-king

Fontanier, Victor *see* Voyages en orient, entrepris par ordre du gouvernement francais, de l'annee 1821 a l'annee 1829

Fontanus, J *see*
- De bello rhodio, libri tres, clementi 7 pont max dedicati...
- Della gverra di rhodi libri 3 aggiunta la discrittione dell 'isola di malta concessa a caualieri, dopo che rhodi fu preso

Fonte invisivel / Schmidt, Augusto Frederico – Rio de Janeiro, Brazil. 1949 – 1r – us UF Libraries [972]

Fontenella. historia de los oraculos – 1868 – 9 – sp Bibl Santa Ana [946]

Fontenelle *see* Historia de los oraculos

Fontenelle, M de *see* Histoire des oracles

Fontenot, M E *see* The effectiveness of acupressure in the treatment of primary dysmenorrhea

Fontes, Amando *see* Corumbas, romance

Fontes brasileiras do panamericanismo / Maul, Carlos – Rio de Janeiro, Brazil. 1941 – 1r – us UF Libraries [972]

Fontes do latim vulgar / Silva Neto, Serafim Da – Rio de Janeiro, Brazil. 1946 – 1r – us UF Libraries [972]

Fonti per la storia d'italia / Italy, Instituto Storico Italiano – v1-50 – 9 – $600.00 – (v10 never publ) – mf#0299; 0300 – us Brook [945]

Fonton, Feliks *see* Histoire de louis 11

Fontoura, Joao Neves Da *see* Voz das opposicoes brasileiras

Fontoynont, Antoine Maurice *see* La grande comore

Fonvielle, Bernard F de *see*
- Lucifer, ou la contre-revolution
- Voyage en espagne en 1798 [mil sept cent quatre-vingt-dix-huit]

Foochow messenger – Foochow, China: American Board of Commissioners for Foreign Missions. v1-(?) nov 1903-oct 1917; ns: apr 1922-spring 1940 (frequency varies) [all publ?] – (= ser Missionary periodicals from the china mainland) – 1r – 1 – $165.00 – us UPA [242]

Food – Toronto. 1985-1995 (1,5,9) – (cont: food in canada) – ISSN: 0829-643X – mf#15479,01 – us UMI ProQuest [660]

Food and Agricultural Organization of the United Nations. World Reporting Service on Plant Diseases and Pests *see* Fao plant protection bulletin

Food and Agriculture Organization of the United Nations *see* Monthly bulletin of agricultural economics and statistics

Food and chemical toxicology – Oxford. 1963+ (1,5,9) – ISSN: 0278-6915 – mf#49072 – us UMI ProQuest [615]

Food and drug law journal – v1- 1946 - – 9 – (filming in process. title varies: v1-4 1946-49) as: food, drug, cosmetic law quarterly. v5-46 1950-91 as: food, drug, cosmetic law journal) – ISSN: 1064-590X – mf#102791 – us Hein [344]

Food and drug packaging – New York. 1973-1984 (1) 1979-1984 (5) 1979-1984 (9) – ISSN: 0015-6272 – mf#9672 – us UMI ProQuest [660]

Food and drug review / U.S. Food and Drug Administration – Washington, DC. v1-50. 1917-66 – 1 – us L of C Photodup [615]

Food and nutrition – Alexandria. 1972-1992 (1) 1972-1992 (5) 1975-1992 (9) – ISSN: 0046-4384 – mf#7356 – us UMI ProQuest [660]

Food and Service Workers of Canada *see*
- Faswoc newsletter
- Faswoc union news

Food and wine : the guide to good taste – New York. 1986+ (1,5,9) – ISSN: 0741-9015 – mf#16126,02 – us UMI ProQuest [640]

Food chemistry – London. 1976+ (1) 1976+ (5) 1987+ (9) – ISSN: 0308-8146 – mf#42194 – us UMI ProQuest [660]

Food control – Kidlington. 1990+ (1,5,9) – mf#17229 – us UMI ProQuest [630]

Food development – Chicago. 1981-1982 (1) 1981-1982 (5) 1981-1982 (9) – (cont: food product development) – mf#9610,01 – us UMI ProQuest [660]

Food, drug, cosmetic law journal – Chicago. 1946-1981 (1) 1971-1981 (5) 1977-1981 (9) – ISSN: 0015-6361 – mf#846 – us UMI ProQuest [640]

Food, drug, cosmetic law journal *see* Food and drug law journal

Food, drug, cosmetic law quarterly *see* Food and drug law journal

Food, drugs and disinfectants act, 1929 (n13 of 1929) = Wet op voedingsmiddels, medisyne en ontsmettingsmiddels, 1929 (n13 van 1929) – Pretoria: Govt Printer 1959 [mf ed Pretoria, RSA: State Library [199-]] – 57p on 1r with other items – 5 – (incl inserted amendments) – mf#op 11155 r23 – us CRL [344]

Food, drugs and disinfectants act (n13 of 1929) and regulations = Wet op voedingsmiddels, medisyne en ontsmettingsmiddels, 1929 (n13 van 1929) en regulasies – Johannesburg, Lex-Patria Publishers [1966?]- [mf ed Pretoria, RSA: State Library [199-]] – 1v on 1r with other items – 5 – (in english & afrikaans) – mf#op 03469 r23 – us CRL [344]

Food engineering – Troy. 1998+ (1,5,9) – (cont: chilton's food engineering) – ISSN: 1522-2292 – mf#23,01 – us UMI ProQuest [660]

Food engineering and ingredients – Sutton. 2000+ (1) – ISSN: 1471-2806 – mf#11781,02 – us UMI ProQuest [660]

Food engineering international – Sutton. 1998-2000 (1,5,9) – (cont: chilton's food engineering international; cont by: food engineering and ingredients) – ISSN: 1521-6004 – mf#11781,01 – us UMI ProQuest [660]

Food for thought – Toronto. 1940-1961 (1) – ISSN: 0383-9540 – mf#342 – us UMI ProQuest [100]

Food hydrocolloids – Kidlington. 1986-1996 (1) 1986-1996 (5) 1986-1996 (9) – ISSN: 0268-005X – mf#16450 – us UMI ProQuest [540]

Food in canada – Toronto. 1985-1985 (1,5,9) – (cont by: food) – ISSN: 0015-6442 – mf#15479 – us UMI ProQuest [660]

Food irradiation alert! : the newsletter of the national coalitionto stop food irradiation / National Coalition to Stop Food Irradiation – 1985 sep/oct-1990 jun – 1r – 1 – mf#1656652 – us WHS [660]

Food management – Cleveland. 1983+ (1,5,9) – ISSN: 0091-018X – mf#12958,02 – us UMI ProQuest [660]

Food manufacture – London. 1927+ (1) 1972+ (5) 1974+ (9) – ISSN: 0015-6477 – mf#2723 – us UMI ProQuest [660]

Food monitor – New York. 1977-1989 (1,5,9) – (cont by: why: challenging hunger and poverty) – ISSN: 0162-0045 – mf#11883 – us UMI ProQuest [660]

Food packer – Pontiac. 1953-1958 (1) – ISSN: 0095-9227 – mf#16474,02 – us UMI ProQuest [660]

Food planning for four hundred millions / Mukerjee, Radhakamal – London: Macmillan and Co, 1938 – (= ser Samp: indian books) – us CRL [350]

Food policy – Kidlington. 1984-1996 (1,5,9) – ISSN: 0306-9192 – mf#17230 – us UMI ProQuest [630]

Food processing – Chicago. 1940+ (1) 1973+ (5) 1973+ (9) – ISSN: 0015-6523 – mf#8801 – us UMI ProQuest [660]

Food processing – Chicago: Putman Pub Co. [v47 n7-13 jul-dec 1986] – 1r – 1 – us CRL [660]

Food product development – New York. 1966-1981 (1) 1973-1981 (5) 1976-1981 (9) – (cont by: food development) – ISSN: 0015-654X – mf#9610 – us UMI ProQuest [660]

Food production management – Timonium. 1904+ (1) 1967+ (5) 1967+ (9) – ISSN: 0191-6181 – mf#2401 – us UMI ProQuest [660]

Food quality and preference – Harlow. 1990-1996 (1,5,9) – ISSN: 0950-3293 – mf#42581 – us UMI ProQuest [660]

Food research institute studies / Stanford University Food Research Institute – Stanford. 1975-1993 (1) 1975-1993 (5) 1975-1993 (9) – (cont: food research institute studies in agricultural economics, trade and development) – ISSN: 0193-9025 – mf#6398,01 – us UMI ProQuest [660]

Food research institute studies in agricultural economics, trade and development / Stanford University Food Research Institute – Stanford. 1960-1974 (1) 1972-1973 (5) (9) – (cont by: food research institute studies) – ISSN: 0015-6566 – mf#6398 – us UMI ProQuest [630]

Food research international – Ottawa. 1999+ (1,5,9) – (cont: canadian institute of food science and technology journal) – ISSN: 0963-9969 – mf#42678 – us UMI ProQuest [660]

Food reviews international – New York. 1993-96 (1,5,9) – ISSN: 8755-9129 – mf#14533 – us UMI ProQuest [660]

Food sciences and nutrition – London. 1988-1990 (1,5,9) – (cont by: international journal of food sciences and nutrition) – ISSN: 0954-3465 – mf#18127,03 – us UMI ProQuest [613]

Food shortage and agriculture / Gandhi, Mahatma – Ahmedabad: Navajivan Pub House, 1949 – (= ser Samp: indian books) – us CRL [630]

Food Store Employees Union, Local 347 *see* Local 347 journal

Food technology – Chicago. 1947+ (1) 1965+ (5) 1970+ (9) – ISSN: 0015-6639 – mf#802 – us UMI ProQuest [660]

Food, Tobacco, Agricultural and Allied Workers Union of America *see* Fta news

Food, Tobacco, Agricultural, and Allied Workers Union of America *see*
- Proceedings of the...national convention of the food, tobacco, agricultural and allied workers of america
- Reports of general executive officers to food, tobacco, agricultural and allied workers unionof america, cio,...national convention

Foodreview – Washington. 1991+ (1,5,9) – (cont: national food review) – ISSN: 1056-327X – mf#11815,01 – us UMI ProQuest [630]

Foods and feeding habits of the pedi... / Quin, P J – Johannesburg, South Africa. 1959 – 1r – us UF Libraries [390]

Foodservice equipment and supplies: fe&s – Newton. 1997+ (1,5,9) – (cont: foodservice equipment and supplies specialist) – ISSN: 1097-2994 – mf#14880,04 – us UMI ProQuest [640]

Foodservice equipment and supplies specialist – Newton. 1986-1997 (1,5,9) – (cont by: foodservice equipment and supplies: fe&s) – ISSN: 0888-8515 – mf#14880,03 – us UMI ProQuest [640]

Fool : by thomas brainless, jester to his majesty the public – Salem. 1807-1807 (1) – mf#3577 – us UMI ProQuest [870]

The fool of quality : or, the history of henry, earl of moreland / Brooke, Henry – a new and rev ed. New York; Derby & Jackson, 1860 – 2mf – 9 – 0-524-05356-1 – mf#1990-5107 – us ATLA [920]

Foolishness of god wiser than the wisdom of men / Trotter, William – s.l, England. 1841 – 1r – us UF Libraries [240]

Foolishness of truth / Parks, William – London, England. 1860 – 1r – us UF Libraries [240]

Fools of fortune : or, gambling and gamblers / Quinn, John Philip – Chicago: The Anti-Gambling Association, 1892 – (comprehending a history of the vice in ancient and modern times, and in both hemispheres, an exposition of its alarming prevalence and destructive effects, with an unreserved and exhaustive disclosure of such frauds, tricks and devices as are practiced by "professional" gamblers, "confidence men" and "bunko steerers") – us CRL [360]

Foon tsite tsoo tsite – New York. N.Y. 1925 – 1 – us AJPC [071]

Foot and ankle international – v1-17. 1980-96 – 1,5,6,9 – $80.00r – us Lippincott [617]

Foot, Jesse *see* The life of john hunter (1728-93)/ john hunter's "directions for preserving animals and parts of animals for examination"

Foot, Lionel R *see* The gold coast and the fantis

The foot of the cross : or, the sorrows of mary / Faber, Frederick William – new ed. London: Burns & Oates, [1857?] – 2mf – 9 – 0-524-06245-5 – mf#1990-5200 – us ATLA [240]

Football and cycling news *see* Ulster cyclist and football news

Football argus [bradford] – Bradford, England 5 sep 1908-11 dec 1926 [mf 1911, 1912] – 1 – (discontinued) – uk British Libr Newspaper [790]

Football argus [newport] – Newport, Wales 30 aug 1919-2 sep 1939, 1 may 1951-29 apr 1967 – 1 – (football ed of: south wales argus; cont by: south wales argus (football ed) 19 aug-2 sep 1967) – uk British Libr Newspaper [790]

Football digest – Evanston. 1971+ (1) 1971+ (5) 1973+ (9) – ISSN: 0015-6760 – mf#6276 – us UMI ProQuest [790]

Football field and sports telegraph – [NW England] Bolton 20 sep 1884-14 mar 1885, 1886-30 apr 1887 [mf ed 2004] – 1 – (title change: cricket & football field [may 1887-1915]; football edition buff [1916-2 sep 1939]; saturday finals; or, white specials [9 sep 1939-1952]; green finals [1953-55]; buff [1956]) – uk Newsplan; uk MLA [072]

Football gazette and telegraph – South Shields, England. Sports Gazette. -w. March 1910-April 1915, Aug 1919-Sept 1939, Sept 1946-Dec 1959. 12 reels – 1 – uk British Libr Newspaper [072]

Football post – [East Midlands] Nottingham, Nottinghamshire 5 sep 1903-2 sep 1939, 31 aug 1946-dec 1950 [mf ed 2002] – 62r – 1 – (football ed of evening post; not publ between 9 sep 1939-aug 1946; coverage wider than sport, incl weekend entertainments, reviews etc) – uk Newsplan [790]

Foote, Arthur *see* The life and times of henry wilder foote

Foote, Charles C *see* Woman's rights and duties

Foote, G W *see* Crimes of christianity pts 1-2

Foote, George William *see* The freethinker, 1881-1919

Foote, Henry Wilder *see*
– James freeman and king's chapel, 1782-87
– Thy kingdom come
Foote, LeRoy *see*
– A defence and exposition of truth
– Scriptural discourses and essays designed to promote growth in grace...
Foote, Shelby *see* Shiloh
Foote, William Henry *see*
– The huguenots
– Sketches of north carolina
– Sketches of virginia [first series]
– Sketches of virginia [second series]
Footfalls of indian history / Noble, Margaret E – London: Longmans, Green, 1915 – 1mf – 9 – 0-524-03369-2 – (incl bibl ref) – mf#1990-3203 – us ATLA [915]
Footman, Henry *see* Reasonable apprehensions and reassuring hints
Footner, Hulbert *see*
– New rivers of the north
– The sealed valley
– Thieves' wit
A footnote to history : eight years of trouble in samoa / Stevenson, Robert Louis – NY: Chas Scribner's Sons, 1892 – 4mf – 9 – $6.00 – mf#LLMC 82-100C Title 22 – us LLMC [980]
Footnotes / East Madison Community Center [Madison, WI] – 1980 oct-dec, 1981 jan-1985 dec, 1986 jan-jul/aug, nov/dec, 1987 jan/feb-mar/apr, nov/dec, 1988 jan/feb-sep/oct – 1r – 1 – (cont: emcc news & activities; cont by: center news [madison, wi]) – mf#1830320 – us WHS [362]
Footnotes : milwaukee area track newsletter / Milwaukee Area Track – preliminaries, 1972 jul-1977 sep – 1r – 1 – (cont by: badgerland striders) – mf#1670608 – us WHS [071]
Footprints in marion county : official publication of marion county genealogical and historical society / Marion County Genealogical & Historical Society – 1978 summer-1983 spring, 1983 summer-1987 spring – 2r – 1 – mf#689663 – us WHS [929]
Footprints of italian reformers / Stoughton, John – [London]: Religious Tract Society, [1881?] – 1mf – 9 – 0-7905-5978-1 – mf#1988-1978 – us ATLA [240]
Footprints of sorrow / Reid, John – [2d ed.] New York: Wilbur B Ketcham [c1869] [mf ed 1984] – 5mf – 9 – 0-8370-0821-2 – (incl bibl ref) – mf#1984-4176 – us ATLA [230]
Footprints of the apostles as traced by saint luke in the acts : being sixty portions for private study and instruction in church / Luckock, Herbert Mortimer – London: Longmans, Green, 1897 – 2mf – 9 – 0-524-05045-7 – (incl bibl ref) – mf#1992-0298 – us ATLA [226]
Foot-prints of the creator : or, the asterolepis of stromness / Miller, Hugh – London, 1849 – (= ser 19th c evolution & creation) – 4mf – 9 – mf#1.1.11582 – uk Chadwyck [210]
The footprints of time : a complete analysis of our american system of government / Bancroft, Charles – Burlington, IA: T T Root, 1879 – 8mf – 9 – $12.00 – mf#LLMC 95-057 – us LLMC [323]
Footsteps in the path of life : meditations and prayers for every sunday in the year / Dods, Marcus – New York: Hodder and Stoughton, 1909 – 1mf – 9 – 0-7905-1651-9 – mf#1987-1651 – us ATLA [240]
Footsteps of freedom : essays / Cousins, James Henry – Madras: Ganesh & Co, 1919 – (= ser Samp: indian books) – us CRL [954]
The footsteps of st paul in rome : an historical memoir / Forbes, S. Russell – 3rd rev and enl ed. London, New York Thomas Nelson, [ca. 1891] – 1mf – 9 – 0-8370-3164-8 – mf#1985-1164 – us ATLA [240]
Footsteps of the flock : origins of louisiana baptists / Wise, Ivan M – 1910 – 1 reel – 1 – $5.36 – (vol 2, part 1. 134p) – us Southern Baptist [242]
Footsteps of the flock : origins of louisiana baptists / Wise, Ivan M – 1 reel – 1 – $7.84 – (vol 2, part 1. 2nd edition) – us Southern Baptist [242]
Footville hustler – Footville WI. 1912 feb 8 – 1r – 1 – mf#916143 – us WHS [071]
Foppens, J F *see*
– Diplomatum belgicorum nova collectio
– Opera diplomatica et historica
For a lasting peace, for a people's democracy – Bucharest, Romania. [s.n.], nov 10 1947-apr 17 1956 – 7r – 1 – us CRL [949]
For a lasting peace, for a people's democracy – Bucharest, [s.n.]. nov 10, 1947-apr 17, 1956 – 7r – 1 – us CRL [320]
For actual settlers : winter in the country of clear days and bright suns, sleeping through a blizzard on the prairie! three weeks travelling in winter through southern manitoba, the turtle mountain country / Armstrong, Louis Olivier – S.l: s.n, 188-? – 1mf – 9 – mf#54633 – cn Canadiana [917]
For better relations with our latin american neighbors : a journey to south america / Bacon, R – Washington, D C 1916 – 3mf – 9 – mf#ILM-4066 – ne IDC [918]
For christ and city! : liverpool sermons and addresses / Stubbs, Charles William – London; New York: Macmillan, 1890 – 1mf – 9 – 0-7905-9696-2 – mf#1989-1421 – us ATLA [240]
For christ in fuh-kien : being a new edition (the fourth) of the story of the fuh-kien mission of the church missionary society / McClelland, T – London: Church Missionary Society, 1904 – 1mf – 9 – 0-7905-6935-3 – mf#1988-2935 – us ATLA [240]
For couples planning to get married = Vir diegene wat gaan trou / South Africa. Department of Health [Departement van Gesondheid] [Departement van Gesondheid] – [Pretoria: Dept of Health 1976?] [mf ed Pretoria, RSA: State Library [199-]] – 20p [ill] on 1r with other items – 5 – (in english & afrikaans) – mf#op 06671 r23 – us CRL [613]
For de fierstunnen : vergnoegte doentjes un vertellsels / Droste, Georg – 2. veraennerte upl. Bremen: F Leuwer, 1922 – 86p – 1 – (foreword by john brinkmann) – mf#7185 – us Wisconsin U Libr [430]
For england's sake / Henley, William Ernest – London, England. 1900 – 1r – us UF Libraries [960]
For ever and ever / Rogers, George – London, England. 1866 – 1r – us UF Libraries [240]
For family worship – New York: Dodd, Mead, c1883 – 2mf – 9 – 0-7905-1566-0 – mf#1987-1566 – us ATLA [240]
For gammel og ung / Homme Children's Home [Wittenberg, WI] – 1894-96, 1897-99 – 2r – 1 – (cont by: christian home) – mf#4800058 – us WHS [362]
For god and the people : prayers of the social awakening / Rauschenbusch, Walter – Boston: Pilgrim Press, c1910 – 1mf – 9 – 0-7905-9602-4 – mf#1989-1327 – us ATLA [240]
'For his sake' : a record of a life consecrated to god and devoted to china. extracts from letters of elsie marshall martyred at hwa-sang...1895 / Marshall, E – London, 1896 – 3mf – 9 – mf#HTM-114 – ne IDC [920]
For immediate release – 1979 jan 3-1986 dec 17 – 1r – 1 – mf#1221162 – us WHS [071]
For india's uplift : a collection of speeches and writings on indian questions / Besant, Annie Wood – Madras: G A Natesan & Co, 1913 [mf ed 1984] – 283p – 1 – mf#1187 – us Wisconsin U Libr [306]
For india's uplift : a collection of speeches and writings on indian questions / Besant, Annie Wood – Madras: GA Natesan & Co, [19–] – (= ser Samp: indian books) – us CRL [954]
For kaempeviserne – Kjobenhavn: Gyldendal 1847 [mf ed Bloomington IN: Indiana Uni Lib, Preservation Dept 1984] – 1r – 1 – us Indiana Preservation [390]
For king and kingdom / Christie, Jas – Carlisle, England. 1897 – 1r – us UF Libraries [240]
For maimie's sake : a tale of love and dynamite / Allen, Grant – New York: F M Lupton, [188-?] – 3mf – 9 – 0-665-90880-6 – mf#90880 – cn Canadiana [830]
For my people – 1973 nov – 1r – 1 – mf#4851974 – us WHS [241]
For now – nos. 1-8. 1966-68 – 1 – us AMS Press [800]
For pacifists / Gandhi, Mahatma – Ahmedabad: Navajivan Pub House, 1949 – (= ser Samp: indian books) – us CRL [320]
For peace and solidarity : bulletin of the peace and solidarity project / Peace and Solidarity Alliance [US] – 1985 mar-1987 nov/dec – 1r – 1 – mf#1568628 – us WHS [320]
For real – v1 n2-v4 n3 [1971 feb-1974 may/jun] – 1r – 1 – mf#1056420 – us WHS [071]
For soldiers and sailors : an abridgment of the book of common worship, published for the national service commission of the presbyterian church in the united states of america – Philadelphia: Presbyterian Board of Publication, 1917 – 2mf – 9 – 0-524-07211-6 – mf#1990-5369 – us ATLA [242]
For the best things / Miller, James Russell – New York: Thomas Y Crowell, c1907 – 1mf – 9 – 0-8370-7312-X – mf#1986-1312 – us ATLA [240]
For the independence of spain, for liberty, for the republic, union of all spaniards / Ibarruri, Dolores – Complete text of the report to the plenary session of the Central Committee of the Communist Party of Spain. Madrid, 1938. Fiche W1757. (Blodgett Collection of Spanish Civil War Pamphlets) – 9 – us Harvard [946]
For the leader company, limited, et al : a speech by nicholas flood davin, mp, delivered in the supreme court of the north-west territories...on the 7th july, 1890 / Davin, Nicholas Flood – Regina: Leader, 1890 – 1mf – 9 – mf#30153 – cn Canadiana [340]
For the people – 1970 jul 4-1975 dec, 1976 jan-1981 mar – 2r – 1 – mf#367516 – us WHS [071]
For the people / Congressional Black Caucus – 1975 apr-1979 – 1r – 1 – (cont by: congressional black caucus reports for the people) – mf#627704 – us WHS [325]
For the record : newsletter of the illinois state archives, office of the secretary of state / Illinois State Archives – v1 n1-v2 n2 [1975 apr-1976 apr], v3 n2 [1979 winter], v4 n2 [1980 winter] – 1r – 1 – mf#599513 – us WHS [324]
For the right = Kampf um's recht / Franzos, Karl Emil – New York: Harper. 2v. [19–?] (mf ed 1990) – 1r – 1 – (filmed with: die juden von barnow) – us Wisconsin U Libr [830]
For the work of the ministry : a manual of homiletical and pastoral theology / Blaikie, William Garden – 6th rev ed. London: J Nisbet, 1896 – 1mf – 9 – 0-8370-6023-0 – (incl ind) – mf#1986-0023 – us ATLA [240]
For your information – 1981 1 [jan]-1982 6 [dec] – 1r – 1 – (cont by: ada world) – mf#670084 – us WHS [071]
Forage crops / Conner, C M – Lake City, FL. 1905 – 1r – us UF Libraries [630]
Foran, Joseph F *see* Predicting muscle fiber type through self-reporting
Foran, Joseph Kearney *see*
– Beauties of the st lawrence
– An essay on obligations
– Irish-canadian representatives
– Poems and canadian lyrics
– The spirit of the age, or, faith and infidelity
– Thomas d'arcy mcgee as an empire builder
Foran, Thomas Patrick *see*
– The code of civil procedure of lower canada
– Digest of reported cases touching the criminal law of canada.
– Trial of ambrose lepine at winnipeg for the wilful murder of thomas scott
Foran, William Robert *see* A cuckoo in kenya
Forastero / Gallegos, Romulo – Buenos Aires, Argentina. 1952 – 1r – us UF Libraries [972]
Forastieri De Flores, Marines *see* Crucificado
Forbes – New York. 1917+ (1) 1964+ (5) 1960+ (9) – ISSN: 0015-6914 – mf#921 – us UMI ProQuest [650]
Forbes, A P *see*
– Charge delivered to the clergy of the diocese of brechin in synod a...
– Jesus our worship
– Primary charge delivered to the clergy of his diocese
Forbes advocate – Forbes, dec 1911-dec 1930 – 10r – A$702.99 vesicular A$757.99 silver – at Pascoe [079]
Forbes advocate – Forbes, jan 1969-jun 1995 – 57r – at Pascoe [079]
Forbes, Alex Penrose *see* Deepening of the spiritual life
Forbes, Alexander Kinloch *see* Ras mala [in roman]
Forbes, Alexander Penrose *see*
– An explanation of the thirty-nine articles
– Kalendars of scottish saints
– Lives of s ninian and s kentigern
– Remains of the late rev arthur west haddan, b d
Forbes, Archibald *see* William of germany: a succinct biography of william 1, german emperor and king of prussia
Forbes, C J F S *see* British burmah and its people
Forbes, Duncan *see*
– Clavis orientalis, pt 2
– An essay on the origin and structure of the hindoostanee tongue, or general language of british india
– A grammar of the hindustani language in the oriental and roman character
– A new persian grammar
Forbes, E *see* Travels in lycia, milyas and cibyratis
Forbes, F E *see*
– Dahomey and the dahomans
– Five years in china
The Forbes Family *see* The forbes papers, 1723-1931
Forbes, Frances Alice *see* A scottish knight-errant
Forbes, Francis et al *see* Patent and trademark laws commission
Forbes, George Henry *see* Doctrinal errors and practical scandals of the english prayer book
Forbes, Henry Prentiss *see* The johannine literature and the acts of the apostles
Forbes, J *see* Oriental memoirs
Forbes, James *see*
– Chorus lady
– Norwegen und seine gletscher
Forbes, John *see*
– Letters of general john forbes relating to the expedition against fort duquesne in 1758
– Predestination and freewill and the westminster confession of faith
– The servant of the lord in isaiah 40-66
– Studies on the book of psalms
– The symmetrical structure of scripture
Forbes, Jonathan *see* Eleven years in ceylon
The forbes papers, 1723-1931 – [mf ed 1969] – 57r – 1 – (with p/g) – us MA Hist [380]
Forbes & parkes gazette – Forbes, oct 1880-oct 1881 – 1r – A$33.66 vesicular A$39.16 silver – at Pascoe [079]
Forbes, Rosita Torr *see* Unicorn in the bahamas
Forbes, S. Russell *see* The footsteps of st paul in rome
Forbes times – Forbes, jan 1899-mar 1920 – 10r – A$640.73 vesicular A$695.73 silver – at Pascoe [079]
Forbes' tourist's directory – Rome, Italy. 1874-15 apr 1882 – 1 – (cont as: forbes' directory and archeological bulletin 1 dec 1882-16 apr 1883. cont as: forbes's directory and bulletin 15 nov 1883-15 apr 1888. cont as: forbes' rome directory and bulletin 1 dec 1888-summer n1889. cont as: rome directory and bulletin 15 nov 1889-15 apr 1890. cont as: the roman news and directory 15 nov 1890-15 apr 1898) – mf#m.f.820 – uk British Libr Newspaper [910]
Forbes, Vernon Siegfried *see* Pioneer travellers of south africa
Forbes, William Cameron *see* Present conditions in spain, january 1938
Forbes, William Trowbridge Merrifield *see* Field tables of lepidoptera
Forbes-Leith, William *see*
– Narratives of scottish catholics under mary stuart and james 6
– Narratives of scottish catholics under queen mary stuart and king james 6th
Forbes-Lindsay, Charles Harcourt Ainslie *see* Panama and the canal to-day
Forbes-Lindsey, Charles Harcourt Ainslie *see* Cuba and her people of to-day
Forbid him not / Minton, Samuel – London, England. 1860 – 1r – us UF Libraries [240]
Forbidden books from the library of a.i. ostroglazov : from the state historical library in moscow – 694mf – 1 – $3,500.00 coll – us UMI ProQuest [020]
Forbidden tree / Gilbert, Nathaniel – London, England. 1805 – 1r – us UF Libraries [240]
Forbiger, Albert *see* Hellas und rom
Forbin, Auguste de *see*
– Souvenirs de la sicile
– Voyage dans le levant, en 1817 et 1818
Forbin, Victor *see* Moeurs haitiennes
Forbundets veckotidning / Evangelical Mission Covenant Church of America – 1915 dec 7-1917 nov 27, 1917 dec 4-1919 jul 29, 1919 sep 16-1921 jun 14, 1921 jun 21-1922 dec 5, 1922 dec 12-1924 may 20, 1924 may 27-1925 nov 17, 1925 nov 24-1927 may 3, 1927 may 10-1928 oct 2, 1928 oct 9-1930 mar 11, 1930 mar 18-1931 aug 18, 1931 aug 25-1932 dec 27, 1933 jan 3-1934 jan 9 – 12r – 1 – (cont by: covenant weekly) – mf#927173 – us WHS [242]
Forbus, W R *see* The suitability and reliability of the physical best fitness test with selected special populations
Forbush, William Byron *see*
– Child study and child training
– The travel lessons on the life of jesus
Forca, cultura e liberdade / Andrade, Almir De – Rio de Janeiro, Brazil. 1940 – 1r – us UF Libraries [972]
Forca nacionalizadora do estado novo / Dantas, Mercedes – Rio de Janeiro, Brazil. 1942 – 1r – us UF Libraries [972]
Forcas armadas em face do momento politico / Pinto, Heraclito Sobral – Rio de Janeiro, Brazil. 1945 – 1r – us UF Libraries [972]
Le forcat : organe socialiste de la region nord. – Lille. juil 1882-juil 1883 – 1 – fr ACRPP [325]
Force and energy : a theory of dynamics / Allen, Grant – London, New York: Longmans, Green and Co, 1888 – 3mf – 9 – mf#05038 – cn Canadiana [530]
La force des choses – Bruxelles 1849 [mf ed Hildesheim 1995-98] – 1v on 1mf – 9 – €40.00 – 3-487-26007-7 – gw Olms [914]
Force for progress / California Labor Federation, AFL-CIO – 1971-73, 1977-81, 1984, 1986-89, 1992-94 – 1r – 1 – (cont: sacramento story) – mf#1723246 – us WHS [331]
Force, Fred P *see* The life story of l. r. millican
La force magique : du mana des primitifs au dynamisme scientifique / Saintyves, Pierre – Paris: E Nourry, 1914 – (= ser Coll science et magie) – 1mf – 9 – 0-524-01871-5 – (incl bibl ref) – mf#1990-2706 – us ATLA [130]
Force no remedy / Besant, Annie (Wood) – [London, 1882] – (= ser 19th c ireland) – 1mf – 9 – mf#1.1.1950 – uk Chadwyck [941]
Force of truth / Scott, Thomas – London, England. 1808 – 1r – us UF Libraries [240]
Force ouvriere / C G T -F O – Paris, 1950-1993 – 1 – (puis hebdomadaire de la confederation force ouvriere) – fr ACRPP [073]
Force ouvriere – Paris: Imp cent de la presse, [1945-66]. 1953-nov 23 1966 – us CRL [071]
Force, Peter *see* Collection
Forcellini, A *see* Totius latinitatis lexicon
Les forces de l'europe / Fer, N – Paris, 1693-1696. 7v – 7mf – 9 – mf#OA-259 – ne IDC [720]
Forces francaises – Paris, 1944 – 1 – (in french) – us UMI ProQuest [934]

FOREIGN

Forces nouvelles / Mouvement Republicain Populaire – Paris. n1-106. 10 fevr 1945-14 fevr 1947 [wkly] – 1 – (mq n99) – fr ACRPP [325]
The forces of the universe / Berwick, George – London: Longmans, Green, & Co 1870 [mf ed 1998] – 1r [ill] – 1 – mf#film mas 28210 – us Harvard [530]
Forchhammer, Ejnar *see* Om richard wagner og hans tannhaeuser
Forcier, Pierre *see* La rupture des accords de cooperation entre le cambodge et les usa
Ford, Abbie A *see* John pierpont
Ford bulletin – 1977 mar 1-may, oct-dec – 1r – 1 – mf#637879 – us WHS [071]
Ford, Charles T *see* From coast to coast
Ford County. Kansas. Board of Commissioners *see* Journals
Ford County. Kansas. District Court *see*
– Records
– Selected case records
Ford County. Kansas. Register of Deeds *see* Records
Ford, David B *see* New england's struggles for religious liberty
Ford, David Barnes *see*
– History of hanover academy
– Studies on the baptismal question
Ford, Eng (Buckinghamshire) Baptists *see* Church books of ford or cuddington and amersham in the county of...
Ford facts / International Union, United Automobile, Aircraft, and Agricultural Implement Workers of America – n8-35 [1941 jan 22-dec 3], v6 n3-v8 n19 [1942 feb 1-1944 oct 1], v9 n24-48 [1945 jun 16-dec 1], v10 n23-v11 [1946 jun 8-1947], 1948-1949 aug 6, 1949 aug 13-1951, 1952-1954, 1955-1960 jul, 1960 aug 6-1963 dec 14, 1964-69, 1967-69, 1970-72, 1970-1984 oct 29, 1970-72 – 10r – 1 – (cont by: uaw facts) – mf#1007645 – us WHS [331]
Ford, Ford Madox, 1873-1939 *see* Ladies whose bright eyes
Ford foundation annual reports – 1950-70 – 20mf – 9 – $90.00 – mf#LLMC 84-465 – us LLMC [360]
Ford foundation letter – New York. 1970-1991 (1) 1975-1991 (5) 1976-1991 (9) – (cont by: ford foundation report) – ISSN: 0015-699X – mf#10572 – us UMI ProQuest [370]
Ford foundation report – New York. 1992+ – 1,5,9 – (cont: ford foundation letter) – ISSN: 1063-7281 – mf#10572,01 – us UMI ProQuest [370]
Ford, Henry Jones *see*
– The rise and growth of american politics
– The scotch-irish in america
Ford, James *see*
– The gospel of s luke
– Holy communion at a visitation
– S paul's epistle to the romans
Ford, James W *see*
– The negroes in a soviet america
– Negro's struggle against imperialism
Ford, John *see* Memoir of william tanner
Ford, Lawrence Carroll *see* Triangular struggle for spanish pensacola, 1689-17...
Ford madox brown : a record of his life and works – London 1896 – (= ser 19th c art & architecture) – 7mf – 9 – mf#4.2.1332 – uk Chadwyck [420]
Ford, Mary Hanford *see* The oriental rose
Ford, Melbourne Haddock *see* The student's legal analysis
Ford Motor Co of Canada *see* Ford sur la ferme
Ford, Paul *see* Our national pie and what it contained
Ford, Paul L *see*
– Bibliography and reference list of the history and literature
– Essays on the constitution of the united states
– Pamphlets on the constitution of the united states
– The writings of thomas jefferson
Ford, Paul Leicester *see*
– His version of it
– A warning to lovers
Ford, Richard *see* Apsley house and walmer castle
Ford, Samuel Howard *see*
– Baptist waymarks
– Brief baptist history
– Origin of the baptists
Ford, Seabury *see* Ohio governor's correspondence
Ford, Sewell *see*
– Horses nine
– Torchy
– Torchy and vee
Ford sur la ferme / Ford Motor Co of Canada – Ontario: Ford, [entre 1904 et 1924] [mf ed 1994] – 1mf – 9 – 0-665-72779-8 – mf#72779 – cn Canadiana [630]
Ford worker / Communist Party of the United States of America – v1-2 n9 [1926 apr-1927 aug] – 1r – 1 – mf#1056429 – us WHS [335]
Forda, Balduinus de *see* Sermones. de commendatione fidei
Les fordcats pour la foi : etude historique (1684-1775) / Coquerel, Athanase – Paris: Michel Levy, 1866 – (= ser Bibliotheque contemporaine) – 1mf – 9 – 0-7905-5593-X – (incl bibl ref) – mf#1988-1593 – us ATLA [944]
Forder, A *see* Ventures among the arabs in desert, tent and town
Forder, Archibald *see* With the arabs in tent and town
Forder, Winter Rand *see* The formation of a ministry to the parents of infants. 1981
Fordham 1557-1950 – (= ser Cambridgeshire parish register transcript) – 13mf – 9 – £16.25 – uk CambsFHS [929]
Fordham entertainment media and intellectual property law forum *see* Fordham intellectual property, media and entertainment law journal
Fordham environmental law journal – v1-12. 1989-2001 – 9 – $216.00 set – (title varies: v1-4 1989-93 as fordham environmental law report) – mf#112391 – us Hein [344]
Fordham environmental law report *see* Fordham environmental law journal
Fordham finance, securities, and tax law forum *see* Fordham journal of corporate and financial law
Fordham, Frieda *see* An introduction to jung's psychology
Fordham intellectual property, media and entertainment law journal – v1-11. 1990-2001 – 9 – $209.00 set – (title varies: v1-3 1990-93 as fordham entertainment media and intellectual property law forum) – mf#113431 – us Hein [346]
Fordham international law forum *see* Fordham international law journal
Fordham international law journal – v1-24. 1977-2001 – 9 – $621.00 set – (title varies: v1-3 1977-80 as fordham international law forum) – ISSN: 0747-9395 – mf#112801 – us Hein [341]
Fordham journal of corporate and financial law – New York. 2000+ (1,5,9) – ISSN: 1532-303X – mf#32209,01 – us UMI ProQuest [346]
Fordham journal of corporate and financial law – v1-5. 1997-2000 – 9 – $41.00 set – (title varies: v1-3 1997-98 as: fordham finance, securities, and tax law forum) – mf#118211 – us Hein [346]
Fordham law review – v1-3. 1914-1917 (all offered) – 5mf – 9 – $7.50 – (this journal ceased publication after v3, but was revived in 1935 with a v4 to date. for copyright reasons only v1-3 can be offered at this time) – mf#LLMC 95-106 – us LLMC [340]
Fordham law review – v1-69. 1914-2001 – 5,6,9 – $1358.00 set – (v1-53 1914-85 on reel or mf $784. v54-69 1985-2001 on mf $574. suspended 1917-34) – ISSN: 0015-704X – mf#102811 – us Hein [340]
Fordham, Reginald Sydney Walter *see* Income tax appeal board practice
Fordham urban law journal – New York. 1979+ (1,5,9) – ISSN: 0199-4646 – mf#12005 – us UMI ProQuest [340]
Ford's christian repository – 1852-Sep 1905 – 1 – us Southern Baptist [242]
Ford's sauk county democrat – Baraboo WI. 1883 may 19-26, 1883 jun 2-1885 dec 12, 1885 dec 19-1886 jun 26 – 3r – 1 – (cont: sauk county democrat [baraboo, wi: 1880]; cont by: sauk county democrat [baraboo, wi: 1886]) – mf#953918 – us WHS [071]
Fordyce, Alexander Dingwall *see*
– Apples of gold in pictures of silver
– The auld kirkyard, fergus
– Extracts from a teacher's observations on school government
– Family record of the name of dingwall fordyce in aberdeenshire, vol 1
– Gleanings from the church-yard
– Letters of a pioneer
– The monumental inscriptions in the cemetary at belleside, fergus
– On co-operation in school matters
– Our sabbath school for ten more years
– The senses considered in their relation to the school
Fordyce, Alexander Dingwall [comp] *see*
– Family record of the name of dingwall fordyce in aberdeenshire, vol 2
– Family record of the name of dingwall fordyce in aberdeenshire, vols 1 and 2
– Memorials of the late hugh mair
Fordyce, James *see* Sermons to young women
Fordyce, John *see* Aspects of scepticism
The fordyce press – Fordyce, NE: R O Barlett, jan 1914-v2 n11. mar 17 1915 (wkly) [mf ed v1 n5. feb 3 [ie] 1914)-15 filmed [1965?]] – 1r – 1 – (absorbed by: hartington herald) – us NE Hist [071]
Forecast for home economics – Dayton. 1956-1986 (1) 1968-1986 (5) 1975-1986 (9) – (cont by: forecast for the home economist) – ISSN: 0015-7090 – mf#2719 – us UMI ProQuest [640]
Forecast for the home economist – New York. 1986-1990 (1) 1986-1990 (5) 1986-1990 (9) – (cont: forecast for home economics) – ISSN: 0890-9849 – mf#2719,01 – us UMI ProQuest [640]
Foredrag mod det humanistiske og saakaldte kristelige frimureri holdte i kristiania og drammen : som tillaeg, odd-fellowordenen, druidernes orden, vidnesbyrd mod hemmelige selskaber / Stub, Hans Gerhard – Kristiania: EC Bjoernstad, 1882 – 1mf – 9 – 0-524-05200-X – mf#1991-2236 – us ATLA [200]
The foregleams of christianity : an essay on the religious history of antiquity / Scott, Charles Newton – rev enl ed. London: Smith, Elder, 1893 – 1mf – 9 – 0-524-01297-0 – mf#1990-2333 – us ATLA [200]
Foreign affairs – New York. 1922+ (1) 1968+ (5) 1960+ (9) – ISSN: 0015-7120 – mf#6 – us UMI ProQuest [327]
Foreign affairs – v1-80. 1922-2001 – 9 – $1734.00 set – ISSN: 0015-7120 – mf#102831 – us Hein [977]
Foreign agriculture – Washington. 1963-1988 [1]; 1971-1988 [5]; 1976-1988 [9] – (cont by: agexporter) – ISSN: 0015-7163 – mf#1727 – us UMI ProQuest [630]
Foreign agriculture – Washington. 1937-1962 (1) – mf#471 – us UMI ProQuest [630]
Foreign Chaplain *see*
– Efficacy of prayer
– Everlasting punishment
Foreign claims settlement commission decisions – Washington: GPO. bk1 1955; bk2 1968; 2 pamphlets 1964, 1967 – 20mf – 9 – $30.00 – (bk 1: settlement of claims by the foreign claims settlement commission of the us and its predecessors from 14 sept 1949-31 mar 1955. bk 2: decisions and annotations, 1950-1967. index/digest of decisions, 1949-1977) – mf#LLMC 94-356B – us LLMC [340]
Foreign commerce.. / Philippines. Dept of Finance and Justice. Bureau of Customs – Manila: Bureau of Printing, 1913 14 – 1 – us Wisconsin U Libr [324]
Foreign conspiracy against the liberties of the united states : the numbers under the signature of brutus, originally published in the new york observer / Morse, Samuel Finley Breese – 7th ed. New York: American and Foreign Christian Union, 1855, c1835 – 1mf – 9 – 0-8370-8363-X – mf#1986-2363 – us ATLA [240]
Foreign economic trends and their implications for the united states – Washington. 1972-1993 (1) 1972-1993 (5) 1976-1993 (9) – ISSN: 0090-9467 – mf#7357 – us UMI ProQuest [337]
Foreign gazetteers of the us board on geographic names on microfiche / U.S. Defense Mapping Agency – 2 grps. 1987-90 – 168mf – 9 – $4,555.00 set – (printed guide incl) – us CIS [910]
Foreign governments / Morstein Marx, Fritz – New York, NY. 1949 – 1r – us UF Libraries [025]
Foreign intelligence literary scene / National Intelligence Study Center et al – v1 n1-v8 n4 [1982 feb-1989 dec] – 1r – 1 – (cont by: world intelligence review [washington, dc]) – mf#644391 – us WHS [150]
Foreign investment law 2-5 and regulations, 1981-1986 / Federated States of Micronesia – n.p., n.d. – 1mf – 9 – $1.50 – mf#LLMC 82-100H Title 10 – us LLMC [324]
Foreign investment laws and regulations / Trust Territory of the Pacific – Saipan: Foreign Investment Branch, Dep of the Attorney General, sep 1976 – 1mf – 9 – $1.50 – mf#LLMC 82-100F Title 99 – us LLMC [332]
Foreign journalists under franco's terror – London, 1937. Fiche W 889. (Blodgett Collection of Civil War Pamphlets) – 9 – us Harvard [946]
Foreign language annals – Yonkers. 1967+ (1) 1975+ (5) 1975+ (9) – ISSN: 0015-718X – mf#10596 – us UMI ProQuest [400]
Foreign letters of the continental congress and the department of state, 1785-1790 / U.S. Dept of State – (= ser General records of the department of state) – 1r – 1 – (with printed guide) – mf#M61 – us Nat Archives [324]
Foreign mission executive minute books / Methodist Church of New Zealand, Methodist Overseas Mission – 8 jan 1925-15 dec 1931 – 1r – 1 – mf#pmb1097 – at Pacific Mss [242]
Foreign mission executive minute books / New Zealand. Methodist Church. Overseas Mission – 8 jan 1925-15 dec 1931 – 1r – 1 – mf#PMB1097 – at Pacific Mss [242]
The foreign mission journal *see* Periodicals
Foreign mission work of american friends : a brief history of their work from the beginning to the year 1912 – [s.l]: American Friends Board of Foreign Missions, 1912 [mf ed 1993] – (= ser Society of friends (quakers) coll) – 1mf – 9 – 0-524-07532-8 – mf#1991-3162 – us ATLA [243]
The foreign mission work of pastor louis harms, and the church at hermansburg / Greenwald, Emanuel – Philadelphia: Lutheran Board of Publication, 1867 – 1mf – 9 – 0-524-07820-3 – mf#1991-3367 – us ATLA [240]
Foreign missionary : an incarnation of a world movement / Brown, Arthur Judson – New York, Chicago: Fleming H Revell c1907 [mf ed 1995] – 1r – 1 – (selection fr correspondence of the late thomas chalmers) – mf#8846 – us Wisconsin U Libr [240]
Foreign missionary / Presbyterian Church in the USA – v9/13-v44/45 n7 – 11r – 1 – (cont by: church at home and abroad) – mf#156179 – us WHS [242]
The foreign missionary : an incarnation of a world movement / Brown, Arthur Judson – New York: FH Revell, c1907 – 1mf – 9 – 0-524-08316-9 – (incl bibl ref and ind) – mf#1993-1011 – us ATLA [240]
The foreign missionary and his work / Cunnyngham, William George Etler – Nashville, Tenn: Pub House of the ME Church, South: Baker & Smith, agents, 1899 – 1mf – 9 – 0-8370-7377-4 – mf#1986-1377 – us ATLA [240]
Foreign missionary chronicle – v. 1-7. 1833-39 – 1 – $50.00 – us Presbyterian [240]
The foreign missionary chronicle – Pittsburgh, Pa. v1-9. 1833-1841 – 3r – 1 – $150.00 – us Presbyterian [240]
Foreign missionary tidings / Presbyterian Church in Canada. Woman's Foreign Missionary Society. Western Divison – [S.l: Arbuthnot Bros, 1897-1914] – 9 – (cont: presbyterian church in canada. woman's foreign missionary society) – mf#P04416 – cn Canadiana [242]
Foreign missions : being a study of some principles and methods in the expansion of the christian church / Malden, Richard Henry – London, New York: Longmans, Green, 1910 – 1mf – 9 – 0-8370-6215-2 – (incl ind) – mf#1986-0215 – us ATLA [240]
Foreign missions / Churton, Edward Townson – London: Longmans, Green, 1901 – 1mf – 9 – 0-524-04766-9 – (incl bibl ref) – mf#1991-2152 – us ATLA [240]
Foreign missions / Duff, Alexander – Edinburgh, Scotland. 1872 – 1r – us UF Libraries [240]
Foreign missions / Martin, George Currie – London: National Council of Evangelical Free Churches, 1905 – 1mf – 9 – 0-8370-6757-X – (incl ind) – mf#1986-0757 – us ATLA [240]
Foreign missions : their place in the pastorate, in prayer, in conferences: ten lectures / Thompson, Augustus Charles – New York: Charles Scribner, 1889 – 2mf – 9 – 0-8370-6422-8 – (incl ind) – mf#1986-0422 – us ATLA [240]
Foreign missions : their relations and claims / Anderson, Rufus – New York: Charles Scribner, 1869 – 1mf – 9 – 0-8370-6083-4 – (incl bibl ref & index) – mf#1986-0083 – us ATLA [240]
Foreign missions after a century / Dennis, James Shepard – New York: Fleming H Revell, c1893 [mf ed 1986] – (= ser Students' lectures on missions 1893) – 1mf – 9 – 0-8370-6253-5 – (incl ind) – mf#1986-0253 – us ATLA [240]
Foreign missions, and mosaic traditions : a lecture / Colenso, John William, bishop of Natal – London, 1865 – (= ser 19th c evolution & creation) – 1mf – 9 – mf#1.1.496 – uk Chadwyck [240]
Foreign Missions Board *see* Methodist church
Foreign missions board methodist church – 1912-1949 – 311r – 1 – $35,765.00 – (regions: africa 52r $5980. asia-central province 9r $1035. south asia 103r $11,845: bengal 8r $920 bombay 7r $805 burma 6r $690 gujarat 5r $575 hyderbad 3r $345 lucknow 12r $1380 southern asia/india 19r $2185 north india 14r $1610 india 11r $1265 india & pakistan / indus river 7r $805 south india 11r $1265. europe 22r $2530. latin america 83r $9545: mexico 8r $920 chile 17r $1955 bolivia 19r $2185 eastern south america 14r $1610 peru/ north andes 11r $1265 brazil 9r $1035 cuba 5r $575. malaya 25r $2875. philippines 17r $1955. with guide. also sold separately $40 d3464.g) – mf#D3464 – us Commission [242]
Foreign Missions Committee of the Presbyterian Church in Victoria *see* New hebrides magazine
Foreign missions conference of north america *see* Interdenominational conference of foreign missionary boards
Foreign Missions Division. Board of Missions and Church Extension *see* Methodist church
Foreign missions of the protestant churches / Baldwin, Stephen Livingstone – New York: Eaton & Mains; Cincinnati: Jennings & Pye c1900 [mf ed 1986] – 1mf – 9 – 0-8370-6005-2 – (incl tables & ind) – mf#1986-0005 – us ATLA [242]
Foreign missions of the protestant churches : their state and prospects / Mitchell, John Murray – Toronto, Canada: Toronto Willard Tract Depot 1888 [mf ed 1986] – 1mf – 9 – 0-8370-6761-8 – (incl ind) – mf#1986-0761 – us ATLA [242]

The foreign missions of the southern baptist convention / Tupper, Henry Allen – Philadelphia: American Baptist Publication Society; Richmond, Va: Foreign Mission Board of the Southern Baptist Convention, [1880?] – 2mf – 9 – 0-7905-8161-2 – mf#1988-6108 – us ATLA [242]

The foreign missions of the united church, 1890-1915 : a brief historical summary / Saeterlie, Martin – Minneapolis, Minn: Augsburg, 1917 – 1mf – 9 – 0-524-06190-4 – mf#1991-2446 – us ATLA [240]

Foreign nationalities branch files, 1942-1945 / U.S. Office of Strategic Services – 2427mf (24:1-29:1) – 9 – $13,085.00 coll – (europe (general & misc) $1355. eastern europe $4245. central europe $4310. southeastern europe $5620. western europe $2365. jewish groups $770. ind only $1210) – us UPA [327]

Foreign nations : africa – (= ser The special studies series) – $10,295.00 coll – (1962-80 7r isbn 0-89093-382-0 $1340. 1980-85 suppl 11r isbn 0-89093-679-X $2135. 1985-88 suppl 10r isbn 1-55655-114-2 $1935. 1989-91 suppl 10r isbn 1-55655-428-1 $1935. 1992-94 suppl 10r isbn 1-55655-534-2 $1935. 1995-97 suppl 8r isbn 1-55655-731-0 $1550. with p/g) – us UPA [327]

Foreign nations : asia – (= ser The special studies series) – 1 – $18,775.00 coll – (asia, 1980-82 suppl 5r isbn 0-89093-434-7 $970. asia, 1982-85 suppl 12r isbn 0-89093-643-9 $2330. asia, 1985-88 suppl 12r isbn 1-55655-112-6 $2330. asia, 1989-91 suppl 9r isbn 1-55655-427-3 $1740. asia, 1992-94 suppl 12r isbn 1-55655-535-0 $2330. asia, 1995-97 suppl 13r isbn 1-55655-722-1 $2520. asia, 1998-2002 suppl 14r* isbn 1-55655-963-1 $2705. china, 1970-80 8r isbn 0-89093-386-3 $1550. japan, korea, & the security of asia, 1970-80 4r isbn 0-89093-384-7 $770. vietnam & southeast asia, 1960-80 13r isbn 0-89093-383-9 $2520. with p/g) – us UPA [321]

Foreign nations : europe and nato – (= ser The special studies series) – 1 – $13,880.00 coll – (1970-80 11r isbn 0-89093-393-6 $1940. 1980-85 suppl 12r isbn 0-89093-681-1 $2120. 1985-88 suppl 8r sbn 1-55655-113-4 $1410. 1989-91 suppl 10r isbn 1-55655-429-x $1760. 1992-94 suppl 14r isbn 1-55655-536-9 $2460. 1995-97 suppl 14r isbn 1-55655-729-9 $2460. 1998-2002 suppl 14r* isbn 1-55655-964-x $2460. with p/g) – us UPA [341]

Foreign nations : latin america – (= ser The special studies series) – 1 – $13,245.00 coll – (1962-80 10r $1935 isbn 0-89093-454-1. suppl: 1980-82 3r $570 isbn 0-89093-435-5. 1982-85 7r $1340 isbn 0-89093-655-2. 1985-88 12r $2330 isbn 0-55655-115-0. 1989-91 14r $2705 isbn 1-55655-426-5. 1992-94 14r $2705 isbn 1-55655-537-7. 1995-97 12r isbn 1-55655-730-2 $2330. with p/g) – us UPA [327]

Foreign nations : the soviet union and republics of the former ussr – (= ser The special studies series) – 1 – (the soviet union, 1970-80 9r isbn 0-89093-385-5 $1740. 1980-82 suppl 8r isbn 0-89093-432-0 $1550. 1982-85 suppl 9r isbn 0-89093-645-5 $1740. 1985-88 suppl 10r isbn 1-55655-111-8 $1935. 1989-91 suppl 13r isbn 1-55655-415-X $2520. the soviet union & republics of the former ussr, 1992-94 suppl 15r isbn 1-55655-533-4 $2905. 1995-97 suppl 12r isbn 1-55655-725-6 $2330. with p/g) – us UPA [327]

Foreign office confidential print : africa – 71r – 1 – $2,790.00 – us Trans-Media [960]

Foreign office confidential print: america – 26r – 1 – $980.00 – us Trans-Media [970]

Foreign office files for china, 1949-1976 : (public record office classes fo 371 and fco 21) – 5pt – 1 – (pt1: complete files for 1949 (pro class fo 371/75731-75957) [31r] £2950; pt2: complete files for 1950 (pro class fo 371/83230-83579) [33r] £3200 [mf ed aug 2000]; pt3: complete files for 1951 (pro class fo 371/92188-92395) [19r] £1800; pt4: complete files for 1952 (pro class fo 371/99229-99387) [22r] £2,100; pt5: complete files for 1953 (pro class fo 371/105188-105355) [18r] £1700 [mf ed spring 2004]; with d/g) – uk Matthew [327]

Foreign office files for cuba : (public record office class fo 371) – 3pt – 1 – (pt1: revolution in cuba 1959-60 (pro classes fo 371/139396-139521, 148178-148345 & prem 11/2622) [13r] £1250; pt2: cuba & the bay of pigs invasion 1961 (pro classes fo 371/156137-156255 & prem 11/3316, 3321 & 3328) [9r] £875; pt3: cuban missile crisis 1962 (pro classes fo371/162308-162436, 168135 & prem 11/3689-3691) [15r] £1450 [mf ed summer 2003]; with d/g) – uk Matthew [972]

Foreign office files for japan and the far east, series 1 : embassy and consular archives – japan (1905-1940) (public record office class fo 262) – [mf ed Marlborough 1994] – 6pt – 1 – (pt1: correspondence to & from japan 1905-20 (pro class fo 262/1466-1511, 2033-34) [18r] £1750; pt2: detailed correspondence for 1921-23 (pro class fo 262/1512-1601) [44r] £4250; pt3: detailed correspondence for 1924-26 (pro class fo 262/1602-72) [44r] £4250; pt4: detailed correspondence for 1927-29 (pro class fo 262/1673-1741) [44r] £4250; pt5: detailed correspondence for 1930-33 (pro class fo 262/1742-1860, 1989-2003+2035) [25r] £2400; pt6: detailed correspondence for 1934-40 (pro class fo 262/1861-1988, 2004-32 + 2036-39) [13r] £1250; with d/g) – uk Matthew [950]

Foreign office files for japan and the far east, series 2 : british foreign office files for post-war japan, 1952-1980 (public record office classes fo 371 and fco 21) – 8pt – 1 – (pt1: complete files for 1952-53 (pro class fo 371/98985-98992, 99013, 99198-99200, 99218, 99227, 99264, 99315, 99388-99542, 99560 & 105361-105464) [38r] £3650; pt2: complete files for 1954-56 (pro class fo 371/110400-110530, 115220-115306 & 121030-121101) [39r] £3750; pt3: complete files for 1957-59 (pro class fo 371/127521-127598, 133577-133659 & 141415-141530) [16r] £1550; pt4: complete files for 1960-62 (pro class fo 371/150561-150654, 158477-158541 & 164958-165033) [25r] £2400; pt5: complete files for 1963-65 (pro class fo 371/170743-170800, 175999-176054 & 181067-181112) [18r] £1750; pt6: complete files for 1966-1968 (pro classes fo 371/187076-187142 & fco 21/238-299) [11r] £1075; pt7: complete files for 1969-1971 (pro class fco 21/555-593, 636-639, 720-769, 798-800 & 877-926) [15r]£1450; pt8: complete files for 1972-1974 (pro class fco 21/959-960, 1026-1054, 1082-1085, 1144-1172, 1212-1217 & 1275-1306) [20r] £1925; with d/g) – uk Matthew [327]

Foreign office files for japan and the far east, series 3 : embassy & consular archives – japan (post 1945) (public record office class fo 262) – 1 – (detailed correspondence for 1945-57 (pro class fo 262/2040-2132) [7r] £675; with d/g) – uk Matthew [327]

Foreign office files for post-war europe, series 1 : the schuman plan and the european coal and steel community, 1950-1957 – [mf ed Marlborough 1995] – 3pt – 1 – (pt1: complete fo 371 files for 1950-53 (pro class fo 371/85841-85869, 86977, 87168, 93826-93844, 94101-94107, 94356, 100247-100265, 100267-100272, 104012-104019, 105951-105961, 106069-106075 & 106077) [20r] £1925; pt2: complete fo 371 files for 1954-55 (pro class fo 371/ 109621, 111250-111264, 111321-111330, 115990-115998, 116036-116057 & 116100-116105) [13r] £1250; pt3: complete fo 371 files for 1956-57 (pro class fo 371/ 120815, 121918-121922, 121925-121928, 121932, 121949-121976, 121984-122005, 122014, 122018-122046, 122050-122061, 124380, 124418, 124451, 124519, 124543-124550, 124559, 124561-124573, 124587, 124590, 124733, 128292-128293, 128315-128324, 128327 & 128329-128330) [28r] £2700; with d/g) – uk Matthew [327]

Foreign office files for post-war europe, series 2 : the treaty of rome and european integration, 1957-1960 – 3pt – 1 – (pt1: files for 1957 (public record office class fo 371/128308-128314, 128325-128326, 128328, 128331-128396, 130988-130991, 131000, bt 241/1700-1701, cab 130/176, t 237/196-197 & t 299/112-115 & 126) [23r] £2200; pt2: files for 1958-59 (public record office class fo 371/134482-134545, 137145, 141134-141139, 142425, 142504, 142561-142569, 142588-142600, 142609-142636) [27r] £2600; pt3: files for 1960 (public record office class fo 371/150217-150227, 150263-150380 and t230/502) [23r] £2200; with d/g) – uk Matthew [327]

Foreign office files for the soviet union : [public record office classes fo 371 and fco 28] – 4pt – 1 – (pt1: complete files for 1960 (pro class fo 371/151908-152050) [11r] £1075; with d/g) – uk Matthew [947]

Foreign office files: united states of america, series 1 : usa – politics and diplomacy, 1960-1974 (public record office classes fo 371 and fco 7: american department – united states) – 2pt – 1 – (pt1: john f kennedy years 1960-63 (pro class fo 371/148576-148649, 155476-155516, 162578-162648 & 168405-168491) [26r] £2500; pt2: lyndon b johnson years 1964-68 (pro class fo 371/174260-174346, 179557-179622 & 184995-185056 & pro class fco 7/738-884) [30r] £2900; with d/g) – uk Matthew [327]

Foreign office files: united states of america, series 2 : vietnam, 1959-1975 (public record office classes fo 371 and fco 15: south east asia department) – 6pt – 1 – (pt1: vietnam 1959-63 (pro class fo 371/144387-144461, 152737-152798, 160107-160175, 166697-166763 & 170088-170153) [33r] £3200; pt2: laos 1959-63 (pro class fo 371/ 143956-144064, 152317-152428, 159811-159956, 166423-166504 & 169802-169876) [58r] £5500 [mf ed jun 2000]; pt3: cambodia 1959-63 (pro class fo 371/ 144344-144386, 152684-152736, 160085-160106, 166664-166696 & 170057-170087) [14r] £1350 [mf ed jun 2000]; pt4: seato, se asia general & thailand 1959-63 – complete files on the vietnam conflict (pro class fo 371/ 143721-143725, 143727-143747, 143769-143774, 143782, 144293,144296-144297, 150381, 152136-152181, 152639-152642, 152644, 152646-152647, 152671, 158379-158380, 159701-159702, 159712-159713, 159715, 159722, 159728-159747, 159756-159758, 160069-160076, 160079-160080, 160083, 164871, 166353-166355, 166359-166360, 166363, 166616-166619, 166622, 166629-166634, 166644-166663, 169678-169679, 169681, 169684, 169686, 169689, 169728-169729, 170016-170020, 170022, 170031-170032, 170038, 170042-170056 and 170634) [25r] £2400; pt5: vietnam, 1964-1966 (pro class fo 371/175464-175533, 180510-180643 & 186279-186419) [44r] £4200 [mf ed fall 2004]; pt6: vietnam, 1967-1968 (pro class fco 15/481-782) [31r] £3000; with d/g) – uk Matthew [327]

Foreign office files: united states of america, series 3 : the cold war (public record office class fo 371 & related files) – 2pt – 1 – (pt1: berlin crisis 1947-50 (pro class fo 371 – germany/70489-70528, 76537-76562, 84977-84994 & related air, cab, defe, do, fo, prem, t & wo files) [30r] £2900; pt2: prague spring & soviet intervention in czechoslovakia, 1967-68 (pro classes prem 13/1373, 1993-1994 & fco 28/38-57, 68-70, 73-75, 87-145, 571-579 & 615-619) [21r] £2000; with d/g) – uk Matthew [327]

Foreign office registers and indexes of correspondence, 1793-1919 / Great Britain. Foreign Office. Public Record Office; ed by Palmer, Greg – [mf ed Chadwyck-Healey] – 3 sets – 9 – (records entire correspondence between british govt and its agents abroad fr the napoleonic period to the treaty of versailles. registers and ind publ in foll groups of countries: usa 55v on 660mf. asia and the pacific 94v on 1078mf. russia, persia and central asia 96v on 1211mf) – uk Chadwyck [327]

Foreign plant diseases / Stevenson, John Albert – Washington, DC. 1926 – 1r – us UF Libraries [630]

Foreign policy / Grant Duff, Mountstuart Elphinstone – London, 1880 – (= ser 19th c books on british colonization) – 1mf – 9 – mf#1.1.8269 – uk Chadwyck [327]

Foreign policy – Washington. 1970+ (1) 1970+ (5) 1975+ (9) – ISSN: 0015-7228 – mf#6812 – us UMI ProQuest [327]

Foreign Policy Association Commission On Cuban A... *see* Problems of the new cuba

Foreign Policy Association Commission On Cuban Af... *see* Problems of the new cuba

Foreign Policy Association, Inc *see* Problemas de la nueva cuba

Foreign policy briefs – 1951 aug 17-1869 jun 30 – 1r – 1 – mf#153517 – us WHS [327]

Foreign policy bulletin – New York. 1921-1961 (1) – mf#934 – us UMI ProQuest [327]

Foreign policy in the far east / Das, Taraknath – New York: Longmans, Green and Co, 1936 – (= ser Samp: indian books) – (foreword by herbert wright) – us CRL [327]

The foreign policy of tanzania, 1961-68 / Shaw, T M – Kampala, 1969 – us CRL [327]

The foreign policy of the indian union / Puntambekar, S V – Baroda: Padmaja Publications, 1948 – (= ser Samp: indian books) – us CRL [327]

Foreign policy reports – New York. 1925-1951 (1) – mf#3493 – us UMI ProQuest [327]

Foreign post world war 2 newspapers – Collection of Newspapers for Various Countries during 1945-1954 – 1 – us NY Public [070]

Foreign press report on west irian / Permanent Mission to the United Nations – New York, 1961-1962. v1-17 – 38mf – 9 – (missing: 1961 v1-9) – mf#SE-1686 – ne IDC [959]

Foreign protestantism within the church of england : the story of an alien theology and its present outcome / Wirgman, Augustus Theodore – London: Catholic Literary Association, 1911 – 1mf – 9 – 0-7905-6974-4 – mf#1988-2974 – us ATLA [242]

Foreign quarterly and westminster review – London. 1827-1847 – 1 – mf#3908 – us UMI ProQuest [073]

Foreign registers and returns / Great Britain. General Register Office – 1627-1960 – uk National [324]

Foreign relations of the united states under the articles of confederation, 1780-1789 (fruac-m) : the complete documentary edition / ed by Giunta, Mary A – ca 39r – 1 – ca $5070.00 – (guide also sold separately $40 s3525.g) – mf#S3525 – National Historical Publications and Records Commission (NHPRC) – us Scholarly Res [327]

Foreign relations papers of the united states – 9 – $7471.00 set – mf#402330 – us Hein [327]

Foreign relations papers of the united states / U.S. Dept of State – 400bks. 1861-1964/68 – 4981mf – 9 – $7471.00 – (the official state dept comp of letters and documents generated in the course of diplomatic relations between us and other countries) – mf#llmc 79-444 – us LLMC [327]

Foreign review : and continental miscellany – London. 1828-1830 (1) – mf#4190 – us UMI ProQuest [920]

Foreign scenes and travelling recreations / Howison, John – Edinburgh 1825 [mf ed Hildesheim 1995-98] – 2v on 4mf – 9 – €120.00 – ISBN-10: 3-487-26683-0 – ISBN-13: 978-3-487-26683-1 – gw Olms [910]

Foreign Service Institute (US) *see*
- Hausa
- Kituba
- Shona
- Twi basic course

Foreign Service Institute [US] *see* Cambodian

Foreign service journal – Washington. 1924+ (1) 1971+ (5) 1976+ (9) – ISSN: 0146-3543 – mf#2445 – us UMI ProQuest [327]

Foreign student-athletes and their motives for attending north carolina ncaa division 1 institutions / Berry, James R – 1999 – 1mf – 9 – $4.00 – mf#PE 4019 – us Kinesology [306]

Foreign trade of india, 1900-1940 : a statistical analysis / Venkatasubbiah, H – New Delhi: Indian Council of World Affairs; Bombay: Oxford University Press, 1946 – (= ser Samp: indian books) – us CRL [380]

Foreign trade of nigeria / Ihaza, Daniel E – 1954 – us CRL [380]

The foreign vocabulary of the qur'an / Jeffery, Arthur – Oriental Institute Baroda, 1938 – 6mf – 8 – €14.00 – ne Slangenburg [260]

The foreigner in china / Wheeler, Lucius N – Chicago: S C Griggs, 1881 [mf ed 1995] – (= ser Yale coll) – 268p – 1 – 0-524-09383-0 – (int by w c sawyer) – mf#1995-0383 – us ATLA [951]

Foreigners in turkey / Brown, P M – Princeton, 1914 – 2mf – 9 – mf#ILM-656 – ne IDC [956]

The foreknowledge of god : and cognate themes in theology and philosophy / McCabe, Lorenzo Dow – Cincinnati: Hitchcock & Walden, 1878 – 2mf – 9 – 0-7905-1360-9 – (incl ind) – mf#1987-1360 – us ATLA [210]

The foreknowledge of god : or, the omniscience of god consistent with his own holiness and man's free agency / Hayes, Joel S – Nashville, Tenn: Pub House of the ME Church, South, 1890 – 1mf – 9 – 0-524-08632-X – (incl bibl ref) – mf#1993-2092 – us ATLA [210]

Forem, Leon *see* Buzshe raynarski, 1897-1914

Den forenede kirke : fred og strid, eller, lidt foreningshistorie / Dahl, Theodor H – Stoughton, Wis: Normannen, 1894 – 1mf – 9 – 0-524-01936-3 – mf#1990-4160 – us ATLA [240]

Den forenede norsk lutherske kirke i amerika / Norlie, Olaf Morgan – Minneapolis, Minn: Augsburg Pub House, 1914 – 2mf – 9 – 0-524-01637-2 – mf#1990-4101 – us ATLA [242]

Forensic engineering – Elmsford. 1987-1991 (1,5,9) – ISSN: 0888-8817 – mf#49501 – us UMI ProQuest [614]

Forensic of pi kappa delta – Brookings. 1915+ (1) 1970+ (5) 1976+ (9) – ISSN: 0015-735X – mf#6552 – us UMI ProQuest [370]

Forensic reports – New York. 1988-1992 (1) 1988-1992 (5) 1988-1992 (9) – ISSN: 0888-692X – mf#16654 – us UMI ProQuest [614]

Forensic science international – Lausanne. 1972+ (1) 1972+ (5) 1987+ (9) – ISSN: 0379-0738 – mf#42195 – us UMI ProQuest [614]

Forensic sciences gazette – Dallas. 1970-1986 (1) 1972-1986 (5) 1974-1986 (9) – ISSN: 0046-4570 – mf#7674 – us UMI ProQuest [614]

Forensischer und kriminologischer umgang mit dem delikt der vergewaltigung und seinen opfern / Schmitz, Christina – (mf ed 1997) – 3mf – 9 – €49.00 – 3-8267-2469-0 – mf#DHS 2469 – gw Frankfurter [345]

Forero Benavides, Abelardo *see* Testimonio contra la barbarie politica

Forero F, Jose Ignacio *see* Historia de la aviacion en colombia

Forero, Manuel Jose *see*
- Camilo torres
- Paginas de la vida colonial
- Proceres y estadistas de colombia

Forerunner / Forerunners Institute [Philadelphia, PA] – 1993 spring – 1r – 1 – mf#5265365 – us WHS [071]

Forerunner / Marantha Campus Ministries International et al – 1981 may-1984 – 1r – 1 – mf#966066 – us WHS [242]

Forerunner – v1-7. 1909-16 [all publ] – (= ser Radical periodicals in the united states, 1881-1960. series 1; Periodicals on women and women's rights, series 1) – 29mf – 9 – $280.00 – us UPA [322]

Forerunner on campus / Maranatha Campus Ministries – v1 n1-v2 n2 [1986 sep-1987 oct] – 1r – 1 – mf#1268046 – us WHS [242]
Forerunners and rivals of christianity : being studies in religious history from 330 b.c. to 330 a.d. / Legge, F – Cambridge: University Press; New York: Putnam [distributor], 1915 – 2mf – us ATLA [230]
Forerunners and rivals of christianity : being studies in religious history from 330 b.c. to 330 a.d / Legge, Francis – Cambridge: University Press; New York: Putnam [distributor], 1915 – 2mf – 9 – 0-7905-5476-3 – (incl bibl ref) – mf#1988-1476 – us ATLA [230]
Forerunners Institute [Philadelphia, PA] *see* Forerunner
Forerunners of modern malawi / Henderson, James – Alice, South Africa. 1968 – 1r – us UF Libraries [960]
Foreshadowings : a proposal for the settlement of the irish land question / Ignotus [pseud] – Dublin, 1870 – (= ser 19th c ireland) – 5mf – 9 – mf#1.1.8186 – uk Chadwyck [333]
Foreshadows : lectures on our lord's parables / Cumming, John – Philadelphia: Lindsay & Blakiston, 1863 – 1mf – 9 – 0-8370-2787-X – mf#1985-0787 – us ATLA [240]
Foresight / Black Teachers Workshop – 1968 nov – 1r – 1 – mf#4851946 – us WHS [370]
Forest advance – Three Lakes WI. 1908 oct 1, 1911 may 4-dec 28 – 1r – 1 – (cont by: new north) – mf#935699 – us WHS [071]
Forest and conservation history – Durham. 1990-1995 (1) 1990-1995 (5) 1990-1995 (9) – (cont: journal of forest history) – ISSN: 1046-7009 – mf#5895,02 – us UMI ProQuest [634]
Forest and stream : a journal of outdooe life, travel, nature study. shooting, fishing, yachting – New York: [Forest and Stream Publ Co,] v1-77. aug 14 1873-dec 30 1911 – 1 – us CRL [790]
Forest and stream : a journal of outdoor life, travel, nature study, shooting, fishing, yachting – New York. 1873-1930 [1] – mf#5551 – us UMI ProQuest [639]
Forest board annual report and financial statement for... / South Australia. Forest Board [-1882] – Adelaide: E Spiller, acting govt printer 1879 (annual) [1878/79] [mf ed 2005] – 1v on 1r – 1 – (cont by: south australia. forest board [-1882]. woods and forests; incl: conservator's annual progress report and appendices) – mf#film mas 36321 – us Harvard [634]
Forest county chief – Crandon WI. 1901 jun 15-29 – 1r – 1 – mf#960673 – us WHS [071]
Forest county spy – Pelican Lake WI. 1885 may 12,26 – 1r – 1 – mf#957194 – us WHS [071]
La forest des hermites et hermitesses d'egypte, et de la palestine... / Blommaert, A – Antwerp: Hierosme Verdussen, 1619 – 2mf – 9 – mf#O-1793 – ne IDC [956]
Forest echo – Crandon WI. 1906 aug 28-1908 may 8, 1908 may 15-1909 dec 24, 1909 dec 31-1911 jul 28, 1911 aug 4-1913 apr 15, 1913 apr 18-1914 dec 25 – 5r – 1 – mf#963543 – us WHS [071]
Forest ecology and management – Amsterdam. 1976+ (1) 1976+ (5) 1987+ (9) – ISSN: 0378-1127 – mf#42068 – us UMI ProQuest [634]
Forest fires in florida – Jacksonville? FL. 1926 – 1r – us UF Libraries [634]
Forest fires in northern canada / Bell, Robert – Washington, DC: Gibson, 1889 – 1mf – 9 – mf#03552 – cn Canadiana [634]
Forest gate gazette and stratford and upton chronicle *see* Forest gate gazette and upon chronicle
Forest gate gazette and upon chronicle – London, UK. 15 dec 1888-27 sep 1902 – 6r – 1 – (incorp with: county borough of west ham gazette. aka: forest gate gazette and stratford and upton chronicle) – uk British Libr Newspaper [072]
The forest grange : a series of twelve letters / Addon, Esther – London: Hamilton, Adams & Co; Birmingham: Hudson & Son. 2v. 1861-62 – (= ser 19th c women writers) – 5mf – 9 – mf#5.1.6 – uk Chadwyck [420]
Forest grove express – Forest Grove OR: W C Benfer, 1916- [wkly] – 1 – (ceased in 1918. absorbed by: washington county news-times (1911-81)) – us Oregon Lib [071]
Forest grove independent – Forest Grove OR: Wheeler & Myers, -1874 [wkly] – 1 – (began in 1873; cont by: washington independent (1874-)) – us Oregon Lib [071]
Forest grove news-times – Forest Grove OR: Times Pub Co, 1981-85 [wkly] – 1 – (cont: washington county news-times. merged with: cornelius times to form: news-times (forest grove, or: 1985-)) – us Oregon Lib [071]
Forest grove news-times *see* News-times (forest grove, or)
Forest grove press – Forest Grove OR: Press Pub Co, -1914 [wkly] – 1 – (began in 1909. absorbed by: washington county news-times (1911-81)) – us Oregon Lib [071]
Forest grove times – Forest Grove OR: Forest Grove Print Co, [wkly] – 1 – (merged with: washington county hatchet, to form: washington county hatchet and forest grove times (1896-97)) – us Oregon Lib [071]
Forest grove times – Forest Grove, Washington County, OR: Forest Grove Printing Co. v3 n1-v6 n47. feb 13 1891-dec 27 1894 – 1 – (merged with: washington county hatchet, to form: washington county hatchet and forest grove times. ceased in 1896) – us Oregon Hist [071]
Forest grove times *see*
– Washington county hatchet
– Washington county hatchet and forest grove times
Forest grove times (forest grove, or) – Forest Grove OR: J B Eddy, [wkly] – 1 – (ceased in 1909. related to: washington county hatchet and forest grove times. absorbed by: washington county news (forest grove, or)) – us Oregon Lib [071]
Forest heights baptist church. athens, georgia : church records – 1984-87 – 1 – 5.00 – us Southern Baptist [242]
Forest hill and sydenham examiner – London. 9 aug 1895-1960 [wkly] – 42r – 1 – (incorp with: south london advertiser fr mar 1933. aka: forest hill sydenham & penge examiner) – uk British Libr Newspaper [072]
Forest hill baptist church. forest hill, louisiana : church records – 1899-1985 – 1 – 60.03 – us Southern Baptist [242]
Forest hill sydenham and penge examiner *see* Forest hill and sydenham examiner
Forest hills journal series / Hamilton Co. Cincinnati – jan 1971-jun 1984, jun 1986-mar 1991 – 27r – 1 – mf#B35655-35681 – us Ohio Hist [071]
Forest history – Santa Cruz. 1957-1974 (1) 1957-1974 (5) (9) – (cont by: journal of forest history) – ISSN: 0015-7422 – mf#5895 – us UMI ProQuest [634]
Forest history cruiser / Forest History Society – v1 n3-4 [1978 mar-jul] – 1r – 1 – (cont: forest history newsletter; cont by: cruiser, forest history) – mf#1160974 – us WHS [978]
Forest history cruiser : a quarterly newsletter of the forest history society / Forest History Society – v4 n1-v8 n1 [1981 mar-1985 spring] – 1r – 1 – (cont: cruiser [santa cruz, ca]; cont by: cruiser [durham, nc]) – mf#1168591 – us WHS [978]
Forest history newsletter / Forest History Society – v1 n1-2 [1977 summer-fall] – 1r – 1 – (cont by: forest history cruiser) – mf#1161211 – us WHS [978]
Forest History Society *see*
– Cruiser
– Forest history cruiser
– Forest history newsletter
Forest improvements by the ccc – Washington, DC: US GPO, 1939 – us CRL [634]
Forest industries – New York. 1950-1992 (1) 1971-1992 (5) 1977-1992 (9) – (cont by: wood technology) – ISSN: 0015-7430 – mf#234 – us UMI ProQuest [634]
Forest industries *see* Western timber industry
Forest industries review *see* Western timber industry
Forest, John H de *see* Sunrise in the sunrise kingdom
Forest leaves – Ayr, Crandon, North Crandon WI. 1885 apr 8-1888 jun 28, 1886 aug 5, sep 16, 1888 jul 5-1890 feb 27, 1890 mar 6-1898 aug 25 – 4r – 1 – mf#957200 – us WHS [071]
Forest life and forest trees : comprising winter camp-life among the loggers, and wild-wood adventure / Springer, John S – New York: Harper, 1851 [mf ed 1982] – 3mf – 9 – mf#35204 – cn Canadiana [634]
Forest life and forest trees : comprising winter camp-life among the loggers, and wild-wood adventure / Springer, John S – New York: Harper, 1856 [mf ed 1984] – 3mf – 9 – 0-665-47610-8 – mf#47610 – cn Canadiana [634]
Forest life in canada west / Moodie, Susanna – [s.l: s.n, 1852?] [mf ed 1984] – 1mf – 9 – 0-665-44671-3 – mf#44671 – cn Canadiana [920]
Forest management grapevine – 1973 dec.-1974 nov – 1r – 1 – (cont by: forestry grapevine) – mf#3578066 – us WHS [071]
Forest measurement / Belyea, Harold Cahill – New York, NY. 1931 – 1r – us UF Libraries [634]
The forest of bourg-marie / Harrison, Susie Frances – Toronto: G N Morang, 1898 – 4mf – 9 – mf#05388 – cn Canadiana [830]
Forest of dean examiner – Cinderford and Blakeney. England. -w. 2 Aug 1873-5 Oct 1877. (3 reels) – 1 – uk British Libr Newspaper [072]
Forest of hermanstadt : a melodrama, in two acts / Jouve, Joseph – London: Clementi, Banger, Hyde, Collard & Davis [180-?] [mf ed 1989] – 1r – 1 – (trans fr french by thomas dibdin) – mf#pres. film 50 – us Sibley [780]
Forest park baptist church. gainesville, florida : church records – 1947-74. Formerly Immanuel Baptist Church, 1947-54. Disbanded in 1974. 1126p – 1 – 50.67 – us Southern Baptist [242]
Forest policy and economics – Amsterdam, 2000+ [1,5,9] – ISSN: 1389-9341 – mf#42835 – us UMI ProQuest [634]
Forest press – Tionesta, PA., 1962-1973 – 13 – $25.00r – us IMR [071]
Forest products journal – Madison. 1951+ (1) 1947+ (5) 1947+ (9) – ISSN: 0015-7473 – mf#2237 – us UMI ProQuest [634]
Forest republican – Crandon WI. 1928 apr 26/1930-2002 jul/dec [gaps] – 61r – 1 – (cont: forest republican, wabeno-soperton advertiser; northern wisconsin news and forest county tribune) – mf#1005579 – us WHS [071]
Forest republican – Crandon WI. 1886 oct 14-1890, 1891-92, 1893-96, 1897-1900, 1901-04, 1905-08, 1909-1913 mar 14, 1915-18, 1921 dec 28, 1922-23 – 9r – 1 – mf#1005584 – us WHS [071]
Forest republican – Tionesta, PA. -w 1889-1894; 1895-1921; 1924-1932 – 13 – $25.00r – us IMR [071]
Forest republican, wabeno-soperton advertiser – Crandon, Soperton, Wabeno WI. 1925-26, 1927-1928 apr 19 – 2r – 1 – (cont: wabeno-soperton advertiser; cont by: forest republican) – mf#1005582 – us WHS [071]
Forest resources of northeastern florida / Ineson, Frank A – Washington, DC. 1938 – 1r – us UF Libraries [634]
Forest scenes and incidents, in the wilds of north america : being a diary of a winter's route from halifax to the canadas, and during four months' residence in the woods on the borders of lakes huron and simcoe / Head, George – 2nd ed. London: J Murray, 1838 [mf ed 1984] – 5mf – 9 – 0-665-08238-X – mf#08238 – cn Canadiana [917]
Forest scenes and incidents, in the wilds of north america : being a diary of a winter's route from halifax to the canadas, and during four months' residence in the woods on the borders of lakes huron and simcoe / Head, George – London 1829 [mf ed Hildesheim 1995-98] – 1v on 3mf – 9 – €90.00 – 3-487-27101-X – gw Olms [880]
Forest scenes and incidents, in the wilds of north america : being a diary of a winter's route from halifax to the canadas, and during four months' residence in the woods on the borders of lakes huron and simcoe / Head, George – London: J Murray, 1829 – 4mf – 9 – mf#35433 – cn Canadiana [917]
Forest science – Bethesda. 1955+ (1) 1972+ (5) 1977+ (9) – ISSN: 0015-749X – mf#6813 – us UMI ProQuest [634]
Foresta, Marie J de *see* Lettres sur la sicile
Forester – Manitowish, Mercer WI. 1931 may 2-aug – 1r – 1 – (cont: mercer-manitowish forester) – mf#1097263 – us WHS [634]
Forester, Alvirda *see*
– Leon county history
– Liberty county
Forestry : a magazine for the country – London: W Rider & Son 1883-86 (mthly) [v7-11 (may 1883-apr 1886)] [mf ed 2005] – 5v on 2r [ill] – 1 – (cont: journal of forestry and estates management; subtitle varies: a journal of forest and estate management; imprint varies: edinburgh: c & r anderson 1885-86) – mf#film mas 99999 – us Harvard [634]
Forestry – Oxford. 1958-1996 (1) 1971-1996 (5) 1974-1996 (9) – ISSN: 0015-752X – mf#1262 – us UMI ProQuest [634]
Forestry grapevine – 1974 dec.-1989 fall – 1r – 1 – (cont: forest management grapevine; cont by: timberline [tomahawk, wi]) – mf#3578077 – us WHS [634]
The forests and the people – [Ottawa?: s.n, 1908?] [mf ed 1994] – 1mf – 9 – 0-665-72922-7 – mf#72922 – cn Canadiana [634]
The forests of canada / Bell, Robert – Montreal: Gazette Print Co, 1886 – 1mf – 9 – mf#00102 – cn Canadiana [634]
Forests protected by the ccc – Washington, DC: US GPO, 1938 – us CRL [634]
La foret du haut-niger / Cousturier, Lucie – Bruges: Les Cahiers d'Aujourd'hui, 1923 – 1 – us CRL [916]
La foret et le cultivateur : conference donnee devant la societe pomologique de la province de quebec au college mcdonald, le 10 decembre 1909 / Chapais, Jean Charles – Quebec: [s.n.], 1910 – 1mf – 9 – 0-665-73100-0 – mf#73100 – cn Canadiana [634]
Forever His Ministries International [Jacksonville FL] *see* Bright side
Forever india / Venkatachalam, Govindraj – Bombay: Nalanda Publications, c1948 – (= ser Samp: indian books) – (int by svetoslav roerich) – us CRL [954]
Forewarned against fascism – London: Anti-fascist Democratic Action. v2-9. 1978-81 – 1 – us Wisconsin U Libr [325]
Forewarnings of bank failure / Dolbeare, Harwood B – Gainesville, FL. 1931 – 1r – us UF Libraries [332]
Forfar dispatch – Forfar: O McPherson 1884-(wkly) [mf ed 2 jan 1992-] – 1 – (not publ: 9 jul-13 aug 1981; imprint varies) – uk Scotland NatLib [072]
Forfar herald – (Angus Herald). Forfar. Scotland. -w. 4 Apr 1884-29 Sep 1933. (50 reels) – 1 – uk British Libr Newspaper [072]
Forfar reformer – Forfar. Scotland. -w. 24 Feb 1883-5 Jan 1885. (27 ft) – 1 – uk British Libr Newspaper [072]
Forfar review – Forfar. Scotland. -w. 5 Jan 1912-16 Apr 1926. (5 reels) – 1 – uk British Libr Newspaper [072]
Forgas armadas e o destino historico do brasil / Moura, Almerio Lourival De – Sao Paulo, Brazil. 1937 – 1r – us UF Libraries [972]
Forge : the bigelow society quarterly / Bigelow Society – 1972 jan-1979 oct – 1r – 1 – (cont: bigelow society quarterly) – mf#500520 – us WHS [366]
Forge / Canadian Communist League [Marxist-Leninist] et al – 1976 jun-1978 dec, 1979-80, 1981-1982 n5, 1983 apr – 3r – 1 – mf#627985 – us WHS [335]
The forge in the forest : being the narrative of the acadian ranger, jean de mer, seigneur de briart: and how he crossed the black abbe: and of his adventures in a strange fellowship / Roberts, Charles George Douglas – Boston, New York: Lamson, Wolffe; Toronto: W Briggs, 1896 – 4mf – 9 – 0-665-12400-7 – mf#12400 – cn Canadiana [910]
The forge in the forest : being the narrative of the acadian ranger, jean de mer, seigneur de briart; and how he crossed the black abbe, and of his adventures in a strange fellowship / Roberts, Charles George Douglas – New York: Grosset & Dunlap, 1896 – 4mf – 9 – mf#50971 – cn Canadiana [910]
The forge in the forest : being the narrative of the acadian ranger, jean de mer, seigneur de briart: and how he crossed the black abbe: and of his adventures in a strange fellowship / Roberts, Charles George Douglas – Toronto: W Briggs; Montreal: C W Coates, 1897? – 4mf – 9 – mf#32480 – cn Canadiana [910]
Forgeron, Jean Baptiste *see* Le protectorat en afrique occidentale francaise et les chefs indigenes
Forget, J *see* De vita et scriptis aphraatis, sapientis persae
Forget, Jacques *see* De vita et scriptis aphraatis, sapientis persae
Forget, Jean-Urgel *see* Histoire de saint-jacques d'embrun, russell, ontario
Forget-me-not – 1823-47 – (= ser English gift books and literary annuals, 1823-1857) – 112mf – 9 – uk Chadwyck [800]
Forging his chains : the autobiography of george bidwell... / Bidwell, George – Hartford: S S Scranton & Co, 1888 – 7mf – 9 – $10.50 – mf#LLMC 92-172 – us LLMC [340]
Forgiveness – London, England. 18– – 1r – us UF Libraries [240]
Forgiveness and law : grounded in principles, interpreted by human analogies / Bushnell, Horace – New York:Scribner, Armstrong, 1874 [mf ed 1985] – 1mf – 9 – 0-8370-3066-8 – mf#1985-1066 – us ATLA [240]
Forgiveness of sins : what is it? – Toronto: Gospel Tract Depository, [187-?] [mf ed 1995] – 1mf – 9 – 0-665-94804-2 – mf#94804 – cn Canadiana [240]
The forgiveness of sins : and other sermons / Smith, George Adam – New York: A C Armstrong, 1904 [mf ed 1986] – 1mf – 9 – 0-8370-9745-2 – mf#1986-3745 – us ATLA [242]
The forgiveness of sins : a study in the apostles' creed / Swete, Henry Barclay – London: Macmillan, 1916 – 2mf – 9 – 0-7905-9699-7 – mf#1989-1424 – us ATLA [240]
Forgotten fantasy – Hollywood. 1970-1971 (1) – ISSN: 0015-7643 – mf#6959 – us UMI ProQuest [400]
A forgotten friend of india : sir charles forbes 1st bart / Wadia, Ruttonjee Ardeshir – Baroda: Padmaja Publ, 1946 – (= ser Samp: indian books) – us CRL [920]
Forgotten people – Canada. jan 1972-dec 1974 – 1r – 1 – cn Commonwealth Imaging [971]
Forgotten people / Native Council of Canada – [v1]: 1st ed [i.e. n1]-v8 n2 [1972 summer-1981] – 1r – 1 – mf#351550 – us WHS [305]
The forgotten war : an appeal from the republic of the south moluccas / Nikijuluw, J P; ed by Dept of Public Information of the RMS – Rotterdam, 1950 – 1mf – 8 – mf#SE-1606 – ne IDC [959]
Forgues, Emile *see* La chine ouverte
[Forgues, P E D] *see* La chine ouverte

Forindien : et udvalg af op opbyggelige missions-fortaellinger / Flood, Johan – Laurvig: J Preutz, 1868 [mf ed 1995] – (= ser Yale coll) – 88p – 1 – 0-524-10269-4 – (in norwegian) – mf#1996-1269 – us ATLA [240]

Forjando patria (por nacionalismo) / Gamio, Manuel – Mexico City? Mexico. 1916 – 1r – us UF Libraries [972]

Forjando vidas / Zapata Castaneda, Adrian – Guatemala, 1949 – 1r – us UF Libraries [972]

Forjett, Charles *see* Our real danger in india

Fork shoals baptist church – Greenville Co. SC. 2046p. 1813-1919, 1929-49, 1951-90 – 1 – $92.07 – (historical highlights 1777-1983. wmu minutes 1978-86. formerly horse creek 1789-99) – mf#0900-8 – us Southern Baptist [242]

Fork shoals church. fork shoals, south carolina : church records – 1857-99 – 1 – us Southern Baptist [242]

Forkel, Johann Nikolaus *see*
– Allgemeine geschichte der musik
– Genauere bestimmung einiger musikalischer begriffe
– Ueber die theorie der musik

Forklaring over fadervor / Rosenius, Carl Olof – Madison: Lisbons norsk-lutherske forening til udgivelse af christelige underviisnings og andagtsboeger for the norske folk i Amerika, 1860 – 1mf – 9 – 0-524-05195-X – mf#1991-2231 – us ATLA [240]

Forks of elkhorn church / Darnell, Ermina Jett – 340p – 1 – us Southern Baptist [242]

Forks of otter creek baptist church. hardin county. kentucky : church records – July 1827-1904 – 1 – $14.04 – us Southern Baptist [242]

Forlong, James George Roche *see*
– Faiths of man
– Rivers of life
– Short studies in the science of comparative religions
– Short studies in the science of comparative religions, embracing all the religions of asia

Forlorn hope – New York NY. 1800 mar 24 – 1r – 1 – mf#305812 – us WHS [071]

Form – Cambridge. 1966-1967 (1) – ISSN: 0532-1697 – mf#3038 – us UMI ProQuest [600]

Form – London, 1921-22 [mf ed Chadwyck-Healey] – (= ser Art periodicals on microform) – 1r – 1 – uk Chadwyck [760]

Form and function of the pauline thanksgivings / Schubert, Paul – Berlin, Germany. 1939 – 1r – us UF Libraries [975]

The form and origin of milton's antitrinitarian conception / Wood, Louis Aubrey – London, Ont: Advertiser Print Co, 1911 [mf ed 1993] – (= ser Unitarian/universalist coll) – 1mf – 9 – 0-524-07780-0 – (incl bibl ref) – mf#1991-3348 – us ATLA [242]

Form, content and technique of traditional literature / Guma, Samson Mbizo – Pretoria, South Africa. 1967 – 1r – us UF Libraries [470]

Die form der hebraeischen poesie / Meier, Ernst Heinrich – Tuebingen: Osiander, 1853 – 1mf – 9 – 0-8370-9296-5 – mf#1986-3296 – us ATLA [470]

Form der sacramenten bruch wie sy zuo basel gebrucht werden, mit sampt eynem kurtzen kinder bericht / Oecolampadius, J – Basel, Lux Schouber, 1537 – 1mf – 9 – mf#PBU-395 – ne IDC [240]

A form for receiving such as have been in schism into the communion of the church of england : and for reconciling those who have lapsed – London: SPCK 1898 [mf ed 1992] – (= ser Church historical society (series) 49) – 1mf – 9 – 0-524-05492-4 – mf#1990-1487 – us ATLA [242]

The form of baptism : an argument designed to prove conclusively that immersion is the only baptism authorized by the bible / Briney, John Benton – St Louis: Christian Pub Co, 1892 – 1mf – 9 – 0-524-02109-0 – mf#1990-4175 – us ATLA [242]

A form of prayer and thanksgiving to almighty god : for the safe delivery of the queen, and the happy birth of a princess / United Church of England and Ireland. Diocese of Toronto – [Toronto: s.n, 1840?] [mf ed 1984] – 1mf – 9 – 0-665-45708-1 – mf#45708 – cn Canadiana [242]

A form of prayer and thanksgiving to be used on monday, the 26th of february, 1838 : being the day appointed by proclamation, for a general thanksgiving to almighty god... / United Church of England and Ireland – [Quebec?: s.n.] 1838 [mf ed 1994] – 1mf – 9 – 0-665-94567-1 – mf#94567 – cn Canadiana [242]

Form of prayer to be used at the religious service in connection with the free christian union – [London?: s.n., 1869?] (London: Woodfall and Kinder) – 1mf – 9 – 0-524-00261-4 – mf#1989-2961 – us ATLA [240]

A form of prayer to be used on friday, the fourth of may 1832 : being the day appointed by proclamation for a general fast and humiliation before almighty god... / United Church of England and Ireland – [Quebec?: s.n, 1832?] [mf ed 1994] – 1mf – 9 – 0-665-94566-3 – mf#94566 – cn Canadiana [242]

Form of the brahms double concerto / Beach, Thelma – U of Rochester 1941 – 1r – 1 – mf#film 884 – us Sibley [780]

The form of the christian temple : being a treatise on the constitution of the new testament church / Witherow, Thomas – Edinburgh: T & T Clark, 1889 – 2mf – 9 – 0-7905-0466-9 – (incl indes) – mf#1987-0466 – us ATLA [240]

Form und geschichte : studie zu einigen methodisch zentralen elementen der kritik der politischen oekonomie von karl marx / Mueller, Ulrich – Heidelberg, 1977 – 3mf – 9 – 3-89349-387-5 – gw Frankfurter [320]

Form und innerlichkeit : beitraege zur geschichte und wirkung der deutschen klassik und romantik / Kohlschmidt, Werner – Bern: Francke, c1955 [mf ed 1993] – (= ser Sammlung dalp v81) – 268p – 1 – (incl bibl ref and ind) – mf#8236 – us Wisconsin U Libr [430]

Form und ueberlieferung der lukas-homilien des origenes / Rauer, M – Leipzig, 1932 – (= ser Tugal 4-47/3) – 2mf – 9 – €5.00 – ne Slangenburg [220]

Form und weise einer visitation : fur die graff und herschafft mansfelt / Sarcerius, E – [Eisleben], 1554 – 1mf – 9 – mf#TH-1 mf 1331 – ne IDC [242]

Form vnd gstalt wie der kinder tauff...dess herren nachtmal...zuo basel...gehalten werden / Oecolampadius, J – [Basel,] 1526 – 1mf – 9 – mf#PBU-373 – ne IDC [240]

Form, Wolfgang *see*
– Widerstand und verfolgung in hessen 1933-1945
– Widerstand und verfolgung in oesterreich 1938 bis 1945

Forma ac ratio tota ecclesiastici ministerij... / Lasco, J – [Francofurti a.M., 1555] – 8mf – 9 – mf#PBA-221 – ne IDC [240]

Forma e expressao no romance brasileiro / Bezerra De Freitas, Jose – Rio de Janeiro, Brazil. 1947 – 1r – us UF Libraries [972]

Forma electionis prioris in ordine praedicatorum / Dominicans. Province of the Holy Name – Sancti Francisci: PJ Thomas, 1887 – 1mf – 9 – 0-524-07196-9 – mf#1990-5354 – us ATLA [240]

Forma y modo de fundar las cofradias del cordon de nro p s fra[n]cisco : y admitir los cofrades della, co[n] el su[m]mario d[e]as gras e i[n]dulge[n]cias q[ue] gana[n] co[n]cedidas por nro s p sixto v – en Mexico: En casa d[e] Pedro Ocharte 1589 – (= ser Books on religion...1543/44-c1800: franciscanos) – 2mf – 9 – mf#crl-199 – ne IDC [241]

Formacao da sociedade brasileira / Sodre, Nelson Werneck – Rio de Janeiro, Brazil. 1944 – 1r – us UF Libraries [306]

Formacao do brasil contemporaneo / Prado Junior, Caio – Sao Paulo, Brazil. 1945 – 1r – us UF Libraries [972]

Formacao do pcb, 1922/1928 / Pereira, Astrojildo – Rio de Janeiro, Brazil. 1962 – 1r – us UF Libraries [972]

Formacao e selecao dos funcionarios locais / Mello, Manoel Caetano Bandeira De – s.l, s.l? 1959 – 1r – us UF Libraries [972]

Formacao historica da nacionalidade brasileira / Oliveira Lima, Manuel De – Rio de Janeiro, Brazil. 1944 – 1r – us UF Libraries [972]

Formacao historica de sao paulo (de comunidade a m... / Morse, Richard M – Sao Paulo, Brazil. 1970 – 1r – us UF Libraries [972]

Formacao historica do brasil / Calogeras, Joao Pandia – Sao Paulo, Brazil. 1938 – 1r – us UF Libraries [972]

Formacao territorial do brasil; origem e evolucao / Dias, Demosthenes de Oliveira – Rio de Janeiro: No Loja Carlos Ribeiro, 1956. 137p. illus. With: Para e Amazonas by J. Mattos; Los Capitales Yanquis en la Argentina by L.V. Sommi; Passos dos Lusiadas Estuados by G. Vasconcel los Abreu.Incl. bibliog. 1 reel. 1273 – 1 – us Wisconsin U Libr [972]

Formacion de la nacionalidadcchilena. santiago de chile, 1943 / Amunategui Solar, Domingo – Madrid: Razon y Fe, 1944 – 1 – sp Bibl Santa Ana [972]

Formacion de la sociedad cubana / Gay-Calbo, Enrique – Habana, Cuba. 1948 – 1r – us UF Libraries [972]

Formacion de las falanges juveniles de franco / Montes, Jose – Caceres: Tip. El Noticiero – sp Bibl Santa Ana [946]

Formacion del pueblo verezolano / Siso, Carlos – New York, NY. 1941 – 1r – us UF Libraries [972]

La formacion profesional del jurista / Cuellar Grajera, Antonio – Badajoz: Imp. Dip. Provincial, 1965. Sep. REE – sp Bibl Santa Ana [340]

Formacion religiosa de jovenes. 2nd ed. trad. de antonio sancho nebot / Toth, Tihamez – Madrid: Razon y Fe, 2v. 1944 – 1 – sp Bibl Santa Ana [240]

Formaciones vegetales de colombia / Instituto Geografico 'Agustin Codazzi' Departamen – Bogota, Colombia. 1963 – 1r – us UF Libraries [972]

The formal garden in england / Blomfield, Reginald Theodore – London 1892 – (= ser 19th c art & architecture) – 3mf – 9 – mf#4.2.1794 – uk Chadwyck [710]

Formal opening of franklin and marshall college in the city of lancaster, june 7, 1853 : together with addresses / Hayes, Alexander L et al – Lancaster, Pa: Pub by order of the Board of Trustees, 1853 – 1mf – 9 – 0-524-08717-2 – mf#1993-1087 – us ATLA [378]

Formal opinions – 1979 jul 23-aug 6 – 1r – 1 – (cont by: summary of requests for formal opinions) – mf#2654538 – us WHS [071]

Formal opinions (a-g) / Pennsylvania – 1889-1950; 1931-42. 7 reels – 1 – $35.00r – us Trans-Media [340]

Formalites a remplir pour obtenir un decret des commissaires nommes pour l'erection des paroisses et la construction et reparation des eglises etc – [s.l: s.n, 186-?] [mf ed 1985] – 1mf – 9 – 0-665-32396-4 – mf#32396 – cn Canadiana [240]

Das formalprinzip des protestantismus : neue prolegomena zu einer evangelischen dogmatik / Resch, Alfred – Berlin: F Berggold, 1876 – 1mf – 9 – 0-8370-8782-1 – mf#1986-2782 – us ATLA [242]

Forman, Henry *see* An account of the work of the north india mission of the presbyterian church of america for the year 1906-1907

Forman, Henry James *see* Pony express

Formation and growth of society out of christian marriage and its c... / Belaney, Robert – London, England. 1881 – 1r – us UF Libraries [306]

Formation ethnique : folk-lore et culture du peuple / Price-Mars, Jean – Port-Au-Prince, Haiti. 1956 – 1r – us UF Libraries [972]

The formation of a ministry to the parents of infants. 1981 / Forder, Winter Rand – 1 – $5.28 – us Southern Baptist [242]

Formation of character : twelve lectures / Palmer, Benjamin Morgan – New Orleans: ES Upton, c1889 – 1mf – 9 – 0-7905-9556-7 – mf#1989-1281 – us ATLA [170]

The formation of christian character : a contribution to christian ethics / Bruce, William Straton – Edinburgh: T & T Clark, 1908 – 1mf – 9 – 0-8370-6090-7 – (incl bibl ref & index) – mf#1986-0090 – us ATLA [240]

Formation of konkani / Katre, Sumitra Mangesh – Bombay: Karnatak Pub House, 1942 – (= ser Samp: indian books) – us CRL [490]

The formation of the gospel tradition / Taylor, Vincent – 1935 – 9 – $10.00 – us IRC [240]

Formation of the union, 1750-1829 / Hart, Albert Bushnell – New York, NY. 1910 – 1r – us UF Libraries [025]

Formations – Madison. 1984-1991 (1,5,9) – ISSN: 0741-5702 – mf#13540 – us UMI ProQuest [400]

Formative influences of legal development / Kocourek, Albert & Wigmore, John H – Boston: Little-Brown, 1918 – 8mf – 9 – $12.00 – mf#LLMC 95-171 – us LLMC [340]

Formby, H *see* Plea of conscience for retiring from pastoral duty

Formby, Henry *see*
– The book of the holy rosary
– Monotheism, in the main derived from the hebrew nation and the law of moses, the primitive religion of the city of rome

Formby times and ainsdale and altcar advertiser – [NW England] Formby 1895-99, 1904, 1907, 1910, 1912-15, 1930, 1940-42, 1952-66, 1975-92, 1993 – 1 – us MLA; uk Newsplan [072]

La forme des prieres et chants ecclesiastiques / Calvin, J – Geneve, 1542 – 3mf – 8 – €7.00 – (facs of original ed located at bibliothek stuttgart, kassel-bale, 1959) – ne Slangenburg [240]

Forme d'oraison pour demander...dieu la saincte predication de l'evangile / Farel, Guillaume – Geneve, Girard, 1545 – 2mf – 9 – mf#PFA-156 – ne IDC [240]

Forme generale et particuliere de la convocation et de la tenue des assemblees nationales ou etats generaux de france, justifiee par pieces authentiques / Lalource & Duval – Recueilli par Lalource et Duval. Paris (1-3). 1789 – 1 – fr ACRPP [324]

Formen der regionalen zusammenarbeit am schwarzen meer und in zentralasien vor dem hintergrund europaeischer integrationserfahrungen : geostrategische implikationen fuer die aussenpolitik der tuerkei / Waltmann, Frank – (mf ed 1995) – 3mf – 9 – €49.00 – 3-8267-2276-0 – mf#DHS 2276 – gw Frankfurter [327]

Formen und motive in den apokryphen apostelgeschichten / Blumenthal, M – Leipzig, 1933 – (= ser Tugal 4-48/1) – 3mf – 9 – €7.00 – ne Slangenburg [220]

Die formenlehre bei john lyly / Kneile, Karl – Tuebingen, 1914 (mf ed 1994) – 2mf – 9 – €31.00 – 3-8267-3038-0 – mf#DHS-AR 3038 – gw Frankfurter [420]

Former and the latter rain / Dow, William – Edinburgh, Scotland. 1866 – 1r – us UF Libraries [240]

Former days / Woodford, James Russell – Cambridge, England. 1885 – 1r – us UF Libraries [240]

Formes surcomposees en francais / Cornu, Maurice – Bern, Switzerland. 1953 – 1r – us UF Libraries [440]

A formiga : pamphleto humoristico, litterario e scientifico – Rio de Janeiro, RJ: Typ Camoes, out 1883 – (= ser Ps 19) – mf#P17,01,135 – bl Biblioteca [079]

Formirovanie burzhuazii v politicheskuiu silu v sibiri / Mosina, I G – Tomsk, 1978 – 2mf – 9 – mf#RPP-32 – ne IDC [325]

Formirovanie finansovogo kapitala v rossii, konets 19 v – 1908 g / Bovykin, V I – M, 1984 – 6mf – 9 – mf#REF-161 – ne IDC [332]

Formosa : internal affairs and foreign affairs, 1945-1954 / U.S. State Dept – (= ser Confidential u s state department central files) – 1 – $3,675.00 coll – (internal affairs, 1945-49 3r isbn 0-89093-733-8 $570. internal affairs & foreign affairs, 1950-54 17r isbn 0-89093-773-7 $3290. with p/g) – us UPA [951]

Formosa under the dutch : described from contemporary records, with explanatory notes and a bibliography of the island / Campbell, William – London: Kegan Paul, 1903 [mf ed 1995] – (= ser Yale coll) – xiv/629p (ill) – 1 – 0-524-09673-2 – mf#1995-0673 – us ATLA [951]

Forms and ritual of the purple order : to be observed in private lodges of the orange association of british north america / Loyal Orange Association of British America – [Toronto?: s.n.] 1848 [mf ed 1984] – 1mf – 9 – 0-665-46141-0 – mf#46141 – cn Canadiana [366]

Forms and systems professional – Philadelphia. 1988-1988 (1,5,9) – ISSN: 0899-7004 – mf#16913 – us UMI ProQuest [680]

Forms for missouri pleading / Pattison, Everett Wilson – St. Louis, Mo.: Gilbert, 1891. 383p. LL-1310 – 1 – us L of C Photodup [340]

Forms for virginia and west virginia annotated, including statutory, common law and equity, commercial, corporation, and criminal forms / Gregory, George Craghead – 2d ed. Charlottesville, VA: Michie, 1925. 1390p. LL-697 – 1 – us L of C Photodup [345]

Forms in civil actions and proceedings in the courts of record of wisconsin / Bryant, Edwin Eustace – 2d ed. Madison, Democrat, 1892. 408 p. LL-819 – 1 – us L of C Photodup [347]

Forms in conveyancing, comprising precedents for ordinary use, and clauses adapted to special and unusual cases / Jones, Leonard Augustus – Boston, Houghton, Mifflin, 1886. 826 p. LL-89 – 1 – us L of C Photodup [340]

Forms of acknowledgments for deeds and other instruments used in the states and territories of the united states. / Reardon, George Evett – Baltimore, Cox, 1882. 140 p. LL-1328 – 1 – us L of C Photodup [340]

Forms of civil procedure adapted to practice and pleading under the code of civil procedure of the state of new york / Lansing, William – Albany, Banks, 1885-88. 3v LL-501 – 1 – us L of C Photodup [347]

Forms of code pleading for nebraska, kansas and oklahoma, fully annotated / Campbell, William S – St. Louis: Thomas, 1912-27. 3v. LL-1278 – 1 – us L of C Photodup [348]

The forms of hebrew poetry : considered with special reference to the criticism and interpretation of the old testament / Gray, George Buchanan – London; New York: Hodder and Stoughton, 1915 – 1mf – 9 – 0-7905-3198-4 – mf#1987-3198 – us ATLA [470]

The forms of hebrew poetry / Gray, George Buchanan – 1915 – 9 – $12.00 – us IRC [470]

Forms of oaths for use in the u.s. district courts / Crawford, Kenneth C – Washington: FJC, 1976 – 1mf – 9 – $1.50 – mf#LLMC 95-387 – us LLMC [347]

Forms of pleadings in civil and criminal cases. / Matthews, William Baynham – Richmond, Va.: Randolph & English, 1873. 395p. LL-700 – 1 – us L of C Photodup [345]

Forms of prayer for public and private use in time of war – London: SPCK, 1918 – 1mf – 9 – 0-524-06697-3 – mf#1990-5268 – us ATLA [240]

Forms of procedure in the courts of admiralty of the united states of america / Pugh, Edward Fox – 2nd, rev. ed. Philadelphia: Johnson, 1903. 376p. LL-1417 – 1 – us L of C Photodup [347]
Forms of religion : as seen in the light of the methods of christ and of the spirit, by following the divine order of development / Coutts, John – London: G Lyal, 1909 – 2mf – 9 – 0-524-04327-2 – mf#1990-3311 – us ATLA [240]
Formula da civilisacao brasileira / Falcao, Annibal – Rio de Janeiro, Brazil. 1934 – 1r – us UF Libraries [972]
The formula of concord : its origin and contents / Fritschel, George John – Philadelphia: Lutheran Pub Soc, 1916 – 1mf – 9 – 0-524-05191-7 – (incl bibl ref) – mf#1991-2227 – us ATLA [240]
Formulae medicae / Llorens y Masdevall, F – Madrid, 1789 – 1mf – 9 – sp Cultura [610]
Formulaire de la priere en famille / Plante, Omer [comp] – ed Sans Renvois, 32e Mille. Quebec: Editions de l'Action Sociale Catholique, 1924 [mf ed 1990] – 2mf – 9 – mf#SEM105P1214 – cn Bibl Nat [241]
Formulaire de prieres, a l'usage des pensionnaires des religieuses ursulines – nouv corr augm ed. Quebec: Impr a la Nouvelle Imprimerie, 1811 [mf ed 1991] – 6mf – 9 – 0-665-90294-8 – (in french and english) – mf#90294 – cn Canadiana [241]
Formularies of faith put forth by authority during the reign of henry 8 : viz articles about religion (1536), the institution of a christian man (1537), a necessary doctrine and erudition for any christian man (1543) / ed by Lloyd, Charles – Oxford: Clarendon Press 1825 [mf ed 1993] – 1mf – 9 – 0-524-05708-7 – mf#1991-2322 – us ATLA [242]
Formularios de procedimiento civil / Rengel-Romberg, Aristides – Caracas, Venezuela. 1962 – 1r – us UF Libraries [350]
Formularios para los fiscales municipales y comarcales / Agundez, Antonio – Caceres: Tip. Extremadura, s.a. – sp Bibl Santa Ana [330]
Formulary – Cleveland. 1995-1996 (1,5,9) – (cont: hospital formulary) – ISSN: 1082-801X – mf#2395,02 – us UMI ProQuest [360]
A formulary of the papal penitentiary in the thirteenth century / ed by Lea, Henry Charles – Philadelphia: Lea Brothers, 1892 [mf ed 1990] – 1mf – 9 – 0-7905-6222-7 – (text in latin, int in english) – mf#1988-2222 – us ATLA [241]
The formulation and use of a staff policy manual within the greene county baptist association / Joslin, James Elliott – 1981 – 1 – 5.04 – us Southern Baptist [242]
La formule bouddhique des douze causes : son sens originel et son interpretation theologique / Oltramare, Paul – Geneve: Georg, 1909 – 1mf – 9 – 0-524-03308-0 – (incl bibl ref) – mf#1990-3193 – us ATLA [280]
Formy ekonomicheskoi samopomoshchi v oblasti remeslennogo truda / Zak, L S – 1912 – 155p 2mf – 9 – mf#COR-29 – ne IDC [335]
Formy i metody kollektivnogo instruktirovaniia v kooperatsii : metod rukovodstvo po org i provedeniiu kollektiv instruktirovaniia / Toranskii, M G – Kharkov, 1929 – 108p 2mf – 9 – mf#COR-276 – ne IDC [335]
Fornaris, Jose *see* Poesias
Fornaro, Carlo de *see* What the catholic church has done to mexico
Fornells deaths – Minorca, Spain. v1. 1783-1888 – 1r – us UF Libraries [324]
Fornells marriages – Minorca, Spain. v1. 1783-1889 – 1r – us UF Libraries [324]
Forner Segarra, Juan Pablo *see*
– Amor de la patria
– El asno erudito
– Carta de bartolo sobrino de don fernando perez
– Conversaciones familiares entre el censor
– Cotejo de las eglogas que ha premiado la real academia de la lengua
– Defensa de don fernando perez...paracuellos
– Discurso antisofistico extractado dcl hombre
– Discursos filosoficos sobre el hombre
– La escuela de la amistad o el filosofo enamorado
– Introduccion...que se recito...sevilla
– Obras
– Oracion apologetica por la espana y su merito literario
– Oracion apologetica por la espana...literario
– Pasatiempo de...respuesta a su oracion apologetica
– La paz. canto heroico
– Reflexiones...escribir la historia de espana
Forner y Segarra, Agustin F *see* Antiguedades de merida
Forneri, Richard Sykes *see* The united empire loyalists of canada
Fornet, Ambrosio *see* En tres y dos
Foro del pueblo – v1 n4-8 [1980 jun/jul-1980 nov], v2 n1-3 [1981 may-sep/oct] – 1r – 1 – mf#668862 – us WHS [071]
Forord : morskabslaesning for enkelte staender efter tid og leilighed / Kierkegaard, Soeren – Kobenhavn: C A Reitzel, 1844 – (= ser Himmelstrup) – 1mf – 9 – 0-7905-3790-7 – mf#1989-0283 – us ATLA [190]
Forort vast – Lerum, Sweden. 1979-80 – 1 – sw Kungliga [078]
Forpost rossii / Primorskaia kraevaia organizatsiia LDPR – Vladivostok, Russia. n2(3) (fev 1997), n5[6](apr 1997)-n6[7] (mai 1997), n10(1997), n13(1997)-n14(1998), n18 (okt 1998) – 1 – mf#mf-12248 (reel 4) – us CRL [077]
Forposten – Marinette WI. 1892 aug-dec, 1894-95, 1896-1900, 1901-05, 1906-1909 jun 25 – 5r – 1 – (cont by: marinette tribunen) – mf#1099105 – us WHS [071]
Forposten – Goteborg, Sweden. 1865-1919 – 20r – 1 – sw Kungliga [078]
Forradalom kiadja a kulfoldi proletarok, nemzetkoszi szocialdemokrata munkas partjanak magyar es roman csoportja – Omsk, Russia, 1918 – 1r – 1 – (in hungarian) – us UMI ProQuest [077]
Forrer, E *see* Die provinzeinteilung des assyrischen reiches
Forres, elgin & nairn gazette – Scotland. -w. 5 Jan 1844-Dec 1883. 11 reels – 1 – uk British Libr Newspaper [072]
Forres, elgin & nairn gazette, northern review and advertiser – Forres: J D Miller 1837-1996 (wkly 1855-]; former frequency: mthly 1837-51; biwkly 1851-55 [mf ed 1997] – 1 – (cont by: forres gazette; imprint varies) – ISSN: 1354-9650 – uk Scotland NatLib [072]
Forres gazette – Forres: Moray & Nairn Newspaper Co 1996- (wkly) [mf ed 1995] – 1 – (cont: forres, elgin & nairn gazette, northern review and advertiser) – ISSN: 1354-9650 – uk Scotland NatLib [072]
Forrest, Benjamin J *see* Bridge launching
Forrest, David William *see*
– The authority of christ
– The christ of history and of experience
Forrest, Edmund William *see* Ned fortescue
Forrest, George William *see* The famine in india
Forrest, James *see*
– Some account of the origin and progress of trinitarian theology
– Studies on the book of psalms
Forrest, T *see* A voyage to new guinea, and the moluccas, from balambangan
Forrest, Thomas *see* Voyage aux moluques et a la nouvelle guinee
Forrest, William Mentzel *see* India's hurt
Forrestal, James V *see* Diaries of james v forrestal, 1944-1949
Forrester, Alexander *see* The object, benefits and history of normal schools
Forrester, Alfred Henry (Alfred Crowquill pseud.) *see* The pictorial grammar
Forrester, Charles Robert *see* The pictorial grammar
Forrester, Henry *see* Christian unity and the historic episcopate
Forrester's boys' and girls' magazine, and fireside companion – Boston. 1848-1857 – 1 – mf#4373 – us UMI ProQuest [370]
Fors, Andrew Peter *see* The ethical world-conception of the norse people
Forsaken idea / Crankshaw, Edward – London, England. 1952 – 1r – us UF Libraries [890]
Forsander, Nils *see*
– Grundlinier till foerelaesningar oefver augsburgiska bekaennelsen
– Den ofoeraendrade augsburgiska bekaennelsen med inledning och foerklaring
Forschende frauen und frauenforschung – Dortmund: projekt vlg, 1993 (mf ed 1996) – (= ser Schriftenreihe der uni dortmund) – 2mf – 3-8267-9714-0 – mf#DHS 9714 – gw Frankfurter [376]
Forschreitende rabbinismus! / Schreiber, Emanuel – Konigsberg i.Pr., Russia. 1877 – 1r – us UF Libraries [939]
Forschung in ephesos / Oesterreichisches Archaeologisches Institut (Vienna, Austria) – Wien: A. Hoelder, 1906 – 1r – 1 – 0-8370-1143-4 – mf#1984-B490 – us ATLA [930]
Forschung und technologische entwicklung (fte) in europa : von einer nationalen zu einer europaeischen fte-politik? / ed by Mueller-Boeling, Detlef & Szyperski, Norbert – Dortmund: projekt vlg, 1994 (mf ed 1996) – 1mf – 9 – €24.00 – 3-8267-9712-4 – mf#DHS 9712 – gw Frankfurter [327]
Forschungen nach einer volksbibel zur zeit jesu : und deren zusammenhang mit der septuaginta-uebersetzung / Boehl, Eduard – Wien: Wilhelm Braumueller, 1873 – 1mf – 9 – 0-7905-0253-4 – (incl bibl ref) – mf#1987-0253 – us ATLA [220]
Forschungen ueber die lateinischen aristoteles-uebersetzungen des 13. jahrhunderts / Grabmann, M – 1916 – (= ser Bgphma 17/5-6) – €12.00 – ne Slangenburg [180]
Forschungen und quellen zur geschichte des konstanzer konzils / Finke, Heinrich – Paderborn: Ferdinand Schoeningh, 1889 – 1mf – 9 – 0-8370-6899-1 – (incl bibl ref and name index to the sources) – mf#1986-0899 – us ATLA [240]
Forschungen zur brandenburgischen und preussischen geschichte – v1-55. 1888-1943 – 9 – $780.00 – mf#0207 – us Brook [943]
Forschungen zur geschichte der fruehmittelalterlichen philosophie / Endres, J A – 1915 – (= ser Bgphma 17/2-3) – €7.00 – ne Slangenburg [100]
Forschungen zur geschichte des neutestamentlichen kanons und der altkirchlichen literatur / Zahn, Theodor – Erlangen: A Deichert 1881-1929 [mf ed 1979] – 10v on 2r – 1 – (v4 & 5 publ by a deichert in erlangen & leipzig; v6-10 in leipzig; incl ind) – mf#film mas c480 – us Harvard [225]
Forschungen zur kultur- und litteraturgeschichte bayerns – Muenchen: G Franzscher Verlag. bk1-5(1893-97) (annual) [v1-16] [mf ed 1978] – 5v on 3r [ill] – 1 – (cont by: forschungen zur geschichte bayerns; imprint varies) – mf#film mas c272 – us Harvard [943]
Forschungsbericht 1990-1991 / ed by Presse- und Informationsstelle Dortmund – Dortmund: projekt vlg, 1993 – (= ser Schriftenreihe der uni dortmund) – 6mf – 3-8267-9716-7 – mf#DHS 9716 – gw Frankfurter [378]
Forschungsbericht 1992-1993 / ed by Presse- und Informationsstelle Dortmund – Dortmund: projekt vlg, 1995 (mf ed 1996) – (= ser Schriftenreihe der uni dortmund) – 8mf – 3-8267-9711-6 – mf#DHS 9711 – gw Frankfurter [378]
Forschungsfahrten im suedlichen eismeer 1819-1821 / Bellingshausen, F von – Leipzig, 1902 – 3mf – 9 – mf#H-6184 – ne IDC [919]
Die forschungsreise s m s gazelle in den jahren 1874 bis 1876 unter kommando des kapitaen zur see freiherrn von schleinitz / Gazelle – North York. 1931+ (1) 1972+ (5) 1975+ (9) – 23mf – 9 – mf#8525 – ne IDC [910]
Forschungsreise s m s planet 1906/07 : vol 3: ozeanographie / Brennecke, W – London. 1862-1881 (1) – 9mf – 9 – mf#2826 – ne IDC [910]
Forschungsreise s m s planet 1906/07 : vol 4: biologie / Graef – Berlin, 1909 – 8mf – 9 – mf#2827 – ne IDC [910]
Forschungsreisen in sued-arabien : bis zum auftreten eduard glasers / Weber, Otto – Leipzig: JC Hinrichs, 1907 [mf ed 1989] – (= ser Der alte orient 8/4) – 1mf – 9 – 0-7905-2095-8 – mf#1987-2095 – us ATLA [915]
Forshall, Josiah et al *see* The holy bible
Forsog til en forbedret gronlandsk grammatica / Fabricius, Otho – andet oplag. Kiobenhavn: Schubart 1801 – (= ser Whsb) – 5mf – 9 – €60.00 – mf#Hu 443 – gw Fischer [430]
Forsog til en oversaettelse af forste bog af davids psalmer : i en verseform, hvorved de kunne synges efter bekjendte psalmemelodier, a soren dahl = Bible. o.t. psalms norwegian paraphrases 1854 – Christiania: B C Fabritius, 1854 – 1r – 1 – us Wisconsin U Libr [240]
Forssk *see* Icones rerum naturalium quas, in itinere orientali...
Forsskal, P *see* Flora aegyptiaco-arabica
Forssman, Julius *see* J k lavater und die religioesen stroemungen des achtzehnten jahrhunderts
Forst de Battaglia, Otto *see* Der kampf mit dem drachen
Forst old school baptist church of roxbury. delaware county – Roxbury, NY – 1 – $13.59 – (brief historical sketch, church minutes sep 1868-may 1952, membership records, letters to lexington-roxbury old school primitive baptist association 1915-66) – mf#7078 – us Southern Baptist [242]
Forst- und jagd-archiv von und fuer preussen / ed by Hartig, Georg Ludwig – Berlin: In Commission bei der Real-Schulbuchhandlung 1816-20 (4n/yr) [1-5 jahrg(1816-20)] [mf ed 2005] – 5v on 2r [ill] – 1 – (cont by: allgemeines forst- und jagd-archiv; imprint varies) – mf#film mas c5699 – us Harvard [634]
The forster and dyce collections : from the national art library, victoria and albert museum, london – 77r – 1 – (coll includes extensive representation of 16th, 17th and 18th century british writers. includes a printed guide) – mf#C35-22400 – us Primary [420]
Forster, Brix *see* Deutsch-ostafrika
Forster, Charles *see*
– The life of john jebb, d.d., f.r.s., bishop of limerick, ardfert, and aghadoe
– A new plea for the authenticity of the text of the three heavenly witnesses
– The one primeval language
Forster, E M [Edward Morgan] *see* Howards end
Forster first (south forster; scituate) baptist church. rhode island : church records; 1769-1837 – Merged with Johnston First (Graniteville) General Six Principle in 1837. 24p – 1 – us Southern Baptist [242]
Forster, G *see* Geschichte der reisen
Forster, Georg *see*
– Ausgewaehlte kleine schriften
– Georg forsters briefe an christian friedrich voss
– Georg forster's saemmtliche schriften
– Lichtstrahlen aus seinen briefen
Forster, Georg et al *see* Hessische beytraege zur gelehrsamkeit und kunst
Forster, George *see* Voyage du bengale a petersbourg a travers les provinces septentrionales de l'inde, le kachmyr, la perse, sur la mer caspienne, etc
Forster, Hans Walter *see* Co-operation with employees
Forster, Henry L *see* Biblical psychology
Forster, Henry Pitt *see* Essay on the principles of sanskrit grammar
Forster, Henry Rumsey *see* The stowe catalogue priced and annotated
Forster, J G *see* Die neue mainzer zeitung oder der volksfreund
Forster, Johann Reinhold *see* Natural history and description of the tyger-car of the cape...
Forster, John *see* Brief memorial of the lord's dealings with george picknell of chalv...
Forster, Jonathan Langstaff *see* Biblical psychology
Forster, Josiah *see* Reflections on the gospel of christ in connexion with the principles and practices of the religious society of friends
Forster volkszeitung – Forst, Lausitz DE, 1962-65 – 1r – 1 – (publ in cottbus) – gw Misc Inst [074]
Forster, William Edward *see*
– Imperial federation
– Our colonial empire
Forster, William Rabbeth *see* The kingdom of god and life therein
Das forsthaus / Lorm, Hieronymus – Wien: Selbstverlag des Verfassers 1864 [mf ed 1995] – 1r – 1 – (filmed with: reichsstaedtische erzaehlungen / hermann kurz) – mf#3679p – us Wisconsin U Libr [820]
Forsthoff, Heinrich *see* Schleiermachers religionstheorie und die motive seiner grundanschauung
Forstliche Blaetter fuer Wuertemberg / ed by Widenmann, W – Tuebingen: H Laupp 1828-[1.-7.heft(1828-34)] [mf ed 2005] – 1r – 1 – (frequency varies; ceased with heft8 (1842)?) – mf#film mas 36989 – us Harvard [634]
Forstliche zeitschrift / ed by Bernhardt, August – Berlin: J Springer 1879- (mthly) [1 jahrg 1.-6. heft (1879 jan-jun)] [mf ed 2005] – 1r [ill] – 1 – (ceased with 1 jahrg 6. heft (jun 1879)?) – mf#film mas 36989 – us Harvard [634]
Forstliche zeitschrift fuer das grossherzogthum baden – Carlsruhe: G Braun'schen Hofbuchh 1838- (irreg) [mf ed bd1-2 heft2s(1838-42)] [mf ed 2005] – 1r – 1 – (v2 n2- incl: verhandlungen des forstlichen vereins im badischen oberlande, later publ separately; ceased in 1843) – mf#film mas 99999 – us Harvard [634]
Forstlicher jahresbericht fuer das jahr... / ed by Weber, Heinrich – Tuebingen: H Laupp'schen Buchhh 1926 (annual) [1924-25] [mf ed 2005] – 1 – (cont: jahresbericht ueber die fortschritte, veroeffentlichungen und wichtigeren ereignisse im gebiete des forst-, jagd- und fischereiwesens; cont by: forstliche rundschau) – mf#film mas 99999 – us Harvard [634]
Forstmann, Max Dieter *see* Der rechtsschutz im schwedischen verwaltungsverfahren
Forstverein fuer das Grossherzogthum Hessen. Jahres-Versammlung *see*
– Bericht ueber die...jahre-versammlung des forstverein fuer das grossherzogthum hessen zu...
– Die jahre-versammlung des forstverein fuer das grossherzogthum hessen zu...
Forstverein fuer das Grossherzogthum Hessen. Versammlung *see*
– Bericht ueber die...versammlung des forstverein fuer das grossherzogthum hessen zu...
Forst-Verein fuer Oesterreich und Salzburg *see* Berichte des forst-vereines fuer oesterreich und salzburg
Forsyth County Defense League *see* All the way
Forsyth county genealogical society journal – v1 n1-v4 n4 [1982 fall-1986 summer] – 1r – 1 – mf#1573055 – us WHS [929]
Forsyth Grant, Minnie *see* Scenes in hawaii
Forsyth, J *see* Remarks on dr heugh's irenicum
Forsyth, James Bell *see* Brief remarks on the waste lands of the crown in the canadas
Forsyth, John R *see* Journal
Forsyth, Joseph *see*
– Remarks on antiquities, arts, and letters
– Remarks on antiquities, arts, and letters during an excursion in italy
Forsyth, Peter Taylor *see*
– The charter of the church
– Christ on parnassus
– The christian ethic of war
– Christian perfection
– The cruciality of the cross
– Faith, freedom, and the future

- The holy father and the living christ
- The justification of god
- Lectures on the church and the sacraments
- Marriage
- Missions in state and church
- The power of prayer
- The principle of authority
- Problems of to-morrow
- Religion in recent art
- Rome, reform and reaction
- The soul of prayer
- Theology in church and state
- The work of christ

Forsyth, Robert Coventry *see*
- The china martyrs of 1900
- Shantung, the sacred province of china in some of its aspects

Forsyth, William *see*
- History of trial by jury
- Hortensius the advocate
- Life of marcus tullius cicero

Forsyth's chips *see* Miscellaneous newspapers of las animas county, reel 2

Fort apache scout / White Mountain Apache Indian Tribe – 1964 apr, 1970 sep-1977 aug 19, 1977 sep-1980 dec, 1981, 1984-87 – 4r – 1 – mf#368283 – us WHS [307]

Fort atkinson chronicle – Fort Atkinson WI. 1895 oct 31-1897 feb 18, 1897 feb 25-1898 aug 18, 1898 aug 25-1899 dec 19 – 3r – 1 – mf#936308 – us WHS [071]

Fort atkinson herald – Fort Atkinson WI. 1866 sep 1,29, 1868 may 7, 1869 apr 1-sep 9, 1871 jun 22-29, 1872 jun 20-1873 sep 4 – 2r – 1 – (cont by: sharon inquirer) – mf#916573 – us WHS [071]

Fort atkinson news – Fort Atkinson WI. 1929 mar 7-1930 feb 28, 1930 mar 7-1931 feb 27, 1931 mar 16-1932 jul 29, 1932 aug 5-1933 nov 24, 1933 dec 1-1935 jun 28, 1935 jul 5-1936 nov 3, 1936 nov 5-1937 may 6, 1937 may 11-1938 may 26, 1938 jun 2-1939 oct 26, 1939 nov 2-1941 apr 4, 1941 apr 11-1942 jan 23 – 11r – 1 – (cont: jefferson county democrat) – mf#936309 – us WHS [071]

Fort atkinson standard – Fort Atkinson WI. 1859 sep 1-1860 jul 5, sep 13-1863 aug 13 – 1r – 1 – mf#916577 – us WHS [071]

Fort bargains – Fort Atkinson WI. 1942 oct 20, dec 29, 1943 oct 20 – 1r – 1 – mf#1212236 – us WHS [071]

The fort beaufort advocate and general advertiser – Fort Beaufort SA, 23 jul 1859-1974 – 1 – sa National [079]

Fort belknap camp crier – 1975 dec 26-1979 sep 6, 1980 mar 27-1981 may 21 – 2r – 1 – (cont: camp crier) – mf#611549 – us WHS [071]

Fort benning bayonet – Columbus, GA. 1942-1991 (1) – mf#62463 – us UMI ProQuest [071]

[Fort bidwell-] bidwell gold nugget (bidwell news) – CA. jan 1907-sep 1917 (broken file) – 3r – 1 – $180.00 – mf#B02237 – us Library Micro [071]

Fort Bliss G I 's for Peace *see* Gigline

Fort bragg advocate news – Fort Bragg, CA. 1984-1999 (1) – mf#61016 – us UMI ProQuest [071]

[Fort bragg-] advocate news – CA. 1971-83 – 18r – 1 – $1080.00 – mf#B02238 – us Library Micro [071]

Fort buchanan sentinel – v10 n20-v11 n37 [1983 may 20-1984 sep 12] – 1r – 1 – (cont by: sentinel [fort buchanan, pr]) – mf#1043333 – us WHS [071]

Fort calhoun chronicle – Fort Calhoun, NE: Frank C Adams. v1 n1. jul 1 1915-n30. nov 28 1946 (wkly) [mf ed with gaps filmed [1972?]] – 7r – 1 – (absorbed by: enterprise (kennard ne). some irregularities in numbering. iss for mar 2 1933-nov 28 1946 lack vol numbering but retain number designations) – us NE Hist [071]

Fort campbell courier – 1984 may-1985 apr, 1985 may-jun, 1985 jul-1986 may – 6r – 1 – mf#550780 – us WHS [071]

Fort capron / Sim, Edith – s.l, s.l? 1936 – 1r – us UF Libraries [978]

Fort caroline / Corse, Carita Doggett – s.l, s.l? 1937 – 1r – us UF Libraries [978]

Fort clinch state park – s.l, s.l? 193-? – 1r – us UF Libraries [978]

Fort clinch state park – s.l, s.l? 193-? – 1r – us UF Libraries [978]

Fort collins argus *see* Miscellaneous newspapers of larimer county

Fort collins courier *see* Miscellaneous newspapers of larimer county

Fort collins mountain and plains weekly *see* Miscellaneous newspapers of larimer county

Fort collins prospectus *see* Miscellaneous newspapers of larimer county

Fort Concho National Historic Landmark et al *see* News from the frontier fort on the conchos

Fort daily news – Fort Atkinson WI. 1942 jan 12-1943 mar 15, 1943 mar 17-dec 31, 1943 oct 18-1945 mar 7, 1945 mar 8-1946 feb 15, 1946 feb 18-1946 oct 27, 1946 oct 28-1947 jun 30, 1947 jul-dec – 7r – 1 – mf#936311 – us WHS [071]

Fort dallas barracks : miami, dade county, florida – s.l, s.l? 193-? – 1r – us UF Libraries [978]

Fort delaware notes / Fort Delaware Society – 1970 jan-1992 feb – 1r – 1 – mf#452908 – us WHS [978]

Fort Delaware Society *see* Fort delaware notes

Fort denaud, hendry county, florida / Huss, Veronica E – s.l, s.l? no date – 1r – us UF Libraries [978]

Fort detrick standard – 1982 may 28-1989 mar 24 – 1r – 1 – (cont by: standard [fort detrick [frederick, md]) – mf#1056459 – us WHS [071]

Fort Dix Free Speech Movement *see* Ultimate weapon of the fort dix free speech movement

Fort Dodge, KS *see* Reports and journals of scouts and marches

Fort erie times review – Ontario, CN. jan 1930-dec 1944 – 15r – 1 – cn Commonwealth Imaging [071]

Fort fest chief – 1977 mar-1982 jul – 1r – 1 – mf#652978 – us WHS [071]

Fort george herald – Fort George, British Columbia, CN. aug 1910-jan 1913 – 1r – 1 – cn Commonwealth Imaging [071]

Fort george island / Corse, Carita Doggett – s.l, s.l? 193-? – 1r – us UF Libraries [978]

Fort george island / Diddell, Mary W – s.l, s.l? 193-? – 1r – us UF Libraries [978]

Fort george island / Duncan, W T – s.l, s.l? 193-? – 1r – us UF Libraries [978]

Fort george island – s.l, s.l? 193-? – 1r – us UF Libraries [978]

Fort george island – s.l, s.l 193-? – 1r – 1 – us UF Libraries [978]

Fort george island / Wilson, Gertrude Rollins – s.l, s.l? 193-? – 1r – us UF Libraries [978]

Fort George Island Company *see* Winter at fort george

Fort george tribune – Fort George, British Columbia, CN. nov 1909-may 1915 – 1r – 1 – cn Commonwealth Imaging [071]

Fort, Gertrude von *see* Le pape du gheto. paris

Fort gordon rambler – v20 n1-17 [1981 jan 9-may 1] – 1r – 1 – (cont: rambler; cont by: semaphore) – mf#580306 – us WHS [071]

Fort greene news – v3 iss 13 [1995 feb 16] – 1r – 1 – mf#5327602 – us WHS [071]

Fort greene samaritan / Hanson Place SDA Church [New York, NY] – 1998 dec – 1r – 1 – mf#5294655 – us WHS [242]

Fort Harker. Kansas *see* Journal of post sutler

Fort howard [city directory : listing] – 1874 – 1r – 1 – mf#3196410 – us WHS [917]

Fort howard herald, and general advertiser for brown county – Green Bay WI. 1873 apr 15, 1874 jan 13-1877 mar 1 – 1r – 1 – (cont by: brown county herald [green bay, wi]) – mf#918916 – us WHS [071]

Fort howard journal – Green Bay WI. 1878 nov22-1880 apr 24 – 1r – 1 – mf#920477 – us WHS [071]

Fort howard monitor – Green Bay WI. 1873 jan 23, 1876 jan 13-1877 jan 4 – 1r – 1 – (cont by: fort howard herald) – mf#918911 – us WHS [071]

Fort howard review – Green Bay WI. 1877 feb 13-1880 aug 18, 1880 aug 25-1883 oct 3, 1883 oct 11-1886 may 29, 1886 jun 5-1887 dec 10, 1887 dec 16-1889 apr 27, 1889 may 4-1890 oct 11, 1890 oct 18-1892 apr 2, 1892 apr 9-1893 sep 23, 1893 sep 30-1895 mar 30 – 9r – 1 – (cont by: green bay review) – mf#947487 – us WHS [071]

Fort jefferson national monument – s.l, s.l? 193-? – 1r – us UF Libraries [978]

[Fort jones-] farmer and miner – CA. 1902-16 – 6r – 1 – $360.00 – mf#B02240 – us Library Micro [071]

[Fort jones-] pioneer press – CA. 1983- – 6r – 1 – $360.00 (subs $50y) – mf#B03220 – us Library Micro [071]

[Fort jones-] scott valley advance – CA. 1897-1908; 1911; 1912; 1913-1916 – 9r – 1 – $540.00 – (see etna) – mf#B02241 – us Library Micro [071]

[Fort jones-] scott valley news – CA. 1878-95 [wkly] – 3r – 1 – $180.00 – mf#B02242 – us Library Micro [071]

[Fort jones-] siskiyou standard – CA. 1917-22 – 2r – 1 – $120.00 – mf#B02243 – us Library Micro [071]

[Fort jones-] the county reporter – CA. 1895-98 – 1r – 1 – $60.00 – mf#B02239 – us Library Micro [071]

[Fort jones-] weekly scott valley news – CA. 1881-86 – 1r – 1 – $60.00 – mf#B02244 – us Library Micro [071]

Fort Larned. Kansas *see* Records

Fort Lauderdale Historical Society *see* New river news

Fort lauderdale history : a report to the members of the historical society of fort lauderdale / Historical Society of Fort Lauderdale – 1962 jul – 1r – 1 – (cont by: new river news) – mf#2332389 – us WHS [978]

Fort lauderdale middle river reclamation district – Ft Lauderdale, FL. 192-? – 1r – us UF Libraries [630]

Fort lauderdale westside gazette – Fort Lauderdale FL. 1979 mar 15/dec 27-1995 jan/jun 22 [gaps] – 28r – 1 – (cont by: westside gazette) – mf#1214516 – us WHS [071]

Fort Leavenworth. Kansas *see* Records

Fort lee trumpeter – Hopewell, VA. 1995+ (1) – mf#69198 – us UMI ProQuest [071]

Fort leonard wood guidon / United States Army Training Center [Fort Leonard Wood, MO] – Fort Leonard Wood WI. v15 n10, 28 [1980 aug 21, 1981 jan 15 – 1r – 1 – (cont by: guidon [fort leonard wood, mo]) – mf#714509 – us WHS [355]

Fort leonard wood guidon / United States Army Training Center [Fort Leonard Wood, MO] – Fort Leonard Wood WI. v18 n28-v19 n5 [1984 jan 5-sep 24] – 1r – 1 – (cont: guidon [fort leonard wood, mo: 1981]; cont by: guidon [fort leonard wood, mo: 1986]) – mf#1345493 – us WHS [355]

Fort leonard wood guidon / United States Army Training Center [Fort Leonard Wood, MO] – Fort Leonard Wood WI. 1987 jan 22-aug 27, 1987 sep 3-1988 may 26 – 2r – 1 – (cont: guidon [fort leonard wood, mo: 1986]; cont by: essayons) – mf#1345246 – us WHS [355]

Fort Leslie J McNair [Washington DC] *see* Pentagram

Fort Lewis College *see* Wambidiota news

Fort loramie progress / Shelby Co. Sidney – v1 n1. (may 1915-jun 1917) [wkly] – 1r – 1 – mf#B11289 – us Ohio Hist [071]

Fort lupton spirit *see* Miscellaneous newspapers of weld county

Fort mcleod collection – Alberta, CN. jan 1897-dec 1919 – 1r – 1 – cn Commonwealth Imaging [071]

Fort mcmurray today – Fort McMurray, Alberta, CN. 1974- – 6r/y – 1 – Can$93.00r – cn Commonwealth Imaging [071]

Fort meade leader – Ft Meade, FL. 1933-1939 – 2r – (gaps) – us UF Libraries [071]

Fort meade post – 1943 nov5-12 – 1r – 1 – mf#928419 – us WHS [071]

Fort meigs memorial commission minute book, 1908-1921 – 1r – 1 – mf#B29155 – us Ohio Hist [355]

Fort mill first baptist church. york baptist association. fort mill, south carolina : church records – 1966-73 – 1 – $79.83 – us Southern Baptist [242]

Fort mitchell baptist church. kentucky : church records – 1924-67 – 1 – us Southern Baptist [242]

Fort myers and lee county, florida – s.l, s.l? 1938 – 1r – us UF Libraries [978]

Fort myers and lower west coast – s.l, s.l? 193-? – 1r – us UF Libraries [978]

Fort myers press – Ft Myers, FL. 1923-1929 – 10r – (gaps) – us UF Libraries [071]

Fort myers section has steady growth – s.l, s.l? 1936 – 1r – us UF Libraries [978]

Fort myers to sanibel island / Frost, J A – s.l, s.l? 1936 – 1r – us UF Libraries [978]

Fort nelson and hudson's bay / Read, David Breakenridge – [S.l: s.n, 1893?] – 1mf – 9 – 0-665-17464-0 – (fr: the canadian magazine) – mf#17464 – cn Canadiana [380]

Fort Ord (CA) *see* Every gi is a p.o.w.

Fort Peck Tribal Executive Board *see* Wotanin wowapi

Fort pierce news – Fort Pierce, FL. 1906 nov 16-191 – 4r – (gaps) – us UF Libraries [071]

Fort pierce news-tribune – Fort Pierce, FL. 1920 jan 14-1952 jan – 106r – (gaps) – us UF Libraries [071]

Fort Point Museum Association *see* Fort point salvo

Fort point salvo / Fort Point Museum Association – v1 n1-v5 n4 [1969 jul-1982 apr] – 1r – 1 – mf#618185 – us WHS [060]

Fort recovery journal / Mercer Co. Fort Recovery – 1893-96/1903,15,24,29-34 (scattered) – 3r – 1 – mf#B37526-37528 – us Ohio Hist [071]

Fort recovery record / Mercer Co. Fort Recovery – 10/15/1919-1/7/1920.7/21-10/1924 – 1r – 1 – mf#B37529 – us Ohio Hist [071]

Fort recovery times / Mercer Co. Fort Recovery – jul 25-sep 26 1924 – 1r – 1 – mf#B37529 – us Ohio Hist [071]

Fort recovery tribune / Mercer Co. Fort Recovery – apr 29 1915 (1 iss only) – 1r – 1 – mf#B37529 – us Ohio Hist [071]

Fort richardson pioneer – Fort Richardson AK. 1984 jan 6-1985 jul 3, 1985 nov15-1986 sep 5, 1986 apr 18-1988 jul 1 – 2r – 1 – (cont: pioneer [fort richardson, ak]; cont by: yukon sentinel; buffalo [greeley, ak]; arctic star) – mf#1138061 – us WHS [071]

Fort, Robert Edwin *see* Analysis of l'ascension for organ by olivier messiaen

Fort Smith Historical Society *see* Journal

Fort smith new era – Fort Smith AR. 1864 jun 4 – 1r – 1 – mf#853690 – us WHS [071]

Fort st george gazette / Madras. Presidency – 1962-1966 – 1 – us NY Public [324]

Fort thompson, hendry county, florida / Huss, Veronica E – s.l, s.l? 193-? – 1r – us UF Libraries [978]

Fort Wayne Federation of Labor *see* Worker

Fort Wayne Women's Bureau *see* Womensword

Fort william free press – Breakish [Isle of Skye]: West Highland Publ Co 1975-76 (wkly) – 1 – uk Scotland NatLib [072]

Fort william-india house correspondence : and other contemporary papers relating thereto / ed by Sinha, Narendra Krishna – Delhi: Publ for the National Archives of India, by the Manager of Publ, Govt of India, 1949- – (= ser Samp: indian books) – us CRL [954]

Fort worth mind – Fort Worth TX. 1943 nov 13, 1947 jan 18, sep 13 [p5-6 only] – 1r – 1 – mf#5019900 – us WHS [071]

Fort worth star-telegram – Fort Worth, TX. 1925+ (1) – mf#60597 – us UMI ProQuest [071]

Fort Y Roldan, Nicolas *see* Cuba indigena

Fortaellinger af danmarks kirkehistorie fra 1517 til 1848 / Koch, Ludvig – Kobenhavn: GEC Gad, 1889 [mf ed 1990] – 1mf – 9 – 0-7905-5357-0 – (in danish) – mf#1988-1357 – us ATLA [240]

Forteckning over talare och material till de lokala spanienkommitteerna, augusti 1937 / Svenska Hjaelpkommitten for Spanien – Stockholm, 1937. Fiche W1218. (Blodgett Collection of Spanish Civil War Pamphlets) – 9 – us Harvard [946]

Fortes, Amyr Borges *see* Historia administrativa, judiciaria e eclesiastica

Fortes, Herbert Parentes *see* Questao da lingua brasileira

Fortes, Meyer *see*
- African political systems
- African political sytems
- Plural society in africa

Fortescue, Adrian *see*
- The greek fathers
- The lesser eastern churches
- The mass
- The orthodox eastern church

Fortescue, John William *see* The writing of history

Fortescue, Thomas Knox *see* General remarks on steam communication

Fortgesetzte betrachtungen ueber die neuesten historischen schriften *see* Betrachtungen ueber die neuesten historischen schriften

Fortgesetzte reise nach hammelburg : oder meine harten schicksale in kautzen-land / Lang, Karl H von – Muenchen 1822 [mf ed Hildesheim 1995-98] – 1mf – 9 – €40.00 – 3-487-29569-5 – gw Olms [914]

Fortgesetzte schubartsche chronik – Stuttgart 1792 – (= ser Dz. historisch-politische abt) – 8mf – 9 – €160.00 – mf#k/n1266 – gw Olms [943]

The forth bridge / Baker, Benjamin – London: s.n, 1884 – 1mf – 9 – mf#00842 – cn Canadiana [624]

Fortia de Piles, Alphonse T de *see* Voyage de deux francais en allemagne, danemarck, suede, russie et pologne

Fortie, Marius *see* Black and beautiful

Fortier, Adelard *see* The economics of war

Fortier, Alcee *see*
- The acadians of louisiana and their dialect
- French literature in louisiana

Fortier, Auguste *see* Les mysteres de montreal

Fortier, de la Broquerie *see* Au service de l'enfance

Fortier, Louis *see*
- Manuel pratique de vocation
- Moyens de connaitre sa vocation

Fortier, Marie-Marthe *see* Bibliographie analytique de l'oeuvre de i-w jones

Fortier, Onesime Laurent *see* Le r pere vincent routier de l'ordre des freres precheurs

Fortier, Suzanne *see* Bibliographie de la poesie canadienne-francaise 1935-1958

Fortieth report...1875 / Diocesan Church Society of New Brunswick – [Fredericton, NB?: s.n.] 1875 [mf ed 1983] – 2mf – 9 – 0-665-43872-9 – mf#43872 – cn Canadiana [242]

Fortificatie, dat is sterckteboowing... / Marolois, S – Amsterdam, 1628 – 2mf – 9 – mf#OA-156 – ne IDC [720]

Fortificatie ofte sterckten-bouwinghe... / Metius, A – Franeker, 1626 – 1mf – 9 – mf#OA-159 – ne IDC [720]

Fortificatio, das ist kuenstliche und wolgegruendte demonstration / Errard, J – Frankfurt a M, 1604 – 2mf – 9 – mf#OA-214 – ne IDC [720]

Fortification uvelle, ou recueil de differantes manieres de fortifier en europe : avec des figures en taille-douce / Pfeffinger, [J F] – uv ed. La Haye, 1740 – 4mf – 9 – mf#OA-272 – ne IDC [720]

Le fortificationi di...nuovamente ristampate... / Lorini, B – Venetia, 1609 – 6mf – 9 – mf#OA-205 – ne IDC [720]
Fortifications and garrison forces in the mandates before pearl harbor, 1934-1941 / U.S. Army. Office of Military History – 1952 – 1 – us L of C Photodup [355]
La fortificazione, guardia, difesa et espugnazione delle fortezze / Tensini, F – Venetia, 1624 – 8mf – 9 – mf#OA-207 – ne IDC [720]
Fortin, Alphonse *see* Les saints martyrs canadiens
Fortin, Charles-Henri *see* Les aventures de pierre
Fortin, Dale A *see* L-carnitine supplementation and the lactate/pyruvate ratio
Fortin, Isabelle *see* Bibliographie analytique des travaux de paul-edouard gagnon
Fortin Magana, Romeo *see*
– Democracia y socialismo, seguido de otros breves e...
– Inquietudes de un ano memorable
Fortin, Pierre *see*
– Le detroit de belle-isle
– Reponse a une adresse de l'assemblee legislative, datee du 19 courant
– Reports of pierre fortin and theophile tetu, stipendiary magistrates
– The straits of belle isle
Fortin-Roussel, Robert *see* Circuit litteraire sur voltaire et rousseau
Fortis, Alberto *see* Voyage en dalmatie
Fortis, Francois M de *see* Voyage pittoresque et historique a lyon
Fortitudine : newsletter of the marine corps historical / Marine Corps Historical Center [US] – 1972 summer-1985 spring – 1r – 1 – mf#969912 – us WHS [355]
Fortlage, C et al *see* Die luecken des hegelschen systems der philosophie
Das fortleben des heidentums in der altchristlichen kirche / Soltau, Wilhelm – Berlin: G. Reimer, 1906 – 1mf – 9 – 0-7905-5967-6 – (incl bibl ref) – mf#1988-1967 – us ATLA [240]
Fort-liberte d'hier et d'aujourd'hui / Calixte, Nyll F – Port-Au-Prince, Haiti. 1960 – 1r – us UF Libraries [972]
Fortnightly – Arlington. 1993-1994 (1,5,9) – (cont: public utilities fortnightly. cont by: public utilities fortnightly) – ISSN: 1074-6099 – mf#2256,01 – us UMI ProQuest [350]
Fortnightly Club. Topeka, Kansas *see* Records
Fortnightly fogbank : for the students of dli-sf / Defense Language Institute [US] – v1 n2-v2 n26 [1984 feb 17-1985 dec 17] – 1r – 1 – mf#1497753 – us WHS [355]
Fortnightly law journal – Toronto, Canada. v1-17. 1931-48 – 1 – $216.00 – mf#0208 – us Brook [340]
Fortnightly review – London. 1865-1934 (1) – mf#2795 – us UMI ProQuest [400]
Fortnightly review of the chicago dental society / Chicago Dental Society – Chicago. 1941-1972 (1) 1971-1972 (5) – (cont by: cds review) – ISSN: 0009-353X – mf#3307 – us UMI ProQuest [617]
Fortnum, Charles Drury Edward *see*
– A descriptive catalogue of the bronzes of european origin in the south kensington museum
– A descriptive catalogue of the maiolica...in the south kensington museum
Fortoul, H *see* De l'art en allemagne
Fortran 4 / Vickers, Frank D – Dubuque, IA. 1978 – 1r – us UF Libraries [500]
Fortress – v4 n1 [1998 feb/mar] – 1r – 1 – mf#4327444 – us WHS [071]
The fortress of quebec 1608-1903 : with ill / Doughty, Arthur George – Quebec: Dussault & Proulx, 1904 [mf ed 1982] – 3mf – 9 – mf#SEM105P81 – cn Bibl Nat [720]
Fortresses et villages desertes du touat-gouara (sahara algerien) : essai d'application de l'analyse photographique aerienne a la recherche archelolgique en regions desertiques – Paris, Ecole pratique des hautes etudes, 1970 – us CRL [930]
Forts and historic points of interest / Fuller, Russell L – s.l, s.l? 1936 – 1r – us UF Libraries [978]
Forts established in florida prior to 1700 / Bird, Phyllis T – s.l, s.l? 193-? – 1r – us UF Libraries [720]
Fortschrift und vollendung bei philo von alexandrien / Voelker, W – Leipzig, 1938 – (= ser Tugal 4-49/1) – 6mf – 9 – €14.00 – ne Slangenburg [270]
Der fortschritt – New York. N.Y. Progress. 1915-32 – 1 – us AJPC [071]
Der fortschritt – Deuben b. Weissenfels DE, 1949 jun-1967 27 dec [gaps] – 3r – 1 – (notes: bkk zeitz / bkw deuben) – gw Misc Inst [074]
Der fortschritt = The progress – New York, NY. 1865-66 – 1 – us AJPC [071]
Der fortschritt – Duesseldorf, Essen DE, 1950 – 1r – 1 – (title varies: 15 apr 1960: deutsche allgemeine zeitung; filmed by misc inst: 1958-62 [3r], 1951-57 [4r], 1960-61) – gw Mikrofilm; gw Misc Inst [074]
Der fortschritt – Bunzlau [Boleslawiec PL], 1855-58 [gaps], 1908 1 jul-1910 30 jun [gaps] – 5r – 1 – (title varies: 1857: niederschlesischer courir) – gw Misc Inst [077]
Fortschritte auf dem gebiete der roentgenstrahlen – Stuttgart. 1932-1952 (1) – mf#129 – us UMI ProQuest [610]
Fortschritte der hebraischen sprachwissenschaft von jehuda chajjug / Rosenak, Leopold – s.l, s.l? 1898 – 1r – us UF Libraries [470]
Fortschritte der neurologie, psychiatrie und ihrer grenzgebiete – Stuttgart. 1976-1976 (1) 1976-1976 (5) 1976-1976 (9) – ISSN: 0015-8194 – mf#10161 – us UMI ProQuest [617]
Fortschritte der ophthalmologie – Berlin. 1983-1983 (1,5,9) – ISSN: 0723-8045 – mf#13255,04 – us UMI ProQuest [617]
Fortschritte der physik – v1-46. 1845-90 – 1 – $1296.00 – (v47-74 1891-1918 $1170 [0209]) – mf#0210 – us Brook [530]
Die fortschritte des zivilrechts im 19. jahrhundert : ein ueberblick ueber die entfaltung des privatrechts in deutschland, oesterreich, frankreich und der schweiz / Hedemann, Justus Wilhelm – Berlin: C Heymann. v1 only. 1910- – (= ser Civil law 3 coll) – 2mf – 9 – mf#LLMC 96-614 – us LLMC [346]
Fortschrittschacht *see* Mansfeld-echo
Fortt, J M *see* The distribution of african population, native and immigrant, in buganda
Fortuin, Foppe *see* Waarom ik tot de christelijk gereformeerde kerk terugkeerde
Fortuin, K W *see* Afgezet naar recht en waarheid
Fortuin, Stephen Daniel *see* Die groepsgebiedewet (no. 41 van 1950) met spesiale verwysing na die kaapse skiereiland (1950-1953)
Fortun Y Fortun, Joaquin *see*
– Etapas
– Pentagrama
[Fortuna-] eel valley advance – CA. Mar 1970-Feb 1975 – 1r – 1 – $60.00 – mf#B02245 – us Library Micro [071]
[Fortuna-] humboldt beacon and fortuna advance – CA. 1907- – 63r – 1 – $3780.00 (subs $120y) – mf#C02246 – us Library Micro [071]
Fortunat : dramatisches maerchen in fuenf acten / Bauernfeld, Eduard von – [Wien?: s.n, 1871?] [mf ed 1993] – (= ser Bauernfeld, eduard von, works 3) – 1 – mf#8509 – us Wisconsin U Libr [820]
Fortunat : etude sur un dernier representant de la poesie latine dans la gaule merovingienne / Tardi, D – Paris, 1927 – 6m – 8 – €14.00 – ne Slangenburg [931]
Fortunat, Dantes *see* Abrege de la geographie de l'ile d'haiti
Fortunate islands / Defries, Amelia Dorothy – London, England. 1929 – 1r – us UF Libraries [972]
The fortunate union = Hao chiu chuan / ed by Baller, Frederick William – Shanghai: American Presbyterian Mission Press, 1911 [mf ed 1996] – (= ser Yale coll) – 371p – 1 – 0-524-10214-7 – (text in chinese, notes in english) – mf#1996-1214 – us ATLA [240]
Fortunati glueckseckel und wuenschhuetlein : ein spiel / Chamisso, Adelbert von; ed by Kossmann, E F – Stuttgart: G J Goeschen, 1895 [mf ed 1993] – (= ser Deutsche litteraturdenkmale des 18. und 19. jahrhunderts 54-55, n f n4-5) – xxxvi/68p – 1 – (incl bibl ref) – mf#8676 reel 5 – us Wisconsin U Libr [820]
Fortunatianus et al *see* Varia rhetorica
Fortunatov, A F *see* Ob izuchenii kooperatsii
Fortunatus / ed by Guenther, Hans – Halle: M Niemeyer, 1914 – 11r – 1 – us Wisconsin U Libr [430]
Fortunatus : hier ist gar kurzweilig zu lesen, was fortunatus und nach ihm seine zwei soehne mit dem glueckssackel und wunschhuetlein fuer wunder vollbracht und erfahren...ein volksbuch aus dem jahre 1509 / ed by Schneider, Gerhard & Arndt, Erwin – [s.l.]: Verlag Mueller & Kiepenheuer [1964?] – 246p (ill) – 1 – us Wisconsin U Libr [390]
Fortune – Chicago. 1930+ (1) 1966+ (5) 1956+ (9) – ISSN: 0015-8259 – mf#1128 – us UMI ProQuest [650]
Fortune, G (George) *see*
– Ideophones in shona
– Ndevo yenombe luvizho and other lilima texts
– Ndevo, yenombe, luvizho and other lilima texts
Fortune, George *see*
– Analytical grammar of shona
– Elements of shona (zezuru dialect)
– Ideophones in shona
Fortune, Marie M *see* Journal of religion and abuse
Fortune news / Fortune Society [New York, NY] – 1971 jun-1973 nov, 1973 dec-1975 jun, 1975 aug-1989 spring – 3r – 1 – mf#1112117 – us WHS [071]
Fortune, R *see*
– A residence among the chinese
– Three years' wandering in the northern provinces of china
Fortune, Robert *see*
– Three years' wanderings in the northern provinces of china
– Two visits to the tea countries of china and the british tea plantations in the himalaya; with a narrative of adventures, and...description of the tea plant.
Fortune Society [New York, NY] *see* Fortune news
Fortune. the struggle in spain – NY, 1937. Fiche W 1215. (Blodgett Collection of Spanish Civil War Pamphlets) – 9 – us Harvard [946]
Fortune-Barthelemy de Felice *see* Encyclopedie (ael1/7)
The fortunes of primitive tribes / Majumdar, Dhirendra Nath – Lucknow: Published for the Lucknow University by the Universal Publishers, 1944 – (= ser Samp: indian books) – us CRL [307]
The fortunes of the landrays / Kester, Vaughan, 1869-1911 – Toronto: McLeod & Allen, c1905, repr 1912 [mf ed 1995] – 6mf – 9 – 0-665-74718-7 – mf#74718 – cn Canadiana [830]
Fortunio, Giovanni Francesco *see* Extracts from regole brievi della volgare grammatica
Forty and eighter : the news magazine of la societe des quarante hommes et huit chevaux / Societe des quarante hommes et huit chevaux – 1964 jul-1974 dec, 1975 jan-1980 dec, 1981-85 – 3r – 1 – mf#709003 – us WHS [366]
The forty days after our lord's resurrection / Hanna, William – New York: Robert Carter, 1864 [mf ed 1985] – (= ser The life of our lord 6) – 1mf – 9 – 0-8370-3463-9 – (incl app) – mf#1985-1463 – us ATLA [220]
Forty days with the master / Huntington, Frederic Dan – New York: EP Dutton, 1891 – 1mf – 9 – 0-7905-3972-1 – mf#1989-0465 – us ATLA [240]
Forty first report...1876 / Diocesan Church Society of New Brunswick – [St John, NB?: s.n.] 1876 [mf ed 1983] – 2mf – 9 – 0-665-43873-7 – mf#43873 – cn Canadiana [242]
The forty martyrs of the sinai desert : and the story of eulogios: from a palestinian syriac and arabic palimpsest / ed by Lewis, Agnes Smith – Cambridge: University Press, 1912 – 1mf – 9 – 0-8370-7402-9 – (text in syriac and arabic; introduction in english) – mf#1986-1402 – us ATLA [240]
Forty years among the zulus / Tyler, Josiah – Boston: Congregational Sunday-School & Pub Society, c1891 [mf ed 1986] – 1mf – 9 – 0-8370-6532-1 – (incl app) – mf#1986-0532 – us ATLA [306]
Forty years at raritan : eight memorial sermons, with notes for a history of the reformed dutch churches in somerset county, n j / Messler, Abraham – New York: A Lloyd, 1873 [mf ed 1990] – 1mf – 9 – 0-7905-5123-3 – mf#1988-1123 – us ATLA [242]
Forty years' experience in sunday-schools / Tyng, Stephen Higginson – New York: Sheldon, 1863 – 1mf – 9 – 0-524-04444-9 – mf#1991-2109 – us ATLA [240]
Forty years' familiar letters of james w alexander : constituting, with the notes, a memoir of his life / Alexander, James W; ed by Hall, John – New York: Scribner; London: Sampson Low, 1860 – 2mf – us ATLA [240]
Forty years' familiar letters of james w. alexander, d.d : constituting, with the notes, a memoir of his life = Correspondence / Alexander, James Waddel; ed by Hall, John – 3rd ed. New York: Scribner; London: Sampson Low, 1860 – 1mf – 9 – 0-7905-4423-7 – mf#1988-0423 – us ATLA [920]
Forty years in brazil / Bennett, Frank – London, England. 1914 – 1r – us UF Libraries [972]
Forty years in burma / Marks, John Ebenezer; ed by Purser, William Charles Bertrand – London: Hutchinson, 1917 – 1mf – 9 – 0-524-03906-2 – mf#1990-1165 – us ATLA [240]
Forty years in china : or, china in transition / Graves, Rosswell Hobart – Baltimore: RH Woodward, 1895 – 5mf – 9 – 0-524-07877-7 – mf#1991-3422 – us ATLA [915]
Forty years in new zealand : including a personal narrative, an account of maoridom, and of the christianization and colonization of the country / Buller, James – London: Hodder and Stoughton, 1878. Chicago: Dep of Photodup, U of Chicago Lib, 1967 (1r); Evanston: American Theol Lib Assoc, 1984 (1r) – 1 – 0-8370-0391-1 – mf#1984-B057 – us ATLA [920]
Forty years in south china, the life of rev. john van nest talmage, d.d / Fagg, John Gerardus – New York: Anson DF Randolph, c1894 – 1mf – 9 – 0-524-06535-7 – mf#1991-2619 – us ATLA [240]
Forty years in the mormon church : why i left it / Evans, Richard C – Toronto: [s.n, 1920?] [mf ed 1995] – 2mf – 9 – 0-665-74163-4 – mf#74163 – cn Canadiana [243]
Forty years in the turkish empire : or, memoirs of rev william goodell / Prime, E D G – Boston, 1891 – 6mf – 9 – mf#HT-170 – ne IDC [920]
Forty years in the turkish empire : or, memoirs of rev william goodell / Prime, Edward Dorr Griffin – New York: Robert Carter, 1876, c1875 – 2mf – 9 – 0-8370-6151-2 – mf#1986-0151 – us ATLA [240]
Forty years' mission work in polynesia and new guinea : from 1835 to 1875 / Murray, Archibald Wright – London: J Nisbet, 1876 – 2mf – 9 – 0-524-04892-4 – mf#1991-2174 – us ATLA [240]
Forty years of official and unofficial life in an oriental crown colony : being the life of sir richard f morgan, kt., queen's advocate and acting chief justice of ceylon – Madras: Higginbotham, 1879 – 1 – us CRL [920]
Forty years of pioneer life : memoir of john mason peck / Peck, John Mason; ed by Babcock, Rufus – Philadelphia: American Baptist Publication Society, c1864 – 1mf – 9 – 0-7905-8246-5 – mf#1988-8109 – us ATLA [975]
Forty years of service : a history of the christian woman's board of missions, 1874-1914 / Harrison, Ida Withers – 2nd ed. [S.l.: s.n., 1915?] – 1mf – 9 – 0-524-06541-1 – mf#1991-2625 – us ATLA [240]
Forty years of the panjab mission of the church of scotland, 1855-1895 / Youngson, J F W – Edinburgh, 1896 – 4mf – 9 – mf#HTM-223 – ne IDC [915]
Forty years' residence in america : or, the doctrine of a particular providence; exemplified in the life of grant thorburn, (the original lawrie todd,) seedsman, new york / Thorburn, Grant – London 1834 [mf ed Hildesheim 1995-98] – 1v on 4mf – 9 – €120.00 – 3-487-27110-9 – gw Olms [920]
Forty-eighth report...1883 / Diocesan Church Society of New Brunswick – [St John NB: s.n.] 1883 [mf ed 1983] – 3mf – 9 – 0-665-43880-X – mf#43880 – cn Canadiana [242]
Forty-fifth report...1880 / Diocesan Church Society of New Brunswick – [St John, NB?: s.n.] 1880 [mf ed 1983] – 2mf – 9 – 0-665-43877-X – mf#43877 – cn Canadiana [242]
Forty-first report of the london society for promoting christianity / London Society For Promoting Christianity Amongst The Jews – London, England. 1849 – 1r – us UF Libraries [240]
Forty-five years in china : reminiscences / Richard, T – London, 1916 – 5mf – 9 – mf#HT-121 – ne IDC [915]
Forty-five years in china : reminiscences / Richard, Timothy – London: T Fisher Unwin, 1916 – 1mf – 9 – 0-524-08503-X – mf#1993-3148 – us ATLA [240]
Forty-four years a public servant / Kincaid, Charles Augustus – Edinburgh: William Blackwood & Sons, 1934 – (= ser Samp: indian books) – us CRL [915]
Forty-fourth report...1879 / Diocesan Church Society of New Brunswick – [St John, NB?: s.n.] 1879 [mf ed 1983] – 2mf – 9 – 0-665-43876-1 – mf#43876 – cn Canadiana [242]
Forty-ninth report...1884 / Diocesan Church Society of New Brunswick – [St John NB: s.n.] 1884 [mf ed 1983] – 3mf – 9 – 0-665-43881-8 – mf#43881 – cn Canadiana [242]
Forty-one years in india from subaltern to commander-in-chief / Roberts, Frederick Sleigh Roberts, 1st earl – [6th ed.] London 1897 – (= ser 19th c british colonization) – 2v on 13mf – 9 – mf#1.1.6876 – uk Chadwyck [954]
Forty-second report...1877 / Diocesan Church Society of New Brunswick – [Fredericton, NB?: s.n.] 1877 [mf ed 1983] – 2mf – 9 – 0-665-43874-5 – mf#43874 – cn Canadiana [242]
Forty-seven identifications of the lost british nation and the uw with the ten lost tribes / Hine, Edward – London: W.E. Guest; Utica: S.E. Lawrence, 1879. 82p – 1 – us Wisconsin U Libr [300]
Forty-seventh report...1882 : incorporated by act of assembly, 16 victoria, cap 4, 14th apr 1853 / Diocesan Church Society of New Brunswick – [St John, NB?: s.n.] 1882 [mf ed 1983] – 2mf – 9 – 0-665-43879-6 – mf#43879 – cn Canadiana [242]
Forty-sixth report...1881 / Diocesan Church Society of New Brunswick – [St John, NB?: s.n.] 1881 [mf ed 1983] – 3mf – 9 – 0-665-43878-8 – mf#43878 – cn Canadiana [242]
Forty-third report...1878 / Diocesan Church Society of New Brunswick – [St John, NB?: s.n.] 1878 [mf ed 1983] – 2mf – 9 – 0-665-43875-3 – mf#43875 – cn Canadiana [242]

FORTY-TWO

Forty-two years amongst the indians and eskimo : pictures from the life of the right reverend john horden, first bishop of moosonee / Batty, Beatrice – London: Religious Tract Society, 1893 – 3mf – 9 – mf#02998 – cn Canadiana [920]
The Forum *see* Dickinson law review
Forum – 1933 oct 19-1936, 1940 jan 4-1941 dec 25 – 2r – 1 – mf#716306 – us WHS [071]
Forum – Amherst. 1973-1977 (1) 1977-1977 (5) 1977-1977 (9) – ISSN: 0034-0162 – mf#7395 – us UMI ProQuest [303]
Forum – New York. 1985-1986 (1) 1985-1986 (5) 1985-1986 (9) – (cont by: penthouse forum) – ISSN: 0160-2195 – mf#15078 – us UMI ProQuest [073]
Forum – Chicago. 1971-1985 (1) 1974-1985 (5) 1975-1985 (9) – (cont by: tort and insurance law journal) – ISSN: 0015-8356 – mf#8156 – us UMI ProQuest [340]
Forum – New York. v2, n1 (dec.1980)-v9, n6 (apr. 1987) – (continues: nyu forum) – us AJPC [073]
Forum – Fargo, ND. 1891+ (1) – mf#60552 – us UMI ProQuest [071]
Forum / Federated Union of Black Arts [South Africa] – 1980 [may?]-oct – 1r – 1 – (cont: fuba forum) – mf#4872416 – us WHS [700]
Forum / Franklin Co. Dublin – v1 n1. may 1970-sep 1984 [mthly, semimthly, wkly, semiwkly] – 8r – 1 – mf#B30040-30047 – us Ohio Hist [071]
Forum – Halesworth, Suffolk. 1981 may 1-1983 nov 11, 1984 feb 10-1985 nov 15, 1985 nov 22-1987 jun, 1987 jul-1988 nov 18, 1988 nov 25-1989 dec 15 – 5r – 1 – mf#1045624 – us WHS [071]
Forum – Houston. 1956-1980 (1) 1972-1980 (5) 1972-1980 (9) – ISSN: 0015-8410 – mf#6794 – us UMI ProQuest [700]
Forum / Michigan Federation of Teachers – 1986 sep-1988 jan – 1r – 1 – (cont: federation forum [detroit, mi: 1982]; cont by: federation forum [detroit, mi: 1988]) – mf#2899805 – us WHS [370]
Forum / Montgomery Co. Dayton – jun 1918-oct 19, mar 37-oct 1946 [wkly, irreg] – 2r – 1 – mf#B5307-5308 – us Ohio Hist [976]
Forum : a national newsletter serving black ministries / American Baptist Churches in the USA – 1992 fall-winter, 1993 summer, 1993 fall-1993/1994 winter – 1r – 1 – mf#3055974 – us WHS [242]
Forum / Ontario Secondary School Teachers' Federation – v8 n2-v12 n4 [1982 mar/apr-1986 dec/1987 jan], v14 n1-2 [1988 mar/apr-may/jun] – 1r – 1 – (cont: bulletin [ontario secondary school teachers' federation]; cont by: education forum [toronto, ont: 1988]) – mf#1532630 – us WHS [373]
Forum / People's Party of the United States – Denver CO. 1904 nov 3 – 1r – 1 – mf#854373 – us WHS [325]
Forum – Providence, RI. 1946-1950 (1) – mf#66305 – us UMI ProQuest [071]
Forum / Ray-O-Vac Corporation – 1980 feb/mar-1985 jun – 1r – 1 – (cont: ray-o-vac forum; cont by: inside rayovaC) – mf#962161 – us WHS [071]
Forum – Sheridan, MT. 1911-1943 (1) – mf#64648 – us UMI ProQuest [071]
Forum – Tacoma, WA. 1903-1918 (1) – mf#67148 – us UMI ProQuest [071]
Forum / US Military Community Aschaffenburg [Germany] – Aschaffenburg. 1982 dec 22, 1983 feb 16, aug 3, sep 14, oct 12-nov 9, 1984 feb 29 – 1r – 1 – (cont by: aschaffenburg forum) – mf#1000936 – us WHS [355]
Forum – Washington. 1977-1981 (1,5,9) – mf#11533 – us UMI ProQuest [610]
Forum : zeitung fuer geistige probleme der jugend – Berlin DE, 1957-1983 31 mar – 1 – (filmed by other misc inst: 1950 27 jan-1983 31 mar [15r]) – gw Misc Inst [360]
O forum : folha judiciaria e accidentalmente politica e litteraria – Recife, PE: Typ do Forum, 12 mar-09 maio 1868 – (= ser Ps 19) – bl Biblioteca [073]
The forum (aba) *see* Tort and insurance law journal (aba)
Forum and century – New York. 1886-1940 (1) – mf#5325 – us UMI ProQuest [900]
Forum and daily tribune : (morning edition) – Fargo, ND. 1925-1966 (1) – mf#65358 – us UMI ProQuest [071]
The forum, bench and bar review – n1-8. 1874-75 (all publ) – 19mf – 9 – $28.50 – mf#LLMC 82-923 – us LLMC [340]
Forum for applied research and public policy – Knoxville. 1989+ (1,5,9) – ISSN: 0887-8218 – mf#17651 – us UMI ProQuest [333]
Forum for changing men – n18-58 [1975 sep-1980 mar/apr] – 1r – 1 – (cont: changing men) – mf#528842 – us WHS [071]
Forum for the Evolution of Progressive Arts *see* Blowin
Forum letter – Delhi. 1973-1996 (1) 1975-1996 (5) 1975-1996 (9) – ISSN: 0046-4732 – mf#8613 – us UMI ProQuest [240]
Forum. madjalah umum mahasiswa *see* Gerakan mahasiswa
Forum peninsula herald – Forks, WA. 1970-1979 (1) – mf#67001 – us UMI ProQuest [071]
Forumeer / American G I Forum – 1970 may-1978 apr, v30 n5-11/12 [1978 may-nov /dec, 1979 jun iss, [1979 jul]-dec, [v31 n2]-v31 n6 [1979 feb-1980 jun] – 1r – 1 – mf#633443 – us WHS [355]
Forumeer – El Paso. 1972-1976 (1) – ISSN: 0015-1287 – mf#7380 – us UMI ProQuest [305]
Forverts = Forward – New York: Jewish Socialist Press Federation, jul 5-dec 1951 – us CRL [071]
Forverts = Jewish daily forward – Chicago: Forward Association, mar 1919-jun 1951 – us CRL [071]
Forverts = Jewish daily forward – Chicago IL. 1934 sep-oct, 1934 nov-dec, 1935 jan 1-feb 28, 1935 mar 1-apr 30, 1936 jul 1-aug 31, 1936 nov 15-dec 31, 1936 oct 1-nov 14, 1936 sep 1-oct 31, 1937 apr 1-may 31, 1937 aug 1-31, 1937 jun 1-jul 31, 1938 jul 1-aug 31, 1938 may 1-jun 30, 1939 jul 1-aug 31, 1939 nov 1-dec 31, 1939 sep 1-oct 31, 1940 jan 1-feb 29, 1940 jul 1-aug 31, 1940 mar 1-apr 30, 1940 may 1-jun 30 – 20r – 1 – (cont by: forverts) – mf#846251 – us WHS [071]
Forward – Glasgow. Scotland. -w. 13 Oct 1906-Dec 1921; 1923. (10 reels) – 1 – uk British Libr Newspaper [072]
Forward – Johannesburg: Commercial Print Co Ltd, 1924-oct 17-dec 5 1952; jan 2-9 1953 – (issues filmed as pt of: st clair drake collection of africana) – us CRL [071]
Forward – Johannesburg. South Africa. -w. Dec 1924-Dec 1926. (1 reel) – 1 – uk British Libr Newspaper [072]
Forward / League of Women Voters of Wisconsin – 1921 jun, sep-1922 mar – 1r – 1 – (cont by: forward [milwaukee, wi: 1922]) – mf#904914 – us WHS [322]
Forward / League of Women Voters of Wisconsin – 1952 jan-1985 jul – 1r – 1 – (cont: forward [milwaukee, wi]) – mf#844780 – us WHS [322]
Forward – n2-60 [1971 may-1977 aug] – 1r – 1 – mf#384136 – us WHS [071]
Forward – Philadelphia. Pa. Vorwarts. 1903-67 – 1 – us AJPC [071]
Forward – Racine WI. 1933 oct 13,20, nov 3 – 1r – 1 – mf#966618 – us WHS [071]
Forward – v1 n1-v8 n5=51 [1974 jul-1984 nov/dec] – 1r – 1 – mf#647044 – us WHS [071]
Forward – v1 n2 [1995 mar], v1 n10 [1995 dec] – 1r – 1 – mf#3297638 – us WHS [071]
Forward *see* Independent labour party newspapers
Forward, 1904-1908 / west bradford gazette, 1905-1906 : from bradford central library – 1r – 1 – (filmed with: west bradford gazette, 1905-06) – mf#97017 – uk Microform Academic [072]
Forward bloc – Calcutta, India. Aug 1939-Jul 1940 – 1r – 1 – us L of C Photodup [079]
The forward movement in religious thought as interpreted by unitarians : five lectures / Herford, Brooke – London: Philip Green, 1895 – 1mf – 9 – 0-524-07688-X – mf#1991-3273 – us ATLA [243]
Forward press – 1928 sep-1943 jan – 1r – 1 – (cont: campaigner [madison, wi]) – mf#3500384 – us WHS [071]
Forward times – Houston, TX. 1960-1983 (1) – mf#66618 – us UMI ProQuest [071]
Forward to freedom : constitutional proposals for a united nigeria – Lagos, 1957 – (filmed with: united labour congress, nigeria. a program for the future and trades union congress of nigeria. social security committee. report to the first annual conference.) – us CRL [321]
Forwerg, Walter *see* Darstellung und eigenschaften der metagermanate des mangans, eisens und kobalts
Forwood, George *see* An examination of the corn returns for the years, 1826, 1827, 1828, and 1829
Fory, M R *see* Premature church-membership
Forzanini, G P *see* Canzone nella nativita di nostro signor giesv christo
Forzano, Giovacchino *see* Cents jours
Fosdick, Harry Emerson *see*
– The assurance of immortality
– The manhood of the master
– The meaning of faith
– The meaning of prayer
– The second mile
Fosdick, Raymond Blaine *see* Public service, international affairs and rockefeller philanthropy
Fosforescencias / Henriquez Urena, Max – Santiago, Cuba. 1930 – 1r – us UF Libraries [972]
Foshee families of america newsletter – v1 n1-v3 n4 [1980 aug-1983 may] – 1r – 1 – mf#1209350 – us WHS [929]
The fo-sho-hing-tsan-king (stbe19) : a life of buddha / Bodhisattva, Asvaghosha – 1883 – (= ser Sacred book of the east (sbte)) – 8mf – 8 – €17.00 – (trans fr sanskrit into chinese by dharmaraksha, a d 420, and fr chinese into english by samuel beal) – ne Slangenburg [280]
Foss, A T *see* Facts for baptist churches
Foss, Claude William *see* Glimpses of three continents
Foss, Cyrus David *see* From the himalayas to the equator
Foss, James Henry *see* Gentleman from everywhere
Foss, Rudolf *see*
– Lebensbilder aus dem zeitalter der reformation
– Die tage trajans und hadrians – leben und schriften agobards, erzbischofs von lyon
Fossatis, D *see* Currus triumphales adventum clarissimorum moschoviae principum paul petrovitz et mariae theodorownae conjugis...in divi marci venetiarum foro die 22 januarii an 1782
Fossatis, G *see* Currus triumphales adventum clarissimorum moschoviae principum paul petrovitz et mariae theodorownae conjugis...in divi marci venetiarum foro die 22 januarii an 1782
Fosse, J C de la *see*
– Nouvelle iconologie historique ou attributs hierogliphyques
– Nouvelle iconologie historique ou attributs hieroglyphiques
Fossey, C *see*
– La magie assyrienne
– Textes assyriens et babyloniens relatifs a la divination
Fossey, Charles *see* La magie assyrienne
Fossi-Barroeta, Luis *see* Politica en tono menor
Fossil algae from guatemala / Johnson, Jesse Harlan – Golden, CO. 1965 – 1r – us UF Libraries [560]
Fossil journal – Fossil OR: S P Shutt, 1886-75 [wkly] – 1 – (merged with: condon globe-times (1919-75) to form: times-journal (1975-).1927-34 incl newspaper publ during school terms by wheeler county high school) – us Oregon Lib [071]
Fossil journal *see*
– Condon globe-times
– Times-journal
Fossil men and their modern representatives : an attempt to illustrate the characters and condition of pre-historic men in europe, by those of the american races / Dawson, John William – [2nd ed] London, 1883 – (= ser 19th c evolution & creation) – 4mf – 9 – mf#1.1.6479 – uk Chadwyck [573]
Fossil sponges : and other organic remains from the quebec group at little metis / Dawson, John William – Montreal: s.n, 1897 – 1mf – 9 – mf#06384 – cn Canadiana [560]
Fossile hoelzer aus ostasien und aegypten / Schenk, A – 1mf – 8 – mf#7670 mf. 223 – ne IDC [930]
Fossiles / Curel, Francois De – Paris, France. 1893 – 1r – us UF Libraries [440]
Fossiles / Curel, Francois De – Paris, France. 1900 – 1r – us UF Libraries [440]
Foster, Addison Pinneo *see* Four pastorates
Foster, Alfred Edye Manning *see* Anglo-catholicism
Foster and Co, Allen *see* Autumn and winter season sketchbook
Foster and mackenzie's canadian farmer's almanac for the year of our lord... – Richmond: Foster & MacKenzie, [1869]- (yrly) [mf ed 1988] – 1mf – 9 – (cont: foster and macleay's canadian farmer's almanac for the year of our lord...; ceased 187-?) – mf#A00654 – cn Canadiana [630]
Foster and macleay's canadian farmer's almanac for the year of our lord... – Richmond: Foster & MacLeay, [186–1868] (yrly) [mf ed 1988] – 1mf – 9 – (cont by: foster and mackenzie's canadian farmer's almanac for the year of our lord...) – mf#A00653 – cn Canadiana [630]
Foster, Arnold *see* Christian progress in china
Foster, Arthur C *see* Celery diseases in florida
Foster, Bernadette L *see* An examination of the bone mineral density status of women with an intellectual disability and risk factors associated with the acquisition of osteoporosis
[Foster city-] foster city progress – CA. 1966-1977 (incomplete); 1978- – 23r – 1 – $1380.00 (subs $90y) – mf#B02247 – us Library Micro [071]
Foster, Finley Milligan *see* The witnessing church
Foster, Frances Allen *see* The northern passion
Foster, Frank Hugh *see*
– Christian life and theology
– The fundamental ideas of the roman catholic church
– A genetic history of the new england theology
– Outline of lectures in systematic theology
– The seminary method of original study in the historical sciences
– The teaching of jesus concerning his own mission
Foster, George Burman *see*
– The finality of the christian religion
– The function of religion in man's struggle for existence
Foster, George Eulas *see*
– The canada temperance manual and prohibitionist's handbook
– Expose budgetaire
– Expose budgetaire par l'hon george e foster... ministere des finances
– The onward march of fifty years
– Some problems of empire
Foster, Harry La Tourette *see*
– A beachcomber in the orient
– Combing the caribbees
– If you go to south america
Foster, Henry La Tourette *see* Caribbean cruise
Foster, James Mitchell *see*
– Christ the king
– Reformation principles stated and applied
Foster, John *see*
– Aux directeurs de la compagnie du chemin de fer de phillipsburg, farnham et yamaska
– Critical essays
– Essay on the evils of popular ignorance
– An essay on the improvement of time
– Fragmentary notes of village sermons
– Glory of the age
– A new system of wooden railways
– Report on the phillipsburg, farnham and yamaska railway
Foster, John Onesimus *see* Life and labors of mrs. maggie newton van cott
Foster, John W *see* A century of american diplomacy
Foster, Joseph *see*
– A pedigree of the forsters and fosters, of the north of england
– Pedigrees of the county families of yorkshire... and authenticated by the members of each family
– The visitation of yorkshire, made in the years 1584-85, by robert glover, somerset herald..
Foster, Michael *see* Examination of the scheme of church-power laid down
Foster, Mulford Bateman *see* Brazil
Foster, Ora Delmer *see* The literary relations of "the first epistle of peter"
Foster, Randolph Sinks *see*
– Centenary thoughts for the pew and pulpit of methodism in eighteen hundred and eighty-four
– Creation
– Evidences of christianity
– God
– Objections to calvinism as it is
– Philosophic basis of theology
– Philosophy of christian experience
– Sin
– Theism
– Union of episcopal methodisms
Foster, Robert Frederick *see* Foster's complete bridge
Foster, Robert Verrell *see*
– A brief introduction to the study of theology
– A commentary on the epistle to the romans
– The lord's prayer
– Old testament studies
– Systematic theology
Foster, Roger *see*
– The federal judiciary acts of 1875 and 1887
– A treatise on pleading and practice in equity in the courts of the united states.
Foster, W *see* Letters received by the east india company
Foster, William *see*
– Early travels in india, 1583-1619
– The embassy of sir thomas roe to india, 1615-19
– The founding of fort st george, madras
Fosterlandet – Chicago IL. 1898 jan 5-dec 28 – 1r – 1 – (cont: framat [chicago, il]) – mf#4019768 – us WHS [071]
Foster's complete bridge / Foster, Robert Frederick – Toronto: Musson, c1905 [mf ed 1996] – 4mf – 9 – 0-665-77563-6 – mf#77563 – cn Canadiana [790]
Foster's monthly reference list – Providence. 1881-1884 – 1 – mf#4614 – us UMI ProQuest [073]
Fothergill, Gerald *see* A list of emigrant ministers to america, 1690-1811
Fotheringham, L Monteith *see* Adventures in nyasaland.
Fotheringham, Thomas Francis *see*
– Atlas of the presbytery of st john, n b
– Church union as affected by the question of valid orders
Foto – Antwerp. 1950-1953 (1) – ISSN: 0015-8682 – mf#465 – us UMI ProQuest [770]
Foto news – Merrill WI. 1982 dec 21/1983 jun-2003 jan/jun [gaps] – 40r – 1 – (cont: merrill shopper & foto news) – mf#1131398 – us WHS [770]
Fouard, Constant *see*
– The christ, the son of god
– Saint paul, ses missions
– Saint peter and the first years of christianity
Foucart, George *see*
– Histoire des religions et methode comparative
– La methode comparative dans l'histoire des religions

Foucart, Paul Francois *see*
– Des associations religeuses chez les grecs
– Les mysteres d'eleusis

Fouchard, Jean *see*
– Aftistes et repertoire des scenes de saint-domingu
– Marrons du syllabaire
– Plaisirs de saint-domingue
– Theatre a saint-domingue

Fouche, Joseph *see*
– Memoirs of joseph fouche...minister of the general police of france
– Memoirs relating to fouche, minister of police under napoleon 1

Foucher, Alfred *see* On the iconography of the buddha's nativity

Foucher, Paul *see*
– Don sebastien de portugal
– Redgauntlet
– Yseult raimbaud

Foucher, Paul Henri *see* Pacte de famine

Fouere, Rene *see* Krishnamurti

Le fouet *see* Le heraut d'armes

Le fouet national – 20 no. Paris. sept 1789-mai 1790 – 1 – (puis tableau de l'europe et de la france corrigee.) – fr ACRPP [073]

Le fouet theatral et litteraire – puis Le Fouet. Art, theatre, litterature. Paris. mars 1868-avr 1869. suivi de: Le Heraut d'armes voir a ce titre – 1 – fr ACRPP [790]

Fougeret, W A *see* Histoire generale de la bastille

Fouillee, Alfred *see*
– La morale, l'art et la religion d'apres m guyau
– La philosophie de socrate

Les fouilles de merida / Melida, Jose Ramon – Archaeciogischen Instituts – 1 – sp Bibl Santa Ana [930]

Le foulbe du nord-cameroun / Etia, Abel Moume – Bergerac: H Trillaud, 1948 – 1 – us CRL [960]

Foulche Delbosc, Isabel *see* Bibliografia de r. foulche-delbosc new york

Foulche-Delbosc, Raymond *see* Essai sur les origines du romancero

Foulis, Christa *see* Predicting the outcome of mild closed head injury using the glasgow coma scale-extended

Foulke, Hugh *see* Memoranda and reflections of rebecca price

Foulkes, Edmund Salisbury *see* The church's creed, or the crown's creed?

Foulks, Theodore *see* Eighteen months in jamaica

Foulon, J-b *see* Itineraire general topographique et hydraulique de la france

Foulston, John *see* The public buildings, erected in the west of england

The foundacion of rhetorike / Reynolds, Richard – 1563 – 9 – us Scholars Facs [420]

Foundation : catalogue de la bibliotheque / Teyler – Philadelphia. 1973-1973 (1) – 21mf – 9 – mf#8578 – ne IDC [020]

The foundation – Atlanta GA: Stewart Missionary Foundation for Africa 1911- [irreg, mthly, bimthly, qrtly] [mf ed 2004] – v1-82 (1911-1977/78) on 3r – 1 – (lacks v47 n1-4; v48 n2-4; v49 n1-2,4; v50 n2-4; v51 n1,3-4; v52 n1 p5-8; v56 n4]; numbering irreg fr 1968-fall 1977-78; publ by the stewart missionary foundation for africa of gammon theological seminary) – ISSN: 0363-6992 – mf#2004-s057 – us ATLA [242]

Foundation action / National Right to Work Legal Defense Foundation – 1981 apr-1982 jun – 1r – 1 – (cont: foundation action in the courts) – mf#654866 – us WHS [322]

Foundation action in the courts / National Right to Work Legal Defense Foundation – 1974 1st qtr-1980 3rd qtr – 1r – 1 – (cont by: foundation action) – mf#674008 – us WHS [322]

Foundation and endowment money management – London. 1998+ (1,5,9) – mf#32369 – us UMI ProQuest [332]

Foundation characters *see* Ping min tsien tzu ko (ccm128)

Foundation Church of the Millennium *see*
– Foundation, visions of the millennium
– Founders

Foundation for Global Peace *see* Bits and peaces

Foundation for P.E.A.C.E. *see* Peace in action

Foundation for reformation research : bulletin of the library – 1966-73 [complete] – 1r – 1 – mf#ATLA S0599 – us ATLA [240]

Foundation for reformation research : monthly newsletter – 1967-88 [complete] – 2r – 1 – mf#ATLA S0600 – us ATLA [240]

Foundation for Reformation Research Library *see* Bulletin of the library

Foundation for reformation research newsletter – St. Louis. 1972-1973 (1) 1967-1973 (5) (9) – (cont by: center for reformation research newsletter) – ISSN: 0360-9707 – mf#6400 – us UMI ProQuest [242]

Foundation for Study of Treaty Law *see* Treaty law manual

Foundation news – Washington. 1990-1994 (1,5,9) – (cont by: foundation news and commentary) – ISSN: 0015-8976 – mf#18360 – us UMI ProQuest [650]

Foundation news and commentary – Washington. 1994-1996 (1,5,9) – (cont: foundation news) – ISSN: 1076-3961 – mf#18360,01 – us UMI ProQuest [650]

The foundation of christian hope / Herndon, Eugene Wallace – Nashville, Tenn: McQuiddy Print Co, 1904 – 1mf – 9 – 0-524-06421-0 – mf#1991-2543 – us ATLA [240]

Foundation of montreal, 250th anniversary celebration 1892 : organizing committees, list of members – S.l: Gazette Printing Co, 1892? – 1mf – 9 – mf#03159 – cn Canadiana [971]

The foundation of professional success / Pryor, Roger Atkinson – Chicago: La Salle, Extension University 1911. 18p. LL-1193 – 1 – us L of C Photodup [340]

The foundation of the university of cambridge *see*
– Papers relating to the bohemian loan, 1620-1622
– The religion of a christian

Foundation rites, with some kindred ceremonies : a contribution to the study of beliefs, customs, and legends connected with buildings, locations, landmarks, etc., etc / Burdick, Lewis Dayton – New York: Abbey Press, c1901 – 1mf – 9 – 0-524-00702-0 – mf#1990-2030 – us ATLA [390]

Foundation truths of scripture as to sin and salvation : in twelve lessons / Laidlaw, John – Edinburgh: T & T Clark; New York: Scribner [distributor], [1897] – 1mf – 9 – 0-7905-1219-X – mf#1987-1219 – us ATLA [220]

Foundation truths of the gospel : essays contributed to the christian – London: Morgan and Scott, [18–?] – 1mf – 9 – 0-7905-0113-9 – (incl bibl ref) – mf#1987-0113 – us ATLA [240]

Foundation, visions of the millennium / Foundation Church of the Millennium – v1-3 [1974 oct/dec-1976 spring] – 1r – 1 – mf#368282 – us WHS [242]

Foundations – Valley Forge. 1958-1982 (1) 1969-1982 (5) 1976-1982 (9) – (cont by: american baptist quarterly) – ISSN: 0015-8992 – mf#2110 – us UMI ProQuest [242]

Foundations : a statement of christian belief in terms of modern thought / Streeter, Burnett Hillman et al – London; New York: Macmillan, 1912 – 2mf – 9 – 0-7905-2154-7 – (incl ind) – mf#1987-2154 – us ATLA [240]

Foundations : a survey of 25 years of activity of the palestine foundation fund-keren hayesod. facts and figures, 1921-1946 / Ulitzer, A – Jerusalem, 1947 – 3mf – 9 – mf#J-28-41 – ne IDC [956]

Foundations / Ulitzur, A – Jerusalem, Israel. 1946 – 1r – us UF Libraries [939]

The foundations : a series of lectures on the evidences of christianity / Gibson, John Monro – 2nd ed. Chicago: Jansen, McClurg, 1880 – 1mf – 9 – 0-8370-3265-2 – mf#1985-1265 – us ATLA [230]

The foundations of a creed / Lewes, George Henry – 3rd ed. London: Truebner, 1874-1875 – (= ser Problems of Life and Mind) – 3mf – 9 – 0-524-00280-0 – mf#1989-2980 – us ATLA [100]

The foundations of belief : being notes introductory to the study of theology / Balfour, Arthur James, 1st earl of – London, 1895 – (= ser 19th c evolution & creation) – 4mf – 9 – mf#1.1.11292 – uk Chadwyck [110]

The foundations of faith : considered in eight sermons / Wace, Henry – 2nd ed. London: John Murray, 1886 – 1mf – 9 – 0-8370-5678-0 – (incl bibl ref) – mf#1985-3678 – us ATLA [210]

Foundations of freedom / Cowen, Denis Victor – Cape Town, South Africa. 1961 – 1r – us UF Libraries [960]

The foundations of history : a series of first things / Schieffelin, Samuel Bradhurst – 3rd ed. New York: Anson D F Randolph, 1864, c1863 – 1mf – 9 – 0-8370-5089-8 – mf#1985-3089 – us ATLA [900]

The foundations of legal liability. / Street, Thomas Atkins – Northport, N.Y., Thompson, 1906. 3 v. LL-1502 – 1 – us L of C Photodup [340]

The foundations of living faiths : an introduction to comparative religion / Bhattacaryya, Haridasa – [Calcutta]: University of Calcutta, 1938- – (= ser Samp: indian books) – us CRL [230]

The foundations of modern europe – 2313mf – 9 – (series 1: the archives of the european movement – from the archives of the european university institute, florence ca 2100mf. with printed guide) – us Primary [940]

The foundations of nationality : a discourse preached...after the great railway celebration, november 1856 / Cordner, John – Montreal: H Rose, 1856 – 1mf – 9 – mf#35475 – cn Canadiana [240]

The foundations of our faith : papers read before a mixed audience of men / Auberlen, Carl August et al – London:Alexander Strahan, 1867 – 1mf – 9 – 0-8370-2459-5 – mf#1985-0459 – us ATLA [240]

Foundations of photography : a tutorial series / Keeling, Derek – 16mf – 9 – £60.00 – (incl text) – mf#FOP – uk World [770]

Foundations of physics – New York. 1970-1996 (1) 1970-1978 (5) – ISSN: 0015-9018 – mf#10855 – us UMI ProQuest [530]

The foundations of religion / Cook, Stanley Arthur – London: T C & E C Jack; New York: Dodge [1914?] [mf ed 1989] – (= ser The people's books 119) – 1mf – 9 – 0-7905-0876-1 – (incl ind) – mf#1987-0876 – us ATLA [200]

The foundations of religion : a study of first principles / Slicer, Thomas Roberts – New York: T R Slicer 1902 [mf ed 1985] – 1mf – 9 – 0-8370-5270-X – mf#1985-3270 – us ATLA [200]

The foundations of religious belief : the methods of natural theology vindicated against modern objections / Wilson, William Dexter – New York: D Appleton 1883 [mf ed 1985] – (= ser The bishop paddock lectures 1883) – 1mf – 9 – 0-8370-5866-X – (incl bibl ref) – mf#1985-3866 – us ATLA [210]

Foundations of success and laws of trade : book devoted to business and its successful prosecution... – London, Ont: S Smith, 1880 – 6mf – 9 – (incl ind) – mf#33638 – cn Canadiana [650]

The foundations of the bible : studies in old testament criticism / Girdlestone, Robert Baker – 4th rev ed. London: Eyre & Spottiswoode, 1892 – 1mf – 9 – 0-8370-2533-8 – (includes appendix & indexes) – mf#1985-0533 – us ATLA [220]

The foundations of the christian faith / Rishell, Charles Wesley – New York: Eaton & Mains; Cincinnati: Curts & Jennings, c1899 – (= ser Library of Biblical and Theological Literature) – 2mf – 9 – 0-7905-8567-7 – mf#1989-1792 – us ATLA [240]

The foundations of the english church / Maude, Joseph Hooper – London: Methuen, 1909 – (= ser Handbooks of english church history) – 1mf – 9 – 0-7905-5013-X – (incl bibl ref) – mf#1988-1013 – us ATLA [240]

Foundations of the nineteenth century = Grundlagen des neunzehnten jahrhunderts / Chamberlain, Houston Stewart – New York: John Lane, 1914 – 1mf – 9 – 0-524-03391-9 – (in english) – mf#1990-0945 – us ATLA [900]

The founder of christendom : an address delivered before the unitarian club of toronto / Smith, Goldwin – Toronto: G N Morang, 1903 – 1mf – 9 – 0-665-76470-7 – mf#76470 – cn Canadiana [240]

Founder of Divine Science Organization *see* [San francisco-] harmony

The founder of mormonism : a psychological study of joseph smith, jr / Riley, Woodbridge – New York: Dodd, Mead, 1903, c1902 – 2mf – 9 – 0-7905-6355-X – (incl bibl ref) – mf#1988-2355 – us ATLA [243]

Founders / Foundation Church of the Millennium – v1-4 n2 [1974 jul-1977 [feb?]] – 1r – 1 – mf#368279 – us WHS [242]

The founders and first three presidents of the bible society / Morris, Henry – London: Religious Tract Society, [1890?] – 1mf – 9 – 0-8370-6587-9 – mf#1986-0587 – us ATLA [920]

Founders and foundations of florida horticulture / Rolfs, P H – Tallahassee, FL. 1935 – 1r – us UF Libraries [630]

The founders and rulers of united israel : from the death of moses to the division of the hebrew kingdom / Kent, Charles Foster – New York: Charles Scribner, 1913, c1908 – 1mf – 9 – 0-7905-0042-6 – mf#1987-0042 – us ATLA [939]

Founders of canadian banking : the hon wm allan, merchant and banker / Shortt, Adam – S.l: s.n, 18–? – 1mf – 9 – mf#18856 – cn Canadiana [332]

Founders of old testament criticism : biographical, descriptive and critical studies / Cheyne, Thomas Kelly – New York: Scribner's, 1893 [mf ed 1984] – (= ser Biblical crit – us & gb 65) – 5mf – 9 – 0-8370-0173-0 – mf#1984-1065 – us ATLA [221]

Founders of the empire / Gibbs, Philip Hamilton – [London], 1899 – (= ser 19th c books on british colonization) – 3mf – 9 – (with four coloured plates and ill) – mf#1.1.7334 – uk Chadwyck [941]

Founders of vijayanagara / Srikantaya, Saklespur – Bangalore City: Mythic Society, 1938 – (= ser Samp: indian books) – us CRL [954]

Founding a protectorate / Sillery, Anthony – London, England. 1965 – 1r – us UF Libraries [960]

The founding of fort st george, madras / Foster, William – London: printed by Eyre and Spottiswoode, 1902 – (= ser Samp: indian books) – us CRL [720]

The founding of the church / Bacon, Benjamin Wisner – Boston: Houghton Mifflin, 1909 – (= ser Modern Religious Problems) – 1mf – 9 – 0-7905-1501-6 – mf#1987-1501 – us ATLA [240]

Founding of the hampton institute / Armstrong, Samuel Chapman – Boston: Directors of the Old South Work, [1904?] – 1mf – 9 – 0-524-04306-X – mf#1990-1232 – us ATLA [370]

The founding of the kashmir state : a biography of maharajah gulab singh, 1792-1858 / Panikkar, Kavalam Madhava – London: George Allen & Unwin, 1953 – (= ser Samp: indian books) – us CRL [920]

Foundry – Cleveland. 1892-1974 (1) 1965-1974 (5) – (cont by: foundry management and technology) – ISSN: 0015-9034 – mf#1076 – us UMI ProQuest [660]

Foundry cub / United Farm Equipment and Metal Workers of America – v1 n1 [1951 mar 1], v3 n5-6 [1951 aug 9-14] – 1r – 1 – mf#3629210 – us WHS [331]

Foundry management and technology – Cleveland. 1974+ (1) 1974+ (5) 1977+ (9) – (cont: foundry) – ISSN: 0360-8999 – mf#1076,01 – us UMI ProQuest [660]

Foundry trade journal – London. 1904-1990 (1) 1977-1990 (5) 1977-1990 (9) – ISSN: 0015-9042 – mf#1294 – us UMI ProQuest [660]

Fountain city beacon – Fountain City WI. 1856 aug 1,15,27, sep 10, dec 13 – 1r – 1 – (cont by: fountain city beacon, and buffalo, dunn, chippewa, trempealeau and clark counties advertiser) – mf#918287 – us WHS [071]

Fountain city daily herald – Fond Du Lac WI. 1856 mar 31-sep 2 – 1r – 1 – (cont by: western freeman; fond du lac weekly commonwealth) – mf#941387 – us WHS [071]

Fountain city daily herald – Fond Du Lac WI. 1854 jul 24, 28, aug -7, 10, 16-sep 4, 7-26, 29-oct 10, 14-20 – 1r – 1 – mf#941382 – us WHS [071]

Fountain city herald – Fond Du Lac WI. 1852 nov 9-1956 sep 2, 1854 mar 7 – 2r – 1 – (cont by: western freeman; fond du lac weekly commonwealth) – mf#941389 – us WHS [071]

Fountain county neighbor – Attica, IN. 1998+ (1) – mf#69526 – us UMI ProQuest [071]

Fountain herald *see* El paso county miscellaneous newspapers, reel 2

Fountain ledger – Attica, IN. 1864-1866 (1) – mf#62716 – us UMI ProQuest [071]

Fountain of light / Rainbow Publications – n2, 8 [1969 jun, nov] v1 n7 [1969 oct] – 1r – 1 – mf#1583332 – us WHS [071]

Fountain, Paul *see* River amazon from its source to the sea

[Fountain valley-] fountain valley daily pilot – CA. 1966 (Scats) – 8r – 1 – $480.00 – mf#H04011 – us Library Micro [071]

Fountain valley news *see* El paso county miscellaneous newspapers

The fountain valley news *see* El paso county miscellaneous newspapers

Fountain warren democrat – Attica, IN. 1898-1966 (1) – mf#62717 – us UMI ProQuest [071]

Fouquet, Karl *see* Jakob ayrers "sidea", shakespeares "tempest" und das maerchen

Fouquet, Leon *see* Luciferianism

Fouquet, Leon Charles *see* Diary

Four advent lectures on concordats / Wiseman, Nicholas Patrick – London, England. 1855 – 1r – us UF Libraries [240]

Four african political systems / Potholm, Christian P – Englewood Cliffs, NJ. 1970 – 1r – us UF Libraries [325]

Four aspects of the music of the beatles : instrumentation, harmony, form, and album unity / Stetzer, Charles William – U of Rochester 1976 [mf ed 19–] – 3mf – 9 – (with bibl) – mf#fiche 1124 – us Sibley [780]

The four canadian highwaymen : or, the robbers of markham swamp / Collins, Joseph Edmund – Toronto: Rose, 1886 – 2mf – 9 – mf#08597 – cn Canadiana [830]

Four centuries of florida ranching / Dacy, George H – St Louis, MO. 1940 – 1r – us UF Libraries [636]

Four centuries of nonconformist disabilities, 1509-1912 / Edwards, William – London: National Council of Evangelical Free Churches, [1912?] – 1mf – 9 – 0-524-05177-1 – mf#1990-5096 – us ATLA [240]

Four centuries of portuguese expansion, 1415-1825 / Boxer, Charles Ralph – Johannesburg, South Africa. 1965, c1961 – 1r – us UF Libraries [960]

Four centuries of scottish psalmody / Patrick, M – London, 1949 – 4mf – 8 – €11.00 – ne Slangenburg [242]

Four centuries of shakespeare : the prompt books : basic documentary sources concerning text and performance – 4 series – (= ser Shakespeare and the Stage) – 215r coll – 1 – (previous title: shakespeare and the stage) – mf#C35-12800 – us Primary [790]

Four centuries of shakespeare: the prompt books, series 1 : basic documentary sources concerning text and performance fromthe folger shakespeare library collection, washington d.c. – 4pt-coll – 86r – 1 – (pt 1: all's well that ends well – julius caesar 21r. pt 2: king henry 4 pt 1 – macbeth 25r. pt 3: measure for measure – pericles 18r. pt 4: romeo and juliet – the winter's tale and misc vols 22r. with guide) – mf#C35-12810 – us Primary [790]

Four centuries of shakespeare: the prompt books, series 2 : basic documentary sources concerning text and performance from the harvard theatre collection – 2pt-coll – 34r – 1 – (pt 1: all's well that ends well – love's labours lost 16r. pt 2: macbeth – the winter's tale and misc vols 18r. with guide) – mf#C35-12820 – us Primary [790]

Four centuries of shakespeare: the prompt books, series 3 : basic documentary sources concerning text and performance from the shakespeare library collection, birmingham public library – 10r – 1 – (covers a great range of productions in england between 1811 and 1929. with guide) – mf#C35-12830 – us Primary [790]

Four centuries of shakespeare: the prompt books, series 4 : basic documentary sources concerning text and performance from the shakespeare centre library collection, stratford-upon-avon – 4pt-coll – 85r – 1 – (pt 1:all's well that ends well – henry 4 pt 2 21r. pt 2: henry 5 – measure for measure 24r. pt 3: the merchant of venice – richard 3 21r. pt 4: romeo and juliet – the winter's tale 19r. with guide) – mf#C35-12840 – us Primary [790]

Four centuries of silence, or, from malachi to christ / Redford, Robert Ainslie – 2nd ed. Chicago: AC McClurg, 1887 – 1mf – 9 – 0-8370-4851-6 – mf#1985-2851 – us ATLA [270]

Four centuries of the panama canal / Johnson, Willis Fletcher – New York, NY. 1906 – 1r – us UF Libraries [972]

Four chapters / Tagore, Rabindranath – Calcutta: Visva-bharati, 1950 – (= ser Samp: indian books) – us CRL [490]

Four commemorative discourses : delivered on his sixty-third birth-day, february 19th, 1865, on the fortieth anniversary of his installation, march 12th, 1865, and on his retirement from pastoral duties, september 9th, 1866 / Bacon, Leonard – New Haven: TJ Stafford, 1866 – 1mf – 9 – 0-524-03313-7 – mf#1990-4673 – us ATLA [240]

Four conferences touching the operation of the holy spirit : delivered at newark, n.j / Ewer, Ferdinand Cartwright – New York: GP Putnam, 1880 – 1mf – 9 – 0-524-08294-4 – mf#1993-3049 – us ATLA [240]

Four conquest of england / St John, James Augustine – London: Smith Elder & Co., 1862. 448p – 1 – us Wisconsin U Libr [941]

Four constituents of the christian character / Wicksteed, Charles – London, England. 1848 – 1r – us UF Libraries [240]

The four days of mayaguez / Rowan, Roy – 1st ed. New York: Norton c1975 [mf ed 1989] – 1r with other items – 1 – mf#mf-10289 seam reel 020/01 [§] – us CRL [327]

Four devils / Fowler, Guy – New York, NY. 1928 – 1r – us UF Libraries [025]

Four early pamphlets, 1783-84 / Godwin, William – Reprint Gainesville, 1966; Delmar, 1977.Introd. by Burton R. Pollin. Essays on politics, education and literature – 9 – us Scholars Facs [320]

Four essays on colonial slavery / Jeremie, John – London: Printed for J Hatchard, 1831 – 2mf – 9 – mf#21353 – cn Canadiana [306]

The four evangelists : with the distinctive characteristics of their gospels / Thomson, Edward A – Edinburgh: T & T Clark, 1868 – 1mf – 9 – 0-8370-9316-3 – mf#1986-3316 – us ATLA [225]

Four gospels : irgizskaia kollektsiia [irgiz collection] – late 1400s-early 1500s – 11mf – 9 – (russian version) – us UMI ProQuest [090]

Four gospels – 1470s – 12mf – 9 – (russian version) – us UMI ProQuest [090]

Four gospels – late 1300s – 9mf – 9 – (russian version) – us UMI ProQuest [090]

Four gospels – 1450s-90s – 13mf – 9 – (russian version with south slavic characteristics) – us UMI ProQuest [090]

Four gospels : vetkovskoe sobranie [vetka collection] – 1470s-90s – 10mf – 9 – (russian version, with south slavic characteristics) – us UMI ProQuest [090]

The four gospels / Fisher, Robert Howie – London: Hodder and Stoughton, 1899 – (= ser Little books on religion) – 1mf – 9 – 0-7905-3130-5 – mf#1987-3130 – us ATLA [226]

The four gospels : from the latin text of the irish codex harleianus numbered harl. 1023 in the british museum library / Buchanan, Edgar Simmons – London: Heath Cranton & Ouseley, 1914 – 1mf – 9 – 0-8370-1857-9 – mf#1987-6244 – us ATLA [226]

The four gospels : from the munich ms. (q) now numbered lat. 6224 in the royal library at munich: with a fragment from st. john in the hof-bibliothek at vienna (cod. lat. 502) / ed by White, Henry Julian – Oxford: Clarendon Press, 1888 – (= ser Old-Latin Biblical Texts) – 1mf – 9 – 0-8370-1801-3 – (includes additions and corrections to old-latin biblical texts, nos. 1 and 2) – mf#1987-6189 – us ATLA [226]

The four gospels : a new translation from the greek text direct, with reference to the vulgate and the ancient syriac version / Spencer, Francis Aloysius – New York: William H Young, 1898 – 1mf – 9 – 0-524-05903-9 – mf#1992-0660 – us ATLA [226]

The four gospels : their age and authorship traced from the fourth century into the first / Kennedy, John; ed by Rice, Edwin Wilbur – Philadelphia: American Sunday-School Union, [1880?] – 1mf – 9 – 0-524-05221-2 – mf#1992-0354 – us ATLA [226]

The four gospels : vol 2, mark, luke, and john: with a commentary / Livermore, Abiel Abbot – Boston: J Munroe, 1842 – 1mf – 9 – 0-524-08605-2 – mf#1993-0040 – us ATLA [226]

The four gospels examined and vindicated on catholic principles / Heiss, Michael – Milwaukee: Hoffmann Bros, 1863 – 1mf – 9 – 0-524-06921-2 – mf#1992-1014 – us ATLA [226]

Four gospels [fragment] : sobornik dvenadtsati mesiatsem [the church calendar for 12 months] – early 1400s – 1mf – 9 – (middle bulgarian version; most likely mss is moldavian by origin) – us UMI ProQuest [090]

The four gospels from a lawyer's standpoint / Bennett, Edmund Hatch – Boston: Houghton, Mifflin, 1899 – 1mf – 9 – 0-8370-2265-7 – mf#1985-0265 – us ATLA [226]

The four gospels from the codex corbeiensis (ff [or ff2]) : being the first complete edition of the ms. now numbered lat. 17225 in the national library at paris / ed by Buchanan, Edgar Simmons – Oxford: Clarendon Press, 1907 – (= ser Old-Latin Biblical Texts) – 1mf – 9 – 0-8370-1799-8 – mf#1987-6187 – us ATLA [226]

The four gospels from the codex veronensis (b) : being the first complete edition of the evangeliarium purpureum in the cathedral library at verona – Oxford: Clarendon Press, 1911 – (= ser Old-Latin Biblical Texts) – 1mf – 9 – 0-8370-1800-5 – mf#1987-6188 – us ATLA [226]

The four gospels in syriac / Bensly, Robert Lubbock et al – Cambridge: University Press, 1894 – 4mf – 9 – 0-8370-1377-1 – mf#1987-6058 – us ATLA [226]

The four gospels in the earliest church history / Nicol, Thomas – Edinburgh: William Blackwood 1908 [mf ed 1985] – (= ser The baird lecture 1907) – 1mf – 9 – 0-8370-4586-X – (incl ind) – mf#1985-2586 – us ATLA [226]

The four gospels translated into the slave language for the indians of north-west america – London: Printed for the British and Foreign Bible Society by Gilbert and Rivington, 1883 – 4mf – 9 – (text in slave. trans by william carpenter bompas) – mf#14248 – cn Canadiana [290]

Four great religions / Besant, Annie Wood – London; New York: Theosophical Pub Society, 1897 – (= ser Samp: indian books) – us CRL [280]

Four great religions : four lectures / Besant, Annie Wood – London: Theosophical Pub Society, 1897 [mf ed 1992] – 1mf – 9 – 0-524-02344-1 – mf#1990-2955 – us ATLA [200]

Four great works of martin gerbert (1720-1793), prince-abbot of st blasien monastery / Gerbert, Martin – repr, 1774-84 – 11 – $180.00 set – (de cantu et musica sacra a prima ecclesiae: 2v 1774. vetus liturgia alemannica: 2v 1776. monumenta veteris liturgiae alemannicae: 2v 1777-79. scriptores ecclesiastici de musica sacra potissimum: 3v 1784) – us Univ Music [780]

Four heatons digest – [North West] Stockport jun 1984-oct/nov 1985 – 1 – uk MLA; uk Newsplan [072]

Four holes baptist church – Orangeburg, SC. 920p. 1820-1864, 1867-1935, 1944-49, 1950-sep 1957 (incomplete) – 1 – $41.40 – (membership rolls 1968, 1979) – mf#5003-8a – us Southern Baptist [242]

Four hundred years : commemorative essays on the reformation of dr. martin luther... / Abbetmeyer, Charles et al; ed by Dau, William Herman Theodore – St Louis, MO: Concordia, 1916 – 1mf – 9 – 0-524-00986-4 – mf#1990-0263 – us ATLA [242]

Four hundred years of freethought / Putnam, Samuel Porter – New York: The Truth Seeker Company, 1894 – 874p – 1 – us Wisconsin U Libr [210]

Four hundred years of world presbyterianism / ed by Drury, Clifford Merrill – San Anselmo, Calif.: [s.n.], c1961. Lib Photographic Service, U of California, 1961 (1r); Evanston: American Theol Lib Assoc, 1984 (1r) – 1 – 0-8370-0010-6 – (includes bibliographies) – mf#1984-B377 – us ATLA [242]

The four irish policies : or, have we no alternative? / Raleigh, Thomas – London, [1886] – (= ser 19th c ireland) – 1mf – 9 – mf#1.1.252 – uk Chadwyck [941]

Four keys to guatemala / Kelsey, Vera – New York, NY. 1939 – 1r – us UF Libraries [972]

Four keys to guatemala / Kelsey, Vera – New York, NY. 1961 – 1r – us UF Libraries [972]

Four key-words of religion : an essay in unsystematic divinity / Huntington, William Reed – 2nd ed. New York: Thomas Whittaker, c1899 [mf ed 1985] – (= ser Briefs on religion 5) – 1mf – 9 – 0-8370-4850-8 – (incl bibl ref) – mf#1985-2850 – us ATLA [240]

Four Lakes Indian Council *see* Four lakes news

Four lakes indian council newsletter – n1, 2-3 [1977 sep, nov-dec], n4, 7-8 [1978 mar, apr-may] – 1r – 1 – (cont by: four lakes news) – mf#634555 – us WHS [305]

Four lakes news / Four Lakes Indian Council – 1978 jun-1981 jun – 1r – 1 – (cont: four lakes indian council newsletter) – mf#634558 – us WHS [305]

Four lectures : delivered in the church of the holy trinity, philadelphia, in the year 1877... / Vinton, Alexander Hamilton – Boston: Alfred Mudge, 1877 – (= ser The bohlen lectures) – 1mf – 9 – 0-8370-5674-8 – mf#1985-3674 – us ATLA [240]

Four lectures : delivered...to the brahmos in bombay and poona in apr and jul 1875 / Goreh, Nehemiah – Bombay: printed at the Education Society's Press, 1875 [mf ed 1991] – 1mf – 9 – 0-524-01511-2 – mf#1990-2487 – us ATLA [230]

Four lectures on some epochs of early church history : delivered in ely cathedral / Merivale, Charles – New York: Anson DF Randolph, [1879?] – 1mf – 9 – 0-524-05325-1 – mf#1990-1443 – us ATLA [240]

Four lectures on the early history of the gospels delivered at milborne port, somerset, advent, 1897 / Wilkinson, J H – London, New York: Macmillan, 1898 – 1mf – 9 – 0-8370-5851-1 – mf#1985-3851 – us ATLA [226]

Four lectures on the western text of the new testament / Harris, James Rendel – London: C J Clay, 1894 – 1mf – 9 – 0-8370-9953-6 – (in english and greek. incl bibl ref) – mf#1986-3953 – us ATLA [225]

Four letters of the apostle paul : a short course / Burton, Ernest De Witt – Chicago, IL: University of Chicago Press, 1908 [mf ed 1990] – (= ser Outline bible study courses of the american institute of sacred literature) – 1mf – 9 – 0-7905-3368-5 – mf#1987-3368 – us ATLA [227]

Four letters to the rev e b elliott on some passages in his hora... / Candlish, Robert S – London, England. 1846 – 1r – us UF Libraries [240]

Four lights / Women's International League for Peace and Freedom – 1945 may-1970 jan – 1r – 1 – (cont: four lights [new york, ny]; cont by: peace and freedom [philadelphia, pa]) – mf#1005916 – us WHS [341]

Four masses of flaminio tresti / Richardson, Jane Marie – U of Rochester 1951 [mf ed 19–] – 4mf – 9 – (with bibl) – mf#fiche 1042 – us Sibley [780]

Four masters of etching... : with original etchings by haden, jacquemart, whistler, and legros / Wedmore, Frederick – London 1883 – (= ser 19th c art & architecture) – 2mf – 9 – mf#4.2.946 – uk Chadwyck [760]

Four men / Belloc, Hilaire – London, England. 1912 – 1r – us UF Libraries [420]

Four methods of teaching english to maswina / Biehler, E – Chishawasha, Zimbabwe. 1906 – 1r – us UF Libraries [420]

Four middle school physical education teachers' experiences during a collaborative action research staff development project / Butt, K L – 1989 – 2mf – 9 – $8.00 – us Kinesology [613]

Four mile baptist church. campbell county. kentucky : church records – 1819-74 – 336p – 1 – $15.12 – us Southern Baptist [242]

Four operas of antonio vivaldi / Rowell, Lewis Eugene – U of Rochester 1958 [mf ed 19–] – 1r – 1 – (with app & bibl) – mf#film 698 – us Sibley [780]

Four pastorates : glimpses of the life and thoughts of eden b. foster, d.d / ed by Foster, Addison Pinneo – Lowell, Mass: George M Elliott, 1883 – 2mf – 9 – 0-524-06484-9 – mf#1991-2584 – us ATLA [240]

Four phases of morals : socrates, aristotle, christianity, utilitarianism / Blackie, John Stuart – New York: Scribner, Armstrong, 1872 – 1mf – 9 – 0-8370-6022-2 – mf#1986-0022 – us ATLA [170]

Four Power Commission of Investigation for the Former Italian Colonies *see* [Report]

Four psalms, 23, 36, 52, 121 : interpreted for practical use / Smith, George Adam – London: Hodder & Stoughton, 1896 [mf ed 1992] – 1mf – 9 – 0-524-05292-1 – mf#1992-0393 – us ATLA [221]

Four quarters – Philadelphia. 1972-1995 (1) 1972-1995 (5) 1976-1995 (9) – ISSN: 0015-9107 – mf#7552 – us UMI ProQuest [073]

Four recent pronouncements / Collins, William Edward – London: SPCK 1899 [mf ed 1993] – (= ser Church historical society (series) 56) – 1mf – 9 – 0-524-05536-X – mf#1990-5140 – us ATLA [242]

Four rivers advocate / Four Rivers Indian Legal Services – 1981 apr 15-1982 oct 25 – 1r – 1 – mf#644022 – us WHS [340]

Four Rivers Indian Legal Services *see* Four rivers advocate

Four sermons / Gruger, Hugo – Manchester, England. 1897 – 1r – us UF Libraries [240]

Four sermons preached before the university of cambridge, in may... / Thorp, Thomas – Cambridge, England. 1838 – 1r – us UF Libraries [240]

Four short lectures on the book of revelation / Douglass, Benjamin – Chicago: [s.n], 1866 – 1mf – 9 – 0-8370-2958-9 – mf#1985-0958 – us ATLA [221]

Four sonatas and two duetts for the harpsichord or piano forte : with an accompanyment for a german flute or violin. opera 18 / Bach, Johann Christian – London: Welcker [1775?] [mf ed 19–] – 1r – 1 – mf#film 80 – us Sibley [780]

Four sonatas for the piano / Reinagle, Alexander – Composer's manuscript, not dated. MUSIC 136, Item 1 – 1 – us L of C Photodup [780]

Four sonatas or duets for two performers on one pianoforte or harpsichord / Burney, Charles – London: printed for aut & sold by R Bremner [pref 1777] [mf ed 19–] – 2mf – 9 – mf#fiche 187 – us Sibley [780]

Four sonnets / Carman, Bliss – Boston: Small, Maynard, c1916 – 1mf – 9 – 0-665-77795-7 – mf#77795 – cn Canadiana [810]

Four speeches delivered in guild-hall, 1643 / Calamy, Edmund – 1646 – 1 – $50.00 – us Presbyterian [941]

Four stages of greek religion : studies based on a course of lectures / Murray, Gilbert – London: Publ for the Columbia University Press by Oxford University Press, 1912 – (= ser Columbia university lectures) – 1mf – 9 – 0-524-00944-9 – mf#1990-2167 – us ATLA [250]

Four state review – 1983 mar-dec – 1r – 1 – mf#649069 – us WHS [071]

Four steps to better government fo new york city / New York Temporary State Commission – New York, NY. 1953-54 – 1r – us UF Libraries [350]

The four streams : newsletter of the diocesan association for western china – Ashford, Kent, etc: The Diocesan Association for Western China. n1-178, (?)-jan 1951; ns: n1 jul 1951 (frequency varies) – (= ser Missionary periodicals from the china mainland) – 1r – 1 – $165.00 – (title varies) – us UPA [242]

Four talks on theology *see*
– Shen hsueh 4 chiang

The four temperaments / Whyte, Alexander – London: Hodder and Stoughton; New York: Dodd, Mead, 1895 – (= ser Little books on religion) – 1mf – 9 – 0-7905-2215-2 – mf#1987-2215 – us ATLA [240]

The four theories of visible church unity : an address / Huntington, William Reed – [S.l.: s.n., 1909?] – 1mf – 9 – 0-7905-6183-2 – mf#1988-2183 – us ATLA [240]

Four tudor books on education / ed by Pepper, Robert D – 1533-1588 – 9 – us Scholars Facs [370]

Four Wheel Drive Auto Co *see* Drive news

Four wing news – Holden, WV. 1942-1949 (1) – mf#67321 – us UMI ProQuest [071]

The four witnesses : being a harmony of the gospels on a new principle / Costa, Isaac da – London: James Nisbet, 1851 – 2mf – 9 – 0-7905-1645-4 – mf#1987-1645 – us ATLA [226]

Four year bummer – Champaign IL. v1-v2 n10 [1969 may-1971 jan] – 1r – 1 – mf#785913 – us WHS [071]

Four years' campaign in india / Taylor, William – 3rd ed. New York: Nelson & Phillips, 1875 – 1mf – 9 – 0-7905-6737-7 – mf#1988-2737 – us ATLA [240]

Four years in ashantee = Vier jahre in asante / Ramseyer, Friedrich August & Kuehne, Johannes; ed by Weitbrecht, Mary – New York: R. Carter, 1875 – 1mf – 9 – 0-7905-5915-3 – (in english) – mf#1988-1915 – us ATLA [240]

Four years in ashantee by the missionaries ramseyer and kuehne. / Ramseyer, Friedrich August & Kuhne; ed by Weitbrecht, Mrs – New York: R Carter, 1875 – 1 – us CRL [960]

Four years in france : or narrative of an english family's residence there during that period; preceded by some account of the conversion of the author to the catholic faith / Beste, Henry – London 1826 [mf ed Hildesheim 1995-98] – 3mf – 9 – €90.00 – 3-487-29703-5 – gw Olms [920]

Four years in southern africa / Rose, Cowper – London 1829 [mf ed Hildesheim 1995-98] – 1v on 2mf – 9 – €60.00 – 3-487-27191-5 – gw Olms [916]

Four years in the old world : comprising the travels, incidents, and evangelistic labors of dr and mrs palmer in england, ireland, scotland and wales / Palmer, Phoebe – 10th ed. New York: Foster & Palmer, Jr, 1866 [mf ed 1984] – (= ser Women & the church in america 175) – 2mf – 9 – 0-8370-1446-8 – mf#1984-2175 – us ATLA [242]

Four years in the old world : comprising the travels, incidents, and evangelistic labors of dr and mrs palmer in england, ireland, scotland and wales / Palmer, Phoebe – Toronto: S Rose, 1866 – 8mf – 9 – 0-665-93545-5 – mf#93545 – cn Canadiana [914]

Four years in the white north / MacMillan, D B – Boston, New York, 1925 – 10mf – 9 – mf#N-306 – ne IDC [919]

Four years of irish history, 1845-1849 / Duffy, Charles Gavan – London, New York: Cassell, Petter, Galpin, (1883). xv,780p. With: Dawn Ginsbergh's Revenge by S.J. Perelman. 1 reel. 1285 – 1 – us Wisconsin U Libr [941]

Four years' residence in the west indies / Bayley, Frederick W – London 1830 [mf ed Hildesheim 1995-98] – 1v on 8mf – 9 – €160.00 – 3-487-26963-5 – gw Olms [918]

Un fourbe demasque – [Montreal?: s.n, 1867?] [mf ed 1994] – 1mf – 9 – 0-665-94736-4 – mf#94736 – cn Canadiana [350]

The four-fold gospel / Simpson, Albert B – New York: Alliance Press, c1890 – 1mf – 9 – 0-8370-7337-5 – mf#1986-1337 – us ATLA [240]

The fourfold gospel : sect 3: the proclamation of the new kingdom / Abbott, Edwin Abbott – Cambridge: University Press; New York: G P Putnam [distributor], 1915 – 2mf – 9 – 0-7905-3420-7 – mf#1987-3420 – us ATLA [226]

The fourfold gospel : sect 4: the law of the new kingdom / Abbott, Edwin Abbott – Cambridge: University Press, 1916 – 2mf – 9 – 0-524-03957-7 – (incl bibl ref) – mf#1992-0000 – us ATLA [226]

The fourfold gospel : sect 5: the founding of the new kingdom / Abbott, Edwin Abbott – Cambridge: University Press, 1917 – 2mf – 9 – 0-524-03958-5 – (incl bibl ref) – mf#1992-0001 – us ATLA [226]

The fourfold gospel : section 1, introduction / Abbott, Edwin Abbott – Cambridge: University Press; New York: G P Putnam [distributor], 1913 – (= ser Diatessarica) – 1mf – 9 – 0-7905-1680-2 – mf#1987-1680 – us ATLA [226]

The fourfold gospel : section 2, the beginning / Abbott, Edwin Abbott – Cambridge: University Press; New York: G P Putnam [distributor], 1914 – (= ser Diatessarica) – 1mf – 9 – 0-7905-1681-0 – mf#1987-1681 – us ATLA [226]

The fourfold sovereignty of god / Manning, Henry Edward – London: Burns, Oates; New York: Benziger Bros, [1871?] – 1mf – 9 – 0-7905-9330-0 – mf#1989-2555 – us ATLA [210]

The fourfold story : a study of the gospels / Genung, George Frederick – Boston: Congregational Sunday-School & Publ Society, c1891 – 1mf – 9 – 0-8370-3251-2 – mf#1985-1251 – us ATLA [220]

Fourfold view of the spiritual life / Moore, Daniel – London, England. 1859 – 1r – us UF Libraries [240]

Fourie, Francois Paul *see* N geskiedenis van die kerklied in die nederduitse gereformeerde kerk

Fourie, Jacob Andries Cornelis *see* Die identifisering van adolessente wat groepdruk moeilik hanteer

Fourie, Pieter P *see* Politics of north-south technology transfer

Fourie, Renata *see* 'N ondersoek na die verband tussen persoonlikheid stipes en 'n sin vir koherensie

Fourie, Roelof Jacobus *see* Evaluation of losses in lv feeders using the beta load model

La fourmi – Port-au-Prince: Impr de L'Oeuvre. 1ere annee: n2-n8. 3 mai-22 mai 1902 – 1 sheet – 9 – us CRL [079]

Fournier, Georges *see* Hydrographie

Fournier, Jules *see*
– Les assurances au canada
– Le canada, son present et son avenir
– Sir lomer gouin
– Souvenirs de prison

Fournier, Narcisse *see*
– Au bord de l'abime
– Celine,
– Davis
– Deux soeurs
– Eleves ensemble
– Souvenirs de la marquise de v...

Fournier, Narcisse) *see* Roman intime, ou, les lettres du mari

Fournier, Pierre-Simon *see* Traite historique et critique sur l'origine et les progres des caracteres de fonte pour l'impression de la musique

Fournier-Verneuil *see*
– Le huron de mont-rouge
– Paris

Four-square, or, the cardinal virtues : addresses to young men / Rickaby, Joseph – New York: Joseph F Wagner, c1908 – 1mf – 9 – 0-7905-9459-5 – mf#1989-2684 – us ATLA [170]

Foursquare world advance / International Church of the Foursquare Gospel – 1977 jan/feb-1985 dec – 1r – 1 – mf#1582412 – us WHS [242]

Fourteen nuts for sceptics to crack / Hastings, H L – London, England. 18– – 1r – us UF Libraries [240]

Fourteen questions : article listing and responding to the fourteen most often asked questions encountered by the joint committee on future status subcommittees in their district hearings / Joint Committee on Future Status Subcommittees [TTPI (U.S.)] – n.a, n.p, n.d. – (= ser Micronesia: prelude to the constitutional convention) – 1mf – 9 – $1.50 – mf#LLMC 82-100F, Title 53 – us LLMC [323]

Fourteen years in basutoland : a sketch of african mission life / Widdicombe, John – London: Church Printing, [pref. 1891]. Chicago: Dep of Photodup, U of Chicago Lib, 1971 (1r); Evanston: American Theol Lib Assoc, 1984 (1r) – 1 – 0-8370-0525-6 – (incl ind) – mf#1984-B239 – us ATLA [240]

Fourteen years in basutoland; a sketch of african mission life / Widdicombe, John – London. 1891 – 1 – us CRL [960]

A fourteenth century english biblical version / ed by Paues, Anna Carolina – Cambridge: University Press, 1904 [mf ed 1990] – 1mf – 9 – 0-8370-1684-3 – (incl bibl ref) – mf#1987-6111 – us ATLA [225]

The fourth anniversary of the spring garden road home of the first baptist church, halifax, ns, lord's day, april 12, 1891 : have we a mission? are the baptists needed to-day?: address / Adams, Henry – Halifax, NS: s.n, 1891 – 1mf – 9 – mf#06153 – cn Canadiana [242]

Fourth annual convention of teachers : in connection with the provincial association of protestant teachers of lower canada / Provincial Association of Protestant Teachers Convention (4e: 1867: Montreal, Quebec) – [Montreal?: s.n, 1867?] [mf ed 1986] – 1mf – 9 – 0-665-62183-3 – mf#62183 – cn Canadiana [366]

Fourth annual lecture and sermon : delivered june, 1882 – St John, NB: J & A McMillan, 1883 – 1mf – 9 – mf#29795 – cn Canadiana [230]

The fourth book of ezra / Bensly, Robert Lubbock – Cambridge: University Press 1895 [mf ed 1989] – (= ser Texts and studies (cambridge, england) 3/2) – 1mf – 9 – 0-7905-3245-X – (in latin; english int by montague rhode james) – mf#1987-3245 – us ATLA [221]

The fourth book of ezra : the latin version from the mss / ed by Bensly, Robert L – 1895 – (= ser Texts and studies (ts)) – 4mf – 9 – €11.00 – (int by m r james) – ne Slangenburg [221]

The fourth book of maccabees – London: SPCK, 1918 – (= ser Translations of Early Documents) – 1mf – 9 – 0-524-06130-0 – (incl bibl ref) – mf#1992-0797 – us ATLA [221]

The fourth book of maccabees and kindred documents in syriac / ed by Bensly, Robert Lubbock – Cambridge: University Press, 1895 – 1mf – 9 – 0-8370-1802-1 – mf#1987-6190 – us ATLA [221]

The fourth book of moses called numbers / ed by Gray, George Buchanan – London: J M Dent; Philadelphia: J B Lippincott 1902 [mf ed 1989] – (= ser The temple bible) – 1mf – 9 – 0-7905-1820-1 – mf#1987-1820 – us ATLA [221]

Fourth brigade account book, 1861 *see* Account book, 1861

Fourth census of the state of florida taken in the year 1915 / Florida Dept Of Agriculture – Tallahassee, FL. 1915 – 1r – us UF Libraries [630]

Fourth census of the united states, 1820 / U.S. Bureau of the Census – (= ser 1790-1840 census schedules) – 142r – 1 – mf#M33 – us Nat Archives [317]

Fourth census of the u.s. – 1820 – 1r – 1 – $50.00 – mf#B50001 – us Library Micro [975]

Fourth commandment, not ceremonial, but moral / Mcneile, Hugh – Liverpool, England. 1856 – 1r – us UF Libraries [240]

The fourth gospel : evidences external and internal of its johannean authorship / Abbot, Ezra et al – New York: Scribner's, 1891 – 1mf – 9 – 0-8370-2006-9 – (incl indes) – mf#1985-0006 – us ATLA [226]

The fourth gospel : the heart of christ / Sears, Edmund Hamilton – 9th ed. Boston: American Unitarian Association, 1890 – 2mf – 9 – 0-524-04925-4 – mf#1992-0268 – us ATLA [226]

The fourth gospel / Hoskyns, Ed – 1947 – 9 – $21.00 – us IRC [240]

The fourth gospel : its purpose and theology / Scott, Ernest Findlay – 2nd ed. Edinburgh: T & T Clark, 1908 – 1mf – 9 – 0-7905-0329-8 – (incl ind) – mf#1987-0329 – us ATLA [226]

The fourth gospel : the question of its origin stated and discussed / Clarke, James Freeman – Boston: Geo H Ellis, 1886 – 1mf – 9 – 0-8370-2671-7 – mf#1985-0671 – us ATLA [226]

The fourth gospel / Schuerer, Emil – London: Francis Griffiths, [1905?] – (= ser Essays for the Times) – 1mf – 9 – 0-524-06159-9 – (incl bibl ref) – mf#1992-0826 – us ATLA [226]

The fourth gospel and some recent german criticism / Jackson, Henry Latimer – Cambridge: University Press, 1906 – 1mf – 9 – 0-8370-3741-7 – (incl bibl ref) – mf#1985-1741 – us ATLA [226]

The fourth gospel in research and debate : a series of essays... / Bacon, Benjamin Wisner – New York: Moffat, Yard, 1910 – 2mf – 9 – 0-7905-0543-6 – (incl bibl ref and index) – mf#1987-0543 – us ATLA [226]

Fourth International *see*
– Spartacist 4

Fourth international – New York. v. 1-17 n1-135. may 1940-spring 1956 – 1 – us NY Public [073]

Fourth Internationalist Tendency [Group] *see* Bulletin in defense of marxism

Fourth letter to n wiseman / Palmer, William – Oxford, England. 1841 – 1r – us UF Libraries [240]

A fourth letter to the people of england : on the conduct of the mrs in alliances, fleets, and armies, since the first differences on the ohio, to the taking of minorca by the french / Shebbeare, John – London: printed for M Collier...1756 [mf ed 1983] – 2mf – 9 – 0-665-40754-8 – mf#40754 – cn Canadiana [320]

Fourth of july raids / Brokensha, Miles – Cape Town, South Africa. 1965 – 1r – us UF Libraries [960]

Fourth report – Manchester? England. 1849? – 1r – us UF Libraries [240]

Fourth report of the committee of the general assembly / Church Of Scotland General Assembly Commmittee On Church Extension – Edinburgh, Scotland. 1838 – 1r – us UF Libraries [240]

Fourth report of the proceedings of the church... / Church Society of the Archdeaconry of New Brunswick – [St John, NB?: s.n.] 1840 [mf ed 1983] – 1mf – 9 – 0-665-43836-2 – mf#43836 – cn Canadiana [240]

Fourth report of the standing committee of grievances / Bas-Canada. Parlement. Chambre d'Assemblee – [S.l.]: [s.n.], [1836?] (mf ed 1990) – 7mf – 9 – mf#SEM105P1225 – cn Bibl Nat [324]

Foury,B *see* Maudave et la colonisation de madagascar

Fout, Henry Harness *see* Our heroes

Fouts, E L *see*
– Manufacture of cultured buttermilk and cottage cheese
– Preparation and use of invert sirup in the manufacture of ice cream

Fovitskii, A L *see* Vestnik vremennogo vserossiiskogo pravitel'stva

Fowkes, Dudley [comp] *see* Catalogue of pre-1650 manuscript maps held by county record offices in england and wales

Fowle, Thomas Welbank *see* The reconciliation of religion and science

[Fowle, Thomas Welbank] *see* A new analogy between revealed religion and the course and constitution of nature

Fowle, William Bentley *see* The bible, the rod, and religion, in common schools

The fowler collection of early architectural books – 86r – 1 – (previous title: fowler collection of early architectural books. based on the fowler architectural collection of the johns hopkins university. coll covers works in architecture to the end of the 18th century. with printed guide) – us Primary [720]

Fowler, Ellen Thorneycroft *see*
– In subjection
– Sirius

Fowler ensign *see* [Fowler-] fowler courier

Fowler family folio – v1 n1-4 [1978 jul-1979 dec], 1983, v1 n5-6 – 1r – 1 – mf#637820 – us WHS [929]

[Fowler-] fowler courier – CA. 1894-1897 – 1r – 1 – $60.00 – (aka: fowler ensign) – mf#B06026 – us Library Micro [071]

[Fowler-] fowler ensign – CA. 1898-1956; 1976-82 – 28r – 1 – $1680.00 – mf#BC02248 – us Library Micro [071]

Fowler, Guy *see* Four devils

Fowler, Harvey *see* Methodist chapel-property case

Fowler, Henry *see* The american pulpit

Fowler, Henry T *see* The american pulpit: sketches, biographical

Fowler, Henry Thatcher *see*
– A history of the literature of ancient israel
– The prophets as statesmen and preachers

Fowler, John *see*
– The history of a railroad difficulty
– Journal of a tour in the state of new york, in the year 1830

Fowler, John A *see* The pennsylvania insurance digest

Fowler, Josiah *see* An analysis of texts of scripture

Fowler, LN *see* New illustrated self-instructor in phrenology and physiology

Fowler, Montague *see*
– Christian egypt
– Church history in queen victoria's reign
– Some notable archbishops of canterbury

Fowler, OS *see* New illustrated self-instructor in phrenology and physiology

Fowler, Philemon H *see* Historical sketch... synod of central new york

Fowler, Philemon Halsted *see* Historical sketch of presbyterianism within the bounds of the synod of central new york

Fowler, Robert Ludlow *see*
– The personal property law of the state of new york being chapter forty-seven of the general laws.
– The real property law of the state of new york

Fowler, Samuel Page *see* An account of the life, character etc of the rev samuel parris, of salem village

Fowler, W J *see* Grace and gold

Fowler, William Chauncey *see*
– The clergy and popular education
– Memorials of the chaunceys

Fowler, William Chauncey et al *see* Centennial papers

Fowler, William Warde *see*
– The city-state of the greeks and romans
– The religious experience of the roman people
– The roman festivals of the period of the republic
– Roman ideas of deity in the last century before the christian era
– Social life at rome in the age of cicero

Fowles, James Henry *see* The necessity of personal communion with christ

Fowlmere 1561-1950 – 6mf – 9 – £7.50 – uk CambsFHS [929]

Fowls of the air / Long, William Joseph – Boston, MA. 1901 – 1r – us UF Libraries [590]

Fox, Bryan D *see* Strategies to enhance self-efficacy to improve exercise adherence in a worksite fitness center

Fox, Charles Donald *see* Truth about florida

Fox cities valley sun – Wisconsin.. v1 n10-19 [1990 oct 18-dec 27], v2 n1-4 [1991 jan 3-24] – 1r – 1 – (cont: valleysun) – mf#1787374 – us WHS [071]

Fox, Cyndy M *see* Psychosocial factors in the development of breast cancer

Fox, Fannie Ferber *see* Fannie fox's cook book

Fox, Francis *see* Introduction to spelling and reading...

Fox, G T *see* A memoir of the rev henry watson fox...

Fox, George *see*
– The journal of george fox
– Passages from the life and writings of george fox, taken from his journal
– Selections from the epistles of george fox

Fox, George Townsend *see* American journals of george townsend fox, 1831-68

Fox, George Townshend *see* The nature and evidences of regeneration

Fox, Harry Halton *see*
– General report on the commercial, industrial, & economic situation of china in june, 1921
– Report on the commercial, industrial and economic situation of china in july 1922

Fox, Henry Watson *see*
– Chapters on missions in south india
– Church missionary sociaety jubilee address no 2

Fox, I *see* Juta's first zulu manual with vocabulary

Fox, J A *see* A key to the irish question

Fox, James Joseph *see* Religion and morality

Fox, John *see*
– Erskine dale, pioneer
– The heart of the hills
– The little shepherd of kingdom come
– The trail of the lonesome pine

Fox, John R *see* Interview transcripts

Fox lake breeze – Fox Lake WI. 1886 jan 6-1887 oct – 1r – 1 – (cont: fox lake breeze and representative; cont by: fox lake representative [fox lake, wi: 1887]) – mf#943230 – us WHS [071]
Fox lake breeze and representative – Fox Lake WI. 1885 dec 2-29 – 1r – 1 – (cont: fox lake representative [fox lake, wi: 1866]; cont by: fox lake breeze) – mf#943228 – us WHS [071]
Fox Lake gazette – Fox Lake WI. 1858 apr 1-1861 may 31, 1861 jun 1-1863 dec 31, 1864 jan 1-1865 mar 8 – 3r – 1 – mf#916566 – us WHS [071]
Fox lake representative – Fox Lake WI. 1866 oct 5-1869 oct 15, 1869 oct 22-1871 jun 9, 1871 jun 16-1874 nov 27, 1874 dec 4-1878 apr 12, 1878 apr 19-1881 apr 29, 1881 may 6-1884 mar 25, 1884 apr 1-1885 nov 25 – 7r – 1 – (cont by: fox lake breeze and representative) – mf#943226 – us WHS [071]
Fox lake representative – Fox Lake WI. 1887 nov/1889 may 10-1962/1964 oct 15 [gaps] – 28r – 1 – (cont: fox lake breeze; cont by: representative [fox lake, wi: 1964]) – mf#943238 – us WHS [071]
Fox lake representative – Berlin, Fox Lake WI. 1969 jan 2-dec 25, 1970 jan-1972 sep 7, 1972 sep 14-1975 apr 24, 1975 may 1-1977 sep 29, 1977 oct 6-1978 dec 28, 1979-80, 1981 jan-sep 17 – 7r – 1 – (cont: representative [fox lake, wi]; cont by: representative [berlin, wi]) – mf#943250 – us WHS [071]
Fox, M A S Columba *see* The life of...john bapti..new york, 1925
Fox, Mary *see* The country house
Fox nation – iss n1-15 [1980 jul 1-1985 may 1] – 1r – 1 – mf#1048148 – us WHS [071]
Fox Novel, M *see* Vease forner y segarra, juan pablo
Fox, Paul *see* Essentials of polish
Fox point [city directory : listing] – 1933 pt1 [p1-1526], 1933 pt2 [p1527-2886], 1936 pt1 [p2-1058], 1936 pt2 [p1059-end], 1939 pt1 [p149-1118], 1939 pt2 [p1119-2172], 1941 pt1 [p87 1520], 1941 pt2 [p1521-end], 1942 pt1 [p49-1110], 1942 pt2 [p1111-2262], 1944 pt1 [p?-1212 [a-s]], 1944 pt2 [p1213-end [s-end]], 1959 pt1 [a-l]], 1959 pt2 [m-z]] – 14r – 1 – mf#3196263 – us WHS [917]
Fox point-bayside herald – Bayside, Fox Point, River Hills, Shorewood WI. 1960 oct 20-1961, 1962 may 3 – 2r – 1 – (cont by: fox point-bayside-river hills herald [shorewood, wi: 1962]) – mf#942269 – us WHS [071]
Fox point-bayside herald – Bayside, Fox Point, River Hills, Shorewood WI. 1963 aug 15-dec, 1964, 1965 jan-jun, 1965 jul-nov 11 – 4r – 1 – (cont: fox point-bayside-river hills herald [shorewood, wi: 1962]; cont by: fox point-bayside-river hills herald [shorewood, wi: 1965]) – mf#942275 – us WHS [071]
Fox point-bayside-river hills herald – Bayside, Fox Point, River Hills, Shorewood WI. 1962 may 10-dec, 1963 aug 8 – 2r – 1 – (cont: fox point-bayside herald [shorewood, wi: 1959]; cont by: fox point-bayside herald [shorewood, wi: 1963]) – mf#942274 – us WHS [071]
Fox point-bayside-river hills herald – Shorewood WI. 1965 nov 18/dec-2002 oct/dec [gaps] – 76r – 1 – (cont: fox point-bayside herald [shorewood, wi: 1963]; cont by: brown deer herald [new berlin, wi: 1997]; glendale herald [new berlin, wi]; shorewood herald [west allis, wi]; whitefish bay herald [west allis, wi]; north shore herald [west allis, wi]) – mf#942278 – us WHS [071]
Fox river journal – Appleton WI. 1905 feb 27-1906 jun 29, 1906 jul 5-1907 dec 27, 1908 jan 3-1908 jun 3, 1909 jun 10-1910 oct 27, 1910 nov 3-1912 mar 28, 1912 apr 4-1913 sep 18, 1913 sep 25-1915 mar 4, 1915 mar 11-1917 mar 1 – 8r – 1 – (cont by: daily fox river journal) – mf#917774 – us WHS [071]
Fox river leader – Aurora, IL. 1909-1922 (1) – mf#62500 – us UMI ProQuest [071]
Fox river patriot – Princeton WI. 1976 nov 17-1978 jun 12, 1978 jun-1980 jan 7, 1980 jan 7-12, 1981-82, 1983-1984 jun – 5r – 1 – (cont by: fox river patriot & flyer) – mf#368288 – us WHS [071]
Fox river patriot – Princeton WI. 1985 may 16-sep 26 – 1r – 1 – (cont: fox river patriot & flyer) – mf#1048734 – us WHS [071]
Fox river patriot and flyer – Princeton WI. iss n146-165 [1984 apr 12-jun, 1984 jul 11-1985 feb 7] – 1r – 1 – mf#1029198 – us WHS [071]
Fox, Sarah E *see* Edwin octavius tregelles, civil engineer and minister of the gospel
Fox Strangways, Arthur Henry *see* The music of hindostan
Fox valley countryside north – Barrington, IL. 1976-1982 (1) – mf#62509 – us UMI ProQuest [071]
Fox valley countryside reader – Barrington, IL. 1983-1984 (1) – mf#68645 – us UMI ProQuest [071]
Fox valley countryside south – Barrington, IL. 1974-1982 (1) – mf#62510 – us UMI ProQuest [071]
Fox valley free press – Oshkosh WI. 1932 aug 5-1936? – 1r – 1 – (cont: fox valley square deal; cont by: oshkosh free press [oshkosh, wi: 1937]) – us WHS [071]
Fox valley free press – Oshkosh WI. [1937/8-1939 nov 10] – 1r – 1 – (cont: oshkosh free press [oshkosh, wi: 1937]; cont by: oshkosh free press [oshkosh, wi: 1939]) – us WHS [071]
Fox Valley Genealogical Society *see*
- Genealogical gems
- Nuggets
Fox valley kaleidoscope – v1 n1-5, v2 n2-3 [1970 mar 6-may 19/jun 9, jun 25/jul?-aug 1/20] – 1r – 1 – mf#679245 – us WHS [071]
Fox valley labor news – 1940 mar 21/1941 dec 25-1996 [gaps] – 26r – 1 – mf#1056502 – us WHS [331]
Fox valley square deal – Neenah, Oshkosh WI. 1931 dec 19-1932 may 21 – 1r – 1 – (cont by: fox valley free press [oshkosh, wi: 1932]) – mf#959490 – us WHS [071]
Fox valley view / Parents Without Partners – v2 n1-?. [1977 jan-1978 apr/jun] – 1r – 1 – mf#621442 – us WHS [305]
Fox, W J *see* History of christ
Fox, William Johnson *see* Church establishment inconsistent with the spirit of christianity a...
Fox, William W *see* In the shadow of the arctic
Foxborough 1720-1849 – Oxford, MA (mf ed 1996) – (= ser Massachusetts vital record transcripts to 1850) – 9mf – 9 – 0-87623-249-7 – (mf 1t-2t: births 1753-1818. mf 2t-3t: marriages & intents 1773-1819. mf 3t: deaths1775-1819. mf 3t-6t: vital records 1720-1849. mf 5t: transient births 1785-1834; deaths 1819-44. mf 6t-7t: marriage intentions 1819-49. mf 7t-8t: marriages 1842-49; births 1843-49. mf 8t-9t: deaths 1843-49. mf 9t: out-of-town marriages 1779-98) – us Archive [978]
Foxborough 1720-1897 – Oxford, MA (mf ed 1996) – (= ser Massachusetts vital records) – 56mf – 9 – 0-87623-383-3 – (mf 1-7: vital records 1720-1860. mf 8-10: town records 1778-92. mf 10-16: town records 1792-1816. mf 17-25: town records 1817-49. mf 26-35: town records 1849-69. mf 36-37: boundaries 1834-1927. mf 38-40: voters 1884-1915. mf 41-42: vital records 1843-54. mf 43-44: birth index 1849-97. mf 44-45: marriage index 1848-97. mf 454-46: death index 1848-97. mf 47-50: births 1855-97. mf 50-53: marriages 1851-98. mf 54-56: deaths 1852-97) – us Archive [978]
Foxcroft, Edmund John Buchanan *see* Australian native policy
Foxe and the english reformation, c1539-1587 : collected manuscript sources from the british library, london – 9r – 1 – £875.00 – uk Matthew [941]
Foxe, J *see* Actes and monuments of matters most speciall and memorable, happenyng in the church
Foxe, John *see*
- The book of martyrs
- Ridley, latimer, cranmer
Foxes of the desert / Carell, Paul – New York: Dutton 1961 [mf ed 1987] – 1r – 1 – (trans fr german by mervyn savill; with: ancient egyptian dances / lexova, i) – mf#7015 – us Wisconsin U Libr [934]
Foxfire – Mountain City. 1967+ (1) 1972+ (5) 1976+ (9) – ISSN: 0015-9220 – mf#7272 – us UMI ProQuest [400]
Foxport baptist church – Los Angeles. 1969-1971 (1) – 1r – 1 – $65.88 – (lacking: jul 1953-oct 1954, sep 1958-may 1960, 1981-nov 1983 (formerly pleasant valley baptist church, 1873-jul 1940); 1991-1995) – mf#6615 – us Southern Baptist [242]
Fox's decisions *see* Haskell's judgements of the honorable edward fox for the maine district and first circuit, 1866-1881
Fox's weekly – Bradford, England. -w. 11 Jan-8 Nov 1883. 30 ft – 1 – uk British Libr Newspaper [072]
Foxton 1599-1950 – (= ser Cambridgeshire parish register transcript) – 5mf – 9 – £6.25 – uk CambsFHS [929]
La foy devoilee par la raison, dans la connaissance de dieu, de ses mysteres, et de la nature / Parisot, Jean Patrocle – 1. ed. Paris: Parisot, 1681 – 1 – us Wisconsin U Libr [240]
Foye, Edward M *see* How to make abstracts of title and searches
Foye, Martin Wilson *see* Antiquity of the church of england
Le foyer des familles illustre – Montreal: C A Marchand, v1 n1(23 oct 1896)- – 9 – ISSN: 1190-7754 – mf#P04137 – cn Canadiana [440]
Le foyer domestique – Ottawa: Bureaux du Foyer domestique, [1876?-1879?] [mf ed v1 n1 1er mars 1876-v4 n6 1er dec 1877; 3e annee n1 3 janv 1878-5e annee n1 1er janv 1880] – 9 – mf#P04008 – cn Canadiana [440]
Foy-Vaillant, Jo *see* Nusimata aerea imperatorum, augustarum et caesarum, in coloniis, municipiis
FPB *see* Exercices orthographiques
Fr alonso de la cruz (alonso de sotomayor) dota a su hija felipa / Meseguer Fernandez, Juan – Madrid: Graf. Calleja, 1970 – 1 – sp Bibl Santa Ana [946]
Fr alonso guerro en la inmaculada en la literatura franciscano-espanola / Uribe, Angel – Archivo Ibero Americano, 1955 – 1 – sp Bibl Santa Ana [440]
Fr hebbels verhaeltnis zu den politischen und sozialen fragen / Steves, Heinrich – Greifswald: F W Kunike, 1909 – 1r – 1 – (incl bibl ref) – us Wisconsin U Libr [430]
Fr hernando de santiago. santander, 1929 / Perez, Luintin – Madrid: Razon y Fe, 1930 – 1 – sp Bibl Santa Ana [240]
Fr hornemanns tagebuch seiner reise von cairo nach murzuck, der hauptstadt des koenigreichs fessan in afrika : in den jahren 1797 und 1798 / Hornemann, Friedrich K – Weimar 1802 [mf ed Hildesheim 1995-98] – 1v on 2mf – 9 – €60.00 – ISBN-10: 3-487-26602-4 – ISBN-13: 978-3-487-26602-2 – gw Olms [916]
Fr joan duns scotus...per universam philosophiam,...contra adversantes defensus / Baro, B – Coloniae Agrippinae, 1664 – 17mf – 9 – mf#CA-6 – ne IDC [241]
Fr juan de cartagena / Vazquez, Isaac – Madrid: Archivo Ibero Americano, 1965 – 1 – sp Bibl Santa Ana [240]
Fr luis de granada, verdadero y unico autor del libro de la oracion / Cuervo, Justo OP – Madrid: Imprenta Revista Archivos, Bibliotecas y Museos, 1918 – 1 – sp Bibl Santa Ana [240]
Fr pedro de jerez en los custodios y provinciales de la provincia de san jose / Perez, Lorenzo – Madrid: Archivo Ibero Americano, 1924 – 1 – (tambien fr jose de santa maria en...; fr baltasar de los angeles...; fr francisco de montemayor...; fr fco de la oliva...) – sp Bibl Santa Ana [240]
Fr schleiermacher's briefwechsel mit j chr gass : mit einer biographischen vorrede / ed by Gass, Wilhelm – Berlin: Georg Reimer 1852 [mf ed 1991] – 1mf – 9 – 0-524-00465-X – mf#1989-3165 – us ATLA [140]
Fr Strehlke *see* Goethe's gedichte
Fra angelico / Douglas, Robert Langton – London 1900 – (= ser 19th c art & architecture) – 4mf – 9 – mf#4.2.1046 – uk Chadwyck [750]
Fra diavolo, oder gasthaus von terracino, romantische oper in drey aufzuegen : vollstaendiger auszug fuer das pianoforte auf 4 haende / Auber, D F E – [18–] – 1r – 1 – mf#pres. film 33 – us Sibley [790]
Fra forst til sidst: gamle og ny fortaellinger / Nielsen, Anton – Odense: Hans Jensens Forlag, 1883. 361p – 1 – us Wisconsin U Libr [430]
Fra gronland till stillehavet / Rasmussen, K – Kobenhavn, 1925-1926. 2v – 18mf – 9 – mf#N-361 – ne IDC [919]
Fra grundtvigianismens og den indre missions tid 1848-1898 / Koch, Ludvig – Kybenhavn: G.E.C. Gad, 1898 – 1mf – 9 – 0-7905-5358-9 – (incl bibl ref) – mf#1988-1358 – us ATLA [240]
Fra kirkens arbeidsmark : afhandlinger, taler og foredrag / Sverdrup, Georg; ed by Helland, Andreas – Minneapolis, MN: Frikirkens Boghandels Forlag, 1911 [mf ed 1993] – (= ser Professor georg sverdrups samlede skrifter i udvalg 4; Lutheran coll) – 1mf – 9 – 0-524-06324-9 – mf#1991-2497 – us ATLA [242]
Fra laaland: ny fortaellinger / Henningsen, Emanuel – Kobenhavn: NC Rom, 1880. 295p – 1 – us Wisconsin U Libr [390]
Fra manchuriet : rejseindtryk / Nyholm, J – Kobenhavn: Det Danske Missionsselskabs Forlag, 1913 [mf ed 1995] – (= ser Yale coll) – 185p (ill) – 1 – 0-524-09427-6 – (in danish) – mf#1995-0427 – us ATLA [951]
Fra paolo sarpi : the greatest of the venetians / Robertson, Alexander – 3rd ed. London: G. Allen, 1911 – 1mf – 9 – 0-7905-6316-9 – mf#1988-2316 – us ATLA [240]
Fra pol til pol / Bauditz, Sophus – Kobenhavn: Gyldendalske boghandel, 1897 [mf ed 1987] – 2v in 1 (ill) – 1 – mf#10695 – us Wisconsin U Libr [430]
Fra ungdomsaar : en oversigt over den forenede norsk lutherske kirkes historie og fremskridt i de svundne femogtyve aar / Bergh, Johan Arndt et al – Minneapolis, Minn, USA: Augsburg Pub House, 1915 – 1mf – 9 – 0-524-02451-0 – mf#1990-4310 – us ATLA [242]
Frachetta, G *see*
- Il primo libro delle orationi nel genere deliberativo...scritte da lui a diuersi principi per la guerra contra il turco
- Il raggvaglio delle marauigliose pompe con le quali mehemet settergi generale di mehemet 3 imperator de' turchi e vscito fuori di constatinopoli
Fractio panis : die aelteste darstellung des eucharistischen opfers in der "capella graeca" / Wilpert, J – Freiburg i.Br., 1895 – €21.00 – ne Slangenburg [241]
Fractio panis / Wilpert, J – Freiburg im Breisgau, 1895 – 4mf – 9 – mf#H-3054 – ne IDC [700]
Fractionated components of resisted reaction time in men and women / Watkinson, Jeffrey – 1996 – 2mf – 9 – $8.00 – mf#PSY 1966 – us Kinesology [612]
O frade – Recife, PE. 13 mar-06 maio 1876 – (= ser Ps 19) – bl Biblioteca [079]
Fradenburgh, Jason Nelson *see*
- Departed gods
- Fire from strange altars
- Light from egypt
- Living religions
- Old heroes
Fradkin, Il'ia Moiseevich *see* Restavratory orla i svastiki
Fradryssa : the ex-monk and imposter / Blenk, James H – [New Orleans, LA: Morning Star, 1910] – 1mf – 9 – 0-8370-8004-5 – mf#1986-2004 – us ATLA [240]
Fradryssa, G V *see* Roman catholicism capitulating before protestantism
Fraedrich, Gustav *see* Ferdinand christian baur, der begruender der tuebinger schule
Fraemmande religionsurkunder / Johansson, Karl Ferdinand et al; ed by Soederblom, Nathan – Stockholm: H Geber, 1908 – 1mf – 9 – 0-524-01965-7 – (incl bibl ref) – mf#1990-2756 – us ATLA [200]
Fraenckische acta erudita et curiosa *see* Nova literaria circuli franconici
Fraenkel, Jonas *see*
- Goethes briefe an charlotte von stein
- Marginalien zu goethes briefen an charlotte von stein
- Zacharias werners weihe der kraft
Fraenkel, Meir *see* Likute lashon
Fraenkel, Peter J *see* Wayaleshi
Fraenkische landeszeitung – Ansbach DE, 18 sep 1946-23 nov 1948* – 1r – 1 – gw Mikrofilm [074]
Der fraenkische merkur : oder unterhaltungen gemeinnuetzigen inhalts fuer die fraenkischen kreislande / ed by Bundschuh, Joh Kaspar – Schweinfurt 1794-1800 – (= ser Dz. historisch-geographische abt) – 44mf – 9 – €440.00 – mf#k/n1283 – gw Olms [074]
Fraenkische nachrichten – Tauberbischofsheim DE, 1977- – ca 8r/yr – 1 – gw Misc Inst [074]
Der fraenkische republikaner / ed by Hartmann, K et al – Mainz 1792-93 [mf ed Hildesheim 1992-98] – (= ser Dz. historisch-politische abt) – 2mf – 9 – €60.00 – mf#k/n1726a – gw Olms [320]
Fraenkische sammlungen von anmerkungen aus der naturlehre arzneygelahrtheit oekonomie und den damit verwandten wissenschaften / ed by Delius, Heinrich Friedrich – Nuernberg 1756-68 – (= ser Dz. abteilung naturwissenschaft) – 8v on 32mf – 9 – €320.00 – mf#k/n3223 – gw Olms [615]
Fraenkische tageszeitung – Nuernberg DE, 1933 1 jun-1945 15 apr – 44r – 1 – mf#4803 – gw Mikropress [074]
Der fraenkische volksfreund – Schwabach DE, 1793 mar-aug – 1r – 1 – gw Misc Inst [074]
Die fraenkischen zuschauer, bey gegenwaertigen besseren aussichten fuer die wissenschaften und das schulwesen im vaterlande / ed by Sprenger, Placidus – Wuerzburg – (= ser Dz) – 3mf – 9 – €90.00 – mf#k/n5389 – gw Olms [370]
Fraenkischer kurier *see* Mittelfraenkische zeitung fuer recht, freiheit und vaterland
Fraenkischer merkur – Schweinfurt DE, 1794-95 – 1r – 1 – gw Mikrofilm [074]
Fraenkischer merkur *see* Rheinische kronik
Fraenkischer tag – Bamberg DE, 1946 8 jan-1948 16 dec [many gaps] – 1r – 1 – (filmed by misc inst: 1968- [ca 9r/yr]) – gw Mikrofilm; gw Misc Inst [074]
Fraenkischer tagespost – Nuremberg, Germany. -d. 1 Sept 1916-6 Aug 1919. Imperfect. 8 reels – 1 – uk British Libr Newspaper [072]
Fraenkisches archiv / ed by Buettner, Heinrich Christoph et al – Ansbach 1790-91 – (= ser Dz. historisch-geographische abt) – 3v on 7mf – 9 – €140.00 – mf#k/n1226 – gw Olms [943]
Fraenkisches magazin fuer statistik, naturkunde und geschichte / ed by Sprengseysen, Chr Fr Kessler von – Sonnenberg 1791 – (= ser Dz. historisch-geographische abt) – 1v on 2mf – 9 – €60.00 – mf#k/n1257 – gw Olms [943]
Fraenkisches volk – Hof DE, 1933 2 jan-31 aug [gaps], 1934-1945 15 apr – 34r – 1 – (title varies: 1 oct 1934: bayerische ostmark [main ed in bayreuth]; 1 aug 1942: hofer tageblatt; 1 mar 1943: hofer ns-zeitung) – gw Misc Inst [074]

Fraenkisches volk [main edition] – Bayreuth DE, 1932 1 oct-1934 24 oct – 8r – 1 – (title varies: 23 jun 1934: bayerische ostwacht; 1 oct 1934: bayerische ostmark; 1 aug 1942: bayreuther kurier. regional ed: nuernberg 1933 19 apr-31 may (gaps) [1r]) – mf#4890 – gw Mikropress [074]
Fraenkisches volksblatt – Wuerzburg DE, 1952 5 apr-1971 22 jun – 67r – 1 – (title varies: 16 mar 1994: volksblatt) – gw Mikrofilm; gw Misc Inst [074]
Fraenzchens lieder / Hoffmann von Fallersleben, August Heinrich – Luebeck: Dittmer, 1859 – 1 – us Wisconsin U Libr [810]
Das fraeulein von scuderi : erzaehlung aus dem zeitalter ludwigs 14 / Hoffmann, E T A [Ernst Theodor Amadeus] – Leipzig: Insel-Verlag [191-?] [mf ed 1995] – 1r – 1 – (filmed with: stunde der entscheidung / bernd hofmann) – mf#3878p – us Wisconsin U Libr [830]
Eine frage : idyll zu einem gemaelde seines freundes alma tadema / Ebers, Georg – Stuttgart: Deutsche Verlags-Anstalt, [1893-97?] [mf ed 1993] – (= ser Georg ebers gesammelte werke 20) – 120/vii/107p (ill) – 1 – mf#8554 reel 4 – us Wisconsin U Libr [880]
Die frage nach dem sinn des daseins / Lauth, R – Muenchen, 1953 – €15.00 – ne Slangenburg [120]
Die frage nach makkabischen psalmen / Goossens, Eduard – Muenster i W: Aschendorff, 1914 [mf ed 1989] – (= ser Alttestamentliche abhandlungen 5/4) – 1mf – 9 – 0-7905-2775-8 – mf#1987-2775 – us ATLA [221]
Fragen der arbeitsoekonomik – Berlin: Die Wirtschaft, 1954-60.Ceased with v19 (1960?). Each no. has distinctive title – 1 – us Wisconsin U Libr [330]
Fragmens de quelques poesies et sentiments d'esprit de m.l. / Labadie, Jean de – Amsterdam, 1678 – 2mf – 9 – mf#PPE-196 – ne IDC [240]
Fragmens de sibaris : ballet ajoute aux surprises de l'amour / Rameau, Jean Philippe – Paris: Bayard, Daumont, Castagnery [1757] [mf ed 1989] – 1r – 1 – (french words, by marmontel) – mf#pres. film 55 – us Sibley [790]
Fragmens d'un voyage en afrique : fait pendant les annees 1785, 1786 et 1787... / Golberry, S M X – Paris, 1802. 2v – 20mf – 9 – mf#A-308 – ne IDC [916]
Fragmens d'un voyage en afrique : fait pendant les annees 1785, 1786 et 1787, dans les contrees occidentales de ce continent, comprises entre le cap blanc de barbarie, par 20 degres, 47 minutes, et le cap de palmes, par 4 degres, 30 minutes, latitude boreale / Golbery, Silvain – Paris [u a] 1802 [mf ed Hildesheim 1995-98] – 2v on 12mf – 9 – €120.00 – 3-487-27203-2 – gw Olms [916]
Fragmens d'un voyage sentimental & pittoresque dans les pyrenees : ou lettre ecrite de ces montagnes / Saint-Amans, J F B de – Farmington. 1975-1981 (1,5,9) – 3mf – 9 – mf#12179 – ne IDC [914]
Das fragment des demetrius / Schiller, Friedrich von – Wien: K Graeser [1893?] [mf ed 1991] – (= ser Graesers schulausgaben classischer werke 48) – 1r – 1 – (incl bibl ref; with cont fr freiherr franz von maltiz; int & ann by adolf lichtenheld. filmed with: schillers demetrius / martin greif) – mf#2872p – us Wisconsin U Libr [820]
Fragment einer schrift des maertyrer-bischofs petrus von alexandrien / Schmidt, Carl – Leipzig, 1901 – (= ser Tugal 2-20/4b) – 1mf – 9 – €3.00 – ne Slangenburg [240]
Fragment of a prajnaparamita manuscript from central asia / Bidyabinod, B B – Calcutta: Govt of India, Central Publication Branch, 1927 – (= ser Samp: indian books) – us CRL [090]
Fragment of an uncanonical gospel from oxyrhynchus / ed by Grenfell, Bernard Pyne & Hunt, Arthur Surridge – London; New York: Published for the Egypt Exploration Fund by Oxford University Press, 1908 – 1mf – 9 – 0-8370-1835-8 – mf#1987-6223 – us ATLA [220]
A fragment of the babylonian "dibbarra" epic / Jastrow, Morris – New York: N D C Hodges, agent; Philadelphia: University of Pennsylvania Press, 1891 [mf ed 1986] – (= ser Publications of the university of pennsylvania. series in philology, literature and archaeology 1/2) – 1mf – 9 – 0-8370-7223-9 – (text in english and akkadian, comm in english. incl bibl ref) – mf#1986-1223 – us ATLA [470]
A fragment of the oliver hart diary : aug 4 1754-oct 27 1754 – 1 – $5.00 – us Southern Baptist [920]
A fragment on the organization of the world : containing observations on the mosaic history of the creation / Snelson, Thomas W – London, 1824 – (= ser 19th c evolution & creation) – 1mf – 9 – mf#1.1.3052 – uk Chadwyck [500]
Fragmenta evangelica : quae ex antiqua recensione versionis syriacae novi testamenti (peshito dictae) a gul curetono vulgata sunt – Lond[on]: Williams & Norgate, 1870 [mf ed 1990] – 2v on 2mf – 9 – 0-8370-1756-4 – mf#1987-6152 – us ATLA [226]
Fragmenta evangelii lucae et libri genesis : ex tribus codicibus graecis quinti, sexti, octavi saeculi, uno palimpsesto ex libya in museum britannicum advecto... / ed by Tischendorf, Constantin von – Lipsiae: JC Hinrichs, 1857 [mf ed 1986] – (= ser Monumenta sacra inedita. nova collectio 2) – 4mf – 9 – 0-8370-9430-5 – mf#1986-3430 – us ATLA [090]
Fragmenta evangelii lucae et libri genesis (msi2) / ed by Tischendorf, G F C – Lipsiae, 1857 – (= ser Monumenta sacra inedita. nova collectio v2) – €52.00 – ne Slangenburg [221]
Fragmenta florulae aethiopico-aegyptiacae ex plantis praecipue ab antonio figari m d musaeo i r florentino missis / Webb, P B – Parisiis, 1854 – 2mf – 8 – mf#1050 – ne IDC [580]
Fragmenta latina evangelii s lucae, parvae genesis et assumptionis mosis, baruch, threni et epistola jeremiae versionis syriacae pauli telensis : cum notis et initio prolegomenon in integram ejusdem versionis editionem / ed by Ceriani, Antonio Maria – Mediolani [Milan]: Typis et impensis Bibliothecae Ambrosianae, 1861 – (= ser Monumenta sacra et profana) – 3mf – 9 – 0-524-02769-2 – mf#1987-6463 – us ATLA [221]
Fragmenta liturgica : documents illustrative of the liturgy of the church of england; exhibiting the several emendations of it, and substitutions for it, that have been proposed, from time to time, and partially adopted, whether at home or abroad / ed by Hall, Peter – Bath: Printed by Binns and Goodwin, 1848. Chicago: Dep of Photodup, U of Chicago Lib, 1978 (1r); Evanston: American Theol Lib Assoc, 1984 (1r) – 1 – 0-8370-0750-X – mf#1984-T067 – us ATLA [241]
Fragmenta origenianae octateuchi editionis : cum fragmentis evangeliorum graecis palimpsestis ex codice leidensi folioque petropolitano quarti vel quinti, guelferbytano codice quinti, sangallensi octavi fere saeculi / ed by Tischendorf, Constantin von – Lipsiae: JC Hinrichs, 1860 – (= ser Monumenta sacra inedita) – 4mf – 9 – 0-8370-9431-3 – mf#1986-3431 – us ATLA [090]
Fragmenta origenianae octateuchi editionis (msi3) / ed by Tischendorf, G F C – Lipsiae, 1860 – (= ser Monumenta sacra inedita. nova collectio v3) – €49.00 – ne Slangenburg [221]
Fragmenta sacra palimpsesta : sive, fragmenta cum novi tum veteris testamenti / ed by Tischendorf, Constantin von – Lipsiae [Leipzig]: J C Hinrichs, 1855 [mf ed 1991] – (= ser Monumenta sacra inedita. nova collectio 1) – 1mf – 9 – 0-7905-8348-8 – (text in greek) – mf#1987-6447 – us ATLA [220]
Fragmenta sacra palimpsesta (msi1) / ed by Tischendorf, G F C – Lipsiae, 1855 – (= ser Monumenta sacra inedita. nova collectio v1) – €46.00 – ne Slangenburg [220]
Fragmenta vaticana / Hollweg, A Bethmann; ed by Maio, Angelo – Bonn, 1833 – €7.00 – ne Slangenburg [241]
Fragmenta versionis latinae antehieronymianae prophetarum hoseae, amosi et michae : e codice fuldensi – Marburgi [Marburg]: Typis et sumptibus Joannis Augusti Kochii, 1856-1858 – 3mf – 9 – 0-524-07960-9 – mf#1992-1115 – us ATLA [221]
Fragmentarische mittheilungen ueber eine reise durch holland und einen theil von belgien im herbste 1834 / Steltzer, C F – Koeln am Rhein 1835 [mf ed Hildesheim 1995-98] – 2mf – 9 – €60.00 – 3-487-29619-5 – gw Olms [914]
Fragmentary notes of village sermons / Foster, John – Leeds, England. 1853 – 1r – us UF Libraries [240]
Fragmentary records of miscellaneous reich ministries and offices, 1919-1945 / Germany. Miscellaneous Reich Ministries and Offices – (= ser National archives coll of foreign records seized, 1941-) – 28r – 1 – mf#T178 – us Nat Archives [943]
Fragmente : neue gedichte / Benn, Gottfried – Wiesbaden: Limes Verlag, c1951 [mf ed 1995] – 32p – 1 – mf#8976 – us Wisconsin U Libr [810]
Fragmente der homilien des cyrill von alexandrien zum lukasevangelium / Sickenberger, J – Leipzig, 1909 – (= ser Tugal 3-34/1b) – 1mf – 9 – €3.00 – ne Slangenburg [240]
Fragmente einer griechischen uebersetzung des samaritanischen pentateuchs / Glaue, Paul & Rahlfs – [S.l: s.n, 1911?] – (= ser Mitteilungen des septuaginta-unternehmens) – 1mf – 9 – 0-7905-3020-1 – (incl bibl ref) – mf#1987-3020 – us ATLA [221]
Fragmente einer lederhandschrift enthaltend mose's letzte rede an die kinder israel mitgetheilt und geprueft / Guthe, Hermann – Leipzig: Breitkopf & Haertel, 1883 – 1mf – 9 – 0-8370-3428-0 – mf#1985-1428 – us ATLA [221]
Fragmente syrischer und arabischer historiker / ed by Baethgen, Friedrich – Leipzig: F A Brockhaus, 1884 – 1mf – 9 – 0-8370-7629-3 – (in german, syriac, arabic. incl bibl ref and index) – mf#1986-1629 – us ATLA [470]
Die fragmente ueber die wirkung des tragischen in wilhelm meisters theatralischer sendung : abhandlung... / ed by Mossdorf-Hasenfratz, Eugenie Helene – Zuerich: Dissertationsdruckerei AG Gebr Leemann, 1945 [mf ed 1993] – 54/[2]p – 1 – (incl bibl ref) – mf#8607 – us Wisconsin U Libr [790]
Fragmente ueber menschenbildung / Arndt, Ernst Moritz – Altona: J F Hammerich, 1805 [mf ed 1988] – 2v (ill) – 1 – mf#7070 – us Wisconsin U Libr [370]
Fragmente ueber verschiedene gegenstaende der neuesten zeitgeschichte / ed by Cranz, August Fr – Berlin 1790 – (= ser Dz. historisch-politische abt) – 10iss on 10mf – 9 – €100.00 – mf#k/n1704 – gw Olms [074]
Fragmente vornicaenischer kirchenvaeter aus den sacra parallela / John of Damascus, Saint; ed by Holl, Karl – Leipzig: J C Hinrichs 1899 [mf ed 1989] – (= ser Tugal 20/2) – 1mf – 9 – 0-7905-4040-1 – (text in greek. notes in german) – mf#1988-0040 – us ATLA; ne Slangenburg [240]
Das fragmententhargum : (thargum jeruschalmi zum pentateuch) / ed by Ginsburger, Moses – Berlin: S Calvary, 1899. Chicago: Dep of Photodup, U of Chicago Lib, 1978 (1r); Evanston: American Theol Lib Assoc, 1984 (1r) – 1 – 0-8370-0609-0 – (incl bibl ref) – mf#1984-T097 – us ATLA [220]
Fragmentos de la vida y virtudes del v ilmo y rumo sr dr don vasco de quiroga : primer obispo de la sta. iglesia catedral de michoacan y fundador del real y primitivo colegio de san nicolas obispo de valladolid / Moreno, Juan Joseph – 1st ed. Mexico:...Colegio de S Ildefonso, ano de 1766 – (= ser Books on religion...1543/44-c1800: biografias de religiosos) – 3mf – 9 – mf#crl-155 – ne IDC [241]
Fragmentos musicos, repartidos en quatro tratados : en que se hallan reglas generales, y muy necessarias para canto llano, canto de organo, contrapunto, y composicion / Nassarre, Pablo – Madrid: en su impr de musica 1700 [mf ed 19–] – 6mf – 9 – mf#fiche 628, 923 – us Sibley [780]
Fragments – v5 n1 [1967 jul/sep], v6-13 n1 [1968-1975], v14-17 n1 [1976-1979], v18 n1-v20 n3 [1980 jan/mar-1982 sep], v20- v23 n1/2 [1982 oct/1985 jun], v23 n3/4 [1985 jul/dec], v24 n1/2 [1986 jan/dec] – 1r – 1 – mf#620988 – us WHS [071]
Fragments de geographes et historiens arabes et persans inedits, relatifs aux anciens peuples du caucase et de la russie meridionale / Defremery – Paris, 1849. v13 – 2mf – 9 – mf#U-537 – ne IDC [915]
Fragments d'une flore de l'arabie petree / Delile, A R – Paris, 1833 – 1mf – 8 – mf#727 – ne IDC [956]
Fragments from graeco-jewish writers / ed by Stearns, Wallace Nelson – Chicago: Uni of Chicago Press, 1908 – 1mf – 9 – 0-8370-5380-3 – mf#1985-3380 – us ATLA [450]
Fragments from holy scripture / Hawker, Robert – London, England. 1819 – 1r – us UF Libraries [220]
Fragments from reimarus : consisting of brief critical remarks on the object of jesus and his disciples as seen in the new testament / ed by Voysey, Charles – Lexington, KY: ATLA, Cttee on Reprinting, 1962 [mf ed 1993] – 1mf – 9 – 0-524-07657-X – (trans by g e lessing) – mf#1992-1098 – us ATLA [240]
Fragments in philosophy and science : being collected essays and addresses / Baldwin, James Mark – New York: Scribner, 1902 – 1mf – 9 – 0-7905-3635-8 – mf#1989-0128 – us ATLA [100]
Fragments of a faith forgotten : some short sketches among the gnostics, mainly of the first two centuries / Mead, George Robert Stow – London: Theosophical Pub Society, 1900 [mf ed 1992] – 1mf – 9 – 0-524-03059-6 – (incl bibl ref) – mf#1990-3162 – us ATLA [290]
Fragments of a world mind / Lohia, Rammanohar – Calcutta: Maitrayani Publ & Booksellers; Allahabad: Distributors, Bookland, 1949 – (= ser Samp: indian books) – us CRL [327]
Fragments of a zadokite work / ed by Schechter, Solomon – Cambridge: University Press, 1910 – 1mf – 9 – 0-7905-0289-5 – (incl ind) – mf#1987-0289 – us ATLA [270]
Fragments of fifty years : some lights and shadows of the work of the japan – [s.l]: [s.n] [c1919] [mf ed 1995] – (= ser Yale coll) – 131p – 1 – 0-524-09328-8 – mf#1995-0328 – us ATLA [240]
The fragments of heracleon (ts1/4) / ed by Brooke, Alan E – 1891 – (= ser Texts and studies (ts)) – 3mf – 9 – €7.00 – ne Slangenburg [240]
Fragments of science : a series of detached essays, addresses, and reviews / Tyndall, John – [10th ed] London, 1899 – (= ser 19th c evolution & creation) – 11mf – 9 – mf#1.1.10859 – uk Chadwyck [500]
Fragments of the books of kings : according to the translation of aquila / Crawford Burkitt, F – Cambridge, 1987 – 2mf – 8 – €5.00 – ne Slangenburg [221]
Fragments of the commentary of ephrem syrus upon the diatessaron / Harris, James Rendel – London: C J Clay, 1895 – 1mf – 9 – 0-8370-9701-0 – (incl bibl ref) – mf#1986-3701 – us ATLA [221]
Fragments of the negro economics division files, 1919-21 – Washington, DC, National Archives and Records Service [19–] – us CRL [330]
Fragments of voyages and travels / Hall, Basil – Edinburgh 1831-33 [mf ed Hildesheim 1995-98] – 9v on 21mf – 9 – €210.00 – 3-487-29923-2 – gw Olms [910]
Fragments on ethical subjects / Grote, George – London: J Murray, 1876 [mf ed 1990] – 1mf – 9 – 0-7905-7636-8 – mf#1989-0861 – us ATLA [170]
Fragments on india / Voltaire, [Francois-Marie Arouet de] – Lahore: Contemporary India Publ, 1937 – (= ser Samp: indian books) – (trans by freda bedi) – us CRL [954]
Fragments politiques et litteraires / Boerne, Ludwig – [Paris]: Pagnerre, 1842 [mf ed 1993] – xxxix/243/[1]pl – 1 – mf#8524 – us Wisconsin U Libr [840]
Fragments, religious and theological : a collection of independent papers relating to various points of christian life and doctrine / Curry, Daniel – New York: Phillips & Hunt, 1880 – 1mf – 9 – 0-524-04068-0 – mf#1991-2013 – us ATLA [240]
Fragoso, Augusto Tasso *see*
– Franceses no rio de janeiro
– Historia da guerra entre a triplice alianca e o pa...
Fragoso, J *see* Discurso de cosas aromaticas...de las indias...para uso de medicinas
Fragrant memories of the tuesday meeting : and the guide to holiness, and their fifty years' work for jesus / Hughes, George – New York: Palmer & Hughes, 1886 [mf ed 1984] – (= ser Women & the church in america 163) – 1mf – 9 – 0-8370-1442-5 – mf#1984-2163 – us ATLA [242]
Fragua y fuelle / Brenes La Roche, Santos – San Juan, Puerto Rico. 1964 – 1r – us UF Libraries [972]
Frahm, Ludwig *see*
– As noch de tankruesel brenn'
– Minschen bi hamborg ruem
– Von morgen bet abend
– Wenn de scharrnbulln brummt
Fraidl, Franz *see* Die exegese der siebzig wochen daniels in der alten und mittleren zeit
Fraie arbaiter velt – London, UK. Heroisgegeben fun Anarchistishen Press Farein. 10 Nov 1905-28 Jun 1906 – 1 – uk British Libr Newspaper [072]
Fraie presse – London, UK. 30 Sept 1937 – 1 – uk British Libr Newspaper [072]
Fraie velt (the free world) – London, UK. May-Jul 1891 – 1 – uk British Libr Newspaper [072]
Fraie vort – London, UK. 15 Sept 1933-26 Jul 1935 – 1 – uk British Libr Newspaper [072]
Fraie vort – London, UK. Aroisgegeben fun Anarchistishen Propaganda Komitet.Sept 1925 – 1 – uk British Libr Newspaper [072]
Fraie yidishe tribune (free jewish tribune) – London, UK. Jun/Jul 1946-Jan/Mar 1948 – 1 – uk British Libr Newspaper [072]
Frailas, A *see* Conocimiento, curacion y preservacion de la peste
Un fraile extremeno en filipinas / Munoz de San Pedro, Miguel – Badajoz: Dip.Prov., 1952 – 1 – sp Bibl Santa Ana [959]
Fraile procer y una fabula poema / Guillen, Flavio – Guatemala, 1932 – 1r – us UF Libraries [972]
Fraiman, Me'ir Ben Ze'ev *see* Torat me'ir
Der fraind – Spb., Warsaw. v1-12. 1903-1913 (oct 13) – 11r – 1 – mf#J-92-8 – ne IDC [077]
Der fraind – St Petersburg, Warsaw. Jan 1903-Dec 1908; Jan 1910-Oct 1913 – 1 – us NY Public [077]
Fraind (the family friend weekly) – London, UK. 6 Jan 1922-27 Jun 1924; 13 Feb 1925-20 Apr 1926 – 1 – uk British Libr Newspaper [072]
Fraktsiia narodnoi svobody v period s 15 oktiabria 1913 g po 14 iiunia 1914 g : pt 1-3: otchet fraktsii. rechi chlenov fraktsii. zakonodatelnye predlozheniia, vnesennye fraktsiei vo vtoruiu sessiiu. chetvertaia gosudarstvennaia duma. sessiia 2-ia – 1914 – 256p 5mf – 9 – mf#RPP-109 – ne IDC [325]

Fraktsiia soiuza 17-go oktiabria v 4-i gosudarstvennoi dume : obzor deiatelnosti na 1-oi sessii, 15 noiabria 1912 g – 25 iiulia 1913 g – 1914 – 90p 1mf – 9 – mf#RPP-183 – ne IDC [325]
Fram – Milwaukee WI. 1895 nov 6-1897 jul 2 – 1r – 1 – (cont by: la crosse tidende; la crosse tidende or milwaukee fram) – mf#1126548 – us WHS [071]
Fram – Fargo ND, Moorhead MN. 1898 may 18-nov 30, dec 28 – 1r – 1 – (cont: fjerde juli og dakota; rodhuggeren; cont by: folkets blad; fram and folkets blad) – mf#543436 – us WHS [071]
Fram – London, UK. 5 Dec 1942; 20 Feb, 20 Mar-30 Oct 1943; 8, 22 Jan, 4 Mar-23 Dec 1944; 6 Jan-23 Apr, 12 May 1945 – 1 – uk British Libr Newspaper [072]
Framat – Malmo, Sweden. 1871-83 – 7r – 1 – sw Kungliga [078]
Frame, Elizabeth *see* Descriptive sketches of nova scotia in prose and verse
Frame, Esther Gordon *see* Reminiscences of nathan t. frame and esther g. frame
Frame, Nathan T *see* Reminiscences of nathan t. frame and esther g. frame
Framery, Nicolas Etienne *see* Le barbier de seville
Framework for caribbean studies / Smith, M G – Mona, Jamaica. 195- – 1r – us UF Libraries [972]
The framework of the church : a treatise on church government / Killen, W D – Edinburgh: T.& T. Clark, 1890 – 1mf – us ATLA [240]
The framework of the church : a treatise on church government / Killen, William Dool – Edinburgh: T & T Clark, 1890 – 1mf – 9 – 0-7905-5238-8 – (incl bibl ref) – mf#1988-1238 – us ATLA [240]
The framework of the economic and financial policies of the government of basutoland, 1967-72 – Mazenod: Mazenod Institute, [197-?] – us CRL [330]
The framework of the future / Amery, Leopold Stennett – London, New York: Oxford University Press, 1944 – (= ser Samp: indian books) – us CRL [954]
Framfari – Canada. sept 1877-dec 1880 – 1r – 1 – (in icelandic. some iss missing) – cn Commonwealth Imaging [071]
The framing of the constitution of the united states / Farrand, Max – New Haven, London: Yale University Press, 1913 – 4mf – 9 – $6.00 – mf#LLMC 95-077 – us LLMC [323]
Framing the constitution of japan, 1944-1949 : primary sources in english, 1944-1949 – (= ser The occupation of japan) – 420mf [20:1, 29:1] – 9 – $5110.00 – 0-88692-155-4 – (guide only $605) – us UPA [323]
Framingham 1687-1849 – Oxford, MA (mf ed 1996) – (= ser Massachusetts vital record transcripts to 1850) – 16mf – 9 – 0-87623-250-0 – (mf 1t-2t: births & deaths 1687-1734. mf 1t-4t: marriages 1700-55. mf 3t: vital records 1708-49. mf 4t: births 1738-65. mf 5t-7t: vital records 1738-1808. mf 7t-10t: births & deaths 1761-1849+. mf 10t-12t: marriage intentions 1807-49. mf 12t-14t: marriages 1806-44. mf 14t-15t: births 1844-49. mf 15t-16t: marriages & deaths 1844-49. mf 16t: out-of-town marriages 1695-1799) – us Archive [978]
Framingham 1687-1905 – Oxford, MA (mf ed 2001) – (= ser Massachusetts vital records) – 137mf – 9 – 0-87623-415-5 – (mf 1-3: births 1687-1809. mf 3: deaths 1694-1809. mf 3-5: marriages 1697-1808. mf 5,10: vital records 1687-1734. mf 5-10: town records 1679-1736. mf 11-15: vital records 1708-1808. mf 16-21: births & deaths 1739-1872. mf 18: deaths 1882-1905. mf 21-25: intentions 1807-1855. mf 23-24: marriages 1806-1843. mf 25: town records 1816-1836. mf 26-34: town records 1736-1789. mf 35-86: town records 1788-1885. mf 87-93: birth index 1687-1900. mf 94-98: marriage index 1700-1900. mf 99-103: death index 1700-1900. mf 104-113: vital records 1843-1868. mf 114-118: births 1869-1901. mf 119-123: marriages 1853-1886. mf 124-128: marriages 1886-1905. mf 129-134: deaths 1865-1904. mf 135-137: deaths 1905-1914) – us Archive [978]
Framingham, Massachusetts. First Baptist Church and Society *see* Records
Framlington and eye mercury – Suffolk, 9 mar 1972-7 oct 1983 – 1 – (cont: suffolk chronicle framlingham & eye edition; cont by: suffolk mercury extra; ceased publ in 1983) – uk Newsplan [072]
Framtiden – Gaevle, Sweden. 1887-88 – 1r – 1 – sw Kungliga [078]
Fran / Ellis, John Breckenridge – Toronto: McLeod & Allen, c1912 [mf ed 1998] – 5mf – 9 – 0-665-65802-8 – mf#65802 – cn Canadiana [830]
Fran de svartas vaerldsdel : taresadd i uganda / Kolmodin, Adolf – Stockholm: Evang. Fosterlands-Stiftelsens Foerlags-Expedition, 189l. Chicago: Dep of Photodup, U of Chicago Lib, 1972 (1r); Evanston: American Theol Lib Assoc, 1984 (1r) – (= ser Missions-bibliotek foer folket) – 1 – 0-8370-0094-7 – (incl bibl ref) – mf#1984-B307 – us ATLA [240]
Fran "soluppgangens land" : japan forr och nu; fornamligast fran missionshistorisk synpunkt / Kolmodin, Adolf – Stockholm: Fosterlands-Stiftelsens Forlags, [1887] [mf ed 1995] – (= ser Yale coll) – 98p (ill) – 1 – 0-524-10174-4 – (in swedish) – mf#1995-1174 – us ATLA [240]
Fran tra till stal – Stockholm: A Bonnier 1953 [mf ed Bloomington IN: Indiana Uni Lib, Preservation Dept 1984] – 1r – 1 – us Indiana Preservation [390]
Franc tireur – Paris, France. 4 sep 1944; 1945-12 oct 1946 – 2 1/2r – 1 – uk British Libr Newspaper [072]
Franca, Antonio *see* Modernismo brasileiro
Le francais – Ed. du soir du journal: Le Matin. paris. 3 dec 1900-juin 1901, 1902-13 sept 1903 – 1 – fr ACRPP [074]
Le francais – Paris: Impr centrale du Chemins du fer, mar 24, apr 19 1871 – (filmed as pt of: commune de paris newspapers) – us CRL [074]
Le francais – Paris. Journal du soir.2 aout 1868-1 nov 1887, 12 oct 1890, 10 oct 1892, 30 mars 1895, 5 juin 1897, 16 mars 1898.Contient egalement l'ed. de Bordeaux, 9-13 mars 1871 – 1 – fr ACRPP [074]
Le francais a l'universite d'ottawa : deux memoires... – Montreal: [Revue franco-americaine, 1911?] [mf ed 1996] – 1mf – 9 – 0-665-78440-6 – mf#78440 – cn Canadiana [378]
Les francais au canada et en acadie / Gourmont, Remy de – Paris: Firmin-Didot, 1888 – 3mf – 9 – mf#03492 – cn Canadiana [971]
Les francais au canada et en acadie : ouvrage illustre de 50 gravures / Gourmont, Remy de – Paris: Firmin-Didot & cie, 1888 [mf ed 1976] – 1r – 5 – mf#SEM16P266 – cn Bibl Nat [971]
Les francais au niger : voyage et combats... / Pietri, Camille – Paris: Hatchette, 1885 – 1 – (filmed with: les explorations au senegal et dans les contrees voisines par j ancelle) – us CRL [960]
Les francais dans l'amerique de nord : ou, histoire des principales familles du canada. supplement / Daniel, Francois – Montreal: E Senecal, 1868 – 1mf – 9 – mf#04202 – cn Canadiana [920]
Francais dans le monde – Paris. 1961+ (1) 1976+ (5) 1976+ (9) – ISSN: 0015-9395 – mf#9781 – us UMI ProQuest [370]
Les francais du canada / Derouet, Camille – Paris: s.n, 1892 – 1mf – 9 – mf#04228 – cn Canadiana [305]
Les francais en amerique : canada, acadie, louisiane / Feyrol, Jacques – Paris: H Lecene et H Oudin, 1886? – 3mf – 9 – mf#05463 – cn Canadiana [305]
Francais et allemands, histoire de leurs relations intellectuelles et sentimentales / Reynaud, Louis – Paris: A. Fayard et Cie., 1930. 386p – 1 – us Wisconsin U Libr [944]
Francais, Gabriel *see* Le complot franc-maconnique et le droit d'accroissement
Francais, J *see*
- L'eglise et la science
- L'eglise et la sorcellerie

Francais, malgaches, bantous, arabes, turcs, chinois, canaques...parlons-nous une meme langue? : essai de semantique comparee, les grands themes universels du langage / Auber, Jacques – Tananarive: Impr officielle, 1958 – us CRL [410]
Le francais moderne : revue de linguistique francaise – Paris. 1933-72 – 1 – fr ACRPP [440]
Une francaise au soudan sur la route de tombouctou, du senegal au niger / Bonnetain, Paul (Madame) – 2. ed. Paris: Librairies-Impr Reunies, 1894 – us CRL [916]
Une francaise au soudan sur la route de tombouctou, du senegal au niger / Bonnetain, Paul (Madame) – 2nd ed. Paris, Librairies-Imprimeries Reunies, 1894 – us CRL [960]
Francaus parle / Leist, Ludovic – Bucharest, Romania. 1908 – 1r – us UF Libraries [960]
France : christianisme et civilisation / Bonet-Maury, Gaston – Paris: Hachette, 1907 – 1mf – 9 – 0-7905-6984-1 – (incl bibl ref) – mf#1988-2984 – us ATLA [240]
France / Eyquem, Marie-Therese – Paris, France. 1940? – 1r – us UF Libraries [025]
France : internal affairs and foreign affairs, 1945-1954 / U.S. State Dept – (= ser Confidential u s state department central files) – 1 – $18,740.00 coll – (internal affairs, 1945-49: pt1: political, governmental & national defense affairs 22r $4245 isbn 0-89093-919-5; pt2: social, economic & industrial affairs 46r $8900 isbn 0-89093-920-9. foreign affairs, 1945-49 4r $770 isbn 0-89093-921-7. internal affairs, 1950-54: pt1: political, governmental & national defense affairs 25r $4840 isbn 0-89093-956-X. foreign affairs, 1950-54 5r 970 isbn 0-89093-955-1. with p/g) – us UPA [944]
France – London, UK. 6 Oct 1921-9 Jul 1969 – 1 – uk British Libr Newspaper [072]
France / Morgan, Sydney – London 1818 [mf ed Hildesheim 1995-98] – 2v on 7mf – 9 – €140.00 – 3-487-29802-3 – gw Olms [914]
France *see*
- Almanach national
- Bulletin annexe au journal officiel
- Bulletin des annonces legales obligatoires
- Bulletin officiel des annonces commerciales
- Bulletin officiel des forces francaises libres
- Documents administratifs
- Elections de 1958
- Les evenements de mai-juin 1968
- Le journal officiel
- Journal officiel de la republique francaise
- Journal officiel de l'empire francais
- Periodiques clandestins 1939-1945
- Recueil des actes administratifs en algerie
- Repertoire de legislation et de jurisprudence forestieres
- Selection de journaux de la periode de la commune mars a mai 1871
- Selection de journaux ephemeres 1848-1849
- Selection de journaux ephemeres 1848-1849, 1
- Selection de journaux ephemeres 1869-1871
- Selection de journaux ephemeres 1869-1871, 1
- Selection de journaux ephemeres 1869-1871, 3
- Selection de journaux ephemeres de la periode de la revolution
- Selection de journaux ephemeres de la periode de la revolution, 1
- Selection de journaux ephemeres de la periode de la revolution, bobine 6
- Selection de journaux ephemeres de la periode de la revolution, bobines 1-7
- Tables annuelles
- Textes d'interet general

La france : description geographique, statistique et topographique / Loriol, V A – Paris 1834 [mf ed Hildesheim 1995-98] – 4v on 8mf – 9 – €160.00 – 3-487-29797-3 – gw Olms [914]
La france : journal des interets monarchiques et religieux de l'Europe – Paris. janv-15 mars 1837 – 1 – fr ACRPP [073]
La france : journal quotien du matin – Sigmaringen DE, 1944 26 oct-1945 21 may? [gaps] – 1r – 1 – (publ by the vihy govt (marschall petain) during the last months of the war) – gw Mikropress [074]
La france – Paris. janv-avr 1863, janv-juin 1870, janv-juin 1871, 1881-82, 1884, juil-dec 1885, janv-juin 1890 – 1 – (politique, scientifique et litteraire puis organe du parti republicain progressiste) – fr ACRPP [073]
France, 1919-1941 – (= ser U s military intelligence reports) – 12r – 1 – $1895.00 – 0-89093-664-1 – (with p/g) – us UPA [355]
France africaine, sahara et soudan : essai sur la mise en valeur du sahara et sur les communications du centre africaine avec l'europe / Pavard, C – Paris: F Leve, 1905 – 1 – us CRL [960]
France. Agence Generale des Colonies *see* Bulletin
France Ambassade (Us) Service De Presse Et D'information *see* Hour of independence
France, Anatole *see* Elm-tree on the mall
France. Ancien Regime *see*
- Arrets de la chambre des comptes de paris
- Arrets de la cour des aides
- Arrets de la cour des monnaies
- Arrets de la cour du parlement
- Arrets du conseil d'etat du roi
- Arrets du grand conseil du roi
- Arrets royaux

La france ancienne et moderne / Carel, Auguste – Paris 1820 [mf ed Hildesheim 1995-98] – 2v on 6mf – 9 – €120.00 – ISBN-10: 3-487-25892-7 – ISBN-13: 978-3-487-25892-8 – gw Olms [944]
France and belgium, 1848-1900 *see* The papers of queen victoria on foreign affairs
France. Annam *see* Moniteur du protectorat de l'annam et du tonkin
France as it is, not lady morgan's france / Playfair, William – London 1819-20 [mf ed Hildesheim 1995-98] – 2v on 5mf – 9 – €100.00 – 3-487-29775-2 – gw Olms [914]
France. Assemblee consultative provisoire *see*
- Debats
- Documents

France. Assemblee de l'Union francaise *see*
- Debats
- Documents

France. Assemblee Nationale *see*
- Annales
- Annales de l'assemblee nationale
- Annales du senat et de la chambre des deputes du 8 mars 1876-28 dec 1880
- Debats de l'assemblee nationale
- Debats parlementaires
- Les documents de l'assemblee nationale
- Documents parlementaires

France. Assemblee nationale *see* Impressions
France. Assemblee nationale. Chambre des Deputes *see* Etat des travaux legislatifs...
France. Assemblee Nationale Constituante *see* Compte rendu des seances
France. Assemblee nationale constituante *see* Proces-verbal de l'assemblee nationale
France. Assemblee nationale legislative *see* Compte rendu des seances
France. Assemblee nationale. Senat *see* [Impressions]
France. Assemblee puis Nationale Constituante *see* Debats
France au combat – Paris, France. 30 nov 1944-1 nov 1945; 3 jan 1946 – 1/2r – 1 – uk British Libr Newspaper [072]
France automobile etc – Paris, France. 1897-24 dec 1898; 1899-1911; 10 jan 1912-14 jul 1914 – 13 1/2r – 1 – uk British Libr Newspaper [072]
France before europe / Michelet, Jules – Boston: Roberts Brothers, 1871 – xxiv/111p – 1 – (trans fr french) – us Wisconsin U Libr [944]
France. Bureau des longitudes *see* Annales de l'observatoire de nice
France. Bureau des longitudes et al *see* Annales du bureau des longitudes
La france catholique : l'hebdomadaire d'information et de culture chretiennes – 1957 – 1 – fr ACRPP [241]
France. Chambre des Deputes *see*
- Debats parlementaires
- Documents parlementaires
- Impressions
- Notices et portraits

France. Chambres francaises *see* Archives parlementaires de 1787-1860
France. Commission archeologique de l'Indo-chine *see* Bulletin de la commission archeologique de l'indochine
France. Commission d'enquete sur les actes du gouvernement de la defense nationale *see*
- Enquete parlementaire sur les actes du gouvernement de la defense nationale...depositions des temoins
- Rapports faits au nom de la commission d'enquete

France. Commission des Archives Diplomatiques *see* Recueil des instructions donnes aux ambassadeurs et ministers de france depuis les traites de westphalie jusqu'a la revolution francaise
France. Commission des Monuments d'Egypte *see* Description de l'egypte
France. Conseil de la Republique *see*
- Debats parlementaires
- Documents parlementaires

France. Conseil d'Etat *see*
- Collection complete des lois, decrets, ordonnances, reglemens avis du conseil d'etat
- Etudes et documents
- Recueil des arrets

France. Conseil Economique *see*
- Avis et rapports
- Bulletin

France. Convention Nationale *see*
- Bulletin
- Collection complete des decrets de la convention nationale

France. Convention Nationale. Comite de Salut Public *see*
- Collection complete des lois et decrets nationale de leur
- Recueil des actes du comite de salut public

France. Cour de Cassation. Chambre criminelle *see* Bulletin des arrets
France. Cour de Cassation. Chambres civiles *see* Bulletin des arrets
France d'amerique / Revert, Eugene – Paris, France. 1949 – 1r – us UF Libraries [972]
La france dans l'afrique du nord : algerie et tunisie / Vignon, Louis Valery – 2e. ed. Paris: Guillaumin, 1887 – 1 – us CRL [960]
La france d'asie – Saigon. 5 nov 1901-30 juin 1906 – 1 – fr ACRPP [073]
La france d'asie – Saigon: Impr de l'Opinion 1901-06 [1901 nov 4-14,21-26, nov 30-dec 10, dec 18-28, 1902 jan 4,9-11, jan 21-oct 11, nov 18-dec 20, 1904: jan 5-26, jan 30-feb 11, mar 5-apr 12, oct 13, dec 8-27/28, 1905 jan 3/4,12, mar 14,30, oct 7, oct 12-dec 13, dec 16-30, 1906 jan 4-18, jan 30-mar 1, mar 8-apr 12, apr 19-may 10,12-29, jun 2,7-23 – 1r – 1 – mf#mf-11767 seam – us CRL [079]
France de demain – Paris, France. 15 jun-15 dec 1898; 1899-1913; 20 jan-20 jul 1914; apr 1920 – 17 1/2r – 1 – uk British Libr Newspaper [072]
France de demain – Paris, France. 30 sep 1914-10 jan 1916 – 2r – 1 – uk British Libr Newspaper [072]
La france de demain – Paris, France. -d. 30 Sept 1914-10 Jan 1916. 2 reels – 1 – uk British Libr Newspaper [072]
La france de demain – Paris. -m June 1898-Jul 1914. 18 reels – 1 – uk British Libr Newspaper [072]
France de marseille et du sud est – Marseilles, France. 10 oct 1944-14 aug 1945 – 1/2r – 1 – uk British Libr Newspaper [072]
France. Direction Generale des Douanes *see* Tableau general du commerce exterieur de la france
France dramatique au dix-neuvieme siecle – Paris: C Tresse. v1-20. 1841 – 1 – $324.00 – (lacks v1&2) – mf#0211 – us Brook [440]

La france en danger / Philalethe – Moscou: Impr francaise [ca 1890] – 1mf – 9 – mf#vrl-110 – ne IDC [366]
La france en ethiopie : histoire des relations de la france avec l'abyssinie chretienne sous les regnes de louis 8 et de louis 14 (1634-1706) d'apres les documents inedits des archives du ministere des affaires etrangeres / Caix de Saint-Aymour, A de – Paris, 1886 – 5mf – 9 – mf#NE-20190 – ne IDC [960]
La france en tunisie – Tunis, Impr Rapide, 1920-33. v1-2, 5-6 – us CRL [079]
France et canada : dieppe-quebec (1639), quebec-dieppe (1912) / Gosselin, Auguste – Ottawa: [s.n], 1914 – 1mf – 9 – 0-665-72781-X – (incl bibl ref) – mf#72781 – cn Canadiana [360]
France et haiti / Firmin, Antenor – Paris, France. 1901 – 1r – us UF Libraries [972]
La france et la prusse avant la guerre / Gramont, Agenor A de – Paris 1872 [mf ed Hildesheim 1995-98] – 1v on 3mf – 9 – €90.00 – 3-487-26029-8 – gw Olms [940]
La france et le canada : rapport au syndicat maritime et fluvial de france / Agostini, Enzo – Paris: s.n, 1886 – 2mf – 9 – mf#00016 – cn Canadiana [338]
La france et le grand schisme d'occident / Valois, Noel – Paris: A Picard, 1896-1902 – 6mf – 9 – 0-7905-8165-5 – (incl bibl ref) – mf#1988-6112 – us ATLA [240]
La france et rome de 1700 a 1715 : histoire diplomatique de la bulle unigenitus jusqu'a la mort de louis 14 d'apres des documents inedits / Le Roy, Albert – Paris: Perrin, 1892 – 2mf – 9 – 0-8370-9006-7 – (incl bibl ref and ind) – mf#1986-3006 – us ATLA [240]
La france et sa politique exterieure en 1867 / Rothan, Gustave – Paris 1887 [mf ed Hildesheim 1995-98] – 2v on 6mf – 9 – €120.00 – ISBN-10: 3-487-26000-X – ISBN-13: 978-3-487-26000-6 – gw Olms [327]
France exterieure et coloniale – Paris, France. 3 dec 1937-31 may 1940 – 1r – 1 – uk British Libr Newspaper [072]
La france exterieure et coloniale – Paris, France. -f. 3 Dec 1937-31 May 1940. 1 reel – 1 – uk British Libr Newspaper [072]
France for the last seven years : or, the bourbons / Ireland, William Henry – London 1822 [mf ed Hildesheim 1995-98] – 1v on 3mf – 9 – €90.00 – ISBN-10: 3-487-26334-3 – ISBN-13: 978-3-487-26334-2 – gw Olms [944]
France in 1829-30 / Morgan, Sydney – London 1830 [mf ed Hildesheim 1995-98] – 2v on 8mf – 9 – €160.00 – 3-487-29801-5 – gw Olms [944]
France independante : l'hebdomadaire des independants et des paysans – Paris, 1950-nov 1962 [wkly] – 1 – fr ACRPP [325]
France independante – Paris: Imp parisiennes reunies, mar 19-apr 2, may 7-21 1962 – us CRL [074]
France. Institut National de la Statistique et des Etudes Economiques *see*
– Annuaire statistique de la france
– Annuaire statistique de la france 1878-1965
– Annuaire statistique de la guadeloupe 1949/1953-1967/1970
– Annuaire statistique de la guyane 1947/1952-1961/1970
– Annuaire statistique de la martinique 1952-1969/1972
France. Journal officiel *see* Liste de beneficiaires de citations
France juive / Drumont, Edouard Adolphe – Paris, France. v1-2. 1887 – 2r – us UF Libraries [939]
France. L'Assemblee nationale *see* Bulletin
France. Laws, Statutes, etc *see*
– Ordonnances des rois de france de la troisieme race
– Recueil des decisions des tribunaux arbitraux mixtes institues par les traites de paix
– Recueil des textes reglementant l'enseignement prive en cochinchine.
– Recueil general des anciennes lois francaises depuis l'an 420 jusqu'a la revolution de 1789
La france, l'emigration, et les colons / Pradt, Dominique Georges Frederic de Riom de Prolhiac de – Paris 1824 [mf ed Hildesheim 1995-98] – 2v on 4mf – 9 – €120.00 – ISBN-10: 3-487-26144-8 – ISBN-13: 978-3-487-26144-7 – gw Olms [944]
France libre – Paris, France. 1 sep 1918-12 aug 1919; 5 sep 1944-6 dec 1945 (imperfect) – 2 1/2r – 1 – uk British Libr Newspaper [072]
La france libre : journal socialiste – Paris. 2 juil 1918-14 15 mars 1931 – 1 – fr ACRPP [325]
La france libre – Paris, France. -w. 1 Sept 1918-11 Aug 1919. Imperfect. 2 reels – 1 – uk British Libr Newspaper [072]
La france litteraire : annales universelles des lettres, des arts et des sciences – Paris. 1832-5 aout 1843 – 5 – fr ACRPP [073]
La france litteraire – Contenant les auteurs francais de 1771 a 1796. Ersche, J. S. Hambourg (I-IV). 1797-1802 – 1 – fr ACRPP [440]
La france litteraire / Querard, Joseph M – 12 v. 1827-64 – 1,9 – us AMS Press [800]
France militaire – Paris, France. 3, 4 jan, 10 aug-30 dec 1897; 1917-11 sep 1918; 9 apr, 13 aug, 19 nov-17 dec 1941 – 4 1/4r – 1 – uk British Libr Newspaper [072]
La france militaire : Journal non politique des armees de terre et de mer – Limoges. 1893, 1911-13, 1919 – 1 – fr ACRPP [355]
France. Ministere de la Marine et des Colonies *see* Senegal et niger
France. Ministere de l'Education nationale *see* Verdun, argonne-metz (1914-1918)
France. Ministere de l'Instruction Publique *see*
– Collections de documents inedits sur l'histoire de france (guizot collection)
– Etat de l'instruction primaire en 1864, d'apres les rapports officiels des inspecteurs d'academie
France. Ministere de l'Instruction Publique et des Beaux-Arts *see*
– Direction de l'enseignement primaire. rapport sur l'organisation et la situation de l'enseignement primaire public en france
– Recueil des monographies pedagogiques publiees a l'occasion de l'exposition universelle de 1889
France. Ministere des Affaires Etrangeres *see*
– Les origines diplomatiques de la guerre de 1870-71, recueil de documents
– Les origines diplomatiques de la guerre de 1870-71. recueil de documents par le ministere des affaires etrangeres
France. Ministere des Affairs Etrangeres *see* Recueil des traites de la france
France. Ministere du Commerce et de l'Industrie, des Postes et des Telegraphes. Statistique generale *see* Resultats statistiques du denombrement de 1896
France missionnaire aux antilles / Noussanne, Henri De – Paris, France. 1936 – 1r – us UF Libraries [972]
La france nouvelle – Lavalle (Argentine) 1943 – 1 – (contient un suppl. en castillano) – fr ACRPP [073]
La france nouvelle – Paris. n1-115. 7 dec 1870-1er avr 1871 – 1 – (mq n63, 89) – fr ACRPP [073]
La france nouvelle – Paris. 22 mai-23 dec 1911 [daily] – 1 – (quotidien, independant, politique, litteraire et financier) – fr ACRPP [073]
France. Office Colonial *see* Les productions de l'afrique occidentale francaise
France (Ordonnances) *see* Code militaire
La france orientale – Tamatave. n2, 5, 7, 35, 37, 40-46, 50; 2e s., n7-11 12. avr 1891-juil 1892 – 1 – (organe independante. journal politico-satirique) – fr ACRPP [073]
France – outre-mer. *see* La depeche coloniale
France. Parlement. Assemblee Constituante *see* Table des matieres des noms de lieux et des noms de personnes contenus daus les proces-verbaux des seances depuis le 5 mai 1789 jusqu'au 30 septembre 1791
France. Parlement. Assemblee des Etats generaux *see* Proces-verbal historique des actes du clerge dispute a l'assemblee des etats generaux des annees 1789 et 1790
France. Parlement. Assemblee Nationale *see*
– Proces-verbal
– Table generale des matieres du proces-verbal
France. Parlement. Assemblee nationale constituante *see* Recueil des rapports, discours et autres pieces
France. Parlement. Chambre de l'Ordre de la noblesse aux Etats Generaux a Versailles *see* Proces-verbal des seances
France. Parlement. Convention Nationale *see*
– Pieces imprimees par ordre de la convention
– Proces-verbal
La france protestante / Haag, Eugene & Haag, Emile – Paris: Bureau de la publication. v1-10. 1846-59 – 1 – $120.00 – mf#0253 – us Brook [242]
La france republicaine – Paris: E Marc-Aurel, 1848. apr ?-29 1848 – us CRL [074]
France. Senat *see*
– Les debats du senat
– Debats parlementaires
– Les documents du senat
– Documents parlementaires
– Impressions
– Notices et portraits
France Service De Coordination De L'enseignement *see* Guadeloupe
La france socialiste : bulletin officiel / La Federation des Travailleurs Socialistes de France – n2-57. Paris. 8 sept 1894-9 nov 1895 – 1 – fr ACRPP [325]
France soir – 1988- – 3r per y – 5 – us UMI ProQuest [074]
La france sous le regne de la convention / Conny de LaFay, Felix J de – Paris 1820 [mf ed Hildesheim 1995-98] – 1v on 3mf – 9 – €90.00 – ISBN-10: 3-487-26282-7 – ISBN-13: 978-3-487-26282-6 – gw Olms [944]
France. Supplement au Journal officiel *see* Liste officielle d'ennemis
France, Thaddeus J *see* The impact of project adventure activities on self-perception
France to-day : its religious orientation = L'orientation religieuse de la france actuelle / Sabatier, Paul – London: J.M. Dent; New York: E.P. Dutton, 1913 – 1mf – 9 – 0-7905-6255-3 – (incl bibl ref. in english) – mf#1988-2255 – us ATLA [200]
France-amerique : hebdomadaire d'information pour les francais aux etats-unis – New York. juil 1943-avr 1947 – 1 – fr ACRPP [073]
France-amerique – New York, NY. -w. 23 May 1943-12 Aug 1945; 19 May 1946-28 Dec 1947; 12 Sept-26 Dec 1948; 13 Feb 1949-13 Dec 1953. 7 reels – 1 – uk British Libr Newspaper [071]
France-equateur – Brazzaville. sept 1952-oct 1959 – 1 – fr ACRPP [073]
France-equateur l'avenir – Brazzaville: R Mahe & F Senez, jan 3 1956-mar 3 1960 – 7r – 1 – us CRL [079]
France-luxembourg : revue politique economique et litteraire – Paris. n1-2, 5-7, 10. mars 1919-mai 1921 – 1 – fr ACRPP [073]
France-nouvelle / Communist Party. France – Paris.24 nov 1945-1980 [wkly] – 1 – fr ACRPP [073]
France-outre-mer – Paris, France. 1948-oct 1961; feb 1962-1976 – 26 1/2r – 1 – (aka: europe france outremer) – uk British Libr Newspaper [073]
Frances b hogan-professional educator, coach and director : of intercollegiate athletics for women at the university of north carolina at chapel hill / Hancock, Elizabeth A – 2000 – 97p on 1mf – 9 – $5.00 – mf#PE 4105 – us Kinesology [370]
Frances, Madeleine *see* Spionza dans les pays neerlandais dans le seconde moitie du 17e siecle
Franceschi, Gustavo Juan *see* El movimiento espanol y el criterio catolico
Franceschini, Giovanni *see* Six sonatas for two violins and violoncello with a thorough bass for the harpsichord, opera 2
Francesco crispi, der advokat italiens : historischer roman / Zeidler, Paul Gerhard – Berlin-Schoeneberg: P J Oestergaard, c1939 – 1r – 1 – us Wisconsin U Libr [830]
Francesco de hollanda : vier gespraeche ueber die malerei, gefuehrt zu rom, 1538 / [Hollanda, F de] Joaquim de Vasconcellos – Wien, 1899. v9 – 5mf – 9 – mf#O-517 – ne IDC [700]
Francesco Maria da Lecce *see* Osservazioni grammaticali nella lingua albanese
Franceses no rio de janeiro / Fragoso, Augusto Tasso – Rio de Janeiro, Brazil. 1950 – 1r – us UF Libraries [972]
France-soir – Paris: France Editions et Publications, 1953-69 – us CRL [074]
France-soir – Paris. Toutes eds. sept 1944-86 – 1 – (toute derniere ed. 1969-oct 1992. suite de: defense de la france) – fr ACRPP [074]
Franchassin, L *see* Des conflits de lois en matiere de mariage au maroc
La franche-maconnerie rendue a sa veritable origine : ou, l'antiquite de la franche-maconnerie: prouvee par l'explication des mysteres anciens et modernes / Lenoir, Alexandre – Paris: Fournier 1814 – 4mf – 9 – mf#vrl-107 – ne IDC [366]
Franchere, Gabriel *see* Relation d'un voyage a la cote du nord-ouest de l'amerique septentrionale
Franchetti, R *see* Nella dancalia etiopica
Franchi, Antonino *see* Il concilio 2 di leone...
Franchi de' Cavalieri, Pio *see* Specimina codicum graecorum vaticanorum
Franchi, Giovan Pietro *see* La cetra sonora
Franchise law journal (aba) – v1-20. 1980-2001 – 9 – $240.00 set – (title varies: v1-3 1980-84 as journal of the forum committee on franchising) – ISSN: 8756-7962 – mf#112091 – us Hein [340]
Franchising world – Washington. 1991+ (1,5,9) – ISSN: 1041-7311 – mf#18305,04 – us UMI ProQuest [650]
La francia y la monarquia en el plata (1818-1820). la politica del duque de richelieu. misiones...buenos aires, 1933 / Belgrano, Mario – Madrid: Razon y Fe, 1935 – 1 – sp Bibl Santa Ana [972]
Francis and dominic : and the mendicant orders / Herkless, John – New York: Scribner 1901 [mf ed 1990] – (= ser The world's epoch-makers) – 1mf – 9 – 0-7905-4753-8 – (incl bibl ref) – mf#1988-0753 – us ATLA [241]
Francis asbury / Mains, George Preston – New York: Eaton & Mains, c1909 – 1mf – 9 – 0-524-06187-4 – mf#1991-2443 – us ATLA [242]
Francis bacon und seine nachfolger / Fischer, Kuno – Leipzig, Germany. 1875 – 1r – us UF Libraries [420]
Francis bedford topographical photographs *see* Photography as art and social history
Francis david, founder and martyr of unitarianism in hungary / Gannett, William Channing – London: Lindsey Press, 1914 – 1mf – 9 – 0-524-08760-1 – mf#1993-3265 – us ATLA [243]
Francis, Edmund *see* Papers
Francis, Frederick J *see* A series of original designs for churches and chapels
Francis, Henry Thomas *see* Jataka tales
Francis hopkinson: his book / Hopkinson, Francis – Autograph songbook. Contains songs and part-songs with keyboard accomp., some with figured bass by Hopkinson and others. Pages 151-152 and 177-178 missing. music 626, music 1045 – 1 – us L of C Photodup [780]
Francis hopkinson, the first american poet-composer (1737-1791) : and james lyon, patriot, preacher, psalmodist (1735-1794): two studies in early american music / Sonneck, Oscar George Theodore – Washington, DC: HL McQueen, 1905 [mf ed 1991] – 1mf – 9 – 0-524-01016-1 – mf#1990-0293 – us ATLA [780]
Francis hutcheson : his life, teaching and position in the history of philosophy / Scott, William Robert – Cambridge: University Press, 1900 – 1mf – 9 – 0-7905-8884-6 – (incl bibl ref) – mf#1989-2109 – us ATLA [100]
Francis j attig, senate service 1952-1974 : reporter of senate debates – (= ser Us senate historical office oral history coll) – 1mf – 9 – $5.00 – us Scholarly Res [323]
Francis, Jabez *see* Printing at home, with full instructions for amateurs
Francis, James Bicheno *see* Address of james bicheno francis, president of the american society of civil engineers
Francis, John Junkin *see* Mills' meetings memorial volume
Francis longe collection of theatrical works / U.S. Library of Congress. Rare Book and Special Collections Division – 2269 English plays, satires, musical dramas, pastorals, masques, etc. 1607-1812. 326v – 56r – 1 – $1,131.00 – us L of C Photodup [820]
Francis, Lynette Crysta-Lee *see* "Housing an illegitimate aristocracy"
Francis, M *see*
– Chevilles de maitre adam
– Famille du porteur d'eau
– Moissonneurs de la beauce
Francis, Mabel B *see*
– History of fort dallas
– Lee county
Francis, Mark E *see* Coverage of african american basketball athletes in sports illustrated (1954 to 1986)
Francis, Nicholas C *see* Collegiate soccer players' perceptions of sport psychology, sport psychologists and sport psychological services
Francis o wilcox, senate service 1947-1955 : first chief of staff, senate foreign relations committee – (= ser Us senate historical office oral history coll) – 3mf – 9 – $15.00 – us Scholarly Res [323]
Francis of Assisi, Saint *see*
– Legend. speculum perfectionis. s. francis of assisi: the mirror of perfection
– Opuscula sancti patris francisci assisiensis
Francis patrick kenrick's opinion on slavery / Brokhage, J D – Washington DC, 1955 – 7mf – 8 – €15.00 – ne Slangenburg [230]
Francis, Phil *see* Beautiful santa cruz county
Francis the first and his times / Coignet, Clarisse Gauthier – From the French by Fanny Twemlow. London: R. Bentley and Son, 1888. iv,371p – 1 – us Wisconsin U Libr [944]
Francis-Boeuf, Jean *see*
– La soudanaise et son amant
– La soudanaise et son amant; roman
Franciscan friars / Gardner, May F – s.l, s.l? 193-? – 1r – us UF Libraries [241]
Franciscan friars – s.l, s.l? 193-? – 1r – us UF Libraries [241]
Franciscan Friars of Marytown et al *see* Newsletter
Franciscan legends in italian art: pictures in italian churches and galleries / Gurney-Salter, Emma – London: J.M. Dent, 1905.20 illus. Bibliography p214-216 – 1 – us Wisconsin U Libr [240]
Franciscan martyrs in england / Hope, Anne Fulton – London: Burns and Oates, 1878 – 1mf – 9 – 0-8370-6983-1 – (incl bibl ref) – mf#1986-0983 – us ATLA [920]
Franciscan message – Pulaski. 1971-1973 (1) 1947-1973 (5) (9) – ISSN: 0015-9824 – mf#6563 – us UMI ProQuest [240]
Franciscan Missionaries of Mary *see* Waifs and strays, vol 1
Franciscan studies – ns: 1(1941)-30(1970) – 231mf – 9 – €440.00 – ne Slangenburg [241]
Franciscan studies – St. Bonaventure. 1941+ (1) 1941+ (5) 1941+ (9) – mf#9813 – us UMI ProQuest [241]
Franciscanismo de cortes y cortesanismo de los franciscanos / Lejarza, Fidel de – Madrid: Missionalia Hispanica, 1948 – 1 – sp Bibl Santa Ana [241]
[Franciscano de la provincia de san miguel] en notas de bibliografia franciscana / Lopez, Atanasio & Blazquez del Barco, Juan – Archivo Ibero Americano, 1926 – 1 – sp Bibl Santa Ana [240]

FRANCISCANOS

Los franciscanos capuchinos en venezuela / Lodores, Baltasar de – Caracas, 1929-1931; Madrid: Razon y Fe, 1933. 3v – 1 – sp Bibl Santa Ana [240]
Los franciscanos y la imprenta en mexico / Zulaica Garate, Roman – Madrid: Razon y Fe, 1940 – sp Bibl Santa Ana [946]
Franciscans *see* Renvncia qve hizo la religion de nuestro padre s...
The franciscans in arizona / Engelhardt, Zephyrin – Harbor Springs, Mich.: Holy Childhood Indian School, 1899 – 1mf – 9 – 0-7905-6523-4 – mf#1988-2523 – us ATLA [240]
The franciscans in california / Engelhardt, Zephyrin – Harbor Springs, Mich: Holy Childhood Indian School, 1897 – 6mf – 9 – 0-7905-6642-7 – mf#1988-2642 – us ATLA [240]
Franciscans. Provincia de San Diego de Mexico *see* Constituciones de la provincia de san diego de mexico / de los menores descalcos de la mas estrecha observancia regular de n s p s francisco en esta nueva-espana
Francisci blanchini veronensis...de tribus generibus instrumentorum musicae veterum organicae dissertatio / Bianchini, Francesco – Romae: impensis Fausti Amidei 1742 [mf ed 19–] – 2mf – 9 – mf#fiche 696 – us Sibley [780]
Francisci salinae...de musica libri septem : in quibus eius doctrinae veritas tam quae ad harmoniam... / Salinas, Francisco – Salmanticae: excudebat M Gastius 1577 [mf ed 19–] – 1r – 1 – mf#film 93 – us Sibley [780]
Francisci xaverii patritii e societate iesu in actus apostolorum commentarium / Patrizi, Francesco Saverio – Romae: Civilitatis Catholicae, 1867 [mf ed 1993] – 1mf – 9 – 0-524-06523-3 – mf#1992-0907 – us ATLA [226]
Francisci xaverii patritii e societate iesu in ioannem commentarium / Patrizi, Francesco Saverio – Romae: B Morini, 1857 – 1mf – 9 – 0-524-07185-3 – mf#1992-1055 – us ATLA [220]
Francisci xaverii patritii e societate iesu in marcum commentarium : cum duabus appendicibus / Patrizi, Francesco Saverio – Romae: Apud Iosephum Spithoever, 1862 – 1mf – 9 – 0-524-06524-1 – mf#1992-0908 – us ATLA [220]
Francisci...alciati...ilustrata / Sanchez de las Brozas, Francisco – 1573 – 9 – sp Bibl Santa Ana [440]
Franciscii sanctii brocensis...coment. in and. alciati emblemata...figuris ilustrata / Sanchez de las Brozas, Francisco – Lugduni: Guliel Ruvillum, 1573 – 1 – sp Bibl Santa Ana [946]
Francisco benegas galvan : obispo de queretaro... – Madrid: Razon y Fe, 1939 – 1 – sp Bibl Santa Ana [240]
Francisco, de Ajofrin, fray *see* Carta familiar de un sacerdote, respuesta a un colegial amigo suyo
Francisco de aldana. el divino capitan / Rivers, Elias L – Badajoz: Institucion de – Servicios Culturales de la Excma. Diputacion Provincial, 1955 – 1 – sp Bibl Santa Ana [946]
Francisco de asis busca al hombre... / Anasagasti, Pedro de – Madrid: Arch. Ibero Americano, 1965 – 1 – sp Bibl Santa Ana [240]
Francisco de hinojosa : el personaje inedito de un drama historico / Munoz de San Pedro, Miguel – Badajoz: Diput. prov. de Badajoz, 1946. Sep. Rev. Est. Ex. – 1 – sp Bibl Santa Ana [946]
Francisco de la Encarnacion *see* La mujer fuerte
Francisco de lizaur (1477-1535) / Munoz de San Pedro, Miguel – Madrid: Juan Bravo, 1962 – 1 – sp Bibl Santa Ana [920]
Francisco de lizaur. hidalgo indiano de principios de siglo 16 / Munoz de San Pedro, Miguel – Madrid: Imp. y Ed Maestre, 1948 – 1 – sp Bibl Santa Ana [946]
Francisco de miranda et alexandre petion / Dalencour, Franciois Stanislas Ranier – Port-Au-Prince, Haiti. 1955 – 1r – us UF Libraries [972]
Francisco de paula santander / Perez Cabrera, Jose Manuel – Habana, Cuba. 1940 – 1r – us UF Libraries [972]
Francisco de pizarro o el pais del oro / Escofet, Jose – Barcelona: S.A. IG.S.Barral Herms, 1929. Col. Los grandes exploradores espanoles. vol 5 – sp Bibl Santa Ana [350]
Francisco de quevedo / Quevedo, Francisco De – Mexico City? Mexico. 1945 – 1r – us UF Libraries [960]
Francisco de zurbaran / Pantorba, Bernardino de – Barcelona: Iberia. Joaquin Gil Editores, S.A., 1946 – 1 – sp Bibl Santa Ana [946]
Francisco de zurbaran. su epoca, su vida y sus obras / Cascales Munoz, Jose – Madrid: C.I.A.P. (S.A.), 1931 – sp Bibl Santa Ana [920]
Francisco gavidia y ruben dario / Ibarra, Cristobal Humberto – San Salvador, El Salvador. 1958 – 1r – us UF Libraries [440]
Francisco isnardi / Venezuela, (Capitania General) Real Audiencia – Caracas, Venezuela. 1960 – 1r – us UF Libraries [972]
Francisco pizarro / Orellana-Pizarro Perez-Aloe, Antonio. Vizconde de Amaya – Madrid: Editorial Saturnino Calleja, S.A., 1928 – 1 – sp Bibl Santa Ana [920]
Francisco pizarro / Quintana, Manuel Jose – Budapest, 1962 – sp Bibl Santa Ana [350]
Francisco pizarro 2_0 edicion / Tena Fernandez, Juan – Plasencia: Editorial Sanchez Rodrigo, 1955 – 1 – sp Bibl Santa Ana [946]
Francisco pizarro. biografia del conquistador del peru / Arciniega, Rosa – Santiago de Chile: Editorial Nascimento, 2nd ed 1941 – sp Bibl Santa Ana [350]
Francisco pizarro debio llamarse diaz o hinojosa / Munoz de San Pedro, Miguel – Badajoz: Imprenta Diputacion Provincial, 1951. Sep. Revista de Estudios Extremenos – sp Bibl Santa Ana [946]
Francisco pizarro, largo en vida y en hazanas / Marquerie, Alfredo – Madrid: Boris. Barrena, Ediciones, 1954 – sp Bibl Santa Ana [910]
Francisco pizarro y el tesoro de atahualpa / Pereyra, Carlos – Madrid: Editorial America, s.a. – sp Bibl Santa Ana [920]
Francisco pizarroso, ofm, en notasde bibliografia franciscana... / Castro, Manuel – Madrid: Graf. Calleja, 1968 – 1 – sp Bibl Santa Ana [240]
Francisco, Ramon *see* Superficies sordidas
Francisco sanchez el brocense / Bell, Aubrey F G – Oxford University Press, 1925 – 1 – sp Bibl Santa Ana [920]
Francisco, Tonolo *see* Manual del catequista. trad. del italiano por d. felix merino revuelta. barcelona, 1943
Franciscus, J *see* ...Oratio ad pium quintum pont
Franciscus junius : een levensbeeld uit den eersten tijd der kerkhervorming / Reitsma, Johannes – Groningen: J B Huber, 1864 – 1mf – 9 – 0-7905-6550-1 – (incl bibl ref) – mf#1988-2550 – us ATLA [240]
Francisi, E *see* Neu-polirter geschicht- kunst- und sittenspiegel auslaendischer voelcker, fuernemlich der sineser...armenier, tuerken, russen...
Le franciste : offizielles organ des franzismus, auflage fuer elsass und lothringen – Metz (F), 1934 aug-1935 apr – 1 – fr ACRPP [073]
Le franciste d'alsace et de lorraine : faschistisches kampfblatt – Strassburg (Strasbourg F), 1937 jul-1 aug, 1938 apr – 1 – fr ACRPP [320]
Le franciste d'alsace et de lorraine = Faszistisches kampfblatt. – 1er juil-1er aout 1937, avr 1938 – 1 – fr ACRPP [320]
Francistown, New Hampshire. Francistown Baptist Church *see* Records
Franck, Adolphe *see*
– Kabbalah
– La kabbale
– Philosophie et religion
Franck, Hans *see*
– Geschlagen!
– Godiva
– Opfernacht
– Das pentagramm der liebe
– Recht ist unrecht
– Der regenbogen
– Die schicksalsuhr
– Ein stueck erde
– Totaliter aliter
– Zeitenprisma
Franck, Harry Alverson *see*
– Mexico and central america
– Pan american highway from the rio grande to...
– Roaming through the west indies
– Trailing cortez through mexico
– Tramping through mexico, guatemala and honduras
– Vagabonding down the andes
– Working north from patagonia: being the narrative of a journey, earned on the way, through southern and eastern south america
Franck, Lic Theol *see* Unbeachtet gebliebene fragmente des pelagius-kommentars zu den paulinischen briefen
Franck, Ludwig *see* Statistische untersuchungen ueber die verwendung der farben in den dichtungen goethe's
Franck, Melchior *see*
– Dialogus metricus
– Melodiarum sacrarum 5, 6, 7, 8, 9, 10. 11.12 vocibus
Der francke – Strassburg (Strasbourg F), 1791 – 1 – fr ACRPP [074]
Francke, August Hermann *see*
– A history of western tibet
– Die mitarbeit der bruedermission bei der erforschung zentral-asiens
Francke, August Hermann [comp] *see* Lower ladakhi version of the kesar saga
Francke, Kuno *see*
– Deutsche arbeit in amerika
– Deutsches schicksal
– The german classics of the nineteenth and twentieth centuries
– Goethes vermaechtnis an amerika
– Die kulturwerte der deutschen literatur des mittelalters
– Die kulturwerte der deutschen literatur in ihrer geschichtlichen entwicklung
– Die kulturwerte der deutschen literatur von der reformation bis zur aufklaerung
– Weltbuergertum in der deutschen literatur von herder bis nietzsche
Franckfurthische gelehrte zeitungen : darinnen die merckwuerdigsten neuigkeiten der gelehrten welt – Frankfurt am Main 1736-71 – (= ser Dz) – 36jg on 178mf – 9 – €1068.00 – mf#k/n109 – gw Olms [074]
Francklin, William *see* The history of the reign of shah-aulum, the present emperor of hindustaun
La franc-maconnerie : histoire authentique des societes secretes depuis les temps les plus recules jusqu'a nos jours... – 2e rev augm ed. Paris: Bloud et Barral, [1886?] – 6mf – 9 – mf#vrl-179 – ne IDC [366]
La franc-maconnerie : son caractere, son organisation, son extension, ses sources, ses affluents, son but et ses secrets / Dechamps, Victor Auguste – Paris: P Lethielleux 1863 – 2mf – 9 – mf#vrl-132 – ne IDC [366]
La franc-maconnerie : synagogue de satan / Meurin, Leo – Paris: Victor Retaux 1893 – 6mf – 9 – mf#vrl-155 – ne IDC [366]
La franc-maconnerie a troyes [1751-1820] / Socard, Emile – Troyes: Dufour-Bouquot 1877 – 1mf – 9 – mf#vrl-111 – ne IDC [366]
La franc-maconnerie allemande pendant la guerre : les loges militaires de campagne / Ligue franc-catholique – Paris: Revue internationale des societes secretes 1918 – 1mf – 9 – mf#vrl-182 – ne IDC [366]
La franc-maconnerie dans l'etat – Bruxelles: H Goemaere 1859 – 1mf – 9 – mf#vrl-180 – ne IDC [366]
La franc-maconnerie dans sa veritable signification : ou, son organisation, son but et son histoire / Eckert, Eduard Emil – Liege: J-G Lardinois 1854 – 2v on 10mf – 9 – mf#vrl-121 – ne IDC [366]
La franc-maconnerie demasquee : son organisation, son but, ses doctrines revolutionnaires & sataniques d'apres les documents authentiques – Tournai: Typ Casterman 1886 – 1mf – 9 – mf#vrl-145 – ne IDC [366]
La franc-maconnerie des femmes / Monselet, Charles – nouv ed. Paris: Michel Levy 1873 – 3mf – 9 – mf#vrl-78 – ne IDC [366]
La franc-maconnerie en elle meme et dans ses rapports avec les autres societes secretes de l'europe : notamment avec la carbonaire italienne / Gyr, Jean-Guillaume, abbe – Liege: J-G Lardinois 1859 – 5mf – 9 – mf#vrl-191 – ne IDC [366]
La franc-maconnerie et la magistrature en france a la vieille de la revolution / Amiable, Louis – Aix: J Remondet-Aubin 1894 – 2mf – 9 – mf#vrl-5 – ne IDC [366]
La franc-maconnerie et la revolution / Estampes, Louis d' – Avignon: Seguin freres 1884 – 6mf – 9 – mf#vrl-42 – ne IDC [366]
La franc-maconnerie et les moyens pour arreter ses ravages / Rosset, Michel – Paris: Victor Lecoffre 1882 – 4mf – 9 – mf#vrl-159 – ne IDC [366]
La franc-maconnerie expliquee par un ami de la verite – Metz: Verronnais 1833 – 3mf – 9 – mf#vrl-181 – ne IDC [366]
La franc-maconnerie feminine : l'ordre mac mixte internationale "le droit humain" / Switkow, N – Brunoy: nouv editions nationales 1933 – 2mf – 9 – mf#vrl-112 – ne IDC [366]
La franc-maconnerie mieux connue : ou, expose abrege de l'origine, des principaux developpements, des constitutions, des resources, des ceremonies, des doctrines et des oeuvres de la franc-maconnerie / Schilt, L de – Lille: I Lefort 1841 – 2mf – 9 – mf#vrl-174 – ne IDC [366]
La franc-maconnerie, religion sociale du principe republicain : premiere partie, la science des interets materiels / Mazaroz, Jean Paul – Paris: L'auteur 1880 – (= ser La revanche de la France par le travail et les interets organises t3 chapitre 2) – 4mf – 9 – mf#vrl-76 – ne IDC [366]
La franc-maconnerie: sa politique et son oeuvre... / Dumont, Auguste – Paris: Courrier litteraire de la presse [ca 1900] – 6mf – 9 – mf#vrl-176 – ne IDC [366]
La franc-maconnerie sous la 3me republique : d'apres les discours maconniques prononces dans les loges par les ff brisson, jules ferry, albert ferry, le royer, floquet... / Leroux, Adrien – Paris: Letouzey & Ane [188-?] – 2v on 11mf – 9 – mf#vrl-108 – ne IDC [366]
De francmacons : van in hunne opkomst tot op den tegenwoordigen tyd... – Rousselaere: David Vanhee 1838 – 2v on 2mf – 9 – mf#vrl-184 – ne IDC [366]
Los francmasones : lo que son – lo que hacen – lo que quieren / Segur, Louis Gaston de – Santiago de Chile: Impr Chilena 1868 – 2mf – 9 – mf#vrl-175 – ne IDC [366]
Francmesnil, Ludovic De *see* Grillon du foyer
Franc-Nohain *see*
– Belle eveillee
– Chapeau chinois
– Salles d'attente
Franco, Afonso Arinos De Melo *see* Conceito de civilisacao brasileira
Franco ami de la france? – Paris, 1938. Fiche W 893. (Blodgett Collection of Spanish Civil War Pamphlets) – 9 – us Harvard [946]
Franco, Cid *see* Independencia economica do brasil
Franco De Almeida, Tito *see*
– Conselheiro francisco jose furtado
Franco, Francisco *see* Habla el caudillo
Franco, Giovanni Giuseppe *see* Simon peter and simon magus
Franco in barcelona – London, 1939. Fiche W 894. (Blodgett Collection of Spanish Civil War Pamphlets) – 9 – us Harvard [946]
Franco Isaza, Eduardo *see* Guerrillas del llano
Franco, Jose *see* Panama defendida
Franco, Jose L *see* Afroamerica
Franco, Jose Luciano *see*
– Placido
– Politica continental americana de espana en cuba
Franco, Jose Ulises *see* Petalos de lealtad
Franco, M *see* Discurso medicinal...en el que se declara la horden...para preservarse de la peste
Franco Oppenheimer, Felix *see*
– Del tiempo y su figura
– Hombre y su angustia
– Lirios del testimonio
Franco Ornes, Pericles *see* Tragedia dominica
Franco R, Ramon *see* Colombia
Franco spain...america's enemy / White, David McKelvy – N.Y., 1945. Fiche W1012. (Blodgett Collection of Spanish Civil War Pamphlets) – 9 – us Harvard [946]
Franco: who is he, what does he fight for? / Curran, Edward Lodge – NY, 1937. Fiche W 822. (Blodgett Collection of Spanish Civil War Pamphlets) – 9 – us Harvard [946]
Franco y Lozano, Francisco *see*
– Antologia latina
– Dialogos de los muertos de luciano
– Geografia de...badajoz
– Homilia a...san juan crisostomo
Franco y Lozano, Francisco et al *see* Trozos selectos...clasicos latinos
Franco-british convention of december 23, 1920 : on certain points connected with the mandates for syria and the lebanon, palestine and mesopotamia – London, 1921 – 1mf – 9 – mf#J-28-168 – ne IDC [956]
Franco-californien – San Francisco CA. 1917 dec 7-1918 jun, 1918 jul-dec, 1919 jan-dec, 1920 jan-dec, 1921 jan-dec – 8r – 1 – (cont: courrier de san francisco [san francisco, ca: daily]; cont by: echo de l'ouest [san francisco, ca]; courrier du pacifique) – mf#1421462 – us WHS [071]
Le franco-californien – San Francisco: Le Franco-Californien Pub Co, dec 7 1917-26] – 18r – us CRL [071]
The franco-canadian annexionists of elmira, ny, to gen benjamin f butler : on canado-american annexation – [Elmira, NY?: s.n, 1866?] [mf ed 1983] – 1mf – 9 – 0-665-44415-X – mf#44415 – cn Canadiana [971]
Francoeur, Lucien *see* A propos de l'ete du serpent
Franco-Franco, Tulio *see* Situation internationale de la republic dominicain
Francois d'Assise, Saint *see* Fioretti, petites fleurs de s francois d'assise
Francois de fenelon / St Cyres, Stafford Harry Northcote, Viscount – London: Methuen, 1901 – 1mf – 9 – 0-524-00109-X – mf#1989-2809 – us ATLA [240]
Francois de foix : opera en trois actes / Berton, Henri – Paris: Duhan [c1809] [mf ed 1990] – 1r – 1 – mf#pres. film 78 – us Sibley [780]
Francois de Sales, Saint *see* Souhaits de bonne annee
Francois, Georges Alphonse Florent Octave *see* L'afrique occidentale francaise
Francois hotman, sa vie et sa correspondance / Dareste, R – Paris, 1876. v2 (p 1-59) – (= ser Revue historique) – 1mf – 9 – mf#PBU-434 – ne IDC [240]
Francois, Jean *see* Vocabulaire austrasien
Francois le champi / ed by Searles, Colbert – New York, Toronto: Oxford UP, 1914 – (= ser Oxford french series) – 4mf – 9 – 0-665-87682-3 – (text in french. int. notes and vocabulary in english) – mf#87682 – cn Canadiana [830]
Francois, Louis von *see* Die akte louise von francois

Francois, Louise von *see*
– Gesammelte werke in fuenf baenden
– Die letzte reckenburgerin
Francois, M G *see* Les productions de l'afrique occidentale francaise
Francois pierrefeu : catholic writer and scientist and former rebel sympathiser speaks of the fascist repression – [s.l: s.n. 193-?] [mf ed 1977] – (= ser Blodgett coll) – 1mf – 9 – mf#w895 – us Harvard [320]
Francois suarez de la compagnie de jesus : d'apres ses lettres, ses autres ecrits inedits et un grand nombre de documents nouveaux / Scorraille, Raoul de – Paris: P Lethielleux, c1912-1913 – 3mf – 9 – 0-524-00103-0 – mf#1989-2803 – us ATLA [240]
Francoise *see* Chroniques de lundi de francoise
Francois-Poncet, Andre *see* Les affinites electives de goethe
Franco's mein kampf; the fascist state in rebel spain / Spanish Information Bureau. New York – N.Y., 1939. Fiche W896. (Blodgett Collection of Spanish Civil War Pamphlets) – 9 – us Harvard [946]
Le franc-parleur – Montreal, QC. 1870-78 – 4r – 1 – cn Library Assoc [071]
Les francs – [s l] 1785 [mf ed Hildesheim 1995-98] – 1v on 1mf – 9 – €40.00 – 3-487-26180-4 – gw Olms [944]
Les francs-macons : initiation a tous leurs mysteres – Paris: Charles Waree [18-?] – 2mf – 9 – mf#vrl-50 – ne IDC [366]
Francs-macons. Grande loge de Quebec *see*
– The book of constitution of the grand lodge of quebec, ancient, free and accepted masons
– The book of constitution of the grand lodge of quebec, ancient free and accepted masons
Les francs-macons peints par euxmemes : scenes de leur vie privee: a bas les masques! – Bruxelles: C-J Fonteyn 1854 – 1mf – 9 – mf#vrl-190 – ne IDC [366]
Francs-macons septembriseurs / Bouchez, E – Paris: Tequi 1892 – 2mf – 9 – mf#vrl-126 – ne IDC [366]
Le franc-tireur *see* Voices from wartime france, 1939-45
Francus, D *see* Gli illvstri et gloriosi gesti, et vittoriose impresse, fatte contra turchi...
Francus, I *see* Historicae relationis continvatio
Frangepan, W *see* Oratio ad serenissimvm carolvm v sacri romani imperij caesarem inclytum...
Frangula, oder, die himmlischen weiber im wald / Jahn, Moritz – Leipzig: P Reclam, 1943, c1933 – 1r – 1 – us Wisconsin U Libr [830]
Frank, Ashley Gavin *see* Investment styles and the asean stock market cycle
The frank b kellogg papers, 1923-1937 – 34r* – 1 – $6580.00 – 1-55655-967-4 – (with p/g) – us UPA [327]
Frank, Bruno *see*
– Cervantes
– Der himmel der enttaeuschten
– The magician
– Politische novelle
– Die schatten der dinge
– Strophen im krieg
– Trenck
Frank, C *see* Lamastu, pazuzu und andere daemonen
Frank, Dirk *see* Das paradox der metafikation
Frank duveneck / Heermann, Norbert – Boston, MA. 1918 – 1r – us UF Libraries [750]
Frank E *see* The representative men of the philippines
Frank, Eli *see* Title to real and leasehold estates and liens
Frank, Ernst *see*
– Kinder in sonne
– Not haemmert menschen
Frank, Felix *see* Marguerite d'angouleme, queen of navarre, 1492-1549
Frank field ellinwood : his life and work / Ellinwood, Mary Gridley – New York: FH Revell, c1911 – 1mf – 9 – 0-524-06534-9 – mf#1991-2618 – us ATLA [240]
Frank, Fr H R *see* Geschichte und kritik der neueren theologie
Frank, Franz Hermann Reinhold *see*
– Dogmatische studien
– System der christlichen gewissheit
– System der christlichen wahrheit
– System of the christian certainty
– Ueber die kirchliche bedeutung der theologie a ritschl's
– Vademecum fuer angehende theologen
– Zur theologie a ritschl's
Frank, Friedrich *see*
– Die bussdisciplin der kirche von den apostelzeiten bis zum siebenten jahrhundert
– Der ritualmord vor den gerichtshoefen der wahrheit und der gerechtigkeit
Frank, Gustav *see* Geschichte der protestantischen theologie
Frank, Heinrich *see* Der oberst
Frank, Helena *see* Yiddish tales
Frank, Henry *see* The doom of dogma and the dawn of truth
Frank, K *see*
– Babylonische beschwoerungreliefs
– Bilder und symbole babylonisch-assyrischer goetter
Frank, Laura B *see* The "heart at work" program
Frank, Lawrence Stroup *see* Carl philipp bach and the growth of the sonata form
Frank, Leonhard *see*
– Die entgleisten
– Im letzten wagen
– Mathilde
– Der mensch ist gut
– Das ochsenfurter maennerquartett
– Die ursache
Frank leslie's boy's and girl's weekly – An illustrated record of outdoor and home amusements. v. 1-36. 13 Oct 1866-9 Feb 1884. 4 nos. wanting – 1 – $260.00 – us L of C Photodup [790]
Frank leslie's budget of humorous and sparkling stories, tales of heroism, adventure and satire – 1878-1896 – 1 – (may-jun 1891; mar 1892, and dec 1894 wanting) – us L of C Photodup [800]
Frank leslie's chimney corner – v1-39, n1-1018. 3 jun 1865-29 nov 1884 – 1 – (2 nos wanting) – us L of C Photodup [073]
Frank leslie's fact and fiction for the chimney corner – v1-2, n1-28. 6 dec 1884-6 jun 1885 – 1 – us L of C Photodup [073]
Frank leslie's illustrated newspaper – New York, NY. 1855-1922 (1) – mf#65072 – us UMI ProQuest [071]
Frank leslie's illustrated newspaper *see* Leslie's illustrated weekly
Frank leslie's illustrierte zeitung – New York NY (USA), 1875 jul-1876 jun, 1877 jan-aug, 1878 aug-dec, 1879 jul-1880 jun – 2r – 1 – gw Misc Inst [071]
Frank leslie's lady's journal : devoted to fashion and choice literature – v1-20, n1-517. 18 nov 1871-8 oct 1881. – 1 – (n517 wanting.) – us L of C Photodup [073]
Frank leslie's lady's magazine – v1-51. sep 1857-dec 1882 – 1 – (5 nos. wanting.) – us L of C Photodup [073]
Frank leslie's new monthly – 1881 – 1 – us L of C Photodup [073]
Frank leslie's new monthly : devoted to light and entertaining literature – v1-6. aug 1863-jul 1866 – 1 – us L of C Photodup [073]
Frank leslie's pleasant hours : devoted to light and entertaining literature – v1-60, n3. Aug 1866-Apr 1896. – 1 – (scattered issues wanting) – us L of C Photodup [073]
Frank leslie's sunday magazine – v. 1-25. Jan 1877-Jun 1889 – 1 – us L of C Photodup [073]
Frank murphy in world war i / Fine, Sidney – Ann Arbor: Michigan Historical Collections 1968. 44p. LL-2279 – 1 – us L of C Photodup [340]
Frank, Othmar *see* Vjasa
Frank, Paul *see* Kampf dem tode
Frank, Robyn *see* Journal of agricultural and food information
Frank, Rudolf *see* Wie der faust entstand
Frank, Shlomo *see* Togbukh fun lodzsher geto
Frank, Ulrich *see* Simon eichelkatz; the patriarch
Frank und die frankisten / Graetz, Heinrich – Breslau, Germany. 1868 – 1r – us UF Libraries [939]
Frank, Uwe *see*
– Die photoakustische bestimmung absoluter optischer extinktionskoeffizienten von adsorbierten farbstoffen
– Photoakustische untersuchung des energietransfers auf fraktalen oberflaechen
Frank ve-adato 1726-1816 / Kraushar, Alexander – Warsaw, Poland. 1895 – 1r – us UF Libraries [939]
Frank, Viktor *see* Russisches christenthum
Frank wedekind / Friedrich, Paul – Berlin: W Borngraeber Verlag Neues Leben, [1913] – 1r – 1 – us Wisconsin U Libr [430]
Frank wedekind / Pissin, Raimund – Berlin: Gose und Tetzlaff, [1905?] – 1r – 1 – us Wisconsin U Libr [430]
Frank, Zevi Pesah *see* Keter torah
Frankby, st john the divine – (= ser Cheshire monumental inscriptions) – 2mf – 9 – £4.00 – mf#393 – uk CheshireFHS [929]
Franke, A H *see* Das alte testament bei johannes
Franke, Carl *see*
– Grundzuege der schriftsprache luthers in allgemeinverstaendlicher darstellung
– Luthers lautlehre
– Luthers satzlehre
– Luthers wortlehre
Franke, Eberhard *see* Das ruhrgebiet und ostpreussen
Franke, Martin *see* Johann friedrich august tischbein
Frankel, A Steven *see* Journal of child abuse and the law
Frankel, Menahem Mordecai *see* Derush ve-hidush 'al ha-torah
Frankel, Mira *see* Press digest index, 1948-1952
Frankel, Sally Herbert *see* Economic impact on under-developed societies
Frankel, Z *see* Zeitschrift fuer die religioesen interessen des judenthums
Frankel, Zacharias *see*
– Grundlinien des mosaisch-talmudischen eherechts
– Historisch-kritische studien zu der septuaginta
Frankel, Zacharias et al *see* Monatsschrift fuer geschichte und wissenschaft des judenthums
Franken, Johan Lambertus Machiel *see* Duminy-dagboeke, duminy diaries
Franken, Richard B *see*
– The attention value of advertisements in a leading periodical
– Newspaper reading habits of business executives and professional men in new york
Frankenberg, W *see*
– Evagrius ponticus
– Die syrischen clementinen
Frankenberg, Wilhelm *see*
– Die composition des deuteronomischen richterbuches (richter 2, 6-16)
– Die datierung der psalmen salomos
– Der organismus der semitischen wortbildung
– Die sprueche
– Das verstaendnis der oden salomos
Frankenberger, Julius *see* Walpurgis
Frankenberger nachrichtsblatt und bezirksanzeiger *see* Intelligenz- und wochenblatt fuer frankenberg mit sachsenburg und umgegend
Frankenberger tageblatt *see* Intelligenz- und wochenblatt fuer frankenberg mit sachsenburg und umgegend
Frankenberger zeitung *see* Kreisblatt fuer den kreis frankenberg-voehl
Frankenburg, Robert *see* Der einsiedler am starnberger see
Frankenheim, Moritz *see* Voelkerkunde
Frankenpost [main edition] – Hof DE, 1945 12 oct-1967 3 aug – 1 – (title varies: 12 oct 1945- in hof fr 1968 as hofer anzeiger. fr 1968 nur der mantel der frankenpost. regional ed: arzberg later: sechsaemter neueste nachrichten, ab 4 jan 1972 als ausg fuer arzberg, wunsiedel, selb 23 may 1953-73, 1992-2000 [13r]; rehau 1951 1 nov-1967 (local pgs); kulmbach 1 may 1959: kulmbacher tagblatt, 2 jul 1973: frankenpost / ksb ausg ksb = kulmbach, stadtsteinach, bayreuth, 2 jan 1996: ausg kulmbach / stadtsteinach 1947 4 jan-1964, 1966-2000 (nur lokalseiten); naila 1988- [ca 7r/yr], 23 apr-31 dec 1996; bayreuth 16 jun 1949-71 [4r], 2 jul 1973-90 [52r], 1992-95 (nur lokalseiten); kronach 1950 6 jul-1972 [5r], 1974 1 apr-1984 30 apr (nur lokalseiten); wunsiedel later: sechsaemterbote 1949 2 jul-1973 (lokalseiten), 1991-1996 7 oct, 1997-2000; muenchberg (for all ed only local pp 1950: ed muenchberg, 1972: ed muenchberg/helmbrechts, 1 jul 1976: ed muenchberg/helmbrechts/frankenwald, 1 jul 1980: ed frankenwald, 1 oct 1981: ed frankenwald/muenchberg-helmbrechtser zeitung, 9 jul 1984: muenchberg-helmbrechtser zeitung) 1 jul 1950-2000 [57r]; stadtsteinach 1951 3 jan-1971 [3r] (nur lokalseiten); ausg f=fernausg 1969-87 [ca 7r/yr]; hof-land 1951 30 jan-1962, 1965-1971 18 dec [gaps], 1978 3 may-1979 28 feb (nur lokalseiten); stiftland=kemnath 1950 1 jul-1987 (nur lokalseiten)) – gw Mikrofilm [074]
Frankenstein, Johannes von *see*
– Der kreuziger
Frankenthaler wochenblatt – Frankenthal, Pfalz DE, 1848 – 1r – 1 – (title varies: 1879: frankenthaler zeitung) – gw Misc Inst [074]
Frankenthaler zeitung *see* Frankenthaler wochenblatt
Frankfort Area Genealogy Society *see* Facts and findings
Frankfort, Henri *see*
– Preliminary reports of the iraq expeditions, 1-5
– The problem of similarity in ancient near eastern religions
Frankfort/mokena star – Chicago Heights, IL. 1976-1999 (1) – mf#62577 – us UMI ProQuest [071]
Frankfurt am main wie es ist : in historisch-statistischer, scientifisch und artistischer, spekulativer und volksthuemlich-charakteristischer beleuchtung und darstellung; ernst und humoristisch gehalten, freisinnig bearbeitet / Wild, Karl – Leipzig 1831 [mf ed Hildesheim 1995-98] – 2mf – 9 – €60.00 – 3-487-29498-2 – gw Olms [914]
Frankfurt chronicle – Frankfurt, Hessen. 1980 dec 5-1983 jun 30, 1983 jul 7-1986 mar 27 – 2r – 1 – mf#1044795 – us WHS [071]
Frankfurt im biedermeier : autobiographische und reiseberichte aus dem leben in der stadt am main zur zeit der bundesversammlung (1816-1866) / Haensel-Hohenhausen, Markus – (mf ed 1992) – 3mf – 9 – €49.00 – 3-89349-540-1 – mf#DHS 540 – gw Frankfurter [943]
Frankfurt. Stadt- und Universitaetsbibliothek. Flugschriftensammlung Gustav Freytag *see* Flugschriftensammlung gustav freytag
Frankfurter allgemeine – Frankfurt: Frankfurter Allgemeine Zeitung, [1949-jun 1 1953-] – 1 – us CRL [074]
Frankfurter allgemeine sonntagszeitung – Frankfurt/M DE, 1990 4 mar- – ca 2r/yr – 1 – gw Mikropress [074]
Frankfurter allgemeine zeitung : [stadtausgabe] – Frankfurt/M DE, 1949 nov-1964 – 90r – 1 – (1965-75 [129r], 1976-86 [152r], 1987-97 [244r]. with suppl) – mf#1815 – gw Mikropress [074]
Frankfurter allgemeine zeitung / d – Frankfurt/M DE, 1975 2 jun-1978 22 may, 1978 21 jul-1979 (tw. ausg. s) – 1 – (filmed by misc inst: 1949 nov-1953 26 may; 1982- [15r/yr]. with suppl) – gw Mikrofilm; gw Misc Inst [074]
Frankfurter allgemeine zeitung fuer deutschland – Frankfurt/M, Germany 1 nov 1949- – 1 – (incl magazine [7 mar 1980-24 dec 1997]; cont: frankfurter zeitung und handelsblatt [16 nov 1866-31 aug 1943]) – uk British Libr Newspaper [074]
Frankfurter allgemeine zeitung / r *see* Frankfurter allgemeine zeitung / s
Frankfurter allgemeine zeitung / s – Frankfurt/M DE, 1976 mai-1981 – 75r – 1 – (fr 8 jan 1988: frankfurter allgemeine zeitung / r [ausg r=rhein-main gebiet]) – gw Misc Inst [074]
Die frankfurter berichte gustavs v. meyern-hohenberg : mit einem lebensbild und einer wuerdigung seines literarischen schaffens / Kummer, Rolf – [S.l.: s.n.] (Leipzig: Druck von Gerhardt), 1934 – 1 – (incl bibl ref) – us Wisconsin U Libr [943]
Frankfurter buecherfreund : mittheilungen aus dem antiquariate von joseph baer und co – Frankfurt. v1-11 1900-13: nf: v1(= 12) 1914-19 -v 4(=15) 1922 [mf ed 1994] – 50mf – 9 – €220.00 – 3-89131-170-2 – gw Fischer [070]
Frankfurter, Felix *see*
– The business of the supreme court: a study in the federal judicial system
– The felix frankfurter papers
– Papers
– A selection of cases under the interstate commerce act
Frankfurter gelehrte anzeigen / ed by Merck, Johann Heinrich et al – Frankfurt am Main 1772-90 – (= ser Dz) – 19v on 226mf – 9 – €1356.00 – mf#k/n288 – gw Olms [074]
Frankfurter gelehrte anzeigen vom jahr 1772 – Heilbronn: Henninger, 1882-83 [mf ed 1993] – (= ser Deutsche litteraturdenkmale des 18. und 19. jahrhunderts 7-8) – 2v – 1 – (incl bibl ref and ind, int by wilhelm scherer, ind by b seuffert) – mf#8676 reel 1 – us Wisconsin U Libr [410]
Frankfurter geschaeftsbericht – Frankfurt/M DE, 1864-1889 jun, 1918-1943 31 aug – 176r – 1 – (cont by: frankfurter handelszeitung, aug 27, 1856. neue frankfurter zeitung, sep 1, 1859. neue deutsche zeitung, nov 2, 1866. frankfurter zeitung, nov 16, 1866) – gw Mikropress [074]
Frankfurter geschaeftsbericht – Frankfurt/M, Stuttgart DE, 1889 jul-1920 12 mar, 1920 25 apr-1922 30 jun, 1923 1 oct-1928, 1935-1943 31 aug [146r until 1917] – 1 – (title varies: 27 aug 1856: frankfurter handelszeitung; 1 sep 1859: neue frankfurter zeitung; 2 nov 1866: neue deutsche zeitung [in stuttgart until 16 nov 1866]; 16 nov 1866: frankfurter zeitung also: frankfurter zeitung und handelsblatt. filmed by misc inst: 1913 feb-mar, 1924 sep & 1930 jan [4r]; 1915, 1942 1 aug-1943 31 aug [4r]. filmed by misc inst: fuer die frau 1926 14 mar-1930; literaturblatt zur frankfurter zeitung 1919-23 [lacking: 1921]) – gw Misc Inst [074]
Der frankfurter goethe / Mentzel, Elisabeth Schippel – Frankfurt a.M: Ruetten & Loening 1900 [mf ed 1990] – 1r – 1 – (filmed with: goethes leipziger krankheit und don sassafras / adolph hansen) – mf#2665p – us Wisconsin U Libr [430]
Frankfurter handelszeitung *see* Frankfurter geschaeftsbericht
Frankfurter journal – Frankfurt Main, 1 Jan-30 Jun 1832 – 1 – (1840-60 62r dm4,650.00) – gw Mikropress [074]
Frankfurter journal *see* Journal
Frankfurter latern (sz2) – Frankfurt/M 1860-65 [mf ed 1998] – (= ser Satirische zeitschriften (sz) 2) – 65mf – 9 – diazo €350 silver €500 – 3-89131-279-2 – (filmed with: friedrich stoltze's frankfurter latern frankfurt/m. 1865-66; neue frankfurter leuchte frankfurt/m. 1868; frankfurter latern frankfurt/m. 1870; deutsche latern frankfurt/m. 1870; frankfurter latern frankfurt/m. 1871; neue frankfurter latern frankfurt/m. 1871; frankfurter latern frankfurt/m. 1872-93) – gw Fischer [870]
Frankfurter nachrichten und intelligenz-blatt *see* Intelligenz-blatt der freyen stadt frankfurt
Frankfurter nachtausgabe *see* Frankfurter neue presse / nachtausgabe
Frankfurter neue presse – Frankfurt/M DE, 1949 3 jun-1953 31 mar – 9r – 1 – (1949 with gaps. filmed by misc inst: 1958-1963 19 jan, 1963 17 apr-1966; 1946 15 apr-1949 jun, 1953 31 mar-1976 okt; 1951 jan, 1976- [ca 7r/yr]. beginning: allgemeine frankfurter neue presse) – gw Mikrofilm; gw Misc Inst [074]

Frankfurter neue presse – Frankfurt/M DE, 1976 nov-1977 apr – 1 – gw Mikropress [074]
Frankfurter neue presse / nachtausgabe – Frankfurt/M DE, 1951-1967 31 mar – 31r – 1 – (title varies: 1951: frankfurter nachtausgabe; 2 may 1966: abendpost frankfurt, nachtausgabe; 1 jul 1966: abendpost nachtausgabe) – gw Mikrofilm [074]
Frankfurter ober-post-amts-zeitung *see* Unvergreifliche postzeitungen
Frankfurter oder-zeitung – Frankfurt/O DE, 1916 apr-jun, 1920-1921 jun, 1923 apr-1924, 1925 jul-1926, aug, 1928 mar, apr, nov, dez – 22r – 1 – gw Misc Inst [074]
Frankfurter postzeitung *see* Unvergreiffliche postzeitungen
Frankfurter presse / ed by Der Amerikanischen 12. Heeresgruppe fuer die deutsche Zivilbevoelkerung – Frankfurt/M DE, 1945 21 apr-26 jul – 1r – 1 – mf#6402 – gw Mikropress [074]
Frankfurter rundschau – Frankfurt/M DE, 1945 1 aug-1957 11 nov – 30r – 1 – (1957 12 nov-1969 [75r], 1970-79 [87r] (through mfa, dortmund), 1980-91 [112r], 1992-94 [36r], 1995-97 [34r], 1998 subsc) – mf#7099 – gw Mikropress [074]
Frankfurter rundschau – Frankfurt/Main, Germany aug 1945-1 jul 1947; 4 mar-15 sep, 3-29 dec 1948; 7 jan-31 dec 1949; 3 may-28 dec 1950; 19 jul-5 aug, 8-14 oct, 11 dec 1951; 10-16 jan, 1 may-29 sep, 3 oct 1952-31 dec 1981 (imperfect) – 1 – uk British Libr Newspaper; us CRL [074]
Frankfurter rundschau / 0 : abendausgabe – Frankfurt/M DE, 1962 25 feb-1975 17 nov – 1 – gw Mikrofilm [074]
Frankfurter rundschau / 1 : deutschland-ausgabe – Frankfurt/M DE, 1947 2 aug-1979 – 1 – (filmed by misc inst: 1945 1 aug-1969 [44r], 1980-97. with suppl) – gw Mikrofilm; gw Misc Inst [074]
Frankfurter rundschau / 2 : stadtausgabe – Frankfurt/M DE, 1945 1 aug-1975, 1982- – 1 – (filmed by misc inst: 1970-81, 1992- [fr 1992 12r/yr]) – gw Mikrofilm; gw Misc Inst [074]
Frankfurter rundschau / land 2 – Frankfurt/M DE, 1972 28 nov-1974 – 1 – gw Mikrofilm [074]
Frankfurter rundschau / land 3 – 5 – Frankfurt/M DE, 1968 8 oct-2002 – 1 – gw Mikrofilm [074]
Frankfurter rundschau / land 6 – Frankfurt/M DE, 1968 8 oct-1974, 1982-1998 26 jan – 1 – gw Mikrofilm [074]
Frankfurter rundschau / land 7 – Frankfurt/M DE, 1968 28 oct-1974, 1988 19 jan-1998 26 jan – 1 – gw Mikrofilm [074]
Frankfurter rundschau / land 8 – 11 – Frankfurt/M DE, 1968 8 oct-1974 – 1 – gw Mikrofilm [074]
Frankfurter rundschau / land 12 – Frankfurt/M DE, 1970 3 aug-1974 – 1 – gw Mikrofilm [074]
Frankfurter rundschau / land 13 – Frankfurt/M DE, 1970 2 nov-1974 – 1 – gw Mikrofilm [074]
Frankfurter rundschau / landausg 2 – Frankfurt/M DE, 2002 27 aug-31 dec – 1r – 1 – (not identical to land 2) – gw Mikrofilm [074]
Frankfurter schulzeitung – Frankfurt/M DE, 1884-92, 1897-1906, 1912-24 – 1 – gw Misc Inst [370]
Frankfurter staats-ristretto – Frankfurt/M DE, 1783-1800 – 8r – 1 – mf#6191 – gw Mikropress [074]
Frankfurter volksblatt 1848 – Frankfurt/M DE, 1851-1852 30 jun – 1 – gw Misc Inst [074]
Frankfurter volksbote – Frankfurt/M DE, 1849 4 apr-1856 – 1r – 1 – gw Misc Inst [074]
Frankfurter wohlfahrtsblaetter – Frankfurt/M DE, 1925, 1927, 1928 – 1 – gw Misc Inst [360]
Frankfurter zeitung – Frankfurt/M DE, 1864-1889 may – 76r – 1 – (1918-43 [98r]; 2r with ind) – mf#7401 – gw Mikropress [074]
Frankfurter zeitung – Frankfurt, Germany. 1924-1929 (1) – mf#67711 – us UMI ProQuest [074]
Frankfurter zeitung *see* Frankfurter geschaeftsbericht
Frankfurter zeitung 1848 – Frankfurt/M DE, 1849 2 feb-30 sep – 1r – 1 – gw Misc Inst [074]
Frankfurter zeitung / stadtausg – Frankfurt/M DE, 1936 jan-jun, 1936 nov-1937 apr, 1937 sep-1939 30 jun – 12r – 1 – gw Misc Inst [074]
Frankfurter zeitung / stadtblatt – Frankfurt/M DE, 1923 3 jan-1933 [gaps] – 16r – 1 – gw Misc Inst [074]
Frankfurter zeitung und handelsblatt – Frankfurt/Main, Germany 16 nov 1866-31 aug 1943 – 1 – (cont: neue frankfurter zeitung [1 jan 1864-17 jul 1866; cont by: frankfurter allgemeine zeitung fuer deutschland [1 nov 1949-]) – uk British Libr Newspaper [074]
Frankfurter zeitung und handelsblatt *see* Frankfurter geschaeftsbericht
Frankl, August *see* Andreas hofer im liede
Frankl, L A *see* Nach jerusalem!
Frankl, Ludwig August *see*
– Anastasius gruen's gesammelte werke
– Cristoforo colombo
– Libanon
– Zur biographie ferdinand raimunds
– Zur biographie franz grillparzer's
Frankl, Oskar *see* Jude in den deutschen dichtungen des 15, 16 und 17 jahrhundertes
Frankl, Victor *see* Espiritu y camino de hispanoamerica
Frankland, Benjamin *see* Recollections of my own life and times
Frankland, Charles *see*
– Narrative of a visit to the courts of russia and sweden, in the years 1830 and 1831
– Travels to and from constantinople, in the years 1827 and 1828
Frankland, William Barrett *see* The early eucharist (a.d. 30-180)
Frankl-Grun, Adolf *see* Vier reden
Franklin and pukekohe times – jan 1921-jun 1940 – 64r – 1 – (aka: franklin times) – mf#15.42 – nz Nat Libr [079]
Franklin auditor / Nathan Hale Society – v1 n1-v5 n12 [1971 mar 25-[1976?]] – 1r – 1 – mf#1111724 – us WHS [071]
Franklin, B *see* Christian experience
Franklin, Benjamin *see*
– Autobiography of benjamin franklin
– Benjamin franklin's account books
– Biographical sketch and writings of elder benjamin franklin
– Christian experience
– The gospel preacher
– On war and peace
– Papers of benjamin franklin
– The pennsylvania gazette, 1728-1815
– Poor richard's almanack
– Predestination and the foreknowledge of god
– Quatuor pour trois violons et violoncelle
– Works of benjamin franklin
Franklin, Christine Ladd *see* Colour and colour theories
Franklin club records, ms 445 – 1895-1901 – 1r – 1 – (minutes of meetings of this free thinker club, at which leon czolgosz heard anarchist emma goldman lecture on 5 may 1901. the lecture contributed to czolgosz's assassination of president wm. mckinley) – us Western Res [320]
Franklin Co. Bexley *see* Spectator
Franklin Co. Canal Winches *see*
– Buckeye news
– Times
– Times series
Franklin Co. Columbus *see*
– American federation of government employees newsletter
– Bohemian
– Booster
– Catholic columbian
– Catholic times
– Christian endeavor world
– Christian witness
– Cio news
– Citizen
– Columbus free press series
– Crisis
– Cross and journal
– Daily capitol fact
– Daily ohio state democrat
– Daily ohio statesman
– Daily reporter
– Daily times
– Democrat
– Democratic call
– Dispatch (1986 edition)
– Eastern review / spectator
– Express
– Express und westbote
– Focus
– Franklin county legal record
– Franklin county news
– Freeman
– Gazette
– Glass workers news
– Good times
– Herold
– Hilltop record
– Hilltop spectator
– Labor news
– Liberal advocate
– Magician
– Messenger
– Metropolitan spectator
– Monitor series
– Morning journal series
– Mutes chronicle
– News
– News (linden-north east)
– News (northland edition)
– Ohio american legion news
– Ohio association public schools employees newsletter
– Ohio christian news
– Ohio chronicle
– Ohio columbian
– Ohio confederate and old school republican
– Ohio coon catcher
– Ohio jewish chronicle
– Ohio jewish chronicle series
– Ohio monitor
– Ohio mute's chronicle series
– Ohio penitentiary news / harbinger news
– Ohio press
– Ohio sentinel
– Ohio sonntagsgast
– Ohio state journal
– Ohio state journal index
– Ohio state journal series
– Ohio state monitor
– Ohio state news
– Ohio statesman
– Ohio waisenfreund
– Ohio/daily ohio press
– Old school republican
– Palladium of liberty
– Post / evening post
– Press post series
– Professional guild of ohio
– Public employees news
– Record
– Rural-urban news / spectator (south east)
– Rural-urban news (west jefferson)
– Rural-urban news (western edition)
– Rural-urban spectator
– Sentinel
– Socialist
– South side booster
– South side leader
– South side spectator
– Southern light
– Spectator (hilltop/w side)
– Spectator
– Spectator (south east rural-urban)
– Spectator (south rural-urban)
– Spectator
– Spectator (west frankln co)
– Star
– State capital fact
– Straight-out harrisonian
– Sunday morning news
– Swan's elevator
– Tagliche westbote
– Telegram
– West side news
– West village spectator
– Westbote
– Western hemisphere
– Whip-poor-will
Franklin Co. Dublin *see*
– Forum
– Villager
Franklin Co. Franklinton *see* Freeman's chronicle
Franklin Co. Gahanna *see*
– Independent news
– Rocky fork enterprise
– Tri-community news
Franklin Co. Grove City *see*
– Grove city record
– Record
Franklin Co. Hilliard *see*
– Northwest news
– Times
– Weekly review
Franklin Co. New Albany *see* Plain news
Franklin Co, OH *see* 1860 census index
Franklin Co. Reynoldsburg *see*
– Little weekly
– News gazette
– Press
– Record
– Reporter
– Spectator
Franklin Co. Westerville *see*
– Public opinion
– Republican gazette
– Review series
– Tan and cardinal
Franklin Co. Whitehall *see*
– Reporter
– Spectator
Franklin Co. Worthington *see*
– Spectator
– Western intelligencer
Franklin county / Woltz, Larry – s.l, s.l? 193-? – 1r – us UF Libraries [978]
Franklin County and Douglas County, KS *see* Funeral registers
Franklin county atlas, 1872 : by caldwell and gould – 1r – 1 – mf#B30575 – us Ohio Hist [978]
Franklin county chronicle – Franklin, NE: Kim L Naden. v1 n1. jun 19 1990- (wkly) [mf ed filmed 1992-] – 1 – (issues for v1 also known as 1st yr) – us NE Hist [071]
Franklin county guard : [consolidated guard, argus and banner] – Bloomington, NE: Huffman & Bower, 1881 (wkly) [mf ed v9 n37. jul 30 1881-dec 2 1892] – 1r – 1 – (cont: banner-guard. cont by: riverton review. publ in riverton ne, dec 2 1892-) – us NE Hist [071]
Franklin county legal record / Franklin Co. Columbus – v1 n1. jan 1879-jan 1880 [wkly] – 1r – 1 – mf#B11782 – us Ohio Hist [071]
Franklin county news – Franklin, NE: Karl L Spence, feb 1910-v5 n483. sep 23 1914 [mf ed with gaps] – 3r – 1 – (cont by: twice-a-week franklin county news) – us NE Hist [071]
Franklin county news – Franklin, NE: Karl L Spence. 14v. v7 n42. nov 30 1916-v20 n22. may 17 1929 (semiwkly) [mf ed with gaps] – 5r – 1 – (formed by the union of: twice-a-week franklin county news and: franklin county progress. absorbed: franklin county times. absorbed by: franklin county sentinel) – us NE Hist [071]
Franklin county news / Franklin Co. Columbus – apr-aug 1955 [wkly] – 1r – 1 – mf#B6672 – us Ohio Hist [071]
Franklin county news – Pukekohe, NZ. jan 1977-dec 1987 – 30r – 1 – mf#15.25 – nz Nat Libr [079]
Franklin county progress – Franklin, NE: John A Barker, oct 31 1912-nov 23 1916// (wkly) [mf ed v2 n47. sep 17 -oct 1 1914 filmed [1980]] – 1r – 1 – (merged with: twice-a-week franklin county news to form: franklin county news (1916)) – us NE Hist [071]
Franklin county sentinel – Franklin, NE: G W Joy & Anne Porter. 75v. v31 n48. oct 28 1920-105th yr n34. jul 6 1994 (wkly) [mf ed with gaps filmed -1994] – 38r – 1 – (cont: franklin sentinel. absorbed: franklin county news (1916), 1929, upland eagle 1932, riverton review 1940, campbell citizen 1944. issue for nov 11 1920 incorrectly dated oct 11 1920) – us NE Hist [071]
Franklin county times – Franklin, NE: Karl L Spence. 1v. v1 n1-5. feb 12-mar 12 1929 (wkly) – 1r – 1 – (absorbed by: franklin county news (1916)) – us NE Hist [071]
Franklin county times – Rocky Mount, VA. 1968-1981 (1) – mf#61897 – us UMI ProQuest [071]
Franklin county tribune – Bloomington, NE: H B Holmes. -v6 n32. oct 5 1922 (wkly) [mf ed dec 27 1917-oct 5 1922 (gaps)] – 1r – 1 – (merged with: bloomington advocate to form: advocate-tribune) – us NE Hist [071]
Franklin courier – jul 1979-mar 1981 – 6r – 1 – mf#15.34 – nz Nat Libr [079]
Franklin d roosevelt : diary and itineraries/usher books – Clearwater Publ Co – 78mf – 9 – $605.00 – us UPA [920]
Franklin d roosevelt and foreign affairs : second series: 1937-1939 / ed by Scheve, Donald – Clearwater Publ Co – 66mf (24:1) – 9 – $690.00 – (guide only $175) – us UPA [327]
Franklin D. Roosevelt Philatelic Society *see* Fireside chats
Franklin democrat – Brookville, IN. 1852-1895 (1) – mf#62736 – us UMI ProQuest [071]
Franklin evening news – Franklin, PA., 1886-1920 – 13 – $25.00r – us IMR [071]
Franklin free press – Franklin, NE: A A Hadden. 7v. apr 1900-v7 n8. may 25 1906 (wkly) [mf ed with gaps filmed [1972?]] – 2r – 1 – (absorbed by: sentinel) – us NE Hist [071]
Franklin, G *see* Energy information database
Franklin gazette – Rocky Mount, VA. 1958-1967 (1) – mf#66859 – us UMI ProQuest [071]
The franklin institute and the making of industrial america – 1824-1950 – 536mf (24:1) – 9 – $3835.00 – (with p/g) – us UPA [600]
Franklin, J *see*
– Narrative of a journey to the shores of the polar sea
– Narrative of a second expedition to the shores of the polar sea
Franklin, James *see* The present state of hayti, (saint domingo)
Franklin, Jodi L *see* A comparison of the yellow springs instruments
Franklin, John *see* Reise an die kuesten des polarmeeres in den jahren 1819, 1820, 1821 und 1822
The franklin kentucky baptist – May 17, 1866 – Jun 22, 1867 – 1 – 6.72 – us Southern Baptist [242]
Franklin minerva – Chambersburg. 1799-1800 (1) – mf#4374 – us UMI ProQuest [900]
Franklin, New Hampshire.East Franklin Baptist Church *see* Records
Franklin news herald – Franklin, PA. -d 1970 – 13 – $25.00r – us IMR [071]
Franklin news post – Rocky Mount, VA. 1984-1999 (1) – mf#61890 – us UMI ProQuest [071]
Franklin. Ohio. First Congregational Church *see* Church records, ms 421
Franklin, ohio, taxes, ms v.f.o. – 1804 – 1r – 1 – (list of the land given in by residents) – us Western Res [978]
Franklin pierce papers – 1820-69 (mf ed 1959) – (= ser Presidential papers microfilm) – 7r – 1 – us L of C Photodup [975]
Franklin pierce papers – 7r – 1 – $245.00 – (with guide) – Dist. us Scholarly Res – us L of C Photodup [975]
Franklin. Presbytery (Pres. Ch. in the USA New School) *see* Minutes, 1846-1870
Franklin repository – Chambersburg, PA. 1808-1865 (1) – mf#65858 – us UMI ProQuest [071]
Franklin repository – Chambersburg, PA. 1884-1901 (1) – mf#65859 – us UMI ProQuest [071]

Franklin, Samuel *see* A critical review of wesleyan perfection
Franklin sentinel – Franklin, NE: N H Miles. 9v. v23 n47. sep 26 1912-v31 n47. oct 21 1920 (wkly) [mf ed with gaps] – 5r – 1 – (cont: sentinel. cont by: franklin county sentinel) – us NE Hist [071]
Franklin times *see* Franklin and pukekohe times
Franklin times advertiser – Pukekohe, NZ. 16 -23 aug; 4 oct 1963-27, may 1964; jan-jul 1979 – 1r – 1 – mf#15.1 – nz Nat Libr [079]
Franklin times (pukekohe) – jan 1921-jun 1940 – 1 – mf#15.42 – nz Nat Libr [079]
Franklin-hales corners hub – Hales Corners WI. 1969 dec 4/1970 aug 27-1996 oct/dec – 79r – 1 – (cont: tri-town hub [hales corners, wi: 1967 : franklin-hales corners ed]; cont by: franklin hub hales corners village hub) – mf#945607 – us WHS [071]
Franklin-Lakeside Neighborhood Association (Madison (WI)) *see* Grapevine
Franklinton baptist church. pleasureville, kentucky : church records – 1848-1940; 1956-66. Formerly Drennons Ridge – 1 – us Southern Baptist [242]
Franklintonian / Ohio Genealogical Society – v10 n1-v15 n10 [1982 jan-1987 nov/dec] – 1r – 1 – (cont: franklin county chapter of the ohio genealogical society) – mf#1111726 – us WHS [929]
Franko, Ivan *see* Zvierinyi biudzhet
Frankreich : eine monatsschrift aus den briefen deutscher maenner in paris / ed by Poel, Peter – Altona 1795-1805 – (= ser Dz. historisch-politische abt) – 96mf – 9 – €960.00 – mf#k/n1741 – gw Olms [860]
Frankreich immer das alte unter der neuen republik : oder eindruecke und erinnerungen aus frankreich im jahre 1850 und der kurz vorhergehenden zeit – Berlin 1851 [mf ed Hildesheim 1995-98] – 1v on 2mf – 9 – €60.00 – 3-487-26003-4 – gw Olms [944]
Frankreich und die franzosen / Schmidt-Weissenfels, Eduard – Berlin 1868 [mf ed Hildesheim 1995-98] – 2v on 4mf – 9 – €120.00 – ISBN-10: 3-487-26002-6 – ISBN-13: 978-3-487-26002-0 – gw Olms [944]
Frankreichs religions- und buergerkriege im sechzehnten jahrhunderte / Herrmann, August Leberecht – Leipzig 1828 [mf ed Hildesheim 1995-98] – 1v on 4mf – 9 – €120.00 – ISBN-10: 3-487-26127-8 – ISBN-13: 978-3-487-26127-0 – gw Olms [944]
Franks, Augustus Wollaston *see* Japanese pottery
Franks, Cyril M *see* Child and family behavior therapy
Franks, Dorothy W *see* Self-esteem and adolescent pregnancy
Franks, Ian *see* Preprogramming vs. on-line preparation in simple movement sequences
Franks, Robert Sleightholme *see* A history of the doctrine of the work of christ in its ecclesiastical development
Franky Vasquez, Pablo *see* Analisis de la poblacion protegida por el seguro s
Frans essink : sin leben un driben olt muenstersch kind / Giese, Franz – 3. Aufl. Leipzig: O Lenz, 1911 (mf ed 1990) – 1r – 1 – (filmed with: zwischen den kriegen) – us Wisconsin U Libr [430]
Frans overbeck : versuch einer wuerdigung / Nigg, W – Muenchen, 1931 – €12.00 – ne Slangenburg [225]
La franscatana : [opera buffa in tre atti di livigni] / Paisiello, Giovanni – [n.p. 1774?] [mf ed 1988] – 1r – 1 – mf#pres. film 17 – us Sibley [780]
Franscini, Stefano *see* Stefano franscini's statistik der schweiz
Fransen van Eck, Cornelis *see* Oratio de praecipuis caussis ethicae christianae, a plurimis christianis nimis neglectae
Frantz, Adolph *see* Lehrbuch des kirchenrechts
Frantz, Clamor *see* Versuch einer geschichte des marien- und annen-cultus in der katholischen kirche
Frantzen, W *see* Die "leben jesu"-bewegung seit strauss
Frantzky, Fr Jos Th *see* Allgemeines europaeisches journal
FRANZ *see* Mon tenor chez les riches
Franz, A *see* Rituale florian
Franz, Albert *see*
- Die kirchlichen benediktionen im mittelalter, vol 1-2
- Die messe im deutschen mittelalter
- Das rituale von st florian aus dem 12. jahrhundert
- Der soziale katholizismus in deutschland bis zum tode kettelers

Franz, Albin *see* Johann klaj
Franz der erste, koenig von frankreich : ein sittengemaelde aus dem sechszehnten jahrhundert / Herrmann, August Leberecht – Leipzig 1824 [mf ed Hildesheim 1995-98] – 1v on 3mf – 9 – €90.00 – ISBN-10: 3-487-26133-2 – ISBN-13: 978-3-487-26133-1 – gw Olms [944]
Franz freiherrn gaudys poetische werke / Gaudy, Franz, Freiherr von – Berlin: Bibliographische Anstalt. 5v in 1. [1891?] – 1r – 1 – us Wisconsin U Libr [430]
Franz, Friedr Christian et al *see* Annalen der teutschen akademien
Franz grillparzer : eine biographische studie / Faeulhammer, Adalbert – Graz: Leuschner & Lubensky, 1884 – 1r – 1 – (incl bibl ref) – us Wisconsin U Libr [920]
Franz grillparzer : eine charakteristik / Koch, Max – Frankfurt am Main: Knauer, 1891 – 1r – 1 – us Wisconsin U Libr [430]
Franz grillparzer : sein leben, dichten und denken / Lange, E – Guetersloh: C Bertelsmann, 1894 – 1r – 1 – (incl bibl ref) – us Wisconsin U Libr [920]
Franz grillparzer : sein leben und schaffen: im hinblick auf den 100. geburtstag / Mahrenholtz, Richard – Leipzig: Renger, 1890 – 1r – 1 – us Wisconsin U Libr [920]
Franz grillparzer : eine studie / Vancsa, Kurt – 1r – 1 – us Wisconsin U Libr [430]
Franz grillparzers lebensgeschichte / Laube, Heinrich – Stuttgart: J G Cotta, 1884 – 1r – 1 – us Wisconsin U Libr [920]
Franz grillprarzer : ein kampf um leben und kunst / Alker, Ernst – Marburg a.L: N G Elwert, 1930 [mf ed 1992] – (= ser Beitraege zur deutschen literaturwissenschaft 36) – 256p – 1 – (incl bibl ref) – mf#8004 reel 3 – us Wisconsin U Libr [430]
Franz kafka : die inszenierung unmoeglicher ueberschreitung in "der process" und anderen schriften / Wuelfingen, Klaus Bock von – (mf ed 1995) – 2mf – 9 – €40.00 – 3-8267-2135-7 – mf#DHS 2135 – gw Frankfurter [430]
Franz lambert von avignon : nach seinen schriften und den gleichzeitigen quellen / Baum, Johann Wilhelm – Strassburg: Treuttel und Wuertz, 1840 – 1mf – 9 – 0-7905-5021-0 – (incl bibl ref) – mf#1988-1021 – us ATLA [920]
Franz liszt (1811-1886) : collected works – Leipzig: Breitkopf & Haertel. 34v. 1907-36 – 11 – $235.00 set – us Univ Music [780]
Franz liszt's briefe / Liszt, Franz; ed by La Mara – Leipzig: Breitkopf & Haertel 1893-1905 [mf ed 1994] – 8v in 5 on 1r – 1 – (letters in german or french; incl ind) – mf#pres. film 139 – us Sibley [920]
Franz, Martin *see* Postnatale entwicklung des mittelohrs bei pachyuromys duprasi
Franz peter schubert (1797-1828) : complete works. critical edition / ed by Brahms, Johannes et al – Leipzig: Breitkopf & Haertel. 21 ser. 1884-97 – 11 – $405.00 set – (incl one revisions report) – us Univ Music [780]
Franz, Rudolf *see* Grillparzers werke
Franz suarez und die scholastik der letzten jahrhunderte / Werner, Karl – Regensburg: GJ Manz, 1861 – 3mf – 9 – 0-7905-8971-0 – (incl bibl ref) – mf#1989-2196 – us ATLA [240]
Franz von assisi und die nachahmung christi / Walter, Johannes von – Lichterfelde-Berlin: E Runge 1910 [mf ed 1990] – (= ser Biblische zeit- und streitfragen 6/5) – 1mf – 9 – 0-7905-6391-6 – (incl bibl ref) – mf#1988-2391 – us ATLA [241]
Franz von baader als begruender der philosophie der zukunf : sammlung der vom jahre 1851 bis 1856 erschienenen recensionen und literarischen notizen ueber franz von baader's saemmtliche werke / ed by Hoffmann, Franz et al – Leipzig: Herrmann Bethmann, 1856 – 1mf – 9 – 0-524-08635-4 – mf#1993-2095 – us ATLA [190]
Franz von kobell : sein leben und seine werke: 1. teil, lebens- und entwicklungsgang: 1. periode (1803-1845) / Dreyer, Aloys; ed by Dreyer, Aloys – Freising: F P Datterer, 1903 – 1r – 1 – us Wisconsin U Libr [920]
Franz von sickingen : dramatisches gedicht in fuenf abtheilungen / Duller, Eduard – Frankfurt/M: J D Sauerlaender, 1833 – 1r – 1 – us Wisconsin U Libr [820]
Franz von sickingen / Maenss, Johannes – Berlin: C Habel, 1877 – 1r – 1 – us Wisconsin U Libr [943]
Franz von sickingen : nach meistens ungedruckten quellen / Ullmann, Heinrich – Leipzig: S Hirzel, 1872 (Grimme & Troemel) – 1r – 1 – (incl bibl ref) – us Wisconsin U Libr [430]
Franz von sickingens fehde gegen trier : und ein gutachten claudius cantiunculas ueber die rechtsansprueche der sickingenschen erben / Bremer, Franz Peter – Strassburg: J H E Heitz 1885 [mf ed 1991] – 1r – 1 – (incl bibl ref; filmed with: freiheit und recht / johann gottfried seume) – mf#2943p – us Wisconsin U Libr [943]
Franz werfel / Braselmann, Werner – Wuppertal-Barmen: Emil Mueller 1960 [mf ed 1996] – (= ser Dichtung und deutung 7) – 1r – 1 – (incl bibl ref; filmed with: der vogel im kaefig / lisa wenger) – mf#4060p – us Wisconsin U Libr [430]
Franz, Wilhelm *see* Britannien und der krieg
Franz x. von zottmann, bischof der dioezese tiraspol : zuege katholischen und deutschen lebens aus russland / Zottmann, Al – Muenchen: Jos Roth, 1904 – 1mf – 9 – 0-524-08890-X – mf#1993-3354 – us ATLA [241]
Franz xavier : ein weltgeschichtliches missionsbild / Venn, Henry & Hoffmann, W – Wiesbaden: Julius Niedner, 1869 [mf ed 1995] – (= ser Yale coll) – 418p – 1 – 0-524-09611-2 – (in german) – mf#1995-0611 – us ATLA [241]
Franzelin, Bernhard *see* Die neueste lehre geysers ueber das kausalitaetsprinzip
Franzelin, Johannes Baptist *see*
- Examen doctrinae macarii bulgakow, episcopi russi schismatici, et iosephi langen, neoprotestantis bonnensis, de processione spiritus sancti
- Theses de ecclesia christi
- Tractatus de divina traditione et scriptura
- Tractatus de sacramentis in genere

Franzen, Raymond H *see*
- The ach index of nutritional status
- Physical defects
- Physical measures of growth and nutrition

Franzen, Stefan *see* Gadolinium-dtpa- und temperaturstudien zur interstitiellen tumortherapie fuer die interventionelle kernspintomographie
Franzero, Charles Marie *see* The life and times of cleopatra
Franzisca hernandez und frai francisco ortiz : anfaenge reformatorischer bewegungen in spanien unter kaiser karl 5 / Boehmer, Eduard – Leipzig: H. Haessel, 1865 – 1mf – 9 – 0-7905-6045-3 – mf#1988-2045 – us ATLA [240]
Die franziskaner in japan einst und jetzt / Boehlen, Hippolytus – Trier: Paulinus-druckerei, 1912 [mf ed 1995] – (= ser Yale coll; Aus allen zonen 13) – 147p (ill) – 1 – 0-524-09948-0 – (in german) – mf#1995-0948 – us ATLA [241]
Franzke, Joachim *see* Pseudoschallwellen-laserspektroskopie an niederdruckplasmen
Der franzl : fuenf bilder eines guten mannes / Bahr, Hermann – 2. aufl. Berlin: Wiener Verlag, 1901 [mf ed 1989] – 375p (ill) – 1 – mf#6973 – us Wisconsin U Libr [820]
Der franzl : fuenf bilder eines guten mannes / Bahr, Hermann – 2. aufl. Berlin: Wiener Verlag, 1901 [mf ed 1989] – 375p (ill) – 1 – mf#6973 – us Wisconsin U Libr [820]
Die franzoesische aufklaerung im spiegel der deutschen literatur des 18. jahrhunderts / ed by Krauss, Werner – Berlin: Akademie-Verlag, 1963 [mf ed 1993] – (= ser Deutsche akademie der wissenschaften zu berlin. schriftenreihe der arbeitsgruppe zur geschichte der deutschen und franzoesischen aufklaerung v10) – clxxxvii/484p – 1 – (incl bibl ref) – mf#8236 – us Wisconsin U Libr [190]
Franzoesische einfluesse bei schiller / Schanzenbach, Otto – Stuttgart: C Liebich, 1885 – 1r – 1 – us Wisconsin U Libr [430]
Der franzoesische einfluss im zweiten teil von gottscheds critischer dichtkunst / Blanck, Karl – Goettingen: W F Kaestner, 1910 [mf ed 1990] – 149p – 1 – (incl bibl ref) – mf#7408 – us Wisconsin U Libr [410]
Franzoesische geschichte vornehmlich im sechzehnten und siebzehnten jahrhundert / Ranke, Leopold von – Stuttgart [u a] 1852-61 [mf ed Hildesheim 1995-98] – 5v on 19mf – 9 – €190.00 – 3-487-26216-9 – gw Olms [944]
Die franzoesische revolution im deutschen drama und epos nach 1815 / Hirschstein, Hans – Stuttgart: J B Metzler, 1912 [mf ed 1992] – (= ser Breslauer beitraege zur literaturgeschichte. neue folge 31) – vii/384p – 1 – (incl bibl ref and ind) – mf#8014 reel 3 – us Wisconsin U Libr [430]
Franzoesische staatsanzeigen : gesammelt und herausgegeben zur geschichte der franzoesischen revolution – Leipzig 1790 – (= ser Dz. historisch-politische abt) – 3mf – 9 – €90.00 – mf#k/n1245 – gw Olms [944]
Franzoesische staatsverwaltung in den rheinischen departementern – Strassburg (Strasbourg F), 1791 3 may-13 jul [gaps] – 1 – fr ACRPP [350]
Die franzoesische tragoedie der ersten haelfte des 17. jahrhunderts im urteile ihrer zeitgenossen / Pizzo, Piero – Zuerich, 1914 (mf ed 1994) – 2mf – 9 – €31.00 – 3-8267-3042-9 – mf#DHS-AR 3042 – gw Frankfurter [440]
Franzoesische zustaende / Heine, Heinrich – Hamburg 1833 [mf ed Hildesheim 1995-98] – 3mf – 9 – €90.00 – 3-487-29794-9 – gw Olms [944]
Franzoesischen Informationsdienst *see* Welt am montag
Die franzoesischen uebertragungen von goethes faust : ein beitrag zur geschichte der franzoesischen uebersetzungskunst / Langkavel, Martha – Strassburg: K J Truebner, 1902 – 1r – 1 – (incl bibl ref) – us Wisconsin U Libr [430]
Die franzoesisch-reformierte gemeinde in frankfurt am main, 1554-1904 / Ebrard, Friedrich Clemens – Frankfurt a M: R Ecklin, 1906 – 1mf – 9 – 0-524-08320-7 – mf#1993-1015 – us ATLA [242]
Franzos, Karl Emil *see*
- For the right
- Heines geburtstag
- Die juden von barnow
- Der kleine martin
- Moschko von parma
- Neue novellen
- Der pojaz
- Saemmtliche werke und handschriftlicher nachlass

Der franzose im deutschen drama / Schilling, Helmut – Heidelberg, 1930 (mf ed 1994) – 2mf – 9 – €31.00 – 3-89349-775-7 – mf#DHS-AR 775 – gw Frankfurter [430]
Franzsen, D G *see* Economic growth and stability in a developing economy
Frapan, Ilse *see* "Fluegel auf!"
Frapie, Leon *see* L'institutrice de province
Frary, George S *see* Diary, ms 3079
Fraseologia de cervantes / Sune Benagages, Juan – Barcelona. 1929 – 1 – us CRL [440]
Fraser, Agnes [pseud] *see*
- British columbia for settlers
- On veldt and farm in bechuanaland

Fraser, Alexander Campbell *see*
- Berkeley
- Berkeley and spiritual realism
- Life and letters of george berkeley...formerly bishop of cloyne
- Locke
- Philosophy of theism
- Thomas reid
- The works of george berkeley

Fraser, Andrew Henderson Leith *see* William carey
Fraser banquet : magnificent tribute of respect and confidence tendered to the honorable c f fraser, commissioner of public works – Toronto?: s.n, 1879? – 1mf – 9 – mf#05608 – cn Canadiana [320]
Fraser, Christine *see* Der austritt deutschlands aus dem volkerbund, seine vorgeschichte und seine nachwirkungen
Fraser, Christopher Findlay *see*
- Speech delivered by hon c f fraser, commissioner of public works
- Speech of the honourable c f fraser

Fraser, Donald *see*
- Autobiography of the late donald fraser
- Blind man of jerusalem
- The future of africa
- The speeches of the holy apostles
- Winning a primitive people

Fraser, Duncan *see*
- Choir of the future
- Church praise

Fraser family papers : writings on the new hebrides by various members of the fraser family – 1881-1921 – 1r – 1 – mf#PMB1037 – at Pacific Mss [920]
Fraser, George A *see*
- British history notes
- Geography notes
- Geography notes for 3rd, 4th, and 5th classes

Fraser herald – Warren, MI. 1926-1931 (1) – mf#63823 – us UMI ProQuest [071]
The fraser institute case : court of queen's bench for lower canada: john fraser et al, appellants and the hon j j c abbott et al., respondents: judgement rendered june 24th, 1873 – Montreal?: J C Becket, 1873 [mf ed 1985] – 1mf – 9 – 0-665-10852-4 – mf#10852 – cn Canadiana [346]
Fraser, J Alban *see* Spain and the west country. london, 1935
Fraser, J B *see* Journal of a tour through part of the snowy range of the himalaya mountains and to the sources of the rivers jumna and ganges
Fraser, J P Munro *see* [Contra costa county-] history of contra costa county
Fraser, James *see*
- Journal of a tour through part of the snowy range of the himala mountains
- Narrative of a journey into khorasan, in the years 1821 and 1822
- Travels and adventures in the persian provinces on the southern banks of the caspian sea
- Treatise [on] justifying faith

Fraser, James B *see* An historical and descriptive account of persia
Fraser, James Nelson *see* The life and teaching of tukaram
Fraser, James, of Dublin *see* Guide through ireland
Fraser, John *see*
- Address by rev john fraser, of kincardine
- Canadian pen and ink sketches
- Death of god's saints
- Erromanga
- Tale of the sea

Fraser, John Foster *see*
- The amazing argentine
- Australia
- Panama and what it means

Fraser, John James *see* Report upon charges relating to the bathurst schools and other schools in gloucester county

Fraser, Lovat *see* Iron and steel in india

Fraser, Mackenzie *see* Correspondence etc, between colonel mackenzie fraser, major magrath and mr maitland

The fraser mines vindicated : or, the history of four months / Waddington, Alfred – Victoria, BC?: s.n, 1888 – 1mf – 9 – mf#42693 – cn Canadiana [622]

Fraser, Philadelphius Bain [comp] *see* A brief statement of the reformed faith

Fraser, Robert G *see* The effect of relaxation training on sport climbing performance of college students

Fraser, Thomas Gamble *see* Records of sport and military life in western india

Fraser times *see* Grand county miscellaneous newspapers

Fraser, W A *see* How the french captured fort nelson

Fraser, William *see*
- Blending lights
- Candid reasons for declining to become a member of temperance socie...
- The state of our educational enterprises

Fraser, William R *see* Church of scotland and its assailants

Fraser, Wm *see* Present agitation for disestamblishment inconsistent with free chur...

Fraserburgh herald – Peterhead: P Scrogie Ltd (wkly) [mf ed 1 jul 1994-] – 1 – (cont: fraserburgh herald and northern counties' advertiser) – uk Scotland NatLib [072]

Fraser-Chamberlain, Isabel *see* Abdul baha on divine philosophy

Fraser's magazine – London. 1830-1882 – 1 – mf#4176 – us UMI ProQuest [073]

Fraser-Winthrop *see* [Fraser-winthrop papers]

[Fraser-winthrop papers] – c1700-c1905 [mf ed Spartanburg SC: Reprint Co 1979] – 4mf – 9 – mf#50-06 – us South Carolina Historical [978]

Frases historicas. vulganizaciones / Oteyza, Luis de – Madrid: Renacimiento, 2th ed 1930 – sp Bibl Santa Ana [946]

Frasquito / Armas Y Cespedes, Jose De – Habana, Cuba. 1894 – 1r – us UF Libraries [972]

Frassen, C *see* Philosophia academica...

Frate angelo da chiarino...osimo 1964 / Berardini, Lorenzo – Madrid: Graf. Calleja, 1966 – 1 – sp Bibl Santa Ana [946]

Frater thomerl : original-zeitbild aus tyrol mit gesang, in einem acte / Berla, Alois – Wien: J Schoenwetter, 1874 [mf ed 1993] – 29p – 1 – mf#8512 – us Wisconsin U Libr [820]

Die fraterherren im luechtenhofe zu hildesheim / Bruggeboes, W – Hildesheim, 1939 – 2mf – 8 – €5.00 – ne Slangenburg [241]

Fraternal age and the fraternal field – Rochester. 1955-1956 (1) – mf#990 – us UMI ProQuest [360]

Fraternal Order of Eagles *see*
- Eagle leader
- Proceedings of the...annual convention

Fraternal review – Helena, MT. 1891-1892 (1) – mf#64456 – us UMI ProQuest [071]

Fraternal society law association proceedings – 1920-41 (all publ) – 48mf – 9 – $72.00 – mf#LLMC 84-467 – us LLMC [340]

Fraternally yours zenska jednota / First Catholic Slovak Ladies Association [US] – 1976 aug-1981 dec – 1r – 1 – mf#922507 – us WHS [366]

Fraternita – (L'Incontro). Turin. Italy. -m. Mar 1949-Dec 1969. (2 reels) – 1 – uk British Libr Newspaper [072]

Fraternite – Algiers. 23 dec 1943-11 apr 1946 – 1/2r – 1 – uk British Libr Newspaper [072]

La fraternite : journal de l'aude – Carcassonne. n1-113. 6 sept 1848-5 juil 1850 – 1 – (lacking: n111) – fr ACRPP [073]

La fraternite : journal moral et politique – 1-23. Paris. mai 1841-mars 1843 – 1 – fr ACRPP [073]

La fraternite – Paris. 20 nov 1869-8 jan 1870 – 1 – (philosophie, sciences, beaux-arts, inventions, musique, theatres) – fr ACRPP [073]

La fraternite de 1845 : organe des interets du peuple – Paris. 1845-jan 1848 – 1 – (journal de reorganisation sociale et de politique generale) – fr ACRPP [073]

La fraternite universelle – Paris: Pommeret et Moreau. v1 n1. dec 1848 – us CRL [074]

Fraternite-revue : revue laique et chretienne de questions sociales, morales, economiques, politiques, artistiques et litteraires – Alencon puis Chartres. oct 1904-janv 1906 – 1 – fr ACRPP [073]

Fraternity / Gent, George William – London, England. 1886 – 1r – us UF Libraries [240]

Frati, Carlo *see* Mapa mas antiguo de la isla de santo domingo

Fratris felicis fabri evagatorium in terrae sanctae, arabiae et egypti peregrinationem / ed by Hassler, Cunradus Dietericus – Stuttgardiae: Sumtibus Societatis Literariae Stuttgardiensis, 1843-49 [mf ed 1993] – (= ser Blvs 2-4) – 3v – 1 – mf#8470 reel 1 – us Wisconsin U Libr [450]

Fratris felicis fabri tractatus de civitate ulmensi, de eius origine, ordine, regimine, de civibus eius et statu / Fabri, Felix; ed by Veesenmeyer, Gustav – Stuttgart: Litterarischer Verein, 1889 [mf ed 1989] (Tuebingen: H Laupp) [mf ed 1993] – (= ser Blvs 186) – xii/251p – 1 – mf#8470 reel 39 – us Wisconsin U Libr [930]

Fratris felicis fabri tractatus de civitate ulmensi, de eius origine, ordine, regimine, de civibus eius et statu / ed by Veesenmeyer, Gustav – Stuttgart: Litterarischer Verein, 1889 (Tuebingen: H Laupp) – us Wisconsin U Libr [450]

Fratris pauli waltheri guglingensis itinerarium in terram sanctam et ad sanctam catharinam / ed by Sollweck, M – Stuttgart: Litterarischer Verein, 1892 (Tuebingen: H Laupp) [mf ed 1993] – (= ser Blvs 189) – 1 – (incl bibl ref and ind) – mf#70 – us Wisconsin U Libr [880]

Fratris pauli waltheri guglingensis itinerarium in terram sanctam et ad sanctam catharinam / Walther, Paulus; ed by Sollweck, M – Stuttgart: Litterarischer Verein, 1892 (Tuebingen: H Laupp) – (incl bibl ref and ind) – us Wisconsin U Libr [450]

Die frau / ed by Lange, Helene – Berlin DE, 1893 oct-1943/44 – 17r – 1 – gw Misc Inst [305]

Frau aja : goethes mutter in ihren briefen und in den erzaehlungen der bettina brentano – Ebenhausen bei Muenchen: Langewiesche-Brandt, 1914 [mf ed 2000] – (= ser Buecher der rose 20) – 377p (ill) – 1 – (incl bibl ref) – mf#10479 – us Wisconsin U Libr [920]

Frau aventiure : lieder aus heinrich von ofterdingens zeit / Scheffel, Joseph Viktor von – 19. Aufl. Stuttgart: A Bonz, 1902 – 1r – 1 – (incl bibl ref) – us Wisconsin U Libr [780]

Die frau buergemeisterin : roman / Ebers, Georg – Stuttgart: Deutsche Verlags-Anstalt, [1893-97?] [mf ed 1993] – (= ser Georg ebers gesammelte werke 7) – 424p – 1 – mf#8554 reel 2 – us Wisconsin U Libr [830]

Die frau buergermeisterin : roman / Ebers, Georg – Stuttgart: Deutsche Verlags-Anstalt, [1893-1897?] [mf ed 1993] – (= ser Georg ebers gesammelte werke 7) – 424p – 1 – mf#8554 reel 2 – us Wisconsin U Libr [830]

Frau geske auf trubernes : eine saga / Tuegel, Ludwig – Muenchen: A Langen, G Mueller 1936 [mf ed 1991] – 1r – 1 – (filmed with: pferdemusik & other titles) – mf#2918p – us Wisconsin U Libr [390]

Frau holde : ein gedicht / Baumbach, Rudolf – New York: H Holt, c1894 [mf ed 1989] – 105p – 1 – (int and notes by laurence fossler) – mf#6983 – us Wisconsin U Libr [810]

Die frau (hq1) : monatsschrift fuer das gesamte frauenleben unserer zeit / ed by Lange, Helene & Baeumer, Gertrud – Berlin 1893/94-1943/44 [mf ed 1991] – (= ser Hq 1) – 51v on 368mf – 9 – €1360.00 coll €30.00y – 3-89131-042-0 – gw Fischer [305]

Die frau im gemeinnuetzigen leben : archiv fuer die gesamtinteressen des deutschen frauen-, arbeits-, erwerbs- und vereinslebens im deutschen reiche und im auslande – Strassburg 1886-89 [mf ed 1998] – (= ser Hq 37) – 4v on 16mf – 9 – €130.00 – 3-89131-297-0 – gw Fischer [305]

Die frau im roemischen christenprocess / Augar, F – Leipzig, 1905 – (= ser Tugal 2-28/4c) – 2mf – 9 – €5.00 – ne Slangenburg [240]

Die frau im staat / ed by Archiv der deutschen Frauenbewegung Kassel – 1919-33 [mf ed 1993] – (= ser Hq 9) – 15v on 34mf – 9 – €200.00 – 3-89131-121-4 – (ind by gilla doelle & cornelia wenzel) – gw Fischer [305]

Frau im wirbel : roman / Bley, Wulf – Leipzig: F Rothbarth, [1942?] [mf ed 1989] – 318p – 1 – mf#7032 – us Wisconsin U Libr [830]

Frau in arbeit – London (GB), 1941 n17 – 1 – gw Misc Inst [331]

Die frau in der dichtung conrad ferdinand meyers / Clauss, Gertrud – Stuttgart: J B Metzler, 1934 – 1r – 1 – (incl bibl ref (p.[v-vi])) – us Wisconsin U Libr [430]

Frau Magdlene : roman / Berens, Josefa – Jena: E Diederichs, 1943, c1935 [mf ed 1989] – 281p – 1 – mf#7006 – us Wisconsin U Libr [830]

Frau meisterin *see* Werksmeister-zeitung

Frau meseck : eine dorfgeschichte / Halbe, Max – Berlin: G Bondi, 1897 – 1r – 1 – us Wisconsin U Libr [430]

Die frau mit den karfunkelsteinen : roman / Marlitt, Eugenie [pseud] – Stuttgart: Union Deutsche Verlagsgesellschaft [1919] [mf ed 1978] – (= ser Romane und novellen 6) – 2r – 1 – mf#film mas c376 – us Harvard; us Wisconsin U Libr [830]

Frau rat : elisabeth goethe, geb. textor / Hoeffner, Johannes – 4. Aufl. Bielefeld: Velhagen & Klasing, 1926 – 1r – 1 – us Wisconsin U Libr [920]

Frau rat goethe und ihre welt : eine farbenskizze / Paquet, Alfons – Frankfurt am Main: Kommissionsverlag von Englert und Schlosser, 1931 – 1r – 1 – us Wisconsin U Libr [920]

Frau sorge : roman / Sudermann, Hermann – Stuttgart: Cotta, 1923 – 1r – 1 – us Wisconsin U Libr [830]

Die frau und ihre zeit – 1908-10 [mf ed 2005] – (= ser 60) – 27mf – 9 – €180.00122=Historische quellen zur frauenbewegung und geschlechterproblematik (hq) – 3-89131-466-3 – gw Fischer [640]

Frau und leben *see* Freiburger tagespost

Frau von stein : goethes freundin und feindin / Nobel, Alphons – [Muenchen]: Muenchner Verlag, c1939 – 1r – 1 – us Wisconsin U Libr [920]

"Fraud" and civil liability under the federal securities laws / Loss, Louis – Washington: FJC, Aug 1983 – 1mf – 9 – $1.50 – mf#LLMC 95-805 – us LLMC [344]

Fraude nos tribunais eclesiasticos / Aradillas Agudo, Antonio – Lisboa: Liber, 1975 – 1 – sp Bibl Santa Ana [240]

Fraudulent mortgages of merchandise / Jones, Leonard Augustus – St. Louis, Jones, 1879. 46 p. LL-434 – 1 – us L of C Photodup [346]

Fraudulent mortgages of merchandise / Pierce, James Oscar – St. Louis: Thomas, 1884. 310p. LL-1060 – 1 – us L of C Photodup [346]

Frauen im garten : eine erzaehlung / Blunck, Hans Friedrich – Hamburg: Hanseatische Verlagsanstalt, c1939 [mf ed 1989] – 221p – 1 – mf#7036 – us Wisconsin U Libr [880]

Frauen im management : forschungsbericht und analyse ausgewaehlter amerikanischer publikationen (1977-1993) / Bauer, Ina – (mf ed 1995) – 2mf – 9 – €40.00 – 3-8267-2249-3 – mf#DHS 2249 – gw Frankfurter [650]

Frauen und film – Berlin DE, 1974-93 – 3r – 1 – mf#12782 – gw Mikropress [305]

Frauenanwalt : organ des verbandes deutscher frauenbildungs- und erwerbvereine – Berlin 1870/71-1875/76 [mf ed 1997] – (= ser Hq 29) – 6v on 46mf – 9 – €240.00 – 3-89131-141-9 – (with: deutscher frauenanwalt 1878-81) – gw Fischer [305]

Frauenberuf : monatsschrift fuer die interessen der gebildeten frauenwelt – Weimar 1887-92 [mf ed 1996] – (= ser Hq 18) – 6v on 29mf – 9 – €220.00 – 3-89131-130-3 – (fr v4 with subtitle: monatsschrift fuer die interessen der frauenfrage) – gw Fischer [305]

Die frauenbestrebungen unserer zeit *see* Allgemeiner frauenkalender

Die frauenbewegung / ed by Archiv der deutschen Frauenbewegung Kassel – 1895-1919 [mf ed 1994] – (= ser Hq 10) – 94mf – 9 – €420.00 – 3-89131-122-2 – (founded by minna cauer; ind by gilla doelle & cornelia wenzel) – gw Fischer [305]

Frauenbilder aus goethe's jugendzeit : studien zum leben des dichters / Duentzer, Heinrich – Stuttgart: J G Cotta, 1852 [mf ed 1992] – xiv/592p – 1 – (incl bibl ref) – mf#7553 – us Wisconsin U Libr [430]

Frauenbildung : zeitschrift fuer die gesamten interessen des weiblichen unterrichtswesens / ed by Wychgram, J – Leipzig/Berlin 1902-23 [mf ed 1996] – (= ser Hq 23) – 22v on 136mf – 9 – €660.00 – 3-89131-135-4 – gw Fischer [305]

Frauenblatt der christlichen gewerkschaften / ed by Gesamtverband der christlichen Gewerkschaften Deutschlands – Moenchen-Gladbach 1920-1933,6 [mf ed 2003] – (= ser Hq 56) – 20mf – 9 – €95.00 – 3-89131-442-6 – gw Fischer [331]

Frauendienst : zeitschrift fuer das gesamtgebiet der wohlfahrtspflege an und durch frauen – 1902-05 – (= ser Hq 52) – 4v on 25mf – 9 – €160.00 – 3-89131-387-X – gw Fischer [240]

Frauenelend und frauenmission in indien / Greundler, O – 5. verm aufl. Basel: Basler Missionsbuchhandlung, 1908 [mf ed 1995] – (= ser Yale coll) – 92p (ill) – 1 – 0-524-10210-4 – (in german) – mf#1996-1210 – us ATLA [305]

Die frauenfrage / Cathrein, Victor – 3., umgearb u verm Aufl. Freiburg i.B.; St Louis, MO: Herder, 1909 – 1mf – 9 – 0-8370-6890-8 – (incl bibl ref and index of names) – mf#1986-0890 – us ATLA [305]

Die frauengestalten der heiligen schrift in der dichtung / ed by Eckart, Rudolf – Langensalza: H Beyer, 1907 [mf ed 1993] – 143p – 1 – (incl ind) – mf#8361 – us Wisconsin U Libr [430]

Das frauenhaus von brescia : [a novel] / Strobl, Karl Hans – Leipzig: L Staackmann c1911 [mf ed 1991] – 1r – 1 – (filmed with: neue balladen und lieder / lulu von strauss und torney) – mf#2906p – us Wisconsin U Libr [830]

Frauenkalender / ed by Deutsch-Evangelischen Frauenbund – 1904-17 – (= ser Hq 38) – 9 – in prep – gw Fischer [305]

Frauenkapital – eine werdende macht : wochenschrift fuer volkswirtschaft, frauenbewegung und kultur / ed by Raschke, Marie et al – 1914-15 [mf ed 1998] – (= ser Hq 36) – 2v on 21mf – 9 – €150.00 – 3-89131-295-4 – (int & ind by gilla doelle) – gw Fischer [305]

Frauenliebe : wochenschrift fuer freundschaft, liebe und sexuelle aufklaerung – Berlin DE, 1926-1931 n48 – 2r – 1 – gw Misc Inst [305]

Frauenlist / ed by Henschel, Erich – Leipzig: S Hirzel, 1937 [mf ed 1993] – (= ser Altdeutsche quellen heft 2) – 31p – 1 – (middle high german text. int in german) – mf#8377 – us Wisconsin U Libr [430]

Frauenlobs streitgedicht zwischen minne und welt : text und untersuchungen / Hildebrand, Alexander – Frankfurt a.M., 1970 – 2mf – 9 – 3-89349-691-2 – gw Frankfurter [430]

Frauenlos und frauenarbeit in der geschichte des christentums : vortraege gehalten auf dem "vierten apologetischen instruktionskursus" zu berlin am 17.-21. oktober 1910 und auf dem "instruktionskursus fuer christliche weibliche liebestaetigkeit" zu breslau am 24.-28. oktober 1910 / Walter, Johannes von – Berlin: Trowitzsch, 1911 – 1mf – 9 – 0-7905-6795-4 – (incl bibl ref) – mf#1988-2795 – us ATLA [240]

Frauen-mission in indien / Weitbrecht, Mary – Geutersloh: C Bertelsmann, 1875 [mf ed 1995] – (= ser Yale coll; Lebensbilder aus der heiden-mission 4) – 122p – 1 – 0-524-09560-4 – (in german) – mf#1995-0560 – us ATLA [240]

Frauenrecht : novelle / Frenzel, Karl – Berlin: Gebrueder Paetel, 1892 – 1r – 1 – us Wisconsin U Libr [430]

Frauen-reich *see* Deutsche hausfrauen-zeitung

Frauen-rundschau : illustrierte wochenschrift fuer die gesamte kultur der frau / ed by Stoecker, Helene et al – 1903-1922 [mf ed 1998] – (= ser Hq 34) – v4-16 on 104mf – 9 – €510.00 – 3-89131-292-X – (previously: dokumente der frauen; with various subtitles fr v10 1909) – gw Fischer [305]

Frauenstaedt, J *see* Schellings vorlesungen in berlin

Frauenstimmrecht : monatshefte des deutschen verbandes fuer frauenstimmrecht / ed by Augspurg, Anita – Berlin, Leipzig. v1-2. 1912/13-1913/14 [mf ed 1998] – (= ser Hq 24) – 15mf – 9 – €130.00 – 3-89131-136-2 – (filmed with: die staatsbuergerin: monatsschrift des deutschen verbandes fuer frauenstimmrecht ed by adele schreiber [v3-8 1914/15-1919]) – gw Fischer [305]

Frauentaschenbuch / ed by de la Motte Fouque & Doering, Georg – Nuernberg 1817-31 [mf ed 1996] – (= ser Hq 22) – 17v on 40mf – 9 – €420.00 – 3-89131-134-6 – gw Fischer [305]

Frauenwelt : eine halbmonatsschrift – Berlin: J H W Dietz Nachf 1924-33 [mf ed 1998] – (= ser Hq 32) – 10v on 59mf – 9 – €310.00 – 3-89131-290-3 – gw Fischer [305]

Frauenwirtschaft – 1910-1928/29 [mf ed 2003] – (= ser Hq 57) – 19v on 57mf – 9 – €360.00 – 3-89131-447-7 – gw Fischer [305]

Frauen-zeitung – Koeln DE, 1848 27 sep – 1 – gw Misc Inst [305]

Frauen-zeitung – Stuttgart DE, 1855-58 – 1 – (with suppl: salon) – gw Misc Inst [305]

Frauenzimmer gesprechspiele : so bey ehr- und tugendliebenden gesellschaften... / Harsdoerffer, G P – Nuernberg: Gedruckt und verlegt bey Wolffgang Endtern, 1644[-49]. 8v – 56mf – 9 – mf#O-25 – ne IDC [090]

Frauenzimmer-zeitung – Kempten DE, 1787 – 1r – 1 – gw Misc Inst [640]

Frauen-zukunft : eine monatsschrift / ed by Lieber, Gabriele von et al – Muenchen, Leipzig 1910-13 [mf ed 1995] – (= ser Hq 21) – 3v on 18mf – 9 – €160.00 – 3-89131-133-8 – gw Fischer [305]

Les fravashis : etude sur les traces dans le mazdeisme d'une ancienne conception sur la survivance des morts / Soederblom, Nathan – Paris: Ernest Leroux, 1899 – 1mf – 9 – 0-7905-6011-9 – (incl bibl ref) – mf#1988-2011 – us ATLA [210]

Fray bartolome de las casas, sus tiempos y su aposislado / Gutierres, Carlos – Madrid, 1878. xxxix, 460p – 1 – us Wisconsin U Libr [240]

Fray d. alonso de valencia y bravo, del orden y caballeria de alcantara (1723-1778) / Velo Nieto, Gervasio – Badajoz: Imprenta de la Diputacion Provincial, 1952 – sp Bibl Santa Ana [240]

Fray ignacio marino, o p / Tisnes Jimenez, Roberto Maria – Bogota, Colombia. 1963 – 1r – us UF Libraries [972]

Fray juan antonio abasolo de la regular observancia de nuestro seraphico padre san francisco... : hazemos saber a vv pp y rr como hemos recibido unas letras patentes en idioma latino... / Abasolo, Juan Antonio – Mexico: [s.n.] 1753 – (= ser Books on religion...1543/44-c1800: franciscanos) – 1mf – 9 – mf#crl-209 – ne IDC [241]
Fray luis bolanos, apostol del paraguay y del rio de la plata. cordoba (tucuman). 1934 / Oro, Buenaventura – Madrid: Razon y Fe, 1935 – 1 – sp Bibl Santa Ana [240]
Fray luis de leon : eine biographie aus der geschichte der spanischen inquisition und kirche im sechszehnten jahrhundert / Wilkens, Cornelius August – Halle: CEM Pfeffer, 1866 – 1mf – 9 – 0-7905-6970-1 – (includes poetry and translations by luis de leon) – mf#1988-2970 – us ATLA [946]
Fray luis de leon y benito arias montano / Lopez de Toro, Jose – Madrid: Rev. Arch. Bibl. y Mus., 1955. pp. 531-548 – 1 – sp Bibl Santa Ana [946]
Fray pedro nunez machado (zafra 1550-burgos 1609) / Guede, Lisardo – Badajoz: Dip. Provincial, 1969. Sep. Ree – 1 – sp Bibl Santa Ana [240]
Fraye erd – Varsha, Poland. 1910 or 1911 – 1r – us UF Libraries [939]
Di fraye shtunde – [New York]: A Hillman 1904 – 1v – 1 – us AJPC [939]
Fraye welt – London. v1-3. 1891-1893 – 1 – us NY Public [073]
Frazer, Ian [comp] *see* Solomon islands political party manifestos, policy statements and programmes of action
Frazer, James George *see*
– Anthologia anthropologica
– Balder the beautiful
– The dying god
– Golden bough
– The golden bough, vol 12
– Leaves from the golden bovgh
– Lectures on the early history of the kingship
– The magic art and the evolution of kings
– Psyche's task
– Spirits of the corn and of the wild
– Taboo and the perils of the soul
– Totemism and exogamy
Frazer, Robert Watson *see*
– British india
– Indian thought past and present
– A literary history of india
Frazier, Edwin Ray *see* Leading teachers of adults and youth to study the historical contexts of scripture
Frazier-lemke farm debt moratorium act. / Skeels, William O – Oklahoma City: Leader, 1937. 100p. LL-1098 – 1 – (1938 supplement. oklahoma city, 1938. 46p. ll-1098) – us L of C Photodup [340]
Frb chicago economic perspectives : a review from the federal reserve bank of chicago / Federal Reserve Bank of Chicago – Chicago. 1983-1988 (1,5,9) – (cont: economic perspectives. cont by: economic perspectives) – ISSN: 0884-7576 – mf#11375,01 – us UMI ProQuest [332]
Fream, William *see*
– Canadian agriculture, pt 1
– Canadian agriculture, pt 2
Frearson's monthly illustrated adelaide news – Adelaide, Australia. Pictorial Australian. -m. 1881-95. 5 reels – 1 – uk British Libr Newspaper [072]
Frearson's monthly illustrated adelaide news – Adelaide, Australia 1881-84 – 1 – (cont by: pictorial australian [ns]: jan 1885-dec 1895) – uk British Libr Newspaper [079]
Freart, R *see* An idea of the perfection of painting...
Freart, R Sieur de Chambray *see* Parallele de l'architectvre antiqve et de la moderne
Frechette, Louis *see*
– A mme honore mercier, fils (ma fille jeanne) a l'occasion de son mariage, 21 avril 1903
– A propos d'education
– A sa majeste victoria 1ere, reine d'angleterre et imperatrice des indes
– Bienvenue a son altesse royale le duc d'york et de cornwall, sept 1901
– Les calomniateurs confondus
– Christmas in french canada
– Un colomniateur demasque par lui-meme
– Le drapeau fantome
– Epaves poetiques / veronica
– Felix poutre
– Feuilles volantes
– Les fleurs boreales / les oiseaux de neige
– Le heros de st-eustache
– Le heros de st-eustache, jean olivier chenier
– Les hommes du jour
– In memoriam
– L'iroquoise du lac saint-pierre
– Jean-baptiste de la salle, fondateur des ecoles chretiennes
– La legende d'un peuple
– Mes loisirs
– Les oiseaux de neige
– Pele-mele
– Pensees d'hiver
– Petite histoire des rois de france
– Philippe-n pacaud
– Poesies choisies, vols 1-3
– Sainte-anne d'auray et ses environs
– Spes ultima
– Stances
– The united states for french canadians
Frechette, Louis-Honore *see*
– La legende d'un peuple
– Mes loisirs
– La voix d'un exile premiere et seconde annee
Freckelton, Thomas Wesley et al *see* Religion and modern thought, and other essays
Fred : the socialist press service – v1 n1-24 [1969 feb 7[?]-jul 28] – 1r – 1 – mf#1056114 – us WHS [335]
Fred burry's journal – Toronto: F W Burry, [1899 or 1900-19–] – 9 – (cont: fred burry's journal of new thought) – mf#P04408 – cn Canadiana [100]
Fred burry's journal of new thought – Toronto: F W Burry, [1898-1899 or 1900] – 9 – (cont by: fred burry's journal) – mf#P04407 – cn Canadiana [100]
Frede, Pierre *see* Aventures lointaines
Fredegarii et aliorum chronica (mgh2:2.bd) : vitae sanctorum / ed by Krusch, B – 1888 – (= ser Monumenta germaniae historica 2: scriptores merovingiarum (mgh2)) – €31.00 – ne Slangenburg [240]
Frederic chopin's (1810-1849) works : first critical edition / ed by Bargiel, Woldemar et al – Leipzig. 14v. 1878-80 – 11 – $110.00 set – (includes suppls and revisions report) – us Univ Music [780]
Frederic, de Ghyvelde, pere *see*
– Album de terre-sainte
– La bonne ste-anne
– Le ciel, sejour des elus
– Saint francois d'assise
– Vie de saint antoine de padoue
– La vierge immaculee
Frederic godet (1812-1900) : d'apres sa correspondance et d'autres documents inedits / Godet, Phillippe – Neuchatel: Attinger, 1913 [mf ed 1984] – 7mf – 9 – 0-8370-0125-0 – (incl bibl ref) – mf#1984-0012 – us ATLA [242]
Frederic ozanam : sa vie et ses oeuvres / Chauveau, Pierre J O – Montreal: C O Beauchemin, 1887 – 7mf – 9 – (int by aut) – mf#03588 – cn Canadiana [241]
Frederic ozanam, professor at the sorbonne : his life and works / O'Meara, Kathleen – 1st American ed. New York: Catholic Publication Society, 1878 – 1mf – 9 – 0-8370-6926-2 – (incl bibl ref) – mf#1986-0926 – us ATLA [920]
Frederic star – Frederic WI. 1903 feb 26/1904-1949 jan 1/1951 jul 26 [gaps] – 24r – 1 – (cont by: inter-county leader [centuria, wi]; inter-county leader and the frederic star) – mf#936249 – us WHS [071]
Frederic william maitland, downing professor of the laws of england; a biographical sketch / Fisher, Herbert Albert Laurens – Cambridge: University Press, 1910. 179p – 1 – us Wisconsin U Libr [920]
Frederica Academy *see* Ebbtide
Frederichs, Julius *see* De secte der loisten, of, antwerpsche libertijnen (1525-1545)
Frederick 2, King of Prussia *see* De la litterature allemande
Frederick denison maurice / Masterman, Charles Frederick Gurney – London: A.R. Mowbray, 1907 – (= ser Leaders Of The Church) – 1mf – 9 – 0-7905-5012-1 – mf#1988-1012 – us ATLA [240]
Frederick Douglass Creative Arts Center *see*
– Fdcac newsletter
– Wordstuff
Frederick douglass' paper – Rochester, N.Y.. 1847-1851 (1) – mf#3111 – us UMI ProQuest [976]
Frederick douglass voice – Rochester NY. 1993 may 24-1996 jun 17 – 1r – 1 – (cont: rochester voice) – mf#2882902 – us WHS [071]
Frederick GIs United *see* Pawn
Frederick II, King of Prussia *see* Friedrich der grosse
Frederick, John Hutchinson *see* The development of american commerce
Frederick, Kurt *see*
– Friedrich wilhelm marpurg's abhandlung von der fuge
– Fugal writing from 1750 to 1827
Frederick l. olmsted papers / Olmsted, Frederick Law – 1777-1952 – 1 – $1,362.00 – us L of C Photodup [920]
Frederick walker : an essay...to which is appended a catalogue / Carr, Joseph William Comyns – [London] 1885 – (= ser 19th c art & architecture) – 1mf – 9 – mf#4.2.372 – uk Chadwyck [740]
The fredericksburg news – 1r – 1 – (issues for feb 4-jun 1848 filmed with: semi-weekly news) – us CRL [071]
Fredericksburger wochenblatt – Fredericksburg TX (USA), 1922 6 apr-1926 [gaps], 1928, 1930-1940 21 aug – 7r – 1 – gw Misc Inst [071]
Fredericq, P *see* Corpus documentorum inquisitionis haereticae
Fredericq, Paul *see* The study of history in holland and belgium
Fredericton telegraph – Fredericton, NB. 1806-07 – 1r – 1 – cn Library Assoc [071]
Frederique, Pierre Frederius *see* Monsieur fenelon duplessis and les protestataires
Fredholm, Karl August *see* Om meteorstenfallet vid hessle den 1 januari 1869
Fredigundis : historischer roman aus der voelkerwanderung / Dahn, Felix – Leipzig: Breitkopf & Haertel, 1899 – 1 – us Wisconsin U Libr [830]
Fredigundis : historischer roman aus der voelkerwanderung (ende des 6. jh) / Dahn, Felix – Leipzig: Breitkopf & Haertel, 1899 – 6r – 1 – us Wisconsin U Libr [830]
Fredmans epistlar / Bellman, Carl Michael – [Stockholm: tryckt hos Anders Zetterberg 1790] [mf ed 1990] – 1r – 1 – mf#pres. film 75 – us Sibley [780]
Fredonian / Ross Co. Chillicothe – sep 1811-oct 1813, apr 1814 [wkly] – 1r – 1 – mf#B1223 – us Ohio Hist [071]
The fredonian – Chillicothe, [OH : Hinde & Richardson]. v1 n52 mar 30 1808-july 20 1815 – 1r – 1 – (scattered issues of this weekly national republican newspaper. with sep 8 1809, shortly after the fredonian's cessation, a new paper, the independent republican, published by p. parcels, began; in 1811, richardson bought the republican and renamed it the fredonian) – mf#M 34 R2.1 002 – us Western Res [071]
Fredro, A M *see* Scriptorum seu togae et belli notationum fragmenta
Free american – Boston MA. 1841 mar 1-dec 2 – 1r – 1 – (cont: massachusetts abolitionist; cont by: emancipator [new york, ny: 1835]; emancipator and free american) – mf#880972 – us WHS [071]
The free american – Columbus, OH. v1 n9. mar 1887 [mf ed 1947] – (= ser Negro Newspapers on Microfilm) – 1r – 1 – us L of C Photodup [071]
The free american and deutscher weckruf und beobachter – 5 Jul 1935-11 Dec 1941 – 2r – 1 – $70.00 – us L of C Photodup [073]
Free and easy club – London, England. 18– – 1r – us UF Libraries [240]
The free and liberal ventilation of sewers in its relation to the sanitation of our buildings / Baillairge, Charles P Florent – S.l: s.n, 1892? – 1mf – 9 – mf#00051 – cn Canadiana [628]
Free and united presbyterian union opposed to the principles / Tyndal, John – Edinburgh, Scotland. 1864 – 1r – us UF Libraries [242]
Free angela and all political prisoners : the newsletter of the united committee to free angela davis / National United Committee to Free Angela Davis et al – Los Angeles CA. 1970 dec 8-29, 1971 jan 15-mar 26, jun 29-aug 1, nov 8 – 1r – 1 – mf#640043 – us WHS [320]
Free angola – New York, Angola Office. n1. 1965; n3 1965 – us CRL [071]
Free baptist cyclopaedia / Burgess, G A – 1889 – 1 – $25.55 – us Southern Baptist [242]
Free campus news – v1-v3 n1 [[1970 mar ?-1971 dec ?]] – 1r – 1 – mf#1056584 – us WHS [071]
A free catholic church / Thomas, Joseph Morgan Lloyd – Boston: American Unitarian Assoc, 1907 [mf ed 1993] – (= ser Unitarian/universalist coll) – 1mf – 9 – 0-524-07766-5 – mf#1991-3334 – us ATLA [241]
Free china journal = Tzu yu chung-kuo chi shih pao – Taipei. 1984-1999 (1,5,9) – (cont by: taipei journal) – ISSN: 0255-9870 – mf#14390 – us UMI ProQuest [070]
Free china review – Taipei. 1951-2000 (1) 1972-2000 (5) 1972-2000 (9) – (cont by: taipei review) – ISSN: 0016-030X – mf#8154 – us UMI ProQuest [073]
Free china weekly – Taipei. 1973-1983 (1) 1977-1983 (5) 1977-1983 (9) – (cont by: free china journal=tzu yu chung-kuo chi shih pao) – ISSN: 0016-0318 – mf#9542 – us UMI ProQuest [079]
Free china weekly – Taipei, Taiwan. -w. 19 July 1964-29 Dec 1974. 3 reels – 1 – uk British Libr Newspaper [072]
Free choice / National Right to Work Committee [US] – v1 n1-2nd quarter 1981 [1965 mar-1981] – 1r – 1 – mf#384134 – us WHS [322]
Free church and american slavery / Macnaughtan, J – Paisley, Scotland. 1846 – 1r – us UF Libraries [230]
Free church circular – v3-v4 n16 [1850 jan 28-1851 jun 28] – 1r – 1 – (cont: spiritual magazine [putney, vt]; cont by: circular [brooklyn, ny]) – mf#1111730 – us WHS [071]
Free church claims / Macgeorge, Andrew – Glasgow, Scotland. 1877 – 1r – us UF Libraries [240]
Free Church Of Scotland *see*
– Appeal for the sustentation fund
– Report by the committee on the destitution in the highlands and isl...
Free church of scotland : her ancestry, her claims, and her conflicts / McCrie, Charles Greig – Edinburgh: T & T Clark [1896?] – (= ser Bible class primers) – 2mf – 9 – 0-7905-7118-8 – mf#1988-3118 – us ATLA [242]
Free church of scotland : monthly record – 1886-1900; 1959-89 – 13r – 1 – (lacking: 1892 p2-3. title varies) – ISSN: 0016-0334 – mf#ATLA S0366 – us ATLA [242]
Free church of scotland / West, John Otho – Edinburgh, Scotland. 1843 – 1r – us UF Libraries [242]
The free church of scotland : her origin, founders and testimony / Bayne, Peter – Edinburgh: T & T Clark; New York: Scribner, [distributor], 1893 – 1mf – 9 – 0-7905-4070-3 – mf#1988-0070 – us ATLA [242]
The free church of scotland, 1843-1910 : a vindication / Stewart, Alexander & Cameron, John Kennedy – Edinburgh: W Hodge, [1910?] – 1mf – 9 – 0-524-02065-5 – mf#1990-0562 – us ATLA [242]
Free church of scotland and the act abolishing patronage / Makellar, William – Edinburgh, Scotland. 1874 – 1r – us UF Libraries [242]
Free church of scotland appeals, 1903-4 : united free church authorised report / ed by Orr, Robert Low – 3rd ed. Edinburgh: Macniven & Wallace, 1904 – 2mf – 9 – 0-524-03640-3 – mf#1990-1068 – us ATLA [242]
Free Church Of Scotland College Committee *see* Report of sub-committee on representations regarding dr dods
Free Church Of Scotland General Assembly *see*
– Pastoral address of the general assembly
– Report on sabbath schools and the young...
– Womens foreign missionary society
Free Church Of Scotland General Assembly (1862) *see* Disruption
The free church principle : its character and history / Moncreiff, Henry Wellwood, Sir – Edinburgh: Macniven & Wallace, 1883 – (= ser Chalmers Lectures) – 1mf – 9 – 0-7905-5850-5 – mf#1988-1850 – us ATLA [240]
Free church principles / Laidlaw, John – Aberdeen, Scotland. 1875 – 1r – us UF Libraries [240]
Free church record : devoted to the advancement of a religion, free fro dogma, superstition and sectarianism – Tacoma, Washington. 1893-1900 [mf ed 2001] – (= ser Christianity's encounter with world religions, 1850-1950) – 1r – 1 – mf#2001-s178 – us ATLA [230]
Free churchman – [Scotland] Glasgow: J R Macnair 23 jun-14 jul 1855 (wkly) [mf ed 2004] – 1r – 1 – uk Newsplan [242]
Free communion shown to be unscriptural / Smellie, James – Edinburgh, Scotland. 1892 – 1r – us UF Libraries [240]
The free communionist : or, unrestricted communion of the lord's supper with all true believers, advocated, and objections of restricted communionists, considered – Dover, NH: Trustees of the Freewill Baptist Connection, 1841. Beltsville, Md: NCR Corp, 1978 (3mf); Evanston: American Theol Lib Assoc, 1978 (3mf) – 9 – 0-8370-1103-5 – mf#1984-4483 – us ATLA [240]
Free Congress Research and Education Foundation *see* Initiative and referendum report
Free course of the word / Sumner, Charles Richard – London, England. 1835 – 1r – us UF Libraries [240]
Free cuba / Guiteras, Juan – Philadelphia, PA. 1896 – 1r – us UF Libraries [972]
Free democrat – Chardon, [OH] : J S Wright, 1852- (weekly) – (other titles: chardon democrat; jeffersonian democrat (chardon, ohio)) – us Western Res [071]
Free democrat : weekly free soil newspaper – Chardon, OH: Brown & Canfield, dec 22 1849-oct 15 1850; mar 30 1852-dec 20 1853 – 4r – 1 – (other titles: chardon democrat) – mf#34 G2.1 006 – us Western Res [071]
Free democratic standard / Summit Co. Akron – nov 1849-oct 1850 [wkly] – 1r – 1 – mf#B27950 – us Ohio Hist [071]
Free enquirer – ser1: v1-3 1825-28 [all publ]. ser2: v1-5 1828-33 [all publ]. ser3: v1-2 1833-35 [all publ] – (= ser Radical periodicals of great britain, 1794-1914. period 1) – 48mf – 9 – $210.00 – us UPA [320]
A free enquiry into the origin of the fourth gospel / Sense, P C – London: Williams & Norgate, 1899 [mf ed 1989] – 2mf – 9 – 0-7905-0516-9 – (incl bibl ref and ind) – mf#1987-0516 – us ATLA [226]
Free following – v2 n10-[v5, n3] [1975 nov 1975-1978 may] – 1r – 1 – mf#391558 – us WHS [071]
Free for all / Better World Educational Corporation – Madison WI. 1973 mar 1-1977 nov 23, 1977 nov 23-1981 jan 14, 1981 jan 14-dec 3 – 3r – 1 – mf#235360 – us WHS [370]

Free hungary see Keresztje southern cross

Free india in asia / Levi, Werner – Minneapolis: University of Minnesota, c1952 – (= ser Samp: indian books) – us CRL [327]

Free indonesia / Central Committee of Indonesian Independence – Brisbane, 1946 – 2mf – 9 – (missing: 1946(aug-sep)) – mf#SE-1479 – ne IDC [959]

Free inquiry – Buffalo. 1980+ (1,5,9) – ISSN: 0272-0701 – mf#12784 – us UMI ProQuest [240]

A free inquiry into the nature and origin of evil : in six letters to... / Jenyns, Soame – London: printed for T Cadell, 1790 [mf ed 1993] – (= ser The works of soame jenyns, esq 3) – 1mf – 9 – 0-524-08637-0 – mf#1993-2097 – us ATLA [210]

Free labour gazette, 1894-96 : the organ of the national free labour association – 1r – 1 – mf#95946 – uk Microform Academic [072]

Free labour press and industrial review, 1899-1907 – 2r – 1 – mf#95947 – uk Microform Academic [338]

Free Lance see Church of scotland and the clerical scandals in old greyfriars' chu...

Free lance see Miscellaneous newspapers of lake county

The free lance – Butte, NE: John C Santee (wkly) [mf ed v1 n49. may 13 1892 filmed [1973]] – 1r – 1 – us NE Hist [071]

The free lance – Omaha, NE: Zook-Quinby Co. v1 n1. jul 14 1899- (wkly)// [mf ed jul 14-aug 25 1899 filmed 1974] – 1r – 1 – us NE Hist [071]

The free lance – Ottawa: Free Lance Print Co, [1893-1896] – 9 – mf#P04929 – cn Canadiana [071]

The free lance – Schuyler, NE: John C Sprecher. 7v. v1 n1. may 8 1903-v7 n46. mar 11 1910 (wkly) [mf ed 1903-04,1910 (gaps) filmed [1974?]-86] – 2r – 1 – us NE Hist [071]

Free lance star – Fredericksburg, VA. 1994-2000 (1) – mf#61883 – us UMI ProQuest [071]

Free life – London. 8 Aug 1890-Aug 1901.-w,-m. 1mqnreels – 1 – uk British Libr Newspaper [072]

Free Lutheran Diet (2nd: 1878: Philadelphia, Pa) see The essays, debates, and proceedings

Free magazine – Lincoln, NE: Free Magazine, [1988]- [mf ed [1990?]] – 1r – 1 – (issues for 1988 lack enumeration and chronology) – us NE Hist [071]

Free man's press – Austin, TX. jul 25 1868- [mf ed 1947] – (= ser Negro newspapers on microfilm) – 1r – 1 – (publ in galveston, tx oct 24 1868. cont: freedman's press. cont by: free man's press) – us L of C Photodup [071]

Free market antitrust immunity reform (fair) act of 2001 : hearing...house of representatives, 107th congress, 2nd session, on h.r. 1253, june 5 2002 / United States. Congress. House. Committee on the Judiciary – Washington: US GPO 2002 [mf ed 2002] – 3mf – 9 – (incl bibl ref) – us GPO [343]

Free masonry in north america from the colonial period to the beginning of the present century : also the history of masonry in new york from 1730 to 1888 / Whittemore, Henry [comp] – New York: Artotype Print & Pub Co 1889 – 6mf – 9 – mf#vrl-92 – ne IDC [366]

Free Media, Inc see Natural bridge

Free men speak – New Orleans LA. 1955 jan 15-1957 oct/nov – 1r – 1 – (cont by: independent american [new orleans, la]) – mf#976318 – us WHS [071]

Free men speak – New Orleans, LA. 1955-1957 (1) – mf#63512 – us UMI ProQuest [071]

The free methodist see Light and life

Free Methodist Church of North America see Light and life

Free, Montague see All about african violets

Free passage / Anarchist-Communist Federation [ACF-NA] – n1 [1979] – 1r – 1 – mf#634164 – us WHS [335]

The free people : official organ of the farmers' and workers' party – Johannesburg: Farmers' and Workers' Party, [-1951]. v2 n2 aug 1938; v2 n12 jun 1939; v3 n7-v4 n2 jan 1940-jan 1941; v4 n39-v13 n79 apr 1941-apr 1951 – 1r – 1 – us CRL [366]

Free presbyterian – Yellow Springs, OH. 10 Aug 1853-30 Sept 1857 – 2r – 1 – us Western Res [242]

Free press – Grand Island, NE: Ed J Hall, 1894 (wkly) [mf ed 1895-1915 (gaps) filmed 1978] – 4r – 1 – (absorbed: central nebraska republican. daily ed: morning free press) – us NE Hist [071]

Free press – Ontario, Canada. 20 nov 1912-25 apr 1918; may-22 oct 1918; nov 1918-12 may 1922 – 67 1/2r – 1 – (aka: london free press) – uk British Libr Newspaper [071]

Free press / Ashtabula Co. Geneva – apr 1 1978-jul 7 1979 (damaged) [daily] – 6r – 1 – mf#B31407-31412 – us Ohio Hist [071]

Free press / Ashtabula Co. Geneva – jan 1930-jun 1938 [daily] – 13r – 1 – mf#B30117-30129 – us Ohio Hist [071]

Free press / Ashtabula Co. Geneva – jan 1953-mar 1971 [daily] – 62r – 1 – mf#B30976-31037 – us Ohio Hist [071]

Free press / Ashtabula Co. Geneva – jul 1938-dec 1952 [daily] – 28r – 1 – mf#B30663-30690 – us Ohio Hist [071]

Free press / Ashtabula Co. Geneva – mar 1971-mar 1978 [daily] – 26r – 1 – mf#B25805-25830 – us Ohio Hist [071]

Free press – Burlington, VT. 1848-2000 (1) – mf#60601 – us UMI ProQuest [071]

Free press – Chattanooga, TN. 1933-1998 (1) – mf#60654 – us UMI ProQuest [071]

Free press – Monroe LA. 1993 may 8-dec, 1994 jan 15-aug 27, 1995 mar 4-dec 23 [v26 n28-v27 n22], 1996 jan 13-dec 21 [v27 n23-v28 n16], 1997 jan 11-dec 27 [v28 n17-v29 n21], 1998 jan-dec – 6r – 1 – (cont: black free press) – mf#2687202 – us WHS [071]

Free press – Green Bay WI. 1914 may 14-jul 14, 1914 jul 15-sep 2, 1914 sep 3-oct 26, 1914 oct 27-dec 22, 1914 dec 23-1915 feb 18, 1915 feb 19-apr 19, 1915 apr 20-jun 28 – 7r – 1 – (cont by: green bay gazette; green bay gazette; green bay press-gazette) – mf#947491 – us WHS [071]

Free press – St Paul, NE: W C Ellis, 1881 (wkly) [mf ed v1 n40. jan 25 1882 filmed [1983]] – 1r – 1 – (cont by: saint paul press) – us NE Hist [071]

Free press – DeWitt, NE: Wm H Stout. v3 n2-32. may 10-dec 6 1879; v1 n1. dec 12 1879- (wkly) [mf ed -dec 31 1880 (gaps) filmed [1974?] – 2r – 1 – (cont: de witt free press. publ in de witt ne, may 10-dec 6 1879; in wilber ne, dec 12 1879- . daily ed: daily free press nov 4 1879-) – us NE Hist [071]

Free press – Oconomowoc WI. 1861 feb 15-1862 aug 16 – 1r – 1 – (cont: oconomowoc free press [oconomowoc, wi: 1858]) – mf#959342 – us WHS [071]

Free press – Tecumseh, NE: Alexis Schumacher, Joe M Hartley (wkly) [mf ed 1954 (gaps) filmed [1980]] – 1r – 1 – (cont: trading post) – us NE Hist [071]

Free press – Detroit, MI. 1837-2000 (1) – mf#60495 – us UMI ProQuest [071]

Free press – Dublin, Ireland. 31 mar-5 jul 1862 – 1r – 1 – uk British Libr Newspaper [072]

Free press – Easton, PA. 1859-1913 (1) – mf#65885 – us UMI ProQuest [071]

Free press – Eau Claire, WI. 1858-1888 (1) – mf#67550 – us UMI ProQuest [071]

Free press – Eau Claire, WI. 1873-1889 (1) – mf#67551 – us UMI ProQuest [071]

Free press – Elwood, IN. 1893-1909 (1) – mf#62775 – us UMI ProQuest [071]

Free press – Green Bay, WI. 1914-1915 (1) – mf#67560 – us UMI ProQuest [071]

Free press – Hope Valley, RI. 1900-1904 (1) – mf#66208 – us UMI ProQuest [071]

Free press – Kinston, NC. 1888-1902 (1) – mf#65319 – us UMI ProQuest [071]

Free press – Kinston, NC. 1990-2000 (1) – mf#61676 – us UMI ProQuest [071]

Free press – Lafayette, IN. 1833-1840 (1) – mf#62866 – us UMI ProQuest [071]

Free press / Lorain Co. Amherst – v1 n1. aug 1875-jun 1879 [wkly] – 1r – 1 – mf#B33274 – us Ohio Hist [071]

Free press – Los Angeles, CA. 1965-1974 (1) – mf#62181 – us UMI ProQuest [071]

Free press – Milwaukee, WI. 1901-1918 (1) – mf#67589 – us UMI ProQuest [071]

Free press – Mount Pleasant, IA. 1951-1969 (1) – mf#63323 – us UMI ProQuest [071]

Free press – Naples, NY. 1833-1891 (1) – mf#65044 – us UMI ProQuest [071]

Free press : newfoundland's home paper – St. John's, Newfoundland, 1901-aug 20 1935 – 1r – 1 – Can$35.00 – (on microfilm (seemingly the only copies extant): nov 26, dec 3,10,24 and 31, 1929. the 5 issues incl accounts of the tsunami that swept the coast of the burin peninsula on nov 18 1929) – cn McLaren [071]

Free press – Oswego, NY. 1830-1834 (1) – mf#65148 – us UMI ProQuest [071]

Free press : published for the benefit of the depositors of the two closed banks / Door County State Bank [Sturgeon Bay, WI] et al – 1933 jan-1934 dec 8 – 1r – 1 – mf#3558329 – us WHS [332]

Free press / Putnum Co. Leipsic – jul 1947-jun 1949 [wkly] – 1r – 1 – mf#B2551 – us Ohio Hist [071]

Free press – Redding, CA. 1895-1906 (1) – mf#62241 – us UMI ProQuest [071]

Free press – Redding, CA. 1904-1904 (1) – mf#62242 – us UMI ProQuest [071]

Free press – Rotorua, NZ. sep-dec 1977 – 1r – 1 – mf#17.10 – nz Nat Libr [079]

Free press – St. John's, Canada. -w. 15 mar 1904-27 dec 1921 – 17r – 1 – uk British Libr Newspaper [071]

Free press – Streator, IL. 1873-1910 (1) – mf#62700 – us UMI ProQuest [071]

Free press – Taft, CA. 1940-1941 (1) – mf#62289 – us UMI ProQuest [071]

Free press – Trumansburg, NY. 1987-2000 (1) – mf#65243 – us UMI ProQuest [071]

Free press – Urbana, OH. 1858-1860 (1) – mf#65698 – us UMI ProQuest [071]

Free press – Vermillion, OH. 1981-1987 (1) – mf#65703 – us UMI ProQuest [071]

Free press : (west suffolk and north essex free press – suffolk and essex press – suffolk free press) – England. Jul 1855-Dec 1869; Apr 1884-Dec 1886; Apr 1888-1975; 1978-80.-w. 109 reels – 1 – uk British Libr Newspaper [072]

Free press – Weston, WV. 1915-1917 (1) – mf#67509 – us UMI ProQuest [071]

Free press – Wexford 1923-71 – 49r – 1 – ie National [072]

Free press – Wexford, Ireland. 14 jan-30 dec 1922; 1950 – 2r – 1 – uk British Libr Newspaper [072]

Free press see
- Bend free press
- London free press
- Miscellaneous newspapers of las animas county, reel 3

The free press – O'Neill, NE: W D Mathews, oct 15 1886 (wkly) [mf ed v1 n2. oct 22 1886 filmed 1973] – 1r – 1 – us NE Hist [071]

The free press – Quakertown, PA. 1972 [daily] – 4r – 13 – $25.00r – us IMR [071]

The free press see
- Gunnison county miscellaneous newspapers

Free press and advertiser – South Shields, England. -w. Jan 1895-Jan 1904. 6 reels – 1 – uk British Libr Newspaper [072]

Free press and area editions / Star Co. Canton – v1 n1. oct 1982-oct 1989 [wkly], semiwkly] – 37r – 1 – mf#B31089-31125 – us Ohio Hist [071]

Free press and pred – Beloit, WI. 1848-1903 (1) – mf#67542 – us UMI ProQuest [071]

Free press (burns, or: 1930) – Burns OR: S D Pierce, -1931 [wkly] [mf ed 1965] – 1r – 1 – (began in 1930; cont by: burns press) – us Oregon Lib [071]

Free press (burns, or: 1932) – Burns OR: S D Pierce, 1932-40 [wkly] [mf ed 1965-66] – 2r – 1 – (cont: burns press (1931-32). absorbed: bend free press; cont by: bend pilot (1940-50). publ dec 9 1938-oct 27 1939 as: deschutes county advertiser) – us Oregon Lib [071]

Free press evening bulletin – Manitoba, CN. oct 1894-apr 1935 – 256r – 1 – cn Commonwealth Imaging [071]

Free press herald / Star Co. Canton – v1 n1. mar-dec 1989 [semiwkly] – 4r – 1 – mf#B31126-31129 – us Ohio Hist [071]

Free press journal and bharat iyoti – Bombay, India. Apr 1944-1993 – 182r – 1 – us L of C Photodup [079]

The free press of springfield – Springfield, MA. v1 n1- dec 1968- [mf ed 19–] – 1 – (a publ of the religion & arts committee of the first unitarian-universalist church of springfield. v1 n1 publ in collaboration with grass roots) – Grass Roots – us Bell [071]

Free press (redmond, or) – Redmond OR: Syd D Pierce, 1910s – 1r – 1 – us Oregon Lib [071]

Free press report on farming – Winnipeg, CN. 1872-83 – 237r – 1 – cn Commonwealth Imaging [630]

Free press series / Greene Co. Xenia – v1 n1. oct 1831-apr 1837, nov 1837-mar 1843 [wkly] – 3r – 1 – mf#B5578-5580 – us Ohio Hist [071]

Free press series / Star Co. Canton – jan 1990-dec 1994 – 10r – 1 – mf#B36316-36325 – us Ohio Hist [071]

Free press standard / Carroll Co. Carrollton – jan 1946-dec 1956 [wkly] – 5r – 1 – mf#B11272-11276 – us Ohio Hist [071]

Free press standard / Carroll Co. Carrollton – jan 1957-dec 1966 [wkly] – 8r – 1 – mf#B11573-11580 – us Ohio Hist [071]

Free press standard / Carroll Co. Carrollton – jan 1967-dec 1983 [wkly] – 28r – 1 – mf#B13368-13395 – us Ohio Hist [071]

Free press standard / Carroll Co. Carrollton – jan 5 1984-dec 26 1991 [wkly] – 13r – 1 – mf#B31636-31648 – us Ohio Hist [071]

Free press standard series / Carroll Co. Carrollton – (1906-16, 19, 20-31, 33-1945) [wkly] – 17r – 1 – mf#B9017-9033 – us Ohio Hist [071]

Free press tribune – Colby, KS. 1971-1980 (1) – mf#61449 – us UMI ProQuest [071]

Free press underground – Columbia, MO: Columbia Free Press, 196- [mf ed 19–] – 1 – us Bell [071]

Free press-times / Ashtabula Co. Geneva – jan 1909-dec 1919 [daily] – 14r – 1 – mf#B30087-30100 – us Ohio Hist [071]

Free press-times / Ashtabula Co. Geneva – jan 1920-dec 1929 [daily] – 16r – 1 – mf#B30101-30116 – us Ohio Hist [071]

Free press-times series / Ashtabula Co. Geneva – mar 1901-apr 1909 very poor quality [wkly] – 14r – 1 – mf#B32749-32762 – us Ohio Hist [071]

Free public libraries for canada : working-men's prize essays – Toronto: Citizen, 1882 – 1mf – 9 – mf#05534 – cn Canadiana [020]

Free Public Library, Ottawa, KS see Board of directors minute books

Free radical biology and medicine – New York. 1987-1995 (1,5,9) – ISSN: 0891-5849 – mf#49520 – us UMI ProQuest [574]

The free religious association : its twenty-five years and their meaning. an address for the twenty-fifth anniversary of the association, at tremont temple, boston... / Potter, William James – Boston: Free Religious Association of America, 1892 – 1mf – 9 – 0-8370-4787-0 – mf#1985-2787 – us ATLA [243]

Free Religious Association (Boston, Mass) Meeting (47th: 1914: Boston, MA) see World religion and world brotherhood

Free review series, 1893/4-1900 – (= ser Periodicals connected with owenite socialism and its successors in secularist, freethought and allied movements, 1834-1916) – 6r – 1 – (cont by: university magazine and free review from v8, 1897) – mf#97175 – uk Microform Academic [073]

Free, Richard W see Lux benigna

Free russia : the organ of the english society of friends of russian freedom – London. jun 1890-oct 1914 jan 1915 [mnthly] – 2r – 1 – uk British Libr Newspaper [073]

Free russia see Russian journals

Free Sons of Israel see Reporter

Free Southern Theater see
- Nkombo
- Records of 1963-1978

Free spaghetti dinner – v1 n16 [1970 jul 31-aug 21] – 1r – 1 – mf#1583381 – us WHS [071]

Free speech – Annandale. 1974-1980 (1) 1974-1980 (5) 1979-1980 (9) – mf#7866 – us UMI ProQuest [380]

Free speech and free thought in america / Haldeman-Julius, Emanuel – Girard, KS: Haldeman-Julius Publications, 1927. 128p. (Big Blue Book No. B-37) – 1 – us Wisconsin U Libr [323]

The free state – Brandon, MS: Free State Pub. Co. v2 n11. jan 20 1900 (wkly) [mf ed 1947] – (= ser Negro Newspapers on Microfilm) – 1r – 1 – us L of C Photodup [071]

Free statia / Collins, William W – Cape Town, South Africa. 1965 – 1r – us UF Libraries [960]

Free Street Theatre see Milwaukee folk

Free sunday advocate and national sunday league record – London. -w. 3 Jul 1869-1 Dec 1890. (2 reels) – 1 – uk British Libr Newspaper [072]

Free testosterone/cortisol responses to short term high-intensity resistance exercise overtraining / Bailey, Jeffery T – 2000 – 69p on 1mf – 9 – $5.00 – mf#PH 1714 – us Kinesology [617]

Free texas / Libertarian Party of Texas – v9 n1-v12 n1 [1980 jan/feb-1983 feb] – 1r – 1 – mf#1092876 – us WHS [325]

Free the land! : republic of new afrika newsletter / Republic of New Afrika [Organization] – n2 [[1971?]] – 1r – 1 – mf#4868200 – us WHS [320]

Free thinkers magazine – London, UK. 1850-51 [irr] – 16ft – 1 – uk British Libr Newspaper [210]

Free Thought League of North America see
- Biron and bruckers sonntags-blatt
- Freidenker

Free thoughts on the probable consequences of the decision in the c... / Gressington, Gilbert – London, England. 1850 – 1r – us UF Libraries [240]

The free trade advocate and journal of political economy – Philadelphia. v. 1-2. 3 Jan-28 Nov 1829 – 1 – us NY Public [380]

Free trade in corn the real interest of the landlord, and the true policy of the state / Rooke, John – London: printed for James Ridgway, 1828 – (= ser 19th c economics) – 1mf – 9 – mf#1.1.219 – uk Chadwyck [380]

Free trader – London. -w. and m. 31 Jul 1903-Dec 1905. (1 reel) – 1 – uk British Libr Newspaper [380]

Free trader – Natchez, MS. 1858-1860 (1) – mf#61082 – us UMI ProQuest [071]

Free trader – Ottawa, IL. 1840-1881 (1) – mf#62671 – us UMI ProQuest [071]

Free trader journal – Ottawa, IL. 1916-1927 (1) – mf#62672 – us UMI ProQuest [071]

Free universal magazine – Baltimore. 1793-1793 (1) – mf#4451 – us UMI ProQuest [200]

Free venice beachhead / Beachhead Collective – Los Angeles CA. 1968 dec 1-1977 jul, 1977 aug-1981, 1982-86 – 8r – 1 – mf#628804 – us WHS [071]

Free voice / International Workers in the Amalgamated Food Industries – 1920 oct 15-1921 jul 15] – 1r – 1 – (cont: hotel worker; cont by: free voice of the amalgamated food workers) – mf#1056597 – us WHS [331]

Free voice of the amalgamated food workers / Amalgamated Food Workers et al – v2 n15-v9 n5 [1921 aug 1-1928 may 1], v9 n6-v16 n4 [1928 jun 1-1935 apr] – 2r – 1 – (cont: free voice [new york, ny]) – mf#3178639 – us WHS [331]

Free voice of the amalgamated food workers / Amalgamated Food Workers of America – 1920-35 – (= ser Labor union periodicals, pt 3: food and agricultural industries) – 2r – 1 – $405.00 – 1-55655-619-5 – us UPA [660]

Free way : a power/line paper – 1972 dec-1977 oct, 1977 dec-1981 dec, 1982 feb-1985 feb – 3r – 1 – (cont: power life; cont by: student magazine) – mf#167579 – us WHS [071]

Free weekly – Marrickville, jan 1963-dec 1968 – 9r – at Pascoe [079]

Free wesleyan church of tonga : miscellaneous printed documents – 1869-1982 – 1r – 1 – mf#pmb doc390 – at Pacific Mss [242]

Free will : the greatest of the seven world-riddles / Gruender, Hubert – St Louis, Mo: B Herder, 1911 – 1mf – 9 – 0-7905-9949-X – mf#1989-1674 – us ATLA [100]

Free will and four english philosophers : hobbes, locke, hume and mill / Rickaby, Joseph – London: Burns and Oates; New York: Benziger, 1906 – 1mf – 9 – 0-7905-9460-9 – (incl bibl ref) – mf#1989-2685 – us ATLA [120]

Free will and human responsibility : a philosophical argument / Horne, Herman Harrell – New York: The Macmillan Company, 1912 – xvi/197p – 1 – us Wisconsin U Libr [120]

Free will baptist – Abstract of the former Articles of Faith confessed by the original Baptist Church, an. 1912. 25p – 1 – 5.00 – us Southern Baptist [242]

Free Will Baptist Foreign Mission Society *see* Records

Free will in relation to statistics / Drummond, Robert Blackley – London, England. 1860 – 1r – us UF Libraries [240]

The freebooter (london) – 11 oct 1823-3 apr 1824 – (= ser 19th c british periodicals) – r51 – 1 – (filmed with: the humming bird (leicester), dec 1824-sep 1825; the london weekly review (london), 16 oct 1839-jan 1840) – us Primary [073]

Freed, Augustus Toplady *see* Iron and steel

Freed, Louis Franklin *see* Sex education in transvaal school

Freedley, Angelo Tillinghast *see*
– The corporation laws of 1883; being a supplement to the general corporation laws of pennsylvania
– The general corporation law of pennsylvania, approved 29 april, 1874.

Freedley, Edwin Troxell *see* The secret of success in life

Freed-man : a monthly magazine devoted to the interests of the freed coloured people – London. 1865-1868 – 1 – mf#4751 – us UMI ProQuest [305]

Freedman / American Tract Society – Boston. v1-6 n3. 1864-69 [all publ] – (= ser Black journals, series 2) – 1r – 1 – $200.00 – us UPA [976]

Freedman, Joseph S [comp] *see* Ethics in the early modern period

Freedman's advocate / National Freedmen's Relief Association – New York. v1-2 n1. 1864-65 [all publ] – (= ser Black journals, series 2) – 1r – 1 – $200.00 – us UPA [976]

Freedman's friend – Philadelphia PA: Friends' Association of Philadelphia and Its Vicinity for the Relief of Colored Freedmen, v1 n14 (1866) [mf ed 2006] – 1r – 1 – (reel incl other titles: 2005-s130 & 2005-s131) – mf#2005-s132 – us ATLA [366]

The freedman's friend – Philadelphia PA: Friends' Association of Philadelphia and Its Vicinity for the Relief of Colored Freedmen 1864- [mf ed 2006] – v1 n14 (1866) on 1r – 1 – (began with v1 n1 [6th month, 1864?] damaged: v1 n14 (1866) p99; reel also incl: catalogue and announcement of the texas deaf and blind institute for colored youths [2005-s130] and: catalogue and announcement of the texas deaf and dumb and blind institute for colored youths [2005-S131]) – mf#2005-s132 – us ATLA [366]

Freedman's journal / American Tract Society – Boston. v1-2. 1865-66 [all publ] – (= ser Black journals, series 2) – 1r – 1 – $200.00 – us UPA [976]

Freedman's press – Austin, TX. jul 18 1868 [mf ed 1947] – (= ser Negro newspapers on microfilm) – 1 – (cont by: free man's press) – us L of C Photodup [071]

The freedman's torchlight – Brooklyn NY: [African Civilization Society] v1 (1866) [mf ed 2005] – 1r – 1 – (no more publ? on reel with 2005-s076) – mf#2005-s075 – us ATLA [360]

Freedmen's aid society records, 1866-1932 – 119r – 1 – $130.00r – (incl guide wh may be purchased separately $10. coll divided as foll: correspondence 1875-1932 112r. annual reports 1866-1924 2r. records of board and committee meetings 1866-1924 5r) – mf#D3472 – us Scholarly Res [366]

Freedmen's record – Boston. 1865-1874 (1) – mf#3091 – us UMI ProQuest [305]

Freedmen's record / New England Freedmen's Aid Society – v1-v5 n9 [1865 jan-1869 feb] – 1r – 1 – mf#1056601 – us WHS [071]

Freedom – 1893 nov 23, 1894 jan 1,15, mar 22-apr 29, may 14,28, jun 4,18, jul 3-16 – 1r – 1 – mf#3236358 – us WHS [071]

Freedom – A journal of anarchist socialism (Communism). London. -m. Oct 1866-Nov Dec 1927, May 1930-Jul Sep 1936. (3 reels) – 1 – uk British Libr Newspaper [072]

Freedom : an anarchist monthly – v1-2 n2,4. 1933-34 [all publ] – (= ser Radical periodicals in the united states, 1881-1960. series 1) – 3mf – 9 – $85.00 – us UPA [335]

Freedom / Church of Scientology International – v4 n1-v17 n5=n7-62 [1971 dec/1972 jan-1984 oct], 1984 dec, v17 n1-12 [1985 jan-jul], v18 n1,2-4 [1985 aug, oct-dec], v18 n5-8 [1986 jan-apr] – 2r – 1 – mf#1033660 – us WHS [290]

Freedom / International Working People's Association – v1 n1-7 [1890 nov-1891 may 1], v2 n3-6 [1892 feb-may] – 1r – 1 – mf#1111735 – us WHS [335]

Freedom : a journal of anarchist communism – London: Freedom Publ Comm. v20-30. 1906-jun 1916 – 1 – us CRL [335]

Freedom – London: J Turner, v20-30 n203-326. 1906-jun 1916 – 1r – 1 – us CRL [072]

Freedom – Mzuzu: [Friendly Publ Ltd] [sep 10/22 1996-dec 15/21 1997] – 1r – 1 – us CRL [079]

Freedom – New York. N.Y. 1913 – 1 – us AJPC [071]

Freedom / Smuts, Jan C – London: Alexander Maclehouse & Co, 1934 – 1mf – 9 – $1.50 – mf#LLMC 92-203 – us LLMC [320]

Freedom – Wellington, NZ. 1953 – 1r – 1 – mf#41.2 – nz Nat Libr [079]

Freedom and culture / Radhakrishnan, Sarvepalli – Madras: GA Natesan & Co, 1936 – (= ser Samp: indian books) – us CRL [303]

Freedom and fellowship in religion : a collection of essays and addresses – Boston:Roberts, 1875 – 1mf – 9 – 0-8370-3186-9 – mf#1985-1186 – us ATLA [240]

Freedom and fellowship in religion : proceedings and papers of the 4th international congress of religious liberals...sep 22-27 1907 / ed by Wendte, Charles William – Boston, Mass: International Council, [1907?] [mf ed 1986] – 2mf – 9 – 0-8370-9000-8 – (incl bibl ref) – mf#1986-3000 – us ATLA [240]

Freedom and friendship : the call of theosophy and the theosophical society / Arundale, George Sydney – Madras: Theosophical Pub House, 1935 – (= ser Samp: indian books) – us CRL [230]

Freedom and independence for the golden lands of australia : the right of the colonies, and the interest of britain and of the world / Lang, John Dunmore – London, 1852 – (= ser 19th c british colonization) – 4mf – 9 – mf#1.1.3491 – uk Chadwyck [320]

Freedom and plenty – 1940 oct – 1r – 1 – (cont: free-economy) – mf#2218594 – us WHS [071]

Freedom and the churches : the contributions of american churches to religious and civil liberty / Rauschenbusch, Walter et al; ed by Wendte, Charles William – Boston: American Unitarian Assoc 1913 [mf ed 1990] – 1mf – 9 – 0-7905-6854-3 – mf#1988-2854 – us ATLA [342]

Freedom at issue – New Brunswick. 1975-1990 (1) 1975-1990 (5) 1975-1990 (9) – (cont by: freedom review) – ISSN: 0016-0520 – mf#10546 – us UMI ProQuest [320]

Freedom call : official newsletter of the milwaukee naacp / National Association for the Advancement of Colored People – 1960 aug-1963 jun – 1r – 1 – (cont: your milwaukee naacp newsletter) – mf#691343 – us WHS [322]

Freedom club bulletin / First Congregational Church, Los Angeles – v17-25 n2 [1966 oct 11-1975 jun 10] – 1r – 1 – mf#368286 – us WHS [242]

Freedom Fellowship Church *see* Freedom today

Freedom first – London. 1972-1973 (1) 1972-1972 (5) (9) – ISSN: 0016-0539 – mf#7082 – us UMI ProQuest [320]

Freedom for spain – London, 1945. Fiche W 897. (Blodgett Collection of Spanish Civil War Pamphlets) – 9 – us Harvard [946]

Freedom for the church of god / Mossman, Thomas Wimberley – London, England. 1876 – 1r – us UF Libraries [240]

Freedom From Religion Foundation [US] *see* Freethought today

Freedom from Religion Foundation [US] *see* Newsletter

Freedom in the church : or, the doctrine of christ as the lord hath commanded and as this church hath received the same according to the commandments of god / Allen, Alexander Viets Griswold – New York: Macmillan, 1907 – 1mf – 9 – 0-8370-8561-6 – (incl bibl ref and ind) – mf#1986-2561 – us ATLA [240]

Freedom in the church of england : six sermons suggested by the voysey judgment / Brooke, Stopford Augustus – 2nd ed London: Henry S King, 1871 – 1mf – 9 – 0-7905-3645-5 – mf#1989-0138 – us ATLA [241]

Freedom information service *see* Mississippi newsletter

Freedom Leadership Foundation *see* Rising tide

Freedom ledger / National Association for the Advancement of Colored People – 1977 sep-1989 apr – 1r – 1 – (cont: naacp racine branch : [newsletter]) – mf#3924379 – us WHS [322]

Freedom magazine / Liberty Amendment Committee of the USA – v17 n1-v18 n2 [1973 spring-1974 summer] – 1r – 1 – (cont: american progress) – mf#1111738 – us WHS [320]

Freedom news – 1968-1971, 1972-1974 mar – 2r – 1 – mf#1111740 – us WHS [320]

Freedom newsdigest – v6 n1-2 [1987 aug-oct] – 1r – 1 – (cont: epia society digest) – mf#1567835 – us WHS [320]

The freedom of authority : essays in apologetics / Sterrett, James Macbride – New York: Macmillan 1905 [mf ed 1985] – 1mf – 9 – 0-8370-5402-8 – (incl ind) – mf#1985-3402 – us ATLA [240]

Freedom of information caselist / Dept of Justice, Office of Information and Privacy – 1985-98 – 70mf – 9 – $105.00 – (add vols planned) – mf#llmc 94-358 – us LLMC [342]

Freedom of information center report / University of Missouri-Columbia – Columbia. 1977-1985 (1) 1977-1985 (5) 1977-1985 (9) – ISSN: 0014-603X – mf#11669 – us UMI ProQuest [322]

Freedom of land / Eversley, George John Shaw-Lefevre, Baron – London, 1880 – (= ser 19th c books on british colonization) – 2mf – 9 – mf#1.1.8270 – uk Chadwyck [323]

Freedom of mind in willing : or, every being that wills a creative first cause / Hazard, Rowland Gibson – New York, London: D. Appleton and Co., 1864 – xviii/465p – 1 – us Wisconsin U Libr [190]

Freedom of mind in willing, or, every being that wills a creative first cause / Hazard, Rowland Gibson; ed by Hazard, Caroline – Boston: Houghton, Mifflin, 1889 – 2mf – 9 – 0-7905-8657-6 – mf#1989-1882 – us ATLA [100]

Freedom of the press : the struggle for copyright and the laws of libel, 1660-1821; a catalog / ed by Parks, Stephen – NY: Garland Publishing Inc, 1974? – 1mf – 9 – $1.50 – mf#LLMC 91-079 – us LLMC [340]

Freedom of the will / Taylor, William – London: Hamilton, Adams, 1881 – (= ser Evangelical Union Doctrinal Series) – 1mf – 9 – 0-7905-8929-X – mf#1989-2154 – us ATLA [100]

The freedom of the will : as a basis of human responsibility and a divine government / Whedon, Daniel Denison – New York: Carlton & Lanahan, c1864 – 1mf – 9 – 0-8370-6538-0 – (incl ind of authors cited) – mf#1986-0538 – us ATLA [120]

Freedom rag – 1994 spring-1995 summer – 1r – 1 – mf#3059603 – us WHS [071]

Freedom reader – v2 n3-5 [1972 sep 1,28, nov] – 1r – 1 – mf#1583404 – us WHS [071]

Freedom review – New York. 1991-1996 (1,5,9) – (cont: freedom at issue) – ISSN: 1054-3090 – mf#10546,01 – us UMI ProQuest [320]

Freedom socialist : voice of the freedom socialist party of washington / Freedom Socialist Party [US] – Seattle WA. v1 n1-2, v2 n1-1982 summer [1966 aug 5-nov 5, 1976 summer-1982 summer] – 1r – 1 – mf#618421 – us WHS [335]

Freedom Socialist Party [US] *see* Freedom socialist

Freedom talks / Life Line Foundation – v1-v2 n11 [1975 jun 29-1976 mar 22] – 1r – 1 – (cont: lifeline freedom talk) – mf#368285 – us WHS [071]

Freedom to express – Grants Pass OR: [s.n.] 1982- [mthly] – 1 – us Oregon Lib [071]

Freedom today / Freedom Fellowship Church – v1 n1-v2 n17 [1975 jul-1978 jan] – 1r – 1 – mf#405106 – us WHS [242]

Freedom unlimited – 1972 oct – 1r – 1 – mf#4717661 – us WHS [071]

Freedom writer / Institute for First Amendment Studies et al – 1985 mar-1987 jan – 1r – 1 – (cont: control q) – mf#1145619 – us WHS [071]

Freedom's battle : being a comprehensive collection of writings and speeches on the present situation / Gandhi, Mahatma – Madras: Ganesh & Co, 1922 – (= ser Samp: indian books) – us CRL [954]

Freedom's facts – Washington. 1954-1971 (1) 1971-1971 (5) (9) – ISSN: 0016-0601 – mf#5862 – us UMI ProQuest [320]

Freedomways – New York. 1961-1985 (1) 1969-1985 (5) 1975-1985 (9) – ISSN: 0016-061X – mf#1609 – us UMI ProQuest [073]

Freehold land times building news – London, UK. 1 apr 1854-1855; jul-dec 1862; jul 1869-1892 [wkly] – 67r – 1 – (aka: building news 1862-; land and building news 1855) – uk British Libr Newspaper [690]

Freeholder : or political essays – London. 1715-1716 (1) – mf#4253 – us UMI ProQuest [320]

Freeholder's journal – London. n1-76. jan 31 1721-may 18 1723 – 1 – us NY Public [073]

Freel, Michael J *see* Survey of indiana high school athletic directors regarding qualifications, education, time obligation and salary

The freelance – Wellington, NZ. jul 1900-dec 1910; jan 1914-2 nov 1960 – 182r – 1 – mf#41.33 – nz Nat Libr [079]

Freeland – New York. v. 1-8. dec 1944-dec 1955 (incomplete). – 1 – us NY Public [073]

Freeland and allied families – v2:iss 5-12 [1972-1986] – 1r – 1 – (cont: freeland quarterly and allied families) – mf#635224 – us WHS [071]

Freeland, Elizabeth M *see* Preceptions of collegiate coaches on coach education1

Freeland quarterly and allied families – v1 n4 [p. 53-72] [1971 oct ?] – 1r – 1 – (cont by: freeland and allied families) – mf#635216 – us WHS [071]

Freeman – Irvington-on-Hudson. 1950-1999 (1) 1968-1999 (5) 1977-1999 (9) – (cont by: ideas on liberty) – ISSN: 0016-0652 – mf#1491 – us UMI ProQuest [320]

Freeman / Franklin Co. Columbus – (jan 1840-feb 1842) – 1r – 1 – mf#B4310 – us Ohio Hist [071]

Freeman / Indianapolis IN – 1915-18, jul 1919-20 [wkly] – 3r – 1 – mf#B5006-5008 – us Ohio Hist [976]

Freeman – Indianapolis, IN. 1888-1916 (1) – mf#62840 – us UMI ProQuest [071]

Freeman / Sandusky Co. Fremont – oct 1849-may 1850 [wkly] – 1r – 1 – mf#B33272 – us Ohio Hist [071]

Freeman / Sandusky Co. Lower Sandusk – v1 n1. feb-oct 1849 [wkly] – 1r – 1 – mf#B33278 – us Ohio Hist [071]

The freeman – Cebu: Freeman Pub Co, jul 10 1925-nov 9 1928 – us CRL [079]

Freeman, Abraham Clark *see* A treatise on the law of executions in civil cases, and of proceedings in aid and restraint thereof

Freeman, and irish-american review – Philadelphia PA. 1890 mar 15-1891 feb 5 – 1r – 1 – mf#1225206 – us WHS [071]

Freeman Education Association *see* Freemanletter

Freeman education association newsletter – 1985 may/jun-1986 jun – 1r – 1 – (cont by: freeman letter) – mf#2840098 – us WHS [370]

Freeman, Edward A *see*
– The chief periods of european history
– Western europe in the fifth century: an aftermath

Freeman, Edward Augustus *see*
– The chief periods of european history
– An essay on the origin and development of window tracery
– The history and conquests of the saracens
– A history of architecture
– The preservation and restoration of ancient monuments
– Principles of church restoration

Freeman, George *see* Sketches in wales

Freeman, H Dwight *see* Strength of irrigation pipe as influenced by coarse aggregate

Freeman, Harrup A *see* Palau's constitutional convention. an informal report

Freeman Institute *see* Behind the scenes

Freeman, J J *see* A dictionary of the malagasy language in two parts

Freeman, James Edward *see* The man and the master

Freeman, James Edward et al *see* A nation-wide preaching mission

Freeman, James Midwinter *see* A short history of the english bible

Freeman, John *see* Church of england schoolmaster

Freeman, John Haskell *see* An oriental land of the free

Freeman, John, of Bhaugulpore *see* A reply to the memorandum of the east india company

Freeman, Jonathan *see* Sermons, 1798-1821

Freeman, Joseph John *see*
– A dictionary of the malagasy language
– The kaffir war

Freeman, Michael D *see* Journal of whiplash and related disorders

Freeman, Paul Douglas *see* Use of the tape recorder as a means of musical expression

Freeman, Phillip *see* Plea for the education of the clergy

Freeman, T R *see*
– Manufacture of ice cream with limited mild solids
– Storing frozen cream

Freeman, Z *see* Manual of american colleges and theological seminaries

Freeman-Grenville, Greville Stewart Parker *see* French at kilwa island
Freemanletter / Freeman Education Association – 1986 aug-sep, 1987 sep, dec, 1988 mar [mutilated]-aug, oct, 1989 feb-1991 dec – 1r – 1 – (cont: freeman education association newsletter) – mf#2686342 – us WHS [370]
Freeman's chancery cases / Mississippi. Supreme Court – 1v. 1839-1843 (all publ) – (= ser Mississippi supreme court reports) – 4mf – 9 – $6.00 – (not a pre-nrs title) – mf#LLMC 95-100 – us LLMC [347]
Freeman's chronicle / Franklin Co. Franklinton – sep 1812-aug 1813, jan-sep 1814 [wkly] – 1r – 1 – mf#B6741 – us Ohio Hist [071]
Freeman's exmouth journal – England, 24 jul 1869-1933 – 57r – 1 – (aka: exmouth journal, 27 apr 1907-) – uk British Libr Newspaper [072]
Freeman's journal – 1881; may-dec 1885 – (= ser New Zealand freeman's Journal (Auckland)) – 4r – (aka: new zealand freeman's journal (auckland)) – mf#11.26 – nz Nat Libr [079]
Freeman's journal – Dublin 1763-1859 – 150r – 1 – ie National [072]
Freeman's journal – Dublin, Ireland. -w. May-June 1892. 1 2 reel – 1 – uk British Libr Newspaper [072]
Freeman's journal – Galveston, TX: R Nelson. v1 n1 (1887)-1891? (wkly) [mf ed 1947] – (= ser Negro Newspapers on Microfilm) – 1r – 1 – us L of C Photodup [071]
Freeman's journal – Sydney, jun 1850-feb 1942 – 116r – 1 – A$6706.74 vesicular A$7344.74 silver – at Pascoe [073]
Freemans journal *see*
- National press
- Weekly national press
Freemans journal and daily commercial advertiser *see* Public register
Freemans journal and daily commercial advertsier (evening ed) – Dublin, Ireland. 1874; sep 1881-dec 1885 – 23r – 1 – uk British Libr Newspaper [072]
Freeman's journal and miscellaneous / Hamilton Co. Cincinnati – (mar 1796-feb 1813) spotty – 1r – 1 – mf#B1222 – us Ohio Hist [071]
Freeman's journal & chillicothe advertiser – Chillicothe, OH. july 11-sept 26, 1800 – 1r – 1 – (second newspaper to be publ in ohio, weekly, national republican) – us Western Res [071]
The freemason – Toronto: Cowan, [1881?]- – 9 – mf#P04467 – cn Canadiana [366]
Free-masonry / Roberts, George – Monmouth, England. 1843 – 1r – us UF Libraries [240]
Freemasonry in china / Giles, Herbert Allen – Amoy, 1880 – (= ser 19th c books on china) – 1mf – 9 – mf#7.1.31 – uk Chadwyck [366]
Freemasons *see*
- Annual communication
- Masonic family
- Masonic journal
- Masonic mirror
- Masonic tidings
Freemasons and spain; struggle of masonic liberalism against reactionism in spain – NY, 1938? Fiche W 898. (Blodgett Collection of Spanish Civil War Pamphlets) – 9 – us Harvard [946]
Freemasons. Barton Lodge, No 6 (Hamilton, Ont) *see*
- Historical sketch of the barton lodge no.6, grc, af and am
- A short historical sketch of the barton lodge of a f and a masons
Freemasons [Belgium] *see* La belgique maconnique
Freemasons. Civil Service Lodge, No 148 (Ottawa, Ont) *see* List of members of the civil service lodge, af and am, n148, grc, ottawa
Freemasons. Constitution *see* The old constitutions belonging to the ancient and honourable society of free and accepted masons of england and ireland
Freemasons. Grand Lodge (Canada) *see* The book of constitution of the grand lodge of ancient and free and accepted masons of canada
Freemasons. Grand Lodge of British Columbia. Communication (9th: 1880: Victoria, BC) *see* Proceedings of the...
Freemasons. Grand Lodge of Ontario *see* The book of constitution of the grand lodge of ancient, free and accepted masons of canada
Freemasons. Grand Lodge of Quebec *see* Proceedings...at an emergent communication held at the village of coteau landing, 6th june, a d 1883, a l 5883
Freemason's magazine and general miscellany – Philadelphia. 1811-1812 (1) – mf#4452 – us UMI ProQuest [975]
Freemasons. Maple Leaf Lodge (St Catharines, Ont) *see* By-laws of maple leaf lodge of ancient free and accepted masons, no, st catharines, c w
Freemasons repository – Providence, RI. 1871-1873 (1) – mf#66306 – us UMI ProQuest [071]
Freemasons United Grand Lodge of England *see* United grand lodge of ancient free and accepted masons of england
Freemen Center for Global Studies [Provo, UT] et al *see* Freemen digest
Freemen digest / Freemen Center for Global Studies [Provo, UT] et al – 1977 dec-1979 dec, 1979 apr-1981 – 2r – 1 – (cont by: freemen digest [salt lake city, ut]) – mf#577313 – us WHS [327]
Freeport journal – Freeport, PA. -w 1885-1932 – 13 – $25.00r – us IMR [071]
Freer, Frederick Ash *see* Edward white, his life and work
The freer gospels / Goodspeed, Edgar Johnson – Chicago: University of Chicago Press, 1914 – (= ser Historical and linguistic studies in literature related to the new testament) – 1mf – 9 – 0-7905-1663-2 – mf#1987-1663 – us ATLA [226]
Freer, James *see* Case of the rev walter c smith
Freer, Marjorie (Mueller) *see* Orchids for april
Das freer-logion / Gregory, Caspar Rene – Leipzig: J C Hinrichs, 1908 – 1mf – 9 – 0-8370-3385-3 – mf#1985-1385 – us ATLA [240]
Freeth, Zahra Dickson *see* Run softly demerara
The freethinker, 1881-1919 : from the national secular society / ed by Foote, George William – (= ser Periodicals connected with owenite socialism and its successors in secularist, freethought and allied movements, 1834-1916) – 33r – 1 – mf#97152 – uk Microform Academic [120]
The freethinker's catechism / Monteil, Edgar – Translated by Frederic W. Mitchell. New York: The Truth Seeker Company, 1880? – 1 – us Wisconsin U Libr [210]
Free-thinking / Sanday, W – London, England. 1886 – 1r – us UF Libraries [240]
Freethought and modern progress : a lecture / Watts, Charles – London, England. 18– – 1r – us UF Libraries [240]
Freethought journal – Toronto: Ontario Free Thought Print and Pub. Co, v1 no 1(Sept 14 1877)- – 9 – mf#P04919 – cn Canadiana [200]
Freethought magazine *see* Torch of reason
Freethought today – Freedom from Religion Foundation. jan 1984- – 1 – us Wisconsin U Libr [210]
Freethought today / Freedom From Religion Foundation [US] – intr iss [1983 sep], v1 n1-v10 n10 [1984 jan-1988 dec] – 1r – 1 – (cont: newsletter [freedom from religion foundation [us]) – mf#683233 – us WHS [290]
Freethought vindicated, or, infidel christianity v. honest unbelief : a lecture delivered in the temperance hall, sydney, june 25 1880, in reply to the revs. a c gillies, j a dowie, the catholic express, and the presbyterian witness by j tyerman / Tyerman, J – Sydney: R.W. Skinner, 1880 – 1mf – 9 – 0-8370-5590-3 – (with an appendix containing correspondence between the rev. a.c. gillies, and j. tyerman) – mf#1985-3590 – us ATLA [210]
Freetown 1686-1890 – Oxford, MA (mf ed 1985) – (= ser Massachusetts vital records) – 37mf – 9 – 0-931248-91-4 – (mf 1-4: records 1686-1764 bk 1. mf 5-10: records 1759-95 bk 2. mf 11-13: records 1795-1847 bk 3. mf 14-15: records 1836-71 bk 5. mf 16-18: records 1822-92 bk 6. mf 19-20: b,m,d 1843-55. mf 21-22: birth records 1856-90 bk 12. mf 23-24: marriages 1854-90 bk 11. mf 25-26: deaths 1852-90 bk 10. mf 27-28: birth index to bks 1,2,3,5,6,9,12. mf 29-30: marriage index to bks 1,2,3,5,9,11. mf 31-32: marriage intention index to bks 1,2,3,5,6. mf 33-34: death index to bks 1,2,3,5,6,9,10. mf 35-37: b,m,d index 1686-1844) – us Archive [978]
Freetown express and christian observer – Freetown. Sierra Leone. -m. Jul 1882-Nov 1884. (33 ft) – 1 – uk British Libr Newspaper [072]
Freewater times (freewater, or) – Freewater OR: Dodd & Kennedy, [wkly] [mf ed 1969] – 1r – 1 – us Oregon Lib [071]
Freewater times (milton-freewater, or) – Freewater OR: Sanderson & Bean, -1951 [wkly] [mf ed 1969] – 8r – 1 – (merged with: milton eagle (1887-1951) to form: eagle times (milton-freewater, or)) – us Oregon Lib [071]
Freeway – Reed City, MI. 1995-1996 (1) – mf#69170 – us UMI ProQuest [071]
Freewill Baptist Associations. Yearly Conference. Maine. 1799-1814 *see* Papers
Freewill baptist magazine – Providence, RI. May 1826-May 1830 – 1 – us ABHS [242]
Freewill baptist quarterly magazine – jun 1839-mar 1841 – 1r – 1 – mf#ATLA R0100B – us ATLA [242]
Freewill baptist register – 1831-55, 1863-69 – 1 – 954.45 – us Southern Baptist [242]
Free-will controversy / Hutchison, Thomas Dancer – London, England. 18– – 1r – us UF Libraries [240]
The freewoman – A weekly feminist review. London. -w. 23 Nov 1911-10 Oct 1912. (1 reel) – 1 – uk British Libr Newspaper [305]
Freeze / Jackson County Citizens for a Nuclear Arms Freeze [NC] – 1983 feb-1987 jan 20 – 1r – 1 – (cont by: newsletter [jackson county peace network [nc]) – mf#1223984 – us WHS [327]
Freeze focus / Nuclear Weapons Freeze Campaign – v4 n2-10 [1984 mar-dec] – 1r – 1 – (cont: freeze newsletter) – mf#958118 – us WHS [327]
Freeze newsletter / Nuclear Weapons Freeze Campaign – v2 n4-v4 n1 [1982 apr-1984 jan] – 1r – 1 – (cont by: freeze focus) – mf#708545 – us WHS [327]
Freeze update / Wisconsin Nuclear Weapons Freeze Campaign – 1983 feb/mar-1984 nov/dec, 1985 mar/apr, jun/jul – 1r – 1 – (cont by: in our hands [madison, wi]) – mf#1098884 – us WHS [327]
Freezing fruits and vegetables on florida farms / Stout, G J – Gainesville, FL. 1948 – 1r – us UF Libraries [634]
The freezing of northern rivers : dances in the far north / Ogilvie, William – [S.l: s.n, 1894?] [mf ed 1982] – 1mf – 9 – 0-665-17537-X – (fr: the canadian magazine) – mf#17537 – cn Canadiana [390]
Frege, W H *see* The leofric collectar, vol 2 (hbs56)
Fregenal de la Sierra. Spain *see*
- Algunas paginas del expediente...construccion de un cementerio
- Proyecto de ordenanzas municipales
Fregeville, Antoine de *see* Palinodie chimique: ov les errevrs de cest art
Frei, Lazar' Isaevich et al *see* Finansirovanie vneshnei torgovli
Frei nach goethe : parodien nach klassischen dichtungen goethes und schillers / ed by Hecht, Wolfgang – Berlin: Ruetten & Loening, 1965 – 1r – 1 – (incl bibl ref) – us Wisconsin U Libr [430]
Freiberger anzeiger – Freiberg, Sachsen DE, 1848 2 mar-1919 – 150r – 1 – (title varies: 16 may 1849: freiberger anzeiger und tageblatt) – gw Misc Inst [074]
Freiberger anzeiger und tageblatt *see* Freiberger anzeiger
Freibrief nebukadnezar's 1, koenigs von babylonien c 1130 v chr / Nebuchadnezzar 1, King of Babylonia – Leipzig: August Pries 1883 [mf ed 1986] – 1mf [ill] – 9 – 0-8370-7816-4 – (text in german & akkadian; comm in german) – mf#1986-1816 – us ATLA [470]
Freiburger bote fuer stadt und land – Freiburg Br DE, 1865 22 apr-1868 20 nov, 1896-1921 – 1 – gw Misc Inst [074]
Freiburger nachrichten – Freiburg Br DE, 1945 5 sep-1946 29 jan – 1 – gw Misc Inst [074]
Freiburger studentenzeitung – Freiburg Br DE, 1930-1937/38, 1951-70 [gaps] – 1 – (filmed with special iss) – gw Misc Inst [378]
Freiburger tageblatt *see* Verkuendigungs-blatt
Freiburger tagespost – Freiburg Br DE, 1949 17 oct-1950 31 mar – 1 – (title varies: 8 jan 1934: tagespost. filmed by other misc inst: 1911-1940 29 feb, 1949 17 oct-1950 31 mar. with suppl: deutsche jugendkraft [fr 1930: sport und volk] 1925-1931; frau und leben 1931-1939 24 aug; im herrgottswinkel 1920-37; jugend und volk 1932-1934 jun; der oberbadische landwirt (fr 19 sep 1935: der oberbadische bauer) 1931-1938 2 mar. regional ed: bue=buehl 1949 17 oct-28 oct; e=emmendingen 1949 17 oct-31 oct; la=lahr 1949 17 oct-31 oct; loe=loerrach 1949 17 oct-1950 31 mar; ra=rastatt 1949 17 oct-28 oct; s=[titisee-] neustadt 1949 16 nov-1950 31 mar; sae=bad saeckingen 1949 17 oct-31 oct. all with gaps) – gw Misc Inst [074]
Freiburger wochenbericht – Freiburg Br DE, 1952 21 mar & 24 dec, 1953 8 jan-1955 23 sep – 1r – 1 – gw Misc Inst [074]
Freiburger wochenblatt *see* Freyburger zeitung
Freiburger zeitung *see* Freyburger zeitung
Freiburger zeitung und handelsblatt *see* Freyburger zeitung
Der freidenker – Leipzig DE, 1918-21 – 1r – 1 – gw Misc Inst [074]
Freidenker – Berlin DE. 1925-33 [gaps] – 1r – 1 – gw Misc Inst [100]
Freidenker – 1875-76, 1877-78, 1878 jun 16, jul 21-28, aug 18-25, 1879 jan 26, apr 30, may 11, sep 21, 1879 – 4r – 1 – (cont: biron & bruckers sonntags-blatt; cont by: pionier [louisville, ky: 1854]; freidenker [milwaukee, wi: 1880]) – mf#1295900 – us WHS [071]
Freidenker – 1880-81, 1882-84, 1884 jan 6-dec 28 – 3r – 1 – (cont: pionier [louisville, ky: 1854]; freidenker [milwaukee, wi: 1875]; cont by: freidenker [milwaukee, wi: 1885]) – mf#1295911 – us WHS [071]
Freidenker / Free Thought League of North America – 1885/1887-1941 jan 19/1942 oct 25 [gaps] – 26r – 1 – (cont: freidenker [milwaukee, wis: 1880]) – mf#1295914 – us WHS [071]
Freidenker : organ der freigeistigen vereinigung der schweiz – v39-41 1956-58. v45-74 1962-91 – 4r – 1 – (lacking: v42-44) – mf#ATLA S0438 – us ATLA [210]
Freideutsche jugend – Hamburg DE, 1916-19 – 1 – gw Misc Inst [943]
Frei-deutschland *see* Rheinisch-westfaelische montagspost
Freidrich, Paul *see* Das grabbe-buch
Der freie angestellte : zeitschrift des zentralverbandes der angestellten, sitz berlin – Berlin: Zentralverband der Angestellten [O Urban] [semimthly] [mf ed 1981] – 2r – 1 – (began: 23 jahrg n20 [1 okt 1919]; ceased: 37 jahrg n11 [1933]; cont: zentralverband der handlungsgehilfen handlungsgehilfen-zeitung; filmed with: fachzeitung fuer schneider & other titles) – mf#7703 reel 56-57 – us Wisconsin U Libr [331]
Der freie arbeiter – Berlin. v1-6. 1904-09 – 1 – us CRL [074]
Der freie arbeiter – Porto Alegre (BR), 1920 15 may-1930 may [gaps] – 1r – 1 – gw Misc Inst [331]
Der freie arbeiter – Berlin DE, 1904-1933 feb [gaps] – 1 – (with suppls: antimilitarismus 1905 oct-1906 oct; die canaille 1905 oct-1906 nov [gaps]; freie literatur 1905 nov-1906 [gaps]; generalstreik 1905-06) – gw Misc Inst [331]
Freie arbeiter – 1911-13 – 1r – 1 – (cont: neues leben) – mf#1430837 – us WHS [331]
Freie arbeiter stimme – New York, NY. In Yiddish. -w. 6 Jan 1950-15 Dec 1961. 2 reels – 1 – uk British Libr Newspaper [071]
Freie arbeiter stimme – New York.Free voice of labor. 1943-45 – 1 – us AJPC [331]
Der freie bauer – Berlin DE, 1945 1 nov-1960 30 oct – 8r – 1 – (cont by: neue deutsche bauernzeitung) – gw Misc Inst [630]
Freie blaetter / ed by Glassbrenner, Adolf – Berlin DE, 1848 6 may-dec – 1 – gw Mikrofilm [074]
Freie blaetter – Zittau DE, 1848 27 oct-1849 19 oct – 1r – 1 – gw Misc Inst [074]
Freie blaetter *see* Freie volksblaetter
Freie bodezeitung – Oschersleben DE, 1962 11 jan-1966 17 nov – 1r – 1 – gw Misc Inst [074]
Freie buehne fuer modernes leben – Berlin. v1-2. 1890-1891 – 324mf – 8 – (cont as: freie buehne fuer entwicklungskampf der zeit, berlin v3-4 1892-1893; cont as: neue deutsche rundschau, berlin v5-14 1894-1903) – mf#H-441 – ne IDC [320]
Freie buehne fuer modernes leben *see* Die neue rundschau
Der freie bund – Leipzig DE, 1898-1902 [gaps] – 1 – gw Misc Inst [074]
Der freie demokrat – Hannover DE, 1946 16 may-24 sep, 1947 1 jan-15 apr & 26 jun-30 dec, 1948 16 jan & 29 jan-6 aug – 1 – gw Misc Inst [074]
Der freie demokrat – Stuttgart DE, 1950 16 aug-1951 12 dec – 1 – gw Misc Inst [320]
Freie demokratische korrespondenz (fdk) – Bonn DE, 1950 17 jan-1957 9 mar – 5r – 1 – mf#4869 – gw Mikropress [074]
Freie Deutsche Hochschule, Paris *see* Zeitschrift fuer freie deutsche forschung
Freie deutsche jugend – Troppau (Opava CZ), 1937 jun/jul-1939 jan/feb – 1r – 1 – (since n5 1938 publ in paris) – gw Misc Inst [320]
Freie deutsche kultur – London (GB), 1941 oct & 1943 oct, 1943 dec-1944 jan, 1945 jul-aug – 1 – gw Misc Inst [072]
Freie deutsche presse : ausgabe nordfranken – Coburg DE, dec 24 1948-may 1960 – 1 – gw Misc Inst [074]
Freie deutsche presse *see* Freisinnige zeitung
Freie deutsche schulzeitung – Leipzig DE, 1881, 1886 – 1r – 1 – (with suppl) – gw Misc Inst [370]
Das freie deutschland / mitteilungen der... *see* Mitteilungen der deutschen freiheitsbibliothek
Freie erde – Neubrandenburg DE, 1976-1990 31 mar – 29r – 1 – (filmed by other misc inst: 1991 [gaps], 1992-; 1990 2 apr-10 nov [2r]. title varies: 2 apr 1990: nordkurier. regional ed (only local pp): altentreptow, apr 2 1958-oct 31 1960, jan 4 1966-70 (incomplete); anklam, apr 5 1958-oct 31 1961, jan 3-aug 25 1967, jun 11 1968-jun 4 1969 (incomplete); demmin, apr 1 1958-feb 15 1962 (incomplete), apr 13-sep 29 1963, apr 25 1967-77 (incomplete); malchin, apr 2 1958-aug 18 1984 (incomplete); neustrelitz 1952 15 aug-1975 [77r], 1954 1 aug-1990 31 mar (gaps) [5r]; pasewalk, apr 1 1958-mar 19 1969 (incomplete); prenzlau, apr 3 1958-dec 25 1964 (incomplete); roebel, apr 3 1958-may 6 1979, aug 2-dec 29 1983 (all incomplete); strasburg, apr 3 1958-oct 31 1961, apr 11 1963-mar 13 1969 (incomplete), jan 3-apr 30 1980; templin, apr 2 1958-oct 31 1961, apr 5-30 1966; teterow, 1963-mar 19 1971, 1973-80 (incomplete); ueckermuende, apr 5 1958-oct 31 1961, mar 9-dec 29 1962; waren, apr 4 1958-70, jan 6 1976-oct 29 1977, feb 17 1979-84 (all incomplete)) – gw Misc Inst [074]

Der freie formelhafte infinitiv der limitation im griechischen / Gruenenwald, L – Wuerzburg: A. Stuber, 1888 – (= ser Beitraege zur historischen syntax der griechischen sprache) – 1mf – 9 – 0-8370-1529-4 – (incl bibl ref) – mf#1987-6068 – us ATLA [450]

Freie Gemeinde *see* Voice of freedom

Freie Gemeinde [Milwaukee, WI] *see* Freie Wort

Die freie generation – Berlin. v1-2. July 1906-June 1908 – 1 – us NY Public [073]

Die freie gewerkschaft – Berlin DE, 1947-1949 26 oct – 9r – 1 – (filmed by mikropress: 1945 9 oct-1946, 1991 1 jan-20 sep [5r] order#3049; filmed by misc inst: 1949 may-aug [1r]; 1945 9 oct-1990 [88r]) – uk British Libr Newspaper; gw Mikropress; gw Misc Inst [331]

Die freie gewerkschaft : organ des fdgb – Berlin DE – 1 – (mai-aug 1949 [1r]; 9 oct 1945-90 [88r]. 1947: tribuene) – Dist. gw Mikrofilm – gw Misc Inst [074]

Der freie hanseat – Bremen DE, 5 sep-23 dez 1947, jan-sep 1948 – 1r – 1 – gw Misc Inst [074]

Freie hessische zeitung – Kassel DE, 1875 [gaps], 1876 feb-jun, 1877 apr-jun, dec [gaps], 1878 jan-26 mar – 2r – 1 – (filmed with suppl) – gw Misc Inst [074]

Freie juedische lehrerstimme : monatsschrift fuer die pflege der interessen des judenthums in schule und haus – Vienna. v1-9. 1912-20 [complete] – (= ser German-jewish periodicals... 1768-1945, pt 2) – 1r – 1 – $125.00 – mf#B77 – us UPA [270]

Freie juedische lehrerstimme – Wien (A), 1912-20 – 1r – 1 – gw Misc Inst [270]

Freie jugend – Berlin DE, 1919 n4-1920 n18 – 1 – gw Misc Inst [305]

Freie klaenge : gedichte / Dorschner-Lanz, Friedrich – Zwodau: Verlag "Freie Worte", 1905 – 1r – 1 – us Wisconsin U Libr [810]

Freie kunst und literatur – Paris (F), 1938 sep-1939 jul – 1r – 1 – gw Misc Inst [700]

Die freie lutherische volkskirche : der lutherischen kirche deutschlands zur pruefung und verstaendigung / Harnack, Theodosius – Erlangen: A Deichert, 1870 – 1mf – 9 – 0-524-01085-4 – mf#1990-4050 – us ATLA [242]

Die freie meinung – Duesseldorf DE, 1958-59 – 1r – 1 – gw Misc Inst [074]

Freie meinung – Bremen DE, 1919 n1-2 – 1r – 1 – gw Misc Inst [074]

Freie meinung – Duesseldorf DE, may 4 1919-dec 28 1920 – 1r – 1 – gw Misc Inst [074]

Die freie presse – Chicago: Free Press Print Co, feb 5-may 26 1872 – 1r – 1 – us CRL [071]

Die freie presse – Kassel DE, 1848 14 mar-29 jun – 1r – 1 – gw Misc Inst [074]

Freie presse – Amsterdam NL, 15 jul 1933-27 jan 1934 – 1r – 1 – gw Misc Inst [074]

Freie presse – Council Bluffs: [s.n.],jul 27 1917-19 – 2r – 1 – us CRL [074]

Freie presse – Elberfeld Barmen DE, 1918; 1925-28 – 14r – 1 – mf#3573 – gw Mikropress [074]

Freie presse – Lodz, Poland. 28 Nov 1928; 10-23 Sept 1939 – 1r – 1 – us L of C Photodup [947]

Freie presse – Berlin DE, 1952 6 jan-1961 28 oct – 3r – 1 – (oppositionelle sozialdemokratie) – gw Mikrofilm [320]

Freie presse : tageszeitung – Buenos Aires (RA), 1972-1976/77 – 15r – 1 – gw Misc Inst [079]

Freie presse – Temeschburg (Timisoara RO), 1930 15 jan-25 may – 1r – 1 – gw Misc Inst [077]

Freie presse *see* Lodzer freie presse

Freie presse and woechentliche tribuene – Omaha, NE. 4v. 29 mar 1923-25 aug 1926 (wkly) [mf ed 1924-26 lacks 29 jul 1925 filmed 1972] – 2r – 1 – (in german. formed by the union of: council bluffs beilage and: woechentliche tribuene. absorbed: carroll demokrat and: weser-nachrichten. cont by wochenblatt der omaha tribuene) – us NE Hist [071]

Freie presse fuer elsass lothringen – Strassburg (Strasbourg F), 1898 2 nov-1916, 1939 22 jan-22 aug – 1 – (filmed by misc inst: 1917-18 (gaps) [3r]. began in schittigheim) – fr ACRPP; gw Misc Inst [074]

Freie presse fuer ober-elsass – Muelhausen / Elsass (Mulhouse F), 1902-18 [gaps] – 1 – (title varies: 1 oct 1904: muelhauser volkszeitung) – fr ACRPP [074]

Freie presse fuer texas – San Antonio TX (USA), 1915 12 may, 1919 29 oct-1929, 1931-1935 9 aug, 1937 5 feb-1938 – 10r – 1 – (with gaps) – gw Misc Inst [071]

Freie presse fur texas – San Antonio: H Pollmar, [1865-]. aug 1917-jul 1918 – 1r – 1 – us CRL [071]

Freie presse [main edition] : chemnitzer zeitung – Chemnitz DE, 1964 2 apr-1967 14 feb – 6r – 1 – (publ started as regional ed of freie presse, zwickau, & became main ed in 1963, while the zwickau-ed became regional ed. today the main ed is called freie presse: chemnitzer zeitung. regional ed: zwickau 1984 2 jan-21 mar, 1984 2 jul-1988, 1992 1 jun-1996 [48r proj 1950-52]; 1952 1 apr-1962 [21r]) – gw Mikrofilm; gw Misc Inst [074]

Freie presse [main edition] – Bielefeld DE, 1958-1963 17 apr; 1966 16 jun-16 jul [small gaps] – 24r – 1 – (merged with: westfaelische zeitung to form: neue westfaelische, bielefeld. filmed by misc inst: (1958-1963 27 mar, 1963 18 apr-1966 15 jun, 1966 18 jul-1967 1 jul); (1947-57) [27r]; filmed by mikropress: 1946 3 apr-1950 [6r] order#6571. regional ed: bueren 1960-1967 1 jul; bersenbrueck, melle, wittlage 1952-1 jul 1967; buende/westf 1952 3 jun-1967 1 jul; detmold 1952-1967 1 jul; guetersloh, rheda-wiedenbrueck (ausg c = guetersloh, (rheda-) wiedenbrueck) 1952-1967 1 jul [gaps]; halle westf 1952-1967 1 jul [gaps]; hoexter 1952 3 jun-1967 1 jul; luebbecke 1952-62 [gaps], 1964-1967 1 jul; minden 1952 3 jun-1967 1 jul [gaps]; osnabrueck 1952 3 jun-1967 1 jul; paderborn 1947-1967 30 jun; warburg 1952 3 jun-1967 1 jul) – gw Mikrofilm; gw Misc Inst; gw Mikropress [074]

Freie presse staatszeitung – Fort Wayne IN (USA), oct 28 1919-jan 27 1927 – 13r – 1 – gw Misc Inst [071]

Freie presse und wochentliche tribune – Omaha. mar 29 1923-24 – us CRL [074]

Freie presse und woechentliche tribuene – Omaha NE (USA), 1923 29 mar-1926 18 aug – 2r – 1 – gw Misc Inst [071]

Freie presse von indiana – Indianapolis, IN. 1856-1880 (1) – mf#62841 – us UMI ProQuest [071]

Freie pressekorrespondenz – Muenchen DE, 1959 jul-1966 – 1 – gw Misc Inst [074]

Das freie rheinland – Duesseldorf DE, 1923 31 aug-13 oct – 1r – 1 – (title varies: wochenblatt v. rheinische landeszeitung) – gw Misc Inst [074]

Freie shriftn – Wilna. n1-17. sept 1926-oct 1935 – 1 – us NY Public [073]

Der freie staatsbuerger – Nuernberg DE, 1848 1 jul-1850 11 apr – 1 – gw Misc Inst [074]

Die freie theologie, oder, philosophie und christenthum in streit und frieden / Biedermann, Alois Emanuel – Tuebingen: LF Fues, 1844 – 1mf – 9 – 0-524-00006-9 – mf#1989-2706 – us ATLA [240]

Freie tribuene – Duesseldorf DE, 1950 aug-1951 – 1r – 1 – gw Misc Inst [074]

Freie tribuene – London (GB), 1944-1946 27 jul – 1r – 1 – gw Misc Inst [072]

Freie Vereinigung Deutscher Gewerkschaften *see* Protokoll ueber die verhandlungen vom kongress..

Das freie volk : demokratisches wochenblatt / ed by Breitscheid, Rudolf – Berlin DE, 1910-1914 8 aug – 2r – 1 – mf#5494 – gw Mikropress [074]

Freie volksblaetter – Koeln DE, 1848 12 apr-1849 7 jan – 1r – 1 – (title varies: 29 oct 1848: freie blaetter (koeln-) muelheim, fr oct 1848 in koeln) – gw Mikrofilm [074]

Freie volkstimme – Zabreh, Czechoslovakia. Aug 1906-1908 – 1r – 1 – us L of C Photodup [077]

Die freie welt : illustrierte wochenschrift der uspd – Berlin DE, 1919-1922 n36 – 1r – 1 – (title varies: 1920: freie welt; also suppl to: freiheit) – gw Misc Inst [074]

Freie welt – 1970-1983 – 1,134mf – 1 – gw Mikropress [320]

Freie wohlfahrtspflege *see* Mitteilungen des reichsverbandes der privaten gemeinnuetzigen kranken- und pflegeanstalten deutschlands

Das freie wort – Bern (CH), 1917-18 [gaps] – 1r – 1 – gw Misc Inst [074]

Das freie wort – Bielitz-Biala (Bielsko-Biala PL), jan-jul 1927 – 1r – 1 – gw Misc Inst [077]

Das freie wort – Bonn DE, 1959-60 (ausg c); 4 jan-27 jun 1964 (ausg l) – 2r – 1 – (filmed by misc inst: oct 1 1960-jun 27 1964) – gw Mikrofilm; gw Misc Inst [074]

Das freie wort – Montevideo (ROU), 1943 jan/feb – 1 – gw Misc Inst [079]

Das freie wort – Rostock DE, 1919 1 mar-1933 12 may – 33r – 1 – (originally in schwerin) – gw Misc Inst [074]

Das freie wort – Reschitza (Resita RO), 1932-33 – 1 – gw Misc Inst [077]

Das freie wort – Essen DE, 8 jan 1919-22; 4 jan 1925-7 aug 1927 – 2r – 1 – (with gaps) – gw Mikrofilm [074]

Das freie wort : zeitung der deutschen kriegsgefangenen in der sowjetunion – Moskau [RUS], 1945 16 aug-27 sep – 1 – (title varies: 19 jul 1943: freies deutschland: organ des nationalkomitees "freies deutschland"; filmed by misc inst: 1941 nov-1945 4 nov [2r]; missing: 1942 n26 & 32) – uk British Libr Newspaper; gw Misc Inst [077]

Das freie wort *see* Freies wort

Freie Wort / Freie Gemeinde [Milwaukee, WI] – v1 n3-v10 n11 [1932 apr-1941 dec] – 1r – 1 – (cont by: voice of freedom [milwaukee, wis]) – mf#3504785 – us WHS [242]

Die freie zeitung – Bern (CH), 1917 4 apr-1920 27 mar – 1r – 1 – gw Misc Inst [074]

Freie zionistische blaetter / ed by Klatzkin, Jakob & Goldmann, Nachum – Heidelberg. n1-4. 1921 [complete] – (= ser German-jewish periodicals...1768-1945, pt 3) – 1r – 1 – $165.00 – mf#B79 – us UPA [270]

Die freien bauern : erzaehlung aus dem norwegischen volksleben / Muegge, Theodor – Berlin: C Flemming und C T Wiskott, [1924?] – 1r – 1 – us Wisconsin U Libr [390]

Die freien rhythmen in der deutschen lyrik : versuch einer uebersichtlichen zusammenfassung ihrer entwicklungsgeschichtlichen eigengesetzlichkeit / Closs, August – Bern: A Francke A G Verlag, c1947 – 198p – 1 – (incl bibl ref) – us Wisconsin U Libr [430]

Freienwalder blick – Bad Freienwalde DE, 1963 8 may-1964 5 dec – 1r – 1 – gw Misc Inst [074]

Freier Deutscher Gewerkschaftsbund. Berlin *see* Geschaeftsbericht

Freier deutscher kulturbund in schweden : mitteilungsblatt des fdkb – Stockholm (S), 1945 feb-1946 apr – 1r – 1 – gw Misc Inst [074]

Freier geist zwischen oder und elbe : dokumente des widerstandes seit 1945 in vers und prosa / ed by Kongress fuer die Freiheit der Kultur – Darmstadt: Montana Verlag, 1954 – 172p – 1 – us Wisconsin U Libr [800]

Freies blatt : organ zur abwehr des antisemitismus – Vienna: Ernst Viktor Zenker. v1-5? 1892-96? [complete] – (= ser German-jewish periodicals...1768-1945, pt 1) – 1r – 1 – $125.00 – mf#B80 – us UPA [939]

Freies blatt – Wien (A), 1892 apr-1896 jun – 1r – 1 – gw Misc Inst [074]

Freies deutschland – Moscow. USSR. -w. 3 Jan 1943-14 May, 16 Aug-27 Sep 1945. (1 reel) – 1 – uk British Libr Newspaper [072]

Freies deutschland : organ der deutschen opposition – Antwerpen & Creil (Oise) B, 1937 14 jan-1939 24 aug – 1r – 1 – gw Misc Inst [074]

Freies deutschland : organ im sinne des nationalkomitees "freies deutschland" – Zuerich (CH), 1943 3 sep-1946 jan – 1r – 1 – gw Misc Inst [074]

Freies deutschland : revista antinazi/antinazi monthly – Mexiko-Stadt (MEX), 1941 nov-1946 jun – 2r – 1 – (cont by: neues deutschland (nueva alemania), 1946. incl suppl: alemania libre 1942 24 jan-1943 1 aug [gaps]) – gw Misc Inst [079]

Freies deutschland – Santiago de Chile (RCH), 1944 apr, aug – 1 – gw Misc Inst [079]

Freies deutschland – Uppsala, Stockholm (S), 1944 may – 1r – 1 – gw Misc Inst [074]

Freies deutschland *see* Das freie wort

Freies deutschland im bild – Moskau (RUS), 1944-1945 jan – 1 – gw Misc Inst [077]

Freies europa – Duesseldorf DE, 1949 – 1r – 1 – gw Misc Inst [074]

Freies heim – Zagreb, Yugoslavia. Oct 1923-1924 – 1r – 1 – us L of C Photodup [949]

Freies hessisches volksblatt – Darmstadt DE, 1848 12 mar-1849 22 sep – 1r – 1 – gw Misc Inst [074]

Freies volk – Duesseldorf. Germany. -d. 21 Nov 1949-2 Jan 1952. (7 reels) – 1 – uk British Libr Newspaper [072]

Freies volk : zentralorgan der kpd – Duesseldorf, 1949-56 – 17r – 1 – gw Mikropress [074]

Freies volk / d *see* Der kampf

Freies wort – Suhl DE, 1954 2 aug-1990 – 75r – 1 – (title varies: 8 mar 1956: das freie wort. filmed by other misc inst: 1992-) – gw Misc Inst [074]

Die freifrauen : ein possenspiel in drei aufzuegen / Hoellrigl, Franz – Leipzig: August Schulze, 1896 – 1 – us Wisconsin U Libr [820]

Freigeist – Liberec, Czechoslovakia. sept 1921-1924; oct 1929-sept 1938 – 9r – 1 – us L of C Photodup [077]

Freight management international – London. 1977-1989 (1) 1977-1989 (5) 1977-1989 (9) – mf#8451 – us UMI ProQuest [380]

Der freihafen – (Hamburg-) Altona DE, 1838-1844 – 1 – gw Misc Inst [074]

Die freiheit – Braunschweig DE, 1919 1 oct-1922 31 oct – 4r – 1 – mf#4769 – gw Mikropress [074]

Die freiheit – Temeschburg (Timisoara RO), 1945-48 – 1 – gw Misc Inst [077]

Die freiheit – Berlin DE, 1918 15 nov-1922 30 sep – 10r – 1 – (title varies: 1 mar 1919: freiheit) – mf#178 – gw Mikropress [074]

Freiheit – 1879 jan 4-1881 jan 22, 1884-1897 [scattered iss], 1900 jan 20-1901 dec, 1902 jan-1904 may 14, 1904 may 21-1910 aug 13 – 3r – 1 – mf#1056623 – us WHS [071]

Freiheit – London (GB), Chicago IL (USA), New York NY (USA), 1879 4 jan-1907 7 dec [many iss missing] – 1r – 1 – (1885 n27: freiheit / amerikanische ausg ed by john most) – gw Misc Inst [072]

Freiheit – Breslau (Wroclaw PL), 1929 jan 10-feb 7 – 1r – 1 – gw Misc Inst [077]

Freiheit – Halle DE, 1946 5 jul-1947 [gaps] – 2r – 1 – mf#6516 – gw Mikropress [074]

Freiheit – Halle, Germany. 1950-Feb 1953; 1962-Apr 1976 – 47r – 1 – us L of C Photodup [074]

Freiheit – Hanau DE, 1919 3 jun-1922 30 sep – 1 – gw Misc Inst [074]

Freiheit – Koenigsberg (Kaliningrad RUS), 1919 2 apr-31 dec, 1921-1922 30 sep – 4r – 1 – gw Misc Inst [077]

Freiheit – New York (etc.). jahrg. 5-31. 1883-1909 – 1 – us NY Public [073]

Freiheit – New York NY. 1890 nov 1-1901 dec 28, 1908 jan 4-1910 aug 13 – 2r – 1 – mf#3423496 – us WHS [071]

Freiheit : niederrheinische tageszeitung der kpd – Duesseldorf DE, 1921-1933 18 feb – 27r – 1 – mf#3907 – gw Mikropress [074]

Freiheit : organ des arbeitenden – Linz, Austria. 8 feb 1946-7 feb 1948 – 1r – 1 – uk British Libr Newspaper [072]

Freiheit – Teplice-Sanov, Czechoslovakia. 1937 – 1r – 1 – us L of C Photodup [077]

Freiheit – Halle S DE, 1950 10 may-1951 20 jun, 1953 7 nov-31 dec, 1967 14 feb – 3r – 1 – (title varies: until 16 jul 1958: mitteldeutsche tageszeitung "freiheit"; 17 mar 1990: mitteldeutsche zeitung. regional ed available (nur kreisseiten): artern, apr 1 1958-oct 31 1961, may 17 1967-nov 12 1969 [gaps]; aschersleben, apr 2 1958-oct 31 1961, mar 4-dec 2 1967, 1969 [gaps], sep 20 1985-90; bernburg, apr 2 1958-61, 1967, 1969 [gaps]; bitterfeld, apr 1 1958-oct 31 1961, feb 27 1967-dec 5 1970 [gaps], jul 19 1983-90; dessau, apr 12 1958-jun 9 1972, jul 19 1983-90 (very scattered); eisleben, apr 1 1958-oct 31 1961, may 19 1988-dec 14 1990; graeffenhainichen, apr 1 1958-oct 31 1961; hohenmoelsen, apr 1 1958-oct 31 1961; koethen, mar 31 1958-nov 20 1964 [gaps]; merseburg, apr 3 1958-apr 10 1970 [gaps], jul 3 1984-90; naumburg, apr 1 1958-oct 31 1961, may 25 1964-sep 17 1966 [gaps], may 19 1988-dec 17 1990; nebra, apr 1 1958-61, apr 11 1964-jun 30 1965 [gaps]; quedlinburg, apr 1 1958-jul 3 1965 [gaps], jan 30 1967-68 [gaps], sep 19 1985-90; querfurt, apr 2 1958-oct 31 1961; rosslau, apr 2 1958-oct 31 1961, jan 7-nov 20 1963; saalkreis, apr 2 1958-66 [gaps]; weissenfels, apr 1 1958-dec 18 1969 [gaps]; wittenberg, apr 8 1958-apr 7 1970 [gaps]; zeitz, apr 1 1958-dec 4 1965 [gaps]. filmed by misc inst: 1946 16 apr-1975 [gaps], 1991-94; 1992- [7r/yr]; 1946 5 jul-1947, 1953 mar-1990 [88r]) – gw Misc Inst [074]

Freiheit *see*
- Der kampf
- Rote tribuene

Freiheit, autoritaet und kirche : eroerterungen ueber die grossen probleme der gegenwart / Ketteler, Wilhelm Emmanuel, Freiherr von – 3. Aufl. Mainz: F Kirchheim, 1862 – 1mf – 9 – 0-7905-7309-1 – (incl bibl ref) – mf#1989-0534 – us ATLA [240]

Freiheit, bruederlichkeit, arbeit *see* Zeitung des arbeiter-vereins zu koeln

Der freiheit eine gasse : aus dem leben und werk georg herweghs / ed by Kaiser, Bruno – Berlin: Verlag Volk & Welt 1948 [mf ed 1995] – 1r – 1 – (incl bibl ref; filmed with: herderbuch: reisejournal / j loeber [ed]) – mf#3636p – us Wisconsin U Libr [800]

Freiheit und brot – Muenchen, DE. Jan-Dec 1933 – 1 – gw Mikropress [074]

Freiheit und recht : eine auswahl aus seinen werken mit einer biographie / Seume, Johann Gottfried – Hildesheim: Verlag Jugend und Volk, [1947?] – 1 – (incl bibl ref) – us Wisconsin U Libr [943]

Freiheit und recht *see* Franz von sickingens fehde gegen trier

Die freiheit und unabhaengigkeit der kirche / Schneemann, G – Freiburg im Breisgau: Herder, 1867 – (= ser Die Encyclica Papst Pius' 9. Vom 8. Dezember 1864) – 1mf – 9 – 0-8370-8304-4 – (incl bibl ref) – mf#1986-2304 – us ATLA [240]

Freiheit-korrespondenz – Muelhausen / Elsass (Mulhouse F), 1933 may-1940 6 mar – 3r – 1 – mf#11005 – gw Mikropress [074]

Freiheits freund – Pittsburgh, PA. 1853-1900 (1) – mf#66038 – us UMI ProQuest [071]

Freiheitsbund deutscher sozialisten : londoner arbeitskreis des freiheitsbundes deutscher sozialisten – London (GB), 1942-47 – 1r – 1 – gw Misc Inst [335]

Der freiheitskampf – Dresden DE, aug 1 1930-apr 1936, jul 1936-feb 1937, may 1937-44, feb-may 8 1945 – 70r – 1 – gw Misc Inst [074]

Der freiheitskampf *see* Budissinische woechentliche nachrichten

Der freiin annette elisabeth von droste-huelshoff gesammelte werke / ed by Droste-Huelshoff, Elisabeth, Freiin von – Muenster: F Schoeningh, 1885-1901 [mf ed 1989] – 4v in 5 [ill] – 1r – 1 – (with biogr int & ann) – mf#7187 – us Wisconsin U Libr [802]

Freikugeln : satirische zeitschrift – Leipzig DE, 1842, 1846 – 1r – 1 – gw Misc Inst [870]

Freiligrath : eine erscheinung aus der stilgeschichte / Klein, Georgette – [S.l.: s.n.], 1919 (Zuerich: Diss-Druckerei Leemann) (mf ed 1990) – 1r – 1 – (filmed with: ferdinand freiligrath) – us Wisconsin U Libr [430]

Freiligrath, Ferdinand *see*
- Die akten ferdinand freiligrath und georg herwegh
- Ferdinand freiligrath
- Ferdinand freiligrath's gesammelte dichtungen
- Ferdinand freiligraths saemtliche werke in zehn bznden
- Gedichte
- Ein glaubensbekenntnis
- Neue gedichte
- Neuere politische und sociale gedichte
- Wir sind die kraft
- Zwischen den garben

Freiligraths einfluss auf die lyriker der muenchener dichterschule / Hallermann, Josef – [S.l.: s.n.], 1917 (Essen: Druck von Fredebeul & Koenen) (mf ed 1990) – 1r – 1 – (filmed with: ferdinand freiligrath) – us Wisconsin U Libr [430]

Freiligraths uebersetzungen englischer dichtungen / Roeschen, Friedrich August – Giessen: Im Selbstverlag des Englischen Seminars der Universit"t Giessen, 1923 (mf ed 1990) – 1r – 1 – (filmed with: ferdinand freiligrath) – us Wisconsin U Libr [430]

Freiling, Howard P *see* An analysis of the factors that influence fan attendance at minor league baseball games

Freimann, Jacob *see* Des gregorius abulfarag, gen bar-hebraeus, scholien zum buche daniel

Der freimaurer! : neue beitraege zur kritik des logenlebens, seiner freunde und feinde / Conrad, Michael Georg – Leipzig: O Heinrichs 1885 – 3mf – 9 – mf#vrl-53 – ne IDC [366]

Die freimaurerei : ein beitrag zur geschichte der politischen geheimbuende / Walther, Hugo – 1. aufl. Wien: J Roller 1910 – 4mf – 9 – mf#vrl-204 – ne IDC [366]

[Freimaurerei] bundesgesetze der grossen national-mutterloge 'zu den drei weltkugeln' : als handschrift gedruckt fuer br freimaurer – Berlin 1928 [mf ed 1992] – 2mf – 9 – gw Frankfurter [943]

Die freimaurerei in ihrem zusammenhang : mit den religionen der alten aegypter, der juden und der christen / Acerrellos, R S – 2. aufl. Leipzig: J J Weber 1836 – 4v in 2 on 16mf – 9 – mf#vrl-35 – ne IDC [366]

Die freimaurerei und das evangelische pfarramt : aus der evangelischen kirchenzeitung – Berlin: Gustav Schlawitz 1854-55 – 3v on 2mf – 9 – mf#vrl-186 – ne IDC [366]

Freimaurer-zeitung : manuscript fuer brueder – 1847-1919 [mf ed 2002] – 73v on 352mf – 9 – €1800 diazo €2160 silver – 3-89131-378-0 – gw Fischer [366]

Freimaurerzeitung *see* Wochenblatt fuer freunde der weisheit und litteratur

Der freimuethige – 1803-11 [mf ed 1997] – (= ser Die zeitschriften des august von kotzebue) – 50mf – 9 – €620.00 – 3-89131-231-8 – gw Fischer [430]

Der freimuethige : oder berlinisches unterhaltungsblatt fuer gebildete, unbefangene leser – Berlin DE, 1810-11, 1815 – 2r – 1 – (several title changes) – gw Misc Inst [074]

Freimuethiges abendblatt – Schwerin DE, 1818 9 jan-1849 29 jun – 10r – 1 – gw Misc Inst [074]

Der freimuetige – Temeschburg (Timisoara RO), 1915 9 dec-1918 18 jul [gaps] – 2r – 1 – gw Misc Inst [077]

Der freimuetige an der haar – Werl DE, 1849-50 – 1 – gw Misc Inst [074]

Freire, Felisbello *see* Historia constitucional da republica dos estados...

Freire, Josue Justiniano *see* Odyssea do 12 regimento

Freire, Laudelino De Oliveira *see* Sonetos brasileiros, seculo 17-20

Freischaerler-reminiscenzen : zwoelf gedichte / Aston, Louise – Leipzig: E O Weller, 1850 [mf ed 1989] – 27p – 1 – mf#6979 – us Wisconsin U Libr [810]

Der freischuetz *see* Gemeinnuetzige unterhaltungs-blaetter

Freise, Otto *see*
- Die drei fassungen von wielands agathon

Freisen, J *see* Liber agendarum ecclesie et diocesis sleszwicensis

Freisen, Joseph *see* Geschichte des canonischen eherechts

Freisinger tagblatt – Freising DE, 1952 3 nov-1968 30 apr – 42r – 1 – (since 23 apr 1968: bezirksausgabe von muenchner merkur, muenchen. filmed by misc inst: 1981- [15r/yr]) – gw Mikrofilm; gw Misc Inst [074]

Der freisinnige – Freiburg Br DE, 1832 1 mar-25 jul – 1 – gw Misc Inst [074]

Freisinnige zeitung – Berlin DE, 1885 20 aug [probe-nr], 1889 5 nov – 1r – 1 – (filmed by mikropress: 1899-1906 [19r] order#4692; filmed by misc inst: 1885 20, 27 aug, 1885 1 sep-1887 aug, 1888-1891 apr, 1891 sep-1897 apr, 1897 sep-1898. title varies: 15 mar 1904-30 jun 1906: freie deutsche presse) – gw Mikrofilm; gw Mikropress; gw Misc Inst [074]

Freisleben, Hans-Joachim *see* Reinigung der mitochondrialen atp-synthase aus rinderherzen funktionelle rekonstitution und rekoppelung synthetisierender f1-partikel an den membranintetralen f0-teil neue medizinische bibliothek

Freistaedter kreisblatt – Freistadt (Kozuchow PL), 1834-39, 1846-47, 1849, 1853-54, 1857, 1864, 1867 – 1 – (title varies: 25 may 1839: kreis-wochenblatt fuer freistadt und neusalz; 25 may 1846: kreis-wochenblatt fuer den gesammten freistaedter kreis; 13 apr 1864: freistaedter wochenblatt fuer stadt und land) – gw Misc Inst [077]

Freistaedter wochenblatt fuer stadt und land *see* Freistaedter kreisblatt

Freit, Lori K *see* A psychosocial assessment on the effects of different evaluation methods on women with a first-time ascus pap smear

Freitag – Berlin DE, 1991- – 1 – (filmed by misc inst: 1990 9 sep-21 dec [bei sonntag mitverfilmt]) – gw Mikropress; gw Misc Inst [074]

Freitag, A *see*
- Architectura militaris...
- Mythologia ethica

Freitag, Britta *see* Tourismusentwicklung in nizza

Der freitagabend : eine familienschrift – Frankfurt a.Main: Franz Benjamin Auffarth. v1. 1859 [complete] – (= ser German-jewish periodicals... 1768-1945, pt 3) – 1r – 1 – $165.00 – mf#B82 – us UPA [939]

Freitagskind : roman / Flake, Otto – Berlin: S Fischer, 1919 – 1r – 1 – us Wisconsin U Libr [830]

Freitas, Affonso Antonio De *see* Vocabulario nheengatu

Freitas, Caio De *see* George canning e o brasil

Freitas, Joao De *see* Umbanda

Freitas, Newton *see*
- Ensaios americanos
- Ensayos americanos (critica literaria)

Freitas, Octavio *see* Domencas africana no brasil

Der freiwillige feuertod in indien und die somaweihe / Hillebrandt, Alfred – Muenchen: Koeniglich Bayerische Akademie der Wissenschaften, 1917 – (= ser Sitzungsberichte der philosophisch-philologischen und der historischen classe der k.b. akademie der wissenschaften zu muenchen) – 1mf – 9 – 0-524-01605-4 – mf#1990-2544 – us ATLA [280]

Der freiwirt – Koeln DE, 1931-33 [gaps] – 1r – 1 – gw Misc Inst [640]

Freiwirtschaft durch freiland und freigeld – Leipzig DE, 1919-27, 1929-33 – 1 – gw Misc Inst [330]

Freiwirtschaftliche zeitung – Hamburg DE, 1924-25 [gaps] – 1 – gw Misc Inst [330]

Freiwirtschaftliche zeitung – Schwarzenburg (CH), 1924 n20-1926 – 1 – gw Misc Inst [330]

Freke, Henry *see* On the origin of species by means of organic affinity

Frelimo information – Alger: Representation en Algerie du Front de liberation du Mozambique, jul 1968?; jun 1970/may 1971 – us CRL [079]

Frelinghuysen, Theodorus Jacobus *see* Sermons

Fremantle, John Morton *see* Gazetteer of muri province, up to december 1919

Fremantle, Stephen James *see* The state of morals and of society in the eastern church in the time of s. chrysostom

Fremantle, William H *see* A collection of the judgments of the judicial committee of the privy council

Fremantle, William Henry *see*
- The gospel of the secular life
- Influence of commerce upon christianity
- Natural christianity

Fremaux, Paul *see* Der sterbende napoleon

Fremde kultur – fremdes geschlecht : studie zur literarischen verarbeitung von fremderfahrungen in frankophonen romanen der letzten beiden jahrzehnte / Mayer, Friederike – Mainz: Gardez, 1995 (mf ed 1995) – (= ser Komparatistik im Gardez) – 4mf – 9 – €45.00 – 3-8267-9666-7 – mf#DHS 9666 – gw Frankfurter [410]

Fremde und fremdes : erotische begegnungen / Kassimis, Johanna – (mf ed 2001) – 416p – 9 – €59.00 – 3-8267-2758-4 – mf#DHS 2758 – gw Frankfurter [410]

Die fremden : ein roman aus der gegenwart / Domanig, Karl – 3., verb. Aufl. Klagenfurt: Verlag der St Josefbuecherbruderschaft, 1911 – 1 – us Wisconsin U Libr [830]

Die fremden : ein roman aus der gegenwart / Domanig, Karl – Klagenfurt: Verlag der St Josefbuecherbruderschaft, 1911 – 1r – 1 – us Wisconsin U Libr [830]

Fremden-blatt – Wien (A), 1915 jun-1919 mar [gaps] – 22r – 1 – (filmed by misc inst: 1864 apr-jun, 1881 apr-jun [3r]; filmed with suppls) – uk British Libr Newspaper; gw Misc Inst [074]

Fremdenblatt – Vienna. jan 1848-dec 1919 – 268r – 1 – us UMI ProQuest [074]

Fremdenbuch fuer heidelberg und die umgegend / Leonhard, Karl C von – Heidelberg 1834 [mf ed Hildesheim 1995-98] – 3mf – 9 – €90.00 – 3-487-29470-2 – gw Olms [914]

Fremden-verkehrs-zeitung – Kassel DE, 1904 14 may-1907 29 sep, 1908 9 may-1913 28 jun – 3r – 1 – (later: casseler fremden-verkehrs-zeitung) – gw Misc Inst [338]

Fremdverstehen : eine untersuchung ueber das verhaeltnis von eigenem und fremdem im hinblick auf bedingungen und moeglichkeiten des verstehens / Hammerschmidt, Anette C – (mf ed 1995) – 4mf – 9 – €56.00 – 3-8267-2152-7 – mf#DHS 2152 – gw Frankfurter [400]

Fremdwoerter im griechischen und lateinischen / Vanicek, Alois – Leipzig: B G Teubner, 1878 – 1mf – 9 – 0-8370-9325-2 – mf#1986-3325 – us ATLA [450]

Fremdwoerterbuch : ein handweiser zur entwelschung fuer amt, schule, haus, leben / Engel, Eduard – Leipzig, 1922 (mf ed 1994) – 2mf – 9 – €31.00 – 3-89349-781-1 – mf#DHS-AR 781 – gw Frankfurter [040]

Das fremdwort bei grimmelshausen : ein beitrag zur fremdworterfrage des 17. jahrhunderts / Hechtenberg, Klara – [s.l: s.n.] 1901 [mf ed 1990] – 1r – 1 – (incl bibl ref. filmed with: friedrich melchior grimm als kritiker.../ karl august georges) – mf#2692p – us Wisconsin U Libr [430]

Das fremdwort bei theodor fontane (briefe, grete minde, l'adultera, irrungen, wirrungen) : ein beitrag zur charakteristik des modernen realistischen romans / Schultz, Albin – [S.l.: s.n.], 1912 (Greifswald: Druck von J Abel) [mf ed 1989] – 116p – 1 – mf#7075 – us Wisconsin U Libr [430]

Fremont d'Ablancourt, Nicolas *see* Suite du neptune francois

Fremont daily herald – Fremont, NE: Smails & Toncray, 1874-v39 n181. jul 31 1910 (daily ex mon) [mf ed 1877-1910 (gaps) filmed 1976] – 18r – 1 – (cont: fremont herald (daily). cont by: fremont herald (weekly 1910). numbering very irregular. one issue no. assigned for each wk's issues, 1877-jul 22 1894) – us NE Hist [071]

Fremont daily tribune – Fremont, NE: Hammond Bros, may 1883-sep 28 1904 (daily ex sun) [mf ed 1887,1890-1904 (gaps) filmed 1972] – 14r – 1 – (cont by: fremont evening tribune) – us NE Hist [071]

Fremont daily tribune – Fremont, NE: Hammond Print Co. v54 n243. feb 24 1937-75th yr n62. jul 25 1942 (daily ex sun) [mf ed lacks aug 30 1937 and jan 24 1938 filmed 1972] – 1r – 1 – (cont: fremont evening tribune. merged with: fremont morning guide to form: fremont guide and tribune. v56 n88-71st yr n87 not publ) – us NE Hist [071]

Fremont evening tribune – Fremont, NE: Hammond Print Co, sep 29 1904-v54 n242. feb 23 1937 (daily ex sun) [mf ed with gaps filmed 1972] – 53r – 1 – (cont: fremont daily tribune. absorbed: fremont tri-weekly tribune. cont by: fremont daily tribune (1937). numbering began with v39 n1 may 14 1921) – us NE Hist [071]

Fremont guide and tribune – Fremont, NE: Fremont Newspapers, Inc. 22v. 75th yr n63. jul 27 1942-96th yr n137. oct 19 1963 (daily ex sun) [mf ed with gaps filmed 1972] – 79r – 1 – (formed by the union of: fremont morning guide and: fremont daily tribune (1937). cont by: fremont tribune (1963). cont the numbering of: fremont daily tribune (1937)) – us NE Hist [071]

Fremont herald – Fremont, NE: R D Kelly, 1871-74// (wkly) [mf ed v1 n26. jan 24 1872] – 1r – 1 – (cont by: fremont weekly herald) – us NE Hist [071]

Fremont herald – Fremont, NE: Herald Co. v39 n1. aug 5 1910-v59 n14. oct 27 1928 (wkly) [mf ed with gaps filmed 1976] – 8r – 1 – (cont: fremont daily herald. vol numbering irregular: v43 repeated, v46 omitted in numbering. v57 n1-v58 not publ) – us NE Hist [071]

Fremont herald-leader : [tri-weekly edition] – Fremont, NE: Dodge County Pub Co. n86. feb 27 1904 (3 times/wk) – 1r – 1 – (cont: fremont tri-weekly herald-leader. cont by: fremont tri-weekly herald. no more publ) – us NE Hist [071]

Fremont, J C *see* Narrative of the exploring expedition to the rocky mountains...and to oregon and north carolina...

Fremont, Jessie Benton *see* Memoirs of my life

Fremont, John Charles *see*
- The exploring expedition to the rocky mountains, oregon and california
- Die felsengebirge oregon und nordcalifornien
- Geographical memoir upon upper california
- Memoirs of my life
- Narrative of the exploring expedition to the rocky mountains in the year 1842
- Reise nach dem felsengebirge im jahre 1842
- Reisen durch die vereinigten staaten von nordamerica nebst einem ausfluge nach canada
- Report of the exploring expedition to the rocky mountains in the year 1842
- A report on an exploration of the country lying between the missouri river and the rocky mountains

Fremont, Joseph *see* Le divorce et la separation de corps

Fremont morning eagle – Fremont, NE: Fremont Tribune. v1 n1. dec 2 1940 (daily ex sun) [mf ed -mar 31 1941 filmed 1999] – 2r – 1 – us NE Hist [071]

Fremont morning guide – Fremont, NE: Morning Guide, apr 21 1933-v9 n260. jul 25 1942 daily ex sun & mon) [mf ed 1939-42 (gaps) filmed 1972] – 8r – 1 – (merged with: fremont daily tribune (1937) to form: fremont guide and tribune) – us NE Hist [071]

[Fremont-] news-register – CA. 1955-72 [daily] – 65r – 1 – $3900.00 – mf#B02250 – us Library Micro [071]

[Fremont-] niles township register – CA. Apr 1927-1954 – 14r – 1 – $840.00 – mf#B02251 – us Library Micro [071]

[Fremont-] ohlone college monitor – CA: ohlone college, 1970-71; 1973-80 – 6r – 1 – $360.00 – mf#B02252 – us Library Micro [378]

Fremont semi-weekly herald – Fremont, NE: N W Smails, jul 2 1897-jul 24 1903 (semiwkly) [mf ed with gaps] – 5r – 1 – (formed by the union of: fremont weekly herald (friday's ed 1892) and: fremont weekly herald (tuesday's ed 1892). merged with: tri-weekly leader to form: fremont tri-weekly herald-leader. issues for jul 2-16 1897 called 27th yr. issues for jul 20 1897-dec 29 1899 called v28 n6-260. issues for jan 2 1900-jun 29 1900 called v28 n261) – us NE Hist [071]

Fremont semi-weekly tribune – Fremont, NE: Hammond Bros. 2v. v23 n27. jan 6 1891-v24 n85. apr 23 1892 (semiwkly) – 2r – 1 – (cont: fremont weekly tribune (1883); cont by: fremont tri-weekly tribune) – us Bell [071]

[Fremont-] the argus – CA. 1960-67; 1967- – 513r – 1 – $30,780.00 (subs $600y) – mf#B02249 – us Library Micro [071]

Fremont tribune – Fremont, NE: J N Hays. v1 n1. jul 24 1868-77// (wkly) – 2r – 1 – (cont by: fremont weekly tribune) – us Bell [071]

Fremont tribune – Fremont, NE: Lester A Walker. 96th yr n138. oct 21 1963- (daily ex sun) – 95r – 1 – (cont: fremont guide and tribune) – us Bell [071]

Fremont tribune – Fremont, NE: Lester A Walker, 96th yr n138. oct 21 1963- (daily ex sun) [mf ed -1968 (lacks oct 13 1964, jul 22 1964, jan 31 1966)] – 24r – 1 – (cont: fremont guide and tribune) – us NE Hist [071]

Fremont tribune – Fremont, NE: Lester A Walker, 96th yr n138 oct 21 1963)- (daily ex sun) [mf ed 1986] – 1 – (cont: fremont guide and tribune) – us Misc Inst [071]

Fremont tribune – Fremont, NE: Hammond Brothers. 2v. v15 n1. jul 14 1882-v16 n12. sep 27 1883 (wkly) – 2r – 1 – (cont: fremont weekly tribune; cont by: fremont weekly tribune (1883)) – us Bell [071]

Fremont tri-weekly herald – Fremont, NE: Dodge County Pub Co. n87. mar 1 1904-05// (3 times/wk) [mf ed -apr 11 1905 (gaps)] – 2r – 1 – (cont: fremont herald-leader. issue for apr 7 1904 called n224 but constitutes n105. issues for jun 28 1904-apr 11 1905 called v32 n143-v34 n90) – us NE Hist [071]

Fremont tri-weekly herald-leader – Fremont, NE: Dodge County Pub Co, jul 28 1903-v1 n84. feb 25 1904 (3 times/wk) – 2r – 1 – (formed by the union of: fremont semi-weekly herald and: tri-weekly leader. cont by: fremont herald-leader. numbering begins with v1 n10 aug 18 1903) – us NE Hist [071]

Fremont tri-weekly tribune – Fremont, NE: Hammond Bros. 27v. v24 n86. apr 26 1892-v50 n39. aug 4 1917 (3 times per wk) – 17r – 1 – (cont: fremont semi-weekly tribune. absorbed by: fremont evening tribune. daily ed: fremont daily tribune 1892-1904 and: fremont evening tribune 1904-17) – us Misc Inst [071]

Fremont tri-weekly tribune – Fremont, NE: Hammond Bros. 27v. v24 n86. apr 26 1892-v50 n39. aug 4 1917 (3 times per wk) [mf ed 1954] – 17r – 1 – (cont: fremont semi-weekly tribune; absorbed by: fremont evening tribune; daily ed: fremont daily tribune 1892-1904 and: fremont evening tribune 1904-17) – us Bell [071]

[Fremont-] washington press – CA. 1909-1910 – 1r – 1 – $60.00 – mf#B02253 – us Library Micro [071]

Fremont weekly herald – Fremont, NE: R D Kelly, 1874-v22 n18. nov 10 1892 (wkly) [mf ed 1876-92 (gaps)] – 7r – 1 – (cont: fremont herald (weekly). split into: fremont weekly herald (friday's ed 1892) and: fremont weekly herald (tuesday's ed 1892)) – us NE Hist [071]

Fremont weekly herald : [friday's edition] – Fremont, NE: N W Smails. v22 n18. nov 12 1892-jun 29 1897 (wkly) [mf ed lacks feb 9 1894, nov 27 1896] – 5r – 1 – (cont: fremont weekly herald. merged with: fremont weekly herald (tuesday's ed 1892) to form: fremont semi-weekly herald) – us NE Hist [071]
Fremont weekly herald : [tuesday's edition] – Fremont, NE: N W Smails. v22 n19. nov 15 1892-jun 25 1897 (wkly) [mf ed lacks feb 12 1895, may 28 1895, dec 29 1896)] – 5r – 1 – (cont: fremont weekly herald. merged with: fremont weekly herald (friday's ed) to form: fremont semi-weekly herald) – us NE Hist [071]
Fremont weekly tribune – Fremont, NE: [Hammond Bros] 8v. v16 n13. oct 4 1883-v23 n26. jan 1 1891 (wkly) [mf ed 1954] – 3r – 1 – (cont: fremont tribune (1882); cont by: fremont semi-weekly tribune. daily ed: fremont daily tribune 1883-91) – us Bell [071]
Fremont weekly tribune – Fremont, NE: W H Michael and Fred Nye, 1877-v14 n52. jul 6 1882 (wkly) – 1r – 1 – (cont: fremont tribune; cont by: fremont tribune (1882)) – us Bell [071]
Fremont/hayward – 1992- – (= ser California telephone directory coll) – 6r – 1 – $300.00 – mf#P00028 – us Library Micro [917]
Fremskridsforening et al *see* Dagslyset
French, Alfred J *see* Life, light, and love
French and english in canada and across the sea / Herridge, William Thomas – [Ottawa?: s.n.], 1917 – 1mf – 9 – 0-665-86518-X – mf#86518 – cn Canadiana [240]
French and indian war orderly books at the massachusetts historical society – 1755-1763 [mf ed 1992] – 1r – 1 – (with p/g) – us MA Hist [355]
French architects and sculptors of the 18th century / Dilke, Emilia Frances (Strong) – London 1900 – (= ser 19th c art & architecture) – 4mf – 9 – mf#4.2.310 – uk Chadwyck [700]
French art and english morals / Trevor, John – London [1886] – (= ser 19th c art & architecture) – 1mf – 9 – mf#4.2.230 – uk Chadwyck [700]
French at kilwa island / Freeman-Grenville, Greville Stewart Parker – Oxford, England. 1965 – 1r – us UF Libraries [960]
French Battery and Carbon Co *see* Ray-o-lite news
French Battery & Carbon Co *see* French flasher
French Battery Co *see* Ray-o-vac news
French bibles – 13th, 14th c – (= ser Holkham library manuscript books 10,13) – 2 col r – 14 – mf#C511,512 – uk Microform Academic [090]
French, C *see* Ireland
French, C J *see* Journal of a tour in upper india
The french canadian cattle / Couture, Joseph-Alphonse – Quebec?: L Brousseau, 1900 – 1mf – 9 – mf#05400 – cn Canadiana [636]
French Canadian Genealogists of Wisconsin *see* Quarterly
The french canadian, imperium in imperio : a lecture on our creed and race problem / Burton, John – Toronto: Copp, Clark, 1887 – 1mf – 9 – mf#00357 – cn Canadiana [377]
French canadian life and character : with historical and descriptive sketches of the scenery and life in quebec, montreal, ottawa, and surrounding country / ed by Grant, George Munro – Chicago: A Belford, 1899 [mf ed 1980] – 3mf – 9 – 0-665-05117-4 – (ill by f b schell et al) – mf#05117 – cn Canadiana [917]
The french canadians in new england / Bender, Prosper – Boston?: s.n, 1892? – 1mf – 9 – (repr fr the new england magazine) – mf#15950 – cn Canadiana [305]
French cathedrals / Winkles, Benjamin – London 1837 – (= ser 19th c art & architecture) – 3mf – 9 – mf#4.2.723 – uk Chadwyck [720]
French drawings and sketchbooks of the nineteenth century, vol 1 / Olsen, Sandra Haller – 1978 – 5 color mf – 15 – $115.00f – 0-226-68796-1 – (122p accompanying text) – us Chicago U Pr [740]
French drawings and sketchbooks of the nineteenth century, vol 2 / Olsen, Sandra Haller – 1978 – 7 color mf – 15 – $120.00f – 0-226-68798-8 – (140p accompanying text) – us Chicago U Pr [740]
French drawings of the sixteenth and seventeenth centuries / Olsen, Sandra Haller – 1977 – 1 color mf – 15 – $35.00f – 0-226-68794-5 – (32p accompanying text) – us Chicago U Pr [740]
The french element in the canadian northwest / Drummond, Lewis Henry – Winnipeg: Northwest Review, 1887 – 1mf – 9 – (in dble clms) – mf#30251 – cn Canadiana [917]
French Equatorial Africa *see* Journal officiel
French Equatorial Africa. Haut Commissariat *see* Annuaire statistique de l'afrique equatoriale francaise 1936-1955
French flasher / French Battery & Carbon Co – 1919 jul-1922 apr – 1r – 1 – (cont by: ray-o-lite news) – mf#3521227 – us WHS [660]
The french forces of the interior : their organization and participation in the liberation of france, 1944 / U.S. Army. European Theater of Operations – 1945 – 1 – $67.00 – us L of C Photodup [355]
French, George *see* Advertising
French, Gilbert James *see* Descriptive catalogue of articles of church decoration
French Guiana *see* Annuaire
French Guinea *see* Journal officiel
French historical studies – Baton Rouge. 1989+ (1,5,9) – ISSN: 0016-1071 – mf#17619 – us UMI ProQuest [944]
French horae : clare college, cambridge, ms kk3.2 – 15th c – 1r – 1 – mf#96817 – uk Microform Academic [390]
The french in africa : (algiers and morocco.) from the portfolio for aug 1844 / Urquhart, David – London 1844 – (= ser 19th c british colonization) – 1mf – 9 – mf#1.1.9034 – uk Chadwyck [322]
French in the west indies / Roberts, Walter Adolphe – Indianapolis, IN. 1942 – 1r – us UF Libraries [972]
French indochina. Conseil superieur *see* Session ordinaire de...
French, J C *see* Himalayan art
French, James *see* A defence of gospel baptism
French jansenists / Tollemache, Marguerite – London: Kegan Paul, Trench, Truebner, 1893 – 1mf – 9 – 0-8370-8553-5 – (incl bibl ref) – mf#1986-2553 – us ATLA [920]
French, John Calvin *see* The art of the pal empire
French, Jonathan *see* Sermons, delivered on the 20th of august, 1812
French, L *see* Reports on agriculture development and land settlement in palestine
French literature in louisiana / Fortier, Alcee – S.l: s.n, 1886? – 1mf – 9 – (incl some french text. incl bibl ref) – mf#54220 – cn Canadiana [440]
French mission life : or, sketches of remarkable conversions and other events among french romanists in the city of detroit / Carter, Thomas – New York: Carlton & Porter, 1857, c1856 [mf ed 1991] – 1mf – 9 – 0-524-00864-7 – mf#1990-0249 – us ATLA [241]
French museum and school for research – s.l, s.l? 193-? – 1r – us UF Libraries [978]
French news – New York. 1957-1973 (1) 1971-1972 (5) – ISSN: 0013-1369 – mf#1513 – us UMI ProQuest [073]
French nineteenth century art (including salon catalogues) : subject collections – (= ser Art exhibition catalogues on microfiche) – 172 catalogues on 269mf – 9 – £1,410.00 – (individual titles not listed separately) – uk Chadwyck [700]
French organ mass in the sixteenth and seventeenth centuries / Howell, Almonte Charles, Jr – U of North Carolina at Chapel Hill 1953 [mf ed 196-] – 8mf – 9 – (musical suppl of examples at end; incl bibl ref & ind) – mf#fiche 402 – us Sibley [780]
French painters of the 18th century / Dilke, Emilia Frances (Strong) – London 1899 – (= ser 19th c art & architecture) – 5mf – 9 – mf#4.2.319 – uk Chadwyck [750]
French political pamphlets, 1547-1648 – 86r – 1 – mf#C39-24100 – us Primary [944]
French pottery / Gasnault, Paul & Garnier, Edouard – [London] 1884 – (= ser 19th c art & architecture) – 3mf – 9 – mf#4.1.402 – uk Chadwyck [730]
French pronouncing book : a new and infallible method of learning and teaching a correct pronouncing of the french language / Rudelle, Lucien de – London, Leicester, 1835 – (= ser 19th c children's literature) – 2mf – 9 – (in 4pt: the last containing a 2nd rev impr ed, of the "art of reading at sight" by lucien de rudelle) – mf#6.1.15 – uk Chadwyck [440]
French prophets of yesterday : a study of religious thought under the second empire / Guerard, Albert Leon – London: T F Unwin, 1913 – 1mf – 9 – 0-7905-6526-9 – (incl bibl ref) – mf#1988-2526 – us ATLA [240]
French prophets of yesterday; a study of religious thought under the second empire / Guerard, Albert Leon – New York: D. Appleton, 1913. 288p – 1 – us Wisconsin U Libr [240]
French, R B *see* Levels of carotene and ascorbic acid in florida-grown foods
French review – Carbondale. 1971+ [1]; 1927+ [5]; 1976+ [9] – ISSN: 0016-111X – mf#6125 – us UMI ProQuest [440]
French revolution / Carlyle, Thomas – London, England. v1-3. 1839 – 1r – us UF Libraries [944]
The french revolution – (= ser The Eighteenth Century coll) – 11r – 1 – (representing works of major and lesser known authors. a special subset of the eighteenth century collection) – us Primary [944]
The french revolution : a sketch / Mathews, Shailer – 2nd ed., rev. New York: Longmans, Green, 1906, c1901 – 1mf – 9 – 0-7905-6243-X – mf#1988-2243 – us ATLA [944]
French revolution, 1789-1941 / Mathews, Shailer – New York, NY. 1925 – 1r – us UF Libraries [025]
The french revolution and religious reform : an account of ecclesiastical legislation and its influence on affairs in france from 1789 to 1804 / Sloane, William Milligan – New York: Scribner, 1901 – (= ser Morse Lectures) – 1mf – 9 – 0-7905-5961-7 – mf#1988-1961 – us ATLA [944]
French revolution in san domingo / Stoddard, Theodore Lothrop – Boston, MA. 1914 – 1r – us UF Libraries [972]
French revolution of 1848 / Arthur, William – London, England. 1849? – 1r – us UF Libraries [240]
The french revolution research collection : 1787-1799 – [mf ed Micro Graphix/Maxwell Communications Corp] – 9 – (in french. with p/g for each sect with int in both french & english. with ind) – us UMI ProQuest [944]
French, Richard Valpy *see* Lex mosaica
French royal and administrative acts, 1256-1794 – 59r – 1 – (approx 16,000 pamphlets on the financial and political administration of france fr the late 13th century to the end of the monarchy. includes printed guide) – mf#C39-27790 – us Primary [944]
French Socialist Federation *see* Socialiste
French Socialist Federation [Johnston City, IL] *see* Germinal
French struggle for the west indies / Crouse, Nellis Maynard – New York, NY. 1943 – 1r – us UF Libraries [972]
French Sudan *see*
– Journal officiel
French, Thomas P *see* Anweisung fuer ansiedler an die ottawa und opeongo strasse und umgegend
French, Warren G *see* Companion to the grapes of wrath
French West Africa *see*
– Journal officiel
– La mauritanie
French West Africa. Direction des Services de la Statistique Generale et de la Mecanographie *see* Annuaire statistique de l'afrique occidentale francaise 1949-1954
French, William Riley *see* Gospel doctrines for the use of sunday schools
French writer andree viollis speaks in paris about the admirable defense of the spanish capital / Ardenne de Tizac, Andree Francoise Cardine d' – [s.l: s.n. 193-] – (= ser Blodgett coll) – 9 – mf#w724 – us Harvard [946]
French-american military relations in the caribbean theatre in world war 2 / U.S. Army. Caribbean Defense Command – 1 – $26.00 – us L of C Photodup [977]
The french-canadian conteur of the olden days / Bender, Prosper – Quebec: [s.n.], 1910 – 1mf – 9 – 0-665-98906-7 – mf#98906 – cn Canadiana [390]
French-Canadian Genealogical Society of Connecticut, Inc *see* Connecticut maple leaf
The french-canadian peasantry : language, customs, mode of life, food, dress / Bender, Prosper – Boston?: s.n, 1890? – 1mf – 9 – mf#54190 – cn Canadiana [306]
French-english, english-french law dictionary / Langstaff, Annie Macdonald – Montreal Wilson and Lafleur, 1937. 141p. LL-2320 – 1 – us L of C Photodup [340]
French-language book of hours manuscripts in the koninklijke bibliotheek [national library of the netherlands], the hague – [mf ed 2007] – 37 titles on 252mf – 9 – €2195.00 – (with p/g) – mf#mmp133 – ne Moran [090]
French-language medieval manuscripts in the koninklijke bibliotheek [national library of the netherlands], the hague – 58r or 910mf – 1,9 – €6895.00 – (with p/g & int by anne s korteweg) – mf#mmp113 – ne Moran [090]
A frenchman in america : (the anglo-saxon race revisited) / O'Rell, Max – Bristol: J W Arrowsmith; London: Simpkin, Marshall, Hamilton, Kent, 1891? – 4mf – 9 – (ill by e w kemble) – mf#33256 – cn Canadiana [917]
A frenchman in america : (the anglo-saxon race revisited) / O'Rell, Max – S.l: s.n, 189-? – 3mf – 9 – mf#39023 – cn Canadiana [917]
A frenchman in america : (the anglo-saxon race revisited) / O'Rell, Max – S.l: s.n, 1891? – 3mf – 9 – mf#29080 – cn Canadiana [917]
A frenchman in america : (the anglo-saxon race revisited) / O'Rell, Max – Toronto: W Bryce, 1891 – (= ser Bryce's home series) – 3mf – 9 – mf#00149 – cn Canadiana [917]
A frenchman in america : recollections of men and things / O'Rell, Max – New York: Cassell Pub Co, c1891 – 5mf – 9 – (ill by e w kemble) – mf#26518 – cn Canadiana [917]
Frenchman valley times – Palisade, NE: W T Brickey. -v9 n13. nov 26 1896 (wkly) [mf ed with gaps] – 1r – 1 – (cont by: palisade press) – us NE Hist [071]
French's sunday forum – Duluth, Superior MN. 1894 sep 16-dec 16 – 1r – 1 – (cont by: sunday morning forum [superior, wis]) – mf#933713 – us WHS [071]
The french-war papers of the marechal de levis / Casgrain, Henri Raymond – Cambridge MA: J Wilson, 1888 – 1mf – 9 – (in english and french. comm by francis parkman and justin winsor) – mf#02545 – cn Canadiana [971]
Freneau, Philip *see* Letters on various interesting and important subjects
Frenk, Azriel Nathan *see*
– Familie dawidsohn
– Meshumadim in poiln
– Yehude polin
Frenkel, Jacob *see*
– G galilei
Frenken, Goswin *see* Die exempla des jacob von vitry
Frensdorf, F *see* Muenchhausens berichte ueber seine mission nach berlin in juni 1740
Frensdorff, F Ferdinand *see* Grundriss zu vorlesungen ueber das deutsche privatrecht
Frensdorff, Salomon *see* Die massora magna
Frenssen, Gustav *see*
– The anvil
– Der brennende baum
– Die brueder
– Die chronik von barlete
– Dorfpredigten
– Die drei getreuen
– Dummhans
– Gruebeleien
– Hilligenlei
– Holyland
– Joern uhl
– Klaus hinrichs baas
– Land an der nordsee
– Das leben des heilands dargestellt
– Lebensbericht
– Lebenskunde
– Luette witt
– Meino der prahler
– Moewen und maeuse
– Otto babendiek
– Der pastor von poggsee
– Peter moors fahrt nach sudwest
– Peter moors fahrt nach suedwest
– Peter moor's journey to southwest africa
– Die sandgraefin
– Vorland
Frente a frente / Midwest Hispanic Catholic Commission et al – 1981 mar-1982 feb – 1r – 1 – (cont: cara a cara) – mf#620801 – us WHS [071]
Frente al futuro / Martinez Conde, Jose – s.l, s.l? 1925 – 1r – us UF Libraries [972]
Frente al silencio / Isla De Rodriguez, Antonia – Habana, Cuba. 1938 – 1r – us UF Libraries [972]
Frente de Libertacao de Mocambique *see* Mozambican revolution
Frente nacional / Vazquez Cobo Carrizosa, Camilo – Cali, Colombia. 196- – 1r – us UF Libraries [972]
Frente nacional y los partidos politicos / Cardenas Garcia, Jorge – Tunja, Colombia. 1958 – 1r – us UF Libraries [972]
Frente popular, boletin de las organizaciones antifascistas hispanas de los e. u. de norte america – Brooklyn, 1937. Fiche W 899. (Blodgett Collection of Spanish Civil War Pamphlets) – 9 – us Harvard [946]
Frente Sandinista de Liberacion Nacional. Nicaragua *see* Barricada
Frentz, Hans *see* Der adjutant
Frenzel, Elisabeth *see* Daten deutscher dichtung
Frenzel, Herbert Alfred *see*
– Aufforderung zum laecheln
– Daten deutscher dichtung
Frenzel, Karl *see*
– Deutsche kaempfe
– Frauenrecht
– Zwei novellen
Freppel, Charles *see*
– Les apologistes chretiens au 2e siecle
– Clement d'alexandrie
– Examen critique de la vie de jesus de m renan
– Origene
– Les peres apostoliques et leur epoque
– Saint cyprien et l'eglise d'afrique au 3e siecle
– Saint irenee et l'eloquence chretienne dans la gaule pendant les deux premiers siecles
– Saint justin
Frequency and quantity of alcohol use of ncaa division 3 student-athletes participating in the minnesota intercollegiate athletic conference / Storsved, John R, II – 1996 – 2mf – 9 – $8.00 – mf#HE 595 – us Kinesology [362]
Frequency technology – Boston. 1962-1970 (1) – ISSN: 0532-6923 – mf#1530 – us UMI ProQuest [621]
Frere, Bartle *see*
– Afghanistan and south africa
– Indian missions
Frere et mari / Humbert, Auguste – Paris, France. 1841 – 1r – us UF Libraries [440]

Frere, Henry Bartle Edward *see* On the impending bengal famine
Frere, John *see* Fasting
Frere, Mary *see* Old deccan days
Frere robert sylvain : docteur es lettres: bibliographie analytique / Vigneault-Page, Celine – 1964 [mf ed 1979] – (= ser Bibliographies du cours...1947-66) – 1mf – 9 – (with ind; pref by maurice lebel) – mf#SEM105P4 – cn Bibl Nat [400]
Frere, W *see* A new history of the book of common prayer
Frere, Walter H *see*
- The anaphora or great eucharistic prayer
- The hereford breviary, vol 1
- The hereford breviary, vol 2
- The hereford breviary, vol 3
- The winchester troper

Frere, Walter Howard *see*
- The church of our fathers
- The english church in the reigns of elizabeth and james 1
- English church ways
- Lancelot andrewes as a representative of anglican principles
- The marian reaction in its relation to the english clergy
- A new history of the book of common prayer
- The principles of religious ceremonial
- Puritan manifestoes
- Some principles of liturgical reform
- Studies in early roman liturgy
- The use of sarum

Freres des ecoles chretiennes *see* Noviciat preparatoire
Les freres des ecoles chretiennes : conference prononcee a l'institut canadien de quebec le 19 avril 1877 / Jolicur, Philippe Jacques – Quebec?: s.n, 1877 – 1mf – 9 – mf#58138 – cn Canadiana [240]
Les freres des ecoles chretiennes et l'enseignement primaire : apres la revolution 1797-1830 / Chevalier, Alexis – Paris: Poussielgue Freres, 1887 – 2mf – 9 – 0-8370-7534-3 – (incl bibl ref) – mf#1986-1534 – us ATLA [377]
Freres feroces : ou, m bonardin a la repetition / Jouslin De La Salle, Armand-Francois – Paris, France. 1825 – 1r – us UF Libraries [440]
Les freres grimm / Tonnelat, Ernest – Paris: A Colin 1912 [mf ed Bloomington IN: Indiana Uni Lib, Preservation Dept 1984] – xii 438p on 1r – 1 – us Indiana Preservation [390]
Fresco decorations and stuccoes of churches and palaces, in italy / Gruner, Ludwig – London 1854 – (= ser 19th c art & architecture) – 3mf – 9 – mf#4.2.121 – uk Chadwyck [720]
Fresenius, J B G W *see* Beitraege zur flora von aegypten und arabien
Fresenius' zeitschrift fuer analytische chemie – Muenchen. 1981-1989 (1) 1981-1989 (5) 1975-1989 (9) – ISSN: 0016-1152 – mf#13168,01 – us UMI ProQuest [540]
Fresh! – 1987 may 23, aug 15, dec 19, 1988 jan 9, feb 20, may 14, jun 25, aug 27, 1989 dec 12, 1998 jan 9-10, 30, apr 26, may 17, jun 7, jul 19[n201], jul 19[n202], aug 30, sep 20, oct 11, nov 22, dec 13, 1999 feb 22, mar 15, apr 5,26, may 17, jun 7,28, jul 19, aug 30, sep 20 – 3r – 1 – (cont: record review) – mf#2337286 – us WHS [071]
A fresh approach to the new testament / Dibelius, Martin – 1936 – 9 – $10.00 – us IRC [242]
A fresh approach to the psalms / Oesterley, William Oscar Emil – New York: Scribner, 1937 [mf ed 1993] – (= ser The international library of christian knowledge) – 1mf – 9 – 0-524-08126-3 – (incl bibl ref) – mf#1993-9032 – us ATLA [221]
Fresh fruit / Brown Daily Herald Voluntary Publishing Association – v4 n16-17 [1976 may 21-jul 4] – 1r – 1 – mf#1583408 – us WHS [071]
Fresh fruit – Providence, RI. 1973-1976 (1) – mf#66307 – us UMI ProQuest [071]
Fresh gleanings / Mitchell, Donald Grant – New York, NY. 1847 – 1r – us UF Libraries [025]
Fresh laurels for the sunday school / Bradbury, William Batchelder – 1867 – 1 – 5.95 – us Southern Baptist [242]
Fresh light from the ancient monuments : a sketch of the most striking confirmations of the bible from recent discoveries in egypt, assyria, palestine, babylonia, asia minor / Sayce, Archibald Henry – 5th ed. [London]: Religious Tract Soc 1890 [mf ed 1990] – (= ser By-paths of bible knowledge 2) – 1mf – 9 – 0-7905-3408-8 – mf#1987-3408 – us ATLA [220]
A fresh study of the fourth gospel / Hitchcock, Francis Ryan Montgomery – London: SPCK; New York: E S Gorham, 1911 [mf ed 1989] – 1mf – 9 – 0-7905-1992-5 – mf#1987-1992 – us ATLA [226]
Fresh tracks in the belgian congo from the uganda border to the mouth of the congo / Norden, Hermann – 57 photographs and two maps. London: H.F. & G. Witherby, 1924. 303p – 1 – us Wisconsin U Libr [960]
Fresh voyages on unfrequented waters / Cheyne, Thomas Kelly – London: A and C Black; New York: Macmillan [distributor], 1914 – 1mf – 9 – 0-7905-0925-3 – (incl bibl ref and indexes) – mf#1987-0925 – us ATLA [220]
Freshfield, D W *see* The exploration of the caucasus
Freshman, Charles *see*
- The autobiography of the rev charles freshman
- The jews and the israelites
- The pentateuch

Freshman eligibility in intercollegiate athletics / Hick, Barbara A & Parkhouse, Bonnie L – 1992 – 1mf – 9 – $4.00 – us Kinesology [378]
Freshman english news – Fort Worth. 1975-1991 (1) 1977-1991 (5) 1977-1991 (9) – (cont by: composition studies/freshman english news) – ISSN: 0739-4713 – mf#10315 – us UMI ProQuest [420]
Freshmen athletes' perceptions of adjustment to intercollegiate athletics / Armenth-Brothers, Francine R – Ball State University, 1995 – 2mf – 9 – $8.00 – mf#PSY1837 – us Kinesology [150]
Freshwater biology – Oxford. 1980+ (1,5,9) – ISSN: 0046-5070 – mf#15527 – us UMI ProQuest [574]
Fresken : neue dichtungen / Vierordt, Heinrich – Heidelberg: C Winter, 1901 – 1r – 1 – us Wisconsin U Libr [810]
Fresn labor citizen / Federated Trades and Building Council [Fresno, CA] – 1945 jun 1-dec, 1946-50, 1951-1953 jun 26 – 3r – 1 – (cont: tri-county labor news; cont by: valley labor citizen) – mf#907974 – us WHS [071]
Fresn resistance newsletter – 1969 oct-dec – 1r – 1 – mf#1056636 – us WHS [071]
Fresno assembly center directory – Fresno Co, CA. 1942 – 5r – 1 – $250.00 – mf#B06085 – us Library Micro [917]
[Fresno-] california farmer : central edition – CA. 1959-jun 1991 – (= ser Wine & agriculture coll) – 54r – 1 – $3240.00 – (see: los angeles, san francisco) – mf#B02256 – us Library Micro [071]
Fresno county "blue book" / Walker, Ben – Fresno Co, CA. 1941 – 1r – 1 – $50.00 – mf#B40215 – us Library Micro [978]
Fresno county, california and the evolution of the fruit vale, 1930 – Fresno Co, CA. 1930 – 1r – 1 – $50.00 – mf#B40216 – us Library Micro [978]
[Fresno county-] coalinga city directories – CA. 1910-1914 – 4r – 1 – $200.00 – mf#D022 – us Library Micro [917]
[Fresno county-] fresno assembly center directory – CA. 1942 – 1r – 1 – $50.00 – mf#D025 – us Library Micro [978]
[Fresno county-] fresno city directories – CA. 1871-1873; 1881-1882; 1886-1990 – 120r – 1 – $6000.00 – mf#D019 – us Library Micro [917]
[Fresno county-] fresno county – 1881-82; 1891; 1900-19; 1955-56 – 22r – 1 – $1100.00 – mf#D017 – us Library Micro [978]
[Fresno county-] fresno county assessor's rolls – CA. 1860; 1862 – 2r – 1 – $100.00 – mf#D018 – us Library Micro [978]
[Fresno county-] fresno, inyo, kern, merced, san bernardino, stanislaus and tulare counties – CA. 1884-1885 – 2r – 1 – $100.00 – mf#D020 – us Library Micro [978]
[Fresno county-] fresno negro directory – CA. 1936 – 1r – 1 – $50.00 – mf#D024 – us Library Micro [978]
[Fresno county-] fresno police annuals – CA. 1951-1990 – 6r – 1 – $300.00 – mf#D026 – us Library Micro [350]
[Fresno county-] fresno sanborn maps – CA. 1888-1919 – 11r – 1 – $550.00 – mf#D021 – us Library Micro [978]
[Fresno county-] fresno sheriff's review – CA. 1958-1990 – 6r – 1 – $300.00 – mf#D027 – us Library Micro [350]
[Fresno county-] history of fresno county – CA: Elliot & Co Publ, 1881 – 1r – 1 – $50.00 – mf#B40217 – us Library Micro [978]
[Fresno county-] la cucaracha records – 1946-50 – (= ser Chicano studies library serial) – 1 – 1 – $50.00 – mf#B06089 – us Library Micro [978]
Fresno county negro directory – Fresno Co, CA. 1936 – 1r – 1 – $50.00 – mf#B06090 – us Library Micro [978]
[Fresno county-] reedley city directories – CA. 1933-1936; 1940 – 4r – 1 – $200.00 – mf#D023 – us Library Micro [917]
[Fresno county-] selma, kingsburg and fowler city directories – CA. 1962-1991 – 13r – 1 – $650.00 – mf#D028 – us Library Micro [917]
[Fresno-] daily evening expositor – CA. 1883-1889; 1892-98 – 29r – 1 – $1740.00 – mf#B02260 – us Library Micro [071]
[Fresno-] fresno asbarez – CA. 1908-12; 1952-73 – 10r – 1 – $600.00 – mf#C02254 – us Library Micro [071]
[Fresno-] fresno bee and guide index – CA. 1870-1974 – 5r; 1mf set – 1,9 – $300.00 1 $500.00 set 9 – mf#B06001 – us Library Micro [071]
[Fresno-] fresno business – CA. 1967-1988 – 3r – 1 – $180.00 – mf#B03599 – us Library Micro [071]
[Fresno-] fresno county and city chamber of commerce business – CA. 1955-67 – 1r – 1 – $60.00 – (cont by: fresno business) – mf#B03598 – us Library Micro [338]
[Fresno-] fresno daily morning republican – CA. 1887-99 – 16r – 1 – $960.00 – mf#R04026 – us Library Micro [071]
[Fresno-] fresno topics – CA. 1946-1955 – 1r – 1 – $60.00 – (cont. by: fresno county and city chamber of commerce business) – mf#B03597 – us Library Micro [071]
[Fresno-] fresno tribune – CA. 1932-1933 – 6 – 1 – $360.00 – mf#C03223 – us Library Micro [071]
[Fresno-] fresno weekly republican – CA. 1876-1899 – 11r – 1 – $660.00 – mf#R04027 – us Library Micro [071]
[Fresno-] guide – CA. 1932-46, 1969-78 – 72r – 1 – $4320.00 – (aka: downtown shopping guide) – mf#B02257 – us Library Micro [071]
[Fresno-] insight – CA. csu fresno: 1969- – 7r – 1 – $420.00 (subs $50y) – mf#B07453 – us Library Micro [302]
Fresno Labor Council *see* Valley labor citizen
[Fresno-] morning republican – CA. 1887-99, 1892-95, jul-dec 1896, jan-jun 1897 (scats), 1897-1932 – 298r – 1 – $17,880.00 – (aka: daily republican) – mf#B02258 – us Library Micro [071]
[Fresno-] parlier progress – CA. 1927-1931 – 5r – 1 – $300.00 – mf#B06027 – us Library Micro [071]
Fresno police annuals – Fresno Co, CA. 1951-90 – 6r – 1 – 300.00 – mf#B06091 – us Library Micro [360]
Fresno Resistance *see* Newsletter
Fresno sanborn maps – Fresno Co, CA. 1888-1919 – 11r – 1 – $550.00 – mf#B06092 – us Library Micro [917]
Fresno sheriff's review – Fresno Co, CA. 1958-90 – 6r – 1 – $300.00 – mf#B06093 – us Library Micro [978]
[Fresno-] shoppers extra : fresno bee – CA. 1983 – 1r – 1 – $60.00 – mf#R04028 – us Library Micro [071]
Fresno State College *see* Raices
[Fresno-] the beam – CA. oct 1943-apr 1944 – 1r – 1 – $50.00 – mf#B06087 – us Library Micro [978]
[Fresno-] the collegian – CA. csu fresno: 1922- – 41r – 1 – $2460.00 – (aka: the daily collegian) – mf#B03715 – us Library Micro [378]
[Fresno-] the cooperative california – CA. 1912-1923 – 1r – 1 – $60.00 – mf#B07452 – us Library Micro [370]
[Fresno-] the express line : the fresno bee – CA. 1993 – 4r – 1 – $240.00 (subs $90y) – mf#R04025 – us Library Micro [071]
[Fresno-] the fresno bee – CA. 1922- [wkly] – 1497r – 1 – $89,820.00 (subs $1440y) – mf#B02255 – us Library Micro [071]
[Fresno-] the fresno news – CA. 1936-1939 – 1r – 1 – $60.00 – mf#B07454 – us Library Micro [370]
[Fresno-] this week – CA. 1983-1993 – 61r – 1 – $3660.00 – mf#R03222 – us Library Micro [071]
[Fresno-] times – CA. 1865 – 1r – 1 – $60.00 – mf#B06028 – us Library Micro [071]
[Fresno-] weekly expositor – CA. 1872-98 – 12r – 1 – $720.00 – mf#BC02259 – us Library Micro [071]
Fresno/clovis – 1966-67; 1980; 1989- – (= ser California telephone directory coll) – 15r – 1 – $750.00 – mf#P00029 – us Library Micro [917]
Fresno/madera – 1965-66; 1980; 1989- – (= ser California telephone directory coll) – 9r – 1 – $450.00 – mf#P00030 – us Library Micro [917]
Freson, Jules G *see* L'esthetique de richard wagner
Fresquet, Fresquito *see* De tonto que soy
Freston, Anthony *see* Collection of evidences for the divinity of our lord jesus christ
Frets – Cupertino. 1979-1989 (1) 1979-1989 (5) 1979-1989 (9) – ISSN: 0162-0401 – mf#12011 – us UMI ProQuest [780]
Fretwell, John *see*
- The christian in hungarian romance
- Three centuries of unitarianism in transylvania and hungary

Freud : his life and his mind / Puner, Helen Walker – New York, NY. 1947 – 1r – us UF Libraries [150]
Freud, Anna *see* The ego and the mechanisms of defence
Freud or jung / Glover, Edward – New York: W.W. Norton, 1950. 207p. Includes index – 1 – us Wisconsin U Libr [616]
Freud, Sigmund *see*
- Gesammelte schriften
- Imago
- Kleine beitraege zur traumlehre
- Le mot d'esprit et ses rapports avec l'inconscient
- The problem of lay-analyses
- Die traumdeutung

Freude der kinderjahre / Schubert, Franz – Juli 1816 [mf ed 1988] – 1r – 1 – (purchased fr: walter schatzki [new york, 1940 nov]; publ in: franz schubert: neue ausgabe saemtliche werke, 4, 10) – mf#pres. film 37, 7 – us Sibley [780]
Die freude in den schriften des alten bundes : eine religionswissenschaftliche studie / Wuensche, August – Weimar: Emil Felber, 1896 – 1mf – 9 – 0-8370-5930-5 – (incl bibl ref) – mf#1985-3930 – us ATLA [240]
Freude und arbeit – Berlin DE, 1936-42 [gaps] – 4r – 1 – gw Misc Inst [074]
Freudenstaedter kreisbote *see* Schwarzwaelder bote [main edition]
Freudenthal *see* Ueber die theologie des xenophanes
Freudenthal, Friedrich *see*
- Bi'n fueer
- In de fierabendstied
- Uenneren strohdack

Freudenthal, Jacob *see*
- Die flavius josephus beigelegte schrift ueber die herrschaft der vernunft (4 makkabaeerbuch)
- Hellenistische studien

Freudenthaler zeitung – Freudenthal (Bruntal CZ), 1920 29 may-1921 5 feb – 1r – 1 – gw Misc Inst [077]
The freudian wish and its place in ethics / Holt, Edwin Bissell – New York: H Holt, 1915 – 1mf – 9 – 0-7905-3961-6 – mf#1989-0454 – us ATLA [150]
Freund aller welt : roman / Flake, Otto – Berlin: S Fischer, 1928 (mf ed 1990) – 1r – 1 – (filmed with: freitagskind) – us Wisconsin U Libr [830]
Freund, Anna *see* Annette von droste-huelshoff in ihren beziehungen zu goethe und schiller und in der poetischen eigenart ihrer gereiften kunst
Der freund der aufklaerung und menschenglueckseligkeit : eine monatsschrift fuer denkende leserinnen und leser aus allen religionen und staenden / ed by Koenig, Joh Christoph – Nuernberg 1785-86 – (= ser Dz) – 2v(=12st) on 6mf – 9 – €120.00 – mf#k/n5131 – gw Olms [230]
Der freund der wahrheit und des volkes – Bamberg DE, 1848 18 aug-1849 29 jan – 1r – 1 – gw Misc Inst [943]
Freund hein : eine leidensgeschichte / Strauss, Emil – Berlin: S Fischer, 1921 – 1r – 1 – us Wisconsin U Libr [830]
Freund hein / Soerensen, Wulf – Berlin: Nordland-Verlag, c1943 (mf ed 1990) – 1r – 1 – (filmed with: der tod vor dem spiegel) – us Wisconsin U Libr [430]
Der freund in der not / Schupp, Johann Balthasar; ed by Braune, Wilhelm – Halle a/S: M Niemeyer 1878 [mf ed 1993] – (= ser Neudrucke deutscher literaturwerke des 16. und 17. jahrhunderts 9) – 11r – 1 – (int by ed) – mf#3387p – us Wisconsin U Libr [830]
Freund, Lothar *see* Die mannschaft der 'samoa'
Freund, Max *see* Israelitischer landes-lehrer-verein in boehmen
Freund, Miriam K *see* Jewish merchants in colonial america
Freund, Wilhelm *see* Triennium philologicum
Freund, William *see* Zur judenfrage in deutschland
Die freunde machen den philosophen *see*
- Ein beitrag zur kritik von lessings laokoon
- Lessings laokoon

Die freunde machen den philosophen, der englaender, der waldbruder / Kaiser, Ilse – Erlangen: E T Jacob, 1917 – 1r – 1 – (incl bibl ref) – us Wisconsin U Libr [430]
Freundesbilder aus goethes leben : studien zum leben des dichters / Duentzer, Heinrich – 2. wohlfeile ausg. Leipzig: Dyk, [1853] [mf ed 1992] – xiv/623p – 1 – (incl bibl ref) – mf#7656 – us Wisconsin U Libr [430]
Die freundin – Berlin DE, 1924-1933 8 mar – 2r – 1 – gw Misc Inst [074]
Die freundin : roman / Hoppe, Ingeborg Marei – Berlin: E Schmidt, 1944 – 1r – 1 – us Wisconsin U Libr [830]
Freundlicher wett-streit...sturm, vauban, coehoorn und rimpler / Sturm, L C – Augspurg, 1718-1721 – 12mf – 9 – mf#OA-222 – ne IDC [720]
Freundlicher wett-streit...sturm, vauban, coehoorn und rimpler / Sturm, L C – Augspurg, 1740 – 2mf – 9 – mf#OA-221 – ne IDC [720]
Die freundschaft : [short stories] / Tuegel, Ludwig – Hamburg: Hanseatische Verlagsanstalt c1939 [mf ed 1991] – 1r – 1 – (filmed with: leuchtendes land / luis trenker) – mf#2917p – us Wisconsin U Libr [830]
Freundschaft – Berlin DE, 1919-1933 n3 – 5r – 1 – gw Misc Inst [074]

Freundschaft – Alma Ata (Kazakhstan), USSR. 1970-1981 – 12r – 1 – $600.00 – (german language) – mf#B63584 – us Library Micro [077]
Freundschaft : mitteilungen der treugemeinschaft sudetendeutscher sozialdemokraten in england – London (GB), 1941 jan-nov – 1r – 1 – gw Misc Inst [072]
Freundschaft : tageszeitung der sowjetdeutschen bevoelkerung kasachstans – Tselinograd (Celinograd KZ), Alma-Ata (Almaty), 1967-69, 1972-90, 1991 1 feb- – 1 – (title varies: 1 feb 1991: deutsche allgemeine zeitung der russlanddeutschen, alma-ata) – gw Misc Inst [077]
O freundschaft! : zeitgemaelde in drei abtheilungen / Tremler, Wenzel – Wien: U Klopf Senior und Alex Eurich, [18-?] – 1r – 1 – us Wisconsin U Libr [820]
Freundschaft und freiheit : ein blatt fuer maennerrechte gegen spiessbuergermoral, pfaffenherrschaft und weiberwirtschaft – Berlin DE, 1921 n1-11 – 1r – 1 – gw Misc Inst [074]
Freundschaftliche lieder / Pyra, Immanuel Jakob & Lange, Samuel Gotthold; ed by Sauer, August – Heilbronn: Henninger, 1885 [mf ed 1993] – l/167p – 1 – (incl bibl ref. ed fr 2nd ed: halle, c h hemmerde, 1749) – mf#8676 reel 2 – us Wisconsin U Libr [780]
Das freundschaftsblatt *see* Die insel der einsamen
Freuntliche ermanung zur grechtigheit... / Bullinger, Heinrich – [Zuerich, Hanns Hager, 1526] – 1mf – 9 – mf#PBU-101 – ne IDC [240]
Frey, Adolf *see*
– Briefe von adolf frey und carl spitteler
– Erinnerungen an gottfried keller
– Gedichte
– J gaudenz von salis-seewis
– Eine untersuchung ueber die bedeutung der empirischen religionspsychologie fuer die glaubenslehre
Frey, Alexander Moritz *see* The stout-heartet cat
Frey, August Emil *see*
– Bartholomaeus ziegenbalg
– Geschichte der reformation
Frey, Axel [comp] *see*
– Baltic biographical archive
– Baltic biographical archive. series 2
– Bibliothek der deutschen literatur. zweites supplement
– Biographisches archiv der sowjetunion (1917-1991)
– Korean biographical archive
– Russisches biographisches archiv
Frey, Axel comp] *see* Russisches biographisches archiv (rba). supplement
Frey, Bernd *see* Comparison of body composition between german and american adults with mental retardation
Frey, Franz Andreas *see* An die souveraine der rheinischen konfoederation
Frey, Henri Nicolas *see* Campagne dans le haut senegal et dans le haut niger, 1885-1886
Frey, Hermann *see*
– Martin greif
– Martin greif in seinen werken
– Martin greifs dramen
– Martin greifs jugenddramen
Frey, J A *see* Baltijas baptisti un bunde, jeb baltijas baptistu tagadejs stahsoklis
Frey, Jakob *see* Gartengesellschaft
Frey, Johannes *see*
– Die letzten lebensjahre des paulus
– Tod, seelenglaube und seelenkult im alten israel
Frey, Joseph Samuel Christian Frederick *see*
– Essays on the passover
– The scripture types
– The theological lectures of rev david bogue
Frey, Lina *see* Briefe von adolf frey und carl spitteler
Frey, Peter *see* Die philosophie der kunst denkt sich zu ende
Frey, Winfried *see* Textkritische untersuchungen zu ottes "eraclius"
Freya – Stuttgart 1861-67 [mf ed 1993] – (= ser Hq 15) – 7v on 38mf – 9 – €170.00 – 3-89131-127-3 – gw Fischer [305]
Freybe, A *see* Faust
Freybe, Albert *see*
– Der deutsche volksaberglaube in seinem verhaeltnis zum christentum und im unterschiede von der zauberei
– Der ethische gehalt in grillparzers werken
Freyberg, Hermann *see*
– Die letzte heuer
– Raetsel um herta
Freyberger gemeinnuetzige nachrichten fuer das chursaechsische erzgebirge – Freiberg, Sachsen DE, 1800-48 – 20r – 1 – (title varies: 1807: freiberger gemeinnuetzige nachrichten fuer das koeniglich saechsische erzgebirge) – gw Misc Inst [943]
[Freyburger] beitraege zur befoerderung des aeltesten christenthums und der neuesten philosophie / ed by Ruef, Kaspar – Ulm 1788-93 – (= ser Dz. abt theologie) – 8v on 30mf – 9 – €300.00 – mf#k/n2226 – gw Olms [230]
Freyburger zeitung – Freiburg Br DE, 1848-49 – 3r – 1 – (title varies: 1802: allgemeines intelligenz- oder wochenblatt fuer das land breisgau und die ortenau; 1808: grossherzoglich-badische privilegirte freyburger zeitung; 3 nov 1810: freiburger wochenblatt; 1848: neue freiburger zeitung; 1852: freiburger zeitung; 1934 n5: freiburger zeitung und handelsblatt; 1935 n276: freiburger zeitung und wirtschaftsblatt; 1940 n60: freiburger zeitung. filmed by other misc inst: 1784, 1793 21 jan-1798, 1800-1943 28 feb. with suppl: unterhaltungs-blatt 1835-48; unterhaltungs-blatt; 1835-48; beilage zur unterhaltung 1851 6 jul-1852) – gw Misc Inst [074]
Der freydenker – Danzig (Gdansk PL), 1741-43 [gaps] – 2r – 1 – gw Misc Inst [077]
Freye urtheile und nachrichten zum aufnehmen der wissenschaften und historie ueberhaupt / ed by Zincke, E H – Hamburg 1744-59 – (= ser Dz) – 16jge on 80mf – 9 – €800.00 – mf#k/n142 – gw Olms [900]
Freygang, Frederika von *see*
– Letters from the caucasus and georgia
– Wilhelm von freygangs...briefe ueber den kaukasus und georgien, nebst angehaengtem reiseberícht ueber persien vom jahre 1812
"Freygeister, naturalisten, atheisten" : ein aufsatz lessings im wahrsager / Consentius, Ernst – Leipzig: E Avenarius, 1899 [mf ed 1992] – 86p – 1 – (incl bibl ref) – mf#7592 – us Wisconsin U Libr [140]
Freymond, Jacques *see* La premiere internationale, recueil de documents
Der freymuethige : eine monatsschrift von einer gesellschaft zu freyburg i b / ed by Kaspar Ruef et al – Ulm, Freyburg 1782-88 – (= ser Dz. abt theologie) – 4v+ 3suppl on 20mf – 9 – €200.00 – mf#k/n2163 – gw Olms [430]
Freymuethige nachrichten von neuen buechern und andern zur gelehrtheit gehoerigen sachen – Zuerich 1744-63 [mf ed 1995] – (= ser Aus den anfaengen des zeitschriftenwesens: fruehe deutsche zeitschriften) – 90mf – 9 – €720.00 – 3-89131-209-1 – (filmed with: woechentliche anzeigen zum vortheil der liebhaber der wissenschaften und kuenste [1764-66] 3v) – gw Fischer; gw Olms [073]
Freyre, Gilberto *see*
– Brasis, brasil e brasilia
– Brazil
– Casa-grande and senzala
– Cultura amenazada
– Em torno de alguns tumulos afro-cristaos de uma area africana
– Guia pratico, historica e sentimental da cidade do...
– Ingleses no brasil
– Interpretacao do brasil
– Interpretacion del brasil
– Masters and the slaves
– Mucambos do nordeste
– Mundo que o portugues criou
– Nacao e exercito
– Nordeste
– Nordeste, aspectos da influencia da canna sobre
– Olinda
– Oliveira lima, don quixoto gordo
– Quase politica
– Regiao e tradicao
– Sobrados e mucambos
Freyre, Gliberto *see* Engenheiro frances no brasil
Freytag, Georg Wilhelm *see* Lexicon arabico-latinum
Freytag, Georg Wilhelm [comp] *see* Darstellung der arabischen verskunst
Freytag, Georg Wilhelm Friedrich *see* Al-muntahab min ta'ri-h halab
Freytag, Gustav *see*
– Die ahnen
– Aus dem jahrhundert des grossen krieges
– Aus einer kleinen stadt
– Briefe an seine gattin
– Dramatische werke
– Erinnerungen aus meinem leben
– Gesammelte werke
– Gustav freytag als politiker, journalist und mensch
– Gustav freytag und herzog ernst von coburg im briefwechsel 1853-1893
– Gustav freytags briefe an albrecht von stosch
– Ingo
– Die journalisten
– Lesebuch aus gustav freytags werken
– The lost manuscript
– Martin luther
– Soll und haben
– Die valentine
– Die verlorene handschrift
– Verlorene handschrift
Freywillige beytraege zu den hamburgischen nachrichten aus dem reiche der gelehrsamkeit / ed by Ziegra, Ch – Hamburg 1772-78 – (= ser Dz) – 6v on 30mf – 9 – €300.00 – mf#k/n289 – gw Olms [000]
Frezier, Amedee *see* Relation du voyage de la mer du sud aux cotes du chili, du perou, et du bresil
Frezier, Amedee F *see* A voyage to the south-sea
Fria ordet – Stockholm, 1880-81 – 2r – 1 – sw Kungliga [078]
Friar magazine – Butler. 1967-1979 (1) 1971-1979 (5) 1978-1979 (9) – ISSN: 0016-1225 – mf#2463 – us UMI ProQuest [240]
The friar of wittenberg / Davis, William Stearns – New York: Macmillan, 1912 – 1mf – 9 – 0-524-02180-5 – mf#1990-0565 – us ATLA [240]
Frick, Alfons *see* Ueber popes einfluss auf hagedorn
Frick, G *see* Goetz von berlichingen
Frick, Otto *see* Mythus und evangelium
Frick, William Keller *see* Henry melchior muhlenberg
Fricke, G A *see* Der paulinische grundbegriff der dikaiosyne theou
Fricke, Gerhard *see* Briefwechsel zwischen goethe und zelter
Fricke, Gustav Adolf *see*
– Ist gott persoenlich?
– Metaphysik und dogmatik
Fricke, Heiko *see* Untersuchung des bei der verformung von metalltraegern entstehenden koerperschall
Fricke, Hermann *see* The odor fontanes letzter romanentwurf
Fricke, Theodore P *see* We found them waiting
Fricker, Karl Viktor *see* Gebiet und gebietshoheit
Die frickes : a trilogy / Wilhelm, Hans Hermann – Berlin: Brunnen-Verlag. 3v. 1918-41 – 1 – us Wisconsin U Libr [830]
Friday caller – Plainfield, IN. 1904-1915 (1) – mf#62937 – us UMI ProQuest [071]
Friday evening post – Black River Falls WI. 1891 dec 18-1895 aug 30 – 1r – 1 – (cont by: fairchild graphic) – mf#1313024 – us WHS [071]
Friday journal – Geneva, NE: Frank O Edgecombe. 2v. 19th yr whole n978. jul 20 1894-20th yr whole n1112. nov 1 1895; 21st yr n7. nov 8 1895-21st yr n51. apr 10 1896 (wkly) [mf ed 1894-96 (gaps)] – 2r – 1 – (cont: republican-journal. merged with: nebraska signal to form: friday journal and the nebraska signal. cont the numbering of republican-journal. issue numbering alternates with tuesday republican and tuesday journal. tuesday ed: tuesday republican 1894-apr 9 1895 and: tuesday journal apr 16 1895-96) – us NE Hist [071]
Friday journal and the nebraska signal – Geneva, NE: Frank O Edgecombe. 1v. 21st yr n53. apr 17 1896 (wkly) – 1r – 1 – (formed by the union of: friday journal and: nebraska signal. merged with: tuesday journal and the nebraska signal to form: nebraska signal. cont the numbering of friday journal. issue numbering alternates with tuesday journal and the nebraska signal. tuesday ed: tuesday journal and the nebraska signal) – us NE Hist [071]
Friday leader – Lafayette, IN. 1951-1951 (1) – mf#62871 – us UMI ProQuest [071]
Friday meeting talks : or, divine prescriptions for the sick and suffering / Simpson, Albert B – New York: Christian Alliance, c1894 [mf ed 1992] – (= ser Christian & missionary alliance coll) – 3mf – 9 – 0-524-02269-0 – mf#1990-4276 – us ATLA [230]
Friday noon – Middletown, OH. 1929-1929 (1) – mf#65586 – us UMI ProQuest [071]
Fridell, Egon *see* Novalis als philosoph
Friderici, Daniel *see*
– Bicinia sacra...
– Deliciae juveniles
– Deliciarum juvenilium ander theil
– Musica figuralis oder neue klaeriche richtige und verstaendliche unterweisung der singe kunst
Fridericus – Berlin 1923 [single iss], 1924-36 [gaps], 1937 n33 [mf ed 2004] – 5r – 1 – (publ in various places: berlin, muenchen, hamburg, köln, bad oeynhausen; filmed by misc inst: 1923 jul-dez; 1922 aug-dez [1r]; 1924-39 [7r]) – gw Mikrofilm; gw Misc Inst [074]
Fridman, M *see* Nasha finansovaia sistema
Fridolin, S P *see* Molochnye tovarichestva moskovskoi gub v 1914 godu
Fridolin sichers chronik / Sicher, F – St Gallen, 1885 – (= ser Mittheilungen der vaterlaendischen Geschichte) – 4mf – 9 – (mittheilungen der vaterlaendischen geschichte. n s v10) – mf#ZWI-51 – ne IDC [240]
Friebe, Freimut *see* Das risorgimento im roman bei george meredith und antonio fogazzaro
Friebe, Karl *see*
– Christian hofmann von hofmannswaldaus grabschriften
– Ueber c. hofmann von hofmannswaldau und die umarbeitung seines getreuen schaefers
Friebe, Michael Horst *see* Ortsaufloesende temperaturmessung im niedrigfeld- und hochfeld kernspintomographen und untersuchung der moeglichkeit zur anwendung in der intra-tumoralen lasertherapie
Fried, Constance R *see* An examination of the test characteristics of the 12 minute aerobic swim test
Fried, John H E *see* United states military intervention in cambodia in the light of international law
Frieda, Juan *see* Quimbayas bajo la dominacion espanola
Friedberg, A *see* Corpus iuris canonici 1-2
Friedberg, Bernhard *see* Luhoth zikaron
Friedberg, Emil *see*
– Aktenstuecke die altkatholische bewegung betreffend
– Aus deutschen bussbuechern
– Lehrbuch des katholischen und evangelischen kirchenrechts
– Sammlung der aktenstuecke zum ersten vaticanischen concil
– Der staat und die katholische kirche im grossherzogthum baden
Friedberger anzeigenblatt – Friedberg, Hessen DE, 1949 6 may-23 jul – 1 – gw Mikrofilm [074]
Friedberger intelligenzblatt *see* Intelligenzblatt fuer die provinz oberhessen
Friede auf erden! : oder die ausweisung am weihnachtsabend; soziales bild in zwei aufzuegen / Lipinski, Richard – 8. Aufl. Berlin: A Hoffmann, 1921 – 1r – 1 – us Wisconsin U Libr [820]
Friede fuer babel und bibel / Giesebrecht, Friedrich – Koenigsberg i Pr: Thomas & Opermann 1903 [mf ed 1993] – 1mf – 9 – 0-524-06677-9 – mf#1992-0930 – us ATLA [220]
Friede, Juan *see*
– Vida y viajes de nicolas federman, conquistador
– Welser en la conquista de venezuela
Friedeberg, S *see* Yehoshua
Friedeberger kreisblatt – Friedeberg (Strzelce Krajenskie PL), 1842 jul-dec, 1847-48, 1935 jan-jun, 1935 oct-1936, 1937 apr-sep, 1938 oct-1939 mar, 1939 jul-dec, 1940 jul-1941 jun, 1941 oct-1942 jun, 1943 jan-jun – 1 – (title varies: 6 feb 1935: die grosse heimatzeitung) – gw Misc Inst [077]
[Friedel, J] *see* Briefe ueber die galanterien von berlin
Friedel und oswald : roman aus der tiroler geschichte / Schmid, Herman – 2.aufl. Leipzig: Keil [18-?] – (= ser Gesammelte schriften. volks- und familien-ausgabe 21-23) – 3pt in 1v – 1 – mf#film mas c438 – us Harvard [830]
Friedell, Egon *see*
– Friedell-brevier
– Die reise mit der zeitmaschine
Friedell-brevier / Friedell, Egon – Wien: E Mueller, 1947 (mf ed 1990) – 1r – 1 – (filmed with: gustav freytag) – us Wisconsin U Libr [880]
Friedemann, Edmond *see* Das judenthum und richard wagner
Frieden / Glaeser, Ernst – 2. Aufl. Berlin: G Kiepenheuer, 1930 (mf ed 1990) – 1r – 1 – (filmed with: hermann von gilm) – us Wisconsin U Libr [430]
Friedens sieg : ein freudenspiel / Schottelius, Justus Georg; ed by Koldewey, Friedrich E – Halle: M Niemeyer, 1900 – us Wisconsin U Libr [430]
Friedens- und kriegs-courier *see* Teutscher kriegs-courier
Friedensbewegung – Berlin, Germany. 1922 – 1r – us UF Libraries [943]
Der friedensbote : amtliche zeitschrift der evangelischen und reformierten kirche – v1-96. n1-13. 1850-1958 [complete] – 32r – 1 – mf#ATLA S0091 – us ATLA [242]
Der friedensplan des leibniz zur wiedervereinigung der getrennten christlichen kirchen : aus seinen verhandlungen mit dem hofe ludwigs 14., leopolds 1. und peters des grossen / Kiefl, Franz Xaver – Paderborn: F Schoeningh, 1903 – 1mf – 9 – 0-524-00047-6 – (incl bibl) – mf#1989-2747 – us ATLA [240]
Friedenspraeliminarien – Leipzig 1809 – (= ser Dz. historisch-politische abt) – v2 on 6mf – 9 – €120.00 – mf#k/n1842 – gw Olms [327]
Die friedenstat – Schoenebeck DE, 1967-1989 23 nov [gaps] – 1r – 1 – (sprengstoffwerk schoenebeck) – gw Misc Inst [621]
Friedenstimme – Halbstadt, Russia. (German Mennonites.) 1906-Nov. 1914 – 1 – us Southern Baptist [242]
Friedenthal, Herbert *see* Reichsverband der juedischen kulturbuende in deutschland
Friedenwald, Julius *see* Diet in health and disease
Friederich carl casimir freiherr von creuz und seine dichtungen : ein beitrag zur litteraturgeschichte des 18. jahrhunderts / Hartmann, Carl – Heidelberg: J Hoerning, 1891 [mf ed 1989] – 88p – 1 – (incl bibl ref) – mf#7160 – us Wisconsin U Libr [430]
Friederich, G *see* De vrijmetselarij en de jezuitisch-hierarchische propaganda
Friederich, G WE *see* Orchestral journal
Friederici, C *see* Bibliotheca orientalis
Friederike brion : ein beitrag zu goethes elsaessischer schuld und zur psychologie seiner liebe / List, Friedrich – Giessen: Ferber, 1923 – 1r – 1 – us Wisconsin U Libr [430]

Friederike von sesenheim im lichte der wahrheit / Duentzer, Heinrich – Stuttgart: J G Cotta, 1893 [mf ed 1990] – 152p – 1 – mf#7372 – us Wisconsin U Libr [430]

Friedl, Peter *see* Cell migration in three-dimensional collagen lattices. integrins, cell-matrix-interactions and migration strategies

Friedlaender, G *see* Beitraege zur reformationsgeschichte

Friedlaender, Heinrich *see*
- Historia economica de cuba

Friedlaender, I *see* Arabisch-deutsches lexikon zum sprachgebrauch des maimonides

Friedlaender, Ludwig *see* Roman life and manners under the early empire

Friedlaender, Ludwig H *see* Ansichten von italien

Friedlaender, Max *see* Gedichte von goethe in compositionen

Friedlaender, Moritz *see*
- Geschichte der juedischen apologetik als vorgeschichte des christenthums
- Griechische philosophie im alten testament
- Das judenthum in der vorchristlichen griechischen welt
- Die religioesen bewegungen innerhalb des judentums im zeitalter jesu
- Synagoge und kirche in ihren anfaengen
- Der vorchristliche juedische gnosticismus
- Zur entstehungsgeschichte des christenthums

Friedlaender, Moriz *see* Der antichrist in den vorchristlichen juedischen quellen

Friedlaender, Solomon Judah *see* Mavo la-tosefta

Friedlaender, Victor *see* Von hueben und drueben

Friedlaender, Walter F *see* Claude lorrain

Friedlaender zeitung – Friedland, Isergebirge (Frydlandt CZ), 1938 5 jan-31 dec – 1r – 1 – gw Misc Inst [077]

Friedlander, Gerald *see*
- Hellenism and christianity
- The jewish sources of the sermon on the mount

Friedlander, Heinrich *see* Historia economica de cuba

Friedlander, Joy L *see* Curricular and pedagogical vision in dance teacher preparation programs in higher education

Friedlander, M H *see* Moses mendelssohn und zeine zeit

Friedlander, Max Hermann *see* Tiferet jisrael

Friedlander, Moriz *see* Vorchristliche judische gnosticismus

Friedlieb, J H *see* Schrift, tradition und kirchliche schriftauslegung

Friedman, Amy A *see* The efficacy of water displacement as a potential tool for assessing total body composition

Friedmann, Hermann *see*
- Deutsche literatur im 20. jahrhundert
- Deutsche literatur im zwanzigsten jahrhundert
- Expressionismus

Friedmann, Joseph *see* Mikhtav 'oz

Friedmann, Sigismund *see* Ludwig anzengruber

Friedrich 2 von Preussen *see* Die morgen stunden eines koeniges an seinen bruder sohn 1766

Friedrich, A *see*
- Emblemata nova
- Emblemes nouveaux

Friedrich bodenstedt's gesammelte schriften : gesammt-ausgaben in zwoelf baenden = Works – Berlin: Verlag der Koeniglichen Hofbuchdruckerei, 1965-69 [mf ed 1989] – (= ser Bibliothek der deutschen literatur fiche 182-191) – 12v in 4 – 1 – mf#7041 – us Wisconsin U Libr [802]

Friedrich creutzer's deutsche schriften : neue und verbesserte: 5. abtheilung – Leipzig, Darmstadt: Leske, 1848 [mf ed 1989] – 2v in 1 – 1 – (v2: frankfurt/main j baer 1854) – mf#7160 – us Wisconsin U Libr [430]

Friedrich creuzer und karoline von guenderode : briefe und dichtungen / ed by Rohde, Erwin – Heidelberg: C Winter, 1896 [mf ed 2001] – xv/142p – 1 – (incl bibl ref) – mf#10506 – us Wisconsin U Libr [860]

Friedrich de la motte fouque als erzaehler / Jeuthe, Lothar – Breslau: F Hirt, 1910 [mf ed 1992] – (= ser Breslauer beitraege zur literaturgeschichte. neue folge 11) – 163p – 1 – (incl bibl ref) – mf#8014 reel 2 – us Wisconsin U Libr [430]

Friedrich dedekinds grobianus / ed by Milchsack, Gustav – Halle: Max Niemeyer, 1882 – us Wisconsin U Libr [430]

Friedrich der grosse : auswahl aus seinen schriften und briefen: nebst einigen gespraechen mit de catt / Frederick II, King of Prussia; ed by Lienhard, F – Stuttgart: Greiner und Pfeiffer, [1907?] (mf ed 1990) – 1r – 1 – (filmed with: gustav freytag, ein publizist) – us Wisconsin U Libr [880]

Friedrich der grosse : dramatische bilder: nach franz [theodor] kugler / Wesendonk, Mathilde – Berlin: Lipperheide 1871 – 1mf – 9 – mf#mw-4 – ne IDC [820]

Friedrich der weise und die schlosskirche zu wittenberg : festschrift zur einweihung der wittenberger schlosskirche, am tage des reformationsfestes, den 31. oktober 1892 / Koestlin, Julius – Wittenberg: R. Herrose, 1892 – 1mf – 9 – 0-7905-6199-9 – mf#1988-2199 – us ATLA [240]

Friedrich, Erich *see* Die siegfried-tragoedie im nibelungenring

Friedrich gerhard's deutsch-amerikanische gewerbe zeitung – New York NY (USA), 1859-60 – 1 – uk British Libr Newspaper [338]

Friedrich gottlieb klopstock / ed by Luetcke, Heinrich – Bielefeld: Velhagen & Klasing, 1931 – 1 – us Wisconsin U Libr [810]

Friedrich gottlieb klopstock : rede gehalten zur klopstockfeier der universitaet marburg am 6. juli 1924 / Elster, Ernst – Marburg a.L: N G Elwert, 1924 [mf ed 1991] – (= ser Marburger akademische reden 41) – 30p – 1 – mf#7519 – us Wisconsin U Libr [850]

Friedrich griese / Melcher, Kurt – Berlin: Junker & Duennhaupt 1936 [mf ed 1992] – (= ser Neue deutsche forschungen. abteilung neuere deutsche literaturgeschichte 7) – 2r – 1 – (incl bibl ref) – mf#3185p – us Wisconsin U Libr [430]

Friedrich haugs epigramme und ihre quellen / Steiner, Emil – [S.l.: s.n.], 1907 (Borna-Leipzig: Buchdruckerei R Noske) – 1r – 1 – (incl bibl ref) – us Wisconsin U Libr [430]

Friedrich hebbel : denker, dichter, mensch / Schuder, Kurt – Leipzig: O Weber, [1909?] [mf ed 1990] – 68p – 1 – mf#7454 – us Wisconsin U Libr [430]

Friedrich hebbel / Fassbinder, Franz – Koeln: Kommissionsverlag von J P Bachem, 1913 [mf ed 1994] – (= ser Vereinsgaben der goerres-gesellschaft 1913 2. vereinsschrift) – 131p – 1 – mf#8750 – us Wisconsin U Libr [430]

Friedrich hebbel / Federn, Etta [Etta Federn-Kohlhaas] – Muenchen: Delphin-Verlag, c1920 [mf ed 2001] – 347p/18pl – 1 – (incl ind) – mf#10587 – us Wisconsin U Libr [430]

Friedrich hebbel als denker / Muenz, Bernhard – 2. aufl. Wien: W Braumueller, 1907 [mf ed 1990] – 119p – 1 – (incl bibl ref) – mf#7454 – us Wisconsin U Libr [140]

Friedrich hebbel als lyriker / Engelhard, Karl – Leipzig: Verlag fuer Literature, Kunst und Musik, 1907 [mf ed 1990] – (= ser Beitraege zur literaturgeschichte 44) – 43p – 1 – mf#7452 – us Wisconsin U Libr [430]

Friedrich hebbel und die gegenwart : die tragische situation des nordischen menschen / Tideman, Wilhelm – [Heidelberg]: Kampmann & Schnabel, 1922 [mf ed 1990] – (= ser Philosophische schriften) – 90p – 1 – mf#7455 – us Wisconsin U Libr [190]

Friedrich hebbel und otto ludwig / Bartels, Adolf – [S.l: s.n.], 1895 [mf ed 1990] – (= ser Grenzboten 3) – 1 – mf#7449 – us Wisconsin U Libr [430]

Friedrich hebbel und seine dramen : ein versuch / Walzel, Oskar Franz – Leipzig: B G Teubner, 1913 [mf ed 1990] – (= ser Aus natur und geisteswelt 408) – 115p/1pl – 1 – mf#7455 – us Wisconsin U Libr [430]

Friedrich hebbels demetrius / Hebbel, Friedrich – Stuttgart: J G Cotta, [1910] [mf ed 1990] – (= ser Cotta'sche handbibliothek) – 130p – 1 – (completed by otto harnack) – mf#7446 – us Wisconsin U Libr [820]

Friedrich hebbels dramatischer stil / Wagner, Albert Malte – Hamburg: L Voss, 1910 [mf ed 1990] – 51p – 1 – (incl bibl ref) – mf#7455 – us Wisconsin U Libr [430]

Friedrich hebbels genoveva : eine monographie / Meszleny, Richard – Berlin: B Behr, 1910 [mf ed 2001] – (= ser Hebbel-forschungen 4) – 174p – 1 – mf#10627 – us Wisconsin U Libr [430]

Friedrich hebbels tagebuecher – Berlin: G Grote, 1885-1887 [mf ed 2001] – 2v – 1 – (pref by felix bamberg) – mf#10585 – us Wisconsin U Libr [880]

Friedrich hebbels und richard wagners nibelungen-trilogien : ein kritischer beitrag zur geschichte der neueren nibelungendichtung / Meinck, Ernst – Leipzig: M Hesse, 1905 [mf ed 1992] – (= ser Breslauer beitraege zur literaturgeschichte 5) – 94p – 1 – mf#8014 reel 1 – us Wisconsin U Libr [430]

Friedrich hebel und otto ludwig : ein vergleich ihrer ansichten ueber das drama / Bruns, Friedrich – Berlin-Steglitz: B Behr (F Feddersen), 1913 [mf ed 2001] – (= ser Hebbel-forschungen 5) – 1r – 1 – (filmed with: friedrich hebbel als kritiker des dramas / arthur kutscher (1907). incl bibl ref) – mf#10627 – us Wisconsin U Libr [430]

Friedrich heinrich jacobi : a study in the origin of german realism / Wilde, Norman – New York: Columbia College, 1894 – 1r – 1 – (incl bibl ref) – us Wisconsin U Libr [430]

Friedrich heinrich jacobi im verhaeltnis zu seinen zeitgenossen, besonders zu goethe : ein beitrag zur entwicklungsgeschichte der neuen deutschen literatur / Deycks, Ferdinand – Frankfurt a.M.: Verlag der Joh. Christ. Hermann'schen Buchhandlung, F E Suchsland, 1848 – 1r – 1 – (incl bibl ref) – us Wisconsin U Libr [430]

Friedrich heinrich jacobi und die fruehromantik / Bossert, Theodor Adolf – Giessen: [s.n.] 1926 [mf ed 1991] – 1r – 1 – (incl bibl ref; filmed with: tiefgluth / pedro ilgen) – mf#7500 – us Wisconsin U Libr [430]

Friedrich hoelderlin : leben und vermaechtnis / Thiele, Herbert – Metz: H Pfleger, 1943 – 1r – 1 – us Wisconsin U Libr [920]

Friedrich hoelderlin / Michel, Wilhelm – Weimar: E Lichtenstein, c1925 – 1r – 1 – us Wisconsin U Libr [430]

Friedrich hoelderlin / Unruh, Friedrich Franz von – Stuttgart: G Truckenmueller, 1942 – 1r – 1 – us Wisconsin U Libr [920]

Friedrich hoelderlin – fritz reuter : zwei biographien / Wilbrandt, Adolf von – Berlin: E Hoffmann, 1894 – 1r – 1 – us Wisconsin U Libr [920]

Friedrich hoelderlin und john keats als geistesverwandte dichter / Wenzel, Guido – Magdeburg: E Baensch, 1896 – 1r – 1 – (incl bibl ref) – us Wisconsin U Libr [410]

Friedrich, Johann *see*
- Beitraege zur geschichte des jesuiten-ordens
- Die constantinische schenkung
- Geschichte des vatikanischen konzils
- Ignaz von doellinger
- Johann adam moehler, der symboliker
- Johann adam moehler der symboliker
- Johann hus
- Johann wessel
- Das lukasevangelium und die apostelgeschichte
- Der mechanismus der vatikanischen religion
- Die merovingerzeit
- Die papst-fabeln des mittelalters
- Die roemerzeit
- Tagebuch, waehrend des vaticanischen concils
- Ueber wahrheit und gerechtigkeit
- Zur aeltesten geschichte des primates in der kirche

Friedrich, Julius *see* Die entstehung der reformatio ecclesiarum hassiae von 1526

Friedrich Karl, Freiherr von Moser *see* Woechentliche frankfurtische abhandlungen, zu erweiterungen der nothwendigen, brauchbaren und angenehmen wissenschaften

Friedrich, Karl Josef [comp] *see* Das hans thoma-buch

Friedrich leopold stolbergs jugendpoesie / Keiper, Wilhelm – Berlin: Mayer & Mueller, 1893 – 1r – 1 – (incl bibl ref) – us Wisconsin U Libr [430]

Friedrich leopolds grafen zu stolberg erste gattin agnes geb. von witzleben : ein lebensbild aus der zeit der empfindlichkeit / Hellinghaus, Otto – Koeln, 1919 (mf ed 1993) – 1mf – 9 – €12.00 – 3-89349-120-1 – mf#DHS-AR 89 – gw Frankfurter [943]

Friedrich ludwig schroeder : ein beitrag zur deutschen litterartur- und theatergeschichte / Litzmann, Berthold – Hamburg: L Voss, 1890-94 – 1r – 1 – (incl bibl ref) – us Wisconsin U Libr [430]

Friedrich ludwig stamm's ulfilas : oder die uns erhaltenen denkmaeler der gotischen sprache / ed by Heyne, Moritz & Wrede, Ferdinand – 9.aufl. Paderborn: F Schoeningh, 1896 [mf ed 1993] – (= ser Bibliothek der aeltesten deutschen litteratur-denkmaeler v1) – xv/443p – 1 – (incl bibl ref) – mf#8437 reel 1 – us Wisconsin U Libr [430]

Friedrich, Martinus *see* Encomium musicae vocalis et instrumentalis

Friedrich matthissons gedichte / ed by Boelsing, Gottfried – Stuttgart: Litterarischer Verein, 1912-13 (Tuebingen: H Laupp, Jr) [mf ed 1993] – (= ser Blvs 257, 261) – 2v – 1 – (incl bibl ref and ind) – mf#8470 reels 52-53 – us Wisconsin U Libr [810]

Friedrich matthissons gedichte / Matthisson, Friedrich von; ed by Boelsing, Gottfried – Stuttgart: Litterarischer Verein. 2v. 1912-13 (Tuebingen: H Laupp, Jr) – (incl bibl ref and ind) – us Wisconsin U Libr [810]

Friedrich melchior grimm als kritiker... *see* Der seltsame springinsfeld

Friedrich melchior grimm als kritiker der zeitgenoessischen literatur in seiner "correspondance litteraire" (1753-1770) : eine literarhistorische studie / Georges, Karl August – 1r – 1 – (incl bibl ref) – us Wisconsin U Libr [430]

Friedrich nietzsche : das subversive als denkansatz in seiner philosophie. ein beitrag zur interpretation / Schart, Franz Friedrich – Mainz: Gardez, 1994 (mf ed 1996) – (= ser Philosophie im Gardez) – 2mf – 3-8267-9668-3 – mf#DHS 9668 – gw Frankfurter [190]

Friedrich nietzsche und die kulturprobleme unserer zeit : vortraege / Kalthoff, Albert – Berlin: CA Schwetschke, 1900 – 1mf – 9 – 0-7905-9293-2 – mf#1989-2518 – us ATLA [190]

Friedrich nietzsches kulturphilosophie und umwertungslehre / Bubnov, Nikolai Mikhailovich – Leipzig: A Kroener 1924 [mf ed 1991] – 1r – 1 – (incl bibl ref; filmed with: die balladen und ritterlichen lieder des frieherrn borries von munchhausen) – mf#2850p – us Wisconsin U Libr [190]

Friedrich, Paul *see*
- Der fall hebbel
- Frank wedekind
- Die hebraischen conditionalsaetze

Friedrich rudolf ludwig von canitz / Lutz, Valentin – [S.l.: s.n.], 1887 (Neustadt: Aktien-Druckerei) – 1r – 1 – (incl bibl ref) – us Wisconsin U Libr [920]

Friedrich rueckert : ein lebens- und charakterbild fuer haus und schule / Beyer, Konrad – Frankfurt a. M: J D Sauerlaender, 1888 [mf ed 1995] – xiii/156/[1]pl – 1 – (incl bibl ref) – mf#8856 – us Wisconsin U Libr [920]

Friedrich schiller : leben, werk und wirkung / Kleinschmidt, Karl – Berlin: Kongress-Verlag, 1955 – 1r – 1 – us Wisconsin U Libr [920]

Friedrich schiller : stuermende jugend. lebenswerk / Scholz, Wilhelm – Muenchen: Deutsches Verlagshaus Bong, c1956 – 1r – 1 – us Wisconsin U Libr [920]

Friedrich schiller, sein leben und seine dichtungen : mit 701 abbildungen nach zeitgenoessischen bildern und illustrationen / Guentter, Otto – Leipzig: J J Weber, c1925 – 1r – 1 – us Wisconsin U Libr [920]

Friedrich schlegel 1794-1802 : seine prosaischen jugendschriften / ed by Minor, J – Wien: C Konegen, 1882 – 1r – 1 – us Wisconsin U Libr [430]

Friedrich schlegels briefe an frau christine von stransky – Wien: Literarischer Verein, 1907 – 2v – 1 – (incl ind) – us Wisconsin U Libr [860]

Friedrich schleiermacher : eine akademische rede / Schenkel, Daniel – Heidelberg: J C B Mohr, 1868 – 1mf – 9 – 0-7905-3623-4 – mf#1989-0116 – us ATLA [140]

Friedrich schleiermacher : ein lebens- und charakterbild / Schenkel, Daniel – Elberfeld: R L Friderichs, 1868 – 1mf – 9 – 0-524-03426-5 – (incl bibl ref) – mf#1990-0980 – us ATLA [240]

Friedrich schleiermacher als religioeser genius deutschlands / Hanne, Johann Wilhelm – Braunschweig: Oehme und Mueller, 1840 – 1mf – 9 – 0-524-00436-6 – mf#1989-3136 – us ATLA [240]

Friedrich schleiermachers grundriss der philosophischen ethik – Berlin: G Reimer 1841 [mf ed 1991] – 1mf – 9 – 0-524-00390-4 – (pref by august twesten) – mf#1989-3090 – us ATLA [170]

Friedrich schleiermacher's philosophische sittenlehre / ed by Kirchmann, Julius Hermann von – Berlin: L Heimann 1870 [mf ed 1991] – (= ser Philosophische bibliothek 24) – 2mf – 9 – 0-524-00333-5 – mf#1989-3033 – us ATLA [170]

Friedrich schleiermacher's reden ueber die religion : kritische ausgabe mit zugrundelegung des textes der ersten auflage / ed by Puenjer, Georg Christian Bernhard – Braunschweig: C A Schwetschke 1879 [mf ed 1990] – 1mf – 9 – 0-7905-4598-5 – mf#1988-0598 – us ATLA [210]

Friedrich spees trutz nachtigall – Heilbronn: G Henninger, 1876 [mf ed 1993] – vii/280p – 1 – (revived by karl simrock) – mf#8456 – us Wisconsin U Libr [810]

Friedrich stoltze's frankfurter latern *see* Frankfurter latern (sz2)

Friedrich und caroline perthes / Adler, Ottilie – Leipzig, 1900 (mf ed 1992) – 2mf – 9 – €24.00 – 3-89349-076-0 – mf#DHS-AR 41 – gw Frankfurter [070]

Friedrich vischer und der zweite teil von goethes faust : rede...der techn. hochschule. stgt. 5.5.1926 / Meyer, Theodor A – Stuttgart: A Bonz, 1927 – 1r – 1 – us Wisconsin U Libr [430]

Friedrich von hagedorns jugendgedichte : eine literarhistorische skizze / Badstueber, Hubert – Wien: A Pichler, 1904 [mf ed 1990] – iv/44p – 1 – (incl bibl ref) – mf#7428 – us Wisconsin U Libr [430]

Friedrich von hardenberg (genannt novalis) : eine nachlese aus den quellen des familienarchivs – 2. aufl. Gotha: F A Perthes, 1883 [mf ed 2001] – vi/278p/1pl (ill) – 1 – mf#10551 – us Wisconsin U Libr [860]

Friedrich von hardenbergs "christenheit oder europa" / Hederer, Edgar – Zeulenroda: B Sporn, 1936 [mf ed 1990] – 99p – 1 – (incl bibl ref) – mf#7435 – us Wisconsin U Libr [430]

Friedrich, Wilhelm *see* Ueber lessings lehre von der seelenwanderung

Friedrich wilhelm joseph schelling : gedaechtnissrede zur feier seines secular-jubilaeums am 27. januar 1875 im akademischen rosensaal zu jena / Pfleiderer, Otto – Stuttgart: JG Cotta, 1875 – 1mf – 9 – 0-7905-9576-1 – mf#1989-1301 – us ATLA [190]

Friedrich wilhelm marpurg (1718-1795) : eight of his works – repr Berlin. 13v – 11 – $220.00 set – (der kritische musikus 1750. abhandlung von der fuge 1753-54. historisch-kritisch beytraege 5v 1754-78. handbuch bey dem generalbasse 1757-62. kritische briefe ueber die tonkunst 2v 1760-64. die kunst das clavier zu spielen 1762. anleitung zum clavierspiel 1765. neue methode allerley arten von temperaturen dem claviere 1790) – us Univ Music [780]

Friedrich wilhelm marpurg's abhandlung von der fuge / Frederick, Kurt – U of Rochester 1951 [mf ed 19–] – 12mf / 3mf – 9 – mf#fiche 212 / fiche 464 – us Sibley [780]

Friedrich wilhelm webers jugendlyrik : auf ihre literarischen quellen und vorbilder untersucht und kritisch gewuerdigt mit benutzung seines ungedruckten nachlasses / Peters, Maria – [S.l: s.n, 1916?] – 1 – (incl bibl ref) – us Wisconsin U Libr [810]

Friedrich wilhelm zachariae in braunschweig / Zimmermann, Paul – Wolfenbuettel: J Zwissler, 1896 – 1r – 1 – (incl bibl ref and index) – us Wisconsin U Libr [430]

Friedrich wilhelme / Rosenberg, Alfred – Muenchen: Zentralverlag der NSDAP, F Eher, 1944 – 1r – 1 – us Wisconsin U Libr [190]

Friedrich, Wolfgang *see* Im klassenkampf

Friedrichs, Hermann *see*
- An der pforte der zukunft
- Gedichte
- Gestalten und leidenschaften
- Lebensbilder
- Liebeskaempfe

Friedrichs von logau saemmtliche sinngedichte / ed by Eitner, Gustav – Stuttgart: Litterarischer Verein, 1872 (Tuebingen: L F Fues) [mf ed 1993] – (= ser Blvs 113) – 817p – 1 – (incl bibl ref and ind) – mf#8470 reel 24 – us Wisconsin U Libr [810]

Friedrichs von logau saemmtliche sinngedichte / ed by Eitner, Gustav – Stuttgart: Litterarischer Verein, 1872 (Tuebingen: L F Fuess) – 1 – (incl bibl ref and ind) – us Wisconsin U Libr [430]

Friedrichstaedter intelligenzblatt *see* Der dittmarser und eiderstedter bote

Frielendorfer zeitung – Frielendorf DE, 1921 1 oct-1933 [gaps], 1935-1937 20 mar – 12r – 1 – (incl suppl: das leben im bild) – gw Misc Inst [074]

Friend / Coleridge, Samuel Taylor – London, England. v1-3. 1850 – 1r – us UF Libraries [420]

Friend / Hawaiian Evangelical Association – 1870 may 1 – 1r – 1 – (cont: friend, of temperance and seamen; cont by: news sheet [hawaiian evangelical association of the congregational christian churches]) – mf#1206041 – us WHS [242]

Friend / Hawaiian Evangelical Association of the Congregational Christian Churches et al – v19 n8-32:10 [1968 dec-1981 dec], v33-42 [1982-1990] – 2r – 1 – (cont: news from hawaiian evangelical association) – mf#618361 – us WHS [071]

Friend : a periodical work devoted to religion, literature and useful miscellany – Albany. 1815-1816 – 1 – mf#3806 – us UMI ProQuest [073]

Friend : a religious and literary journal – Philadelphia. 1827-1906 – 1 – mf#4453 – us UMI ProQuest [073]

The friend – Bloemfontein SA, 1896-1949 – 283r – 1 – sa National [079]

The friend : journal of the british quaker movement / Friends House Library. The Religious Society of Friends – 1843-1999+ – 115r – 1 – £5280.00 – mf#FRI – uk World [243]

The friend *see* The british friend

Friend and guide / Equitable Fraternal Union [Neenah, WI] – v1 n1-v21 n4 [1897 sep-1917 dec], v23 n12-v33 n5 [1920 aug-1930 jan] – 2r – 1 – (cont by: friend & guide, the messenger) – mf#1097331 – us WHS [366]

Friend and guide, the messenger / Equitable Fraternal Union [Neenah, WI] et al – 1930 feb-1933 dec, 1934 jan-1942 aug – 2r – 1 – (cont: friend and guide; cont by: equitable reserve guide) – mf#1097328 – us WHS [366]

A friend at court – v1-2. 1896-98 – 4mf – 9 – $6.00 – mf#LLMC 82-924 – us LLMC [340]

Friend [bloemfontein, south africa] – South Africa [mf ed National Library of South Africa] – jan 1937-dec 1939 – 1 – sa State Libr [079]

The friend daily telegraph – Friend, NE: E Whitcomb (daily) [mf ed v3 n2. oct 1 1884 filmed [1973]] – 1r – 1 – us NE Hist [071]

Friend free press – Friend, NE: H G Vines, 1890-v5 n19. jan 2 1891 (wkly) [mf ed with gaps filmed 1980-[85]] – 2r – 1 – (cont: weekly free press. cont by: people's rip-saw) – us NE Hist [071]

Friend, George William [comp] *see* An alphabetical list of engravings declared at the office of the printsellers' association, london

The "friend" in his family : or, a familiar exposition of some of the religious principles of the society of friends – London: Alfred W Bennett, 1865 – 1mf – 9 – 0-8370-8886-0 – (with brief biographical notices of a few of its early members) – mf#1986-2886 – us ATLA [240]

Friend indeed : for women in the prime of life – intro, 1984 apr-1993 mar – 1r – 1 – mf#2579629 – us WHS [305]

Friend, Margaret L *see* Without fear or favour

Friend of australia : or, a plan for exploring the interior, and for carrying on a survey of the whole continent of australia / Gardiner, Allen – London 1830 [mf ed Hildesheim 1995-98] – 1v on 3mf – 9 – €90.00 – ISBN-10: 3-487-26821-3 – ISBN-13: 978-3-487-26821-7 – gw Olms [919]

The friend of china – Canton: William Tarrant, oct 6 1860-dec 21 1861; feb 18 1863 – us CRL [079]

The friend of china – Hong Kong: Richard Oswald, mar 17 1842 – us CRL [079]

The friend of china and hong kong gazette – Victoria, HK: [Richard Oswald], mar 24 1842-1858 – 10r – 1 – us CRL [079]

Friend of india – Serampore, India: Mission Press, 1818-26 [mf ed 2001] – (= ser Christianity's encounter with world religions, 1850-1950) – 3r – 1 – (cont by the mthly & quarterly ser both of wh ran fr 1820-26) – mf#2001-s061-063 – us ATLA [230]

Friend of india – Serampore, India. 1852.-w. 1 reel – 1 – uk British Libr Newspaper [072]

Friend of man – Providence, RI. 1842-1843 (1) – mf#66308 – us UMI ProQuest [071]

Friend of man – Utica. 1836-1842 (1) – mf#5554 – us UMI ProQuest [976]

The friend of moses : or, a defence of the pentateuch as the production of moses and an inspired document, against the objections of modern skepticism / Hamilton, William Thomas – New York: M W Dodd, 1852 [mf ed 1989] – 2mf – 9 – 0-7905-1449-4 – (publ also in edinburgh as: the pentateuch and its assailants) – mf#1987-1449 – us ATLA [221]

Friend of peace – Boston. 1815-1827 (1) – mf#4454 – us UMI ProQuest [978]

The friend of peace – 1816-28 – (= ser The library of world peace studies) – 21mf – 9 – $155.00 – us UPA [320]

Friend of russians – Chicago, IL. n12-44. 1942-48 [complete] – 1r – 1 – (cont: russian millions) – mf#ATLA S0713B – us ATLA [073]

Friend of the free state – Bloemfontein SA, 1857-90 – 1 – sa National [079]

A friend of the queen (marie antoinette-count de fersen) / Gaulot, Paul – aut ed. New York: Appleton, 1893 [mf ed 1987] – xii/371p – 1 – mf#6878 – us Wisconsin U Libr [920]

The friend of the sovereignty – Bloemfontein SA, 1850-80 – 16r – 1 – sa National [960]

The friend sentinel – Friend, NE: W I Brundage, dec 1897 (wkly) [mf ed 1898-1907,1911 (gaps)] – 1 – (some irregularities in numbering) – us NE Hist [071]

Friend telegraph – Friend, NE: E Whitcomb. 38v. v28 n37. oct 20 1905-v65 n35. aug 28 1942 (wkly) [mf ed with gaps] – 9r – 1 – (cont: friend weekly telegraph) – us NE Hist [071]

Friend weekly telegraph – Friend, NE: E Whitcomb. 23v. v6 n23. jul 6 1883-v28 n36. oct 13 1905 (wkly) [mf ed with gaps] – 5r – 1 – (cont: friendville telegraph. cont by: friend telegraph) – us NE Hist [071]

Friendly address to jews / Gaius – London, England. 18– – 1r – us UF Libraries [240]

Friendly address to the episcopalians of scotland / Jolly, Alexander – Aberdeen, Scotland. 1826 – 1r – us UF Libraries [240]

Friendly address to the receivers of the doctrines of the new jerus... / Jones, Richard – Manchester, England. 1807? – 1r – us UF Libraries [240]

Friendly advice from a minister to the servants of his parish – London, England. 1814 – 1r – us UF Libraries [240]

Friendly agitator / Philadelphia Yearly Meeting of the Religious Society of Friends – 1971 apr/may, sep/oct-nov, 1972 feb, jul/aug-nov/dec, 1973 jan/feb-1974 mar, may/jun-1975 may/jun, 1976 feb, apr, 1977 feb,jun, nov, 1978 may-1983 sep – 1r – 1 – mf#500063 – us WHS [366]

The friendly arctic : the story of five years in polar regions / Stef nsson, V – New York, 1922 – 16mf – 9 – mf#N-406 – ne IDC [919]

A friendly exchange of views between quebec and ontario / Unity Publicity Bureau – Quebec: the Telegraph Printing Co, 1917 [mf ed 1991] – 1mf – 9 – mf#SEM105P1438 – cn Bibl Nat [971]

Friendly hints to candid sceptics / Hastings, H L – London, England. 1882 – 1r – us UF Libraries [240]

Friendly hints to female servants / Watkins, Henry George – London, England. 18– – 1r – us UF Libraries [240]

Friendly letters to a universalist on divine rewards and punishments / Whitman, Bernard – Cambridge: Brown, Shattuck, 1833 – 1mf – 9 – 0-524-07066-0 – mf#1992-1029 – us ATLA [240]

Friendly letters to the society of friends on some of their distinguishing principles / Wardlaw, Ralph – Glasgow: A Fullarton, 1836 – 5mf – 9 – 0-524-07472-0 – mf#1991-3132 – us ATLA [240]

The friendly messenger – Philadelphia: [s n] n440-489 (jul 1931-sep 1945) [gaps] [qrtly] [mf ed 2005] – 1r – 1 – ("for free distribution among the colored people of the southern states of the united states of america, and liberia in africa"; publ by: trustees under the will of charles l willits, deceased; cont: african's friend; lacks several iss; some pgs damaged) – mf#2005-s107 – us ATLA [200]

Friendly relations between great britain and germany : souvenir volume of the visit to germany of representatives of the british christian churches, june 7th to 20th, 1909 / ed by Siegmund-Schultze, Friedrich – Berlin: Printed by HS Hermann, [1909?] – 1mf – 9 – 0-524-00105-7 – mf#1989-2805 – us ATLA [240]

The friendly stars / Martin, Martha Evans – New York, London: Harper & Bros 1907 [mf ed 1998] – 1r [pl/ill] – 1 – (incl ind; int by harold jacoby) – mf#film mas 28407 – us Harvard [520]

A friendly talk about revision / Morris, Edward D – 1891 – 9 – $50.00 – us Presbyterian [240]

A friendly visit to the house of mourning / Cecil, Richard – New Brunswick: printed by Abraham Blauvelt, 1801 [mf ed 1994] – 1mf – 9 – 0-665-94669-4 – mf#94669 – cn Canadiana [240]

Friendly visit to the house of mourning / Cecil, Richard – London, England. 18– – 1r – us UF Libraries [240]

Friendly visitor : being a collection of select and original pieces, instructive and entertaining, suitable to read in all families – New York. 1825-1825 (1) – mf#3807 – us UMI ProQuest [240]

Friendly woman – 1978 spring, 1980 winter, fall, 1982 spring-fall, 1983 winter-fall, 1984 winter-1988 fall, 1989 winter-fall – 1r – 1 – mf#1056645 – us WHS [305]

The friends : who they are, what they have done / Beck, William – 1st ed. New York: Friends' Book and Tract Committee, 1897 – 1mf – 9 – 0-8370-8648-5 – (incl ind) – mf#1986-2648 – us ATLA [240]

Friends and foes in the transkei : an englishwoman's experience during the cape frontier war of 1877-78 / Pritchard, Helen M – London: Low, Marston, Searle & Rivington, 1880 – 1 – us CRL [960]

Friends association for abolishing state regulation of vice (1873-1910) / friends association for the promotion of social purity (1910-1926) : papers from the moral reform movement / Religious Society of Friends (Quakers) – 1864-1926 – 4r – 1 – 1-897955-29-4 – uk Academic [366]

Friends association for the promotion of social purity (1910-1926) *see* Friends association for abolishing state regulation of vice (1873-1910) / friends association for the promotion of social purity (1910-1926)

Friends' Association of Philadelphia and Its Vicinity, for the Relief of Colored Freedmen. Executive Board *see*
- Annual report of the executive board of the friends' association of philadelphia and its vicinity
- Report of the executive board of the friends' association of philadelphia and its vicinity

Friends beyond seas / Hodgkin, Henry Theodore – London: Headley Bros 1916 [mf ed 1992] – 1mf – 9 – 0-524-02828-1 – (incl bibl ref) – mf#1990-4449 – us ATLA [240]

Friends Committee on National Legislation [US] *see*
- Fcnl action
- Fcnl memo
- Fcnl washington newsletter
- Report on indian legislation

Friends Committee on National Legislation, Washington, DC *see* Fcnl washington newsletter

Friends for Jamaica *see* Caribbean newsletter

Friends for jamaica newsletter – iss 14 [1981 nov] – 1r – 1 – (cont by: caribbean newsletter [new york, ny]) – mf#4872494 – us WHS [071]

Friends House Library. The Religious Society of Friends *see*
- Anti-slavery collection
- The british friend
- Early quaker writings 17th-18th centuries
- The friend
- The friends' quarterly
- The great book of sufferings
- The london two weeks meeting
- Quaker digest registers of births, marriages and burials for england and wales
- The quaker manuscripts collection, pt a

Friends in the seventeenth century / Evans, Charles – new and rev ed. Philadelphia: Friends' Book-Store [distributor], 1885 – 2mf – 9 – 0-524-02464-2 – mf#1990-4323 – us ATLA [240]

Friends' intelligencer – Philadelphia. 1844-1910 – 1 – mf#4143 – us UMI ProQuest [073]

Friends journal – 1980-81, 1982 feb-1983 dec, 1984-85, 1986-1988 jun – 4r – 1 – (cont: friend; friends intelligencer) – mf#597187 – us WHS [071]

Friends journal – Philadelphia. 1955+ (1) 1971+ (5) 1977+ (9) – ISSN: 0016-1322 – mf#2675 – us UMI ProQuest [301]

The friends' library : comprising journals, doctrinal treatises, and other writings of members of the religious society of friends / ed by Evans, William & Evans, Thomas – Philadelphia, 1837-1850 – 16mf – 9 – 0-524-03150-9 – mf#1990-4599 – us ATLA [240]

The friends' meeting-house, fourth and arch streets, philadelphia : a centennial celebration, sixth month fourth, 1904 / Vaux, George et al – Philadelphia, PA: JC Winston, [1904?] – 1mf – 9 – 0-524-02967-9 – mf#1990-4519 – us ATLA [240]

Friends of a half century : fifty memorials with portraits of members of the society of friends, 1840-1890 / ed by Robinson, William – 1st ed. London: Edward Hicks, 1891 [mf ed 1992] – (= ser Society of friends (quakers) coll) – 1mf – 9 – 0-524-04124-5 – mf#1992-2010 – us ATLA [243]

Friends of allensworth, los angeles chapter newsletter – 1994 apr/jun – 1r – 1 – mf#4877104 – us WHS [366]

Friends of Channel 21 *see* TwentyOne magazine

The friends of christ keep his commandments or obedience the test of discipleship / Fittz, Hervey – (A sermon). 1834 – 1 – 5.00 – us Southern Baptist [242]

Friends of Democracy and Independence in Spain *see* Italians in spain

Friends of europe publications – London (GB), n.d, n1-75 – 1 – gw Misc Inst [940]

Friends of Haiti *see* Haiti report

Friends of Janet *see* Design of the times

Friends of Micronesia *see* Newsletter

Friends of Old Sturbridge Village *see* Rural visitor

Friends of raton anthropology – v4 n4 [1979 fall], v6 n1 [1981 jan], v6 n3-v8 n4 [1981 jan-1983 jul] – 1r – 1 – mf#1420813 – us WHS [301]

Friends of Rural Education *see* Watch dog

Friends of russia *see* Gospel in russia

Friends of russians – London. n12-44. 1937-41 [complete] – 1r – 1 – (filmed with: gospel in russia) – mf#ATLA S0702A – us ATLA [240]

Friends of Spanish Democracy *see* Spain

Friends of the children of lascahobas, haiti, inc – 1991 jan, sep, 1994 apr – 1r – 1 – mf#5308511 – us WHS [071]

Friends of the city of new york in the nineteenth century / Wood, William H S – New York: [s.n.], 1904 – 1mf – 9 – 0-524-03629-2 – mf#1990-4789 – us ATLA [240]

The friends' quarterly / Friends House Library. The Religious Society of Friends – 1867-1999+ – 41r – 1 – £1800.00 – mf#FQE – uk World [240]

Friends review : a religious, literary and miscellaneous journal – Philadelphia. 1847-1894 – 1 – mf#3984 – us UMI ProQuest [073]

Friends service council. annual reports – London: Friends Service Council (Quakers), 1927-78 [mf ed 2001] – (= ser Christianity's encounter with world religions, 1850-1950) – 9r – mf#2001-s073-081 – us ATLA [243]

Friendship / Black, Hugh – Chicago: Fleming H Revell c1903 [mf ed 1986] – 1mf – 9 – 0-8370-6021-4 – mf#1986-0021 – us ATLA [240]

Friendship *see* Peng yu (ccm288)

Friendship baptist church fishville (pollock), louisiana : church minutes – Pollock, LA. 2246p. 1870-1947; 1948-99 – 1 – $101.07 – mf#6981 – us Southern Baptist [242]

Friendship baptist church. jefferson county. tennessee : church records – 12 Mar 1819-97 – 1 – us Southern Baptist [242]

Friendship baptist church. jefferson county. tennessee : church records – 1858-1902 – 1 – 8.28 – us Southern Baptist [242]

Friendship baptist church. spartanburg county. south carolina : church records – 1801-1933, 1919-74 – 1 – 72.68 – us Southern Baptist [242]

Friendship Community *see* Communist

Friendship Force of Wisconsin *see* Newsletter

Friendship House *see* Community

The friendship of art / Carman, Bliss – Toronto: Copp, Clark Co, 1904 – 4mf – 9 – 0-665-73651-7 – mf#73651 – cn Canadiana [840]

The friendship of books, and other lectures / Maurice, Frederick Denison; ed by Hughes, Thomas – London: Macmillan, 1874 – 1mf – 9 – 0-7905-7448-9 – mf#1989-0673 – us ATLA [000]

Friendship reporter – Friendship WI. 1908 aug 21/sep 10-1995 jul/dec [gaps] – 59r – 1 – (cont: dells reporter) – mf#1139501 – us WHS [071]

Friendship village / Gale, Zona – New York, NY. 1909, 1908 – 1r – us UF Libraries [960]

Friendship's offering – 1824-44 – (= ser English gift books and literary annuals, 1823-1857) – 98mf – 9 – uk Chadwyck [800]

Friendville advocate – Friendville, NE: Wm A Connell, 1877-78// (wkly) [mf ed v1 n8. may 4-nov 16 1877 (gaps)] – 1r – 1 – (cont by: friendville telegraph) – us NE Hist [071]

Friendville telegraph – Friendville, NE: Wells and Allen, 1878-v6 n22. jun 29 1883 (wkly) [mf ed with gaps] – 2r – 1 – (cont: friendville advocate. cont by: friend weekly telegraph) – us NE Hist [071]

Fries, Adelaide Lisetta *see* The moravians in georgia, 1735-1740

Fries, J *see*
- Dictionariolum puerorum tribus linguis latina, gallica et germanica conscriptum...
- Dictionarium latinogermanicum...
- Novum latinogermanicum et germanicolatinum lexicon... ex probatis auctoribus digestum...

Fries, Kena *see* Orlando in the long, long ago and now

Fries, S A *see* Moderne darstellungen der geschichte israels

Fries, Samuel Andreas *see* Die gesetzesschrift des koenigs josia

Friesche lust-hof... / Starter, J J – Amsterdam: Dirck Pietersz Voscuyl, 1621 – 4mf – 9 – mf#0-3274 – ne IDC [090]

Friese, Philip Christopher *see* Semitic philosophy

Friesen-courier – Bredstedt DE, 1903 11 apr-1942 – 1 – gw Misc Inst [074]

Frieze, Henry Simmons *see* Giovanni dupre

Frigid facts / Montgomery Co. Dayton – aug 1946-mar 1949, dec 1950, dec 1951 [mthly] – 1r – 1 – mf#B10190 – us Ohio Hist [331]

Frihetsvaennen – Stockholm, 1860-64 – 9 – sw Kungliga [078]

Friis, A *see* Danmark ekspeditionen til gronlands nordostkyst

Frimmel, T *see* Der anonimo morellia (marcanton michiel's notizia d'opere del disegno)

Frimorgn – Riga. Apr 16 1926-May 13 1934. Incomplete – 1 – us NY Public [077]

The fringe of the east : a journey through past and present provinces of turkey / Lukach, H C – London, 1913 – 5mf – 9 – mf#AR-2029 – ne IDC [915]

Frings, Ketti *see* Hold back the dawn

Frings, Theodor *see*
- Eneide
- Morant und galie

Frisancho Pineda, David *see* Jatun rijchari-h

Frisbie-Frisbee Family Association of America *see* Bulletin of the frisbie-frisbee...

Frisch, Christian *see* Joannis kepleri astronomi opera omnia

Frisch, Daniel *see* Heymland

Frischauer, Paul *see*
- A great lord
- Presidente vargas

Frische clavier fruechte : oder sieben suonaten von guter invention und manier auff dem claviere zu spielen / Kuhnau, Johann – Dresden, Leipzig: Joh Christoph Zimmermann 1700 – 9 – mf#pres. film 174 – us Sibley [780]

Frischlin, Nicodeums *see* Deutsche dichtungen

Frischmann, David *see* Ba-arets

Das frisierte testament : eine komoedie in prosa / Pfannenschmidt, Heinz – Berlin: P Neff c1940 [mf ed 1991] – 1r – 1 – (filmed with: die leute aus der mohrenapotheke / ernst penzoldt) – mf#2862p – us Wisconsin U Libr [830]

Die frist / Ehrler, Hans Heinrich – Muenchen: G Mueller c1931 [mf ed 1989] – 1r – 1 – (filmed with: menschen und affen / albert ehrenstein) – mf#7207 – us Wisconsin U Libr [830]

Fristoe, William *see* A concise history of the ketocton baptist association

Frit danmark – London, UK. 16 Dec 1940-29 Jun 1945 – 1 – uk British Libr Newspaper [072]

Fritaenkaren – Rock Island, IL: Lutheran Augustana Book Concern, 1893 – 1mf – 9 – 0-524-05707-9 – mf#1991-2321 – us ATLA [240]

Frith, Francis *see* The quaker ideal

[Frith, Francis et al] *see* A reasonable faith

Frith, William Powell *see*
- John leech
- My autobiography and reminiscences
- The railway station painted

Fritsch, Erdmann *see* Islam und christentum im mittelalter

Fritsch, Gustav *see*
- Die eingeborenen sued-afrika's

Fritsch, Johann *see* Taschenbuch fuer reisende ins riesengebirge

Fritsch, Otto *see* Martin opitzen's buch von der deutschen poeterei

Fritschel, George John *see*
- The formula of concord
- Geschichte der lutherischen kirche in amerika
- Die schriftlehre von der gnadenwahl

Fritschel, Gottfried *see*
- Geschichte der christlichen missionen unter den indianern nordamerikas im 17. und 18. jahrhundert
- Die religion der geheimen gesellschaften
- Traktat von der gnadenwahl

Fritschel, Siegmund *see* Die unterscheidungslehren der synoden von iowa und missouri

Fritsches new ulmer wochenblatt – New Ulm MN (USA), 1928 16 jun-1929 23 feb [gaps] – 1r – 1 – gw Misc Inst [071]

Fritts, Suzanne M *see* Psychological factors that predispose athletes to injury

Fritz ellrodt : roman / Gutzkow, Karl – 2. aufl. Jena: H Costenoble, 1874 [mf ed 2001] – 3v in 2 – 1 – mf#10520 – us Wisconsin U Libr [830]

Fritz flesch collection on jews in south africa / Flesch, Fritz, collector – 1946-72 – 1 – us CRL [939]

Fritz, Georg *see* Ad majorem dei gloriam!

Fritz, Josef *see* Das wagnervolksbuch im 18. jahrhundert

Fritz mauthners ausgewaehlte schriften / Mauthner, Fritz – Stuttgart: Deutsche Verlags-anstalt [c1919] – 6v on 1r – 1 – mf#film mas 8595 – us Harvard [802]

Fritz, Otto *see* Johann peter hebels ausgewaehlte erzaehlungen u gedichte

Fritz reuter / Griese, Friedrich – Stuttgart: J G Cotta 1938 [mf ed 1992] – 1r – 1 – (filmed with: heiterer guckkasten / bruno wolfgang) – mf#2865p – us Wisconsin U Libr [430]

Fritz reuter, heinrich seidel und der humor in der neueren deutschen dichtung / Biese, Alfred – Kiel: Lipsius und Tischer, 1891 [mf ed 1993] – (= ser Deutsche schriften fuer litteratur und kunst. 1. reihe 5) – 55p – 1 – mf#8035 – us Wisconsin U Libr [430]

Fritz stavenhagens "mudder mews" / Stolle, Carl – Marburg a.L.: N G Elwert, 1926 – 1r – 1 – (incl bibl ref) – us Wisconsin U Libr [430]

Fritz und die soldatenstiefel : geschichten vom kriege / Wiemer, Rudolf Otto – Muenchen: Deutscher Volksverlag, 1943 – 1r – 1 – us Wisconsin U Libr [943]

Fritze, Franz *see* Der nordwesten des thueringer waldes oder zehn tage in ruhla

Fritze, J G *see* Medizinische annalen fuer aerzte und gesundheitsliebende

Fritzlar, Herbert von *see* Herbort's von fritslar liet von troye

Fritzlarer kreis-anzeiger *see* Kreis-anzeiger

Fritzlarer zeitung – Fritzlar DE, 1888 4 sep-1920 – 20r – 1 – (incl suppl) – gw Misc Inst [074]

Fritzlar-homberger allgemeine *see* Hessische allgemeine (hna)

Fritzsch, Robert *see* Ueber wolframs von eschenbach religiositaet

Fritzsche, Gerhard *see* Schwert und kelle

Fritzsche, O *see* Glarean, sein leben und seine schriften

Fritzsche, Otto Fridolin *see*
- Kurzgefasstes exegetisches handbuch zu den apokryphen des alten testamentes
- Libri apocryphi veteris testamenti graece

Fritzsche, Volkmar *see* Das berufsbewusstsein jesu mit beruecksichtigung geschichtlicher analogien untersucht

Frizen, F *see* Pod gnetom religii

Frizen, I I *see* Drug naroda

Frobenius, Leo *see*
- African genesis
- Unbekannte afrika
- Das zeitalter des sonnengottes

Frobisher, M *see* The three voyages of...in search of a passage to cathaia and india by the north-west, a d 1576-1578

Frodsham and helsby weekly news and kingsley reporter – [NW England] Frodsham, Cheshire Record Off 11 jun-31 dec 1909 – 1 – uk MLA; uk Newsplan [072]

Froeb, Hermann *see* Ernst kochs "prinz rosa-stramin"

Froebel journal, the... 1965-74 : the journal of the national froebel foundation – n1-30 – 46mf – 9 – mf#86823 – uk Microform Academic [073]

Froebel, Julius *see*
- Aus amerika
- Reise in die weniger bekannten thaeler auf der nordseite der penninischen alpen
- Die republikaner

Ein froehlich herz / Kinau, Rudolf – Hamburg: Quickborn-Verlag, 1941 – 1r – 1 – us Wisconsin U Libr [430]

Der froehliche botschafter – 1841-1901 [complete] – 17r – 1 – (title varies) – mf#atla s0115 – us ATLA [242]

Das froehliche buch : aus deutscher dichter und maler kunst / Avenarius, Ferdinand [comp]; ed by Kunstwart – Muenchen: G D W Callwey im Kunstwart-Verlage, 1909 [mf ed 1993] – ix/422p/12pl (ill) – 1 – mf#8360 – us Wisconsin U Libr [810]

Der froehliche goethe / Bode, Wilhelm – Berlin: E S Mittler, 1912 [mf ed 1990] – xi/383p – 1 – mf#7350 – us Wisconsin U Libr [430]

Der froehliche weinberg : lustspiel in drei akten / Zuckmayer, Carl – Berlin: Propylaeen-Verlag c1925 [mf ed 1991] – 1r – 1 – (filmed with: untersuchungen zur biographie philipp zesens / max gebhardt) – mf#2966p – us Wisconsin U Libr [820]

Froehling, Fritz *see* Die meldertasche

Froehling, Lori A *see* Effectiveness of exercise versus exercise plus tape in the management of females with patellofemoral pain

Froelich, J-C *see* La tribu konkomba du nord togo

Froelich, Jean Claude *see* La tribu konkomba du nord togo

Froelich, Marcus *see* Sefer ha-madrikh

Froelich, Richard *see* Tamulische volksreligion

Die froemmigkeit des grafen ludwig von zinzendorf : ein psychoanalytischer beitrag zur kenntnis der religioesen sublimierungsprozesse und zur erklaerung des pietismus / Pfister, Oskar Robert – Leipzig: F Deuticke, 1910 – (= ser Schriften zur angewandten seelenkunde) – 1mf – 9 – 0-7905-6875-6 – (incl bibl ref) – mf#1988-2875 – us ATLA [150]

Die froemmigkeit philos und ihre bedeutung fuer das christentum : eine religionsgeschichtliche studie / Windisch, Hans – Leipzig: J C Hinrichs, 1909 – 1mf – 9 – 0-7905-0460-X – (incl bibl ref and indexes) – mf#1987-0460 – us ATLA [240]

Froereisen, Isaac *see*
- Griechische dramen in deutschen bearbeitungen

Die froesche des aristophanes : mit ausgewaehlten antiken scholien / ed by Suess, Wilhelm – Bonn: A Marcus & E Weber 1911 [mf ed 1992] – (= ser Kleine texte fuer theologische und philologische vorlesungen und uebungen 66) – 2mf – 9 – 0-524-04692-1 – (in greek; int in german) – mf#1990-3401 – us ATLA [450]

Frog farms near orlando / Harold, William G – s.l, s.l? 1936 – 1r – us UF Libraries [639]

Froger, Francois *see* Relation d'un voyage fait en 1695, 1696 et 1697 aux cotes d'afrique, detroit de magellan, brezil, cayenne et isles antilles

Froger, J *see* Les origines de la prime

Froger, L *see* Les chants de la messe aux 8th et 9th siecle

Froget, Barthelemy *see* De l'habitation du saint-esprit dans les aames justes d'apres la doctrine de saint thomas d'aquin

Frohes leben : geschichten / Steguweit, Heinz – feldpostausg. Muenchen: A Langen, G Mueller 1943, c1934 [mf ed 1991] – (= ser Die kleine buecherei 34) – 1r – 1 – (filmed with: ins volle menschenleben & other titles) – mf#2897p – us Wisconsin U Libr [830]

Frohlich, Louis D *see* The law of motion pictures, including the law of the theatre treating of the various rights of the author, actor.

Frohman, Daniel *see* Memories of a manager; reminiscences of the old lyceum and of some players of the last quarter century

Frohn, K von *see* Neueste staatskunde von deutschland

Frohnmeyer, Ludwig Johannes *see*
- Bilderatlas zur bibelkunde
- A progressive grammar of the malayalam language for europeans

Frohschammer, J *see* Athenaeum

Frohschammer, Jakob *see*
- Das neue wissen und der neue glaube
- Die philosophie des thomas von aquino
- The reality of romanism
- The romance of romanism
- Ueber den ursprung der menschlichen seelen
- Ueber die bedeutung der einbildungskraft in der philosophie kant's und spinoza's
- Ueber die genesis der menschheit und deren geistige entwicklung in religion, sittlichkeit und sprache
- Zur wuerdigung der unfehlbarkeit des papstes und der kirche

Froidmont, Libert *see* Labyrinthvs sive de compositione continvi liber vnvs

Froissart, and his times / Saint Leger, Francis – London 1832 [mf ed Hildesheim 1995-98] – 3v on 8mf – 9 – €160.00 – ISBN-10: 3-487-26306-8 – ISBN-13: 978-3-487-26306-9 – gw Olms [440]

Froissart, Jean *see*
- Chroniques de jean froissart
- Sir john froissart's chronicles of england, france, spain, and the adjoining countries

Froitzheim, Johann *see*
- Goethe und heinrich leopold wagner
- Lenz und goethe

Das frolockende augspurg : wie solches wegen der hoechst-begleuckten geburt desz durchleuchtigsten ertz-herzogen und printzen von asturien leopoldi 2 / Kolb, J Chr – Augspurg: Gedruckt bey Andreas Maschenbauer, 1716 – 3mf – 9 – mf#O-17 – ne IDC [090]

From a colonial governor's note-book / St Johnston, Thomas Reginald – London, England. 1936 – 1r – us UF Libraries [972]

From abraham to david : the story of their country and times / Harper, Henry Andrew – New York: Macmillan, 1892 [mf ed 1986] – 1mf – 9 – 0-8370-9872-6 – (incl ind) – mf#1986-3872 – us ATLA [221]

From advent to advent : sermons / Moore, Aubrey Lackington – London: Percival, 1892 [mf ed 1991] – 1mf – 9 – 0-7905-9415-3 – mf#1989-2640 – us ATLA [242]

From akbar to aurangzeb : a study in indian economic history / Moreland, William Harrison – London: Macmillan, 1923 – (= ser Samp: indian books) – us CRL [330]

From an indian zenana : the story of lydia muttulakshmi / Picken, W H Jackson – London: Charles H Kelly, 1892 [mf ed 1995] – (= ser Yale coll) – 64p (ill) – 1 – 0-524-09977-4 – (pref by mrs wiseman) – mf#1995-0977 – us ATLA [920]

From apollyonville to the holy city : a poem / Allen, John Slater – Halifax, NS: Printed for the author at the Wesleyan Office, 1880 – 4mf – 9 – mf#05825 – cn Canadiana [810]

From atheism to christianity / Porter, George P – New York: Nelson & Phillips; Cincinnati: Hitchcock & Walden, 1873, c1872 [mf ed 1985] – 1mf – 9 – 0-8370-4781-1 – mf#1985-2781 – us ATLA [210]

From baca to beulah : sequel to "valley of baca" / Smith, Jennie – Philadelphia: Garringues Bros, 1880 [mf ed 1984] – (= ser Women & the church in america 185) – 1mf – 9 – 0-8370-1448-4 – mf#1984-2185 – us ATLA [305]

From bad to worse / hard to beat / and, a terrible christmas : three stories of montreal life / Phillips, John Arthur – Montreal: Lovell, 1877 – 4mf – 9 – mf#12030 – cn Canadiana [880]

From benguella to the territory of yacca / Capello, Hermenegildo Carlos De Brito – New York, NY. v1-2. 1969 – 1r – us UF Libraries [960]

From benguella to the territory of yacca : description of a journey into central and west africa....in the years 1877-1880 / Capello, H & Ivens, R – London, 1882. 2v – 11mf – 9 – mf#HT-24 – ne IDC [916]

From bombay to bushire, and bussora : including an account of the present state of persia, and notes on the persian war / Shepherd, William Ashton – London: R. Bentley, 1857. xi,236p – 1 – us Wisconsin U Libr [950]

From bortkiewics to kolmogorov : yet more russian papers on probability and statistics. aus dem russischen von oscar sheynin / Sheynin, Oscar – (mf ed 1999) – 3mf – 9 – €49.00 – 3-8267-2656-1 – mf#DHS 2656 – gw Frankfurter [510]

From boston to bareilly and back / Butler, William – New York: Phillips & Hunt; Cincinnati: Cranston & Stowe 1885 [mf ed 1993] – 2mf [ill] – 9 – 0-524-08352-5 – mf#1993-3052 – us ATLA [242]

From buddhism to christianity *see* Seng lu hsin chu chi (ccm207)

From buddhist priest to christian evangelist / Vories, William Merrell – Hachiman, Omi, Japan: Omi mission, [1917?] [mf ed 1995] – (= ser Yale coll) – 51p (ill) – 1 – 0-524-09829-8 – mf#1995-0829 – us ATLA [920]

From capetown to ladysmith / Steevens, George Warrington – Edinburgh, Scotland. 1900 – 1r – us UF Libraries [960]

From coast to coast : a farmer's ramble through canada, and the canadian pacific railway system / Ford, Charles T – [Exeter, England?: s.n.], 1899 [mf ed 1981] – 1mf – 9 – mf#14989 – cn Canadiana [917]

From columbus to bolivar / Arias Larreta, Abraham – Los Angeles, CA. 1965 – 1r – us UF Libraries [972]

From comte to benjamin kidd : the appeal to biology or evolution for human guidance / Mackintosh, Robert – New York: Macmillan, 1899 – 1mf – 9 – 0-7905-9326-2 – mf#1989-2551 – us ATLA [574]

From constantinople to the home of omar khayyam : travels in transcaucasia and northern persia for historic and literary research / Jackson, Abraham Valentine Williams – New York: Macmillan, 1911 – 2mf – 9 – 0-524-03673-X – (incl bibl ref) – mf#1990-3251 – us ATLA [915]

From 'coolie location' to group area : a brief account of johannesburg's indian community / Randall, Peter & Desai, Yunus – Johannesburg, South African Institute of Race Relations, 1967 – us CRL [960]

From daniel bernoulli to urlanis : still more russian papers on probability and statistics. aus dem russischen von oscar sheynin / Sheynin, Oscar – (mf ed 2000) – 3 mf – 9 – €49.00 – 3-8267-2696-0 – mf#DHS 2696 – gw Frankfurter [510]

From dark to dawn : being a second series of night scenes in the bible / March, Daniel – Philadelphia: McCurdy, 1878 – 2mf – 9 – 0-524-03982-8 – mf#1992-0025 – us ATLA [220]

From darkness to light : history of the eight prisons which have been, or are now, in montreal, from a d 1760 to a d 1907 "civil and military"... / Borthwick, John Douglas – Montreal: Gazette Print, 1907 – 2mf – 9 – 0-665-71602-8 – mf#71602 – cn Canadiana [365]

From darkness to light : a series of autobiographical sketches relating to religious experiences / Sanders, B B et al – Cincinnati, O[hio]: Standard Pub Co, c1907 – 1mf – 9 – 0-524-07041-5 – mf#1991-2894 – us ATLA [240]

From darkness to light : a sketch of the life of james t quinlan / Ellis, Joseph J – Baltimore: American Job Print Office, 1897 – 1mf – 9 – 0-524-04209-8 – mf#1990-5000 – us ATLA [240]

From darkness to light: the story of a telugu convert / Clough, John Everett – 3rd ed. Boston: W.G. Corthell, 1882. 288p. illus – 1 – us Wisconsin U Libr [240]

From davidov to romanovsky : more russian papers on probability and statistics / ed by Sheynin, Oscar – (mf ed 1998) – 3mf – 9 – €49.00 – 3-8267-2579-4 – mf#DHS 2579 – gw Frankfurter [510]

From dawn to sunrise : a review, historical and philosophical, of the religious ideas of mankind / Smith, J Gregory, Mrs – Rouses Point, NY: Lovell, 1876 – 1mf – 9 – 0-524-02373-5 – mf#1990-2984 – us ATLA [200]

From early hinduism to neo-vedanta : paradigm shifts in sacred psychology and mysticism their implications for south african hindus / Saradananda, swami – Uni of South Africa 2001 [mf ed Johannesburg 2001] – 5mf [ill] – 9 – (incl bibl ref) – mf#mfm15017 – sa Unisa [280]

From eden to sahara : florida's tragedy / Small, John Kunkel – Lancaster, PA. 1929 – 1r – us UF Libraries [500]

From egypt to palestine through sinai, the wilderness and the south country : observations of a journey made with special reference to the history of the israelites / Bartlett, Samuel Colcord – New York: Harper, 1879 – 2mf – 9 – 0-7905-3422-3 – mf#1987-3422 – us ATLA [916]

From egyptian rubbish-heaps : five popular lectures on the new testament with a sermon delivered at northfield, massachusetts, in august, 1914 / Moulton, James Hope – 2nd ed. London: Charles H Kelly 1917 – 1mf – 9 – 0-524-08087-9 – mf#1992-1147 – us ATLA [220]

From emory's bar at the west end of contract 60 to port moody (burrard inlet), british columbia : specification for the construction of the work / Compagnie du chemin de fer canadien du Pacifique – [Ottawa?: s.n, 1881?] [mf ed 1983] – 1mf – 9 – mf#13610 – cn Canadiana [380]

From epworth to london with john wesley : being fifty photo-engravings of the sacred places of methodism... / Edmondson, George W [comp] – Toronto: W Briggs, c1890 – 3mf – 9 – 0-665-89022-2 – mf#89022 – cn Canadiana [242]

From fact to faith / Gibson, John Monro – New York: Fleming H Revell, [1898?] – 1mf – 9 – 0-7905-3134-8 – mf#1987-3134 – us ATLA [240]

From faith to faith : sermons / Bernard, John Henry – London: Isbister, 1895 – 1mf – 9 – 0-7905-0798-6 – (incl bibl ref) – mf#1987-0798 – us ATLA [240]

From far formosa : the island, its people and missions / Mackay, G L; ed by Macdonald, J A – Edinburgh, London, 1896 – 5mf – 9 – mf#HTM-108 – ne IDC [915]

From far formosa : the island, its people and missions / Mackay, George Leslie; ed by Macdonald, James Alexander – New York: F H Revell, c1895 – 1mf – 9 – 0-7905-4832-1 – mf#1988-0832 – us ATLA [915]

From fundamental to accessory in the development of the nervous system and of movements / Burk, Frederic – S.l: s.n, 1898? – 1mf – 9 – mf#00872 – cn Canadiana [611]

From gottsched to hebbel / Mason, Gabriel Richard – London: G Harrap, 1961 [mf ed 1993] – 286p (ill) – 1 – (incl ind) – mf#8219 – us Wisconsin U Libr [430]

From grace to grace / Capel-Cure, Edward – London, England. 18– – 1r – us UF Libraries [240]

From his birth, 9 sept 1826 until aug 1876 / Spencer, John Henderson – Incomplete transcription by Mrs. J. Henry Simpson, 1947. 230p – 1 – 8.05 – us Southern Baptist [242]

From individual to collective voice : the overlapping roles of choreographer, performer, and designer during the creation of the three ligeti etudes / Savino, Cynthia – 2000 – 38p on 1mf – 9 – $5.00 – mf#PE 4164 – us Kinesology [790]

From island to island in the south seas : or, the work of a missionary ship / Cousins, George [comp] – 3rd rev ed. London: London missionary Society; John Snow, 1894 – (= ser Yale coll) – viii/124p (ill) – 1 – 0-524-09618-X – mf#1995-0618 – us ATLA [240]

From jerusalem to antioch : sketches of the primitive church / Dykes, James Oswald – London: Hodder and Stoughton, 1875 – 2mf – 9 – 0-7905-1754-X – mf#1987-1754 – us ATLA [221]

From jerusalem to nicaea : the church in the first three centuries / Moxom, Philip Stafford – Boston: Roberts, 1895 – (= ser Lowell institute lectures) – 2mf – 9 – 0-7905-8858-7 – mf#1989-2083 – us ATLA [240]

From joseph to joshua / Rowley, H H – Oxford, 1948 – 9 – $10.00 – us IRC [221]

From leopoldville to lagos / Congo (Democratic Republic) – Leopoldville, Congo. 1962 – 1r – us UF Libraries [960]

From letter to spirit : an attempt to reach through varying voices the abiding word / Abbott, Edwin Abbott – London: Adam and Charles Black, 1903 – 2mf – 9 – 0-8370-9520-4 – (incl bibl ref and indexes) – mf#1986-3520 – us ATLA [220]

From liberal ulster to england – [London, 1886] – (= ser 19th c ireland) – 1mf – 9 – mf#1.1.253 – uk Chadwyck [941]

From magic to science : essays on the scientific twilight / Singer, Ch – New York, 1928 – €15.00 – ne Slangenburg [080]

From malachi to matthew : outlines of the history of judea from 440 to 4 b.c. / Moss, Richard Waddy – London: Charles H Kelly, 1899 – 1mf – 9 – 0-8370-9406-2 – (incl ind) – mf#1986-3406 – us ATLA [939]

From markov to kolmogorov : russian papers on probability and statistics / ed by Sheynin, Oscar – (mf ed 1998) – 3mf – 9 – €49.00 – 3-8267-2514-X – mf#DHS 2514 – gw Frankfurter [510]

From memory's shrine : the reminiscences of carmen sylva = Mein penatenwinkel / Elisabeth, Queen – Philadelphia: J B Lippincott Company, 1911 (mf ed 1990) – 1r – 1 – (filmed with: astra. in english) – us Wisconsin U Libr [920]

From montreal, in appeal : kerr, esquire, appellant, and la croix, esquire, respondent: appellant's case, 1821 / Kerr, James – [Montreal?: s.n, 1821?] [mf ed 1993] – 1mf – 9 – 0-665-94587-6 – (incl french text) – mf#94587 – cn Canadiana [345]

From naboth's vineyard : being impressions formed during a fourth visit to south africa undertaken at the request of the tribune newspaper / Butler, William Francis – London: Chapman & Hall Ltd 1907 – (= ser [Travel descriptions from south africa, 1711-1938]) – 3mf – 9 – mf#zah-87 – ne IDC [916]

From naptown to sportstown : growth politics, urban development, and economic change in indianapolis / Schimmel, Kimberly S – 1994 – 3mf – $12.00 – us Kinesology [303]

From nowhere to beulahland : a personal narrative / Brown, Elijah P – 2nd ed. Chicago: Winona, 1904 – 1mf – 9 – 0-8370-7368-5 – mf#1986-1368 – us ATLA [920]

From ocean to ocean : a record of the work of the woman's american baptist home mission society – [s.l: s.n.] c1921- [annual] [mf ed 2003] – 2r – 1 – (iss for 1936-37 lacks collective title; has distinctive title. 1931/32 and 1933/34 not publ) – mf#2003-s046 – us ATLA [242]

From olivet to patmos : the first christian century in picture and story / Houghton, Louise Seymour – New York: American Tract Society, c1893 – 1mf – 9 – 0-8370-3674-7 – mf#1985-1674 – us ATLA [221]

From one business to another : fobta / B/E Productions [Oconomowoc, WI] – 1987-88, v3 n2 [feb 1984], 1985 jan-1986 dec – 2r – 1 – (cont by: fobta) – mf#1268980 – us WHS [338]

From opitz to lessing : a study of pseudo-classicism in literature / Perry, Thomas Sergeant – Boston: J R Osgood, 1885 [mf ed 1993] – vi/207p – 1 – (incl ind) – mf#8175 – us Wisconsin U Libr [430]

From pillar to post / Bangs, John Kendrick – New York, NY. 1916 – 1r – us UF Libraries [025]

From pioneer home to the white house / Thayer, William Makepeace – Boston, MA. 1882 – 1r – us UF Libraries [025]

From prophecy to exorcism : the premisses of modern german literature / Hamburger, Michael – London: Longmans, c1965 [mf ed 1993] – vii/167p – 1 – (incl bibl ref and ind) – mf#8271 – us Wisconsin U Libr [430]

From rome to protestantism / McGerald, Samuel – Buffalo, NY: Christian Literature Co, 1900 – 1mf – 9 – 0-8370-8770-8 – mf#1986-2770 – us ATLA [240]

From savagery to civilisation / Roy, Manabendra Nath – Calcutta: Digest Book House, 1940 – (= ser Samp: indian books) – us CRL [301]

From schola to cathedral / Brown, Gerard Baldwin – Edinburgh 1886 – (= ser 19th c art & architecture) – 3mf – 9 – mf#4.2.93 – uk Chadwyck [720]

From silence to personal voice : the journey of an emerging artist / Priest, Jill G – 1999 – 30p on 1mf – 9 – $5.00 – mf#PE 4123 – us Kinesology [790]

From slavery to a bishopric : or, the life of bishop walter hawkins of the british methodist episcopal church, canada / Edwards, S J Celestine – London: J Kensit, 1891 – 1mf – 9 – 0-524-06995-6 – mf#1991-2848 – us ATLA [242]

From solomon to the captivity : the story of the two hebrew kingdoms / Gregg, David & Mudge, Lewis W – New York: American Tract Soc, 1890 (mf ed 1995) – 1r – 1 – mf#ZZ-34512 – us NY Public [221]

From solomon to the captivity : the story of the two hebrew kingdoms / Gregg, David & Mudge, Lewis Ward – New York: American Tract Society, c1890 – 1mf – 9 – 0-8370-9474-7 – mf#1986-3474 – us ATLA [221]

From space lab to space station – 1984 – (= ser Advances in the astronautical sciences 56) – 9 – $30.00 – us Univelt [629]

From st francis to dante : translations from the chronicle of the franciscan salimbene, 1221-1288 / Coulton, George Gordon – 2nd rev enl ed. London: David Nutt, 1907 – 2mf – 9 – 0-7905-4265-X – mf#1988-0265 – us ATLA [240]

From strength to strength / Jowett, John Henry – London: Hodder and Stoughton, [1898?] – (= ser Little books on religion) – 1mf – 9 – 0-7905-0957-1 – mf#1987-0957 – us ATLA [240]

From strength to strength : three sermons on stages in a consecrated life / Westcott, Brooke Foss – London; New York: Macmillan, 1890 – 1mf – 9 – 0-7905-9753-5 – mf#1989-1478 – us ATLA [240]

From sunrise land : letters from japan / Carmichael, Amy – 2nd ed. London: Marshall Bros, 1895 [mf ed 1995] – (= ser Yale coll) – vii/180p (ill) – 1 – 0-524-09674-0 – (pref by c a fox) – mf#1995-0674 – us ATLA [950]

From sunrise land, letters from japan / Carmichael, A Wilson- – London, 1895 – 2mf – 9 – mf#HT-25 – ne IDC [915]

From talk to text : or, a likely story!– likely enough / Ballard, Addison – New York:Longmans, Green, 1904 [mf ed 1985] – 1mf – 9 – 0-8370-2172-3 – mf#1985-0172 – us ATLA [210]

From the baroque library nuenning *see* Aus der barockbibliothek nuenning

From the book of myths / Carman, Bliss – Boston: L C Page, 1902 – (= ser Pipes of Pan) – 2mf – 9 – 0-665-78048-6 – mf#78048 – cn Canadiana [810]

From the book of myths *see* Pipes of pan

From the book of valentines *see* Pipes of pan

From the cam to the cays / Carr, John David – London, England. 1961 – 1r – us UF Libraries [972]

From the caves and jungles of hindostan = Iz peshcher i debrei indii / Blavatsky, Helena Petrovna – London: Theosophical Publishing Society, 1892 – 1mf – 9 – 0-524-02416-2 – (in english) – mf#1990-3000 – us ATLA [915]

From the cradle to the grave : life of eld. solon a. howenstine / Howenstine, Lydia – Fort Wayne, Ind: DW Underwood, printer, 1894 – 2mf – 9 – 0-524-08793-8 – mf#1993-3285 – us ATLA [240]

From the cradle to the grave (index) : reminiscences of wm. bull meek camptonville, california – 1r – 5,9 – $50.00 – mf#B40150 – us Library Micro [920]

From the exile to the advent / Fairweather, William – Edinburgh: T & T Clark, 1895 [mf ed 1988] – (= ser Handbooks for bible classes and private students) – 1mf – 9 – 0-7905-0011-6 – (incl bibl ref & ind) – mf#1987-0011 – us ATLA [221]

From the fight / Carmichael, Amy – London: Church of England Zenana Missionary Society; Marshall Bros, [1900] [mf ed 1995] – (= ser Yale coll) – 62p (ill) – 1 – 0-524-09305-9 – (original drawings by f a baker) – mf#1995-0305 – us ATLA [242]

From the gold mine to the pulpit : the story of the rev t l jones, backwoods methodist preacher in the pacific northwest during the closing years of the 19th century / Jones, Thomas Lewis – Cincinnati: printed...by Jennings & Pye [c1904] [mf ed 1984] – 3mf – 9 – 0-8370-0986-3 – mf#1984-4330 – us ATLA [242]

From the green book of the bards *see* Pipes of pan

From the himalayas to the equator : letters, sketches and addresses, giving some account of a tour in india and malaysia / Foss, Cyrus David – New York: Eaton & Mains, c1899 – 1mf – 9 – 0-524-07680-4 – mf#1991-3265 – us ATLA [910]

From the horse's mouth / Western Pennsylvania Teamsters for a Democratic Union et al – 1976 jul-1977 feb,v3 n3-v7 n7 [1977 feb i.e. mar-1981 nov/dec] – 1r – 1 – mf#647085 – us WHS [331]

From the new york evening express, tuesday, april 22, 1873 canada correspondence : the series of letters we are publishing from canada...the rebellion of 1837, interesting reminiscences, progress of events, the ministers sent out from england... / Brown, Thomas Storrow – [Montreal?: s.n, 1873?] – 1mf – 9 – 0-665-94530-2 – mf#94530 – cn Canadiana [971]

From the reformation to the puritan revolution : papers of the york court of high commission, c1560-1641 – 14r – 1 – (from the borthwick institute for historical research, university of york. incl printed guide) – mf#C39-17200 – us Primary [941]

From the restoration of 1660 to the revolution of 1688 / Brown, John – London: National Council of Evangelical Free Churches, 1904 – (= ser Eras Of Nonconformity) – 1mf – 9 – 0-7905-4608-6 – mf#1988-0608 – us ATLA [941]

From the shores / Parents Without Partners – v4 n6 [1976 jun], v6 n5-12 [1978 may-dec] – 2r – 1 – mf#619919 – us WHS [305]

From the tablets of sumer / Kramer, SN – Falcons Wing Press, 1956 – 9 – $12.00 – us IRC [930]

From the uttermost to the uttermost : the life story of josephus pulis / Simpson, Albert B – New York City: Christian Alliance, c1914 [mf ed 1992] – (= ser Christian & missionary alliance coll) – 1mf – 9 – 0-524-02146-5 – mf#1990-4212 – us ATLA [240]

From third to fourth : a report on the activities of the yci since its third world congress / Young Communist International. Executive Committee – Stockholm: Publ by The Committee, 1924 (mf ed 19–) – 84p – mf#ZT-SFC pv124 n9 – us NY Public [335]

From tribal rule to modern government / Rhodes-Livingstone Institute Conference (13th : 1960) – Lusaka, Zambia. 1960 – 1r – us UF Libraries [960]

From trusteeship to ... ? : micronesia and its future / Micronesia Support Committee and the Pacific Concerns Research Center – 2nd ed. Honolulu: Maka'ainana Media, aug 1982 – (= ser Micronesia: evolution to separate political entities) – 1mf – 9 – $1.50 – mf#LLMC 82-100F, Title 102 – us LLMC [323]

From union to apartheid / Ballinger, Margaret – New York, NY. 1969 – 1r – us UF Libraries [960]

From vision to reality / Yaari, M – Tel Aviv, 1963 – 2mf – 9 – mf#J-28-25 – ne IDC [956]

From west to east : being the story of a recent visit to indian missions / Weatherley, Ella M – London: Zenana Bible and Medical Mission, [1910] [mf ed 1995] – (= ser Yale coll) – 128p (ill) – 1 – 0-524-09326-1 – (int by e g ingham) – mf#1995-0326 – us ATLA [240]

From whose bourne / Barr, Robert – New York, London: F A Stokes, c1896 – 3mf – 9 – (ill by frank m gregory) – mf#27189 – cn Canadiana [830]

From "winner" to "sign" : the unchanged understanding of the church-world relation in 20th century ecumenical thought / Fubara-Manuel, Benebo Fubara – 2002 [mf ed 2003] – 1r – 1 – mf#d00005 – us ATLA [240]

From "winner" to sign : the changed understanding of the church-world relation in 20th-century ecumenical thought / Fubara-Manuel, Benebo Fubara – [mf ed 2003] – 1r – 1 – (with bibl) – mf#d00005 – us ATLA [230]

From yeravda mandir : ashram observances / Gandhi, Mahatma – Ahmedabad: Navajivan Pub House, 1945 – (= ser Samp: indian books) – (trans fr original gujarati by valji govindji desai) – us CRL [180]

Frome and north somerset labour party records, 1918-1983 – (= ser Labour party in britain, origins and development at local level. series 2) – 5r – 1 – (with p/g. int by andrew thorpe) – mf#97561 – uk Microform Academic [325]

Fromm und frei : eine ostergabe in religioesen dichtungen / Allmers, Hermann – Oldenburg: Schulzesche Hof-Buchhandlung und Hof-Buchdruckerei (A Schwartz), [1889?] [mf ed 1996] – 62p – 1 – mf#9580 – us Wisconsin U Libr [810]

Fromm und frei! : wahre worte fuer tapfere juenglinge / Pfennigsdorf, Emil – Dessau: Buchh des Evangelischen Vereinshauses, 1900 – 1mf – 9 – 0-7905-9572-9 – mf#1989-1297 – us ATLA [170]

Frommann, Karl *see*
– Herbort's von fritslar liet von troye
– Der johanneische lehrbegriff in seinem verhaeltnisse zur gesammten biblisch-christlichen lehre

FROMME

Der fromme naturkundige – Danzig (Gdansk PL), 1738 jun-1739 nov – 1 – gw Misc Inst [240]
Fromme-Bechem, Annemarie *see* Die grosse ordnung
Frommel, Gaston *see*
– Etudes de theologie moderne
– Etudes religieuses et sociales
– Lettres et pensees
Frommel, Wilhelm *see* Sammlung von vortraegen fuer das deutsche volk, erster band
Frommett, B R *see*
– Desiat let sovetskoi promyslovoi kooperatsii, 1917-1927
– Kooperativnoe vospitanie detei i vozrozhdenie chelovechestva
– Russkii sotsializm i kooperatsiia
La fronde : journal feministe – Paris. 9 dec 1897-1er mars 1905, 3-24 juil 1914, 26 mai-dec 1926, 26 avr, 4 mai 1929 – 1 – (Quot. jusqu'en 1903 et a partir de 1926.) – fr ACRPP [305]
Le frondeur – Liege, Belgium. -w. 24 April 1880-30 Sept 1888. 3 reels – 1 – uk British Libr Newspaper [949]
Frondizi, Arturo *see* Tratado de rio de janeiro
Frondizi, Josefina B De *see* Nuestra america
Fronduer – Liege Belgium, 24 apr 1880-30 sep 1888 – 2 1/2r – 1 – uk British Libr Newspaper [074]
Froneman, Helouise *see* Being an asperger vs being real
Fronemann, Wilhelm *see*
– Das erbe wolgasts
– Hammerschlaege
Fronimo, dialogo di vincentio galilei... : sopra l'arte del bene intavolare, et rettamente sonare la mvsica negli strumenti artificali si di corde come di fiato, & in particulare nel liuto / Galilei, Vincenzo – In Vineggia: Appresso l'herede di G Scotto 1584 [mf ed 19–] – 1r – 1 – mf#film 812 – us Sibley [780]
Fronmueller, G F C *see*
– Die briefe petri und der brief judae
– The epistle general of jude
Fronmueller, Petri *see* The epistles general of peter
Die front – Zuerich (CH), 1934-37 [gaps] – 1 – gw Misc Inst [074]
Front – The Hague. v1 n1-4. dec. 1930-june 1931 – 1 – us NY Public [073]
Le front – Paris. nov 1935-38 – 1 – fr ACRPP [073]
Front and forward / Gimbels Midwest – 1981 n3-1986 n3 – 1r – 1 – mf#679972 – us WHS [071]
Front commun / Mouvement Frontiste – Paris. n1-5. dec 1933-mai juin 1934 [bimnthly] – 1 – fr ACRPP [073]
Front d'action Politique *see* Journal du frap
Front de Liberation Nationale Algerienne *see* El-moudjahid
Front der herzen : roman / Hohlbaum, Robert – Berlin: K H Bischoff 1944 [mf ed 1990] – 1r – 1 – (filmed with: gestern / hugo von hofmannsthal) – mf#2729p – us Wisconsin U Libr [830]
Front der Sozialistischen Demokratie und Einheit *see* Neuer weg
Front line / New Century Policies Educational Programs – v1 n1-v4 n3 [1982 jan-1985 nov] – 1r – 1 – mf#998633 – us WHS [370]
Front line : a temple university student publication / Temple University – iss n1-3 [1980 dec-1981 apr] – 1r – 1 – mf#634397 – us WHS [378]
Front line – Washington DC. n1-4 [1980 spring-1982 spring] – 1r – 1 – mf#669896 – us WHS [071]
Front mondial = Weltfront /Worldfront – n1-9. janv-sept 1933 – 1 – (devenu: front mondial. contre la guerre et le fascisme. n.s., no. 1-52. oct 1933-nov 1935. fond. henri barbusse. paris. devenu: paix et liberte voir a ce titre) – fr ACRPP [073]
Front mondial *see* Weltfront gegen imperialistischen krieg und faschismus
Front national – Paris, France. 6 sep 1944-16 nov 1946 – 2r – 1 – uk British Libr Newspaper [072]
Front national-syndicaliste *see* L'assault
Front Natsional-revoliutsionnogo deistviia *see* Nash marsh
Front nauki i tekhniki – Moscow. Jan 15 1930-July 1938 – 1 – us NY Public [500]
Front nouveau *see* Die neue front
Le front ouvrier – Laprairie: [s.n.] ([s.l.]: [s.n.]) v1 n1 2 dec 1944-v10 n16 20 mars 1954 (wkly) [mf ed 1983] – 9r – 1 – (merger of: la jeunesse ouvriere and: le mouvement ouvrier; becomes: la jeunesse ouvriere) – mf#SEM35P180 – cn Bibl Nat [073]
Le front ouvrier *see* La jeunesse ouvriere
Front page – v1 n1-v2 n4 [1975 jul-1977 nov], n9-14 [1978 feb-1979 aug] – 1r – 1 – mf#665989 – us WHS [071]
Front patriotique pour le progres (central african republic) – s.l, s.l? 19–? – 1r – us UF Libraries [960]
Front range people's press – n10-11, 15-16 [1972 apr-may, Sep- oct], n23 [1973 may] – 1r – 1 – (cont: snake ranch news) – mf#1583415 – us WHS [071]
Front social : bulletin mensuel de la troisieme force. ni fascisme, ni bolchevisme, contre le capitalisme – Paris. n7, 13-15.1933-oct 1934 [mnthly] – 1 – fr ACRPP [325]
Front uni national du Kampuchea *see*
– Declaration des intellectuels patriotes
– Political programme of the national united front of kampuchea [nufk]
– Programme politique du front uni national du kampuchea [funk]
Fronteira brasileo-boliviana pelo amazonas / Lopes Goncalves, Augusto Cezar – Lisboa, Portugal. 1901 – 1r – us UF Libraries [972]
Fronteiras do brasil no regime colonial / Macedo Soares, Jose Carlos De – Rio de Janeiro, Brazil. 1939 – 1r – us UF Libraries [972]
Fronteiras e fronteiros / Goycochea, Luis Felipe De Castilhos – Sao Paulo, Brazil. 1943 – 1r – us UF Libraries [972]
Frontera de la republica dominicana con haiti – Ciudad Trujillo, Dominican Republic. 1946 – 1r – us UF Libraries [972]
Frontera dominico-haitiana / Rodriguez, Cayetano Armando – Santo Domingo, Dominican Republic. 1929 – 1r – us UF Libraries [972]
Frontera sur de mexico / Trens Esquinca, Leonor – Mexico City? Mexico. 1953 – 1r – us UF Libraries [972]
Fronteras / Arrivi, Francisco – San Juan, Puerto Rico. 1960 – 1r – us UF Libraries [972]
Frontier – O'Neill City, NE: W D Mathews. 85v. 1880-v85 n6. may 27 1965 (wkly) [mf ed with gaps filmed [1966]-1970] – 39r – 1 – (absorbed: item (o'neill, ne) 1892, verdigre eagle (1931) 1964, stuart advocate 1965. merged with: holt county independent (o'neill, ne 1897) to form: frontier and holt county independent. some sects titled: frontier and verdigre eagle may 14-nov 26 1964) – us NE Hist [071]
Frontier – Chicago. 1961-1970 (1) – ISSN: 0016-2086 – mf#1551 – us UMI ProQuest [500]
Frontier – London. 1958-1976 (1) 1967-1976 (5) 1967-1967 (9) – ISSN: 0016-2078 – mf#2673 – us UMI ProQuest [240]
The frontier – Missoula, Montana. v1-19. may 1920-summer 1939 [incomplete] – 1 – us NY Public [073]
The frontier / Platt, Ward – New York City: Literature Dept, Presbyterian Home Missions 1908 [mf ed 1986] – (= ser Forward mission study courses) – 1mf – 9 – 0-8370-6308-6 – (incl ind) – mf#1986-0308 – us ATLA [240]
Frontier and holt county independent – O'Neill, NE: Miles Pub Co. v99 n48. nov 26 1987- (wkly) [mf ed nov 26 1987- filmed 1988-] – 1 – (cont: holt county independent (1984)) – us NE Hist [071]
Frontier and holt county independent – O'Neill, NE: G E Miles. 20v. v77 n22. jun 3 1965-v96 n22. may 31 1984 (wkly) [mf ed filmed 1970-85] – 34r – 1 – (formed by the union of: frontier and: holt county independent (1897). cont by: holt county independent (1984)) – us NE Hist [071]
Frontier and midland – Missoula. 1920-1939 (1) – mf#4657 – us UMI ProQuest [400]
Frontier by air / Hager, Alice Rogers – New York, NY. 1942 – 1r – us UF Libraries [972]
Frontier county faber – Stockville, NE: A E Powers, may 1884-1895// (wkly) [mf ed 1887-95 (gaps) filmed 1970] – 1r – 1 – (cont by: faber) – us NE Hist [071]
Frontier county journal – Eustis, NE: D Ralph Lee. v1 n1. jan 8 1897- (wkly) [mf ed with gaps] – 1r – 1 – us NE Hist [071]
Frontier county republican – Stockville, NE: Frontier County Print Co, jul 1892-v11 n52. jul 2 1903 (wkly) [mf ed 1892,1895-1903 (gaps) filmed 1970] – 4r – 1 – (cont by: republican) – us NE Hist [071]
Frontier missionary problems : their character and solution / Kinney, Bruce – New York: FH Revell Co, c1918 – 3mf – 9 – 0-524-07892-0 – mf#1991-3437 – us ATLA [240]
Frontier news – 1952 jun 19-1955 apr 5, 1955 may 5-1957 mar, 1957 apr-1958 dec, 1959 jan-1961 jul 14 – 4r – 1 – mf#3475050 – us WHS [071]
Frontier nursing service quarterly bulletin – Lexington. 1974+ (1) 1976+ (5) 1976+ (9) – ISSN: 0016-2116 – mf#8580 – us UMI ProQuest [610]
Frontier sentinel – Newry, Ireland. 15 Oct 1904-26 Aug 1972 – 58 1/2r – 1 – (lacking: dec 1930) – uk British Libr Newspaper [072]
Frontier sketches : illustrations of the pioneer work of home missionaries as told by the workers or drawn from their experience / ed by Grose, Howard Benjamin – New York: American Baptist Home Mission Society, [1908?] – 1mf – 9 – 0-524-06620-5 – mf#1991-2675 – us ATLA [240]
Frontier standard and east london gazette – East London SA, jan 1890-sep 1891 [mf ed Cape Town: SA Library 1983] – 1r – 1 – (cont by: east london standard) – sa National [079]
Frontier times – 1923 oct-1925 aug, 1925 sep-1927 apr, 1927 may-1929 sep, 1968 aug-1971 nov, 1972 jan-1973 nov, 1974 jan-1976 jul, 1976 sep-1979 mar, 1979 jul-1981 may – 8r – 1 – mf#429903 – us WHS [071]
Frontier times – 1984 oct-1985 oct – 1r – 1 – mf#1277659 – us WHS [071]
La frontiere nord de la province de quebec / Cazes, Paul de – Quebec: A Cote, 1886 – 1mf – 9 – mf#02119 – cn Canadiana [917]
Les frontieres de la cote d'ivoire, de la cote d'or et du soudan / Delafosse, Maurice – Paris: Masson, 1908 – 1 – us CRL [960]
Frontieres du bresil et de la guyane anglaise – Paris, France. v1-3. 1903 – 1r – us UF Libraries [972]
Les frontieres du cambodge... / Chhak, Sarin – Paris: Dalloz 1966- [mf ed 1989] – (= ser Centre d'etude des pays d'extreme -orient, asie du sud-est 1) – 1r with other items – 1 – (with bibl; t1: les frontieres du cambodge avec les anciens pays de la federation indochinoise: le laos et le vietnamn [cochinchine et annam]) – mf#mf-10289 seam reel 002/20 [§] – us CRL [959]
Frontiers – Boulder. 1975+(1,5,9) – ISSN: 0160-9009 – mf#13578 – us UMI ProQuest [305]
Frontiers International *see* Frontiersman
Frontiers of health services management – Michigan. 1984+ (1,5,9) – ISSN: 0748-8157 – mf#14619 – us UMI ProQuest [360]
Frontiersman / Frontiers International – 1954 jan – 1r – 1 – mf#5132312 – us WHS [071]
Frontisme *see* La fleche
Frontline / Common Cause [US] – v2 n1-v6 n3 [1976 jan/feb-1980 may/jun] – 1r – 1 – (cont by: in common [washington, dc: 1975]; common cause [washington, dc]) – mf#701251 – us WHS [071]
Frontline / Institute for Scientific Socialism [Oakland, CA] – 1983 apr 11, v1 n1-v4 n2 [1983 jun 27-1986 jul 21], v4 n4-v7 n10 [1986 aug 4-1989 nov 27], transition iss n1-2 [1990 feb-mar] – 2r – 1 – (cont by: crossroads [oakland, ca]) – mf#1027775 – us WHS [335]
Frontline solutions – Duluth. 2000+ (1) – ISSN: 1528-6363 – mf#29401 – us UMI ProQuest [629]
Frontline solutions. pan-european edition – Duluth. 2000+ (1,5,9) – mf#29402 – us UMI ProQuest [629]
Frontlines : reason's newsletter for libertarians / Reason Foundation – 1978 dec-1983 may – 1r – 1 – mf#669887 – us WHS [140]
Frontpage / Newspaper Guild of New York – v4 n9 [1946 sep], v23 i.e. 24 n8-? [1967 sep 19-1977 nov] – 2r – 1 – (cont: guildpaper) – mf#368911 – us WHS [071]
Froriep, Just Friedrich *see* Erfurtische gelehrte zeitung
Der frosch : familiendrama in einem act nach henrik ipsen / Hartleben, Otto Erich – 3. aufl. Berlin: S Fischer 1901 [mf ed 1990] – 1r – 1 – (filmed with: mei erich / selma hartleben & other titles) – mf#2699p – us Wisconsin U Libr [820]
Frossard, Charles Louis *see*
– Deux sermons
– L'eglise sous la croix pendant la domination espagnole
– Etude historique et bibliographique sur la discipline ecclesiastique des eglises reformees de France
– Les origines de la faculte de theologie protestante de montauban
Frossard, Edouard *see* Catalogue of the important historical collection of coins and medals made by gerald e hart, esq
Frossard, LO *see*
– Mon journal de voyage en russie
– Rapport sur les negociations conduites a moscou, suivi des theses presentees au 2eme congres de l'internationale communiste
Frost, A F *see* Evolution
Frost, Adelaide Gail *see* By waysides in india
The frost genealogy : descendants of william frost of oyster bay, new york / Frost, Josephine C – New York: Frederick H Hitchcock, 1912 – 1r – 1 – us Western Res [920]
Frost, George E *see* The patent system and the modern economy.
Frost, Gilman Dubois *see* Microfilm edition of dr gilman frost's genealogical records of hanover, new hampshire
Frost, Henry Weston *see* Men who prayed
Frost illustrated – Fort Wayne IN. 1995 aug 2/8-dec 27/1996 jan 2, 1996 jan 3/9-dec 25/31, 1997 jan 1/7-dec 31/1998 jan 6, 1998 jan 7/13-dec 30/jan 5, 1999 jan 6/12-dec 2000 jan 4, 2000 jan 5/11-jun 28/jul 4, 2000 jul 5/11-dec 27/2001 jan 2 – 7r – 1 – (cont: coffee break [fort wayne, in]) – mf#2891925 – us WHS [071]
Frost, J A *see* Fort myers to sanibel island
Frost, J M *see* Correspondence files of corresponding secretaries, baptist sunday school board
Frost, J M et al *see* Baptists' why and why not
Frost, James Marion *see*
– The memorial supper of our lord
– The moral dignity of baptism
– Pedobaptism, is it from heaven or of men?
– The school of the church
Frost, John *see* History of the united states
Frost, Josephine C *see* The frost genealogy
Frost, Jules A *see*
– Chronological history
– Famous edison fish story
– Points of interest
– Saddlebag bank
– Tarpon tournaments
– Thomas a edison
Frost, Mary *see* Zambian oral literature
Frost, Maurice *see* English and scottish psalm and hymn tunes, 1543-1677
Frost, R *see* Letter to the very reverend the warden of manchester
Frost, Robert *see* West-running brook
Frost, Thomas *see* The old showman and the old london fairs
Frost, Thomas et al *see* Old church life
Frost, Thomas Gold *see* A treatise on the federal corporation tax law
Frostproof news – Frostproof, FL. 1988 may 19-1997 – 10r – (gaps) – us UF Libraries [071]
Frothingham, A L *see* Stephen bar sudali
Frothingham, Arthur Lincoln *see*
– Christian philosophy
– Stephen bar sudhaili, the syrian mystic, and the book of hierotheos
Frothingham, Ephraim Langdon *see*
– Christian philosophy
– A statement of the trinitarian principle
Frothingham, Frederick *see* The lord's song and other sermons
Frothingham, Octavius Brooks *see*
– Boston unitarianism, 1820-1850
– George ripley
– Gerrit smith
– Recollections and impressions, 1822-1890
– The religion of humanity
– Transcendentalism in new england
Frothingham, Richard *see* A tribute to thomas starr king
Frotola...per la vittuoria de i nuostri segnore contra i turchi / Magagno, G B – Np, [1571] – 1mf – 9 – mf#H-8320 – ne IDC [956]
D'frou kaetheli und ihri buebe : des "staern vo buebebaerg" zweiter teil: berndeutsche erzaehlung / Tavel, Rudolf von – 3. aufl. Bern: A Francke, 1938 [mf ed 1996] – 431p – 1 – mf#9305 – us Wisconsin U Libr [880]
Froude, James Anthony *see*
– Book of job
– Bunyan
– Calvinism
– The divorce of catherine of aragon
– The english in ireland in the eighteenth century
– English in the west indies
– The english in the west indies
– Lectures on the council of trent
– Life and letters of erasmus
– Luther
– The nemesis of faith
– Oceana, or, england and her colonies
– The pilgrim
– Short studies on great subjects. fourth series
– Short studies on great subjects. third series
– Theological unrest
Froude, Richard Hurrell *see* Remains of the late reverend richard hurrell froude
Frou-frou – Hull [Quebec: s.n. 1896] – 9 – ISSN: 1190-7665 – mf#P04093 – cn Canadiana [440]
Frou-frou – Juiz de Fora, MG. 31 out 1897 – (= ser Ps 19) – mf#P17,02,107 – bl Biblioteca [079]
Froufrou / Meilhac, Henri – Paris, France. 1884 – 1r – us UF Libraries [440]
Frowein, Eberhard *see* Mein eignes propres geld
The frozen pirate / Russell, William Clark – Toronto: W Bryce, 1887 – 5mf – 9 – mf#32814 – cn Canadiana [830]
Frucht, Else *see* Goethes vermaechtnis
Die fruchtbringende gesellschaft und johann valentin andreae / Begemann, Wilhelm – Berlin: E S Mittler und Sohn, 1911 [mf ed 1993] – xii/79p – 1 – (incl bibl ref) – mf#8175 – us Wisconsin U Libr [430]
Fructucso y Tristancho, Gonzalo *see* Excursiones briologicas por, la provincia de badajoz
Fructuoso, Gonzalo *see* Industrias rurales
Die fruechte ihrer haende : materialien zum wandel kultureller leitbilder der frau im christlichen abendland / Heubach, Helga – (mf ed 1996) – 6mf – 9 – €62.50 – 3-8267-2226-4 – mf#DHS 2226 – gw Frankfurter [305]

Frueh- und spaetmorbiditaet und -letalitaet nach interventioneller therapie bei diabetikern mit instabiler angina pectoris / Gaudesius, Giedrius – (mf ed 1999) – 2mf – 9 – €40.00 – 3-8267-2649-9 – mf#DHS 2649 – gw Frankfurter [616]
Fruehbrodt, Gerhard *see* Der impressionismus in der lyrik der annette von droste-huelshoff
Fruehchristliche apologeten, 1. bd (bdk12 1.reihe) – (= ser Bibliothek der kirchenvaeter. 1. reihe (bdk 1.reihe)) – €18.00 – ne Slangenburg [230]
Fruehchristliche apologeten, 2. bd (bdk14 1.reihe) – (= ser Bibliothek der kirchenvaeter. 1. reihe (bdk 1.reihe)) – €18.00 – ne Slangenburg [230]
Das fruehchristliche und merowingische mains : nach den bodenfunden dargestellt / Behrens, G – Mainz, 1950 – €5.00 – ne Slangenburg [240]
Fruehe griechische sagenbilder in boeotien / Hampe, Roland – Athen: Deutsches Archaeologisches Institut 1936 [mf ed 1981] – 1r – 1 – (imperfect; with bibl footnotes) – us UW Libraries [930]
Die fruehen kraenze / Zweig, Stefan – Leipzig: Insel-Verlag, 1917 [mf ed 1992] – 84p – 1 – mf#7801 – us Wisconsin U Libr [810]
Fruehererkennung und vermeidung von arbeitsbedingten erkrankungen : arbeitsbericht des zentrums arbeit und gesundheit dortmund-wuppertal (zag) / Bolt, Hermann M et al – (mf ed 1998) – 1mf – 9 – €30.00 – 3-8267-2522-0 – mf#DHS 2522 – gw Frankfurter [616]
Fruehgeschichte des deutschen schrifttums / Baesecke, Georg – Halle/S: M Niemeyer, 1953 [mf ed 1993] – (= ser Georg baesecke. vor- und fruehgeschichte des deutschen schrifttums 2) – xii/203/[18]p – 1 – (incl bibl ref) – mf#8166 – us Wisconsin U Libr [430]
Ein fruehling / Raabe, Wilhelm Karl – Berlin: Otto Janke, 1872 – 1r – 1 – us Wisconsin U Libr [830]
Fruehling eines deutschen menschen : die geschichte des jungen goethe / Hofer, Klara – Leipzig: Hesse & Becker, [1932] – 1r – 1 – (incl bibl ref) – us Wisconsin U Libr [920]
Fruehlingsboten : gedichte / Berens, August – St Charles, MO: Evangelische Synode von Nord-Amerika; Hamburg: Agentur des Ruahen Hauses, 1889 [mf ed 1989] – vii/167p – 1 – mf#7008 – us Wisconsin U Libr [810]
Eine fruehlingsfahrt nach den canarischen inseln / Christ, Hermann – Basel [u a] 1886 [mf ed Hildesheim 1995-98] – 1v on 2mf [ill] – 9 – €60.00 – 3-487-27281-4 – gw Olms [914]
Fruehlings-lieder / Ehrler, Hans Heinrich – Muenchen: A Langen c1913 [mf ed 1989] – 1r – 1 – (filmed with: gesicht und antlitz & other titles) – mf#7210 – us Wisconsin U Libr [810]
Fruehlingssturm / charfreitag / der gang nach emmaus / pfingsten in weimar : die geschichte einer deutschen familie in zwei jahrhunderten: romantrilogie / Hohlbaum, Robert – Berlin: Vier Falken Verlag c1926 [mf ed 1995] – 1r – 1 – (filmed with: der tor und der tod / hugo von hofmannsthal) – mf#3879p – us Wisconsin U Libr [830]
Der fruehlingswalzer : a short novel / Hohlbaum, Robert – Reichenberg: Gebr Stiepel [c1925] [mf ed 1990] – 1r – 1 – (filmed with: getrennt marschieren / robert hohlbaum & other titles) – mf#2730p – us Wisconsin U Libr [830]
Fruehneuhochdeutsches glossar / Goetze, Alfred – Bonn: A Marcus und E Weber, 1912 – (= ser Kleine texte fuer vorlesungen und uebungen) – 1mf – 9 – 0-524-05282-4 – mf#1992-0383 – us ATLA [430]
Fruehpost – Strassburg (Strasbourg F), 1789 13-30 aug – 1 – fr ACRPP [074]
Das fruehroemische lager bei hofheim im taunus / Ritterling, E – Wiesbaden, 1913 – €25.00 – ne Slangenburg [930]
Fruehrot : ein buch von heimat und jugend / Winnig, August – Stuttgart: J G Cotta, 1926 – 1r – 1 – us Wisconsin U Libr [920]
Fruentlich verglimpfung vnd ableynung ueber die predig des treffenlichen, martini luthers wider die schwermer... / Zwingli, H – Zuerich, 1544 – 1mf – 9 – mf#ME-1218 – ne IDC [242]
Frug, Semen Grigor'evich *see*
– Oysgeveylte shriften
– Shire
Fruges, G-M, de *see* J-j olier, 1608-1657
Fruhgeschichte des judischen volkes / Helling, Fritz – Frankfurt am Main, Germany. 1947 – 1r – us UF Libraries [939]
Fruhlingsnacht [f von eichendorf] r schumann, op 39, no 12 / Schumann, Robert – [184-?] [mf ed 19–] – 1mf – 9 – mf#fiche 1115 – us Sibley [780]
Die fruhschriften / Marx, Karl – Stuttgart, 1953. 588p – 1 – us Wisconsin U Libr [335]
Fruin, T A *see* De economische politiek van het nieuwe indonesie
Fruit and truck farms on the east coast of florida : in the famous indian river district – Chicago, IL. 1910? – 1r – us UF Libraries [630]
Fruit and vegetable growing in manatee county, florida – Norfolk, VA. 1911 – 1r – us UF Libraries [634]
Fruit farming at kelowna : the orchard city of british columbia / Central Ikanagon Lands Ltd – [Kelowna, BC?: s.n, 1910 or 1911?] – 1mf – 9 – 0-665-75263-6 – mf#75263 – cn Canadiana [634]
Fruit from the jungle / Wood, M D – Mountain View CA: Pacific Press Publ Assn [1919] [mf ed 1995] – (= ser Yale coll) – 331p (ill) – 1 – 0-524-10173-6 – mf#1995-1173 – us ATLA [954]
Fruit grower – Wenatchee, WA. 1926-1931 (1) – mf#69281 – us UMI ProQuest [071]
Fruit Growers' Association of Upper Canada *see* Constitution and by-laws of the fruit growers' association of upper canada
Fruit outlook and situation – Washington. 1981-1983 (1) 1981-1983 (5) 1981-1983 (9) – (cont: fruit situation) – ISSN: 0277-6073 – mf#9158,01 – us UMI ProQuest [634]
Fruit situation – Washington. 1974-1980 (1) 1976-1980 (5) 1976-1980 (9) – (cont by: fruit outlook and situation) – ISSN: 0364-8648 – mf#9158 – us UMI ProQuest [634]
Fruit varieties journal – University Park. 1963-1999 (1) 1971-1999 (5) 1975-1999 (9) – (cont by: journal of american pomological society) – ISSN: 0091-3642 – mf#2530 – us UMI ProQuest [634]
A fruitful life / Drury, B Paxson – 1 – $8.26 – (a narrative of the experiences and missionary labors of stephen paxson 1882) – us Southern Baptist [242]
A fruitful life : a narrative of the experiences and missionary labors of stephen paxson / Drury, Belle Paxson – Philadelphia: American Sunday-School Union, 1882 [mf ed 1991] – 1mf – 9 – 0-524-00537-0 – (int by c l goodell) – mf#1990-0037 – us ATLA [240]
Fruitful manitoba : homes for millions, the best wheat land and the richest grazing country under the sun – S.l: s.n, 1891? – 1mf – 9 – mf#56110 – cn Canadiana [630]
Fruitman – v5 n1-10 [1902 jan-oct] – 1r – 1 – (cont: northern fruits; cont by: fruitman and garden guest) – mf#955003 – us WHS [634]
Fruitman and garden guest / Iowa Horticultural Society – v5 n11-v8 n7/8 [1902 nov-1905 jul/aug] – 1r – 1 – (cont: fruitman; cont by: fruitman and gardener) – mf#955008 – us WHS [634]
Fruitman and gardener – v8 n9-v10 n10 [1905 sep-1907 oct] – 1r – 1 – (cont: fruitman and garden guest; cont by: fruitman and gardener including the strawberry magazine) – mf#955011 – us WHS [634]
Fruitman and gardener including the strawberry magazine – 1907 nov-1911 nov, 1911 dec-1916 jul – 2r – 1 – (cont: fruitman and gardener; strawberry magazine) – mf#955014 – us WHS [634]
Fruits of christianity / Besant, Annie Wood – London, England. 18– – 1r – us UF Libraries [240]
Fruits of the spirit / Beecher, Henry Ward – London, England. 1886? – 1r – us UF Libraries [240]
Fruits of the spirit, the ornaments of christians / Burnside, Robert – London, England. 1805 – 1r – us UF Libraries [240]
[Fruitvale-] fruitvale shopping news – CA. 1934-1939 – 1r – 1 – $60.00 – mf#B03132 – us Library Micro [071]
Frumnorraen malfraedi / Johannesson, Alexander – Reykjavik, Iceland . 1920 – 1r – us UF Libraries [960]
Frusciano, Thomas J *see* Journal of archival organization
Frutaz, A P *see* Contributo alla storia della riforma del messale promulgato da san pio 5th nel 1570
Frutos Cortes, Eugenio *see*
– Antropologia filosofica 1. preliminares y cuestiones basicas
– Antropologia filosofica 2. dimensiones entitativas del hombre
– Balmes en la encrucijada filosofica
– Calderon de la barca
– Calderon de la barca. autos sacramentales. antologia
– Contribucion a una antologia de la realidad historica
– Creacion filosofica y creacion poetica
– Creacion poetica (j. guillen, salinas, a. machado, d. alonso, s.j. de la cruz, m. pinillos)
– De la caracterologia individual a la colectiva
– La esencial heterogeneidad del ser en antonio machado
– Etica elemental
– La filosofia de calderon en sus autos sacramentales
– Historia de la filosofia y de las ciencias
– El humanismo y la moral de juan pablo sartre
– Idea del teatro de ortega y gasset
– Immanencia y trascendencia del ser y del conocer en heidegger
– Introduccion a la filosofia 1. introduccion y logica
– Introduccion a la filosofia 2. psicologia y etica
– El nuevo humanismo
– Origen, naturaleza y destino del hombre en los autos sacramentales de calderon
– La persona humana
– El primer bergson en antonio machado
– Teoria del conocimineto y ontologia
– La vina destruide. a hungria, en su martirio
– La voluntad y el libre albedrio en los autos sacramentales de calderon
Fry, Benjamin St James *see*
– The life of rev. enoch george
– The life of rev. richard whatcoat
– The life of rev. william m'kendree
– Woman's work in the church
Fry, Francis *see*
– The bible by coverdale 1805
– A bibliographical description of the editions of the new testament, tyndale's version in english
– A description of the great bible, 1539
Fry, Henry *see*
– Atlantic steam navigation
– The history of north atlantic steam navigation
– Lloyd's
Fry, Jacob *see* The sin of adultery
Fry, Joan Mary *see*
– Christ and peace
– The communion of life
– The way of peace
Fry, Joseph Storrs *see* Concise history of tithes
Fry, Lucius G *see* Archidiaconal functions
Fry, Roger Eliot *see* Giovanni bellini
Fry, Thomas *see* Necessity of religious knowledge to salvation
Fryburg pioneer : [official billings county paper 1916-1919] – Fryburg, Billings Co, ND: Thurston & Tharalson. v1 n1 oct 9 1913-v6 n44 aug 8 1919 (wkly) – 1 – (some cols in german. merged with: billings county herald to form: billings county pioneer) – mf#02450-02452 – us North Dakota [071]
Frye, William *see* In whitest africa
Fryeburg post – Hampden Highlands, ME. 1958-1975 (1) – mf#63559 – us UMI ProQuest [071]
Fryer, Alfred Cooper *see* Cuthberht of lindisfarne
Frying pan – 1976 jul-1979 jun, 1979 jul-1982 jun, 1982 jul-1986 jul – 3r – 1 – mf#1078001 – us WHS [071]
Fryksdalens tidning – Arvika, 1916-20 – 3r – 1 – sw Kungliga [078]
Fryksdalingen – Filipstad, Sweden. 1885-87 – 1r – 1 – sw Kungliga [078]
Fryksdalsbygden – Sunne, Sweden. 1979- – 1 – (nordvarmland, 1979) – sw Kungliga [078]
Fryksdalsbygden – Sunne, Sweden. 1909-31 – 27r – 1 – sw Kungliga [078]
Frysk en frij – Leeuwarden, Netherlands. 25 May 1945-28 Oct 1966 – 9r – 1 – uk British Libr Newspaper [949]
Fsj novena...virgen de la coronada...vca de los barros – 1878 – 9 – sp Bibl Santa Ana [240]
Fsw-echo – Tangermuende DE, 1974 6 aug-1990 apr [gaps] – 2r – 1 – (faser- und spanplattenwerk) – gw Misc Inst [074]
Ft caroline / Corse, Carita Doggett – s.l, s.l? 193-? – 1r – us UF Libraries [978]
Ft hood sentinel – Temple TX. 1980 aug 21-1981 jan 9, 1981 jun 18-dec, 1982 jan-dec, 1983 jan-dec, 1984 apr-dec, 1985 apr-dec, 1986 jan 9-apr 3, 1986 apr-jul – 17r – 1 – (cont: armored sentinel; cont by: sentinel [temple, tx]) – mf#583534 – us WHS [071]
Ft ogelthorne press – Rome, GA. 1991-2000 (1) – mf#68776 – us UMI ProQuest [071]
Ft ord panorama – 1980 sep 11-1981 jun, 1981 jul-1982 mar, 1982 apr-dec, 1983 jan-jun, 1983 jun 30-dec 22, 1984 jan-dec, 1985 jan-dec 19, 1986 jan-jun, 1986 jul 3-dec 18, 1987 jan-dec – 13r – 1 – (cont by: panorama [fort ord, ca]) – mf#579494 – us WHS [071]
Ft riley post – 1980 aug 22-1981 jun, 1981 jul-1982 apr, 1982 may-dec, 1983 jan-dec, 1984 jan-dec, 1985 jan-dec, 1986 jan-dec, 1987 aug 14-dec 31, 1987 jan 9-aug 7, 1988 jan 8-jun 24, 1988 jul-dec, 1989 jan-dec – 17r – 1 – mf#584048 – us WHS [071]
Fta – includes on reel: fta with pride. v1 iss 1 [1971 jun], v2 iss 1 [1971 oct], v1 n1=1-v4 n6=31 [1968 jun 23-1972 jan] – 1r – 1 – mf#964719 – us WHS [320]
Fta news / Food, Tobacco, Agricultural and Allied Workers Union of America – v1-v9 n7 [1939 jul-1950 sep] – 1r – 1 – mf#1056115 – us WHS [331]
Fta with pride / Heidelberg Liberation Front – v1:iss 1-v3:iss 8 [1971 may-1973 may] – 1r – 1 – mf#918891 – us WHS [320]
Ftd misal breve... – Madrid: Razon y Fe, 1927 – 1 – sp Bibl Santa Ana [946]
Fto vida y virtudes del venerable siervo de dios marcelino champagnat. barcelona, fto. 1929 / Bayle, Constantino – Madrid: Razon y Fe, 1929 – 1 – sp Bibl Santa Ana [240]
Fu, Chen-ch'uan *see* Kai liang chien so i chien shu
Fu, Chiao-chin *see* T'u ti hsing cheng, tu ti shih yung
Fu, Chi-t'ui *see* Ou chan i lai shih chieh ching chi ta shih
Fu ch'ou / Pa, Chin – Shang-hai: Hsin Chung-kuo shu chu, Min kuo 24 [1935] – (= ser P-k&k period) – us CRL [480]
Fu fu; fu, chang chung shuoo fa / Chang, Ch'ang-jen – Shang-hai: Ch'ang ch'eng shu chu, Min kuo 24 [1935] – (= ser P-k&k period) – us CRL [306]
Fu hsing chung-hua / Shih, Min – Shang-hai: Chung-kuo tzu ch'iang hsueh she, Min kuo 24 [1935] – (= ser P-k&k period) – us CRL [951]
Fu hsing pi chiao yen chiu ti 1 chi : i-ta-li fu hsing yu chung -kuo / Liu, Wen-tao & Hsueh, Kuang-ch'ien – [China: sn, Min kuo 24 ie 1935] – (= ser P-k&k period) – us CRL [951]
Fu huo / Hsia, Yen – Ch'ung-ch'ing: Mei hsueh ch'u pan she, Min kuo 33 [1944] – (= ser P-k&k period) – us CRL [951]
Fu huo ti mei kuei / Hou, Yao – Shang-hai: Shang wu yin shu kuan, Min kuo 21 [1932] – (= ser P-k&k period) – us CRL [820]
Fu jen chi / Ch'en, Wei-sung – Shang-hai: Ta tung shu chu, 1932 – (= ser P-k&k period) – us CRL [305]
Fu, Jen-ta *see* T'ai-p'ing yang chu kuo ti ching chi tou cheng yu erh tz'u ta chan
Fu, Jo-yu *see* Ping min tsien tzu ko (ccm128)
Fu kuei fu yun : [san mu hsi chu] / Chang, Chun-hsiang – Shang-hai: Shih chieh shu chu, 1944 (1946 printing) – (= ser P-k&k period) – us CRL [820]
Fu, Lan *see* Su sung shih wu hsing shih pien
Fu lu: tu mu chu chi / K'ang, Min – Shang-hai: Chung-kuo t'u shu tsa chih kung ssu, 1940 – (= ser P-k&k period) – us CRL [820]
Il fu mattia pascal, romanzo / Pirandello, Luigi – Firenze: R. Bemporad & Figlio, 1921. 1 reel. 1268 – 1 – us Wisconsin U Libr [820]
Fu nu erh t'ung pao hu wen t'i / Wang, Yun-wu & Li, Sheng-wu – Shang-hai: Shang wu yin shu kuan, Min kuo 22 [1933] – (= ser P-k&k period) – us CRL [305]
Fu nu she hui k'o hsueh ch'ang shih tu pen / Shen, Chih-yuan – Shang-hai: Sheng huo shu tien, Min kuo 26 [1937] – (= ser P-k&k period) – us CRL [300]
Fu nu t'an sou / Chin, Chung-hua – Shang-hai: Nu tzu shu tien, Min kuo 22 [1933] – (= ser P-k&k period) – us CRL [305]
Fu nu wen t'i / Chang, P'ei-fen – Shang-hai: Shang wu yin shu kuan, Min kuo 22 [1933] – (= ser P-k&k period) – us CRL [305]
Fu nu wen t'i / Ch'u, Yun – Shang-hai: Tu shu sheng huo she, Min kuo 25 [1936] – (= ser P-k&k period) – us CRL [305]
Fu nu wen t'i chung yao yen lun chi / Chung-kuo kuo min tang Hsuan ch'uan pu – Nan-ching: Kuo min tang chung yang hsuan ch'uan pu, Min kuo 18 [1929] – (= ser P-k&k period) – us CRL [305]
Fu nu wen t'i ti ko fang mien / Chin, Chung-hua – Shang-hai: K'ai ming shu tien, Min kuo 23 [1934] – (= ser P-k&k period) – us CRL [305]
Fu nu yun tung / Key, Ellen – Shang-hai: Shang wu yin shu kuan, 1936 – (= ser P-k&k period) – us CRL [305]
Fu, Sheng *see* Fu sheng kuo chi lun wen chi ti i chi
Fu sheng kuo chi lun wen chi ti i chi / Fu, Sheng – Shang-hai: Sheng huo shu tien, Min kuo 22 [1933] – (= ser P-k&k period) – us CRL [327]
Fu shih hua chi ch'i t'a : ming chia hsiao shuo chi / [Yao] P'eng-tzu et al – Shang-hai: Liang yu fu hsing t'u shu kung ssu, 1940 – (= ser P-k&k period) – us CRL [830]
Fu shih ta yao / Suzuki, Torao – [China]: Cheng chung shu chu, Min kuo 36 [1947] – (= ser P-k&k period) – us CRL [951]
Fu, Shuang-chi *see* Min tsu chan cheng ch'uan chun chan chi shih liao ts'un yao ch'u p'ein
Fu tan mirror – Shanghai. 1934-37 – 1/4r – 1 – uk British Libr Newspaper [072]
Fu tan ta hsueh fu chung san shih chou chi nien ts'e – Shang-hai: Kai hsiao, [Min kuo 24 [1935]] – (= ser P-k&k period) – us CRL [951]
Fu tan ta hsueh san shih chou nien kung shang kuan li hsueh hsi chi nien k'an – Shang-hai: Fu tan ta hsueh, Min kuo 24 [1935] – (= ser P-k&k period) – us CRL [951]
Fu, Ts'an-yen *see* Jih-pen chan shih mao i cheng ts'e
Fu, T'ung-hsien *see* Hsien tai che hsueh chih k'o hsueh chih ch'u
Fu, Tung-hua *see* Han yu sheng niu pien chuan chih ting lu
Fu wu tao te / P'an, Wen-an – Shang-hai: Shang wu yin shu kuan, [1939] – (= ser P-k&k period) – us CRL [170]
Fu, Wu-kang *see* Li shih ts'ung t'an

Fu yin chih chen kuei (ccm134) / Gutzlaff, Karl Friedrich August – Hsin-chia-p'o, 1837 [mf ed 198?] – (= ser Ccm 134) – 1 – mf#1984-b500 – us ATLA [225]
Fu yin ho ts'an (ccm234) / Luce, Henry Winters – Shanghai, 1902 [mf ed 198?] – (= ser Ccm 234) – 1 – (pref in english) – mf#1984-b500 – us ATLA [226]
Fu yin hsin pao (ccs) = Gospel news – Fu-chou. n32. 1877 [complete] [mf ed 198?] – (= ser Chinese christian serials coll) – 1 – mf0296i – us ATLA [240]
Fu yin te chun pei (ccc86) = The preparation for the gospel / Ch'en, Chi-yun – Shanghai, 1949 [mf ed 198?] – (= ser Ccm 86) – 1 – mf#1984-b500 – us ATLA [226]
Fu yin te chun pei (ccm87) = The gospel for all / Chen, Chi-yun – Shanghai, 1949 [mf ed 198?] – (= ser Ccm 87) – 1 – mf#1984-b500 – us ATLA [226]
Fu yu sheng ching (ccm141) = The children's bible: scripture selections – Hong Kong, 1952 [mf ed 198?] – (= ser Ccm 141) – 1 – mf#1984-b500 – us ATLA [220]
Fuad, Koeprueluezade Mehmed *see* Yeni osmanli tarih edebiyati
Fuba forum / Federated Union of Black Arts [South Africa] – 1979 dec – 1r – 1 – mf#4872381 – us WHS [700]
Fubara-Manuel, Benebo Fubara *see*
– From "winner" to "sign"
– From "winner" to sign
Fu-chien chan shih ching chi ti li / Hsu, T'ien-t'ai – Nan-p'ing: Fu-chien jen wen ch'u pan she, Min kuo 32 [1943] – (= ser P-k&k period) – us CRL [339]
Fu-chien hsi nan lu k'uang chi hua / Cheng, Hua – [China: sn, 1933] – (= ser P-k&k period) – us CRL [380]
Fu-chien hsiang-shih lu – List of successful candidates in the imperial examination in Fukien province: 1751, 1752, 1875, 1876, 1879, 1882, 1885, 1888, 1889, 1891, 1893, 1897. 1 reel – 1 – us Chinese Res [951]
Fu-chien jih-pao – Foochow, Fukien. March 1961-Dec 11, 1963. Scattered issues missing. 6 reels – 1 – us Chinese Res [079]
Fu-chien li nien tui wai mao i t'ung chi – Fu-chien: Sheng cheng fu mi shu ch'u kung pao shih, Min kuo 24 [1935] – (= ser P-k&k period) – us CRL [380]
Fu-chien sheng ching chi chien she hui lan – [China]: Fu-chien sheng ching chi chien she chi hua wei yuan hui hsuan ch'uan ch'u, Min kuo 30 [1941] – (= ser P-k&k period) – us CRL [339]
Fu-chien sheng ch'u pu cheng li t'u ti kai k'uang – [Fu-chien: Fu-chien sheng, 1939] – (= ser P-k&k period) – us CRL [630]
Fu-chien sheng hsien cheng jen yuan hsun lien so *see* Hsien cheng jen yuan hsun lien
Fu-chien sheng jen shih hsing cheng yu hsun lien – [China: Fu-chien sheng ti fang hsing cheng kan pu hsun lien t'uan, 1940] – (= ser P-k&k period) – us CRL [350]
Fu-chien sheng ti cheng kai k'uang – [China: sn, 1939] – (= ser P-k&k period) – us CRL [630]
Fu-chien sheng ti fang hsing cheng kan pu hsun lien t'uan k'o ch'eng – [China: Fu-chien sheng ti fang hsing cheng kan pu hsun lien t'uan, 1940] – (= ser P-k&k period) – us CRL [350]
Fu-chien sheng wu nien lai chung teng chiao yu – [np, nd] – (= ser P-k&k period) – us CRL [370]
Fu-chien sheng yin hang *see* Fu-chien sheng yin hang chang tse hui pien
Fu-chien sheng yin hang chang tse hui pien / Fu-chien sheng yin hang – [Fu-chou shih?: Fu-chien sheng yin hang tsung kuan li ch'u mi shu shih], Min kuo 29 [1940] – (= ser P-k&k period) – us CRL [951]
Fu-chien sheng yin hang Tsung kuan li ch'u *see* Liu nien lai ti fu-chien sheng yin hang
Fu-chien sheng yin hang wu chou nien chi nien ts'e – [Fu-chien: Kai yin hang, 1940?] – (= ser P-k&k period) – us CRL [951]
Fu-chien ti i yun – [China]: Fu-chien yun shu kung ssu, 1941 – (= ser P-k&k period) – us CRL [380]
Fu-chien t'i yu – [Fu-chien: Sheng li Fu-chou kung kung t'i yu ch'ang yen chiu pu, Min kuo 26 [1937]] – (= ser P-k&k period) – us CRL [790]
Fu-chou chi-tu chiao ch'ing nien hui li shih (ccm348) : ko hsiang kuei tse fu – [s.l: s.n, 19–] [mf ed 198?] – (= ser Ccm 348) – 1 – mf#1984-b500 – us ATLA [360]
Fu-chou t'u ti teng chi – [China: sn, 1939] – (= ser P-k&k period) – us CRL [630]
Fuchs, A *see* Der kosmopolitische beobachter
Fuchs, Albert *see*
– Les debuts de la litterature allemande du 8e au 12e siecles
– Goethe, un homme face a la vie
– Die temporalsaetze mit den konjunktionen "bis" und "so lange als"
Fuchs, Eduard et al *see* Aus dem klassenkampf
Fuchs, Emil *see*
– Schleiermachers religionsbegriff und religioese stellung zur zeit der ersten ausgabe der reden
– Vom werden dreier denker
Fuchs, Friedrich *see* Die central-karpathen mit den naechsten voralpen
Fuchs, Georg *see* Des burschen heimkehr
Fuchs, Gerd *see* Literatur und wirklichkeit
Fuchs, Hans *see*
– Als seekadett nach fernost
– Eine insel im la plata
Fuchs, Karl *see* Johann gabriel seidl
Fuchs, M et al *see* Apologetische vortraege
Fuchs, Meik *see* Dreiguds un noschens
Fudge, B R *see* Relation of magnesium deficiency in grapefruit leaves to yield and chemical composition of fruit
Fueeterer, Ulrich *see* Ulrich fueeterers prosaroman von lanzelot
Fueetrer, Ulrich *see*
– Die gralepen in ulrich fueetrers bearbeitung (buch der abenteuer)
– Poytislier
Fueggetlen magyarorszag *see* Del kereszt je or southern cross
Fuehmann, Franz *see* Erfahrungen und widersprueche
Der fuehrer – Berlin DE, 1921-23 – 1r – 1 – gw Misc Inst [074]
Der fuehrer : offizielles organ des unabhaengigen ordens der sonderbaren bruder in den v st von amerika – New York NY 18 sep 1874 – 1 – uk British Libr Newspaper [366]
Der fuehrer : weckrufe nationalsozialistischen glaubens und wollens – Karlsruhe DE, 1927 5 nov-1928, 1930-1945 30 mar – 69r – 1 – (filmed by misc inst: 1927 5 nov-1928 29 dec, 1930 4 jan-1945 30 mar [12r]) – gw Mikropress; gw Misc Inst [320]
Der fuehrer *see* Niederrheinischer bote
Fuehrer conferences on matters dealing with the german navy, 1939-1945 / Germany. Kriegsmarine Oberkommando – (= ser World war 2 research colls) – 1r – 1 – $175.00 – 0-89093-198-4 – us UPA [934]
Fuehrer conferences on matters dealing with the german navy, sept 1939-april 1945 – 1984 – 1r – 1 – $130.00 – mf#S1654 – U.S. Naval Historical Center – us Scholarly Res [355]
Der fuehrer durch den harz – Quedlinburg [1834] [mf ed Hildesheim 1995-98] – (= ser Fbc) – 2mf [ill] – 9 – €60.00 – 3-487-29591-1 – gw Olms [914]
Fuehrer durch die sammlungen des museums fuer voelkerkunde / Berlin. Staatliche Museen. Museum fuer Voelkerkunde – Berlin: Koenigliche Museum 1887 [mf ed 1979] – 1r – 1 – mf#film mas 9263 – us Harvard [390]
Ein fuehrer durch goethes faust : 1. und 2. teil / Kaempfer, August Hermann – Halle (Saale): Verlag der Buchhandlung des Waisenhauses, 1920 – 1r – 1 – us Wisconsin U Libr [430]
Fuehrer durch koeln / Vulpius, J E – Koeln 1836 [mf ed Hildesheim 1995-98] – 1mf – 9 – €40.00 – 3-487-29544-X – gw Olms [914]
Fuehrer- und amtsblatt des gaues mecklenburg-luebeck – Schwerin 1935 1 apr-1945 apr [mf ed 2005] – 4r – 1 – (1 apr 1937: fuehrer- und amtsblatt des gaues mecklenburg) – gw Mikrofilm [350]
Fuehrung und geleit : ein lebensgedenkbuch / Carossa, Hans – Leipzig: Insel-Verlag 1934, c1933 [mf ed 1989] – 1r – 1 – (filmed with: georg buchners drama "dantons tod" / hans landsberg) – mf#7143 – us Wisconsin U Libr [880]
Fuel – Kidlington. 1982+ (1) 1982+ (5) 1982+ (9) – ISSN: 0016-2361 – mf#1260,01 – us UMI ProQuest [333]
Fuel and energy abstracts – Kidlington. 1978-1994 (1) 1978-1994 (5) 1979-1994 (9) – ISSN: 0140-6701 – mf#17231,01 – us UMI ProQuest [550]
Fuel for the flame / Waugh, Alec – New York, NY. 1960 – 1r – us UF Libraries [025]
Fuel processing technology – Amsterdam. 1978+ (1) 1978+ (5) 1987+ (9) – ISSN: 0378-3820 – mf#42069 – us UMI ProQuest [660]
Fuelleborn, Georg Gustav *see* Beytraege zur geschichte der philosophie
Fueller, Franziska *see* Das psychologische problem der frau in kleists dramen und novellen
Das fuellhorn : ein zeitblatt zunaechst fuer und ueber israeliten / ed by Rosenfeld, S W – Bamberg, Dinkelsbuehl. v1-2. 1835-36 – (= ser German-jewish periodicals...1768-1945, pt 3) – 1r – 1 – $165.00 – (lacking: n23 v2) – mf#B83 – us UPA [939]
Fuellkrug, Gerhard *see* Gottesknecht des deuterojesaja
Fueloep-Miller, Rene *see* The mind and face of bolshevism
Fueloil and oil heat – Cedar Grove. 1942-1977 (1) 1971-1977 (5) 1977-1977 (9) – (cont by: fueloil and oil heat and solar systems) – ISSN: 0016-2418 – mf#1103 – us UMI ProQuest [690]
Fueloil and oil heat – Fairfield. 1990-1991 (1) 1990-1991 (5) 1990-1991 (9) – (cont: fueloil and oil heat magazine. cont by: fueloil and oil heat with air conditioning) – ISSN: 1061-141X – mf#1103,03 – us UMI ProQuest [690]
Fueloil and oil heat and solar systems – Cedar Grove. 1977-1985 (1) 1977-1985 (5) 1977-1985 (9) – (cont: fueloil and oil heat. cont by: fueloil and oil heat magazine) – ISSN: 0148-9801 – mf#1103,01 – us UMI ProQuest [690]
Fueloil and oil heat magazine – Cedar Grove. 1985-1990 (1) 1985-1990 (5) 1985-1990 (9) – (cont: fueloil and oil heat and solar systems. cont by: fueloil and oil heat) – ISSN: 0888-0735 – mf#1103,02 – us UMI ProQuest [690]
Fueloil and oil heat with air conditioning – Fairfield. 1991-1997 (1) 1991-1997 (5) 1991-1997 (9) – (cont: fueloil and oil heat. cont by: oilheating) – ISSN: 1060-9725 – mf#1103,04 – us UMI ProQuest [690]
Die fuenf buecher mosis : in uebersichtlicher nebeneinanderstellung des urtextes, der septuaginta, vulgata und luther-uebersetzung, so wie der wichtigsten varianten der vornehmsten deutschen uebersetzungen fuer den praktischen handgebrauch – 4. aufl. Bielefeld: Velhagen & Klasing, 1875 – 10mf – 9 – 0-524-08207-3 – mf#1993-0002 – us ATLA [221]
Fuenf festpredigten augustins in gereimter prosa / Augustine, Saint, Bishop of Hippo; ed by Lietzmann, Hans – Bonn: A Marcus & E Weber 1905 [mf ed 1992] – (= ser Kleine texte fuer theologische vorlesungen und uebungen 13/2) – 1mf – 9 – 0-524-04667-0 – (text in latin; notes in german) – mf#1990-1294 – us ATLA [241]
Fuenf jahre unter den staemmen des kongo-staates / Ward, H – Leipzig, 1891. 3v – 5mf – 9 – mf#H-6177 – ne IDC [916]
Fuenf madchen *see* Die wassernot im emmenthal / fuenf maedchen / durch die branntweinlaeuter
Fuenf neue arabische landschaftsnamen im alten testament : mit einem exkurs ueber die paradiesesfrage / Koenig, Eduard – Berlin: Reuther & Reichard, 1901 – 1mf – 9 – 0-8370-1945-1 – (incl bibl ref) – mf#1987-6332 – us ATLA [221]
Fuenf schloesser : altes und neues aus mark brandenburg / Fontane, Theodor – 5-6.aufl. Stuttgart: J G Cotta, 1920 [mf ed 1989] – viii/454p – 1 – mf#7248 – us Wisconsin U Libr [914]
Fuenff predigen : von dem wercke der concordien / Andreae d A, J – Dreszden, etc, nd – 3mf – 9 – mf#TH-1 mf 55-57 – ne IDC [242]
Fuenff und zwanzig bedenckliche figuren mit erbaulichen erinnerungen : dem tugend und kunstliebenden zu guter gedechtnus in kupffer gebracht / Meyer, C – Zuerich, 1674 – 1mf – 9 – mf#O-1459 – ne IDC [090]
Fuenfftzig impresen : oder sinnbilder sambt einem dem alphabet nach eingerichtem indice... – Muenchen: Johann Juecklin, 1678 – 2mf – 9 – mf#O-11 – ne IDC [090]
Der fuenffuessige jambus bei christian dietrich grabbe : ein beitrag zur metrik / Kessler, Hugo – Muenster: Westfaelische Vereinsdruckerei 1913 [mf ed 1990] – 1r – 1 – (incl bibl ref. filmed with: christian dietrich grabbe in der nachschillerischen entwickelung / joseph gieben) – mf#2687p – us Wisconsin U Libr [430]
Der fuenffuessige jambus in den dramen friedrichs halms : eine metrische untersuchung / Lambertz, Paul – Muenster, 1914 (mf ed 1994) – 1mf – 9 – €24.00 – 3-8267-3108-5 – mf#DHS-AR 3108 – gw Frankfurter [430]
Fuenffzehen symbola : welche zu der hoechst betaurlichen begraebnus auch durch thomas bernhardt de lillis seynd componirt worden / Lillis, Thomas Bernhardt de – n.p, 1685 – 1mf – 9 – mf#O-1923 – ne IDC [090]
Das fuenfte evangelium (das heilige land) / Brueckner, Martin – Tuebingen: J C B Mohr 1910 [mf ed 1993] – (= ser Religionsgeschichtliche volksbuecher fuer die deutsche christliche gegenwart 1/21) – 1mf – 9 – 0-524-06118-1 – (incl bibl ref) – mf#1992-0785 – us ATLA [220]
Fuenfundsiebzig punkte zur beantwortung der frage, absolute oder relative wahrheit der hl schrift? : eine kritik der schrift dr fr eggers / Holzhey, Carl – Muenchen: J J Lentner, 1909 [mf ed 1989] – 1mf – 9 – 0-7905-0430-8 – mf#1987-0430 – us ATLA [220]
Fuenfzig feldpostbriefe eines frankfurters : aus den jahren 1870 und 1871 / Wuelker, Richard Paul – 2. Aufl. Halle (Saale): M Niemeyer, 1876 – 1 – us Wisconsin U Libr [860]
Fuenfzig feuilletons : mit einem praeludium in versen / Kuernberger, Ferdinand – Wien: T Daberkow, [1905?] – 1 – us Wisconsin U Libr [840]
Fuenfzig jahre goethe-gesellschaft / Goetz, Wolfgang – Weimar: Goethe-Gesellschaft, 1936 [mf ed 1993] – (= ser Schriften der goethe-gesellschaft 49) – vi/102p – 1 – mf#8657 reel 11 – us Wisconsin U Libr [430]
Fuenfzig jahre im predigtamte : freud' und leid aus dem leben und der fuenfzigjaehrigen dienstzeit eines evangelischen predigers in amerika / Hoehn, M – Chicago. Ill: M Hoehn, 1913 – 1mf – 9 – 0-524-07324-4 – mf#1991-3039 – us ATLA [240]
Fuenfzig unterdrueckte balladen und liebeslieder des 16. jahrhunderts : mit den alten singweisen / ed by Ditfurth, Franz Wilh Freiherr v – Heilbronn: G Henninger, 1877 – 1r – 1 – (incl bibl ref and title index) – us Wisconsin U Libr [780]
Fuente de Cantos. Ayuntamiento *see* Ordenanzas municipales
Fuente del Maestre *see*
– Fiestas patronales. septiembre 1975
– Reglamento y ordenanzas de la comunidad de labradores de fuente del maestre
– Revista de las fiestas que se celebran en fuente del maestre en honor del santisimo cristo de las misericordias
Fuente sellada / Torrens De Garmendia, Mercedes – Habana, Cuba. 1956 – 1r – us UF Libraries [972]
Fuentes, Eduardo Sanchez *see* Fok-lor en la musica cubana
Fuentes historicas sobre colon y america / Angleria, Pedro Martir – 1892. v. 1-4 – 9 – sp Bibl Santa Ana [970]
Fuentes, Milton *see* Discurso de santos
Las fuentes para la historia de la pedagogia espanola / Floriano Cumbreno, Antonio C – Madrid, 1943 – 1 – sp Bibl Santa Ana [370]
Fuentes, Patricia De *see* Conquistadors
Fuentes y Guzman, Francisco A *see* Historia de guatemala
Fuentes Y Guzman, Francisco Antonio De *see* Recordacion florida
Fuentes y Guzman, Francisco Antonio de *see* Recordacion florida. discurso historial y de demostracion natural, material, militar y politica del reyno de guatemala
Fuentes-Figueroa Rodriguez, Julian *see* Historia de venezuela
Fuenzig geistliche homilien (bdk10 1.reihe) / Makarius der Aegypter – (= ser Bibliothek der kirchenvaeter. 1. reihe (bdk 1.reihe)) – €17.00 – ne Slangenburg [240]
Fuer aeltere litteratur und neuere lectuere / ed by Canzler, Karl Christian et al – Leipzig 1783-85 – (= ser Dz. abt literatur) – 3jge on 18mf – 9 – €180.00 – mf#k/n4537 – gw Olms [430]
Fuer das voelkerrecht / Mueller, Max Ludwig – Tuebingen: Kloeres, 1916. 32p – 1 – us Wisconsin U Libr [341]
Fuer die frau *see* Frankfurter geschaeftsbericht
Fuer die frau meisterin – Duesseldorf DE, 1912-14 – 1r – 1 – gw Misc Inst [074]
Fuer die jesuiten : kurzgefasste geschichte der gesellschaft jesu im gegensatze zum protestantismus und zum freimaurerthum / Ruetjes, Heinrich – Emmerich: JL Romen, 1872 – 1mf – 9 – 0-524-05893-8 – mf#1991-2343 – us ATLA [241]
Fuer die jugend *see* Hoefer intelligenz-blatt
Fuer freunde der tonkunst / Rochlitz, Friedrich – Leipzig: C Cnobloch 1830-45 [mf ed 1979] – 4v on 1r – 1 – mf#film mas 9112 – us Harvard [780]
Fuer geist und herz – 1786 [mf ed 1997] – (= ser Die zeitschriften des august von kotzebue) – 13mf – 9 – €120.00 – 3-89131-237-7 – gw Fischer [430]
Fuer polens freiheit : achthundert jahre deutsch-polnische freundschaft in der deutschen literatur / Haeckel, Manfred [comp] – Berlin: Verlag Blick nach Polen [1952?] – 404p – 1 – (int by rudolf leonhard and leo kruczkowski; incl bibl ref) – us Wisconsin U Libr [430]
Fuer staat und volk – Melsungen DE, 1929 7 apr-12 may – 1r – 1 – gw Misc Inst [320]
Fuer und wider kahnis : kritik der dogmatik von kahnis mit bezug auf dessen vertheidigungsschrift / Delitzsch, Franz – Leipzig: Doerffling und Franke, 1863 – 1mf – 9 – 0-8370-9613-8 – mf#1986-3613 – us ATLA [240]
Fuer unsere kleinen : illustrierte monatsschrift fuer kinder von 4 bis 10 jahren – Gotha DE, 1884-85, 1887-88 [all incomplete], 1891 n7-9 – 1 – gw Misc Inst [801]
Fuera de acta / Sancho, Alfredo – Mexico City? Mexico. 1968 – 1r – us UF Libraries [972]
Fuerbringer, Hermann *see* Die kuenstlerischen voraussetzungen des genter altars der brueder van eyck
Fuerer-Haimendorf, Christoph von *see*
– The aboriginal tribes of hyderabad...
– The naked nagas
Die fuernemsten heupstueck der christlichen lehre / Chemnitz d A, M – Wulffenbuettel, 1569 – 4mf – 9 – mf#TH-1 mf 211-214 – ne IDC [242]
Fuero de albarracin (siecle 13) – Albarracin – 1r – 5,6 – sp Cultura [340]
Fuero de caceres (anno 1229-1231) – Caceres – 1r – 5,6 – sp Cultura [340]
Fuero de cordoba. expedientes sobre el fuero (anno 1241-sieclo 1900) – Cordoba – 1r – 5,6 – sp Cultura [340]

El fuero de plasencia / Benavides Checa, Jose – 1896 – 9 – sp Bibl Santa Ana [946]
Fuero de poblacion otorgado por...don carlos 3 a las localidades formadas en la sierra morena por la llamada "colonizacion interior"... / Cotta y Marquez de Prado, Ventura de – s.1, s.i, s.a. Sep. a Mancha. pp. 1-30 – 1 – sp Bibl Santa Ana [946]
Fuero de poblacion otorgado...carlos 3 a las localidades formadas en la sierra morena por la llamada "colonizacion interior" de espana, que afecto a parte de la provincia de ciudad real – Badajoz, 1961. Sep.Rev. de Est. Regionales. La Mancha – sp Bibl Santa Ana [946]
Fuero de usagre (siglo 18) / Urena y Semenjaud, Rafael & Bonilla y San Martin, Adolfo; ed by Hijos de Reus – Madrid, 1907 – 1 – sp Bibl Santa Ana [946]
Fuero indigena venezolano / Venezuela Comision Indigenista – Caracas, Venezuela. pt1-2. 1954 – 1r – us UF Libraries [972]
El fuero real de espana hecho por alfonso 9 / Diaz de Montalvo, A – Burgos, 1533 – 11mf – 9 – sp Cultura [946]
Fueros de la iglesia ante el liberalismo y el cons... / Cadavid G, J Ivan – Medellin, Colombia. 1955 – 1r – us UF Libraries [972]
Fuers publikum gewaehlt-erzaehlt : prosa aus sechs jahrzehnten kabarett / ed by Bemmann, Helga – Berlin: Henschelverlag, 1971 – 1r – 1 – (incl ind) – us Wisconsin U Libr [430]
Die fuersorge : zeitschrift fuer alle zweige der oeffentlichen und freien wohlfahrtspflege – Berlin [1] 1924-[2] 1925 [mf ed 2005] – (= ser Freie wohlfahrtspflege 6) – 140mf – 9 – €880.00 – 3-89131-469-8 – (with: deutsche zeitschrift fuer wohlfahrtspflege [1] 1925/26 apr-[20] 1944/45; with suppl: das fuersorgerecht [1] 1932/33) – gw Fischer [360]
Fuerst bismarck : sein politisches leben und wirken urkundlich in thatsachen und des fuersten eigenen kundgebungen / Hahn, Ludwig Ernst [comp] – Berlin: W Hertz 1878-91 [mf ed 1979] – 5v on 2r – 1 – (v1: bis 1870, v2: 1870 bis 1877, v3: bis 1879, v4: 1879-85, v5: 1885-90; v5 is a cont by karl wippermann) – mf#film mas c701 – us Harvard [943]
Fuerst bismarck und der antisemitismus / Popper-Lynkeus, Josef – Wien: Engel 1886 [mf ed 19–] – 1r – 1 – mf#film mas 572 – us Harvard [939]
Fuerst bismarck und die kaiserin augusta / Bosbach, Heinz – Koeln, 1936 (mf ed 1993) – 1mf – 9 – €24.00 – 3-89349-335-2 – mf#DHS-AR 188 – gw Frankfurter [943]
Fuerst ganzgott und saenger halbgott : [short story] / Arnim, Ludwig Achim, Freiherr von – Wien: Herz-Verlag, 1922 [mf ed 1996] – 46p (ill) – 1 – (ill by karl harmos) – mf#9605 – us Wisconsin U Libr [830]
Fuerst, Julius *see*
- Geschichte der biblischen literatur und des juedisch-hellenistischen schriftthums
- Hebraeisches und chaldaeisches schul-woerterbuch ueber das alte testament
- Der kanon des alten testaments
- Librorum sacrorum veteris testamenti concordantiae hebraicae atque chaldaicae
- Der orient

Fuerst, Max *see* Biographisches lexikon fuer das gebiet zwischen inn und salzach
Fuerst, R H *see* Neues modelbuch von unterschiedlicher art...
Fuerst, Rudolf *see* Deutsche erzaehler des achtzehnten jahrhunderts
Die fuersten fallen : roman aus hundert jahren anarchie / Euringer, Richard – Leipzig: Grethlein 1935 [mf ed 1989] – 1r – 1 – (filmed with: die arbeitslosen) – mf#7226 – us Wisconsin U Libr [830]
Der fuersten- und volksfreund – Kassel DE, 1831 4 mar-6 aug – 1r – 1 – gw Misc Inst [074]
Fuerstenfeldbrucker tagblatt – Fuerstenfeldbruck DE, 1987- – 15r/yr – 1 – (bezirksausgabe von muenchner merkur, muenchen) – gw Misc Inst [074]
Fuersten-postkarten : sammlung fruehsorge = Royal postcards : the fruehsorge collection / Behrens, Christoph [comp] – [mf ed 1988] – 86mf (1:24) – 9,15 – silver €1260.00 – ISBN-10: 3-598-32532-0 – ISBN-13: 978-3-598-32532-8 – (incl printed guide) – gw Saur [900]
Die fuerstenspiegel des hohen und spaeten mittelalters (mgh schriften:2.bd) / Berges, W – 1938 – (= ser Monumenta germaniae historica. schriften (mgh schriften)) – €19.00 – ne Slangenburg [931]
Fuerstenthal, Johann August Ludwig *see* Repertorium ueber saemmtliche, durch die gesetz-sammlung und die amts-blaetter der koeniglichen regierungen
Fuerstenthumer zeitung – Koeslin (Koszalin, PL), 1912 jul-dec – 1r – 1 – gw Misc Inst [077]
Das fuerstentum mentesche : studie zur geschichte westkleinasiens im 13-15 jh / Wittek, P – Istanbul, 1934 – 4mf – 8 – mf#U-659 – ne IDC [956]
Die fuerstin : erzaehlungen / Heuschele, Otto – Stuttgart: J F Steinkopf, 1945 – 1r – 1 – us Wisconsin U Libr [830]
Die fuerstin amalie von galitzin und friedrich leopold graf zu stolberg-stolberg : ein beitrag zur stellung des gallitzin-kreises in der deutschen literatur- und geistesgeschichte / Wolf, Otmar – Wuerzburg, 1952 (mf ed 1992) – 3mf – 9 – €49.00 – 3-89349-243-7 – mf#DHS-AR 34 – gw Frankfurter [943]
Fuerstlich lippischer kalender auf das jahr... – Lemgo DE, 1848-50 – 1r – 1 – gw Misc Inst [074]
Fuerstlich lippisches intelligenzblatt *see* Lippische intelligenzblaetter
Fuerstlich lippisches regierungs- und anzeigeblatt *see* Lippische intelligenzblaetter
Fuerstlich schwarzburgisch-rudolstaedtisches gnaedigst privilegiertes wochenblatt *see* Rudolstaedtische woechentliche anzeigen und nachrichten
Fuerstlicher baumeister : oder architectura civilis, wie grosse fuersten und herren pallaeste... flaeglich anzulegen und nach heutiger art auszuzieren / Decker, P – Augsburg. 2v. 1711-1716 – 10mf – 9 – mf#OA-102 – ne IDC [720]
Fuerstlich-oranien-nassau-fuldaische woechentliche polizei-, kommerzien- und zeitungsanzeigen *see* Fuldaische wochentlich policey- und commercien-anzeigen
Fuertes Acevedo, Maximo *see*
- El ano meteorologico 1879
- El ano meteorologico 1881
- Bosquejo...literatura de asturias...
- Bosquejos cientificos
- Discurso...pedro calderon
- Mineralogia asturiana

Fuertes, Damaso *see* Cronica de la primera asamblea misional diocesana de badajoz celebrada en zafra...
Fuertes, Jose V *see* Lecciones de ritos romanos
Fuerther nachrichten – Fuerth DE, 1978 1 jun-2002 – 1 – (filmed by misc inst: 1977-[ca 13r/yr]. ba v. nuernberger nachrichten, nuernberg) – gw Mikrofilm; gw Misc Inst [074]
Fuerther tagblatt – Fuerth DE, 1848-49 – 1r – 1 – gw Misc Inst [074]
La fuerza mayor en el derecho mercantil. / Hewstone Burotto, Luis – Santiago de Chile: Cultura 1945. 109p. LL-4073 – 1 – us L of C Photodup [346]
Fuerzas de pasion / Robiou, Jose Ramon – Ciudad Trujillo, Dominican Republic. 1958 – 1r – us UF Libraries [972]
Fuessli, Hans Heinrich *see*
- Neues schweitzersches museum
- Schweitzersches museum

Fuesslin, C *see* Theatrum gloriae sanctorum, erectum a venerando...
Fuesslin, J C *see*
- Beytraege zur erlaeuterung der kirchen-reformations-geschichten
- Epistolae ab ecclesiae helveticae reformatoribus vel ad eos scriptae

Fuessly, J C *see* Magazin fuer die liebhaber der entomologie
Fueter, Eduard *see* Religion und kirche in england im fuenfzehnten jahrhundert
Fueyuzat – Selanik: Yeni Asir Matbaasi. Sahib-i Imtiyaz: Ali Riza; Mueduer ve Sermuharriri: Hakki Baha, 1908-09. n3. 15 mart 1325 [1909] – (= ser O & t journals) – 2mf – 9 – $40.00 – us MEDOC [956]
Fuga / Ozores, Rentato – Panama, 1959 – 1r – us UF Libraries [972]
Fugaku hiyaku-kei : or a hundred views of fuji (fusiyama) / Dickins, Frederick Victor – London 1880 – (= ser 19th c art & architecture) – 2mf – 9 – mf#4.1.301 – uk Chadwyck [700]
Fugal writing from 1750 to 1827 / Frederick, Kurt – U of Rochester 1957 [mf ed 19–] – 2v on 10mf – 9 – (with app) – mf#fiche 217 – us Sibley [780]
Fuggetlen magyarorszag *see* Keresztje southern cross
Fughe e capriccj : per clavicembalo o per l'organo: opera prima / Marpurg, Friedrich Wilhelm – Berlin: J J Hummel [1777] [mf ed 199-] – 1r – 1 – mf#pres. film 134 – us Sibley [780]
The fugitive – Nashville. v1-4. apr 1922-dec 1925 – 1 – us NY Public [810]
The fugitive / Tagore, Rabindranath – London: Macmillan and Co, 1925 – (= ser Samp: indian books) – us CRL [490]
Fugitive from spain : a lecture given in lima, peru / Sassone, Felipe – [San Francisco] 1937 – 9 – (extracts trans into english; publ in spanish in "el comercio de lima", peru...nov 8 & 9, 1936) – mf#w1156 – us Harvard [946]
The fugitive slave law / Hall, B M – Schenectady, [NY]: Riggs, 1850 (mf ed 1989) – 1r – 1 – mf#ZZ-30406 – us NY Public [976]
The fugitive slave law and its victims / [May, Samuel] – rev enl ed. New York: American Anti-Slavery Society, 1861 [mf ed 1992] – (= ser Anti-slavery tracts. new series) – 1mf – 9 – 0-524-01949-5 – (1st publ 1856) – mf#1990-0538 – us ATLA [976]
Fugitt, Stephen Mark *see* Biblical philistines
Fugue before bach / Presser, William – U of Rochester 1947 [mf ed 19–] – 1r / 12mf – 1,9 – (with bibl) – mf#film 177 / fiche 174 – us Sibley [780]
Fugue in the orchestral works of bartok / Berleant, Arnold – U of Rochester 1955 [mf ed 1955] – 2mf – 9 – mf#fiche978 – us Sibley [780]
Fuhlrott, Joseph *see* Marien-predigten fuer die vorzueglichsten feste der allerseligsten mutter gottes
Fuhn, S J *see* Ha-karmel
Führer, Alois Anton *see* The sharqi architecture of jaunpur
Fuhrer zur eroffnungsfeier des nord-ostsee-kanals: mit lagenplan der kriegsschiffe, stadtplan und-ansichten; eisenbahn-und dampfschiffs-fahrplanen; programm der kieler wocho.. – Kiel: Verlag der "Nord-Ostsee-Zeitung", 1895 – 1 – us Wisconsin U Libr [949]
Fuhrmann, J *see* Irish medieval monasteries on the continent
Fuji flyer – Tokota Air Base. 1981 may-1983 apr, 1983 may-1987 may – 2r – 1 – mf#1278385 – us WHS [355]
Fujii, Sadafumi *see* Shaji torishirabe ruisan
Fujii, Sensho *see* Bukkyo shoshi
Fujimori, Seikichi *see* Wen i hsin lun
Fujimoto-Kanatani, Koichiro *see* Determining the essential elements of golf swings used by elite golfers
Fujishima, Ryauon *see* Le bouddhisme japonais
Fujita, Tokutaro *see* Nihon kinsei kayo shiryoshu
Fukai, Kanichiro *see* [Japanese commentaries on the "four shoo" or the books of the four philosophers]
Fuken tokeisho shusei : series 1: meiji nenkan fuken tokeisho shusei (statistical annuals of the respective prefectures of japan in the meiji era, 1873-1912 – (= ser Complete coll of the statistical annuals of the respective prefectures of japan, 1873-1972) – 2050v on 504r – 1 – Y3,478,000 – (with 64p guide. in japanese. all of the annual statistical reports of the 47 prefectures) – ja Yushodo [315]
Fuken tokeisho shusei : series 2: taisho showa nenkan fuken tokeisho shusei (statistical annuals of the respective prefectures of japan in the taisho and showa eras, 1913-1945) – (= ser Complete coll of the statistical annuals of the respective prefectures of japan, 1873-1972) – 4550v on 636r – 1 – Y4,140,000 – (with 32p guide. in japanese. all of the annual statistical reports of the 47 prefectures) – ja Yushodo [315]
Fuken tokeisho shusei : series 3: todofuken tokeisho shusei, sengo-hen (statistical annuals of the respective prefectures of japan in the showa era, 1947-1972) – (= ser Complete coll of the statistical annuals of the respective prefectures of japan, 1873-1972) – 1220v on 320r – 1 – Y2,850,000 – (with 24p guide. in japanese. all of the annual statistical reports of the 47 prefectures) – ja Yushodo [315]
Fukien Mission *see* Fukien mission. general report. women's auxiliary
Fukien mission. general report. women's auxiliary : china's children's helping band / Fukien Mission – 1919-35 – 1r – 1 – (lacking: 1924. 1928-29) – mf#ATLA S0724A – us ATLA [240]
Fukien Province (China) Mi shu ch'u Pien i shih *see* Min cheng i nien
Fuks, Lajb *see* Het leven der joden in de sowjet-unie
Fuks, Sh L *see* Podol'skii krai
Fukuzawa kankei monjo : records concerning yukichi fukuzawa and keio gijuki / Keio Gijuku Fukuzawa Memorial Center [comp] – 240r – 1 – Y3600,000 – (in japanese) – ja Yushodo [950]
Fukyo Taikwan *see* Buddhist meditations from the japanese
A fulani grammar / Leith-Ross, Sylvia – [Lagos: Govt Printing Office, 1919?] – 1 – us CRL [490]
The fulani of northern nigeria : some general notes / St Croix, F W de – [Lagos: Govt Printer, 1944; 1945] – 1 – us CRL [305]
Fulbourn all saints & st vigor 1538-1950 – (= ser Cambridgeshire parish register transcript) – 13mf – 9 – £16.25 – uk CambsFHS [929]
Fulcher, George Williams *see* Life of thomas gainsborough
Fulcrum / Socialist Party of Canada – v1 n1 [i.e. 1st ed]-v10 n2 [1968 may-1977] – 1r – 1 – (cont by: socialist fulcrum) – mf#615612 – us WHS [071]
Fulda, Fuerchtegott Christian *see* Trogalien zur verdauung der xenien
Fulda, Hermann *see* Das kreuz und die kreuzigung
Fulda, Ludwig *see*
- Aus der werkstatt
- Gedichte
- Der heimliche koenig
- Hoehensonne
- Jugendfreunde
- Lebensfragmente
- Neue gedichte
- Die rueckkehr zur natur
- Schlaraffenland
- Der seeraeuber
- Der sohn des kalifen

Fulda-eder-bote – Koerle DE, 1933 1 feb-may 31 – 1r – 1 – gw Misc Inst [074]
Fuldaer kreisblatt *see* Fuldaische wochentlich policey- und commercien-anzeigen
Fuldaer politische zeitung – Fulda DE, 1831 16 sep-1832 20 jun [gaps] – 1 – (title change: 1832: fuldaer zeitung. incl suppl: kastalia) – gw Misc Inst [074]
Fuldaer volkszeitung – Fulda DE, 1945, 31 oct-22 dec, 1946-48, 1949 13 jan-1974 29 jun – 85r – 1 – gw Mikrofilm [074]
Fuldaer zeitung *see* Fuldaer politische zeitung
Fuldaer zeitung 1874 – Fulda DE, 1951 17 mar-1968 – 57r – 1 – (filmed by misc inst: 1874-1945 28 mar, 1969- [ca 9r/yr]) – gw Mikrofilm; gw Misc Inst [074]
Fuldaische wochentlich policey- und commercien-anzeigen – Fulda DE, 1765-1922 30 aug [gaps] – 1 – (title varies: 1771: fuldaische wochentliche polizei-, kommerzien- und zeitungsanzeigen; 1802 n43: fuerstlich-oranien-nassau-fuldaische woechentliche polizei-, kommerzien- und zeitungsanzeigen; 1804: fuldaisches intelligenzblatt; 1811: intelligenzblatt fuer das departement fulda; 1814: fuldaisches intelligenzblatt; 1 aug 1815: provinzial-blatt fuer das grossherzogthum fulda; 1822: wochenblatt fuer die provinz fulda; 1849 n14: wochenblatt fuer den verwaltungsbezirk fulda; 1851 n38: wochenblatt fuer die provinz fulda; 1866 n87: wochenblatt des vorhinnigen regierungsbezirkes fulda; 1869: kreisblatt; 1873: fuldaer kreisblatt; 30 apr 1920: fuldaer tageblatt) – gw Misc Inst [074]
Fuldaische wochentliche polizei-, kommerzien- und zeitungsanzeigen *see* Fuldaische wochentlich policey- und commercien-anzeigen
Fuldaisches intelligenzblatt *see* Fuldaische wochentlich policey- und commercien-anzeigen
Fulda-werra-zeitung – Eschwege DE, 1888-1923 31 aug, 1924 2 feb-1935 25 oct – 58r – 1 – (title varies: 7 nov 1902: eschweger zeitung. with suppl) – gw Misc Inst [074]
Fulda-werra-zeitung *see* Deutsches familienblatt
Fulfillment of the prophecies concerning ammon, moab, and philistia – London, England. 18– – 1r – us UF Libraries [240]
The fulfilment of a dream of pastor hsi's : the story of the work in hwochow / Cable, A Mildred – London: Morgan & Scott; China Inland Mission, 1917 [mf ed 1995] – (= ser Yale coll) – xx/268p (ill) – 1 – 0-524-09321-0 – mf#1995-0321 – us ATLA [920]
Fulfilment of the christian ministry / Jackson, Thomas – London, England. 1839 – 1r – us UF Libraries [240]
Fulford, Francis *see*
- An address delivered in the chapel of the general theological seminary of the protestant episcopal church in the united states on friday, nov 13th 1852
- A letter to the bishops, clergy and laity of the united church of england and ireland in the province of canada
- A sermon, preached on sunday, 5th jan 1862

Fulford, Francis Woodbury *see* Soeren aabye kierkegaard
Fulford, Henry William *see* The general epistle of st james
Fulgentius von Ruspe (Fulgentius of Ruspe, Saint) *see* Ausgewaehlte schriften (bdk9 2.reihe)
Fulham advertiser – [London & SE] Hammersmith 3 feb 1893 [mf ed 2003] – 1r – 1 – uk Newsplan [072]
Fulham and hammersmith chronicle *see* Fulham chronicle
Fulham and hammersmith guardian – London, UK. 2 jan 1986-21 dec 1990; 4 jan-20 dec 1991; jan-24 dec 1992; 8 jan-24 dec 1993 – 12r – 1 – (aka: hammersmith and fulham guardian; hammersmith fulham and chiswick guardian) – uk British Libr Newspaper [072]
Fulham and walham green news – London, UK. 1889; 1890; 1892-25 mar 1904 – 14r – 1 – uk British Libr Newspaper [072]
Fulham and west london observer – London, UK. 1896; 1914; 1915 – 3r – 1 – (aka: fulham observer; west london observer) – uk British Libr Newspaper [072]
Fulham chronicle – London, 6 apr 1888-1979; 11 jan-13 jun 1980; 11 jul 1980-1989; 1992-jun 1998 – 131 1/2r – 1 – (aka: fulham and hammersmith chronicle) – uk British Libr Newspaper [072]
Fulham observer *see* Fulham and west london observer
The fulham papers : 17th-18th centuries / Lambeth Palace Library – 20r – 1 – £950.00 – mf#FPA – uk World [242]

Fulham post – London, UK. 1 may-24 dec 1987; 1988-19 dec 1991 – 8r – 1 – uk British Libr Newspaper [072]
Fulham times *see* Hammersmith and fulham times
Fu-liang chiang ti hei yeh / Yu, Feng – Kuei-lin: Tso che shu fang, Min kuo 32 [1943] – (= ser P-k&k period) – us CRL [830]
Fulke, W *see*
– A briefe and plaine declaration
– A briefe confutation
A full account and collation of the greek cursive codex evangelium 604 (egerton 2610 in the british museum) : with two facsimiles: together with ten appendices... / Hoskier, Herman Charles – London: David Nutt, 1890 [mf ed 1986] – 1mf – 9 – 0-8370-9249-3 – mf#1986-3249 – us ATLA [226]
A full account of the trial of james suiter, sen, william suiter, jun, and jame suiter, jun : for the murder of living lane: with portraits of the three individuals and an account of the execution of william suiter, and their three declarations – Quebec: [s.n.] 1834 [mf ed 1983] – 1mf – 9 – 0-665-41700-4 – mf#41700 – cn Canadiana [345]
Full and authentic report of the discussion on church establishment / Leckie, Charles – Edinburgh, Scotland. 1838 – 1r – us UF Libraries [240]
Full and authentic report of the tilak trial, 1908 : being the only authorised verbatim account of the whole proceedings with introduction and character sketch of bal gangadhar tilak together with press opinion – Bombay: NC Kelkar, 1908 – (= ser Samp: indian books) – us CRL [954]
A full and circumstantial account of the trial of the rev. doctor dodd, at the sessions house in the old bailey, on saturday the 22nd of february, 1777. / Dodd, William – London, Richardson and Urquhart 1777? 55 p. LL-2249 – 1 – us L of C Photodup [340]
A full and plaine declaration of ecclesiasticall discipline owt off the word off god... / Cartwright, T & Travers, W – n.p., 1574 – 4mf – 9 – mf#PW-66 – ne IDC [240]
A full collation of the codex sinaiticus with the received text of the new testament : to which is prefixed a critical introduction / Scrivener, Frederick Henry Ambrose [coll] – Cambridge: Deighton, Bell & Co 1864 [mf ed 2005] – 1r – 1 – 0-524-10534-0 – ("with scrivener's reprint (1862) of stephens' text of the new testament") – mf#b00744 – us ATLA [225]
A full description of the soil, water, timber, and prairies of each lot, or quarter section of the military land between the mississippi and illinois rivers / Van Zandt, Nicholas B – 1818. 127p. 1r – 1 – $8.00 – us Minn Hist [355]
A full description of the two historical paintings of the funeral of the late sir john s d thompson... / Bell-Smith, Frederick Marlett – S.l.: s.n, 1895? – 1mf – 9 – mf#02323 – cn Canadiana [750]
Full employment action news – 1978 jan-1980 jun – 1r – 1 – mf#672540 – us WHS [331]
Full employment newsreporter / National Committee for Full Employment – v1 n1-2 [1975 apr-sep/oct] – 1r – 1 – mf#700874 – us WHS [331]
A full exposure of the c b s or dark lantern association : containing the proceedings of this secret political society, letters, correspondence etc etc – [Brockville, Ont?: s.n, 1861?] [mf ed 1983] – 1mf – 9 – 0-665-44528-8 – mf#44528 – cn Canadiana [366]
The "full faith and credit clause" of the united states constitution; an instrument of federalism / Virginia. Commission on Constitutional Government – Richmond, 1966 25 p. LL-4 – 1 – us L of C Photodup [342]
Full gospel baptist times – 1994 dec, 1996 mar, jun – 1r – 1 – mf#4027705 – us WHS [071]
A full history of the wonderful career of moody and sankey in great britain and america : embracing, also, the best portions of mr. moody's sermons... / Goodspeed, Edgar Johnson – New York: H S Goodspeed, c1877 [mf ed 1990] – 2mf – 9 – 0-7905-8008-X – mf#1988-8008 – us ATLA [240]
Full life magazine / American Mutuality Foundation – 1979 feb-1985 dec, 1986 jan-1995 jul – 2r – 1 – (cont: full life [los angeles, ca]) – mf#1043026 – us WHS [368]
Full report of a conference of working men, clergy and others – Liverpool, England. 1868 – 1r – us UF Libraries [240]
A full report of the venning vs hunter trial : at the circuit court, st john, before his honor justice ritchie, mar 1863, as reported for the "morning telegraph" / Venning, W N – [Saint John NB?: s.n.] 1863 [mf ed 1983] – 1mf – 9 – 0-665-41718-7 – mf#41718 – cn Canadiana [347]
Full service ca (microform) – All abstracts in CA. -w. 1907-. Printed volume indexes – 6,9 – us Chemical [540]
Fuller, Andrew *see*
– The atonement of christ and the justification of the sinner
– Backslider
– The care of the soul
– Christian patriotism
– The complete works of the rev. andrew fuller
– The complete works of the rev. andrew fuller, with a memoir of his life by andrew gunton fuller
– Householder and the labourers
– Importance of a deep and intimate knowledge of divine truth
– Jesus the true messiah
– The last remains of the rev. andrew fuller
Fuller, Benjamin Apthorp Gould *see* The problem of evil in plotinus
Fuller, Catherine J *see* A descriptive analysis of aerobic instructor behaviors and related student responses
Fuller, Edgar I *see* The visible of the invisible empire
Fuller findings – 1983 oct-1988 apr – 1r – 1 – mf#1322380 – us WHS [071]
Fuller, Francis *see* Medicina gymnastica
Fuller, Halsey Oakley *see* Halsey in the west indies
Fuller, J G *see* Reasons for christian communion
Fuller, Jennie *see* [Missionaries and missions to jews]
Fuller, John Channing *see* Reinhold niebuhr's theological ethics
Fuller, John Frederick Charles *see* The conquest of red spain
Fuller, Margaret *see* Memoirs of margaret fuller ossoli
Fuller, R Buckminster *see* Preview of building
Fuller restoration of the diaconate : a means of strengthening the ch... / Mackenzie, Henry – London, England. 1845 – 1r – us UF Libraries [240]
Fuller, Richard *see*
– The benevolence of the gospel toward the poor
– The cross
Fuller, Robert Hart *see* South africa at home
Fuller, Russell L *see*
– Forts and historic points of interest
– Health
Fuller, Samuel *see* Education in the two andovers
Fuller, Tamela G *see* The effects of a weight training course on stress levels and locus of control in college females
Fuller, Thomas Brock *see*
– The roman catholic church, not the mother church of england
– Systematic beneficence
Fuller, William Henry *see*
– The colonial question
– Flapdoodle
– H m s parliament
Fuller-Eliott-Drake, Elizabeth Douglas *see* The family and heirs of sir francis drake
Fullerenes, nanotubes, and carbon nanostructures – New York. 2002+ (1,5,9) – ISSN: 1536-383X – mf#20891,01 – us UMI ProQuest [540]
Fuller's practice reports / Michigan. Supreme Court – 1v. 1896 – (= ser Michigan supreme court reports) – 1mf – 9 – $4.50 – mf#LLMC 84-161 – us LLMC [347]
Fullerton, Alexander *see* The wilkesbarre letters on theosophy
[Fullerton-] california spirit and corridors – CA: csu fullerton, 1979-81 – 2r – 1 – $120.00 – mf#R02261b – us Library Micro [370]
[Fullerton-] daily titan – CA: CSU Fullerton, 1994 – 20r – 1 – $1200.00 – (aka: titan times) – mf#R02263 – us Library Micro [378]
[Fullerton-] fullerton news tribune – CA. 1895-– 425r – 1 – $25,500.00 (subs $440y) – mf#R02262 – us Library Micro [071]
Fullerton, George Stuart *see*
– The conception of the infinite and the solution of the mathematical antinomies
– On sameness and identity
– A plain argument for god
– The world we live in
Fullerton, George Stuart et al *see*
– Essays, philosophical and psychological
– On the perception of small differences
Fullerton, Georgiana *see* The life of luisa de carvajal
The fullerton news – Fullerton, NE: W T Hastings, 1893-v4 n25. dec 25 1896 (wkly) [mf ed 1895-96 (gaps)] – 1r – 1 – (merged with: nance county journal to form: fullerton news-journal) – us NE Hist [071]
Fullerton news and nance county journal – Fullerton, NE: W H Totten. v17 n11. jan 8 1897-v20 n39. jul 27 1900; v8 n4 aug 2 1900-v8 n38. mar 28 1901 (wkly) [mf ed 1897-1901] – 2r – 1 – (cont: fullerton news-journal. cont by: news-journal. some irregularities in numbering) – us NE Hist [071]
Fullerton news-journal – Fullerton, NE: W H Totten. v17 n10. jan 1 1897=v4 n26 (wkly) – 1r – 1 – (formed by the union of: fullerton news and: nance county journal. cont by: fullerton news and nance county journal. no more publ) – us NE Hist [071]
Fullerton post – Fullerton, NE: Post Pub Co, 1888-in the yr of existence, going on 43 yrs n21. sep 13 1928 (wkly) [mf ed 1892,1895-28 (gaps)] – 11r – 1 – (merged with: news-journal to form: nance county journal. issues for nov 18 1892-mar 24 1905 also called whole n233-857. some irregularities in numbering) – us NE Hist [071]
Fullerton, Robert Stewart *see* Memoir
Fullerton, William Young *see*
– C h spurgeon
– New china
Fulleylove, John *see* The holy land
Fulliquet, Georges *see* La justification par la foi
Fully, freely and entirely / Delaware Heritage Commission – v1 n1-v3 n4 [1985 spring-1987 winter] – 1r – 1 – mf#1344730 – us WHS [978]
Fulness and freeness of the gospel message / Chalmers, Thomas – Brighton, England. 1847 – 1r – us UF Libraries [226]
The fulness of blessing : or, the gospel of christ: as illustrated from the book of joshua / Smiley, Sarah Frances – London: Hodder & Stoughton, 1876 – 1mf – 9 – 0-8370-5276-9 – (incl ind) – mf#1985-3276 – us ATLA [220]
The fulness of christ : an essay / Weston, Frank – London, New York: Longmans, Green, 1916 [mf ed 1991] – 1mf – 9 – 0-7905-9760-8 – mf#1989-1485 – us ATLA [242]
Fuloep-Miller, Rene *see* Gandhi, the holy man
Fulton, Alvin W *see* Ernest bloch's sacred service
Fulton, Chester Alan *see* Mining practice in the florida pebble phosphate field
Fulton Co. Archbold *see*
– Advocate
– Buckeye
– Farmland news
– Herald
Fulton Co. Fayette *see* Review
Fulton Co. Pettisville *see* Progress
Fulton county historical and genealogical society – 1977 jan-1987 oct – 1r – 1 – (cont: fulton county historical society newsletter) – mf#1685326 – us WHS [978]
Fulton county sun – Rochester, IN. 1913-1921 (1) – mf#62967 – us UMI ProQuest [071]
Fulton democrat – McConnellsburg, PA. -w 1889-1981 – 13 – $25.00r – us IMR [071]
Fulton first baptist church : church records – Fulton, KY. 1734p. 1891-1971 – 1 – $78.03 – (incl minutes of buckingham united baptist church, fulton, kentucky, dec 1865-75) – us Southern Baptist [242]
Fulton, John *see*
– The chalcedonian decree
– Index canonum
Fulton, Justin Dewey *see*
– Charles h spurgeon
– Is it mary or the lady of the jesuits?
– The outlook of freedom
– The true woman
Fulton, Levi S *see*
– A practical system of book-keeping by single and double entry
Fulton, Maurice G *see* Bryce on american democracy
Fulton, William *see* The theology of the reformed church in its fundamental principles
Fumagalli, Camillo *see* Il diritto di fraterna nella giurisprudenza da accursio alla codificazione
Fumagalli, Guiseppe *see* Bibliografia etiopica
O fumante : orgam indispensavel – Sao Joao del Rei, MG: Typ Commercial, 20 ago 1898 – (= ser Ps 19) – mf#P17,02,97 – bl Biblioteca [079]
Fumet, Stanislas *see* Le bienhereux martin de porres...paris
Fun – London. 1861-1901 (1) – mf#4254 – us UMI ProQuest [420]
Fun dervaytns / Hirschkan, Zevi – Berlin, Germany. 1922 – 1r – us UF Libraries [939]
Fun journal – Indianapolis. 1971-1974 (1) 1971-1973 (5) (9) – ISSN: 0016-2647 – mf#6187 – us UMI ProQuest [370]
Fun onzog tsu fatvirklekhung / Kaplansky, Solomon – Warsaw, Poland. 1932 – 1r – us UF Libraries [939]
Fun shpanie biz holand / Malach, Leib – Warszawa, Poland. 1937 – 1r – us UF Libraries [939]
Fun "zshargon" tsu yidish / Dubnow, Simon – Wilno, Lithuania. 1929 – 1r – us UF Libraries [939]
Funafuti, or three months on a coral island : an unscientific account of a scientific expedition / David, [C M] – London, 1899 – 4mf – 9 – mf#HT-87 – ne IDC [919]
Funchal, Agostinho de Sousa Coutinho *see* O conde de linhares
Funck, Friedrich *see* 1793 beitrag zur geheimen geschichte der franzoesischen revolution
Funck, Heinrich *see*
– Beitraege zur wieland-biographie
– Goethe und lavater
Funck, J *see* Auszug vnd kurtzer bericht
The function of religion in man's struggle for existence / Foster, George Burman – Chicago:University of Chicago Press, 1909 – 1mf – 9 – 0-8370-3166-4 – mf#1985-1166 – us ATLA [210]
The function of the church / Fairchild, Edwin Milton – Chicago: University of Chicago; London: Luzac, [1869?] – 1mf – 9 – 0-8370-7859-8 – mf#1986-1859 – us ATLA [240]
Function of the vibrissae in the behavior of the white rat / Vincent, Stella Burnham – Cambridge, MA. 1912 – 1r – us UF Libraries [590]
The function of universities in religion : an address...san francisco, april 26 1897 / Howison, George Holmes – San Francisco: C A Murdock, 1897 – 1mf – 9 – 0-7905-3862-8 – mf#1989-0355 – us ATLA [240]
Functional analysis and its applications – New York. 1967-1995 (1) 1967-1977 (5) – ISSN: 0016-2663 – mf#10826 – us UMI ProQuest [530]
The functional development of christian education in the church of the united brethren in christ / Boyer, John Neely – BD: Bonebrake Theological Seminary 1930 – 1r – 1 – $35.00 – mf$-15 – us Commission [242]
Functional ecology – Oxford. 1987-1996 (1,5,9) – ISSN: 0269-8463 – mf#15616 – us UMI ProQuest [574]
Functional motor skills and the developmentally disabled / Young, Marnie J & Fisher, Janet M – 1991 – 2mf – 9 – $8.00 – us Kinesology [150]
Functional relationships between boron and various anions in the nutrition of the tomato / Beckenbach, J R – Gainesville, FL. 1944 – 1r – us UF Libraries [574]
The functions of dance in social gatherings in seattle's arab-american community / Wartluft, Elizabeth M – 1994 – 2mf – $8.00 – us Kinesology [790]
Fund for Labor Defense [US] *see* New york labor punch out
Fund for Peace *see* In the public interest
Fund raising management – Garden City. 1969+ (1) 1974+ (5) 1975+ (9) – ISSN: 0016-268X – mf#7930 – us UMI ProQuest [650]
Der fund von tell-amarna und die bibel / Vogel, August – Braunschweig: H Wollermann, 1898 – (= ser Veroeffentlichungen des Bibelbundes) – 1mf – 9 – 0-7905-3238-7 – mf#1987-3238 – us ATLA [220]
Fundacao da cidade paraense / Porto, Arthur – Rio de Janeiro, Brazil. 1938 – 1r – us UF Libraries [972]
Fundacion de capellania de missas : y dote para religiosas perpetuamente annuales / Vizarron y Eguiarreta, Juan Antonio de – [Mexico: s.n. 1743] – (= ser Books on religion...1543/44-c1800: miscelanea) – 1mf – 9 – mf#crl-415 – ne IDC [241]
Fundacion de la ciudad de gracias a dios / Lunardi, Federico – Tegucigalpa, Mexico. 1946 – 1r – us UF Libraries [972]
Fundacion de la universidad en guatemala (1548-168 / Mata Gavidia, Jose – Guatemala, 1954 – 1r – us UF Libraries [378]
La fundacion de merida / Alvarez Saenz de Buruaga, Jose – Sep. de "Emerita Augusta" – 1 – sp Bibl Santa Ana [946]
Fundacion del colegio de coyoacan hernan y martin cortes / Mateos, S F – Madrid: Missionalia Hispanica, 1947 – 1 – sp Bibl Santa Ana [060]
Fundacion del colegio real de san ildefonso en diez y siete de enero de mil seiscientos y diez y ocho / Mexico [City]. Colegio real de Ildefonso – [Mexico City: s.n. 1618] – (= ser Books on religion...1543/44-c1800: jesuitas) – 1mf – 9 – mf#crl-215 – ne IDC [241]
Fundacion espanola...por el virrey del peru, don francisco de toledo. lima, 1926 / Urteaga, H H & Romero, Carlos A – Madrid: Razon y Fe, 1929 – 1 – sp Bibl Santa Ana [946]
Fundacion "Fernando Valhondo Calaff" *see* Fundacion fernando valhondo calaff
Fundacion fernando valhondo calaff / Fundacion "Fernando Valhondo Calaff" – Caceres: Tip. La Minerva, 1960 – 1 – sp Bibl Santa Ana [060]
Fundacion Para El Progreso De Colombia *see* Politica urbana para los paises en desarrollo
Fundacion y fabrica del convento de san antonio de padua de almendralejo en la provincia franciscana de san gabriel / Barrado Manzano, Arcangel – Madrid: Archivo Ibero-Americano, 1960 – 1 – sp Bibl Santa Ana [240]
Fundacion, y primero siglo, del muy religioso convento de sr s joseph de religiosas carmelitas descalzas de la ciudad de la puebla de los angeles, en la nueva espana : el primero que se fundo en la america septemtrional, en 27. de diziembre de 1604 / Gomez de la Parra, Jose – en la Puebla de los Angeles:...M de Ortega, ano de 1732 – (= ser Books on religion...1543/44-c1800: ordenes, etc: carmelitas descalzas) – 7mf – 9 – mf#crl-174 – ne IDC [241]

Fundaciones benficas de la provincia de caceres anteriores a 1850 / Orti Belmonte, Miguel Angel – Caceres: Imp. Sanguino, 1949 – 1 – sp Bibl Santa Ana [946]

Fundaciones testamentarias de fr. miguel de medina, jeronimo de guadalupe en el convento de san francisco de medina de pomar, burgos, 1915 / Gracia Villacampa, Carlos – Madrid: Archivo Ibero Americano, 1916 – 1 – sp Bibl Santa Ana [240]

El fundador de montevideo : montevideo, 1928 / Sallaberry, Juan Faustino – Madrid: Razon y Fe, 1929 – 1 – sp Bibl Santa Ana [946]

Los fundadores de bogota / Bayle, Constantino – Madrid: Razon y Fe, 1924 – 1 – sp Bibl Santa Ana [946]

Fundamenta partiturae in compendio data : das ist: kurtzer und gruendlicher unterricht, den general-bass, oder partitur, nach denen reglen recht und wohl schlagen zu lehrnen / Gugl, Matthaeus – Augspurg & Insprugg: J Wolff 1757 [mf ed 198-?] – 2mf – 9 – mf#fiche 870 – us Sibley [780]

Fundamental and clinical pharmacology – Paris. 1989-1992 (1,5,9) – ISSN: 0767-3981 – mf#42605 – us UMI ProQuest [615]

Fundamental Baptist Fellowship *see* Information bulletin

The fundamental christian faith : the origin, history and interpretation of the apostles' and nicene creeds / Briggs, Charles Augustus – New York: Charles Scribner, 1913 – 1mf – 9 – 0-7905-4152-1 – (incl bibl ref) – mf#1988-0152 – us ATLA [240]

Fundamental christology : a discussion of foundation doctrines concerning the christ / Young, George Lindley – Boston: Advent Christian Publication Society, 1906 – 1mf – 9 – 0-524-00233-9 – mf#1989-2933 – us ATLA [240]

Fundamental constitution / Carolina State – 1669 [mf ed 1981] – 3mf – 9 – mf#51-501 – us South Carolina Historical [323]

The fundamental error of christendom / Moore, William Thomas – St Louis: Christian Publishing, c1902 – 1mf – 9 – 0-524-02130-9 – mf#1990-4196 – us ATLA [240]

The fundamental fallacy of socialism : an exposition of the question of landownership: comprising an authentic account of the famous mcglynn case / ed by Preuss, Arthur – St Louis, MO: B Herder, 1908 – 1mf – 9 – 0-8370-7010-4 – (incl bibl ref) – mf#1986-1010 – us ATLA [333]

Fundamental gymnastics / Bukh, Niels E – 1938 – 3mf – 9 – $9.00 – us Kinesology [790]

The fundamental ideas of christianity / Caird, John – Glasgow: J MacLehose; New York: Macmillan, 1899 – (= ser Gifford lectures) – 2mf – 9 – 0-7905-3650-1 – mf#1989-0143 – us ATLA [240]

The fundamental ideas of the roman catholic church : explained and discussed for protestants and catholics / Foster, Frank Hugh – Philadelphia: Presbyterian Board of Publ & Sabbath-School Work, 1899, c1898 [mf ed 1986] – 1mf – 9 – 0-8370-8108-4 – (incl ind) – mf#1986-2108 – us ATLA [241]

The fundamental principle of the word of god, the legitimate basis of temperance societies : a discourse delivered in the congregational chapel, montreal...13th october, 1840 / Atkinson, Timothy – [Montreal?: s.n.], 1840 (Montreal: Campbell & Becket) – 1mf – 9 – 0-665-89963-7 – mf#89963 – cn Canadiana [210]

The fundamental principles of christian ethics : five lectures / Conway, James Joseph – Chicago: D H McBride, 1896 – 1mf – 9 – 0-8370-6728-6 – mf#1986-0728 – us ATLA [230]

The fundamental principles of christian unity : lectures delivered in lent, 1902 / Parks, James Lewis et al – Washington: Church Militant, 1902 – (= ser Churchman's League Lectures) – 1mf – 9 – 0-524-02705-6 – mf#1990-0686 – us ATLA [240]

Fundamental principles of the metaphysic of ethics / Kant, Immanuel – Trans. by Thomas Kingsmill Abbott. 10th ed. London, New York: Longmans, Green and Co., 1926.102p – 1 – us Wisconsin U Libr [190]

Fundamental principles of the organization, management, and teaching of the school band / Cecil, Herbert Myron – U of Rochester 1953 [mf ed 19–] – 2mf – 9 – mf#fiche389 – us Sibley [780]

The fundamental problems of metaphysics / Lindsay, James – Edinburgh: W Blackwood, 1910 – 1mf – 9 – 0-7905-9019-0 – mf#1989-2244 – us ATLA [110]

Fundamental questions : chiefly relating to the book of genesis and the hebrew scriptures / Clark, Edson Lyman – New York: G P Putnam, 1882 – 1mf – 9 – 0-8370-2665-2 – mf#1985-0665 – us ATLA [220]

Fundamental rights : a constitutional and juridical study with particular reference to india in the light of the experience of the united states of america and the united kingdom / Ramaswamy, M – New Delhi: Indian Council of World Affairs, 1946 – (= ser Samp: indian books) – (foreword by maurice gwyer) – us CRL [323]

Fundamental rights and constitutional remedies / Aggarawala, Om Prakash – Delhi: Metropolitan Book Co, 1953-1954 – (= ser Samp: indian books) – us CRL [323]

The fundamental truths of the christian religion : sixteen lectures delivered in the university of berlin during the winter term 1901-02 = Grundwahrheiten der christlichen religion / Seeberg, Reinhold; ed by Morrison, W D – New York: G P Putnam; London: Williams & Norgate, 1908 – 1mf – 9 – 0-8370-7426-6 – mf#1986-1426 – us ATLA [240]

The fundamental unity of india : from hindu sources / Mookerji, Radhakumud – London, New York: Longmans, Green and Co, 1914 – (= ser Samp: indian books) – (int by j ramsay macdonald) – us CRL [954]

Die fundamentale glaubenslehre der katholischen kirche : vorgelegt und gegen die modernen sozialen irrtuemer verteidigt von papst leo 13. – Paderborn: Ferdinand Schoeningh, 1903 – 2mf – 9 – 0-8370-8492-X – (incl ind) – mf#1986-2492 – us ATLA [241]

Fundamentalist – 1917-53. (Name change: The Searchlight, 1917-27) – 1 – 525.73 – us Southern Baptist [242]

Fundamentals : or, bases of belief concerning man, god, and the correlation of god and men: a handbook of mental, moral, and religious philosophy / Griffith, Thomas – London: Longmans, Green 1871 [mf ed 1985] – 1mf – 9 – 0-8370-3395-0 – (incl app) – mf#1985-1395 – us ATLA [210]

The fundamentals – A Testimony to the Truth. Chicago: Testimony Publishing Co., n.d. v1-12.1660p – 1 – 66.40 – us Southern Baptist [242]

The fundamentals : a testimony to the truth / Orr, James – Chicago, IL: Testimony Pub [1910] [mf ed 1986] – 12v on 12mf – 9 – 0-8370-6772-3 – (incl ind) – mf#1986-0772 – us ATLA [240]

The fundamentals and their contrasts / Buckley, James Monroe – Nashville: Publishing House of the Methodist Episcopal Church, South, 1906 – (= ser The Quillian Lectures) – 1mf – 9 – 0-8370-2507-9 – mf#1985-0507 – us ATLA [240]

Fundamentals of civil legislation of the u.s.s.r. and the union republics. fundamentals of civil procedure of the u.s.s.r. and the union republics / Russia. (1923-U.S.S.R.). Laws, Statutes, etc – Official texts. Translated from the Russian. Moscow: Progress Publishers 1968. 130p. LL-4211 – 1 – us L of C Photodup [348]

Fundamento juridico del nuevo ideal nacional / Cova Garcia, Luis – Caracas, Venezuela. 1955 – 1r – us UF Libraries [340]

Fundamentos da cultura catarinese – Rio de Janeiro, Brazil. 1970 – 1r – us UF Libraries [972]

Fundamentos da poesia brasileira / Lima, Silvio Julio De Albuquerque – Rio de Janeiro, Brazil. 1930 – 1r – us UF Libraries [972]

Fundamentos de religion / Marquez, Gabino – Madrid: Apostolado de la Prensa, 2nd ed 1915 – 1 – sp Bibl Santa Ana [200]

Fundamentos de religion / Marquez, Gabino – Madrid: Apostolado de la Prensa, 3rd ed 1920 – 1 – sp Bibl Santa Ana [240]

Fundamentos de sociologia / Colorado, Vicente – 1883. Prologo pur Urbano Gonzalez Serrano – 9 – sp Bibl Santa Ana [300]

Fundamentos de uma filosofia brasileira / Netto, Aben Attar – Rio de Janeiro, Brazil. 1947 – 1r – us UF Libraries [972]

Los fundamentos del battlismo / Giudici, Roberto B – Montevideo: (Prometeo), 1946. 142p – 1 – us Wisconsin U Libr [972]

Fundamentos del mundo nuevo / Stefanich, Juan – Buenos Aires, Argentina. 1944 – 1r – us UF Libraries [025]

Fundamentos nacionais da politica do acucar / Lima Sobrinho, Barbosa – Rio de Janeiro, Brazil. 1943 – 1r – us UF Libraries [972]

Funde und forschungen : eine festgabe fuer julius wahle zum 15. februar 1921 / Deetjen, Werner et al – Leipzig: Insel Verlag, 1921 – 1r – 1 – us Wisconsin U Libr [430]

Fundgruben des orients / ed by Hammer-Purgstall, Josef von – Wien 1810-19 [mf ed Hildesheim 1994] – 6v on 34mf – 9 – diazo €148.00 silver €188.00 – gw Olms [950]

A funding plan for the renovation of the a.e. finley golf course / Howden, Jeffrey B – 1997 – 2mf – 9 – $8.00 – mf#RC 517 – us Kinesology [790]

Die fundlandschaft guspini, provinz cagliari, sardinien : archaeologische studien in suedwestsardinien mit einer einfuehrung zur sardischen vorgeschichte / Koberstein, Astrid Beate – (mf ed 1993) – 4mf – 9 – €49.00 – 3-89349-704-8 – mf#DHS 704 – gw Frankfurter [930]

The fundraising process for the mccaskill soccer center / Foels, Tracie L – 1999 – 1mf – 9 – $4.00 – mf#PE 3953 – us Kinesology [790]

Funebris pompa serenissimi ranutii farnesii parmae et placentiae ducis 4... / Caprara, A – Parmae: Typis Anthaei Viothi, 1622 – 1mf – 9 – mf#O-2002 – ne IDC [090]

Funeral addresses / Roberts, Samuel – Conway, Wales. 1880? – 1r – us UF Libraries [240]

Funeral agenda / Stolz, Joseph – [s.l: s.n, 1897?] [mf ed 1985] – 1mf – 9 – 0-8370-5427-3 – (incl bibl ref) – mf#1985-3427 – us ATLA [270]

Funeral de un sueno / Altamirano, Carlos Luis – San Jose, Costa Rica. 1958 – 1r – us UF Libraries [972]

The funeral director – Montreal: D R Nelson, [1887?-18–] – 9 – mf#P04305 – cn Canadiana [390]

The funeral discourse, occasioned by the death of the rev robert morrison : delivered before the london missionary society, at the poultry chapel feb 19 1835 / Fletcher, Joseph – London: Frederick Westley & A H Davis, 1835 [mf ed 1995] – (= ser Yale coll) – 75p – 1 – 0-524-09781-X – mf#1995-0781 – us ATLA [240]

Funeral discourse on the death of the rev william orme / Fletcher, Joseph – London, England. 1830 – 1r – us UF Libraries [240]

Funeral furniture and stone vases / Petrie, W M – London, 1937. 2pts – 4mf – 9 – mf#NE-20378 – ne IDC [930]

Funeral Home, Havana, KS *see* Records

The funeral of the late lieutenant colonel cameron : will take place on monday, the 17th instant, at 3 o'clock... – [Toronto?: s.n. 1842?] [mf ed 1983] – 1mf – 9 – 0-665-38686-9 – mf#38686 – cn Canadiana [090]

The funeral of the late mrs george hamilton : will proceed from the residence of mr samuel p jarvis, to the place of interment on friday next, at 4 o'clock – [Toronto?: s.n. 1829?] [mf ed 1983] – 1mf – 9 – 0-665-38691-5 – mf#38691 – cn Canadiana [090]

The funeral of the late mrs henry scadding : ...spadina toronto, sept 26 1843 – [Toronto?: s.n. 1843?] [mf ed 1983] – 1mf – 9 – 0-665-38685-0 – mf#38685 – cn Canadiana [090]

The funeral of the late mrs steers : ...toronto, 29th july 1839 – [Toronto?: s.n. 1839?] [mf ed 1983] – 1mf – 9 – 0-665-38687-7 – mf#38687 – cn Canadiana [090]

Funeral oration : preached by the rev archdeacon o'keeffe, on pope... / O'keeffe, Thomas – Cork, Ireland. 1868 – 1r – us UF Libraries [240]

Funeral oration of his eminence cardinal weld / Wiseman, Nicholas Patrick – London, England. 1837 – 1r – us UF Libraries [240]

Funeral oration of pius 7 / Soulacroix, Abbe – London, England. 1824 – 1r – us UF Libraries [240]

Funeral records / Gibbons Mortuary (GSU), Coffey County, KS – undated – 1 – us Kansas [920]

Funeral registers / Franklin County and Douglas County, KS – undated – 1 – us Kansas [920]

Funeral registers / Labette County, KS – 1889-1926 – 1 – us Kansas [920]

A funeral sermon *see* Life of the rev alex mathieson...minister of st. andrew's church, montreal

Funeral sermon : on the death of mr i i jun... / Woodd, Basil – London, England. 1794 – 1r – us UF Libraries [240]

Funeral sermon : preached at spa-fields chapel, july 3, 1791 / Jones, David – London, England. 1791 – 1r – us UF Libraries [240]

Funeral sermon of the rev joseph brown : rector of christ's c... / Curling, W – London, England. 1867 – 1r – us UF Libraries [240]

Funeral sermon on sir john thompson / O'Brien, Cornelius – Halifax [NS]: E P Meagher, 1906 – 1mf – 9 – (with app) – mf#73673 – cn Canadiana [240]

Funeral sermon on the death of john harrison / Proud, J – London, England. 1798 – 1r – us UF Libraries [240]

Funeral services at williamstown, mass., in love and honor of rev. mark hopkins, d.d., ll.d : tuesday, june 21st, 1887 – [New Haven: Press of Tuttle, Morehouse & Taylor, 1887?] – 1mf – 9 – 0-524-08363-0 – mf#1993-3063 – us ATLA [240]

Funerale fatto nel duomo di torino alla gloriosa memoria... / Giuglaris, L – Torino: Appresso gl'Heredi di Gio, 1638 – 3mf – 9 – mf#O-1589 – ne IDC [090]

Funes, Jorge Ernesto *see* Caballeros de espuela dorada (descubrimiento y conquista del peru)

Funes Peraza, Berta *see* Mensaje en el tiempo

Funfzig jahre lebenserfahrungen eines judischen lehrers und... / Wolff, Lion – Leipzig, Germany. 1919 – 1r – us UF Libraries [939]

Fungi parasitic upon aleyrodes citri... / Fawcett, H S – s.l, s.l? 1908 – 1r – us UF Libraries [630]

Fungus disease of the san jose scale (sphaerostilbe coccophila, tul) / Rolfs, P H – Lake City, FL. 1897 – 1r – us UF Libraries [630]

Fungus diseases of scale insects and whitefly / Rolfs, P H – Gainesville, FL. 1908 – 1r – us UF Libraries [630]

Fungus diseases of scale insects and whitefly / Rolfs, P H – Gainesville, FL. 1913 – 1r – us UF Libraries [630]

Funk – Berlin DE, 1924-1944 jun – 5r – 1 – (with suppl: die funk-bastler) – gw Mikrofilm [380]

Funk alle tage *see* Die mirag

Funk, Chr B et al *see* Leipziger magazin zur naturkunde, mathematik und oekonomie

Funk, Daniel C *see* Fan loyalty

The funk enterprise – Funk, NE: L T Brooking. v1 n1. sep 30 1898- (wkly) [mf ed -mar 3 1899 (gaps) filmed 1979] – 1r – 1 – us NE Hist [071]

Funk, F X *see* Opera patrum apostolocorum

Funk, Franz Xaver von *see*
- Die apostolischen konstitutionen
- Die echtheit der ignatianischen briefe
- Kirchengeschichtliche abhandlungen und untersuchungen
- Manual of church history
- Das testament unseres herrn und die verwandten schriften

Funk, Isaac Kaufman *see*
- The complete preacher
- The next step in evolution

Funk, Jacob *see* War versus peace

Funk, Joseph *see* The reviewer reviewed

Funk, Samuel *see* Akiba

Funk und bewegung – Berlin DE, 1933-36 – 1r – 1 – gw Misc Inst [790]

Funk und schall – Frankfurt/M DE, 1929 15 oct-1935 28 aug [gaps] – 2r – 1 – gw Mikrofilm [790]

Funk, W P *see* Die zweite apokalypse des jacobus aus nag-mammadi-codex 5

Funk, Wendy W *see* The effects of creative dance on movement creativity in third grade children

Der funke – Berlin DE? n.d. – 1 – gw Misc Inst [074]

Der funke – Landsberg (Bez Halle) DE, 1959 may-1960 jul [gaps] – 1r – 1 – gw Misc Inst [074]

Der funke – Magdeburg DE, 1968-90 [gaps] – 3r – 1 – gw Misc Inst [074]

Der funke – Dessau DE, 1965 15 jan-1974 nov [gaps], 1975 jan-nov, 1976-1992 11 dec [gaps] – 5r – 1 – (notes: raw dessau) – gw Misc Inst [530]

Der funke – Paris (F), Wien/Prag, 1933 – 1 – gw Misc Inst [074]

Funke, Erich *see* Modern german prose

Funke, Odilia *see* Meister eckehart

Funkhouser, George Absalom *see* The divinity of our lord

Funk-korrespondenz – Koeln DE, 2 dec 1953-2003 – 48r – 1 – gw Mikrofilm [380]

Funkschau – Muenchen DE, 1929 1 jan-24 dec – 1r – 1 – (with suppls: neues vom funk / der bastler / der fernempfang) – gw Mikrofilm [380]

Funk-spiegel *see* Westfunk

Funkstunde – Berlin DE, 1924 16 nov-1937 26 dec – 9mf=16df – 9 – gw Mikrofilm [380]

Die funktion der gtpase von elongationsfaktor g aus escherichia coli bei der translokation am ribosom / Savelsbergh, Andreas – (mf ed 1997) – 1mf – 9 – €30.00 – 3-8267-2489-5 – mf#DHS 2489 – gw Frankfurter [574]

Die funktion der nebenfiguren in fontanes romanen / Buscher, Heide – Bonn, 1969 – 1 – gw Mikropress [440]

Funktion magazine – 1996 sep – 1r – 1 – mf#4851573 – us WHS [071]

Funktionelle charakterisierung eines regulativen elementes zwischen transkriptions- und translationsstart im e-kristallipromotor / Krauss, Eberhard – (mf ed 1997) – 2mf – 9 – €40.00 – 3-8267-2430-5 – mf#DHS 2430 – gw Frankfurter [574]

Funktionelle charakterisierung von domaenen des elongationsfaktors g / Borowski, Christian – (mf ed 1995) – 1mf – 9 – €30.00 – 3-8267-2192-6 – mf#DHS 2192 – gw Frankfurter [574]

Funktionelle morphologie und cytochemie osteoblastaerer rattenosteosarkomzellen unter einwirkung von calcium; parathormon oder aluminium / Niemann, Frank-Michael – Hamburg 1985 (mf ed 1995) – 2mf – 9 – €40.00 – 3-8267-2269-8 – mf#DHS-AR 2269 – gw Frankfurter [616]

Funk-wacht – Hamburg DE, 1933 31 dec-1941 31 may – 13r – 1 – gw Mikrofilm [790]

Funk-woche – Berlin DE, 1926 9 may-1937 23 oct – 4r – 1 – gw Mikrofilm [380]

Funnell, William *see* A collection of voyages in four volumes, vol 4

Funny pages – iss n6-42. oct 1936-oct 1940 – 15 – (issues 36-42 apr-oct 1940 set of 7mf $50.55) – mf#002CM-008CM – us MicroColour [740]

Funnyworld – Alexandria. 1966-1980 (1) 1966-1980 (5) 1966-1980 (9) – ISSN: 0071-9943 – mf#10615 – us UMI ProQuest [790]

Funston, Frederick *see*
- Memories of two wars
- Papers

The fur country : or, seventy degrees north latitude / Verne, Jules – New York: Lovell, 1876 [mf ed 1987] – 6mf – 9 – 0-665-41558-3 – (trans fr french by n d'anvers) – mf#41558 – cn Canadiana [440]

Fur farming in the province of quebec : describing the most approved methods of propagating foxes and other fur-bearing animals in captivity / Chambers, Edward Thomas Davies – Quebec: [s.n.], 1920 – 1mf – 9 – 0-659-90682-1 – (also available in french) – mf#9-90682 – cn Canadiana [639]

Fur, fin, and feather : a compilation of the game laws of the principal states and provinces of the united states and canada: together with a list of hunting and fishing localities and other useful information for gunners and anglers – New York: Charles Suydam, Publisher, 1875 – us CRL [340]

The fur worker – Long Island City, N.Y. v. 1-14 no. 7. Oct 3 1916-Apr 1931 – 1 – us NY Public [331]

Furber, Daniel Little et al *see* Professor park and his pupils

Furber, Holden *see* John company at work

Die furche – Wien (A), 1946 19 jan-1975 27 dec – 1 – gw Misc Inst [074]

Die furcht vor dem denken – occam und luther / Schlatter, Adolf von & Kropatschek, Friedrich – Guetersloh: C. Bertelsmann, 1900 – (= ser Beitraege zur foerderung christlicher theologie) – 1mf – 9 – 0-7905-3283-2 – (incl bibl ref) – mf#1987-3283 – us ATLA [242]

Das furchtmotiv in der katholischen busslehre von augustin bis petrus lombardus / Hunzinger, A W – Naumburg, 1906 (mf ed 1993) – 1mf – 9 – €24.00 – 3-89349-340-9 – mf#DHS-AR 193 – gw Frankfurter [720]

Das furchtproblem in der katholischen lehre von augustin bis luther / Hunzinger, August Wilhelm – Leipzig: A. Deichert 1906 [mf ed 1990] – (= ser Lutherstudien 2/1) – 1mf – 9 – 0-7905-6184-0 – (incl bibl ref) – mf#1988-2184 – us ATLA [241]

Furer-Haimendorf, Christoph von *see* The aboriginal tribes of hyderabad

Furet de londres – London, UK. 8-19 Jan 1942 – 1 – uk British Libr Newspaper [072]

Le furet du vesinet : journal illustre des concerts et fetes – Saint-Germain. 1862-63 – 1 – fr ACRPP [073]

Furetiere, Antoine *see*
- Dictionaire universel
- Dictionnaire universel

Furey, Francis Thomas *see* Life of leo 13. and history of his pontificate

Fur-farming in canada / Canada. Commission of Conservation. Committee on Fisheries, Game and Fur-Bearing Animals – By J. Walter Jones. Montreal: Gazette Printing Co. Ltd., 1913. viii, 166p. plates, maps, tables – 1 – us Wisconsin U Libr [636]

Furies – v1-v2 n3 [1972 jan-1973 may/jun] – 1r – 1 – mf#781713 – us WHS [071]

Furlong Cardiff, Guillermo *see*
- Cartografia jesuistica del rio de la plata...
- El padre jose quiroga
- La personalidad y la obra de tomas falkner. buenos aires, 1929
- Tradicion religiosa en la escuela argentina

Furlong, John William *see* Military notes on cuba

Furman, Hart, Botsford et al *see* Sermons and miscellaneous pamphlets

Furman history – Publ. in the Greenville News, Greenville, SC, 15 Apr 1951 and 11 Nov 1958. 48p – 1 – 5.00 – us Southern Baptist [242]

Furman, Richard *see* Correspondence

Furman, Wood *see* A history of charleston association of baptist churches in the state of south carolina

Furmanov, Dmitrii *see* Shar zemli

Furmer, Bernardo *see* De rerum usu et abusu

Furneaux, Rupert *see*
- Zulu war

Furness, William Henry *see*
- Discourses
- Genius of christianity
- The gospels, historical
- Jesus, the heart of christianity
- The power of spirit manifest in jesus of nazareth
- Remarks on the four gospels
- The story of the resurrection of christ told once more
- The unconscious truth of the four gospels
- The veil partly lifted and jesus becoming visible

Furnishing world – London. 1950-1954 (1) – ISSN: 0016-3015 – mf#560 – us UMI ProQuest [740]

Furniss, Harry *see* The works of charles burton barber

Furniture : architecture, applied arts, studio arts – (= ser Art exhibition catalogues on microfiche) – 29 catalogues on 33mf – 9 – £275.00 – (individual titles not listed separately) – uk Chadwyck [740]

Furniture – (= ser Christie's pictorial archive: painting and graphic art) – 165mf – 9 – $1235.00 – 0-907006-22-1 – (incl ormulu & musical instruments; over 9500 reproductions) – uk Mindata [740]

Furniture and decorative accessories *see* The index of american design (tiam)

Furniture and upholstery journal and undertakers gazette – Toronto: J. Acton, [1895?-18– or 19–] – 9 – mf#P05031 – cn Canadiana [680]

Furniture and woodwork collection / Victoria and Albert Museum. London – (= ser Decorative art in the victoria and albert museum) – 110mf – 9 – $810.00 – 0-907006-20-5 – (over 6500 reproductions) – uk Mindata [740]

Furniture production – Franklin. 1964-1987 (1) 1971-1987 (5) 1974-1987 (9) – ISSN: 0532-8942 – mf#2219 – us UMI ProQuest [740]

Furniture with candelabra and interior decoration / Bridgens, Richard – London 1838 – (= ser 19th c art & architecture) – 2mf – 9 – mf#4.2.853 – uk Chadwyck [740]

Furniture worker / North Central District Council of Furniture and Woodenware Workers et al – v4-v8 n2 [1940 feb-1944 jun] – 1r – 1 – mf#1111793 – us WHS [680]

Furniture workers' journal / International Furniture Workers' Union of America – v1 n1, v2-v9 n4 [1883 feb 1, 1884 jan 4-1891 apr 14] – 1r – 1 – mf#1056706 – us WHS [331]

Furniture workers press : official organ of the united furniture workers of america, cio / United Furniture Workers of America – 1965-70, 1939-43, 1944-55, 1956-70, 1971-77, 1978-86 – 6r – 1 – (cont: people's press; cont by: iue news) – mf#1206389 – us WHS [331]

Furniture/today – High Point. 1985+ (1,5,9) – ISSN: 0194-360X – mf#15030 – us UMI ProQuest [740]

Furnivall, F J *see*
- Emblemes and epigrames...a d 1600
- The story of england by robert manning of brunne

Furrer, Reinhold *see* Die haftung des kommanditisten im vergleich mit der haftung des komplementars auf grundlage des franzosischen, schweizerischen und deutschen handelsrechtes

Furrow – Maynooth. 1950+ (1) 1971+ (5) 1976+ (9) – ISSN: 0016-3120 – mf#3037 – us UMI ProQuest [240]

Furrow corn belt edition – Moline. 1975-1980(1,5,9) – ISSN: 0016-3112 – mf#10573 – us UMI ProQuest [630]

Furs and fur garments / Davey, Richard Patrick Boyle – London [1895] – (= ser 19th c art & architecture) – 3mf – 9 – mf#4.2.1484 – uk Chadwyck [740]

Furst bismarck und der antisemitismus – Wien, Austria. 1886 – 1r – us UF Libraries [939]

Furst, Julius *see* Geschichte des karaerthums

Furstenberg, Dorylee *see* Influence of water therapy on selected physiological variables in pregnant women

Furtado, Celso *see*
- Brasil en la encrucijada historica
- Diagnosis of the brazilian crisis
- Dialectica del desarrollo
- Perspectiva da economia brasileira

Further account of discoveries in natural history, in the western states / Rafinesque-Schmaltz, C S – Oxford. 1973+ (1,5,9) – 1mf – 9 – mf#8119 – ne IDC [910]

Further animadversions on dr haweis' misquotations and misrepresentations / Milner, Isaac – Cambridge, England. 1801 – 1r – us UF Libraries [240]

Further comments on dr pusey's renewed explanation / Dodsworth, William – London, England. 1851 – 1r – us UF Libraries [240]

Further copies or extracts of correspondence relative to the affairs of lower canada and upper canada : lower canada, upper canada, nova scotia, new brunswick, prince edward island – London, England: s.n, 1838?] (mf ed 1982) – 1mf – 9 – mf#SEM105P109 – cn Bibl Nat [324]

Further copies or extracts of correspondence relative to the affairs of lower canada and upper canada : lower canada, upper canada, nova scotia, new brunswick, prince edward island – [London, England: s.n, 1838?] (mf ed 1982) – 1mf – 9 – mf#SEM105P109 – cn Bibl Nat [323]

Further correspondence relative to the projected railway from halifax to quebec : (in continuation of papers presented by command of her majesty 16th june 1851) / Canada. Gouverneur general – London: printed by George Edward Eyre & William Spottiswoode, 1852 [mf ed 1982] – 2mf – 9 – mf#SEM105P120 – cn Bibl Nat [380]

Further correspondence with the governments of canada, prince edward island and newfoundland : respecting the treaty of washington and canadian pacific railway: in continuation of papers presented march, 1873 – London: printed by W Clowes for HMSO, 1873 [mf ed 1984] – 1mf – 9 – 0-665-45726-X – mf#45726 – cn Canadiana [380]

Further education – London. 1950-1951 – 1 – mf#676 – us UMI ProQuest [374]

Further excavations at mohenjo-daro : being an official account of archaeological excavations at mohenjo-daro carried out by 1927 and 1931 / Mackay, Ernest John Henry – Delhi: Manager of Publications, 1938- – (= ser Samp: indian books) – us CRL [930]

Further leaves from assam : a continuation of my journal "twenty years in assam" / ed by Moore, P H [Mrs] – Nowgong: [s.n] 1907 [mf ed 1995] – (= ser Yale coll) – xi/191p – 1 – 0-524-09172-2 – mf#1995-0172 – us ATLA [920]

Further papers relative to the affairs of lower canada – [London, England: s.n, 1837] (mf ed 1984) – 1mf – 9 – mf#SEM105P405 – cn Bibl Nat [336]

Further recollections of an indian missionary / Leupolt, C B – London, 1884 – 5mf – 9 – mf#HTM-100 – ne IDC [915]

Further recollections of an indian missionary / Leupolt, C B – London: James Nisbet, 1884 [mf ed 1995] – (= ser Yale coll) – xi/403p (ill) – 1 – 0-524-10140-X – mf#1995-1140 – us ATLA [920]

Further remarks on the voyages of john meares, esq : in which several important facts misrepresented in the said voyages, relative to geography and commerce, are fully substantiated / Dixon, George – London: Printed by John Stockdale...and George Goulding... 1791 – 1mf – 9 – mf#38598 – cn Canadiana [917]

A further report from the committee of secrecy appointed to inquire into the conduct of robert, earl of orford... / Grande-Bretagne. Parliament. House of Commons – London: T Leech, 1742 [mf ed 1974] – 1r – 5 – mf#SEM16P3 – cn Bibl Nat [324]

Further researches into the history of the ferrar-group / Harris, James Rendel – London: C J Clay, 1900 – 2mf – 9 – 0-8370-3492-2 – mf#1985-1492 – us ATLA [220]

Further studies in the prayer book / Dowden, John – London: Methuen, 1908 – 1mf – 9 – 0-7905-4401-6 – mf#1988-0401 – us ATLA [240]

Furttenbach, J *see*
- Architectura civilis...
- Architectura martialis...
- Architectura recreationis...
- Architectura universalis...
- Mannhaffter kunst-spiegel...

[Furttenbach, J] *see* Architectura privata...

Furtwaengler, Adolf *see* Masterpieces of greek sculpture

Furugh – Rasht. sal-i 1, shumarah-i 1-12. day 1306-aban va azar 1308 [dec 1927-oct, nov 1938] – 1r – 1 – $80.00 – us MEDOC [079]

Furukawa, Takeji *see* Hsueh yeh hsing yu min tsu hsing

Fusarium wilt of watermelons / Walker, M N – Gainesville, FL. 1941 – 1r – us UF Libraries [634]

Fusco, Federico *see* Diamantina

Fuse / Arton's Cultural Affairs Society & Publishing – v7 n5-v11 n3 [1984 feb-1987 fall] – 1r – 1 – (cont: centerfold; cont by: fuse magazine) – mf#1582379 – us WHS [071]

Fusion – Boston. 1973-1973 – 1 – ISSN: 0016-3163 – mf#6188 – us UMI ProQuest [073]

Fusion – New York. 1987-1987 (1,5,9) – ISSN: 0148-0537 – mf#15628,01 – us UMI ProQuest [333]

Fusion of sacred and secular elements in benjamin britten's vocal and choral literature / Damp, Alice Bancroft – U of Rochester 1973 [mf ed 1973] – 4mf – 9 – mf#fiche987 – us Sibley [780]

Fussell, L *see* A journey round the coast of kent

Fusslin, Johann C *see* Works

Fussreise durch russland und die sibirische tartarei : und von der chinesischen graenze nach dem eismeer und kamtschatka / Cochrane, John – Weimar 1825 [mf ed Hildesheim 1995-98] – 1v on 3mf – 9 – €90.00 – ISBN-10: 3-487-26485-4 – ISBN-13: 978-3-487-26485-1 – gw Olms [910]

Fussreise vom brocken auf den vesuv und rueckkehr in die heimath / Ife, August – Leipzig 1820 [mf ed Hildesheim 1995-98] – 1v on 2mf – 9 – €60.00 – 3-487-27747-6 – gw Olms [914]

Fussreise von wien nach dem schneeberge : mit historischen nachrichten von der entstehung und den aeltesten bewohnern der in dieser gegend liegenden schloesser und ortschaften / Embel, Franz – Wien 1801 [mf ed Hildesheim 1995-98] – 2mf – 9 – €60.00 – 3-487-29422-2 – gw Olms [914]

Fussreise zweyer schlesier durch italien und ihre begebenheiten in neapel / Wehrhan, Otto – Breslau 1821 [mf ed Hildesheim 1995-98] – 3mf – 9 – €90.00 – 3-487-29275-0 – gw Olms [914]

Die fusswassung in monastischen brauchtum und in der lateinischen liturgie (tab47) / Schaefer, Th. – 1956 – (= ser Texte und arbeiten. beuron (tab). beitraege zur ergruendung des aelteren lateinischen und christlichen schrifttums und gottesdienstes) – €7.00 – ne Slangenburg [241]

Fu-tan t'ung hsueh hui *see* Fu-tan t'ung hsueh hui hui yuan lu

Fu-tan t'ung hsueh hui hui yuan lu / Fu-tan t'ung hsueh hui – Shang-hai: Fu-tan t'ung hsueh hui, [1933] – (= ser P-k&k period) – us CRL [030]

Futch, Merrill Charles *see* Vitamin a assay of one type of dried citrus pulp

The futility of technical, industrial, vocational and continuation schools : a paper. / Crane, Richard Teller – Chicago, 1911 – 13p – 1 – us Wisconsin U Libr [370]

Future – Tulsa. 1971-1986 (1) 1971-1986 (5) 1975-1986 (9) – (cont by: jaycees magazine) – ISSN: 0016-3260 – mf#6066 – us UMI ProQuest [380]

The future capital of the british empire : a possible solution of the suez canal and eastern questions. a political study by a conservative-radical – 2nd ed. London, 1884 – (= ser 19th c books on british colonization) – 1mf – 9 – mf#1.1.2069 – uk Chadwyck [330]

Future Club of New Orleans *see* Black river journal

Future generations computer systems: fgcs – Amsterdam. 1986-1995 (1,5,9) – ISSN: 0167-739X – mf#42540 – us UMI ProQuest [000]

Future is ours, comrade / Novak, Joseph – New York, NY. 1964 – 1r – us UF Libraries [025]

The future leadership of the church / Mott, John Raleigh – New York: Student Dept., Young Men's Christian Association, 1908 – 1mf – 9 – 0-7905-5074-1 – (incl bibl ref) – mf#1988-1074 – us ATLA [240]

The future life : four sermons. preached at st. john's notting hill... / Dudden, Frederick Homes – London: Longmans, Green, 1915 – 1mf – 9 – 0-524-04903-3 – mf#1992-0246 – us ATLA [240]

The future marquis / Aldrich, Annie Charlotte Catharine – London: Hurst & Blackett Publ. 3v. 1881 – (= ser 19th c women writers) – 10mf – 9 – mf#5.1.111 – uk Chadwyck [420]

The future of africa / Fraser, Donald – London: Student Volunteer Missionary Union, 1911 – 1mf – 9 – 0-7905-4577-2 – (incl bibl ref) – mf#1988-0577 – us ATLA [960]

The future of british america : independence! how to prepare for it / Tickle, Paul I – [Toronto?: s.n.], 1865 [mf ed 1983] – 1mf – 9 – mf#23225 – cn Canadiana [971]

The future of canada : address delivered by j m clark before the mulock club, toronto / Clark, John Murray – [Toronto?: s.n, 1903?] – 1mf – 9 – 0-665-72091-2 – mf#72091 – cn Canadiana [320]

The future of canada / a perplexed imperialist / the canadian flag etc / Ewart, John Skirving – [Ottawa?: s.n, 1908?] [mf ed 1994] – 1mf – 9 – 0-665-73196-5 – (incl bibl ref) – mf#73196 – cn Canadiana [320]

The future of catholic peoples : protestant and catholic civilization compared / Haulleville, Prosper Charles Alexander – New York: Hickey, c1878 – 1mf – 9 – 0-8370-8116-5 – (incl bibl ref) – mf#1986-2116 – us ATLA [241]

The future of exchange and the indian currency / Jevons, Herbert Stanley – London: Oxford University Press, 1922 – (= ser Samp: indian books) – us CRL [332]

The future of india – London, 1859 – (= ser 19th c british colonization) – 1mf – 9 – mf#1.1.8979 – uk Chadwyck [954]

The future of india / Moon, Penderel – London: Pilot Press, 1945 – (= ser Samp: indian books) – us CRL [954]

The future of india and south-east asia / Panikkar, Kavalam Madhava – London: George Allen & Unwin Ltd; Bombay: Allied Publishers, 1945 – (= ser Samp: indian books) – us CRL [954]

The future of islam / Blunt, Wilfrid Scawen – London: Kegan Paul, Trench, 1882 – 1mf – 9 – 0-524-01681-X – mf#1990-2583 – us ATLA [260]

The future of mcgill university : annual university lecture, session 1880-81 / Dawson, John William – Montreal?: s.n, 1881? – 1mf – 9 – mf#03666 – cn Canadiana [378]

Future of our schools / Bentwich, Herbert – London, England. 1908 – 1r – us UF Libraries [939]
The future of religion, and other essays / Momerie, Alfred Williams – Edinburgh: William Blackwood, 1893 – 1mf – 9 – 0-7905-9818-3 – mf#1989-1543 – us ATLA [240]
The future of science = Avenir de la science / Renan, Ernest – Boston: Roberts Brothers, 1891 – 2mf – 9 – 0-7905-3275-1 – (incl bibl ref. in english) – mf#1987-3275 – us ATLA [210]
The future of the co-operative movement in india / Qureshi, Anwar Iqbal – London, New York: Oxford University Press, 1947 – (= ser Samp: indian books) – (forewords by j coatman, vera anstey) – us CRL [334]
The future of the dominion of canada : an address delivered before the canadian club of new york / Collins, Joseph Edmund – S.l: s.n, 1877? – 1mf – 9 – mf#56103 – cn Canadiana [971]
The future of the kanaka / Jacomb, Edward – Westminster: P S King & Son, 1919 [mf ed 1995] – (= ser Yale coll) – 222p – 1 – 0-524-09536-1 – mf#1995-0536 – us ATLA [980]
The future of u s-saudi relations : hearing... house of representatives, 107th congress, 2nd session, may 22 2002 / United States. Congress. House. Committee on International Relations. Subcommittee on the Middle East and South Asia – Washington: US GPO 2002 [mf ed 2002] – 1mf – 9 – 0-16-068881-7 – (incl bibl ref) – us GPO [327]
Future Political Status Commission [TTPI (US)] *see*
– Interim report to the congress of micronesia
– Papers of the 2nd meeting
– Report to the congress of micronesia
Future probation : a symposium on the question "is salvation possible after death?" / Leathes, Stanley et al – London: James Nisbet 1886 [mf ed 1986] – (= ser Nisbet's theological library) – 1mf – 9 – 0-8370-8835-6 – mf#1986-2835 – us ATLA [240]
Future probation and foreign missions : certain duties and usages at the rooms of the american board / Thompson, Augustus Charles – Boston: Beacon Press, 1886 [mf ed 1990] – 1mf – 9 – 0-7905-6510-2 – (incl bibl ref) – mf#1988-2510 – us ATLA [240]
Future probation examined / Love, William De Loss – New York: Funk & Wagnalls, 1888 [mf ed 1986] – 1mf – 9 – 0-8370-8762-7 – (incl ind) – mf#1986-2762 – us ATLA [240]
Future punishment : comprising four parochial sermons, with an introduction on the scriptural doctrine of retribution, and an essay on prayers for the dead / McKim, Randolph Harrison – New York: T Whittaker, 1883 – 1mf – 9 – 0-524-04408-2 – (incl bibl ref) – mf#1992-0101 – us ATLA [240]
Future religious policy of america : a discussion of eleven great living questions / Halstead, William Riley – Cincinnati: Hitchcock and Walden, 1877, c1876 – 1mf – 9 – 0-7905-7939-1 – mf#1989-1164 – us ATLA [200]
Future retribution : viewed in the light of reason and revelation / Row, Charles Adolphus – new and enl ed. London: Isbister, 1889 – 2mf – 9 – 0-7905-3471-1 – (incl ind) – mf#1987-3471 – us ATLA [240]
Future space activities – 1976 – 9 – $20.00 – us Univelt [629]
The future state / Gayford, Sydney Charles – 2nd ed. London: Rivingtons, 1904 – (= ser Oxford church text books) – 1mf – 9 – 0-524-04904-1 – (incl bibl ref) – mf#1992-0247 – us ATLA [240]
The future state and free discussion : four sermons. preached in the first presbyterian church of oakland / Hamilton, Laurentine – San Francisco: JH Carmany, 1869 – 1mf – 9 – 0-524-06137-8 – mf#1992-0804 – us ATLA [242]
The future tenses of the blessed life / Meyer, Frederick Brotherton – New York: Fleming H Revell, c1892 – 1mf – 9 – 0-8370-7174-7 – mf#1986-1174 – us ATLA [240]
Future u.s. space program [aasms30] – 1979 – (= ser Aasms 1968) – 5papers and 60 abstracts on 6mf – 9 – $15.00 – 0-87703-129-0 – (suppl to v38, advances) – us Univelt [350]
FutureBanker – New York. 1997+ (1,5,9) – ISSN: 1092-9061 – mf#29628 – us UMI ProQuest [332]
Futures / Aroni, Julius – New Orleans, Gresham, 1882. 101 p. LL-439 – 1 – us L of C Photodup [340]
Futures – Chicago. 1983+ (1) 1983+ (5) 1983+ (9) – (cont: commodities) – ISSN: 0746-2468 – mf#8097,01 – us UMI ProQuest [332]
Futures – Kidlington. 1968+ (1) 1968+ (5) 1968+ (9) – ISSN: 0016-3287 – mf#13330 – us UMI ProQuest [600]
Futures research quarterly – Bethesda. 1985+ (1,5,9) – (cont: world future society bulletin) – ISSN: 8755-3317 – mf#15296 – us UMI ProQuest [500]
Futurism : subject collections – (= ser Art exhibition catalogues on microfiche) – 10 catalogues on 12mf – 9 – £100.00 – (individual titles not listed separately) – uk Chadwyck [700]
The futurism of young asia / Sarkar, Benoy Kumar – Leipzig: Verlog Von Markert & Petters, 1922 – (= ser Samp: indian books) – us CRL [950]
Futurist – Washington. 1967+ (1) 1967+ (5) 1967+ (9) – ISSN: 0016-3317 – mf#6998 – us UMI ProQuest [500]
Futurista : semanario critico e noticioso – Itajai, SC: Typ de Itajahy, 01 jan 1927 – (= ser Ps 19) – mf#UFSC/BPESC – bl Biblioteca [079]
O futuro : folha litteraria, noticiosa e commercial – Senhor do Bonfim, BA. 24 mar 1878; 14 fev 1880 – (= ser Ps 19) – mf#P18B,02,56 – bl Biblioteca [079]
O futuro : hebdomadario academico-republicano – Rio de Janeiro, RJ: Typ Uniao Academica, 06-13 ago 1881 – (= ser Ps 19) – mf#P18A,07,49 – bl Biblioteca [320]
O futuro : orgao das ideas republicanas – Belem, PA: Typ Republicana, 20 abr 1872 – (= ser Ps 19) – bl Biblioteca [320]
O futuro : periodico scientifico e litterario – Recife, PE: Typ Commercial, 10 jun, 30 ago 1864 – (= ser Ps 19) – mf#P17,02,164 – bl Biblioteca [079]
O futuro : periodico scientifico e litterario – Recife, PE: Typ Industrial, 01-15 jun, ago, 01 set 1878 – (= ser Ps 19) – mf#P17,2,184 – bl Biblioteca [079]
El futuro de la agricultura nacional-sindicalista / Falange Espanola Tradicionalista y de las Juntas Ofensivas nacional-Sindicalistas – n.p., 1937? Fiche W 870. (Blodgett Collection of Spanish Civil War Pamphlets) – 9 – us Harvard [946]
Fux, Johann Joseph *see*
– Gradus ad parnassum
– Practical rules for learning composition
– Traite de composition musicale
Fux, Manfred *see* Ein beitrag zum entwurf von zeitreihenreglern
Fuxhoffer, D *see* Monasteriologiae regni hungariae
Fuyuzat – Baku, 1910- . shumarah-'i 3-10 [1 nov 1910-15 mar 1911] – 1r – 1 – $53.00 – us MEDOC [079]
Fuyuzat – Istanbul. 1 sene n1-3. 19 tesrinisani 1324-? [2 dec 1908-?] – (= ser O & t journals) – 4mf – 9 – $60.00 – us MEDOC [956]
Fuzilamentos de 1894 no parana / Carneiro, David – Rio de Janeiro, Brazil. 1937 – 1r – us UF Libraries [972]
Fuzuli / Sueleyman, Nazif – Istanbul, 1927 – (= ser Ottoman literature, writers and the arts) – 3mf – 9 – $55.00 – us MEDOC [470]
Fuzuli *see*
– The divan project
– Hadikat uel-su'ada
Fuzuli, Kulliyat-i (Divan-i) *see* The divan project
Fuzzy sets and systems – Amsterdam. 1978+ (1) 1978+ (5) 1987+ (9) – ISSN: 0165-0114 – mf#42070 – us UMI ProQuest [510]
Fv. pedro de feria y su doctrina zapoteca. estudio bibliografico / Salvador y Conde, Jose – Madrid: Missionalia Hispanica, 1947 – 1 – sp Bibl Santa Ana [240]
Fve miles high : the story of an attack on the second highest mountain in the world / Bates, Robert H et al – New York: Dodd, Mead & Co, 1939 – (= ser Samp: indian books) – us CRL [790]
Fvndamenta lvtheranae doctrinae de ubiquitate : ...corporis christi in eucharistia, ad orthodoxae fidei normam expensa. pt 1 / [Hardesheim, C] – n.p, 1579 – 3mf – 9 – mf#PBU-597 – ne IDC [240]
Fvndamentvm firmvm / Bullinger, Heinrich – Tigvri, Christoph Froschouer, 1563 – 4mf – 9 – mf#PBU-222 – ne IDC [240]
Fwp journal – Braamfontein. 1973-1973 (1) – ISSN: 0015-9026 – mf#7493 – us UMI ProQuest [020]
Fy 1997 application for grants under the fund for the improvement of education : assessment development grants – Washington DC: US Dept of Education, Office of Educational Research & Improvement [1997] [mf ed 1998] – 1mf – 9 – us GPO [350]
Fyfe, James Hamilton *see* British enterprise beyond the seas
Fyfe, Robert Alexander *see* The teaching of the new testament in regard to the soul, and the nature of christ's kingdom
Fyffe, David *see* The essentials of christian belief
Fyi : the news bulletin of the exhibits round table – St. Louis. 1991-1993 (1) – mf#12517,04 – us UMI ProQuest [020]
Fynney, Fred B *see* Zululand and the zulus
Fyns tidende – Odense, Denmark. 1945 – 3r – 1 – uk British Libr Newspaper [074]
Fyrtiofemte psalmen af psaltaren fversttning med anmrkningar / Swartling, Karl – Upsala: Akademiska boktryckeriet, 1872 [mf ed 1985] – 1mf – 9 – 0-8370-5470-2 – mf#1985-3470 – us ATLA [221]
Fyshe, Thomas Maxwell *see* Discussion, design, and specifications for a reinforced concrete bridge abutment
Fysk : facts you should know / Virginia Penitentiary – v10 n1-2 [1981 dec-1982 mar] – 1r – 1 – (cont by: fysk magazine) – mf#1781682 – us WHS [365]
Fysk magazine / Virginia Penitentiary – v10 n3-v13 n2 [1982 jun-1986 summer] – 1r – 1 – mf#1781688 – us WHS [365]
Fysy, Frederic *see* Coming of christ, ad 1947
Fytche, A *see* Burma past and present with personal reminiscences of the country
Fyvie, William *see* A vocabulary english and goojurattee
Fyzee Rahamin, Atiya Begum *see* Sangit of india
Fz [freiwirtschaftliche zeitung] – Hamburg DE, 1924-33 [gaps] – 1 – gw Misc Inst [330]
G : good news / Elf-Lords of Rivendell (Whitewater, WI) – 1969 oct 7-1970 dec 15 – 1r – 1 – mf#1112084 – us WHS [071]
G a buergers ausgewaehlte werke : in zwei baenden – Stuttgart: J G Cotta. 2v. [1885] [mf ed 1989] – 2v – 1 – (biogr int by richard maria werner) – mf#7094 – us Wisconsin U Libr [802]
G A O journal – Washington DC 1988-92 – 1,5,9 – (cont: gao review) – ISSN: 1045-3261 – mf#16498 – us UMI ProQuest [350]
G A Ogle and Co *see* Standard atlas of phillips county, colorado
G B *see* Nothing against the commandments
G b u reporter – Pittsburgh PA. 1893-1980 – 8r – 1 – us IHRC [073]
G c lichtenberg's briefe an dieterich, 1770-1798 : zum 100. todestage lichtenberg's / Lichtenberg, Georg Christoph; ed by Grisebach, Eduard – Leipzig: Dieterich, T Weicher, 1898 [mf ed 1995] – ix/145p/1pl (ill) – 1 – mf#8817 – us Wisconsin U Libr [920]
G c t – Waco TX 1978-86 – 1,5,9 – (cont by: gifted child today) – ISSN: 0164-9728 – mf#12496.03 – us UMI ProQuest [640]
G E lessings uebersetzungen aus dem franzoesischen friedrichs des grossen und voltaires : im auftrag der gesellschaft fuer deutsche litteratur in berlin / Lessing, Gotthold Ephraim; ed by Schmidt, Erich – Berlin: W Hertz, 1892 – 1r – 1 – us Wisconsin U Libr [430]
G f news letter / Girl Friends, Inc – 1964 – 1r – 1 – (cont: girl friend newsletter) – mf#5012742 – us WHS [305]
G FD *see* Captain lightfoot
G G gervinus : ein kapitel ueber literaturgeschichte / Rychner, Max – Bern: Seldwyla, 1922 (mf ed 1990) – 1r – 1 – (filmed with: zwischen den kriegen. incl bibl ref) – us Wisconsin U Libr [430]
G galilei / Frenkel, Jacob – Warsaw, Poland. 1900 – 1r – us UF Libraries [939]
G galilei / Frenkel, Jacob – Warsaw, Poland. no date – 1r – us UF Libraries [939]
O g h monitor / Operation Going Home [Committee] – 1979 nov/dec, 1980 mar/apr, nov/dec, 1981 may/jun, 1982 jan/feb – 1r – 1 – mf#5297745 – us WHS [355]
G I Rights and Information Center *see* Fight back!
G LP *see* La voix paysanne
G Maria de Alboraya, Domingo de *see* Historia del monestario de yuste
G p [general practice] – Leawood KS 1950-69 – 1 – ISSN: 0016-3600 – mf#2269 – us UMI ProQuest [610]
G Q campus and career annual – New York NY 1965-68 – 1 – ISSN: 0433-051X – mf#6762 – us UMI ProQuest [331]
G Q Candaridddhi *see* Sandana khmaer-russi
G Q scene – New York NY 1966-68 – 1 – ISSN: 0434-9997 – mf#6760 – us UMI ProQuest [740]
Gaa paa! – Girard KS, Minneapolis MN. 1909 nov 13-1912 dec 28, 1913 jan 4-1915 dec 25, 1916 jan 1-1918 oct 26 – 3r – 1 – (continued by: folkets rhost) – mf#880922 – us WHS [071]
Gaab, Ernst *see* Der hirte des hermas
Gaagskaia konferentsiia, iiun'-iiul' 1922 g : sobranie dokumentov / Conference on Russia, (1922 : Hague, Netherlands); ed by Lashkevich, Georgii Nikolaevich – Moskva: Nar komissariat po inostrannym delam, 1922 [mf ed 2002] – 1r – 1 – (filmed with: cheshskiia glossy v mater verborum / razbor a o patery (1878)) – mf#5256 – us Wisconsin U Libr [336]
Gaastra, Dieuke *see* Bijdrage tot de kennis van hat vedische ritueel, jaiminiyasrautasutra
Gabaldon Marquez, Joaquin *see*
– Misiones venezolanas en los archivos europeos
– Muestrario de historiadores coloniales de venezuela
Gabaldon Marquez,Joaquin *see* Don gerardo patrullo y otros desmayos
Gabatshwane, S M *see*
– Introduction to the bechaunaland protectorate history
– Seretse khama and botswana
– Tshekedi khama of bechuanaland
Gabbai, Meir Ben Ezekiel Ibn *see* Avodat ha-kodesh
Gabela, Raymoth Vika *see* Parental involvement as an administrative component of educational administration for the black people in south africa
Gabele, Anton *see* In einem kuehlen grunde
Gabelentz, Georg von der *see* Confucius und seine lehre
Gabelentz, Hans Conon von der *see* Elemens de la grammaire mandchoue
Gabinete de leitura, seroes das familias brasileiras : jornal para todas as classes, sexos e idades – Rio de Janeiro, RJ: Typ Commercial de J de N Silva, 13 ago 1837-08 abr 1838 – (= ser Ps 19) – mf#P02,05,41 – bl Biblioteca [073]
Gabinete Psicopedagogico Sanchez Rodrigo *see* Manual de la bateria de test gesar
Gabinetto armonico pieno d'istromenti sonori : indicati, e spiegati dal padre filippo bonanni / Buonanni, Filippo – Roma: Nella Stamperia di Giorgio Placho 1722 [mf ed 19–] – 7mf – 9 – (apparently a later iss of 1722 ed: "aggiunta" [p171-177] lacking in original) – mf#fiche 368 – us Sibley [780]
Gabino marquez, s.j. deberes patrioticos / Meseguer, Pedro – Madrid: Razon y Fe, 1941 – sp Bibl Santa Ana [240]
Gabino tejado / Blanco Garcia, Francisco – Madrid: Saenz de Jubera, 1909 – sp Bibl Santa Ana [440]
Gabler, Karl *see* Faust-mephisto, der deutsche mensch
Gablonzer tagblatt – Gablonz, Neisse (Jablonec nad Nisou CZ), 1935 1 may-1938 – 5r – 1 – gw Misc Inst [077]
Gabon *see*
– Bulletin officiel
– Journal officiel
– Journal officiel de la republique gabonaise
Gabon. Direction de la Statistique et des Etudes Economiques *see* Rapport annuel sur la situation economique, financiere et sociale de la republique gabonaise 1961-1971
Gabon-Congo *see* Journal officiel
Gaboon and Corisco Mission. (Pres. Ch. in the USA) *see* Minutes and records
Gaborit, Prosper *see* Vie de m francois mabileau
Gabriel *see* Etude du tshiluba
Gabriel charland et sa descendance : dictionnaire genealogique / Charland, Maurice – Quebec: Librairie Pruneau, 1933 [mf ed 1990] – 3mf – 9 – mf#SEM105P1215 – cn Bibl Nat [929]
Gabriel, Charles Nicolas *see* Le marechal de camp desandrouins, 1729-1792
Gabriel de Jesus, C D *see* Sermon predicado en la misa nueva del presbitero don constantino lancho solana celebrada en la iglesia parroquial de santa marina, en canaveral (caceres) por el fr
Gabriel de Talavera, Fray *see* Historia...guadalupe
Gabriel, M *see*
– Argentine
– Chaine electrique
– Dejeuner d'employes
– Eau de javelle
– Jacquot
– Tambour et la vivandiere
Gabriel marcel et karl jaspers / Ricoeur, P – Paris, 1947 – 8mf – 8 – €17.00 – ne Slangenburg [140]
Gabriel marcel et karl jaspers / Ricoeur, Paul – Paris, France. 1947 – 1r – us UF Libraries [140]
Gabriel west : and other poems / Currie, Margaret Gill – Fredericton, NB: H A Cropley, 1866 – 2mf – 9 – mf#36943 – cn Canadiana [810]
Gabriel y Galan, Jose Maria *see*
– Nuevas castellanas
– Obras completas
– Solo para mi lugar
Gabriel y Ruiz de Apodaca, Fernando de *see* Poesias
Gabriel-de-l'Annonciation, soeur *see* Bibliographie analytique sur la methodologie de l'histoire du canada (1950-1962)
Gabrieli, Andrea *see*
– Concerti di andrea et gio. gabrieli...
– Il primo libro de madrigali a sei voci
– Primus liber missarum sex vocum
– Il secondo libro de madrigali a sei voci
Gabrieli, Giovanni *see*
– Canzoni et sonate...
– Sacrae symphoniae 6, 7, 8, 10, 12, 14, 15 & 16 vocibus
– Symphoniae sacrae liber secundus...
Gabrielle : ou, les aides-de-camp / Ancelot, Francois – Paris, France. 1839 – 1r – us UF Libraries [440]
Gabulli, Plorio A *see* Cantos de amor y de dolor
Gabungan koperasi konsumsi djakarta-raya / Laporan umum Pengurus Lembaga Administrasi Negara – Djakarta, 1964 – 3mf – 9 – mf#SE-1481 – ne IDC [959]
Gabungan perindustrian / Laporan tahun – Djakarta, 1950-1952 – 2mf – 9 – mf#SE-1482 – ne IDC [959]

Gacelas de hafiz / Hafiz – Mexico City? Mexico. 1944 – 1r – us UF Libraries [025]

Gaceta – Tampa, FL. 1922 oct 6-1998 apr – 110r – (gaps) – us UF Libraries [071]

La gaceta / Costa Rica – San Jose. 1892-1948 – 1 – us NY Public [972]

Gaceta algodonera – Buenos Aires. v18, n213-v. 37, n433. oct 1941-29 feb 1960 – 1 – us NY Public [073]

Gaceta de buenos aires – Buenos Aires. 21 jul-24 nov 1934 – 1/4r – 1 – uk British Libr Newspaper [072]

Gaceta de colombia – Bogota, Colombia. 1822-1831 – 1r – us UF Libraries [079]

Gaceta de la habana / Cuba – Sept 4, 1800-June 1902 – 1 – us L of C Photodup [324]

Gaceta de los tribunales / Chile – 1841-50 – 1 – us L of C Photodup [972]

Gaceta de los tribunales. – Santiago de Chile. On film: 1841-1950; indexes 1898-1902, 1905-12. LL-02005 – 1 – us L of C Photodup [340]

Gaceta de madrid / Spain – Boletin oficiel. 1856-99; 1936-38 – 1 – us L of C Photodup [324]

Gaceta de mexico, compendio de noticias de nueva espana / Mexico – 1796-1809 – 1 – 54.00 – us L of C Photodup [340]

Gaceta de paz; diario de la justicia de paz y de informacion juridica general – v. 1-147. 1935-65. Includes indexes for v. 1-89. v. 132 33, 140-141, 144-145 wanting – 1 – 700.00 – us L of C Photodup [972]

Gaceta del foro – Buenos Aires, Argentina. 3 feb 1916-16 apr 1964 – 104r – 1 – (including "diccionario de jurisprudencia") – uk British Libr Newspaper [073]

Gaceta del foro : jurisprudencia, legislacion, doctrina – Buenos aires. 3 feb 1916-16 apr 1964 – n2-17414 – 1 – (incl: diccionario de jurisprudencia 1916-23 (-1932-38). incl ind by ricardo victorica) – uk British Libr Newspaper [340]

Gaceta del foro. jurisprudencia-legislacion-doctrina – v v. 1-250. 1916-67 – 1 – us L of C Photodup [972]

Gaceta del gobierno / Bolivia – 1841-62 – 1 – 81.00 – us L of C Photodup [972]

Gaceta del gobierno / Mexico. (State) – Toluca. On film: 1851-1924. LL-02025 – 1 – us L of C Photodup [340]

Gaceta del gobierno del ecuador / Ecuador – 1843-45 – (el 21 de junio. 1845-46. el nacional. 1846-49. 121.00; 1) – us L of C Photodup [340]

Gaceta diario oficial / Costa Rica – 1877-1957 – 1 – us L of C Photodup [324]

Gaceta diario oficial / Costa Rica – 1970- – 1 – us L of C Photodup [340]

La gaceta diario oficial / Honduras – 1876-1949 – 1 – us L of C Photodup [324]

La gaceta diario oficial / Honduras – 1971- – 1 – us L of C Photodup [340]

La gaceta diario oficial / Nicaragua – 1845-1946, 1962-69 – 1 – us L of C Photodup [324]

La gaceta diario oficial / Nicaragua – 1970- – 1 – (subject index. 1970-74. 1) – us L of C Photodup [340]

Gaceta espanola – London, UK. Nov 1884; 7 Jan 1885-10 Mar 1886; 5 Jan-23 Mar 1887; 2 Jan 1889-20 May 1891; 11 Nov 1891 – 1 – uk British Libr Newspaper [072]

Gaceta judicial – Tegucigalpa. Honduras. -w. 10 Feb 1900-12 Mar 1904, 2 Jul 1907-11 Apr 1911. (Imperfect). (2 reels) – 1 – uk British Libr Newspaper [072]

Gaceta municipal / Sucre. Venezuela (District) – Jan.-Sept. 1950 – 1 – us NY Public [324]

Gaceta oficial / Bolivia – 1911-15, 24-28, 43-44. 1970- – 1 – us L of C Photodup [972]

Gaceta oficial / Cuba – 1964-Feb 1969 very incomplete. Feb 1970- – 1 – us L of C Photodup [972]

Gaceta oficial / Cuba – Havana. 1902-1966 – 1 – us NY Public [972]

Gaceta oficial / Dominican Republic – 1865-1969 – 1 – $1,909.00 – us L of C Photodup [972]

Gaceta oficial / Dominican Republic – 1970-. LL-02087 – 1 – us L of C Photodup [340]

Gaceta oficial / Dominican Republic – Ciudad Trujillo. 1945-1966 – 1 – us NY Public [972]

Gaceta oficial / Mexico. (Federal district) – 1943-1947 – 1 – us NY Public [324]

Gaceta oficial / Panama – 1970- – 1 – us L of C Photodup [340]

Gaceta oficial / Paraguay – Asuncion. 1901-1903, 1918-1942, 1945, 1959-May 29, 1963 – 1 – us NY Public [324]

Gaceta oficial / Venezuela – 1827-1969 – 1 – 4623.00 – us L of C Photodup [324]

Gaceta oficial / Venezuela – 1970- – 1 – us L of C Photodup [972]

Gaceta oficial / Vera Cruz. Mexico (State) – Jalapa-Enriquez. On film: 1853-1965. LL-02039 – 1 – us L of C Photodup [340]

Gaceta oficial americana – London, UK. 17 Jul 1873; 17 Apr 1874-29 Aug 1875 – 1 – uk British Libr Newspaper [072]

Gaceta oficial del estado / Sucre. Venezuela (State) – Cumana. Jan.-Nov. 1950 – 1 – us NY Public [972]

Gaceta oficial del estado miranda / Miranda. Venezuela – Los Teques. 1949-Apr. 1960 – 1 – us NY Public [324]

Gaceta oficial del estado zulia / Zulia. Venezuela. (State) – Maracaibo. 1950-1966 – 1 – us NY Public [972]

La gaceta portena – Buenos Aires. 9 mar 1984- [semimnthly] – 1 – us Wisconsin U Libr [073]

Gaceta sud americana y de espana – London, UK. 8 Dec 1885-15 Dec 1886 – 1 – uk British Libr Newspaper [072]

Gache, Roberto *see* Paris, glosario argentino

Gachon, Paul *see* Quelques preliminaires de la revocation de l'edit de nantes en languedoc (1661-1685)

Gaddi, G B *see*
- Roma bilitata nelle sue fabbriche dalla santit...di stro sigre clemente xii
- Roma nobilitata nelle sue fabbriche dalla santit...di nostro signore clemente xii

Gaddi, Giuseppe *see* Poveri bimbi; la barbarie dei ribelli spagnuoli

Gaddum, L W *see*
- Pectic constituents of citrus fruits
- Study of some trace elements in fertilizer materials

Gadea Pico, Ramon A *see* Pequeno poema de la aldea y otros poemas

Gadgil, Dhananjaya Ramchandra *see*
- Economic effects of irrigation
- Regulation of wages
- War and indian economic policy

Gadjah-Mada *see* Jajasan badan penerbit godjah mada

Gadolinium-dtpa- und temperaturstudien zur interstitiellen tumortherapie fuer die interventionelle kernspintomographie / Franzen, Stefan – (mf ed 1995) – 2mf – 9 – €40.00 – 3-8267-2200-0 – mf#DHS 2200 – gw Frankfurter [616]

Gadsby and Arnold *see* Our catalogue

Gadsby, John *see* Memoirs of the principal hymn-writers and compilers of the 17th, 18th and 19th centuries

Gadsby, William *see* The works of the late william gadsby, manchester

Gadsden county, florida – Tallahassee, FL. 192- – 1r – us UF Libraries [630]

Gadsden county times – Quincy, FL. 1907 mar 08-1997 – 74r – (gaps) – us UF Libraries [071]

Gadsen (sic) county / Woltz, Larry – s.l, s.l? 193-? – 1r – us UF Libraries [978]

Gadzhieva, S *see* Dinamika izmeneniia polozheniia dagestanskoi zhenshchiny i semia

Gae update – Tucker GA 1975-81 – 1 – (cont by: update) – ISSN: 0164-467X – mf#10466.01 – us UMI ProQuest [370]

Gaebelein, Arno Clemens *see*
- The acts of the apostles
- The harmony of the prophetic word
- Hath god cast away his people?
- The lord of glory
- The prophet daniel
- The prophet joel
- Studies in zechariah
- The work of christ

Gaebert, Hans Walter *see*
- Flucht zum fakir von ipi
- Flugmeldehelferin inge berger

Gaebler, Joachim *see* Internationale freizuegigkeit des kapitals und unterentwickelte laender

Gaede, Werner *see* Goethes torquato tasso im urteil von mit- und nachwelt

Gaedertz, Karl Theodor *see*
- Bei goethe zu gaste
- Emanuel geibel, saenger der liebe, herold des reiches

Gaegenbericht...vff den bericht herren johansen brentzen / Bullinger, Heinrich – [Zuerich, Christoph Froschauer, 1562] – 3mf – 9 – mf#PBU-221 – ne IDC [240]

Gaegensatz vnnd kurtzer begriff der euangelischen vnd baepstischen leer / Bullinger, Heinrich – Zuerych, Christoffel Froschouer, 1551 – 1mf – 9 – mf#PBU-165 – ne IDC [240]

Gael – Dublin, Ireland. 6 jan-24 may 1924 – 1/4r – 1 – uk British Libr Newspaper [072]

The gaelic american – New York. Sept 19 1903-Dec 15 1951 – 1 – us NY Public [071]

Gaelic club, cleveland, ohio, records – Cleveland, Cuyahoga, OH. 1930-39 – 1r – 1 – (minutes, membership rosters, and financial balance reports of this social and athletic club, located on cleveland's west side, serving irish immigrants and irish americans) – us Western Res [790]

Gaelic folk tales / O'Sheridan, Mary Grant – Boston: R G Badger c1926 [mf ed Bloomington IN: Indiana Uni Lib, Preservation Dept 1984] – 167p on 1r – 1 – us Indiana Preservation [390]

Gaeltacht – Gweedore, Donegal feb-jun 1973 – 0.25r – 1 – ie National [072]

Gaenssle, Carl *see* The hebrew particle asher

Gaertner, K Chr *see* Sammlung vermischter schriften von den verfassern der bremischen beytraege

Gaertner, Karl Christian *see* Neue beytraege zum vergnuegen des verstandes und des witzes

Gaertnerpost *see* Deutsche gaertnerpost

Gaeste im paradies / Andres, Stefan Paul – Muenchen: Paul List, c1937 [mf ed 1995] – 299p – 1 – (first publ in 1937 under title: mosellaendische novellen) – mf#8919 – us Wisconsin U Libr [830]

Gaetane de montreuil et ses oeuvres : conference donnee a trois-rivieres, le 18 juillet 1916 / Lacerte, Adele Bourgeois – [Ottawa?: s.n.] 1916 [mf ed 1995] – 1mf – 9 – 0-665-74770-5 – mf#74770 – cn Canadiana [810]

Gaeterbock, F *see* Ottonis morenae et continuatorum historia frederici 1 (mgh6:7.bd)

Gaetzinger, E *see* Joachim von watt

Gaf – 1969 jul 4-dec 13 – 1r – 1 – mf#901056 – us WHS [071]

Gaffarel, Paul *see*
- Decouvertes des portugais en amerique au temps de christophe colomb
- Les decouvreurs francais du 14e au 16e siecle
- Etude sur les rapports de l'amerique et de l'ancien continent avant christophe colomb
- Histoire de la decouverte de l'amerique depuis les origines jusqu'a la mort de christophe colomb, vol 1
- Histoire de la decouverte de l'amerique depuis les origines jusqu'a la mort de christophe colomb, vol 2
- Histoire de la decouverte de l'amerique depuis les origines jusqu'a la mort de christophe colomb, vols 1 and 2
- Nunez de balboa

Gaffat ethiopia : amharic and english monthly newspaper – Adelphi MD. 1992 may-1993 mar, may-1994 jan/feb, 1995 mar 12, dec – 1r – 1 – mf#2681049 – us WHS [071]

Gaffney first baptist church. cherokee county. south carolina : church records – Oct 1980-Sept 1986 – 1 reel – 1 – $11.25 – (wmu minutes, 1977, 1981-1988. 250p) – us Southern Baptist [242]

Gaffurio, Franchino *see*
- Angelicum ac divinum opus musice
- De harmonia musicorum instrumentorum opus
- Musice utriusque cantus practica
- Practica musice

Gaffurius, Franchinus *see*
- Angelicum ac divinum opus musice...
- Practica musice franchini gafori laudensis...
- Theorica musice

Gag airwaves / Gospel Music Workshop of America – v3 n1 [1997 feb] – 1r – 1 – mf#4023604 – us WHS [780]

Gagalis, Zisis *see* Effects of leg exercise and insulin injection sites on blood glucose in persons with insulin dependent diabetes mellitus (iddm)

Gagarin, Jean [comp] *see* L'eglise russe et l'eglise catholique

Gagarin, S P *see* Vseobshchii geograficheskii i statisticheskii slovar'

Gage county democrat – Beatrice, NE: G P Marvin. v1 n1. nov [ie dec] 19 1879-dec 2 1909 (wkly) – 3r – 1 – (v2 n1 dec 17 1880-v10 n52 dec 6 1888 called also whole n53-whole n467. iss for dec 13 1888- called 10th yr n1-) – us NE Hist [071]

Gage county farm journal – Liberty, NE: Ivan D Long, -v48 n52. may 7 1931 (wkly) [mf ed v42 n40. feb 19 1925-may 7 1931 (gaps) filmed 1977] – 2r – 1 – (cont: liberty journal. cont by: liberty journal (1931)) – us NE Hist [071]

The gage county herald – Beatrice, NE: H T Wilson (wkly) [mf ed jan 18 1901-apr 28 1905 (gaps)] – 1r – 1 – us NE Hist [071]

Le gage des divines fiancailles (de arrha animae) / Hugh of Saint-Victor – Bruges, 1923 – €5.00 – (trans and ann by m ledrus) – ne Slangenburg [241]

Gage, Matilda Joslyn *see* Woman, church and state

The gage of the two civilizations: shall christendom waver? : being an inquiry into the causes of the rupture of the english and french treaties of tien-tsin... / [Nye, Gideon] – Macao, [s.n.], 1860 [mf ed 1995] – (= ser Yale coll) – 1 – 0-524-09075-0 – mf#1995-0075 – us ATLA [950]

Gage papers, american manuscripts in the..., 1731-1874 : from sussex archaeological society, lewes – (= ser British records relating to america in microform) – 3r – 1 – (with int by julian gwyn) – mf#96826 – uk Microform Academic [025]

Gage, Sandra L *see* Marketing structures, activities and outcomes amongst selected national sport organizations

Gage, Thomas *see*
- English-american
- Nueva relacion que contiene los viajes de tomas ga...
- Travels in the new world

Gage, William Leonard *see*
- German rationalism
- The home of god's people
- Light in darkness
- Trinitarian sermons

Gager, William *see* Ulysses redux: tragoedia nova

Gagern, Friedrich, Freiherr von *see* Der marterpfahl

Gagern, Hans C E, Freiherr von *see*
- Mein antheil an der politik

Gage's school examiner and monthly review – [S.l: s.n, 1881-18– or 19–] – 9 – mf#P04403 – cn Canadiana [370]

Gage's standard book-keeping, by single and double entry : designed for use in the public and high schools, formerly called beatty and clare's book-keeping – 9th ed. Toronto; Winnipeg: W J Gage, 1883 [mf ed 1987] – (= ser W J Gage and Co's educational series) – 3mf – 9 – 0-665-10411-1 – (incl ind and publ list) – mf#10411 – cn Canadiana [650]

Gageure imprevue / Scribe, Eugene – Paris, France. 1809 – 1r – us UF Libraries [440]

Gagin, P *see* Sacramentaire gelasien d'angouleme'

Gagini, Carlos *see*
- Los aborigenes de costa rica
- Cuentos
- Teatro

Gagnebin, Ferdinand Henri *see* Liste des eglises wallonnes des pays-bas et des pasteurs qui les ont desservies

Gagneius, Ioan. *see* In quatuor sacro-sancta iesu christi evangelia..scholia

Gagnol, abbe *see* Le jansenisme convulsionnaire et l'affaire de la planchette

Gagnon, Charles-Octave *see*
- Chronologie de l'histoire des etats-unis d'amerique
- Mandements, lettres pastorales et circulaires des eveques de quebec

Gagnon, Ernest *see*
- Canadiens, mefiez-vous
- Chansons populaires du canada
- Choses d'autrefois
- Famille charles-edouard gagnon
- Feuilles volantes
- Louis jolliet, decouvreur du mississipi [sic] et du pays des illinois, premier seigneur de l'ile d'anticosti
- Pages choisies
- Reponse a la brochure de monsieur l'abbe h-r casgrain

Gagnon, Francoise *see* Bibliographie analytique de simone bussieres

Gagnon, Gilberte *see* Bibliographie analytique de la litterature pedagogique canadienne francaise de 1790 a 1900

Gagnon, Huguette *see* Bibliographie analytique de alain grandbois

Gagnon, Jacques Etiennette *see* Bibliographie analytique de l'oeuvre de olivette lamontagne

Gagnon, Jeff L *see* Mechanical work and kinematic differences between overground and treadmill walking

Gagnon, Marcelle *see* Bibliographie analytique de madame helene b beausejour

Gagnon, Phileas *see*
- Essai de bibliographie canadienne,. vol 1
- Essai de bibliographie canadienne, vol 2
- Essai de bibliographie canadienne, vols 1-2

Gahan, James Joseph *see*
- Canada
- The immaculate mary
- Lecture on "ireland in sunshine and shadow"
- Lecture [on "ireland in sunshine and shadow"]

Gahan, William *see* Sermons and moral discourses for all the sundays and principle festivals of the year

O gahucho na corte : jornal politico e joco-serio – Rio de Janeiro, RJ: Typ de Silva Lima, 31 mar 1849 – (= ser Ps 19) – mf#P14,02,33 n04 – bl Biblioteca [320]

Gai sallusti crispi bellum catilinae / Sallust – Sallust's Catiline with parallel passages from Cicero's orations against Cataline. Introd., notes, and vocabulary by Jared W. Scudder.Boston: Allyn and Bacon, (c1900). xx,126p – 1 – us Wisconsin U Libr [900]

Gai sollii apollinaris sidonii epistulae et carmina (mgh1:8.bd) / ed by Luetjohann, Ch – 1887 – (= ser Monumenta germaniae historica 1: scriptores – auctores antiquissimi) – €29.00 – (accedunt fausti aliorumque epistulae ad ruricium aliosque. ruricii epistulae rec et emend b krusch) – ne Slangenburg [240]

Gaidukevich, L F *see* Rabochaia zhizn'

Gaiko geppo : foreign affairs monthly, republic of china from 1932-1936 = Wai-chaio-yueh-pao – v1-9. 1932-36 – 10r – 1 – Y99,000 – (in chinese) – ja Yushodo [951]

Gaiko koho : official gazette of the foreign affairs, china from 1921-1928 = Wai-chaio kung-pao – n1-82. 1921- 28 – 14r – 1 – Y120,000 – (in chinese; lacking: n14) – ja Yushodo [951]

Gaiko-iho furoku geppo *see* Gaimu-sho ho

The gaikwads of baroda : english documents / ed by Gense, J H & Banaji, D R – Bombay: DB Taraporevala Sons & Co, [1936]-1945 – (= ser Samp: indian books) – us CRL [920]

Gail, Jean Francois *see* Reflexions sur le gout musical en france
[Gailkircher, W] *see* Quadriga aeternitatis
Gailland, Maurice, S J *see* Diary
Gaillard, Gabriel Henri *see* Histoire de francois premier, roi de france
Gaillard, Thomas *see*
– The principles of surveying
Gaillardot, C *see* Catalogue de l'herbier de syrie
Gailly de Taurines, Charles *see* L'avenir politique du canada et des canadiens francais
Gailor, Thomas Frank et al *see* Christian unity and the bishops' declaration
Gailus-Doering, Sigrid *see* Die imaginaere und die reale hexe
Gaimar, Geoffrey *see* Lestorie des engles solum la translacion maistre geffrei gaimar (rs91)
Gaimu-sho geppo *see* Gaimu-sho ho
Gaimu-sho ho : journal of the ministry of foreign affairs, 1892 to 1946 – 24r – 1 – Y360,000 – (in japanese. title change: "gaiko-iho furoku geppo" "gaimu-sho geppo" "gaimu-sho ho") – ja Yushodo [327]
Gaimu-sho ho *see* Gaimu-sho ho
Gaimu-sho teikoku gikai chosho : research papers compiled for use in the diet – 44th 1920/21-69th 1936 – 114r – 1 – Y1,710,000 – (with 44p guide. in japanese) – ja Yushodo [324]
The gaina-sutras, pt 1 (stbe22) : the akaranga-sutra and the kalpa-sutra – 1884 – (= ser Sacred book of the east (sbte)) – 7mf – 8 – €15.00 – (trans fr prakrit by herman jacobi) – ne Slangenburg [280]
The gaina-sutras, pt 2 (stbe45) : the uttaradhyayana sutra, the sutrakritanga sutra – 1895 – (= ser Sacred book of the east (sbte)) – 9mf – 8 – €18.00 – (trans fr prakrit by hermann jacobi) – ne Slangenburg [280]
Gaines, A G et al *see* The latest word of universalism
Gainesville advocate – Gainesville, FL. 1890 feb 15; mar 09 – 11r – us UF Libraries [071]
Gainesville, alachua county, florida – s.l, s.l? 1930 – 1r – us UF Libraries [630]
Gainesville evening news – Gainesville, FL. 1895 feb 10 – 1r – us UF Libraries [071]
Gainesville first baptist church. gainesville, georgia : wmu records – 1896-1954 – 1 – $39.84 – us Southern Baptist [242]
Gainesville independent – Gainesville, FL. 1965-1976 – 10r – (gaps) – us UF Libraries [071]
Gainesville independent – Gainesville, FL. 1970 and 1975 – 2r – (gaps) – us UF Libraries [071]
Gainesville local guide : gainesville – s.l, s.l? 193-? – 1r – us UF Libraries [978]
Gainesville star – Gainesville, FL. 1904 may-oct 8 – 1r – us UF Libraries [071]
Gainesville sun – Gainesville, FL. 1879 feb 19-1982 feb – 345r – (gaps) – us UF Libraries [071]
Gainesville times – Gainesville, FL. 1876 jul 6-1879 feb 12 – 1r – us UF Libraries [071]
Gainesville times – Gainesville, FL. 1877 apr 14 – 1r – us UF Libraries [071]
Gainesville voice – Gainesville FL: 1986 nov 3, dec 19, 1987 jan 16, feb 20, jun 5 – 1r – 1 – mf#4027738 – us WHS [071]
Gainesville weekly bee – Gainesville, FL. 1882 may 12-1884 apr – 1r – us UF Libraries [071]
Gains to the bible from modern criticism : and other essays / Smith, John Frederick et al – London: British & Foreign Unitarian Assoc 1913 [mf ed 1993] – 1mf – 9 – 0-524-07762-2 – mf#1991-3330 – us ATLA [243]
Gainsboro' evening news – [East Midlands] Lincolnshire 31 may 1880-19 sep 1882 [mf ed 2004] – 5r – 1 – uk Newsplan [072]
Gainsborough evening news – Lincolnshire 1923-75 – 1 – (called gainsborough news 1867-1953; then retford,worksop,isle of axholme & gainsborough news 1954-65; then gainsborough evening news 1966-97) – uk Newsplan [072]
Gainsborough news [midweek ed] – Gainsborough, Lincolnshire [1883]-to date [mf ed 1986-] – 1 – (retford, worksop, isle of axholme and gainsborough news: evening ed to 30 oct 1923 then gainsborough evening news 6 nov 1923-7 oct 1975. never a true evening paper: it has always been the tuesday afternoon ed of a friday paper; the gainsborough news) – uk Newsplan [072]
Gainsborough standard – Gainsborough, Lincolnshire, 1986-2006 – 1 – uk Newsplan [072]
Gainsborough Stock and Subscription Library *see* Laws of the gainsburgh stock and subscription library
Gainsford, Robert John *see* Reformatory schools
Gainza, Francisco *see* Milagros y novena de la santisima virgen del rosario, patrona universal de las islas filipinas, que se venera in el templo de santo domingo de manila
Gairdner, J *see*
– Letters and papers illustrative of the reign of richard 3 and henry 7
– Memorials of henry 7
Gairdner, James *see*
– England
– The english church in the 16th century
– Lollardy and the reformation in england
– The reign of henry 8 from his accession to the death of wolsey
Gairdner, William Henry Temple *see*
– D M thornton
– Echoes from edinburgh, 1910
Gairdner, William Henry Temple et al *see* The vital forces of christianity and islam
Gairloch & district times – [Gairloch: s.n.] 1978- (biwkly) – 1 – uk Scotland NatLib [072]
Gaiser, Gerd *see* Reiter am himmel
Gaismaier, Josef *see* Justinus kerners saemtliche poetische werke
Gaistliche lieder un[d] psalmen, d mart luthers und andern frommen christen : nach ordnung der jarzeiten – Augspurg: Schoenigk 1580 – (= ser Hqab. literatur des 16. jahrh.) – 6mf – 9 – €70.00 – mf#1580c – gw Fischer [780]
Gait and posture – Oxford, England 1993+ – 1,5,9 – ISSN: 0966-6362 – mf#20195 – us UMI ProQuest [613]
Gait, Edward *see* A history of assam
Gait perturbation response in anterior cruciate ligament deficiency and surgery / Ferber, Ronald R – 2001 – 215p on 31mf – 9 – $15.00 – mf#PE 4203 – us Kinesology [617]
Gaitan / Osorio Lizarazo, Jose Antonio – Buenos Aires, Argentina. 1952 – 1r – us UF Libraries [972]
Gaitan, Luis Alejandro *see* Jurisprudencia de la corte suprema de justicia
Gaitan P, Aquilino *see* Por que cayo el partido conservador
Gaitana / Vargas Villamil, Luis Hernando – Bogota, Colombia. 1959 – 1r – us UF Libraries [972]
Gaitanides, Johannes *see* Georg rudolf weckherlin
Gaither reporter – 1993 oct, 1994 dec, 1995 jul – 1r – 1 – mf#3912685 – us WHS [071]
Gaitskell, Charles D *see* Art and crafts in our schools
Gaius *see* Friendly address to jews
Gaja baru – Djakarta, 1959-1960 – 4mf – 9 – (missing: 1959, v1(2-end); 1959/1960, v2(1-5)) – mf#SE-882 – ne IDC [959]
Gal, Hans *see* Johannes brahms (1833-1897)
Galaal, Muusa H I *see* The terminology and practise of somali weather lore, astronomy, and astrology
Galaay : ou, la quete de l'epouse: conte wolof / Ndiaye, Moussa – 1987 – us CRL [390]
Galactic and extragalactic studies : first collection / Shapley, Harlow – [Cambridge MA]: Harvard Observatory 1940 [mf ed 1998] – 1r [pl/ill] – 1 – (incl bibl ref) – mf#film mas 28415 – us Harvard [520]
Galagauan ni dona maria na na-agom nin saua ibago iyo si don juan rey del mundo de austria – [Nueva Caceres]: Libreria Mariana [190-?] [mf ed Bloomington IN: Indiana Uni Lib, Preservation Dept 1984] – (= ser Coll...in the bikol language) – 1r – 1 – us Indiana Preservation [490]
Galan, Leocadio *see*
– Cien razones (aunque sea tarde)
– Dirsos...!tengo madre! cosas de un recluta
Galan Saval, Ricardo *see* Escuela elemantal de trabajo y de capatazes agricolas de caceres
Galan y Galan, FG *see* De mi vieja extremadura (paginas montanchegas)
Galan y los comuneros / Gutierrez, Jose Fulgencio – Bucaramanga, Colombia. 1939 – 1r – us UF Libraries [972]
Galand, A *see* Bibliotheque orientale
Galand, C *see* Bibliotheque orientale
Das galante wien : sittengemaelde / Gross-Hoffinger, Anton – Leipzig 1846 [mf ed Hildesheim: 1995-98] – (= ser Fbc) – 2v on 5mf – 9 – €100.00 – 3-487-29466-4 – gw Olms [914]
Galanter, Marc *see* Substance abuse
Galanti, Raphael Maria *see* Historia do brasil
Galanus, Clemens *see* Conciliationis ecclesiae armenae cum romana
Galaripsos / Deligne, Gaston Fernando – Ciudad Trujillo, Dominican Republic. 1946 – 1r – us UF Libraries [972]
Galarreta, Luis Adam *see* Bocetos y recuerdos
Galashiels telegraph – [Scotland] Galashiels: A Walker & Son 3 nov 1896-dec 1950 (wkly) [mf ed 2004] – 309v on 51r – 1 – (cont by: border telegraph [oct 1902-]) – uk Newsplan [072]
Galashiels weekly journal – [Scotland] Galashiels: J Collie 29 jan-23 apr 1842 (wkly) [mf ed 2004] – 13v on 1r – 1 – uk Newsplan [072]
La galatea : edicion publicada por rodolfo schevill y adolfo bonilla / Cervantes Saavedra, Miguel de – Madrid: Rodriguez 1914 [mf ed 1987] – 2v on 1r – 1 – mf#1936 – us Wisconsin U Libr [830]
Der galaterbrief : nach seiner echtheit untersucht: nebst kritischen bemerkungen zu den paulinischen hauptbriefen / Steck, Rudolf – Berlin: Georg Reimer, 1888 – 1mf – 9 – 0-8370-9421-6 – (incl bibl ref) – mf#1986-3421 – us ATLA [227]
Der galaterbrief im feuer der neuesten kritik : besonders des prof dr loman in amsterdam sowie des prof rudolf steck in berlin / Schmidt, Paul Viktor – Leipzig: August Neumann, 1892 – 2mf – 9 – 0-8370-9653-7 – (incl bibl ref) – mf#1986-3653 – us ATLA [227]
Galathee / Barbier, Jules – Paris, France. 1875 – 1r – us UF Libraries [440]
Galatians 3:28 Press *see* Perspective
Galatians and romans : with introduction and notes / Mackenzie, William Douglas – New York: Fleming H Revell; London: Andrew Melrose, 1912 – 1mf – 9 – 0-7905-1430-3 – (incl ind) – mf#1987-1430 – us ATLA [227]
Galatians, ephesians, philippians, colossians, 1 and 2 thessalonians, 1 and 2 timothy, titus and philemon : a popular commentary upon a critical basis... / Clark, George Whitefield – Philadelphia: American Baptist Publ Soc 1903 [mf ed 1993] – (= ser Clark's peoples commentary) – 2mf [ill] – 9 – 0-524-06571-3 – mf#1992-0914 – us ATLA [227]
Galatinus, P *see* De arcanis catholici veritatis libri 12
The galax gatherers : the gospel among the highlanders / Guerrant, Edward Owings; ed by Guerrant, Grace – Richmond, VA: Onward Press, c1910 – 1mf – 9 – 0-8370-6059-1 – (incl ind) – mf#1986-0059 – us ATLA [240]
Galaxy / Altus Air Force Base (OK) – 1981 may 1-1983 jul, 1983 aug-1985 jun, 1985 jul-1987 apr 14, 1987 may 1-1988 oct 28, 1988 nov 4-1989 dec 31 – 5r – 1 – (continued by: patriot (altus, ok)) – mf#999067 – us WHS [355]
Galaxy – La Salle IL 1866-78 – 1 – mf#4133 – us UMI ProQuest [071]
Galbraith, John *see*
– In the new capital
– Technical education
Galbraith, John S *see* Reluctant empire
Galbraith, Richard *see* Blow the trumpet
Galbraith, Thomas *see*
– Bensalem, or, the new economy
– General financial and trade review of the city of toronto for 1880
– A new chapter added to political economy
– New monetary theory
Galbraith, W O *see* Colombia
Gale, Charles James *see* A treatise on the law of casements
Gale, George *see* Historic tales of old quebec
Gale, James Scarth *see*
– Korea in transition
– Korean sketches
– Korean-english dictionary
– The vanguard
Gale, John T *see* Account, guide and form book for administrators and executors in the state of ohio
Gale, Martha *see* Woman's high calling
Gale, Samuel *see* Nerva
Gale, Thomas *see* De mysteriis liber
Gale, William Daniel *see* Deserve to be great
Gale, Zona *see* Friendship village
Galeana, Juan *see* Pedimento y replica del promotor fiscal del tribunal de circiuto
Galeazzi, Francesco *see* Elementi teorico-practici de musica by francesco galeazzi
Galectin-3 im nervensystem : neue einblicke in die bedeutung eines lektins / Probstmeier, Rainer – (mf ed 2000) – 2mf – 9 – €40.00 – 3-8267-2692-8 – mf#DHS 2692 – gw Frankfurter [574]
Gale-morant papers, 1731-1925 – University of Exeter Library – (= ser BRRAM series) – 2r – 1 – £134 / $268 – (with p/g; int by r b sheridan) – mf#r97047 – uk Microform Academic [972]
Galen, Philipp *see* Der irre von st james
Galena daily advertiser – Galena IL. 1852 oct 20, 1853 sep 3 – 1r – 1 – (continued by: galena daily gazette) – mf#976081 – us WHS [071]
Galena democrat and public advertiser – Galena IL 1838 dec 6 – 2r – 1 – mf#874096 – us WHS [071]
Galena democrat for the country – Galena IL. 1840 may 30, sep 12 – 1r – 1 – mf#874097 – us WHS [071]
Galena gazette and advertiser – Galena IL. 1841 jan 11 – 1r – 1 – mf#872734 – us WHS [071]
Galena industrial press – Galena IL. 1882 oct 5-1883 sep 13, 1883 sep 30-1885 jan 29, 1885 feb 5-dec 31 – 3r – 1 – (cont: industrial press; cont by: galena press) – mf#887621 – us WHS [071]
Galena press – Galena IL. 1886 jan-aug 26, 1886 sep 2-1888 apr 12, 1888 apr 19-1890 jan 29 – 3r – 1 – (cont: galena industrial press) – mf#887626 – us WHS [071]
Galena sentinel – Galena IL. 1843 sep 2, 1844 nov 23, dec 7-14 – 2r – 1 – mf#846231 – us WHS [071]
Galenian – Galena IL. [1834 mar 14-1836 sep 20] – 1r – 1 – (cont: miners' journal; cont by: galena democrat) – mf#870122 – us WHS [071]
Galeno *see* Opera medica
Galeno ilustrado : avicena explicado y doctores defendidos / Lopez Cornejo, A – Sevilla, 1699 – 6mf – 9 – sp Cultura [610]
Galenus *see* De placitis hippocratis et platonis libri novem
A galeria – Rio de Janeiro, RJ: Typ do Diario de N L Vianna, 15 abr-30 maio 1945 – (= ser Ps 19) – mf#P14,04,29 – bl Biblioteca [079]
Galeria cearense – Fortaleza, CE: Typ Universal, Lith Cearense, 29 set-nov 1895; jan-fev 1896; 10 abr 1897 – (= ser Ps 19) – mf#60A,01,11 n01 – bl Biblioteca [079]
Galeria dos presidentes de sao paulo / Egas, Eugenio – Sao Paulo, Brazil. v1-3. 1926-1927 – 2r – us UF Libraries [972]
Galeria heroica de mexico / Moreno, Pablo C – Torreon, Mexico. 1954 – 1r – us UF Libraries [972]
Galeria nacional de hombres ilustres o notables / Samper, Jose Maria – Bogota, Colombia. 1879 – 1r – us UF Libraries [972]
Galeria puertorriquena / Fernandez Juncos, Manuel – San Juan, Puerto Rico. 1958 – 1r – us UF Libraries [972]
Galeria universitaria. tomo 1. caracas, 1934 / Dominguez, Rafael – Madrid: Razon y Fe, 1936 – 1 – sp Bibl Santa Ana [946]
Galerie chalette catalogue – New York, 1954-1970 – (= ser Art exhibition catalogues on microfiche) – 30 catalogues on 31mf – 9 – £260.00 – (individual titles not listed separately) – uk Chadwyck [700]
Galerie de l'ancienne cour : ou memoires anecdotes pour servir a l'histoire des regnes de louis 14 et de louis 15 – [s l] 1786 [mf ed Hildesheim 1995-98] – 3v on 16mf – 9 – €160.00 – 3-487-26097-2 – gw Olms [944]
Galerie des contemporains illustres : par un homme de rien. 3e livraison (3e du 10e vol) m spontini / Lomenie, Louis de – Paris: A Rene & Cie [1847] [mf ed 1993] – 1r – 1 – mf#pres. film 127 – us Sibley [780]
Galerie historique / Dionne, Narcisse Eutrope – Quebec: Laflamme & Proulx. 8v. 1909-13 [mf ed 1985] – 25mf – 9 – mf#SEM105P497 – cn Bibl Nat [971]
Galerie l'Art francais *see* "L'art francais" presente...[...du nouveau avec marc-aurele fortin, arca]
Galerie Nationale du Canada *see* Exhibition catalogues, 1919-1959
Galerie photographique des eveques de quebec depuis mgr de laval jusqu'a nos jours : dediee a monseigneur c f baillargeon, administrateur du diocese / Livernois, J B – Quebec: [s.n.] 1863 [mf ed 1983] – 1mf – 9 – 0-665-38219-7 – mf#38219 – cn Canadiana [241]
Galeries martiniquaises / Philemon, Cesaire – Fort-de-France, Martinique. 1930 – 1r – us UF Libraries [972]
Galesburg labor news / Galesburg Trade and Labor Assembly – Galesburg IL. v2 n38 [1897 may 29]-1900, 1901/02-1984/v92 n4 [1986 sep 18] [gaps] – 36r – 1 – mf#1289927 – us WHS [331]
Galesburg Trade and Labor Assembly *see* Galesburg labor news
Galesville independent – Galesville WI. 1894 mar 2-1895 may 31, 1895 jun 7-1896 dec 26, 1897 jan 1-1900 sep 7 – 3r – 1 – (cont: independent [galesville, wi: 1891]) – mf#944155 – us WHS [071]
Galesville independent – Galesville WI. 1890 oct 17-1891 feb 20 – 1r – 1 – (cont: independent [galesville, wi]; cont by: independent [galesville, wi: 1891]) – mf#944151 – us WHS [071]
Galesville independent – Galesville WI. [1874 nov5-1887 mar 3], 1876 jan 13-1878 dec 26, 1879 jan 2-1882 jun 8, 1882 jun 15-1885 jun 11, 1885 jun 18-1887 feb 10 – 5r – 1 – (continued by: independent (galesville, wi)) – mf#944144 – us WHS [071]
Galesville journal – Galesville WI. 1870 may 6-1870 dec 30, 1971 jan 1-1973 jan 2 – 2r – 1 – (continued by: trempealeau county record; journal & record) – mf#943513 – us WHS [071]
Galesville republican – Centerville, Ettrick, Galesville, Holmen, Trempealeau WI. 1897 sep 30/1904 oct 27 [scattered]-1995 [gaps] – 61r – 1 – (cont: galesville independent; galesville independent; galesville independent) – mf#1022564 – us WHS [071]
Galesville transcript – Galesville WI. 1860 mar 16-1861 dec 27, 1862 jan 3-1863 nov 27, 1863 dec 4-1865 dec 29, 1866 jan 1-1867 nov 8 – 4r – 1 – (continued by: trempealeau county record and galesville transcript) – mf#943410 – us WHS [071]
Der galgenstrick : eine komoedie in drei aufzuegen / Erler, Otto – Leipzig: H Haessel 1924 [mf ed 1989] – 1r – 1 – (filmed with: das leben des hartwig bruckner / otto hans engstler) – mf#7218 – us Wisconsin U Libr [820]

Galiani, Ferdinando *see*
- Del dialetto napoletano
- Vocabolario delle parole del dialetto napoletano, che piu si scostano dal dialetto toscano

Galib, Ismail *see*
- Takvim-i meskukat-i osmaniye
- Takvim-i meskukat-i selcukiye

Galib, Seyh *see* The divan project
Galibert, Leon *see* L'algerie ancienne et moderne
Galich, Iurii *see* Volchii smekh
Galich, Manuel *see*
- Del panico al ataque
- Pescado indigesto
- Por que lucha guatemala
- Tren amarillo

Galicia, reyno de christo sacramentado, y primogenito de la iglesia entre las gentes : santiago, principe hereditario de este reyno en que sentado a la diestra de su divino primo, maestro, y rev como se lo avia pedido... / Seguin, Paschasio de – en Mexico: En la Imprenta del Nuevo Rezado de dona Maria de Rivera...ano santo de 1750 – (= ser Books on religion...1543/44-c1800: vidas y cultos de santos) – 2v on 2mf – 9 – mf#crl-123 – ne IDC [241]
Galicia. Sejm *see* Stenograficzne sprawozdania
Galiffe, Jacques *see* Italy and its inhabitants
Galiffe, John Barthelemy Gaifre *see*
- Quelques pages d'histoire exacte
- Le refuge italien de geneve

Galiffe, John-Barthelemy-Gaifre *see* La chaine symbolique
Galignani's messenger – Paris. 13 oct 1824-31 mar 1825; 1 oct 1828-31 aug 1831; 1 jun-31 dec 1832; 1834; jan-2 nov 1835; feb 1836-jun 1837; jan-29 sep 1838; 1 jul-30 sep 1839; 12 nov-30 nov 1840; 16 mar 1848-16 mar 1849; 15 apr 1853-jun 1865; 1 jul, 21 sep 1865; 24 nov 1865-jun 1870; 1 jul 1870-1872; 1890-jul 1904 (imperfect) – 91 1/2r – 1 – (aka: daily messenger galignanis messenger (afternoon ed)) – uk British Libr Newspaper [072]
Galii, M *see* Volia [vladivostok: 1919]
Galilaea auf dem oelberg : wohin jesus seine juenger nach der auferstehung beschied / Hofmann, Rudolph Hugo – Leipzig: Alexander Edelmann, 1896 – 1mf – 9 – 0-7905-1996-8 – (incl bibl ref) – mf#1987-1996 – us ATLA [220]
The galilean : or, jesus the world's savior / Lorimer, George Claude – Boston: Silver, Burdett, 1892 – 2mf – 9 – 0-7905-8834-X – mf#1989-2059 – us ATLA [221]
Galilean christianity / Elliott-Binns, L E – S.C.M. Press, 1956 – 9 – $10.00 – us IRC [240]
The galilean gospel / Bruce, Alexander Balmain – Cincinnati: Jennings & Graham [18–?] [mf ed 1985] – (= ser Library of standard religious authors) – 1mf – 9 – 0-8370-2483-8 – mf#1985-0483 – us ATLA [240]
Galilee : sa vie, ses decouvertes et ses travaux / Parchappe, Maximien – Paris: L Hachette 1866 [mf ed 1998] – 1r – 1 – (incl bibl ref) – mf#film mas 28210 – us Harvard [520]
Galilee in the time of christ / Merrill, Selah – Boston: Congregational Publishing Society, c1881 – 1mf – 9 – 0-7905-0377-8 – (incl ind) – mf#1987-0377 – us ATLA [930]
Galilee in the time of christ / Merrill, Selah – London, England. 1891 – 1r – us UF Libraries [939]
Galilee, ses travaux scientifiques et sa condamnation : lecture publique faite devant l'institut-canadien / Dessaulles, L A – Montreal: L'Avenir, 1856 – 1mf – 9 – mf#34082 – cn Canadiana [520]
Galilee-Belfer, Adam *see* The effect of modified pnf trunk strengthening on functional performance in female rowers
Galilei e kant, o, l'esperienza e la critica nella filosofia moderna / Dominicis, Saverio F de – Bologna: N Zanichelli, 1874 – 1mf – 9 – 0-7905-7439-X – (incl bibl ref) – mf#1989-0664 – us ATLA [190]
Galilei, Galileo *see* Dialogues concerning two new sciences
Galilei, Vincenzo *see*
- Dialogo di vincentio galilei nobile fiorentino della musica antica, et della moderna
- Fronimo, dialogo di vincentio galilei...

Galileistudien : historisch-theologische untersuchungen ueber die urtheile der roemischen congregationen im galileiprocess / Grisar, Hartmann – Regensburg; New York: F. Pustet, 1882 – 1mf – 9 – 0-7905-6229-4 – (incl bibl ref) – mf#1988-2229 – us ATLA [210]
Galileo galilei and the roman curia : from authentic sources = Galileo galilei und die roemische curie / Gebler, Karl von – London: C Kegan Paul, 1879 – 1mf – 9 – 0-8370-7294-8 – (incl ind) – mf#1986-1294 – us ATLA [945]
Galindo Herrero, Santiago *see*
- Donoso cortes
- Donoso cortes en la ultima etapa de su vida

Galindo Lena, Carlos *see*
- Hablo de tierra conocida
- Ser en el tiempo

Galinee, Rene de Brehant de *see* Exploration of the great lakes, 1669-1670
Galinier *see* Voyage en abyssinie dans les provinces du tigre
Galische sprachlehre *see* Vergleichungstafeln der europaeischen stamm-sprachen und sued-, west-asiatischer
Galisteo. Ayuntamiento *see*
- Ferias y fiestas de nuestra senora de la asuncion durante...agosto 1962
- Ferias y fiestas de nuestra senora de la asuncion los dias...agosto 1960
- Ferias y fiestas en honor de ntra. sra. de la asuncion. 1971

Galitsye un ir bafelkerung / Rubsztein, Ben Zion – Varshe, Poland. 1923 – 1r – us UF Libraries [939]
Gall, August, Freiherr von *see*
- Altisraelitische kultstaetten
- Die einheitlichkeit des buches daniel

Gall, August, Frieherr von *see* Die herrlichkeit gottes
Gall, Frederick Beckles *see* Gazetteer of bauchi province
Gall, James *see*
- Primeval man unveiled
- Second initiatory cathechism, with paraphrase exercises, and proofs on the lesson system

Gall, Ludwig *see* Meine auswanderung nach den vereinigten staaten in nord-amerika, im fruehjahr 1819 und meine rueckkehr nach der heimath im winter 1820
Gall, M M *see* Maternal and fetal responses to maximal exercise during swimming and cycling
Gallagher, Charles Wesley *see* God revealed, or, nature's best word
Gallagher, Mason *see* The regard due to the virgin mary
Gallagher, Michael *see* Psychological skills training programs of successful division 1 women's swim programs
Galland, Antoine *see*
- Ali-baba
- Reise nach aegypten

Galland, Henri *see* Essai sur les motazelites
Galland, Julien *see* Recueil des rits et ceremonies du pelerinage de la mecque
Gallant, Abraham Nephtali *see* Midrash veha-mishneh
Gallardo, B J *see* Ensayo de una biblioteca espanola de libros raros y curiosos
Gallardo, Bartolome Jose *see*
- Al zurriagazo zurribanda
- Catalogo formado por...de los principales articulos que componen la selecta libreria de d.j. boehl de faber
- Ensayo...biblioteca...libros raros

Gallardo Bonilla, Leandro *see* Descripcion proclama...badajoz...al trono...rey d. fernando 6
Gallardo de Alvarez, Isabel *see*
- Cuentos de la abuelita
- Nuestra senora de fatima

Gallardo Diaz, Fernando *see* Cronicas de ayer
Gallardo, Luis F *see* Despues del brocal
Gallardo y Gomez, Manuela *see* Muchachas en flor. madrid, 1946
Gallardo y Victor, Manuel *see* Documentos historicos sobre las persecuciones sufridas por la masoneria en espana
Gallas, Heidrun *see* Die synthetische evolutionstheorie und ihre kritiker
Gallatin, Albert *see*
- General land office correspondence
- Letters of albert gallatin, on the oregon question
- The papers of albert gallatin
- Suggestions on the banks and currency of the several united states

The gallatin debate / Lipscomb-Griffin – 1872 – 1 – 5.00 – us Southern Baptist [242]
Gallatin first baptist church. gallatin, tennessee : church records – 1878-Sept 1987. 1354p – 1 – 60.93 – us Southern Baptist [242]
[Gallaup de Chastueil] *see* Discours sur les arcs triomphaux dresses en la ville d'aix, a l'heureuse arriv, e de monseigneur le duc de bourgogne, et de monseigneur le duc de berry
Gallay, P *see* Briefe (gcsej6)
A gallegada : folha reaccionaria – Rio de Janeiro, RJ, 04 maio-jun 1883; 05 ago 1886 – (= ser Ps 19) – mf#P05,04,190 – bl Biblioteca [079]
Gallego, Ignacio *see* El problema campesino in andalucia
Gallego, Laura *see* Presencia
Gallego Rojas, Gilberto *see* Tratado de las cosas humildes
Gallego Y Garcia, Tesifonte *see* Cuba por fuera
Gallegos, Anibal *see* Belice mexicano
Gallegos, Gerardo *see* Americas
Os gallegos no brazil : jornal regenerador – Rio de Janeiro, RJ. 28 jul 1886 – (= ser Ps 19) – mf#P17,03,99 – bl Biblioteca [079]
Gallegos Rocafull, Jose Manuel *see*
- Crusade or class war?
- Crusade or class war? the spanish military revolt

Gallegos, Romulo *see*
- Canaima
- Forastero
- Pobre negro

Gallegos Valdes, Luis *see* Panorama de la literatura salvadorena
Gallenga, Antonio Carlo Napoleone *see* Pearl of the antilles
Galleon – 1986 nov-1987 jul – 1r – 1 – mf#1803742 – us WHS [071]
Gallerani, Alessandro *see* Jesus all good
Gallereia petra velikago v imperatorskoi publichnoi biblioteke / Stasov, V – 1903 – 3mf – 9 – mf#R-11257 – ne IDC [947]
Galleria civica d'arte moderna catalogue – Turin, 1939-74 – (= ser Art exhibition catalogues on microfiche) – ca 27mf – 9 – uk Chadwyck [700]
Galleria di pitture dell'...tommaso ruffo, vescovo di palestrina, e di ferrara... / Agnelli, J – Ferrara, 1734 – 4mf – 9 – mf#O-1186 – ne IDC [700]
Galleria Pesaro. Milan *see* Libero andreotti
Gallerie der neuesten reisen, von russen durch russland und fremde laender unternommen / Budberg, Leonhard G von – Zerbst 1832 [mf ed Hildesheim 1995-98] – 2mf – 9 – €60.00 – 3-487-28965-2 – gw Olms [910]
Galleries and cabinets of art in great britain / Waagen, Gustav Friedrich – London 1857 – (= ser 19th c art & architecture) – 6mf – 9 – mf#4.2.813 – uk Chadwyck [700]
Gallery / Afro-American Museum of Detroit – 1983 jan, may, 1984 feb – 1r – 1 – (continued by: gallery (museum of african american history (detroit, mi))) – mf#2535549 – us WHS [071]
The gallery at castle howard / Tatham, Charles Heathcote – London 1811 – (= ser 19th c art & architecture) – 1mf – 9 – mf#4.2.1448 – uk Chadwyck [700]
Gallery of antiquities selected from the british museum / Arundale, Francis – London [1842,43] – (= ser 19th c art & architecture) – 3mf – 9 – mf#4.1.362 – uk Chadwyck [740]
The gallery of british artists illustrated... : from their most popular works / Sherer, John – London [1879,80?] – (= ser 19th c art & architecture) – 6mf – 9 – mf#4.1.213 – uk Chadwyck [700]
The gallery of modern sculpture / Hall, Samuel Carter – London [1849]-54 – (= ser 19th c art & architecture) – 5mf – 9 – mf#4.1.186 – uk Chadwyck [730]
The gallery of modern sculpture / ed by Hall, Samuel Carter – London: G Virtue, [1849]-1854 – (= ser 19th c art & architecture) – 5mf – 9 – mf#4.1.186 – uk Chadwyck [730]
The gallery of pictures painted by benjamin west – [London? 1811] – (= ser 19th c art & architecture) – 1mf – 9 – mf#4.2.1690 – uk Chadwyck [750]
Gallery of the society of painters in watercolours – London 1833 – (= ser 19th c art & architecture) – 1mf – 9 – mf#4.2.1770 – uk Chadwyck [750]
Galletly, Alexander *see* Ancient towers and doorways
Galletti, Johann *see*
- Johann georg august galletti's allgemeine weltkunde
- Reise nach paris im sommer 1808 [achtzehnhundertacht]

Galley – Frankfurt. 1977 dec 20-1983 aug – 1r – 1 – mf#1240238 – us WHS [071]
Galley, Alfred *see* Die busslehre luthers und ihre darstellung in neuster zeit
Galley, Suzi-Lyn *see* The effectiveness of microcurent electrical nerve stimulation (m.e.n.s.) in the treatment of post acute lymphedema in ankle injuries
Galli, Angelo *see* Sull'opportunita delle strade ferrate nello stato pontificio e sui modi per adottarle
Gallia christiana : in provincias ecclesiasticas distributa – Lutetiae Parisiorum: excudebat Johannes-Baptista Coignard. 16v. 1715-1865 [mf ed 1984] – 10r – 1 – (with ind) – mf#SEM35P206 – cn Bibl Nat [240]
Gallia christiana (gc) : opera et studio monachorum congregationis s mauri osb – Parisiis. v1-16. 1739-1865 – 16mf on 726mf – 8 – €1384.00 – (cont by b haureau; vols listed individually) – ne Slangenburg [240]
Gallia Co. Gallipolis *see*
- Bulletin
- Daily tribune
- Gallia gazette
- Gallia times
- Journal
- Journal series
- Tribune

Gallia Co. Vinton *see*
- Gallia republican
- Vinton leader

Gallia deplorata : sive relatio, de luctuoso bello, quod rex christianissimus contra vicinos populos molitur – n.p, 1641 – 1mf – 9 – mf#O-1578 – ne IDC [090]
Gallia gazette / Gallia Co. Gallipolis – (apr 1819-jun 1821) – 1r – 1 – mf#B28766 – us Ohio Hist [071]
Gallia republican / Gallia Co. Vinton – v1 n1. oct 1855-oct 1857 [wkly] – 1r – 1 – mf#B6625 – us Ohio Hist [071]
Gallia times / Gallia Co. Gallipolis – 1-12/1930, 1-12/32, 1/34-12/1946 [wkly] – 7r – 1 – mf#B33843-33849 – us Ohio Hist [071]
Gallia times / Gallia Co. Gallipolis – jun 1913-22, aug 1924-29 [wkly] – 7r – 1 – mf#B10555-10561 – us Ohio Hist [071]
The gallican church : a history of the church of france from the concordat of bologna, a.d. 1516, to the revolution / Jervis, William Henley – London: John Murray, 1872 – 3mf – 9 – 0-7905-4874-7 – (incl bibl ref) – mf#1988-0874 – us ATLA [240]
The gallican church : sketches of church history in france / Lloyd, Julius – London: SPCK, [1879?] – 1mf – 9 – 0-7905-5373-2 – mf#1988-1373 – us ATLA [240]
The gallican church and the revolution : a sequel to the "history of the church of france from the concordat of bologna to the revolution" / Jervis, William Henley – London: Kegan Paul, Trench, 1882 – 2mf – 9 – 0-7905-4875-5 – (incl bibl ref) – mf#1988-0875 – us ATLA [944]
Gallicanism and ultramontanism in catholic europe in the 18th century : foreign correspondence and other documents from the archive of the jansenist archbishops of utrecht, 1723-1807 – [mf ed 2003] – 192mf – 9 – €2500.00 – (fr utrecht archives, the netherlands; with p/g) – mf#mmp108 – ne Moran [241]
Galliculus, Johannes *see*
- Isagoge de composicione cantus
- Libellus de compositione cantus

Gallieni, Joseph Simon *see* Neuf ans a madagascar
Gallieni, Joseph-Simon *see*
- Deux campagnes au soudan francais, 1886-1888
- Rapport d'ensemble sur la pacification, l'organisation et la colonisation de madagascar (octobre 1896 a mars 1899)

Gallienne, Richard le *see* The religion of a literary man (religio scriptoris)
Gallions reach. leipzig, 1928 / Tomlinson, H M; ed by Bayle, Constantino – Madrid: Razon y Fe, 1928 – 9 – sp Bibl Santa Ana [830]
Gallison, John *see* Gallison's reports of cases in the first circuit, 1812-1815
Gallison's reports of cases in the first circuit, 1812-1815 / Gallison, John – 2nd ed. Boston: Little-Brown. v1-2. 1845 (all publ) – (= ser Early federal nominative reports) – 15mf – 9 – $22.50 – mf#LLMC 81-452 – us LLMC [324]
El gallo *see* Denver county miscellaneous newspapers, reel 3
Gallo – v2 n3-v? n? [1969 jan-1980 mar/may] – 1r – 1 – mf#690423 – us WHS [071]
Gallo – Washington DC 1978-81 – 1,5,9 – ISSN: 0016-416X – mf#7947 – us UMI ProQuest [240]
Gallo, A *see* Declaracion breve...y sumaria del valor del oro...
Gallo canto / Marquez Y De La Cerra, Miguel F – Rio Piedras, Puerto Rico. 1972 – 1r – us UF Libraries [972]
Gallo pinto / Carballido Rey, Jose M – La Habana, Cuba. 1965 – 1r – us UF Libraries [972]
Gallo, Ugo *see* Storia della letteratura ispano-americana
Gallois, Durmart le *see* Li romans de durmart le galois
Les gallois en amerique au 12e siecle / Beauvois, Eugene – S.l: s.n, 189- – 1mf – 9 – mf#04067 – cn Canadiana [917]
Gallois, M-Aug *see* L'apocalypse de s jean
Galloway, Alexander *see* Skeletal remains of bambandyanalo
Galloway, Charles B *see* The editor-bishop, linus parker
Galloway, Charles Betts *see*
- Christianity and the american commonwealth
- A circuit of the globe
- The editor-bishop, linus parker
- Great men and great movements
- Modern missions

Galloway express and general advertiser for kirkcudbrightshire and wigtownshire – [Scotland] Castle-Douglas: D Halliday 23 mar-26 sep 1872; 5 jun-14 aug 1873 (wkly) [mf ed 2004] – 1r – 1 – (began on oct 6 1870) – uk Newsplan [072]
Galloway gazette – Newton-Stewart: W M Leslie 1870- (wkly) [mf ed 1999-] – 1 – (absorbed: newton-stewart journal and wigtownshire times; imprint varies) – uk Scotland NatLib [072]
Galloway, George *see*
- The philosophy of religion
- The principles of religious development
- Studies in the philosophy of religion

Galloway, Joseph *see* Political reflections on the late colonial governments

GANAPAKS

Galloway news – Castle Douglas: J H Maxwell Ltd 1931- (wkly) [mf ed 1999-2001] – 1 – (cont: galloway news and kirkudbrightshire advertiser; absorbed: dumfries news; imprint varies) – ISSN: 0016-4178 – uk Scotland NatLib [072]
Galloway post and county advertiser – [Scotland] Stranraer: R Dick 10 feb 1866 & 27 apr, 4 may 1867 (wkly) [mf ed 2004] – 1r – 1 – (ceased in 1879?) – uk Newsplan [072]
Galloway register and stranraer advertiser – [Scotland] Stranraer: H Wylie 1 mar 1836-1 nov 1837 (mthly) [mf ed 2004] – 1r – 1 – (cont by: galloway register and stranraer advertiser) – uk Newsplan [072]
Galloway, Shayne P *see* The use of assessment by wilderness orientation programs
Galloway, Thomas Walton *see* Zoology
Galloway, William Brown *see* The testimony of science to the deluge
Galloway, William Johnson *see* Advanced australia
Gall's newsletter – Kingston, Jamaica. 7 Mar 1895-27 Mar 1899 (imperfect).-d. 14 reels – 1 – uk British Libr Newspaper [072]
Gall-stone surgery : with a report of a successful case of choledochotomy / Armstrong, George E – S.l: s.n, 1895? – 1mf – 9 – mf#40972 – cn Canadiana [617]
Gallup looks at the movies : audience research reports, 1940-1953 – 1979 – 4r – 1 – $520.00 – mf#S1853 – us Scholarly Res [790]
Gallup Organization *see* 1988 presidential election polls
Gallus : oder, roemische scenen aus der zeit augusts: zur genaueren kenntniss des roemischen privatlebens / Becker, Wilhelm Adolph – Berlin: S Calvary 1880-82 [mf ed 1992] – 3v on 3mf [ill] – 9 – 0-524-04858-4 – mf#1990-3420 – us ATLA [930]
Gallus A Koenigsaal *see* Dialogus malegranatum
Gallus, Heidrun *see* Untersuchungen zur anthocyanbiosynthese an genetisch definierten blueten und zellkulturen der dahlie und der sommeraster
Gallus, N *see*
– Catechismvs predigsweise gestelt fuer die kirche zu regenspurg
– Eine dispvtation von mitteldingen vnd von den itzigen verenderungen in kirchen die christlich vnd wol geordent sind
– Waechterstimme nic galli wo vnd in was stuecken vnter dem namen lutheri der augspurgischen confession vnd h schrifft
[Gallus, N] *see* Qvaestio de libero arbitrio, qvatenvs illa qvibvsdam nunc disceptatur in ecclesijs augustanae confessionis
Gallus oheims chronik von reichenau / ed by Barack, K A – Stuttgart: Litterarischer Verein, 1866 – (incl bibl ref and ind) – us Wisconsin U Libr [430]
Gallus oheims chronik von reichenau / ed by Barack, K A – Stuttgart: Litterarischer Verein, 1866 [mf ed 1993] – (= ser Blvs 84) – 246p – 1 – (incl bibl ref and ind) – mf#8470 reel 17 – us Wisconsin U Libr [880]
Galluzzi, Francesco Maria *see* Vida del venerable padre antonio baldinucci
Galluzzi, Tarquinio *see* [In aristotelis libros quinque priores moralium ad nicomachum]
Gallyon, R N *see* Vocabulary of kwara'ae, solomon islands
Galoa ou edongo d'antan / M'Beye, Ogoula – Port-Gentil, Gabon, 1957 – us CRL [960]
Galope de astros / Echevers, Malin De – Guatemala, 1936 – 1r – us UF Libraries [972]
Galoppe D'onquaire, Jean Hyacinthe Adonis *see* Femme de quarante ans
Galpin, Barrie *see* Computers, topic work and young children
Galpin society journal – West Sussex, England 1948+ – 1,5,9 – ISSN: 0072-0127 – mf#3126 – us UMI ProQuest [780]
Galt, Alexander Tilloch *see*
– The canadian tariff
– Church and state
– Civil liberty in lower canada
– The saint lawrence and atlantic railroad
[Galt-] galt herald – CA. 1912- – 70r – 1 – $4200.00 (subs $100y) – mf#C02264 – us Library Micro [071]
[Galt-] galt weekly gazette – CA. 1882-1911 – 7r – 1 – $420.00 – mf#C03225 – us Library Micro [071]
Galt, John *see*
– Letters from the levant
– The life of benjamin west
– The life of lord byron
– The life, studies, and works of benjamin west
– Voyages and travels, in the years 1809, 1810, and 1811
Galt summer carnival : thursday and friday, june 12th and 13th, 1890 – [London ON?: s.n. 1890?] [mf ed 1987] – 1mf – 9 – 0-665-53027-7 – mf#53027 – cn Canadiana [390]
Galtier, P *see* L'eglise et la remission des peches aux premiere siecles
Galton, Arthur *see*
– Church and state in france, 1300-1907
– The message and position of the church of england
– Our attitude towards english roman catholics and the papal court
Galton, Francis *see*
– Bericht eines forschers im tropischen suedafrika
– English men of science
– Hereditary genius
– Natural inheritance
– Vacation tourists and notes of travel in 1860 1861, 1862-3
Galton, John Lincoln *see* Notes of lectures on the book of revelation
Galtruchius, P *see* Philosophiae, ac mathematicae totius...institutio
Galum pyam dipani tika / Mhuin, Sa khan Kuiy to – Ran kun: U Tan Mon Kri mi sa cu Ca pe: Mui jo Ca pe Phran khyi re 1975 [mf ed 1994] – on pt of 1r – 1 – mf#11052 r1716 n1 – us Cornell [959]
Galura – Bandung: Yayasan Penerbit Galura (wkly) [mf ed Honolulu, HI: University of Hawaii at Manoa, University Libraries 1990] – 2r – 1 – (in sundanese; lacking several iss) – mf#mf-6894 seam – us CRL [079]
Galuth / Alpersohn, Marcos – Buenos Ayres, Argentina. 1929 – 1r – us UF Libraries [939]
Galvan, Manuel De Jesus *see*
– Enriquillo
Galvanometer : and its uses / Haskins, Charles Hamilton – New York, NY. 1873 – 1r – us UF Libraries [621]
Galvao, Eduardo Eneas *see* Santos e visagens
Galvao, Francisco *see* Diretrizes do estado novo
Galvao, Henrique *see* Huila
Galvao, Jesus Bello *see* Programacao do ensino e desenvolvimento economico
Galvas pilsetas avize – St Petersburg, Russia 1906-19 [mf ed Norman Ross] – 1 – mf#nrp-1697 – us UMI ProQuest [077]
Galveston daily news – Galveston, Houston TX. 1943 jun 20-21 – 1r – 1 – (cont: news bulletin [houston, tx]; galveston tribune) – mf#4267571 – us WHS [071]
Galveston Historical Foundation *see* Saccarappa
Galveston tri-weekly news – Galveston, Houston TX. 1863 mar2, 16-20,27, apr 17,22-26, aug 7,26, sep 4 – 1r – 1 – (cont: tri-weekly news [houston, tx: 1861]) – mf#4267628 – us WHS [071]
Galvez en la encrucijada / Arriola, Jorge Luis – Mexico City? Mexico. 1961 – 1r – us UF Libraries [972]
Galvez G, Maria Albertina *see* Emblemas nacionales
Galvez, Manuel *see*
– Espana y algunos espanoles
– In a forshtadt
– Miercoles santo
Galvez Y Del Monte, Wenceslao *see* De lo mas hondo
Galvis Salazar, Fernando *see* Jose eusebio caro
Galway advrtiser – Galway, Ireland. 1986-19 dec 1991; 1992; 1993 – 23r – 1 – uk British Libr Newspaper [072]
Galway american – Galway, Ireland. 12 apr 1862-27 jun 1863 – 1r – 1 – uk British Libr Newspaper [072]
The galway american – Galway. Ireland. -w. 12 Apr 1862-27 Jun 1863 – 1r – 1 – uk British Libr Newspaper [072]
Galway express etc – Galway, Ireland. -w. 29 jun 1853-4 sep 1920 – 30 1/2r – 1 – uk British Libr Newspaper [072]
Galway free press – Galway, Ireland. -w. 1832-28 mar 1835 – 1 1/2r – 1 – (publ 1832-28 mar 1835) – uk British Libr Newspaper [072]
Galway independent paper – Galway, Ireland. -w. 1829-31 mar 1832 – 3 1/4r – 1 – uk British Libr Newspaper [072]
Galway Industrial Society *see* Report of the galway industrial society...domestic industry during the famine in connaught
Galway mercury and connaught weekly advertiser – Galway, Ireland. -w. 16 oct 1844-29 jul 1846; 2 sep 1846-10 mar 1860 – 6 1/2r – 1 – uk British Libr Newspaper [072]
Galway observer – Galway, Ireland. 12 jul 1989-30 jun 1991; 9 sep 1991-23 dec 1992 – 6r – 1 – uk British Libr Newspaper [072]
Galway observer – Galway, Ireland. 18 nov, 23 dec 1882; 20 jan, 24 feb, 3 mar-2 jun, 7, 28 jul, 8 sep, 20 oct 1883; 17 may, 7, 14 jun, 19 jul 1884; 8 jun 1889-1921; 11 feb 1922-25 aug 1923; 28 feb 1925-1 oct 1966 (1918 imperfect).-d – 25 1/4r – 1 – uk British Libr Newspaper [072]
Galway packet and connaught advocate – Galway. Ireland. -w. 24 apr 1852-20 dec 1854 – 2r – 1 – uk British Libr Newspaper [072]
Galway patriot – Galway, Ireland. -w. 18 jul 1835-23 oct 1839 – 3r – 1 – uk British Libr Newspaper [072]
Galway pilot and connaught advertiser – Galway, Ireland. -w. 1905-12 oct 1918 – 6r – 1 – uk British Libr Newspaper [072]
Galway press – Galway, Ireland. -w. 17 mar 1860-28 dec 1861 – 1 1/2r – 1 – uk British Libr Newspaper [072]
Galway standard – Galway, Ireland. -w. 21 oct 1842-1 dec 1843 – 1/2r – 1 – uk British Libr Newspaper [072]
Galway vindicator and connaught advertiser – Galway, Ireland. -w. 10 jul 1841-4 nov 1899 (imperfect) – 55 1/2r – 1 – uk British Libr Newspaper [072]
Galway weekly advertiser – Galway, Ireland. -w. 1823-20 may 1843 – 7r – 1 – uk British Libr Newspaper [072]
Galzy, Jeanne *see* George sand
Gama, Annibal *see* D pedro na regencia
Gamache, Sylvie *see* Jouons avec les livres
Gamaliel / Community for Creative Non-violence (Washington, DC) – v2 n2-v4 n1 [summer 1976-spring 1978] – 1r – 1 – mf#645249 – us WHS [071]
Gamarra, Abelardo M *see* Rasgos de pluma
Gamazo, German *see* Recurso de casacion... ferrocarriles extremenos
Gamba, [J F] *see* Voyage dans la russie meridionale et particulierement dans les provinces situees au del...du caucase, fait depuis 1820-1824
Gamba, Jean *see* Voyage dans la russie meridionale, et particulierement dans les provinces situees au-dela du caucase
Gamba, Pietro *see* Narrative of lord byron's last journey to greece
Gambach, Nesim *see* Januca
Gambalo, Francesco *see* Sacred works of franciscus gambalo
Gambarini, Elisabetta de *see* Lessons for the harpsichord
Gambart, Ernest *see* On piracy of artistic copyright
Gamber, Stanislas *see* Le livre de la genese dans la poesie latine au 5me siecle
Gamberti, D *see* L'idea di un prencipe et eroe christiano in francesco i d'este...
Gambetta, C *see* Selected characteristics of u.s. wheelchair basketball players, their sport participation and their attitudes towards their coaches
Gambetta, Leon M *see* Discours et plaidoyers politiques
Gambia *see* Statistical summary 1964-1968
The gambia : a physical, regional and economic geography / Jarrett, Harold Reginald – London, 1947 – 1 – us CRL [960]
Gambia, annual departmental reports relating to the... 1881-1966 – (= ser Annual departmental reports relating to african countries prior to independence) – 31r – 1 – (with guide. int by d c dorward) – mf#97077 – uk Microform Academic [960]
The gambia echo – Bathurst, Gambia: The Gambia Echo Newspaper Syndicate, dec 1 1947; jan 12, mar 15-22 1948 – us CRL [079]
Gambia, government publications relating to the... 1822-1965 – (= ser Government publications relating to african countries prior to independence) – 71r – 1 – (int by d c dorward and alan butler) – mf#96979 – uk Microform Academic [960]
The gambia outlook – Bathurst: Senegambian Press, mar 1961-jan 1962 – us CRL [079]
Gambia Peoples' Party *see* Gambia peoples' party (gpp) manifesto
Gambia peoples' party (gpp) manifesto / Gambia Peoples' Party – [Banjul, Gambia]: The Party, [1987] (mf ed 1995) – 1mf – 9 – mf#FSN-019,574 – us NY Public [960]
Gambier-Parry, T R *see* The colbertine breviary, vol 1-2 (hbs43-44)
Gambir vajirasar : tipannasadhikadvisatagatha / Qaem, Brah Nanapavaravijja – Bhnam Ben: Buddhasasana Pandity 2494 [1952] [mf ed New Haven CT: SEAsia Coll, Yale Uni Library 1992] – 1r with other items – 1 – (in khmer & pali [in khmer script]; added t.p. in french: vajirasara / traduit du pali en cambodgien par preas sirisammativong em) – mf#mf-10392/3 – us CRL [280]
Gambla, Michael *see* Einfluesse impliziter eignungstheorien auf die beobachtungsgenauigkeit in assessment-centern
Gamble, D P *see* Wolof-english dictionary
Gamble, John *see*
– The spiritual sequence of the bible
– A view of society and manners, in the north of ireland, in the summer and autumn of 1812
– Views of society and manners in the north of ireland
Gamble mansion / Sponenbarger, Lilliam B – s.l, s.l? 193-? – 1r – us UF Libraries [978]
Gamble, W H *see* Trinidad
Gambler, Kl *see*
– Das sakramentar von jena
– Sakramentartypen
– Wege zum urgregorianum
Gambling / Walters, W – London, England. 18– – 1r – us UF Libraries [360]
Gambling behavior among college student-athletes, non-athletes, and former athletes / Bourn, Drew F – 1998 – 2mf – 9 – $8.00 – mf#PSY 2009 – us Kinesology [150]
Gamboa road gang / Beleno C, Joaquin – Panama, 1960 – 1r – us UF Libraries [972]
Gamboian, Nancy *see* The use of somatic training to improve pelvit tilt and lumbar lordosis alignment during quiet stance and dynamic dance movement
Gambrall, Theodore Charles *see*
– Church life in colonial maryland
– Studies in the civil, social and ecclesiastical history of early maryland
Gambrell, James Bruton *see*
– The baptist general convention and its work
– Recollections of confederate scout service
– Ten years in texas
The game birds of india, burmah and ceylon / Hume, Allan O & Marshall, C H T – 24mf – 15 – $480.00 – us UMI ProQuest [590]
The game birds of manitoba / Atkinson, George E – Winnipeg: Manitoba Free Press, 1898 – 1mf – 9 – mf#30243 – cn Canadiana [639]
Game of billiards and how to play it / Roberts, John – London, England. 1913 – 1r – us UF Libraries [790]
Gamersfelder, Hans *see* Der gantz psalter davids
Gamertsfelder, Solomon Jacob *see*
– A bible study on prayer
– Systematic theology
Games and puzzles – Luton, England 1972-80 – 1,5,9 – mf#9847 – us UMI ProQuest [790]
Games for the playground, home, school and gymnasium / Bancroft, Jessie Hubbell – New York, NY. 1909 – 1r – us UF Libraries [790]
Gamewell, Mary Ninde *see*
– The gateway to china
– New life currents in china
Gamez, A *see*
– Censura sencilla del papel que publico en esta corte el reverendo fray buenaventura angeleres
– Discurso del cometa inocente...
– Discurso filosofico, medico e historial...en defensa de la medicina dogmatica y su sangria...
Gamin de paris / Bayard, Jean-Francois-Alfred – Paris, France. 1837 – 1r – us UF Libraries [440]
Gamio, Manuel *see* Forjando patria (por nacionalismo)
Gamio, Manuel [comp] *see* The mexican immigrant
Det gamle groenlands nye perlustration / Egede, H – Kobenhavn. v54. 1741 – 4mf – 9 – mf#N-196 – ne IDC [919]
Gamlingay 1602-1950 – 12mf – 9 – £16.00 – uk CambsFHS [929]
Gammack, James *see* Lecture on the hagiology and parochial dedications of scotland
Gammage, Kimberley L *see* Validation of the revised exercise motivation questionnaire and examination of the relationship between motivation and adherence
Gamma-rays from aluminum due to proton bombardment / Plain, Gilbert John et al – [Lancaster PA] 1940 [mf ed 1987] – 1r – 1 – (repr fr: physical review, v57 n3 (1 feb 1940). incl bibl ref) – mf#9469 – us Wisconsin U Libr [530]
Gammell, William *see* A history of american baptist missions in asia, africa, europe and north america
Gammon, Samuel R *see* The evangelical invasion of brazil
Gammon Theological Seminary *see*
– Catalog...of gammon theological seminary
– Catalogue of gammon school of theology
– Circular of the gammon school of theology
– Quarterly bulletin
Gamnit nayopay / Gnac Gi Len – Bhnam Ben: Ron Bumb Ma Yoen 2515 [1971]- [mf ed 1990] – 1r with other items – 1 – (in khmer) – mf#mf-10289 seam reel 120/3 [§] – us CRL [320]
Gamp tree – v1-6 n4 [1972 jan/feb/mar-1977 oct/nov/dec] – 1r – 1 – mf#368291 – us WHS [071]
Gams, Pius Bonifatius *see* Die kirchengeschichte von spanien
Gan Sukhum *see* Cas min dan dum
Ganaches / Sardou, Victorien – Paris, France. 1863 – 1r – us UF Libraries [440]
Ganado karakul / Carbonero Bravo, D – Madrid: Mag, 1954 – 1 – sp Bibl Santa Ana [946]
Ganado lanar / Junta Provincial de Fomento Pecuario. Badajoz – Badajoz: Tip. La Minerva Extremena, 1963. Publ. no 14 – sp Bibl Santa Ana [946]
Ganado news bulletin – 1947-48 – (= ser American indian periodicals... 2) – 1mf – 9 – $95.00 – us UPA [305]
Ganado porcino extremeno / Calles Mariscal, Juan & Calles Mariscal, Alfredo – Madrid: Artes Graficas J. San Martin, 1946 – sp Bibl Santa Ana [946]
Ganado today / College of Ganado – 1976 winter [v1 n2], 1977 winter [v2 n3] – 1r – 1 – mf#819196 – us WHS [378]
Ganan Gonzalez, Felix *see* Ejemplos de ortografia espanola...
Ganapaks prajadhipateyy 1972-73 / Cau Sau – Bhnam Ben: Ron Bumb Khmaer 1974 [mf ed 1990] – 1r with other items – 1 – (in khmer) – mf#mf-10289 seam reel 128/7 [§] – us CRL [320]

Ganapati Lakshmana *see* Essay on the promotion of domestic reform among the natives of india

Ganavarta – Calcutta, India. 1957-62 – 6r – 1 – us L of C Photodup [079]

Gandara, F *see* Armas i triunfos...de los hijos de galicia...

Gandara, Raul *see* Padre damian

Gandara, Salvador de la *see* Carta, que sobre la vida

Gandarias de Colmenares, Eusebio *see* Resena de varios...minas de valdelayegua

Gandavo, Henricus A *see* Summa theologica ab hieronymus scarpario

Gandee, B F *see* The artist

Ganderias, Perfecto *see* Tardes de la quinta... cristiano

Gandersheimer kreisblatt – Bad Gandersheim DE, 1983 1 jun- – 5r/yr – 1 – gw Misc Inst [074]

Die gandersheimer reimchronik des priesters eberhard / ed by Wolff, Ludwig – Halle/S: Niemeyer Verlag, 1927 [mf ed 1993] – (= ser Altdeutsche textbibliothek 25) – xlii/79p – 1 – mf#8193 reel 3 – us Wisconsin U Libr [430]

Gandhi : world citizen / Lester, Muriel – Allahabad: Kitab Mahal, 1945 – (= ser Samp: indian books) – us CRL [920]

Gandhi, a biographical study / Sen, Ela – Calcutta: Susil Gupta, 1945 – (= ser Samp: indian books) – us CRL [920]

Gandhi against fascism / ed by Chander, Jag Parvesh – Lahore: Free India Publications, [1938] – (= ser Samp: indian books) – us CRL [954]

Gandhi and anarchy / Sankaran Nair, Chettur – 2nd ed Madras: Tagore, (1922) 262p 1 reel.1247 – 1 – us Wisconsin U Libr [335]

Gandhi and gandhism / Gupta, Nagendranatha – Bombay: Hind Kitabs, 1945 – (= ser Samp: indian books) – (foreword by k natarajan) – us CRL [954]

Gandhi and gandhism, a study / Pattabhi Sitaramayya, Bhogaraju – Allahabad: Kitabistan, 1942- – (= ser Samp: indian books) – us CRL [954]

Gandhi and non-violent resistance : the non-co-operation movement of india: gleanings from the american press / Watson, Blanche [comp] – Madras: Ganesh & Co, 1923 – (= ser Samp: indian books) – us CRL [321]

Gandhi and stalin : two signs at the world's crossroads / Fischer, Louis – London: Victor Gollancz Ltd, 1948 – (= ser Samp: indian books) – us CRL [327]

A gandhi anthology / Desai, Valaji Govindaji [comp] – Allahabad: Navajivan Pub House, 1952 – (= ser Samp: indian books) – us CRL [954]

Gandhi as i know him / Yajnik, Indulal Kanaiyalal – Delhi: Danish Mahal, 1943 – (= ser Samp: indian books) – us CRL [920]

Gandhi, Dhiren *see* Prayer and other sketches of mahatma gandhi

Gandhi era in world politics / Krishnamurti, Y G – Bombay: Popular Book Depot, 1943 – (= ser Samp: indian books) – (foreword by s radhakrishnan) – us CRL [327]

Gandhi, fighter without a sword / Eaton, Jeanette – New York: William Morrow & Co, 1950 – (= ser Samp: indian books) – (ill by ralph ray) – us CRL [920]

Gandhi, Mahatma *see*
- Bapu's letters
- Bapu's letters to mira, 1924-1948
- Basic education
- Cent per cent swadeshi
- Delhi diary
- Diet and diet reform
- Drink, drugs and gambling
- Economics of khadi
- Ethical religion
- Ethics of fasting
- Food shortage and agriculture
- For pacifists
- Freedom's battle
- From yeravda mandir
- Gandhigrams
- Gandhiji's correspondence with the government, 1942-44
- Gita the mother
- The good life
- A guide to health
- Hindu dharma
- India of my dreams
- Indian home rule
- The indian states' problem
- Key to health
- Mahatma gandhi at work
- Mahatma gandhi, his own story
- Mohan-mala
- My early life, 1869-1914
- The nation's voice
- Non-violence in peace and war
- Ramanama
- Rebuilding our villages
- Satyagraha
- Satyagraha in south africa
- Selected letters
- Selected writings of mahatma gandhi
- Self-restraint v self-indulgence
- The story of my experiments with truth
- To a gandhian capitalist
- To the hindus and muslims
- To the princes and their people
- To the students
- Towards new education
- Towards non-violent socialism
- The unseen power
- The wheel of fortune
- The wisdom of gandhi
- Women and social injustice
- Young india, 1924-1926

Gandhi, Manmohan Purushottam *see* How to compete with foreign cloth

Gandhi marg : [english ed] – New Delhi, India 1957-77 – 1,5,9 – ISSN: 0016-4437 – mf#8043 – us UMI ProQuest [079]

Gandhi Memorial Leprosy Foundation *see* Report of the drug screening trials of ayurvedic therapy in the cure of leprosy

Gandhi memorial peace number / ed by Roy, Kshitis – Santiniketan: Visva-Bharati, 1949 – (= ser Samp: indian books) – us CRL [954]

Gandhi, Namita *see* Effect of an exercise program on quality of life of women with fibromyalgia

Gandhi sahitya suci = Gandhiana: a bibliography of gandhian literature / Desapande, Panduranga Ganesa [comp] – Ahmedabad: Navajivan Pub House, 1948 – (= ser Samp: indian books) – us CRL [010]

Gandhi, tagore, and nehru / Kripalani, Krishna – Bombay: Hind Kitabs, 1947 – (= ser Samp: indian books) – us CRL [954]

Gandhi, the apostle : his trial and his message / Muzumdar, Haridas Thakordas – Chicago: Universal Pub Co, 1923 – (= ser Samp: indian books) – us CRL [920]

Gandhi, the holy man / Fuloep-Miller, Rene – London; New York: GP Putnam & Sons, 1931 – (= ser Samp: indian books) – (trans fr german by f s flint and d f tait) – us CRL [920]

Gandhi, the man of destiny : a passion play / Thadani, T V – [Karachi: TV Thadani], 1930 – (= ser Samp: indian books) – us CRL [820]

Gandhi, the master / Munshi, Kanaiyalal Maneklal – Delhi: Rajkamal Publications, 1948 – (= ser Samp: indian books) – us CRL [920]

Gandhi, the statesman / Kripalani, Jiwatram Bhagwandas – Delhi: Ranjit Printers & Publishers, 1951 – (= ser Samp: indian books) – us CRL [920]

Gandhi triumphant! : the inside story of the historic fast / Muzumdar, Haridas Thakordas – New York: Universal Pub Co, 1939 – (= ser Samp: indian books) – us CRL [954]

Gandhi versus the empire / Muzumdar, Haridas Thakordas – New York: Universal Pub Co, c1932 – (= ser Samp: indian books) – (foreword by will durant) – us CRL [327]

Gandhi, Virchand Raghavji *see* The karma philosophy

Gandhian economic thought / Kumarappa, Jagadisacandra – Bombay: Vora & Co Publishers, 1951 – (= ser Samp: indian books) – us CRL [330]

The gandhian economy and other essays / Kumarappa, Joseph Cornelius – Wardha: All India Village Industries Association, 1949 – (= ser Samp: indian books) – us CRL [330]

Gandhian outlook and techniques – New Delhi: Ministry of Education, Govt of India, 1953 – (= ser Samp: indian books) – (foreword by maulana abul kalam azad) – us CRL [320]

The gandhian plan of economic development for india / Shriman Narayan – Bombay: Padma Publications, 1944 – (= ser Samp: indian books) – (foreword by mahatma gandhi) – us CRL [330]

Gandhian plan reaffirmed / Shriman Narayan – Bombay: Padma Publications, 1948 – (= ser Samp: indian books) – (foreword by rajendra prasad) – us CRL [330]

The gandhian way / Kripalani, Jiwatram Bhagwandas – Bombay: Vora & Co Publishers, 1938 – (= ser Samp: indian books) – us CRL [320]

Gandhigrams / Tikekar, S R [comp] – Bombay: Hind Kitabs, 1947 – (= ser Samp: indian books) – us CRL [080]

Gandhiji as we know him : by seventeen contributors / ed by Shukla, Chandrashanker – Bombay: Vora & Co, 1945 – (= ser Samp: indian books) – (foreword by sarojini naidu) – us CRL [920]

Gandhiji, his life and work / ed by Tendulkar, D G et al – Bombay: Karnatak Pub House, [1944] – (= ser Samp: indian books) – us CRL [920]

Gandhiji in indian villages / Desai, Mahadev Haribhai – Madras: S Ganesan, 1927 – (= ser Samp: indian books) – us CRL [954]

Gandhiji's correspondence with the government, 1942-44 / Gandhi, Mahatma – Ahmedabad: Navajivan Pub House, 1945 – (= ser Samp: indian books) – us CRL [954]

Gandhism : an analysis / Spratt, Philip – Madras: Huxley Press, 1939 – (= ser Samp: indian books) – us CRL [320]

Gandhism : a socialistic approach / Agarwala, Amar Narain – Allahabad: Kitab-Mahal, [1944?] – (= ser Samp: indian books) – us CRL [320]

Gandhism, nationalism, socialism / Roy, Manabendra Nath – Calcutta: Bengal Radical Club, [1940] – (= ser Samp: indian books) – (int by benoyendra nath banerjee) – us CRL [320]

Gandhism reconsidered / Dantwala, Mohanlal Lalloobhai – Bombay: Padma Publications, 1945 – (= ser Samp: indian books) – us CRL [320]

Gandhism versus socialism / Gregg, Richard Bartlett – New York: John Day Co, c1932 – (= ser Samp: indian books) – us CRL [320]

Gandia, Enrique De *see* Don ramiro en america

Gandia, Enrique de *see* La revision de la historia argentina

Gandier, Alfred *see* A sermon for the new year

Gandini, Francesco *see* Itineraire de l'europe

Gandolphy, Peter *see*
- Congratulatory letter to the rev herbert marsh...
- Second letter to the rev herbert marsh

Gandy, Joseph *see* Designs for cottages, cottage farms

Gandy, Joseph Michael *see* The rural architect

Gandy, Michael *see* Architectural illustrations of windsor castle

Gan-eden / Hurlbert, William Henry – New York, NY. 1854 – 1r – us UF Libraries [972]

Ganesa : a monograph on the elephant-faced god / Getty, Alice – Oxford: Clarendon Press, 1936 – (= ser Samp: indian books) – (int by alfred foucher) – us CRL [280]

Ganeshan, Vridhagiri *see* Das indienbild deutscher dichter um 1900

Der gang der handlung in goethes faust / Harnack, Otto – Darmstadt: A Bergstraesser [1902?] [mf ed 1990] – 1r – 1 – (filmed with: goethe's faust / albert gruen) – mf#7347 – us Wisconsin U Libr [430]

Der gang der kirche in lebensbildern / Kahnis, Karl Friedrich August – Leipzig: Doerffling und Franke, 1881 – 2mf – 9 – 0-7905-4881-X – mf#1988-0881 – us ATLA [240]

Ein gang durch die christliche welt : studien ueber die entwicklung des christlichen geistes in briefen an einen laien / Lang, Heinrich – Berlin: G Reimer, 1859 – 1mf – 9 – 0-7905-9295-9 – mf#1989-2520 – us ATLA [240]

Der gang nach emmaus *see* Fruehlingssturm / charfreitag / der gang nach emmaus / pfingsten in weimar

Gang Pun Jhyan *see*
- Khdam kamadeb
- Panji lohit
- Qantat bhloen knun tuan

Ganga dass : a tale of hindustan / Calkins, Harvey Reeves – New York: Abingdon Press, [c1917] [mf ed 1995] – (= ser Yale coll; Books on christian stewardship) – 79p – 1 – 0-524-09905-7 – mf#1995-0905 – us ATLA [954]

Gangaa – Nigeria: s.n., apr 1979 – 1r – 1 – us CRL [079]

Gangadin *see* Europeans' guide and medical companion in India

The gangas of talkad: a monograph on the history of mysore from the fourth to the close of the eleventh century / Krishna Rao, MV – Madras: B.G. Paul, 1936.306p. ill. Bibliography – 1 – us Wisconsin U Libr [954]

Gangauf, Theodor *see*
- Des heiligen augustinus speculative lehre von gott dem dreieinigen
- Metaphysische psychologie des heiligen augustinus

Gang-box news / International Brotherhood of Electrical Workers – 1979 feb-1986 jul – 1r – 1 – mf#1477260 – us WHS [331]

Gange, C du *see* Descriptio magnae ecclesiae seu sanctae sophiae (cbh11,2)

Ganghofer, Ludwig *see*
- Gewitter im mai; der besondere
- Das grosse jagen
- Lebenslauf eines optimisten

Gangoly, Ordhendra Coomar *see*
- The art of java
- Indian architecture
- Love-poems in hindi
- Southern indian bronzes (first series)

Gangooly, J C *see* Life and religion of the hindoos

Gangooly, Joguth Chunder *see* Life and religion of the hindoos

Gangopadhyaya, Tarakanatha *see* The brothers

Gangstead, Sandra K *see*
- Effects of instruction on the analytical proficiency of physical education majors in fundamental sport skills analysis
- The effects of two instructional conditions on sport skill specific analytic proficiency of physical education majors

Ganguli, Taraknath *see*
- Svarnalata

Ganguly, Dhirendra Chandra *see* The eastern calukyas

Ganguly, Manomohan *see* Handbook to the sculptures in the museum of the bangiya sahitya parishad

Ganguly, Nalin C *see* Raja ram mohun roy

[Ganichev, S I et al] *see* Ot grivny do rublia

Ganilh, Charles *see* De la contre-revolution en france

Ganit vidya : thnak di 3 / Quy van thun et al – Bhnam Ben: Krasuan Qap Ram Jati 1974 [mf ed 1990] – 1r with other items – 1 – (in khmer) – mf#mf-10289 seam reel 120/8 [§] – us CRL [510]

Ganjinah-i ma'arif – Tabriz. sal-i 1, shumarah-i 1-8. 3 rabi al-avval 1341-8 zu'l qa'dah 1341 [24 oct 1922-23 jun 1923] – 1r – 1 – $90.00 – us MEDOC [079]

Gann – 1907-84 – 1,5,9 – (cont by: japanese journal of cancer research: gann) – ISSN: 0016-450X – mf#7995.02 – us UMI ProQuest [616]

Gann : the japanese journal of cancer research – v1-60. 1907-69 – 1 – us AMS Press [616]

Gann, Lewis H *see*
- Birth of a plural society
- History of northern rhodesia, early days to 1953
- Huggins of rhodesia

Gann, Thomas William Francis *see* Discoveries and adventures in central america

Gannett, Betty *see* Papers of betty gannett, 1929-1970

Gannett, Ezra Stiles *see* The faith of the unitarian christian explained, justified and distinguished

Gannett, William Channing *see*
- The childhood of jesus
- Ezra stiles gannett, unitarian minister in boston, 1824-1871
- Francis david, founder and martyr of unitarianism in hungary

Gannon, Edward K *see* Correlation of abdominal accessory expiratory muscle strength and pulmonary functions in older adults

Gano, Darwin Curtis *see* Commercial law

Gano, John *see*
- Evidence of george washington's religion
- Memoirs

Gano, W *see* Florida

The ganoid fishes of the british carboniferous formations / Traquair, R H – London, 1877-1914. 7pts – 8mf – 9 – mf#G-218 – ne IDC [590]

Ganong, William Francis *see* The itinerary of jacques cartier's first voyage

Gano-Overway, Lori A *see* Goal perspectives and their relationship to beliefs, affective responses and coping strategies among african and anglo american athletes

Gans, David Ben Solomon *see* Zemah dawid

Die gans der fuchs : drei dutzend fabeln / La Fontaine et al – Zuerich: Diogenes Verlag, c1957 – 1r – 1 – (german text in part translated from french, russian and spanish) – us Wisconsin U Libr [390]

Gans, Edgar Hilary *see* Digest of maryland statutes and decisions on criminal law

Ganschinietz, R *see* Hippolytos' capitel gegen die magier

Ganschinietz, Richard *see* Hippolytos' capitel gegen die magier, refut. haer. 4 28-42

Ganshof Van Der Meersch, W J *see* Fin de la souverainete belge au congo

Ganss, Henry George *see* Mariolatry, new phases of an old fallacy

Gansser, Georgine *see* Enzymologische aspekte der homofermativen milchsaeuregaerung in mutans-streptokokken

Gantillon, Simon *see* Maya

Gants jaunes / Bayard, Jean-Francois-Alfred – Paris, France. 1835 – 1r – us UF Libraries [440]

Gantz, J L *see* Development of the attitudes toward the disabled in physical education scale

Der gantz psalter davids : in gesangs weyse gestelt durch hansen gamersfelder. also, das sich die psalmen alle durch aus, in mannigfeltiger melodey hernach angezeicht, fein un[d] lieblich singen lassen... / Gamersfelder, Hans – Nuernberg: Berg & Neuber 1542 – (= ser Hqab. literatur des 16. jahrh.) – 4mf – 9 – €50.00 – mf#1542b – gw Fischer [780]

Der gantz psalter davids: mit angehenckten lobgesengen des alten und neuen testaments... : nach der gemeinen alten kirchischen latinischen edition auff verss und reimweiss gar trewlich, verstendlich und geschicklich gestellet durch rutgerum edingium – Coelln: Cholinus 1574 – (= ser Hqab. literatur des 16. jahrh.) – 6mf – 9 – €70.00 – mf#1574b – gw Fischer [780]

Die gantze bibel : das ist: alle buecher allts unnd news testaments...verteutschet... / Jud, L – Zuerich: Christoffel Froschouer, [1539]-1540 – 15mf – 9 – mf#PBU-490 – ne IDC [240]

Der gantze psalter : das ist alle psalmen dess koeniglichen propheten dauids, an der zahl 150.: mit hoechstem fleiss vnnd treuwen verteutschet – Franckfurt am Mayn: Bassaeus 1582 – (= ser Hqab. literatur des 16. jahrh.) – 5mf – 9 – €60.00 – mf#1582a – gw Fischer [780]

GARCIA

Der gantze psalter davids : darneben alle andern psalmen vnd geistliche lieder, im alten vnd newen testament: gesangsweise gefasset / Spangenberg, Cyriacus – Frankfurt: Rab 1582 – (= ser Hqab. literatur des 16. jahrh.) – 7mf – 9 – €75.00 – mf#1582c – gw Fischer [780]
Ganz neue biblische bilder-ergoetzung dem alter und der jugend zur beschauung und erbauung, aus dem alten testament angestellet und mitgetheilet – Nuernberg, n d – 5mf – 9 – mf#0-1136 – ne IDC [700]
Die ganze aesthetik in einer nuss : oder, neologisches woerterbuch / Schoenaich, Christoph Otto, Freiherr von; ed by Koester, Albert – Berlin: B Behr (E Bock), 1900 [1993] – (= ser Deutsche litteraturdenkmale des 18. und 19. jahrhunderts 70-81) – xxviii/612p – 1 – (incl bibl ref and ind; with reprod of original t p) – mf#8676 reel 5 – us Wisconsin U Libr [430]
Der ganze prolog des johannesevangeliums in satzfolge und -gliederung woertliches citat aus jesaia : eine studie des christusbildes nach der aneinanderhaltung beider testamente / Steinfuehrer, W – Leipzig: Doerffling & Franke, 1904 – 1mf – 9 – 0-8370-9307-4 – mf#1986-3307 – us ATLA [226]
Gao auditing standards : 1972, 1981 and 1988 revisions / General Accounting Office – Washington: GAO. 1v ea. 1972; 1881; 1988 – 4mf – 9 – $6.00 – mf#llmc 90-375A – us LLMC [336]
Gao, Jiaping *see* In vivo insulin action on whole body and individual tissues in obese shhf/ mcc-cp rats with or without acute exercise
Gao review – Washington DC 1972-87 – 1,5,9 – (cont by: gao journal) – ISSN: 0016-3414 – mf#7911 – us UMI ProQuest [350]
Gaona, Juan de *see* Coloquio de la paz y tranquilidad christiana en lengua mexicana: con licencia, y priuilegio
Gaos, Jose *see* Pensamiento espanol
Gap hill baptist church. pickens county. south carolina : church records – 1954-72 – 1 – 7.38 – us Southern Baptist [242]
Gaponenko, L S et al *see* Revoliutsionnoe dvizhenie v rossii posle sverzheniia samoderzhaviia
Gaposchkin, Cecilia Helena Payne *see*
– Radio talks from the harvard observatory
– The universe of stars
– Variable stars
Gaposchkin, Sergei *see* Variable stars
Gappmaier, Eduard *see* Maximal exercise testing and aerobic training in multiple sclerosis
Gapura : almanak nasional – Djakarta, 1951-1956 – 44mf – 9 – mf#SE-660 – ne IDC [959]
Gapura : madjalah bulanan gema kehidupan kota / Pemerintah daerah Kotamadya – Surabaja, 1968-1972 – 47mf – 9 – mf#SE-1483 – ne IDC [959]
Gapway baptist church. marion county. south carolina : church records – 1833-67 – 1 – 5.92 – us Southern Baptist [242]
Garabedian, Richard A *see* The relationship between health-promoting attitudes and exercise in the elderly
Garach, Hematlal *see* Auditor switching
Garage and motor agent – London, England 1950-54 – 1 – ISSN: 0016-4526 – mf#504 – us UMI ProQuest [380]
Garahan, Melbourne *see* Stiffs
Garant, J-Honorat *see* Bio-bibliographie de monsieur elphege bois
Garantias y los principios sociales en la constitu... / Paz Barnica, Edgardo – Tegucigalpa, Mexico. 1963 – 1r – us UF Libraries [972]
Garasse, F *see*
– La doctrine curieuse des beaux esprits de ce temps...
– Le rabelais reforme par les ministres...
Garat, Dominique-Pierre-Jean *see* Trois romances avec accompagnement de piano ou harpe... oeuvre 5
Garau, F *see*
– El sabio instruido de la gracia...
– El sabio instruido de la naturaleza, en quarenta maximas politicas, y morales...
Garaud, Louis *see* Trois ans a la martinique
Garavini Di Turno, Sadio *see* Diamond river
The garb law : an argument on the pennsylvania garb law in relation to public school teachers / Bucher, George & McGuire, F W – Quarryville, PA: G Bucher, 1908 – 1mf – 9 – 0-524-04710-3 – mf#1990-5062 – us ATLA [344]
Garbage – Gloucester, England 1993-94 – 1,5,9 – ISSN: 1044-3061 – mf#19240 – us UMI ProQuest [333]
Garbalosa, Graziella *see* Narkis
Garbe, Richard *see*
– Indien und das christentum
– The philosophy of ancient india
– Die saamkhya-philosophie
– Samkhya und yoga
Garber, Klaus *see*
– Deutsche dichtung von der aeltesten bis auf die neueste zeit
– Handbuch des personalen gelegenheitsschrifttums in europaeischen bibliotheken und archiven
– Innere geschichte der entwicklung der deutschen national-litteratur
Garber-bund un bershter-bund / Dubnova-Erlikh, Sofiia – Varshe, Poland. 1937 – 1r – us UF Libraries [939]
[Garberville-] southern humbolt life and times – CA. 1987-1991 – 3r – 1 – $180.00 – mf#B05037 – us Library Micro [071]
[Garberville-] star root – CA. 1977-1982 – 1r – 1 – $60.00 – mf#B03226 – us Library Micro [071]
[Garberville-] the redwood record – CA. 1962-91 – 19r – 1 – $1140.00 (subs $50y) – mf#B02265 – us Library Micro [071]
Garbett, Edward *see*
– The bible and its critics
– The dogmatic faith
– God's word written
Garbett, Edward Lacy *see* Rudimentary treatise on...design in architecture
Garbett, G Kingsley *see* Growth and change in a shona ward
Garbett, James *see*
– Church and the age
– Church of england and the church of rome
– Communion of saints
– Diocesan synods and convocation
Garbett, John *see* Church defended
Garborg, Arne *see* Straumdrag
Garborg, Hulda *see* Helenes historie
Garca Calderon, Ventura *see* Semblanzas de america
Garceran, Rafael *see* Falange desde febrero de 1936 al gobierno nacional
Garces y Gonzales, Valeriano *see* Vocabulario descriptivo...diccionario
Garcia A, J Luis *see* Corazon de indio
Garcia, Alf I *see*
– La administracion de sacramentos en toledo despues del cambio de rito
– Edicion tridentina del manual toledano y su incorporacion al ritual romano
– El manual tolendano para la administration de sacramentos a travers de los siglos
Garcia Alonso, Francisco *see* Flores de heroismo. sevilla, 1939
Garcia Alzola, Ernesto *see* Marti va con nosotros
Garcia, Andres Javier *see* Carta edificativa
Garcia Angulo, Efrain *see* Puerto rico
Garcia, Antonio *see* Necrologio de la provincia serafica de cartagena desde su restauracion en 1878, murcia, 1967
Garcia Arias, Luis *see* Las embajadas de don juan antonio de vera y zuniga en italia
Garcia Barcena, Rafael *see*
– Aforismos de luz y caballero
– Responso heroico
Garcia, Bartholome *see* Manual para administrar los santos sacramentos de penitencia, eucharistia, extrema-uncion, y matrimonio
Garcia Bauer, Carlos *see*
– Controversia sobre el territorio de belice
– En el amanecer de una nueva era
Garcia Bermejo, Antonio *see*
– Oracion funebre...exequias
– Oracion funebre...san isidro
Garcia Blanco, Manuel *see* De las andanzas de unamuno por tierras extremenas
Garcia Bote, Eduardo *see* Serie de conferencias explicadas en dicha corporacion durante el curso de 1908-1909 (federacion taquigrafa espanola)
Garcia Calderon K, Manuel *see* La capacidad cambiaria en el derecho internacional privado
Garcia Calderon, Ventura *see*
– Mejores cuentos americanos
– Vale un peru
Garcia Camino, Victor Gerardo *see* Aproximacion a un estudio de antonio hurtado como poeta
Garcia Caminos Burgos, Luis F *see* Aproximacion a un estudio de antonio hurtado como poeta
Garcia Cantero, Gabriel *see* El vinculo del matrimonio...
Garcia Carrasco, Florencio *see*
– Cuaderno de unidades didacticas no 4. naturaleza 1
– Cuaderno de unidades didacticas no 4 naturaleza 2
– Cuaderno de unidades didacticas no 4. vida social 1
– Cuaderno de unidades didacticas no 4 vida social 2
– Cuaderno de unidades didacticas no 4 vida social 3
Garcia Carrasco, Florenico *see* Cuaderno de unidades didacticas no 4 naturaleza 3
Garcia Carrasco, Francisco A *see*
– Cuaderno de unidades didacticas no 4. naturaleza 1
– Cuaderno de unidades didacticas no 4 naturaleza 2
– Cuaderno de unidades didacticas no 4 naturaleza 3
– Cuaderno de unidades didacticas no 4. vida social 1
– Cuaderno de unidades didacticas no 4 vida social 2
– Cuaderno de unidades didacticas no 4 vida social 3
Garcia Carrasco, Francisco A et al *see* Yo soy... lengua espanola curso no 6
Garcia Carrasco, Francisco Andres *see* Extracto de la tesis de sobre la accion cultural de espana en marruecos
Garcia Chuecos, Hector *see*
– Relatos y comentarios sobre temas de historia vene...
– Siglo dieciocho venezolano
Garcia, Clersida *see* A fieldwork study of how young children learn fundamental motor skills and how they progress in the development of striking
Garcia de Atocha, Jose *see*
– Memoria sobre...guadalupe
– Respuestas...d. felipe rosado de belalcazar
Garcia de Diego, Jose A *see* Escarceos de toponimia extremena
Garcia de Diego, V *see* Idea de un principe politico christiano representada en cien empresas
Garcia de la Cuesta, Gregorio *see* Manifiesto que presenta a europa...extremadura
Garcia de la Fuente, P Arturo *see*
– El caso del obispo marcial de merida
– El concilio 3 emeritense
Garcia de la Huerta, Vicente *see*
– Biblioteca militar espanola
– La escena espanola...teatro
– La fe triunfante del amor
– Leccion critica a los lectores de la memoria de cosme damian sobre el teatro
– Leccion critica a los lectores del papel intitulado
– Leccion historica al profesor paredes
– Poesias
– Raquel
Garcia de Medrano *see* La regla y el...santiago del espada
Garcia de paredes / Llano y Persi, Manuel de – 1848 – 9 – sp Bibl Santa Ana [830]
Garcia De Paredes, Carlos M *see* Minotauro
Garcia de Soto, Jesus *see* Nuevas aportaciones al estudio de la necropolis oriental de merida
Garcia Del Rio, Juan *see* Meditaciones colombianas
Garcia Diaz, Manuel *see* Neoclasicos en puerto rico
Garcia Ensenat, Ezequiel *see* Escudo oficial del municipo de la habana
Garcia Espinosa, Juan Manuel *see* Primer director del centro regional de la unesco
Garcia, Fernandez *see* Lexicon scholasticum philosophico-theologicum...
Garcia Figueroa, Fernando *see* Plano-guia-callejero de caceres
Garcia Flores, Juan Felipe (Don Felipe) *see*
– Resumen de la temporada taurina 1968 en extremadura
– Toros en extremadura. amplios detalles de los espectaculos taurinos celebrados en las plazas de esta region durante la temporada de 1969
Garcia Fox, Leonardo *see* Reflejos en el agua
Garcia, Francesc V *see* La armonia del parnas, mes nvmerosa en las poesias varias del atlant del cel poetic
Garcia Garces, Narciso *see*
– Exposicion del dogma catolico
– Titulos y grandezas de maria
Garcia Garcia, A *see* Manuel gercia garrido
Garcia Garcia, Casimiro *see* Elementos de religion. la doctrina de nuestro senor jesucristo
Garcia Garcia, Juan *see* Los benficios del telefono
Garcia Garcia, Rafael *see*
– Arias montano y la politica de felipe 2
– Las "elucidationesin evangelia" de benito arias montano
Garcia Garfalo Y Mesa, Manuel *see* Vida de jose maria heredia en mexico, 1825-1839
Garcia Garofalo Y Mesa, Manuel *see* Marta abreu arencibia y el dr luis estevez y rome
Garcia, Genaro *see* Don juan de palafox y mendoza
Garcia Gil, Manuel *see*
– Carta pastoral
– Carta pastoral. confirmacion
– Pastoral sobre el sacramento de la confirmacion
Garcia Godoy, Federico *see*
– Al margen del plan peynado
– Alma dominicana
– Americanismo literario
– Antologia
– De aqui y de alla
– Guanuma
– Literatura americana de nuestros dias
– Rufinito
Garcia Gomez, Aristides *see* Todo un poco
Garcia Gutierrez, Jesus *see* Apuntes para la historia del origen y desenvolvimiento del regio patronato indiano hasta 1857
Garcia, Helen M *see*
– Authors in miami
– Concerning dade county
Garcia Hernandez, F *see* Tratado de fiebres malignas...
Garcia Hernandez, Manuel *see* Estampas venezolanas
Garcia, Huecos *see* Hector estudios de historia colonial venezolana...
Garcia Icazbalceta, Joaquin *see* Conquista y colonizacion de mejico
Garcia Jalon, Miguel *see* A. por el excmo. sr. conde de montijo...sra. dona mariana enriquez
Garcia Jimeno, Fernando *see*
– Nota del dia
– Ripios. poesias
Garcia Jimeno, Fernando y Jesus *see* El vestido largo
Garcia, John L *see* Journal of creativity in mental health
Garcia, Jose *see* Proyecto...pagar las contribuciones
Garcia, Jose Gabriel *see* Compendio de la historia
Garcia, Juan Francisco *see* Suite de impresiones para piano
Garcia, Juan Justo *see*
– Elementos de aritmetica
– Elementos de aritmetica, algebra y geometria
– Elementos de aritmetica, tomo 1
– Elementos de...logica
– Nuevos elementos de geografia general astronomica
Garcia Kohly, Mario *see*
– Alma cubana a traves de sus poetas
– Politica internacional cubana
Garcia Laguardia, Jorge Mario *see* Antecedentes del seguro social en guatemala
Garcia laso de la vega / Salazar, Antonio – Badajoz: Imprenta Dipt. Provincial, 1963 – sp Bibl Santa Ana [946]
Garcia Lluberes, Alcides *see* Americo lugo
Garcia Lopez, Rafael *see* Manual para el cultivo y beneficio del tabaco en filipinas
Garcia Lorca, Federico *see*
– Alocucion a los actores argentinos. manuscrito
– Apuntes sobre la escenificacion de los romances. manuscrito
– Canciones y otros poemas
– La casa de bernarda alba
– Comedia sin titulo. ms.
– La comedianta. ms.
– Conferencia recital sobre el romancero gitano
– Cuadernillo de miguel picazo, manuscrito
– Libro de poemas
– El maleficio de la mariposa
– Mariana pineda
– Mariana pineda, fragmento manuscrito
– Poemas del cante jondo. fragmento manuscrito
– Poeta en nueva york. manuscrito
– Los suenos de mi prima aurelia
– Titeres de cachiporra
– Yerma
– La zapatera prodigiosa
Garcia, Manuel *see* Depechos legales...tierra santa
Garcia, Marcellan, Jose *see* Catalogo del archivo de musica de la real capilla de palacio
Garcia Mari, Raul *see* Marti
Garcia Marquez, Gabriel *see* Hojarasca
Garcia Matos, Manuel *see*
– L'anthologie du folklore musical d'espague
– Lirica popular de la alta extremadura
Garcia Mejia, Rene *see* Golpe a las 2 am
Garcia, Miguel *see* Campana de portugal... extremadura
Garcia, Miguel Angel *see*
– Anecdotas centroamericanos
– Asamblea nacional constituyente de 1885
Garcia Miranda, Vicenta *see* Flores del valle
Garcia Monge, Eduardo *see* Mis pasatiempos
Garcia Montes Y Angulo, Jose *see* Cuba y su futuro
Garcia Mora, Jose *see* Alegacion en derecho... virgen del puerto...plasencia
Garcia moreno / Berthe, Augustine – Paris, France. v1-2. 1903 – 1r – us UF Libraries [972]
Garcia moreno : un gobernante modelo / Bayle, Constantino – Madrid: Razon y Fe, 1921 – 1 – sp Bibl Santa Ana [350]
Garcia moreno y la instruccion publica por julio tobar donoso / Bayle, Constantino – Madrid: Razon y Fe, 1924 – 1 – sp Bibl Santa Ana [946]
Garcia Morente, Manuel *see* Origenes del nacionalismo espanol
Garcia Morgado, Jose *see* Por los fueros de la verdad
Garcia Nieto, Jose *see* Pregon pronunciado por d jose garcia nieto...
Garcia Oliver, Juan *see* El fascismo internacional y la guerra antifascista espanola
Garcia Oro, J *see* San pedro de alcantara... madrid, 1965
Garcia Pedrosa, Jose R *see* Legislacion social de cuba
Garcia Pelaez, Francisco De Paula *see* Memorias para la historia del antiguo reino
Garcia Perez, Juan *see* La desamortizacion de las propiedades...valencia de alcantara
Garcia Porras-Pita, Armando *see* Materia penal para los estudiantes de derecho pena

Garcia Prada, Carlos *see* Personalidda historica de colombia
Garcia, R *see* A comparison of grip strength in young athletes and non-athletes
Garcia Rios, Miguel A *see* Aquellos tiempos
Garcia Rodriguez, Jose Maria *see* Espanol en la espanola
Garcia Rojo, D *see* Madrid. biblioteca nacional. catalogo de incunables
Garcia Romero de Tejada, Jose *see*
– El libro del jurado
– La psicologia con la ontologia
Garcia Romero de Tejada, Julio *see* La psicologia y su relacion con la ontologia
Garcia Romero, Diego *see* Procedimientos practicos modernos para lafabricacion de vinos en extremadura
Garcia rovira / Gomez L, Efrain – Bogota, Colombia. 1946 – 1r – us UF Libraries [972]
Garcia Rubio, Manuel *see* Sencillos...geografia
Garcia, S Ismael *see* Medio siglo de poesia panamena
Garcia Salas, Jose Maria *see* Parnaso centroamericano
Garcia Salinero, Fernando *see* Una omision en la polemica forneriana
Garcia Samudio, Nicolas *see* Cronica del muy magnifico capitan d gonzalo suare
Garcia Sanchez, Federico *see* Te demu landanus
Garcia Sanchez, Francisco *see*
– El castillo de medellin en la ruta del turismo
– Medellin, ruta de turismo
Garcia Sanchez, Miguel A *see* Proceso de cerete
Garcia Sandoval, Eugenio *see* Excavaciones arqueologicas en la zona de merida
Garcia Santillan, Juan Carlos *see* Legislacion sobre indios del rio de la plata en el siglo 16th. madrid. 1928
Garcia, Susan C *see* Validity of the sit-and-reach test for male and female adolescents
Garcia, Telesforo *see* Navegamtes y descubridores espanoles del mar pacifico. vasco nunez de balboa
Garcia Troncoso, Parmenio Constantino *see* Payeyo garcia troncoso
Garcia Vazquez, Sebastian *see* El pintor eugenio hermoso
Garcia Vega, Lorenzo *see*
– Cetreria del titere
– Suite para la espera
Garcia Vidal, Ceferino *see* Seminario de plasencia, el. apuntes historicos
Garcia y Bellido, Antonio *see*
– El culto a ma-bellona en la espana romana
– El jarro ritual lusitano de la coleccion calzadilla
– Nombres de artistas en la espana romana
Garcia Y Carcia, Jose Antonio *see* Relaciones de los vireyes del...
Garcia y Garcia, Antonio *see* Los manuscritos juridicos medievales de...
Garcia y Garcia, Casimiro *see*
– Programa de 2nd curso de religion (3rd del bachillerato) historia de la iglesia y liturgia 1937-1938
– Programa de religion y moral para los cursos 2nd, 4th, 5th y 6th de bachillerato 1937-1938
Garcia y Romero de Tejada, Jose *see* El libro del jurado, 2nd parte
Garcia-Pelayo, Manuel *see* Derecho constitucional comparado
Garcin de Tassy *see*
– Histoire de la litterature hindoui et hindoustani
– Memoire sur les particularites de la religion musulmane dans l'inde d'apres les ouvrages hindoustanis
Garcini, Maria Del Carmen *see* Antologia del cuento hispanoamericano
Garcke, A *see* Die botanischen ergebnisse der reise seiner koenigl hoheit des prinzen waldemar von preussen in den jahren 1845 und 1846
Garcon, l'addition? / Davanne, J B – Paris, France. 1871 – 1r – us UF Libraries [440]
Garcon sans-souci / Perin, Rene – Paris, France. 1818 – 1r – us UF Libraries [440]
Garconne : junggesellin – Berlin DE, 1930-1932 n20/21 – 1r – 1 – gw Misc Inst [074]
La garconne : roman / Margueritte, Victor – Paris: E Flammarion, c1922 [mf ed 2000] – 1r – 1 – (filmed with: les caves du vatican / andre gide) – mf#10471 – us Wisconsin U Libr [830]
Gard, Anson Albert *see* How to see montreal
Gardane, Ange *see* Ange's von gardane kaiserl franz gesandtschafts-sekretaers tagebuch
[Gardane, P A M de] *see* Journal d'un voyage dans la turquie-d'asie et la perse, fait en 1807 et 1808
Gardariki : ein stufenbuch aus russischem raum / Brandt, Dagmar – 3. aufl. Berlin: Wiking Verlag, 1944 [mf ed 1989] – 937p – 1 – (incl bibl) – mf#7062 – us Wisconsin U Libr [890]
La garde blanche : Organe national, contre-revolutionnaire – n1-8. Paris. mars-avr 1919 – 1 – fr ACRPP [073]
Garde d'haiti / Mccrocklin, James H – Annapolis, MD. 1956 – 1r – us UF Libraries [972]
Garde forestier / Leuven, Adolphe De – Paris, France. 1845? – 1r – us UF Libraries [440]
Le garde mobile – Paris: Leautey, aug 1848 – us CRL [074]
Garde, V *see* Den danske konebaads-expedition til gronlands ostkyst
Gardeleger altmark echo – Gardelegen DE, 1962 3 feb-1967 30 mar – 1r – 1 – gw Misc Inst [074]
Garden – Bronx NY 1977-90 – 1,5,9 – ISSN: 0191-3999 – mf#11785 – us UMI ProQuest [635]
Garden – London, England 1975+ – 1,5,9 – (cont: journal of the royal horticultural society) – ISSN: 0308-5457 – mf#500,01 – us UMI ProQuest [635]
Garden and field – Adelaide, Australia. -m. April 1894-Aug 1901; 12 Sept 1903-Sept 1905 – 5r – 1 – uk British Libr Newspaper [635]
Garden and forest : a journal of horticulture, landscape art and forestry – Killen TX 1888-97 – 1 – mf#2893 – us UMI ProQuest [634]
Garden architecture and landscape gardening / Hughes, John Arthur – London 1866 – (= ser 19th c art & architecture) – 3mf – 9 – mf#4.2.1647 – uk Chadwyck [710]
Garden book of barbados – Bridgetown? Barbados. 1935 – 1r – us UF Libraries [630]
Garden City, Kansas. Community Congregational Church *see* Records
Garden county news – Lewellen, NE: L M Warner, jan 1910 (wkly) [mf ed v2 n31. jan 29 1910- (gaps) filmed 1970-] – 1 – (cont: deuel county news. absorbed: west nebraska beacon. publ in lewellen jan 29 1910-sep 12 1913; in oshkosh sep 19 1913- . issues for jul 26 1913- called v5 n4-) – us NE Hist [071]
Garden journal – Bronx NY 1951-76 – 1,5 – ISSN: 0016-4585 – mf#1905 – us UMI ProQuest [635]
The garden of eloquence / Peacham, Henry – 1593 – 9 – us Scholars Facs [410]
Garden of spices – 1899 – 1 – $50.00 – us Presbyterian [780]
Garden of the glades – Jacksonville, FL. 1914 – 1r – us UF Libraries [630]
Garden party in aid of st mark's church, parkdale : to be held...parkdale on wednesday, september 5th 1883 – [Toronto?: s.n. 1883?] [mf ed 1983] – 1mf – 9 – 0-665-39741-0 – mf#39741 – cn Canadiana [090]
Garden supply retailer – Minnetonka MN 1979-82 – 1,5,9 – (cont: home & garden supply merchandiser) – ISSN: 0195-1386 – mf#1668.01 – us UMI ProQuest [635]
[Gardena-] gardena valley news – CA. 1950-Aug 1964; Nov 1964-65 – 36r – 1 – $2160.00 – mf#B02266 – us Library Micro [071]
The gardener / Tagore, Rabindranath – London: Macmillan and Co, 1929 – (= ser Samp: indian books) – us CRL [490]
Gardener's gazette – London, UK. 1837. -m. 1 reel – 1 – uk British Libr Newspaper [635]
Gardeners magazine – Melbourne, jun 1855-may 1856 – 1r – A$27.50 vesicular A$33.00 silver – at Pascoe [079]
Gardening in florida / Whitner, J N – Jacksonville, FL. 1885 – 1r – us UF Libraries [630]
Gardens of the caribbees / Starr, Ida May Hill – Boston, MA. v1-2. 1904 (1903) – 1r – us UF Libraries [630]
Gardiennes / Perochon, Ernest – Paris, France. 1924 – 1r – us UF Libraries [025]
Gardin, John Emile *see*
– Liberty bonds and civilization
– Liberty bonds for the business woman
Gardiner, A F *see* Narrative of a journey to the zoolu country in south africa
Gardiner, A H *see*
– The admonitions of an egyptian sage
– Egyptian letters to the dead
– A topographical catalogue of the private tombs of thebes
Gardiner, Allen *see* Friend of australia
Gardiner, C Harvey (Clinton Harvey) *see* Naval power in the conquest of mexico
Gardiner, Clinton Harvey *see* Constant captain, gonzalo de sandoval
Gardiner, Frederic *see*
– Aids to scripture study
– A harmony of the four gospels in greek
– The last of the epistles
– Leviticus
– The old and new testaments in their mutual relations
– The principles of textual criticism
Gardiner gazette – Gardiner City OR: O L Williams, [wkly] – 1 – us Oregon Lib [071]
Gardiner, Grace Anne Marie Louise (Napier) *see* The complete indian housekeeper and cook
Gardiner, Harry Norman *see* Jonathan edwards
Gardiner, J S *see* The fauna and geography of the maldive and laccadive archipelagoes
Gardiner, John Hays *see* The bible as english literature
Gardiner, Maine.West Gardiner Free Will Baptist Church/Spears Corner Free Will Baptist Church *see* Records
Gardiner, Richard *see*
– Memoirs of the siege of quebec
Gardiner, Robert *see* Observations on the prospective benefits derivable from the incorporation of the artillery with the cavalry and infantry of the army
Gardiner, Robert Hallowell *see*
– Correspondence, 1910-1924
– [Letters]
Gardiner, Robert William *see* Report on gibraltar considered as a fortress and a colony
Gardiner, Samuel R *see* History of england
Gardiner, Samuel Rawson *see*
– The constitutional documents of the puritan revolution, 1625-1660
– Cromwell's place in history
– History of the commonwealth and protectorate, 1649-1656
– History of the great civil war, 1642-1649
– What gunpowder plot was
Gardinier, David E *see* Cameroon
Gardista – Bratislava, Czechoslovakia. Jul 1942-Mar 1945 (scattered issues) – 4r – 1 – us L of C Photodup [077]
Gardner 1737-1849 – Oxford, MA (mf ed 1996) – (= ser Massachusetts vital record transcripts to 1850) – 6mf – 9 – 0-87623-251-9 – (mf 1t: marriage intentions 1786-1832. mf 1t-2t: marriages 1792-1829. mf 2t-4t: births & deaths 1737-1847. mf 4t: marriages & intentions 1836-42. mf 4t-5t: births 1842-49. mf 5t: marriages 1843-49. mf 6t: deaths 1843-49) – us Archive [978]
Gardner 1762-1892 – Oxford, MA (mf ed 1986) – (= ser Massachusetts vital records) – 31mf – 9 – 0-87623-037-0 – (mf 1-4: births & deaths 1762-1843. mf 5: births & deaths 1785-96. mf 6-7: publishments 1786-1832. mf 8-10: intentions of marriages 1839-79. mf 11-12: b,m,d 1844-52. mf 13-17: births 1857-91. mf 18-19: index to births 1857-91. mf 20-23: marriages 1853-92. mf 24: index to marriages 1853-92. mf 25-29: deaths 1858-92. mf 30-31: index to deaths 1858-92) – us Archive [978]
Gardner, Alice *see*
– The conflict of duties and other essays
– Julian, philosopher and emperor
– The lascarids of nicaea
– The odore of studium
– Rome, the middle of the world
– Studies in john the scot
– Synesius of cyrene
– Within our limits
Gardner, Brian *see*
– Lion's cage
– Mafeking
Gardner, Charles Edwyn *see* Life of father goreh
Gardner, Christopher Thomas *see* Simple truths
Gardner, E A *see* Naukratis
Gardner, Edmund Garratt *see*
– Dante and the mystics
– The dialogues of saint gregory, surnamed the great
– The king of court poets
– Saint catherine of siena
Gardner, Ella *see* Life in japan
Gardner, Ernest Arthur *see* Religion and art in ancient greece
Gardner, G W *see* Bull swamp baptist church
Gardner, George *see* Viagens no brasil
Gardner, George W *see* Echoes from the fleeting years
Gardner, Gregory A *see* Clinical instruction in athletic training
Gardner, Henry A *see* Questions and answers on tung oil production in america
Gardner, J K *see* A comparison of occupational stress and related variables among salespersons, clerical staff, service technicians, and managers of the mid-ohio district of the xerox corporation
Gardner, James *see*
– The faiths of the world
– Memoirs of christian missionaries
Gardner, John M *see* American negligence reports, current series
Gardner, M B *see* Shall the sword devour forever?
Gardner, Mary Tracy *see* Winners of the world during twenty centuries
Gardner, May F *see* Franciscan friars
Gardner, Percy *see*
– The ephesian gospel
– The growth of christianity
– A historic view of the new testament
– Modernity and the churches
– The origin of the lord's supper
– The religious experience of saint paul
Gardner, Percy et al *see* The faith and the war
Gardner, Robert G *see* Jeremiah walker: georgia general baptist
Gardner, Samuel A *see* Latest interpretations
Gardner, W R W *see*
– The qurranic doctrine of god
– The qurranic doctrine of salvation
– The qurranic doctrine of sin
Gardner, W W *see* Missiles of truth
Gardner, William Edward *see* Winners of the world during twenty centuries
Gardner, William James *see* History of jamaica from its discovery
Gardner's baptist church. tar river association. warren county. north carolina : church records – 1844-1930 – 1 – 9.90 – us Southern Baptist [242]
[Gardnerville-] nevada lutheran – NV. 10 jun 1918 (only known issue) – 1r – 1 – $60.00 – mf#U04533 – us Library Micro [242]
[Gardnerville-] nevada magazine – NV. 1889-1900; 1945-1949 – 4r – 1 – $240.00 – mf#U04843 – us Library Micro [071]
[Gardnerville-] record – NV. 1898-1904 [wkly] – 1r – 1 – $60.00 – mf#U04534 – us Library Micro [071]
[Gardnerville-] the courier – NV. 1901-09 [wkly] – 4r – 1 – $240.00 – mf#U04532 – us Library Micro [071]
[Gardnerville-] the record-courier – NV. 1909-[wkly] – 88r – 1 – $5280.00 (subs $275y) – mf#UN04535 – us Library Micro [071]
Gardthausen, V *see*
– Catalogus codicum graecorum sinaiticorum
– Die griechischen schreiber des mittelalters und der renaissance
Gardthausen, Viktor *see*
– Das buchwesen im altertum und im byzantinischen mittelalter
– Die schrift, unterschriften und chronologie im altertum und im byzantinischen mittelalter
Garduna / Zeno Gandia, Manuel – San Juan, Puerto Rico. 1955 – 1r – us UF Libraries [972]
Gareau, Tony *see* Hyperbaric oxygen therapy in the treatment of sports injuries
Gareis, Karl *see* Introduction to the science of law
Gareis, Reinhold *see* Geschichte der evangelischen heidenmission
Garenganze : or, seven years' pioneer mission work in central africa / Arnot, Frederick Stanley – Chicago: Fleming H Revell, [1889?] – 1mf – 9 – 0-8370-6321-3 – mf#1986-0321 – us ATLA [920]
Garenganze : or, seven years' pioneer mission work in central africa / Arnot, Frederick Stanley – London, [1889] – 4mf – 9 – mf#HTM-6 – ne IDC [916]
Garezera raksti – v25-26 [1947-1948] – 1r – 1 – (continued by: garezera raksti (1965)) – mf#681880 – us WHS [071]
Garfias, Robert *see* Basic melody of the togaku pieces of the gagaku repertoire
Garfield County Genealogists *see* Garfield county roots and branches
Garfield county miscellaneous newspapers – Denver, CO – (crystal river empire (jan 24 1924); daily post reminder (feb 16 1933, aug 27 1935, sep 1 1936-dec 31 1936); gazette shopper (1969-74); glenwood high country gazette (1969-70); glenwood gazette (june 15 1968-sep 17 1969); glenwood springs reminder record (1964-67); morning reminder (mar 12 1954); the pool (oct 11 1890); the ute chief (july 9 1887-aug 24 1888); new castle nonpareil (feb 21 1896); rifle reveille (sep 6 1890, oct 30 1908); rifle telegram (jul 23 1925); garfield county news (aug 11 1911-oct 27 1911); silt searchlight (nov 4 1913) – mf#MF Z99 G18 – us Colorado Hist [071]
Garfield county news *see* Garfield county miscellaneous newspapers
[Garfield county quaver] – Burwell, NE: [W T Hastings], -1891// (wkly) [mf ed apr 12-jul 12 1888 (gaps) filmed [1973] – 1r – 1 – (absorbed by: garfield enterprise) – us NE Hist [071]
Garfield county roots and branches / Garfield County Genealogists – v1 n1-8 n4 [1978 feb.-1985 fall] – 1r – 1 – (continued by: roots & branches (enid, ok)) – mf#1685725 – us WHS [929]
Garfield enterprise – Burwell, NE: Todd Bros, 1888-94// (wkly) [mf ed jun 23 1892] – 1r – 1 – (absorbed: garfield county quaver. cont by: burwell progress) – us NE Hist [071]
Garfield heights leader – Garfield Heights, OH: August E. Kleinschmidt, feb 25 1971-nov 20 1986 – 13r – 1 – (weekly cleveland suburban newspaper) – mf#(M) 34 C9.3 211 – us Western Res [071]
Garfield Heights. Ohio. St. John Evangelical Lutheran Church *see* Church records
Garfield heights tribune *see* Neighborhood news / the garfield heights tribune
Garfield, James A *see* Papers
Garfinkel, Marian S *see* The effect of yoga and relaxation techniques on outcome variables associated with osteoarthritis of the hands and finger joints
Gargan, Denis *see* Ancient church of ireland
Garganta la Olla. Ayuntamiento *see* Fiestas en garganta la olla 1974
Gargasz, Kimberly L *see* Participation in white water rafting instruction by adults with developmental disabilities
Gargata manju gisun-i bithe = Dan qing yu – [Jingzhou]: Jingzhou zhu fang fan yi zong xue, Guangxu xin mao [1891] [mf ed 1966] – (= ser Tenri coll of manchu-books in manchu-characters. series 1, linguistics 47; Mango bunkenshu. 1, gogaku hen) – 8v on 1r – 1 – (in manchu and chinese) – ja Yushodo [480]

Garibaldi, Anita see Garibaldi en america
Garibaldi en america / Garibaldi, Anita – Buenos Aires, Argentina. 1930 – 1r – us UF Libraries [972]
Garibaldi news – Garibaldi OR: [s.n.] -1932 [wkly] – 1 – (cont by: garibaldi-rockaway news (1932-36)) – us Oregon Lib [071]
Garibaldi-rockaway news – Garibaldi OR: F C Baker, 1932-36 [wkly] – 1 – (cont: garibaldi news (-1932); cont by: north tillamook county news (1936-47)) – us Oregon Lib [071]
Gariepy, Charles-Napoleon see
– De jure et justitia
– Nouveau code de droit canonique et theologie morale
Garimpeiro / Guimaraes, Bernardo – Rio de Janeiro, Brazil. 1945 – 1r – us UF Libraries [972]
Garimpeiro / Guimaraes, Bernardo – Sao Paulo, Brazil. 1962 – 1r – us UF Libraries [972]
Garimpeiro : orgao litterario e noticioso – Bagagem, MG: Typ do Garimpeiro, 02-16 out 1886 – (= ser Ps 19) – mf#P31,03,20 – bl Biblioteca [440]
O garimpeiro : publicacao hebdomadaria – Rio de Janeiro, RJ: Typ Central, 06-20 nov 1881 – (= ser Ps 19) – mf#P17,01,144 – bl Biblioteca [079]
Garin, Andre-Marie see Chemin de la croix et autres prieres
Gariod, Charles see Decouvertes des portugais en amerique au temps de christophe colomb
Garisa, Rupert C see Papers relating to nauru
Garlake, Peter S see The kingdoms of africa
Garland : or new general repository of fugitive poetry – La Salle IL 1825 – 1 – mf#3808 – us UMI ProQuest [810]
Garland, Augustus Hill see
– Experience in the supreme court of the united states, with some reflections and suggestions as to that tribunal
– A treatise on the constitution and jurisdiction of the united states courts on pleading, practice and procedure therein.
Garland, Hamlin see
– Boy life on the prairie
– Crumbling idols: twelve essays on art and literature
Garland herald – Lincoln, NE: Interstate Newspaper Co, 1917-v27 n2. apr 25 1934 (wkly) [mf ed 1920-34 (gaps) filmed 1976] – 4r – 1 – (cont: germantown herald. absorbed by: lancaster county weekly. publ in havelock ne, mar 1927-jul 22 1931. issue for apr 13 1927 incorrectly dated mar 13 1927) – us NE Hist [071]
Garland, Landon Cabell et al see Discussions in theology
Garland, Nicholas Surrey see
– Compilation of the laws and amendments thereto relating to building societies, loan companies, joint stock companies, and interest on mortgages and other acts pertaining to monetary institutions
– Garland's banks, bankers and banking and financial directory of canada
– Garland's banks, bankers and banking in canada
– Parliamentary directory and statistical guide
Garland, Nicholas Surrey [comp] see Parliamentary directory and statistical guide
The garland of life : poems, west and east / Cousins, James Henry – Madras: Ganesh & Co, 1917 – (= ser Samp: indian books) – us CRL [810]
Garlandia, Johannes de (John Garland) see Morale scolarium
Garland's banks, bankers and banking and financial directory of canada : with a list of bank solicitors and commercial lawyers, and a brief analysis of the commercial laws of the several provinces of the dominion / ed by Garland, Nicholas Surrey – Ottawa: Mortimer, 1895 – 6mf – 9 – (incl ind) – mf#07718 – cn Canadiana [332]
Garland's banks, bankers and banking in canada : with list of bank solicitors and commercial lawyers – Ottawa: s.n, 1890? – 1mf – 9 – mf#55373 – cn Canadiana [332]
Garland's banks, bankers and banking in canada : with list of bank solicitors and commercial lawyers: to which has been added statistics of the dominion / ed by Garland, Nicholas Surrey – Ottawa: Mortimer, 1890 – 4mf – 9 – (incl ind) – mf#07719 – cn Canadiana [332]
Garlick, Don see Technical problems for the violoncello in contemporary string quartet literature
Garlick, Peter C see African traders in kumasi
Garma c c chang tibetan collection – 528mf – 9 – $320.00 – (over 130 tibetan works from the xylograph and mss coll; title list on 1mf available $1) – us IASWR [280]
GARMENT WORKER / United Garment Workers of America – 1893 apr-1903 aug – 1r – 1 – (continued by: weekly bulletin of the clothing trades) – mf#3389101 – us WHS [680]
Garment worker / International Ladies' Garment Workers' Union – v15 n8 [1972 oct], v16 n4 [1974 aug], v17, n1 (1975 jan) – 1r – 1 – mf#368290 – us WHS [680]
Garment worker : official organ of the united garment workers of america / United Garment Workers of America – 1945 sep 7-1954 dec, 1955-58, 1959-64, 1965 jan-1967 dec, 1968 jan 5-apr 5, may 3-1977 dec – 5r – 1 – (cont: weekly bulletin of the clothing trades) – mf#1111820 – us WHS [331]
Garment workers in action : history of the garment workers of south africa to 1952 / Sachs, E S – Johannesburg: Eagle Press, 1957 – us CRL [960]
Garmisch-partenkirchener tagblatt – Garmisch-Partenkirchen DE, 1945 2 jan-14 apr – 1r – 1 – (bezirksausgabe von muenchner merkur, muenchen; filmed by misc inst: 1977- [13r/yr]) – gw Mikrofilm; gw Misc Inst [074]
Garmon newsletter – n1-26 [1985 apr-1989 jun] – 1r – 1 – mf#1542259 – us WHS [071]
Garneau, Alfred see Poesies
Garneau, Francois-Xavier see Histoire du canada
Garneau, Marthe see Bibliographie de l'oeuvre de monsieur gerard morisset
Garneau, Robert see Bibliographie de l'oeuvre de monsieur rolland dumais
Garner, J Dianne see Journal of women and aging
A garner of saints : being a collection of the legends and emblems usually represented in art / Hinds, Allen Banks – New York: EP Dutton; London: JM Dent, 1900 [mf ed 1990] – 1mf – 9 – 0-7905-5708-8 – mf#1988-1708 – us ATLA [700]
Garnett, Lucy Mary Jane see Mysticism and magic in turkey
Garnett, Richard see Life of ralph waldo emerson
Garnier, Edouard see French pottery
Garnier, Francis see Chronique royale du cambodge
Garnier, Ilse see L'expressionnisme allemand
Garnier, L see Premier grand quatuor, pour flute, violon, alto et basse, oeuvre 2
Garnier, Pierre see Gottfried benn
Garnier, T D see Zur entwicklungsgeschichte der novellendichtung ludwig tieck's
Garnier, Thomas see Christ the world's peace
Garo jungle book : or, the mission to the garos of assam / Carey, William et al – Philadelphia: Judson Press, [1919] [mf ed 1995] – (= ser Yale coll) – 283p (ill) – 1 – 0-524-09139-0 – mf#1995-0139 – us ATLA [954]
Garonne – Toulouse, France. 29 jun 1940-10 jul 1944 – 5 1/2r – 1 – uk British Libr Newspaper [072]
O garoto : critico, desopilante, molieresco, rabelaiseano – Fortaleza, CE. 03 nov 1907-12 dez 1908 – (= ser Ps 19) – bl Biblioteca [079]
Garra de luz / Henriquez Urena, Max – Habana, Cuba. 1958 – 1r – us UF Libraries [972]
Garran, Robert Randolph see The coming commonwealth
Garrat, Geoffrey Theodore see An indian commentary
Garratt, G T see
– The legacy of india
– Rise and fulfilment of british rule in india
Garrazin, General see Geschichte des kriegs in spanien
Garreau, Albert see Saint albert le grand
Garrett a morgan papers see Morgan, garrett a, papers, ms 3534
Garrett, Agnes see Suggestions for house decoration in painting, woodwork
Garrett Biblical Institute, Evanston, Ill. Bureau of Social and Religious Research see Indiana racial study
Garrett, Candi L see Heat distribution in the lower leg from pulsed short wave diathermy and ultrasound treatments
Garrett e o romantismo / Braga, Teofilo – Porto, Portugal. 1903 – 1r – us UF Libraries [025]
Garrett, H L O see Mughal rule in india
Garrett, Harold see Studies in the propagation of the tung tree, aleurites fordi, hemsl...
Garrett, Herbert G see The life insurance act
Garrett, Herbert Leonard Offley see
– Events at the court of ranjit singh, 1810-1817
– The punjab a hundred year ago as described by v jacquemont (1831) and a soltykoff
Garrett, James L, Jr see Baptist church discipline, broadman press
Garrett, John see A classical dictionary of india
Garrett, John C see
– The centennial
Garrett, Rhoda see Suggestions for house decoration in painting, woodwork
Garrettsville. Ohio. Baptist Church see Church records, ms 2087
Garrettsville. Ohio. Church of Christ see Church records, ms 2843
Garrick Club. London see Kemble prompt books in the garrick club, london
Garrick, David see The papers of david garrick, 1717-79
Garrido, Eduardo see Razones contra ultrajes
Garrido, Pablo see Esoteria y fervor populares de puerto rico
Garrido Santiago, Manuel see Arquitectura religiosa del s. 16 de la tierra de barros
Garrigo, Roque E see
– Discoursos leidos en la recepcion
– Historia documentada de la conspiracion
Garrigo, Roque E see Misoneismo politico-ornamental
Garris, Edward Walter see Special methods in teaching vocational agriculture
Garrison argus – Garrison, NE: Roseleta J Clark. v1 n1. may 22 1902-v6 n23. feb 6 1908 (wkly) [mf ed 1905-08 (gaps) filmed [1974?]] – 2r – 1 – (suspended with may 30 1907; resumed with sep 5 1907. iss for sep 5-oct 10 1907 called v1 n1-6. iss for oct 17 1907-feb 6 1908 called v6 n7-23) – us NE Hist [071]
Garrison, DR see Occupational stress and job satisfaction related to management styles of american- and japanese-owned companies in america
Garrison fork baptist church. beech grove, tennessee : church records – 1809-1933 – 1 – us Southern Baptist [242]
Garrison, James Harvey see
– Christian union
– Half-hour studies at the cross
– Helps to faith
– The holy spirit
– A nineteenth century movement
– The old faith restated
– Our first congress
– The reformation of the nineteenth century
– The story of a century
– The witness of jesus, and other sermons
A garrison romance / Laffan, Bertha Jane (Grundy) – London: Eden, Remington & Co Publ Co, 1892 – (= ser 19th c women writers) – 4mf – 9 – mf#5.1.42 – uk Chadwyck [810]
Garrison tribune see Alamosa county miscellaneous newspapers
Garrison, William Lloyd see
– Abolition and emancipation
– The abolition of slavery
– The loyalty and devotion of colored americans in the revolution and war of 1812
– William lloyd garrison papers, 1833-1882
Garrison, Winfred Ernest see Alexander campbell's theology
Garrod, George Watts see
– The epistle to the colossians
– The first epistle to the thessalonians
Garrod, Heathcote William see
– Einhard's life of charlemagne
– The religion of all good men
Garron De Doryan, Victoria see Aire, el agua y el arbol
Garrow, David see Centers of the southern struggle
Garrow, David J see The martin luther king, jr fbi file
Garrucci, R see Storia della arte cristiana nei primi otto secoli della chiesa
Garry, Robert see La renaissance du cambodge
Garsault see Arte del barbero-peluquero-banero
Garside, Charles Brierley see The sacrifice of the eucharist, and other doctrines of the catholic church explained and vindicated
Garst, Laura DeLany see
– In the shadow of the drum tower
– A west-pointer in the land of the mikado
Garstang courier – [NW England] Garstang 1965-26 jan 1974 – 1 – (title change: courier [2 feb 1974-1977]) – uk MLA; uk Newsplan [072]
Garstang guardian – [NW England] Garstang Lib 1987 – 1 – uk MLA; uk Newsplan [072]
Garstang, John see
– A short history of ancient egypt
– The syrian goddess
Garston and woolton weekly news, and south liverpool reporter – [NW England] Liverpool jan 1913-dec 1950 [mf ed 2004] – 42r – 1 – (missing: 1940; title change: liverpool weekly news [jan 1961-dec 1974, feb-dec 1975, jan 1978]) – uk Newsplan; uk MLA [072]
Garston & woolton reporter and district observer – [NW England] Liverpool 9 jun 1888-dec 1920 [mf ed 2004] – 33r – 1 – (missing: 1897, 1912) – uk Newsplan [072]
Der garten von vaux-le-vicomte / Bechter, Barbara – (mf ed 1993) – 3mf – 9 – €49.00 – 3-89349-850-8 – mf#DHS 850 – gw Frankfurter [710]
Die gartenbauwirtschaft – Berlin DE, 1933-40 – 2r – 1 – gw Misc Inst [635]
Gartengesellschaft / Frey, Jakob; ed by Bolte, Johannes – Stuttgart: Litterarischer Verein, 1896 (Tuebingen: H Laupp, Jr) – (latin text with an introduction and commentaries in german) – us Wisconsin U Libr [830]
Gartenhof, Kaspar see Die bedeutendsten romane philipps von zesen und ihre literargeschichtliche stellung
Gartenkalender [auf das jahr 1782-89] / ed by Hirschfeld,Christian Cay Lorenz – Kiel & Dessau 1782/83, Kiel 1784-86, Braunschweig 1787-89 – (= ser Dz) – 7jge on 15mf – 9 – €150.00 – mf#k/n2936 – gw Olms [635]
Die gartenlaube : illustrierte familienzeitschrift – Leipzig DE, 1853-1933 n50, 1934-1941 n50, 1942-1944 sep – 84r – 1 – (title varies: 1935?: die neue gartenlaube: illustrierte familienzeitschrift. filmed by misc inst: 1853-55, 1857-58, 1897 [only suppl], 1915-16, 1917 [gaps], 1918, ind: 1853-1902, contents ind: 1915-16 [10r]. incl suppl: deutsche blaetter 1862 oct-dec) – gw Mikrofilm [640]
Die gartenlaube (iz1) / ed by Estermann, Alfred – Leipzig, Berlin 1853-1944 [mf ed 1999] – (= ser Illustrierte zeitschriften (iz) 1) – 1382mf – 9 – diazo €6090 silver €7880 – 3-89131-346-2 – gw Fischer [306]
Gartner's notes to the interstate commerce commission reports : vols 1-41 / U.S. Interstate Commerce Commission – Louisville: Banks. 3v. 1915-17 (all publ) – (= ser Interstate commerce commission reports) – 5mf – 9 – $22.50 – mf#LLMC 84-116 – us LLMC [324]
Garuda / Suara pembaruan – Djakarta, 1970-1971(1-21) – 22mf – 9 – mf#SE-1937 – ne IDC [959]
Garuda indonesian airways – Djakarta, 1956-1958 – 2mf – 9 – mf#SE-682 – ne IDC [959]
The garuda purana (saroddhara) / Naunidhirama – Allahabad: PGnini Office, 1911 – (= ser Samp: indian books) – (with english trans by ernest wood and s v subrahmanyam; int fr sris chandra vasu) – us CRL [280]
The garuda puranam / ed by Dutt, Manmatha Nath – Calcutta: Society for the Resuscitation of Indian Literature, 1908 – 2mf – 9 – 0-524-07080-6 – mf#1991-0062 – us ATLA [280]
Garuda visi – [Fei ying zhou bao] – Bandung: Garuda Giriharja Abdimedia (wkly) [mf ed Chicago IL: filmed by Preservation Resources, Bethlehem, PA for SEAsian MF Project, CRL 2002] – (= ser Indonesian political tabloids microfilm coll) – 1r with other items – 1 – (began in 1999; in indonesian) – mf#mf-13314 seam – us CRL [079]
Garver, Earl Simeon see Puerto rico
Garvey's voice / Universal Negro Improvement Association – 1973 apr, 1977 mar/apr-1990 dec/jan] – 1r – 1 – (cont: negro world) – mf#1826131 – us WHS [305]
Garvie, Alfred E see A course of bible study for adolescents
Garvie, Alfred Ernest see
– Can we still follow jesus?
– The christian certainty amid the modern perplexity
– The evangelical type of christianity
– The gospel according to st luke
– The gospel for to-day
– A guide to preachers
– A handbook of christian apologetics
– The missionary obligation in the light of the changes of modern thought
– The ritschlian theology
– Romans
– Studies in the inner life of jesus
– Studies of paul and his gospel
Garvin, Hugh Carson see What the bible teaches
Garvin, John E see The centenary of the society of mary
Garvin, John William see The collected poems of isabella valancy crawford
Gary see Eudore et cymodocee
Gary american – Gary IN. 1971 feb 20/26, mar 13/19, 1973 dec 17/27, 1974 jan 19/26, feb 16/23, 1975 nov 15 – 1r – 1 – (cont: gary colored american; gary sun [gary, in: 1929]) – mf#2704374 – us WHS [305]
Gary business monthly : a publication of the small business development council / Gary Small Business Development Council (Gary, IN) – 1986 aug /sep – 1r – 1 – mf#4877121 – us WHS [338]
Gary crusader – Gary IN. 1968 sep 7-1970 apr 25, 1970 may 2-1971 oct 30, 1971 nov6-1974 apr 27, 1974 may 4-1976 jun 26, 1976 jul 3-1978 jul 29, 1978 aug-1979 dec, 1980-92, 1993 jan 2-jun 26, 1993 jul 3-dec 25, 1997 oct 11, 1998 jan 3-jun 27, 1998 jul 4-dec 26, 1999 jan 2-dec 25, 2000 jan 1-jun 24 – 26r – 1 – mf#857064 – us WHS [071]
Gary post-tribune – Gary IN. 1929 may 6 – 1r – 1 – (cont: gary evening post and daily tribune; cont by: post-tribune) – mf#4364639 – us WHS [071]
Gary, Roberta Sexton see Practical organ accompaniment for the magnificat in d major and for the cantata no 106 of johann sebastian bach
Gary Small Business Development Council (Gary, IN) see Gary business monthly
Garzarella, L see Predicting body composition of healthy females by b-mode ultrasound
Garzia, Gabriele see Comento filologico-esegetico sul primo salmo
Garzoni, Maurizio see Grammatica e vocabulario della lingua kurda
Gas / Kaiser, Georg – Potsdam, Germany. c1918 – 1r – us UF Libraries [025]
Gas – New York NY 1925-73 – 1,5 – mf#1039 – us UMI ProQuest [550]

Gas abstracts – Des Plaines IL 1945-94 – 1,5,9 – ISSN: 0016-4844 – mf#253 – us UMI ProQuest [550]

Gas engine troubles and remedies / Stritmatter, Albert – Cincinnati, OH. 1903 – 1r – us UF Libraries [621]

Gas engineering and management – London, England 1974-96 – 1,5,9 – (cont: institution of gas engineers (london, england) journal; cont by: international gas engineering & management) – ISSN: 0306-6444 – mf#7129.02 – us UMI ProQuest [622]

Gas journal – Tonbridge, England 1849-1972 – 1,5 – mf#1390 – us UMI ProQuest [550]

Gas separation & purification – Oxford, England 1989-96 – 1,5,9 – ISSN: 0950-4214 – mf#17232 – us UMI ProQuest [660]

Gas times – London, England 1950-56 – 1 – mf#475 – us UMI ProQuest [550]

Gas turbine international – Norwalk CT 1960-75 – 1,5,9 – (cont by: sawyer's gas turbine international) – ISSN: 0435-1312 – mf#2101.02 – us UMI ProQuest [621]

Gas warfare / Farrow, Edward Samuel – New York, NY. c1920 – 1r – us UF Libraries [025]

Gas world – London, England 1976-87 – 1,5,9 – (cont: gas world & gas journal) – ISSN: 0308-7654 – mf#5925.01 – us UMI ProQuest [550]

Gas world and gas journal – London, England 1968-74 – 1,5 – (cont by: gas world) – ISSN: 0308-1654 – mf#5925.01 – us UMI ProQuest [550]

Gasa, Velisiwe Goldencia *see* The impact of disrupted family life and school climate on the self-concept of the adolescent

Gasco y Navarro, J M *see* Asserta aphoristica et chirurgica ex libris...

Gascoigne, Thomas *see* Loci e libro veritatum

Gascon y Miramon, Antonio *see* Los criaderos de hierro de burguillos (badajos)

Gascoyne advance – Gascoyne, Bowman Co, ND: L Pitsor. v1 n1 feb 10 1910-jan 1915?/ / (wkly) – 1 – (cont by: mineral springs tribune) – mf#11453 – us North Dakota [071]

Gascoyne gazette – Gascoyne, ND: W C Smith. v1 n1 apr 7 1915-v3 n31 nov 7 1917 (wkly) – 1 – (merged with: farmers leader to form: the farmers leader and gascoyne gazette. missing: 1916 dec 27; 1917 mar 21, may 23, aug 15) – mf#08217 – us North Dakota [630]

Gascoyne news – Gascoyne, Bowman Co, ND: Roy L Johnston, sep 5, 1918?-dec 1934?// (wkly) [mf ed with gaps] – 1 – (many iss misnumbered and/or misdated; cont: gascoyne pennant) – mf#10366-10369 – us North Dakota [071]

Gascoyne pennant – Gascoyne, Bowman Co, ND: W C Smith. v1 n2 nov 21 1917-v1 n40 aug 15 1918 (wkly) – 1 – (cont: yellowstone trail pennant; missing: 1918 aug 1) – mf#08377 – us North Dakota [071]

Das gasel in der deutschen dichtung und das gasel bei platen / Tschersig, Hubert – Leipzig: Quelle & Meyer, 1907 [mf ed 1992] – (= ser Breslauer beitraege zur literaturgeschichte. neue folge 1) – xii/229p – 1 – (incl bibl ref and ind) – mf#8014 reel 2 – us Wisconsin U Libr [430]

Gas-Faucher, F *see* En sampan sur les lacs du cambodge et a angkor

Gaskell and the brontes : literary manuscripts of elizabeth gaskell (1810-1865) & the brontes from the brotherton library, university of leeds – [mf ed Marlborough 2003] – 7r – 1 – £675.00 – uk Matthew [420]

Gaskell, Elizabeth Cleghorn *see* Cranford

Gaskell, William *see*
- Duties of the individual to society
- Faithful religious teacher
- God manifest in christ

Gaskill, Jackson *see* The printing-machine manager's complete practical handbook

Gaskiya ta fi kwabo – Zaria: M. Abubakar Iman Kagara, sep 1974-dec 1982 – us CRL [960]

Gaskiya ta fi kwabo – Zaria, Nigeria: M Abubakar Iman Kagara, [aug 23 1950-dec 1957; aug 1959-dec 1972; 1973-nov 1981] – 1 – us CRL [079]

Gaskoin, Herman (Mrs) *see* Children's treasury of bible stories

Gasnault, Paul *see* French pottery

Gaspar de San Augustin *see*
- Compendio de la arte de la lengua tagala
- Confessionario copioso en lengua espanola, y tangela

Gaspar berse, of de nederlandsche franciscus xaverius : eene bijdrage tot de geschiedenis der societeit van jezus in indie van 1546-53 / Nieuwenhoff, Willem Frederik van – Rotterdam: G W van Belle, 1870 [mf ed 1995] – (= ser Yale coll) – viii/410p (ill) – 1 – 0-524-10107-8 – (in dutch) – mf#1995-1107 – us ATLA [241]

Gaspar Da Madre De Deos *see* Memorias para a historia da capitania de s vicent

Gaspar de Segovia, Jose M *see* Por don juan de ovando...lavadero lanas de almendralejo

Gaspar de Segovia, Joseph Manuel *see* Por los...monasterios de...

Gaspar octavio hernandez / Pena, Concha – Panama, 1953 – 1r – us UF Libraries [972]

Gaspard de coligny : admiral of france / Whitehead, Arthur Whiston – London: Methuen, 1904 – 1mf – 9 – 0-7905-6334-7 – mf#1988-2334 – us ATLA [944]

Gaspard l'avise / Barre, M – Paris, France. 1812 – 1r – us UF Libraries [440]

Gaspard l'avise / Barre, M – Paris, France. 1814 – 1r – us UF Libraries [440]

Gaspard l'avise / Barre, M – Paris, France. 1818 – 1r – us UF Libraries [440]

O gasparense : orgam independente e noticioso – Gaspar, SC: Typ Gasparense, 23 dez 1923 – (= ser Ps 19) – mf#P33,06,09 – bl Biblioteca [079]

Gasparian, Fernando *see* Defesa da economia nacional

Gasparin, Agenor, comte de *see*
- The doctrine of plenary inspiration
- L'eglise selon l'evangile

Gasparini, Graziano *see*
- Arquitectura colonial en venezuela
- Casa colonial venezolana
- Promesa de venezuela
- Templos coloniales de venezuela

Gasparini, Joseph *see* The attributes of christ

Gasparis megandri...in epistolam pauli ad ephesios comentarius... / Grossmann, K – Basileae, Henricus Petrus, [1534] – 3mf – 9 – mf#PBU-608 – ne IDC [240]

Gasparis megandri...in epistolam pauli ad galatas commentarius... / Grossmann, K – Tigvri, rhellicani epistola... / Grossmann, K – Tigvri, officina Froschoviana, 1533 – 1mf – 9 – mf#PBU-495 – ne IDC [240]

Gasparo contarini, 1483-1542 : eine monographie / Dittrich, Franz – Braunsberg: Ermlaendischen Zeitungs- und Verlagsdruckerei, 1885 – 3mf – 9 – 0-7905-4559-4 – (incl bibl ref) – mf#1988-0559 – us ATLA [920]

Gasparo contarini und das regensburger concordienwerk des jahres 1541 / Brieger, Theodor – Gotha: Perthes, 1870 – 1mf – 9 – 0-524-03273-4 – (incl bibl ref) – mf#1990-0884 – us ATLA [240]

Gasparri, Pietro *see*
- Tractatus canonicus de sacra ordinatione
- Tractatus canonicus de sanctissima eucharistia

The gaspe magazine and instructive miscellany – [New Carlisle, Quebec?: s.n. 1849-] – 9 – mf#P05119 – cn Canadiana [420]

Gasperini, Auguste de *see* La nouvelle allemagne musicale

Gaspesiana / Saint-Denis, soeur – 1963 [mf ed 1979] – (= ser Bibliographies du cours...1947-66) – 2mf – 9 – (with ind; pref by guy fortier) – mf#SEM105P4 – cn Bibl Nat [010]

Gaspesie : textes, photos: guide touristique / Parise, Marie et al – [1ere ed]. Montreal: Editions 0.25, [1966?] (mf ed 1998) – 3mf – 9 – mf#SEM105P2909 – cn Bibl Nat [917]

La gaspesie : esquisse generale, terres a coloniser, etc, etc / Pelland, Alfred – Quebec: [Dept de la colonisation, des mines et des pecheries], 1914 [mf ed 1995] – 9 – cn Bibl Nat [971]

La gaspesie / Ferland, Jean-Baptiste-Antoine – Quebec?: A Cote, 1877 – 4mf – 9 – mf#09173 – cn Canadiana [917]

La gaspesie : promenades dans le golfe saint-laurent: nouvelle-ecosse, ile du prince-edouard, nouveau-brunswick, la baie des chaleurs, la gaspesie / Faucher de Saint-Maurice – Montreal: Cadieux & Derome, 1886? – 3mf – 9 – mf#27314 – cn Canadiana [917]

La gaspesie en 1888 / Bechard, Auguste – [Quebec?: Nationale], 1918 – 2mf – 9 – 0-665-71655-9 – mf#71655 – cn Canadiana [917]

La gaspesie, la suisse canadienne / ed by Blais, Isidore – Rimouski: [1938?] (mf ed 1994) – 1mf – 9 – (in french and english) – mf#SEM105P2119 – cn Bibl Nat [917]

La gaspesie, promenades dans le golfe saint-laurent : nouvelle ecosse, ile du prince edouard, nouveau brunswick, la baie des chaleurs, la gaspesie / Faucher de Saint-Maurice – 3e ed. Montreal: Librairie Saint-Joseph: Cadieux & Derome, [s.d] (mf ed 1974) – 1r – 5 – mf#SEM16P38 – cn Bibl Nat [917]

Gasquet, A *see* De l'autorite imperiale en matiere religieuse a byzance

Gasquet, Abbot *see* Edward 6th and the book of common prayer

Gasquet, Francis Aidan *see*
- The bosworth psalter
- Collectanea anglo-premonstratensia
- Edward 6 and the book of common prayer
- England under the old religion
- The eve of the reformation
- Henry 8 and the english monasteries
- Henry the third and the church
- The last abbot of glastonbury and other essays
- The old english bible
- Parish life in mediaeval england

Gasquet, Francis Aidan, Cardinal *see*
- Codex vercellensis
- English monastic life
- The first divorce of henry 8
- The great pestilence (a.d. 1348-9)

Gasquet, Francis Aiden, Cardinal *see* Lord acton and his circle

Gass, J *see* Une ordonnance curieuse

Gass, Joachim Christian *see* Fr schleiermacher's briefwechsel mit j chr gass

Gass, Patrick *see*
- A journal of the voyages and travels of a corps of discovery
- Tagebuch einer entdeckungs-reise durch nord-amercia

Gass, Wilhelm *see*
- Dr. e.l. th. henke's neuere kirchengeschichte
- Fr schleiermacher's briefwechsel mit j chr gass
- Gennadius und pletho
- Georg calixt und der synkretismus
- Geschichte der christlichen ethik
- Geschichte der protestantischen dogmatik
- Symbolik der griechischen kirche

Die gasselbuben : geschichte aus den bayerischen vorbergen / Schmid, Herman – 2.aufl. Leipzig: Keil [18–?] – (= ser Gesammelte schriften. volks- und familien-ausgabe 28) – 1 – (bound with: almenrausch und edelweiss and das muenchener kindeln) – mf#film mas c438 – us Harvard [830]

Gassendi, Pierre *see* Opera omnia

Gasser, J C *see* Vierhundert jahre zwingli-bibel 1524-1924

Gasser, Johann Conrad *see* Das alte testament und die kritik

Gasseren : [a periodical devoted to the interests of foreign missions especially in the island of madagascar [later expanded to africa, china, and other countries]] – Minneapolis MN: Lutheran Board of Missions (Folkebladets trykkeri)[1900-16] [mthly. biwkly] [mf ed 2005] – 16v on 2r – 1 – (ed by georg sverdrup [nov 1900-jun 1907]; andreas helland [jul 1907-16]; some pgs damaged) – mf#2005c-s056 – us ATLA [242]

Gassol, Ventura *see* Les tombes flamejants

Gast, E R *see* Emilia galotti

Gast, J *see* De anabaptismi exordio, erroribvs, historijs abominandis, confutationibus adiectis libri duo...

Gastfahrten : reise-erfahrungen und studien / Rossmann, Wilhelm – Leipzig 1880 [mf ed Hildesheim 1995-98] – 3mf – 9 – €90.00 – 3-487-29909-7 – gw Olms [910]

Das gastmahl : novelle / Klaehn, Friedrich Joachim – Muenchen: F Eher 1943 [mf ed 1990] – 1r [ill] – 1 – (filmed with: benno papentrigk's schuttelreime) – mf#2758p – us Wisconsin U Libr [830]

Gastoldi, Giovanni Giacomo *see*
- Baletti a tre voci
- Concenti musicali con le sue sinfonie a otto voci
- Il primo libro de madrigali a cinque voci
- Il secondo libro de madrigali a cinque voci

Gaston et bayard / Belloy, Pierre-Laurent Buyrette De – Paris, France. 1801 – 1r – us UF Libraries [440]

Gaston, Hugh *see* A scripture account of the faith and practice of christians

Gaston news *see*
- Cornelius news
- West washington county news

Gaston, William *see* William gaston papers

Gastonia first baptist church. gastonia, north carolina : church records – 1912-57 – 1 – 87.84 – us Southern Baptist [242]

Gastriklands tidning – Borlange, Sweden. 1979- – 1 – sw Kungliga [078]

Gastroenterologia – Basel, Switzerland 1966-67 – 1 – (cont by: digestion) – mf#2058 – us UMI ProQuest [574]

Gastroenterologie clinique et biologique – Paris, France 1977-81 – 1,5,9 – ISSN: 0399-8320 – mf#11317 – us UMI ProQuest [616]

Gastroenterology – Philadelphia PA 1979+ – 1,5,9 – ISSN: 0016-5085 – mf#77 – us UMI ProQuest [616]

Gastroenterology abstracts and citations – Washington DC 1975-78 – 1,5,9 – ISSN: 0016-5093 – mf#7358 – us UMI ProQuest [616]

Gastroenterology and endoscopy news – New York NY 1986-91 – 1,5,9 – (cont: american journal of proctology, gastroenterology & colon & rectal surgery) – ISSN: 0162-6566 – mf#1996,02 – us UMI ProQuest [616]

Gastroenterology clinics of north america – Philadelphia PA 1987+ – 1,5,9 – ISSN: 0889-8553 – mf#12719,01 – us UMI ProQuest [616]

Gastroenterology nursing – v7-19. 1985-96 – (= ser Sga Journal) – 1,5,6,9 – $65.00r – (formerly: sga journal) – us Lippincott [616]

Gastrointestinal endoscopy – Oxford, England 1949+ – 1,5,9 – ISSN: 0016-5107 – mf#6217 – us UMI ProQuest [616]

Gastrointestinal radiology – Heidelberg, Germany 1981-92 – 1,5,9 – ISSN: 0364-2356 – mf#13169 – us UMI ProQuest [616]

Gastronome sans argent / Scribe, Eugene – Paris, France. 1821 – 1r – us UF Libraries [440]

Gasunas, Geraldine *see* The physiological responses of the addition of hand held weights to stationary bicycling

Gataker, T *see* True contentment in the gaine of godlines, with its self-sufficiencie

Gatch, Asbury P *see* 9th regiment, ov cavalry

Gatcomb's musical gazette, devoted to the interest of banjo, mandolin and guitar – v.1-12. 1887-99 – 1 – us L of C Photodup [780]

The gate and the cross : or, pilgrim's progress in romans: an excursus and parallelism / Peck, George Bacheler – Boston: Watchword Publ Co, 1889 – 1mf – 9 – 0-8370-4692-0 – mf#1985-2692 – us ATLA [220]

Gate city journal – Nyssa OR: Vahl & Megorden, -1937 [wkly] – 1 – (began in 1910. cont: nyssa sun; cont by: nyssa gate city journal (1937-)) – us Oregon Lib [071]

Gate city news / National Association of Letter Carriers – nov 1979-feb/mar 1985 – 1r – 1 – (cont: branch 5 newsletter) – mf#1209351 – us WHS [380]

The gate of peace / Carman, Bliss – New Canaan [CT: s.n.], 1909 – 1mf – 9 – 0-665-78239-X – mf#78239 – cn Canadiana [810]

Gate of the pacific / Pim, Bedford Clapperton Trevelyan – London, England. 1863 – 1r – us UF Libraries [972]

The gates ajar / Phelps, Elizabeth Stuart – London, Ont: E A Taylor, 1869 [mf ed 1984] – 3mf – 9 – 0-665-32331-X – mf#32331 – cn Canadiana [830]

Gates county journal – Ladysmith WI. 1901 dec 28-1902 may 31, 1902 jun 7-1904 jan 9, 1904 jan 16-1905 aug 26, 1905 sep 2-1905 nov 18 – 4r – 1 – (cont: weekly journal [ladysmith, wi]; cont by: rusk county journal) – mf#932423 – us WHS [071]

Gates, Errett *see*
- Disciples of christ
- The early relation and separation of baptists and disciples

Gates, Hartley Baxter *see*
- The dominion of canada, its interests, prospects and policy

Gates, Helen Dunn *see* A consecrated life

Gates, Henry Louis Jr *see* Black literature, 1827-1940

Gates, Otis H *see* Federal food and drug act decisions

Gates researcher – v1 n1-v4 n4 [1982 sep-1985 dec] – 1r – 1 – mf#979778 – us WHS [071]

Gates vs burgoyne : a plea in behalf of gates: "anchor" i e john watts de peyster draws out an answer – S.l: s.n, 1883? – 1mf – 9 – mf#32175 – cn Canadiana [975]

Gateshead and tyneside echo – England. -w. 24 Apr 1879-19 Jan 1880. (2 reels) – 1 – uk British Libr Newspaper [072]

Gateshead guardian – England. -w. 6 Jul 1895-23 Jun 1900. (Wanting 1896, 1897). (3 reels) – 1 – uk British Libr Newspaper [072]

Gateshead observer – England. -w. 18 Nov 1837-6 Oct 1886. (Wanting Jul 1870-Jul 1871). (59 reels) – 1 – uk British Libr Newspaper [072]

Gateshead post – 1986-Jun 1988; Jul 7-Dec 29 1988; Jan-Jun 1989; Jul 6-Dec 28 1989; 1990-96 – 1 – uk British Libr Newspaper [072]

Gateway – Frankfurt. 1981 may 1-1986 mar 28 – 1r – 1 – mf#1045644 – us WHS [071]

Gateway diggers / Gateway Genealogical Society (Iowa) – 1983 mar-1988 jan/mar – 1r – 1 – (continued by: clinton county gateway genealogical society newsletter) – mf#1496844 – us WHS [929]

Gateway Genealogical Society (Iowa) *see* Gateway diggers

Gateway heritage – St Louis MO 1980+ – 1,5,9 – ISSN: 0198-9375 – mf#12372.01 – us UMI ProQuest [071]

The gateway to china : pictures of shanghai / Gamewell, Mary Ninde – New York: Fleming H Revell, [1916] [mf ed 1995] – (= ser Yale coll) – 252p (ill) – 1 – 0-524-09149-8 – mf#1995-0149 – us ATLA [915]

Gateway Transportation Co Inc *see* Gateways

Gateways / Gateway Transportation Co Inc – sep 1976-sep 1979 – 1r – 1 – mf#645615 – us WHS [380]

Gatfworld – Sewickley PA 1989+ – 1,5,9 – (cont: graphic arts abstracts) – ISSN: 1048-0293 – mf#17375 – us UMI ProQuest [740]

Gathered fragments / Woodard, Luke – Columbus, O[hio]: Jos H Miller, 1883 – 1mf – 9 – 0-524-07172-1 – mf#1991-2961 – us ATLA [240]

Gathering place – v1 n5-v2 n4 [1971 jun-1972 apr] – 1r – 1 – mf#1056831 – us WHS [071]

Gathering post – may 1982-oct 1984 – 1r – 1 – mf#1081910 – us WHS [071]

GAYRAUD

Gati sacca : [a novel] / Citta – Ran Kun: Tan On Ca pe tuik 1954 [mf ed 1990] – 1r with other items – 1 – (in burmese) – mf#mf-10289 seam reel 192/2 [§] – us CRL [830]
Gatien, Felix X *see* Histoire du cap-sante
Gatilok : ou, l'art de bien se conduire / Ukana Suttantaprija, Ind – Phnom-Penh: Editions de l'Institut bouddhique 1959-1973 [mf ed 1990] – 1r with other items – 1 – (title in khmer on added t.p: gatilok r cpap dunman khluan; text in khmer) – mf#mf-10289 seam reel 127/1 [§] – us CRL [170]
The gatineau beaver : official organ of the forty-third battalion bazaar – Ottawa: Sergeant's Mess at the [D]rill Hall, [1899] – 9 – ISSN: 1190-7355 – mf#P04317 – cn Canadiana [355]
Gatineau, Pean *see* Leben und wunderthaten des heiligen martin
Gatley, congregational: burials 1777-1967 – [North Cheshire FHS] – (= ser Cheshire church registers) – 1mf – 9 – £2.50 – mf#181 – uk CheshireFHS [929]
Gatling gun : a periodical of the period – v1 n1-5, n7-10 [1898 jun-oct, dec-1899 mar] – 1r – 1 – mf#1056833 – us WHS [071]
Gatn [german-american trade news] – New York NY 1973-80 – 1,5,9 – ISSN: 0192-0103 – mf#7214 – us UMI ProQuest [337]
O gato : album de caricaturas – Rio de Janeiro, RJ. 1911-27 set 1913 – (= ser Ps 19) – mf#P03,02,18-22 – bl Biblioteca [870]
Gato azul / Pozo Seiglie, Orlando Del – Havana, Cuba. 1964 – 1r – us UF Libraries [972]
El gato negro – Caceres, 1923, 1924 y 1932 – 5 – sp Bibl Santa Ana [073]
Gaton Richiez, Carlos *see* Jurisprudencia en las republica dominicana
Gator – Helsingfors: Svenska litteratursallskapet i Finland forlag 1949 [mf ed Bloomington IN: Indiana Uni Lib, Preservation Dept 1984] – 1r – 1 – us Indiana Preservation [390]
Los gatos mail *see* Miscellaneous saratoga newspapers
[Los gatos-] mail – CA. 1893-1953 – 35r – 1 – $2100.00 – (aka: los gatos mail news) – mf#BC02406 – us Library Micro [071]
[Los gatos-] news – CA. 1881-1904; 1906-15 [wkly] – 7r – 1 – $420.00 – mf#B02407 – us Library Micro [071]
[Los gatos-] times – CA. 1936-59 – 27r – 1 – $1620.00 – (aka: daily times) – mf#BC02405 – us Library Micro [071]
[Los gatos-] weekly times – CA. 1982 – 25r – 1 – $1500.00 (subs $120y) – mf#B02410 – us Library Micro [071]
Gatsiskii, A S *see* Nizhegorodskii sbornik, izdavaemyi nizhegorodskim gubernskim statisticheskim komitetom
Gattenberger, K *see* Vliianie russkogo zakonodatel'stva na proizvoditel'nost' torgovogo bankovogo kredita
Gatter, Christoph Wilh Jakob *see* Neues technologisches magazin
Gatterer, Christoph Wilhelm Jakob *see* Technologisches magazin
Gatterer, Christoph Wilhelm Jakob et al *see* Annalen der forst- und jagd-wissenschaft
Gatterer, Johann Christoph *see*
– Allgemeine historische bibliothek
– Historisches journal von mitgliedern des kgl historischen instituts zu goettingen
Gatti, Attilio *see* Sangoma
Gattine, M A de *see* Relation curieuse et nouvelle d'un voyage de congo fait es annees 1666 et 1667
Gattula, Erasmus *see*
– Ad historiam cassinensis accessiones
– Historia abbatiae cassinensis
Gatty, Alfred *see* Baptism misunderstood
Gaubil, A *see*
– Memoire sur le thibet & sur la royaume des eleuthes
– Memoire sur les isles que les chinois appellent isles de lieou-kieou...
Gaubil, J *see* Catalogue synonymique des coleopteres d'europe et d'algerie
Gaubote : nachrichtenblatt fuer den gau teutoburger wald weser-bergland des touristenvereins "die naturfreunde" – Bielefeld DE, 1926-1933 n4 – 1r – 1 – gw Misc Inst [790]
La gauche r.d.r : Journal du rassemblement democratique revolutionnaire – n1-13. Paris. mai 1948-mars 1949 – 1 – (mq no. 9-11) – fr ACRPP [325]
La gauche revolutionnaire – n1-14. Paris. oct 1935-janv 1937 – 1 – fr ACRPP [325]
Gaucho / Coni, Emilio Angel – Buenos Aires, Argentina. 1945 – 1r – us UF Libraries [972]
Gauchos / Azevedo, Thales De – Salvador, Brazil. 1958 – 1r – us UF Libraries [972]
Gaud, Fernand *see* Les mandja (congo francais)
Gaud Rodriguez, Santos *see* Joven cadete
Gaudapada Acarya *see* The agamasastra of gaudapada
Gaudeamus! : lieder aus dem engeren und weiteren / Scheffel, Joseph Viktor von – Stuttgart: A Bonz, 1903 – 1r – 1 – us Wisconsin U Libr [780]
Gaudesius, Giedrius *see* Frueh- und spaetmorbiditaet und -letalitaet nach interventioneller therapie bei diabetikern mit instabiler angina pectoris
Gaudette Peace and Justice Center *see* Harvest of justice
Gaudier, Benito *see* Nuestro mayaguez de ayer y el verdadero origen de...
Gaudig, H *see* Prinz friedrich von homburg
Gaudij paschales iesv christi redivivi : in gloriosissimae resurrectionis ejus laetam celebrationem relatio historia, a qvatuor evangelistis consignata, & melodia harmonica adornata / Besler, Samuel – Breslae [i.e. Wroclaw] in officina typographica Baumanniana [1612] [mf ed 19–] – 1r – 1 – mf#film 1379 – us Sibley [780]
Gaudin, Jacques *see* Voyage en corse, et vues politiques sur l'amelioration de cette isle
Gaudy, Franz, Freiherr von *see* Franz freiherrn gaudys poetische werke
Gauguin, Paul *see* Paul gauguin's intimate jounrals
Gauken – Madison, Stoughton WI. 1890 jul 23 – 1r – 1 – (continued by: normannen) – mf#940199 – us WHS [071]
Das gauklerzelt : roman / Bruees, Otto – Guetersloh: C Bertelsmann, [1943] [mf ed 1989] – 224p – 1 – mf#7092 – us Wisconsin U Libr [830]
Gaul, L *see* Alberts des grossen verhaeltnis zu plato
Gaul, Sabine Hildegard Elisabeth *see* Vergleich ausgewaehlter mineralstoffe und spurenelemente in baerlauch- und knoblauchpflanzen unterschiedlicher standorte
Gauldin, Robert *see* Historical development of scoring for the wind ensemble
La gaule chretienne : d'apres les ecrivains et les monuments anciens – Paris: Librairie Hachette, 1879 – 1mf – 9 – 0-524-04610-7 – mf#1990-1270 – us ATLA [240]
Gaulke, Johannes *see* Hagenow und sohn
Gaulois – Brussels Belgium, 11 oct 1944-5 apr 1945 – 1/2r – 1 – uk British Libr Newspaper [074]
Le gaulois – Paris: Impr de G Kugelman, [1868-]. jan 16-21,23,26-31, feb 1-4,7-8,15,20,22-25, mar 24-25,28-30 1871 – (issues filmed as pt of: commune de paris newspapers) – us CRL [074]
Le gaulois – Paris. 5 juil 1868-30 mars 1929 – 1 – (litteraire et politique) – fr ACRPP [073]
Le gaulois – Petite gazette, critique, satirique et anecdotique puis Journal hebdomadaire biographique illustre. Dir. Jean Dolent et Alfred Sirven. no. 1-140; n.s., no. 1-35. Paris. 10 nov 1857-1er sept 1861 – 1 – fr ACRPP [073]
Le gaulois du dimanche – Paris. 20 juin 1897-aout 1914 [wkly] – 1 – (supplement litteraire et illustre du gaulois quotidien) – fr ACRPP [440]
Gaulot, Paul *see* A friend of the queen (marie antoinette-count de fersen)
Gault, Robert *see* Popery
Gaultier, D Z *see* L'alienation mentale devant la justice criminelle
Gaultier, J *see*
– L'anatomie du calvinisme...
– Table chronographique de l'estat du christianisme
Gault's decisions / Georgia – 1v. 1820-1846 (all publ) – (= ser Georgia appellate reports) – 1mf – 9 – $1.50 – (no pre-nrs cases) – mf#LLMC 94-001 – us LLMC [340]
Gaume, Jean *see* Abrege du catechisme de perseverance
Gaumont – Berlin DE, n.d. [1906/07?] – 1 – gw Mikrofilm [074]
Gaumont *see* Kinematographische wochenschau
Gaunt, Mary Eliza Bakewell *see* Reflection in jamaica
Gauntlet – (Carlile). v1-60. 1833-34 [all publ] – (= ser Radical periodicals of great britain, 1794-1914. period 1) – 10mf – 9 – $175.00 – us UPA [900]
Gauntlet / University of Wisconsin Center – Baraboo/Sauk County – v10 n5-v12 n2 [1978 feb 6-1985 nov 18] – 1r – 1 – mf#1331229 – us WHS [071]
Gauss, Carl Friedrich *see* Theoria motus corporum coelestium in sectionibus conicis solem ambientium
Gauss, J H *see* The bible's authority supported by the bible's history
Gauss, Karl *see* Reformationsversuche in der basler bischofsstadt pruntrut
Gaussen, L *see* Theopneustic
Gaussen, S R L *see* Theopneusty
Gaussen, Samuel Robert Louis *see*
– The canon of the holy scriptures from the double point of view of science and of faith
– The prophet daniel
Gaussy, Fernand *see* Laclos 1741-1803.
Gautama, the buddha / Radhakrishnan, Sarvepalli – London: Humphrey Milford, [1938] – (= ser Samp: indian books) – us CRL [280]
Los gautes espanoles. madrid...1954 / Menendez Pidal Navascues, Fautismo – Hidalguia 11, Abril-junio, 1954 – 1 – sp Bibl Santa Ana [946]
Gautherot, Gustave *see* Septembre 1792. histoire politique des massacres...
Gauthier de Brecy, Charles *see* Revolution de toulon, en 1793
Gauthier, Georges *see* Bibliographie de madame gabrielle roy
Gauthier, H *see* Dictionnaire des noms geographiques contenus dans les textes hieroglyphiques
Gauthier, Yves *see* Alcool, alcoolisme, milieu de travail
Gauthier-Chasse, Helene *see* A diable-vent
Gautier Benitez, Jose *see* Poesias
Gautier Dapena, Jose A *see* Trayectoria del pensamiento liberal puertorriqueno
Gautier, Emile Felix *see*
– L'afrique noire occidentale
– Missions au sahara
Gautier, J *see* Le voyageur dans le royaume des pays-bas
Gautier, Judith *see* Richard wagner et son oeuvre poetique depuis rienzi jusqu'a parsifal
Gautier, L *see* Oeuvres poetiques d'adam de saint-victor
Gautier, Lucien *see*
– Au dela du jourdain
– La mission du proph ete ezechiel
– Souvenirs de terre-sainte
Gauttier d'Arc, Edouard *see*
– Ceylan
– Histoire des conquetes des normands
Gauvreau, Joseph *see*
– Entretien au peuple
– Proces-verbaux des assemblees
Gauvreau, Louis N *see* Petit traite sur la culture du tabac
Gavagan, Joseph Andrew *see* Powers of the united states supreme court
Gavakari – Manamade, India. Jul-Sept 1966 – 1r – 1 – us L of C Photodup [079]
Gavalda y Cabre, Jose Maria de *see* Reparacion y ejemplaridad...
Gavan Duffy, Thomas *see* Hope
Gavault, Paul *see* Petite chocolatiere
Gavaut, minard and cie / Gondinet, Edmond – Paris, France. 1879 – 1r – us UF Libraries [440]
Gavazzi *see* Orations
Gavazzi, Alessandro *see* The lectures complete of father gavazzi
Gavel / Milwaukee Bar Association – v1-35 and 1966 index. 1938-76 – 69mf – 9 – $103.00 – mf#LLMC 84-607 – us LLMC [340]
Gavel (milwaukee) – v1-35. 1938-76 (all publ) – 9 – $220.00 set – ISSN: 0093-1845 – mf#102881 – us Hein [340]
Gavel (north dakota) – v15-47. 1970-99 – 9 – $281.00 set – ISSN: 0093-1845 – mf#400900 – us Hein [340]
Gaventa William C *see* Journal of religion, disability and health
Gavidia, Francisco *see*
– Cuentos y narraciones
– Discursos
– Encomendero
– Historia moderna de el salvador
Gavillet, Andre *see* Litterature au defi aragon surrealiste
The gavimath and palkigundu inscriptions of asoka / ed by Turner, R L – Hyderabad: Dept of Archaeology, Govt of Hyderabad, 1952 – (= ser Samp: indian books) – us CRL [700]
Gavin, Antonio *see* The great red dragon
Gavin, Timothy P *see* Mechanisms for the decline in arterial oxygenation during exercise in normoxia and acute hypoxia
Gavino Rodriguez, Martin *see* Exposicion del... ayuntamiento...badajoz
Gaviota / Caballero, Fernan – Madrid, Spain. 1881 – 1r – us UF Libraries [025]
Gaviota / Caballero, Fernan – Madrid, Spain. v1-2. 1928 – 1r – us UF Libraries [025]
Gavrilov, A V *see* Ocherki istorii s peterburgskoi sinodalnoi tipografii vyp 1
Gavroche – Paris. nov 1944-mai 1948 [wkly] – 1 – (hebdomadaire litteraire, artistique, politique et social) – fr ACRPP [073]
Gavroche – Paris, France. 9 nov 1944-8 nov 1945; 17 jan-26 sep 1946 – 1r – 1 – uk British Libr Newspaper [072]
Gawan : ein mysterium / Stucken, Eduard – 4. Aufl. Berlin: E Reiss, [1911] – 1r – 1 – us Wisconsin U Libr [430]
Gawrila sarytschew's russisch-kaiserlichen generalmajors von der flotte achtjaehrige reise im nordoestlichen sibirien, auf dem eismeere und dem nordoestlichen ozean – Leipzig 1805-15 [mf ed Hildesheim 1995-98] – 3v on 6mf – 9 – €120.00 – 3-487-28968-7 – gw Olms [910]
Gawsworth, st james – [North Cheshire FHS] – (= ser Cheshire monumental inscriptions) – 2mf – 9 – £3.25 – mf#120 – uk CheshireFHS [929]
Gawsworth, st james: baptisms 1557-1885 – [North Cheshire FHS] – (= ser Cheshire church registers) – 3mf – 9 – £4.00 – mf#282 – uk CheshireFHS [929]
Gawsworth, st james: burials 1557-1997 – [North Cheshire FHS] – (= ser Cheshire church registers) – 7mf – 9 – £8.00 – mf#209 – uk CheshireFHS [929]
Gay activists alliance records, 1970-1983 – 21r – 1 – (contains minutes of meetings, international and general correspondence of group members and a large number of ephemera) – us Primary [305]
Gay Calbo, Enrique *see* Bandera
Gay community news – -w. Boston, G.C.N., Inc – 1 – us Wisconsin U Libr [305]
Gay, Ebenezer *see* A call from macedonia
Gay hoa cuc : tho / Mong Tuyet – Paris: Rung Truc 1974 [mf ed 1992] – on pt of 1r – 1 – mf#11052 r219 n3 – us Cornell [810]
Gay liberator – v1,no.11-. Sept 1971-. Detroit: Pansy Press.ill. Continues: Detroit Gay Liberator – 1 – us Wisconsin U Libr [305]
Gay life : the midwest gay weekly – Chicago. v1-11 n31/32. jun 20 1975-jan 30 1986// – 11r – 1 – Can$845.00 – (suppl incl: escape (for wisconsin), and sister spirit. documents the beginning of the aids crisis as it affected the gay community in chicago and the midwest) – cn McLaren [305]
Gay, Luz *see* Poesias
Gay madison / United (Madison, WI: Organization) – feb/mar 1979-sep 1982 – 1r – 1 – (continued by: out (madison, wi)) – mf#626193 – us WHS [305]
Gay news – -bw. 1972-. London, England: Gay News Ltd – 1 – us Wisconsin U Libr [305]
Gay news *see* Sexual politics in britain
Gay Peoples Union *see* U w s p – gay peoples union
Gay renaissance newsletter – feb 1976-feb 1978 – 1r – 1 – mf#402516 – us WHS [305]
Gay renaissance newsletter – Gay Center, Madison. 1979? -m. Ceased publ – 1 – us Wisconsin U Libr [305]
The gay rights movement : from the new york public library's international gay information center archives – 2 sects – (= ser Gay and lesbian studies) – 45r coll – 1 – (sect 1: the mattachine society of new york records, 1951-76 24r c39-28891. sect 2: gay activists alliance records, 1970-83 21r c39-28892) – mf#C39-28890 – us Primary [305]
Gay rights movement, series 3 : act up: the aids coalition to unleash power – 159r – 1 – us Primary [305]
Gay rights movement, series 4 : national gay and lesbian task force records, 1973-2000 – [mf ed 2002] – 298r – 1 – us Primary [305]
Gay rights movement, series 5 : gay activism in britain from 1958: the hall-carpenter archives from the london school of economics / London School of Economics. The Hall-Carpenter Archives – [mf ed 2002] – 1 – (pt1: the albany trust c92r. pt2: the campaign for homosexual equality: forthcoming) – us Primary [305]
Gay rights movement, series 6 : atlanta lesbian feminist alliance archives, c1972-1994 – [mf ed 2003] – ca 190r in 4pts – 1 – (pt1: administrative files 14r. pt2: subject files 42r. pt3: archives 11r. pt4: periodical collections 123r) – us Primary [305]
Gay, Sophie *see* Physiologie du ridicule
Gay, Sydney Howard *see* James madison
Gay, Teofilo *see* Histoire des vaudois
Gaya De Garcia, Maria Cristina *see* Raiz y cielo
Gaya, L de *see* L'art de la guerre...
Gaya Nuno, Juan Antonio *see* Zurbaran
Gayacao, D Juan *see* Nuevo vocabulario y guia de conversaciones espanol-panayano
Gayarre, Charles *see* Essai historique sur la louisiane
Gay-Calbo, Enrique *see*
– Discoursos leidos en la recepcion
– En el centenario de ayestaran
– Formacion de la sociedad cubana
– Nuestro problema constitucional
Gaye, J W *see* Carteggio inedito d'artisti dei secoli 14, 15, 16...
Gaye-i milliye – Sivas, 1922. Sahib-i Imtiyaz: Maksud Azmi. n2. 3 mart 1338 [1922], 3,6. 10 mart 1338 [1922] – (= ser O & t journals) – 1mf – 9 – $25.00 – us MEDOC [956]
Gayer, Arthur Edward *see* Fallacies and fictions relating to the irish church establishment...
Gayer, Christina D *see* Physiological discriminators of rowing performance in male, club rowers
Gayet, Louis *see* Le grand schisme d'occident
Gayford, Sydney Charles *see* The future state
Gayley, Charles Mills *see* The classic myths
Gayot, Gerard G *see*
– Clerge indigene
– Titans de 1804
Gayraud, Hippolyte *see* Thomisme et molinisme

Gaz – Paris, France. 10 feb-15 oct 1860; 15 feb 1861-1868; 31 jan 1869-15 sep 1870; 15 jul 1871-15 dec 1872; 1873-80; 15 jan 1881-15 jun 1886 – 7r – 1 – uk British Libr Newspaper [072]

Le gaz – Paris. -w.10 Feb 1857-15 Jun 1886 – 7r – 1 – uk British Libr Newspaper [622]

Gaza, a city of many battles : from the family of noah to the present day / Dowling, Theodore Edward – London: SPCK; New York: E S Gorham, 1913 – 1mf – 9 – 0-7905-6990-6 – (incl bibl ref) – mf#1988-2990 – us ATLA [956]

Gazali / Carra de Vaux, Bernard, Baron – Paris: Felix Alcan, 1902 – (= ser Les Grands Philosophes) – 1mf – 9 – 0-524-02295-X – mf#1990-2918 – us ATLA [260]

Gazavat – Berlin DE, 1943 20 oct-1944 23 jun [gaps] – 1r – 1 – (in cyrillic) – gw Misc Inst [074]

Gazelle *see* Die forschungsreise s m s gazelle in den jahren 1874 bis 1876 unter kommando des kapitaen zur see freiherrn von schleinitz

Gazet – Antwerp Belgium, 22 sep 1944-12 jul 1945 – 1/2r – 1 – uk British Libr Newspaper [074]

Gazet van antwerpen – Antwerp Belgium, 19 sep 1944-11 jul 1945; 17 dec 1946 – 1r – 1 – uk British Libr Newspaper [074]

Gazet van brussel – Brussels Belgium, 26 aug 1916-3 apr 1917 – 1r – 1 – uk British Libr Newspaper [074]

A gazeta – Laguna, SC: Typ Central, 23 mar, jul, 02 nov 1930 – (= ser Ps 19) – bl Biblioteca [079]

A gazeta : a voz do povo – Florianopolis, SC. 16 ago 1934; jun 1935; mar-abr 1936; set-out 1937; fev 1939; abr 1940; fev, 30 maio 1945 – (= ser Ps 19) – bl Biblioteca [079]

Gazeta : organ, posviashchennyi voprosam khlebnoi torgovli – Spb., 1907-1914 – 12mf – 9 – mf#R-8130 – ne IDC [077]

Gazeta academica : periodico dedicado as sciencias, artes e lettras – Rio de Janeiro, RJ: Typ Cosmopolita, 22 jan-02 fev 1876 – (= ser Ps 19) – mf#P19A,04,47 – bl Biblioteca [073]

Gazeta academica : periodico dos alumnos da faculdade de medicina – Rio de Janeiro, RJ: Typ de A Marques & C, 07-19 jun 1884 – (= ser Ps 19) – mf#DIPER – bl Biblioteca [610]

Gazeta academica : periodico dos alumnos da faculdade de medicina – Rio de Janeiro, RJ: Typ de A Marques & C, 15 jun-out 1883; maio-15 out 1884 – (= ser Ps 19) – bl Biblioteca [610]

Gazeta academica de sciencias & lettras – Recife, PE: Typ do Correio da Noite, jun 1879 – (= ser Ps 19) – mf#P17,02,183 – bl Biblioteca [073]

Gazeta artistica : revista quinzenal musica,litteratura e bellas artes – Sao Paulo, SP. 11 nov 1909-out 1911; nov 1913-fev 1914 – (= ser Ps 19) – mf#P18,02,07 – bl Biblioteca [700]

Gazeta clinica : revista trimensal de medicina e cirurgia – Rio de Janeiro, RJ. 04 maio 1888 – (= ser Ps 19) – mf#DIPER – bl Biblioteca [610]

Gazeta colonial – Caxias do Sul, RS: Typ da Gazeta Colonial, 30 jun 1906; 10 out 1908-17 maio 1909 – (= ser Ps 19) – bl Biblioteca [079]

Gazeta da bahia – Bahia, 08 jan 1879-31 dez 1886 – (= ser Ps 19) – mf#P11,02,68 – bl Biblioteca [321]

Gazeta da tarde – Bahia, 23 jun 1881; maio 1882; set 1884; out, 25 nov 1885 – (= ser Ps 19) – mf#P11,02,20 – bl Biblioteca [321]

Gazeta da tarde – Para, 21 abr 1890 – (= ser Ps 19) – bl Biblioteca [079]

Gazeta da varginha – Varginha, MG. 26 abr 1894 – (= ser Ps 19) – bl Biblioteca [079]

Gazeta das petas – Rio de Janeiro, RJ. 03 out 1881 – (= ser Ps 19) – mf#P17,01,147 – bl Biblioteca [079]

Gazeta de alemquer – Alemquer, PA. 20-30 jan,abr-maio 1885; jan, mar-maio, jul-ago, out-dez 1890; maio 1891; jan-mar, jul-dez 1894; jan-abr 1895; out-02 dez 1908 – (= ser Ps 19) – mf#DIPER – bl Biblioteca [079]

Gazeta de botafogo : orgao independente, noticioso e litterario – Rio de Janeiro, RJ. 08-22 ago 1909 – (= ser Ps 19) – mf#DIPER – bl Biblioteca [079]

Gazeta de cataguazes : consagrada aos interesses da lavoura e commercio – Cataguases, MG: Typ da Gazeta de Cataguazes, 13 jan 1884 – (= ser Ps 19) – bl Biblioteca [380]

Gazeta de leopoldina – Leopoldina, MG. 27 set 1896; fev-jul 1898; jan, mar, maio, jul-set, dez 1913; fev, set 1915; abr 1918; jan, dez 1922; dez 1926; jan-abr 1927; fev, out 1932; abr, jul-dez 1933; out 1934;18 set 1960 – (= ser Ps 19) – mf#P11B,3,54 – bl Biblioteca [440]

Gazeta de macao – Macao: M M D Pegado, jan-aug 1839 – us CRL [079]

Gazeta de macao – Macao: Typographia do Governo, 1825-26 – 1r – 1 – us CRL [079]

Gazeta de macau e timor = Ao-men hsin-wen chih – Macau: Typographia mercantil, [1872-]. sep 20 1872-apr 20 1874 – 1r – 1 – us CRL [079]

Gazeta de mage – Rio de Janeiro, RJ. 01 jan-15 fev 1903 – (= ser Ps 19) – mf#DIPER – bl Biblioteca [079]

Gazeta de manaos : orgao imparcial – Manaus, AM. 13 jun 1886 – (= ser Ps 19) – bl Biblioteca [079]

Gazeta de manicore : orgao do partido conservador – Manaus, AM: Typ da Gazeta de Manicore, 16 jan 1887 – (= ser Ps 19) – bl Biblioteca [079]

Gazeta de noticias – Maceio, AL. 22 jul 1879; abr-maio 1880; nov 1881; jan-mar,maio-jul 1882; 15 fev 1883 – (= ser Ps 19) – mf#P18B,01,32 – bl Biblioteca [079]

Gazeta de noticias – Rio de Janeiro Brazil, feb-oct 1916 – 4r – 1 – uk British Libr Newspaper [079]

Gazeta de ouro fino : orgam official dos poderes municipaes – Ouro Fino, MG. 31 jan 1892-dez 1894; jan, mar-out, dez 1895; jan-dez 1896; jul 1897; fev, mar 1898; dez 1910; fev, maio-out, dez 1914; jan 1915; 11 jul 1925 – (= ser Ps 19) – mf#P11B,03,61 – bl Biblioteca [321]

Gazeta de paracatu – Paracatu, MG. 25 mar 1894 – (= ser Ps 19) – bl Biblioteca [079]

Gazeta de petropolis – Petropolis, RJ: Typ da Gazeta de Petropolis, 02 jun 1892-28 dez 1904 – (= ser Ps 19) – mf#P18A,04,32 – bl Biblioteca [321]

Gazeta de porto novo – Porto Novo da Cunha, MG. 12 mar 1896; 03 ago 1899 – (= ser Ps 19) – bl Biblioteca [079]

Gazeta de propria : orgao dos interesses sociaes, commercio e lavoura do baixo s francisco – Propria, SE. Typ da Gazeta de Propria, 30 mar-maio, ago 1884; 24 jun 1885 – (= ser Ps 19) – bl Biblioteca [079]

Gazeta de uberaba – Uberaba, MG: [s.n.] 19 mar 1888; abr 1889; nov 1901; jan-fev 1902; mar, dez 1906; maio 1908; fev, maio, jul 1912; jan, set 1917; 19 out 1938 – (= ser Ps 19) – mf#P11B,03,48 – bl Biblioteca [079]

Gazeta dlia russkikh rabochikh / Rabotnik – Geneva, 1875-1876. v1-2. nos 1-14/15 – 3mf – 9 – mf#R-18146 – ne IDC [077]

Gazeta do banho : orgao dedicado aos banhistas – Rio de Janeiro, RJ. 25 dez 1881 – (= ser Ps 19) – mf#P17,01,148 – bl Biblioteca [079]

Gazeta do commercio – Paraiba: [s.n.] 01 maio 1894 – (= ser Ps 19) – bl Biblioteca [380]

Gazeta do instituto hahnemanniano do brasil – Rio de Janeiro, RJ: Typ Teixeira & Comp, ago-set, dez 1859; jan 1860 – (= ser Ps 19) – mf#P17,01,139 – bl Biblioteca [610]

Gazeta do rio de janeiro – Rio de Janeiro, RJ: Impressao Regia, 10 set 1808-31 dez 1822 – (= ser Ps 19) – mf#P4,1,1-28 – bl Biblioteca [321]

Gazeta do sertao – Campina Grande, PB: Typ da Gazeta do Sertao, 01 set 1888-06 maio 1891 – (= ser Ps 19) – mf#P11B,04,05 – bl Biblioteca [321]

Gazeta do sul – Quilimane: [Alfredo de Aguiar, may 22-jul 20, dec 21 1889 – us CRL [079]

Gazeta dos domingos : revista encyclopedica semanal do rio de janeiro – Rio de Janeiro, RJ: Typ Americana, 06 jan-03 fev 1839 – (= ser Ps 19) – mf#P02,04,30 – bl Biblioteca [073]

Gazeta dos estados – Rio de Janeiro, RJ. 30 ago-15 nov 1908 – (= ser Ps 19) – mf#DIPER – bl Biblioteca [079]

Gazeta dos hospitaes do rio de janeiro – Rio de Janeiro, RJ: Typ Guanabarense de L A F Menezes, 01 mar-dez 1850; jan, mar-dez 1851; jan-15 fev 1852 – (= ser Ps 19) – mf#P14,04,39 – bl Biblioteca [362]

Gazeta fluminense – Petropolis, RJ: [s.n.] 03 fev-31 dez 1905 – (= ser Ps 19) – bl Biblioteca [079]

Gazeta handlowa – Warsaw, Poland. 1950-70; 1972-78 – 17r – 1 – us L of C Photodup [943]

Gazeta kabare : organ noveishego tipa / ed by Tomskii, I S – Pg[Petrograd]: T-vo "Gazetnykh rabotnikov" 1918 [1918-] – (= ser Asn 1-3) – n1,2 [1918] item 8, on reel n1 – 1 – mf#asn-2 008 – ne IDC [077]

Gazeta krakowska – Krakow, Poland. Jun 1952-Apr 1976; 1978-1992 – 76r – 1 – us L of C Photodup [943]

Gazeta kujawska – Inowroclaw, Poland. 1953-Jan 1960 – 12r – 1 – (some missing issues) – us L of C Photodup [943]

Gazeta ludowa – Warsaw. Poland. -d. Jul 1946-Nov 1949. (14 reels) – 1 – uk British Libr Newspaper [943]

Gazeta ludowa – Warsaw, Poland. Nov 1945-Nov 1949 – 5r – 1 – us L of C Photodup [943]

Gazeta maritima : orgao da marinha mercante, navegacao, commercio e industrias maritimas – Rio de Janeiro, RJ. 12 nov-20 dez 1903 – (= ser Ps 19) – mf#P19A,04,140 – bl Biblioteca [380]

Gazeta medica brazileira : revista quinzenal de medicina, cirurgia e pharmacologia – Rio de Janeiro, RJ: Typ de Oliveira & Silva, 15 mar-31 ago 1882 – (= ser Ps 19) – mf#DIPER – bl Biblioteca [610]

Gazeta mod i novostei – M., 1831-1835 – 79mf – 9 – mf#R-3351 – ne IDC [077]

Gazeta morska – Gdansk, Poland. Sept 1-30 1945; July 14-31 1946 – 1r – 1 – us L of C Photodup [077]

Gazeta naval – Rio de Janeiro, RJ: Imprensa Industrial, 01-15 dez 1877 – (= ser Ps 19) – bl Biblioteca [355]

Gazeta niedzielna – London, UK. May 1949- – 1 – uk British Libr Newspaper [072]

Gazeta official – Belem, PA. Typ Commercial, 06 set-out 1858; jan 1859-30 jun 1860 – (= ser Ps 19) – bl Biblioteca [350]

Gazeta oficial / Panama – 1876-1969. Incomplete – 1 – us L of C Photodup [324]

Gazeta olsztynska – Olsztyn, Poland. Dec 1952-1957 (scattered issues); Feb 1959-1976 – 35r – 1 – us L of C Photodup [943]

Gazeta operaria : orgam dedicado especialmente aos interesses dos artistas e operarios – Rio de Janeiro, RJ. 08-22 jan 1881 – (= ser Ps 19) – mf#P05,04,191 – bl Biblioteca [780]

Gazeta operaria : orgam proletariado do rio de janeiro – Rio de Janeiro, RJ. 09 dez 1884-25 fev 1885 – (= ser Ps 19) – mf#P19A,04,91 – bl Biblioteca [780]

Gazeta operaria – Rio de Janeiro, RJ. 20 set 1902-fev 1903; set-08 dez 1906 – (= ser Ps 19) – mf#P11A,01,11 – bl Biblioteca [790]

Gazeta politicheskaia i literaturnaia : izdanie russkikh politicheskikh emigrantov – Geneva, 1877-1890. nos 1-112 – 30mf – 9 – mf#R-18123 – ne IDC [077]

Gazeta politicheskaia i literaturnaia – M., 1859-1860 – 29mf – 9 – mf#1787 – ne IDC [077]

Gazeta politicheskaia i literaturnaia – Spb., 1825-1863 – 3369mf – 9 – (missing: 1836(43); 1861(41-43, 49, 250, 281, 298-299); 1862(292); 1863(169-171)) – mf#R-4240 – ne IDC [077]

Gazeta politicheskaia i literaturnaia / ed by Pogodin, M P – M., 1867-1868 – 28mf – 9 – mf#R-8371 – ne IDC [077]

Gazeta polska – Warsaw, Poland. 25 Jan 1919; Apr-Dec 1934; Feb 1939; Jun-Sept 1939 – 4r – 1 – us L of C Photodup [943]

Gazeta polska w chicago – Chicago, IL: W Dyniewicz, 1888; 1890; jan 5 1905; 1907; 1909-10; 1912-jan 20 1917 – 14r – 1 – us CRL [071]

Gazeta pomorska – Bydgoszcz, Poland. Jul-Dec 1950; 1952-92 – 84r – 1 – us L of C Photodup [943]

Gazeta popular – Macae, RJ: Typ da Gazeta Popular, 05 dez 1877; mar 1878; jun, nov 1880; set 1883; abr 1884; 12 ago 1885 – (= ser Ps 19) – mf#P18A,04,33 – bl Biblioteca [321]

Gazeta poranna – Warsaw, Poland. Jan-Feb 1919 – 1r – 1 – us L of C Photodup [943]

Gazeta poznanska (gazeta zachodnia) – Poznan, Poland. nov 1950-1973; apr 1975-92 – 80r – 1 – us L of C Photodup [077]

Gazeta promyshlennosti : severnyi muravei – Spb., 1830-1833 – 51mf – 9 – mf#R-3769 – ne IDC [077]

Gazeta promyshlennosti i torgovli – M., 1862. v1-51 – 23mf – 9 – (missing: 1862(6-8, p 53-57; 29-32; 35-39)) – mf#R-1501 – ne IDC [077]

Gazeta promyshlennosti i torgovli : pribavlenie k gazete "den" – M., 1863(1-52) – 15mf – 9 – (missing: 1863(6, 13, 22-24)) – mf#R-1502 – ne IDC [077]

Gazeta rabochego i krest'ianskogo / Russia – St. Petersburg. 1917-18 – 1 – $19.00 – us L of C Photodup [324]

Gazeta robotnicza – London, UK. Aug 1948-Apr 1952 – 1 – uk British Libr Newspaper [072]

Gazeta robotnicza – Wroclaw, Poland. Apr 1951-1992 – 90r – 1 – (missing: 1952) – us L of C Photodup [943]

Gazeta romaneasca = La gazette roumaine – Paris. 1 no. 1935 – 1 – fr ACRPP [073]

Gazeta sadowa warszawska – Warsaw. On film: v2-65; 1874-1938. LL-0239 – 1 – us L of C Photodup [340]

Gazeta tarnowska – Tarnow, Poland. 1953 – 1r – 1 – us L of C Photodup [943]

Gazeta torunska – Torun, Poland. 1953; Feb-Mar 1954; 1955-59 – 13r – 1 – us L of C Photodup [943]

Gazeta ufficiale / Sicily – Palermo. 1958-66 – 1 – us NY Public [945]

Gazeta voprosov / Russkoe natsional'no-osvoboditel'noe dvizhenie – Moscow, Russia. n1(sen 1996)-n2[3](1997), n4(1997) – 1 – (cont by: russkaia gazeta voprosov) – mf#mf-12248 (reel 4) – us CRL [077]

Gazeta vserossiiskogo soiuza soldat i matrosov : izdanie tsentralnogo komiteta partii sotsialistov-revoliutsionerov – Paris, 1907-1914, nos 1-60; 1921, nos 1-3 – 20mf – 9 – (missing: 1907, nos 1, 4/5, 7/8, 11/12; 1907-1908, no 17) – mf#R-18044 – ne IDC [077]

Gazeta warszawska – Warsaw, Poland. Jan-Feb 1919 – 1r – 1 – us L of C Photodup [943]

Gazeta wspolczesna – Bialystok, Poland. Feb-Sept 1916; 1953-90; 1992 – 65r – 1 – us L of C Photodup [943]

Gazeta wyborcza – Warsaw, Poland. May 17 1989-Jan 1 1993 – 25r – 1 – us L of C Photodup [077]

Gazeta zachodnia *see* Gazeta poznanska (gazeta zachodnia)

Gazeta zielonogorska – Zielona Gora, Poland. 1953-Jul 1954; Oct 1954-Feb 1955 (scattered issues); Jun 1955-Jun 1970 – 26r – 1 – us L of C Photodup [943]

Gazeta zyrtare e rps te shqiperise – Tirane, 1944-1991. jan 18 1955-83 – 13r – us CRL [079]

Gazeteci lisani / Sait Pasa, Kuecuek – Dersaadet: Sabah Matbaasi, 1327 – (= ser Ottoman literature, writers and the arts) – 2mf – 9 – $40.00 – us MEDOC [470]

Gazetinha – orgam litterario, noticioso, recreativo, critico, humoristico... – Juiz de Fora, MG: Typ do Pharol, 01 out 1886 – (= ser Ps 19) – mf#P17,02,76 – bl Biblioteca [079]

Gazetinha : orgam provisorio do club quatro de marco – Uberaba, MG: Typ da Gazetinha, 03-06 set 1896 – (= ser Ps 19) – mf#P11B,03,93 – bl Biblioteca [321]

Gazetinha – Porto Alegre, RS: Typ da Gazetinha, 03,17,21 jan 1897 – (= ser Ps 19) – bl Biblioteca [079]

Gazetinha – Uba, MG. 12 nov 1896 – (= ser Ps 19) – bl Biblioteca [079]

Gazetta da matta – Carangola, MG. 15 nov 1896 – (= ser Ps 19) – bl Biblioteca [079]

Gazetta italiana di londra – London, UK. 20 Sept 1896-22 Sept 1900 – 1 – uk British Libr Newspaper [072]

Gazette / Adams Co. Manchester – v1 n1. aug 1867-jul 1869 [wkly] – 1r – 1 – mf#B6623 – us Ohio Hist [071]

Gazette – Paris. 31 jan 1855-64; 1869; jan-3 feb, 22 apr 1870; 1 sep 1877-oct 1900; 8 jan 1906-19 jun 1915 [daily] – 89 1/4r – 1 – (aka: gazette de france) – uk British Libr Newspaper [074]

Gazette / Ashtabula Co. Conneaut – 12/1834-5/36, 8-12/36, 9/41-3/1843 [wkly] – 1r – 1 – mf#B5523 – us Ohio Hist [071]

Gazette / Ashtabula Co. Jefferson – (1/1878-06,6/14-2/16,12/25-10/1926) damaged [wkly] – 4r – 1 – mf#B8576-8579 – us Ohio Hist [071]

Gazette – Wedderburn OR: E M M Bogardus, [wkly] – 1 – (began in 1895. cont: gold beach gazette (-1895)) – us Oregon Lib [071]

Gazette / Belmont Co. Saint Clairsv – (1931-55), 1971-73 [wkly] – 10r – 1 – mf#B25319-25328 – us Ohio Hist [071]

Gazette / Belmont Co. Saint Clairsv – (9/1825-98,2/03-06,3/10-1917) [wkly] – 20r – 1 – mf#B4340-4359 – us Ohio Hist [071]

Gazette – Blantyre, Lanarkshire, Scotland, UK. 1935-7 Feb 1953; 13 Jun 1959-31 Jan 1964. -w.8 reels – 1 – uk British Libr Newspaper [072]

Gazette / Brown Co. Georgetown – oct 1905-sep 1922,nov 1922-sep 1925 [wkly] – 9r – 1 – mf#B7409-7417 – us Ohio Hist [071]

Gazette / Clark Co. Springfield – jan-jun 1908 [daily] – 2r – 1 – mf#B34660-34661 – us Ohio Hist [071]

Gazette / Columbiana Co. East Liverpool – dec 1871-nov 1875 [wkly] – 1r – 1 – mf#B1796 – us Ohio Hist [071]

Gazette – Plum Creek, NE: C B Signor. v1 [n1] mar 12 1884- (wkly) [mf ed -1887 (gaps) filmed 1979] – 2r – 1 – (cont by: plum creek semi-weekly gazette) – us NE Hist [071]

Gazette – Stevens Point WI. 1885 sep 23/1886 jun 9-1919 sep 2/1921 jan 11 [gaps] – 32r – 1 – (cont: portage county gazette; cont by: stevens point journal [stevens point, wi: 1872]; gazette and stevens point journal) – mf#954776 – us WHS [071]

Gazette – Whitewater WI. 1892 jul 13-1897 dec 31; 1898-1899 jan 18 – 2r – 1 – (cont: weekly gazette [whitewater, wi]; cont by: whitewater gazette [whitewater, wi: 1899]) – mf#949820 – us WHS [071]

Gazette – Monroe WI. 1882 mar 3-nov 30, 1883 may 10 – 2r – 1 – (continued by: monroe sun (monroe, wi: 1881); monroe sun-gazette) – mf#1108632 – us WHS [071]

Gazette – Watertown WI. 1879 jul 22-1880 feb 3 – 1r – 1 – (continued by: watertown gazette) – mf#945679 – us WHS [071]

Gazette / Delaware Co. Delaware – jan 1858-dec 1874 [wkly] – 7r – 1 – mf#B131-137 – us Ohio Hist [071]

Gazette – mai 1631-1761. – 1 – (devenu: gazette de france. 1762-15 aout 1792. a paru ensuite sous des titres divers. paris. mai 1631-1792. a partir de sept 1805 voir a gazette de france) – fr ACRPP [073]

Gazette – Eau Claire WI. 1896 oct 2-1898 jul 15 – 1r – 1 – mf#961927 – us WHS [071]

Gazette / Fairfield Co. Lancaster – (8/1826-2/1856, 3/1858-11/1910) [wkly, semiwkly, wkly] – 26r – 1 – mf#B11187-11212 – us Ohio Hist [071]

Gazette : feuille officielle d'annonces judiciaires et legales / L'Hotel Drouot – Paris, 1929 – 1 – fr ACRPP [340]

Gazette / Franklin Co. Columbus – jan 1868-jul 1869 [wkly] – 2r – 1 – mf#B1219-1220 – us Ohio Hist [071]

Gazette / Great Britain. Western Pacific High Commission – Sura. 1943-1967 – 1 – us NY Public [980]

Gazette / Greene Co. Xenia – jan 1903-dec 1912 [semiwkly] – 9r – 1 – mf#B10440-10448 – us Ohio Hist [071]

Gazette / Highland Co. Hillsboro – 1879-1910,1912-15,1917-24 [wkly] – 21r – 1 – mf#B8988-9008 – us Ohio Hist [071]

Gazette / Highland Co. Hillsboro – jan 1910-dec 1911, jan-dec 1916 [wkly] – 2r – 1 – mf#B33944-33945 – us Ohio Hist [071]

Gazette / Highland Co. Hillsboro – sep 1857-mar 1861, (feb-aug 1868) [wkly] – 2r – 1 – mf#B407-408 – us Ohio Hist [071]

Gazette / Huron Co. Bellevue – (1869, 1871, 1873) scattered [wkly] – 1r – 1 – mf#B13209 – us Ohio Hist [071]

Gazette / Huron Co. Bellevue – 1877-82, 94-98, 1900-aug 1919 [wkly] – 14r – 1 – mf#B11829-11842 – us Ohio Hist [071]

Gazette / Huron Co. Bellevue – 1940-jun 1950 [daily] – 21r – 1 – mf#B11843-11863 – us Ohio Hist [071]

Gazette / Huron Co. Bellevue – jan 1906-dec 1975 (fire damaged) [daily] – 126r – 1 – mf#B937-1062 – us Ohio Hist [071]

Gazette / Huron Co. Bellevue – jan 1976-dec 1984 [daily] – 27r – 1 – mf#B13425-13451 – us Ohio Hist [071]

Gazette / Huron Co. Bellevue – jan 1985-dec 1993 [daily] – 27r – 1 – mf#B34781-34807 – us Ohio Hist [071]

Gazette / Knox Co. Centerburg – may 1935-dec 1983 [wkly] – 21r – 1 – mf#B13215-13235 – us Ohio Hist [071]

Gazette / Mid-Continent Railway Historical Society – 1971 nov-1973 mar, 1973 apr-1975 aug – 2r – 1 – (cont: railway gazette [north freedom, wi]; cont by: railway gazette [north freedom, wi: 1975]) – mf#635694 – us WHS [380]

Gazette / Montgomery Co. Far/Germ. – aug 27 1993-dec 26 1996 – 3r – 1 – mf#B36996-36998 – us Ohio Hist [071]

Gazette / Muskingum Co. Zanesville – (jan 1852-apr 1856) [wkly] – 1r – 1 – mf#B5614 – us Ohio Hist [071]

Gazette / Nigeria. Central-West State – Kaduna. v. 1, no. 1-v. 2. no. 5. Supplement. Jun. 15, 1967-Feb. 29, 1968 – 1 – us NY Public [324]

Gazette / Nigeria. North-Eastern State – Kaduna. v. 1, no. 1-v. 3, no. 50. Suppl. 15 Jun 1967-18 Dec 1969 – 1 – us NY Public [960]

Gazette / Nigeria. North-Western State – Kaduna. v. 1, no. 1-v. 3, no. 51. Suppl. 15 Jun 1967-26 Dec 1969 – 1 – us NY Public [960]

Gazette / Nigeria. Western Region – Ibadan. 1957-1966 – 1 – us NY Public [960]

Gazette – Oshkosh WI. 1896 feb 26 – 1r – 1 – mf#959797 – us WHS [071]

Gazette / Preble Co. Camden – (1894-97, 1899-oct 1902) [wkly] – 3r – 1 – mf#B3948-3950 – us Ohio Hist [071]

Gazette / Star Co. Massillon – mar 1843-aug 1844 [wkly] – 1r – 1 – mf#B6624 – us Ohio Hist [071]

Gazette – Sturtevant WI. 1985 jun 6-nov 14 – 1r – 1 – mf#1049878 – us WHS [071]

Gazette / Trumbull Co. Cortland – v1 n1. may 1876-79, 1882 [wkly] – 2r – 1 – mf#B11295-11296 – us Ohio Hist [071]

Gazette / Warren Co. Lebanon – jan 1884-feb 1893 [semiwkly, wkly] – 4r – 1 – mf#B10807-10810 – us Ohio Hist [071]

Gazette / Washington Co. Marietta – v1 n1. jul 1833-oct 1840, dec 1840-feb 1843 [wkly] – 2r – 1 – mf#B12085-12086 – us Ohio Hist [071]

The gazette – Parnell, NZ. 1981-83 – 1 – (aka: parnell-remuera gazette) – mf#11.40 – nz Nat Libr [079]

The gazette – Basingstoke, England. Extra edition. -w. Aug-Dec 1980 – 39ft – 1 – uk British Libr Newspaper [072]

The gazette – Basingstoke, England. Mid-week edition. 1976-80. Lacking 1979 – 8r – 1 – uk British Libr Newspaper [072]

The gazette – Basingstoke, England. Week-end edition. -w. 1976-81. Lacking 1979. 20 reels – 1 – uk British Libr Newspaper [072]

The gazette – Beatrice, NE: [s.n.] v1 n1. mar 31 1928- [mf ed [1993]] – 1r – 1 – us NE Hist [071]

The gazette – Montreal, Quebec, CN. 1878- – 36r/y – 1 – Can$2660.00 silver Can$2500.00 vesicular – cn Commonwealth Imaging [071]

The gazette – Portland, ME. apr 16 1798-apr 22 1799 – 1 – us CRL [071]

The gazette and bankruptcy court reporter – v1 no 1-21. 1867-68 – 2mf – 9 – $3.00 – mf#LLMC 95-290 – us LLMC [340]

Gazette and bulletin – Williamsport, PA., 1911 – 13 – $25.00r – us IMR [071]

Gazette and daily – York, PA., 1933-1970 – 13 – $25.00r – us IMR [071]

Gazette and democrat – New Castle, PA. -w 1870-1874 – 13 – $25.00r – us IMR [071]

Gazette and east florida herald – St Andrews, FL. 1821 jul-dec; 1823-1826 – 1r – us UF Libraries [071]

Gazette and echo (beeston and west notts gazette echo) – England. 1913-38, 1940, 1949-50 – 24r – 1 – uk British Libr Newspaper [072]

Gazette and herald dushore – Dushore, PA., 1905-1925 – 13 – $25.00r – us IMR [071]

Gazette and inner city news – Auckland, NZ. 1987-89 – 3r – 1 – mf#11.70 – nz Nat Libr [079]

The gazette and land bulletin – Waycross, GA; Brunswick, GA; Tampa, FL: Gazette Pub Co, 1896 (wkly) [mf ed 1947] – (= ser Negro Newspapers on Microfilm) – 1r – 1 – us L of C Photodup [071]

Gazette and miami register / Butler Co. Hamilton – oct 1819-dec 1820 [wkly] – 2r – 1 – mf#B13149-13150 – us Ohio Hist [071]

Gazette and miami valley advertiser / Montgomery Co. Germantown – mar 1845-dec 1848 [wkly] – 1r – 1 – mf#B5458 – us Ohio Hist [071]

Gazette and stevens point journal – Stevens Point WI. 1921 jan 19-may 18, 1921 may 25-1923 mar 21 – 2r – 1 – (cont: gazette [stevens point, wi]; stevens point journal) – mf#954786 – us WHS [071]

Gazette and universal daily advocate – Philadelphia, PA., 1826-1834 – 13 – $25.00r – us IMR [071]

Gazette anecdotique, litteraire, artistique et bibliographique – Paris. 1876-99 – 1 – fr ACRPP [073]

Gazette [basingstoke & district ed] – Basingstoke, England 16 may 1975-9 jan 1981 – 1 – (iss twice wkly fr 1981 to aug; cont: basingstoke gazette [3 oct 1969-9 may 1975]; cont by: basingstoke & north hants gazette [16 jan 1981-21 feb 1986]) – uk British Libr Newspaper [072]

Gazette [basingstoke & north hampshire] – Basingstoke, England 12 jan 1998-31 aug 2001 [mf 1979-] – 1 – (cont: basingstoke & north hampshire gazette [24 feb 1986-9 jan 1998]; cont by: basingstoke & north hampshire gazette [7 sep 2001-5 jun 2005]) – uk British Libr Newspaper [072]

Gazette [basingstoke, tadley] – Basingstoke, England 9 jun 2005- [mf 1979-] – 1 – (cont: basingstoke & north hampshire gazette [7 sep 2001-5 jun 2005]) – uk British Libr Newspaper [072]

Gazette constitutionnelle de l'arrondissement de cambrai – Cambrai. 4 dec 1838-12 fevr 1839, 1848-27 mars 1852, avr 1863, 18 nov 1865-25 janv 1866. – 1 – (devenu: gazette de cambrai) – fr ACRPP [073]

Gazette d'agriculture, commerce, arts et finances *see* Gazette du commerce

Gazette de bonn – Bonn DE, 1789 n1-209 – 1r – 1 – gw Misc Inst [323]

Gazette de bruxelles – (Gazette francoise des Pays-Bas Gazette des Pays-Bas). Brussels. Belgium. -w. 30 Jul 1756-3 Feb 1763. (6 reels) – 1 – uk British Libr Newspaper [949]

Gazette de champfleury – Paris. no1-10. nov-dec 1856 – 1 – fr ACRPP [073]

Gazette de charleroi – Charleroi Belgium, 27 nov 1941; jan-5 jul 1944 – 1r – 1 – uk British Libr Newspaper [074]

Gazette de france – Paris. 23 sept 1805-juin 1841, 1842-sept 1919 – 1 – (a paru sous des titres divers) – fr ACRPP [073]

Gazette de france *see*
– Gazette

Gazette de grande-bretagne – London, UK. 19 Jul 1924-3 Aug 1929; 1931-30 Jul 1932 – 1 – uk British Libr Newspaper [072]

Gazette de guernesey – Saint Peter Port. 14 Oct 1837; 20 Apr 1839; 6 Jan 1894-28 Mar 1936 – 1 – uk British Libr Newspaper [072]

La gazette de hollande – The Hague, Netherlands. -w. 1 Aug 1914-28 June 1922. 14 reels – 1 – uk British Libr Newspaper [949]

La gazette de hue – Hue: Dac-Lap [1935-[1935 mar 1-dec 27, 1936 jan 10-jun 7, 1939 jan 7-dec 30] – 1r – 1 – mf#mf-11762 seam – us CRL [079]

Gazette de la guadeloupe – La Guadeloupe. I, n22-32, 39-52; II, n1. mai 1788-janv 1789 – 1 – fr ACRPP [073]

Gazette de la louisiane – New Orleans LA. 1818 nov 19-21 – 1r – 1 – (cont: louisiana gazette and new orleans mercantile advertizer; cont by: gazette de l'etat) – mf#861798 – us WHS [071]

Gazette de lausanne et journal suisse – Lausanne: P Seguin, jan 3 1966-dec 31 1966/jan 1 1967 – 4r – 1 – us CRL [074]

Gazette de l'equateur – Coquilhatville: L'Avenir belge, aug 1952-sep 1957 – 1r – 1 – us CRL [079]

La gazette de l'etat / Pondicherry. India – Pondicherry. 1964-1966 – 1 – us NY Public [324]

La gazette de l'hotel drouot : l'hebdomadaire des ventes publiques – 1891-1988 – 106r – 1 – $15,950.00 – us UPA [700]

Gazette de liege – Liege Belgium, 26 oct 1944-8 aug 1946 – 3r – 1 – uk British Libr Newspaper [074]

Gazette de l'ile de jersey – Saint Helier. 5 Aug 1786-27 Dec 1788 – 1 – uk British Libr Newspaper [072]

Gazette de lyon – Lyon. 5 avr 1845-27 juil 1853 – 1 – fr ACRPP [073]

Gazette de paris – Paris. oct 1871-janv 1873 – 1 – (hebdomadaire puis journal financier hebdomadaire) – fr ACRPP [073]

Gazette de paris – Paris. oct 1789-aout 1792 – 1 – fr ACRPP [073]

La gazette de paris : non politique – Paris. 6 avr 1856-1er avr 1860 avec un spec. de sept 1853. Interrompu du 29 sept au 30 nov 1859 – 1 – fr ACRPP [073]

Gazette de prague – Prague, Czechoslovakia. Apr 1920-Sept 1926 – 4r – 1 – us L of C Photodup [077]

Gazette de prague – Prague, Czechoslovakia. -w. 4 jan 1922-29 sep 1926 – 2 1/4r – 1 – uk British Libr Newspaper [077]

Gazette de sante : oder gemeinnuetziges medicinisches magazin / ed by Rahn, Johann Heinrich – Zuerich 1782-85 – (= ser Dz) – 4jge on 18mf – 9 – €180.00 – mf#k/n3596 – gw Olms [610]

Gazette de sante : oder gemeinnuetziges medicinisches magazin /.../ – Zuerich (CH), 1782-86 – 2r – 1 – gw Misc Inst [610]

Gazette de sorel – Sorel, QC: G I Barthe, 1862-73 – 6r – 1 – ISSN: 1204-2870 – cn Library Assoc [971]

Gazette des architectes et du batiment – Paris, 1853-86 – (= ser Architectural periodicals at avery library, columbia university) – 11r – 1 – $1490.00 – us UPA [720]

Gazette des ardennes – Charleville (F), Rethel (F), 1914 15 nov-1918 1 oct [gaps] – 3r – 1 – gw Misc Inst [074]

Gazette des ardennes : journal des pays occupes paraissant quatre fois par semaine – Charleville. nov 1914-nov 1918 – 1 – fr ACRPP [073]

Gazette des ardennes / edition illustree – Charleville (F), 1914 1 nov-1918 8 nov – 1r – 1 – gw Misc Inst [074]

Gazette des beaux arts – Paris, 1859-68 v1-25; 1869-88 v1-38; 1889-1908 v1-40; 1909-19 v1-15; 1920-28 v1-18+ind 1859-63; 1864-68; 1859-1908 – 1160mf – 9 – mf#0-501 – ne IDC [700]

Gazette des beaux-arts – Paris. 1866-67, 1962 – 1 – fr ACRPP [700]

Gazette des cours de l'europe : le royaliste, ami de l'humanite – Paris. 46 no. sept 1790-avr 1792 – 1 – fr ACRPP [944]

Gazette des deux-ponts / journal – Zweibruecken, Mannheim DE, 1795 18 mar-1795 13 oct, 1795 13 nov-1796 9 jul, 1796 29 jul-1798, 1809-1810 31 oct – 4r – 1 – (title varies: 29.7.1796?: gazette des deux ponts / nouvelles. local ed: zweibruecken 1774-75, 1796-1797 30 sep) – gw Misc Inst [074]

Gazette des etrangers – Paris. no. 2531-2750. juil-13 dec 1868 – 1 – (litterature, beaux-arts, sport, bourse, industrie, chronique de la cour, de la ville et du theatre) – fr ACRPP [073]

Gazette des femmes – Quebec [Province]. 1979 oct [v1 n1]-1985 mar/apr – 1r – 1 – (cont: bulletin du c s f) – mf#1046447 – us WHS [305]

Gazette des lettres – Paris, France. 31 may 1947-22 jul 1950 – 1r – 1 – uk British Libr Newspaper [072]

Gazette des saints-simoniens *see* L'organisateur

Gazette des travaux publics *see* Le locateur

Gazette des tribunaux – Paris, France. -w. 1850-1939; 6 jan 1940-1 mar 1943; 11 july-31 dec 1943; 1 jan, 23 apr 1944; 22 apr-dec 1945; 1947-1955 – 172 3/4r – 1 – uk British Libr Newspaper [072]

Gazette des tribunaux – Port-au-Prince: Charles Heraux, 3eme annee n2-4e annee n16 15 janv 1887-15 aout 1888; 5e annee n1-n24 1 janv-15 dec 1889; 13e annee n1-n16 1 janv-15 aout 1902; 14e annee n1-16e annee n14 1 janv 1903-15 juil 1905; 17e annee n1-19e annee n21 1 janv 1906-1 nov 1908; 20e annee n1-n9; 22e annee n1-24e annee n51 – 30 sheets – 9 – us CRL [079]

Gazette des tribunaux *see* Index to gazette des tribunaux

La gazette des trois-rivieres – Trois Rivieres, QC: Ludger Duvernay, 1817-21 – 1r – 1 – cn Library Assoc [971]

Gazette du bas-languedoc – Nimes. 24 aout 1848-49, 1951 – 1 – (biwkly, daily after1er dec 1849) – fr ACRPP [073]

Gazette du bon ton – Paris, 1913-25 [mf ed Chadwyck-Healey] – (= ser Art periodicals on microform) – 4r – 14 – uk Chadwyck [740]

Gazette du cinema – Dir. Maurice Scherer, Eric Rohmer. no. 1-5. Paris. mai-nov 1950 – 1 – fr ACRPP [790]

Gazette du commerce – Paris. 1763-83 – 1 – (puis gazette du commerce, de l'agriculture et des finances; gazette d'agriculture, commerce, arts et finances) – fr ACRPP [380]

Gazette du commerce, de l'agriculture et des finances *see* Gazette du commerce

Gazette du commerce; et d'agriculture – Saint Helier. 29 Apr 1837 – 1 – uk British Libr Newspaper [072]

Gazette du commerce et litteraire – Montreal, QC: Chez F Nesplet & C Berger, 1778-79 – 1r – 1 – ISSN: 0826-0583 – cn Library Assoc [971]

La gazette du franc : le conseiller de la famille francaise – Nos 93-145. Paris. 1927 – 1 – fr ACRPP [640]

Gazette du grand-duchy de francfort *see* Journal de francfort 1794

La gazette du languedoc – Memorial de Toulouse. Journal des interets provinciaux. Toulouse. juin 1831-52 – 1 – fr ACRPP [073]

Gazette du midi : journal du soir – Marseille. 1848-51 – 1 – fr ACRPP [073]

Gazette du palais – Paris, France. -d. 5 jun, 31 dec 1943; 1, 11 jan, 3, 9 may 1944; 31 dec 1955-1962; jul 1963-1977 – 83r – 1 – uk British Libr Newspaper [074]

Gazette ealing borough *see* Middlesex county times

Gazette extra – Basingstoke, England 20 aug 1980-6 may 1981 – 1 – (cont by: basingstoke gazette extra [13 may 1981-17 nov 1999]) – uk British Libr Newspaper [072]

Gazette extra [basingstoke & north hampshire ed] – Basingstoke, England 16 feb 2000-5 sep 2001 – 1 – (cont: extra [24 nov 1999-9 feb 2000]; cont by: basingstoke extra-12 sep 2001-) – uk British Libr Newspaper [072]

La gazette financiere – v1 n1-52. 1909-10 – (= ser O & t journals) – 10mf – 9 – $165.00 – us MEDOC [332]

Gazette (hammersmith and fulham ed) *see* Shepherds bush gazette and west london post

Gazette (hemel hempstead) *see* Hemel hempstead gazette and west herts advertiser

Gazette (hitchin) *see* Hitchin and royston express

Gazette judiciaire et commerciale de lyon – Lyons, France. 23 jun 1943-4 jul 1944 – 1/2r – 1 – uk British Libr Newspaper [072]

La gazette litteraire de l'europe – Paris. Par Arnaud et Suard. Mars 1764-1 mars 1766 (I-VIII) – 1 – fr ACRPP [400]

Gazette [london] : international journal for mass communication studies – London, England 1991+ – 1,5,9 – ISSN: 0016-5492 – mf#16787.01 – us UMI ProQuest [302]

Gazette (manitoba) – 1870-1900 – 8r – 1 – cn Library Assoc [971]

Gazette medicale de france – Paris, France 1977-80 – 1,5,9 – ISSN: 0016-5557 – mf#8625 – us UMI ProQuest [610]

Gazette [midweek ed] – Basingstoke, England 20 may 1975-18 dec 1979 – 1 – (cont: basingstoke midweek gazette [27 feb 1973-13 may 1975]) – uk British Libr Newspaper [072]

Gazette national : ou, monitor universel – 1789-1810 – 1 – $1320.00 – mf#0230 – us Brook [074]

Gazette nationale – Paris: Impr de A Gutot, jul 16-aug 13 1848 – us CRL [074]

Gazette nationale ou le moniteur universel *see* Le moniteur universel

Gazette news – Daytona Beach, FL. 1901-1915 – 8r – (gaps) – us UF Libraries [071]

Gazette [northern ed] – Basingstoke, England 7 jan 2000-31 aug 2001) – 1 – (cont by: basingstoke & north hampshire gazette (northern ed) 7 sep 2001-22 feb 2002) – uk British Libr Newspaper [072]

Gazette oakdale – Oakdale, NE: S M Figge, 1893-93// (wkly) [mf ed v1 n7. nov 17 1893 filmed 1973] – 1r – 1 – (merged with: public opinion (neligh ne) to form: yeoman (nelgih ne)) – us NE Hist [071]

Gazette of india / India – New Delhi. 1912-1942 – 1 – us NY Public [954]

Gazette of pakistan / Pakistan – 1967- – 1 – 65.00y – us L of C Photodup [954]

Gazette of the state of georgia – Savannah GA. 1783 feb 13 – 1r – 1 – (continued by: georgia gazette (savannah, ga: 1788)) – mf#855547 – us WHS [978]

Gazette of the state of south-carolina – Charleston SC. 1777 dec 23 – 1r – 1 – (cont: south-carolina gazette; cont by: state gazette of south carolina) – mf#780700 – us WHS [978]

Gazette of the united states – New York and Philadelphia. -d. 15 Apr 1789-14 Sep 1793, 23 Dec 1793-17 May 1797, 1 Aug-27 Dec 1798, 21-26 Nov 1803. (Imperfect). (4 reels) – 1 – uk British Libr Newspaper [071]

Gazette of the united states – Philadelphia. Sept 1 1791-Dec 31 1792 – 1 – us NY Public [071]

Gazette rancher-farmer : serving the ranchers, farmers and homemakers of southwest nebraska and northwest kansas – McCook, NE: [Harry D Strunk] feb 16 1954-feb 5 1957 (wkly) – 2r – 1 – (cont: mccook daily gazette; cont by: rancher and farmer) – us Bell [636]

Gazette series / Clark Co. Springfield – apr-dec 1905, jul 1906-jun 1908 [daily] – 7r – 1 – mf#B10820-10826 – us Ohio Hist [071]

Gazette series / Delaware Co. Delaware – 1875-89, 1891-1914 [wkly, semiwkly] – 32r – 1 – mf#B10407-10438 – us Ohio Hist [071]

Gazette series / Jefferson Co. Steubenville – (1874-87,90-00,05-06,10,19-3/1925) [daily] – 59r – 1 – mf#B3991-4049 – us Ohio Hist [071]

Gazette series / Jefferson Co. Steubenville – jan 1870-feb 1875 [wkly, semiwkly] – 2r – 1 – mf#B5543-5544 – us Ohio Hist [071]

Gazette shopper *see* Garfield county miscellaneous newspapers

The gazette (stevenage) *see* Stevenage gazette

Gazette thru register / Miami Co. Piqua – (1820-62) scattered [wkly] – 2r – 1 – mf#B8558-8559 – us Ohio Hist [071]

Gazette universelle : ou papiers-nouvelles de tous les pays et de tous les jours – Paris. dec 1789-aout 1792 – 1 – fr ACRPP [073]

Gazette van detroit – Detroit, Roseville MI. 1984 jun 21-1985 dec, 1986 jan 2-1987 sep 10 – 2r – 1 – mf#851851 – us WHS [071]

Gazette (yiewsley and w drayton ed) *see* Yiewsley and w drayton gazette

Gazette-advertiser – Giltner, NE: J C Bierbower. 8v. v38 n34. jun 20 1940-v45 n15. feb 13 1947 (wkly) [mf ed filmed 1975] – 2r – 1 – (formed by the union of: giltner gazette and: phillips advertiser. cont by: giltner gazette (1947)) – us NE Hist [071]

Gazette-chronicle / Belmont Co. Saint Clairsv – jan 1974-jul 1983 [wkly] – 5r – 1 – mf#B23278-23282 – us Ohio Hist [071]

Gazetteer – La Salle IL 1824 – 1 – mf#4455 – us UMI ProQuest [240]

Gazetteer of bauchi province / Gall, Frederick Beckles – London: Waterlow, 1920 – 1 – (filmed with: gazetteer of plateau province by c g ames) – us CRL [960]

Gazetteer of ilorin province – London: Waterlow, 1921 – 1 – (filmed with: gazetteer of bauchi province by f b gall) – us CRL [960]

Gazetteer of kano province / Gowers, William Frederick – London: Waterlow, 1921 – 1 – (filmed with: gazetteer of bauchi province by f b gall) – us CRL [960]

Gazetteer of muri province, up to december 1919 / ed by Fremantle, John Morton – [London: Waterlow, 1920] – 1 – (filmed with: gazetteer of bauchi province by f b gall) – us CRL [960]

Gazetteer of nupe province – London: Waterlow, 1920 – 1 – (filmed with: gazetteer of bauchi province by f b gall) – us CRL [960]

Gazetteer of plateau province – Lagos, 1933 – 1 – (filmed with: gazetteer of bauchi province by f b gall) – us CRL [550]

Gazetteer of sokoto province / Arnett, Edward John – London: Waterlow, 1920 – 1 – (filmed with: gazetteer of bauchi province by f b gall) – us CRL [910]

Gazetteer of the kontagora province : compiled from the provincial records / Duff, E C – London: Waterlow, 1920 – 1 – (filmed with: gazetteer of bauchi province by f b gall) – us CRL [960]

A gazetteer of the province of upper canada : to which is added, an appendix, describing the principal towns, fortifications and rivers in lower canada – New-York: publ by Prior and Dunning...: Pelsue & Gould, Print, 1813 [mf ed 1983] – 2mf – 9 – 0-665-44538-5 – mf#44538 – cn Canadiana [917]

Gazetteer of yola province – [Lagos: Govt Press, 1927] – 1 – (filmed with: gazetteer of bauchi province by f b gall) – us CRL [960]

Gazetteer of zaria province / Arnett, Edward John – [s.l: s.n.], 1920 – 1 – (filmed with: gazetteer of bauchi province by f b gall) – us CRL [910]

The gazette-journal – Hastings, NE: Wigton Bros. v8 n31. dec 9 1880- =v9 n49-1883// (wkly) [mf ed -nov 30 1882 filmed 1973] – 2r – 1 – (formed by the union of: adams county gazette and: hastings journal. split into: hastings daily gazette-journal and: hastings weekly gazette-journal) – us NE Hist [071]

Gazette-times – Corvallis OR: N R Moore, 1909 [wkly] – 1 – (formed by the union of: corvallis times and: corvallis weekly gazette; cont by: weekly gazette-times) – us Oregon Hist [071]

Gazette-times *see*
- Corvallis times
- Heppner herald
- Heppner times

Gazette-times (corvallis, or) – Corvallis OR: N R Moore, 1909 [wkly] – 1r – 1 – (related to daily ed: daily gazette-times, 1909. merger of: corvallis times; corvallis weekly gazette; cont by: weekly gazette-times) – us Oregon Lib [071]

Gazette-times (heppner, or) – Heppner OR: V Crawford, 1912-25 [wkly] – 1 – (merger of: heppner gazette (heppner, or); heppner times. absorbed: heppner herald; cont by: heppner gazette-times (1925-)) – us Oregon Lib [071]

Gazier, Augustin Louis *see* Histoire generale du mouvement janseniste depuis ses origines jusqu'a nos jours

Gazin, V *see* Moia gazeta [ostrov: 1918]

Gazophylacium medico-physicum oder schatzkammer (ael3/2) / Woyt, Johann Jacob – Leipzig 1709 [mf ed 1993] – (= ser Archiv der europaeischen lexikographie: fach-enzyklopaedien) – 13mf – 9 – €110.00 – 3-89131-151-6 – (filmed with: medicinisch- und natuerlicher dinge...; int by michael stolberg) – gw Fischer [610]

Gaztelu, Teodoro *see* Establecimiento de banos mineromedicinales de alange

Gazul, Arturo *see* El libro gris

Gazzam, Audley William *see* Gazzam's treatise on the bankrupt law

Gazzam's treatise on the bankrupt law / Gazzam, Audley William – 4th ed. Albany, N.Y.: Little, 1872. 789p. LL-579 – 1 – us L of C Photodup [346]

Gazzetta – 1974 feb-1976 oct 29, 1976 nov 5-1978 sep 29, 1978 oct 6-1980 dec, 1981-84 – 4r – 1 – mf#368293 – us WHS [071]

Gazzetta del massachusetts – Boston, MA: J M Gubitosi & Co, [aug 8 1903-1905; mar 23 1907-1911; 1915-1938] – 26r – 1 – us CRL [071]

Gazzetta delle campagne – Turin, Italy. 1876-80.-m. 1 reel – 1 – uk British Libr Newspaper [074]

Gazzetta di napoli – Naples, Italy. -w. 5 Jan 1877-31 Dec 1887. 22 reels – 1 – uk British Libr Newspaper [074]

Gazzetta di palermo – Palermo, Italy. Gazzetta provinciale La Nuova gazzetta. -w. 1 April 1869-31 Dec 1887. 31 reels – 1 – uk British Libr Newspaper [072]

La gazzetta di syracuse – Syracuse, NY: Ray Pub Co Inc, [mar 8 1918-mar 24 1939; jul 1940-feb 1952] – 1 – us CRL [071]

Gazzetta italiano di londra – London, UK. 23 May-31 Dec 1871 – 1 – uk British Libr Newspaper [072]

Gazzetta musicale di milano – Milan. 1842-1902. Lacking: v7-20; 24-26 – 24r – 1 – $459.00 – us L of C Photodup [780]

Gazzetta privilegiata di venezia – Venice, Italy. Gazzetta di Venezia. Jan. 1847-Jun. 1920 and supplement – 1 – us NY Public [074]

Gazzetta ufficiale della repubblica italiana / Italy – 1861-1894. 1899-1944. Turin, etc – 1 – us NY Public [945]

La gazzetta ufficiale dell'unione europee *see* The official journal of the european union

Gazzettino – London, UK. 3 May 1902 – 1 – uk British Libr Newspaper [072]

Gbenedio, Nelson A *see* Effect of flexibility exercises on range of motion and physical performance of developmentally disabled adults

Gciu 777 bulletin / Graphic Communications International Union – v37 n10-11 [1985 feb 8-mar 8], v38 n2-3, 5-9 [1985 jun 14-jul 12, sep 13-1986 jan 10], v39 n1 (1986 may 9] – 1r – 1 – mf#1056716 – us WHS [740]

Gciu news local 583 / Graphic Communications International Union – v38 n6-12 [1985 oct 11-1986 apr 1], v39 n1,3,5-8,10,12 [1986 may 9, jul 11, sep 12-dec 12, 1987 feb 13, apr 10], v40 n5 [1987 sep 11] – 1r – 1 – (cont: bindery news labor) – mf#1573610 – us WHS [331]

Gda : zeitschrift des gewerkschaftsbundes der angestellten – Berlin DE, 1920-1933 30 jun – 4r – 1 – gw Mikrofilm [331]

Gda : zeitschrift des gewerkschaftsbundes der angestellten – Leipzig DE, 1925, 1927-28 – 1r – 1 – gw Mikrofilm [331]

Gde vsenlenskaia tserkov? : k voprosu o soedinenii tserkvei i k ucheniiu o tserkvi / Svetlov, P – Sviato – Troitskaia Sergieva Lavra 1905 – 200p 4mf – 8 – (kritiko-bibliograficheskoe obozrenie literatury po starokato licheskomu voprosu v dukhovnoi pechati za 1904 god) – mf#R-7283 – ne IDC [243]

Gdr bulletin – -q. vol.1, no.1-2 (Apr. 1975)-. St. Louis, Mo.: Dept of Germanic Languages and Literature, Washington Univesity, 1975- . Newletter for literature and culture in the German Democratic Republic – 1 – us Wisconsin U Libr [943]

Geada, Rita *see*
- Cuando cantan las pisadas
- Desvelado silencio

Der geaechtete : die bruedersage / Kremer, Hannes – 2.aufl. Muenchen: F Eher, 1943 [mf ed 1992] – 153p – 1 – mf#7524 – us Wisconsin U Libr [390]

Gear, Hiram Lewis *see* A treatise on the law of landlord and tenant with special reference to the american law

Gear, James Henderson Sutherland et al *see*
- Bilharzia in south africa
- Bilharzia in south Africa
- Malaria in southern Africa

Geary County, KS *see* Births, deaths, and marriage record cards

Geauga Baptist Association *see* Printed minutes of the annual session of the geauga baptist association, 1835-1861

Geauga Co. Burton *see*
- Geauga leader series
- Geauga times leader
- Independent

Geauga Co. Chardon *see*
- Geauga county news
- Geauga county record
- Geauga democrat
- Geauga freeman
- Geauga record
- Geauga republic
- Geauga republican
- Geauga republican-record
- Geauga times leader
- Jeffersonian democrat
- Times
- Western reserve times

Geauga Co. Chesterland *see*
- News
- West geauga communicator

Geauga Co. Historical Society *see* Pioneer and general history of geauga co, [ohio]

Geauga Co. Middlefield *see*
- Geauga independent
- Messenger

Geauga county engineer records – Geauga, OH. 1802-1951 – 6r – 1 – us Western Res [620]

Geauga County Genealogical Society *see* Raconteur

Geauga county military history of ohio, 1669-1865 – 1r – 1 – mf#B27279 – us Ohio Hist [355]

Geauga county news / Geauga Co. Chardon – jan 1937-dec 1938 [wkly] – 1r – 1 – mf#B29879 – us Ohio Hist [071]

Geauga county news / Geauga Co. Chardon – jul 1922-36, 1939-jul 1944 [wkly] – 10r – 1 – mf#B12863-12872 – us Ohio Hist [071]

Geauga County. Ohio. Russell Township *see* Records, ms 3106

Geauga county record / Geauga Co. Chardon – jan 1903-dec 1918 [wkly] – 8r – 1 – mf#B11398-11405 – us Ohio Hist [071]

Geauga county record / Geauga Co. Chardon – jan 1919-dec 1921 [wkly] – 1r – 1 – mf#B12873 – us Ohio Hist [071]

Geauga county record : weekly democratic newspaper – Chardon, OH. 22 Apr 1892-30 Dec 1904 – 2r – 1 – us Western Res [071]

Geauga democrat / Geauga Co. Chardon – jan 1866-dec 1871 [wkly] – 2r – 1 – mf#B273-274 – us Ohio Hist [071]

Geauga democratic record : weekly democratic newspaper – Chardon, OH. 13 Jan 1887-28 Apr 1888 – 1r – 1 – us Western Res [071]

Geauga freeman / Geauga Co. Chardon – (may 1840-nov 1842) scattered (irreg] – 1r – 1 – mf#B32788 – us Ohio Hist [071]

Geauga gazette / Lake Co. Painesville – v1 n1. (aug 1828-dec 1832) [wkly] – 1r – 1 – mf#B32899 – us Ohio Hist [071]

Geauga gazette *see* Chardon spectator and geauga gazette

Geauga independent / Geauga Co. Middlefield – (oct 1884-apr 1885) [wkly] – 1r – 1 – mf#B32897 – us Ohio Hist [071]

Geauga leader series / Geauga Co. Burton – v1 n1. (dec 1874-dec 1943) [wkly] – 27r – 1 – mf#B32793-32819 – us Ohio Hist [071]

Geauga record / Geauga Co. Chardon – jan 1953-jul 1962 [wkly] – 7r – 1 – mf#B6450-6456 – us Ohio Hist [071]

Geauga record – Chardon, OH: Geauga Publ, Inc, 1952-62 (weekly) – 1 – (titled merged: geauga times-leader (burton, ohio), and geauga times-leader and record) – mf#34 G2.1 028 – us Western Res [071]

Geauga republic / Geauga Co. Chardon – apr 1850-apr 1851 [wkly] – 1r – 1 – mf#B275 – us Ohio Hist [071]

Geauga republic : weekly whig newspaper – Chardon, OH. 25 Dec 1849-17 Jan 1854 – 1r – 1 – us Western Res [071]

Geauga republican / Geauga Co. Chardon – jan-dec 1873 [wkly] – 1r – 1 – mf#B1220 – us Ohio Hist [071]

Geauga republican / Geauga Co. Chardon – v1 n1. jan-dec 1872, jan 1874-dec 1921 [wkly] – 23r – 1 – mf#B249-271 – us Ohio Hist [071]

Geauga republican : weekly republican newspaper – Chardon, OH. 1 Jan 1873-31 Dec 1873 – 1r – 1 – us Western Res [071]

Geauga republican and whig : weekly whig newspaper – Chardon, OH. 20 May 1843-16 Mar 1847 – 1r – 1 – us Western Res [071]

Geauga republican-record / Geauga Co. Chardon – 1923-25, 1927-42 [wkly] – 8r – 1 – mf#B12878-12885 – us Ohio Hist [071]

Geauga republican-record / Geauga Co. Chardon – jan 1943-dec 1952 [wkly] – 6r – 1 – mf#B6152-6157 – us Ohio Hist [071]

Geauga republican-record / Geauga Co. Chardon – jan-dec 1922, jan 1926-dec 1926 [wkly] – 2r – 1 – mf#B271-272 – us Ohio Hist [071]

Geauga times leader / Geauga Co. Burton – 1955-58 [wkly] – 4r – 1 – mf#B2031-2034 – us Ohio Hist [071]

Geauga times leader / Geauga Co. Burton – 1959-apr 1961 [wkly] – 3r – 1 – mf#B6109-6111 – us Ohio Hist [071]

Geauga times leader / Geauga Co. Chardon – may 1961-apr 1963 [wkly, semiwkly] – 3r – 1 – mf#B6111-6113 – us Ohio Hist [071]

Geauga times leader – Burton, OH: Geauga Times-Leader Co, 1942-mar 29 1962 (weekly); apr 2-jul 12 1962 (semiweekly) – 1 – (merger of: geauga leader (burton, ohio: 1924) and, middlefield times (middlefield, ohio: 1917). previous title: geauga record (chardon, ohio: 1952). cont by: geauga times-leader and record) – us Western Res [071]

Die gebaerde; der fremde / Wiechert, Ernst Emil – Zuerich: Im Verlag der Arche, 1947 – 1r – 1 – us Wisconsin U Libr [430]

Gebannt und erloest / Werner, E – Philadelphia: Morwitz. 2v in 1. [1889?] – 1 – us Wisconsin U Libr [830]

Gebannt und erloest / Werner, E – Philadelphia: Morwitz. 2v in 1. [1889?] – 1r – 1 – us Wisconsin U Libr [830]

Gebeda, Linda Albert *see* Mentoring system as a tool for classroom transformation

Gebel, Bjoern *see* Die leistungspolitik von dienstanbietern fuer mobile commerce im rahmen einer one-to-one marketing-konzeption

Geben und nehmen : schauspiel in fuenf aufzuegen / Langen, Martin – Paris: A Langen [1902?] – 1r – 1 – us Wisconsin U Libr [820]

Gebertus, M *see* Monumenta veteris liturgiae alemannicae

Das gebet bei paulus / Juncker, Alfred – Berlin: Edwin Runge, 1905 – 1mf – 9 – 0-8370-9554-9 – mf#1986-3554 – us ATLA [240]

Das gebet des herrn / Kamphausen, Adolf – Elberfeld: RL Friderichs, 1866 – 1mf – 9 – 0-524-08450-5 – mf#1993-2055 – us ATLA [240]

Das gebet hieremie des propheten : ausslegung diss gebets in gesang weyss / Kress, Johann – [Augsburg]: [Ulhart] 1525 – (= ser Hqab. literatur des 16. jahrh.) – 1mf – 9 – €20.00 – mf#1525b – gw Fischer [780]

Das gebet im judentum / Perles, Felix – Frankfurt a. M: J Kauffmann, 1904 – 1mf – 9 – 0-7905-3047-3 – mf#1987-3047 – us ATLA [270]

Das gebet in der aeltesten christenheit : eine geschichtliche untersuchung / Goltz, Eduard – Leipzig: J C Hinrichs, 1901 – 1mf – 9 – 0-7905-4531-4 – (incl bibl ref) – mf#1988-0531 – us ATLA [240]

Das gebet in der aeltesten christenheit : eine geschichtliche untersuchung / Goltz, Eduard von der – Leipzig: J.C. Hinrichs, 1901 – 1mf – us ATLA [240]

Gebet un erbauungsbuch fur israeliten / Prager, M – Brilon, Germany. 1860 – 1r – us UF Libraries [939]

Gebet und gottesdienst im neuen testament / Nielen, J M – Freiburg i.Br., 1937 – 7mf – 8 – €15.00 – ne Slangenburg [225]

Gebete und hymnen an nergal / Boellenruecher, J – Leipzig, 1904 – (= ser Leipziger semitistische studien) – 1mf – 9 – (leipziger semitistische studien. v1, pt 6)) – mf#NE-20108 – ne IDC [956]

Das gebetsproblem im anschluss an schleiermachers predigten und glaubenslehre / Menegoz, Fernand – Leipzig: JC Hinrichs, 1911 – 1mf – 9 – 0-7905-9813-2 – (bibliographical footnotes) – mf#1989-1538 – us ATLA [240]

Gebeurtenisse uit di kaffer-oorloge fan 1834 / Coetser, Paulus Petrus Johannes – Kaapstad, South Africa. 1963 – 1r – us UF Libraries [960]

Gebhardt, Bruno *see* Die gravamina der deutschen nation gegen den roemischen hof

Gebhardt, Hermann *see* The doctrine of the apocalypse

Gebhardt, Max *see* Untersuchungen zur biographie philipp zesens

Gebhardt, Oscar von *see*
- Acta martyrum selecta
- Die akten der edessenischen bekenner curjas, samonas und abibos
- Die altercatio simonis iudaei et theophili christiani
- Die evangelien des matthaeus und des marcus
- Das evangelium und die apokalypse des petrus
- Graecus venetus
- Hermae pastor graece
- Hieronymus liber de viris inlustribus; gennadius liber de viris inlustribus – der sogenannte sophronius
- Passio s theclae virginis
- Passio s. theclae virginis
- Die psalmen salomo's
- Der sogenannte sophronius [tugal1-14/1b]

Gebhart, Emile *see*
- Autour d'un tiare
- L'italie mystique

Gebiet und gebietshoheit / Fricker, Karl Viktor – Tubingen: Verlage der H. Laupp'schen Buchhandlung, 1901. 112p. LL-4071 – 1 – us L of C Photodup [341]

Gebildete leser *see* Morgenblatt fuer gebildete staende / gebildete leser (1807-1865)

Gebirgsfreund *see* Lusatia 1885

Gebler, Carl *see* Zum religionsunterricht im schullehrerseminar

Gebler, Karl von *see* Galileo galilei and the roman curia

Gebler, Tobias Philipp, Freiherr von *see* Aus dem josephinischen wien

Der gebrauch des alten testaments in den neutestamentlichen schriften / Clemen, August – Guetersloh: C Bertelsmann, 1895 – 1mf – 9 – 0-8370-2679-2 – mf#1985-0679 – us ATLA [220]

Der gebrauch des imperativischen infinitivs im griechischen / Wagner, Richard – [S.l.: s.n., 1891?] (Leipzig: Hesse & Becker) – 1mf – 9 – 0-8370-9197-7 – mf#1986-3197 – us ATLA [450]

Gebrauch und miszbrauch des lateinischen singens und betens beym offentlichen gottes-dienst : aus gottes wort der kirchen-geschichte und der reinen theologen schrifften zur treuhertzige vermahnung...durch johannem musicovium / Muscovius, Johann – Wittenberg: druckts M Schulze [169-?] [mf ed 19–] – 3mf – 9 – mf#fiche 50 – us Sibley [780]

Gebrueder hagedorn : schauspiel in fuenf akten / Petersen, Johannes – Leipzig: Oswald Mutze, 1881 – 1r – 1 – us Wisconsin U Libr [820]

Die gebruik van die ontwikkelingsgefasiliteerde groepmodel vir egskeidingsgetraumatiseerde adolessente / Jakobsen, Marikje – Uni of South Africa 2000 [mf ed Johannesburg 2000] – 7mf [ill] – 9 – (incl bibl ref) – mf#mfm14846 – sa Unisa [370]

Die gebruik van die opvoedkundig-sielkundige relasieteorie in die identifisering van'n middeljarekrisis / Botha, Susanna Petronella Wilhelmina – Uni of South Africa 2000 [mf ed Johannesburg 2000] – 5mf [ill] – 9 – (text in afrikaans; abstract in afrikaans and english; incl bibl ref) – mf#mfm14839 – sa Unisa [370]

Die gebruik van hipnoterapie in die hantering van depressie en angs by adolessente en volwassenes / Geer, Lorna Francis – Uni of South Africa 2000 [mf ed Johannesburg 2000] – 4mf – 9 – (text in afrikaans) – mf#mfm14901 – sa Unisa [615]

Die geburt des jahrtausends : [essays and poems] / Eggers, Kurt – Leipzig: Schwartzhaeupter-Verlag c1936 [mf ed 1990] – 1r – 1 – (filmed with: der junge hutten) – mf#7205 – us Wisconsin U Libr [800]

Die geburtsgeschichte jesu christi / Soltau, Wilhelm – Leipzig: Dieterich, 1902 – 1mf – 9 – 0-8370-5327-7 – (includes bibliographical references inschriften zu ehren des augustus) – mf#1985-3327 – us ATLA [220]

Geburtshilfe und frauenheilkunde – Stuttgart, Germany 1976+ – 1,5,9 – ISSN: 0016-5751 – mf#10162 – us UMI ProQuest [618]

Das geburtsjahr christi : geschichtlich-chronologische untersuchungen / Zumpt, August Wilhelm – Leipzig: B G Teubner, 1869 – 1mf – 9 – 0-8370-9359-7 – (incl bibl ref) – mf#1986-3359 – us ATLA [220]

Die geburtsstunde...de heyelte / Elias de Tejada, Francisco – Madrid: Arbor, 1953 – 1 – sp Bibl Santa Ana [946]

Gebweiler anzeiger *see* Die belchenstimme

Gebweiler neueste nachrichten – Gebweiler, Elsass (Guebwiller F), 1935 jan-29 jun – 1 – fr ACRPP [074]

Gebweiler tagblatt – Gebweiler, Elsass (Guebwiller F), 1906-1918 16 nov – 1 – fr ACRPP [074]

Gebweiler volksblatt *see* Die belchenstimme

Gec journal of research – Chelmsford, England 1983-96 – 1,5,9 – (cont by: gec journal of technology) – ISSN: 0264-9187 – mf#14903.01 – us UMI ProQuest [600]

Gec journal of science and technology – Chelmsford, England 1930-83 – 1,5,9 – ISSN: 0302-2587 – mf#1258 – us UMI ProQuest [621]

Geck, Rudolf *see* So war das

Gecmis guenler / Uenaydin, Rusen Esref – [Istanbul]: Matbaa-yi Orhaniye, 1919 – (= ser Ottoman literature, writers and the arts) – 3mf – 9 – $55.00 – us MEDOC [470]

Gedachtnisrede auf rabbiner dr meier hildescheimer / Grunberg, Samuel – Berlin, Germany. 1935 – 1r – us UF Libraries [939]

Gedaechtniseffekte bei nichtlinearen oszillatoren / Suenner, Tobias – (mf ed 1995) – 1mf – 9 – €30.00 – 3-8267-2096-2 – mf#DHS 2096 – gw Frankfurter [530]

Gedancken von der grossen landes-wirthschaft – Dresden, Gotha, Frankfurt/M, Leipzig DE, 1756-57 – 1r – 1 – gw Misc Inst [333]

Der gedanke / ed by Michelet, Ludwig – 1861-84 [mf ed 1994] – 9v on 28mf – 9 – €190.00 – 3-89131-160-5 – gw Fischer [074]

Der gedanke der paepstlichen weltherrschaft bis auf bonifaz 8 / Hauck, Albert – Leipzig: A Edelmann, 1904 – 1mf – 9 – 0-524-00757-8 – (incl bibl ref) – mf#1990-0189 – us ATLA [240]

Gedanken / Moltke, Helmuth Carl Bernhard; ed by Cochenhausen, Friedrich von – Berlin: Atlantis-Verlag, c1941 – 1r – 1 – us Wisconsin U Libr [943]

Gedanken goethes in der neuzeitlichen biologie / Loesche, Martin – Bremen: J Storm, 1949 [mf ed 1990] – 26p – 1 – mf#7390 – us Wisconsin U Libr [430]

Gedanken nach zwei uhr nachts / Waescher, Aribert – Berlin: Buchwarte-Verlag L Blanvalet, 1939 – 1r – 1 – us Wisconsin U Libr [830]

Gedanken ueber die nachahmung der griechischen werke in der malerei und bildhauerkunst / Winckelmann, Johann Joachim; ed by Seuffert, Bernhard – Heilbronn: Henninger, 1885 [mf ed 1993] – (= ser Deutsche litteraturdenkmale des 18. und 19. jahrhunderts 20) – ix/44p (ill) – 1 – (repr of: "erste ausgabe 1755 mit oesers vignetten". int by von ulrichs) – mf#8676 reel 2 – us Wisconsin U Libr [700]

Gedanken ueber faust 2 / Ziegler, Konrat – Stuttgart: J B Metzler, 1919 – 1r – 1 – us Wisconsin U Libr [430]

Gedanken ueber goethe / Hehn, Victor – 3., verm. Aufl. Berlin: Borntraeger, 1895 – 1r – 1 – (incl ind) – us Wisconsin U Libr [430]

Gedanken von bestimmung moralischen werths / Dalberg, Karl Theodor von – Erfurt: Bey Georg Adam Keyser 1782 – (= ser Ethics in the early modern period) – 1mf – 9 – mf#pl-29 – ne IDC [170]

Gedanken, vorschlaege und wuensche zur verbesserung der oeffentlichen erziehung als materialien zur paedagogik / ed by Resewitz, Friedr Gabriel – Berlin, Stettin 1778-86 – (= ser Dz) – 5v[zu je 4st] on 15mf – 9 – €150.00 – mf#k/n635 – gw Olms [370]

Gedanken zum sexualproblem : mit einem geleitwort von dr. placzek / Bang, Herman – Bonn, 1922 (mf ed 1994) – 1mf – 9 – €24.00 – 3-8267-3025-9 – mf#DHS-AR 3025 – gw Frankfurter [150]

Gedankendichtung der fruehromantik / Boehm, Hans [comp] – Muenchen: G D W Callwey, 1925 [mf ed 1993] – (= ser Kunstwart buecherei 27) – 109p – 1 – (int by comp) – mf#8317 – us Wisconsin U Libr [430]

Die gedankeneinheit des ersten briefes petri : ein beitrag zur neutestamentlichen theologie / Koegel, Julius – Guetersloh: C Bertelsmann, 1902 – (= ser Beitraege zur foerderung christlicher theologie) – 1mf – 9 – 0-524-05917-9 – (incl bibl ref) – mf#1992-0674 – us ATLA [227]

Gedankengang des v frank'schen systems der christlichen wahrheit / Vollert, Wilhelm – Leipzig: A Deichert (Georg Boehme), 1895 [mf ed 1985] – 1mf – 9 – 0-8370-5676-4 – mf#1985-3676 – us ATLA [240]

Gedankenharmonie aus goethe und schiller : lebens- und weisheitssprueche aus deren werken: ein fuehrer durch das leben und die sittliche welt / Goethe, Johann Wolfgang von; ed by Gottschall, Rudolf – Leipzig: C F Amelang, 1866 – us Wisconsin U Libr [430]

Gedankenlyrik / Goethe, Johann Wolfgang von – Muenchen: G D W Callwey, 1923 (mf ed 1990) – 1r – 1 – (filmed with: poetry and truth) – us Wisconsin U Libr [810]

Gedankenwelt der halacha / Breuer, Raphael – Frankfurt am Main, Germany. 1913 – 1r – us UF Libraries [939]

Geddes, Alexander *see*
- Apology for slavery
- Critical remarks on the hebrew scriptures

Geddes, M *see* The church history of ethiopia

Geddes, Patrick *see*
- Every man his own art critic at the manchester exhibition, 1887
- The life and work of sir jagadis c bose

Geddes, Thomas Edward *see* La resurrection de jesu-christo, nuestro senor

Geden, Alfred Shenington *see*
- Comparative religion
- A concordance to the greek testament
- The massoretic and other notes
- Outlines of introduction to the hebrew bible
- Select passages illustrating mithraism...with an introduction
- Studies in the religions of the east

Geden, John Dury *see* The doctrine of a future life as contained in the old testament scriptures

Gedenkblatt an professor a bearliner / Grun, Oscar – Zuerich, Switzerland. 1915 – 1r – us UF Libraries [939]

Gedenkblatt der neuendettelsauer heidenmission in queensland und neu-guinea, 1885-1910 / Flierl, Johann – 2. verb aufl. Neuendettelsau: Missionhauses, 1910 [mf ed 1995] – (= ser Yale coll) – 94p (ill) – 1 – 0-524-10277-5 – (in german) – mf#1996-1277 – us ATLA [240]

Gedenkblatter / Kayserling, Meyer – Leipzig, Germany. 1892 – 1r – us UF Libraries [939]

Gedenkblatter fur die bruder ehrenvizegrossprasident hugo kuznitzky / B'nai B'rith Grossloge Fur Deutschland 8 – Berlin, Germany. 1934 – 1r – us UF Libraries [939]

Gedenkblatter fur oberrabbiner salomon kutna / Nobel, Israel – Filehne, Poland. 1910 – 1r – us UF Libraries [939]

Gedenkblatter zum 25 jahrigen bestehen der synagoge friedberger – Frankfurt am Main, Germany. 1932 – 1r – us UF Libraries [939]

Gedenkblatter zur erinnerung an das 175 jahrige jubilaum / Mainzer, Moritz – Frankfurt am Main, Germany. 1914 – 1r – us UF Libraries [939]

Gedenkblatter zur erinnerung an die samson raphael mirsch-feier der... – Frankfurt am Main, Germany. 1908? – 1r – us UF Libraries [939]

Gedenkblatter zur erinnerung an rabbiner dr a m goldschmidt / Israelitische Religionsgemeinde (Leipgiz, Germany) – Leipzig, Germany. 1889 – 1r – us UF Libraries [939]

Gedenkboek 1150-1940 : studien en essais onder leiding van p clerinx cssr / Christina de Wonderbare – Leuven, 1950 – €5.00 – ne Slangenburg [240]

Gedenkboek van een vijf-en-twintigjarig zendelingsleven op nieuw-guinea (1862-1887) / Hasselt, J L van – Utrecht: Kemink & Zoon, 1888. Chicago: Dep of Photodup, U of Chicago Lib, 1975 (1r); Evanston: American Theol Lib Assoc, 1984 (1r) – 1 – 0-8370-0425-X – mf#1984-B448 – us ATLA [240]

Gedenkboek van het vijftigjarig jubileum der christelijke gereformeerde kerk, a.d. 1857-1907 – 2. vermeerderde oplage. Grand Rapids, Mich: JB Hulst, [1907?] – 1mf – 9 – 0-524-07520-4 – mf#1991-3150 – us ATLA [240]

Gedenkbuch "chewra kadischa" – Wien, Austria. 1905? – 1r – us UF Libraries [939]

Gedenkbuch des metzer buergers philippe de vigneulles : aus den jahren 1471-1522 / ed by Michelant, Heinrich – Stuttgart: Literarischer Verein, 1852 [mf ed 1993] – (= ser Blvs 24) – xxxv/444p – 1 – (incl bibl ref and ind. french text. int in german) – mf#8470 reel 5 – us Wisconsin U Libr [880]

Gedenkbuch, erinnerung an karl heinzen und an die enthuellungsfeier des heinzen-denkmals am 12. juni 1886 in boston, mass. / Schmemann, Karl – Milwaukee, Wis.: Freidenker Pub. Co., 1887 – 1r – 1 – (incl bibl ref) – us Wisconsin U Libr [430]

Gedenkrede auf stefan george : gehalten am 13. dezember 1933 in der aula der universitaet bonn / Clemen, Paul – Bonn: Scheur, 1934 (mf ed 1990) – 1r – 1 – (filmed with: das werk georges) – us Wisconsin U Libr [430]

Gedenkschrift fuer ferdinand josef schneider, 1879-1954 / ed by Bischoff, Karl – Weimar: H. Boehlaus Nachfolger, 1956 [mf ed 1993] – 377p – 1 – (incl bibl ref) – mf#8046 – us Wisconsin U Libr [430]

Gedenkschrift zur feier des 25 jahr bestandes des israel / B'nai B'rith District No 10 "Moravia" – Brunn, Czechoslovakia. 1921 – 1r – us UF Libraries [939]

Gedenkwaardigheden uit de geschiedenis van gelderland / Nijhoff, I A – Amsterdam. v1-6. 1830-1862 – €155.00 – ne Slangenburg [242]

Gedenkwaerdig bedryf der nederlandsche oost-indische maetschappye, op de kuste en het keizerrijk van taising of sina... / Dapper, O – Amsterdam: Jacob van Meurs, 1670 – 18mf – 9 – mf#HT-780 – ne IDC [915]

Gedeon, Manouel Io *see* Kanonikai diataxeis, epistolai, lyseis, thespismata ton hagiotaton patriarchon konstantinoupoleos

Gederici, V *see* Deh numi pietosi

Das gedicht (mme5) : blaetter fuer die dichtung / ed by Ellermann, Heinrich – 1934/35-1943/44 [mf ed 1999] – (= ser Marbacher mikrofiche-editionen (mme) 5; Kultur – literatur – politik: deutsche zeitschriften des 19./20. jahrhunderts (klp)) – 10v on 43mf – 9 – €250.00 – 3-89131-354-3 – gw Fischer [410]

Gedicht und gedanke : auslegungen deutscher gedichte / ed by Burger, Heinz Otto – Halle (Saale): M Niemeyer, c1942 [mf ed 1996] – 434p – (incl bibl ref) – mf#9498 – us Wisconsin U Libr [430]

Gedichtbuechelchen / Asmus, Georg – Leipzig: E H Mayer [1891] [mf ed 1988] – 163p – 1 – (incl comedy: der blumenstrauss) – mf#6968 – us Wisconsin U Libr [810]

Gedichte / Ambrosius, Johanna – 28. aufl. Koenigsberg i. Pr: Thomas & Oppermann 1896 [mf ed 1991] – 1r – 1 – (filmed with: die wanderung des herrn ulrich von hutten / will vesper) – mf#2945p – us Wisconsin U Libr [810]

Gedichte / Beck, Karl Isidor – 4. der neuen ausg. 3. aufl. Berlin: Voss, 1846 [mf ed 1989] – 344p – 1 – mf#7002 – us Wisconsin U Libr [810]

Gedichte / Bielfeld, H A – Milwaukee: Freidenker Pub Co, 1889 [mf ed 1989] – 196p – 1 – mf#7020 – us Wisconsin U Libr [810]

Gedichte / Brachvogel, Udo – Leipzig: B Westermann; New York: Lemcke & Buechner, 1912 [mf ed 1989] – 286p – 1 – mf#7061 – us Wisconsin U Libr [810]

Gedichte / Busse, Carl – 4. aufl. Stuttgart: A G Liebeskind 1899 [mf ed 1989] – 1r – 1 – (filmed with: zu guter letzt / wilhelm busch) – mf#7097 – us Wisconsin U Libr [810]

Gedichte / Carossa, Hans – 2nd enl ed. Leipzig: Insel-Verlag 1912 [mf ed 1995] – 1r – 1 – (filmed with: graefin lassbergs enkelin / fr lehne) – mf#3812p – us Wisconsin U Libr [810]

Gedichte / Castelhun, Friedrich Karl – 3. verb verm aufl. Zuerich: Buchhandlung des Schweizerischen Gruetlivereins; Milwaukee: Freidenker Publ Co 1901 [mf ed 1989] – 1r – 1 – (filmed with: die poesie, ihr wesen und ihre formen / moriz carriere) – mf#7146 – us Wisconsin U Libr [810]

Gedichte / Conde, Jean de [Jehan de Condet]; ed by Tobler, Adolf – Stuttgart: Litterarischer Verein, 1860 [mf ed 1993] – (= ser Blvs 54) – 186p – 1 – mf#8470 reel 11 – us Wisconsin U Libr [810]

Gedichte / Cornelius, Peter – Leipzig: C F Kahnt, 1890 – 1r – 1 – us Wisconsin U Libr [810]

Gedichte / Cornelius, Peter – Leipzig: C F Kahnt, 1890 – 1 – us Wisconsin U Libr [810]

Gedichte / Dach, Simon; ed by Ziesemer, Walther – Halle/S: M Niemeyer, 1936-38 [mf ed 1989] – (= ser Schriften der koenigsberger gelehrten gesellschaft. sonderreihe 4-7) – 4v – 1 – mf#7162 – us Wisconsin U Libr [810]

Gedichte / Dach, Simon; ed by Ziesemer, Walther – Halle/Saale : M. Niemeyer, 1936-38 [mf ed 1989] – (= ser Schriften der koenigsberger gelehrten gesellschaft. sonderreihe 4-7) – 4v – 1 – us Wisconsin U Libr [810]

Gedichte / Dahn, Felix – Leipzig: Breitkopf & Haertel. 3v. 1898-99 – us Wisconsin U Libr [810]

Gedichte / Dahn, Felix – Leipzig: Breitkopf & Haertel. 3v. 1898-99 – 1 – us Wisconsin U Libr [810]

Gedichte / Deinhardstein, Johann Ludwig – Berlin: Duncker und Humblot, 1844 – us Wisconsin U Libr [810]

Gedichte / Deinhardstein, Johann Ludwig – Berlin: Duncker und Humblot, 1844 – 1 – us Wisconsin U Libr [810]

Gedichte / Dilg, William – Milwaukee, WI: J B Hoeger, 1866 [mf ed 1989] – xii/237p – 1 – mf#7177 – us Wisconsin U Libr [810]

Gedichte / Drescher, Martin – [Chicago, IL]: Columbia Printing Co, 1909 [mf ed 1989] – 219p – 1 – mf#7185 – us Wisconsin U Libr [810]

Gedichte / Droste-Huelshoff, Annette von – 3. Aufl. Paderborn: F Schoeningh, 1887 – 1 – us Wisconsin U Libr [810]

Gedichte / Droste-Huelshoff, Annette von – 6. Aufl. Paderborn: F Schoeningh, [1900] – 1 – us Wisconsin U Libr [810]

Gedichte / Droste-Huelshoff, Annette von – 7. Aufl. Paderborn: F Schoeningh, [1907?] – 1 – us Wisconsin U Libr [810]

Gedichte / Droste-Huelshoff, Annette von – Paderborn: F Schoeningh, 1887 – 1r – 1 – us Wisconsin U Libr [810]

Gedichte / Droste-Huelshoff, Annette von – Paderborn: F Schoeningh, [1900] – 1r – 1 – us Wisconsin U Libr [810]

Gedichte / Droste-Huelshoff, Annette von – Paderborn: F Schoeningh, [1907?] – 1r – 1 – us Wisconsin U Libr [810]

Gedichte / Dulk, Albert Friedrich Benno – Stuttgart: J H W Dietz, 1887 [mf ed 1989] – 96p – 1 – mf#7190 – us Wisconsin U Libr [810]

Gedichte / Elze, Karl – Halle: M Niemeyer, 1878 (mf ed 1990) – 1r – 1 – (filmed with: astra) – us Wisconsin U Libr [810]

Gedichte : faust i und ii / Goethe, Johann Wolfgang von – Cambridge, MA: Foreign Books, Inc, [19–?] (mf ed 1990) – 1r – 1 – (filmed with: acht lieder von goethe) – us Wisconsin U Libr [800]

Gedichte / Fischer, Johann Georg; ed by Lissauer, Ernst – Stuttgart: J G Cotta, 1923 – 1r – 1 – us Wisconsin U Libr [810]

Gedichte / Fontane, Theodor – 12-14.aufl. Stuttgart: J G Cotta, 1908 [mf ed 1990] – xii/418p – 1 – mf#7073 – us Wisconsin U Libr [810]

Gedichte / Freiligrath, Ferdinand – Stuttgart: J G Cotta, 1848 – 1r – 1 – us Wisconsin U Libr [810]

Gedichte / Frey, Adolf – 2., verm Aufl. Leipzig: H Haessel, 1908 (mf ed 1990) – 1r – 1 – (filmed with: unneren strohdack) – us Wisconsin U Libr [810]

Gedichte / Friedrichs, Hermann – Leipzig: W Friedrich, [18–] (mf ed 1990) – 1r – 1 – (filmed with: gustav freytag) – us Wisconsin U Libr [810]

Gedichte / Fulda, Ludwig – Stuttgart: J G Cotta, [19–?] (mf ed 1990) – 1r – 1 – (filmed with: aus der werkstatt) – us Wisconsin U Libr [810]

Gedichte / Geib, August – 2., verm Ausg. Leipzig: Druck und Verlag der Genossenschaftsdruckerei, 1876 (mf ed 1990) – 1r – 1 – (filmed with: liebe, leben, kampf) – us Wisconsin U Libr [810]
Gedichte / Geibel, Emanuel – 119. Aufl. Stuttgart: J G Cotta, 1893 (mf ed 1990) – 1r – 1 – (filmed with: gedichte) – us Wisconsin U Libr [810]
Gedichte : gesamtausgabe mit einem lebensbilde des dichters von karl weinhold / Strachwitz, Moritz Graf – 8. Aufl. Breslau: E Trewendt, 1891 – 1r – 1 – us Wisconsin U Libr [810]
Gedichte / Gilm, Hermann von; ed by Greinz, Rudolf Heinrich – Leipzig: P Reclam, 1894 (mf ed 1990) – 1r – 1 – (filmed with: zwischen den kriegen) – us Wisconsin U Libr [810]
Gedichte / Gilm, Hermann von – Leipzig: A G Liebeskind, 1894 – 1r – 1 – us Wisconsin U Libr [810]
Gedichte / Goethe, Johann Wolfgang von – Berlin: G Grote, 1871 – 1r – 1 – us Wisconsin U Libr [810]
Gedichte / Goethe, Johann Wolfgang von; ed by Korff, H A – Leipzig: S Hirzel, 1949, c1947 – 1r – 1 – us Wisconsin U Libr [810]
Gedichte / Gruen, Anastasius – 2. Aufl. Leipzig: Weidmann, 1838 – 1 – us Wisconsin U Libr [810]
Gedichte / Gruen, Anastasius – Leipzig: Weidmann, 1838 – 1r – 1 – us Wisconsin U Libr [810]
Gedichte / Guenther, Johann Christian; ed by Litzmann, Berthold – Leipzig: P Reclam, [1880] – 1r – 1 – (incl bibl ref) – us Wisconsin U Libr [810]
Gedichte / Hassaurek, Friedrich – Cincinnati: M & R Burgheim, c1877 – 1r – 1 – us Wisconsin U Libr [810]
Gedichte / Hebbel, Friedrich – gesammt-ausg stark verm verb. Stuttgart: J G Cotta, 1857 [mf ed 1993] – x/474p – 1 – mf#8660 – us Wisconsin U Libr [810]
Gedichte / Hebbel, Friedrich – gesammt-ausg, stark vermehrt und verbessert. Stuttgart: J G Cotta, 1857 [mf ed 1993] – x/474p – 1 – mf#8660 – us Wisconsin U Libr [810]
Gedichte / Huch, Ricarda – Leipzig: H Haessel, 1894 – 1r – 1 – us Wisconsin U Libr [810]
Gedichte / Lenau, Nikolaus – Berlin: G Hempel, [1879] [mf ed 1993] – viii/200p – 1 – mf#7617 – us Wisconsin U Libr [810]
Gedichte / Lexow, Friedrich – New York: E Steiger, 1872 – 1r – 1 – us Wisconsin U Libr [810]
Gedichte / Liliencron, Detlev, Freiherr von – Wiesbaden: Verlag des Volksbildungsvereins, 1909 – 1r – 1 – us Wisconsin U Libr [810]
Gedichte / Loeben, Otto Heinrich, Graf von; ed by Pissin, Raimund – Berlin: B Behr, 1905 [mf ed 1994] – (= ser Deutsche litteraturdenkmale des 18. und 19. jahrhunderts 135, 3 folge n15) – xvii/171p – 1 – (incl bibl ref) – mf#8676 reel 8 – us Wisconsin U Libr [810]
Gedichte / Meissner, Alfred – 2., stark vermehrte Aufl. Leipzig: F L Herbig, 1846 – 1r – 1 – us Wisconsin U Libr [810]
Gedichte / Miegel, Agnes – Stuttgart: J G Cotta, 1935 – 1r – 1 – us Wisconsin U Libr [810]
Gedichte / Mueller, Wilhelm; ed by Hatfield, James Taft – vollst krit ausg. Berlin: B Behr, 1906 [mf ed 1993] – (= ser Deutsche litteraturdenkmale des 18. und 19. jahrhunderts 137, 3. folge n17) – xxxi/513p/3pl – 1 – (incl bibl ref and ind) – mf#8676 reel 8 – us Wisconsin U Libr [810]
Gedichte : neue auswahl / Hartmann, Moritz – Stuttgart: J G Cotta, 1874 [mf ed 2001] – 342p – 1 – mf#10556 – us Wisconsin U Libr [810]
Gedichte = Poems / Becker, Nicolaus – Koeln: M DuMont-Schauberg, 1841 [mf ed 1993] – (= ser Bibliothek der deutschen literatur 58) – 218p – 1 – mf#8512 – us Wisconsin U Libr [810]
Gedichte = Poems / Chamisso, Adelbert von; ed by Rauschenbusch, Wilhelm – 2. aufl. Berlin: G Grote 1876 [mf ed 1993] – 1r – 1 – (filmed with: der wandsbecker bote / karl gerok [comp]; & other titles) – mf#8537 – us Wisconsin U Libr [810]
Gedichte = Poems / Ehrler, Hans Heinrich – Stuttgart: Greiner & Pfeiffer [1919?] [mf ed 1989] – 1r [ill] – 1 – (filmed with: fruehlings-lieder) – mf#7210 – us Wisconsin U Libr [810]
Gedichte = Poems / Hopfen, Hans – Berlin: A Hofmann 1883 [mf ed 1995] – 1r – 1 – (filmed with: fraenzchens lieder / hoffmann von fallersleben) – mf#3757p – us Wisconsin U Libr [810]
Gedichte : jubilaeums-ausgabe zum hundertsten geburtstage des dichters (1791-1891) = Poems / Grillparzer, Franz; ed by Sauer, August – Stuttgart: J G Cotta, 1891 [mf ed 1996] – xiv/612p/1pl – 1 – mf#9660 – us Wisconsin U Libr [810]
Gedichte = Poems / Meyer, Conrad Ferdinand – 16. aufl. Leipzig: H Haessel, 1900 [mf ed 1995] – xv/397p – 1 – mf#8823 – us Wisconsin U Libr [810]
Gedichte = Poems / Moericke, Eduard Friedrich – 13. mit einem Nachtrag verm. Aufl. Leipzig: G J Goeschen, 1898 – 1r – 1 – us Wisconsin U Libr [810]
Gedichte / Prutz, Robert Eduard – Zuerich: Druck und Verlag des literarischen Comptoirs, 1843 – 1r – 1 – us Wisconsin U Libr [810]
Gedichte / Schack, Adolf Friedrich von – 3. Aufl. Stuttgart: J G Cotta, 1874 – 1r – 1 – us Wisconsin U Libr [810]
Gedichte / Scriba, Carl – 2. Aufl. Butzbach: M Kuhl, 1850 – 1r – 1 – us Wisconsin U Libr [810]
Gedichte / Simrock, Karl Joseph – Leipzig: Hahn, 1844 – 1r – 1 – us Wisconsin U Libr [810]
Gedichte / Storm, Theodor – 13. Auflage. Berlin: Gebrueder Paetel, 1903 – 1r – 1 – us Wisconsin U Libr [810]
Gedichte / Strodtmann, Adolf – 3. und verm. Gesammt-Ausg. Leipzig: P Reclam, [18–?] – 1r – 1 – us Wisconsin U Libr [810]
Gedichte / Strodtmann, Adolf – 3., verm. Ges.-Ausg. Leipzig: P Reclam, [18–?] – 1 – us Wisconsin U Libr [810]
Gedichte / Viereck, George Sylvester – Leipzig: Hesse & Becker, 1922 – 1r – 1 – (appendix: george sylvester viereck: an appreciation) – us Wisconsin U Libr [810]
Gedichte / Weckherlin, Georg Rodolf; ed by Goedeke, Karl – Leipzig: Brockhaus, 1873 – 1 – us Wisconsin U Libr [810]
Gedichte / Weckherlin, Georg Rudolf; ed by Fischer, Hermann – Stuttgart: Litterarischer Verein. 3v. 1894- (Tuebingen: H Laupp, Jr) – us Wisconsin U Libr [810]
Gedichte / Weckherlin, Georg Rudolf; ed by Goedeke – Leipzig: Brockhaus, 1873 – 1r – 1 – us Wisconsin U Libr [810]
Gedichte / Werfel, Franz – Berlin: P Zsolnay, 1927 – 1r – 1 – us Wisconsin U Libr [810]
Gedichte / Werfel, Franz – Berlin: P Zsolnay, 1927 – 1 – us Wisconsin U Libr [810]
Gedichte / Wilbrandt, Adolf von – Wien: L Rosner, 1874 – 1r – 1 – us Wisconsin U Libr [810]
Gedichte *see*
– Briefe an und von johann heinrich merck
– Joseph von eichendorff
– Moses mendelssohn
Gedichte aus dreissig jahren / Werfel, Franz – Stockholm: Bermann-Fischer Verlag, 1939 – 1r – 1 – us Wisconsin U Libr [810]
Gedichte der gebrueder wolf : eine auswahl / Wolf, Johann Theobald; ed by Mueller, Eugen – Strassburg: K J Truebner, 1916 – 1r – 1 – us Wisconsin U Libr [810]
Gedichte der gefangenen : ein sonettenkreis / Toller, Ernst – 2. Aufl. Muenchen: K Wolff, 1923 – 1r – 1 – us Wisconsin U Libr [810]
Gedichte des 12. und 13. jahrhunderts / ed by Hahn, K A – Quedlinburg, Leipzig: G Basse, 1840 [mf ed 1993] – (= ser Bibliothek der gesammten deutschen national-literatur von der aeltesten bis auf die neuere zeit sect1/20) – vii/152p – 1 – mf#8438 reel 5 – us Wisconsin U Libr [810]
Gedichte des deuterojesaias / Praetorius, Franz – Berlin, Germany. 1922 – 1r – us UF Libraries [939]
Gedichte des koenigsberger dichterkreises aus heinrich alberts arien und musicalischer kuerbshuette, 1638-1650 / ed by Fischer, L H – Halle: Max Niemeyer, 1883 – us Wisconsin U Libr [810]
Die gedichte des michel beheim / ed by Gille, Hans & Spriewald, Ingeborg – Berlin: Akademie-Verlag, 1968-72 [mf ed 1993] – (= ser Deutsche texte des mittelalters 60, 64, 65) – 3v in 4 – 1 – (incl bibl ref and ind) – mf#8623 reel 18 – us Wisconsin U Libr [810]
Die gedichte des wilden mannes / ed by Standring, Bernard – Tuebingen: Max Niemeyer, 1963 [mf ed 1993] – (= ser Altdeutsche textbibliothek n59) – xiv/62p – 1 – (incl bibl ref) – mf#8193 reel 5 – us Wisconsin U Libr [810]
Gedichte fuer ein Volk / Becher, Johannes Robert – Leipzig: Insel-Verlag, 1919 [mf ed 1989] – 107p – 1 – mf#6994 – us Wisconsin U Libr [810]
Gedichte goethes an frau v stein : in faksimilenachbildung / ed by Wahle, Julius – Weimar: Verlag der Goethe-Gesellschaft, 1924 [mf ed 1993] – (= ser Schriften der goethe-gesellschaft 37) – 16p/8pl/12lea – 1 – mf#8657 reel 9 – us Wisconsin U Libr [810]
Gedichte goethes veranschaulicht nach form- und strukturwandel / Meschke, Waltraut – Berlin. Akademie-Verlag, 1957 – 1r – 1 – (incl bibl ref) – us Wisconsin U Libr [430]
Die gedichte heinrichs des teichners / ed by Niewoehner, Heinrich – Berlin: Akademie-Verlag, 1953-56 [mf ed 1994] – (= ser Deutsche texte des mittelalters 44, 46, 48) – 3v – 1 – (incl bibl ref) – mf#8623 reel 11 – us Wisconsin U Libr [810]
Gedichte in hochdeutscher mundart / Stoltze, Friedrich – Frankfurt a.M.: H Keller, 1862 – 1r – 1 – us Wisconsin U Libr [390]
Gedichte in schwaebischer mundart / Heerbrandt, Gustav – New York: Selbstverlag von G Heerbrandt, 1892 – 1r – 1 – us Wisconsin U Libr [810]
Gedichte und aufsaetze : erschienen in der 'syracuse union', syracuse, ny, 1903 / Benignus, Wilhelm – New York, NY: W Benignus, c1903 [mf ed 1989] – 38p – 1 – mf#7005 – us Wisconsin U Libr [810]
Gedichte und gedanken / Thoma, Hans; ed by Eberlein, Karl – Konstanz: Reuss & Itta, 1919 – 1r – 1 – (includes bibliographical references) – us Wisconsin U Libr [800]
Gedichte, volkslieder, legenden, sagen / Wesendonk, Mathilde – Zuerich: [s.n. 1864?] – 2mf – 9 – mf#mw-5 – ne IDC [802]
Gedichte vom hausrat aus dem 15. und 16. jahrhundert : in facsimiledruck / ed by Hampe, Theodor – Strassburg: J H E Heitz (Heitz & Muendel), 1899 [mf ed 1993] – (= ser Drucke und holzschnitte des 15. und 16. jahrhunderts in getreuer nachbildung 2) – 109p/1pl (ill) – 1 – mf#8376 – us Wisconsin U Libr [810]
Die gedichte von albert ehrenstein / Ehrenstein, Albert – Leipzig: Ed Strache, [1920] (mf ed 1990) – 1r – 1 – (filmed with: die geburt des jahrtausends) – us Wisconsin U Libr [810]
Gedichte von goethe in compositionen / ed by Friedlaender, Max – Weimar: Goethe-Gesellschaft, 1896-1916 [mf ed 1994] – (= ser Schriften der goethe-gesellschaft 11, 31) – 2v – 1 – (in german. incl bibl ref and ind) – mf#8657 reel 3 – us Wisconsin U Libr [430]
Gedichte von johann nicolaus goetz : aus den jahren 1745-65, in urspruenglicher gestalt / ed by Schueddekopf, Carl – Stuttgart: G J Goeschen, 1893 [mf ed 1993] – (= ser Deutsche litteraturdenkmale des 18. und 19. jahrhunderts 42) – xxxvi/89p – 1 – (incl bibl ref) – mf#8676 reel 4 – us Wisconsin U Libr [810]
Gedichte von joseph freiherrn von eichendorff / Eichendorff, Joseph, Freiherr von – 16. Aufl. Leipzig: C F Amelang, 1892 – 1r – 1 – (incl ind) – us Wisconsin U Libr [810]
Die gedichte walthers von der vogelweide / Vogelweide, Walther von der – 6. ausg. Berlin: G Reimer, 1891 [mf ed 1993] – xviii/234p – 1 – mf#8444 – us Wisconsin U Libr [810]
Gedichte walthers von der vogelweide – Jena: H Coftenoble, 1881 – 1r – 1 – (incl bibl ref. german translations of middle high german poems) – us Wisconsin U Libr [810]
Gedichten van jacob zeeus – Delf: Reinier Boitet, 1721 – 7mf – 9 – mf#O-3200 – ne IDC [090]
Gedichten van jakob zeeus – Amsterdam: Antoni Schoonenburg, 1737 – 11mf – 9 – mf#O-807 – ne IDC [090]
Gedik, S *see*
– Antipistorius
– Calviniana religio
– Drey christliche vnd in gottes wort vnd der alten lehrer schrifften wolgegruendte predigten
– Postilla das ist avszlegung der euangelien durchs gantze jahr
– Von bildern vnd altarn jn den euangelischen kirchen augspurgischer confession
– Von den ceremonien bey dem heiligen abendmahl
Gedike, Friedrich *see* Berlinische monatsschrift 1783-96 / berlinische blaetter 1797-98 / neue berlinische monatsschrift 1799-1811
Gedike, Friedrich et al *see* Berlinische monatsschrift
Das gedoppelte tun : beitraege zur analyse des verhaeltnisses zwischen maennern und frauen in einer antagonistischen gesellschaft / Mueller, Hannelore – 1995 – 3mf – 9 – 3-8267-2103-9 – mf#DHS 2103 – gw Frankfurter [305]
Gedud ha-'avodah – Tel-Aviv, Israel. 1931 – 1r – us UF Libraries [939]
"Gedung buku nasional" / Daftar buku Indonesia – Djakarta, 1955-1958 – 16mf – 9 – mf#SE-632 – ne IDC [959]
Gee, Henry *see*
– The elizabethan clergy and the settlement of religion, 1558-1564
– The elizabethan prayer-book and ornaments
– The reformation period
Gee, Henry et al *see* Typical english churchmen from parker to maurice
The geelong advertiser – Victoria, australia. 23 jan-15 nov 1841; 19 sep 1844-31 dec 1860; 2 oct 1866; 25, 27 jul, 26 oct 1867; 28 mar 1868; 5 sep 1892-30 oct 1918; 21 jun 1919-30 apr, 26 dec 1940-6 oct 1942 – 1 – (imperfect) – mf#M.C.838 – uk British Libr Newspaper [079]
Geen, M S *see* Making of south africa
Geenzier, Enrique *see* Viejo y nuevo
Geer, C de *see* Memoires pour servir a l'histoire des insectes
Geer, George Jarvis *see*
– Conversion of st paul: three discourses
– The conversion of st paul
Geer, Lorna Francis *see* Die gebruik van hipnotherapie in die hantering van depressie en angs by adolessente en volwassenes
Geer memorial baptist church. easley, south carolina : church records – Pickens County, Nov 1902-Sept 1985. Piedmont Assoc. Membership Rolls to Feb 1960. 1306p – 1 – us Southern Baptist [978]
Geere, Henry Valentine *see* By nile and euphrates
Geesaman cousins / Geesaman Family Association – v1 n1-v4 n2, v5 n1 [1975 jan-1978 jul, 1980 oct] – 1r – 1 – mf#500056 – us WHS [929]
Geesaman Family Association *see* Geesaman cousins
Geessel om uyt te dryven den arminiaenschen quel-geest / Trigland, J – Ed 3. Amstelredam, 1628 – 1mf – 9 – mf#PBA-348 – ne IDC [240]
Geest en vorm der vrijmetselarij : een hand-en leerboek voor bb vv, naar het hoogduitsch van j g findel / Findel, Joseph Gabriel – Zwolle: Van Hoogstraten & Gorter 1875 – 3mf – 9 – mf#vrl-101 – ne IDC [366]
Het geestelijck cieraet van christi bruylofts-kinderen ofte de praktijcke des heylighen avontmaels... / Teelinck, W – Amstelredam, 1644 – 2mf – 9 – mf#PBA-318 – ne IDC [240]
'T geestelijck roer van 't coopmans schip... / Udemans, G C – Dordrecht, 1640 – 9mf – 9 – mf#PBA-362 – ne IDC [240]
Het geestelyck jubilee van het jaer o.h. m.dc.l... / Sambeeck, J – t'Antwerpen: Philips van Eyck, 1663 – 5mf – 9 – mf#O-3161 – ne IDC [090]
Het geestelyck kaert-spel met herten troef : oft het spel der liefde... / Joseph...Sancta Barbara – t'Antwerpen: Jacobus van Gaesbeeck, [c1666] – 6mf – 9 – mf#O-3097 – ne IDC [090]
Het geestelyck kaert-spel met herten troef... / Joseph...Sancta Barbara – t'Antwerpen: Franciscus Muller, 1712 – 6mf – 9 – mf#O-3096 – ne IDC [090]
Geestesgesondheidsfasiliteite in die republiek van suid-afrika : 1972 = Mental health facilities in the republic of south africa: 1972 – Johannesburg: Suid-Afrikaanse Nasionale Raad vir Geestesgesondheid [1972] [mf ed Pretoria, RSA: State Library [199-]] – 36lea on 1r with other items – 5 – (in afrikaans & english) – mf#A70/0500 r25 – us CRL [362]
The geeta : as a chaitanyite reads it / Bhakti Hridaya Bon, swami – Bombay: Popular Book Depot, 1938 – (= ser Samp: indian books) – us CRL [280]
Gefaehrtin meines sommers : [a novel] / Boerner, Klaus Erich – Berlin: Holle, 1938 [mf ed 1989] – 192p – 1 – mf#7050 – us Wisconsin U Libr [830]
Der gefaelschte brief des bischofs theonas an den oberkammerherrn lucian / Harnack, Adolf von – Leipzig, 1903 – (= ser Tugal 2-24/3c) – 1mf – 9 – €3.00 – ne Slangenburg [240]
Der gefaelschte brief des bishofs theonas an den oberkammerherrn lucian *see* Der pseudocyprianische traktat de singularitate clericorum
Das gefaengnis zum preussischen adler : eine selbsterlebte schildbuergerei / Wille, Bruno – Jena: E Diederichs 1914 [mf ed 1991] – 1r [ill] – 1 – (filmed with: prisoner halm / karl wilke) – mf#3054p – us Wisconsin U Libr [830]
Der gefangene von metz : vaterlaendisches lustspiel in fuenf aufzuegen / Gutzkow, Karl – Berlin: G Bernstein, 1871 [mf ed 1993] – 108p – 1 – mf#8668 – us Wisconsin U Libr [820]
Die gefangenschaftsbriefe / Haupt, Erich – 8. bezw. 7. aufl. Goettingen: Vandenhoeck und Ruprecht, 1902 – (= ser Kritisch Exegetischer Kommentar Ueber Das Neue Testament) – 2mf – 9 – 0-7905-3448-7 – (incl bibl ref) – mf#1987-3448 – us ATLA [227]
Geffcken, Friedrich Heinrich *see*
– Church and state
– Die voelkerrechtliche stellung des papstes
Geffcken, J *see*
– Kompositon und entstehungszeit der oracula sibyllina
– Die oracula sibyllina
Geffcken, Johannes *see*
– Aus der werdezeit des christentums
– Der bildercatechismus des fuenfzehnten jahrhunderts
– Christliche apokryphen
– Kaiser julianus
– Sokrates und das alte christentum
– Zwei griechische apologeten
Geffner, Robert *see* Journal of aggression, maltreatment and trauma
Geffner, Robert A *see*
– Journal of child sexual abuse
– Journal of emotional abuse
Geffrei Gaimar *see* Lestorie des engles solum la translacion maistre geffrei gaimar (rs91)
Gefle dagblad – Gavle, Sweden. 1895-1978 – 450r – 1 – sw Kungliga [078]
Gefle dagblad – Gavle, Sweden. 1979- – 1 – sw Kungliga [078]

Gefleborgs laens tidning – Gaevle, Sweden. 1889-95 – 10r – 1 – sw Kungliga [078]

Gefleposten – Gaevle, Sweden. 1859 – 1r – 1 – sw Kungliga [078]

Gefleposten – Gavle, Sweden. 1864-1941 – 1 – sw Kungliga [078]

Der gefrorene dionysos *see* Die liebesschaukel

Der gefrorene dionysus : erzaehlung / Andres, Stefan Paul – Berlin: Ulrich Riemerschmidt, c1942 [mf ed 1995] – (= ser Die liebesschaukel) – 239p – 1 – (later publ under title: die liebesschaukel) – mf#8919 – us Wisconsin U Libr [880]

Der gefrorene kuss *see* Auf wache

Gefuege und mechanische eigenschaften von keramiken im system spinell-aluminiumoxid / Vollweiler, Lutz – (mf ed 1996) – 1mf – 9 – €30.00 – 3-8267-2294-9 – mf#DHS 2294 – gw Frankfurter [660]

Gegen den haeretiker : buch 4 u. 5 = Adversus haereses / Irenaeus – Leipzig: J C Hinrichs, 1910 – (= ser Tugal) – 1mf – 9 – 0-7905-1718-3 – (incl ind. in armenian) – mf#1987-1718 – us ATLA [240]

Gegen den militarismus und gegen die neuen steuern / Liebknecht, Wilhelm & Bebel, A – Berlin, 1893 – 1 – gw Mikropress [943]

Gegen den strom : lyrisches und satyrisches / Palmer, Albert – 2. Aufl. Leipzig: O Wigand, 1884 – 1r – 1 – us Wisconsin U Libr [430]

Gegen den strom – New York NY (USA), 1938 mar-1939 oct/nov – 1r – 1 – gw Misc Inst [071]

Gegen den strom – Breslau (WrocLaw PL), Berlin DE, Paris (F), 1928 17 nov-1935 n5 [gaps] – 4r – 1 – (title varies: 1936: der internationale klassenkampf kpo; fr 1929 in berlin, fr may 1933 in paris) – gw Misc Inst [077]

Gegen die arianer, 1. bd (bdk13 1.reihe) : briefe an serapion und epiktet / Athanasius – (= ser Bibliothek der kirchenvaeter. 1. reihe (bdk 1.reihe)) – €19.00 – ne Slangenburg [240]

Gegen die blutbeschuldigung / Horovicz, Jonathen Benjamin – Wien, Austria. 1903 – 1r – us UF Libraries [939]

Gegen die haeresien, 1. bd (bdk3 1.reihe) / Irenaeus – (= ser Bibliothek der kirchenvaeter. 1. reihe (bdk 1.reihe)) – €14.00 – ne Slangenburg [240]

Gegen die haeresien, 2. bd (bdk4 1.reihe) / Irenaeus – (= ser Bibliothek der kirchenvaeter. 1. reihe (bdk 1.reihe)) – €14.00 – ne Slangenburg [240]

Gegen die heiden / ueber die menschwerdung / leben des hl antonius und pachomius, 2. bd (bdk31 1.reihe) / Athanasius – (= ser Bibliothek der kirchenvaeter. 1. reihe (bdk 1.reihe)) – €15.00 – ne Slangenburg [240]

Gegen renan, leben jesu / Gerlach, Hermann – Berlin: G Schlawitz, 1864 – 1mf – 9 – 0-7905-3376-6 – (incl bibl ref) – mf#1987-3376 – us ATLA [240]

Der gegen-angriff : antifaschistische wochenschrift – Berlin DE, Prag [CZ], Zuerich [CH], Paris [F], 1933 apr-1936 14 mar – 3r – 1 – (cont: deutsche volkszeitung, prag; with suppl: roter pfeffer 1933-34 [gaps]) – gw Misc Inst [320]

Der gegen-angriff – Koeln DE, 1932-33 [gaps] – 1r – 1 – gw Misc Inst [074]

Der gegenangriff : anti-faschistische zeitschrift – Prague, Zurich, Paris. avr 1933-mars 1936 – 1 – fr ACRPP [325]

Gegenbaur, C von *see* Morphologisches jahrbuch

Die gegenreformation in schlesien / Ziegler, Heinrich – Halle a. S.: Verein fuer Reformationsgeschichte, [1888?] – (= ser Schriften Des Vereins Fuer Reformationsgeschichte) – 1mf – 9 – 0-7905-5079-2 – (incl bibl ref) – mf#1988-1079 – us ATLA [943]

Der gegensatz des classischen und des romantischen in der neueren philosophie / Hermann, Conrad – Leipzig: M Schaefer, 1877 – 1mf – 9 – 0-524-00271-1 – mf#1989-2971 – us ATLA [190]

Der gegensatz des katholicismus und protestantismus : nach den principien und hauptdogmen der beiden lehrbegriffe / Baur, Ferdinand Christian – 2. verb. Tuebingen: L F Fues, 1836 [mf ed 1989] – 2mf – 9 – 0-7905-4069-X – (incl bibl ref) – mf#1988-0069 – us ATLA [230]

Gegenseitigen beziehungen zwischen der modernen mission und cultur = Modern missions and culture: their mutual relations / Warneck, Gustav – New ed. Edinburgh: James Gemmell, 1888 – 1mf – 9 – 0-8370-6446-5 – (in english) – mf#1986-0446 – us ATLA [240]

Der gegenstoss – Prag (CZ), 1933 4 aug-22 dec [gaps] – 1 – gw Misc Inst [077]

Der gegenwaertige kampf um das alte testament : vortrag / Oettli, Samuel – Guetersloh: C Bertelsmann, 1896 [mf ed 1989] – 1mf – 9 – 0-7905-1015-4 – mf#1987-1015 – us ATLA [221]

Gegenwaertiger zustand von tunkin, cochinchina und der koenigreiche camboja, laos und lac-tho / Bissachere, Pierre J de la – Weimar 1813 [mf ed Hildesheim 1995-98] – 1v on 3mf – 9 – €90.00 – ISBN-10: 3-487-26535-4 – ISBN-13: 978-3-487-26535-3 – gw Olms [959]

Die gegenwart : berliner wochenschrift fuer juedische angelegenheiten / ed by Hirsch, Carl – Berlin: Julius Benzian. v1-2. 1867-68 – (= ser German-jewish periodicals...1768-1945, pt 3) – 1r – 1 – $165.00 – mf#B84 – us UPA [939]

Die gegenwart – Vienna, Austria jan 1859-dec 1870 [mf ed Norman Ross] – 5r – 1 – (foederalistisch) – mf#nrp-1956 – us UMI ProQuest [074]

Die gegenwart : eine halbmonatsschrift – Freiburg Br, Frankfurt/M DE, 1946-58 – 8r – 1 – (1-13th yr with yrly ind of past publ) – mf#7553 – gw Mikropress [073]

Die gegenwart : organ fuer die interessen des judentums – Prague. v1-3. 1867-70 – (= ser German-jewish periodicals...1768-1945, pt 2) – 1 – $75.00 – (lacking: n2-12 in v1) – mf#B-HUC – us UPA [270]

Die gegenwart – Prag (CZ), 1867 28 nov-1870 – 1r – 1 – gw Misc Inst [077]

Gegenwart – Appleton WI. 1898 nov 7-1899, 1900-14, 1915-1916 jun 12 – 10r – 1 – mf#917762 – us WHS [071]

Die gegenwart (klp13) : wochenschrift: literatur, kunst und oeffentliches leben / ed by Lindau, Paul – Berlin: v1 1872-v52 1897 [=yr1-26]; yr27 1898-60 1931 [=v53-120] [mf ed 2003] – (= ser Marbacher mikrofiche-editionen (mme) 13; Kultur – literatur – politik: deutsche zeitschriften des 19./20. jahrhunderts (klp)) – 510mf – 9 – €2300.00 – 3-89131-444-2 – gw Fischer [074]

Die gegenwart, leipzig 1848-1856 : eine enzyklopaedische darstellung der neuesten zeitgeschichte fuer alle staende – [mf ed 1985] – (= ser Enzyklopaedische information im 19. jahrhundert) – 26mf (1:42) – 9 – diazo €265.00 – ISBN-10: 3-598-30671-7 – ISBN-13: 978-3-598-30671-6 – gw Saur [943]

Gegenwart und zukunft der philosophie in deutschland / Gruppe, O F & Roser, Andreas; ed by Roser, Andreas – Berlin: 1855 (mf ed 1996) – (= ser Passauer Texte zur Philosophie) – 3mf – 9 – €49.00 – 3-8267-2334-1 – mf#DHS 2334 – gw Frankfurter [100]

Ejn gegenwurff vnd widerweer huldrych zuinglins, wider hieronymum emser... / Zwingli, H – Zuerich: Christoph Froschouer, 1525 – 1mf – 9 – mf#PBU-512 – ne IDC [242]

Geggie, Robert *see* A practical guide to a right understanding of the prefixes and affixes in the english language

Der gegner : blaetter zur kritik der zeit – Berlin DE, 1919 1 apr-1922 24 mar – 1 – gw Misc Inst [074]

Die gegner "edgars" : und ihre leistungen / Hammerstein, Ludwig von – Trier: Paulinus-Druckerei, 1887 – 1mf – 9 – 0-8370-7152-6 – mf#1986-1152 – us ATLA [230]

Die gegner zwinglis am grossmuensterstift in zuerich / Pestalozzi, T – Zuerich, 1918 – (= ser Schweizer studien zur geschichtswissenschaft) – 3mf – 9 – mf#ZWI-72 – ne IDC [242]

Geharnschte venus / Stieler, Kaspar von; ed by Raehse, Th – Halle: Max Niemeyer, 1888 – us Wisconsin U Libr [430]

Een geheiligd leven / David, V D – Nijmegen: P J Milborn, 1898 [mf ed 1995] – (= ser Yale coll) – vii/230p (ill) – 1 – 0-524-09902-2 – (in dutch) – mf#1995-0902 – us ATLA [240]

Het geheim der vrijmetselarij opengelegd / Geysbeek, Pieter Gerardus Witsen – Amsterdam: Lodewijk van Es 1831 – 2mf – 9 – mf#vrl-47 – ne IDC [366]

Geheime geschichte des neuen franzoesischen hofes : in briefen waehrend der monate august, september und oktober 1805 / Goldsmith, Lewis – St Petersburg [i.e. Leipzig 1806-1807 [mf ed Hildesheim 1995-98] – 2v on 5mf – 9 – €100.00 – ISBN-10: 3-487-26399-8 – ISBN-13: 978-3-487-26399-1 – gw Olms [944]

Geheime gesellschaften in alter und neuer zeit, ihre organisation, ihre zwecke und ziele : mit besonderer beruecksichtigung der freimaurer- und odd-fellow-logen, des druiden- und illuminaten-ordens / Hein, Erich – Leipzig: Raimund Gerhard 1913 – 2mf – 9 – mf#vrl-201 – ne IDC [366]

Geheime liebschaften heinrich's des vierten : aus original-manuscripten gezogen, und gesammelt waehrend der anwesenheit der armeen in frankreich im jahre 1815 / Pappenheim, Albert von – Nuernberg 1824 [mf ed Hildesheim 1995-98] – 2v on 4mf – 9 – €120.00 – ISBN-10: 3-487-26104-9 – ISBN-13: 978-3-487-26104-1 – gw Olms [920]

Geheime nachrichten ueber napoleon bonaparte. von einem manne, der ihn seit funfzehn jahren nicht verlassen hat – Leipzig 1815 (mf ed 1992) – 2mf – 9 – 3-89349-111-2 – mf#DHS-AR 80 – gw Frankfurter [944]

Die geheime offenbarung des apostels johannes : und zwar die ersten drei kapitel derselben in zehn vortraegen / Paulhuber, Xav – Schaffhausen: Hurter, 1851 – 1mf – 9 – 0-7905-0437-5 – (incl bibl ref) – mf#1987-0437 – us ATLA [225]

Die geheime offenbarung und die zukunstserwartungen des urchristentums / Rohr, Ignaz – 1. & 2. aufl. Muenster i W: Aschendorff 1911 [mf ed 1992] – (= ser Biblische zeitfragen 4/5) – 1mf – 9 – 0-524-05630-7 – (incl bibl ref) – mf#1992-0485 – us ATLA [220]

Geheime sekte van 't kimpasi / Wing, Joseph Van – Brussels, Belgium. 1921? – 1r – us UF Libraries [960]

Geheime unterredungen zwischen zweyen vertrauten freunden, einem theologo philosophizante und philosopho theologizante, von magia naturali. zum druck gegeben vom collegio curiosorum in deutschland – Cosmopoli, 1702 – 1 – us Wisconsin U Libr [240]

Die geheimen gesellschaften mit vollem rechte verurtheilt von der katholischen kirche / Becker, Wilhelm – St Louis, MO: B Herder, [18–?] – 1mf – 9 – 0-524-02518-5 – mf#1990-0618 – us ATLA [241]

Geheimer briefwechsel zwischen dem kaiser napoleon und dem papst pius 7 : aus den urkundlichen akten gezogen, nebst dem bericht ueber die gewaltthaetige entfuehrung des sr. paepst. heiligkeit nach frankreich / [Bourges, Charles Doris des] – O.O. 1814 (mf ed 1993) – 2mf – 9 – €24.00 – 3-89349-255-0 – mf#DHS-AR 112 – gw Frankfurter [240]

Geheimes kinder-spiel-buch mit vielen bildern / Ringelnatz, Joachim [Hans Boetticher] – Potsdam: G Kiepenheuer, 1924 [mf ed 1989] – 48p (ill) – 1 – mf#7055 – us Wisconsin U Libr [880]

Das geheimnis der alten mamsell : roman / Marlitt, Eugenie [pseud] – Stuttgart: Union Deutsche Verlagsgesellschaft [1919] [mf ed 1978] – (= ser Romane und novellen 1) – 2r – 1 – mf#film mas c376 – us Harvard [830]

Das geheimnis der froemmigkeit und die gottmenschheit christi : ein beitrag zur deutung des schluffes von 1. tim. 3 / Bleibtreu, Walther – Guetersloh: C Bertelsmann, 1906 – (= ser Beitraege zur foerderung christlicher theologie) – 1mf – 9 – 0-524-08030-5 – mf#1992-1123 – us ATLA [220]

Das geheimnis der universitaet / Rosenstock-Huessy, Eugen – Stuttgart, 1958 – 6mf – 8 – €14.00 – ne Slangenburg [378]

Das geheimnis in der religion : vortrag / Duhm, Bernhard – Freiburg i. B.: J C B Mohr, 1896 – 1mf – 9 – 0-7905-1594-6 – mf#1987-1594 – us ATLA [210]

Geheimniss des gnaden-bunds / Lampe, F A – Bremen, 1729-37 – 47mf – 9 – mf#PBA-215 – ne IDC [240]

Das geheimniss und die innere einheit der drey goethe'schen balladen : der fischer, der erlkoenig und der todtentanz / Schrader, Hermann – Berlin: H Dolfuss 1881 [mf ed 1990] – 1r – 1 – (filmed with: goethe's sprache und die antike / carl olbrich) – mf#7396 – us Wisconsin U Libr [430]

Die geheimnisse des glaubens / Schoeberlein, Ludwig – Heidelberg: Carl Winter, 1872 – 1mf – 9 – 0-8370-5312-9 – mf#1985-3312 – us ATLA [240]

Geheimnisse des reifen lebens : aus den aufzeichnungen angermanns / Carossa, Hans – Leipzig: Insel-Verlag 1937 [mf ed 1989] – 1r – 1 – (filmed with: georg buchners drama dantons tod / hans landsberg) – mf#7143 – us Wisconsin U Libr [830]

Geheimnisse einiger philosophen und adepten: aus der verlassenschaft eines alten mannes – Erster theil. Leipzig: C.G. Hilscher, 1780. 1 reel. 1203 – 1 – us Wisconsin U Libr [540]

Der geheimnisvolle hof : eine erzaehlung aus dem norden / Asbeck, Wilhelm Ernst – Dresden: Meinhold, c1940 [mf ed 1988] – 136p – 1 – mf#6958 – us Wisconsin U Libr [880]

Geheimnisvolle inseln tropen-afrikas / Bernatzik, Hugo Adolf – Berlin, Germany. 1933 – 1r – us UF Libraries [960]

Gehetzt uebers meer / Korn, Heinz – 5. und 6. Aufl. Berlin: Junge Generation Verlag, [1944] – 1r – 1 – us Wisconsin U Libr [830]

Gehirne : novellen / Benn, Gottfried – Leipzig: K Wolff, 1916 – (= ser Buecherei "der juengste tag" 35) – 52p – 1 – mf#7006 – us Wisconsin U Libr [830]

Gehler, Johann Samuel Traugott *see* Johann samuel traugott gehlers physikalisches woerterbuch

Gehre, Horst *see* Die entwicklung der amtshaftung in deutschland seit dem 19. jahrhundert

Gehring, Albert *see* Racial contrasts

Gehring, Alwin *see*
- Bartholomaeus ziegenbalg
- Braune christen im hause des herrn

Gehring, Friedrich Wilhelm *see* Die volksdeutsche dichtung in unserer zeit

Geht dir da nicht ein auge auf : gedichte / ed by Schramm, Godehard et al – Frankfurt/M: Fischer Taschenbuch Verlag, 1974 – 1r – 1 – us Wisconsin U Libr [810]

Geib, August *see* Gedichte

Geibel, Emanuel *see*
- Emanuel geibel's briefe an karl freiherrn von der malsburg und mitglieder seiner familie
- Emanuel geibels gesammelte werke
- Gedichte
- Meister andrea
- Ein ruf von der trave
- Zeitstimmen

Geider, Stefan *see* Die historische entwicklung der interdependenz von atmung und herz-kreislaufsystem und der einflu.. der atmung auf die herzzeitintervalle unter besonderer beruecksichtigung der koerperposition

Geiermann, P *see* A manual of theology for the laity

Die geige : vier novellen / Binding, Rudolf Georg – Potsdam: Ruetten & Loening, 1941 [mf ed 1989] [mf ed 1989] – 212p – 1 – mf#7024 – us Wisconsin U Libr [830]

Geigel Polanco, Vicente *see*
- Canto de tierra adentro
- Canto del amor infinito
- Despertar de un pueblo
- Independencia de puerto rico

Geigel Sabat, Fernando Jose *see* Balduino enrico

Geigel Y Zenon, Jose *see* Articulos politico-humoristicos y literarios

Geiger, Abraham *see*
- Judaism and its history
- Juedische zeitschrift fuer wissenschaft und leben
- Sadducaeer und pharisaeer
- Urschrift und uebersetzungen der bibel

Geiger, Albert James *see* Study of the farm shop instruction in the vocational agricultural s...

Geiger, Bernhard *see* Die amesa spentas

Geiger, Eugen *see* Der meistergesang des hans sachs

Geiger, L-B *see*
- La participation dans la philosophie de s thomas d'aquin

Geiger, Ludwig *see*
- Anton reiser
- Aus chamissos fruehzeit
- Charlotte von schiller und ihre freunde
- De la litterature allemande
- Deutsche satiriker des 16. jahrhunderts
- Goethe in frankfurt am main 1797
- Goethe und die seinen
- Johann reuchlin
- Johann reuchlins briefwechsel
- Renaissance und humanismus in italien und deutschland
- Das studium der hebraeischen sprache in deutschland
- Unbekannte aufsaetze und gedichte
- Zeitschrift fuer die geschichte der juden in deutschland

Geiger, Ludwig et al *see* Boernes werke

Geiger, Paul *see* Deutsches volkstum in sitte und brauch

Geiger, W *see* Die pehleviversion des ersten capitels des vendidaed

Geiger, Wilhelm *see*
- The dipavamsa and mahavamsa and their historical development in ceylon
- Zarathushtra in the gathas and in the greek and roman classics

Geijer, Erik Gustaf *see* The history of the swedes

Geikie, Cunningham *see*
- Holy land and the bible
- Hours with the bible
- The life and words of christ
- Our new religions

Geikie, James *see* The great ice age and its relation to the antiquity of man

Geikie, John Cunningham *see*
- The english reformation
- The holy land and the bible
- The precious promises
- Reply to a special report of the superintendent of education

Geikie's literary news – Toronto: J.C. Geikie, [1856-18–] – 9 – mf#P05593 – cn Canadiana [400]

Geil, W E *see* A yankee in pigmy land

Geil, William Edgar *see* Yankee on the yangtze

Geilenkirchener volkszeitung – Geilenkirchen DE, 1957 2 nov-1959 30 jun – 1 – (bezirksausgabe von aachener volkszeitung) – gw Misc Inst [074]

Geiler, J *see* Navicula sive speculum fatuorum...

Geilfus, G *see* Erzaehlung des sempacher krieges...

Geilinger, Walter *see* Kilimandjaro

Geisel, Judith Charlotte *see* Tasso und sein gefolge

Geiselmann, Josef Rupert *see* Geist des christentums und die katholizismus

Geiseltal-echo *see* Unser grundstoff

Geiseltal-kurier *see* Unser grundstoff

Die geisha o-sen : geisha-lieder nach japanischen motiven / Henschke, Alfred (pseud. Klabund) – Muenchen: Roland-Verlag A Mundt, 1918 – 1r – 1 – us Wisconsin U Libr [480]
Geisler *see* Ueber die schriftstellerische thaetigkeit thomas abbt's
Geisler, PR *see* Tissue degradation markers and subjective reports of pain as a result of eccentric muscular contractions
Geisler, S A *see* Eccentric peak torque and maximal repetition work percentages of the dominant external rotators in college division 1 baseball players
Geisler, Victor *see* Hermann siebeck's religionsphilosophie dargestellt und beurteilt
Geislinger zeitung – Geislingen a.d. Steige DE, 1980- – 6r/yr until 1994 – 1 – (bezirksausgabe von suedwest-presse, ulm; filmed by other misc inst: 1987- [6r/yr]) – gw Misc Inst [074]
Die geissel / ed by Rebmann, sp. Vollmer, Georg Friedr – Uppsala 1797, Paris 1797/98, Mainz 1797 – (= ser Dz. historisch-politische abt) – 3jge[zu je 12iss] on 26mf – 9 – €260.00 – mf#k/n1758 – gw Olms [933]
Geissler, Horst Wolfram *see* Grillparzer und schopenhauer
Geissler, Mortiz *see* Am i a christian?
Geissler, Rolf *see* Dekadenz und heroismus
Geist der goethezeit : versuch einer ideellen entwicklung der klassisch-romantischen literaturgeschichte / Korff, Hermann August – Leipzig: Koehler & Amelag, 1964-66 [mf ed 1993] – 5v on 2r – 1 – mf#7847 – us Wisconsin U Libr [430]
Der geist der lutherischen theologen wittenbergs im verlaufe des 17. jahrhunderts : theilweise nach handschriftlichen quellen / Tholuck, August – Hamburg: F. und A. Perthes, 1852 – 1mf – us ATLA [242]
Der geist der lutherischen theologen wittenbergs im verlaufe des 17. jahrhunderts : theilweise nach handschriftlichen quellen / Tholuck, August – Hamburg: F und A Perthes, 1852 – 1mf – 9 – 0-7905-6698-2 – mf#1988-2698 – us ATLA [242]
Der geist des christenthums : seine entwickelung und sein verhaeltnis zu kirche und cultur der gegenwart / Hanne, Johann Wilhelm – Elberfeld: RL Friderichs, 1867 – 1mf – 9 – 0-8370-3466-3 – mf#1985-1466 – us ATLA [240]
Geist des christentums und die katholizismus / Geiselmann, Josef Rupert – Mainz: Mattias Gruenewald, 1940 – 1r – 1 – 0-8370-1145-0 – mf#1984-B491 – us ATLA [241]
Der geist des hohen liedes : geschichte, kritik und uebersetzung / Altschul, Jakob – Wien: Wilhelm BraumUeller, 1874 – 1mf – 9 – 0-8370-2082-4 – mf#1985-0082 – us ATLA [220]
Geist des musikalischen kunstmagazins / Reichardt, Johann Friedrich; ed by Alberti, I – Berlin: gedruckt & in Commission bey J F Unger 1791 [mf ed 19–] – 3mf – 9 – mf#fiche 932, 652 – us Sibley [780]
Geist des ostens – Munich, 1913-15 [mf ed 2001] – (= ser Christianity's encounter with world religions, 1850-1950) – 1r – 1 – (in german) – mf#2001-s154 – us ATLA [073]
Der geist gottes und die verwandten erscheinungen im alten testament und im anschliessenden judentum / Volz, Paul – Tuebingen: J C B Mohr (Paul Siebeck), 1910 – 1mf – 9 – 0-7905-0408-1 – (incl indes) – mf#1987-0408 – us ATLA [221]
Geist, Hermann *see* Wie fuehrt goethe sein titanisches faustproblem, das bild seines eigenen lebenskampfes, vollkommen einheitlich durch?
Geist und buchstabe der dichtung : Goethe, Schiller, Kleist, Hoelderlin / Kommerell, Max – 3rd rev enl ed. Frankfurt/Main: V Klostermann, 1944 [mf ed 1993] – 357p – 1 – mf#8219 – us Wisconsin U Libr [410]
Geist und form : aufsaetze zur deutschen literaturgeschichte / Vietor, Karl – Bern: A Francke c1952 [mf ed 1992] – 1r – 1 – (incl bibl ref & ind. filmed with: saggi di letteratura tedesca / leonello vincenti & other title) – mf#3140p – us Wisconsin U Libr [430]
Geist und freiheit : allgemeine kritik des gesetzesbegriffes in natur- und geisteswissenschaft / Koehler, Walther – Tuebingen: JCB Mohr, 1914 – 1mf – 9 – 0-7905-9402-1 – (incl bibl ref) – mf#1989-2627 – us ATLA [100]
Geist und gesellschaft : ueber die aufloesung der staendischen gesellschaft im epischen werk von karl gutzkow / Kramp, Willy – Wuerzburg: K Triltsch, 1937 – 1r – 1 – (incl bibl ref (p. 68-70)) – us Wisconsin U Libr [430]
Geist und leben – 20(1947)-31(1958) – 105mf – 9 – €252.00 – (cont: zeitschrift fuer ascese und mystik) – ne Slangenburg [230]
Geist und leben : vortrage und aufsaetze / Koch, Franz – Hamburg: Hanseatische Verlagsanstalt, [c1939] [mf ed 1993] – 239p – 1 – (incl bibl ref) – mf#8085 – us Wisconsin U Libr [430]
Geist und schrift bei sebastian franck : eine studie zur geschichte des spiritualismus in der reformationszeit / Hegler, Alfred – Freiburg i. B: J C B Mohr, 1892 – 1mf – 9 – 0-7905-4231-5 – (incl bibl ref) – mf#1988-0231 – us ATLA [140]
Der geisterbeschwoerer : volksschauspiel in 4 akten / Zwerenz, Carl – Wien: F S Hummel 1869 [mf ed 1995] – 1r – 1 – (filmed with: addrich im moos / heinrich zschokke) – mf#3766p – us Wisconsin U Libr [820]
Der geisterseher : aus den papieren des grafen o– / Schiller, Friedrich von – Muenchen: Georg Mueller 1922 [mf ed 1995] – 1r – 1 – (1st pt ed by friedrich schiller. 2nd pt ed by hanns heinz ewers. filmed with: schiller, don carlos / rudolf ibel [ed]) – mf#3731p – us Wisconsin U Libr [430]
Die geisterseher : humoristischer roman / Mauthner, Fritz – Berlin: Verlag des Vereins der Buecherfreunde 1894 [mf ed 1996] – 1r – 1 – (filmed with: zwischen sumpf und firmament / kurt martens) – mf#3950p – us Wisconsin U Libr [830]
Die geisterwelt im glauben des paulus / Dibelius, Martin – Goettingen: Vandenhoeck & Ruprecht, 1909 – 1mf – 9 – 0-8370-2903-1 – (incl ind of subjects, of greek words, and of citations from biblical, and extra-biblical literature) – mf#1985-0903 – us ATLA [225]
Geistesgeschichtliche aspekte des genossenschaftlichen bildungsgedankens : unter besonderer beruecksichtigung von v a hubers schriften / Jansen, Brigitte E S – (mf ed 1995) – 3mf – 9 – €49.00 – 3-8267-2100-4 – mf#DHS 2100 – gw Frankfurter [370]
Die geistesgeschichtliche bedeutung der bibel / Eucken, Rudolf – Leipzig: A Kroener, 1917 – 1mf – 9 – 0-524-04397-3 – mf#1992-0090 – us ATLA [220]
Der geisteskampf des christentums gegen den islam bis zur zeit der kreuzzuege / Keller, Adolf – Leipzig: W Faber, 1896 – 1mf – 9 – 0-524-01562-7 – (incl bibl ref) – mf#1990-2516 – us ATLA [230]
Die geisteskultur von tarsos im augusteischen zeitalter : mit beruecksichtigung der paulinischen schriften / Boehlig, Hans – Goettingen: Vandenhoeck & Ruprecht, 1913 – 1mf – 9 – 0-7905-0812-5 – (incl bibl ref and indexes) – mf#1987-0812 – us ATLA [260]
Die geisteswelt ulrich zwinglis / Koehler, W – Gotha, 1920 – 2mf – 9 – mf#ZWI-76 – ne IDC [242]
Geisthardt, Hans-Juergen *see* Literatur im blickpunkt
Geisthardt, Hans-Juergen et al *see* Literatur im blickpunkt
Die geistige einwirkung der person jesu auf paulus : eine historische untersuchung / Koelbing, Paul – Goettingen: Vandenhoeck & Ruprecht, 1906 – 1mf – 9 – 0-8370-3959-2 – mf#1985-1959 – us ATLA [920]
Die geistige offenbarung gottes in der geschichtlichen person jesu / Steinmann, Theophil – Goettingen: Vandenhoeck & Ruprecht, 1903 – 1mf – 9 – 0-8370-55431 – (incl bibl ref) – mf#1985-3543 – us ATLA [240]
Geistiges vermaechtnis *see* Nikolaus lenaus geistiges vermaechtnis
Geistliche gesenge – Bresslaw: Winckler 1541 – (= ser Hqab. literatur des 16. jahrh.) – 1mf – 9 – €20.00 – mf#1541b – gw Fischer [780]
Die geistliche gestalt eines evangelischen lehrers : nach dem sinn und exempel der alten / Arnold, Gottfried – Frankfurt: Johann Georg Boehmen, 1723 – 1r – 1 – 0-8370-0469-1 – mf#1984-B259 – us ATLA [150]
Geistliche herzens einbildungen inn zweihundert und fuenfzig biblischen figur-spruechen angedeutet... / [Mattsperger, M] – Augstburg, 1685 – 2mf – 9 – mf#0-1135 – ne IDC [700]
Das geistliche jahr / geistliche lieder / Droste-Huelshoff, Annette von; ed by Arens, Eduard – Leipzig: M Hesse [between 1900 und 1920] [mf ed 1995] – 1r [ill] – 1 – (filmed with: gesammelte werke in drei baenden / richard dehmel) – mf#3824p – us Wisconsin U Libr [800]
Die geistliche leiter zum himmelreich... : reimweiss beschrieben... / Grezel, Wolfgang – s.l. 1594 – (= ser Hqab. literatur des 16. jahrh.) – 1mf – 9 – €20.00 – mf#1594 – gw Fischer [780]
Geistliche lieder : dere etliche von alters her in der kirchen eintrechtiglich gebraucht, und etliche zu unser zeit, von erleuchteten, fromen christen und gottseligen lerern new zugericht sind... – s.l. 1566 – (= ser Hqab. literatur des 16. jahrh.) – 2mf – 9 – €30.00 – mf#1566a – gw Fischer [780]
Geistliche lieder : mit einer newen vorrede d mart luth; warnung d mart luth – Leipzig: Berwald 1556 – (= ser Hqab. literatur des 16. jahrh.) – 7mf – 9 – €75.00 – mf#1556 – gw Fischer [780]
Geistliche lieder *see* Das geistliche jahr / geistliche lieder
Geistliche lieder d martini lutheri : vnd anderer fromen christen: nach ordnung der jarzeit, mit collecten vnd gebeten – Leipzig: Beyer 1583 – (= ser Hqab. literatur des 16. jahrh.) – 7mf – 9 – €75.00 – mf#1583 – gw Fischer [780]
[Geistliche lieder und psalmen] / Luther, Martin – Magdeburg: Lotther 1546 – (= ser Hqab. literatur des 16. jahrh.) – 4mf – 9 – €50.00 – mf#1546b – gw Fischer [780]
[Geistliche lieder und psalmen] geystlike leder un[d] psalmen, uppet nye gebetert : dyth sint twee gesanck-boekelin, und mit velen andern gesengen...vormeret unde gebetert / Luther, Martin – Magdeborch: Walther c1540 – (= ser Hqab. literatur des 16. jahrh.) – 4mf – 9 – €50.00 – mf#1540b – gw Fischer [780]
Geistliche lieder vnd psalmen : durch d mart luther, vnd andere fromme christen auffs new zusamenbracht – Leipzig: Berwaldt 1553 – (= ser Hqab. literatur des 16. jahrh.) – 6mf – 9 – €70.00 – mf#1553b – gw Fischer [780]
Geistliche lieder vnd psalmen : welche von fromen christen gemacht, vnd zusamen gelesen sind; auffs new vbersehen, gebessert vnd gemehret / Luther, Martin – Erffurdt: Sachsse [nach 1546] – (= ser Hqab. literatur des 16. jahrh.) – 5mf – 9 – €60.00 – mf#1546c – gw Fischer [780]
Geistliche lieder vnd psalmen der alten apostolischer recht vnd wargleubiger christlicher kirchen... : auffs fleissigste und christlichste zusamen bracht gemehret vnd gebessert... / Leisentrit, Johann – Bautzen: [Wolrab] 1573 – (= ser Hqab. literatur des 16. jahrh.) – 8mf – 9 – €80.00 – mf#1573c – gw Fischer [780]
[Geistliche lieder von der allerheiligisten jungfraw maria] das ander theil geistlicher lieder von der allerheiligisten jungfrawen maria...mit...fleis zusammen bracht gemehret vnd gebessert / Leisentrit, Johann – Bautzen 1573 – (= ser Hqab. literatur des 16. jahrh.) – 2mf – 9 – €30.00 – mf#1573d – gw Fischer [780]
Die geistliche lyrik der juden in nachdichtungen / Wiener, Meir – Wien; Leipzig: R Loewit, 1920 [mf ed 1993] – 188p – 1 – (incl bibl ref. no more publ) – mf#8146 – us Wisconsin U Libr [430]
Geistliche selbstbekenntnisse : ueber das wesen und leben der evangelisch-lutherischen kirche / Appelius, Karl Theodor – Leipzig: Eduard Kummer, 1867 [mf ed 1986] – 1mf – 9 – 0-8370-8644-2 – mf#1986-2644 – us ATLA [242]
Geistliche todts-gedancken bey allerhand gemaehlden und schildereyen... / [Rentz, M] – Passau: Gedruckt bey Friderich Gabriel Mangold; Linz: Verlegts Franz Anton Ilger, 1753 – 5mf – 9 – mf#0-1848 – ne IDC [090]
Geistliche volkslieder : ans alter und neuerer Zeit / ed by Hommel, Friedrich – Leipzig: B G Teubner, 1864 [mf ed 1993] – xviii/308p (ill) – 1 – (incl bibl ref and ind) – mf#8190 – us Wisconsin U Libr [780]
Die geistlichen uebungen des ignatius von loyola = Exercitia spiritualia / Ignatius of Loyola, Saint; ed by Schickele, Rene – Berlin: H. Seemann Nachfolger, [19–?]. Chicago: Dep of Photodup, U of Chicago Lib, 1972 (1r); Evanston: American Theol Lib Assoc, 1984 (1r) – (= ser Kultur-Documente) – 1 – 0-8370-0430-6 – mf#1984-B309 – us ATLA [241]
Die geistlichen uebungen des ignatius von loyola : eine psychologische studie / Holl, Karl – Tuebingen: J C B Mohr 1905 [mf ed 1990] – (= ser Sammlung gemeinverstaendlicher vortraege und schriften aus dem gebiet der theologie und religionsgeschichte 41) – 1mf – 9 – 0-7905-5843-2 – mf#1988-1843 – us ATLA [241]
Geistliches jahr : in liedern auf alle sonn- u festtage / Droste-Huelshoff, Annette von – Muenster/W: Aschendorff, 1913 [mf ed 1989] – 232p – 1 – mf#7186 – us Wisconsin U Libr [810]
Geistliches magazien – La Salle IL 1764-71 – 1 – mf#4456 – us UMI ProQuest [200]
Ein geistliches spiel von s meinrads leben und sterben / ed by Morel, P Gall – Stuttgart: Litterarischer Verein, 1863 [mf ed 1993] – (= ser Blvs 69) – 126p – 1 – mf#8470 reel 14 – us Wisconsin U Libr [241]
Geistliches und weltliches aus dem tuerkisch-griechischen orient : selbsterlebtes und selbstgesehenes / Gelzer, Heinrich – Leipzig: B G Teubner 1900 [mf ed 1986] – 1mf [ill] – 9 – 0-8370-8021-5 – (incl bibl ref) – mf#1986-2021 – us ATLA [956]
Geistreiche gesaenge und lieder auf alle sonntags-evangelien und episteln so in dem christlichen jahre enthalten seyn : der 1. theil (-11 theil) – Schweidnitz: Johann Christian Mueller, 1725-26. 11v – 39mf – 9 – mf#0-1866 – ne IDC [090]
Geistreiches fast- und nachtmahlbuechlein... – Zuerich: Buercklischer Truckerey – 5mf – 8 – €12.00 – (trans fr drelincourt's french version) – ne Slangenburg [880]
Geistweit, William Henry *see* The young christian and his bible
Geitmann, Anja *see* Growth and formation of the cell wall in pollen tubes of nicotiana tabacum and petunia hybrida
Geklibene shriftn / Kacyzne, Alter – Varshe, Poland. 1951 – 1r – us UF Libraries [939]
Geklibene shriftn / Rozobski, Mordekhai – Buenos Aires, Argentina. 1947 – 1r – us UF Libraries [939]
Gelaehmte schwingen : lustspiel in einem aufzuge / Thoma, Ludwig – Muenchen: A Langen, c1918 – 1r – 1 – us Wisconsin U Libr [820]
Das gelahrte preussen /.../ – Thorn (Torun PL), 1722 oct-1724 sep – 1 – gw Misc Inst [077]
Gelasius kirchengeschichte (gcsej5) / ed by Loeschcke, G & Heinemann, M – 1918 – (= ser Griechische christlichen schriftsteller der ersten jahr- hunderte (gcsej)) – €15.00 – ne Slangenburg [241]
Das gelbe ahornblatt : ein leben in geschichten / Brehm, Bruno – Karlsbad-Drahowitz: A Kraft [1943?] [mf ed 1989] – 1r – 1 – (filmed with other titles) – mf#7066 – us Wisconsin U Libr [830]
Das gelbe buch : novellen und gedichte / Gleichen-Russwurm, Alexander von et al – Stuttgart: Verlagsgesellschaft "Das gelbe Blatt" 1919 [mf ed 1993] – 1r – 1 – (filmed with: die zeit traegt einen roten stern) – mf#3339p – us Wisconsin U Libr [800]
Gelbe hefte : historische und politische zeitschrift fuer das christliche deutschland – Muenchen: Verlag der Gelben Hefte. jahrg1-17 1924/25-[1980] – 18v on 7r – 1 – (cont: historisch-politische blaetter fuer das katholische deutschland; ceased in 1942 with v18 n3. cf. gesamtverzeichnis der deutschsprachigen schriftums) – mf#film mas c721 – us Harvard [241]
Gelbe hefte : historische und politische zeitschrift fuer das katholische deutschland – Muenchen [1] 1924/25-[18] 1941/42 [mf ed 2005] – (= ser Kultur – literatur – politik 19) – 165mf – 9 – €820.00 – 3-89131-461-2 – gw Fischer [241]
Die gelbe post : ostasiatische halbmonatsschrift – Schanghai (VR), 1939 1 may-1 nov – 1r – 1 – gw Misc Inst [079]
Der gelbe seedieb : roman / Seeliger, Ewald Gerhard – Berlin: Ullstein 1915 [mf ed 1991] – 1r – 1 – (filmed with: charles sealsfield (carl postl) / albert b faust) – mf#2941p – us Wisconsin U Libr [830]
Gelbhaus, Sigmund *see* Rabbi jehuda hanassi und die redaction der mischna
Gelbmann, Gerhard *see* Die pragmatische kommunikationstheorie rekonstruktion, wissenschaftsphilosophischer hintergrund, kritik
Geld und erfahrung / Eyth, Max – Hamburg-Grossborstel: Verlag der Deutschen Dichter-Gedaechtnis-Stiftung 1916 [mf ed 1993] – 1r – 1 – (with portrait of eyth; int by carl mueller-rastatt; ill by theodor herrmann. filmed with: murillo / ernst eckstein) – mf#8574 – us Wisconsin U Libr [830]
Geld und geist / Gotthelf, Jeremias [Albert Bitzius]; ed by Bloesch, Hans – Muenchen, Bern: E Rentsch, 1930 [mf ed 1993] – (= ser Saemtliche werke in 24 baenden 7) – 437p – 1 – mf#8522 reel 2 – us Wisconsin U Libr [890]
Geld und werthpapiere : eine besprechung der fuer den bankverkehr erheblichen bestimmungen des entwurfes eines buergerlichen gesetzbuches fuer das deutsche reich / Koch, Richard – Berlin, Leipzig: J Guttentag, 1889 – (= ser Civil law 3 coll; Beitraege zur erlaeuterung und beurtheilung des entwurfes eines buergerlichen gesetzbuches fuer das deutsche reich) – 1mf – 9 – (includes bibliographic references) – mf#LLMC 96-600 – us LLMC [346]
Geldard, Sarah R *see* Acceptable service (what it really is)
Geldart, Ernest *see* A manual of church decoration and symbolism
[Gelder, A de] Lilienfeld, K *see* Arent de gelder
Gelder, Elias Van *see* Volksschule des judischen alterthums nach talmudischen
Geldern'sche zeitung *see* Kreis-blatt
Gelders, V *see* Quelques aspects de l'evolution des colonies en 1938
Der geldstag : oder, die wirtschaft nach der neuen mode / Gotthelf, Jeremias [Albert Bitzius]; ed by Hunziker, Rudolf – Erlenbach, Zuerich: E Rentsch, 1923 [mf ed 1993] – (= ser Saemtliche werke in 24 baenden 8) – 416px – 1 – mf#8522 reel 2 – us Wisconsin U Libr [830]
Geldtheorie und wirtschaftswachstum : neuere methodik und deutungen / Gruner, Hans – Heidelberg, 1961 – 2mf – 9 – 3-89349-391-3 – gw Frankfurter [332]
Gelee, Claude *see* Claude lorrain
Ein gelegenheitsgedicht von brockes / ed by Gundolf, Friedrich – Heidelberg: C Winter, 1931 [mf ed 1989] – 5/[12]p – 1 – mf#7089 – us Wisconsin U Libr [810]
Ein gelegenheitsgedicht von brockes *see* Die ravensburger fahnentraeger

Gelegenheitsgedichte und prologe fuer arbeiterfeste : mit einem anhang, winke fuer redner / Wittich, Manfred – 2. durchgesehene und verm. Aufl. Muenchen: M Ernst, 1894 – 1r – 1 – us Wisconsin U Libr [810]

Gelehrte abhandlungen und nachrichten aus und von russland : geliefert von der schule der sprache, kuenste... / ed by Buesching, Ant Friedr – Leipzig, Koenigsberg, Mitau 1764-65 – (= ser Dz) – 2st on 4mf – 9 – €120.00 – mf#k/n239 – gw Olms [947]

Gelehrte abhandlungen und nachrichten aus und von russland – Leipzig, Koenigsberg, Mitau DE, 1764-65 – 1 – gw Misc Inst [074]

Gelehrte beytraege zu den mecklenburg-schwerinschen nachrichten *see* Mecklenburgische nachrichten, fragen und anzeigungen

Das gelehrte hannover : oder lexikon von schriftstellern und schriftstellerinnen, gelehrten geschaeftsmaennern und kuenstlern die seit der reformation in und ausserhalb den saemtlichen zum jetzigen koenigreich hannover gehoerigen provinzen gelebt haben und noch leben / Rotermund, Heinrich Wilhelm – Bremen 1823 [mf ed Hildesheim 1983] – (= ser Die schriftsteller- und gelehrtenlexika des 17., 18., und 19. jahrhunderts) – 2v on 19mf – 9 – diazo €88.00 silver €106.00 – gw Olms [030]

Das gelehrte schwaben : oder lexicon der jetzt lebenden schwaebischen schriftsteller / Gradmann, Johann Jacob – Ravensburg 1802 [mf ed Hildesheim 1983] – (= ser Die schriftsteller- und gelehrtenlexika des 17., 18., und 19. jahrhunderts) – 1v on 10mf – 9 – diazo €52.80 silver €69.80 – gw Olms [430]

Gelehrtes fuerstenthum baireut / Fikenscher, Georg Wolfgang Augustin – Erlangen, Nuernberg 1801-05 [mf ed Hildesheim 1983] – (= ser Die schriftsteller- und gelehrtenlexika des 17., 18., und 19. jahrhunderts) – 12v on 34mf – 9 – diazo €158.00 silver €188.00 – gw Olms [430]

Die geleise des kirchenjahres und der wege gottes in der gruendung seines reiches auf erden : in der anordnung und dem zusammenhange der sonn- und festtaeglichen evangelien (de tempore) / Seiss, Martin – Regensburg; New York: Friedrich Pustet, 1875 – 1mf – 9 – 0-8370-7507-6 – mf#1986-1507 – us ATLA [240]

Gelesnoff (Zhelieznov), Vladimir *see* The pathway of faith

Gelfand, Mark I *see* The war on poverty, 1964-1968

Gelfand, Michael *see*
- African background
- African crucible
- African's religion
- Gubulawayo and beyond
- Medicine and custom in africa
- Mother patrick and her nursing sisters
- Northern rhodesia in the days of the charter
- Shona religion
- Sick african
- Witch doctor

Geliebeter jan : briefe gehen nach dem osten / Loeff, Friedel – Berlin: M Warneck, 1944 – 1r – 1 – us Wisconsin U Libr [860]

An einen geliebten soldaten : neue verse / Doehrn, Gisela – Berlin: H von Hugo, 1941 [mf ed 1989] – 74p – 1 – mf#7180 – us Wisconsin U Libr [810]

Gelilot ha-arets / Kahane, Hillel – Bucharest, Romania. 1880 – 1r – us UF Libraries [939]

Gelineau, J *see* Chant et musique dans le culte chretien

Gell, William *see*
- The itinerary of greece
- Narrative of a journey in the morea
- Pompeiana

Gella Iturriaga, Jose *see* Refranero del mar. madrid, 1944

Gellert als romanschriftsteller / Kretschmer, Elisabeth – Breslau: Breslauer Genossenschafts-Buchdruckerei, 1902 – 1r – 1 – us Wisconsin U Libr [430]

Gellert, Christian Fuerchtegott *see*
- C f gellert's saemtliche schriften
- Gellerts aelteste fabeln
- Gellerts dichtungen
- Poetische werke

Gellert und holland : ein beitrag zu...der geistigen und literarischen beziehungen zwischen deutschland und holland... / Noordhoek, Willem Johannes – Amsterdam: H J Paris, 1928 – 1r – 1 – (incl bibl ref) – us Wisconsin U Libr [430]

Gellerts aelteste fabeln / Gellert, Christian Fuerchtegott; ed by Handwerck, Hugo – Marburg: R Friedrichs Universitaets-Buchdruckerei. 2v in 1. 1904-1907 (mf ed 1990) – 1r – 1 – (filmed with: emanuel geibel. incl school reports) – us Wisconsin U Libr [430]

Gellerts dichtungen / Gellert, Christian Fuerchtegott; ed by Schullerus, A – Krit durchges und erl Ausg. Leipzig; Wien: Bibliographisches Institut, [1891] (mf ed 1990) – 1r – 1 – (filmed with: emanuel geibel) – us Wisconsin U Libr [810]

Gellerts lustspiele : ein beitrag zur deutschen litteraturgeschichte des 18. jahrhunderts / Haynel, Woldemar Claudius – Emden; Borkum: W Haynel, 1896 – 1 – us Wisconsin U Libr [430]

Gellerts lustspiele / Capt, Louis – Zuerich: Kommerzdruck & Verlags AG 1949 [mf ed 1989] – 1r – 1 – (incl bibl ref; filmed with: emanuel geibel / arno holz [ed]) – mf#7287 – us Wisconsin U Libr [430]

Gellerts schwedische graefin : der roman der welt- und lebensanschauung des vorsubjektivischen buergertums: eine entwicklungsgeschichtliche analyse / Brueggemann, Fritz – [Aachen]: Aachener Verlags- und Druckerei-Gesellschaft 1925 [mf ed 1989] – 1r – 1 – (incl bibl ref; filmed with: emanuel geibel / arno holz [ed]) – mf#7287 – us Wisconsin U Libr [430]

Gellhorn, Eleanor Cowles *see* Mckay's guide to bermuda, the bahamas, and the car

Gelligaer – a collection of registers – 2mf – 9 – £2.50 – (grave registers for st cattwg church 1883-92, graig bargoed ind births & baptisms 1832-38; ystrad mynach siloah membership list 1916-20; indentures, bonds & removal orders 1791-93) – uk Glamorgan FHS [929]

Gelligaer, glamorgan, parish church of hengoed baptist chapel : baptisms 1801-1839, burials 1738-1871 & members registration 1861-1872 – 2mf – 9 – £2.50 – uk Glamorgan FHS [929]

Gelligaer, glamorgan, parish church of st cattwg : baptisms 1696-1909, burials 1696-1908, marriages 1709-1837 – 3mf – 9 – £3.75 – uk Glamorgan FHS [929]

Gelligaer, st cattwg, monumental inscriptions – 5mf – 9 – £6.25 – uk Glamorgan FHS [929]

Gellivarebladet – Lulea, Sweden. 1899-1901 – 1 reel – 1 – sw Kungliga [078]

Gelmintozy cheloveka, zhivotnykh, rastenii i mery borby s nimi : tezisy dokladov konferentsii vsesoiuznogo obshchestva gelmintologov an sssr, moskva, 27-29 ianvaria 1981 g / ed by Filippov, V V – Moskva: Obshchestvo, 1980 – us CRL [077]

Gelnhaeuser anzeiger – Gelnhausen DE, 1933 2 nov-1934 16 jan, 1934 19 feb-1935 18 oct – 5r – 1 – gw Misc Inst [074]

Gelnhaeuser nachrichten *see* Tages-zeitung fuer den kreis gelnhausen

Gelnhaeuser tageblatt *see* Kreis-blatt

Gelnhaeuser zeitung *see* Kinzig-bote

Gelombang *see* Kementerian penerangan republik indonesia

Het geloofsbegrip van calvijn / Dee, Simon Pieter – Kampen: JH Kok, [1918?] – 1mf – 9 – 0-524-06403-2 – (incl bibliographic references) – mf#1991-2525 – us ATLA [242]

"Geloofskrisis as gesigsbedrog" : n postmodernistiese siening van spiritualiteit en pastoraat / Greyling, J P – Stellenbosch: U of Stellenbosch 1998 [mf ed Gauteng: MGX 1998] – 2mf – 9 – (mf 1/2 is missing; incl bibl; text in afrikaans) – mf#mf.1358 – sa Stellenbosch [207]

Gelora KIAA *see* Ec kiaa seksi penprop

Gelora nusantara – Jogjakarta, 1964-1965 – 6mf – 9 – mf#SE-884 – ne IDC [950]

Gelora teknologi / Dewan Mahasiswa ITB – Bandung, 1964 – 2mf – 9 – mf#SE-714 – ne IDC [959]

Gelpi, Roberto Zoilo *see* Phytochemical study of the florida oil of sweet orange

Gelpi Y Ferro, Gil *see* Historia de la revolucion y guerra de cuba

Gelre : bijdragen en mededeelingen – 1(1898)-49(1949) – 328mf – 9 – €625.00 – (index: 1-40 (1897-1937) 5mf €12) – ne Slangenburg [073]

Gelsenkirchener allgemeine zeitung – Gelsenkirchen 1914 21 jul-31 dez, 1915 1 apr-1916 31 mar, 1916 2 okt-1917 31 mar, 1931 1 apr-30 jun, 1932 1 jul-30 sep, 1933 1 jan-30 jun, 1943 1 okt-31 dec [mf ed 2004] – 8r – 1 – (filmed by misc inst: 1921 feb-1928 mar) – gw Mikrofilm; gw Misc Inst [074]

Gelsenkirchener nachrichten – Gelsenkirchen 1952 1 apr-30 jun, 1953 1 apr-30 jun [mf ed 2004] – 1r – 1 – (vlg in gelsenkirchen-buer) – gw Mikrofilm [074]

Gelsenkirchener volkszeitung – Gelsenkirchen DE, 1903 1 may-1905 31 mar – 2r – 1 – (nov? 1904: volkszeitung) – gw Misc Inst [074]

Gelsenkirchener zeitung – Gelsenkirchen DE, 1902 2 jan-1940 30 jun – 51r (mit ergaenzungen auf mpf) – 1 – gw Misc Inst [074]

Die geltenden papstwahlgesetze / ed by Giese, Friedrich – Bonn: A Marcus und E Weber, 1912 – (= ser Kleine texte fuer vorlesungen und uebungen) – 1mf – 9 – 0-524-04677-8 – (incl bibl ref) – mf#1990-1304 – us ATLA [240]

Gelzer, H *see* Texte der notitiae episcopatuum

Gelzer, Heinrich *see*
- Ausgewaehlte kleine schriften
- Byzantinische kulturgeschichte
- Geistliches und weltliches aus dem tuerkisch-griechischen orient
- Die neuere deutsche national-literatur nach ihren ethischen und religioesen gesichtspunkten

Gelzer, Heinrich et al *see* Patrum nicaenorum nomina latine, graece, coptice, syriace, arabice, armeniace

Gem – 1829-32 – (= ser English gift books and literary annuals, 1823-1857) – 12mf – 9 – uk Chadwyck [800]

Gem / International Jewelry Workers' Union (founded 1916) – 1952 nov-1965, 1966 mar-1974 may – 2r – 1 – mf#1056875 – us WHS [331]

The gem : a selection of the most popular and choice hymns and tunes for sabbath schools – Toronto: W C Chewett, 1868 – 2mf – 9 – mf#12580 – cn Canadiana [780]

The gem : a weekly journal devoted to pleasant and instructive home reading – St John, NB: R and E Armstrong, [1879-1886?] – 9 – mf#P04527 – cn Canadiana [640]

Gema alma mater / Institut Keguruan dan Ilmu Pendidikan – Medan, 1968. v1-2(4) – 2mf – 9 – (missing: 1968 v1(1)) – mf#SE-1486 – ne IDC [950]

Gema bukit barisan / Semdam-II/BB – Medan, 1970(1) – 2mf – 9 – mf#SE-1487 – ne IDC [950]

Gema harapan – Manado: Yayasan Berkat Karunia (biwkly) [mf ed Chicago IL: filmed by Preservation Resources, Bethlehem, PA for SEAsian MF Project, CRL 2002] – (= ser Indonesian political tabloids microfilm coll) – 1r with other items [v2 n33 (week 3 feb 2000) – 1 – (began in 1999?) – mf#mf-13314 seam – us CRL [320]

Gema pembangunan irian barat / Sekretariat Koordinator Urusan Irian Barat – Djakarta, 1965 – 1mf – 9 – (missing: 1965 v1(1-4)) – mf#SE-1488 – ne IDC [959]

Gema pemuda al-irsjad / Pemuda Al-Irsjad – Djakarta, 1954-1956 – 14mf – 9 – (missing: 1954, v1(1-2); 1955, v2(1, 4-5, 7-8); 1956, v2(10)) – mf#SE-718 – ne IDC [959]

Gema press – Jogjakarta, 1962-1967(5) – 11mf – 9 – (missing: 1962-1964, v1-3(1); 1965, v4(12); 1966, v5(3-12)) – mf#SE-715 – ne IDC [959]

Gema reformasi : gr: tabloid berita mingguan – Jakarta: Media Citra Mahkotapos 1998 [wkly] [mf ed Chicago IL: filmed by Preservation Resources, Bethlehem, PA for SEAsian MF Project, CRL 2002] – (= ser Indonesian political tabloids microfilm coll) – 1r with other items – 1 – (Began in 1998) – mf#mf-13314 seam – us CRL [079]

Gema SSBRI *see* Madjalah resmi serikat sekerdja bank rakjat indonesia

Gemaehlde aus dem naturreich beyder sicilien / Spallanzani, Lazzaro – Wien 1824 [mf ed Hildesheim 1995-98] – 2mf – 9 – €60.00 – 3-487-29190-8 – gw Olms [914]

Gemaehlde von neapel und seinen umgebungen / Rehfues, Philipp J von – Zuerich 1808 [mf ed Hildesheim 1995-98] – 3v on 5mf – 9 – €100.00 – 3-487-29217-3 – gw Olms [914]

Gemaelde der kueste von guinea und der einwohner derselben : wie auch der daenischen colonien auf dieser kueste, entworfen waehrend meines aufenthaltes in afrika in den jahren 1805 bis 1809 / Monrad, Hans – Weimar 1824 [mf ed Hildesheim 1995-98] – 1v on 3mf – 9 – €90.00 – ISBN-10: 3-487-26486-2 – ISBN-13: 978-3-487-26486-8 – gw Olms [916]

Gemaelde der ostsee in physischer, geographischer, historischer und merkantilischer ruecksicht / Catteau-Calleville, Jean – Weimar 1815 [mf ed Hildesheim 1995-98] – 1v on 4mf – 9 – €120.00 – gw Olms [914]

Gemaelde des griechischen archipelagus / Murhard, Friedrich – Berlin 1807-08 [mf ed Hildesheim 1995-98] – 2v on 6mf – 9 – €120.00 – 3-487-29074-X – gw Olms [949]

Gemaelde von konstantinopel / Murhard, Friedrich – Penig 1804 [mf ed Hildesheim 1995-98] – 3v on 9mf – 9 – €180.00 – 3-487-29120-7 – gw Olms [915]

Gemaelde von madrid / Fischer, Christian – Berlin 1802 [mf ed Hildesheim 1995-98] – 3mf – 9 – €90.00 – 3-487-29875-9 – gw Olms [914]

Gemaelde von sardinien : in historischer, politischer, geographischer und naturhistorischer hinsicht / Azuni, Domenico A – Leipzig 1803 [mf ed Hildesheim 1995-98] – (= ser Fbc) – 2v on 5mf – 9 – €100.00 – 3-487-29182-7 – gw Olms [914]

Gemaelde von ungern / Csaplovics, Janos – Pesth 1829 [mf ed Hildesheim 1995-98] – 2v on 5mf – 9 – €100.00 – 3-487-29160-6 – gw Olms [914]

Gemaelde von valencia / Fischer, Christian – Leipzig 1803 [mf ed Hildesheim 1995-98] – 2v on 4mf – 9 – €120.00 – 3-487-29804-X – gw Olms [914]

Gemaelde-Galerie, Dresden *see* The dresden gallery

Gemah ripah / Bank Koperasi Tani dan Nelajan – Djakarta, 1963-1969 – 36mf – 9 – (1963, v1(nov-dec); 1964, v2(jan-apr); 1966, v4(12); 1967, v5(3-8, 9); 1969, v7(2-3)) – mf#SE-683 – ne IDC [959]

Gemblacensis, Guibertus *see* Epistolae

Die gemeinde : organ des bundes evangelisch-freikirchlicher gemeinden – 1965-92 [complete] – Inquire – 1 – ISSN: 0016-6073 – mf#ATLA S0379 – us ATLA [242]

Gemeinde – Vienna, Austria. 1958-jul 1974 – 3r – 1 – uk British Libr Newspaper [072]

Die gemeinde in der apostolischen zeit und im missionsgebiet; das wunder in der synagoge / Schlatter, Adolf von – Guetersloh: C Bertelsmann, 1912 – (= ser Beitraege zur foerderung christlicher theologie) – 1mf – 9 – 0-524-00599-0 – mf#1990-0099 – us ATLA [240]

Die gemeinde. (the church) – German Baptist. 1946-60 – 1 – us Southern Baptist [242]

Die gemeinde unterm kreuz : oder, botschafter des heils in christo – v1-4. 1885-88 [complete] – (= ser Mennonite serials coll) – 1r – 1 – mf#ATLA 1994-S019 – us ATLA [242]

Gemeindeblatt / Der Juedischen Gemeinde zu Berlin – Berlin. Jarg. 1-28. no. 45. Jan. 13, 1911-Nov. 6, 1938 – 1 – us NY Public [074]

Gemeindeblatt der deutsch-israelitischen gemeinde zu hamburg – Hamburg DE, 1925 10 may-1938 12 aug – 1r – 1 – gw Misc Inst [270]

Gemeindeblatt der deutsch-israelitischen gemeinde zu hamburg : juedisches gemeindeblatt fuer das gebiet der hansestadt hamburg – Hamburg. v1-14. 1925-38 – (= ser German-jewish periodicals...1768-1945, pt 1) – 1r – 1 – $125.00 – (lacking: misc iss) – mf#B444 – us UPA [939]

Gemeindeblatt der israelitischen religionsgemeinde dresden : amtliches organ des dresden / ed by Ploemacher, L & Anschel, Leo – v1-14. 1925-38 – (= ser German-jewish periodicals...1768-1945, pt 1) – 2r – 1 – $220.00 – (lacking: v9 1933 and misc iss) – mf#B438 – us UPA [270]

Gemeindeblatt der israelitischen religionsgemeinde dresden – Dresden DE, may 25 1925-nov 1932, 1934-oct 22 1938 – 2r – 1 – gw Misc Inst [270]

Gemeindeblatt der israelitischen religionsgemeinde zu leipzig / ed by Cohn, Gustav – Leipzig. v1-14. 1925-38 – (= ser German-jewish periodicals...1768-1945, pt 1) – 2r – 1 – $220.00 – (lacking: v13 1937 and v14 1938, all except n34) – mf#B453 – us UPA [270]

Gemeindeblatt der juedischen gemeinde zu berlin – Berlin. v. 1-28 no. 45. Jan 13 1911-Nov 6 1938 – 1 – us NY Public [939]

Gemeindeblatt fuer die juedischen gemeinden preussens *see* Verwaltungsblatt des preussischen landesverbandes juedischer gemeinden

Gemeindeblatt fur die israelitische gemeinde frankfurt a main – Frankfurt a.Main. v1-16. 1922/23-1937/38* – (= ser German-jewish periodicals...1768-1945, pt 1) – 2r – 1 – $220.00 – (cont as: (1) frankfurter israelitisches gemeindeblatt: organ der israelitischen gemeinde. (2) juedisches gemeindeblatt fuer die israelitische gemeinde zu frankfurt a main) – mf#B442 – us UPA [270]

Der gemeindebote – Berlin, Germany 1890-93 [mf ed Norman Ross] – 7r – 1 – mf#nrp-296 – us UMI ProQuest [939]

Gemeinde-bote – London (GB), 1938-39 [gaps] – 1 – gw Misc Inst [072]

Die gemeindegesaenge der heiligen messe / Winterswyl, L A & Messerschmid, F – Wuerzburg, 1940 – 1mf – 8 – €3.00 – ne Slangenburg [240]

Das gemeindekind : erzaehlung / Ebner-Eschenbach, Marie von – Berlin: Gebrueder Paetel 1887 [mf ed 1993] – 2v on 1r – 1 – (filmed with: glaubenslos?) – mf#8573 – us Wisconsin U Libr [830]

Gemeindezeitung fuer den regierungsbezirk kassel – Kassel DE, 1868 11 jan-31 dec – 1r – 1 – gw Misc Inst [074]

Gemeindezeitung fuer den synagogenbezirk duesseldorf – Duesseldorf DE, 1930-31 [gaps], 1932-nov 5 1938 – 2r – 1 – (title varies: 14 aug 1937: juedisches gemeindeblatt fuer den synagogenbezirk duesseldorf) – gw Misc Inst [270]

Gemeinde-zeitung fuer die israelitischen gemeinden wuerttembergs – Stuttgart DE, 1924 15 apr-1938 1 nov – 2r – 1 – (filmed by other misc inst: 1937 2 may-1938 1 nov [gaps]. after 16 apr 1937: juedisches gemeindeblatt fuer die israelitischen gemeinden wuerttembergs) – gw Misc Inst [939]

Gemeindezeitung fuer die israelitischen gemeinden wuerttembergs / ed by Rieger, Dr & Sternheim, Hans – Stuttgart. v1-15. 1924-38 – (= ser German-jewish periodicals...1768-1945, pt 1) – 2r – 1 – $220.00 – (lacking: v2,13) – mf#B475 – us UPA [939]

Der gemeinnuetzige – Hagen, Westf DE, 1951-57, 1963-92 [gaps] – 175r – 1 – (gaps in 1990 & 1991 replaced through westfaelische rundschau; title varies: 24 nov 1949: neue hohenlimburger zeitung, 1 oct 1975: westfalenpost/wp /hohenlimburg [ut: neue hohenlimburger zeitung]; filmed by misc inst: 1958-62, 1993-) – gw Mikrofilm; gw Misc Inst [074]

Gemeinnuetzige betrachtung[en] der neuesten schriften : welche religion, sitten und besserung des menschlichen geschlechts betreffen / ed by Seiler, Georg Friedrich – Erlangen 1776-1800 – (= ser Dz. abt theologie) – 25v on 200mf – 9 – €1200.00 – mf#k/n2130 – gw Olms [200]

Gemeinnuetzige briefe – Goettingen DE, 1739 – 1r – 1 – gw Misc Inst [074]

Gemeinnuetzige nachrichten fuer die provinz ostfriesland – Aurich DE, 1805-08 – 2r – 1 – gw Misc Inst [943]

Gemeinnuetzige stadt- und landzeitung – Kahla, Thuer DE, 1799-1800 [gaps], 1801 – 1r – 1 – (title varies: 1801: gemeinnuetzige zeitung fuers volk) – gw Misc Inst [074]

Gemeinnuetzige unterhaltungen aus der arzneykunde, naturgeschichte und oekonomie / ed by Reyher, Joh Georg – Kiel 1790-91 – (= ser Dz. abteilung naturwissenschaft) – 2jge on 6mf – 9 – €120.00 – mf#k/n3342 – gw Olms [615]

Gemeinnuetzige unterhaltungs-blaetter – Hamburg, Berlin DE, 1811-13 – 1r – 1 – (filmed by other misc inst: 1831-48, 1850-51, 1853, 1861, 1864, 1867-73 [11r]. title varies: 10 apr 1811: privilegirte gemeinnuetzige unterhaltungsblaetter; 18 jan 1812: hamburgisches unterhaltungsblatt; 25 jan 1812: hamburgische unterhaltungs-blaetter; 6 jan 1813: hamburgisches unterhaltungsblatt; 1826: der freischuetz; ab 1861? in berlin) – gw Misc Inst [074]

Gemeinnuetzige zeitung fuers volk *see* Gemeinnuetzige stadt- und landzeitung

Gemeinnuetziger anzeiger – Rottweil DE, 1836 jan-mar – 1r – 1 – gw Misc Inst [074]

Gemeinnuetziger anzeiger zum breslauer intelligenz-blatt *see* Breslausche auf das interesse der commerzien der schl. lande eingerichtete frag- und anzeigungs-nachrichten

Gemeinnuetziges anhaltisches wochenblatt – Koethen DE, 1784-86 – 1r – 1 – gw Misc Inst [074]

Gemeinnuetziges colberger wochenblatt – Kolberg (Kolobrzeg PL), 1832, 1939 may-dec, 1940 may-aug, 1941 may-aug, 1942 [gaps], 1943 apr-dec – 9r – 1 – (title varies: 1 jul 1933: kolberger zeitung) – gw Misc Inst [077]

Gemeinnuetziges, unterhaltendes neustaedter wochenblatt *see* Vaterland

Gemeinnuetziges wochenblatt – Bernkastel DE, 1849 – 1r – 1 – (title varies: 3 jan 1847: bernkast'ler wochenblatt; 3 jan 1849: bernkast'ler tageblatt; 3 jul 1850: bernkast'ler zeitung; 1 jan 1885: bernkasteler zeitung) – gw Misc Inst [074]

Gemeinnuetziges wochenblatt – Schwaebisch Gmuend DE, 1825 13 jul-1944 3 jul – 59r – 1 – (title varies: 4 aug 1840: intelligenzblatt; 6 dec 1842: der bote vom remsthale; 2 jan 1864: der remsthalbote; 1873: rems-zeitung; 1 jul 1936: schwaebische rundschau. filmed by misc inst: 1977- [ca 7r/yr]) – gw Misc Inst [074]

Gemeinnuetziges wochenblatt – Paderborn, Hamm (Westf), Dortmund DE, 1848-1850 3 sep, 1851-59, 1861-1869 28 sep [gaps], 1870 1 jan-16 apr, 1871-1872 20 jun, 1872 2 jul-1877, 1878 1 jul-1881 3 dec – 1 – (title varies: 6 apr 1848: westfaelische zeitung; fr 30 mar 1850 publ in hamm, fr 10 oct 1850 in paderborn, fr 9 sep 1855 in dortmund) – gw Misc Inst [074]

Gemeinnuetziges wochenblatt – Tilsit (Sowjetsk RUS), 1826 [gaps] – 1r – 1 – (with gaps. filmed by other misc inst: 1818-21, 1823-1825 4 nov [gaps], 1827 [gaps] & 1830, 1832-34, 1836-40, 1845-47, 1860-61 [gaps], 1864 [gaps] & 1884 [gaps], 1889 1 jan-28 apr [gaps] (21r); filmed with suppl: intelligenznachrichten) – gw Misc Inst [077]

Gemeinnuetziges wochenblatt fuer den buerger und landmann – Luebeck DE, 1794 4 oct-1795 26 sep – 1r – 1 – gw Misc Inst [074]

Gemeinnuetziges wochenblatt fuer friedberg und die gegend *see* Allgemeines friedberger wochenblatt fuer stadt- und landleute

Gemeinnuetziges wochenblatt fuer geilenkirchen, heinsberg und die umgebung – Geilenkirchen, Heinsberg DE, 1845 – 1 – gw Misc Inst [074]

Gemeinnuetziges wochenblatt fuer rendsburg und umliegende gegend – Rendsburg DE, 1808, 1810, 1817-1944, 1950-92 – 1 – (filmed by other misc inst: 1976- [ca 7r/yr]. title varies: 2 aug 1848: rendsburger wochenblatt, later: rendsburger tageblatt, schleswig-holsteinische landeszeitung, schleswig-holsteinische tagespost) – gw Misc Inst [074]

Gemeinnutzige betrachtung der neuesten schriften – 1776-78 [complete] – 2r – 1 – mf#ATLA S0888 – us ATLA [073]

Gemeinnutzige betrachtungen der neuesten schriften – 1779-1800 [incomplete] – 8r – 1 – mf#ATLA S0888A – us ATLA [073]

Die gemeinschaft : hefte fuer die religioese erstaerkung des judentums – Berlin: Liberaler Synagogenverein Norden in Berlin. v1-22. 1925-33 – (= ser German-jewish periodicals...1768-1945, pt 2) – 1r – 1 – $115.00 – (lacking: n10,16-18) – mf#B85 – us UPA [270]

Die gemeinschaft : hefte fuer die religioese erstarkung des judentums – Berlin DE, 1925-33 [gaps] – 1r – 1 – gw Misc Inst [270]

Die gemeinschaft – Berlin DE, 1917-23 – 1 – (title varies: 1922 n14: der beamtenbund) – gw Misc Inst [350]

Die gemeinschaft der eigenen – Berlin DE, 1919-1924/25 n7 – 1r – 1 – gw Misc Inst [074]

Gemeinschaft der heiligen und heiligungs-gemeinschaften / Arnold, Carl Franklin – Gr Lichterfelde-Berlin: E Runge 1909 [mf ed 1989] – (= ser Biblische zeit- und streitfragen 5/1) – 1mf – 9 – 0-7905-3001-5 – (incl bibl ref) – mf#1987-3001 – us ATLA [240]

Gemeinschaft und einzelmensch / Welty, E – Salzburg/Leipzig, 1935 – 6mf – 8 – €14.00 – ne Slangenburg [140]

Die gemeinschaften und sekten wuerttembergs / Palmer, Christian; ed by Jetter – Tuebingen: H Laupp, 1877 – 1mf – 9 – 0-524-05159-3 – (incl bibl ref) – mf#1990-1415 – us ATLA [240]

Gemeinschaftliche grammatik der arischen und der semitischen sprachen : voran eine darlegung der entstehung des alfabets / Raabe, Andreas – Leipzig: Julius Klinkhardt, 1874 [mf ed 1986] – 1mf – 9 – 0-8370-8217-X – mf#1986-2217 – us ATLA [470]

Gemeinschaftsblatt *see* Der komet

Gemella philosophia intellectus et voluntatis seu scientiarum et morum quam... / Scharz, J J – Lincij: Typis Joannis Jacobi Mayr, [1676] – 1mf – 9 – mf#O-1901 – ne IDC [090]

Gemelli Careri, G F *see*
- Voyage du docteur jean-francesco gemelli careri...la chine
- A voyage round the world...

Das gemerkbuechlein des hans sachs, 1555-1561 : nebst einem anhange, die nuernberger meistersinger-protokolle von 1595-1605 / ed by Drescher, Karl – Halle: M Niemeyer 1898 [mf ed 1993] – (= ser Neudrucke deutscher literaturwerke des 16. und 17. jahrhunderts 149-152) – 11r – 1 – (incl bibl ref & ind) – mf#3387p – us Wisconsin U Libr [880]

Geminiani, Francesco *see*
- L'art de jouer le violon
- Concerti grossi con due violini, viola e violoncello di concertino obligati, e due altri violini e basso di concerto grosso...
- Concerti grossi con due violini, viola e violoncello di concertino obligati, e due altri violini e basso di concerto grosso...no 1-6
- Concerti grossi con due violini, viola e violoncello di concertino, obligati, e due altri violini e basso di concerto grosso...opera terza
- Concerti grossi con due violini, violoncello e viola di concertino obligati
- Concerti grossi con due violini, violoncello e viola di concertino obligati, e due altri violini, e basso di concerto grosso ad arbitrio...opera seconda
- Dictionaire harmonique
- Inchanted forrest
- New and compleat instructions for the violin
- Rules for playing in a true taste on the violin, german flute, violoncello, and harpsichord, particularly the thorough-bass
- Treatise of good taste in the art of musick

Gemma musicalis...liber secundus – 1589 – (= ser Mssa) – 8mf – 9 – €100.00 – mfchl 138 – gw Fischer [780]

Gemma musicalis...liber tertius – 1590 – (= ser Mssa) – 6mf – 9 – €80.00 – mfchl 139 – gw Fischer [780]

Gemma musicalis...quatuor, quinque, sex et plurium vocum – 1588 – (= ser Mssa) – 8mf – 9 – €100.00 – mfchl 137 – gw Fischer [780]

Gemmill, John Alexander *see*
- The canadian parliamentary companion, 1883
- Note on the probable origin of the scottish surname of gemmill or gemmell
- Notes on parliamentary divorce in canada
- The ogilvies of montreal
- The practice of the parliament of canada upon bills of divorce

Gempar – Djakarta, 1964 – 3mf – 9 – mf#SE-719 – ne IDC [950]

Gems for christian ministers – London, England. 18– – 1r – us UF Libraries [240]

Gems of art from the great exhibition : being a series of drawings of...statuary / Concanen, E – London [1852] – (= ser 19th c art & architecture) – 1mf – 9 – mf#4.2.1561 – uk Chadwyck [740]

Gems of chinese literature / Giles, Herbert Allen – London: Bernard Quaritch; Shanghai: Kelly & Walsh, 1884 – (= ser 19th c books on china) – 3mf – 9 – mf#7.1.17 – uk Chadwyck [480]

Gems of fancy cookery : a collection of reliable and useful household recipes – Goderich, Ont: F Jordan, [1890?] [mf ed 1984] – 1mf – 9 – 0-665-01604-2 – mf#01604 – cn Canadiana [640]

Gems of french art : a series of carbon photographs from the pictures of eminent modern artists / Scott, William Bell – London 1871 – (= ser 19th c art & architecture) – 2mf – 9 – mf#4.2.1267 – uk Chadwyck [700]

Gems of hope in memory of the faithful departed / Bate, Fanny – Guelph, Ont?: s.n, c1899 – 4mf – 9 – mf#27766 – cn Canadiana [240]

Gems of modern belgian art : a series of carbon photographs / Scott, William Bell – London 1872 – (= ser 19th c art & architecture) – 2mf – 9 – mf#4.2.244 – uk Chadwyck [700]

Gems of modern german art : a series of carbon-photographs / Scott, William Bell – London 1873 – (= ser 19th c art & architecture) – 2mf – 9 – mf#4.2.245 – uk Chadwyck [700]

Gems of the bog : a tale of the Irish peasantry / Chaplin, Jane Dunbar – Boston: American Tract Society, 1869? – 5mf – 9 – mf#27435 – cn Canadiana [390]

Gems of thought from noble thinkers – sep 1894-jan 1896; sep 1894-jan 1896 – 1r – mf#ZB 33 – nz Nat Libr [079]

Gemueths-schaetze : hinterlassene gedichte / Hiller, Louise – New York: [s.n.], 1883 – 1r – 1 – us Wisconsin U Libr [810]

Die gemutsart jesu : nach jetziger wissenschaftlicher, insbesondere jetziger psychologischer methode / Baumann, Julius – Leipzig: Alfred Kroner, 1908 – 1mf – 9 – 0-8370-2211-8 – mf#1985-0211 – us ATLA [150]

Gen, B A *see* Shona structure

Gen grant's visit – s.l, s.l? 193-? – 1r – us UF Libraries [978]

Gen jose mario jernandez – s.l, s.l? 1938 – 1r – us UF Libraries [978]

Genand, J A *see* Notes de voyage

Genathliacon serenissimo neo-nato archiduci austriae leopoldo, augustissimi, romanorum imperatoris leopoldi primi... / Bischoff, Erich – Viennae: Apud Susannam Christinam, Matthaei Cosmerovij, [1701] – 1mf – 9 – mf#O-90 – ne IDC [090]

Genaue darlegung des orthodoxen glaubens (bdk44 1.reihe) / Johannes von Damascus (John of Damascus, Saint) – (= ser Bibliothek der kirchenvaeter. 1. reihe (bdk 1.reihe)) – €15.00 – ne Slangenburg [243]

Genauere bestimmung einiger musikalischer begriffe : zur ankuendigung des akademischen winter-concerts von michaelis 1780 bis ostern 1781 / Forkel, Johann Nikolaus – Goettingen: J C Dieterich 1780 [mf ed 19–] – 1r – 1 – mf#film 1030, 1709 – us Sibley [780]

Genau'r glyn hundred : parishes covered: aberystwyth st michael, eglwys fach, llanbadarn fawr, llanfihangel genau'r glyn, llangynfelyn, ysbyty cynfyn – (= ser Index to burials 1813-1837, 1838-1865 and 1866-1920) – 9 – (v1: 1813-37 [2mf] £3; v7: 1838-65 [3mf] £4; v13: 1866-1920 [3mf] £3) – uk Cardiganshire [941]

Genc kalemler – Selanik. Mueduer-i Mes'ul ve Sahib-i Imtiyaz: Nesimi Sarim, 1910-12. 2 cilt n1. 8 nisan 1327 [1911], 5. 19 haziran 1327 [1911] – (= ser O & t journals) – 1mf – 9 – $25.00 – us MEDOC [956]

Gence, comtesse de *see* Le cabinet de toilette d'une honnete femme

Gench, Barbara E *see*
- Bone density patterns in adult females with a history of anorexia nervosa
- Effect of age on reaction and movement times in girls and women
- Motor ability testing of speech handicapped preschool children

Genclik – Ankara: Resimli Ay Matbaasi. Sahib-i Imtiyaz: ve Mueduer-i Mes'ul: Cemal. Sayi 1-2 1928 [mthly] – (= ser O & t journals) – 2mf – 9 – $40.00 – us MEDOC [956]

Gendang budaja / Penguasa Darurat Militer Daerah Sumatra Utara – Medan, 1962 – 6mf – 9 – mf#SE-1489 – ne IDC [950]

Gendarme de Bevotte, Georges *see* La legende de don juan

Gendarme par telephone / Agno, Jehan D' – Paris, France. 1912? – 1r – us UF Libraries [440]

Gender and education – Abingdon, Oxfordshire 1997+ – 1,5,9 – ISSN: 0954-0253 – mf#20940 – us UMI ProQuest [370]

Gender and leadership : a comparison of division 1 athletic directors / Richhart, Christina L – 1988 – 1mf – 9 – $4.00 – mf#PE 3857 – us Kinesology [305]

Gender and sexuality at play : women professional athletes and the people who watch them / Nelson, Kelly – 2000 – 240p on 3mf – 9 – $15.00 – mf#PE 4197 – us Kinesology [305]

Gender classifications of female athletes and nonathletes at women's and coeducational colleges / Chick, Susan A & Murray, Mimi – 1992 – 2mf – $8.00 – us Kinesology [150]

Gender differences in overt coaching behaviors of high school soccer coaches / Millard, Linda & Brockmeyer, Gretchen A – 1992 – 2mf – 9 – $8.00 – us Kinesology [150]

Gender differences in peak blood lactate concentration and blood lactate removal following strenuous exercise / Zhang, Q – 1991 – 2mf – 9 – $8.00 – us Kinesology [612]

Gender differences in running economy / Davies, Michael J & Mahar, Matthew T – 1992 – 2mf – $8.00 – us Kinesology [612]

Gender differences in sport centrality / Allan, Diane E & Boydell, Cary – 1992 – 2mf – $8.00 – us Kinesology [150]

Gender differences in sport orientation and goal orientation / Kleppinger, Alison – Springfield College, 1995 – 2mf – 9 – $8.00 – mf#PSY1848 – us Kinesology [150]

Gender differences in sport participation and incentive motivation of former college basketball players and swimmers / Schwartz, Diana L – Springfield College, 1995 – 2mf – 9 – $8.00 – mf#PSY1862 – us Kinesology [150]

Gender differences in substrate metabolism and thermoregulation during rest and exercise in cold and warm water / Shea, Kyla – 1993 – 2mf – $8.00 – us Kinesology [612]

Gender differences in the relationships among self-confidence, gender-appropriateness, and value / Clifton, Robert T & Gill, Diane L – 1992 – 2mf – 9 – $8.00 – us Kinesology [150]

Gender differences in walking with respect to movement of the pelvis / Johansen, Michelle K – University of British Columbia, 1996 – 1mf – 9 – mf#PE 3656 – us Kinesology [612]

Gender differences regarding knowledge of child health and development among high school students / Roeschlein, Debra L & Gilbert, Kathleen R – 1992 – 2mf – 9 – $8.00 – us Kinesology [613]

Gender & history – Oxford, England 1989+ – 1,5,9 – mf#17391 – us UMI ProQuest [305]

Gender issues – Piscataway NJ 1998+ – 1,5,9 – (cont: feminist issues) – ISSN: 1098-092X – mf#12222,01 – us UMI ProQuest [305]

Gender related differences in performance levels of triathletes / Santanello, T – 1992 – 2mf – 9 – $8.00 – us Kinesology [150]

Gender & society – Newbury Park CA 1987+ – 1,5,9 – ISSN: 0891-2432 – mf#17053 – us UMI ProQuest [301]

Genders – Austin TX 1988-93 – 1,5,9 – ISSN: 0894-9832 – mf#16465 – us UMI ProQuest [000]

Gendlin, E I *see*
- Zapiski riadovogo revoliutsionera
- Zemskie izvestiia

Gendre d'un millionnaire / Leonce – Paris, France. 1845 – 1r – us UF Libraries [440]

Gendrikov, V B *see* Kratkii putevoditel po fondam lichnogo proiskhozhdeniia rukopisnogo otdela muzeia istorii religii i ateizma

Gendron, Pierre-Saul *see* La famille de nicolas gendron

Gendry, Jules *see* Pie 6

Gene – Oxford, England 1977+ – 1,5,9 – ISSN: 0378-1119 – mf#42076 – us UMI ProQuest [575]

Gene analysis techniques – Oxford, England 1989 – 1,5,9 – (cont by: genetic analysis, techniques & applications) – ISSN: 0735-0651 – mf#42450.03 – us UMI ProQuest [575]

Gene structure and expression – Oxford, England 1982+ – 1,5,9 – (cont: nucleic acids & protein synthesis) – ISSN: 0167-4781 – mf#42170 – us UMI ProQuest [574]

Genealogia de la casa de urries, del marques de velilla de ebro / T'Serclaes, Duque de – Madrid: Tip. Arch., Bibl. y Mus, 1930. B.R.A.H. 97, pp. 12-13 – sp Bibl Santa Ana [920]

Genealogia de los conquistadores de cuyo / Morales Guinazu, Fernando – Buenos Aires, 1932; Madrid: Razon y Fe, 1933 – 1 – sp Bibl Santa Ana [920]

Genealogia serenissimae domus austriacae a philippo primo rege hispaniarum... – Graz: Fred. Widmanstetter, 1666 – 2mf – 9 – mf#O-1992 – ne IDC [090]

Genealogias del nuevo reyno de granada. 2 vol. bogota, 1943 / Florez de Ocariz, Juan – Madrid: Razon y Fe, 1947 – 1 – sp Bibl Santa Ana [920]

Genealogical advertiser : a quarterly magazine of family – v1-4 [1898 mar-1901] – 1r – 1 – mf#4927868 – us WHS [929]

Genealogical aids bulletin : gab / Miami Valley Genealogical Society – 1971 jun-1972 feb – 1r – 1 – (continued by: genealogical aids bulletin of the miami valley genealogical society) – mf#500714 – us WHS [929]

Genealogical aids bulletin / Miami Valley Genealogical Society – 1978 fall-1985 fall – 1r – 1 – (cont: genealogical aids bulletin of the miami valley genealogical society) – mf#1725016 – us WHS [929]

Genealogical aids bulletin of the miami valley genealogical society / Miami Valley Genealogical Society – 1972 jun-1978 winter, 1978 spring-summer – 2r – 1 – (cont: genealogical aids bulletin [dayton, oh]; cont by: genealogical aids bulletin [dayton, oh: 1978]) – mf#1725000 – us WHS [929]

A genealogical and biographical record of the savery families (savory and savary) and of the severy family (severit, savery, savory, savary) : descended from early immigrants... and extracts from english, new england, and barbadoes records relating to families of both names / Savary, Alfred William – Boston: Collins, 1893 – 4mf – 9 – (incl ind) – mf#35129 – cn Canadiana [929]

Genealogical and family history of the state of new hampshire / Stearns, Ezra Scollay – 4v. 1908 – 1 – us L of C Photodup [920]

Genealogical and personal history of fayette county pennsylvania, vols 1-3 / Jordan, John W & Hadden, James – New York: Lewis Historical Publishing Co, 1912 – 1r – 1 – us Western Res [978]

Genealogical Association for Uncommon Surnames *see* Drady, drawdy, droddy, drody, drude and variants [o'grady, a variant of draddy]

Genealogical Center (Atlanta, GA) *see* Genealogy tomorrow

A genealogical chart of the male descendants of allen perley / Perley, George Augustus [comp] – Fredericton NB: G A Perley 1877 [mf ed 1987] – 1mf – 9 – 0-665-53097-8 – mf#53097 – cn Canadiana [929]

Genealogical computer pioneer – v1 n101-v4 n105/6 [1982 nov-1986 jun/jul] – 1r – 1 – mf#1026965 – us WHS [929]

Genealogical data relating to women in the western reserve before 1840 (1850) / Cleveland Centennial Commission. Women's Dept – 1r – 1 – (original coll has title: names of women who were born in...county, ohio, or who came to it prior to 1840) – us Western Res [929]

Genealogical gems : publication of the fox valley genealogical society / Fox Valley Genealogical Society – v1 n1-v4 n1 [1982 summer-1985 summer] – 1r – 1 – mf#932527 – us WHS [929]

Genealogical helper – Logan UT 1947-91 – 1,5,9 – (cont by: everton's genealogical helper) – ISSN: 0016-6359 – mf#6817.03 – us UMI ProQuest [929]

Genealogical & Historical News [Organization] *see* Jones of virginia letter

A genealogical history of the milesian families of ireland : with the monument to brian boroimhe: the chart of the armorial bearings of the same families / Courcy, B W De [comp] – Cincinnati: W F Overdiek & M L Riegel, 1880 [mf ed 1987] – 77p – 1 – mf#7599 – us Wisconsin U Libr [929]

Genealogical journal / Genealogical Society of Davidson County, North Carolina – 1981 spring-1986 fall; 1987 winter-1993 fall – 2r – 1 – mf#1140580 – us WHS [929]

Genealogical journal – Salt Lake City UT 1972+ – 1,5,9 – ISSN: 0146-2229 – mf#7644 – us UMI ProQuest [929]

Genealogical library quarterly / Augustan Society – 1976 sep-1981 – 1r – 1 – (continued by: genealogical library journal) – mf#368294 – us WHS [929]

Genealogical magazine of new jersey – v1-3 [1925-1928] – 1r – 1 – mf#1111836 – us WHS [929]

Genealogical material, ms 1826 / Spooner Family – 1879-1935 – 4r – 1 – us Western Res [920]

Genealogical material, ms 2500 / Oberholtzer Family – Excerpts – 3r – 1 – us Western Res [920]

Genealogical memoranda : snively / Snively, William Andrew – Brooklyn, NY, 1883 – 1r – 1 – (original in the library of congress) – us Western Res [920]

Genealogical news / Milholan and Associates – v1 n6-v1 n10 [1983 jun-1983 nov] – 1r – 1 – mf#697517 – us WHS [929]

Genealogical newsletter of the nova scotia historical society / Nova Scotia Historical Society – v2 n103-v2 n109 [n24-30] [summer 1978-winter 1980] – 1r – 1 – (continued by: genealogical newsletter of the royal nova scotia historical society) – mf#657173 – us WHS [929]

Genealogical newsletter of the royal nova scotia historical society / Royal Nova Scotia Historical Society – v2 n1010-v4 n102 [i e 3][n30-n40], [spring 1980-autumn 1982] – 1r – 1 – (cont: genealogical newsletter of the nova scotia historical society; cont by: nova scotia genealogist) – mf#657171 – us WHS [929]

Genealogical notes on cape cod families, 1620-1901 / Brownson, Lydia B et al – 8r – 1 – $1040.00 – mf#S1831 – us Scholarly Res [978]

Genealogical records of austin bearse (or bearce) of barnstable, cape cod, massachusetts, usa, a.d. 1638 to a.d. 1933 / Meadows, Fannie L & Ames, Jennie M – 1r – 1 – (manuscript genealogy) – us Western Res [978]

Genealogical records of the pioneers of tampa and... / Harrison, Charles Edward – Tampa, FL. 1915 – 1r – us UF Libraries [920]

Genealogical reference builders newsletter – 1967 jan-1970 nov, 1970 index, 1971 feb-1974 sep, 1974 oct-1977 nov, 1982 nov-1984 apr – 4r – 1 – mf#1048048 – us WHS [929]

Genealogical register of the descendants : of john scanton of guildford, connecticut, who died in the year 1651 / Scranton, Erastus – Hartford: Press of Case, Tiffany & Co, 1855 – 1r – 1 – us Western Res [920]

Genealogical Society of Central Missouri *see* Reporter: quarterly of genealogical society of central missouri

Genealogical Society of Davidson County, North Carolina *see* Genealogical journal

Genealogical Society of DeKalb County, Illinois *see*
- Cornsilk from dekalb county, il
- Cornsilk newsletter from dekalb co il

Genealogical Society of Greater Miami *see* Heritage

Genealogical society of greater miami : [newsletter] – 1977 sep-1980 mar – 1r – 1 – (continued by: heritage (miami, fl)) – mf#1685339 – us WHS [929]

Genealogical Society of Harlan County, Kentucky et al *see* Harlan footprints

Genealogical Society of Okaloosa County, Florida *see* Journal of the genealogical society of okaloosa county, florida

Genealogical society of okaloosa county journal – 1978 spring-1986 spring – 1r – 1 – (cont: journal of the genealogical society of okaloosa county, florida; cont by: journal of northwest florida) – mf#1033903 – us WHS [071]

Genealogical Society of South Brevard *see* Bulletin of the genealogical...

Genealogical Society of the Northern Territory *see*
- Annual single numbers series alphabetical index, 1949-1958
- Annual single numbers series alphabetical index, 1965-1969
- Annual single numbers series alphabetical index, 1969-1976
- Annual single numbers series alphabetical index, 1971-1976
- Annual single numbers series alphabetical index, 1977-1978
- Annual single numbers series alphabetical index, 1979-1980
- Annual single numbers series alphabetical index, 1981-1983
- Annual single numbers series alphabetical index, 1984-1986
- Annual single numbers series alphabetical index, 1987-1989
- Annual single numbers series alphabetical index, 1990-1991
- Annual single numbers series alphabetical index, 1992-1993

Genealogical Society of Washtenaw Co. Michigan *see* Family history capers

Genealogical society of weld county / Weld County Genealogical Society – v1 n101-v8 n104 [1975 aug-1982 may] – 1r – 1 – (continued by: weld county, colorado) – mf#978985 – us WHS [929]

Genealogical Text, Research & Equipment [Proprietorship : Sainte Genevieve, MO] *see* Genealogists' exchange

Genealogical tips / Tip-O'-Texas Genealogical Society – 1977 apr/jun-1986 jan/mar – 1r – 1 – mf#1023247 – us WHS [929]

La genealogie des familles gouin et allard : avec arbre des familles richer-lafleche, fugere, guillet, methot, chapdelaine, pinard-lauziere, bibaud / Desaulniers, Francois Lesieur – [Montreal?: A-P Pigeon], 1909 – 2mf – 9 – 0-665-73930-3 – mf#73930 – cn Canadiana [929]

La genealogie des familles richer de la fleche et hamelin : avec notes historiques sur sainte-anne-de-la-perade, les grondines, etc / Desaulniers, Francois Lesieur – [Montreal?: A P Pigeon], 1909 – 4mf – 9 – 0-665-73929-X – mf#73929 – cn Canadiana [929]

Die genealogie des koenigs jojachin und seiner nachkommen (1 chron. 3, 17-24) in geschichtlicher beleuchtung : eine kritische studie zur juedischen geschichte und litteratur / Rothstein, Johann Wilhelm – Berlin: Reuther & Reichard, 1902 – 1mf – 9 – 0-7905-2691-3 – mf#1987-2691 – us ATLA [220]

La genealogie du grant turc a present regnant – N p, 1519 – 2mf – 9 – mf#H-8135 – ne IDC [950]

Genealogie et notes historiques, etc : famille baillairge, ses ancetres, ses descendants et ses allies, au canada et a l'etranger, 1605-1895 / Baillairge, George Frederick – Joliette, PQ: Impr du bon combat, du couvent et de la famille, 1894 – 3mf – 9 – mf#06817 – cn Canadiana [929]

Genealogie vorstelijke familie van djohor door l c van ranzow, 1827 – 1mf – 8 – mf#SD-102 mf 45 – ne IDC [929]

Les genealogies de soixante et sept tres nobles et tres illustres maisons, partie de france, partie estrageres, yssues de merouee... / Lusignano, S di – Paris, 1587 – 3mf – 9 – mf#H-8369 – ne IDC [956]

The genealogies of our lord and saviour jesus christ : as contained in the gospels of st matthew and st luke / Hervey, A C – Cambridge: Macmillan; London: T Hatchard, 1853 – 1mf – 9 – 0-7905-1097-9 – mf#1987-1097 – us ATLA [220]

Genealogies of the shepherd islands, vanuatu : (by island, village and family), 1930s – 1r – 1 – (restricted access) – mf#PMB1132 – at Pacific Mss [980]

Genealogies of the tribes of british somaliland and mijertein – [Hargeisa?]: publ under the auspices of the Military Govt, Somaliland Protectorate, 1944 – 1 – us CRL [960]

Genealogisches taschenbuch der adeligen haeuser – Brunn. v. 1-19. 1870-94. (Wanting v. 2, 12) – 1 – 60.00 – us L of C Photodup [920]

Genealogist : official journal of the american-canadian genealogical society of new hampshire / American-Canadian Genealogical Society of New Hampshire et al – 1975 may-1985 fall – 1r – 1 – (continued by: american-canadian genealogist) – mf#1206362 – us WHS [929]

Genealogist – v1 n104-v2 n102 [aug 1971-oct/dec 1972] – 1r – 1 – mf#1111839 – us WHS [929]

Genealogists' exchange / Genealogical Text, Research & Equipment [Proprietorship : Sainte Genevieve, MO] – ed 20-24[1983 mar 10-jul], v3 n101-v6 n103[1983 aug-1986 nov/dec], v6 n104, n64 [1987 spring] – 1r – 1 – (cont: genealogists' information exchange) – mf#1056908 – us WHS [929]

The genealogist's guide / Marshall, G W – 1903 – 1r – 1 – mf#129 – uk Microform Academic [920]

Genealogists' magazine – London, England 1969+ – 1,5,9 – ISSN: 0016-6391 – mf#10693 – us UMI ProQuest [929]

Genealogy / Bobo and Ray Families. South Carolina – 1744-1946. 251p – 1 – us Southern Baptist [920]

Genealogy / Indiana Historical Society – n1-100 [1973 oct-1986 apr] – 1r – 1 – mf#605674 – us WHS [929]

Genealogy america – v1 n1-v2 n11 [1984 jan-1985 nov] – 1r – 1 – mf#1009265 – us WHS [929]

Genealogy and history of the hatch family : descendants of thomas and grace hatch of dorchester, yarmouth, and barnstable, massachusetts / Hatch Genealogical Society – pt1-pt8 [1925-31] – 1r – 1 – mf#2896248 – us WHS [929]

Genealogy bug – v5 n102/3-v5 n104 [1976 jul/aug-dec] – 1r – 1 – (cont: martin genealogist [1976]) – mf#923074 – us WHS [929]

Genealogy club of hemet-san jacinto : [newsletter] / Hemet-San Jacinto Genealogical Society – v3 n1-v4 n1,3-4,5,7 [1978 jan-1979 jan, mar-apr, aug, oct – 1r – 1 – (continued by: valley genealogist of hemet-san jacinto) – mf#1575494 – us WHS [929]

Genealogy Division, Indiana State Library *see* 1840 federal population census, indiana

Genealogy news of the blalock-blaylock clans – v1 n101-v8 n105 [1980 sep-1987 sep/oct] – 1r – 1 – (continued by: bla(y)lock genealogy news) – mf#1344663 – us WHS [929]

Genealogy newsclippings, 1935 – 1r – 1 – mf#B31147 – us Ohio Hist [978]

Genealogy notes on frisby family – s.l, s.l? 193-? – 1r – us UF Libraries [920]

The genealogy of the cushing family : an account of the ancestors and descendants of matthew cushing, who came to america in 1638 / Cushing, James Stevenson – [Montreal?: Perrault Print], 1905 – 8mf – 9 – 0-659-91083-7 – (1st ed 1877) – mf#9-91083 – cn Canadiana [929]

Genealogy of the family normandeau : dit deslauriers / Filteau, Louis Honore – Ottawa: A Bureau, 1894 – 2mf – 9 – mf#05458 – cn Canadiana [929]

Genealogy of the lyman family / Coleman, Lyman – 1872 – 1 – $50.00 – us Presbyterian [920]

Genealogy of the merrick-mirick-myrick family / Merrick, George B – Madison, WI: Tracy, Gibb & Co, 1902 – 1r – 1 – us Western Res [920]

Genealogy of the south-indian gods : a manual of the mythology and religion of the people of southern india, including a description of popular hinduism = Genealogie der malabarischen goetter / Ziegenbalg, Bartholomaeus; ed by Germann, Wilhelm – Madras: Higginbotham, 1869 – 1mf – 9 – 0-524-03128-2 – (in english) – mf#1990-3181 – us ATLA [280]

Genealogy today – 1983 jan 1-1984 jun – 1r – 1 – mf#857028 – us WHS [929]

Genealogy tomorrow / Genealogical Center (Atlanta, GA) – v1 n1-v2 n12 [1984 oct-1985 dec] – 1r – 1 – mf#1085471 – us WHS [929]

Genebrardus, Gilb. *see*
- Chronographiae libri quatuor
- Psalmi davidis

Die geneesheer in die gesondheidsdiens = The medical practitioner in the health service – Pretoria: Dept van Gesondheid [1979?] [mf ed Pretoria, RSA: State Library [199-]] – 1v [ill] on 1r with other items – 5 – (also available in english) – mf#op 06798 r24 – us CRL [362]

Genelli, Christoph *see* The life of st. ignatius of loyola

Gener, J B *see* Theologia dogmatico-scholastica...

Genera insectorum... / ed by Wytsman, P A G – Bruxelles 1902-35 – v1-203 on 684mf – 9 – (1924 not publ) – mf#z-1736/2 – ne IDC [590]

Genera lichenum : an arrangement of the north american lichens / Tuckerman, Edward – Amherst [MA]: E Nelson, 1872 [mf ed 1985] – 4mf – 9 – 0-665-33144-4 – mf#33144 – cn Canadiana [580]

Generacion asesinada / Marrero, Levi – Habana, Cuba. 1934 – 1r – us UF Libraries [972]

Generacion del medio siglo – Bogota, Colombia. 1955 – 1r – us UF Libraries [972]

Generacion del treinta / Melendez, Concha – San Juan, Puerto Rico. 1960 – 1r – us UF Libraries [972]

La generacion espanola que vivio la derrota / Sanchez Montes, Juan – Madrid: Arbor, 1949 – 1 – sp Bibl Santa Ana [946]

Generaciones y semblanzas / Perez De Guzman, Fernan – Madrid, Spain. 1941 – 1r – us UF Libraries [972]

A general abridgement of law and equity : alphabetically digested under the proper titles; with notes and references to the whole / Viner, Charles – 2nd ed. London: G G J, J Robinson etc, Dublin. v1-124. 1791-94 (all publ) – 176mf – 9 – $264.00 – mf#LLMC 95-245 – us LLMC [324]

A general abridgement of law and equity / Viner, Charles – v1-24. 1791-95 – 9 – $1199.00 – mf#0670 – us Brook [342]

General account of the first settlement / Hall, Richard – Bridgetown? Barbados. 1924 – 1r – us UF Libraries [972]

General Accounting Office *see* Gao auditing standards

General advertiser – Dublin, Ireland. 21, 28 jan, 4 feb-25 mar, 13 may-16 sep, 14 oct-30 dec 1837; 1838-11 jan 1840; 4, 11 apr, 2 may, 29 aug, 21 nov, 19, 26 dec 1840; 27 mar 1841-24 dec 1847; 1848-26 nov 1859; 1860-1896; 5 jan-22 mar 1924 – 26 1/4r – 1 – (aka: general advertiser for dublin and all ireland; general advertiser and local government and legal record) – uk British Libr Newspaper [072]

General advertiser : or limerick gazette – Limerick, Ireland. 17 sep 1804-10 nov 1820 – 15 1/2r – 1 – uk British Libr Newspaper [072]

General advertiser *see*
- Cape town daily news / the general advertiser
- Durban advocate / general advertiser
- The independence / general advertiser

General advertiser for dublin and all ireland *see* General advertiser

General advertiser for dublin etc – Dublin, jan-mar 1924 – 1r – 1 – ie National [072]

General advertiser for liverpool and surrounding districts – [NW England] Liverpool 15 nov 1902-18 apr 1903 – 1 – uk MLA; uk Newsplan [072]

General Agreement on Tariffs and Trade [Organization] *see* Protocol for the accession of cambodia to the general agreement on tariffs and trade

A general and bibliographical dictionary of the fine arts / Elmes, James – London 1826 – (= ser 19th c art & architecture) – 8mf – 9 – mf#4.2.1482 – uk Chadwyck [057]

General and special indexes to the general correspondence of the office of the secretary of the navy, july 1897-aug 1926 / U.S. Secretary of the Navy – (= ser General records of the department of the navy, 1798-1947) – 119r – 1 – (with printed guide) – mf#M1052 – us Nat Archives [355]

General and special orders, department of texas, 1878-93 / United States Adjutant-General's Office – Washington: National Archives and Records Service, 1974. reel 8: v530, 532-533; reel 9: v534-537; reel 10: v538-540; reel 11: v541-542 – 4r – 1 – us CRL [324]

General andre fontanges chevallier / Benoit, Julien – Paris, France. 1924 – 1r – us UF Libraries [972]
General assembly journal of the presbyterian church in the united states / Presbyterian Church in the USA – n1 [may 19 1871], n3 [may 22 1871], n6 [may 26 1871], n8 [may 29 1871]-n10 [may 31 1871], v8 n1 [may 21 1880]-n10 [jun 1 1880], v9 n1 [may 20 1881]-n10 [may 31 1881] – 1r – 1 – (cont: reunion assembly reporter) – mf#933524 – us WHS [242]
The general assembly of 1866 / Boardman, Henry Augustus – Philadelphia: JB Lippincott, 1867 – 2mf – 9 – 0-524-07852-1 – mf#1991-3397 – us ATLA [240]
General assembly record *see* Kung pao (ccs)
General Associate Synod (Scotland) *see* Narrative of the state of religion in britain and ireland
General Association of General Baptists *see* General baptist messenger
General baptist history / ed by Montgomery, David B – Evansville: Courier Co, 1882 [mf ed 1993] – 1mf – 9 – 0-524-08487-4 – mf#1993-3132 – us ATLA [242]
General baptist messenger / General Association of General Baptists – v92 n15 [1977 aug], v94 n1-v97 n7 [1979 jan-1982 jul], v97 n8-v104 n9 [1982 aug-1989 sep] – 2r – 1 – mf#618168 – us WHS [242]
General baptist messenger – Owensville, IN/ Poplar Bluff,MO. 1914-15, 1928-66. – 1 – (single reels available) – us ABHS [242]
The general baptist repository (london) – 1802-14 – (= ser 19th c british periodicals) – r52 – 1 – us Primary [242]
The general baptist repository (london) – 1815-21 – (= ser 19th c british periodicals) – r53 – 1 – us Primary [242]
General biographical dictionary : containing an historical and critical account of the lives and writings of the most eminent persons in every nation / Chalmers, Alexander – London. v1-32. 1812-17 – 9 – $803.00 – mf#0145 – us Brook [929]
General calleja / Barrios Y Carrion, Leopoldo – Madrid, Spain. 1896 – 1r – us UF Libraries [972]
General catalogue of all publications of the government of bombay (including sind) – Bombay, Govt Print and Stationery. n4 1927; n9 1933; n12 1938; 1958 – us CRL [324]
General catalogue of all publications of the government of india and local governments and administrations – Calcutta. n23-37. 1915-1922 – us CRL [324]
General catalogue of double stars within 121° of the north pole / Burnham, Sherburne Wesley – [Washington]: Carnegie Institution of Washington 1906 [mf ed 2005] – (= ser Publication / carnegie institution of washington 5) – 2v on 2r – 1 – mf#film mas c6032 – us Harvard [520]
General catalogue of the mccormick theological seminary of the presbyterian church, 1830-1900 / Mccormick Theological Seminary – Chicago: Rogerson Press, 1900 [mf ed 2004] – 1r – 1 – 0-524-10467-0 – mf#b00684 – us ATLA [242]
The general catechism : revised, corrected, and enlarged, and prescribed to be taught, throughout the dioceses of kingston and toronto – [Kingston, Ont?: s.n.] 1844 [mf ed 1983] – 1mf – 9 – 0-665-44536-9 – mf#44536 – cn Canadiana [241]
Le general cavaignac devant l'assemblee nationale – Paris [1848?] – us CRL [324]
General circular / British Chamber of Commerce in Indonesia – Djakarta, 1959-1964 – 187mf – 9 – (missing: 1959(2387, 2392, 2399, 2415-2417, 2510, 2538, 2586, 2592)) – mf#SE-673 – ne IDC [959]
A general collection of the best and most interesting voyages and travels in all parts of the world : many of which are now first translated into english; digested on a new plan / ed by Pinkerton, John – London 1808-14 [mf ed Hildesheim 1995-98] – (= ser Fbc) – 17v on 160mf – 9 – €960.00 – 3-487-29805-8 – gw Olms [910]
General conference mennonite church : handbook of information – 1951-56 [complete] – (= ser Mennonite serials coll) – 1r – 1 – mf#ATLA 1993-S008 – us ATLA [242]
General Conference of Protestant Missionaries in Japan (1900: Tokyo) *see* Proceedings of the general conference of protestant missionaeries in japan
General conference of the african m e church : [reports] / African Methodist Episcopal Church. General Conference – Philadelphia: W S Young, printer. 11th-13th (1856-64) [mf ed 2005] – 1 – (cont: african methodist episcopal church. general conference. proceedings of the...general conference of the african m e church; cont by: african methodist episcopal church. general conference. quadrennial session of the general conference of the african methodist episcopal church; some pgs damaged) – mf#2005-s087 – us ATLA [242]
General conference of the african m e church : [reports] / African Methodist Episcopal Church. General Conference – Philadelphia: W S Young, printer. 11th-13th (1856-64) [mf ed 2005] – 1 – (cont: african methodist episcopal church. general conference. proceedings of the...general conference of the african m e church; cont by: african methodist episcopal church. general conference. quadrennial session of the general conference of the african methodist episcopal church; some pgs damaged) – mf#2005-s086 – us ATLA [242]
General Conference of the Congregational Churches of Connecticut *see* Centennial papers
General Conference of the Protestant Missionaries of China (1877: Shanghai, China) *see* Resolutions and appeal unanimously adopted by the conference of protestant missionaries at shanghai, may 16th 1877
The general conferences of the methodist episcopal church from 1792 to 1896 – Cincinnati: Curts & Jennings, 1900 – 1mf – 9 – 0-524-05362-6 – mf#1990-5113 – us ATLA [242]
General Construction Workers Industrial Union No 310 *see* Minutes of the convention of the construction workers industrial union no 310, iww
General Convention for Religious Liberty *see* Minutes, 1766-75; presbyterian church in the u.s.a., records, 1706-88
General convention of congregational ministers and churches of vermont : minutes / Congregational Ministers and Churches of Vermont – 1855-98 [complete] – 3r – 1 – mf#ATLA S0608 – us ATLA [242]
The general convention of the baptists of north america : history of events leading to the organization of the convention and minutes of the first convention, saint louis, mo., may 16, 1905 – [S.l.]: printed by the order of the Executive Committee, [1905?] – 1mf – 9 – 0-524-08864-0 – mf#1993-3328 – us ATLA [242]
The general corporation act of new jersey. / New Jersey. Laws, Statutes, etc – 4th ed., with the amendments of 1902. Trenton? 1902. 240 (i.e. 250)p. LL-962 – 1 – us L of C Photodup [348]
The general corporation law of pennsylvania, approved 29 april, 1874. / Freedley, Angelo Tillinghast – Philadelphia: Johnson, 1882. 141p. LL-75 – 1 – us L of C Photodup [348]
The general corporation law, the stock corporation law, the transportation corporations law, and the business corporation law, of the state of new york. / Haviland, Charles Tappan – New York: Diossy, 1891. 155p. LL-694 – 1 – (new york: diossy, 1892. 187p. ll-693) – us L of C Photodup [346]
General correspondence of the alaskan territorial governor, 1909-1958 / U.S. State Dept. – 378r – 1 – (with printed guide) – mf#M939 – us Nat Archives [324]
General correspondence of the office of the secretary of commerce, 1929-1933 / U.S. Dept of Commerce – (= ser General records of the department of commerce) – 16r – 1 – (with printed guide) – mf#M838 – us Nat Archives [380]
General Council of the Congregational and Christian Churches of the United States *see* Social action
General Council of the Evangelical Lutheran Church in North America. Board of Foreign Missions *see*
– India files 1880-1911
– Minutes, reports, and publications
– Reports, publications, and minutes
General Council of the Evangelical Lutheran Church in North America. Women's Missionary Society *see*
– Women's missionary society constitutions and bylaws n d
– Women's missionary society history 1902-1919, 1902-1911
– Women's missionary society pamphlets 1894-1918, 1912-1916
– Women's missionary society records 1913-1918
General court martial of general george armstrong custer, 1867 / U.S. Army. Judge Advocate General – (= ser Records of the office of the judge advocate general (army)) – 1r – 1 – mf#T1103 – us Nat Archives [355]
General d jose n rodriguez ante sus compatriotas / Rodriguez, Jose N – Guatemala, 1898 – 1r – us UF Libraries [972]
General decorative and applied art : incl sculpture, arms & armour, clocks, antiquities, tribal art, tapestries & icons – (= ser Christie's pictorial archive: decorative and applied art) – 171mf – 9 – $1285.00 – 0-907006-42-6 – (over 15,000 reproductions) – uk Mindata [700]
General dentistry – Chicago IL 1976+ – 1,5,9 – (cont: journal – academy of general dentistry) – ISSN: 0363-6771 – mf#8038,01 – us UMI ProQuest [617]
General description (monroe county, florida) / Saunders, H J – s.l, s.l? 1936 – 1r – us UF Libraries [978]
A general description of china... / Grosier, J B G A – London, 1788. 2v – 13mf – 9 – mf#HT-522 – ne IDC [590]
General description of orange county, florida / Mason, Z H – Apopka, FL. 1881 – 1r – us UF Libraries [630]
A general description of scotland : containing an account of its situation, extent, rivers...; to which is prefixed a copious travelling guide...; forming an itinerary of scotland / Cooke, George A – London [circa 1820] [mf ed Hildesheim 1995-98] – 1v on 1mf – 9 – €40.00 – 3-487-28836-2 – gw Olms [914]
General description of shangae and its environs : extracted from native authorities / Medhurst, W H – Shangae: Mission Press, 1850 – 3mf – 9 – mf#HT-659 – ne IDC [915]
A general dictionary, historical and critical... (ael1/45.5) – London 1734-41 [mf ed 1999] – 10v on 74mf – 9 – €720.00 – 3-89131-334-9 – gw Fischer [059]
General documentation, 1969-73 : study project on christianity in apartheid society – Johannesburg, [19–?] – us CRL [240]
General Drivers, Helpers, Petroleum and Inside Workers Union *see* Organizer
The general east india guide and vade mecum : for the public functionary, government officer, private agent, trader of foreign sojourner, in british india, and the adjacent parts of asia immediately connected with the honourable the east india company... / Gilchrist, John – London 1825 [mf ed Hildesheim 1995-98] – 1v on 4mf – 9 – €120.00 – 3-487-27530-9 – gw Olms [380]
The general ecclesiastical constitution of the american church : its history and rationale / Perry, William Stevens – New York: T. Whittaker, 1891 – (= ser The bohlen lectures) – 1mf – 9 – 0-7905-5730-4 – mf#1988-1730 – us ATLA [240]
General education and technical education and developmental research / ed by Shah, K T – Bombay: Vora & Co, 1948 – (= ser Samp: indian books) – us CRL [370]
The general education board : series 1: appropriations; subseries 1: the early southern program – 159r – 1 – $130.00r $20,7670.00 coll – (states in coll also listed separately. printed guide for entire collection. also sold separately $25 s3298.g) – mf#S3298 – Rockefeller Archive Centre, North Tarrytown – us Scholarly Res [370]
General education board archive : series 1: appropriations subseries 3: new southern program and related programs, 1931-1961 – 201r – 1 – $26,130.00 – mf#S3350 – us Scholarly Res [370]
The general education board archives : series 1: appropriations; subseries 3: new southern program and related programs, 1931-1961 – 201r – 1 – $26,130.00 – (guide available dec 1996 and only sold separately unless entire collection is purchased s3350.g $35) – mf#S3350 – Rockefeller Archive Center, North Tarrytown, NY, A Division of the Rockefeller univ – us Scholarly Res [370]
The general effect of mandatory minimum prison terms / Meierhoefer, Barbara S – Washington: FJC, 1992 – 1mf – 9 – $1.50 – mf#LLMC 95-830 – us LLMC [345]
General electric forum – 1958-63 – 1 – us L of C Photodup [621]
General electric forum – Schenectady NY 1958-69 – 1 – ISSN: 0435-2572 – mf#1527 – us UMI ProQuest [621]
General electric review – Killen TX 1952-58 – 1 – ISSN: 0095-9480 – mf#839 – us UMI ProQuest [621]
General eliseo payan, vicepresidente de la republi... / Gonzalez Toledo, Aureliano – Bogota, Colombia. 1887 – 1r – us UF Libraries [972]
The general epistle of james : with notes and introduction / Carr, Arthur – Cambridge: University Press; New York: Macmillan [distributor], 1896 – (= ser Cambridge Greek Testament For Schools And Colleges) – 1mf – 9 – 0-7905-3307-3 – mf#1987-3307 – us ATLA [227]
The general epistle of st james / Fulford, Henry William – London: Methuen 1901 [mf ed 1989] – (= ser The churchman's bible) – 1mf – 9 – 0-7905-0944-X – (incl bibl ref) – mf#1987-0944 – us ATLA [227]
The general epistles : james, peter, john, and jude. introduction, authorized version, revised version, with notes, index and map / ed by Bennett, William Henry – NY: Henry Frowde, 1901 – 1mf – 9 – 0-8370-2268-1 – (incl ind) – mf#1985-0268 – us ATLA [227]
The general epistles of ss james, peter, john, and jude : with notes critical and practical / Sadler, Michael Ferrebee – London: George Bell, 1891 – 1mf – 9 – 0-8370-5024-3 – mf#1985-3024 – us ATLA [227]
Un general espanol del siglo 27, don jose de garro / Arrillaga, Enrique de – Madrid: Razon y Fe, 1935 – 1 – sp Bibl Santa Ana [920]
General evening post – Dublin, Ireland. 1782-31 jul 1784 – 1 1/2r – 1 – uk British Libr Newspaper [072]
General evening post – Dublin. Ireland. -sw. 3 Jan 1782-31 May, 24 Jun-10 Jul, 30 Sep, 30 Oct, 13 Nov-30 Dec 1783. (1 reel) – 1 – uk British Libr Newspaper [072]
Le general faidherbe / Brunel, Ismael-Matthieu – Paris: C Delagrave, 1890 – 1 – us CRL [920]
General federation clubwoman – v50 n7,n9, v51 n2-n9 [1972 mar, may/jun, oct-1973 may/jun] – 1r – 1 – (cont: clubwoman [washington, dc]; cont by: general federation clubwoman news) – mf#803469 – us WHS [366]
General federation clubwoman – v57 n5-v58 n9 [1979 jan/feb-1980 may] – 1r – 1 – (cont: general federation clubwoman news; cont by: gfwc clubwoman news) – mf#898412 – us WHS [366]
General Federation clubwoman news – 1976 sep – 1r – 1 – (cont: general federation clubwoman; cont by: general federaton clubwoman [1978]) – mf#867467 – us WHS [366]
General federation magazine – v16 n9-v17 n9, v18, n10 [1917 dec-1918 sep, 1919 dec] – 1r – 1 – (cont: general federation of women's clubs magazine; cont by: general federation news) – mf#783052 – us WHS [366]
General federation news – v3 n3-4, v9 n1, v9 n12 [1922 nov-dec, 1928 jul, 1929 jun] – 1r – 1 – (cont: general federation magazine; cont by: clubwoman [washington, dc]) – mf#897064 – us WHS [366]
General Federation of Women's Clubs *see* GFWC clubwoman
General financial and trade review of the city of toronto for 1880 / Galbraith, Thomas – Toronto?: s.n, 1881 – 2mf – 9 – mf#01460 – cn Canadiana [380]
General francisco morazan, articulos publicados en... / Montufar, Lorenzo – Guatemala, 1896 – 1r – us UF Libraries [972]
General goes depoe / Coutinho, Lourival – Rio de Janeiro, Brazil. 1956 – 1r – us UF Libraries [972]
General grant, the lessons of his life and death : sermon preached by request in zion presbyterian church, brantford, ont, sabbath ev'g, sept 13, 1885 / Cochrane, William – Brantford, Ont?: s.n, 1885 – 1mf – 9 – mf#01119 – cn Canadiana [920]
General headquarters, southwest pacific area, 1941-1945 : chronological index and summary of communications – (= ser World war 2 research colls) – 12r – 1 – $2085.00 – 0-89093-738-9 – (with p/g) – us UPA [355]
General hints to emigrants : containing notices of the various fields for emigration, with practical hints on preparation for emigrating... – London, 1866 – (= ser 19th c books on british colonization) – 3mf – 9 – mf#1.1.2926 – uk Chadwyck [304]
A general historico-critical introduction to the old testament = Handbuch der historisch-kritischen einleitung in das alte testament / Haevernick, Heinrich Andreas Christoph – Edinburgh: T & T Clark 1852 [mf ed 1989] – (= ser Clark's foreign theological library 28) – 1mf – 9 – 0-7905-1717-5 – (incl bibl ref. in english) – mf#1987-1717 – us ATLA [221]
General history for colleges and high schools / Myers, Philip Van Ness – Boston, MA. 1906 – 1r – us UF Libraries [900]
The general history of china : containing a geographical, historical, chronological...of the empire of china, chinese tartary, corea and thibet / Du Halde, J B – London: J Watts, 1741. 4v – 24mf – 9 – mf#HT-510 – ne IDC [915]
A general history of the baptist denomination in america and other parts of the world / Benedict, David – New York: Lewis Colby, 1848 [mf ed 1989] – 3mf – 9 – 0-7905-4137-8 – (incl bibl ref) – mf#1988-0137 – us ATLA [242]
A general history of the catholic church : from the commencement of the christian era until the present time = Histoire generale de l'eglise / Darras, Joseph Epiphane – New York: P O'Shea, 1868, c1865 [mf ed 1986] – 4v on 8mf – 9 – 0-8370-9052-0 – (incl bibl ref and ind) – mf#1986-3052 – us ATLA [241]
The general history of the christian church : from her birth to her final triumphant state in heaven / Walmesley, Charles – 5th American ed. New York: D & J Sadlier, 1860 – 1mf – 9 – 0-524-03471-0 – mf#1990-1014 – us ATLA [240]
The general history of the mogol empire : from its foundation by tamerlane, to the late emperor orangzeb; extracted from the memoirs of m manouchi, a venetian, and chief physitian to orangzeb for above forty years / Catrou, Francois – London: printed for Jonah Bowyer, 1709 [mf ed 1995] – (= ser Yale coll) – 366p – 1 – 0-524-09400-4 – mf#1995-0400 – us ATLA [954]

General history of the principal discoveries and improvements in useful art – La Salle IL 1726-27 – 1 – mf#5556 – us UMI ProQuest [700]

A general history of the sabbatarian churches : embracing accounts of the armenian, east indian, and abyssinian episcopacies in asia and africa... / Davis, Tamar – Philadelphia: Lindsay & Blakiston, 1851 [mf ed 1991] – 1mf – 9 – 0-524-01108-7 – mf#1990-0322 – us ATLA [242]

General history of the science and practice of music / Hawkins, John – London: T Payne & son 1776 [mf ed 19–] – 5v on 60mf – 9 – mf#fiche 492 – us Sibley [780]

A general history of the world / Barth, Christian Gottlob – New York: Lane & Scott, 1853 [mf ed 1992] – 1mf – 9 – 0-524-04606-9 – (rev by daniel parish kidder) – mf#1990-1266 – us ATLA [900]

General hospital psychiatry – Oxford, England 1979+ – 1,5,9 – ISSN: 0163-8343 – mf#42077 – us UMI ProQuest [616]

General index to compiled military service records of revolutionary war soliders / U.S. War Dept – (= ser War department coll of revolutionary war records) – 58r – 1 – (with printed guide. reproduces the most comprehensive name ind to american soldiers who served during the revolution) – mf#M860 – us Nat Archives [355]

General index to compiled service records of volunteer soldiers who served during the war with spain / U.S. War Dept. Adjutant General's Office – (= ser Records of volunteer soldiers who served during the war with spain) – 126r – 5 – (with printed guide) – mf#M871 – us Nat Archives [355]

General index to pension files, 1861-1934 / U.S. Veterans Administration – (= ser Records of the veterans administration) – 544r – 5 – mf#T288 – us Nat Archives [355]

General index to the historical and biographical works of john strype – Oxford: Clarendon Press 1828 [mf ed 1959] – 2v on 1r – 1 – mf#1984-s016h – us ATLA [242]

A general index to the statutes of new brunswick now in force, other than those contained in the consolidated statutes : being chiefly local and private acts with a table of acts which since 1854 have expired or become obsolete or have been repealed otherwise then by chapter 120 of the consolidated statutes / Burbidge, George Wheelock – Fredericton, NB: s.n, 1878 – 2mf – 9 – mf#55302 – cn Canadiana [348]

General index to the statutes...,public and private, passed during the years 1869-1919 (both included) / Prince Edward Island. Laws, Statutes, etc – Charlottetown: Dillon, 1918. 116p. LL-2331 – 1 – us L of C Photodup [348]

General information index to names and subjects ("the de grange index"), 1789-1889 / U.S. War Dept. Office of the Chief of Engineers – (= ser Records of the office of the chief of engineers) – 7r – 1 – mf#M1703 – us Nat Archives [355]

General information regarding the virgin islands / United States Dept Of The Interior – Washington, DC. 1939 – 1r – us UF Libraries [972]

General instructions for making the meteorological observations at the senior county grammar schools in upper canada – Toronto: printed for the Dept of Public Instruction for Upper Canada by Lovell & Gibson, 1857 [mf ed 1983] – 1mf – 9 – mf#SEM105P337 – cn Bibl Nat [550]

General instructions in music : containing precepts and examples in every branch of the science / Bemetzrieder, Anton – London: printed & sold by aut...[1790] [mf ed 19–] – 6mf – 9 – (in french & english) – mf#fiche 347 – us Sibley [780]

General introduction to statistical account of upper canada : compiled with a view to a grand system of emigration, in connexion with a reform of the poor laws / Gourlay, Robert – London 1822 [mf ed Hildesheim 1995-98] – 1v on 3mf – 9 – €90.00 – 3-487-27090-0 – gw Olms [317]

General introduction to statistical account of upper canada : compiled with a view to a grand system of emigration, in connexion with a reform of the poor laws / Gourlay, Robert – London: Publ by Simpkin and Marshall...and J M Richardson, 1822 [mf ed 1984] – 6mf – 9 – (with ind) – mf#SEM105P374 – cn Bibl Nat [317]

General introduction to statistical account of upper canada : compiled with a view to a grand system of emigration, in connexion with a reform of the poor laws / Gourlay, Robert Fleming – London, 1822 – (= ser 19th c books on british colonization) – 6mf – 9 – mf#1.1.9893 – uk Chadwyck [971]

General introduction to the prophetic writings of the old testament : and especially to the minor prophets / Elliott, Charles – New York: Charles Scribner, c1874 [mf ed 1986] – (= ser A commentary on the holy scriptures. old testament 14/1) – 1mf – 9 – 0-8370-6040-0 – mf#1986-0040 – us ATLA [221]

A general introduction to the study of holy scripture : in a series of dissertations... / Breen, Andrew Edward – 2nd rev enl ed. Rochester, NY: J P Smith, 1908 [mf ed 1989] – 2mf – 9 – 0-7905-0675-0 – (incl ind) – mf#1987-0675 – us ATLA [220]

General introduction to the study of holy scripture : the principles, methods, history, and results of its several departments and of the whole / Briggs, Charles Augustus – [2nd rev ed] New York: Scribner's, 1899 [mf ed 1985] – 2mf – 9 – 0-8370-2451-X – (incl ind) – mf#1985-0451 – us ATLA [220]

General introduction to the study of the holy scriptures / Gigot, Francis Ernest – 3rd rev ed. New York: Benziger, 1903, c1900 [mf ed 1989] – (= ser Introduction to the study of the holy scriptures 1) – 2mf – 9 – 0-7905-1398-6 – mf#1987-1398 – us ATLA [220]

General jackson's fine : an examination into the question of martial law / Ingersoll, Charles Jared – Washington, Blair and Rives, 1843. LL-380 – 88p – 1 – us L of C Photodup [340]

General james wilkinson's order book, december 31, 1796-march 8, 1808 / U.S. War Dept. Adjutant General's Office – (= ser Records of the adjutant general's office, 1780's-1917) – 3r – 5 – (with printed guide) – mf#M654 – us Nat Archives [355]

General james wolfe, his life and death : a lecture...on tuesday, september 13, 1859, being the anniversary day of the battle of quebec... / Bell, Andrew – Montreal, Quebec: J Lovell, 1859 – 1mf – 9 – (incl bibl ref) – mf#44161 – cn Canadiana [355]

General justo rufino barrios / Carranza, Jesus E – Guatemala, 1956 – 1r – us UF Libraries [972]

General land office circulars and general regulations / Fisher, CG – Washington: GPO, 1930 – 18mf – 9 – $27.00 – mf#llmc 85-303 – us LLMC [348]

General land office correspondence / Gallatin, Albert – 1r – 1 – mf#B26321 – us Ohio Hist [324]

The general law of partnership as applied to commercial and business liabilities / Button, Charles P – New York, Hassel, 1881. 16 p. LL-574 – 1 – us L of C Photodup [346]

General legal forms and precedents, for ordinary use and with explanatory changes adapted to special cases. / Jones, James – Chicago, Myers, 1894. 929 p. LL-1410 – 1 – us L of C Photodup [340]

General letter : architecture / Davis, Mary Irene – s.l, s.l? 1936 – 1r – us UF Libraries [720]

General linguistics – Ann Arbor MI 1955-66 – 1 – ISSN: 0016-6553 – mf#8505 – us UMI ProQuest [400]

General magazine and historical chronicle : for all the british plantations in america – La Salle IL 1741 – 1 – mf#3523 – us UMI ProQuest [071]

General magazine and impartial review [1787] – La Salle IL 1787-92 – 1 – mf#4718 – us UMI ProQuest [071]

General magazine and impartial review [1798] – La Salle IL 1798 – 1 – mf#4457 – us UMI ProQuest [071]

General magloire ambroise a-t-il ete tue ou s'est-... / Ambroise, Fernand – Port-Au-Prince, Haiti. 1937 – 1r – us UF Libraries [972]

General maximo gomez – Habana, Cuba. 1900 – 1r – us UF Libraries [972]

General mayor de la universidad de san carlos en g... / Ferrus Roig, Francisco – Guatemala, 1962 – 1r – us UF Libraries [972]

General Mining Association of Quebec *see* The journal of the general mining association of the province of quebec

General Missionary Convention of the Methodist Episcopal Church. 1st *see* The open door

General Motors Corporation *see*
– Delco electronics broadcaster
– Rock river review

General Motors Corporation et al *see*
– Local 717 news
– Local 717 union news

General murgueitio / Tascon, Tulio Enrique – Bogota, Colombia. 1915 – 1r – us UF Libraries [972]

General news / General Telephone Company of Illinois – v1 n1-v2 n12 [1981 jun 26-1982 dec 24] – 1r – 1 – (cont: wisconsin news [sun prairie, wi]; cont by: mto news) – mf#656904 – us WHS [380]

General news / General Telephone Company of Wisconsin – v15, n103-v20, n1012 [1974 mar-1979 dec 21] – 1r – 1 – (continued by: wisconsin news (sun prairie, wi)) – mf#577304 – us WHS [380]

General news *see* Miscellaneous newspapers of las animas county, reel 1

General news and timpas times *see* Miscellaneous newspapers of las animas county, reel 1

General news letter – Dublin, Ireland. 3 oct 1744 – 1/4r – 1 – uk British Libr Newspaper [072]

General notice of a reply by major robinson... dated 30th march, 1849 : to observations by mr wilkinson on his report of the exploratory survey for the halifax and quebec railway / Wilkinson, John – Fredericton NB: J Simpson, 1852 – 1mf – 9 – mf#22318 – cn Canadiana [380]

General office subject files, 1966-1972 – 1ser – (= ser Papers of the naacp 30) – 1 – (ser a: subject files 22r isbn 1-55655-902-X $4840. with p/g) – us UPA [322]

General officers' washington report / United Association of Journeymen and Apprentices of the Plumbing and Pipe Fitting Industry – 1976 nov-1979 dec, 1980 jan-1982 jun, 1982 jul-1985 dec, 1986 jan-1990 dec – 4r – 1 – (cont: general officers' weekly newsletter) – mf#2516398 – us WHS [680]

General officers' weekly newsletter / United Association of Journeymen and Apprentices of the Plumbing and Pipe Fitting Industry – 1964 oct 28-1968 feb 29, 1968 mar 6-1972 jan 28 – 2r – 1 – (continued by: general officers' washington report) – mf#661612 – us WHS [680]

General orders : his excellency the governor in chief and commander of the forces, has received a despatch from major general de rottenburgh, transmitting a letter... / Great Britain. Army – [Quebec?]: John Neilson... [1813?] [mf ed 1993] – 1mf – 9 – 0-665-91309-5 – (in english and french) – mf#91309 – cn Canadiana [971]

General orders / Legion of Valor of the United States of America – 1965 apr 23 [v75 n8], 1968-1977 dec – 1r – 1 – (cont: general orders, army and navy legion of valor of the united states of america) – mf#379640 – us WHS [355]

General orders and circulars / Cuba Military Governor, 1899 (John R Brooke) – Havana, Cuba. 1900 – 1r – us UF Libraries [355]

General orders and circulars of the confederate war department, 1861-1865 / U.S. War Dept. Confederate Records – (= ser War department coll of confederate records) – 1r – 1 – (with printed guide) – mf#M901 – us Nat Archives [355]

General orders and circulars of the war department and headquarters of the army, 1809-1860 / U.S. War Dept. Adjutant General's Office – (= ser Records of the adjutant general's office, 1780's-1917) – 8r – 1 – (with printed guide) – mf#M1094 – us Nat Archives [355]

General orders, general field orders and circulars, 1861-62 / United States Army Dept of Missouri and Mississippi – Washington, DC: National Archives and Records Service, 1977 – us CRL [355]

General orders kept by general william heath, may 23, 1777-oct 20, 1778 / U.S. War Dept. – (= ser War department coll of revolutionary war records) – 1r – 1 – mf#T42 – us Nat Archives [355]

General ordinances of the city of st augustine... / Saint Augustine (Fla) Ordinances, Etc – St Augustine, FL. 1928 – 1r – us UF Libraries [978]

General organization bulletin for ... / Industrial Workers of the World – (1971 feb-1980 n5] – 1r – 1 – mf#605587 – us WHS [331]

General ospina / Sanchez Camacho, Jorge – Bogota, Colombia. 1960 – 1r – us UF Libraries [972]

General pharmacology – Oxford, England 1970+ – 1,5,9 – ISSN: 0306-3623 – mf#49075.01 – us UMI ProQuest [615]

General photographic file of the u.s. forest service, 1886- / U.S. Forest Service – (= ser Records Of The Forest Service) – 121r – 1 – mf#M1127 – us Nat Archives [634]

General photographs of the bureau of ships, 1914 – (= ser Records of the bureau of ships) – 1100r – 1 – mf#M1222 – us Nat Archives [355]

General post office advertiser – Dublin, Ireland. 22 jul, 26 oct 1741 – 1/4r – 1 – uk British Libr Newspaper [072]

General practice clinics – Washington DC 1949-50 – 1 – ISSN: 0097-1634 – mf#148 – us UMI ProQuest [610]

General price current, mercantile register, and shipping list – Hong Kong: John Cairns, jan 3-aug 22, 1845 – (filmed consecutively with: hongkong register, jan 3-aug 22 1845) – us CRL [380]

General Primitive Baptist State Convention of Texas *see* Proceedings of the organization of the general primitive baptist state convention of texas

General principles of christian ethics : the first part of the system of christian ethics = Christliche sittenlehre. selections / Schmid, Christian Friedrich – Philadelphia: Lutheran Bookstore, 1872 – 1mf – 9 – 0-524-05090-2 – (in english) – mf#1991-2214 – us ATLA [170]

The general principles of constitutional law in the united states / Cooley, Thomas McIntyre – 3rd ed. Boston: Little, Brown, 1898 – 5mf – 9 – $7.50 – mf#LLMC 95-064 – us LLMC [323]

The general principles of the law of contract / Hammon, Louis Lougee – Saint Paul, Keefe-Davidson, 1902. 1233 p. LL-340 – 1 – us L of C Photodup [346]

The general principles of the law of insurance / Peele, Stanton Canfield – Washington, D.C.: Tibbetts, 1901. 69 1p. LL-1204 – 1 – us L of C Photodup [346]

General principles of the philosophy of nature / Stallo, John Bernard – 1848 – 1 – us CRL [140]

General properties of some tropical and sub-tropical fruits of florida / Abbott, Ouida Davis – Gainesville, FL. 1931 – 1r – us UF Libraries [634]

General public acts of congress respecting the sale and disposition of public lands : with a g opinions, 1776-1838 – Washington: Gales & Seaton. 2v. 1838 – (= ser Land laws of the united states, 1776-1938) – 19mf – 9 – $28.50 – mf#LLMC 82-101-1 – us LLMC [343]

General publications of the venice biennale with la biennale di venezia *see* Publications of the venice biennale, 1895-1977 (pvb)

General radio experimenter – Concord MA 1926-70 – 1 – mf#2082 – us UMI ProQuest [380]

General ramon leocadio bonachea / Carbonell, Nestor – Habana, Cuba. 1947 – 1r – us UF Libraries [972]

General records of the american commission to negotiate peace, 1918-1931 / U.S. Commission to Negotiate Peace – (= ser Records of the american commission to negotiate peace) – 563r – 1 – (with printed guide) – mf#M820 – us Nat Archives [327]

General register office: birth certificates from the presbyterian, independent and baptist registry and from the wesleyan methodist metropolitan registry – 1742-1840 – 207 file(s) – 1 – uk National [242]

General register office: foreign registers and returns – 1627-1960 – 163 files + vols – 1 – uk National [324]

General register office: miscellaneous foreign death returns – 1830-1921 – 69 papers + vols – 1 – uk National [933]

General register office: miscellaneous foreign returns – 1831-1969 – 59v – 1 – uk National [350]

General register office: registers and returns of births, marriages and deaths in the protectorates etc of africa and asia – 1895-1965 – 15v – 1 – uk National [929]

General register office: registers of births, marriages and deaths surrendered to the non parochial registers commission of 1857 : and other registers and church records – 1646-1970 – 314 microform – 1 – uk National [929]

General register office: registers of births, marriages and deaths surrendered to the non-parochial registers commissions of 1837 and 1857 – 1567-1858 – 4680 file(s) – 1 – uk National [350]

General register office: registers of clandestine marriages and of baptisms in the fleet prison, king's bench prison, the mint and the may fair chapel – 1667-c1777 – 835v – 1 – uk National [350]

General register office: society of friends' registers, notes and certificates of births, marriages and burials – 1578-1841 – 1674v – 1 – uk National [243]

General regulations and prize list... : september 28th, 29th, 30th and october 1st, 1880 / Agricultural and Industrial Exhibition (1880: Kentville, Nova Scotia) – Kentville, NS: Western Chronicle Office, 1880 – 1mf – 9 – mf#06915 – cn Canadiana [630]

General regulations and prize list... : september 30th, october 1st, 2nd, 3rd and 4th 1878 / Agricultural and Industrial Exhibition (1878: Truro, Nova Scotia) – Truro, NS?: W B Alley, 1878 – 1mf – 9 – mf#06917 – cn Canadiana [630]

General regulations and prize list... : october 1st, 2nd, 3rd, 4th and 5th, 1877 / Provincial Agricultural Exhibition (1877 : Kentville, NS) – Kentville, NS?: Western Chronicle, 1877 – 1mf – 9 – mf#67242 – cn Canadiana [630]

General regulations and prize list... : october 7th, 8th and 9th, fair day 10th, 1878 / King's County Agricultural and Industrial Exhibition (1878 : Kentville, NS) – Halifax, NS?: Nova Scotia Print, 1878 – 1mf – 9 – mf#67239 – cn Canadiana [630]

General regulations, and prize list... : october 9th, 10th, 11th, 12th and 13th, 1876 / Provincial Exhibition (1876: Truro, Nova Scotia) – Truro, NS?: W B Alley, 1876 – 1mf – 9 – mf#63627 – cn Canadiana [630]

General regulations for the execution of the mortgage laws for cuba, puerto rico and the philippines, 1893 – Washington: GPO, 1899 – 2mf – 9 – $3.00 – mf#LLMC 92-313 – us LLMC [348]

General relativity and gravitation – Dordrecht, Netherlands 1970+ – 1,5,9 – ISSN: 0001-7701 – mf#10854 – us UMI ProQuest [530]

General Relief Committee (Montreal, Quebec) *see* Proceedings of the general relief committee: appointed by the citizens of montreal

General remarks on steam communication : with reference to the united kingdoms as the centre / Fortescue, Thomas Knox – Dublin: S J Machen, 1845 [mf ed 1983] – 1mf – 9 – 0-665-44505-9 – mf#44505 – cn Canadiana [380]

General report : with the addresses at the devotional meetings = a / Pan-Anglican Congress 1908 – London: Society for Promoting Christian Knowledge; New York: E S Gorham, 1908 – 1mf – 9 – 0-8370-9090-3 – mf#1986-3090 – us ATLA [240]

General report and minutes of evidence of the royal commission on the state of universities and colleges of scotland, 1832-37 : command n310 – 32mf – 9 – mf#86967 – uk Microform Academic [324]

General report in regard to the share and loan capital, the traffic in passengers and goods : and the working expenditure of the railway companies in the united kingdom 1871-1901 – [mf ed Chadwyck-Healey] – (= ser British government publications...1801-1977) – 1r – 1 – uk Chadwyck [380]

General report of the deputation sent by the american board to china in 1907 / American Board of Commissioners for Foreign Missions – Boston: The Board, 1907 [mf ed 1995] – (= ser Yale coll) – 57p (ill) – 1 – 0-524-10021-7 – mf#1995-1021 – us ATLA [951]

General report on the commercial, industrial, & economic situation of china in june, 1921 / Fox, Harry Halton – London: His Majesty's Stationery Office, 1921. 64p. illus – 1 – us Wisconsin U Libr [951]

General repository and review – La Salle IL 1812-13 – 1 – mf#3809 – us UMI ProQuest [240]

General review of british and foreign literature – La Salle IL 1806 – 1 – mf#4255 – us UMI ProQuest [410]

General richepanse / Sainte-Croix De La Ronciere, Georges – Paris, France. 1933 – 1r – us UF Libraries [972]

General rigby, zanzibar, and the slave trade, with journals / Rigby, Christopher Palmer – London, England. 1935 – 1r – us UF Libraries [960]

General rule of the apostolic union of secular priests / Apostolic Union of Secular Priests – S.l: Office of the Messenger of the Sacred Heart, 1880? – 1mf – 9 – (trans fr french) – mf#07062 – cn Canadiana [241]

General rules and orders. / New Brunswick. Canada. Supreme Court – Toronto and Edinburgh: Carswell, 1881. 287, 1p. LL-2381 – 1 – us L of C Photodup [340]

General rules and regulations for the government of the common gaols of canada – [Quebec?: s.n.] 1861 [mf ed 1983] – 1mf – 9 – 0-665-44516-4 – mf#44516 – cn Canadiana [365]

General rules of the quebec benevolent society, passed 7th august, 1805 = Regles generales de la societe bienveillante de quebec, faites le 7e aout, 1805 / Societe bienveillante de Quebec – Quebec: John Neilson, 1805 [mf ed 1974] – 1r – 5 – mf#SEM16P197 – cn Bibl Nat [366]

General section circular u / Commercial Advisory Foundation in Indonesia – Djakarta, 1965-1972(67) – 27mf – 9 – (missing: 1965(22); 1966(31)-1970(1-2, 4, 8, 10, 12, 14, 17, 19); 1971(23, 27, 29, 32, 36, 37, 40, 43, 45, 48); 1972(53-55, 57, 62-64)) – mf#SE-1381 – ne IDC [959]

General sketch of the history of pantheism / Plumptre, Constance E – London: WW Gibbings, 1878 – 2mf – 9 – 0-524-00846-9 – mf#1990-2092 – us ATLA [210]

General sociology : an exposition of the main development in sociological theory from spencer to ratzenhofer / Small, Albion Woodbury – Chicago: University of Chicago Press [etc] 1905 [mf ed 1970] – (= ser Library of american civilization 10354) – xiii/739p on 1mf – 9 – us Chicago U Pr [301]

General sociology / Small, Albion Woodbury – Chicago, IL. 1905 – 1r – us UF Libraries [301]

General statement of the annual revenue and expenditure of the province of canada : from the period of the union of the late province of upper and lower canada, to the end of the year 1849 – Toronto: printed by Rollo Campbell, 1850 [mf ed 1984] – 1mf – 9 – mf#SEM105P434 – cn Bibl Nat [336]

The general statutes of the state of minnesota as amended by subsequent legislation. / Minnesota. Laws, Statutes, etc – St. Paul: West, 1894. 2v. L.C. set incomplete: v.1, p. 1273-1274 wanting. LL-967 – 1 – us L of C Photodup [348]

General strike – London. Issued by the International Libertarian Group of Correspondence. 1 Oct 1903. 3 ft – 1 – uk British Libr Newspaper [072]

General strike newspapers – [NW England] Lancashire may 1926 [mf ed 2002] – 1r – 1 – uk Newsplan [331]

General strike newspapers etc – [NW England] may 1926 [mf ed 2002] – 1 – uk Newsplan [331]

General subjects – Oxford, England 1964+ – 1,5,9 – ISSN: 0304-4165 – mf#42175 – us UMI ProQuest [574]

A general subscription for the relief of sufferers by the late destructive fires throughout the province of new brunswick – Fredericton [NB]: G K Lugrin, 1825 [mf ed 1983] – 1mf – 9 – 0-665-44517-2 – mf#44517 – cn Canadiana [360]

General suggestions for leaders of mission study classes / Sailer, T H P – rev ed. New York: Student Volunteer Movement, c1911 – 1mf – 9 – 0-524-02988-1 – mf#1990-0775 – us ATLA [240]

General summation / International Military Tribunal for the Far East – Tokyo. LL-028 – 1 – us L of C Photodup [340]

A general survey of the history of the canon of the new testament / Westcott, Brooke Foss – 7th ed. London, New York: Macmillan, 1896 [mf ed 1989] – 2mf – 9 – 0-7905-2622-0 – (incl ind) – mf#1987-2622 – us ATLA [225]

General synod of the evangelical lutheran church in the united states : woman's home and foreign missionary society. minutes 1879-1919 / General Synod of the Evangelical Lutheran Church in the United States. Woman's Home and Foreign Missionary Society – 1r – 1 – (with ind; forms pt of: subgroup gs 16/4 executive committee) – mf#xa0095r – us ATLA [242]

General synod of the evangelical lutheran church in the united states : woman's home and foreign missionary society. minutes 1887-1916 / General Synod of the Evangelical Lutheran Church in the United States. Woman's Home and Foreign Missionary Society – 1r – 1 – (with ind; forms pt of: subgroup gs 16/5 general literature committee) – mf#xa0096r – us ATLA [242]

General Synod of the Evangelical Lutheran Church in the United States. Board of Foreign Missions *see*
- American evangelical lutheran mission files 1875-1919
- Incorporation papers 1911

General Synod of the Evangelical Lutheran Church in the United States. Woman's Home and Foreign Missionary Society *see*
- Board of trustees correspondence 1918
- Board of trustees minutes 1884-1916
- General synod of the evangelical lutheran church in the united states
- Woman's home and foreign missionary society convention programs 1893-1919, 1899-1919
- Woman's home and foreign missionary society minutes 1879-1919

General Synod of the Evangelical Lutheran Church in the United States. Women's Home and Foreign Missionary Society *see* Women's home and foreign missionary society correspondence 1899, 1909, 1915, 1918

General Telephone Company of Illinois *see* General news

General Telephone Company of Wisconsin *see*
- General news
- Wisconsin area news
- Wisconsin news

General theatrical programme – (Theatrical Programme). London. -w. Dec 1883-Feb 1886. (2 reels) – 1 – uk British Libr Newspaper [790]

General theological library : bulletin – v1-77. 1908-83 [complete] – 3r – 1 – mf#ATLA S0631 – us ATLA [020]

General theory of law / Korkunov, N M – N.Y.: The Macmillan Co, 1909 (reprint 1922) – 6mf – 9 – $9.00 – mf#LLMC 95-180 – us LLMC [340]

General topography : martin county / Lyons, Isabel J – s.l, s.l? 1936 – 1r – us UF Libraries [978]

General topography / Saunders, H J – s.l, s.l? 193-? – 1r – us UF Libraries [978]

A general treatise on kansas pleading and practice under the code of civil procedure / Taylor, Irwin – Topeka, Kan., Crane, 1888. 662 p. LL-568 – 1 – us L of C Photodup [347]

General u s grant's tour around the world – Chicago, IL. 1879 – 1r – us UF Libraries [910]

A general view of the criminal law of england / Stephen, James Fitzjames – 1st ed. London: Macmillan, 1863 – 6mf – 9 – $9.00 – (2nd ed. london: macmillan, 1890 5mf $7.50 84-807b) – mf#LLMC 84-807A – us LLMC [345]

A general view of the history of the english bible / Westcott, Brooke Foss – 2nd ed. London, New York: Macmillan, 1872 [mf ed 1985] – 1mf – 9 – 0-8370-5805-8 – (incl bibl ref, app & ind) – mf#1985-3805 – us ATLA [220]

A general view of the rise, progress, and corruptions of christianity / Whately, Richard – New York: William Gowans, 1860 [mf ed 1985] – 1mf – 9 – 0-8370-5763-9 – (with sketch of life of aut & catalogue of his writings) – mf#1985-3763 – us ATLA [240]

General wolfe : a favorite song / Smart, Thomas – Covent Garden, [London]: printed for G Goulding, [ca 1770] (mf ed 1988) – 1mf – 9 – mf#SEM105P909 – cn Bibl Nat [780]

General wood workers' journal / International Furniture Workers' Union of America – v1 n1-52 [1891 jun 15-1896 jan 15] – 1r – 1 – mf#1056922 – us WHS [680]

General works and fiction from india from the british library, london *see* Colonial discourses, series 3

General y natural historia de las indias / Fernandez de Oviedo Valdes, Gonzalo – Sevilla: Juan Cromberger, 1535 – 1 – sp Bibl Santa Ana [970]

General york : vaterlaendisches schaupspiel in fuenf akten / Greif, Martin – Leipzig: C F Amelang, 1912 – 1r – 1 – us Wisconsin U Libr [430]

General-anzeiger : fuer gross-oberhausen und das nordwestliche industrie-gebiet – Oberhausen DE, 1958 3 jan-1960 [gaps] (nur lokalteil) – 7r – 1 – gw Misc Inst [074]

General-anzeiger – Berlin-Reinickendorf DE, 1933 2 oct-1936 9 dec – 7r – 1 – (incl suppls) – gw Misc Inst [074]

General-anzeiger – Joliet, IL: [R Zintzch & Co], feb 1896-feb 1897; sep 1917; oct 1917-sep 1937 – 20r – 1 – us CRL [071]

General-anzeiger – Joliet IL (USA), 1922 18 feb-1933 8 jul [gaps], 1933 30 sep-7 oct [gaps] – 5r – 1 – gw Misc Inst [071]

General-anzeiger : tageszeitung fuer ostfriesland, oldenburger land und emsland – Rhauderfehn DE, 1988- – ca 8r/yr – 1 – gw Misc Inst [074]

General-anzeiger – Luebeck DE, 1890 1 jul-30 sep, 1894 3 jan-1896, 1897 1 jul-1898 30 jun, 1899 1 jan-30 jun, 1900 1 jul-1901 30 jun, 1902-07, 1909 1 jan-30 jun, 1910-1941 1 apr – 186r – 1 – (title varies: 1 jan 1899: luebecker general-anzeiger) – gw Mikrofilm [074]

General-anzeiger – Halle S DE, 1918 8 nov-30 nov – 1r – 1 – (title varies: 1 jun 1918: hallische nachrichten. banned fr 15-29 mar 1920. filmed by other misc inst: 1889 22 mar-1919 30 sep, 1920 2 jan-13 mar & 30 mar-30 apr, 1920 1 jul-1926, 1927 1 jul-1944 31 mar) – gw Misc Inst [074]

General-anzeiger – Werder DE, 1936 jun-sep – 1r – 1 – gw Misc Inst [074]

General-anzeiger *see*
- Dortmunder nachrichten [main edition]
- Duerener zeitung 1875
- Duisburger tageblatt 1881
- Dürener zeitung 1875
- General-anzeiger fuer elberfeld-barmen
- General-anzeiger fuer wesel
- Generalanzeiger fur marburg und umgebung 1887
- Der pforzheimer beobachter
- Woechentliches frag- und kundschaffts-blath

Generalanzeiger – Muenchen DE, 1892 29 sep-1943 31 mar – 1 – (title varies: 16 sep 1898: muenchener zeitung) – gw Misc Inst [074]

General-anzeiger der stadt bad nauheim – Bad Nauheim DE, 1906 4 jan-1913 29 dec [gaps] – 6r – 1 – gw Mikrofilm [074]

General-anzeiger der stadt magdeburg und provinz sachsen – Magdeburg DE, 1914 1 aug-30 sep, 1915 1 jul-30 sep, 1916 1 jul-30 sep, 1931 sep – 2r – 1 – (title varies: 24 sep 1908: magdeburger general-anzeiger. filmed by misc inst: 1883 3 jul-1884 30 sep, 1885 2 jul-1888 30 jun, 1889 7 jul-1912 31 mar, 1916 1 oct-1920 31 mar, 1920 1 oct-1921 30 sep, 1923-1930 31 jan, 1930 1 jun-1931 30 aug, 1931 1 nov-1932 30 apr, 1932 1 jul-1933 31 aug, 1934 1 jul-1937 31 jan, 1937 2 mar-1939 30 jun, 1939 aug, 1941 1 jan-11 may [171r]. with suppl) – gw Mikrofilm; gw Misc Inst [074]

General-anzeiger der stadt Wuppertal *see* General-anzeiger fuer elberfeld-barmen

General-anzeiger des amtsgerichtsbezirks koetzschenbroda *see* Koetzschenbrodaer zeitung

General-anzeiger fuer chemnitz und umgegend *see* Chemnitzer anzeiger und stadtbote

General-anzeiger fuer die gesamten interessen des judentums – Berlin DE, 1902-03, 1906, 1907 [gaps], 1908-10, 1911 [gaps] – 4r – 1 – gw Misc Inst [939]

General-anzeiger fuer die kreise grevenbroich, moenchen-gladbach und bergheim *see* Rheinische landeszeitung

Generalanzeiger fuer die oberaemter reutlingen, tuebingen, rottenburg, herrenberg, nuertingen, urach und muensingen – Reutlingen DE, 1976- – ca 9r/yr – 1 – (title varies: 13 jun 1916: reutlinger general-anzeiger) – gw Misc Inst [074]

General-anzeiger fuer die stadt und den bezirk ludwigshafen am rhein – Ludwigshafen DE, 1876 [single iss], 1877, 1881 [single iss], 1884-87 [single iss], 1888-1920 [gaps], 1922-1941 18 apr – 1 – gw Misc Inst [074]

General-anzeiger fuer duesseldorf und umgegend – Duesseldorf DE, 1958 3 jan-1961 3 may, 1961 29 jun-1974 25 may – 1 – (title varies: 29 nov 1906: duesseldorfer general-anzeiger; 1 jan 1918: duesseldorfer nachrichten; 8 jan 1919: rote fahne am niederrhein; 9 jan 1919: duesseldorfer nachrichten; 2 jul 1948: westdeutsche zeitung; 22 oct 1949: duesseldorfer nachrichten; 27 oct 1979: westdeutsche zeitung. filmed by bnl: duesseldorf de, 1916 sep-1919 6 aug [16r]. filmed by other misc inst: 1925 feb [1r]; 1876 8 oct-1917, 1951 apr-1977 [more than 267r]; 1951 2 apr-1957, 1978- [ca 10r/yr until 1957 30r]. with suppl: am rhein fr 13 jul 1911: rhein und duessel 1909 12 apr-1917 6 oct [4r] publ in duesseldorf & essen; illustrirtes sonntags-blatt 1890, 1892-93, 1895, 1900-01 (all with gaps) [2r]; sonderausgabe fuer das besetzte gebiet 1919 3 jul-1920 28 jan [3r] publ in duesseldorf-oberkassel; sonntagsblatt 1904 3 jul-1909 28 nov [2r]; westdeutsche sportzeitung 1924-1931 25 feb [2r]) – gw Misc Inst [074]

General-anzeiger fuer elberfeld-barmen – Wuppertal DE, 1949 1 oct-1957 – 37r – 1 – (title varies: 1 jan 1929: general-anzeiger der stadt wuppertal; 2 oct 1944: wuppertaler nachrichten; 1 oct 1949: general-anzeiger der stadt wuppertal / w, a; 16 aug 1971: general-anzeiger / a; 30 dec 1972: wz. general-anzeiger / a; 16 oct 1978: wz. westdeutsche zeitung; publ in wuppertal, fr 1 jul 1971 in duesseldorf. filmed by misc inst: 1969- [ca 10r/yr]; 1958-1966 12 oct, 1967 24 may-1969 5 aug, 1969 8 dec-1980 11 jan, 1980 8 may-31 dec [206r]) – gw Mikrofilm; gw Misc Inst [074]

Generalanzeiger fuer fabrikbedarf *see* Generalanzeiger fuer fabrikbedarf 1912

Generalanzeiger fuer fabrikbedarf 1912 – Duesseldorf DE, 1912-1922 23 sep – 5r – 1 – (title varies: 1915: allgemeine industriezeitung; 1920: generalanzeiger fuer fabrikbedarf) – gw Misc Inst [074]

General-anzeiger fuer hamburg-altona – Hamburg DE, 1888 2 sep-1945 3 may, 1952 13 sep-1957 31 mar – 199r – 1 – (title varies: 28 aug 1922: hamburger anzeiger. with suppl: die deutsche arbeitsfront 1933-39; die frau von heute in heim und beruf [fr 1937: die frau unserer zeit] 1932-40; fuer jungens und deerns 1925-40; fuer unsere frauen 1907-23; handels- und schiffahrtsblatt 1924-30; haus, hof und garten 1922-23 [gaps]; illustrierte wochenbeilage 1924 21 feb-1941 30 aug [gaps]; kleingarten und siedlung 1924-32; prima 1954-57; reisen und wandern 1929-38 [gaps]; der siedler 1922-23 [gaps]; der sonntag 1922-24; sport, spiele, turnen [fr 1937 sport]; 1925-39 [gaps]; ulenspegel 1922-37; vom sonntag zum alltag 1925-32 [gaps]) – gw Misc Inst [074]

General-anzeiger fuer kassel und umgegend – Kassel DE, 1888-1892 30 jun – 5r – 1 – (incl suppl) – gw Misc Inst [074]

Generalanzeiger fuer koeslin und umgebung – Koeslin (Koszalin PL), 1891 [gaps], 1893 n152-306 – 1 – (missing: n180) – gw Misc Inst [077]

General-anzeiger fuer leipzig und umgegend – Leipzig DE, 1903-18 – 55r – 1 – (later: leipziger abendzeitung) – gw Misc Inst [074]

Generalanzeiger fuer marburg und umgebung *see* Generalanzeiger fur marburg und umgebung 1887

General-anzeiger fuer solingen und umgegend *see* Tages-anzeiger

General-anzeiger fuer wesel – Wesel DE, 1961-1964 31 jul – 13r – 1 – (title varies: 1 apr 1944: volksfreund; 1 sep 1952: general-anzeiger. filmed by misc inst: 1958-60; 1952 1 sep-1957 [15r]. with suppl: bulletin pour les prisonniers francais en allemagne 1914 23 sep-1915 6 feb [1r]) – gw Mikrofilm [074]

General-anzeiger fuer wilster, st margarethen – Wilster DE, 1892-94 – 3r – 1 – (title varies: 17 mar 1893: tageblatt fuer glueckstadt und umgegend) – gw Misc Inst [074]

Generalanzeiger fur marburg und umgebung 1887 – Marburg DE, 1887 23 jan-1922 30 nov – 68r – 1 – (title varies: 1 apr 1888: annoncen-blatt (general-anzeiger) fuer marburg und umgebung; 1 oct 1890: general-anzeiger; 27 sep 1891: generalanzeiger fuer marburg und umgebung; 1 nov 1893: hessische landeszeitung. incl suppls) – gw Misc Inst [074]

General-anzeiger / rote erde *see* Dortmunder nachrichten [main edition]

General-anzeiger und fremdenblatt – Posen (Poznan PL), 1852 1 jan-30 jun – 1 – gw Misc Inst [077]

Der general-bass in der composition : oder: neue und grueendliche anweisung, wie ein music-liebender mit besonderm vortheil, durch die principia der composition... / Heinichen, Johann David – Dressden: Bey dem autore 1728 [mf ed 19–] – 21mf – 9 – mf#fiche 731 – us Sibley [780]

General-commission [commisie-generaal] for the netherlands indies, 1946-1947 : pts 3.1.-3.3: papers of the members – [mf ed 2007] – (= ser Dutch political conflict with the republic of indonesia, 1945-1949 3) – 178mf – 9 – €2800.00 – (in dutch, indonesian, english; with p/g & concordance) – mf#mmp126-128 – ne Moran [350]

Generale kerckelycke historie van den gheboorte onzes h iesu christi tot het iaer 1624 / Baronius, C & Spondanus, H – Antwerpen, 1623 – (= ser Kerckelycke historie van neder-landt) – €229.00 – (bound with: h rosweydus: kerckelycke historie van neder-landt) – ne Slangenburg [242]

Generale legende der heylighen met het leven iesu christi ende marie : vergadert wt de h schrifture, oude vaders ende registers der h kercke / Ribadineira, P & Rosweydus, H – 2e druck. Antwerpen. v1-2. 1629 – €177.00 – ne Slangenburg [240]

"Generale repetitsye" / Verber, H [comp] – Kharkov, Ukraine 1931 – 1r – 1 – us UF Libraries [939]

Generales y doctores / Loveira, Carlos – Havana, Cuba. 1962 – 1r – us UF Libraries [972]

General-geschichte der fuerchterlichsten giftmischerinn gesche margarethe gottfried, gebornen timm : aus den besten quellen geschoepft; in gemaessheit beschlusses des hohen senats vom 17. december 1830 zur herausgabe bestaetigt / Feilner, Franz – Bremen 1831 [mf ed Hildesheim 1995-98] – 1mf – 9 – €40.00 – 3-487-29357-9 – gw Olms [880]

General-gouvernement – Krakow, Poland 1940-44 [mf ed Norman Ross] – 2r – 1 – mf#nrp-784 – us UMI ProQuest [939]

Die generalin : erzaehlung / Heuschele, Otto – Leipzig: Bong, 1943 – 1r – 1 – us Wisconsin U Libr [830]

Die generalin und andere geschichten / Best, Walter – 6. aufl. Muenchen: F Eher, 1943 [mf ed 1989] – (= ser Soldaten-kameraden 28) – 69p – 1 – mf#7014 – us Wisconsin U Libr [830]

General-inspektion des militaer-verkehrswesens (bestand ph 9 5) / inspektion des militaer-luft- und kraftfahrwesens (bestand ph 9 20) bd 25 / ed by Fleischer, Hans-Heinrich – 1986 – xv/89p – €3.50 – 978-3-89192-005-3 – gw Bundesarchiv [355]

Generalis collectionis : omnium operum illustrissimi ac reverendissimi domini, domini jacobi thomae josephi wellens... / Wellens, J – t'Antwerpen: C M Spanoghe, 1784 – 4mf – 9 – mf#O-3268 – ne IDC [090]

Generalis totius ordinis clericorum canonicorum historia / Pennotto, Gabr – Romae, 1624 – €95.00 – ne Slangenburg [240]

Generalisimo maximo gomez / Henriquez Y Carvajal, Federico – Santo Domingo, Dominican Republic. 1932 – 1r – us UF Libraries [972]

El generalisimo trujillo / Fernandex Mato, Romas – Ciudad Trujillo, 1944 – 1 – us CRL [920]

Generalisimo trujillo molina / Diaz Valdeparez, J – Santiago, Dominican Republic. 1946 – 1r – us UF Libraries [972]

Generalisimo trujillo molina / Perez Leyba, Salvador A – Ciudad Trujillo, Dominican Republic. 1940 – 1r – us UF Libraries [972]

Generalkommission der Gewerkschaften Deutschlands *see* Rechenschaftsbericht

Das generalkonzil im grossen abendlaendischen schisma / Bliemetzrieder, Franz – Paderborn: Ferdinand Schoeningh, 1904 – 1mf – 9 – 0-8370-8165-3 – (incl bibl ref) – mf#1986-2165 – us ATLA [240]

Generalmajor v. stille und friedrich der grosse contra lessing / Fisch, Richard – Berlin: Weidmann, 1885 – 1r – 1 – us Wisconsin U Libr [430]

General-post-office-advertiser – Dublin, Ireland. 22 July, 26 Oct 1741. 3 ft – 1 – uk British Libr Newspaper [072]

Generalregister zu band 1-61 (bdk62/63 1.reihe) – (= ser Bibliothek der kirchenvaeter. 1. reihe (bdk 1.reihe)) – €27.00 – ne Slangenburg [240]

The general's letters, 1885 : being a reprint from the war cry of letters to soldiers and friends scattered through out the world / Booth, William – London: Salvation Army, 1890 [mf ed 1987] – 204p – 1 – mf#2047 – us Wisconsin U Libr [860]

Generasi baru / Dewan Nasional Permuda Rakjat – Djakarta, 1963-1964. v1(1-6) – 3mf – 9 – (missing: 1964(3-4)) – mf#SE-366 – ne IDC [959]

Generation harmonique : ou traite de musique theorique et pratique / Rameau, Jean Philippe – Paris: Prault fils 1737 [mf ed 19–] – 5mf – 9 – mf#fiche 509 – us Sibley [780]

A generation of religious progress : issued in commemoration of the 21st anniversary of the union of ethical societies / Johnston, Harry Hamilton et al; ed by Spiller, Gustav – London: publ for the Union of Ethical Societies and the Rationalist Press Assoc Ltd by Watts, 1916 [mf ed 1991] – 1mf – 9 – 0-7905-9287-8 – mf#1989-2512 – us ATLA [200]

Generations / Jewish Historical Society of Maryland – v1 n1-v5 n2 [1978 dec-1985 apr] – 1r – 1 – (continued by: generations (baltimore, md: 1986)) – mf#1265571 – us WHS [939]

Generations – San Francisco CA 1993+ – 1,5,9 – ISSN: 0738-7806 – mf#19371 – us UMI ProQuest [618]

Generations sanpete / Manti High School (UT) – v1 n1-n7 [1978 summer-1982] – 1r – 1 – mf#1224279 – us WHS [373]

Genes & development – Cold Spring Harbor NY 1990+ – 1,5,9 – ISSN: 0890-9369 – mf#18232 – us UMI ProQuest [575]

La genese : traduction d'apres l'hebreu, avec distinction des elements constitutifs du texte, suivie d'un essai de restitution des livres primitifs dont s'est servi le dernier redacteur / Lenormant, Francois – Paris: Maisonneuve, 1883 – 1mf – 9 – 0-524-04786-3 – mf#1992-0206 – us ATLA [221]

La genese des mythes / Krappe, Alexander Haggerty – Paris: Payot 1938 [mf ed Bloomington IN: Indiana Uni Lib, Preservation Dept 1984] – 1r – 1 – us Indiana Preservation [390]

Genese social da gente bandeirante / Ferreira, Tito Livio – Sao Paulo, Brazil. 1944 – 1r – us UF Libraries [972]

Genese und kritik des subjektbegriffs : zur selbstthematisierung der menschen als subjekte / Guttandin, Friedhelm – (mf ed 1997) – 4mf – 9 – €56.00 – 3-8267-2480-1 – mf#DHS 2480 – gw Frankfurter [120]

Genesee farmer – La Salle IL 1840-65 – 1 – mf#3783 – us UMI ProQuest [630]

Genesee farmer and gardener's journal – La Salle IL 1831-39 – 1 – mf#3985 – us UMI ProQuest [634]

Genesee Synod. (Pres. Church in the USA) *see* Minutes, 1821-1870

Genesee Valley. Presbytery. (Pres. Church in the USA) *see* Minutes, 1858-1886

La genesi : con discussioni critiche / Minocchi, Salvatore – Firenze: Biblioteca scientifico-religiosa, 1908 – 1mf – 9 – 0-8370-4445-6 – mf#1985-2445 – us ATLA [221]

Die genesis / Delitzsch, Franz – Leipzig: Doerffling und Franke, 1852 – 1mf – 9 – 0-8370-9374-0 – mf#1986-3374 – us ATLA [221]

Die genesis / Dillmann, August – 4. aufl. Leipzig: S Hirzel, 1882 – 2mf – 9 – 0-8370-9461-5 – mf#1986-3461 – us ATLA [221]

Die genesis / Hoberg, Gottfried – 2. verm und verb aufl. Freiburg i B: Herder, 1908 – (= ser Exegetisches Handbuch zum Pentateuch) – 2mf – 9 – 0-524-08081-X – mf#1992-1141 – us ATLA [221]

Genesis = Genesis / Dillmann, August – Edinburgh: T & T Clark, 1897 – 3mf – 9 – 0-8370-2916-3 – (includes subject and lexical indexes. in english) – mf#1985-0916 – us ATLA [221]

Genesis / Gunkel, Hermann – 2. verb aufl. Goettingen: Vandenhoeck und Ruprecht, 1902 – 2mf – 9 – 0-8370-9476-3 – (incl ind) – mf#1986-3476 – us ATLA [221]

Genesis / Holzinger, Heinrich – Freiburg i.B: J C B Mohr, 1898 – 1mf – 9 – 0-8370-3647-X – (includes appendix) – mf#1985-1647 – us ATLA [221]

Genesis / Jaume, Adela – Habana, Cuba. 1954 – 1r – us UF Libraries [972]

Genesis / ed by Kalisch, Marcus Moritz – english ed. London: Longman, Brown, Green, Longmans, and Roberts 1858 [mf ed 1989] – (= ser A historical and critical commentary on the old testament) – 2mf – 9 – 0-7905-2417-1 – mf#1987-2417 – us ATLA [221]

Genesis / Mitchell, Hinckley G T – New York: Macmillan, 1909 – (= ser The bible for home and school) – 1mf – 9 – 0-8370-4454-5 – (incl ind and appendixes) – mf#1985-2454 – us ATLA [221]

Genesis : or, the first book of moses: together with a general theological and homiletical introduction to the old testament / Lange, Johann Peter – New York: Charles Scribner, 1884, c1868 [mf ed 1985] – (= ser A commentary on the holy scriptures. old testament 1) – 2mf – 9 – 0-8370-5630-6 – (english trans by tayler lewis & abraham gosman. incl bibl) – mf#1985-3630 – us ATLA [221]

Genesis / Weidner, Revere Franklin – Chicago: Fleming H Revell, c1892 – 1mf – 9 – 0-524-04418-X – mf#1992-0111 – us ATLA [221]

Genesis 2 – Cherry Plain NY 1979-80 – 1 – ISSN: 0016-6669 – mf#12031 – us UMI ProQuest [320]

Genesis and exodus / Simpson, Albert B – New York: Word, Work & World Pub Co, 1888 [mf ed 1992] – 1mf – 9 – 0-524-02147-3 – mf#1990-4213 – us ATLA [221]

Genesis and exodus / Terry, Milton S & Newhall, Fales H – New York: Hunt & Eaton; Cincinnati: Cranston & Stowe c1889 [mf ed 1990] – (= ser Commentary on the old testament 1) – 2mf [ill] – 9 – 0-8370-1655-X – mf#1987-6085 – us ATLA [221]

Genesis and geology : or, an investigation into the reconciliation of the modern doctrines of geology with the declarations of scripture / Crofton, Denis – Boston: Phillips, Sampson, 1853 [mf ed 1986] – 1mf – 9 – 0-8370-9928-5 – mf#1986-3928 – us ATLA [221]

Genesis and geology : a plea for the doctrine of evolution: being a sermon preached nov 5th 1871... / Henslow, George – London, 1871 – (= ser 19th c evolution & creation) – 1mf – 9 – mf#1.1.11604 – uk Chadwyck [210]

The genesis and growth of religion : the l.p. stone lectures for 1892, at princeton theological seminary, new jersey / Kellogg, S H – New York: Macmillan, 1892 – 1mf – 9 – 0-7905-1123-1 – (incl bibl ref) – mf#1987-1123 – us ATLA [210]

Genesis and its authorship : two dissertations / Quarry, John – rev ed. London: Williams and Norgate, 1873 – 2mf – 9 – 0-7905-1374-9 – (incl bibl ref) – mf#1987-1374 – us ATLA [221]

Genesis and modern science / Perce, Warren Raymond – New York: J. Pott, 1897. 362p. illus., plates, diagrs – 1 – us Wisconsin U Libr [500]

Genesis and prophets / Palfrey, John Gorham – Boston: James Monroe, 1840-52 [mf ed 1989] – (= ser Academical lectures on the jewish scriptures and antiquities 2-3) – 2v on 3mf – 9 – 0-7905-2258-6 – mf#1987-2258 – us ATLA [221]

Genesis and science : or, the first leaves of the bible / Arnold, John Muehleisen – 2nd ed. London: Longmans, Green, 1875 – 1mf – 9 – 0-7905-3069-4 – mf#1987-3069 – us ATLA [221]

Genesis and semitic tradition / Davis, John D – New York: Charles Scribner, 1894 – 1mf – 9 – 0-8370-6105-9 – (incl bibl ref) – mf#1986-0105 – us ATLA [221]

Genesis de estado moderno en espana / Sanchez Bella, Ismael – Pamplona. 1956 – 1 – us CRL [946]

Genesis de la convencion dominico-americana / Troncoso De La Concha, M De J – Santiago, Dominican Republic. 1946 – 1r – us UF Libraries [972]

Genesis disclosed : being the discovery of a stupendous error which changes the entire nature of the account of the creation of mankind / Davies, Thomas Alfred – New York: G W Carleton, 1874 [mf ed 1985] – 1mf – 9 – 0-8370-2844-2 – mf#1985-0844 – us ATLA [221]

Genesis graece : e fide editionis sixtinae... / ed by Lagarde, Paul de – Lipsiae [Leipzig]: BG Teubneri, 1868 [mf ed 1989] – 1mf – 9 – 0-7905-1962-3 – (in greek. int in latin, greek & hebrew) – mf#1987-1962 – us ATLA [221]

The genesis of american anti-missionism / Carroll, Benajah Harvey – Louisville, KY: Baptist Book Concern, 1902 [mf ed 1990] – 1mf – 9 – 0-7905-5927-7 – (incl bibl ref) – mf#1988-1927 – us ATLA [242]

The genesis of american anti-missionism / Carroll, BH – 1902 – 1 – 8.26 – us Southern Baptist [242]

The genesis of california's first constitution, 1846-1849 / Rockwell et al – CA. 1846-49 – 1 – $50.00 – mf#B50525 – us Library Micro [323]

Genesis of churches in the united states of america, in newfoundland and the dominion of canada / Croil, James – Montreal: F Brown, 1907 – 4mf – 9 – 0-665-71277-4 – (incl ind) – mf#71277 – cn Canadiana [240]

The genesis of genesis : a study of the documentary sources of the first book of moses in accordance with the results of critical science / Bacon, Benjamin Wisner – Hartford: Student Pub Co, 1892, c1891 – 1mf – 9 – 0-8370-2141-3 – (incl bibl ref) – mf#1985-0141 – us ATLA [221]

Genesis of new port richey / Avery, Elroy Mckendree – New Port Richey, FL. 1924 – 1r – us UF Libraries [978]

The genesis of the chirala station of the guntur, india, mission of the evangelical lutheran church (general synod) in the united states of america / Harris, E C – Sterling, IL: Lutheran Brotherhood, 1918 – 1mf – 9 – 0-524-08374-6 – mf#1993-3074 – us ATLA [242]

Genesis of the federal judiciary system; address before virginia state bar association, august 2, 1904 / Richards, Walter Buck – Richmond: Everett Waddey, 1904. 36p LL-11217 – 1 – us L of C Photodup [340]

The genesis of the heavens and the earth and all the host of them / Dana, James Dwight – Hartford: Student Publ Co, 1890 – 1mf – 9 – 0-8370-2821-3 – mf#1985-0821 – us ATLA [220]

The genesis of the new england churches / Bacon, Leonard – New York: Harper 1874 [mf ed 1990] – 2mf – 9 – 0-7905-5020-2 – (incl bibl ref) – mf#1988-1020 – us ATLA [243]

Genesis of the social conscience : the relation between the establishment of christianity in europe and the social question / Nash, Henry Sylvester – New York: Macmillan, 1897 – 1mf – 9 – 0-7905-9042-5 – mf#1989-2267 – us ATLA [301]

Genesis of the united states / Brown Alexander – Boston, MA. v1-2. 1897 – 2r – us UF Libraries [025]

Genesis und keilschriftforschung : ein beitrag zum verstaendnis der biblischen ur- und patriarchengeschichte / Nikel, Johannes – Freiburg i B: Herder, 1903 – 1mf – 9 – 0-524-06850-X – mf#1992-0992 – us ATLA [221]

Genesius *see* Historiarum libri 8 (cshb22)

Genesius, G *see* Thesaurus philologicus criticus linguae hebraeae et chaldaeae v t

Genesius, J *see* De rebus constantinopolitanis libri 4 (cbh30,1)

Genest, J *see* Some accounts of the english stage, 1660-1830

Genet (Carpentras), Eleazar *see* Liber cantici magnificat

Genet, Edmond C *see* Papers

Genetic analysis, techniques and applications – Oxford, England 1990 – 1,5,9 – (cont: gene analysis techniques) – ISSN: 1050-3862 – mf#42450.03 – us UMI ProQuest [575]

Genetic counselling = Genetiese raadgewing / South Africa. Department of Health [Departement van Gesondheid] [Departement van Gesondheid] – Pretoria: Dept of Health Genetics Div [1975?] [mf ed Pretoria, RSA: State Library [199-]] – 12p [ill] on 1r with other items – 5 – (in english & afrikaans) – mf#op 06447 r23 – us CRL [362]

Genetic disorders and high-risk populations = Genetiese afwykings en hoerisikobevolkings – [Pretoria: Dept of Health 1975?] [mf ed Pretoria, RSA: State Library [199-]] – 10p on 1r with other items – 5 – (in english & afrikaans) – mf#op 06432 r23 – us CRL [616]

A genetic history of the new england theology / Foster, Frank Hugh – Chicago: The University of Chicago Press, 1907 [mf ed 1970] – (= ser Library of american civilization 11003) – xv/568p on 1mf – 9 – us Chicago U Pr [242]

A genetic history of the new england theology / Foster, Frank Hugh – Chicago: University of Chicago Press, 1907 [mf ed 1989] – 2mf – 9 – 0-7905-4572-1 – (incl bibl ref) – mf#1988-0572 – us ATLA [240]

Genetic history of the problems of philosophy / Banerjee, Muraly Dhar – Calcutta: University of Calcutta, 1935 – (= ser Samp: indian books) – (developed & completed by hiranmay banerjee) – us CRL [100]

Genetic resources and crop evolution – Dordrecht, Netherlands 1995+ – 1,5,9 – ISSN: 0925-9864 – mf#18656 – us UMI ProQuest [580]

Genetic, social, and general psychology monographs – Washington DC 1985+ – 1,5,9 – ISSN: 8756-7547 – mf#16463,01 – us UMI ProQuest [150]

Genetic theory of reality : being the outcome of genetic logic as issuing in the aesthetic theory of reality aalled pancalism / Baldwin, James Mark – New York: GP Putnam, 1915 – (= ser The Science Series) – 1mf – 9 – 0-7905-3529-7 – (incl bibl ref) – mf#1989-0022 – us ATLA [100]

Genetic theory of reality : being the outcome of genetic logic as issuing in the aesthetic theory of reality called pancalism; with an extended glossary of terms / Baldwin, James Mark – New York: G.P. Putnam, 1915 – 1mf – us ATLA [160]

Genetica – Dordrecht, Netherlands 1989+ – 1,5,9 – ISSN: 0016-6707 – mf#16788 – us UMI ProQuest [575]

Genetica iberica – Madrid, Spain 1974-80 – 1,5,9 – ISSN: 0016-6693 – mf#8749 – us UMI ProQuest [575]

Genetical research – Cambridge, England 1960+ – 1,5,9 – ISSN: 0016-6723 – mf#2847 – us UMI ProQuest [590]

Genetics – Bethesda MD 1916+ – 1,5,9 – ISSN: 0016-6731 – mf#12982 – us UMI ProQuest [575]

Genetics / White, Edith Grace – New York, NY. 1962 – 1r – us UF Libraries [575]

Genetics in plant and animal improvement / Jones, Donald Forsha – New York, NY. 1925 – 1r – us UF Libraries [575]

Genetic-speculative philosophy of religion = Genetisch-spekulative religionsphilosophie / Pfleiderer, Otto – London: Williams and Norgate, 1888 – (= ser Religionsphilosophie auf geschichtlicher Grundlage) – 1mf – 9 – 0-7905-8719-X – (incl bibl ref. in english) – mf#1989-1944 – us ATLA [200]

Die genetische anlage des christlichen lebens / Beck, Johann Tobias; ed by Lindenmeyer, Julius – Guetersloh: Bertelsmann, 1882 – (= ser Vorlesungen ueber christliche Ethik) – 1mf – 9 – 0-7905-3538-6 – mf#1989-0031 – us ATLA [240]

Genetische entwickelung der vornehmsten gnostischen systeme / Neander, August – Berlin: F Duemmler, 1818 – 1mf – 9 – 0-7905-7060-2 – mf#1988-3060 – us ATLA [290]

Die genetische entwicklung der sog ordines minores in den drei ersten jahrhunderten / Wieland, F – Rom, 1897 – €12.00 – ne Slangenburg [240]

Geneva : a monthly review of world affairs – 1930-36 – (= ser The library of world peace studies) – 12mf – 9 – $105.00 – (formerly: league of nations in review, jun 1930-31; league of nations, mar-may 1930) – us UPA [321]

Geneva express – Lake Geneva WI. 1856 mar 29-aug 9, oct 25, nov 1 – 1r – 1 – (cont: geneva weekly express; cont by: elkhorn independent [elkhorn, wi: 1855]) – mf#929714 – us WHS [071]

Geneva gazette – Geneva, NE: Edith M Pray. 30v. -v16 n25. dec 29 1899; 19th yr n26. jan 5 1900-30th yr n32. apr 20 1911 (wkly) [mf ed v12 n25. nov 15 1895-1911 (gaps)] – 8r – 1 – (merged with: nebraska signal and the grafton sun and the exeter enterprise to form: nebraska signal and the geneva gazette) – us NE Hist [071]

Geneva journal – Geneva, NE: J A Loudermilch, 1885-feb 1894// (wkly) [mf ed v5 n7. nov 2 1889] – 1r – 1 – (merged with: fillmore county republican to form: republican-journal) – us NE Hist [071]

Geneva lake herald – Lake Geneva WI. 1872 mar 9, apr 20-1873 apr 19, also nov 22, 1873 suppl, 1874 jan 10-1876 dec 20, 1877 jan 6-1879 may 17 – 3r – 1 – (continued by: lake geneva herald (lake geneva, wi: 1879)) – mf#929790 – us WHS [071]

Geneva lake mirror – Lake Geneva WI. 1860 feb 9-1861 feb 14, 1861 feb 7, mar 14 – 2r – 1 – mf#929717 – us WHS [071]

Geneva papers on risk and insurance. issues and practice – Oxford, England 1981+ – 1,5,9 – ISSN: 1018-5895 – mf#12881 – us UMI ProQuest [368]

Geneva, past and present : an historical and descriptive guide for the use of foreign visitors in geneva / Doumergue, Emile – Geneva: Atar, [1909?] – 1mf – 9 – 0-7905-5385-6 – mf#1988-1385 – us ATLA [915]

Geneva. Presbytery (Pres. Church in the USA) *see* Minutes

Geneva review – Geneva, NE: A T and E J Scott. v8 n11. jan 4 1883-jun 1889// (wkly) [mf ed 1883-85 (gaps)] – 2r – 1 – (cont: fillmore county review. cont by: fillmore county republican) – us NE Hist [071]

Geneva. Synod (Pres. Church in the USA) *see* Minutes, 1812-1881

Geneva weekly express – Lake Geneva WI. 1855 oct 20-1856 mar 15 – 1r – 1 – (continued by: geneva express) – mf#929711 – us WHS [071]

Genevan – Lake Geneva WI. v1 n22 1858 oct 28 – 1r – 1 – mf#932228 – us WHS [071]

Geneve plagiaire... / Coton, P – Paris, 1618 – 21mf – 9 – mf#CA-121 – ne IDC [240]

Genevieve : ou, la jalousie paternelle / Scribe, Eugene – Paris, France. 1846 – 1r – us UF Libraries [440]

Geng, Lizhong *see*
– A perspective of sport consumer behavior in the people's republic of china
– Sports sponsorship in china

Gengangaren – Vadstena, Sweden. 1844-54 – 1 – sw Kungliga [078]

Genie du christianisme : ou beautes de la religion chretienne / Chateaubriand, Francois-Rene, vicomte de – 7e ed. Paris: Le Normant. 1823 [mf ed 1985] – 5v on 1mf – 9 – 0-665-50538-8 – mf#50538 – cn Canadiana [240]

Genie francais et l'ame haitienne / Coicou, Massillon – Paris, France. 1904 – 1r – us UF Libraries [972]

Genie und charakter / Ludwig, Emil – Berlin, Germany. 1927, c1924 – 1r – us UF Libraries [025]

Genin, Francois *see* Les jesuites et l'universite

El genio literario de extremadura : apuntes de literatura regional / Lopez Prudencio, Jose – Badajoz: imp vicente rodriguez, 1912 – 1 – sp Bibl Santa Ana [440]

Genio y figura / Corretjer, Juan Antonio – Guaynabo, Puerto Rico. 1961 – 1r – us UF Libraries [972]

Genitourinary medicine – London, England 1985-97 – 1,5,9 – (cont: british journal of venereal diseases; cont by: sexually transmitted infections) – ISSN: 0266-4348 – mf#1354.02 – us UMI ProQuest [616]

Genius – Munich, 1919-21 [mf ed Chadwyck-Healey] – (= ser Art periodicals on microform) – 1r + 23 col slides – 1 – uk Chadwyck [700]

The genius and character of emerson : lectures at the concord school of philosophy / ed by Sanborn, Franklin Benjamin – Boston: James R Osgood, 1885 – 2mf – 9 – 0-524-00317-3 – mf#1989-3017 – us ATLA [420]

The genius and mission of methodism : embracing what is peculiar in doctrine, government, modes of worship, etc / Strickland, William Peter – Boston: C.H. Peirce, 1851 – 1mf – 9 – 0-7905-6090-9 – mf#1988-2090 – us ATLA [242]

The genius and mission of the protestant episcopal church in the united states / Colton, Calvin – New York: Stanford & Swords, 1853 – 1mf – 9 – 0-7905-4727-9 – mf#1988-0727 – us ATLA [242]

Der genius der zeit : ein journal / ed by Hennings, August – Altona 1794-1800 – (= ser Dz. historisch-politische abt) – 21v on 83mf – 9 – €830.00 – mf#k/n1732 – gw Olms [074]

Der genius des neunzehnten jahrhunderts / ed by Hennings, August – Altona 1801-02 – (= ser Dz. historisch-politische abt) – 6v on 18mf – 9 – €180.00 – mf#k/n1790 – gw Olms [074]

Genius im wort : von deutschem dichten und denken / Benz, Richard – Jena: E Diederichs, 1941, c1936 [mf ed 1989] – (= ser Deutsche reihe 42) – 78/[2]p – 1 – mf#7006 – us Wisconsin U Libr [430]

Genius of christianity / Furness, William Henry – London, England. 1830 – 1r – us UF Libraries [240]

The genius of christianity, or, the spirit and beauty of the christian religion = Genie du christianisme / Chateaubriand, Francois-Rene – 2nd rev. ed. Baltimore: John Murray, 1856 – 2mf – 9 – 0-7905-4546-2 – (in english) – mf#1988-0546 – us ATLA [240]

The genius of israel : a reading of hebrew scriptures prior to the exile / Noyes, Carleton Eldredge – Boston: Houghton Mifflin, 1924 (mf ed 1995) – 1r – 1 – mf#ZZ-34512 – us NY Public [939]

Genius of liberty – Fredericksburg VA. 1799 apr 12, may 3 – 1r – 1 – (cont: genius of liberty; and fredericksburg & falmouth advertiser; cont by: courier [fredericksburg, va: 1800]) – mf#885331 – us WHS [071]

Genius of liberty, and fredericksburg and falmouth – Fredericksburg VA. 1798 nov 13 – 1r – 1 – (continued by: genius of liberty (fredericksburg, va: 179?)) – mf#885329 – us WHS [071]

The genius of protestantism : a book for the times / Edgar, Robert McCheyne – 2nd ed. Edinburgh: Oliphant, Anderson & Ferrier, 1900 – 1mf – 9 – 0-8370-8666-3 – (incl bibl ref and index) – mf#1986-2666 – us ATLA [242]

The genius of shakespeare : and other essays / Osborne, William Frederick – Toronto: W Briggs, 1908 – 2mf – 9 – 0-659-91290-2 – mf#9-91290 – cn Canadiana [420]

The genius of the roman rite : being a paper. read at the meeting of the historical research society at archbishop's house, westminster... / Bishop, Edmund – 2nd ed. London: FE Robinson, 1902 – 1mf – 9 – 0-524-04037-0 – mf#1990-4945 – us ATLA [241]

Genius of universal emancipation – La Salle IL 1821-39 – 1 – mf#4458 – us UMI ProQuest [320]

Genkin, D M *see*
– Khozraschet v promkooperatsii
– Novyi zakon o promyslovoi kooperatsii 11 maia 1927 g
– Sbornik postanovlenii o promyslovoi kooperatsii i kustarnoi promyshlennosti
– Zakonodatelstvo o promyslovoi kreditnoi kooperatsii

Genlis, Stephanie F de *see* Dictionnaire critique et raisonne des etiquettes de la cour, des usages du monde, des amusemens, des modes, des moeurs, etc, des francois

Gennadi, G *see* Spisok knig o russkikh monastyriakh i tservakh

Gennadius *see* Contre les doutes de plethon sur aristote

Gennadius und pletho : aristotelismus und platonismus in der griechischen kirche / Gass, Wilhelm – Breslau: A Gosohorsky, 1844 [mf ed 1990] – 2v on 1mf – 9 – 0-7905-6702-4 – (greek texts in v2. incl bibl ref) – mf#1988-2702 – us ATLA [243]

Gennep, Arnold van *see* Le folklore

Gennrich, Paul *see*
– Der kampf um die schrift in der deutsch-evangelischen kirche des neunzehnten jahrhunderts
– Die lehre von der wiedergeburt

Gennuso, Giuseppo *see* La question siciliana

Gennuso, Guiseppe *see* La questione siciliana

The genoa banner – Genoa, NE: Banner Print Co, 1892-95// (wkly) [mf ed v1 n27. jul 28 1892] – 1r – 1 – us NE Hist [071]

Genoa city broadcaster – Genoa City, Lake Geneva WI. 1944 jan 20-1945, 1946-48, 1949-51, 1952-1954 may 6 – 4r – 1 – (continued by: lake geneva broadcaster) – mf#929913 – us WHS [071]

Genoa city marquee – Genoa City WI. 1955 jun 9-oct 13 – 1r – 1 – (cont: marquee [genoa city, wi: 1955]) – mf#918088 – us WHS [071]

[Genoa-] courier – NV. 1881-99 – 7r – 1 – $420.00 – (aka: weekly courier; cont by: courier, gardnerville) – mf#U04539 – us Library Micro [071]

[Genoa-] douglas county banner – NV. oct-dec 1865 [wkly] – 1r – 1 – $60.00 – mf#U04537 – us Library Micro [071]

[Genoa-] genoa enterprise – NV. 1992- – 2r – 1 – $120.00 (subs $50y) – mf#U04844 – us Library Micro [071]

Genoa junctionjournal – Genoa Junction WI. 1895 feb 22-sep 4 – 1r – 1 – (continued by: lake geneva news (lake geneva, wi: 1879)) – mf#918286 – us WHS [071]

Genoa leader – Genoa, NE: F H Young, 1881-v48 n1. jun 25 1926 (wkly) [mf ed 1895-1906 (gaps)] – 9r – 1 – (cont: genoa banner. merged with: genoa times to form: genoa leader times) – us NE Hist [071]

Genoa leader times – Genoa, NE: J E Tesarek. v48 n2. jul 2 1926- (wkly) [mf ed with gaps] – 1 – (formed by the union of: genoa leader and: genoa times. publ as: genoa leader-times jul 16 1926- . cont the numbering of: genoa leader. some irregularities in numbering. 84th year not publ) – us NE Hist [071]

[Genoa-] the nevada prohibitionist – NV. 1 apr 1889 (only known issue) – 1r – 1 – $60.00 – mf#U04538 – us Library Micro [071]

Genoa times – Genoa, NE: C J Stockwell. 25v. v1 n1. jan 17 1902-v25 n4. jun 25 1926 (wkly) [mf ed with gaps] – 13r – 1 – (merged with: genoa leader to form: genoa leader times) – us NE Hist [071]

Genome – Ottawa, Canada 1989+ – 1,5,9 – ISSN: 0831-2796 – mf#17209,01 – us UMI ProQuest [575]

Genossenschaft Deutscher Buehnen-Angehoeriger *see*
– Gettke's buehnen-almanach
– Neuer theater-almanach fuer das jahr...

Der genossenschaftsbauer – Berlin DE, 1961 6 jan-1964 25 dec – 8r – 1 – (filmed by misc inst: 1955 Aug-1960 29 oct, 1961-90 [49r]; title varies: 4 nov 1960: neue deutsche bauernzeitung) – uk British Libr Newspaper; gw Misc Inst [630]

Die genossin – Berlin, Hannover DE, 1924-27, 1928 [gaps], 1929-1931 oct, 1932-33 – 2r – 1 – gw Misc Inst [074]

Genoude, Antoine E de *see* Voyage dans la vendee et dans le midi de la france

Genouillac, Henri de *see* L'eglise chretienne au temps de saint ignace d'antioche

Genouy, Oswald *see* Vers la religion eternelle

Genovefa : trauerspiel in 3 aufzuegen / Wesendonk, Mathilde – Zuerich: Buerkli 1866 – 2mf – 9 – mf#mw-10 – ne IDC [820]

Genoveva : eine tragoedie in fuenf acten / Hebbel, Friedrich – Hamburg: Hoffmann und Campe, 1843 [mf ed 1995] – 234p – 1 – (in verse) – mf#8764 – us Wisconsin U Libr [820]

Genrebilder / Ebner-Eschenbach, Marie von – Leipzig: H Fikentscher, H Schmidt & H Guenther, [1928] – 2r – 1 – us Wisconsin U Libr [430]

Gens, G F *see* Nachrichten ueber chiwa, buchara, chokand, und den nordwestlichen theil des chinesischen staates...

Gense, J H *see* The gaikwads of baroda

Gensel, Reinhold *see* Gutzkows werke

Gensichen, Otto Franz *see* Das haideroeslein von sesenheim

Gent, George William *see* Fraternity

Gent, R A *see* Sermon on the death of the duke of wellington

"Genta" / Madjalah taman sari – Pontianak, 1964 – 1mf – 9 – mf#SE-919 – ne IDC [959]

Genta – Solo, 1964-1965 – 26mf – 9 – mf#SE-885 – ne IDC [950]

Genta Islam *see* Madjlis ulama dst i djawa barat

Genta kedjaksaan *see* Persatuan djaksa-djaksa

Genta massa *see* Jajasan serba/guna

Genta pemuda – Djakarta, 1964-1965 – 3mf – 9 – mf#SE-721 – ne IDC [959]

Gente / Associated Students of UCLA – v13 n1-v16 n3 [1982 oct-1985 feb] – 1r – 1 – (cont: gente de aztlan [los angeles, ca: 1973]; cont by: gente de aztlan [los angeles, ca: 1986]) – mf#1278054 – us WHS [378]

Gente / Associated Students of UCLA – v1 n6 [1971 may 31], v2 n3 [1972 feb 7], v3 n1-2 [1972 oct 3-31] – 1r – 1 – (continued by: gente de aztlan (los angeles, ca: 1973)) – mf#1269282 – us WHS [378]

Gente de aztlan / Associated Students of UCLA – v6 n5-v12 n2 [1976 may-1981 nov] – 1r – 1 – (cont: gente [los angeles, ca: 1971]; cont by: gente [los angeles, ca: 1982]) – mf#615814 – us WHS [378]

Gente de playa giron / Gonzalez De Cascorro, Raul – Habana, Cuba. 1962 – 1r – us UF Libraries [972]

Gente de portal / Roman, Miguel Alberto – Ciudad Trujillo, Dominican Republic. 1954 – 1r – us UF Libraries [972]

Gente nostra – Rome, Italy. 1 jan 1933-1 dec 1935 – 1 – mf#m.f.829 – uk British Libr Newspaper [074]

Gente nueva – Montijo, 1920 – 5 – sp Bibl Santa Ana [073]

Gentes herbarum – Ithaca NY 1920-84 – 1,5,9 – ISSN: 0072-0879 – mf#2687 – us UMI ProQuest [580]

Gentil, Emile *see* La chute de l'empire de rabah

Gentil, Robert *see*
– Grande geographie de l'ile d'haiti
– Haiti a l'exposition colombienne de chicago

The gentile and the jew in the courts of the temple of christ : an introduction to the history of christianity = Heidenthum und judenthum / Doellinger, Johann Joseph Ignaz von – London: Longman, Green, Longman, Roberts, and Green, 1862 – 3mf – 9 – 0-7905-4626-4 – (incl bibl ref. in english) – mf#1988-0626 – us ATLA [240]

Gentile, Dina *see* The perceptions of sport management students toward the market ability of professional athletes

Gentile, Giovanni *see* Il modernismo

The gentile nations : or, the history and religion of the egyptians, assyrians, babylonians, medes, persians, greeks, and romans / Smith, George – New York: Carlton & Phillips, 1855 [mf ed 1992] – (= ser Sacred annals 3) – 2mf – 9 – 0-524-03993-3 – mf#1992-0036 – us ATLA [221]

Gentili, Tommaso Maria *see* Memorie di un missionaerio domenicano nella cina

Gentilism : religion previous to christianity / Thebaud, Augustus J – New York: D and J Sadlier, 1876 – 1mf – 9 – 0-524-01310-1 – mf#1990-2346 – us ATLA [290]

The gentle art of making enemies : [correspondence published by j a m whistler] / Whistler, James (Abbott) McNeill – [new ed]. London 1892 – (= ser 19th c art & architecture) – 4mf – 9 – mf#4.1.355 – uk Chadwyck [860]

The gentle reader / Crothers, Samuel McChord – Boston: Houghton, Mifflin, 1904 – 1mf – 9 – 0-524-07812-2 – mf#1991-3359 – us ATLA [410]

The gentle skeptic : or, essays and conversations of a country justice on the authenticity and truthfulness of the old testament records / ed by Walworth, Clarence Augustus – 2nd rev ed. New York: D Appleton, 1863 – 1mf – 9 – 0-8370-7435-5 – mf#1986-1435 – us ATLA [220]

Gentleman dick of the greys : and other poems / Cockin, Hereward Kirby – Toronto: C B Robinson, 1889 – 2mf – 9 – mf#32224 – cn Canadiana [810]

Gentleman from everywhere / Foss, James Henry – Boston, MA. 1902 – 1r – us UF Libraries [978]

A gentleman of leisure : a novel / Fawcett, Edgar – Toronto: Rose-Belford, 1881 [mf ed 1993] – 3mf – 9 – 0-665-91452-0 – mf#91452 – cn Canadiana [830]

Gentleman-adventurer, botefuhr / Scoville, Dorothy R – s.l, s.l? 1936 – 1r – us UF Libraries [978]

The gentleman's house / Kerr, Robert – 3rd ed. London 1871 – (= ser 19th c art & architecture) – 7mf – 9 – mf#4.2.568 – uk Chadwyck [720]

Gentleman's journal : or, the monthly miscellany – La Salle IL 1692-94 – 1 – mf#4256 – us UMI ProQuest [071]

Gentleman's journal for the war : being an historical account and geographical description of several strong cities, towns and ports of europe – La Salle IL 1693-94 – 1 – mf#4257 – us UMI ProQuest [073]

Gentleman's magazine, index to the... 1731-1818 – 1r – 1 – mf#96589 – uk Microform Academic [073]

Gentleman's vade-mecum : or the sporting and dramatic companion – La Salle IL 1836 – 1 – mf#5557 – us UMI ProQuest [790]

Gentlemans vode mecum – Philadelphia, PA., 1827-1836 – 13 – $25.00r – us IMR [071]
Gentlemen and ladies' town and country magazine – La Salle IL 1789-90 – 1 – mf#4459 – us UMI ProQuest [640]
The gentlemens amusement / Shaw, R & Carr, B – A Selection of solos, duetts, overtures, arranged as duetts, rondos & romances. Philadelphia: B. Carr 1794-96. MUSIC 123, Item 14 – 1 – us L of C Photodup [780]
The gentlemen's book of etiquette; and manuel of politeness. / Hartley, Cecil B – Boston: DeWolfe, Fiske & Co., 1873. 332p – 1 – us Wisconsin U Libr [390]
Gentlemen's fancy dress : how to choose it / Holt, Ardern – [4th ed?] London [1898?] – (= ser 19th c art & architecture) – 2mf – 9 – mf#4.2.168 – uk Chadwyck [740]
The gentlemen's handbook on poker / Florence, William James – New York: G. Routledge, 1892. xi,195p. illus – 1 – us Wisconsin U Libr [790]
Gentry – 1983 jan – 1r – 1 – mf#4852842 – us WHS [929]
Gentry, Deborah Barnes *see* Journal of teaching in marriage and family
Gentry family gazette and genealogy exchange – n22-23 [1983 oct-dec], n24-46 [1984 feb-1990 apr], indexes v4-5 – 1r – 1 – (cont: gentry family gazette and genealogy) – mf#1712345 – us WHS [929]
Gentry, Grier B *see* Coaching motivation and efficiency
Genty, Louis *see* L'influence de la decouverte de l'amerique sur le bonheur du genre-humain
Gentz, Friedrich von *see*
– Historisches journal
– Neue deutsche monatsschrift
Gentz, Friedrich von et al *see* Deutsche monatsschrift
Gentz, G *see* Die kirchengeschichte der nicephorus callistus xanthopupos und ihre quellen
Gentz, Leslie M E *see* An offensive seasonal analysis of girl's high school fast-pitch softball in lowa and Michigan (1994-1998)
Genuina labor periodistica de enrique jose varona / Entralgo, Elias Jose – Habana, Cuba. 1949 – 1r – us UF Libraries [972]
Genuinae relationes inter sedem apostolicam et assyriorum orientalium seu chaldaeorum ecclesiam : nunc majori ex parte primum editae historicisque adnotationibus illustratae – Roma: Ermanno Loescher, 1902 – 2mf – 9 – 0-8370-7865-2 – (incl bibl ref and index) – mf#1986-1865 – us ATLA [240]
Der genuine ablauf der motivik im musical the new starlight express / Petri, Hasso Gottfried – (mf ed 2001) – 158p 2mf – 9 – €40.00 – 3-8267-2768-1 – mf#DHS 2768 – gw Frankfurter [780]
A genuine account of nova scotia : containing a description of its situation, air, climate, soil and its produce... – [Dublin]: London printed: and, Dublin, repr for Philip Bowes...1750 [mf ed 1984] – 1mf – 9 – 0-665-44277-7 – mf#44277 – cn Canadiana [917]
Genuine character of the gospel stated and illustrated / Thomson, Andrew – Edinburgh, Scotland. 1815 – 1r – us UF Libraries [220]
The genuineness and authorship of the pastoral epistles / James, J D – London: Longmans, Green, 1906 – 1mf – 9 – 0-8370-3763-8 – (incl ind) – mf#1985-1763 – us ATLA [227]
The genuineness, authenticity, and inspiration of the word of god / Greenfield, William – New York: R Carter, 1853 – 1mf – 9 – 0-524-08077-1 – mf#1992-1137 – us ATLA [220]
Genung, George Frederick *see*
– Book of leviticus
– Book of numbers
– The fourfold story
– The magna charta of the kingdom of god
Genung, John Franklin *see*
– Ecclesiastes
– The epic of the inner life
– The hebrew literature of wisdom in the light of to-day
Genzel', P P *see* Finansovaia reforma v rossii
Genzmer, Felix et al *see* Geschichte der deutschen literatur
Geo batten and co's directory of the religious press of the united states : a list of nearly all religious periodicals with their denomination or class, frequency of issue, number of pages, size of pages, whether illustrated, subscription price, circulation, distribution, editor and publisher – 3rd ed. New York: Geo Batten, 1897 – 1mf – 9 – 0-7905-4129-7 – mf#1988-0129 – us ATLA [070]
Geo info systems – Cleveland OH 1997+ – 1,5,9 – (cont by: geospatial solutions) – ISSN: 1051-9858 – mf#19720.01 – us UMI ProQuest [000]
Geo p rowell and co's american newspaper directory – 11th n4 [1879 oct], 14th [1881] – 2r – 1 – (continued by: american newspaper directory (new york, ny: 1886)) – mf#1727553 – us WHS [070]
Geoarchaeology – Hoboken NY 1986+ – 1,5,9 – ISSN: 0883-6353 – mf#18106 – us UMI ProQuest [930]
Geobyte – Tulsa OK 1985-93 – 1,5,9 – ISSN: 0885-6362 – mf#16050 – us UMI ProQuest [550]
Geochemistry international – Hoboken NJ 1964-96 – 1,5,9 – ISSN: 0016-7029 – mf#14352 – us UMI ProQuest [550]
Geochimica et cosmochimica acta – Oxford, England 1950+ – 1,5,9 – ISSN: 0016-7037 – mf#49076 – us UMI ProQuest [550]
Geode / Wisconsin State College and Institute of Technology – v37 n4-5 [1961 oct 30-dec 18] – 1r – 1 – (cont: w i t geode) – mf#1885214 – us WHS [378]
Geoderma – Oxford, England 1967+ – 1,5,9 – ISSN: 0016-7061 – mf#42078 – us UMI ProQuest [630]
Geodesy, mapping and photogrammetry – v1-3, 1959-61 bimonthly as Geodesy and Cartography; v4-14, 1962-72 bimonthly as Geodesy and Aerophotography. Quarterly under present title. In English – 1,5,6 – (v1-3 1959-61 $50.00 set. v4-15 1962-73 $20.00y. v16 1974 $30.00. v17 1975 $30.00. v18 1976 $35.00. v19 1977 $50.00. v20 1978 $40.00) – us AGU [550]
Geodinamica acta – Oxford, England 2001+ – 1,5,9 – ISSN: 0985-3111 – mf#16580 – us UMI ProQuest [550]
Geoexploration – Oxford, England 1963-91 – 1,5,9 – (cont by: journal of applied geophysics) – ISSN: 0016-7142 – mf#42079.01 – us UMI ProQuest [622]
Geoffrey chaucer of england / Chute, Marchette Gaylord – New York, NY. 1946 – 1r – us UF Libraries [420]
Geoffrey stirling : a novel / Laffan, Bertha Jane (Grundy) – London: Chapman & Hall Ltd. 3v. 1883 – (= ser 19th c women writers) – 12mf – 9 – mf#5.1.43 – uk Chadwyck [810]
Geoffrion, Louis Philippe *see* Reglement annote de l'assemblee legislative
Geoffroy de Villeneuve, Rene *see* L'afrique
Geoffroy, Julien Louis *see* Manuel dramatique: a l'usage des auteurs et des acteurs, et necessaire aux gens du monde qui aiment les idees toutes trouvees, et les jugemens tout faits
Geoforum – Oxford, England 1970+ – 1,5,9 – ISSN: 0016-7185 – mf#49077 – us UMI ProQuest [900]
Geognosia. componentes de la corteza terrestre / Gil Calvo, Joaquin – Villafranca de los Barros (Badajoz): Graficas Crasferv, 1964 – 1 – sp Bibl Santa Ana [240]
Geognosy : or, the facts and principles of geology against theories / Lord, David Nevins – 2nd ed. New-York: Franklin Knight, 1857, c1855 – 2mf – 9 – 0-8370-9964-1 – (incl bibl ref) – mf#1986-3964 – us ATLA [220]
Geografia : guia y plan para su estudio, con aplicacion especial a la economia politica, tomo 1 / Beltran y Rozpide, Ricardo – Madrid: Razon y Fe, 1926 – 1 – sp Bibl Santa Ana [330]
Geografia antigua / Ramirez de Arellano, Rafael – Madrid: Fortanet, 1915. B.R.A.H. lxvi/pp. 110-115 – 1 – sp Bibl Santa Ana [900]
Geografia da fome / Castro, Josue De – Rio de Janeiro, Brazil. 1948 – 1r – us UF Libraries [972]
Geografia de bolivia y peru / Sievers, Wilhelm – Madrid: Razon y Fe, 1931 – 1 – (tambien de ecuador, colombia y venezuela) – sp Bibl Santa Ana [972]
Geografia de costa rica / Montero Barrantes, Francisco – Barcelona, Spain. 1892 – 1r – us UF Libraries [918]
Geografia de costa rica / Quiros Amador, Tulia – San Jose, Costa Rica. 1954 – 1r – us UF Libraries [918]
Geografia de costa rica / Trejos, Jose Francisco – San Jose, Costa Rica. 1937 – 1r – us UF Libraries [918]
Geografia de costa rica / Vincenzi, Moises – San Jose, Costa Rica. 1936 – 1r – us UF Libraries [918]
Geografia de costa rica para 1 ano / Urena Morales, Gabriel – San Jose, Costa Rica. 1965 – 1r – us UF Libraries [918]
Geografia de cuba / Marrero, Levi – Habana, Cuba. 1950 – 1r – us UF Libraries [918]
Geografia de cuba / Mestre Llano, Eloy – Habana, Cuba. 1948 – 1r – us UF Libraries [918]
Geografia de cuba / Nunez Jimenez, Antonio – Habana, Cuba. 1959 – 1r – us UF Libraries [918]
Geografia de cuba para uso de las escuelas / Aguayo, Alfredo Miguel – Habana, Cuba. 1928 – 1r – us UF Libraries [972]
Geografia de espana... / Martin Echevarria, L; ed by Bayle, Constantino – Madrid: Razon y Fe, 1928 – 9 – sp Bibl Santa Ana [914]
Geografia de la isla de cuba / Pichardo Y Tapia, Estaban – Habana, Cuba. 1854-55 – 1r – us UF Libraries [918]
Geografia de la isla de puerto rico / Asenjo, Conrado – San Juan, Puerto Rico. 1927 – 1r – us UF Libraries [918]
Geografia de las islas britanicas. barcelona, 1929 / Moscheles, J – Madrid: Razon y Fe, 1930 – 1 – sp Bibl Santa Ana [941]
Geografia de panama / Crespo, Jose D – Boston, MA. 1928 – 1r – us UF Libraries [918]
Geografia de santo domingo / Cucurullo, Oscar – Ciudad Trujillo, Dominican Republic. 1956 – 1r – us UF Libraries [918]
Geografia de suiza. barcelona, 1929 / Walser, H – Madrid: Razon y Fe, 1930 – 1 – sp Bibl Santa Ana [946]
Geografia de venezuela / Alvarez, Ramon – Caracas, Venezuela. 1962 – 1r – us UF Libraries [918]
Geografia de venezuela / Vila, Marco Aurelio – Caracas, Venezuela. 1956 – 1r – us UF Libraries [918]
Geografia de venezuela / Vila, Marco Aurelio – Caracas, Venezuela. 1961 – 1r – us UF Libraries [918]
Geografia de venezuela / Vila, Marco Aurelio – Caracas, Venezuela. 1962 – 1r – us UF Libraries [918]
Geografia de...badajoz / Franco y Lozano, Francisco – 1894 – 9 – sp Bibl Santa Ana [914]
Geografia de...badajoz / Munoz de Rivera, Antonio – 1894 – 9 – sp Bibl Santa Ana [914]
Geografia del arte en colombia, 1960 / Barney Cabrera, Eugenio – Bogota, Colombia. 1963 – 1r – us UF Libraries [700]
Geografia del atlantico / Escalante, Aquiles – Barranquilla, Colombia. 1961 – 1r – us UF Libraries [910]
Geografia del japon. barcelona, 1929 – Madrid: Razon y Fe, 1930 – 1 – sp Bibl Santa Ana [915]
Geografia del tachira / Vila, Marco Aurelio – Caracas, Venezuela. 1957 – 1r – us UF Libraries [972]
Geografia do acucar / Varzea, Afonso – Rio de Janeiro, Brazil. 1943 – 1r – us UF Libraries [972]
Geografia dos transportes no brasil / Silva, Moacir Malheiros Fernandes – Rio de Janeiro, Brazil. 1949 – 1r – us UF Libraries [380]
Geografia e historia de la republica dominicana / Inchaustegui Cabral, Joaquin Marino – Santiago, Dominican Republic. 1939 – 1r – us UF Libraries [972]
Geografia e historia del departamento del valle de... / Camacho Perea, Miguel – Cali, Colombia. 1964 – 1r – us UF Libraries [972]
Geografia economica / Schmidt, Walter; ed by Bayle, Constantino – Madrid: Razon y Fe, 1928 – 9 – sp Bibl Santa Ana [330]
Geografia elemental de la republica del salvador / Castro, Juan Francisco – San Salvador, El Salvador. 1905 – 1r – us UF Libraries [972]
Geografia espiritual / Massiani, Felipe – Caracas, Venezuela. 1949 – 1r – us UF Libraries [200]
Geografia fisica y de la republica de colombia / Botero M, Jose Manuel – Medellin, Colombia. 1939 – 1r – us UF Libraries [972]
Geografia fisica y de la republica de colombia / Botero M, Jose Manuel – Medellin, Colombia. 1960 – 1r – us UF Libraries [972]
Geografia fisica y economica de colombia / Arango Cano, Jesus – Bogota, Colombia. 1964 – 1r – us UF Libraries [972]
Geografia general nacionalista de la america del c... / Castillo, Jose Leon – Guatemala, 1924 – 1r – us UF Libraries [972]
Geografia guerrera de colombia / Riascos Grueso, Eduardo – Cali, Colombia. 1949 – 1r – us UF Libraries [972]
Geografia humana do brasil para o terceiro ano / Azevedo, Aroldo De – Sao Paulo, Brazil. no date – 1r – us UF Libraries [972]
Geografia ilustrada de costa rica / Trejos, Juan – San Jose, Costa Rica. 1941 – 1r – us UF Libraries [972]
Geografia ilustrada de el salvador, c a / Fonseca, Pedro S – Barcelona, Spain. 1926 – 1r – us UF Libraries [972]
Geografia politica. barcelona, 1929 / Dix, Arthur – Madrid: Razon y Fe, 1930 – 1 – sp Bibl Santa Ana [320]
Geografia...distinta in 12 libri : ne' quali; oltra l'esplicatione di molti luoghi di tolomeo, e della bussola, e dell' aguglia... / Sanuto, L – Vinegia, 1588 – 10mf – 9 – mf#H-8411 – ne IDC [956]
Geografica descripcion / Burgoa, Francisco de – Mexico: Talleres graficos de la nacion 1934 [mf ed 1981] – (= ser Publicaciones del archivo general de la nacion 25-26) – 2v on 1r – 1 – (incl repr of t.p. of original ed; sequel to aut's palestra historial) – mf#6097 – us Wisconsin U Libr [918]
Geografica descripcion de la parte septentrional del polo artico de la america : y nueva iglesia de las indias occidentales y sitio astronomico de esta provincia de predicadores de antequera, valle de oaxaca... / Burgoa, Francisco de – En Mexico: impr Iuan Ruyz, ano de 1674 – (= ser Books on religion...1543/44-c1800: historia ecclesiastical) – 10mf – 9 – mf#crl-163 – ne IDC [241]
Geografica descripcion de la parte septentrional del polo artico de la america y nueva iglesia de las indias occidentales...2 vol. mexico, 1934 / Burgoa, Francisco de – Madrid: Razon y Fe, 1936 – 1 – sp Bibl Santa Ana [970]
Geograficheskie izvestiia, vydavaemye ot russkogo geograficheskogo obshchestva – Spb., 1848-1850. 17 pts – 10mf – 9 – (missing: 1848(3); 1850(1-4)) – mf#1732 – ne IDC [077]
Geograficheskie karty rossii 15-19 stoletii – 1892 – 2mf – 8 – mf#R-7105 – ne IDC [947]
Geografichesko-statisticheskii slovar' amurskoi i primorskoi oblastei / Kirillov, A – Blagoveshchensk, 1894 – 10mf – 8 – mf#R-8394 – ne IDC [314]
Geografichesko-statisticheskii slovar' rossiiskoi imperii / Semenov, P P – Spb, 1863-1885. 5v – 114mf – 8 – mf#1279 – ne IDC [314]
Geografiese aspekte van rekreasie en vryetydsbesteding in bellville-suid / McPherson, Elsworth Adam – U of the Western Cape 1988 [mf ed S.l: s.n. between 1987 & 1991] – 3mf – 9 – (summary in english) – sa Misc Inst [306]
Geographia historica : tomo 7: de persia, del mogol, de la india; y sus reynos, de la china... / Velarde, P M – Madrid: Manuel de Moya, 1752 – 3mf – 9 – mf#HT-614 – ne IDC [915]
Geographia historica palestinae antiquae / Szczepanski, L – Romae, 1926 – 5mf – 9 – mf#H-2945 – ne IDC [956]
Geographic cutters / U.S. Library of Congress – 24mf – 9 – $50.00 – (lists of geographic cutters for u.s. cities, counties, and regions) – us L of C Photodup [020]
Geographic index to correspondence of the military intelligence division of the war department general staff, 1917-1941 / U.S. War Dept. Military Intelligence Division – (= ser Records of the war department general and special staffs) – 17r – 1 – (with printed guide) – mf#M1474 – us Nat Archives [355]
Geographica helvetica – Bern, Switzerland 1950-55 – 1 – ISSN: 0016-7312 – mf#535 – us UMI ProQuest [301]
Geographical abstracts: human geography – Oxford, England 1989+ – 1,5,9 – ISSN: 0953-9611 – mf#42590 – us UMI ProQuest [900]
Geographical abstracts: physical geography – Oxford, England 1989+ – 1,5,9 – ISSN: 0954-0504 – mf#42589 – us UMI ProQuest [900]
A geographical, agricultural and mineralogical sketch / Hunt, Thomas Sterry – Quebec: printed at "Le Canadien" Office, 1865 [mf ed 1990] – 1mf – 9 – mf#SEM105P1219 – cn Bibl Nat [917]
Geographical analysis – Oxford, England 1969+ – 1,5,9 – ISSN: 0016-7363 – mf#5752 – us UMI ProQuest [900]
A geographical and commercial view of northern central africa : containing a particular account of the course and termination of the great river niger in the atlantic ocean / MacQueen, James – Edinburgh 1821 [mf ed Hildesheim 1995-98] – 1v on 2mf – 9 – €60.00 – 3-487-27335-7 – gw Olms [916]
Geographical and historical observations upon the map of thibet : containing the territories of the grand lama, and the neighbouring countries... / Du Halde, J B – London, 1741. v4 – 2mf – 9 – mf#HT-510 – ne IDC [915]
A geographical and statistical description of scotland : containing a general survey of that kingdom, its climate, mountains, lakes, rivers, products, population... / Playfair, James – Edinburgh 1819 [mf ed Hildesheim 1995-98] – 2v on 7mf – 9 – €140.00 – 3-487-27903-7 – gw Olms [914]
Geographical annual : or family cabinet atlas / Starling, Thomas – London 1832 [mf ed Hildesheim 1995-98] – 1mf – 9 – €40.00 – 3-487-29974-7 – gw Olms [912]
A geographical description of southampton island and notes on the eskimo / Comer, G – New York, 1910. v42 – 1mf – 9 – mf#N-174 – ne IDC [919]
A geographical description of the four parts of the world : taken from the notes and workes of the famous monsieur sanson...and other eminent travellers and authors... / Blome, Richard – London: printed by T N for R Blome...1670 [mf ed 1984] – 7mf – 9 – 0-665-32096-5 – (incl ind) – mf#32096 – cn Canadiana [910]
Geographical essays / Law, Bimala Churn – London: Luzac & Co, 1937- – (= ser Samp: indian books) – us CRL [954]

Geographical handbook: palestine and transjordan / U.S. Naval Intelligence – 1943 – 9 – $24.00 – us IRC [915]

Geographical, historical and statistical repository – La Salle IL 1824 – 1 – mf#3778 – us UMI ProQuest [900]

A geographical, historical, and topographical description of van diemen's land : with important hints to emigrants, and useful information respecting the application for grants of land / Evans, George W – London 1822 [mf ed Hildesheim 1995-98] – 1v on 1mf – 9 – €40.00 – 3-487-26795-0 – gw Olms [919]

Geographical, historical, political, philosophical and mechanical essays, no 2 : containing a letter representing the impropriety of sending forces to virginia; the importance of taking fontenac... / Evans, Lewis – London: printed for R & J Dodsley...1756 [mf ed 1984] – 1mf – 9 – 0-665-44274-2 – mf#44274 – cn Canadiana [975]

A geographical historie of africa : before which out of the best ancient and modern writers is prefixed a general description of africa, and also a particular treatise of all the maine lands and isles undescribed / Leo, John – Londini: G Bishop, 1600 – (in arabicke and italian. filmed with: blyden, e w christianity, islam and the negro race) – us CRL [960]

Geographical journal – Oxford, England 1893+ – 1,5,9 – ISSN: 0016-7398 – mf#548 – us UMI ProQuest [900]

Geographical magazine – Killen TX 1874-78 – 1 – mf#2796 – us UMI ProQuest [900]

The geographical magazine – London, 1935-99+ – 59r (49 col) – 1,14 – £3950.00 – mf#GMZ – uk World [910]

A geographical memoir of the persian empire, accompanied by a map / Macdonald, Kinneir, J – London, 1813 – 6mf – 9 – mf#AR-2041 – ne IDC [956]

Geographical memoir upon upper california : addressed to the senate of the united states in 1848 / Fremont, John Charles; ed by McCarty, William – Philadelphia: W McCarty, 1849 – 1mf – 9 – mf#16686 – cn Canadiana [917]

Geographical memoirs on new south wales : containing an account of the surveyor general's late expedition to two new ports / Field, Barron – London 1825 [mf ed Hildesheim 1995-98] – 1v on 4mf – 9 – €120.00 – ISBN-10: 3-487-26801-9 – ISBN-13: 978-3-487-26801-9 – gw Olms [919]

Geographical, natural, and civil history of chili / Molina, Giovanni – [London 1809 [mf ed Hildesheim 1995-98] – 2v on 5mf – 9 – €100.00 – ISBN-10: 3-487-26826-4 – ISBN-13: 978-3-487-26826-2 – gw Olms [918]

Geographical review – New York NY 1916+ – 1,5,9 – ISSN: 0016-7428 – mf#124 – us UMI ProQuest [900]

Geographical sketch of st domingo, cuba and nicar... / Clark, Benjamin C – Boston, MA. 1850 – 1r – us UF Libraries [972]

A geographical, statistical, and historical description of hindostan : and the adjacent countries / Hamilton, Walter – London 1820 [mf ed Hildesheim 1995-98] – 2v on 18mf – 9 – €180.00 – 3-487-27264-4 – gw Olms [915]

A geographical study of human settlement in the eastern province of the gold coast colony west of the volta delta / Boateng, Ernest Amano – Oxford, 1954 – us CRL [960]

La geographie – Paris. v1-32; 62-72. 1900-18-19; 1934-39 – 1 – us L of C Photodup [910]

Geographie a l'usage des ecoliers du petit seminaire de quebec – a Quebec: Chez J Neilson...1804 [mf ed 1985] – 1mf – 9 – 0-665-01723-5 – mf#01723 – cn Canadiana [370]

Geographie a l'usage des ecoliers du petit seminaire de quebec – Quebec: J Neilson, 1804 [mf ed 1974] – 1r – 5 – mf#SEM16P142 – cn Bibl Nat [900]

Geographie abregee : par demandes et par reponses: divisee par lecons, pour l'instruction de la jeunesse... / Lenglet Dufresnoy, abbe – 8th corr enl ed. Paris: Chez la Veuve Tilliard... 1774 [mf ed 1984] – 4mf – 9 – 0-665-44881-3 – mf#44881 – cn Canadiana [910]

Geographie ancienne et moderne de la chine / Couvreur, Seraphin – Hien Hien: Mission Catholique, 1917 [mf ed 1995] – (= ser Yale coll) – 424p – 1 – 0-524-09073-4 – (in chinese) – mf#1995-0073 – us ATLA [915]

Geographie de la jeunesse, ou nouveau manuel de geographie : contenant la description detaillee des empires, des royaumes, et d'autres etats... / Depping, Georges – Paris 1824 [mf ed Hildesheim 1995-98] – 2v on 8mf – 9 – €160.00 – 3-487-29964-X – gw Olms [910]

Geographie de la palestine, vols 1-2 (etb) / Abel, Felix-Marie – Paris. v1-2. 1933-1938 – €38.00 – ne Slangenburg [956]

Geographie de la republique d'haiti / Chauvet, Henri – Port-Au-Prince, Haiti. 1929 – 1r – us UF Libraries [972]

Geographie de l'afrique chretienne : mauretanies / Toulotte, Anatole, Monseigneur – Montreuil-sur-mer: Imprimerie Notre-Dame des Pres, 1894 – 1mf – 9 – 0-8370-7674-9 – mf#1986-1674 – us ATLA [240]

La geographie de l'egypte a l'epoque copte / Amelineau, Emile – Paris: Imprimerie nationale, 1893 – 1mf – 9 – 0-524-04633-6 – (incl bibl ref) – mf#1990-3376 – us ATLA [930]

Geographie de l'empire de russie : contenant la russie d'europe et la russie d'asie / Rabbe, Alphonse – Paris 1828 [mf ed Hildesheim 1995-98] – 2v on 5mf – 9 – €100.00 – 3-487-29028-6 – gw Olms [910]

Geographie de l'ethiopie. ce que j'ai entendu, faisant suite...ce que j'ai vu / Abbadie, A d' – Paris, 1890. 1v – 6mf – 9 – mf#NE-20175 – ne IDC [916]

Geographie de l'ile d'haiti / Chauvet, Henri – Port-Au-Prince, Haiti. 1912 – 1r – us UF Libraries [972]

Geographie des alten palaestina / Buhl, Frants – Freiburg i. B.: J.C.B. Mohr (Paul Siebeck), 1896 – (= ser Grundriss der theologischen wissenschaften) – 1mf – 9 – 0-7905-0749-8 – (incl bibl ref and ind) – mf#1987-0749 – us ATLA [900]

Geographie du cambodge / Aymonier, Etienne – Paris: E Leroux 1876 [mf ed 1989] – 1r with other items – 1 – mf#mf-10289 seam reel 001/12 [§] – us CRL [915]

Geographie du cambodge / Delvert, Jean – [Phnom Penh]: Service francais d'information au Cambodge [195-?] [mf ed 1989] – 2v in 1 on 1r with other items – 1 – mf#mf-10289 seam reel 007/03 [§] – us CRL [330]

Geographie du cambodge & de l'asie des moussons / Tan-kim-Huon – Phnom-Penh: Impr Henry 1957 [mf ed 1989] – 1r with other items – 1 – ("a l'usage des maitres de l'enseignement primaire; des professeurs et elves de l'enseignement secondaire national") – mf#mf-10289 seam reel 004/12 [§] – us CRL [915]

Geographie du cours elementaire : ou primaire a l'usage des ecoles chretiennes / Adelbertus, frere – Montreal: freres des ecoles chretiennes, 1876 [mf ed 1994] – 9 – cn Bibl Nat [900]

Geographie du cours elementaire ou inferieur : a l'usage des ecoles chretiennes – Montreal: C O Beauchemin & Valois, 1873 [mf ed 1984] – 1mf – 9 – 0-665-46147-X – mf#46147 – cn Canadiana [370]

Geographie economique : lere lecon: les regions geographiques de la province de quebec / Brouillette, Benoit – Montreal: Federation des Chambres...1943 [mf ed 1992] – 1mf – 9 – mf#SEM105P1497 – cn Bibl Nat [330]

Geographie elementaire descriptive : ou, lecons graduees de geographie: a l'usage des colleges, des maisons d'education et des ecoles normales / Boniface, Alexandre – 5e rev corr ed. Paris: J Delalain, 1844 [mf ed 1984] – 5mf – 9 – 0-665-43135-X – mf#43135 – cn Canadiana [910]

Geographie generale : contenant la geographie physique, politique, administrative, historique, agricole, industrielle et commerciale de chaque pays avec des notions sur le climat... / Dussieux, Louis – Paris, Lyon: J Lecoffre, 1866 – 11mf – 9 – (incl bibl ref) – mf#49040 – cn Canadiana [900]

Geographie historique et les droits territoriaux / Peralta, Manuel Maria De – Paris, France. 1900 – 1r – us UF Libraries [972]

Geographie historique, physique et statistique du royaume des pays-bas et de ses colonies / Cloet, Jean J de – Bruxelles 1822 [mf ed Hildesheim 1995-98] – 2v on 3mf – 9 – €90.00 – 3-487-29650-0 – gw Olms [910]

Geographie locale / Dartigue, Maurice – Port-Au-Prince, Haiti. v1. 1931- – 1r – us UF Libraries [972]

Geographie moderne : precedee d'un petit traite de la sphere et du globe, ornee de traits d'histoire naturelle et politique... / Croix, Louis Antoine Nicolle de la – nouv rev augm ed. Paris: Chez Auguste Delalain...1812 [mf ed 1985] – 2v on 1mf – 9 – mf#51314 – cn Canadiana [910]

Geographie physique et politique de l'espagne et du portugal : suivie d'un itineraire detaille de ces deux royaumes / Antillon, Isidoro de – Paris 1823 [mf ed Hildesheim 1995-98] – 3mf – 9 – €90.00 – 3-487-29814-7 – gw Olms [914]

Geographie physique, historique et statistique de la france : avec deux tables des matieres, dont l'une, par ordre alphabetique, peut servir de dictionnaire geographique abrege de la france, vol 4 / Mentelle, Edme – Paris 1821 [mf ed Hildesheim 1995-98] – 5mf – 9 – €100.00 – 3-487-29967-4 – gw Olms [944]

Geographie und statistik des grossherzogthums baden nach den neuesten bestimmungen bis zum 1 maerz 1820 / Demian, Johann – Heidelberg 1820 [mf ed Hildesheim 1995-98] – 2mf – 9 – €60.00 – 3-487-29479-6 – gw Olms [914]

Geographie universelle : ou description de toutes les parties du monde / Malte-Brun, Conrad – Paris: E & V Penaud. 8v. 1851? [mf ed 1984] – 8v on 1mf – 9 – mf#46059 – cn Canadiana [910]

Geographie universelle : traduite de l'allemand de mr buesching sur la cinquieme edition nouvellement revue & fort augmentee / Buesching, Anton – Strasbourg 1772-79 [mf ed Hildesheim 1995-98] – 9v on 40mf – 9 – €40.000 – 3-487-29975-5 – gw Olms [910]

La geographie universelle : ou l'on donne une idee abregee des quatre parties du monde, et des differens lieux qu'elles renferment; traduite de l'allemand / Huebner, Johann – Basel 1746 [mf ed Hildesheim 1995-98] – 6v on 19mf – 9 – €190 – 3-487-29985-2 – gw Olms [910]

Geographie universelle, ancienne et moderne, mathematique, physique, statistique, politique et historique, des cinq parties du monde : redigee d'apres ce qui a ete publie d'exact et de nouveau par les geographes, les naturalistes... / ed by Mentelle, Edme – Paris 1816 [mf ed Hildesheim 1995-98] – 16v on 59mf – 9 – €590.00 – 3-487-29979-8 – gw Olms [900]

Geographische beschreibung brasiliens / Macedo, Joaquim Manuel De – Leipzig, Germany. 1873 – 1r – us UF Libraries [918]

Geographische beschreibung von island / Gliemann, Theodor – Altona 1824 [mf ed Hildesheim 1995-98] – 2mf – 9 – €60.00 – 3-487-28935-0 – gw Olms [914]

Geographische Gesellschaft. Munich *see* Mitteilungen der geographischen gesellschaft in muenchen

Geographische und ethnographische studien zum 3. und 4. buche der koenige / Doeller, Johannes – Wien: Mayer, 1904 – (= ser Theologische Studien der Leo-Gesellschaft) – 1mf – 9 – 0-8370-2945-7 – mf#1985-0945 – us ATLA [220]

Die geographischen und voelkerkundlichen quellen und anschauungen in herders "ideen zur geschichte der menschheit" / Grundmann, Johannes – Berlin: Weidmann, 1900 – 1r – 1 – (incl bibl ref) – us Wisconsin U Libr [100]

Geographischer buechersaal zum nutzen und vergnuegen eroeffnet / ed by Hager, Joh Georg – Chemnitz 1766-78 – (= ser Dz. historisch-geographische abt) – 3v on 15mf – 9 – €150.00 – mf#k/n1025 – gw Olms [910]

Geographisches lesebuch zum nutzen und vergnuegen / ed by Fabri, Johann Ernst – Halle 1782-84 – (= ser Dz. historisch-geographische abt) – 4v on 8mf – 9 – €160.00 – mf#k/n1128 – gw Olms [910]

Geographisches magazin / ed by Fabri, Johann Ernst – Dessau, Leipzig 1783-84 – (= ser Dz. historisch-geographische abt) – 4v on 12mf – 9 – €120.00 – mf#k/n1142 – gw Olms [910]

Geographisches statistisch-topographisches lexikon von schwaben : oder vollstaendige alphabetische beschreibung aller im schwaebischen kreis liegenden staedte, kloester, schloesser, doerfer, flecken, hoefe, berge, thaeler, fluesse, seen, merkwuerdiger gegenden usw / Roeder, Philipp – Ulm 1791/92 [mf ed Hildesheim 1995-98] – 7mf – 9 – €140.00 – 3-487-29504-0 – gw Olms [914]

Geographisch-historische beschreibung des oestlichen kaukasus : zwischen den fluessen terek, aragwi, kur und dem kaspischen meere / Klaproth, Julius – Weimar 1814 [mf ed Hildesheim 1995-98] – 1v on 2mf – 9 – €60.00 – ISBN-10: 3-487-26530-3 – ISBN-13: 978-3-487-26530-8 – gw Olms [914]

Geographisch-historisches wochenblatt : zur erlaeuterung der begebenheiten des tages – Bremen DE, 1798-1801 – 1r – 1 – gw Misc Inst [900]

Geographisch-statistische beschreibung von californien : aufschluesse ueber die lage, den boden und das clima des landes, ueber seine bewohner, ihr leben, sitten und gebraeuche, ueber staatsverfassung, religion, ueber bodenerzeugnisse und handel; mit besonderer beruecksichtigung seines mineralreichthums... / Hartmann, Carl – Weimar 1849 [mf ed Hildesheim 1995-98] – 1v on 2mf – 9 – €60.00 – 3-487-27157-5 – gw Olms [917]

Geographisch-statistisches zeitungs-, post- und comptoir-lexicon : nach den neuesten bestimmungen fuer studierende, zeitungsleser, reisende und geschaeftsleute jeder art / Stein, Christian – Leipzig 1811 [mf ed Hildesheim 1995-98] – 2v on 7mf – 9 – €140.00 – 3-487-29914-3 – gw Olms [059]

Geographisch-statistisch-topographische uebersicht des regierungsbezirks minden / Seemann, W – Muenster [u a] 1845 [mf ed Hildesheim 1995-98] – 2mf – 9 – €60.00 – 3-487-29519-9 – gw Olms [914]

Geographisch-statistisch-topographisches lexikon von wuertemberg : oder: alphabetische beschreibung aller staedte, doerfer, weiler, schloesser, baeder, berge, fluesse, seen usw in hinsicht der lage, anzahl der bewohner, nahrungsquellen, merkwuerdigkeiten, wichtigsten ereignisse der aeltern und neuern zeit / Korsinsky, Bernhard – Stuttgart 1833 [mf ed Hildesheim 1995-98] – 3mf – 9 – €90.00 – 3-487-29492-3 – gw Olms [914]

Geography / Marie de l'Incarnation, mere – Quebec: C Darveau, 1886 [mf ed 1995] – 1mf – 9 – 0-665-94782-8 – (also available in french) – mf#94782 – cn Canadiana [910]

Geography : martin county / Lyons, Isabel J – s.l, s.l? 193-? – 1r – us UF Libraries [978]

Geography anatomiz'd : or, the geographical grammar: being a short and exact analysis of the whole body of modern geography... / Gordon, Patrick – 20th corr enl ed. London: printed for J & P Knapton, J Brotherton, J Clarke...1754 [mf ed 1984] – 6mf – 9 – 0-665-44934-8 – mf#44934 – cn Canadiana [910]

A geography and atlas of protestant missions : their environment, forces, distribution, methods, problems, results and prospects at the opening of the 20th century / Beach, Harlan P – New York: Student Volunteer Movt for Foreign Missions, 1901 [mf ed 1989] – 2mf – 9 – 0-7905-4072-X – mf#1988-0072 – us ATLA [242]

The geography and history of nova scotia : with a general outline of geography, and a sketch of the british possessions in north america / Calkin, John Burgess – Halifax, NS: A W Mackinlay, 1864 – 2mf – 9 – (incl: "vocabulary of geographical terms") – mf#37415 – cn Canadiana [917]

The geography, history, and statistics, of america, and the west indies : exhibiting a correct account of the discovery, settlement, and progress of the various kingdoms, states, and provinces of the western hemisphere, to the year 1822; with additions relative to the new states of south america, etc / Carey, Henry – London 1823 [mf ed Hildesheim 1995-98] – 1v on 4mf – 9 – €120.00 – 3-487-27171-0 – gw Olms [970]

Geography notes / Henderson, George E & Fraser, George A – Toronto: Educational Publishing, 1897 – 2mf – 9 – mf#28355 – cn Canadiana [900]

Geography notes for 3rd, 4th, and 5th classes / Henderson, George E & Fraser, George A – Toronto: Educational Pub Co, 1898 – (= ser School helps series) – 2mf – 9 – mf#16793 – cn Canadiana [917]

Geography of bermuda / Watson, J Wreford – London, England. 1965 – 1r – us UF Libraries [919]

Geography of dade county, florida / Sanderson, Isabelle – s.l, s.l? 1936 – 1r – us UF Libraries [978]

Geography of early buddhism / Law, Bimala Churn – London: Kegan Paul, Trench, Trubner & Co, 1932 – (= ser Samp: indian books) – (foreword by f w thomas) – us CRL [280]

The geography of hudson's bay – (= ser Hakluyt society. extra series 11) – 3mf – 7 – mf#448 – uk Microform Academic [917]

The geography of hudson's bay / Coats, W; ed by Barrow, J – London, 1852 – 3mf – 9 – mf#N-165 – ne IDC [919]

Geography of latin america / Carlson, Fred Albert – New York, NY. 1943 – 1r – us UF Libraries [918]

Geography of middle america / Seeman, Albert L – Seattle, WA. 1941 – 1r – us UF Libraries [918]

A geography of the bible : compiled for the american sunday school union / Alexander, James Waddel & Alexander, Joseph Addison – Philadelphia: American Sunday School Union, 1830 [mf ed 1984] – 3mf – 9 – 0-8370-1018-7 – (incl ind) – mf#1984-4401 – us ATLA [220]

Geography of the coffee industry of puerto rico / Campbell, David Stephen – Chicago, IL. 1947 – 1r – us UF Libraries [338]

The geography of the heavens and class book of astronomy : accompanied by a celestial atlas / Burritt, Elijah Hinsdale – 5th ed. New York: Huntington & Savage 1843 [mf ed 1998] – 1r [ill] – 1 – (incl ind; int by thomas dick) – mf#film mas 28425 – us Harvard [520]

Geojournal – Dordrecht, Netherlands 1989+ – 1,5,9 – ISSN: 0343-2521 – mf#14749 – us UMI ProQuest [550]

Geokhimicheskie issledovaniia – Moskva: IMGRE, [1970-] v2. 1972 – us CRL [947]

Geologia historica do brazil / Brazil. Divisao de Geologia e Mineralogia – Rio de Janeiro, Brazil. 1930 – 1r – us UF Libraries [972]

Geologia salvadorena / Larde Y Larin, Jorge – San Salvador, El Salvador. 1952 – 1r – us UF Libraries [972]

Geological abstracts – Oxford, England 1990+ – 1,5,9 – ISSN: 0954-0512 – mf#42592 – us UMI ProQuest [550]

Geological and ground water conditions in florida / United States Army Corps Of Engineers – Washington, DC. 1936 – 1r – us UF Libraries [500]

Geological bibliography of mid-continent basement usa / Sheahan, Patricia – Boulder CO: Geological Soc of America, c1984 – 1mf – 9 – 0-8137-6015-1 – us GPO [550]

Geological bulletin / Florida Geological Survey – Tallahassee, FL. n12-53. 1935-1971 – 6r – us UF Libraries [550]

The geological evidences of the antiquity of man : with remarks on theories of the origin of species by variation / Lyell, Charles – London, 1863 – (= ser 19th c evolution & creation) – 6mf – 9 – mf#1.1.9070 – uk Chadwyck [573]

The geological history of lake superior / Bell, Robert – Toronto: Murray, 1899 – 1mf – 9 – mf#03554 – cn Canadiana [550]

Geological journal – Hoboken NJ 1979+ – 1,5,9 – ISSN: 0072-1050 – mf#11997,01 – us UMI ProQuest [550]

Geological magazine – Cambridge, England 1949+ – 1,5,9 – ISSN: 0016-7568 – mf#462 – us UMI ProQuest [550]

Geological magazine – London: Dulau & Co. v23-71. 1886-1934 – 1 – $540.00 – (ind 1864-1903) – mf#0231 – us Brook [550]

The geological observer / De la Beche, Henry Thomas – London, 1851 – (= ser 19th c evolution & creation) – 9mf – 9 – mf#1.1.1073 – uk Chadwyck [550]

The geological relations of the principal nova scotia minerals / Gilpin, Edwin – S.l: s.n, 18–? – 1mf – 9 – mf#06902 – cn Canadiana [550]

Geological report : review of the oil and gas possi... / Hill, Edward Allison – Tallahassee, FL. 1927 – 1r – us UF Libraries [550]

Geological society of america bulletin – Boulder CO 1890+ – 1,5,9 – ISSN: 0016-7606 – mf#10629 – us UMI ProQuest [550]

Geological Society of Australia *see* Journal of the geological society of australia

Geological society of india. journal – Bangalore, India 1973 – 1 – ISSN: 0016-7622 – mf#8800 – us UMI ProQuest [550]

Geological Society of London *see* Proceedings of the geological society of london

Geological society of london. transactions – 1811-42 – 1 – $216.00 – mf#0232 – us Brook [550]

Geological society of south africa. transactions – Marshalltown, South Africa 1979 – 1,5,9 – ISSN: 0371-7208 – mf#11729.01 – us UMI ProQuest [550]

Geological society of south africa. transactions – v1-51. 1915-49 – 9 – $660.00 – mf#0233 – us Brook [550]

Geological survey and marine corps surveys and maps of the dominican republic, 1919-1923 / U.S. Geological Survey – (= ser Records Of The U.S. Geological Survey) – 6r – 1 – mf#T282 – us Nat Archives [550]

Geological survey of canada – Montreal: printed by John Lovell, 1863 [mf ed 1992] – 1mf – 9 – mf#SEM105P1684 – cn Bibl Nat [550]

Geologie en geohydrologie van het eiland curacao / Molengraaff, Gerard Johan Hendrik – Delft, Netherlands. 1929 – 1r – us UF Libraries [550]

Geologische rundschau – Dordrecht, Netherlands 1910-74 – 1 – (cont by: international journal of earth sciences: geologische rundschau) – ISSN: 0016-7835 – mf#10146.01 – us UMI ProQuest [550]

Geologische studien ueber niederlaendisch west-ind... / Martin, Karl – Leiden, Netherlands. 1888 – 1r – us UF Libraries [550]

Geology – Boulder CO 1973+ – 1,5,9 – ISSN: 0091-7613 – mf#13403 – us UMI ProQuest [550]

Geology : united states exploring expedition. during the years 1838-1842 under the command of charles wilkes / Dana, J D – London. 1992-1996 (1) – 28mf – 9 – mf#2835 – ne IDC [910]

Geology and its teaching : especially as it relates to the development theory as propounded in "vestiges of creation," and darwin's "origin of species." an original essay – London, 1861 – (= ser 19th c evolution & creation) – 1mf – 9 – mf#1.1.9069 – uk Chadwyck [550]

Geology and mineralogy considered with reference to natural theology / Buckland, William – London, 1836, 1837 – (= ser 19th c evolution & creation) – 10mf – 9 – mf#1.1.10844 – uk Chadwyck [210]

Geology and revelation : or, the ancient history of the earth / Molloy, Gerald – London: Longmans, Green, Reader & Dyer, 1870 [mf ed 1989] – 2mf – 9 – 0-7905-2930-0 – mf#1987-2930 – us ATLA [221]

Geology of a portion of the laurentian area to the north of montreal / Adams, Frank Dawson – Montreal: s.n, 1897 – 3mf – 9 – mf#36906 – cn Canadiana [550]

The geology of genesis : an inquiry into the credentials of the mosaic record of creation / Robinson, E Colpitt – London, 1885 – (= ser 19th c evolution & creation) – 2mf – 9 – mf#1.1.7150 – uk Chadwyck [210]

The geology of michipicoten island / Burwash, Edward Moore – [Toronto]: University Library, 1905 – (= ser University of toronto studies. geological series 3) – 1mf – 9 – 0-659-91343-7 – mf#9-91343 – cn Canadiana [550]

Geology of the goldfields of british guiana / Harrison, John Burchmore – London, England. 1908 – 1r – us UF Libraries [550]

Geology of the province of camaguey, cuba / Macgillavry, Henry James – Utrecht, Netherlands. 1937 – 1r – us UF Libraries [550]

The geology of vancouver and vicinity / Burwash, Edward Moore – Chicago: University of Chicago Press, [1918] – 2mf – 9 – 0-659-91388-7 – mf#9-91388 – cn Canadiana [550]

Geology of venezuela and trinidad / Liddle, Ralph Alexander – Ithaca, NY. 1946 – 1r – us UF Libraries [550]

Geology today – Oxford, England 1985+ – 1,5,9 – ISSN: 0266-6979 – mf#15545 – us UMI ProQuest [550]

Geo-marine technology – Washington DC 1964-67 – 1 – mf#2172 – us UMI ProQuest [550]

Geomechanics abstracts – Oxford, England 1997+ – 1,5,9 – ISSN: 1365-1617 – mf#42785 – us UMI ProQuest [622]

Geometrie of te meet-const / Marolois, S – Amsterdam, 1629 – 2mf – 9 – mf#OA-155 – ne IDC [720]

Geometrie, toise et le tableau stereometrique : lecture faite...du quebec 20 mars 1872 / Baillairge, Charles P Florent – Quebec: C Darveau, 1873 – 1mf – 9 – mf#00055 – cn Canadiana [510]

Geometrische zuordnung sequentieller roentgenbilder mit hilfe drehungs- und masstabsinvarianter bildmuster-merkmale / Harendt, Norbert – (mf ed 1995) – 2mf – 9 – €40.00 – 3-8267-2204-3 – mf#DHS 2204 – gw Frankfurter [530]

Geometry and faith : a supplement to the ninth bridgewater treatise / Hill, Thomas – 3rd ed, greatly enl. Boston: Lee and Shepard; New York: CT Dillingham, 1882 – 1mf – 9 – 0-7905-8660-6 – mf#1989-1885 – us ATLA [240]

Geometry, mensuration and the stereometrical tableau : lecture read...20th march 1872 / Baillairge, Charles P Florent – Quebec: C Darveau, 1873 – 1mf – 9 – mf#00809 – cn Canadiana [510]

Geometry of the zeros of a polynomial in a complex variable / Marden, Morris – New York, NY. 1949 – 1r – us UF Libraries [510]

Geomicrobiology journal – Abingdon, Oxfordshire 1988+ – 1,5,9 – ISSN: 0149-0451 – mf#11689 – us UMI ProQuest [576]

Geomorphology – Oxford, England 1987+ – 1,5,9 – ISSN: 0169-555X – mf#42551 – us UMI ProQuest [550]

Geon yits hak / Lipschitz, Jacob Lipmann – Vilna, Lithuania. 1899 – 1r – us UF Libraries [939]

'Tgeopende : en bereidwillige herte, na den heere jesus... / [Boekholt, J] – t'Amsterdam: Johannes Boekholt, 1693 – 3mf – 9 – mf#O-3035 – ne IDC [090]

Geophysical journal – Oxford, England 1988 – 1,5,9 – (cont by: geophysical journal of the ras, dgg, and egs) – ISSN: 0952-4592 – mf#16701.02 – us UMI ProQuest [550]

Geophysical journal international – Oxford, England 1989+ – 1,5,9 – (cont: geophysical journal of the ras, dgg, and egs) – ISSN: 0956-540X – mf#16701,02 – us UMI ProQuest [550]

Geophysical journal of the royal astronomical society / Royal Astronomical Society – Oxford, England 1958-87 – 1,5,9 – ISSN: 0016-8009 – mf#1298 – us UMI ProQuest [550]

Geophysical prospecting – Oxford, England 1980+ – 1,5,9 – ISSN: 0016-8025 – mf#15546 – us UMI ProQuest [550]

Geophysical research letters – v1- 1974-. Semi-monthly since 1992 – (v1-3 1974-76 $25.00y 1,5,6,9,13. v4 1977 $25.00 1,5,6,13 $35.00 9. v5 1978 $30.00 1,5,6,13 $40.00 9. v6 1979 $35.00 1,5,6,1,3 $50.00 9. v7 1980 $35.00 1,5,6,13 $65.00 9. v8 1981 $35.00 1,5,6,13 $105.00 9. v9 1982 $70.00 1,5,6,13 $145.00 9 v19. 1992 $480.00. v20 1993 $498.00. v21 $590.00 10. v22 1995 $680.00) – us AGU [550]

Geophysics – Tulsa OK 1936+ – 1,5,9 – ISSN: 0016-8033 – mf#1005 – us UMI ProQuest [550]

Geopolitiek asia timoer raja : koetipan pidato / Bekki, A – (Djakarta): Djawa Shimbun Sha, 2604 – 48p 1mf – 9 – mf#SE-2002 mf24 – ne IDC [959]

Georg : eine dorfgeschichte aus dem ries / Meyr, Melchior – Bayreuth: Gauverlag Bayreuth, 1944 – 1r – 1 – us Wisconsin U Libr [390]

Georg benedict winer's grammatik des neutestamentlichen sprachidioms / Winer, Georg Benedikt – 8. aufl. Goettingen: Vandenhoeck & Ruprecht 1894-98 [mf ed 1991] – 3mf – 9 – 0-7905-8352-6 – (in german & greek; incl bibl ref) – mf#1987-6451 – us ATLA [450]

Georg buechner : eine biographische erzaehlung / Bauer, Franz – Berlin: Neues Leben, c1949 [mf ed 1993] – 163p – 1 – mf#8526 – us Wisconsin U Libr [830]

Georg buechner : versuch ueber die tragische existenz / Schmid, Peter – Bern: P Haupt, 1940 – 1r – 1 – us Wisconsin U Libr [430]

Georg buechner in selbstzeugnissen und bilddokumenten / Johann, Ernst – Hamburg: Rowohlt, 1958 – 1r – 1 – (incl bibl ref) – us Wisconsin U Libr [430]

Georg buechner in selbstzeugnissen und bilddokumenten / Johann, Ernst – Hamburg: Rowohlt, 1958 – 1 – (incl bibl ref) – us Wisconsin U Libr [920]

Georg buechner und der dandysmus / Gunkel, Richard – Utrecht: R Kemink, 1953 – 1r – 1 – (incl bibl ref and ind) – us Wisconsin U Libr [430]

Georg buechner und der dandysmus / Gunkel, Richard – Utrecht: Kemink, 1953 – 1 – (incl bibl ref and index) – us Wisconsin U Libr [430]

Georg buechner und die romantik / Lipmann, Heinz – Muenchen: M Hueber, 1923 – 1r – 1 – us Wisconsin U Libr [430]

Georg buechner und seine zeit / Mayer, Hans – Berlin: Volk und Welt, [195-?] [mf ed 1989] – 397p – 1 – mf#7151 – us Wisconsin U Libr [430]

Georg buechner und shakespeare / Vogeley, Heinrich – Marburg; Wuerzburg: Dissertationsdruckerei und Verlag K Triltsch, 1934 – 1r – 1 – us Wisconsin U Libr [410]

Georg buechners aesthetische anschauungen / Nahke, Heinz – Dresden: Verlag der Kunst, 1955 – 1 – us Wisconsin U Libr [430]

Georg buechners aesthetische anschauungen / Nahke, Heinz – Dresden: Verlag der Kunst, 1955 – 1r – 1 – us Wisconsin U Libr [430]

Georg buechners "danton" / Koenig, Fritz – Halle (Saale): M Niemeyer, 1924 – 1r – 1 – us Wisconsin U Libr [430]

Georg buechners drama "dantons tod" / Landsberg, Hans – [S.l.: s.n.], 1900 (Berlin: Druck von E Ebering) – 1r – 1 – (incl bibl ref) – us Wisconsin U Libr [430]

Georg buechners saemtliche poetische werke : nebst einer auswahl seiner briefe / Buechner, Georg; ed by Zweig, Arnold – Muenchen: Roesl 1923 [mf ed 1993] – 1r – 1 – (filmed with: der erzieherische gehalt in j j breitingers "critischer dichtkunst" / jakob braeker) – mf#8525 – us Wisconsin U Libr [800]

Georg calixt und der synkretismus : eine dogmenhistorische abhandlung / Gass, Wilhelm – Breslau: A Gosohorsky, 1846 – 1mf – 9 – 0-524-01226-1 – (incl bibl ref) – mf#1990-0365 – us ATLA [240]

Georg calixtus und seine zeit / Henke, Ernst Ludwig Theodor – Halle: Buchh des Waisenhauses, 1853-1860 – 3mf – 9 – 0-7905-8199-X – (incl bibl ref) – mf#1988-8082 – us ATLA [943]

Georg christoph lichtenbergs aphorismen / ed by Leitzmann, Albert – Berlin: B Behr, 1902-1908 [mf ed 1993] – (= ser Deutsche litteraturdenkmale des 18. und 19. jahrhunderts 123, 131, 136, 140-141) – 5v (ill) – 1 – mf#8676 reel 6, 7, 8 – us Wisconsin U Libr [390]

Georg christoph lichtenberg's gedanken und maximen : lichtstrahlen aus seinen werken – Leipzig: F A Brockhaus, 1871 [mf ed 1996] – 226p – 1 – (biogr int by eduard grisebach) – mf#9708 – us Wisconsin U Libr [390]

Georg ebers gesammelte werke – Stuttgart: Deutsche Verlags-Anstalt, [1893-1897?] [mf ed 1993] – 5r – 1 – mf#8554 – us Wisconsin U Libr [802]

Georg enach : iz epoki bor by shveitsarskogo naroda za nezavisimost = Juerg jenatch / Meyer, Conrad Ferdinand – Moskva: Gos izd-vo; Petrograd: Tipografiia im N Bukharina, 1923 [mf ed 1995] – 299p – 1 – mf#8824 – us Wisconsin U Libr [830]

Georg forster : das abenteuer seines lebens unter wiedergabe vieler briefe und tagebucheintragungen / Langewiesche, Wilhelm – Ebenhausen im Isartal: W Langewiesche-Brandt, [1923] (mf ed 1990) – 1r – 1 – (filmed with: theodor fontane) – us Wisconsin U Libr [430]

Georg forster, der naturforscher des volks / Moleschott, Jacob – Volksausg. Frankfurt (Main): Meidinger, 1857 – 1r – 1 – us Wisconsin U Libr [920]

Georg forster nach seinen originalbriefen / Zincke, Paul – Dortmund: F W Ruhfus. 2v. 1915 – 1r – 1 – us Wisconsin U Libr [920]

Georg forsters briefe an christian friedrich voss / Forster, Georg; ed by Zincke, Paul – Dortmund: F W Ruhfus, 1915 (mf ed 1990) – 1r – 1 – (filmed with: theodor fontane. incl ind) – us Wisconsin U Libr [860]

Georg forster's saemmtliche schriften / Forster, Georg – Leipzig: F A Brockhaus, 1843 [mf ed 1986] – 9v on 1mf – 9 – 0-665-38846-2 – (ed by his daughter. with portrait of forster by g g gervinus. incl bibl ref) – mf#38846 – cn Canadiana [910]

Georg forsters tagebuecher / ed by Zincke, Paul & Leitzmann, Albert – Berlin: B Behr (F Feddersen), 1914 [mf ed 1993] – xlv/436p/1pl – 1 – (incl bibl ref and ind) – mf#8676 reel 9 – us Wisconsin U Libr [430]

Georg friedrich haendels lebensbeschreibung : nebst einem verzeichnisse seiner ausuebungswerke und deren beurtheilung / Mainwaring, John – Hamburg: Auf kosten des uebersetzers 1761 [mf ed 19–] – 3mf – 9 – (trans of memoirs of the life of the late george frederic handel by johann mattheson) – mf#fiche 435 – us Sibley [780]

Georg friedrich handel's (1685-1759) works : edition of the deutschen haendelgesellschaft / ed by Chrysander, Friedrich – Leipzig, Bergdorf-bei-Hamburg: Breitkopf & Haertel, Chrysander. 96v + 6 suppls. 1858-1894; 1902 – 11 – $625.00 set – (v49 not publ) – us Univ Music [780]

Georg herwegh und seine deutschen vorbilder / Hensold, Karl – Ansbach: C Bruegel, 1916 (mf ed 1990) – 1r – 1 – (filmed with: goethes faust in urspruenglicher gestalt. incl bibl ref) – us Wisconsin U Libr [430]

Georg leonhard hartmanns versuch einer beschreibung des bodensee's – St Gallen 1808 [mf ed Hildesheim 1995-98] – 2mf – 9 – €60.00 – 3-487-29487-7 – gw Olms [914]

Georg, Manfred *see* Grabbes doppeltes gesicht

Georg melchior kraus / Schweinsberg, Eberhard Schenk zu, Freiherr – Weimar: Goethe-Gesellschaft, 1930 [mf ed 1993] – (= ser Schriften der goethe-gesellschaft 43) – 40p/52pl (ill) – 1 – (incl bibl ref) – mf#8657 reel 10 – us Wisconsin U Libr [430]

Die georg philipp telemann-sammlung / ed by Staatsbibliothek zu Berlin – Preussischer Kulturbesitz – [mf ed 2001-03] – (= ser Musikhandschriften der staatsbibliothek zu berlin – preussischer kulturbesitz 2; Music manuscripts of the staatsbibliothek zu berlin – preussischer kulturbesitz) – 356mf (1:24) – 9 – silver €3100.00 – ISBN-10: 3-598-34433-3 – ISBN-13: 978-3-598-34433-6 – (incl guide) – gw Saur [780]

Die georg philipp telemann-sammlung : supplement 2 / Fischer, Axel [comp]; ed by Singakademie zu Berlin – [mf ed 2003] – (= ser Musikhandschriften der staatsbibliothek zu berlin – preussischer kulturbesitz) – 122mf – 9 – silver €890.00 – ISBN-10: 3-598-34441-4 – ISBN-13: 978-3-598-34441-1 – (incl guide with int by ralph-j reipsch) – gw Saur [780]

Die georg philipp telemann-sammlung der stadt- und universitaetsbibliothek frankfurt am main / ed by Stadt- und Universitaetsbibliothek Frankfurt am Main – [mf ed 2002-2003] – 405mf (1:24) in 3 installments – 9 – silver €3750.00 – ISBN-10: 3-598-34980-7 – ISBN-13: 978-3-598-34980-5 – (incl guide) – gw Saur [780]

Georg rollenhagens spiel vom reichen manne und armen lazaro / ed by Bolte, Johannes – Halle: M Niemeyer, 1929 – us Wisconsin U Libr [430]

Georg rollenhagens spiel von tobias / ed by Bolte, Johannes – Halle: M Niemeyer, 1930 – (incl bibl ref) – us Wisconsin U Libr [430]

Georg rudolf weckherlin : the embodiment of a transitional stage in german metrics / Schaffer, Aaron – Baltimore: The Johns Hopkins Press, 1918 – 1 – (incl bibl ref) – us Wisconsin U Libr [430]

Georg rudolf weckherlin : versuch einer physiognomischen stilanalyse / Gaitanides, Johannes – [S.l.: s.n.], 1936 – 1r – 1 – (incl bibl ref) – us Wisconsin U Libr [430]

Georg rudolf weckherlins gedichte / ed by Fischer, Hermann – Stuttgart: Litterarischer Verein, 1894-. (Tuebingen: H Laupp, Jr) [mf ed 1993] – (= ser Blvs 199-200, 245) – 3v – 1 – mf#8470 reels 41, 50 – us Wisconsin U Libr [810]

Georg schwartzerdt, der bruder melanchthons und schultheiss zu bretten : festschrift zur feier des 25jaehrigen bestehens des vereins fuer reformationsgeschichte / Mueller, Nikolaus – Leipzig: Verein fuer Reformationsgeschichte 1908 [mf ed 1990] – (= ser Schriften des vereins fuer reformationsgeschichte 25/96-97) – 1mf – 9 – 0-7905-5125-X – mf#1988-1125 – us ATLA [242]

Georg schwartzerdt, der bruder melanchthons und schultheiss zu bretten : festschrift zur feier des 25jaehrigen bestehens des vereins fuer reformationsgeschichte / Mueller, Nikolaus – Leipzig: Verein fuer Reformationsgeschichte; 25. Jahrg., Schrift 96/97) – 1mf – us ATLA [242]

Georg thyms gedicht thedel von wallmoden [1558] / ed by Zimmermann, Paul – Halle: Max Niemeyer, 1887 – 11r – 1 – (incl bibl ref) – us Wisconsin U Libr [430]

GEORG

Georg trakl / Jaspersen, Ursula – Hamburg: Hansischer Gildenverlag, 1947 – 1r – 1 – (incl bibl ref) – us Wisconsin U Libr [430]

Georg viscount valentia's und heinrich salt's reisen nach indien, ceylon, dem rothen meere, abyssinien und aegypten : in den jahren 1802, 1803, 1804, 1805 und 1806 – Weimar 1811 [mf ed Hildesheim 1995-98] – 9mf – 9 – €180.00 – ISBN-10: 3-487-26538-9 – ISBN-13: 978-3-487-26538-4 – gw Olms [910]

Georg wickrams werke / ed by Bolte, Johannes & Scheel, Willy – Stuttgart: Litterarischer Verein. 8v. 1901-06 (Tuebingen: H Laupp, Jr) – (incl bibl ref) – us Wisconsin U Libr [430]

Georg wickrams werke / ed by Bolte, Johannes & Scheel, Willy – Stuttgart: Litterarischer Verein, 1901-06 (Tuebingen: H Laupp, Jr) [mf ed 1993] – (= ser Blvs 222-223, 229-230, 232, 236-237, 241) – 8v – 1 – (incl bibl ref. v2-8 ed by johannes bolte alone) – mf#8470 reels 46-49 – us Wisconsin U Libr [802]

Georg witzel, ein altkatholik des 16. jahrhunderts / Schmidt, Gustav Lebrecht – Wien: W Braumueller, 1876 – 1mf – 9 – 0-7905-6948-5 – mf#1988-2948 – us ATLA [241]

George / Gundolf, Friedrich – Berlin: G Bondi, 1920 – 1r – 1 – us Wisconsin U Libr [430]

George – New York NY 1998-2001 – 1,5,9 – ISSN: 1084-662X – mf#21841 – us UMI ProQuest [320]

George 3, King *see* King's proclamation, for the encouragement of piety and virtue

George a smathers : us senator from florida 1951-1969 – (= ser Us senate historical office oral history coll) – 4mf – 9 – $20.00 – us Scholarly Res [323]

The george advertiser – George SA, dec 1 1864-dec 28 1870 (wkly) [mf ed Cape Town: SA library 1985] – 3r – 1 – mf#MS00373 – sa National [079]

George and bessie derrick / Diggs, Paul – s.l, s.l? 1939 – 1r – us UF Libraries [978]

George and knysna herald – George SA, 21 dec 1881-1971 – 1 – (cont as: the south western herald, 1971-) – sa National [079]

George, Betty Grace (Stein) *see* Education for africans in tanganyika

The george brinton mcclellan papers – 82r – 1 – $2,870.00 – Dist. us Scholarly Res – us L of C Photodup [355]

George Brown / Lewis, John – Toronto: Morang, 1906 [mf ed 1995] – (= ser The makers of canada) – 4mf – 9 – 0-665-77453-2 – mf#77453 – cn Canadiana [971]

George brown : pioneer-missionary and explorer / Brown, George – London: Hodder and Stoughton, 1908 – 2mf – 9 – 0-7905-6284-7 – mf#1988-2284 – us ATLA [240]

George buchanan / Wallace, Robert & Smith, John Campbell – Edinburgh: Oliphant, Anderson & Ferrier, [1899?] – (= ser Famous Scots Series) – 1mf – 9 – 0-7905-6033-X – mf#1988-2033 – us ATLA [941]

George buchanan, humanist and reformer : a biography / Brown, Peter Hume – Edinburgh: D Douglas 1890 [mf ed 1988] – 1r – 1 – mf#1553 – us Wisconsin U Libr [140]

George buchanan, humanist and reformer : a biography / Brown, Peter Hume – Edinburgh: D Douglas, 1890 – 1mf – 9 – 0-7905-4441-5 – (incl bibl ref) – mf#1988-0441 – us ATLA [920]

George, C *see* Chronica del esforcado principe y capitan iorge castriote rey de epiro, o albania...

George calvert and cecilius calvert : barons baltimore of baltimore / Browne, William Hand – New York: Dodd, Mead c1890 [mf ed 1991] – (= ser Makers of america) – 1mf – 9 – 0-524-00516-8 – mf#1990-0016 – us ATLA [241]

George canning e o brasil / Freitas, Caio De – Sao Paulo, Brazil. v1-2. 1958 – 1r – us UF Libraries [972]

The george chalmers collection – 8r – 1 – $280.00 – Dist. us Scholarly Res – us L of C Photodup [975]

George cruikshank : the artist, the humourist, and the man / Bates, William – London 1878 – (= ser 19th c art & architecture) – 1mf – 9 – mf#4.2.1165 – uk Chadwyck [740]

George cruikshank's omnibus : illustrated / Blanchard, Samuel Laman – [new ed] London 1869 – (= ser 19th c art & architecture) – 4mf – 9 – mf#4.2.1653 – uk Chadwyck [740]

George fox / Hodgkin, Thomas – London: Methuen, 1896 – (= ser Leaders Of Religion) – 1mf – 9 – 0-7905-5709-6 – mf#1988-1709 – us ATLA [240]

George fox and the early quakers / Bickley, Augustus Charles – London: Hodder and Stoughton, 1884 – 2mf – 9 – 0-524-02811-7 – (incl bibliographic references and ind) – mf#1990-4432 – us ATLA [243]

George gershwin / Armitage, Merle – New York, NY. 1938 – 1r – us UF Libraries [780]

The george gordon meade collection, 1793-1896 – ca 14r – 1 – ca $1820.00 – (with guide) – mf#S3360 – Historical Society of Pennsylvania. Available Spring, 1997 – us Scholarly Res [975]

George grenfell and the congo : a history and description of the congo independent state and adjoining districts of congoland... / Johnston, Harry Hamilton – London: Hutchinson, 1908 – 3mf – 9 – 0-524-03719-1 – mf#1990-4824 – us ATLA [916]

George H LaBarre Galleries, Inc *see* Labarre newsletter

George h.c. macgregor, m.a : a biography / Macgregor, Duncan Campbell – New York: Revell, 1901 – 1r – 1 – 0-8370-0312-1 – mf#1984-B133 – us ATLA [920]

George, Henry *see* Progress and poverty

George herbert / Nichols, William – London, England. 1891 – 1r – us UF Libraries [420]

George, Hereford Brooke *see* Historical evidence

George, hofmannsthal, rilke / ed by Sommerfeld, Martin – New York: W W Norton, c1938 – 1r – 1 – (incl bibl ref) – us Wisconsin U Libr [430]

George hubbard pepper papers, 1895-1918 : documents from an archaeologist famous for fieldwork in the american southwest – [mf ed Norman Ross Publ] – 8r – 1 – (with p/g) – us UMI ProQuest [933]

George j pinwell and his works / Williamson, George Charles – London 1900 – (= ser 19th c art & architecture) – 3mf – 9 – mf#4.2.525 – uk Chadwyck [750]

George, James *see*
- An address delivered at the opening of queens sic college, 1853
- An address delivered on the 5th april, 1855
- An address to those who have been baptized in infancy
- A brief inquiry into causes of the poetic element in the scottish mind
- Christ crucified
- The duties of subjects to their rulers
- A few comments upon mr macaulay's remarks on the internal water communications of the canadas
- A few remarks on internal improvements in the canadas
- The field and the men for it
- The good old way
- The mission of great britain to the world
- Moral courage
- Prospectus of the saint lawrence company
- The relation between piety and intellectual labor
- The sabbath school of the fireside and the sabbath school of the congregation as it ought to be
- What is civilization?

George jameson : the scottish vandyck / Bulloch, John – Edinburgh 1885 – (= ser 19th c art & architecture) – 3mf – 9 – mf#4.2.1331 – uk Chadwyck [750]

George, Jody et al *see* Handbook on jury use in the federal courts

George, Joelle *see* Effect of cd-rom enhanced lectures on substance abuse test scores

George, Johann Friedrich Leopold *see*
- Die aelteren juedischen feste
- Mythus und sage

George, L *see* Psychologie

George leatrim : or, the mother's test / Moodie, Susanna – Edinburgh: Oliphant, Anderson & Ferrier, 1882 [mf ed 1994] – 1mf – 9 – 0-665-94667-8 – mf#94667 – cn Canadiana [830]

George magoffin humphrey papers, 1912-1970 / Humphrey, George Magoffin – [mf ed 1981] – 26r – 1 – (incl 24p guide. correspondence, speeches, official documents & reports...and other materials relating primarily to humphrey's service as treasury secretary in the eisenhower administration) – mf#ms3132 – us Western Res [975]

George Mason University *see* Federal one

George mason university civil rights law journal – v1-11. 1990-2001 – 9 – $166.00 set – ISSN: 1049-4766 – mf#112921 – us Hein [322]

George maxwell gordon : the pilgrim missionary of the punjab: a history of his life and work 1839-1880 / Lewis, Arthur – London, 1889 [i.e. 1888] – (= ser 19th c british colonization) – 5mf – 9 – mf#1.1.8958 – uk Chadwyck [240]

George mueller of bristol and his witness to a prayer-hearing god / Pierson, Arthur Tappan – New York: Baker and Taylor, c1899 – 2mf – 9 – 0-8370-6303-5 – mf#1986-0303 – us ATLA [240]

George, Nathan Dow *see*
- Annihilationism not of the bible
- Universalism not of the bible

George Nathaniel Curzon, Marquis Curzon of Kedleston *see* Curzon, india and empire

George phoenix – George Town, Cape of Good Hope: Meyer & Gaugain. n1 [7 apr 1869]-n18 [3 aug 1869] (wkly) [mf ed Cape Town: SA Lib 1992] – 1r – 1 – mf#mp.1242 – sa National [079]

George ripley / Frothingham, Octavius Brooks – Boston: Houghton, Mifflin, c1882 [mf ed 1990] – (= ser American men of letters) – 1mf – 9 – 0-7905-4582-9 – mf#1988-0582 – us ATLA [070]

George, Robert James *see*
- The covenanter pastor
- The covenanter vision
- Pastor and people

George s frary diary *see* Diary, ms 3079

George sand / Galzy, Jeanne – Paris, France. 1950 – 1r – us UF Libraries [440]

George, Stefan *see*
- Deutsche dichtung
- Zeitgenoessische dichter

George, Stefan Anton *see*
- Briefwechsel zwischen george und hofmannsthal
- Die buecher der hirten- und preisgedichte, der sagen und saenge, und der haengenden gaerten
- Die buecher der hirten- und preisgedichte, der sagen und saenge und der haengenden gaerten
- Die fibel
- Hymnen; pilgerfahrten; algabal
- Das jahr der seele
- Der krieg
- Der siebente ring
- Tage und taten
- Der teppich des lebens und die lieder vom traum und tod

George stephens, piano forte and melodeon ware rooms, division street, cobourg, c w : pianos from the best new york, boston and albany makers for sale or to let... – [s.l: s.n. 187-?] [mf ed 1983] – 1mf – 9 – 0-665-39733-X – mf#39733 – cn Canadiana [780]

George, T J *see* The briton in india

George tames : professional photographer who specialized in capitol hill – (= ser Us senate historical office oral history coll) – 3mf – 9 – $15.00 – us Scholarly Res [770]

George w chadwick : his life and works / Campbell, Douglas Graves – U of Rochester 1957 [mf ed 19–] – 7mf – 9 – mf#fiche705 – us Sibley [780]

George w crile papers *see* Crile, george w, papers, ms 2806

George washington bicentennial news / United States George Washington Bicentennial Commission (US) – Alexandria VA. 1930 oct-1931 dec – 1r – 1 – mf#3282664 – us WHS [975]

George washington international law review – Washington DC 2002+ – 1,5,9 – (cont: george washington journal of international law & economics) – mf#8766,02 – us UMI ProQuest [341]

George washington journal of international law and economics – Washington DC 1982-2000 – 1,5,9 – (cont: journal of international law & economics) – ISSN: 0748-4305 – mf#8766.02 – us UMI ProQuest [341]

George washington journal of international law and economics – v1-33. 1966-2001 – 5,6,9 – $641.00 set – (v1-18 1966-84 on reel $253. v19-33 1985-2001 on mf $388. title varies: v1-2 1966-68 as studies in law and economic development. v3-5 1968 -71 as journal of law and economic development. v6-15 1971-81 as journal of international law and economics) – ISSN: 0748-4305 – mf#104041 – us Hein [341]

George washington law review – v1-68. 1932-2000 – 5,6,9 – $1636.00 set – (v1-53 1932-85 on reel or mf $1144.00. v54-68 1985-2000 on mf $492) – ISSN: 0016-8076 – mf#102891 – us Hein [340]

George washington letters, 1781-1798 – Charleston SC: Printed by C C Sebring [mf ed 1993] – 1mf – 9 – mf#51-178 – us South Carolina Historical [327]

George washington papers – 124r – 1 – $4,340.00 – Dist. us Scholarly Res – us L of C Photodup [975]

George Washington University. Dept of Law *see* Examination questions given in the law school of columbian university

George Washington University Seminar Conference *see*
- Argentina, brazil and chile since independence
- Caribbean area

George Washington University Seminar Conference... *see* Colonial hispanic america

George whitefield : a biography, with special reference to his labors in america / Belcher, Joseph – New York: American Tract Society, [1857?] – 2mf – 9 – 0-7905-4079-7 – mf#1988-0079 – us ATLA [920]

Sir George Williams University *see* Direction one

George wilson / Macaulay, James – London, England. 18– – 1r – us UF Libraries [240]

Georgekreis und literaturwissenschaft : zur wuerdigung und kritik der geistigen bewegung stefan georges / Roessner, Hans – Frankfurt a/M: M Diesterweg, 1938 (mf ed 1990) – 1r – 1 – (filmed with: der dritte humanismus im werke stefan georges und thomas manns...) – us Wisconsin U Libr [430]

Georgel, Jean *see* Voyage a saint-petersbourg, en 1799-1800

George's creek baptist church. pickens county. south carolina : church records – 1859-1909, 1916, 1941-58, 1965-69 – 1 – $17.82 – us Southern Baptist [242]

Georges, Karl August *see* Friedrich melchior grimm als kritiker der zeitgenoessischen literatur in seiner "correspondance litteraire" (1753-1770)

Georges Scholarus [Gennadius] *see* Contre les doutes de plethon sur aristote

Georges-Andre, soeur *see* Bio-bibliographie analytique, 1941-1957

Georgetown 1639-1905 – Oxford, MA (mf ed 1998) – (= ser Massachusetts vital records) – 53mf – 9 – 0-87623-395-7 – (mf 1-6: births 1639-1888 a-y. mf 6-8: marriage 1836-88 a-y. mf 8-13: paupers & accounts 1841-1915. mf 14-18: tax assessments 1863-65. mf 18-20: rebellion record 1861-65. mf 21-22: birth index 1825-97. mf 22-24: birth index 1898-1996. mf 25-28: groom index 1836-97. mf 28-30: marriage index 1898-1996. mf 31-33: death index 1836-1996. mf 34-35,44: publishments 1838-1907. mf 36: births 1825-51. mf 37: deaths 1836-49, 1859. mf 37: marriages 1836-49+. mf 38-39: births 1837-70. mf 39: marriages 1844-51; deaths 1844-58. mf 40-41: births 1870-1905. mf 42-44: marriages 1852-97. mf 45: marriages 1898-1905. mf 46-50: deaths 1860-94. mf 51-53: deaths 1895-1909) – us Archive [978]

Georgetown 1779-1849 – Oxford, MA (mf ed 1996) – (= ser Massachusetts vital record transcripts to 1850) – 3mf – 9 – 0-87623-252-7 – (mf 1t: births 1780-1851; marriage 1779; deaths 1838-49. mf 2t: intentions 1838-49; marriages 1836-49. mf 3t. births 1837-49; marriages 1843-49; deaths 1844-49) – us Archive [978]

Georgetown advocate – Georgetown DC. 1845 jul 3 [v4 n700] – 1r – 1 – mf#845948 – us WHS [071]

Georgetown baptist church – Washington. 1972+ (1) 1946+ (5) 1972+ (9) – 1r – 1 – $10.00 – (incl history 1843-1964) – mf#6532 – us Southern Baptist [242]

Georgetown baptist church. georgetown county. south carolina : church records – 1909-21, 1953-72 – 1 – us Southern Baptist [242]

Georgetown College *see* Historical catalog and history

Georgetown College. Georgetown, Ky *see* Catalog

Georgetown Colloquium On Africa 1st, Georgetown University *see* New forces in africa

The georgetown courier – Georgetown, [i.e. Washington] DC: J D McGill, 1865-76 – 1 – us CRL [071]

Georgetown first baptist church. georgetown, south carolina : church records – 1805-21 – 1 – 5.00 – us Southern Baptist [242]

[Georgetown-] georgetown gazette – CA. 1880-1924; 1933-35 – 10r – 1 – $600.00 – mf#BC02267 – us Library Micro [071]

[Georgetown-] georgetown news – CA. 1854-56 – 1r – 1 – $60.00 – mf#C02268 – us Library Micro [071]

Georgetown immigration law journal – v1-15. 1985-2001 – 9 – $404.00 set – ISSN: 0891-4370 – mf#111151 – us Hein [340]

Georgetown international environmental law review – Washington DC 1999+ – 1,5,9 – ISSN: 1042-1858 – mf#29971 – us UMI ProQuest [341]

Georgetown journal of gender and the law – v1-2. 1999-2001 – 9 – $72.00 – ISSN: 1525-6146 – mf#118021 – us Hein [342]

Georgetown journal of legal ethics – Washington DC 1999+ – 1,5,9 – ISSN: 1041-5548 – mf#29074 – us UMI ProQuest [170]

Georgetown journal on fighting poverty *see* Georgetown journal on poverty law and policy

Georgetown journal on poverty law and policy – v1-8. 1993-2001 – 9 – $171.00 set – (title varies: v1-5 1993-98 as) – ISSN: 1075-0827 – mf#115481 – us Hein [362]

Georgetown law journal – v1-14. 1912/13-1925/26 – 40mf – 9 – $60.00 – (v7 1918/19 was never published. add vols as copyright expires) – mf#LLMC 95-105 – us LLMC [340]

Georgetown law journal – v1-89. 1912-2001 – 1,5,6,9 – $2294.00 – (v7 never publ. v1-84 1912-96 in reel or mf $2068. v85-89 1997-2001 on mf $226.) – ISSN: 0016-8092 – mf#102901 – us Hein [340]

Georgetown law journal – WashingtonDC 1990+ – 1 – ISSN: 0016-8092 – mf#14456 – us UMI ProQuest [340]

The georgetown library society of south carolina : and the book borrowing habits of ten of its antebellum members / McInvaill, Dwight Emlyn Huger – 1978 [mf ed 1981] – 2mf – 9 – mf#50-02 – us South Carolina Historical [020]

[Georgetown-] rural review – CA. 1963-65 – (= ser Wine & agriculture coll) – 1r – 1 – $60.00 – mf#C02270 – us Library Micro [071]

[Georgetown-] town crier – CA. 1961-72 – 4r – 1 – $240.00 – mf#C02269 – us Library Micro [071]

George-town weekly ledger – Georgetown DC. 1792 jul 28 – 1r – 1 – mf#850820 – us WHS [071]
Georgi, J G *see* Bemerkungen einer reise im russischen reich im jahre 1772
Georgia – (= ser General education board: the early southern program) – 28r – 1 – $3640.00 – us Scholarly Res [370]
Georgia : code of georgia annotated – Norcross: The Harrison Co, 1936-apr 2002 update – 9 – $5146.00 set – mf#401760 – us Hein [348]
Georgia : session laws of american states and territories – 1787-2001 – 9 – $4139.00 set – mf#402600 – us Hein [348]
Georgia *see*
– Gault's decisions
– Georgia appellate reports
– Reports and opinions
– Reports, post-nrs
– Reports, pre-nrs
– Superior court decisions
Georgia, 1752-67 : from the public record office, london – (= ser Naval office shipping lists; British records relating to america in microform) – 1r – 1 – mf#96557 – uk Microform Academic [975]
Georgia Academy of Science *see* Bulletin of the georgia academy of science
Georgia african methodist of the african methodist church : official organ of the 6th episcopal district, ame church – Atlanta GA, mar 1950 [mf ed 2004] – 1r – 1 – (filmed with: christ methodist episcopal church (pittsburgh pa) year book [order032] and: negro commission bulletin [order034]) – mf#2004-s033 – us ATLA [242]
Georgia analytical repository – Ed. by Henry Holcombe. 1802-03 – 1 – us Southern Baptist [242]
Georgia analytical repository – La Salle IL 1802-03 – 1 – mf#3578 – us UMI ProQuest [240]
Georgia analytical repository – Savannah. 1802-03 – 1 – us ABHS [073]
Georgia appellate reports / Georgia – v1-24. 1907-20 – 37mf (1:42) 273mf (1:24) – 9 – $576.00 – (no pre-nrs vols. updates planned as copyright ends) – mf#LLMC 84-130 – us LLMC [340]
Georgia armchair researcher – v4, n101-v5, n104 [1983 spring-[1984/85 winter]] – 1r – 1 – (cont: armchair researcher) – mf#1207082 – us WHS [071]
Georgia attorney general reports and opinions – 1878-2000 + tables and indexes – 6,9 – $599.00set – (1878-81, 1885-1979 on reel $245. 1980-2000 + tables and indexes on mf $354) – mf#408190 – us Hein [348]
Georgia baptist – Augusta/Atlanta.General Missionary Baptist Convention of Georgia. 1898-1900, 1902, 1918, 1927-31, 1936-49. Single reels available – 1 – us ABHS [242]
The georgia baptist – Washington. 1972-1978 (1) 1950-1978 (5) 1974-1978 (9) – 1r – 1 – mf#6525 – us Southern Baptist [242]
Georgia baptist association : woman's missionary union scrapbook and history / Davison, Annie – 1 reel – 1 – $13.12 – (also includes church histories, biographies, and an historical table, 1890-1936) – us Southern Baptist [242]
Georgia Baptist Associations *see* Historical materials
Georgia baptist convention board minutes – 1816-64. 598p – 1 – us Southern Baptist [242]
Georgia bar association annual reports – 1884-1962 – 301mf – 9 – $451.00 – (lacking: 1959) – mf#LLMC 84-469 – us LLMC [340]
Georgia bar journal (macon) – v1-26. 1938-64 (all publ) – 9 – $336.00 set – (cont by: georgia state bar journal) – mf#102911 – us Hein [340]
Georgia bar journal (ns) – v1-6. 1995-2001 – 9 – $124.00v – (cont: georgia state bar journal) – mf#116451 – us Hein [340]
Georgia. Black Baptist Associations *see* Minutes
Georgia citizen – Macon GA. 1856 jul 19 – 1r – 1 – mf#859926 – us WHS [071]
Georgia Commission for the National Bicentennial Celebration *see* Georgia gazette
Georgia decisions / Georgia. Supreme Court – pts1-2. 1842-1843 (all publ) – (= ser Georgia supreme court reports) – 3mf – 9 – $4.50 – (a pre-nrs title) – mf#LLMC 94-003 – us LLMC [347]
Georgia educator – Tucker GA 1970-74 – 1 – mf#10465 – us UMI ProQuest [370]
Georgia enterprise – Covington GA. 1866 apr 27, may 11, jul 6 – 1r – 1 – (continued by: covington news) – mf#868696 – us WHS [071]
The georgia form book. / Silman, James B – Atlanta: Harrison, 1882. 396p. LL-127 – 1 – us L of C Photodup [348]
Georgia gazette / Georgia Commission for the National Bicentennial Celebration – v1 n1-v4 n1 [1973 jun-1976 jun] – 1r – 1 – mf#522769 – us WHS [975]
The georgia gazette, 1763-1770 – feb 7 1763-may 23 1770 [mf ed 1974] – 2r – 1 – (with p/g) – us MA Hist [071]
Georgia genealogical survey – v1, n101-v2, n104 [1981-1982] – 1r – 1 – mf#697560 – us WHS [978]
Georgia. General Assembly *see* The revolutionary records...1769-1784
Georgia historical chronicle – 1975 jan 6-1977 jan 24 – 1r – 1 – mf#1057012 – us WHS [978]
Georgia historical quarterly – Savannah GA 1917+ – 1,5,9 – ISSN: 0016-8297 – mf#229 – us UMI ProQuest [978]
Georgia historical quarterly – v7 n2-v8 n4 [1917 mar-1924 dec] – 1r – 1 – mf#770720 – us WHS [978]
Georgia informer – v12 n1-11 [1994 oct-1995 aug] – 1r – 1 – mf#3143885 – us WHS [071]
Georgia journal of international and comparative law – Athens GA 1970+ – 1,5,9 – ISSN: 0046-578X – mf#10075 – us UMI ProQuest [341]
The georgia jurist – v1 no 1-3. 1938 (all publ) – 1mf – 9 – $1.50 – mf#LLMC 84-470 – us LLMC [340]
The georgia law reporter – v1. 1885-86 (all publ) – 3mf – 9 – $13.50 – mf#LLMC 84-472 – us LLMC [340]
Georgia law review – v1 no 1-3. 1927-28 – 1mf – 9 – $1.50 – mf#LLMC 84-471 – us LLMC [340]
Georgia law review – v1-35. 1966-2001 – 1,5,6 – $980.00 set – (v1-29 1966-95 in reel $792. v30-35 1995-2001 in mf $188) – ISSN: 0016-8300 – mf#102941 – us Hein [340]
Georgia lawyer (macon) – v1-2. 1930-32 (all publ) – 9 – $18.00 set – mf#102951 – us Hein [340]
Georgia librarian – McDonough GA 1964-97 – 1,5,9 – (cont by: georgia library quarterly) – ISSN: 0016-8319 – mf#6712.01 – us UMI ProQuest [020]
Georgia library quarterly – McDonough GA 1998+ – 1,5,9 – (cont: georgia librarian) – mf#6712,01 – us UMI ProQuest [020]
Georgia loan office records relating to the loan of 1790 / U.S. Treasury Dept. Bureau of the Public Debt – (= ser Records of the bureau of the public debt) – 2r – 1 – mf#T694 – us Nat Archives [336]
Georgia loan office records relating to various loans, 1804-1818 / U.S. Treasury Dept. Bureau of the Public Debt – (= ser Records of the bureau of the public debt) – 1r – 1 – mf#T788 – us Nat Archives [336]
Georgia. New Sunbury Baptist Association. Savannah Extension Board *see* Record book, 1935-49
Georgia peace and justice report / Atlanta Clergy and Laity Concerned – 1984 sep/oct-1985 sep/oct – 1r – 1 – (cont: atlanta report) – mf#1108256 – us WHS [071]
Georgia. Presbytery (Cum. Pres. Ch.) *see* Minutes, 1880-1899
Georgia review – Athens GA 1947+ – 1,5,9 – ISSN: 0016-8386 – mf#765 – us UMI ProQuest [400]
Georgia social science journal – Ann Arbor MI 1986-92 – 1,5,9 – ISSN: 0016-8408 – mf#12692.01 – us UMI ProQuest [300]
Georgia. State Bar Association *see* Proceedings, 1884-1962
Georgia state bar journal – v1-31. 1964-95 (all publ) – 9 – $325.00 set – (cont: georgia bar journal (macon)) – ISSN: 0016-8416 – mf#102961 – us Hein [340]
Georgia state university law review – v1-18. 1984-2002 – 9 – $446.00 – ISSN: 8755-6847 – mf#110251 – us Hein [344]
Georgia straight – 1967 may 19-1969 dec 3, 1969 dec 3-1970 dec 29, 1971 jan 6-sep 28, 1971 sep 28-1972 dec 28, 1973 jan-1974 jan 3, 1974 jan 3-1975 jan 20, 1975 jan-dec, 1976 dec-oct 27, 1976 jan-1977 jan, 1977 oct-1978 aug, 1978 apr-1979 may 30 – 11r – 1 – (continued by: georgia grape; vancouver free press) – mf#770722 – us WHS [071]
Georgia. Supreme Court *see*
– Dudley's reports
– Georgia decisions
– Georgia supreme court reports
– Robert m charlton's reports
– Thomas t u p charlton's reports
Georgia supreme court reports / Georgia. Supreme Court – v1-161. 1846-1925 – 1543mf – 9 – $2314.00 – (pre-nrs: v1-22 1846-86 168mf $252.00. updates not poss before year 2006) – mf#LLMC 82-982 – us LLMC [347]
Georgia supreme court. reports of cases in law and equity argued and determined in the supreme court of georgia – Ann Arbor MI 1805-87 – 1 – mf#1788 – us UMI ProQuest [348]
Georgia trend – Norcross GA 1985+ – 1,5,9 – ISSN: 0882-5971 – mf#14905 – us UMI ProQuest [338]
Georgia. University *see* Phelps-stokes fellowship fund
Georgian – Savannah GA. 1818 nov28 – 1r – 1 – mf#859948 – us WHS [071]
The georgian bay : an account of its position, inhabitants, mineral interests, fish, timber and other resources, with map and illustrations / Hamilton, James Cleland – Toronto: J Bain; London: E Marbrough, 1893 – 2mf – 9 – mf#33399 – cn Canadiana [917]
The georgian era: memoirs of the most eminent persons, who have flourished in great britain, from the accession of george the first to the demise of george the fourth. – London: Vizetelly, Branston and Co., 1832-34. 4v. By ....Clarke, Brit. Museum – 1 – us Wisconsin U Libr [920]
Georgiana, Charlotte *see* Aristocratic women
Georgievskij gor sovet rk i kd *see* Izvestiia georgievskogo soveta rabochikh, soldatskikh i krest'ianskikh deputatov
Georgii Acropolitae *see* Historia (cbh14)
Georgii acropolitae annales *see* Breviarium historiae metricum (cshb29)
Georgii Cedreni *see* Compendium historiarum (cbh8)
Georgii Codini *see*
– De officiis magnae ecclesiae et aulae constantinopo-litanae
– Excerpta de antiquitatibus constantinopolitani
– Excerpta de antiquitatibus constantinopolitanis
Georgii Pachymeris *see*
– De michael et andronico palaeologis libri 13
– Historia rerum a michaele palaeologo
– Historia rerum ab andronico seniore
Georgii Phranzae *see* Chronicon (cbh30,2)
Georgii Syncelli *see* Chronographia [cbh5]
Georgii vallae placentini viri clariss[imi] de expetendis, et fvgiendis rebvs opvs... / Valla, Giorgio – Venetiis: In aedibvs Aldi Romani...1501 [mf ed 19–] – 2v on 1r – 1 – mf#film 2578 – us Sibley [780]
Les georgiques chretiennes: poeme couronne par l'academie francaise / Jammes, Francis – 5th ed. Paris: Mercure de France, 1914.216p – 1 – us Wisconsin U Libr [810]
Georgius Cedrenus *see* Ioannis scylitzae ope ab imm bekkero suppletus et emendatus (cshb34,35)
Georgius phrantzes, ioannes cananus, ioannes anagnostes (cshb36) / ed by Bekkeri, Imm – Bonnae, 1838 – (= ser Corpus scriptorum historiae byzantinae (cshb)) – €21.00 – ne Slangenburg [243]
Georgius syncellus et nicephorus cp (cshb12,13) / ed by Dindorfii, Guil – Bonnae. v1. 1829 – (= ser Corpus scriptorum historiae byzantinae (cshb)) – €27.00 – (v2 bonnae 1829 €21) – ne Slangenburg [243]
Geoscience canada – St John's NL 1974+ – 1,5,9 – ISSN: 0315-0941 – mf#10496 – us UMI ProQuest [550]
Geospatial solutions – Cleveland OH 2000+ – 1,5,9 – ISSN: 1529-7403 – mf#19720,01 – us UMI ProQuest [000]
Geotechnica – Praha: Ceskoslovenske akademie ved. v25. 1959 – us CRL [077]
Geotechnical testing journal – Conshohocken PA 1978+ – 1,5,9 – ISSN: 0149-6115 – mf#11879 – us UMI ProQuest [620]
Geotextiles and geomembranes – Oxford, England 1988+ – 1,5,9 – ISSN: 0266-1144 – mf#42516 – us UMI ProQuest [612]
Geothals, genius of the panama canal / Bishop, Joseph Bucklin – New York, NY. 1930 – 1r – us UF Libraries [972]
Geothermics – Oxford, England 1972+ – 1,5,9 – ISSN: 0375-6505 – mf#49261 – us UMI ProQuest [550]
Geotimes – Alexandria VA 1956+ – 1,5,9 – ISSN: 0016-8556 – mf#6668 – us UMI ProQuest [550]
Geoynim un gdoylim / Lunski, Hayim-Haikl – Vilna, Lithuania. 1931 – 1r – us UF Libraries [939]
Gepraegte form : goethes morphologie und die muenzkunst / Kuhn, Hermann – Weimar: H Boehlaus Nachf, 1949 [mf ed 1993] – (= ser Schriften der hallischen goethe-gesellschaft) – 62p/1lea/24pl (ill) – 1 – (incl bibl ref) – mf#8656 – us Wisconsin U Libr [430]
Gepshtein, D *see* Evreiskie avtonomisty
Ger y Lobez, Florencio *see*
– Tratado de construccion civil
– Tratado de construccion civil. atlas
Der gerade weg/illustrierter sonntag : deutsche zeitung fuer wahrheit und recht – Muenchen DE, 1929 31 mar-1938 8 mar – 4r – 1 – (filmed with: romanblaetter) – mf#4417 – gw Mikropress [074]
Gerakan mahasiswa / Forum. madjalah umum mahasiswa – Djakarta, 1954-1959 – 26mf – 9 – (missing: 1954, v1(2-3, 5-end); 1955, v2(2-3); 1956, v2(10-end)-v3(1, 7)) – mf#SE-363 – ne IDC [959]
Gerakan mahasiswa djakarta – Djakarta, 1948-1952 – 4mf – 9 – (missing: 1948-1951 v1-4(4)) – mf#SE-447 – ne IDC [950]
Geraldine : a souvenir of the st lawrence / Hopkins, Alphonso Alva – Boston, New York: Houghton, Mifflin, 1894 – 4mf – 9 – 0-665-91065-7 – mf#91065 – cn Canadiana [917]
Geraldine county chronicle – 31 aug 1878-1880; 13 apr 1895-1904; 26 jan-sep 1905; jan 1906-dec 1907; jan 1909-aug 1915 – 1 – (title changes to: geraldine guardian fr apr 1895) – mf#75.3 – nz Nat Libr [079]
Geraldine guardian *see* Geraldine county chronicle
Geraldton express and murchison gold fields news – Geraldton, Australia 2 nov 1894-3 mar 1916 (imperfect) – 1 – (cont: victorian express, west australian advertiser & pastoral, agricultural & mining register [21 jan 1885-26 oct 1894]) – uk British Libr Newspaper [079]
Geraldton-murchison telegraph – Geraldton, Australia. Sep 1892-Jan 1899.-w. 6 reels – 1 – uk British Libr Newspaper [072]
Geralton-longlac times-star – 1937-1986 – cn Commonwealth Imaging [071]
Gerard, Alexander *see* An essay on taste 1759...
Gerard david : painter and illuminator / Weale, William Henry James – London 1895 – (= ser 19th c art & architecture) – 1mf – 9 – mf#4.2.1389 – uk Chadwyck [750]
Gerard, Frances A *see* Angelica kauffmann
Gerard grote (1340-1384) : et les debuts de la devotion moderne / Epiney-Burgard, G – Wiesbaden, 1970 – 9mf – 8 – €18.00 – ne Slangenburg [240]
Gerard, Guillaume Samsoen de *see* Irene von starenburg
Gerard, James Watson *see* Titles to real estate in the state of new york
Gerard, John *see* What was the gunpowder plot?
Gerard manley hopkins : the man and the poet / Srinivasa Iyengar, K R – London; New York: Oxford University Press, 1948 – (= ser Samp: indian books) – (foreword by jerome d'souza) – us CRL [420]
Gerard, Pierre Auguste Florent *see* Histoire des francs d'austrasie
Gerasimov, M N *see* Nasha tribuna
Geraskin, S V *see* Nekropol donskogo monastyria
Gerathewohl, Fritz *see*
– Das deutsche vortragsbuch
– Goethe in heutiger sicht
Geraud, Jules *see* Patent laws of latin america (south and central america)
La gerbe : hebdomadaire de la volonte francaise – n1-214. Paris. 11 juil 1940-17 aout 1944 [wkly] – 1 – fr ACRPP [320]
La gerbe : petite revue artistique, litteraire et scientifique – n1-2. nov 1885 – 1 – (devenu: le sans-titre. petite revue artistique. n3-5. dec 1885-janv 1886. devenu: la revue blanche, artistique, litteraire et scientifique. n6-14. paris. fevr-juin 1886) – fr ACRPP [073]
Gerbe de sang / Depestre, Rene – Port-Au-Prince, Haiti. 1946 – 1r – us UF Libraries [972]
Gerbe pour deux amis / Camille, Roussan – Port-Au-Prince, Haiti. 1945 – 1r – us UF Libraries [972]
Gerber, Carl Friedrich von *see* System des deutschen privatrechts
Gerber, Ernst Ludwig [comp] *see* Historisch-biographisches lexicon der tonkuenstler
Der gerber etc – Vienna, Austria. jan 1887-15 dec 1889; 1890-15 dec 1909 – 8r – 1 – uk British Libr Newspaper [072]
[Gerber-] tehama county reporter – CA. 1953-63 [wkly] – 3r – 1 – $180.00 – mf#B02272 – us Library Micro [071]
[Gerber-] the gerber star – CA. 1924-39 – 5r – 1 – $300.00 – mf#BC02271 – us Library Micro [071]
Gerberding, George Henry *see*
– Life and letters of w. a. passavant, d.d
– The lutheran catechist
– Problems and possibilities
– The way of salvation in the lutheran church
Gerberg, Eugene J *see* Florida butterflies
Gerberon, Gabriel *see* Histoire general du jansenisme
Gerbert, Camill *see* Geschichte der strassburger sectenbewegung zur zeit der reformation 1524-1534
Gerbert, Martin *see*
– Four great works of martin gerbert (1720-1793), prince-abbot of st blasien monastery
– Scriptores ecclesiastici de musica sacra potissimum
Gerbert, une pape philosophe : d'apres l'histoire et d'apres la legende / Picavet, Francois – Paris: Ernest Leroux, 1897 – (= ser Bibliotheque de l'ecole des hautes etudes) – 1mf – 9 – 0-8370-8212-9 – (incl bibl ref) – mf#1986-2212 – us ATLA [240]
Gerbertus, M *see* Vetus liturgia alemanica
Gerbier, B *see*
– A brief discourse concerning the three chief principles of magnificent building
– Crijghs-architecture ende fortificatien
Gerbillon, J F *see*
– Historical observations on grand tartary
– Observations historiques sur la grande tartarie
– Travels into western tartary
Gerbrandy, P S *see*
– Ambon en de a r-partij
– Ambon en de ar-partij de vrijheidsstrijd van de republiek der zuid-molukken
Gerchunoff, Alberto *see* Pino y la palmera

Gerdener, G B A *see* Studies in the evangelisation of south africa
Gerdes, D *see* Epistolarum fasciculus
Gerdes, Dan *see* Florilegium historico-criticum librorum rariorum
Gerdesius, D *see* Introductio in historiam evangelii seculo 16 passim per europam renovati doctrinaeque reformatae
Gerding *see* Die bahn swakopmund-windhoek
Gere, V *see* Pervaia russkaia gsudarstvennaia duma
Gere, V I *see* Vtoraia gosudarstvennaia duma
Die gerechten von kummerow : roman / Welk, Ehm – Berlin: Deutscher Verlag c1943 [mf ed 1991] – 1r – 1 – (filmed with: vereinsamtes herz / josef weinheber) – mf#2982p – us Wisconsin U Libr [830]
Gerechtigkeit / International Ladies Garment Workers' Union – New York, N.Y. Yiddish edition of Justice. v. 1-40. 1919-Jan 1958 – 1 – us NY Public [071]
Die gerechtigkeit gottes im roemerbrief / Hellegers, Frederick Riker – Tuebingen: Christian Gulde, 1939 – 1mf – 9 – 0-524-08106-9 – (incl bibl ref) – mf#1993-9012 – us ATLA [227]
Geredja Katolik di Indonesia. buku tahunan *see* Kantor waligeredja indonesia
Gereformeerde geloofsleer : voor de literarische klassen van de theologische school en calvin college te grand rapids, mich / Heyns, William – Grand Rapids, MI: Eerdmans Sevensma, 1916 – 1mf – 9 – 0-524-06544-6 – mf#1991-2628 – us ATLA [242]
De gereformeerde kerk aan den arbeid 1657-1672 / Knappert, Laurentius – Leiden: E J Brill, 1913 [mf ed 1992] – 72p on 1mf – 9 – 0-524-05437-1 – (incl bibl ref) – mf#1990-1469 – us ATLA [242]
Geregtelike nadoodse ondersoek na die dood van stephen bantu biko gehou in die sinagoge, pretoria : extracts from the evidence [...] at the old synagogue, pretoria, during the inquest into the death of stephen bantu biko, nov-dec 1, 1977 – Johannesburg,[19–?] – us CRL [960]
Het gereinigt herte door 't geloof : aangetoond in verscheide dichtkundige uitbreidinge... / Dulken, G van – Amsteldam: Mercelis van Heems, 1715 – 3mf – 9 – mf#O-3063 – ne IDC [090]
Het gereinigt herte door 't geloof : aangetoond in verscheide dichtkundige uitbreidinge... / Dulken, G van – Amsterdam: Arend Hennebo, 1739 – 3mf – 9 – mf#O-3064 – ne IDC [090]
Gereke, Paul *see*
- Engelhard
- Die legenden
- Seifrits alexander

Gerena Bras, Gaspar *see* Aljibe
Gerest, Regis *see* Veritas. la vie chretienne raisonnee et meditee 4. la maison paternelle 1933
Le gerfaut : revue de la societe ornithologique du centre de la belgique – Bruxelles 1911-46 – v1-36 on 106mf – 9 – mf#z-2019c12 – ne IDC [590]
Gerhaeusser, Wilhelm *see* Muenchener septuaginta-fragmente
Gerhard, Adele *see*
- Das bild meines lebens
- Pflueger

Gerhard, Hans Ferdinand *see* In der jodutenstrasse
Gerhard, J *see*
- Commentarius super genesin
- Homiliarum sacrarum in pericopas evangeliorum dominicalium
- Locorum theologicorum cum pro adstruenda veritate
- Ein vnd fuenfftzig gottselige, christliche evangelische andachten oder geistreiche betrachtungen

Gerhard, Johann *see* Erklaerung der historie des leidens und sterbens unsers herrn christi jesu
Gerhard, Paul *see* Geschichte und beschreibung der mission unter den kolhs in ostindien
Gerhard, Wilhelm *see*
- Historien der alden e
- Das weib

Gerhards, Marty D *see* Handrail assisted versus nonhandrail assisted stairmaster gauntlet ergometry
Gerhardt, Dagobert von *see*
- Caritas
- Gerke suteminne
- Gewissensqualen
- Hypochondrische plaudereien

Gerhardt, Paul *see*
- Lyra gerhardti, or, a selection of paul gerhardt's spiritual songs
- Paul gerhardt's geistliche lieder
- Wach auf, mein herz

Gerhardt, Peter *see* Materielles scheidungsrecht
Gerhart, Emanuel Vogel *see*
- The education of woman
- Institutes of the christian religion
- An introduction to the study of philosophy

Gerhart hauptmann : kritische studien / Kutscher, Artur – Hirschberg: O Reier, [1909?] – 1r – 1 – (incl bibl ref) – us Wisconsin U Libr [430]
Gerhart hauptmann : eine studie / Behl, Carl Friedrich Wilhelm – Berlin: W Borngraeber, [1913?] [mf ed 1990] – 31p/[1]pl – 1 – mf#7445 – us Wisconsin U Libr [430]
Gerhart hauptmann / Sulger-Gebing, Emil – 3. verb. & verm. Aufl. Leipzig: B G Teubner, 1922 – 1r – 1 – (incl bibl ref) – us Wisconsin U Libr [430]
Gerhart hauptmann / Woerner, U C – 2. verb. & verm. Aufl. Berlin: A Duncker, 1901 – 1r – 1 – us Wisconsin U Libr [430]
Gerhart hauptmann and john galsworthy : a parallel / Trumbauer, Walter H R – Philadelphia, PA: [s.n.], 1917 – 1r – 1 – (incl bibl ref) – us Wisconsin U Libr [410]
Gerhart hauptmann – aus dem leben des deutschen geistes in der gegenwart : fuenf reden / Kuehnemann, Eugen – Muenchen: Beck, 1922 – 1r – 1 – us Wisconsin U Libr [430]
Gerhart hauptmann und seine besten buehnenwerke : eine einfuehrung / Bab, Julius – Berlin: F Schneider, c1922 [mf ed 2001] – 203p/1pl – 1 – mf#10576 – us Wisconsin U Libr [430]
Gerhart hauptmanns "emanuel quint" : eine studie / Faesi, Robert – Zuerich: Schulthess, 1912 – 1r – 1 – us Wisconsin U Libr [430]
Gerhart hauptmanns leben chronik und bild / Behl, Carl Friedrich Wilhelm & Voigt, Felix A [comp] – Berlin: Suhrkamp Verlag, 1942 [mf ed 1995] – 175p (ill) – 1 – mf#9087 – us Wisconsin U Libr [920]
Gerhart hauptmanns "till eulenspiegel" : ein studie / Enking, Ottomar – 1. und 2. Aufl. Berlin: S Fischer, 1930, c1929 – 1r – 1 – us Wisconsin U Libr [430]
Gerhart hauptmanns "till eulenspiegel" : versuch einer ideengehaltlichen, formalaesthetischen und literarhistorischen wuerdigung / Schwager, Lothar Helmut – Leipzig: Schwarzenberg & Schumann, [1930] – 1 – (incl bibl ref) – us Wisconsin U Libr [430]
Gerhart hauptmanns veland : seine entstehung und deutung / Hemmerich, Karl – Wuerzburg: Druck von Memminger, 1935 – 1r – 1 – (incl bibl ref) – us Wisconsin U Libr [430]
Gerhart, Isaac *see* Choral harmonie. enthaltend kirchen-melodien
Geriatric nursing – Oxford, England 1980+ – 1,5,9 – ISSN: 0197-4572 – mf#12205 – us UMI ProQuest [618]
Geriatrics – Cleveland OH 1946+ – 1,5,9 – ISSN: 0016-867X – mf#12940 – us UMI ProQuest [618]
Geriatrisches assessment im altersheim unter besonderer beruecksichtigung psychotroper medikation / Jost, Holger Wilfried – (mf ed 1997) – 2mf – 9 – €40.00 – 3-8267-2497-6 – mf#DHS 2497 – gw Frankfurter [618]
Der gerichtssaal; zeitschrift fuer zivil-und militarstrafrecht und strafprozessrecht sowie die erganzenden disziplinen – Stuttgart. On film: v1-116; 1849-1942. LL-0224 – 1 – us L of C Photodup [340]
Der gerichtstag : in fuenf buechern / Werfel, Franz – Leipzig: K Wolff 1923, c1919 [mf ed 1991] – 1r – 1 – (filmed with: der weg ins licht / gisela wenz-hartmann) – mf#3011p – us Wisconsin U Libr [830]
Die gerichtsverhandlung als literarisches motiv in der deutschen literatur des ausgehenden mittelalters / Strothmann, Friedrich Wilhelm – Jena: E Diederich, 1930 [mf ed 1993] – (= ser Deutsche arbeiten der universitaet koeln 2) – 75p – 1 – (incl bibl ref) – mf#8215 reel 1 – us Wisconsin U Libr [430]
Gerichts-zeitung – Wien, Manz etc. v.1-82; 1850-1931 – 1 – (missing: v76-78; 1925-27.) – mf#LL-0200 – us L of C Photodup [340]
Gerichtszeitung *see* Hamburg-altonaer volksblatt
Gerifaltes extremenos : angel marina, el juglar de la virgen morena / Gutierrez Macias, Valeriano – Badajoz: Dip. Provincial, 1974. Sep. REE – 1 – sp Bibl Santa Ana [946]
Los gerifaltes marxistas por... / Gonzalez Alvarez, Claudio – Caceres: Tip. Floriano, 1939 – 1 – sp Bibl Santa Ana [335]
Gerin, Elzear *see* Le saint-maurice
Gering courier – Gering, NE: Wood & Wisner. 14th yr n16. aug 3 1900- (wkly) – 1 – (cont: gering weekly courier. absorbed: gering midwest 1927, banner county news 1955 and: minatare free press 1964) – us NE Hist [071]
The gering courier – Gering, NE: A B Wood. 1v. v1 n1. apr 27 1887-v1 n42. feb 9 1888 (wkly) [mf ed lacks dec 22 1887] – 1 – (cont by: gering weekly courier) – us NE Hist [071]
Gering, Hugo *see* Glossar zu den liedern der edda
The gering midwest – Gering, NE: [Will M Maupin] 1918-v9 n34. jul 1 1927 (wkly) [mf ed 1919-27 (gaps)] – 3r – 1 – (cont: midwest magazine. absorbed: lyman enterprise. absorbed by: gering courier (1900)) – us NE Hist [071]
Gering weekly courier – Gering, NE: Wood & Bristol. 14v. v1 n43. feb 16 1888-14th yr n15. jul 27 1900 (wkly) [mf ed with gaps] – 3r – 1 – (cont: gering courier. absorbed: our home (minatare ne). cont by: gering courier (1900)) – us NE Hist [071]
Gerin-Lajoie, Antoine *see*
- Le catechisme des electeurs d'apres l'ouvrage de a gerin-lajoie
- Catechisme politique ou elemens du droit public et constitutionnel du canada
- Dix ans au canada de 1840 a 1850
- Jean rivard
- Jean rivard, economiste
- Jean rivard, le defricheur
- Le jeune latour

Gerin-Lajoie, Marie *see*
- Traite de droit usuel
- A treatise on everyday law

Gerke, F *see* Die stellung des ersten clemensbriefes, innerhalb der entwicklung der altchristlichen gemeindeverfassung und des kirchenrechts
Gerke suteminne : ein maerkisches kulturbild aus der zeit des ersten hohenzollern / Gerhardt, Dagobert von – Breslau: S Schottlaender. 3v in 2. 1906 – 1r – 1 – us Wisconsin U Libr [430]
Gerken, IU S *see*
- Russkaia armiia [omsk: 1918]
- Russkii voin

Gerlach, Andreas *see* Die wechselwirkung organischer molekuele mit supra-molekularen verbindungen
Gerlach, Fritz *see* Auf neuer scholle
[Gerlach-] gerlach express – NV. 19 jun 1964 – 1r – 1 – $60.00 – mf#U04572 – us Library Micro [071]
Gerlach, Hans Egon *see* Goethe erzaehlt sein leben
Gerlach, Hellmut von *see* Der zusammenbruch der deutschen polenpolitik
Gerlach, Hermann *see*
- Die dotationsansprueche und der nothstand der evangelischen kirche im koenigreich preussen
- Gegen renan, leben jesu
- Die letzten dinge

Gerlach, Kurt *see*
- Die birken in den steinen
- Die strasse nach prag

Gerlach, Martin von *see* Der mensch im stande der schuld nach dem buche jesaja
Gerlach, Otto von *see* Commentary on the pentateuch
[Gerlach-] valley press – NV. 1961-1962 – 1r – 1 – $60.00 – mf#U04540 – us Library Micro [071]
Gerlache, Eugene de *see* De laatste dagen van het pauselijk leger
Gerland, Georg *see* Der mythus von der sintflut
Gerle, Wolfgang *see*
- Boehmen
- Prag und seine merkwuerdigkeiten

Gerler, Jr, Edwin R *see* Journal of school violence
The germ : thoughts toward nature in poetry, literarure and art – no. 1-4. 1901 – 1 – us AMS Press [420]
Germain, Alexandre *see* Une loge maconnique d'etudiants a montpellier
Germain, Andre *see* Goethe et bettina
Germain, M *see* Museum italicum
Germain, Michel *see* Monasticon gallicanum
Germain, Simone *see* Pedagogie partique
Germain, Victorin *see*
- Allo!...allo! ici la creche
- Les recits de la creche

Germaine-Marie, soeur *see* Bibliographie de la psychologie rationnelle au canada francais, 1945-1963
German administration official records / Nauru – 1887-1916 – 1r – 1 – mf#pmb16 – at Pacific Mss [324]
German affairs – London (GB), 1947 6 nov-1948 14 jun – 1r – 1 – gw Misc Inst [943]
German air force reports : luftgaukommandos, flak, deutsche luftwaffenmission in rumanien / Germany. Air Force – (= ser National archives coll of foreign records seized, 1941-) – 64r – 1 – mf#T405 – us Nat Archives [355]
German american / German American Emergency Conference – 1943 feb-1946 feb 15, 1951 apr-1960 apr, 1960 jun-1968 jun – 3r – 1 – mf#701895 – us WHS [305]
The german american – New York NY (USA), 1942 1 may-1952 31 mar – 2r – 1 – gw Misc Inst [305]
German American Emergency Conference *see* German american
German american law journal – v1-5. 1991-96 – 9 – $50.00 set – ISSN: 1083-1894 – mf#116581 – us Hein [340]
German and japanese surrender documents of world war 2 and the korean armistice agreements / U.S. Joint Chiefs of Staff – (= ser Records Of The U.S. Joint Chiefs Of Staff) – 1r – 1 – mf#T826 – us Nat Archives [355]
The german and swiss settlements of colonial pennsylvania : a study of the so-called pennsylvania dutch / Kuhns, Oscar – New York: H Holt, 1901, c1900 – 1mf – 9 – 0-7905-5415-1 – (incl bibl ref) – mf#1988-1415 – us ATLA [975]
German anti-supernaturalism : six lectures on strauss's life of jesus / Harwood, Philip – London: C Fox, 1841 [mf ed 1993] – 1mf – 9 – 0-524-06572-1 – mf#1992-0915 – us ATLA [240]
The german army high command, 1938-1945 – (= ser World war 2 research colls) – 4r – 1 – $710.00 – 0-89093-107-0 – (with p/g. coll of 42 archival titles written by former oberkommando officers under allied supervision at the interrogation enclosure in neustadt, germany, giving the inside history of the german army command structure. trans into english) – us UPA [355]
German Baptist Brethren (US) *see* Missionary hymns
German baptist brethren (u.s.). annual meeting. proceedings of the annual meeting – Myersdale PA: J Quinter 1876-79 (Huntingdon PA: J R Durborrow) [mf ed 2007] – 1 – (imprint varies; minutes first publ in 1778; damaged: 1877 p17; cont by: german baptist brethren (u.s.). annual meeting. report of the proceedings of the brethren's annual meeting for... [1880-84] 1r [1090]; german baptist brethren (u.s.). annual meeting. report of the brethren's annual meeting [1885-88] 1r [1091]; german baptist brethren (u.s.). annual meeting. report of the brethren's annual meeting proceedings of the brethren's annual meeting [1889-1902] 1r [1092]; german baptist brethren (u.s.). annual meeting. full report of the proceedings of the brethren's annual meeting [1903-07] 1r [1093]) – mf1089 – us ATLA [242]
German Baptist Brethren (US). General Church Erection and Missionary Committee *see* Annual report of the general missionary committee and the book and tract work...
German Baptist Convention *see*
- Annual
- Jahrbuch des bundes

German baroque literature : from the the yale university collection and the private collection of harold hantz – 1280r – 1 – (yale coll: 669r c35-28341. harold jantz coll: 611r c35-28342. guide available) – us Primary [430]
German books : a selection from the most important publications of the years 1914-1925: exhibited at new york...earl hall, columbia university, autumn, 1925 = Deutsche buecher: eine auswahl der wichtigsten erscheinungen aus den jahren 1914-1925 – Leipzig: Boersenverein der Deutschen Buchhaendler 1925 [mf ed 2005] – 1r – 1 – (preceded by essay in german & english: deutschlands anteil an der internationalen wissenschaftlichen arbeit = germany's work for international science [34p]) – mf#11476 – us Wisconsin U Libr [430]
German books and periodicals : from the wilhelm scherer collection – 123r – 1 – (monographs and periodicals on german philosophy, aesthetics, art, history and literature from the german baroque period to the 19th century. title listing available) – mf#C35-15000 – us Primary [430]
German books on china from the late 15th century to 1920, pt 1 : history = Deutschsprachige schriften zu china vom spaeten 15.jahrhundert bis 1920, teil 1: geschichte / Vonderstein, Mirko [comp]; ed by Walravens, Hartmut – [mf ed 2003-04] – 685mf (1:24) in 3 installments – 9 – diazo €4770.00 (silver €5670 isbn: 978-3-598-35386-4) – ISBN-10: 3-598-35382-0 – ISBN-13: 978-3-598-35382-6 – gw Saur [951]
German books on islam from the 16th century to 1900, pt 1 : religion and theology, law and customs = Deutschsprachige schriften zum islam vom 16. jahrhundert bis 1900, teil 1: religion und theologie, recht und sitte / Cikar, Jutta [comp] – [mf ed 2002-03] – 788mf (1:24) in 3 installments – 9 – diazo €5100.00 (silver €6000 isbn: 978-3-598-35193-8) – ISBN-10: 3-598-35192-5 – ISBN-13: 978-3-598-35192-1 – (with guide) – gw Saur [260]
German books on islam from the 16th century to 1900, pt 2 : history of the arab world and persia = Deutschsprachige schriften zum islam vom 16. jahrhundert bis 1900, teil 2: geschichte der arabischen welt und persiens / Cikar, Jutta [comp] – [mf ed 2003-04] – 1087mf (1:24) in 5 installments – 9 – diazo €8500.00 (silver €9750 isbn: 978-3-598-35291-1) – ISBN-10: 3-598-35290-5 – ISBN-13: 978-3-598-35290-4 – (with guide) – gw Saur [260]
German books on islam from the 16th century to 1900, pt 3 : history of the ottoman empire = Deutschsprachige schriften zum islam vom 16. jahrhundert bis 1900, teil 3: geschichte des osmanischen reiches – [mf ed 2005-07] – 1189mf (1:24) in 5 installments – 9 – diazo €8500.00 (silver €9750 isbn: 978-3-598-35305-5) – ISBN-10: 3-598-35304-9 –

ISBN-13: 978-3-598-35304-8 – (incl guide) – gw Saur [260]

German books on japan 1477 to 1945, pt 1 : history = Deutschsprachige schriften zu japan 1477 bis 1945, teil 1: geschichte / ed by Walravens, Hartmut – [mf ed 2002-04] – 1425mf (1:24) in 6 installments – 9 – diazo €9900.00 (silver €11,700 isbn: 978-3-598-35173-0) – ISBN-10: 3-598-35172-0 – ISBN-13: 978-3-598-35172-3 – (with guide) – Dist. ja Yushodo – gw Saur [950]

German books on japan 1477 to 1945, pt 2 : literature, music and fine arts = Deutschsprachige schriften zu japan 1477 bis 1945, teil 2: literatur, musik und bildende kunst / ed by Walravens, Hartmut – [mf ed 2004] – 649mf (1:24) in 3 installments – 9 – diazo €4950.00 (silver €5850 isbn: 978-3-598-35231-7) – ISBN-10: 3-598-35230-1 – ISBN-13: 978-3-598-35230-0 – (with guide) – gw Saur [700]

A german buddhist, oberpraesidialrat theodor schultze : a biographical sketch / Pfungst, Arthur – [2nd ed] London: Luzac, 1902 [mf ed 1993] – 1mf – 9 – 0-524-07793-2 – (trans fr german by l f de wilde) – mf#1991-0170 – us ATLA [280]

German, Christian *see* Der zeitgeist und die kirche

The german church on the american frontier : a study in the rise of religion among the germans of the west, based on the history of the evangelischer kirchenverein des westens (evangelical church society of the west) 1840-1866 / Schneider, Carl Edward – St. Louis, Mo.: Eden Pub. House, 1939. Chicago: Dep of Photodup, U of Chicago Lib, 1975 (1r); Evanston: American Theol Lib Assoc, 1984 (1r) – 1 – 0-8370-1530-8 – (incl bibl ref) – mf#1984-B465 – us ATLA [240]

The german church struggle : tribulation and promise / Barth, Karl – 2nd ed. [London]: Kulturkampf Association, [1938?] – 1mf – 9 – 0-524-08094-1 – mf#1993-9000 – us ATLA [240]

The german classics of the nineteenth and twentieth centuries : masterpieces of german literature, translated into english / ed by Francke, Kuno – Albany: J B Lyon c1913-c1914 [mf ed 1993] – 20v (ill) – 1 – (some vols publ: new york: german publ soc. these have title: the german classics : masterpieces of german literature translated into english) – mf#8630 – us Wisconsin U Libr [430]

German "colonisation" begun in spain – NY, 1937. Fiche W 910. (Blodgett Collection of Spanish Civil War Pamphlets) – 9 – us Harvard [946]

German correspondent – La Salle IL 1820-21 – 1 – mf#3777 – us UMI ProQuest [071]

The german demand for colonies / Glahn, Gerhard Ernst Ludwig Von – Evanston, 1939 – us CRL [943]

German documents among the war crimes records of the judge advocate division, headquarters, us army, europe / U.S. Army Commands – (= ser Records of united states army commands, 1942-) – 20r – 1 – mf#T1021 – us Nat Archives [345]

German domination of st andrews bay / Writers' Program (Fla) – s.l, s.l? 1918 – 1r – us UF Libraries [978]

The german drama of the nineteenth century = Deutsche drama des neunzehnten jahrhunderts in seiner entwicklung dargestellt / Witkowski, Georg – New York: H Holt, 1909 [mf ed 1993] – x/230p – 1 – (incl ind. authorized trans fr 2nd german ed by l e horning) – mf#8282 – us Wisconsin U Libr [790]

The german drama of the nineteenth century = Deutsche drama des neunzehnten jahrhunderts in seiner entwicklung dargestellt / Witkowski, Georg – New York: H Holt, 1909 [mf ed 1993] – x/230p – 1 – (trans fr 2d german ed by l e horning. incl ind) – mf#8282 – us Wisconsin U Libr [430]

German dramatists of the 19th century / Kaufmann, Friedrich Wilhelm – Los Angeles: Lymanhouse, c1940 [mf ed 1993] – 8282 – 1 – mf#vi/215p – us Wisconsin U Libr [430]

German economic review – Stuttgart, Germany 1963-77 – 1,5,9 – ISSN: 0016-8734 – mf#5157 – us UMI ProQuest [338]

German education : past and present = Deutsche bildungswesen in seiner geschichtlichen entwickelung / Paulsen, Friedrich – New York: Charles Scribner, 1908 – 1mf – 9 – 0-524-02649-1 – (incl bibl ref. in english) – mf#1990-0673 – us ATLA [370]

The german element in the united states : with special reference to its political, moral, social, and educational influence / Faust, Albert Bernhardt – Boston: Houghton Mifflin, 1909 – 3mf – 9 – 0-7905-6587-0 – (incl bibl ref) – mf#1988-2587 – us ATLA [305]

German foreign ministry archives : filmed by the american historical association, 1867-1920 / Germany. Foreign Ministry – (= ser National archives coll of foreign records seized, 1941-) – 434r – 1 – mf#T149 – us Nat Archives [943]

German Fusilier Society. South Carolina *see* Minutes, 1905-1914

German history – Newbury Park CA 2002+ – 1,5,9 – ISSN: 0266-3554 – mf#17497 – us UMI ProQuest [943]

German immigration into pennsylvania through the port of philadelphia / Diffenderffer, Frank Ried – Lancaster, PA. 1900 – 1r – us UF Libraries [304]

German influence on samuel taylor coleridge / Goodman, Hardin Mcdonald – s.l, s.l? 1957 – 1r – us UF Libraries [420]

German international – Bonn-Lengsdorf, Germany 1974-79 – 1,5,9 – ISSN: 0016-8769 – mf#9772 – us UMI ProQuest [338]

German liberty authors / Florer, Warren Washburn – Boston: R G Badger c1918 [mf ed 1993] – 1r – 1 – (incl ind. filmed with: the susanna theme in german literature / paul f casey) – mf#8145 – us Wisconsin U Libr [430]

German life and letters – Oxford, England 1936+ – 1,5,9 – ISSN: 0016-8777 – mf#1241 – us UMI ProQuest [430]

German literature of the mid-nineteenth century in england and america : as reflected in the journals, 1840-1914 / Hathaway, Lillie Vinal – Boston: Chapman & Grimes, c1935 – 1 – us Wisconsin U Libr [430]

German literature through nazi eyes / Atkins, Henry Gibson – London: Methuen, 1941 [mf ed 1993] – vii/136p – 1 – (incl bibl ref and ind) – mf#8119 – us Wisconsin U Libr [430]

German lyric poetry : a critical analysis of selected poems from klopstock to rilke / Prawer, Siegbert Salomon – London: Routledge & K Paul, 1952 – 1r – 1 – (incl bibl ref) – us Wisconsin U Libr [810]

German lyric poetry / Macleod, Norman – New York: Harcourt, Brace & Co [1930?] – 158p – 1 – (incl ind) – us Wisconsin U Libr [430]

The german lyrik / Lees, John – London, Toronto: J M Dent & sons Ltd; New York: E P Dutton & Co, 1914 [mf ed 1993] – vii/266p – 1 – mf#8282 – us Wisconsin U Libr [430]

German men of letters : twelve literary essays / ed by Natan, Alex – London: O Wolff, 1961- [mf ed 1992] – 1r – 1 – (v4 ed by brian keith-smith. v1-3 have subtitle: twelve literary essays. v4 has distinctive title: essays on contemporary german literature. incl bibl ref) – mf#8249 – us Wisconsin U Libr [840]

German military and technical manuals, 1910-1945 – (= ser National archives coll of foreign records seized, 1941-) – 126r – 1 – mf#T283 – us Nat Archives [355]

The german novelists – London: F Warne, [1880?] [mf ed 1993] – (= ser Chandos library) – xv/623p – 1 – (trans fr originals with critical and biogr notices by thomas roscoe) – mf#8190 – us Wisconsin U Libr [390]

The german peace offer : address by colonel george t denison before the empire club of toronto, 28th december, 1916 / Denison, George Taylor – [Toronto?: s.n, 1916?] – 1mf – 9 – 0-665-73922-2 – mf#73922 – cn Canadiana [933]

German, Pedro M *see* Mujer y patria

German philosophy and politics / Dewey, John – New York: H Holt, 1915 – (= ser John Calvin McNair Lectures) – 1mf – 9 – 0-7905-7287-7 – mf#1989-0512 – us ATLA [190]

The german pietists of provincial pennsylvania, 1694-1708 / Sachse, Julius Friedrich – Philadelphia: Printed for the author, 1895 (Philadelphia: PC Stockhausen) – 2mf – 9 – 0-7905-6947-7 – mf#1988-2947 – us ATLA [243]

German pow camp newspapers – L110093 – 15r – 1 – $525.00 – us L of C Photodup [350]

German propaganda of the first world war *see* First world war: a documentary record, series 1

German protestantism and the right of private judgement in the interpretation of holy scripture : a brief history of german theology from the reformation to the present time: in a series of letters to a layman / Dewart, Edward Hartley – Oxford: J H Parker; London: J G F and J Rivington, 1844 – 1mf – 9 – 0-7905-1038-3 – (incl bibl ref) – mf#1987-1038 – us ATLA [242]

German psychology of today, the empirical school / Ribot, Theodule Armand – Trans. from 2nd French ed. by James Mark Baldwin. New York: Scribner, c1886. xxi,307p – 1 – us Wisconsin U Libr [150]

German publications on alcoholism – Berlin, 1-99. 1891-1958? – 7r – us UF Libraries [360]

German quarterly – Cherry Hill NJ 1928+ – 1,5,9 – ISSN: 0016-8831 – mf#12356 – us UMI ProQuest [430]

German rationalism : in its rise, progress, and decline, in relation to theologians, scholars, poets, philosophers, and the people = Kirchengeschichte des 18. und 19. jahrhunderts / Hagenbach, Karl Rudolf; ed by Gage, William Leonard & Stuckenberg, John Henry Wilbrandt – Edinburgh: T & T Clark, 1865 [mf ed 1990] – (= ser Vorlesungen ueber wesen und geschichte der reformation 5-6) – 1mf – 9 – 0-7905-4966-2 – (english trans fr german by ed) – mf#1988-0966 – us ATLA [230]

The german reformation / Schaff, Philip – 2nd rev ed. New York: Scribner, 1892 – 2mf – 9 – 0-524-01891-X – (incl bibl ref) – mf#1990-0518 – us ATLA [242]

German Reformed Church (US). Liturgical Committee *see* The liturgical question with reference to the provisional liturgy of the german reformed church

German reformed messenger *see* Reformed church messenger

German religious life in colonial times / Bittinger, Lucy Forney – Philadelphia: JB Lippincott, 1906 [mf ed 1989] – 1mf – 9 – 0-7905-4091-6 – (incl bibl ref) – mf#1988-0091 – us ATLA [240]

German reports on synthetic rubber, 1937-1945 / U.S. Reconstruction Finance Corporation – (= ser Records Of The Reconstruction Finance Corporation) – 13r – 1 – mf#T948 – us Nat Archives [338]

A german scholar in the east : travel scenes and reflections = Welt des ostens / Hackmann, Heinrich Friedrich – London: K Paul, Trench, Truebner; New York: James Pott, 1914 [mf ed 1991] – 1mf – 9 – 0-524-01444-2 – (in english) – mf#1990-2439 – us ATLA [915]

The german sectarians of pennsylvania, 1708-1800 : a critical and legendary history of the ephrata cloister and the dunkers / Sachse, Julius Friedrich – Philadelphia: Printed for the author, 1899-1900 (Philadelphia: PC Stockhausen) – 3mf – 9 – 0-7905-7141-2 – mf#1988-3141 – us ATLA [243]

German sixteenth century imprints from the reformation period – 1520-1666 – 1 – us Southern Baptist [242]

German social report – Germany 1963-70 – 1 – mf#10705 – us UMI ProQuest [300]

The german soul in its attitude towards ethics and christianity, the state, and war : two studies / Huegel, Friedrich, Freiherr von – London: JM Dent; New York: EP Dutton, 1916 – 1mf – 9 – 0-7905-3916-0 – mf#1989-0409 – us ATLA [943]

German studies presented to leonard ashley willougby : by pupils, colleagues and friends on his retirement – Oxford: B. Blackwell, 1952 – 1 – (incl bibl ref) – us Wisconsin U Libr [430]

German studies presented to professor h.g. fiedler, m.v.o., by pupils, colleagues, and friends on his seventy-fifth birthday, 28 april 1937 – Oxford: Clarendon Press, 1938 – 1r – 1 – (incl bibl ref) – us Wisconsin U Libr [943]

German Texan Heritage Society *see* Newsletter

German tribune – Hamburg DE, 10 feb 1970-dec 1976 – 7r – 1 – uk British Libr Newspaper [074]

German universities : a narrative of personal experience, together with recent statistical information, practical suggestions, and a comparison of the german, english and american systems of higher education / Hart, James Morgan – New York: Putnam, 1874 – 1mf – 9 – 0-7905-4970-0 – mf#1988-0970 – us ATLA [378]

The german universities : their character and historical development = Wesen und geschichtliche entwicklung der deutschen universitaeten / Paulsen, Friedrich – New York: Macmillan, 1895 – 1mf – 9 – 0-7905-6871-3 – (incl bibl ref. in english) – mf#1988-2871 – us ATLA [378]

The german universities and university study = deutschen universitaeten und das universitaetsstudium / Paulsen, Friedrich – London: Longmans, Green, 1908 – 2mf – 9 – 0-524-03657-8 – (incl bibl ref. in english) – mf#1990-1085 – us ATLA [378]

The german universities for the last fifty years = das universitaetsstudium in deutschland waehrend der letzten 50 jahre / Conrad, Johannes – Glasgow: D. Bryce, 1885 – 1mf – 9 – 0-7905-8022-5 – (in english) – mf#1988-6003 – us ATLA [378]

German-Acadian Coast Historical and Genealogical Society *see* Voyageurs

German-american bulletin / Montgomery Co. Dayton – jul 1933-jun 1941 [wkly] – 3r – 1 – mf#B5195-5199 – us Ohio Hist [071]

German-American National Congress *see* Deutsch-amerikaner

German-Canadian Historical Association *see* Canadiana germanica

German-english dictionary for foresters – 1939 – 16mf – 7 – mf#474 – uk Microform Academic [430]

Germanenzug / Gmelin, Otto – Jena: E Diederichs, 1940, c1934 (mf ed 1990) – 1r – 1 – (filmed with: sommerwind ueber tormohlenhof) – us Wisconsin U Libr [830]

Germaneren / Germanske S S Norge – Kamporgan for Germanske SS. Norge. Oslo, Norway. -w. Oct. 1942-Apr. 1945. (very imperfect, 1 reel) – 1 – uk British Libr Newspaper [072]

Germania – Cleveland, Ohio. -d. Jan 1889-Jun 1889 – 1 – us Western Res [071]

Germania – Milwaukee WI. 1873 nov 8-dec, 1874 jan-sep 19, 1876 sep 19 – 2r – 1 – (continued by: germania und abend=post) – mf#1165019 – us WHS [071]

Germania – Milwaukee WI. 1873 jun 18, 1876 jan 26, 1873 jun 18-dec 10, 1875 jun 16-1876 jun 7, 1878 jun 12-1879 jun 4 – 2r – 1 – (continued by: milwaukee herold; milwaukee america) – mf#1093690 – us WHS [071]

Germania – Berlin DE, 1870 28 dec-1938 [gaps] – 196r – 1 – (filmed by misc inst: 1919 jul-sep, 1920 jan-mar, 1922 jan-mar, jul-sep, 1923 jan-1 jul, 1924 jan-apr [8r]; 1918 1 oct-1919 30 jun, 1920-1925 30 jun, 1927 1 jul-31 dec, 1930 1 jul-31 aug. incl suppls: akademische stimmen 1922 jan-mar; blaetter fuer rechtspflege 1922 jan-mar; kirche und welt 1911 24 sep-1914 2 aug; sonntagsblatt 1897 30 may-1898 11 sep; wissenschaftliche beilage [ab 1924?: wissenschaftliche blaetter] 1924 apr, 1925 mar-apr) – mf#207 – gw Mikropress; gw Misc Inst [074]

Germania / Jefferson Co. Steubenville – 8/1876-99,01-02,05-1916 [wkly] – 20r – 1 – (in german) – mf#B8821-8840 – us Ohio Hist [071]

Germania : korrespondentblatt der roemisch-germanischen kommission – Frankfurt/M DE, 1917-63 – 6r – 1 – mf#4660 – gw Mikropress [943]

Germania – Leipzig DE, 1851, 1852 – 1r – 1 – gw Misc Inst [943]

Germania – London, UK. 9 Apr-7 May 1859 – 1 – uk British Libr Newspaper [072]

Germania : neueste nachrichten aus deutschland, oesterreich, der schweiz und amerika – New York NY 23 feb 1874 – 1 – uk British Libr Newspaper [071]

Germania / Summit Co. Akron – (1876-82), 1888-90, 1893 [wkly, semiwkly] – 6r – 1 – (in german) – mf#B10549-10554 – us Ohio Hist [071]

Germania / Summit Co. Akron – jan 1919-jul 1920 [irreg, twice wkly] – 2r – 1 – (in german) – mf#B3935-3936 – us Ohio Hist [071]

Germania / Wyandot Co. Upper Sandusk – v1 n1. jun 1886-dec 1895 [wkly] – 4r – 1 – (in german) – mf#B8493-8496 – us Ohio Hist [071]

Germania *see* Milwaukee-germania-abend-post

The germania; a collection of the most favorite operatic airs, marches, polkas, waltzes, dances, and melodies of the day / Burditt, B A – Arranged in an easy and familiar style, for four, five, and six instruments. Boston: Oliver Ditson, 1855. MUSIC 1991, Item 1 – 1 – us L of C Photodup [780]

Germania deplorata : sive relatio, qua pragmatica momenta belli pacisque expenduntur – [Paris], 1641 – 1mf – 9 – mf#O-00 – ne IDC [090]

Germania Judaica – Koelner Bibliothek zur Geschichte des deutschen Judentums e.V. *see*
- Deutschsprachige zeitungen aus palaestina und israel
- Deutschsprachige zeitungen aus palaestina und israel, abt 1
- Deutschsprachige zeitungen aus palaestina und israel, abt 2

Germania und abend=post – 1897 may-dec, 1898 jan-dec, 1899 jan-dec, 1900 jan-dec, 1901 jan-nov 2 – 21r – 1 – (cont: germania [milwaukee, wi: 1873 : daily]; abend post; cont by: milwaukee germania abendpost) – mf#1166680 – us WHS [071]

Germania-herold – Milwaukee WI. 1913 jan 2/mar 6-1918 apr 25-may 25 [gaps] – 32r – 1 – (cont: milwaukee germania abendpost; milwaukee herold und seebote; cont by: milwaukee-herold [milwaukee, wi: 1918 : daily]) – mf#855189 – us WHS [071]

Germania-herold *see* Milwaukee-germania-abend-post

Germanic review – Washington DC 1955+ – 1,5,9 – ISSN: 0016-8890 – mf#958 – us UMI ProQuest [430]

Germanicus / Arnault, Antoine-Vincent – Paris, France. 1817 – 1r – us UF Libraries [440]

Germanien – Berlin DE, 1933 n1, 1937, 1939 n4 – 1 – gw Misc Inst [074]

Germanien, heilig herz europas : der germanische gedanke bei ernst moritz arndt, 1769-1860 / Zander, Alfred – Berlin-Grunewald: Der Reichsfuehrer SS, SS-Hauptamt, [194-?] – 1r – 1 – us Wisconsin U Libr [943]

Germaniia / Heine, Heinrich – Moskva: Gos. Izd-vo "Khudozhestvennaia Literatura", 1938 – 1 – us Wisconsin U Libr [830]

Germanische alterthuemer. mit text, uebersetzung und erklaerung von tacitus germania / Holtzmann, Adolf – Ed. by Alfred Holder. Leipzig: B.G. Teubner, 1873. iv,313p – 1 – us Wisconsin U Libr [943]

Germanische Fuerstentochter *see* Eine germanische fuerstentochter – die liebe siegt

Eine germanische fuerstentochter – die liebe siegt : zwei erzaehlungen / Germanische Fuerstentochter – Milwaukee, Wis.: Germania, 1913 – 1r – 1 – us Wisconsin U Libr [830]

Germanische heldensage / Schneider, Hermann – Berlin: W de Gruyter, 1933-1962 [mf ed 1993] – (= ser Grundriss der germanischen philologie 10:1-3) – 2v in 3 on 1r – 1 – (publ out of sequence. incl bibl ref and ind to all vols) – mf#7846 – us Wisconsin U Libr [390]

Germanische mythen : forschungen / Mannhardt, Wilhelm – Berlin: F Schneider, 1858 – 1mf – 9 – 0-524-04525-9 – (incl bibl ref) – mf#1990-3359 – us ATLA [290]

Germanische mythologie / Mogk, Eugen – Leipzig: C.J. Goeschen, 1906. 129p – 1 – us Wisconsin U Libr [390]

Germanisch-romanische monatsschrift – Heidelberg. v1-8. 1909-1920 – 61mf – 9 – mf#H-10027 – ne IDC [430]

Germanistische abhandlungen – Breslau. v1-68. 1882-1934 – 1 – $432.00 – mf#0234 – us Brook [430]

German-jewish periodicals from the leo baeck institute in new york, 1768-1945 / Leo Baeck Institute, New York – (= ser Research colls in judaica) – 3pt on 251r – 1 – $27,025.00 coll – (pt1 116r $11,975. pt2 54r $5410 originally publ by clearwater publ co. pt3 81r $11,065. individual titles listed separately) – us UPA [939]

Germann, Wilhelm *see*
- Genealogy of the south-indian gods
- Die kirche der thomaschristen
- Missionar christian friedrich schwartz
- Ziegenbalg und pleutschau

The germans / Wylie, Ida Alena Ross – Indianapolis: The Bobbs-Merrill Company, 1911. 4+361p. Plates. Illus.Publ. 1910 by Mills & Boon, London, under title: My German Year. With a different set of illus.1 reel. 1260 – 1 – us Wisconsin U Libr [390]

Germans in the conquest of america / Arciniegas, German – New York, NY. 1943 – 1r – us UF Libraries [972]

Germans who never lost / Hoyt, Edwin Palmer – New York, NY. 1968 – 1r – us UF Libraries [910]

Germanske S S Norge *see* Germaneren

Germanton times – Holbrook, jul 1884-jan 1885 – 1r – A$32.78 vesicular A$38.28 silver – at Pascoe [079]

Germantown banner-press – Colgate, Germantown, Hubertus etc WI. 1979 mar 15/jun-1995 oct/dec – 49r – 1 – (cont: germantown news-banner; germantown press) – mf#944167 – us WHS [071]

Germantown gleaner – Germantown, NE: C A Fetterman, 1894 (wkly) [mf ed -1897 (gaps) filmed 1979] – 1r – 1 – us NE Hist [071]

Germantown, OH *see* Cemetery records, 1849-1929

Germantown press – Colgate, Germantown, Hubertus etc WI. 1975 nov 5-1978 mar 8, 1978 mar 15-dec 28, 1979 jan 1-mar 8 – 3r – 1 – (continued by: germantown news-banner; germantown banner-press) – mf#944165 – us WHS [071]

Germanus *see* Lerchensang und schwerterklang

Germany : internal affairs and foreign affairs, 1930-1966 / U.S. State Dept – (= ser Confidential u s state department central files) – 1 – $101,440.00 coll – (internal affairs, 1930-41 59r isbn 0-89093-637-4 $11,430. foreign affairs, 1930-39 12r isbn 0-89093-427-4 $2330. internal affairs, 1942-44 35r isbn 0-89093-429-0 $6765. foreign affairs, 1940-44 5r isbn 0-89093-428-2 $970. internal affairs, 1945-49: pt1: political, governmental, & national defense affairs 41r isbn 0-89093-430-4 $7930; pt2: social, economic, & industrial affairs 68r isbn 0-89093-431-2 $13,140. internal affairs & foreign affairs: 1950-54 45r isbn 1-55655-193-2 $8710; 1955-59 37r isbn 1-55655-194-0 $7160; 1960-jan 1963 35r isbn 1-55655-750-7 $6765. federal republic of germany: internal affairs, 1950-54: pt1: political, governmental, & national defense affairs 56r isbn 0-89093-911-X $10,820. foreign affairs, 1950-54 9r isbn 0-89093-910-1 $1740. internal affairs, 1955-59: pt1: political, governmental, & national defense affairs 29r isbn 1-55655-447-8 $5610; pt2: social, economic, & industrial affairs 40r isbn 1-55655-448-6 $7745. federal republic of germany: foreign affairs, 1955-59 5r isbn 1-55655-446-X $970. internal affairs, 1960-jan 1963 46r isbn 1-55655-752-3 $8900. foreign affairs, 1960-jan 1963 5r isbn 1-55655-753-1 $970. subject-numeric files, feb 1963-66: pt1: political, governmental, & national defense affairs 34r* isbn 1-55655-976-3 $6580. with p/g) – us UPA [943]

Germany : its universities, theology and religion / Schaff, Philip – Philadelphia: Lindsay & Blakiston; New York: Sheldon, Blakeman, 1857 – 1mf – 9 – 0-7905-6731-8 – mf#1988-2731 – us ATLA [943]

Germany : monthly reports by the executive committee of the social democratic party of germany – Prag (CZ), Paris (F), 1937 jun-1940 mar – 3r – 1 – (english ed of: deutschlandberichte der sopade) – gw Misc Inst [943]

Germany : series 1: 1906-1925 – 4pt – (= ser Confidential british foreign office political correspondence) – 1 – $87,415.00 coll – (pt1: 1906-19 105r $20,895 isbn 1-55655-530-X. pt 2: 1920-21 139r $27,660 isbn 1-55655-630-6. pt3: 1922-23 150r $29,850. pt4: 1924-25 94r $18.700. with p/g) – us UPA [327]

Germany *see*
- Deutscher reichs-anzeiger und preussischer staats-anzeiger
- Journaux publies par les prisonniers de guerre allemands en france 1946-1948

Germany, 1919-1941 – (= ser U s military intelligence reports) – 28r – 1 – $4430.00 – 0-89093-426-6 – (with p/g) – us UPA [355]

Germany, 1941-1944 : reports – 2pt – (= ser U s military intelligence reports) – 1 – (pt1: geography, population & social conditions, politics & govt, economy & finance 42r $6635 isbn 0-89093-475-4. pt2: national defense, army, navy & military aviation 37r $5850 isbn 0-89093-476-2. with p/g) – us UPA [355]

Germany. Air Force *see* German air force reports

Germany. Air Force High Command *see* Records of the headquarters of the german air force high command

Germany and central europe, 1841-1900 *see* The papers of queen victoria on foreign affairs

Germany and its occupied territories during world war 2 / U.S. Office of Strategic Services & U.S. State Dept – (= ser Oss/state department intelligence and research reports 4) – 22r – 1 – $3405.00 – 0-89093-120-8 – (with p/g) – us UPA [943]

Germany. Armed Forces High Command. Headquarters *see* Records of the headquarters of the german armed forces high command

Germany. Army *see* Records of german army areas

Germany. Army Field Commands *see* Records of german field commands

Germany. Army High Command. Headquarters *see* Records of the headquarters of the german army high command

Germany. Auswaertiges Amt *see*
- Die grosse politik der europaischen kabinette
- La politique exterieure de l'allemagne

Germany. Berlin Documents Center *see*
- Documents concerning jews in the berlin document center
- Name index of jews whose german nationality was annulled by the nazi regime

Germany. Bundesrat *see*
- Denkschrift zum entwurf eines buergerlichen gesetzbuchs
- Entwurf eines buergerlichen gesetzbuches fuer das deutsche reich, erste lesung

Germany. Democratic Republic *see* Statutes

Germany. Democratic Republic. Volkskammer *see* Volkskammer..

Germany. Embassy at Washington, DC *see* Archives of the german embassy at washington

Germany. Federal Republic *see* Deutscher wetterdienst

Germany. Federal Republic. Bundesministerium fuer Verkehr *see* Verkehrsblatt

Germany. Federal Republic. Bundesrat *see* Verhandlungen des deutschen bundesrates, 1949-2009 ff

Germany. Federal Republic. Bundestag *see*
- Verhandlungen des deutschen bundestages, 1949-1999
- Verhandlungen, stenographische berichte

Germany. Foreign Ministry *see*
- A catalog of files and microfilms of the german foreign ministry archives, 1867-1920
- German foreign ministry archives
- Index of microfilmed records of the german foreign ministry and the reich's chancellery covering the weimar period
- Miscellaneous records of the german foreign office received by the department of state
- Records of the german foreign ministry pertaining to china, 1919-1935
- Records of the german foreign office filmed for the university of london
- Records of the german foreign office received by the department of state
- Records of the german foreign office received by the department of state from st antony's college
- Records of the german foreign office received by the department of state from the university of california (project 1)
- Records of the german foreign officer received by the department of state from the british museum

Germany. (formerly Democratic Republic). Staatliche Zentralverwaltung fuer Statistik *see* Statistisches jahrbuch der deutschen demokratischen republik 1955-1965

Germany. (formerly Federal Republic). Statistische Bundesamt *see* Statistisches jahrbuch fuer die bundesrepublik deutschland 1952-1965

Germany. (formerly Federal Republic). Statistisches Reichsamt *see* Statistisches jahrbuch fuer das deutsche reich 1880-1942

Germany. German Navy *see* Guide to the records of the german navy, 1850-1945

Germany. Heer. Oberkommando *see* Oberkommando des heeres

Germany in the later middle ages, 1200-1500 / Stubbs, William; ed by Hassall, Arthur – London, New York: Longmans, Green, 1908 [mf ed 1991] – 1mf – 9 – 0-524-01132-X – (incl bibl ref) – mf#1990-0346 – us ATLA [931]

Germany. Kingdom of Prussia *see* State archives-baptists

Germany. Kommission fuer die zweite Lesung des Entwurfs des Buergerlichen Gesetzbuchs *see* Protokolle der kommission fuer die zweite lesung des entwurfs des buergerlichen gesetzbuchs, im auftrage des reich-justizamts

Germany. Kommission zur Ausarbeitung des Entwurfes eines Buergerlichen Gesetzbuchs *see*
- Entwurf einer grundbuchordnung fuer das deutsche reich
- Entwurf eines buergerlichen gesetzbuches fuer das deutsche reich
- Entwurf eines buergerlichen gesetzbuches fuer das deutsche reich, erste berathung
- Entwurf eines einfuehrungsgesetzes zum buergerlichen gesetzbuche fuer das deutsche reich, erste lesung
- Entwurf eines familienrechts fuer das deutsche reich
- Entwurf eines gesetzes fuer das deutsche reich
- Entwurf eines rechtes der erbfolge fuer das deutsche reich

Germany. Kriegsmarine Oberkommando *see* Fuehrer conferences on matters dealing with the german navy, 1939-1945

Germany. Laws, Statutes, etc *see* Reichs-gesetzblatt

Germany. Miscellaneous Reich Ministries and Offices *see* Fragmentary records of miscellaneous reich ministries and offices, 1919-1945

Germany. National Socialist German Labor Party *see* Records of the national socialist german labor party

Germany. Navy *see* Records of the german navy, 1850-1945, received from the united states naval history division

Germany. Navy High Command. Headquarters *see* Records of the headquarters of the german navy high command (okm)

Germany. Navy. Operations Division. Staff *see* War diary, operations division, german naval staff, 1939-45

Germany. Nazi Cultural and Research Institutions *see* Records of nazi cultural and research institutions

Germany. North German Confederation, 1866-70. Reichstag *see*
- Stenographische berichte und anlagen ueber die verhandlungen des reichstages des norddeutschen bundes
- Stenographische berichte und anlagen ueber die verhandlungen des reichstags des nord-deutschen bundes

Germany. Office of the Deputy for Serbian Economy *see* Records of the office of the deputy for serbian economy

Germany. Office of the Reich Commissioner *see* Records of the office of the reich commissioner for the strengthening of germandom

Germany. Reich Air Ministry *see* Records of the reich air ministry (reichsluftfahrtministerium)

Germany. Reich Commissioner *see* Records of the reich commissioner for the baltic states, 1941-1945

Germany. Reich Ministry *see*
- Records of the reich ministry for armaments and war production
- Records of the reich ministry for public enlightenment and propaganda, 1936-1944
- Records of the reich ministry for the occupied eastern territories, 1941-1945
- Records of the reich ministry of economics

Germany. Reich Office for Soil Exploration *see* Reich office for soil exploration

Germany. Reich SS. and German Police *see* Records of the reich leader of the ss and chief of the german police

Germany. Reichsarbeitsministerium. Hauptabteilung *see* Der arbeitseinsatz im deutschen reich

Germany. Reichsjustizamt und Reichsjustizministerium *see*
- Denkschrift zum entwurf eines buergerlichen gesetzbuchs nebst drei anlagen
- Zusammenstellung der gutachtlichen aeusserungen zu dem entwurf eines zum buergerlichen gesetzbuch gefertigt im reichsjustizamt

Germany. Reichskanzler-Amt *see* Handbuch fuer das deutsche reiche auf das jahr...

Germany. Reichsministerium fuer Wissenschaft, Erziehung und Volksbildung *see*
- Deutsche wissenschaft, erziehung, und volksbildung
- Richtlinien fuer die leiberserziehung in jungenschulen

Germany. Reichstag *see*
- Erste, zweite und dritte berathung des entwurfs eines buergerlichen gesetzbuchs im reichstage
- Verhandlungen. legislaturperiode
- Verhandlungen. stenographische

Germany. Statistiches Reichsamt *see* Statistisches jahrbuch fuer das deutsche reich

Germany (territory under allied occupation, 1945-1955, russian zone) hauptverwaltung arbeit und sozialfursorge – Jahrbuch: arbeit und sozialfursorge. Berlin – 1 – us Wisconsin U Libr [943]

Germany (territory under allied occupation 1945-1955, russian zone) statistisches zentralamt. volks-und berufszahlung vom 29. oktober 1946 in der sowjetischen besatzungszone deutschlands – Berlin: Deutscher Zentralverlag, 1948 – 1 – us Wisconsin U Libr [943]

Germany to-day – London (GB), 1938-1940 mar – 2r – 1 – (oct 1939: inside nazi-germany, suppl to: austria to-day) – gw Misc Inst [943]

Germany today : newsletter – New York NY (USA), 1945 21 jun-1946 7 dec – 1r – 1 – gw Misc Inst [943]

Germany. Todt Organization *see* Records of the todt organization

Germany, turkey and armenia – London, 1917 – 2mf – 9 – mf#AR-1439 – ne IDC [327]

Germany (West) Bundesministerium Fur Gesamtdeutsche Fragen *see* Schein und wirklichkeit

Germany (West) Constitution *see* Grundgesetz

Germany's business leaders, 1400-1917 – 426mf (28:1) – 9 – $3790.00 – (with p/g) – us UPA [650]

Germar, E R *see* Magazin der entomologie

Germar, Ernst *see* Reise nach dalmatien und in das gebiet von ragusa

Gemelshausen / Gerstaecker, Friedrich; ed by Lewis, Orlando F – Boston: D C Heath, c1902 – 1r – 1 – us Wisconsin U Libr [430]

Germenes incorruptibles / Ycaza, Jorge Enrique De – Panama, 1944 – 1r – us UF Libraries [972]

Germinal / French Socialist Federation [Johnston City, IL] – v1 n1-v2 n2 [1922 jan-1923 feb] – 1r – 1 – (cont: socialiste [johnston city, il]) – mf#964211 – us WHS [335]

Germinal – Chicago IL, 1913* – 1r – 1 – (italian periodical) – us IHRC [073]

Germinal : tout pour la republique – Paris. v1, n1; v2,n1-220. 18 dec 1892, 29 janv-6 sept 1893 – 1 – fr ACRPP [073]

Germinal, cuentos / Carias Reyes, Marcos – Tegucigalpa, Mexico. 1936 – 1r – us UF Libraries [972]

Germiny, Charles *see* La politique de leon 13

Germiston advocate – Germiston, SA. 1923-79 – 45r – 1 – sa National [079]

Germogen, Bishop of Pskov and Porkhov *see* O bogosluzhenii pravoslavnoi tserkvi

Germonik, Ludwig *see* Die brandschatzung zur franzosenzeit 1809-13 in illyrien, oder, die gestoerte see-idylle

Gernandt, Leon *see* The protection of water during armed conflict

Gerning, Johann I von *see* Reise durch oestreich und italien

Gernler, L *see* Disputationes exegeticae in confessionem helveticam

Gero, Ihan *see* Il primo libro de madrigali italiani et canzon francese a due voci

Gerock, C F *see* Versuch einer darstellung der christologie des koran

Gerodontology – Oxford, England 1988-90 – 1 – ISSN: 0734-0664 – mf#12770 – us UMI ProQuest [618]

Geroi nashego vremeni / Lermontov, Mikhail – Zheneva, Russia. 1945 – 1r – us UF Libraries [025]

Geroi rodiny – (city unknown), Russia 1941-45 [mf ed Norman Ross] – 1 – mf#nrp-98 – us UMI ProQuest [934]

Geroicheskaia krasnoarmeiskaia – (city unknown), Russia 1939-45 [mf ed Norman Ross] – 1 – mf#nrp-99 – us UMI ProQuest [934]

Geroicheskii pokhod – (city unknown), Russia 1939 [mf ed Norman Ross] – 1 – mf#nrp-100 – us UMI ProQuest [934]

Gerok, Karl *see*
- Die apostelgeschichte in bibelstunden
- Hirtenstimmen
- Pilgerbrod
- Predigten auf alle fest-, sonn- und feiertage des kirchenjahrs
- Der wandsbecker bote

Gerold, Rosa von *see* Eine herbstfahrt nach spanien

Gerolstein : epilogue des mysteres de paris / Sue, Eugene – Bruxelles 1843 [mf ed Hildesheim 1995-98] – 1v on 1mf – 9 – €40.00 – ISBN-10: 3-487-25864-1 – ISBN-13: 978-3-487-25864-5 – gw Olms [914]
Geronimo castillo de bovadilla / Elias de Tejada Spinola, Francisco – Madrid: Grafica Universal, 1939 – sp Bibl Santa Ana [940]
Gerontologia – Basel, Switzerland 1966-74 – 1,5,9 – ISSN: 0016-898X – mf#2059 – us UMI ProQuest [618]
Gerontologia clinica – Basel, Switzerland 1966-74 – 1,5,9 – ISSN: 0016-8998 – mf#2060 – us UMI ProQuest [618]
Gerontologist – Washington DC 1961+ – 1,5,9 – ISSN: 0016-9013 – mf#2281 – us UMI ProQuest [618]
Gerontology and geriatrics education / ed by Dawson, Grace D – v1- . 1980- . q – 1, 9 ($200.00 in US $280.00 outside hardcopy subsc) – us Haworth [618]
Gerrard, Thomas John *see* Bergson
Gerretsen, Jan Hendrik *see* Micronius
Gerretson, Frederik Carel *see* Coens eerherstel
Gerrit smith : a biography / Frothingham, Octavius Brooks – New York: G.P. Putnam, 1878, c1877 – 1mf – 9 – 0-7905-6063-1 – mf#1988-2063 – us ATLA [920]
Gerritsz, Hessell *see* Descriptio ac delineatio geographica detectionis freti, sive, transitus ad occasum, sufra terras americanas, in chinam atq
Gerry, Elbridge *see* Elbridge gerry papers 1744-1895
Gerses, D *see* Epistolae ad henr bullingerum
Gersfelder kreisblatt – Gersfeld DE, 1883 22 aug-1931, 1934-1941 28 mar – 30r – 1 – (incl suppl) – gw Misc Inst [074]
Gershuni, G *see* Iz nedavnego proshlogo
Gershuni, Grigorii Andreevich *see* Heldn fun der revolutsye fun noentn 'over...
Gersin, M *see* Chercheuse d'esprit
Gerson, Adolf *see* Der chacham kohelet als philosoph und politiker
Gerson, Brasil *see* Ouro, o cafe e o rio
Gerson, Jean *see* Tripartito del christianissimo y consolatorio doctor juan gerson de doctrina christiana a cualquier muy prechosa
Gerson, Jean de (Jean Charlier) *see*
– Alphabetum divini amoris
– Opera omnia
– Traktat ueber die hinfuehrung der kleinen zu christus
Gerson, Menachem *see* Werkleute
Gerstaecker, Friedrich *see*
– Die flusspiraten des mississippi
– Germelshausen
– Gesammelte schriften
– Gold
– Im busch
– In den pampas
– The little whaler; or, the adventures of charles hollberg
– Les pirates du mississipi
– Die regulatoren in arkansas
Gerstell, Richard *see* How to survive an atomic bomb
Gerstenberg, Heinrich *see*
– Deutschland, deutschland ueber alles!
– Hoffmann von fallersleben und sein deutsches vaterland
Gerstenberg, Heinrich Wilhelm von *see* H w v gerstenbergs rezensionen in der hamburgischen neuen zeitung
Gerstenberg, Heinrich Wilhelm von et al *see*
– Briefe ueber merkwuerdigkeiten der litteratur
– Der hypochondrist
Gerstenhauer, Arthurius *see* De alcaei et sapphonis copia vocabulorum
Gerstner, Hermann *see*
– Auf grosser fahrt
– Es war in einer sommernacht
– Faehnrich charlotte
– Der graue rock
– Mit helge suedwaerts
– Requiem fuer einen gefallenen
– Die strasse ins waldland
– Zwischen den kriegen
Gertrud von loden : eine erzaehlung aus der schwedenzeit / Quandt, Clara – Braunschweig: Benno Goeritz, 1891 – 1r – 1 – us Wisconsin U Libr [830]
Gertrude anoda kuramba enock / Rubio, J – Gwelo, Zimbabwe. 1962 – 1r – us UF Libraries [960]
Gertrude et mon ceur / Acremant, Albert – Paris, France. 1937 – 1r – us UF Libraries [440]
Gertrude van rensselaer wickham papers *see* Wickham, gertrude van rensselaer, papers, ms 1085
Gertrudis gomez de avellaneda / Cuba Servicio Femenino Para La Defensa Civil – Habana, Cuba. 1947 – 1r – us UF Libraries [972]
Gertrudis gomez de avellaneda / Figarola-Caneda, Domingo – Madrid, Spain. 1929 – 1r – us UF Libraries [972]
Gertrudis gomez de avellaneda / Gomez De Avellaneda Y Arteaga, Gertrudis – Barcelona, Spain. 1953 – 1r – us UF Libraries [972]
Gertsen, A *see* [Sbornik]
Gertsenshtein, M I *see* Zemelnaia reforma v programme partii narodnoi svobody
Gertsenshtein, M Ia *see* Khar'kovskii krakh
Gervais, Albert *see*
– Guide des adresses de la ville de joliette pour l'annee 1900
– Joliette illustre
Gervais, F L P *see* Voyage autour du monde par les mers de l'inde et de chine execute sur la corvette de l'etat la favorite pendant les annees 1830, 1831, 1832
Gervais, Paul *see* Histoire naturelle des mammiferes
Gervais, Pierre D *see* Golf putting and preferences for cognitive training
Gervais star – Gervais OR: P P Hassler, 1926-[wkly] – 1 – (cont: gervais weekly star (-1926)) – us Oregon Lib [071]
Gervais, Villius *see* Developpement de l'enseignement populaire a cuba
Gervais weekly star – Gervais, OR: W J Clarke. v9 n31-v35 n50. jan 11 1901-may 28 1926 – 1 – (cont by: gervais star) – us Oregon Hist [071]
Gervais weekly star – Gervais OR: W J Clarke, -1926 [wkly] – 1 – (cont by: gervais star (1926-)) – us Oregon Lib [071]
Gervaix, Jean Francois Regis *see* Zephyrin guillemin
Gervase of Canterbury *see* Historical works (rs73)
Gerwerkschaftsbund der Angestellten *see* Ordentlicher bundestag
Gerz, Alfred *see*
– Der carnaval und die somnambuele
– Mathilde moehring
– Meister martin der kuefner und seine gesellen
Ges u cristo nella litteratura contemporanea straniera e italiana : studio storico-scientifico / Labanca, Baldassare – Torino: Fratelli Bocca, 1903 – 2mf – 9 – 0-8370-4026-4 – (includes bibliographies and indexes of biblical books cited and of names) – mf#1985-2026 – us ATLA [240]
Gesaenge aus den drei reichen : ausgewaehlte gedichte / Werfel, Franz – 2. Aufl. Leipzig: K Wolff, c1917 – 1r – 1 – us Wisconsin U Libr [810]
Gesaenge mit begleitung des fortepiano : zweyte sammlung / Schroeter, Corona – Weimar: in commission bey dem Industrie Comptoir 1794 [mf ed 1992] – 1r – 1 – mf#pres. film 112 – us Sibley [780]
Gesaenge unter der fahne : vier kantaten / Boehme, Herbert – Muenchen: Zentralverlag der NSDAP, F Eher, [1935?] [mf ed 1989] – (= ser Kameraden 4) – 72p – 1 – mf#7043 – us Wisconsin U Libr [810]
Gesammelte abhandlungen / Lagarde, Paul de – Anastatischer Neudruck. Goettingen: In Commission bei Lueder Horstmann, 1896 – 1mf – 9 – 0-8370-7803-2 – (incl bibl ref and indexes) – mf#1986-1803 – us ATLA [470]
Gesammelte abhandlungen zur biblischen wissenschaft / Kuenen, Abraham – Freiburg i.B: J C B Mohr (Paul Siebeck), 1894 [mf ed 1989] – 2mf – 9 – 0-7905-1343-9 – mf#1987-1343 – us ATLA [221]
Gesammelte abhandlungen zur roemischen religions- und stadtgeschichte : ergaengungsband zu des verfassers religion und kultus der roemer / Wissowa, Georg – Muenchen: CH Beck, 1904 – 1mf – 9 – 0-524-02679-3 – (incl bibl ref) – mf#1990-3109 – us ATLA [250]
Gesammelte aufsaetze / Danzel, Theodor Wilhelm; ed by Jahn, Otto – Leipzig: Dyk, 1855 [mf ed 1989] – xxxiv/244p – 1 – mf#7169 – us Wisconsin U Libr [802]
Gesammelte aufsaetze : neue folge / Ritschl, Albrecht – Freiburg i B: JCB Mohr, 1896 [mf ed 1990] – 1mf – 9 – 0-7905-7363-6 – mf#1989-0588 – us ATLA [240]
Gesammelte briefe des heiligen franciscus xaverius : des grossen indianerapostels aus der gesellschaft jesu. als grundlage der missionsgeschichte spaterer zeiten... – Augsburg: Nicolaus Doll, 1794 [mf ed 1996] – (= ser Yale coll) – 3v (ill) – 1 – 0-524-10253-8 – (in german) – mf#1996-1253 – us ATLA [241]
Gesammelte civilistische schriften / Arndts von Arnesberg, Karl Ludwig, Ritter – Stuttgart, Cotta. v1-3. 1873-74 – (= ser Civil law 3 coll) – 19mf – 9 – (incl bibl ref and index) – mf#LLMC 96-544 – us LLMC [346]
Gesammelte dichtungen / Guenderode, Karoline von; ed by Salomon, Elisabeth – Muenchen: Drei Masken Verlag, 1923 [mf ed 2001] – xxi/492p/1pl – 1 – (incl bibl ref) – mf#10506 – us Wisconsin U Libr [810]
Gesammelte dichtungen / Wagner, Christian; ed by Guentter, Otto – 1. Ausg. Stuttgart: Strecker und Schroeder, 1918 – 1r – 1 – us Wisconsin U Libr [810]
Gesammelte erzaehlungen = Short stories / Raabe, Wilhelm Karl – Berlin: O. Janke, 1896-1900 – 1 – us Wisconsin U Libr [830]
Gesammelte gedichte / Carossa, Hans – Leipzig: Insel-Verlag 1943 [mf ed 1989] – 1r – 1 – (filmed with: georg buchners drama dantons tod / hans landsberg) – mf#7143 – us Wisconsin U Libr [810]
Gesammelte gedichte / Carossa, Hans – Zuerich: Verlag der Arche c1949 [mf ed 1989] – 1r – 1 – (filmed with: georg buchners drama dantons tod / hans landsberg) – mf#7143 – us Wisconsin U Libr [810]
Gesammelte gedichte / Keller, Gottfried – Berlin: W Hertz, 1883 – 1 – us Wisconsin U Libr [810]
Gesammelte gedichte / Sallet, Friedrich von – Leipzig: P Reclam, [18–?] – 1r – 1 – us Wisconsin U Libr [810]
Gesammelte geschichten und novellen / Riehl, Wilhelm Heinrich – Stuttgart, Germany. v1. 1879 – 1r – us UF Libraries [830]
Gesammelte geschichten und novellen = Short stories / Riehl, Wilhelm Heinrich – Stuttgart: J G Cotta 1879 [mf ed 1995] – 1r – 1 – (filmed with: der fluch der schoenheit) – mf#3716p – us Wisconsin U Libr [830]
Gesammelte novellen und erzaehlungen / Birch-Pfeiffer, Charlotte – Leipzig: Reclam, 1863-65 [mf ed 1993] – 3v on 1r – 1 – mf#8535 – us Wisconsin U Libr [800]
Gesammelte patristische untersuchungen / Draeseke, Johannes – Altona: A.C. Reher, 1889 – 1mf – 9 – 0-7905-6165-4 – mf#1988-2165 – us ATLA [240]
Gesammelte schriften / Berger, Alfred, Freiherr von; ed by Bettelheim, Anton & Glossy, Karl – Wien; Leipzig: Deutsch-Oesterreichischer Verlag, 1913 [mf ed 1989] – 3v – 1 – mf#7009 – us Wisconsin U Libr [802]
Gesammelte schriften / Birmann, Martin – Basel: Reich 1894 [mf ed 1979] – 2v in 1 on 1r [ill] – 1 – (v1: lebensbild und aufsaetze meist biographischen inhalts; v2: zur geschichte der landschaft basel) – mf#film mas 8765 – us Harvard [802]
Gesammelte schriften / Boerne, Ludwig – Wien: Tendler & Comp 1868?- [mf ed 1979] – 12v on 1r – 1 – mf#film mas 8592 – us Harvard [802]
Gesammelte schriften / Dreisel, Hermann O – Milwaukee (WI): Freidenker Pub Co, 1905 – 1r – 1 – us Wisconsin U Libr [800]
Gesammelte schriften / Freud, Sigmund – t. I-XII). Wien – 1 – fr ACRPP [150]
Gesammelte schriften / Gerstaecker, Friedrich – Jena. 43v. 1872-79 – 1 – $184.00 – us L of C Photodup [800]
Gesammelte schriften / Goerres, Joseph von; ed by Schellberg, Wilhelm et al – Koeln: Gilde-Verlag, 1926- [mf ed 1993] – 4r (ill) – 1 – (incl bibl ref and ind) – mf#8181 – us Wisconsin U Libr [802]
Gesammelte schriften / Meissner, Alfred – Leipzig: F W Grunow 1871-76 [mf ed 1979] – 18v on 2r – 1 – mf#film mas c472 – us Harvard [802]
Gesammelte schriften / ed by Mommsen, T – Berlin, 1905-1913. v1-8 – 90mf – 8 – mf#H-329 – ne IDC [956]
Gesammelte schriften / Strauss, David Friedrich – Bonn. v1-12. 1876-77 – 1 – $108.00 – (in german) – mf#0572 – us Brook [802]
Gesammelte schriften / Zschokke, Heinrich – Aarau: H R Sauerlaender 1865 [mf ed 1979] – 10v in 5 on 1r – 1 – mf#film mas 8653 – us Harvard [802]
Gesammelte schriften in 15 (16) baenden = Works / Laube, Heinrich – Wien: W Braumueller, 1875-82 – 1 – us Wisconsin U Libr [800]
Gesammelte vortraege und abhandlungen dr. richard rothe's aus seinen letzten lebensjahren = Selections. 1886 / Rothe, Richard – Elberfeld: RL Friderichs, 1886 – 1mf – 9 – 0-524-00090-5 – mf#1989-2790 – us ATLA [240]
Gesammelte vortraege verschiedenen inhalts = Lectures. selections / Luthardt, Christoph Ernst – Leipzig: Doerffling und Franke, 1876 – 1mf – 9 – 0-7905-9790-X – mf#1989-1515 – us ATLA [240]
Gesammelte werke : anfaenge und ziele des jahrhunderts / Gutzkow, Karl – Jena [1874]-[1876] [mf ed Hildesheim 1995-98] – 4v on 3mf – 9 – €90.00 – 3-487-26013-1 – gw Olms [802]
Gesammelte werke / Anzengruber, Ludwig; ed by Bettelheim, Anton et al – Stuttgart: J G Cotta 1890 [mf ed 1978] – 10v on 1r – 1 – mf#film mas 8276 – us Harvard [802]
Gesammelte werke : eine auswahl in fuenf baenden / Fontane, Theodor – Berlin: Fischer, 1920 [mf ed 1993] – 5v on 2r – 1 – (int by paul schlenther) – mf#8200 – us Wisconsin U Libr [800]
Gesammelte werke : band 1-5 / Braun, Lily – Berlin-Grunewald [1922] [mf ed 1995] – 29mf – 9 – €190.00 – 3-8267-3178-6 – (v1: im schatten der titanen und: julie vogelstein: lily braun. ein lebensbild 6mf €45 isbn 3-8267-3173-5 dhs-ar 3173. v2: memoiren einer sozialistin. lehrjahre. roman 6mf isbn 3-8267-3174-3 dhs-ar 3174. v3: memoiren einer sozialistin. kampfjahre. roman 6mf €45.00 isbn 3-8267-3175-1 dhs-ar 3175. v4: lebensucher. roman und: mutter maria. eine tragoedie in fuenf akten 6mf €45 isbn 3-8267-3176-x dhs-ar 3176. v5: die liebesbriefe der marquise. madeleine guimard. eine lyrische oper in drei akten 5mf €40 isbn 3-8267-3177-8 dhs-ar 3177) – mf#DHS-AR 3178 – gw Frankfurter [430]
Gesammelte werke / Bierbaum, Otto Julius – Muenchen: G Mueller, [c1922] [mf ed 1989] – 7v – 1 – mf#7022 – us Wisconsin U Libr [820]
Gesammelte werke / Braun, Lily – Berlin-Grunewald: H Klemm, [1923] [mf ed 1989] – 5v – 1 – (each vol has also special t p) – mf#7064 – us Wisconsin U Libr [802]
Gesammelte werke / Busch, Wilhelm – Muenchen: Braun & Schneider c1921-23 [mf ed 1995] – 2v on 1r [ill] – 1 – mf#3805p – us Wisconsin U Libr [802]
Gesammelte werke / Ernst, Paul – Muenchen: A Langen, G Mueller, 1928-37 [mf ed 1989] – 19v – 1 – mf#6984 – us Wisconsin U Libr [802]
Gesammelte werke : erste serie / Zahn, Ernst – Stuttgart: Deutsche Verlags-Anstalt. 10v. [192-?] – 1 – us Wisconsin U Libr [800]
Gesammelte werke / Freytag, Gustav – Leipzig: S Hirzel 1887-88 [mf ed 1979] – 22v on 3r – 1 – mf#film mas c541 – us Harvard [802]
Gesammelte werke / Greif, Martin – Leipzig: C F Amelang. 3v. 1895-1896 (mf ed 1990) – 1r – 1 – (filmed with: deutsche schriften) – us Wisconsin U Libr [802]
Gesammelte werke / Heyse, Paul – Berlin: W Hertz. 10v. 1872-74 – 6r – 1 – us Wisconsin U Libr [800]
Gesammelte werke / Hoelderlin, Friedrich; ed by Boehm, Wilhelm – Jena: E Diederichs. 3v. 1905 – 1 – us Wisconsin U Libr [800]
Gesammelte werke / MacKay, John Henry – Berlin: B Zack 1911 [mf ed 1978] – 8v on 1r – 1 – mf#film mas 8193 – us Harvard [802]
Gesammelte werke : neue serie / Heyse, Paul – Berlin: W Hertz. 28v. 1884-1914 – 1 – us Wisconsin U Libr [800]
Gesammelte werke / Stehr, Hermann; ed by Tau, Max – Trier: F Lintz, 1924 – 2r – 1 – us Wisconsin U Libr [800]
Gesammelte werke = Works / Rosegger, Peter – Neub. und neueingeleit. Ausg. Leipzig: L Staackmann, 1922-1924 – 1 – (incl ind) – us Wisconsin U Libr [800]
Gesammelte werke = Works. 1911 / Wildenbruch, Ernst von; ed by Litzmann, Berthold – Berlin: G. Grote. 6v. 1911-1924 – 5r – 1 – (incl bibl ref) – us Wisconsin U Libr [800]
Gesammelte werke in fuenf baenden / Francois, Louise von – Leipzig: Insel-Verlag. 5v. [1918?] – 1r – 1 – us Wisconsin U Libr [800]
Die gesammelten gedichte / Lasker-Schueler, Else – 2. Aufl. Leipzig: K Wolff, [19–?] – 1r – 1 – us Wisconsin U Libr [810]
Die gesammelten werke / Bismarck, Otto, Fuerst von – Berlin. 15v in 19. 1924-35 – 9 – $264.00 – (in german) – mf#0107 – us Brook [943]
Gesammeltes werk / Binding, Rudolf Georg – Frankfurt/M: Ruetten & Loening, c1927 [mf ed 1989] – 5v – 1 – (each vol has sep t p) – mf#7023 – us Wisconsin U Libr [802]
Gesammtgeschichte des neuen testaments : oder neutestamentliche isagogik / Guericke, H E F – 2nd ed. Leipzig, 1854 – 13mf – 8 – €25.00 – ne Slangenburg [225]
Gesammt-register zu amtsblatt and gesetzessammlungen des kantons zurich / Zurich. (Canton). Laws, Statutes, etc. (Indexes) – Zurich. Genossenschaftsbuchdruckerei, 1882. 182 p. LL-4085 – 1 – us L of C Photodup [348]
Gesammt-verlags-katalog des deutschen buchhandels und des mit ihm im direkten verkehr stehenden auslandes : mikrofiche edition der ausgabe 1881-1893 = Complete publishers' catalogue of the german book trade and foreign countries in contact with it / ed by Russell, Adolph & Basch, Johannes – [mf ed 1986] – 217mf (1:24) – 9 – silver €840.00 – ISBN-10: 3-598-10625-4 – ISBN-13: 978-3-598-10625-5 – gw Saur [070]
Gesamtausgabe der erzaehlenden schriften : in neun baenden / Fontane, Theodor – Berlin: S Fischer, 1925 [mf ed 1989] – 9v on 2r – 1 – mf#7247 – us Wisconsin U Libr [802]
Gesamtdeutsche rundschau – Dortmund, Bonn DE, 1954 26 feb-1959 6 mar – 1r – 1 – gw Misc Inst [380]
Gesamtueberblick ueber die polnische presse *see* Gesamtueberblick ueber die polnische tagesliteratur
Gesamtueberblick ueber die polnische tagesliteratur – Posen (Poznan PL), Berlin DE, 1908-1909 10 aug, 1912-1915 23 feb, 1916-1918 13 nov, 1919 & 1924, 1928-30, 1932-36, 1938-1939 2 sep – 1 – (title varies: 1920: gesamtueberblick ueber die polnische presse; publ in berlin fr 1920) – gw Misc Inst [077]

Gesamtverantwortung und spezifik: die cdu-presse in der entwicklung des ddr-journalismus 1957 bis 1961 / Fischer, Gerhard – Berlin: Union Verlag, 1971. 175p – apply for prices – us Wisconsin U Libr [070]

Gesamtverband der Arbeitnehmer der offentlichen betriebe und des Personen- und verkehrs *see* Jahrbuch

Gesamtverband der Christlichen Gewerkschaften Deutschlands *see* Jahrbuch...

Gesamtverband der christlichen Gewerkschaften Deutschlands *see* Frauenblatt der christlichen gewerkschaften

Gesamtverzeichnis des deutschsprachigen schrifttums 1700-1910 = Bibliography of german-language publications / ed by Schmuck, Hilmar – [mf ed 1986] – 795mf (1:24) – 9 – diazo €2950.00 (silver €3250 isbn: 978-3-598-30595-5) – ISBN-10: 3-598-30590-7 – ISBN-13: 978-3-598-30590-0 – gw Saur [014]

Gesamtverzeichnis des deutschsprachigen schrifttums 1911-1965 = Bibliography of german-language publications, 1911-1965 / ed by Oberschlep, Reinhard – [mf ed 1984] – 400mf (1:42) – 9 – diazo €2950.00 (silver €3250 isbn: 978-3-598-30456-9) – ISBN-10: 3-598-30455-2 – ISBN-13: 978-3-598-30455-2 – gw Saur [010]

Die gesandtschafts-reise nach siam und hue, der hauptstadt von cochinchina, in den jahren 1821 bis 1822 : aus dem tagebuch des verstorbenen george finlayson, wundarzt und naturforscher der gesandtschaft; mit einem vorwort ueber den verfasser – Weimar 1827 [mf ed Hildesheim 1995-98] – 1v on 3mf – 9 – €90.00 – 3-487-26481-1 – gw Olms [915]

Gesang buechlin, darinn der gantze psalter davids : sampt andern gaistlichen gesaengen mit jren melodeyen begriffen – corr aufl, Augspurg: Ulhart [ca 1570] – (= ser Hqab. literatur des 16. jahrh.) – 6mf/6mf – 9 – €70.00/€70.00 – mf#1570b/1570c – gw Fischer [780]

Gesang des deutschen / Hoelderlin, Friedrich – Frankfurt (Main): Hausdruckerei der Bauerschen, [1925?] – 1 – us Wisconsin U Libr [810]

Ein gesangbuch der brueder inn behemen unnd merherrn, die man auss hass & neyd, pickharden, waldenses, &c nennet : von jhnen auff ein newes (sonderlich vom sacrament des nachtmals) gebessert, unnd etliche schoene newe geseng hinzu gethan – Nuernberg: Berg & Newber [c1560] – (= ser Hqab. literatur des 16. jahrh.) – 6mf – 9 – €70.00 – mf#1560a – gw Fischer [780]

Das geschaeft dess menschen... : die zweite auflage / Schoenberg, M von – St Gallen: Gedruckt, und zu finden im Fuerstlichen Gotteshaus, 1776 – 4mf – 9 – mf#0-2036 – ne IDC [090]

Die geschaeftige martha *see* Der froehliche botschafter

Geschaefts- und kassenbericht / Zentralverband christlicher Lederarbeiter Deutschlands – 1907/09. Includes Protokoll der V. Generalversammlung, 1909. (Serial publications of German trade unions in the Memorial Library, University of Wisconsin-Madison) – 1 – us Wisconsin U Libr [330]

Geschaefts- und unterhaltungsblatt fuer den kreis gladbach und umgebung – Moenchengladbach DE, 1837-38 – 1r – 1 – (title varies: 3 jan 1847: gladbacher kreis-blatt fuer geschaefte, politik und unterhaltung; 3 jan 1864: gladbacher zeitung. filmed by misc inst: 1843-44, 1847-1848 may) – gw Mikropress; gw Misc Inst [074]

Geschaeftsbericht / Allgemeiner deutscher Gewerkschaftsbund. Ortsausschuss Berlin – Berlin. v15, 20, 27, 29-40. 1903-31 – (= ser Serial publications of german trade unions) – 1 – (incl: bericht des arbeitersekretariats) – us Wisconsin U Libr [331]

Geschaeftsbericht / ed by Club der Filmindustrie – Berlin DE, 1916, 1918 – 1 – gw Mikrofilm [650]

Geschaeftsbericht / Deutscher Arbeitgeberbund fuer das Baugewerbe – Berlin. 1926/27, 1929/30 – (= ser Serial publications of german trade unions) – 1 – us Wisconsin U Libr [331]

Geschaeftsbericht / Deutscher Handels- und Industrieangestellten-Verband – 1927-28. (Serial publications of German trade unions in the Memorial Library, University of Wisconsin-Madison.) – 1 – us Wisconsin U Libr [330]

Geschaeftsbericht / Deutscher Werkmeister-Verband – Duesseldorf. 1924-25, 1926-27, 1928-29, 1930-31. (Serial publications of German trade unions in the Memorial Library, University of Wisconsin-Madison.) – 1 – us Wisconsin U Libr [330]

Geschaeftsbericht / Deutscher Wirtschaftsbund fuer das Baugewerbe – Berlin. 1928-29, 1930-31. (Serial publications of German trade unions in the Memorial Library, University of Wisconsin-Madison.) – 1 – us Wisconsin U Libr [330]

Geschaeftsbericht / Freier Deutscher Gewerkschaftsbund. Berlin – 1946. 431p. illus. map, ports – 1 – us Wisconsin U Libr [330]

Geschaeftsbericht / Verband der Bergarbeiter Deutschlands – Bochum. 1913/14. (Serial publications of German trade unions in the Memorial Library, University of Wisconsin-Madison) – 1 – us Wisconsin U Libr [330]

Geschaeftsbericht / Verband der Buchbinder und Papierverarbeiter Deutschlands – Berlin. 1906, 1921, 1925-31. Title varies: Bericht. (Serial publications of German trade unions in the Memorial Library, University of Wisconsin-Madison.) – 1 – us Wisconsin U Libr [330]

Geschaeftsbericht / Verband der Gaertner und Gaertnereiarbeiter – 1920/25. s.l.: s.n. Until 1919, organization called Allgemeiner deutscher Gaertnerverein. Some vols. also include ITS: Verhandlungsbericht des Verbandstages. (Serial publications of German trade unions in the Memorial Library, University of Wisconsin-Madison.) – 1 – us Wisconsin U Libr [330]

Geschaeftsbericht / Verband der Gemeinde- und Staatsarbeiter – Berlin. 1922/23, 1926, 1928. (Serial publications of German trade unions in the Memorial Library, University of Wisconsin-Madison.) – 1 – us Wisconsin U Libr [330]

Geschaeftsbericht / Verband der Graphischen Hilfsarbeiter und -arbeiterinnen Deutschlands – 1927. Berlin. (Serial publications of German trade unions in the Memorial Library, University of Wisconsin-Madison.) – 1 – us Wisconsin U Libr [330]

Geschaeftsbericht / Verband deutscher Arbeitsnachweise – 1898/99. Berlin. (Serial publications of German trade unions in the Memorial Library, University of Wisconsin-Madison) – 1 – us Wisconsin U Libr [330]

Geschaeftsbericht / Verband deutscher Textilarbeiter – 1906/07. Berlin. (Serial publications of German trade unions in the Memorial Library, University of Wisconsin-Madison.) – 1 – us Wisconsin U Libr [330]

Geschaeftsbericht / Vereinigung der deutschen Arbeitgeberverbaende – Berlin. 1921/22, 1927/29. (Serial publications of German trade unions in the Memorial Library, University of Wisconsin-Madison.) – 1 – us Wisconsin U Libr [330]

Geschaeftsbericht / Zentralverband der Angestellten – 1919/20, 1924/25, 1928/30. Berlin. (Serial publications of German trade unions in the Memorial Library, University of Wisconsin-Madison.) – 1 – us Wisconsin U Libr [330]

Geschaeftsbericht / Zentralverband der Arbeitnehmer oeffentlicher Betriebe und Verwaltungen – Koeln. 1925 27. (Serial publications of German trade unions in the Memorial Library, University of Wisconsin-Madison.) – 1 – us Wisconsin U Libr [330]

Geschaeftsbericht / Zentralverband der Handlungsgehilfen – 1912. (Serial publications of German trade unions in the Memorial Library, University of Wisconsin-Madison.) – 1 – us Wisconsin U Libr [330]

Geschaeftsbericht / Zentralverband der Lederarbeiter und Arbeiterinnen Deutschlands – 1905-07. Berlin. (Serial publications of German trade unions in the Memorial Library, University of Wisconsin-Madison.) – 1 – us Wisconsin U Libr [330]

Geschaeftsbericht.. / Gewerkschaft deutscher Eisenbahner – v4-5. 1928-31. (Serial publications of German trade unions in the Memorial Library, University of Wisconsin-Madison.) – 1 – us Wisconsin U Libr [330]

Geschaeftsbericht.. / Gewerkverein christlicher Bergarbeiter Deutschlands – Essen. 1911/12, 1921/25, 1928/29. Title varies. (Serial publications of German trade unions in the Memorial Library, University of Wisconsin-Madison) – 1 – us Wisconsin U Libr [330]

Geschaeftsbericht... / Arbeiter-Sekretariat. Halle – Halle. v2,4. 1901-03. (Serial publications of German trade unions in the Memorial Library, University of Wisconson-Madison) – 1 – us Wisconsin U Libr [330]

Geschaeftsbericht... / Arbeiterwohlfahrt Hauptausschuss – 1926 – (= ser Serial publications of german trade unions) – 1 – us Wisconsin U Libr [331]

Geschaeftsbericht der konzentration – Berlin DE, 1925-29 – 1 – gw Misc Inst [650]

Geschaeftsbericht des hilfsvereins der deutschen juden – Berlin, Germany 1909-14 [mf ed Norman Ross] – 2r – 1 – mf#nrp-297 – us UMI ProQuest [939]

Geschaeftsbericht des rheinisch-wesfaelischen kohlensyndikats – Essen-Ruhr, 1893-1939/40 – 2r – 1 – gw Mikropress [074]

Geschaeftsbericht des vorstandes / Zentralverband aller in der Schmiederei beschaeftigten Personen – 1906. Hamburg. (Serial publications of German trade unions in the Memorial Library, University of Wisconsin-Madison.) – 1 – us Wisconsin U Libr [330]

Geschaeftsbericht des westdeutschen rundfunks [...] – Koeln DE, 1924-36 – 1 – gw Misc Inst [380]

Geschaeftsbericht des zentralvorstandes / Zentralverband der Schuhmacher Deutschlands – 1908/09, 1910/11. Nuernberg. (Serial publications of German trade unions in the Memorial Library, University of Wisconsin-Madison.) – 1 – us Wisconsin U Libr [330]

Geschaeftsberichte der bezirke/ landesverbaende / Sozialdemokratische Partei Deutschlands – 1946-1968 – 1 – gw Mikropress [943]

Geschaeftsbriefe schiller's / ed by Goedeke, Karl – Leipzig: Veit, 1875 – 1r – 1 – us Wisconsin U Libr [920]

Die geschaefts-ordnung des concils von trient : aus einer handschrift des vaticanischen archives zum erstenmale genau und vollstaendig an's licht gestellt: sammt einem vorberichte – Wien: Carl Gerold, 1871 – 1mf – 9 – 0-8370-8331-1 – mf#1986-2331 – us ATLA [240]

Geschcihte des christlichen gottesdienstes : ein handbuch fuer vorlesungen und uebungen im seminar / Koestlin, Heinrich Adolf – Freiburg i. B.: J C B Mohr, 1887 – 1mf – 9 – 0-7905-5241-8 – (incl bibl ref) – mf#1988-1241 – us ATLA [240]

Geschichtchen fuer meine tochter / Bouilly, Jean Nicolas – Leipzig: Hartmann, 1816 [mf ed 1989] – 2v in 1 – 1 – (trans by kotzebue) – mf#7061 – us Wisconsin U Libr [830]

Geschichte aegypten's unter den pharaonen *see* Egypt under the pharaohs

Geschichte alexanders des dritten und der kirche seiner zeit / Reuter, Hermann – Leipzig: B G Teubner 1860-64 [mf ed 1993] – 3v on 5mf – 9 – 0-524-05883-0 – (incl bibl ref) – mf#1990-5177 – us ATLA [240]

Geschichte babyloniens und assyriens / Hommel, Fritz – Berlin: G Grote 1885-[88] – (= ser Allgemeine geschichte in einzeldarstellungen 1/2) – [ill/pl] – 1 – (iss in 5pt [1885-88]) – mf#film mas c604 – us Harvard [930]

Geschichte babyloniens und assyriens / Winckler, Hugo – Leipzig: Eduard Pfeiffer 1892 [mf ed 1989] – (= ser Voelker und staaten des alten orients 1) – 1mf – 9 – 0-7905-2813-4 – (incl bibl ref & ind) – mf#1987-2813 – us ATLA [930]

Die geschichte bileams und seine weissagungen / Hengstenberg, Ernst Wilhelm – Berlin: L Oehmigke, 1842 – 1mf – 9 – 0-524-05215-8 – mf#1992-0348 – us ATLA [220]

Geschichte der allgemeinen evang.-lutherischen synode von ohio und anderen staaten / Peter, Philip Adam & Schmidt, William – Columbus, Ohio: Verlagshandlung der Synode, 1900 – 2mf – 9 – 0-524-02489-8 – mf#1990-4348 – us ATLA [242]

Geschichte der altchristkiche litteratur bis eusebius / Harnack, Adolf von – Leipzig. v1-3. 1893-1904 – €80.00 – ne Slangenburg [240]

Geschichte der alt-'ebraeischen literatur : fuer denkende bibelleser / Schultze, Martin – Thorn: Ernst Lambeck, 1870 [mf ed 1989] – 1mf – 9 – 0-7905-2132-6 – mf#1987-2132 – us ATLA [221]

Geschichte der alten welt / Rostovtzeff, Michael Ivanovcitch – Wiesbaden, Germany. v1-2. 1941 – 1r – us UF Libraries [830]

Geschichte der althebraeischen litteratur : apokryphen und pseudoepigraphen / Budde, Karl – Leipzig: C F Amelang, 1906 – 2mf – 9 – 0-8370-2511-7 – (incl ind) – mf#1985-0511 – us ATLA [221]

Geschichte der altkirchlichen litteratur / Bardenhewer, Otto – Freiburg. v1-4. 1902-1924 – 69mf – 8 – €132.00 – ne Slangenburg [240]

Geschichte der alttestamentlichen literatur in aufsaetzen / Hausrath, Adolf – Heidelberg: J C B Mohr, 1864 – 1mf – 9 – 0-7905-1004-9 – mf#1987-1004 – us ATLA [221]

Geschichte der alttestamentlichen weissagung, theile 1 : die vorgeschichte der alttestamentlichen weissagung / Baur, Gustav – Giessen: J Ricker, 1861 – 1mf – 9 – 0-8370-2216-9 – (no further parts were published) – mf#1985-0216 – us ATLA [221]

Geschichte der amerikanischen urreligionen / Mueller, Johann Georg – Basel: Schweighauser, 1855 – 2mf – 9 – 0-524-06931-X – mf#1990-3557 – us ATLA [290]

Geschichte der angelsachsen bis zum tode koenig aelfreds / Winkelmann, Eduard August – Berlin: G Grote 1883 – (= ser Allgemeine geschichte in einzeldarstellungen 2/3) – [ill/pl/facs] – 1 – (with dahn, felix: urgeschichte der germanischen und romanischen voelker [berlin 1881-89] 4. bd) – mf#film mas c604 – us Harvard [941]

Geschichte der apologie des christentums / Zoeckler, Otto – Guetersloh: Bertelsmann, 1907 [mf ed 1991] – 2mf – 9 – 0-7905-8992-3 – (incl bibl ref & ind) – mf#1989-2217 – us ATLA [240]

Geschichte der apostolichen verkuendigung / Noesgen, Karl Friedrich – Muenchen: CH Beck, 1893 – (= ser Geschichte der neutestamentlichen offenbarung) – 2mf – 9 – 0-7905-1543-1 – (incl bibl ref and indexes) – mf#1987-1543 – us ATLA [240]

Geschichte der arabischen litteratur / Brockelmann, Carl – Weimar: E. Felber, 1898-1902 – 3mf – 9 – 0-7905-8020-9 – (incl bibl ref) – mf#1988-6001 – us ATLA [470]

Geschichte der architekur von den ealtesten zeiten bis zur... / Lubke, Wilhelm – Leipzig, Germany. v1-2. 1884-1886 – 1r – us UF Libraries [720]

Geschichte der arianischen haeresie : bis zur entscheidng von nikaea 325 / Koelling, W – Guetersloh. v1-2. 1874-1883 – €27.00 – ne Slangenburg [240]

Geschichte der attributenlehre : in der juedischen religionsphilosophie des mittelalters von saadja bis maimauni / Kaufmann, David – Gotha: F A Perthes, 1877 [mf ed 1990] – 2mf – 9 – 0-7905-5291-4 – (incl bibl ref) – mf#1988-1291 – us ATLA [270]

Die geschichte der auferweckung des lazarus / Steinmeyer, Franz Ludwig – Berlin: Wiegandt und Grieben, 1888 – 1mf – 9 – 0-8370-5389-7 – (incl bibl ref) – mf#1985-3389 – us ATLA [240]

Geschichte der aufloesung der alten gottesdienstlichen kirche deutschlands / Graff, P – Goettingen. v1-2. 1937-1939 – €29.00 – ne Slangenburg [240]

Geschichte der ausgestorbenen alten friesischen oder saechsischen sprache / Wiarda, Tileman Dothias – Aurich, Bremen: Winter, Foerster [in kommission] 1784 – (= ser Whsb) – 1mf – 9 – €20.00 – mf#Hu 158 – gw Fischer [430]

Geschichte der basler mission, 1815-1899 / Eppler, Paul – Basel: Verlag der Missionsbuchh, 1900 – 1mf – 9 – 0-524-00750-0 – (incl bibl ref) – mf#1990-0182 – us ATLA [240]

Geschichte der berliner missionsgesellschaft : und ihrer arbeiten in suedafrica, mit einer uebersichtskarte und vielen bildern / Wangemann, Hermann Theodor – Berlin: Im Selbstverlag des Ev. Missionshauses in Berlin, 1872-1877. Chicago: Dep of Photodup, U of Chicago Lib, 1971 (1r); Evanston: American Theol Lib Assoc, 1984 (1r) – 1 – 0-8370-0536-1 – (incl bibl ref and ind) – mf#1984-B219 – us ATLA [240]

Geschichte der bernischen taeufer / Mueller, Ernst – Frauenfeld: J Hueber, 1895 – 1mf – 9 – 0-524-01232-6 – mf#1990-0371 – us ATLA [240]

Geschichte der beziehungen zwischen theologie und naturwissenschaft : mit besondrer ruecksicht auf schoepfungsgeschichte / Zoeckler, Otto – Guetersloh: C Bertelsmann, 1877-1879 – 4mf – 9 – 0-7905-9554-0 – (incl bibl ref) – mf#1989-1259 – us ATLA [210]

Geschichte der biblischen literatur und des juedisch-hellenistischen schriftthums / Fuerst, Julius – Leipzig: Bernhard Tauchnitz. 2v. 1867-70 – 4mf – 9 – 0-7905-0890-7 – (incl bibl ref) – mf#1987-0890 – us ATLA [221]

Geschichte der biblischen offenbarung : als einleitung in's alte und neue testament / Haneberg, Daniel Bonifacius von – 4. Aufl. Regensburg: Georg Joseph Manz, 1876 – 3mf – 9 – 0-7905-0999-7 – (incl bibl ref and index) – mf#1987-0999 – us ATLA [220]

Geschichte der boehmischen reformation im fuenfzehnten jahrhundert / Krummel, Leopold – Gotha: FA Perthes, 1866 – 2mf – 9 – 0-7905-5769-X – (incl bibl ref) – mf#1988-1769 – us ATLA [242]

Geschichte der byzantinischen litteratur : von justinian bis zum ende des ost-roemischen reiches (527-1453) / Krumbacher, K – Muenchen, 1891 – €21.00 – ne Slangenburg [241]

Geschichte der byzantinischen litteratur : von justinian bis zum ende des ostroemischen reiches (527-1453) / Krumbacher, Karl – 2. Aufl. Muenchen: C.H. Beck, 1897 – (= ser Handbuch der klassischen Altertumswissenschaft) – 3mf – 9 – 0-7905-8040-3 – (incl bibl ref) – mf#1988-6021 – us ATLA [930]

Geschichte der cansteinschen bibelanstalt in halle / Bertram, Oswald – Halle: Verlag der Buchh. des Waisenhauses, 1863 – 1mf – 9 – 0-8370-1785-8 – mf#1987-6173 – us ATLA [220]

Geschichte der censur in zuerich : monatsschrift des wissenschaftlichen vereins in zuerich / Meyer von Knonau, G – Zuerich, 1859. v4, p 1-16 – 1mf – 9 – mf#ZWI-44 – ne IDC [242]

Geschichte der chinesischen mission : unter der leitung des pater johann adam schall, priesters aus der gesellschaft jesu / Schall von Bell, Johann Adam – Wien: Mechitaristen-Congregations-Buchhandlung, 1834 [mf ed 1995] – (= ser Yale coll) – 461p – 1 – 0-524-09768-2 – (trans fr latin into german. ann by ig sch von mannsegg) – mf#1995-0768 – us ATLA [241]

Geschichte der chinesischen mission... / Schall von Bell, J A – Wien, 1834 – 5mf – 9 – mf#HT-915 – ne IDC [915]

Geschichte der christlichen arabischen literatur / Graf, G – Citta del Vaticano. v1-4. 1944-1951 – €107.00 – ne Slangenburg [240]

Geschichte der christlichen dogmen in pragmatischer entwickelung / Lentz, Carl Georg Heinrich – Helmstedt: Verlag der C G Fleckeisen, 1834-1835 – 2mf – 9 – 0-524-08848-9 – (incl bibl ref) – mf#1993-2133 – us ATLA [240]

Geschichte der christlichen eschatologie innerhalb der vornicaenischen zeit : mit theilweiser einbeziehung der lehre vom christlichen heile ueberhaupt / Atzberger, Leonhard – Freiburg im Breisgau; St Louis, Mo: Herder, 1896 – 7mf – 9 – 0-524-00002-6 – (incl bibl ref) – mf#1989-2702 – us ATLA [240]

Geschichte der christlichen ethik / Gass, Wilhelm – Berlin: G Reimer, 1881-87 – 3mf – 9 – 0-7905-4681-7 – (incl bibl ref) – mf#1988-0681 – us ATLA [170]

Geschichte der christlichen ethik seit der reformation / Luthardt, Christoph Ernst – Leipzig: Doerffling & Franke, 1893 – (= ser Geschichte der christlichen ethik) – 2mf – 9 – 0-7905-9318-1 – (incl bibl ref and ind to all vols of geschichte der christlichen ethik) – mf#1989-2543 – us ATLA [170]

Geschichte der christlichen ethik vor der reformation / Luthardt, Christoph Ernst – Leipzig: Doerffling & Franke, 1888 – (= ser Geschichte der christlichen ethik) – 1mf – 9 – 0-7905-9319-X – (incl bibl ref) – mf#1989-2544 – us ATLA [170]

Geschichte der christlichen kirche / Schleiermacher, Friedrich [Ernst Daniel]; ed by Bonnell, E – Berlin: G Reimer 1840 [mf ed 1990] – 2mf – 9 – 0-7905-3982-9 – mf#1989-0475 – us ATLA [240]

Geschichte der christlichen kirche / Zeller, Eduard – Stuttgart: Franckh, 1848 – 1mf – 9 – 0-524-00426-9 – mf#1989-3126 – us ATLA [240]

Geschichte der christlichen kirche *see* History of the church

Geschichte der christlichen litteraturen des orients / Brockelmann, Carl et al – Leipzig: CF Amelang, 1907 – (= ser Die Litteraturen des ostens in Einzeldarstellungen) – 1mf – 9 – 0-524-08315-0 – (incl bibl ref) – mf#1993-1010 – us ATLA [470]

Geschichte der christlichen missionen unter den indianern nordamerikas im 17. und 18. jahrhundert : nebst einer beschreibung der religion der indianer / Fritschel, Gottfried – Nuernberg: G Loehe, 1870 – 1mf – 9 – 0-524-04549-6 – mf#1991-2113 – us ATLA [240]

Geschichte der christlichen philosophie zur zeit der kirchenvaeter / Stoeckl, Albert – Mainz: F Kirchheim, 1891 – 1mf – 9 – 0-7905-9684-9 – (incl bibl ref) – mf#1989-1409 – us ATLA [240]

Die geschichte der dalailamas / Schulemann, Guenther – Heidelberg: C Winter, 1911 – (= ser Religionswissenschaftliche Bibliothek) – 1mf – 9 – 0-524-02542-8 – (incl bibl ref) – mf#1990-3037 – us ATLA [280]

Geschichte der deutschen arbeitsbewegung in 15 kapiteln / Berlin. Institut fuer Marxismus-Leninismus – Autorenkollektiv: Walter Ulbricht et al.1966-69. 15v – 1 – us Wisconsin U Libr [335]

Geschichte der deutschen bibeluebersetzungen in der schweizerisch-reformirten kirche : von der reformation bis zur gegenwart / Mezger, J J – Basel: Bahnmaier, 1876 – 1mf – 9 – 0-8370-9169-1 – (incl bibl ref) – mf#1986-3169 – us ATLA [220]

Geschichte der deutschen bibeluebersetzungen in der schweizerisch-reformirten kirche von der reformation bis zur gegenwart / Mezger, J J – Basel, 1876 – 5mf – 9 – mf#ZWI-45 – ne IDC [242]

Geschichte der deutschen dichtung / Pfeiffer-Belli, Wolfgang – Freiburg: Herder, 1954 – 1r – 1 – (incl bibl ref and index) – us Wisconsin U Libr [430]

Geschichte der deutschen dichtung / Roehl, Hans – Leipzig: B G Teubner, 1914 – 1r – 1 – (incl bibl ref and index) – us Wisconsin U Libr [430]

Geschichte der deutschen dichtung im elften und zwolften jahrhundert / Scherer, Wilhelm – Strassburg, Germany. 1875 – 1r – us UF Libraries [430]

Geschichte der deutschen dichtung nach ihren epochen dargestellt / Schneider, Hermann – Bonn: Athenaeum-Verlag, 1949-50 [mf ed 1993] – 2v – 1 – mf#8056 – us Wisconsin U Libr [430]

Geschichte der deutschen dichtung nach ihren epochen dargestellt / Schneider, Hermann – Bochum: Deutscher Buchklub, [1952?] [mf ed 1993] – 776p – 1 – (incl bibl ref and ind) – mf#8056 – us Wisconsin U Libr [430]

Geschichte der deutschen dichtung von den aeltesten denkmaelern bis auf die neuzeit / Roquette, Otto – 3. durchgesehene Aufl, neue unveraenderte Ausg. Frankfurt a.M.: Literarische Anstalt Ruetten & Loening, 1882 – 1 – (incl ind) – us Wisconsin U Libr [430]

Geschichte der deutschen elegie / Beissner, Friedrich – Berlin: de Gruyter, 1941 [mf ed 1993] – (= ser Grundriss der germanischen philologie 14) – xvi/246p – 1 – (incl bibl ref and ind) – mf#8188 – us Wisconsin U Libr [430]

Geschichte der deutschen evangelischen synode von nord-amerika : im auftrage der synode zu ihrem fuenfundsiebzigjaehrigen jubilaeum / Muecke, Albert – St Louis, MO: Eden Publ House, 1915 [mf ed 1990] – 1mf – 9 – 0-7905-6073-9 – (in german) – mf#1988-2073 – us ATLA [242]

Geschichte der deutschen frauendichtung seit 1800 / Spiero, Heinrich – Leipzig, 1913 [mf ed 1992] – 1mf – 9 – €24.00 – 3-89349-109-0 – mf#DHS-AR 78 – gw Frankfurter [430]

Geschichte der deutschen goethe-biographie : ein kritischer abriss / Maync, Harry Wilhelm – 2. Abdruck. Leipzig: H Haessel, 1914 – 1 – us Wisconsin U Libr [430]

Geschichte der deutschen historiographie : seit dem auftreten des humanismus / Wegele, Franz X von; ed by Koeniglich Bayerische Akademie der Wissenschaften. Historische Kommission – Muenchen: R Oldenbourg, 1885 – (= ser Geschichte der wissenschaften in Deutschland) – 3mf – 9 – 0-7905-8234-1 – (incl bibl ref) – mf#1988-6134 – us ATLA [900]

Geschichte der deutschen im staate new york / Kapp, Friedrich – New York, NY. 1867 – 1r – us UF Libraries [305]

Geschichte der deutschen literatur / Bartels, Adolf – Leipzig: H Haessel, 1924-1928 [mf ed 1993] – 3v – 1 – (incl ind) – mf#8114 – us Wisconsin U Libr [430]

Geschichte der deutschen literatur / Fechter, Paul – [Guetersloh]: C Bertelsmann, 1952 – 1r – 1 – (incl ind) – us Wisconsin U Libr [430]

Geschichte der deutschen literatur / Fechter, Paul – [Guetersloh]: S Mohn, [1960] – 1r – 1 – (incl ind) – us Wisconsin U Libr [430]

Geschichte der deutschen literatur / Grabert, Willy – Muenchen: Bayerischer Schulbuch-Verlag, 1953 – 1 – (incl ind) – us Wisconsin U Libr [430]

Geschichte der deutschen literatur / Howald, Johann – Konstanz: C Hirsch, [1903] – 1r – 1 – (incl ind) – us Wisconsin U Libr [430]

Geschichte der deutschen literatur : mit einem abriss der geschichte der deutschen sprache und metrik / ed by Boetticher, Gotthold & Kinzel, Karl – 2. verb aufl. Halle a.S: Buchhandlung des Waisenhauses, 1896 [mf ed 1993] – xii/178p – 1 – (incl bibl ref and ind) – mf#8071 – us Wisconsin U Libr [430]

Geschichte der deutschen literatur / Scherer, Wilhelm & Walzel, Oskar – 4. aufl. Berlin: Askanischer Verlag, C A Kindle, 1928 [mf ed 1993] – xvi/942p – 1 – (bibl by josef koerner) – mf#7837 – us Wisconsin U Libr [430]

Geschichte der deutschen literatur / Scherer, Wilhelm & Walzel, Oskar – 4. Aufl. Berlin: Askanischer Verlag, C A Kindle, 1928 [mf ed 1993] – xvi/942p – 1 – (incl bibl ref and ind. bibl by josef koerner) – mf#7837 – us Wisconsin U Libr [430]

Geschichte der deutschen literatur : von den anfaengen bis zum ende des spaetmittelalters (1490) / Genzmer, Felix et al – Stuttgart: J B Metzler, 1962 [mf ed 1993] – 286p – 1 – (incl bibl ref and ind) – mf#8156 – us Wisconsin U Libr [430]

Geschichte der deutschen literatur : von goethes tod bis zur gegenwart / Alker, Ernst – Stuttgart, Cotta [1949-50] [mf ed 1993] – 2v on 1r – 1 – mf#8212 – us Wisconsin U Libr [430]

Geschichte der deutschen literatur 1789 bis 1806 / Dahnke, Hans-Dietrich – 2nd rev ed. Berlin: VEB Deutscher Verlag, 1958 [mf ed 1993] – (= ser Lehrbriefe für das fernstudium der oberstufenlehrer) – 395p – 1 – (incl bibl ref and ind) – mf#8243 – us Wisconsin U Libr [430]

Geschichte der deutschen literatur bis zur mitte des elften jahrhunderts / Unwerth, Wolf von – Berlin: Vereinigung wissenschaftlicher Verleger, 1920 – 1r – 1 – (incl bibl ref and index) – us Wisconsin U Libr [430]

Geschichte der deutschen literatur im achtzehnten jahrhundert / Hettner, Hermann – Berlin: Aufbau-Verlag, 1961- [mf ed 1993] – 1r – 1 – (text revision by gotthard erler. incl bibl ref) – mf#8222 – us Wisconsin U Libr [430]

Geschichte der deutschen literatur in der schweiz / Baechtold, J – Frauenfeld, 1892 – 1mf0mf – 9 – mf#ZWI-80 – ne IDC [430]

Geschichte der deutschen literatur von den anfaengen bis zur gegenwart / Linden, Walther – 5. Aufl. Leipzig: P Reclam, 1944, c1937 – 1r – 1 – (incl ind) – us Wisconsin U Libr [430]

Geschichte der deutschen literatur von goethes tod bis zur gegenwart *see* Die deutsche literatur im 19. jahrhundert, 1832-1914

Geschichte der deutschen litteratur von den aeltesten zeiten bis zur gegenwart / Vogt, Friedrich Hermann Traugott & Koch, Max – Leipzig: Bibliographisches Institut, 1897 – 1 – (incl ind) – us Wisconsin U Libr [430]

Geschichte der deutschen litteratur von ihren anfaengen bis auf die neueste zeit / Hirsch, Franz – Leipzig: W Friedrich [1883-85] [mf ed 1979] – (= ser Geschichte der weltlitteratur in einzeldarstellungen 5) – 3v on 1r – 1 – (v1: das mittelalter; v2: von luther bis lessing; v3: von goethe bis zur gegenwart) – mf#film mas 8652 – us Harvard [430]

Geschichte der deutschen mystik im mittelalter / Preger, W – Leipzig. v1-3. 1874-1893 – 8 – €48.00 – (pt1 leipzig 1874 9mf. pt2 leipzig 1881 8mf. pt3 leipzig 1893 8mf) – ne Slangenburg [230]

Geschichte der deutschen national-literatur : zum gebrauch an hoeheren unterrichtsanstalten und zum selbststudium / Kluge, Hermann – 21. verb. Aufl. Altenburg: O Bonde, 1890 – 1r – 1 – (incl bibl ref and index) – us Wisconsin U Libr [430]

Geschichte der deutschen ode / Vietor, Karl – Muenchen: Drei Masken Verlag 1923 [mf ed 1993] – (= ser Geschichte der deutschen literatur nach gattungen 1) – 1r – 1 – (incl bibl ref & ind. filmed with: christliches erbe und lyrisches gestaltung / hans giesecke) – mf#8297 – us Wisconsin U Libr [430]

Geschichte der deutschen poetik / Markwardt, Bruno – Berlin: W de Gruyter & Co, 1937-1967 – 5v – 1 – (incl bibl ref and ind) – us Wisconsin U Libr [430]

Geschichte der deutschen rechtswissenschaft / Stintzing, Roderich – Muenchen: R Oldenbourg. 3v in 4. 1880-1910 – (= ser Civil law 3 coll; Geschichte der wissenschaften in deutschland, neuere zeit) – 23mf – 9 – (part 2 was edited after the author's death by ernst landsberg (1860-1927), who completed the work. part 3 has the title: geschichte der deutschen rechtswissenschaft...fortsetzung zu der geschichte der deutschen rechtswissenschaft, erste und zweite abtheilung) – mf#LLMC 96-534 – us LLMC [340]

Geschichte der deutschen reformation / Bezold, Friedrich von – Berlin: Historischer Verlag Baumgaertel 1890 [mf ed 1991] – (= ser Allgemeine geschichte in einzeldarstellungen 3/1) – 3mf – 9 – 0-524-00741-1 – mf#1990-0173 – us ATLA [242]

Geschichte der deutschen sprache / Bach, Adolf – Heidelberg, Germany. 1949 – 1r – us UF Libraries [430]

Geschichte der deutschen sprache – Mit Texten und Uebersetzungshilfen. Berlin: Verlag Volk und Wissen, 1969. 428p. with illus. and 4 folding tables (in pocket) – 1 – us Wisconsin U Libr [430]

Geschichte der deutschen sprache und poesie : vorlesungen, gehalten an der universitaet bonn seit dem wintersemester 1818/19 / Schlegel, August Wilhelm von; ed by Koerner, Josef – Berlin: B Behr (F Feddersen), 1913 [mf ed 1993] – (= ser Deutsche litteraturdenkmale des 18. und 19. jahrhunderts 147, 3. folge n27) – xxxviii/184p – 1 – (incl bibl ref) – mf#8676 reel 9 – us Wisconsin U Libr [430]

Geschichte der deutsch-lutherischen kirche / Uhlhorn, Friedrich – Leipzig: Doerffling & Franke, 1911 – 2mf – 9 – 0-7905-8079-9 – mf#1988-6060 – us ATLA [242]

Geschichte der dogmatik in russischer darstellung – Guetersloh: C Bertelsmann, 1902 – 1mf – 9 – 0-8370-7463-0 – (incl bibl ref) – mf#1986-1463 – us ATLA [200]

Die geschichte der dogmatischen florilegien vom 5.-8. jahrhundert / Schermann, Theodor – Leipzig: J C Hinrichs, 1904 [mf ed 1989] – (= ser Tugal 28/1 [nf 13/1]) – 1mf – 9 – 0-7905-1732-9 – (in german & greek. incl bibl ref & ind) – mf#1987-1732 – us ATLA [240]

Die geschichte der dogmatischen florilegien vom 5.-8. jahrhundert / Schermann, Theodor – Leipzig, 1904 – (= ser Tugal 2-28/1) – 2mf – 9 – €5.00 – ne Slangenburg [240]

Geschichte der dogmen, oder, darstellung der glaubenslehren des christenthums : von seiner stiftung bis auf die neuern zeiten, insbesondere fuer studierende der theologie und zur vorbereitung auf ihre pruefung / Ruperti, F A – Berlin: FA Herbig, 1831 – 1mf – 9 – 0-7905-9098-0 – mf#1989-2323 – us ATLA [240]

Geschichte der domschule zu reval : 1319-1939 / Thomson, Erik – Wuerzburg: Holzner Verlag, 1969 – 1 – (incl bibl ref and index) – us Wisconsin U Libr [377]

Geschichte der einfuehrung des christenthums im suedwestlichen deutschland : besonders in wuertemberg / Hefele, Karl Joseph von – Tuebingen: H Laupp, 1837 – 1mf – 9 – 0-7905-7241-9 – (incl bibl ref) – mf#1988-3241 – us ATLA [240]

Geschichte der englischen gesandtschaft an den hof von kabul, im jahre 1808 : nebst ausfuehrlichen nachrichten ueber das koenigreich kabul, den dazu gehoerigen laendern und voelkerschaften / Elphinstone, Mountstuart – Weimar 1817 [mf ed Hildesheim 1995-98] – 2v on 7mf – 9 – €140.00 – 3-487-26520-6 – gw Olms [956]

Geschichte der englischen litteratur / Brink, Bernhard Aegidius Konrad Ten – Strassburg, Germany. v1-2. 1899 – 1r – us UF Libraries [420]

Geschichte der entdeckung und eroberung peru's : nebst ergaenzung aus augustins de zarate und garcilasso's de la vega berichten / Xeres, Francisco de – Stuttgart [u.a. 1843 [mf ed Hildesheim 1995-98] – 1v on 2mf – 9 – €60.00 – ISBN-10: 3-487-26831-0 – ISBN-13: 978-3-487-26831-6 – gw Olms [972]

Geschichte der entstehung und ausbildung des kirchenstaates / Sugenheim, Samuel – Leipzig: FA Brockhaus, 1854 – 1mf – 9 – 0-7905-6686-9 – (incl bibl ref) – mf#1988-2686 – us ATLA [240]

Geschichte der eroberung von mesopotamien und armenien von mohammed ben omar el wakedi / Mordtmann, A D – Hamburg, 1847 – 3mf – 9 – mf#AR-1812 – ne IDC [956]

Geschichte der ersten deutschen lutherischen ansiedlung in altenburg, perry co., mo : mit besonderer beruecksichtigung der dortigen kirchlichen bewegungen / Schieferdecker, Georg Albert – Clayton Co, Iowa: Seminars Wartburg, 1865 – 2mf – 9 – 0-524-07972-2 – mf#1990-5417 – us ATLA [242]

Geschichte der erziehung und bildung des israelitischen volkes und entwicklung der goettlichen heilsidee : durch die propheten bis zur ankunft des messias / Westermayer, Anton – Schaffhausen: Friedr Hurter, 1861 [mf ed 1993] – (= ser Das alte testament und seine bedeutung 4) – 2mf – 9 – 0-524-06224-2 – mf#1992-0862 – us ATLA [221]

Geschichte der ethik als philosophischer wissenschaft / Jodl, Friedrich – 2., neu bearb und verm Aufl. Stuttgart: Cotta, 1906-1912 – 4mf – 9 – 0-524-08347-9 – mf#1993-2037 – us ATLA [170]

Geschichte der evangel. kirche deutschlands : in der ersten haelfte des 19. jahrhunderts / Tischhauser, Christian – Basel: R Reich, 1900 – 2mf – 9 – 0-7905-7152-8 – (incl bibl ref) – mf#1988-3152 – us ATLA [242]

Geschichte der evangelischen fluechtlinge in der schweiz / Moerikofer, J C – Leipzig, S Hirzel, 1876 – 5mf – 9 – mf#PBU-436 – ne IDC [242]

Geschichte der evangelischen gemeinschaft / Orwig, Wilhelm W – 1. Aufl. Cleveland, Ohio: Verlegt von C Hammer fuer die Evang Gemeinschaft, 1857 – 2mf – 9 – 0-524-03447-8 – mf#1990-4707 – us ATLA [242]

Geschichte der evangelischen heidenmission : mit besonderer beruecksichtigung der deutschen / Gareis, Reinhold – Konstanz: Carl Kirsch, [1901?] – 2mf – 9 – 0-8370-7141-0 – (incl ind) – mf#1986-1141 – us ATLA [242]

Geschichte der evangelischen kirche in deutschland / Rocholl, Rudolf – Leipzig: A Deichert, 1897 – 2mf – 9 – 0-524-01531-7 – (incl bibl ref) – mf#1990-0437 – us ATLA [242]

Geschichte der evangelischen kirchenverfassung in deutschland / Richter, Aemilius Ludwig – Leipzig: Bernh Tauchnitz, 1851 – 1mf – 9 – 0-7905-6945-0 – (incl bibl ref) – mf#1988-2945 – us ATLA [242]

Die geschichte der evangelisch-lutherischen missouri-synode in nord-amerika : und ihrer lehrkaempfe von der saechsischen auswanderung im jahre 1838 an bis zum jahre 1884 / Hochstetter, Chr – Dresden: H J Naumann 1885 [mf ed 1992] – 2mf – 9 – 0-524-02473-1 – mf#1990-4332 – us ATLA [242]

Geschichte der evangelisch-lutherischen st. lorenz-gemeinde u.a.c. zu frankenmuth, mich / Mayer, Emanuel A – St Louis, MO: Concordia Pub House, 1895 – 1mf – 9 – 0-524-06876-3 – mf#1990-5295 – us ATLA [242]

Geschichte der evangel.-luth. synode von iowa und anderen staaten / Deindoerfer, Johannes – Chicago, Ill.: Wartburg Pub. House, 1897 – 2mf – 9 – 0-7905-5203-5 – mf#1988-1203 – us ATLA [242]

Die geschichte der familie lessing / Buchholtz, Arend [comp]; ed by Lessing, Carl Robert – Berlin: Druck von O v Holten 1909 [mf ed 1992] – 2v on 1r – 1 – mf#3080p – us Wisconsin U Libr [929]

Geschichte der fanatischen und enthusiastischen wiedertaeufer vornehmlich in niederdeutschland : melchior hofmann und die secte der hofmannianer / Krohn, Barthold Nicolaus – Leipzig: Bey B C Breitkopf, 1758 – 1r – 1 – 0-8370-0323-7 – mf#1984-B472 – us ATLA [943]

Geschichte der fatimiden-chalifen / Wuestenfeld, F – Goettingen, 1881 – 5mf – 9 – mf#NE-190 – ne IDC [956]

Die geschichte der festung glatz / Koehl, Eduard – Wuerzburg: Holzner Verlag, 1972 – 10r – 1 – (incl bibl ref and index) – us Wisconsin U Libr [943]

Die geschichte der festung koenigsberg/pr., 1257-1945 / Ehrhardt, Traugott – Wuerzburg: Holzner, 1960 – 10r – 1 – (incl bibl ref) – us Wisconsin U Libr [943]

Geschichte der freimaurerei : ein beitrag zur kultur- und literatur-geschichte des 18. jahrhunderts / Boos, Heinrich – 2. rev aufl. Aarau: Druck und Verlag von Sauerlaender & Co 1906 – 5mf – 9 – mf#vrl-37 – ne IDC [366]

Geschichte der freimaurerei in oesterreich-ungarn / Abafi, Lajos [Ludwig] – Budapest: L Aigner [1890-99] – 5v on 4mf – 9 – (filmed: bd1-4 [1890-93]) – mf#vrl-31 – ne IDC [366]

Geschichte der friedrichsschule zu gumbinnen : ein beitrag zur kultur- und bildungsgeschichte ostpreussens / Kirrinnis, Herbert – Wuerzburg: Holzner Verlag, 1963 – 1 – (incl bibl ref) – us Wisconsin U Libr [370]

Die geschichte der geburt des herrn und seiner ersten schritte im leben : in bezug auf die neueste kritik / Steinmeyer, Franz Ludwig – Berlin: Wiegandt & Grieben, 1873 [mf ed 1988] – (= ser Apologetische beitraege 4) – 1mf – 9 – 0-7905-0390-5 – mf#1987-0390 – us ATLA [240]

Geschichte der gefangenschaft auf st helena / Montholon, Charles J de – Leipzig 1846 [mf ed Hildesheim 1995-98] – 1v on 3mf – 9 – €90.00 – ISBN-10: 3-487-26365-3 – ISBN-13: 978-3-487-26365-6 – gw Olms [920]

Geschichte der gegenreformation / Droysen, Gustav – Berlin: G Grote 1893 – (= ser Allgemeine geschichte in einzeldarstellungen 3/3/1) – [ill/facs] – 1 – mf#film mas c604 – us Harvard [940]

Geschichte der gegenwaertigen zeit [...] – Strassburg (Strasbourg F), 1790 1 oct-1793 1 jan – 1 – fr ACRPP [933]

Geschichte der geheimen gesellschaften und der republikanischen partei in frankreich : vom regierungsantritt louis philipps bis zur februarrevolution, 1830-1848 / LaHodde, Lucien de – Basel 1851 [mf ed Hildesheim 1995-98] – 1v on 3mf – 9 – €90.00 – ISBN-10: 3-487-26066-2 – ISBN-13: 978-3-487-26066-2 – gw Olms [944]

Geschichte der gelehrtheit / Wieland, Christoph Martin; ed by Hirzel, Ludwig – Frauenfeld: J Huber, 1891 – 1r – 1 – (incl bibl ref) – us Wisconsin U Libr [000]

Geschichte der gnostisch-manichaeischen sekten im frueheren mittelalter / Doellinger, Johann Joseph Ignaz von – Muenchen: CH Beck, 1890 [mf ed 1990] – (= ser Beitraege zur sektengeschichte des mittelalters 1) – 1mf – 9 – 0-7905-5268-X – (incl bibl ref) – mf#1988-1268 – us ATLA [290]

Geschichte der gottesbeweise im mittelalter bis zum ausgang der hochscholastik / Grunwald, G – Muenster, 1907 – (= ser Bgphma 6/3) – 3mf – 8 – €7.00 – ne Slangenburg [110]

Geschichte der griechischen literatur / Aly, Wolfgang – Bielefeld, Germany. 1925 – 1r – us UF Libraries [450]

Geschichte der gruendung der armenisch-evangelischen gemeinde in schamachi : ein lebensbild aus der armenischen kirche und basler mission / Eppler, Christoph Friedrich – Basel: Verlag des Missionskomptoirs, 1873 – 1mf – 9 – 0-7905-6524-2 – mf#1988-2524 – us ATLA [240]

Geschichte der gruendung und ausbreitung der zur synode von missouri, ohio und andern staaten gehoerenden evangelisch-lutherischen gemeinden u.a.c. zu chicago, illinois : zur erinnerung an die am trinitatis-sonntag, den 31. mai 1896, stattgefundene feier des fuenfzigjaehrigen bestehens der ev. luth. kirche zu chicago – [S.l.: s.n.], 1896 (Chicago: Lange) – 2mf – 9 – 0-524-02253-4 – mf#1990-4260 – us ATLA [242]

Geschichte der hauen'schen erziehungsanstalt zu berlin / Gotz, Oscar – Berlin, Germany. 1909 – 1r – us UF Libraries [939]

Geschichte der hebraeischen sprache und schrift / Gesenius, W – Leipzig, 1815 – 3mf – 9 – mf#NE-483 – ne IDC [470]

Geschichte der hebrer *see* History of the hebrews

Geschichte der heiligen monika / Bougaud, Emile – Mainz: Franz Kirchheim; Milwaukee: Hoffman [distributor], 1870 – 1mf – 9 – 0-8370-6887-8 – mf#1986-0887 – us ATLA [920]

Die geschichte der heiligen schriften, alten testaments / Reuss, Eduard – 2. verm u verb Ausg. Braunschweig: C A Schwetschke, 1890 – 2mf – 9 – 0-7905-0213-5 – (includes bibliographies) – mf#1987-0213 – us ATLA [221]

Die geschichte der heiligen schriften neuen testaments / Reuss, Eduard – 6th rev enl ed. Braunschweig, 1887 – 12mf – 8 – €23.00 – ne Slangenburg [225]

Geschichte der heiligen schriften, neuen testaments = History of the sacred scriptures of the new testament / Reuss, Eduard – Boston: Houghton, Mifflin. 2v. 1884 – 2mf – 9 – 0-7905-0215-1 – (in english. includes bibliographies and index) – mf#1987-0215 – us ATLA [225]

Geschichte der herrschenden ideen des islams : der gottesbegriff, die prophetie und staatsidee / Kremer, Alfred, Freiherr von – Leipzig: FA Brockhaus, 1868 – 2mf – 9 – 0-524-04522-4 – (incl bibl ref) – mf#1990-3356 – us ATLA [260]

Geschichte der herzoglichen franzschule in dessau, 1799-1849 / Horwitz, Ludwig – Dessau, Germany. 1894 – 1r – us UF Libraries [939]

Geschichte der hexenprozesse in bayern : im lichte der allgemeinen entwicklung / Riezler, Sigmund – Stuttgart: JG Cotta, 1896 – 1mf – 9 – 0-524-02189-9 – (incl bibl ref) – mf#1990-0574 – us ATLA [943]

Geschichte der indischen litteratur / Winternitz, Moritz – Leipzig: CF Amelang, 1909-[1922?] – (= ser Die Litteraturen des Ostens in Einzeldarstellungen) – 4mf – 9 – 0-524-07504-2 – mf#1991-0125 – us ATLA [490]

Geschichte der insel hayti / Handelmann, Heinrich – Kiel, Germany. 1856 – 1r – us UF Libraries [972]

Geschichte der irischen kirche von anfang bis zum 12. jahrhundert / Delius, W – Basel, 1954 – €11.00 – ne Slangenburg [241]

Geschichte der israel / Buxbaum, Heinrich – Pressburg, Czechoslovakia. 1884 – 1r – us UF Libraries [939]

Geschichte der israelitischen religion / Marti, Karl – 5. aufl. Strassburg: Friedrich Bull, 1907 – 1mf – 9 – 0-8370-9488-7 – (incl bibl ref and indexes) – mf#1986-3488 – us ATLA [221]

Geschichte der jesuiten in den laendern deutscher zunge im 16. jahrhundert / Duhr, Bernhard – Freiburg im Breisgau; St. Louis, Mo.: Herder, 1907 – 9mf – 9 – 0-7905-8054-3 – (incl bibl ref) – mf#1988-6035 – us ATLA [241]

Geschichte der jesuiten in den laendern deutscher zunge in der ersten haelfte des 17. jahrhunderts / Duhr, Bernhard – Freiburg im Breisgau; St. Louis, Mo.: Herder, 1913 – 14mf – 9 – 0-7905-8055-1 – (incl bibl ref) – mf#1988-6036 – us ATLA [241]

Geschichte der juden : von den aeltesten zeiten bis auf die gegenwart / Graetz, H – Leipzig. v1-2. 1874-1876 – 8 – €63.00 – (v1 leipzig 1874 10mf. v2/1 leipzig 1875 9mf. v2/2 leipzig 1876 9mf) – ne Slangenburg [939]

Geschichte der juden in lubeck / Carlebach, Salomon – Lubeck, Germany. 1898 – 1r – us UF Libraries [939]

Geschichte der juden in mahran und oesterr-schlesien / Elvert, Christian – Brunn, Czechoslovakia. 1895 – 1r – us UF Libraries [939]

Geschichte der juden in oedenburg / Pollack, Miksa – Wien, Austria. 1929 – 1r – us UF Libraries [939]

Die geschichte der juden in palaestina seit dem jahre 70 nach chr : eine skizze / Hoelscher, Gustav – Leipzig: J C Hinrichs, 1909 – (= ser Schriften Des Institutum Delitzschianum Zu Leipzig) – 1mf – 9 – 0-7905-1998-4 – mf#1987-1998 – us ATLA [270]

Geschichte der juden in stadt und stift essen bis zur sakularisatio... / Samuel, Salomon – Essen-Ruhr, Germany. 1905 – 1r – us UF Libraries [939]

Geschichte der juden in wien / Wolf, Gerson – Wien, Austria. 1876 – 1r – us UF Libraries [939]

Geschichte der judischen reformbewegung von mendelssohn / Seligmann, Caesar – Frankfurt am Main, Germany. 1922 – 1r – us UF Libraries [939]

Geschichte der juedischen apologetik als vorgeschichte des christenthums / Friedlaender, Moritz – Zuerich: Caesar Schmidt, 1903 – 2mf – 9 – 0-8370-9864-5 – (incl bibl ref) – mf#1986-3864 – us ATLA [270]

Geschichte der juedischen literatur / Karpeles, Gustav – Berlin: Robert Oppenheim, 1886 – 3mf – 9 – 0-524-08050-X – mf#1991-0266 – us ATLA [410]

Geschichte der katholischen katachese / Probst, Ferdinand – Breslau: Franz Goerlich, 1886 – 1mf – 9 – 0-524-01526-0 – (incl bibl ref) – mf#1990-0432 – us ATLA [241]

Geschichte der katholischen kirche : von der mitte des 18. jahrhunderts bis zum vatikanischen konzil / Mirbt, Carl – Berlin: G J Goeschen 1913 [mf ed 1990] – (= ser Sammlung goeschen) – 1mf – 9 – 0-7905-5665-0 – (incl bibl ref) – mf#1988-1665 – us ATLA [241]

Geschichte der katholischen kirche in deutschland im neunzehnten jahrhundert : 1. band: vom beginne des 19. jahrhunderts bis zu den concordatsverhandlungen / Brueck, Heinrich – Mainz, 1902 (mf ed 1994) – 6mf – 9 – €250.00 – 3-89349-184-8 – gw Frankfurter [241]

Geschichte der katholischen kirche in irland von der einfuehrung des christenthums bis auf die gegenwart / Bellesheim, Alphons – Mainz: F Kirchheim, 1890-1891 – 6mf – 9 – 0-524-02880-X – (incl bibl ref) – mf#1990-4471 – us ATLA [241]

Geschichte der katholischen kirche in schottland : von der einfuehrung des christentums bis auf die gegenwart / Bellesheim, Alphons – Mainz: Franz Kirchheim. 2v. 1883 – 4mf – 9 – 0-8370-6800-2 – (incl ind) – mf#1986-0800 – us ATLA [241]

Geschichte der katholischen literatur deutschlands vom 17. jahrhundert bis zur gegenwart / Bruehl, I A – Wien, Leipzig, 1861 (mf ed 1993) – 5mf – 9 – €74.00 – 3-89349-124-4 – mf#DHS-AR 91 – gw Frankfurter [430]

Geschichte der katholischen missionen in ostindien von der zeit vasco da gama's bis zur mitte des achtzehnten jahrhunderts / Muellbauer, M – Muenchen, 1851 – 7mf – 9 – mf#787 – ne IDC [915]

Geschichte der katholischen reformation : erster band / Maurenbrecher, Wilhelm – Noerdlingen: C H Beck, 1880 [mf ed 1986] – 1mf – 9 – 0-8370-8279-X – (no more publ) – mf#1986-2279 – us ATLA [241]

Geschichte der kirche russlands = Istoriia russkoi tserkvi / Filaret, Archbishop of Chernigov – Frankfurt a. M.: Joseph Baer, Sotheran, 1872 – 2mf – 9 – 0-8370-7695-1 – mf#1986-1695 – us ATLA [240]

Geschichte der kirchenverfassung deutschlands im mittelalter / Werminghoff, Albert – Leipzig: BG Teubner, 1905 – 1mf – 9 – 0-7905-6966-3 – (incl bibl ref) – mf#1988-2966 – us ATLA [240]

Die geschichte der kirchenweihe vom 1.-7. jahrhundert / Stiefenhofer, D – Muenchen, 1909 – €7.00 – ne Slangenburg [241]

Geschichte der kirchlichen armenpflege / Ratzinger, Georg – Freiburg im Breisgau: Herder, 1868 – 2mf – 9 – 0-8370-7095-3 – (incl bibl ref) – mf#1986-1095 – us ATLA [240]

Geschichte der kirchlichen liturgie des bisthums augsburg / Hoeynck, F A – Augsburg, 1889 – 8mf – 8 – €17.00 – ne Slangenburg [241]

Geschichte der kirchlichen revolution oder protestantischen reform des kantons bern und umliegender gegenden / Haller, K L – Luzern, Raeber, 1836 – 4mf – 9 – mf#PBU-453 – ne IDC [242]

Geschichte der kirchlichen trennung zwischen dem orient und occident : von den ersten anfaengen bis zur juengsten gegenwart / Pichler, Aloys – Muenchen: M. Rieger, 1864-1865 – 4mf – 9 – 0-7905-5789-4 – (incl bibl ref) – mf#1988-1789 – us ATLA [240]

Geschichte der koelnischen minoriten-ordensprovinz / Eubel, Konrad – Koeln: J & W Boisseree, 1906 – (= ser Veroeffentlichungen des historischen Vereins fuer den Niederrhein) – 1mf – 9 – 0-524-04488-0 – (incl bibl ref) – mf#1990-1250 – us ATLA [243]

Geschichte der kosmologie in der griechischen kirche bis auf origenes / Moeller, E W – Halle, 1860 – €19.00 – ne Slangenburg [240]

Geschichte der kosmologie in der griechischen kirche bis auf origines : mit specialuntersuchungen ueber die gnostischen systeme / Moeller, Wilhelm – Halle: Fricke, 1860 [mf ed 1990] – 2mf – 9 – 0-7905-6767-9 – (incl bibl ref) – mf#1988-2767 – us ATLA [290]

Geschichte der kreuzzuege / Kugler, B – Berlin, 1880 – 6mf – 9 – mf#ILM-609 – ne IDC [931]

Geschichte der kreuzzuege / Kugler, Bernhard – 2nd ed. Berlin: G Grote 1891 – (= ser Allgemeine geschichte in einzeldarstellungen 2/5) – [pl/ill/facs] – 1 – (incl bibl ref) – us Harvard [931]

Geschichte der kreuzzuege im umriss / Roehricht, R – Innsbruck, 1898 – 3mf – 9 – mf#H-2908 – ne IDC [931]

Geschichte der lateinischen sprache / Stolze, Friedrich – Leipzig, 1910 (mf ed 1992) – 1mf – 9 – €24.00 – 3-89349-114-7 – mf#DHS-AR 83 – gw Frankfurter [450]

Geschichte der lehre vom heiligen geiste : in zwei buechern / Noesgen, Karl Friedrich – Guetersloh: C Bertelsmann, 1899 – 1mf – 9 – 0-7905-9543-5 – (incl bibl ref) – mf#1989-1248 – us ATLA [240]

Geschichte der letzten systeme der philosophie in deutschland von kant bis hegel / Michelet, Karl Ludwig – Berlin: Duncker und Humblot, 1837-1838 – 4mf – 9 – 0-524-08546-3 – mf#1993-2071 – us ATLA [190]

Geschichte der logik im abendlande / Prantl, Carl – Leipzig: S Hirzel, 1855-1870 – 5mf – 9 – 0-7905-3560-2 – (incl bibl ref) – mf#1989-0053 – us ATLA [160]

Geschichte der logosidee in der christlichen litteratur / Aall, Anathon – Leipzig: OR Reisland, 1899 – (= ser Logos) – 2mf – 9 – 0-7905-7795-X – (incl bibl ref) – mf#1989-1020 – us ATLA [240]

Geschichte der logosidee in der griechischen philosophie / Aall, Anathon – Leipzig: OR Reisland, 1896 [mf ed 1991] – (= ser Der logos 1) – 1mf – 9 – 0-7905-7796-8 – (incl bibl ref) – mf#1989-1021 – us ATLA [180]

Geschichte der lutherischen kirche in america / Graebner, Augustus Lawrence – St Louis, MO: Concordia Publishing House, 1892 – 2mf – 9 – 0-7905-7108-0 – mf#1988-3108 – us ATLA [242]

Geschichte der lutherischen kirche in amerika = History of the evangelical lutheran church in the united states / Jacobs, Henry Eyster; ed by Fritschel, George John – Guetersloh: C Bertelsmann 1896-97 [mf ed 1992] – 2v on 2mf – 9 – 0-524-02474-X – (in german. incl bibl ref) – mf#1990-4333 – us ATLA [242]

Geschichte der lutherischen mission / Plitt, Gustav Leopold; ed by Hardeland, Otto – Berlin: A Deichert, 1894-1895 – 2mf – 9 – 0-524-06438-5 – mf#1991-2560 – us ATLA [242]

Geschichte der maronen-negern auf jamaika : nebst einer schilderung des vormaligen und jetzigen zustandes dieser insel / Dallas, Robert – Weimar 1805 [mf ed Hildesheim 1995-98] – 1v on 3mf – 9 – €90.00 – ISBN-10: 3-487-26575-3 – ISBN-13: 978-3-487-26575-9 – gw Olms [918]

Geschichte der mennoniten : von menno simons' austritt aus der roemisch-katholischen kirche in 1536 bis zu deren auswanderung nach amerika in 1683 / Cassel, Daniel Kolb – Philadelphia: J Kohler, 1890 – 2mf – 9 – 0-8370-8970-0 – (incl ind) – mf#1986-2970 – us ATLA [243]

Die geschichte der menschheit / Rohrbach, Paul – Koenigstein im Taunus: KR Langewiesche, c1914 – 1mf – 9 – 0-524-04590-9 – mf#1992-0178 – us ATLA [900]

Geschichte der messgewaender / Papas, T – Muenchen, 1965 – 6mf – 8 – €14.00 – ne Slangenburg [240]

Geschichte der mission der evangelischen brueder auf den caraibischen inseln s thomas, s croix und s jan / Oldendorp, C G A; ed by Bossart, J J – Barby, 1777. 2v – 14mf – 9 – mf#HTM-152 – ne IDC [919]

Geschichte der mittelhochdeutschen literatur / Vogt, Friedrich Hermann Traugott – 3. umgearb. Aufl. Berlin; Leipzig: Vereinigung Wissenschaftlicher Verleger, 1922 – 1 – (incl bibl ref) – us Wisconsin U Libr [430]

Geschichte der moralstreitigkeiten in der roemisch-katholische kirche / Doellinger, Ignaz von & Reusch, H – Noerdlingen. v1-2. 1889 – €37.00 – ne Slangenburg [241]

Geschichte der moralstreitigkeiten in der roemisch-katholischen kirche : seit dem sechzehnten jahrhundert. mit beitraegen zur geschichte und charakteristik des jesuitenordens / ed by Doellinger, Johann Joseph Ignaz von & Reusch, Franz Heinrich – Noerdlingen: CH Beck, 1889 – 3mf – 9 – 0-524-04172-5 – (incl bibl ref) – mf#1990-4976 – us ATLA [241]

Geschichte der nauenschen erziehungsanstalt zu berlin / Gotz, Oscar – Berlin, Germany. 1932? – 1r – us UF Libraries [939]

Geschichte der neueren philosophie von baco und cartesius bis zur gegenwart / Stoeckl, Albert – Mainz: F Kirchheim, 1883 – 3mf – 9 – 0-7905-8156-6 – (incl bibl ref) – mf#1988-6103 – us ATLA [100]

Geschichte der neuesten reformen der judischen gemeinde berlin's / Pinner, M – Berlin, Germany. 1857 – 1r – us UF Libraries [939]

Geschichte der neuhebraischen literatur / Klausner, Joseph – Berlin, Germany. 1921 – 1r – us UF Libraries [470]

Geschichte der niederfraenkischen geschaeftssprache / Heinzel, Richard – Paderborn: Schoeningh, 1874. 464p – 1 – us Wisconsin U Libr [430]

Geschichte der optik : from aristotle to newton / Wilde, E – Berlin. v1. 1838 – 1r – 1 – mf#96366 – uk Microform Academic [530]

Geschichte der orientalischen angelegenheit im zeitraume des pariser und des berliner friedens / Bamberg, Felix – Berlin: G Grote 1892 – (= ser Allgemeine geschichte in einzeldarstellungen 4/5) – [ill] – 1 – mf#film mas c604 – us Harvard [940]

Geschichte der orientalischen kirchen von 1453-1898 / Kyriakos, A Diomedes – Leipzig: A Deichert, 1902 – 1mf – 9 – 0-7905-4994-8 – (incl bibl ref) – mf#1988-0994 – us ATLA [240]

Geschichte der patriarchalischen und mosaischen offenbarung bis zur zeit der richter / Westermayer, Anton – Schaffhausen: Friedr Hurter, 1861 [mf ed 1993] – (= ser Das alte testament und seine bedeutung 2) – 2mf – 9 – 0-524-06225-0 – mf#1992-0863 – us ATLA [221]

Geschichte der paulinischen forschung von der reformation bis auf die gegenwart : a critical history = Paul and his interpreters / Schweitzer, Albert – London: Adam and Charles Black, 1912 – 1mf – 9 – 0-8370-9577-8 – (in english. includes bibliographies and index) – mf#1986-3577 – us ATLA [240]

Geschichte der perser und araber zur zeit der sasaniden / Noeldeke, T – Leyden, 1879 – 10mf – 8 – mf#U-584 – ne IDC [956]

Geschichte der pflanzung und leitung der christlichen kirche durch die apostel = History of the planting and training of the christian church by the apostles / Neander, August – London: Henry G Bohn. 2v. 1851 – 4mf – 9 – 0-7905-0047-7 – (incl indes) – mf#1987-0047 – us ATLA [225]

Geschichte der pflanzung und leitung der christlichen kirche durch die apostel / Neander, August – 5. aufl. Gotha: Friedrich Andreas Perthes, 1862 – 2mf – 9 – 0-8370-9890-4 – (incl bibl ref and indexes) – mf#1986-3890 – us ATLA [225]

Geschichte der pflanzung und leitung der christlichen kirche durch die apostel / Neander, August – Hamburg- v1-2. 1832-33 – 8 – €29.00 – (v1 hamburg 1832 (8mf). v2 hamburg 1833 (7mf)) – ne Slangenburg [226]

Geschichte der philosophie / Schleiermacher, Friedrich [Ernst Daniel]; ed by Ritter, Heinrich – Berlin: G Reimer 1839 [mf ed 1990] – 1mf – 9 – 0-7905-3983-7 – mf#1989-0476 – us ATLA [100]

Geschichte der philosophie / Schwegler, A – Tuebingen, 1870 – 8mf – 8 – €17.00 – ne Slangenburg [100]

Geschichte der philosophie / Stoeckle, A – Mainz, 1875 – 24mf – 8 – €46.00 – ne Slangenburg [100]

Geschichte der philosophie des mittelalters / Marbach, G O – Leipzig, 1841 – €15.00 – ne Slangenburg [100]

Geschichte der philosophie des mittelalters / Stoeckl, Albert – Mainz: F Kirchheim, 1864-1866 – 6mf – 9 – 0-7905-8919-2 – (incl bibl ref) – mf#1989-2144 – us ATLA [180]

Geschichte der philosophie im umriss *see* History of philosophy in epitome

Geschichte der philosophischen terminologie im umriss / Eucken, Rudolf – Leipzig: Veit, 1879 – 1mf – 9 – 0-7905-7566-3 – (incl bibl ref) – mf#1989-0791 – us ATLA [100]

Geschichte der phoenizier / Pietschmann, Richard – Berlin: G Grote 1889 – (= ser Allgemeine geschichte in einzeldarstellungen 1/4/2) – [ill/pl] – 1 – (with: justi, f: geschichte des alten persiens [berlin 1879]) – mf#film mas c604 – us Harvard [930]

Geschichte der poetischen litteratur der deutschen / Hahn, Werner; ed by Kreyenberg, Gotthold – Berlin: W Hertz, 1897 – 1r – 1 – (incl ind) – us Wisconsin U Libr [430]

Geschichte der poetischen national-literatur der hebraeer / Meier, Ernst Heinrich – Leipzig: W Engelmann, 1856 – 2mf – 9 – 0-524-04913-0 – (incl bibl ref) – mf#1992-0256 – us ATLA [470]

Die geschichte der predigt in deutschland bis auf karl den grossen, 600-814 : lateinische predigten von verfassern fremdlaendischer herkunft / Albert, Felix Richard – Guetersloh: C Bertelsmann, 1892 [mf ed 1989] – (= ser Die geschichte der predigt in deutschland bis luther 1) – 1mf – 9 – 0-7905-4365-6 – (incl bibl ref) – mf#1988-0365 – us ATLA [240]

Geschichte der presbyterial- und synodalverfassung : seit der reformation / Lechler, Gotthard Victor – Leiden: D. Noothoven van Goor, 1854 – 1mf – 9 – 0-7905-6072-0 – (incl bibl ref) – mf#1988-2072 – us ATLA [242]

Geschichte der protestantischen dogmatik : in ihrem zusammenhange mit theologie ueberhaupt / Gass, Wilhelm – Berlin: G Reimer, 1854-1867 – 5mf – 9 – 0-7905-7107-2 – (incl bibl ref and indexes) – mf#1988-3107 – us ATLA [242]

Geschichte der protestantischen dogmatik von melanchthon bis schleiermacher / Herrmann, Wilhelm – Leipzig: Breitkopf und Haertel, 1842 – 1mf – 9 – 0-7905-9388-2 – (incl bibl ref) – mf#1989-2613 – us ATLA [242]

Geschichte der protestantischen kirchenverfassung / Sehling, Emil – 2. aufl. Leipzig: B G Teubner 1914 [mf ed 1990] – (= ser Grundriss der geschichtswissenschaft zur einfuehrung in das studium der deutschen geschichte des mittelalters und der neuzeit 2/8) – 1mf – 9 – 0-7905-6497-1 – (incl bibl ref) – mf#1988-2497 – us ATLA [242]

Geschichte der protestantischen theologie : besonders in deutschland / Dorner, Isaak August – Muenchen: JG Cotta, 1867 – (= ser Geschichte der wissenschaften in Deutschland) – 3mf – 9 – 0-8370-8503-9 – (incl bibl ref) – mf#1986-2503 – us ATLA [242]

Geschichte der protestantischen theologie / Frank, Gustav – Leipzig: Breitkopf und Haertel, 1862-1905 – 1mf – 9 – 0-7905-4576-4 – (incl bibl ref) – mf#1988-0576 – us ATLA [242]

Geschichte der protestantischen theologie : von der konkordienformel an bis in die mitte des achtzehnten jahrhunderts / Planck, Gottlieb Jakob – Goettingen: Vandenhoeck und Ruprecht, 1831 – 1mf – 9 – 0-7905-6717-2 – mf#1988-2717 – us ATLA [242]

Geschichte der quietistischen mystik in der katholischen kirche / Heppe, Heinrich – Berlin: Wilhelm Hertz, 1875 – 2mf – 9 – 0-7905-7000-9 – (incl bibl ref) – mf#1988-3000 – us ATLA [241]

Geschichte der reaction kaiser julians gegen die christliche kirche / Rode, Friedrich – Jena: H Dabis, 1877 – 1mf – 9 – 0-524-04148-2 – (incl bibl ref) – mf#1990-1218 – us ATLA [240]

Geschichte der reformation : bis zur vollendung der konkordienformel und dem erstmaligen erscheinen des konkordienbuches am 25. juni 1580 fortgefuehrt / Frey, August Emil – Neueste Aufl. Lahr (Baden); New York, NY: E Kaufmann, [1908?] – 1mf – 9 – 0-524-00990-2 – mf#1990-0267 – us ATLA [242]

Geschichte der reformation / Myconius, Friedrich; ed by Clemen, Otto – Leipzig: R Voigtlaender, [1914?] – (= ser Voigtlaenders Quellenbuecher) – 1mf – 9 – 0-524-02916-4 – (incl bibl ref) – mf#1990-0732 – us ATLA [242]

Geschichte der reformation des sechszehnten jahrhunderts / Merle d'Aubigne, J H – Stuttgart, 1861-1862. 5 v – 29mf – 9 – mf#ZWI-92 – ne IDC [242]

Geschichte der reformation im elsass und besonders in strasburg : nach gleichzeitigen quellen / Roehrich, Timotheus Wilhelm – Strasburg: F C Heitz, 1830-1832. Chicago: Dep of Photodup, U of Chicago Lib, 1966 (1r); Evanston: American Theol Lib Assoc, 1984 (1r) – 1 – 0-8370-0447-0 – mf#1984-B094 – us ATLA [242]

Geschichte der reformation in der grafschaft oettingen, 1522-1569 / Herold, Reinhold – Halle: Verein fuer Reformationsgeschichte 1902 [mf ed 1990] – (= ser Schriften des vereins fuer reformationsgeschichte 20/75) – 1mf – 9 – 0-7905-5094-6 – (incl bibl ref) – mf#1988-1094 – us ATLA [242]

Geschichte der reformation in polen / Wotschke, Theodor – Leipzig: Verein fuer Reformationsgeschichte durch R Haupt, 1911 – (= ser Studien zur Kultur und Geschichte der Reformation) – 1mf – 9 – 0-524-01034-X – mf#1990-0311 – us ATLA [242]

Geschichte der reformation in venedig / Benrath, Karl – Halle: Verein fuer Reformationsgeschichte, 1887 – (= ser [Schriften des Vereins fuer Reformationsgeschichte]) – 1mf – 9 – 0-7905-4604-3 – (incl bibl ref) – mf#1988-0604 – us ATLA [945]

Geschichte der reformation und gegenreformation in der ehemaligen freien reichstadt dinkelsbuehl (1524-1648) / Buerckstuemmer, Christian – Leipzig: Verein fuer Reformationsgeschichte 1914-15 [mf ed 1990] – (= ser Schriften des vereins fuer reformationsgeschichte 14/57) – 1mf – 9 – 0-524-02277-1 – (incl bibl ref) – mf#1990-0582 – us ATLA [242]

Geschichte der reformirten kirche in russland : kirchenhistorische studie / Dalton, Hermann – Gotha: R Besser, 1865 – 1mf – 9 – 0-524-02002-7 – mf#1990-0547 – us ATLA [242]

Geschichte der reformirten kirchen in lithauen / Lukaszewicz, Jozef – Leipzig: Dyk, 1848-1850 – 2mf – 9 – 0-7905-7176-5 – (incl bibl ref) – mf#1988-3176 – us ATLA [242]

Geschichte der reisen : die seit cook an der nordwest- und nordost-kueste von amerika und in dem noerdlichsten amerika...unternommen worden sind / Forster, G – Berlin, 1791. 3v – 17mf – 9 – mf#H-6157 – ne IDC [910]

Die geschichte der religion / Pfleiderer, Otto – Leipzig: Fues, 1869 – (= ser Religion, ihr Wesen und ihre Geschichte) – 2mf – 9 – 0-7905-7451-9 – mf#1989-0676 – us ATLA [200]

Geschichte der religion : darstellung der inneren entwicklung und aeusseren gestaltung der religioesen idee / Scherr, Johannes – 2. Aufl. Leipzig: O Wigand, 1860 – 3mf – 9 – 0-524-04652-2 – (incl bibl ref) – mf#1990-3395 – us ATLA [200]

Geschichte der religionsphilosophie von spinoza bis auf die gegenwart / Pfleiderer, Otto – 3., erw Aufl. Berlin: G Reimer, 1893 – 2mf – 9 – 0-7905-9060-3 – (incl bibl ref) – mf#1989-2285 – us ATLA [200]

Geschichte der revolution in england / Stern, Alfred – Berlin: G Grote 1881 – (= ser Allgemeine geschichte in einzeldarstellungen 3/4) – [ill/pl] – 1 – mf#film mas c604 – us Harvard [941]

Geschichte der ritter des ordens pour la merite im weltkrieg / ed by Moeller-Witten, Hanns – Berlin: Bernard & Graefe, 1935. 2v – 1 – us Wisconsin U Libr [943]

Geschichte der roemischen kirche / Langen, Joseph – Bonn: Max Cohen, 1881-93 [mf ed 1986] – 4v on 8mf – 9 – 0-8370-7883-0 – (incl bibl ref & ind) – mf#1986-1883 – us ATLA [241]

Geschichte der romanschen sprache : abgelesen in der koeniglichen gesellschaft der wissenschaften, den 10 nov 1775 / Planta, Joseph – Chur: Typographische Gesellschaft 1776 – (= ser Whsb) – 1mf – 9 – €20.00 – (trans fr english) – mf#Hu 142 – gw Fischer [450]

Geschichte der schweizerisch-reformierten kirchen / Bloesch, Emil – Bern: Schmid & Francke, 1898-1899 – 3mf – 9 – 0-7905-5512-3 – (incl bibl ref) – mf#1988-1512 – us ATLA [242]

Geschichte der see-reisen und entdeckungen im sued-meerwelche : auf befehl sr grossbrittannischen majestaet [george des dritten] unternommen...worden sind; aus den tagebuechern der [schiffs-]befehlshaber und den handschriften... – Berlin 1774-88 [mf ed Hildesheim 1995-98] – 7v on 41mf – 9 – €410.00 – 3-487-26628-8 – gw Olms [919]

Geschichte der slavenapostel konstantinus (kyrillus) und methodius / Goetz, Leopold Karl – Gotha: F A Perthes, 1897 – 1mf – 9 – 0-7905-4853-4 – (incl bibl ref) – mf#1988-0853 – us ATLA [240]

Geschichte der slawenapostel cyrill und method und der slawischen liturgie / Ginzel, J A – Wien, 1861 – €14.00 – ne Slangenburg [243]

Geschichte der stadt babylon / Winckler, Hugo – Leipzig: JC Hinrichs, 1904 [mf ed 1989] – (= ser Der alte orient 6/1) – 1mf – 9 – 0-7905-2814-2 – mf#1987-2814 – us ATLA [241]

Geschichte der stadt hohenstein in ostpreussen / Hartmann, Ernst of Osterode – Wuerzburg: Holzner-Verlag, 1959 – 10r – 1 – us Wisconsin U Libr [914]

Geschichte der stadt liebemuehl / Hartmann, Ernst of Osterode – Wuerzburg: Holzner, 1964 – 1r – 1 – (incl bibl ref and index) – us Wisconsin U Libr [914]

Geschichte der stadt schneidemuehl / Boese, Karl – 2. voll umgearb erw aufl. Wuerzburg: Holzner Verlag, 1965 [mf ed 1993] – (= ser Ostdeutsche beitraege aus dem goettinger arbeitskreis 30 [i e 34]) – 234p (ill) – 1 – mf#8098 reel 5 – us Wisconsin U Libr [943]

Geschichte der strassburger sectenbewegung zur zeit der reformation 1524-1534 / Gerbert, Camill – Strassburg: JH Ed Heitz (Heitz & Muendel), 1889 – 1mf – 9 – 0-8370-8902-6 – (incl bibl ref) – mf#1986-2902 – us ATLA [944]

Die geschichte der synagogen-gemeinde zu stettin : eine studie zur geschichte des pommerschen judentums / Peiser, Jacob – 2., bearb. und erw. Aufl. Wuerzburg: Holzner Verlag, 1965 – 1 – (incl bibl ref) – us Wisconsin U Libr [270]

Geschichte der synkretistischen streitigkeiten in der zeit des georg calixt / Schmid, Heinrich – Erlangen: C Heyder, 1846 – 1mf – 9 – 0-524-00099-9 – (incl bibl ref) – mf#1989-2799 – us ATLA [240]

Geschichte der technologie seit der wiederherstellung der wissenschaften am ende des 18. jhdts – Band 1-3. Gottingen. 1807, 1810, 1811 – 1 – gw Mikropress [500]

Geschichte der uebersiedlung von vierzig tausend armeniern : welche im jahre 1828 aus der persischen provinz aderbaischan nach russland auswanderten / Glinka, Sergej – Leipzig 1834 [mf ed Hildesheim 1995-98] – 1v on 1mf – 9 – €40.00 – 3-487-27640-2 – gw Olms [947]

Geschichte der union der ruthenischen kirche mit rom : von den aeltesten zeiten bis auf die gegenwart / Pelesh, Julian – Wien: Mechitharisten-Buchdr. 2v. 1878-81 – 5mf – 9 – 0-8370-9017-2 – (incl bibl ref) – mf#1986-3017 – us ATLA [243]

Geschichte der union der ruthenischen kirche mit rom von den altesten zeiten bis auf die gegenwart / Pelesh, Iulian – Wien: Mechitharisté Buchdruckerei, 1878.Includes bibliog. references – 1 – us Wisconsin U Libr [243]

Geschichte der wiedertaeufer in der schweiz zur reformationszeit / Nitsche, Richard – Einsiedeln, New York, Cincinnati, St Louis, 1885 (mf ed 1993) – 2mf – 9 – €31.00 – 3-89349-296-8 – mf#DHS-AR 156 – gw Frankfurter [242]

Geschichte der wiedertaeufer und ihres reichs zu muenster / Keller, Ludwig – Muenster: Coppenrath, 1880 – 1mf – 9 – 0-7905-4534-9 – (incl bibl ref) – mf#1988-0534 – us ATLA [240]

Geschichte der wiedertaeufer zu muenster : nach urkunden und berichten von zeitgenossen dem deutschen volke / Faesser, Johann Chr – 2., gaenzlich umgearb Aufl. Muenster: E C Brunn, [1861?] – 1mf – 9 – 0-8370-8898-4 – mf#1986-2898 – us ATLA [243]

Geschichte der zeichnenden kuenste von ihrer wiederauflebung bis auf die neuesten zeiten / Fiorillo, J D – Goettingen, 1789-1806. 4v – 31mf – 9 – mf#O-1195 – ne IDC [700]

Geschichte des allmaeligen verfalls der unirten ruthenischen kirche im 18. und 19. jahrhundert unter polnischem und russischem scepter / Likowski, Eduard – Posen: A Toczynski. 2v. 1885-87 – 2mf – 9 – 0-8370-8271-4 – mf#1986-2271 – us ATLA [240]

Geschichte des alten aegyptens / Meyer, Eduard – Berlin: G Grote 1887 – (= ser Allgemeine geschichte in einzeldarstellungen 1/1) – [ill/pl/facs] – 1 – (with int: geographie des alten aegyptens, schrift und sprache seiner bewohner by johannes duemichen) – mf#film mas c604 – us Harvard [930]

Geschichte des alten bundes / Hasse, Friedrich Rudolf – Leipzig: Wilhelm Engelmann, 1863 – 1mf – 9 – 0-7905-0895-8 – (incl bibl ref) – mf#1987-0895 – us ATLA [220]

Geschichte des alten indiens / Lefmann, Salomon – Berlin: G Grote 1890 – (= ser Allgemeine geschichte in einzeldarstellungen 1/3) – [ill/pl] – 1 – mf#film mas c604 – us Harvard [954]

Geschichte des alten persiens / Justi, Ferdinand – Berlin: G Grote 1879 – (= ser Allgemeine geschichte in einzeldarstellungen 1/4/1) – [ill/pl/facs] – 1 – mf#film mas c604 – us Harvard [930]

Geschichte des alterthums / Duncker, Maximilian Wolfgang – 4th impr ed. Berlin: Duncker & Humblot 1874- [mf ed 1979] – 9v in 8 on 3r – 1 – (incl various ed of some vols) – mf#film mas c665 – us Harvard [900]

Geschichte des apostolischen zeitalters zur zerstoerung jerusalems / Ewald, Heinrich – 3rd ed. Goettingen, 1868 – 14mf – 8 – €27.00 – ne Slangenburg [220]

Die geschichte des bisthums bamberg : nach den quellen / Looshorn, Johann – Muenchen: P Zipperer 1886-1910 [mf ed 2005] – 8v in 7 on 4r [ill] – 1 – 0-524-10528-6 – (each vol has also special t p; v5-8 impr reads: bamberg, verlag und druck der handels-druckerei; incl "orts- und personen-register"; v1: die gruendung und das erste jahrhundert des bisthums bamberg, oder die heiligen kaiser heinrich und kunigunda; v2: das bisthum bamberg von 1102-1303; v3: 1303-1399; v4: 1400-1556; v5: 1556-1622; v6: 1623-1729; v7: 1729-1808; lacks: v2 p351-366) – mf#b00740 – us ATLA [241]

Geschichte des breviers : versuch einer quellenmaessigen darstellung der entwicklung des altkirchlichen und des roemischen officiums bis auf unsere tage / Baeumer, Suitbert – Freiburg i B: Herder, 1895 – 2mf – 9 – 0-524-06239-0 – (incl bibl ref) – mf#1990-5194 – us ATLA [240]

Geschichte des buddhismus in der mongolei / Jigs-med nam-mka; ed by Huth, Georg – Strassburg: KJ Truebner, 1892-1896 – 2mf – 9 – 0-524-05859-8 – mf#1990-3523 – us ATLA [280]

Geschichte des bundesgedankens im alten testament, 1. haelfte / Karge, Paul – Muenster i W: Aschendorff, 1919 [mf ed 1989] – (= ser Alttestamentliche abhandlungen 2/1-4) – 2mf – 9 – 0-7905-2478-3 – (no more publ. incl bibl ref & ind) – mf#1987-2478 – us ATLA [221]

Geschichte des canonischen eherechts : bis zum verfall der glossenlitteratur / Freisen, Joseph – 2. verm Ausg. Paderborn: F Schoeningh, 1893 – 3mf – 9 – 0-524-04542-9 – mf#1990-5049 – us ATLA [240]

Geschichte des chinesischen reiches *see* Geutzlaff's geschiedenis van het chinesche rijk

Geschichte des deutschen kirchenliedes bis auf luthers zeit / Fallersleben, Hoffmann von – Hannover, 1861 – 10mf – 8 – €19.00 – ne Slangenburg [780]

Geschichte des deutschen privatrechts / Thudichum, Friedrich von – Stuttgart: F Enke, 1894 – (= ser Civil law 3 coll) – 5mf – 9 – (incl bibl ref and index) – mf#LLMC 96-527 – us LLMC [346]

Geschichte des deutschen volkes seit dem ausgang des mittelalters / Jannsen, Johannes – St. Louis, MO: Herder, (1893-1901). 8v.3 reels. 1259 – 1 – us Wisconsin U Libr [943]

Geschichte des deutschen volksschulwesens / Heppe, Heinrich – Gotha: FA Perthes, 1858-1860 – 5mf – 9 – 0-7905-8142-6 – mf#1988-6089 – us ATLA [370]

Geschichte des dorfes schmiehel einschliesslich einer kurzen... / Neu, Heinrich – Ettenheim, Germany. 1902 – 1r – us UF Libraries [943]

Geschichte des dreissigjaehrigen krieges / Winter, Georg – Berlin: G Grote 1893 – (= ser Allgemeine geschichte in einzeldarstellungen 3/3/2) – [ill] – 1 – mf#film mas c604 – us Harvard [932]

Geschichte des elsasses von den aeltesten zeiten bis auf die gegenwart : bilder aus dem politischen und geistigen leben der deutschen westmark / Lorenz, Ottokar – Berlin 1871 [mf ed Hildesheim 1995-98] – 2v on 4mf – 9 – €120.00 – 3-487-25942-7 – gw Olms [943]

Geschichte des englischen deismus / Lechler, Gotthard Victor – Stuttgart: J.G. Cotta, 1841 – 2mf – 9 – 0-7905-6601-X – (incl bibl ref) – mf#1988-2601 – us ATLA [210]

Geschichte des ennsthales : 16. august bis 24. december 1881; (als manuscript gedruckt) / Hohenlohe, Philipp – Wien 1882 [mf ed Hildesheim 1995-98] – 1mf – 9 – €40.00 – 3-487-29425-7 – gw Olms [943]

Geschichte des ersten kreuzzuges / R"hricht, R – Innsbruck, 1901 – 3mf – 9 – mf#H-2907 – ne IDC [931]

Geschichte des ersten kreuzzuges / Roehricht, Reinhold – Innsbruck: Wagner, 1901 – 1mf – 9 – 0-524-04317-5 – (incl bibl ref) – mf#1990-1243 – us ATLA [940]

Geschichte des erziehungswesens und der cultur der juden in frankreich und deutschland : von der begruendung der juedischen wissenschaft in diesen laendern bis zur vertreibung der juden aus frankreich (10.-14. jahrhundert) / Guedemann, Moritz – Wien: Alfred Hoelder, 1880 – 1mf – 9 – 0-8370-7634-X – (incl bibl ref) – mf#1986-1634 – us ATLA [370]

Geschichte des erzienhungswesens und der cultur / Gudemann, Mortiz – Berlin, Germany. 1922 – 1r – us UF Libraries [939]

Geschichte des fraeuleins von sternheim / La Roche, Sophie von; ed by Ridderhoff, Kuno – Berlin: B Behr, 1907 [mf ed 1994] – (= ser Deutsche litteraturdenkmale des 18. und 19. jahrhunderts 138, 3. folge n18) – xxxix/345p – 1 – (incl bibl ref) – mf#8676 reel 8 – us Wisconsin U Libr [830]

Geschichte des franzoesischen calvinismus in seiner bluethe : bis zum aufstande von amboise i. j. 1560 / Polenz, Gottlob von – Gotha: FA Perthes, 1857 – 8mf – 9 – 0-524-08805-5 – (incl bibl ref) – mf#1993-3297 – us ATLA [944]

Geschichte des gelehrten unterrichts auf den deutschen schulen und universitaeten : vom ausgang des mittelaters bis zur gegenwart / Paulsen, Friedrich – 2., umgearb. und sehr erw. Aufl. Leipzig: Veit, 1896-1897 – 4mf – 9 – 0-7905-8063-2 – mf#1988-6044 – us ATLA [370]

Die geschichte des gregorianischen gesanges in den protestantischen gottesdiensten / Schrems, Th – Freiburg Schw., 1930 – €7.00 – ne Slangenburg [240]

Die geschichte des griechischen skeptizismus / Goedeckemeyer, Albert – Leipzig: Dietrich, 1905 – 1mf – 9 – 0-7905-7629-5 – (incl bibl ref) – mf#1989-0854 – us ATLA [180]

Geschichte des heiligen thomas von aquin / Mettenleiter, Dominikus – Regensburg: Friedrich Pustet, 1856 – 1mf – 9 – 0-524-02282-8 – mf#1990-0587 – us ATLA [241]

Geschichte des hellenismus / Droysen, Johann Gustav – 2nd ed. Gotha: F A Perthes 1877-78 [mf ed 1979] – 6v in 3 on 1r – 1 – (pt1: geschichte alexanders des grossen; pt2: geschichte der diadochen; pt3: geschichte der epigonen; with app: ueber die hellenischen staedtergruendungen; ind by alfred schulz) – mf#film mas 9043 – us Harvard [930]

Geschichte des herrn c le beau, advocat im parlament : oder, merckwuerdige und neue reise zu denen wilden des nordlichen theils von america... – Erfurt [Germany]: Druckts und verlegts Joh David Jungnicol, 1752 [mf ed 1982] – 2v on 1mf – 9 – 0-665-18221-X – mf#18221 – cn Canadiana [917]

Die geschichte des hussitenthums und prof. constantin hoefler : kritische studien / Palacky, Frantisek – Prag: F Tempsky, 1868 – 1mf – 9 – 0-7905-7068-8 – mf#1988-3068 – us ATLA [240]

Die geschichte des juedischen volkes und seiner literatur vom babylonischen exile bis auf die gegenwart mit einem anhange : proben der juedischen literatur / Baeck, Samuel – 3. verb. Aufl. Frankfurt a.M. 19906 (mf ed 1996) – (= ser Monographien zur wissenschaft des judentums) – 8mf – 9 – €73.00 – 3-8267-3187-5 – mf#DHS 50002 – gw Frankfurter [939]

Geschichte des kaisers napoleon / Laurent, Paul – Leipzig 1840 [mf ed Hildesheim 1995-98] – 1v on 10mf – 9 – €100.00 – ISBN-10: 3-487-26232-0 – ISBN-13: 978-3-487-26232-1 – gw Olms [944]

Geschichte des karaerthums / Furst, Julius – Leipzig, Germany. v1-3. 1862-1869 – 1r – us UF Libraries [939]

Geschichte des katholischen modernismus / Kuebel, Johannes – Tuebingen: J C B Mohr, 1909 – 1mf – 9 – 0-8370-8916-6 – (incl ind) – mf#1986-2916 – us ATLA [241]

Geschichte des kirchengesanges in der deutschen reformirten schweiz seit der reformation / Weber, H – Zuerich, 1876 – 3mf – 9 – mf#ZWI-103 – ne IDC [242]

Geschichte des kirchenlieds und kirchengesangs der christlichen : inbesondere der deutschen evangelischen kirche = History of church songs and hymns for christians, particularly of the german evangelical church / Koch, Eduard Emil – repr 3rd ed. Stuggart. 8v. 1866-77 – 11 – $145.00 set – (with index vol) – us Univ Music [780]

Geschichte des kriegs in spanien / Garrazin, General – 1815 – 9 – sp Bibl Santa Ana [946]

Geschichte des kulturkampfes im deutschen reiche : im auftrage des zentralkomitees fuer die generalversammlungen der katholiken deutschlands / Kissling, Johannes Baptist – Freiburg i B: Herdersche Verlagshandlung, 1911-1916 – 4mf – 9 – 0-524-08328-2 – (incl bibl ref and ind) – mf#1993-1023 – us ATLA [241]

Die geschichte des leidens und sterbens, der auferstehung und himmelfahrt des herrn / Belser, Johannes Evangelist – Freiburg i B: Herder, 1903 – 2mf – 9 – 0-524-06032-0 – mf#1992-0745 – us ATLA [220]

Geschichte des machtverfalls der tuerkei bis ende des 19. jahrhunderts : und die phasen der "orientalischen frage" bis auf die gegegenwart / Sax, C Ritter von – Wien, 1913 – 8mf – 9 – mf#AR-1836 – ne IDC [956]

Geschichte des materialismus und kritik seiner bedeutung in der gegenwart / Lange, Friedrich Albert – 7. Aufl. Leipzig: J Baedeker, 1902 – 3mf – 9 – 0-7905-9998-8 – (incl bibl ref) – mf#1989-1723 – us ATLA [100]

Geschichte des monismus / Eisler, Rudolf – Leipzig: Alfred Kroener, 1910 – 1mf – 9 – 0-7905-7726-7 – (incl bibl ref) – mf#1989-0951 – us ATLA [100]

Die geschichte des montanismus / Bonwetsch, G N – Erlangen, 1881 – €11.00 – ne Slangenburg [240]

Die geschichte des montanismus / Bonwetsch, Gottlieb Nathanael – Erlangen: A Deichert, 1881 [mf ed 1989] – 1mf – 9 – 0-7905-4094-0 – (incl bibl ref) – mf#1988-0094 – us ATLA [240]

Geschichte des muensterischen aufruhrs : in drei buechern / Cornelius, Carl Adolf – Leipzig: TO Weigel, 1855-1860 – 2mf – 9 – 0-524-01882-0 – (incl bibl ref) – mf#1990-0509 – us ATLA [943]

Geschichte des neutestamentlichen kanon / Credner, Karl August – Berlin: G Reimer, 1860 – 1mf – 9 – 0-7905-3323-5 – (incl bibl ref) – mf#1987-3323 – us ATLA [225]

Die geschichte des pfarrers von kalenberg / ed by Dollmayr, Viktor – Halle: M Niemeyer, 1906 – (incl bibl ref) – us Wisconsin U Libr [430]

Die geschichte des pietismus / Schmid, Heinrich – Noerdlingen: C H Beck, 1863 – 2mf – 9 – 0-8370-8865-8 – (incl bibl ref) – mf#1986-2865 – us ATLA [240]

Geschichte des pietismus in den schweizerischen reformierten kirchen / Hadorn, Wilhelm – Konstanz, Deutschland: Carl Hirsch, [1901?] – 2mf – 9 – 0-524-07289-2 – mf#1990-5376 – us ATLA [242]

Geschichte des pietismus in der lutherischen kirche des 17. und 18. jahrhunderts / Ritschl, Albrecht – Bonn: A Marcus, 1884-86 [mf ed 1990] – 2v on 3mf – 9 – 0-7905-3561-0 – (incl bibl ref) – mf#1989-0054 – us ATLA [242]

Geschichte des pietismus in der reformirten kirche / Ritschl, Albrecht – Bonn: A Marcus, 1880 [mf ed 1990] – 2mf – 9 – 0-7905-3562-9 – (incl bibl ref) – mf#1989-0055 – us ATLA [242]

Geschichte des pietismus und der mystik in der reformierten kirche : namentlich der niederlande / Heppe, Heinrich – Leiden: E J Brill, 1879 – 2mf – 9 – 0-7905-6650-8 – (incl bibl ref) – mf#1988-2650 – us ATLA [242]

Geschichte des politischen franzoesischen calvinismus : vom aufstand von amboise i. j. 1560 bis zum gnadenedict von nimes i. j. 1629 / Polenz, Gottlob von – Gotha: FA Perthes, 1859-1869 – 27mf – 9 – 0-524-08806-3 – (incl bibl ref) – mf#1993-3298 – us ATLA [944]

Geschichte des prinzen biribinker / Wieland, Christoph Martin – Leipzig: G H Wigand, [1902?] – 1r – 1 – us Wisconsin U Libr [830]

Geschichte des pronomen reflexivum / Dyroff, Adolf – Wuerzburg: A Stuber, 1892-1893 – (= ser Beitraege zur historischen syntax der griechischen sprache) – 1mf – 9 – 0-8370-1427-1 – mf#1987-6064 – us ATLA [450]

Die geschichte des propheten jona : nach einer karschunischen handschrift der kgl. bibliothek zu berlin: ein beitrag zur jona-exegese / Wolf, Benedict – 2. aufl. Berlin: M Poppelauer, 1899 – 1mf – 9 – 0-7905-0118-X – (incl bibl ref) – mf#1987-0118 – us ATLA [221]

Geschichte des protestantismus in oesterreich : in umrissen / Loesche, Georg – Tuebingen: Mohr, 1902 – 1mf – 9 – 0-7905-6541-2 – (incl bibl ref) – mf#1988-2541 – us ATLA [242]

Die geschichte des rabbi jesus von nazareth / Delff, Heinrich Karl Hugo – Leipzig: Wilhelm Friedrich, [1889?] – 2mf – 9 – 0-524-04451-1 – (incl bibl ref) – mf#1992-0120 – us ATLA [221]

Geschichte des rationalismus. erste abtheilung, geschichte des pietismus und des ersten stadiums der aufklaerung / Tholuck, August – Berlin: Wiegandt und Grieben, 1865 – 1mf – 9 – 0-8370-8870-4 – (no more published. incl bibl ref) – mf#1986-2870 – us ATLA [210]

Geschichte des reiches gottes bis auf jesus christus / Koenig, Eduard – Berlin: Martin Warneck, 1908 – 1mf – 9 – 0-8370-3968-1 – (incl bibl ref and index) – mf#1985-1968 – us ATLA [220]

Die geschichte des reiches gottes im alten bunde : zum studium und zur unterrichtlichen behandlung der biblischen geschichte fuer praeparanden, seminaristen und lehrer / Kahle, F Hermann – 10. verb Aufl. Breslau [Wroclaw]: C Duelfer, 1900 – (= ser Hilfsbuch beim evangelischen Religions-Unterricht) – 1mf – 9 – 0-524-05218-2 – mf#1992-0351 – us ATLA [220]

Geschichte des reiches gottes unter dem alten bunde *see* History of the kingdom of god under the old testament

Geschichte des religionsunterrichts in der evangelischen volksschule wuerttembergs / Weisenboehler, Oskar – Marbach: A Remppis, 1903 – 1mf – 9 – 0-8370-7598-X – mf#1986-1598 – us ATLA [377]

Die geschichte des richters von orb / Weismantel, Leo – Freiburg im Breisgau: Herder, 1927 – 1r – 1 – (cover title: johann christin: der richter von orb) – us Wisconsin U Libr [830]

Geschichte des roemischen kaiserreiches / Hertzberg, Gustav Friedrich – Berlin: G Grote 1880 – (= ser Allgemeine geschichte in einzeldarstellungen 2/1) – [ill/pl] – 1 – mf#film mas c604 – us Harvard [930]

Geschichte des roemischen katechismus / Corvin von Skibniewski, Stephan Leo, Ritter – Rom; New York: Friedrich Pustet, 1903 – 1mf – 9 – 0-8370-8414-8 – mf#1986-2414 – us ATLA [240]

Geschichte des romans und der novelle in deutschland / Borcherdt, Hans Heinrich – Leipzig: J J Weber, 1926- [mf ed 1993] – v1 – 1 – (projected later vols never publ; cont by: der roman der goethezeit) – mf#8192 – us Wisconsin U Libr [430]

Geschichte des schiksals der freymaeurer zu neapel – Frankfurt: [s.n.] 1779 – 3mf – 9 – mf#vrl-58 – ne IDC [366]

Geschichte des schweizerischen bundesrechtes von den ersten ewigen buenden bis auf die gegenwart / Bluntschli, J R – Zuerich, 1849-1852. 2 v – 11mf – 9 – mf#ZWI-22 – ne IDC [240]

Geschichte des semitischen altertums in tabellen / Floigl, Victor – Leipzig:Wilhelm Friedrich, 1882 – 1mf – 9 – 0-8370-3155-9 – mf#1985-1155 – us ATLA [939]

Geschichte des sonettes in der deutschen dichtung : mit einer einleitung ueber heimat, entstehung und wesen der sonettform / Welti, Heinrich – Leipzig: Veit, 1884 – 1r – 1 – (incl bibl ref and index) – us Wisconsin U Libr [430]

Geschichte des spanischen protestantismus im sechszehnten jahrhundert / Wilkens, Cornelius August – Guetersloh: C. Bertelsmann, 1888 – 1mf – 9 – 0-7905-6150-6 – (incl bibl ref) – mf#1988-2150 – us ATLA [242]

Geschichte des untergangs der antiken welt / Seeck, Otto – Stuttgart: Metzler 1920-23 [mf ed 1979] – 6v in 12 on 2r – 1 – (with app; incl various ed of some vols) – mf#film mas c670 – us Harvard [930]

Geschichte des untergangs des griechisch-roemischen heidentums / Schultze, Viktor – Jena: H Costenoble, 1887-1892 – 2mf – 9 – 0-524-08056-9 – (incl bibl ref) – mf#1991-0272 – us ATLA [250]

Geschichte des vatikanischen konzils / Friedrich, Johann – Bonn: P Neusser, 1877-1887 – 7mf – 9 – 0-524-04358-2 – (incl bibl ref) – mf#1990-5041 – us ATLA [241]

Geschichte des vatikanischen konzils : von seiner ersten ankuendigung bis zu seiner vertagung: nach den authentischen dokumenten / Granderath, Theodor – Freiburg i.B.; St Louis, MO: Herder. 3v. 1903-06 – 6mf – 9 – 0-8370-9063-6 – (incl bibl ref and index) – mf#1986-3063 – us ATLA [241]

Geschichte des volkes israel : 1. band: einleitung in die geschichte des volkes israel / Ewald, Heinrich – 3. ausg. Goettingen, 1864 – 11mf – 8 – €21.00 – ne Slangenburg [939]

Geschichte des volkes israel / Guthe, Hermann – Freiburg i.B.: J C B Mohr (Paul Siebeck), 1899 – 1mf – 9 – 0-8370-9951-X – (incl bibl and index) – mf#1986-3951 – us ATLA [939]

Geschichte des volkes israel / Stade, Bernhard et al – 2. aufl. Berlin: G Grote 1888-89 [mf ed 1993] – (= ser Allgemeine geschichte in einzeldarstellungen 1/6) – 2v on 4mf – 9 – 0-524-06053-3 – (incl bibl ref) – mf#1992-0766 – us ATLA; us Harvard [939]

Geschichte des volkes israel : von anbeginn bis zur eroberung masada's im jahre 72 nach christus / Hitzig, Ferdinand – Leipzig: S Hirzel, 1869 – 2mf – 9 – 0-7905-1008-1 – (incl bibl ref and index) – mf#1987-1008 – us ATLA [939]

Geschichte des volkes jisrael von zerstoerung des ersten tempels bis zur einsetzung des mackabaeers schimon zum hohen priester und fuersten / Herzfeld, Levi – Leipzig: Oskar Leiner, 1870 – (= ser Schriften (Institut zur Foerderung der israelitschen Literatur)) – 1mf – 9 – 0-524-08078-X – mf#1992-1138 – us ATLA [939]

Geschichte des wachstums und der erfindungen der chemie, in der neuern zeit / Wiegleb, Johann Christian – Berlin, 1790-91. 2v – 1 – us Wisconsin U Libr [540]

Geschichte des zeitalters der entdeckungen / Ruge, Sophus – Berlin: G Grote 1881 – (= ser Allgemeine geschichte in einzeldarstellungen 2/9) – [ill/pl/facs] – 1 – mf#film mas c604 – us Harvard [900]

Geschichte des zuercherischen schulwesens : bis gegen das ende des sechzehnten jahrhunderts / Ernst, Ulrich – Winterthur: Bleuler-Hausheer, 1879 – 1mf – 9 – 0-8370-7630-7 – mf#1986-1630 – us ATLA [370]

Geschichte des zuercherischen schulwesens... / Ernst, U – Winterthur, Bleuler-Hausheer, 1879 – 3mf – 9 – mf#PBU-437 – ne IDC [242]

Geschichte des zweiten kaiserreiches und des koenigreiches italien / Bulle, Konstantin – Berlin: G Grot'sche Verlagsbuchhandlung 1890 – (= ser Allgemeine geschichte in einzeldarstellungen 4/4) – [pl/ill/facs] – 1 – mf#film mas c604 – us Harvard [940]

Die geschichte deutsch-juedischer refugees in schottland / Koelmel, Rainer – Heidelberg, 1979 – 4mf – 3-89349-837-0 – gw Frankfurter [941]

Die geschichte eines genies : novelle / Schubin, Ossip – 2. Aufl. Berlin: Paetel, 1890 – 1 – us Wisconsin U Libr [830]

Geschichte eines knaben / Wiechert, Ernst Emil – Tuebingen: R Wunderlich, [194-?] – 1r – 1 – us Wisconsin U Libr [830]

Geschichte frankreichs im revolutionszeitalter / Wachsmuth, Wilhelm – Hamburg 1840-1844 [mf ed Hildesheim 1995-98] – 5v on 35mf – 9 – €350.00 – ISBN-10: 3-487-26203-7 – ISBN-13: 978-3-487-26203-1 – gw Olms [944]

Geschichte, geographie und bedeutung de insel trin... / Gommersbach, Wilhelm – Bonn, Germany. 1907 – 1r – us UF Libraries [972]

Geschichte israels bis auf alexander den grossen / Oettli, Samuel – Calw: Verlag der Vereinsbuchh., 1905 – (= ser Die Geschichte Israels) – 2mf – 9 – 0-7905-1137-1 – (incl bibl ref and index) – mf#1987-1137 – us ATLA [939]

Geschichte israels bis auf die griechische zeit / Benzinger, Immanuel – Leipzig: G J Goeschen 1904 [mf ed 1989] – (= ser Sammlung goeschen) – 1mf – 9 – 0-7905-0724-2 – mf#1987-0724 – us ATLA [939]

Geschichte israels in einzeldarstellungen / Winckler, Hugo – Leipzig: E Pfeiffer, 1895-1900 – 2mf – 9 – 0-524-05763-X – (incl bibl ref) – mf#1992-0606 – us ATLA [930]

Geschichte israels unter den richtern und koenigen / Westermayer, Anton – Schaffhausen: Friedr Hurter, 1861 [mf ed 1993] – (= ser Das alte testament und seine bedeutung 3) – 2mf – 9 – 0-524-06285-4 – mf#1992-0866 – us ATLA [221]

Geschichte israels von josua bis zum ende des exils / Nikel, Johannes – 1. & 2. aufl. Muenster i W: Aschendorff 1910 [mf ed 1992] – (= ser Biblische zeitfragen 3/3-4) – 1mf – 9 – 0-524-04109-1 – (incl bibl ref) – mf#1992-0067 – us ATLA [221]

Die geschichte jesu : auf grund freier geschichtlicher untersuchungen ueber das evangelium und die evangelien / Noack, Ludwig – Mannheim: J Schneider, 1876 – 3mf – 9 – 0-7905-3460-6 – (incl bibl ref) – mf#1987-3460 – us ATLA [220]

Die geschichte jesu / Schmidt, Paul Wilhelm – Freiburg i. B: J C B Mohr, 1899 – 1mf – 9 – 0-8370-9819-X – mf#1986-3819 – us ATLA [220]

Geschichte jesu christi / Noesgen, Karl Friedrich – Muenchen: C H Beck, 1891 – (= ser Geschichte der neutestamentlichen offenbarung) – 2mf – 9 – 0-7905-1544-X – (incl bibl ref and indexes) – mf#1987-1544 – us ATLA [220]

Geschichte masurens / Toeppen, Max – Danzig, 1870 – 1r – 1 – gw Mikropress [943]

Die geschichte meines lebens : vom kind bis zum manne / Ebers, Georg – Stuttgart: Deutsche Verlags-Anstalt, [1893-97?] [mf ed 1993] – (= ser Georg ebers gesammelte werke 25) – 522p – 1 – mf#8554 reel 4 – us Wisconsin U Libr [920]

Geschichte napoleon bonaparte's / Buchholz, Paul Ferdinand Friedrich – Berlin 1827-1829 [mf ed Hildesheim 1995-98] – 3v on 13mf – 9 – €130.00 – ISBN-10: 3-487-26241-X – ISBN-13: 978-3-487-26241-3 – gw Olms [944]

Geschichte, quellen und literatur des wuerttembergischen privatrechts / Waechter, Carl Georg von – Stuttgart: J B Metzler. 2v in 1. 1839-42 – (= ser Civil law 3 coll; Handbuch des im koenigreiche wuerttemberg geltenden privatrechts) – 12mf – 9 – mf#LLMC 96-523 – us LLMC [346]

Geschichte roms und der paepste im mittelalter *see* History of rome and the popes in the middle ages

Geschichte rueckwaerts? / Wolff, Eugen – Kiel: Lipsius und Tischer, 1892 – 1r – 1 – us Wisconsin U Libr [943]

Geschichte siciliens im alterthum / Holm, Adolf – Leipzig: W Engelmann 1880-98 [mf ed 1979] – 3v on 1r [ill/pl] – 1 – mf#film mas 9096 – us Harvard [945]

Geschichte und beschreibung der kanarieninseln aus dem franzoesischen / Bory de Saint-Vincent, Jean B – Weimar 1804 [mf ed Hildesheim 1995-98] – 1v on 3mf – 9 – €90.00 – 3-487-26590-7 – gw Olms [914]

Geschichte und beschreibung der mission unter den kolhs in ostindien / Gerhard, Paul – Berlin: Buchhandlung der Gossnerischen Mission, 1883 [mf ed 1995] – (= ser Yale coll) – 140p – 1 – 0-524-10182-5 – (in german) – mf#1995-1182 – us ATLA [240]

Geschichte und beschreibung des muensters zu freiburg im breisgau / Schreiber, Heinrich – Freiburg im Breisgau 1820 [mf ed Hildesheim 1995-98] – 2mf – 9 – €60.00 – 3-487-29495-8 – gw Olms [943]

Geschichte und beschreibung von newfoundland und der kueste labrador / 917 – Weimar 1822 [mf ed Hildesheim 1995-98] – 1v on 2mf – 9 – €60.00 – 3-487-26495-1 – gw Olms [971]

Geschichte und besiedlung des ratiborer landes : mit beitraegen von georg raschke und ferdinand huetteroth / Hyckel, Georg – Wuerzburg: Holzner, 1961 – 1 – (incl bibl ref) – us Wisconsin U Libr [943]

Geschichte und dogmatik : eine erkenntnistheoretische untersuchung / Vowinckel, Ernst – Leipzig: A Deichert (Georg Boehme), 1898 – 1mf – 9 – 0-8370-6444-9 – (incl bibl ref) – mf#1986-0444 – us ATLA [110]

Geschichte und erzaehlungen – Danzig 1771-78 – (= ser Dz) – 10pt on 22mf – 9 – €220.00 – mf#k/n5367 – gw Olms [390]

Geschichte und geschichtswissenschaft in der methodologischen diskussion der gegenwart / Rhein, Monika – Frankfurt a.M., 1983 – 3mf – 9 – 3-89349-393-X – gw Frankfurter [943]

Geschichte und historie in der religionswissenschaft : ueber die notwendigkeit in der religionswissenschaft zwischen geschichte und historie strenger zu unterscheiden / Wobbermin, Georg – Tuebingen: J C B Mohr 1911 [mf ed 1991] – (= ser Zeitschrift fuer theologie und kirche 1911/2) – 1mf – 9 – 0-7905-8750-5 – (incl bibl ref) – mf#1989-1975 – us ATLA [240]

Geschichte und kirche / Doellinger, Johann Joseph Ignaz von – Muenchen o. J. (mf ed 1995) – 2mf – 9 – €31.00 – 3-8267-3152-2 – mf#DHS-AR 3152 – gw Frankfurter [240]

Geschichte und kritik der neueren theologie : insbesondere der systematischen / Frank, Fr H R – 4. Aufl. Leipzig: A Deichert, 1908 – 2mf – 9 – 0-8370-8670-1 – (incl bibl ref and index) – mf#1986-2670 – us ATLA [210]

Geschichte und kritischer katalog des deutschen, niederlaendischen und franzoesischen kupferstichs im 15 jahrhundert / Lehrs, M – Wien. v1-9. 1908-1934 – 75mf – 9 – mf#0-340 – ne IDC [700]

Geschichte und literatur der kirchengeschichte / Staeudlin, Carl Friedrich; ed by Hemsen, Johannes Tychsen – Hannover: Hahn, 1827 – 1mf – 9 – 0-524-01897-9 – (incl bibl ref) – mf#1990-0-524 – us ATLA [012]

Geschichte und offenbarung im alten testament / Lotz, Wilhelm – Leipzig: J C Hinrichs, 1891 – 1mf – 9 – 0-8370-4183-X – mf#1985-2183 – us ATLA [221]

Geschichte und politik – Berlin DE, 1802-04 – 3r – 1 – gw Misc Inst [320]

Geschichte und politik / ed by Woltmann, Karl Ludwig – Berlin 1800-05 – (= ser Dz. historisch-politische abt) – 6jge[zu je 3bdn] on 49mf – 9 – €490.00 – mf#k/n1783 – gw Olms [900]

Geschichte und rechtliche stellung der juden in pommern / Grotesend, Ulrich – Marburg, Germany. 1931 – 1r – us UF Libraries [939]

Geschichte und system der mittelalterlichen weltanschauung / Eicken, Heinrich von – Stuttgart: J.G. Cotta, 1887 – 2mf – 9 – 0-7905-8056-X – (incl bibl ref) – mf#1988-6037 – us ATLA [940]

Geschichte und system des iranischen strafrechts / Daftary, Ali Akkbar Khan – Halle-Wittenberg, 1935 (mf ed 1994) – 2mf – 9 – €31.00 – 3-8267-3000-3 – mf#DHS-AR 3000 – gw Frankfurter [345]

Geschichte unserer missionsstation kotapad in jeypur (vorderindien) / Gloyer, E – Breklum: Missionshauses, 1907 [mf ed 1995] – (= ser Yale coll) – 135p (ill) – 1 – 0-524-09019-X – (in german) – mf#1995-0019 – us ATLA [240]

Die geschichte unseres volks. bilder aus der vergangenheit und gegenwart der deutschen in rumanien / Mueller-Langenthal – Hermannstadt – 1 – gw Mikropress [943]

Die geschichte vom alten blute und von der ungeheuren verlassenheit : erzaehlung / Grimm, Hans – Berlin: Deutsche Buch-Gemeinschaft, c1931 – 1r – 1 – us Wisconsin U Libr [830]

Geschichte von halitsch und wladimir bis 1772.. / Engel, Johann Christian von – Wien: Franz Jakob Kaiserer, 1792. 2v in 1. 1 reel. 1245 – 1 – us Wisconsin U Libr [947]

Geschichte von hellas und rom / Hertzberg, Gustav Friedrich – Berlin: G Grote 1879 – (= ser Allgemeine geschichte in einzeldarstellungen 1/5) – 2v [ill/pl] – 1 – mf#film mas c604 – us Harvard [930]

Die geschichte von joseph dem zimmermann / Morenz, S – Berlin, 1951 – (= ser Tugal 5-56) – 3mf – 9 – €7.00 – ne Slangenburg [240]

Geschichte von sudafrika / Hintrager, Oscar – Muenchen, Germany. 1952 – 1r – us UF Libraries [960]

Geschichten / ed by Kollektiv Eulenspiegel Verlag – Berlin: Eulenspiegel Verlag, 1967 – 1r – 1 – (incl bibl ref) – us Wisconsin U Libr [830]

Die geschichten des majors / Hopfen, Hans – 3. Aufl. Berlin: Richard Wilhelmi, 1882 – 1r – 1 – us Wisconsin U Libr [830]

Die geschichten des majors / Hopfen, Hans – 3. aufl. Berlin: Richard Wilhelmi 1882 [mf ed 1995] – 1r – 1 – (filmed with: fraenzchens lieder / hoffmann von fallersleben) – mf#3757p – us Wisconsin U Libr [830]

Geschichten und bilder zur foerderung der inneren mission *see* Fliegende blaetter aus dem rauhen hause zu horn bei hamburg (fw1)

Geschichten und novellen : gesamtausgabe / Riehl, Wilhelm Heinrich – Stuttgart: J G Cotta 1923 [mf ed 1979] – 7v on 1r – 1 – mf#film mas 8596 – us Harvard [830]

Geschichten und skizzen aus der heimath / Seidel, Heinrich – 2nd rev ed. Leipzig: A G Liebeskind 1889 – 1 – mf#film mas c419 – us Harvard [880]

Die geschichten und taten wilwolts von schaumburg / ed by Keller, Adelbert von – Stuttgart: Literarischer Verein, 1859 [mf ed 1993] – (= ser Blvs 50) – 208p – 1 – mf#8470 reel 11 – us Wisconsin U Libr [920]

Geschichte-schreiber der deutschen vorzeit – v1-94. 1894-1914 – 9 – $498.00 – mf#0239; 0240 – us Brook [943]

Der geschichtforscher / ed by Meusel, Johann Georg – Halle 1775-79 [mf ed Hildesheim 1992-98] – (= ser Dz. historisch-geographische abt) – 7pt on 15mf – 9 – €150.00 – mf#k/n1079 – gw Olms [931]

Geschichtliche aufsaetze und gedichte / Leibniz, Gottfried Wilhelm von; ed by Pertz, G H – Hannover, 1847 – €15.00 – ne Slangenburg [430]

Der geschichtliche christus : eine reihe von vortraegen mit quellenbeweis und chronologie des lebens jesu / Keim, Theodor – 3., vielfach erw Aufl. Zuerich: Orell, Fuessli, 1866 – 1mf – 9 – 0-8370-3869-3 – (incl bibl ref) – mf#1985-1869 – us ATLA [240]

Der geschichtliche christus und die christliche glaubenslehre / Schnedermann, Georg – Leipzig: Bernhard Richter, 1902 – 1mf – 9 – 0-524-06000-2 – mf#1992-0737 – us ATLA [240]

Der geschichtliche christus und die moderne philosophie : eine genetische darlegung der philosophischen voraussetzungen im streit um die christusmythe / Kiefl, Franz Xaver – Mainz: Kirchheim, 1911 – 1mf – 9 – 0-524-00048-4 – mf#1989-2748 – us ATLA [240]

Geschichtliche entwickelung der constructionen mit prin / Sturm, Josef – Wurzburg: A. Stuber, 1882 – (= ser Beitraege zur historischen syntax der griechischen sprache) – 1mf – 9 – 0-8370-1654-1 – (incl bibl ref) – mf#1987-6084 – us ATLA [450]

Die geschichtliche entwickelung der kirche im 19. jahrhundert und die ihr dadurch gestellte aufgabe – die forschungen ueber die paulinischen briefe : ihr gegenwaertiger stand und ihre aufgaben / Sell, Karl & Heinrici, Carl Friedrich Georg – Giessen: J Ricker, 1887 – (= ser Vortraege der theologischen Konferenz zu Giessen) – 1mf – 9 – 0-524-02651-3 – mf#1990-0675 – us ATLA [240]

Die geschichtliche entwicklung des deutschen realschulwesens : abschnitt 6: bestrebungen auf dem gebiet der schulreform in den jahren 1882-1890 / Wetzstein, O – Neustrelitz, 1911 (mf ed 1993) – 1mf – 9 – €19.00 – 3-89349-312-3 – mf#DHS-AR 168 – gw Frankfurter [373]

Die geschichtliche entwicklung des realschulwesens in deutschland : abschnitt 1: die entstehung deutscher realschulen im 18. jahrhundert / Wetzstein, O – Neustrelitz, 1906 (mf ed 1993) – 1mf – 9 – €19.00 – 3-89349-313-1 – mf#DHS-AR 169 – gw Frankfurter [373]

Der geschichtliche jesus : eine allgemeinverstaendliche untersuchung der frage: hat jesus gelebt, und was wollte er? / Clemen, Carl – Giessen: Alfred Toepelmann, 1911 – 1mf – 9 – 0-7905-0874-5 – mf#1987-0874 – us ATLA [240]

Geschichtliche studien : albert hauck zum 70. geburtstage / Dobschuetz, Ernst von et al – Leipzig: JC Hinrichs, 1916 – 1mf – 9 – 0-524-03577-6 – (incl bibl ref) – mf#1990-1037 – us ATLA [240]

Geschichtliches der israelitischen kultusgemeinde altenstadt / Rose, Hermann – Altenstadt, Germany. 1931 – 1r – us UF Libraries [939]

Die geschichtlichkeit des markusevangeliums / Weiss, Bernhard – Berlin: Edwin Runge 1905 [mf ed 1989] – (= ser Biblische zeit- und streitfragen 1/3) – 1mf – 9 – 0-7905-0525-8 – mf#1987-0525 – us ATLA [225]

Die geschichtlichkeit des sinaibundes / Giesebrecht, Friedrich – Koenigsburg i. Pr: Thomas & Oppermann, 1900 – 1mf – 9 – 0-8370-3273-3 – (incl bibl ref) – mf#1985-1273 – us ATLA [240]

Geschichts- und Altertumforschender Verein. Eisenberg *see* Mitteilungen des geschichts- und altertumsforschenden vereins zu eisenberg

Geschichts- und lebensbilder aus der erneuerung der religioesen lebens in den deutschen befreiungskriegen *see* Religious life in germany during the wars of independence

Die geschichtsschreibung der reformation und gegenreformation : bodin und die begruendung der geschichtsmethodologie durch bartholomaeus keckermann / Menke-Glueckert, Emil – Leipzig: J C Hinrichs, 1912 – 1mf – 9 – 0-7905-5434-8 – (incl bibl ref) – mf#1988-1434 – us ATLA [242]

Die geschichtschreibung im alten testament / Schmidt, Hans – Tuebingen: J C B Mohr 1911 [mf ed 1989] – (= ser Religionsgeschichtliche volksbuecher fuer die deutsche christliche gegenwart 2/16) – 1mf – 9 – 0-7905-3223-9 – (incl bibl ref) – mf#1987-3223 – us ATLA [221]

Geschichtsconstruction oder wissenschaft? : ein wort zur verstaendigung ueber die wellhausensche geschichtsauffassung mit besonderer beziehung auf die vorprophetische stufe der religion israels und die religionsgeschichtliche stellung davids / Baentsch, Bruno – Halle a S: J Krause, 1896 – 1mf – 9 – 0-524-06827-5 – (incl bibl ref) – mf#1992-0969 – us ATLA [270]

Die geschichtsphilosophie des heiligen augustinus : mit einer kritik der beweisfuehrung des materialismus gegen die existenz des geistes / Reinkens, Joseph Hubert – Schaffhausen: Fr Hurter, 1866 – 1mf – 9 – 0-7905-9605-9 – mf#1989-1330 – us ATLA [100]

Die geschichtsphilosophie in den historischen dramen julius mosens : beitrag zur entwicklungsgeschichte der dichterpersoenlichkeit / Fehn, Andreas – Bamberg: Buchner, 1915 – 1r – 1 – (incl bibl ref) – us Wisconsin U Libr [430]

Geschichtsphilosophische ansaetze in der fruehromantik / Klawon, Dieter – Frankfurt a.M., 1977 – 3mf – 9 – 3-89349-673-4 – gw Frankfurter [190]

Geschichtsschreibung und geschichtsauffassung im elsass zur zeit der reformation : vortrag / Lenz, Max – Halle: Verein fuer Reformationsgeschichte, 1895 – (= ser [Schriften des Vereins fuer Reformationsgeschichte]) – 1mf – 9 – 0-7905-4705-8 – mf#1988-0705 – us ATLA [944]

Die geschichtstreue theologie und ihre gegner : oder neues licht und neues leben / Volkmar, G – Zuerich, 1858 – €5.00 – ne Slangenburg [240]

Die geschicke judas und israels im rahmen der weltgeschichte / Lehmann-Haupt, Carl Friedrich – Tuebingen: J C B Mohr (Paul Siebeck) 1911 [mf ed 1989] – (= ser Religionsgeschichtliche volksbuecher fuer die deutsche christliche gegenwart 2/1,6) – 1mf – 9 – 0-7905-1012-X – (incl bibl ref) – mf#1987-1012 – us ATLA [939]

Geschied- en tijdrekenkundig overzicht van het gouvernement makasar / Ligtvoet, A – 3v. 1669-1846 – 18mf – 8 – mf#SD-103 mf 1-18 – ne IDC [959]

Geschiedenis, constitutie en bij-wetten van de "holland union benevolent association" te grand rapids, mich : opgericht den 10den maart, 1892 / ed by Holland Union Benevolent Association – [Grand Rapids, Mich]: Paul Hugenholtz, [1892?] – 1mf – 9 – 0-524-06623-X – mf#1991-2678 – us ATLA [240]

De geschiedenis der christelijke gereformeerde kerk in nederland : aan het volk verhaald / Verhagen, J – Kampen: G P Zalsman, 1886 [mf ed 1993] – (= ser Presbyterian coll) – 554 [ie 454p] on 5mf – 9 – 0-524-07367-8 – mf#1990-5404 – us ATLA [242]

Geschiedenis der doopsgezinden in friesland : van derzelver ontstaan tot dezen tyd / Cate, Steven Blaupot ten – Leeuwarden: W Eekhoff, 1839 – 1mf – 9 – 0-7905-7209-5 – (incl bibl ref) – mf#1988-3209 – us ATLA [240]

Geschiedenis der doopsgezinden in holland, zeeland, utrecht en gelderland : van derzelver ontstaan tot op dezen tijd, uit oorspronkelijke stukken en echte berigten / Cate, Steven Blaupot ten – Amsterdam: P N van Kampen, 1847 – 2mf – 9 – 0-7905-4785-6 – (incl bibl ref) – mf#1988-0785 – us ATLA [240]

De geschiedenis der kerkmuziek in de nederlanden sedert de hervorming / Kat, A I M – Hilversum, 1939 – 7mf – 8 – €15.00 – ne Slangenburg [780]

Geschiedenis der nederduitsche gereformeerde kerk in zuid-afrika / McCarter, John – Amsterdam: Hoeveker & Zoon, [1875?] – (= ser [Travel descriptions from south africa, 1711-1938]) – 2mf – 9 – mf#zah-1 – ne IDC [242]

Geschiedenis der nederlandsche hervormde kerk gedurende de 16e en 17e eeuw / Knappert, Laurentius – Amsterdam: Meulenhoff, 1911 – 1mf – 9 – 0-524-02402-2 – (incl bibl ref) – mf#1990-0605 – us ATLA [240]

Geschiedenis der nederlandsche hervormde kerk gedurende de 18e en 19e eeuw / Knappert, Laurentius – Amsterdam: Meulenhoff, 1912 – 1mf – 9 – 0-524-02403-0 – (incl bibl ref) – mf#1990-0606 – us ATLA [240]

Geschiedenis der nederlandsche taal / Winkel, Jan Te – Culemborg, Netherlands. 1901 – 1r – us UF Libraries [960]

Geschiedenis en kritiek der hedendaagsche oud-katholieke beweging in duitschland van juli 1870 tot mei 1877 / Knuttel, Willem Pieter Cornelis – Leiden: S C van Doesburgh, 1877 [mf ed 1992] – 1mf – 9 – 0-524-04136-9 – mf#1990-1206 – us ATLA [241]

Geschiedenis van curacao / Amelunxen, C P – Hillegom, Netherlands. 1929 – 1r – us UF Libraries [972]

Geschiedenis van de doopsgezinden te straatsburg van 1525 tot 1557 / Hulshof, Abraham – Amsterdam: J Clausen, 1905 [mf ed 1990] – 1mf – 9 – 0-7905-4533-0 – (in dutch & latin. incl bibl ref) – mf#1988-0533 – us ATLA [242]

Geschiedenis van de hervorming en de hervormde kerk der nederlanden / Reitsma, J – Groningen, 1916 – €31.00 – ne Slangenburg [242]

Geschiedenis van de kolonien essequebo / Netscher, P M – Gravenhage, Netherlands. 1888 – 1r – us UF Libraries [972]

De geschiedenis van de liturgische geschriften der nederlandsch hervormde kerk op nieuw onderzocht / Sart, Joan Willem Frederik Gobius du – Utrecht: A J van Huffel, 1886 [mf ed 1993] – (= ser Presbyterian coll) – ix/180p on 3mf – 9 – 0-524-07365-1 – (incl bibl ref) – mf#1990-5402 – us ATLA [242]

Geschiedenis van de orde der vrijmetselaren in nederland : onderhoorige kolonien en landen / Maarschalk, H – Breda: P B Nieuwenhuijs 1872 – 5mf – 9 – mf#vrl-109 – ne IDC [366]

Geschiedenis van de oud-katholieke kerk van nederland / Kleef, B A van – Rotterdam, 1937 – €13.00 – ne Slangenburg [241]

Geschiedenis van de vorsten van palembang – n.d. – 2mf – 8 – mf#SD-102 mf 1-2 – ne IDC [959]

Geschiedenis van de vroomheid in de nederlanden / Axters, St – Antwerpen. v1-4. 1950-1960 – €103.00 – (v1: de vroomheid tot rond het jaar 1300, 1950 €27. v2: de eeuw van ruusbroec, 1953 €31. v3: de moderne devotie 1380-1550, 1956 €25. v4: na trente, 1960 €21) – ne Slangenburg [242]

Geschiedenis van den godsdienst *see* Tiele's kompendium der religionsgeschichte

Geschiedenis van den oorsprong, de invoering en de lotgevallen van den heidelbergschen catechismus / Schotel, Gilles Dionysius Jacobus – Amsterdam: W.H. Kirberger, 1863 – 1mf – 9 – 0-7905-6364-9 – (incl bibl ref) – mf#1988-2364 – us ATLA [240]

Geschiedenis van het buddhisme in indie / Kern, H – Haarlem. v1-2. 1882-1884 – €48.00 – ne Slangenburg [280]

Geschiedenis van het lutheranisme in de nederlanden tot 1618 / Pont, Johannes Wilhelm – Haarlem: E F Bohn, 1911 – (= ser Verhandelingen Rakende den Natuurlijken en Geopenbaarden Godsdienst) – 2mf – 9 – 0-7905-8222-8 – (incl bibl ref) – mf#1988-6122 – us ATLA [242]

Geschiedenis van het nederlandsche zendelinggenootschap en zijne zendingsposten / Kruijf, Ernst Frederik – Groningen: JB Wolters, 1894 – 8mf – 9 – 0-524-07436-4 – (incl bibl ref and ind) – mf#1991-3096 – us ATLA [240]

Geschiedenis van het protestantisme van den munsterschen vrede tot de fransche revolutie, 1648-1789 / Maronier, Jan Hendrick – Nieuwe met een aanhangsel vermeerderde uitg. Leiden: EJ Brill, [1901] – 2mf – 9 – 0-524-01115-X – (incl errata, ind and bibl notes) – mf#1990-0329 – us ATLA [242]

De geschiedenis van het socinianisme in de nederlanden / Slee, Jacob Cornelius van – Haarlem: F Bohn, 1914 [mf ed 1993] – (= ser Unitarian/universalist coll) – viii/336p/1pl on 1mf – 9 – 0-524-07718-5 – mf#1991-3303 – us ATLA [242]

Geschiedenis van kalilah en daminah : uit het maleisch in het madureesch / Adi Koro, Raden Pandji – Batavia: Landsdrukkerij, 1879 [mf ed 1975] – 329p – 1 – mf#4549 – us Wisconsin U Libr [390]

Geschiedkundig onderzoek naar den waldenzischen oorsprong van de nederlandsche doopsgezinden / Cate, Steven Blaupot ten – Amsterdam: F. Muller, 1844 – 1mf – 9 – 0-7905-5928-5 – (incl bibl ref) – mf#1988-1928 – us ATLA [240]

Geschiedkundige tijdtafel van suriname / Oudschans Dentz, Frederik – Amsterdam, Netherlands. 1949 – 1r – us UF Libraries [972]

Geschlagen! : deutsche tragoedie in sieben stationen / Franck, Hans – Stuttgart: W Seifert, c1923 (mf ed 1990) – 1r – 1 – (filmed with: von morgen bet abend) – us Wisconsin U Libr [820]

Ein geschlecht : tragoedie / Unruh, Fritz von – Muenchen: Kurt Wolff Verlag, 1922, c1917 – 1r – 1 – us Wisconsin U Libr [430]

Geschlecht und gesellschaft, vol 1 / Vanselow, Karl et al; ed by Vanselow, Karl – Berlin, Leipzig, Wien: Verlag der Schoenheit 1906-1926/27 [mf ed 1991] – (= ser Hq 3) – 14v on 94mf – 9 – €390.00 – 3-89131-044-7 – (later publ in : muenchen, dresden, leipzig) – gw Fischer [305]

Geschlechtskunde / Hirschfeld, Magnus – Stuttgart 1926-30 [mf ed 2001] – (= ser Hq 46) – 5v on 37mf – 9 – €230.00 – 3-89131-372-1 – (v1: die koerperseelischen grundlagen. v2: folgen und folgerungen. v3: einblicke und ausblicke. v4: bilderteil. v5: registerteil) – gw Fischer [305]

Der geschwinde rechner : oder des haendlers nuetzlicher gehuelfe – Chesnut-Hill, 1793 (mf ed 1994) – 1mf – 9 – 3-8267-3028-3 – mf#DHS-AR 3028 – gw Frankfurter [510]

Die geschwister : [a novel] / Bertsch, Hugo – 12. aufl. Stuttgart; Berlin: J G Cotta, 1912, c1903 [mf ed 1989] – 218p – 1 – (pref by adolf wilbrandt) – mf#7014 – us Wisconsin U Libr [830]

Geschwister : roman / Huch, Friedrich – Berling: S Fischer, [1921] – 1r – 1 – us Wisconsin U Libr [830]

Der gesellige – Graudenz (Grudziadz PL), 1837-40, 1914 1 apr-1915 31 mar, 1924 1 apr-30 sep [gaps], 1925 1 jan-31 mar [gaps], 1927 1 jan-31 mar, 1928 1 jul-30 sep [gaps], 1929 3 apr-1930 30 sep, 1931 1 jul-1932 30 jun, 1932 1 oct-1933 30 jun, 1934-1935 31 mar, 1935 1 jul-31 dec, 1936 oct-1937 1 jan, 1938 1 apr-31 dec – 35r – 1 – gw Misc Inst [077]

Die gesellschaft : internationale revue fuer sozialismus und politik – Berlin DE, 1924-mar 1933 – 5r – 1 – gw Mikropress [074]

Die gesellschaft – Muenchen DE, 1885 1 jan-31 mar – 1r – 1 – gw Misc Inst [074]

Gesellschaft fuer Deutsche Philologie in Berlin *see* Jahresbericht ueber die erscheinungen auf dem gebiete der germanischen philologie

Gesellschaft fuer Erdkunde *see*
– Verhandlungen der gesellschaft fuer erdkunde zu berlin
– Zeitschrift der gesellschaft fuer erdkunde zu berlin

Gesellschaft fuer Geschichte des Landvolks und der Landwirtschaft *see* Zur ostdeutschen agrargeschichte

Gesellschaft fuer musikforschung – v. 1-29. 1870-1905 – 1 – 55.00 – us L of C Photodup [780]

Gesellschaft fuer Pommersche Geschichte und Alterthumskunde *see* Monatsblaetter

Gesellschaft fuer romanische literatur – Dresden. v1-50. 1903-38 – 9 – $1083.00 – mf#0241 – us Brook [440]

Gesellschaft fuer Salzburger Landeskunde *see* Mitteilungen der gesellschaft fuer salzburger landeskunde

Gesellschaft fuer Unternehmensgeschichte *see* Privatbanken in der ns-zeit

Gesellschaft fuer Wiener Theaterforschung *see* Festschrift fuer eduard castle zum achtzigsten geburtstag

Gesellschaft Naturforschender Freunde. Berlin *see* Sitzungsberichte

Der gesellschaft naturforschender freunde zu berlin : neue schriften – Berlin 1795-1803 – (= ser Dz. abteilung naturwissenschaft) – 4v on 23mf – 9 – €230.00 – mf#k/n3364 – gw Olms [500]

Gesellschaft rheinlaendischer Gelehrter *see* Rheinisches conversations-lexicon (ael1/20)

Gesellschaft und organisation : zur soziologischen theorie von organisationen / Herrmann, Peter – 1993 – 7mf – 9 – 3-89349-684-X – mf#DHS 684 – gw Frankfurter [303]

Gesellschaft zur erforschung juedischer kunstdenkmaeler e v zu frankfurt a main : mitteilungen – Duesseldorf: Heinrich Frauberger. v1-8. 1900-15 – (= ser German-jewish periodicals...1768-1945, pt 2) – 1r – 1 – $125.00 – mf#B88 – us UPA [939]

Gesellschaft zur Foerderung der Wissenschaft des Judentums (Germany) *see* Moses ben maimon

Der gesellschafter – Nagold DE, 1973 2 jul-1979 – 44r – 1 – (bezirksausgabe von schwarzwaelder bote, oberndorf) – gw Misc Inst [074]

Der gesellschafter – Hamburg: R Wald & M Beyer 1894- [jahrg1-3(1894-97)] [mf ed 198-] – 4v – 1 – (ceased in 1898; n1-3, 1. jahrg, have sub-title: monatsschrift fuer vornehme unterhaltung; eidtors: 1. jahrg n1-3, roderich wald & max beyer; n4-8, roderich wald and g a mueller; n9-2, jarhg n3, roderich wald; 2. jahrg n3-3. jahrg n6, eduard moos; imprint varies: 1. jahrg n1-6, hamburg, verlag des "gesellschafters"; n7-9, hamburg, verlag von n nachum; n10-3. jahrg n6, erfurt, druck und verlag von eduard moos) – mf#film mas 8776 – us Harvard [430]

Der gesellschafter : oder blaetter fuer geist und herz – Berlin DE, 1817 1 jan-30 jun, 1820-1823 31 may, 1825 n7, 1828 11 aug, 1848 – 3r – 1 – (missing: 1820 20 sep-27 dec, 1822 1 apr-29 jun, 1822 2 oct-1823 2 may. filmed with suppls) – gw Misc Inst [074]

Gesellschafter – London, UK. 1890-91 – 1 – uk British Libr Newspaper [072]

Gesellschaftliche bemuehungen, der welt die christliche religion anzupreisen – Goettingen, Gotha DE, 1772-73 – 1r – 1 – gw Misc Inst [240]

Gesellschaftsspiegel – Wuppertal-Elberfeld DE, 1845-46 – 1r – 1 – gw Misc Inst [074]

Gesenhoff, Georgia *see* Untersuchungen zum griechischen schmuck an beispielen des 7. und 6. jahrhunderts v. chr

Gesenius, H *see* Hebrew grammar

Gesenius' hebrew grammar = Hebraeische grammatik / Gesenius, Wilhelm; ed by Kautzsch, Emil – 2nd English ed. Oxford: Clarendon Press, 1910 – 6mf – 9 – 0-8370-1666-5 – (incl bibl ref. in english) – mf#1987-6096 – us ATLA [470]

Gesenius' hebrew grammar : seventeenth edition, with numerous corrections and additions = Hebraeische grammatik / Gesenius, Wilhelm – New and rev. ed. New York: American Book Company, c1855 – 1mf – 9 – 0-7905-1047-2 – (incl ind) – mf#1987-1047 – us ATLA [470]

Gesenius, W *see* Geschichte der hebraeischen sprache und schrift

Gesenius, Wilhelm *see*
– Gesenius' hebrew grammar
– Guilielmi gesenii, philosophiae et theologiae doctoris . . . thesaurus philologicus criticus linguae hebraeae et chaldaeae veteris testamenti
– Hebraeisch-deutsches handwoerterbuch ueber die schriften des alten testaments
– A hebrew and english lexicon of the old testament
– Lexicon manuale hebraicum et chaldaicum in veteris testamenti libros
– Versuch ueber die maltesische sprache

Das gesetz chammurabis und moses : eine skizze / Grimme, Hubert – Koeln: JP Bachem, 1903 – 1mf – 9 – 0-8370-7699-4 – (incl bibl ref) – mf#1986-1699 – us ATLA [340]

Das gesetz der form : briefe an tote / Hefele, Herman – Jena: E Diederichs 1921 [mf ed 1990] – 1r – 1 – (filmed with: friedrich hebbel und die gegenwart / wilhelm tideman) – mf#2706p – us Wisconsin U Libr [840]

Das gesetz der liebe / Ehrler, Hans Heinrich – Gotha: Klotz 1928 [mf ed 1989] – 1r – 1 – (filmed with: fruehlings-lieder) – mf#7210 – us Wisconsin U Libr [890]

Das gesetz des herrn : oder, die heiligen zehn gebote / Seeberg, P – 2. verm. Aufl. Berlin:Eduard Beck, 1867 – 1mf – 9 – 0-8370-5206-8 – mf#1985-3206 – us ATLA [220]

Das gesetz hammurabis und die thora israels : eine religions- und rechtsgeschichtliche parallele / Oettli, Samuel – Leipzig: A Deichert, 1903 – 1mf – 9 – 0-8370-7573-4 – (incl bibl ref) – mf#1986-1573 – us ATLA [340]

Gesetz uber die verwaltungsgerichtsbarkeit in bayern, wurttemberg-baden und hessen, mit kommentar von paulus van husen / Bavaria. Laws, Statutes, etc – Stuttgart, Poeschel 1947 177 p. LL-4088 – 1 – us L of C Photodup [348]

Gesetz ueber den vaterlandischen hilfsdienst / Konferenz von Vertretern der gewerkschaftlichen Organisationen und Angestelltenverbande – 1917. 78p. Generalkommission der Gewerkschaften Deutschlands. (Serial publications of German trade unions in the Memorial Library, University of Wisconsin-Madison.) – 1 – us Wisconsin U Libr [331]

Das gesetz ueber die presse vom 12. mai 1851, aus der entstehungsgeschichte, der rechtslehre und der entscheidungen des koeniglichen ober-tribunals erlautert. / Hartmann, L – Berlin: Decker, 1865 2 347 1p. LL-4104 – 1 – us L of C Photodup [340]

Das gesetz und christus im evangelium : zur revision der kirklichen lehre "de lege et evangelio" / Bugge, Christian August – Christiania [Oslo]: in Commission bei Jacob Dybwad, AW Broeggers, 1903 – (= ser Videnskapsselskapets Skrifter) – 1mf – 9 – 0-7905-0748-X – (incl bibl ref) – mf#1987-0748 – us ATLA [220]

Gesetz und evangelium / Walther, Carl Ferdinand Wilhelm – St Louis, Mo: Concordia Pub House, 1893 – 1mf – 9 – 0-524-08692-3 – mf#1993-3217 – us ATLA [220]

Gesetz- und verordnungsblatt land sachsen – Dresden DE, 1948-49 – 2r – 1 – gw Misc Inst [350]

Gesetz zur befreiung von nationalsozialismus und militarismus mit den ausfuehrungsvorschriften und formularen / Schullze, Erich – Muenchen, 1946 (mf ed 1995) – 3mf – 9 – €74.00 – 3-8267-3147-6 – mf#DHS-AR 3147 – gw Frankfurter [348]

Gesetzblatt der ddr – 1949-1989 – 1,797mf – 1 – gw Mikropress [348]

Die gesetze hammurabis : rektoratsrede. gehalten am stiftungsfeste der hochschule zuerich... / Cohn, Georg – Zuerich: Art Institut Orell Fuessli, I903 – 1mf – 9 – 0-8370-7685-4 – (incl bibl ref) – mf#1986-1685 – us ATLA [340]

Die gesetze hammurabis : und ihr verhaeltnis zur mosaischen gesetzgebung sowie zu den 12 tafeln / Mueller, David Heinrich – Wien: Alfred Hoelder, 1903 – 1mf – 9 – 0-8370-7571-8 – (incl bibl ref) – mf#1986-1571 – us ATLA [340]

Die gesetze hammurabis in umschrift und uebersetzung : dazu einleitung, woerter-, eigennamen-verzeichnis die sog. sumerischen familiengesetze und die gesetztafel brit. mus. 82-7-14, 988 / ed by Winckler, Hugo – Leipzig: JC Hinrichs, 1904 – 1mf – 9 – 0-524-05665-X – mf#1992-0515 – us ATLA [340]

Die gesetze hammurabis koenigs von babylon um 2250 v chr : das aelteste gesetzbuch der welt – Leipzig: JC Hinrichs, 1902 [mf ed 1986] – (= ser Der alte orient 4/4) – 1mf – 9 – 0-8370-7465-7 – mf#1986-1465 – us ATLA [340]

Gesetze, parlamentarische regeln und geschafts-ordnung / B'nai B'rith District No8 Silesia-Loge 36, Nr 477 – Liegnitz, Poland. 19-? – 1r – us UF Libraries [939]

Die gesetzessammlung (ulozenie) von 1649 und ihre auswirkungen auf die kirche in der aera nikons : untersuchungen zum verhaeltnis zwischen staat und kirche im moskauer russland / Heise, Christoph Ulrich – Heidelberg, 1972 – 2mf – 9 – 3-89349-769-2 – gw Frankfurter [348]

Die gesetzesschrift des koenigs josia : eine kritische untersuchung / Fries, Samuel Andreas – Leipzig: A Deichert, 1903 – 1mf – 9 – 0-8370-3206-7 – mf#1985-1206 – us ATLA [221]

Die gesetzgebung mosis im lande moab : ein beitrag zur einleitung in's alte testament / Riehm, Eduard – Gotha: Friedrich Andreas Perthes, 1854 – 1mf – 9 – 0-7905-2034-6 – (incl bibl ref) – mf#1987-2034 – us ATLA [270]

Geshikhte fun der yidisher arbeter-bavegung in lodzsh / Jasny, A Wolf – Lodz, Poland. 1937 – 1r – us UF Libraries [939]

Geshikhte fun idn in brazil / Raizman, Itzhak Z – Sao Paulo, Brazil. 1935 – 1r – us UF Libraries [939]

Geshikhte fun yidishen legyon / Jabotinsky, Vladimir – Varsha, Poland. 1929 – 1r – us UF Libraries [939]

Geshikhtes / Fininberg, Ezra – Moskve, Russia. 1939 – 1r – us UF Libraries [939]

Die gesicherten ergebnisse der bibelkritik und das von uns verkuendete gotteswort / Loeber, Richard – Gotha: Gustav Schloessmann, 1889 – 1mf – 9 – 0-7905-2052-4 – mf#1987-2052 – us ATLA [220]

Das gesicht im nebel : erzaehlung / Doerfler, Peter – Leipzig: P Reclam 1944 [mf ed 1990] – 1r – 1 – (aft by josef magnus wehner. filmed with: zwischen den garben / ferdinand freiligrath & other titles) – mf#7322 – us Wisconsin U Libr [830]

Das gesicht im nebel *see* Goethes egmont und schillers wallenstein

Gesicht und antlitz : neue gedichte / Ehrler, Hans Heinrich – Gotha: L Klotz 1928 [mf ed 1989] – 1r – 1 – (filmed with: fruhlings-lieder) – mf#7210 – us Wisconsin U Libr [810]

Gesichter und gesichte / Usinger, Fritz – Darmstadt: E Roether, 1965 – 1r – 1 – us Wisconsin U Libr [430]

Gesinsbeplanning : 'n beskouing uit die oogpunt van demografiese ontwikkeling en bevolkingsbeheer = Family planning / Roux, J P – Pretoria: Dept van Gesondheid [1972?] [mf ed Pretoria, RSA: State Library [199-]] – 7p on 1r with other items – 5 – (incl bibl ref) – mf#op 06529 r24 – us CRL [613]

N geskiedenis van die kerklied in die nederduitse gereformeerde kerk / Fourie, Francois Paul – Uni of South Africa 2000 [mf ed Pretoria: UNISA 2000] – 10mf – 9 – (incl bibl; text in afrikaans; incl abstract in afrikaans, english & german) – mf#mfm14794 – sa Unisa [242]

Die geskiedenis van die lutherse kerk aan die kaap / Hoge, J – Kaapstad: Cape Times Ltd 1938 – (= ser [Travel descriptions from south africa, 1711-1938]) – 3mf – 9 – (in: argief-jaarboek vir suid-afrikaanse geskiedenis) – mf#zah-31 – ne IDC [242]

Geskiedenis van die nederduitse gereformeerde sendingkerk oranje vrystaat (1910-1963) / May, John Mosoeu – U of the Western Cape 1986 [mf ed S.l: s.n. 1986] – 3mf – 9 – (summary in afrikaans & english; incl bibl) – sa Misc Inst [960]

Die geskiedenis van die ng sendinggemeente beaufort-wes 1818-1860 / Wyk, Albertus Benjamin van – U of the Western Cape 1990 [mf ed S.l: s.n. 1990] – 2mf – 9 – (summary in afrikaans & english; incl bibl) – sa Misc Inst [960]

Geskiedenis van die suid-afrikaanse republiek / Pelzer, Auguste – Kaapstad: A A Balkema, 1950 – 1 – us CRL [960]

Geslachtkundige aanteekeningen : t.a.v.d. gecommiteerden ten landdage van overijssel zedert 1610-1794 / Doorninck, J van – Deventer, 1871 – €32.00 – ne Slangenburg [949]

Geslison, Jeannette E K *see* A comparison of village and staged versions of selected hungarian dance styles

Gesnaldo dello spirito sancto : santa gema galgani / Bayle, Constantino – Madrid: Razon y Fe, 1944 – 1 – sp Bibl Santa Ana [240]

Gesner, B C *see* Will the old book stand?

Gesner, C *see*
– Historia animalium. libri 1 de quadrupedibus viviparis...
– Historia animalium. libri 2 de quadrupedibus viviparis...
– Historia animalium. libri 3 de de avium natura...
– Historia animalium. libri 4 de piscium et aquatilium...
– Historia animalium. libri 5 de serpentium natura...
– Historiae insectorum libellus qui est de scorpione

Gesner, Joh Albr et al [comp] *see* Selecta physico-oeconomica

Gesner, S *see*
– Controversiae inter theologos vvittenbergenses de regeneratione et electione dilvcida explicatio dd egidii hvnnii, polycarpi leyseri, salomonis gesneri
– Pro sanctissimo libro christianae concordiae dispvtatio prima sex capitvm

Gesolei – Duesseldorf DE, 1926 n1-162 – 1r – 1 – gw Misc Inst [074]

Gesondheidsbevorderende beweging = Health promoting movement / South Africa. Department of Health [Departement van Gesondheid] [Departement van Gesondheid] [Departement van Gesondheid] – [Pretoria: Dept van Gesondheid 1980?] [mf ed Pretoria, RSA: State Library [199-]] – 112p [ill] on 1r with other items – 5 – (in afrikaans & english) – mf#op 06882 r26 – us CRL [613]

Gesprache in dem reiche derer welt-weissen, in acht verschiedenen theilen zusammen gefasset, und mit einer vorrede von dem vorzuge der neuern.. – Halle, 1722 – 1 – us Wisconsin U Libr [400]

Gespraech eines lebensmueden mit seiner seele / Erman, A – Berlin, 1896 – 2mf – 9 – mf#NE-20386 – ne IDC [400]

Das gespraech jesu mit dem samariterin / Steinmeyer, Franz Ludwig – Berlin: Weigandt und Grieben, 1887 – 1mf – 9 – 0-524-05636-6 – mf#1992-0491 – us ATLA [220]

Das gespraech ueber formen : und, platons lysis deutsch / Borchardt, Rudolf – Leipzig: J Zeitler, 1905 [mf ed 1989] – 78p – 1 – mf#7052 – us Wisconsin U Libr [880]

Das gespraech ueber gedicht / Hofmannsthal, Hugo von – Berlin: Hyperionverlag 1918 [mf ed 1992] – 1r – 1 – (filmed with: die goethe-bildnesse / hermann rollett) – mf#3079p – us Wisconsin U Libr [430]

Gespraech von der musik : zwischen einem organisten und adjuvanten, darinnen nicht nur von verschiedenen missbraeuchen, so bey der musik eingerissen gehandelt... / Voigt, C – Erfurth: J D Jungnicol 1742 [mf ed 19–] – 4mf – 9 – mf#fiche 858 – us Sibley [780]
Gespraech zwischen einem musico theoretico : und einem studioso musices von der praetorianischen / printzischen / werckmeisterischen...zu befoerderung reiner harmonie / Sorge, Georg Andreas – Lobenstein: Im Verlag des Autoris [1748] [mf ed 19–] – 2mf – 9 – mf#fiche 944, 667 – us Sibley [780]
Gespraeche am abend : aus dem tagebuch des andreas thorstetten / Gmelin, Otto – Feldpostausg. Jena: E Diederichs, 1943, c1941 (mf ed 1990) – 1r – 1 – (filmed with: sommerwind ueber tormohlenhof) – us Wisconsin U Libr [430]
Gespraeche in dem reiche der todten... – Leipzig DE, 1720 n11-1739 n240 [gaps] – 19r – 1 – gw Mikrofilm [130]
Gespraeche jesu mit seinen juengern nach der auferstehung : ein katholisch-apostolisches senderschreiben des 2. jahrhunderts / Schmidt, Carl – Leipzig, 1919 – (= ser Tugal 3-43) – 12mf – 9 – €23.00 – ne Slangenburg [240]
Gespraeche mit daemonen : gedichte, maerchen, briefe / Arnim, Bettina von – Berlin: im Propylaeen-Verlag, c1922 [mf ed 1993] – (= ser Saemtliche werke v7) – 562p/pl – 1 – (incl bibl ref and ind for entire set of coll works (v1-7)) – mf#8196 reel 2 – us Wisconsin U Libr [802]
Gespraeche mit einem grobian – Leipzig: F A Brockhaus, 1866 [mf ed 1993] – xii/383p – 1 – mf#8459 – us Wisconsin U Libr [080]
Gespraeche mit goethe in den letzten jahren seines lebens : 1823-1832 / Eckermann, Johann Peter; ed by Castle, Eduard – Berlin: Bong & Co, c1916 [mf ed1999] – 2v in 1 (ill) – 1 – (incl ind by hans erich neumann) – mf#10130 – us Wisconsin U Libr [080]
Der gespraechige [...] – Danzig (Gdansk PL), 1828-1829 mar – 1 – gw Misc Inst [077]
Gespraechsbefaehigung im berufsbezogenen portugiesischunterricht / Groeschl, Juergen – (mf ed 1993) – 3mf – 9 – €49.00 – 3-89349-760-9 – mf#DHS 760 – gw Frankfurter [440]
Ain gesprech etlicher predicanten zu basel : gehalten mit etlichen bekennern des widertauffs / [Oecolampadius, J] – [Basel: Valentin Curio, 1525] – 1mf – 9 – mf#ME-89 – ne IDC [242]
Gesprechbiechlin neuew karsthans / Bucer, Martin; ed by Lehmann, Ernst – Halle: M Niemeyer 1930 [mf ed 1993] – 1r – 1 – (filmed with: neudrucke deutscher literaturwerke des 16. und 17. jahrhunderts) – mf#3387p – us Wisconsin U Libr [202]
Gesprekvoering met die pasiente in die verskillende suid-afrikaanse tale = Conversations with patients in the different south african languages – [Pretoria: Dept of Health 1979?] [mf ed Pretoria, RSA: State Library [199-]] – 140p on 1r with other items – 5 – (in afrikaans, english, northern sotho, sotho, tsonga, tswana, venda, xhosa & zulu) – mf#op 06822 r26 – us CRL [362]
Gesproken woordkunst van de nkundo / Rop, Albert De – Tervuren, Belgium. 1956 – 1r – us UF Libraries [960]
Gess, Friedrich Wilhelm *see* Deutliche und moeglichst vollstaendige uebersicht ueber das theologische system dr. friedrich schleiermachers
Gess, T W *see* The revelation of god in his word
Gess, Wolfgang Friedrich *see*
– Das apostolische zeugniss von christi person und werk
– Christi zeugniss von seiner person und seinem werk
– Das dogma von christi person und werk
– Die inspiration der helden der bibel und der schriften der bibel
Gessa *see* Manual de aplicacion y correcion de la bateria "c"
Gessen, I V (Iosif Vladimirovich) *see* Nakanunie probuzhdeniia
Gessie, Berlingiero *see* La spada di honore
Gessner, C *see*
– De omni rerum fossilium genere, gemmis, lapidibus, metallis,...
– De piscibus et aquatilibus omnibus libelli 3. novi
– De piscinis...
– Historiae animalium...
– Icones animalium quadrupedum viviparorum et oviparorum, quae in historia animalium c. gesneri describuntur,...
– Icones animalium quadrupedum viviparorum et oviparorum, quae in historia animalium c. gesneri...describuntur,...
– Icones avium omnium, quae in historia avium c. gesneri describuntur,...
– Lexicon graecolatinum denuo impressum... novissime per adrianum iunium...locupletatum...
– Libellus de lacte et operibus lactariis philologus pariter ac medicus...
– Nomenclator aquatilium animantium
– Opera
Gessner, K *see* Bibliotheca universalis
[Gessner, K] *see* Appendix bibliothecae conradi gesneri
Gessner, K et al *see* Bibliotheca...
Gessner, Theodor *see* Das hohe lied salomonis
Gesta abbatum fontanellensium (mgh7:28.bd) – 1886 – (= ser Monumenta germaniae historica 7: scriptores rerum germanicarum in usum scholarum (mgh7)) – €3.00 – ne Slangenburg [240]
Gesta abbatum monasterii s albani *see* Chronia monasterii s albani 4 [rs28]
Gesta christi : or, a history of humane progress under christianity / Brace, Charles Loring – 4th ed. New York: A C Armstrong, 1884, c1882 [mf ed 1990] – 2mf – 9 – 0-7905-5577-8 – (1st ed publ 1882) – mf#1988-1577 – us ATLA [240]
Gesta de heroes / Ibarzabal, Federico De – Habana, Cuba. 1918 – 1r – us UF Libraries [972]
Gesta episcoporum (mgh5:14.bd) : historiae (suppl tom 1-12 pars 2) suppl tomi 13 – 1883 – (= ser Monumenta germaniae historica 5: scriptores in folio (mgh5)) – €35.00 – ne Slangenburg [240]
Gesta frederici 1. imperatoris in lombardia auctore cive mediolanensi (mgh7:27.bd) – 1892 – (= ser Monumenta germaniae historica 7: scriptores rerum germanicarum in usum scholarum (mgh7)) – €5.00 – (accedunt gesta frederici 1 in expeditione sacra) – ne Slangenburg [240]
Gesta greyorum... : henry, prince of purpoole / England. Inns of Court – London, 1688 – 2mf – 9 – $3.00 – (account of gray's inn revels under henry helmes, "prince of purpoole") – mf#LLMC 84-282 – us LLMC [941]
Gesta hammaburgensis ecclesiae pontificum *see* Adam's von bremen hamburgische kirchengeschichte
Die gesta innocentii 3. im verhaeltniss zu den regesten desselben papstes / Elkan, Hugo – Heidelberg: J. Hoerning, 1876 – 1mf – 9 – 0-8370-7939-X – (incl bibl ref) – mf#1986-1939 – us ATLA [240]
Gesta pontificum romanorum / Monumenta Germaniae Historica. Scriptores – 14mf – 8 – mf#367 – ne IDC [700]
Gesta romanorum : das ist der roemer tat / ed by Keller, Adelbert von – Quedlinburg, Leipzig: G Basse, 1841 [mf ed 1993] – (= ser Bibliothek der gesammten deutschen national-literatur von der aeltesten bis auf die neuere zeit sect1/23) – viii/174p – 1 – mf#8438 reel 5 – us Wisconsin U Libr [430]
Gesta saec 13 (mgh5:25.bd) – 1880 – (= ser Monumenta germaniae historica 5: scriptores in folio (mgh5)) – €48.00 – ne Slangenburg [240]
Gesta stephani regis anglorum (rs82/3) – 1886 – (= ser The rolls series (rs)) – €19.00 – (incl: richard of hexham: historia de gestis regis stephani et de bello de standard (1135-1139). aelred of rievaulx: relatio de standardo. jordan fantosme: chronique de la guerre entre les anglois et les ecossais (1173-1174), with a trans. richard of devizes: de rebus gestis ricardi primi (1189-1192)) – ne Slangenburg [931]
Die gestalt der wortform und des satzes unter einwirkung des rhythmus bei chaucer und gower / Bihl, Josef – Tuebingen, 1915 (mf ed 1994) – 2mf – 9 – €31.00 – 3-8267-3043-7 – mf#DHS-AR 3043 – gw Frankfurter [420]
Die gestalt des bildenden kuenstlers in der dichtung / Laserstein, Kaete – Berlin: W de Gruyter & Co, 1931 – 2r – 1 – (incl bibl ref and index) – us Wisconsin U Libr [430]
Gestalten des christlichen abendlandes – Munich: Koesel-Pustet, 1937-40 [mf ed 2001] – (= ser Christianity's encounter with world religions, 1850-1950) – 1r – 1 – (in german. contains biogr of: anselm von canterbury (1033-1109), johannes von ruysbroeck (1293-1381), katharina von genua (1447-1510), bernard overberg (1754-1826) and martin deutinger (1815-64)) – mf#2001-s006 – us ATLA [920]
Gestalten deutscher dichtung : eine literaturgeschichte / Denecke, Rolf – Frankfurt am Main: Hirschgraben-Verlag, 1965 – 1 – 1 – (incl ind) – us Wisconsin U Libr [430]
Gestalten und leidenschaften : dichtungen / Friedrichs, Hermann – Hamburg: Verlagsanstalt und Druckerei Actien-Gesellschaft, 1889 (mf ed 1990) – 1r – 1 – (filmed with: gustav freytag) – us Wisconsin U Libr [810]
Gestalten und probleme / Winkler, Eugen Gottlob – Leipzig-Markkleeberg: Karl Rauch Verlag, c1937 – 1 – us Wisconsin U Libr [840]
Die gestaltung der feste im jahres- und lebenslauf in der ss-familie *see* Nsdap (national socialist german workers party) nazi publications
Die gestaltung des kuenstlerischen kaleidoskops : zur filmaesthetik von jaen-luc godard / Prassel, Caroline – (mf ed 1996) – 3mf – 9 – €49.00 – 3-8267-2323-6 – mf#DHS 2323 – gw Frankfurter [440]
Gestaltung, umgestaltung : festschrift zum 75. geburtstag von hermann august korff / ed by Mueller, Joachim – Leipzig: Koehler & Amelang, 1957 – 1 – (in german; one article in english. incl bibl ref) – us Wisconsin U Libr [430]
Gestaltungen des frauen-bildes in deutscher lyrik / Schukart, Hanns – Bonn a. Rh.: L Roehrscheid, 1933 – 1r – 1 – (incl bibl ref) – us Wisconsin U Libr [430]
Gestaltungselemente im bildwerk von otto mueller / Decker, Marlene – Dortmund: projekt vlg. 1993 (mf ed 1996) – 3mf – 9 – €38.00 – 3-8267-9706-X – mf#DHS 9706 – gw Frankfurter [700]
Gestaltungsmittelanalyse bei dienstleistungen / Willms, Jens-Peter – (mf ed 1995) – 2mf – 9 – €40.00 – 3-8267-2085-7 – mf#DHS 2085 – gw Frankfurter [650]
Gestas heroicas / Perez Ortiz, Ramon – Ciudad Trujillo, Dominican Republic. 1952 – 1r – us UF Libraries [972]
Gestationsdiabetes in nicaragua : untersuchungen zur praevalenz des gestationsdiabetes und zur risikoabschaetzung in der geburtshilfe des hospital aleman nicaraguense in managua / Moeller, Christoph – (mf ed 1998) – 2mf – 9 – €40.00 – 3-8267-2584-0 – mf#DHS 2584 – gw Frankfurter [618]
La geste tiedo / Sy, Amadou Abel – 1980 – us CRL [390]
Die gesteigerte effizienz durch das arrangement : dargestellt an dem werk "der tod und das maedchen" (franz schubert / gustav mahler) / Petri, Hasso Gottfried – [mf ed 2002] – 2mf – 9 – €40.00 – 3-8267-2784-3 – mf#DHS2784 – gw Frankfurter [780]
Gestern : dramatische studie in einem akt in versen / Hofmannsthal, Hugo von – 3. Aufl. Berlin: S Fischer, 1909 – 1r – 1 – us Wisconsin U Libr [810]
Der gestiefelte kater / das rotkaeppchen : zwei maerchenspiele mit musik und reigen / Herrmann, Emil Alfred; ed by Benz, Richard – Jena: E Diederichs 1911 [mf ed 1990] – 1r – 1 – (filmed with: das problem "volkstum und dichtung" bei herder / reta schmitz) – mf#2724p – us Wisconsin U Libr [790]
Gestion agricola despues de 1898 / Colon, Edmundo Dimas – San Juan, Puerto Rico. 1948 – 1r – us UF Libraries [972]
Gestion oficial en agricultura / Cabal Cabal, Camilo J – Bogota, Colombia. 1952 – 1r – us UF Libraries [630]
Gestirn des krieges : gedichte / Schuett, Bodo – Jena: E Diederichs, 1941 – 1 – us Wisconsin U Libr [810]
Gestirn des krieges *see* Inquiry into the sources of charles sealsfield's novel morton
Der gestohlene mond / Barlach, Ernst; ed by Dross, Friedrich – Berlin: Suhrkamp, 1948 [mf ed 1989] – 270p – 1 – mf#6979 – us Wisconsin U Libr [830]
Die gestohlene seele : eine erzaehlung aus chile / Vegesack, Siegfried von – Muehlacker: E Haendle [194-?] [mf ed 1991] – 1r – 1 – (filmed with: blumbergshof) – mf#2944p – us Wisconsin U Libr [880]
Der gestorbene / Becher, Johannes Robert – Regensburg: F L Habbel, 1921 [mf ed 1989] – 43p/[1]pl (ill) – 1 – mf#6994 – us Wisconsin U Libr [890]
Gestos / Sarduy, Severo – Barcelona, Spain. 1963 – 1r – us UF Libraries [972]
Gestoso y Perez, Jose *see* De sevilla a guadalupe. breves apuntes tomados a vuela pluma
Gestrikland – Gaevle, Sweden. 1877-80 – 2r – 1 – sw Kungliga [078]
Gestuehl der alten : [short stories] / Blunck, Hans Friedrich – Leipzig: Insel-Verlag, 1943 [mf ed 1989] – (= ser Insel-buecherei 538) – 78p – 1 – mf#7036 – us Wisconsin U Libr [830]
Gesundheitsfoerderung als arbeitsfeld sozialer therapie / Weil, Thomas – (mf ed 1995) – 3mf – 9 – €49.00 – 3-8267-2164-0 – mf#DHS 2164 – gw Frankfurter [360]
Die gesundheitsfuehrung – Berlin DE, 1939 n1-3, 1940-1945 n2 – 1 – gw Misc Inst [613]
Gesundheitsfuersorge der inneren mission *see* Mitteilungen des deutschen evangelischen krankenhausverbandes (fw3)
Gethsemane and after : a new setting of an old story / Brady, Cyrus Townsend – New York: Moffat, Yard, 1907 [mf ed 1985] – 1mf – 9 – 0-8370-2436-6 – mf#1985-0436 – us ATLA [790]
Gethsemane baptist church – Sacramento. 1972-1979 (1) 1972-1979 (5) 1976-1979 (9) – 1r – 1 – $41.13 – mf#6505 – us Southern Baptist [242]
Gethsemani : ou notice sur l'eglise de l'agonie ou de la priere / Orfali, G – Paris, 1924 – 2mf – 9 – mf#H-2857 – ne IDC [956]
Getilgte paulus-und psalmtexte (tab14) : unter getilgten ambr liturgiestuecken cod sang 908 / Dold, Alban – 1928 – (= ser Texte und arbeiten. beuron (tab). beitraege zur ergruendung des aelteren lateinischen und christlichen schrifttums und gottesdienstes) – €5.00 – ne Slangenburg [240]
Getino, Luis A *see* Manipuelo de flores del maestro fr. francisco de vitoria
Getino, Luis G *see* Justicia y caracter de la guerra nacional espanola
Getreide-zeitung – Bremen DE, 1852-53 – 2r – 1 – gw Misc Inst [074]
Getrennt marschieren : erzaehlung / Hohlbaum, Robert – Muenchen: A Langen/G Mueller 1935 [mf ed 1990] – (= ser Die kleine buecherei 52) – 1r – 1 – (filmed with: der fruehlingswalzer / robert hohlbaum) – mf#2730p – us Wisconsin U Libr [830]
Die getrennten : gedichte / Gstettner, Hans – Berlin: Suhrkamp, 1944 – 1r – 1 – us Wisconsin U Libr [810]
Das getriebe – Dessau DE, 1953 8 jan-1990 [gaps] – 6r – 1 – (notes: maschinenfabrik) – gw Misc Inst [621]
Gettell, Raymond Garfield *see* Political science
Getting late – n1 [1967] – 1r – 1 – mf#901081 – us WHS [071]
Getting results- for the hands-on manager – New York NY 1997 – 1,5,9 – (cont: getting results- for the hands-on manager c [office ed]) – ISSN: 1088-4343 – mf#1115,04 – us UMI ProQuest [650]
Getting results- for the hands-on manager. a [plant ed] – New York NY 1997 – 1,5,9 – ISSN: 1088-4343 – mf#25469 – us UMI ProQuest [650]
Getting results- for the hands-on manager. c [office ed] – New York NY 1996-97 – 1,5,9 – (cont by: getting results- for the hands-on manager) – ISSN: 1088-4343 – mf#1115.04 – us UMI ProQuest [650]
Getting together – 1970 aug-1975 jan 15, 1976 may-1978 aug – 2r – 1 – mf#1112014 – us WHS [071]
Getting together – v2 iss 3 [1972 may 19] – 1r – 1 – mf#901087 – us WHS [071]
Getting what we want; how to apply psychoanalysis to your own problems / Edson, David Orr, M D – New York, London: Harper & Brothers, (c1921). 4p,l,286,(1)p. Fold. form – 1 – us Wisconsin U Libr [150]
Gettke's buehnen-almanach / ed by Genossenschaft Deutscher Buehnen-Angehoeriger – Leipzig: Luckhardt [etc 1872-88] [2.-17. jahrg (1874-89)] (annual) [mf ed 2001] – 17v on 5r – 1 – (cont by: neuer theater-almanach fuer das jahr...; imprint varies) – mf#film mas c4850 – us Harvard [790]
Gettleman; Marvin et al *see* Conflict in indo-china
Gettorfer nachrichten – Gettorf DE, 1895-1939 15 sep – 1 – gw Misc Inst [074]
Getty, Alice *see*
– Ganesa
– The gods of northern buddhism
Gettysburg College *see* Gettysburg intercultural advancement news
Gettysburg intercultural advancement news / Gettysburg College – 1990 feb – 1r – 1 – mf#4876450 – us WHS [378]
Getuigenis der 25jarige evangelie-bediening / Beij, B de – Milwaukee: Houtkamp, 1872 – 1mf – 9 – 0-524-06597-7 – mf#1991-2652 – us ATLA [240]
Getulio, este desconhecido / Josefsohn, Leon – Rio de Janeiro, Brazil. 1957 – 1r – us UF Libraries [972]
Getulio vargas / Carrazzoni, Andre – Buenos Aires, Argentina. 1941 – 1r – us UF Libraries [972]
Getulio vargas : (esboco de biografia) / Pessoa Cavalcanti De Albuquerque, Epitacio – Rio de Janeiro, Brazil. 1938 – 1r – us UF Libraries [972]
Getulio vargas / Luis, Pedro – Sao Paulo, Brazil. 1946 – 1r – us UF Libraries [972]
Getulio vargas e o direito social trabalhista / Pimpao, Hirose – Rio de Janeiro, Brazil. 1942 – 1r – us UF Libraries [972]
Getulio vargas, meu pai / Peixoto, Alzira Vargas do Amaral – Rio de Janeiro, Editora Globo 1960 – 1 – us Wisconsin U Libr [920]
Getz, Feivel Meir *see* Dat veha-hinukn
Getz, Pierce A *see* Organ mixtures in comtemporary american practice
Getze, J A *see* Young organist
Geuer, F *see* Die kirchenpolitik des kanzlers michel de l'hospital
Geuffroy, A *see*
– Aulae turcicae, othomanniciq've imperii descriptio...pars 1. solymanni 12 and selymi 13 tvrcar impp contra christianos...pars 2
– Briefue descriptio de la covrt dv grant tvrc et vng sommaire du regne des othmans auec vn abrege de leurs folles superstitions...
Geunta unsjiah : fakultas keguruan universitas sjiah kuala – Banda, 1970(aug)-1971(may) – 3mf – 9 – (missing: 1970, v1(dec); 1971, v2(jan-feb)) – mf#SE-1492 – ne IDC [378]

Geus, Christoph *see* Klinische untersuchungen zur thermoregulation am beispiel des temperaturmusters der haut und ihre moegliche bedeutung fuer zahnaerztliche diagnostische und therapeutische fragestellungen
Geutzlaff, de apostel der chinezen : in zijn leven en zijne werkzaamheid / Erdbrink, Gerhard Rudolf – Rotterdam: M Wijt & Zonen, 1850 [mf ed 1995] – (= ser Yale coll) – 53p – 1 – 0-524-09801-8 – (in dutch) – mf#1995-0801 – us ATLA [240]
Geutzlaff, Karl Friedrich August *see*
– Geutzlaff's geschiedenis van het chinesche rijk
– Journal of three voyages along the coast of china, in 1831, 1832 and 1833
Geutzlaff's geschiedenis van het chinesche rijk : van de oudste tijden tot op den vrede van nanking = Geschichte des chinesischen reiches von den aeltesten zeiten bis auf den frieden von nanking / ed by Neumann, Karl Friedrich & Meppen, K N – 's Gravenhage: K Fuhri, 1852 [mf ed 1995] – (= ser Yale coll) – 2v in 1 (ill) – 1 – 0-524-09737-2 – (trans fr german into dutch) – mf#1995-0737 – us ATLA [951]
Gevaert, F A et al *see* Andre ernest modeste gretry
De gevallen van een hedendaagsch vry-metselaar : uit het hoogduitsch vertaald – Amsterdam: G Roos 1805 – 2v on 8mf – 9 – mf#vrl-52 – ne IDC [366]
Gevatter tod : fuer die maerchenspiele der kuenstlerischen volksbuehne nach grimms maerchen in rede und handlung gesetzt / Guembel-Seiling, Max – Leipzig: Breitkopf & Haertel, 1918 – 1r – 1 – us Wisconsin U Libr [820]
Gevelsberger zeitung *see* Ennepethal-zeitung
Gevelsberger zeitung milsper-voerder zeitung *see* Ennepethal-zeitung
Gevet / Prilutski, NOah – Varshe, Poland. 1923 – 1r – us UF Libraries [939]
Geveze – Istanbul: Asir Matbassi. Sahib-i Imtiyaz: Kirkor Faik, 1908-09. n6-36,38,41,43. 23 tesrinisani 1324-2 nisan 1325 [1908-09] – (= ser O & t journals) – 3mf – 9 – $65.00 – us MEDOC [956]
Gevule tsiyon / Motskevitch, Shabbetai – Vilna, Lithuania. 1899 – 1r – us UF Libraries [939]
Gewalt an frauen : zur sozialpaedagogischen handlungskompetenz in der arbeit mit betroffenen als notwendige grundlage zukunftsweisender psycho-sozialer/therapeutischer/ gesundheitsfoerdernder ausbildung / Ragoss, Silvia – (mf ed 1995) – 3mf – 9 – €49.00 – 3-8267-2132-2 – mf#DHS 2132 – gw Frankfurter [360]
Gewalt ueber das feuer : eine erzaehlung aus der urzeit / Blunck, Hans Friedrich – jugendausg. Reutlingen: Ensslin & Laiblin, c1955 [mf ed 1989] – 128p (ill) – 1 – mf#8984 – us Wisconsin U Libr [390]
Gewalt ueber das feuer : eine sage von gott und mensch / Blunck, Hans Friedrich – Jena: E Diederichs, 1928 [mf ed 1989] – 225p – 1 – mf#7036 – us Wisconsin U Libr [390]
Gewaltakte und greueltaten der polen waehrend des 3. aufstandes in. . . *see* Nsdap (national socialist german workers party) nazi publications
Die gewalten : ein band balladen / Csokor, Franz Theodor – Berlin-Charlottenburg: A Juncker, [19–] [mf ed 1989] – 71p – 1 – mf#7161 – us Wisconsin U Libr [780]
Gewappnetes herz : gedichte vom krieg / Ehrke, Hans – Braunschweig: G Westermann, c1943 (mf ed 1990) – 1r – 1 – (filmed with: menschen und affen) – us Wisconsin U Libr [810]
Gewerbe- und handelszeitung – Prag (CZ), 1930 sep-1938 31 mar – 4r – 1 – gw Misc Inst [380]
Gewerbe-blatt fuer sachsen – Leipzig DE, 1834-44 – 5r – 1 – gw Misc Inst [338]
Gewerbe-zeitung – Leipzig DE, 1846 apr-1847 mar – 1r – 1 – gw Misc Inst [330]
Gewerbliche rundschau – Prag (CZ), 1938 16 apr-10 sep – 1r – 1 – gw Misc Inst [380]
Die gewerbliche stellung der frau im mittelalterlichen koeln / Behaghel, Wilhelm – Berlin: W Rothschild 1910 – (= ser Abhandlungen zur mittleren und neueren geschichte) – 1mf – 9 – 0-524-08515-3 – (incl bibl ref) – mf#1993-1045 – us ATLA [943]
Gewerbliches tageblatt fuer kassel und die umgegend – Kassel DE, 1853 5 dec-1871, 1881-1932 29 sep – 155r – 1 – (with gaps. later: kasseler tageblatt) – gw Misc Inst [338]
Gewerkschaft der Angestellten *see* Zeitschrift
Gewerkschaft deutscher Eisenbahner *see* Geschaeftsbericht..
Der gewerkschaftler : mitteilungsblatt der gewerkschaft hessen-pfalz – Neustadt [Haardt / a.d. Weinstr] DE, 1948 jun & jul – 1r – 1 – (filmed by mka: 1946-49 [1r] [3869]; cont: welt der arbeit, koeln) – gw Mikrofilm; gw Mikropress [331]
Gewerkschaftliche frauenzeitung – Berlin DE, 1916-27 – 3r – 1 – gw Misc Inst [331]
Gewerkschaftliche frauenzeitung – Berlin. v7. 1922. (Serial publications of German trade unions in the Memorial Library, University of Wisonsin-Madison.) – 1 – us Wisconsin U Libr [331]
Gewerkschaftliche informationen – Koeln, Bielefeld, Duesseldorf DE, 1946 24 oct-1949 [gaps] – 1r – 1 – (title varies: gewerkschaftliche praxis 1949) – mf#3874 – gw Mikropress [331]
Gewerkschaftliche praxis *see* Gewerkschaftliche informationen
Gewerkschafts-archiv – Jena. v1-18. 1924-33. (Serial publications of German trade unions in the Memorial Library, University of Wisconsin-Madison.) – 1 – us Wisconsin U Libr [331]
Gewerkschaftseinheit : informationsblatt der einheitsgewerkschaft fuer den bezirk koblenz-trier – Koblenz, Trier DE, 1945 5 sep-12 dec – 1r – 1 – mf#3876 – gw Mikropress [331]
Gewerkschaftskommission. Berlin *see* Rechenschafts-bericht
Gewerkschaftsleben – No.1-. Jan 1981-. -m. Monatsschrift des FDGB – 1 – us Wisconsin U Libr [331]
Die gewerkschaftsstimme – Aschaffenburg, Muenchen DE, 1910-1933 n13 – 16r – 1 – (with suppl) – gw Misc Inst [331]
Gewerkschafts-zeitung : informationsblatt der einheitsgewerkschaften rheinland-hessen-nassau – Koblenz DE, 1946 1 jun-1949 25 dec [gaps] – 1r – 1 – mf#3873 – gw Mikropress [331]
Gewerkschafts-zeitung : organ der bayerischen gewerkschaften – Muenchen DE, 1947 10 jan-25 dec – 1r – 1 – mf#3875 – gw Mikropress [331]
Gewerkschafts-zeitung : zeitschrift der freien gewerkschaft in der britischen zone – Hamburg DE, 1946 12 feb-1947 1 apr – 1r – 1 – mf#3872 – gw Mikropress [331]
Gewerkschafts-zeitung fuer das gebiet suedwuerttenberg und hohenzollern – Tuttlingen DE, 1946 15 sep-1948 31 jan – 1r – 1 – (title change with n3 onwards: die schaffenden 1948 18 feb-1949 31 dec) – mf#3878 – gw Mikropress [331]
Der gewerksvereinsbote – Duesseldorf DE, 1901-05, 1907-sep 1912 – 1 – (title varies: 1906: westdeutsche arbeiterpost, jun 1907: duesseldorfer post, jul 1912: westdeutsche post) – gw Misc Inst [331]
Der gewerkverein – Berlin DE, 1908-18 – 3r – 1 – (filmed by misc inst: 1877-89, 1906-19 [1r]) – gw Mikropress; gw Misc Inst [331]
Gewerkverein christlicher Bergarbeiter Deutschlands *see*
– Geschaeftsbericht..
– Protokoll...
Das gewissen : die entwickelung seiner namen und seines begriffes / Kaehler, Martin – Halle: J Fricke, 1878 – 1mf – 9 – 0-7905-9290-8 – (incl bibl ref) – mf#1989-2515 – us ATLA [170]
Gewissen – Berlin DE, 1919 1 apr-1920 – 1r – 1 – (filmed by mikropress: 1928-1929 30 mar (gaps) [1r] order#5184; filmed by misc inst: 1921-22 [1r]; 1921-mar 1929 [3r]) – gw Mikrofilm; gw Mikropress; gw Misc Inst [074]
Das gewissen und die gewissensfreiheit : zehn vortraege / Simar, Hubert Theophil – Freiburg i.B.; St Louis, MO: Herder, 1874 – 1mf – 9 – 0-8370-7264-6 – (incl bibl ref) – mf#1986-1264 – us ATLA [240]
Die gewissenhaften : eine komoedie / Erler, Otto – Wiemar: F Fink c1938 [mf ed 1989] – 1r – 1 – (filmed with: die kleine weltlaterne / peter bamm) – mf#7216 – us Wisconsin U Libr [820]
Gewissensfragen : religioese briefe aus der gegenwart fuer die gegenwart / Wimmer, Richard – Tuebingen: JCB Mohr (Paul Siebeck), 1902 [mf ed 1985] – 1mf – 9 – 0-8370-5868-6 – mf#1985-3868 – us ATLA [210]
Gewissensqualen : zwei novellen / Gerhardt, Dagobert von – Berlin: Verlag des Vereins der Buecherfreunde, Schall & Grund, [1894?] – 1r – 1 – us Wisconsin U Libr [430]
Die gewissheit des glaubens und die freiheit der theologie / Herrmann, Wilhelm – 2. neu bearb Aufl. Freiburg i B: JCB Mohr, 1889 – 1mf – 9 – 0-7905-7759-3 – mf#1989-0984 – us ATLA [240]
Gewissheit des siegs und sicht auf grosse tage : gesammelte sonette, 1935-1938 / Becher, Johannes Robert – Moskau: Meshdunarodnaja Kniga (Das Internationale Buch), 1939 [mf ed 1989] – 151p – 1 – mf#6994 – us Wisconsin U Libr [810]
Das gewissheitsproblem in der systematischen theologie bis zu schleiermacher / Heim, Karl – Leipzig: JC Hinrichs, 1911 – 1mf – 9 – 0-7905-3903-9 – (includes bibliographical footnotes and index) – mf#1989-0396 – us ATLA [240]
Gewitter im mai; der besondere / Ganghofer, Ludwig – Berlin: Ullstein, [1912?] (mf ed 1990) – 1r – 1 – (filmed with: gustav freytag, ein publizist) – us Wisconsin U Libr [830]
Gewordene liturgie / Jungmann, J A – Innsbruck, 1941 – 7mf – 8 – €15.00 – ne Slangenburg [240]
Geyer, B *see*
– D thomae aquinatis. de essentia et potentiis animae in generali (ia, q 75-77). una cum guilelmi de la mare correctorii art 28
– Magistri echardi quaestiones et sermo parisienses
– Peter abaelardus philosophische schriften
– S thomae de aquino quaestiones de trinitate divina. summa theologica 1, q 27-32
– Die sententiae divinitatis
Geyer, Carl-Friedrich *see* Kritische theorie und metaphysik
Geyer, Mauritius *see* Observationes epigraphicae de praepositionum graecarum
Geyer, P *see* Itenera hierosolymitana saecvli 4-8
Geyer, Ursula *see* Der adlerflug im romischen konserkrationszeremoniell
Geymet, Enrico *see* E la casa un paradiso
Geymonat, Jean *see* Michel servet et ses idees religieuses
Geysbeek, Pieter Gerardus Witsen *see* Het geheim der vrijmetselarij opengelegd
[Geyserville-] geyserville press – CA. 1940-60 – 7r – 1 – $420.00 – mf#B02273 – us Library Micro [071]
Geystliche lieder : mit e newen vorrede, d m luth – Leipzig: Babst 1567 – (= ser Hqab. literatur des 16. jahrh.) – 5mf – 9 – €60.00 – mf#1567a – gw Fischer [780]
Geystliche lieder : mit einer newen vorrede d mart luthers; warnung d m luth – [Nuernberg]: [Neuber] [vor 1564] – (= ser Hqab. literatur des 16. jahrh.) – 4mf – 9 – €50.00 – mf#1564b – gw Fischer [780]
Geystliche lieder d martin luthers : von newem zugericht, mit vil schoenen psalmen vnd liedern gemehret (psalmen vnd geystliche lieder, welche von frommen christen gemacht... sindt) – [Nuernberg]: [Neuber] 1567 – (= ser Hqab. literatur des 16. jahrh.) – 8mf – 9 – €80.00 – mf#1567b – gw Fischer [780]
Geystliche lieder d martini lutheri – Leipzig: Steinmann 1588 – (= ser Hqab. literatur des 16. jahrh.) – 6mf – 9 – €70.00 – mf#1588c – gw Fischer [780]
De gezaghebbers der oost-indische compagnie op hare buiten-comptoiren in azie / Wijnaendts van Resandt, Willem – Amsterdam: Liebaert, 1944 [mf ed 1989] – (= ser Centrale dienst voor sibbekunde. genealogische bibliotheek 2) – 316p – 1 – mf#2763 – us Wisconsin U Libr [380]
Gezamelte shriften / Shneour, Zalman – Warsaw, Poland. 1910 or 11 – 1r – us UF Libraries [939]
Gezang un deklamatyse / Kassel, David – Warszawa, Poland. 192-? – 1r – us UF Libraries [939]
Gezangboek voor vrijmetselaaren – Amsterdam: J S Van Esveldt Holtrop 1806 – 6mf – 9 – mf#vrl-120 – ne IDC [366]
Gezangen voor de a loge louisa augusta : in het oosten van purmerende – [s.l: s.n] 5834 [1834] – 2mf – 9 – mf#vrl-48 – ne IDC [366]
Gfroerer, August Friedrich *see*
– Die heilige sage
– Das heiligthum und die wahrheit
– Das jahrhundert des heils
– Philo und die alexandrinische theosophie
– Untersuchung ueber alter, ursprung, zweck der dekretalen des falschen isidorus
GFWC clubwoman / General Federation of Women's Clubs – 1980 sep-1983 nov – 1r – 1 – (cont: general federation clubwoman [1978]) – mf#898420 – us WHS [366]
The ghadr directory : containing the names of persons who have taken part in the ghadr movement in america, europe, africa and afghanistan as well as india / India. Intelligence Bureau – New Delhi: Govt of India Press, 1934 – 1 – us CRL [954]
Ghana : internal affairs and foreign affairs, 1960-jan 1963 / U.S. State Dept – (= ser Confidential u s state department central files) – 12r – 1 – $2320.00 – 1-55655-910-0 – (with p/g) – us UPA [327]
Ghana *see* Subsidiary legislation – supplement
Ghana agriculture, 1890-1962 : a bibliography of crop and stock,co-operation and forestry, food and fishery / Tetteh, S N – Accra, Government Printing Office [1962] – us CRL [960]
The ghana archive of the basel mission, 1829-1918 – 153r – 1 – (with guide. int by paul jenkins) – mf#97040 – uk Microform Academic [240]
Ghana Commissioner for Local Government Enquiries *see* Report
Ghana daily express – Accra: Ausco Press and Pub Co, feb 1954-mar 1955] – us CRL [079]
Ghana daily mail – Accra: Amalgamated Press Ltd, jul 1957-apr 1958] – us CRL [079]
The ghana evening news – Accra: Heal Press, mar 1954-aug 1958 – us CRL [079]
Ghana (formerly Gold Coast). Central Bureau of Statistics *see* Statistical yearbook 1961-1970
Ghana journal of science – Legon, Ghana 1961-75 – 1 – ISSN: 0016-9544 – mf#6941 – us UMI ProQuest [500]
Ghana nationalist – Accra: Fortitude Press & Pub Co, mar 1953-apr 1955 – us CRL [079]
The ghana report : economic development and investment opportunities, legal problems relative to investment, sociological factors relative to general economic development – New York, 1959 – us CRL [338]
Ghana statesman – Accra. Ghana. Sept. 10, 17; Oct. 1, 1948 – 1 – us NY Public [079]
Ghana times – Accra: Star Pub Co, oct 4, 1958-jun 30, 1960 – us CRL [079]
Ghanaian language materials and miscellanea, 1966-68 – Chicago, University of Chicago, Photodup Dept, 1973 – us CRL [470]
The ghanaian times – Accra: Star Pub Co, jul 1960-sep 20 1962 – us CRL [079]
Ghani, Muhammad 'Abdu'l *see* Pre-mughal persian in hindustan
Ghassemlou, Abdul Rahman *see* Kurdistan and the kurds
Ghatakarparam oder das zerbrochene gefaess : ein sanskritisches gedicht / Kalidasa; ed by Dursch, Georg Martin – Berlin: Duemmler 1828 – (= ser Whsb) – 1mf – 9 – €20.00 – (trans & ann by ed; sanskrit in devanagari script) – mf#Hu 246 – gw Fischer [810]
Ghazarian, M *see* Armenien unter der arabischen herrschaft...
Ghazzali *see*
– Ad-dourra al-faakhira
– Ihya' 'ulum al-din
– Von der ehe
The ghebers of hebron : an introduction to the gheborim in the lands of sethim... / Dunlap, Samuel Fales – new rev ed. New York: JW Bouton, 1898 – 3mf – 9 – 0-524-06376-1 – (incl bibl ref) – mf#1990-3543 – us ATLA [200]
Het gheestelijck kaertspel met herten troef... / Joseph...Sancta Barbara – Antwerpen: M. Cnobbaert, 1676 – 6mf – 9 – mf#O-650 – ne IDC [090]
Gheestelycke sermoonen : de naevolghinghe des armen leven christi – het merch der zielen / Tauler, Johann – Antwerpen, 1707 – 28mf – 8 – €54.00 – (trans by p j de lixbona) – ne Slangenburg [240]
Gheestelycke sermoonen / Tauler, Johann – Antwerpen, 1683 – €60.00 – (trans by p j de lixbona) – ne Slangenburg [240]
De gheestelycke vryagie waer christus de ziele is vryende : seer schoon ende profytelijck om den mensch inde liefde godts t'ontsteken – Brussel: Schoevaerdts, 1649 – 10mf – 9 – mf#O-3276 – ne IDC [090]
Ghellinck, J de *see*
– L'essor de la litterature latine au 12th siecle
– Litterature latine au moyen age
– Patristique et moyen-ayge
Gheon, Henri *see*
– Prodigue de londres
– Quete heroique du graal
Gherardi, Alessandro *see*
– Le lettere di santa caterina de'ricci
– Nuovi documenti e studi intorno a girolamo savonarola
Gherardius, P *see* In foedvs et victoriam contra tvrcas...
De gheschiedenisse ende den doodt der vromer martelaren... / Haemstede, A C – n.p, 1559 – 5mf – 9 – mf#PBA-182 – ne IDC [240]
Ghesquiere, J *see* Acta sanctorum belgii selecta
Il ghetto di mantova / Carnevali, Luigi – Mantova, Italy. 1884 – 1r – us UF Libraries [939]
Il ghetto di mantova / Carnevali, Luigi – Mantova, Italy. 1884 – 1r – us UF Libraries [939]
Il ghetto di roma / Natali, Ettore – Roma: Stab tip della Tribuna, 1887- (mf ed 1995) – 1r – 1 – mf#ZZ-34373 – us NY Public [241]
Der ghetto und die juden in rom / Gregorovius, Ferdinand – Berlin: Im Schocken Verlag, 1935 (mf ed 1995) – (= ser Buecherei des schocken verlags) – 1r – 1 – (incl bibl ref) – mf#ZZ-34373 – us NY Public [939]
Gheyn, J *see* Maniement d'armes...
Ghiano, Juan Carlos *see* Testimonio de la novela argentina
Ghiraldo, Alberto *see*
– Antologia americana
– Archivo de ruben dario
Ghirardi, Alfred A *see* Radio trouble-shooter's handbook
Ghirardini, G *see* Relation du voyage fait...la chine sur le vaisseau l'amphitrite, en l'annee 1698
Ghirlanda di varii fiori : overo, intavolatvra di ghitarra spagnvola, doue che da se stesso ciascuno potra imparare con grandissima facilita, e breuita / Abbatessa, Giovanni Battista. – In Milano, appresso Lodouico Monza [1650?] [mf ed 1988] – 1r – 1 – mf#pres. film 35 – us Sibley [780]
Ghisi, Andrea *see* Laberinto dato novamente in luce dal clarissimo signor andrea ghisi
Ghisletti, Louis V *see* Mwiskas
Ghitzis, Moisey *see* Mame erd

GIDEON

Ghond, the hunter / Mukerji, Dhan Gopal – New York: EP Dutton & Co, 1928 – (= ser Samp: indian books) – (ill by boris artzybasheff) – us CRL [954]
Ghose, Aurobindo *see*
– Baji prabhou
– Bal gangadhar tilak
– Bankim-tilak-dayananda
– Bases of yoga
– The brain of india
– The century of life
– The doctrine of passive resistance
– The human cycle
– Ideal and progress
– Ideal and progress; essays
– The ideal of human unity
– The ideal of the karmayogin
– Isha upanishad
– Kalidasa
– Letters of sri aurobindo
– The life divine
– Lights on yoga
– The message of the gita
– The mother
– The national value of art
– The renaissance in india
– The riddle of this world
– Rishi bunkim chandra
– Savitri
– Six poems of sri aurobindo
– Songs to myrtilla
– The spirit and form of indian polity
– A system of national education
– Thoughts and glimpses
– The uttarapara speech of sri aurobindo ghose
– War and self-determination
– Yoga and its objects
Ghose, Bimal Comar *see* Planning for india
Ghose, Girish Chunder *see* Selections from the writings of girish chunder ghose
Ghose, Jogendra Chunder *see* The english works of raja ram mohun roy
Ghose, Manmohan *see* Songs of love and death
Ghose, Moti Lal *see* Speeches and writings
Ghose, Rashbehary *see* Speeches delivered on various occasions
Ghose, Sarat Chandra *see* Life of dr mahendra lal sircar
Ghose, Subhendu *see* Netaji bose
Ghose, Sudhindra Nath *see*
– And gazelles leaping
– The vermilion boat
Ghosh, Aurobindo *see* Views and reviews
Ghosh, D *see* Pressure of population and economic efficiency in india
Ghosh, D P *see* Designs from orissan temples
Ghosh, J *see* Higher education in bengal under british rule
Ghosh, Jajneswar *see* Samkhya and modern thought
Ghosh, Jamini Mohan *see*
– Sannyasi and fakir raiders in bengal
– The sannyasis of mymensingh
Ghosh, Jitendra Nath *see* Netaji subhas chandra
Ghosh, Jyotish Chandra *see* Bengali literature
Ghosh, Kali Charan *see* Famines in bengal, 1770-1943
Ghosh, Krishnachandra *see* An epitome of jainism
Ghosh, Manmathanath *see*
– The life of grish chunder ghose
– Selections from the writings of girish chunder ghose
Ghosh, Manomohan *see*
– A history of cambodia from the earliest times to the end of the french protectorate
– Paniniya siksa
Ghosh, Manoranjan, Rai Sahib *see* Rock-paintings
Ghosh, Nagendra Nath *see* Early history of india
Ghosh, Praphullachandra *see* India as known to ancient and mediaeval europe
Ghosh, S L *see* Urban morals in ancient india
Ghosh, Tushar Kanti *see* The bengal tragedy
Ghosha, Binaya *see* Primitive indian architecture
Ghosha, Ramachandra *see* History of hindu civilisation
Ghosha, Sisirakumara *see* Pictures of indian life
Ghoshal, Sarat Chandra *see* Pariksamukham
Ghoshal, Subodh Krishna *see* Sarkarism
Ghoshal, Upendra Nath *see*
– The beginnings of indian historiography and other essays
– A history of hindu political theories
Ghost – La Salle IL 1796 – 1 – mf#4719 – us UMI ProQuest [071]
Ghost at noon / Moravia, Alberto – New York, NY. 1955 – 1r – us UF Libraries [025]
Ghost buildings / Zimmerman, Louis P – s.l, s.l? 1936 – 1r – us UF Libraries [978]
Ghost dance – East Lansing MI 1968-83 – 1,5,9 – ISSN: 0016-9633 – mf#8322 – us UMI ProQuest [810]
Ghost land : or, researches into the mysteries of occultism. illustrated in a series of autobiographical sketches / Britten, William; ed by Britten, Emma Hardinge – Chicago, IL: Progressive Thinker Pub House, 1897 – 1mf – 9 – 0-524-01169-9 – mf#1990-2245 – us ATLA [130]
Ghosts and their relations : pen and ink sketches of men and noted places, tales, essays, etc etc / Clark, Daniel – Toronto: W Warwick, 1874 – 4mf – 9 – (incl ind) – mf#11858 – cn Canadiana [130]
Ghulam Husain Khan, Tabatabai *see* A translation of seir mutaqherin
Ghun Srun *see* Siavbhau jamdan bhag sneha
Ghurye, Govind Sadashiv *see*
– The aborigines – "so called" – and their future
– Caste and race in india
– Indian costumes
– Indian sadhus
– Occidental civilization
– Race relations in negro africa
GI news / Vietnam Veterans Against the War/ Winter Soldier Organization – 1974 apr, aug-1975 jul – 1r – 1 – mf#1111799 – us WHS [355]
GI organizer – 1969 apr 2-aug 5 – 1r – 1 – mf#892642 – us WHS [355]
GI press service / Student Mobilization Committee to End the War in Vietnam – 1969 jun 26-1971 sep – 1r – 1 – mf#766212 – us WHS [355]
Gi stories of the ground, air and service forces in the european theater of operations – 53 unit histories – 1 – us NY Public [940]
GI voice – 1969 mar – 1r – 1 – mf#903275 – us WHS [355]
Gia – 1969 nov – 1r – 1 – mf#901075 – us WHS [071]
Gia dinh ma bay : tieu thuyet / Phan, Tu – Ha-noi: Van Hoc 1975 [mf ed 1992] – on pt of 1r – mf#11052 r45 n1 – us Cornell [480]
Gia-dinh bao – Saigon. 1865, 1872, 1874-76, 1893, 1895, 1897-1909. LO. Per. 552 – 1 – fr ACRPP [959]
Giai pha m xuan at mao 1975 suong nguyet anh / Truong Nu- Trung Hoc Suong Nguyet anh – [Saigon: Suong Nguyet anh, 1975] [mf ed 1993] – on pt of 1r – 1 – mf#11052 r478 n12 – us Cornell [959]
Giai phong : co quan cu mat tran dan toc giai phong mien nam viet nam – [Hanoi: s.n, sep 1975-jun 1976] – 1 – us CRL [079]
Giambelluca, Christopher *see* Reducing anterior shear during knee extension
Giamberandini, G *see* La consacrazione eucharistica nella chiesa copta
Giang son – Ha Noi: [s.n. jan 1 1953-jan 31 1954, mar 1-jun 30 1954 – 3r – 1 – mf#mf-5516 seam – us CRL [079]
Giang-son viet-nam, day non-nuo'c ninh thuan / Nguyen, Di'nh Tu – Saigon: So'ng Mo'i 1974 [mf ed 1976] – 1r – 1 – us L of C Photodup [959]
Giang-van de-tam : ban van-chuong va khoa-hoc / Thuan Phong – Saigon: a-Chau 1958 [mf ed 1992] – on pt of 1r – 1 – mf#11052 r11 n10 – us Cornell [959]
Giannini, Amedeo *see* Lo stato giuridico della gente dell'aria, diritto internazionale ed interno
Giannotti, Donati *see* Opere politiche e letterarie.
The giant cities of bashan : and syria's holy places / Porter, Josias Leslie – New York: Thomas Nelson, 1884. Beltsville, Md: NCR Corp, 1978 (5mf); Evanston: American Theol Lib Assoc, 1984 (5mf) – 9 – 0-8370-0883-2 – (incl ind) – mf#1984-4184 – us ATLA [915]
The giant judge : or, the story of samson, the hebrew hercules / Scott, William Anderson – San Francisco: Whitton, Towne, 1858 – 1mf – 9 – 0-524-05633-1 – mf#1992-0488 – us ATLA [220]
Giard, Gilles *see* Rapport du comite profil sur les enquetes sociologiques aupres des etudiants du secondaire de la region et aupres des etudiants et etudiantes du collegial du college de l'assomption, hiver 1986
Il giardiniere avviato nellesercizio della sua professione dal cav... – Milano, 1812. 2v – 9mf – 9 – mf#GDI-22 – ne IDC [710]
Il giardiniere avviato nell'esercizio della sua professione dal cav f re / Re, F – Milano, 1812. 2v – 9mf – 9 – mf#GDI-22 – ne IDC [700]
Il giardiniero francese, ovvero trattato del tagliare gl'alberi... / Dahauron, R – Venetia, 1723 – 55p 2mf – 9 – mf#GDI-7 – ne IDC [710]
Giauque, Florien *see*
– The laws relating to roads and ditches, bridges and watercourses in the state of ohio.
– A manual for guardians and trustees of minors, insane persons, imbeciles, idiots, drunkards, and for guardians ad litem, resident and non-resident, affected by the laws of ohio
– A manual for notaries public, general conveyancers, commissioners, justices, mayors, consuls.
– The settlement of estates of deceased persons, including the subjects of wills, executors, administrators, testamentary trustees...with such estates in ohio
Giavi *see* De la carpa a la gloria
Gibb, Elias John Wilkinson et al *see* The sacred books and early literature of the east, vol 6
Gibb, H A R *see*
– The arab conquests in central asia
Gibb, H O *see* "Torheit" und "raetsel" im neuen testament
Gibb, Heather *see* The relationship of selected health screenings to elementary teachers' attitude and intent to teach cardiovascular education
Gibb, John *see* The confessions of augustine
Gibbes, James Shoolbred *see* Letterbooks
Gibbes, Lewis R *see* Papers
Gibbes, Robert Wilson *see* Cuba for invalids
Gibbings, Richard *see* Report of the trial and martyrdom of pietro carnesecchi
Gibbins, Henry de Beltgens *see* British commerce and colonies
Gibbon / Morison, James Augustus Cotter – New York, NY. 1901 – 1r – us UF Libraries [025]
Gibbon / Morison, James Cotter – New York: Harper 1879 [mf ed 1990] – (= ser English men of letters (new york, ny)) – 1mf – 9 – 0-7905-5954-4 – mf#1988-1954 – us ATLA [930]
Gibbon, Edward *see*
– Autobiography of edward gibbon
– History of christianity
The gibbon gazette – Gibbon, NE: C Putnam. v1 n1. oct 18 1900- (wkly) [mf ed oct 18, nov 22 1900 filmed [1985] – 1r – 1 – us NE Hist [071]
Gibbon, Guy E *see* The mississippian occupation of the red wing area
Gibbon, Perceval *see* Vrouw grobelaar's leading cases
Gibbon reporter – Gibbon, NE: W H Carson (wkly) [mf ed v6 n40. mar 12 1896-1912, 1914 (gaps)] – 1 – (cont: gibbon reporter and farmers' alliance advocate) – us NE Hist [071]
Gibbon reporter and farmers' alliance advocate – Gibbon, NE: S Watson, 1890 (wkly) [mf ed jun 23, dec 8 1892] – 1r – 1 – (cont by: gibbon reporter) – us NE Hist [071]
Gibbons, David et al *see* The metropolitan building act
Gibbons, James *see*
– The causes and cure of unbelief
– The faith of our fathers
– Our christian heritage
Gibbons Mortuary (GSU), Coffey County, KS *see* Funeral records
Gibbons, Ruth E G *see* A prismatic approach to the analysis of style in dance
Gibbons, Simon *see* Baptism, the sacrament of regeneration
Gibbons, Thomas *see* Memoirs of eminently pious women
The gibbs archive : the papers of antony gibbs and sons, 1744-1953 – 319r – 1 – £16,995.00 – (complete archive of this famous london merchant and banking house includes substantial source material related to latin america and australia. reels may be purchased separately) – mf#GAR – uk World [380]
Gibbs, E Nathan *see* Career mobility patterns of head coaches in the national basketball association
Gibbs, Edward J *see* England and south africa
Gibbs, Ellen *see* The bible references of john ruskin
Gibbs, Frederick Waymouth *see* English law and Irish tenure
Gibbs, George *see* The judicial chronicle, being a list of the judges of the courts of common law and chancery in england and america and of the contemporary reports
Gibbs, Henry *see*
– Twilight in south africa
Gibbs, John *see* Designs for gothic ornaments and furniture
Gibbs, Joseph *see* Confirmation sermon
Gibbs, Josiah Willard *see* Philological studies with english illustrations
Gibbs, Mary *see* The bible references of john ruskin
Gibbs memorial baptist church : articles of faith, church directories, church histories, church minutes 150=bostwick, ga. 1902-1961 – 1 – $27.54 – mf#6805 – us Southern Baptist [242]
Gibbs, Nanette [comp] *see* American biographical archive (aba1)
Gibbs, Peter *see* Avalanche in central africa
Gibbs, Philip *see* Wings of adventure and other little novels
Gibbs, Philip Hamilton *see* Founders of the empire
Gibbs, W H *see* No interest for money, except to the government, then not to exceed 3 percent
Gibbs, William *see*
– The handbook of architectural ornament...
– The universal decorator
Gibbud, H B *see* [Sermons on christian life]
Gibernau, Jose *see*
– Solution for the iberian puzzle
– Spain and the world
Giberne, Agnes *see* Val and his friends
Giberne, C C *see* Portraits of schoolgirls and other persons in south india
Gibier, abbe *see* Le catholicisme dans les temps modernes
Gibner, N P *see* Sistema kooperatsii
Gibraltar *see*
– Gibraltar chronicle and official gazette
– Statistical blue books 1828-1947
Gibraltar chronicle – Gibraltar. -w. Jan 1944-June 1945; Jan 1948-June 1949; Jan 1950-Dec 1972. 32 1 2 reels – 1 – uk British Libr Newspaper [072]
Gibraltar chronicle and official gazette / Gibraltar – no. 27344-32231. 1926-41. nos. 31193-31218, 31659-31687, 31788-31869, 31926-32164 And Other Scattered numbers missing – 1 – us L of C Photodup [324]
Gibraltar e olivenoa / Veiga, Estacio – 1863 – 9 – sp Bibl Santa Ana [946]
Gibraltar post – Gibraltar. -w. Jan 1958-Dec 1968. 4 reels – 1 – uk British Libr Newspaper [072]
Gibson, Daniel B *see* The lord's supper
Gibson, Edgar C S *see* The book of job
Gibson, Edgar Charles Sumner *see*
– Messages from the old testament
– The old testament in the new
– The three creeds
Gibson, Edmund [comp] *see* A preservative against popery in several select discourses upon the principal heads of controversy between protestants and papists
Gibson, Emily Cooper *see* Study of the major organ works of paul hindemith
Gibson, J D *see* Addresses
Gibson, James *see*
– Christian sabbath
– Plain but friendly remonstrance, addressed to the glasgow memomoria
– Poor in the land
– Poor man's enemies exposed
– Principle of voluntary churches and not the principle of an establi...
– Remarks on the speech of ac dick
– Sermon preached in the parish church of mildenhall, suffolk
Gibson, Jeremy Sumner Wycherley [comp] *see* Census returns, 1841-1881, on microfilm
Gibson, John *see*
– The botany of the eastern coast of lake huron
– Infant baptism a true sacrament
– Lord's supper
– Reasons for not joining the primitive methodists
Gibson, John Campbell *see* Mission problems and mission methods in south china
Gibson, John E *see* Nonlinear automatic control
Gibson, John Monro *see*
– The ages before moses
– Christianity according to christ
– The devotional use of the holy scriptures
– The foundations
– From fact to faith
– The gospel of st matthew
– The mosaic era
– Protestant principles
– Rock versus sand
– The unity and symmetry of the bible
Gibson, Margaret Dunlop *see*
– Apocrypha arabica
– Apocrypha sinaitica
– An arabic version of the acts of the apostles and the seven catholic epistles
– An arabic version of the epistles of st paul to the romans, corinthians, galatians
– How the codex was found
– The palestinian syriac lectionary of the gospels
– Palestinian syriac texts from palimpsest fragments in the taylor-schechter collection
– Studia sinaitica 2
Gibson, Margaret Dunlop [comp] *see* Catalogue of the arabic mss in the convent of s catharine on mount sinai
Gibson, Oscar Lee *see* Analytical study of the timbre of the clarinet
Gibson, William *see*
– The abbe de lamennais and the liberal catholic movement in france
– The year of grace
Gibson, William Ralph Boyce *see*
– God with us
– A philosophical introduction to ethics
– The problem of logic
– Rudolf eucken's philosophy of life
GI-Civilian Alliance for Peace *see* Counterpoint
Gicovate, Bernard *see* Conceptos fundamentales de literatura comparada
Giddings deutsches wochenblatt – Giddings, TX (USA), 1921 15-19 dec, 1923-24, 1926 5 aug-1933 21 dec, 1935-1938 22 dec [gaps] – 6r – 1 – gw Misc Inst [071]
Giddings, Edward Jonathan *see* American christian rulers
Giddins, Kevin J *see* Influencing a broader understanding of jazz dance
Gide, Andre *see* Theseus
Gide, Andre et al *see* Rainer maria rilke (1875-1926)
Gide, Andre Paul Guillaume *see* Lettres a angele
Gideon and the judges : a study, historical and practical / Lang, John Marshall – New York: Fleming H Revell, [189-?] – 1mf – 9 – 0-8370-9961-7 – (incl bibl ref) – mf#1986-3961 – us ATLA [221]

GIDEON'S

Gideon's faith / Wilson, William – London, England. 1833 – 1r – us UF Libraries [240]
Gidik – Istanbul: Selanik Matbassi. Sahib-i Imtiyaz: H Seyfeddin; Mueduer-i Mes'ul: M Hazim, Arif Hikmet. n1. 21 tesrinievvel 1326 [1910] – (= ser O & t journals) – 1mf – 9 – $25.00 – us MEDOC [956]
Gidney, William Thomas *see*
– The history of the london society for promoting christianity amongst the jews
– The jews and their evangelization
Gidney, Williams Thomas *see* At home and abroad
Gidra – v1-6 n3 apr 1969-mar 1974 – 1r – 1 – mf#1057212 – us WHS [071]
Gidrogeologiia sssr / ed by Sidorenko, A V – Moskva: Nedra, [1966-] – us CRL [550]
'n Gids tot die wet op gesondheid, wet n63 van 1977. – Pretoria: Govt Printer [1977?] [mf ed Pretoria, RSA: State Library [199-]] – 52p on 1r with other items – 5 – mf#op 09464 r23 – us CRL [344]
De gids voor indonesie – Djakarta, 1952-1953 – 8mf – 9 – mf#SE-1493 – ne IDC [959]
Gidulianov, P V *see* Otdelenie tserkvi ot gosudarstva v sssr
Gidumal, Dayaram *see* The status of woman in india
Gieben, Joseph *see* Christian dietrich grabbe in der nachschillerischen entwickelung
Giefang rhbao *see*
– Jie fang ri bao
Gierach, Erich *see* Das maerterbuch
Gierke, Otto Friedrich von *see*
– Deutsches privatrecht
– Der entwurf eines buergerlichen gesetzbuchs und das deutsche recht
– Personengemeinschaften und vermoegenseinbegriffe in dem entwurfe eines buergerlichen gesetzbuches fuer das deutsche reich
– Political theories of the middle age
Gierloff-Emden, Hans Gunter *see* Kuste von el salvador
Giertych, Jedrzej *see* Tragizm losow polski
Gies, L *see* Elten, land und leute
Giese, Erich *see* Wie erschliessen wir unsere kolonien?
Giese, Franz *see* Frans essink
Giese, Friedrich *see*
– Altosmanischen anonymen chroniken
– Die geltenden papstwahlgesetze
– Tevarih-i al-i osman [asikpasazade tarihi]
Giese, Wilhelm *see* Pueblos romanicos y su cultura popular
Giesebrecht, Franz *see* Ein deutscher kolonialheld
Giesebrecht, Friedrich *see*
– Die alttestamentliche schaetzung des gottesnamens und ihre religionsgeschichtliche grundlage
– Beitraege zur jesaiakritik
– Die berufsbegabung der alttestamentlichen propheten
– Das buch jeremia
– Die degradationshypothese und die alttestamentliche geschichte
– Friede fuer babel und bibel
– Die geschichtlichkeit des sinaibundes
– Die grundzuege der israelitischen religionsgeschichte
– Die hebraeische praeposition lamed
– Jeremias metrik
– Der knecht jahves des deuterojesaia
– Der wendepunkt des buches hiob, capitel 27 und 28
Giesebrecht, Friedrich et al *see* The ologische studien
Giesecke, Hans Heinrich *see* Christliches erbe und lyrische gestaltung
Gieseler, Johann Carl Ludwig *see* A text-book of church history
Gieseler, Johann Karl Ludwig *see* Die protestantische kirche frankreichs von 1787 bis 1846
Giesen, Adolf *see*
– Eberhard von groote
– Die gottesbeweise bei franz brentano
Giessener allgemeine *see* Giessener freie presse
Giessener anzeiger *see* Anzeigeblatt fuer die stadt giessen
Giessener freie presse – Giessen, Lahn DE, 1946 25 jan-1947 25 nov, 1948 5 oct-1949 6 oct – 2r – 1 – (title varies: 3 jan 1966: giessener allgemeine. filmed by misc inst: 1976 4 may- [ca 11r/yr]) – gw Misc Inst [074]
Giesserei-zeitung – Berlin DE, 1913 & 1921, 1923-26 – 4r – 1 – (filmed by bnl: 1904-15 dec 1909 [6r]) – gw Mikrofilm; uk British Libr Newspaper [670]
Giessler, Klaus-Volker *see*
– Nachlass alfred von tirpitz (bestand n 253) bd 65
– Nachlass wilhelm groener (bestand n 46) bd 30
Gietmann, Gerhard *see*
– Commentarius in ecclesiasten et canticum canticorum
– De re metrica hebraeorum
Giffard, Chris *see* "The hour of youth has struck"
Giffen, Robert *see* The progress of the working classes in the last half century
Giffoni, Maria Amalia Correa *see* O registro das dancas e folguedos populares
Gifford, Archer *see* Unison of the liturgy
Gifford, Edwin Hamilton *see*
– The incarnation
– Voices of the prophets
Gifford, John *see* Orange, a political rhapsody in three cantos
Gifford, John C *see* Living by the land
Gifford, John Clayton *see*
– Billy bowlegs and the seminole war
– Tropical subsistence homestead
Gifford, Miram Wentworth *see* Laws of the soul, or, the science of religion and the future life
Gifhorner tageszeitung – Gifhorn DE, 1907 15 dec-1916, 1917 jul-1921 jun, 1922-1931 jun, 1932-1935 29 jun – 38r – 1 – gw Misc Inst [074]
Gift from the ministry of foreign affairs and external commerce of belgium / Belgium. Ministry of Foreign Affairs – 30r – 1 – mf#T1113 – us Nat Archives [327]
Gift of god / Noel, Baptist Wriothesley – London, England. 1851 – 1r – us UF Libraries [240]
The gift of immortality : a study in responsibility / Slattery, Charles Lewis – Boston: Houghton Mifflin, 1916 – (= ser Raymond F. West Memorial Lectures on Immortality) – 1mf – 9 – 0-7905-8586-3 – mf#1989-1811 – us ATLA [240]
The gift of tongues / Hayes, Doremus Almy – New York: Methodist Book Concern, c1913 – 1mf – 9 – 0-7905-1150-9 – mf#1987-1150 – us ATLA [400]
Gift & stationery business [gsb] – New York NY 1989-95 – 1 – (cont by: giftware business) – ISSN: 0896-4092 – mf#11940.04 – us UMI ProQuest [640]
Gifted child quarterly – 1957+ – 1,5,9 – ISSN: 0016-9862 – mf#1649 – us UMI ProQuest [640]
Gifted child today – Waco TX 1986-93 – 1,5,9 – (cont: g/c/t; cont by: gifted child today magazine) – ISSN: 0892-9580 – mf#12496.03 – us UMI ProQuest [640]
Gifted child today – Waco TX 2000+ – 1,5,9 – (cont: gifted child today magazine) – mf#12496.03 – us UMI ProQuest [640]
Gifted child today magazine – Waco TX 1993-99 – 1,5,9 – (cont: gifted child today; cont by: gifted child today) – ISSN: 1076-2175 – mf#12496.03 – us UMI ProQuest [640]
Giganov, Josif *see* Slovar' rossijsko-tatarskij
Gigante e o rio / Soares, Alvaro Teixeira – Rio de Janeiro, Brazil. 1957 – 1r – us UF Libraries [972]
Gigantes y cabezudos / Luque Lobos, Jorge – Buenos Aires, Argentina. 1929 – 1r – us UF Libraries [972]
Giger, George Musgrave *see* Sermons
Gigiena i sanitarnoe delo – Pg., 1914-1916 – 49mf – 9 – mf#R-9275 – ne IDC [077]
Gigiena truda i professionalnye zabolevaniia / Ministerstvo zdravookhraneniia Soiuza SSR – Moskva: Medgiz, [1957-]. v10. 1966 – 1r – 1 – us CRL [947]
Gigline / Fort Bliss G I 's for Peace – feb 1970 p15-19 – 1r – 1 – mf#767841 – us WHS [355]
Gigmanag = Our people / Native Council of Prince Edward Island et al – 1976 jan-1984 dec, 1985 jan-1989 jul – 2r – 1 – mf#621668 – us WHS [305]
Gignac, Francoise *see* Bibliographie analytique de mademoiselle simone pare
Gignoux, Claude-Joseph *see* Bourges pendant la guerre
Gigot, Francis Ernest *see*
– Biblical lectures
– Christ's teaching concerning divorce in the new testament
– Didactic books and prophetical writings
– General introduction to the study of the holy scriptures
– The historical books
– Outlines of jewish history
– Outlines of new testament history
Gihi pratipatti bises bistar / Brah Silasamvara [Ras] – Bhnam Ben: Pannagar Gim Sen, 2509 [1966] [mf ed 1990] – 1r with other items – 1 – (in khmer) – mf#mf-10289 seam reel 113/8 [§] – us CRL [170]
Gihr, Nikolaus *see* Die sequenzen des roemischen messbuches
Gijig-anang mekateokonaie, s j o gagikwewinan / Artus, Gaston Andre – Tours France: Impr A Mame, 1898 – 3mf – 9 – mf#00033 – cn Canadiana [490]
Gil Ayuso, F *see* Noticias bibliograficas de textos y disposiones legales de los reinos de castilla...
Gil Becerra, Benito *see*
– Asserta theo-subtitulia...efficacia
– Ave maria
Gil becerra, benito, en la inmaculada en la literatura franciscano-espanola / Uribe, Angel – Archivo Ibero Americano, 1955 – 1 – sp Bibl Santa Ana [240]
Gil blas – Paris. 30 may 1891-23 sep 1898; 6 jan-20 jan 1899 [wkly] – 4r – 1 – (aka: gil blas illustre) – uk British Libr Newspaper [074]
Gil blas – Madrid, Spain. -w. 3 Nov 1864-29 Sept 1872. 3 reels – 1 – uk British Libr Newspaper [074]
Gil blas – Paris. nov 1879-4 aout 1914, 20 janv, 9 juil, 31 dec 1921, 1er juil 1922, juin 1931, 8 nov 1937-12 janv 1938 – 1 – fr ACRPP [073]
Le gil blas : jornal politique, satyrique et artistique – Rio de Janeiro, RJ: Typ da Gazeta de Noticias, 14 out 1877-01 set 1878 – (= ser Ps 19) – mf#P19A,04,146 – bl Biblioteca [320]
Le gil blas *see* Le messager du bresil
Gil blas illustre – Paris, France. -w. 28 June 1891-25 Dec 1896. 3 reels – 1 – uk British Libr Newspaper [072]
Gil blas illustre : supplement – Paris. mai 1891-aout 1903 – 1 – fr ACRPP [073]
Gil blas illustre *see* Gil blas
Gil Calvo, Joaquin *see* Geognosia. componentes de la corteza terrestre
Gil de Godoy, Juan *see* El mejor guzman de los buenos
Gil De Rubio, Victor M *see*
– Matices
– Perfiles
– Redobles
Gil, F *see* Disertacion fisico-medica..., para preservar...de viruela
Gil Farres, Octavio *see*
– Extension urbana de la merida romana
– Lucernas romanas decoradas del museo emeritense
Gil Fortoul, Jose *see*
– Humo de mi pipa
– Paginas de ayer
Gil Garcia, Bonifacio *see*
– Cancionero popular de extremadura. contribucion al folklore musical de la region
– El canto de relacion en el folklore infantil de extremadura
– Las flores en la tradicion extremena
– Folklore extremeno. extremadura y la posible regionalizacion de su musica popular. la tradicion en la cancion extremena y su evolucion
– Hallazgo de veintiocho canciones populares de extremadura, recogidas en los anos 1884-85
– El pajarillo de la tradicion extremena
Gil Julian, Juan *see* Pergaminos del museo arqueologico provincial de badajoz
Gil Ramirez, Jose *see* Novena de sta quiteria
Gil Sanz, J *see* El triumpho vindicado de la calumnia, impostura e ignorancia contra la medicina...
Gil y de Pina, J *see* Tratado breve de la curacion del garrotillo
Gila River Indian Community [Arizona: Association] *see* Pima maricopa echo
Gilabert, A G *see* Durruti un anarquista integro
Gilbacher sonntags-blatt – Juechen DE, 1913 apr-1914 2 aug – 1r – 1 – (with suppl: thomas a kempis 1913 apr-dec [gaps]; covers kempen [1r]) – gw Misc Inst [074]
Gilbert and ellice islands colony advisory council : minutes of meetings – 1963-67 – 1r – 1 – mf#pmb doc25 – at Pacific Mss [980]
Gilbert and ellice islands colony advisory council : proceedings of the colony conferences – 1956-62 – 1r – 1 – mf#pmb doc26 – at Pacific Mss [980]
Gilbert and sullivan, part 1 : the correspondence, diaries, literary manuscripts & prompt copies of w s gilbert (1836-1911) from the british library, london – [mf ed Marlborough 2003] – 19r – 1 – £1800.00 – uk Matthew [790]
Gilbert, Ashhurst Turner *see* Pictorial crucifixes
Gilbert, Ashurst Turner *see* Commandment of god made of none effect by the traditions of men
Gilbert, Carole M *see* Journal of hospital librarianship
Gilbert, D H *see* Florida
Gilbert, David W *see* Memoranda and notebooks
Gilbert, Dorie J *see* Journal of hiv/aids and social services
Gilbert, E W *see* Piety honored after death...
Gilbert, Frank *see* Railway law in illinois
Gilbert, Frank G *see* Street railway reports, annotated
Gilbert, G T *see* Chartularies of st mary's abbey, dublin (rs80)
Gilbert, George Holley *see*
– The first interpreters of jesus
– Interpretation of the bible
– The poetry of job
– A primer of the christian religion
– The revelation of jesus
– A short history of christianity in the apostolic age
– The student's life of jesus
– The student's life of paul
Gilbert, Glen Alexander *see*
– Aviacion civil en el salvador
– Civil aviation in el salvador
Gilbert h. grosvenor collection of photographs of the alexander graham bell family / U.S. Library of Congress. Prints and Photographs Division – 28,000 images. 3 reels. P&P11533 – 1 – $23.00r – us L of C Photodup [770]
Gilbert herald – Gilbert MN. 1922 jun 23-1924 oct 30, 1924 jul 4-1925 oct 30 – 1r – 1 – mf#770734 – us WHS [071]
Gilbert, Hubert E *see* Landsknechte
Gilbert, J T *see* Register of the abbey of st thomas the martyr, dublin (rs94)
Gilbert, James Stanley *see* Panama patchwork
Gilbert, Jesse Samuel *see* Blessed are they
Gilbert, John Th *see* Historic and municipal documents, ireland, ad 1172-1320 (rs53)
Gilbert, John Thomas *see* Calendar of ancient records of dublin
Gilbert, Joseph *see* The christian atonement, its basis, nature, and bearings
Gilbert, Juliet Francis *see* An investigation of internet usage among a group of professionals in south africa
Gilbert, Kathleen R *see*
– Gender differences regarding knowledge of child health and development among high school students
– The relationship between identity status and contraceptive practices of college students
Gilbert, Nathaniel *see* Forbidden tree
Gilbert, Otto *see* Griechische religionsphilosophie
Gilbert, Pamela *see* British museum entomological literature, 1800-1864
Gilbert, Paul James *see* The king's greatest business
Gilbert, Paul S ET A L *see* Beginning somali history
[Gilbert-] record – NV. 1925-27 [wkly] – 1r – 1 – $60.00 – mf#U04541 – us Library Micro [071]
Gilbert, Stephen *see* Cleveland, ohio, taxes, ms v.f. o
Gilbert, T *see* Voyage from new south wales to canton, in the year 1788
Gilbert, W S *see* Best known works of w s gilbert
Gilbertese myths, legends and oral traditions / Grimble, Arthur – n.d. – 1r – mf#pmb69 – at Pacific Mss [390]
Gilberti, Maturino *see* Thesoro spiritual de pobres en le[n]gua de michuaca[n]
The gilbertine rite, vol 1-2 (hbs59-60) / Woolley, R M – 1921-1923 – (= ser Henry bradshaw society (hbs)) – 2v on 9mf – 8 – €18.00 – ne Slangenburg [241]
Gilberto freyre / Meneses, Diogo De Melo – Rio de Janeiro, Brazil. 1944 – 1r – us UF Libraries [972]
Gilberto freyre : sua ciencia, sua filosofia, sua ar... – Rio de Janeiro, Brazil. 1962 – 1r – us UF Libraries [972]
Gilbey, Geoffrey *see* Saturday smile
Gilbey, Walter. 1st Baronet *see*
– Animal painters of england from the year 1650
– Life of george stubbs
Gilbreath, Frank A *see* Story of old fort myers
Gilburth, Kenneth Riley *see* An investigation of the process of change in the major contemporary schools of psychotherapy
Gilchrist, Alexander *see*
– Life of william blake
– Life of william etty, r a
Gilchrist, Beth Bradford *see* The life of mary lyon
Gilchrist county journal – Trenton, FL. 1934-1997 – 46r – (gaps) – us UF Libraries [071]
Gilchrist, John *see* The general east india guide and vade mecum
Gilchrist, John Borthwick *see*
– The hindee-roman orthoepigraphical ultimatum
– The strangers east indian guide to the hindoostanee
Gilchrist, Robert Niven *see*
– Indian nationality
– The separation of executive and judicial functions
Gilcrease gazette / Thomas Gilcrease Institute of American History and Art – v4 n2-v16 n3, 1967 may-1978 may – 1r – 1 – mf#483328 – us WHS [975]
Gildea, George Robert *see*
– Reproductive relief spinning in the west of ireland
Gildener hon / Weinper, Zishe – New York, NY. 1927 – 1r – us UF Libraries [939]
Gilder, William *see* In eis und schnee
Gildersleeve, Basil L *see* Latin exercise-book
Gilead baptist church. hettick, illinois : church records – 1869-Jun 1986. History, 1869-1969. 1752p – 1 – 78.84 – us Southern Baptist [242]
Gilead baptist church. union county. johnsville, south carolina : church records – 1838-Feb 1953 – 1 – us Southern Baptist [242]
Gilead, Zerubavel *see* Pirke palmah (mi-pi lohamim)
Giles, Chauncey *see* The true and false theory of evolution
Giles County Historical Society *see* Bulletin of the giles county...

Giles, Edward Bowyer *see* The history of the art of cutting in england
Giles, Henry *see*
– Christian view of retribution hereafter
– Creeds
– Lectures and essays on irish and other subjects
– Man, the image of god
Giles, Herbert Allen *see*
– China and the chinese
– Confucianism and its rivals
– Freemasonry in china
– Gems of chinese literature
– A glossary of reference on subjects connected with the far east
– Glossary of reference on subjects connected with the far east
– History of chinese literature
– Religions of ancient china
Giles, Herbert Allen et al *see* Great religions of the world
Giles, John Allen *see* Heathen records to the jewish scripture history
Giles, Lionel *see* Christians at chen-chiang fu
Giles of Assisi *see* Dicta beati aegidii assisiensis
Giles of rome on boethius : "diversum est esse et id quod est" / Nash, P W – Toronto, 1950 – 1mf – 8 – €3.00 – ne Slangenburg [180]
Giles, Scott L *see* An investigation of the career mobility patterns of ncaa division 1-a head football coaches
Gileston, st giles; and st athan, st tathan, monumental inscriptions – 1mf – 9 – £1.25 – uk Glamorgan FHS [929]
Gilfield baptist church, petersburg, virginia, 1803-1903 / Kennard, Richard – 1 – 5.00 – us Southern Baptist [242]
Gilfillan, John Alexander *see* Singing-schools in america
Gilfillan, Samuel *see*
– Essay on brotherly love
– Essay on the sanctification of the lord's day
Das gilgamesch-epos – Goettingen: Vandenhoeck & Ruprecht, 1911 – (= ser Forschungen zur religion und literatur des alten und neuen testaments) – 1mf – 9 – 0-7905-2525-9 – mf#1987-2525 – us ATLA [470]
Gilgamesh *see* Das babylonische nimrodepos
Gilgandra weekly – Gilgandra, jan 1969-dec 1992 – 26r – at Pascoe [079]
Gilhodes, C *see* The kachins
Giliarov-Platonov, N P *see*
– Iz perezhitogo
– Universitetskii vopros
– Voprosy very i tserkvi
Giliberto, Vincenzo *see* La citt...d'iddio incarnato
Gilii, Filippo Salvadore *see* Ensayo de historia americana
Giliomee, Yolande *see* Teenagers interviewing problems
Gilkey *see* Ohio hundred year book, 1787-1901
Gilkey, Langdon *see* Reaping the whirlwind
Gill 1719-1849 – Oxford, MA (mf ed 1996) – (= ser Massachusetts vital record transcripts to 1850) – 4mf – 9 – 0-87623-253-5 – (mf 1t-2t: births & deaths 1719-1845. mf 2t: marriages & intentions 1794-1843. mf 3t: intentions 1795-1805, 1837-49; out-of-town marriages 1794-98. mf 4t: vital records 1843-49) – us Archive [978]
Gill 1754-1895 – Oxford, MA (mf ed 1987) – (= ser Massachusetts vital records) – 19mf – 9 – 0-87623-057-5 – (mf 1-4: births & deaths 1754-1810. mf 5-7: births & deaths 1804-44. mf 8-9: town records 1793-1824. mf 10-11: index to births 1843-1930. mf 12-13: index to marriages 1843-1930. mf 14: index to deaths 1843-1930. mf 15-16: b,m,d 1843-63. mf 17-19: b,m,d 1860-92) – us Archive [978]
Gill and johnson's law reports / Maryland. Court of Appeals – v1-12. 1829-43 (all publ) – (= ser Maryland supreme court reports) – 24mf – 9 – $108.00 – (a pre-nrs title) – mf#LLMC 84-149 – us LLMC [347]
Gill, Charles *see*
– Le cap eternite
– Les soirees du chateau de ramezay
Gill, Diane L *see*
– Effect of an active attentional strategy on running economy of low economical runners
– Gender differences in the relationships among self-confidence, gender-appropriateness, and value
– An investigation of self-efficacy and control theory with elite distance runners
– Psychological and physiological changes associated with a period of increased training
Gill, Everett *see* Protestants of the east
Gill, George *see* The oxford and cambridge geography
Gill, John *see*
– The cause of god and truth
– An exposition of the book of solomon's song
– Notices of the jews
Gill, Stephen Romney *see*
– Letters
Gill, W W *see*
– Historical sketches of savage life in polynesia
– Life in the southern isles
– Work and adventure in new guinea 1877 to 1885
Gill, William Hugh *see*
– Esther
– The incarnate word
Gill, William Icrin *see*
– Christian conception and experience
– Evolution and progress
Gill, William Wyatt *see*
– Life in the southern isles
– Myths and songs from the south pacific
Gille, Frank *see* Spaetblutungen nach tonsillektomie
Gille, Hans *see* Die gedichte des michel beheim
Gille, Hans Hermann Karl *see* Das neue deutschland im gedicht
Gille, Philippe *see* Charbonniers
Gill[e/i]land researcher's aide – v1 n1-v5 n4 [1981/1982 1st iss-1986/1987 4th iss] – 1r – 1 – mf#1832810 – us WHS [071]
Gillen, Francis James *see* The native tribes of central australia
Gillentine, John A *see* A comparison of the sportsmanship attitudes and/or moral reasoning of interscholastic coaches
Gilles, P *see*
– De bosporo thracio libri 3
– De topographia constantinopoleos, et de illvis antiqvitatibvs libri qvatvor
Gillespie, Alexander *see* Gleanings and remarks, collected during many months of residence at buenos ayres, and within the upper country
Gillespie, Charles George Knox *see* The sanitary code of the pentateuch
Gillespie, David F *see* Journal of social service research
Gillespie, William *see* Rebellion of absalom
Gillespie, William Honyman *see* The argument
Gillet, Joseph E *see*
– Propalladia and other works of...
Gillet, Joseph H E *see* Propalladia and other works of...
Gillet, Martin Stanislaus *see* L'education du caractere
Gillett, Ezra Hall *see*
– Ancient cities and empires
– England two hundred years ago
– God in human thought
– History and literature of the unitarian controversy
– History of the presbyterian church in the united states of america
– The life and times of john huss
Gillett forum *see* Miscellaneous newspapers of teller county
Gillett times – Gillett WI. 1918 may 18, jun 1, 1926 jul 29, 1927 jan 6-1928 dec 20, 1929 jan 3-1930 may 29, 1930 jun 5-1930 jul 31, 1931 jan 1-1932 sep 29, 1932 oct 6-1934 mar 29, 1934 apr 5-1935 jun 27, 1936 jan 2-1936 may 28, 1936 jun 4-1939 mar 23, 1939 apr 6-1940 jun 27, 1941 jan 2-1941 dec 18, 1943 jan 7-1943 aug 26, 1943 nov 4-1945 dec 27, 1943 sep 2-1944 dec 28, 1946-48, 1949-51 – 12r – 1 – (continued by: oconto falls herald (oconto falls, wi: 1920); oconto county times-herald) – mf#944157 – us WHS [071]
Gillette, Abram Dunn *see* Minutes of the philadelphia baptist association from a.d. 1707 to a.d. 1807
Gillette, Charles J *see* Perceptions of discrimination in athletic training education programs
Gillhoff, Johannes *see* Juernjakob swehn
Gillies, Archibald *see* Secularist man-trap
Gillies, Hugh Cameron *see* Elements of gaelic grammar
Gillies, James Robertson *see* Jeremiah
Gillies, Samuel *see* Lecture on the rise, institution, object, and progress of the princ...
Gillin, John Lewis *see* The dunkers
Gillin, John Philip *see* San luis jilotepeque
Gillingham, James *see* Eight days with the spiritualists
Gilliodts-Van Severen, Louis *see* Cartulaire de l'ancien consulat d'espagne a bruges
Gillis, James Donald *see* The great election
Gillisonville baptist church (called coosawhatchie, 1832-85). jasper county. south carolina : church records – 1873-98; 1950-81 – 1 – us Southern Baptist [242]
Gillmore, Parker *see*
– All round the world
– A hunter's adventures in the great west
– Lone life
– Lone life, vol 1
– Lone life, vol 2
– Prairie and forest
Gillot, Hubert *see* Denis diderot
Gillow, Joseph *see* The haydock papers
Gillow, Thomas *see* Catholic principles of allegiance illustrated
The gillows' archive : 'patterns of elegance' – 103r (1 col) – 1 – £4300.00 – mf#GIL – uk World [640]
Gill's law reports / Maryland. Court of Appeals – v1-9. 1843-51 (all publ) – (= ser Maryland supreme court reports) – 18mf – 9 – $81.00 – (a pre-nrs title) – mf#LLMC 84-150 – us LLMC [347]
Gilly, D *see* Handbuch der land-bau-kunst
Gilly, William Stephen *see*
– Hora catechetica
– Our protestant forefathers
– Vigilantius and his times
Gilm, Hermann von *see*
– Gedichte
– Hermann von gilms familien – und freundesbriefe
Gilman, Arthur *see*
– A library of religious poetry
– The story of the saracens
Gilman, Charlotte Perkins *see* Women and economics; a study of the economic relation between men and women as a factor in social evolution
Gilman, Gorham Dummer *see* Journal of a canoe voyage along the kauai palis, made in 1845
Gilman herald – Gilman WI. 1943 dec 10-1946, 1947-63,1964-1965 sep 17 – 7r – 1 – mf#944141 – us WHS [071]
Gilman, Samuel C *see* The conquest of the sioux
Gilmantown, New Hampshire. Gilmantown Baptist Church *see* Records
Gilmartin, Aron Seymour *see* Some collegations [sic] of the unitarian and ethical culture movements in america
Gilmer first baptist church. gilmer, texas : church records – 1899-1988 – 2r – 1 – $133.56 – (2,968p) – us Southern Baptist [242]
Gilmer first baptist church. gilmer, texas : church records – Newsletter, Vols 1-17, 19-35, Apr 1955-1989 – 3r – 1 – $163.35 – (3,630p) – us Southern Baptist [242]
Gilmore, Andrew *see* The historic garrison at annapolis royal, n s
Gilmore, Carole A *see* An examination of the academic performance of student-athlete admission exceptions at a divison 1-a institution
Gilmore, David Chandler *see* The end of the law
Gilmore, Eugene Allen *see* Handbook on the law of partnership, including limited partnerships
Gilmore, Gary D *see* The impact of the la crosse wellness project on the health promotion involvement of college students residing on the campus of the university of wisconsin-la crosse
Gilmore genealogical newsletter – 1985 1st qtr-1988 4th qtr – 1r – 1 – mf#2256839 – us WHS [929]
Gilmore, George W *see* Korea from its capital
Gilmore, George William *see*
– The church, the people, and the age
– The johannean problem
– Korea from its capital
Gilmore, James Houston *see* Notes of a course of lectures on smith's mercantile law
Gilmore, Robert L *see* Caudillism and militarism in venezuela, 1810-1910
Gilmorehill globe – [Scotland] Glasgow: Students' Representative Council, Glasgow University 10 oct 1932-20 feb 1935 (wkly during session) – 1r – 1 – uk Newsplan [378]
Gilmour, J *see*
– Among the mongols
– James gilmour of mongolia
Gilmour, James *see*
– Among the mongols
– James gilmour of mongolia
Gilmour, R *see* Buku re masoko anoyera e chirangano che kare ne chipswa
Gilow, Hermann *see* Die grundgedanken in heinrich von kleists "prinz friedrich von homburg"
Gilpin county miscellaneous newspapers – Denver, CO (mf ed 1991) – 1r – 1 – (the pine cone (jul 3 1897-oct 20 1900); black hawk advetiser (apr 14 1888); black hawk independent (jul 23 1898-oct 1 1898); black hawk times (may 4 1887-aug 17 1887); colorado miner (jul 4 1863-aug 29 1863); daily black hawk journal (aug 5 1873); daily colorado miner (nov 16 1863); the post (may 17 1879, may 31 1879); weekly mining journal (jan 3 1865-jul 11 1865); colorado herald (may 20 1871); daily colorado herald (jun 18 1868-feb 3 1872); gilpin daily graphic (oct 18 1882); little kingdom come (scattered issues feb 16 1970-dec 21 1970); weekly colorado herald (apr 13 1870); pine creek gold belt (may 1 1896-jul 24 1896); tolland herald (sep 1 1905)) – mf#MF Z99 G427 – us Colorado Hist [071]
Gilpin daily graphic *see* Gilpin county miscellaneous newspapers
Gilpin, Edwin *see*
– The geological relations of the principal nova scotia minerals
– The iron ores of pictou county, nova scotia
Gilpin, H D *see* Gilpin's reports of cases in the eastern district of pennsylvania, 1828-1826
Gilpin, John Bernard *see* Sable island
Gilpin, Richard *see* Daemonologia sacra
Gilpin, W *see* An essay upon prints
Gilpin, William *see*
– Address of gov william gilpin of colorado territory
– Explanation of the duties of religion
– Observations on several parts of the counties of cambridge, norfolk, suffolk, and essex
– Observations on the coasts of hampshire, sussex, and kent
– Observations on the western parts of england
– Remarks on forest scenery, and other woodland views
– Two essays
– Voyage en differentes parties de l'angleterre
Gilpin, William Sawrey *see* Practical hints upon landscape gardening
Gilpin's reports of cases in the eastern district of pennsylvania, 1828-1826 / Gilpin, H D – Philadelphia: Nicklin. 1v. 1837 (all publ) – (= ser Early federal nominative reports) – 7mf – 9 – $10.50 – mf#LLMC 81-453 – us LLMC [345]
Gil-Robles y Quinones, Jose Maria *see* Spain in chains
[Gilroy-] california weekly leader – CA. feb 19 1875-dec 17 1875 – 1r – 1 – $110.00 – mf#R03227 – us Library Micro [071]
[Gilroy-] gilroy advocate – CA. 1886-1946 – 24r – 1 – $1440.00 – mf#RC02274 – us Library Micro [071]
[Gilroy-] gilroy dispatch – CA. 1981-1983 – 1r – 1 – $60.00 – mf#R04029 – us Library Micro [071]
[Gilroy-] the gilroy dispatch – CA. 1925 – 202r – 1 – $12,120.00 (subs $360y) – mf#RC02275 – us Library Micro [071]
Gilruth, James *see* Ironton of sweet long ago, 1872-1874
Gilson, Etienne *see* La liberte chez descartes et la theologie
Gilson, J P *see* The mozarabic psalter (hbs30)
Gilson, Richard P *see*
– Samoa 1830-1900 and other research materials
– Samoa 1830-1900 drafts and research materials and cook islands, fiji, french polynesia, gilbert and ellice islands colony (kiribati), niue and png
Giltebrandt, P A *see* Rukopisnoe otdelenie vilenskoi publichnoi biblioteki...
Giltner gazette – Giltner, NE: [J C Bierbower] 17v. v45 n16. feb 20 1947-v61 n17. dec 27 1962 (wkly) [mf ed with gaps filmed 1975] – 5r – 1 – (cont: gazette-advertiser) – us NE Hist [071]
Giltner gazette – Giltner, NE: [J C Bierbower] 1901-v38 n33. jun 13 1940 (wkly) [mf ed 1918-40 (gaps) filmed 1975] – 7r – 1 – (merged with: phillips advertiser to form: gazette-advertiser) – us NE Hist [071]
Gim Lian [Vaddhanaprita], Hluan Vicitravadakar *see* Vijja prampi prakar
Gim Saet *see*
– Kanna kamsat
– Siavbhau pravattisastr
Gimbeletter – Gimbels, Milwaukee WI. v16 n1-v22 n20 [1974 jan 1-1980 oct] – 1r – 1 – mf#500709 – us WHS [071]
Gimbelite – 1976 summer-1981 – 1r – 1 – mf#354463 – us WHS [071]
Gimbels Midwest *see* Front and forward
Gimenez Caballero, Ernesto *see* Espana y franco
Gimenez de la Torre, Pelayo *see* Tratado practico sobre el mal rojo del cerdo
Gimenez, Joseph Patrick *see* Deep waters
Gimi : thnak di 1 / Toy Khiav Kumar et al – Bhnam Ben: Ron Bumb Mahaladh [mf ed 1990] – 1r with other items – 1 – (in khmer) – mf#mf-10289 seam reel 120/7 [§] – us CRL [500]
Gimli balour – Canada. jan 1903-dec 1909 – 1r – 1 – (in icelandic. some iss missing) – cn Commonwealth Imaging [071]
Gimli gimlungur – Canada. jan 1909-dec 1911 – 1r – 1 – (in icelandic) – cn Commonwealth Imaging [071]
Gimmerthal, Armin *see* Hinter der maske
Gimson and Barnsley : designs and drawings in cheltenham art gallery and museum – 4r (1 col) – 1 – £250.00 – (incl printed guide) – mf#GBR – uk World [740]
The gin mill primer : a book of easy reading lessons for children of all ages, especially for boys who have votes / Bengough, John Wilson – Toronto: W Briggs; Montreal: C W Coates, 1898 – 1mf – 9 – mf#10227 – cn Canadiana [362]
Ginal, J R *see* Die unbefleckte empfaengniss der seligsten jungfrau maria
Gindely, Antonin *see* History of the thirty years' war
Gindert, Christine *see* Textverwaltung als aufgabe der mensch-computer-interaktion
Gindin, I F *see*
– Banki i promyshlennost' v rossii
– Gosudarstvennyi bank i ekonomicheskaia politika tsarskogo pravitel'stva, 1861-1892
– Russkie kommercheskie banki
Gindraux, Jules *see* Histoire du christianisme dans le monde paien

Gingras, Apollinaire *see*
– Au foyer de mon presbytere
– Le bas-canada entre le moyen-age et l'age moderne
– L'echo des coeurs
– L'emballement
Gingras, Isaie *see* Preuve testimoniale dans l'affaire des syndics de la ville de longueuil...
Gingras, Jean-Jules *see* Bibliographie analytique de l'oeuvre de monsieur le chanoine paul-emile crepeault, 1944-1964
Gingras, Jules Fabien *see* Manuel des expressions vicieuses les plus frequentes
Gingras, Leon *see* L'orient
Gingras, Paul-Emile *see* Rapport d'une etude confiee au cadre par la digec sur l'evaluation des colleges
Giniewski, Paul *see*
– Bantustans
– Two faces of apartheid
Ginisty, Paul *see* Chartreuse de parme
Ginius, L *see* Ad christianos principes de svscepto pro christiana rep contra turcas bello communiter conficiendo...
Ginko soran : annual directory of banks in japan: the banking bureau of the ministry of finance – 1st-49th. 1895-1942 – 18r – 1 – Y178,000 – (in japanese) – ja Yushodo [332]
Ginouvier, J F *see* Tableau de l'interieur des prisons de france
Ginsberg, Hyman *see* Questions and answers to past bar examination questions
Ginsberg, Leon *see* Administration in social work
Ginsberg, Morris *see* She hui hsueh tao yen
Ginsbourg, Anna *see* Jewish refugees in shanghai
Ginsburg, Christian David *see*
– Coheleth, commonly called the book of ecclesiastes
– The essenes
– Introduction to the massoretico-critical edition of the hebrew bible
– The kabbalah
– The massoreth ha-massoreth of elias levita
– The moabite stone
– The song of songs
Ginsburg, S L *see* A wandering jew in brazil
Ginsburger, M *see*
– Israelitische friedhof in jungholz
– Thargum jonathan ben usiel zum pentateuch
Ginsburger, Moses *see* Das fragmententhargum
Ginseng / Kains, Maurice Grenville – New York, NY. 1903 – 1r – us UF Libraries [025]
Ginther, A *see* Mater amoris et doloris
Ginza *see* Mandaeische schriften
Ginzberg, Louis *see*
– Teshuvah bi-devar yenot ha-kesherim
– Yerushalmi fragments from the genizah
Ginzburg, Isidor *see* Talmud
Ginzburg, Vul'f Veniaminovich *see* Gornye tadzhiki
Ginzel, J A *see* Geschichte der slawenapostel cyrill und method und der slawischen liturgie
Ginzkey, Franz Karl *see*
– Meistererzaehlungen
– Rositta
Il gior31 dicembre 1815 : descrizione del solenne ingresso in miladelle loro maesta ii err francesco i e maria luigia d'austria – Milano, 1816 – 1mf – 9 – mf#O-1096 – ne IDC [700]
Giorda, Joseph *see* Lu tel kaimintis kolinzuten kuitlt smiimii
Giordani, Giuseppe *see* Caro mio ben
Giordano bruno / McIntyre, James Lewis – London, New York: The Macmillan Co., 1903 – xvi/365p – 1 – us Wisconsin U Libr [190]
Giordano bruno : philosopher and martyr / Brinton, Daniel Garrison – Philadelphia: D McKay 1890 [mf ed 1987] – 1r – 1 – (with: giordano bruno / holyoake, g j) – mf#1969 – us Wisconsin U Libr [130]
Giordano bruno : trauerspiel in drei aufzuegen / Wilbrandt, Adolf von – Wien: L Rosner, 1874 – 1r – 1 – us Wisconsin U Libr [820]
Giordano bruno and the relation of his philosophy to free thought; a lecture...new york liberal club, oct. 30, 1885 / Davidson, Thomas – Boston: Index Assoc., 1886. 45p – 1 – us Wisconsin U Libr [180]
Giordano bruno und nicolaus von cusa : eine philosophische abhandlung / Clemens, Franz Jakob – Bonn: J Wittmann, 1847 – 1mf – 9 – 0-524-00014-X – (incl bibl ref) – mf#1989-2714 – us ATLA [100]
Giordano brunos einfluss auf goethe und schiller : vortrag gehalten in der richard wagner-gesellschaft zu berlin am 25. oktober 1906 / Kuhlenbeck, L – Leipzig: T Thomas, 1907 – 1r – 1 – (incl bibl ref) – us Wisconsin U Libr [430]
Giordano, Luca *see* Extracts from regole brievi della volgare grammatica
Giorg, Kara *see*
– Abendglocken
– Poesien des alltags
Giorgi, A A *see* Alphabetum tibetanum missionum apostolicorum commodo editum
Giornale araldico-genealogico-diplomatico – Fermo. v1-28 1874-1905 – 1 – $161.00 – us L of C Photodup [920]
Giornale delle arti e delle industrie – Turin, Florence, Italy. jan 1861-dec 1865; 1871-89 [wkly] – 20r – 1 – uk British Libr Newspaper [073]
Giornale di malacologia / Stobel, E [comp] – Pavia 1853-54 – 2v on 7mf – 9 – mf#z-900/2 – ne IDC [590]
Il giornale di toronto : il settimanale per tutta la famiglia – Toronto. v1-13. feb 10 1967-oct 1979//? (wkly) – 13r – 1 – Can$862.00 – (in italian and english) – cn McLaren [071]
Il giornale d'italia – Roma: L'Unione publicita italiana SA, jun 1939; dec 1938-jun 7 1944 – 1 – us CRL [074]
Il giornale d'italia – Roma: Societa per la pubblicita la Italia, 1949-52 – us CRL [074]
Giornale italiano – Sydney, mar 1932-jun 1940 – 3r – A$196.28 vesicular A$212.78 silver – at Pascoe [079]
Il giornale italiano – New York: Italian Press Pub. Association, jan-sep 27 1919 – us CRL [074]
Giornale italiano di chemioterapia – Torino Caselle, Italy 1975-79 – 1,5,9 – ISSN: 0017-0445 – mf#10029 – us UMI ProQuest [615]
Giornale patriottico di corsica / ed by Buonarroti, Filippo – Bastia (Corsica). v1-2. apr-nov 1790 – (= ser Rare or unique periodicals and volumes from the feltrinelli archives) – 4mf – 9 – $70.00 – us UPA [362]
Giornale storico della letteratura italiana – v. 1-140, SUPPS. 1-28. Index v. 1-100. 1883-1963 – 1 – us L of C Photodup [440]
Il giorno – 1987- – 4r per y – 5 – enquire for prices – us UMI ProQuest [074]
Il giorno. (l'elettricita) – Milan, Italy. 1 may 1882-24 dec 1908 [wkly] – 29r – 1 – uk British Libr Newspaper [073]
Giot nang hong / Duy-Nguyen – [Saigon] May Hong [1974] [mf ed 1992] – (= ser Tu sach may hong, loai da c biet) – on pt of 1r – 1 – mf#11052 r178 n7 – us Cornell [959]
Giosue carducci, l'homme et le poete / Jeanroy, A – Paris. 1911 – 1 – us CRL [920]
Giovagnoli, Antonio Francesco *see* The life of saint margaret of cortona
Giovanelli, Ruggiero *see*
– Motecta partim quinis, partim octonis vocibus
– Motecta quinque vocum liber secundus
Giovanni and matteo villani – 14th c – (= ser Holkham library manuscript books 552) – 1r – 1 – mf#97106 – uk Microform Academic [900]
Giovanni bellini / Fry, Roger Eliot – London 1899 – (= ser 19th c art & architecture) – 2mf – 9 – mf#4.2.186 – uk Chadwyck [750]
Giovanni cassiano ed evagrio pontico / Marsili, S – Roma, 1936 – 4mf – 8 – €11.00 – ne Slangenburg [241]
Giovanni da Rovigo *see* Lexicon bonaventurianum
Giovanni di Fidanza *see* Die psychologie bonaventura's nach den quellen dargestellt
Giovanni dupre / Frieze, Henry Simmons – London 1886 – (= ser 19th c art & architecture) – 3mf – 9 – mf#4.2.616 – uk Chadwyck [730]
Giovanni gentiles philosophie und paedagogik / Baur, Johannes – Muenchen, 1935 (mf ed 1992) – 3mf – 9 – €24.00 – 3-89349-009-4 – mf#DHS-AR 12 – gw Frankfurter [140]
Giovanni pierluigi da palestrina (1525-1594) : first critical edition / ed by Witt, Theodor de et al – Leipzig: Breitkopf & Haertel. 33v. 1862-1907 – 11 – $325.00 set – us Univ Music [780]
La giovinezza di hamann / Accolti Gil Vitale, Nicola – Varese: Editrice Magenta, c1957 [mf ed 1993] – (= ser Saggi e ricerche 3) – ixiv/169p – 1 – (incl bibl ref) – mf#8670 – us Wisconsin U Libr [190]
Giovio, P *see*
– Commentario de le cose de tvrchi, di pavlo iovio, vescovo di nocera, a carlo qvinto imperadore avgvsto
– Commentario de le cose de tvrchi, di pavlo iovio, vescovo di nocere, a carlo qvinto imperadore avgvsto
– Commentario de le cose de tvrchi, et del s georgio scanderbeg, principe di epyrro
– Commentarius captae vrbis, dvctorecarolo borbonio, ad exquisitum modum confectus...
– Histoire...svr les choses faictes et auenues de son temps en toutes les parties du monde...
– ...Opera qvotqvot extant omnia
– Turcicarum rervm commentarivs...
– Vrsprung des turkischen reichs bis auff denitzigen solyman...
Giovio, Paolo *see*
– Dialogo de las empresas militares
– Dialogo dell' imprese militari et amorose
– Dialogo dell' imprese militari et amorose di monsignor giovio vescovo di nocera
– Dialogo dell' imprese militari et amorose di monsignor paolo giovio vescovo di nucera...
– Dialogo dell'imprese militari et amorose
– Dialogo dell'imprese militari et amorose di monsignor giovio vescovo di nocera
– Ragionamento...sopra i motti e disegni d'arme e d'amore, che communemente chiamano
– Le sententiose imprese di monsignor paulo giovio
Gipa news – [may? 1971-may? 1974] – 1r – 1 – mf#1056718 – us WHS [071]
Gips als historischer aussenbaustoff in der windsheimer bucht : verbreitung, gewinnung und bestaendigkeit im vergleich zu anderen oertlichen naturwerksteinen / Lucas, Hans Guenter – (mf ed 1994) – 4mf – 9 – €56.00 – 3-89349-871-0 – mf#DHS 871 – gw Frankfurter [550]
The gipsy smith missions in america : a volume commemorative of his sixth evangelistic campaign in the united states 1906-1907 – Boston, Mass.: Interdenominational Pub., 1907, c1906 – 1mf – 9 – 0-8370-6243-8 – mf#1986-0243 – us ATLA [240]
Giraffe / Hamilton Co. Cincinnati – aug-oct 1842 [daily] – 1r – 1 – mf#B3318 – us Ohio Hist [320]
Giral Pereira, Jose *see* Problemas de la alimentacion en la post-guerra
Giraldi cambrensis opera, vol 8 (rs21) : de principis instructione liber / ed by Warner, G F – (= ser The rolls series (rs)) – (with ind to v1-4 and 8 1891 €18) – ne Slangenburg [931]
Giraldi cambrensis opera, vols 1-4 (rs21) / ed by Brewer, J S – (= ser The rolls series (rs)) – (v1: invectionum libellus; symbolum elect 1861 €19. v2: gemma ecclesiastica 1862 €17. v3: de invectionibus, lib 4, de menevensi eccl dialogus; vita s david 1863 €18. v4: speculum ecclesiae; de vita galfridi archiepiscopi eboracensis sive certamina galfridi eb archiepisc 1873 €18. v5: topographia hibernica, et expugnati hibernica 1867 €19. v6: itinerarium kambriae, et descriptio kambriae 1868 €15. v7: vita s remigii, et vita s hugonis 1877 €17. v8: de principis instructione liber. with ind to v1-4 and 8 1891 €18) – ne Slangenburg [240]
Giraldi cambrensis opera, vols 5-7 (rs21) / ed by Dimock, J F – (= ser The rolls series (rs)) – (v5: topographia hibernica, et expugnati hibernica 1867 €19. v6: itinerarium kambriae, et descriptio kambriae 1868 €15. v7: vita s remigii, et vita s hugonis 1877 €17) – ne Slangenburg [931]
Giraldo Jaramillo, Gabriel *see*
– Estudios historicos
– Grabado en colombia
– Notas y documentos sobre el arte en colombia
– Pinacotecas bogotanas
Giraldo, Juan Manuel *see* Vida y...don diego de arce reynoso
Giraldo Londono, Pedronel *see* Don fernando
Giraldus *see*
– Giraldi cambrensis opera, vol 8
– Giraldi cambrensis opera, vols 1-4
– Giraldi cambrensis opera, vols 5-7
Giralt Thovar, Arturo *see* Normas de cultivo para obtener beneficio con los maices...
Giran, Etienne *see*
– Christianisme et liberte
– Jesus of nazareth
– Sebastien castellion et la reforme calviniste
Girard cosmopolite – Girard, PA. -w 1899-1910 – 13 – $25.00r – us IMR [071]
Girard herald – Girard KS. 1886 jan 7-1888 sep 29, 1888 oct 5-1891 mar 28, 1891 apr 4-jun 28 – 3r – 1 – (continued by: western herald (girard, ks)) – mf#880427 – us WHS [071]
Girard, Jules *see* Le sentiment religieux en grece
Girard, Louis *see* La politique des travaux publics du second empire
Girard weekly journal : weekly republican newspaper – Girard, OH, 4 Jan 1910-24 Mar 1911 – 1r – 1 – us Western Res [071]
Girardeau, John Lafayette *see*
– Calvinism and evangelical arminianism
– Discussions of philosophical questions
– Discussions of theological questions
– Instrumental music in the public worship of the church
– Sermons
– Theology as a science, involving an infinite element
– The will in its theological relations
Girardey, Ferreol *see* The practical catechist
Girardin, Emile De *see*
– Ecole des journalistes
– Joie fait peur
– Supplice d'une femme
Girardin, Emile de, Mme *see* La joie fait peur
Girasol / Rosa-Nieves, Cesareo – San Juan, Puerto Rico. 1960 – 1r – us UF Libraries [972]
Giraud, Leon *see* Le roman de la femme chretienne
Giraud, Philippe *see* Est-ce st paul a athenes?... au milieu de l'areopage?
Giraud, Victor *see* Pascal, l'homme, l'oeuvre, l'influence
Giraudet, Eugene *see* Traite de la danse
Giraudier, Antonio *see*
– Bordes
– Cerca
– Green against linen
– Mano en el espacio
– Piedras magicas
– Piensame
– Schnecke am ufer und andere gedichte
Girauld de Montpellier, A *see* Utilissim, promt...y facil remey e memorial para preservarse y curar de la peste
Girault de Saint-Fargeau, Eusebe *see*
– Dictionnaire de la geographie physique et politique de la france
– Guide pittoresque du voyageur en france
– Histoire nationale
Girdlestone, Charles *see*
– God's word and ministers
– Letter on church reform
Girdlestone, Henry *see* Notes on the apocalypse
Girdlestone, Robert Baker *see*
– The anatomy of scepticism
– The building up of the old testament
– Deuterographs
– The foundations of the bible
– How to study the english bible
– Old testament theology and modern ideas
– Outlines of bible chronology
– The student's deuteronomy
– Synonyms of the old testament
Girdling the globe : from the land of the midnight sun to the golden gate, a record of a tour around the world / Miller, Daniel Long – Mount Morris, Ill: Brethren Pub House, 1898 – 2mf – 9 – 0-524-02835-4 – mf#1990-4456 – us ATLA [910]
Girerd, Sylvain *see* L'oeuvre militaire de la galissoniere au canada
Girgensohn, Karl *see*
– Die religion, ihre psychischen formen und ihre zentralidee
– Seele und leib
Giri, Mahadebananda, swami *see* Vedic culture
Girid – 1310 [1892] – (= ser Vilayet salnames) – 4mf – 9 – $60.00 – us MEDOC [956]
Girl friend newsletter – 1963 – 1r – 1 – (cont: girl friends incorporated news letter; cont by: g f news letter) – mf#5012735 – us WHS [071]
Girl Friends, Inc *see* G f news letter
Girl friends incorporated news letter – 1962 feb 28 – 1r – 1 – (continued by: girl friend newsletter) – mf#5012732 – us WHS [071]
The girl i can't forget = Celle qu'on n'oublie pas / Carbonneau, Fred – Montreal: The Popular Music Pub, 1926 [mf ed 1988] – 1mf – 9 – (in english and french) – mf#SEM105P926 – cn Bibl Nat [780]
The girl in the case / Barr, Robert – London, Toronto: Hodder & Stoughton, [1914?] – 2mf – 9 – 0-665-71693-1 – mf#71693 – cn Canadiana [830]
The girl of the new day / Knox, Ellen Mary – Toronto: McClelland & Stewart, c1919 [mf ed 1995] – 3mf – 9 – 0-665-74721-7 – (incl app) – mf#74721 – cn Canadiana [305]
A girl of to-day / Adams, Ellinor Davenport – London, Glasgow, Dublin: Blackie & Son Ltd, 1898 – (= ser 19th c women writers) – 4mf – 9 – mf#5.1.59 – uk Chadwyck [640]
Girl Scouts of Milwaukee County *see* Soundings
Girlhood's hand-book of woman / ed by Donnelly, Eleanor C – 3rd ed. St Louis, MO: B Herder, 1914 [mf ed 1986] – 1mf – 9 – 0-8370-6897-5 – mf#1986-0897 – us ATLA [305]
Girls' education in india : in the secondary and collegiate stages / Dasgupta, Jyotiprova – [Calcutta]: University of Calcutta, 1938 – (= ser Samp: indian books) – us CRL [376]
The girls of miss clevelands' / Embree, Beatrice – Toronto: Musson, c1920 [mf ed 1998] – 3mf – 9 – 0-665-98441-3 – mf#98441 – cn Canadiana [830]
The girl's own paper – Toronto: Warwick Bro's and Rutter, [1880-1907] – 9 – mf#P04991 – cn Canadiana [073]
Il giro del mondo : giornale di viaggi, geografia e costumi – Milano 1879-80 [mf ed Hildesheim 1995-98] – 10mf – 9 – €80.00 – 3-487-26624-5 – gw Olms [945]
Girod, Amury *see* Notes diverses sur le bas-canada
Girodon, P *see* Expose de la doctrine catholique
Giron – Miami, FL. 1965 dec-1995 mar – 1r – us UF Libraries [071]
Giron, Jose Eduardo *see* Notario practico, tratado de notaria
Giron, Manuel Antonio *see* Pediatria social
Giron Mena, Manuel Antonio *see* Medicina social
Giron, Pedro *see* Cronica el emperador carlos 5th. edicion de juan sanchez montes. prologo de peter rasow. madrid, 1964
Gironcourt, George R de *see* Missions de gironcourt en afrique occidentale, 1908/1909/1912
La gironde – Bordeaux. 1858-1926 – 1 – fr ACRPP [073]
Le girondin – Paris: Impr de E Briere, feb 29-mar 1 1848 – us CRL [074]
Gironella, Jose Maria *see* Cipreses creen en dios
Girouard, Desire *see*
– L'album de la famille girouard
– Les anciennes cotes du lac saint-louis
– Les anciens postes du lac saint-louis
– The beauharnois canal question
– The bills of exchange act, 1890

- Considerations sur les lois civiles du mariage
- Essai sur les lettres de change et les billets promissoires
- La famille cousineau
- La famille girouard
- La famille girouard en france
- Lake st louis old and new, illustrated
- The old settlements of lake st louis
- Une page sombre de notre histoire
- The royal commission
- Supplement to lake st louis, etc etc

Girouard, Desire [comp] *see* Un tableau interessant pour les electeurs de la province de quebec

Giroux, Henri *see*
- Guide illustre de montreal et de ses institutions catholiques
- Une heroine du canada
- Histoire de la communaute de notre dame de charite du bon-pasteur de montreal
- Histoire et miracles de ste anne de beaupre
- La misericorde ou 50 annees de devouement et d'agnegation des religieuses de misericorde a montreal

Giroux, Pauline *see* Bibliographie analytique du docteur louis-georges godin (1897-1932)

Giroux, Yvette *see* Bibliographie analytique de monsieur andre giroux

Girst, Judah Loeb *see* Bi-netivot ha-zeman veha-netsah

Girtanner, Christoph *see* Politische annalen

Girton 1599-1950 – 6mf – 9 – £7.50 – uk CambsFHS [929]

Girvan courier and carrick district news – [Scotland] South Ayrshire, Girvan: W Stevans 18 may 1904-27 oct 1909 (wkly) [mf ed 2004] – 6r – 1 – (cont by: carrick courier, circulating in the kingdom of carrick [30 nov 1904-oct 1909]) – uk Newsplan [072]

Girvan gazette and carrick advertiser – [Scotland] Glasgow: Hay Nisbet & Co jan-sep 1901 (mthly) [mf ed 2004] – 9v on 1r – 1 – uk Newsplan [072]

Girvan town crier and pavilion news – [Scotland] South Ayrshire, Girvan: printed at the Carrick Herald Office 15 aug 1925-25 apr 1931 (wkly) [mf ed 2004] – 3r – 1 – uk Newsplan [072]

Giry, A *see* Manuel de diplomatique

Giry, Arthur *see* Notices bibliographiques sur les archives des eglises et des monasteres de l'epoque carolingienne

Giry, F *see* La vie de m jean-jacques olier

GI's Against Fascism *see* Duck power

Gisberti voetii tractatus selecti de politica ecclesiastica. series prima = Tractatus selecti de politica ecclesiastica. series prima / Voet, Gijsbert; ed by Hoedemaker, Phillipus Jacobus – Amstelodami: JH Kruyt, 1885 – (= ser Bibliotheca reformata) – 5mf – 9 – 0-524-07469-0 – mf#1991-3129 – us ATLA [240]

Gisberti voetii tractatus selecti de politica ecclesiastica. series secunda = Tractatus selecti de politica ecclesiastica. series secunda / Voet, Gisberti; ed by Hoedemaker, Phillipus Jacobus – Amstelodami: JH Kruyt, 1886 – (= ser Bibliotheca reformata) – 5mf – 9 – 0-524-07470-4 – mf#1991-3130 – us ATLA [240]

Gisbertus voetius / Duker, A C – Leiden. v1-3. 1897-1914 – 3v on 23mf – 8 – €52.00 – ne Slangenburg [242]

Gisbertus voetius / Duker, Arnoldus Cornelius – Leiden: Brill, 1897-1915 – 4mf – 9 – 0-7905-5085-7 – (incl ind) – mf#1988-1085 – us ATLA [920]

Gisborne herald – aug 1939-dec 1952; 8 may-24 sep 1953; 2 jan-9 apr 1954; 23 jan-28 jul 1962; 5 dec 1963-5 jan 1965; jan-jun 1970; may 1976; oct 1976; dec 1976; jan 1977-oct 1998 – 1 – mf#18.1 – nz Nat Libr [079]

Gisborne, Thomas *see* Considerations on modern theories of geology

Gisborne times – 4 jan 1905-5 oct 1906; 2 jan 1907-nov 1912; 17 apr 1913-17 dec 1913; 3 jan-31 mar 1927; 15 feb 1933-31 mar – 1 – mf#18.2 – nz Nat Libr [079]

Gisborne, William *see* The colony of new zealand

Gisholt Machine Co *see* News crib

Gisleberti chronicon hanoniense (mgh7:29.bd) – 1869 – (= ser Monumenta germaniae historica 7: scriptores rerum germanicarum in usum scholarum (mgh7)) – €14.00 – ne Slangenburg [240]

Gislenius, A *see* Initera constantinopolitanvm et amasianvm...

Gismondi, Enrico *see* Linguae syriacae

Gisneros y Sevillano, Juan *see* Extirpacion total de la laringe por cacinoma

Gissing, George *see* Workers in the dawn

Gist – 1974-90 – 1r – 1 – (cont: foreign policy outlines) – mf#1937844 – us WHS [327]

The gist of japan : the islands, their people, and missions / Peery, Rufus Benton – New York: Fleming H Revell, 1897 [mf ed 1995] – (= ser Yale coll) – 317p (ill) – 1 – 0-524-09648-1 – mf#1995-0648 – us ATLA [950]

GIs-WACs United Against the War *see* Left face

G-i-t : glas und instrumenten-technik – Darmstadt: Hoppenstedt Wirtschaftsverlag GmbH. v32 n6. jun 1988 – 1r – 1 – us CRL [660]

Git Han *see*
- Siksa qanubaky
- Upakarn qanubaky

Gita : meditations / Vaswani, Thanwardas Lilaram – Poona: Gita Pub House, [1900?]- – (= ser Samp: indian books) – us CRL [280]

Gita and gospel / Farquhar, John Nicol – 3rd ed. Madras: Christian Literature Society for India, 1917 – 1mf – 9 – 0-524-01958-4 – (incl bibl ref) – mf#1990-2749 – us ATLA [280]

Gita lankara dipani, kho, gita sippam a khre kham – Ran Kun: Kkhuccha saya pita kat ca pum nhip tuik 1974 [mf ed 1990] – 1r with other items – 1 – (in burmese; incl bibl ref & ind) – mf#mf-10289 seam reel 157/5 [§] – us CRL [780]

Gita the mother / Gandhi, Mahatma; ed by Chander, Jag Parvesh – Lahore: Indian Print Works, [19–] – (= ser Samp: indian books) – us CRL [280]

Gita vesantara / Tan Aon, Raisihakui – Ran Kun: CapeU 1975 [mf ed 1990] – [ill] 1r with other items – 1 – (in burmese) – mf#mf-10289 seam reel 145/3 [§] – us CRL [780]

La gitana extremena y otros poemas / Alvarez Joven, Arturo – Salamanca: Imprenta Comercial Salmantina, 1949 – sp Bibl Santa Ana [810]

Gitanjali = Song offerings / Tagore, Rabindranath – London: Macmillan & Co, 1917 – (= ser Samp: indian books) – (int by w b yeats) – us CRL [780]

Gitanos, gitanerias. cantos, bailes y cosas de los gitanos de espana / Lopez Martinez, Antonio – Merida, 1950 – sp Bibl Santa Ana [946]

Gitarist – Moscow. n1-12 1904; n1-12 1905; n1-12 1906 [mthly] – 18mf – 9 – us UMI ProQuest [780]

Giteau, Madeleine *see*
- Histoire d'angkor
- Histoire du cambodge

Gitsham, Ernest *see* A first account of labour organisation in south africa

Gitto, Anita T *see* Relationship of excess post-exercise oxygen consumption to vo2max and recovery rate

Giudici, Roberto B *see* Los fundamentos del battislmo

Giuglaris, L *see* Funerale fatto nel duomo di torino alla gloriosa memoria...

Giugno, Carl *see* Ein schmetterling

Giuliani, G B *see* Descrittione dell' apparato fatto nella festa di s. giovanni dal fedelissimo popolo napolitano

Giunta, Mary A *see* Foreign relations of the united states under the articles of confederation, 1780-1789 (fruac-m)

Giuseppe, Maria *see* The lives of the blessed leonard of port maurice and of the blessed nicholas fattore

Giustina : a spanish tale of real life. a poem in three cantos / Law, Elizabeth Susan, Baroness Colchester – [London]: 1833 – (= ser 19th c women writers) – 1mf – 9 – mf#5.1.11 – uk Chadwyck [810]

Giustiniani, A *see* Castigatissimi annali con la loro copiosa tavola della eccelse and illustrissima repubii di genoa, da fideli and approuati scrittore...

Giustizia / International Ladies' Garment Workers' Union – 1977 oct-1986 dec – 1r – 1 – mf#1214361 – us WHS [680]

Giustizia / International Ladies Garment Workers' Union – Organo Ufficiale. New York etc. v1-29 Jan 18 1919-1946 – 1 – us NY Public [331]

Giustizia e liberta – Paris, France. -w. 18 may 1934-8 sep 1939; 22 apr 1940 – 2r – 1 – uk British Libr Newspaper [072]

La giustizia fra intrighi e tradimenti / Conti, Giovanni – 2nd ed. Roma, Casa, 1952. 115 p. LL-4069 – 1 – us L of C Photodup [340]

La giustizia nella somalia, guglielmo ciamarra. raccolta di giurisprudenza coloniale. / Ciamarra, Guglielmo – Napoli: Giannini, 1914. 421p. LL-12011 – 1 – us L of C Photodup [340]

Giuzal'an, L T *see* Rukopisi shakh-name v leningradskikh sobraniiakh

Giv'at pinhas / Dembitzer, Phinehas Elijah – Krakow, Poland. 1925 – 1r – us UF Libraries [939]

Given, John James *see* The truth of scripture

Givens, Nick K *see* Echoes from hell

Giver and his gifts / Moore, Daniel – London, England. 1860 – 1r – us UF Libraries [240]

Givon, Talmy *see*
- Si-luyana language
- Siluyana language

Givstino historico... – Vinegia, 1561 – 3mf – 9 – mf#H-8408 – ne IDC [956]

Gizeh and rifeh / Petrie, W M – London, 1907. 2 v – 5mf – 9 – mf#NE-20355 – ne IDC [956]

Gizetti, Aleksandr *see*
- Etiudy o zapadnoi literature
- Svetlyi dukhom

Gizli el / Guentekin, Resat Nuri – Dersaadet: Sems Matbaasi, 1925 – (= ser Ottoman literature, writers and the arts) – 3mf – 9 – $55.00 – us MEDOC [470]

Gjellerup, Karl *see* Die huegelmuehle

Gjellerup, Karl Adolph *see*
- An der grenze
- Der pilger kamanita
- Die weltwanderer

Gjentagelsen : et forsog i den experimenterende psychologi / Constantius, Constantin – Kobenhavn: CA Reitzel, 1843 – 1mf – us ATLA [150]

Gjentagelsen et forsoeg i den experimenterende psychologi / Kierkegaard, Soeren – Kobenhavn: C A Reitzel, 1843 – (= ser Himmelstrup) – 1mf – 9 – 0-7905-3791-5 – mf#1989-0284 – us ATLA [150]

Gjertsen, Melchior Falk *see* Referat af forhandlingerne i en fri conferents i decorah, iowa

Gjorabok *see* Arsping hins...

Gjorabok...arsping hins... / Evangeliska luterska kirkjufelag islendinga i vesturheimi – [Winnipeg MB?: s.n.] 1910- [annual] [mf n26-51 1910-35] – 2r – 1 – (lacks: n47-50. ceased in 1950? filmed with earlier titles: arsfundr hins...and: arsping hins...) – mf#2003-s502c – us ATLA [242]

Gk's weekly – London. March 21, 1925-March 12, 1927 – 1 – us NY Public [073]

Gla, Dietrich *see* Die originalsprache des matthaeusevangeliums

Glacial and inter-glacial deposits near toronto / Coleman, Arthur Philemon – Chicago: University of Chicago Press, 1895? – 1mf – 9 – mf#00704 – cn Canadiana [550]

The glacial nightmare and the flood : a second appeal to common sense from the extravagance of some recent geology / Howorth, Henry Hoyle – London, 1893 – (= ser 19th c evolution & creation) – 2v on 12mf – 9 – mf#1.1.7591 – uk Chadwyck [550]

Glaciation of high points in the southern interior of british columbia / Dawson, George M – London?: s.n, 1889 – 1mf – 9 – mf#02278 – cn Canadiana [550]

Glacier / Tlinkit Training Academy – v1 n9, 1886 aug – 1r – 1 – mf#639839 – us WHS [071]

Glackemeyer, Edouard Claude *see*
- An alphabetical index to the laws of canada
- Tableau alphabetique des cites, villes, villages, paroisses et cantons dans chaque comte de la province de quebec...

Glad tidings : comprising sermons and prayer-meeting talks delivered at the new york hippodrome / Moody, Dwight Lyman – New York: E B Treat 1877, c1876 [mf ed 2002] – 1r – 1 – 0-524-10368-2 – (rev, corr with biogr sketch & ind arr by h h birkins) – mf#b00618 – us ATLA [240]

Glad tidings – [S.1: s.n. 1864?-18–] – 9 – mf#P06023 – cn Canadiana [200]

Gladbacher kreis-blatt fuer geschaefte, politik und unterhaltung *see* Geschaefts- und unterhaltungsblatt fuer den kreis gladbach und umgebung

Gladbacher zeitung *see* Geschaefts- und unterhaltungsblatt fuer den kreis gladbach und umgebung

Gladbecker stadtanzeiger *see* Gladbecker volkszeitung

Gladbecker volkszeitung – Gladbeck, Herne DE, 1957 2 oct-30 dec, 1958 1 apr-1960 – 1 – (title varies: 2 jan 1951: gladbecker stadtanzeiger; 10 oct 1953: ruhr-nachrichten. main ed in dortmund; ed for gladbeck, herne, wanne) – gw Misc Inst [074]

Gladden, Ted *see* The north dakota judicial education plan

Gladden, Washington *see*
- Applied christianity
- Burning questions of the life that now is and of that which is to come
- The christian pastor and the working church
- The christian way
- Christianity and socialism
- The church and modern life
- The church and the kingdom
- The great war
- How much is left for the old doctrines?
- The interpreter
- The labor question
- The lord's prayer
- Myrrh and cassia
- The practice of immortality
- Recollections
- Ruling ideas of the present age
- Seven puzzling bible books
- Social facts and forces
- Social salvation
- Where does the sky begin
- Who wrote the bible?
- Witnesses of the light
- Working people and their employers
- The young men and the churches

Gladdening river : twenty-five years' guild influence among the himalayas / Manuel, David Gilmour – London: A & C Black, 1914 [mf ed 1995] – (= ser Yale coll) – xxiii/260p (ill) – 1 – 0-524-09385-7 – (with foreword by baron carmichael of skirling) – mf#1995-0385 – us ATLA [242]

Glades county democrat – Moore Haven, FL. 1920 jun 21-1997 – 52r – (gaps) – us UF Libraries [071]

O gladiador – Meces do Pombal, MG. 03 jun 1894 – (= ser Ps 19) – bl Biblioteca [079]

Gladiolus thrips in florida / Wilson, J W – Gainesville, FL. 1941 – 1r – us UF Libraries [630]

Gladius ecclesiae : church lessons for young churchmen / Titcomb, Jonathan Holt – 6th ed. London: Church of England Sunday School Institute, [18–?] – 1mf – 9 – 0-8370-8795-3 – (incl ind) – mf#1986-2795 – us ATLA [240]

Gladness in jesus / Boardman, William Edwin – New and rev ed. Boston: Willard Tract Repository, c1870 – 1mf – 9 – 0-7905-3591-2 – mf#1989-0084 – us ATLA [240]

Gladstone age press – Manitoba, CN. oct 1981-jun 1988 – 8r – 1 – cn Commonwealth Imaging [071]

Gladstone and other addresses / Tupper, Kerr B – 1898 – 1 – 9.55 – us Southern Baptist [242]

Gladstone observer – Gladstone. jan 1974-dec 1979, jan 1981-dec 1982 – 24r – at Pascoe [079]

Gladstone on macleod and macaulay : two essays – Toronto: Beldford, 1876 – 1mf – 9 – mf#13329 – cn Canadiana [070]

Gladstone review – Milwaukie OR: North Clackamas Publ & Printers Inc, 1965- [wkly] – 1 – us Oregon Lib [071]

Gladstone, W E *see*
- Church of england and ritualism
- Ecce homo
- Letter to the right rev william skinner
- Parliamentary oaths
- Rome and the newest fashions of religion

Gladstone, William Ewart *see*
- Church in wales
- Contemporary estimates of his life and character
- Correspondance on church and religion
- Correspondence on church and religion of william ewart gladstone
- The creation story
- Gladstone on macleod and macaulay
- The impregnable rock of holy scripture
- The irish church
- Juventus mundi
- Rome and the newest fashions in religion
- The state in its relations with the church
- Studies subsidiary to the works of bishop butler
- Vatican decrees in their bearing on civil allegiance
- The vatican decrees in their bearing on civil allegiance

Glaeser, Ernst *see*
- Frieden
- Los que teniamos doce anos
- El ultimo civil

Glage, Max *see* Der grundfehler der ritschlschen theologie

Glagolitica : ueber die glagolitische literatur: das alter der bukwitza, ihr muster nach dem sie gebildet worden, den ursprung der roemisch-slawischen liturgie...ein anhang zum slavin; mit zwey kupfertafeln / Dobrovsky, Josef – Prag 1807 – (= ser Whsb) – 2mf – 9 – €30.00 – mf#Hu 178 – gw Fischer [410]

Glahn, Gerhard Ernst Ludwig Von *see* The german demand for colonies

Glaisher, James *see* Travels in the air

Glaister, Elizabeth *see*
- Art embroidery
- Needlework

Le glaive de la parolle veritable / Farel, Guillaume – Geneve, Jean Girard, 1550 – 6mf – 9 – mf#PFA-159 – ne IDC [240]

Glaive de l'esprit – London, UK. 20 Dec 1940; 17 Jan, 14 Mar-Aug 1941 – 1 – uk British Libr Newspaper [072]

Le glaive de l'esprit – Londres. n8-10. 28 juil-sept 1941 – 1 – fr ACRPP [240]

Glamorgan cemeteries, crematoria & graveyard lists – 1mf – 9 – £1.25 – uk Glamorgan FHS [929]

Glamorgan County Council *see* County hall records, 1719-1890

Glamorgan county times – [Wales] Treorchy jan 1909-dec 1950 [mf ed 2004] – 42r – 1 – (cont by: glamorgan county times & free press [jul 1949-dec 1950]) – uk Newsplan [072]

Glamorgan families (members' interests directory) – 1991 – 1mf – 9 – £1.25 – uk Glamorgan FHS [929]

Glamorgan free press #1 – [Wales] LLGC 9 may 1891-dec 1909 [mf ed 2004] – 16r – 1 – (missing: 1901-02; cont by: glamorgan free press, pontypridd & rhonddas chronicle [jan 1906-dec 1909]) – uk Newsplan [072]

Glamorgan free press, pontypridd and rhondda chronicle – [Wales] LLGC jan 1910-dec 1943 [mf ed 2003] – 40r – 1 – (cont by: glamorgan free press and rhondda chronicle [jan 1911-dec 1912]; glamorgan free press, pontypridd, rhonda and caerphilly chronicles [jan 1913-dec 1921]; glamorgan free press and rhondda leader, pontypridd, rhondda and caerphilly chronicles and maesteg, garw and ogmore telegraph [jan 1922-dec 1929]; free press and rhondda leader [jan 1930-dec 1943]) – uk Newsplan [072]

Glamorgan gazette – Bridgend, Wales. -w. April 1894-Dec 1977. Lacking 1896; July-Dec 1897; 1899. 77 1 2 reels – 1 – uk British Libr Newspaper [072]

Glamorgan gazette – Maesteg, Wales. -w. Jan-Dec 1979. 3 reels – 1 – uk British Libr Newspaper [072]

Glamorgan marriage index, grooms & brides – pre-1837 – 15mf – 9 – £18.75 – uk Glamorgan FHS [929]

Glamorgan militia lists – 1mf – 9 – £11.25 set – (contains: caerphilly hundred (incl merthyr tydfil) 1819-21; cowbridge & ogmore hundreds 1819-21; dinas powis & kibbor (cardiff) hundreds 1819-21; llangyfelach hundred 1819-20; miskin hundred 1819-21; neath hundred 1819-20; newcastle (bridgend) hundred 1819-21; swansea hundred 1819-20) – uk Glamorgan FHS [929]

Glamorgan & monmouthshire mining accident book – 1933-34 – 1mf – 9 – £1.25 – uk Glamorgan FHS [622]

Glamorgan monumental inscriptions master index 90,000 names – 7mf – 9 – £8.75 – uk Glamorgan FHS [929]

Glamorgan register of electors – 1845/6 – 2mf – 9 – £2.50 – uk Glamorgan FHS [325]

Glamorgan strays collection – n1 (april 2000) – 2mf – 9 – £2.50 – uk Glamorgan FHS [929]

Glamorgan testators pcc wills – 1601-1770 – 1mf – 9 – £1.25 – uk Glamorgan FHS [929]

Glamour – New York NY 1939+ – 1,5,9 – ISSN: 0017-0747 – mf#695 – us UMI ProQuest [740]

The glamour and tragedy of the zulu war / Clements, W H – London: J Lane, [1936] – 1 – us CRL [960]

A glance at london, brussels, and paris – Edinburgh 1829 [mf ed Hildesheim 1995-98] – 1v on 2mf – 9 – €60.00 – 3-487-27801-4 – gw Olms [914]

A glance at some of the beauties and sublimities of switzerland : with excursive remarks on the various objects of interest, presented during a tour through its picturesque scenery / Murray, John – London 1829 [mf ed Hildesheim 1995-98] – (= ser Fbc) – 2mf – 9 – €60.00 – 3-487-29333-1 – gw Olms [914]

A glance at the ecclesiastical councils of new england / Dexter, Henry Martyn – Boston: Wiggin & Lunt, 1867 [mf ed 1990] – 1mf – 9 – 0-7905-7220-6 – (incl bibl ref) – mf#1988-3220 – us ATLA [240]

Glance at the events of 1848 / Habershon, Matthew – London, England. 1849 – 1r – us UF Libraries [240]

Glance at the religious progress of the country in a hundred years : a paper...in rochester, ny, oct 19 1876 / Atwood, Isaac Morgan – Boston: Universalist Pub House, 1876 [mf ed 1992] – (= ser Unitarian/universalist coll) – 1mf – 9 – 0-524-04724-3 – mf#1991-2129 – us ATLA [240]

A glance backward at fifteen years of missionary life in north india / Warren, Joseph – Philadelphia: Presbyterian Board of Publ, 1856 [mf ed 1986] – 1mf – 9 – 0-8370-6630-1 – mf#1986-0630 – us ATLA [920]

Glances over the field of faith and reason, or, christianity in its idea and development : its connection with human progress and unity / Ashley, R K – Boston: Crocker and Brewster, 1855 – 1mf – 9 – 0-7905-9117-0 – mf#1989-2342 – us ATLA [240]

The glands of destiny (a study of the personality) / Cobb, Ivo Geikie – New York: MacMillan, 1928. vii,295p. Bibliog – 1 – us Wisconsin U Libr [150]

Glanes paleolithiques anciennes dans le bassin du guadiana / Brenil, H – 1917 – 1 – sp Bibl Santa Ana [560]

Le glaneur – Port-au-Prince: Imp H Amblard, jan 24, jan 30-feb 3 1900 – 1 sheet – 9 – us CRL [079]

Le glaneur – Port-au-Prince: [Impr V Pierre-Noel, 1ere annee, n1-n5. sep 1925-janv 1926 – 1 sheet – 9 – us CRL [079]

Le glaneur : recueil litteraire des jeunes – Montreal: P J Bedard. 2e annee 1re livraison 10 juin 1892-2e annee 9e livraison 10 oct 1892 (bimthly) [mf ed 1986] – 1r – 5 – (cont by: ecrin litteraire 1181-1714; merged with: le glaneur (levis, quebec) to become: le recueil litteraire) – mf#SEM16P360 – cn Bibl Nat [073]

Le glaneur du haut-rhin – Colmar. 1848-53 – 1 – fr ACRPP [073]

Le glaneur (levis, quebec) – Levis: [s.n.] v1 n1 nov 1890-v1 n12 avril 1892 (irreg) [mf ed 1986] – 1r – 5 – (merged with: le recueil litteraire to become: le glaneur: recueil litteraire des jeunes) – mf#SEM16P361 – cn Bibl Nat [073]

Le glaneur: recueil litteraire des jeunes *see* Le glaneur (levis, quebec)

La glaneuse – Journal des salons et des theatres. puis Journal populaire. Red. en chef J.-A. Granier. no. 1-318. Lyon. juin 1831-mars 1834 – 1 – fr ACRPP [410]

Glanures : les aspirations: poesies canadiennes de w chapman / Lesage, Jules Simeon – [Quebec?: L Brousseau], 1904 – 1mf – 9 – 0-665-85323-8 – mf#85323 – cn Canadiana [410]

Glanvill, Joseph *see*
- Plus ultra
- Saducismus triumphatus; or, full and plain evidence concerning witches and apparitions

Glardon, Auguste *see* Missions dans l'inde

Glarean, Heinrich *see*
- Glariani dokekachordon
- Musicae epitome ex glareani dodecachordo

Glarean, sein leben und seine schriften / Fritzsche, O – Frauenfeld, 1890 – 2mf – 9 – mf#ZWI-88 – ne IDC [240]

Glareani dodekachordon...basileae / Glareanus, Henricus – [Colophon: Basileae...1547] [mf ed 19–] – 9mf – 9 – mf#fiche 480 – us Sibley [780]

Glareanus, H *see* Descriptio de situ helvetiae et vicinis gentibus...de quatuor helvetiorum pagis...cum commentariis osualdi myconii...ad maximilianum augustum...panegyricon

Glareanus, Henricus *see*
- Auss glareani musick ein vsszug
- Glareani dodekachordon...basileae

Glareanus, Henricus Loriti *see* Liber ecclesiasticorum carminum

Glariani dokekachordon / Glarean, Heinrich – 1547 – (= ser Mssa) – 6mf – 9 – €80.00 – mfchl 56b – gw Fischer [780]

Glas – Belgrade, Yugoslavia. Sept 1944-Feb 1952 – 8r – 1 – us L of C Photodup [949]

Glas crnogorca – Cetinje, Yugoslavia. -d. Jan 1893-Dec 1913. 4 reels – 1 – uk British Libr Newspaper [949]

Glas istre – Pula, Yugoslavia. 7 Nov 1947; Mar 1955-Jun 1957 (scattered issues) – 2r – 1 – us L of C Photodup [949]

Glas juga – Skopje, Yugoslavia. Mar 1941 (scattered issues) – 1r – 1 – us L of C Photodup [949]

Glas kanadskih – Windsor Ontario, Canada. 15 may 1952; 17 sep 1953; 1 oct 1954-20 dec 1956; 1957-16 dec 1971; 1972-20 dec 1973; 3 jan-19 dec 1974; 2 jan-11 dec 1975 – 12 3/4r – 1 – uk British Libr Newspaper [072]

Glas naroda – New York: Slovenic Pub Co, 1924-1940; 1946-oct 24 1963 – us CRL [071]

Glas naroda – New York NY, 1893-1950* – 1r – 1 – (slovenian newspaper) – us IHRC [071]

Glas naroda – New York NY, 1912, 1916-21 – 10r – 1 – (slovenian newspaper) – us IHRC [071]

Glas sdz : official organ of the slovenian mutual benefit association = Sdz voice – Cleveland, OH: The Slovenian Mutual Benefit Assoc. v7 n47. dec 2 1948-1959 (weekly) – (in slovenian and english. between 1946-48, title changed from: glas slovenske dobrodelne zveze, to: glas sdz, and between 1956-59, it changed to: our voice. numbering irregular) – us Western Res [071]

Glas slavonije – Osijek, Yugoslavia. Jan 1955; 1956-Jun 1957 – 4r – 1 – us L of C Photodup [949]

Glas sv. antuna – Buenos Aires, Argentina. -m. Oct 1947-June 1955. 2 reels – 1 – uk British Libr Newspaper [072]

Glas svobode – Chicago: M V Konda, aug 1917-apr 1927 – 9r – 1 – us CRL [071]

Glas svobode – Chicago IL, 1907-11, 1918, 1020, 1922* – 1r – 1 – (slovenian newspaper) – us IHRC [071]

Glas svobode – Pueblo CO, 1902-07 – 3r – 1 – (slovenian newspaper) – us IHRC [071]

Der glasberg : roman einer jugend, die hinauf wollte / Wille, Bruno – Berlin, Ullstein c1920 [mf ed 1991] – 1r – 1 – (filmed with: prisoner halm / karl wilke) – mf#3054p – us Wisconsin U Libr [830]

Glasenapp, Carl Friedrich *see* Richard wagner's leben und wirken

Glasenapp, Carl Friedrich [comp] *see* Wagner-lexikon

Glasenapp, Helmuth von *see*
- The doctrine of karman in jain philosophy
- Die lehre vom karman in der philosophie der jainas nach den karmagranthas

Glaser, Adolf *see* Schlitzwang

Glaser, Eduard *see* Jehowah-jovis und die drei soehne noah's

Glaser, Waldemar *see* Ein trupp sa

Glasgow advertiser [glasgow, scotland : 1825] – [Scotland] Glasgow: printed...by A & J M Duncan University 24 may-21 jun 1825 (twice/wk, wkly) [mf ed 2004] – 1r – 1 – uk Newsplan [072]

Glasgow argus – [Scotland] Glasgow: G Mackay 31 jan 1857-feb 1858 (wkly) [mf ed 2003] – 57v on 1r – 1 – (incorp with: national penny press) – uk Newsplan [072]

Glasgow chronicle – [Scotland] Glasgow: D Prentice mar 1811-dec 1832, jan 1838-9 aug 1843 (3/wk) [mf ed 2003] – 39r – 1 – (missing: 1833, 1840-jan 1842; absorbed: glasgow journal (glasgow, scotland : 1741)) – uk Newsplan [072]

Glasgow citizen : or, west of scotland journal & advertiser – [Scotland] Glasgow: J Hedderwick & Sons 6 jan 1844-13 jul 1912 (wkly) [mf ed 2004] – 89r – 1 – (missing: 1847; cont by: glasgow citizen [6 jul 1861-jun 1865]; glasgow weekly citizen [jul 1865-jun 1891]; weekly citizen [jul 1891-jun 1896]; saturday weekly citizen [jul 1896-dec 1906]; weekly citizen [jan 1907-jul 1912]) – uk Newsplan [072]

Glasgow clincher – [Scotland] Glasgow: [A Petrie jul 1897-1903 (mthly) [mf ed 2003] – 1r – 1 – uk Newsplan [072]

Glasgow commonweal – [Scotland] Glasgow: printed & publ for the "Commonweal" Society by the Labour Literature Society apr-nov 1896 (mthly) [mf ed 2004] – 1r – 1 – (cont by: common weal [oct-nov 1896]) – uk Newsplan [072]

Glasgow constitutional – [Scotland] Glasgow: James M'Nab 31 oct 1835-dec 1845 (semiwkly) [mf ed 2003] – 10r – 1 – (cont: constitutional; absorbed: renfrewshire advertiser) – uk Newsplan [072]

Glasgow cornucopia – [Scotland] Glasgow: G Nicholls 19 nov 1831 (wkly) [mf ed 2003] – 1r – 1 – (cont: edinburgh cornucopia; cont by: cornucopia britannica) – uk Newsplan [072]

Glasgow courant / ed by Simpson, Matthew – [Scotland] Glasgow 11 nov 1715-may 1716, 18 nov 1745-oct 1760 (wkly) [mf ed 2003] – 8r – 1 – (cont by: west country intelligence [31 jan-may 1716]; glasgow courant [18 nov-oct 1760]; ceased in 1760; absorbed by: glasgow journal) – uk Newsplan [072]

Glasgow courant and west highland advertiser – [Scotland] Glasgow: Pare, George & Co 7 jul 1855 [mf ed 2004] – 1v on 1r – 1 – uk Newsplan [072]

Glasgow courier *see* Voice of poland

Glasgow daily stock and share list – [Scotland] Glasgow: by authority of the Committee of the Glasgow Stock Exchange 1 jul 1904-2 jan 1948 (daily) [mf ed 2004] – 188r – 1 – (began with 3rd ser n1 on 29 mar 1873; cont: glasgow daily share list; absorbed by: glasgow stock exchange official list) – uk Newsplan [332]

Glasgow evening news – Scotland, UK. 1893. -d. 2 reels – 1 – uk British Libr Newspaper [072]

Glasgow examiner – Scotland. -w. Jan 1858-3 Sept 1864. 6 1 reels – 1 – uk British Libr Newspaper [072]

Glasgow. Faculty of Procurators *see* Report by committee of the faculty of procurators in glasgow appointed to consider and report upon a bill (as amended in committee) to consolidate and amend the laws relating to procedure in the court of session in scotland...

Glasgow free press – [Scotland] Glasgow: William Bennett 1 jan 1823-24 jun 1835 [mf ed 2003] – 13v on 9r – 1 – (missing: 1827-31 & various) – uk Newsplan [072]

Glasgow harlequin : a weekly record of glasgow pantomime & pantomimists – [Scotland] Glasgow: Hodge, Wm and Co 17 dec 1895-14 jan 1896 (wkly) [mf ed 2003] – 1r – 1 – uk Newsplan [790]

Glasgow herald – Glasgow, Scotland 26 aug 1805-30 dec 1850 (imperfect) – 1 – (cont: herald & advertiser [ns] 1 nov 1802-23 aug 1805) – uk British Libr Newspaper [072]

Glasgow highlander : a weekly newspaper for gaels at home and abroad = Gaidheal ghlaschu – [Scotland] Glasgow: A Sinclair 23 sep-25 nov 1933 (wkly) [mf ed 2003] – 1v on 1r – 1 – (in english & gaelic) – uk Newsplan [072]

Glasgow journal (various) / ed by Stalker, Andrew – [Scotland] Glasgow: printed...for Andrew Stalker & Alexander Carlile 27 jul 1741-dec 1817 (wkly) [mf ed 2003] – 16r – 1 – (cont: glasgow weekly journal [jul 1741-jul 1743]; absorbed: glasgow courant in 1760; glasgow chronicle in 1779) – uk Newsplan [072]

Glasgow mercantile advertiser – Scotland. 1853, 1858, 1863, 1868, 1873, 1878 [wkly] – 6r – 1 – (aka: mercantile advertiser & shipping gazette) – uk British Libr Newspaper [072]

Glasgow mercantile advertiser – [Scotland] Glasgow: A Moody 5 oct 1852-dec 1878, jan 1893-jun 1896 (wkly) [mf ed 2003] – 26r – 1 – (cont by: merchantile advertiser and shipping list [jan 1870-jun 1896]) – uk Newsplan [072]

Glasgow Public Library *see* The william smeal collection

Glasgow saturday post – Scotland. -w. May 1840-Apr 1843, Feb 1844-Jan 1845. (2 reels) – 1 – uk British Libr Newspaper [072]

Glasgow sentinel – Scotland. -w. 1850-77. (40 reels) – 1 – uk British Libr Newspaper [072]

Glasgow sentinel (glasgow, scotland: 1809) – [Scotland] Glasgow: W Lane & Co 6 jun 1809-26 feb 1811 (three times/wk) [mf ed 2003] – 2v on 2r – 1 – uk Newsplan [072]

Glasgow sentinel (glasgow, scotland: 1821) – [Scotland] Glasgow: Alexander & Borthwick 10 oct 1821-22 jan 1823 (three times/wk) [mf ed 2003] – 2v on 1r – 1 – (cont: clydesdale journal (hamilton, scotland)) – uk Newsplan [072]

Glasgow south & eastwood extra – Glasgow: Newtext Composition -1997 (wkly) [mf ed 6 jan 1994-] – 1 – (cont by: extra. glasgow south & eastwood) – uk Scotland NatLib [072]

Glasgow sportsman – [Scotland] Glasgow: F Wicks 25 mar-18 apr 1889 (daily) [mf ed 2003] – 22v on 1r – 1 – uk Newsplan [790]

Glasgow times – [Scotland] Glasgow 25 jun 1855-9 jun 1869 – 760v on 12r – 1 – (preceded by a specimen number dated may 17, 1855 with title: the glasgow times and scottish daily advertiser; cont as: glasgow times and western counties chronicle [jan 1860-jun 1869]) – uk Newsplan [072]

Glasgow times and advocate of social, political and ecclesiastical reform – [Scotland] Glasgow: W J Paterson 23 jan-17 apr 1847 (wkly) [mf ed 2003] – 13v on 1r – 1 – uk Newsplan [072]

Glasgow trades council, 1858-1951 – (= ser Labour party in britain, origins and development at local level. series 1) – 14r – 1 – (int by w hamish fraser) – mf#97133 – uk Microform Academic [331]

Glasgow "unique" advertiser and literary journal – [Scotland] Glasgow: Dunlop & Co feb-oct 1892 (mthly) [mf ed 2004] – 1r – 1 – uk Newsplan [072]

Glasgow volunteer gazette : the organ of the citizen training force in glasgow and district, and later, the glasgow volunteer force – [Scotland] Glasgow: W Hodge & Co 5 jun 1915-15 mar 1919 (wkly) [mf ed 2003] – 4r – 1 – (cont by: scottish volunteer gazette) – uk Newsplan [072]

Glasgow, W. Melanchthon *see* History of the reformed presbyterian church

Glasgow weekly herald – [Scotland] Glasgow: H Munro 12 nov 1864-29 apr 1938 (wkly) [mf ed 2003] – 3782v on 124r – 1 – (cont by: weekly herald (glasgow, scotland) [sep 1937-apr 1938]) – uk Newsplan [072]

Glasgow weekly mail – [Scotland] Glasgow: printed by C Gunn 1 mar 1862-dec 1869, jan 1871-78, 1894-9 may 1931 (wkly) [mf ed 2003] – 93r – 1 – (missing: 1870, 1882, 1885, 1891, 1893, 1921; merged with: scottish weekly record to form: weekly mail and record [jul 1915-jun 1920]; weekly record [jul 1920-may 1931]) – uk Newsplan [072]

Glasgow weekly star – [Scotland] Glasgow: R Gillespie 3 jan 1874-13 mar 1875 (wkly) [mf ed 2003] – 1r – 1 – (a weekly literary supplement to the evening star) – uk Newsplan [072]

Glasgow west-end mercury and partick advertiser – [Scotland] Glasgow: W Forbes 7 oct-8 nov 1857 (wkly) [mf ed 2004] – 1r – 1 – uk Newsplan [072]

Glasgow, William Melanchton *see* History of the reformed presbyterian church in america

Glasgow, William Melancthon *see* Cyclopedic manual of the united presbyterian church of north america

Das glashaus – Berlin 1920 n1-17 [mf ed 2004] – 1 – gw Mikrofilm [074]

Glasilo k s k jednote – Chicago IL, 1915* – 1r – 1 – (slovenian newspaper) – us IHRC [071]

Glasilo k s k jednote – Chicago IL, 1915-45 – 16r – 1 – (slovenian newspaper) – us IHRC [071]

Glasilo snpj – Chicago IL, 1910-11, 1913-14* – 1r – 1 – (slovenian newspaper) – us IHRC [071]

Glasilo snpj – Chicago IL, 1910-15 – 2r – 1 – (slovenian newspaper) – us IHRC [071]

Glasmacher : fuenf erzaehlungen / Leutelt, Gustav – Karlsbad: A Kraft, [1944] – 1r – 1 – us Wisconsin U Libr [830]

Glasnik – Calumet MI, 1901-15* – 1r – 1 – (slovenian newspaper) – us IHRC [071]

Glasow, Catharina von *see* Ein beitrag zur kenntnis der genstruktur der phosphoenolpyruvat-carboxylase hoeherer pflanzen

Glass – Redhill, England 1977-91 – 1,5,9 – ISSN: 0017-0984 – mf#3074 – us UMI ProQuest [740]

Glass age – London, England 1968-90 – 1,5,9 – ISSN: 0017-0992 – mf#2752.01 – us UMI ProQuest [720]

Glass and ceramics – Dordrecht, Netherlands 1970+ – ISSN: 0361-7610 – mf#10884 – us UMI ProQuest [740]

GLEANINGS

Glass Bottle Blowers Association of the United States and Canada *see*
– Blowers' report of the proceedings of the final wage conference
– Blowers' report of the sessions of the final wage conference
– Proceedings of the glass bottle blowers' association of the united states and canada
– Proceedings of the...annual convention of the glass bottle blowers' associationof the united states and canada
– Report of proceedings of the glass bottle blowers' association of the united states and canada
– Workers' report of the sessions of the final wage conference between the national vial and bottle manufacturers' association and the glass bottle blow
Glass Bottle Blowers' Association of the United States and Canada *see* Secretary's quarterly statement
Glass, Charles Gordon *see* Stray leaves from scotch and english history
Glass cutter / Window Glass Cutters' League of America – 1938 jul-1946 oct, 1946 nov-1958 aug – 2r – 1 – mf#1057256 – us WHS [680]
Glass, David *see* The alaskan boundary line
Glass digest – New York NY 1981+ – 1,5,9 – ISSN: 0017-1018 – mf#11915,01 – us UMI ProQuest [740]
Glass fiberboard srm for thermal resistance / Hust, Jerome G – Gaithersburg MD: US Dept of Commerce, National Bureau of Standards [mf ed 1985] – 9 – (with bibl) – us GPO [660]
Glass, Harold Maurice *see* South african policy towards basutoland
Glass, Henry Alexander *see*
– The barbone parliament
– The story of the psalters
Glass industry – New York NY 1920+ – 1,5,9 – ISSN: 0017-1026 – mf#6401 – us UMI ProQuest [740]
Glass, mosaics and jewelry : architecture, applied arts, studio arts – (= ser Art exhibition catalogues on microfiche) – 8 catalogues on 12mf – 9 – £100.00 – (individual titles not listed separately) – uk Chadwyck [740]
The glass of fashion up to date – Toronto: Delineator Pub. Co, 18– -189- or 19–] – 9 – mf#P05156 – cn Canadiana [740]
Glass paperweights *see* History of glass
Glass, Paul *see* Der kreis sensburg
Glass workers news / Franklin Co. Columbus – jan 1956-nov 1979 [mthly, bimthly] – 7r – 1 – mf#B9761-9767 – us Ohio Hist [331]
Glass workers news / United Glass and Ceramic Workers of North America – 1956-57, 1958-60, 1961-62, 1963-65, 1966-68, 1969-71, 1972-1982 nov – 7r – 1 – (cont: cio news [glass workers ed: 1944]; cont by: aluminum light) – mf#659716 – us WHS [660]
Glassberg, Abraham *see* Die beschneidung in ihrer geschichtlichen, ethnographischen, religioesen und medicinischen bedeutung
Glassbrenner, Adolf *see*
– Freie blaetter
– Neuer reineke fuchs
– Unterm brennglas
Glasscock, R S *see* Selecting, fitting and showing the beef steer
Glasse, Samuel *see* Sennacherib defeated, and his army destroyed...
Glassworker / Brotherhood of Painters, Decorators, and Paperhangers of America – winter 1946-winter 1960 – 1r – 1 – mf#1057258 – us WHS [640]
Glassworker : official organof the amalgamated glassworkers' international association / Amalgamated Glass Workers' International Association of America – 1903 aug 15-1915 sep 15 – 1r – 1 – mf#1428990 – us WHS [331]
Glastonbury : an address...by the bishop of stepney...on tuesday, aug 3 1897... / Browne, George Forrest – London: SPCK 1897 [mf ed 1993] – (= ser Church historical society (series) 30) – 1mf – 9 – 0-524-05490-8 – (incl ind) – mf#1990-1485 – us ATLA [240]
Glastonbury abbey documents – 13th-16th c – (= ser Archives of the marquess of bath, longleat house, warminster, wiltshire) – 30r – 1 – (with p/g) – uk Microform Academic [025]
Glaswegian – [Scotland] Glasgow 17 jun-23 sep 1886 [mf ed 2003] – 1r – 1 – uk Newsplan [072]
Glatthaar, Joseph T *see* Confederate military manuscripts
Glatz, Karl Jordan *see* Chronik des bickenklosters zu villingen 1238 bis 1614
Glatzel, Max *see* Julius leopold klein als dramatiker
Glatzer kreisblatt – Glatz (Klodzko PL), 1848 8 aug-1846, 1928 – 1 – gw Misc Inst [077]
Der glaube an die gottheit christi : eine studie zur theologie ritchls und kaftan / Lechler, Paul – Berlin: Reuther & Reichard, 1895 [mf ed 1985] – 1mf – 9 – 0-8370-4384-0 – mf#1985-2384 – us ATLA [240]
Der glaube an jesus christus : ein vortrag / Bassermann, Heinrich – Frankfurt a.M: M Diesterweg, 1881 – 1mf – 9 – 0-8370-2573-7 – mf#1985-0573 – us ATLA [240]
Der glaube der modernen wissenschaft gegenueber / Segur, Louis Gaston – Mainz: Franz Kirchheim, 1874 [mf ed 1985] – 1mf – 9 – 0-8370-5219-X – mf#1985-3219 – us ATLA [210]
Glaube + gewissen – Halle (S [an der Saale]): Max Niemeyer [1955-72] [mf ed 1987] – 18v on 5r – 1 – (merged with: evangelisches pfarrerblatt to form: standpunkt) – mf0640b – us ATLA [242]
Glaube, historie und sittlichkeit : eine systematische untersuchung ueber die theologischen prinzipien im denken albert schweitzers / Browarzik, Ulrich – [s.l.: s.n., 1959?] Chicago: Dep of Photodup, U of Chicago Lib, 1968 (1r); Evanston: American Theol Lib Assoc, 1984 (1r) – 1 – 0-8370-0105-6 – mf#1984-B091 – us ATLA [240]
Der glaube lebt : rufe der zeit / Boehme, Herbert – Muenchen: Zentralverlag der NSDAP, F Eher, [1935] [mf ed 1989] – 56p – 1 – mf#7043 – us Wisconsin U Libr [810]
Glaube, liebe und gute werke : eine untersuchung der prinzipiellen eigentuemlichkeit der evangelisch-lutherischen ethik / Bensow, Oscar – Guetersloh: C Bertelsmann, 1906 – (= ser Beitraege zur foerderung christlicher theologie) – 1mf – 9 – 0-524-06079-7 – mf#1991-2392 – us ATLA [242]
Der glaube luthers in seiner freiheit von menschlichen autoritaeten : rede. gehalten bei dem antritt des rectorates der universitaet leipzig... / Brieger, Theodor – Leipzig: A Edelmann, [1892?] – 1mf – 9 – 0-524-05430-4 – mf#1990-1462 – us ATLA [242]
Der glaube nach der anschauung des alten testamentes : eine untersuchung ueber bedeutung von he-remin im alttestamentlichen sprachgebrauch – die ehe nach der lehre des roemischen katechismus / Bach, Ludwig & Sommer, Christian – Guetersloh: C Bertelsmann, 1900 – (= ser Beitraege zur foerderung christlicher theologie) – 1mf – 9 – 0-524-07722-3 – (incl bibl ref) – mf#1992-1105 – us ATLA [221]
Glaube und erfahrung : saetze aus den werken / Grimm, Hans – Muenchen: A Langen/ G Mueller, 1937 – 1r – 1 – us Wisconsin U Libr [430]
Glaube und heimat : evangelisches sonntagsblatt fuer thueringen – v21-46. 1966-91 – Inquire – 1 – (lacks some iss) – mf#ATLA S0429 – us ATLA [242]
Glaube und heimat – Jena DE, 1958 4 may-1986 – 6r – 1 – gw Misc Inst [074]
Glaube und lehre : theologische streitschriften / Lipsius, Richard Adelbert – Kiel, Hadersleben: Schwers'sche Buchhandlung, 1871 [mf ed 2004] – 1r – 1 – 0-524-10486-7 – (incl bibl ref) – mf#b00701 – us ATLA [210]
Der glaube und seine bedeutung fuer erkenntnis, leben und kirche : mit ruecksicht auf die hauptfragen der gegenwart / Koestlin, Julius – Berlin:Reuther & Reichard, 1895 – 1mf – 9 – 0-8370-4422-7 – (incl bibl ref) – mf#1985-2422 – us ATLA [210]
Glaube und unglaube in der weltgeschichte : ein kommentar zu augustins de civitate dei: mit einem exkurs, fruitio dei, ein beitrag zur geschichte der theologie und der mystik / Scholz, Heinrich – Leipzig: JC Hinrichs, 1911 – 1mf – 9 – 0-7905-9630-X – (incl bibl ref) – mf#1989-1355 – us ATLA [240]
Der glaube unserer vaeter : als der germanen ureigenes altes testament und grundlage einer kraeftigeren, nationalen volkserziehung allen vaterlands-freunden "so weit die deutsche zunge klingt" / Hoffmeister, Hermann – Berlin: Kogge & Fritze, 1882 – 2mf – 9 – 0-524-05855-5 – mf#1990-3519 – us ATLA [290]
Glauben und wissen : ausgewaehlte vortraege und aufsaetze / Lipsius, Richard Adelbert – Berlin: CA Schwetschke, 1897 – 2mf – 9 – 0-524-08543-9 – mf#1993-2068 – us ATLA [240]
Glauben und wissen bei den grossen denkern des mittelalters / Betzendoerfer, Walter – Gotha, 1931 – 5mf – 8 – €12.00 – ne Slangenburg [230]
Der glaubensact des christen : nach begriff und fundament / Koenig, Eduard – Erlangen:Andr. Deichert (Georg Boehme), 1891 – 1mf – 9 – 0-8370-4433-2 – (incl bibl ref) – mf#1985-2433 – us ATLA [210]
Ein glaubensbekenntnis : zeitgedichte / Freiligrath, Ferdinand – Mainz: V von Zabern, 1844 (mf ed 1990) – 1r – 1 – (filmed with: neuere politische und sociale gedichte) – us Wisconsin U Libr [810]
Glaubensgewissheit : eine untersuchung ueber die lebensfrage der religion / Heim, Karl – Leipzig: JC Hinrichs, 1916 – 1mf – 9 – 0-7905-7757-7 – mf#1989-0982 – us ATLA [240]
Die glaubenslehre / Ewald, Heinrich – Leipzig: F C W Vogel, 1873-1874 – (= ser Lehre der Bibel von Gott) – 2mf – 9 – 0-8370-1966-4 – mf#1987-6353 – us ATLA [220]
Glaubenslehre / Stephan, Horst – Berlin, Germany. 1941 – 1r – us UF Libraries [943]
Die glaubenslehre der evangelisch-protestantischen kirche : nach ihrer guten begruendung, mit ruecksicht auf das beduerfniss der zeit / Steudel, Johann Christian Friedrich – Tuebingen: CF Osiander, 1834 – 2mf – 9 – 0-524-08651-6 – mf#1993-2111 – us ATLA [242]
Die glaubenslehre der evangelisch-reformirten kirche / Schweizer, A – Zuerich, 1844-1847. 2 v – 14mf – 9 – mf#ZWI-49 – ne IDC [242]
Glaubenslehre und gebraeuche der aelteren abessinischen kirche / Kromrei, Ernst – Leipzig: A Th Engelhardt, 1895 – 1mf – 9 – 0-8370-7641-2 – mf#1986-1641 – us ATLA [240]
Glaubenslos? : erzaehlung / Ebner-Eschenbach, Marie von – Berlin: Gebrueder Paetel, 1893 – 2r – 1 – us Wisconsin U Libr [430]
Glaubenslos? : erzaehlung / Ebner-Eschenbach, Marie von – Berlin: Gebrueder Paetel, 1911 – 1r – 1 – us Wisconsin U Libr [430]
Glaubenslos?; unsuehnbar / Ebner-Eschenbach, Marie von – Leipzig: H Fikentscher, H Schmidt & H Guenther, [1928] – 2r – 1 – us Wisconsin U Libr [430]
Die glaubensparteien in der eidgenossenschaft und ihre beziehungen zum ausland, vornehmlich zum hause habsburg und zu den deutschen protestanten 1527-1531 / Escher, H – Frauenfeld, 1882 – 4mf – 9 – mf#ZWI-87 – ne IDC [240]
Glaubensregel, heilige schrift und taufbekenntnis : untersuchungen ueber die dogmatische autoritaet, ihr werden und ihre geschichte, vornehmlich in der alten kirche / Kunze, Johannes – Leipzig: Doerffling & Franke, 1899 – 2mf – 9 – 0-524-00761-6 – mf#1990-0193 – us ATLA [240]
Glauber, Johann *see* The works of the highly experienced and famous chymis
Die glaubwuerdigkeit der evangelischen geschichte : mit bezug auf dav. friedr. strauss und bruno bauer und die durch dieselben angeregten streitigkeiten / Grimm, Wilibald – Jena: C Hochhausen, 1845 – 1mf – 9 – 0-7905-1602-0 – (incl bibl ref) – mf#1987-1602 – us ATLA [225]
Die glaubwuerdigkeit der evangelischen geschichte : zugleich eine kritik des lebens jesu von strauss / Tholuck, August – 2. aufl. Hamburg: Friedrich Perthes, 1838 – 2mf – 9 – 0-7905-2436-8 – mf#1987-2436 – us ATLA [225]
Die glaubwuerdigkeit des alten testamentes im lichte der inspirationslehre und der literarkritik / Nikel, Johannes – 1. & 2. aufl. Muenster i W: Aschendorff 1908 [mf ed 1992] – (= ser Biblische zeitfragen 1/8) – 1mf – 9 – 0-524-05624-2 – (incl bibl ref) – mf#1992-0479 – us ATLA [221]
Die glaubwuerdigkeit des irenaeischen zeugnisses ueber die abfassung des vierten kanonischen evangeliums / Gutjahr, F S – Graz, 1904 – 4mf – 8 – €11.00 – ne Slangenburg [240]
Die glaubwuerdigkeit des markusevangeliums / Rohr, Ignaz – 1. & 2. aufl. Muenster i W: Aschendorff 1909 [mf ed 1993] – (= ser Biblische zeitfragen 2/4) – 1mf – 9 – 0-524-06156-4 – mf#1992-0823 – us ATLA [225]
Die glaubwuerdigkeit unserer evangelien : ein beitrage zur apologetik / Boese, Heinrich – Freiburg i B: Herder, 1895 – (= ser Ergaenzungshefte zu den "Stimmen aus Maria-Laach") – 1mf – 9 – 0-524-05971-3 – mf#1992-0708 – us ATLA [220]
Glaue, Paul *see*
– Fragmente einer griechischen uebersetzung des samaritanischen pentateuchs
– Die vorlesung heiliger schriften im gottesdienste
Glavnaia astronomicheskaia observatoriia [Soviet Union] *see* Izvestiia glavnoi astronomicheskoi observatorii
Glavneishie reformy, provedennye n kh bunge v finansovoi sisteme rossii : opyt kriticheskoi otsenki deiatel'nosti n kh bunge, kak ministra finansov, (1881-1887 gg) / Kovan'ko, P – Kiev, 1901 – 9mf – 9 – mf#REF-474 – ne IDC [332]
Glavnoe Pravlenie Gosudarstvennogo Strakhovaniia (Gosstrakh) *see*
– Dekrety o gosudarstvennom strakhovanii
– Polozhenie o gosudarstvennom strakhovanii soiuza ssr
Glavnoe Pravlenie Gosudarstvennogo Strakhovaniia (Gosstrakh Soiuza SSR) *see* Svod rasporiazhenii glavnogo pravleniia, deistvuiushchikh na 1-oe ianvaria 1926 goda
Glavnoe Pravlenie Gosudarstvennogo Strakhovaniia SSSR (GOSSTRAKH) *see* Shest'desiat let sel'skogo obiazatel'nogo strakhovaniia
Glavnoe Pravlenie Gosudarstvennogo Strakhovaniia SSSR (Gosstrakh) *see* Piat' let gosudarstvennogo strakhovaniia v sssr
Glavnoe Pravlenie Gosudarstvennogo Strakhovaniia. Tarifno-statisticheskii otdel *see* Statisticheskie svedeniia po strakhovaniiam ot ognia za 1924-25 operatsionnyi god
Glawe, Walther *see* Die beziehung des christentums zum griechischen heidentum
Glaze – 1980 jul, dec, 1981 dec, 1982 may, oct – 1r – 1 – mf#802607 – us WHS [071]
Glazebrook, Michael George *see*
– The end of the law
– Studies in the book of isaiah
Glazenap, Sergei Pavlovich *see*
– Mesures micrometriques d'etoiles doubles faites a st-petersbourg et a domkino
Glazer, A R *see* Effects of submaximal exercise on the mood of female bulimics
Glazier, Richard *see*
– Historical and descriptive notes on ornament
– A manual of historic ornament
Glazman, Ari *see* Fentster tsu der velt
Gle annali avero le vite de' principi et signori della casa othomana / Sansovino, F – Venetia, 1571 – 3mf – 9 – mf#H-8318 – ne IDC [956]
The gleam / Younghusband, Francis Edward – London: John Murray, 1923 – (= ser Samp: indian books) – us CRL [280]
Gleams from paul's prison : or, studies for the daily life in the epistle to the philippians / Hoyt, Wayland – Philadelphia: Griffith and Rowland, 1903 – 2mf – 9 – 0-524-03975-5 – (incl bibl ref) – mf#1992-0018 – us ATLA [220]
Gleams of sunshine, optimistic poems / Chant, Joseph Horatio – Toronto: Printed...by W Briggs, 1915 – 3mf – 9 – 0-665-73080-2 – mf#73080 – cn Canadiana [810]
Gleaner – Arena WI. 1894 nov 8, dec 6 – 1r – 1 – mf#1794707 – us WHS [071]
Gleaner – Chatham, NB. 1829-80 – 17r – 1 – cn Library Assoc [079]
The gleaner – Brantford, Ont: T Somerville, [1886?-19–] – 9 – mf#P04275 – cn Canadiana [240]
The gleaner – Kingston: The Gleaner Co Ltd, dec 7 1992- – us CRL [070]
The gleaner – San Francisco, CA; Portland, OR.1865 – 1 – us AJPC [071]
The gleaner and de cordova's advertising sheet – Kingston. Jamaica. -sw. 8 Sep, 8, 24 Nov, 25 Dec 1866, 8, 24 Jan, 9, 25 Feb, 11, 13 Mar, 24 Apr, 24 Oct 1867, 26 Mar 1868, 27 May, 9, 18 Jun 1875. (13 ft) – 1 – uk British Libr Newspaper [072]
Gleanings – Kikungshan, Honan: Lutheran United Mission. v2-12 (1920-1935) [mf ed 2005] – 1r [ill] – 1 – (cont by: china gleanings; lacks: v2 n3-4, v3 [entire vol], v4 n1,2 [entire iss], v5 [entire vol], v6 n1, v7 n3) – mf#2005C-S016 – us ATLA [242]
Gleanings : gathered at bapu's feet / Mirabehn – Ahmedabad: Navajivan Pub House, 1949 – (= ser Samp: indian books) – us CRL [320]
Gleanings and remarks, collected during many months of residence at buenos ayres, and within the upper country : with a prefatory account of the expedition from england... / Gillespie, Alexander – Leeds 1818 [mf ed Hildesheim 1995-98] – 1v on 3mf – 9 – €90.00 – ISBN-10: 3-487-26851-5 – ISBN-13: 978-3-487-26851-4 – gw Olms [880]
Gleanings from fifty years in china / Little, Archibald John – London: Sampson Low, Marston [1910] [mf ed 1995] – (= ser Yale coll) – xvi/335p (ill) – 1 – 0-524-09499-3 – (rev by mrs archibald little) – mf#1995-0499 – us ATLA [915]
Gleanings from quebec / Fairchild, George Moore – Quebec: F Carrel, 1908 [mf ed 1995] – 3mf – 9 – 0-665-74211-8 – mf#74211 – cn Canadiana [917]
Gleanings from the church-yard : a selection of old inscriptions / Fordyce, Alexander Dingwall – S.l: s.n, 1880 – 1mf – 9 – mf#03147 – cn Canadiana [929]
Gleanings from the nineteenth century / Croil, James – [Montreal?: Mitchell & Wilson], 1913 – 3mf – 9 – 0-665-73008-X – mf#73008 – cn Canadiana [900]
Gleanings from the public press relating to the appointment of the honorable j r gowan to the senate of canada – [Toronto?: s.n, 1885?] [mf ed 1993] – 1mf – 9 – 0-665-91415-6 – mf#91415 – cn Canadiana [325]
Gleanings from westminster abbey : with appendices, supplying further particulars, and completing the history of the abbey buildings... / Scott, George Gilbert – Oxford, London: J H & Jas Parker, 1861 – (= ser 19th c art & architecture) – 2mf – 9 – (ill by numerous plates & woodcuts) – mf#4.1.209 – uk Chadwyck [720]
Gleanings in africa : exhibiting a faithful and correct view of the manners and customs of the inhabitants of the cape of good hope, and surrounding country... – London 1805 [mf ed Hildesheim 1995-98] – 1v on 3mf [ill] – 9 – €90.00 – 3-487-27292-X – gw Olms [960]
Gleanings in africa – New York, NY. 1969 – 1r – us UF Libraries [960]
Gleanings in bee culture – Medina OH 1873-1992 – 1,5,9 – (cont by: bee culture) – ISSN: 0017-114X – mf#1894 – us UMI ProQuest [630]

Gleanings in buddha-fields : studies of hand and soul in the far east / Hearn, L – Boston, New York, 1897 – 4mf – 9 – mf#HTM-80 – ne IDC [915]

Gleanings in buddha-fields : studies of hand and soul in the far east / Hearn, Lafcadio – Boston: Houghton, Mifflin, 1898 – 1mf – 9 – 0-524-00892-2 – mf#1990-2115 – us ATLA [915]

Gleanings in harvest fields, 1889-1907 / Methodist New Connexion Missions – 15mf – 9 – (missing: 1889(p1-36); 1890(p49-128); 1891(p161-224, 237-260); 1892/1893(p285-344); 1893/1894(p369-396); 1899(p97-112); 1903(p161-176); 1904(p81-96); 1905(p113-128); 1906(p161-176)) – mf#H-2742 – ne IDC [956]

Gleanings in holy fields / Macmillan, Hugh – London: Macmillan 1899 [mf ed 1984] – 2mf – 9 – 0-8370-0752-6 – mf#1984-6247 – us ATLA [240]

Gleanings in the italian field of celtic epigraphy / Rhys, John – s.l, s.l? 1919 – 1r – us UF Libraries [490]

Gleanings in the west of ireland / Osborne, Sidney Godolphin. Lord – London, 1850 – (= ser 19th c ireland) – 3mf – 9 – mf#1.1.3180 – uk Chadwyck [941]

Gleanings of a wanderer : in various parts of england, scotland, and north wales – London 1805 [erschienen] 1806 [mf ed Hildesheim 1995-98] – 1v on 2mf [ill] – 9 – €60.00 – 3-487-26455-2 – gw Olms [914]

Gleason, Arthur H *see* Papers

Gleason, William J *see* History of soldiers' and sailors' monument

Glebe – New York. v. 1-2. sept 1913-nov 1914 (ncomplete) – 1 – us NY Public [073]

Glebe, Jean de la *see*
- Le diable est aux vaches
- L'industrie avicole dans la province de quebec

Glebe weekly – Glebe, jan 1972-aug 1989 – 23r – at Pascoe [079]

Gledhill, Alan *see* The republic of india

Gleichen-Russwurm, Alexander, Freiherr von *see* Im gruenen salon

Gleichen-Russwurm, Alexander von *see* Klassische schoenheit

Gleichen-Russwurm, Alexander von et al *see* Das gelbe buch

Gleiches recht : soziales drama in vier akten / Grelling, Richard – Berlin: H Steinitz, 1892 – 1r – 1 – us Wisconsin U Libr [820]

Die gleichheit – New York. N.Y. (Equality). 1913-19 – 1 – us AJPC [071]

Die gleichheit *see* Die arbeiterin

Gleichheit – New York. v. 1-5. Dec 5 1913-Jan 11 1919 – 1 – us NY Public [071]

Gleichheit : official organ of the social democrats – Wiener Neustadt, Austria jan 1870-sep 1877 [mf ed Norman Ross] – 1r – 1 – mf#nrp-2156 – us UMI ProQuest [074]

Gleichheit – Wien (A), 1874-75, 1877 [gaps] – 1r – 1 – gw Misc Inst [074]

Das gleichnis vom verlorenen sohn : lukas 15, 11-32 / Koegel, Julius – Berlin: Edwin Runge 1909 [mf ed 1989] – (= ser Biblische zeit- und streitfragen 5/9) – 1mf – 9 – 0-7905-2725-1 – mf#1987-2725 – us ATLA [220]

Die gleichnisreden jesu im allgemeinen / Juelicher, Adolf – 2. neubearb aufl. Freiburg i. B: J C B Mohr, 1899 – 1mf – 9 – 0-7905-2123-7 – (incl bibl ref and ind) – mf#1987-2123 – us ATLA [220]

Die gleichnisreden jesu im lichte der rabbinischen gleichnisse des neutestamentlichen zeitalters : ein beitrag zum streit um die "christusmythe" und eine widerlegung der gleichnistheorie juelichers / Fiebig, Paul – Tuebingen: J C B Mohr (Paul Siebeck), 1912 – 1mf – 9 – 0-7905-0884-2 – (incl ind) – mf#1987-0884 – us ATLA [225]

Die gleichnisse jesu : zugleich eine anleitung zu einem quellenmaessigen verstaendnis der evangelien / Weinel, Heinrich – Leipzig: B G Teubner, 1904 – (= ser Aus natur und geisteswelt) – 1mf – 9 – 0-524-04815-0 – (incl bibl ref) – mf#1992-0235 – us ATLA [220]

Gleig, George Robert *see*
- The great problem
- The history of the british empire in india
- A narrative of the campaigns of the british army, at washington, baltimore, and new orleans

Gleim, Johann Wilhelm Ludewig *see*
- Briefwechsel zwischen gleim und heinse
- Briefwechsel zwischen gleim und ramler

Gleim, Johann Wilhelm Ludwig *see*
- Briefwechsel zwischen gleim und uz
- Preussische kriegslieder von einem grenadier

Gleim und die klassiker goethe, schiller, herder : ein beitrag zur literaturgeschichte des 18. jahrhunderts / Kozlowski, Felix von – Halle a/S: Verlag der Buchhandlung des Waisenhauses, 1906 (mf ed 1990) – 1r – 1 – (filmed with: stilprobleme in gessners kunst und dichtung. incl bibl ref) – us Wisconsin U Libr [430]

Glen Covenant, New Hampshire. Glen Covenant Free Will Baptist Church *see* Records

Glen flora star – Glen Flora WI. 1901 apr 11-1903 apr 2 – 1r – 1 – mf#917913 – us WHS [071]

Glen, Francis Wayland *see*
- Annexation
- Continental union versus reciprocity

Glen innes examiner – Glen Innes. 1874-78, 1880-81, 1883-88, 1890-1901, 1911, 1931-35, 1946-68 – at Pascoe [079]

Glen innes examiner – Glen Innes, jan 1931-dec 1947 – 14r – A$967.60 vesicular A$1044.60 silver – at Pascoe [079]

Glen innes examiner – Glen Innes, jan 1948-1955 – 4r – A$176.00 vesicular A$198.00 silver – at Pascoe [079]

Glen innes examiner – Glen Innes, jan 1969-dec 1996 – at Pascoe [079]

Glen innes examiner – Glen Innes, jul 1897-dec 1907 – 4r – A$154.00 vesicular A$176.00 silver – at Pascoe [079]

Glen innes guardian – Glen Innes, jan 1899-dec 1906 – 3r – A$204.56 vesicular A$221.06 silver – at Pascoe [079]

Glen, William *see* Journal of a tour from astrachan to karass, north of the mountains of caucasus

The glenbow collection – Canada. 1885-1938 – 25r – 1 – (a coll of significant publ related to the history and economic development of alberta) – cn Commonwealth Imaging [971]

Glencoe baptist church. glencoe, kentucky : church records – 1878-Nov 1956 – 1 – us Southern Baptist [242]

Glencoe transcript – Ontario Prov., CN. apr 1873-1914 – 1 – cn Commonwealth Imaging [071]

Glendale – 1923-50; 1986; 1988 – (= ser California telephone directory coll) – 19r – 1 – $950.00 – mf#P00031 – us Library Micro [917]

[Glendale-] asbarez – CA. 1974- – 48r – 1 – $2880.00 (subs $160y) – mf#C02368 – us Library Micro [071]

Glendale baptist church. nashville, tennessee : church records – 1950-69. (Includes bulletins and newsletters) – 1 – us Southern Baptist [242]

[Glendale-] california magyarsag – CA. 1957- – 14r – 1 – $60.00 – mf#C02374 – us Library Micro [071]

[Glendale-] glendale evening news – CA. 1914; Anniversary Issue-Fall of 1914; 1915; 1921-1928 – 73r – 1 – $4380.00 – mf#H04012 – us Library Micro [071]

[Glendale-] glendale news – CA. 1905; 1906-1913; 1920-1928 – 63r – 1 – $3780.00 – mf#H03229 – us Library Micro [071]

[Glendale-] glendale news-press – CA. 1928-1931; 1933-1981; 1983 – 842r – 1 – $50,520.00 (subs $600y) – mf#H04013 – us Library Micro [071]

[Glendale-] glendale scene – CA. 1964 – 1r – 1 – $60.00 – mf#H03228 – us Library Micro [071]

Glendale herald – Glendale, Shorewood, Whitefish Bay WI. 1955 aug 7/1955 dec 31-1984 jan/feb 2 [gaps] – 39r – 1 – (cont: glendale-town times; cont by: herald [brown deer, wis ed: 1981]; brown deer/glendale herald) – mf#946720 – us WHS [071]

Glendale log – Glendale OR: C J Shorb, [wkly] – 1 – (successor to: glendale news (1902-26). 1928-30 incl newspaper pub by glendale high school students) – us Oregon Lib [071]

The glendale log *see* Glendale news

Glendale news – Glendale OR: H W Hulbert, 1902-26 [wkly] – 1 – (succeeded by: glendale log) – us Oregon Lib [071]

Glendale news *see* Glendale log

Glendale-brown deer news – Bayside, Brown Deer, Glendale, River Hills WI. 1958 oct-1964 may – 1r – 1 – (cont: town of milwaukee news) – mf#920585 – us WHS [071]

Glendale/burbank – 1987; 1991- – (= ser California telephone directory coll) – 7r – 1 – $350.00 – mf#P00032 – us Library Micro [917]

Glendale-town times – Glendale, Shorewood WI. 1953 aug 7-1953 dec 25, 1953 sep 18-1954, 1955 jan 1-1955 aug 5 – 3r – 1 – (continued by: glendale herald) – mf#946716 – us WHS [071]

Glendive times – Montana. 1882 jan 19-1884 dec 27 – 1r – 1 – mf#949243 – us WHS [071]

[Glendora-] glendora magazine – CA. 1983- – 3r – 1 – $180.00 (subs $50y) – mf#R04030 – us Library Micro [071]

[Glendora-] glendora press – CA. 1926- – 5r – 1 – $300.00 (subs $50y) – mf#RH03230 – us Library Micro [071]

[Glendora-] glendora signal – CA. 1887- – 2r – 1 – $120.00 (subs $50y) – mf#RC03600 – us Library Micro [071]

Glenelg, Charles Grant, Baron *see*
- Copies or extracts of despatches from sir f b head
- Extract of a despatch from lord glenelg to the earl of durham
- Lord glenelg's despatches to sir f b head
- Lower canada
- Papers relative to the affairs of lower canada
- Return to an address of the honourable the house of commons, dated 5 march 1839

Glenesk, Algernon Borthwick, Baron *see* The origin and objects of the primrose league

Glenn, Cherie A *see* Attitudes of therapeutic recreation professionals toward persons with aids and the relationship of their attitude to their knowledge of aids

[Glenn county-] butte, colusa, glenn, nevada, placer, shasta, sutter, tehama and yuba counties – CA. 1892-1894 – 1r – 1 – $50.00 – mf#D008 – us Library Micro [978]

Glenn/tehama counties – 1916-33; 1992- – (= ser California telephone directory coll) – 20r – 1 – $1000.00 – mf#P00033 – us Library Micro [917]

Glenrock item – Glenrock, PA. -w 1874-1886; 1886-1943 – 13 – $25.00r – us IMR [071]

Glenrothes gazette, leslie and markinch news – Kirkcaldy: Strachan & Livingston Ltd 1962- (wkly) [mf ed 6 jan 1994-] – 1 – (not publ: 13 mar 1980, 9 jul-13 aug 1981) – ISSN: 1354-6074 – uk Scotland NatLib [072]

Glensharrold in 1888 / Pease, Alfred Edward – [London], 1888 – (= ser 19th c ireland) – 1mf – 9 – mf#1.1.1942 – uk Chadwyck [339]

The glenvil globe – Glenvil, NE: A D Scott. v1 n1. feb 12 1915-18// (wkly) [mf ed with gaps] – 1r – 1 – (occasional articles in german) – us NE Hist [071]

The glenville bee – Glenville, NE: S Lounsbury, 1899 (wkly) [mf ed -jan 11 1901 (gaps)] – 1r – 1 – us NE Hist [071]

Glenville surprise – Glenville, NE: N F Kletzing, 1895-v1 n40. dec 6 1895 (wkly) [mf ed v1 n40. dec 6 1895 filmed [1973]] – 1r – 1 – (absorbed by: fairfield tribune) – us NE Hist [071]

Glenwood gazette *see* Garfield county miscellaneous newspapers

Glenwood high country gazette *see* Garfield county miscellaneous newspapers

Glenwood springs reminder record *see* Garfield county miscellaneous newspapers

Gletscherfahrten in den berner alpen / Roth, Abraham – Berlin 1861 [mf ed Hildesheim 1995-98] – 2mf – 9 – €60.00 – 3-487-29360-9 – gw Olms [914]

Gley, Gerard *see*
- Langue et litterature des anciens francs
- Voyage en allemagne et en pologne

Glick, Jeffrey *see* A case study of selected effects of an organized summer residential camp upon staff memebers

Glickson, Moshe *see* 'Olamenu

Glidden [enterprise] – Glidden WI. 1904 jun 10/1906-1995 [gaps] – 42r – 1 – mf#1043714 – us WHS [071]

Gliederung der sudafrikanischen bantusprachen / Van Warmelo, Nicolaas Jacobus – s.l, s.l? 1927 – 1r – us UF Libraries [470]

Gliemann, Theodor *see* Geographische beschreibung von island

Glienke, Wolfgang *see* Der einfluss einer hiv-1-infektion humaner monozyten/makrophagen auf die genexpression immunregulatorischer proteine in vitro

Gliksman, Baruch Bendet *see* Sefer vikuah

Gliksman, Pinhas Zelig *see*
- Rabenu elyakim gets
- Rav shel simhah

Glim / Inns of Court Students Union. London – v1-4 nos 11-31 (all publ?) – 54mf – 9 – $81.00 – mf#LLMC 84-283 – us LLMC [340]

Glimpse at the indian mission-field and leper asylums in 1886-1887 [microform] / Bailey, Wellesley Crosby – London: John F Shaw, [1888] [mf ed 1995] – (= ser Yale coll) – iv/188p (ill) – 1 – 0-524-09415-2 – mf#1995-0415 – us ATLA [362]

Glimpses from central honan / ed by Carlberg, Gustav – [s.l: s.n.] 1922-23 [mthly ex jul & aug] [mf ed 2006] – v1-2 (1922-23) on 1r – 1 – (some pgs damaged; publ by the augustana synod mission, mar-nov 1923; cont by: honan glimpses) – mf#2006c-s002 – us ATLA [242]

Glimpses in pioneer life on puget sound / Atwood, A – Seattle: Denny-Coryell, 1903 [mf ed 1992] – (= ser Methodist coll) – 2mf – 9 – 0-524-05175-5 – mf#1990-5094 – us ATLA [242]

Glimpses of africa, west and southwest coast : containing the author's impressions and observations during a voyage of six thousand miles from sierra leone to st. paul de loanda and return, including the rio del ray and cameroons rivers, and the congo river, from its mouth to matadi / Smith, Charles Spencer – Nashville, TN: Publ House AME Church Sunday School Union, 1895 – 1mf – 9 – 0-7905-6446-7 – mf#1988-2446 – us ATLA [916]

Glimpses of alaska : a collection of views of the interior of alaska and the klondike district / Wilson, Veazie – Chicago: Rand, McNally, 1897 [mf ed 1980] – 3mf – 9 – mf#09154 – cn Canadiana [770]

Glimpses of bengal : selected from the letters of sir rabindranath tagore, 1885-1895 – London: Macmillan and Co, 1921 – (= ser Samp: indian books) – us CRL [954]

Glimpses of dakkan history / Rama Rao, M – Bombay; New York: Orient Longmans, 1951 – (= ser Samp: indian books) – us CRL [954]

Glimpses of destiny from the book / Chisholm, Murdoch – Halifax [NS]: T C Allen, [1917?] – 2mf – 9 – 0-665-74157-X – (incl app) – mf#74157 – cn Canadiana [220]

Glimpses of fifty years : the autobiography of an american woman / Willard, Frances Elizabeth – Boston: Woman's Temperance Publication Association, 1889 – (= ser Women & the church in america) – 2mf – 9 – 0-8370-1414-X – mf#1984-2189 – us ATLA [240]

Glimpses of gandhiji / Diwakar, Ranganath Ramachandra – Bombay: Hind Kitabs, 1949 – (= ser Samp: indian books) – (foreword by sardar vallabhbhai patel) – us CRL [920]

Glimpses of george fox and his friends / Budge, Jane – London: SW Partridge, [18–?] – 1mf – 9 – 0-7905-9908-2 – mf#1989-1633 – us ATLA [242]

Glimpses of glory : or, incentives to holy living: an antidote to weariness in well-doing, and comfort for the afflicted and bereaved / ed by Zethar – Toronto: W Briggs; Montreal: C W Coates, 1890 – 2mf – 9 – mf#26070 – cn Canadiana [240]

Glimpses of god : and other sermons / Newton, Benjamin Gwernydd – Cleveland: Franklin Avenue Congregational Church, 1897 – 1mf – 9 – 0-8370-7414-2 – mf#1986-1414 – us ATLA [240]

Glimpses of india : a unique collection of landscapes and architectural beauties / Thakur Singh, S G – Calcutta: Punjab Fine Art Association, [19–]- – (= ser Samp: indian books) – (foreword by rabindra nath tagore; int by abanindra nath tagore) – us CRL [750]

Glimpses of indian life / Streatfeild, Henrietta S – London: Marshall Brothers [1908] [mf ed 1995] – (= ser Yale coll) – x/171p (ill) – 1 – 0-524-09396-2 – mf#1995-0396 – us ATLA [954]

Glimpses of jesus : or, letters of c.h. balsbaugh. containing also his autobiography / Balsbaugh, Christian Hervey – Mt Morris, IL: James M Neff, 1895 – 2mf – 9 – 0-524-03488-5 – mf#1990-4710 – us ATLA [240]

Glimpses of life in bermuda and the tropics / Newton, Margaret – London, England. 1897 – 1r – us UF Libraries [972]

Glimpses of mughal architecture / Saraswati, Sarasi Kumar; ed by Goswami, A – [Sl: sn, 1953] (Calcutta: Gossain & Co) – (= ser Samp: indian books) – (int with historical analysis by jadunath sarkar) – us CRL [720]

Glimpses of old english homes / Balch, Elizabeth – London 1890 – (= ser 19th c art & architecture) – 3mf – 9 – mf#4.1.294 – uk Chadwyck [720]

Glimpses of sunshine and shade in the far north : or, my travels in the land of the midnight sun / Craig, Lulu Alice – Cincinnati: Editor Pub Co, 1900 – 2mf – 9 – mf#14824 – cn Canadiana [810]

Glimpses of the ages / Scholes, Theophilus E Samuel – or, The "superior" and "inferior" races, so-called discussed in the light of science and history – (filmed with: fournier, g la raza negra es la mas antigua de las razas humanas) – us CRL [573]

Glimpses of the past / Wordsworth, Elizabeth – London: AR Mowbray, [1912?] – 1mf – 9 – 0-524-05389-8 – mf#1991-2295 – us ATLA [378]

Glimpses of the past in the red river settlement : from letters of mr john pritchard, 1805-1836 – Middlechurch, Man: Rupert's Land Indian Industrial School Press, 1892 – 1mf – 9 – (notes by george bryce) – mf#30452 – cn Canadiana [240]

Glimpses of the supernatural : being facts, records and traditions relating to dreams, omens, miraculous occurrences, apparitions, wraiths, warnings, second-sight, witchcraft, necromancy, etc / ed by Lee, Frederick George – New York: G W Carleton, 1875 – 1mf – 9 – 0-7905-6532-3 – mf#1988-2532 – us ATLA [130]

Glimpses of the unseen : a study of dreams, premonitions, prayer and remarkable answers, hypnotism, spiritualism... / Austin, Benjamin Fish – Toronto, Brantford Ont: Bradley-Garretson, 1898? – 6mf – 9 – (int by e i badgley) – mf#32089 – cn Canadiana [130]

Glimpses of three continents : a series of travels in india, the bible lands, and europe / Foss, Claude William – Rock Island, Ill: Augustana Book Concern, 1912 – 2mf – 9 – 0-524-04548-8 – mf#1991-2112 – us ATLA [910]

Glimpses of world history : being further letters to his daughter, written in prison, and containing a rambling account of history for young people / Nehru, Jawaharlal – Allahabad: Kitabistan, 1934-1935 – (= ser Samp: indian books) – us CRL [900]

Glinka, Sergej *see* Geschichte der uebersiedlung von vierzig tausend armeniern

Glisson, Silas Nease *see* Cultural nationalism and colonialism in nineteenth-century irish horror fiction

Gloag, Paton J *see*
- A critical and exegetical commentary on the acts of the apostles
- National religion

Gloag, Paton James *see*
- Closing address
- Death, gain to the believer
- Introduction to the catholic epistles
- Introduction to the johannine writings
- Introduction to the pauline epistles
- Introduction to the synoptic gospels
- Life of paul
- The messianic prophecies
- The primeval world
- A treatise on justification by faith

Global and planetary change – Oxford, England 1989+ – 1,5,9 – ISSN: 0921-8181 – mf#42562 – us UMI ProQuest [550]

Global biogeochemical cycles – Quarterly. – $148.00 – (v6 1992 $135.00. v8 1994 $185.00. v9 1995 $250.00) – us AGU [550]

Global communications – Shawnee Mission KS 1991 – 1 – ISSN: 0195-2250 – mf#12335 – us UMI ProQuest [380]

Global cosmetic industry – Carol Stream IL 1999+ – 1,5,9 – (cont: dci) – ISSN: 1523-9470 – mf#2551.02 – us UMI ProQuest [640]

Global economic outlook – Toronto, Canada 1990-99 – 1,5,9 – ISSN: 0820-5167 – mf#18213.03 – us UMI ProQuest [330]

Global environmental change – Oxford, England 1990+ – 1,5,9 – ISSN: 0959-3780 – mf#18298 – us UMI ProQuest [333]

Global environmental change [aasms60] : the role of space in understanding earth – 1990 – (= ser Aasms 1968) – 1paper on 1mf – 9 – $10.00 – 0-87703-324-2 – (suppl to vol 76, science and technology) – us Univelt [550]

Global environmental change, pt b : environmental hazards – Oxford, England 1999+ – 1,5,9 – ISSN: 1464-2867 – mf#42843 – us UMI ProQuest [333]

Global finance – New York NY 1990+ – 1,5,9 – ISSN: 0896-4181 – mf#18357 – us UMI ProQuest [332]

Global fund news – London, England 2001+ – 1,5,9 – ISSN: 1529-5710 – mf#32365 – us UMI ProQuest [332]

Global governance – v1-7. 1995-2001 – 9 – $259.00 set – ISSN: 1075-2846 – mf#116981 – us Hein [340]

Global outlook – Toronto, Canada 2000+ – 1,5,9 – ISSN: 1495-6764 – mf#18213,03 – us UMI ProQuest [650]

Global power report – New York NY 1999+ – 1,5,9 – ISSN: 1095-6441 – mf#22916,02 – us UMI ProQuest [333]

Global trade – Philadelphia PA 1989-92 – 1,5,9 – (cont by: global trade & transportation; cont: american import/export global trade) – ISSN: 1060-0906 – mf#311.08 – us UMI ProQuest [337]

Global trade & transportation – Philadelphia PA 1993-94 – 1,5,9 – (cont: global trade) – ISSN: 1069-2843 – mf#311.08 – us UMI ProQuest [337]

Globale konvergenz von zufallsstrategien in der optimierung : eine analyse der uebergangswahrscheinlichkeiten in markov-ketten auf die bedingung der global asymptomischen konvergenz und die entwicklung einer neuen optimierungsstrategie evolutionary accepting / Krimphove, Frank – (mf ed 1995) – 1mf – 9 – €30.00 – 3-8267-2272-8 – mf#DHS 2272 – gw Frankfurter [510]

Globalization and the russian far east prospects for intergration / Anders, Rainer-Elk – 2001 – 3mf – 9 – €50.11 – 3-8267-2766-5 – mf#DHS 2766 – gw Frankfurter [327]

Globe – Toronto, Canada. 2 jan 1863-30 sep 1864; 1870-21 dec 1877; 1878-24 jun 1881; 15 jun 1883-jun 1885; 1893-3 feb 1909 – 55 1/2r – 1 – (aka: weekly globe and canada farmer; globe and canada farmer) – uk British Libr Newspaper [072]

Globe – Camp Lejeune NC. 1948 sep 9-dec, 1949-51, 1951-54, 1955 – 4r – 1 – (cont: camp lejeune globe; camp lejeune globe; camp lejeune globe; cont by: camp lejeune globe [camp lejeune, nc: 1956]) – mf#3626105 – us WHS [071]

Globe – Camp Lejeune NC. 1976 jul 22-1977, 1978, 1979, 1980, 1983 – 5r – 1 – (cont: camp lejeune globe [camp lejeune, nc: 1956]; cont by: camp lejeune globe [camp lejeune, nc: 1985]) – mf#3627493 – us WHS [071]

Globe – Washington DC. 1831 dec 17, 1836 jan 23, apr 2, nov 12, 1841 jan 11-14, feb 15, mar 4 – 1r – 1 – (continued by: weekly union (washington, dc)) – mf#845764 – us WHS [071]

Globe / Defense Language Institute (US) – 1979 jun 7-1983 dec 15 – 1r – 1 – mf#1002317 – us WHS [327]

Globe – La Salle IL 1819 – 1,5,9 – mf#3810 – us UMI ProQuest [410]

Globe / Montgomery Co. Dayton – oct 1941-jan 1942, jun-aug 1942 [wkly] – 1r – 1 – mf#B5002 – us Ohio Hist [071]

Globe – Sydney, Australia. 28 jan 1886-14 jun 1887 – 6r – 1 – uk British Libr Newspaper [072]

Globe – Toronto, ON: G Brown, 1844-49, 1858-69 – 50r – 1 – ISSN: 0839-3680 – cn Library Assoc [071]

Le globe – Journal des interets economiques, puis Revue economique hebdomadaire. Dir., A. Coste, puis A. Burdeau. Paris. quot., puis hebd. 1871-85 – 1 – fr ACRPP [330]

Le globe : journal philosophique et litteraire – Paris. v1-8. 1824-1830 – 210mf – 8 – (missing: v8(161, 254, 255, 283, 297, 298, 303, 310) 1830) – mf#H-1380 – ne IDC [073]

Le globe – Paris. n1-33. 15 janv-16 fevr 1868 – 1 – (politique, litteraire et financier) – fr ACRPP [944]

The globe – London. -d. 1822-99, 228mqn reels – 1 – uk British Libr Newspaper [072]

The globe – Sydney, Australia. 28 Jan 1886-14 Jun 1887.-d. 6 reels – 1 – uk British Libr Newspaper [072]

The globe – Toronto, Canada. -d. 1870-73 – 8r – 1 – uk British Libr Newspaper [072]

The globe – Toronto, Canada. Weekly Globe and Canada Farmer. -w. Jan 1871-Feb 1909 – 44r – 1 – uk British Libr Newspaper [072]

Globe and canada farmer *see* Globe

The globe and mail – Toronto: The Globe Print Co, [1936-]. jun 1938-feb 1946 – us CRL [071]

The globe and mail – 1849- – 1 – (yrly reel count varies) – us UMI ProQuest [072]

Globe christmas numbers – 1885, 1888-89, 1897-1912// – 2r – 1 – Can$185.00 – (19 nos (all publ?) of a lavishly-ill literary suppl to the toronto globe) – cn McLaren [400]

Le globe: journal philosophique et litteraire – Paris. Biweekly, later daily. 15 Sep 1824-20 Apr 1832 – 1 – fr ACRPP [100]

The globe=citizen – South Omaha, NE: Citizen Print Co. v4 n38. aug 13 1909- (wkly) [mf ed aug 13 1909-mar 11 1910 (gaps) filmed 1980] – 1r – 1 – us NE Hist [071]

Globe-journal – Falls City, NE: Jacob Baily, jul 15 1875-v15 n733. feb 4 1882 (wkly) [mf ed with gaps] – 3r – 1 – (formed by the union of: nemaha valley journal (1868) and: little globe. cont by: falls city journal. v14 not publ) – us NE Hist [071]

Globensky, Charles Auguste Maximilien *see* La rebellion de 1837-38

Globescope / Art, Business & Culture Exchange International – 1983 jan-feb/mar, 1984 jul – 1r – 1 – mf#4877753 – us WHS [071]

Globetrotter – 1981 may-1982 jun, 1982 jul-1983 jun, 1983 jul 22-1985 aug, 1985 sep-1986 aug, 1986 sep-1987 aug 21, 1987 sep-1988, 1989 – 7r – 1 – mf#651522 – us WHS [071]

Il globo – Rome, Italy. 1 jun 1946-18 may 1947; 1 aug 1950-30 dec 1951; 2 jan-31 dec 1957 – 1 – (imperfect) – mf#m.f.878.l – uk British Libr Newspaper [074]

O globo : jornal philosophico, literario, industrial e scientifico – Rio de Janeiro, RJ: Typ J R da Costa, 13 out 1844 – (= ser Ps 19) – mf#P03A,03,17 n02 – bl Biblioteca [073]

O globo – Maranhao: Typ de J O M da Cunha Torres, 20 jan 1852-set 1855; jul 1858-30 dez 1859 – (= ser Ps 19) – mf#P06,03,16 n01P06,03,17-19 – bl Biblioteca [079]

Der globus – Berlin DE, 1941 n5, 7, 9-18, 1942 n19-27 – 1 – gw Misc Inst [074]

Globus – Warsaw. v. 1-2. July 1932-Nov 1933 – 1 – us NY Public [947]

Globus – Zagreb. Yugoslavia. -w. 3 Jan 1960-12 May 1963. (7 reels) – 1 – uk British Libr Newspaper [949]

Y gloch = The bell – [Wales] Gwynedd 30 rhagfyr 1903-7 ebrill 1937 [mf ed 2004] – 15r – 1 – uk Newsplan [072]

Die glocke – Oelde DE, 1 jul 1927-30 sep 1935 [gaps]; 2 nov 1949-57; 18 oct-9 nov 1960; 18 feb-1may 1961; 1951-80 [gaps] – 25mf – 9 – (filmed by misc inst: 1976- [ca 8r/yr]; 1958-1960 19 oct, 1960 10 nov-1961 17 feb, 1961 3 may-1980 24 feb, 1980 21 may-30 dec. regional & local ed: ahlen=ahlener tageblatt 1951-70 [only local sect]; a=rheda-wiedenbrueck 1951-70; b=beckum 1951-70; c,d,e=guetersloh 1951-70, 1978 1 sep- [8r/yr]; f=warendorf 1940 11 jun-26 jun, 1951-70 [gaps]) – gw Mikrofilm; gw Misc Inst [074]

Die glocke – Leipzig, Berlin, Dresden DE, Wien (A), 1861-62 – 1r – 1 – (filmed by other misc inst: 1859 26 mar-24 dec, 1860 jan-23 jun, 1861 n105-156, 1862-1864 sep, undated: iss11-13; 1859-60 [2r]) – gw Misc Inst [074]

Die glocke : wochenschrift fuer politik, finanzwirtschaft und kultur – Muenchen, Berlin DE, 1915 1 sep-1925 26 sep – 7r – 1 – mf#2267 – gw Mikropress [073]

Glocke – London, UK. 15 May 1881 – 1 – uk British Libr Newspaper [072]

Die glocken von danzig : eine geschichte aus danzigs grosser zeit / Enderling, Paul – Stuttgart: K Thienemanns Verlag, [1943] (mf ed 1990) – 1r – 1 – (filmed with: die kleine weltlaterne) – us Wisconsin U Libr [830]

Glockenspiel, gesammelte gedichte / Seidel, Heinrich – Leipzig: Liebeskind 1889 – 1 – mf#film mas c419 – us Harvard [810]

Glockmann, G *see* Homer in der fruehchristlichen literatur bis justinin

Gloeckner, Karl *see* Brentano als maerchenerzaehler

Die gloecknerstochter / Hahn-Hahn, Ida, Graefin – Regensburg: J Habbel, [19–?] [mf ed 1993] – (= ser Graefin ida hahn-hahn works. 1. serie 17-18) – 2v – 1 – mf#8669 – us Wisconsin U Libr [430]

Gloel, Heinrich *see* Der wetzlarer goethe

Gloel, Johannes *see*
- Der heilige geist in der heilsverkuendigung des paulus
- Die juengste kritik des galaterbriefes

Glogau, Gustav *see* Die ideale der socialdemokratie und die aufgabe des zeitalters

Glogauer kreisblatt – Glogau (Glogow, PL), 1847 – 1 – gw Misc Inst [077]

Gloomy summer / Gutch, Charles – London, England. 1860 – 1r – us UF Libraries [240]

Gloria : utopischer roman / Bade, Wilfrid – Berlin: W Andermann, 1939 [mf ed 1989] – 344p – 1 – mf#6971 – us Wisconsin U Libr [830]

Gloria bellica serenissimi et potentissimi principis maximiliani...quam heroi maximo heroes et heroides... / [Stengel, G] – [Ingolstadii: Ex officina typographica Gregorii Haenlini], 1623 – 5mf – 9 – mf#O-1993 – ne IDC [090]

Gloria christi : an outline study of missions and social progress / Lindsay, Anna Robertson Brown – New York: Macmillan 1907 [mf ed 1986] – (= ser United study of missions) – 1mf – 9 – 0-8370-6141-5 – (incl ind) – mf#1986-0141 – us ATLA [240]

Gloria llamo dos veces / Gonzalez Herrera, Julio – Ciudad Trujillo, Dominican Republic. 1944 – 1r – us UF Libraries [972]

Gloria patri : our talks about the trinity and the new trinitarianism / Whiton, James Morris – 2nd ed. New York: Thomas Whittaker, 1904, c1892 – 1mf – 9 – 0-8370-5760-4 – mf#1985-3760 – us ATLA [240]

Gloria ueber der welt : roman / Bade, Wilfrid – Berlin: Ullstein, c1937 [mf ed 1989] – 225p – 1 – mf#6971 – us Wisconsin U Libr [830]

Y glorian [casnewydd-ar-wysg] : newyddiadur wythnosol cenedlaethol – [Wales] LLGC 1 mawrth-27 gorff 1867 [mf ed 2003] – 1r – 1 – uk Newsplan [072]

Y glorian newyddiadur wythnosol – [Wales] Gwynedd ion 1899-28 medi 1916 [mf ed 2003] – 18r – 1 – uk Newsplan [072]

Glorias de asturias / Leon Gutierrez, Florencio – 1896 – 9 – sp Bibl Santa Ana [946]

Glorias de espana. hernando cortes / Sandoval, M de – 1898 – 9 – sp Bibl Santa Ana [946]

Glorias del sacerdocio reveladas a santa brigida juntamente con sus obligaciones y danos : sacadas del libro autentico de las revelaciones de la santa, y ponderdadas por el padre antonio nadal, de la compania de jesus – en Lima: En la Plazuela de San Christoval, ano 1751 – (= ser Books on religion...1543/44-c1800: mistica y meditacion) – 2mf – 9 – mf#crl-285 – ne IDC [241]

Glorias dominicanas en su esclarecido, e ilustre militar tercer orden : tomo 1o. contiene el origen de este venerable instituto, su antiguedad... / Hidalgo y Costilla, Miguel – en Mexico: impr D Joseph Fernandez de Jaurequi, ano de 1795 – (= ser Books on religion...1543/44-c1800: ordenes, etc: dominicos) – 2v on 3mf – 9 – mf#crl-196 – ne IDC [241]

Glorias...hernando cortes / Sandoval, M de – 18-8 (sic) – 9 – sp Bibl Santa Ana [946]

Glories of christ's kingdom / Dewar, Daniel – London, England. 1820 – 1r – us UF Libraries [240]

The glories of divine grace : a free rendering of the original treatise of p. eusebius nieremberg, s.j = Herrlichkeiten der goettlichen gnade / Scheeben, Matthias Joseph – 2nd ed. New York: Benziger Brothers, c1886 – 2mf – 9 – 0-7905-7460-8 – (in english) – mf#1989-0685 – us ATLA [240]

The glories of hindustan / Nawrath, Ernst Alfred – London: Methuen & Co, 1935 – (= ser Samp: indian books) – us CRL [915]

Glories of india on indian culture and civilization / Acharya, Prasanna Kumar – Allahabad: Jay Shankar Bros, 1952 – (= ser Samp: indian books) – us CRL [954]

The glories of magadha / Samaddar, Jogindra Nath – [Patna: Patna University, 1924] – (= ser Samp: indian books) – (foreword by a b keith) – us CRL [954]

Glorieux, P *see*
- Aux origines de la sorbonne
- La litterature quodlibetique 2
- La litterature quodlibetique de 1260 a 1320
- Repertoire des maitres en theologie de paris au 13th siecle

La gloriosa et felice vittoria consegvita dall' armata christiana contra quella del turcho – Venetia, 1571 – 1mf – 9 – mf#H-8177 – ne IDC [950]

Gloriosa sotaina do primeira imperio / Britto, Jose Gabriel De Lemos – Sao Paulo, Brazil. 1937 – 1r – us UF Libraries [972]

Glorioso pasado historico de camaguey, 1868-1878 y... / Acosta Leon, Raul D – Camaguey? Cuba. 1951 – 1r – us UF Libraries [972]

The glorious company of the apostles : being studies in the characters of the twelve / Jones, John Daniel – London: J Clarke; New York: [distr by] Thomas Whittaker, 1904 – 1mf – 9 – 0-8370-3798-0 – mf#1985-1798 – us ATLA [225]

The "glorious enterprise" : the plan of campaign for the conquest of new france, its origin, history and connections with the invasions of canada / Lighthall, William Douw – Montreal: C.A. Marchand, Printer to the Numismatic and Antiquarian Society, [1902?] [mf ed 1997] – 1mf – 9 – 0-665-83544-2 – (repr fr: canadian antiquarian and numismatic journal. 3rd series v3 n5) – mf#83544 – cn Canadiana [971]

A glorious future for australia : or, the freeman's guide-book: being a letter respectfully dedicated to the electors of victoria under its new constitution / Nemo – Melbourne 1856 – (= ser 19th c british colonization) – 1mf – 9 – mf#1.1.4935 – uk Chadwyck [325]

The glorious gospel : the center of christianity / Adcock, Adam Kennedy – Cincinnati: Standard Pub Co, c1916 [mf ed 1993] – (= ser Christian church (disciples of christ) coll) – 1mf – 9 – 0-524-06006-1 – mf#1991-2366 – us ATLA [240]

The glorious land : short chapters on china, and missionary work there / Moule, Arthur Evans – London: Church Missionary Society, 1891 [mf ed 1995] – (= ser Yale coll) – 108p (ill) – 1 – 0-524-09285-0 – mf#1995-0285 – us ATLA [951]

The glorious lord / Meyer, Frederick Brotherton – Chicago: Fleming H Revell, c1896 – 1mf – 9 – 0-8370-7175-5 – mf#1986-1175 – us ATLA [240]

Glorious recovery by the vaudois of their valleys, from the original with a compendious history of that people, previous and subsequent to that event, by hugh dyke acland / Arnaud, Henri & Acland, Hugh Dyke – London: John Murray, Albemarle-Street, 1827 – 1r – 1 – $16.72 – us Southern Baptist [242]

Glorious victories of amda seyon, king of ethiopia / Royal Chronicle Of Abyssinia – Oxford, England. 1965? – 1r – us UF Libraries [960]

The glory after the passion : a study of the events in the life of our lord from his descent into hell to his enthronement in heaven / Stone, James Samuel – London: Longmans, Green, 1913 [mf ed 1992] – 1mf – 9 – 0-524-05423-1 – mf#1992-0433 – us ATLA [240]

The glory and divinity of the holy bible and its spiritual sense : a lecture / Hyde, John – London: James Speirs, 1889 – 1mf – 9 – 0-8370-3712-3 – mf#1985-1712 – us ATLA [220]

The glory and joy of the resurrection / Paton, James – New York: American Tract Society, 1902 – 1mf – 9 – 0-524-05107-0 – mf#1992-0328 – us ATLA [220]

Glory dead / Calder-Marshall, Arthur – London, England. 1939 – 1r – us UF Libraries [972]

Glory of christ as the risen saviour and future judge / Scholefield, James – Cambridge, England. 1839 – 1r – us UF Libraries [240]

Glory of god displayed in the building up of zion / Davidson, Thomas – Edinburgh, Scotland. 1802 – 1r – us UF Libraries [240]

Glory of god in gathering his people to himself / Hawker, Robert – London, England. 1823 – 1r – us UF Libraries [240]

Glory of good deeds / Harrison, J C – London, England. 1870 – 1r – us UF Libraries [240]

Glory of the age : john foster on missions: with an essay on the skepticism of the church / Foster, John – New York: Edward H Fletcher, 1851 – 1mf – 9 – 0-8370-6814-2 – mf#1986-0814 – us ATLA [240]

Glory of the immortal life : embracing the prophecies and proofs of the great doctrine of immortality in the analogies of nature, the longings and demands of the soul, the clear and sufficient assurances of divine revelation . . . / Stebbins, Jane E – Norwich, CT: JH Jewett, 1873, c1871 [mf ed 1991] – 2mf – 9 – 0-7905-8738-6 – (1st ed 1867 publ under title: our departed friends) – mf#1989-1963 – us ATLA [240]

The glory of the ministry : paul's exultation in preaching / Robertson, A T – New York: Fleming H Revell, c1911 – 1mf – 9 – 0-7905-0221-6 – (incl bibl ref) – mf#1987-0221 – us ATLA [220]

The glory of the redeemer in his person and work / Winslow, Octavius – 4th ed. Philadelphia: Lindsay & Blakiston, 1859 – 1mf – 9 – 0-8370-5438-9 – (incl ind) – mf#1985-3438 – us ATLA [240]

Glory of the two crown's heads, adam and christ, unveill'd / Culy, David – Spilsby, England. 1820 – 1r – us UF Libraries [240]

Glory of young men / Coats, Walter William – Girthon, Scotland. 1891 – 1r – us UF Libraries [240]

The glory that was gurjaradesa / ed by Munshi, K M – Bombay: Bharatiya Vidya Bhavan, 1943- – (= ser Samp: indian books) – us CRL [954]

Glorying in the cross of christ / Maclaurin, John – Edinburgh, Scotland. 1816 – 1r – us UF Libraries [240]

Glos anglii – Cracow. Poland. -w. 9 Nov 1946-2 Apr 1949. (Imperfect). 3 reels – 1 – uk British Libr Newspaper [943]

Glos gminy zydowskiej – Warsaw. v. 1-3. 1937-1939 – 1 – us NY Public [939]

Glos konfederacji – London, UK. Sept 1953-Mar/ Oct 1960 – 1 – uk British Libr Newspaper [072]

Glos ludu – Lublin, Poland. Nov 1944-Jan 1945 – 1r – 1 – us L of C Photodup [943]

Glos ludu – Warsaw, Poland. Feb 1945-1948 – 9r – 1 – us L of C Photodup [943]

Glos ludu – Warsaw. Poland. -w. 28 May 1946-12 Jul 1947. (12 reels) – 1 – uk British Libr Newspaper [943]

Glos narodu – Czestochowa, Poland. Aug-Sept 1945 – 1r – 1 – us L of C Photodup [943]

Glos narodu – Jersey City, NJ. -w. 19 Oct 1950-23 June 1955. Imperfect. 3 reels – 1 – uk British Libr Newspaper [071]

Glos polek – Chicago, IL: Women's Voice Pub Co, 1902-03; 1910-73 – 26r – 1 – us CRL [071]

Glos polek : the polish women's voice / Polish Women's Alliance of America – 1976 apr 15, 1977 may 19-1981, 1982-85 – 2r – 1 – mf#622862 – us WHS [305]

Glos polski – Canada. jan 1938-dec 1990 – 41r – 1 – (in polish) – cn Commonwealth Imaging [071]

Glos polski – Liverpool. 1911-Jun 1912 – 1 – uk British Libr Newspaper [077]

Glos polski – Paris, France. -w. 28 nov 1939-9 jun 1940 – 1r – 1 – uk British Libr Newspaper [077]

Glos polski w londynie – London, UK. Sept 1940; 1944 – 1 – uk British Libr Newspaper [072]

Glos pomorza – Koszalin, Poland. 1953-79 – (= ser Glos Koszalinski) – 43r – 1 – (previously: glos koszalinski) – us L of C Photodup [943]

Glos pracy – Krakow, Poland. Aug 19-Nov 1945 – 1r – 1 – us L of C Photodup [077]

Glos pracy – Warsaw, Poland. Feb 1951-1981 – 61r – 1 – us L of C Photodup [943]

Glos robotniczy – Lodz, Poland. Jul 1950-1992 – 91r – 1 – (cont as: glos poranny as of 30 jan 1990) – us L of C Photodup [943]

Glos szczecinski – Szczecin, Poland. 31 Dec 1952-1992 – 79r – 1 – us L of C Photodup [943]

Glos wielkopolski – Poznan, Poland. 18 Aug 1945-Jun 1970 – 32r – 1 – us L of C Photodup [943]

Glos wybrzeza – Gdansk, Poland. 14,18,27 Sept 1951; 1953-92 – 84r – 1 – us L of C Photodup [943]

La glosa que...codigo...partidas / Lopez de Tovar, Gregorio – 1878 – 9 – (tomos 2-6 1878) – sp Bibl Santa Ana [946]

Glosando a benito arias montano. la funcion social del ingenio / Goni, Blas – Malaga: Revista Espanola de Estudios Biblicos, 1928 – 1 – sp Bibl Santa Ana [946]

Glosarios del bajo espanola en venezuela / Alvarado, Lisandro – Caracas, Venezuela. v1-2. 1954 – 1r – us UF Libraries [972]

Glosas / Sanchez-Arjona, Vicente – Sevilla: Editorial Franciscana de San Antonio, 1950 – 1 – sp Bibl Santa Ana [810]

Glosas martianas / Le Riverend Bruzone, Pablo – Miami, FL. 1964 – 1r – us UF Libraries [972]

Glossa – Burnaby, Canada 1967-82 – 1,5,9 – ISSN: 0017-1271 – mf#6562 – us UMI ProQuest [400]

Glossae in matthaeum : formae tplila 146 / Otfridus Wizanburgensis – [mf ed 2003] – (= ser ILL – ser a; Cccm 200) – 9mf+vii/104p – 9 – €62.00 – 2-503-65002-3 – be Brepols [400]

Glossaire franco-canadien et vocabulaire de locutions vicieuses usitees au canada / Dunn, Oscar – Quebec: impr A Cote, 1880 [mf ed 1974] – 1r – 5 – (int by m frechette) – mf#SEM16P136 – cn Bibl Nat [440]

Glossar zu den liedern der edda / Gering, Hugo – Paderborn: F Schoeningh, 1923 – 3r – 1 – us Wisconsin U Libr [430]

Glossarium ad scriptores mediae et infimae graecitatis duos in tomos digestum / Cange, Ch Dufresne Du – Lugduni, 1688 – €103.00 – ne Slangenburg [221]

Glossarium mediae et infimae latinatis conditum a carclo du fresne / Du Cange, Charles d. Fresne – 10v. 1883-87 – 1 – 95.00 – us L of C Photodup [450]

Glossarium mediae et infimae latinitatis / Cange, Ch. Dufresne Du – new ed. Parisiis. v1-9. 1883-1887 – 9v on 156mf – 8 – €298.00 – ne Slangenburg [450]

Glossarium novum ad scriptores medii aevi – Parisiis. Tomi 1-4. 1766 – 9 – €195.00 – ne Slangenburg [220]

A glossary, bengali and english : to explain the tota-itihas, the batris singhasan, the history of raja krishna chandra, the purusha-parikhya, the hitopadesa / Haughton, Graves Chamney – London: printed by Cox & Baylis, 1825 – (= ser 19th c books on linguistics) – 2mf – 9 – mf#2.1.54 – uk Chadwyck [040]

Glossary of ecclesiastical ornament and costume : with extracts from the works of durandus, georgius, bona, catalini, gerbert, martene, molanus, thiers, mabillon, ducange, etc / Pugin, Augustus Welby Northmore [comp] – London: Henry G Bohn, 1844 – (= ser 19th c art & architecture) – 5mf – 9 – mf#4.1.185;c.4.1.234;c.4.1.235 – uk Chadwyck [740]

A glossary of liturgical and ecclesiastical terms / Lee, Frederick George – London: Bernard Quaritch, 1877 [mf ed 1990] – 2mf – 9 – 0-7905-4888-7 – (incl bibl ref) – mf#1988-0888 – us ATLA [052]

Glossary of physical education terms, part 1 / College Physical Education Association Committee on Terminology – 1937 – 9 – us Kinesology [790]

A glossary of reference on subjects connected with the far east / Giles, Herbert Allen – Hongkong: Lane, Crawford, c1878 [mf ed 1987] – iii/182p – 1 – mf#10220 – us Wisconsin U Libr [959]

Glossary of reference on subjects connected with the far east / Giles, Herbert Allen – 3rd ed. Shanghai: Kelly & Walsh, 1900 [mf ed 1995] – (= ser Yale coll) – 328p – 1 – 0-524-09136-6 – mf#1995-0136 – us ATLA [950]

Glossary of technical terms, phrases, and maxims of the common law / Stimson, Frederic Jesup – Boston, Little, Brown, 1881. 305 p. LL-318 – 1 – us L of C Photodup [346]

A glossary of the aramaic inscriptions / Cook, Stanley Arthur – Cambridge: University Press; New York: Macmillan [dist], 1898 [mf ed 1986] – 1mf – 9 – 0-8370-8091-6 – (incl bibl ref) – mf#1986-2091 – us ATLA [470]

A glossary of the west saxon gospels : latin-west saxon and west-saxon latin / Harris, Mattie Anstice – Boston: Lamson, Wolffe, 1899 [mf ed 1985] – (= ser Yales studies in english 6) – 1mf – 9 – 0-8370-3498-1 – mf#1985-1498 – us ATLA [226]

A glossary of words and terms used in the constitution – Saipan: (= Education for Self-Government, 1976 – (= ser Micronesian constitutional convention, 1975) – 1mf – 9 – $1.50 – mf#LLMC 82-100F, Title 104 – us LLMC [323]

Glossay, Karl *see* Wien 1840-1848

Glossed texts, aldhelmiana, psalms / ed by Pulsiano, Philip – [mf ed Binghamton NY, 1996] – (= ser ASMMF) – 862 folios – 8 – $120.00v / £76.00v [institution] ($96v / £60v if part of subsc) – 0-86698-210-8 – mf#mr169 – us MRTS [090]

Glossolalia : a critical study of alleged origins, the new testament and the early church / Lovekin, Arthur Adams – [Sewanee, Tenn.] 1962 – 1r – 1 – 0-8370-1484-0 – mf#1984-B027 – us ATLA [240]

Glossolalia / Vivier, Lincoln Morse Van Eetveldt – 1960 – 1r – 1 – 0-8370-0457-8 – mf#1984-B011 – us ATLA [240]

Die glossolalie in der alten kirche : in dem zusammenhang der geistesgaben und des geisteslebens des alten christenthums / Hilgenfeld, Adolf – Leipzig: Breitkopf und Haertel, 1850 – 1mf – 9 – 0-7905-1954-2 – (incl bibl ref) – mf#1987-1954 – us ATLA [220]

Glossop chronicle and advertiser etc – Glossop, Derbyshire, 5 nov 1859-to date (wkly) – 1 – (1861 incomplete, 1911 mf only) – uk Newsplan [072]

Glossop record *see* The record

Glossop-dale chronicle and north derbyshire reporter – [NW England] Stalybridge 1859-9 feb 1923 – 1 – (title change: glossop chronicle [16 feb 1923-2 jul 1937]; glossop chronicle & advertiser, north derbyshire & north cheshire reporter [9 jul 1937]) – uk MLA; uk Newsplan [072]

Glossy, Karl *see* Gesammelte schriften

Gloster, Archibald *see* A letter to the right honourable the earl of buckinghamshire, late secretary of the colonial department

Gloucester 1634-1895 – Oxford, MA (mf ed 1989) – (= ser Massachusetts vital records) – 375mf – 9 – 0-87623-093-1 – (mf 1-27: church records 1703-1835. mf 28-47: town records 1642-1752. mf 48-68: land grants 1707-1820. mf 69-93: b,m,d 1643-1794. mf 94-107: b,i,m,d 1772-1807. mf 108-116: b,i,m,d 1797-1828. mf 117-140: b,i,m,d 1740-1848. mf 141-154: b,m,d 1843-50. mf 154-156: marriages & intentions 1839-61. mf 157-164: deaths 1851-64. mf 165-169: deaths 1865-73. mf 170-180: deaths 1874-83. mf 181-199: deaths 1884-93. mf 200-220: deaths 1894-1903. mf 221-228: marriages 1851-69. mf 229-231: marriages & intents 1861-74. mf 232-235: intentions 1861-74. mf 236-286: marriages & intents 1874-1903. mf 287-294: intentions 1890-95. mf 295-315: births 1851-68. mf 316-324: births 1868-73. mf 325-330: births 1873-76. mf 331-344: births 1874-1883. mf 345-361: births 1884-93. mf 362-369: births 1893-98. mf 370-375: births 1894-95) – us Archive [978]

Gloucester 1641-1849 – Oxford, MA (mf ed 1996) – (= ser Massachusetts vital record transcripts to 1850) – 54mf – 9 – 0-87623-254-3 – (mf 1t-5t: vital records 1641-1727. mf 5t-10t: vital records 1716-39. mf 10t-20t: vital records 1722-99. mf 20t-26t: vital records 1767-1840. mf 27t-32t: vital records 1775-1842. mf 32t-36t: marriages 1827-39. mf 36t-40t: vital records 1764-1858. mf 40t-42t: marriages 1839-49. mf 43t: out-of-town marriages 1676-1799. mf 43t-44t: marriages 1844-50. mf 45t-49t: births 1843-49. mf 50t-53t: deaths 1843-50. mf 54t: vital records 1751-1849) – us Archive [978]

Gloucester advocate – Gloucester, jan 1969-dec 1996 – 9 – at Pascoe [079]

Gloucester. England *see* Ward lists and other records of the city of gloucester, 1843-86

Gloucester examiner – Raymond terrace, nov 1893-jun 1912 – 3r – A$99.00 vesicular A$115.50 silver – at Pascoe [079]

Gloucester free press and weekly advertiser – [SW England] Gloucestershire 14 jul 1855-dec 1873 [mf ed 2003] – 12r – 1 – (missing: 1861; cont as: gloucester mercury [jan 1856-dec 1860]; gloucester mercury and weekly advertiser [jan 1862-dec 1864]; gloucester mercury [jan 1865-dec 1873]) – uk Newsplan [072]

Gloucester journal – [SW England] Gloucestershire jun 1730-dec 1792 [mf ed 2004] – 32r – 1 – uk Newsplan [072]

Gloucester labour party records, 1899-1951 – (= ser Labour party in britain, origins and development at local level. series 1) – 3r – 1 – (int by roger eatwell) – mf#97129 – uk Microform Academic [325]

Gloucester mercury – [SW England] Gloucestershire aug 1828-mar 1829 [mf ed 2004] – 1r – 1 – uk Newsplan [072]

Gloucester standard – [SW England] Gloucestershire jan 1873-aug 1902 [mf ed 2004] – 29r – 1 – (missing: 1898; cont by: gloucester standard & gloucestshire news [jan 1878-aug 1902]) – uk Newsplan [072]

Gloucestershire chronicle – England.6 Jul 1833-1848. -w. 6 reels – 1 – uk British Libr Newspaper [072]

Gloucestershire news – [SW England] Gloucester 12 nov 1870-17 feb 1872 [mf ed 2004] – 2r – 1 – uk Newsplan [072]

Gloucster advocate – Gloucester. jan 1905-dec 1912, jan 1925-dec 1946 – 16r – A$1109.77 vesicular A$1197.77 silver – at Pascoe [079]

Glove workers' bulletin – v1-24 n9 [1937 nov-1961 jul] – 1r – 1 – mf#1057277 – us WHS [680]

Glove workers' journal : official organ, international glove workers union of america / International Glove Workers Union of America – 1905 dec – 1r – 1 – mf#4967027 – us WHS [331]

Glover, Archibald Edward, 1860?- *see* Thousand miles of miracle in china

Glover, Edward *see* Freud or jung

Glover, Elizabeth Rosetta Scott *see* Life of sir john hawley glover

Glover, J *see* Le livere de reis de brittanie e le livere de reis de engleterre (rs42)

Glover, John Corbett *see* "The flying priest"

Glover, Octavius *see*
- Doctrine of the person of christ
- A short treatise on sin

Glover, Richard *see* Herbert stanley jenkins, m.d., f.r.c.s., medical missionary, shensi, china

Glover, Robert *see*
- Ebenezer
- The visitation of yorkshire, made in the years 1584-85, by robert glover, somerset herald..

Glover, Robert Hall *see* The real heart of the missionary problem

Glover, Terrot Reaveley *see*
- The christian tradition and its verification
- The jesus of history
- Life and letters in the fourth century
- Virgil

Glover, W B *see* Evangelical nonconformists and higher criticism in the nineteenth century

The glow-worm : and other beetles / Fabre, Jean Henri – Toronto: McClelland & Stewart, 1919 [mf ed 1994] – 6mf – 9 – 0-665-72829-8 – (trans by alexander teixeira de mattos) – mf#72829 – cn Canadiana [590]

The glowworm – Toronto: [s.n. 1891-189-?] – 9 – mf#P04265 – cn Canadiana [420]

Gloy, Albert *see* Sommerwind ueber tormoehlenhof

Gloyer, E *see* Geschichte unserer missionsstation kotapad in jeypur (vorderindien)

Glt – Goteborg, Sweden. 1997 – sw Kungliga [078]

Gluck, Christophe Willibald *see* Armide

Gluck, James Fraser *see* The law of receivers of corporations including national banks.

Gluckman, M *see*
- Analysis of a social situation in modern zululand
- The economy of the central barotse plain
- Essays on lozi land and royal property
- Malinowski's sociological theories

Glucoregulation and work performance in gluconeogenesis-inhibited iron deficient rats / Linderman, J K – 1991 – 2mf – 9 – $8.00 – us Kinesology [590]

Glueck auf – Halle S DE, 1967, 13mar-22 jul [gaps] – 1r – 1 – (bkw mulde-nord) – gw Misc Inst [074]

Glueck auf! / Werner, E – Philadelphia: Morwitz. 2v. [1889?] – 1 – us Wisconsin U Libr [830]

Das glueck dieses sommers : novelle / Sturm, Stefan – 5. aufl. Guetersloh: C Bertelsmann 1943 [mf ed 1991] – 1r – 1 – (filmed with: totenhorn-sudwand / karl hans strobl.) – mf#2907p – us Wisconsin U Libr [830]

Glueck, Hermann *see* Der dialekt in den dorfgeschichten berthold auerbachs und melchior meyrs

Glueck, Sheldon *see* The sheldon glueck papers

Glueckauf : berg- und huettenmaennische zeitschrift – Essen DE, 1865-1918 – 48r – 1 – (1865-16 may 1883 as suppl of: essener zeitung) – mf#3727 – gw Mikropress [622]

Glueckauf – Marienberg DE, 1881-1943 n3 [mpf] – 1 – (tw. auch: glueck auf; previously publ in schneeberg & schwarzenberg) – gw Misc Inst [074]

Glueckauf! – Bochum, Dortmund, Gelsenkirchen DE, 1889 16 mar-1907 – 8r – 1 – (several title changes: deutsche berg- und huettenarbeiter-zeitung; 1903: deutsche bergarbeiter-zeitung; 1905: bergarbeiter-zeitung; 1931: die bergbau-industrie; 1933: der deutsche bergknappe. filmed by mikropress: 1889 16 mar-1933 24 jun [13r]) – gw Misc Inst [622]

Das glueckhafte schiff von zuerich : [abdruck der ausgabe von 1577] = [Lucky ship of zurich] / Fischart, Johann; ed by Baesecke, Georg – Halle a/S: M Niemeyer 1901 [mf ed 1993] – (= ser Neudrucke deutscher literaturwerke des 16. und 17. jahrhunderts 182) – 11r – 1 – (incl bibl ref) – mf#3387p – us Wisconsin U Libr [810]

Glueckliche insel : erzaehlungen / Blunck, Hans Friedrich – feldpostausg. Bayreuth : Gauverlag Bayreuth, 1942 [mf ed 1989] – (= ser Die kleine glockenbuecherei 2) – 111p (ill) – 1 – mf#7037 – us Wisconsin U Libr [880]

Glueckliche menschen : roman / Polenz, Wilhelm von – 3. Aufl. Berlin: F Fontane, 1905 – 1r – 1 – us Wisconsin U Libr [830]

Das glueckskind : fuer die maerchenspiele der kuenstlerischen volksbuehne schlicht und getreu in rede und handlung gebracht nach dem maerchen der gebrueder grimm "der teufel mit den drei goldenen haaren" / Guembel-Seiling, Max – Leipzig: Breitkopf & Haertel 1918 [mf ed 1990] – (= ser Deutsche maerchenspiele) – 1r – 1 – (filmed with: guerillaskrieg: versprengte lieder) – mf#2694p – us Wisconsin U Libr [790]

Der gluecksspilz : roman / Berend, Alice – Muenchen: Albert Langen, c1919 [mf ed 1995] – 219p – 1 – mf#8976 – us Wisconsin U Libr [830]

Glueckstaedter anzeiger – Glueckstadt DE, 1960 23 jan-1965 [gaps] – 1r – 1 – gw Misc Inst [074]

Glueckstaedtische fortuna – Glueckstadt DE, 1794-1945 5 may, 1946 4 jan-1970 – 1 – (title varies: 3 jan 1801: glueckstaedtsche fortuna; 2 jan 1858: glueckstaedter fortuna. with suppl: extrablaetter 1914 25 jul-1918 10 nov) – gw Misc Inst [074]

Glueckstaedtsche fortuna *see* Glueckstaedtische fortuna

Der gluecksucher und die sieben lasten : ein hohes lied / Becher, Johannes Robert – Moskau: Verlagsgenossenschaft Auslaendischer Arbeiter in der UdSSR, 1938 [mf ed 1989] – 149p – 1 – mf#6994 – us Wisconsin U Libr [810]

Die gluecks-woche – Duesseldorf DE, 1955 5 nov-1956 31 mar – 1r – 1 – gw Misc Inst [074]

Gluehender tag : maenner in der bewaehrung / Goote, Thor – 7. Aufl. Guetersloh: C Bertelsmann, 1943 – 1r – 1 – us Wisconsin U Libr [830]

Gluth, Oskar *see*
– Lenz als dramatiker
– Panks lachende erben
– Der verhexte spitzweg

Glutz-Blotzheim, Robert *see* Handbuch fuer reisende in der schweiz

Glycerol-induced hyperhydration during long term exercise in a heated environment / Swan, Jacob G – 1997 – 2mf – 9 – $8.00 – mf#PH 1592 – us Kinesology [612]

Glycosylated hemoglobin and the oxygen kinetics in individuals with type 2 diabetes / Dwyer, Gregory B & Wallace, Janet P, – 1992 – 2mf – 9 – $8.00 – us Kinesology [613]

Glyncorrwg, glamorgan, parish church of st john : baptisms 1702-1905, burials 1702-1908, marriages 1724-1837 – 1mf – 9 – £1.25 – uk Glamorgan FHS [929]

Glynneath & aberpergwm, st cadoc, monumental inscriptions – 1mf – 9 – £1.25 – uk Glamorgan FHS [929]

Glynneath, addoldy, monumental inscriptions – 2mf – 9 – £2.50 – uk Glamorgan FHS [929]

Glyntaff, glamorgan, parish church of st mary : baptisms 1848-1926, burials 1848-1884 – [Glamorgan]: GFHS [mf ed c2003] – 2mf – 9 – £2.50 – uk Glamorgan FHS [929]

Glyptodon news – v1 n1,2,4 [1973 mar 19/apr 2, apr 3/apr 17, may 7/may 21] – 1r – 1 – mf#1583426 – us WHS [071]

Gmb journal – General, Municipal, Boilermakers & Allied Trades Union, 1983-. -m. Continues: GMW Journal – 1 – us Wisconsin U Libr [330]

Gmelin, J G *see* Reise durch sibirien von dem jahre 1733 bis 1743

Gmelin, Johann *see* Voyage en siberie

Gmelin, Otto *see*
– Germanenzug
– Gespraeche am abend
– Die gralsburg
– Das gruene glas
– Konradin reitet
– Das neue reich
– Der ruf zum reich

Gmina wyznaniowa zydowska w lodzi : kronika gminy wyznaniowj zydowska w lodzi – Lodz: Gmina Wyznaniowa Zydowska. v1 n1-3 aug-dec 1929; v4 n4-5 apr-may 1932; v5 n6-7 dec 1934 – us CRL [943]

Gmina wyznaniowa zydowska w warszawie i jej instytucje – Warszawa: Gmina, 1927 – us CRL [077]

Gmuender tagespost – Schwaebisch Gmuend DE, 1980 1 jan- – 79r until 1990 – 1 – (bezirksausgabe von suedwest-presse, ulm) – gw Misc Inst [074]

Gmw journal – v34, no.12 (Dec 1971)-v37, no.2 (Feb 1974); n.s. Sept 1981; tabloid Oct Nov 1981-Nov Dec 1982. Esher, Surrey: General and Municipal Workers' Union. illus. semimonthly. Continued by: GMB Journal – 1 – us Wisconsin U Libr [330]

Gnac Gi Len *see* Gamnit nayopay

Gnad, Ernst *see* Ueber robert hamerlings lyrik

Gnade und wahrheit : eine lyrische dichtung / Berens, August – St Louis, MO: Verlag der Deutschen Evangelischen Synode von Nord-Amerika, 1890 [mf ed 1989] – viii/64p – 1 – mf#7008 – us Wisconsin U Libr [810]

Das gnadenbild der mater ter admirabilis von ingolstadt in bayern : geschichtlicher bericht und gebete / Hattler, Franz – 2. Aufl. Freiburg i.B.: Herder, 1880 – 1mf – 9 – 0-8370-9789-4 – mf#1986-3789 – us ATLA [240]

Die gnadenlehre des petrus lombardus / Schupp, J – Freiburg, 1932 – 6mf – 8 – €14.00 – ne Slangenburg [240]

Die gnaedige frau von paretz : dramolet in einem aufzug / Wichert, Ernst – 2. Aufl. Leipzig: P Reclam jun., [1894?] – 1r – 1 – us Wisconsin U Libr [820]

Gnaedigst privilegiertes leipziger intelligenz-blatt in frag- und anzeigen, vor stadt- und landwirthe – Leipzig DE, 1769-70, 1787-91 – 1 – gw Misc Inst [074]

Gnaedigst privilegirt onolzbachische wochentliche frag- und anzeigungsnachrichten – Ansbach DE, 1773-78 – 1r – 1 – gw Misc Inst [943]

Gnaedigst privilegirte braunschweigische zeitung fuer staedte, flecken und doerfer besonders fuer den deutschen landmann – Braunschweig DE, 1848-49 – 1r – 1 – (filmed by other misc inst: 1786-1809, 1811-21, 1823, 1825, 1828-47 [26r]) – gw Misc Inst [074]

Gnagnalasan na agui ni jazmina asim mirasole sa cahadean nin antioquia asin francia – [Nueva Caceres?: Libreria Mariana? 190-?] [mf ed Bloomington IN: Indiana Uni Lib, Preservation Dept 1984] – (= ser Coll...in the bikol language) – 1r [ill] – 1 – us Indiana Preservation [490]

The gnaosis of the light : a translation of the untitled apocalypse contained in the codex brucianus – London: John M Watkins, 1918 – 1mf – 9 – 0-524-07607-3 – mf#1991-0133 – us ATLA [240]

Gneisenau, August von *see* Papers of august von gneisenau, ca. 1785-1831

Gneisse, Karl *see* Der begriff des kunstwerks in goethes aufsatz von deutscher baukunst (1772) und in schillers aesthetik

Gnerich, Ernst *see* Andreas gryphius und seine herodes-epen

Gnesin, Mikhail Fabianovich *see* Variationen ueber ein juedisches volksthema

Gnevushev, A M *see* Volia sibiri

Gnistan – Goteborg, Sweden. 1895-97 – sw Kungliga [078]

Gnistan – Stockholm, Sweden. 1967-78 – 7r – 1 – sw Kungliga [078]

Gnistan – Stockholm, Sweden. 1979-86 – 1 – sw Kungliga [078]

Der gnom : wochenschrift fuer den gesamten bergbau – Duesseldorf DE, 1898 jul-1901 jun – 2r – 1 – uk British Libr Newspaper [622]

Gnomic literature in bible and apocyrpha : with special reference to the gnomic fragments and their bearing on the proverb collections / Levi, Gerson Baruch – 1mf – 9 – 0-8370-9714-2 – (incl ind of biblical citations) – mf#1986-3714 – us ATLA [270]

Die gnosis / Koehler, Walther – Tuebingen: J C B Mohr (Paul Siebeck) 1911 [mf ed 1986] – (= ser Religionsgeschichtliche volksbuecher fuer die deutsche christliche gegenwart 4/16) – 1mf – 9 – 0-8370-9879-3 – mf#1986-3879 – us ATLA [290]

Die gnosis – [Wien: Manz] v1. 1903 [biwkly] [mf ed 2003] – 1v on 1r – 1 – (merged with: lucifer to form: lucifer gnosis) – mf#2003-s500 – us ATLA [290]

Die gnosis *see*
– Lucifer mit der gnosis
– Luzifer

Gnosis en evangelie : eene historische studie / Bolland, Gerardus Johannes Petrus Josephus – Amsterdam: W Versluys, 1906 – 1mf – 9 – 0-524-06832-1 – mf#1992-0974 – us ATLA [221]

The gnostic heresies of the first and second centuries / Mansel, Henry Longueville; ed by Lightfoot, Joseph Barber – London: John Murray, 1875 – 1mf – 9 – 0-7905-1230-0 – (incl bibl ref and indexes) – mf#1987-1230 – us ATLA [290]

Gnostica – St Paul MN 1976-79 – 1,5,9 – (cont: gnostica news) – ISSN: 0145-885X – mf#8376.01 – us UMI ProQuest [290]

Gnostica news – St Paul MN 1971-74 – (cont by: gnostica) – ISSN: 0362-8922 – mf#8376.01 – us UMI ProQuest [290]

Gnosticism and agnosticism : and other sermons / Salmon, George – London; New York: Macmillan, 1887 – 1mf – 9 – 0-7905-3217-4 – mf#1987-3217 – us ATLA [290]

Le gnosticisme et la franc-maconnerie : consideree dans son origine, son organisation, ses bases, son but, les moyens employes pour attendre le but propose et ses destinees / Haus, Edouard – Bruxelles: H Goemaere 1875 – 5mf – 9 – mf#vrl-192 – ne IDC [366]

The gnostics and their remains, ancient and medieval / King, Charles William – 2nd ed. New York: G.P. Putnam, 1887 – 2mf – 9 – 0-7905-6236-7 – (includes annotated bibliographical appendix by joseph jacobs) – mf#1988-2236 – us ATLA [210]

Le gnostique de clement d'alexandrie / ed by Fenelon – Paris, 1930 – 6mf – 8 – €14.00 – (int by p dudon) – ne Slangenburg [240]

Gnostiques et gnosticisme : etude critique des documents du gnosticisme chretien aux 2e et 3e siecles / Faye, Eugene de – Paris: E Leroux, 1913 – (= ser Bibliotheque de l'ecole des hautes etudes) – 2mf – 9 – 0-7905-4635-3 – (incl bibl ref) – mf#1988-0635 – us ATLA [240]

Gnostische schriften in koptischer sprache aus dem codex brucianus / ed by Schmidt, Carl – Leipzig: J C Hinrichs 1892 [mf ed 1989] – (= ser Tugal 8/1-2) – 2mf – 9 – 0-7905-1914-3 – (in coptic, german & greek; incl bibl ref & ind) – mf#1987-1914 – us ATLA; ne Slangenburg [290]

Die gnostischen quellen hipployts / Staehelin, H – Leipzig, 1890 – (= ser Tugal 1-6/3a) – 2mf – 9 – €5.00 – ne Slangenburg [240]

Die gnostischen quellen hippolyts in seiner hauptschrift gegen die haeretiker – sieben neue bruckstücke der syllogismen des apelles : die gwynn'schen cajus- und hippolytus-fragmente / Staehelin, Hans & Harnack, Adolf von – Leipzig: JC Hinrichs, 1890 – (= ser Tugal) – 1mf – 9 – 0-7905-1917-8 – mf#1987-1917 – us ATLA [240]

Die gnostischen schriften des koptischen papyrus berolinensis 8502 / Till, W – Berlin, 1955 – (= ser Tugal 5-60) – 6mf – 9 – €14.00 – ne Slangenburg [240]

Gnothi seauton : oder magazin zur erfahrungsseelenkunde / ed by Moritz, Karl Philipp et al – Berlin 1783-93 – (= ser Dz. abt philosophie) – 10v on 30mf – 9 – €300.00 – mf#k/n549 – gw Olms [100]

Gnueg, Hiltrud *see* Literarische utopie-entwuerfe

Go and tell john : a sketch of the medical and philanthropic work of the board of foreign missions of the presbyterian church in the u.s.a / Halsey, Abram Woodruff – New York: Board of Foreign Missions of the Presbyterian Church in the USA, 1914 – 1mf – 9 – 0-524-06306-0 – mf#1991-2479 – us ATLA [242]

Go to joseph / Cracknell, J E – London, England. 1865 – 1r – us UF Libraries [240]

Go up higher : or, religion in common life / Clarke, James Freeman – Boston: Lee & Shepard, c1877 [mf ed 1992] – (= ser Unitarian/universalist coll) – 1mf – 9 – 0-524-02458-8 – mf#1990-4317 – us ATLA [243]

Goa *see* Boletim oficial do estado da india

Goad, Charles Edward *see*
– Atlas of the city of montreal
– Insurance plan of the city of montreal, quebec, canada

Goad, Chas E *see* Atlas of the city of montreal

Goad, George Washington *see* Instructions to juries especially adapted to the laws of texas

Goadby, Edwin *see* The present depression in trade

Goadby, Joseph Jackson *see* Bye-paths in baptist history

Goal involvements, goal orientation, and perceptions of parent- and coach-initiated motivational climates among youth sport participants / Daw, Jessica L – 1999 – 2mf – 9 – $8.00 – mf#PSY 2119 – us Kinesology [150]

The goal of india / Holland, William Edward Sladen – London: United Council for Missionary Education, 1917 [mf ed 1995] – (= ser Yale coll) – 256p (ill) – 1 – 0-524-09874-3 – mf#1995-0874 – us ATLA [954]

The goal of india / Holland, William Edward Sladen – Madras: Christian Literature Society for India, 1919 [mf ed 1995] – (= ser Yale coll) – iv/178p – 1 – 0-524-10062-4 – (with chapter (the other half of india) by mrs w s urquhart) – mf#1995-1062 – us ATLA [954]

The goal of the human race, or, the development of civilisation, its origin and issue = Urspruenge und ziele unserer kulturentwicklung / Grau, Rudolf Friedrich – London: Simpkin, Marshall, Hamilton, Kent, 1892 – 1mf – 9 – 0-7905-3380-4 – (in english) – mf#1987-3380 – us ATLA [100]

Goal orientation and level of satisfaction in runners / Maday, Kristen M – 2000 – 146p on 2mf – 9 – $10.00 – mf#PSY 2140 – us Kinesology [150]

Goal orientation and moral atmosphere in youth sport : an examination of lying, hurting, and cheating behavior in girls' soccer / Stephens, Dawn E & Bredemeier, Brenda Jo Light – 1993 – 3mf – 9 – $12.00 – us Kinesology [150]

Goal perspectives and their relationship to beliefs, affective responses and coping strategies among african and anglo american athletes / Gano-Overway, Lori A – Purdue University, 1995 – 2mf – 9 – $8.00 – mf#PSY1844 – us Kinesology [150]

Goar, J *see*
– Chronographia [cbh5]
– Chronographia [cbh6]
– De officiis magnae ecclesiae et aulae constantinopo-litanae
– Rituale graecorum

Goat without horns / Davis, Beale – New York, NY. 1925 – 1r – us UF Libraries [972]

Gobankata daidaiki : directory of military officials in the tokugawa regime, 1584-1819. in the holdings of national archives / Okano, Magojuro [comp] – 394bks on 52r – 1 – Y540,000 – (with 20p guide. in japanese) – ja Yushodo [950]

Gobat, S *see* Journal of a three years' residence in abyssinia

Gobat, Samuel *see*
– Journal of a three years' residence in abyssinia
– Journal of a three years' residence in abyssinia: in furtherance of the objects of the church missionary society – a brief history of the church of abyssinia

Gobernador de nicaragua en el siglo 16 / Molina Arguello, Carlos – Sevilla, Spain. 1949 – 1r – us UF Libraries [972]

Gobernadores de antioquia 1571-1819. bogota, 1932 / Restrepo Saenz, Jose Ma. – Madrid: Razon y Fe, 1934 – 1 – sp Bibl Santa Ana [946]

Gobernantes de caldas / Jimenez Tobon, Gerardo – Manizales, Colombia. 1955 – 1r – us UF Libraries [972]

Gobernantes de columbia (1810-1957) / Mendoza Velez, Jorge – Bogota, Colombia. 1957 – 1r – us UF Libraries [972]

Los gobernantes de mexico : hernando cortes / Rivera Cambas, Manuel – Tabucaya: (Mexico. Editorial Cttp apte), 1962 – sp Bibl Santa Ana [320]

Gobierno Civil *see* Protocolo

Gobierno de puerto rico / Ramos De Santiago, Carmen – Rio Piedras, Puerto Rico. v1-2. 1965 – 1r – us UF Libraries [972]

Gobierno universal y la solucion integral del problema judio / Pacifico, Justo – Buenos Aires, Argentina. 1945? – 1r – us UF Libraries [939]

Gobillon, N *see* La vie de la venerable louise de marillac, veuve de m legras

Gobineau, Arthur *see*
– D pedro 2 e o conde de gobineau
– Essai sur l'inegalite des races humaines

Gobineau, Arthur, comte de *see* Souvenirs de voyage

Goblet D'alviella, Eugene *see* Contemporary evolution of religious thought in england, america

Goblet d'Alviella, Eugene, Comte *see*
– Hierographie
– Hierologie
– Hierosophie
– Lectures on the origin and growth of the conception of god

Goblet d'Alviella, Eugene, comte *see*
– The contemporary evolution of religious thought in england, america and india
– Introduction a l'histoire generale des religions
– The migration of symbols

Goblin : canada's only humourous publication, containing...satire, caricature and dependable literary criticism – Toronto: The Goblins Ltd. v1-8 n2. feb 1921-oct 1927 (mthly) – 3r – 1 – Can$245.00 – cn McLaren [400]

Goby, Emile *see* Confitures de ma tante

Gochet, Alexis Marie *see*
– Soldats et missionaires au congo de 1891 a 1894
– La traite des negres et la croisade africaine.

Gockel, hinkel und gackeleia : ein maerchen / Brentano, Clemens – 2. aufl. Berlin: Morawe & Scheffelt 1912 [mf ed 1993] – 1r – 1 – (filmed with: der erzieherische gehalt in j j breitingers "critischer dichtkunst" / jakob braeker) – mf#8525 – us Wisconsin U Libr [430]

God : conferences delivered at notre dame in paris... = Conferences de notre-dame de paris / Lacordaire, Henri-Dominique – New York: P O'Shea, c1871 – 1mf – 9 – 0-8370-7558-0 – (in english) – mf#1986-1558 – us ATLA [240]

God : an enquiry into the nature of man's highest ideal and a solution of the problem from the standpoint of science / Carus, Paul – Chicago: Open Court, 1908 – (= ser Christianity of To-day Series) – 1mf – 9 – 0-8370-2597-4 – mf#1985-0597 – us ATLA [210]

God : his knowability, essence and attributes: a dogmatic treatise / Pohle, Joseph – St Louis MO: B Herder 1911 [mf ed 1985] – (= ser Dogmatic theology 1) – 2mf – 9 – 0-8370-2753-5 – (incl bibl ref & ind; aut english version with some abridgement & add ref by arthur preuss) – mf#1985-0753 – us ATLA [210]

God : nature and attributes / Foster, Randolph Sinks – New York: Eaton & Mains 1897 [mf ed 1991] – (= ser Studies in theology) – 1mf – 9 – 0-7905-9202-9 – mf#1989-2427 – us ATLA [210]

God *see* Shang ti lun (ccm163)

God able- and willing to save / Molyneux, Capel – London, England. 1870 – 1r – us UF Libraries [240]

God against slavery : and the freedom and duty of the pulpit to rebuke it as a sin against god / Cheever, George Barrell – New York: J H Ladd, 1857 – 1mf – 9 – 0-7905-5864-5 – mf#1988-1864 – us ATLA [976]

God and amn : philosophy of the higher life / Shumaker, Elmer Ellsworth – New York: G P Putnam, 1909 – 1mf – 9 – 0-8370-5603-9 – mf#1985-3603 – us ATLA [210]

God and bread : with other sermons / Vincent, Marvin Richardson – New York: Dodd, Mead, 1884 [mf ed 1988] – 1mf – 9 – 0-7905-0407-3 – mf#1987-0407 – us ATLA [240]

God and christ : sermons preached in bedford chapel / Brooke, Stopford Augustus – London: P Green, 1894 – 1mf – 9 – 0-7905-3696-X – mf#1989-0189 – us ATLA [240]

God and freedom in human experience / d'Arcy, Charles Frederick – London: E Arnold, 1915 – (= ser Donnellan Lectures) – 1mf – 9 – 0-7905-3779-6 – mf#1989-0272 – us ATLA [210]

God and his book / Saladin – London: W Stewart [18–] [mf ed 1985] – 1mf – 9 – 0-8370-4974-1 – mf#1985-2974 – us ATLA [210]

God and little children : the blessed state of all who die in childhood / Van Dyke, Henry – New York: Anson DF Randolph, c1890 – 1mf – 9 – 0-8370-2909-0 – mf#1985-0909 – us ATLA [240]

God and the bible : a review of objections to "literature and dogma" / Arnold, Matthew – New York: Macmillan, 1903 [mf ed 1986] – 1mf – 9 – 0-8370-9440-2 – (incl bibl ref) – mf#1986-3440 – us ATLA [240]

God and the future life : the reasonableness of christianity / Nordhoff, Charles – NY: Harper, c1883 – 1mf – 9 – 0-8370-4598-3 – mf#1985-2598 – us ATLA [210]

God and the individual / Strong, Thomas Banks – London; New York: Longmans, Green, 1903 – 1mf – 9 – 0-7905-9691-1 – mf#1989-1416 – us ATLA [210]
God and the soul : an essay towards fundamental religion / Armstrong, Richard Acland – London: Philip Green, 1896 – 1mf – 9 – 0-8370-2113-8 – mf#1985-0113 – us ATLA [210]
God and the state = Dieu et l'etat / Bakunin, Mikhail Aleksandrovich – new rev ed. London: [s.n.], 1910 – 1mf – 9 – 0-524-04311-6 – (in english) – mf#1990-1237 – us ATLA [190]
God bless my soul! – London, England. 18– – 1r – us UF Libraries [240]
God borby : publitsistich khronika, 1905-1906 / Miliukov, P N – 1907 – 550p on 6mf – 9 – mf#RPP-121 – ne IDC [325]
God first – go forward : the impact of the south africa general mission/africa evangelical fellowship on the africa evangelical church, 1962-1994 / Kopp, Thomas Joseph – Uni of South Africa 2001 [mf ed Johannesburg 2001] – 4mf – 9 – (incl bibl ref) – mf#mfm14979 – sa Unisa [242]
God in christ : three discourses delivered at new haven, cambridge and andover / Bushnell, Horace – centenary ed. New York: Charles Scribner's, 1903 [mf ed 1984] – 4mf – 9 – 0-8370-0217-6 – mf#1984-0044 – us ATLA [240]
God in christ jesus : a study of st. paul's epistle to the ephesians / Lidgett, John Scott – London: Charles H Kelly, 1915 – 1mf – 9 – 0-524-03980-1 – mf#1992-0023 – us ATLA [220]
God in evolution : a pragmatic study of theology / Johnson, Francis Howe – New York: Longmans, Green, 1911 – 1mf – 9 – 0-7905-7794-1 – mf#1989-1019 – us ATLA [210]
God in freedom / Luzzatti, Luigi – New York, NY. 1930 – 1r – us UF Libraries [025]
God in his providence : a comprehensive view of the priciples and particulars of an active divine providence over man... / Fernald, Woodbury Melcher – Boston: Otis Clapp, 1859 [mf ed 1984] – 5mf – 9 – 0-8370-0930-8 – (incl bibl ref) – mf#1984-4295 – us ATLA [240]
God in history / Cumming, J – London, England. 1848? – 1r – us UF Libraries [240]
God in history : or, the progress of man's faith in the moral order of the world = Gott in der geschichte / Bunsen, Christian Karl Josias, Freiherr von – London: Longmans, Green, and Co, 1868-1870 – 4mf – 9 – 0-524-05454-1 – (in english) – mf#1990-3480 – us ATLA [210]
God in history : or, facts illustrative of the presence and providence of god in the affairs of men / Cumming, John – New York: Lane & Scott, 1852 [mf ed 1984] – 2mf – 9 – 0-8370-0961-8 – mf#1984-4314 – us ATLA [210]
God in human thought : or, natural theology traced in literature, ancient and modern, to the time of bishop butler... / Gillett, Ezra Hall – New York: Scribner, Armstrong, 1874 [mf ed 1991] – 2mf – 9 – 0-7905-7741-0 – (incl bibl ref) – mf#1989-0966 – us ATLA [210]
God in nature and life : selections from the sermons and writings of walter r brooks / Brooks, Walter Rollin – New York: Anson DF Randolph, c1889 [mf ed 1993] – 1mf – 9 – 0-524-08262-6 – mf#1993-3017 – us ATLA [242]
God in the constitution : "a review" of col robert g ingersoll / Pearne, Thomas Hall – Cincinnati: Geo P Houston, 1890 [mf ed 1984] – 1mf – 9 – 0-8370-1065-9 – mf#1984-4407 – us ATLA [240]
God incarnate / Kingdon, Hollingworth Tully – New York: T Whittaker 1890 [mf ed 1991] – (= ser The bishop paddock lectures 1890) – 1mf – 9 – 0-7905-9995-3 – (incl bibl ref) – mf#1989-1720 – us ATLA [210]
God is in heaven : an investigation of the concept of god in the book of ecclesiastes / Estes, Henry B 2 – 1982 – 1 – $7.84 – us Southern Baptist [242]
God is love – London, England. 1800 – 1r – us UF Libraries [240]
God is love : a supplement to the author's discourse on the reasonableness of future, endless punishment / Adams, Nehemiah – Boston: Gould and Lincoln, 1858 – (= ser Truths for the Times) – 1mf – 9 – 0-524-05591-2 – mf#1992-0446 – us ATLA [240]
God is love / Whytehead, R – London, England. 1836 – 1r – us UF Libraries [240]
God is spirit, god is love : a treatise on spiritual unitarianism / Elliot, George – London: Philip Green, 1895 – 1mf – 9 – 0-8370-8811-9 – mf#1986-2811 – us ATLA [210]
The god juggernaut and hinduism in india : from a study of their sacred books and more than 5,000 miles of travel in india / Zimmerman, Jeremiah – New York: Fleming H Revell, c1914 – 1mf – 9 – 0-524-01327-6 – mf#1990-2363 – us ATLA [280]
The god man / Dixon, Amzi Clarence – Baltimore: Wharton, Barron, c1891 – 1mf – 9 – 0-8370-2929-5 – mf#1985-0929 – us ATLA [240]
God, man, and the bible / Baylee, Joseph – London, England. 1860 – 1r – us UF Libraries [210]
God manifest in christ / Gaskell, William – London, England. 1854 – 1r – us UF Libraries [240]
The god of israel : a paper read before the international positivist congress at naples, 27th april-3rd may 1908 / Levy, Joseph Hiam – London: Lawrence Nelson, [1908?] – 1mf – 9 – 0-8370-4094-9 – (incl bibl ref) – mf#1985-2094 – us ATLA [210]
God of the matopo hills / Daneel, M L – Hague, Netherlands. 1970 – 1r – us UF Libraries [960]
God out and man in : or, replies to robert g. ingersoll / Platt, William Henry – Rochester, NY: Steele & Avery, 1883 [mf ed 1990] – 1mf – 9 – 0-7905-9582-6 – mf#1989-1307 – us ATLA [210]
God revealed in the process of creation and by the manifestation of jesus christ : including an examination of the development theory contained in the "vestiges of the natural history of creation" / Walker, James Barr – Boston: Gould and Lincoln; New York: Sheldon, Lamport & Blakeman, 1856, c1855 – 1mf – 9 – 0-7905-7488-8 – mf#1989-0713 – us ATLA [210]
God revealed, or, nature's best word / Gallagher, Charles Wesley – New York: Eaton & Mains, 1899 – 1mf – 9 – 0-7905-9935-X – mf#1989-1660 – us ATLA [210]
God sluzhby sotsialistov kapitalistam: : ocherki po istorii kontrrevoliutsii v 1918 godu / Vladimirova, V; ed by Iakovlev, I A – 1927 – 386p on 5mf – 9 – mf#RPP-11 – ne IDC [325]
God spake all these words / Brookes, James Hall – [s.l.]: J T Smith; St Louis: Rev J W Allen, Presbyterian Bd of Publ [distributor], c1895 – 1mf – 9 – 0-8370-2468-4 – mf#1985-0468 – us ATLA [220]
God, the creator and lord of all / Harris, Samuel – New York: Scribner, 1896 – 3mf – 9 – 0-7905-7828-X – mf#1989-1053 – us ATLA [210]
God to be obeyed rather than men / Buchanan, Robert – Glasgow, Scotland. 1839 – 1r – us UF Libraries [240]
The god varuna in the rig-veda : a paper / Griswold, Hervey De Witt – Ithaca, NY: Taylor and Carpenter, 1910 – 1mf – 9 – 0-524-01491-4 – (incl bibl ref) – mf#1990-2467 – us ATLA [280]
The god we trust : studies in the devotional use of the apostles' creed / Ross, George Alexander Johnston – New York: FH Revell, c1913 – (= ser Cole lectures) – 1mf – 9 – 0-7905-9620-2 – mf#1989-1345 – us ATLA [240]
God with us : a study in religious idealism / Gibson, William Ralph Boyce – London: Adam and Charles Black, 1909 – 1mf – 9 – 0-8370-3267-9 – mf#1985-1267 – us ATLA [200]
God with us, or, the person and work of christ : with an examination of "the vicarious sacrifice" of dr. bushnell / Hovey, Alvah – 2nd ed. Boston: Gould and Lincoln; New York: Sheldon, 1872 – 1mf – 9 – 0-8370-4855-9 – mf#1985-2855 – us ATLA [240]
God x / Sel'sko-khoziaistvennyi obzor Chernigovskoi gubernii za 1913 god – Chernigov, 1915 – 9mf – 8 – mf#RZ-198 – ne IDC [314]
Godard, Benjamin *see* Chant et baiser
Godard, Charles *see* Le brahmanisme
Godard, John George *see* Poverty
Godart, Felix *see* La reforme judiciaire de 1789 a 1794 d'apres camille desmoulins et la legislation
Godbey, Geoffrey C *see*
– The relation of self-esteem to constraints on leisure among adolescents
– Tourist travel motivations and satisfaction study of china/yangzte river adventure tours
Godbey, William Baxter *see*
– Commentary on the new testament
– Translation of the new testament from the original greek
Godbout, Leopold *see* Ecole populaire de cooperation
Goddard, Dwight *see*
– A buddhist bible
– The divine urge to missionary service
– Was jesus influenced by buddhism
Goddard, Evelyn Lily *see* Harmonic technique of ernest chausson as exemplified in his chamber music
Goddard, Harold Clarke *see* Studies in new england transcendentalism
Goddard, Pliny E *see* Ethnology of the north california coast indian tribes hupa wyot pomo miwak
Het goddelyck herte : ofte de woonste godts in het herte / Bottens, Fulg – Ghendt. v1-3. 1716 – 3v on 33mf – 8 – €63.00 – ne Slangenburg [240]
Goddelycke aandachten ofte vlammende begeerten eens boetvaerdige geheijligd en lief-rijcke ziele / [Serrarius, P] – t'Amsterdam: Salomon Savrij, [1653] – 2mf – 9 – mf#O-3203 – ne IDC [090]
De goddelycke voorsienigheydt uytgebeeldt in joseph... / Nerrincq, Franciscus – t'Antwerpen: Ignatius Leyssens, 1710 – 6mf – 9 – mf#O-701 – ne IDC [090]
Goddelycke wenschen verlicht : met sinnebeelden en vierige uytspraken der out-vaders... / Hugo, H – [Antwerpen]: P I Paets, 1645 – 5mf – 9 – mf#O-3091 – ne IDC [090]
Goddelycke wenschen verlicht : met sinnebeelden, ghedichten en vierighe uyt-spraecken der oudvaders... / Hugo, H – t'Hantwerpen: Hendrick Aertssens, 1629 – 8mf – 9 – mf#O-640 – ne IDC [090]
Goddelyke liefde-vlammen... / Luyken, Jan – t'Amsteldam: Joannes Boekholt, 1691 – 3mf – 9 – mf#O-3033 – ne IDC [090]
Goddelyke liefde-vlammen of vervolg van jesus en de ziele... / Luyken, Jan – Amsterdam: de Verheydens, 1736 – 3mf – 9 – (titlepg missing) – mf#O-3032 – ne IDC [090]
Godden, G M *see* Communist attack in great britain, london 1938
Godden, Gertrude M *see*
– Communism in spain, 1931-1936
– Communist operations in spain
– New spain; its people, its ruler
Godden, John *see* Notes and reminiscences of a journey to england
Godden, Rumer *see*
– Bengal journey
– Rungli-rungliot
Godding, Marc Antone *see* Eighteenth-century english organ voluntary
Gode, Antonio *see* Asedio de huesca, 18 julio 1936, 25 marzo 1938
Godefridus, abbas Admontensis olim Weingarttensis *see* Homiliae
Godefroy, Fredric *see* Dictionnaire de l'ancienne langue francaise
Godefroy, J *see* Bibliotheque des benedictins de la congregation de saint-vanne et saint-hydulphe (afm29)
Godefroy, Jean *see* Chronique metrique de godefroy de paris
Goderich, Frederick John Robinson, Viscount *see*
– Letter to the right hon lord viscount goderich
– Message
Goderich, Lord Viscount *see* Copy of a memorial from james stuart, esquire, his majesty's attorney general for lower canada
Goderich star – Ontario, Canada. 9 may 1913-22 dec 1921 – 9r – 1 – uk British Libr Newspaper [071]
Godesberger volkszeitung – Bonn DE, 1913 jul-dez, 1920 jan-jun, 1921-1923 jun, 1924-27, 1928 jul-1941 mar – 49r – 1 – (with suppl: drachenfelser echo, wochenblatt fuer mehlem und umgebung apr 1-nov 27 1931, 1932-sep 22 1933, oct 20 1933 [2r]. title varies: deutsche reichszeitung, jan 2 1934; mittelrheinische landes-zeitung, oct 1 1934) – gw Misc Inst [074]
Godet, Frederic Louis *see*
– The atonement in modern religious thought
– Bibliska studier [1st series]
– Bibliska studier [2nd series]
– Commentary on st paul's epistle to the romans
– Commentary on the gospel of john
– Etudes bibliques
– Etudes bibliques. deuxieme serie
– Examen des principales questions critiques soulevees de nos jours au sujet du quatri eme evangile
– Histoire de la reformation et du refuge
– Introduction au nouveau testament
– Introduction to the new testament
– Lectures in defence of the christian faith
– Studies of creation and life
– Studies on the epistles
– Studies on the new testament
Godet, Fredric Louis *see* Commentary on the gospel of st luke
Godet, Phillippe *see* Frederic godet (1812-1900)
Godey's lady's book / Cairns Collection of American Women Writers – v90/91 [1875] – 1r – 1 – (cont: lady's book; cont by: godey's magazine) – mf#2593367 – us WHS [305]
Godey's magazine – La Salle IL 1830-98 – 1 – mf#3756 – us UMI ProQuest [780]
Godfrey, John Blennerhassett *see* The means of preventing the downfall of the british empire
Godfrey, L M *see* Indirect domestic influence
Godfrey, Walter Hindes *see* History of architecture in london
The godhead of jesus : four sermons / Perowne, Edward Henry – Cambridge: Deighton, Bell, 1867 – (= ser Hulsean Lectures) – 1mf – 9 – 0-7905-9056-5 – mf#1989-2281 – us ATLA [240]
Godinet, Adolphe *see* Voyage en normandie et en bretagne
Godinez de la Paz, Carlos *see* Consideraciones... ferrocarriles...caceres
Goding, MW *see* Statement of objectives and policies of the trust territory of the pacific islands (ttpi)
Godinho, Manuel *see* Relacao do novo caminho que fez por terra e mar vindo de india para portugal...
[Godinho, N] *see* De abassionorum rebus...
Godism / Robertson, John M – Bradford, England. 18– – 1r – us UF Libraries [210]
Godiva : drama in fuenf akten / Franck, Hans – Muenchen: Delphin-Verlag, c1919 (mf ed 1990) – 1r – 1 – (filmed with: von morgen bet abend) – us Wisconsin U Libr [820]
Godkin, Edwin L *see* Unforeseen tendencies of democracy
Godkin, James *see*
– Church principles of the new testament
– Ireland and her churches
– The land-war in ireland
– The religious history of ireland, primitive, papal and protestant, including the evangelical missions, catholic agitations, and church progress of the last half-century
– The rights of ireland
De godlievende ziel vertoont in zinnebeelden / Hugo, Hermannus & Vaenius, Othon – Utrecht: H & J Besseling, 1749 – 3mf – 9 – mf#O-642 – ne IDC [090]
De godlievende ziel vertoont in zinnebeelden met dichtkunstige verklaringen van Jan Suderman / Hugo, Hermannus & Vaenius, Othon – t'Amsterdam: H. Wetstein, 1724 – 3mf – 9 – mf#O-641 – ne IDC [090]
Godlonton, Robert *see*
– Case of the colonists of the eastern frontier of the cape of good hope
– Narrative of the irruption of the kafir hordes
– Narrative of the keffir war 1850-1851-1852
De godloosheyd der achthiende eeuw : of, kort-begryp van de saemenzweeringen der philosofiten, vry-metzelaers [franc-macons,] zoogezeyde verlichte [illumines,] en jacobins, tegen den godsdienst en den troon, ontdekt door haer-zelven... / Haspeslagh, Louis – [Belgium?: s.n.] 1804 – 3mf – 9 – mf#vrl-200 – ne IDC [366]
Godloosheyd der achttiende eeuw : of, kort begryp van de zamenzweeringen der philosophisten, vry-metzelaers, [franc-macons] zoo-gezeyde verlichte, [illumines] en jacobins, tegen den godsdienst en den troon, ontdekt door haer zelven... / Haspeslagh, Louis – Brussel: G Cuelens 1816 – 4mf – 9 – mf#vrl-183 – ne IDC [366]
The godly pastor : life of the rev. jeremiah hallock, of canton, conn.: to which is added a sketch of the life of the rev. moses hallock, of plainfield, mass / Yale, Cyrus – A new ed rev and enl. New York: American Tract Society, [1854?] – 1mf – 9 – 0-7905-6978-7 – mf#1988-2978 – us ATLA [240]
Godly union and concord : sermons preached mainly in westminster abbey in the interest of christian fraternity / Henson, Hensley – London: J Murray, 1902 – 1mf – 9 – 0-7905-4750-3 – mf#1988-0750 – us ATLA [240]
The god-man / Edwards, Thomas Charles – London: Hodder and Stoughton, 1895 – (= ser Davies Lecture) – 1mf – 9 – 0-8370-3708-5 – (incl bibl ref and indexes) – mf#1985-1708 – us ATLA [240]
The godman collection : persian ceramic art... with examples from other collections / Wallis, Henry – London 1894 – (= ser 19th c art & architecture) – 3mf – 9 – mf#4.2.1536 – uk Chadwyck [730]
Godolin, Pierre *see* Las obros de pierre goudelin
Godolphin / Lytton, Edward Bulwer Lytton, Baron – Boston, MA. 189– – 1r – us UF Libraries [420]
Godolphin : mr isaac's new dances performed at court on her majesty's birthday 1714 / Paisible, James – London: printed for & engraved by D Wright...[1714] [mf ed 1988] – 1r – 1 – mf#pres. film 47 – us Sibley [790]
Godolphin, Gregory *see* The unique
Godon, Julien *see* Painted tapestry
Godovoi otchet za 1923-1924 god : (za pervyi operatsionnyi god: s 1 iiunia 1923 g. po 30 sentiabria 1924 goda) / Ural'skii oblastnoi Sel'sko-Khoziastvennyi Bank £Uralsel'khozbank" – Sverdlovsk, 1924 – 5mf – 9 – mf#REF-112 – ne IDC [332]
Godoy / Izquierdo Hernandez, Manuel – Badajoz: Dip. Provincial, 1967 – sp Bibl Santa Ana [946]
Godoy, Gustavo *see* Semper et ubique
Godoy, Manuel *see*
– Carlos 4 y maria luisa, de juan perez de guzman y gallo
– Carta al marques de san simon
– Cuenta...memorias criticas y apologeticas...carlos 4 de borbon
– Disputationes...thomae eidem angelico
– Manifiesto...aranjuez, madrid y bayona
– Memorias

Godoy, Pedro de *see*
- Disputationes theologicae
- Disputationes theologicae in tertiam partem divi thomae
- Disputationes theologicae intertiam partem divi thonae...
- Disputationes theologicae intertian pastem divi thonae...

Godoy. principe de la paz y de bassano / Taxonera, Luciano de – Barcelona: Edit. Juventud, 1946 – 9 – sp Bibl Santa Ana [940]

Godrycz, John A *see* The doctrine of modernism and its refutation

God's ancient people not cast away / Blomfield, Charles James – London, England. 1843? – 1r – us UF Libraries [240]

The gods and other lectures / Ingersoll, Robert Green – Washington, DC: C P Farrell, 1879, c1874 – 1mf – 9 – 0-8370-3724-7 – mf#1985-1724 – us ATLA [210]

Gods and rituals / Middleton, John – Garden City, NY. 1967 – 1r – us UF Libraries [025]

God's balance of faith and freedom / Waterman, Lucius – Milwaukee: Young Churchman, c1911 – (= ser Mary Fitch Page Lectures) – 1mf – 9 – 0-7905-6396-7 – mf#1988-2396 – us ATLA [240]

God's champion, man's example : a study of the conflict of our divine deliverer / Birks, H A – London: Religious Tract Society, 1891 – 1mf – 9 – 0-7905-3010-4 – mf#1987-3010 – us ATLA [220]

God's chief mercy / Maclaurin, John – London, England. 18-- – 1r – us UF Libraries [240]

God's co-operative society : suggestions on the strategy of the church / Marson, Charles Latimer – London: Longmans, Green, 1914 – 1mf – 9 – 0-524-04879-7 – mf#1990-5077 – us ATLA [240]

God's hand in america / Cheever, George Barrell – New-York: M.W. Dodd; London: Wiley & Putnam, 1841 – 1mf – 9 – 0-7905-4202-1 – mf#1988-0202 – us ATLA [240]

God's image in man : some intuitive perceptions of truth / Wood, Henry – 6th ed. Boston: Lee and Shepard, 1894, c1892 – 1mf – 9 – 0-8370-5749-3 – mf#1985-3749 – us ATLA [210]

God's image in man and its defacement in the light of modern denials / Orr, James – New York: A C Armstrong; London: Hodder and Stoughton, 1905 – (= ser Stone Lectures) – 1mf – 9 – 0-7905-3153-4 – (incl bibl ref) – mf#1987-3153 – us ATLA [240]

God's iron : a life of the prophet jeremiah / Birmingham, G A – London, 1956 – 5mf – 8 – €12.00 – ne Slangenburg [240]

God's jester / Blount, Melesina Mary – New York, NY. 1930 – 1r – us UF Libraries [972]

God's living oracles : being the exeter hall lectures on the bible / Pierson, Arthur T – New York: Baker & Taylor, c1904 – 1mf – 9 – 0-7905-1784-1 – mf#1987-1784 – us ATLA [220]

God's means of grace : a discussion of the various helps divinely given as aids to christian character, and a plea for fidelity to their scriptural form and purpose / Yoder, Charles Francis – Elgin IL: Brethren Pub House 1908 [mf ed 1992] – 2mf – 9 – 0-524-03809-0 – mf#1990-4881 – us ATLA [240]

Gods mercie mixed with his justice, or his peoples deliverance in times of danger / Cotton, John – 1641 – 9 – us Scholars Facs [240]

God's method with man : or, sacred scenes along the path to heaven / Gorham, Barlow Weed – Cincinnati: Hitchcock and Walden, 1879. Beltsville, Md: NCR Corp, 1978 (3mf); Evanston: American Theol Lib Assoc, 1984 (3mf) – 9 – 0-8370-0947-2 – mf#1984-4298 – us ATLA [240]

God's missionary plan for the world / Bashford, James Whitford – New York: Young People's Missionary Movt, 1908 – 1mf – 9 – 0-8370-6011-7 – (with index) – mf#1986-0011 – us ATLA [240]

God's oath : a study of an unfulfilled promise of god / Ottman, Ford Cyrinde – New York City: Pub office "Our Hope" [1911?] [mf ed 1986] – 1mf – 9 – 0-8370-6291-8 – (incl ind) – mf#1986-0291 – us ATLA [240]

The gods of india : a brief description of their history, character and worship / Martin, Edward Osborn – London: JM Dent; New York: EP Dutton, 1914 – 1mf – 9 – 0-524-01287-3 – mf#1990-2323 – us ATLA [280]

The gods of india : a brief description of their history, character and worship / Martin, Edward Osborn – London: J.M. Dent; New York: E.P. Dutton, 1914. 330p. ill – 1 – us Wisconsin U Libr [280]

Gods of modern grub street : impressions of contemporary authors / Adcock, Arthur St John – New York: Frederick A Stokes Co, 1923 [mf ed 1985] – viii/326p – 1 – mf#1308 – us Wisconsin U Libr [420]

The gods of northern buddhism : their history, iconography, and progressive evolution through the northern buddhist countries / Getty, Alice – Oxford: Clarendon Press, 1928 – (= ser Samp: indian books) – (int on buddhism trans fr french of j deniker; ill fr coll of henry h getty) – us CRL [280]

The gods of olympos : or, mythology of the greeks and romans = Olymp / Petiscus, August Heinrich; ed by Raleigh, Katherine A – New York: Cassell Pub Co, c1892 – 1mf – 9 – 0-524-04348-5 – (incl bibl ref. in english) – mf#1990-3332 – us ATLA [250]

The gods of our fathers : a study of saxon mythology / Stern, Herman Isidore – New York: Harper, 1898 – 1mf – 9 – 0-524-01303-9 – mf#1990-2339 – us ATLA [290]

The gods of the egyptians : or, studies in egyptian mythology / Budge, Ernest Alfred Wallis – Chicago: Open Court. 2v. 1904 – 4mf – 9 – 0-7905-0867-2 – (incl bibl ref and index. text in english and egyptian) – mf#1987-0867 – us ATLA [390]

God's other children and other sketches / Manganyi, M C – New Haven, 1973 – (filmed with: the politics of separate freedoms) – us CRL [240]

God's plan for soul-winning / Hogben, Thomas – [2nd ed]. Toronto: A Sims, 1907 – 1mf – 9 – 0-8370-6264-0 – mf#1986-0264 – us ATLA [240]

God's plan for world redemption : an outline study of the bible and missions / Watson, Charles Roger – Philadelphia, PA: Board of Foreign Missions of the United Presbyterian Church of NA, c1911 – 1mf – 9 – 0-7905-6397-5 – mf#1988-2397 – us ATLA [240]

God's predestination the confidence of his saints / Harding, Thomas – London, England. 1850 – 1r – us UF Libraries [240]

God's requirements : and other sermons / Chapin, Edwin Hubbell – New York: J Miller, 1881 – 1mf – 9 – 0-524-08356-8 – mf#1993-3056 – us ATLA [240]

God's rescues : or, the lost sheep, the lost coin, and the lost son: three discourses on luke 15 / Williams, William R – New York: Anson D F Randolph, 1871 – 1mf – 9 – 0-8370-7353-7 – mf#1986-1353 – us ATLA [220]

God's revelations of himself to men : as successively made in the patriarchal, jewish, and christian dispensations and in the messianic kingdom / Andrews, Samuel James – 2nd rev enl ed. New York: Putnam's, 1901 [mf ed 1985] – 1mf – 9 – 0-8370-2103-0 – (incl app & notes) – mf#1985-0103 – us ATLA [220]

God's songs and the singer : four sermons / Bain, John Wallace – Pittsburgh: United Presbyterian Board of Publication, 1871 – 1mf – 9 – 0-524-01075-7 – mf#1990-4040 – us ATLA [220]

God's thoughts fit bread for children : a sermon preached before the connecticut sunday-school teachers' convention, at the pearl-street congregational church, hartford, conn... / Bushnell, Horace – Boston: Nichols and Noyes, 1869 – 1mf – 9 – 0-524-08568-4 – mf#1993-3153 – us ATLA [242]

God's timepiece for man's eternity : its purpose of love and mercy, its plenary infallible inspiration, and its personal experiment of forgiveness and eternal life in christ / Cheever, George Barrell – [2nd ed]. New York: A C Armstrong, 1888, c1883 – 2mf – 9 – 0-7905-0923-7 – mf#1987-0923 – us ATLA [220]

God's two books : or, nature and the bible have one author / Balfour, Thomas Alexander Goldie – London: James Nisbet, 1861 – 1mf – 9 – 0-8370-2166-9 – mf#1985-0166 – us ATLA [210]

God's way of holiness / Bonar, Horatius – New York: Robert Carter, 1867 – 1mf – 9 – 0-8370-2769-1 – mf#1985-0769 – us ATLA [240]

God's whip *see* Bicz bozy

God's white throne : a rational evangelical theodicy / Palmer, Byron – 3rd ed. Cincinnati: Jennings and Graham, 1904 – 1mf – 9 – 0-8370-3973-8 – mf#1985-1973 – us ATLA [210]

God's will and man's shall / Hawker, Robert – London, England. 1823 – 1r – us UF Libraries [240]

God's witness / Hawker, Robert – London, England. 1820 – 1r – us UF Libraries [240]

God's witness in prophecy and history : bible studies on the historical fulfilments of jacob's prophetic blessings on the twelve tribes contained in genesis 49: with a supplementary enquiry into the history of the lost tribes / Bellett, John Crosthwaite – London: J Masters, 1884 – 1mf – 9 – 0-8370-6162-8 – (incl bibl ref and index and appendixes containing different versions of genesis 49 and deuteronomy 33, 6-25) – mf#1986-0162 – us ATLA [220]

God's witness to his own word / Bryan, R G – London: Skeffington, 1892 – 1mf – 9 – 0-8370-2502-8 – mf#1985-0502 – us ATLA [220]

God's wonderful work in france : an introduction of the deputation from the protestants of france to the american churches / Bacon, Leonard Woolsey – New York: Special Commission on the Work of the Deputation, [188-?] – 1mf – 9 – 0-524-03211-4 – mf#1990-0839 – us ATLA [242]

God's word : man's light and guide: a course of lectures / Taylor, William M et al – New York: American Tract Society, c1877 – 1mf – 9 – 0-8370-3316-0 – mf#1985-1316 – us ATLA [220]

God's word and ministers / Girdlestone, Charles – London, England. 1838 – 1r – us UF Libraries [240]

God's word in man's language / Nida, E – New York, 1952 – 3mf – 8 – €7.00 – ne Slangenburg [210]

God's word in man's language / Nida, Eugene Albert – New York, NY. 1952 – 1r – us UF Libraries [240]

God's word written : the doctrine of the inspiration of holy scripture / Garbett, Edward – Boston: American Tract Society, [1868] – 1mf – 9 – 0-8370-3230-X – mf#1985-1230 – us ATLA [220]

God's works made to be remembered / Calthrop, Gordon – London, England. 18-- – 1r – us UF Libraries [240]

God's world and other sermons / Mills, Benjamin Fay – New York: Revell, [c1894] Beltsville, Md: NCR Corp, 1978 (4mf); Evanston: American Theol Lib Assoc, 1984 (4mf) – (= ser Revivalism and revival preachers in america) – 9 – 0-8370-1226-0 – mf#1984-3011 – us ATLA [242]

Godsdienst : volgens de beginselen der ethische richting onder de modernen / Hooykaas, Isaac et al – 's Hertogenbosch: GH van der Schuyt, 1876 – 1mf – 9 – 0-8370-3324-1 – mf#1985-1324 – us ATLA [240]

Degodsdienst uit plichtbesef en de geloofsvoorstelling uit dichtende verbeelding geboren? : bedenkingen tegen dr l w e rauwenhoff's wijsbegeerte van den godsdienst / Cannegieter, Tjeerd – Leiden: E J Brill, 1890 [mf ed 1985] – 284p on 1mf – 9 – 0-8370-2582-6 – (in dutch) – mf#1985-0582 – us ATLA [230]

Godsdienst van israel tot den ondergang van den joodschen staat *see* The religion of israel to the fall of the jewish state

Godsdienstige bewegingen preanger regentschappen / Mailrapport. n262. 1886 – 1mf – 8 – mf#SD-101 mf 4 – ne IDC [959]

De godsleer der middeleeuwsche joden : bijdrage tot de geschiedenis der leer aangaande god / Muller, Pieter Johannes – Groningen: J B Wolters, 1898 [mf ed 1985] – 1mf – 9 – 0-8370-4538-X – (incl list of hebrew words and ind) – mf#1985-2538 – us ATLA [270]

De godsleer van calvijn : uit religieus oogpunt beschouwd en gewaardeerd / Muller, Pieter Johannes – Groningen: B J Wolters, 1881 [mf ed 1993] – (= ser Presbyterian coll) – 126p on 3mf – 9 – 0-524-07443-7 – (in dutch) – mf#1991-3103 – us ATLA [242]

De godsleer van zwingli en calvijn / Muller, Pieter Johannes – Sneek: J Campen, 1883 [mf ed 1992] – (= ser Presbyterian coll) – 115p on 1mf – 9 – 0-524-05088-0 – (incl bibl ref) – mf#1991-2212 – us ATLA [242]

De godspraken van amos / Gunning, J H – Leiden: E J Brill, 1885 [mf ed 1985] – 1mf – 9 – 0-8370-3423-X – mf#1985-1423 – us ATLA [221]

Godts, G M *see* Why do protestants not invoke the virgin?

Het godtvruchtich herte : den koninghlijcken throon van jesys den vreedsamighen salomon / Luzvic, S – t'Antwerpen: Henricus Aertsens, 1627 – 3mf – 9 – mf#0-3118 – ne IDC [090]

Godwi : ein kapitel deutscher romantik / Kerr, Alfred – Berlin: G Bondi, 1898 [mf ed 1989] – xi/136p – 1 – (incl bibl ref) – mf#7084 – us Wisconsin U Libr [430]

Godwin, Benjamin *see*
- The jubilee memorial of horton college, bradford, containing the sermon preached at the jubilee service, 2 aug 1854; also
- Lectures on the atheistic controversy, 1836, delivered at sion chapel, bradford, yorkshire

Godwin, Edward William et al *see* Artistic conservatories

Godwin, George *see*
- Buildings and monuments
- The churches of london

Godwin, John Henry *see*
- Christian faith
- The epistle of the apostle paul to romans

Godwin, Parke *see*
- Biography of william cullen bryant
- The history of france
- Tales from the german of heinrich zschokke

Godwin, William *see*
- Essays
- Four early pamphlets, 1783-84
- Lives of the necromancers: or, an account of the most eminent persons in successive ages who have claimed...or to whom has been imputed...the exercise of magical power

Goe, F F *see* Disciples indeed

Goebbels, Joseph *see*
- Un discurso del dr. goebbels, ministro de propaganda de alemania
- The truth about spain
- La verdad sobre espana
- Die wahrheit ueber spanien

Goebel, BM *see* The effects of supervised cardiac rehabilitation on selected coronary artery disease risk factors following coronary artery bypass graft surgery

Goebel, Julius *see*
- Goethes faust
- Hoffmann von fallerslebens "texanische lieder"

Goebel, K C T F *see* Reise in die steppen des suedlichen russlands...

Goebel, K von *see* Wilhelm hofmeister – the works and life of a nineteenth century botanist

Goebel, Kaete *see* Die quellen und die entstehungszeit von thomas heywood's "iron age"

Goebel, Rubye K *see*
- Agricultural data
- Annual florida events
- Army and navy
- Boat trips
- City government of daytona beach
- Daytona beach art school
- District or neighborhood architecture and housing
- Drama no 666
- Educational centers
- Ethnography
- Flora
- Florida preparatory school
- Manufacturing and industry : n632 bunnell
- Manufacturing and industry n632 (daytona beach)
- Manufacturing and industry : n632
- Museums and collections
- Music
- Music, n664
- Points of interest
- Recreation and amusement centers
- Recreation : n680
- Recreation no 680
- Religious institutions and structures
- St paul's catholic parochial school
- Societies and associations
- Tourist camps
- Watermelon feast
- Waterways

Goebel, Siegfried *see*
- Die parabeln jesu
- The parables of jesus

Goebel's probate reports / Ohio. Supreme Court – Hamilton County. 1v. 1885-90 (all publ) – (= ser Ohio appellate decisions) – 4mf – 9 – $6.00 – mf#llmc 84-183 – us LLMC [348]

Die goechhausen : briefe einer hofdame aus dem klassischen weimar / Goechhausen, Luise von; ed by Deetjen, Werner – Berlin: E S Mittler, 1923 (mf ed 1990) – 1r – 1 – (filmed with: golowin. incl bibl ref) – us Wisconsin U Libr [860]

Goechhausen, E A A von *see* Eudaemonia

Goechhausen, Luise von *see* Die goechhausen

Goeckingk, Ludwig Friedrich Guenter von *see* Journal von und fuer deutschland

Goede, Arian de *see* Oud-nederlandsch procesrecht

Goede, Christian A *see* England, wales, irland und schottland

Goedeckemeyer, Albert *see* Die geschichte des griechischen skeptizismus

Goedeke *see* Gedichte

Goedeke, Karl *see*
- Emanuel geibel
- Gedichte
- Geschaeftsbriefe schiller's
- Schillers saemmtliche schriften
- Spiegel des regiments

Goedeke, Karl, 1814-87 *see* Goethes leben..

Goedings, Peter *see* Der mineralstoffhaushalt in viscum album l (weissbeerige mistel)

Goedsche, Friedrich W *see* Die weltumsegler oder abenteuer und seltsame schicksale der familie neander auf ihrer reise durch die welt

Goegrafia descriptiva de la republica dominicana / Inchaustegui Cabral, Joaquin Marino – Republica Dominicana, Dominican Republic. 1957 – 1r – us UF Libraries [918]

Goehler, Georg *see* Cornelius freundt

The goehner news – Goehner, NE: L H Warner. v1 n1. sep 22 1899- (wkly) [mf ed -apr 20 1900 (gaps) filmed 1979] – 1r – 1 – (publ in milford ne, oct 13 1899-apr 20 1900) – us NE Hist [071]

Goehringer, Diane M *see* A study of the perceived effects of the repeal of the pennsylvania interscholastic athletic association constitutional bylaw article 11, section 2

Goeje, Claudius Henricus De *see* Zondvloed en zondeval bij de indianen van west-ind...
Goeje, M J de *see* Historia khalifatus omari 2, jazidi 2 et hischami
Goeje, Michael Jan de et al *see* Orientalische studien
Goekalp, Ziya *see* Tuerk medeniyet tarihi
Goeken, Walther *see* Herder als deutscher
Goelzer, Henri *see* Etude lexicographique et grammaticale de la latinite de saint jeraome
Goenuel yuvasi / Cahid, Burhan [Morkaya] – Istanbul: Burhan Cahid ve Suerekasi, 1926 – (= ser Ottoman literature, writers and the arts) – 4mf – 9 – $60.00 – us MEDOC [470]
Goepfert, A et al *see* Methods of multicriteria decision theory
Goepp, Edouard *see* Les grands hommes de la france
Goeransson, Nils Johan *see* Undersoekning af religionen
Goeree, W *see*
– D'algemeene bouwkunde
– Inleydinge tot de algemeene teyken-konst...
Goeres-gesellschaft zur pflege der wissenschaft im katholischen Deutschland *see* Historisches jahrbuch
Goerges, E *see* Wegweiser durch das wesergebiet von muenden bis minden, nebst teutoburgerwald u deister
Goering, Hermann *see* Selected research documents relating to hermann goering
Goering, Reinhard *see*
– Seeschlacht
– Die suedpolexpedition des kapitaens scott
Goering, Theodor *see* Der messias von bayreuth
Goerlitzer anzeiger *see* Der anzeiger
Goerlitzer anzeiger 1945 – Goerlitz DE, 1945 7 jun-1948 16 jan – 2r – 1 – gw Misc Inst [074]
Goerlitzer anzeiger und vergnuegungsblatt – Goerlitz DE, 1885 – 1r – 1 – gw Misc Inst [074]
Goerlitzer fama – Goerlitz DE, 1842, 1847-49 – 2r – 1 – gw Misc Inst [074]
Goerlitzer haus- und grundbesitzerzeitung – Goerlitz DE, 1924-27, 1929 & 1939 – 1r – 1 – gw Misc Inst [640]
Goerlitzer landbund – Goerlitz DE, 1926 – 1r – 1 – gw Misc Inst [074]
Goerlitzer tageblatt – Goerlitz DE, 1856 1 oct-1862, 1864 31 may-1928 – 153r – 1 – (title varies: 1 jan 1863: niederschlesische zeitung. with suppl) – gw Misc Inst [074]
Goerlitzer volkszeitung – Goerlitz DE, 1899 28 jan-1933 4 mar – 71r – 1 – gw Misc Inst [074]
Goerlitzer vororts-zeitung – Goerlitz DE, 1898-1900 – 2r – 1 – gw Misc Inst [074]
Goerlitzer zeitung – Goerlitz DE, 1891 26 nov-1892 30 sep – 2r – 1 – gw Misc Inst [074]
Goerner, Karl August *see* Rothkaeppchen
Goerres / Sepp, Johann Nepomuk – Berlin: E Hofmann, 1896 – 1r – 1 – us Wisconsin U Libr [920]
Goerres, Guido *see*
– Der heilige stuhl
– Historisch-politische blaetter
Goerres, J *see*
– Das rothe blatt
– Der ruebezahl
Goerres, Joseph *see*
– Die legende der hl jungfrau und maertyrin sankt katharina
– Teutschland und die revolution
Goerres, Joseph von *see*
– Athanasius
– Charakteristiken und kritiken von joseph goerres
– Charakteristiken und kritiken von joseph goerres aus den jahren 1804 und 1805
– Der dom von koeln und das muenster von strassburg
– Europa und die revolution
– Gesammelte schriften
– Joseph von goerres gesammelte schriften
– Die triarier
Goerres, Marie *see* Joseph von goerres gesammelte schriften
Goerres und sein "rothes blatt" / Nothardt, Fritz – [s.l.: s.n.] 1932 – 1r – 1 – (incl bibl ref) – us Wisconsin U Libr [943]
Die goerres-gesellschaft, 1876-1901 : denkschrift zur feier ihres 25jaehrigen bestehens nebst jahresbericht fuer 1900 / Cardauns, Hermann – Koeln: JP Bachem, 1901 – 1mf – 9 – 0-524-08222-7 – mf#1993-1007 – us ATLA [430]
Goertz, Carl von *see* Reise um die welt in den jahren 1844-1847
Goertz, Hartmann *see* Vom wesen der deutschen lyrik
Goerz *see* Boekan impian, boekan lamoenan
Goes, Albrecht *see*
– Moerike
– Ueber das gespraech
– Unquiet night
– Unruhige nacht
Goes, D de *see* Fides, religio, moresque aethiopum
Goes, Gustav *see* Die trommel schlug zum streite
Goesbriand, Louis de *see*
– Les canadiens des etats-unis
– A relation of the first pilgrimage from the diocese of burlington to st anne of beaupre, june 20, 1882
Goeschel, Carl *see* Unterhaltungen auf einer reise von und nach naumburg an der saale ueber jena, rudolfstadt, saalfeld, gera, altenburg und zeitz
Goeschel, Karl Friedrich *see* Der mensch nach leib, seele und geist diesseits und jenseits
Goessl, Manfred M *see* Die wirtschaftlichkeit von ersatzinvestitionen. investitionsrechnung in der betriebswirtschaftlichen praxis
Goessling, Tobias *see* Entscheidung in modellen
Goesswein, G *see* Schriftgemaesse und erbauliche erklaerung der offenbarung st johannis
Goeta lejon – Goeteborg, Sweden. 1890-91 – 1r – 1 – sw Kungliga [078]
Goeteborgs allehanda – Goeteborg, Sweden. 1880 – 1r – 1 – sw Kungliga [078]
Goeteborgs annonstidning – Goeteborg, Sweden. 1881-82 – 1r – 1 – sw Kungliga [078]
Goeteborgs fria tidning – Goeteborg, Sweden. 2005- – 1 – sw Kungliga [078]
Goeteborgs nyheter – Goeteborg, Sweden. 1892-1921 – 8r – 1 – sw Kungliga [078]
Goeteborgs stiftstidning – Goeteborg, Sweden. 1907-34 – 26r – 1 – sw Kungliga [078]
Goeteborgs veckotidning – Goeteborg, Sweden. 1893-1977 – 37r – 1 – sw Kungliga [078]
Goeteborgs weckoblad – Goeteborg, Sweden. 1875-92 – 6r – 1 – sw Kungliga [078]
Goeteborgstidningen dagbladet – Goeteborg, Sweden. 1884-85 – 1r – 1 – sw Kungliga [078]
Goeteborgstrakten – Goeteborg, Moelndal, Sweden. 1923-70 – 42r – 1 – sw Kungliga [078]
Goeters, Wilhelm *see*
– Die vorbereitung des pietismus
– Die vorbereitung des pietismus in der reformierter kirche der niederlaende
Goethe / Amann, Paul – Paris: Rieder, [1932] [mf ed 1993] – (= ser Maitres des litteratures 12) – 120p/[60]pl – 1 – (incl bibl ref) – mf#8637 – us Wisconsin U Libr [430]
Goethe / Angelloz, Joseph-Francois – Paris: Mercure de France, 1949 [mf ed 1990] – 384p – 1 – (incl bibl ref) – mf#7349 – us Wisconsin U Libr [430]
Goethe / Carlyle, Thomas – Berlin : Oesterheld, 1910 [mf ed 1999] – 176p – 1 – mf#10131 – us Wisconsin U Libr [170]
Goethe / Charpentier, John – Paris: J Tallandier 1943 [mf ed 1990] – 1r – 1 – (incl bibl ref; filmed with: goethe in vertraulichen briefen seiner zeitgenossen / wilhelm bode [comp]) – mf#2795p – us Wisconsin U Libr [430]
Goethe – Weimar: H Boehlaus Nachf, 1936-71 [mf ed 1994] – 33v on 7r (ill) – 1 – (cont: goethe-gesellschaft (weimar, germany). jahrbuch der goethe-gesellschaft; cont by: goethe-jahrbuch. publ suspended 1945-46. suppl with title: goethe-bibliographie accompany ea iss, 14/15 52-) – mf#8665 – us Wisconsin U Libr [430]
Goethe / Croce, Benedetto – London: Methuen, 1923 [mf ed 1993] – xxi/208p – 1 – (int by by douglas ainslie) – mf#8654 – us Wisconsin U Libr [430]
Goethe : discorsi pronunciati a weimar e a roma 24 marzo e 2 aprile 1932-x / Farinelli, Arturo – Roma: Reale Accademia d'Italia, 1933 – 1r – 1 – us Wisconsin U Libr [430]
Goethe : drei reden / Schweitzer, Albert – Muenchen: Biederstein, 1949 – 1r – 1 – us Wisconsin U Libr [430]
Goethe : eine einfuehrung in leben und werk unter besonderer beruecksichtigung seiner jugendzeit / Goethe, Johann Wolfgang von; ed by Victor, Walther – Berlin: Verlag Neues Leben, 1960 – 1 – (incl bibl ref) – us Wisconsin U Libr [430]
Goethe : ensayos poeticos / Gonzalez Serrano, Urbano – Madrid: Lib. Int. de Fdez. Villegary Cia s.a. 3rd ed – 9 – sp Bibl Santa Ana [430]
Goethe : gedenkrede gehalten bei der feier der 100. wiederkehr seines todestages in seiner vaterstadt frankfurt a.m. am 22. maerz 1932 / Schweitzer, Albert – Muenchen: C H Beck, 1933 – 1r – 1 – us Wisconsin U Libr [430]
Goethe / ed by Graef, Hans Gerhard – Leipzig: Inselverlag, 1919-20 [mf ed 1994] – 16v on 5r (ill) – 1 – (v6 dated 1919, all other vols dated 1920) – mf#8618 – us Wisconsin U Libr [802]
Goethe / Grimm, Herman Friedrich; ed by Hansen, Wilhelm – vollstaend ausg. Detmold-Hiddesen: Maximilian-Verlag, 1948 [mf ed 1993] – (= ser Maximilian-buecherei 3) – 568p – 1 – (incl bibl ref and ind. originally issued in 2v 1876-77, under title: goethe: vorlesungen gehalten an der kgl universitaet zu berlin) – mf#8640 – us Wisconsin U Libr [430]
Goethe / Gundolf, Friedrich – 7. unveraend aufl. 14-16. Berlin: G Bondi, 1920, c1916 [mf ed 1993] – viii/795p – 1 – mf#8641 – us Wisconsin U Libr [430]
Goethe / Gundolf, Friedrich – Berlin: G Bondi, 1918, c1916 [mf ed 1996] – (= ser Werke der wissenschaft aus dem kreise der blaetter fuer die kunst) – viii/795p – 1 – (incl ind) – mf#9653 – us Wisconsin U Libr [430]
Goethe / Hauptmann, Gerhart – New York: Columbia University Press, 1932 – 1r – 1 – us Wisconsin U Libr [430]
Goethe / Heinemann, Karl – Leipzig: E A Seemann, 1895 – 1r – 1 – us Wisconsin U Libr [430]
Goethe / Herford, Charles Harold – London: T C & E C Jack, [19–?] – 1r – 1 – us Wisconsin U Libr [430]
Goethe : leben, gedanken, bildnisse / Langewiesche, Wilhelm & Langewiesche, Karl Robert – Koenigstein im Taunus: Verlag der eiserne Hammer, c1932 – 1r – 1 – us Wisconsin U Libr [430]
Goethe : ein lesebuch fuer unsere zeit / Goethe, Johann Wolfgang von; ed by Victor, Walther – Weimar: Volksverlag, 1962 – 1r – 1 – (incl bibl ref) – us Wisconsin U Libr [430]
Goethe : maximen und reflexionen / ed by Hecker, Max – Weimar: Goethe-Gesellschaft, 1907 [mf ed 1993] – xxxviii/411p/1pl – 1 – (incl bibl ref and ind) – mf#8657 reel 6 – us Wisconsin U Libr [880]
Goethe / Meyer, Richard Moritz – 2. aufl. Berlin: E Hofmann & Co, 1898 [mf ed 1999] – (= ser Geisteshelden (fuehrende geister) 3. sammlung 1-3) – xxxii/747p/3pl (ill) – 1 – mf#10177 – us Wisconsin U Libr [430]
Goethe : notas apressadas de um jornalista / Ribeiro, Joao – Rio de Janeiro: Revista de Lingua Portuguesa, 1932 [mf ed 1993] – 118p – 1 – (portuguese trans fr german) – mf#8652 – us Wisconsin U Libr [430]
Goethe : ein profil / Paasche, Fredrik – Stuttgart: Greiner und Pfeiffer, 1922 – 1r – 1 – us Wisconsin U Libr [430]
Goethe / Robertson, J G – London: G Routledge, New York: E P Dutton, 1927 [mf ed 2000] – (= ser Republic of letters) – 1 – (incl bibl ref & ind) – mf#10472 – us Wisconsin U Libr [430]
Goethe : der roman von seiner erweckung / Trentini, Albert – Muenchen: G D W Callwey, 1923 – 1r – 1 – us Wisconsin U Libr [430]
Goethe : sein leben / Federn, Etta [Etta Federn-Kohlhaas] – Stuttgart: Union Deutsche Verlagsgesellschaft, [1922?] [mf ed 1993] – (= ser Lichter am weg) – 253p (ill) – 1 – (incl bibl ref and ind) – mf#8652 – us Wisconsin U Libr [430]
Goethe : sein leben und seine werke / Baumgartner, Alexander; ed by Stockmann, Alois – 4. aufl. Freiburg i.B: Herder, 1923-25 [mf ed 1990] – 2v on 1r – 1 – mf#7350 – us Wisconsin U Libr [430]
Goethe : skizzen zu des dichters leben und werken / Graef, Hans Gerhard – Leipzig: H Haessel, 1924 [mf ed 1990] – x/487p/12pl (ill) – 1 – (incl bibl ref) – mf#7555 – us Wisconsin U Libr [920]
Goethe : vermaechtnis und aufruf: eine einfuehrung / Resch, Johannes – Berlin: Verlag Neues Leben, c1949 [mf ed 1993] – 333p – 1 – (incl bibl ref) – mf#8641 – us Wisconsin U Libr [430]
Goethe = Vie de goethe / Carre, Jean Marie – New York: Coward-McCann, 1929 [mf ed 1993] – ix/308p – 1 – (trans fr french by eleanor hard) – mf#8654 – us Wisconsin U Libr [430]
Goethe : vier reden / Schweitzer, Albert – 3. erw. Aufl. Muenchen: C H Beck, 1950 – 1 – us Wisconsin U Libr [430]
Goethe : vorlesungen gehalten an der kgl universitaet zu berlin / Grimm, Herman Friedrich – 3. durchges aufl. Berlin: W Hertz, 1882 [mf ed 1993] – viii/524p – 1 – (incl bibl ref and ind) – mf#8640 – us Wisconsin U Libr [430]
Goethe / Witkowski, Georg – Leipzig: E A Seemann: Gesellschaft fuer graph Industrie, 1899 [mf ed 1999] – (= ser Dichter und darsteller 1) – 270p/6/7pl (ill) – 1 – (incl ind) – mf#10177 – us Wisconsin U Libr [430]
Goethe / Wolff, Max Josef – Leipzig: B G Teubner, 1921 – 1r – 1 – us Wisconsin U Libr [430]
Goethe als bildender kuenstler : mit 60 lichtdruckbildern / Federmann, Arnold – Stuttgart: J G Cotta, 1932 – 1r – 1 – (incl bibl ref) – us Wisconsin U Libr [700]
Goethe als denker / Siebeck, Hermann – Stuttgart: F Frommann (E Hauff), 1902] [mf ed 2000] – (= ser Frommanns klassiker der philosophie 15) – 244p – 1 – (incl bibl ref) – mf#10473 – us Wisconsin U Libr [100]
Goethe als erbe seiner ahnen / Bradish, Joseph Arno von – Berlin, New York: B Westermann 1933 [mf ed 1990] – 1r – 1 – (filmed with: goethe als mensch und deutscher / gunther heyd) – mf#7372 – us Wisconsin U Libr [430]
Goethe als erzieher nietzsches / Saleski, Maria Agnes – Leipzig: Schwarzenberg & Schumann, [1929] – 1r – 1 – (incl bibl ref) – us Wisconsin U Libr [943]
Goethe als freimaurer / Deile, Gotthold – Berlin: E S Mittler 1908 – (= ser Stunden mit goethe) – 4mf – 9 – mf#vrl-54 – ne IDC [366]
Goethe als freimaurer / Deile, Gotthold – Berlin: E S Mittler, 1908 – 1r – 1 – (incl bibl ref) – us Wisconsin U Libr [430]
Goethe als geschichtsphilosoph und die geschichtsphilosophische bewegung seiner zeit / Menke-Glueckert, Emil – Leipzig: R Voigtlaender, 1907 – 1r – 1 – us Wisconsin U Libr [430]
Goethe als kabbalist in der "faust"-tragoedie / Louvier, Ferdinand August – Berlin: Verlag des Bibliographischen Bureaus, 1892 – 1r – 1 – (incl bibl ref) – us Wisconsin U Libr [430]
Goethe als kuender des lebens / Horneffer, Ernst – Muenchen: Erasmus-Verlag, 1947 [mf ed 1993] – (= ser Ernst reinhardt buecherreihe) – 415p – 1 – mf#8654 – us Wisconsin U Libr [430]
Goethe als mensch und deutscher / ed by Heyd, Guenther – Potsdam: Ruetten & Loening, c1937 – 1r – 1 – us Wisconsin U Libr [920]
Goethe als mensch und deutscher *see* Goethe als erbe seiner ahnen
Goethe als naturforscher / Strunz, Franz – Wien: Verlag des Volksbildungshauses Wiener Urania, 1917 – 1r – 1 – us Wisconsin U Libr [500]
Goethe als naturforscher und herr du bois-reymond als sein kritiker : eine antikritik / Kalischer, Salomon – Berlin: G Hempel, 1883 – 1r – 1 – us Wisconsin U Libr [430]
Goethe als naturforscher und in besonderer beziehung auf schiller : eine rede / Virchow, Rudolf – Berlin, 1861 (mf ed 1993) – 1mf – 9 – €24.00 – 3-89349-260-7 – mf#DHS-AR 117 – gw Frankfurter [430]
Goethe als paedagog : vortrag gehalten im bruenner lehrervereine am 22. maerz 1880 – Leipzig: H Pfeil, 1880 – 1r – 1 – us Wisconsin U Libr [370]
Goethe als patient / Veil, Wolfgang Heinrich – Jena: G Fischer, 1939 – 1r – 1 – (incl bibl ref) – us Wisconsin U Libr [920]
Goethe als persoenlichkeit : berichte und briefe von zeitgenossen / Amelung, Heinz [comp] – Muenchen: G Mueller; Berlin: Propylaeen-Verlag, 1914-25 – (= ser Propylaeen-ausgabe von goethes saemtlichen werken suppl 1-3) – 3v – 1 – mf#6975 – us Wisconsin U Libr [430]
Goethe als raetseldichter / Biedermann, Flodoard, Freiherr von – Berlin: H Berthold AG, Abt Privatdrucke, 1924 [mf ed 1992] – 45p – 1 – (portrait of goethe by ferdinand jagemann) – mf#7985 – us Wisconsin U Libr [430]
Goethe als rechtsanwalt / Wieruszowski, Alfred – Coeln a. Rhein: P Neubner, [1909?] – 1r – 1 – (incl bibl ref) – us Wisconsin U Libr [340]
Goethe als religioeser charakter / Loew, Wilhelm – Muenchen: Ch Kaiser, 1924 – 1r – 1 – us Wisconsin U Libr [430]
Goethe als repraesentant des buergerlichen zeitalters : rede zum 100. todestag goethes gehalten am 18. maerz 1932 in der preussischen akademie der kuenste zu berlin / Mann, Thomas – Berlin: S Fischer, c1932 – 1r – 1 – us Wisconsin U Libr [430]
Goethe als seelenforscher / Klages, Ludwig – Leipzig: Verlag von J A Barth, 1932 – 1r – 1 – us Wisconsin U Libr [430]
Goethe als vater einer neuen aesthetik / Steiner, Rudolf – Duesseldorf: Deibele & Teubig, 1948 – 1r – 1 – us Wisconsin U Libr [110]
Goethe als zeichner : ein beitrag zum bilde seiner persoenlichkeit / Drost, Willi – 2. verm. Aufl. Potsdam: Akademische Verlagsgesellschaft Athenaion, [1938] – 1r – 1 – (incl bibl ref) – us Wisconsin U Libr [430]
Goethe and august von koetzebue / Stenger, Gerhard – Breslau: F Hirt, 1910 [mf ed 1992] – (= ser Breslauer beitraege zur literaturgeschichte. neue folge 12) – 176p – 1 – (incl bibl ref) – mf#8014 reel 2 – us Wisconsin U Libr [430]
Goethe and democracy : address delivered in the coolidge auditorium in the library of congress on may 2, 1949 / Mann, Thomas – Washington, D.C.: [Library of Congress], 1950 – 1r – 1 – us Wisconsin U Libr [430]
Goethe and his woman friends / Crawford, Mary Caroline – Boston: Little, Brown, 1911 [mf ed 1993] – xiii/452p/1lea/60pl (ill) – 1 – (incl bibl ref and ind) – mf#8653 – us Wisconsin U Libr [920]
Goethe and schiller's xenions / ed by Carus, Paul – Chicago: Open Court, 1896 – 1r – 1 – (in english/german) – us Wisconsin U Libr [430]
Goethe and the conduct of life / Thomas, Calvin – Ann Arbor: Andrews & Witherby, 1886 [mf ed 1970] – (= ser University of michigan. philosophical papers. first series 2) – 28p – 1 – mf#3425 – us Wisconsin U Libr [170]
Goethe and the twentieth century / Robertson, John George – Cambridge [England]: The University Press; New York: G P Putnam, 1912 [mf ed 1990] – 1 – (incl ind) – mf#7396 – us Wisconsin U Libr [430]

Goethe as revealed in his poetry / Fairley, Barker – London: J M Dent, 1932 [mf ed 1994] – ix/210p – 1 – mf#8660 – us Wisconsin U Libr [430]

Goethe, Catharina Elisabeth *see*
– Die briefe der frau rath goethe
– Briefe von goethes mutter an die herzogin anna amalia

The goethe centenary at the university of wisconsin : a memorial volume of addresses and some other contributions / ed by Hohlfeld, A R – Madison, Wis.: University of Wisconsin, 1932 – 1r – 1 – us Wisconsin U Libr [430]

Goethe, das sinnbild deutscher kultur / Barthel, Ernst – Darmstadt: Ernst Hofmann, 1930 [mf ed 1999] – vii/348p/[1]pl – 1 – (incl ind) – mf#10172 – us Wisconsin U Libr [430]

Goethe der bildner / Thode, Henry – Heidelberg: C Winter, 1906 – 1r – 1 – us Wisconsin U Libr [430]

Goethe der deutsche / Bartels, Adolf – Frankfurt a/M: M Diesterweg, 1932 [mf ed 1990] – 192p – 1 – (incl ind) – mf#7341 – us Wisconsin U Libr [430]

Goethe, der deutsche prophet in der faust- und meisterdichtung : mit einem anhang der benuetzten, teilweise erst neu aufgefundenen quellen in goethes werken, korrespondenzen etc / Umfrid, Otto Ludwig – Stuttgart: A Bonz, 1893 – 1r – 1 – (incl bibl ref) – us Wisconsin U Libr [430]

Goethe der grosse humanist / Fischer, Ernst – Wien: Globus-Verlag, 1949 – 1r – 1 – us Wisconsin U Libr [430]

Goethe en angleterre / Carre, Jean Marie – Paris: Plon-Nourrit, c1920 [mf ed 1999] – xviii/300p – 1 – (incl bibl ref and ind) – mf#10184 – us Wisconsin U Libr [410]

Goethe en france : etude de litterature comparee / Baldensperger, Fernand – 2. rev ed. Paris: Hachette, 1920 [mf ed 1993] – 398p – 1 – mf#8654 – us Wisconsin U Libr [410]

Goethe erzaehlt sein leben / ed by Gerlach, Hans Egon & Herrmann, Otto – Hamburg: Hamburger Buchring, C Wegner, c1949 [mf ed 1993] – 533p – 1 – (incl bibl ref) – mf#8607 – us Wisconsin U Libr [920]

Goethe et bettina : le vieillard et la jeune fille / Germain, Andre – Paris: Les Editions de France, 1939 [mf ed 1993] – iv/252p/1pl – 1 – mf#8464 – us Wisconsin U Libr [920]

Goethe et la france : ce qu'il en a connu, pense et dit / Loiseau, Hippolyte – Paris: V Attinger, 1930 [mf ed 2000] – (= ser Occident 12) – 362p – 1 – (incl ind) – mf#10458 – us Wisconsin U Libr [410]

Goethe et la litterature francaise / Caumont, Armand – Frankfurt a.M: Mahlau & Waldschmidt 1885 [mf ed 1990] – 1r – 1 – (filmed with: goethe und pestalozzi / gottfried bohnenblust) – mf#7384 – us Wisconsin U Libr [410]

Goethe et schiller : la litterature allemande a weimar, la jeunesse de schiller, l'union de goethe et de schiller, la vieillesse de goethe / Bossert, Adolphe – 5 rev ed. Paris: Hachette 1903 [mf ed 1993] – 1r – 1 – (incl bibl ref; filmed with: ruckblicke auf mein leben / karl gutzkow) – mf#7467 – us Wisconsin U Libr [430]

Goethe gegen kant : goethes wissenschaftliche leistung als naturforscher und philosoph / Seidel, Fritz – Berlin: L Grosser, 1948 [mf ed 1993] – 118p/2pl (ill) – 1 – mf#8642 – us Wisconsin U Libr [100]

Goethe im gespraech / ed by Deibel, Franz & Gundelfinger, Friedrich – Leipzig: Insel-Verlag, 1906 [mf ed 1991] – xiv/365p – 1 – (incl bibl ref) – mf#7542 – us Wisconsin U Libr [080]

Goethe in berka an der ilm / Graef, Hans Gerhard – Weimar: Gustav Kiepenheuer, 1911 [mf ed 1999] – 92p – 1 – mf#10137 – us Wisconsin U Libr [920]

Goethe in briefen und gespraechen = Correspondence. selections / Goethe, Johann Wolfgang von; ed by Beutler, Ernst – Leipzig: P Reclam jun, 1942 – (incl bibl ref) – us Wisconsin U Libr [860]

Goethe in den jahren 1771 bis 1775 / Abeken, Bernhard Rudolf – Hannover: C Ruempler, 1861 [mf ed 1996] – 434p – mf#9653 – us Wisconsin U Libr [920]

Goethe in dornburg / Sternaux, Ludwig Friedrich – Berlin: E Runge, 1919 – 1r – 1 – us Wisconsin U Libr [430]

Goethe in frankfurt am main 1797 : aktenstuecke und darstellung / Geiger, Ludwig – Frankfurt a.M.: Ruetten & Loening, 1899 – 1r – 1 – us Wisconsin U Libr [430]

Goethe in heutiger sicht / Gerathewohl, Fritz – Muenchen, Berlin: R Oldenbourg, 1942 – 1r – 1 – (incl bibl ref (p. 51-52)) – us Wisconsin U Libr [430]

Goethe in meinem leben : erinnerungen und betrachtungen / Abeken, Bernhard Rudolf; ed by Heuermann, Adolf – Weimar: H Boehlau, 1904 [mf ed 1990] – vii/278p – 1 – (incl ind) – mf#7349 – us Wisconsin U Libr [920]

Goethe in vertraulichen briefen seiner zeitgenossen : auch eine lebensgeschichte / ed by Bode, Wilhelm – Berlin: E S Mittler, 1921 – 1r – 1 – (incl bibl ref) – us Wisconsin U Libr [920]

Goethe in zuerich / Zollinger, Friedrich – Zuerich: Atlantis-Verlag, 1932 – 1r – 1 – us Wisconsin U Libr [920]

Goethe, Johann Wolfgang von *see*
– Acht lieder
– Adalbert von weislingen
– Aus goethes tagebuechern
– The auto-biography of goethe
– Der befreier
– Campagne in frankreich
– Clavijo
– Conversations of goethe with eckermann and soret
– Conversations with eckermann
– Dichtung und wahrheit
– Die drei aeltesten bearbeitungen von goethe's iphigenie
– Elegie, september 1823
– Ephemerides
– Faust
– Le faust
– Faust
– Il faust
– Faust
– Faust, a tragedy
– Le faust de goethe
– Faust de goethe
– Le faust de goethe
– Faust, first part
– Faust und urfaust
– Fausto
– Gedankenharmonie aus goethe und schiller
– Gedankenlyrik
– Gedichte
– Goethe
– Goethe in briefen und gespraechen
– Goethe, schiller
– Goethe, the story of a man
– Goethe ueber seinen faust
– Goethe und uvarov und ihr briefwechsel
– Goethe-briefe
– Goethes briefe
– Goethes clavigo
– Goethes egmont
– Goethes erste weimarer gedichtsammlung mit varianten
– Goethes faust
– Goethe's faust
– Goethes faust
– Goethe's faust
– Goethes faust
– Goethes 'faust'
– Goethe's faust
– Goethes faust
– Goethes faust am hofe des kaisers
– Goethes faust in urspruenglicher gestalt
– Goethe's gedichte
– Goethes gedichte
– Goethe's gedichte
– Goethes gedichte in zeitgeschichtlicher auswahl
– Goethes goetz von berlichingen
– Goethe's iphigenie auf tauris
– Goethes iphigenie auf tauris
– Goethes kleinere aufsaetze
– Goethes liebesgedichte
– Goethes lyrik
– Goethe's poems
– Goethe's poems and aphorisms
– Goethes roemische elegien
– Goethes saemtliche werke
– Goethe's saemtliche werke
– Goethes tagebuecher der sechs ersten weimarischen jahre
– Goethes tasso
– Goethe's theory of colours
– Goethes werke
– Goethes Werke
– Goethes werke
– Goethes Werke
– Goetz von berlichingen
– Goetz von berlichingen mit der eisernen hand
– The gretchen episode from goethe's faust
– Die guten frauen
– Hermann and dorothea
– Hermann und dorothea
– Iphigenia in tauris
– Iphigenie auf tauris
– Italienische reise
– Juristische abhandlung ueber die floehe
– Die leiden des jungen werthers
– Letters from goethe
– Die lyrischen meisterstuecke von johann wolfgang von goethe
– Das maerchen
– Maximen und reflexionen
– The maxims and reflections of goethe
– Memoirs of goethe
– Miscellaneous travels of j w goethe
– Mit goethe durch die schweiz
– Novellen und maerchen
– Poems and ballads of goethe
– The poems of goethe
– Poesia e verita
– Poetry and truth
– Propylaeen
– Select minor poems
– Stirb und werde
– Das tagebuch
– Torquato tasso
– Le tragedie du docteur faust de goethe en vers francais
– Ueber den bologneser spat
– Unterhaltungen deutscher ausgewanderten
– Vergangenheit und gegenwart in eins
– Voyage en italie
– Werke
– Wert und wuerde
– West-oestlicher divan
– Wilhelm meisters theatralische sendung
– Wilhelm meisters wanderjahre
– Xenien 1796
– Zur morphologie

Goethe, karl august und ottokar lorenz : ein denkmal / Duentzer, Heinrich – Dresden: Dresdener Verlagsanstalt, 1895 [mf ed 1990] – 126p – 1 – mf#7372 – us Wisconsin U Libr [430]

Goethe og werther / Enault, Louis – Kobenhavn: V Pontoppidan, 1889 – 1r – 1 – (incl bibl ref) – us Wisconsin U Libr [430]

Goethe on nature and on science / Sherrington, Charles Scott – 2nd ed. Cambridge: University Press, 1949 [mf ed 1993] – 53p – 1 – (incl bibl ref) – mf#8656 – us Wisconsin U Libr [110]

Goethe, Ottilie von *see* Aus ottilie von goethes nachlass

Goethe, schiller : ueber das theater; eine auswahl aus ihren schriften / Goethe, Johann Wolfgang von; ed by Eggebrecht, Axel – Berlin: B. Henschel, 1949. vii,495p – 1 – us Wisconsin U Libr [430]

Goethe, sein leben und seine werke / Bielschowsky, Albert – 28. aufl. Muenchen: C H Beck, 1914 [mf ed 1996] – 2v – 1 – (incl bibl ref and ind. v2 completed after aut's death by t ziegler) – mf#9652 – us Wisconsin U Libr [430]

Goethe, the story of a man : being the life of johann wolfgang goethe as told in his own words and the words of his contemporaries / ed by Lewisohn, Ludwig – New York: Farrar, Straus, 1949 [mf ed 1998] – 2v – 1 – (trans by ed. incl bibl ref) – mf#9941 – us Wisconsin U Libr [860]

Goethe ueber freunde und feinde : zwei kapitel aus goethes lebenskunst / Bode, Wilhelm – Berlin: E S Mittler, 1913 [mf ed 1990] – 35p – 1 – mf#7350 – us Wisconsin U Libr [920]

Goethe ueber seinen faust / Goethe, Johann Wolfgang von – Leipzig: Insel-Verlag, [19–] [mf ed 1990] – (= ser Insel-Buecherei 44) – 83p – 1 – (int by hans heinrich borcherdt) – mf#7346 – us Wisconsin U Libr [840]

Goethe ueberzeitlich / Baeumer, Gertrud – Berlin: F A Herbig, 1932 [mf ed 1990] – 91p/1pl – 1 – mf#7383 – us Wisconsin U Libr [430]

Goethe, un homme face a la vie : essai de biographie interieure / Fuchs, Albert – Paris: Aubier, 1946 – 1r – 1 – (incl bibl ref) – us Wisconsin U Libr [920]

Goethe und aristoteles / Petersen, Peter – Braunschweig: G Westermann, 1914 – 1r – 1 – (incl bibl ref) – us Wisconsin U Libr [410]

Goethe und august von kotzebue / Stenger, Gerhard – Breslau: F Hirt, 1910 – 1 – (incl bibl ref) – us Wisconsin U Libr [430]

Goethe und beethoven / Engelsmann, Walter – Augsburg: B Filser, c1931 – 1r – 1 – us Wisconsin U Libr [430]

Goethe und byron : eine darstellung des persoenlichen und litterarischen verhaeltnisses mit besonderer beruecksichtigung des "faust" und "manfred" / Sinzheimer, Siegfried – [s.l: s.n.] 1894; Muenchen: Kgl Hof und Universitaets-Buchdruckerei C Wolf – 1 – (incl bibl ref) – us Wisconsin U Libr [430]

Goethe und charlotte v. stein / Hoefer, Edmund – 2. Aufl. Leipzig: Xenien-Verlag, 1911 – 1r – 1 – us Wisconsin U Libr [430]

Goethe und christentum : die religion und ethik goethes und der hauptvertreter des christentums / Fliedner, Wilhelm – Gotha: L Klotz, 1930 – 1r – 1 – us Wisconsin U Libr [430]

Goethe und christiane : von wesen und sinn ihrer lebensgemeinschaft / Martin, Bernhard – Kassel: Baerenreiter-Verlag, 1949 [mf ed 1993] – 98p – 1 – mf#8653 – us Wisconsin U Libr [920]

Goethe und das christentum / Kahle, Wilhelm – Duelmen i.Westf.: A Laumann, 1949 – 1r – 1 – us Wisconsin U Libr [430]

Goethe und das classische alterthum : die einwirkung der antike auf goethes dichtungen im zusammenhange mit dem lebensgangs des dichters / Thalmayr, Franz – Leipzig: G Fock, 1897 – 1r – 1 – (incl bibl ref) – us Wisconsin U Libr [430]

Goethe und das volkslied / Waldberg, Max, Freiherr von – Berlin: W Hertz, 1889 [mf ed 1993] – 31p – 1 – mf#8655 – us Wisconsin U Libr [430]

Goethe und das weimarer hoftheater : mit vielen bildern nach alten vorlagen / Hoeffner, Johannes – Weimar: G Kiepenheuer, 1913 – 1r – 1 – us Wisconsin U Libr [430]

Goethe und der arzt von heute / Oehme, Curt – Stuttgart: Hippokrates-Verlag, 1950 [mf ed 1993] – 47p – 1 – (incl bibl ref) – mf#8656 – us Wisconsin U Libr [140]

Goethe und der katholizismus / Hammerschmidt, Ferdinand – Breslau: O Borgmeyer, [1932] – 1r – 1 – us Wisconsin U Libr [430]

Goethe und der okkultismus / Seiling, Max – Leipzig: O Mutze, [1901?] – 1r – 1 – us Wisconsin U Libr [130]

Goethe und der orient / Krueger-Westend, Herman – Weimar: H Boehlau, 1903 – 1r – 1 – us Wisconsin U Libr [430]

Goethe und die bedeutung des gegenstandes fuer die bildende kunst / Schulz-Uellenberg, Gisela – Muenchen: Filser-Verlag, 1947 [mf ed 1993] – 380p – 1 – (incl bibl ref) – mf#8656 – us Wisconsin U Libr [700]

Goethe und die bildende kunst : festrede, gehalten in der oeffentlichen sitzung der bayerischen akademie der wissenschaften zur feier des 173. stiftungstages am 11. mai 1932 / Pinder, Wilhelm – Muenchen: Verlag der Bayerischen Akademie der Wissenschaften: C H Beck, 1933 – 1r – 1 – us Wisconsin U Libr [700]

Goethe und die bildende kunst / Stelzer, Otto – Braunschweig: Vieweg, 1949 – 1r – 1 – us Wisconsin U Libr [700]

Goethe und die descendenzlehre / Wasielewski, Waldemar von – Frankfurt a.M.: Ruetten & Loening, 1903 – 1r – 1 – (inclues bibliographical references) – us Wisconsin U Libr [920]

Goethe und die deutsche gegenwart / Linden, Walther – Berlin: Bong, c1932 – 1r – 1 – us Wisconsin U Libr [430]

Goethe und die deutschen : vom nachruhm eines dichters = German image of goethe / Leppmann, Wolfgang – Stuttgart: W Kohlhammer, 1962 [mf ed 1993] – (= ser Sprache und literatur 3) – 296p – 1 – (incl bibl ref and ind; german trans fr english) – mf#8653 – us Wisconsin U Libr [430]

Goethe und die juden / Bab, Julius – Berlin: Philo Verlag, 1926 [mf ed 1990] – (= ser Die morgen-reihe 3. schrift) – 36p – 1 – (incl bibl ref) – mf#7383 – us Wisconsin U Libr [430]

Goethe und die juden / Koch, Franz – Hamburg: Hanseatische Verlagsanstalt, c1937 [mf ed 2000] – (= ser Schriften des reichinstituts fuer geschichte des neuen deutschlands 12) – 37p – 1 – mf#10455 – us Wisconsin U Libr [320]

Goethe und die juden / Teweles, Heinrich – Hamburg: W Gente, 1925 – 1r – 1 – (the "register" incl brief biogr notices) – us Wisconsin U Libr [943]

Goethe und die kirche seiner zeit / Blanckmeister, Franz – Dresden: F Sturm, 1923 [mf ed 1990] – 187p – 1 – mf#7351 – us Wisconsin U Libr [170]

Goethe und die liebe : zwei vortraege / Schroeer, Karl Julius – Heilbronn: Gebr Henninger, 1884 [mf ed 2000] – xi/78p – 1 – mf#10463 – us Wisconsin U Libr [920]

Goethe und die musik / Abert, Hermann – Stuttgart: J Engelhorn, 1922 [mf ed 1990] – (= ser Musikalische volksbuecher) – 127p – 1 – (incl bibl ref) – mf#7378 – us Wisconsin U Libr [780]

Goethe und die naturwissenschaften / Benn, Gottfried – Zuerich: Verlag der Arche, 1949 [mf ed 1990] – 57p – 1 – mf#7383 – us Wisconsin U Libr [430]

Goethe und die naturwissenschaften / Walden, Paul – Bremen: G A v Halem, c1933 – 1r – 1 – us Wisconsin U Libr [500]

Goethe und die philosophie / Bauch, Bruno – Tuebingen: J C B Mohr (P Siebeck), 1928 [mf ed 1990] – (= ser Philosophie und geschichte 20) – 36p – 1 – (incl bibl ref) – mf#7383 – us Wisconsin U Libr [410]

Goethe und die physik : vortrag gehalten in der muenchner universitaet am 9. mai 1923 / Wien, Wilhelm – Leipzig: J A Barth, 1923 – 1r – 1 – (incl bibl ref) – us Wisconsin U Libr [530]

Goethe und die romantik : briefe mit erlaeuterungen / ed by Schueddekopf, Carl & Walzel, Oskar – Weimar: Goethe-Gesellschaft, 1898-99 [mf ed 1993] – (= ser Schriften der goethe-gesellschaft 13-14) – 2v – 1 – (incl bibl ref and ind) – mf#8657 reel 4 – us Wisconsin U Libr [860]

Goethe und die schweiz / Bohnenblust, Gottfried – Frauenfeld: Huber, c1932 [mf ed 1999] – (= ser Schweiz im deutschen geistesleben 72-73) – 264p – 1 – (incl bibl ref and ind) – mf#10179 – us Wisconsin U Libr [914]

Goethe und die seinen : quellenmaessige darstellungen ueber goethes haus / Geiger, Ludwig – Leipzig: R Voigtlaender, 1908 – 1r – 1 – (incl bibl ref) – us Wisconsin U Libr [430]

Goethe und die urpflanze / Bliedner, Arno – Frankfurt (Main): Ruetten & Loening, 1901 [mf ed 1990] – iv/75p/[2]pl (ill) – 1 – (incl bibl ref) – mf#7383 – us Wisconsin U Libr [580]

Goethe und die wertherzeit : ein vortrag / Knortz, Karl – Zuerich: Verlags-Magazin (J Schabelitz), 1885 – 1r – 1 – us Wisconsin U Libr [430]

Goethe und die wirtschaft / Dantz, Antonie – Borna-Leipzig: R Noske, 1935 – 1r – 1 – (incl bibl ref) – us Wisconsin U Libr [430]

Goethe und dresden / Biedermann, Flodoard, Freiherr von – Berlin: G Hempel, 1875 [mf ed 1990] – 172p – 1 – (incl ind) – mf#7351 – us Wisconsin U Libr [920]

Goethe und frau v stein / Adler, Emma – Leipzig: Toeplitz & Deuticke, 1887 [mf ed 1990] – 16p – 1 – mf#7378 – us Wisconsin U Libr [920]

Goethe und grossbritannien / Vollrath, Wilhelm – Erlangen: Palm & Enke, 1932 – 1r – 1 – (includes bibliobraphical references) – us Wisconsin U Libr [430]

Goethe und hebbel : eine antithese: festvortrag zur dezennarfeier des wuerttembergischen goethebundes am 22. november 1910 im buergermuseum zu stuttgart / Zinkernagel, Franz – Tuebingen: J C B Mohr, 1911 – 1r – 1 – us Wisconsin U Libr [430]

Goethe und heinrich leopold wagner : ein wort der kritik an unsere goethe-forscher / Froitzheim, Johann – Strassburg: J H E Heitz (Heitz & Muendel), 1889 – 1r – 1 – us Wisconsin U Libr [430]

Goethe und johann peter hebel / Rehm, Walther – Freiburg i. Br: Selbstverlag der Universitaet, 1949 [mf ed 1996] – (= ser Freiburger universitaetsreden. n f 7) – 34p – 1 – (incl bibl ref) – mf#9654 – us Wisconsin U Libr [920]

Goethe und karl august : studien zu goethes leben / Duentzer, Heinrich – 2. neubearb und vollend aufl. Leipzig: Dyk, 1888 [mf ed 1992] – 3pts in 1v – 1 – (incl ind) – mf#7554 – us Wisconsin U Libr [920]

Goethe und kein ende / Du Bois-Reymond, Emil Heinrich – Leipzig: Veit, 1883 – 1r – 1 – (incl bibl ref) – us Wisconsin U Libr [430]

Goethe und lavater : briefe und tagebuecher / ed by Funck, Heinrich – Weimar: Goethe-Gesellschaft, 1901 [mf ed 1993] – (= ser Schriften der goethe-gesellschaft 16) – 1 – (incl bibl ref & ind) – mf#8657 reel 4 – us Wisconsin U Libr [880]

Goethe und lavater : zeugnisse ihrer freundschaft – Zuerich: Rascher, 1918 [mf ed 1990] – (= ser Schweizerische bibliothek 2) – 96p – 1 – mf#7372 – us Wisconsin U Libr [920]

Goethe und luther / Benrath, Paul – Tuebingen: J C B Mohr, 1919 [mf ed 1990] – p76-96 – 1 – (incl bibl ref) – mf#7383 – us Wisconsin U Libr [430]

Goethe und marianne von willemer : eine biographische studie / Pyritz, Hans Werner – 3. aufl. Stuttgart: J B Metzler, 1948 [mf ed 1993] – 131p – 1 – (incl bibl ref) – mf#8653 – us Wisconsin U Libr [920]

Goethe und oesterreich : briefe mit erlaeuterungen / ed by Sauer, August – Weimar: Goethe-Gesellschaft, 1902-04 [mf ed 1993] – (= ser Schriften der goethe-gesellschaft 17-18) – 2v – 1 – (incl bibl ref) – mf#8657 reel 5 – us Wisconsin U Libr [430]

Goethe und pestalozzi / Bohnenblust, Gottfried – Bern: E Bircher, [1923?] [mf ed 1990] – (= ser Schriften der freistudentenschaft bern 1) – 23p – 1 – mf#7384 – us Wisconsin U Libr [430]

Goethe und pestalozzi *see*
– Goethekult und goethephilologie
– Wandlungen in goethes religion
– Wir heissen's fromm sein

Goethe und plotin / Koch, Franz – Leipzig: J J Weber, 1925 [mf ed 2000] – 263p – 1 – mf#10454 – us Wisconsin U Libr [140]

Goethe und schiller in briefen / Voss, Heinrich – Leipzig: Philipp Reclam, [1895] [mf ed 1990] – 1 – (incl bibl ref) – mf#7401 – us Wisconsin U Libr [860]

Goethe und schopenhauer : ein beitrag zur entwicklungsgeschichte der schopenhauerschen philosophie / Doell, Heinrich – Berlin: E Hofmann, 1904 – 1r – 1 – us Wisconsin U Libr [430]

Goethe und sein sohn : weimarer erlebnisse in den jahren 1827-1831 / Holtei, Karl von – 1. Ausgabe in Auswahl nach Holteis Lebenserinnerungen "Vierzig Jahre". Hamburg: Vera-Verlag, c1924 – 1r – 1 – us Wisconsin U Libr [430]

Goethe und seine auslaendischen besucher / Landgraf, Hugo – Muenchen: Deutsche akademie, 1932 [mf ed 2000] – 158p/6lea (ill) – 1 – (incl bibl ref) – mf#10456 – us Wisconsin U Libr [920]

Goethe und seine eltern / Krueger-Westend, Herman – Weimar: H Boehlau, 1904 – 1r – 1 – us Wisconsin U Libr [430]

Goethe und seine eltern *see* Herder und die deutsche kulturanschauung

Goethe und seine freunde im briefwechsel / ed by Meyer, Richard M – Berlin: G Bondi, 1909-11 [mf ed 1991] – 3v – 1 – mf#7534 – us Wisconsin U Libr [920]

Goethe und seine welt : 580 abbildungen / ed by Wahl, Hans & Kippenberg, Anton – Leipzig: Insel-Verlag, 1932 [mf ed 1993] – 306p (ill) – 1 – mf#8653 – us Wisconsin U Libr [740]

Goethe und seine zeit / Alt, Karl Hermann – Leipzig: Quelle & Meyer, 1911 [mf ed 1990] – (= ser Wissenschaft und bildung 99) – 155p – 1 – (incl bibl ref) – mf#7383 – us Wisconsin U Libr [430]

Goethe und tischbein / Oettingen, Wolfgang von – Weimar: Goethe-Gesellschaft, 1910 [mf ed 1993] – (= ser Schriften der goethe-gesellschaft 25) – 40p/25pl (ill) – 1 – (incl bibl ref) – mf#8657 reel 6 – us Wisconsin U Libr [430]

Goethe und tolstoi / Mann, Thomas – Aachen: Verlag "Die Kuppel", 1923 – 1r – 1 – us Wisconsin U Libr [430]

Goethe und uvarov und ihr briefwechsel / Goethe, Johann Wolfgang von – St. Petersburg: H Schmitzdorff, 1888 – 1r – 1 – (incl bibl ref) – us Wisconsin U Libr [920]

Goethe und weimar : mit einem goethebildnis von karl bauer / Schrumpf, Ernst – Muenchen: C H Beck, 1912 – 1r – 1 – us Wisconsin U Libr [430]

Goethe von schiller in briefen / Voss, Heinrich; ed by Graf, Hans Gerhard – Leipzig: Philipp Reclam, [1895] – 1r – 1 – (incl bibl ref (p.[126])) – us Wisconsin U Libr [430]

Goethe, weimar und jena im jahre 1806 : nach goethes privatacten: am fuenfzigjaehrigen todestage goethes / Keil, Robert; ed by Keil, Richard & Keil, Robert – Leipzig: E Schloemp, 1882 – 1r – 1 – (incl bibl ref) – us Wisconsin U Libr [430]

Die goethe-bildnisse : biographisch-kunstgeschichtlich dargestellt; mit 78 holzschnitten, 8 radierungen und 2 heliogravuren / Rollett, Hermann – Wien: W Braumueller, 1883 – 1 – (incl bibl ref) – us Wisconsin U Libr [760]

Goetheborgs weckolista – Goeteborg, Sweden. 1750-1758 – 3r – 1 – sw Kungliga [078]

Goethe-briefe : mit einleitungen und erlaeuterungen / Goethe, Johann Wolfgang von; ed by Stein, Philipp – Berlin: O Elsner, 1902-05 – 2r – 1 – (incl bibl ref) – us Wisconsin U Libr [920]

Goethe-festwoche 1946 in bremen : veranstaltet von der bremer ortsvereinigung der goethe-gesellschaft in weimar vom 25. bis 31. august 1946 – Bremen: F Truejen, 1947 – 1r – 1 – us Wisconsin U Libr [430]

Goethe-handbuch – 3v. Stuttgart: J.B. Metzler, 1916-18. 1 reel. 1221 – 1 – us Wisconsin U Libr [430]

Das goethehaus am frauenplan : die geschichte des hauses von der erbauung bis zu goethes zeit / Weichberger, Alexander – Weimar: H Boehlau 1932 [mf ed 1990] – 1r [ill] – 1 – (incl bibl ref. filmed with: goethe und weimar / ernst schrumpf) – mf#7378 – us Wisconsin U Libr [430]

Goethe-kalender : 1906 – Leipzig: Dieterich'sche Verlagsbuchhandlung, 1906- [mf ed 1994] – (ill) – 1 – (no evidence that vols 1915 and 1916 were ever publ. issues 1906-14, 1917-28 ed by karl heinemann. issues 1929- ed by frankfurter goethe-museum) – mf#8636 – us Wisconsin U Libr [390]

Goethekult und goethephilologie : eine streitschrift / Braitmaier, Friedrich – Leipzig: A Weigel 1892 [mf ed 1990] – 1r – 1 – (filmed with: goethe und pestalozzi / gottfried bohnenblust) – mf#7384 – us Wisconsin U Libr [430]

Ein goethepreis / Bewer, Max – 3. aufl. Dresden: Gloess, 1900 [mf ed 1993] – 80p (ill) – 1 – mf#8652 – us Wisconsin U Libr [943]

Goethe-probleme / Wukadinovic, Spiridion – Halle (Saale): M Niemeyer, 1926 – 1r – 1 – (incl bibl ref) – us Wisconsin U Libr [430]

Goethes aeltere zeitgenossen / ed by Sevin, Ludwig – Karlsruhe: J J Reiff, 1902 [mf ed 1990] – 1r – 1 – (filmed with: gedankenlyrik) – us Wisconsin U Libr [430]

Goethes anschauung vom menschen / Vietor, Karl – Bern: Francke c1960 [mf ed 1996] – 1r – 1 – (incl bibl ref. filmed with: goethe; et, la synthese / leon daudet) – mf#4291p – us Wisconsin U Libr [120]

Goethes ballader – Stockholm: Hugo Geber, 1900 [mf ed 1993] – 77p – 1 – (swedish trans by carl snoilsky) – mf#8614 – us Wisconsin U Libr [810]

Goethes bedeutung fuer die gegenwart : zwei vortraege gehalten zur feier des 150. geburtstages in der aula des kgl. gymnasiums zu neuwied / Biese, Alfred – Neuwied: L Heuser, 1900 [mf ed 1990] – 39p – 1 – mf#7383 – us Wisconsin U Libr [850]

Goethes beitraege zu den frankfurter gelehrten anzeigen von 1772 : zugleich beitrag zur kenntnis der sprache des jungen goethe / Modick, Otto – Borna-Leipzig: R Noske, 1913 – 1r – 1 – (incl bibl ref (p.[126])) – us Wisconsin U Libr [430]

Goethes berliner beziehungen / Arnhold, Erna – Gotha: L Klotz, 1925 [mf ed 1990] – vii/455p – 1 – mf#7349 – us Wisconsin U Libr [920]

Goethes briefe / Goethe, Johann Wolfgang von; ed by Hellen, Eduard von der – Stuttgart: J G Cotta, [1901-13] [mf ed 1990] – (= ser Cotta'sche bibliothek der weltliteratur) – 6v on 1r – 1 – (incl bibl ref) – mf#7462 – us Wisconsin U Libr [860]

Goethes briefe an charlotte von stein / ed by Borcherdt, Hans Heinrich – Berlin: Deutsche bibliothek, [19–] [mf ed 1991] – 2v – 1 – mf#7539 – us Wisconsin U Libr [860]

Goethes briefe an charlotte von stein / ed by Fraenkel, Jonas – Jena: E Diederichs, 1908 – 1r – 1 – us Wisconsin U Libr [920]

Goethes briefe an e. th. langer / ed by Zimmermann, Paul – Wolfenbuettel: J Zwissler, 1922 – 1r – 1 – (incl bibl ref) – us Wisconsin U Libr [920]

Goethes briefe an frau von stein : nebst dem tagebuch aus italien und briefen der frau von stein / Heinemann, K [comp] – Stuttgart: J G Cotta, [1894] [mf ed 1993] – (= ser Cotta'sche bibliothek der weltlitteratur) – 4v in 2 – 1 – mf#8622 – us Wisconsin U Libr [920]

Goethes briefe an frau von stein / ed by Schoell, Adolf & Wahle, Julius – Frankfurt a.M.: Ruetten & Loening, 1899 – 1r – 1 – us Wisconsin U Libr [920]

Goethes briefwechsel mit antonie brentano 1814-1821 / ed by Jung, Rudolf – Weimar: H Boehlaus Nachf., 1896 – 1r – 1 – us Wisconsin U Libr [920]

Goethes briefwechsel mit christian gottlob voigt / ed by Tuemmler, Hans – Weimar: H Boehlaus Nachfolger, 1949-62 [mf ed 1993] – (= ser Schriften der goethe-gesellschaft 53-56) – 4v – 1 – (incl bibl ref and ind) – mf#8657 reel 12 – us Wisconsin U Libr [430]

Goethe's briefwechsel mit den gebruedern von humboldt / ed by Bratranek, Franz Thomas – Leipzig: F A Brockhaus, 1876 [mf ed 1993] – (= ser Neue mittheilungen aus johann wolfgang von goethe's handschriftlichem nachlasse 3) – xlix/443p – 1 – (incl bibl ref and ind) – mf#8607 – us Wisconsin U Libr [920]

Goethes briefwechsel mit einem kinde / Arnim, Bettina von – Berlin: im Propylaeen-Verlag, c1920 [mf ed 1993] – (= ser Saemtliche werke 3-4) – 2v/pl (ill) – 1 – mf#8196 reel 1 – us Wisconsin U Libr [860]

Goethes briefwechsel mit einem kinde : seinem denkmal / ed by Grimm, Herman Friedrich – 4. aufl. Berlin: W Hertz, 1890 [mf ed 1993] – (= ser Cotta'sche handbibliothek 132-134) – 3v in 1 (ill) – 1 – mf#8464 – us Wisconsin U Libr [920]

Goethes briefwechsel mit friedrich rochlitz / ed by Biedermann, Woldemar, Freiherr von – Leipzig: F W v Biedermann, 1887 – 1r – 1 – (incl bibl ref) – us Wisconsin U Libr [920]

Goethes briefwechsel mit heinrich meyer / ed by Hecker, Max – Weimar: Goethe-Gesellschaft, 1917-32 [mf ed 1993] – (= ser Schriften der goethe-gesellschaft 32, 34-35, 35/2) – 4v – 1 – (incl bibl ref and ind) – mf#8657 reel 8 – us Wisconsin U Libr [430]

Goethes briefwechsel mit marianne von willemer / ed by Stein, Philipp – Leipzig: Insel-Verlag, 1908 [mf ed 1993] – lx/338p/2pl (ill) – 1 – (incl bibl ref and ind) – mf#8604 – us Wisconsin U Libr [920]

Goethes briefwechsel mit seiner frau / ed by Graf, Hans Gerhard – Frankfurt a.M.: Ruetten & Loening, 1916 – 1r – 1 – us Wisconsin U Libr [920]

Goethes briefwechsel mit thomas carlyle / ed by Hecht, Georg – Dachau: Einhorn-Verlag, [1913] – 1 – (incl the english text of carlyle's letters to goethe) – us Wisconsin U Libr [920]

Goethes buehnenbearbeitung von "romeo und julia" / Wendling, Emil – Zabern: [s.n.], 1907 – 1 – 1 – (incl bibl ref) – us Wisconsin U Libr [790]

Goethes campagne in frankreich, 1792 : eine philologische untersuchung aus dem weltkriege / Roethe, Gustav – Berlin: Weidmann, 1919 [mf ed 1990] – 1r – 1 – (filmed with: campagne in frankreich. incl bibl ref and ind) – us Wisconsin U Libr [430]

Goethes charakter : eine seelenschilderung / Saitschick, Robert – Stuttgart: F Frommann (E Hauff), 1898 – 1r – 1 – us Wisconsin U Libr [430]

Goethes clavigo : edited with the variants of all of the older editions / Goethe, Johann Wolfgang von; ed by Strube, Claire M M – Tuebingen: H Laupp, 1923 [mf ed 1990] – 1r – 1 – (filmed with: das volkslied und sein einfluss auf goethe's lyrik. incl bibl ref) – us Wisconsin U Libr [820]

Goethes deutsche gesinnung : ein beitrag zur geschichte seiner entwicklung / Winter, Friedrich Gotthard – [S.l.: s.n.], 1880 Leipzig: Druck der Rossberg'schen Buchdruckerei – 1r – 1 – (incl bibl ref) – us Wisconsin U Libr [430]

Goethes deutsche sendung : eine festrede / Korff, Hermann August – Leipzig: J J Weber, 1932 [mf ed 1990] – 24p – 1 – mf#7390 – us Wisconsin U Libr [080]

Goethe's dichtung und wahrheit : selections from books 1-11 / ed by Jagemann, Hans C G von – New York: H Holt, 1901, c1890 [mf ed 1993] – 373p – 1 – (incl bibl ref. german text. int and notes in english) – mf#8607 – us Wisconsin U Libr [430]

Goethes 'die aufgeregten' : zur frage der politischen dichtung in deutschland / Demetz, Peter – Hann M.nden: F Nowack, c1952 – 1r – 1 – us Wisconsin U Libr [810]

Goethes egmont : ein trauerspiel / Goethe, Johann Wolfgang von – Wien: K Graeser, [1893?] (mf ed 1990) – 1r – 1 – (filmed with: das gesicht im nebel) – us Wisconsin U Libr [820]

Goethes egmont : ein trauerspiel in fuenf aufzuegen / Goethe, Johann Wolfgang von – 3., verb Aufl. Paderborn: F Schoeningh, 1895 (mf ed 1990) – 1r – 1 – (filmed with: das gesicht im nebel. incl bibl ref) – us Wisconsin U Libr [820]

Goethes egmont / Vollmer, Friedrich – Leipzig: H Bredt, 1895 (mf ed 1990) – 1r – 1 – (filmed with: studien zu goethes egmont. incl bibl ref) – us Wisconsin U Libr [430]

Goethes egmont / Zimmermann, Ernst – Halle a/S: M Niemeyer, 1909 (mf ed 1990) – 1r – 1 – (filmed with: studien zu goethes egmont. incl bibl ref) – us Wisconsin U Libr [430]

Goethes egmont und schillers wallenstein : eine parallele der dichter / Bratranek, Franz Thomas – Stuttgart: Cotta 1862 [mf ed 1990] – 1r – 1 – (filmed with: das gesicht im nebel / peter doerfler) – mf#7322 – us Wisconsin U Libr [430]

Goethes ehe / Hofer, Klara – 1.-3. Aufl. Stuttgart, Berlin: J G Cotta, 1920 – 1r – 1 – us Wisconsin U Libr [920]

Goethes eigenhaendige reinschrift des west-oestlichen divan : eine auswahl von 28 blaettern im faksimile-nachbildung / ed by Burdach, Konrad – Weimar: Goethe-Gesellschaft, 1911 [mf ed 1993] – (= ser Schriften der goethe-gesellschaft 26) – 37p/28pl – 1 – (incl bibl ref) – mf#8657 reel 6 – us Wisconsin U Libr [430]

Goethes einfluss auf george meredith / Wilcox, Richard – Frankfurt a.M., 1976 (mf ed 1994) – 3mf – 9 – 3-89349-941-5 – mf#DHS-AR 941 – gw Frankfurter [410]

Goethes einfluss auf novalis heinrich von ofterdingen / Woltereck, Kaete A – Weida i.Th.: Thomas & Hubert, 1914 – 1 – (incl bibl ref) – us Wisconsin U Libr [430]

Goethes einfluss auf uhland / Sintenis, Franz – Dorpat: C Mattiesen, 1871 – 1r – 1 – (incl bibl ref) – us Wisconsin U Libr [430]

Goethes erste weimarer gedichtsammlung mit varianten / Goethe, Johann Wolfgang von; ed by Leitzmann, Albert – Bonn: A Marcus & E Weber, 1910 [mf ed 1990] – (= ser Kleine texte fuer theologische und philologische vorlesungen und uebungen 63) – 34/[1]p – 1 – (incl bibl ref) – mf#7318 – us Wisconsin U Libr [810]

Goethes ethische ansichten : ein beitrag zur geschichte der philosophie unserer dichterheroen / Melzer, Ernst – Neisse: J Graveur, 1890 – 1r – 1 – (incl bibl ref) – us Wisconsin U Libr [430]

Goethe's faust : a commentary / Snider, Denton Jaques – St Louis: William Harvey Miner Co, c1886 [mf ed 1996] – 2v – 1 – mf#9643 – us Wisconsin U Libr [430]

Goethe's faust : first part / Goethe, Johann Wolfgang von – Oxford: B Blackwell, 1924 [mf ed 1990] – 1r – 1 – (trans by john todhunter; int by j g robertson. filmed with: faust und urfaust / goethe ann by ernst beutler) – mf#7330 – us Wisconsin U Libr [820]

Goethe's faust : part one and selected sections of part two in the german original with an english translation / ed by Raschen, J F L – Ithaka, NY: Thrift Press, 1949 [mf ed 1993] – xviii/360p – 1 – (incl bibl ref. parallel german and english text. int and notes in english. trans by ed) – mf#8611 – us Wisconsin U Libr [820]

Goethe's faust : six essays / Fairley, Barker – Oxford: Clarendon Press, 1953 [mf ed 1993] – vi/132p – 1 – mf#8604 – us Wisconsin U Libr [430]

Goethe's faust – London: J M Dent; New York: E P Dutton, 1926 [mf ed 1993] – xxiv/594p/8pl (ill) – 1 – (trans into english by w h van der smissen. comm and notes by w h van der smissen. int by robert falconer) – mf#8615 – us Wisconsin U Libr [810]

Goethes "faust" : eine evangelische auslegung / Melzer, Friso – Berlin: Furche-Verlag, 1932 – 1 – us Wisconsin U Libr [430]

Goethes "faust" : eine freimaurertragoedie: versuch einer klaerung, kein kommentar / Rost, Else – Neue, erw Aufl. Muenchen: Ludendorff, [1936] – 1r – 1 – us Wisconsin U Libr [430]

Goethes "faust" : eine historische erlaeuterung / Riemann, Robert – Leipzig: Dieterich, 1911 – 1r – 1 – us Wisconsin U Libr [430]

Goethes 'faust' : eine tragoedie in zwei teilen / Goethe, Johann Wolfgang von – Leipzig: Breitkopf & Haertel. 2v in 1. 1908 (mf ed 1990) – 1r – 1 – (filmed with: goethes faust) – us Wisconsin U Libr [820]

Goethes faust : eine analyse der dichtung / Buechner, Wilhelm – Leipzig: B G Teubner, 1921 [mf ed 1993] – v/128p – 1 – (incl bibl ref) – mf#8604 – us Wisconsin U Libr [430]

Goethes faust : eine analyse der dichtung / Buechner, Wilhelm – Leipzig: B G Teubner 1921 [mf ed 1993] – 1r – 1 – (incl bibl ref; filmed with: goethe's faust / by barker fairley; & other titles) – mf#8604 – us Wisconsin U Libr [430]

Goethes faust : andeutungen ueber sinn und zusammenhang des ersten und zweiten theiles der tragoedie / Deycks, Ferdinand – 2., stark verm u verb Ausg. Frankfurt a/M: J C Hermann, 1855 (mf ed 1990) – 1r – 1 – (filmed with: goethes faust in seiner haltesten gestalt. incl bibl ref and ind) – us Wisconsin U Libr [430]

Goethes faust / Aster, Ernst von – Muenchen: Roesl, 1923 [mf ed 1993] – (= ser Philosophische reihe 75) – 155p – 1 – (incl bibl ref) – mf#8603 – us Wisconsin U Libr [110]

Goethes faust : briefwechsel mit einer dame [i.e. emilie lichtenberger, geb. burkhardt] / ed by Gruen, Albert – Gotha: H Scheube, 1856 – 1 – us Wisconsin U Libr [430]

Goethes faust : ein buch des lebens / Tolle, Hugo – Leipzig: O Hillmann, 1922 – 1r – 1 – us Wisconsin U Libr [430]

Goethes faust : ein deutscher mythus / Gruetzmacher, Richard Heinrich – Berlin: G Stilke. 2v in 1. 1936 – 1r – 1 – us Wisconsin U Libr [430]

Goethes faust : die dramatische einheit der dichtung / Rickert, Heinrich – Tuebingen: J C B Mohr, 1932 – 1r – 1 – us Wisconsin U Libr [430]

Goethes faust : eine einfuehrung / Litzmann, Berthold – Berlin: E Fleischel, 1904 – 1r – 1 – us Wisconsin U Libr [430]

Goethes faust : eine einfuehrung / Pfeiffer, Johannes – 2., verb Aufl. Bremen: J Storm, 1947 – 1r – 1 – (incl bibl ref) – us Wisconsin U Libr [430]

Goethes faust : einfuehrung und deutung / Gramsch, Alfred – Braunschweig: G Westermann, 1949 – 1r – 1 – us Wisconsin U Libr [430]

Goethes faust : entstehungsgeschichte und erklaerung / Minor, Jakob – Stuttgart: Cotta. 2v. 1901 – 1r – 1 – us Wisconsin U Libr [430]

Goethes faust : erster teil / ed by Duentzer, Heinrich – 7. Aufl. Leipzig: E Wartig, [1909] – 1r – 1 – (incl bibl ref) – us Wisconsin U Libr [430]

Goethes faust : erster teil / Duentzer, Heinrich – 7. aufl. Leipzig: E Wartig, [1909] [mf ed 1990] – 240p – 1 – (incl bibl ref) – mf#7344 – us Wisconsin U Libr [430]

Goethes faust : erster teil / Goethe, Johann Wolfgang von; ed by Goebel, Julius – 2nd rev ed. New York: H Holt, 1927, c1907 (mf ed 1990) – 1r – 1 – (filmed with: goethes faust. incl bibl ref) – us Wisconsin U Libr [820]

Goethes faust : erster und zweiter teil / Goethe, Johann Wolfgang von; ed by Mecker, Max – 2. Aufl. Leipzig: J J Weber, 1923 (mf ed 1990) – 1r – 1 – (filmed with: goethes faust) – us Wisconsin U Libr [820]

Goethes faust : erster und zweiter theil / Duentzer, Heinrich – 6. neubearb aufl. Leipzig: E Wartig, E Hoppe, 1899-1900 [mf ed 1990] – (= ser Erlaeuterungen zu den deutschen klassikern. abt 1, erlaeuterungen zu goethes werken 12-14) – 2v in 1 – 1 – (incl bibl ref) – mf#7344 – us Wisconsin U Libr [430]

Goethes faust : erster und zweiter theil / Marbach, Gotthard Oswald – Stuttgart: J G Goeschen, 1881 – 1r – 1 – us Wisconsin U Libr [430]

Goethes faust : erster und zweiter theil: text u. erlaeuterung in vorlesungen / ed by Oettingen, Alexander von – Erlangen: A Deichert. 2v. 1880 – 1r – 1 – us Wisconsin U Libr [430]

Goethes faust / Federn, Etta [Etta Federn-Kohlhaas] – Berlin: Horodisch & Marx, 1927 [mf ed 1990] – 92/1p – 1 – (incl bibl ref) – mf#7353 – us Wisconsin U Libr [430]

Goethes faust : first part / Goethe, Johann Wolfgang von – Oxford: B Blackwell, 1924 (mf ed 1990) – 1r – 1 – (filmed with: faust und urfaust) – us Wisconsin U Libr [820]

Goethes faust : the first part / Goethe, Johann Wolfgang von – London: Rivington, [1882?] (mf ed 1990) – 1r – 1 – (filmed with: goethes faust) – us Wisconsin U Libr [820]

Goethes faust / Fischer, Kuno – 3. durchges u verm Aufl. Stuttgart: J C Cotta. 4v in 2. 1893 – 1r – 1 – us Wisconsin U Libr [430]

Goethes faust / Fischer, Kuno; ed by Michels, Victor – 7. Aufl. Heidelberg: C Winter. 4v in 3. 1913 – 1r – 1 – us Wisconsin U Libr [430]

Goethes faust : a fragment of socialist criticism / Hitch, Marcus – Chicago: C H Kerr, 1908 – 1r – 1 – us Wisconsin U Libr [430]

Goethes faust : fuenfzehn vortraege / Hauri, Johannes – Berlin-Zehlendorf (Wsb): C Skopnik, 1910 – 1r – 1 – us Wisconsin U Libr [430]

Goethes faust : fuer die auffuehrung als mysterium in zwei tagewerken eingerichtet von otto devrient / Goethe, Johann Wolfgang von – 4., unver"nd Aufl. Leipzig: Breitkopf & Haertel. 2v in 1. 1896 (mf ed 1990) – 1 – (filmed with: goethes faust) – us Wisconsin U Libr [430]

Goethes faust / Goethe, Johann Wolfgang von – 7., durchgearb Aufl. Leipzig: Hesse & Becker. 2v. 1924 (mf ed 1990) – 1 – (filmed with: das spiel vom doktor faust von goethe) – us Wisconsin U Libr [430]

Goethes faust / Goethe, Johann Wolfgang von – Berlin: Kuehling & Guettner, 1908 (mf ed 1990) – 1r – 1 – (filmed with: goethes faust) – us Wisconsin U Libr [430]

Goethes faust / Goethe, Johann Wolfgang von – Leipzig: K W Hiersemann, 1907 (mf ed 1990) – 1r – 1 – (filmed with: goethes faust) – us Wisconsin U Libr [430]

Goethes faust / Goethe, Johann Wolfgang von – Gotha: F A Perthes, 1888 (mf ed 1990) – 1r – 1 – (filmed with: studien zu goethes egmont. incl bibl ref) – us Wisconsin U Libr [430]

Goethes faust / Goethe, Johann Wolfgang von; ed by Hellen, Eduard von der – Stuttgart: J G Cotta, [19–?] (mf ed 1990) – 1 – (filmed with: le faust) – us Wisconsin U Libr [430]

Goethes faust / Goethe, Johann Wolfgang von; ed by Petsch, Robert – 2. Ausg. Leipzig: Bibliographisches Institut, [1925?] – 1 – (incl bibl ref) – us Wisconsin U Libr [820]

Goethes faust / Hefele, Herman – 3. Aufl. Stuttgart: F Frommann, 1946 – 1r – 1 – us Wisconsin U Libr [430]

Goethes faust / Hefele, Herman – Stuttgart: F Frommann (H Kurtz), 1931 – 1r – 1 – us Wisconsin U Libr [430]

Goethes faust : i aarene 1788-89 / Sarauw, Christian Preben Emil – Kobenhavn: A F Host, 1919 – 1r – 1 – (incl bibl ref) – us Wisconsin U Libr [430]

Goethes faust : in saemtlichen fassungen mit den bruchstuecken und entwuerfen des nachlasses / Goethe, Johann Wolfgang von; ed by Alt, Karl – Berlin: Bong, [1909?] (mf ed 1990) – 1r – 1 – (filmed with: goethes faust. incl bibl ref) – us Wisconsin U Libr [820]

Goethes faust : indledning og forklaring / Koch, Carl – Kobenhavn: K Schonberg, 1901 – 1r – 1 – (incl bibl ref) – us Wisconsin U Libr [430]

Goethes faust : nach seiner entstehung, idee und composition / Fischer, Kuno – 2. neu bearb u verm Aufl. Stuttgart: b J. G. Cotta, 1887 – 1r – 1 – us Wisconsin U Libr [430]

Goethes faust : pt i / Goethe, Johann Wolfgang von; ed by Priest, George M – Princeton, NJ: Princeton University Press, 1929 (mf ed 1990) – 1 – (filmed with: le faust de goethe. incl bibl ref) – us Wisconsin U Libr [820]

Goethes faust / Saupe, Ernst Julius – Leipzig: F Fleischer, 1856 – 1r – 1 – us Wisconsin U Libr [430]

Goethes faust : seine kritiker und ausleger / Koestlin, Karl – Tuebingen: H Laupp, 1860 – 1r – 1 – (incl bibl ref) – us Wisconsin U Libr [430]

Goethes faust : teil i nebst urfaust / Goethe, Johann Wolfgang von; ed by Reh, Hans – Langensalza: J Beltz, 1937 (mf ed 1990) – 1r – 1 – (filmed with: goethes faust) – us Wisconsin U Libr [820]

Goethes faust / Trendelenburg, Adolf – Berlin: W de Gruyter, 1921-22 [mf ed 1993] – 2v – 1 – (incl bibl ref) – mf#8615 – us Wisconsin U Libr [430]

Goethes faust / Vischer, Friedrich Theodor – 3rd ed. Stuttgart: J G Cotta, 1921 – 1r – 1 – (incl bibl ref) – us Wisconsin U Libr [430]

Goethes faust : weg und sinn seines lebens, seiner rettung / Buehlmann, Heinrich – Zuerich: Amalthea-Verlag 1931 [mf ed 1990] – 1r – 1 – (filmed with: a passage in the night / sholem asch) – mf#7342 – us Wisconsin U Libr [430]

Goethes faust : zeugnisse und excurse zu seiner entstehungsgeschichte / Pniower, Otto – Berlin: Weidmann, 1899 – 1r – 1 – (incl bibl ref and indexes) – us Wisconsin U Libr [430]

Goethes faust : zur ersten einfuehrung in das verstaendnis der dichtung / Wagner, Kurt – Bielefeld; Leipzig: Velhagen & Klasing, 1926 – 1r – 1 – (incl bibl ref) – us Wisconsin U Libr [430]

Goethes faust : zweiter theil: tragoedie in fuenf akten / Goethe, Johann Wolfgang von – Dresden: E Pierson, 1880 (mf ed 1990) – 1r – 1 – (filmed with: goethes faust) – us Wisconsin U Libr [820]

Goethes faust als einheitliche dichtung / Baumgart, Hermann – Koenigsberg i/Pr: W Koch, 1893-1902 [mf ed 1990] – 2v – 1 – mf#7340 – us Wisconsin U Libr [430]

Goethes faust als einheitliche dichtung / Schreyer, Hermann – Halle (Saale): Verlag der Buchhandlung des Waisenhauses, 1881 – 1r – 1 – us Wisconsin U Libr [430]

Goethes faust als erzaehlung : zur einfuehrung in das verstaendnis des originals / Kupffer, Julius – Naumburg (Saale): A Schirmer, 1892 – 1r – 1 – us Wisconsin U Libr [430]

Goethes faust als levensbeeld / Bruinwold Riedel, J – Utrecht: W Leijdenroth 1905 [mf ed 1990] – 1r – 1 – (filmed with: a passage in the night / sholem asch) – mf#7342 – us Wisconsin U Libr [430]

Goethes faust am hofe des kaisers : [des zweyten theiles erste abteilung] in drei akten / Goethe, Johann Wolfgang von; ed by Tewes, Friedrich – Berlin: G Reimer, 1901 (mf ed 1990) – 1 – (filmed with: goethes faust in urspruenglicher gestalt. incl new scene between faust and mephisto written by eckermann) – us Wisconsin U Libr [820]

Goethes faust auf der buehne : beitraege zum probleme der auffuehrung und inszenierung des gedichtes / Kilian, Eugen – Muenchen; Leipzig: G Mueller, 1907 – 1r – 1 – us Wisconsin U Libr [430]

Goethes faust auf der deutschen buehne : eine jahrhundertbetrachtung / Petersen, Julius – Leipzig: Quelle & Meyer, 1929 – 1r – 1 – (incl bibl ref) – us Wisconsin U Libr [430]

Goethes faust, ein geheimbuch : nachweise aus des dichters briefen, tagebuechern etc / ed by Steinzaenger, O – Hamburg: C Boysen, 1906 – 1r – 1 – us Wisconsin U Libr [430]

Goethes faust erster und zweiter theil / Sengler, J – Berlin: F Henschel, 1873 – 1r – 1 – us Wisconsin U Libr [430]

Goethes faust im blickfeld des 20. jahrhunderts : eine weltanschauliche darstellung / Bertram, Johannes – 4. aufl. Hamburg: Hamburger Kulturverlag (Produktion Dreizack), 1949 [mf ed 1990] – 384p – 1 – (incl ind and bibl ref) – mf#7340 – us Wisconsin U Libr [430]

Goethes faust im lichte der kulturphilosophie spenglers / Jacobskoetter, Ludwig – Berlin: E S Mittler, 1924 – 1r – 1 – us Wisconsin U Libr [430]

Goethes faust im zwanzigsten jahrhundert / Hohlenberg, Johannes – Basel: R Geering, 1931 – 1r – 1 – us Wisconsin U Libr [430]

Goethes faust in seiner aeltesten gestalt / Collin, Joseph – Frankfurt a/M: Ruetten & Loening, 1896 (mf ed 1990) – 1 – (filmed with: versuch einer geschichte des volksschauspiels vom doctor faust. incl bibl ref) – us Wisconsin U Libr [430]

Goethes faust in urspruenglicher gestalt / Goethe, Johann Wolfgang von – 4. Abdruck. Weimar: Boehlau, 1899 (mf ed 1990) – 1 – (filmed with: faust) – us Wisconsin U Libr [820]

Goethes faust und der geist der magie / Birven, Henri Clemens – Leipzig: Talisverlag, 1923 [mf ed 1990] – viii/168p/[1]pl – 1 – (incl bibl ref) – mf#7341 – us Wisconsin U Libr [430]

Goethes faust und die bildende kunst / Storck, Willy F – Leipzig: Xenien-Verlag, 1912 – 1r – 1 – (incl bibl ref) – us Wisconsin U Libr [430]

Goethes faust und die vollendung des menschen / Levinstein, Kurt – Berlin: W de Gruyter, 1948 [mf ed 1993] – 132p – 1 – mf#8604 – us Wisconsin U Libr [430]

Goethes faustdichtung : in ihrer kuenstlerischen einheit / Valentin, Veit – Berlin: E Felber, 1894 – 1r – 1 – us Wisconsin U Libr [430]

Goethes faustdichtung : ein neuer originalkommentar / Wilhelmi, Rudolf – Hamburg: C Boysen, 1908 – 1r – 1 – us Wisconsin U Libr [430]

Goethes faustidee nach der urspruenglichen conception / Gwinner, Wilhelm von – Frankfurt (Main): J Baer, 1892 – 1 – us Wisconsin U Libr [430]

Goethes "fischer" : das wasser rauscht', das wasser schwoll: eine poetische studie / Kuester, Rudolf – Breslau: Priebatsch, 1918 – 1r – 1 – (incl bibl ref) – us Wisconsin U Libr [430]

Goethes freundinnen : briefe zu ihrer charakteristik / ed by Baeumer, Gertrud – Leipzig; Berlin: B G Teubner, 1909 [mf ed 1990] – 318p/12pl – 1 – mf#7351 – us Wisconsin U Libr [920]

Goethe's gedichte / ed by Fr Strehlke – Berlin: F Duemmler. 3v. [1886-88] – 1r – 1 – (incl bibl ref and ind) – us Wisconsin U Libr [810]

Goethe's gedichte / Goethe, Johann Wolfgang von – Berlin: G Hempel. 3v. 1882-84 – 1r – 1 – (incl bibl ref and ind) – us Wisconsin U Libr [810]

Goethes gedichte = Poems / Goethe, Johann Wolfgang von – Berlin: S Fischer. 2v. [1905?] – 1r – 1 – (incl bibl ref and ind) – us Wisconsin U Libr [810]

Goethes gedichte in zeitgeschichtlicher auswahl / Goethe, Johann Wolfgang von; ed by Witkop, Philipp – Stuttgart: Strecker und Schroeder, 1932 (mf ed 1990) – 1r – 1 – (filmed with: acht lieder von goethe) – us Wisconsin U Libr [810]

Goethes "geheimnisse" und seine "indischen legenden" / Baumgart, Hermann – Stuttgart: J G Cotta, 1895 [mf ed 1990] – vi/110p – 1 – mf#7383 – us Wisconsin U Libr [430]

Goethes geistesart in ihrer offenbarung durch seinen faust und durch das maerchen "von der schlange und der lilie" / Steiner, Rudolf – Dornach, Switzerland: Philosophisch-anthroposophischer Verlag am Goetheanum, 1926 [mf ed 1990] – 95p – 1 – mf#7360 – us Wisconsin U Libr [110]

Goethes geistesgestalt / Steffen, Albert – Dornach: Verlag fuer Schoene Wissenschaften, 1932 [mf ed 1993] – 392p – 1 – (incl bibl ref) – mf#8652 – us Wisconsin U Libr [430]

Goethes geistige welt / Loesche, Martin – Stuttgart: S Hirzel, 1948 [mf ed 1993] – 379p – 1 – (incl bibl ref and ind) – mf#8655 – us Wisconsin U Libr [430]

Goethes geschichtlicher sinn / Tellenbach, Gerd – [Freiburg i.B.]: Selbstverlag der Universitaet, 1949 – 1r – 1 – (incl bibl ref) – us Wisconsin U Libr [430]

Goethes gespraeche / ed by Biedermann, Woldemar, Freiherr von – 2. durchges stark verm aufl, gesamtausg. Leipzig: F W v Biedermann, 1909-1911 [mf ed 1999] – 5v – 1 – (incl bibl ref and ind) – mf#10147 – us Wisconsin U Libr [080]

Goethes gesundheitspflege : essen und trinken – zwei kapitel aus "goethes lebenskunst" / Bode, Wilhelm – Berlin: E S Mittler, 1913 [mf ed 1990] – 48p – 1 – mf#7486 – us Wisconsin U Libr [920]

Goethes goetz von berlichingen / Duentzer, Heinrich – 5. durchgesehene und verm aufl. Leipzig: E Wartig, 1894 [mf ed 1990] – (= ser Erlaeuterungen zu den deutschen klassikern. abt 1, erlaeuterungen zu goethes werken 6) – 181p – 1 – (incl bibl ref) – mf#7362 – us Wisconsin U Libr [430]

Goethes goetz von berlichingen in zeichnungen von franz pforr / ed by Benz, Richard – Weimar: Goethe-Gesellschaft, 1941 [mf ed 1993] – (= ser Schriften der goethe-gesellschaft 52) – 42p/24pl (ill) – 1 – mf#8657 reel 11 – us Wisconsin U Libr [740]

Goethes goetz von berlichingen und shakespeares historische dramen : abhandlung des oberlehrers august huther / Huther, August – [S.l.: s.n.], 1893; Cottbus: Druck von A Hein – 1r – 1 – us Wisconsin U Libr [430]

Goethes gottfried von berlichingen / Schregle, Hans – Halle (Saale): M Niemeyer, 1923 – 1r – 1 – (incl bibl ref) – us Wisconsin U Libr [430]

Goethes gotz von berlichingen mit der eisernen hand – New York, NY. 1896 – 1r – us UF Libraries [430]

Goethes intuition / Emrich, Hermann – Tuebingen: J C B Mohr, 1928 – 1r – 1 – us Wisconsin U Libr [430]

Goethes iphigenie : ihr verhaeltniss zur griechischen tragoedie und zum christentum / Mueller, H F – Heilbronn: Henninger, 1882 – 1r – 1 – us Wisconsin U Libr [430]

Goethes iphigenie : ein vortrag / Heinzelmann, W – Erfurt: H Neumann, 1891 – 1r – 1 – us Wisconsin U Libr [430]

Goethe's iphigenie auf tauris – London, New York: Macmillan, 1904, c1898 [mf ed 1993] – lxi/180p – 1 – (incl bibl ref and ind. german text. int and notes in english by charles a eggert) – mf#8606 – us Wisconsin U Libr [820]

Goethes iphigenie auf tauris : in vierfacher gestalt / ed by Baechtold, Jakob – Freiburg i B [Germany]: J C B Mohr, 1883 [mf ed 1993] – viii/125p – 1 – mf#8605 – us Wisconsin U Libr [820]

Goethes iphigenie auf tauris – 3. aufl. Leipzig: H Bredt, 1906 [mf ed 1998] – (= ser Die deutschen klassiker 5) – x/236p – 1 – (incl bibl ref. ann by m evers) – mf#9981 – us Wisconsin U Libr [820]

Goethes iphigenie auf tauris / Stoffel, J – Langensalza: H Beyer, 1899 [mf ed 1998] – (= ser Deutsche dramen und epische dichtungen fuer den schulgebrauch erlaeutert 6) – 73p – 1 – mf#9981 – us Wisconsin U Libr [430]

Goethes kleinere aufsaetze / Goethe, Johann Wolfgang von; ed by Seidlitz, Woldemar von – Muenchen: F Bruckmann, 1904 – 1r – 1 – (incl bibl ref) – us Wisconsin U Libr [430]

Goethe's knowledge of english literature / Boyd, James – Oxford: The Clarendon Press, 1932 [mf ed 1993] – (= ser Oxford studies in modern languages and literature) – xvii/310p – 1 – (incl bibl ref and ind) – mf#8655 – us Wisconsin U Libr [410]

Goethe's knowledge of french literature / Barnes, Bertram – Oxford: The Clarendon Press, 1937 [mf ed 1993] – (= ser Oxford studies in modern languages and literature) – viii/172p – 1 – (incl bibl ref and ind) – mf#8655 – us Wisconsin U Libr [410]

Goethes kopf und gestalt / Bauer, Karl – Berlin: E S Mittler, 1908 [mf ed 1990] – xi/62p/[23]pl (ill) – 1 – mf#7353 – us Wisconsin U Libr [430]

Goethes laufbahn als schriftsteller / Mann, Thomas – Muenchen: R Oldenbourg, c1933 – 1r – 1 – us Wisconsin U Libr [430]

Goethes leben / Bode, Wilhelm – Berlin: E S Mittler, 1920-27 [mf ed 1992] – 9v on 2r (ill) – 1 – mf#7637 – us Wisconsin U Libr [920]

Goethes leben / Duentzer, Heinrich – 2. durchgesehene verm aufl. Leipzig: Fues (R Reisland), 1883 [mf ed 1999] – xii/707p/5pl (ill) – 1 – (incl ill and ind) – mf#10131 – us Wisconsin U Libr [920]

Goethes leben : eine krankengeschichte / Lorenz, Friedrich – Jena: G Neuenhahn, 1938 – 1 – (incl bibl ref) – us Wisconsin U Libr [920]

Goethes leben.. / Goedeke, Karl, 1814-87 – Suppl. zu den Werken des Dichters. Stuttgart: J.G. Cotta, (1883). 187p. Illus. With: Laerdal og Borgund by J.A. Laberg. 1 reel. 1291 – 1 – us Wisconsin U Libr [430]

Goethes leben im garten am stern / Bach, Adolf – Berlin, 1917 (mf ed 1994) – 4mf – 9 – €45.00 – 3-8267-3015-1 – mf#DHS-AR 3015 – gw Frankfurter [430]

Goethes leben in seinen briefen / ed by Bab, Julius – Berlin: Deutsche Buch-Gemeinschaft, [1929-31] [mf ed 1992] – 3v (ill) – 1 – mf#7464 – us Wisconsin U Libr [920]

Goethes leben, leisten und leiden : in goethe's bildersprache / Schauffler, Theodor – Heidelberg: C Winter, 1913 – 1r – 1 – (incl ind) – us Wisconsin U Libr [430]

Goethes lebenskunst / Bode, Wilhelm – 3. aufl. Berlin: E S Mittler, 1902 [mf ed 1990] – vi/267p – 1 – (portrait of goethe by c a schwerdtgeburth) – mf#7351 – us Wisconsin U Libr [430]

Goethes leipziger krankheit und "don sassafras" / Hansen, Adolfph – Leipzig: J Woerner, 1911 – 1r – 1 – (incl bibl ref) – us Wisconsin U Libr [430]

Goethes leipziger liederbuch : i. (einleitung und gedicht i-iv) / Strack, Adolf – Giessen: J Ricker, 1893 (mf ed 1990) – 1r – 1 – (filmed with: das volkslied und sein einfluss auf goethe's lyrik. incl bibl ref) – us Wisconsin U Libr [430]

Goethes leipziger liederbuch / Strack, Adolf – Giessen: J Ricker, 1893 (mf ed 1990) – 1r – 1 – (filmed with: das volkslied und sein einfluss auf goethe's lyrik. incl bibl ref) – us Wisconsin U Libr [430]

Goethes liebesgedichte / Goethe, Johann Wolfgang von; ed by Graf, Hans Gerhardt – Leipzig: Insel-Verlag, 1912 (mf ed 1990) – 1r – 1 – (filmed with: acht lieder von goethe) – us Wisconsin U Libr [810]

Goethes lyrik : ausgewaehlt und erklaert fuer die oberen klassen hoeherer schulen von franz kern – Berlin: Nicolai, 1889 – 1r – 1 – (incl bibl ref and ind) – us Wisconsin U Libr [810]

Goethes lyrik in weisen deutscher tonsetzer bis zur gegenwart / Holle, Hugo – Muenchen: Wunderhorn-Verlag, 1914 (mf ed 1990) – 1r – 1 – (filmed with: das volkslied und sein einfluss auf goethe's lyrik. incl bibl ref) – us Wisconsin U Libr [430]

Goethes lyrische dichtung in ihrer entwicklung und bedeutung / Baumgart, Hermann; ed by Baumgart, Gertrud – Heidelberg: C Winter, 1931-39 [mf ed 1993] – 3v – 1 – (incl bibl ref and ind) – mf#8597 – us Wisconsin U Libr [430]

Goethe's lyrische gedichte / Duentzer, Heinrich – 3. neubearb aufl. Leipzig: E Wartig (E Hoppe), 1896-98 [mf ed 1993] – (= ser Erlaeuterungen zu den deutschen klassikern. abt 1, erlaeuterungen zu goethes werken 22-28) – 12v in 10 – 1 – mf#8594 – us Wisconsin U Libr [430]

Goethe's maerchen : ein politisch-nationales glaubensbekenntniss des dichters / Baumgart, Hermann – Koenigsberg: Hartung, 1875 [mf ed 1999] – 131p – 1 – mf#10173 – us Wisconsin U Libr [430]

Goethes morphologie / Bergmann, Wolfgang – Freiburg: K Alber, 1949 [mf ed 1993] – 26p/2pl (ill) – 1 – (incl bibl ref) – mf#8656 – us Wisconsin U Libr [430]

Goethes morphologie : (metamorphose der pflanzen und osteologie): ein beitrag zum sachlichen und philosophischen verstaendnis und zur kritik der morphologischen begriffsbildung / Hansen, Adolph – Giessen: Alfred Töpelmann, 1919 [mf ed 1999] – 1 – mf#10197 – us Wisconsin U Libr [430]

Goethes naturphilosophie im faust : ein beitrag zur erklaerung der dichtung / Hertz, Gottfried Wilhelm – Berlin: E S Mittler, 1913, c1912 – 1r – 1 – (incl bibl ref) – us Wisconsin U Libr [430]

Goethes naturwissenschaftliche correspondenz : (1812-32) / ed by Bratranek, F T H – Leipzig: F A Brockhaus, 1874-76 – 1r – 1 – (incl indes) – us Wisconsin U Libr [920]

Goethes nausikaa / Kettner, Gustav – Berlin: Weidmann, 1912 – 1r – 1 – us Wisconsin U Libr [430]

Goethes "novelle" : der schauplatz: coopersche einfluesse / Wukadinovic, Spiridion – Halle (Saale): M Niemeyer, 1909 – 1 – (incl bibl ref) – us Wisconsin U Libr [430]

Goethe's novelle 'die wahlverwandtschaften' : ein rekonstruktionsversuch / Wolff, Hans Matthias – Bern: Francke, c1955 [mf ed 1993] – 86p – 1 – mf#8606 – us Wisconsin U Libr [830]

Goethes paedagogik : vortrag gehalten zum besten der wilhelm-augusta-stiftung fuer frankfurter lehrerkinder am 7. februar 1881 / Eiselen, F – Frankfurt a.M.: M Diesterweg, 1881 – 1r – 1 – us Wisconsin U Libr [430]

Goethes paedagogische provinz : eine deutung von goethes erziehungsbild / Nitschke, Otfried – Wuerzburg: K Triltsch, 1937 – 1r – 1 – (incl bibl ref (p. 80-84)) – us Wisconsin U Libr [430]

Goethes persoenlichkeit : drei reden des kanzlers friedrich v. mueller, gehalten in den jahren 1830 und 1832 / Mueller, Friedrich von; ed by Bode, Wilhelm – Berlin: Mittler, 1901 – 1r – 1 – us Wisconsin U Libr [430]

Goethe's poems / Goethe, Johann Wolfgang von – Boston, MA: D C Heath, 1911 – 1 – (german text with an introduction and notes in english. includes bibliographical references) – us Wisconsin U Libr [810]

Goethe's poems / Goethe, Johann Wolfgang von – New York: H Holt, c1901 – 1r – 1 – (german text with introduction and notes in english. includes bibliographical references) – us Wisconsin U Libr [810]

Goethe's poems and aphorisms / Goethe, Johann Wolfgang von; ed by Bruns, Friedrich – New York: Oxford University Press, 1932 (mf ed 1990) – 1r – 1 – (filmed with: acht lieder von goethe. in english and german) – us Wisconsin U Libr [430]

Goethes preisaufgaben fuer bildende kuenstler : 1799-1805 / Scheidig, Walther – Weimar: H Boehlaus Nachfolger, 1958 [mf ed 1993] – xi/535p/1lea/46pl (ill) – 1 – (incl bibl ref and ind) – mf#8657 reel 13 – us Wisconsin U Libr [700]

Goethes propylaeen / Boehlich, Ernst – Stuttgart: Metzler, 1915 [mf ed 1992] – (= ser Breslauer beitraege zur literaturgeschichte. neue folge 44) – viii/170p – 1 – mf#8014 reel 5 – us Wisconsin U Libr [430]

Goethes rede zum schaekespears tag – Weimar: [Goethe-Gesellschaft], 1938 [mf ed 1993] – (= ser Schriften der goethe-gesellschaft 50) – 8pl – 1 – mf#8657 reel 11 – us Wisconsin U Libr [850]

Goethes reim / Wehnert, Bruno – Berlin: B Paul, 1899 – 1r – 1 – us Wisconsin U Libr [430]

Goethes relativitaetstheorie der farbe : nebst einer musikaesthetischen parallele / Barthel, Ernst – Bonn: F Cohen, 1923 [mf ed 1990] – 71p (ill) – 1 – mf#7383 – us Wisconsin U Libr [430]

Goethes religioese jugendentwicklung / Schubert, Hans von – Leipzig: Quelle & Meyer, 1925 – 1r – 1 – us Wisconsin U Libr [240]

Goethes religioeses erleben im zusammenhang seiner intuitiv-organischen weltanschauung / Neubauer, Ernst – Tuebingen: J C B Mohr (P Siebeck), 1925 – 1r – 1 – (incl bibl ref (p. 79-84)) – us Wisconsin U Libr [430]

Goethes religiositaet / Aner, Karl – Tuebingen: J C B Mohr, 1910 [mf ed 1990] – 32p – 1 – mf#7383 – us Wisconsin U Libr [430]

Goethes rheinreise mit lavater und basedow im sommer 1774 / ed by Bach, Adolf – Zuerich 1923 (mf ed 1993) – 2mf – 9 – €24.00 – 3-89349-263-1 – mf#DHS-AR 120 – gw Frankfurter [914]

Goethes roemische elegien : nach der aeltesten reinschrift / Goethe, Johann Wolfgang von; ed by Leitzmann, Albert – Bonn: A Marcus und E Weber, 1912 [mf ed 1990] – (= ser Kleine texte fuer theologische und philologische vorlesungen und uebungen 100) – 56p – 1 – mf#7371 – us Wisconsin U Libr [810]

Goethe's saemtliche werke = Works / Goethe, Johann Wolfgang von – Stuttgart: J G Cotta. 6v. 1854-55 – 2r – 1 – us Wisconsin U Libr [430]

Goethes saemtliche werke – Propylaeen-Ausg. Muenchen: G Mueller, 1909 – 45v – 1 – (incl ind) – mf#6974 – us Wisconsin U Libr [802]

Goethes saemtliche werke : jubilaeumsausgabe in 40 baenden / ed by Hellen, Eduard von der – Stuttgart: J G Cotta [1902-12] [mf ed 1994] – 40v – 1 – (incl bibl ref and ind. int in v1 dated 1902, int in ind to set dated 1912) – mf#8619 – us Wisconsin U Libr [802]

Goethes "satyros" und der urfaust / Schneider, Ferdinand Josef – Halle/S: M Niemeyer, 1949 [mf ed 1993] – (= ser Hallische monographien 12) – 33p – 1 – (incl bibl ref) – mf#8606 – us Wisconsin U Libr [430]

Goethes schoene seele susanna katharina v. klettenberg : ein lebensbild im anschlusse an eine sonderausgabe der bekenntnisse einer schoenen seele / Dechent, Hermann – Gotha: F A Perthes, 1896 – 1r – 1 – us Wisconsin U Libr [920]

Goethes schweizer reise 1775 : zeichnungen und niederschriften / ed by Koetschau, Karl & Morris, Max – Weimar: Goethe-Gesellschaft, 1907 [mf ed 1993] – (= ser Schriften der goethe-gesellschaft 22) – 49p/16pl (ill) – 1 – (incl bibl ref) – mf#8657 reel 6 – us Wisconsin U Libr [430]

Goethes "selige sehnsucht" : ein gespraech um die moeglichkeit einer christlichen deutung / Rang, Florens Christian & Rang, Bernhard – Freiburg: Herder, 1949 – 1r – 1 – us Wisconsin U Libr [430]

Goethes sprache und die antike : studien zum einfluss der klassischen sprachen auf goethe's poetischen stil / Olbrich, Carl – Leipzig: F W v Biedermann, 1891 – 1r – 1 – us Wisconsin U Libr [430]

Goethes sprache und stil im alter / Knauth, Paul – Leipzig: E Avenarius, 1898 – 1r – 1 – (incl bibl ref and index) – us Wisconsin U Libr [430]

Goethes stammbaeume : eine genealogische darstellung / Duentzer, Heinrich – Gotha: F A Perthes, 1894 [mf ed 1996] – 168p – 1 – mf#9654 – us Wisconsin U Libr [929]

Goethes stellung zu tod und unsterblichkeit / Koch, Franz – Weimar: Goethe-Gesellschaft, 1932 [mf ed 1993] – vi/336p – 1 – (incl bibl ref) – mf#8657 reel 11 – us Wisconsin U Libr [110]

Goethes stellung zum christenthum : ein literarischer beitrag / Oosterzee, Johannes Jacobus van – Bielefeld: Velhagen und Klasing, 1858 – 1r – 1 – (incl bibl ref) – us Wisconsin U Libr [210]

Goethes stellung zur franzoesischen romantik / Wadepuhl, Walter – [S.l.: s.n.], 1924 – 1r – 1 – (incl bibl ref) – us Wisconsin U Libr [410]

Goethes stellung zur religion / Filtsch, Eugen – Langensalza, 1879 (mf ed 1992) – 1mf – 9 – €24.00 – 3-89349-116-3 – mf#DHS-AR 85 – gw Frankfurter [200]

Goethes tagebuecher der sechs ersten weimarischen jahre (1776-82) / Goethe, Johann Wolfgang von; ed by Duentzer, Heinrich – Leipzig: Dyk, 1889 [mf ed 1990] – 261p – 1 – (incl bibl ref and ind) – mf#7366 – us Wisconsin U Libr [880]

Goethes tasso / Duentzer, Heinrich – 5. neu durchgesehene verm aufl. Leipzig: E Wartig, E Hoppe, 1898 [mf ed 1998] – (= ser Erlaeuterungen zu den deutschen klassikern. abt 1, erlaeuterungen zu goethes werken 10) – 200p – 1 – (incl bibl ref) – mf#9990 – us Wisconsin U Libr [430]

Goethes tasso / Fischer, Kuno – Heidelberg: C Winter, [18–?] – 1 – us Wisconsin U Libr [430]

Goethes tasso und kuno fischer : nebst einem anhange: goethes tasso und goldonis tasso / Kern, Franz – Berlin: Nicolai, 1892 – 1r – 1 – us Wisconsin U Libr [430]

Goethes testament : die loesung des faust-raetsels: der deutung erstes bis drittes buch: euphorismen, eilebeute, hexenkueche / Ullrich, Albert – Dessau: Faust-Verlag, 1921 – 1r – 1 – us Wisconsin U Libr [430]

Goethes theater-roman : festtagsgruss an konrad zwierina / Seuffert, Bernhard – Graz: Leuschner und Lubensky, 1924 – 1r – 1 – us Wisconsin U Libr [430]

Goethe's theory of colours / Goethe, Johann Wolfgang von – London 1840 – (= ser 19th c art & architecture) – 5mf – 9 – (trans fr german) – mf#4.2.1597 – uk Chadwyck [700]

Goethes torquato tasso im urteil von mit- und nachwelt / Gaede, Werner – Essen: National-Zeitungs-Verlag, 1931 – 1r – 1 – (incl bibl ref) – us Wisconsin U Libr [430]

Goethes typusbegriff / Spinner, Heinrich – Horgen-Zuerich: Verlag der Muenster-Presse, 1933 – 1r – 1 – (incl bibl ref) – us Wisconsin U Libr [500]

Goethes und stifters nausikaa-tragoedie : ueber die urphaenomene / Augustin, Hermann – Basel: B Schwabe, 1941 [mf ed 1990] – 91p – 1 – mf#7383 – us Wisconsin U Libr [410]

Goethes unterhaltungen deutscher ausgewanderten im spannungsfeld von franzoesischer revolution und aesthetischer erziehung / Schieferer, Robert – (mf ed 1997) – 1mf – 9 – €30.00 – 3-8267-2426-7 – mf#DHS 2426 – gw Frankfurter [430]

Goethes urteile ueber shakespeare aus seiner persoenlichkeit erklaert / Eckert, Georg Heinrich – [S.l.: s.n.] Goettingen: Druck von L Hofer, 1918 – 1r – 1 – (incl bibl ref) – us Wisconsin U Libr [410]

Goethes vater : eine studie / Ewart, Felicie – Hamburg: L Voss, 1899 – 1r – 1 – us Wisconsin U Libr [920]

Goethes verhaeltnis zu den organischen naturwissenschaften : vortrag gehalten im wissenschaftlichen verein zu berlin / Schmidt, Oscar – Berlin: W Hertz, 1853 – 1r – 1 – us Wisconsin U Libr [500]

Goethes verhaeltnis zu hans sachs : beilage zum 30. jahresbericht des hohenzollern-gymnasiums zu schwedt a.o. / Cleve, Karl – Schwedt a.O.: F Freyhoff, 1911 – 1r – 1 – us Wisconsin U Libr [430]

Goethes verhaeltnis zu klopstock : ihre geistigen, litterarischen und persoenlichen beziehungen / Lyon, Otto – Leipzig: T Grieben (L Fernau), 1882 – 1r – 1 – (incl bibl ref) – us Wisconsin U Libr [430]

Goethes vermaechtnis / Frucht, Else – 2. Aufl. Muenchen; Leipzig: Delphin-Verlag, c1913 – 1r – 1 – us Wisconsin U Libr [430]

Goethes vermaechtnis an amerika : vortrag gehalten im deutschen gesellig-wissenschaftlichen verein von new york am 12. oktober 1899 / Francke, Kuno – [S.l.: s.n.], 1899 – 1r – 1 – us Wisconsin U Libr [430]

Goethes weg zur hoehe : neue bearbeitung von "goethes bestem rat" / Bode, Wilhelm – Berlin: E S Mittler, 1912 [mf ed 1990] – 56p – 1 – mf#7372 – us Wisconsin U Libr [430]

Goethes welt- und lebensanschauung / Menzel, Alfred – Hamburg: Pfadweiser-Verlag, 1919 – 1r – 1 – us Wisconsin U Libr [430]

Goethes welt- und lebensanschauung / Ziegler, Theobald – Berlin: G Reimer, 1914 – 1r – 1 – us Wisconsin U Libr [140]

Goethes weltanschauung : reden und aufsaetze / Spranger, Eduard – Wiesbaden: Insel-Verlag, 1949 [mf ed 1993] – 255p – 1 – (incl bibl ref) – mf#8656 – us Wisconsin U Libr [080]

Goethes weltwende-schicksal : festvortrag zur feier der 100. wiederkehr von goethes todestag gehalten in der aula der vereinigten friedrichs-universitaet halle-wittenberg am 18. februar 1932 / Schneider, Ferdinand Josef – Halle (Saale): M Niemeyer, 1932 [mf ed 1990] – (= ser Hallische universitaetsreden 55) – 23p – 1 – mf#7396 – us Wisconsin U Libr [430]

Goethes Werke / ed by Kippenberg, Anton et al – [Mainz: Druck der Mainzer Presse, 1936] [mf ed 1989] – 22v – 1 – (each vol also has a special t p) – mf#6986 – us Wisconsin U Libr [802]

Goethes Werke / Trunz, Erich et al [comp] – Hamburger ausg. Hamburg: C Wegner, 1948-60 [mf ed 1993] – 14v on 4r – 1 – (incl bibl ref and ind) – mf#8627 – us Wisconsin U Libr [802]

Goethes werke / Goethe, Johann Wolfgang von – Stuttgart: J G Cotta. 36v. 1866-68 (mf ed 1990) – 6r – 1 – us Wisconsin U Libr [802]

Goethes werke : illustrirt von ersten deutschen kuenstlern / Goethe, Johann Wolfgang von – 3. aufl. Stuttgart: Deutsche Verlags-Anstalt, [between 1884-1894?] [mf ed 1996] – 5v (ill) – 1 – mf#9691 – us Wisconsin U Libr [802]

Goethes werke / ed by Kurz, Heinrich – Hildburghausen: Verlag des Bibliographischen Instituts, 1868-69 [mf ed 1989] – (= ser Bibliothek der deutschen nationalliteratur) – 12v – 1 – mf#6985 – us Wisconsin U Libr [802]

Goethes werke in sechs baenden / ed by Schmidt, Erich – Leipzig: Insel-Verlag, [1925?] [mf ed 1996] – 6v – 1 – (incl bibl ref and ind. int by by gustav roethe) – mf#9637 – us Wisconsin U Libr [802]

Goethes "werther" im urteil des 19. jahrhunderts : romantik bis naturalismus 1830-1880 / ed by Bickelmann, Ingeborg – [S.l: s.n.], 1937 Gelnhausen: Dissertationsdruckerei F W Kalbfleisch [mf ed 1990] – 68p – 1 – (incl bibl ref) – mf#7371 – us Wisconsin U Libr [430]

Goethes 'werther' in der niederlaendischen literatur : ein beitrag zur vergleichenden literaturgeschichte / Menne, Karl Johannes Joseph – Leipzig: M Hesse, 1905 [mf ed 1992] – (= ser Breslauer beitraege zur literaturgeschichte 6) – 94p – 1 – (incl bibl ref and ind) – mf#8014 reel 1 – us Wisconsin U Libr [410]

Goethes wilhelm meister und die aesthetische doctrin der aelteren romantik / Prodnigg, Heinrich – Graz: Verlag der Steierm. Landes-Oberrealschule, 1891 – 1 – us Wisconsin U Libr [430]

Goethes wissenschaftslehre in ihrer modernen tragweite / Barthel, Ernst – Bonn: F Cohen, 1922 [mf ed 1999] – 119p – 1 – (incl bibl ref) – mf#10172 – us Wisconsin U Libr [430]

Goethe-studien – Tokyo: Japanisch-Deutsches Kultur-Institut, [1933?] [mf ed 1993] – (= ser Japanisch-deutscher geistesaustausch 4) – 85p – 1 – mf#8637 – us Wisconsin U Libr [430]

Die goethezeit / Martini, Fritz – Stuttgart: C E Schwab, c1949 [mf ed 1993] – (= ser Ces-buecherei v11) – 177p – 1 – (incl bibl ref) – mf#8207 – us Wisconsin U Libr [430]

Die goethezeit in deutschland : sieben skizzen / Herse, Wilhelm – Hannover: Wissenschaftliche Verlagsanstalt, 1949 – 1 – us Wisconsin U Libr [430]
Goethezeit und katholizismus im werk ida hahn-hahns : ein beitrag zur geistesgeschichte des 19. jahrhunderts / Guntli, Lucie – Emsdetten, 1931 (mf ed 1992) – 1mf – 9 – €24.00 – 3-89349-043-4 – mf#DHS-AR 6 – gw Frankfurter [430]
Das goetterdekret ueber das abaton / Junker, H – Wien, 1913 – (= ser Kaiserliche Akademie Wissenschaften Wien, Philosophisch-historische Klasse. Denkschriften) – 2mf – 9 – (kaiserliche akademie wissenschaften wien. philosophisch-historische klasse. denkschriften v56) – mf#NE-20024 – ne IDC [956]
Die goetterfamilie : kosmopolitische komoedie / Dehmel, Richard – Berlin: S Fischer, 1921 [mf ed 1989] – 108p – 1 – mf#7170 – us Wisconsin U Libr [820]
Goetternamen : versuch einer lehre von der religioesen begriffsbildung / Usener, Hermann Karl – Bonn: F Cohen, 1896 – 1mf – 9 – 0-7905-6381-9 – (incl bibl ref) – mf#1988-2381 – us ATLA [210]
Die goetternamen in den babylonischen siegelcylinder-legenden / Krausz, Joseph & Hommel, Fritz – Leipzig: Otto Harrassowitz, 1911 – 1mf – 9 – 0-7905-1340-4 – (incl ind) – mf#1987-1340 – us ATLA [930]
Die goetterwelt der alten deutschen / Kaiser, W – Prag: Verlag des deutschen Vereines zur Verbreitung gemeinnuetziger Kenntnisse, 1880 – 1r – 1 – us Wisconsin U Libr [390]
Goetti und gotteli : berndeutsche novelle / Tavel, Rudolf von – 6. aufl. Bern: A Francke, 1931 [mf ed 1996] – (= ser Familie landorfer v3) – 332p/1pl (ill) – 1 – mf#7743 – us Wisconsin U Libr [830]
Die goettin laechelt : erzaehlung / Berger, Siegfried – Leipzig: P Reclam, 1942 [mf ed 1989] – 70p – 1 – mf#7010 – us Wisconsin U Libr [880]
Die goettin psyche in der hellenistischen und fruehchristlichen literatur / Reitzenstein, Richard – Heidelberg: C Winter, 1917 – (= ser Sitzungsberichte der Heidelberger Akademie der Wissenschaften, Philosophisch-Historische Klasse) – 1mf – 9 – 0-524-02230-5 – (incl bibl ref) – mf#1990-2904 – us ATLA [250]
Goettingen und seine umgebungen : ein taschenbuch vorzueglich fuer studirende und reisende / ed by Veldeck, Heinrich – Goettingen 1824 [mf ed Hildesheim 1995-98] – 2v on 6mf – 9 – €120.00 – 3-487-29532-6 – gw Olms [914]
Goettingensche wochenzeitung für stadt und land – Goettingen DE, 1848, 4 apr-4 oct – 1r – 1 – gw Mikrofilm [074]
Goettingensches wochenblatt – Goettingen DE, 1870 2 jan-4 nov [gaps] – 1r – 1 – (title varies: 1819: goettingensches wochenblatt; 2 jan 1868: goettinger tageblatt) – gw Mikrofilm [074]
Goettingensches wochenblatt *see* Goettingisches wochenblatt
Goettinger anzeigenblaetter – Goettingen DE, 1931 15 nov-1933 25 feb – 1r – 1 – gw Mikrofilm [074]
Goettinger Arbeitskreis *see* Ein ostpreussisches pfarrerleben
Goettinger blick – Goettingen DE, 1969 15 jan-1992 – 20r – 1 – gw Mikrofilm [074]
Goettinger deutscher bote – Goettingen DE, 1907 26 jan & 6 feb, 1 aug-31 dec – 1r – 1 – (with suppl) – gw Mikrofilm [074]
Goettinger echo am freitag – Goettingen DE, 1970 10 apr-2 oct – 1r – 1 – gw Mikrofilm [074]
Goettinger fanfare *see* Fanfare
Goettinger freie presse *see* Göttinger anzeiger 1881
Goettinger freizeit-magazin – Goettingen DE, 1983 30 jul-1991 12 jun – 4r – 1 – gw Mikrofilm [790]
Goettinger leben – Goettingen DE, 1925 1 oct-1936 28 jun – 2r – 1 – gw Mikrofilm [074]
Goettinger musenalmanach auf 1770 / ed by Redlich, Carl – Stuttgart: G J Goeschen, 1894 [mf ed 1993] – (= ser Deutsche litteraturdenkmale des 18. und 19. jahrhunderts 49-50) – 110p – 1 – (title of almanach as originally ed by heinrich christian boie: musenalmanach fuer das jahr 1770) – mf#8676 reel 4 – us Wisconsin U Libr [810]
Goettinger musenalmanach auf 1771 / ed by Redlich, Carl – Stuttgart: G J Goeschen, 1895 [mf ed 1993] – (= ser Deutsche litteraturdenkmale des 18. und 19. jahrhunderts 52-53, n f n2-3) – 100p – 1 – mf#8676 reel 5 – us Wisconsin U Libr [810]
Goettinger musenalmanach auf 1772 / ed by Redlich, Carl – Leipzig: G J Goeschen, 1897 – (= ser Deutsche litteraturdenkmale des 18. und 19. jahrhunderts 64-65, n f n14-15) – 122p – 1 – mf#8676 reel 5 – us Wisconsin U Libr [810]
Goettinger tageblatt *see*
– Goettingensches wochenblatt
– Goettingisches wochenblatt
Goettinger tageblatt 1889/1949 – Goettingen DE, 1889 6 aug-1905, 1949 27 oct-1975 – 1 – (with suppls. filmed by misc inst: 1976- [ca 8r/yr]) – gw Mikrofilm; gw Misc Inst [074]
Goettinger universitaets-zeitung – Goettingen DE, 1945 11 dec-1949 28 jan [gaps] – 1 – (cont: deutsche universitaets-zeitung) – uk British Libr Newspaper [378]
Goettinger woche – Goettingen DE, 1985 28 jun-1990 12 oct – 2r – 1 – gw Mikrofilm [074]
Goettinger zeitung – Goettingen DE, 1869 1 jul-27 dec, 1870 1 jul-31 dec, 1874 1 jul-1878, 1922 – 9r – 1 – (with suppls) – gw Mikrofilm [074]
Goettingische anzeigen von gelehrten sachen – Goettingen 1739-1801 [mf ed Hildesheim 1980] – 1327mf – 9 – diazo €4380.00 silver €4980.00 – (began as: goettingische zeitungen von gelehrten sachen, goettingen 1739-52 [mf ed 1980] 183mf) – gw Olms [430]
Goettingische anzeigen von gelehrten sachen – Goettingen 1753-1801 – (= ser Dz) – 214v+ind on 1144mf – 9 – €6864.00 – mf#k/n187 – gw Olms [500]
Goettingische gelehrte anzeigen : unter aufsicht der akademie der wissenschaften – Goettingen 1802-92 [mf ed Hildesheim 2003] – 1243mf – 9 – diazo €4980.00 silver €5800.00 – gw Olms [943]
Goettingische [gelehrte] beytraege zum nutzen und vergnuegen – Goettingen 1768 – (= ser Dz) – 1jg[=1-39st] on 2mf – 9 – €60.00 – mf#k/n5321 – gw Olms [870]
Goettingische philosophische bibliothek : worinnen nachrichten von den neuesten schriften der heutigen weltweisen...wie auch kurze untersuchungen mitgeteilt werden / ed by Windheim, Christian Ernst von – Hannover [v9: Nuernberg] 1749-57 – (= ser Dz) – 9v[zu je 6st] on 36mf – 9 – €360.00 – mf#k/n538 – gw Olms [100]
Goettingische policey-amts nachrichten – Goettingen DE, 1755 [gaps], 1756, 1757 [gaps] – 1r – 1 – gw Misc Inst [350]
Goettingische policey-amts-nachrichten : oder vermischte abhandlungen zum vortheil des nahrungsstandes aus allen theilen der oekonomischen wissenschaften / ed by Justi, Johann Heinrich Gottlieb von – Goettingen 1755-57 – (= ser Dz) – quartal 1-8 on 7mf – 9 – €140.00 – mf#k/n2868 – gw Olms [500]
Goettingische zeitungen von gelehrten sachen – Goettingen 1739-52 – (= ser Dz) – 13jge+ind on 183mf – 9 – €1098.00 – mf#k/n122 – gw Olms [500]
Goettingische zeitungen von gelehrten sachen *see* Goettingische anzeigen von gelehrten sachen
Goettingisches historisches magazin / ed by Meiners, Christoph et al – Hannover 1787-91 – (= ser Dz. historisch-geographische abt) – 8v on 41mf – 9 – €410.00 – mf#k/n1190 – gw Olms [943]
Goettingisches historisches magazin – Hannover DE, 1787-88, 1790-94 – 8r – 1 – (title varies: 1792: neues goettingisches historisches magazin) – gw Misc Inst [943]
Goettingisches magazin der wissenschaften und litteratur / ed by Lichtenberg, Georg Christoph et al – Goettingen 1780-85 – (= ser Dz) – 4jge on 20mf – 9 – €200.00 – mf#k/n346 – gw Olms [500]
Goettingisches magazin fuer industrie und armenpflege / ed by Wagemann, Ludwig Gerhard – Goettingen 1789-1803 – (= ser Dz) – 6v on 16mf – 9 – €160.00 – mf#k/n2757 – gw Olms [330]
Goettingisches philosophisches museum / ed by Buhle, Joh Gottlieb Gerh et al – Goettingen 1798-99 – (= ser Dz. abt philosophie) – 2v[zu je 2st] on 6mf – 9 – €120.00 – mf#k/n583 – gw Olms [100]
Goettingisches wochenblatt – Goettingen DE, 1870 2 jan-4 nov [gaps] – 1r – 1 – (title varies: 1819: goettingensches wochenblatt; 2 jan 1868: goettinger tageblatt. with suppls) – gw Mikrofilm [074]
Die goettliche komoedie : entwicklungsgeschichte und erklaerung / Vossler, Karl – Heidelberg: C Winter, 1907-1910 – 3mf – 9 – 0-524-01321-7 – (incl bibl ref) – mf#1990-2357 – us ATLA [440]
Das goettliche "noch nicht!" : ein beitrag zur lehre vom heiligen geist / Oettingen, Alexander von – Erlangen: A Deichert, 1895 – 1mf – 9 – 0-7905-9829-9 – mf#1989-1554 – us ATLA [240]
Der goettliche ruf : leben und werk von robert mayer: roman / Finckh, Ludwig – Muenchen: Deutscher Volksverlag 1931 [mf ed 1989] – 1r – 1 – (filmed with: der deutsche finckh) – mf#7241 – us Wisconsin U Libr [830]
Das goettliche selbstbewusstsein jesu : nach dem zeugnis der synoptiker / Steinbeck, Johannes – Leipzig: A Deichert, 1908 – 1mf – 9 – 0-524-05749-4 – (incl bibl ref) – mf#1992-0592 – us ATLA [220]
Die goettliche vorherbestimmung bei paulus und in der posidonianischen philosophie / Liechtenhan, R – Goettingen, 1922 – 3mf – 8 – €7.00 – ne Slangenburg [226]
Die goettliche vorsehung und das selbstleben der welt / Schmidt, Wilhelm – Berlin: Wiegandt & Grieben, 1887 – 1mf – 9 – 0-7905-7463-2 – mf#1989-0688 – us ATLA [210]
Die goettliche weisheit als persoenlichkeit im alten testamente / Goettsberger, Johann – 1.& 2. aufl. Muenster in Westfalen: Aschendorff 1919 [mf ed 2005] – (= ser Biblische zeitfragen; 9. folge 1/2) – 1r with other items – 1 – 0-524-10543-X – mf1074a – us ATLA [221]
Die goettliche zuvorersehung und erwaehlung in ihrer bedeutung fuer den heilsstand des einzelnen glaeubigen nach dem evangelium des paulus : eine biblisch-theologische untersuchung / Mueller, Karl – Halle (Saale): Max Niemeyer, 1892 – 1mf – 9 – 0-8370-9641-3 – (incl indes) – mf#1986-3641 – us ATLA [240]
Goettsberger, Johann *see*
– Adam und eva
– Barhebrus und seine scholien zur heiligen schrift
– Die goettliche weisheit als persoenlichkeit im alten testamente
Der goettweiger trojanerkrieg / ed by Koppitz, Alfred – Berlin: Weidmann, 1926 [mf ed 1993] – (= ser Deutsche texte des mittelalters 29) – xxviii/483p/[1]pl – 1 – (incl bibl ref and ind) – mf#8623 reel 6 – us Wisconsin U Libr [810]
Goetz, Delia *see* Neighbors to the south
Goetz, Hermann *see*
– The art and architecture of bikaner state
– The crisis of indian civilisation in the eighteenth and early nineteenth centuries
Goetz, Johann Nikolaus *see* Gedichte von johann nicolaus goetz
Goetz, K G *see* Der alte anfang und die urspruengliche form von cyprians schrift ad donatum
Goetz, Karl *see* Die heimstaetter
Goetz, Karl Gerold *see* Die todestage der apostel paulus und petrus
Goetz, Leopold Karl *see*
– Geschichte der slavenapostel konstantinus (kyrillus) und methodius
– Gosudarstvo i tserkov v drevnei rossii
– Ignatius von loyola und der protestantismus
– Leo 13.
– Der ultramontanismus als weltanschauung
Goetz von berlichingen : ein schauspiel / Goethe, Johann Wolfgang von – Stuttgart: J G Cotta, 1846 [mf ed 1992] – xcv/191p – 1 – (incl bibl ref. int and comm by a chuquet) – mf#7632 – us Wisconsin U Libr [820]
Goetz von berlichingen : schauspiel in 5 aufzuegen fuer schulgebrauch und selbstunterricht / Goethe, Johann Wolfgang von; ed by Frick, G – Leipzig: B G Teubner, 1905 [mf ed 1990] – (= ser Deutsche schulausgaben) – 138p – 1 – mf#7362 – us Wisconsin U Libr [820]
Goetz von berlichingen mit der eisernen hand : ein schauspiel / Goethe, Johann Wolfgang von – new ed. Paris: L Cerf, 1885 [mf ed 1991] – xcv/191p – 1 – (incl bibl ref. int and comm by a chuquet) – mf#7632 – us Wisconsin U Libr [820]
Goetz, Walther *see*
– Johann calvin
– Die quellen zur geschichte des hl. franz von assisi
Goetz, Wolfgang *see*
– Fuenfzig jahre goethe-gesellschaft
– Robert emmet
– Schiller
Goetze, A *see* Aus dem sozialen und politischen kampf
Goetze, Alfred *see* Fruehneuhochdeutsches glossar
Goetze, E *see* Hans Sachs
Goetze, Edmond *see* Saemmtliche fastnachtspiele
Goetze, Edmund *see*
– Der huernen seufrid
– Saemtliche fabeln und schwaenke
Goetze, Robert *see* H heine's 'buch der lieder' und sein verhaeltnis zum deutschen volkslied
Der goetzendienst goethe / Lutz, Joseph Maria – Muenchen: Drei Eulen-Verlag 1925 [mf ed 1990] – 1r – 1 – (incl bibl ref. filmed with: goethe als religioser charakter / wilhelm loew) – mf#7395 – us Wisconsin U Libr [430]
Goetzinger, Ernst *see* Joachim vadian
Goetzinger, Wilhelm *see* Schandau und seine umgebungen
Goez, J F de *see* Exercises d'imagination de differens caracteres et formes humaines...
Goezes streitschriften gegen lessing / ed by Schmidt, Erich – Stuttgart: G J Goeschen, 1893 [mf ed 1993] – (= ser Deutsche litteraturdenkmale des 18. und 19. jahrhunderts 43-45) – v/208p – 1 – mf#8676 reel 4 – us Wisconsin U Libr [430]
Goezmann, Louis V de *see* Histoire politique des grandes querelles entre l'empereur charles 5, et francois 1, roi de france
Goff, C C *see*
– Melon aphid, aphis gossypii glover
– Pepper weevil
– Relative susceptibility of some annual ornamentals to root-know
Goffin, Herbert J *see* At grips
Goffinet, Hippolyte *see* Cartulaire de l'abbaye d'orval
Goffstown, New Hampshire. Goffstown Baptist Church *see* Records
Goforth, Jonathan *see* Chinese christian general
Goforth, Rosalind *see* Chinese diamonds for the king of kings
Goforth, William Wallace *see* Report on the economic conditions of the canadian plumbing and heating industry
Gogarten, F *see* Fichte als religioeser denker
Gogebic iron tribune – Hurley WI. 1886 may 8-1887 oct 8, 1887 oct 15-1889 apr 20, 1889 apr 27-1890 oct 11, 1890 oct 18-1892 feb 28, 1892 mar 5-1893 dec 2 – 5r – 1 – (continued by: montreal river miner (hurley, wi: 1885); montreal river miner and gogebic iron tribune) – mf#923586 – us WHS [071]
Gogerly, Daniel John *see* Ceylon buddhism
Go-getter / Milwaukee Junior Association of Commerce – v2 n25, v2 n27-v3 n14 1922 oct 27, 1922 nov 10-1923 jun 29 – 1r – 1 – mf#690235 – us WHS [380]
Go-getter letter / U((nited States Army Recruiting Battalion (Kansas City, MO) – v4 n11 [1983 dec], v5 n5 [1984 may], v6 n10 [1985 dec], v7 n1,4-11 [1986 jan, apr-dec], v8 n3-4,6-7,11-12 [1987 mar-apr, jun-jul, nov-dec], v9 n4-9,11-12 [1988 apr-sep, nov-dec] – 1r – 1 – (continued by: bugle call (kansas city, mo)) – mf#1520634 – us WHS [355]
Gogol, Nikolai Vasilevich *see*
– Inspector general
– Sochineniia
Gogol, Nikolai Vasilievich *see* Gogolevskie teksty
Gogol, Nikolai Vasiltevich *see* Evenings on a farm near dikanka
Gogol', Nikolai Vesil'evich *see* Mertvye dushi
Gogolevskie teksty / Gogol, Nikolai Vasilievich – S. Petersburg: Izd. Otdniia russkago iazyka i slovesnosti Imp. akademii nauk, 1910 – 1 – us Wisconsin U Libr [460]
Gogolok, Kirstin *see* Die verstaendlichkeit in der informationsgesellschaft
Gogou, Constantin *see* 1re these: sur une inegalite lunaire a longue periode due a l'action perturbatrice de mars
Goguel, Maurice *see*
– L'apotre paul et jesus-christ
– Les chretiens et l'empire romain a l'epoque du nouveau testament
– L'eucharistie
– L'evangile de marc et ses rapports avec ceux de mathieu et de luc
– Introduction au nouveau testament
– La notion johannique de l'esprit et ses antecedents historiques
Goguelat, Francois de *see* Memoire de m le baron de goguelat, lieutenant-general, sur les evenemens relatifs au voyage de louis 16 a varennes
Gohiet, Francois *see*
– Conferences sur la question ouvriere
– A night with the philosophers
Gohin, Ferdinand *see* Transformations de la langue francaise pendant
Goias / Lisita Junior – Goiania, Brazil. 1965 – 1r – us UF Libraries [972]
Goias (Brazil) Governor *see* Relatorios dos presidentes, 1a republica, 1891-1929
Goias (Brazil) President *see* Relatorios dos presidentes, epoca do imperio, 1835-1889
Going public : the ipo reporter – New York NY 1990-93 – 1,5,9 – ISSN: 0278-0038 – mf#18457,02 – us UMI ProQuest [332]
Going to the sun / Lindsay, Vachel – New York, NY. 1923 – 1r – us UF Libraries [025]
Goings forth of jehovah in his trinity of persons, in acts of perso... / Hawker, Robert – London, England. 1821 – 1r – us UF Libraries [240]
Goings on – v1 n1-2 [1987 jun-jul] – 1r – 1 – mf#1567833 – us WHS [071]
Goinsiday telling the news – v1 n1-v4 n3, 1972 aug-1975 mar – 1r – 1 – mf#606181 – us WHS [071]
Goitein, Hirsch *see* Das problem der theodicee in der aelteren juedischen religionsphilosophie
Goitein, S D *see* Von den juden jemens
Gokhale, Balkrishna Govind *see* Buddhism and asoka
Gokhale, Gopal Krishna *see* Speeches of gopal krishna gokhale
Gol de letra / Pedrosa, Milton – Rio de Janeiro, Brazil. 1967 – 1r – us UF Libraries [972]
Golan, Ron *see* Effects of thymopentin on the responses of hypothalamic-pituitary-adrenal axis to a high intensity dynamic exercise protocol
Golash, Deirdre *see* The bail reform act of 1984
Golat-teman / Tabib, Avraham – Tel-Aviv, Israel. 1931 – 1r – us UF Libraries [939]
Golberry, S M X *see* Fragmens d'un voyage en afrique

Golbery, Silvain *see*
- Fragmens d'un voyage en afrique
- Reise nach senegal
- Travels in africa

Golborne reporter and star – [NW England] Golborne 1967-68, 1972-11 jul 1979 – 1 – uk MLA; uk Newsplan [072]

[Golconda-] nevada miner – NV. feb-sep 1902 [wkly] – 2r – 1 – $120.00 – mf#U04542 – us Library Micro [071]

[Golconda-] the news – NV. jul-dec 1899 [wkly] – 1r – 1 – $60.00 – mf#U04543 – us Library Micro [071]

Gold : aus der goldgraeberzeit kaliforniens / Gerstaecker, Friedrich; ed by Menny, Rudolf – Reutlingen: Ensslin & Laiblin, [19–?] (mf ed 1990) – 1r – 1 – (filmed with: die regulatoren in arkansas) – us Wisconsin U Libr [430]

Gold : roman aus den minenfeldern kaliforniens / Gerstaecker, Friedrich – Wilhelmshaven: Hera-Verlag, 1950 (mf ed 1990) – 1r – 1 – (filmed with: die regulatoren in arkansas) – us Wisconsin U Libr [830]

Gold! – v1 n1-n18, 1969-77 summer – 1r – 1 – mf#1112075 – us WHS [071]

Gold am pazifik : eine erzaehlung aus kaliforniens grossen tagen / Eberle, Josef – Stuttgart: Silberburg, 1935 [mf ed 1989] – 214p – 1 – mf#7193 – us Wisconsin U Libr [830]

Gold and incense : a west country story / Pearse, Mark Guy – New York: Eaton & Mains, [c1895] Beltsville, Maryland: NCR Corp, 1978 (1mf); Evanston: American Theol Lib Assoc, 1984 (1mf) – 9 – 0-8370-0838-7 – mf#1984-4232 – us ATLA [975]

Gold and the gospel in mashonaland, 1888 / Knight-Bruce, George Wyndham Hamilton – London, England. 1949 – 1r – us UF Libraries [960]

Gold and the south african economy / Katzen, Leo – Cape Town, South Africa. 1964 – 1r – us UF Libraries [330]

Gold beach gazette – Ellensburg OR: W Sutton, [wkly] – 1 – (ceased in 1895; cont by: gazette (wedderburn, or)) – us Oregon Lib [071]

Gold beach globe – Gold Beach OR: I N Muncy, [wkly] – 1 – (ceased in 1916. absorbed by: gold beach reporter (-1923)) – us Oregon Lib [071]

Gold beach reporter – Gold Beach OR: Reporter Pub Co, -1923 [wkly] – 1 – (absorbed gold beach globe (-1916); cont by: curry county reporter (1923-). 1917-18 & 1921-22 incl newspaper publ during school terms by gold beach school students. 1922-23 by brookings school students. 1921-23 incl newspaper devoted to brookings, or and southern curry county) – us Oregon Lib [071]

Gold beach reporter *see* Gold beach globe

[Gold center-] news – NV. 29 sep 1906; 12 jan 1907 [wkly] – 1r – 1 – $60.00 – mf#U04544 – us Library Micro [071]

Gold, Charles E *see* A study of the gospel song

[Gold circle-] miner – NV. 11 apr 1908 [wkly] – 1r – 1 – $60.00 – mf#U04545 – us Library Micro [071]

[Gold circle-] news – NV. 26 sep 1908 [wkly] – 1r – 1 – $60.00 – mf#U04546 – us Library Micro [071]

[Gold circle-] porcupine – NV. 20 may 1914 [wkly] – 1r – 1 – $60.00 – mf#U04547 – us Library Micro [071]

Gold coast aborigine – Cape Coast. Ghana. -w. Jan 1898-Jun 1902. (48 ft) – 1 – uk British Libr Newspaper [072]

Gold coast africa : [catalog] – 1998 winter – 1r – 1 – mf#4888664 – us WHS [071]

Gold coast and british togoland, annual departmental reports relating to the... 1843-1956 – (= ser Annual departmental reports relating to african countries prior to independence) – 110r – 1 – (with guide. int by d c dorward) – mf#97004 – uk Microform Academic [960]

The gold coast and the fantis : a complete compendium for miners, traders, and students of native life / Foot, Lionel R & Jones, T F E – London: Gold Coast Globe Pub Co, 1903 – us CRL [960]

Gold coast assize – Cape Coast. Ghana. -m. Nov 1883-Feb 1884. (3 ft) – 1 – uk British Libr Newspaper [072]

Gold coast chronicle – Accra. Ghana. -f. Jun 1894-Dec 1901 – 1r – 1 – uk British Libr Newspaper [079]

Gold Coast. (Colony). Legislative Council *see* Minutes

Gold coast daily mail – Accra: Amalgamated Press Ltd, jul 18, 21, 23, 1956 – us CRL [079]

Gold coast echo – Cape Coast. Ghana. -irr. Jan, Jul-Dec 1888. (12 ft) – 1 – uk British Libr Newspaper [072]

Gold coast express – Accra. Ghana. -w. Jul-Dec 1897, Sep 1899-May 1900. (1 reel) – 1 – uk British Libr Newspaper [079]

Gold coast free press – Accra. Ghana. -m. Aug-Nov 1899. (6 ft) – 1 – uk British Libr Newspaper [079]

Gold coast, government publications relating to the... 1846-1957 – (= ser Government publications relating to african countries prior to independence) – 151r – 1 – (with int by r j a r rathbone) – mf#96929 – uk Microform Academic [960]

Gold coast independent – Accra. Ghana. -w. Jun 1922-Jan 1940, Feb-Mar 1941, Mar 1942-Sep 1947, Jun 1948. (Imperfect). (32 reels) – 1 – uk British Libr Newspaper [079]

The gold coast independent – Accra: Independent press Ltd, apr 9, aug 9, 21, sep 11 1954 – us CRL [079]

Gold coast independenti – Accra. Ghana. -w. 14, 21 Dec 1895, 6 Jun 1896-19 Feb, 30 Sep 1898. (Imperfect). (36 ft) – 1 – uk British Libr Newspaper [079]

Gold coast leader – Cape Coast. Ghana. -w. Jun 1902-Aug 1929. (9 reels) – 1 – uk British Libr Newspaper [072]

Gold coast methodist – Cape Coast. Ghana. -m. Dec 1886. (2 ft) – 1 – uk British Libr Newspaper [072]

Gold coast methodist times – Cape Coast. Ghana. -m. Jul-Dec 1897. (9 ft) – 1 – uk British Libr Newspaper [072]

Gold coast nation – Cape Coast. Ghana. -w. Mar 1912-Dec 1920. (Imperfect). (3 reels) – 1 – uk British Libr Newspaper [072]

Gold coast news – Cape Coast. Ghana. -w. Mar-Aug 1885. (Imperfect). (8 ft) – 1 – uk British Libr Newspaper [072]

The gold coast observer – Cape Coast. Ghana. may 16-23, 30; June 27; July 4, 11, 1941 – 1 – us NY Public [079]

Gold coast observer and weekly advertiser – Cape Coast: F G Mensah, sep 19-27 1952; apr 9 1954 – us CRL [079]

Gold coast people – Cape Coast. Ghana. -w. 26 Oct 1891. (4 ft) – 1 – uk British Libr Newspaper [072]

Gold coast pioneer – Accra. Ghana. -m. Feb 1921 – 1 – uk British Libr Newspaper [079]

Gold coast records of the united society for the propagation of the gospel, 1753-1951 – 12r – 1 – mf#96923 – uk Microform Academic [220]

Gold coast review – Accra, Govt Printer. v2-5 n2. 1926-31 – us CRL [079]

Gold coast spectator – Accra. Ghana. -w. Oct 1937-Sep 1938. (2 reels) – 1 – uk British Libr Newspaper [072]

Gold coast times – Cape Coast. Ghana. may 22; June 19, 26; July 3, 17; Aug. 7, 14; Sept. 11, 18, 25; Oct. 2, 9, 23; Nov. 6, 13, 20, 27, 1926 – 1 – us NY Public [079]

Gold coast times – Cape Coast. Ghana. -w. 28 Mar 1874-31 Mar 1875, 22 May-11 Sep 1875, 17 Nov 1877-25 Jan 1878, 11 Feb, 9, 30 Jul 1881-12 Feb 1885. (Imperfect). (1 reel) – 1 – uk British Libr Newspaper [072]

Gold coast times – Cape Coast. Ghana. -w. 4 Jan 1930-23 May 1936, 28 Nov 1936-6 Jul 1940. (6 reels) – 1 – uk British Libr Newspaper [072]

[Gold creek-] news – NV. 1896-97 [wkly] – 1r – 1 – $60.00 – mf#U04548 – us Library Micro [071]

Gold, diamonds and orchids / Lavallre, William Johanne – New York, NY. 1935 – 1r – us UF Libraries [972]

The gold dollar – Austin TX: Jacob Fontaine [mf ed 2004] – aug 1876 [complete] on 1r – 1 – (began in 1876?) – mf#2004-s109 – us ATLA [071]

The gold fields mercury – Pilgrim's Rest SA, 30 jun 1876-7 feb 1878 – 1r – 1 – sa National [960]

Gold fields of alaska : klondike gold fields and northwest territory – [S.I.]: North American Transportation and Trading Co, [1897?] [mf ed 1981] – 2mf – 9 – mf#15730 – cn Canadiana [622]

Gold fields of alaska : klondike gold fields and northwest territory – [S.I: North American Transportation & Trading Co?, 1898?] [mf ed 1981] – 2mf – 9 – mf#15731 – cn Canadiana [622]

The gold fields of canada : a paper read before the literary and historical society of quebec, 18th november, 1863 / Douglas, James – Quebec?: Hunter, Rose, 1863 – 1mf – 9 – mf#23082 – cn Canadiana [622]

The gold fields of new ontario / Baelz, Walter – [Ontario?: s.n, 1912?] – 1mf – 9 – 0-665-87890-7 – (trans fr german "zeitschrift f prakt, geologie" by t l walker) – mf#87890 – cn Canadiana [622]

The gold fields of new ontario : comprising the lake of the woods, rainy lake, seine river, the manitou and michipicoton districts... – S.l: CPR?, 1899 – 1mf – 9 – mf#00456 – cn Canadiana [622]

Gold fields of st domingo / Courtney, Wilshire S – New York, NY. 1860 – 1r – us UF Libraries [550]

The gold fields of the klondike : fortune seekers' guide to the yukon region of alaska and british america: the story as told by ladue, berry, phiscator and other gold finders / Leonard, John William – London: T F Unwin; Chicago: A N Marquis, c1897 [mf ed 1980] – 3mf – 9 – mf#15512 – cn Canadiana [622]

Gold fields of the klondike and the wonders of alaska : a masterly and fascinating description of the newly-discovered gold mines, how they were found, how worked... / Ingersoll, Ernest – [S.I.]: Edgewood Pub Co, c1897 [mf ed 1982] – 6mf – 9 – (int by henry w elliott) – mf#15271 – cn Canadiana [622]

Gold fields of the klondike and the wonders of alaska : a masterly and fascinating description of the newly-discovered gold mines, how they were found, how worked... / Ingersoll, Ernest – slightly enl ed. [S.I.]: Edgewood Pub Co, c1897 [mf ed 1982] – 6mf – 9 – 0-665-16472-6 – (int by henry w elliott) – mf#16472 – cn Canadiana [622]

Gold fields of the klondike and the wonders of alaska : a masterly and fascinating description of the newly-discovered gold mines, how they were found, how worked... / Ingersoll, Ernest – St John, NB: Earle, c1897 [mf ed 1984] – 6mf – 9 – 0-665-16471-8 – (int by henry w elliott) – mf#16471 – cn Canadiana [622]

The gold fields of the world : our knowledge of them and its application to the gold fields of canada / Anderson, William James [comp] – Quebec?: G & G E Desbarats, 1864 – 1mf – 9 – mf#23185 – cn Canadiana [622]

Gold fields revisited : being further glimpses of the gold fields / Mathers, Edward Peter – Pretoria, South Africa. 1970 – 1r – us UF Libraries [550]

Gold flower – Minneapolis, Saint Paul MN. 1971 nov-1979 apr/may – 1r – 1 – mf#491016 – us WHS [071]

Gold hill news – Gold Hill OR: News Print Co, 1897- [wkly] – 1 – us Oregon Lib [071]

Gold hill nugget – Gold Hill OR: J Campbell, 1949 [wkly] – 1 – (merged with: rogue river record and: grants pass bulletin (1927-49) to form: bulletin (grants pass or)) – us Oregon Lib [071]

Gold in canada : the chaudiere valley and its mineral wealth – Quebec: Morning Chronicle, 1880 – 2mf – 9 – mf#00586 – cn Canadiana [622]

Gold in the east : being observations on a practical method of establishing a gold currency in india, and its influence on the trade and finance of that country / Daniell, Clarmont – [London], [1879] – (= ser 19th c books on british colonization) – 1mf – 9 – mf#1.1.885 – uk Chadwyck [332]

Gold, irving, scrapbook – Cleveland, Cuyahoga, OH. 1930-36 – 1r – 1 – (scrapbook, consisting of posters, clippings, photographs, telegrams, and sheet music relating to gold's early career as a dance band musician and booking agent in around cleveland) – us Western Res [920]

The gold measures of nova scotia and deep mining / Faribault, Eugene Rodolphe; ed by Nova Scotia. Mining Society – Halifax, NS: The Society, 1900? – 1mf – 9 – (together with other papers bearing upon nova scotia gold mines) – mf#34046 – cn Canadiana [622]

Gold, Michael *see* Jews without money

Gold or laurel : the olympic tradition in a changing world – Emory University, 1990 – (= ser United states olympic academy 14) – 4mf – 9 – $16.00 – us Kinesology [790]

Gold paved the way / Cartwright, Alan Patrick – London, England. 1967 – 1r – us UF Libraries [960]

Gold, Pleasant Daniel *see*
- History of volusia county, florida
- In florida's dawn

Gold rush gazette *see* Clear creek county miscellaneous newspapers

Gold, sport, and coffee planting in mysore : with chapters on coffee planting in coorg, the mysore representative assembly, the indian congress, caste, and the indian silver question / Elliot, Robert Henry – Westminster, 1894 – (= ser 19th c books on british colonization) – 6mf – 9 – mf#1.1.8134 – uk Chadwyck [954]

Gold spring diary : diary of john jolly in goldfields / Jolly, John – 1854-55 – 1r – 1 – $50.00 – mf#B40014 – us Library Micro [917]

Gold star dust : official publication of gold star post – 1978 feb,jun-1985 dec – 1r – 1 – mf#1012647 – us WHS [071]

Gold star mother / American Gold Star Mothers, Inc – v21 n11-v36 n4 [1968 nov-1983 jul/aug] – 1r – 1 – mf#572098 – us WHS [071]

Gold, Steven N *see* Journal of trauma practice

Gold und reiherfedern : gestalten und erlebnisse aus ecuador / Uden, Horst – Salzburg: Verlag "Das Bergland-Buch", 1942 – 1r – 1 – us Wisconsin U Libr [390]

Goldammer, Kurt *see* Novalis und die welt des ostens

Goldammer, Peter *see* Begegnungen und wuerdigungen

The gold-bearing veins of bag bay, near lake of the woods / McKellar, Peter – S.l: s.n, 1899? – 1mf – 9 – mf#09570 – cn Canadiana [622]

Goldbeck, Eduard *see* Der rote leutnant

Goldberg, Efraim *see* Yerushe fun doyre's

Goldberg, Isaac *see*
- Havelock ellis: a biographical and critical survey
- Literatura hispanoamericana
- Man mencken

Goldberg, K N *see* Otdelenie tserkvi ot gosudarstva i shkoly ot tserkvi

Goldberg, Leah *see* Sifrut yafah 'olamit be-targumeha le-'ivrit

Goldberger, Michael *see* The effects of contextual interference and three levels of difficulty on the acquisition, retention, and transfer of hockey striking skills by second grade children

Goldberger, Philipp *see* Die allegorie in ihrer exegetischen anwendung bei maimonides

Goldblatt, I *see* Mandated territory of south west africa

Goldblum, N *see* Recommendation for a national laboratory service in ethiopia

Goldelman, Salomon *see* In goles bay di ukrainer

Goldelse : roman / Marlitt, Eugenie [pseud] – Stuttgart-Berlin: Union Deutsche Verlagsgesellschaft [1919] [mf ed 1978] – (= ser Romane und novellen 8) – 2r – 1 – mf#film mas c376 – us Harvard [830]

Golden : the newsletter of the national caucus and center on black aged, inc / National Caucus and Center on Black Aged [US] – 1987 winter-spring – 1r – 1 – (cont: golden page; cont by: golden page [1992]) – mf#4866975 – us WHS [305]

Golden age / Cartwright, Alan Patrick – Cape Town, South Africa. 1968 – 1r – us UF Libraries [960]

The golden age : or, the new and happy, commercial, all-abundant, and pacific era of mankind! / Edwards, George – [London, 1815] – (= ser 19th c economics) – 1mf – 9 – mf#1.1.111 – uk Chadwyck [330]

The golden age of german literature / Lohan, Robert – 2nd rev ed. New York: F Ungar, 1948, c1945 [mf ed 1993] – 1r – 1 – (incl bibl ref and ind) – us Wisconsin U Libr [430]

The golden aphroditis and granges garden of verse / Grange, John – 1577 – 9 – us Scholars Facs [810]

Golden bay argus – Collingwood, NZ. 1833-may 1911 – 16r – 1 – mf#50.3 – nz Nat Libr [079]

Golden belt – Red Cloud, NE: Willcox & McMillan. v1 n1. apr 28 1893 (wkly) [mf ed may 5 1893-oct 23 1896 (gaps) filmed 1968-74] – 2r – 1 – us NE Hist [071]

The golden boat / Tagore, Rabindranath – London: George Allen & Unwin Ltd, 1932 – (= ser Samp: indian books) – (trans by bhabani bhattacharya) – us CRL [490]

Golden bough / Frazer, James George – New York, NY. 1922 – 1r – us UF Libraries [420]

Golden bough / Frazer, James George – New York, NY. 1930, 1922 – 1r – us UF Libraries [420]

The golden bough, vol 12 : bibliography and general index: a study in magic and religion / Frazer, James George – 3rd rev enl ed. London: Macmillan 1915 [mf ed 1992] – 2mf – 9 – 0-524-05847-4 – mf#1990-3511 – us ATLA [230]

The golden breath : studies in five poets of the new india / Anand, Mulk Raj – London: John Murray, 1933 – (= ser Samp: indian books) – us CRL [490]

Golden caribbean / Blaney, Henry Robertson – Boston, MA. 1900 – 1r – us UF Libraries [972]

A golden chaine : or, the description of theologie, containing the order of the causes of salvation and damnation, according to gods word / Perkins, W – Cambridge: Iohn Legat, 1600 – 20mf – 9 – mf#PW-78 – ne IDC [240]

Golden complex / Dodd, Lee Wilson – New York, NY. 1927 – 1r – us UF Libraries [025]

Golden crescent *see* Miscellaneous newspapers of teller county

Golden days : a tale of girls' school life in germany / Adams-Acton, Marion (Hamilton) & Hering, Jeanie – [London], Paris, New York: Cassell, Petter, & Galpin, 1873 – (= ser 19th c children's literature) – 4mf – 9 – mf#6.1.38 – uk Chadwyck [305]

Golden days for boys and girls – Philadelphia. 6 mar 1880-28 Nov 1885; 13 Mar 1886-13 Nov 1897; 18 Nov 1899-8 Nov 1902.-w. 21 reels – 1 – uk British Libr Newspaper [800]

Golden Gate Baptist Theological Seminary *see* Bulletins, catalogs and other material

Golden gate bridge : newspaper clippings – San Francisco, CA. 1932-58 – 1r – 1 – $50.00 – mf#B40307 – us Library Micro [978]

Golden gate law review *see* Golden gate university law review

The golden gate; sparkling jewels; gospel trumpet; shining pearls / Shaw, Knowles – 1 – us Southern Baptist [242]

Golden gate university law review – v1-31. 1971-2001 – 5,6,9 – $549.00 set – (v1-14 1971-984 on reel $203. v15-31 1985-2001 on mf $347. title varies: v1-5 1971-75 as golden gate law review) – ISSN: 0363-0307 – mf#102991 – us Hein [340]
Golden gate university. school of law. golden gate university law review – San Francisco CA 1976+ – 1,5,9 – ISSN: 0363-0307 – mf#10614.01 – us UMI ProQuest [378]
Golden, George Fuller *see* My lady vaudeville and her white rats
Golden gleanings of poetry and song with choice selections of prose : containing the best productions of the most celebrated authors of all ages and countries / Northrop, Henry Davenport [comp] – St John, NB: R A H Morrow, [1897?] [mf ed 1983] – 8mf – 9 – (publ also under title: the brightest gems of poetry, prose and song) – mf#32672 – cn Canadiana [810]
Golden globe *see* Jefferson county miscellaneous newspapers
Golden griffon : published inthe interest of 108th div[ision] [tng] personnel – v6 n3 [1983 fall], v7 n1-3 [1984 spring-fall], v8 n4 [1985 winter], v9 n1 [1986 spring] – 1r – 1 – (continued by: griffon (charlotte, nc)) – mf#1057322 – us WHS [071]
Golden harvest / Dean, Nina Oliver – Ocala, FL. 1937 – 1r – us UF Libraries [978]
Golden hind – London, 1922-24 [mf ed Chadwyck-Healey] – (= ser Art periodicals on microform) – 1r – 1 – uk Chadwyck [760]
Golden hours : a magazine for sunday reading – La Salle IL 1866-84 – 1 – mf#4720 – us UMI ProQuest [071]
Golden, Jane E et al *see* Responses to graded exercise testing in normal children and adolescents with high and low left ventricular mass
Golden jubilee of the general association of colored baptists in kentucky : the story of 50 years' work from 1865-1915, including many photos and sketches / ed by Parrish, Charles Henry – Louisville, KY: Mayes Print Co, 1915 [mf ed 1993] – 1mf – 9 – 0-524-08494-7 – mf#1993-3139 – us ATLA [242]
Golden jubilee of the montreal trades and labor council, 1897-1947 : commemorative banquet held october 24th 1947...montreal – [Montreal?: s.n, 1947?] (mf ed 1998) – 1mf – 9 – mf#SEM105P3025 – cn Bibl Nat [380]
Golden jubilee of the reverend fathers dowd and toupin : with historical sketch of irish community of montreal, biographies of pastors of "recollet" and "st patrick's", etc / ed by Curran, John Joseph – Montreal?: J Lovell, 1887 – 2mf – 9 – (incl ind) – mf#02243 – cn Canadiana [920]
Golden land / De Onis, Harriet – New York, NY. 1948 – 1r – us UF Libraries [972]
Golden links / United Steelworkers of America – v8 n9-v11 n4 [1976 aug-1982 may] – 1r – 1 – mf#615825 – us WHS [660]
Golden lodge news / United Steelworkers of America – [1975 jun-1987 mar] – 1r – 1 – (cont: golden lodge news [1941]) – mf#1330918 – us WHS [660]
Golden lotus / Hsiao-Hsiao-Sheng – London, England. v1-4. 1939 – 1r – us UF Libraries [480]
Golden lotus : a periodical on buddhism – Philadelphia: Golden Lotus Press, 1944-67 [mf ed 2001] – (= ser Christianity's encounter with world religions, 1850-1950) – 4r – 1 – mf#2001-s100 – us ATLA [280]
Golden Mean Society *see*
– Double eagle report
– Journal from the golden mean society
– Journal-report from the golden mean society
Golden page / National Caucus and Center on Black Aged (US) – v1 n5-v4 n6 [1977 mar-1981 apr/may] – 1r – 1 – mf#606290 – us WHS [305]
Golden republic / Bulpin, Thomas Victor – Cape Town, South Africa. 1953 – 1r – us UF Libraries [960]
Golden roots of the mother lode / Tuolumne County Genealogical Society – v1 n4/v2 n1-v6 n4/v7 n1 [1981 winter/1982 spring-1986 fall/1987 winter] – 1r – 1 – (cont: mother lode-ore) – mf#1582354 – us WHS [929]
Golden rule / United Society of Christian Endeavor – 1889 dec 12-1891 sep 24, missing/mutilated 1989 dec 26, 1890 feb 6-mar 6, v7 [1892 oct-1893 sep] – 2r – 1 – (continued by: christian endeavor world) – mf#1387003 – us WHS [240]
The golden rule – Vicksburg, MS: Golden Rule Pub Co. v2 n38. jan 27 1900 (wkly) [mf ed 1947] – (= ser Negro Newspapers on Microfilm) – 1r – 1 – us L of C Photodup [071]
Golden star – British Columbia, CN. oct 1963-dec 1980 – 24r – 1 – cn Commonwealth Imaging [071]
Golden thoughts on mother, home and heaven : from poetic and prose literature of all ages and all lands – rev enl ed. Montreal: A J Cleveland, 1882 [mf ed 1993] – 5mf (ill) – 9 – 0-665-91442-3 – mf#91442 – cn Canadiana [230]
The golden threshold / Naidu, Sarojini – London: William Heinemann, 1914 – (= ser Samp: indian books) – (int by arthur symons) – us CRL [490]
The golden trade : or, a discovery of the river gambra, and the golden trade of the ethiopians / Jobson, R – London, 1933 – 5mf – 9 – mf#A-319 – ne IDC [916]
The golden treasury of indian literature / ed by Shah, Sirdar Ikbal Ali – London: Sampson Low, Marston & Co [1938] [mf ed 1996] – (= ser Samp: indian books) – 1r – 1 – (filmed with other items) – mf#mf-10448 r189 – us CRL [490]
Golden vials full of odours / Blakely, John – Glasgow, Scotland. 1861 – 1r – us UF Libraries [240]
Golden weekly globe *see* Jefferson county miscellaneous newspapers
The golden years 1929-41 *see* Kine weekly
Goldenberg, Marni A *see* Understanding the benefits of ropes course experiences using means-end analysis
Goldene buch – 1929 jan-1931 sep, 1931 oct-1934 apr, 1934 may-1938 jan, 1938 feb-1939 nov, 1939 dec-1942 may, 1942 jun-1944 nov, 1944 dec-1946 nov, 1946 dec-1948 feb – 8r – 1 – mf#568475 – us WHS [071]
Die goldene bulle karls 4. vom jahre 1356 (mgh leges 4:11.bd) – (= ser Monumenta germaniae historica leges 4. fontes iuris germanici antiqui in usum scholarum separatim editi (mgh leges 4)) – €7.00 – ne Slangenburg [342]
Der goldene chersones / Bishop, [J F] – Leipzig: F Hirt & Sohn, 1884 – 5mf – 9 – mf#SE-20203 – ne IDC [915]
Das goldene erbe : roman / Finckh, Ludwig – Muenchen: Deutscher Volksverlag 1944, c1943 [mf ed 1989] – 1r – 1 – (filmed with: der deutsche finckh) – mf#7241 – us Wisconsin U Libr [830]
Die goldene gans : weihnachtsmaerchenspiel in 5 bildern / Daehnhardt, Oskar – Leipzig: B Liebisch, 1910 [mf ed 1989] – 787p – 1 – mf#7169 – us Wisconsin U Libr [390]
Der goldene kelch *see* Aus nacht zum licht
Das goldene vlies : dramatisches gedicht in drei abteilungen / Grillparzer, Franz – school ed. Stuttgart, Berlin: Cotta 1906 [mf ed 1990] – 1r – 1 – (int & ann by adolph lichtenheld. filmed with: das kind des torfmachers / friedrich griese) – mf#2688p – us Wisconsin U Libr [810]
Die goldene zeit neue geschichten aus der heimath / Seidel, Heinrich – Leipzig: A G Liebeskind 1889 – 1 – mf#film mas c419 – us Harvard [830]
Goldenseal / Science and Culture Center (Charleston, W VA) – (1977 apr-1982 summer], [1982 fall-1985 winter], [1986 spring-1988 winter] – 3r – 1 – mf#609579 – us WHS [071]
Golder, Christian *see* History of the deaconess movement in the christian church
Golder, F A *see* Bering's voyages
[Goldfield-] chronicle – NV. 1907-09 [daily; wkly] – 4r – 1 – $240.00 – mf#U04549 – us Library Micro [071]
[Goldfield-] daily news – NV. 1909-10; 8 mar 1911 [daily] – 2r – 1 – $120.00 – mf#U04550 – us Library Micro [071]
[Goldfield-] daily tribune – NV. 1906-30 [daily] – 27r – 1 – $1620.00 – mf#U04551 – us Library Micro [071]
[Goldfield-] enterprise – NV. feb-nov 1958 [wkly] – 1r – 1 – $60.00 – mf#U04552 – us Library Micro [071]
[Goldfield-] goldfield news – NV. 1904-10; 1946-47 (broken series) – 3r – 1 – $180.00 – mf#U04556 – us Library Micro [071]
[Goldfield-] gossip – NV. 1906-07 – 2r – 1 – $120.00 – (aka: goldfield gossip weekly) – mf#U04553 – us Library Micro [071]
Goldfield gossip weekly *see* [Goldfield-] gossip
[Goldfield-] nevada mining bulletin – NV. dec 1906 – 1r – 1 – $60.00 – mf#U04554 – us Library Micro [071]
[Goldfield-] nevada mining news – NV. jun 1906 – 1r – 1 – $60.00 – mf#U04555 – us Library Micro [071]
[Goldfield-] nevada workman – NV. 7 sep 1907 – 1r – 1 – $60.00 – mf#U04566 – us Library Micro [071]
Goldfield news – Barberton. SA. 1888-1947 – 2r – 1 – sa National [079]
[Goldfield-] news and beatty bulletin – NV. 1947-56 [wkly] – 4r – 1 – $300.00 – mf#U04557 – us Library Micro [071]
[Goldfield-] news and weekly tribune – NV. 1914-46 [wkly] – 7r – 1 – $420.00 – mf#U04558 – us Library Micro [071]
[Goldfield-] sporting bulletin – NV. 29 aug 1906 – 1r – 1 – $60.00 – mf#U04561 – us Library Micro [071]
[Goldfield-] sun – NV. 1905-06 [daily] – 1r – 1 – $60.00 – mf#U04562 – us Library Micro [071]
[Goldfield-] the post – NV. 25 may 1912 – 1r – 1 – $60.00 – mf#U04559 – us Library Micro [071]
[Goldfield-] the review – NV. 1905-09 [wkly] – 5r – 1 – $300.00 – mf#U04560 – us Library Micro [071]
Goldfield times – Barberton. South Africa. 1886-88 – 2r – 1 – sa National [079]
[Goldfield-] vigilant – NV. 30 jan 1905 [wkly] – 1r – 1 – $60.00 – mf#U04563 – us Library Micro [071]
[Goldfield-] weekly market letter – NV. 1909 – 1r – 1 – $60.00 – mf#U04564 – us Library Micro [071]
[Goldfield-] weekly news – NV. 1905-1910; 1929; 1931 [wkly] – 7r – 1 – $420.00 – mf#U04565 – us Library Micro [071]
The goldfields of mashonaland / Sawyer, Arthur Robert – Manchester: J. Heywood, 1894. 99p. fold. maps – 1 – us Wisconsin U Libr [622]
Goldflies, David G *see* The relationship between habitual diet and blood pressure among black and white adults
Gold-foil : hammered from popular proverbs / Holland, Josiah Gilbert – New York: Charles Scribner, 1883, c1881 – 1mf – 9 – 0-8370-7513-0 – mf#1986-1513 – us ATLA [240]
Goldfuss, G A *see* Petrefacta germaniae...
Goldhammer, Leopold *see* Reichsbote
Goldhan, A H *see* Ueber die einwirkung des goethischen werthers und wilhelm meisters auf die entwicklung edward bulwers
[Goldhill-] daily news – NV. 1867-82 (scats) – 1r – 1 – $60.00 – (1978 nevada day edition) – mf#U04567 – us Library Micro [071]
Goldie and McCulloch Co Ltd *see* Banker's safes
Goldie and mcculloch manufacturers of steam engines, boilers, mill gearing, and furnishings of every description : also the well-known wheelock automatic engines, high pressure, compound and compound condensing – Galt, Ont: [s.n, 188-?] (Toronto: Bingham & Webber) – 1mf – 9 – 0-665-90256-5 – mf#90256 – cn Canadiana [621]
Goldie, Fay *see* Lost city of the kalahari
Goldie, Francis *see*
– The first christian mission to the great mogul
– Life of saint aloysius gonzaga
– The life of st. alonso rodriguez
– The life of the blessed antony baldinucci
Goldie, T W *see* The mosaic account of the creation of the world and the noachian deluge
Goldin, L *see* Hantbukh fun der velt-literatur
Golding, Charles *see* Suffolk scarce tracts, 1595 to 1684
Goldington – (= ser Bedfordshire parish register series) – 1mf – 9 – £3.00 – uk BedsFHS [929]
Goldman, Cheryl L *see* An assessment of the relationship between participation in intercollegiate athletics and the dynamics of romantic relationships
Goldman, E *see* Anarkhizm
Goldman, Emma *see* Emma goldman papers
Goldman, Isaac *see* Shivat tsiyon
Goldman, Jerry *see* The seventh circuit preappeal program
Goldman, L I *see* Politicheskie protsessy v rossii (1901-1917)
Goldman, Norma *see* The big apple
Goldmann, Karl *see* Richard voss
Goldmann, N *see* La uvelle fortification
Goldmann, Nachum *see* Freie zionistische blaetter
Goldmark, Carl *see* Die koenigen von saba
Goldmine – Iola WI 1993+ – 1,5,9 – ISSN: 1055-2685 – mf#20539,02 – us UMI ProQuest [622]
Goldmining leases for the northern *see* Northern territory gold mining/mineral lands/pastoral leases/pearling
Der goldne topf : ein maerchen aus der neuen zeit / Hoffmann, E T A [Ernst Theodor Amadeus] – Wiesbaden: Herta Hartmanshenn [1947?] [mf ed 1994] – 1r – 1 – (filmed with: hoelderlin: choix de textes, bibliographie, dessins...) – mf#3641p – us Wisconsin U Libr [390]
Goldondrinas / Grau Archilla, Raul – San Juan, Puerto Rico. 1956 – 1r – us UF Libraries [972]
Goldpanner – Fairbanks AK. 1981 may-1983 sep, 1983 oct-1986 mar – 2r – 1 – mf#1045638 – us WHS [071]
Goldrush gazette *see* Jacksonville goldrush gazette
Goldsack, William *see* Muhammad and the bible
Goldschmidt, A M *see* Gedenkblatter zur erinnerung an rabbiner dr a m goldschmidt
Goldschmidt, L *see*
– Babylonische talmud
– Bibliotheca aethiopica
Goldschmidt, Lazarus *see* Sefer yetsirah
Goldschmidt, Meir *see*
– Evrei
– Livs erindringer og resultater
Goldschmidt, Moritz *see*
– Sone von nausay
Goldschmit, Rudolf K *see* Heidelberg als stoff und motiv der deutschen dichtung
Goldshtein, M L *see* Pechat pered sudom
Goldshtein, ML *see* Rechi i stat'i
Goldsmith, F H *see* John ainsworth, pioneer kenya administrator, 1864-1946
Goldsmith, Lewis *see*
– Geheime geschichte des neuen franzoesischen hofes
– Historia secreta del gabinete de napoleon (anno 1811)
– Statistics of france
Goldsmith, Oliver *see*
– Oliver goldsmith and thomas gray
– The poems of oliver goldsmith
– Vicar of wakefield
Goldsmiths' Alliance, Ltd *see* [Illustrated catalogue]
The goldsmiths'-kress library of economic literature – ongoing – units of ca 50r ea – 4309r – 1 – (combines the pre-1850 monographic and pre-1906 serials holdings of the goldsmiths' library of economic literature at the university of london and the kress library of business and economics at the harvard graduate school of business administration. suppl with material from seligman collection, butler library, columbia univ. and sterling library at yale. subject break-outs available: trade and manufacturers; colonies; transport; general and misc; corn laws; commerce; finance; social conditions; politics; agriculture) – us Primary [330]
The goldsmiths'-kress library of economic literature : supplements to segments 1-3 (goldsmiths' library of economic literature, (printed books, periodicals and manuscripts to 1850) – 46r unit 86, reel n427-472 – 1 – us Primary [330]
Goldson, William *see* Observations on the passage between the atlantic and pacific oceans
Goldstein, Julius et al *see* Der morgen
Goldstein, Walter Benjamin *see* Wassermann
Goldthwaite, Vere *see* The philosophy of ingersoll
Goldwin smith and the jews / Bendavid, Isaac Besht – S.l: s.n, 1891? – 1mf – 9 – mf#02322 – cn Canadiana [939]
Goldwin smith papers, 1823-1910 – [mf ed ProQuest] – 28r – 1 – (with p/g) – us UMI ProQuest [080]
Goldziher, Ignac *see*
– Le culte des ancetres et le culte des morts chez les arabes
– Mohammed and islam
– Muhammedanische studien
– Mythology among the hebrews and its historical development
– Stellung der alten islamischen orthodoxie zu den antiken wissenschaften
– Translation of the chapter on hadaith and the new testament
– Vorlesungen ueber den islam
Golenischeff, W *see*
– Les papyrus hieratiques nos 1115, 1116a et 1116b de l'ermitage imperial a st petersbourg
– Papyrus hieratiques, nos 58001-58036
Golf business – Middleburg OH 1976 – 1 – (cont: golfdom) – ISSN: 0148-3706 – mf#10205.01 – us UMI ProQuest [790]
Golf championships / Harold, William G – s.l, s.l? 1936 – 1r – us UF Libraries [790]
Golf collectors' society bulletin – 1970 dec-1984 nov – 1r – 1 – (cont: bulletin [golfiana collectors club]; cont by: bulletin [golf collectors society]) – mf#956165 – us WHS [790]
Golf courses / Trainor, A W – s.l, s.l? 1936 – 1r – us UF Libraries [790]
Golf digest – New York NY 1971+ – 1,5,9 – ISSN: 0017-176X – mf#60015 – us UMI ProQuest [790]
Golf journal – Far Hills NJ 1980+ – 1,5,9 – ISSN: 0017-1794 – mf#11237 – us UMI ProQuest [790]
Golf magazine – New York NY 1974+ – 1,5,9 – ISSN: 1056-5493 – mf#9316 – us UMI ProQuest [790]
Golf putting and preferences for cognitive training / Gervais, Pierre D – 2000 – 115p on 2mf – 9 – $10.00 – mf#PSY 2137 – us Kinesology [150]
Golfdom – Middleburg OH 1975-76 – 1,5,9 – (cont by: golf business) – ISSN: 0017-1905 – mf#10205.01 – us UMI ProQuest [790]
Golfiana Collectors Club *see* Bulletin of the golfiana...
Golgotas / Echeverria Rodriguez, Roberto – Barranquilla, Colombia. 1944 – 1r – us UF Libraries [972]
Golgotha and the church of the holy sepulchre / Parrot, Andre – S.C.M. Press, 1957 – 9 – $10.00 – us IRC [930]
Golgotha and the holy sepulchre / Wilson, Charles William; ed by Watson, Charles Moore – London: Committee of the Palestine Exploration Fund, 1906 – 1mf – 9 – 0-7905-0458-8 – mf#1987-0458 – us ATLA [915]

Goliad messenger – Goliad TX. 1863 sep 10 – 1r – 1 – mf#858518 – us WHS [071]

Golibart Gonzalez, Porfirio *see* Ritmos de la montana

Golightly, C P *see* Look at home

Golightly, Charles Portales *see* The position of the right rev. samuel wilberforce, d.d., lord bishop of oxford, in reference to ritualism

Golikov, I I *see* Deianiia petra velikogo, mudrogo preobrazitelia rossii, sobrannye iz dostovernykh istochnikov i raspolozhennye po godam

Golikov, P I *see* Chto takoe proizvoditelno-trudovaia artel i kak ee organizovat

Gollancz, Hermann *see* Mafteah shelomoh

Gollenperger, Dr. *see* Radikale lieder

Gollock, Georgina Anne *see*
- Lives of eminent africans – tonga
- A winter's mails from ceylon, india and egypt

Golos – St Petersburg, Russia 1863-67 [mf ed Norman Ross] – 1 – mf#nrp-1594 – us UMI ProQuest [077]

Golos – Yaroslavl', Russia 1917 [mf ed Norman Ross] – 5r – 1 – mf#nrp-2191 – us UMI ProQuest [077]

Golos. 1863-1883 (incomplete) – 53r – 1 – mf#RF-3 – ne IDC [077]

Golos armii : gaz voen, obshchestv i lit – Omsk [Akmol obl]: Osvedom otd Shtaba 2-oi Armii 1919 [1919 18 sent-] – (= ser Asn 1-3) – n1-9 [1919] [gaps] item 82, on reel n17 – 1 – mf#asn-1 082 – ne IDC [077]

Golos bashkir : organ pravitel'stva bashkirii – Orenburg: [s n] 1918 [1918 [?]-14 avg] – (= ser Asn 1-3) – n2-3 [1918] item 83, on reel n17 – 1 – (cont by: vestnik pravitel'stva bashkirii) – mf#asn-1 083 – ne IDC [077]

Golos bednoty : organ kaliazinskogo komiteta rkp(b) i uezdnogo soveta rabochikh i krest'ianskikh deputatov. fraktsiia rkp(b) pri ispolkome – Kalyazin, Russia 1918 [mf ed Norman Ross] – 1r – 1 – mf#nrp-607 – us UMI ProQuest [077]

Golos checheno-ingushetii : respublikanskaia obshchestvenno-politicheskaia gazeta – Groznyi: Golos Checheno-Ingushetii -1992 [1991-92] [Minneapolis MN: East View Publ [199-] – 3r – 1 – (iss for jan 1-may 20 1992 filmed with: golos chechenskoi respubliki may 21-jun 27 1992) – mf#mf-11996 seemp – us CRL [077]

Golos chechenskoi respubliki : respublikanskaia obshchestvenno-politicheskaia gazeta – Groznyi: Golos Chechenskoi Respubliki 1992- (mf ed Minneapolis MN: East View Publ [199-] – 2r – 1 – (iss for may 20-jun 27 1992 filmed with: golos checheno-ingushetii jan 1-may 20 1992) – mf#mf-11996 seemp (reel 3) may 20-jun 27 1992; mf-11997 seemp (1r) jul 2-dec 31 1992 – us CRL [077]

Golos dagestana : organ dagestanskogo oblastnogo ispolnitel'nogo komiteta – Temir-Khan-Shura, Russia 1917 [mf ed Norman Ross] – 1r – 1 – mf#nrp-1808 – us UMI ProQuest [077]

Golos fronta : ezhedn obshchestv -polit i lit gaz / ed by Bekarevich, A – Melitopol' [Zaporozh gub]: [s n] 1920 [1920-] – (= ser Asn 1-3) – n115-143 [1920] [gaps] item 100, on reel n21 – 1 – mf#asn-1 100 – ne IDC [077]

Golos iuga [elisavetgrad: 1918-1919] / ed by Gorshkov, D S – Elisavetgrad [Kherson gub]: [s n] 1918-19 [1904 5 dek-1919 [?]] – (= ser Asn 1-3) – n13 [1918] n55-58 [1919] item 101, on reel n21 – 1 – mf#asn-1 101 – ne IDC [077]

Golos iuga [poltava: 1919] : ezhedn obshchestv -polit i lit gaz / ed by Popovskii, A I – Poltava: B A Bartoshevich 1919 [1919 4 avg-] – (= ser Asn 1-3) – n79-80 [1919] item 102, on reel n21 – 1 – mf#asn-1 102 – ne IDC [077]

Golos iunoshestva : vykhodit po povodu sozdaniia kluba uchashcheisia molodezhi – Chita [Zabaik obl]: [s n] 1918 [1918 2 maia [19 apr]-] – (= ser Asn 1-3) – n1 [1918] item 103, on reel n21 – 1 – mf#asn-1 103 – ne IDC [077]

Golos kavkaza – Tbilisi, Russia 1906-17 [mf ed Norman Ross] – 1 – (reel contains short runs of multiple titles. for complete listing of titles on a reel, please inquire) – mf#nrp-1794 – us UMI ProQuest [077]

Golos kooperatora – Stavropol, 1916-1917(4) – 3 – 9 – mf#COR-576 – ne IDC [335]

Golos kooperatsii – Kiev, 1918-1919(7) – 5 – 9 – mf#COR-577 – ne IDC [335]

Golos krest'ianina : organ ispolnitel'nogo komiteta turkestanskogo kraevogo soveta krest'ianskikh deputatov – Tashkent, Uzbekistan 1918 [mf ed Norman Ross] – 1r – 1 – mf#nrp-1759 – us UMI ProQuest [077]

Golos krest'ianina – Vladivostok, Russia 1919 [mf ed Norman Ross] – 1r – 1 – mf#nrp-2072 – us UMI ProQuest [077]

Golos krest'ianstva : gaz sotsial'nodemokrat, vnepartiin, otstaivaiushchaia interesy krest'ianstva i derevni / ed by Zeriunov, A I – Vladivostok [Primor obl]: [s n] 1919 [1919 [9 marta]-[1921]] – (= ser Asn 1-3) – n11-20 [1919] [gaps] item 85, on reel n17 – 1 – mf#asn-1 085 – ne IDC [077]

Golos kubantsa : ezhedn vnepartiin gaz – Ekaterinodar [Kuban obl]: [s n] 1919 [1919 26 okt-] – (= ser Asn 1-3) – n1-3 [1919] item 86, on reel n17 – 1 – mf#asn-1 086 – ne IDC [077]

Golos maksimalista : ezhenedel'nik: organ omsk soiuza sots -rev maksimalistov – Omsk [Akmol obl]: [s n] 1918 [1918 25 apr-] – (= ser Asn 1-3) – n1 [1918] item 87, on reel n17 – 1 – mf#asn-1 087 – ne IDC [077]

Golos minuvshago : zhurnal istorii i istorii literatury – v1-11. 1913-oct 1923 – 1 – $420.00 – mf#0243 – us Brook [460]

Golos minuvshago na chuzho-istoronie : zhurnal istorii i istorii literatury. – v1-19. 1923-28 – 1 – $108.00 – mf#0244 – us Brook [460]

Golos momenta : ezhened gaz / ed by Mil'shtein, L M – Eniseisk: Nar -resp gruppa 1918 [1917 18 apr-] – (= ser Asn 1-3) – n38-49 [1918] item 88, on reel n17 – 1 – mf#asn-1 088 – ne IDC [077]

Golos moskvy – Moscow, Russia 1906-15 [mf ed Norman Ross] – 1 – mf#nrp-1069 – us UMI ProQuest [077]

Golos moskvy – Moskva: Moskovskoe tovarishchestvo dlia izdaniia knig i gazet. n2-25 jan 3-31 1912; n26-49 feb 1-29 1912 – 1 – us CRL [947]

Golos naroda / Chitinskaia regional'naia organizatsii Liberal'no-Demokraticheskoi partii Rossii – Chita, Russia. n4[21](noi 1997), n1[23](fev 1998)-2[24](mar 1998), n4[26](iiun 1998)-n7[29](sen 1998) – 1 – mf#mf-12248 (reel 4) – us CRL [077]

Golos naroda : ezhedn polit gaz otv / ed by Fel'dman, M S – Tomsk: [s n] 1918 [1918 [1 iiunia]-] – (= ser Asn 1-3) – n3-157 [1918] [gaps] item 89, on reel n18 – 1 – mf#asn-1 089 – ne IDC [077]

Golos naroda : organ kustanajskikh obshchestvennykh organizatsij – Kustanaj, Kazakhstan, 1917 [mf ed Norman Ross] – 1r – 1 – mf#nrp-815 – us UMI ProQuest [077]

Golos nizhegorodtsa – Nizhny-Novgorod, Russia 1917 [mf ed Norman Ross] – 1r – 1 – mf#nrp-1211 – us UMI ProQuest [077]

Golos pravdy – New York: Golos Pravdy Pub Co. v1-8. mar 1938-dec 1945 – 1 – us CRL [073]

Golos pravdy : organ kronshtadskogo komiteta rsdrp – Penza, Russia 1919 [mf ed Norman Ross] – 3r – 1 – mf#nrp-1369 – us UMI ProQuest [077]

Golos priural'ya – Chelyabinsk, Russia 1909-15 [mf ed Norman Ross] – 1 – mf#nrp-404 – us UMI ProQuest [077]

Golos rabochego [gomel': 1920] : organ gomel gub i gor kom "bunda" / Vseobshch evr rabochii soiuz "Bund" – Gomel': [s n] 1920 [1920 7 iiunia-] – (= ser Asn 1-3) – n2-4 [1920] item 9, on reel n1 – 1 – mf#asn-2 009 – ne IDC [077]

Golos rabochego [ufa: 1918] : organ ufim gub biuro rsdrp – Ufa: [s n] 1918 [1918 1 [14] fevr-] – (= ser Asn 1-3) – n1-218 [1918] [gaps] item 90, on reel n18,19 – 1 – mf#asn-1 090 – ne IDC [077]

Golos rodiny / ed by Petelin, I I – Vladivostok [Primor obl]: T-vo "Svobodnaia Rossiia" 1918-19 [1917 12 [25] avg-1923 [?]] – (= ser Asn 1-3) – n313 [1918]-n556, 1-39 [1919] [gaps] item 91, on reel n19,20 – 1 – mf#asn-1 091 – ne IDC [077]

Golos rossii : ezhedn vnepartiin obshchestv -polit i lit gaz / ed by Markova, E IA – Baku: P G Bulgakov 1919-20 [1919-] – (= ser Asn 1-3) – n42-71 [1919]-n6-71 [1920] [gaps] item 92, on reel n20 – 1 – mf#asn-1 092 – ne IDC [077]

Golos rossii / Natsional'no-Respublikanskaia partiia Rossii – St Petersburg, Russia. n2(1991), n3(1991), n5(1991), n2[7](1991), n3[8](1992)-5[10](1993) – 1 – mf#mf-12248 (reel 4) – us CRL [077]

Golos rossii : organ russkoj demokraticheskoj mysli – Berlin, Germany 1919-20 [mf ed Norman Ross] – 2r – 1 – mf#nrp-324 – us UMI ProQuest [074]

Golos rossii – Sofia, Bulgaria. Jun 1936-Aug 1938 – 1r – 1 – us L of C Photodup [949]

Golos severnogo uchitelia : organ arkhang gub uchit soiuza / ed by Sobolev, I M – Arkhangel'sk: [s n] 1918-1919 [1917 23 marta-1919 [?]] – (= ser Asn 1-3) – n28-30 [1918]-n14-30 [1919] item 93, on reel n20 – 1 – mf#asn-1 093 – ne IDC [077]

Golos sibiri : ezhedn obshchestv -polit, demokrat gaz / ed by Arkhipova, M I – Tomsk: G M Gal'chenko 1918-19 [1918 15 dek-1919 28 fevr] – (= ser Asn 1-3) – n3-12 [1918]-n2-43 [1919] [gaps] item 94, on reel n20 – 1 – mf#asn-1 094 – ne IDC [077]

Golos sibiri – Omsk, Russia 1909-11 [mf ed Norman Ross] – 1 – mf#nrp-1255 – us UMI ProQuest [077]

Golos sibiriaka : voen -lit gaz / ed by Chaban, D – Ekaterinburg: [s n] 1919 [1918 [?]-1919 [iiul']] – (= ser Asn 1-3) – n7-44 [1919] [gaps] item 96, on reel n20 – 1 – mf#asn-1 096 – ne IDC [077]

Golos sibirskoi armii : gaz voen, obshchestv i lit – [Ekaterinburg]: Inform otd Shtaba Sib armii 1919 [1919 26 marta-[iiul']] – (= ser Asn 1-3) – n1-32 [1919] [gaps] item 95, on reel n20 – 1 – mf#asn-1 095 – ne IDC [077]

Golos simbirskoi kooperatsii – Simbirsk, 1918-1919(5/6) – 4 – 9 – (missing:1918(1-4, 6-8),1919(1-4)) – mf#COR-579 – ne IDC [335]

Golos soldata : izdaetsia armejskim komitetom / Armiia chetvertaia – (Russia) 1917 [mf ed Norman Ross] – 1r – 1 – mf#nrp-83 – us UMI ProQuest [077]

Golos soldata : organ pri sovete rabochikh i soldatskikh deputatov gor ekaterinoslava – Dnipropetrovsk, Ukraine 1917-18 [mf ed Norman Ross] – 1r – 1 – mf#nrp-466 – us UMI ProQuest [077]

Golos sotsial'demokrata – Geneva etc. v. 1-5 no. 1-26. Feb 1908-Dec 1911 – 1 – us NY Public [335]

Golos stepi : koop i obshchestv -polit gaz / ed by Kustov, A V – Pavlodar [Semipalat obl]: T-vo kooperativov 1919 [1919 1 ianv-] – (= ser Asn 1-3) – n1-43 [1919] [gaps] item 97, on reel n21 – 1 – (suppl: biulleteni gazety "golos stepi" [asn-1 020]; cont: pavlodarskii telegraf) – mf#asn-1 097 – ne IDC [077]

Golos stepi – Omsk, Russia 1907 [mf ed Norman Ross] – 1 – mf#nrp-1256 – us UMI ProQuest [077]

Golos taigi : gaz vnepartiin, polit, ekon i lit / ed by Sokolov, A K – Mariinsk [Tom gub : Mariin osvedpunkt 1919 [1919 29 noiab-] – (= ser Asn 1-3) – n1 [1919] item 98, on reel n21 – 1 – mf#asn-1 098 – ne IDC [077]

Golos tatar : ezhenedel'nyi organ / Vremennago krymsko-musul'manskago ispolnitel'nago komiteta – Simferopol': Tip "Millet" Vremennago krymsko-musul'manskago ispolnitel'nago komiteta [1917- g izd 1-i n1-i-] – item 4 on reel of 7 titles – 1 – (filmed: g1 n1-g1 n13 (1917) [lacks n3-4,6,10-11]) – mf#mr-7 – ne IDC [305]

Golos truda – Barnaul, Russia 1917-18 [mf ed Norman Ross] – 1 – mf#nrp-264 – us UMI ProQuest [077]

Golos truda : organ iakut kom partii sotsialistov-revoliutsionerov / ed by Klingof, D F – IAkutsk: [s n] 1918 [1917 6 dek-1920 5 iiunia] – (= ser Asn 1-3) – n8-32 [1918] [gaps] item 99, on reel n21 – 1 – mf#asn-1 099 – ne IDC [335]

Golos trudovogo krest'ianstva : organ ispolnitel'nogo komiteta vserossijskogo soveta krest'ianskikh deputatov 2-go sozyva – St Petersburg, Russia 1917 [mf ed Norman Ross] – 2r – 1 – mf#nrp-1595 – us UMI ProQuest [077]

Golos truzhenika / Industrial Workers of the World – v2 n76-v9 n25/26 [1919 dec 27-1927 apr/may] – 1r – 1 – mf#369063 – us WHS [331]

Golos ukrainy – Kiev: Verkhovnyi Sovet Ukrainy, 1991- – 1 – us East View [077]

Golos zabaikal'ia : nar obshchestv -lit gaz / ed by IAkovlev, R I – Chita [Zabaik obl]: [s n] 1919 [1919 27 noiab-] – (= ser Asn 1-3) – n1-11 [1919] [gaps] item 84, on reel n17 – 1 – mf#asn-1 084 – ne IDC [077]

Golosa iz Rossii *see* [Sbornik]

Golosa iz rossii – (2nd Edition). London. v. 1-9. 1858-1859 – 1 – us NY Public [947]

Golovnin, Vasilii M *see* Begebenheiten des capitains von der russisch-kaiserlichen marine golownin

Golovnin, Vasily Mikhaylovich *see* Narrative of my captivity in japan, during the years 1811, 1812 and 1813

Golowin : novelle / Wassermann, Jakob – Berlin: S Fischer, 1930, c1920 (mf ed 1990) – 1r – 1 – (filmed with: die gochhausen) – us Wisconsin U Libr [830]

Golpe a las 2 am / Garcia Mejia, Rene – Guatemala, 1958 – 1r – us UF Libraries [972]

Golpe de abril / Moniz, Edmundo – Rio de Janeiro, Brazil. 1965 – 1r – us UF Libraries [972]

Golpe y porrazo / Martinez Herrera, Alberto – Habana, Cuba. 1964 – 1r – us UF Libraries [972]

Golson, Jack *see* Papers on cultural policy in papua new guinea

Golther, Wolfgang *see*
- Die deutsche dichtung im mittelalter, 800 bis 1500
- Das lied vom huernen seyfrid
- Der nibelunge not
- Parzival in der deutschen literatur
- Religion und mythus der germanen
- Die sagengeschichtlichen grundlagen der ringdichtung richard wagners
- Tristan und isolde in der franzoesischen und deutschen dichtung des mittelalters und der neuzeit
- Zur deutschen sage und dichtung

Gol'tsman, A *see* Khoziaistvennyi rost i khoziaistvennye zatrudneniia

Gol'tsman, M T *see* Sostav sovetskikh i torgovykh sluzhashchikh

Goltz, Eduard *see*
- Das gebet in der aeltesten christenheit
- Ignatius von antiochien als christ und theologe – griechische excerpte aus homilien des origenes
- Eine textkritische arbeit des zehnten bezw sechsten jahrhunderts
- Tischgebete und abendmahlsgebete in der altchristlichen und in der griechischen kirche

Goltz, Eduard, Freiherr von der *see*
- Der dienst der frau in der christlichen kirche
- Logos soterias pros ten parthenon

Goltz, Eduard von der *see*
- Das gebet in der aeltesten christenheit
- Ignatius von antiochien als christ und theologe
- Logos soterias pros ten parthenon – de virginitate
- Logos soterias pros ten parthenon (de virginitate
- Eine textkritische arbeit des zehnten bezw sechsten jahrhunderts
- Tischgebete und abendmahlsgebete in der altchristlichen und in der griechischen kirche

Goltz, Hermann, Freiherr von der *see*
- Gottes offenbarung durch heilige geschichte
- Die reformirte kirche genf's im neunzehnten jahrhundert

Goltz, Hermann von der *see* Die christlichen grundwahrheiten

Goltz, Joachim, Freiherr von der *see*
- Einst auf der lorettohoehe
- Ewig wiederkehrt die freude
- Vater und sohn
- Von mancherlei hoelle und seligkeit

[Goltzius] Hirschmann, O *see* Hendrick goltzius als maler, 1600-1617

Golubev, A *see*
- Otchet ego prevoskhoditel'stvu g upravliaiushchemu ministerstvom finansov o revizii ekaterinodarskogo gorodskogo obshchestvennogo banka, proizvedennoi s 9 po 15 oktiabria 1903 g
- Otchet ego prevoskhoditel'stvu gospodinu ministru finansov o revizii taganrogskogo obshchestva vzaimnogo kredita, proizvedennoi po vysochaishemu poveleniiu, s 24 oktiabria po 3 noiabria 1902 g
- Otchet ego vysokoprevoskhoditel'stvu gospodinu ministru finansov o revizii sel'sko-khoziaistvennogo i promyshlennogo banka v rostove na donu, proizvedennoi po vysochaishemu poveleniiu, s 3 po 20 oktiabria 1902 g

Golubev, A K *see*
- Russkie banki
- Sbornik materialov po voprosam, podlezhavshim obsuzhdeniiu s"ezda
- Statistika kratkosrochnogo kredita

Golubinskii, Evgenii Evstignieevich *see* K nashei polemikie s staroobriadtsami

Golubinskii, G G *see* Istoria kanonizatsii sviatykh v russkoi tserkvi

Golubkova, S N *see* Kachestvo i standart v promkooperatsii

Golubovich, G *see* Bibliotheca bio-bibliografica della terra santa e dell' oriente francescano

Golunskii, Sergei Aleksandrovich *see* Sudoustroistvo sssr; uchebnik dlia iuridicheskikh shkol

Golus : zeitschrift der juedischen emigration aus deutschland – Prag (CZ), 1933 n1 – 1r – 1 – (publ once only) – gw Misc Inst [939]

Golyshenko, V S *see* Sinaiskii paterik

Golz, Bruno *see* Zwei schwaebische erzaehler

Goma, Isidoro *see*
- La eucaristia y la vida cristiana. 2nd ed. 2 tomos. barcelona, 1934
- La familia, segun el derecho natural y cristiano...
- Jesucristo redentor
- Las modas y el lujo, ante la ley cristiana, la sociedad y el arte. 3rd ed san sebastian, 1935

Goma Tomas, Isidro *see* Antilaicismo

Goma y Tomas, Isidro *see* El evangelio explicado, vol 1

Gomantak – Panjim, India. Jul 1966-1993 – 86r – 1 – us L of C Photodup [079]

Gomara, Francisco Lopez de *see* The conquest of the weast india

Gomarus, F *see* Opera theologica omnia...

Gomarus, Franciscus (Francis Gomar) *see* Opera theologica omnia...

Gomery, Douglas *see* The will hays papers

Gomes carneiro : o general da republica / Calmon, Pedro – Rio de Janeiro, Brazil. 1933 – 1r – us UF Libraries [972]

Gomes Da Silva, Francisco *see* Memorias do chalaca

Gomes, Eduardo *see* Campanha de libertacao [discursos]

Gomes, Henrique de Barros *see* O padroado da coroa de portugal nas indias orientaes e a concordata de 23 de junho de 1886

Gomes, Lindolfo *see* Contos populares brasileiros

Gomes, Ruy Cinatti Vaz Monteiro *see* Exploracoes botanicas em timor

Gomez Acedo, Francisco *see* Novena de nuestra senora de sopetran ordenada de nueve preciosas piedras de que se compone el nombre sopetrana...

Gomez Alfau, Luis Emilio *see* Ayer o el santo domingo de hace 50 anos

Gomez Alvarez, Roberto *see* Poemas profanos

Gomez, Antonio *see*
– Discursos evangelicos
– Sermon del doctor seraphico s. buenaventura...

Gomez, Antonio J *see* Monografias eclesiastica y civil de medelllin

Gomez Aristizabal, Horacio *see* Teoria gorgona

Gomez Ballesteros, Francisco *see* El liberalismo el socialismo y la solucion nacionalsindicalista

Gomez Barrientos, Estanislao *see* 25 anos a traves del estado de antioquia

Gomez Bravo, Juan *see*
– Advertencias a la istoria (sic) de merida
– Catalogo de los obispos de cordoba. 1st parte
– Catalogo... obispos de cordoba

Gomez Bravo, Vicente *see*
– Aesthetica. nociones de la belleza y de las artes
– Aquaviva
– Espana en america
– Gramatica historica espanola y antologias
– Una hija de maria (maria perez de guzman y sanjuan)
– Lyra hispana...
– Lyra hispana. cuestomatia escolar para lectura y analisis literario
– Lyra hispana. prosa selecta
– Lyra hispana. prosa selecta. silva dramatica
– Noche-buena en familia
– Prosa selecta de autores espanoles para lectura y analisis literario
– Silva dramatica...
– Silva dramatica. asuntos del teatro espanol dispuestos para estudio literario
– Subplemento al tesoro poetico castellano del siglo 19
– Tesoro poetico castellano del siglo 19

Gomez, C R A *see* Caso palmer

Gomez Canedo, Lino *see* Don juan de carvajal. un espanol al servicio de la santa sede. madrid, 1947

Gomez Canedo, Luis *see* Un espanol al servicio de la santa sede. don juan de carvajal, cardenal de sant'angelo, lugado en alemania y hungria (1.399-1.469)

Gomez Carbonell, Maria *see*
– Estudio critico biografico de juan clemente zenea
– Homenaje a bernarda toro de gomez

Gomez carrillo 30 anos despues / Barreintos, Alfonso Enrique – Barcelona, Spain. 1959 – 1r – us UF Libraries [972]

Gomez Carrillo, Augustin *see* Comprendio de historia de la america central

Gomez Carrillo, Enrique *see*
– Cultos profanos
– Del amor, del dolor, y del vicio
– En el reino de la frivolidad
– Entre encajes
– Evangelio del amor
– Grecia eterna
– Paginas escogidas
– Por tierras lejanas
– Primer libro de las cronicas
– Treinta anos de mi vida
– Tres novelas inmorales

Gomez Castanos, Marcial *see* Mocion presentada al ayuntamiento de olivenza

Gomez Costa, Arturo *see* San juan, ciudad fantastica de america

Gomez de Arteche, Jose *see* La conquista de mejico

Gomez de Avellaneda, Gertrudis *see* Guatimozin. tomo 2

Gomez De Avellaneda Y Arteaga, Gertrudis *see*
– Antologia
– Baltasar
– Gertrudis gomez de avellaneda
– Love letters
– Sab

Gomez De Avellaneda Y Arteaga, Gertrudos *see* Obras de la avellaneda

Gomez de Cervantes, Gonzalo *see* La vida economica social de nueva espana al finalizar el siglo 16. prologo y notas de alberto ma carreno. mexico, 1944

Gomez de la Parra, A *see* Polyanthea medicis speciosa chirurgis mirifica myrepsicis valde utilis et necessaria

Gomez de la Parra, Jose *see* Fundacion, y primero siglo, del muy religioso convento de sr s joseph de religiosas carmelitas descalzas de la ciudad de la puebla de los angeles, en la nueva espana

Gomez De La Serna, Ramon *see* Goya

Gomez de la Serna, Ramon *see* Mi tia carolina coronado

Gomez de Mercado y Miguel, F *see* Isabel 1, reina de espana y madre de america. madrid, 1943

Gomez de Santana, Indalecio *see*
– Memoria leida el dia 16 de septiembre en la apertura del curso de 1867 a 1868 en el instituto de segunda ensenanza de caceres, por...
– Memoria...23 de septiembre...de 1867 a 1868... colegio de isabel 2...
– Memoria...caceres
– Memoria...instituto de 2a ensenanza

Gomez Efe *see* Tragedie delminero

Gomez, Florencio *see* Discurso-politico...colero norbo

Gomez, Francisco Gregorio *see* Burguesito recien pescado

Gomez Guillen, Roman *see* Juan vazquez en la catedral de plasencia

Gomez Hermosilla, Jose *see*
– Arte de hablar en prosa y verso
– Arte de hablar...verso

Gomez Hoyos, Rafael *see* Revolucion granadina de 1810

Gomez Hurtado, Alvaro *see* Revolucion en america

Gomez Jara, Francisco *see* Discurso pronunciado en la colaboracion de grados de licenciados en jurisprudencia...

Gomez Jara y Herrera, Juan de la Cruz *see* Apuntes historicos...villa de fuente del maestre desde...

Gomez, Jose Jorge *see* Corteza y la savia

Gomez, Juan Gualberto *see*
– Cuestion de cuba en 1884
– Por cuba libre

Gomez L, Efrain *see* Garcia rovira

Gomez, Laureano *see*
– Comentarios a un regimen
– Crimen de la magdalena
– Cuadrilatero
– Desde el exilio

Gomez, M *see* ...De que el aforismo primero de hipocrates...sirve a la milicia como a la medicina...

Gomez, Madeleine A *see* Jornadas divertidas, politicas sentencias y hechos memorables de reyes y heroes de la antiguedad

Gomez, Maximo *see*
– Cartas de maximo gomez
– Horas de tregua

Gomez Miedes, B *see*
– Enchiridion o manual instrumento de salud contra el morbo articular que llaman gota...
– Enquiridion contra el morbo articular

Gomez monsegu c.p., bernardo y elias de tejada, f. la riqueza espiritual de espana / Elordury, E – Madrid: Razon y Fe, 1944 – 1 – sp Bibl Santa Ana [240]

Gomez Naranjo, Pedro Alejandro *see* Sal de la historia

Gomez Nogales, Salvador *see*
– La filosofia de la naturaleza y la psicologia segun ibn hazm
– Horizonte de la metafisica aristotelica

Gomez, P *see* Satisfaccion al publico contra la adicion apologetica que a su dissertacion medico-moral del primer tomo de la palestra critica medica

Gomez Picon, Alirio *see* Semblanza de antonio jose restrepo

Gomez Picon, Rafael *see*
– Magdalena, rio de colombia
– Timana

Gomez Restrepo, Antonio *see*
– Bogota
– Historia de la literatura colombiana
– Oraciones academicas

Gomez Robledo, Antonio *see*
– Filosofia en el brasil
– Mexico y el arbitraje internacional

Gomez Robles, Julio *see*
– Statement of the laws of guatemala in matters affe...

Gomez Rodeles, Cecilio *see* Historia de la publicacion monumenta historica societatis jesu

Gomez Takiwah, Mariano *see* Chonbilal ch'ulelal

Gomez Villafranca, Ramon *see*
– Cooperacion de roman gomez villafranca a la bibliografiade arias montano, nº 1
– Extremadura en la guerra de la independencia espanola
– Los extremanos en las cortes de cadiz
– Seminario conciliar de san aton. gabinete numismatico. catalogo

Gomez-Bravo, Vicente *see*
– En familia
– Estetica siemprevivas y ensayos

Gomez-Moreno, M *see* Iglesias mozarabes

Gomis, Juan B *see* Magisterio de amor fray juan de los angeles (1536-1610)

Gomme, George Laurence Gomme *see* Seven sages of rome

Gommersbach, Wilhelm *see* Geschichte, geographie und bedeutung de insel trin...

Gomperz, Th *see* Griechische denker

Gonaecierum libri 3. de morbis mulierum communibus... / Mercado, Luys de – Basilea, 1588 – 9 – sp Cultura [610]

Gonardiya *see* The study of patanjali

Goncalves, Alpheu Dinix *see* Observacoes astronomicas sobre o cometa de halley

Goncalves, Carlos Alberto *see* Modern brazil

Goncalves de magalhaes / Magalhaes, Domingos Jose Goncalves De – Sao Paulo, Brazil. 1946 – 1r – us UF Libraries [972]

O goncalvismo em pitangui / Diniz, Silvio Gabriel – Belo Horizonte, Brazil. 1969 – 1r – us UF Libraries [972]

Goncharov, Ivan *see* The precipice

Goncourt, Edmond *see* La femme au dix-huitieme siecle

Goncourt, Jules *see* La femme au dix-huitieme siecle

Gonda, Jan *see*
– Notes on brahman
– Remarks on similes in sanskrit literature

Gondinet, Edmond *see*
– Gavaut, minard and cie
– Revoltees
– Voyage d'agrement

Gondolier – Venice, FL. 1946 mar 6-1977 – 16r – (gaps) – us UF Libraries [071]

Gondon, G *see* L'imitateur de jesus-christ

Gondon, Jules *see* Notice biographique sur le r. p. newman de l'oratoire de saint-philippe de neri

The gondwana and the gonds / Singh, Indrajit – Lucknow: Universal Publishers, 1944 – (= ser Samp: indian books) – (foreword by radhakamal mukerjee) – us CRL [305]

Gonet, J B *see*
– Clypeus theologiae thomisticae contra novos eius impugnatorus
– Manuale thomistarum seu brevis theologiae cursus...

Gonfle / Martin Du Gard, Roger – Paris, France. 1928 – 1r – us UF Libraries [440]

Gong *see* Pendidikan umum

Gongora Echenique, Manuel *see* Que he visto en cuba

Gongora Y Argote, Luis De *see* Gongora y el 'polifemo'

Gongora y el 'polifemo' / Gongora Y Argote, Luis De – Madrid, Spain. v1-2. 1961 – 1r – us UF Libraries [960]

Gongren ribao – Peking. Mar 1953; 25 jul 1957; nov 1928; 15 nov 1958; jan-jun 1962; jan-may 1963; nos 6099-6203 1980 – 2 1/2r – 1 – uk British Libr Newspaper [072]

Goni, Blas *see* Glosando a benito arias montano. la funcion social del ingenio

Gonidec, Jean-Francois-Marie-Maurice-Agatha le *see*
– Dictionnaire celto-breton
– Grammaire celto-bretonne

Goniec krakowski – Krakow, Poland. Aug 1940-Jan 1945 – 6r – 1 – us L of C Photodup [943]

Goniec polski – South Bend: Goniec Polski Pub Co, jun 27 1896-1927; 1928-64 – 57r – 1 – us CRL [071]

Gonino, I *see* History of art in sardinia, judaea, syria, and asia minor

Gonkong v sisteme mirovykh ekonomicheskikh sviazei / Kukolevskii, A G; ed by Medovogo, A I – Moskva: Izd-vo "Mezhdunarodnye otnosheniia", 1972 – us CRL [947]

Gonnelieu, R P de *see* L'imitation de jesus-christ

Gontaut Biron, R de *see* Sur les routes de syrie apres neuf ans de mandat

Gontery, J *see* La pierre de touche, ou la vraye methode pour desabuser les esprits

Gontijo De Carvalho, Antonio *see*
– Calogeras
– Ensaios biograficos
– Raul fernandes

Gonzaga law review – v1-36. 1966-2001 – 5,6,9 – $750.00 set – (v1-20 1966-85 on reel $330. v21-36 1985-2001 on mf $420) – ISSN: 0046-6115 – mf#103001 – us Hein [340]

Gonzaga, Mary *see* The mysticism of johann joseph von goerres as a reaction against rationalism

Gonzaga, Norberto *see*
– Angola
– Historia de angola, 1482-1963

Gonzaga, Tomas Antonio *see* Marilia de dirceu e mais poesias

Gonzague, Elisabetta R de *see* Lettres de madame la princesse de g...

Gonzales Borreguero, G *see* La caza de la perdiz con reclamo

Gonzales de la Calle, Pedro Urbano *see* Varias notas y apuntes sobre temas de letras clasicas

Gonzales de Mendoza, Juan *see* Historia de las cosas mas notables del gran reyno de china

Gonzales, Jose Maria *see*
– El dia de colon y de la paz
– El dia de colon y la paz

Gonzales Llubera, Ignacio *see* Viajes de benjamin de tudela. 1160-1173

Gonzales, Narciso Gener *see* In darkest cuba

Gonzales Pintado, Gaspar *see* Los martires de las misiones del paraguay. bilbao, 1934. madrid, 1934

Gonzales Ruiz, Ricardo [Editor] *see* Guatemala de hoy

Gonzales y Gomez de Soto, Juan Jose *see* Breve historial de las sagradas reliquias que se veneran en las parroquias de santa eulalia y sta. marta la mayor de merida

Gonzalez Alberty, Fernando *see* Grito, poemario de vanguardia

Gonzalez Alcorta, Leandro *see* Que pasa en cuba

Gonzalez Alonso, Diego *see* El templo de ammon y los pitagoricos

Gonzalez Alvarez, Claudio *see* Los gerifaltes marxistas por...

Gonzalez, Antonio *see* Capsulas gelatinosas

Gonzalez, Antonio E *see* Guerra del chaco

Gonzalez Arrili, Bernardo *see* Deliciosa jujuy

Gonzalez Bazan, Carlos R *see*
– Al son de mi mejorana
– Canto y saloma

Gonzalez Castell, Rafael *see* Entre mis cuatro paredes

Gonzalez, Castro *see* Compendaria in graecam via

Gonzalez Castro, Jose *see* Medico de guijo de santa barbara. tratamiento de la neuralgia ciatica por la cauterizacion del helix

Gonzalez, Ceferino *see*
– La rebellion militaire en espagne et l'incomprehension des democraties europeennes devant un aussi grave probleme
– Voie libre a la verite

Gonzalez Concepcion, Felipe *see* Lejania

Gonzalez Cuadrado, Antonio *see*
– Instituto general tecnico de badajoz. memoria del curso de 1904 a 1905...
– Memoria...curso de 1889 a 1890...
– Memoria...instituto de badajoz
– Programa de geografia

Gonzalez de Agueros, Pedro *see* Arbol cronologico que manifiesta los comisarios generales de indias del orden de san francisco y plan de todas las provincias...

Gonzalez de Carreras, Diego *see* Tertius articulus que...

Gonzalez De Cascorro, Raul *see*
– Arboles sin raices
– Concentracion publica
– Gente de playa giron
– Semilla
– Vidas sin domingo

Gonzalez de Cellorigo, M *see* Memorial de la politica necesario y util restauracion a la republica de espana

Gonzalez De La Calle, Pedro Urbano *see* Contribucion al estudio del bogotano

Gonzalez de la Calle, Pedro Urbano *see*
– Arias montano
– Consideraciones acerca de la segunda paradoja de el brocense
– Vida profesional y academica de francisco de sanchez de las brozas

Gonzalez de Manuel, Thomas *see* Verdadera relacion...las batuecas

Gonzalez de Mendoza, J *see*
– Del' historia della chinas...
– Histoire du grand royaume de la chine...
– Historia delas cosas mas notables, ritos y costvmbres...
– Nova et succincta, vera tamen historia de amplissimo, potentissimo china...

Gonzalez De Padrino, Flor *see* Escuela rural

Gonzalez De Soto, Cristobal M *see* Noticia historica de la republica de venezuela

Gonzalez Del Valle Y Carvaval, Emilio Martin *see* Poesia lirica en cuba

Gonzalez Del Valle Y Ramirez, Francisco *see*
– Cronologia herediana
– Del epistolario de heredia
– Heredia en la habana

Gonzalez Echegaray, Carlos *see* Morfologia y sintaxis de la lengua bujeba

Gonzalez, Edelmira *see* Gris mayor

Gonzalez, Eloy Guillermo *see* Historia estadistica de cojedes (desde 1771)

Gonzalez, F et al *see* Caracterizacion y propiedades de una vermiculita de badajoz

Gonzalez Fernandez, Hector *see* Historia de colombia

Gonzalez, Fernando *see* Santander

Gonzalez Garcia, Matias *see*
– Cuentos
– Tesoro del ausubal

Gonzalez Ginorio, Jose *see*
– Descubrimiento de puerto rico
– Tanama

Gonzalez Gonzalez, Valeriano *see* Rosa, la de villambro

Gonzalez, Graciela *see* Carne y alma

Gonzalez Guinan, Francisco *see* Historia contemporanea de venezuela

Gonzalez Hernandez, Juan *see*
– La s sociedades de credito y los ferrocarriles extremenos
– Las sociedades de credito y los ferrocarriles extremenos

Gonzalez Herrera, Edelmira *see* Alma llanera

Gonzalez Herrera, Julio *see*
– Gloria llamo dos veces
– Trementina cleren i bongo

Gonzalez Holguin, Diego *see*
– Gramatica y arte nueva de la lengva general de todo el peru, llamada lingua qquichua, o lengua del inca
– Gramatica y...el peru
– Vocabulario de la lengua general de todo el peru llamada lengua qquichua, o del inca

Gonzalez, J de M *see* Los hijos de la fortuna. novela original de costumbres espanolas

Gonzalez, Jose Emilio *see* Poetas puertorriquenos de la decada de 1930

Gonzalez, Jose Luis *see*
– 5 cuentos de sangre
– Paisa

Gonzalez, Josemilio *see* Cantico mortal a julia de burgos

Gonzalez, Juan Natalicio *see* Solano lopez y otros ensayos
Gonzalez, Juan Vicente *see* Presencia de juan vicente gonzalez
Gonzalez, julio. alfonso 9th. madrid, 1944 / Cereceda, F – Madrid: Razon y Fe, 1946 – 1 – sp Bibl Santa Ana [946]
Gonzalez Lopez, Felipe *see* Leyendas y tradiciones portoplantenas
Gonzalez, Luis Felipe *see*
– Obra cultural de don miguel obregon
– Origen y desarrollo de las poblaciones de heredia
Gonzalez, Manuel Dionisio *see*
– Apendice de la memoria historica
– Memoria historica de la villa
Gonzalez, Manuel Pedro *see*
– Estudios sobre literaturas hispano-americanas
– Indagaciones martianas
– Revaloracion de marti
– Rosalia de castro en ingles
Gonzalez Martinez, Enrique *see* Preludios
Gonzalez Menendez-Reigada, Albino *see* Catecismo patriotico espanol
Gonzalez, Miguel *see* Sangre en cuba
Gonzalez Montalvo, Ramon *see*
– Barbasco
– Tinajas
Gonzalez Montero, Belisario *see* Miscelanea
Gonzalez Olmedilla, Juan *see* Ofrenda de espana a ruben dario
Gonzalez Orellana, Carlos *see* Historia de la educacion en guatemala
Gonzalez, Otto Raul *see* Sombras era
Gonzalez Palencia, A *see*
– Madrid. archivo historico nacional. extracto del catalogo de documentos del consejo de indias...
– Moros y cristianos en espana medieval
Gonzalez, Palencia Angel *see* The flame of hispanicism
Gonzalez Patino, Julian *see* Nociones de geologia y prehistoria de colombia
Gonzalez Pintado, Gaspar *see* Paje, misionero y martir
Gonzalez R, Mario Gilberto *see* Pequena resena bio-bibliografica de carlos wyld os...
Gonzalez, Rafael *see* Sociedad extremena de fomento-reforma de proyecto banco hipitecario
Gonzalez Ramos, Vicente *see* Pregon de la santisima virgen de la victoria
Gonzalez Ricardo, Rogelio *see* Aguinaldos
Gonzalez Rios, Policarpo *see* Studies on the histology and cause of storage pitting of citrus fruits
Gonzalez Ruiz, Nicolas *see* Hernando cortes y francisco pizarro
Gonzalez Ruiz, Ricardo *see* El salvador de hoy
Gonzalez Serrano, Urbano *see*
– En pro y en contra
– Estudios criticos
– Goethe
– Preocupaciones sociales
– La psicologia del amor
– La psicologia fisiologica
Gonzalez, T *see* Censo de poblacion de las provincias y partidos de la corona de aragon en el siglo 16
Gonzalez, Thomas A *see* Caloosahatchee
Gonzalez Toledo, Aureliano *see* General eliseo payan, vicepresidente de la republi...
Gonzalez Valcarcel, Jose Manuel *see* Treinta anos de restauracion monumental en caceres
Gonzalez Valencia, Jose Maria *see* Separation of panama from colombia
Gonzalez, Valentin *see* El campesino
Gonzalez Velez, Francis *see* Remanso
Gonzalez Viquez, Cleto *see* Capitulos de un libro sobre historia financiera de
Gonzalez y Centeno, V *see* Memorias academicas de la real sociedad de medicina...
Gonzalez Y Contreras, Gilberto *see*
– Ausencia pura
– Cristal de epoca
– Ultimo cuadillo
Gonzalez y Diaz Tunon, Ceferino *see* Estudios sobre la filosofia de santo tomas
Gonzalez y Gomez de Soto, Juan Jose *see*
– Correspondiente en merida de la r.a. de la historia
– Emerita augusta. apuntes monograficos acerca de la catedral metropolitana de santa maria jerusalen
– Epitome historico de merida
Gonzalez y Gonzalez, Valeriano *see* Primicial (la desgracia pudo mas...odisea...el doctor amores...!pero no al axilo!)
Gonzalez y Grez, Juan J *see*
– Estudio...virgen maria...almoharin
– El obrero
Gonzalez Y Gutierrez, Diego *see* Nobles pasiones del 68
Gonzalez Zeledon, Manuel *see* Cuentos
Gonzalez-Blanco, Pedro *see*
– Presidente machado
– Problema de belice y sus alivios
– Trujillo
Gonzalez-Doria, Fernando De *see*
– Don mirocletes
– Mi compadre
Gonzaliz, Manuel Pedro *see* Antologia critica de jose marti
Gonze, Collin *see* South african crisis and united states policy
Gonzenbach Freire, Carlos *see* Study of resistance of some potato hybrids to infection
Gonzolo Picon, Febres *see* Obras completas...
Gooch, G P *see* British documents on the origin of the war, 1898-1914
Gooch, George Peabody *see*
– History and historians in the nineteenth century
– The history of english democratic ideas in the seventeenth century
Good, Adolphus Clemens *see* Journal
Good and faithful service / Capel-Cure, Edward – London, England. 18– – 1r – us UF Libraries [240]
Good, DL *see* College women, alcohol consumption, and negative sexual outcomes
Good earth of terrebonne – 1976 sep-1977 apr, 1977 apr/may – 2r – 1 – mf#4848495 – us WHS [071]
Good fight / Mahabane, Zaccheus R – Evanston, IL. 1966 – 1r – us UF Libraries [960]
Good form for all occasions; a manual of manners, dress and entertainment for both men and women / Hall, Florence Marion Howe – New York, London: Harper, 1914. vii,228p – 1 – us Wisconsin U Libr [390]
Good fortune – Brooklyn, NY. 1981-86 – 1 – us AJPC [071]
Good friday / Holland, Henry Scott – London, New York: Longmans, Green, 1899 [mf ed 1991] – 1mf – 9 – 0-7905-9967-8 – mf#1989-1692 – us ATLA [240]
Good gestes / Wren, Percival Christopher – New York, NY. 1929 – 1r – us UF Libraries [025]
Good government – Killen TX 1881-1906 – 1 – ISSN: 0017-2065 – mf#2894 – us UMI ProQuest [350]
Good grandmother / Hughes, Mary – London, England. 1823 – 1r – us UF Libraries [240]
Good hope news – Dallas TX. 1992 mar – 1r – 1 – (continued by: african herald (dallas, tx)) – mf#2569918 – us WHS [071]
Good housekeeping – New York NY 1885+ – 1,5,9 – ISSN: 0017-209X – mf#3118 – us UMI ProQuest [640]
Good, J I *see* The reformed reformation
Good, James I *see* Life of rev benjamin schneider
Good, James Isaac *see*
– The early fathers of the reformed church in the united states
– Famous missionaries of the reformed church
– Famous places of the reformed churches
– Famous reformers of the reformed and presbyterian churches
– The heidelberg catechism in its newest light
– Historical hand-book of the reformed church in the united states
– History of the reformed church in the united states, 1725-1792
– History of the reformed church in the u.s. in the nineteenth century
– History of the reformed church of germany, 1620-1890
– History of the swiss reformed church since the reformation
– John huss and the presbyterians and reformed
– Life of rev. benjamin schneider, d.d
– Life pictures of john calvin
– Minutes and letters of the coetus of the german reformed congregations in pennsylvania, 1747-1792
– The origin of the reformed church in germany
– Rambles round reformed lands
– Women of the reformed church
Good, John Booth *see*
– A vocabulary and outlines of grammar of the nitlakapamuk or thompson tongue
– Work in british columbia
Good, John Mason *see* Memoirs of the life and writings of the reverend alexander geddes, ll.d
The good life / Gandhi, Mahatma; ed by Chander, Jag Parvesh – Lahore: Free India Publications, [between 1900 and 1950] – (= ser Samp: indian books) – us CRL [170]
Good man / Alexander, William Lindsay – Edinburgh, Scotland. 1873 – 1r – us UF Libraries [240]
Good man is hard to find / O'connor, Flannery – New York, NY. 1955 – 1r – us UF Libraries [025]
Good morning – v1 n6-v2 n3 [1976 oct. 26-1977 may 3] – 1r – 1 – (cont: good morning in shelburne county) – mf#300950 – us WHS [071]
Good morning – v1-3. 1919-21 [all publ] – (= ser Radical periodicals in the united states, 1881-1960. series 1) – 7mf – 9 – $105.00 – us UPA [073]
Good morning in shelburne county – v1 n1-v1 n4 [1976 aug 24-sep 28] – 1r – 1 – (continued by: good morning (shelburne (ns))) – mf#310969 – us WHS [071]
Good neighbors : argentina, brazil, chile and / Herring, Hubert Clinton – New Haven, CT. 1941 – 1r – us UF Libraries [972]
Good neighbourship among nations / Ikatan Indonesia Untuk Perserikatan Bangsa-Bangsa – Djakarta, 1955-1957 – 3mf – 9 – (missing: 1955(1-2); 1956/1957(6/8)) – mf#SE-1917 – ne IDC [327]
Good newes from virginia / Whitaker, Alexander – 1613 – 9 – 5.00 – us Scholars Facs [978]
Good news / Church of God in Christ – v1 n1 [[1980?]] – 1r – 1 – mf#4114807 – us WHS [240]
Good news – Monona, Madison. 1982 feb 9-dec 21, 1983 jan 4-dec 27, 1984, 1985, 1986-1987 mar 30, 1987 mar 31-1988 jul 4, 1988 jul-1989 sep 26, 1989 sep 27-1990 dec – 8r – 1 – (cont: tri-m) – mf#702061 – us WHS [071]
Good news – Nashville. 1970-76. 2 reels – 1 – us L of C Photodup [780]
The good news : a semi-monthly undenominational religious periodical – Prescott, CW [Ont: Printed and pub at the "Evangelizer Office", [1861?-18– or 19–] – 9 – (incl ind) – mf#P06044 – cn Canadiana [240]
The good news *see* Chi-tu yen hsing lu hsin pien [ccm323]
Good news acid – v1 n1-3 [1971] – 1r – 1 – mf#1583442 – us WHS [071]
Good news express – 1996 mar, 1997 feb – 1r – 1 – mf#4027729 – us WHS [071]
Good news from a far country / Hawker, Robert – London, England. 1824 – 1r – us UF Libraries [240]
Good news herald – 1993 sep-1998 dec – 1r – 1 – mf#3118964 – us WHS [071]
Good news journal : a newspaper of positivity, the key to success – 1992 nov-1994 dec 27 – 1r – 1 – mf#2847490 – us WHS [150]
The good old days of honorable john company : being curious reminiscences illustrating manners and customs of the british in india during the rule of the east india company from 1600 to 1858 with brief notices of places and people of those times etc etc etc / Carey, W H [comp] – Calcutta: R Cambray & Co, 1906-1907 – (= ser Samp: indian books) – us CRL [306]
The good old times : the story of the manchester rebels of '45 / Ainsworth, William Harrison – London: Tinsley, 1873 [mf ed 1988] – 3v – 1 – mf#2169 – us Wisconsin U Libr [830]
The good old way : a sermon preached at the opening of the new presbyterian church, scarborough, on sabbath, 3rd february, 1850 / George, James – Toronto : A H Armour, 1850 – 1mf – 9 – mf#41528 – cn Canadiana [240]
Good one christian may do – London, England. 18– – 1r – us UF Libraries [240]
Good out of africa / Culwick, A T – 2mf – 9 – mf#363/12 – uk Microform Academic [960]
Good reasons for not being a congregationalist or the answer to the question, why am i not a congregationalist? – 1 – 5.00 – us Southern Baptist [242]
The good red earth : [novel] / Phillpotts, Eden – Toronto: W Briggs, 1901 – 4mf – 9 – 0-659-90446-2 – mf#9-90446 – cn Canadiana [830]
Good, Reynolds E *see* Brazil
Good roads for wisconsin – (1916 jul-1921 may], [1921 aug-1925 feb] – 2r – 1 – mf#918306 – us WHS [366]
Good samaritan – 1947-51* – (= ser Chinese christian coll 52) – 1r – 1 – (in chinese) – mf#ATLA S0296H – us ATLA [240]
The good samaritan : and other bible stories dramatized / Cole, Edna Earle – Boston: R G Badger; Toronto: Copp, Clark, c1915 – 2mf – 9 – 0-659-91891-9 – (ill by harold wagner) – mf#9-91891 – cn Canadiana [820]
Good Samaritan Medical Center [Milwaukee WI] *see* Center scope
Good shepherd / Chin, John – London, England. 1815 – 1r – us UF Libraries [240]
Good shepherd / Church of God in Christ – 1992 may/jun – 1r – 1 – mf#4027715 – us WHS [240]
Good shepherd / Parish, H – London, England. 1824 – 1r – us UF Libraries [240]
A good speed to virginia / Gray, Robert – 1609. Bound with Richard Rich, News from Virginia. 1610 – 9 – 5.00 – us Scholars Facs [978]
Good templar – Glasgow, Scotland, UK. 1892-1907. -w. 4 reels – 1 – uk British Libr Newspaper [072]
Good Templars, International Order of Wisconsin *see* North-western advance
Good tidings / Bisbee, Frederick Adelbert et al – Boston: Universalist Pub House, 1900 – 1mf – 9 – 0-524-07717-7 – mf#1991-3302 – us ATLA [240]
Good tidings – Ferndale WA: J B Boulet. v3-9. 1908-14 [mthly] – 1r – 1 – us Oregon Lib [241]
Good tidings, pertaining to the earth and the race as disclosed in the scriptures – New York: Published by an Association of Believers, 1871 (Hartford, Conn: Press of Case, Lockwood & Brainard) – 1mf – 9 – 0-524-05280-8 – mf#1992-0381 – us ATLA [220]
Good times / Franklin Co. Columbus – v1 n1. jun 1901-oct 1902 [mthly] – 1r – 1 – mf#B9750 – us Ohio Hist [073]
Good times : universal life. bulletin of the church of the times – 1968 jan 24-1969 oct 9, 1969 apr 2-1970 sep 18, 1970 sep 25-1972 aug 7, 1970 sep 25-1972 aug 9 – 4r – 1 – mf#1057360 – us WHS [071]
Good will : a collection of christmas stories / Pearse, Mark Guy – London: Wesleyan Conference Office, [187-?]Beltsville, Md: NCR Corp, 1978 (3mf); Evanston: American Theol Lib Assoc, 1984 (3mf) – 9 – 0-8370-0835-2 – mf#1984-4233 – us ATLA [240]
Good work for ontario : splendid record of the whitney government: success in every branch of the public service... – Toronto: Southam Press, [1911] – 1mf – 9 – 0-665-86473-6 – mf#86473 – cn Canadiana [325]
Good years *see* Hua nien (ccs30)
Goodall, Edward Basil Herbert *see* Some wemba words
Goodall, Elizabeth *see* Prehistoric rock art of the federation of rh0desia and nyasaland
Good-bye dolly gray / Kruger, Rayne – Philadelphia, PA. 1960 – 1r – us UF Libraries [960]
Goodbye to all that – v1 n1-40 [1970 sep 15-1973 jun] – 1r – 1 – mf#1112087 – us WHS [071]
Goodchild, R G *see* The limes tripolitanvs in the light of recent discoveries
Goode, Christine Mary *see* Preparation and negotiation
Goode, F *see* Christ
Goode, Francis *see* The better covenant practically considered
Goode, J F *see* History of tugalo baptist association, georgia
Goode, S W *see* Municipal calcutta
Goode, William *see*
– Aged christian's hope
– Altars prohibited by the church of england
– Case as it is
– The divine rule of faith and practice
– Is the reformation a blessing?
– Letter to sir w p wood
– Reply to the letter and declaration respecting the royal supremacy...
Goode, William H *see*
– Outposts of zion
– Outposts of zion with limnings of mission life
Goodell, Amelia *see* Haven family history and reunions, 1896
Goodell, Charles Le Roy *see*
– Pastoral and personal evangelism
– The pastor's vade mecum
Goodell, Thomas Dwight *see* A school grammar of attic greek
Goodell, William *see* The american slave code in theory and practice
Gooden, G M *see* Conflict in spain 1920-1937
Goodenow, Smith Bartlett *see*
– Bible chronology carefully unfolded
– Woman's voice in the church
Goodeve, Joseph *see* The law of evidence
Goodfellow, Clement Francis *see* Great britain and south african confederation, 1870-1881
Goodfellow, David Martin *see* Principles of economic sociology
Goodfellow review of crafts – Berkeley CA 1978-83 – 1,5,9 – ISSN: 0162-2765 – mf#11748,01 – us UMI ProQuest [740]
Goodger, D R *see* Papua new guinea patrol reports and related correspondence
Goodhart, C J *see* Lawfulness of marriage with a deceased wife's sister
Goodhue county historical news – (1967 jun-1982 jun] – 1r – 1 – mf#620716 – us WHS [071]
Goodhue, J A *see* The crucible
Goodison, William *see* A historical and topographical essay upon the islands of corfu, leucadia, cephalonia, ithaca, and zante
Goodland banner *see* Miscellaneous kansas state newspapers
Goodluck, W R *see* A letter, to the citizens of london
Goodman, George *see* The church in victoria during the episcopate of the right reverend charles perry
Goodman, Hardin Mcdonald *see* German influence on samuel taylor coleridge
Goodman, Paul *see*
– The synagogue and the church
– Zionism and the jewish diaspora
Goodman Service Club *see* Service news
Goodmann, Gustav *see* Probate proceedings and administration of estates
The goodness of god / Bascom, John – New York: G P Putnam; London: Knickerbocker Press, 1901 – 1mf – 9 – 0-8370-2563-X – mf#1985-0563 – us ATLA [210]
The goodness of god : a thanksgiving sermon preached in st alban's church, ottawa, on october 22nd, 1868 / Bedford-Jones, T – Ottawa?: Bell & Woodburn, 1868 – 1mf – 9 – mf#10253 – cn Canadiana [242]
Goodness of god acknowledged in recovery from sickness / Turner, William – Birmingham, England. 1808 – 1r – us UF Libraries [240]

Goodnow, Frank Johnson *see*
– Selected cases on government and administration
– Social reform and the constitution
Goodnow, Josephine A B *see* Great missionaries of the church
Goodpaster, Stacee *see* Perceptions of competence and control as predictors of athletes'interpretation of parental involvement in gymnastics
Goodrich, Arthur et al *see* The story of the welsh revival
Goodrich baptist church. charleston county, south carolina : church records – 1963-1980 – 1r – 1 – $5.00 – us Southern Baptist [242]
Goodrich, Charles Augustus *see* The ecclesiastical class book
Goodrich, Charles B *see* The science of government as exhibited in the institutions of the u.s.
Goodrich, Chauncey *see*
– Do missions pay?
– Pocket dictionary (chinese-english) and pekingese syllabary
Goodrich, Joseph King *see* The coming china
Goodrich, Samuel G *see* Popular biography: embracing the most eminent characters of every age, nation, and profession.
Goodrich, Samuel Griswold *see* A pictorial history of america
Goodrich, William Winton *see* Legal ethics, the duty of the hour
Goodrick, Alfred Thomas Scrope *see* The book of wisdom
Goodridge, Richard E W *see* On the proposed change of time marking to a decimal system
Goodsell, Charles T *see* Administration of a revolution
Goodsell, Fred Field *see* American board in china, 1830-1950
Goodsir, Joseph Taylor *see* The westminster confession of faith examined on the basis of the other protestant confessions
Goodspeed, Calvin *see*
– Book of genesis
– Some unsolved problems of the higher criticism
Goodspeed, E J *see* The conflict of severus patriarch of antioch by athanasius
Goodspeed, Edgar Johnson *see*
– Die aeltesten apologeten
– The bixby gospels
– The book of thekla
– The epistle to the hebrews
– The freer gospels
– A full history of the wonderful career of moody and sankey in great britain and america
– The harvard gospels
– The haskell gospels
– Index apologeticus
– Index patristicus
– The martyrdom of cyprian and justa
– The newberry gospels
– The story of the new testament
Goodspeed, George Stephen *see*
– A history of the ancient world
– A history of the babylonians and assyrians
– Israel's messianic hope to the time of jesus
– The world's first parliament of religions
Goodspeed's Book Shop *see* Flying quill
[Goodspring-] gazette – NV. 29 jul 1916; 11 may 1918; 1919-21 [wkly] – 2r – 1 – $120.00 – mf#U04568 – us Library Micro [071]
Goodway hymns and songs for general use in all holiness meetings – 1 – $50.00 – us Presbyterian [780]
Goodwill Industries of Madison *see* Goodwill news
Good-will messenger – Valley City, ND: Peoples Opinion Printing Co, apr 1928-aug 1935// (wkly) – 1 – (merged with: peoples opinion to form: peoples opinion and good-will messenger) – mf#06831 – us North Dakota [071]
Goodwill news / Goodwill Industries of Madison – 1975 sep/oct-1977 mar/apr – 1r – 1 – mf#369065 – us WHS [338]
Goodwin, Daniel Raynes *see* Notes on the late revision of the new testament version
Goodwin, E P et al *see* Jew and gentile
Goodwin, Edward Lewis *see* The colonial church in virginia
Goodwin, Ernst C *see* An evaluation of the square and staggered stance
Goodwin, Francis *see*
– Domestic architecture
– Rural architecture
Goodwin, Frank Judson *see* A harmony of the life of st paul
Goodwin, Gwendoline *see* Anthology of modern indian poetry
Goodwin, Harvey *see*
– Church of england admonished by the examples of former times
– Church of england past and present
– A commentary on the gospel of s matthew
– Creation
– The doctrines and difficulties of the christian faith contemplated from the standing ground afforded by the catholic doctrine of the being of our lord jesus christ
– Lesson taught by human suffering
– Memoir of bishop mackenzie
– Plain thoughts concerning the meaning of holy baptism
– Reasonable service
Goodwin, Harvey et al *see* Credentials of christianity
Goodwin, Henry Martyn *see* Christ and humanity
Goodwin, Jeff E *see* Bandwidth knowledge of results in motor skill performance and learning
Goodwin, John Albert Reed *see* Long range planning in the face of change
Goodwin, Nolan W *see* Winterton remembered
Goodwin, T *see* A short account of the art of polychrome
Goodwin, T S *see* Congregationalism
Goodwin, Thomas *see*
– Indispensible and absolute necessity of regeneration
– Sermons and notes of sermons
Goodwin, Thomas A *see* Lovers three thousand years ago
Goodwin, William Brownell *see* Spanish and english ruins in jamaica
Goodwin, William W *see* Greek reader
Goodwin's town officer / Thomas, Benjamin F – 4th ed. Worchester (Mass.): Dorr, Howland & Co, 1837 – 4mf – 9 – $6.00 – (covers the range of duties and legal responsibilities of various municipal officers in early 19th century massachusetts) – mf#LLMC 96-065 – us LLMC [350]
Goody, Jack *see* Ethnography of the northern territories of the gold coast
Gool, J van *see* De nieuwe schouburgh der nederlantsche kunstschilders en schilderessen...
Goold, W H *see*
– Maynooth endowment
– Patronage opposed to the independence of the church
– Speech delivered at the second annual meeting of the scottish refor...
Gooldy grapevine – v1-3 n3/4 [1974 may 15-1976 summer/fall] – 1r – 1 – mf#369061 – us WHS [071]
Goole and marshland gazette, and monthly miscellany – [Yorkshire & Humberside] feb 1854-mar 1870 [mf ed 2004] – 1 – uk Newsplan [072]
Goole, marshland and howden gazette and snaith and rawcliffe advertiser – [Yorkshire & Humberside] East Riding 12 mar 1870-mar 1873 [mf ed 2003] – 6r – 1 – uk Newsplan [072]
Goole saturday journal – [Yorkshire & Humberside] East Riding 11 may 1889-26 mar 1941 [mf ed 2003] – 47r – 1 – (missing: 1890, 1897-98, 1900; cont by: goole saturday journal and goole wednesday journal [jan 1893-dec 1913]; goole journal [jan 1914-mar 1941]) – uk Newsplan [072]
Goondiwindi argus – Goondiwindi – at Pascoe [079]
Goonewardene, T W *see* Sinhalese diary, 1893
Goor, Yehudah *see* Kitsur divre ha-yamin le-'am yisra'el me-reshit heyoto 'ad ha-yom h...
Goossens, Eduard *see* Die frage nach makkabischen psalmen
Gooszen, M A *see*
– Aanteekeningen ter toelichting van den strijd over de praedestinatie in het gereformeerd protestantisme
– Bijdrage tot de kennis van het gereformeerd protestantisme
Gooszen, Maurits Albrecht *see* De heidelbergsche catechismus en het boekje van de breking des broods
Goote, Thor *see*
– Gluehender tag
– 'Rangehen Ist Alles!
– Sie werden auferstehen!
– Wir tragen das leben
Gopal krishna gokhale : his life and speeches / Hoyland, John Somervell – Calcutta: YMCA Pub House, 1948 – (= ser Samp: indian books) – us CRL [954]
Gopal, Mysore Hatti *see*
– Mauryan public finance
– Towards a realistic tax policy for india
GOPaper / Nevada Republican Central Committee – v1-4 [1974 jul-1977 aug /sep/oct] – 1r – 1 – mf#369064 – us WHS [325]
Gopinatha, Diksita, Bhatta *see* Bhattagopinathadiksitaviracita samskararatnamala...
Goppelzrieder, Casparus *see* Psalmodia sacra cum litaniis et aliis precationibus singulis hebdomadae diebus accomodata
Gora / Tagore, Rabindranath – London: Macmillan and Co, 1924 – (= ser Samp: indian books) – us CRL [490]
Gorakhnath and the kanphata yogis / Briggs, George Weston – Calcutta: YMCA Pub House; New York: Oxford University Press, 1938 – (= ser Samp: indian books) – us CRL [280]
Goral yehude romaniyah / Kuperstein, Leib – Tel-Aviv, Israel. 1943/44 – 1r – us UF Libraries [939]
Gorbachev, I A *see* Tovarichestva polnye, na vere, kreditnye, ssudo-sberegatelnye, trudovye i s peremennym kapitalom
Gorbunova, L *see* Chernaia sotnia
Gorce, Paul-Marie de la *see* Tendances de la politique francaise et europeenne vis-a-vis du conflit israelo-arabe
Gorchakov, M *see*
– Monastyrskii prikaz
– O zemelnykh vladeniiakh vserossiiskikh mitropolitov, patriarkhov i sv sinoda
Gorchakovskii, P L *see* 8 vsesoiuznoe soveshchanie "izuchenie i osvoenie flory i rastitelnosti vysokogorii"
De gordel der waarheid : leerrede over efezen 6:14a. afscheids-preek...12 maart, 1876, te grand rapids, mich / Boer, Geert Egberts – Grand Rapids, MI: Standaard Drukkerij, 1876 [mf ed 1993] – (= ser Reformed church coll) – 51p on 1mf – 9 – 0-524-06706-6 – mf#1991-2736 – us ATLA [227]
The gordian knot : or, the problem which baffles infidelity / Pierson, Arthur Tappan – New York: Funk & Wagnalls, 1902 [mf ed 1985] – 1mf – 9 – 0-8370-4749-8 – mf#1985-2749 – us ATLA [230]
Gordils, Jose *see* Violetas
Gordimer, Nadine *see*
– The black interpreters
– Lying days
– South african writing today
Gordo Moreno, Angel *see*
– A la legion cantos de amor y de dolor de espana por...
– Mis cantares
The Gordoa Family *see* The gordoa family papers
The gordoa family papers : personal papers covering an important period in mexican history – 1822-46 [mf ed Norman Ross Publ] – 4r – 1 – (with p/g) – us UMI ProQuest [972]
Gordon, A *see* Itinerarium septentrionale
Gordon, A D *see*
– Briefe aus palaestina
– Selected essays
Gordon, Aaron *see* Teshuvot milu'ot even
Gordon, Adele *see* Transformation of south african farm schools
Gordon, Adoniram Judson *see*
– Ecce venit
– The holy spirit in missions
– In christ
– The ministry of healing
– The ministry of women
– The twofold life
Gordon, Aharon Ben Me'ir *see* Even me'ir
Gordon, Alexander *see*
– Heads of english unitarian history
– Heresy
– The personality of michael servetus (1511-1553
– Report of an official visit to transylvania
Gordon, Alexander Reid *see*
– The early traditions of genesis
– The poets of the old testament
Gordon, Andrew *see* Our india mission
Gordon, Andrew Robertson *see* Rapport sur l'expedition a la baie d'hudson en 1886
Gordon, Ann *see* The papers of elizabeth cady stanton and susan b. anthony
Gordon, Anna Adams *see* The beautiful life of francis e willard
Gordon, Arthur *see*
– Fijian pamphlets collected by sir arthur gordon
– Fijian pamphlets collected by sir arthur gordon, vols 1-5
– High commission fiji pamphlets
– High commission, fiji, pamphlets
Gordon, Bert *see* Journal of chromatographic science
Gordon Cumming, Constance Frederica *see*
– Two happy years in ceylon
– Work for the blind in china
Gordon, Douglas Hamilton *see* An evaluation of water and soil resources for irrigation in the wasbank drainage basin
Gordon, Elizabeth Anna *see*
– World-healers
– World-healers; or, the lotus gospel and its bodhisattvas, compared with early christianity.
Gordon, Ernest B *see* Adoniram judson gordon
Gordon, George A *see*
– Through man to god
– Ultimate conceptions of faith
Gordon, George Angier *see*
– Aspects of the infinite mystery
– The christ of to-day
– The mission of the prophet
– The new epoch for faith
– Religion and miracle
– Revelation and the ideal
– The witnesses to immortality in literature, philosophy and life
Gordon, George Angier et al *see* The claims and opportunities of the christian ministry
Gordon, George Hamilton *see* Earl of aberdeen's correspondence with the rev dr chalmers
Gordon, George J R *see* New cathedral for aberdeen
Gordon, Hanford Lennox *see* A lecture on the harper's ferry tragedy
Gordon, Helen Cameron *see* Syria as it is
Gordon, I *see* Liturgia et potestas in re liturgica
Gordon, J *see* Controversiarum epitomes...
Gordon, James *see* Appeal to unionists
Gordon, James B *see* An historical and geographical memoir of the north-american continent
Gordon, James Bentley *see* An historical and geographical memoir of the north-american continent, its nations and tribes
Gordon, James Edward *see* British legislature
Gordon, John *see*
– A letter to the subscribers to the 8th edition of the encyclopaedia britannica
– Nonconformity and liberty
– Power of faith
– Thomas aikenhead
Gordon, John James Hood *see* The sikhs
Gordon journal – Gordon, NE: H G Lyon, 1891-v26 n17 [ie 19] mar 29 1917 (wkly) [mf ed 1892-1917 (gaps)] – 8r – 1 – (occasional articles in german. merged with: sheridan county democrat to form: sheridan county democrat and gordon journal. some irregularities in numbering) – us NE Hist [071]
Gordon, Judah Leib *see*
– Mivhar shirav
– 'Olam Ke-Minhago
– Shire-'allah
Gordon, King *see* United nations in the congo
Gordon, L et al *see* Etapy zhiznennogo tsikla i byt rabotaiushchei zhenshchiny
Gordon, Patrick *see* Geography anatomiz'd
Gordon, Robert Winslow *see*
– American folksong texts
– Folk-songs of america
Gordon, Samuel *see* Birobidzshaner toyshvim
Gordon, Samuel Dickey *see*
– Quiet talks about calvary
– Quiet talks about jesus
– Quiet talks about our lord's return
Gordon, Saul *see* Gordon's annotated forms of agreement
Gordon, Stewart *see* Cecile staub genhart
Gordon, Thomas *see* Craftsmen
The gordon w prange collection : the most comprehensive collection of publications issued in japan from 1945-1949 / Supreme Commander of the Allied Powers. Civil Censorship Detachment – [mf ed Norman Ross Publ 1996] – 13,743 titles on 62,976mf [magazine subsets] 44r [justin williams & charles kades papers] – 9,1 – (most comprehensive coll of magazines publ in japan during the early post-war years. most magazines are in japanese, with a few in english and other languages. a 3v p/g (in japanese) for the prange magazine coll. ind on microfilm & p/g available for the justin williams papers. the charles l kades papers incl a finding aid) – University of Maryland – us UMI ProQuest [950]
Gordon, William N *see* The leisure activity selection process
Gordon, William Robert *see* The science of revealed truth impregnable
Gordon, Wolff von *see* Die dramatische handlung in sophokles' "koenig oidipus" und kleists "der zerbrochene krug"
Gordon's annotated forms of agreement / Gordon, Saul – New York: Prentice-Hall, 1923. 904p. LL-401 – 1 – us L of C Photodup [340]
Gordonvale burial register – 1mf – 9 – A$6.00 – at Cairns [929]
Gordonvale monumental inscriptions – 2mf – 9 – A$12.00 – at Cairns [929]
Gore, Catherine *see* The historical traveller
Gore, Charles *see*
– The basis of anglican fellowship in faith and organization
– The body of christ
– The church and the ministry
– The clergy and the creeds
– The creed of the christian
– The incarnation of the son of god
– Leo the great
– Lux mundi
– The ministry of the christian church
– The mission of the church
– Orders and unity
– The permanent creed and the christian idea of sin
– The question of divorce
– St paul's epistle to the romans
– Spiritual efficiency
– Thoughts on religion
– William law's defence of church principles
Gore, Charles et al *see* Oxford house papers. third series
Gore gazette and ancaster, hamilton, dundas and flamborough advertiser – Ancaster, ON. 1827-29 – 1r – 1 – ISSN: 1490-7224 – cn Library Assoc [071]
Gore, John *see* King charles 5
Gore, John Ellard *see* The visible universe
Gore, Montague *see*
– A few brief hints on the causes of the present distress
– Lecture on the products and resources of british india...
– Letter to his grace the duke of wellington etc on the present state of affairs in india
– A postscript to the third edition of suggestions on the amelioration of the present condition of ireland

– Suggestions for the amelioration of the present condition of ireland
– Thoughts on the present state of ireland
Gore, N A *see* A bibliography of ramayana
Gore, Richard T *see* Instrumental works of georg muffat
Gore rinoyera re sangano – Mariannhill, South Africa. 1918 – 1r – us UF Libraries [960]
Gore standard – jan-jun 1907; sep 1907-dec 1910 – 12r – 1 – (previous title: southern standard) – mf#85.11 – nz Nat Libr [079]
Gore standard *see* Southern standard
Gore, T J et al *see* That they all may be one
Gore, Willard Clark *see* The imagination in spinoza and hume
Goreh, Nehemiah *see* Four lectures
Gorelik, Aaron *see* Shturemdike yorn
Gorelik, Schmarja *see* Yidishe kep
Goren, Charles Henry *see* Point count bidding in contract bridge
Gorenie i vzryv : materialy chetvertogo vsesoiuznogo simpoziuma po goreniiu i vzryvu, 23-27 sentiabria 1974 g. / Akademiia nauk SSSR, Otdelenie ordena Lenina Instituta khimicheskoi fiziki; ed by Stesik, L N – Moskva: Nauka, 1977 – us CRL [947]
Gorets – Vladikavkaz, Russia 1906 [mf ed Norman Ross] – 1 – (reel contains short runs of multiple titles; for complete listing of titles on a reel, please inquire) – mf#nrp-2223 – us UMI ProQuest [077]
Gorev, B *see* Anarkhizm v rossii
Gorev, B I *see* Anarkhisty, maksimalisty i makhaevtsy
Gorey correspondent – Gorey, Ireland. 2 feb 1861-19 dec 1863; 1864-1892 – 22 1/4r – 1 – (aka: gorey correspondent and arklow standard. incorp with: enniscorthy recorder from 1893) – uk British Libr Newspaper [072]
Gorey correspondent and arklow standard *see* Gorey correspondent
Gorgeous gallery of gallant inventions (1578) / Proctor, Thomas – Cambridge, MA. 1926 – 1r – us UF Libraries [025]
Gorgon – v1-49. 1818-19 [all publ] – (= ser Radical periodicals of great britain, 1794-1914. period 1) – 5mf – 9 – $95.00 – us UPA [335]
Gorgoniev, IUrii Aleksandrovich *see* The khmer language
Gorham, Barlow Weed *see*
– Camp meeting manual
– Concerning them that are asleep
– God's method with man
Gorham case briefly considered / Dodsworth, William – London, England. 1850 – 1r – us UF Libraries [240]
Gorham, Charles Turner *see*
– Ethics of the great religions
– The first easter dawn
Gorham controversy briefly noticed / Macdonnell, Eneas – London, England. 1850 – 1r – us UF Libraries [240]
Gorham, George Cornelius *see*
– The case of the rev g c gorham against the bishop of exeter
– Examination before admission to a benefice by the bishop of exeter
Gorham v the bishop of exeter / Hook, Walter Farquhar – London, England. 1850 – 1r – us UF Libraries [240]
Gori, A F *see* Symbolae litterariae opvscvla varia philologica scientifica antiqvaria...et monumenta medii aevi
Gorilas / Fonseca, Gondin Da – Sao Paulo, Brazil. 1963 – 1r – us UF Libraries [972]
Gorin, I *see* Ishimskii krai
Gorinov, M M *see* 20-e gody
Gorinov, M M et al *see* Istoricheskoe znachenie nepa sbornik nauchnykh trudov
Gor'kovskaia kommuna – Nizhnij Novgorod, Russia 1973 [mf ed Norman Ross] – 4r – 1 – mf#nrp-1212 – us UMI ProQuest [077]
Gor'kovskaia pravda – Nizhnij Novgorod, Russia 1974-88 [mf ed Norman Ross] – 1r – 1 – mf#nrp-1213 – us UMI ProQuest [077]
Gor'kovskii rabochii – Nizhnij Novgorod, Russia 1975-85 [mf ed Norman Ross] – 1r – 1 – mf#nrp-1214 – us UMI ProQuest [077]
Gorman, Kathleen *see* Recommendations for the process of producing an initial season of the nutcracker ballet
Gorman, Michael James *see* A manual of county court practice in ontario
Gorman, Monica E *see* Test-retest study of the biodex closed kinetic chain attachment for the upper extremity
Gorman, Robert A *see* Copyright law
Gorman, W A R *see* Simple silozi
Gornik codzienny = The miner's daily – Wilkes-Barre, PA: Gornik Pub Co, apr 1-jun 19 1922 – 1r – 1 – us CRL [071]
Gornye tadzhiki / Ginzburg, Vul'f Veniaminovich – Moskva: Izd-vo Akademii nauk SSSR 1937 [mf ed Bloomington IN: Indiana Uni Lib, Preservation Dept 1984] – 1r – 1 – us Indiana Preservation [390]
Gorod i derevnia – 1923-1927(13/14) – 91mf – 9 – mf#COR-580 – ne IDC [335]
Gorodiski, Jonah *see* Aroysgevorfene reyd
Gorodovikov, Oka Ivanovich *see* V riadakh pervoi konnoi
Gorodskaia duma : biul / ed by Zaitsev, I I – Perm': Gor uprava, 1919 [1919 6 ianv-] – (= ser Asn 1-3) – n1-14 [1919] item 104, on reel n21 – 1 – mf#asn-1 104 – ne IDC [077]
Gorodskie lombardy v rossii / Serebriakov, Ia A – Spb, 1907 – 1mf – 9 – mf#REF-463 – ne IDC [332]
Gorodskie obshchestvennye banki rossii : obzor ikh deiatel'nosti po 1 ianvaria 1871 goda / Ososov, Vla – Spb, 1872. 3v – 4mf – 9 – mf#REF-341 – ne IDC [332]
Gorodskoe delo – Spb., Pg., 1909-1917 – 412mf – 9 – (missing: 1909(24); 1914(21); 1915(20); 1916(2-3); 1917(4-24)) – mf#R-2348 – ne IDC [077]
Gorodskoe ot ognia strakhovanie so vzaimnoiu mezhdu gorodami garantiei : doklad saratov dume glasnogo va. korobkova. (s proektom ustava). reshenie pozharnogo voprosa v gorodakh zavisit ot organizatsii strakhovaniia / Korobkov, VA – Spb, 1887 – 1mf – 9 – mf#REF-446 – ne IDC [332]
Gorodskoi vestnik : [ezhedn munitsip gaz: izd samar gor dumy] / ed by Vozdvizhenskii, V N – Samara: Gor upr, 1918 [1906 17 dek-1918 27 sent] – (= ser Asn 1-3) – n1-96 [1918] [gaps] item 105, on reel n21 – 1 – mf#asn-1 105 – ne IDC [077]
Goron, Marie Francois *see*
– L'amour a paris, nouveaux memoires, 1
– L'amour a paris, nouveaux memoires, 2
– L'amour a paris, nouveaux memoires, 3
– L'amour a paris, nouveaux memoires, 4
Goropius, J *see* Opera
Gorospe, Manuel Ignacio de *see* El dr d manuel ignacio de gorospe y padilla
Gorozhanskii, I I *see* Damaskin semenov-rudnev, episkop nizhegorodskii [1737-1795]
Gorr, Adolph *see* The influence of greek antiquity on modern german drama
Gorraeus, Ioannis *see* Definitionum medicarum libri 24 literis graecis distincti (ael3/1)
Gorrell's history of the american expeditionary forces air service, 1917-1919 / U.S. Army. American Expeditionary Forces – (= ser Records of the american expeditionary forces (world war 1), 1917-1923) – 58r – 1 – (with printed guide) – mf#M990 – us Nat Archives [355]
Gorrie, P Douglass *see* The lives of eminent methodist ministers
Gorrie, Peter Douglass *see*
– The churches and sects of the united states
– Episcopal methodism as it was and is
– History of the methodist episcopal church in the united states
– The lives of eminent methodist ministers
Gorriti, Juana Manuela *see* Oasis en la vida
El gorro de dormir / Vargas, Adolfo de – 1876 – 9 – sp Bibl Santa Ana [830]
Gorse, Jean Eugene *see* Territoire des comores
Gorshkov, D S *see* Golos iuga [elisavetgrad: 1918-1919]
Gorshkov, G P et al *see* Seismotektonika alpiiskogo skladchatogo poiasa i
Gorskaia pravda – Vladikavkaz, Russia 1922-24 [mf ed Norman Ross] – 11r – 1 – mf#nrp-2065 – us UMI ProQuest [077]
Gorskii, A V *see*
– Opisanie slavianskikh rukopisei moskovskoi sinodalnoi (patriarshei) biblioteki
– Opisanie velikikh chetikh-minei makariia mitropolita vserossiiskago
Gorst, Harold Edward *see* China
Gorst, John Eldon *see* The maori king
Gorter, James Polk *see* Law of evidence
Gorter, Richard *see* "Mehr licht"
Gorton, openshaw and bradford reporter : a localised edition of the ashton reporter – [NW England] Manchester 30 may 1874-7 jun 1924 [mf 1874,1876,1900] – 1 – (wanting 1896,1911; cont by: gorton & openshaw reporter [14 jun 1924-19 jun 1936]; gorton & openshaw reporter & droylsden & clayton herald [26 jun 1936-1969]) – uk Newsplan [072]
Gorvie, Max *see* Our people of the sierra leone protectorate
Gorzny, Willi [comp] *see*
– Biografisch archief van de benelux
– Deutsches biographisches archiv
'Gos Lo-tsa-ba Gzon-nu-dpal *see* The blue annals
Gosche, Richard *see* Ueber ghazzaalais leben und werke
Goschen, George Joachim *see* Reports and speeches on local taxation
Goschen, George Joachim Goschen, 1st viscount *see* The cry of "justice to ireland"
Gosford, Archibald Acheson, Earl of *see* Papers relative to the affairs of lower canada
Gosford star – Gosford. jul 1973-dec 1979, jan 1982-dec 1983 – 19r – at Pascoe [079]
Gosford times – Gosford. jul 1897-dec 1907, jul-dec 1911, jan 1929-dec 1940 (misc periods), jan 1952-dec 1958, jan 1960-dec 1961 – 20r – A$1307.90 vesicular A$1417.90 silver – at Pascoe [079]
Gosford times – jan 1915-dec 1928 – 14r – 9 – A$462.00 vesicular 539.00 silver – at Pascoe [079]
Goshen 1781-1898 – Oxford, MA (mf ed 1983) – (= ser Massachusetts vital records) – 6mf – 9 – 0-931248-34-5 – (mf 1-2: early records 1781-1820. mf 3: early records 1821-43. mf 4: m,b,d,intention 1822-53. mf 5-6: b,m,d 1844-98) – us Archive [978]
Goshen and the shrine of saft el henneh : (1885) / Naville, E – London, 1887 – (= ser Mees 4) – 3mf – 8 – €7.00 – ne Slangenburg [930]
Goshen baptist church. lincoln county. georgia : church records – 1802-67 – 1 – 9.72 – us Southern Baptist [242]
Goshen, New Hampshire. Goshen Baptist Church *see* Records
Goshen trails hamilton county historical society news / Hamilton County Historical Society (IL) – v1 n1-v18 n4 [1965 apr-1982 oct] – 1r – 1 – mf#369059 – us WHS [978]
Gosky, Martino *see* Arbustum vel arboretum augustaeum
Goslarsche zeitung – Goslar DE, 1976- – ca 8r/yr – 1 – gw Misc Inst [074]
Goslarsche zeitung *see* Harzburger zeitung
Gosling, W G *see* Labrador
Gosman, Abraham *see* Historical sketches of the missions in japan, korea
Goson kensetsu riron, ichimei, chugoku minzoku no zento / Liang, Shu-ming – Tokyo: Dai Asea Kensetsusha, Showa 17 [1942] – (= ser P-k&k period) – us CRL [338]
Gospel / Tyerman, Luke – London, England. 1851 – 1r – us UF Libraries [226]
The gospel according to darwin / Hutchinson, Woods – Chicago: Open Court, 1898 – 1mf – 9 – 0-8370-3702-6 – mf#1985-1702 – us ATLA [210]
The gospel according to darwin / Hutchinson, Woods – London, 1898 – (= ser 19th c evolution & creation) – 3mf – 9 – mf#1.1.4274 – uk Chadwyck [575]
The gospel according to john = Das evangelium nach johannes / Lange, Johann Peter; ed by Schaff, Philip – New York: Charles Scribner, 1871 [mf ed 1985] – (= ser A commentary on the holy scriptures. new testament 3) – 2mf – 9 – 0-8370-5977-1 – (trans fr german by ed) – mf#1985-3977 – us ATLA [226]
The gospel according to luke = Das evangelium nach lukas / Oosterzee, Johannes Jacobus van – New York: Scribner, Armstrong, 1886, c1868 [mf ed 1986] – (= ser A commentary on the holy scriptures. new testament 2/2) – 1mf – 9 – 0-8370-6768-5 – (trans fr german by philip schaff & charles casey starbuck) – mf#1986-0768 – us ATLA [227]
The gospel according to luke / Luce, H K – 1936 – 9 – $15.00 – us IRC [240]
The gospel according to luke / Riddle, Matthew Brown – New York: Charles Scribner, 1882, c1881 – 1mf – 9 – 0-8370-9731-2 – mf#1986-3731 – us ATLA [226]
The gospel according to luke : with notes, comments, maps, and illustrations / Abbott, Lyman – New York: A S Barnes, 1878 – 1mf – 9 – 0-8370-2020-4 – mf#1985-0020 – us ATLA [226]
The gospel according to mark / Alexander, Joseph Addison – New York: Charles Scribner, 1858 – 2mf – 9 – 0-7905-0060-4 – mf#1987-0060 – us ATLA [226]
The gospel according to mark = Evangelium nach markus / Lange, Johann Peter – 4th ed. New York: Charles Scribner, 1886, c1866 [mf ed 1986] – (= ser A commentary on the holy scriptures. new testament 2/1) – 1mf – 9 – 0-8370-6749-9 – (english trans fr german with additions by william g t shedd) – mf#1986-0749 – us ATLA [226]
The gospel according to mark / Hort, A F – 1914 – 9 – $10.00 – us IRC [240]
The gospel according to mark / Riddle, Matthew Brown – New York: Charles Scribner, 1881 – 1mf – 9 – 0-8370-9732-0 – mf#1986-3732 – us ATLA [226]
The gospel according to matthew / Alexander, Joseph Addison – New York: Scribner, Armstrong, 1873, c1860 – 2mf – 9 – 0-7905-1623-3 – mf#1987-1623 – us ATLA [226]
The gospel according to matthew : together with a general theological and homiletical introduction to the new testament = Das evangelium nach matthaeus / Lange, Johann Peter – New York: Charles Scribner, 1865, c1864 [mf ed 1986] – (= ser A commentary on the holy scriptures. old testament 1) – 2mf – 9 – 0-8370-6750-2 – (trans fr 3rd german ed into english, with additions original and selected by philip schaff. incl bibl) – mf#1986-0750 – us ATLA [226]
The gospel according to matthew / McNeile, Alan Hugh – 1915 – 9 – $18.00 – us IRC [240]
The gospel according to matthew / Schaff, Philip – New York: Charles Scribner, 1882, c1881 – 1mf – 9 – 0-8370-9740-1 – mf#1986-3740 – us ATLA [226]
The gospel according to matthew / Williams, Nathaniel Marshman – Boston: Gould & Lincoln, 1873, c1870 – 1mf – 9 – 0-8370-5858-9 – mf#1985-3858 – us ATLA [226]
The gospel according to paul : a sermon delivered sep 17 1828... / Beecher, Lyman – Boston: Publ by request of the Church [by] T R Marvin, Printer, 1829 – 1mf – 9 – 0-7905-3302-2 – mf#1987-3302 – us ATLA [240]
The gospel according to peter, and the revelation of peter : two lectures on the newly recovered fragments, together with the greek texts / Robinson, Joseph Armitage – 2nd ed. London: C J Clay; New York: Macmillan [dist] 1892 [mf ed 1985] – 1mf – 9 – 0-8370-5844-9 – (incl bibl ref) – mf#1985-3844 – us ATLA [226]
The gospel according to s john : translated from the eleven oldest versions except the latin, and compared with the english bible – London: Joseph Masters, 1862 – 5mf – 9 – 0-8370-1687-8 – (incl bibl ref) – mf#1987-6114 – us ATLA [226]
The gospel according to s luke : in greek: after the westcott and hort text / ed by Wright, Arthur – London, New York: Macmillan, 1900 [mf ed 1986] – 3mf – 9 – 0-8370-9439-9 – (int & ann in english. text in greek. incl ind) – mf#1986-3439 – us ATLA [226]
The gospel according to s mark : illustrated (chiefly in the doctrinal and moral sense) from ancient and modern authors – 2nd ed. London: J. Masters, 1864 – 2mf – 9 – 0-8370-1194-9 – (incl ind) – mf#1987-6024 – us ATLA [226]
The gospel according to saint matthew : and part of the first chapter of the gospel according to saint mark / Cheke, John – Cambridge [UK]: J and J Deighton, 1843. Chicago: Dep of Photodup, U of Chicago Lib, 1968 (1r); Evanston: American Theol Lib Assoc, 1984 (1r) – 1 – 0-8370-0395-4 – (trans into english fr greek) – mf#1984-B076 – us ATLA [226]
The gospel according to saint matthew – St Louis, MO: B Herder, 1898 – 1mf – 9 – 0-524-07338-4 – mf#1992-1069 – us ATLA [226]
The gospel according to saint matthew : with maps, notes and introduction / Carr, Arthur – Cambridge: University Press, 1894 – 1mf – 9 – 0-8370-9768-1 – (incl ind) – mf#1986-3768 – us ATLA [226]
The gospel according to saint matthew : with notes critical and practical / Sadler, Michael Ferrebee – 5th ed. London: G Bell, 1890 – 2mf – 9 – 0-524-05693-5 – mf#1992-0543 – us ATLA [226]
The gospel according to satan / Grey, Standish – 4th ed. London:Kerby & Endean; New York:James Pott, 1881 – 1mf – 9 – 0-8370-3393-4 – mf#1985-1393 – us ATLA [230]
Gospel according to st john – London: Cassell, Petter & Galpin, [18–?] – (= ser New testament commentary) – 2mf – 9 – 0-524-05393-6 – (incl bibl ref) – mf#1992-0403 – us ATLA [226]
The gospel according to st john : authorised version: with introduction and notes / Clark, Henry William – New York: Fleming H Revell, [1907?] – 1mf – 9 – 0-7905-1322-6 – (incl ind) – mf#1987-1322 – us ATLA [226]
The gospel according to st john : the authorised version with introduction and notes – London: John Murray, 1902. Beltsville, Md: NCR Corp, 1978 (5mf); Evanston: American Theol Lib Assoc, 1984 (5mf) – (= ser Biblical crit – us & gb) – 9 – 0-8370-0214-1 – mf#1984-1100 – us ATLA [226]
The gospel according to st john : the greek text – London: John Murray, 1908 – 10mf – 9 – 0-7905-2452-X – (incl ind) – mf#1987-2452 – us ATLA [226]
The gospel according to st john / Rickaby, Joseph – London: Burns & Oates, [1898?] – (= ser Scripture Manuals for Catholic Schools) – 1mf – 9 – 0-524-06035-5 – mf#1992-0748 – us ATLA [226]
The gospel according to st john : with notes critical and practical / Sadler, Michael Ferrebee – 6th ed. London: G Bell, 1893 – 2mf – 9 – 0-524-05297-2 – mf#1992-0398 – us ATLA [226]
The gospel according to st luke / Carr, Arthur – London: Rivingtons, 1875 – 1mf – 9 – 0-7905-0072-8 – (incl indes) – mf#1987-0072 – us ATLA [226]
The gospel according to st luke / Plumptre, Edward Hayes; ed by Ellicott, Charles John – London, New York: Cassell, [18–?] – 2mf – 9 – 0-8370-6834-7 – (incl indes) – mf#1986-0834 – us ATLA [226]
The gospel according to st luke / ed by Vincent, Marvin Richardson – London: J M Dent; Philadelphia: J B Lippincott 1902 [mf ed 1989] – (= ser The temple bible) – 1mf – 9 – 0-7905-1854-6 – mf#1987-1854 – us ATLA [226]

The gospel according to st luke : with introduction and notes / Garvie, Alfred Ernest – New York: Fleming H Revell; London: Andrew Melrose, 1911 – 1mf – 9 – 0-7905-1395-1 – (incl ind) – mf#1987-1395 – us ATLA [226]

The gospel according to st luke : with introduction, notes and maps / Lindsay, Thomas M – New York: Scribner & Welford, [1887?] – (= ser Handbooks for bible classes and private students) – 1mf – 9 – 0-8370-4138-4 – (incl ind) – mf#1985-2138 – us ATLA [226]

The gospel according to st luke : with maps, notes and introduction / Farrar, Frederic William – Cambridge: University Press, 1893 – 2mf – 9 – 0-8370-9778-9 – (discussion in english; text in greek. incl indes) – mf#1986-3778 – us ATLA [226]

The gospel according to st luke : with maps, notes and introduction / Farrar, Frederic William – Cambridge: University Press; London: C J Clay, 1891 – 1mf – 9 – 0-8370-3097-8 – (incl indes) – mf#1985-1097 – us ATLA [226]

Gospel according to st mark : the greek text with introduction, notes, and indices / Swete, Henry Barclay – 3d ed. London: Macmillan, 1909 – 1r – 1 – 0-8370-0434-9 – mf#1984-B361 – us ATLA [226]

The gospel according to st mark / Buisson, J C du – London: Methuen 1906 [mf ed 1989] – (= ser The churchman's bible) – 1mf – 9 – 0-7905-0823-0 – (incl bibl ref) – mf#1987-0823 – us ATLA [226]

The gospel according to st mark : the greek text / ed by Hort, Arthur – Cambridge: University Press; New York: Macmillan [distributor], 1902 – 1mf – 9 – 0-8370-7157-7 – mf#1986-1157 – us ATLA [226]

The gospel according to st mark : in the original greek, with a digest of notes from various commentators, for the use of schools and students generally / Major, John Richardson – London: Longmans, Green, 1871 – 1mf – 9 – 0-524-08508-0 – mf#1993-0033 – us ATLA [226]

The gospel according to st mark / ed by Plummer, Alfred – Cambridge:University Press, 1915 – 1mf – 9 – 0-8370-5448-6 – (incl ind) – mf#1985-3448 – us ATLA [226]

The gospel according to st mark / Smith, Sydney Fenn – London: Burns, Oates & Washbourne, 1915 – (= ser Scripture manuals for catholic schools) – 1mf – 9 – 0-524-06568-3 – mf#1992-0911 – us ATLA [226]

The gospel according to st mark : with commentary / Plumptre, Edward Hayes – London: Cassell, Petter & Galpin [1879?] [mf ed 1992] – (= ser New testament commentary for schools) – 1mf – 9 – 0-524-05289-1 – mf#1992-0390 – us ATLA [226]

The gospel according to st mark : with introduction and notes / Green, S Walter – New York: Fleming H Revell; London: Andrew Melrose, [1908?] – 1mf – 9 – 0-7905-1325-0 – (incl ind) – mf#1987-1325 – us ATLA [226]

The gospel according to st mark / with introduction, notes and maps – Edinburgh: T & T Clark, [1883?] – (= ser Handbooks for bible classes and private students) – 1mf – 9 – 0-7905-2017-6 – mf#1987-2017 – us ATLA [226]

The gospel according to st mark : with maps, notes and introduction / Maclear, George Frederick – 1st ed. Cambridge: University Press, 1883 – 1mf – 9 – 0-8370-9803-3 – (discussion in english; text in greek. incl ind) – mf#1986-3803 – us ATLA [226]

The gospel according to st matthew : english and ojibway versions in parallel readings – Toronto, Rochester, NY: Int'l Evangelical & Colportage Mission of Algoma and the North-West, 1897 – 2mf – 9 – (trans into ojibway by jones brothers) – mf#09713 – cn Canadiana [226]

The gospel according to st matthew : the greek text with introduction, notes, and indices – London: Macmillan, 1915 – 2mf – 9 – 0-524-05596-3 – mf#1992-0451 – us ATLA [226]

The gospel according to st matthew – London: Burns Oates & Washbbourne, [1899?] – (= ser Scripture manuals for catholic schools) – 1mf – 9 – 0-524-06569-1 – mf#1992-0912 – us ATLA [226]

The gospel according to st matthew : with commentary / Plumptre, Edward Hayes – London: Cassell, Petter, Galpin, [1879?] – (= ser New testament commentary for schools) – 2mf – 9 – 0-524-04919-X – mf#1992-0262 – us ATLA [226]

The gospel according to st paul : studies in the first eight chapters of his epistle to the romans / Dykes, James Oswald – New York: Robert Carter 1888 [mf ed 1985] – 1mf – 9 – 0-8370-3012-9 – mf#1985-1012 – us ATLA [227]

The gospel according to the jews and pagans : the historical character of the gospel established from non-christian sources / Stokes, Samuel E; ed by Murray, John Owen Farquhar – London, New York: Longmans, Green, 1913 [mf ed 1991] – 1mf – 9 – 0-7905-8920-6 – mf#1989-2145 – us ATLA [226]

Gospel advocate : conducted by a society of gentlemen – La Salle IL 1821-25 – 1 – mf#4461 – us UMI ProQuest [240]

Gospel advocate and impartial investigator – La Salle IL 1823-29 – 1 – mf#4462 – us UMI ProQuest [240]

The gospel among the slaves : a short account of missionary operations among the african slaves of the southern states / ed by Harrison, William Pope – Nashville, TN: Pub House of the M E Church, South, 1893 [mf ed 1990] – 1mf – 9 – 0-7905-4918-2 – mf#1988-0918 – us ATLA [240]

The gospel and frontier peoples : a report of a consultation december 1972 / ed by Pierce Beaver, R – South Passadena, 1973 – 9mf – 8 – €18.00 – ne Slangenburg [220]

The gospel and human needs / Figgis, John Neville – London, New York: Longmans, Green, 1909 – (= ser Hulsean Lectures) – 1mf – 9 – 0-7905-4519-5 – (incl bibl ref) – mf#1988-0519 – us ATLA [240]

The gospel and its elements / Challen, James – Philadelphia: J Challen, 1856 – 1mf – 9 – 0-524-06013-4 – mf#1991-2373 – us ATLA [226]

The gospel and its ministry / Anderson, Robert – 2nd ed. Bosoton: Believers' Book-Rooms, [18–] – 1mf – 9 – 0-8370-2497-8 – mf#1985-0497 – us ATLA [240]

The gospel and its witnesses : some of the chief facts in the life of our lord and the authority of the evangelical narratives considered in lectures chiefly preached at st. james's westminster / Wace, Henry – 2nd ed. London: John Murray, 1884 – 1mf – 9 – 0-8370-9336-8 – (incl bibl ref) – mf#1986-3336 – us ATLA [226]

The gospel and modern life : sermons on some of the difficulties of the day, with a preface on a recent phase of deism / Davies, John Llewelyn – London: Macmillan, 1869 – 1mf – 9 – 0-7905-9177-4 – mf#1989-2402 – us ATLA [240]

The gospel and philosophy : six lectures preached in trinity chapel, new york / Dix, Morgan – New York: E & J B Young, 1886 – 1mf – 9 – 0-8370-2927-9 – mf#1985-0927 – us ATLA [230]

The gospel, and romanism in canada : an historical sketch of the grande ligne mission in lower canada / Lafleur, Theodore – [Montreal?: J Starke], 1866 – 1mf – 9 – 0-665-92956-0 – mf#92956 – cn Canadiana [242]

The gospel, and romanism in canada : an historical sketch of the grande ligne mission in lower canada / Lafleur, Theodore – [Montreal?: s.n.] 1866 [mf ed 1993] – 1mf – 9 – 0-665-92956-0 – mf#92956 – cn Canadiana [242]

The gospel and the church = Evangile et l'eglise / Loisy, Alfred Firmin – new ed. London: I Pitman, 1908 [mf ed 1989] – 1mf – 9 – 0-7905-3147-X – (trans by christopher home. with pref memoir by g tyrrell) – mf#1987-3147 – us ATLA [241]

Gospel and the doctrine which is not the gospel / Curling, W – Oxford, England. 1842 – 1r – us UF Libraries [220]

The gospel and the mala : the story of the hyderabad wesleyan mission / Lamb, Frederick – Mysore: Wesleyan Mission Press, 1913 [mf ed 1995] – (= ser Yale coll) – 120p (ill) – 1 – 0-524-09122-6 – mf#1995-0122 – us ATLA [242]

The gospel and the modern man / Mathews, Shailer – New York: Macmillan, 1910 – (= ser Haverford Library Lectures) – 1mf – 9 – 0-7905-7995-2 – mf#1989-1200 – us ATLA [240]

The gospel and the new world / Speer, Robert Elliott – New York: Fleming H Revell, c1919 – 1mf – 9 – 0-524-06321-4 – mf#1991-2494 – us ATLA [240]

The gospel as taught by calvin / Reed, Richard Clark – Jackson, Miss: Presbyterian Reformation Society – 2mf – 9 – 0-524-07912-9 – mf#1991-3457 – us ATLA [242]

The gospel awakening : comprising the sermons and addresses, prayer meeting talks and bible readings of the great revival meetings... / Moody, Dwight Lyman et al – 3rd ed. Chicago: J Fairbanks, 1878 – 2mf – 9 – 0-524-05956-X – mf#1991-2356 – us ATLA [240]

Gospel banner – v1-92. 1878-1969 [gaps] – (= ser Mennonite serials coll) – 43r – 1 – mf#ATLA 1994-S004 – us ATLA [242]

The gospel banner : gospel worker edition – v23 n13- . v24, v26 n2-24. 1900-01, 1903 [gaps] – (= ser Mennonite serials coll) – 2r – 1 – mf#ATLA 1994-S030 – us ATLA [242]

Gospel best promulgated by national schools / Wrangham, Francis – York, England. 1808 – 1r – us UF Libraries [226]

The gospel by moses in the book of genesis : or, the old testament unveiled / Putnam, Catherine H – New York: Edward H Fletcher, 1854 – 2mf – 9 – 0-7905-3424-X – mf#1987-3424 – us ATLA [221]

Gospel communicator : or, philanthropist's journal – Glasgow, Scotland, 1823-27 (fortnightly) [mf ed 2001] – (= ser Christianity's encounter with world religions, 1850-1950) – 1r – 1 – mf#2001-s195 – us ATLA [220]

Gospel communicator / Pentecostal Word Explosion (Chicago, IL) – 1995 apr – 1r – 1 – mf#4024314 – us WHS [242]

The gospel developed through the government and order of the churches of jesus christ / Johnson, William B – 1846 – 1 – 8.05 – us Southern Baptist [242]

Gospel doctrines for the use of sunday schools / French, William Riley – Boston: Universalist Pub House, c1865 – 1mf – 9 – 0-524-06408-3 – mf#1991-2530 – us ATLA [240]

Gospel ethnology / Pattison, Samuel Rowles – new ed. London: Religious Tract Society [1887?] [mf ed 1989] – 1mf – 9 – 0-7905-1975-5 – (incl bibl ref) – mf#1987-1975 – us ATLA [240]

The gospel for a world of sin : a companion-volume to "the gospel for an age of doubt" / Van Dyke, Henry – New York: Macmillan, 1900, c1899 – 1mf – 9 – 0-8370-5642-X – (incl bibl ref) – mf#1985-3642 – us ATLA [240]

The gospel for all *see* Fu yin te chun pei (ccm87)

The gospel for an age of doubt : the yale lectures on preaching 1896 / Dyke, Henry van – New York: Macmillan, 1897, c1896 [mf ed 1985] – 2mf – 9 – 0-8370-5645-4 – (incl bibl ref & ind) – mf#1985-3645 – us ATLA [230]

The gospel for to-day / Garvie, Alfred Ernest – London: James Clarke; Glasgow: Inglis Ker, 1904 – 1mf – 9 – 0-524-00265-7 – mf#1989-2965 – us ATLA [226]

Gospel from two testaments : sermons on the international sunday-school lessons for 1893 / Burnham, Sylvester et al; ed by Andrews, Elisha Benjamin – Providence: E A Johnson, 1892 – 2mf – 9 – 0-524-08205-7 – mf#1993-0000 – us ATLA [220]

Gospel glass / Bloomfield, John – London, England. 1860 – 1r – us UF Libraries [220]

The gospel hammer and highway grader : or, rubbish cleaned from the way of life / Bashor, Stephen Henry – Lanark, IL: Brethren at Work Steam Printing House, 1878 – 1mf – 9 – 0-524-03537-7 – mf#1990-4732 – us ATLA [240]

Gospel herald – Beamsville, CN. 1937-92 – 16r – 1 – cn Commonwealth Imaging [240]

Gospel herald / Church of God of the Mountain Assembly – 1977-82, 1983-89 – 2r – 1 – mf#654756 – us WHS [071]

Gospel herald – Voree WI. [1847 sep 23-1850 may 9] – 1r – 1 – (cont: zion's reveille) – us WHS [071]

Gospel herald – v1-48. 1908-55 [complete] – (= ser Mennonite serials coll) – 48r – 1 – ISSN: 0017-2340 – mf#ATLA 1990-S000 – us ATLA [242]

The gospel herald – Ontario, CN. 1937-91 – 15r – 1 – (cumulative ind 1969-84) – cn Commonwealth Imaging [240]

Gospel herald [1820] – La Salle IL 1820-29 – 1 – mf#3986 – us UMI ProQuest [220]

Gospel herald [1989] – Beamsville, Canada 1989 – 1 – ISSN: 0829-4666 – mf#16084 – us UMI ProQuest [220]

Gospel history : a syllabus of professor c.w. hodge's gospel history / Hodge, Caspar Wistar – Princeton: Charles S Robinson, 1879 – 1mf – 9 – 0-8370-3605-4 – mf#1985-1605 – us ATLA [220]

The gospel history : a compendium of critical investigations in support of the historical character of the four gospels = Wissenschaftliche kritik der evangelischen geschichte / Ebrard, Johannes Heinrich August; ed by Bruce, Alexander B – Edinburgh: T & T Clark 1863 [mf ed 1989] – (= ser Clark's foreign theological library 19) – 2mf – 9 – 0-7905-0708-0 – (english trans fr german by james martin. incl bibl ref) – mf#1987-0708 – us ATLA [226]

The gospel history and its transmission / Burkitt, Francis Crawford – Edinburgh: T & T Clark, 1906 – 1mf – 9 – 0-8370-2538-9 – mf#1985-0538 – us ATLA [226]

Gospel holiness / Dayton, A C – 1879 – 1 – 5.00 – us Southern Baptist [242]

Gospel hymns and sacred songs / Bliss, PO & Sankey, Ira D – 1875 – 1 – 5.00 – us Southern Baptist [242]

Gospel hymns nos. 5 and 6, combined : for use in gospel meetings and other religious services / Sankey, Ira David et al – New York: Biglow & Main, c1892 – 5mf – 9 – 0-524-08785-7 – mf#1993-1093 – us ATLA [780]

The gospel in burmah : the story of its introduction and marvellous progress among the burmese and karens / Wylie, Macleod (Mrs) – New York: Sheldon; Boston: Gould & Lincoln, 1860 – 1mf – 9 – 0-8370-6553-4 – mf#1986-0553 – us ATLA [240]

The gospel in canada : and its relation to huron college in addresses by the lord bishop of huron; right rev dr mcilvaine... – London: W Hunt, [1865?] [mf ed 1984] – 2mf – 9 – 0-665-38608-7 – (incl bibl ref; int by t r birks) – mf#38608 – cn Canadiana [226]

The gospel in central america, published in london in 1850 / Crowe, Frederick – 1 – us Southern Baptist [242]

The gospel in enoch, or, truth in the concrete : a doctrinal and biographical sketch / Tucker, Henry Holcombe – Philadelphia: J B Lippincott, 1869 – 1mf – 9 – 0-8370-5583-0 – (includes appendix) – mf#1985-2289 – us ATLA [226]

The gospel in futuna : with chapters on the islands of the new hebrides, the people, their customs, religious beliefs, etc / Gunn, W – London, 1914 – 4mf – 9 – mf#HTM-75 – ne IDC [919]

The gospel in isaiah : illustrated in a series of expositions, topical and practical founded upon the sixth chapter / Robinson, Charles S – Chicago: Fleming H Revell, c1895 – 1mf – 9 – 0-8370-4929-6 – mf#1985-2929 – us ATLA [226]

The gospel in its native land / Carnapas, Anna Macdonald – Chicago: The Gospel...c1909 [mf ed 1989] – 1mf – 9 – 0-7905-0627-0 – mf#1987-0627 – us ATLA [915]

The gospel in latin lands : outline studies of protestant work in the latin countries / Clark, Francis Edward – New York: Macmillan, 1909 – 1mf – 9 – 0-8370-6095-8 – (includes bibliographies & index) – mf#1986-0095 – us ATLA [242]

The gospel in latin lands : outline studies of protestant work in the latin countries of europe and america / Clark, Francis Edward & Clark, Harriet Elizabeth – New York, Toronto: Macmillan, 1909 – 4mf – 9 – 0-665-98252-6 – mf#98252 – cn Canadiana [242]

The gospel in leviticus : or, an exposition of the hebrew ritual / Seiss, Joseph Augustus – Philadelphia:Lindsay & Blakiston, 1860, c1859 [mf ed 1985] – 1mf – 9 – 0-8370-5221-1 – mf#1985-3221 – us ATLA [221]

The gospel in pagan religions : some thoughts suggested by the world's parliament of religions to an orthodox christian – Boston: Arena Publ Co, 1894 – 1mf – 9 – 0-8370-3348-9 – mf#1985-1348 – us ATLA [230]

Gospel in russia – 1936-46 [complete] – 1r – 1 – (filmed with: friends of russia) – mf#ATLA S0702B – us ATLA [240]

Gospel in russia *see* Friends of russians

The gospel in russia – New York – 1r – 1 – $5.00 – (printed by the all-russian evangelical christian union, u.s.a. nov. 1926, oct. 1928, july-oct. 1930, jan. & apr. 1931, oct. 1933, jan. 1935) – us Southern Baptist [242]

The gospel in santhalistan : [by an old indian] – London: James Nisbet, 1875 [mf ed 1995] – (= ser Yale coll) – x/98p – 1 – 0-524-09336-9 – (pref by horatius bonar) – mf#1995-0336 – us ATLA [220]

The gospel in south india : or, the religious life, experience, and character of the hindu christians / Mateer, S – London, [1880] – 3mf – 9 – mf#HTM-120 – ne IDC [915]

The gospel in the gospels / Dubose, William Porcher – New York: Longmans, Green, 1906 – 1mf – 9 – 0-7905-7338-5 – mf#1989-0563 – us ATLA [226]

The gospel in the psalms : a series of expositions based on revised translation / M'Lean, Daniel – Edinburgh: Andrew Elliot, 1875 – 1mf – 9 – us ATLA [240]

The gospel in water, or campbellism / Jarrel, Willis Anselm – 1886 – 1 – us Southern Baptist [242]

Gospel inquirer – La Salle IL 1823-24 – 1 – mf#3811 – us UMI ProQuest [220]

Gospel invitation to the throne of grace / Macdonald, T M – London, England. 1874 – 1r – us UF Libraries [226]

The gospel its own advocate / Griffin, George – New York: D Appleton, 1850 – 1mf – 9 – 0-524-05804-0 – mf#1992-0631 – us ATLA [240]

The gospel its own witness / Leathes, Stanley – London: Henry S. King, 1874. Beltsville, Md: NCR Corp, 1978 (2mf); Evanston: American Theol Lib Assoc, 1984 (2mf) – (= ser Hulsean Lectures) – 9 – 0-8370-0860-3 – mf#1984-4208 – us ATLA [220]

Gospel letters / Kirk, John – Glasgow: "Christian News" Office, 1862 – 1mf – 9 – 0-8370-4456-1 – mf#1985-2456 – us ATLA [220]

The gospel liturgy : a prayer-book for churches, congregations, and families / Universalist Church of America – Philadelphia: G Collins, 1857 – 1mf – 9 – 0-524-07486-0 – (incl ind) – mf#1990-5412 – us ATLA [240]

The gospel manual : an arrangement of the four gospels blended into one continuous record of the life and ministry of jesus christ – San Francisco: Occident Print and Pub Co, 1886 – 1mf – 9 – 0-524-06829-1 – mf#1992-0971 – us ATLA [226]
Gospel message / Dealtry, W – London, England. 1829 – 1r – us UF Libraries [226]
The gospel message : or, essays, addresses, suggestions, and warnings on the different aspects of christian missions to non-christian races and peoples / Cust, Robert Needham – London: Luzac, 1896 [mf ed 1986] – 2mf – 9 – 0-8370-6661-1 – (incl ind) – mf#1986-0661 – us ATLA [240]
Gospel messenger – 1827 jul 28-1836 jan 30, 1847 jan 22-1848 dec 29, 1849 jan 5-1853 nov 4, 1854 jan 20-1859 dec 30, 1860 jan 6-1861 dec 26, 1862 jan 2-1863 dec 31, 1864 jan 7-1865 dec 28 – 7r – 1 – mf#1112094 – us WHS [071]
Gospel messenger : official organ of the congregational holiness church – 1977 jun-1985 dec – 1r – 1 – mf#1668521 – us WHS [242]
Gospel messenger : or, universalist advocate – London, C.W. [Ont]: J R Lavell, [1849?-18–] – (= ser Universalist advocate) – 9 – mf#P05032 – cn Canadiana [240]
Gospel ministry : as instituted by christ, a good work / Taylor, John – Edinburgh, Scotland. 1831 – 1r – us UF Libraries [220]
Gospel ministry / Wylie, James A – London, England. 1857 – 1r – us UF Libraries [220]
The gospel miracles : an essay / Illingworth, John Richardson – London: Macmillan, 1915 – 1mf – 9 – 0-7905-7343-1 – mf#1989-0568 – us ATLA [226]
The gospel miracles in their relation to christ and christianity / Taylor, William Mackergo – New York: A D F Randolph, 1880 – 1mf – 9 – 0-8370-5487-7 – (incl bibl ref) – mf#1985-3487 – us ATLA [240]
Gospel music and ministry connection : gmmc – v2 n3 [1994apr] – 1r – 1 – mf#4023522 – us WHS [780]
Gospel music magazine – v1 n1 [1985 mar] – 1r – 1 – mf#4023539 – us WHS [780]
Gospel Music Workshop of America *see* Gag airwaves
The gospel narratives : their origin, peculiarities and transmission / Miles, Henry Adolphus – Boston: Wm Crosby and H P Nichols, 1848 – 1mf – 9 – 0-524-08509-9 – mf#1993-0034 – us ATLA [226]
Gospel news *see* Fu yin hsin pao (ccs)
Gospel news report = Fu yin hsin pao – n32. Jul 1877* – (= ser Chinese christian coll 53) – 1r – 1 – (in chinese) – mf#ATLA S0296l – us ATLA [240]
The gospel of barnabas / Ragg, Lonsdale & Ragg, Lara – Oxford, 1907 – 12mf – 8 – €23.00 – ne Slangenburg [226]
The gospel of buddha according to old records / Carus, Paul – Chicago: Open Court, 1894 – 1mf – 9 – 0-524-01173-7 – mf#1990-2249 – us ATLA [280]
The gospel of common sense : as contained in the canonical epistle of james / Deems, Charles Force – New York: Wilber B Ketcham, c1888 – 1mf – 9 – 0-8370-2859-0 – mf#1985-0859 – us ATLA [220]
Gospel of evolution / Adamson, William – Edinburgh, Scotland. 1885 – 1r – us UF Libraries [220]
The gospel of experience, or, the witness of human life to the truth of revelation / Newbolt, William Charles Edmund – London; New York: Longmans, Green, 1896 – (= ser The boyle lectures) – 1mf – 9 – 0-7905-9825-6 – mf#1989-1550 – us ATLA [240]
The gospel of forgiveness : a series of discourses / Candlish, Robert Smith – Edinburgh: A and C Black, 1878 – 2mf – 9 – 0-7905-0872-9 – mf#1987-0872 – us ATLA [240]
The gospel of gladness : and its meaning for us / Clifford, John – Edinburgh: T & T Clark, 1912 – (= ser The Scholar as Preacher) – 1mf – 9 – 0-7905-7328-8 – mf#1989-0553 – us ATLA [240]
The gospel of good will : as revealed in contemporary scriptures / Hyde, William De Witt – New York: Macmillan, 1916 – (= ser Lyman Beecher Lectures) – 1mf – 9 – 0-7905-7837-9 – mf#1989-1062 – us ATLA [240]
The gospel of healing / Simpson, Albert B – rev ed. New York: Christian Alliance, c1915 [mf ed 1992] – 1mf – 9 – 0-524-02148-1 – mf#1990-4214 – us ATLA [230]
The gospel of industry : a survey of industrial training on baptist mission fields / Lipphard, William Benjamin – Philadelphia: Pub for American Baptist Foreign Mission Society and Woman's American Baptist Foreign Mission Society by the American Baptist Publ Soc, 1918 – 1mf – 9 – 0-524-06910-7 – (incl bibl ref) – mf#1991-2823 – us ATLA [242]
The gospel of jesus the son of god : an interpretation for the modern man / Knox, George William – Boston: Houghton, Mifflin, 1909 – 1mf – 9 – 0-8370-4442-1 – mf#1985-2442 – us ATLA [240]
The gospel of john : an exposition / Erdman, Charles Rosenbury – Philadelphia: Westminister Press, 1917 – 1mf – 9 – 0-524-08074-7 – mf#1992-1134 – us ATLA [226]
The gospel of john / Simpson, Albert B – New York: Christian Alliance Pub Co, c1904 [mf ed 1992] – (= ser Christian & missionary alliance coll) – 1mf – 9 – 0-524-02149-X – mf#1990-4215 – us ATLA [226]
Gospel of john and the acts of the apostles / Simpson, Albert B – New York: Christian Alliance Pub Co, 1891 [mf ed 1992] – (= ser Christian & missionary alliance coll) – 1mf – 9 – 0-524-02150-3 – mf#1990-4216 – us ATLA [226]
The gospel of joy / Brooke, Stopford Augustus – London: Isbister, 1898 – 1mf – 9 – 0-7905-3613-7 – mf#1989-0106 – us ATLA [240]
The gospel of labor / Stelzle, Charles – New York: F H Revell, c1912 – 1mf – 9 – 0-7905-6087-9 – mf#1988-2087 – us ATLA [240]
The gospel of law : a series of discourses upon fundamental church doctrines / Stewart, Samuel James – Boston: George H Ellis, 1882 – 1mf – 9 – 0-8370-5544-X – mf#1985-3544 – us ATLA [210]
The gospel of life : thoughts introductory to the study of christian doctrine / Westcott, Brooke Foss – 2nd ed. London; New York: Macmillan, 1895 – 1mf – 9 – 0-8370-5806-6 – mf#1985-3806 – us ATLA [230]
The gospel of life : thoughts introductory to the study of christian doctrine / Westcott, Brooke Foss – London; New York: Macmillan, 1892. Beltsville, Md: NCR Corp, 1978 (4mf); Evanston: American Theol Lib Assoc, 1984 (4mf) – (= ser Biblical crit – us & gb) – 9 – 0-8370-0252-4 – (incl bibl ref) – mf#1984-1101 – us ATLA [220]
The gospel of life in the syriac new testament : the syriac, peshito, contrasted with the greek, with respect to the following words, viz.: sozo, soteria, soter / Pettingell, John Hancock – Yarmouth, Me: Scriptual Publication Society, c1886 – 1mf – 9 – 0-524-06151-3 – mf#1992-0818 – us ATLA [225]
The gospel of luke the apostles' creed / Craig, Austin – Boston: Wm Crosby and HP Nichols, [18–?] – 1mf – 9 – 0-524-08272-3 – mf#1993-3027 – us ATLA [226]
The gospel of matthew / Robinson, Theodore Henry – London: Hodder and Stoughton, 1928 – (= ser The Moffatt New Testament Commentary) – 1mf – 9 – 0-524-08135-2 – mf#1993-9041 – us ATLA [226]
The gospel of narada / ed by Greenlees, Duncan – Madras: Theosophical Pub House, 1951 – (= ser Samp: indian books) – (trans fr sanskrit of narada pancaratra, the narada bhakti sutras, and the narada gita, with a running comm and int by) – us CRL [280]
The gospel of paul / Everett, Charles Carroll – Boston:Houghton, Mifflin, 1893 – 1mf – 9 – 0-8370-3083-8 – (incl ind) – mf#1985-1083 – us ATLA [226]
The gospel of reconciliation : or, at-one-ment / Walker, William Lowe – Edinburgh: T & T Clarke 1909 [mf ed 1985] – 1mf – 9 – 0-8370-2930-9 – (incl ind) – mf#1985-0930 – us ATLA [240]
The gospel of s john, register of fragments, etc, facsimiles / ed by Horner, George – Oxford: Clarendon Press, 1911 – (= ser The Coptic Version of the New Testament in the Southern Dialect) – 1mf – 9 – 0-524-02767-6 – mf#1987-6461 – us ATLA [226]
The gospel of s luke / Ford, James – London: J Masters, 1851 – 2mf – 9 – 0-8370-1499-9 – mf#1987-6065 – us ATLA [226]
The gospel of s luke / ed by Horner, George – Oxford: Clarendon Press, 1911 – (= ser The Coptic Version of the New Testament in the Southern Dialect) – 2mf – 9 – 0-524-02766-8 – mf#1987-6460 – us ATLA [226]
The gospel of s matthew – 2nd ed. London: J Masters, 1859 – 2mf – 9 – 0-8370-1833-1 – mf#1987-6221 – us ATLA [226]
The gospel of saint john in west-saxon / ed by Bright, James Wilson – Boston, MA: D C Heath, 1904 – (= ser The belles-lettres series) – 1mf – 9 – 0-8370-1304-6 – mf#1987-6039 – us ATLA [226]
The gospel of saint mark in gothic / ed by Skeat, Walter William – Oxford: Clarendon Press, 1882 – (= ser Clarendon press series) – 1mf – 9 – 0-8370-1980-X – mf#1987-6367 – us ATLA [226]
The gospel of saint matthew in west-saxon / ed by Bright, James Wilson – Boston, MA: D C Heath, 1904 – (= ser The belles-lettres series) – 1mf – 9 – 0-8370-1305-4 – mf#1987-6040 – us ATLA [226]
The gospel of selfless action : or, the gita according to gandhi – Ahmedabad: Navajivan Pub House, 1946 – (= ser Samp: indian books) – (trans of original in gujarati; add int and comm by mahadev desai) – us CRL [280]
The gospel of spiritual insight : being studies in the gospel of st. john / Deems, Charles Force – New York: Wilbur B Ketcham, c1891 – 2mf – 9 – 0-8370-9935-8 – mf#1986-3935 – us ATLA [220]
Gospel of sri ramakrishna / Ramakrishna – New York, NY. 1942 – 1r – us UF Libraries [280]
The gospel of sri ramakrishna – Madras: Sri Ramakrishna Math, 1947 – (= ser Samp: indian books) – (trans by swami nikhilananda) – us CRL [280]
The gospel of st john : an exposition, exegetical and homiletical... / Whitelaw, Thomas – Glasgow: James Maclehose, 1888 – 2mf – 9 – 0-7905-0529-0 – (incl ind) – mf#1987-0529 – us ATLA [226]
The gospel of st john / Maclaren, Alexander – New York: A C Armstrong, 1894 [mf ed 1989] – (= ser Bible class expositions) – 1mf – 9 – 0-7905-2851-7 – mf#1987-2851 – us ATLA [226]
The gospel of st john and the synoptic gospels / Barth, Fritz – New York: Eaton & Mains, c1907 – 1mf – 9 – 0-8370-2194-4 – mf#1985-0194 – us ATLA [226]
The gospel of st luke translated into the slave language for indians of north-west america – London: Printed for the British and Foreign Bible Society, 1890 – 2mf – 9 – (in the syllabic character. trans by w c bompas and w d reeve) – mf#06639 – cn Canadiana [290]
The gospel of st mark / Maclaren, Alexander – London: Hodder & Stoughton, 1893 [mf ed 1989] – (= ser Bible class expositions) – 1mf – 9 – 0-7905-1291-2 – mf#1987-1291 – us ATLA [226]
Gospel of st matthew / Simpson, Albert B – 2nd rev ed. New York: Alliance Press, c1904 [mf ed 1991] – (= ser Christ in the bible 13; Christian and missionary alliance coll) – 1mf – 9 – 0-524-01818-9 – mf#1990-4156 – us ATLA [226]
The gospel of st matthew / Gibson, John Monro – New York: A C Armstrong, [1890] – 2mf – 9 – 0-8370-2289-4 – mf#1985-0289 – us ATLA [226]
The gospel of st matthew translated into the slave language for the indians of north-west america : in the syllabic character – London: Printed for the British and Foreign Bible Society, 1886 – 1mf – 9 – mf#14680 – cn Canadiana [290]
The gospel of the atonement : being the hulsean lectures for 1898-99 / Wilson, James Maurice – London, New York: Macmillan 1899 [mf ed 1985] – 1mf – 9 – 0-8370-2995-3 – mf#1985-0995 – us ATLA [240]
The gospel of the divine sacrifice : a study in evangelical belief: with some conclusions touching life / Hall, Charles Cuthbert – New York: Dodd, Mead, 1897, c1896 – 1mf – 9 – 0-8370-4728-5 – mf#1985-2728 – us ATLA [240]
The gospel of the hereafter / Smyth, John Paterson – New York: Fleming H Revell, c1910 – 1mf – 9 – 0-524-05698-6 – mf#1992-0548 – us ATLA [240]
Gospel of the incarnation / Punshon, William Morley – London, England. 1877? – 1r – us UF Libraries [220]
Gospel of the kingdom *see* Social progress
The gospel of the kingdom : a popular exposition of the gospel according to matthew / Spurgeon, Charles Haddon – New York: Fleming H Revell, c1913 – 2mf – 9 – 0-524-04416-3 – mf#1992-0109 – us ATLA [226]
Gospel of the kingdom to be universally preached / Johnston, John – London, England. 1818? – 1r – us UF Libraries [220]
The gospel of the lord : an early version which was circulated by marcion of sinope as the original gospel / Marcion of Sinope – Guernsey: Publ for John Whitehead: Printed and sold by T M Bichard, [1891?] – 1mf – 9 – 0-8370-1745-9 – mf#1987-6141 – us ATLA [220]
The gospel of the old testament : an explanation of the types and figures by which christ was exhibited under thhe legal dispensation / Mather, Samuel – Philadelphia: A Towar, 1834 [mf ed 2003] – 1r – 1 – 0-524-10460-3 – (rewritten fr original work of samuel mather by aut of "the listener", "christ our example" etc. originally publ under title: the figures or types of the old testament) – mf#b00680 – us ATLA [221]
The gospel of the resurrection : thoughts on its relation to reason and history / Westcott, Brooke Foss – 4th ed. London: Macmillan, 1879 – 1mf – 9 – 0-7905-2210-1 – mf#1987-2210 – us ATLA [226]
The gospel of the secular life : sermons preached at oxford, with a prefatory essay / Fremantle, William Henry – New York: Scribner, 1883 – 1mf – 9 – 0-7905-5653-7 – mf#1988-1653 – us ATLA [240]
The gospel of the twelve apostles : together with the apocalypses of each one of them / ed by Harris, James Rendel – Cambridge: University Press; New York: Macmillan [dist], 1900 [mf ed 1989] – 1v on 1mf – 9 – 0-7905-2467-8 – (in english & syriac, int also in greek) – mf#1987-2467 – us ATLA [226]
The gospel on the banks of the niger : journal and notices of the native missionaries accompanying the niger expedition of 1857-1859 / Crowther, S & Taylor, J C – London, 1859 – 5mf – 9 – mf#HTM-47 – ne IDC [916]
The gospel on the banks of the niger : journals and notices of the native missionaries accompanying the niger expedition of 1857-1859 / Crowther, Samuel & Taylor, John Christopher – London: Church Missionary House, 1859. Chicago: Dep of Photodup, U of Chicago Lib, 1978 (1r); Evanston: American Theol Lib Assoc, 1984 (1r) – 1 – 0-8370-0764-X – mf#1984-T126 – us ATLA [240]
Gospel palladium – La Salle IL 1823-24 – 1 – mf#3812 – us UMI ProQuest [240]
Gospel plan and method of salvation / Smyth, T S – St Austell, England. 1818 – 1r – us UF Libraries [240]
The gospel plan of salvation / Brents, Thomas Wesley – Cincinnati: Bosworth, Chase & Hall, 1874 – 2mf – 9 – 0-524-06983-2 – mf#1991-2836 – us ATLA [240]
The gospel preacher / Franklin, Benjamin – Indianapolis IN: Daniel Sommer, 1909 [mf ed 1993] – (= ser Christian church (disciples of christ) coll) – 2v on 3mf – 9 – 0-524-07565-4 – mf#1991-3185 – us ATLA [240]
The gospel preacher – Ashland OH: [s.n] 1879-82 (wkly) [mf ed 2003] – 4v on 2r – 1 – (not publ jun 7-14 1881. title iss jun 7-10 1881: daily gospel preacher incl in 2nd reel. merged with: progressive christian (berlin pa) to form: progressive christian and gospel preacher) – mf1039a – us ATLA [242]
The gospel preacher *see* Daily gospel preacher
Gospel questions and answers / Denney, James – London: Hodder & Stoughton [1896?] [mf ed 1991] – (= ser Little books on religion) – 1mf – 9 – 0-7905-9183-9 – mf#1989-2408 – us ATLA [242]
The gospel records : their genuineness, authenticity, historic verity, and inspiration, with some preliminary remarks on the gospel history = Allgemeine einleitung in die schriften des neuen testaments / Nast, Wilhelm – Cincinnati: Curts & Jennings; New York: Eaton & Mains, c1866 [mf ed 1989] – 1mf – 9 – 0-7905-2250-0 – (in english) – mf#1987-2250 – us ATLA [226]
The gospel reflector – Philadelphia. v. 1 no. 1-12. June 15 1845 – 1 – us NY Public [240]
Gospel registry – 1993 apr-nov /dec [v4 n4-8] – 1r – 1 – mf#2858491 – us WHS [220]
Gospel routes – 1980 apr/may – 1r – 1 – mf#4983088 – us WHS [071]
Gospel sermons / McCosh, James – New York: Scribner, 1890, c1888 – 1mf – 9 – 0-7905-9413-7 – mf#1989-2638 – us ATLA [240]
Gospel song : a historical and analytical study / Boyer, Horace Clarence – U of Rochester 1964 [mf ed 19–] – 1r – 1 – mf#film 1227 – us Sibley [780]
Gospel songs and hymns. no. 1 : for the sunday school, prayer meeting, social meeting, general song service / Holsinger, George Blackburn – Mt Morris, Ill: Brethren Pub House, 1898 – 3mf – 9 – 0-524-07200-0 – (incl ind) – mf#1990-5358 – us ATLA [780]
The gospel sources of christian-jewish prejudice : teaching church school teachers to apply contemporary biblical studies to the task of interpreting the problematic gospel texts / Kaminsky, David Cyril – Princeton, New Jersey, 1976. Chicago: Dep of Photodup, U of Chicago Lib, 1976 (1r); Evanston: American Theol Lib Assoc, 1984 (1r) – 1 – 0-8370-1288-0 – mf#1984-T014 – us ATLA [240]
Gospel speaks / LADA's Music Shoppe (Marietta, GA) – v2 n9 [1995 may/jun] – 1r – 1 – mf#4114520 – us WHS [780]
Gospel spreading church newsletter – 1994 jan, apr, jul, oct, 1995 jan, apr, jul, oct, 1996 jan, apr, jul, 1997 apr – 1r – 1 – mf#3074452 – us WHS [220]
Gospel studies = Etudes evangeliques / Vinet, Alexandre Rodolphe – New York: MW Dodd, 1849 [mf ed 1990] – 1mf – 9 – 0-7905-7614-7 – (in english. int by robert baird) – mf#1989-0839 – us ATLA [240]
Gospel teacher *see* Progressive christian [berlin pa]
The gospel teacher / ed by Kaufman, P J – Foraker IN: L S Hostetler. v1 n11-v37 1891-1931 [mf ed 1994] – (= ser Mennonite serials coll) – 2r – 1 – (publ suspended oct 1892-96. several iss lacking or missing) – mf#1994-s016 – us ATLA [242]

Gospel temperance / Van Buren, J M – New York, NY. 1878 – 1r – us UF Libraries [220]

Gospel temperance herald and blue ribbon official gazette – London. 5 apr 1882-9 feb 1910 [wkly] – 11r – 1 – (wanting 12 jun 1901-29 jun 1902) – uk British Libr Newspaper [073]

Gospel tidings / Evangelical Mennonite Brethren Church et al – 1977-79, 1980-83 – 2r – 1 – (continued by: fellowship focus) – mf#977305 – us WHS [220]

Gospel tidings – v31-32. 1941-1953 [complete] – (= ser Mennonite serials coll) – Inquire – 1 – mf#ATLA 1994-S032 – us ATLA [242]

Gospel time – 1997 apr-sep/oct, 1998 feb/mar-may/jun – 1r – 1 – mf#3907702 – us WHS [220]

Gospel times – 1994 nov 18, dec 18, 1995 jan 15, feb 12, mar 10, apr 2, may 1,25, jun 18, jul 21, aug 13, sep 18, nov 20, 1996 may 5, sep 20, oct 20 – 1r – 1 – mf#3177852 – us WHS [220]

Gospel times / Christian Tabernacle Baptist Church (Chicago, IL) – 1980 aug – 1r – 1 – mf#4027693 – us WHS [242]

The gospel to be published and applied against all sin : a discourse. delivered at the anniversary of the congregational board of publication, in central church, boston... / Cheever, George Barrell – Boston: Congregational Board of Publication, 1856 – 1mf – 9 – 0-524-08742-3 – mf#1993-3247 – us ATLA [242]

The gospel to the africans : a narrative of the life and labours of the rev william jameson in jamaica and old calabar / Robb, A – Edinburgh, [1861] – 4mf – 9 – mf#HTM-163 – ne IDC [916]

The gospel to the poor versus pew rents / Austin, Benjamin Fish – Toronto: W Briggs; Montreal: C W Coates, 1884 – 2mf – 9 – (int by bishop carman. papers on the pew system by newman hall et al) – mf#06703 – cn Canadiana [230]

Gospel tribune – 1993 apr, 1996 feb, may – 1r – 1 – mf#4027671 – us WHS [071]

The gospel tribune and christian communionist : a monthly inter-denominational journal – Toronto: R Dick, 1856-[1858?] – 9 – (incl ind) – mf#P04290 – cn Canadiana [240]

The gospel tribune for alliance and intercommunion throughout evangelical christendom – Toronto: R Dick, 1854-1856 – 9 – (incl ind) – mf#P04289 – cn Canadiana [240]

Gospel trumpet – La Salle IL 1822-23 – 1 – mf#3776 – us UMI ProQuest [240]

Gospel truth – Oklahoma City OK 1967-73 – 1,5 – ISSN: 0017-2383 – mf#2295 – us UMI ProQuest [220]

Gospel truths – v1 n7-n8, n13-n14. 1902 [complete] – (= ser Mennonite serials coll) – 1r – 1 – mf#ATLA 1994-S020 – us ATLA [242]

Gospel visitant – La Salle IL 1811-18 – 1 – mf#3813 – us UMI ProQuest [220]

Gospel voice – 1979 oct/nov – 1r – 1 – mf#4878600 – us WHS [071]

Gospel witness – Toronto. v1-58. 1922-80 [complete] – 23r – 1 – (title varies: gospel witness and protestant advocate) – mf#ATLA RS0148 – us ATLA [242]

The gospel witness : organ of the federation of evangelical lutheran churches in india – Guntur: printed at the AELM Press 1905- [qrtly, mnthly] [mf ed 2006] – v1-65 (1905-70) [gaps] on 13r – 1 – (imprint varies; publ in: guntur 1905-jan 1923, 1926-aug 1927, mar 1930-may 1957; rajahmundry feb 1923-25; madras sep 1927-feb 1930; jun 1957-1970; some iss missing; some pgs damaged) – ISSN: 0017-2391 – mf#2006c-s001 – us ATLA [242]

The gospel witness – Toronto, April 9, 1931, February 11 and 18, 1932. Publ. No. 2204-4 g. One of several items on a reel – 1 – us Southern Baptist [242]

The gospel witness – v1-3. 1905-08 [complete] – (= ser Mennonite serials coll) – 2r – 1 – mf#ATLA 1991-S000 – us ATLA [242]

Gospel witness and protestant advocate *see* Gospel witness

Gospel world news – v5 n65 [1997apr] – 1r – 1 – mf#4023597 – us WHS [220]

Gospel-criticism and historical christianity : a study of the gospels and of the history of the gospel-canon during the second century with a consideration of the results of modern criticism / Cone, Orello – New York: G P Putnam, 1891 – 1mf – 9 – 0-8370-2723-3 – (incl ind) – mf#1985-0723 – us ATLA [226]

The gospels = Evangiles et la seconde generation chretienne / Renan, Ernest – London: Mathieson, [189-?] – (= ser Histoire des Origines du Christianisme) – 1mf – 9 – 0-7905-2686-7 – (in english) – mf#1987-2686 – us ATLA [226]

The gospels / Pullan, Leighton – London, New York: Longmans, Green 1912 [mf ed 1989] – (= ser The oxford library of practical theology) – 1mf – 9 – 0-7905-1842-2 – (incl bibl ref & ind) – mf#1987-1842 – us ATLA [226]

The gospels : with moral reflections on each verse / Quesnel, Pasquier – Philadelphia: Parry & McMillan. 2v. 1855 – 4mf – 9 – 0-8370-9020-2 – mf#1986-3020 – us ATLA [226]

The gospels according to s matthew and s mark / ed by Stubbs, Charles William – London: J M Dent; Philadelphia: J B Lippincott 1901 [mf ed 1989] – (= ser The temple bible) – 1mf – 9 – 0-7905-1853-8 – mf#1987-1853 – us ATLA [226]

The gospels as historical documents / Stanton, Vincent Henry – Cambridge: The University Press, 1903-20 – 1r – 1 – 0-8370-1535-9 – mf#1984-B282 – us ATLA [226]

The gospels, historical : address delivered at the unitarian conference in washington d.c... and other sermons / Furness, William Henry – [S.l.: s.n.], 1895 – 1mf – 9 – 0-7905-3438-X – mf#1987-3438 – us ATLA [243]

The gospels in greek *see* Psalterium

The gospels in greek/psalterium – 11th, 12th c – (= ser Holkham library manuscript books 4,25) – 1r – 1 – mf#95967 – uk Microform Academic [090]

The gospels in the light of historical criticism : with a preface on (1) the obligations of the clergy, (2) the resurrection of our lord / Chase, Frederic Henry – London, New York: Macmillan, 1914 – 1mf – 9 – 0-7905-0630-0 – mf#1987-0630 – us ATLA [226]

The gospels in the second century : an examination of the critical part of a work entitled "supernatural religion" / Sanday, William – London: Macmillan, 1876 – 1mf – 9 – 0-8370-5040-5 – (incl bibl ref, indexes of biblical passages cited and chronology) – mf#1985-3040 – us ATLA [226]

Gospels of anarchy : and other contemporary studies / Lee, Vernon – London, England. 1909 – 1r – us UF Libraries [960]

The gospels of s matthew and s mark / ed by Horner, George – Oxford: Clarendon Press, 1911 – (= ser The Coptic Version of the New Testament in the Southern Dialect) – 2mf – 9 – 0-524-02765-X – mf#1987-6459 – us ATLA [226]

Gospels of yesterday : drummond, spencer, arnold / Watson, Robert Alexander – 3rd ed. London: James Nisbet, 1889 – 1mf – 9 – 0-8370-5706-X – (incl bibl ref) – mf#1985-3706 – us ATLA [210]

The gospel-visitor – Poland OH, 1857-73 [mthly] [mf ed 2003] – 23v on 4r – 1 – (merged with: christian family companion to form: christian family companion and gospel visitor) – mf1026 – us ATLA [242]

Gosper county citizen – Elwood, NE: John W Thomas, 1883-96// (wkly) [mf ed 1892] – 1r – 1 – (absorbed by: elwood republican) – us NE Hist [071]

Gosper county citizen – Elwood, NE: John W Thomas, 1883-96// (wkly) [mf ed v9 n26. jun 23 1892] – 1r – 1 – (absorbed by: elwood republican) – us NE Hist [071]

Gosper county enterprise – Elwood, NE: A L Squire. v1 n1. jun 15 1899-v2 n42. mar 28 1901 (wkly) [mf ed withgaps filmed 1970] – 1r – 1 – (has occasional suppls) – us NE Hist [071]

Gospodarstvo – Buenos Aires, Argentina, 1926* – 1r – 1 – (slovenian newspaper) – us IHRC [079]

Gosport & fareham express & standard – Ennisworth, England. Aug-Dec 1982; 1983-May 1986. -w. 3 reels – 1 – uk British Libr Newspaper [072]

Gosport journal – Portsmouth, England. 3 Jan-28 Mar 1958; 4 Jul-24 Dec 1958; 2 Jan-19 Jun 1959; 14 Aug-23 Dec 1959; 1960-64. -w. 10 reels – 1 – uk British Libr Newspaper [072]

Gosport standard – England.Nov-Dec 1976; 17, 24 Mar, 12 May-22 Dec 1977; Jan-Aug, 21 Sept-21 Dec 1978; 1979-24 Dec 1980; 8 Jan 1981-5 Aug 1982 – 1 – uk British Libr Newspaper [072]

Goss, Benjamin D *see* Not as simple as black and white

Goss, C C *see* Statistical history of the first century of american methodism

Goss, Charles Chaucer *see* Statistical history of the first century of american methodism

Goss, Charles Frederic *see* The redemption of david corson

Goss, Charles W F *see* London directories from the guildhall library, 1677-1855

Goss, CW F *see* The london directories, 1677-1855

Goss, L Allan *see* The story of we-than-da-ya

Gossalbes, Jose *see* Informacion que da al publico el dr. jose gossalbes...sobre la ultima enfermedad...

Gossard, H A *see*
- Cottony cushion scale
- Insecticides and fungicides
- Insects of the pecan
- Some common florida scales
- Two peach scales
- White fly

Gosse, Edmund *see* Jeremy taylor

Gosse, P H *see*
- Anonymous letter
- Dying postman
- New forest
- Reapers

Gosse, P H, Mrs *see* Consumptive death-bed

Gosse, Philip Henry *see*
- The canadian naturalist
- Creation
- Dollar's worth
- Omphalos

Gosse, Philip Henry (Mrs) *see* Christian soldier

Gossel, J *see* Grabreden

Gosselin, Auguste *see*
- Le 19e siecle tableaux des premieres annees
- Un bon patriote d'autrefois
- Le docteur labrie
- L'eglise et l'etat au canada apres la conquete du pays par les anglais
- Un episode de l'histoire de la dime au canada 1705-1707
- France et canada
- Juridiction exercee par l'archeveque de rouen
- M jean le sueur
- Le venerable francois de laval, premier eveque de quebec et apotre du canada
- Vie de mgr de laval
- Le vrai monument de champlain

Gosselin, Charles *see* Petit traite de grammaire anglaise

Gosselin, David *see*
- Abrege complet de l'histoire sainte
- Le catechime des commencants
- Catechisme populaire de la lettre encyclique de notre t saint-pere leon 13
- Chronological and alphabetical tables of the principal facts of the history of canada, 1492-1887
- Le code catholique
- Dictionnaire genealogique des familles de charlesbourg
- Les etapes d'une classe au petit seminaire de quebec, 1859-1868
- Figures d'hier et d'aujourd'hui a travers saint-laurent, i o, vol 1
- Figures d'hier et d'aujourd'hui a travers saint-laurent, i o, vol 2
- Figures d'hier et d'aujourd'hui a travers saint-laurent, i o, vol 3
- Histoire du cap-sante
- Histoire populaire de l'eglise du canada
- Manuel du pelerin a la bonne sainte-anne de beaupre
- Melanges historiques
- Nouvelle heure des congreganistes
- Pages d'histoire ancienne et contemporaine de ma paroisse natale, saint-laurent, ile d'orleans
- Tablettes chronologiques et alphabetiques des principaux evenements de l'histoire du canada

Gosselin, M *see* The power of the pope during the middle ages

Gosselman, Carl *see* Reise in columbien in den jahren 1825 und 1826

The gossip : a canadian society journal with philatelic and numismatic departments – Ottawa: Gossip Pub Co, [1887] – 9 – ISSN: 1190-6316 – mf#P04690 – cn Canadiana [760]

Gossman, Stephen J *see* Farm labor camp design in rural marion county

Gossner Mission, Berlin *see* 'Die biene auf dem missionsfelde'

Gossners mission unter hindus und kolhs um neujahr 1878 / Plath, Karl Heinrich Christian – Berlin: Gossnerischen Mission, 1879 – 1mf – 9 – 0-7905-6665-6 – mf#1988-2665 – us ATLA [240]

Gossners segensspuren in nordindien : eine geschichtliche und missionstheoretische reisebeschreibung / Plath, Karl Heinrich Christian – Friedenau-Berlin: Gossnerschen Mission, 1896 [mf ed 1995] – (= ser Yale coll) – 162p – 1 – 0 524-10144-2 – (in german) – mf#1995-1144 – us ATLA [240]

Gost' = The guest – Riga, Warsaw – 1r – 1 – $12.80 – (from years 1923, 1925, 1929, 1930, 1935 (incomplete)) – us Southern Baptist [242]

Gosta berlings saga...2 upplagan / Lagerlof, Selma – Stockholm, 1895. 2v – 1 – us Wisconsin U Libr [830]

Gostling, William *see* Music part-books

Gostwick, Joseph *see* Outlines of german literature

Gosudarev pechatnyi dvor i sinodalnaia tipografiia v Moskve *see* Istoricheskaia spravka

Gosudarstvennaia duma *see*
- Stenograficheskie otchety

Gosudarstvennaia duma. stenogramma zasedanii – Moscow: Parlament Rossiiskoi Federatsii, 1994- – 9 – us East View [077]

Gosudarstvennaia duma v rossii : sbornik dokumentov i materialov / Kalinychev, F I – 1957 – 646p 7mf – 9 – mf#RPP-18 – ne IDC [325]

Gosudarstvennoe khoziaistvo rossii v pervoi chetverti 18 stoletiia i reforma petra velikogo / Miliukov, P N – 1892 – 16mf – 8 – mf#R-7698 – ne IDC [947]

O gosudarstvennoi promyshlennosti / Losev, V N – Moskva: "Moskovskii rabochii," [1925?] [mf ed 2004] – (= ser K itogam 14 s'ezda vkp(b) 3) – 1r – 1 – (filmed with: pis'mo tovarishchu emigrantu / v sukhomlinov (v1-2 1919-20)) – us Wisconsin U Libr [338]

O gosudarstvennoi strakhovoi monopolii : protokol zasedaniia iuridicheskoi komissii petrogradskogo otdeleniia instituta finansovo-ekonomicheskikh issledovanii / Pergament, M Ia – M, 1923 – 1mf – 9 – mf#REF-119 – ne IDC [332]

Gosudarstvennye finansy tsarskoi rossii v epokhu imperializma / Pogrebinskii, A P – M, 1968 – 4mf – 9 – mf#REF-159 – ne IDC [332]

Gosudarstvennyi astronomicheskii institut im P K Shternberga *see*
- Biulleten' gosudarstvennogo astronomicheskogo instituta imeni p k shternberga pri moskovskom ordena lenina gosudarstvennom universitete imeni m v lomonosova
- Trudy gosudarstvennogo astronomicheskogo instituta im p k shternberga

Gosudarstvennyi bank : dannye po kontoram i otdeleniiam za 1903-1912 gg / ed by Slanskii, E N – Spb, 1913 – 9mf – 9 – mf#REF-242 – ne IDC [332]

Gosudarstvennyi bank : issledovanie ego ustroistva, ekonomicheskogo i finansovogo znacheniia / ed by Sudeikin, VT – Spb, 1891 – 10mf – 9 – mf#REF-246 – ne IDC [332]

Gosudarstvennyi bank : kratkii ocherk deiatel'nosti za 1860 -1910 gody / ed by Slanskii, E N – Spb, 1910 – 5mf – 9 – mf#REF-245 – ne IDC [332]

Gosudarstvennyi bank i ekonomicheskaia politika tsarskogo pravitel'stva, 1861-1892 / Gindin, I F – M, 1960 – 8mf – 9 – mf#REF-249 – ne IDC [332]

Gosudarstvennyi Bank R S F S R *see*
- Pervyi god deiatel'nosti
- Voprosy bankovoi politiki

Gosudarstvennyi Bank RSFSR *see* Mesiats raboty pravleniia gosudarstvennogo banka rsfsr (16 noiabria-15 dekabria 1921 g)

Gosudarstvennyi Bank SSSR *see* Sbornik tsirkuliarov rasporiazhenii pravleniia gosudarstvennogo banka sssr posledovavshikh s 1 oktiabria 1927 g po 31 dekabria 1927

Gosudarstvennyi Bank. SSSR *see* Economic survey

Gosudarstvennyi Bank. Upravlenie po delam melkogo kredita *see* Otchet po melkomu kreditu za 1913 g

Gosudarstvennyi dvorianskii zemel'nyi bank, 1885-1910 gg – Spb, 1910 – 2mf – 9 – mf#REF-273 – ne IDC [332]

Gosudarstvennyi finansovyi kontrol' : sbornik zakonov, instruktsii, pravil, tsirkuliarov i t p materialov po gosudarstvennomu finansovomu kontroliu / Landa, I I & Lukashevker, D S – M, 1927 – 7mf – 9 – mf#REF-46 – ne IDC [332]

Gosudarstvennyi Istoricheski Muzei. Moscow *see* Trudy

Gosudarstvennyi kontrol', 1811-1911 – Spb, [1911] – 8mf – 9 – mf#REF-223 – ne IDC [332]

Gosudarstvennyi kontrol' i raskhodovanie narodnykh deneg *see* Kratkoe rukovodstvo

Gosudarstvennyi kontrol' v rossii : ego istoriia i sovremennoe ustroistvo v sviazi s izlozheniem smetnoi sistemy, kassovogo poriadka i ustroistva gosudarstvennoi otchetnosti / Sakovich, VA – Ed 2. Spb, 1897-1898. 2v – 13mf – 9 – mf#REF-224 – ne IDC [332]

Gosudarstvennyi kredit v sovetskoi rossii / Sokol'nikov, G – M, 1923 – 1mf – 9 – mf#REF-30 – ne IDC [332]

Gosudarstvo i narodnoe obrazovanie v rossii 18 veka / Vladimirskii-Budanov, M – Iaroslavl, 1874 – 6mf – 8 – mf#R-7957 – ne IDC [947]

Gosudarstvo i pravo (opyt izlozheniia marksistskogo ucheniia o sushchestve gosudarstva i prava). s predisloviem n. v. krylenko / Ksenofontov, F – Moskva, IUridicheskoe izdatel'stvo N. K. IU., 1924. 171 p. LL-4002 – 1 – us L of C Photodup [340]

Gosudarstvo i tserkov v drevnei rossii : kievskii period 988-1240 = Staat und kirche in altrussland: kiever periode 988-1240 / Goetz, Leopold Karl – Berlin: Duncker, 1908 – 1mf – 9 – 0-7905-5829-7 – (incl bibl ref) – mf#1988-1829 – us ATLA [240]

Gosudarstvo-gorod antichnago mira : opyt istoricheskago postroeniia politicheskoi i sotsial'noi evoliutsii antichnykh grazhdanskikh obshchin / Kareev, Nikolai Ivanovich – S-Peterburg: Tip M Stasiulevicha, 1903 [mf ed 2002] – 1r – 1 – (filmed with: monarkhii drevniago vostoka i greko-rimskago mira / n karieev (1904) & other titles. incl bibl ref) – mf#5233 – us Wisconsin U Libr [930]

Goswami, A *see* Glimpses of mughal architecture

Goszczy'nski, Seweryn *see* Dziennik podrozy do tatrow

Gota – Vadstena, Sweden. 1854-56 – 1 – sw Kungliga [078]

Gota alvdalsnyheterna *see* Alingsas tidning

Gota de tiempo / Andreu Iglesias, Cesar – San Juan, Puerto Rico. 1958 – 1r – us UF Libraries [972]

Gotama bhu ra rhan e buddhuppatti / E Mon, Buil – Ran kun: E nnvan Chve Ca pe 1974 [mf ed 1994] – on pt of 1r – 1 – (in burmese) – mf#11052 r1944 n7 – us Cornell [280]

Gotama the man / Davids, Caroline Augusta Foley Rhys – London: Luzac & Co, 1928 – (= ser Samp: indian books) – us CRL [280]

Gotas de sangre / Blanck Y Menocal, Guillermo De – Habana, Cuba. 1920 – 1r – us UF Libraries [972]

Gotch, John Alfred *see* Architecture of the renaissance in england

Goteborg – Goteborg, Sweden. 1886 – 1 – sw Kungliga [078]

Goteborgs aftonblad – Goteborg, Sweden. 1888-95, 1902 – 21r – 1 – sw Kungliga [078]

Goteborgs handels- och sjofartstidning – Goteborg, Sweden. 1832-1985 – 1 – sw Kungliga [078]

Goteborgs marknadsberaettelse – Goteborg, 1880-84, 1886-1913 – 9 – sw Kungliga [078]

Goteborgs morgonpost – Goteborg, Sweden. 1896-1950 – 1 – (suppl: vy och revy, 1910-11) – sw Kungliga [078]

Goteborgs nyheter – Goeteborg, 1884-85 – 9 – sw Kungliga [078]

Goteborgskuriren – Uddevalla, Sweden. 1890-1900, 1979-86 – 1 – sw Kungliga [078]

Goteborgsnytt – Molndal, Sweden. 1979-84 – 1 – sw Kungliga [078]

Goteborgsposten – Goteborg, Sweden. 1858-1978 – 976r – 1 – sw Kungliga [078]

Goteborgsposten – Goteborg, Sweden. 1979- – 1 – sw Kungliga [078]

Goteborgstidningen – Goteborg, Sweden. 1967-89 – 1 – (aka: gt) – sw Kungliga [078]

Goteborgstidningen – Goteborg, Sweden. 1902-67. GT, 1967-78 – 515r – 1 – sw Kungliga [078]

Der gotha : der "oesterreich-gotha". mit ergaenzungswerken zum deutschen adel – [mf ed 1997] – 140mf (1:24) – 9 – silver €2348.00 – ISBN-10: 3-598-30359-9 – ISBN-13: 978-3-598-30359-3 – gw Saur [920]

Der gotha, 1763-1944 – [mf ed 1982-84] – 794mf (1:24) – 9 – silver €6648.00 – ISBN-10: 3-598-30330-0 – ISBN-13: 978-3-598-30330-2 – gw Saur [920]

Gothaer tagespost – Gotha DE, 1993- – 2r/yr – 1 – (bezirksausgabe von thueringische landeszeitung, weimar) – gw Misc Inst [074]

Gothaische gelehrte zeitungen / Kluepfel, Emil Chr – Gotha 1774-1804 – (= ser Dz) – 31jge[zu je 52st] on 180mf – 9 – €1080.00 – mf#k/n308 – gw Olms [430]

Gothaisches genealogisches taschenbuch der freiherrlichen hauser – 1920-25 – 1 – us L of C Photodup [920]

Gothaisches magazin der kuenste und wissenschaften / ed by Schack, Hermann Ewald – Gotha 1776-77 – (= ser Dz) – 2v[zu je 4st] on 6mf – 9 – €120.00 – mf#k/n316 – gw Olms [700]

Gotheborgska nyheter – Goteborg, Sweden. 1765-1848 – 16r – 1 – sw Kungliga [078]

Gotheborgsposten – Goteborg, Sweden. 1813-32, 1850-51 – 5r – 1 – sw Kungliga [078]

Gothein, Eberhard *see*
- Ignatius von loyola
- Ignatius von loyola und die gegenreformation
- Politische und religioese volksbewegungen vor der reformation

Gothenburg independent – Gothenburg, NE: H C Booker. v21 n32. nov 16 1905-v41 n43. jan 31 1925 (wkly) [mf ed with gaps] – 8r – 1 – (cont: gothenburg independent and gothenburg sun. merged with: gothenburg times (1908) to form: gothenburg times and gothenburg independent) – us NE Hist [071]

Gothenburg independent – Gothenburg, NE: Willard & Springsteen. v1 n1. may 9 1885-v20 n38. dec 29 1904 (wkly) [mf ed with gaps filmed – 1984] – 8r – 1 – (merged with: gothenburg sun to form: gothenburg independent and gothenburg sun) – us NE Hist [071]

Gothenburg independent and gothenburg sun – Gothenburg, NE: H C Booker. v20 n[39]. jan 5 1905-v21 n31. nov 9 1905 (wkly) [mf ed jan 5 1905] – 1r – 1 – (formed by the union of: gothenburg independent and: gothenburg sun. cont by: gothenburg independent (1908)) – us NE Hist [071]

The gothenburg sun – Gothenburg, NE: J B McKnight, 1900-v5 n39. dec 30 1904 (wkly) [mf ed 1901-04 (gaps) filmed 1978] – 2r – 1 – (merged with: gothenburg independent to form: gothenburg independent and gothenburg sun) – us NE Hist [071]

The gothenburg times – Gothenburg, NE: R D & D P Holmes. 40th yr n5. aug 14 1947- (wkly) – 1 – (cont: gothenburg times, gothenburg independent and farnam echo) – us NE Hist [071]

The gothenburg times – Gothenburg, NE: J C Holmes. 17v. v1 n1. jul 17 1908-v17 n33. jan 28 1925 (wkly) [mf ed with gaps] – 13r – 1 – (merged with: gothenburg independent (1908) to form: gothenburg times and gothenburg independent) – us NE Hist [071]

Gothenburg times and gothenburg independent – Gothenburg, NE: J C Holmes. 16v. v17 n34. feb 4 1925-32nd yr n31. feb 1 1940 (wkly) [mf ed with gaps] – 15r – 1 – (formed by the union of: gothenburg times (1908) and: gothenburg independent (1908). merged with: farnam echo to form: gothenburg times, gothenburg independent and farnam echo) – us NE Hist [071]

Gothenburg times, gothenburg independent and farnam echo – Gothenburg, NE: R D and D P Holmes. 9v. 32nd yr n32. feb 8 1940-40th yr n4. aug 7 1947 (wkly) – 7r – 1 – (formed by the union of: gothenburg times and gothenburg independent and: farnam echo. cont by: gothenburg times (1947)) – us NE Hist [071]

Gothic architecture in france, england, and italy / Jackson, Thomas Graham, Sir – Cambridge: University Press; Chicago: University of Chicago Press [distributor], 1915 – 3mf – 9 – 0-7905-8039-X – mf#1988-6020 – us ATLA [720]

Gothic blimp works – n3-6 [1969 aug-nov] – 1r – 1 – mf#1057384 – us WHS [071]

Gothic fiction : rare printed works from the sadleir-black collection of gothic fiction at the alderman library, university of virginia – 6pt – 1 – (pt1: matthew lewis & gothic horror – beckford to lewis [23r] £2150; pt2:... – mackenzie to zschokke [21r] £1950; pt3: gothic terror: radcliffe & her imitators – boaden to meeke [26r] £2450; pt4:... – pickard to wilkinson [23r] £2050; pt5: domestic & sentimental gothic – bennett to lamb [21r] £1950; pt6:... – lathom to warner [21r] £1950; with d/g) – uk Matthew [420]

Gothic forms applied to furniture metal work and decoration / Talbert, Bruce J – Birmingham 1867 – (= ser 19th c art & architecture) – 3mf – 9 – mf#4.2.191 – uk Chadwyck [740]

Gothic furniture in the style of the 15th century / Pugin, Augustus Welby Northmore – London: Ackermann & Co, 1835 – (= ser 19th c art & architecture) – 1mf – 9 – mf#4.1.10 – uk Chadwyck [740]

The gothic model-book : the architecture of the middle ages / Statz, Vincenz & Ungewitter, Georg Gottlob – London [1862] – (= ser 19th c art & architecture) – 8mf – 9 – mf#4.2.1325 – uk Chadwyck [720]

Gothic ornaments drawn from existing authorities / Colling, James Kellaway – London [1848-50] – (= ser 19th c art & architecture) – 6mf – 9 – mf#4.2.1028 – uk Chadwyck [740]

Gotifredi viterbiensis gesta friderici 1 et heinrici 4 imperatorum metrice scripta (mgh7:30.bd) – 1870 – (= ser Monumenta germaniae historica 7: scriptores rerum germanicarum in usum scholarum (mgh7)) – €5.00 – ne Slangenburg [240]

Gotland – Visby, Sweden. 1886-99 – 1 – sw Kungliga [078]

Gotlands allehanda – Visby, Sweden. 1873-1953, 1979- – 1 – sw Kungliga [078]

Gotlands folkblad – Visby, Sweden. 1928-83 – 1 – (aka: gotlands tidningar) – sw Kungliga [078]

Gotlands folkblad – Visby, Sweden. 1928-83 – 1 – (aka: gotlanningen tidningar) – sw Kungliga [078]

Gotlands laens tidning – Visby, 1849-58 – 9 – sw Kungliga [078]

Gotlands tidning – Visby, 1859-65, 1867-88 – 9 – sw Kungliga [078]

Gotlands tidningar – Visby, Sweden. 1983- – 1 – sw Kungliga [078]

Gotlands tidningar *see*
- Gotlands folkblad
- Gotlanningen

Gotlanningen – Visby, Sweden. 1884-1978 – 1 – (title changes to: gotlands tidningar 1983) – sw Kungliga [078]

Gotlanningen tidningar *see* Gotlands folkblad

Goto, Shimpei *see* Goto shimpei monjo

Goto shimpei monjo : count shimpei goto records. in the holdings of the goto shimpei memorial museum, iwate prefecture – 1857-1929 – 130,000p on 88r – 1 – Y880,000 – (with 106p guide. in japanese) – ja Yushodo [950]

Gots folk / Birnbaum, Nathan – Berlin, Germany. 1921 – 1r – us UF Libraries [939]

Gotschlich, Emil *see* Lessing's aristotelische studien und der einfluss derselben auf seine werke

Gotsmakhs kremil t.e. 25 raznikh evreiskikh narodnikh piecen i kuletov pietykh y evreiskom narodnam teatr goldfadena *see* "Hotsmah's kremil" fun fershidene antiken

Ein gott! / Grundwald, Max – Hamburg, Germany. 1896 – 1r – us UF Libraries [939]

Gott in der natur / Flammarion, Camille – Halle: O Hendel [1902?] [mf ed 1998] – (= ser Bibliothek der gesamt-literatur 1612-1616) – 1r [pl/ill] – 1 – (incl bibl ref) – mf#film mas 28211 – us Harvard [210]

Gott in frankreich? / Sieburg, Friedrich – Frankfurt am Main, Germany. 1929 – 1r – us UF Libraries [025]

Gott und die naturgesetze / Pfaff, Friedrich – Heidelberg: Carl Winter, 1881 – (= ser Sammlung von Vortraegen fuer das deutsche Volk) – 1mf – 9 – 0-524-03469-9 – mf#1990-1012 – us ATLA [210]

Gott und die wissenschaft / Buechner, Ludwig – Leipzig: Theod. Thomas, 1897 – 1mf – 9 – 0-8370-2519-2 – mf#1985-0519 – us ATLA [210]

Gott und goetter : eine studie zur vergleichenden religionswissenschaft / Pesch, Christian – Freiburg i B: Herder, 1890 – (= ser Ergaenzungshefte zu den "Stimmen aus Maria-Laach") – 1mf – 9 – 0-524-05867-9 – (incl bibl ref) – mf#1990-3531 – us ATLA [230]

Gott und sein reich : philosophische darlegung der freien goettlichen selbstentwicklung zum allumfassenden organismus / Meyr, Melchior – Stuttgart:Gebrueder Maentler (A. Kroener), 1860 – 1mf – 9 – 0-8370-4421-9 – mf#1985-2421 – us ATLA [210]

Gott und seine offenbarungen in natur und geschichte / Hamberger, Julius – 2. verb Aufl. Guetersloh: C Bertelsmann, 1882 – 1mf – 9 – 0-8370-3456-6 – mf#1985-1456 – us ATLA [240]

Gotte, Jules *see* Sponge culture

Gotteberg, J A O *see* Ti aar i hunan

Das gotteerlebnis des germanischen menschen : weltanschauliches in der dichtung von hans fr blunck / Etscheid, Lisel – Bonn a/Rh: L Roehrscheid 1932 [mf ed 1992] – 1r – 1 – (incl bibl ref. filmed with: hoelderlins deutung des "oedipus" und der "antigone" / hans schrader & other titles) – mf#3114p – us Wisconsin U Libr [430]

Gotterthrone im urwald : auf den spuren altindo-malaiischer kultur / Krug, Hans-Joachim – Berlin, Germany. 1943 – 1r – us UF Libraries [939]

Gottes dasein bewiesen am wissen und sein : ein beitrag zur rechenschaft unsres glaubens / Doederlein, Julius – Erlangen: Eduard Besold, 1871 – 1mf – 9 – 0-8370-2943-0 – mf#1985-0943 – us ATLA [210]

Gottes fuehrung im alten testament / Moser, Friedrich Carl, Freiherr von – 2. aufl. Calw: Verlag der Vereinsbuchh, [18–?] – (= ser Brosamen) – 1mf – 9 – 0-524-07843-2 – mf#1992-1109 – us ATLA [221]

Das gottes kind : ein weihnachtsspiel / Herrmann, Emil Alfred – Jena: E Diederichs 1912 [mf ed 1991] – 1r [ill] – 1 – (filmed with: das problem "volkstum und dichtung" bei herder / reta schmitz) – mf#7474 – us Wisconsin U Libr [820]

Gottes offenbarung durch heilige geschichte : nach ihrem wesen beleuchtet in einer reihe oeffentlicher vortraege / Goltz, Hermann, Freiherr von der – Basel: Felix Schneider, 1868 – 1mf – 9 – 0-524-04573-9 – mf#1992-0161 – us ATLA [220]

Die gottesanrede im ante-sanctus / Beckman, S – Muenster, 1932 – 2mf – 8 – €5.00 – ne Slangenburg [240]

Der gottesbegriff in den heidnischen religionen der neuzeit : eine studie zur vergleichenden religionswissenschaft / Pesch, Christian – Freiburg i B: Herder, 1888 – (= ser Ergaenzungshefte zu den "Stimmen aus Maria-Laach") – 1mf – 9 – 0-524-05869-5 – (incl bibl ref) – mf#1990-3533 – us ATLA [230]

Der gottesbegriff in den heidnischen religionen des alterthums : eine studie zur vergleichenden religionswissenschaft / Pesch, Christian – Freiburg i B: Herder, 1885 – (= ser Ergaenzungshefte zu den "Stimmen aus Maria-Laach") – 1mf – 9 – 0-524-05868-7 – (incl bibl ref) – mf#1990-3532 – us ATLA [230]

Der gottesbeweis aus der bewegung bei thomas von aquin : auf seinen wortlaut untersucht: ein beitrag zur textkritik und erklaerung der summa contra gentiles / Weber, Simon – Freiburg i.B, St Louis, MO: Herder 1902 [mf ed 1991] – 1mf – 9 – 0-524-00202-9 – (in german & latin) – mf#1989-2902 – us ATLA [240]

Die gottesbeweise bei franz brentano / Seiterich, Eugen & Giesen, Adolf – Freiburg, 1936 (mf ed 1992) – 2mf – 9 – €24.00 – 3-89349-067-1 – mf#DHS-AR 30 – gw Frankfurter [230]

Die gottesbezeichnungen in den liturgien der ostkirchen / Buckel, A – Wuerzburg, 1938 – 3mf – 8 – €7.00 – ne Slangenburg [243]

Das gottesdienstliche jahr der juden / Schaerf, Theodor – Leipzig:J C Hinrichs, 1902 – 1mf – 9 – 0-8370-5067-7 – (incl bibl ref) – mf#1985-3067 – us ATLA [270]

Die gottesdienstliche schriftlesung 1 : stand und aufgaben der perikopenforschung / Kunze, G – Goettingen, 1947 – 4mf – 8 – €11.00 – ne Slangenburg [240]

Der gottesdienstliche volksgesang im juedischen und christlichen altertum / Leitner, Franz – Freiburg i. Br., 1906 – 6mf – 8 – €14.00 – ne Slangenburg [780]

Die gottesdienstlichen gebraeuche der griechen und roemer / Seemann, Otto – Leipzig: Verlag des litterarischen Jahresberichts, 1888 – (= ser Kulturbilder aus dem klassischen Altertume) – 1mf – 9 – 0-524-03483-4 – mf#1990-3225 – us ATLA [250]

Das gotteserlebnis der reformation : eine apologetische rede in erweiterter form / Mandel, Hermann – Guetersloh: C Bertelsmann, 1916 – (= ser Beitraege zur foerderung christlicher theologie) – 1mf – 9 – 0-524-00642-3 – (incl bibl ref) – mf#1990-0142 – us ATLA [242]

Das gotteserlebnis in hebbels dramen / Blaustein, Leopold – Berlin: Reuther & Reichard, 1929 [mf ed 1990] – viii/68p – 1 – (incl bibl ref) – mf#7452 – us Wisconsin U Libr [430]

Gottesfriede und treuga dei (mgh schriften:20.bd) / Hoffmann, H – 1964 – (= ser Monumenta germaniae historica. schriften (mgh schriften)) – €15.00 – ne Slangenburg [241]

Der gottesgedanke in der geschichte der philosophie. erster teil, von heraklit bis jakob boehme / Schwarz, Hermann – Heidelberg: C Winters, 1913 – (= ser Synthesis) – 2mf – 9 – 0-7905-8580-4 – mf#1989-1805 – us ATLA [210]

Gottesgedanken in israels koenigtum / Boehmer, Julius – Guetersloh: C Bertelsmann, 1902 – (= ser Beitraege zur foerderung christlicher theologie) – 1mf – 9 – 0-524-06736-8 – mf#1992-0939 – us ATLA [221]

Die gottesheilige messe von gott allein erstiftet / Murner, Thomas; ed by Pfeiffer-Belli, Wolfgang – Halle: M Niemeyer, 1928 – us Wisconsin U Libr [240]

Gottesknecht des deuterojesaja : eine kritisch-exegetische und biblisch-theologische studie / Fuellkrug, Gerhard – Goettingen: Vandenhoeck & Ruprecht, 1899 – 1mf – 9 – 0-8370-3212-1 – mf#1985-1212 – us ATLA [221]

Die gotteslehre des irenaeus / Kunze, Johannes – Leipzig: Doerffling & Francke, 1891 – 1mf – 9 – 0-7905-6070-4 – (incl bibl ref) – mf#1988-2070 – us ATLA [210]

Die gottesoffenbarung in jesu christo : nach wesen, inhalt und grenzen: unter dem geschichtlichen, psychologischen und dogmatischen gesichtspunkte / Schwartzkopff, Paul – Giessen: J Ricker, 1896 – 1mf – 9 – 0-8370-5610-1 – (incl bibl ref) – mf#1985-3610 – us ATLA [240]

Gottesstaat, 1. bd (bdk1 1.reihe) : buch 1-8 / Augustinus (Augustine, Saint, Bishop of Hippo) – (= ser Bibliothek der kirchenvaeter. 1. reihe (bdk 1.reihe)) – €5.00 – ne Slangenburg [240]

Gottesstaat, 2. bd (bdk16 1.reihe) : buch 9-16 / Augustinus (Augustine, Saint, Bishop of Hippo) – (= ser Bibliothek der kirchenvaeter. 1. reihe (bdk 1.reihe)) – €18.00 – ne Slangenburg [240]

Gottesstaat, 3. bd (bdk28 1.reihe) : buch 17-27 / Augustinus (Augustine, Saint, Bishop of Hippo) – (= ser Bibliothek der kirchenvaeter. 1. reihe (bdk 1.reihe)) – €18.00 – ne Slangenburg [241]

Gottfried arnolds auserlesene send-schreiben derer alten / Arnold, Gottfried – Frankfurt: Theod. Phillipo Calvisio, 1700 – 1r – 1 – 0-8370-0470-5 – mf#1984-B257 – us ATLA [943]

Gottfried august buerger : der roman seines lebens in seinen briefen und gedichten / ed by Mederow, Paul Wolfgang – Berlin: Morawe & Scheffelt, 1912 [mf ed 1989] – 276p – 1 – mf#7095 – us Wisconsin U Libr [430]

Gottfried august buerger und philippine gatterer : ein briefwechsel aus goettingens empfindsamer zeit / ed by Ebstein, Erich – Leipzig: Dieterich, 1921 [mf ed 1989] – 221p (ill) – 1 – (incl bibl) – mf#7095 – us Wisconsin U Libr [860]

Gottfried benn : un demi siecle vecu par un poete allemand / Garnier, Pierre – Paris: Andre Silvaire, c1959 – 1r – 1 – (incl bibl ref) – us Wisconsin U Libr [430]

Gottfried der student : ein moralisches akademisches epos nach alten handschriften / Bimstein, Emanuel – 3. aufl. Heiligenstadt: F W Cordier, [18–?] [mf ed 1989] – 216p – 1 – mf#7023 – us Wisconsin U Libr [810]

Gottfried keller / Enders, Carl Friedrich – Leipzig: P Reclam, c1921 – 1r – 1 – us Wisconsin U Libr [920]

Gottfried keller : festvortrag bei der am 19. juli 1919 von der universitaet bern in ihrer aula veranstalteten keller-hundert-jahrfeier / Maync, Harry Wilhelm – Bern: K J Wysz, 1919 – 1r – 1 – (incl bibl ref) – us Wisconsin U Libr [430]
Gottfried keller : ein literarischer essay / Brahm, Otto – Berlin: A B Auerbach 1883 [mf ed 1995] – 1r – 1 – (filmed with: der gruene heinrich / gottfried keller; ed by emil ermatinger) – mf#3648p – us Wisconsin U Libr [430]
Gottfried keller / Rosenfeld, W – Leipzig: Sphinx-Verlag, [1907?] – 1r – 1 – (incl bibl ref) – us Wisconsin U Libr [920]
Gottfried keller : sechs vortraege / Steiner, Gustav – Basel: Helbing & Lichtenhahn, 1918 – 1r – 1 – us Wisconsin U Libr [430]
Gottfried keller : zu seinem hundertsten geburtstage (19. juli 1919) / Arx, Walther von – Basel: Verein fuer Verbreitung guter Schriften, 1919 [mf ed 1995] – 72p – 1 – mf#8786 – us Wisconsin U Libr [430]
Gottfried keller im europaeischen gedanken / Hochdorf, Max – Zuerich: Rascher, 1919 – 1r – 1 – us Wisconsin U Libr [430]
Gottfried keller im spiegel seiner zeit : urteile und berichte von zeitgenossen ueber den menschen und dichter / ed by Zaech, Alfred – Zuerich: Scientia, 1952 – 1 – us Wisconsin U Libr [430]
Gottfried keller und die frauen : ein stueck herzenstragik / Huber, Walther – 2. Aufl. Bern: F Wyss, 1919 – 1r – 1 – us Wisconsin U Libr [430]
Gottfried keller und j v widmann : briefwechsel / ed by Widmann, Max – Basel: Rhein-Verlag, 1922 – 1 – (incl bibl ref & ind) – us Wisconsin U Libr [920]
Gottfried kellers dramatische bestrebungen / Preitz, Max – Marburg a.L.: N G Elwert, 1909 – 1r – 1 – (incl bibl ref (5th prelim. leaf)) – us Wisconsin U Libr [430]
Gottfried kellers gesammelte werke – Stuttgart, Germany. v1-10. 1912 – 2r – us UF Libraries [430]
Gottfried kellers glaube : ein bekenntnis zu seinem protestantismus / Buri, Fritz – Bern: P Haupt 1944 [mf ed 1995] – 1r – 1 – (filmed with: gottfried keller im spiegel seiner zeit / alfred zaech [ed]) – mf#3649p – us Wisconsin U Libr [430]
Gottfried kellers "gruener heinrich" von 1854/5 und 1879/80 : beitraege zu einer vergleichung / Leppmann, Franz – Berlin: E Ebering [1902] [mf ed 1990] – 1r – 1 – (incl bibl ref. filmed with: gottfried keller und die frauen / walther huber) – mf#2755p – us Wisconsin U Libr [430]
Gottfried kellers politische anschauungen / Schlomer, Harm Henry – Heidelberg-Handschusheim: H Fahrer, 1936 – 1r – 1 – (incl bibl ref) – us Wisconsin U Libr [430]
Gottfried wilhelm sacer's "reime dich, oder ich fresse dich...northausen 1673" / Pfeil, Leopold – Heidelberg, 1914 (mf ed 1995) – 2mf – 9 – €31.00 – 3-8267-3139-5 – mf#DHS-AR 3139 – gw Frankfurter [430]
Gottfried's von strassburg tristan / ed by Bechstein, Reinhold – 3. aufl. Leipzig: F A Brockhaus, 1890-1891 [mf ed 1993] – (= ser Deutsche classiker des mittelalters 7-8) – 2v – 1 – (middle high german text. int in german. incl bibl ref and ind) – mf#8398 – us Wisconsin U Libr [810]
Gottheil, Gustav *see* Hymns and anthems adapted for jewish worship
Gottheil, Gustav et al *see* The message of the world's religions
Gottheil, Richard James Horatio *see* A treatise on syriac grammar
Die gottheit des heiligen geistes : nach den griechischen vaetern des vierten jahrhunderts / Schermann, Theodor – Freiburg i B: Herder, 1901 – (= ser Strassburger theologische Studien) – 3mf – 9 – 0-524-04499-6 – (incl bibl ref) – mf#1990-1261 – us ATLA [240]
Gotthelf : die geheimnisse des erzaehlers / Muschg, Walter – Muenchen: Beck, c1931 [mf ed 1989] – ix/579p – 1 – (incl bibl) – mf#7030 – us Wisconsin U Libr [430]
Gotthelf, Jeremias *see*
– Der bauern-spiegel
– Bauernspiegel
– Familienbriefe jeremias gotthelfs
Gotthelf, Jeremias [Albert Bitzius] *see*
– Albert bitzius
– Die armennot / ein sylvestertraum / eines schweizers wort
– Bilder aus der schweiz
– Erlebnisse eines schuldenbauers
– Geld und geist
– Der geldstag
– Die kaeserei in der vehfreude
– Saemtliche werke in 24 baenden
Gotthelf, Jeremias [pseud: Albert Bitzius] *see*
– Jakobs, des handwerksgesellen wanderungen durch die schweiz
– Jeremias gotthelf's ausgewaehlte werke
– Kaethi, die grossmutter
– Kaethi die grossmutter
– Kalendergeschichten
– Kleinere erzaehlungen
– Leiden und freuden eines schulmeisters
– Uli der paechter
– Ulric the farm servant
– Volksausgabe seiner werke im urtext
– Die wassernot im emmental / die armennot / eines schweizers wort
– Die wassernot im emmenthal / fuenf maedchen / dursli der branntweinlaeuter
– Wie anne baebi jowaeger
– Wie uli der knecht gluecklich wird
– Zeitgeist und bernergeist
Gotthold ephraim lessing : eine einfuehrung in sein leben und werk / Seidel, Siegfried – Berlin: Verlag Neues Leben, 1963 – 1r – 1 – (incl bibl ref & index) – us Wisconsin U Libr [430]
Gotthold ephraim lessing / ed by Vorbrodt, Walther – Berlin: Union Deutsche Verlagsgesellschaft, 1920 – 1r – 1 – us Wisconsin U Libr [920]
Gotthold ephraim lessing, his life and his works / Zimmern, Helen – London: Longmans, Green, 1878 – 2mf – 9 – 0-7905-8990-7 – mf#1989-2215 – us ATLA [430]
Gotthold ephraim lessings saemtliche schriften = Works / Lessing, Gotthold Ephraim; ed by Lachmann, Karl – 3. durchges. verm. Aufl. Stuttgart: G J Goeschen, 1866-1924 – 1 – (incl bibl ref & index) – us Wisconsin U Libr [800]
Göttinger anzeiger 1881 – Goettingen DE, 1893 1 jan-30 jun – 1r – 1 – (title varies: 1 oct 1882: goettinger freie presse; 1 jan 1891: goettinger anzeiger) – gw Mikrofilm [074]
Gottlob david hartmann : ein lebensbild aus der sturm- und drangzeit / Lang, Wilhelm – Stuttgart: W Kohlhammer, 1890 – 1r – 1 – us Wisconsin U Libr [430]
Der gottmensch, das ebenbild des unsichtbaren gottes : beitrag zur christologie / Keerl, Philipp Friedrich – Basel: Bahnmaier, 1866 – 2mf – 9 – 0-7905-3453-3 – (incl bibl ref) – mf#1987-3453 – us ATLA [240]
Gott-natur : goethes naturanschauung im lichte seiner froemmigkeit / Fischer, Paul – Weimar: H Boehlaus Nachfolger, 1932 – 1r – 1 – us Wisconsin U Libr [430]
Gottsaeliger vnd grundtlicher bericht von der hochheit...heiliger goettlicher geschrifft... / Bullinger, Heinrich – Zuerych, Christoffel Froschouer, 1572 – 2mf – 9 – mf#PBU-240 – ne IDC [240]
Gottschalck, Friedrich *see* Taschenbuch fuer reisende in den harz
Gottschald, Max *see* Deutsche namenkunde
Gottschalk, Hanns *see* Schicht und schacht
Gottschalk, Louis Moreau *see* Study and edition of recently discovered works of louis moreau gottschalk
Gottschall, Lori L *see* Training patterns and illness during a men's collegiate basketball season
Gottschall, Rudolf *see* Gedankenharmonie aus goethe und schiller
Gottschall, Rudolf von *see*
– Christian dietrich grabbe
– Die deutsche nationallitteratur des neunzehnten jahrhunderts
– Kaiser napoleon 3
Gottschall's reports / Ohio. Montgomery County. Dayton – 1v. 1865-73 (all publ) – (= ser Ohio appellate decisions) – 6mf – 9 – $9.00 – mf#llmc 84-184 – us LLMC [340]
Gottsched : biographische skizze / Reichel, Eugen – Berlin: Gottsched-Verlag, 1900 – 1r – 1 – (incl bibl ref) – us Wisconsin U Libr [920]
Gottsched, Johann Christoph *see*
– Joh. chr. gottscheds sterbender cato
– Neuer buechersaal der schoenen wissenschaften und freyen kuenste
– Das neueste aus der anmuthigen gelehrsamkeit
Gottsched und die reform der deutschen literatur im achtzehnten jahrhundert / Koch, Max – Hamburg: J F Richter, 1887 – 1r – 1 – (incl bibl ref) – us Wisconsin U Libr [430]
Der gottschedkreis und russland : deutsch-russische literaturbeziehungen im zeitalter der aufklaerung / Lehmann, Ulf – Berlin: Akademie-Verlag 1966 [mf ed 1993] – 1r – 1 – (incl bibl ref. filmed with: deutsch-indische geistesbeziehungen / ludwig alsdorf) – mf#8141 – us Wisconsin U Libr [410]
Gottscheds bedeutung fuer die geschichte der deutschen philologie / Lachmann, Hans – Leipzig: Kommissionsverlag von Alfred Lorentz, 1931 – 1r – 1 – (incl bibl ref) – us Wisconsin U Libr [430]
Gottscheer zeitung – Gottschee (Kocevje SLO), 1938 1 oct-20 dec – 1r – 1 – gw Misc Inst [077]
Gottscheer zeitung – Kocevje, Yugoslavia. 20 Aug 1924; Feb 1925-1941 – 3r – 1 – us L of C Photodup [949]
Gottschick, Johannes *see* Die kirchlichkeit der s.g. kirchlichen theologie
Gottselige begirden r.p. hermanni hugonis s. jes. verteutscht durch r.p.f. carl. stengelium ord. s. ben. / Hugo, H – Colln: In Verlegung Constantini und Johan Muttich, 1636 – 5mf – 9 – mf#0-07 – ne IDC [090]
Gotwald, Luther Alexander *see* Trial of l.a. gotwald, d.d
Gotz, Oscar *see*
– Geschichte der hauen'schen erziehungsanstalt zu berlin
– Geschichte der nauenschen erziehungsanstalt zu berlin
Goubau, Francisci *see* Apostolicarum epistolarum libri quinque
Goubaud, Adolphe *see* Madame goubaud's point lace book
Goucher baptist church – Washington. 1971+ (1) 1971+ (5) 1975+ (9) – 1 – $14.85 – mf#6500 – us Southern Baptist [242]
Goucher, John Franklin *see* Christianity and the united states
Goudar, Ange *see* Le brigandage de la musique italienne
Goudge, Henry Leighton *see*
– The first epistle to the corinthians
– The mind of st paul
Goudge, Henry Leighton et al *see* The place of women in the church
Goudie's perpetual sleigh road supersedes the railway : and is capable of carrying passengers at a rate of eighty to one hundred miles an hour... – [Toronto?: s.n.], 1874 [mf ed 1981] – 2mf – 9 – mf#23942 – cn Canadiana [380]
Goudin, A *see* Philosophia iuxta inconcussa tutissimaque divi thomae dogmata...
Gougaud, L *see*
– Anciennes coutumes claustrales
– Christianity in celtic lands
– Eremites et reclus
Gougaud, Louis *see* Les chretientes celtiques
Gouge, William *see*
– A commentary on the whole epistle to the hebrews
– The saints' sacrifice
Gough, Archibald Edward *see* The philosophy of the upanishads and ancient indian metaphysics
Gough, John B *see* Habit
Gough, John Bartholomew *see*
– Autobiography and personal recollections of john b gough
– Platform echoes
Gough, Robert E *see* Journal of small fruit and viticulture
Gouhier, Henri Gaston *see* Essais sur descartes
Goujon, Alexandre M *see* Bulletins officiels de la grande armee dictes par l'empereur napoleon
Goulart, Gastao *see* Verdades da revolucao paulista
Goulart, Jose Alipio *see*
– Cavalo na formacao do brasil
– Tropas e tropeiros na formacao do brasil
Goulart, S *see*
– 28 discours chrestiens, touchant l'estat du monde et de l'eglise de dieu
– Recueil des choses mémorables ou histoire des cinq rois
[Goulart, S] *see* Memoires de la ligue, 1585-1598
Goulart, Simon *see*
– Memoire de la ligue
– Theatre du monde
Goulburn, Edward Meyrick *see*
– The doctrine of the resurrection of the body
– Everlasting punishment
– John william burgon
– The pursuit of holiness
– Thoughts on personal religion
Goulburn, Edward Meyrick et al *see* Replies to "essays and reviews"
Goulburn evening penny post – Goulborn, Australia. 31 jan 1911-29 jun 1922; 4 jan-aug, 1 sep, 12 dec 1952 – 15r – 1 – (aka: goulburn evening post) – uk British Libr Newspaper [079]
Goulburn evening penny post – Goulburn, jan 1901-dec 1968 – (= ser Goulburn evening post) – 110r – A$7038.28 vesicular A$7643.28 silver – (aka: goulburn evening post) – at Pascoe [079]
Goulburn evening penny post – Goulbourn. oct 1870-dec 1876, jan 1878-dec 1900 – 26r – A$858.00 vesicular A$1001.00 silver – at Pascoe [079]
Goulburn evening post *see* Goulburn evening penny post
Goulburn post – Goulburn, jan 1969-jun 1997 – at Pascoe [079]
Gould, Daniel *see*
– Collegiate soccer players' perceptions of sport psychology, sport psychologists and sport psychological services
– Elite athletes in flow
Gould, Daniel R *see* Social loafing and crew
Gould, E M E Baring *see* With note-book and camera
Gould, Elgin Ralston Lovell *see*
– Popular control of the liquor traffic
– The progress of labour statistics in the united states
– The social condition of labor
Gould, Ezra Palmer *see*
– Commentary on the epistles to the corinthians
– Critical and exegetical commentary on the gospel
– A critical and exegetical commentary on the gospel according to st mark
Gould, F J *see* Agnosticism writ plain
Gould, Frederick James *see*
– A concise history of religion
– New conversion
– Noble path
Gould, George *see*
– Documents relating to the settlement of the church of england by the act of uniformity of 1662
– Open communion and the baptists of norwich
Gould, George Milbry *see* The meaning and method of life
Gould, J *see* The zoology of the voyage of hms beagle...during the years 1832-1836
Gould, James *see* A treatise on the principles of pleading, in civil action
Gould, John M *see* Notes on the revised statutes of the u.s.
Gould, John Melvile *see* A treatise on the law of waters, including riparian rights, and public and private rights in waters tidal and inland
Gould, Joseph *see* The letter-press printer
Gould, Peter R *see* The development of the transportation pattern in ghana
Gould, V Ward *see* Government land in florida
Gouldsbury, Cullen *see* Rhodesian rhymes
Goulet, M *see* Fetes...l'occasion du mariage de s m napoleon...avec marie-louise...
Goulon *see* Memoires pour l'attaque et la defense d'une place
Goumaz, Louis *see*
– Het ambt bij calvijn
– Calvinisme et liberte
– La doctrine du salut (doctrina salutis)
– Timothee, ou, le ministere evangelique...d'apres calvin
Gounod, Charles *see*
– Chants sacres
– Faust
Gour, Claude-Gilles *see* Institutions constitutionnelles et politiques du cambodge
Gour, Hari Singh *see*
– Facts and fancies
– The spirit of buddhism
Gouraige, Ghislain *see* Independance d'haiti devant la france
Gourbillon, Joseph A de *see* Voyage critique a l'etna en 1819
Gourbillon, Joseph Antoine De *see* Marquis de tulipano
Gourcuff, Olivier De *see* Chatiee!
Gourd, Alphonse *see* Les chartes coloniales et les constitutions des etats-unis de l'amerique du nord
Goureaud, P H *see* Au maroc, 1911-1914
Gourgaud, Gaspard *see* Refutation de la vie de napoleon
Gourge, Klaus *see* Oekonomie und psychoanalyse ueber die moeglichkeit einer psychoanalytischen oekonomie als ansatz zur erweiterung der oekonomischen rationalitaet
Gourgue, Gerard *see* Probleme de la delinquance juvenile et l'instituti
Gourick's washington digest – v1-21. 1889-1909 – 21mf – 9 – $31.50 – mf#LLMC 84-473 – us LLMC [240]
Gourlay, Janet *see* The temple of mut in asher
Gourlay, Robert *see*
– General introduction to statistical account of upper canada
– Statistical account of upper canada
Gourlay, Robert Fleming *see*
– General introduction to statistical account of upper canada
– Statistical account of upper canada
Gourmet – New York NY 1981+ – 1,5,9 – ISSN: 0017-2553 – mf#987 – us UMI ProQuest [640]
Gourmont, Remy De *see*
– Culture des idees
– Decadence and other essays on the culture of ideas
Gourmont, Remy de *see*
– Les canadiens de france
– Les francais au canada et en acadie
– A virgin heart
Gourock advertiser – [Scotland] Inverclyde, Gourock: T S Aikman 4 apr 1908-17 sep 1909 (wkly) [mf ed 2004] – 77v on 1r – 1 – uk Newsplan [072]
Goury, Jules *see* Plans, elevations, sections and details of the alhambra
Gousset, cardinal (Thomas Marie Joseph) *see* La croyance generale et constante de l'eglise touchant l'immaculee conception de la bienheureuse vierge marie
Gousset, Thomas Marie Joseph *see* Theologie morale
Gout, Raoul *see*
– Du protestantisme au catholicisme

Les gouts-reunis : ou nouveaux concerts a l'usage de toutes les sortes d'instrumens de musique augmentes d'une grande sonade en trio... / Couperin, Francois – [18-?] [mf ed 19–] – 1mf / 1r – 9,1 – (19th-c mss facs of 1724 publ) – mf#fiche 1218 / pres. film 63 – us Sibley [780]

Les gouttelettes : sonnets / Lemay, Pamphile – Montreal: Beauchemin, 1904 [mf ed 1995] – 3mf – 9 – 0-665-74849-3 – mf#74849 – cn Canadiana [810]

Gouvea, F A de *see* Iornada do arcebispo de goa dom frey aleixo de menezes primaz da india oriental...

Gouvernement haitien / Dautant, Caius – Port-Au-Prince, Haiti. 1910 – 1r – us UF Libraries [972]

Gouvernante, comedie : en cinq actes et en vers / La Chaussee, Nivelle De – Paris, France. 1808 – 1r – us UF Libraries [440]

Le gouvernement de la defense nationale / Simon, Jules – Paris 1874 [mf ed Hildesheim 1995-98] – 1v on 3mf – 9 – €90.00 – ISBN-10: 3-487-25994-x – ISBN-13: 978-3-487-25994-9 – gw Olms [350]

Le gouvernement de la province de quebec : pendant les annees 1875, 1876 et 1877 jusqu'au 2 mars 1878 – Quebec: L'Eclaireur, 1878 [mf ed 1980] – 1mf – 9 – 0-665-02291-3 – mf#02291 – cn Canadiana [971]

Gouvernement du general boisrond-canal / Etheart, Liautaud – Port-Au-Prince, Haiti. 1882 – 1r – us UF Libraries [972]

Gouvernement et legislature de la province de quebec = Gouvernment [sic] and legislature of the province of quebec – [S.l: s.n.], 1888 [mf ed 1986] – 1mf – 9 – 0-665-57753-2 – mf#57753 – cn Canadiana [323]

Gouvernement et legislature de la province de quebec = Government and legislature of the province of quebec – [Levis, Quebec?: s.n.], 1892 [mf ed 1986] – 1mf – 9 – 0-665-57748-6 – mf#57748 – cn Canadiana [323]

Gouvernement et legislature de la province de quebec : government and legislature of the province of quebec – [S.l: s.n.], 1887 [mf ed 1986] – 1mf – 9 – 0-665-57747-8 – mf#57747 – cn Canadiana [323]

Gouvernement et legislature de la province de quebec : government and legislature of the province of quebec – [S.l: s.n.], 1890 [mf ed 1980] – 1mf – 9 – 0-665-06403-9 – mf#06403 – cn Canadiana [323]

Gouvernement et legislature de la province de quebec : government and legislature of the province of quebec – [S.l: s.n.], 1890 [mf ed 1986] – 1mf – 9 – 0-665-56280-2 – mf#56280 – cn Canadiana [323]

Gouvernement et legislature de la province de quebec = Government and legislature of the province of quebec – [S.l: s.n.], 1894 – 1mf – 9 – 0-665-57749-4 – mf#57749 – cn Canadiana [323]

Gouvernement et legislature de la province de quebec = Government and legislature of the province of quebec – [S.l: s.n.], 1895 [mf ed 1986] – 1mf – 9 – 0-665-57750-8 – mf#57750 – cn Canadiana [323]

Gouvernement et legislature de la province de quebec = Government and legislature of the province of quebec – [S.l: s.n.], 1896 [mf ed 1986] – 1mf – 9 – 0-665-57751-6 – mf#57751 – cn Canadiana [323]

Gouvernement et legislature de la province de quebec = Government and legislature of the province of quebec – [S.l: s.n.], 1897 [mf ed 1986] – 1mf – 9 – 0-665-57752-4 – mf#57752 – cn Canadiana [323]

Gouvernement et legislature de la province de quebec = Government and legislature of the province of quebec – [S.l: s.n.], 1900 [mf ed 1986] – 1mf – 9 – 0-665-57754-0 – mf#57754 – cn Canadiana [323]

Gouvernement francais : evenements politiques, 1534, jacques quartier i e cartier decouvre la baie des chaleurs – S.l: s.n, 1859? – 1mf – 9 – mf#57755 – cn Canadiana [971]

Gouvernement General de l'Indochine, Cochinchine *see* Rapports au conseil colonial

Gouvernement general de l'Indochine, Concession francaise de Tourane *see* Budget municipal pour l'exercice...

Le gouvernement mercier : trois annees de progres, de rehabilitation et de revendication – [Quebec: s.n.] 1890 [mf ed 1985] – 4mf – 9 – 0-665-11700-0 – (incl ind) – mf#11700 – cn Canadiana [325]

Le gouvernement provincial devant l'opinion : discours-programme prononce le 6 septembre 1896 a saint-jean-port-joli / Flynn, Edmund James – Quebec: impr generale, 1897 – 1mf – 9 – mf#03117 – cn Canadiana [323]

Gouvernement revolutionnaire de l'Angola en exil *see* Communique

Gouvernementsblad van nieuw-guinea – Hollandia, 1950-1962 – 111mf – 9 – (missing: 1956(25); 1961(6, 71)) – mf#SE-237 – ne IDC [950]

Gouvernementsblad van suriname / British Guiana – Paramaribo (etc.). 1952-1966- – 1 – us NY Public [324]

Le gouverneur et madame de ramezay recoivent chronique des fetes du 250e anniversaire du chateau de ramezay = The governor's reception: chronicle of the 250th birthday of the chateau de ramezay / Morin, Victor – Montreal: la Societe d'archeologie et de numismatique, 1957 [mf ed 1987] – 2mf – 9 – (trans by john d king) – mf#SEM105P762 – cn Bibl Nat [971]

Un gouverneur general de l'algerie, l'amiral de gueydon / Dominique, L C – Alger: A. Jourdan, 1908.viii,563p. Incl. bibliog. references.With: The Germans by I.A.R. Wylie. 1 reel. 1260 – 1 – us Wisconsin U Libr [960]

The gouverneur morris papers – 6r – 1 – $210.00 – Dist. us Scholarly Res – us L of C Photodup [977]

Gouverneurs de la rosee / Roumain, Jacques – Port-Au-Prince, Haiti. 1944 – 1r – us UF Libraries [972]

Gouverneurs, lieutenants-gouverneurs, et administrateurs de la province de quebec, des bas et haut canadas, du canada sous l'union et de la puissance du canada, 1763-1908 / Audet, Francis-Joseph – Ottawa: Impr pour la Societe Royale du Canada – 1mf – 9 – 0-665-73043-8 – (incl english text) – mf#73043 – cn Canadiana [971]

Gouverneursjahre in kamerun / Puttkamer, Jesko Albert Eugen Von – Berlin, Germany. 1912 – 1r – us UF Libraries [960]

Gouzy, Rene *see*
- Au grand soleil d'afrique
- Des gorilles, des nains et meme...des hommes

Gov f w pickens accounts *see* Accounts

Govan chronicle – [Scotland] Glasgow: J Cossar 1 jan 1876-28 jun 1878 (wkly) [mf ed 2004] – 2r – 1 – (cont by: govan chronicle and partick observer [1878]) – uk Newsplan [072]

Goveia, Elsa V *see* Slave society in the british leeward islands at th...

Govender, Saravani *see* The influence of participatory development on the communication patterns of the parachute packing section of the sandf

The governance of empire : being suggestions for the adaptation of the british constitution to the conditions of union among the overseas states / Lighthall, William Douw – Montreal: The author, 1910 [mf ed 1997] – 1mf – 9 – 0-665-83547-7 – mf#83547 – cn Canadiana [327]

The governance of women's intercollegiate athletics : association for intercollegiate athletics for women (aiaw) 1976-1982 / Willey, Suzanne C – 1996 – 4mf – 9 – $16.00 – mf#PE 3782 – us Kinesology [790]

Governantes del nuevo reyno de granada durante el... / Restrepo Tirado, Ernesto – Buenos Aires, Argentina. 1934 – 1r – us UF Libraries [972]

Governing – Washington DC 1987+ – 1,5,9 – ISSN: 0894-3842 – mf#16276 – us UMI ProQuest [350]

Governing palestine / Machover, J M – London, 1936 – 4mf – 9 – mf#ILM-1972 – ne IDC [956]

O governista – Sao Paulo, SP: Typ do Governo, 03 ago 1850-28 jun 1851 – (= ser Ps 19) – mf#P19,03,08 – bl Biblioteca [320]

Government accountants journal – Potomac VA 1976-2000 – 1,5,9 – (cont: federal accountant) – ISSN: 0883-1483 – mf#10381.02 – us UMI ProQuest [336]

Government activities from independence until today, jul 1 1960-dec 31 1963 – Mogadiscio, Presidency of the Council of Ministers, 1964 – us CRL [321]

The government and its care of children / Spain. Embajada. United States – Washington, DC, 193? Fiche W1178. (Blodgett Collection of Spanish Civil War Pamphlets) – 9 – us Harvard [946]

Government and politics in the twentieth century / Carter, Gwendolen Margaret – New York, NY. 1965 – 1r – us UF Libraries [325]

Government and politics in tribal societies / Schapera, Isaac – London, England. 1956 – 1r – us UF Libraries [325]

Government and politics of switzerland / Brooks, Robert Clarkson – Yonkers-on-Hudson, NY. 1921 – 1r – us UF Libraries [325]

Government and religion of the virginia indians / Hendren, Samuel Rivers – Baltimore: Johns Hopkins Press 1895 [mf ed 1991] – (= ser Johns hopkins university studies in historical and political science 13/11-12) – 1mf – 9 – 0-524-01370-5 – (incl bibl ref) – mf#1990-2382 – us ATLA [307]

Government by all the people / Wilcox, Delos F – New York, NY. 1912 – 1r – us UF Libraries [320]

Government civil en haiti / Janvier, Louis Joseph – Lille, France. 1905 – 1r – us UF Libraries [350]

The government class book : a manual of instruction in the principles of constitutional government and law / Young, Andrew W & Clark, Salter S – New York: Maynard, Merrill & Co, 1899 – 4mf – 9 – $6.00 – (with ny: suppl by myron t. scudder) – mf#LLMC 92-162 – us LLMC [342]

The government code of guam, 1952 / Bohn, John A – Agana: Gov of Guam, 1952 – 7mf – 9 – $10.50 – mf#LLMC 82-100B Title 19 – us LLMC [324]

The government code of guam, 1961 : the 1952 code revised / Bohn, John A – Agana: Gov of Guam. 2v. 1960 – 18mf – 9 – $24.00 – (with 1964 suppl) – mf#LLMC 82-100B Title 2 – us LLMC [324]

The government code of guam, 1970 / Bohn, John A – n.p; n.d. 3v – 31mf – 9 – $46.50 – (with 1974 suppl) – mf#LLMC 82-100B Title 15 – us LLMC [324]

The government college record – Lahore, Printed at the Tribune Press. n1. jun 1900 – us CRL [378]

Government contracts digest / U.S. Bureau of the Budget – Washington: GPO, 1925 (all publ) – 2mf – 9 – $3.00 – (digest of the opinions of the us courts and the attorney-general on cases relating to government contracts) – mf#LLMC 84-109 – us LLMC [346]

Government contracts review – v1-2. 1957-58 – 11mf – 9 – $16.50 – mf#LLMC 84-474 – us LLMC [346]

Government control of railways; estimates of the pooled revenue, receipts and expenses and resultant net revenue 1939/40-1947 : [and] british transport commission: annual report, statement of accounts and statistics for the year ended 31st dec...1948-1962 – [mf ed Chadwyck-Healey] – (= ser British government publications...1801-1977) – 2r – 1 – uk Chadwyck [350]

Government data systems – Midland Park NJ 1977-88 – 1,5,9 – ISSN: 0046-6212 – mf#11695 – us UMI ProQuest [000]

Government executive – Washington DC 1968+ – 1,5,9 – ISSN: 0017-2626 – mf#7549 – us UMI ProQuest [350]

Government gazette / Australia. Northern Territory – Darwin. 1958-Jan. 1969- – 1 – us NY Public [324]

Government gazette / Australia. Western – Perth. 1958-1969- – 1 – us NY Public [324]

Government gazette – Darwin, nov 1873-may 1942 – 8r – A$472.41 vesicular A$516.41 silver – at Pascoe [079]

Government gazette / Malta – Commercial edition. 1957-1958. Valetta – 1 – us NY Public [324]

Government gazette / Mauritius – Port Louis. 1958-1966- – 1 – us NY Public [324]

Government gazette – Mbabane. v4-5 n140-269. 1966-1967 – us CRL [324]

Government gazette / Queensland – Brisbane. 1958-1968 – 1 – us NY Public [980]

Government gazette / Selangor – Kuala Lumpur. 1957-1966 – 1 – us NY Public [959]

Government gazette – South Africa. 4 jan 1980- – 10,800mf – 9 – sa State Libr [960]

Government gazette / Transvaal – Pretoria. South Africa. 1900-10 – sa National [960]

Government gazette / Victoria. Australia – Melbourne. 1957-1969- – 1 – us NY Public [980]

Government gazette / Zambia – Lusaka. Zambia Republic. Oct 28, 1964-1968 – 1 – us NY Public [960]

Government in the sunshine act : an interpretive guide / Berg, Richard K & Klitzman, SH – Washington: GPO, 1978 (all publ) – 2mf – 9 – $3.00 – mf#LLMC 94-346 – us LLMC [340]

Government information quarterly – Oxford, England 1992+ – 1,5,9 – ISSN: 0740-624X – mf#19772 – us UMI ProQuest [350]

Government land in florida / Gould, V Ward – Deland, FL. 1915 – 1r – us UF Libraries [333]

Government legislation and the aborigines / Australia. Federal Council for Aboriginal Advancement. Sub-committee on Legislative Reform – Melbourne 1964. 35 p. LL-2292 – 1 – us L of C Photodup [340]

Government news / Connecticut State Employees Association – v10 n1-v13 n8 [1977 jan-1980 aug] – 1r – 1 – (continued by: csea news (hartford, ct)) – mf#815470 – us WHS [350]

The government of american samoa / Greer, Richard A – 2nd rev ed. jan 1954 – 3mf – 9 – $4.50 – mf#LLMC 82-100C Title 43 – us LLMC [323]

The government of india / MacDonald, James Ramsay – London: Swarthmore Press, [1919] [mf ed 1995] – (= ser Yale coll) – ix/291p – 1 – 0-524-09028-9 – mf#1995-0028 – us ATLA [323]

The government of india / Malcolm, John – London 1833 [mf ed Hildesheim 1995-98] – 1v on 4mf – 9 – €120.00 – 3-487-27534-1 – gw Olms [954]

The government of ireland bill / Chamberlain, Joseph. – [London], [1886] – (= ser 19th c ireland) – 1mf – 9 – mf#1.1.245 – uk Chadwyck [323]

The government of ireland bill / Devonshire, Spencer Compton Cavendish, 8th Duke of – London, [1886] – (= ser 19th c ireland) – 1mf – 9 – mf#1.1.244 – uk Chadwyck [323]

Government of Niue, Justice, Lands and Survey Department *see*
- Land court
- Land titling project reports
- Registers of births, deaths and marriages

Government of Niue, Justice, Lands and Survey Department, Land Court *see* Applications to the land court

Government of Rajasthan Appointments (O and M) Department *see* Secretariat manual (as amended up to 31st march, 1959)

Government of switzerland / Rappard, William Emmanuel – New York, NY. 1936 – 1r – us UF Libraries [325]

The government of the city of frankfort-on-the-main / Dodge, Martin Herbert – New York?: s.n., 1920 – us CRL [350]

The government of the east india company, and its monopolies : or, the young india party, and free trade? / Lewin, Malcolm – London, 1857 – (= ser 19th c british colonization) – 1mf – 9 – mf#1.1.7278 – uk Chadwyck [338]

The government of the empire : a consideration of means for the representation of the british colonies in an imperial parliament / Bousfield, William – London, 1877 – (= ser 19th c books on british colonization) – 1mf – 9 – mf#1.1.3719 – uk Chadwyck [954]

Government of the national church – Ramsgate, England. 1870 – 1r – us UF Libraries [240]

Government of the Northern Mariana Islands *see* Constitution of the commonwealth of the northern mariana islands, 1976

Government of the northern mariana islands – briefing booklet – n.p, n.d. – 2mf – 9 – $3.00 – mf#llmc82-100j, title 21 – us LLMC [350]

The government of the people in the state of connecticut / Douglas, Charles H – New York: Hinds, Hayden & Eldredge, Inc., 1917 – 3mf – 9 – $4.50 – mf#LLMC 92-159 – us LLMC [340]

Government of Tonga *see* Premier's (shirley baker's) letterbooks, 1873-74, 1880-90

Government of tonga : letterbooks-out – 1873-74, 1880-83 – 1r – 1 – mf#PMB1089 – at Pacific Mss [980]

Government of trinidad and tobago / Reis, Charles – Port-of-Spain, Trinidad and Tobago. 1947 – 1r – us UF Libraries [972]

Government organization manuals 1900-1980 : the largest collection of government organization manuals in the world / ed by Korman, Richard I – [mf ed Chadwyck-Healey] – 8710mf – 9 – (73 countries listed. in english, french, german, portuguese, russian, spanish. deatils supplied on request) – uk Chadwyck [323]

Government product news – Cleveland OH 1973-94 – 1,5,9 – ISSN: 0017-2642 – mf#6850 – us UMI ProQuest [350]

Government publications relating to the cape of good hope to 1910 – 1 – (group i: votes & proceedings, annexures and select committee reports of the cape house of assembly and the legislative council, 1854-1910, 285r with guide, 97056. group ii: statistical registers of the cape of good hope, 1821-1909, 47r with guide, 97056) – uk Microform Academic [324]

Government publications review – Oxford, England 1973-93 – 1,5,9 – (cont by: journal of government information) – ISSN: 0277-9390 – mf#49078.01 – us UMI ProQuest [350]

Government reports announcements & index – Washington DC 1946-96 – 1,5,9 – ISSN: 0097-9007 – mf#1453 – us UMI ProQuest [350]

Government standard / American Federation of Government Employees – 1933 dec 29-1937 jun 11, 1937 jun 18-1940 dec 20, 1941 jan 3-1945 dec 21, 1946-71, 1972 jan 14-1976 dec, 1977-79, 1980-1984 jun – 12r – 1 – (continued by: afge government standard) – mf#3420280 – us WHS [350]

Government subsidies and the postal services with india, china, and australia – London, 1879 – (= ser 19th c books on british colonization) – 1mf – 9 – mf#1.1.890 – uk Chadwyck [350]

Government technical reports – More than 60,000 available each year in microfiche. Sources include U.S. and foreign government agencies and contractors – 9 – $9.00 – us NTIS [600]

Government union review – Vienna, Austria 1980+ – 1,5,9 – ISSN: 0270-2487 – mf#12538.01 – us UMI ProQuest [331]

Governmental affairs bulletin / Wisconsin State Chamber of Commerce – v1 n2,4-v8 n8 [1955 jan 21, feb 4-1962 aug 14] – 1r – 1 – (continued by: wisconsin governmental affairs) – mf#1345069 – us WHS [350]

Governmental finance – Chicago IL 1972-84 – 1,5,9 – ISSN: 0091-4835 – mf#11310 – us UMI ProQuest [336]

Governmental research notes – Long Beach CA 1938-66 – 1 – mf#9796 – us UMI ProQuest [350]

Governmental system in southern rhodesia / Murray, D J – Oxford, England. 1970 – 1r – us UF Libraries [325]

"Governor fred hall, a study of a political personality" / Green, Robert – undated – 1 – us Kansas [320]
The governor general in ooterparah in 1865 / Lawrence, John Laird Mair Lawrence – [Calcutta, 1866?] – (= ser 19th c british colonization) – 1mf – 9 – mf#1.1.2513 – uk Chadwyck [350]
Governor jonathan belcher letter books, 1723-1754 – [mf ed 1966] – 11r – 1 – (with p/g. unique insight into the career of jonathan belcher as governor of massachusetts and new hampshire (1729-41), and of new jersey (1746-57), and into new england politics and political tumult during this important period in american history) – us MA Hist [978]
Governors' Conference *see*
– Proceedings
– Proceedings of conferences
Governor's correspondence relating to luke short / Kansas. Governor, 1883-85 (Governor Glick) – May-Jun 1883 – 1 – us Kansas [978]
Governors' Interstate Indian Council *see* Report of the...annual meeting
Governors of jamaica in the... / Cundall, Frank – London, England. 1937 – 1r – us UF Libraries [972]
Governors of jamaica in the first half... / Cundall, Frank – London, England. 1937 – 1r – us UF Libraries [972]
Governors of jamaica in the seventeenth century / Cundall, Frank – London, England. 1936 – 1r – us UF Libraries [972]
Governors of Ohio *see* Calendar of official papers, (1803-1878)
Governors' papers – 72r (3r 35mm, 69r 16mm) – 1,5 – mf#B26011-26081 – us Ohio Hist [324]
Governor's report on commonwealth affairs : jan 1978-jan 1982 / Camacho, Carlos S – Saipan: Office of the Governor, 1982? – 2mf – 9 – $3.00 – mf#llmc82-100j, title 23 – us LLMC [324]
Govett, James R *see* The relative importance of proprioception, ligament laxity and strength on functional performance in the acl deficient and acl reconstructed knee
Govett, R *see* Exposition of the gospel of st john
Govett, Robert *see*
– Christians!
– Entrance into the millennial kingdom
Govinda, Anagarika Brahmacari *see* The psychological attitude of early buddhist philosophy
Govinda Tirtha, swami *see* The nectar of grace
Govindacharya, Alkondavilli *see*
– The divine wisdom of the draavida saints
– The life of raamaanujaachaarya
– Mazdaaism in the light of vishnuism
Govinden, Winston Sivalingum *see* The not yet healed
Govion Broglio Solari, Catherine de *see* Venice under the yoke of france and of austria
Gow, Andrew J *see*
– The effect of exercise on glyceraldehyde-3-phosphate dehydrogenase and superoxide dismutase activities in the post-ischemic heart
– Effect of sprint training upon sarcoplasmic reticulum ca2+ atpase and na+-ca2+ exchanger mrna expression in rat myocardium
Gow, William *see* Apocalypse unveiled and a fight with death and slander
Gowansville baptist church. gowansville, south carolina : church records – 1835-83.Formerly: Cross Roads Baptist Church – 1 – 76.50 – us Southern Baptist [242]
Gowen, Herbert Henry *see*
– Church work in british columbia
– A history of indian literature
– The revelation of "the things that are"
Gower, calvinistic methodist chapels, monumental inscriptions – 1mf – 9 – £1.25 – uk Glamorgan FHS [929]
Gower, Elizabeth M *see* Coronary artery disease among young adults under 50 in la crosse county
Gower (incl loughor) (h.o.107/1424) – (= ser 1841 census returns [glamorgan]) – 4mf – 9 – £5.00 – uk Glamorgan FHS [314]
Gower (incl loughor) (r.g.11/5367,8,9,70) – (= ser 1881 census returns [glamorgan]) – 6mf – 9 – £7.50 – uk Glamorgan FHS [314]
Gower, Myrna Zoe *see* Case study approach to some features of cross-cultural social work practice with indian families
Gowers, William Frederick *see* Gazetteer of kano province
Gowlett, D F *see*
– Morphology of the substantive in lozi
– Morphology of the verb in lozi
Gowon, Yakubu *see* Faith in unity
Gowring, John William *see* Doctrines of free and sovereign grace
Goy, Ina *see* Lexikologische untersuchungen zum fachwortschatz der oekologie und des umweltschutzes im neugriechischen
Goya / Gomez De La Serna, Ramon – Madrid, Spain. 1928 – 1r – us UF Libraries [700]
Goya / Poore, Charles Graydon – New York, NY. 1938 – 1r – us UF Libraries [700]
Goya, moratin, melendez valdes y donoso cortes. resena historica de los anteriores enterramientos y traslaciones de sus restos mortales hasta su inhumacion en el mausoleo del cementerio de san isidro el dia 11 de mayo de 1900 / Mesonero Romanos, Manuel – Madrid: Imp. Hijos de M.G. Hernandez, 1900 – sp Bibl Santa Ana [946]
Goyat, Michel *see* Le code du travail malgache
Goyau, Lucie Faure *see* L'evolution feminine
Goyau, Lucie Felix-Faure *see*
– Newman
Goyaz : orgam democratico – Goias. 17 set 1885-nov 1889; fev 1890-jun 1891; set 1892-ago 1893; fev, abr-ago 1894; mar-dez 1900; mar, jul-dez 1901; jan 1902-dez 1903; jan, maio-jun 1904; jan-31 dez 1910 – (= ser Ps 19) – mf#P11B,06,05 – bl Biblioteca [320]
Goycochea, Luis Felipe De Castilhos *see* Fronteiras e fronteiros
Goyette, Armand *see* Histoire genealogique et livre de famille des goyette, 1659-1959
Goyette, Gabriel *see* L'ideologie scolaire du conseil de l'instruction publique de la province de quebec, 1927-1964
Gozoy liturgia de la santa misa / Aradillas Agudo, Antonio – Barcelona: Juan Flors, editor, 1960 – 1 – sp Bibl Santa Ana [240]
Gpo sales publications reference file / U.S. Government Printing Office – Bimonthly and monthly suppl. "Documents in Print" catalog – 9 – $117.00y in US $146.25y outside – 0-16-011178-1 – mf#S-N 721-002-00000-4. Sub-list ID-PRF – us GPO [324]
Gq [gentlemen's quarterly] – New York NY 1931+ – 1,5,9 – ISSN: 0016-6979 – mf#5027 – us UMI ProQuest [071]
Graaf, H J de *see*
– De band tussen ambon en nederland
– De band tussen ambon en nederland
The graaff-reinet herald – 25 aug 1852-27 aug 1884 – 12r – 1 – mf#MS00257 – sa National [079]
De graaff-reinetter – 4 jul 1885-27 nov 1902 [mf ed Cape Town: SA library, 1983] – 13r – 1 – (frequency varies. suspended apr 9 1900-dec 31 1900) – sa National [079]
Graah, W *see* Undersogelses-reise til ostkysten af gronland
Graah, W A *see* Narrative of an expedition to the east coast of greenland...
Graap, Paul-Gerhard *see* Richard wagners dramatischer entwurf jesus von nazareth
Graauwhart, Hendrik *see*
– Christelyke bedenkingen en voorbeeldlyke zeedelessen afgeleid uit 's werelds eerste toestand
– Leerzame zinnebeelden
– Voorbeeldelyke zeede-lessen
Grab- und weihdenkmaeler aus den territorien von augusta traiana und kabyle vom 1. bis 3. jahrhundert n chr : mit katalog abbildungsteil / Conrad, Sven – (mf ed 1994) – 1mf – 9 – €30.00 – mf#DHS 863 – gw Frankfurter [930]
Grab- und weihdenkmaeler aus den territorien von augusta traiana und kabyle vom 1. bis 3. jahrhundert n. chr / Conrad, Sven – (mf ed 1994) – 3mf – 9 – €49.00 – 3-89349-862-1 – (with catalogue. abbildungsteil: 1mf dm60 isbn: 3-89349-863-x) – mf#DHS 862 – gw Frankfurter [930]
Graba, Carl *see* Tagebuch, gefuehrt auf einer reise nach faeroe im jahre 1828
Grabado en colombia / Giraldo Jaramillo, Gabriel – Bogota, Colombia. 1959 – 1r – us UF Libraries [972]
Grabar, A *see* La peinture religieuse en bulgarie
Grabar, Oleg *see* The illustrations of the maqamat
Grabbe / Krack, Otto – Berlin, Leipzig: Schuster & Loeffler, [1904?] – 1r – 1 – us Wisconsin U Libr [430]
Grabbe als kritiker / Els, Hans van – Marburg, 1914 (mf ed 1994) – 2mf – 9 – €31.00 – 3-8267-3096-8 – mf#DHS-AR 3096 – gw Frankfurter [430]
Grabbe, Christian Dietrich *see*
– Aschenbroedel
– Christian dietrich grabbe's saemmtliche werke und handschriftlicher nachlass
– Kaiser friedrich barbarossa
Das grabbe-buch / ed by Friedrich, Paul & Ebers, Fritz – Detmold: Meyer (M Staercke) 1923 [mf ed 1990] – 1r [ill] – 1 – (incl bibl ref. filmed with: rudolf von gottschall / moritz brasch) – mf#2685p – us Wisconsin U Libr [430]
Grabbes doppeltes gesicht / Georg, Manfred – Berlin-Lichterfelde: E Runge, [1922?] – 1r – 1 – (incl bibl ref) – us Wisconsin U Libr [920]
Das grabdenkmal des koenigs chephren / Hoelscher, U – 6mf – 9 – (veroeffentlichungen der ernst von sieglin expedition in egypten, leipzig 1912) – mf#NE-394 – ne IDC [956]
Das grabdenkmal des koenigs ne-user-re : ausgrabungen der deutschen orient-gesellschaft in abusir, 1902-1904 / Borchardt, L – Leipzig, 1907 – 10mf – 8 – mf#H-239 – ne IDC [930]
Grabenhorst, Georg *see*
– Regimentstag
– Spaete heimkehr
Grabenzeitung – Berlin DE, 1945 17 mar-10 apr – 1 – gw Misc Inst [929]
Graber, Gustav Hans *see* Die schwarze spinne
Graber, Lisa M *see* The relationship between rebounding and winning percentage n.c.a.a. division i big sky conference women's basketball
Grabert, Willy *see* Geschichte der deutschen literatur
Die grabesschuld : nachgelassene erzaehlung / Sealsfield, Charles; ed by Meissner, Alfred – Leipzig: E J Guenther, 1873 – 1r – 1 – us Wisconsin U Libr [830]
Grabill, Stephen John *see* Theological foundation for a reformed doctrine of natural law
Grabmann, M *see*
– Forschungen ueber die lateinischen aristoteles-uebersetzungen des 13. jahrhunderts
– Mittelalterlich geistesleben
– Die philosophia pauperum und ihr verfasser albert von orlamuende
– Die theologische erkenntnis- und einleitungslehre des hl thomas von aquin
– Die werke des hl thomas von aquin
Grabmann, Martin *see*
– Die lehre des heiligen thomas von aquin von der kirche als gotteswerk
– P heinrich denifle
– Die philosophische und theologische erkenntnislehre des kardinals matthaeus von aquasparta
Grabowski, Tadeusz *see* Literatura aryanska w polsce, 1560-1660
Grabowsky, Ian *see* A history in diary form of civil aviation in papua and new guinea
Der grabpalast des patuamenap in der thebanischen nekropolis / Duemichen, J – Leipzig, 1884-1894. 3v – 11mf – 9 – mf#NE-371 – ne IDC [956]
Grabreden / Gossel, J – Frankfurt am Main, Germany. 1892 – 1r – us UF Libraries [943]
Grabreden / Gudemann, Moritz – Wien, Austria. 1894 – 1r – us UF Libraries [943]
Die grabschrift des darius zu nakschi rustam / Darius I, King of Persia; ed by Hitzig, Ferdinand – Zuerich: Orell, Fuessli, 1847 – 1mf – 9 – 0-8370-7690-0 – (text in german and old persian; notes in german. incl bibl ref) – mf#1986-1690 – us ATLA [470]
Graca, Arnobio *see* Economic politica e economia brasileira
Grace, actual and habitual : a dogmatic treatise / Pohle, Joseph – St Louis MO: B Herder 1915 [mf ed 1991] – (= ser Dogmatic theology 7) – 2mf – 9 – 0-7905-9443-9 – (incl bibl ref) – mf#1989-2668 – us ATLA [240]
Grace and duty of being spiritually minded / Owen, John – London, England. 1816 – 1r – us UF Libraries [240]
Grace and gold : or, scriptural giving / Fowler, W J – Truro NS: News Pub, 1909 [mf ed 1995] – 3mf – 9 – 0-665-74193-6 – mf#74193 – cn Canadiana [240]
Grace and truth : under twelve different aspects / Mackay, William Paton – Edinburgh: James Taylor; Chicago: FH Revell, 1878 – 1mf – 9 – 0-8370-4254-2 – mf#1985-2254 – us ATLA [240]
Grace baptist church. lexington, kentucky : church records – 1924-73.History, 1924-64 – 1 – us Southern Baptist [242]
Grace church newsletter / Grace Episcopal Church [Madison, WI] – 1982 jan 15-jun 7, aug 27-nov 30, 1983 jan 4-apr 4 – 1r – 1 – (cont: newsletter [grace episcopal church [madison, wi]: 1977]; cont by: newsletter [grace episcopal church [madison, wi]: 1983]) – mf#1534431 – us WHS [071]
Grace church newsletter / Grace Episcopal Church [Madison, WI] – v1 n1-v2 n4 [1973 sep-1975 jun] – 1r – 1 – (cont: parish news letter [madison, wi: 1966]; cont by: parish news letter [madison, wi: 1975]) – mf#1534430 – us WHS [242]
Grace church record / Grace Episcopal Church (Madison, WI) – v1 n1-4, v2 n1-9,11 [1872 jul-1875 apr] – 1r – 1 – mf#587255 – us WHS [242]
Grace Episcopal Church *see*
– Newsletter
– Parish news letter
Grace Episcopal Church (Madison, WI) *see* Grace church record
Grace Episcopal Church [Madison, WI] *see*
– Grace church newsletter
– Grace episcopal church newsletter
Grace episcopal church newsletter / Grace Episcopal Church [Madison, WI] – 1987 sep 1, nov 1-dec 1, 1988 jan 5-dec – 1r – 1 – (cont: newsletter [grace episcopal church [madison, wi]: 1983]; cont by: grace notes [madison, wi: 1989]) – mf#1533872 – us WHS [242]
Grace hospital bulletin – Detroit MI 1916-72 – 1,5 – mf#6282 – us UMI ProQuest [362]
Grace, John *see* Haven of rest
Grace journal : a publication of grace theological seminary – Winona Lake IN: Grace Theological Seminary...with Grace Seminary Alumni Assoc 1960-73 [mf ed 1977] – 14v on 2r – 1 – (v1 had only 2 iss; with ind) – mf0325 – us ATLA [240]
The grace magazine – Washington DC: Grace Publ Assoc. v6 n13 (dec 15 1940), v7 n18 (dec 1941), aug 1948 [mf ed 2005] – 1r [ill] – 1 – (missing: (jan-nov 1941, jan 1942-jul 1948)) – mf#2005-s029 – us ATLA [242]
The grace of christ : or, sinners saved by unmerited kindness / Plumer, William Swan – Philadelphia: Presbyterian Board of Publ, c1853 – 1mf – 9 – 0-7905-9583-4 – mf#1989-1308 – us ATLA [240]
The grace of god magnified : an experimental tract / Taliaferro, Hardin E – Charleston: Southern Baptist Publication Society, 1859 – 2mf – 9 – 0-524-07920-X – mf#1991-3465 – us ATLA [240]
Grace of god that bringeth salvation hath appeared to all men / Ayre, John – London, England. 1839 – 1r – us UF Libraries [240]
Grace of god the cause of ministerial excellence and usefulness / Scales, Thomas – London, England. 1847 – 1r – us UF Libraries [240]
Grace of the apostolic priesthood / Carter, T T – London, England. 1870 – 1r – us UF Libraries [240]
Grace of the gospel : how popery mars it / Stowell, Hugh – Preston, England. 1851? – 1r – us UF Libraries [240]
Grace theological journal – Winona Lake IN 1985-91 – 1,5,9 – ISSN: 0198-666X – mf#15323 – us UMI ProQuest [240]
The graces : a classical allegory, interspersed with poetry, and illustrated by explanatory notes: together with a poetical fragment entitled psyche among the graces / Wieland, Christoph Martin – London: G and W B Whittaker, 1823 – 1 – (incl bibl ref) – us Wisconsin U Libr [810]
The graces of interior prayer : a treatise on mystical theology = des graces d'oraison / Poulain, Augustin – London: K Paul, Trench, Truebner, 1912 – 2mf – 9 – 0-524-07167-5 – (incl bibl ref. in english) – mf#1991-2956 – us ATLA [240]
Graceta del foro – Buenos Aires. 3 feb 1916-aug 1930; 1931-30 apr 1964 – 92r – 1 – uk British Libr Newspaper [072]
Gracey, David *see* Sin and the unfolding of salvation
Gracey, J T *see* Medical work of the woman's foreign missionary society
Gracey, J T [Mrs] *see* Eminent missionary women
Gracey, John Talbot *see* India
Gracey, Kathryn H *see* Effects of elevated muscle temperature on exercise-induced muscle sympathetic nerve activity
Graci Larravide, Mario *see* Sacrificio de amor
Gracia asin desgracia sa buhay ni liberato na villano na nacaagom qui librada asin princesa esteleta sa reino nin francia – [Nueva Caceres: Libreria Mariana 190-?] [mf ed Bloomington IN: Indiana Uni Lib, Preservation Dept 1984] – (= ser Coll...in the bikol language) – 2v on 1r – 1 – us Indiana Preservation [490]
Gracia Villacampa, Carlos *see*
– Una comision de carlos 5th al...p. francisco de los angeles quinones (despues obispo de coria)
– Fundaciones testamentarias de fr. miguel de medina, jeronimo de guadalupe en el convento de san francisco de medina de pomar, burgos, 1915
La gracia y el sacerdocio de cristo – Francisco Navares Merino Lima: Facultad de Teologia Pontificia y Civil de Lima, 1967 – us CRL [972]
Gracian *see*
– De konst des wijsheid
– De konst der wijsheid
Gracias, Joao Baptista Amancio *see* Origens do christianismo na india
Gracias, Louis *see* Eastern clay
Grada, Cormac O *see* Nineteenth century books on ireland collection
Gradaus mein deutsches volk! – Muenchen DE, 1848 30 dec-1849, 1850 n46, 52 – 1 – gw Misc Inst [943]
Grade 1 teachers' involvement in school-based curriculum development in the northern province / Lumadi, Mutendwahothe Walter – Uni of South Africa 2000 [mf ed Johannesburg 2000] – 11mf – 9 – (incl bibl ref) – mf#mfm14914 – sa Unisa [370]
Grade 12 provincial examination papers : with marking keys and suggestions for teachers / British Columbia. Ministry of Education – [Victoria BC: The Ministry 1987 [mf ed 1993] – 1v on 1r – 1 – cn UBC Preservation [370]
Grade 12 teachers' attitude towards mass-media transmitted educational supplements / Mokgathe, Phefo L – Pretoria: Vista University 2000 [mf ed 2000] – 2mf – 9 – mf#mfm15123 – sa Unisa [302]

Grade, Tiffany J *see* Deluxe and illuminated manuscripts
Graded bible lessons and materials / Presbyterian Church in the U.S.A. Board of Christian Education – 1937-1948 – 1 – $50.00 – us Presbyterian [240]
Graded elementary magazine – Oct 1930-Sep 1933 – 1 – us Southern Baptist [242]
Graded junior pupil, years 1-4 – 1943 – 1 – us Southern Baptist [242]
Graded junior teacher, years 1-4 – 1943 – 1 – 49.42 – us Southern Baptist [242]
Graded lesson helper – Apr 1926-Sep 1930 – 1 – 77.35 – us Southern Baptist [242]
The graded school : a graded course of instruction for public schools / Wells, William Harvey – New York: A.S. Barnes & Co., 1867. 200p. fold. forms – 1 – us Wisconsin U Libr [370]
Graded studies in the new testament : designed for christian bible-schools / Pendleton, Huntington King & Pendleton, Philip Yancy – Cincinnati, OH: Standard Pub Co, c1898 – 1mf – 9 – 0-524-06275-7 – mf#1991-2466 – us ATLA [225]
Graded zulu exercises / Doke, Clement Martyn – Lovedale, South Africa. 1931 – 1r – us UF Libraries [470]
Grading, packing and stowing florida produce / Ensign, M R – Gainesville, FL. 1932 – 1r – us UF Libraries [630]
Gradis, Gaston *see* A la recherche du grand-axe
Gradmann, Johann Jacob *see* Das gelehrte schwaben
Graduale ad usum ordinia praedicatorum... : french illuminated manuscript, between 1253 & 1262? dominican gradual, incl scripture lessons, some offices – [between 1253 and 1262?] [mf ed 19–] – 15mf – 9 – mf#fiche1, 914 – us Sibley [780]
Graduale alderspacense (cima61) : color microfiche edition of the manuscript muenchen, bayerische staatsbibliothek, clm 2541/2542 – (mf ed 2001) – (= ser Codices illuminati medii aevi (cima)) – 56p on 10 color mf – 15 – €390.00 – 3-89219-061-5 – (int to the gradual of aldersbach & cistercian plainchant by david hiley) – gw Lengenfelder [090]
Graduale romanum – 1591 – (= ser Mssa) – 8mf – 9 – €100.00 – mfchl 81 – gw Fischer [780]
Graduale romanum : de tempore, [et] sanctis, ad ritum missalis, ex decr. s. conc. trid. restituti... – Venetiis: Iuntas 1596 – (= ser Hqab. literatur des 16. jahrh.) – 6mf – 9 – €70.00 – mf#1596a – gw Fischer [780]
Graduale romanum : proprium de tempere adventus ad sabbatum sanctum – [15–?] [mf ed 1988] – 1r – 1 – mf#pres. film 12 – us Sibley [780]
[Graduale romanum...et in agenda mortuorum] – [15–?] [mf ed 1988] – 1r – 1 – mf#pres. film 15 – us Sibley [241]
Graduale sarisburiense : british musem "additional manuscripts mss 12, 194" – London, 1894 – €44.00 – (with a dissertation and historical ind...by w h frere) – ne Slangenburg [240]
Graduale – sequentiar (cima60) : farbmikrofiche-edition der handschrift salzburg, bibliothek der erzabtei st peter (osb), cod a 9 11 – (mf ed 2001) – (= ser Codices illuminati medii aevi (cima) 60) – 74p on 8 color mf – 15 – €360.00 – 3-89219-060-7 – (int by stefan engels) – gw Lengenfelder [090]
Graduate woman – Washington DC 1979-85 – 1,5,9 – (cont: aauw journal; cont by: outlook american association of university women) – ISSN: 0161-5661 – mf#904.02 – us UMI ProQuest [376]
Graduating engineer – Ann Arbor MI 1984-94 – 1,5,9 – ISSN: 0193-2276 – mf#12318 – us UMI ProQuest [620]
Gradus ad parnassum / Fux, Johann Joseph – 1725 – (= ser Mssa) – 3mf – 9 – €50.00 – mfchl 51a – gw Fischer [780]
Gradus ad parnassum : sive manuductio ad compositionem musicae regularem, methodo nova ac certa, nondum ante tam exacto ordine in lucem edita / Fux, Johann Joseph – Viennae: typis Joannis Petri van Ghelen 1725 [mf ed 19–] – 16mf – 9 – mf#fiche 465 – us Sibley [780]
Gradwell letters and other papers, the... 1777-1833 : from westminster cathedral archives – 1r – 1 – (with ind) – mf#2721 – uk Microform Academic [241]
Gradwell, Robert *see* Dissertation on the fable of papal antichrists
Grae actualites – [s.l.]: Publie par le Departement des relations exterieures du Gouvernement revolutionnaire de l'Angola en exil. n1-3 – us CRL [327]
Graeber, H J *see* Die konfessionelle schule
Graeber, Herm Joh *see* Versuch einer historischen erklaerung der offenbarung des johannes
Graebner, Augustus Lawrence *see*
– "Bis hieher"
– Dr. martin luther
– Geschichte der lutherischen kirche in america
– Half a century of sound lutheranism in america
– Outlines of doctrinal theology
– Protestantischer nachruf zum gedaechtniss papst leo 13
– Trial and self-conviction of pope leo 13
Graeca latina / Schulze, Wilhelm – Gottingae: Officina Academica Dieterichiana, 1901 – 1mf – 9 – 0-8370-9178-0 – (incl bibl ref) – mf#1986-3178 – us ATLA [450]
Graecae grammaticae rudimenta : in usum scholarum – Editio 18 Oxonii [Oxford]: E Typographeo Clarendoniano, 1875 – 1mf – 9 – 0-8370-9211-6 – mf#1986-3211 – us ATLA [450]
Graecus venetus : pentateuchi, proverbiorum, ruth, cantici, ecclesiastae, threnorum, danielis versio graeca / ed by Gebhardt, Oscar von – Lipsiae: F A Brockhaus, 1875 [mf ed 1990] – (= ser Biblioteca nazionale marciana) – 2mf – 9 – 0-8370-1691-6 – (pref by franciscus delitzsch) – mf#1987-6115 – us ATLA [221]
Graef *see* Forschungsreise s m s planet 1906/07
Graef, Hans Gerhard *see*
– Goethe
– Goethe in berka an der ilm
– Jahrbuch der goethe-gesellschaft
Graef, Hans Gerhard [comp] *see* Aus goethes tagebuechern
Graef, Hermann *see* Heinrich von kleist
Graef, Peter Leo *see* Hilfe, die nicht ankommt
Graefe, Alan R *see*
– An examination of environmental attitudes among college students
– The impacts of marine debris, weather conditions, and unexpected events on recreational boater satisfaction on the delaware inland bays
– Outdoor recreation participation of maryland residents in maryland state forests and parks
– A simulation approach to crowding in outdoor recreation
Graefe, C F von *see* Encyclopaedisches woerterbuch der medicinischen wissenschaften (ael3/13)
Graefe, Johann Friedrich *see* Samlung verschiedener und auserlesener oden
Graefe, Johanna *see* Ueber den "zauberberg" von thomas mann
Graefe, Sylvia *see* Zentralkomitee der sozialistischen einheitspartei deutschlands. buero erich honecker 1957-1989 (bestand dy 30) bd 81
Graefe's archive for clinical and experimental ophthalmology – Dordrecht, Netherlands 1980-84 – 1,5,9 – (cont: albrecht von graefes archiv fuer klinische und experimentelle ophthalmologie) – ISSN: 0721-832X – mf#13170.05 – us UMI ProQuest [617]
Graeffer, Franz *see* Oesterreichische nationalenzyklopaedie (ael1/10)
Graefin erika lehr- und wanderjahre / Schubin, Ossip – Braunschweig: G Westermann, c1921 – 1 – us Wisconsin U Libr [830]
Graeflich erbachisches wochen-blatt fuer den landkreis erbach – Erbach / Odenw DE, 1977- – 7r/yr – 1 – (bezirksausgabe von darmstaedter tagblatt; title varies: 5 jan 1828: graeflich erbachisches wochen-blatt; 1 apr 1848: wochen-blatt fuer die bezirke erbach und brenberg; 26 aug 1848: intelligenz-blatt fuer den regierungsbezirk erbach; 2 sep 1848: anzeige-blatt fuer den regierungsbezirk erbach; 21 oct 1848: erbacher anzeige-blatt; 3 aug 1852: anzeige-blatt fuer die kreise erbach und neustadt; 5 jan 1855: anzeige-blatt fuer den kreis erbach; 1867: erbacher kreisblatt; 1900: centralanzeiger fuer den odenwald; 1 aug 1949: odenwaelder heimatzeitung; 2 feb 2002: odenwaelder echo odenwaelder heimatzeitung; fr 1 nov 1972 ba v. darmstaedter tagblatt, ausg erbach-michelstadt) – gw Misc Inst [074]
Graeme, Bruce *see* Arraches aux tenebres
Der graenz-bote – Tuttlingen DE, 1947-1958 10 jan, 1958 9 oct-1959 – 21r – 1 – (bezirksausgabe von schwaebische zeitung, leutkirch; filmed by other misc inst: 1983- [8r/yr]; title varies: 1943-45: tuttlinger zeitung, 1946-49: schwaebisches tagblatt, main ed in tuebingen, fr 1961 regional ed of: schwaebische zeitung, leutkirch) – gw Misc Inst [074]
Graepp, L W et al *see* Aus nacht zum licht
Graesse, J G Th *see* Die beiden aeltesten lateinischen fabelbuecher des mittelalters
Graessner, Gernot Heinrich Willi *see* Deutschland und die nationalsozialisten in den vereinigten staaten von amerika
Graetz, H *see* Geschichte der juden
Graetz, Heinrich *see*
– Emendationes in plerosque sacrae scripturae veteris testamenti libros
– Frank und die frankisten
– History of the jews
– Kohelet
– Kritischer commentar zu den psalmen
– Popular history of the jews
– Schir ha-schirim
Graevenitz, George von *see* Italienische reise
Graf adolf friedrich von schack : ein literarischer essay / Brenning, Emil – Bremen: Carl Rocco 1885 [mf ed 1995] – 1r – 1 – (incl bibl ref; filmed with: zur ehre gottes / sacher-masoch) – mf#3724p – us Wisconsin U Libr [840]
Graf adolf friedrich von schack : ein literarisches portrait / Babel, Eugen – Wien: Carl Gerold's Sohn, 1885 [mf ed 1995] – 82p – 1 – mf#8857 – us Wisconsin U Libr [430]
Graf alfred keyserling erzaehlt / Keyserling, Alfred – Kaunas und Leipzig: Pribacis, 1937. xii,399p – 1 – us Wisconsin U Libr [947]
Graf, Arturo *see* Un monte di pilato in italia
Graf essex / Laube, Heinrich – 8. Aufl. Leipzig: J J Weber, [1898?] – 1 – us Wisconsin U Libr [820]
Graf, Friedrich Hartmann *see* Sonata or trio for two german flutes and a bass
Graf, G *see*
– Ein bisher unbekanntes werk des patriarchen eutychios von alexandrien
– Die christlich-arabische litteratur bis zur fraenkischen zeit...
– Des theodor abu kurra traktat ueber den schoepfer und die wahre religion
– Geschichte der christlichen arabischen literatur
– Die philosophie und gotteslehre des jahja ibn 'adi und spaeteren autoren
– Ein reformversuch innerhalb der koptischen kirche im 12. jahrhundert
Graf, Georg *see*
– Catalogue de manuscrits arabes chretiens con-serves au caire
– Die ueberlieferung der arabischen uebersetzung des diatessarons
Graf, Hans *see* Entwicklung der wahlen und politischen parteien in gross-dortmund
Graf, Hans Gerhard *see*
– Goethe von schiller in briefen
– Goethes briefwechsel mit seiner frau
Graf, Hans Gerhardt *see* Goethes liebesgedichte
Graf, Karl Heinrich *see*
– Eduard reuss' briefwechsel mit seinem schueler und freunde karl heinrich graf
– Der prophet jeremia
– Der segen moses (deuteronomium 100 33)
– Der stamm simeon
Graf s iu vitte / Kleinov, G M – Spb, 1906 – 1mf – 9 – mf#REF-477 – ne IDC [332]
Graf s iu vitte, kak ministr finansov / Lutokhin, D A – Pg, 1915 – 1mf – 9 – mf#REF-478 – ne IDC [332]
Graf simon iv. zur lippe und die weserrenaissance : bautaetigkeit eines lippischen landesherrn (1563-1613) / Hilker, Susanne – (mf ed 2000) – 3mf – 9 – €49.00 – 3-8267-2681-2 – mf#DHS 2681 – gw Frankfurter [720]
Ein graf spielt theater *see* Ein schloss in boehmen
Der graf von charolais : ein trauerspiel / Beer-Hofmann, Richard – 4. aufl. Berlin: S Fischer, 1906 [mf ed 1991] – 264p – 1 – mf#7532 – us Wisconsin U Libr [820]
Der graf von essex im deutschen drama / Baerwolf, Walther – Tuebingen, 1919 (mf ed 1994) – 1mf – 9 – €24.00 – 3-8267-3010-0 – mf#DHS-AR 3010 – gw Frankfurter [430]
Grafe, Eduard *see*
– Die stellung und bedeutung des jakobusbriefes in der entwicklung des urchristentums
– Ueber veranlassung und zweck des roemerbriefes
Grafenstein, Andreas *see* Moeglichkeiten und grenzen einer erhoehung der staatswirtschaftlichen planungsrationalitaet durch cutback-management-techniken und erfolgskontrollen
Graff, Eberhard Gottlieb *see* Deutsche interlinearversionen der psalmen
Graff, P *see* Geschichte der aufloesung der alten gottesdienstlichen kirche deutschlands
Graff, Sigmund *see* Die endlose strasse
Graffenauer, Jean *see* Meine berufsreise durch deutschland, preussen und das herzogthum warschau
Graffin, B *see* Aphraetes
Graffiti Press *see* Nma&m conscience
Grafica – North Hollywood CA 1972-83 – 1,5,9 – ISSN: 0017-2898 – mf#7326 – us UMI ProQuest [305]
Grafrica – 1982 mar 21 – 1r – 1 – mf#4851583 – us WHS [071]
Die grafschaft glatz : ihre natur und deren beziehungen zu geschichte und leben der menschen / Kutzen, Joseph A – Glogau 1873 [mf ed Hildesheim 1995-98] – 2mf [ill] – 9 – €60.00 – 3-487-29502-4 – gw Olms [943]
Der grafschafter *see* Dorf-chronik 1848
Grafschafter nachrichten – Nordhorn DE, 1978- – ca 7r/yr – 1 – gw Misc Inst [074]
Grafton 1693-1849 – Oxford, MA (mf ed 1996) – (= ser Massachusetts vital record transcripts to 1850) – 15mf – 9 – 0-87623-255-1 – (mf 1t: births, intentions, marriages 1693-1766. mf 2t: vital records 1721-82. mf 3t-6t: marriages & intentions 1781-1849. mf 7t: marriages & births 1735-1824. mf 8t-9t: births & deaths 1756-1849. mf 10t: births 1791-1846. mf 11t: births & deaths 1788-1850. mf 12t: births 1846-48. mf 13t: births 1848-49; marriages 1844-45. mf 14t: marriages 1845-49; deaths 1844-47. mf 15t: deaths 1847-49) – us Archive [978]
Grafton 1693-1900 – Oxford, MA (mf ed 1993) – (= ser Massachusetts vital records) – 169mf – 9 – 0-87623-164-4 – (mf 1-5: birth index 1693-1900. mf 6-15: births 1693-1900. mf 16-20: marriage index 1729-1900. mf 21-29: marriages 1729-1905. mf 30-33: death index 1721-1900. mf 34-40: deaths 1721-1900. mf 41-44: birth index 1693-1934. mf 45-48: death index 1721-1934. mf 49-56: birth, death 1693-1851. mf 57-60: birth, death 1744-1843. mf 61-67: birth, death 1779-1856. mf 68-78: proprietors 1728-1775. mf 79-83: town records 1735-1782. mf 84-90: town records 1752-89. mf 84-90: town records 1752-89. mf 91-97: town records 1779-1800. mf 98-107: town records 1799-1820. mf 108-114: town record 1820-39. mf 115-124: town record 1821-48. mf 125-138: intentions 1735-1900. mf 139-141: death index 1844-84. mf 142-150: deaths 1844-1905. mf 151-152: marriage index 1844-85. mf 153-159: marriages 1844-1900. mf 160-161: birth index 1844-62. mf 162-166: births 1844-84. mf 167-169: birth 1885-1900) – us Archive [978]
Grafton and coos counties bar association annual meetings – New Hampshire. 3v in 5 books. 1882-98 – 24mf – 9 – $36.00 – mf#LLMC 84-476 – us LLMC [340]
Grafton argus – Grafton, 1874-75; 1903-20 – at Pascoe [079]
Grafton argus – Grafton, jan 1899-dec 1907 – 5r – 9 – A$321.29 vesicular A$348.79 silver – at Pascoe [079]
Grafton, Charles Chapman *see*
– A catholic atlas
– Difficulties of faith
– A letter about the mission to be held at the church of st john the evangelist, montreal
– Pusey and the church revival
Grafton County. New Hampshire. Bar *see* Proceedings of the grafton county bar, on the death of the hon. harry hibbard.
Grafton daily argus – Grafton, 1921-22 – at Pascoe [079]
Grafton enterprise – Grafton WI. 1927 jul 27-aug 3 – 1r – 1 – (continued by: port washington pilot; port washington star and ozaukee county advertiser; port washington pilot, port washington star and grafton enterprise) – mf#916186 – us WHS [071]
Grafton galleries exhibition of l'art nouveau / Grafton Galleries, London – London [1899?] – (= ser 19th c art & architecture) – 2mf – 9 – mf#4.2.1091 – uk Chadwyck [700]
Grafton Galleries, London *see*
– L'art decoratif francais
– Fair women
– Grafton galleries exhibition of l'art nouveau
Grafton sun – Grafton, NE: Shoff & Lowley, feb 1898-jan 29 1910// (wkly) [mf ed v1 n4. mar 18 1898-1902 (gaps)] – 1r – 1 – (merged with: nebraska signal and the exeter enterprise to form: nebraska signal and the grafton sun and the exeter enterprise) – us NE Hist [071]
Grafton, Thomas William *see*
– Alexander campbell
– Men of yesterday
Gragera Vega, Silverio *see* Procuradores de los tribunales ejercientes en espana
Gragger, Robert *see* Preussen, weimar und die ungarische koenigskrone
Gragnon, Alfred *see* Inspecteur grey
Graham, A D *see* On faith
Graham, Allen D *see* Cruelty and christianity
Graham, Angela H *see* The national endowment for the arts dance program, 1965-1971
Graham baptist church. sumter county. south carolina : church records – 1886-1924. Roll. 1952-59 – 1 – 7.92 – us Southern Baptist [242]
Graham, Bruce J *see* A comparison of hydrostatic weighing and displacement plethysmography for determining body density of young elite female gymnasts
Graham, David *see* A treatise on the law of new trials in cases civil and criminal
Graham, Gabriela Cunninghame *see* Santa teresa
Graham, George Ransom *see* Cost of living on one hundred farms columbia county, florida, year...
Graham, H Q. *see* Pastoral register
Graham, Hugh Davis *see* Civil rights during the nixon administration, 1969-1974
Graham, Isabella *see* The power of faith
Graham, J A *see* On the threshold of three closed lands

Graham, J Miller *see*
- East of the barrier
- East of the barrier, or side lights on the manchuria mission

Graham, J R G *see* Answer by her majesty's government to the memorial transmitted to s...

Graham, James *see* Interviews from villages in the njombe district, tanzania

Graham, James Robert *see* The planting of the presbyterian church in northern virginia prior to the organization of winchester presbytery, december 4, 1794

Graham, John *see*
- Discourse delivered at great st mary's church, cambridge
- Elisha's tribute to the memory of elijah
- Revolution settlement of the church of scotland

Graham, John Anderson *see*
- Missionary expansion since the reformation
- On the threshold of three closed lands

Graham, John William *see* War from a quaker point of view

Graham journal of health and longevity – La Salle IL 1837-39 – 1 – mf#3748 – us UMI ProQuest [613]

Graham, Maria *see*
- Letters on india
- Three months passed in the mountains east of rome

Graham, Ralph E *see* The relationship of aerobic fitness, type a behavior pattern, and hostility to baroreflex responses and cardiovascular reactivity to nonexertional stressors

Graham, Richard *see* Century of brazilian history since 1865

Graham, Robert Douglas *see* Rough passage

Graham, Rose *see* S gilbert of sempringham and the gilbertines

Graham, Sigrid *see* A systemic description of drug rehabilitation programmes

Graham, Stephen *see*
- Changing russia
- Great russian short stories
- In quest of el dorado
- The way of martha and the way of mary
- With poor immigrants to america
- With the russian pilgrims to jerusalem

Graham, Timothy *see* Corpus christi college, cambridge 1

Graham,John *see* Refutation of a number of pernicious errors

Graham's illustrated magazine – La Salle IL 1826-58 – 1 – mf#4580 – us UMI ProQuest [640]

Graham's magazine – La Salle IL 1837-40 – 1 – mf#3953 – us UMI ProQuest [700]

Graham's magazine – v1-53. 1826-58 – (= ser The casket) – 1 – us AMS Press [410]

Grahamstown journal – 1831-1919 – 1 – (title varies: the jounal) – mf#MS00047 – sa National [079]

The grahamstown observer – jul 3 1933-sep 25 1933? (wkly) [mf ed Cape Town: SA Library 1985] – 1r – 1 – mf#MS00208 – sa National [079]

Grahamstown (south africa) diocesan records / United Society for the Propagation of the Gospel. Archives – 19th c – 12r – 1 – (int by isobel pridmore) – mf#96723 – uk Microform Academic [025]

Grahl-Moegelin, Walter *see* Die lieblingsbilder im stil e.t.a. hoffmanns

Grail *see* The irish theosophist

Graillot, Henri *see* Le culte de cybele, mere des dieux

The grain, grass and gold fields of south-western canada : edmonton, alberta, canada, described as a mixed farming and mining country... / Cowie, Isaac [comp] – Edmonton: s.n, 1897 – 1mf – 9 – mf#30156 – cn Canadiana [630]

Grain growers guide – Winnipeg, CN. 1908-28 – 13r – 1 – cn Commonwealth Imaging [660]

Grain millers news / American Federation of Grain Millers – 1983 may-1988 oct – 1r – 1 – mf#1057421 – us WHS [630]

Grain Services Union [CLC] *see* Facts and comments

The grain trade : extract from a paper on "the graphical delineation of statistical facts" / Harvey, Arthur – S.l: s.n, 1863? – 1mf – 9 – mf#41508 – cn Canadiana [380]

A graine of musterd-seede : or, the least measure of grace that is or can be effectuall to salvation / Perkins, W – London: Iohn Legate, 1611 – 2mf – 9 – mf#PW-79 – ne IDC [240]

Grains de bon sens – 15e Mille. Montreal: Impr du Messager, 1918 [mf ed 1990] – 3mf – 9 – (ill by edmond-joseph massicotte) – mf#SEM105P1212 – cn Bibl Nat [230]

Graiul nou – Bucharest. Rumania. -sw. 18 Dec 1944-22 Feb 1948. (Very imperfect). (2 reels) – 1 – uk British Libr Newspaper [949]

Grajevsky Jacob Osher *see* Berit ha-levi

Die gralepen in ulrich fueetrers bearbeitung (buch der abenteuer) / ed by Nyholm, Kurt – Berlin: Akademie-Verlag, 1964 [mf ed 1993] – (= ser Deutsche texte des mittelalters 57) – cxiv/384p/5pl – 1 – (incl bibl ref and ind) – mf#8623 reel 8 – us Wisconsin U Libr [430]

Die gralsburg : erzaehlung / Gmelin, Otto – Leipzig: P List, c1935 (mf ed 1990) – 1r – 1 – (filmed with: sommerwind ueber tormohlenhof) – us Wisconsin U Libr [830]

Gram, Hans *see*
- America – a new march
- The death song of an indian chief, from "ouabi"

Gramatica castellana / Alonso, Amado – Buenos Aires, Argentina. 1939 – 1r – us UF Libraries [440]

Gramatica castellana / Alonso, Amado – Buenos Aires, Argentina. v1-2. 1940-41 – 1r – us UF Libraries [440]

Gramatica castellana / Castillo, Manuel – 1898 – 9 – sp Bibl Santa Ana [440]

Gramatica changana / Ribeiro, Armando – Canicado, Mozambique. 1965 – 1r – us UF Libraries [470]

Gramatica de la lengua arabiga / Moreno Nieto, Jose – 1872 – 9 – sp Bibl Santa Ana [470]

Gramatica de la lengua catalana / Fabra, Pompeu – Barcelona, Spain. 1912 – 1r – us UF Libraries [440]

Gramatica de la lengua espanola / Real Academia Espanola – Nueva ed. reformada. Madrid: Perlado, Paez y Compania, 1924. 564p. 1 reel. 1283 – 1 – us Wisconsin U Libr [440]

Gramatica elemental de la lengua latina y castellana / Lama y Lena, Rafael & Lozano, Francisco Franco – Gijon: Imprenta de Mussel, 2nd ed 1894 – 1 – sp Bibl Santa Ana [440]

Gramatica francesa / Nunez Coronado, J – Badajoz: Imp. Grafica Iberia, 2nd ed 1945 – 1 – sp Bibl Santa Ana [440]

Gramatica francesa / Nunez Coronado, J – Badajoz: La Minerva Extremena, 3rd ed 1946 – 1 – sp Bibl Santa Ana [440]

Gramatica graeca / Sanchez de las Brozas, Francisco – 1592 – 9 – sp Bibl Santa Ana [450]

Gramatica historica espanola y antologias / Gomez Bravo, Vicente – Medieval y del siglo 15. Madrid: Graficas Nebrija, 1957 – 1 – sp Bibl Santa Ana [440]

Gramatica ilocana / Lopez, Francisco, Fr; ed by Carro, P – 3rd ed. Malabon: Estab tip del Asilo de huerfanos de Malabon, a cargo de pp. Agustinos calzados, 1895 – us CRL [400]

Gramatica latina / Santos Coco, Francisco – Badajoz: Tip. Artes Graficas, 1932 – 1 – sp Bibl Santa Ana [440]

Gramatica latina (i. 1468) / Mates, Bartolome – Barcelona – 1r – 5,6 – sp Cultura [450]

Gramatica marastta : a mais vulgar que se pratica nos reinos do nizamaxa e idalxa offerecida aos muitos reverendos padres missionarios dos dittos reinos – Roma: Est da Sag Congreg de propag fide 1778 – (= ser Whsb) – 1mf – 9 – €20.00 – mf#Hu 261 – gw Fischer [410]

Gramatica umbundu / Valente, Jose Francisco – Lisboa, Portugal. 1964 – 1r – us UF Libraries [470]

Gramatica y arte nueva de la lengva general de todo el peru, llamada lingua qquichua, o lengua del inca / Gonzalez Holguin, Diego – Ciudad de los Reyes [i. e. Lima]: Canto 1607 – (= ser Whsb) – 4mf – 9 – €50.00 – mf#Hu 388 – gw Fischer [440]

Gramatica y...el peru / Gonzalez Holguin, Diego – 1842 – 9 – sp Bibl Santa Ana [440]

Gramatices...fallaces e prolixae / Sanchez de las Brozas, Francisco – 1595 – 9 – sp Bibl Santa Ana [450]

Gramaticus latinae institutiones / Sanchez de las Brozas, Francisco – 1576 – 9 – sp Bibl Santa Ana [450]

Gramatik fun der yidisher shprakh / Reisen, Zalman – Vilne, Lithuania. 1920 – 1r – us UF Libraries [470]

Gramatyka historyczna jezyka czeskiego / Lehr-Sptawinski, Tadeusz – Warszawa: Panstwowe Wydawn, Naukowe, 1957-. Map. Bibliog. 1 reel. 1287 – 1 – us Wisconsin U Libr [460]

Gramberg, Gerhard Anton *see* Blaetter vermischten inhalts

Grambling State University *see* Gsu today

Gramera malagasy : dingana faharoa / Rajaobelina, Prosper – [Tananarive]: Edisiona Salohy, 1960 – 1 – us CRL [490]

Graminaeus, Diederich *see* Mysticus aquilo, sive declaration vaticinii jeremie prophetae

La grammaire / Labiche, Eugene; ed by Squair, John – Toronto: Copp, Clark [1906?] [mf ed 1995] – 2mf – 9 – 0-665-74780-2 – (incl english text) – mf#74780 – cn Canadiana [820]

Grammaire arabe : a l'usage des eleves de l'ecole speciale des langues orientales vivantes; avec figures / Sacy, Antoine Isaac Silvestre de – Paris: Impr Imperiale 1810 – (= ser Whsb) – 11mf – 9 – €95.00 – mf#Hu 208 – gw Fischer [410]

Grammaire catalane / Fabra, Pompeu – Paris, France. 1946 – 1r – us UF Libraries [440]

Grammaire categorielle du francais. etude theorique et implantation. le systeme grace (grammaire categorielle etendue) / Segond, Frederique – 1mf – 9 – (10065) – fr Atelier National [440]

Grammaire celto-bretonne : contenant les principes de l'orthographe, de la prononciation, de la construction des mots et des phrases, selon le genie de la langue celto-bretonne / Gonidec, Jean-Francois-Marie-Maurice-Agatha le – Paris: Chez l'auteur 1807 – (= ser Whsb) – 4mf – 9 – €50.00 – mf#Hu 110 – gw Fischer [410]

Grammaire de la langue arameenne : selon les deux dialectes syriaque et chaldaique / Dawud, Yusuf – 2e ed rev corr augm. Mossoul: Impr des peres dominicains 1896-98 [mf ed 1986] – 2v on 3mf – 9 – 0-8370-9053-9 – mf#1986-3053 – us ATLA [470]

Grammaire de la langue armenienne : ou l'on expose les principes et le regles de la langue, d'apres les meilleurs grammairiens, et les auteurs originaux et suivant les usages particuliers de l'idiome haikien / Chahan de Cirbied, Jacques – Paris: Everat 1823 – (= ser Whsb) – 10mf – 9 – €90.00 – mf#Hu 229 – gw Fischer [490]

Grammaire de la langue khmere / Maspero, Georges – Paris: Impr Nationale 1915 [mf ed 1989] – 1r with other items – 1 – ("ouvrage publie sous le patronage de l'ecole francaise d'extreme-orient") – mf#mf-10289 seam reel 024/13 [§] – us CRL [480]

Grammaire de la langue turque : dialecte osmanli / Deny, J – Paris, 1921 – 2mf2 – 8 – mf#U-302 – ne IDC [470]

Grammaire demotique / Lexa, F – Prague, 1949-1951. 7 v – 36mf – 9 – mf#NE-20061 – ne IDC [956]

Grammaire du kiyombe / Clerq, L De – Bruxelles, Belgium. 1921 – 1r – us UF Libraries [470]

Grammaire du lomongo / Rop, Albert De – Leopoldville, Congo. 1958 – 1r – us UF Libraries [470]

Grammaire esquimaude composee en 1928 / Turquetil, Arsene – 2e ed. Outremont: [s.n, 1938?] (mf ed 1999) – 2mf – 9 – mf#SEM105P3105 – cn Bibl Nat [490]

Grammaire esquimaude du sous-dialecte de l'ungava / Schneider, Lucien – [Quebec]: Ministere des richesses naturelles, [1968?] (mf ed 1988) – 2mf – 9 – (with ind) – mf#SEM105P937 – cn Bibl Nat [490]

Grammaire et dictionnaire de lingala / Guthrie, Malcolm – Cambridge, England. 1939 – 1r – us UF Libraries [470]

Grammaire et exercices pratiques – Kiniama, Zaire. 19–? – 1r – us UF Libraries [470]

Grammaire francaise elementaire : suivie d'une methode d'analyse grammaticale raisonnee, a l'usage des ecoles chretiennes – 2e ed. [Montreal: s.n.] 1843 [mf ed 1984] – 3mf – 9 – 0-665-45527-5 – (incl ind) – mf#45527 – cn Canadiana [440]

Grammaire francoise : pour servir d'introduction a la grammaire latine / Houdet, Antoine-Jacques – Montreal: J Brown, 1811 [mf ed 1971] – 1r – 5 – mf#SEM16P45 – cn Bibl Nat [440]

Grammaire francoise : pour servir d'introduction a la grammaire latine / Houdet, Antoine-Jacques & Riviere, Claude – Montreal: J Brown, 1811 [mf ed 1971] – 1r – 5 – mf#SEM16P46 – cn Bibl Nat [440]

Grammaire francoise : pour servir d'introduction a la grammaire latine – Montreal: Impr par J Brown...1811 [mf ed 1984] – 1mf – 9 – 0-665-44866-X – mf#44866 – cn Canadiana [440]

Grammaire grecque du nouveau testament / Combe, Ernest – Lausanne: Georges Bridel; Paris: Librairie Fischbacher, [1894?] – 1mf – 9 – 0-8370-7854-7 – (incl indes) – mf#1986-1854 – us ATLA [450]

Grammaire hebraeique : precedee d'un precis historique sur la langue hebraeique / Preiswerk, Samuel – 3e rev et corr ed. Bale: H Georg 1871 [mf ed 1986] – 1mf – 9 – 0-8370-9263-9 – mf#1986-3263 – us ATLA [470]

Grammaire hebraique : avec paradigmes, exercices de lecture, chrestomathie et indice bibliographique = Hebaeische grammatik / Strack, Hermann Leberecht – ed rev augm. Carlsruhe: H Reuther 1886 [mf ed 1986] – (= ser Porta linguarum orientalium 1) – 1mf – 9 – 0-8370-9273-6 – (trans fr german by anton jean baumgartner) – mf#1986-3273 – us ATLA [470]

Grammaire kinyarwanda / Hurel, Eugene – Kabgayi, Rwanda. 1959 – 1r – us UF Libraries [470]

Grammaire latine : suivie des regles de la versification – Montreal: J Brown, 1811 [mf ed 1971] – 1r – 5 – mf#SEM16P48 – cn Bibl Nat [450]

Grammaire latine : suivie des regles de la versification – Montreal: J Brown, 1811 [mf ed 1971] – 1r – 5 – mf#SEM16P47 – cn Bibl Nat [450]

Grammaire malgache fondee sur les principes de la grammaire javanaise : suivie d'exercices et d'un recueil de cent et un proverbes / Marre-de Marin – Paris: Chez Maisonneuve et Cie, 1876 – 1 – us CRL [490]

Grammaire malgache fondee sur les principes de la grammaire javanaise, suivie d'exercices et d'un recueil de cent et un proverbes / Marin, Marre-de – Paris: Maisonneuve, 1876 – us CRL [470]

Grammar and language : a philosophical study / Starck, Ed L – Boston: W B Clarke, 1887 – 1mf – 9 – 0-8370-8313-3 – mf#1986-2313 – us ATLA [420]

Grammar and vocabulary of the bullom language / Nylaender, Gustavus Reinhold – London: Church Missionary Society 1813 – (= ser Whsb) – 2mf – 9 – €30.00 – mf#Hu 354 – gw Fischer [470]

Grammar and vocabulary of the bullom language / Nylander, Gustavus Reinhold – London: Printed for the Church missionary society, by Ellerton and Henderson, 1814 – us CRL [470]

A grammar and vocabulary of the susco language : to which are added the names of some of the sosoo towns / Brunton, E – Edinburgh: Printed by J Ritchie, 1802 – 1 – us CRL [490]

Grammar la cinyanja – Nyasaland, Malawi. 1930 – 1r – us UF Libraries [470]

Grammar of central karanga / Marconnes, Francisque A – Johannesburg, South Africa. 1931 – 1r – us UF Libraries [470]

A grammar of chinyanja : a language spoken in british central africa on and near the shores of lake nyasa – Aberdeen: G & W Fraser, 1891 – 1 – us CRL [490]

Grammar of gambian mandinka / Rowlands, Evan Celyn – London, England. 1959 – 1r – us UF Libraries [470]

A grammar of high tamil / Beschi, Constantino Giuseppe – Trichinopoly: St Joseph's Industrial School Press, 1917 – 1 – (latin text publ for the first time by I besse) – us CRL [490]

A grammar of japanese ornament and design / Cutler, Thomas William – London 1880 [i.e. 1879-80] – (= ser 19th c art & architecture) – 2mf – 9 – mf#4.2.42 – uk Chadwyck [740]

Grammar of new testament greek = Grammatik des neutestamentlichen griechisch / Blass, Friedrich Wilhelm – 2nd rev enl ed. London; New York: Macmillan, 1905 – 1mf – 9 – 0-8370-9365-1 – (incl indes) – mf#1986-3365 – us ATLA [450]

Grammar of northern transvaal ndebele / Ziervogel, D – Pretoria, South Africa. 1959 – 1r – us UF Libraries [470]

The grammar of ornament / Jones, Owen – London 1868 – (= ser 19th c art & architecture) – 8mf – 9 – mf#4.2.1473 – uk Chadwyck [740]

The grammar of science / Pearson, Karl – 2nd rev enl ed. London: A & C Black, 1900 – xviii/548p – 1 – (incl bibl) – us Wisconsin U Libr [500]

The grammar of science / Pearson, Karl – London: W. Scott, 1892. xvi,493p. (The Contemporary Science Series) – 1 – us Wisconsin U Libr [500]

The grammar of south indian (karnatic) music / Subrahmanya Ayyar, Chandrasekhar – Madras: CS Ayyar, 1951 – (= ser Samp: indian books) – us CRL [780]

Grammar of swazi / Ziervogel, D – Johannesburg, South Africa. 1952 – 1r – us UF Libraries [470]

A grammar of the ancient dialect of the canarese language – 2nd rev enl ed. Bangalore: Basel Mission Book & Tract Depository, 1889 – 1 – us CRL [490]

A grammar of the arabic language : according to the principles taught and maintained in the schools of arabia / Lumsden, Matthew – Calcutta: printed by F Dissent, 1813 – (= ser 19th c books on linguistics) – 8mf – 9 – (v1 only, no more publ) – mf#2.1.5 – uk Chadwyck [470]

A grammar of the arabic language / Caspari, Carl Paul – 1862 – 9 – $27.00 – us IRC [470]

A grammar of the bechuana language / Archbell, James – Graham's Town, Cape of Good Hope: Meurant & Godlonton, 1837 – 1 – us CRL [490]

A grammar of the bechuana language / Archbell, James – Graham's Town: Cape of Good Hope. Meurant & Godlonton, 1837[1838] – (= ser 19th c books on linguistics) – 2mf – 9 – mf#2.1.13 – uk Chadwyck [470]

A grammar of the benga language / Mackey, James L – New York: Mission House, 1855 – 1 – us CRL [490]

A grammar of the chinyanja language as spoken at lake nyassa : with chinyanja-english and english-chinyanja vocabularies / Riddel, Alexander – Edinburgh: J Maclaren, 1880 – 1 – us CRL [490]

A grammar of the cingalese language / Chater, James – Colombo: printed at the Govt Press, 1815 – (= ser 19th c books on linguistics) – 2mf – 9 – mf#2.1.55 – uk Chadwyck [490]

Grammar of the cingalese language / Chater, James – Colombo: Bergman 1815 – (= ser Whsb) – 2mf – 9 – €30.00 – mf#Hu 274 – gw Fischer [490]

Grammar of the classical arabic language / Howell, M S – Allahabad – 3r – 1 – mf#96749 – uk Microform Academic [470]

A grammar of the cree language / Howse, Joseph – London: Truebner 1865 [mf ed Bloomington IN: Indiana Uni Lib, Preservation Dept 1984] – xix 324p on 1r – 1 – us Indiana Preservation [490]

A grammar of the cree language : with which is combined an analysis of the chippeway dialect / Howse, Joseph – London: J G F & J Rivington, 1844 – (= ser 19th c books on linguistics) – 4mf – 9 – mf#2.1.14 – uk Chadwyck [490]

Grammar of the dialects of vernacular syriac as spoken by the eastern syrians of kurdistan, north-west persia, and the plain of mosul : with notices of the vernacular of the jews of azerbaijan and of zakhu near mosul / Maclean, Arthur John – Cambridge: University Press, 1895 – 1mf – 9 – 0-8370-8275-7 – mf#1986-2275 – us ATLA [470]

Grammar of the fulde language / Reichardt, Charles August Ludwig – London: Church Missionary House, 1876 – 1 – (with appendix of some original traditions & portions of scripture transl into fulde, together with 8 chapters of the book of genesis) – us CRL [490]

Grammar of the german language / Curme, George Oliver – New York, NY. 1922 – 1r – us UF Libraries [430]

A grammar of the greek language / Crosby, Alpheus – 23rd ed. Boston: Phillips, Sampson & Co, 1858 [mf ed 1987] – xiv/[9]/464p/viii – 1 – mf#2043 – us Wisconsin U Libr [450]

A grammar of the greek language / Jelf, William Edward – 4th ed. Oxford: James Parker, 1866 [mf ed 1992] – 12mf – 9 – 0-524-03882-1 – (with additions and corr) – mf#1987-6495 – us ATLA [450]

A grammar of the hebrew language / Green, William Henry – 3rd ed. New York: John Wiley, 1865, c1861 [mf ed 1986] – 1mf – 9 – 0-8370-9150-0 – (incl ind) – mf#1986-3150 – us ATLA [470]

A grammar of the hebrew language, comprised in a series of lectures : compiled from the best authorities, and drawn principally from oriental sources, designed for the use of students in the universities / Lee, Samuel – 2nd ed. London: printed for James Duncan, 1832 – (= ser 19th c books on linguistics) – 5mf – 9 – mf#2.1.17 – uk Chadwyck [470]

A grammar of the hebrew language of the old testament = Grammatik der hebraischen sprache des a.t. / Ewald, Georg Heinrich August von – [London]: Whittaker and Co, 1836 – (= ser 19th c books on linguistics) – 5mf – 9 – mf#2.1.16 – uk Chadwyck [470]

Grammar of the high dialect of the tamil language, termed shen-tamil : to which is added an introduction to tamil poetry / Beschi, Costanzo Giuseppe – Madras: College Press 1822 – (= ser Whsb) – 2mf – 9 – €30.00 – (trans fr original latin by benjamin guy babington) – mf#Hu 270 – gw Fischer [490]

A grammar of the hindi language : in which are treated the high hindi, braj, and the eastern hindi of the ramayan of tulsi das... / Kellogg, Samuel Henry – 2nd rev enl ed. London: K Paul, Trench, Truebner, 1893. Chicago: Dep of Photodup, U of Chicago Lib, 1967 (1r); Evanston: American Theol Lib Assoc, 1984 (1r) – 1 – 0-8370-0159-5 – (incl bibl ref) – mf#1984-B085 – us ATLA [490]

Grammar of the hindustami language / Shakespeare, John – London: Cox & Bayles 1813 – (= ser Whsb) – 3mf – 9 – €40.00 – mf#Hu 257 – gw Fischer [490]

A grammar of the hindustani language in the oriental and roman character / Forbes, Duncan – new ed. London: W H Allen, 1855 [mf ed 1990] – viii/148/28/56p/[14]pl (ill) – 1 – mf#7417 – us Wisconsin U Libr [490]

Grammar of the homeric dialect / Monro, David Binning – Oxford, England. 1882 – 1r – us UF Libraries [450]

Grammar of the iberno-celtic, or irish language : to which is prefixed, an essay on the celtic language / Vallancey, Charles – 2nd enl ed. Dublin: Marchbank 1782 – (= ser Whsb) – 4mf – 9 – €50.00 – mf#Hu 086 – gw Fischer [490]

A grammar of the idiom of the new testament = Grammatik des neutestamentlichen sprachidioms / Winer, Georg Benedikt – 7th enl ed. Andover: Warren F Draper, 1877, c1874 [mf ed 1989] – 2mf – 9 – 0-7905-0535-5 – (incl ind. in english) – mf#1987-0535 – us ATLA [450]

A grammar of the idiom of the new testament : prepared as a solid basis for the interpretation of the new testament / Winer, Georg Benedikt; ed by Thayer, J H – 7th enl ed. Andover: W F Draper, 1897 [mf ed 1987] – xviii/728p – 1 – (improved by gottlieb luenemann. rev aut trans) – mf#8662 – us Wisconsin U Libr [450]

Grammar of the irish language / O'donovan, John – Dublin, Ireland. 1845 – 1r – us UF Libraries [490]

Grammar of the kaffir language / Mclaren, James – London, England. 1906 – 1r – us UF Libraries [470]

Grammar of the kafir language / Boyce, William Binnington – Cape Town, South Africa. 1956 – 1r – us UF Libraries [470]

Grammar of the lamba language / Doke, Clement Martyn – London, England. 1922 – 1r – us UF Libraries [470]

Grammar of the mahratta language : to which are added dialogues on familiar subjects / Carey, William: – Serampore: Mission press 1805 – (= ser Whsb) – 3mf – 9 – €40.00 – mf#Hu 263 – gw Fischer [490]

A grammar of the maskoke : or creek language / Bucknew, H F – 140p – 1 – $5.00 – us Southern Baptist [490]

Grammar of the modern syriac language : as spoken in oroomiah, persia, and in koordistan / Stoddard, David Tappan – [S.l.: s.n., 1853?] – 1mf – 9 – 0-8370-7671-4 – mf#1986-1671 – us ATLA [470]

A grammar of the new testament diction : intended as an introduction to the critical study of the greek new testament = Grammatik des neutestamentlichen sprachidioms / Winer, Georg Benedikt – Philadelphia: Smith, English, 1859 [mf ed 1993] – 2mf – 9 – 0-524-07125-X – (incl bibl ref. trans fr german by edward masson) – mf#1992-1041 – us ATLA [225]

A grammar of the new testament greek = Grammatik des neutestamentlichen sprachgebrauchs / Buttmann, Alexander – Andover: W F Draper, 1873 [mf ed 1989] – 1mf – 9 – 0-8370-1184-1 – (trans fr german by j h thayer) – mf#1987-6018 – us ATLA [450]

A grammar of the old testament in greek, according to the septuagint / Thackeray, Henry St John – Cambridge: University Press, 1909 [mf ed 1988] – 1mf – 9 – 0-7905-0170-8 – (no more publ. incl indes) – mf#1987-0170 – us ATLA [450]

A grammar of the persian language : comprising a portion of the elements of arabic inflexion / Lumsden, Matthew – Calcutta: printed by T Watley. 2v. 1810 – (= ser 19th c books on linguistics) – 13mf – 9 – mf#2.1.30 – uk Chadwyck [470]

A grammar of the persian language : to which are subjoined, several dialogues / Muhammad Ibrahim, Mirza – London: Wm H Allen & Co, 1841 – (= ser 19th c books on linguistics) – 3mf – 9 – (with an alphabetical list of the english and persian terms of grammar, and app on the use of arabic words) – mf#2.1.52 – uk Chadwyck [470]

Grammar of the phoenecian language / Harris, Z S – 1936 – 9 – $10.00 – us IRC [470]

Grammar of the punjabee language / Carey, William – Serampore: Mission-Press 1812 – (= ser Whsb) – 2mf – 9 – €30.00 – mf#Hu 259 – gw Fischer [490]

A grammar of the samaritan language : with extracts and vocabulary / Nicholls, G F – London: Samuel Bagster [1858?] [mf ed 1986] – 1mf – 9 – 0-8370-7654-4 – mf#1986-1654 – us ATLA [470]

Grammar of the sanskrita language / Wilkins, Charles – London: Bulmer Kingsbury 1808 – (= ser Whsb) – 8mf – 9 – €80.00 – (parallelsachtitel in sanskrit) – mf#Hu 237 – gw Fischer [490]

A grammar of the septuagint / Thackeray, Henry St John – Cambridge. 1900 – 9 – $21.00 – us IRC [221]

Grammar of the sesuto language / Jacottet, Edouard – Johannesburg, South Africa. 1927 – 1r – us UF Libraries [470]

Grammar of the sindebele dialect of zulu / O'neil, J – Bulawayo, Zimbabwe. 1912 – 1r – us UF Libraries [470]

Grammar of the tahitian dialect of the polynesian language / Davies, John – Tahiti: Mission Pr 1823 – (= ser Whsb) – 1mf – 9 – €20.00 – mf#Hu 449 – gw Fischer [490]

A grammar of the tamil language / Rhenius, Charles Theophilius Ewald – Madras: printed at the Church Mission Press, 1836 – (= ser 19th c books on linguistics) – 9 – (with app) – mf#2.1.44 – uk Chadwyck [490]

A grammar of the vulgate / Plate, W E & White, H J – Oxford, 1926 – 3mf – 8 – €7.00 – ne Slangenburg [221]

The grammar of tiv / Abraham, Roy Clive – Kaduna, Nigeria: Printed by the Govt Printer, 1933 – 1 – us CRL [490]

The grammar school system of ontario : a correspondence between the board of trustees... and the rev e ryerson... / Clinton County Grammar School (Ont). Board of Trustees – Clinton ON: s.n, 1868 – 1mf – 9 – (repr fr: 'clinton new era') – mf#23520 – cn Canadiana [370]

Grammar yoruba / Lijadu, E M – Ode Ondo, 1898 (mf ed 1976) – 1mf – 9 – mf#NYPL FSN SC 015,135 – us NY Public [470]

Grammatica aegyptiaca utriusque dialecti / Scholtz, Christian:; ed by Woide, Charles Godfrey – Oxonii: Clarendon 1778 – (= ser Whsb) – 2mf – 9 – €30.00 – (with handwritten corr & additions by w von humboldt) – mf#Hu 379 – gw Fischer [470]

Grammatica arabica / Postel, G – Parisiis, [1540] – 1mf – 9 – mf#H-8266 – ne IDC [470]

Grammatica critica linguae arabicae : cum brevi metrorum doctrina / Ewald, Heinrich – Lipsiae: Hahn 1831-33 – (= ser Whsb) – 9mf – 9 – €85.00 – mf#Hu 210 – gw Fischer [470]

Grammatica della lingua ebraica / Luzzatto, Samuel David – Padova: A. Bianchi, 1853 – 2mf – 9 – 0-8370-1198-1 – mf#1987-6028 – us ATLA [470]

Grammatica e vocabulario della lingua kurda / Garzoni, Maurizio – Roma: Stamp della Sac Congreg di propag fide 1727 – (= ser Whsb) – 3mf – 9 – €40.00 – mf#Hu 225 – gw Fischer [490]

Die grammatica figurata des mathias ringmann (philesius vogesigena) in faksimiledruck / Ringmann, Matthias; ed by Wieser, Fr R v – Strassburg: J H Ed Heitz, 1905 – (incl bibl ref. latin text with an introduction in german) – us Wisconsin U Libr [430]

Grammatica graecitatis novi testamenti : quam ad georgii wineri ejusdem argumenti librum, germanico idiomate conscriptum / Beelen, Jan Theodor – Lovanii [Louvain]: CJ Fonteyn et apud Vanlinthout, 1857 – 2mf – 9 – 0-524-05968-3 – (incl bibl ref) – mf#1992-0705 – us ATLA [450]

Grammatica latino-hibernica : nunc compendiata / O'Molloy, Franciscus – Romae: Typographia S Cong de Propag 1677 – (= ser Whsb) – 4mf – 9 – €50.00 – mf#Hu 084 – gw Fischer [410]

Grammatica, libri 16 / Priscianus [Priscian: Priscianus Caesariensis] – 14 and 15th c – (= ser Holkham library manuscript books 406,423) – 1r – 1 – (filmed with: galfridus de vinesauf: poetria novella) – mf#96849 – uk Microform Academic [450]

...Grammatica linguae amharicae / Ludolf, Hiob – Frankfort, 1698 – 1 – us CRL [470]

Grammatica linguae amharicae : quae vernacula est habessihorum, in usum eorum qui cum anitqua bac et praeclare natione christiana conversari volent, edita: plura habes in praefatione / Ludolfo, lobo – Francofurti ad Moenum: Prostat apud Johannem David. Zunnerum: Impressit Martinus Jacquetus, 1698 – (filmed with ludolf, hiob. lexicon ambarico-latinum) – us CRL [470]

Grammatica linguae hebraicae cum exercitiis et glossario : studiis academicis / Zapletal, Vincenz – Paderbornae: Sumptibus Ferdinandi Schoeningh, 1902 – 2mf – 9 – 0-524-05923-3 – mf#1992-0680 – us ATLA [470]

Grammatica linguae mauro-arabicae : uxta vernaculi idiomatis usum. accessit vocabularium latino-mauro-arabicum / Dombay, Franz Lorenz von – Vindobonae 1800 – (= ser Whsb) – 2mf – 9 – €30.00 – mf#Hu 361 – gw Fischer [410]

Grammatica syriaca / Hoffmann, Andreas Gottlieb – Halis: Impensis Librariae Orphanotrophei, 1867 [mf ed 1986] – 1mf – 9 – 0-8370-8994-8 – (incl bibl ref) – mf#1986-2994 – us ATLA [470]

Grammaticae latinae institutiones / Sanchez de las Brozas, Francisco – 1595 – 9 – sp Bibl Santa Ana [450]

Grammatical index to the chandogya-upanisad / Little, Charles Edgar – New York, NY. 1900 – 1r – us UF Libraries [490]

Grammatical outline and vocabulary of the oji language : with especial reference to the akwapim-dialect, together with a collection of proverbs of the natives / Riis, Hans Nicolaus – Basel: Bahnmaier, 1854 – 1 – us CRL [490]

Grammatici latini / ed by Keil, H – Lipsiae, 1857-1880 – 95mf – 8 – (incl suppl) – mf#H-214 – ne IDC [450]

Grammaticus *see* Answer to dr kidd's appeal to the public

Grammatik der aramaeischen muttersprache jesu / Schultze, Martin – Berlin: S Calvary, 1899 – 1mf – 9 – 0-8370-7261-1 – (incl bibl ref) – mf#1986-1261 – us ATLA [470]

Grammatik der biblisch-chaldaeischen sprache und des idioms des thalmud babli : ein grundriss = Elementi grammaticali del caldeo biblico e del dialetto talmudico babilonese / Luzzatto, Samuel David – Breslau [Wroclau]: Schletter, 1873 – 1mf – 9 – 0-8370-9294-9 – (in german. incl bibl ref) – mf#1986-3294 – us ATLA [470]

Grammatik der biblisch-chaldaischen sprache und des idioms des thal... / Luzzatto, Samuel David – Breslau, Germany. 1873 – 1r – us UF Libraries [470]

Grammatik der englischen sprache auf wissenschaftlicher grundlage / Deutschbein, Max – Heidelberg, Germany. 1953 – 1r – us UF Libraries [420]

Grammatik der eskimo-sprache : wie sie im bereich der missions-niederlassungen der brudergemeine an der labradorkuste gesprochen wird / Bourquin, Theodor – London: Moravian Mission Agency, 1891 – 5mf – 9 – mf#00217 – cn Canadiana [490]

Grammatik der griechischen papyri aus der ptolomaeerzeit : mit einschluss der gleichzeitigen ostraka und der in aegypten verfassten inschriften / Mayser, Edwin – Leipzig: B G Teubner 1906 [mf ed 1988] – 2mf – 9 – 0-7905-0143-0 – (incl ind) – mf#1987-0143 – us ATLA [450]

Grammatik der griechischen vulgarsprache in historischer entwicklung / Mullach, Friedrich Wilhelm August – Berlin: Ferd Duemmler, 1856 [mf ed 1986] – 1mf – 9 – 0-8370-9299-X – (incl ind) – mf#1986-3299 – us ATLA [450]

Grammatik der mongolischen sprache / Schmidt, Isaak Jacob – St Petersburg: Buchdruckerei der Kaiserlichen Akad der Wiss 1831 – (= ser Whsb) – 2mf – 9 – €30.00 – mf#Hu 304 – gw Fischer [490]

Grammatik der neuhebraeischen sprache / Siegfried, Carl – Karlsruhe: H Reuther; New York: B Westermann, 1884 [mf ed 1986] – (= ser Lehrbuch der neuhebraeischen sprache und literatur 1-2) – 1mf – 9 – 0-8370-9271-X – mf#1986-3271 – us ATLA [470]

Grammatik der septuaginta : laut- und wortlehre / Helbing, Robert – Goettingen: Vandenhoeck und Ruprecht, 1907 – 1mf – 9 – 0-8370-9288-4 – mf#1986-3288 – us ATLA [450]

Grammatik der slavischen sprache in krain, kaernten und steyermark / Kopitar, Bartholomaeus – Laibach: Korn 1808 – (= ser Whsb) – 6mf – 9 – €70.00 – mf#Hu 191 – gw Fischer [460]

Grammatik der syrischen sprache : mit vollstaendigen paradigmen, chrestomathie und woerterbuche / Uhlemann, Friedrich – 2. ueberarbeitete und verm Ausg. Berlin: Jonas, 1857 – 2mf – 9 – 0-8370-7595-5 – (incl ind) – mf#1986-1595 – us ATLA [470]

Grammatik der tigrina-sprache in abessinien, hauptsaechlich in die gegend von aksum und adoa / Praetorius, F – Halle, 1871 – 4mf – 9 – mf#NE-20246 – ne IDC [470]

Grammatik des biblisch-aramaeischen : mit den nach handschriften berichtigten texten und einem woerterbuch / Strack, Hermann Leberecht – 3., grossenteils neubearbeitete Aufl. Leipzig: J C Hinrichs, 1901 – 1mf – 9 – 0-8370-7267-0 – mf#1986-1267 – us ATLA [470]

Grammatik des biblischen und targumischen chaldaismus : fuer akademische vorlesungen / Winer, Georg Benedikt – 2., durchaus verb Aufl. Leipzig: Tr Woeller, 1842 – 1mf – 9 – 0-7905-8321-6 – (incl bibl ref) – mf#1987-6426 – us ATLA [470]

Grammatik des duala (kamerun) / Ittmann, Johannes – Nendeln, LIECHTENSTEIN . 1969 – 1r – us UF Libraries [470]

Grammatik des juedisch-palaestinischen aramaeisch : nach den idiomen des palaestinischen talmud, des onkelostargum und prophetentargum, und der jerusalemischen targume / Dalman, Gustaf – 2. Aufl, verm und vielfach umgearb. Leipzig: J C Hinrichs, 1905 – 1mf – 9 – 0-8370-9220-5 – (incl ind) – mf#1986-3220 – us ATLA [470]

Grammatik des neutestamentlichen sprachidioms *see* Treatise on the grammar of new testament greek

Grammatik des otjiherero nebst worterbuch / Viehe, G – Stuttgart, Germany. 1897 – 1r – us UF Libraries [470]

Grammatische grundmuster der modernen chinesischen umgangssprache und kontrastanalyse chinesisch-deutsch : unter einbezug valenztheoretischer fragestellungen... / Schmidt, Wolfgang G A – (mf ed 1996) – 2mf – 9 – €40.00 – 3-8267-2378-3 – mf#DHS 2378 – gw Frankfurter [410]

Grammatische untersuchungen ueber die biblische graecitaet : ueber die lesezeichen / Lipsius, Karl Heinrich Adelbert – Leipzig: J C Hinrichs, 1863 – 1mf – 9 – 0-8370-9292-2 – (incl bibl ref) – mf#1986-3292 – us ATLA [450]

Die grammatischen schulen der araber : erste abtheilung, die schulen von basra und kufa und die gemischte schule / Fluegel, Gustav – Leipzig: In Commission bei F A Brockhaus, 1862 – 1mf – 9 – 0-8370-7697-8 – (in german and arabic. no more published. incl bibl ref) – mf#1986-1697 – us ATLA [470]

Grammatography : a manual of reference to the alphabets of ancient and modern languages = Alphabete orientalischer und occidentalischer sprachen / Ballhorn, Friedrich – London: Truebner, 1861 [mf ed 1986] – 1mf – 9 – 0-8370-8082-7 – (incl ind. in english) – mf#1986-2082 – us ATLA [400]

Grammofon i fonograf – St Petersburg. n1-3 1902; n1-13, 16-17 1903; n1-12 1904; n1 1907 – 14mf – 9 – (title changed in 1905-06 to: svet i zvuk. in aug-sep 1906 to: grammofon i fotografiia) – us UMI ProQuest [780]

Grammofon i fotografiia – aug-sep 1906 [mthly] – 2mf – 9 – us UMI ProQuest [780]

Grammofon i fotografiia *see* Grammofon i fonograf

Grammofonnaia zhizn' – Moscow. n1-18 1911; n19-31 1912 [biwkly] – 14mf – 9 – us UMI ProQuest [780]

Grammofonnyi mir = Die grammofon-welt – St Petersburg. n1-12 1910; n1-23 1911; n1-20 1912; n1-20 1913; n1-14 1914; from 1915 on in sussian only n1-5,7-8 1915; n1-12 1916; n1-2,6-8 1917 [mthly] – 48mf – 9 – us UMI ProQuest [780]

Grammont, Henri Delmas de *see* Histoire d'alger sous la domination turque

Gramont, Agenor A de *see* La france et la prusse avant la guerre

Gramophone – Teddington, England 1923+ – 1,5,9 – ISSN: 0017-310X – mf#1338 – us UMI ProQuest [780]

Gramsch, Alfred *see* Goethes faust

Gran aj-jum palas / Recinos, J Humberto – Guatemala, 1964 – 1r – us UF Libraries [972]

Gran bretana mercantil e industrial – London, UK. Sept 1926 – 1 – uk British Libr Newspaper [072]

Gran burundun-burunda ha muerto / Zalamea, Jorge – Bogota, Colombia. 1960? – 1r – us UF Libraries [972]

El gran cardinal de espana, don pedro gonzalez de mendoza / La Cadena, Ramon Lacadena y Brualla – 2nd ed. Madrid, 1942 – 1 – us CRL [920]

Gran colombia y espana (1819-1822) / O'leary, Daniel Florencio – Madrid, Spain. 1919 – 1r – us UF Libraries [972]

Gran crisis y la necesidad de una confederacion pa... / Abalo, J L – Habana, Cuba. 1940 – 1r – us UF Libraries [972]

Gran desafio / Ruiz Novoa, Alberto – Bogota, Colombia. 1965 – 1r – us UF Libraries [972]

La gran extremadura, una utopia / Sanchez Morales, Narciso – Badajoz: Imp. Diput. Provincial, 1968 – sp Bibl Santa Ana [946]

Gran incognito / Barahona Jimenez, Luis – San Jose, Costa Rica. 1953 – 1r – us UF Libraries [972]

Gran literatura iberoamericana / Torres-Rioseco, Arturo – Buenos Aires, Argentina. 2nd ed. 1951 – 1r – us UF Libraries [440]

Gran matematico y fecundo poeta. arsenio gallego hernandez / Fernandez Sanchez, Teodoro – Badajoz: Dip. Provincial, 1974. Sep. REE – 1 – sp Bibl Santa Ana [510]

Gran miami – Miami, FL. 1988 apr 07-1989 jan 26 – 1r – us UF Libraries [071]

La gran noche del mundo / Perez Lozano, Jose Maria – Madrid: PPC, 1962 – 1 – sp Bibl Santa Ana [946]

Gran revolucion africana / Bayo, Armando – Habana, Cuba. 1966 – 1r – us UF Libraries [960]

Granada relocation center records, 1943-1945 – [mf ed 1992] – 2r – 1 – (with finding aid) – us UW Libraries [977]

O granadeiro – Rio de Janeiro, RJ: Typ do Diario de N L Vianna, 22 mar-16 maio 1845 – (= ser Ps 19) – mf#P14,04,28 – bl Biblioteca [320]

Granados, Anastasio *see* El cardenal goma, primado de espana. madrid, 1969

Granados, Rafael M *see* Historia de colombia

Granau, Johannes *see* Leise beschwoerung

Granbery, John Cowper *see*
- Experience, the crowning evidence of the christian religion
- Outline of new testament christology

Granby 1710-1900 – Oxford, MA (mf ed 1993) – (= ser Massachusetts vital records) – 36mf – 9 – 0-87623-171-7 – (mf 1-7,9: town records 1765-1845. mf 7-11: vital records 1710-1863. mf 12-17: tax lists 1782-1824. mf 18-29: tax lists 1824-49. mf 30: vital records 1843-60. mf 31: births 1860-93. mf 32: marriages 1860-93. mf 33: deaths 1860-93. mf 34: deaths 1894-1903. mf 35: marriages 1894-1904. mf 36: births 1894-1902) – us Archive [978]

Granby gazette and shefford county advertiser – Granby, QC. 1855-77 – 2r – 1 – ISSN: 1184-5880 – cn Library Assoc [971]

Granby leader and eastern townships record – Granby, QC. 1891-1901 – 1r – 1 – ISSN: 1487-0371 – cn Library Assoc [071]

Granby leader-mail – Granby, QC. 1901-10 – 3r – 1 – cn Library Assoc [071]

Grancini, Michel'Angelo *see* Musica ecclesiastica da capella

Grancolas, M J *see*
- Les anciennes liturgies 2. l'ancien sacramentaire de l'eglise 1
- Les anciennes liturgies 3. l'ancien sacramentaire de l'eglise 2

Le grand alphabet francois : divise par syllabes (sic) – nouv ed. Quebec: impr a la Nouvelle Imprimerie, 1806 [mf ed 1974] – 1r – 5 – mf#SEM16P152 – cn Bibl Nat [220]

Grand Army Home for Veterans [King WI] *see* Courier

Grand army journal – Washington, DC, 30 Apr 1870-15 July 1871 – 1 reel – 1 – us Western Res [355]

Grand Army of Labor *see* Overland for one dollar

Grand Army of the Republic *see*
- Proceedings of...annual encampment of the department of wisconsin, grand army of the republic
- Reveille
- Western veteran

Grand Army of the Republic. Dept of the Potomac *see* Journal of the annual encampment

Grand Army of the Republic. District of Kansas *see* Records

Grand Army of the Republic et al *see* Journal of proceedings...annual encampment

Grand Army of the Republic. Post 332. White Cloud, Kansas *see* Minutes

Grand Army of the Republic. Post No. 69 (O.M. Mitchell), Osborne, KS *see* Minutes and roster

Grand Army of the Republic Post No 187 *see* Records, ms 2758

Grand Canyon College. Phoenix, Arizona *see* Catalogs and college records

Un grand chef berbere : le caid goundafi / Justinard, Leopold Victor – Casablanca: Editions Atlantides, [1951] – 1 – us CRL [920]

Grand cinquantenaire de la st-jean-baptiste : discours des canadiens eminents fait tant au congres national qu'au banquet suivi d'une description complete de toutes les fetes – Lancaster. 1989-1989 (1) – 2mf – 9 – (incl ind) – mf#13823 – cn Canadiana [360]

Grand communication / Order of the Eastern Star – 80th [1994 jun 30/jul 2] – 1r – 1 – mf#5004674 – us WHS [366]

Grand concert pour le clavecin ou forte piano : avec l'accompagnement des plusieurs instruments, liv 1 / Mozart, Wolfgang Amadeus – Amsterdam: J Schmitt [1782?] [mf ed 19–] – 9pt on 1r – 1 – mf#film 2545 – us Sibley [780]

Grand concert under the auspices of the knights of the maccabees of the world in aid of "jabal" tent no 18 : ...petersville... december the 17th, 1878, at eight o'clock, mrs raymond, pianiste... – S.l: s.n, 1878? – 1mf – 9 – mf#58956 – cn Canadiana [790]

Grand concerto a la chasse, op 64 / Steibelt, Daniel – London: printed by Joseph Dale & Son...[180-?] [mf ed 1989] – 1r – 1 – mf#pres. film 53 – us Sibley [780]

Grand concerto for the pianoforte, op 33 / Steibelt, Daniel – London: Muzio Clementi & Co...[180-?] [mf ed 1992] – 1r – 1 – mf#pres. film 114 – us Sibley [780]

Grand county advocate *see* Grand county miscellaneous newspapers

Grand county citizen *see* Grand county miscellaneous newspapers

Grand county miscellaneous newspapers – Denver, CO (mf ed 1991) – 1r – 1 – (grand lake prospector (scattered issues 1882-86); middle park times (scattered issues 1889-1905); grand county news (jan 8 1904); grand county advocate (mar 9 1905-dec 23 1905); grand county citizen (aug 2 1912); kremmling register (may 27 1926); kremmling news (scattered issues 1912-21); fraser times (jan 1 1914)) – mf#MF Z99 G6764 – us Colorado Hist [071]

Grand county news *see* Grand county miscellaneous newspapers

Grand cwa maritime championship meet : labor day sports under the auspices of the st john bicycle and athletic club – S.l: s.n, 1896? – 1mf – 9 – mf#55376 – cn Canadiana [790]

Le grand dictionnaire historique : ou le melange curieux de l'histoire sacree et profane / Moreri, Louis – nouv dern rev corr augm ed. Paris: Chez Jean Baptiste Coignard, 1718 [mf ed 1983] – 4r – 1 – mf#SEM35P186 – cn Bibl Nat [059]

Le grand dictionnaire historique (ael1/44) / Moreri, Louis – [mf ed 1998] – (= ser Archiv der europaeischen lexikographie, abt 1: enzyklopaedien) – 532mf – 9 – €3530.00 set – 3-89131-342-X – (gesamtedition der ausgaben lyon 1681, utrecht 1692, paris 1699, paris 1718, paris 1725, basel 1731-32, amsterdam 1740, paris 1759, deutsche ausgaben: leipzig, 1709, 1730-1732; vols available individually) – gw Fischer [900]

Le grand dictionnaire historique (ael1/44.1) – 2nd ed. Lyon 1681 [mf ed 1998] – (= ser Archiv der europaeischen lexikographie, abt 1: enzyklopaedien) – 2v on 28mf – 9 – €230.00 – 3-89131-320-9 – gw Fischer [900]

Le grand dictionnaire historique (ael1/44.2) / ed by Clerc, Jean le – 6th ed. Utrecht 1692 [mf ed 1998] – (= ser Archiv der europaeischen lexikographie, abt 1: enzyklopaedien) – 4v on 25mf – 9 – €230.00 – 3-89131-321-7 – gw Fischer [900]

Le grand dictionnaire historique (ael1/44.3) / ed by Vaultier – [9th ed] Paris 1699 [mf ed 1998] – (= ser Archiv der europaeischen lexikographie, abt 1: enzyklopaedien) – 4v on 28mf – 9 – €230.00 – 3-89131-322-5 – gw Fischer [900]

Le grand dictionnaire historique (ael1/44.4) / ed by Pin, Louis-Ellies du & le Cointe, abbe – [15th ed] Paris 1718 [mf ed 1998] – (= ser Archiv der europaeischen lexikographie, abt 1: enzyklopaedien) – 5v on 53mf – 9 – €410.00 – 3-89131-323-3 – gw Fischer [900]

Le grand dictionnaire historique (ael1/44.5) / ed by Barre, Louis-Francois-Joseph de la, vailly et al – [17th ed] Paris 1725 [mf ed 1998] – (= ser Archiv der europaeischen lexikographie, abt 1: enzyklopaedien) – 6v on 65mf – 9 – €490.00 – 3-89131-324-1 – gw Fischer [900]

Le grand dictionnaire historique (ael1/44.6) / ed by Roques, Pierre – [18th ed] Basel 1731-32 [mf ed 1998] – (= ser Archiv der europaeischen lexikographie, abt 1: enzyklopaedien) – 6v on 58mf – 9 – €450.00 – 3-89131-325-X – gw Fischer [059]

Le grand dictionnaire historique (ael1/44.7) – 18. et derniere ed [22nd ed] Amsterdam 1740 [mf ed 1998] – (= ser Archiv der europaeischen lexikographie, abt 1: enzyklopaedien) – 8v on 63mf – 9 – €500.00 – 3-89131-326-8 – gw Fischer [059]

Le grand dictionnaire historique (ael1/44.8) : nouvelle edition / ed by Drouet, Etienne-Francois – [24th ed] Paris 1759 [mf ed 1998] – (= ser Archiv der europaeischen lexikographie, abt 1: enzyklopaedien) – 10v on 120mf – 9 – €690.00 – 3-89131-327-6 – gw Fischer [059]

Grand echo du nord de la france *see* Echo du nord

Grand evenement : la fete de l'immaculee-conception de la vierge marie, a rome, le 8 dec 1854 – Montreal: Senecal & Daniel, 1855 [mf ed 1994] – 1mf – 9 – 0-665-94734-8 – mf#94734 – cn Canadiana [241]

Grand forks gazette – Grand Forks, British Columbia, CN. apr 1905-dec 1970 – 26r – 1 – cn Commonwealth Imaging [071]

Grand forks miner and miner gazette – Grand Forks, British Columbia, CN. may 1898-dec 1901 – 1r – 1 – cn Commonwealth Imaging [071]

Grand forks news and news gazette – Grand Forks, British Columbia, CN. jan 1901-dec 1905 – 2r – 1 – cn Commonwealth Imaging [071]

Grand gibier et terres inconnues / Bary, Maxime De – Paris, France. 1910 – 1r – us UF Libraries [960]

Grand island anzeiger – Grand Island, NE: J P Windolph, 1889-jahrg 4 n31. 14 apr 1893 (wkly) [mf ed 1976] – 1r – 1 – (in german. merged with: herold to form grand island anzeiger und herold) – us NE Hist [071]

Grand island anzeiger und herold – Grand Island, NE: J P Windolph. jahrg 4 n32. 21 apr 1893-jahr 21 n27. 8 mar 1901 (wkly) [mf ed lacks 9 okt 1896 and 9 nov 1900 filmed 1976] – 5r – 1 – (in german. formed by the union of: grand island anzeiger and: herold. merged with: nebraska staats-anzeiger to form: nebraska staats-anzeiger und herold. jahrg 5 n17-jahrg 14 n16 not publ. on sundays publ as: sonntags-blatt des anzeiger und herold 5 apr 1896-98) – us NE Hist [071]

Grand island daily independent : [regular edition] – Grand Island, NE: J.W. Liveringhouse. v1 n1. jan 21 1884-v119 n32. nov 10 1989 (daily ex jan 1, jul 4, thanksgiving, dec 25) [mf ed 1884-1944 (gaps) 1970-[1989]] – 240r – 1 – (absorbed: doniphan index and: nebraska courier. merged with: grand island daily independent (mail ed) to form: grand island independent. on sat publ as: the grand island weekender independent jan 21 1967-sep 22 1979 and: the grand island saturday independent sep 29 1979-nov 4 1989. on sun publ as: the grand island sunday independent sep 30 1979-nov 5 1989) – us NE Hist [071]

Grand island daily independent : [regular edition] – Grand Island, NE: J W Liveringhouse. v1 n1. jan 21 1884-v119 n32. nov 10 1989 (daily ex jan 1, jul 4, thanksgiving, dec 25) – 343r – 1 – (absorbed: doniphan index and: nebraska courier; merged with: grand island daily independent (mail ed) to form: grand island independent; on saturdays publ as: the grand island weekender independent jan 21 1967-sep 22 1979 and: the grand island saturday independent sep 29 1979-nov 4 1989; on sundays publ as: the grand island sunday independent sep 30 1979-nov 5 1989) – us Bell [071]

The grand island daily independent : [mail edition] – Grand Island, NE: Grand Island Daily Independent. v82 n233. sep 30 1965-nov 10 1989// (daily ex jan 1, jul 4, thanksgiving, dec 25) – 158r – 1 – (merged with: grand island daily independent (regular ed) to form: grand island independent; on saturdays publ as: the grand island weekender independent jan 21 1967-sep 22 1979; on sundays publ as: the grand island sunday independent sep 30 1979-89) – us Bell [071]

Grand island daily press – Grand Island, NE: Augustine Bros, nov 26 1900-v6 n244. oct 11 1902 (daily ex sun) [mf ed with gaps filmed 1972] – 4r – 1 – (cont: grand island daily republican) – us NE Hist [071]

Grand island daily republican – Grand Island, NE: Seth P Mobley. v1 n1. jul 19 1897-v3 n138. jan 6 1900; ns: v1 n1. jan 8 1900-v4 n111. nov 24 1900 (daily ex sun) [mf ed with gaps filmed 1971-72] – 7r – 1 – (cont by: grand island daily press. issues for jan 8-mar 26 1900 also called old ser v3 n139. weekly ed: central nebraska republican) – us NE Hist [071]

Grand island daily times – Grand Island, NE: J S & C W Stidger, 1886 (daily) [mf ed v1 n34. jun 25 1886 filmed [1996]] – 1r – 1 – us NE Hist [071]

Grand island democrat – Grand Island, NE: Adams & Risley. v 17 n22. dec 6 1901-(wkly) [mf ed 1901-08 (gaps) filmed 1978] – 3r – 1 – (cont: democrat) – us NE Hist [071]

Grand island herald – Grand Island, NE: Herald Pub Co, sep 1930-62nd yr n87. sep 9 1942 (daily ex sun & mon) [mf ed with gaps filmed 1976] – 6r – 1 – (cont: grand island herald combined with the grand island shopping news. publ as: grand island daily herald may 11-jun 23 1933 and: grand island herald & shopper's bulletin dec 30 1938-jan 13 1939 and: grand island herald and shoppers' bulletin jan 17-feb 3 1939 and: grand island daily herald may 12-sep 9 1942. some irregularities in numbering) – us NE Hist [071]

Grand island herald – Grand Island, Hall County, NE: Henry Garn. v2 n7. sep 17 1886-(wkly) [mf ed 1886-87 (gaps) filmed 1976] – 1r – 1 – (cont: herald. german ed: herold) – us NE Hist [071]

Grand island herald – Grand Island, NE: A-H Pub Co, jul 1918-v49 n19. nov 29 1928 (wkly) [mf ed filmed 1976] – 4r – 1 – (cont: nebraska staats-anzeiger und herold. merged with: grand island shopping news, to form: grand island herald combined with the grand island shopping news) – us NE Hist [071]

Grand island herald combined with the grand island shopping news – Grand Island, NE: Herald Pub Co. v49 n20. dec 6 1928-sep 1930// (wkly) [mf ed with gaps filmed 1976] – 2r – 1 – (formed by the union of: grand island herald (1918) and: grand island shopping news. cont by: grand island herald (1930)) – us NE Hist [071]

Grand island independent – Grand Island, NE: Mr & Mrs Mobley, 1882 (semiwkly) [mf ed -1912 (gaps) filmed [1976-[80?]] – 7r – 1 – (cont: platte valley independent. absorbed: anti-monopolist) – us NE Hist [071]

Grand island journal – Grand Island, NE: W M Smith, 1890 (wkly) [mf ed v1 n50. nov 18 1892] – 1r – 1 – us NE Hist [071]

Grand island times – Grand Island, NE: Stevenson & Williams. v1 n1. jul 16 1873-(wkly) [mf ed 1873-82,1887,1889,1891-92 (gaps) filmed 1972] – 4r – 1 – (publ as: semi-weekly grand island times feb 28 1882, times dec 14 1888. daily ed: daily evening times (1873), and: grand island times (1878), and: grand island daily times (1886)) – us NE Hist [071]

Grand island times (daily edition) – Grand Island, NE: [C P R Williams] v1 n1. sep 18 1878- (daily) [mf ed -sep 20 1878 filmed 1972] – 1r – 1 – (weekly ed: grand island times) – us NE Hist [071]

Grand island weekly journal – Grand Island, NE: L J Simmons, feb 1900 (wkly) [mf ed v1 n4. mar 1-jul 19 1900 (gaps) filmed [1979?] – 1r – 1 – us NE Hist [071]

Le grand jeu – Paris. n1-3. ete 1928-automne 1930 – 1 – (reconstitution du no. 4 qui devait paraitre en 1932. collection privee. disponible en reimpression) – fr ACRPP [073]

Le grand journal : moniteur de la semaine. – Paris. 14 fevr, 3 avr 1864-25 nov 1866 – 1 – fr ACRPP [073]

Le grand journal – Montreal: Cie de publ Alpha. v1 n1 2/8 janv 1967-v2 n9 26 fevr/3 mars 1968 (wkly) [mf ed 1974] – 2r – 1 – mf#SEM35P103 – cn Bibl Nat [073]

Le grand journal illustre – Montreal: Publ Quebecor. v2 n10 4/10 mars 1968-v18 n45 12/18 oct 1985 (wkly) [mf ed 1974-85] – 33r – 1 – mf#SEM35P103 – cn Bibl Nat [073]
Grand junction daily news *see* Miscellaneous newspapers of mesa county
Grand junction morning record *see* Miscellaneous newspapers of mesa county
Grand lake prospector *see* Grand county miscellaneous newspapers
Grand march in the romance of blue beard : for the piano-forte with an accompaniment for tambourine & guitar / Kelly, Michael – London: printed for Corri, Dussek & Co [c1798] [mf ed 19–] – 1r – 1 – mf#pres. film 100 – us Sibley [780]
Das "grand ministere" leon gambettas 10. november 1881 – 26. januar 1882 : ein beitrag zur parlamentsgeschichte der dritten republik / Pfeiffer, Peter – Heidelberg, 1974 – 3mf – 9 – 3-89349-376-X – gw Frankfurter [320]
Grand mogol / Chivot, Henri – Paris, France. 1921 – 1r – us UF Libraries [440]
Grand opera house, london, ont, programme : season 1897-98, tuesday, ev'g, march 29... – S.l: s.n, 1898? – 1mf – 9 – mf#57024 – cn Canadiana [790]
Grand opera house, london, ont, programme : three nights and saturday matinee, september 7, 8, 9... – London, Ont?: Advertiser, 1899? – 1mf – 9 – mf#55749 – cn Canadiana [790]
Le grand orient de france : ses doctrines et ses actes: documents inedits / Bidegain, Jean – Paris: Libr Antisemite 1905 – 4mf – 9 – mf#vrl-125 – ne IDC [366]
Un grand peuple de l'afrique equatoriale : elements d'une monographie sur l'urundi et les warundi / Burgt, Johannes Michael M van der – Bois-le-Duc, Netherlands: Societe "L'illustration catholique," 1903 – 1 – us CRL [960]
Un grand politique catholique : carl lueger, bourgmestre de vienne / Liber – Quebec: editions de l'Action sociale catholique, 1912 [mf ed 1996] – 1mf – 9 – 0-665-80840-2 – mf#80840 – cn Canadiana [350]
Grand quatuor pour deux violons, alto & violoncelle, oeuvre 18me / Brandl, Johann Evangelist – Offenbach s/M: J Andre [181-?] [mf ed 19–] – 4pt on 1r – 1 – mf#pres. film 80 – us Sibley [780]
Grand rapids chronicle / Michigan Federation of Labor – 1898 jun 30-1903 dec 31, 1904 jan 1-1908 may 28 – 2r – 1 – mf#1112112 – us WHS [331]
Grand rapids cio news / Kent County Industrial Union Council, CIO – v1 n1-12, v2 n1-2,4-25, v3 n1-4,6-16,19,21, v4 n2-4 [1955 jul 28-dec 29, 1956 jan 12-26, feb 23-dec 24, 1957 jan 7-feb 18, mar 18-sep 16, nov 15, dec 24, 1958 feb 14-apr 25] – 1r – 1 – (continued by: grand valley afl-cio news) – mf#1770084 – us WHS [331]
The grand rapids furniture journal – 1880-1930 – 70r – 1 – $5,600.00 – (with printed index) – us UMI ProQuest [680]
Grand Rapids Historical Society *see* Newsletter
Grand Rapids Inter-Tribal Council *see* Turtle talk
Grand Rapids tribune – Grand Rapids, Wisconsin Rapids WI. 1900 apr 21-1901 dec 21, 1908 aug 26-1910 mar 30, 1910 apr 6-1911 nov 22, 1911 nov 29-1912 dec 25 – 4r – 1 – (cont: centralia enterprise and tribune; cont by: wood county tribune [wisconsin rapids, wi]) – mf#951861 – us WHS [071]
Grand Rapids tribune – Grand Rapids, Wisconsin Rapids WI. 1873 aug 30-1875 mar 31, 1875 apr 1-1876 aug 12, 1876 aug 19-1879 may 3, 1879 may 10-1882 aug 12, 1882 aug 19-1885 nov 7, 1885 nov 14-1887 may 28 – 6r – 1 – (continued by: centralia enterprise (wisconsin rapids, wi); centralia enterprise and tribune) – mf#929898 – us WHS [071]
Grand rapids tribune – Grand Rapids, Wisconsin Rapids WI. 1920 mar 1-1920 mar 20 – 1r – 1 – (cont: grand rapids leader; cont by: grand rapids daily tribune) – mf#1145849 – us WHS [071]
The grand rebel : an impression of shivaji, founder of the maratha empire / Kincaid, Dennis – London: Collins, 1937 – (= ser Samp: indian books) – us CRL [920]
Grand River. Ohio. Grand River Baptist Association *see* Church records
Grand River. Ohio. Presbytery of Grand River *see* Presbyterian church in the usa, presbytery of grand river records, 1814-1818, 1829-1870
Grand river sachem – Caledonia, ON. 1866-68 – 1r – 1 – cn Library Assoc [971]
Un grand roi du cambodge, jayavarman 7 : traduit du francais en cambodgien, par kromokar choum-mau / Coedes, George et al – Phnom-Penh: Editions de la Bibliotheque royale 1935 [mf ed 1989] – 1r with other items – 1 – (title & text in khmer; added t.p. in french) – mf#mf-10289 seam reel 002/18 [§] – us CRL [959]
Un grand romancier d'amour et d'aventure au 13e siecle. chretien de troyes. / Cohen, Gustave – Paris. 1931 – 1 – us CRL [440]
Grand rounds – [Halifax, NS: s.n.], 1876 – 9 – mf#P04282 – cn Canadiana [240]
Le grand schisme d'occident : d'apres les documents contemporains deposes aux archives secretes du vatican / Gayet, Louis – Florence: Loescher et Seeber; Berlin: S. Calvary, 1889 – 4mf – 9 – 0-8370-9062-8 – (incl bibl ref) – mf#1986-3062 – us ATLA [940]
Grand sonate pour piano, violon, violoncelle ou basson, op 69 / Steibelt, Daniel – Paris: Richault [c1819] [mf ed 1989] – 3pt on 1r – 1 – mf#pres. film 53 – us Sibley [780]
Grand trio pour piano, violon et violoncelle, oeuvre 49 / Mendelssohn-Bartholdy, Felix – Leipsic: Breitkopf & Haertel [18–] [mf ed 19–] – 1r – 1 – mf#film 2569 – us Sibley [780]
Grand trunk railway : letter of mr brassey to the hon john ross, president of the company / Brassey, Thomas, Earl – Toronto?: Leader and Patriot Office, 1856 – 1mf – 9 – mf#32349 – cn Canadiana [380]
Grand trunk railway : the shortest and most direct route between all points east and west, montreal, quebec – [s.l: s.n. 1886?] [mf ed 1987] – 1mf – 9 – 0-665-42531-7 – mf#42531 – cn Canadiana [380]
Grand Trunk Railway Company of Canada *see*
- By-laws, rules, special rules, regulations and orders
- Official programme of the tercentenary celebration of the founding by samuel de champlain in 1608 of the ancient capital of canada, historic quebec

Grand trunk railway of canada : the great international route between east and west – S.l: s.n, 1889? – 1mf – 9 – mf#18714 – cn Canadiana [380]
Grand trunk railway of canada, great western division, no 5, time tables, no 5, and special instructions for the exclusive use and guidance of employes : previous time tables to be destroyed, to take effect on monday, november 19th, 1883, at 12.35 am – S.l: s.n, 1883? – 1mf – 9 – mf#56484 – cn Canadiana [380]
Grand United Order of Galilean Fishermen *see* Fishermen's net
Grand United Order of Odd Fellows in America *see* Wisconsin weekly blade
Le grand vaincu / Cauvain, Henri – Paris: Lecoffre, 1878 [mf ed 1980] – (= ser Coll hetzel. bibliotheque d'education et de recreation) – 2v on 1mf – 9 – 0-665-04257-4 – mf#04257 – cn Canadiana [830]
Grand valley afl-cio news / Kent County Labor Council [MI] – v4 n6-8, v5 n2, v6 n3-5, v11 n4,6-8, v12 n3-12, v13 n1-12, v14 n1-5 [1958 aug 29-nov 21, 1959 feb 13, 1960 mar 23-may 25, 1965 apr 22, jun 30-aug 26, 1967 jan 19-dec 21, 1968 jan 31-may 23] – 1r – 1 – (cont: grand rapids cio news; cont by: grand valley labor news) – mf#1770136 – us WHS [331]
Grand Valley labor news / Kent County Labor Council [MI] – v14 n6-12, v15 n1-9 [1968 jun 27-dec 19, 1969 jan 30-sep 30, nov 25-dec 23] – 1r – 1 – (cont: grand valley afl-cio news) – mf#1770178 – us WHS [331]
Grand valley star and vidette – Ontario, CN. 1902- – 1r/y – 1 – Can$93.00 – cn Commonwealth Imaging [071]
La grand victoire dv tresillvstre roy de poloine, contre dayeuode duc de muldauie... – Paris, 1531 – 1mf – 9 – mf#H-8140 – ne IDC [956]
Le grand vocabulaire francois – Paris: C Panckoucke. v1-30 – 1 – $400.00 – mf#0325 – us Brook [440]
Le grand voyage du pays des hurons : situe en l'amerique vers la mer douce, es derniers confins de la nouvelle france dite canada... / Sagard, Gabriel – [s.l: s.n.] 1865. [mf ed 1984] – 8v on 1mf – 9 – mf#41197 – cn Canadiana [971]
Grand voyage, 'journey's end' / Sherriff, Robert Cedric – Paris, France. 1930 – 1r – us UF Libraries [440]
Grand-Carteret, John *see*
- L'enseigne
- La femme en culotte

Le grand-duche de berg (1806-1813) : etude sur la domination francaise en allemagne sous napoleon 1 / Schmidt, Charles – Paris, 1905 (mf ed 1992) – 4mf – 9 – €24.00 – 3-89349-021-3 – mf#DHS-AR 61 – gw Frankfurter [940]
[Le grande-] advocate – CA. 1911-1942 – 10r – 1 – $600.00 – mf#B06033 – us Library Micro [071]
Grande assemblee a levis dimanche prochain, le 15 octobre : m frechette, mp defie de s'y rendre par m l g desjardins / Desjardins, Louis Georges – S.l: s.n, 1876? – 1mf – 9 – mf#03902 – cn Canadiana [323]
La grande bataille – La France aux republicains. Red. en chef Lissagaray (1893) puis Organe republicain. Politique, financier, industriel, commercial. Dir. F. Fourcroy. Paris. Quot. puis irr. 19 janv-7 juin 1893. quelques no. en 1912-16, 1920-21, 1926-30, 1935 – 1 – fr ACRPP [073]
La grande bataille : organe de la ligue des anti-cléricaux francais – Paris. 21 aout 1898-12 avr 1899 – 1 – fr ACRPP [320]
La grande chronique de bomu / Lotar, L – Bruxelles: G van Campenhout, 1940 – 1 – us CRL [960]
La grande chronique de l'ubangi / Lotar, L – Bruxelles: G van Campenhout, 1937 – 1 – us CRL [960]
La grande colere de la mere duchene – Paris: Impr. de Beaule et Maignand, may 1849 – us CRL [944]
La grande comore / Fontoynont, Antoine Maurice & Raomandahy – Tananarive: Impr moderne de l'Emyrne, Pitot de la Beaujardiere, 1937 – 1 – us CRL [960]
La grande comore par le docteur fontoynont et raomandahy, medicin indigene – Tananarive, Imprimerie moderne de l'Emyrne, Pitot de la Beaufardiere, 1937 – us CRL [610]
Grande demonstration religieuse dans l'eglise de notre-dame de montreal... : mardi le 18 fevrier, 1868... – [Montreal: s.n.], 1868 – 1mf – 9 – 0-665-06340-7 – mf#06340 – cn Canadiana [320]
La grande encyclopedie (ael1/3) – Paris 1886-1902 [mf ed 1992] – (= ser Archiv der europaeischen lexikographie, abt 1: enzyklopaedien) – 31v on 404mf – 9 – €1340.00 – 3-89131-054-4 – gw Fischer [030]
La grande et merueilleuse, and trescruelle oppugnatio de la noble cite de rhodes... / Bourbon, J de – [Paris], 1527 – 1mf – 9 – mf#H-8235 – ne IDC [956]
La grande figure de norodom sihanouk : telle qu'elle est depointe par les documents de valeur historique decouverts dans les archives du palais royal / Sam Sary – Phnom-Penh: Impr du Palais Royal 1955 [mf ed 1989] – 1r with other items – 1 – mf#mf-10289 seam reel 005/04 [§] – us CRL [959]
La grande france – Paris. n1-4, 6-7, 13-15, 17-44. 1900-03 – 1 – (art, litterature sociale, colonies) – fr ACRPP [073]
Grande geographie de l'ile d'haiti / Gentil, Robert – Paris, France. 1896 – 1r – us UF Libraries [918]
La grande guerre ecclesiastique : la comedie infernale et les noces d'or, la suprematie ecclesiastique sur l'ordre temporel / Dessaulles, L A – Montreal?: s.n, 1873 – 2mf – 9 – mf#23883 – cn Canadiana [241]
La grande guerre et l'effort britannique : voyage des journalistes canadiens en angleterre et en france, confiance generale en une victoire prochaine... / Robillard, Charles – [Montreal ?: s.n.], 1918 [mf ed 1992] – 1mf – 9 – mf#SEM105P1664 – cn Bibl Nat [914]
[Le grande-] news – CA. 1946-1948 – 1r – 1 – $60.00 – mf#B06034 – us Library Micro [071]
Grande prairie northern tribune – Alberta, CN. jun 1932-dec 1939 – 3r – 1 – cn Commonwealth Imaging [071]
La grande revue – Montreal: Arthur Saint-Pierre. v1 n1 21 avril 1917-v1 n6 26 mai 1917 (wkly) [mf ed 1984] – 1r – 1 – mf#SEM35P194 – cn Bibl Nat [073]
La grande revue – Paris. 1898-1940 – 1 – fr ACRPP [073]
La grande revue *see* Revue de paris et de saint-petersbourg
Grande ronde sentinel – La Grande OR: McComas & Jeffery, 1868- [wkly] – 1 – us Oregon Lib [071]
Une grande route maritime canadienne en territoire canadien : ou, le canal de la baie georgienne: numero de luxe-souvenir du bulletin de la chambre de commerce du district de montreal, juin 1914 – [Montreal?: s.n.], 1914 – 2mf – 9 – 0-665-73857-9 – mf#73857 – cn Canadiana [380]
Grande sinal : revista de espiritualidade e pastoral – Petropolis, Brazil: Editora Vozes Ltda [v27-51 (1973-1997)] (bimthly) – 11r – 1 – us CRL [241]
Grande sonate pour le pianoforte / Cramer, Johann Baptist – Leipzig: C F Peters [182-] [mf ed 19–] – 1r – 1 – mf#film 1491 – us Sibley [780]
Grande sonate pour piano forte : intitule les adieux, de dussek a son ami clementi. oeuvre 4 / Dussek, Johann Ladislaus – Paris: Pleyel [1811-?] [mf ed 19–] – 1r – 1 – mf#pres. film 91 – us Sibley [780]
Grande tombola au profit du monument national : ouverture le 20 mai, 1906 : dans les salles du monument national, angle des rues dalhousie et george – [Ottawa?: Cie d'Impr d'Ottawa, 1906? – 1mf – 9 – 0-665-75771-9 – mf#75771 – cn Canadiana [971]
La grande troniciade ou itineraire de quebec a la riviere-du-loup : poeme badin / Cassegrain, Arthur – Ottawa: G E Desbarats, 1866 – 2mf – 9 – mf#33239 – cn Canadiana [810]
Grande vida de fernao dias pais / Taunay, Afonso De E – Rio de Janeiro, Brazil. 1955 – 1r – us UF Libraries [972]
Grande-Bretagne *see*
- An act to make temporary provision for the government of lower canada
- Anno primo victoriae reginae, magnae britanniae et hiberniae
- Reports of the commissioners appointed to inquire into the grievances complained of in lower canada

Grande-Bretagne. Colonial Department *see* Papers relating to the red river settlement
Grande-Bretagne. Colonial Office *see*
- Canada, papers relating to the removal of the seat of government, and to the annexation movement
- Communications between the colonial office and the governors of upper and lower canada
- Copies of any despatches from the governor-general of canada to her majesty's secretary of state for the colonies in regard to the commercial changes now under the consideration of the imperial legislature
- Copies or extracts of correspondence alluded to in lord glenelg's despatch to sir francis head, 7th september 1837
- Copy of the memorial from the board of trade at toronto to the british government regarding cheap postage
- Hudson bay company
- Returns relating to the legislative councils of lower canada

Grande-Bretagne. Monarque *see* Letters patents of the township of newton granted
Grande-Bretagne. Parliament *see*
- Copies of correspondence relative to the affairs of canada
- Copies or extracts of correspondence relative to the affairs of canada
- Correspondence relative to the affairs of canada, 1841
- Correspondence relative to the affairs of canada, 1846

Grande-Bretagne. Parliament, 1774. House of Commons *see* Debates of the house of commons in the year 1774 on the bill for making more effectual provision for the government of province of quebec
Grande-Bretagne. Parliament. House of Commons *see*
- A bill
- Bill (as amended by the committee) for uniting the legislatures of lower and upper canada
- Copy of the fourth report of the standing committee of grievances made to the assembly of lower canada
- Copy of the minutes of the evidence taken before the select committee appointed in the year 1834
- Correspondence relative to the affairs of canada
- First report from the select committee on emigration, scotland
- A further report from the committee of secrecy appointed to inquire into the conduct of robert, earl of orford...
- Rapport du comite choisi sur le gouvernement civil du canada
- Report from select committee on lower canada
- Report from the select committee on aborigines (british settlements)
- Report from the select committee on the civil government of canada
- Report from the select committee on the hudson's bay company

Grande-Bretagne. Parliament. House of Lords *see* Appendix to minutes of evidence taken before select committee of the house of lords on colonization from ireland
Grande-Bretagne. Privy Council *see* Affaire guibord
Grande-Bretagne. Privy Council Judicial Committee *see* Jugement des lords du comite judiciaire du conseil prive sur l'appel de dame henriette brown vs les cure et marguilliers de l'oeuvre et fabrique de notre-dame de montreal, au canada, prononce le 21 nov 1874
Grande-Bretagne. War Office *see* Drill and rifle instruction for the corps of rifle volunteers
Granderath, Theodor *see* Geschichte des vatikanischen konzils
Grandes amores del poeta luis llorens torres / Llorens, Washington – San Juan, Puerto Rico. 1959 – 1r – us UF Libraries [972]
Grandes de espana. primera serie : capitanes. biografia de orellana / Anonimo – Madrid: Ediciones de la Vicesecretaria de Educacion Popular, 1945 – 1 – sp Bibl Santa Ana [920]
Grandes de mexico / Palavicini, Felix Fulgencio – Mexico City? Mexico. 1948 – 1r – us UF Libraries [972]
Grandes ferias de ganado de todas clases...los dias 18, 19, 20 de junio...1960 / Logrosan. Ayuntamiento – Caceres: Tip. La Minvera, 1960 – 1 – sp Bibl Santa Ana [630]
Grandes ferias de ganados de todas clases en logrosan (caceres) – Caceres: Tip. La Minerva, 1962 – 1 – sp Bibl Santa Ana [946]

Grandes ferias de ganados de todas clases en...junio 1959 / Logrosan. Ayuntamiento – Caceres: Imprenta La Minerva, 1959 – 1 – sp Bibl Santa Ana [390]

Grandes fiestas 1976. dias 12, 13, 14, 15 y 16 de septiembre en honor del santisimo cristo, patrono de esta villa – Caceres: Imp. La Minerva, 176 – 1 – sp Bibl Santa Ana [240]

Grandes fiestas, 1976 en honor del santisimo cristo...caceres... – Caceres: Tip. La Minerva, 1975 – 1 – sp Bibl Santa Ana [240]

Grandes fiestas de san bartolome. agosto 1972 – Trujillo: Imp. Gexme, 1972 – 1 – sp Bibl Santa Ana [390]

Grandes fiestas de santiago de toda clase de ganados y generos de comercio...24, 25 y 26 de julio, 1960 / Casatejada. Ayuntamiento – Caceres: Tip. Minerva, 1960 – 1 – sp Bibl Santa Ana [630]

Grandes fiestas en honor de la santisima virgen del rosario. 1974 / Alcuescar. Ayuntamiento – Caceres: Imp. La Minerva, 1974 – 1 – sp Bibl Santa Ana [390]

Grandes fiestas en honor de la santisima virgen del rosario 1975 – Caceres: Imp. La Minerva, 1975 – 1 – sp Bibl Santa Ana [240]

Grandes fiestas en honor de la santisima virgen del rosario 1976 – Caceres: Tip. La Minerva, 1976 – 1 – sp Bibl Santa Ana [240]

Grandes fiestas en honor de la santisima virgen del rosario. septiembre octubre 1972 – Caceres: Imp. La Minerva, 1972 – 1 – sp Bibl Santa Ana [240]

Grandes fiestas en honor de la virgen de la salud – Merida: Imp. J. Rejas, 1970 – 1 – sp Bibl Santa Ana [240]

Grandes fiestas en honor de la virgen del rosario, 1977 / Alcuesca R. Ayuntamiento – Caceres: Imp. La Minerva, 1977 – sp Bibl Santa Ana [390]

Grandes fiestas en honor de san fernando. estacion de la renfe...julio, 1959 / Valencia de Alcantara. Estacion de Renfe – Caceres: Tip. El Noticiero S.L., 1959 – 1 – sp Bibl Santa Ana [390]

Grandes fiestas en honor de san juan bautista, 1974 / Herreruela. Ayuntamiento – Caceres: Tip. La Minerva, 1974 – 1 – sp Bibl Santa Ana [390]

Grandes fiestas en honor de san juan bautista 1975 – Caceres: Tip. La Minerva, 1975 – 1 – sp Bibl Santa Ana [240]

Grandes fiestas en honor del emigrante 1974 – Caceres: Tip. La Minverva, 1974 – 1 – sp Bibl Santa Ana [390]

Grandes fiestas en honor de...la virgen de la soterrana. 1972 – Trujillo: Imp. Gexme, 1972 – 1 – sp Bibl Santa Ana [240]

Grandes fiestas patronales en honor de ntra. sra. del consuelo. 1972 / Logrosan. Ayuntamiento – Caceres: Imp. La Minerva, 1972 – 1 – sp Bibl Santa Ana [390]

Grandes fiestas patronales en honor de nuestra senora del consuelo. septiembre-octubre 1973 – Caceres: Imp. La Minerva, 1973 – 1 – sp Bibl Santa Ana [240]

Grandes fiestas que la real villa de...celebra en honor de sus patronos san fulgencio y santa florentina. agosto 1973 – Caceres: Imp. La Minerva, 1973 – 1 – sp Bibl Santa Ana [390]

Les grandes lignes des migrations des bantous de la province orientale du congo belge / Moeller, A – Bruxelles: G van Campenhout, 1936 – 1 – us CRL [300]

Grandes momentos de la filosofia en cuba / Agramonte Y Pichardo, Roberto Daniel – Habana, Cuba. 1950 – 1r – us UF Libraries [100]

Grandes poetas romanticos do brasil / Ramos, Frederico Jose Da Silva – Sao Paulo, Brazil. 1949 – 1r – us UF Libraries [440]

Los grandes sabios / Munoz de la Pena, Arsenio – Madrid: Aguilar, 1964 – 1 – sp Bibl Santa Ana [946]

Grandeur et misere de la femme / Nayrac, Jean Paul – Paris: Michalon, 1905 – (= ser Les femmes [coll]) – 2mf – 9 – mf#10595 – fr Bibl Nationale [305]

Grandeza y afirmacion de mexico / Avila Camacho, Manuel – Mexico City? Mexico. 1943 – 1r – us UF Libraries [972]

Grandezas de la ciudad de dios / Moreno Maldonado, Jose – Madrid: Razon y Fe, 1927 – 1 – sp Bibl Santa Ana [910]

Grandezas y miserias de dos victorias / Caycedo, Bernardo J – Bogota, Colombia. 1951 – 1r – us UF Libraries [972]

Grandezas y miserias de la revolucion social espanola / Marti-Ibanez, Felix – Barcelona, 1937? Fiche W 1028. (Blodgett Collection of Spanish Civil War Pamphlets) – 9 – us Harvard [946]

Grandezze della citta di roma... / Mari, G – Roma, 1625 – 2mf – 9 – mf#0-1044 – ne IDC [720]

Grandfield first baptist church. grandfield, oklahoma : church records – 1916-66 – 1 – 63.81 – us Southern Baptist [242]

Grandgeorge, L *see* Saint augustin et le neo-platonisme

Grandidier, A *see*
- Ethnographie de madagascar
- Histoire physique, naturelle et politique de madagascar

Grandidier, Alfred et al *see* Collection des ouvrages anciens concernant madagascar

Grandidier, G *see* Ethnographie de madagascar

Grandidier, Guillaume *see*
- Bibliographie de madagascar
- Expressions figurees de la langue malgache

Grandin court baptist church. roanoke, virginia : church records – 1945-Sep 1966 – 1 – 77.40 – us Southern Baptist [242]

Grandin de L'eprevier, Marie-Louise *see* Prieres d'un petit enfant

Grandin, Leonce *see* Le dahomey, a l'assaut du pays des noirs

Grandin, Vital *see* Un supreme appel l'eveque du nord-ouest supplie tous les amis de la justice au canada de l'aider a proteger ses ouailles contre les tyrans d'ottawa

Grandmougin, Charles Jean *see* Esquisse sur richard wagner

Grandmoulin, J *see* Traite elementaire de droit civil egyptien indigene et mixte compare avec le droit francais, conforme aux programmes de l'ecole khediviale et des facultes francaises de droit

Grand-papa guerin / Laurencin, M – Paris, France. 1838 – 1r – us UF Libraries [440]

Grand-prairie daily herald-tribune – Alberta, CN. 1964- – 12r/y – 1 – Can$1065.00 – cn Commonwealth Imaging [071]

Grandpre, Gustave *see* Promenade au croisic, suivie d'iseul et almanzor, ou la grotte a madame

Grandpre, Louis Marie Joseph O'hier, comte de *see* Voyage a la cote occidentale d'afrique fait dans les annees 1786 et 1787

Les grands convertis : m. paul bourget, m. j.-k. huysmans, m. brunetiere, m. coppee / Sageret, Jules – Paris: Societe du Mercure de France, 1906 – 1mf – 9 – 0-8370-8302-8 – mf#1986-2302 – us ATLA [240]

Les grands ecrivains francais / Sainte-Beuve, Charles Augustin – Paris. v1-23. 1828 – 9 – $605.00 – mf#0527 – us Brook [920]

Grands fiestas en honor del santisimo cristo patrono de esta villa – Caceres: Imp. La Minerva, 1972 – 1 – sp Bibl Santa Ana [240]

Grands germand herald – Philadelphia, PA., 1840 – 13 – $25.00r – us IMR [071]

Les grands hommes de la france : navigateurs / Goepp, Edouard & Cordier, Emile L – Paris: P DuCrocq, 1873 – 5mf – 9 – mf#06328 – cn Canadiana [910]

Grands inities / Schure, Edouard – Paris, France. 1927 – 1r – us UF Libraries [025]

Les grands pretres – S.l: s.n, 1882? – 1mf – 9 – mf#03498 – cn Canadiana [920]

Les grands voyageurs contemporains : ouvrage contenant cent quarante et un dessins gravees sur bois, trente et un portraits de voyageurs et vingt-sept cartes itineraires / Meissas, G – Paris: Hatchette, 1894 – 1 – us CRL [910]

Grandsire, N A *see* Vingt jours de route, et genealogie historique de la famille des coches, messageries, diligences, voitures publiques, malle-postes, etc

Grandval, Marie Felicie Clemence de Reiset, vicomtesse de *see* Mazeppa

Grane, William Leighton *see* Church divisions and christianity

Granet, Marcel *see* La polyginie sororale et le sororat dans la chine feodale

Grange bulletin – Toronto: [s.n. 1881?-18– or 19–] – 9 – mf#1190-657X – cn Canadiana [630]

Grange, Eugene *see* Pauline

Grange, John *see* The golden aphroditis and granges garden of verse

Grangemouth advertiser – Grangemouth: F Johnston & Co Ltd 1979- (wkly) [mf ed 7 jan 1998-] – 1 – (cont: grangemouth advertiser and eastern district chronicle) – uk Scotland NatLib [072]

Grangemouth advertiser and eastern district chronicle – [Scotland] Grangemouth: W Glen & Co 29 dec 1900-dec 1950 (wkly) [mf ed 2003] – 27r – 1 – (missing: 1902; cont by: grangemouth advertiser) – uk Newsplan [072]

Grangemouth gazette and falkirk news – [Scotland] Grangemouth: G G Stephen 10 jun 1876-8 nov 1879 (wkly) [mf ed 2003] – 2r – 1 – (cont: grangemouth gazette and stirling and linlithgowshires advertiser [jun 1876-13 jul 1878]) – uk Newsplan [072]

Granger – Auburn, NE: Dundas & Wheeldon. 24v. v19 n3. jan 15 1892-v42 n52. dec 28 1915 (wkly) – 10r – 1 – (cont: nemaha county granger; absorbed by: nemaha county republican (auburn ne)) – us Bell [071]

Granger : devoted to the interests of patrons of husbandry in canada – London, Ont: [s.n. 1875-1876] – 9 – (cont by: the canadian granger) – mf#P04439 – cn Canadiana [636]

Granger, Frank *see* The worship of the romans

Granger, Frank Stephen *see* The worship of the romans

Grani – Arbeitslager der Staatlosen Moenchehof bei Kassel: Posev 1946- [mf ed Bloomington IN: Indiana Uni Lib, Preservation Dept 1989] – 13r – 1 – us Indiana Preservation [073]

Grani – Zhurnal Literatury, Iskusstvo; Obshchestvennoi Mysli. Limburg-Lahn Western Germany etc. v. 1-62. July 1946-1966 – 1 – us NY Public [700]

Granier De Cassagnac, Adolphe *see* Histoire de la chute du roi louis-philippe

Granier, Gerhard *see*
- Nachlass hans von seeckt (bestand n 247) bd 19
- Nachlass kurt von schleicher (bestand n 42) bd 17
- Nachlass magnus von levetzow (bestand n 239) bd 21

Granier, Jacquelin P *see* Human rights, the helsinki accords and the united states

Granier, Jacqueline P *see* Human rights, european politics, and the helsinki accord

Granite City. Illinois. First Baptist Church *see* Newspaper

Granite cutters' journal – 1877 may, 1884 jul, nov, 1885 apr-1886 apr, 1887 mar-1896 dec, 1897 jan-1903 mar, 1903 apr-1910 mar, 1931 sep-1939 dec – 4r – 1 – mf#780704 – us WHS [670]

Granite mining journal *see* Chaffee county miscellaneous newspapers

Granite monthly : a magazine of literature, history and state progress – v15 [1893] – 1r – 1 – (continued by: new hampshire, the granite state monthly) – mf#2800170 – us WHS [071]

Granite state monthly – Killen TX 1877-1930 – 1 – mf#2895 – us UMI ProQuest [071]

[Granite-] times – NV. 17 apr, 1 may 1908 [wkly] – 1r – 1 – $60.00 – mf#U04569 – us Library Micro [071]

Granitstein, Moses *see*
- Eybike tfise
- Mentsh-simfonie

Grannan, Charles P *see* Questions d'ecriture sainte

Grano de arena / Aristeguieta Rojas, Francisco De Paula – Caracas, Venezuela. 1974 – 1r – us UF Libraries [972]

Granos de arena / Sanchez-Arjona, Vicente – Sevilla: Imp. Carlos Acuna, Tomo 3. 1953 – 1 – sp Bibl Santa Ana [810]

Granos de arenas / Sanchez-Arjona, Vicente – Sevilla: Graficas La Gavidia, Tomo 1. 1952 – 1 – sp Bibl Santa Ana [810]

Granos de arenas / Sanchez-Arjona, Vicente – Sevilla: Graficas Tirvia, Tomo 2. 1953 – 1 – sp Bibl Santa Ana [810]

Granott, Abraham *see* Problemot shel hapolitikah ha-karka'it be-erets-yisrael

Gransden, little – bishop's transcripts 1600-1754 – (= ser Cambridgeshire parish register transcript) – 1mf – 9 – £1.25 – uk CambsFHS [929]

Grant, A J *see* Europe in the nineteenth century (1789-1914)

Grant advocate – Bloomington WI. 1873 jun 18-1874 sep 9 – 1r – 1 – mf#957441 – us WHS [071]

Grant, Alexander Henley *see* The church seasons

Grant, Andrew *see* History of brazil

Grant, Anthony *see* The past and prospective extension of the gospel by missions to the heathen

Grant, Asahel *see* The nestorians

Grant, Chapman *see* Herpetology of the cayman islands

Grant, Charles *see* History of mauritius or the isle of france and the neighbouring islands

Grant county advocate – Lancaster WI. 1875 jan 13-1877 dec 19 – 1r – 1 – (continued by: grant county argus) – mf#930460 – us WHS [071]

Grant county argus – Lancaster WI. 1876 jul 31-1878 apr 15 – 1r – 1 – (cont: grant county advocate; cont by: grant county gazette) – mf#930462 – us WHS [071]

Grant county blue mountain eagle – John Day; Canyon City OR: Eagle Pub Co, 1948-72 [wkly] – 1 – (cont: blue mountain eagle; absorbed by: john day valley ranger (1931-48); cont by: blue mountain eagle (1972-)) – us Oregon Lib; us Oregon Hist [071]

Grant county clarion – Lancaster WI. 1961 sep 11-1962 may 28 – 1r – 1 – mf#933392 – us WHS [071]

Grant county democrat – Muscoda WI. 1906 oct 18-1909 apr 9, 1909 apr 16-1910 oct 28, 1910 nov 4-1912 apr 19, 1912 apr 26-1913 oct 17, 1913 oct 24-1915 apr 16, 1915 apr 23-1917 jan 26, 1917 feb 2-1918 aug 23, 1918 aug 30-1919 jan 10 – 8r – 1 – (cont: valley voice; cont by: muscoda progressive [muscoda, wi: 1913]; muscoda progressive and grant county democrat) – mf#1138893 – us WHS [071]

Grant county democrat – Platteville WI. 1885 jul 16-1887 dec 29 – 1r – 1 – (continued by: grant county news (platteville, wi)) – mf#1001530 – us WHS [071]

Grant county democrat – Prairie Du Chien WI. 1870 oct 7-1873 apr 28 – 1r – 1 – mf#930472 – us WHS [071]

Grant county express – Canyon City OR: Express Pub Co [wkly] – 1 – us Oregon Lib [071]

Grant county gazette – Lancaster WI. 1978 jun 5-1879 dec 19 – 1r – 1 – (cont: grant county argus; cont by: inter county gazette) – mf#930464 – us WHS [071]

Grant county herald – Lancaster WI. 1850 sep 26/dec-1967 jan 4/1968 jun 26 [gaps] – 56r – 1 – (continued by: grant county independent; grant county herald independent) – mf#1133857 – us WHS [071]

Grant county herald – Lancaster WI. [1843 mar 18-1844 oct 5] – 1r – 1 – (continued by: wisconsin herald and grant county advertiser) – mf#1139533 – us WHS [071]

Grant county herald independent – Lancaster WI. 1968 jul 4/1969 apr 3-2001 jul-dec [gaps] – 69r – 1 – (cont: grant county herald [lancaster, wi: 1850]; grant county independent; bloomington record; cassville american) – mf#1107412 – us WHS [071]

Grant county historical society offers you here and there in grant county – 1981 feb.-1983 feb – 1r – 1 – (continued by: here & there in grant county) – mf#944322 – us WHS [978]

Grant county independent – Lancaster, Muscoda WI. 1935 sep 5-1936 aug 27, 1936 sep 3-1937 apr 29, 1937 may 6-1938 mar 31, 1938 aug 7-1939 aug 31, 1939 sep 7-1941 sep 25, 1941 oct 2-1943 oct 14, 1943 oct 21-1945, 1946-65, 1966 nov 24-1967 sep 21, 1966-1966 nov 17, 1967 sep 28-1968 jun 27 – 20r – 1 – (continued by: grant county herald (lancaster, wi: 1850); grant county herald independent) – mf#1107407 – us WHS [071]

Grant county journal – Prairie City OR: Gilman Bros, -1937 [wkly] – 1 – (cont by: journal (1937-1942)) – us Oregon Lib [071]

Grant county news – Platteville WI. 1890 jan 25-dec, 1891-93, 1894-1897 jul 14, 1897 jul 21-1900 sep 19, 1900 sep 26-1903, 1904-48, 1949-1952 jul – 23r – 1 – (cont: grant county democrat [platteville, wi]; cont by: platteville journal [platteville, wi: 1899]; platteville journal and grant county news) – mf#1001529 – us WHS [071]

Grant county news – Canyon City OR: S H Shepherd, 1879-1908 [wkly] – 1 – (cont: grant county times; absorbed by: blue mountain eagle) – us Oregon Lib; us Oregon Hist [071]

Grant county news – Hyannis, NE: John and Gloria Barkley. v1 n1. oct 1 1970- (wkly) [mf ed with gaps filmed 1978-] – 1 – (vol numbering dropped with n12 dec 13 1984; resumed with v99 n21 [ie 22] feb 8 1990 cont the numbering of grant county tribune (1897)) – us NE Hist [071]

Grant county times – Canyon City OR: H G Guild, [wkly] – 1 – (cont by: grant county news) – us Oregon Lib [071]

Grant county tribune – Hyannis, NE: P M Alwood. v8 n48. feb 26 1897-80th yr n13. aug 28 1969 (wkly) [mf ed with gaps filmed 1971-[79]] – 23r – 1 – (cont: grant county tribune and live stock journal. issue numbering dropped with 57th yr feb 6 1946; resumed with 62nd yr n1 may 4 1950) – us NE Hist [071]

The grant county tribune – Hyannis, NE: Cushman & Alwood, 1889-v2 n43. feb 19 1891 (wkly) [mf ed 1890-91 (gaps) filmed 1971] – 1r – 1 – (cont by: grant county tribune and live stock journal) – us NE Hist [071]

Grant county tribune and live stock journal – Hyannis, NE: Cushman & Alwood. 7v. v2 n44. feb 26 1891-v8 n47. feb 18 1897 (wkly) [mf ed with gaps filmed 1971] – 2r – 1 – (cont: grant county tribune. cont by: grant county tribune (1897)) – us NE Hist [071]

Grant county witness – Lancaster, Platteville WI. 1859 may 26-1862 may 8, 1862 may 9-1863, 1864 jan 7-1868 jun 4, 1868 jun 11-1870, 1871-98, 1899-1900, 1901-1902 jun, 1902 jul-1903, 1904-1905 jun, 1905 jul-1906 feb 14 – 17r – 1 – (continued by: platteville witness and mining times) – mf#987249 – us WHS [071]

Grant, Daniel *see* Land tenure in ireland

Grant, David *see* Sermons

Grant Duff, Mountstuart Elphinstone *see* Foreign policy

Grant, Duncan *see* Publication of the gospel, first at jerusalem

Grant, Elihu *see* The peasantry of palestine

Grant family magazine – 1900 feb-1901 – 1r – 1 – mf#2697739 – us WHS [640]

Grant, George *see* Account book

Grant, George Monro *see*
- Advantages of imperial federation
- Ocean to ocean
- Our picturesque northern neighbor
- Picturesque canada
- Principal grant's letters on prohibition
- The religions of the world

Grant, George Munro *see*
- French canadian life and character
- Picturesque quebec microform

Grant, Helen *see* Scottish women's protestant union

Grant, Henry *see*
- Ireland's hour

Grant, J A *see* The botany of the speke and grant expedition

Grant, James *see*
- Impressions of ireland and the irish
- Jakob grant's koenigl grossbritannischen schiffs-lieutenant's bericht von einer entdeckungs-reise nach neu-sued-wallis
- Verhaal van eene ontdekkingsreize, na nieuw-zuid-wales

Grant, James Augustus *see*
- Colonial discourses, series 2
- Private note book, 1858-1863

Grant, James Charles *see* Speech of james charles grant, esquire on the inexpediency of an elective council

Grant, Joanne *see* Black protest

Grant, Jonathan S et al *see* Cambodia

Grant, Kenneth James *see* My missionary memories

Grant, Miles *see* Positive theology

Grant, Paul *see* Kweli times

Grant, Percy Stickney *see* Socialism and christianity

Grant, President *see* The agency of a.b. steinberger in the samoan islands

Grant reception – s.l, s.l? 193-? – 1r – us UF Libraries [978]

Grant, Robert *see*
- History of physical astronomy from the earliest ages to the middle of the 19th century
- A sketch of the history of the east-india company

Grant, Ulysses *see* Papers

Le grant voyage de hierusalem diuise en deux parties / Le Huen, N – Paris, 1522 – 5mf – 9 – mf#H-8229 – ne IDC [915]

Grant, W L *see* Acts of the privy council of england

Grant, William John *see* The spirit of india

Grant, William Lawson *see* Canadian constitutional development

Granta – London, England 1987+ – 1,5,9 – ISSN: 0017-3231 – mf#16634 – us UMI ProQuest [071]

Grantchester st andrew & st mary 1539-1950 – 8mf – 9 – £10.00 – uk CambsFHS [929]

Grantham, David B *see* Parents successfully separating from adolescents: a study in the phenomenon of separation

Grantham journal – Grantham, Lincolnshire, feb 1854-to date [mf ed feb 1854-jun 1855, 1871, jan-jun 1897; 1986-] – 1 – (began as grantham journal of useful instructive and entertaining knowledge and monthly advertiser; then grantham journal and mercantile advertiser n4 may 1854-n18 jul 1855, then grantham journal, wkly, n19 7 jul 1855-to date) – uk Newsplan [072]

Grantham post and lincoln leicester and notts telegraph – [East Midlands] Lincolnshire 25 mar 1882-8 sep 1883 [mf ed 2002] – 2r – 1 – (discontinued) – uk Newsplan [072]

Granton news – Granton WI. 1912 oct 18 – 1r – 1 – (continued by: neillsville times; republican and press; neillsville press) – mf#916189 – us WHS [071]

Grantown supplement – Grantown-on-Spey: Angus Stuart 1881-1914 (wkly) [mf ed 1891-95] – 1 – (cont by: strathspey news and grantown-on-spey supplement) – uk Scotland NatLib [072]

Grant's cases / Pennsylvania. Superior Court – v1-3. 1814-63 (all publ) – (= ser Pre-nrs nominative reports) – 20mf – 9 – $30.00 – mf#LLMC 84-196 – us LLMC [340]

Grant's ontario error and appeal reports / Ontario. Canada – v1-3. 1846-66 (all publ) – 16mf – 9 – $24.00 – mf#LLMC 81-044 – us LLMC [340]

Grants pass bulletin – Grants Pass OR: Bulletin Pub Co, 1960-64 [wkly] – 1 – (cont: bulletin (grants pass, or); cont by: bulletin (grants pass, or: 1964)) – us Oregon Lib [071]

Grants pass bulletin – Grants Pass OR: D L Ewing, 1927-49 [wkly] – 1 – (merged with: rogue river record and: gold hill nugget, to form: bulletin (grants pass or). cont: southern oregon spokesman (1924-27)) – us Oregon Lib [071]

Grants pass bulletin *see* Gold hill nugget

Grants pass courier – Grants Pass OR: A E Voorhies, 1934-41 [daily ex sun] – 1 – (cont: daily grants pass courier (1931-34). absorbed: illinois valley courier (1935); cont by: grants pass daily courier (1941-)) – us Oregon Lib [071]

Grants pass courier *see* Illinois valley courier

Grants pass daily courier – Grants Pass OR: A E Voorhies, 1941- [daily ex sun & hols] – 1 – (cont: grants pass courier (grants pass, or)) – us Oregon Lib [071]

Grants pass daily courier – Grants Pass OR: A E Voorhies, 1919-31 [daily ex sun] – 1 – (cont: rogue river courier (-1918); cont by: daily grants pass courier (1931-34)) – us Oregon Lib [071]

Grants pass daily courier – Grants Pass, OR: A E Voorhies. v9 n83-v21 n207. jan 2 1919-may 21 1931 – 1 – (cont: rogue river courier; cont by: daily grants pass courier; aka: daily courier) – us Oregon Hist [071]

Grant's petersburg progress – Petersburg VA. 1865 apr 3 – 1r – 1 – mf#881664 – us WHS [071]

Grant's upper canada chancery reports / Ontario. Canada – v1-29. 1849-82 (all publ) – 226mf – 9 – $339.00 – mf#LLMC 81-045 – us LLMC [340]

Grantsmanship center news : the grantsmanship center – Los Angeles CA 1979-85 – 1,5,9 – ISSN: 0364-3115 – mf#12526 – us UMI ProQuest [378]

[Grantsville-] bonanza – NV. 7 may, 30 jun 1881 [wkly] – 1r – 1 – $60.00 – mf#U04570 – us Library Micro [071]

Granucci, N *see* ...L'eremita, la carcere, e l' diporto

Granville 1732-1902 – Oxford, MA (mf ed 1988) – (= ser Massachusetts vital records) – 32mf – 9 – 0-87623-089-3 – (mf 1-3: town & vital records 1751-96. mf 4-7: town records 1751-88. mf 7-8: marriages & intentions 1787-99. mf 9-13: town & vital records 1757-1836. mf 14-16: baptisms, marriages, deaths 1739-1861. mf 17: estate valuations in 1853. mf 18: rebellions 1861-65. mf 19-20: index: b,m,d 1855-90. mf 21-22: b,m,d 1843-58. mf 23-27: b,m,d 1855-90. mf 28: intentions of marriage 1906-40. mf 29-31: deaths 1891-1951. mf 32: marriages & births 1891-1902) – us Archive [978]

Granville 1735-1849 – Oxford, MA (mf ed 1996) – (= ser Massachusetts vital record transcripts to 1850) – 6mf – 9 – 0-87623-256-X – (mf 1t-2t: vital records 1735-99. mf 2t-3t: marriage & intentions 1787-99. mf 3t: births & deaths 1757-1836. mf 3t-6t: marriages & intentions 1796-1831. mf 6t: births 1792-1804, 1843-49; marriages 1843-49; deaths 1843-49) – us Archive [978]

Granville, Joseph Mortimer *see* The secret of a good memory

Granville sharp papers : from gloucestershire record office – [mf ed Marlborough, 1996] – (= ser Abolition and emancipation 4) – 30r – 1 – $3850.00 – (with guide) – uk Matthew [976]

Grao de areia e estudos brasileiros / Amado, Gilberto – Sao Paulo, Brazil. 1948 – 1r – us UF Libraries [972]

Grape growing in florida / Dickey, R D – Gainesville, FL. 1938 – 1r – us UF Libraries [634]

Grape growing in florida / Dickey, R D – Gainesville, FL. 1947 – 1r – us UF Libraries [634]

Grapes of wrath – v1-3 n7 [1972 nov20-1975 summer] – 1r – 1 – mf#1112114 – us WHS [071]

Grapevine / Central State Hospital [WI] – [1973-75, scattered iss] – 1r – 1 – (cont: patient's pen; cont by: ideas & insights) – mf#707392 – us WHS [362]

Grapevine – San Vito Dei Normanni As. 1981 may 1-1984 dec 21 – 1r – 1 – (cont: san vito grapevine) – mf#1099033 – us WHS [071]

Grapevine / Franklin-Lakeside Neighborhood Association (Madison (WI)) – iss 1-iss 10, iss 12-iss 19, [1979 mar/apr-1981 nov, [1981 feb]-1982 may, 1983 mar] – 1r – 1 – mf#669876 – us WHS [366]

Grapevine : newsletter of the movement for a new society / Movement for a New Society – 1983 mar-1986 jan, 1986 feb-1988 jun – 2r – 1 – (lacks: 1987 apr; cont: dandelion wine) – mf#1010144 – us WHS [366]

Grapevine : official publication of the state of nevada employee's association / State of Nevada Employee's Association – v8 n2,5-v12 n3 [1976 mar 5, aug 2-1980 may 2] – 1r – 1 – (continued by: state employee news) – mf#1124988 – us WHS [366]

Grapheus, C *see*
- Spectaculorum in susceptione philippi hisp. prin. divi caroli. 5...
- La tres admirable, tres magnificque, et triumphante entree, du treshault et trespuissant prince philipes...

Graphic *see* Wheatland newspapers

The graphic – Chicago, IL. 4 Jul 1891-30 Jun 1894 – 3r – 1 – uk British Libr Newspaper [071]

The graphic – London, England. Jan 1873-Dec 1874; Jan 1877-Jun 1878 – 7r – 1 – uk British Libr Newspaper [072]

Graphic art / Victoria and Albert Museum. London – (= ser Fine art and design in the victoria and albert museum) – 165mf – 9 – $1225.00 – 0-907006-40-X – (with almost 10,000 reproductions; with printed ind) – uk Mindata [760]

Graphic art in british guiana / Roth, Vincent – Georgetown, Guyana. 1949 – 1r – us UF Libraries [740]

The graphic arts / Hamerton, Philip Gilbert – London 1882 – (= ser 19th c art & architecture) – 7mf – 9 – mf#4.2.134 – uk Chadwyck [760]

The graphic arts : a magazine for printers and users of printing – Boston, Mass.: National Arts Pub Co, [1911-]. v4. jul 1912-jun 1913 – us CRL [680]

Graphic arts abstracts – Sewickley PA 1947-88 – 1,5,9 – (cont by: gatfworld) – ISSN: 0017-3282 – mf#9929 – us UMI ProQuest [740]

Graphic Arts International Union *see*
- Bookbinders' bulletin
- Union tabloid

Graphic arts monthly – New York NY 1988+ – 1,5,9 – (cont: graphic arts monthly & the printing industry) – ISSN: 1047-9325 – mf#398,01 – us UMI ProQuest [680]

Graphic arts monthly and the printing industry – New York NY 1929-86 – 1 – (cont by: graphic arts monthly) – ISSN: 0017-3312 – mf#398.01 – us UMI ProQuest [680]

Graphic arts unionist – Washington DC 1964-77 – 1,5,9 – ISSN: 0017-3363 – mf#3465 – us UMI ProQuest [331]

Graphic Communications International Union *see*
- Gciu 777 bulletin
- Gciu news local 583
- Southern california graphic arts unions monthly

Graphic history of the south african war, 1899-1900 / Huyshe, Wentworth – London, England. 1900 – 1r – us UF Libraries [960]

Graphic illustrations : with historical and descriptive accounts, of toddington / Britton, John – London [1840] – (= ser 19th c art & architecture) – 3mf – 9 – mf#4.2.1406 – uk Chadwyck [740]

Graphic (newberg, or) – Newberg OR: Newberg Graphic, 1993-98 [semiwkly] – 1 – (cont: newberg graphic; cont by: newberg graphic (newberg, or)) – us Oregon Lib [071]

Graphic news republican / Harding Co. Kenton – aug 1911-feb 1920 [wkly] – 4r – 1 – mf#B9172-9175 – us Ohio Hist [071]

Graphic scenes in african story : settlers – slavery – missions and missionaries – battle-fields / Bruce, Charles – Edinburgh: W P Nimmo, Hay & Mitchell, [188-?] – 1mf – 9 – 0-8370-6248-9 – mf#1986-0248 – us ATLA [960]

Graphical and literary illustrations of fonthill abbey, wiltshire / Britton, John – London 1823 – (= ser 19th c art & architecture) – 2mf – 9 – mf#4.2.1249 – uk Chadwyck [740]

Graphic-mirror / Preble Co. New Paris – jan 1958-aug 1962 [wkly] – 2r – 1 – mf#B32394-32395 – us Ohio Hist [071]

Graphic-news / Harding Co. Kenton – 1889, 1902, feb 1903-jul 1911 [wkly] – 6r – 1 – mf#B9034-9039 – us Ohio Hist [071]

Graphicommunicator : the newspaper of the graphic communications union – 1983 jul,oct-1985 dec; 1986 feb-1994 nov/dec – 2r – 1 – (cont: union tabloid) – mf#1028207 – us WHS [740]

Graphic-review *see* Newport graphic-review

The graphics : western graphic – San Francisco, CA. v6-7 1899; v9 july-dec 1900; v33-34. 4 jun 1910-27 may 1911 – 3r – 1 – $150.00 – mf#B06102 – us Library Micro [978]

Graphis – v1-26. 1944-70 – 1 – us AMS Press [700]

Die graphischen kuenste – v1-22. 1879-99 – 1 – $252.00 – mf#0247 – us Brook [760]

Graphitologie / Coupal, Louis – Montreal: [s.n], c1920 – 1mf – 9 – 0-659-90909-X – mf#9-90909 – cn Canadiana [810]

Graphs and combinatorics – Dordrecht, Netherlands 1989+ – 1,5,9 – ISSN: 0911-0119 – mf#16998 – us UMI ProQuest [510]

Grapow, H *see* Religioese urkunden

Grapow, Hermann et al *see* Textbuch zur religionsgeschichte

Grappe, Georges *see* J-H newman

Grappenhall, st wilfrid – [North Cheshire FHS] – (= ser Cheshire monumental inscriptions) – 2mf – 9 – £3.25 – mf#121 – uk CheshireFHS [929]

Grappin, Pierre *see* La theorie de genie dans le preclassicisme allemand

Grasberger, Hans *see*
- Die naturgeschichte des schnaderhuepfels
- Die schoene kastellanin / maria buch

Grasedieck, Brunhild *see* "Mary hamilton"

Grases, Pedro *see* Resumen de la historia de venezuela de andres bell

Graslitzer grenzbote – Graslitz (Kraslice CZ), 1933 5 oct-14 dec – 1r – 1 – gw Misc Inst [077]

Grass and forage science : the journal of the british grassland society / British Grassland Society – Oxford, England 1979+ – 1,5,9 – ISSN: 0142-5242 – mf#15548,01 – us UMI ProQuest [630]

Grass, K *see* Vom juedischen kriege

Grass, Karl *see* Sizilische reise

Grass, Karl Konrad *see*
- Die russischen sekten
- Zur lehre von der gottheit jesu christi

Grass, L I *see* Strakhovanie sel'skokhoziaistvennykh posevov ot neurozhaia

Grass roots – 1974 feb-1978 jun/jul – 1r – 1 – mf#369058 – us WHS [071]

Grass roots – v1 n1-2 [1975 sep 10-oct 6], n3-4 [1983 mar-may] – 1r – 1 – (continued by: siskiyou country) – mf#1653065 – us WHS [071]

Grass roots : national newsletter of the people's party / Populist Party (US) – v1 n3-v4 n10 [1972 mar-1975 dec] – 1r – 1 – mf#369066 – us WHS [325]

Grass roots campaigning – 1984 oct-1992 sep – 1r – 1 – mf#2478115 – us WHS [071]

Grass roots forum – 1967 mar 11-1980 mar – 1r – 1 – mf#157529 – us WHS [071]

Grass roots forum – San Gabriel CA 1967-80 – 1 – ISSN: 0017-3517 – mf#7379 – us UMI ProQuest [320]

Grass roots news – v3 n3-5 [1973 aug-1974 apr/may] – 1r – 1 – mf#369062 – us WHS [071]

Grass roots of indianapolis – 1971 mar 31-aug – 1r – 1 – mf#486849 – us WHS [071]

Grass roots pachyderm – v1 n3-v2 n4 [1975 spring-1976 spring] – 1r – 1 – mf#369070 – us WHS [071]

[Grass valley-] foothill weekly tidings – CA. 1877-1904 – 8r – 1 – $480.00 – (aka: evening telegraph; weekly telegraph; weekly tidings) – mf#C03601 – us Library Micro [071]

[Grass valley-] grass valley telegraph – CA. Jul 1853-Oct 1858 (broken file) – 1r – 1 – $60.00 – mf#C02276 – us Library Micro [071]

Grass valley journal – Grass Valley OR: Journal Pub Co, -1931 [wkly] – 1 – (merged with: sherman county observer (1897-1931) to form: sherman county journal (1931-)) – us Oregon Lib [071]

Grass valley journal *see* Sherman county journal

[Grass valley-] nevada nation – CA. 1859-1860; 1861-1862; 1863-1865 – 5r – 1 – $300.00 – (aka: nevada national; grass valley national; daily national) – mf#C03602 – us Library Micro [071]

[Grass valley-] the union – CA. 1865-66 (broken file), 1867-89; 1890- – 442r – 1 – $26,520.00 (subs $350y) – (aka: daily morning union; morning union; morning daily union and herald) – mf#B02277 – us Library Micro [071]

Grass valley/nevada city – 1992- – (= ser California telephone directory coll) – 3r – 1 – $150.00 – mf#P00034 – us Library Micro [917]

Grasses, forage plants, tomato blight / Rolfs, P H – Lake City, FL. 1893 – 1r – us UF Libraries [630]

Grasses of north america, v1 : chapters on the physiology, composition, selection, improving and cultivation of grasses, management of grass lands... / Beal, William James – New York: H Holt, 1896, c1887 – v1 on 6mf – 9 – mf#05301 – cn Canadiana [630]

Grasses of north america, v2 : the grasses classified, described and each genus illustrated... / Beal, William James – New York: H Holt, 1896 – v2 on 8mf – 9 – mf#05302 – cn Canadiana [580]

Grasses of north america, vols 1 and 2 : the grasses classified, described and each genus illustrated, with chapters on their geographical distribution and a bibliography / Beal, William James – New York: H Holt, 1896 – 2v on 1mf – 9 – mf#05300 – cn Canadiana [580]

Grasset S Sauveur *see* Encyclopedie des voyages

Grasse-Tilly, Francois Joseph Paul, marquis de *see* Memoire du comte de grasse sur le combat naval du 12 avril 1782

Grasset-Saint-Sauveur, Andre *see*
- Beschreibung der ehemaligen venetianischen besitzungen auf dem festen lande und an den kuesten von griechenland
- Travels through the balearic and pithiusian islands

Grasset-Saint-Saveur, J *see* Encyclopedie des voyages

Grasshof, Fritz *see* Halunkenpostille

Grassi, Alfio de *see* Addresses

Grassland news – Thedford, NE: Roy & Irene Alleman. 2v. 1953-v2 n45. sep 22 1955 (wkly) [mf ed v1 n9. jan 7 1954-sep 22 1955 filmed 1979] – 1r – 1 – us NE Hist [071]

Grassmann, Gottfried Ludolf *see* Neue berliner beytraege zur landwirtschaftswissenschaft

Grassmann, Robert *see* Die verfluchungen und beschimpfungen des herrn christus und der christen durch die paepste, bischoefe und priester der roemischen kriche und die pflicht jedes christen diesen versuchungen gegenueber

Grasso, Andrew T *see* The relationship of competitive state anxiety and athletic performance in high school basketball players

Grassroots – Berkeley CA 1973-81 – 1 – mf#9759 – us UMI ProQuest [071]

Grassroots – Cape Town SA, 1 mar 1980- 31 dec 1987 – 1r – 1 – sa National [960]
Grassroots / Grassroots Collective (Berkeley, CA) – Berkeley CA. 1974 aug 21/sep 1, oct 16/29, 1978 jan 11-1982 feb 9, 1982 feb 10-1986 jun 10 – 3r – 1 – mf#369067 – us WHS [071]
Grassroots Collective (Berkeley, CA) *see* Grassroots
Grassroots development : journal of the inter-american foundation – Rosslyn VA 1989-95 – 1 – ISSN: 0733-6608 – mf#15022.01 – us UMI ProQuest [327]
Grassroots editor – Joplin MO 1960+ – 1,5,9 – ISSN: 0017-3541 – mf#9760 – us UMI ProQuest [070]
Grassy cove and mt. pleasant baptist church. cumberland county. tennessee : church records – 1833-1924 – 1 – 5.00 – us Southern Baptist [242]
Grassy knoll gazette – v10 n1-v14 n12 [1986:2-1990 12] – 1r – 1 – (continued by: ahimsa news) – mf#1787104 – us WHS [071]
Grassy pond baptist church. gaffney, south carolina : church records – 9Feb 1879-27 Sep 1959, 1965-1987 (Lacking 1944-24 Sep 1950). Roll Book. 1958-Mar 1965 – 1 – $50.58 – us Southern Baptist [242]
Gratacap, Louis Pope *see*
– The analytics of a belief in a future life
– Philosophy of ritual
– The world as intention
Grateful alie – London, England. 18– – 1r – us UF Libraries [240]
Gratiant, Gilbert *see*
– Ile federee francaise de la martinique
Gratiolet, Louis-Pierre *see* Recherches sur l'anatomie de l'hippopotame
Gratiot reporter – Gratiot WI. 1912 jun 13, oct 24, 1913 may 22-oct 9 – 1r – 1 – mf#916339 – us WHS [071]
O gratis : jornal puramente d'anuncios declaracoes, reclamacoes, correspondencia... – Rio de Janeiro, RJ: Typ do Gratis, 26 nov-07 dez 1844 – (= ser Ps 19) – mf#P15,01,71-72 – bl Biblioteca [079]
Graton, John R *see* Correspondence
Gratry, Auguste *see*
– Guide to the knowledge of god
– Henri perreyve
– Mgr. l'evaeque d'orleans et mgr. l'archevaque de malines
– Le mois de marie de l'immaculee conception
– Papal infallibility untenable
– Le rev. p. gratry
– Ueber die erkenntniss der seele
– Ueber die erkenntniss des menschen in seiner denkthaetigkeit
Die gratulanten kommen : edvard grieg zum 150. geburtstag / ed by Oelmann, Klaus Henning – (mf ed 1993) – 2mf – 9 – €40.00 – 3-89349-814-1 – mf#DHS 814 – gw Frankfurter [780]
Gratulatio ad venerabilem presbyterum, dominum gabrielem de saconay, praecentorem ecclesiae lugdunensis, de pulchra et eleganti praefatione quam libro regis angliae inscripsit / [Calvin, J] – [Geneva: Conrad Badius], 1561 – 1mf – 9 – mf#CL-10 – ne IDC [242]
Gratz, L O *see*
– Effect of seed-potato treatment on yield and rhizoctonosis in florida from 1924 to 1929
– Infection of potato tubers by alternaria solani in relation to storage conditions
– Irish potato disease investigations, 1924-1925
– Potato spsraying and dusting experiments in florida, 1924 to 1929
– Tests of cigar-wrapper tobacco varieties resistant to blackshank
– Variety tests of white potatoes
Grau Archilla, Raul *see*
– Goldondrinas
– Vertigo de la nube
Grau, Rudolf Freidrich *see* The goal of the human race, or, the development of civilisation, its origin and issue
Grau, Rudolf Friedrich *see*
– Das selbstbewusstsein jesu
– Semiten und indogermanen in ihrer beziehung zu religion und wissenschaft
Grau San Martin, Ramon *see* Revolucion constructiva
Graubart, Judah Leib *see* Sefer zikaron
Graubner-Scheffler, Annette *see* Psychoedukative gruppenarbeit mit psychoseerfahrenen
Graudenzer anzeiger fuer stadt und land – Graudenz (Grudziadz PL), 1849 & 1852 – 1 – gw Misc Inst [077]
Der graue mann : eine volksschrift / ed by Jung, Joh Heinrich et al – Nuernberg 1795-1833 – (= ser Dz. abt theologie) – 7v on 20mf – 9 – €200.00 – (iss39ff ed by de valenti) – mf#k/n2269 – gw Olms [240]
Der graue rock : novelle / Gerstner, Hermann – 5. aufl. Muenchen: Zentralverlag der NSDAP, F Eher [1941?] [mf ed 1990] – 1r – 1 – (filmed with: die regulatoren in arkansas / friedrich gerstacker) – mf#2609p – us Wisconsin U Libr [830]
Das graue ungeheuer / ed by Wekhrlin, Wilhelm Ludwig – [Nuernberg] 1784-87 [mf ed Hildesheim 1992-98] – (= ser Dz) – 12v on 36mf – 9 – €360.00 – mf#k/n5666 – gw Olms [390]
Grauer, Albert W *see* Vocal style of sixt dietrich and johann eccard
Graul, Karl *see*
– The distinctive doctrines of the different christian confessions in the light of the word of god
– Die unterscheidungslehren der verschiedenen christlichen bekenntnisse im lichte goettlichen worts
Die gravamina der deutschen nation gegen den roemischen hof : ein beitrag zur vorgeschichte der reformation / Gebhardt, Bruno – 2. Aufl. Breslau: Koebner, 1895 – 1mf – 9 – 0-524-04309-4 – (incl bibl ref) – mf#1990-1235 – us ATLA [240]
Grave, and the reverence due to it... / Harston, Edward – Oxford, England. 1856 – 1r – us UF Libraries [240]
Grave, B *see* Kadety v 1905-1906 gg
Grave, Joao *see* Cartas para o brasil
Gravel, Albert *see*
– Notes preliminaires
– "Suagothel"
Gravel and Golddust *see* The southeast asia pamphlet collection at monash university
Gravel, Pierre Bettez *see* Remera a community in eastern ruanda
Gravel ridge first baptist church. north little rock, arkansas : church records – Mar 1949, Feb 1965 – 1 – us Southern Baptist [242]
Graveley 1599-1950 – (= ser Cambridgeshire parish register transcript) – 4mf – 9 – £5.00 – uk CambsFHS [929]
Gravelot, H Fcochin, C N *see* Iconologie par figures o- trait
Grave-mounds and their contents : a manual of archaeology / Jewitt, Llewellynn Frederick William – London 1870 – (= ser 19th c art & architecture) – 4mf – 9 – mf#4.1.334 – uk Chadwyck [930]
Graven image – v1 n5 [1970] – 1r – 1 – mf#1583460 – us WHS [071]
Graven in the rock : or, the historical accuracy of the bible / Kinns, Samuel – London: Cassell, 1891 [mf ed 1992] – 2mf – 9 – 0-524-04404-X – mf#1992-0097 – us ATLA [221]
Graves, Mrs. *see* Personal testimonies of
Graves, Absalom *see* Hymns, psalms and spiritual songs
Graves, Algernon *see*
– Catalogue of the works of the late sir edwin landseer
– A dictionary of artists who have exhibited works in the principal london exhibitions from 1760 to 1893
– A history of the works of sir joshua reynolds
Graves, Anna Melissa *see* Benvenuto cellini had no prejudice against bronze
Graves, Armgaard Karl *see*
– Secrets of the german war office
– Secrets of the hohenzollerns
Graves, Charles *see* A history of unitarianism
Graves, Charles Alfred *see* Real property
Graves creek baptist church. henderson county. kentucky : church records – 1810-1941 – 1 – 61.20 – us Southern Baptist [242]
Graves, F A *see* Personal testimonies of
Graves, Frank Pierrepont *see*
– Great educators of three centuries
– A history of education before the middle ages
– A history of education during the middle ages
– A history of education in modern times
– Peter ramus and the educational reformation of the sixteenth century
Graves, Henry Clinton *see* Handbook of christian doctrine
Graves, J *see* Roll of the proceedings of the king's council in ireland (rs69)
Graves, J R *see*
– The great iron wheel
– The little seraph
– The new great iron wheel
– Old landmarkism, what is it?
– Southern baptist almanac and annual register
– The southern psalmist
– Trial of, before the first baptist church of nashville, tennessee
– The tri-lemma
– The work of christ in the covenant of redemption
Graves, James Robinson *see*
– The bible doctrine of the middle life as opposed to swedenborgianism and spiritism
– The graves-ditzler or great carrollton debate
– Old landmarkism, what is it?
– The tri-lemma
– The trilemma
Graves, Kersey *see* The world's sixteen crucified saviors
The graves of myles standish and other pilgrims / Huiginn, Eugene Joseph Vincent – Beverley, Mass.: EJV Huiginn, 1914 – 1mf – 9 – 0-7905-4929-8 – mf#1988-0929 – us ATLA [243]
Graves, Rosswell Hobart *see* Forty years in china
The graves-ditzler or great carrollton debate : on the mode of baptism, infant baptism, church of christ, the lord's supper, believers' baptism, final perseverance of saints / Graves, James Robinson & Ditzler, Jacob – Memphis: Southern Baptist Publ Soc, 1876 – 3mf – 9 – 0-524-03710-8 – mf#1990-4815 – us ATLA [240]
Gravesend and milton journal – England, 23 Aug 1834-25 Nov 1837 – 1r – 1 – uk British Libr Newspaper [072]
Gravesend argus, kent and essex chronicle and port of london journal – [London & SE] Kent Arts & Lib, Gravesend Lib 15 may 1880-9 may 1885 – 1 – (discontinued) – uk Newsplan [072]
Gravesend journal etc – Gravesend, England. 6 Jul 1864-1876; 1878-1892.-w. 16 reels – 1 – uk British Libr Newspaper [072]
Gravesend & northfleet standard – England.8 Apr 1892-1915. -w. 23 reels – 1 – uk British Libr Newspaper [072]
Gravesend & northfleet standard, etc. (football ed.) – England.5 Jan-6 Apr, 21 Sept-28 Dec 1895; 4 Jan-11 Apr, 26 Sept-26 Dec 1896. -w.1/2 reel – 1 – uk British Libr Newspaper [072]
Gravesend reporter – (Gravesend and Dartford Reporter). England. -w. 2 Feb 1856-Dec 1969. (160 reels) – 1 – uk British Libr Newspaper [072]
The grave-tree / the wind and the tree / seven wind songs / overlord / Carman, Bliss – [New York: s.n, 1892?] – 1mf – 9 – 0-665-07214-7 – mf#07214 – cn Canadiana [810]
Gravette news herald – Gravette AR. 1957 sep 26 – 1r – 1 – mf#853697 – us WHS [071]
Gravier, Gabriel *see*
– Decouverte de l'amerique par les normands au 10e siecle
– Voyage a segou, 1878-1879
Gravissimae quaestionis de christianarum ecclesiarum : in occidentis praesertium partibus... / Ussher, J [Archbishop of Armagh] – Londini: Excudebat Bonham Norto, 1613 – 5mf – 9 – mf#PW-30 – ne IDC [240]
Gravitation : an elementary explanation of the principal perturbations in the solar system:[written for the penny cyclopaedia...] / Airy, George Biddell – London: C Knight 1834 – 1r [ill] – 1 – mf#film mas 28229 – us Harvard [520]
O gravoche : orgao critico-humoristico – Ouro Preto, MG. 01 jan 1900 – (= ser Ps 19) – bl Biblioteca [079]
Grawert, Rolf *see* Verwaltungsabkommen zwischen bund und landern in der bundesrepublik deutschland. eine kritische untersuchung der gegenwartigen staatspraxis mit eine-zusammenstellung der zwischen bund und landern abgeschlossenen abkomren
Gray, A *see* Botany. phanerogamia
Gray, Andrew *see*
– Letter to the inhabitants of aberdeenshire and the neighbouring cou...
– The origin and early history of christianity in britain
– Oxford tractarianism, the scottish episcopal college
– Persecution, the lairds, the lawyers, and the moderate clergy
– Present conflict between the civil and ecclesiastical courts examin...
– Public religious intercourse with the office-bearers of the establi...
Gray, Arlene E *see* Lowell mason's contribution to american church music
Gray, Arthur Romeyn *see* An introduction to the study of christian apologetics
Gray, Asa *see*
– Darwiniana
– Natural science and religion
– Natural selection not inconsistent with natural theology
Gray, Basil *see* Rajput painting
Gray, C N *see*
– Confession as taught by the church of england
– Life of robert gray
Gray, Carol *see* Human experimentation
Gray, Charles Norris *see* Life of robert gray
Gray, Clayton *see* Le vieux montreal (montreal qui disparait)
Gray, Clifton Daggett *see* The samas religious texts
Gray, Edward A *see* A holistic analysis of stress with implications for stress management as a function of pastoral counseling
Gray, Elizabeth (nee McEwen) *see* Journal and correspondence
Gray, G P *see* Morphology of the substantive in herero
Gray, George Buchanan *see*
– A critical introduction to the old testament
– The divine discipline of israel
– The forms of hebrew poetry
– The fourth book of moses called numbers
– Sacrifice in the old testament
– Studies in hebrew proper names
Gray, George Zabriskie *see*
– The children's crusade
– The church's certain faith gray
Gray, Georgina L *see* Oxygen consumption during kayak paddling
Gray, J E *see*
– Catalogue of the specimens of lizards in the collection of the british museum
– The zoology of the voyage of hms erebus and terror
Gray, J M *see* Mutesa of bufanda
Gray, James *see*
– Buddhaghosuppatti
– A dissertation on the coincidence between the priesthoods of jesus christ and melchisedec
– The earth's antiquity in harmony with the mosaic record of creation
– Jinaalankaara
– Last charge of an ascending redeemer
Gray, James Martin *see*
– The bulwarks of the faith
– How to master the english bible
– Synthetic bible studies
Gray, John *see*
– File of correspondence relating to emin pasha extracted from the zanzibar archives
– Irish church establishment
– Obnoxious oaths and catholic disabilities
Gray, John G *see* Effects of limited and expanded rest intervals on the navy physical readiness test
Gray, John Hamilton *see* Churchmen and dissenters
Gray, John Milner *see* History of zanzibar
Gray, John Stanley *see*
– Communicative speaking
Gray, Lc *see* History of agriculture in the southern united states to 1860
Gray leafspot : a new disease of tomatoes / Weber, George F – Gainesville, FL. 1932 – 1r – us UF Libraries [630]
Gray, Marvin *see* Catastrophe model of anxiety and performance
Gray, Nelson Cockburn *see*
– Canada, ontario, the british flag and other poems
– New patriotic poems
Gray panther network; age and youth in action / Gray Panther Project Fund – Philadelphia, PA, [quarterly] 1977-78 [bimonthly] 1979- – 1 – us Wisconsin U Libr [305]
Gray Panther Project Fund *see* Gray panther network; age and youth in action
Gray panthers network – 1972-87 – 1 – $50.00 – us Presbyterian [305]
Gray panthers network – 1976 aug-nov, 1977 apr-dec, 1978 spring-winter, 1979 jan-1983 sep/oct, 1984 spring-1987 fall, v17 n1-5 [1988 jan/feb-nov/dec] – 1r – 1 – (cont: network [gray panthers]; cont by: network [gray panthers project fund]) – mf#971556 – us WHS [071]
Gray, Patrick, 1819?-1876 *see* The new heavens and the new earth
[Gray, R] *see*
– A journal of the bishop's visitation tour through the cape colony, in 1848
– A journal of the bishop's visitation tour through the cape colony, in 1850
Gray, R M *see* Mahatma gandhi
Gray, Richard *see* Two nations
Gray, Robert *see*
– Catechism
– Dialogue between a churchman and a methodist
– A good speed to virginia
– Judgment
– Serious address to seceders and sectarisits, of every description, who exist in separation from...
Gray, Robert A *see* Gray's civil government of florida
Gray, Robert Earl *see* Treatment of the trombone in contemporary chamber liturature
Gray, Robert F *see* Sonjo of tanganyika
Gray, Thomas *see* Poetic commonplace books and manuscripts of thomas gray
Gray, W H *see* Sermon on the disestablishment
Gray, William *see*
– Diaries, correspondence and miscellaneous papers
– Journal, report and letterbook
– Press cuttings, pamphlets and articles on kanaka labour traffic and missionary life in the new hebrides, 1884-1915
– Travels in western africa
Gray, William Scott *see* Guidebook for days and deeds
Graya : a magazine for circulation among members of gray's inn – n1-86. 1927-82 (all publ) – 88mf – 9 – $132.00 – mf#LLMC 84-283 – us LLMC [073]
Gray's civil government of florida / Gray, Robert A – Tallahassee, FL. 1921 – 1r – us UF Libraries [350]
Grays einfluss auf die deutsche lyrik im achtzehnten jahrhundert / Uebel, Otto – Heidelberg, 1914 (mf ed 1994) – 1mf – 9 – €19.00 – 3-8267-3098-4 – mf#DHS-AR 3098 – gw Frankfurter [410]

Gray's inn : commemoration of the tercentenery of francis bacon / Gray's Inn. London – London: Printed by order of the Masters of the Bench, – 1mf – 9 – $1.50 – mf#LLMC 84-286 – us LLMC [920]
Gray's inn : notes illustrative of its history and antiquities / Douthwaite, William R – London: Benson & Page, 1876 – 2mf – 9 – $3.00 – mf#LLMC 84-281 – us LLMC [340]
Gray's inn and lincoln's inn / Bellot, Hugh Hale L – London: Methuen & Co, 1925? – 3mf – 9 – $4.50 – mf#LLMC 84-273 – us LLMC [340]
Gray's inn journal – La Salle IL 1753-54 – 1 – mf#4258 – us UMI ProQuest [340]
The gray's inn journal / Murphy, Arthur – 4mf – 9 – $6.00 – (a repr of 101 papers by charles ranger (pseud. of arthur murray), of which some appeared in the craftsman and some after 1753, were issued separately. on republ in gemtleman's magazine, the 101 nos were extended to 104 with major alterations) – mf#LLMC 84-285 – us LLMC [073]
Gray's inn library : catalogue of books / Severn, M D – London: Witherby & Co, 1906 – 11mf – 9 – $16.50 – mf#LLMC 84-288 – us LLMC [340]
Gray's Inn. London *see*
- Gray's inn
- Gray's inn pension book

Gray's Inn. London. Library *see*
- Gray's inn library
- Mr baldwin's visit

Gray's inn pension book : the records of the honorable society / Gray's Inn. London – London: Chiswick. 2v. 1901-10 – 12mf – 9 – $18.00 – mf#LLMC 84-291 – us LLMC [340]
Gray's monthly record and general advertising index – [Scotland] Edinburgh: J Gray feb-mar 1834 (mthly) [mf ed 2004] – 1r – 1 – uk Newsplan [072]
Grayson baptist church : minutes, financial records, membership – Grayson, LA. 1936-1995 – 1 – $60.75 – mf#6917 – us Southern Baptist [242]
Grayson gateway / Denison Library (TX) – v1 n1-v4 n4 [1981 sep-1985 jun] – 1r – 1 – mf#927105 – us WHS [071]
Grazer tagblatt – Graz (A), 1906 may – 1r – 1 – gw Misc Inst [074]
Grazhdanin – St Petersburg, Russia 1872-1914 [mf ed Norman Ross] – 1 – mf#nrp-101 – us UMI ProQuest [077]
Grazhdanin-voin : a m koliubakin (1868-1915) – 1915 – 179p 2mf – 9 – mf#RPP-113 – ne IDC [325]
Grazhdanskaia aviatsiia – Moskva: Gos obedinennoe nauch-tekhn. izd-vo. 1967-1972 – us CRL [077]
Grazia e giustizia / Italy. Ministero di Giustizia e dei Culti – Annuario. Roma. 1864-1919 – 1 – us NY Public [324]
Great adventure of panama / Bunau-Varilla, Philippe – Garden City, NY. 1920 – 1r – us UF Libraries [972]
Great american lawyers / Lewis, W D – Philadelphia: John C. Winston Co. v1-8 + index. 1907-09 – 54mf – 9 – $81.00 – mf#LLMC 81-407 – us LLMC [340]
The great apostasy : considered in the light of scriptural and secular history / Talmage, James Edward – Salt Lake City, Utah: Deseret News, 1909 – 1mf – 9 – 0-7905-7084-X – (incl bibl ref) – mf#1988-3084 – us ATLA [240]
The great assize : war studies in the light of christian ideals / Rollings, William Swift – London: HR Allenson, [1916?] – 1mf – 9 – 0-524-07714-2 – mf#1991-3299 – us ATLA [240]
The great awakening : a history of the revival of religion in the time of edwards and whitefield / Tracy, Joseph – Boston: Tappan & Dennet, 1842, c1841 – 2mf – 9 – 0-7905-6371-1 – mf#1988-2371 – us ATLA [240]
Great awakening and other revivals in the religious life of... / Mitchell, Mary Hewitt – New Haven, CT. 1934 – 1r – us UF Libraries [025]
The great awakening in columbus, ohio : under the labors of rev b fay mills and his associates / ed by Stauffer, Henry – Columbus, Ohio: W. L. Lemon, [c1895] Beltsville, Md: NCR Corp, 1978 (2mf); Evanston: American Theol Lib Assoc, 1984 (2mf) – (= ser Revivalism and revival preachers in america) – 9 – 0-8370-1225-2 – (incl ind) – mf#1984-3012 – us ATLA [242]
The great awakening of 1740 : lectures delivered before the baptist church of evanston, il, the second baptist church of chicago, and other churches / Chapell, Frederic Leonard – Philadelphia: American Baptist Publication Society, 1903. Chicago: Dep of Photodup, U of Chicago Lib, 1974 (1r); Evanston: American Theol Lib Assoc, 1984 (1r) – 1 – 0-8370-0060-2 – mf#1984-6019 – us ATLA [240]
Great barford – (= ser Bedfordshire parish register series) – 2mf – 9 – £5.00 – uk BedsFHS [929]
Great barrington 1741-1849 – Oxford, MA (mf ed 1996) – (= ser Massachusetts vital record transcripts to 1850) – 3mf – 9 – 0-87623-257-8 – (mf 1t: births 1741-1813; marriage intentions 1761-1802. mf 1t-2t: marriages 1795-1844. mf 2t: out-of-town marriages 1762-99. mf 2t-3t: births 1847-49. mf 3t: marriages 1843-49; deaths 1843-49) – us Archive [978]
Great barrington 1745-1903 – Oxford, MA (mf ed 1990) – (= ser Massachusetts vital records) – 51mf – 9 – 0-87623-104-0 – (mf 1-2: vital & church records 1770-1856. mf 3: sheffield north parish meetings 1745-61. mf 4-8: town minutes 1761-91. mf 9-11: b,d,m 1794-1855. mf 12-15: b,m,d 1843-68. mf 16-17: index to marriage intents 1850-80. mf 18-20: marriage intentions 1850-80. mf 21-25: index to deaths 1850-1960. mf 26-28: deaths 1863-1903. mf 29-34: index to marriages 1848-1959. mf 35-39: marriages 1843-1901. mf 40-42: marriages 1880-94. mf 43-49: index to births 1848-1959. mf 50-51: births 1867-90) – us Archive [978]
Great battle / Macdona, H Victor – London, England. 1871 – 1r – us UF Libraries [240]
The great book of sufferings : 1659-1793 / Friends House Library. The Religious Society of Friends – 29r – 1 – £1350.00 – (accounts of prosecutions against quakers) – mf#GBS – uk World [240]
Great bowl of alachua / Alachua County (FL) Chamber Of Commerce – Gainesville, FL. 1925? – 1r – us UF Libraries [978]
Great Britain *see*
- Board of trade/overseas department economic surveys, 1921-1961
- Carrington and kirwan's reports
- Carrington and marshman's reports
- Carrington and payne's reports
- Documents and correspondence relating to the quest
- The jurist, new series
- The jurist, old series
- Yearbooks

Great Britain. *see* Report of the royal commission on the poor law, 1909
Great britain : england, wales, and scotland / Karl Baedeker (Firm) – Leipsic, Germany. 1887 – 1r – us UF Libraries [941]
Great britain : handbook for travellers / Karl Baedeker (Firm) – Leipsic, Germany. 1897 – 1r – us UF Libraries [914]
Great britain : internal affairs and foreign affairs, 1930-1954 / U.S. State Dept – (= ser Confidential u s state department central files) – 1 – $47,430.00 coll – (internal affairs, 1930-39: pt1: political, governmental, & national defense affairs 37r isbn 0-89093-554-8 $7160; pt2: social, economic, & industrial affairs 42r isbn 0-89093-555-6 $8130. foreign affairs, 1930-39 26r isbn 0-89093-556-4 $5040. internal affairs, 1940-44: pt1: political, governmental, & national defense affairs 31r isbn 0-89093-394-4 $5985; pt2: social, economic, & industrial affairs 63r isbn 0-89093-395-2 $12,210. foreign affairs, 1940-44 9r isbn 0-89093-396-0 $1740. internal affairs, 1945-49: pt1: political, governmental, & national defense affairs 26r isbn 0-89093-457-6 $5040. foreign affairs, 1945-49 10r isbn 0-89093-459-2 $1935. foreign affairs, 1950-54 14r isbn 1-55655-896-1 $2705. with p/g) – us UPA [941]
Great britain, 1300 to 1980 *see* History of glass
Great britain and germany / Wilkinson, Spenser – London, Toronto: Oxford University Press, H Milford, [1914] – 1mf – 9 – 0-665-66904-6 – mf#66904 – cn Canadiana [933]
Great britain and south african confederation, 1870-1881 / Goodfellow, Clement Francis – Cape Town, South Africa. 1966 – 1r – us UF Libraries [960]
Great Britain. Army *see* General orders
Great britain as seen by canadian eyes : inaugural address delivered in convocation hall, november 6th, 1896 / Bryce, George – [Winnipeg?: s.n, 1896] [mf ed 1980] – 1mf – 9 – 0-665-00978-X – mf#00978 – cn Canadiana [941]
Great Britain. Board of Trade *see*
- Registrar general of shipping and seamen
- Registrar general of shipping and seamen and predecessor

Great britain board of trade. board of trade journal – Norwich, England 1886-1970 – 1 – mf#626 – us UMI ProQuest [380]
Great britain british guiana and british honduras report – London, England. 1948 – 1r – us UF Libraries [972]
Great Britain British Honduras Land Use Survey *see* Land in british honduras
Great Britain. Census Office *see* Census of england and wales, 1871
Great Britain. Central Advisory Committee for England *see* Half our future
Great Britain Central Office Of Information Reference Division *see* Botswana
Great Britain. Central Office of Information Reference Division *see* Zambia
Great Britain. Central Office of Information. Reference Division *see* The khmer republic
GREAT BRITAIN COLONIAL OFFICE *see* British territories in east and central africa, 1945-1950
Great Britain Colonial Office *see*
- Basutoland, bechuanaland protectorate and swaziland
- British territories in east and central africa, 1945-1950
- Closer association of the british west indian colonies
- Papers relating to the question of the closer union of kenya

Great Britain. Colonial Office *see*
- Cameroons under united kingdom administration
- Rules and regulations for her majesty's colonial service

Great Britain. Colonial Office. Nigeria *see* Intelligence reports on southern nigeria
Great britain colonial office pamphlets about africa – London: The Office, [1974?] – us CRL [324]
Great Britain Commission Of Inquiry Into Disturbances In Zanzibar *see* Report of a commission of inquiry into disturbances in zanzibar
Great Britain. Commissioners of Health *see* Reports on the state of large towns and populous districts, 1844-45
Great Britain. Commissioners on African Settlements *see* Report on the slave trade
Great Britain. Committee on the Limitation of Actions *see* Report
Great Britain. Court of Common Pleas *see* H blackstone's reports
Great Britain. Court of Exchequer *see* Reports of cases concerning the revenue, argued and determined in the court of exchequer, from easter term 1743, to hilary term 1767
Great Britain. Court of King's Bench *see*
- Barnewell and adolphus' reports
- Barnewell and alderson's reports
- Barnewell and cresswell's reports
- Burrow's reports
- Durnford and east's reports
- East's reports
- Nevile and manning's reports
- Nevile and perry's reports
- Reports of cases argued and adjudged in the king's courts at westminster...1742-1774

Great Britain. Court of Queen's Bench *see* Best and smith's reports
Great Britain. Courts *see*
- British ruling cases
- Central criminal sessional papers
- English ruling cases

Great Britain. Courts of Chancery, King's Bench and Common Pleas *see* The law journal / the law journal reports
Great Britain. Courts of Common Law and Equity *see* The revised reports
Great Britain. Department of Practical Art *see* A catalogue of the articles of ornamental art
Great britain dept of employment. employment gazette – Norwich, England 1974-93 – 1,5,9 – ISSN: 0264-7052 – mf#9889.01 – us UMI ProQuest [350]
Great Britain. Embassy. Prague *see* Czechoslovak republic press review
Great Britain. England *see*
- The bar reports
- Cobbett's state trials / howell's state trials
- Cox's criminal cases
- Cox's magistrates cases
- A general abridgement of law and equity
- The law reporter
- The times law reports
- The weekly notes

Great Britain. England. Court of Common Pleas *see*
- Bingham's new cases
- Bingham's reports
- Bosanquet and puller's new reports
- Bosanquet and puller's reports
- Broderip and bingham's reports
- Common bench reports, new series
- Common bench reports, old series
- Moore and payne's reports
- Moore and scott's reports
- Moore's common pleas reports

Great Britain. England. Courts of Chancery *see*
- Beavan's reports
- Hare's reports
- Vesey's senior's reports

Great Britain. Exchequer *see* Red book of the exchequer (rs99)
Great Britain Foreign Office *see* Correspondence with the united states
Great Britain. Foreign Office *see*
- Collected diplomatic documents relating to the outbreak of the european war
- Papers concerning affairs in liberia
- Recent diplomatic exchanges concerning the proposal for an international conference on the neutrality and territorial integrity of cambodia
- Registers (library series) and indexes of general correspondence

Great Britain. Foreign Office. Public Record Office *see* Foreign office registers and indexes of correspondence, 1793-1919
Great Britain. General Board of Health *see* Urban and rural social conditions in industrial britain, series one
Great Britain. General Register Office *see*
- Birth certificates from the presbyterian, independent and baptist registry
- Foreign registers and returns
- Miscellaneous foreign death returns
- Miscellaneous foreign marriage returns
- Miscellaneous foreign returns
- Registers and returns of births, marriages and deaths in the protectorates etc of africa and asia
- Registers of births, marriages and deaths surrendered to the non parochial registers commission of 1857
- Registers of births, marriages and deaths surrendered to the non-parochial registers commissions of 1837 and 1857
- Registers of clandestine marriages and of baptisms in the fleet prison, king's bench prison, the mint and the may fair chapel
- Society of friends' registers, notes and certificates of births, marriages and burials

Great Britain. High Commissioner for Aden *see* High commissioner's gazette
Great Britain. High commissioner for Basutoland, the Bechuanaland Protectorate, and Swaziland *see* Official gazette
Great Britain. Historical Manuscripts Commission *see* Report on the manuscripts of lord middleton, preserved at wollaton hall, nottinghamshire
Great Britain. House of Commons *see* The journal of the house of commons
Great britain. house of commons journal – Killen TX 1547-1900 – 1 – mf#2587 – us UMI ProQuest [323]
Great Britain. House of Lords *see*
- Bligh's parliamentary reports
- Brown's parliamentary cases
- Clark and finnelly's reports
- House of lords cases (clark and finnelly)

Great britain in 1833 / Haussez, Charles L d' – London 1833 [mf ed Hildesheim 1995-98] – 2v on 4mf – 9 – €120.00 – 3-487-28881-8 – gw Olms [941]
Great Britain. India Office *see* A list of the principal indian government publications on sale in this country and at the various government presses in india
Great Britain. International Dance Teachers Association *see* Ballroom dancing to bronze medal standard
Great Britain. Laws, Statutes, etc *see*
- Combination laws, select committee on the...
- Law reports: pre-1865
- Law reports: pre-1865: english nominatives
- Local, personal, and private acts
- Private acts
- Public general acts, 1714-1933
- The statutes at large (pickering's statutes), 1225-1869
- Statutes of the realm

Great Britain. Local Government Board *see* Urban and rural social conditions in industrial britain, series two
Great Britain. Medical Officers of Health *see* Urban and rural social conditions in industrial britain, series three
Great Britain. Ministry of Labour *see* Report on the establishment and progress of joint industrial councils set up in accordance with the recommendations of the committee on relations between employers and employed "the whitley" committee
Great Britain Ministry Of Overseas Development Economic Survey *see* Development of the bechuanaland economy
Great Britain. Ministry of Reconstruction Civil War Workers' Committee *see* 2nd, 3rd, 4th and 5th interim reports of the civil war workers' committee
Great Britain Naval Intelligence Division *see* Belgian congo
Great britain, palestine and the jews – New York, 1918 – 1mf – 9 – mf#ILM-2481 – ne IDC [956]
Great Britain Parliament *see* First prayer book of edward 6
Great Britain. Parliament *see*
- Debates
- Parliamentary reports and papers relating to the navy, 1801-1900
- Records of the parliament holden at westminster, 28 febr in the 33rd year of the reign of edward 1

Great Britain, Parliament, House Of Commons *see* Church in wales
Great Britain. Parliament. House of Commons *see*
- Division lists
- Population of england and wales
- Report from the select committee of the house of commons on the earl of elgin's collection
- Report on the visit by an all party group of members of parliament to spain
- Sessional papers

Great britain. parliament. house of commons. journal – v1-9. 1547-1685 – 9 – $600.00 – (v34 only 1972-74 $50 [0249]) – mf#0248 – us Brook [324]

Great Britain Parliament House of Commons. Select Committee on the East India Company *see* The fifth report from the select committee of the house of commons on the affairs of the east india company

Great Britain. Parliament. House of Lords *see*
- Cases submitted to the house of lords on appeal from the courts of england, scotland, and northern ireland
- Sessional papers

Great britain parliament. parliamentary debates : authorized edition – London, England 1803-1908 – 1 – mf#5652 – us UMI ProQuest [323]

Great Britain. Privy Council *see*
- Acts of the privy council of england, colonial series
- Acts of the privy council of england, new series
- Law reports: indian appeals
- Moore's privy council cases
- Records and briefs of cases submitted on appeal to the privy council

Great Britain. Public Record Office *see*
- Calendar of entries in the papal registers relating to great britain and ireland. papal letters
- Calendar of state papers, domestic series of the reign of charles i, 1625-1649
- The clarendon papers, 1867-1870
- Colonial office 824 / 1-2. brunei. sessional papers
- Hand-book to the public records

Great britain public record office. calendar of house of commons – Norwich, England 1760-75 – 1 – mf#2584 – us UMI ProQuest [324]

Great britain public record office. calendar of state papers : domestic – charles 2 – Paris, France 1660-85 – 1 – mf#3220 – us UMI ProQuest [324]

Great britain public record office. calendar of state papers : domestic, edward 6, mary, elizabeth 1, and james 1 – Norwich, England 1547-1625 – 1 – mf#2590 – us UMI ProQuest [324]

Great britain public record office. calendar of state papers : domestic series, of the reign of charles 1, 1625-1649 – Norwich, England 1625-49 – 1 – mf#2588 – us UMI ProQuest [324]

Great britain public record office. calendar of state papers : domestic-commonwealth – Norwich, England 1649-60 – 1 – mf#2589 – us UMI ProQuest [324]

Great britain public record office. calendar of state papers colonial series – Paris, France 1661-1733 – 1 – mf#2583 – us UMI ProQuest [324]

Great Britain. Royal Academy of Dancing *see* Ballet in education; children's examinations

Great Britain. Royal Air Force *see* Royal air force final reports on operations

Great Britain. Royal Commission for the Paris Exhibition, 1878 *see*
- The english guide to the paris exhibition, 1878
- An illustrated catalogue of painting and sculpture

Great Britain. Royal Commission of Fine Arts, Westminster Hall *see*
- Clarke's critical catalogue of the works of art sent in
- [Exhibition catalogue 1843]
- [Exhibition catalogue 1844]
- [Exhibition catalogue 1845]
- [Exhibition catalogue 1847]

Great Britain. Royal Commission on Legal Services *see* The public evidence presented to the royal commission on legal services, 1979

Great Britain. Royal Commissions on Ancient and Historical Monuments and Constructions *see*
- Historic buildings in britain
- Wales

Great Britain. Royal Photographic Society *see* The journal of the photographic society

Great Britain. School of Design *see* The drawing book of the government school of design

Great Britain. Secondary School Examinations Council on Secondary School Examinations other than G. C. E *see* Report of the committee on secondary schools examinations other than g c e, 1958

Great Britain. Stationery Office *see*
- Annual catalogues of british official and parliamentary publications, 1894-1909
- Annual catalogues of british official and parliamentary publications, 1910-1919

Great Britain. Superior Courts of Equity and Law *see* The weekly reporter

Great Britain. Unemployment Assistance Board *see* Origins of the welfare state in britain

Great Britain. War Office. General Staff *see*
- Military report on the gold coast, ashanti and the northern territories
- Route book of the gold coast colony, ashanti and the northern territories

Great Britain. War Office. General Staff. Geographical Section *see* The anglo-egyptian sudan

Great Britain. War Office. Intelligence Division *see* Military report on somaliland, 1907

Great Britain West India Royal Commission (1938-) *see* Statement of action taken

Great Britain West India Royal Commission (1938-1... *see* Recommendations

Great Britain. Western Pacific High Commission *see* Gazette

The great canadian north-west : manitoba, keewatin and north west territories – Winnipeg: Chisholm & Dickson, 1881 – 1mf – 9 – mf#30346 – cn Canadiana [917]

The great canal at suez : its political, engineering, and financial history. with an account of the struggles of its projector, ferdinand de lesseps / Fitzgerald, Percy Hetherington – London. 2v. 1876 – (= ser 19th c books on british colonization) – 8mf – 9 – mf#1.1.9579 – uk Chadwyck [380]

Great case of tythes truly stated, clearly opened, and fully resolv... / Pearson, Anthony – London, England. 1835 – 1r – us UF Libraries [240]

The great centre : an astronomical study / Hamilton, James Cleland – S.l: s.n, 1892? – 1mf – 9 – mf#56283 – cn Canadiana [520]

The great charter of christ : being studies in the sermon of the mount / Carpenter, William Boyd – New York: Thomas Whittaker, 1896 – 1mf – 9 – 0-8370-2588-5 – mf#1985-0588 – us ATLA [220]

The great civil war in dorset, 1642-1660 / Bayley, Arthur Rutter – Taunton: Barnicott and Pearce: Wessex Press, 1910 [mf ed 1980] – xx/493p – 1 – mf#93 – us Wisconsin U Libr [941]

A great cloud of witnesses : being a brief treatise... / Wilson, Louis Charles – Cincinnati, OH: Standard Pub Co, 1901 [mf ed 1992] – (= ser Christian church (disciples of christ) coll) – 1mf – 9 – 0-524-02175-9 – mf#1990-4241 – us ATLA [240]

Great cohary baptist church. eastern association. north carolina : church records – 1790-1855 – 1 – 7.92 – us Southern Baptist [242]

The great commission / Kilbourne, Ernest A – 3rd ed. Tokyo: Oriental missionary society, 1913 [mf ed 1996] – (= ser Yale coll) – 134p – 1 – 0-524-10280-5 – mf#1996-1280 – us ATLA [950]

The great companion / Abbott, Lyman – New York: Outlook, 1904 – 1mf – 9 – 0-7905-1567-9 – mf#1987-1567 – us ATLA [240]

The great company (1667-1871), vol 1 : being a history of the honourable company of merchants-adventurers trading into hudson's bay / Willson, Beckles – Toronto: Copp, Clark, 1900 – 1v of 2 on 5mf – 9 – (int by lord strathcona and mount royal) – mf#37689 – cn Canadiana [380]

The great company (1667-1871), vol 2 : being a history of the honourable company of merchants-adventurers trading into hudson's bay / Willson, Beckles – Toronto: Copp, Clark, 1900 – 1v of 2 on 5mf – 9 – (incl ind) – mf#37690 – cn Canadiana [380]

The great company, vols 1 and 2 : being a history of the honourable company of merchants-adventurers trading into hudson's bay / Willson, Beckles – Toronto: Copp, Clark, 1900 – 2v on 1mf – 9 – (int by lord strathcona and mount royal) – mf#37688 – cn Canadiana [380]

The great conflict : a discourse concerning baptists, and religious liberty / Lorimer, George Claude – Boston: Lee & Shepard; New York: Charles T Dillingham, 1877 [mf ed 1986] – 1mf – 9 – 0-8370-8921-2 – (incl bibl ref) – mf#1986-2921 – us ATLA [242]

The great conquest : or, miscellaneous papers on missions / Ellinwood, Frank Field – New York: William Rankin, 1876 – 1mf – 9 – 0-8370-6113-X – mf#1986-0113 – us ATLA [240]

The great consummation and the signs that herald its approach / Taylor, Daniel Thompson – Boston: Advent Christian Publ Soc, 1891 [mf ed 1991] – 2mf – 9 – 0-524-00162-6 – mf#1989-2862 – us ATLA [240]

The great controversy between christ and satan : from the destruction of jerusalem to the end of time / White, Ellen Gould Harmon – 3rd ed. Oakland, Cal.: Pacific Press; Battle Creek, Mich.: Review and Herald, 1885, c1884 – 2mf – 9 – 0-7905-3496-7 – mf#1987-3496 – us ATLA [240]

The great cylinder inscriptions a and b of gudea / Price, I M – Leipzig, 1899; 1927 – (= ser Assyriologische bibliothek) – 8mf – 9 – (assyriologische bibliothek v15, 26) – mf#NE-20032 – ne IDC [956]

The great debate : a verbatim report of the discussion of the meeting of the american board of commissioners for foreign missions. held at des moines, iowa... – Boston: Houghton, Mifflin, 1886 – 2mf – 9 – 0-524-07223-X – mf#1991-2964 – us ATLA [240]

Great decisions – New York NY 1955-77 – 1,5,9 – mf#3481 – us UMI ProQuest [327]

The great didactic of john amos comenius : now for the first time englished = Didactica magna / ed by Keatinge, Maurice Walter – London: Adam & Charles Black, 1896 [mf ed 1986] – 2mf – 9 – 0-8370-7615-3 – (incl bibl ref. int by maurice walter keatinge) – mf#1986-1615 – us ATLA [370]

The great dionysiak myth / Brown, Robert – London: Longmans, Green, 1877-1878 – 3mf – 9 – 0-524-04506-2 – (incl bibl ref) – mf#1990-3340 – us ATLA [250]

The great disciples of the buddha / Myint Hlaing, Maung – Rangoon, Burma: Zeyar Hlaing Literature House 1981 [mf ed 1990] – 1r with other items – 1 – mf#mf-10289 seam reel 189/8 [§] – us CRL [280]

Great disruption principle / Bruce, John – Edinburgh, Scotland. 1859 – 1r – us UF Libraries [240]

The great doctrines of the bible / Evans, William – Chicago: Bible Institute Colportage Association, c1912 – 1mf – 9 – 0-7905-1383-8 – (incl ind) – mf#1987-1383 – us ATLA [220]

The great eastern : being a full description and historical account of this monarch of the ocean – [Quebec?: s.n.] 1861 [mf ed 1984] – 1mf – 9 – 0-665-45461-9 – mf#45461 – cn Canadiana [623]

The great eastern – Grahamstown SA, n1 (feb 2 1864)-n499 (sep 9 1868) (twice/wk) [mf ed Cape Town: SA library 1986] – 8r – 1 – (numbering irregular) – mf#MS00372 – sa National [079]

Great educators of three centuries : their work and its influence on modern education / Graves, Frank Pierrepont – New York: Macmillan, 1912 – 1mf – 9 – 0-7905-5332-5 – (incl bibl ref) – mf#1988-1332 – us ATLA [370]

The great election / Gillis, James Donald – Halifax [NS]: T C Allen, [189-?] [mf ed 1993] – 1mf – 9 – 0-665-91441-5 – mf#91441 – cn Canadiana [971]

Great elephant / Scholefield, Alan – New York, NY. 1968 (c1967) – 1r – us UF Libraries [960]

The great encyclical letters of pope leo 13 : translations from approved sources / Leo 13, Pope – New York: Benziger Bros, 1903 – 2mf – 9 – 0-524-03137-1 – mf#1990-4586 – us ATLA [240]

The great enigma / Lilly, William Samuel – 2nd ed. London: John Murray, 1893 [mf ed 1985] – 1mf – 9 – 0-8370-4132-5 – (incl bibl ref & ind) – mf#1985-2132 – us ATLA [210]

Great epic of india : its character and origin / Hopkins, Edward Washburn – New York: C Scribner, 1901 – (= ser Yale bicentennial publications) – 2mf – 9 – 0-524-04518-6 – mf#1990-3352 – us ATLA [490]

The great epic of israel : the web of myth, legend, history, law, oracle, wisdom and poetry of the ancient hebrews / Fiske, Amos Kidder – New York: Sturgis & Walton, 1911 [mf ed 1992] – 1mf – 9 – 0-524-04795-2 – mf#1992-0215 – us ATLA [221]

The great epics of ancient india : condensed into english verse / Dutt, Romesh Chunder – London: JM Dent & Co, 1900 – (= ser Samp: indian books) – (int by f max muller) – us CRL [810]

Great exhibition / Binney, Thomas – London, England. 1851 – 1r – us UF Libraries [240]

Great exhibition *see*
- Palais de cristal
- Pilote de londres

Great Falls Genealogy Society *see* Treasure state lines

The great famine, 1847 – Cheltenham, [1847] – (= ser 19th c ireland) – 1mf – 9 – mf#1.1.9428 – uk Chadwyck [941]

The great famine and its causes / Nash, Vaughan – [London] 1900 – (= ser 19th c british colonization) – 4mf – 9 – (with photos by aut & map of india showing the famine area) – mf#1.1.6276 – uk Chadwyck [360]

The great famine in ireland / O'Brien, William Patrick – [London], 1896 – (= ser 19th c ireland) – 4mf – 9 – mf#1.1.4638 – uk Chadwyck [941]

The great fire in st john's newfoundland, 8 july, 1892 / Harvey, Moses – Boston?: s.n, 1892 – 1mf – 9 – mf#67815 – cn Canadiana [971]

The great game : a plea for a british imperial policy / Thorburn, Walter Millar – London 1875 – (= ser 19th c british colonization) – 3mf – 9 – mf#1.1.9651 – uk Chadwyck [320]

Great german short novels and stories / ed by Cerf, Bennett A – New York: B A Cerf, D S Klopfer, c1933 – 1r – 1 – us Wisconsin U Libr [830]

The great gold lands of south africa : a vacation run in cape colony, natal, the orange free state, and the transvaal visiting the diamond mines and the gold fields; the scenes of the boer war and the war in zululand... / Smith, Ronald M – [London] 1891 – (= ser 19th c british colonization) – 4mf – 9 – mf#1.1.7031 – uk Chadwyck [916]

The great gorham case : a history in five books, including expositions of the rival baptismal theories / Binney, Thomas – London: Partridge and Oakey, 1850 – 1mf – 9 – 0-524-03608-X – mf#1990-4768 – us ATLA [240]

The great gulf *see* Hong kou (ccm18)

Great harvest – London, England. 18– – 1r – us UF Libraries [240]

Great hatred / Samuel, Maurice – New York, NY. 1941 (c1940) – 1r – us UF Libraries [025]

The great historical process against fascism / Spain. Embajada. United States – Washington, DC, n.d. Fiche W1179. (Blodgett Collection of Spanish Civil War Pamphlets) – 9 – us Harvard [946]

Great hymns and hymn writers *see* Sheng shih tien kao (ccm317)

The great ice age and its relation to the antiquity of man / Geikie, James – [London] 1874 – (= ser 19th c evolution & creation) – 7mf – 9 – mf#1.1.11618 – uk Chadwyck [550]

Great importance of a re-union between the catholics and the protestants... / Spencer, George – Manchester, England. 1839? – 1r – us UF Libraries [240]

The great indian mutiny of 1857 : its causes, features, and results / Kennedy, James – London, [1858] – (= ser 19th c british colonization) – 1mf – 9 – mf#1.1.8884 – uk Chadwyck [954]

Great indians / Radhakrishnan, Sarvepalli – Bombay: Hind Kitabs, 1954 – (= ser Samp: indian books) – (introductory essay on the author by d s sarma) – us CRL [954]

The great iron wheel / Graves, J R – 1856 – 1 – us Southern Baptist [242]

Great issues / Horton, Robert Forman – London: TF Unwin, 1909 – 1mf – 9 – 0-7905-3964-0 – mf#1989-0457 – us ATLA [240]

Great jesuit plot of the nineteenth century / Paterson, William – Edinburgh, Scotland. 1894 – 1r – us UF Libraries [240]

Great jewish short stories / Bellow, Saul – New York, NY. 1963 – 1r – us UF Libraries [420]

The great joy of saints / Simson, John – 1654- – 1 – us Southern Baptist [242]

A great judicial character, roger brooke taney / Gregory, Charles Noble – New Haven, CT: Field 1908. 18p. LL-490 – 1 – us L of C Photodup [340]

The great klondike gold fields : an exhaustive description and full information for prospectors, and up-to-date map of alaska – [New York?]: Alaska Transportation, Trading and Mining C[o], [1897?] [mf ed 1982] – 1mf – 9 – mf#32878 – cn Canadiana [622]

Great lakes bulletin – 1918 feb 27-1918 sep 30, 1918 oct 1-1919 mar 12, 1919 mar 13-dec 31 – 3r – 1 – mf#875039 – us WHS [071]

Great lakes communicator – v8 n6/7-v11 n2 [1978 mar/apr-1980 nov] – 1r – 1 – (cont: communicator [ann arbor, mi]) – mf#338401 – us WHS [071]

Great lakes entomologist – East Lansing MI 1966+ – 1,5,9 – ISSN: 0090-0222 – mf#2131 – us UMI ProQuest [590]

Great lakes gazette – v6 n4-v10 n4 [1978 jan-1980 summer iss, may/aug] – 1r – 1 – (cont: people & places) – mf#644216 – us WHS [071]

Great lakes indian community voice / Great Lakes Inter-Tribal Council – v1 n26-v5 n5 [1968 jul 29-1975 aug/sep] – 1r – 1 – mf#369069 – us WHS [305]

Great Lakes Indian Fish and Wildlife Commission *see* Masinaigan

Great Lakes Inter-Tribal Council *see* Great lakes indian community voice

Great Lakes Movement for a Democratic Military *see*
- Dare to struggle
- Navy times are changin'
- Now here this!
- Out now!
- Out now
- Right-on post

Great lakes news letter – Ann Arbor MI 1956-84 – 1,5,9 – ISSN: 0017-3665 – mf#6402 – us UMI ProQuest [639]

Great lakes research checklist – Ann Arbor MI 1959-95 – 1,5,9 – ISSN: 0072-7326 – mf#6403 – us UMI ProQuest [333]

Great Lakes Schooner Association *see* Schooner

Great Lakes seafarer / Seafarers' International Union of North America – v1-4 n6 [1943 mar-1945 oct] – 1r – 1 – mf#1057511 – us WHS [380]

The great lakes seafarer – Detroit. v. 7-10. Mar 1959-Jan 1964. Incomplete – 1 – us NY Public [978]

The great law : a study of religious origins and of the unity underlying them / Williamson, William – London: Longmans, Green, 1899 – 1mf – 9 – 0-524-01389-6 – (incl bibl ref) – mf#1990-2401 – us ATLA [200]
The great law book : the kingdom and reign of the messiah / Kingsbury, Harmon – New York: William Gowans, 1857. Beltsville, Md: NCR Corp, 1978 (4mf); Evanston: American Theol Lib Assoc, 1984 (4mf) – 9 – 0-8370-0950-2 – (incl ind) – mf#1984-4325 – us ATLA [220]
The great liberation = Mahanirvana tantra – Madras: Ganesh & Co, 1927 – (= ser Samp: indian books) – (trans fr sanskrit, with comm by arthur avalon) – us CRL [280]
The great lone land : a narrative of travel and adventure in the north-west of america / Butler, William Francis – 5th ed. London: S Low, Marston, Low & Searle, 1873 [mf ed 1983] – 5mf – 9 – 0-665-38357-6 – mf#38357 – cn Canadiana [917]
A great lord = Ein grosser herr / Frischauer, Paul – New York: Random House, c1937 [mf ed 1989] – 371p – 1 – (trans fr german by phyllis and trevor blewitt) – mf#7279 – us Wisconsin U Libr [430]
Great luso-brazilian figure / Boxer, Charles Ralph – London, England. 1957 – 1r – us UF Libraries [972]
The great match, and other matches – Boston: Roberts Brothers, 1877. 293p., illus. (No name series) – 1 – us Wisconsin U Libr [999]
The great meaning of metanoia : an undeveloped chapter in the life and teaching of christ / Walden, Treadwell – New York: Thomas Whittaker, 1896 – 1mf – 9 – 0-8370-5683-7 – mf#1985-3683 – us ATLA [240]
Great men and great movements : a volume of addresses / Galloway, Charles Betts; ed by Candler, Warren Akin – Nashville, Tenn: Pub House ME Church, South, 1914 – 1mf – 9 – 0-524-04359-0 – mf#1990-5042 – us ATLA [240]
The great men of god : biographies of patriarchs, prophets, kings and apostles / Guthrie, Thomas et al – New York: Nelson & Phillips, 1877 – 1mf – 9 – 0-524-08507-2 – mf#1993-0032 – us ATLA [220]
Great men of the christian church / Walker, Williston – Chicago: University of Chicago Press, 1908 – (= ser Constructive Bible Studies) – 1mf – 9 – 0-7905-6144-1 – (incl bibl ref) – mf#1988-2144 – us ATLA [240]
Great missionaries of the church / Creegan, Charles Cole & Goodnow, Josephine A B – New York: Thomas Y Crowell, c1895 – 1mf – 9 – 0-8370-6484-8 – mf#1986-0484 – us ATLA [920]
The great mogul / Payne, Robert – London: William Heinemann Ltd, 1950 – (= ser Samp: indian books) – us CRL [830]
The great mother of the gods / Showerman, Grant – 1901 – (= ser Bulletin of the university of wisconsin) – 1mf – 9 – 0-524-01299-7 – (incl bibl ref) – mf#1990-2335 – us ATLA [250]
Great mystery of godliness incontrovertible / Henderson, E – London, England. 1830 – 1r – us UF Libraries [240]
Great north road / Green, Lawrence George – Cape Town, South Africa. 1961 – 1r – us UF Libraries [960]
Great Northern and Northern Pacific Railway Companies *see* Steam locomotive drawings
Great Northern Railway Company *see* Annual reports
Great Northern Railway Company. Advertising and Publicity Dept *see* Magazine and newspaper advertisements and articles and other publicity
Great Northern Railway Company. Personnel Dept *see* Index to personnel files
Great Northern Railway of Canada *see* Prospectus of the great northern railway (of canada)
The great north-west of canada : a paper read at conference, indian and colonial exhibition, london, june 8th, 1886 / Begg, Alexander – London: H Blacklock, [1886?] [mf ed 1982] – 1mf – 9 – mf#30165 – cn Canadiana [971]
Great opinions by great judges / Snyder, William Lamartine – New York, Baker, Voorhis, 1883. 792 p. LL-1082 – 1 – us L of C Photodup [340]
The great parliamentary bore : [the case of prince azeem jah and britain's interests in its indian empire] / Bell, Evans – London, 1869 [i.e. 1868] – (= ser 19th c books on british colonization) – 2mf – 9 – mf#1.1.6072 – uk Chadwyck [954]
The great passion-prophecy vindicated / Maitland, Brownlow – London: Christian Evidence Committee of the SPCK, 1884 – 2mf – 9 – 0-524-06146-7 – mf#1992-0813 – us ATLA [220]
The great pestilence (a.d. 1348-9) : now commonly known as the black death / Gasquet, Francis Aidan, Cardinal – London: Simpkin Marshall, Hamilton, Kent, 1893 – 1mf – 9 – 0-8370-6905-X – (incl bibl ref and index) – mf#1986-0905 – us ATLA [941]
Great plains journal – Lawton OK 1961+ – 1,5,9 – ISSN: 0017-3673 – mf#9660 – us UMI ProQuest [071]
Great plains natural resources journal – v1-6. 1996-2002 – 9 – $127.00 set – mf#117741 – us Hein [344]
Great plains quarterly – Lincoln NE 1989+ – 1,5,9 – ISSN: 0275-7664 – mf#18013 – us UMI ProQuest [071]
The great poets and their theology / Strong, Augustus Hopkins – Philadelphia: American Baptist Publication Society, 1899, c1897 – 2mf – 9 – 0-7905-7543-4 – mf#1989-0768 – us ATLA [810]
The great preparation : or, redemption draweth nigh / Cumming, John – New York: Carlton, 1863. Beltsville, Md: NCR Corp, 1978 (4mf); Evanston: American Theol Lib Assoc, 1984 (4mf) – 9 – 0-8370-0960-X – mf#1984-4315 – us ATLA [240]
The great problem : can it be solved? / Gleig, George Robert – Edinburgh: William Blackwood, 1876 [mf ed 1985] – 1mf – 9 – 0-8370-3308-X – mf#1985-1308 – us ATLA [240]
The great problems = Massimi problemi / Varisco, Bernardino – New York: Macmillan, 1914 – (= ser Library of Philosophy) – 1mf – 9 – 0-7905-8951-6 – (in english) – mf#1989-2176 – us ATLA [190]
The great question, will you consider the subject of personal religion? / Boardman, Henry Augustus – Philadelphia: American Sunday-School Union, c1855 – 3mf – 9 – 0-7905-3588-2 – mf#1989-0081 – us ATLA [240]
The great red dragon : or, the master-key to popery / Gavin, Antonio – Boston: Samuel Jones, 1854 – 1mf – 9 – 0-8370-8985-9 – mf#1986-2985 – us ATLA [230]
Great red island / Stratton, Arthur – New York, NY. 1964 – 1r – us UF Libraries [960]
The great redemption : or, gospel light, under the labors of moody and sankey / Moody, Dwight Lyman – Chicago: The Century Book and Paper Co., 1889. Beltsville, Md: NCR Corp, 1978 (6mf); Evanston: American Theol Lib Assoc, 1984 (6mf) – (= ser Revivalism and revival preachers in america) – 9 – 0-8370-1228-7 – mf#1984-3013 – us ATLA [240]
The great redemption : a treatise on various doctrines of the new testament religion as delivered to us by our lord and saviour jesus christ and by his inspired apostles, and enjoined upon all his followers / Leckrone, Quincy – North Manchester, Ind: Bible Student Pub Co, 1898 – 1mf – 9 – 0-524-04056-7 – mf#1990-4964 – us ATLA [240]
Great register of los angeles county – Los Angeles Co, CA. 1873-86 – 1r – 1 – $50.00 – mf#C40227 – us Library Micro [978]
The great register of placer county – Placer Co, CA. 1867-82 – 1r – 1 – $50.00 – mf#C40250 – us Library Micro [978]
The great register of plumas county – Plumas Co, CA. 1867; 1879; 1896 – 1r – 1 – $50.00 – mf#B40251 – us Library Micro [978]
The great register of sacramento county – Sacramento Co, CA. 1867-79 – 1r – 1 – $50.00 – mf#C40252 – us Library Micro [978]
The great register of san francisco city and county – San Francisco, CA. 1866; 1867; 1869; 1871 – 1r – 1 – $50.00 – mf#C40308 – us Library Micro [978]
Great register of siskiyou county – Siskiyou Co, CA. oct 24 1892 – 1r – 1 – $50.00 – mf#B40271 – us Library Micro [978]
Great register of voters for fresno county – Fresno Co, 1867-77; AA-LY (1890); MA-ZO (1890); A-E (1892-95); F-L (1892-95); M-R (1892-95); S-Z (1892-95) – 8r – 1 – $400.00 – mf#B06094 – us Library Micro [324]
Great register of yuba county – Yuba Co, CA. 1867-96 – 1r – 1 – $50.00 – mf#C40288 – us Library Micro [978]
The great rejected books of the biblical apocrypha – New York: Parke, Austin & Lipscomb c1917 [mf ed 1992] – (= ser The sacred books and early literature of the east 14) – 1mf – 9 – 0-524-03974-7 – (trans by robert henry charles; bibl by edward h johns; incl bibl ref) – mf#1992-0017 – us ATLA [220]
The great religions of india / Mitchell, John Murray – Edinburgh: Oliphant, Anderson and Ferrier, 1906 – (= ser Duff Missionary Lectures) – 1mf – 9 – 0-524-01201-6 – mf#1990-2277 – us ATLA [280]
Great religions of the world / Giles, Herbert Allen et al – New York: Harper, 1901 – 1mf – 9 – 0-524-00723-3 – mf#1990-2051 – us ATLA [200]
Great religious teachers of the east / Martin, Alfred Wilhelm – New York: Macmillan, 1911 – 1mf – 9 – 0-524-00934-1 – mf#1990-2157 – us ATLA [280]
The great revival in the west, 1797-1805 / Cleveland, Catharine Caroline – Chicago, Ill: University of Chicago Press, 1916 – 1mf – 9 – 0-524-06242-0 – (incl bibl ref) – mf#1990-5197 – us ATLA [240]
The great revival of 1800 / Speer, William – Philadelphia: Presbyterian Board of Publication, c1872 – 1mf – 9 – 0-7905-5970-6 – mf#1988-1970 – us ATLA [240]
The great revival of the eighteenth century / Hood, Edwin Paxton – Philadelphia: American Sunday-School Union, [1882] Beltsville, Md: NCR Corp, 1977 (4mf); Evanston: American Theol Lib Assoc, 1984 (4mf) – 9 – 0-8370-1650-9 – mf#1984-6243 – us ATLA [240]
Great revivals and the great republic / Candler, Warren Akin – Nashville: Publishing House of the M. E. Church, South, 1904. Beltsville, Md: NCR Corp, 1978 (4mf); Evanston: American Theol Lib Assoc, 1984 (4mf) – (= ser Revivalism and revival preachers in america) – 9 – 0-8370-0227-3 – (incl ind) – mf#1984-3003 – us ATLA [242]
The great revolt in india : its effects upon the missions of the presbyterian board / Wilson, John Leighton & Lowrie, John Cameron – New York: Printed for the Board of Foreign Missions by Edward O Jenkins, 1857 – 1mf – 9 – 0-524-07278-7 – mf#1991-3019 – us ATLA [954]
The great rift valley : being the narrative of a journey to mount kenya and lake baringo... / Gregory, J W – London, 1896 – 6mf – 9 – mf#HT-53 – ne IDC [590]
Great River Genealogical Society *see*
– Yellowjacket
– Yellowjacket of the great river genealogical society
Great River Road Association et al *see*
– News letter
– Pilot's wheel
Great river road news / Mississippi River Parkway Commission – 1963 dec-1973 sep – 1r – 1 – (cont: news letter [mississippi river parkway commission]; cont by: pilot's wheel) – mf#162005 – us WHS [071]
Great russian plays / Houghton, Norris – New York, NY. 1960 – 1r – us UF Libraries [460]
Great russian revolution / Chernov, Viktor Mikhailovich – New Haven: Yale UP 1936 [mf ed 1986] – 1r – 1 – (trans & abr by philip e mosely; with bibl ref) – mf#1797 – us Wisconsin U Libr [947]
Great russian short stories / Graham, Stephen – New York, NY. 1959 – 1r – us UF Libraries [460]
The great salvation / Zollars, Ely Vaughan – Cincinnati: Standard Pub Co, c1896 – 1mf – 9 – 0-524-06508-X – mf#1991-2608 – us ATLA [240]
Great saviors of the world / Abhedananda, swami – New York: Vedaanta Society, c1911 – 1mf – 9 – 0-524-00674-1 – mf#1990-2002 – us ATLA [280]
Great sayings by great lawyers : immortal thoughts snatched from oblivion / Clark, G J – Kansas City MO: Vernon Law Bk Co, 1926 – 9mf – 9 – $13.50 – mf#llmc92-208 – us LLMC [340]
The great schoolmen of the middle ages : an account of their lives and the services they rendered to the church and the world / Townsend, William John – London: Hodder & Stoughton, 1881 [mf ed 1992] – 1mf – 9 – 0-524-04628-X – (incl bibl ref) – mf#1990-1288 – us ATLA [180]
Great south african christians / Davies, Horton – Westport, CT. 1970 – 1r – us UF Libraries [960]
Great south land / Koebel, William Henry – New York, NY. 1920 – 1r – us UF Libraries [972]
Great southwest *see* Miscellaneous newspapers of la plata county, colorado
Great speckled bird / Atlanta Cooperative News Project et al – Atlanta GA. 1968 mar 15-1969 mar 10, 1970 jul 27-1971 apr 26, 1969 mar 17-1970 jul 20, 1971 may 3-1972 mar 27, 1972 apr 3-dec 25, 1973, 1974, 1975 – 7r – 1 – mf#770752 – us WHS [071]
Great speckled bird – Atlanta GA 1968-76 – 1 – ISSN: 0017-369X – mf#7709 – us UMI ProQuest [590]
The great speckled bird – n1-37. 1968-March 1969 – 1 – us AMS Press [073]
Great speeches – Toronto?: s.n, 1881 – 1mf – 9 – mf#08817 – cn Canadiana [080]
Great speeches by great lawyers : a collection of arguments and speeches before courts and juries by eminent lawyers / Snyder, William Lamartine – New York: Baker, Voorhis & Co, 1892 – 8mf – 9 – $12.00 – mf#LLMC 92-131 – us LLMC [340]
The great standard oil monopoly case : united states of america v standard oil company of new jersey / U.S. Supreme Court – 6r – 1 – $935.00 – 0-89093-018-X – (anti-trust case against john d rockefeller's standard oil co in 1911. with p/g) – us UPA [343]
The great supper not calvinistic : being a reply to the rev. dr. fairchild's discourses on the parable of the great supper / Lee, Leroy Madison; ed by Summers, Thomas Osmond – Nashville, Tenn: Southern Methodist Pub House, 1883 – 1mf – 9 – 0-7905-9301-7 – mf#1989-2526 – us ATLA [240]
The great supper of god : or, discourses on weekly communion = Communion hebdomadaire / Coube, Stephen; ed by Brady, F X – New York: Benziger, 1901 – 1mf – 9 – 0-8370-6811-8 – mf#1986-0811 – us ATLA [240]
Great sutton, st john – (= ser Cheshire monumental inscriptions) – 1mf – 9 – £2.50 – mf#90 – uk CheshireFHS [929]
Great swamp erie da da boom – [v1]-2 n8 [[1970? dec]-1972 apr 26] – 1r – 1 – mf#1057514 – us WHS [071]
The great taboo / Allen, Grant – New York: A L Burt, 1890? – 3mf – 9 – (incl publ list) – mf#18108 – cn Canadiana [890]
The great teacher : characteristics of our lord's ministry / Harris, John – 13th american from the 10th London rev ed. Boston: Gould, Kendall and Lincoln, 1850 – 1mf – 9 – 0-524-08410-6 – mf#1993-0025 – us ATLA [220]
Great thoughts of the bible / Reid, John – New York: Wilbur B Ketcham, c1891 – 1mf – 9 – 0-8370-3986-X – (incl ind) – mf#1985-1986 – us ATLA [220]
Great trek / Walker, Eric Anderson – London, England. 1965 – 1r – us UF Libraries [960]
The great tribulation : or, the things coming on the earth / Cumming, John – London: Richard Bentley, 1860. Beltsville, Md: NCR Corp, 1978 (6mf); Evanston: American Theol Lib Assoc, 1984 (6mf) – 9 – 0-8370-0966-9 – mf#1984-4309 – us ATLA [240]
The great value and success of foreign missions : proved by distinguished witnesses / ed by Liggins, John – New York: Baker & Taylor, c1888 [mf ed 1986] – 1mf – 9 – 0-8370-6139-3 – (incl app & ind. int by arthur tappan pierson) – mf#1986-0139 – us ATLA [240]
The great victory in boston – Boston, Mass.: Committee of One Hundred, 1889 – (= ser Committee of One Hundred Series) – 1mf – 9 – 0-524-02338-7 – mf#1990-0594 – us ATLA [240]
Great victory of cambodian people : warmly congratulating the patriotic cambodian armed forces and people on the liberation of phnom penh and all cambodian. = Chien-p'u-chai jen min ti wei ta sheng li – Peking: Foreign languages Press [mf ed 1990] – 1r with other items – 1 – mf#mf-10289 seam reel 125/2 [§] – us CRL [959]
The great war : six sermons / Gladden, Washington – Columbus, O[hio]: McClelland, [1915?] – 1mf – 9 – 0-7905-5600-6 – mf#1988-1600 – us ATLA [240]
Great warford, baptist chapel – (= ser Cheshire monumental inscriptions) – 1mf – 9 – £2.50 – mf#16 – uk CheshireFHS [929]
The great west *see* Miscellaneous newspapers of pueblo county
Great west series / Hamilton Co. Cincinnati – v1 n1. jan-dec 1850, may 1853-apr 1855 [daily] – 1r – 1 – mf#B6620 – us Ohio Hist [071]
Great Western Railway Co (Canada) *see* Report by the directors...to the shareholders upon the report made by the commission
Great Western Railway Company of Canada *see* Reply of the president and directors to the report of the committee of investigation
Great western railway of canada, and united states mail route : from suspension bridge, n f, to detroit, and branches from hamilton to toronto...: passenger train time table, june 1861 – [Hamilton, Ont?: s.n, 1861?] [mf ed 1984] – 1mf – 9 – 0-665-45632-8 – mf#45632 – cn Canadiana [380]
Great western railway – roll of honour – 1914-18 – 1mf – 9 – £1.25 – uk Glamorgan FHS [380]
The great white banner / Keough, Walter James – [St John NB?: s.n.] 1913 [mf ed 1996] – 2mf – 9 – 0-665-78740-5 – mf#78740 – cn Canadiana [810]
Great white throne – Edinburgh, Scotland. 1867 – 1r – us UF Libraries [240]
Great women of india / ed by Madhavananda, Swami & Majumdar, Ramesh Chandra – Mayavati, Almora: Advaita Ashrama, 1953 – (= ser Samp: indian books) – us CRL [954]
Great zimbabwe, mashonaland, rhodesia / Hall, Richard Nicklin – New York, NY. 1969 – 1r – us UF Libraries [960]
Greater amusements – Sunnyside NY 1975-78 – 1,5,9 – ISSN: 0017-3701 – mf#9678 – us UMI ProQuest [790]
Greater asia – Rangoon, Burma. 7 Feb 1943-21 Apr 1945 – 1r – 1 – us L of C Photodup [079]
Greater baton rouge black pages : business listing and information guide – 1996 – 1r – 1 – mf#3912391 – us WHS [917]

Greater Buffalo Industrial Union Council *see* Union leader

Greater Cleveland Genealogical Society *see* Certified copy

Greater cleveland rapid transit authority records *see* Rta records, lr-ra 0001, 1848-1958

Greater cleveland regional transit authority records, lr-ra 0001 – 1848-1958 – 25r – 1 – (records of several plank toll road and street car companies, predecessors to the cleveland railway co, of the crc, of the shaker heights rapid transit, and copies of the tayler grant negotiations and settlement) – us Western Res [978]

Greater enterprise news – Portland OR: Clarke Pub Co Inc, 1964-66 [wkly] – 1 – (merger of: greater eastside news; enterprise (parkrose, or: 1962); cont by: greater enterprise news north) – us Oregon Lib [071]

Greater enterprise news north – Portland OR: Clarke Pub Co Inc, 1966-67 [wkly] – 1 – (cont: greater enterprise news (1964-1966); cont by: Clarke press (greater enterprise news north ed)) – us Oregon Lib [071]

The greater half of the continent / Wiman, Erastus – Toronto: Hunter, Rose, 1889 – 1mf – 9 – mf#34517 – cn Canadiana [917]

Greater Harrisburg Region Central Labor Council *see* Central penna. labor news

Greater Holy Temple COGIC (Jacksonville, FL) *see* Holy temple good news

Greater Holy Temple COGIC [Jacksonville FL] *see* Daily walk

Greater india / Tagore, Rabindranath – Madras: S Ganesan, 1921 – (= ser Samp: indian books) – us CRL [954]

Greater Johnstown Regional Central Labor Council *see*
- Council's voice
- Labor council newsletter

Greater La Crosse Area Chamber of Commerce *see* Update

Greater La Crosse Chamber of Commerce *see* La crosse roads

Greater Lansing Labor Council *see* Skilled craftsman and lansing industrial news

Greater Lansing Labor Council [MI] et al *see* Lansing industrial news

Greater lawrence jewish news – (Lawrence, Mass.) V34, no. 3 (Dec. 1963); v34, no. 5 (Feb. 1964)-v34, no. 7 (Apr.1964) – us AJPC [939]

The greater life and work of christ : as revealed in scripture, man, and nature / Patterson, Alexander – Chicago: Fleming H Revell, c1896 – 1mf – 9 – 0-8370-3982-7 – mf#1985-1982 – us ATLA [240]

Greater lowell indian cultural association – (1977 feb/may, aug, oct-dec, 1978 nov-1985 nov] – 1r – 1 – mf#1231467 – us WHS [071]

Greater Madison Chamber of Commerce *see* Statistical bulletin

Greater Madison Chamber of Commerce [Madison WI] *see* Issues inreview

Greater Madison Convention and Visitor's Bureau *see* In the bureau spotlight

Greater Madison Convention and Visitors Bureau *see*
- Bureau drawer
- Calendar of events

Greater Menomonie Area Chamber of Commerce *see* You are menomonie news

Greater milwaukee star – Milwaukee WI. 1968 apr 10-oct 12, 1968 apr 10-dec 28, 1968 oct 19-1969 nov 29, 1969 dec 6-1971 mar 13, 1969 jan 4-dec 27, 1970 jan 3-1971 jan 23, 1971 feb 6-sep 11, 1971 mar 20-oct 30, 1971 sep 18-oct 30 – 9r – 1 – (cont: milwaukee star [milwaukee, wi: 1961]; cont by: your greater milwaukee star) – mf#871437 – us WHS [071]

Greater Milwaukee TRIM Committee *see* Trim bulletin

Greater Minneapolis Council of Churches *see*
- Vision on the wind
- Wig-i-wam

Greater minnesota edition / Minnesota Women's Consortium – ed 1-41 [1984 aug 1-1988 jul 6] – 1r – 1 – mf#1057522 – us WHS [305]

Greater new guide baptist church informer – 1991 apr-may – 1r – 1 – mf#3912529 – us WHS [242]

Greater new haven black coalition weekly – New Haven CT. 1972 feb/mar-sep – 1r – 1 – mf#874815 – us WHS [071]

Greater news – 1998 jan 10-dec 10, 1999 feb 13-oct 9 – 2r – 1 – (cont: new jersey greater news) – mf#3641974 – us WHS [071]

Greater News of Northern New Jersey *see* New jersey greater news

Greater oregon – Halsey OR: H F Lake & J F Howard, 1929-78 [wkly] [mf ed 1960-75] – 18r – 1 – (cont: halsey enterprise (1917-1924). merged with: benton county herald (corvallis, or); weekly oregon herald (albany, or: 1978)) – us Oregon Lib [071]

Greater portland news – Portland OR: George A Denfeld & Beth Urba Denfeld, 1955- [wkly] – 1 – (cont: upper sandy news (1947-). pub with: upper sandy news, sep 29 1955-feb 23 1956) – us Oregon Lib [071]

Greater south africa / Smuts, Jan Christiaan – Johannesburg, South Africa. 1940 – 1r – us UF Libraries [960]

Greater two rivers shopper – Denmark WI. 1974 aug 7-dec 18, 1975 dec 3-1976 apr 28, 1975 jan 8-jul 2, 1975 jul 9-nov 26, 1976 may 5-sep 15, 1976 sep 22-1977 feb 7, 1977 dec 8-1978 mar 13, 1977 feb 10-1977 may 23, 1977 may 26-sep 5, 1977 sep 8-dec 5, 1978 jun 15, 1978 mar 16-jun 12 – 12r – 1 – (continued by: two rivers shopper) – mf#956872 – us WHS [071]

The greatest bank in america / Oxley, James Macdonald – [S.l: s.n, 1900?] [mf ed 1982] – 1mf – 9 – 0-665-17912-X – (fr: the canadian magazine) – mf#17912 – cn Canadiana [332]

The "greatest discovery" exploded : or, the death, resurrection and second coming of christ established against the aspersion and claims of mirza ghulam ahmad of qadian / Thakur Dass, G L – 1st ed. Lodiana: American Tract Society, 1903 [mf ed 1992] – 1mf – 9 – 0-524-02617-3 – mf#1990-3067 – us ATLA [260]

The greatest of the judges : principles of church life illustrated in the history of gideon / Miller, William – London: Hodder and Stoughton, 1878 – 1mf – 9 – 0-524-05232-8 – mf#1992-0365 – us ATLA [220]

The greatest of the world's forces applied through a half-day perpetual, industrial, and universal school / Farrar, Ephraim H – [Montreal?: s.n, 1885?] [mf ed 1984] – 1mf – 9 – 0-665-01792-8 – mf#01792 – cn Canadiana [370]

The greatest theme in the world / Marsh, Frederick Edward – New York: Alliance Press, c1908 [mf ed 1992] – (= ser Christian & missionary alliance coll) – 1mf – 9 – 0-524-02123-6 – mf#1990-4189 – us ATLA [240]

The greatest thing in the world / Drummond, Henry – Toronto: Hodder & Stoughton, c1918 – 1mf – 9 – 0-665-66625-X – mf#66625 – cn Canadiana [230]

Greatest treasure in the exhibition / Merle D'aubigne, J H – London, England. 18– – 1r – us UF Libraries [240]

The greatest work in the world : or, the evangelization of all peoples in the present century / Pierson, Arthur Tappan – 4th rev ed. New York: Fleming H Revell, c1891 – 1mf – 9 – 0-8370-7253-0 – mf#1986-1253 – us ATLA [240]

The greatest work in the world : or, the mission of christ's disciples / Titus, Charles Buttz – [Tentative ed] [S.l.: s.n.], c1906 – 1mf – 9 – 0-524-06450-4 – mf#1991-2572 – us ATLA [240]

Greatness at the feet of jesus / Fleming, J – London, England. 18– – 1r – us UF Libraries [240]

The greatness of human nature : the key to the religion of w e channing / Kyper, Ralph Edward – Chicago, 1936. Chicago: Dep of Photodup, U of Chicago Lib, 1971 (1r); Evanston: American Theol Lib Assoc, 1984 (1r) – 1 – 0-8370-0383-0 – mf#1984-B169 – us ATLA [240]

Greatrex, Charles Butler *see* Whittlings from the west

La great-west life – S.l: s.n, 189-? – 1mf – 9 – mf#61341 – cn Canadiana [360]

Greatwood, Edward Albert *see* Die dichterische selbstdarstellung im roman des jungen deutschland

Greaves, Charles Augustus *see* The science of life

Greaves, H L *see* Metropolitan for scotland

Greaves, Joseph *see* Christian researches in syria and the holy land, in 1823 and 1824

Greaves, William Herbert *see* Common pleas act, 1911

Grebe, Walter *see* Die erzaehlungstechnik viktor scheffels

Grebenshchikov, Georgii *see* Kupava

Le grec, le latin : leur utilite pour apprecier la signification des mots actuels de la langue... / Baillairge, Charles P Florent – Ottawa?: s.n, 1899? – 1mf – 9 – mf#04355 – cn Canadiana [450]

Grece, Charles *see* Facts and observations respecting canada, and the united states of america

Grece, Charles Frederick *see*
- Essays on practical husbandry
- Facts and observations respecting canada, and the united states of america

Grechko, A M *see* Organizatsiia narodnogo khoziaistva i kooperatsiia

Grecia eterna / Gomez Carrillo, Enrique – Guatemala, 1964 – 1r – us UF Libraries [972]

Grecian and roman mythology / Dwight, Mary Ann – New York: Putnam, 1849 – 2mf – 9 – 0-524-02016-7 – mf#1990-2791 – us ATLA [250]

Die gred : roman aus dem alten nuernberg / Ebers, Georg – Stuttgart: Deutsche Verlags-Anstalt, [1893-1897?] [mf ed 1993] – (= ser Georg ebers gesammelte werke 14-15) – 2v – 1 – mf#8554 reel 2 – us Wisconsin U Libr [830]

Die gred : roman aus dem alten nuernberg / Ebers, Georg – Stuttgart: Deutsche Verlags-Anstalt. [1893-97?] [mf ed 1993] – (= ser Georg ebers gesammelte werke 14-15) – 2v – 1 – mf#8554 reel 3 – us Wisconsin U Libr [830]

Grediagin, Ann *see* The effect of a 50-km ultramarathon on vitamin b6 metabolism and plasma and urinary urea nitrogen

Greebe, Cornelius Aleidus Arnoldus Ioannes *see* De dioscuris

Greece / Conder, Josiah – London 1826 [mf ed Hildesheim 1995-98] – 2v on 5mf – 9 – €100.00 – 3-487-29082-0 – gw Olms [914]

Greece, 1847-63 *see* The papers of queen victoria on foreign affairs

Greece, ancient and modern / Felton, Cornelius Conway – Lectures before the Lowell Institute. 7th ed. Boston: Houghton, Mifflin, 1886.2v in 1 – 1 – us Wisconsin U Libr [450]

Greece and babylon : a comparative sketch of mesopotamian, anatolian and hellenic religions / Farnell, Lewis Richard – Edinburgh: T & T Clark, 1911 – 1mf – 9 – 0-7905-3738-9 – (incl bibl ref) – mf#1989-0231 – us ATLA [250]

Greece and the golden horn / Olin, Stephen – New York: J. C. Derby, 1854. Beltsville, Md: NCR Corp, 1978 (4mf); Evanston: American Theol Lib Assoc, 1984 (4mf) – 9 – 0-8370-0942-1 – mf#1984-4303 – us ATLA [914]

Greece and the greeks of the present day / About, Edmond – New York: Dix, Edwards & Co, 1857 [mf ed 1987] – xvi/360p – 1 – mf#9111 – us Wisconsin U Libr [949]

Greece. Ethnike Statisstike Hyperesia *see*
- Statistike epeteris tes hellados 1930-1939
- Statistike epeteris tes hellados 1954-1965

Greece, New York. Greece Baptist Church *see* Records

Greece & rome – Cambridge, England 1931+ – 1,5,9 – ISSN: 0017-3835 – mf#1247 – us UMI ProQuest [450]

The greek and eastern churches / Adeney, Walter Frederic – New York: Charles Scribner 1908 [mf ed 1986] – 2mf – 9 – 0-8370-7600-5 – (incl bibl ref & ind) – mf#1986-1600 – us ATLA [243]

The greek and eastern churches : their history, faith, and worship / ed by Kidder, Daniel Parish – New York: Carlton & Phillips 1854 [mf ed 1986] – 1mf – 9 – 0-8370-8265-X – mf#1986-2265 – us ATLA [243]

A greek and english lexicon of the new testament / Robinson, Edward – new rev ed. New York: Harper & Brothers, c1850 [mf ed 1993] – 8mf – 9 – 0-524-08089-5 – mf#1992-1149 – us ATLA [450]

A greek and english lexicon to the new testament : in which the words and phrases... are distinctly explained... / Parkhurst, John – new ed. London: Printed for George Cowie, 1825 [mf ed 1991] – 712p – 1 – mf#7635 – us Wisconsin U Libr [450]

Greek and gothic progress and decay in the three arts / Tyrwhitt, Richard St John – London, England. 1881 – 1r – us UF Libraries [720]

Greek and roman ghost stories / Collison-Morley, Lacy – Oxford: BH Blackwell; London: Simpkin, Marshall, 1912 – 1mf – 9 – 0-524-01686-0 – mf#1990-2588 – us ATLA [450]

Greek and roman stoicism and some of its disciples : epictetus, seneca and marcus aurelius / Davis, Charles Henry Stanley – Boston: Herbert B Turner, 1903 – 1mf – 9 – 0-524-01051-X – mf#1990-2199 – us ATLA [180]

The greek boy and the sunday-school : comprising ceremonies of the greek church... / Castanis, C Plato – 2nd ed. Philadelphia: William S Martien, 1852, c1847 – 1mf – 9 – 0-8370-8248-X – mf#1986-2248 – us ATLA [240]

The greek catholic faith : a homily / Anatolius, Bishop of Mohilew and Mstislaw – New York: E P Dutton, 1873 – 1mf – 9 – 0-8370-7761-3 – mf#1986-1761 – us ATLA [242]

The greek catholic sower – [Joliet, IL: Pittsburgh Byzantine Diocesan Press. v1 n1-n7 mar-sep 1949; v2 n1-6 jan-jun 1950] – 1 – us CRL [241]

Greek catholic union messenger = Viestnik greko kaft: sojedinenija – Homestead, PA: Greek Catholic Union, 1952-76, 1953-75 – 1 – us CRL [071]

Greek church, her doctrines and principles contrasted with those of... / Bardsley, Joseph – London, England. 1870 – 1r – us UF Libraries [240]

Greek coins; poems...with memorabilia by floyd dell, edna kenton and susan glaspell / Cook, George Cram – New York, (c1925). 142p. With: The Royal Woman by H. Mann. 1 reel. 1297 – 1 – us Wisconsin U Libr [810]

Greek divination : a study of its methods and principles / Halliday, William Reginald – London: Macmillan, 1913 – 1mf – 9 – 0-524-01706-9 – (incl bibl ref) – mf#1990-2608 – us ATLA [250]

The greek fathers / Fortescue, Adrian – London: Catholic Truth Society, 1908 – 1mf – 9 – 0-524-03093-6 – (incl bibl ref) – mf#1990-0818 – us ATLA [240]

Greek folk songs from the ottoman provinces of northern hellas – 2nd ed, rev and enl. London: Ward and Downey 1888 [mf ed Bloomington IN: Indiana Uni Lib, Preservation Dept 1984] – xxxix 290p on 1r – 1 – us Indiana Preservation [390]

The greek gospel : an interpretation of the coming faith / Usher, Edward Preston – Grafton, Mass, USA: E Usher, 1909 – 1mf – 9 – 0-7905-8613-4 – mf#1989-1838 – us ATLA [240]

A greek grammar for schools and colleges / Hadley, James – New York: D Appleton, 1877 [mf ed 1993] – 1mf – 9 – 0-524-08407-6 – mf#1993-0022 – us ATLA [450]

A greek grammar for the use of schools and colleges / Sophocles, Evangelinus Apostolides – new ed. Hartford: WJ Hamersley, 1853 [mf ed 1993] – 1mf – 9 – 0-524-07343-0 – mf#1992-1074 – us ATLA [450]

The greek grammar of roger bacon and a fragment of his hebrew grammar / ed by Nolan, Edmond & Hirsch, Samuel Abraham – Cambridge: University Press, 1902 [mf ed 1986] – 1mf – 9 – 0-8370-9682-0 – (in latin, english & greek. incl ind) – mf#1986-3682 – us ATLA [450]

A greek grammar of the new testament = Grammatik des neutestamentlichen sprachidioms / Winer, Georg Benedikt – Andover: printed...by Flagg & Gould, 1825 [mf ed 1989] – 1mf – 9 – 0-7905-2574-7 – (english trans fr german by moses stuart and edward robinson) – mf#1987-2574 – us ATLA [450]

Greek learning in the western church during the seventh and eighth centuries, a.d / Lumby, Joseph Rawson – Cambridge: [s.n.], 1878 (J. Palmer) – 1mf – 9 – 0-7905-4947-6 – mf#1988-0947 – us ATLA [450]

Greek lexicon of the roman and byzantine periods / Sophocles, Evangelinus Apostolides – 1900 – 9 – $39.00 – us IRC [450]

Greek life and thought; a portrayal of greek civilization / Van Hook, Larue – New York: Columbia University Press, 1923. xiv,329p. front., illus., plates, map – 1 – us Wisconsin U Libr [900]

The greek liturgies : chiefly from original authorities / ed by Swainson, Charles Anthony – Cambridge: University Press, 1884 – 2mf – 9 – 0-8370-5839-2 – (with an appendix containing the coptic ordinary canon of the mass from 2 mss in the british museum, ed & transl by c bezold) – mf#1985-3839 – us ATLA [240]

Greek manuals of church doctrine / Duckworth, Henry Thomas Forbes – London: Published for the Eastern Church Association [by] Rivingtons, 1901 – 1mf – 9 – 0-7905-4416-4 – mf#1988-0416 – us ATLA [240]

Greek mythology systematized / Scull, Sarah Amelia – Philadelphia: Porter and Coates, c1880 – 1mf – 9 – 0-524-01298-9 – mf#1990-2334 – us ATLA [250]

The greek new testament / ed by Tregelles, Samuel Prideaux – London: Samuel Bagster: C J Stewart, 1857-1879 – 11mf – 9 – 0-8370-1981-8 – mf#1987-6368 – us ATLA [225]

Greek original of the new testament asserted / Burgess, Thomas – London, England. 1823 – 1r – us UF Libraries [240]

Greek Orthodox Archdiocese of North and South America *see* Orthodox observer

Greek orthodox theological review – Brookline MA 1954+ – 1,5,9 – ISSN: 0017-3894 – mf#2443 – us UMI ProQuest [243]

Greek palimpsest fragments of the gospel of saint luke : obtained in the island of zante, by the late general colin macaulay, and now in the library of the british and foreign bible society / ed by Tregelles, Samuel Prideaux – London: Samuel Bagster, 1861 [mf ed 1989] – 1mf – 9 – 0-8370-1306-2 – mf#1987-6041 – us ATLA [225]

Greek philosophers / Benn, Alfred William – London, England. v1-2. 1882 – 1r – us UF Libraries [025]

Greek piety / Nilsson, Martin Persson – Oxford, England. 1948 – 1r – us UF Libraries [025]

Greek pottery / Lane, Arthur – Faber & Faber. 1948 – 9 – $10.00 – us IRC [930]

Greek proverbs from mrs peter caravasios / Caravasios, Peter – s.l, s.l? 1939 – 1r – us UF Libraries [978]

Greek reader : consisting of selections from xenophon, plato, herodotus, and thucydides: with notes adapted to goodwin's greek grammar, parallel references to crosby's and hadley's grammars, and copperplate maps / ed by Goodwin, William W & Allen, Joseph H – rev ed. Boston: Ginn and Heath, 1877, c1871 – 1mf – 9 – 0-8370-9236-1 – (texts in greek; notes in english) – mf#1986-3236 – us ATLA [450]

Greek, roman and byzantine studies – Durham NC 1994+ – 1,5,9 – ISSN: 0017-3916 – mf#18425,01 – us UMI ProQuest [450]

Greek sculpture and modern art / Walston, Charles – Cambridge, England. 1914 – 1r – us UF Libraries [720]

Greek study : church and religious life / Hargis, Modeste – s.l, s.l? 1939 – 1r – us UF Libraries [978]

Greek study : citizens of us enjoy big heritage / Adallis, Diogenes – s.l, s.l? 1939 – 1r – us UF Libraries [978]

Greek study : daughters of penelope / Hargis, Modeste – s.l, s.l? 1939 – 1r – us UF Libraries [978]

Greek study : emigration to the united states and... / Hargis, Modeste – s.l, s.l? 1939 – 1r – us UF Libraries [978]

Greek study : interview with oldest greek in pensa... / Hargis, Modeste – s.l, s.l? 1939 – 1r – us UF Libraries [978]

Greek study : pensacola florida / Hargis, Modeste – s.l, s.l? 1939 – 1r – us UF Libraries [978]

Greek study : pensacola florida : escambia county / Hargis, Modeste – s.l, s.l? 1939 – 1r – us UF Libraries [978]

Greek terra-cotta statuettes / Huish, Marcus Bourne – London 1900 – (= ser 19th c art & architecture) – 5mf – 9 – mf#4.2.1219 – uk Chadwyck [730]

Greek terracotta statuettes / Hutton, Caroline Amy – London 1899 – (= ser 19th c art & architecture) – 2mf – 9 – mf#4.2.1390 – uk Chadwyck [730]

The greek testament : with a critically revised text, a digest of various readings, marginal references to verbal and idiomatic usage, prolegomena, and a critical and exegetical commentary / Alford, Henry – 7th ed. London: Rivingtons; Cambridge: Deighton, Bell, 1871-1875 – 9mf – 9 – 0-8370-1168-X – (incl bibl ref) – mf#1987-6004 – us ATLA [220]

Greek testament lessons for colleges, schools, and private students : consisting chiefly of the sermon on the mount, and the parables of our lord / Smith, John Hunter – Edinburgh: William Blackwood, 1884 – 1mf – 9 – 0-8370-5292-0 – mf#1985-3292 – us ATLA [220]

The greek testament: with a critically revised text; a digest of various readings: marginal references to verbal and idiomatic usage: prolegomena: and a critical and exegetical commentary – By Henry Alford.London: Rivingtons, 1871-74. 4v. folding tab – 1 – us Wisconsin U Libr [240]

Greek, the language of christ and his apostles / Roberts, Alexander – London; New York: Longmans, Green, 1888 – 2mf – 9 – 0-7905-0219-4 – (incl bibl ref and index) – mf#1987-0219 – us ATLA [450]

Greek thought and the origins of the scientific spirit / Robin, Leon – London: K. Paul, Trench, Trubner & Co; New York: A.A. Knopf, 1928. Trans. from the French by M.R. Dobie – xx/409p – 1 – us Wisconsin U Libr [180]

Greek vase paintings / Harrison, Jane Ellen & MacColl, Dugald Sutherland – London 1894 – (= ser 19th c art & architecture) – 8mf – 9 – mf#4.2.1446 – uk Chadwyck [740]

Greek vases : historical and descriptive / Horner, Susan – London 1897 – (= ser 19th c art & architecture) – 3mf – 9 – mf#4.1.406 – uk Chadwyck [740]

The greek verb : its structure and development = Verbum der griechischen sprache seinem baue nach dargestellt / Curtius, Georg – London: J Murray, 1880 – 2mf – 9 – 0-8370-1830-7 – (in english) – mf#1987-6218 – us ATLA [450]

The greek versions of the testaments of the twelve patriarchs / ed by Charles, Robert Henry – Oxford: Clarendon Press, 1908 – 1mf – 9 – 0-7905-0922-9 – (text in greek; introduction, notes, and appendices in english, greek, and hebrew. incl bibl ref and index) – mf#1987-0922 – us ATLA [450]

Greek votive offerings : an essay in the history of greek religion / Rouse, William Henry Denham – Cambridge: University Press, 1902 – 2mf – 9 – 0-524-04867-3 – (incl bibl ref) – mf#1990-3429 – us ATLA [243]

Greek way / Hamilton, Edith – New York, NY. 1930 – 1r – us UF Libraries [025]

Greek-americans of florida – s.l, s.l? 1936 – 1r – us UF Libraries [978]

A greek-english lexicon of the new testament : being grimm's wilke's clavis novi testamenti = Lexicon graeco-latinum in libros novi testamenti / Grimm, Carl Ludwig Wilibald – New York: Harper, 1887 [mf ed 1990] – 8mf – 9 – 0-8370-1879-X – (trans, rev and enl by joseph henry thayer. in english) – mf#1987-6266 – us ATLA [052]

Greek-english word-list : containing about 1000 most common greek words / Baird, Robert – Boston: Ginn, 1893 [mf ed 1986] – 1mf – 9 – 0-8370-9202-7 – (incl ind) – mf#1986-3202 – us ATLA [450]

Greeks and latins : being a full and connected history of their dissensions and overtures for peace down to the reformation / Ffoulkes, Edmund S – London: Longmans, Green, 1867 – 2mf – 9 – 0-7905-4515-2 – (incl bibl ref) – mf#1988-0515 – us ATLA [240]

The greeks and the persians / Cox, George William – New York: Scribner, Armstrong, 1876 – (= ser Epochs of Ancient History) – 1mf – 9 – 0-524-04449-X – mf#1992-0118 – us ATLA [930]

Greeks in america : an account of their coming, progress, customs, living, and aspirations: with an historical introduction and the stories of some famous american-greeks / Burgess, Thomas – Boston: Sherman, French, 1913 – 1mf – us ATLA [305]

Greeks in america : an account of their coming, progress, customs, living, and aspirations: with an historical introduction and the stories of some famous american-greeks / Burgess, Thomas – Boston: Sherman, French, 1913 – 1mf – 9 – 0-7905-4106-8 – (incl bibl ref) – mf#1988-0106 – us ATLA [975]

Greeks lexicon of the roman and byzantine periods (from b c 146 to a d 1100) / Sophocles, Evangelinus Apostolides – Cambridge, 1914 – 21mf – 8 – €40.00 – ne Slangenburg [930]

Greeks of tarpon springs, florida / Lovejoy, Gordon Williams – s.l, s.l? 1938 – 1r – us UF Libraries [978]

Greeley, A W *see* Three years of arctic service

Greeley center independent – Greeley Center, NE: H L Ganoe, oct 9 1886-88// (wkly) [mf ed lacks jan 28-feb 11, mar 25 1887] – 1r – 1 – (merged with: greeley news to form: greeley leader) – us NE Hist [071]

Greeley citizen – Greeley, NE: Edward P Curran. v41 n21. dec 26 1935- (wkly) – 1 – (cont: greeley citizen and the leader-independent) – us NE Hist [071]

Greeley citizen – Greeley, NE: James B Barry, sep 1892-v33 n19. dec 11 1924 (wkly) [mf ed 1895-1924 (gaps)] – 9r – 1 – (merged with: leader-independent to form: greeley citizen and the leader-independent. monday ed: citizen (1918)) – us NE Hist [071]

Greeley citizen and the leader-independent – Greeley, NE: Edward P Curran. v33 n20. dec 18 1924-v41 n20. dec 19 1935 (wkly) [mf ed with gaps] – 3r – 1 – (formed by the union of: greeley citizen and: leader-independent. cont by: greeley citizen (1935). some irregularities in numbering. cont the numbering of: greeley citizen) – us NE Hist [071]

Greeley county independent – Scotia, NE: H Allnutt, 1889-94// (wkly) [mf ed v1 n23. jun 25 1891, jun 9 1892 filmed [1979-90]] – 2r – 1 – (merged with: greeley leader to form: leader-independent) – us NE Hist [071]

The greeley herald – Greeley, NE: N H Parks. v8 n1. nov 18 1890- (wkly) [mf ed -1892 (lacks apr 30-may 71891) filmed 1979] – 1r – 1 – (cont: scotia herald) – us NE Hist [071]

Greeley home journal *see* Miscellaneous newspapers of weld county

Greeley leader – Greeley Centre, NE: Barngrover & Ganoe, apr 1888-94// (wkly) [mf ed -1892 (gaps)] – 3r – 1 – (formed by the union of: greeley center independent and: greeley news. merged with: greeley county independent to form: leader-independent. publ in greeley center mar 29 1889-92. issues for apr 13-dec 7 1888 called v1 n2-36. issues for jan 11 1889-jul 1 1892 called whole n117-296) – us NE Hist [071]

Greeley morning spokesman *see* Miscellaneous newspapers of weld county

Green against linen / Giraudier, Antonio – New York, NY. 1957 – 1r – us UF Libraries [972]

Green, Anna Elizabeth *see* Riglyne vir die ouer van die chroniese siek kind

Green, Anna Katharine *see* The mystery of the hasty arrow

Green, Arnold *see* Ohio supreme court practice; containing the law, decisions and forms, with full directions for proceeding in mandamus, quo-warranto, habeas corpus.

Green, Arthur George *see* A systematic survey of the organic colouring matters

Green, Ashbel *see*
- A historical sketch
- The life of ashbel green, vdm
- Presbyterian missions

Green bag : an entertaining magazine of the law – Killen TX 1889-1914 – 1 – mf#2896 – us UMI ProQuest [340]

The green bag – v1-26. 1889-1914 (all publ) – 62mf – 9 – $279.00 – (ind incl) – mf#LLMC 82-925 – us LLMC [073]

Green bag: 2nd series : an entertaining journal of law – Cleveland: The Green Bag, Inc. v1-3. 1997-2000 – 9 – $104.00 – ISSN: 1095-5216 – mf#117941 – us Hein [340]

Green bay advocate – Green Bay WI. 1846 aug 13-1850, 1851-58, 1859-1862 jul 31, 1862 aug 21-1864, 1865-91, 1892-1894 may 31, 1894 jun 1-1897, 1899 jan 1-1900 dec 18, 1900 dec 21-1902 may 16, 1902 may 19-1903 jun 12, 1903 jun 15-1904 dec 31, 1905, 1906 jan-dec 18 – 22r – 1 – (continued by: green bay gazette; green bay gazette) – mf#930436 – us WHS [071]

Green bay advocate – Green Bay WI. 1902 oct 8-dec 31, 1903 jan-jun, 1903 jul 16-dec, 1903 jun 15-1904 dec 31, 1904 jan-dec, 1905 jan-dec, 1906 jan-dec – 9r – 1 – (continued by: green bay gazette; green bay gazette) – mf#947513 – us WHS [071]

Green bay banner – Green Bay WI. 1858 nov 19-1859 oct 21 – 1r – 1 – mf#944293 – us WHS [071]

Green Bay Business College [WI] *see* Proclamation

Green bay catholic compass – 1978 oct 7-1979 dec, 1979 sep-1979 dec, 1980-84, 1985 oct-1986 jun, 1986 jul-1987 mar, 1987 apr-dec, 1988 jan-dec, 1989, 1990 – 14r – 1 – (cont: spirit [green bay : 1977]; cont by: compass [green bay, wi]) – mf#500715 – us WHS [071]

Green bay [city directory : listing] – 1872, 1931, 1874, 1892, 1896-97, 1900 – 4r – 1 – mf#3197672 – us WHS [071]

Green bay community center newsletter – Palestine TX. 1983 apr, jun, 1984 may, 1985 jun, 1986 mar-apr, jun, aug, 1987 feb, may, jul, 1988 jan, may, sep, 1989 may, 1990 jan, may-jun, oct, 1991 jan-feb, apr, 1993 jun, 1994 feb – 1r – 1 – mf#4878552 – us WHS [071]

Green bay daily advocate – Green Bay WI. 1896 sep 16-1897 apr 1, 1897 apr 2-1897 oct 12, 1897 oct 14-1898 jan 14 – 3r – 1 – (continued by: evening advocate (green bay, wi)) – mf#947496 – us WHS [071]

Green bay daily herald – Green Bay WI. 1900 jun 12-1901 feb 5 – 1r – 1 – (cont: green bay news and social mirror; daily times [green bay, wi: 1899]) – mf#918908 – us WHS [071]

Green bay data – Green Bay WI. 1882 dec 26-1883 jan 11 – 1r – 1 – (cont: data [green bay, wi: 1882]) – mf#918681 – us WHS [071]

Green Bay Education Association *see* News

Green bay gazette – Green Bay WI. 1914 feb 16-1914 jun 1 – 2r – 1 – (cont: daily state gazette; green bay advocate [green bay, wi: 1902]; cont by: free press [green bay, wi]; green bay press-gazette) – mf#1011873 – us WHS [071]

Green bay gazette – Green Bay WI. 1866 mar 3-1868 nov 28, 1868 dec 5-1870 aug 6 – 2r – 1 – (continued by: state gazette (green bay, wi: 1870)) – mf#1044797 – us WHS [071]

Green bay globe – Green Bay WI. 1876 jul 4-1879 dec 24, 1880-1883 aug 31 – 2r – 1 – (cont: weekly globe [green bay, wi: 1876]) – mf#918696 – us WHS [071]

Green bay herald – Green Bay WI. 1911 aug 17-1912 aug 8, 1913 aug 16-1914 jun 6 – 1r – 1 – mf#920479 – us WHS [071]

Green bay historical bulletin / Brown County Historical Society et al – Green Bay WI. 1925 feb-1934 – 1r – 1 – mf#419839 – us WHS [978]

Green bay news-chronicle – Green Bay WI. 1976 apr 12/may 29-2004 dec 16/31 [gaps] – 449r – 1 – (cont: brown county chronicle; daily news [green bay, wi]) – mf#1144877 – us WHS [071]

Green bay press-gazette – Green Bay WI. 1915 jun 29/30-1953 jul/aug [gaps] – 346r – 1 – (cont: free press [green bay, wi]; green bay gazette) – mf#879641 – us WHS [071]

Green bay register – Green Bay WI. 1957 mar 29-1958, 1959 jan-1960 jun, 1960 jul-1961 dec, 1962-64, 1965 jan 1-sep 3, 1965 sep 10-1966 apr 15, 1966 apr 22-dec 2, 1966 dec 9-1967 jul 21, 1967 jul 28-1968 mar 22, 1968 dec 13-1969 sep 12, 1968 mar 29-dec 6, 1969 sep 19-1970 sep 4 – 14r – 1 – (continued by: spirit (green bay, wi: weekly ed)) – mf#1057531 – us WHS [241]

Green bay republican – Green Bay WI. 1841 oct 16-1843 feb 18, 1841 nov 20, dec 11,18, 1842 sep 10, 1843 jan 28, aug 26, oct 17-1844 nov 5 – 2r – 1 – (continued by: wisconsin republican (green bay, wi: 1844)) – mf#920581 – us WHS [071]

Green bay review – Green Bay WI. 1896 jan 4/1897 jul 3-1917 aug 11/1919 jan 11 [gaps] – 16r – 1 – (cont: fort howard review) – mf#947488 – us WHS [071]

Green bay semi-weekly gazette – Green Bay WI. 1899 may 27/nov 8-1915 apr 27-oct 15 [gaps] – 23r – 1 – (cont: green bay weekly gazette) – mf#1044807 – us WHS [071]

Green bay spectator – Green Bay WI. 1852 jan 3-nov 30, 1852 feb 21, apr 24, aug 3, sep 28 – 2r – 1 – mf#918682 – us WHS [071]

Green bay story / Procter & Gamble Paper Products Co – 1976 oct-1987 dec – 1r – 1 – (cont: charmin story) – mf#1582648 – us WHS [670]

Green bay volks zeitung – Green Bay WI. 1874 may 28-jun 18, jul 16-nov 19, 1876 jan 6, jul 20, aug 31 – 1r – 1 – mf#1011534 – us WHS [071]

Green bay weekly gazette – Green Bay WI. 1894 aug 28-1895 aug 21, 1895 aug 28-1897 feb 24, 1897 mar 3-1898 aug 3, 1898 aug 10-1899 may 24 – 4r – 1 – (cont: state gazette [green bay, wi: 1870]; cont by: green bay semi-weekly gazette) – mf#1044804 – us WHS [071]

Green book magazine – Killen TX 1909-21 – 1 – mf#3302 – us UMI ProQuest [400]

Green book vacation guide – 1949 – 1r – 1 – mf#4879128 – us WHS [640]

Green brier baptist church. claiborne county. mississippi : church records – 1884-Jun 1902 – 1 – 5.09 – us Southern Baptist [242]

Green county herald – Monroe WI. 1933 jan 20-1935, 1936-40 – 2r – 1 – (cont: green county herold; schweizer nachrichten; cont by: amerikanische schweizer nachrichten und green county herold) – mf#1124640 – us WHS [071]

Green county herold – Monroe WI. 1877 oct 3-1882, 1883, 1892 sep-1894, 1895-98, 1899-1902, 1903-31, 1932-1933 jan 18 – 12r – 1 – (cont: wisconsin botschafter; cont by: green county herald) – mf#1124635 – us WHS [071]

Green county real estate journal – Monroe WI. 1889 aug 15-1890 jan 16 – 1r – 1 – (cont: real estate journal; cont by: county journal) – mf#1125621 – us WHS [333]

Green county reformer – Monroe WI. 1874 jan 15-oct 22, 1874 oct 29-1878 jan 24, 1878 jan 31-1880 may 6 – 3r – 1 – (cont: green county republican and liberal press) – mf#1124515 – us WHS [071]

Green county republican – Monroe WI. 1871 jul 25-1873 sep 23 – 1r – 1 – (continued by: liberal press; green county republican and press) – mf#1120621 – us WHS [071]

Green county republican and liberal press – Monroe WI. 1873 nov 28-1874 jan 8 – 1r – 1 – (cont: green county republican and press; cont by: green county reformer) – mf#1124514 – us WHS [071]

Green county republican and press – Monroe WI. 1873 oct 3-nov 21 – 1r – 1 – (cont: green county republican; liberal press; cont by: green county republican and liberal press) – mf#1124512 – us WHS [071]

Green county review – v1 n1-v11 n4 [1977 oct.-1988 jul] – 1r – 1 – mf#1685682 – us WHS [978]

Green cove springs, florida, 1816 / Key, Fanny – s.l, s.l? 193-? – 1r – us UF Libraries [978]

Green, Dennis Howard *see* Konrads 'trojanerkrieg' und gottfrieds 'tristan'

Green, Duff *see* Duff green papers

Green, Edmund S *see*
- Green's digest of annotated cases
- Green's digest of the cases in the american state reports

Green, Edmund Tyrrell *see*
- The book of jeremiah and lamentations
- The thirty-nine articles and the age of the reformation

Green, Edwin Luther *see* School history of florida

Green egg – St Louis MO 1972-75 – ISSN: 0046-6395 – mf#7648 – us UMI ProQuest [130]

Green final – [NW England] Oldham Lib 6 sep 1919, 1959-73 – 1 – uk MLA; uk Newsplan [072]

Green final – [NW England] Douglas, Manx NHL 1935-40, 1947-1971* – 1 – (title change: green final and women's magazine [1970-71]) – uk MLA; uk Newsplan [072]

Green, Francis Marion *see*
- The christian ministers' manual
- Christian missions
- The life and times of john franklin rowe
- The standard manual for sunday school workers

Green, Frederick *see* A brief review of criminal cases in the supreme court for the past year

Green, George *see* Catalogue of the eastlake library in the national gallery

Green, George Alfred Lawrence *see* Editor looks back

Green, H *see*
- The mirrour of maiestie
- Shakespeare and the emblem writers

Green hand / Chapman, Paul Wilbur – Chicago, IL. 1932 – 1r – us UF Libraries [025]

GREENOCK

Green havoc in the lands of the caribbean / Wardlaw, C W – Edinburgh, Scotland. 1935 – 1r – us UF Libraries [972]
Green hill baptist church. mount juliet, tennessee : church records – 1900-Sep 1967 – 1 – us Southern Baptist [242]
Green island gazette – Dunedin, NZ. 13 nov 1975-mar 1985// – 4r – 1 – (only publ 13 nov 1975-mar 1985. ceased publ mar 1985) – mf#81.5 – nz Nat Libr [079]
Green, J Howard *see*
– Henry w chandler and his recollections of the flo...
– Unsolved murder
Green, James *see* The spanish armada
Green, James L *see* Journals and correspondence
Green, John *see* Correspondence
Green, John P *see* John p green papers, ms 3379
Green, John Paterson *see* John paterson green papers, 1869-1910
Green, Joseph Henry *see*
– Spiritual philosophy
Green, Joseph Joshua *see*
– Leaves from the journal of joseph james neave
– Souvenir of the address to king edward 7, 1901
Green, Katharine Rogers et al *see* Ma-li-hsun hsiao chuan (ccm250)
Green, Katherine R *see* Ma-li-hsun hsiao chuan
Green, L P *see* Development in africa
Green lake county democrat – Markesan, Princeton WI. 1879 feb 13-1880 oct 14, 1880 oct 21-1882 jun 15, 1882 jun 22-1883 dec 27, 1884 jan 10-1885 sep 16 – 4r – 1 – (cont: independent [princeton, wi]) – mf#1124676 – us WHS [071]
Green lake county reporter – Green Lake, Markesan WI. 1983 may 12/jun-1999 jan/jun [gaps] – 33r – 1 – (cont: green lake county reporter & markesan regional reporter) – us WHS [071]
Green lake county reporter – 1919-63, 1964/1965 oct 14-1978 may 1/11 [gaps] – 30r – 1 – (continued by: markesan regional reporter; green lake county reporter & markesan regional reporter) – mf#1048110 – us WHS [071]
Green lake county reporter and markesan regional reporter – Green Lake, Markesan WI. 1978 may 18-dec 31, 1979 jan-dec, 1980 jan-sep, 1980 oct-1981 jun, 1981 jul-dec, 1982 jan-sep, 1982 oct-1983 may 5 – 8r – 1 – (cont: green lake county reporter [green lake, wi: 1894]; markesan regional reporter; cont by: green lake county reporter [green lake, wi: 1983]) – mf#1048130 – us WHS [071]
Green lake guide – Green Lake, Princeton WI. 1940 jun 28 – 1r – 1 – mf#916501 – us WHS [071]
Green laker – 1977 may 24-1978 aug 9, sep 3, nov 29-dec 27, 1979 mar 28-1980 aug 20, 1981 may 21-1986 aug 26 – 3r – 1 – mf#1112270 – us WHS [071]
Green, Lawrence George *see*
– Almost forgotten, never told
– Beyond the city lights
– Decent fellow doesn't work
– Great north road
– Grow lovely, growing old
– I heard the old men say
– Like diamond blazing
– Lords of the last frontier
– On wings of fire
– So few are free
– South african beachcomber
– Tavern of the seas
– These wonders to behold
– Thunder on the blaauwberg
– Where men still dream
– White man's grave
Green, Lawrence George) *see* In the land of the afternoon
Green leaves / Sullivan, Timothy Daniel – Dublin, Ireland. 1886 – 1r – us UF Libraries [960]
Green light – Denver, CO: [s.n.] – 3r – 1 – (a system newspaper for employees of the rio grande railroad) – mf#MF Gre823d – us Colorado Hist [380]
Green Mountain Folklore Society *see* Green mountain whittlin's
Green mountain gem : a monthly journal of literature, science, and the arts – La Salle IL 1843-49 – 1 – mf#4558 – us UMI ProQuest [071]
Green mountain repository – La Salle IL 1832 – 1 – mf#4375 – us UMI ProQuest [071]
Green mountain sentinel : your vermont air national guard – 1981 jul-1989 dec – 1r – 1 – mf#1728138 – us WHS [355]
Green mountain whittlin's / Green Mountain Folklore Society – v1-4 – 1r – 1 – mf#216480 – us WHS [390]
Green, Nathan C *see* Story of spain and cuba
Green, Nicholas John *see* Criminal law reports
Green, Oscar Olin *see* Normal evangelism
Green pond baptist church. spartanburg county. woodruff, south carolina : church records – 1804-1971 – 1 – 5.00 – us Southern Baptist [242]
Green quarterly : an anglo-catholic magazine – London, 1924-34 [mf ed 2001] – (= ser Christianity's encounter with world religions, 1850-1950) – 2r – 1 – mf#2001-s099 – us ATLA [241]
Green revolution – Cochranville PA 1963+ – 1,5,9 – ISSN: 0017-3983 – mf#7551 – us UMI ProQuest [333]
Green revolution / School of Living (York, PA) et al – 1963 jan 1-1975 jun, 1975 aug-1977 oct, 1977 dec-1979 early winter – 3r – 1 – mf#165942 – us WHS [333]
Green, Richard *see*
– Anti-methodist publications issued during the eighteenth century
– John wesley, evangelist
– The mission of methodism
– Wesley bibliography
Green, Robert *see* "Governor fred hall, a study of a political personality"
Green, S G *see* Polycarp of smyrna
Green, S Walter *see* The gospel according to st mark
Green, Salmon et al *see* Addresses delivered at richmond, vermont, june 28, 1895
Green, Samuel *see* A discourse...at plymouth, dec 20 1828
Green, Samuel A *see* Piracy off the florida coast and elsewhere
Green, Samuel Gosnell *see*
– The acts of the apostles
– A brief introduction to new testament greek
– The christian creed and the creeds of christendom
– Church establishments considered
– A handbook of church history
– A handbook to old testament hebrew
– Handbook to the grammar of the greek testament
Green, T *see* Instructions for the poor
Green teacher – Toronto, Canada 1998+ – 1,5,9 – ISSN: 1192-1285 – mf#26237 – us UMI ProQuest [370]
Green, Thomas Andre *see* Colorado addendum to green's pleading and practice
Green, Thomas Hill *see*
– Martial law in hawaii
– Prolegomena to ethics
– Works of thomas hill green
Green, Thomas Sheldon *see* Critical greek and english concordance of the new testament
Green tree and the dry... / Guinness, H Grattan – Stirling, Scotland. 1858 – 1r – us UF Libraries [240]
The green vaults dresden illustrations of the choicest works in that museum of art / Gruner, Ludwig – London 1876 – (= ser 19th c art & architecture) – 2mf – 9 – mf#4.1.441 – uk Chadwyck [740]
Green, William Henry *see*
– The argument of the book of job unfolded
– An elementary hebrew grammar
– A grammar of the hebrew language
– A hebrew chrestomathy
– The pentateuch
– Pentateuch analysis
– Prophets and prophecy
– The unity of the book of genesis
Green, William Henry [comp] *see* Moses and the prophets
Green, William Mercer *see* Memoir of rt. rev. james hervey otey, d.d., ll.d
Green year *see* Nu ch'ing nien yueh k'an (ccs)
Greenback standard – Greenback standard WI. 1878 mar 22-oct 11 – 1r – 1 – (continued by: oshkosh standard) – mf#961072 – us WHS [071]
Green-bay intelligencer – Green Bay WI. 1833 dec 11-1835 jun 13 – 1r – 1 – (continued by: green-bay intelligencer, and wisconsin democrat) – mf#920642 – us WHS [071]
Green-bay intelligencer, and wisconsin democrat – Green Bay WI. 1835 jun 27-1836 jun 1 – 1r – 1 – (continued by: wisconsin free press; wisconsin free press; wisconsin democrat) – mf#920647 – us WHS [071]
Green-bay post – Green Bay WI. 1857 oct 3-1858 nov 10, 1858 mar 10 – 2r – 1 – mf#944353 – us WHS [071]
Greenberg & Co, L S *see* [L s greenberg & co's trade catalogue of jewelry]
Greenberg, David Benjamin *see* Shopping guide to mexico, guatemala, and the carib
Greenberg, Doreen L *see* The effect of the paradoxical intervention of symptom prescription on state anxiety levels and performance in young competitive swimmers
Greenbie, Sydney *see*
– Central five
– Fertile land, brazil
Greenbrier first baptist church. greenbrier, tennessee : church records – 1885-1973. Records. 1905-75 – 1 – 73.53 – us Southern Baptist [242]
Greenburg, Joseph Harold *see* Studies in african linguistic classification
Greencastle press – Greencastle, PA. -w 1889-1912 – 13 – $25.00r – us IMR [071]
Greendale bulletin – Greendale WI. 1938 aug 24 – 1r – 1 – (continued by: greendale review) – mf#951297 – us WHS [071]
Greendale review – Greendale, Hales Corners WI. 1938 aug 25-1943 oct 27, 1943 nov 10-1952 jan 31 – 2r – 1 – (cont: greendale bulletin; cont by: tri-town news [hales corners, wi]) – mf#951307 – us WHS [071]
Greendale sun – Greendale WI. 1946 dec 6-1947 feb 21, mar 7-apr 18 – 1r – 1 – mf#916359 – us WHS [071]
Greendale village life – Franklin, Greendale, Greenfield WI. 1960 oct 27/1963 apr 4-1996 nov/dec [gaps] – 87r – 1 – mf#1011872 – us WHS [071]
Greene, Carol *see* A physiological profile of champion level female triathletes
Greene Co. Beavercreek *see*
– Daily news
Greene Co. Cedarville *see* Herald
Greene Co. Fairborn *see* Voice
Greene Co. Xenia *see*
– Daily republican
– Democrat news
– Free press series
– Gazette
– Herald
– Republican series
– Sentinel
– Torch-light
– Western cornet
– Wood construction
Greene Co. Yellow Spring *see*
– American
– Antioch record series
– News
– Review / citizen / torch / news
Greene county atlas, 1874 – 1r – 1 – mf#B27424 – us Ohio Hist [978]
Greene county democrat – Eutaw AL. 1993 jul 7-dec 29, 1993 may 12-1993 jun 30, 1994 jan 5-jun 29, 1994 jul 6-dec 28, 1995 jan 4-jun 28 – 5r – 1 – (cont: eutaw mirror) – mf#2731301 – us WHS [071]
Greene county democrat – Waynesburg, PA. -w 1889-1904; 1905-1912 – 13 – $25.00r – us IMR [071]
Greene County Historical Society *see* Greene hills echo
Greene, Daniel *see* Public land statutes of the united states
Greene, Daniel Crosby *see* The christian movement in japan
Greene, E A *see* Saints and their symbols: a companion in the churches and picture galleries of europe
Greene, E G *see* Pathfinder for the organization and work of the woman's christian temperance union
Greene, F V *see* Report on the russian army and its campaigns in turkey in 1877-1878
Greene, Francis Vinton *see* Sketches of army life in russia
Greene, Graham *see* Quiet american
Greene hills echo / Greene County Historical Society – v8 n1-v10 n3 [1978/79 [winter]-1982 sep] – 1r – 1 – (continued by: greene hills echo (1983)) – mf#697470 – us WHS [978]
Greene, Homer *see* Aus nacht zum licht
Greene, J Milton *see* Historical sketch of the mission in persia under the care of the board of foreign missions of the presbyterian church
Greene, Jesse *see* Life, three sermons, and some of the miscellaneous writings of rev. jesse greene
Greene, JG *see* Tiffany's transcript appeals
Greene, Maine. Free Will Baptist Church *see* Records
Greene, Maria Louise *see* The development of religious liberty in connecticut
Greene, Mary *see* Life, three sermons, and some of the miscellaneous writings of rev. jesse greene
Greene, Maurice *see* Deborah and barak
Greene, Nathanael *see*
– Nathanael greene papers
– The papers of general nathanael greene
Greene, Nelson Lewis *see* An historical chart of german literature for use in schools and colleges
Greene, Robert *see*
– Ciceronis amor: tullies love
– The history of orlando furioso, 1594..
Greene, Thomas *see* Principles of religion explained and proved from the scriptures
Greene, William Brenton *see* Christian doctrine
Greene's reports – v1-4. 1847-54 (all publ) – (= ser Iowa supreme court reports) – 28mf – 9 – $42.00 – (a pre-nrs title) – mf#LLMC 94-006 – us LLMC [347]
Greenfield 1709-1849 – Oxford, MA (mf ed 1996) – (= ser Massachusetts vital record transcripts to 1850) – 7mf – 9 – 0-87623-258-6 – (mf 1t-2t: marriage intentions 1783-1815. mf 2t: marriages 1784-95. mf 2t-3t: births & deaths 1709-1825. mf 3t-4t: births 1749-64. mf 4t: deaths 1753-69. mf 4t-5t: births & deaths 1779-1854+. mf 5t-6t: marriage intentions 1815-49. mf 6t: marriages 1821-46; out-of-town marriages 1768-98. mf 7t: births & marriages 1844-50; deaths 1848-49) – us Archive [978]
Greenfield 1750-1891 – Oxford, MA (mf ed 1987) – (= ser Massachusetts vital records) – 39mf – 9 – 0-87623-054-0 – (mf 1-2: minutes, marriages, intentions 1753-72. mf 3-6: vital records 1750-1815. mf 7-8: b,m,d 1843-56. mf 9-13: births 1857-91. mf 14-18: marriages 1855-91. mf 19-22: deaths 1857-91. mf 23-28: index to births 1849-1925. mf 29-30: index to marriage intentions 1845-85. mf 31-35: index to marriages 1845-1925. mf 36-39: index to deaths 1848-1931) – us Archive [978]
Greenfield, Dominic *see* Perceived adequacy of professional preparation in sport psychology among ncaa division 1a head athletic trainers
Greenfield greendale guardian – Greendale, Greenfield, Hales Corners WI. 1964 apr 16-dec 31, 1965 jan 7-oct 21 – 2r – 1 – (cont: greenfield guardian; cont by: greenfield greendale woman and guardian) – mf#921902 – us WHS [071]
Greenfield greendale woman and guardian – Greendale, Greenfield, Hales Corners WI. 1965 oct 28-1966 apr 28 – 1r – 1 – (cont: greenfield greendale guardian) – mf#921916 – us WHS [071]
Greenfield guardian – Greenfield WI. 1962 oct 19-1964 apr 10 – 1r – 1 – (continued by: greenfield greendale guardian) – mf#921900 – us WHS [071]
Greenfield, M Rose *see* Five years in ludhiana
Greenfield observer – Greenfield WI. 1984 feb 16/jun-2003 nov/dec – 86r – 1 – (cont: observer [greenfield, wi]) – mf#1011869 – us WHS [071]
Greenfield observer – Greenfield WI. 1968 jun 20/1969 apr 17-1979 jan/apr 19 [gaps] – 19r – 1 – (continued by: observer (greenfield, wi)) – mf#1011867 – us WHS [071]
The greenfield papyrus in the british museum / Budge, Ernest Alfred Wallis – London, 1912 – 6mf – 9 – mf#NE-370 – ne IDC [930]
Greenfield, William *see* The genuineness, authenticity, and inspiration of the word of god
Greenford and northolt gazette – 1951; 18 mar 1974-1978; 16 feb 1979-9 sep 1988; 22 dec 1989- 24 dec 1992; 8 jan 1993-dec 1998 83 3/4r – 75 1/4r – 1 – (incorp with gazette (ealing borough edt) from 16 sep 1988-1 sep 1989. aka: middlesex county times and west middlesex gazette; (greenford and northolt edt)) – uk British Libr Newspaper [072]
Greenford and northolt gazette *see* Middlesex county times
Greenford northolt and southall recorder – London UK, 16 oct 1987-88; 13 jan-21 dec 1989; jan-14 sep 1990 – 5r – 1 – uk British Libr Newspaper [072]
Greenhalgh, Heidi A *see* Cross-validation of a quarter-mile walk test for college males and females
Greenhough, John Gersham *see* The conduct of public worship
Greenidge, Abel Hendy Jones *see* Roman public life
GreenLake spectator – Berlin, Green Lake WI. 1861 jul 2, 1862 dec 23-1864 sep 13, 1864 oct 12-1866 aug 22 – 2r – 1 – mf#922773 – us WHS [071]
Greenland missions : with biographical accounts of some of the principal converts / [Cranz, D] – Dublin, 1831 – 4mf – 9 – mf#HTM-46 – ne IDC [919]
Greenland missions : with biographical sketches of some of the principal converts – Dublin 1831 [mf ed Hildesheim 1995-98] – 1v on 2mf – 9 – €60.00 – 3-487-27081-1 – gw Olms [919]
Greenland, the adjacent seas, and the north-west passage to the pacific ocean : illustrated on a voyage to davis's strait during the summer of 1817 / O'Reilly, B – London, 1818 – 12mf – 9 – mf#H-488 – ne IDC [919]
Greenleaf gazette / Greenleaf Grocery – n101-13, [1979 apr-1980 oct] – 1r – 1 – mf#616592 – us WHS [640]
Greenleaf Grocery *see* Greenleaf gazette
Greenleaf, Simon *see*
– An examination of the testimony of the four evangelists
– A treatise on the law of evidence
Greenlees, Duncan *see* The gospel of narada
Greenock advertiser, and clyde and renfrewshire chronicle – Greenock : [s.n.] (semiwkly) – 6v – 1 – (cont: greenock advertiser (greenock, scotland : 1802); cont by: greenock advertiser, and clyde & west country chronicle) – uk Scotland NatLib [072]
Greenock advertiser, and clyde commercial herald [1819-1830] – Greenock: J Mennons 1819-30 (semiwkly) – 13v – 1 – (formed by union of: greenock herald (greenock, scotland : 1813) and: greenock advertiser (greenock, scotland : 1814); cont by: greenock advertiser, and clyde commercial journal) – uk Scotland NatLib [072]

Greenock advertiser, and clyde commercial herald [1830-1877] – Greenock: J Mennons 1830-77 (3 times/wk) – 1 – (cont: greenock advertiser, and clyde commercial herald; cont by: greenock advertiser & clyde commercial and shipping journal) – uk Scotland NatLib [072]

Greenock advertiser and clyde & west country chronicle – Greenock: J Mennons 1807-14 (3 times/wk) – 7v – 1 – (cont: greenock advertiser, and clyde and renfrewshire chronicle; cont by: greenock advertiser (greenock, scotland : 1814)) – uk Scotland NatLib [072]

Greenock advertiser (greenock, scotland : 1802) – Greenock: [s.n.] 1802 (wkly) [mf ed 1994] – 1v on 1r – 1 – (cont by: greenock advertiser, and clyde and renfrewshire chronicle) – uk Scotland NatLib [072]

Greenock advertiser (greenock, scotland : 1814) – Greenock: J Mennons 1814-19 (semiwkly) – 6v – 1 – (cont: greenock advertiser and clyde & west country chronicle; merged with: greenock herald (greenock, scotland: 1813) to form: greenock advertiser, and clyde commercial herald) – uk Scotland NatLib [072]

Greenock advertising gazette – [Scotland] Inverclyde, Greenock: J McCunn 23 jan-19 jun 1866 (wkly) [mf ed 2004] – 22v on 1r – 1 – uk Newsplan [072]

Greenock and port-glasgow argus : or, pickings for all palates – [Scotland] Inverclyde, [Greenock]: printed by J Blair 4 may-1 jun 1850 (mthly) [mf ed 2004] – 2v on 1r – 1 – uk Newsplan [072]

Greenock commonwealth [Greenock, Scotland : 1928] – [Scotland] Inverclyde, Greenock: J Anderson, Labour & Co-operative Joint Committee sep 1928-may 1931 (mthly) [mf ed 2004] – 31v on 1r – 1 – uk Newsplan [072]

Greenock commonwealth [greenock, scotland : 1936] – [Scotland] Inverclyde, Greenock: J Anderson, Labour & Co-operative Joint Committee nov 1936 [mf ed 2003] – 1v on 1r – 1 – uk Newsplan [072]

Greenock daily press and coast advertiser – [Scotland] Inverclyde, Greenock: R Fleming 5 mar-28 dec 1867 (daily) [mf ed 2004] – 1r – 1 – (cont by: greenock daily press weekly review [4 may-13 jul 1867]; cont by: greenock press weekly review [20 jul-dec 1867]) – uk Newsplan [072]

Greenock & district news – [Scotland] Inverclyde, Greenock: J S Thomson 10 oct 1947-3 feb 1950 (wkly, fortnightly, every 3 wks) [mf ed 2004] – 1r – 1 – (not publ between apr 29 1949 & feb 3 1950) – uk Newsplan [072]

Greenock election squib – [Scotland] Inverclyde, Greenock: J M Pollock 1 nov 1905-5 nov 1907 (irreg) [mf ed 2004] – 1r – 1 – (cont by: greenock and district election squib [3 nov 1906-nov 1907]) – uk Newsplan [325]

Greenock herald – Scotland. -w. 1853, 1858, 1863, 1868, 1878. 4 reels – 1 – uk British Libr Newspaper [072]

Greenock herald, and general advertiser – [Scotland] Inverclyde, Greenock: J Blair 19 feb 1852-79, 1893-5 nov 1937 (wkly) [mf ed 2003] – 49r – 1 – (missing: 1853, 1873; cont by: greenock herald [25 jul 1874-8 jun 1928]; greenock herald and west coast courier [15 jun 1928-31 oct 1930]; west coast courier [7 nov 1930-5 jun 1936]; west coast courier and greenock herald [12 jun 1936-5 nov 1937]) – uk Newsplan [072]

Greenock news and public guide – [Scotland] Inverclyde, Greenock: M'Cunn & Fleming 5 aug 1869-25 jun 1870 (daily) [mf ed 2004] – 6r – 1 – (formed by union of: greenock news and weekly press and: public guide) – uk Newsplan [072]

Greenock news-clout : an unstamped periodical, legal successor to young greenock, aurora, and quilp's budget – [Scotland] Inverclyde, Greenock: J Lennox 8 aug 1849-30 nov 1850 (mthly) [mf ed 2004] – 1r – 1 – uk Newsplan [072]

Greenock observer and clyde shipping gazette – [Scotland] Inverclyde, Greenock: J H Kippen 4 jan 1844-7 may 1846 (wkly) [mf ed 2003] – 1r – 1 – uk Newsplan [380]

Greenock protestant – [Scotland] Inverclyde, Greenock: printed by W Campbell aug 1852 (mthly) [mf ed 2004] – 1r – 1 – (a monthly journal for the defence of truth in opposition to popish error and for the discussion of subjects of general local interest) – uk Newsplan [242]

Greenock record and clyde shipping list – [Scotland] Inverclyde, Greenock: A Johnston 21 feb-17 jun 1850 (twice/wk) [mf ed 2003] – 34v on 1r – 1 – uk Newsplan [380]

Greenock telegraph : inverclyde's evening newspaper – Greenock: Orr, Pollock & Co Ltd 1985- (daily ex sun) – 1 – (not publ: 25, 26 dec 2005; 25, 26 dec 2006; 1-2 jan 2007; cont: greenock telegraph and clyde shipping gazette) – ISSN: 0963-3219 – uk Scotland NatLib [072]

Greenough, Charles Pelham *see* A digest of the reported decisions of the courts of the united states of america, and of great britain and her colonies, relating to the rights and liabilities of gas companies

Greenough, Horatio *see* The travels, observations, and experience of a yankee stonecutter

Greenpeace chronicles / Greenpeace Foundation – n13-n21 [1979 feb-1979 nov] – 1r – 1 – mf#628639 – us WHS [333]

Greenpeace examiner / Greenpeace USA – 1980 apr-1986 oct/dec – 1r – 1 – mf#1219311 – us WHS [333]

Greenpeace Foundation *see* Greenpeace chronicles

Greenpeace New Zealand *see* Campaigns protesting against nuclear testing in the pacific

Greenpeace new zealand newsletter – 1974-2004 – 1r – 1 – mf#pmb doc464 – at Pacific Mss [333]

Greenpeace USA *see* Greenpeace examiner

Green's criminal law reports – v1-2. 1995 – 18mf – 9 – $27.00 – mf#LLMC 95-121 – us LLMC [345]

Green's digest of annotated cases / Green, Edmund S – New York, San Francisco: Thompson Co, Bancroft-Whitney. 1v. 1921 (all publ) – 24mf – 9 – $36.00 – mf#LLMC 84-695E – us LLMC [348]

Green's digest of the cases in the american state reports / Green, Edmund S – San Francisco: Bancroft-Whitney. v1-5. 1904-12 (all publ) – (= ser American state reports. trinity series, pt 3) – 82mf – 9 – $123.00 – mf#LLMC 78-038B – us LLMC [348]

Greensboro review – Greensboro NC 1966+ – 1,5,9 – ISSN: 0017-4084 – mf#6711 – us UMI ProQuest [400]

Greensboro sun – 1974 feb-1976 nov, v1 n1-v5 n7 [1972 mar 22-1976 nov] – 2r – 1 – mf#1057546 – us WHS [071]

Greensburg and american herald – Greensburg, PA., 1856-1864 – 13 – $25.00r – us IMR [071]

Greensburg gazette – Greensburg, PA., 1816-1821 – 13 – $25.00r – us IMR [071]

Greensburg herald – Greensburg, PA. -w 1864-1870 – 13 – $25.00r – us IMR [071]

Greensburg. Kansas. Greensburg Town Company *see* History

Greensburg press – Greensburg, PA. -w 1887-1912 – 13 – $25.00r – us IMR [071]

Greenstone, James L *see* Journal of police crisis negotiations

Greenup, Albert William *see*
- The new testament in the revised version of 1881
- A short commentary on the book of lamentations
- Taanit yerushalmi

Greenville black star / South Carolina Black Media Group – Greenville SC. 1988 dec 1/3-29/31, 1989 jan 5/7-aug 17 – 1r – 1 – mf#1663924 – us WHS [071]

[Greenville-] bulletin – CA. 1886-87 (broken file) [wkly] – 1r – 1 – $60.00 – mf#B02278 – us Library Micro [071]

Greenville first baptist church. greenville, south carolina : church records – 1831-1939, 1947-75 – 1 – us Southern Baptist [242]

[Greenville-] indian valley record – CA. 1954-57; 1972; 1977 [wkly] – 4r – 1 – $240.00 – mf#B02279 – us Library Micro [071]

[Greenville-] the record – CA. 1957-59 [wkly] – 2r – 1 – $120.00 – mf#B02280 – us Library Micro [071]

Greenville weekly observer – Greenville AL. 1865 apr 22 [extra] – 1r – 1 – mf#853659 – us WHS [071]

Greenville Woman's College. South Carolina *see* Catalogs and college records

Greenwald, Emanuel *see*
- The foreign mission work of pastor louis harms, and the church at hermansburg
- The lutheran reformation
- Sprinkling, the true mode of baptism
- The true church

Greenwald, Jekuthiel Judah *see* Mekorot le-korot yisrael

Greenwald, Leopold *see* Korot ha-torah veha-emunah be-hungariyah

Greenwall, Harry James *see*
- His highness the aga khan
- Unknown liberia

Greenwell, Dora *see*
- Carmina crucis
- John woolman
- The patience of hope
- The power of prayer
- A present heaven

Greenwell, Scott D *see* Estimating body fat percentage using simple measures

Greenwich 1741-1890 – Oxford, MA (mf ed 1984) – (= ser Massachusetts vital records) – 84mf – 9 – 0-931248-74-4 – (mf 1: b,d,m,intentions 1741-77. mf 2: b,m,d 1748-79. mf 3: b,m,d,intentions 1750-1811. mf 4-5: marriages 1816-34; births 1778-1839. mf 6-9: marriages & births 1792-1850. mf 10-13: b,m,d 1845-90. mf 14-15: marriages & intentions 1868-90. mf 16-33: index to births 1741-1937. mf 34-53: index to marriage intentions 1750-1919. mf 54-71: index to marriages 1749-1937. mf 72-83: index to deaths 1746-1937. mf 84: relocation of voter registrations 1938) – us Archive [978]

Greenwich and deptford chronicle – Greenwich, UK. 18 mar-dec 1871; 1877; 1879 – 2r – 1 – (incorp with: greenwich and deptford observer from oct 1885) – uk British Libr Newspaper [072]

Greenwich and woolwich free press – Greenwich, UK. 27 aug 1987-88; 12 jan-16 mar 1989 – 1 3/4r – 1 – (aka: south london free press (greenwich and woolwich ed)) – uk British Libr Newspaper [072]

Greenwich labour party records, 1920-87 – (= ser Labour party in britain, origins and development at local level. series 2) – 8r – 1 – (with p/g. int by fred lindop) – mf#97567 – uk Microform Academic [325]

Greenwich mercury – London, England. -w. 15 april 1981-jun 1998 56 1/2r – 1 – uk British Libr Newspaper [072]

Greenwich observations in astronomy and magnetism made at the royal observatory, greenwich, the royal greenwich observatory, herstmonceux, and the royal greenwich observatory, abinger, in the year... / Jones, Harold Spencer – London: HMSO 1959- (annual) [mf ed 2000] – 3v [1953-55] on 1r [ill] – 1 – (cont: greenwich observations in astronomy, magnetism and meteorology made at the royal observatory, greenwich, the royal greenwich observatory, herstmonceux and the royal greenwich observatory, abinger in the year...; aka: greenwich observations; "publ by order of the board of admiralty in obedience to his majesty's command") – mf#film mas c4563 – us Harvard [520]

Greenwich observations in astronomy, magnetism and meteorology made at the royal observatory, greenwich, the royal greenwich observatory, herstmonceux, and the royal greenwich observatory, abinger, in the year... / Jones, Harold Spencer – London: HMSO 1957-58 (annual) [mf ed 2000] – 5v [1948-52] on 3r (ill) – 1 – (cont: observations made at the royal observatory, greenwich, in the year...in astronomy, magnetism and meteorology; cont by: greenwich observations in astronomy and magnetism made at the royal observatory, greenwich, the royal greenwich observatory, herstmonceux, and the royal greenwich observatory, abinger, in the year...; iss for 1948 called: greenwich observations in astronomy, magnetism and meteorology made at the royal observatory, greenwich, and the royal greenwich observatory, abinger, in the year...; aka: greenwich astronomical observations; greenwich observations) – mf#film mas c4562 – us Harvard [520]

Greenwich villager – New York. v. 1-2 n15. may 15 1933-feb 1934 – 1 – us NY Public [410]

Greenwich woolwich and deptford gazette – 1834-9 apr 1981 – 186r – 1 – (cont as local eds: deptford & peckham mercury; greenwich mercury; lewisham mercury; woolwich mercury; aka: kentish mercury; south east london mercury; south east london and kentish mercury; lewisham mercury london; mercury lewisham borough; south east london mercury) – uk British Libr Newspaper [072]

Greenwich, woolwich and deptford gazette – [London & SE] Lewisham LHAC 1834-39 – 1 – (cont as: kentish mercury) – uk Newsplan [072]

Greenwood baptist church. lincolnton, georgia : church records – Aug 1909-Aug 1951; Oct 1969-Apr 1970; Sept 1973-Sept 1977 – 1 – us Southern Baptist [242]

Greenwood, F W P *see* Remarks on the duty of observing the lord's supper

Greenwood, Francis William Pitt *see* A history of king's chapel in boston

Greenwood gazette – Lincoln, NE: Inter-State Newspaper Co, 1889-v39 n14. apr 25 1934 (wkly) [mf ed 1892,1895-1934 (gaps)] – 16r – 1 – (absorbed by: lancaster county weekly. publ in havelock ne, may 11 1927-jul 22 1931) – us NE Hist [071]

Greenwood gleaner – Greenwood WI. 1900 feb 2/1901 aug 2-1968 oct 10/1969 jun 12 [gaps] – 35r – 1 – (continued by: loyal tribune; spencer record; tribune record gleaner) – mf#946908 – us WHS [071]

Greenwood, J Michael *see*
- Computer-aided transcription
- Follow-up study of word processing and electronic mail in the 3rd circuit court of appeals
- The impact of word processing and electronic mail on us courts of appeals

Greenwood, Michael et al *see* A comparative evaluation of stenographic and audiotape methods for us district court reporting

Greenwood news – Greenwood, NE: Geo W Brewster. 3v. v4 n29. jun 7 1901-v6 n28. jun 5 1903 (wkly) [mf ed with gaps] – 1r – 1 – (cont: news-record) – us NE Hist [071]

Greenwood record – Greenwood, NE: Geo B Pickett, sep 1897-v3 n43. jun 29 1900 (wkly) [mf ed with gaps] – 1r – 1 – (merged with: greenwood news to form: news-record) – us NE Hist [071]

Greenwood, Thomas *see*
- Cathedra petri
- Latest heresy

Greenwood, Thomas. F R G S *see* Museums and art galleries

Greer first baptist church. greer, south carolina : church records – 1908-24, 1944-68 – 1 – 60.17 – us Southern Baptist [242]

Greer, J R *see* Quakerism

Greer, Richard A *see* The government of american samoa

Greerton gazette – Tauranga, NZ. jun-aug 1981 – 1r – 1 – mf#16.29 – nz Nat Libr [079]

Greg, William *see* Sketches in greece and turkey

Greg, William Rathbone *see* The creed of christendom

Gregg, David *see*
- Between the testaments
- From solomon to the captivity
- Prayer a fact

Gregg, Frank M *see* Anti-slavery notes

Gregg, John A F *see* The wisdom of solomon

Gregg, Mary Kirby *see* Chapters on trees

Gregg, Richard Bartlett *see*
- Economics of khaddar
- Gandhism versus socialism
- The power of non-violence

Gregg, Tresham Daines *see* Apostasy of the roman catholic church clearly demonstrated

Gregg, William *see*
- History of the presbyterian church in the dominion of canada

Gregg, William H *see* Where, when, and how to catch fish on the east coast

Grego, Joseph *see* Rowlandson the caricaturist

Gregoire, Achille *see* Bal champetre au cinquieme etage

Gregoire, H *see*
- An enquiry concerning the intellectual and moral faculties and literature of negroes
- Marc le diacre

Gregoire, Herman *see*
- Histoire du congo pour la jeunesse
- Makako, singe d'afrique
- Masako, singe d'afrique; roman

Gregoire le Grand (Gregory The Great) *see* Le pastorale

Gregoire le grand (he5) : les etats barbares et la conquete arabe (590-757) – Paris, 1938 – (= ser Histoire de l'eglise (he)) – €29.00 – ne Slangenburg [241]

Gregoire, the priest and the revolutionist / Gregory, Caspar Rene – 1876 – 1mf – 9 – 0-7905-6750-4 – (incl bibl ref) – mf#1988-2750 – us ATLA [240]

Gregor 1. der grosse : ein lebensbild / Bonsmann, Th – Paderborn: Junfermann, 1890 – 1mf – 9 – 0-8370-7849-0 – mf#1986-1849 – us ATLA [920]

Gregor 7 : sein leben und wirken / Martens, Wilhelm – Leipzig: Duncker & Humblot. 2v. 1894 – 2mf – 9 – 0-8370-7887-3 – (incl ind) – mf#1986-1887 – us ATLA [240]

Gregor 8. 57taegiges pontifikat / Nadig, Paul – Basel: Allg Schweizer Zeitung, 1890 – 1mf – 9 – 0-8370-7893-8 – mf#1986-1893 – us ATLA [240]

Gregor der Grosse *see*
- Ausgewaehlte schriften, 2.bd (bdk3 2.reihe)
- Buch der pastoralregel (bdk4 2.reihe)

Gregor der grosse : lebensbild zur 1300 jaehrigen wiederkehr seines todestages / Bilguer, Dr von – Berlin: Germania, 1904 – 1mf – 9 – 0-8370-7921-7 – mf#1986-1921 – us ATLA [240]

Gregor der grosse und seine zeit. erster band / Pfahler, Georg – Frankfurt am Main: Carl Bernhard Lizius, 1852 – 1mf – 9 – 0-8370-8138-6 – (incl bibl ref) – mf#1986-2138 – us ATLA [240]

Gregor, Frantiska *see* The story of bohemia

Gregor, Leigh Richmond *see* The new canadian patriotism

Gregor von Nazianz (Gregory of Nazianzus, Saint) *see*
- Briefe
- Reden 1-20, 1. bd (bdk59 1.reihe)

Gregor von Nyssa (Gregory of Nyssa, Saint) *see* Grosse kathechese (bdk56 1.reihe)

Gregorian chants for canticles and psalter / Bedford-Jones, T – S.l: s.n, 1868? – 1mf – 9 – mf#01013 – cn Canadiana [780]

The gregorian sacramentary under charles the great (hbs49) / Wilson, H Austin – London, 1915 – (= ser Henry bradshaw society (hbs)) – 7mf – 8 – €15.00 – ne Slangenburg [241]

Gregorianum – 1(1920)-27(1946) – 238mf – 9 – €454.00 – ne Slangenburg [243]
Gregorii 1 papae registrum epistolarum, libri 8-14 (mgh epistolae 1:2.bd) : libri 8-14 – 1899 – (= ser Monumenta germaniae historica epistolae. 1. epistolae in quarto (mgh epistolae 1)) – €32.00 – ne Slangenburg [227]
Gregorii 1 papae registrum epistolarum, libri 8-14 (mgh epistolae 1:2.bd) : libri 8-14 – 1899 – (= ser Monumenta germaniae historica epistolae. 1. epistolae in quarto (mgh epistolae 1)) – €32.00 – ne Slangenburg [241]
Gregorii 1 papae registrum epistolarum (mgh epistolae 1:1.bd) : libri 1-7 – 1887-1891 – (= ser Monumenta germaniae historica epistolae. 1. epistolae in quarto (mgh epistolae 1)) – €23.00 – ne Slangenburg [227]
Gregorii abulfarag bar ebhraya in evangelium matthaei scholia = Horreum mysteriorum. selections / Bar Hebraeus – Gottingae: In aedibus Dieterichianis, 1879 – 1mf – 9 – 0-524-02759-5 – mf#1987-6453 – us ATLA [226]
Gregorii bar ebhraya in evangelium iohannis commentarius : e thesauro mysteriorum desumptum = Horreum mysteriorum. selections / Bar Hebraeus; ed by Schwartz, R – Gottingae: In aedibus Dieterichianis, 1878 – 1mf – 9 – 0-524-02760-9 – mf#1987-6454 – us ATLA [226]
Gregorii nysseni sententiae de salute adipiscenda / Herrmann, Wilhelm – Halis Saxonum: Formis Karrasianis, 1875 – 1mf – 9 – 0-7905-7402-0 – mf#1989-0627 – us ATLA [240]
Gregorii turonensis opera (mgh2:1.bd) / ed by W Arndt, W & Krusch, B – 1884 – (= ser Monumenta germaniae historica 2: scriptores merovingiarum (mgh2)) – €48.00 – ne Slangenburg [241]
Gregorio 10 *see* Decretales cum glossis (siecle 14)
Gregorio Rocasolano, Antonio del *see* Reflexiones sobre la labro cientifica del r.p. confirma navas s.j. prologo y recopilacion de ignacio sala de castelarnan, s.j.
Gregorio, San *see*
- Homiliae (codice uncial, siecle 6-7)
- Moralia in job (siecle 10)

Gregorio y epifanio / Gutierrez Gonzalez, Gregorio – Lima, Peru. 1926? – 1r – us UF Libraries [972]
Gregorius / Aue, Hartmann von der – 2. aufl. Halle a.S: M Niemeyer, 1900 [mf ed 1993] – (= ser Altdeutsche textbibliothek 2) – xxiii/103p – 1 – (incl bibl ref) – mf#8193 reel 1 – us Wisconsin U Libr [810]
Gregorius 16, Pope *see* Triomphe du st-siege et de l'eglise; ou, les novateurs modernes combattus avec leurs propres armes
Gregorius de Valencia *see*
- Analysis fidei catholicae
- Commentariorum theologicorum...

Gregorius des Grossen, H (Gregory The Great) *see* Saemtliche briefe
Gregorius Magnus *see*
- Homiliae in evangelia (ccsl141
- In canticum canticorum. in librum primum regum
- Registrum epistularum

Gregorius Thaumaturgus *see* Ueber die beiden hierarchien (bdk2 1.reihe)
Gregorius Turonensis *see* Libri miraculorum aliaque opera minora
Gregorius-blatt : organ fuer katholische kirchenmusik – v. 1-29 – 1 – 59.00 – us L of C Photodup [780]
Gregorovius, Ferdinand *see*
- Corsica
- The emperor hadrian
- Der ghetto und die juden in rom
- Lucretia borgia
- The roman journals of ferdinand gregorovius 1852-1874
- Siciliana
- The tombs of the popes
- Urban 8. im widerspruch zu spanien und dem kaiser
- Wanderjahre in italien

Gregors des grossen lehre von den engeln / Kurz, L – Rotenburg, 1938 – 3mf – 8 – €7.00 – ne Slangenburg [241]
Gregory 1, Pope *see*
- Ausgewaehlte schriften des heiligen gregorius des grossen, papstes und kirchenlehrers
- The dialogues of saint gregory, surnamed the great
- Lex levitarum, or, preparation for the cure of souls – regula pastoralis

Gregory 1, Pope Saint (Gregory the Great) *see* Gregory the great's moralia
Gregory 1, Saint *see*
- Le pastorale
- Saemtliche briefe

Gregory 1, Saint [Gregory The Great] *see* Rhetorica ad herennium...
Gregory 7, Pope *see* Monumenta gregoriana
Gregory 9 and greek ordinations / Lacey, Thomas Alexander – London: SPCK 1898 [mf ed 1992] – (= ser Church historical society (series) 33) – 1mf – 9 – 0-524-05507-6 – mf#1990-1502 – us ATLA [243]
Gregory 15, Pope *see* Bulla de su santidad
Gregory 16, Pope *see* Triomphe du st. siege et de l'eglise, ou, les novateurs moderns combattus avec leurs propres armes
Gregory, Alfred *see* Robert raikes, journalist and philanthropist
Gregory, Arden R *see* Cardiac rehabilitaion exercise adherence
Gregory, Arthur Edwin *see* The hymn-book of the modern church
Gregory, Benjamin *see*
- The holy catholic church, the communion of saints
- Side lights on the conflicts of methodism

Gregory, Caspar Rene *see*
- Canon and text of the new testament
- Das freer-logion
- Gregoire, the priest and the revolutionist
- Die griechischen handschriften des neuen testaments
- Die koridethi evangelien th038
- Die schriften von carl wessely
- Textkritik des neuen testamentes
- Vorschlaege fuer eine kritische ausgabe des griechischen neuen testaments
- Wellhausen und johannes

Gregory, Caspar Rene et al *see* The ologische studien
Gregory, Charles Hutton *see* Report of mr charles hutton gregory, c e
Gregory, Charles Noble *see* A great judicial character, roger brooke taney
Gregory, Daniel Seeley *see* The church in america and its baptisms of fire
Gregory, George Craghead *see* Forms for virginia and west virginia annotated, including statutory, common law and equity, commercial, corporation, and criminal forms
Gregory, Isaac *see* Rough notes on the silver crisis and on the debased money of india
Gregory, J W *see* The great rift valley
Gregory, Jc *see* Nature of laughter
Gregory, John *see* Diaghilev's oversight
Gregory, John Burslem *see* The oracles ascribed to matthew by papias of hierapolis
Gregory, John Milton *see* The bible – how to teach the bible
Gregory, John Robinson *see*
- A history of methodism
- The theological student

Gregory, Josephene T *see* Nassau county, florida
Gregory of nazianzum : ho theologos, "the divine" = Gregorius von nazianz, der theologe. selections / Ullmann, Karl – London: John W Parker, 1851 – 1mf – 9 – 0-8370-9909-9 – (in english) – mf#1986-3909 – us ATLA [240]
Gregory of Nazianzus, Saint *see* The five theological orations of gregory of nazianzus
Gregory of Nyssa, Saint *see* The catechetical oration of gregory of nyssa
Gregory, Olinthus *see* The evidences of christianity
Gregory, Saint, Bishop of Tours *see* Histoire des francs, textes de manuscrits de corbie et de bruxelles
Gregory the great / Barmby, James – London: SPCK 1892 [mf ed 1986] – (= ser The fathers for english readers) – 1mf – 9 – 0-8370-7842-3 – (incl bibl ref) – mf#1986-1842 – us ATLA [240]
Gregory the great : his place in history and thought / Dudden, Frederick Homes – London; New York: Longmans, Green. 2v. 1905 – 4mf – 9 – 0-8370-7936-5 – (incl bibl & ind) – mf#1986-1936 – us ATLA [920]
Gregory the great's moralia : emmanuel college ms. 112 – 14th c – 1r – 14 – mf#C594 – uk Microform Academic [240]
Gregory, Theodor Emanuel Gugenheim *see* Ernest oppenheimer and the economic development of southern africa
Gregory, Tom *see* History of solano and napa counties, california
Greif, Martin *see*
- General york
- Gesammelte werke
- Schillers demetrius

Greiff, Max *see* Internationales privatrecht nach dem einfuehrungsgesetze zum buergerlichen gesetzbuche
Greifswalder gemeinnuetziges wochenblatt fuer den bauern und landmann [...] – Greifswald DE, 1794 jun-1975 may – 1r – 1 – gw Misc Inst [630]
Greifswalder studien : theologische abhandlungen hermann cremer zum 25 jaehrigen professorjubilaeum / Oettli, Samuel et al – Guetersloh: C Bertelsmann, 1895 – 1mf – 9 – 0-8370-3389-6 – (incl bibl ref) – mf#1985-1389 – us ATLA [220]
Greifswalder tageblatt 1992 – Greifswald DE, 1992 – 1 – gw Misc Inst [074]
Greifswaldisches wochen-blatt von allerhand gelehrten und nuetzlichen sachen – Greifswald DE, 1743 – 1 – gw Misc Inst [943]
Greig, B F *see* Jesus of nazareth passeth by...
Greig, T Watson *see* Ladies' dress shoes of the nineteenth century
Greiner, Leo *see*
- Arbaces und panthea
- Lenau
- Das tagebuch

Greiner, Martin *see* Die entstehung der modernen unterhaltungsliteratur
Greiner, N Gretchen *see* Like it is
Greiner, Wilhelm *see* Die ersten bnovellen otto ludwigs und ihr verhaeltnis zu ludwig tieck
Greiner-Mai, Herbert *see*
- Die deutsche kriminalerzaehlung von schiller bis zur gegenwart
- Die ursache
- Der verbrecher aus verlorener ehre
- Wer ist schuld?

Greinz, Rudolf *see*
- Der heilige berg athos
- Heinrich heine und das deutsche volkslied

Greinz, Rudolf Heinrich *see* Gedichte
Der greis : eine wochenschrift / ed by Patzke, Johann Samuel – Magdeburg 1763-66 – (= ser Dz) – 16pt on 24mf – 9 – €240.00 – mf#k/n5283 – gw Olms [370]
Greith, Carl *see* Die deutsche mystik im prediger-orden (von 1250-1350)
Greko-kaftoliceske sojedinenije v ssa kalendar' – 1912-1937, 1939-1943, 1945-1977 – 1 – us CRL [520]
Grell, Guenther *see* Wij bouwen zeedijken
Grellet, Henry Robert *see* The case of england and western australia in respect to transportation
Grelling, Richard *see* Gleiches recht
El gremio de plateros en las indias occidentales / Torre Revello, Jose – Buenos Aires, 1932; Madrid: Razon y Fe, 1933 – 1 – sp Bibl Santa Ana [972]
Gremio Oficial de Exportadores de Pimenton *see* El pimenton de la vera en su cocina
Gremio tres de maio – Itajai, SC. 12 out-15 nov 1902 – (= ser Ps 19) – mf#P16,02,54 – bl Biblioteca [972]
Grenada : five year development plan, 1964-1968 / Trinidad And Tobago Development Programme Commiss... – Port-of-Spain, Trinidad and Tobago. 1965 – 1r – us UF Libraries [972]
Grenada *see*
- Grenada government gazette
- Statistical blue books 1860-1938

Grenada government gazette / Grenada – Saint George. 1957-1968- – 1 – us NY Public [972]
Grenaderskij korpus ispolnitel'nyj komitet *see* Izvestiia ispolnitel'nogo komiteta
The grenadian voice – St. Georg's, Grenada. June 13 1981-1991 – 6r – 1 – us L of C Photodup [074]
Grendon, Felix *see* The anglo-saxon charms
Grenfell, Bernard Pyne *see*
- Fragment of an uncanonical gospel from oxyrhynchus
- Logia iesou
- New sayings of jesus; and, fragment of a lost gospel

Grenfell record – Grenfell, jan 1969-jun 1984; jul-dec 1992 – 24r – 9 – at Pascoe [079]
Grenfell record – Grenfell, may 1875-dec 1953 – 29r – A$2101.31 vesicular A$2260.81 silver – at Pascoe [079]
Grenfell, Wilfred Thomason *see* A man's faith
Grenier, F *see* Code recreatif des francs-macons
Grenier, Henry Napoleon *see* Lecons de photographie
Grenier, Jacques R de *see* Memoires de la campagne des decouvertes dans les mers de l'inde
Grenier, Jean *see* Entretiens sur le bon usage de la liberte
Grenier, Jean-Pierre *see*
- Jofroi
- Orion le tueur

Grenier's rubber news – Kuala Lumpur. Malaysia. oct 1909-sep 1918, jan 1919-jun 1920 [wkly] – 9r – 1 – uk British Libr Newspaper [670]
Grenna tidning – Granna, Sweden. 1858-1964 – 1 – sw Kungliga [078]
Grennaposten – Graenna, 1992, 1993 – 9 – sw Kungliga [078]
Grennaposten – Graenna, Sweden. 1891 – 1r – 1 – sw Kungliga [078]
Grennen, Joseph E *see* Review notes and study guide to shakespeare's henry 4, part 1
Grenon, Hector *see* Au temps des "petits chars"
Grenon, P *see* Documentos historicos coleccionados po...seccion geografia...
Der grenzbauer : roman / Boris, Otto – Dresden: Deutscher Literaturverlag, O Melchert, 1943 [mf ed 1989] – 428p (ill) – 1 – mf#7052 – us Wisconsin U Libr [830]
Grenzbote – Bratislava, Czechoslovakia. May-Jul 1941; Nov 1943-Mar 1945 – 5r – 1 – us L of C Photodup [077]
Grenzbote – Pressburg (Bratislava SK), 1934 2 jan-1944 31 mar – 23r – 1 – (missing: 1943) – gw Misc Inst [077]
Die grenze mitten durch das herz / Brehm, Bruno – Muenchen: R Piper c1938 [mf ed 1989] – 1r – 1 – (filmed with other titles) – mf#7066 – us Wisconsin U Libr [830]
Grenz-echo – Eupen (B), 1949 3 jan-11 jan & 6 sep-14 nov, 1950 28 jan-11 mar & 20 jun-26 jul, 1951 [single iss] & jun [gaps], 1952-57 – 14r – 1 – (filmed by misc inst: 1928 8 feb-1940 mai [gaps], 1949 3 jan-1957, 1972-; 1958-66; 2000 1 apr-) – gw Mikrofilm; gw Misc Inst [074]
Grenzflaechennahe gitterstrukturen zur untersuchung von oberflaechen und duennen filmen / Mueller, Klaus Guenther – [mf ed 1996) – 2mf – 9 – €40.00 – 3-8267-2386-4 – mf#DHS 2386 – gw Frankfurter [530]
Die grenzgebiete zwischen privatrecht und strafrecht : kriminalistische bedenken gegen den entwurf eines buergerlichen gesetzbuches fuer das deutsche reich / Liszt, Franz von – Berlin, Leipzig: J Guttentag, 1889 – (= ser Civil law 3 coll; Beitraege zur erlaeuterung und beurtheilung des entwurfes eines buergerlichen gesetzbuches fuer das deutsche reich) – 1mf – 9 – mf#LLMC 96-602 – us LLMC [346]
Grenzland – Saarbruecken. dec 1934 – 1 – (cont. by: westland) – fr ACRPP [073]
Grenzland *see* Westland
Grenzland-bote : wyrzysker zeitung – Wyrzysk (Wirsitz PL), 1932 19 jan-23 apr, 1935 2 may-1939 29 jun – 2r – 1 – gw Misc Inst [077]
Grenzland-kurier *see* Dreistaedte-zeitung
Grenzland-zeitung – Grottau (Hradek nad Nisou CZ), 1929/30 4 jan-1937 – 4r – 1 – gw Misc Inst [077]
Grenzmann, Wilhelm *see* Die jungfrau von orleans in der dichtung
Der grenzmark-rappe : grenzmaerkische sagen, erzaehlungen, balladen und gedichte / Menzel, Herybert – Hamburg: Hanseatische Verlagsanstalt 1933 [mf ed 1990] – 1r – 1 – (filmed with: gedichte / alfred meissner) – mf#2833p – us Wisconsin U Libr [800]
Der grenzpolizist – Koenigs Wusterhausen DE, 1956-61 – 3r – 1 – (title varies: 21 sep 1961: grenzsoldat) – gw Misc Inst [327]
Grenzpost – Zwittau (Svitavy CZ), 1929 17 aug-1938 – 4r – 1 – gw Misc Inst [077]
Grenzsoldat *see* Der grenzpolizist
Die grenzwacht – Schneidemuehl (Pila PL), 1924 jan-jun – 1r – 1 – gw Misc Inst [077]
Grenzwacht – Glatz (Klodzko PL), 1936 2 jan-31 mar & 1 oct-31 dec, 1937 1 oct-31 dec, 1938 1 jul-30 sep, 1939 2 oct-1941 30 jun, 1941 1 oct-30 dec – 7r – 1 – gw Misc Inst [077]
Grenzwacht – Hindenburg (Zabrze), Poland sep 1924-apr 1945 [mf ed Norman Ross] – 3r – 1 – (non-partisan, independent weekly for all border regions of the steiermark, burgenland and corinthia) – mf#nrp-1450 – us UMI ProQuest [074]
Grenzwacht – Osijek, Yugoslavia. 1942-43 – 1r – 1 – us L of C Photodup [949]
Grenzwacht *see* Slawonischer volksbote
Grenzwarte – Bocholt, Borken DE, 1927 1 jan-24 mar & 30 jun-31 dec, 1929 3 jan-30 mar, 1930 25 feb & 1 apr-30 jun, 1930 1 oct-1931 29 apr, 1932 1 apr-30 jun, 1933 2 jan-31 mar, 1933 1 oct-1934 9 feb – 7r – 1 – (title varies: 1 feb 1934: der neue tag. vlg in oberhausen) – gw Mikrofilm [074]
Die grenz-zeitung – Schlawe (SLawno, PL) 1934 1 mar-30 mar, 1935 11 aug-13 sep [gaps], 1936-37 [gaps], 1938 14 jan-7 apr [gaps] & 1938, 1 jul-31 aug [gaps] [mf ed 2004] – 2r – 1 – (vlg in stolp; 6 feb 1936?: schlawer grenzzeitung; 27 jan 1937: die grenzzeitung; vbg: stolp, schlawe, buetow-rummelsburg; filmed by misc inst: 1936/37 (single iss) 1r; 1936 1 jul-30 okt, 1939 21 jul-1940 [gaps], 1942) – gw Mikrofilm [077]
Gresens, Sabine et al *see* Reichsministerium fuer kirchliche angelegenheiten (bestand r 5101) bd 68
Gresham gazette – Gresham, NE: Hugh M McGaffin, 1894-v94 n16. jun 25 1992 (wkly) [mf ed 1896-99,1902,1907-92 (gaps) filmed 1971-92] – 23r – 1 – (cont: gresham review. absorbed by: shelby sun (1904). some irregularities in numbering) – us NE Hist [071]
Gresham outlook – Gresham, Multnomah County, OR: H L St Clair. v1 n1-v68 n77 mar 3 1911-sep 23 1978; 68th yr n78-81st yr n45 sep 27 1978-jun 8 1991 – 1 – (cont by: outlook) – us Oregon Hist [071]
Gresham outlook – Gresham OR: H L St Clair, 1911-91 [semiwkly] – 1 – (cont by: outlook (1991-)) – us Oregon Lib [071]
Gresham review – Gresham, NE: S Rhodes, 1888-94// (wkly) [mf ed 4th yr n10. mar 13, apr 24, jul 17 1891] – 1r – 1 – (cont by: gresham gazette) – us NE Hist [071]
Gresley, Joseph-Edouard le *see* Bibliographie analytique de monsieur alphonse desilets
Gresley, W *see*
- Letter to the dean of bristol on what he considers the "fundamental...
- Real danger of the church of england
- Second word of remonstrnace with the evangelicals

Gresley, William *see* Distinctive tenets of the church of england
Gresset, J-B *see* Discours sur l'harmonie
Gressington, Gilbert *see* Free thoughts on the probable consequences of the decision in the c...
Gressmann, H *see*
- Nonnenspiegel und moenchsspiegel des euagrios pontikos
- Studien zu eusebs theophanie

Gressmann, Hugo *see*
- Albert eichhorn und die religionsgeschichtliche schule
- Die ausgrabungen in palaestina und das alte testament
- Moses und seine zeit
- Studien zu eusebs theophanie
- Der ursprung der israelitisch-juedischen eschatologie
- Das weihnachts-evangelium auf ursprung und geschichte

Gressmann, Hugo et al *see* Altorientalische texte und bilder zum alten testamente
Gresswell, George *see* An examination of the theory of evolution
Greswell, William Henry Parr *see* The growth and administration of the british colonies 1837-1897
Greta and branxton gazette – Greta, jan 1891-jun 1892 – 1r – 9 – A$40.30 vesicular A$45.80 silver – at Pascoe [079]
The gretchen episode from goethe's faust / ed by Heffner, Roe Merrill Secrist et al – Boston: Houghton Mifflin, c1950 [mf ed 1993] – ix/144p – 1 – (german text. int and notes in english) – mf#8622 – us Wisconsin U Libr [820]
Grete herball – The grete herball whiche geueth parfyt knowlege and vnderstandyng of all maner of herbes & there gracyous vertues... London, 1526 – 1 – us Wisconsin U Libr [631]
Grete minde / Fontane, Theodor – Niedersedlitz: Schuster, [19–?] [mf ed 1989] – 95p – 1 – mf#7248 – us Wisconsin U Libr [830]
Grethenbach, Constantine *see* Secular view of the bible
Gretna breeze – Gretna, NE: John Bradford. 1st yr n1. jun 16 1899- (wkly) [mf ed - 1935,1938- (gaps)] – 1 – (absorbed: gretna tribune. publ in papillion may 27 1943-aug 23 1945; gretna ne, aug 30 1945-nov 29 1951; papillion dec 6 1951-) – us NE Hist [071]
Gretna guide and news – Gretna, NE: Ostdiek Publ, 1961 (wkly) [mf ed 1964- filmed 1976-] – 1 – (title varies slightly) – us NE Hist [071]
Gretna news – Gretna, NE: Speedie & Patterson, sep 1896-v1 n36. may 27 1897 (wkly) – 1r – 1 – (merged with: gretna reporter to form: news-reporter) – us NE Hist [071]
Gretna reporter – Gretna, NE: John Bradford, jul 1888-v10 n29. may 28 1897 (wkly) [mf ed 1891-97 (gaps)] – 1r – 1 – (cont: sarpy county democrat. merged with: gretna news to form: news-reporter. issues for aug 4 1892-jan 26 1893 also called whole n196-321) – us NE Hist [071]
Gretna tribune – Gretna, NE: W E Patterson. v1 n1. jul 13 1900-v1 n38. mar 29 1901 (wkly) [mf ed with gaps] – 1r – 1 – (absorbed by: gretna breeze) – us NE Hist [071]
Gretry, Andre Ernest Modeste *see* Ouverture de la caravane
Gretser, J *see* De officiis magnae ecclesiae et aulae constantinopo-litanae (cbh18)
Gretsko, P et al *see* Organ russkikh revoliutsionerov
Greuel in spanien. la terreur en espagne – Olten, 1936? Fiche W930. (Blodgett Collection of Spanish Civil War Pamphlets) – 9 – us Harvard [946]
Greulich, Oskar *see* Platens litteratur-komoedien
Greundler, O *see* Frauenelend und frauenmission in indien
Greuner, Ruth *see* Zeitzuender im eintopf
Greve generale – London, UK. 18 Mar-2 Jun 1902 – 1 – uk British Libr Newspaper [072]
La greve generale / Congres National des Chambres Syndicales et Groupes Corporatifs – Organe du Comite d'organisation de la greve generale. 2e annee, no. 1-7. Paris. 13 janv-18 mars 1894; 3e annee, no. 1-14. mars 1899-sept 1900. mq no. 3, 9, 11 – 1 – fr ACRPP [331]
Greve, H E *see* De bronnen van carel van mander voor "het leven der doorluchtighe nederlandtsche en hoogduytsche schilders"
Greven, Joseph *see* Die anfaenge der beginen
Grevenbroicher zeitung – Grevenbroich DE, 1913 apr-1922 mar, 1923 apr-1925 mar – 1 – gw Misc Inst [074]
Grevener anzeiger – Greven DE, 1953 5 feb-1970 – 1 – gw Misc Inst [074]
Greving, Joseph *see*
- Johann eck als junger gelehrter
- Pauls von bernried vita gregorii 7. papae

Grewe, Norbert *see* Beratungslehrer
Grewingk, Constantin et al *see* Ueber die meteoritenfaelle von pillistfer, buschhof und igast in liv- und kurland
The grey books index 1900-74 : riba members' work illustrated in architectural periodicals 1900-1974 / Royal Institute of British Architects (RIBA) – 2pts – 269mf – 9 – £1420.00 coll – (incl printed guide. pt a: index to members' work in periodicals 1900-19 : building types fiche nos 1-7 architects' names fiche nos 8-12. pt b: index to members' work in periodicals 1920-74: architects' names fiche nos 13-143 building types fiche nos 144-269) – mf#GBI – uk World [720]
Grey, C *see* Travels to tana and persia
Grey, Douglas *see* A practical treatise upon modern printing machinery and letterpress printing
Grey, Henry *see*
- Diffusion of christianity dependent on the exertions of christians
- Remarks relative to his connection with the letters of anglicanus

Grey, Henry George Grey, 3rd earl *see* The colonial policy of lord john russell's administration
Grey, Henry George Grey, Earl *see* A bill
Grey, Jemima, Marchioness *see* Aristocratic women
The grey papers, 1748-1894 : from the archives of the earl of halifax, garrowby – 8r – 1 – mf#97012 – uk Microform Academic [025]
Grey river argus – Greymouth, NZ. jan-dec 1904; jan-jun 1907; jan-dec 1939; 1 jul-13 sep 1966 – 1 – (title changes to: argus leader (greymouth)) – mf#60.1 – nz Nat Libr [079]
Grey river argus *see* Argus leader (greymouth)
Grey, Standish *see* The gospel according to satan
Grey, William Henry *see* Church leases
The grey world / Underhill, Evelyn – London: William Heinemann, 1904 – 1mf – 9 – 0-7905-8611-8 – mf#1989-1836 – us ATLA [100]
Grey, Zane *see*
- Comics and scrapbooks
- Correspondence celebrating 20 years with harper brothers
- Diary

Grey, zane – 1r – 1 – mf#B26960 – us Ohio Hist [790]
Greyerz, Otto von *see* Blumenlese aus den saemmtlichen werken von johann rudolf wyss dem juengern
Greyling, J P *see* "Geloofskrisis as gesigsbedrog"
Greymouth evening star – jan 1901-jun 1904; jan 1905-jun 1914; jul 1915-dec 1939; 24 feb-30 nov 1956; 29 apr-2 oct 1959; 11 dec 1959-30 apr 1960; 3 apr-23 dec 1963; 22 jul-22 aug 1964; jul 1976-oct 1998 – 1 – mf#60.2 – nz Nat Libr [079]
Greytown gazette – Greytown, South Africa. 1903-88 – 24r – 1 – sa National [079]
Greytown news – oct 1977-sep 1980 – 1r – 1 – mf#48.16 – nz Nat Libr [079]
Grezel, Wolfgang *see* Die geistliche leiter zum himmelreich...
Griaule, M *see* Jeux et divertissements abyssins
Gribble, Cecil F *see*
- Collection of pamphlets relating to the methodist church in fiji, 1878-1970
- Miscellaneous papers relating to tonga and fiji

Gribble, Charles Bessly *see* Christ glorified!
Gribble, Francis Henry *see* The comedy of catherine the great
Gribble, Phillip A *see* Effects of static and hold-relax stretching on hamstring range of motion using the flexability le1000
Gribovskii, Viacheslav Mikhailovich *see* Mater ialy dlia istorii vysshago suda i nadzora v pervuiu polovinu..
Gridiron – La Salle IL 1822-23 – 1 – mf#3814 – us UMI ProQuest [073]
Gridiron revivdus / Montgomery Co. Miamisburg – apr-may 1839 very short [wkly] – 1r – 1 – mf#B5001 – us Ohio Hist [071]
Gridley, Albert Leverett *see* The first chapter of genesis as the rock foundation for science and religion
Gridley camp news / Gridley Migratory Labor Camp – v1 n3, 5-12, 14, 16 [1942 apr 24-jul 31] – 1r – 1 – (cont: tent city news) – mf#3547288 – us WHS [305]
[Gridley-] gridley herald – CA. 1880-85; 1888-92; 1902- [wkly] – 65r – 1 – $3900.00 (subs $50y) – mf#B02281 – us Library Micro [071]
Gridley Migratory Labor Camp *see* Gridley camp news
Grieben, Theobald *see* Die sudeten
Griechen und semiten auf dem isthmus von korinth : religionsgeschichtliche untersuchungen / Maass, Ernst – Berlin: G Reimer, 1903 – 1mf – 9 – 0-524-00929-5 – (incl bibl ref) – mf#1990-2152 – us ATLA [250]
Griechenland-index = Bilddokumentation zur kunst in griechenland / ed by Bildarchiv Foto Marburg – Deutsches Dokumentationszentrum fuer Kunstgeschichte Philipps- Universitaet Marburg – [mf 2000-01] – 340mf (1:24) in 2 installments – 9 – silver €5200.00 – ISBN-10: 3-598-34523-2 – ISBN-13: 978-3-598-34523-4 – gw Saur [700]
Griechentum und christentum : zwoelf hibbertvorlesungen ueber den einfluss griechischer ideen und gebraeuche auf die christliche kirche = Influence of greek ideas and usages upon the christian church / Hatch, Edwin – Freiburg i B: J C B Mohr 1892 [mf ed 1990] – (= ser Hibbert lectures (london, england) 1888) – 1mf – 9 – 0-7905-7237-0 – (incl bibl ref; trans fr english into german by erwin preuschen) – mf#1988-3237 – us ATLA [240]
Griechentum und christentum / Rohr, Ignaz – 1. & 2. aufl. Muenster i W: Aschendorff 1912 [mf ed 1993] – (= ser Biblische zeitfragen 5/8) – 1mf – 9 – 0-524-05745-1 – (incl bibl ref) – mf#1992-0588 – us ATLA [250]
Griechentum und judentum im letzten jahrhundert vor christus / Heinisch, Paul – 1. & 2. aufl. Muenster i W: Aschendorff 1908 [mf ed 1993] – (= ser Biblische zeitfragen 1/12) – 1mf – 9 – 0-524-06920-4 – (incl bibl ref) – mf#1992-1013 – us ATLA [270]
Griechisch-byzantinische gespraechsbuecher und verwandtes aus sammelhandschriften / ed by Heinrici, Carl Friedrich Georg – Leipzig: B G Teubner, 1911 [mf ed 1990] – (= ser Abhandlungen der koeniglich saechsischen gesellschaft der wissenschaften 28/8) – 1mf – 9 – 0-7905-4813-5 – (in greek & german. incl bibl ref) – mf#1988-0813 – us ATLA [450]
Griechische christliche schriftsteller der ersten drei jahrhunderte – Leipzig. v1-34. 1897-1926 – 327mf – 8 – mf#1057c – ne IDC [450]
Griechische denker / Gomperz, Th – Leipzig. v1-3. 1896-1909 – 8 – €63.00 – (v1: leipzig 1896 10mf. v2: leipzig 1902 13mf. v3: leipzig 1909 10mf) – ne Slangenburg [180]
Griechische dramen in deutschen bearbeitungen / Spangenberg, Wolfhart & Froereisen, Isaac; ed by Daehnhardt, Oskar – Stuttgart: Litterarischer Verein, 1896 (Tuebingen: H Laupp, Jr) [mf ed 1993] – (= ser Blvs 211-212) – 2v – 1 – (incl bibl ref) – mf#8470 reel 44 – us Wisconsin U Libr [820]
Griechische dramen in deutschen bearbeitungen / Spangenberg, Wolfhart & Froereisen, Isaac; ed by Daehnhardt, Oskar – Stuttgart: Litterarischer Verein. 2v. 1896 (Tuebingen: H Laupp, Jr) – (incl bibl ref) – us Wisconsin U Libr [450]
Griechische epigraphik / Hiller Von Gaertringen, Friedrich – Leipzig, Germany. 1925 – 1r – us UF Libraries [025]
Griechische excerpte aus homilien des origenes / Klostermann, Erich – Leipzig, 1894 – (= ser Tugal 1-12/3b) – 1mf – 9 – €3.00 – ne Slangenburg [450]
Griechische geschichte : von ihrem ursprunge bis zum untergange der selbstaendigkeit des griechischen volkes / Holm, Adolf – Berlin: S Calvary & Co 1886-91 [mf ed 1979] – (= ser Calvary's philologische und archaeologische bibliothek [81-85, 89-99], 107-114, i) – 3v on 1r – 1 – mf#film mas 9111 – us Harvard [930]
Griechische goetterlehre / Welcker, Friedrich Gottlieb – Goettingen: Dieterich, 1857-1862 – 5mf – 9 – 0-524-02439-1 – (incl bibl ref) – mf#1990-3023 – us ATLA [250]
Griechische grammatik / Meyer, Gustav – Leipzig: Breitkopf & Haertel, 1886 – (= ser Bibliothek indogermanischer grammatiken) – 2mf – 9 – 0-524-05926-8 – mf#1992-0683 – us ATLA [450]
Die griechische, griechisch-roemische und alt-christliche lateinische musik / Moehler, A – Rom, 1898 – €7.00 – ne Slangenburg [780]
Griechische inschriften zur griechischen staatenkunde / Bleckmann, Friedrich – Bonn: A Marcus & E Weber, 1913 [mf ed 1992] – (= ser Kleine texte fuer vorlesungen und uebungen 115) – 1mf – 9 – 0-524-04635-2 – (incl bibl ref) – mf#1990-3378 – us ATLA [450]
Griechische kirche / Hasemann, J – [Leipzig: Brockhaus, 1866] [mf ed 1986] – 1mf – 9 – 0-8370-8184-X – (incl bibl) – mf#1986-2184 – us ATLA [243]
Griechische liturgien (bdk5 1.reihe) – (= ser Bibliothek der kirchenvaeter. 1. reihe (bdk 1.reihe)) – €21.00 – ne Slangenburg [243]
Griechische mythologie / Preller, Ludwig – 4th ed. ed by Carl Robert. Berlin: Weidmann, 1894-1926.2v in 3 – 1 – us Wisconsin U Libr [250]
Griechische mythologie / Preller, Ludwig – 2. Aufl. Berlin: Weidmann, 1860-1861 – 3mf – 9 – 0-524-02433-2 – (incl bibl ref) – mf#1990-3017 – us ATLA [250]
Griechische mythologie und religionsgeschichte / Gruppe, Otto – Muenchen: CH Beck, 1902-1906 – (= ser Handbuch der klassischen Altertumswissenschaft) – 21mf – 9 – 0-524-07376-7 – mf#1991-0096 – us ATLA [250]
Griechische papyri / Lietzmann, Hans [comp] – 2. aufl. Bonn: A Marcus & E Weber 1910 [mf ed 1992] – (= ser Kleine texte fuer theologische und philologische vorlesungen und uebungen 14) – 1mf – 9 – 0-524-04700-6 – (text in greek; notes in german; incl bibl ref) – mf#1990-3409 – us ATLA [090]
Griechische philosophie im alten testament : eine einleitung in die psalmen- und weisheitsliteratur / Friedlaender, Moritz – Berlin: Georg Reimer, 1904 – 1mf – 9 – 0-7905-0943-1 – (incl bibl ref) – mf#1987-0943 – us ATLA [221]
Die griechische philosophie im buche der weisheit / Heinisch, Paul – Muenster i.W: Aschendorff, 1908 [mf ed 1989] – (= ser Alttestamentliche abhandlungen 1/4) – 1mf – 9 – 0-7905-0897-4 – (incl bibl ref & ind) – mf#1987-0897 – us ATLA [221]
Griechische religionsphilosophie / Gilbert, Otto – Leipzig: W Engelmann, 1911 [mf ed 1992] – 2mf – 9 – 0-524-02203-8 – (incl bibl ref) – mf#1990-2877 – us ATLA [180]
Griechische schulgrammatik : auf grund der ergebnisse der vergleichenden sprachforschung / Koch, Ernst – 8. Aufl. Leipzig: B G Teubner, 1881 – 1mf – 9 – 0-8370-9253-1 – (includes glossary and index) – mf#1986-3253 – us ATLA [450]
Griechische schulgrammatik / Curtius, Georg – 5. Aufl. Prag: F Tempsky, 1862 – 1mf – 9 – 0-8370-9219-1 – (incl indes) – mf#1986-3219 – us ATLA [450]
Die griechische sprache im zeitalter des hellenismus : beitraege zur geschichte und beurteilung der koine / Thumb, Albert – Strassburg: Karl J Truebner, 1901 – 1mf – 9 – 0-8370-9317-1 – (incl bibl ref and indexes) – mf#1986-3317 – us ATLA [450]
Griechische syntax zum neuen testament : nebst uebungsstuecken zum uebersetzen ins griechische fuer formenlehre und syntax / Heusser, Theodor – Basel: CS Spittler, 1889 – 1mf – 9 – 0-524-08080-1 – mf#1992-1140 – us ATLA [450]
Die griechische uebersetzung der viri inlustres des hieronymus / Wentzel, G – Leipzig, 1895 – (= ser Tugal 1-13/3) – 1mf – 9 – €3.00 – ne Slangenburg [450]
Die griechische uebersetzung der viri inlustres des hieronymus / Wentzel, Georg – Leipzig: J C Hinrichs, 1895 – (= ser Tugal) – 1mf – 9 – 0-7905-1799-X – (incl bibl ref) – mf#1987-1799 – us ATLA [240]
Die griechische uebersetzung des apologeticus tertullian's / medicinisches aus der aeltesten kirchengeschichte / Harnack, Adolf von – Leipzig: J C Hinrichs 1892 [mf ed 1989] – (= ser Tugal 8/4) – 1mf – 9 – 0-7905-1883-X – (in german & greek. incl bibl ref & ind) – mf#1987-1883 – us ATLA; ne Slangenburg [230]
Griechische und lateinische lehnwoerter im talmud, midrasch und targum / Krauss, Samuel – Berlin: S Calvary. 2v. 1898-99 – 3mf – 9 – 0-8370-8267-6 – (incl indes) – mf#1986-2267 – us ATLA [470]
Griechische zauberpapyri und das gemeinde- und dankgebet im 1. klemensbriefe / Schermann, Theodor – Leipzig: J C Hinrichs, 1909 – (= ser Tugal) – 1mf – 9 – 0-7905-1733-7 – (incl bibl ref) – mf#1987-1733 – us ATLA [240]
Griechische zauberpapyri und das gemeinde- und dankgebet im 1. klemensbriefe / Schermann, Theodor – Leipzig, 1909 – (= ser Tugal 3-34/2b) – 2mf – 9 – €5.00 – ne Slangenburg [240]
Griechische-kulturgeschichte / Burckhardt, Jakob; ed by Oeri, Jakob – Berlin, Stuttgart: W Spemann [1898-1902] [mf ed 1979] – 4v on 1r – 1 – (with bibl footnotes) – mf#film mas 9023 – us Harvard [930]
Die griechischen christlichen schriftsteller der ersten jahr- hunderte (gcsej) – Leipzig. v1-49. 1901- – 547mf – 8 – €1043.00 – (vols also listed separately) – ne Slangenburg [240]
Die griechischen culte und mythen in ihren beziehungen zu den orientalischen religionen / Gruppe, Otto – Leipzig: BG Teubner, 1887 – 2mf – 9 – 0-524-04162-8 – (incl bibl ref) – mf#1990-3292 – us ATLA [250]
Die griechischen handschriften des neuen testaments / Gregory, Caspar Rene – Leipzig: J C Hinrichs, 1908 – 1mf – 9 – 0-8370-3386-1 – mf#1985-1386 – us ATLA [225]
Die griechischen handschriften des neuen testaments in der udssr / Treu, K – Berlin, 1966 – (= ser Tugal 5-91) – 7mf – 9 – €15.00 – ne Slangenburg [225]
Die griechischen inschriften der palaestina tertia westlich der "araba"' / Alt, Albrecht – Berlin, Leipzig, 1921 – (= ser Wissenschaftliche Veroeffentlichungen des Deutsch-Tuerkischen Denkmalschutz-Kommandos, Pt 2) – 2mf – 9 – mf#H-2880 – ne IDC [956]
Die griechischen papyrusurkunden : ein vortrag / Wilcken, Ulrich – Berlin: G. Reimer, 1897 – 1mf – 9 – 0-7905-3499-1 – (incl bibl ref) – mf#1987-3499 – us ATLA [450]

Die griechischen schreiber des mittelalters und der renaissance / Vogel, M & Gardthausen, V – Leipzig, 1909 – 9mf – 8 – €18.00 – ne Slangenburg [450]
Griechisches biographisches archiv (gba) = Greek biographical archive (gba) / Schmuck, Hilmar [comp] – [mf ed 1998-2001] – 453mf (1:24) in 12 installments – 9 – diazo €10,060.00 (silver €11,080 isbn: 978-3-598-34181-6) – ISBN-10: 3-598-34180-6 – ISBN-13: 978-3-598-34180-9 – (with printed ind) – gw Saur [949]
Griechisches erbe; das urbild der antike im widerschein des heutigen lebens / Klatt, Fritz – Berlin: W. de Gruyter & Co., 1943. 100p. 20 pl. 1 reel. 1293 – 1 – us Wisconsin U Libr [900]
Der griechisch-syrische text des matthaeus : e 351 im verhaeltnis zu tatian ssc ferrar / Pott, August – Leipzig: B G Teubner, 1912 – 1mf – 9 – 0-7905-3101-1 – mf#1987-3101 – us ATLA [225]
Griechisch-syrisch-hebraeischer index zur weisheit des jesus sirach / Smend, Rudolf – Berlin, G Reimer, 1907 – 1mf – 9 – 0-7905-3169-0 – mf#1987-3169 – us ATLA [221]
Grieco, Agrippino *see* Evolucao da prosa brasileira
Grier, Thomas Graham *see* On the canal zone, panama
Grierson, Elizabeth Wilson *see* Our scottish heritage
Grierson, G A *see* The pisaca languages of north western india
Grierson, George A *see* Hatim's tales
Grierson, George Abraham *see*
– Linguistic survey of india
– Note on recent translations of scripture into hindi and bengali
– Notes on tul'si das
Gries, John Matthew *see* Present home financing methods
Griesbach, Heinz *see* Deutsche sprachlehre fur auslander
Griese, Friedrich *see*
– Baeume im wind
– Fritz reuter
– Der heimliche koenig
– Das kind des torfmachers
– Das letzte gesicht
– Mein leben
– Mensch aus erde gemacht
– Der ruf des schicksals
– Die wagenburg
– Die weisskoepfe
– Winter
Griesel, August F *see* A w griesel's neuestes gemaelde von prag
Grieshaber, Christoph *see* Europafaehigkeit der schweizerischen alters- und hinterlassenenversicherung (ahv)
Griesheim, Christ Ludwig von *see* Beytraege zur aufnahme des bluehenden wohlstands der staaten
Griesheimer, Friedrich *see* Vergleich deutscher sonderstrafnormen mit den entsprechenden oesterreichischen bestimmungen
Griesinger, Theodor *see* The mysteries of the vatican
The grievances between authors and publishers : being the report of the conferences of the incorporated society of authors held at willis's rooms, in march, 1887 / Incorporated Society of Authors – London: various publ, 1887 – (= ser 19th c publishing...) – 3mf – 9 – (with add matter & summary) – mf#3.1.66 – uk Chadwyck [070]
Grieving the holy spirit / Sortain, J – Brighton, England. 1859 – 1r – us UF Libraries [240]
Der griff ins all : anekdoten und kurze geschichten / Lerbs, Karl – feldpostausg. Berlin: T Knaur 1943 [mf ed 1990] – 1r – 1 – (filmed with: die freunde machen den philosophen...von jakob michael reinhold lenz / ilse kaiser [comp]) – mf#2823p – us Wisconsin U Libr [830]
Griffen, A M *see* Soil survey of escambia county, florida
Griffen, Albert *see* The key note
Griffes : two coloristic works / Keyton, Robert – U of Rochester 1952 [mf ed 19–] – 2mf – 9 – mf#fiche 1161 – us Sibley [780]
Griffin, A W *see* Chitonga vocabulary of the zambesi valley
Griffin, Appleton Prentiss Clark [comp] *see* Select list of books (with references to periodicals) relating to the far east
Griffin baptist church. pickens county. south carolina : church records – 1857-1972 – 1 – 46.22 – us Southern Baptist [242]
Griffin, Edward D *see* A plea for africa
Griffin, George *see* The gospel its own advocate
Griffin, George Douglas *see* Important national information, canadian finances examined
Griffin, George Eugene *see* Concerto (no 1) for piano forte, op 1
Griffin, Gilderoy Wells *see* New south wales: her commerce and resources
Griffin, Levi Thomas *see*
– Cases on personal property.
– Sprague's illustrative cases on personal property
Griffin, Lisa M *see* An examination of the relationship between teacher enthusiasm and alt-pe
Griffin, Martin Ignatius Joseph *see* Catholics and the american revolution
Griffin, Sybil N *see*
– Mysterious music of the alafia river
– Strawberry schools in hillsborough county
Griffin, W S *see* Canadian poems
Griffin, Watson *see* The provinces and the states
Griffin, Zebina Flavius *see* Chundra lela
Griffing, James Sayre *see*
– Papers
Griffing, Jane R *see* Letters from florida on the scenery, climate, social and material c...
Griffis, William Elliot *see*
– America in the east
– American in the east
– Corea, the hermit nation
– Corea, without and within
– Hepburn of japan and his wife and helpmates
– The influence of the netherlands in the making of the english commonwealth and the american republic
– The lily among thorns
– A maker of the new orient
– The mikado's empire
– Modern pioneer in korea
– The religions of japan
– Verbeck of japan
Griffis, William Elliott *see* Dux christus
Griffith, Coleman R *see*
– Psychology and athletics
– The psychology of coaching
Griffith, D W *see* D w griffith papers, 1897-1954
Griffith, Elmer Cummings *see* Epochs in baptist history
Griffith, F L *see*
– El eersheh
– Hieratic papyri from kahun and gurob
– The inscriptions of si-t and der rifeh
– The mound of the jew and the city of onias
– Stories of the high priests of memphis
Griffith, Fl *see* El beresh, pt 2
Griffith, FL *see* Beni-hasan, 4: zoological and other details
Griffith, Gareth E *see* An electromyographic comparison of seated and standing up-hill cycling
Griffith john : de apostel van centraal-china / Marang, Gerardus Pieter – [Rotterdam: J M Bredee, 1913] [mf ed 1995] – (= ser Yale coll; Lichtstralen op den akker der wereld [19. jaarg 1913] 4) – 50p (ill) – 1 – 0-524-09602-3 – (in dutch) – mf#1995-0602 – us ATLA [240]
Griffith john : founder of the hankow mission, central china / Robson, William – London: S W Partridge [1901?] [mf ed 1995] – (= ser Yale coll) – 176p (ill) – 1 – 0-524-09930-8 – (last chapter of this ed written by frank b broad of the mission house staff) – mf#1995-0930 – us ATLA [920]
Griffith john : founder of the hankow mission, central china / Robson, William – New York: Fleming H Revell, [1888?] – 1mf – 9 – 0-8370-6355-8 – mf#1986-0355 – us ATLA [240]
Griffith john : the story of fifty years in china / John, Griffith – popular rev ed. New York: A C Armstrong, 1908 – 2mf – 9 – 0-8370-6581-X – (incl ind) – mf#1986-0581 – us ATLA [240]
Griffith john : the story of fifty years in china / Thompson, Ralph Wardlaw – 2nd ed. London: Religious tract society, 1907 [mf ed 1995] – (= ser Yale coll) – xvi/544p (ill) – 1 – 0-524-09783-6 – mf#1995-0783 – us ATLA [920]
Griffith, Joseph B *see* Causal attributions and task persistence of learned-helpless and mastery-oriented sixth graders
Griffith, M E Hume- *see* Behind the veil in persia and turkish arabia
Griffith parks news *see* [Los angeles-] wilshire press – griffith parks news
Griffith, Ralph Thomas Hotchkin *see* Idylls from the sanskrit
Griffith, Thomas *see* Fundamentals
Griffith times – Griffith, jan 1971-dec 1972 – 3r – at Pascoe [079]
Griffith, W *see*
– Itinerary notes of plants collected in the khasyah and bootan mountains, 1837-1838, in affghanistan and neighbouring countries, 1839-1841
– Muscologia itineris assamici
Griffith, William *see* A treatise on the jurisdiction and proceedings of justices of the peace in civil suits
Griffith, William Brandford *see* A digest of and index to the reports of cases decided in the supreme court of the gold coast colony. 1844-1931
Griffith, William Herbert *see* The effects of couple communication training on marital perceptions
Griffith, William J *see* Santo tomas
Griffith-Jones, Ebenezer *see*
– The ascent through christ
– The challenge of christianity to a world at war
– The economics of jesus
– Faith and verification, with other studies in christian thought and life
– Types of christian life
Griffiths, Charles John *see* A narrative of the siege of delhi
Griffiths, D, Jr *see* Two years' residence in the new settlements of ohio, north america
Griffiths, Frederick A *see* Notes on military law
Griffiths, Howard *see* Study of british opinion on the problems and policies of the union of south africa
Griffiths, John *see*
– J griffiths, md, der koenigl gesellschaft zu edimburg und mehrerer auswaertigen gelehrten gesellschaften mitgliede, neue reise in arabien, die europaeische und asiatische tuerkey
– The paintings in the buddhist cave-temples of ajanta
Griffiths, Percival Joseph *see* The british impact on india
Griggs, Edward Howard *see*
– American statesmen
– Moral education
– The new humanism
Griggs, Leverett *see* Letters to a theological student
Griggs, Sutton Elbert *see* Hindered hand
Griggs, William *see* India
Griggs, William Charles *see* Odds and ends from pagoda land
Grignon de Montfort, Louis-Marie, Saint *see*
– Methode pour reciter avec fruit le saint-rosaire
– Le secret de marie devoile a l'ame pieuse
Grignon, Edmond *see* Album historique publie a l'occasion des fetes du cinquantenaire de la paroisse de sainte-agathe-des-monts, 1861-1911
Grignon, Joseph-Jerome *see*
– Un lutrin canadien
– Le luxe de notre epoque
– Le vieux temps
Grignon, Wilfrid *see*
– La culture du ginseng
– Le petit livre d'or du cultivateur et du colon
– Quelle est la meilleure poudre de condition pour tous les animaux de la ferme
Grigor'ev, G G *see* Vecherniaia pochta
Grigor'ev, M I *see* Melkii kredit v iaroslavskoi gubernii
Grigor'ev, N S *see* Otechestvo [perm': 1919]
Grigorev, V V *see* Istoricheskii ocherk russkoi shkoly
Grigor'iants, T S *see* Odnodnevnye gazety sssr 1917-1983
Grigorovich, V I *see* Ocherk puteshestviia po evropeiskoi turtsii
Grigorovitza, Emanuel *see* Die quellen von cl brentanos 'gruendung der stadt prag'
Grigsby, James Edward *see* The criminal law, including the federal penal code.
Grigson, Geoffrey *see* Places
The grihya-sutras (stbe29) : rules of vedic domestic ceremonies. pt 1: sankhayana, asvalayana, paraskara, khadira – 1886 – (= ser Sacred book of the east (sbte)) – 8mf – 8 – €17.00 – (trans by hermann oldenberg) – ne Slangenburg [280]
The grihya-sutras (stbe30) : rules of vedic domestic ceremonies. pt 2: gobhila, hiranyakesin, apastamba – 1892 – (= ser Sacred book of the east (sbte)) – 7mf – 8 – €15.00 – (trans by hermann oldenberg. apastamba, yagna-paribhasha-sutras trans by f max mueller) – ne Slangenburg [280]
Grijalva, Juan de *see* Cronica de la orden de n p s augustin en las prouincias de la nueua espana
Grill, Julius *see*
– Der achtundsechzigste psalm
– Die persische mysterienreligion im roemischen reiche und das christentum
– Der primat der petrus
– Zur kritik der komposition des buchs hiob
Die grille – 1811-12 [mf ed 1997] – (= ser Die zeitschriften des august von kotzebue) – 7mf – 9 – €120.00 – 3-89131-233-4 – gw Fischer [430]
Die grille / ed by Kotzebue, August von – Koenigsberg 1811-12 – (= ser Dz) – 3v on 9mf – 9 – €180.00 – mf#k/n6173 – gw Olms [870]
Grille, Francois *see* Suite au memorial de sainte-helene
Grillen und pillen aus abraham a sancta clara : hundert stuecklein / Abraham a Sancta Clara; ed by Bertsche, Karl – Muenchen-Pasing: Filser-Verlag, 1948 [mf ed 1993] – (= ser Abraham a sancta clara selections, 1948 1. werkchen) – 140p – 1 – (modern german trans of early modern german text) – mf#8451 – us Wisconsin U Libr [830]
Der grillenpfiff : kriminal-roman / Baumgarten, Harald – Berlin: Aufwaerts-Verlag, c1943 [mf ed 1989] – (= ser Der aufwaerts-kriminal-roman 29) – 239p – 1 – mf#6991 – us Wisconsin U Libr [830]
O grillo – Ouro Preto, MG. 10 dez 1905-06 jan 1906 – (= ser Ps 19) – bl Biblioteca [079]
Grillo, Max *see* Hombre de las leyes
Grillo que canto sobre el canal / Korsi, Demetrio – Panama, Panama. 1937 – 1r – us UF Libraries [972]
Grillon du foyer / Francmesnil, Ludovic De – Paris, France. 1905 – 1r – us UF Libraries [440]
Grillparzer / Hohlbaum, Robert – Stuttgart: J G Cotta 1938 [mf ed 1990] – (= ser Die dichter der deutschen) – 1r – 1 – (filmed with: grillparzers verhaeltnis zur politischen tendenzliteratur seiner zeit / konrad beste) – mf#2690p – us Wisconsin U Libr [430]
Grillparzer als archivdirector / Wolf, Gerson – Wien: Winter, 1874 – 1r – 1 – us Wisconsin U Libr [920]
Grillparzer, der tragiker der schuld / Sprengler, Joseph – Lorch: A Buerger, 1947 – 1r – 1 – us Wisconsin U Libr [430]
Grillparzer e i suoi drammi / Vincenti, Leonello – Milano: R Ricciardi, 1958 [mf ed 1993] – vii/290p – 1 – (incl bibl ref) – mf#8701 – us Wisconsin U Libr [430]
Grillparzer, Franz *see*
– Die ahnfrau
– Des meeres und der liebe wellen
– A dream is life
– Esther
– Euripides medea und das goldene vliess von grillparzer
– Gedichte
– Das goldene vlies
– Grillparzers saemmtliche werke in zwanzig baenden
– Grillparzers werke
– Koenig ottokars glueck und ende
– Saemtliche werke
– Sappho
– El sueno es vida
Grillparzer und die wissenschaft : drei vortraege / Redlich, Oswald – Wien: A Hartleben, [1925?] – 1r – 1 – (incl bibl ref) – us Wisconsin U Libr [500]
Grillparzer und schopenhauer / Geissler, Horst Wolfram – [S.l.: s.n.] 1915 (Weimar: G Uschmann) – 1r – 1 – (incl bibl ref) – us Wisconsin U Libr [943]
Grillparzer und seine werke / Paoli, Betty – Stuttgart: J G Cotta, 1875 – 1r – 1 – us Wisconsin U Libr [430]
Grillparzer unter goethe's einfluss / Waniek, Gustav – Bielitz: Im Verlage des k. k. Staats-Obergymnasiums, 1893 – 1r – 1 – (incl bibl ref) – us Wisconsin U Libr [430]
Grillparzers "ahnfrau" und die schicksalsidee / Terlitza, Victor – Bielitz: Verlag der k. k. Staats-Oberrealschule, 1883 – 1r – 1 – us Wisconsin U Libr [430]
Grillparzers "ahnfrau" und die wiener volksdramatik / Mueller, Curt – Leipzig: E Wiegandt, 1911 – 1r – 1 – (incl bibl ref) – us Wisconsin U Libr [430]
Grillparzers gespraeche : und die charakteristiken seiner persoenlichkeit durch die zeitgenossen – Wien: Verlag des Literarischen Vereins, 1904-16 [mf ed 1993] – (= ser Schriften des literarischen vereins in wien 1, 3, 6, 12, 15, 20) – 6v – 1 – (incl bibl ref and ind) – mf#8308 reel 1-5 – us Wisconsin U Libr [080]
Grillparzers kunstphilosophie / Reich, Emil – Wien: Manz, 1890 – 1r – 1 – us Wisconsin U Libr [700]
Grillparzers lyrik als ausdruck seines wesens / Zausmer, Otto – Wien, Leipzig: Deutscher Verlag fuer Jugend und Volk, c1933 – 1r – 1 – (incl bibl ref) – us Wisconsin U Libr [430]
Grillparzers menschenauffassung / Mueller, Joachim – Weimar: H Boehlau, 1934 – 1r – us Wisconsin U Libr [430]
Grillparzers saemmtliche werke in zwanzig baenden / ed by Sauer, August – 5. ausg. Stuttgart: J G Cotta Nachf, [1892] [mf ed 2001] – 20v – 1 – mf#10524 – us Wisconsin U Libr [802]
Grillparzers verhaeltnis zu shakespeare / Braun, Hanns – [s.l: s.n.] 1916 [mf ed 1990] – 1r – 1 – (incl bibl ref; filmed with: grillparzers verhaltnis zur politischen tendenzliteratur seiner zeit / konrad beste) – mf#2690p – us Wisconsin U Libr [410]
Grillparzers verhaeltnis zur politik seiner zeit : ein beitrag zur wuerdigung seines schaffens und seiner persoenlichkeit / Buecher, Wilhelm – Marburg a.L: N G Elwert 1913 [mf ed 1992] – (= ser Beitraege zur deutschen literaturwissenschaft 19) – 1r – 1 – (incl bibl ref & ind; filmed with: johann rist als weltlicher lyricker / von oskar kern) – mf#3098p – us Wisconsin U Libr [430]
Grillparzers verhaeltnis zur politischen tendenzliteratur seiner zeit / Beste, Konrad – [S.l: s.n.] 1915 [mf ed 1990] – 47p – 1 – (incl bibl ref) – mf#7419 – us Wisconsin U Libr [430]
Grillparzers verhaltnis zur politischen tendenzliteratur seiner zeit *see* Grillparzers verhaltnis zu shakespeare

Grillparzers werke / ed by Franz, Rudolf – krit durchges erlaeut ausg. Leipzig, Wien: Bibliographisches Institut [1903?] [mf ed 1993] – (= ser Meyers klassiker-ausgaben) – 5v – 1 – (incl bibl ref and ind) – mf#8699 – us Wisconsin U Libr [802]
Grillparzers werke *see* Saemtliche werke
Grima y Villa-Senor Gabriel de *see* Ceremorias en el colegio militar de nuestra sra. sta. maria de tudia del orden de santiago de la universidad de salamanca
Grima y Villa-Senor, Gabriel de *see* Ceremonia del...santa maria de tudia
Grimarest, Jean Leonor Le Gallois de *see* Traite du recitatif dans la lecture
Grimball, John *see* Log of the shenandoah
Grimberg, Barbara *see* Umweltorientiertes marketing im handel
Grimble, Arthur *see* Gilbertese myths, legends and oral traditions
Grimblot, Paul *see* Letters of william 3 and louis 14 and their ministers
Grime, Elder J H *see* Manuscripts and pamphlets
Grime, J H *see* What is an orthodox baptist?
Grimes bulletin / Bodak, Shirley L – 1977 mar 1-1977 dec – 1r – 1 – mf#637883 – us WHS [071]
Grimes, John *see* Remarks on the practicability and advantage of opening up a communication between the east coast of the peninsula of india and the cotton districts of nagpore
Grimke, Angelina Emily *see*
– Appeal to the christian women of the south
– Letters to catherine e beecher
Grimke, Archibald Henry *see* William lloyd garrison, the abolitionist
Grimke, Francis James *see*
– Next step in racial cooperation
– Phase of the race problem looked at from within the race itself
Grimke, Sarah Moore *see* An epistle to the clergy of the southern states
The grimke sisters : sarah and angelina grimke, the first american women advocates of abolition and woman's rights / Birney, Catherine H – Boston: Lee and Shepard, 1885. Beltsville, Md: NCR Corp, 1978 (4mf); Evanston: American Theol Lib Assoc, 1984 (4mf) – (= ser Women & the church in america) – 9 – 0-8370-0737-2 – mf#1984-2046 – us ATLA [975]
Grimke, Thomas Smith *see* An inquiry into the accordancy of war with the principles of christianity
Grimm, August T von *see* Wanderungen nach suedosten
Grimm, Brueder *see* Deutsche sagen
Grimm, Carl Ludwig Wilibald *see*
– A greek-english lexicon of the new testament
– Kurzgefasstes exegetisches handbuch zu den apokryphen des alten testamentes
Grimm, Eduard *see* Die ethik jesu
Grimm, George *see* The doctrine of the buddha
Grimm, Hans *see*
– Die drei lachenden geschichten
– Die geschichte vom alten blute und von der ungeheuren verlassenheit
– Glaube und erfahrung
– Meine geliebten claudius-gedichte
– Die olewagen saga
– Die olewagen-saga
– Wie grete aufhoerte ein kind zu sein
– Der zug des hauptmanns von erckert
– Zug des hauptmanns von erckert
Grimm, Herman Friedrich *see*
– Achim von arnim und die ihm nahe standen
– Goethe
– Goethes briefwechsel mit einem kinde
– The life and times of goethe
Grimm, Hermann Friedrich *see*
– Das kind
– Das leben goethes
– Life of michael angelo
Grimm, J *see* Deutsche mythologie
Grimm, Jacob *see*
– Deutsche grammatik
– Deutsche mythologie
– Deutsche rechtsalterthuemer
– Deutsche sagen
– Drei reden jakob grimms
– Grimm's household tales
– Kinder- und hausmaerchen
– Rotkaeppchen
– Ueber den ursprung der sprache
Grimm, Jacob Ludwig Carl *see*
– Kinder- und hausmaerchen
– Kinder- und hausmaerchen der brueder grimm
Grimm, Joseph *see* Die samariter und ihre stellung in der weltgeschichte
Grimm, Karl Josef *see* Euphemistic liturgical appendixes in the old testament
Grimm, Ludwig *see* Trogalien zur verdauung der xenien
Grimm, Reinhold *see* Strukturen
Grimm, Wilhelm *see*
– Deutsche sagen
– Kinder- und hausmaerchen
– Kinder- und hausmaerchen der brueder grimm
– Novellen
Grimm, Wilibald *see*
– Die glaubwuerdigkeit der evangelischen geschichte
– Kurzgefasste geschichte der lutherischen bibeluebersetzung bis zur gegenwart
Grimmaer pflege *see* Grimmaisches wochenblatt fuer stadt und land
Grimmaisches wochen- und intelligenzblatt *see* Grimmaisches wochenblatt fuer stadt und land
Grimmaisches wochenblatt *see* Grimmaisches wochenblatt fuer stadt und land
Grimmaisches wochenblatt fuer stadt und land – Grimma DE, 1813-45 – 148r – 1 – (title varies: 8 jan 1820: grimmaisches wochenblatt; 10 jan 1829: grimmaisches wochen- und intelligenzblatt; 4 jan 1834: grimmaisches wochen- und anzeigeblatt; 1 oct 1881: aufg in: nachrichten fuer grimma. with suppl: grimmaer pflege 1922-1944 jul/aug (gaps) [3r]) – gw Misc Inst [074]
Grimme, Hubert *see*
– Das gesetz chammurabis und moses
– Grundzuege der hebraeischen akzent- und vokallehre
– Das israelitische pfingstfest und der plejadenkult
– Mohammed
– Psalmenprobleme
Grimmelshausen / Busse, Hermann Eris – Stuttgart: J G Cotta, 1944, c1939 [mf ed 1993] – 91p – 1 – mf#8452 – us Wisconsin U Libr [430]
Grimmelshausen / Hayens, Kenneth Cochrane – London; New York: Pub for St Andrews University by H Milford, OUP, 1932 [mf ed 1993] – (= ser St andrews university publications 34) – 252p – 1 – (incl bibl ref and ind) – mf#8452 – us Wisconsin U Libr [430]
Grimmelshausen, erloesung und barocker geist / Burkhard, Werner – Frankfurt am Main: M Diesterweg, 1929 [mf ed 1993] – (= ser Deutsche forschungen 22) – 154p – 1 – (incl bibl ref) – mf#8023 reel 4 – us Wisconsin U Libr [430]
Grimmelshausen, Hans Jakob Christoph von *see*
– Der abenteuerliche simplicissimus
– Der abenteuerliche simplicissimus und andere schriften
– Continuatio des abentheurlichen simplicissimi
– Grimmelshausens courasche
– Grimmelshausens simplicissimus teutsch
– Grimmelshausens springinsfeld
– Grimmelshausens werke in vier teilen
– Grimmelshausens wunderbarliches vogelnest
– Der seltsame springinsfeld
Grimmelshausens courasche / ed by Scholte, Jan Hendrik – Halle: Niemeyer, 1923 [mf ed 1993] – (= ser Neudrucke deutscher litteraturwerke des 16. und 16. jahrhunderts 246-248) – lvi/168p (ill) – 1 – (fr oldest ed 0f 1670. incl bibl ref) – mf#8413 reel 9 – us Wisconsin U Libr [830]
Grimmelshausens simplicissimus teutsch / ed by Scholte, Jan Hendrik – Halle: M Niemeyer, 1938 [mf ed 1993] – (= ser Neudrucke deutscher literaturwerke des 16. und 17. jahrhunderts 302-309) – 463p (ill) – 1 – (ed of 1669. incl bibl ref) – mf#8413 reel 11 – us Wisconsin U Libr [830]
Grimmelshausens sprichwoerter und redensarten / Lenschau, Martha – Frankfurt am Main: M Diesterweg, 1924 [mf ed 1993] – (= ser Deutsche forschungen 10) – 154p – 1 – (incl bibl ref) – mf#8023 reel 2 – us Wisconsin U Libr [430]
Grimmelshausens springinsfeld / ed by Scholte, Jan Hendrik – Halle: M Niemeyer, 1928 [mf ed 1993] – (= ser Neudrucke deutscher litteraturwerke des 16. und 16. jahrhunderts 249-252) – xxxix/139p – 1 – (fr 1670 ed. incl bibl ref) – mf#8413 reel 9 – us Wisconsin U Libr [830]
Grimmelshausens und zesens josephsromane : ein vergleich zweier barockdichter / Stucki, Clara – Horgen-Zuerich: Verlag der Muenster-Presse, 1933 [mf ed 1993] – (= ser Wege zur dichtung 15) – 149p – 1 – (incl bibl ref) – mf#8452 – us Wisconsin U Libr [430]
Grimmelshausens werke in vier teilen / ed by Borcherdt, Hans Heinrich – Berlin: Deutsches Verlagshaus Bong & Co, [1921] [mf ed 2001] – 4v in 3 on 1r (ill) – 1 – (incl bibl ref) – mf#10499 – us Wisconsin U Libr [830]
Grimmelshausens wunderbarliches vogelnest : erster teil / ed by Scholte, Jan Hendrik – Halle (Saale): M Niemeyer, 1931 [mf ed 1993] – xvii/148p/1pl (ill) – 1 – (fr 1672 ed. incl bibl ref) – mf#8413 reel 10 – us Wisconsin U Libr [830]
Grimmer kreis-wochenblatt – Grimmen DE, 1855-56 – 1r – 1 – gw Misc Inst [074]
Grimm's household tales / Grimm, Jacob – New York, NY. n d – 1r – us UF Libraries [025]
Grimouard, Henri *see* Amiral de grimouard au port-au-prince
Grimpolario / Labrador Ruiz, Enrique – Habana, Cuba. 1937 – 1r – us UF Libraries [972]
Grimsby advertiser *see* Grimsby independent and advertiser
Grimsby and north lincolnshire advertiser *see* Grimsby independent and advertiser
Grimsby guardian – England. -w. 9 Oct 1857-8 Nov 1867. (4 reels) – 1 – uk British Libr Newspaper [072]
Grimsby herald – England. -w. 31 Jul 1863-1 Oct 1881. (1872 imperfect). (8 reels) – 1 – uk British Libr Newspaper [072]
Grimsby independent and advertiser – England. 7 jun 1861-22 oct 1887 [wkly] – 12r – 1 – (aka: grimsby and north lincolnshire advertiser; grimsby advertiser) – uk British Libr Newspaper [072]
Grimsby independent, lindsey and general advertiser – England. -w. 4 Jun 1858-27 May 1859. (33 ft) – 1 – uk British Libr Newspaper [072]
Grimsby weekly express – [Yorkshire & Humberside] North East Lincs 5 may 1883-dec 1897 [mf ed 2004] – 12r – 1 – (missing: 1898; cont as: grimsby express (weekly edition) [jan 1886-dec 1889]; grimsby weekly express [jan 1890-dec 1897]) – uk Newsplan [072]
Grimshaw, William *see* The merchants' law book; being a treatise on the law of account render, attachment, bailment, bills of exchange and promissory notes.
Grimthorpe, Edmund Beckett *see* Church restoration
Grimthorpe, Edmund Beckett. 1st Baron *see*
– A book on building
– Church restoration
– Lectures on gothic architecture
Grimthorpe, Edmund Beckett, Baron *see* Should the revised new testament be authorised?
Grimwade, Eric Illingworth *see* Religious life and thought in the english novel of the nineteenth century
Grin and barrett – v1 n1-3 [1980 jul 13-1986 jun] – 1r – 1 – mf#1477169 – us WHS [071]
A grinalda : jornal dos domingos – Rio de Janeiro, RJ: Typ de M J Cardozo & C, 23 jul-12 nov 1848 – (= ser Ps 19) – bl Biblioteca [079]
A grinalda : periodico litterario – Rio de Janeiro, RJ: [s.n.] 10 nov 1850 – (= ser Ps 19) – mf#P15,01,46 n02 – bl Biblioteca [440]
A grinalda : revista semanal, litteraria e recreativa – Typ de Francisco de Paula Brito, 02-09 dez 1861 – (= ser Ps 19) – mf#P17,01,140 – bl Biblioteca [073]
Grinberg, B I *see* Ukazatel-spravochnik promyslovoi kooperatsii moskovskoi oblasti
Grinchuckle – Montreal: A Gilbert. sep 23 1869-feb 24 1870// – 1r – 1 – Can$65.00 – (with illustrations by j w bengough) – cn McLaren [400]
Grinding stone : an independent news monthly – v1-2, n5. 1968-november 1969 – 1 – us AMS Press [073]
Grindlay, Robert Melville *see* Scenery costumes and architecture
Grindrod, Edmund *see* Proud abased
Grinell, DeWitt Clinton *see* Diary; papers
Grinfield, Edward William *see* An apology for the septuagint
Grinfield, Thomas *see* Duties and rewards of the christian minister
Gring, Ambrose Daniel [comp] *see* Eclectic chinese-japanese-english dictionary
Gringo lenca / Oqueli, Arturo – Tegucigalpa, Mexico. 1947 – 1r – us UF Libraries [972]
Gringoire – Le grand hebdomadaire parisien, politique, litteraire. Paris. nov 1928-mai 1944 – 1 – fr ACRPP [800]
Gringos / Wise, Henry Augustus – New York, NY. 1850 – 1r – us UF Libraries [972]
Grinnell herald-register – Grinnell IA. 1936 apr 20 – 1r – 1 – (cont: grinnell herald [grinnell, ia: 1889]; grinnell register) – mf#3925837 – us WHS [071]
Grinnell, William Morton *see* Address delivered by william morton grinnell
Grinstead, William *see* Account book
Grinten, Willem Christiaan Leonard van der *see* De verplichtingen van den werkgever
Grinter, Moses *see* Account book
Les griots : la revue scientifique et litteraire d'haiti – Port-au-Prince: [s.n., v1 n1-v2 n2/3 jul/aug/sep 1938- oct/nov/dec 1939/jan/feb/mar 1940 – 15r – 1 – us CRL [073]
Grip : independent journal of humour and caricature / ed by Bengough, John Wilson – Toronto: Grip Printing & Pub Co. v1-42 may 24 1873-dec 29 1894// – 7r – 1 – Can$425.00 – (ill by ed. an unparalleled source of 19thc canadian political satire) – cn McLaren [320]
The grip – Williamsport, PA., 1889 – 13 – $25.00r – us IMR [071]
Grip strength profiles of elementary aged males and females / Svehla, B G – 1991 – 1mf – 9 – $4.00 – us Kinesology [790]
Gripen – Sodertalje, Sweden. 1866, 1880-90 – 1 – sw Kungliga [078]
Grip-sack : a receptacle of light literature, fun and fancy – Toronto: Grip Printing & Pub. Co. v1-2. 1882-83//? – 1r – 1 – Can$110.00 – cn McLaren [071]
Griqualand west advertiser *see* Diamond fields witness and griqualand west advertiser
Gris mayor / Gonzalez, Edelmira – Habana, Cuba. 1937 – 1r – us UF Libraries [972]
Grisanti, Angel *see*
– Proceso contra los asesinos delgran mariscal de ay...
– Resumen historico de la instruccion publica en ven...
– Vargas intimo
Grisar, Erich *see* Monteur klinkhammer
Grisar, Hartmann *see*
– Galileistudien
– History of rome and the popes in the middle ages
– Luther
– Die roemische kapelle sancta sanctorum und ihr schatz
Die grisardis des erhart grosz : nach der breslauer handschrift / Gross, Erhart; ed by Strauch, Philipp – Halle/S: M Niemeyer Verlag, 1931 [mf ed 1993] – (= ser Altdeutsche textbibliothek 29) – xl/74p – 1 – (middle high german. int in german. incl bibl ref) – mf#8193 reel 3 – us Wisconsin U Libr [430]
Die grisardis des erhart grosz : nach der breslauer handschrift / ed by Strauch, Philipp – Halle (Saale): M Niemeyer Verlag, 1931 [mf ed 1993] – (= ser Altdeutsche textbibliothek 29) – xl/74p – 1 – (middle high german text; int in german. incl bibl ref) – mf#8193 reel 3 – us Wisconsin U Libr [430]
Grischa : ein trauerspiel / Heiseler, Henry von – Muenchen: Musarion, c1919 [mf ed 1995] – 54p – 1 – mf#9082 – us Wisconsin U Libr [820]
Griscom, John Hoskins *see* Memoir of john griscom, ll.d., late professor of chemistry and natural philosophy
Grisdale, Joseph Hiram *see* Growing and using mangels, sugar mangels and forage sugar beets
Grisebach, Eduard *see*
– Deutsche literatur 1770-1870
– G c lichtenberg's briefe an dieterich, 1770-1798
– Der neue tanhaeuser
– Tanhaeuser in rom
Griselda / Hauptmann, Gerhart – 2. Aufl. Berlin: S Fischer, 1909 – 1r – 1 – us Wisconsin U Libr [820]
Griselda : the history of patient grisel / ed by Wheatley, Henry B – London: printed for the Villon Society 1885 [mf ed Bloomington IN: Indiana Uni Lib, Preservation Dept 1984] – (= ser Chap-books and folk-lore tracts. 1st ser 4) – 1r – 1 – (aka: the ancient true and admirable history of patient grisel, a poore mans daughter in france) – us Indiana Preservation [390]
Griseldis : dramatische dichtung in einem vorspiel und drei akten / Stach, Ilse von – Kempten: J Koesel & F Pustet, 1921 – 1r – 1 – us Wisconsin U Libr [810]
Griseldis : dramatisches gedicht in fuenf akten / Halm, Friedrich – 11. Aufl. Wien: C Gerold, 1896 – 1r – 1 – us Wisconsin U Libr [810]
Griseldis : ein volksstueck in vier akten / Berger, Ludwig – Muenchen: K Wolff, c1921 [mf ed 1989] – 108p – 1 – mf#7009 – us Wisconsin U Libr [820]
Griselidis / Silvestre, Armand – Paris, France. 1923 – 1r – us UF Libraries [440]
Grishinskii, A S et al *see* Istoriia russkoi armii i flota
Grissom, William Lee *see* History of methodism in north carolina
Grisson, Rudolf Hermann Rulemann *see* Beitraege zur auslegung von richard wagners "ring des nibelungen"
Grist mill / Ballston Spa Area Historical Society et al – v1 n1-v16 n3, v17 n1,3-4, v18 n1-2,4, v19 n1-v21 n4 [1967 mar-1982 sep, 1983 mar, sep-dec, 1984 mar-may, dec, 1985 mar-1987 dec] – 1r – 1 – mf#621004 – us WHS [071]
Grist mill / Cuyahoga Co. Cleveland – feb 1947-jan 1949 [wkly] – 1r – 1 – mf#B1063 – us Ohio Hist [071]
Grist, William Alexander *see* The historic christ in the faith of to-day
Griswold, Hervey De Witt *see*
– Brahman
– The chet rami sect
– The dayanandi interpretation of the word "deva" in the rig veda
– The god varuna in the rig-veda
– Mirza ghulam ahmad
– The radha swami sect
– The religion of the rigveda
Griswold, Hervey DeWitt *see* Village evangelization (n2)
Griswold, Latta *see* The episcopal church
Griswold linkage – v1 n3-v3 n4 [1980-1981/82] – 1r – 1 – mf#599841 – us WHS [071]
Griswold, Luther D *see* Luther d. griswold papers, ms p.p.
Grit – Toronto. v1 n1-13. dec 1-17 1917// (daily) – 1r – 1 – Can$70.00 – (incl "an appeal to women voters." publ during the 1917 federal election campaign) – cn McLaren [325]

Grit city news – v1-[v2? n4] [1970 may 26-[1972 apr]] – 1r – 1 – mf#1057567 – us WHS [071]

O grito da razao na corte do rio de janeiro – Rio de Janeiro, RJ: Imprensa Nacional, 23 fev-22 mar 1825 – (= ser Ps 19) – mf#P01,04,12 – bl Biblioteca [321]

Grito de independencia / Zamora Castellanos, Pedro – Guatemala, 1935 – 1r – us UF Libraries [972]

Grito de independencia en colombia – Medellin, Colombia. 1960 – 1r – us UF Libraries [972]

Grito de paz / Marin, Thelvia – Habana, Cuba. 1964 – 1r – us UF Libraries [972]

Grito del norte – v1-6 n6 [1968 aug 24-1973 aug] – 1r – 1 – mf#806602 – us WHS [071]

El grito del pueblo – Manila: P H Pobiete, jan 13 1900; mar 30 1901; jan 9 1902 – us CRL [079]

O grito do povo – Rio de Janeiro, RJ: Typ Reis, 18 jun-set 1887; set-dez 1888; 25 jan 1889 – (= ser Ps 19) – mf#P17,01,160 – bl Biblioteca [320]

Grito, poemario de vanguardia / Gonzalez Alberty, Fernando – San Juan, Puerto Rico. 1931 – 1r – us UF Libraries [972]

Gritsenko, I F *see* Sistematicheskii ukazatel russkoi literatury po kooperatsii 1856-1924 gg

Grlic, Danko *see* Covek danas

Groat, George Gorham *see* Attitude of american courts in labor cases

Grob, Fritz *see* Jeremias gotthelfs geld und geist

Grob, Johann *see*
– Epigramme

Das grobe hemd : volksstueck in vier acten / Karlweis, C – [Wien]: Wiener Verlag 1901 [mf ed 1990] – 1r – 1 – (filmed with: on the eve / leopold kampf) – mf#2752p – us Wisconsin U Libr [820]

Grobler, Hermien *see* Die terapeutiese begeleiding van die kind na die dood van 'n ouer

Grobler, Johannes Marthinus *see* 'N ondersoek van afrikaanssprekendes se behoeftes aan afrikaanse televisieprogramme

Groce family newsletter – v1 n1-3 [1986 nov-1987 mar] – 1r – 1 – mf#1288493 – us WHS [071]

Grocock, Robert *see* Accoustical phenomena as they relate to the performance and manufacture of the modern valve trumpet

Grocott's mail – Grahamstown SA, 1871 [mf ed Cape Town: SA library 1985] – 1r – 1 – mf#MS00451 – sa National [079]

Grodnenskaia pravda – Grodno, Belarus 1973-88 [mf ed Norman Ross] – 5r – 1 – mf#nrp-519 – us UMI ProQuest [077]

Grodnenskie gubernskie vedomosti – Grodno, Belarus 1838-1915 [mf ed Norman Ross] – 50r – 1 – mf#nrp-517 – us UMI ProQuest [077]

Groeben, Georg Dietrich von der *see* Neue kriegsbibliothek

Groeger, Detlef *see* Ueber eine spezielle art von multiplikativen galois-modulhomomorphismen auf einem relativ-abelschen zahlkoerper

Groen, D H *see* Rusticii helpidii carmina notis criticis versione batava commentarioque exegetico instructa

Groen, J *see* Onze lagere school

Groene, Valentin *see*
– Der ablass
– Compendium der kirchengeschichte

Groener, Wilhelm *see* Papers of general wilhelm groener, 1867-1939

Groenewegen, H *see* Hieroglyphica

Groenewegen, H Ij *see* De evolutieleer en het godsdienstig geloof

Die groeninger theologen : seinen fruehern und jetzigen zuhoerern am 25-jaehrigen jubelfeste seines amtsantrittes = Groninger godgeleerden / Hofstede de Groot, Petrus – Gotha: Friedrich Andreas Perthes, 1863 – 1mf – 9 – 0-8370-8679-5 – (incl bibl ref) – mf#1986-2679 – us ATLA [240]

Groenings, Jakob *see* A catholic catechism for the parochial and sunday schools of the united states

Groeper, Richard *see* Neue beitraege zu heinrich von kleist

Die groepsgebiedewet (no. 41 van 1950) met spesiale verwysing na die kaapse skiereiland (1950-1953) / Fortuin, Stephen Daniel – U of the Western Cape 1990 [mf ed S.l: s.n. 1990] – 2mf – 9 – (incl text of the act; summary in afrikaans & english; incl bibl) – sa Misc Inst [343]

Groes faen, st david, and babell calvinistic methodist, monumental inscriptions – 1mf – 9 – £1.25 – uk Glamorgan FHS [929]

Groesbeck, John *see* The crittenden commercial arithmetic and business manual...

Groeschl, Juergen *see* Gespraechsbefaehigung im berufsbezogenen portugiesischunterricht

Die groesse der musikalische intervalle als grundlage der harmonie / Bellermann, Heinrich – Berlin: J Springer 1873 [mf ed 1991] – 1r – 1 – mf#pres. film 102 – us Sibley [780]

Groesse der natur; ruf des freien landes; vom inhalt des lebens / Binding, Rudolf Georg – Leipzig: Gesellschaft der Freunde der Deutschen Buecherei, 1931 [mf ed 1989] [mf ed 1989] – (= ser Jahresgabe der gesellschaft der freunde der deutschen buecherei 13) – 55p – 1 – mf#7024 – us Wisconsin U Libr [880]

Groesse und grenzen der kleinbuergerlich-demokratischen bewegung in der revolution von 1848-49 in deutschland – Berlin: Sekretariat des Zentralvorstandes der Liberal-Demokratischen Partei Deutschlands, 1968. 153p. Bibliog – 1 – us Wisconsin U Libr [943]

Groessel, Wolfgang *see* Die stellung der lutherischen kirche deutschlands zur mission im 17. jahrhundert

Groessenwahn : patalogischer roman / Bleibtreu, Karl – Leipzig: W Friedrich, 1888 [mf ed 1989] – 3v in 2 – 1 – mf#8108 – us Wisconsin U Libr [830]

Die groesseren dichtungen von james montgomery / Wissmann, Paul – Koenigsberg i. Pr. 1913 (mf ed 1994) – 1mf – 9 – €24.00 – 3-8267-3049-6 – mf#DHS-AR 3049 – gw Frankfurter [420]

Groesste denkwuerdigkeiten der welt : oder sogenannte relationes curiosae / ed by Happel, Eberhard G – Hamburg 1683-91 [mf ed 1993] – (= ser Aus den anfaengen des zeitschriftenwesens: fruehe deutsche zeitschriften) – 5v on 28mf – 9 – €270.00 – 3-89131-094-3 – gw Fischer [900]

Groff, George Weidman *see* Lychee and lungan

Grogan, Jennifer *see* The measurement of perceived-self and perceived-parental gender-role orientation

Grogan, Kevin [comp] *see* A collection in the making

Grohmann, Wilhelm *see* Lutherische metaphysik

Grohmann, Willy *see* Vers und prosa im hohen drama des achtzehnten jahrhunderts

Groizard y Coronado, Carlos *see* Cuartillas

Grolmann, Karl Ludwig Wilhelm von *see* Magazin fuer die philosophie des rechts und der gesetzgebung

Gromada – Warsaw, Poland. 4 Jul 1951; Jan-May 1952 – 1r – 1 – us L of C Photodup [943]

Gromada rolnik polski – Warsaw, Poland. Jun 1952-Jun 1958; 1959; 1961-May 1979 – 29r – 1 – us L of C Photodup [943]

Groman, V G et al *see* Narodnoe khoziaistvo sssr v 1923-1924 g

Gromov, Mikhail *see* Across the north pole to america

Gromov, Nikolai *see* Pered razsvietom

Gronau, K *see* Poseidonios und die juedisch-christliche genesisexegese

Grondal, Bror Leonard *see* Bror leonard grondal papers, 1908-1974 [bulk 1945-1974]

Grondeise van 'n verantwoordbare opvoedkundige sielkundige praktyk : met besondere verwysing na pedoterapie / Kruger, Jacolien – Uni of South Africa 2000 [mf ed Johannesburg 2000] – 4mf – 9 – (text in afrikaans) – mf#mfm14913 – sa Unisa [370]

Grondel, A H *see* Enkele opmerkingen naar aanleiding van de enquete inzake

Grondelicke onderrichtinghe : vande leere ende den geest des hooftketters david joris... / Emmius, [U] – Middelburgh, 1599 – 2mf – 9 – mf#PBA-172 – ne IDC [240]

De gronden, afbeeldingen en beschrijvingen der aldervoornaamste en aldernieuwste gebouwen... / Vingboons, P – Leiden, 1715 – 5mf – 9 – mf#OA-291 – ne IDC [720]

Gronden en afbeeldsels der voornaamste gebouwen... / Vingboons, P – Amsterdam, [1688] – 2mf – 9 – mf#OA-69 – ne IDC [720]

Grondich bericht : van de eerste beghinselen der wederdoopsche seckten... / Moded, H – Middelburgh, 1603 – 4mf – 9 – mf#PBA-272 – ne IDC [240]

Grondich bericht van de wettelijcke beroepinghe der predicanten ofte kerckendienaren... / Helmichius, W – Delft, 1611 – 1mf – 9 – mf#PBA-192 – ne IDC [240]

Grondig onderwys in de fregoriaansche choorzang of choral, nevens eenige aanmerkingen over di zang-konst : bestaande in dertien lessen...en wel byzonderlyk der organisten / Juerrns, J F – Amsterdam: F J van Tetroode, voor rekening van den autheur 1789 [mf ed 19–] – 1mf – 9 – mf#fiche 498 – us Sibley [780]

Grondslag, wezen en openbaring van het godsdienstig geloof : volgens de heilige schrift / Hoekstra, Sytze – Rotterdam: Altmann & Roosenburg, 1861 – 1mf – 9 – 0-7905-7761-5 – mf#1989-0986 – us ATLA [240]

Grondt-regulen der bouw-const : ofte de uytnementheyt van vijf orders... / Schuym, [J] – Amsterdam, 1662 – 2mf – 9 – mf#OA-77 – ne IDC [720]

Gronemann, S *see* Die jonathan'sche pentateuch-uebersetzung in ihrem verhaeltnisse zur halacha

Gronland geographisk og statistisk beskrevet / Rink, H – Kobenhavn. 2v. 1852-1857 – 21mf – 9 – mf#H-497 – ne IDC [919]

Gronland i tohundredaaret for hans egedes landing – Kobenhavn, 1921. v60-61 – 31mf – 9 – mf#H-478 – ne IDC [917]

Gronland langs polhavet / Rasmussen, K – Kobenhavn, Kristiania, 1919 – 18mf – 9 – mf#N-359 – ne IDC [917]

Gronlund, Laurence *see*
– Ca ira!
– The cooperative commonwealth in its outlines

Grontregulen der bow-const : ofte de uytnementheyt van de vyf orders der architectvra / Scamozzi, [V] – Anstelrodamis, 1640 – 4mf – 9 – mf#OA-79 – ne IDC [720]

Groom, Edward *see* The art of transparent painting on glass

Groombridge, Lana *see* A study of the stages of readiness to adopt exercise and strength training behaviors among adults 65 years and older

Groos, Karl *see* The play of man

Groot, J F de *see* De verbo incarnato

Groot, Jan Jakob Maria de *see*
– Le code du mahaayaana en chine
– Les fetes annuellement celebrees a emoui
– Religion in china
– The religion of the chinese
– Sectarianism and religious persecution in china

Groot, Jose Manuel *see*
– Historia eclesiastica y civil de nueva granada
– Historia y cuadros de costumbres

Het groot natuur- en zedekundig werelttoneel... / Poot, H K – Delft, 1743-50. 3v – 56mf – 9 – mf#O-406 – ne IDC [090]

Groot schilderboek... / Lairesse, G de – Haarlem, 1740. 2v – 24mf – 9 – mf#O-334 – ne IDC [700]

Het groote nord-westen : de grootste en vruchtbaarste velden waar de landverhuizers zich kunnen vestign: inlichtingen voor de uitwijkelingen – Ottawa: Dept of Agriculture, 1882 – 1mf – 9 – mf#53720 – cn Canadiana [304]

De groote schouburgh der Nederlantsche konstschilders en schilderessen... / Houbraken, A – Amsterdam, 1718-1721. 3v – 15mf – 9 – mf#O-305 – ne IDC [700]

"Grootlamsiekte" : a specific syndrome of prolonged gestation in sheep caused by a shrub, salsola tuberculata (fenzl ex moq) schinz var tomentosa c a smith ex aellen / Basson, P A et al – Pretoria: Govt Printer [1969] [mf ed Pretoria, RSA: State Library [199-]] – 1r with other items – 5 – (incl bibl ref) – mf#r24 – us CRL [634]

Gropp, I *see* Historia monasterii amorbacensis ord s benedicti

Gros, jules *see* Les voyages et decouvertes de paul soleillet dans le sahara et dans le soudan en vur dun projet dun chemin de fer transsaharien

Gros, L *see* Histoire de notre dame de lourdes

Grosart, Alexander Balloch *see*
– Daemonologia sacra
– Drowned
– The prince of light and the prince of darkness in conflict
– Representative nonconformists

Grosch, Hermann *see* Die echtheit des zweiten briefes petri

Grosch, Rudolf *see* Die jugenddichtung friedrich hoelderlins

Grosclaude, Charles *see* Exposition et critique de l'ecclesiologie de calvin

Grosclaude, E *see* Un parisien...madagascar

Grose, F *see* Rules for drawing caricaturas

Grose, Howard Benjamin *see*
– Advance in the antilles
– Aliens or americans?
– Frontier sketches
– The incoming millions
– The judson centennial, 1814-1914

Groseclose, J Sidney *see* Die althochdeutschen poetischen denkmaeler

Groser, Thomas *see*
– Covenant name of god
– On prayer

Groser, William H *see* A hundred years work for children, 1803-1903

Groser, William Howse *see*
– Scripture natural history
– The teacher's model and the model teacher

Grosh, C *see* Schluffel heileger schrift

Die grosherzogin von gerolstein : oper in 3 acten und 4 bildern / Offenbach, Jacques – [n.p. 1887?] [mf ed 19–] – 1r – 1 – mf#pres. film 21 – us Sibley [780]

Grosier, J B G A *see*
– Description generale de la chine,...
– A general description of china...

Grosier, Jean B *see* De la chine, ou description generale de cet empire

Grosley, Pierre *see* Observations sur l'italie et sur les italiens

Groslier, George *see*
– A l'ombre d'angkor
– Recherches sur les cambodgiens s'apres les textes et les monuments depuis les premiers siecles de notre ere

Grosman, Moisheh *see* In farkishuftn land fun legendarn dzshugashvili

Gross, Abbie Mae *see* History of homestead

Gross, Ch *see* The sources and literature of english history

Gross daytoner zeitung / Montgomery Co. Dayton – aug 1914-jun 1947 [wkly] – 35r – 1 – (in german) – mf#B5069-5103 – us Ohio Hist [071]

Gross, Elke Christiane *see* Lipoprotein (a)

Gross, Erhart *see*
– Die grisardis des erhart grosz

Gross, Fred Louis *see* The law of real estate brokers, with 1917 supplement.

Gross, G *see* Die bedeutung des aesthetischen in der evangelischen religion – noch ein wort ueber den christlichen dienst

Gross, Imke *see* Die rolle des mthfr tt 677 genotyps und weiterer prothrombotischer risikofaktoren fuer die entstehung von sinusvenenthrombosen bei kindern mit all

Gross, J *see* La divinisation du chretien d'apres les peres grecs

Gross, Johannes M *see* Studie zur praezision des celay r-systems

Gross, Tilman *see* Modell fuer die ausbreitung von viralen infektionskrankheiten am beispiel des bakteriophagen phi x 174

Grossalmeroder zeitung [vlg dittmar] – Grossalmerode DE, 1898 6 jan-30 jun – 1r – 1 – (publ in dittmar in grossalmerode) – gw Misc Inst [074]

Grossalmeroder zeitung [vlg vogt] – Hessisch-Lichtenau DE, 1898 26 oct-1910 [gaps] – 1 – (publ in vogt in hessisch-lichtenau) – gw Misc Inst [074]

Gross-daytoner zeitung – Dayton OH (USA), 1923-1937 26 jun [gaps] – 20r – 1 – gw Misc Inst [071]

Das grossdeutsche reich : ein appell auf die befreiung oesterreichs: mit liedern / Boehme, Herbert – Muenchen: Zentralverlag der NSDAP, F Eher, [1938] [mf ed 1989] – (= ser Kameraden 20) – 45p – 1 – mf#7044 – us Wisconsin U Libr [943]

"Grossdeutschland im weltgeschehen." / U.S. Library of Congress. Prints and Photographs Division – 1939-42 – ca 1325photos on 1r – 1 – (4 annual news photo compilations, publ for the reichsministerium fuer volksaufklaerung und propaganda) – us L of C Photodup [943]

Der grosse alexander / Quilichino, da Spoleto; ed by Guth, Gustav – Berlin: Weidmann, 1908 – (= ser Deutsche texte des mittelalters 13) – xii/102p/2pl – 1 – (incl bibl ref and ind. middle high german trans fr latin) – mf#8623 reel 4 – us Wisconsin U Libr [430]

Der grosse baum : erzaehlung / Hoerner, Herbert von – Stuttgart: J Engelhorns Nachf. A Spemann 1940, c1938 – 1r – 1 – mf#2727p – us Wisconsin U Libr [880]

Der grosse bergarbeiter-streik des jahres 1889 im rheinisch-westfaelischen kohlenrevier : ein wort zur abwehr / Lensing, L – Dortmund 1889 – 1 – gw Mikropress [943]

Das grosse bestiarium der modernen literatur / Blei, Franz – 4. aufl. Berlin: E Rowohlt, 1922 [mf ed 1989] – 252p – 1 – (incl ind) – mf#7031 – us Wisconsin U Libr [410]

Grosse brand und der wiederaufbau von hamburg / Faulwasser, Julius – Hamburg, Germany. 1892 – 1r – us UF Libraries [025]

Grosse, Carl *see* Reise nach dem hohen norden durch schweden, norwegen und lappland

Das grosse conversationslexikon fuer gebildete staende (ael1/5) / Meyer, Josef – Hildburghausen 1840-55 [mf ed 1993] – (= ser Archiv der europaeischen lexikographie, abt 1: enzyklopaedien) – 46v+6 suppl vols on 405mf – 9 – €2660.00 – 3-89131-068-4 – (int by otmar seemann) – gw Fischer [030]

Grosse deutsche lexika : aufklaerung und fruehes neunzehntes jahrhundert / ed by Killy, Walther – [mf ed 1993] – 421mf (1:24) – 9 – silver €4700.00 – ISBN-10: 3-598-40620-7 – ISBN-13: 978-3-598-40620-1 – (incl: universal-lexikon oder vollstaendiges encyclopaedisches woerterbuch, bd 1-26 altenburg: literatur-comptoir 1835-36 [211mf] €2458 [isbn 978-3-598-40621-8]; allgemeine deutsche real-encyklopaedie fuer die gebildeten staende: (conversations-lexikon) leipzig: brockhaus, 1827 [113mf] €1548 [isbn 978-3-598-40622-5]; huebner, johann: reales staats-, zeitungs- und conversations-lexicon, leipzig: gleditsch, 1737 [13mf] €198 [isbn 978-3-598-40623-2]; huebner, johann: reales, staats-, zeitungs-, und conversations-lexicon, leipzig: glegitsch, 1795 [15mf] €240 [isbn 978-3-598-40624-9]; hederich, benjamin: reales schul-lexicon, leipzig, gleditsch, 1717 [16mf] €249 [isbn 978-3-598-40625-6]; hederich, benjamin: gruendliches lexicon mythologicum, leipzig, gleditsch, 1741 [11mf] €144 [isbn 978-3-598-40626-3]; hederich, benjamin: gruendliches antiquitaeten-lexicon, leipzig, gleditsch, 1743 [17mf] €268 [isbn 978-3-598-40627-0]; walch, johann georg: philosophisches lexicon, leipzig, gleditsch, 1775 [25mf] €288 [isbn 978-3-598-40628-7]) – gw Saur [430]

Die grosse entscheidung : drama / Klass, Gert von – Wiesbaden: Verlag Der Greif, c1944 – 1r – 1 – us Wisconsin U Libr [820]

Die grosse fahrt : ein roman von seefahrern, entdeckern, bauern und gottesmaennern / Blunck, Hans Friedrich – Muenchen: A Langen/ G Mueller, 1935 [mf ed 1989] – 318p – 1 – mf#7037 – us Wisconsin U Libr [830]

Die grosse heimatzeitung *see* Friedeberger kreisblatt

Der grosse imhoff : ein deutscher kolonisator: roman / Beilelstein, Felix Wilhelm – Darmstadt: L Kichler, 1942 [mf ed 1989] – 366p – 1 – mf#7005 – us Wisconsin U Libr [830]

Das grosse jagen : roman aus dem 18. jahrhundert / Ganghofer, Ludwig – Berlin: G Grote 1920, c1918 [mf ed 1990] – (= ser Grotesche sammlung von werken zeitgenoessischer schriftsteller 133) – 1r [ill] – 1 – (filmed with: lebenslauf eines optimisten & other titles) – mf#7284 – us Wisconsin U Libr [830]

Das grosse jahrzehnt in der kritik seiner zeit : die wesentlichen und die umstrittenen rezensionen aus der periodischen literatur des ueberganges von der klassik zur fruehromantik, begleitet von den stimmen der umwelt, in einzeldarstellungen / ed by Fambach, Oscar – Berlin: Akademie-Verlag, 1958 [mf ed 1993] – (= ser Jahrhundert deutscher literaturkritik, 1750-1850 v4) – xix/684p – 1 – (incl bibl ref) – mf#8223 reel 2 – us Wisconsin U Libr [430]

Der grosse janja : ein kattowitzer roman / Ulitz, Arnold – [Breslau]: W G Korn [1939] [mf ed 1991] – 1r – 1 – (filmed with: ararat) – mf#2932p – us Wisconsin U Libr [830]

Grosse, Julius *see*
- Gudrun
- Gundel vom koenigsee

Grosse kathechese (bdk56 1.reihe) : ueber das gebet des herrn / ueber die 8 seligkeiten / dialog ueber die seele / leben der seligen makrina / Gregor von Nyssa (Gregory of Nyssa, Saint) – (= ser Bibliothek der kirchenvaeter. 1. reihe (bdk 1.reihe)) – €15.00 – ne Slangenburg [240]

Grosse koalition und opposition : die politik der fdp 1966-1969 / Koetteritzsch, Georg A – (mf ed 1999) – 8mf – 9 – €68.00 – 3-8267-2673-1 – mf#DHS 2673 – gw Frankfurter [321]

Der grosse komet von 1556 und seine bevorstehende wiederkehr / Jahn, Gustav Adolph – Leipzig: J J Weber 1856 [mf ed 1998] – 1r [ill] – 1 – mf#film mas 28291 – us Harvard [520]

Grosse kuenstlerlexika vom 16. bis zum fruehen 19. jahrhundert : inst 1: italien; inst 2: deutschland, niederlande; inst 3: frankreich, spanien, england und irland, inst 4: boehmen, maehren und die schweiz = Great dictionaries of artists from 16th to early 19th century / ed by Schuette, Ulrich – [mf ed 2001-03] – (= ser Nachschlagewerke und quellen zur kunst 4) – 342mf (1:24) – 9 – diazo €2689.00 (silver €3289 isbn: 978-3-598-34991-1) – ISBN-10: 3-598-34990-4 – ISBN-13: 978-3-598-34990-4 – (with guide) – gw Saur [700]

Das grosse leben : goethes briefe / ed by Hartung, Ernst – Ebenhausen bei Muenchen: W L Brandt 1924 [mf ed 1990] – 2v on 1r – 1 – (filmed with: briefwechsel zwischen goethe und staatsrath schultz / h duntzer [ed]) – mf#2786p – us Wisconsin U Libr [920]

Die grosse ordnung : roman einer familie am strom / Fromme-Bechem, Annemarie – 2. Aufl. Muenchen: Zentralverlag der NSDAP, F Eher, 1943 (mf ed 1990) – 1r – 1 – (filmed with: liebeskaempfe) – us Wisconsin U Libr [830]

Die grosse politik der europaischen kabinette / Germany. Auswaertiges Amt – v1-40. 1871-1914 – 1 – $432.00 – mf#0235 – us Brook [940]

Das grosse signal : jungen im verratenen rheinland / Lux, Hanns Maria – Berlin: W Limpert c1937 [mf ed 1993] – 1r [ill] – 1 – (incl "geschichtliches nachwort" by hans bellinghausen; ill by willy thomsen. filmed with: die fahrt nach letztesand / martin luserke) – mf#7604 – us Wisconsin U Libr [830]

Die grosse steinplatteninschrift nebukadnezars 2. in transcribiertem babylonischen grundtext : nebst uebersetzung und commentar / Flemming, Johannes – Goettingen: Dieterich, 1883 – 1mf – 9 – 0-7905-3085-6 – mf#1987-3085 – us ATLA [470]

Die grosse stunde : erzaehlungen / Helke, Fritz – Stuttgart: Union Deutsche Verlagesgesellschaft, 1943 – 1r – 1 – us Wisconsin U Libr [830]

Die grosse suende : ein buergerliches trauerspiel / Bahr, Hermann – Zuerich: Verlags-Magazin (J Schabelitz), 1889 [mf ed 1998] – 119p – 1 – mf#9961 – us Wisconsin U Libr [820]

Die grosse veraenderung und unsere literatur / Seghers, Anna – Berlin: Aufbau-Verlag, 1956 [mf ed 1993] – 46p – 1 – mf#8257 – us Wisconsin U Libr [430]

Das grosse wandern : roman / Rainalter, Erwin Herbert – [Berlin]: Volksverband der Buecherfreunde: Wegweiser-Verlag 1945 [mf ed 1996] – 1r – 1 – (filmed with: anna seghers / [heinz neugebauer]) – mf#9241 – us Wisconsin U Libr [830]

Das grosse wandern : ein spiel vom ewigen deutschen schicksal / Eggers, Kurt – Berlin-Schoeneberg: Volkschaft-Verlag fuer Buch, Buehne & Film 1934 [mf ed 1989] – (= ser Spiele vom schaffenden volk 3) – 1r – 1 – (filmed with: die geburt des jahrtausends) – mf#7205 – us Wisconsin U Libr [820]

Die grosse woge : roman / Baudissin, Eva Fanny Bernhardine Tuerk, graefin von – Stuttgart: J Engelhorn 1919 [mf ed 1995] – 1r – 1 – (filmed with: zwuesche tuer u angle / ernst balzli) – mf#3779p – us Wisconsin U Libr [830]

Die grosse wolthat so unser herre gott durch d martinum luther der welt erzeiget : in reimen kuertzlich zusammen gefasset / [Alber, E] – np, [1546] – 1mf – 9 – mf#TH-1 mf 9 – ne IDC [242]

Grosse-Brauckmann, Emil *see* Der psaltertext bei theodoret

The grosse-isle tragedy and the monument to the irish fever victims 1847 reprinted : with additional information and illustrations, from the daily telegraph's commemorative souvenir, issued on the occasion of the unveiling of the national memorial on the 15th august, 1909... / Jordan, John A – Quebec: Telegraph Printing, 1909 [mf ed 1994] – 9 – cn Bibl Nat [971]

Die grossen kappadocier : basilius, gregor von nazianz und gregor von nyssa als exegeten / Weiss, Hermann – Braunsberg: A Martens, 1872 – 1mf – 9 – 0-524-08184-0 – mf#1992-1170 – us ATLA [220]

Die grossen kriege in der geschichte des deutschen volkes / Jacob, Karl – Tuebingen: Kloeres, 1915. 28p – 1 – us Wisconsin U Libr [943]

Die grossen socialen fragen der gegenwart : sechs predigten. gehalten in mainz... / Ketteler, Wilhelm Emmanuel, Freiherr von – Mainz: Kirchheim, 1878 – 1mf – 9 – 0-524-03234-3 – ("mit einem anhange: leichenrede, gehalten im jahre 1848 zu frankfurt am grabe des fuersten lichnowski und des generals von auerswald.") – mf#1990-0862 – us ATLA [240]

Grossenhainer tageblatt *see* Grossenhainer wochenblatt

Grossenhainer wochenblatt – Grossenhain DE, 1807-08, 1812-15, 1817-19, 1821-29, 1831-38, 1840-1945 20 apr [gaps] – 160r – 1 – (title varies: 1893: grossenhainer tageblatt) – gw Misc Inst [074]

Grosser, Guenther *see* Das buendnis der parteien

Grosser tempel in mexico – s.l, s.l? 1755 – 1r – us UF Libraries [025]

Grosses illustriertes frauen-lexikon : sicherster ratgeber und bequemstes nachschlagebuch ueber alle beduerfnisse und angelegenheiten in wohnung, kueche, keller, garten / ed by Krueger, Auguste & Dillon, Franz J – Berlin: Herlet 1900 [mf ed 2002] – (= ser Hq 54) – 15mf – 9 – 3-89131-394-2 – gw Fischer [640]

Grosses vollstaendiges universal-lexikon aller wissenschaften und kuenste (ael1/32) / Zedler, Johann Heinrich – Halle/Leipzig 1732-54 [mf ed 1995] – (= ser Archiv der europaeischen lexikographie, abt 1: enzyklopaedien) – 64v+4 suppl vol on 630mf – 9 – €3960.00 – 3-89131-213-X – gw Fischer [030]

Grosseteste, Robert, Bishop of Lincoln *see*
- Roberti grosseteste episcopi quondam lincolniensi epistolae
- The sermons of robert grosseteste

Gross-flottbeker tageblatt – Hamburg DE, 1911-34 – 48r – 1 – gw Misc Inst [074]

Die grossfuerstin : roman / Samarow, Gregor – 2. Aufl. Stuttgart: Deutsche Verlags-Anstalt – 1r – 1 – us Wisconsin U Libr [830]

Gross-gerauer echo *see* Heimat-zeitung des kreises gross-gerau

Grossherzoglich badische staats-zeitung *see* Carlsruher zeitung

Grossherzoglich badisches allgemeines anzeigenblatt – Karlsruhe DE, 1856-68 – 2r – 1 – gw Misc Inst [074]

Grossherzoglich badisches anzeige-blatt fuer den seekreis – Konstanz DE, 1821-25 – 2r – 1 – gw Misc Inst [074]

Grossherzoglich badisches anzeige-blatt fuer kinzig, murg- und pfinzkreis *see* Allgemeines intelligenz- oder wochenblatt fuer saemtliche hochfuerstliche lande

Grossherzoglich badisches mittelrheinisches provinzial-blatt *see* Allgemeines intelligenz- oder wochenblatt fuer saemtliche hochfuerstliche lande

Grossherzoglich bergisches archiv *see* Bergisches archiv

Grossherzoglich hessische landzeitung *see* Hessen-darmstaedtische privilegirte landes-zeitung

Grossherzoglich mecklenburg-schwerinsches officielles wochenblatt *see* Herzoglich mecklenburg-schwerinsches officielles wochenblatt

Grossherzoglich mecklenburg-strelitzscher officieller anzeiger fuer gesetzgebung und staatsverwaltung fuer das fuerstenthum ratzeburg – Schoenberg DE, 1831-1898 30 sep, 1907-12, 1914-18 – 1 – (filmed by misc inst: 1849, 19.1. – 1919, 31 may. title varies: officielle beilage zu den woechentlichen anzeigen fuer das fuerstenthum ratzeburg 1838 7 feb-1841 20 jan [1r]) – gw Mikrofilm; gw Misc Inst [342]

Grossherzogl[ich] s[achsen] weimar-eisenachisches regierungs-blatt – Eisenach DE, 1819 15 jan-24 dec, 1830 – 2r – 1 – gw Misc Inst [074]

Grossherzoglich-badische privilegirte freyburger zeitung *see* Freyburger zeitung

Grossherzoglich-badisches oberrheinisches provinzial-blatt – Freiburg Br DE, 1803-55 – 23r – 1 – gw Misc Inst [943]

Grossherzogliche Sternwarte zu Heidelberg. Astrometrisches Institut *see* Veroeffentlichungen der grossherzoglichen sternwarte zu heidelberg [astrometrisches institut]

Grossherzogliche Sternwarte zu Heidelberg [Koenigstuhl] *see* Veroeffentlichungen der grossherzoglichen sternwarte zu heidelberg [koenigstuhl]

Gross-Hoffinger, Anton *see*
- Das galante wien
- Oesterreich wie es ist
- Ungarn, das reich, land und volk wie es ist

Gross-kikindaer zeitung – Gross-Kikinda (Velika Kikinda YU), 1902 17 aug-21 dec – 1r – 1 – gw Misc Inst [077]

Gross-kokler bote *see* Schaeszburger zeitung

Grossman, Edgar F *see*
- Determination of the winter survival of the cotton boll weevil by field counts
- Heat treatment for controlling the insect pests of stored corn
- Hibernation of the cotton boll weevil under controlled temperature
- How the boll weevil ingests poison
- Methods for making counts of boll weevil infestation

Grossman, George S *see* Omnibus copyright revision

Grossman, James *see* Black workers in the era of the great migration, 1916-1929

Grossman, Leonid Petrovich *see* Ns leskov

Grossman, Louis *see* Inaugural sermon delivered in the temple beth-el at detroit, mich...

Grossman, mary b, papers, ms 3660 – 1923-72 – 1r – 1 – (grossman was a suffragist and a pioneering woman lawyer. she was the first woman elected to a municipal judgeship in the united states, in 1923) – us Western Res [340]

Grossman, V I *see* Nalogovyi spravochnik dlia kustarei i remeslennikov i ikh pervichnikh obedinenii

Grossmann, K *see*
- Gasparis megandri...in epistolam pauli ad ephesios comentarius...
- Gasparis megandri...in epistolam pauli ad galatas commentarius...
- In divi pauli epistolas tres, ad timotheum et titum...
- Jacob unrest
- Jacob unrest, oesterreichische chronik
- Eyn kurtze aber christenliche vslegung fuer die jugend der gebotten gottes...

Grossmann, Kurt *see* Die menschenrechte

Grossmann, Louis *see*
- Maimonides
- Selected writings of isaac m. wise
- Some chapters on judaism and the science of religion

Gross-new-yorker zeitung – New York NY. 1890 apr 7 – 1r – 1 – mf#910619 – us WHS [071]

Die grosstadtluft : schwank in vier akten / Blumenthal, Oscar & Kadelburg, Gustav – Berlin: E Bloch, [1891] [mf ed 1989] – 141p – 1 – mf#7036 – us Wisconsin U Libr [820]

Grosvenor Gallery, London *see*
- Exhibition of the works of sir john e millais
- The grosvenor gallery winter exhibition
- [Summer exhibition catalogue 1877]
- [Summer exhibition catalogue 1878]
- [Summer exhibition catalogue 1880]
- [Summer exhibition catalogue 1881]
- [Summer exhibition catalogue 1882]
- [Summer exhibition catalogue 1883]
- [Summer exhibition catalogue 1884]
- [Summer exhibition catalogue 1885]
- [Summer exhibition catalogue 1886]
- [Summer exhibition catalogue 1887]
- [Summer exhibition catalogue 1888]
- [Summer exhibition catalogue 1889]
- [Summer exhibition catalogue 1890]

Grosvenor Gallery, London. *see* [Summer exhibition catalogue 1879]

The grosvenor gallery winter exhibition : [catalogues 1877-90] / Grosvenor Gallery, London – London 1877-90 – (= ser 19th c art & architecture) – 19mf – 9 – mf#4.2.1635 – uk Chadwyck [700]

Grosvenor notes : with facsimiles of sketches by the artists / Blackburn, Henry – London 1878-90 – (= ser 19th c art & architecture) – 10mf – 9 – mf#4.2.284 – uk Chadwyck [700]

Grosze passions cantate : [du goettlicher, warum bist du so in des todes schmerz versunken?] / Bach, Carl Philipp Emanuel – [178-?] [mf ed 1988] – 1r – 1 – mf#pres. film 14 – us Sibley [780]

Grot, Nikolaus *see* Nietzsche und tolstoi

Grote, A R *see* North american entomologist

Grote, Augustus Radcliffe *see*
- Check list of the noctuidae of america, north of mexico
- The effect of the glacial epoch upon the distribution of insects in north america
- The hawk moths of north america
- An illustrated essay on the noctuid of north america

Grote, George *see*
- Fragments on ethical subjects
- The minor works of george grote
- Plato and the other companions of sokrates

Grote, Hans Henning, Freiherr *see*
- Der hauptmann
- Die hoehle von beauregard

Grote, Harriet (Lewin) *see* Memoir of the life of ary scheffer

Grote, John *see*
- An examination of the utilitarian philosophy
- Exploratio philosophica

Grote, Ludwig [comp] *see* Martin luther und seine mitstreiter

Grotefend, G A *see* Preussisch-deutsche gesetz-sammlung, 1806-99

Grotefend, Georg Friedrich *see*
- Erlaeuterung der babylonischen keilinschriften aus behistun
- Erlaeuterung der inschrift aus den oberzimmern in nimrud
- Erlaeuterung der keilinschriften babylonischer backsteine
- Erlaeuterung einer inschrift des letzten assyrisch-babylonischen koenigs aus nimrud
- Erlaeuterung zweier ausschreiben des koeniges nebukadnezar in einfacher babylonischer keilschrift
- Die tributverzeichnisse des obelisken aus nimrud

Grotefend, Otto *see* Urkunden der familie von saldern

Grotesend, Ulrich *see* Geschichte und rechtliche stellung der juden in pommern

Das groteske in der fruehen prosa v.a. kaverins (1920-1931) / Suchy, Joerg – (mf ed 1994) – 2mf – 9 – €40.00 – 3-8267-2013-X – mf#DHS 2013 – gw Frankfurter [460]

The grotesque in church art / Wildridge, Thomas Tindall – London 1899 – (= ser 19th c art & architecture) – 3mf – 9 – mf#4.2.1187 – uk Chadwyck [700]

Groth, Andrea W *see* A study of the characteristics of participants at the fedex wellness center

Groth, Klaus *see* Lebenserinnerungen

Grothe, J A *see* Archief voor de geschiedenis der oude hollandsche zending

Grotius, H *see*
- Annales et historiae de rebus belgicis
- Opera omnia theologica

Grotius, Hugo *see* Defence of the catholic faith

Grotius, Hugo de *see*
- Opera omnia theologica...
- The rights of war and peace

Grotkass, Carolin *see* Aktivitaet und aktivierbarkeit von polyphenoloxidasen in embryogenen und nicht-embryogenen suspensionskulturen von euphorbia pulcherrima willd. ex. klotzsch

Groto, L *see* Oratione...fatta in vinegia, per l'allegrezza della uittoria ottenuta contra turchi dalla santissima lega

Groton 1640-1900 – Oxford, MA (mf ed 1995) – (= ser Massachusetts vital records) – 133mf – 9 – 0-87623-381-7 – (mf 1-4: vital records 1656-1829. mf 5-11: vital records 1640-1853. mf 12-14: vital records 1674-1758. mf 15-24: vital records 1674-1758. mf 25-42: proprietors 1711-1829. mf 43-48: land records 1662-1772. mf 49-58: town records 1655-1797. mf 59: births 1829-43. mf 59-78: town records 1740-1825. mf 79-88: town meetings 1813-35. mf 89-101: town orders 1789-1808. mf 101-102: boundaries 1828-29. mf 103-105: town records 1815-48. mf 106-107: paupers 1854-1902. mf 108: dogs 1858-69. mf 108-110: taxes 1863, 1879. mf 110-113: veterans 1861-83. mf 114-115: birth index 1843-1901. mf 115-116: death index 1843-1901. mf 117-118: marriage index 1843-1901. mf 119-120: births 1843-53. mf 120: marriages 1843-55. mf 121: deaths 1843-52. mf 122-126: births 1854-1901. mf 126-129: marriages 1856-1901. mf 130-133: deaths 1853-1901) – us Archive [978]

Groton 1648-1849 – Oxford, MA (mf ed 1996) – (= ser Massachusetts vital record transcripts to 1850) – 19mf – 9 – 0-87623-259-4 – (mf 1t-3t. births & marriages 1674-1753. mf 1t-5t: deaths 1734-1829. mf 3t-5t: births 1689-1795. mf 6t-7t: marriages 1732-72. mf 7t-8t: births & deaths 1648, 1714-1843. mf 9t-12t: births & deaths 1701-1851. mf 12t-13t: marriages 1772-98; births & deaths 1771-1850. mf 13t: out-of-town marriages 1685-1799. mf 14t-16t: marriages & intentions 1795-1849. mf 16t-17t: births 1843-49. mf 18t: marriages 1843-49. mf 19t: deaths 1843-49) – us Archive [978]

Groton landmark – Groton MA. 1895 may 25/1896 oct 17-1916 jan 8/1917 jan 27 [gaps] – 16r – 1 – mf#1231355 – us WHS [071]

Groton, William Mansfield *see* The christian eucharist and the pagan cults

La grotta di trofonio : opera comica in due atti [di giambattista casti]... / Salieri, Antonio – Vienna: Artaria Compagni [1785] [mf ed 19–] – 7mf – 9 – mf#fiche 1108 – us Sibley [780]

Grou, Jean Nicholas *see* Manual for interior souls

Grou, Jean Nicola *see* Characteristics of true devotion

Grou, Jean Nicolas *see* The christian sanctified by the lord's prayer

Ground covers for florida gardens / Crevasse, J M – Gainesville, FL. 1941 – 1r – us UF Libraries [630]

Ground level – n14-17 [1995 aug/sep-1996 apr] – 1r – 1 – mf#3403387 – us WHS [071]

The ground of national grievances examined : with a proposal for the liquidation of the national debt – Newcastle: printed by & for J Clark, 1820 – (= ser 19th c economics) – 1mf – 9 – mf#1.1.66 – uk Chadwyck [339]

The ground of woman's eligibility – [New York: Hunt & Eaton, 1891] Beltsville, Md: NCR Corp, 1978 (1mf); Evanston: American Theol Lib Assoc, 1984 (1mf) – (= ser Women & the church in america) – 9 – 0-8370-1617-7 – mf#1984-2019 – us ATLA [240]

Ground on which sinners can come to the holy god – London, England. 18– – 1r – us UF Libraries [240]

Ground reaction forces analysis of a variety of jumping activites1 : in growing children / Tsang, Garry – 2000 – 136p on 2mf – 9 – $10.00 – mf#PE4121 – us Kinesology [612]

Ground water – Westerville OH 1963+ – 1,5,9 – ISSN: 0017-467X – mf#10126 – us UMI ProQuest [550]

Ground water age – Latham NY 1973-91 – 1,5,9 – ISSN: 0046-645X – mf#8437 – us UMI ProQuest [333]

Grounds and reasons of christian regeneration / Law, William – London, England. 1845 – 1r – us UF Libraries [240]

Grounds for laying before the council of king's college, london / Jelf, R W – Oxford, England. 1853 – 1r – us UF Libraries [240]

Grounds for laying before the council of king's college, london, certain statements contained in a recent publication entitled theological essays by the rev. f.d. maurice, m.a., professor of divinity in king's college / Jelf, Richard William – 2nd ed. Oxford: JH Parker, 1853 – 1mf – 9 – 0-524-05082-1 – mf#1991-2206 – us ATLA [240]

Grounds for remaining in the anglican communion / Faber, Frederick William – London, England. 1846 – 1r – us UF Libraries [241]

Grounds maintenance – Shawnee Mission KS 1966+ – 1,5,9 – ISSN: 0017-4688 – mf#6496 – us UMI ProQuest [790]

The grounds of faith : four lectures / Manning, Henry Edward – London: Burns and Lambert, 1852 – 1mf – 9 – 0-8370-6820-7 – mf#1986-0820 – us ATLA [230]

Grounds of legislative restriction applied to public-houses / Arnot, William – Glasgow, Scotland. 18– – 1r – us UF Libraries [240]

The grounds of theistic and christian belief / Fisher, George Park – New York: Scribner, 1915, c1911 – 2mf – 9 – 0-7905-4521-7 – (incl bibl ref) – mf#1988-0521 – us ATLA [240]

Grounds of union between the churches of england and of rome consid... / Church Of England Diocese Of Durham – London, England. 1811 – 1r – us UF Libraries [240]

Grounds of unitarian dissent / Yates, James – Glasgow, Scotland. 1813 – 1r – us UF Libraries [243]

Grounds on which the church of england separated from the church of... / Church Of England Diocese Of Durham – London, England. 1807 – 1r – us UF Libraries [241]

Grounds on which the church of england separated from the church of... / Church Of England Diocese Of Durham – London, England. 1809 – 1r – us UF Libraries [241]

Ground-water mining in the united states / Sloggett, Gordon & Dickason, Clifford – Washington DC: US Dept of Agriculture, Economic Research Service, 1986 [mf ed 1986] – (= ser Agricultural economic report 555) – 9 – (with bibl) – us GPO [630]

The groundwork of a system of evangelical lutheran theology / Sprecher, Samuel – Philadelphia: Lutheran Pub Soc, 1879 [mf ed 1991] – 2mf – 9 – 0-7905-9677-6 – mf#1989-1402 – us ATLA [242]

Groundwork of economics / Mukerjee, Radhakamal – London; New York: Longmans, Green, and Co, 1925 – (= ser Samp: indian books) – us CRL [330]

Group and organization management – Newbury Park CA 1992+ – 1,5,9 – (cont: group & organization studies) – ISSN: 1059-6011 – mf#10927.01 – us UMI ProQuest [150]

Group and organization studies – Newbury Park CA 1976-91 – 1,5,9 – (cont by: group & organization management) – ISSN: 0364-1082 – mf#10927.01 – us UMI ProQuest [150]

Group areas act / Horrell, Muriel – Johannesburg, South Africa. 1956 – 1r – us UF Libraries [960]

The group areas act – Cape Town: Juta, 1953 – 1 – us CRL [960]

Group cohesion and perceptions of pressure to conform in a university residence / Ramsay, Michael C – 1994 – 1mf – $4.00 – us Kinesology [150]

Group decision and negotiation – Dordrecht, Netherlands 1992+ – 1,5,9 – ISSN: 0926-2644 – mf#18657 – us UMI ProQuest [303]

Group [dordrecht] – Dordrecht, Netherlands 1977+ – 1,5,9 – ISSN: 0362-4021 – mf#11179 – us UMI ProQuest [150]

A group dynamic interpretation of a teambuilding event : a case study / Britton, Mignon – Uni of South Africa 2000 [mf ed Johannesburg 2000] – 5mf – 9 – (incl bibl ref) – mf#mfm15091 – sa Unisa [150]

Group for Advertising Progress *see* Final proof

A group intervention programme for adolescents of divorce / Johnson, Colleen – Uni of South Africa 2000 [mf ed Pretoria: UNISA 2000] – 6mf – 9 – (incl bibl ref) – mf#mfm 15120 – sa Unisa [362]

Group [loveland] – Loveland CO 1985+ – 1,5,9 – ISSN: 0163-8971 – mf#15244 – us UMI ProQuest [240]

Group prejudices in india : a symposium / ed by Nanavati, Manilal B & Vakil, C N – Bombay: Vora & Co, 1951 – (= ser Samp: indian books) – us CRL [305]

Group process in the organizational development of the church / Sidener, Roger Don – Princeton, New Jersey, 1976. Chicago: Dep of Photodup, U of Chicago Lib, 1976 (1r); Evanston: American Theol Lib Assoc, 1984 (1r) – 1 – 0-8370-1286-4 – mf#1984-T017 – us ATLA [240]

Group psychotherapy and psychodrama – Washington DC 1947-75 – 1,5,9 – ISSN: 0096-0586 – mf#6905.04 – us UMI ProQuest [150]

Group psychotherapy, psychodrama & sociometry – Washington DC 1976-80 – 1,5,9 – (cont by: journal of group psychotherapy, psychodrama & sociometry) – ISSN: 0146-6178 – mf#6905.04 – us UMI ProQuest [150]

Group theories of religion and the individual / Webb, Clement Charles Julian – London: G Allen & Unwin; New York: Macmillan, 1916 – (= ser Wilde Lectures in Natural and Comparative Religion) – 1mf – 9 – 0-7905-9745-4 – mf#1989-1470 – us ATLA [301]

Groupe de travail sur la prostitution des mineurs *see* Rapport sur la prostitution chez les mineurs

Groupe des Etudiants Socialistes *see* La jeunesse socialiste

Groupe du Bas-Languedoc de l'Association Sully *see* Bulletin

Groupe Marxiste Revolutionnaire *see* La taupe rouge

Groupement de petit-ekonda / Mune, Pierre – Bruxelles, Belgium. 1959 – 1r – us UF Libraries [960]

Grout, Lewis *see* Zulu-land, or, life among the zulu-kafirs

Grove, Alfred Thomas *see* Africa south of the sahara

Grove, C A *see* The effects of high impact exercise versus low impact exercise on bone density in postmenopausal women

Grove City College (PA). Young Americans for Freedom *see* Entrepreneur

Grove city record / Franklin Co. Grove City – jan 1993-dec 1998 – 7r – 1 – mf#B 41440-41446 – us Ohio Hist [071]

Grove, Frederick Philip *see* Oscar wilde

Grove level baptist church. dalton, georgia : church records – Sept 1853 – Mar 1952. Incomplete – 1 – 49.86 – us Southern Baptist [242]

Groveland farms – Groveland, FL. 1917 – 1r – us UF Libraries [630]

Groveland graphic – Groveland, FL. 1923 sep 3-1942 aug – 8r – (gaps) – us UF Libraries [071]

Groveland, lake county, florida / Allen, L – s.l, s.l? 1936 – 1r – us UF Libraries [978]

[Groveland-] tuolumne prospector – CA. 1901-1904 – 1r – 1 – $60.00 – mf#C03603 – us Library Micro [071]

Grover, Alonzo J *see* Romanism

[Grover city-] five cities times press recorder – CA. 1970-79 – 36r – 1 – $2160.00 – mf#B02282 – us Library Micro [071]

Grover cleveland papers – (mf ed 1958) – (= ser Presidential papers microfilm) – 164r – 1 – (with guide) – Dist. us Scholarly Res – us L of C Photodup [975]

Grover first baptist church. kings mountain association. grover, north carolina : church records – 1897-1963 – 1 – us Southern Baptist [242]

Grover, Frederick Warren *see* Inductance calculations, working formulas and tables

Grover, Thomas Johnson *see* The plough and harrow address to the electors of canada

Grover w ensley, senate service 1949-1957 : executive director, joint economic committee – (= ser Us senate historical office oral history coll) – 2mf – 9 – $10.00 – us Scholarly Res [323]

Groves, A N *see* Journal of a residence at bagdad, during the years 1830 and 1831

Groves, Anthony *see* Journal of a residence at bagdad, during the years 1830 and 1831

Groves, Henry *see* Baptism

Groves, Henry Charles *see* Doctrines and practices of the jesuits

Groves, Michelle D *see* Dynamic functional assessment of the lower extremity in the non-varsity athletic population

Groves, W C *see*
– Ethnographic studies of new ireland
– Report on education in the british solomon islands

Groves, William Charles *see* Papers relating to education in papua new guinea and nauru

Grow lovely, growing old / Green, Lawrence George – Cape Town, South Africa. 1951 – 1r – us UF Libraries [960]

Growing and using mangels, sugar mangels and forage sugar beets / Grisdale, Joseph Hiram – [Ottawa?: s.n, 1911?] – 1mf – 9 – 0-665-99629-2 – (with notes on their chemical composition by frank t shutt; incl ind) – mf#99629 – cn Canadiana [630]

Growing disciples – Nashville TN 1994-97 – 1,5,9 – (cont: discipleship training) – ISSN: 0162-4601 – mf#21596 – us UMI ProQuest [200]

Growing evil / Cox, J – London, England. 18– – 1r – us UF Libraries [240]

Growing narcissi, gladioli and dahlias under florida conditions / Van Cleef, Clinton B – s.l, s.l? 1926 – 1r – us UF Libraries [630]

Growing of easer lily bulbs under florida conditions / Cooke, Alfred Fuller – s.l, s.l? 1927 – 1r – us UF Libraries [630]

Growing plants without soil by the water-culture method / Hoagland, D R – Tallahassee, FL. 1938 – 1r – us UF Libraries [630]

Growing sea island cotton under florida conditions / United States Works Projects Administration (Fla) – Tallahassee, FL. 1938 – 1r – us UF Libraries [630]

Growing up in puerto rico / Mcfadden, Dorothy Loa Mausolff – Morristown, NJ. 1958 – 1r – us UF Libraries [972]

Growler – Toronoto. v1 n1-4. jul 22-aug 19 1864//? [wkly] – 1r – 1 – Can$110.00 – cn McLaren [071]

Growls from uganda / Critolaus – London: E Stock, 1909 – 1 – us CRL [960]

Growth – Hulls Cove ME 1937-87 – 1,5,9 – (cont by: growth, development, and aging) – ISSN: 0017-4793 – mf#2183.01 – us UMI ProQuest [574]

The growth and administration of the british colonies 1837-1897 / Greswell, William Henry Parr – London, 1898 – (= ser 19th c books on british colonization) – 3mf – 9 – mf#1.1.4557 – uk Chadwyck [941]

Growth and change – Oxford, England 1970+ – 1,5,9 – ISSN: 0017-4815 – mf#6171 – us UMI ProQuest [338]

Growth and change in a shona ward / Garbett, G Kingsley – Salisbury, Zimbabwe. 1960 – 1r – us UF Libraries [960]

Growth and education / Tyler, John M – 1907 – 4mf – 9 – $12.00 – us Kinesology [370]

Growth and formation of the cell wall in pollen tubes of nicotiana tabacum and petunia hybrida / Geitmann, Anja – (mf ed 1997) – 2mf – 9 – €40.00 – 3-8267-2395-3 – mf#DHS 2395 – gw Frankfurter [574]

Growth behavior and maintenance of organic foods in bahia grass / Leukel, W A – Gainesville, FL. 1930 – 1r – us UF Libraries [630]

Growth condition of an ice layer in freezing soils under applied loads / Takeda, Kazuo – [Hanover NH]: US Army Corps of Engineers... 1993-94 [mf ed 1994] – 2mf – 9 – us GPO [550]

Growth, development, and aging – Hulls Cove ME 1988+ – 1,5,9 – (cont: growth) – ISSN: 1041-1232 – mf#2183.01 – us UMI ProQuest [574]

The growth hormone response to exercise at different times of the day / Bullard, J M – 1991 – 1mf – 9 – $4.00 – us Kinesology [613]

Growth in grace / Bailey, John – London, England. 1808 – 1r – us UF Libraries [240]

Growth of african civilization : west africa 1000-1800 / Davidson, Basil – London, England. 1965 – 1r – us UF Libraries [960]

The growth of african literature : a survey of the works published by african writers / Joseph, Stanislaus – Montreal, 1952 – (in english and french) – us CRL [470]

The growth of christianity : london lectures / Gardner, Percy – London: Adam and Charles Black, 1907 – 1mf – 9 – 0-7905-0025-6 – (incl bibl ref) – mf#1987-0025 – us ATLA [240]

The growth of church institutions / Hatch, Edwin – New York: T. Whittaker, 1887 – 1mf – 9 – 0-7905-3199-2 – (incl bibl ref) – mf#1987-3199 – us ATLA [240]

The growth of europe / Cole, Grenville Arthur James – London: Williams & Norgate; Toronto: W Briggs, 1914 – (= ser Home university library of modern knowledge) – 3mf – 9 – 0-665-66340-4 – mf#66340 – cn Canadiana [914]

The growth of federal finance in india : being a survey of india's public finances from 1833 to 1939 / Thomas, Parakunnel Joseph – London, New York: Oxford University Press, 1939 – (= ser Samp: indian books) – us CRL [336]

Growth of labor law in the united states : 1962 edition / U.S. Dept of Labor – Washington: GPO, 1962 [all publ] – 4mf – 9 – $6.00 – mf#llmc 95-009A – us LLMC [344]

Growth of labor law in the united states : 1967 edition / U.S. Dept of Labor – Washington: GPO, 1967 [all publ] – 4mf – 9 – $6.00 – mf#LLMC 95-009B – us LLMC [344]

Growth of leaves : proceedings / Easter School In Agricultural Science (3d : 1956... – London, England. 1956 – 1r – us UF Libraries [500]

Growth of nationality in the united states : a social study / Bascom, John – New York: GP Putnam, 1899 – 1mf – 9 – 0-7905-3995-0 – mf#1989-0488 – us ATLA [975]

The growth of sport in a southern city : a study of the organizational evolution of baseball in louisville, kentucky, as an urban phenomenon, 1860-1900 / Sullivan, Dean A – 1989 – 2mf – 9 – $8.00 – mf#PE 4001 – us Kinesology [790]

The growth of the church in its organization and institutions / Cunningham, John – London: Macmillan, 1886 – (= ser Croall Lectures) – 1mf – 9 – 0-7905-5595-6 – (incl bibl ref) – mf#1988-1595 – us ATLA [240]

The growth of the constitution in the federal convention of 1787 : an effort to trace the origin and development of each separate clause from its first suggestion in that body to the form finally approved / Meigs, William M – 2nd ed. Philadelphia/London: J B Lippincott Co, 1900 – 5mf – 9 – $7.50 – mf#LLMC 90-358 – us LLMC [323]

Growth of the gospel / Porter, J Scott – Belfast, Northern Ireland. 1833 – 1r – us UF Libraries [240]

The growth of the manor / Vinogradoff, Paul – 2nd, rev ed. London: George Allen, 1911 – 1mf – 9 – 0-524-05163-1 – (incl bibl ref) – mf#1990-1419 – us ATLA [941]

The growth of the new testament : a study of the books in order / Horton, Robert Forman – Boston: Pilgrim Press, [1913?] – 1mf – 9 – 0-7905-0373-5 – mf#1987-0373 – us ATLA [225]

Growth of the oral method of instructing the deaf : an address delivered november 10, 1894, on the twenty-fifth anniversary of the opening of the horace mann school, boston, mass / Bell, Alexander Graham – Boston: Rockwell and Churchill, 1896 [mf ed 1981] – 1mf – 9 – 0-665-24571-8 – (incl bibl ref) – mf#24571 – cn Canadiana [360]

The growth of the protestant episcopal church in the diocese of massachusetts during the nineteenth century : a brief sketch / Magrath, John T – Cambridge, Mass: [s.n.], 1901 – 1mf – 9 – 0-524-03849-X – mf#1990-4896 – us ATLA [242]

The growth of the soul : a sequel to esoteric buddhism / Sinnett, Alfred Percy – London: Theosophical Pub Society, 1896 – 1mf – 9 – 0-524-02052-3 – mf#1990-2827 – us ATLA [280]

Growth of the spirit of christianity : from the 1st century to the dawn of the lutheran era / Matheson, George – Edinburgh: T & T Clark, 1877 [mf ed 1991] – 2v on 2mf – 9 – 0-7905-9335-1 – mf#1989-2560 – us ATLA [240]

Growth of trade and industry in modern india : an introductory survey / Vakil, Chandulal Nagindas et al – Calcutta; New York: Longmans, Green and Co, 1931 – (= ser Samp: indian books) – us CRL [380]

Growth stock digest – Edwardsburg MI 1972 – 1,5 – ISSN: 0017-484X – mf#6404 – us UMI ProQuest [332]

Growth strategies – Santa Monica CA 2001+ – 1 – mf#23056.01 – us UMI ProQuest [338]

Groyse mener – New York, NY. n d – 1r – us UF Libraries [939]

Groznenskii rabochii – Groznyj, Chechen Republic 1973-88 [mf ed Norman Ross] – 5r – 1 – mf#nrp-520 – us UMI ProQuest [077]

Groznenskii rabochii : obshchestvenno-politicheskaia gazeta – Groznyi: Groznenskii rabochii 1995- (mf ed Minneapolis MN: East View Publ [199-]) – 3r – 1 – mf#mf-11998 seemp (jun 1995-97); mf-12753 seemp (1998-2000) – us CRL [077]

Grrrande peche au goujon electoral : ou, blageurs taisez-vous! – Paris [1848?] – us CRL [944]

Gruau de LaBarre, Modeste *see* Intrigues devoilees

Grub, George *see* An ecclesiastical history of scotland

Grubb, Cassel William *see* Edition of the stabat mater and an analysis of the stabat mater by luigi boccherini

Grubb, Edward *see*
- Authority and the light within
- Social aspects of the quaker faith
- What is quakerism?

Grubb, Eugene H *see* Potato

Grubb, Sarah *see* A selection from the letters of the late sarah grubb (formerly sarah lynes)

Grubb, W Barbrooke *see* Among the indians of the paraguayan chaco

Grube, August W *see* Alpenwanderungen

Grube, K *see*
- Chronicon windeshemense und liber de reformatione monastica des augustinerpropstes joh busch
- Des augustinerpropstes ioannes busch

Grube, Wilhelm *see* Religion und kultus der chinesen

Die grubenlampe – Wernigerode DE, 1967 12 jan-1968 15 jul [gaps] – 1r – 1 – (veb harzer eisenerzgruben) – gw Misc Inst [660]

Grubenmann, Yvonne de Athayde *see* Un cosmopolite suisse

Gruber, Foster M *see* Automotive engine testing

Gruber, Hermann *see* Mazzini, freimaurerei und weltrevolution

Gruber, Johann Gottfried *see* Allgemeine encyclopaedie der wissenschaften und kuenste (ael1/33)

Gruber, Levi Franklin *see* The truth about the so-called luther's testament in english, tyndale's new testament

Grubnuk, N A *see* Trudy vtorogo seminara "akusticheskie statisticheskie modeli okeana"

Grub-street journal – La Salle IL 1730-37 – 1 – mf#4259 – us UMI ProQuest [070]

Grudzien, M N *see* Blutiker onhoyb

Grudzinski, Herbert *see* Shaftesburys einfluss auf chr m wieland

Gruebeleien : [essays and reflections] / Frenssen, Gustav – Berlin: G Grote 1920 [mf ed 1989] – 1r – 1 – (filmed with: dorfpredigten) – mf#7264 – us Wisconsin U Libr [840]

Grueber, C S *see*
- Omission not prohibition
- Plain discourse o "the one faith"
- Second letter to his grace the archbishop of canterbury, being an exposure of the rev w goode's...

Grueber, Charles Stephen *see* Article 29 considered...

Grueber, Herbert Appold *see* Roman medallions in the british museum

Grueber, J *see*
- Travels from china to europe, in 1661
- Voyage...la chine

[Grueber, J] *see* Alcune lettere latine del suddetto padre toccanti l'istesse materie

Grueder, C S *see* Presence, the sacrifice, the adoration

Gruehl, M *see* The citadel of ethiopia

Gruen, Albert *see* Goethes faust

Gruen, Anastasius *see*
- Anton auerspergs (anastasius gruens) politische reden und schriften
- Gedichte
- In der veranda
- Nikolaus lenaus dichterischer nachlass

Gruen, Ferdinand Bernard *see* English grammar in american high schools since 1900

Gruen, Karl *see* Louis napoleon bonaparte

Gruenanger, Carlo *see*
- La letteratura tedesca medievale
- Storia della letteratura tedesca medioevale

Gruenau, W von *see* Die staats- und voelkerrechtliche stellung egyptens

Gruenbaum, Max *see* Neue beitraege zur semitischen sagenkunde

Gruenberg, Benjamin C in collaboration with U.S Bureau *see* High schools and sex education

Gruenberg, Gottfried *see* Wir kaempfen und siegen fuer dich freiheit

Gruenberg, Karl et al *see* Hammer und feder

Gruenberg, Paul *see*
- Die evangelische kirche
- Philipp jakob spener

Gruenberger kreis- und intelligenzblatt – Gruenberg (Zielona Gora PL), 1867-68 – 1 – gw Misc Inst [077]

Gruenberger kreisblatt – Gruenberg (Zielona Gora PL), 1920-1932 24 sep – 1 – gw Misc Inst [077]

Gruenberger wochenblatt – Gruenberg (Zielona Gora PL), 1825 2 jul-1828, 1830-31, 1832 7 apr-1840, 1842-65 [gaps], 1867-71, 1873 1 jan-28 sep, 1939 2 jan-1 oct, 1941 1 oct-1942, 1943 1 jul-1944 29 feb [gaps] – 20r – 1 – gw Misc Inst [077]

Gruender, Hubert *see* Free will

Das gruenderthum in der musik : ein epilog zur bayreuther grundsteinlegung / Mohr, Wilhelm – Koeln: M DuMont-Schauberg 1872 – 1mf – 9 – mf#wa-70 – ne IDC [780]

Gruendliche ableinung fuenffzig statlicher ausserlesener vnd in alle ewige ewigkeit unerweisslicher calvinischer ertz- vnd hauptluegen / Hoe von Hoenegg, M – Leipzig, 1621 – 1mf – 9 – mf#TH-1 mf 675 – ne IDC [242]

Gruendliche einleitung in die anfangslehren der tonkunst : zum gebrauche musikalischer lehrstunden... / Albrecht, Johann Lorenz – Langensalza: Johann Christian Martini 1761 [mf ed 19–] – 3mf – 9 – mf#fiche321 – us Sibley [780]

Gruendliche, summarische, apostolische ausfuehrung, der gantzen reinen catholischen, evangelischen lehre in funffzig predigten verfasset / Hoe von Hoenegg, M – Leipzig, 1611 – 9mf – 9 – mf#TH-1 mf 680-688 – ne IDC [242]

Gruendlicher bericht : auff die von den calvinisten eingegebene klaegliche supplikation, darinnen die himmelsche goettliche wahrheit / Hoe von Hoenegg, M – Leipzig, 1605 – 4mf – 9 – mf#TH-1 mf 676-679 – ne IDC [242]

Gruendlicher bericht des deutschen meistergesangs / Puschman, Adam Zacharias; ed by Jonas, Richard – Halle: Max Niemeyer, 1888 – 11r – 1 – (incl bibl ref) – us Wisconsin U Libr [430]

Gruendlicher und deutlicher unterricht zur verfertigung der vollstaendigen saeulen-ordnung / Schuebler, J J – Nuernberg, n d. 4v – 1mf – 9 – mf#OA-110 – ne IDC [720]

Gruendtlicher bericht auff johann sturmij / Andreae d A, J – Tuebingen, 1581 – 4mf – 9 – mf#TH-1 mf 21-24 – ne IDC [242]

Gruendtlicher warhafftiger vnd bestendiger bericht : von christlicher einigkeit der theologen vnd predicanten / [Andreae d A, J] – Wolffenbuettel, 1570 – 2mf – 9 – mf#TH-1 mf 58-59 – ne IDC [242]

Gruendung und entwicklung der kreditinstitute in mannheim und ludwigshafen von der mitte des 19. jahrhunderts bis 1911 / Wuestenhoff, Conrad Kleefeld von – Heidelberg, 1967 – 3mf – 9 – 3-89349-757-9 – gw Frankfurter [332]

Gruendungsgeschichte der stifter...des alten bistums muenster / Tibus, A – Muenster, 1867 – 5mf – 8 – €12.00 – ne Slangenburg [241]

Das gruene blatt – Dortmund DE, 1963-66 – 1 – gw Misc Inst [631]

Gruene brueche : schilderungen und erzaehlungen aus dem wild- und waidmannsleben des hochgebirges / Achleitner, Arthur – Stuttgart: Adolf Bonz, 1894 [mf ed 1995] – 223p – 1 – mf#8917 – us Wisconsin U Libr [880]

Das gruene glas / Gmelin, Otto – Koeln: Im Staufen-Verlag [194-?] [mf ed 1990] – 1r – 1 – (filmed with: sommerwind ueber tormohlenhof / albert gloy) – mf#7310 – us Wisconsin U Libr [830]

Der gruene heinrich : roman / Keller, Gottfried – new ed. Stuttgart: G J Goeschen 1879-80 [mf ed 1995] – 4v on 1r – 1 – (v1 in w.w. set is a different 1884 ed (3. aufl)) – mf#3904p – us Wisconsin U Libr [830]

Der gruene heinrich *see* Gottfried keller

Die gruene post – Berlin DE, 1927 10 apr-1939, 1941-1942 20 dec – 17r – 1 – (filmed with suppl: 1941 2 nov-28 dec) – gw Misc Inst [074]

Gruenebaum, G E von *see* Modern islam

Grueneisen, Carl *see* Der ahnenkultus und die urreligion israels

Der gruenenden jugend ueberfluessige gedanken : abdruck der ausgabe von 1678 / Weise, Christian – Halle a. S: M Niemeyer 1914 [mf ed 1993] – (= ser Neudrucke deutscher literaturwerke des 16. und 17. jahrhunderts 242-245) – 11r – 1 – (incl bibl ref; the work of editing was begun by r bernfeld & completed & int by max freiherr von waldberg) – mf#3387p – us Wisconsin U Libr [430]

Gruenenwald, L *see* Der freie formelhafte infinitiv der limitation im griechischen

Gruenewald und der edelmann : und andere geschichten / Doderer, Otto – Prag, Leipzig: Noebe, 1944 [mf ed 1989] – (= ser Feldpostreihe noebe 18) – 63p – 1 – mf#7180 – us Wisconsin U Libr [830]

Gruenfeld, A *see* Die lehre vom goettlichen willen bei den juedischen religionsphilosophen des mittelalters von saadja bis maimuni

Gruenstein, Bernard *see* The church and the jew

Gruenwedel, Albert *see* Mythologie des buddhismus in tibet und der mongolei

Gruetzmacher, Richard H *see* Jungfrauengeburt

Gruetzmacher, Richard Heinrich *see*
- Goethes faust
- Ist das liberale jesusbild modern?
- Nietzsche und wir christen
- Wesen und grenzen des christlichen irrationalismus

Gruey, Louis Jules *see* Bulletin chronometrique

Gruger, Hugo *see* Four sermons

Grumbine, Jesse Charles Fremont *see* Evolution and christianity

Grumbler – La Salle IL 1715 – 1 – mf#5560 – us UMI ProQuest [390]

Grumbler – Toronto, mar 20 1858-jan 1 1869// (irreg) – 1r – 1 – Can$110.00 – cn McLaren [320]

Grun, Oscar *see* Gedenkblatt an professor a bearliner

Grunberg, Samuel *see* Gedachtnisrede auf rabbiner dr meier hildesheimer

Grunbergensis ethicorum libri duo : methodice conscripti, et cum priscorum philosophorum demonstrationibus, tum historicorum exemplis illustrati / Scultetus, Abraham – Lugduni: [s.n.] 1593 – (= ser Ethics in the early modern period) – 2mf – 9 – mf#pl-122 – ne IDC [170]

Grundaspekte des mensch-seins bei romano guardini : eine anthropologisch-religions-philosophische untersuchung / Lee, Kyung-Won – (mf ed 1996) – 3mf – 9 – €49.00 – 3-8267-2307-4 – mf#DHS 2307 – gw Frankfurter [120]

Grundbegriffe der naturphilosophie bei wilhelm von ockham / Moser, S – Innsbruck, 1932 – 4mf – 8 – €11.00 – ne Slangenburg [140]

Die grundbegriffe in den kosmogonien der alten voelker / Lukas, Franz – Leipzig: W Friedrich, 1893 [mf ed 1991] – 1mf – 9 – 0-524-01618-6 – (incl bibl ref) – mf#1990-2557 – us ATLA [240]

Das grundbekenntnis der kirche und die modernen geistesstroemungen / Schmidt, Wilhelm – Guetersloh: C Bertelsmann, 1905 – 1mf – 9 – 0-524-00786-1 – (incl bibl ref) – mf#1990-0218 – us ATLA [240]

Das grundbekenntniss der evangelisch-lutherischen kirche : mit einer geschichtlichen einleitung und kurzen erklaerenden anmerkungen versehen / Pieper, Anton – St Louis, Mo: Luth Concordia-Verlag, 1880 – 1mf – 9 – 0-7905-9580-X – mf#1989-1305 – us ATLA [242]

Das grunddogma des romanismus : oder, die lehre von der kirche / Delitzsch, Johannes – Gotha: Rud. Besser, 1875 – 1mf – 9 – 0-8370-5910-0 – mf#1985-3910 – us ATLA [240]

Der grundfehler der ritschlschen theologie / Glage, Max – Kiel: M Liebscher, 1893 [mf ed 1991] – 1mf – 9 – 0-7905-7742-9 – (no more publ? incl bibl ref) – mf#1989-0967 – us ATLA [242]

Grundformen volkstuemlicher erzaehlerkunst in den kinder- und hausmaerchen der brueder grimm : ein stilkritischer versuch / Berendsohn, Walter Arthur – Hamburg: W Gente, 1921 [mf ed 1990] – 143p – 1 – (incl bibl ref) – mf#7423 – us Wisconsin U Libr [430]

Grundfragen der motologie : eine untersuchung der ganzheitlichkeit am beispiel von kindlicher entwicklung und angst / Haegele, Sigurd – (mf ed 1993) – 3mf – 9 – €49.00 – 3-89349-792-7 – mf#DHS 792 – gw Frankfurter [150]

Der grundgedanke in goethes faust / Splettstoesser, Willi – Berlin: G Reimer 1911 [mf ed 1990] – 1r – 1 – (filmed with: urfaust? / hermann schneider) – mf#7360 – us Wisconsin U Libr [430]

Die grundgedanken des jakobusbriefes : verglichen mit den ersten briefen des petrus und johannes / Vowinckel, Ernst – Guetersloh: C Bertelsmann, 1899 – (= ser Beitraege zur foerderung christlicher theologie) – 1mf – 9 – 0-524-06003-7 – mf#1992-0740 – us ATLA [227]

Die grundgedanken in heinrich von kleists "prinz friedrich von homburg" / Gilow, Hermann – Berlin: R Gaertner, 1893 – 1r – 1 – us Wisconsin U Libr [430]

Grundgesetz / Germany (West) Constitution – Berlin, Germany. 1956 – 1r – us UF Libraries [025]

Grundgesetz der grossen loge von preussen : genannt royal york zur freundschaft – Berlin, 1906 (mf ed 1992) – 2mf – 3-89349-107-4 – mf#DHS-AR 76 – gw Frankfurter [943]

Grundideen, erscheinungsformen und ursachen politischen hindu-fundamentalismus in indien von 1875 bis heute aus modernisierungstheoretischer sicht / Schworck, Andreas – (mf ed 1995) – 1mf – 9 – €30.00 – 3-8267-2099-7 – mf#DHS 2099 – gw Frankfurter [320]

Die grundirrthuemer unserer zeit / Roh, Peter – 5. Aufl. Freiburg i.B.; St Louis, MO: Herder, 1890 – 1mf – 9 – 0-8370-7981-0 – (incl bibl ref) – mf#1986-1981 – us ATLA [240]

Grundke, Otto *see* Kant's entwickelung vom realismus aus nach dem subjectiven idealismus hin

Grundlage einer ehren-pforte : woran der tuechtigsten capellmeister, componisten, musikgelehrten, tonkuenstler &c leben, werke, verdienste &c erscheinen sollen / Mattheson, Johann – Hamburg: in Verlegung des Verfassers 1740 [mf ed 19–] – 11mf – 9 – mf#fiche 223 – us Sibley [780]

Grundlage zu einer hessischen gelehrten- und schriftsteller-geschichte / Strieder, Friedrich Wilhelm et al – Kassel, Marburg 1781-1868 [mf ed Hildesheim 1983] – (= ser Die schriftsteller- und gelehrtenlexika des 17., 18., und 19. jahrhunderts) – 21v on 117mf – 9 – diazo €498.00 silver €598.00 – gw Olms [430]

Grundlagen der aegyptisch-semitischen wortvergleichung / Calice, F – Wien, 1936 – (= ser Wiener Zeitschrift fuer Kunde des Morgenlandes) – 4mf – 9 – (beihefte zur "wiener zeitschrift fuer kunde des morgenlandes" v1) – mf#NE-470 – ne IDC [470]

Die grundlagen der christologie schleiermachers : die entwicklung der anschauungsweise schleiermachers bis zur glaubenslehre / Bleek, Hermann – Freiburg i B: JCB Mohr, 1898 – 1mf – 9 – 0-7905-9137-5 – (incl bibl ref) – mf#1989-2362 – us ATLA [240]

Die grundlagen der demokratie *see* Nsdap (national socialist german workers party) nazi publications

Grundlagen der judischen ethik – Berlin, Germany. 1920 – 1r – us UF Libraries [939]

Die grundlagen des lutherischen kirchenregimentes / Mejer, Otto – Rostock: Stiller, 1864 – 1mf – 9 – 0-7905-6815-2 – (incl bibl ref) – mf#1988-2815 – us ATLA [242]

Grundlagen des neunzehnten jahrhunderts / Chamberlain, Houston Stewart – Muenchen, Germany. v1-2. 1942 – 1r – us UF Libraries [025]

Grundlagen fuer eine umgestaltung des alttestamentlichen religionsunterrichts / Meltzer, Hermann – Dresden: Bleyl & Kaemmerer, 1897 – 1mf – 9 – 0-8370-7965-9 – (incl bibl ref) – mf#1986-1965 – us ATLA [221]

Grundlagen, stile, gestalten der deutschen literatur : eine geschichtliche darstellung / Hoffmann, G F & Roesch, Herbert – Frankfurt am Main: Hirschgraben, 1968 – 1r – 1 – (incl ind) – us Wisconsin U Libr [430]

Grundlagen und sinn der griechischen geschichte / Stier, Hans Erich – Stuttgart, Germany. 1945 – 1r – us UF Libraries [025]

Grundlagen und ziele der religioes-liberalen judentums / Norden, Joseph et al – Frankfurt a. M. 1918 (mf ed 1997) – (= ser Monographien zur wissenschaft des judentums) – 1mf – 9 – €24.00 – 3-8267-3215-4 – mf#DHS 10000 – gw Frankfurter [270]

Grundlegung der homiletik : in einigen vorlesungen über den wahren charakter eines protestantischen geistlichen / Marheineke, Philipp – Hamburg: F Perthes 1811 [mf ed 2005] – 1r – 1 – 0-524-10526-X – us ATLA [242]

Grundlegung zur wissenschaftlichen konstruktion des gesammten woerter- und formenschatzes : zunaechst der semitischen, versuchsweise und in grundzuegen auch der indo-germanischen sprachen / Drechsler, Christoph Moritz Bernhard Julius – Erlangen: Palm & Enke 1830 – (= ser Whsb) – 4mf – 9 – €50.00 – mf#Hu 014 – gw Fischer [410]

Grundlehre fur metallwerker, fachkunde, fachrechnen / Schumann, Willy – Frankfurt am Main, Germany. 1943 – 1r – us UF Libraries [500]

Grundlehren der religionsphilosophie / Drobisch, Moritz Wilhelm – Leipzig: L Voss, 1840 – 1mf – 9 – 0-524-02870-2 – (incl bibl ref) – mf#1990-3143 – us ATLA [200]

Die grundlehren des christenthums aus dem bewusstsein des glaubens / Schenkel, Daniel – Leipzig: F A Brockhaus 1877 [mf ed 1993] – 2mf – 9 – 0-524-08561-7 – mf#1993-2086 – us ATLA [240]

Grundler, J E] *see* A letter to the reverend mr geo lewis...

Ein grundlick bericht van der lere und dem geist des ertzketters david joris : uth synen schrifften und wercken flytich und getrouwelick vorvatet... / Emmius, U – n.p, 1597 – 2mf – 9 – mf#PBA-174 – ne IDC [240]

Grundling, Julius *see* Louis napoleon
Grundlinien christlicher irenik : aufruf und beitrag zum frieden unter den christlichen confessionen und nationen / Hasse, Hermann Gustav – Leipzig: J Lehmann, 1882 – 1mf – 9 – 0-7905-7756-9 – mf#1989-0981 – us ATLA [240]
Grundlinien der kirchengeschichte : in der form von dispositionen fuer seine vorlesungen / Loofs, Friedrich – Halle a. S: M. Niemeyer, 1910 – 2mf – 9 – 0-7905-5427-5 – mf#1988-1427 – us ATLA [240]
Grundlinien der systematischen theologie : zum gebrauche bei vorlesungen / Bachmann, Philipp – Leipzig:A. Deichert (Georg Boehme), 1908 – 1mf – 9 – 0-8370-2540-0 – mf#1985-0540 – us ATLA [240]
Grundlinien der theologie martin kaehlers / Zaenker, Otto – Guetersloh: C Bertelsmann, 1914 – (= ser Beitraege zur foerderung christlicher theologie) – 1mf – 9 – 0-7905-8984-2 – (incl bibl ref) – mf#1989-2209 – us ATLA [240]
Grundlinien des mosaisch-talmudischen eherechts / Frankel, Zacharias – Breslau, Germany. 1860 – 1r – us UF Libraries [939]
Grundlinien einer erkenntnistheorie der goetheschen weltanschauung mit besonderer ruecksicht auf schiller : zugleich eine zugabe zu goethes "naturwissenschaftlichen schriften" in kuerschners deutscher national-literatur / Steiner, Rudolf – erw. Aufl. Stuttgart: Der Kommende Tag, 1924 – 1r – 1 – us Wisconsin U Libr [430]
Grundlinien einer theorie des bewusstseins / Bergmann, Julius – Berlin: O Loewenstein, 1870 – 1mf – 9 – 0-7905-7323-7 – mf#1989-0548 – us ATLA [120]
Grundlinien zum religions-unterricht an den oberen klassen gelehrter schulen : nebst einem anhang, die augsburgische confession mit einleitung und erklaerung / Thomasius, Gottfried – 5. aufl. Nuernberg: August Recknagel, 1867 [mf ed 1986] – 1mf – 9 – 0-8370-7743-5 – (in german & latin) – mf#1986-1743 – us ATLA [377]
Grundlinier till foerelaesningar oefver augsburgiska bekaennelsen / Forsander, Nils – Rock Island, Ill: Lutheran Augustana Book Concern, [189-?] – 1mf – 9 – 0-524-07240-X – mf#1991-2981 – us ATLA [240]
Grundmann, Johannes *see* Die geographischen und voelkerkundlichen quellen und anschauungen in herders "ideen zur geschichte der menschheit"
Grund-richtiger, kurtz- leicht und noethiger unterricht der musicalischen kunst : wie man fueglich und in kurtzer zeit choral und figural singen... / ed by Speer, Daniel – Ulm: G W Kuehnen 1687 [mf ed 19–] – 1r – 1 – mf#film 1322 – us Sibley [780]
Grundriss der christlichen apologetik : zum gebrauche bei akademischen vorlesungen / Schultz, Hermann – 2. erw. Aufl. Goettingen: Vandenhoeck & Ruprecht, 1902 [mf ed 1985] – 1mf – 9 – 0-8370-5162-2 – (incl bibl ref) – mf#1985-3162 – us ATLA [240]
Grundriss der christlichen apologetik *see* Outlines of christian apologetics
Grundriss der christlichen dogmengeschichte / Nitzsch, Friedrich – Berlin: E.S. Mittler, 1870 – 1mf – 9 – 0-7905-5498-4 – (incl bibl ref) – mf#1988-1498 – us ATLA [240]
Grundriss der christlichen ethik / Lange, Johann Peter – Heidelberg: Carl Winter, 1878 – 1mf – 9 – 0-8370-2094-8 – (incl bibl ref) – mf#1985-0094 – us ATLA [230]
Grundriss der christlichen glaubens- und sittenlehre : als compendium fuer studirende und als leitfaden fuer den unterricht an hoeheren schulen / ed by Pfleiderer, Otto – 4. Aufl. Berlin:Georg Reimer, 1888 – 1mf – 9 – 0-8370-5450-8 – (incl bibl ref) – mf#1985-3450 – us ATLA [240]
Grundriss der deutschen literaturgeschichte – Berlin: Vereinigung Wissenschaftlicher Verleger. 2v. 1920-22 – 1 – (no more publ? includes bibliographic references and index) – us Wisconsin U Libr [430]
Grundriss der dogmengeschichte : die entstehung des dogmas und seine entwickelung im rahmen der morgenlaendischen kirche / Harnack, Adolf von – Freiburg i B: JCB Mohr, 1889 – 1mf – 9 – 0-524-05320-0 – mf#1990-1438 – us ATLA [240]
Grundriss der dogmengeschichte : entwickelungsgeschichte der christlichen lehrbildungen / Dorner, August – Berlin: G Reimer, 1899 – 2mf – 9 – 0-7905-9264-9 – (incl bibl ref and ind) – mf#1989-2489 – us ATLA [240]
Grundriss der dogmengeschichte / Seeberg, Reinhold – Leipzig: A Deichert, 1901 – 1mf – 9 – 0-7905-9632-6 – (incl bibl ref) – mf#1989-1357 – us ATLA [240]
Grundriss der einleitung in das neue testament / Langen, Joseph – 2. aufl. Bonn: Eduard Weber, 1873 – 1mf – 9 – 0-8370-6911-4 – (includes bibliographies) – mf#1986-0911 – us ATLA [225]
Grundriss der encyclopaedie der theologie / Dorner, Isaak August – Berlin: Georg Reimer, 1901 – 1mf – 9 – 0-8370-2953-8 – mf#1985-0953 – us ATLA [240]
Grundriss der evangelischen dogmatik : zum gebrauche bei akademischen vorlesungen / Schultz, Hermann – 2. erw. Aufl. Goettingen: Vandenhoek & Ruprecht, 1892 – 1mf – 9 – 0-8370-5616-0 – (incl bibl ref) – mf#1985-3616 – us ATLA [242]
Grundriss der evangelischen ethik : zum gebrauche bei akademischen vorlesungen / Schultz, Hermann – Goettingen: Vandenhoeck & Ruprecht, 1891 – 1mf – 9 – 0-8370-6367-1 – (incl bibl ref) – mf#1986-0367 – us ATLA [230]
Grundriss der geschichte der griechischen philosophie *see* Outlines of the history of greek philosophy
Grundriss der geschichte der klassischen philologie / Gudeman, Alfred – Leipzig, Germany. 1909 – 1r – us UF Libraries [450]
Grundriss der geschichte der philosophie *see* History of the ancient and mediaeval philosophy, vol 1
Grundriss der geschichte des neutestamentlichen kanons : eine ergaenzung zu der einleitung in das neue testament / Zahn, Theodor – 2. verm vielfach bearb aufl. Leipzig: A Deichert (Georg Boehme), 1904 – 1mf – 9 – 0-8370-6556-9 – (incl bibl ref) – mf#1986-0556 – us ATLA [225]
Grundriss der griechischen litteratur / Bernhardy, Gottfried – Halle: E Anton 1876-80 [mf ed 1992] – 2v in 3v on 1r – 1 – (v1: innere geschichte der griechischen litteratur [4th rev 1876]; v2/1: epos, elegie, iamben, melik [3rd rev 2nd impr 1877]; v2/2: dramatische poesie, alexandriner, byzantiner, fabel [3rd rev 2nd impr 1880]) – mf#film mas 22577.1 – us Harvard [450]
Grundriss der patrologie : mit besonderer beruecksichtigung der dogmengeschichte / Rauschen, Gerhard – 4. und 5., verm und verb Aufl. Freiburg i B: Herder, 1913 – 1mf – 9 – 0-524-02798-6 – mf#1990-0702 – us ATLA [240]
Grundriss der patrologie / Rauschen, G – Freiburg i. Br, 1906 – 4m – 9 – €11.00 – ne Slangenburg [240]
Grundriss der religionsphilosophie / Dorner, August – Leipzig: Duerr, 1903 – 2mf – 9 – 0-7905-8784-X – (incl bibl ref) – mf#1989-2009 – us ATLA [200]
Grundriss der ritschlschen dogmatik / Kuegelgen, Constantin von – 2. veraend aufl. Leipzig: Richard Woepke, 1903 [mf ed 1990] – 1mf – 9 – 0-7905-7884-0 – (incl bibl ref) – mf#1989-1109 – us ATLA [242]
Grundriss der symbolik fuer vorlesungen / Plitt, Gustav Leopold – Erlangen: A Deichert, 1875 – 2mf – 9 – 0-7905-9064-6 – mf#1989-2289 – us ATLA [240]
Grundriss der theologischen ethik / Kirn, Otto – 3. Aufl. Leipzig: A Deichert, 1912 – 1mf – 9 – 0-524-08452-1 – mf#1993-2057 – us ATLA [170]
Grundriss der vergleichenden grammatik der semitischen sprachen / Brockelmann, Carl – Berlin: Reuther & Reichard; New York: Lemcke & Buechner, 1908-1913 – 4mf – 9 – 0-8370-1875-7 – (incl bibl ref) – mf#1987-6262 – us ATLA [470]
Grundriss einer lautlehre der bantusprachen / Meinhof, Carl – Berlin, Germany. 1910 – 1r – us UF Libraries [470]
Grundriss einer schoenen stadt in absicht ihrer anlage und einrichtung zur bequemlichkeit, zum vergnuegen, zum anwachs und zur erhaltung ihrer einwohner nach bekannten mustern entworfen / Willebrand, J P – Hamburg, Leipzig, 1775-1776. 2v – 9mf – 9 – mf#OA-126 – ne IDC [720]
Grundriss einer systematischen theologie des judentums auf geschichtlicher grundlage / Kohler, Kaufmann – Leipzig: Gustav Fock, 1910 – (= ser Schriften (gesellschaft zur foerderung der wissenschaft des judentums (germany))) – 1mf – 9 – 0-7905-1770-1 – (incl bibl ref & ind) – mf#1987-1770 – us ATLA [270]
Grundriss zu vorlesungen ueber das deutsche privatrecht : mit einschluss des lehn- und handelsrechts nebst beigefuegten quellen / Kraut, Wilhelm Theodor; ed by Frensdorff, F Ferdinand – 6., verm u verb Aufl. Berlin: J Guttentag, 1886 – (= ser Civil law 3 coll) – 7mf – 9 – (incl bibl ref) – mf#LLMC 96-526 – us LLMC [346]
Die grundsaetze der musikalischen komposition / Sechter, Simon – Leipzig: Breitkopf & Haertel 1853-54 [mf ed 1991] – 3v on 1r – 1 – (contents: [1. abt]: die richtige folge der grundharmonien...; 2. abt: von den gesetzen des taktes in der musik. vom einstimmigen satze. die kunst zu einer gegebenen melodie die harmonie zu finden; 3. abt: vom drei- und zweistimmigen satze. rhythmische entwuerfe. vom strengen satze, mit kurzen andeutungen des freien satzes. vom doppelten contrapunkte; with bibl) – mf#pres. film 103 – us Sibley [780]
Grundsaetze des gemeinen deutschen privatrechts mit einschluss des handels-, wechsel- und seerechts / Mittermaier, Carl Joseph Anton – 3., umgearb u sehr verm Aufl. Landshut: P Kruell. 2v in 1. 1827 – (= ser Civil law 3 coll) – 10mf – 9 – (incl bibl ref) – mf#LLMC 96-532 – us LLMC [346]
Grundsaetze des generalbasses : als erste linien zur composition / Kirnberger, Johann Philipp – Berlin: J J Hummel [1781] [mf ed 19–] – 1r / 3mf – 1,9 – mf#film 1050 / fiche 503 – us Sibley [780]
Grundsaetze evangelisch-lutherischer kirchenverfassung / Hoefling, Johann Wilhelm Friedrich – 3., sehr verm. und verb. Aufl. Erlangen: T. Blaesing, 1853 – 1mf – 9 – 0-7905-6232-4 – mf#1988-2232 – us ATLA [242]
Grundsaetze reformierter kirchenverfassung / Rieker, Karl – Leipzig: C.L. Hirschfeld, 1899 – 1mf – 9 – 0-7905-6252-9 – (incl bibl ref and index) – mf#1988-2252 – us ATLA [242]
Das grundschulwesen in den provinzen des osmanischen reiches waehrend der herrschaftsperiode abduelhamids 2 (1876-1908) / Somel, Aksin – (mf ed 1993) – 5mf – 9 – €59.00 – 3-8267-2115-2 – mf#DHS 2115 – gw Frankfurter [956]
Der grundstein – Hamburg DE, 1914-17 – 1r – 1 – gw Misc Inst [074]
Der grundstein / ed by Leber, Georg – Frankfurt/M 1952 6 apr-19 okt [gaps], 1960, 1964-66, 1972 [mf ed 2004] – 3r – 1 – gw Mikrofilm [074]
Der grundstein 1888 bis 1933 : gewerkschaftszeitung des deutschen baugewerbes / ed by IG Bauen – Agrar – Umwelt – [mf ed 2003] – 355mf (1:24) – 9 – silver €1990.00 – ISBN-10: 3-598-35141-0 – ISBN-13: 978-3-598-35141-9 – (with guide bk; int by peter ruetters) – gw Saur [331]
Grundt, Friedrich Immanuel *see* Hebraeische elementargrammatik
Grundtlicher bericht vnnd auszzug... / Vadian, J – [Zuerich, Christoph Froschauer, 1542] – 1mf – 9 – mf#PBU-405 – ne IDC [240]
Der grundton in goethes lebensanschauung / Vogt, Paul – Weimar: H Boehlau 1932 [mf ed 1990] – 1r – 1 – (incl bibl ref. filmed with: faust als tragodie / benno von wiese) – mf#2678p – us Wisconsin U Libr [140]
Grundtvig, Nicolai Frederik Severin *see*
– Nik. fred. sev. grundtvigs udvalgte skrifter
– Skal den lutherske reformation virkelig fortsoettes?
Grundvorstellungen der amerikanischen wirtschafts-ethik : zur ideologie der "prosperity" / Koenigsgarten, Hugo F – Wien, 1934 (mf ed 1995) – 2mf – 9 – €31.00 – 3-8267-3160-3 – mf#DHS-AR 3160 – gw Frankfurter [330]
Die grundwahrheiten der christlichen religion : ein akademisches publikum in sechzehn vorlesungen... / Seeberg, Reinhold – 4. verb. Aufl. Leipzig: A Deichert (Georg Boehme), 1906 – 1mf – 9 – 0-8370-5209-2 – mf#1985-3209 – us ATLA [240]
Grundwald, Max *see* Ein gott!
Grundy county labor news – v1 n10-v2, v7 n3-v12 n8 [1947 feb 12-1948 apr 6, 1952 jul 15-1957 dec 10] – 1r – 1 – mf#1057590 – us WHS [331]
Grundy, John *see*
– Christianity
– Reciprocal duties of ministers and congregations
Grundy, John Clowes *see* Catalogue of...modern pictures and drawings
Grundy, Julia Margaret Kunkle *see* Ten days in the light of acca
Grundy, Robert Caldwell *see* A discussion of the mode and subjects of christian baptism
Die grundzuege der alttestamentlichen weisheit : ein beitrag zur theologie des alten testaments / Oehler, Gust. Fr – Tuebingen: L F Fues, [1854?] – 1mf – 9 – 0-7905-1772-8 – (incl bibl ref) – mf#1987-1772 – us ATLA [221]
Grundzuege der griechischen etymologie / Curtius, Georg – 5. umgearb Aufl. Leipzig: BG Teubner, 1879 – 2mf – 9 – 0-524-08308-8 – mf#1993-0013 – us ATLA [450]
Grundzuege der hebraeischen akzent- und vokallehre : mit einem anhange, ueber die form des namens jahwae / Grimme, Hubert – Freiburg (Schweiz): Commissionsverlag der Universitaetsbuchh, 1896 – (= ser Collectanea friburgensia) – 2mf – 9 – 0-8370-1725-4 – (incl bibl ref) – mf#1987-6121 – us ATLA [470]
Die grundzuege der israelitischen religionsgeschichte / Giesebrecht, Friedrich – Leipzig: B G Teubner, 1904 – (= ser Aus natur und geisteswelt) – 1mf – 9 – 0-524-02715-3 – (incl bibl ref) – mf#1990-3118 – us ATLA [270]
Die grundzuege der lehre von tempus und modus im griechischen / Aken, Adolf Friedrich – Rostock: Stiller, 1861 – 1mf – 9 – 0-8370-9200-0 – (incl indes) – mf#1986-3200 – us ATLA [450]
Grundzuege der literaturgeschichte / Lechner, Hermann – Erw. Aufl. Innsbruck: Tyrolia-Verlag, 1947 – 1r – 1 – (incl ind) – us Wisconsin U Libr [430]
Grundzuege der lyrik goethes / Achelis, Thomas – Bielefeld: Velhagen & Klasing, 1900 [mf ed 1993] – (= ser Velhagen and klasings sammlung deutscher schulausgaben 81) – iv/120p – 1 – (incl bibl ref and ind) – mf#8591 – us Wisconsin U Libr [430]
Grundzuege der neutestamentlichen graecitaet : nach den besten quellen fuer studierende der theologie und philologie / Schirlitz, Samuel Christoph – Giessen: Ferber, 1861 – 1mf – 9 – 0-8370-9177-2 – (incl indes) – mf#1986-3177 – us ATLA [450]
Grundzuege der physiologischen optik / Aubert, Hermann – Leipzig, 1876 – 1 – gw Mikropress [612]
Grundzuege der religionswissenschaft : eine kurzgefasste einfuehrung in das studium der religion und ihrer geschichte / Tiele, Cornelius Petrus – Tuebingen: J C B Mohr, 1904 – 1mf – 9 – 0-8370-5534-2 – mf#1985-3534 – us ATLA [200]
Grundzuege der schriftsprache luthers in allgemeinverstaendlicher darstellung / Franke, Carl – 2nd rev enl ed. Halle (Saale): Verlag der Buchhandlung des Waisenhauses, 1913-1922 [mf ed 1993] – 3v – 1 – (incl bibl ref and ind) – mf#8195 – us Wisconsin U Libr [430]
Grundzuege des rhythmus, des vers- und strophenbaues in der hebraeischen poesie : nebst analyse einer auswahl von psalmen und anderen strophischen dichtungen der verschiedenen vers- und strophenarten, mit vorangehendem abriss der metrik der hebraeischen poesie / Ley, Julius – Halle: Buchh des Waisenhauses, 1875 – 3mf – 9 – 0-524-06741-4 – mf#1992-0944 – us ATLA [470]
Grundzuege einer kontrastiven valenzgrammatik fuer den fremdsprachenunterricht : deskriptive, sprachtypologische und curriculare aspekte am beispiel des deutschen, koreanischen und chinesischen / Schmidt, Wolfgang G A – Bochum, 1990 (mf ed 1997) – 4mf – 9 – €56.00 – 3-8267-2392-9 – mf#DHS 2392 – gw Frankfurter [370]
Grundzuege und chrestomathie der papyruskunde / Mitteis, L – Berlin: B G Teubner, 1912 – 5mf – 9 – 0-7905-3042-2 – (incl bibl ref) – mf#1987-3042 – us ATLA [450]
Grundzuge einer vergleichenden grammatik der bantusprachen / Meinhof, Carl – Hamburg, Germany. 1948 – 1r – us UF Libraries [470]
Grunebaum-Ballin, Paul Frederic Jean *see* Henri gregoire, lami des hommes de toutes les couleurs
Gruner, Christian Gottfried *see*
– Almanach fuer aerzte und nichtaerzte
– Neues taschenbuch fuer aerzte und nichtaerzte
Gruner, Hans *see* Geldtheorie und wirtschaftswachstum
Gruner, Ludwig *see*
– Descriptions of the plates of fresco decorations
– Fresco decorations and stuccoes of churches and palaces, in italy
– The green vaults dresden illustrations of the choicest works in that museum of art
– The terra-cotta architecture of north italy (12th – 15th centuries)
Gruner, Wilhelm Heinrich Ludwig *see* The decorations of the garden-pavilion in the grounds of buckingham palace
Grunewald : reminiscences d'allemagne – Quebec?: C Darveau, 1878 – 1mf – 9 – mf#09021 – cn Canadiana [914]
Grunewald-echo – Berlin DE, 1901 2 jul-1931, 1933-37, 1939 – 14r – 1 – (filmed with suppls) – gw Misc Inst [074]
Grunfeld, Max *see* Leben und lieben im ghetto
Grunt-shtrikhn fun der yidisher filozofye / Finkelstein, Leo – Varshe, Poland. 1937 – 1r – us UF Libraries [939]
Grunt-shtrikhn fun yidishn realizm / Oislender, Nokhum – Wilno, Lithuania. 1928 – 1r – us UF Libraries [939]
Grunwald, G *see* Geschichte der gottesbeweise im mittelalter bis zum ausgang der hochscholastik
Grunzel, Josef *see* Die wirtschaftliche konzentration
Grupe-Loercher, Erica *see* Der weg ueber den vulkan
Grupo de dolmenes en termino de barcarrota, provincia de badajoz / Melida, Jose Ramon – Etnografia y Prehistoria. T.III. Madrid, Museo Antropologico. 1924 – 1 – sp Bibl Santa Ana [946]
Grupo Saker-Ti *see*
– Cuentos de guatemala, 1952
– Doce poemas
– Siete afirmaciones
Grupo sindical de colonization num. 7959 *see* Ordenanzas de regimen interior
Grupp, Georg *see* Kulturgeschichte der roemischen kaiserzeit

"Gruppa osvobozhdenie truda" – Moscow. n1-6. 1924-1928 – 1 – us NY Public [947]
Gruppa zhurnalistov *see* Kievskie novosti
Die gruppe 47 : bericht, kritik, polemik: ein handbuch / ed by Lettau, Reinhard – Neuwied, Berlin: Luchterhand, c1967 [mf ed 1993] – 565p – 1 – (incl bibl ref and ind) – mf#8256 – us Wisconsin U Libr [430]
Gruppe bosemueller : roman / Beumelburg, Werner – Oldenburg: G Stalling, c1930 [mf ed 1989] – 332p – 1 – mf#7017 – us Wisconsin U Libr [830]
Gruppe Internationale Marxisten (GIM) *see* Was tun
Gruppe, O F *see* Gegenwart und zukunft der philosophie in deutschland
Gruppe, Otto *see*
– Griechische mythologie und religionsgeschichte
– Die griechischen culte und mythen in ihren beziehungen zu den orientalischen religionen
– Die rhapsodische theogonie und ihre bedeutung innerhalb der orphischen litteratur
Gruppe, Otto Friedrich *see* Die kosmischen systeme der griechen
Grusinische grammatik *see* Vergleichungstafeln der europaeischen stamm-sprachen und sued-, west-asiatischer
Grussendorf, Hermann *see* Der monolog im drama des sturm und drang
Grutter, Virginia *see* Dame la mano
Grutze, Albert Lewis *see* A collection of wills
Grutzner, Sally J *see* The effects of galvanic current and ice on muscle temperature
Gruver, BM *see* The social construction of leisure for women in academe
Gruziia : ezhedn polit i lit gaz / ed by Dzhaparidze, L S – Tiflis: E D Gordeladze 1919-21 [1918 3 sent-1921 25 fevr] – (= ser Asn 1-3) – n2 [1919]-n38 [1921] [gaps] item 106, on reel n21,22 – 1 – mf#asn-1 106 – ne IDC [077]
Gruzinskaia ssr za 20 let (statisticheskii sbornik) – Tbilisi, 1941 – 2mf – 9 – mf#RHS-18 – ne IDC [314]
Gruzinskii, A S *see* Iz istorii perevoda evangeliia v iuzhnoi rossii v 16 veke
Gry, Leon *see*
– Le millenarisme dans ses origines et son developpement
– Les paraboles d'henoch et leur messianisme
[Grynaeus, S] *see* Novvs orbis regionvm ac insvlarvm veteribvs incognitarvm...
Gryphius, Andreas *see*
– Andreas gryphius lateinische und deutsche jugenddichtungen
– Andreas gryphius lyrische gedichte
– Catharina von georgien
– Horribilicribrifax
– Olivetum
– Peter squenz
– Sonn- und feiertages-sonette
– Sonnette
Gryphon – Florennes Air Base. 1986 jan 10-1989 jan 6 – 1r – 1 – mf#1570862 – us WHS [355]
Gryphon guardian – Wuescheim. 1986 oct 1-1988 nov 10 – 1r – 1 – (continued by: wueschheim warrior) – mf#1711080 – us WHS [071]
Grythytte tidning – Nora, Sweden. 1885-1907 – 1 – sw Kungliga [078]
Grzimek, Bernhard *see* Serengeti shall not die
Gsangbuechlin, darinn der gantze psalter davids : sampt andern gaistlichen gesangen, mit jren melodeyen begriffen – korr aufl. Augspurg: Ulhart 1557 – (= ser Hqab. literatur des 16. jahrh.) – 5mf – 9 – €60.00 – mf#1557a – gw Fischer [780]
Gschwind, Karl *see* Die niederfahrt christi in die unterwelt
Gsell, Stephane *see* Histoire ancienne de l'afrique du nord
Gsn [gay studies newsletter] – Toronto. v1-16 n1. apr 1974-mar 1989// – 1r – 1 – Can$95.00 – (superseded by: lesbian and gay studies newsletter) – cn McLaren [305]
Gstettner, Hans *see* Die getrennten
Gsu today / Grambling State University – 1990 winter – 1r – 1 – mf#4877103 – us WHS [378]
Gt – Goeteborg, Sweden. 1995- – 9 – sw Kungliga [078]
Gte automatic electric : a world-wide communications journal – Phoenix AZ 1977-78 – 1,5,9 – (cont: gte automatic electric technical journal; cont by: gte automatic electric: world-wide communications journal) – ISSN: 0147-3328 – mf#2526.03 – us UMI ProQuest [380]
Gte automatic electric technical journal – Phoenix AZ 1948-76 – 1,5,9 – (cont by: gte automatic electric: a world-wide communications journal) – ISSN: 0147-331X – mf#2526.03 – us UMI ProQuest [380]
Gte automatic electric world-wide communications journal – Phoenix AZ 1980-82 – 1,5,9 – (cont: gte automatic electric: a world-wide communications journal; cont by: gte network systems world-wide communications journal) – ISSN: 0273-141X – mf#2526.03 – us UMI ProQuest [380]
Gu kamm gu gap : pralom lok tam manosancetana / Kuy Yak Hu – Bhnam Ben: Ron Bumb Syn Hen 2508 [1965] [mf ed 1989] – 1r with other items – 1 – mf#mf-10289 seam reel 095/12 [§] – us CRL [959]
Gu kri rhan : [a novel] / Mra Cakra, Cha ra kri – Ran Kun: Tan Mon Kri Ca up tuik [195-?] [mf ed 1990] – 1r with other items – 1 – (in burmese) – mf#mf-10289 seam reel 198/6 [§] – us CRL [830]
Gu ta gu thai khanse nhac kon : [a novel] / Sa Ra, Mon – Ran Kun: Rasa Ca pe 1976 [mf ed 1990] – 1r with other items – 1 – (in burmese) – mf#mf-10289 seam reel 172/7 [§] – us CRL [830]
Guac Gi Len *see* Vijja bal ratth
Guadalajara, 8-23 marzo, 1937 : con i progioniere italiani, dopo la vittoria dell'esertio popolare spagnolo / Tedeschi, Paolo – Parigi: Ed de Coltura Sociale 1937 [mf ed 1977] – (= ser Blodgett coll) – 1mf – 9 – mf#w1222 – us Harvard [946]
Guadalupe – Caceres, 1906-1909 – 5 – sp Bibl Santa Ana [073]
Guadalupe – Caceres, 1914 y 1915 – 5 – sp Bibl Santa Ana [073]
Guadalupe, arte, devocion y... / Alvarez, Arturo – Madrid: Archivo Ibero Americano, 1965 – 1 – sp Bibl Santa Ana [240]
Guadalupe (caceres) / Junta Provincial de Turismo – Vitoria: Tip. Fournier, s.a. – 1 – (fotos mas) – sp Bibl Santa Ana [338]
Guadalupe de extremadura en indias / Bayle, Constantino – Madrid: Razon y Fe, 1928 – 9 – sp Bibl Santa Ana [972]
Guadalupe en la america andina. madrid, 1969 / Meseguer Fernandez, Juan & Alvarez, Arturo – Madrid: Graf. Calleja, 1969 – 1 – sp Bibl Santa Ana [972]
Guadalupe en los siglos 17; 18 / Jimenez Priego, Teresa – Badajoz: Dip. Provincial, 1975. Sep. REE – 1 – sp Bibl Santa Ana [946]
Guadalupe, impresiones artistico-religiosas / Pedrajas y Nunez-Romero, Eloy – Badajoz: Uceda Hermanos, 1902 – 1 – sp Bibl Santa Ana [240]
Guadeloupe / Champon, E – Paris, France. 1902 – 1r – us UF Libraries [972]
Guadeloupe : etude geographie / Lasserre, Guy – Bordeux, France. v1-2. 1961 – 1r – us UF Libraries [972]
Guadeloupe / France Service De Coordination De L'enseignement – Paris, France. 1946 – 1r – us UF Libraries [972]
Guadeloupe... / Lara, Oruna – Paris, France. 1921 – 1r – us UF Libraries [972]
Guadeloupe Commission Locale Du Plan Sur Le Bilan *see* Rapport general de la commission locale du plan su...
Guadeloupe et ses iles / Jorond, Antoine Victor – Basse-Terre, Guadeloupe. 1965 – 1r – us UF Libraries [972]
Guadeloupe, guyane, martinique, saint-pierre et mi... / Paris Exposition Coloniale Internationale, 1931 – Paris, France. 1931 – 1r – us UF Libraries [972]
Guadelupe, Andres *see*
– Commentaria in hossean prophetam
– Historia...de los angeles
– Mystica theologia
Guadiana – Badajoz, 1946 – 5 – sp Bibl Santa Ana [073]
Guadiana. seminario de actualidades extremenas – Badajoz, 1946 (1 a 11) – sp Bibl Santa Ana [946]
Guajira ante el congreso de colombia / Vives Echeverria, Jose Ignacio – Bogota, Colombia. 1965 – 1r – us UF Libraries [972]
[Gualala-] independent coast observer – CA. 1969 – 19r – 1 – $1140.00 (subs $90y) – mf#B03233 – us Library Micro [071]
Gualandi, M *see* Nuova raccolta di lettere sulla pittura, scultura ed architettura
Gualteri burlaei liber de vita et moribus philosophorum / ed by Knust, Hermann – Stuttgart: Litterarischer Verein, 1886 (Tuebingen: H Laupp) – us Wisconsin U Libr [450]
Gualteri burlaei liber de vita et moribus philosophorum / ed by Knust, Hermann – Stuttgart: Litterarischer Verein, 1886 (Tuebingen: H Laupp) [mf ed 1993] – (= ser Blvs 177) – 441p – 1 – mf#8470 reel 37 – us Wisconsin U Libr [110]
Gualtieri, F *see* Corona per la vittoria del sereniss don gio d'avstria
Gualtieri, G *see* Relationi della venvta de gli ambasciatori giaponesi...roma, sino alla partita di lisbona
Guam : proceedings of the constitutional convention, 1977 – Agana: The Territory of Guam, Mar 1979 – 16mf – 9 – $24.00 – mf#LLMC 82-100B Title 35 – us LLMC [323]
Guam : proceedings of the first constitutional convention, 1969-1970 – Agana: Garrison & McCarter Inc, n.d. – 28mf – 9 – $42.00 – mf#LLMC 82-100B Title 32 – us LLMC [342]
Guam : session laws of american states and territories – 1975-96 – 9 – $203.00 set – mf#402610 – us Hein [348]
Guam *see* Guam news letter
Guam administrative rules, 1991 / Guam. Office of the Attorney General, Division of the Compiler of Laws – 1v titles 1-6. Agana: Govt of Guam, 1991 – 8mf – 9 – $12.00 – (updates planned) – mf#LLMC 82-100B Title 36 – us LLMC [324]
Guam. (Commonwealth) *see*
– Administrative rules and regulations of the government of guam, 1975
– Annual reports of the governor of guam, 1938-1981
– Civil regulations with the force and effect of law in guam, 1947
– The guam recorder, old series
Guam. (Commonwealth). 1st Legislature *see* Statutes and amendments to the codes of the territory of guam, 1951-1952
Guam. (Commonwealth). 5th Legislature *see* The government code of guam, 1961
Guam. (Commonwealth). 10th Legislature *see* The civil code of the territory of guam, 1970
Guam. (Commonwealth). Laws, Statutes, etc *see*
– The civil and penal codes of the territory of guam, 1953
– The civil code of guam, 1947
– The code of civil procedure and probate code of guam, 1953
– The code of civil procedure and probate code of guam, 1970
– The code of civil procedure of guam, 1947
– The government code of guam, 1970
– The penal code of guam, 1947
– The penal code of the territory of guam, 1970
– The probate code of guam, 1947
Guam. (Commonwealth). Legislative Counsel *see* The government code of guam, 1952
Guam Federation of Teachers, Local 1581 *see* Union
Guam Govt *see* Guam naval government brief extracts, 1905
Guam Law Revision Commission *see* Guam organic act and related federal laws thru june 5, 1979
Guam Legislature *see* Session laws of guam, 1975/76-1991/92
Guam legislature session laws, executive orders, resolutions – v1-12. 1975-96 – 9 – $203.00 set – mf#306781 – us Hein [348]
Guam naval government brief extracts, 1905 : brief extracts relative to the island of guam from publications, memoranda furnished to congress, general orders and annual reports for 1901-1904 / Guam Govt – Washington: GPO, 1905 – 2mf – 9 – $3.00 – mf#LLMC 82-100B Title 31 – us LLMC [324]
Guam news letter / Guam – Agana. 1909-1922 – 1 – us NY Public [079]
Guam. Office of the Attorney General *see* Guam organic act and related federal laws thru august 31, 1984
Guam. Office of the Attorney General, Division of the Compiler of Laws *see* Guam administrative rules, 1991
Guam organic act and related federal laws thru august 31, 1984 / Guam. Office of the Attorney General – Agana: Office of the Attorney General, Div of Compiler of Laws, 1984 – 1mf – 9 – $1.50 – mf#LLMC 82-100B Title 25 – us LLMC [340]
Guam organic act and related federal laws thru june 5, 1979 / Guam Law Revision Commission – Agana, 1979 – 1mf – 9 – $1.50 – mf#LLMC 82-100B Title 27 – us LLMC [323]
The guam recorder – Agana. v. 1-17, no. 2. 1924-May 1940 – 1 – us NY Public [079]
The guam recorder, new series – Univ of Guam, Micronesia Research Centre. v1-9. oct/dec 1971-79 (all publ?) – 16mf – 9 – $24.00 – mf#LLMC 82-100B Title 22 – us LLMC [972]
The guam recorder, old series / Guam. (Commonwealth) – v12-18. mar 1924-nov 1941 – 48mf – 9 – $72.00 – (incl ind) – mf#LLMC 82-100B Title 21 – us LLMC [972]
Guam. US Congress *see*
– Alien labor program in guam
– Amendment to the organic act of guam
– Non-voting delegates, guam and the virgin islands
– Proposed constitution for guam
– Providing for the establishment of a constitution for guam
Guam-federal digest, 1950-1987 / Compiler of Laws, Office of the Attorney General – Agana n.d. – 4mf – 9 – $6.00 – (digest of all publ guam cases reported in the supreme court reporter, federal reporter 2d and the federal supplement; pref dated jun 1988) – mf#llmc82-100b, title 37 – us LLMC [348]
Guan, Juchuang *see* Fan yi lei bian
Guanabara, Alcindo *see* Presidencia campos salles
Guanacaste / Costa Rica Secretaria De Gobernacion – San Jose, Costa Rica. 1924 – 1r – us UF Libraries [972]
Guanajuato. Mexico (State) *see* Periodico oficial
Guandique, Jose Salvador *see* Roberto edmundo canessa
Guangming ribao – Peking. Mar 1953; mar 1957-59; mar 1961-65 – 13 3/4r – 1 – (in chinese) – uk British Libr Newspaper [072]
Guantanamo Bay Naval Base [Cuba] *see* Daily gazette
Guantanamo gazette – Guantanamo Bay Naval Base [Cuba]. v43 n148-151, 154-155, 158-160, 163-166, 168-171, 173-180 [1987 aug 3-6,11-12,17-19,24-27, 31-sep 3,8,17], v44 n1-19,26-36,38-40,42-77,79-83,85-124,126-230,232-237 [1988 jan 4-feb 1,10-25,29-mar 2,4-apr 28, may 2-6,10, jul 6,8-dec 13,15-30] – 1r – 1 – (cont: daily gazette [guantanamo bay, cuba]; cont by: guantanamo bay gazette) – mf#1726486 – us WHS [071]
Guanuma : novela historica / Garcia Godoy, Federico – Santo Domingo, Dominican Republic. 1914 – 1r – us UF Libraries [972]
O guanumby : dedicado ao bello-sexo cermense – Carmo, RJ. 29 dez 1908-16 mar 1909; 06 abr 1909 – (= ser Ps 19) – mf#DIPER – bl Biblioteca [073]
O guaracyaba : jornal litterario e instructivo – Rio de Janeiro, RJ. set 1850-abr 1851; out 1853-29 jan 1854 – (= ser Ps 19) – mf#P01B,05,14-15 – bl Biblioteca [073]
Guaranteed 4% insurance bonds / Federal Life Assurance Co – [s.l: s.n. 189-?] [mf ed 1987] – 1mf – 9 – 0-665-41048-4 – mf#41048 – cn Canadiana [368]
O guarany – Diamantina, MG: Typ do Monitor do Norte, 31 jan, 02 mar 1878 – (= ser Ps 19) – mf#P17,02,106 – bl Biblioteca [079]
O guarany : jornal politico, litterario e industrial – Rio de Janeiro, RJ: Typ de Paula Brito, 06 ago-08 set 1853 – (= ser Ps 19) – mf#P15,01,50 – bl Biblioteca [079]
O guarany : periodico critico e joco-serio – Maceio, AL: Typ Uniao, 22 set, 25 nov 1879 – (= ser Ps 19) – mf#P18B,01,09 – bl Biblioteca [079]
Guard (eugene, or) – Eugene City OR: J B Alexander, -1870 [wkly] – 1 – (began in 1867; cont by: eugene city guard) – us Oregon Lib [071]
Guard news : official publication of international union, united plant guard workers of america / United Plant Guard Workers of America – 1950-56, 1957-1965 summer, 1966 jan-1970 dec, 1971 jan/feb-1990 mar/may – 4r – 1 – (continued by: security link) – mf#1057597 – us WHS [331]
O guarda nacional : alerta, alerta – Rio de Janeiro, RJ: Typ de L A F Menezes, 13 fev-03 ago 1849 – (= ser Ps 19) – mf#P14,02,35 – bl Biblioteca [320]
O guarda nacional mineiro – Ouro Preto, MG: Typ Patricia do Universal, 01 jan 1838-out 1839; jan-16 ago 1840 – (= ser Ps 19) – mf#P19B,01,02 – bl Biblioteca [320]
Guardia – 1969 aug-1975 aug, 1978 sep-1982 – 2r – 1 – (continued by: nueva guardia) – mf#27700 – us WHS [071]
Guardia, Elpidio De La *see* Historia de guanabacoa
Guardia Quiros, Victor *see* Escarceos literarios
Guardian – Cape Town: Stewart Printing Co. jun 18-dec 1937 (1r); MF-6505 CAMP (12r) 1938-may 22 1952 – us CRL [072]
Guardian – Upper Heyford, England. 1985 oct 4-1986, 1987-1988 feb, 1988 mar-1989 jun, 1989 jul 7-1990 sep 28 – 4r – 1 – (cont: sky king) – mf#1330670 – us WHS [071]
Guardian – Boston MA. 1902 jun 13-1903 aug 8 – 1r – 1 – (continued by: boston guardian) – mf#672841 – us WHS [071]
Guardian – Cowra. mar 1896-dec 1907; mar 1933-dec 1948; nov 1950-feb 1959 (misc iss) – 7r – 1 – A$479.47 vesicular A$517.97 silver – at Pascoe [079]
Guardian : [daily and morning] – Charlottetown, PEI. 1890-1903 – 24r – 1 – ISSN: 0830-2678 – cn Library Assoc [071]
Guardian – Dublin, Ireland. 31 oct 1846-30 jan 1847 – 1r – 1 – uk British Libr Newspaper [072]
Guardian – Fort Polk LA. 1983 jul 8-1984 sep 28, 1985 nov-1986 aug, 1986 sep-1987 aug, 1987 sep-1988 may, 1988 jun 10-1989 jan, 1989 feb-jun, 1989 jul-oct, 1989 nov-1990 feb – 8r – 1 – mf#1000547 – us WHS [071]
Guardian / Institute for Independent Social Journalism, Inc [New York, NY] – New York NY. 1968 feb 10-1969 jun 14, 1969 jun 21-1971 feb 13, 1970 jun 13, 1971 feb 27-dec 15, 1972 jan-1973 jun, 1973 jul-1974 oct – 5r – 1 – (cont: national guardian) – mf#202634 – us WHS [071]
Guardian – La Salle IL 1807-08 – 1 – mf#3579 – us UMI ProQuest [071]
Guardian / Neturei Karta of the USA – n1-2 [1974 apr-jul 30] – 1r – 1 – (continued by: jewish guardian (brooklyn (ny))) – mf#202634 – us WHS [071]
Guardian – nov 1969-nov 1973; jul 1983-dec 1986 – 9r – at Pascoe [079]
Guardian – Rangoon, Burma. 1973-Aug 1988 – 50r – 1 – us L of C Photodup [079]
Guardian – Rangoon Burma, oct 1973-28 oct 1974 – 2 1/2r – 1 – uk British Libr Newspaper [079]

Guardian – Sydney, 1844 – 1r – 9 – A$27.50 vesicular A$33.00 silver – at Pascoe [079]
Guardian – Cape Town SA, 1937-62 – 20r – 1 – (title varies: clarion, 1952; people's world, 1952; advance, 1954-62; new age, 1954-62) – sa National [079]
Guardian – v1 n1-v7 n1 [1978 apr-1987 sep] – 1r – 1 – mf#1221277 – us WHS [071]
Guardian – Wexford, Ireland. 1986-89; 11 jan 1990-1993 – 24 1/2r – 1 – uk British Libr Newspaper [072]
The guardian – 1846-1951 (mf ed 1999) – Lambeth Palace Library – uk World [072]
The guardian – 1864 – 1r – 1 – sa National [079]
The guardian – Dublin. Ireland. -w. 31 Oct 1846-30 Jan 1847. (10 ft) – 1 – uk British Libr Newspaper [072]
The guardian – London. -w. 1852; 1860-65. (13 reels) – 1 – uk British Libr Newspaper [072]
The guardian – Madras. India. v. 21, no. 17-36. 29 Apr-9 Sep 1943 – 1 – us NY Public [079]
The guardian – Manchester: Manchester Guardian and Evening News Ltd, aug 1959-dec 1972 – us CRL [072]
The guardian – New Castle, PA. -w 1889-1901 – 13 – $25.00r – us IMR [071]
The guardian – Palmerston North, NZ. jan 1974-dec 1986 – 34r – 1 – mf#45.8 – nz Nat Libr [079]
The guardian – Rangoon: The Guardian Press, 1959-apr 1973; jan-apr 1975; jan-aug 1977 – 1 – us CRL [079]
The guardian – Wexford. Ireland. -w. 20 Nov 1847-30 Dec 1848. (37 ft) – 1 – uk British Libr Newspaper [072]
Guardian [addison, steele, and others] – Killen TX 1713 – 1 – mf#3201 – us UMI ProQuest [071]
Guardian and constitutional advocate – Belfast Ireland, 19 jun 1827-25 mar 1836 – 8r – 1 – uk British Libr Newspaper [072]
The guardian and constitutional advocate – Belfast. Ireland. -sw. 19 Jun 1827-25 Mar 1936. (8 reels) – 1 – uk British Libr Newspaper [072]
Guardian and gazette extra *see* Waltham forest guardian and independent extra
Guardian and monitor – La Salle 1819-28 – 1 – mf#3815 – us UMI ProQuest [071]
Guardian and times *see* Mortlake and barnes guardian
Guardian and tipperary (north riding) and ormond advertiser *see* Nenagh guardian
Guardian (ealing ed) *see* Ealing and chiswick guardian
The guardian (ellesmere) *see* Malvern record
Guardian extra *see* Waltham forest guardian and independent extra
Guardian gazette – Macksville, 1958-66 – at Pascoe [079]
Guardian index 1842-1928 – [mf ed Marlborough 1994] – 3pt – 1 – (pt1: 1842-80 [18r] £1700; pt2: 1881-1904 [23r] £2150; pt3: 1905-28 [24r] £2250; with d/g) – uk Matthew [072]
Guardian index, 1929-1972 – [mf ed Marlborough 1993] – 5pt – 9 – (pt1: 1929-35 [474mf] £2950; pt2: 1936-45 [479mf] £2950; pt3: 1946-55 [474mf] £2950; pt4: 1956-62 [429mf] £2700; pt5: 1963-72 [513mf] £3200; with d/g) – uk Matthew [072]
Guardian index, 1973-1985 – 2pt – 9 – (pt1: 1973-78 [184mf] £1250; pt2: 1979-85 [153mf] £1050; with d/g) – uk Matthew [072]
Guardian journal : a daily paper – Nottingham, Nottinghamshire, 7 sep 1953-19 jun 1973 – 1 – (formed by merger of: nottingham guardian and: nottingham journal; publ in a variety of ed & had a nottinghamshire guardian weekly suppl) – uk Newsplan [072]
Guardian (kingston borough ed) *see* Kingston borough guardian
Guardian midweek extra (waltham forest and redbridge ed) *see* Waltham forest guardian and independent extra
Guardian (nenagh ed) *see* Nenagh guardian
Guardian of truth = Wei li pao – v7-9. 1956-58* – (= ser Chinese christian coll 62) – 1r – 1 – (in chinese) – mf#ATLA S0296R – us ATLA [240]
Guardian pictoral – Marrickville, apr 1970-jan 1971 (misc iss); jan 1972-dec 1979 – 5r – A$237.76 vesicular A$267.76 silver – at Pascoe [079]
Guardian pictorial – Marrickville – 5r – A$305.54 vesicular A$333.04 silver – at Pascoe [079]
Guardian (surrey and kingston ed) *see* Kingston borough guardian
Guardian today : the voice of the malawi congress party – Lilongwe: The Party, apr 7/13-jun 11/15, jun 25/30, jul 5, 9, 16, 1993; jul 30/aug 5 1993-jan 28/feb 3 1994; feb 11/17-mar 25/31, apr 8/14-apr 29/may 5, may 13/19 1994 – 1r – 1 – us CRL [079]
Guardian (wanstead loughton and buckhurst hill ed) *see* Woodford loughton and buckhurst hill guardian
Guardiola, Esteban *see*
– Historia de la universidad de honduras
– Impugnacion al folleto que, con el titulo de...
Guards / International Guards Union of America – v1 n3-v4 n2 [1952 jan-1955 feb] – 1r – 1 – mf#1057600 – us WHS [331]
Guardsman – Washington DC 1978 – 1,5,9 – (cont by: national guard) – ISSN: 0163-3953 – mf#11942.01 – us UMI ProQuest [355]
Guarena. Ayuntamiento *see* Ordenanzas municipales
Guarini, M A *see* Compendio historico...di ferrara...
Guarnello, A *see* Canzone nella felicissima vittoria christiana contra infideli al sereniss d gio d'avstria
Guaro y champana / Lindo, Hugo – San Salvador, El Salvador. 1961 – 1r – us UF Libraries [972]
O guasca da corte : periodico jocoso, politico e imparcial – Rio de Janeiro, RJ: Typ Brasiliense, 01 maio-19 ago 1851 – (= ser Ps 19) – mf#P10,05,17 – bl Biblioteca [320]
Guasima : cuadros jibaros / Fonfrias, Ernesto Juan – San Juan, Puerto Rico. 1957 – 1r – us UF Libraries [972]
Guasp, Ignacio *see*
– Cronicas y 105 sentencias leves
– Tradicion
Guastaferri, Fabritio *see* Lettera di fabritio guastaferri al sig. gio. francesco salitti..
Guatamala *see* Tarifa para el cobro de los derechos de importacio
Guatemala : ancient and modern / Munoz, Joaquin – New York, NY. 1940 – 1r – us UF Libraries [972]
Guatemala : la democracia y el imperio / Arevalo, Juan Jose – Buenos Aires, Argentina. 1964 – 1r – us UF Libraries [972]
Guatemala : la democracia y el imperio / Arevalo, Juan Jose – Mexico City? Mexico. 1954 – 1r – us UF Libraries [972]
Guatemala / Fergusson, Erna – New York, NY. 1937 – 1r – us UF Libraries [972]
Guatemala : from where the rainbow takes its colors / Munoz, Joaquin – Guatemala City, 1952, c1940 – 1r – us UF Libraries [972]
Guatemala : genio y figura / Guillen, Pedro – Guatemala, 1954 – 1r – us UF Libraries [972]
Guatemala : a historical survey / Jensen, Amy Elizabeth – New York, NY. 1955 – 1r – us UF Libraries [972]
Guatemala : the land of the quetzal / Brigham, William Tufts – Gainesville, FL. 1965 – 1r – us UF Libraries [972]
Guatemala : monografia sociologica / Montefort Toledo, Mario – Mexico City? Mexico. 1959 – 1r – us UF Libraries [972]
Guatemala : monumentos historicos y arqueologicos / Rubin De La Borbolla, Daniel Fernando – Mexico City? Mexico. 1953 – 1r – us UF Libraries [972]
Guatemala : past and present / Jones, Chester Lloyd – Minneapolis, MN. 1940 – 1r – us UF Libraries [972]
Guatemala / Rosenthal, Mario – New York, NY. 1962 – 1r – us UF Libraries [972]
Guatemala / Sapia Martino, Raul – Guatemala, 1965 – 1r – us UF Libraries [972]
Guatemala : volcanic but peaceful / United States Office Of Inter-American Affairs – Washington, DC. 1943 – 1r – us UF Libraries [972]
Guatemala *see*
– Diario de centro america. guatemala
– El guatemalteco diario de centro america
– Legislacion indigenista de guatemala
– Legislacion revolucionaria
– Leyes mas importantes de hacienda y economia, clas...
– Leyes vigentes
– Leyes y reglamentos de hacienda 1926
– Memoria de labores de un ano de gobierno, 30 de ma...
– Ordenanza de la policia nacional
– Ordenanza militar para el regimen
– Recopilacion de leyes agrarias
Guatemala and the states of central america / Domville-Fife, Charles William – London, England. 1913 – 1r – us UF Libraries [972]
Guatemala ante america / Guatemala Secretaria De Relaciones Exteriores – Guatemala, 1951 – 1r – us UF Libraries [972]
Guatemala Asamblea Constituvente (1945) *see* Diario de sesiones
Guatemala city / Pan American Union – Washington, DC. 1949 – 1r – us UF Libraries [972]
Guatemala (City) Cabildo *see* Libro viejo de la fundacion de guatemala
Guatemala Comision De Los Quince *see* Diario de sessiones de la comision de los quince e...
Guatemala Constitution *see* Constitucion de la republica de guatemala
Guatemala de hoy / Gonzales Ruiz, Ricardo [Editor] – Guatemala, 1949 – 1r – us UF Libraries [972]
Guatemala Direccion General De Estadistica *see*
– Demarcacion politica de la republica de guatemala
– Estudio sobre las condiciones de vida de 179 famil...
– Guatemala y los censos de 1950
Guatemala. Direccion General de Estadistica *see*
– Anuario de la direccion general de estadistica 1898
– Anuario estadistico 1970
– Guatemala en cifras 1955-1969
– Republica de guatemala
Guatemala en cifras 1955-1969 / Guatemala. Direccion General de Estadistica – (= ser Latin american & caribbean...1821-1982) – 28mf – 9 – uk Chadwyck [318]
Guatemala Laws, Statutes, Etc *see*
– Codigo de trabajo
– Codigo de trabajo (decreto numero 330 del congreso
– Constitucion y codigos de la republica de guatemala
– Decreto 203
– Decretos-leyes del actual gobierno (emitidos hasta...)
– Ley organica y reglamentaria de instruccion public
Guatemala Laws, Statutes, Etc (Indexes) *see* Catalogo razonado de las leyes de guatemala
Guatemala. Ministerio de Economia y Trabajo *see* Memoria de las labores realizadas durante el ano...
Guatemala Ministerio De Educacion Publica *see* Educacion guatemalteca
Guatemala nuestra / Marinello, Juan – Habana, Cuba. 1961 – 1r – us UF Libraries [972]
Guatemala para el turista / Valle, Jose – Guatemala, 1929 – 1r – us UF Libraries [972]
Guatemala. Secretaria de Educacion Publica *see* Memoria de las labores del poder ejecutivo en el ramo de educacion publica durante el ano administrativo de...presentada a la asamblea legislativa en sus sesiones ordinarias de...
Guatemala Secretaria De Fomento *see* Memoria
Guatemala Secretaria De Relaciones Exteriores *see*
– Controversia sobre belice durante el ano de 1946
– Guatemala ante america
– Libro blanco de guatemala
– Opinion centroamericana a proposito del libro
– Puntos capitales que sostiene el gobierno
– Puntos capitales que sostiene el gobierno de guate...
– White book
Guatemala y los censos de 1950 / Guatemala Direccion General De Estadistica – Guatemala, 1950 – 1r – us UF Libraries [972]
El guatemalteco diario de centro america / Guatemala – 1970- – 1 – (ind 1971-76) – us L of C Photodup [972]
Guateque / Romero Plazas, Elias – Bogota, Colombia. 1962 – 1r – us UF Libraries [972]
Guateque a alfonso camin en decimas de batey / Sanjurjo, Jose – Habana, Cuba. 1953 – 1r – us UF Libraries [972]
Guatimala, or, the republic of central america, in 1827-8 : being sketches and memorandums made during a twelve-months' residence / Dunn, Henry – London 1829 [mf ed Hildesheim 1995-98] – 1v on 2mf – 9 – €60.00 – 3-487-26967-8 – gw Olms [880]
Guatimozin. tomo 2 / Gomez de Avellaneda, Gertrudis – 1846 – 9 – sp Bibl Santa Ana [946]
Guay, Charles *see*
– Chronique de Rimouski
– Recueil de prieres
Guay, Marcel *see* Bibliographie de paul-andre lamontagne
Guayacuya / Corvington, Hermann – Port-Au-Prince, Haiti. 1944 – 1r – us UF Libraries [972]
Guayana y sus problemas / Oxford-Lopez, Eduardo – Caracas, Venezuela. 1942 – 1r – us UF Libraries [972]
Guayaquil – Bogota, Colombia. 1957 – 1r – us UF Libraries [972]
Guazzo, M *see*
– Histoire...di tvtti i fatti degni di memoria nel mondo svccessi dell' anno 1524...
– Historie...conteneno le gverre di mahometto imperatore de turchi...
Guazzo, Stefano *see* Dialoghi piacevoli del sig. stefano guazzo
Gubanov, N F *see* Amurskoe slovo
Gubelman, M I *see* Smychka
Gubener anzeiger – Guben DE, 1884-86 – 1 – gw Misc Inst [074]
Gubener tageblatt – Guben DE,1906 n231-306 – 1 – gw Misc Inst [074]
Gubener zeitung – Guben DE, 1871-73, 1881 jan-jul, 1883 nov-1887 may, 1892 20 nov-1899 jan, 1900 oct-1901 jul, 1902 nov-1905 apr, 1915 jan-jun, 1917 feb-1918 mar, 1923 jun, 1924 jan-aug, 1925 dec-1926 mar, 1927 mar-aug, 1928 15 aug-15 nov, 1929 10 sep-1930 13 mar, 1930 19 jun-30 sep, 1931 jan-13 jul, 1931 28 oct-1932 19 aug, 1933 25 jul-oct, 1936 jul-dec, 1937 20 apr-1938 31 may, 1939 jan-1940 18 mar, 1941 aug-1943 may 1 – (fr jun 1935 ausg a, fr 3 may1 943 ausg fuer guben & fuerstenberg/oder) – gw Misc Inst [074]
Guber, Boris Andreevich *see* Sosedi
Gubernatis, Alessandro de *see* Le tradizioni popolari di s stefano di calcinaia
Gubernatorial inauguration address / Camacho, Carlos S – Saipan: the Commonwealth Government, n.d. – 1mf – 9 – $1.50 – mf#llmc82-100j, title 22 – us LLMC [080]
Gubernskiia viedomosti / Russia – 1838-1906 – 1 – $1531.00 – us L of C Photodup [073]
Gubernur, (rangkajo basa) reshuffle : penambahan dan pengurangan rentjana anggaran belandja tahun dinas 1959 dari dinas2 daerah swatantra tingkat i sumatera barat, west sumatra (province) – [Padang, 1959] – 1mf – 9 – mf#SE-11591 – ne IDC [959]
Gubulawayo and beyond / Gelfand, Michael – London, England. 1968 – 1r – us UF Libraries [960]
Guchkov, a i v tretei gosudarstvennoi dume : (1907-1912 gg) sbornik rechei – 1912 – 248p 3mf – 9 – mf#RPP-186 – ne IDC [325]
Gudari *see* Spanish-basque political periodicals
Gude, Carl *see* Gudes erlaeuterungen deutscher dichtungen
Gude, Ludvig Jacob Mendel *see* Om magister s. kierkegaards forfattervirksomhed
Gudehus, Jonas *see* Meine auswanderung nach amerika im jahre 1822, und meine rueckkehr in die heimath im jahre 1825
Gudeman, Alfred *see* Grundriss der geschichte der klassischen philologie
Gudemann, Moritz *see*
– Grabreden
– Jerusalem, die opfer und die orgel
– Torah veha-hayim be-artsot ha-ma'arav
Gudemann, Mortiz *see* Geschichte des erzienhungswesens und der cultur
Gudensberger zeitung – Gudensberg DE, 1905 5 oct-1909 15 apr, 1913-19, 1922-28, 1930-1936 24 dec – 9r – 1 – (incl suppl) – gw Misc Inst [074]
Gudes erlaeuterungen deutscher dichtungen : ausgefuehrte anleitungen zur aesthetischen wuerdigung und unterrichtlichen behandlung / Gude, Carl; ed by Linde, Ernst – Leipzig: F Brandstetter. 10v. 1910-1928 – 1 – (vol publ out of sequence and in different editions) – us Wisconsin U Libr [430]
Gudgeon, T W *see* Defenders of new zealand
Gudgeon, Walter Edward *see* Papers
Gudlindlu mntanami / Sigogo, Ndabezinhle S – Gwelo, Zimbabwe. 1967 – 1r – us UF Libraries [960]
Gudok – Baku, Azerbaijan 1907-09 [mf ed Norman Ross] – 1 – mf#nrp-246 – us UMI ProQuest [077]
Gudra puce : pulkveza kalpaka latviesu skolas jauniesu zurnals / Colonel Kalpak's Latvian School (Wauwatosa, WI) – n1-n13 [1966 apr-1981 pasavari] – 1r – 1 – mf#1331217 – us WHS [460]
Gudrun = Kudrun / Loeschhorn, H [comp] – Halle/S: Verlag der Buchhandlung des Waisenhauses, 1891 [mf ed 1993] – (= ser Denkmaeler der aelteren deutschen literatur 1/2; Deutsche heldensage 2) – 1 – mf#8185 – us Wisconsin U Libr [810]
Gudrun : ein schauspiel / Strauss und Torney, Viktor Friedrich von – Frankfurt a.M.: K T Voelcker, 1851 – 1r – 1 – us Wisconsin U Libr [820]
Gudrun : schauspiel in 5 akten / Wesendonk, Mathilde – Zuerich: Schabelitz 1868 – 3mf – 9 – mf#mw-11 – ne IDC [820]
Gudrun : schauspiel in fuenf aufzuegen / Caro, Carl – Breslau: E Trewendt 1877 [mf ed 1989] – 1r – 1 – (filmed with: georg buchners drama dantons tod / hans landsberg) – mf#7143 – us Wisconsin U Libr [820]
Gudrun : schauspiel in fuenf aufzuegen / Grosse, Julius – Leipzig: J J Weber, 1870 – 1r – 1 – us Wisconsin U Libr [820]
Gudruns tod : tragoedie / Schumann, Gerhard – Wien: K H Bischoff, 1943 – 1r – 1 – us Wisconsin U Libr [890]
Die gudrunsage : drei vortraege ueber ihre erste gestalt und ihre wiederbelebung, gehalten in schleswig im januar 1867 / Keck, Karl Heinrich – Leipzig: B G Teubner, 1867 [mf ed 1993] – 84p – 1 – mf#8034 – us Wisconsin U Libr [390]
Guds uforandersighed : en tale / Kierkegaard, Soeren – Kobenhavn: C A Reitzel, 1855 – (= ser Himmelstrup) – 1mf – 9 – 0-7905-7415-2 – mf#1989-0640 – us ATLA [210]

Guds veie med et gjenstridigt folk : en historisk beretning / Himle, Thorstein – Red Wing MN: [s.n.] 1902 [mf ed 1992] – 1mf – 9 – 0-524-05192-5 – mf#1991-2228 – us ATLA [951]
Gudstrons uppkomst / Soederblom, Nathan – Stockholm: H. Gebers, 1914 – 1mf – 9 – 0-7905-6012-7 – (incl bibl ref) – mf#1988-2012 – us ATLA [200]
Guede, Lisardo *see*
– Fray pedro nunez machado (zafra 1550-burgos 1609)
– La merced (compendio historico en 12 lecciones)
Guedemann, Moritz *see*
– Geschichte des erziehungswesens und der cultur der juden in frankreich und deutschland
– Das judenthum
Gueder, Eduard *see*
– Die lehre von der erscheinung jesu christi unter den todten
– Vergleichende darstellung des lutherischen und reformirten lehrbegriffs
Guedj, Elijah *see* Zeh ha-shulhan
Gueenaga de Silva, Rosario *see* Relacion de pero lopez. vision de un conquistador del siglo 16
Gueft ue senid / Bey, Mehmet Ata – Istanbul: Mihran Matbaasi, 1304 [1887] – (= ser Ottoman literature, writers and the arts) – 2mf – 9 – $40.00 – us MEDOC [470]
Gueiros, Optato *see* Lampeao
Guel, Conde de (Marques de Comillas) *see* Apuntes de recuerdos
Gueladio ham bodedio : heros de la poulagou a travers deux recits epiques peuls / Wane, Animata – 1980 – us CRL [944]
Gueldene aepfel in silbernen schalen : das ist, worte geredet zu seiner zeit ueber 400. sinnbilder... / Pfeffel, J A – Augspurg, Detleffsen: Gedruckt bey Christoph Peter, 1746 – 2mf – 9 – mf#0-1254 – ne IDC [090]
Gueldene rose, d.i. einfaeltige beschreibung des allergroessesten von.. – Hamburg, 1705 – 1 – us Wisconsin U Libr [800]
Gueldenstaedt, J A [von] *see* Reisen durch russland und im caucasischen gebuerge
Guell Rente, Jose *see* Restos de colon...
Guell Y Ferrer, Juan *see* Rebelion cubana
Guell Y Rente, Juan *see* Ultimos cantos
Guelnihal / Kemal, Namik – Dersaadet, [1875] – (= ser Ottoman literature, writers and the arts) – 3mf – 9 – $55.00 – us MEDOC [470]
Guelzar, Divan-i *see* The divan project
Guelzow, Erich *see*
– Ernst moritz arndt in schweden
– Ernst moritz arndts briefe an eine freundin
– Unserm lehrer gustav ehrismann zum gedaechtnis
Guembel-Seiling, Max *see*
– Bruder lustig
– Gevatter tod
– Das glueckskind
– Die kluge bauerntochter
– Marienkind
– Das tapfere schneiderlein
– Der treue johannes
– Das wasser des lebens
– Die zertanzten schuhe
Guemruek nizamname-i umumiyesi – Dersaadet [Istanbul]: Mahmud bey Matbaasi, 1309 [1893] – 2mf – 9 – $40.00 – us MEDOC [380]
Guemruek siyasetimizin esaslari – [Istanbul]: Tuerk Ocaklari Merkez Heyeti Matbaasi, 1928 – 1mf – 9 – $25.00 – us MEDOC [380]
Guenaydin : tageszeitung fuer tuerken in europa – Frankfurt/M DE, sep 1 1978-aug 31 1979 – 3r – 1 – gw Misc Inst [074]
Die guenderode / Arnim, Bettina von – Berlin: im Propylaeen-Verlag, c1920 [mf ed 1993] – (= ser Saemtliche werke v2) – 602p – 1 – mf#8196 reel 1 – us Wisconsin U Libr [920]
Guenderode, Karoline von *see*
– Correspondence of fraeulein guenderode and bettine von arnim
– Dichtungen
– Friedrich creuzer und karoline von guenderode
– Gesammelte dichtungen
– Karoline von guenderode und ihre freunde
Guenduez, Aka *see* Yarim tuerkler
Guenes – Istanbul: Matbaa-i Ebuezziya, Vatan Matbaasi, 1927. Mueduer-i Mes'ul: Orhan Seyfi [Orhon] n1-17. 1 kanunisani-1 tesrinievvel 1927 – (= ser O & t journals) – 6mf – 9 – $90.00 – us MEDOC [956]
Guenin, Eugene *see* La nouvelle-france
Guenon, Rene *see*
– East and west
– Introduction to the study of the hindu doctrines
– Man and his becoming
Guenser zeitung – Guens (Koeszeg H), 1929 6 jan-1938 – 3r – 1 – gw Misc Inst [077]
Guentekin, Resat Nuri *see*
– Acimak
– Gizli el
Guenter, Heinrich *see* Die christliche legende des abendlandes
Guenter und christiane : roman / Ball, Kurt Herwarth – 1. aufl. Berlin: Buchverlag Der Morgen, 1964 [mf ed 1995] – 285p – 1 – mf#8970 – us Wisconsin U Libr [830]
Guenther, A C L G *see* Reptiles and fishes of the south sea islands
Guenther, Agnes *see* Von der hexe die eine heilige war
Guenther, Carl *see* Heinrich zschokkes jugend- und bildungsjahre (bis 1798)
Guenther, Dandy *see* Die sprache der werbung in den printmedien
Guenther, Ernst *see* Die entwicklung der lehre von der person christi im 19. jahrhundert
Guenther, Hans *see* Fortunatus
Guenther, Hans-Christian *see* Immunhistochemische darstellung peripherer neuraler und neuroendokriner zellelemente bei dysplasien und karzinomata in situ der harnblase
Guenther, Johann Christian *see*
– Gedichte
– Johann christian guenthers saemtliche werke
– Saemtliche werke
Guenther, Johannes von *see* Sonettengarten
Guenther, Kurt Martin *see* Die entwicklung der novellistischen kompositionstechnik kleists bis zur meisterschaft
Guenther, Martin *see* Populaere symbolik
Guenther, Oskar *see* Das verhaeltnis der ethik thomas hill greens zu derjenigen kants
Guentter, Otto *see*
– Friedrich schiller, sein leben und seine dichtungen
– Gesammelte dichtungen
Guenzburg, Johann Eberlin von *see* Ausgewaehlte schriften
Guenzburger zeitung – Guenzburg DE, 1978- – (= ser Bezirksausgabe von augsburger allgemeine) – ca 9r/yr – 1 – gw Misc Inst [074]
Guer, J A *see* Moeurs et usages des turcs, leur religion, leur gouvernement civil, militaire et politique
Gueranger, Prosper *see*
– De la monarchie pontificale a propos du livre de mgr l'eveque de sura
– Defence of the roman church against father gratry
– Die hoechste lehrgewalt des papstes
– Lettre a monseigneur l'archeveque de rheims
Guerard, Albert Leon *see*
– French prophets of yesterday
– French prophets of yesterday; a study of religious thought under the second empire
Guerard, Emile *see* Epoque de 1815
Guerber, Helene Adeline *see*
– Legends of the middle ages
– Legends of the virgin and christ, with special reference to literature and art
– Myths of northern lands
Guericke, H E F *see* Gesammtgeschichte des neuen testaments
Guericke, Heinrich Ernst Ferdinand *see* Manual of church history
Guerilla – Toronto. v1-3 n32. jun 5 1970-jun 30 1973// (wkly) – 3r – 1 – Can$330.00 – cn McLaren [331]
Guerilla [sic] – n12-13 [1970 nov13-27], v2 n25 [1971 dec 1], v2 n32-34,37-38,42-43,45,48 [1972 feb 2-16, mar 8-15, apr 12-19, may 3,26] – 1r – 1 – (continued by: guerilla free press) – mf#1583448 – us WHS [320]
Guerillaskrieg : versprengte lieder – Belle-Vue bei Konstanz: Verlags- und Sortiments-Buchhandlung, 1845 – 1r – 1 – us Wisconsin U Libr [810]
Guerin, Eugenie de *see*
– Journal of eugenie de guerin
– Letters of eugenie de guerin
Guerin, J M F *see* Astronomie indienne d'apres la doctrine et les livres anciens et modernes des brammes sur l'astronomie, l'astrologie et la chronologie
Guerin, M *see* L'encyclique rerum novarum "sur la condition des ouvriers"
Guerin, Marc-Aime *see* Histoire de la pedagogie
Guerin, V *see* Description de l'aile de patmos et de l'aile de samos
Guerin, v *see* Description geographique, historique et archeologique de la palestine
Guerin, Victor *see* La terre sainte
Guerini, Francesco *see* Sonate a violino con viola di gamba o cembalo...opera prima
Gueriniere, Joseph *see* Histoire generale du poitou
Guerin-Meneville, Felix Edouard *see* Iconographie du regne animal de g. cuvier; ou, representation d'apres nature de l'une des especes les plus remarquables et souvent non encore figurees, de chaque genre d'ani-maux
Guerino detto il meschino : storia delle grandi imprese e vittorie de lui riportate contro i turchi durante il regno di carlo magno... / Andrea da Barberino – Milano: Bietti, 1910 [mf ed 1986] – 566p – 1 – mf#7239 – us Wisconsin U Libr [830]
Guerino il meschino : romanzo cavalleresco di andrea da barberino... / Andrea da Barberino – Milano: Bertieri e Vanzetti, 1923 [mf ed 1986] – 5p/15-267/[1]p (ill) – 1 – mf#7240 – us Wisconsin U Libr [830]
Guerinot, Armand *see*
– Recherches sur l'origine de l'idee de dieu d'apres le rig-veda
– Repertoire d'epigraphie jaina
Guerke, Britta *see* Neurohumorale effekte einer therapie mit ramipril
[Guerneville-] the paper – CA. 1979-1989 – 9r – 1 – $540.00 – mf#B05039 – us Library Micro [071]
[Guerneville-] the russian river news – CA. 1970-1993 – 16r – 1 – $960.00 – mf#B06031 – us Library Micro [071]
Guernsey, Alice Margaret *see*
– Citizens of to-morrow
– Under our flag
Guernsey breeders' journal – Reynoldsburg OH 1910+ – 1,5,9 – ISSN: 0017-5110 – mf#227 – us UMI ProQuest [636]
Guernsey, Clark *see* [Clark Guernsey]
Guernsey Co. Cambridge *see*
– Clarion of freedom
– Daily jeffersonian
– Guernsey times series
Guernsey Co. Cumberland *see*
– Echo
Guernsey Co. Pleasant City *see* News
Guernsey Co. Quaker City *see*
– Home towner
– Press-advertiser
Guernsey Co. Senacaville *see* Times
Guernsey county atlas, 1870 – 1r – 1 – mf#B7070 – us Ohio Hist [978]
Guernsey county atlas, 1870 – 1r – 1 – mf#B7070 – us Ohio Hist [978]
Guernsey evening press *see* Evening press
Guernsey jeffersonian : [weekly democratic newspaper] – Washington, OH. jan 3 1844-may 15 1873 – 4r – 1 – $460.00 – mf#D3490P05 – us Western Res [071]
Guernsey, Rocellus Sheridan *see* A key to story's equity jurisprudence
Guernsey times series / Guernsey Co. Cambridge – apr 1840-may 1842 [wkly] – 1r – 1 – mf#B28802 – us Ohio Hist [071]
Guerpinar *see* Nimet sinas
Guerra a muerte / Tosta Garcia, Francisco – Caracas, Venezuela. 1906 – 1r – us UF Libraries [972]
Guerra a muerte! – Buenos Aires, Argentina. 1945 – 1r – us UF Libraries [972]
Guerra, Antonio Teixeira *see* Dicionario geologico-geomorfologico
Guerra, Arthur *see* The effect of sodium citrate ingestion on 1600 meter running performance
Guerra Camacho, Mercedes *see* El colera morbo en badajoz en 1883
Guerra civil. inventario de la documentacion de la generalitat de cataluna – 1989 – 9 – sp Cultura [355]
Guerra! cuba / Burguete, Ricardo – Buenos Aires, Argentina. 1902 – 1r – us UF Libraries [972]
Guerra de 85 / Palacio, Julio H – Bogota, Colombia. 1936 – 1r – us UF Libraries [972]
Guerra de extremadura y sitios de badajoz (1706) / Silva, Alejandro de – Badajoz: Tip.Euc. A.Arqueros, 1945 – 1 – sp Bibl Santa Ana [350]
Guerra de independencia de cuba / Varona Guerrero, Miguel Angel – Habana, Cuba. v1-3. 1946 – 1r – us UF Libraries [972]
Guerra de la liga y la invasion de quijano / Fernandez Guardia, Ricardo – San Jose, Costa Rica. 1930 – 1r – us UF Libraries [972]
Guerra de la liga y la invasion de quijano / Fernandez Guardia, Ricardo – San Jose, Costa Rica. 1950 – 1r – us UF Libraries [972]
Guerra de los mil das en el sur de colombia / Coral, Leonidas – Pasto, Colombia. 1939? – 1r – us UF Libraries [972]
Guerra de s le sao paulo, 1932 / Osorio, Manoel – Sao Paulo, Brazil. 1932 – 1r – us UF Libraries [972]
Guerra del chaco / Gonzalez, Antonio E – Sao Paulo, Brazil. 1941 – 1r – us UF Libraries [972]
Guerra del tiempo / Carpentier, Alejo – Habana, Cuba. 1963 – 1r – us UF Libraries [972]
Guerra do flores / Barroso, Gustavo – Sao Paulo, Brazil. 1930 – 1r – us UF Libraries [972]
Guerra do rosas / Barroso, Gustavo – Sao Paulo, Brazil. 1929 – 1r – us UF Libraries [972]
Guerra dominico-haitiana / Rodriguez Demorizi, Emilio – Ciudad Trujillo, Dominican Republic. 1957 – 1r – us UF Libraries [972]
Guerra dominico-haitiana / Rodriguez Demorizi, Emilio – Santiago, Dominican Republic. 1944 – 1r – us UF Libraries [972]
Guerra, Dora *see* Signos menos
Guerra dos barbaros / Taunay, Affonso De E – Sao Paulo, Brazil. 1936 – 1r – us UF Libraries [972]
Guerra dos mascates / Ferrer, Vicente – Lisboa, Portugal. 1915 – 1r – us UF Libraries [972]
Guerra, Felipe Leon *see*
– Notas a las antiguedades de extremadura
– Notas a las antiguedades e extremadura de jose viu
Guerra fisica, proezas medicales, hazanas de la ignorancia... – [16–] – (on same reel: various authors) – us CRL [946]
Guerra Flores, Jose *see*
– Flecha de sombra
– Poemas del ocaso
Guerra Guerra, Arcadio *see*
– Apuntes bibliograficos de la prensa periodica de la baja extremadura. 1 y 2
– El badajoz del siglo 16
– Carta de privilegio de los reyes catolicos a la ciudad de badajoz, fechada en el campamento real "sobre toro" el dia 21 de julio de 1475
– Cartas a lopez prudencio
– De historia. relaciones de badajoz con la corte y con don manuel godoy durante el valimiento de este
– Festejos en honor del principe de la paz habidos en badajoz en 1807
– Instituto militar pestolozziano de madrid, obra del extremeno manuel godoy
– La mineria en la baja extremadura en la primera mitad del siglo 29
– Precios y salarios en badajoz durante el bienio 1775-76
Guerra Hontiveros, Marcelino *see*
– Apuntes...villa de gata
– Narraciones de un trovador
Guerra, Jorge *see* Nueve cuentos por un peso
Guerra, Jose Joaquin *see* Estudios historicos
La guerra nacional espanola ante la moral y el derecho / Menendez-Reigada, Ignacio G – Bilbao, 1937? Fiche W 1044. (Blodgett Collection of Spanish Civil War Pamphlets) – 9 – us Harvard [946]
Guerra phisica, proezas medicales o hazanas de la ignorancia [4th quarter of 17th cent] – Madrid, Biblioteca Nacional [19–] – us CRL [946]
Guerra, Ramiro *see*
– Azucar y poblacion en las antillas
– Defensa nacional y la escuela
– Historia de cuba
– Historia elemental de cuba
– Mudos testigos
– Primeras crisis economicas de cuba
Guerra revolucionaria / Pinto, Bilac – Rio de Janeiro, Brazil. 1964 – 1r – us UF Libraries [972]
Guerra santa: el sentido catolico de la guerra espanola. burgos, 1938 / Castro Albarran, A de – Burgos: Razon y Fe, 1938 – 1 – sp Bibl Santa Ana [946]
Guerra y los basiliscos / Llopis, Rogelio – Habana, Cuba. 1962 – 1r – us UF Libraries [972]
Guerra y marina, epoca de carlos 1 de espana... / Alvarez, C – Valladolid, 1949 – 9 – sp Cultura [920]
Guerrant, Edward Owings *see*
– The galax gatherers
– The soul winner
Guerrant, Grace *see* The galax gatherers
Guerras civiles del peru / Cieza Leon, Pedro de – Madrid – 1 – sp Bibl Santa Ana [972]
Guerras de bolivar / Rivas Vicuna, Francisco – Caracas, Venezuela. v1-2. 1921 – 1r – us UF Libraries [972]
Guerras de retaguardia / Juez, Antonio – Badajoz: Tip.Espanola, 1937 – 1 – sp Bibl Santa Ana [946]
Guerras piraticas de filipinas / Barrantes Moreno, Vicente – 1878 – 9 – sp Bibl Santa Ana [959]
Guerra-Trigueros, Alberto *see* Minuto de silencio
La guerre / Carred, Henri – Paris: Imp E Bautruche, 1848 – us CRL [944]
La guerre : ...les clercs de st-viateur, a notre-dame de lourdes, le 16 aout 1914 / Emard, Joseph-Medard – Valleyfield [Quebec: s.n.] 1914 [mf ed 1994] – 1mf – 9 – 0-665-73227-9 – mf#73227 – cn Canadiana [241]
La guerre aerienne illustree – Paris. n1-163. 16 nov 1916-25 dec 1919 – 1 – (devenu: la vie aerienne illustree) – fr ACRPP [073]
La guerre americaine, son origine et ses vraies causes : lecture faite a l'institut-canadien, le 14 decembre 1864 / Dessaulles, L A – Montreal?: s.n, 1865 – 6mf – 9 – mf#34768 – cn Canadiana [976]
La guerre anglo-boer : histoire et recits d'apres des documents officiels / Rosny, J H – Paris: Editions de la Revue Blanche, 1902 – 1 – us CRL [960]
La guerre au dahomey, 1888-1893 : d'apres les documents officiels / Aublet, Edouard Edmond – Paris: Berger-Levrault, 1894 – 1 – us CRL [960]
Guerre au sexe / Jouhaud, Auguste – Paris, France. 1856 – 1r – us UF Libraries [440]
Guerre dans l'afrique australe / Doyle, Arthur Conan – Paris, France. 1902 – 1r – us UF Libraries [960]
La guerre de russie : aventures d'un soldat de la grande armee / Vekeman, Gustave – Montreal?: s.n, 1895? – 2mf – 9 – (ill by j-b lagace) – mf#28654 – cn Canadiana [830]
La guerre de sarsa-dengel contre les falachas / Halevy, J – Paris, 1907 – 2mf – 9 – mf#NE-20234 – ne IDC [956]

La guerre d'europe / Champagne, Philias – Nashua, NH: P Champagne, 1915 [mf ed 1988] – 1mf – 9 – mf#SEM105P910 – cn Bibl Nat [780]

Guerre d'independence: quels sont les nationaux – Madrid, 193? Fiche W933. (Blodgett Collection of Spanish Civil War Pamphlets) – 9 – us Harvard [946]

La guerre du dahomey : journal de campagne dun sous-lieutenant d'infanterie de marine / Morienval, Henri – Paris: A Hatier, [1893?] ed illus – 1 – us CRL [960]

La guerre et la condition privee de la femme / Isore, Andre – Paris: de Boccard, 1919 – (= ser Les femmes [coll]) – 6mf – 9 – mf#9687 – fr Bibl Nationale [305]

Guerre et religion / La Souchere, Elena de – Paris, 1938? Fiche W934. (Blodgett Collection of Spanish Civil War Pamphlets) – 9 – us Harvard [946]

Guerre ouverte : ou ruse contre ruse / Dumaniant, Antoine-Jean – Paris, France. 1809 – 1r – us UF Libraries [440]

Guerre sociale – Paris, France. Sep-nov 1914; 1915 – 1r – 1 – uk British Libr Newspaper [072]

La guerre sociale – 19 dec 1906-15 – 1 – (hebd. puis quot. a partir du 6 aout 1914. devenu: la victoire. 1916. paris. dec 1906-18) – fr ACRPP [320]

O guerreiro – Rio de Janeiro, RJ: Typ Guanabarense de L A F de Menezes, 08 jan-set, 05-12 nov 1853 – (= ser Ps 19) – mf#P01B,05,11 – bl Biblioteca [320]

Guerreiro, Fernao *see*
– Jahangir and the jesuits
– Relacao anual das coisas que fizeram os padres da companhia

Guerreiro, Manuel Viegas *see* Bochimanes!

Guerrero / Sarasqueta De Smyth, Acracia – Panama, Panama. 1962 – 1r – us UF Libraries [972]

Guerrero *see* Ricardo wagner, el hombre i el artista

Guerrero, Alonso *see*
– Abismo...y discurso...virgen maria
– Norte y guia para el camino del cielo

Guerrero Castillo, Julian N *see* Managua

Guerrero, E *see*
– Beitia, eugenio. apostolado de los seglares 2nd edicion. madrid, 1939
– Muchachas en flor. madrid, 1946

Guerrero, Eduardo Garcia *see* Ricardo wagner, el hombre i el artista

Guerrero, Francisco *see* Canticum beatae mariae quod magnificat nuncupatur

Guerrero, J *see* Sol de la medicina que alumbra a los que ignoran la verdadera doctrina...

Guerrero. Mexico (State) *see* ...Periodico oficial del gobierno del estado de guerrero

Guerrero Meza, Hector Emilio *see* Deben derogarse los escritos de replica, duplica y extracto de litis del codigo de procedimientos civiles del estado de la baja california

Guerrero, Pedro *see* Constituciones synodales... granada

Guerrero, Placido *see* Tentativa...leccion de don vicente garcia de la huerta

Guerrero, Rafael *see* Cronica de la guerra de cuba (1895)

Guerrero Y Pallares, Teodoro *see* Anatomia del corazon

Guerrero y Pallares, Teodoro *see* La nube negra

Guerrero Yoacham, Cristian *see* Conferencias del niagara falls

Les guerres d'afrique depuis la conquete d'alger par les francais jusqu'a et compris l'expedition de kabylie en 1858 / Ladimir, Jules – Paris: B Renault, 1859 – 1 – us CRL [960]

Les guerres de la revolution / Chuquet, Arthur – Paris. v1-11. 1886-96 – 1 – $60.00 – (in french) – mf#0153 – us Brook [944]

Guerrier de Dumast, Auguste Prosper Francois *see* La maconnerie

Guerriers et sorciers en somalie / Lippmann, Alphonse – [Paris]: Hatchette, [1953] – 1 – us CRL [306]

Guerrilla and counterguerrilla warfare in russia during world war 2 / U.S. Army. Office of the Chief of Military History – 1963 – 1 – $20.00 – us L of C Photodup [947]

The guerrilla resistance movement in the philippines / U.S. Army. Far East Command – 1948. 2v – 1 – us L of C Photodup [959]

Guerrillas del llano / Franco Isaza, Eduardo – Bogota, Colombia. 1959 – 1r – us UF Libraries [972]

Guerrillero / Navarrete Porrata, Gabriel – Puerto Rico, Puerto Rico. 1958 – 1r – us UF Libraries [972]

Guerrilleros intelectuales / Agudelo Ramirez, Luis Eduardo – Medellin, Colombia. 1957 – 1r – us UF Libraries [972]

Guerro, Alonso *see* [Natural de fuente de cantos] en notas de bibliografia franciscana

Guerrra-Trigueros, Alberto *see* Surtidor de estrellas

Guertler N *see* Historia templariorum

Guerttler, Karin R *see* Kuenec artues der guote

Guery, Ch *see* Antiques ceremonies dans l'abbaye de saint-evroult

Guesde, Jules *see*
– Le programme du parti ouvrier
– Le socialisme au jour le jour

Guesped novela / Salazar Dominguez, Jose – Caracas, Venezuela. 1946 – 1r – us UF Libraries [972]

Guesses at purpose in nature : with especial reference to plants / James, William Powell – London [1882] – (= ser 19th c evolution & creation) – 3mf – 9 – mf#1.1.10880 – uk Chadwyck [574]

Guesses at the riddle of existence : and other essays on kindred subjects / Smith, Goldwin – Toronto: Copp, Clark, 1897 – 3mf – 9 – (incl publ list) – mf#34321 – cn Canadiana [210]

Guesses at truth / Hare, Julius Charles & Hare, Augustus William – [3rd ed] New York: EP Dutton, 1877 [mf ed 1993] – (= ser Anglican/ episcopal coll) – 2mf – 9 – 0-524-05950-0 – mf#1991-2350 – us ATLA [240]

Guessfeldt, Paul *see* In den hochalpen

Guest, Shannon M *see* The influence of dispositional goal orientation, perceptions of the motivational climate, and scholarship level on sport commitment in elite level athletes

Guestlings *see* The white and black books of the cinque ports from 1433

Gueterbock, K *see* Byzanz und persien in ihren diplomatisch-voelkerrechtlichen beziehungen im zeitalter justinians

Gueterbock, Karl *see* Der islam im lichte der byzantinischen polemik

Guetersloher zeitung – Guetersloh DE, 1954-1960 30 sep – 1 – (title varies: 25 oct 1921: guetersloher zeitung und tageblatt; 2 dec 1935: westfaelische zeitung / b; main ed in bielefeld; publ in guetersloh, fr 1 nov 1922 in bielefeld) – gw Misc Inst [074]

Guetersloher zeitung und tageblatt *see* Guetersloher zeitung

Gueterwagen – Halle S DE, 1962 18 aug-1970 nov, 1972-1974 aug, 1975-1989 7 nov – 5r – 1 – (with gaps) – gw Misc Inst [074]

Gueterwagen – Magdeburg DE, 1966 19 jan-1968 aug, 1969-1975 nov, 1976-1977 oct, 1978-1992 21 dec – 4r – 1 – (with gaps) – gw Misc Inst [074]

Guettee, Wladimir, abbe *see*
– Exposition de la doctrine de l'eglise catholique orthodoxe
– The papacy

Guettler, Felix *see* Wordsworth's politische entwicklung

Guettler, Wilhelm *see* Die religioese kindererziehung im deutschen reiche

Guetzlaff, Carl *see* Journal of three voyages along the coast of china, in 1831, 1832, and 1833

Guetzlaff, Karl Friedrich August *see*
– The life of taou-kwang, late emperor of china
– On the present state of buddhism in china

Guevara, Antonio *see* Epistolas familiares

Guevara Castaneira, Josefina *see* Del yunque a los andes

Guevara, Miguel Tadeo de *see* Visita sin despedida que hizo maria santisima de guadalupe al reyno

Gueye, Lamine *see* Etapes et perspectives de l'union francaise

Gugeline : ein buehnenspiel in fuenf aufzuegen / Bierbaum, Otto Julius – Berlin: Schuster & Loeffler, [1899] [mf ed 1989] – 105p (ill) – 1 – mf#7020 – us Wisconsin U Libr [820]

Gugl, Matthaeus *see* Fundamenta partiturae in compendio data

Gugler, Julius *see*
– Dramatisches
– Der stern des westens

Guglielmi, Pietro Alessandro *see* Robert und kalliste

Guglieri, A *see* Documentos de la compania de jesus en el archivo historico nacional

Guglieri Navarro, A *see* Madrid. archivo historico nacional. seccion de sigilografia. catalogo de sellos. 1-3

Gugomo, Gottlieb F von *see* Reise von bucharest, der hauptstadt in der wallachei, ueber giurgewo, rustschuk, durch oberbulgarien, bis gegen die graenzen von rumelien, und dann durch unterbulgarien ueber silistria

Gugy, Augustus *see* Some remarks on the pamphlet of william foster coffin, esquire etc

Guha, Praphulla Kumar *see* Tragic relief

Guha-Thakurta, Prabhucharan *see* The bengali drama

Guhl, E *see* Kuenstlerbriefe uebersetzt und erlaeutert von dr ernst guhl

Gui de Cambrai *see* Barlaam und josaphat

Guia artistica, mercantil e industrial de caceres... – Caceres: Imp. Santos Floriano Glez, s.a. – 1 – sp Bibl Santa Ana [338]

Guia Civica De Guatemala *see*
– Imagenes de la revolucion

Guia comercial de ferias y fiestas mayo, 1945 / Caceres. Ayuntamiento – Caceres: Imp. Garcia Floriano – sp Bibl Santa Ana [390]

Guia comercial industrial profesional clasificada – Panama, Panama. 1947 – 1r – us UF Libraries [972]

Guia comercial literaria, mayo 1913 – Caceres: Tip. El Noticiero, S.A. – 1 – sp Bibl Santa Ana [440]

Guia comercial y turistica de la ciudad de guatema... – Guatemala, 1936 – 1r – us UF Libraries [972]

Guia de bibliografia historica portuguesa / Academia Portuhuesa da Historia – Madrid: Archivo Ibero Americano, 1960 – 1 – sp Bibl Santa Ana [946]

Guia de bogota / Hernandez De Alba, Guillermo – Bogota, Colombia. 1948 – 1r – us UF Libraries [972]

Guia de caceres y su provincia / Rosa Roque, Julio – Caceres: Tip. El Noticiero, 1951 – 1 – sp Bibl Santa Ana [946]

Guia de espectaculos. feria y fiestas agosto 1974 / Empresa Ber-Maq – Valencia de Alcantara: Tip. Avila, 1974 – 1 – sp Bibl Santa Ana [390]

Guia de espectaculos ferias y fiestas, 1975 – Valencia de Alcantara: Tip. Avila, 1975 – 1 – sp Bibl Santa Ana [390]

Guia de filipinas para [...] – Manila, 1879, 1884, 1885, 1886, 1889-97,1898 – (issues for 1879, 1884 filmed with: guia de forasteros en las islas filipinas para el ano [...] 1862-1865) – us CRL [959]

Guia de forasteros en filipinas : para el ano de [...] – Manila: Imprenta de los Amigos del Pais, a cargo de M Sanchez, [1858-] 1859, 1861-65 – (issues for 1859, 1861 1879, 1884 filmed with: guia de forasteros en las islas filipinas para el ano [...] 1854-57; issues for 1862-65 filmed with: guia de filipinas para [...] 1879, 1884) – us CRL [959]

Guia de forasteros en filipinas, para el ano de [...] – Manila: Imprenta de los Amigos del Pais, a cargo de M Sanchez, 1858, 1860 – us CRL [959]

Guia de forasteros en las islas filipinas para el ano [...] – Manila: Imprenta de D Miguel Sanchez, [-1857]. 1851 – 1r – 1 – us CRL [980]

Guia de foresteros en las islas filipinas para el ano... – Manila: Imprenta de D Miguel Sanchez, 1842-50; 1852; 1854-57 – 1 – us CRL [959]

Guia de fuentes para la historia de africa subsahariana – Madrid, 1987 – 9 – sp Cultura [960]

Guia de fuentes para la historia de asia en espana – Madrid, 1987 – 9 – sp Cultura [950]

Guia de fuentes para la historia de iberoamerica conservadas en espana, tomo 1 – Madrid, 1966-1969 – 2mf – 9 – sp Cultura [972]

Guia de historia de venezuela, 1492-1945 / Arellano Moreno, Antonio – Caracas, Venezuela. 1955 – 1r – us UF Libraries [972]

Guia de la ciudad de plasencia (caceres) por un placentino – Plasencia: Imprenta Placentina, Ano 1. 1905 – 1 – sp Bibl Santa Ana [946]

Guia de la ciudad de plasencia (caceres) por...ano 1906 / Rosado, Joaquin – Plasencia: Farmacia de Rosado, s.a. – sp Bibl Santa Ana [946]

Guia de los archivos de madrid – Madrid, 1952 – 9 – sp Cultura [020]

Guia de manaus / Correa, Luiz De Miranda – Rio de Janeiro, Brazil. 1969 – 1r – us UF Libraries [972]

Guia de merida / Almagro Basch, Martin – 5th ed corregida y aumentada con los ultimos hallazgos. Valencia: Direccion General de Bellas Artes, 1972 – 1 – sp Bibl Santa Ana [910]

Guia de merida / Almagro Basch, Martin – 7th ed corregida y aumentada con los ultimos hallazgos. Valencia: Direccion General del Patrimonio Artistico y Cultural, 1977 – 1 – sp Bibl Santa Ana [910]

Guia de ouro preto / Bandeira, Manuel – Rio de Janeiro, Brazil. 1938 – 1r – us UF Libraries [972]

Guia de ouro preto / Bandeira, Manuel – Rio de Janeiro, Brazil. 1963 – 1r – us UF Libraries [972]

Guia de ouro preto / Bandeira, Manuel – Rio de Janeiro, Brazil. 1967 – 1r – us UF Libraries [972]

Guia de trujillo / Acedo, Federico – Madrid: Artistica, 1925 – 1 – sp Bibl Santa Ana [946]

Guia de villafranca de los barros / Bogeat y Asuar, Antonio – Villafranca de los Barros, 1919 – 1 – sp Bibl Santa Ana [946]

Guia del archivo de la corona de aragon / Martinez Fernando, J E – Madrid, 1958 – 9 – sp Cultura [946]

Guia del archivo historico nacional / Sanchez Belda, L – Madrid, 1958 – 9 – sp Cultura [946]

Guia del espectador / Publicidad Cos – Caceres: Tip. El Noticiero, 1948 – 1 – (seria a n2 abril 1948) – sp Bibl Santa Ana [946]

Guia del forastero – Caceres.1899 – 9 – sp Bibl Santa Ana [074]

Guia del inversionista / Ayala Munoz, Ruben – Guatemala, 1964 – 1r – us UF Libraries [972]

Guia del maestro costarricense / Vincenzi, Moises – San Jose, Costa Rica. 1941 – 1r – us UF Libraries [972]

Guia deportiva de caceres 1975-76 – Caceres: Delegacion Provincial de Educacion Fisica y Deportes. Imp. Extremadura, 1976 – 1 – sp Bibl Santa Ana [790]

Guia didactica de la escuela nueva / Aguayo, Alfredo Miguel – Habana, Cuba. 1938 – 1r – us UF Libraries [972]

Guia do estado de santa catarina – Florianopolis, Brazil. 194- – 1r – us UF Libraries [972]

Guia general de la republica de panama – Panama, Panama. 1932 – 1r – us UF Libraries [972]

Guia general...badajoz / Sanchez Arjona y Sanchez Arjona, Francisco – 1881 – 9 – sp Bibl Santa Ana [946]

Guia geografica y administrativa de la isla de cub... / Imberno, Pedro Jose – Habana, Cuba. 1891 – 1r – us UF Libraries [972]

Guia higienica y medica del maestro / Delvaille, C – Badajoz: La Minerva Extremena, 1894 – 1 – sp Bibl Santa Ana [610]

Guia historia y descriptiva de los archivos, bibliotecas y museos arqueologicos de espana – Madrid, 1921 – 9 – sp Cultura [946]

Guia historica de el salvador / Larde Y Larin, Jorge – San Salvador, El Salvador. 1958 – 1r – us UF Libraries [972]

Guia historico-geografica de los 126 municipios / Correa, Ramon C – Tunja, Colombia. 1938 – 1r – us UF Libraries [972]

Guia industrial de angola / Associacao Industrial De Angola – Luanda, Angola. 1960 – 1r – us UF Libraries [960]

Guia local comercial de managua / Bravo Aguilera, Francisco – Managua, Nicaragua. 1930 – 1r – us UF Libraries [972]

Guia municipal de colombia / Molina, Roberto – Bogota, Colombia. 1937 – 1r – us UF Libraries [972]

Guia oficial / Philippines – 1834-98 – 1 – us CRL [324]

Guia oficial de espana – Madrid: Imprenta Real, 1770-1927 – 1 – $1872.00 – (lacks: 1773, 1807, 1809-11, 1814, 1905. name varies) – mf#0252 – us Brook [946]

Guia oficial de filipinas / Philippines – 1879-97 – 1 – us CRL [959]

Guia oficial de las ferias y fiestas de miajadas, agosto 1971 / Miajadas. Ayuntamiento – Caceres: Imp. T. Rodriguez, 1971 and 1974 – 1 – sp Bibl Santa Ana [914]

Guia oficial del ilustre colegio notarial de caceres. ano de 1936 – Caceres: Edit. Extremadura, 1936 – (tambien anos 1940, 1941, 1943, 1944, 1945, 1946, 1947, 1948, 1949, 1950, 1951, 1952, 1953, 1954, 1956, 1957, 1958, 1959, 1961, 1962, 1963, 1970) – sp Bibl Santa Ana [946]

Guia organica de la universidad de san carlos de g... – Guatemala, 1952 – 1r – us UF Libraries [972]

Guia para el trabajador espanol en francia, 1977 / Portalin Garcia, A – Caceres: Imp. Garcilaso, 1977 – 1 – sp Bibl Santa Ana [331]

Guia. Plasencia *see* Muy noble, muy leal y muy benefica

Guia poetica da cidade do rio de janeiro / Castro, Luiz Paiva De – Rio de Janeiro, Brazil. 1965 – 1r – us UF Libraries [972]

Guia politica, eclesiastica y militar del virreynato del peru : para el ano de 1794 / Unanue, Jose Hipolito – [Lima]: impr Real de los Ninos Huerfanos [1794] – (= ser Books on religion...1543/44-c1800: historia ecclesiastical) – 5mf – 9 – mf#crl-165 – ne IDC [240]

Guia popular-callejera e historico-turistica de llerena la llana y santiaguista – Badajoz: Imp. INCA, 1965 – sp Bibl Santa Ana [946]

Guia pratico, historica e sentimental da cidade do... / Freyre, Gilberto – Rio de Janeiro, Brazil. 1961 – 1r – us UF Libraries [972]

Guia sociogeografica de guatemala / Valle Matheu, Jorge Del – Guatemala, 1956 – 1r – us UF Libraries [972]

Guia turistica de colombia / Valencia Restrepo, Ricardo – Bogota, Colombia. 1936 – 1r – us UF Libraries [972]

Guia turistica de plasencia / Plasencia. Guia – Plasencia: Sanguino, Impresor, 1961 – 1 – sp Bibl Santa Ana [914]

Guia viaria de ciudad trujillo / Jesus Mejia, Manuel De – Ciudad Trujillo, Dominican Republic. 1944 – 1r – us UF Libraries [972]

Guiana / Rodway, James – London, England. 1912 – 1r – us UF Libraries [972]

Guiana graphic – Georgetown, Guyana. Guyana Graphic. -d. 7 Oct 1951-28 June 1953; 1 Jan 1959-9 Feb 1961; 14 Jan 1963-1 April 1966; 1 Oct 1970-Dec 1972. 37 reels – 1 – uk British Libr Newspaper [079]

Il guiba esplorato / Bottego, Vittorio – Rome, 1895 – 1 – us CRL [910]

Il guiba esporato : sotto gli auspici della societa geografica italiana / Bottego, Vittorio – Roma, E Loescher, 1895 – us CRL [945]
Guibal, R *see* Peut-on fermer le canal de suez?
Guibert, Archdeacon of Toulouse *see* Leben des heiligen papstes leo 9
Guibert de Tournai *see* Tractatus de pace
Guibert, J *see*
– El caracter
– La primavera de la vida
Guibert, Jacques A de *see* Journal d'un voyage en allemagne
Guibert, Michel Claude *see* Memoires pour servir a l'histoire de la ville de dieppe
Guibertus S Mariae de Novigento *see* Opera omnia
Guicciardijn, L *see* Beschrijvinghe van alle de nederlanden...
Guicciardini, L *see* Account of the ancient flemish school of painting
Guichardus, T *see* Oratio habita ab eloquentissimo viro
Guichenon, Samuel *see*
– Histoire de bresse et de bugey
– Histoire genealogique de la royale maison de savoye
Guichenot, A *see* Histoire naturelle des reptiles et des poissons
Guidance and control 1979 [aasms32] – 1979 – (= ser Aasms 1968) – 3papers on 2mf – 9 – $10.00 – 0-87703-128-2 – (suppl to v40, advances) – us Univelt [629]
Guidance and control 1981 [aasms36] – 1981 – (= ser Aasms 1968) – 7papers on 5mf – 9 – $15.00 – 0-87703-156-8 – (suppl to v45, advances) – us Univelt [629]
Guidance and control 1982 [aasms38] – 1982 – (= ser Aasms 1968) – 1paper on 1mf – 9 – $10.00 – 0-87703-180-0 – (suppl to v48, advances) – us Univelt [629]
Guidance and control 1983 [aasms44] – 1983 – (= ser Aasms 1968 44) – 2papers on 2mf – 9 – $10.00 – 0-87703-214-9 – (suppl to v51, advances) – us Univelt [629]
Guidance and control 1984 [aasms48] – 1984 – (= ser Aasms 1968) – 6papers on 4mf – 9 – $15.00 – 0-87703-201-7 – (suppl v55, advances) – us Univelt [629]
Guidance and control 1985 [aasms50] – 1985 – (= ser Aasms 1968) – 7papers on 3mf – 9 – $15.00 – 0-87703-213-0 – (suppl to v57, advances) – us Univelt [629]
Guidance and control 1986 [aasms53] – 1986 – (= ser Aasms 1968) – 7papers on 3mf – 9 – $15.00 – 0-87703-259-9 – (suppl to v61, advances) – us Univelt [629]
Guidance and control 1988 [aasms56] – 1988 – (= ser Aasms 1968) – 3papers on 2mf – 9 – $10.00 – 0-87703-290-4 – (suppl to v66, advances) – us Univelt [629]
Guidance and control 1992 [aasms64] – 1992 – (= ser Aasms 1968) – 6papers on 3mf – 9 – $20.00 – 0-87703-355-2 – (suppl to v78, advances) – us Univelt [629]
Guidance and control 1993 [aasms67] – 1993 – (= ser Aasms 1968) – 6papers on3mf – $20.00 – 0-87703-367-6 – (suppl to v81, advances) – us Univelt [629]
Guidance and control 1997 [aasms75] – 1997 – (= ser Aasms 1968) – 11papers on 6mf – $30.00 – 0-87703-436-2 – (suppl to v94, advances) – us Univelt [629]
Guidance and control conference 1978 [aasms29] – 1978 – (= ser Aasms 1968) – 22papers on 9mf – 9 – $20.00 – 0-87703-179-7 – us Univelt [629]
Guidance and counselling – Toronto, Canada 1985+ – 1,5,9 – (cont: school guidance worker) – ISSN: 0831-5493 – mf#15640 – us UMI ProQuest [370]
Guidance implications related to the eating habits of adolescents / Schnel, Nadine Deboreh – Uni of South Africa 2001 [mf ed Johannesburg 2001] – 4mf – 9 – (incl bibl ref) – mf#mfm15016 – sa Unisa [613]
Guide / Christian Labour Association of Canada – 1977-1984 – 1r – 1 – mf#705215 – us WHS [366]
Guide – Alliance, NE: J S Paradis. 4v. v7 n48. jan 23 1895-v10 n48. jan 13 1898 (wkly) – 2r – 1 – (cont: hemingford guide; absorbed by: alliance times) – us Bell [071]
Guide – Cedarburg, Grafton, Port Washington WI. 1980 mar 5-apr, 1980 may-dec, 1981 jan-dec, 1982 jan-dec, 1983 jan-dec, 1984 jan-aug 8 – 1r – 1 – (cont: ozaukee county guide [port washington, wi]; cont by: ozaukee county guide [grafton, wi]) – mf#946586 – us WHS [071]
Guide : a monthly journal devoted to legal news and public affairs – v1-4. 1892-95 (all publ) – (= ser Historical legal periodical series) – 1 – $50.00 – (available on reel only) – mf#408990 – us Hein [340]
Guide *see* [San francisco-] shipping guide
Le guide : ou nouvelle description d'amsterdam enseignant aux voyageurs, et aux negocians, son origine, ses agrandissemens and son etat actuel... – Amsterdam 1753 [mf ed Hildesheim 1995-98] – (= ser Fbc) – 5mf [ill] – 9 – €100.00 – 3-487-29660-8 – gw Olms [914]
The guide – Battleford, Sask: Indian Industrial School, [1892-189- or 19–] – 9 – mf#P05979 – cn Canadiana [378]
Guide book to the west indies – New York, NY. 1921 – 1r – us UF Libraries [972]
Le guide commercial pour la ville de detroit et ses environs : contenant l'histoire de detroit, montreal et quebec: avec les noms de leurs fondateurs / Bedard, J Alphonse – Detroit: C M Rousseau, 1881 – 1mf – 9 – mf#04066 – cn Canadiana [971]
Guide de berlin, de potsdam et des environs : ou description abregee des choses remarquables qui s'y trouvent; avec un plan de la ville de berlin / Nicolai, Friedrich – Berlin 1793 [mf ed Hildesheim 1995-98] – 2mf – 9 – €60.00 – 3-487-29567-9 – (trans fr german) – gw Olms [914]
Guide de la legislation du travail / Haiti (Republic) Department Du Travail – Port-Au-Prince, Haiti. 1955 – 1r – us UF Libraries [972]
Guide de l'instituteur : contenant une serie de reponses aux questions inserees dans la circulaire numero douze du surintendant de l'education... / Valade, Francois-Xavier – 3e ed. Montreal: J B Rolland, impr-libraire, 1853 [mf ed 1995] – 9 – cn Bibl Nat [370]
Guide des adresses de la ville de joliette pour l'annee 1900 / Gervais, Albert – Joliette, Quebec: A Gervais, 1900? – 1mf – 9 – mf#06379 – cn Canadiana [917]
Guide des jeunes amoureux pour parler et ecrire – Quebec?: L P Normand, 1863 – 1mf – 9 – mf#35511 – cn Canadiana [390]
Guide des voyageurs dans le royaume des pays-bas – Bruxelles 1818 [mf ed Hildesheim 1995-98] – (= ser Fbc) – 2mf – 9 – €60.00 – 3-487-29658-6 – gw Olms [914]
Guide des voyageurs en italie et en suisse / Reichard, Heinrich – Weimar 1810 [mf ed Hildesheim 1995-98] – 2v on 3mf – 9 – €90.00 – 3-487-29328-5 – gw Olms [914]
Guide d'ouro preto / Bandeira, Manuel – Rio de Janeiro, Brazil. 1948 – 1r – us UF Libraries [972]
Guide du cambodge / Cambodia. Departement du tourisme – [Phnom-Penh 1969?] [mf ed 1989] – 1r with other items – 1 – (also available in english [mf-10289 seam reel 002/01]) – mf#mf-10289 seam reel 002/02 [§] – us CRL [915]
Le guide du colon francais au canada – Ottawa?: s.n, 1886 – 1mf – 9 – mf#12223 – cn Canadiana [971]
Guide du colon, province de quebec / Quebec (Province). Departement des terres de la couronne et al – Levis: Mercier & cie, 1877-1944 [mf ed 1987] – 9 – mf#SEM105P618 – cn Bibl Nat [350]
Le guide du concert; hebdomadaire musical illustre – v. 1-32. 1910-52 – 1 – us L of C Photodup [780]
Guide du cotillon : et les danses de salon / Bail, P & Stilb, G – Paris: Maison Bail, 1895 (mf ed 1988) – 1r – 1 – (incl directions for dancing, and floor diagrams) – mf#ZZ-29,063 – us NY Public [790]
La guide du cultivateur : ou cours d'argiculture / Rouleau, Charles-Edmond – [Quebec?: s.n.] 1890 [mf ed 1981] – 5mf – 9 – 0-665-12930-0 – mf#12930 – cn Canadiana [630]
Guide du diplomate guineen – Conakry: Imprimerie Nationale "Patrice Lumumba", 1979 – us CRL [980]
Guide du jeune homme : recueil de prieres suivi du petit office de la sainte-vierge... – Montreal: Cadieux & Derome, 1882 [mf ed 1984] – 6mf – 9 – 0-665-45340-X – (in french and latin) – mf#45340 – cn Canadiana [230]
Guide du jeune pianiste : classifications methodique et graduee d'oeuvres diverses pour piano... / Eschmann, Johann Carl; ed by Dussault, Joseph Daniel – Lotbiniere, 1886 [mf ed 1990] – 3mf – 9 – mf#SEM105P11 – cn Bibl Nat [780]
Guide du pelerin a sainte-anne de beaupre – Montreal: C O Beauchemin & fils, 1900 [mf ed 1992] – 1mf – 9 – mf#SEM105P1741 – cn Bibl Nat [241]
Guide du voyageur a lausanne et dans ses environs – Lausanne 1834 [mf ed Hildesheim 1995-98] – 1mf [ill] – 9 – €40.00 – 3-487-29386-2 – gw Olms [914]
Guide du voyageur au congo belge et au ruanda-urundi / Belgium Office Du Tourisme Du Congo Belge Et Du Ruanda-Urundi – Bruxelles, Belgium. 1954 – 1r – us UF Libraries [960]
Guide du voyageur en abyssinie / Afevork, G J – Rome, Paris, 1908 – 3mf – 9 – mf#NE-20280 – ne IDC [916]
Guide du voyageur en espagne / Bory de Saint-Vincent, Jean B de – Paris 1823 [mf ed Hildesheim 1995-98] – 5mf – 9 – €100.00 – 3-487-29835-X – gw Olms [914]
Guide du voyageur en france / Audin, Jean M – Paris [circa 1824] [mf ed Hildesheim 1995-98] – 4mf – 9 – €120.00 – 3-487-29692-6 – gw Olms [914]
Guide du voyageur en suisse / Martyn, Thomas – Lausanne 1790 [mf ed Hildesheim 1995-98] – 2mf – 9 – €60.00 – 3-487-29391-9 – gw Olms [914]
Guide economique de la republique d'haiti / Institut Haitien De Statistique – Port-Au-Prince, Haiti. 1964 – 1r – us UF Libraries [330]
A guide for notaries public and commissioners. / Barber, Gershom Morse – 3d ed. Cleveland 1894. 151 p. LL-6 – 1 – us L of C Photodup [340]
Guide for prescribed fire in southern forests / Wade, Dale D – Atlanta, GA. 1989 – 1r – us UF Libraries [500]
Guide for the perplexed / Maimonides, Moses – London, England. 1904 – 1r – us UF Libraries [939]
Guide for travelers : time tables of ocean and river steamers, railways, street cars and omnibus lines – [Quebec?: s.n.] 1884 [mf ed 1984] – 1mf – 9 – 0-665-45575-5 – (in english and french) – mf#45575 – cn Canadiana [380]
Guide & gazette : broughty ferry guide, carnoustie gazette, monifieth advertiser – Arbroath: Arbroath Herald Ltd 1983- (wkly) [mf ed 4 jan 1992-] – 1 – (cont: broughty ferry guide and carnoustie gazette, monifieth advertiser; suppl: arbroath & angus today) – uk Scotland NatLib [072]
Guide [hagerstown] – Hagerstown MD 1953-73 – 1 – ISSN: 0017-5226 – mf#3019 – us UMI ProQuest [240]
Guide illustre de montreal et de ses institutions catholiques : avec programme de la st jean baptiste pour 1884 / Giroux, Henri – Montreal: [s.n.], 1884 – 2mf – 9 – mf#03443 – cn Canadiana [971]
Guide illustre du sylviculteur canadien / Chapais, Jean Charles – Quebec: J A Langlais, 1891 – 3mf – 9 – (incl ind) – mf#26921 – cn Canadiana [634]
Guide indispensable au peuple : contenant l'adresse des principales maisons de commerce, liste de membres du conseil de ville et des comites... / Beauchamp, Joseph – [Quebec: s.n.], 1892 – 1mf – 9 – 0-665-03036-3 – mf#03036 – cn Canadiana [917]
Guide indispensable au peuple : contenant l'adresse des principales maisons de commerce, liste de membres du conseil de ville et des comites... / Beauchamp, Joseph – Quebec?: s.n, 1893 – 1mf – 9 – mf#55833 – cn Canadiana [030]
Guide lines for educational psychologists : in the therapeutical application of the medical hypnoanalysis with anxiety clients / Roets, Susanna – Uni of South Africa 2001 [mf ed Johannesburg 2001] – 6mf – 9 – (incl bibl ref) – mf#mfm14860 – sa Unisa [150]
Guide map and history of the klondike, alaska gold fields – Chicago: L M Lord, [1898?] [mf ed 1981] – 1mf – 9 – mf#15072 – cn Canadiana [622]
Guide marks for young churchmen = Recent past. selections / Wilmer, Richard Hooker – New York: Thomas Whittaker, 1889 – 1mf – 9 – 0-524-04545-3 – mf#1990-5052 – us ATLA [240]
Le guide musical. revue internationale de la musique et des theatres lyriques – v. 1-61, no. 38-40. 1855-1917 – 1 – us L of C Photodup [780]
The guide of the perplexed of maimonides = Dalalat al-haririn / Maimonides, Moses – London: Truebner 1885 [mf ed 1993] – (= ser The english and foreign philosophical library 28-30) – 3v on 3mf – 9 – 0-524-08303-7 – (trans into english & ann by michael friedlaender) – mf#1993-4008 – us ATLA [270]
Guide parlementaire historique de la province de quebec, 1792 a 1902 / Desjardins, Joseph – Quebec: [s.n.], 1902 [mf ed 1988] – 5mf – 9 – (with ind) – mf#SEM105P960 – cn Bibl Nat [323]
Guide pittoresque aux eaux d'aix en savoie – Paris 1834 [mf ed Hildesheim 1995-98] – 2mf [ill] – 9 – €60.00 – 3-487-29208-4 – gw Olms [914]
Guide pittoresque du voyageur en belgique / Ferrier de Tourettes, Alexandre – Bruxelles 1839 [mf ed Hildesheim 1995-98] – 2mf – 9 – €60.00 – 3-487-29663-2 – gw Olms [914]
Guide pittoresque du voyageur en france : contenant la statistique et la description complete des 86 departements / Girault de Saint-Fargeau, Eusebe – Paris 1838 [mf ed Hildesheim 1995-98] – 6v on 34mf – 9 – €340.00 – 3-487-29771-X – gw Olms [914]
Guide pour l'application des lois sociales / Indochina. French – Saigon: Impr. de l'Union, 1936. 45p. LL-10007 – 1 – us L of C Photodup [340]
Guide pratique pour la recherche et l'exploitation / Levat, Edouard David – Paris, France. 1898 – 1r – us UF Libraries [972]
Guide pratique pour le choix des professions feminines / Bureau, Helene – Paris: Colin, 1921 – (= ser Les femmes [coll]) – 2mf – 9 – fr Bibl Nationale [331]
Guide rock signal – Guide Rock, NE: S B Newmeyer (wkly) [mf ed v1 n2-v2 n48. feb 19 1883-84 (lacks v2 n5 1884)] – 1r – 1 – (cont by: guide rock signal and republican valley farmer) – us NE Hist [071]
Guide rock signal – Guide Rock, NE: H Vaughan. -v75 n14. apr 8 1957 (wkly) [mf ed v25 n51-v75. jan 3 1908-57 (gaps)] – 15r – 1 – (cont: guide rock weekly signal, republican valley farmer. absorbed by: commercial advertiser) – us NE Hist [071]
Guide rock signal and republican valley farmer – Guide Rock, NE: S B Newmeyer (wkly) [mf ed v4 n1-v52. jan 9 1886-67] – 1r – 1 – (cont: guide rock signal. cont by: guide rock weekly signal, republican valley farmer. publ as: guide rock weekly signal, republican valley farmer, may 1886) – us NE Hist [071]
The guide rock weekly newsletter – Guide Rock, NE: S B Newmeyer. 2v. feb 1 1906-v2 n15. may 9 1907 (wkly) [mf ed with gaps] – 1r – 1 – us NE Hist [071]
Guide through ireland / Fraser, James, of Dublin – Dublin, 1838 – (= ser 19th c ireland) – 7mf – 9 – mf#1.1.6585 – uk Chadwyck [914]
A guide through the royal porcelain works / Worcester Royal Porcelain Co Ltd – [Worcester 1885?] – (= ser 19th c art & architecture) – 1mf – 9 – mf#4.2.537 – uk Chadwyck [730]
A guide to all the watering and sea-bathing places : with a description of the lakes; a sketch of a tour in wales, and various itineraries / Feltham, John – London 1815 [mf ed Hildesheim 1995-98] – (= ser Fbc) – 4mf – 9 – €120.00 – 3-487-28825-7 – gw Olms [914]
Guide to archives and manuscripts at harvard and radcliffe : 3,800 descriptive entries for individual manuscript collections – [mf ed Chadwyck-Healey, 1990] – 17mf – 9 – 0-89887-081-X – (with p/g) – uk Chadwyck [090]
A guide to biblical study / Peake, Arthur Samuel – New York: Dodd Mead, 1897 [mf ed 1986] – 1mf – 9 – 0-8370-9408-9 – (int by a m fairbairn) – mf#1986-3408 – us ATLA [220]
Guide to botswana / Winchester-Gould, G A – Gaberone, Botswana. 1968 – 1r – us UF Libraries [960]
Guide to british west indian archive materials / Bell, Herbert Clifford Francis – Washington, DC. 1926 – 1r – us UF Libraries [972]
Guide to buddhahood : being a standard manual of chinese buddhism / Hsuan Fo Pu – Shanghai: Christian Literature Society, 1907 – 1mf – 9 – 0-524-02350-6 – mf#1990-2961 – us ATLA [240]
A guide to burghley house / Blore, Thomas – Stamford [1815?] – (= ser 19th c art & architecture) – 2mf – 9 – mf#4.1.427 – uk Chadwyck [720]
"Guide to captured german documents" (maxwell air force base, alabama, dec 1952) and "supplement" (national archives, washington, dc, 1959) / U.S. National Archives and Records Service – (= ser Records of the national archives and records administration) – 1r – 1 – mf#T1183 – us Nat Archives [324]
Guide to churchmen about baptism and regeneration / Ryle, J C – Ipswich, England. 1857 – 1r – us UF Libraries [240]
Guide to church-reform / Duncombe, Edward – London, England. 1833 – 1r – us UF Libraries [240]
Guide to commercial shark fishing in the caribbean area / Caribbean Commission – Washington, DC. 1945 (I.E. 1947) – 1r – us UF Libraries [500]
A guide to commissioners in chancery / Matthews, James Muscoe – 2d ed. Richmond, VA: Randolph and English, 1871. 254p. LL-912 – 1 – us L of C Photodup [340]
Guide to current official statistics. / India. Office of the Economic Advisor – Delhi: Manager of Publ, [1945-49] – 1 – (filmed with: india: imperial record dept list of the heads of administrations in india and of the india office in england) – us CRL [954]
The guide to eaton hall – [Manchester 1885] – (= ser 19th c art & architecture) – 1mf – 9 – mf#4.1.219 – uk Chadwyck [720]
A guide to elephanta / Sastri, Hiranand – Delhi: Manager of Publ, 1934 – (= ser Samp: indian books) – us CRL [915]
A guide to federal agency rulemaking / Administrative Conference of the US (ACUS) – 1st ed 1983. Washington: GPO, 1983 (all publ) – 4mf – 9 – $6.00 – mf#LLMC 94-336A – us LLMC [340]
A guide to federal agency rulemaking / Administrative Conference of the US (ACUS) – 2nd ed 1991. Washington: GPO, 1991 (all publ) – 5mf – 9 – $7.50 – mf#LLMC 94-336B – us LLMC [340]
A guide to figure drawing / Hicks, George Edgar – London 1853 – (= ser 19th c art & architecture) – 1mf – 9 – mf#4.1.386 – uk Chadwyck [740]

A guide to figure painting in water-colours / Whiteford, Sydney T – London [1870?] – (= ser 19th c art & architecture) – 2mf – 9 – mf#4.1.418 – uk Chadwyck [750]

Guide to florida – New York, NY. 1873 – 1r – us UF Libraries [630]

Guide to florida / Rambler – New York, NY. 1875 – 1r – us UF Libraries [630]

A guide to health / Gandhi, Mahatma – Madras: S Ganesan, 1921 – (= ser Samp: indian books) – (trans by a rama iyer) – us CRL [613]

Guide to historic quebec and its principal business houses / Le Moine, James MacPherson [comp] – [Quebec?: s.n, between 1890 and 1900] – 1mf – 9 – 0-665-94184-6 – mf#94184 – cn Canadiana [971]

Guide to historic quebec and lower st-lawrence – Quebec: H H Wright, 1892 – 2mf – 9 – mf#04394 – cn Canadiana [917]

Guide to jamaica / Bowen, Calvin – Kingston, Jamaica. 1958? – 1r – us UF Libraries [972]

Guide to jamaica / Olley, Philip Peter – Glasgow, Scotland. 1937 – 1r – us UF Libraries [972]

A guide to magistrates and constables. / Pollard, John Garland – Richmond, Va.: Waddey, 1906. LL-1336 – 1 – us L of C Photodup [340]

A guide to magistrates, with practical forms for the discharge of their duties out of court / Mayo, Joseph – 2d ed. Richmond: Morris, 1860. 726p. LL-291 – 1 – (ibid. richmond: goode, 1892. 711p. ll-965) – us L of C Photodup [347]

Guide to marine insurance : being a handbook on the law and practice of marine insurance with special reference to policies on goods / Keate, Henry – London: Pitman & Sons, (post 1919?) – 3mf – 9 – $4.50 – mf#LLMC 92-120 – us LLMC [368]

Guide to mars / Moore, Patrick – New York, NY. 1957 – 1r – us UF Libraries [500]

Guide to negro periodical literature – Jefferson City, Missouri. v. 1-4. Feb 1941-Feb 1943; Jan-Sept 1946 – 1 – us NY Public [305]

A guide to new brunswick, british north america etc / Atkinson, Christopher William – Edinburgh?: s.n, 1843 (Edinburgh: Anderson & Bryce) – 3mf – 9 – mf#28524 – cn Canadiana [917]

A guide to nizamu-d din / Hasan, Zafar, Khan Bahadur – Calcutta: Supt, Govt Print, India, 1922 – (= ser Samp: indian books) – us CRL [915]

A guide to painting on glass / Bielfeld, H – London 1855 – (= ser 19th c art & architecture) – 1mf – 9 – mf#4.2.1038 – uk Chadwyck [740]

A guide to preachers / Garvie, Alfred Ernest – New York: GH Doran [1906?] [mf ed 1990] – 1mf – 9 – 0-7905-3936-5 – mf#1989-0429 – us ATLA [240]

A guide to punjab government reports and statistics / Fazal, Cyril P K – Lahore, The Civil and Military Gazette, 1939 – us CRL [324]

A guide to rawlinson c745-747 (bodleian library, oxford) : "correspondance from the outforts to cape coast castle, 1681-1699" – Madison. University of Wisconsin, 1972 – (filmed with royal african co of england correspondence) – us CRL [960]

A guide to reading the hebrew text : for the use of beginners / Vibbert, William H – Andover: Warren F Draper, 1872 [mf ed 1986] – 1mf – 9 – 0-8370-9328-7 – mf#1986-3328 – us ATLA [270]

A guide to records relating to ghana in repositories in the u.k. excluding the public record office / Agyei, Samuel Kwasi – 1988 – 7mf – 9 – £35.00 – (with guide) – mf#(Altair) – uk Microform Academic [324]

Guide to salvation : the life and teachings of jesus christ. designed expressly for universalist sunday schools / Fletcher, L J – Boston: Universalist Pub House, c1863 – (= ser Doctrinal Series (Boston, Mass.)) – 1mf – 9 – 0-524-07679-0 – mf#1991-3264 – us ATLA [240]

A guide to sanchi / Marshall, John Hubert – Delhi: Manager of Publ, 1936 – (= ser Samp: indian books) – us CRL [930]

Guide to selected legal sources of mainland china / Hsia, Tao-tai – Washington: Library of Congress, 1967. 357p. LL-10036 – 1 – us L of C Photodup [340]

Guide to standard shona spelling – Chishawasha, Zimbabwe. 1955 – 1r – us UF Libraries [470]

Guide to swaziland / Andrews, Bruce – Johannesburg, South Africa. 1970 – 1r – us UF Libraries [960]

A guide to taxila / Marshall, John Hubert – Delhi: Manager of Publ, 1936 – (= ser Samp: indian books) – us CRL [930]

A guide to the antiquities of upper egypt : from abydos to the sudan frontier / Weigall, Arthur Edward Pearse Brome – New York: Macmillan, 1910 [mf ed 1992] – 2mf – 9 – 0-524-05113-5 – mf#1992-0334 – us ATLA [930]

Guide to the art of illuminating and missal painting / Audsley, William James & Audsley, George Ashdown – [2nd ed] London 1861 – (= ser 19th c art & architecture) – 2mf – 9 – mf#4.1.253 – uk Chadwyck [740]

A guide to the churches and missions in the city of new york / New York. (City) – 1877 – 9 – $50.00 – us Presbyterian [240]

Guide to the city of london, ontario, canada – [London, Ont?: London Print & Litho Co], 1892 – 1mf – 9 – 0-665-91095-9 – mf#91095 – cn Canadiana [917]

Guide to the city of quebec : descriptive and illustrated with map / Carrel, Frank – Quebec: F Carrel, 1915 – 3mf – 9 – (incl some text in french) – mf#98704 – cn Canadiana [917]

Guide to the fishing and hunting resorts in the vicinity of the grand trunk railway of canada : containing particulars of fish, game, hotels, livery and general facilities – S.l: s.n, 1890? – 2mf – 9 – mf#58127 – cn Canadiana [790]

A guide to the french language : consisting of vocabulary, verbs, dialogues, and exercises / Appleton, Elizabeth – London: printed for G & W B Whittaker, 1824 – (= ser 19th c children's literature) – 4mf – 9 – mf#6.1.14 – uk Chadwyck [440]

Guide to the health act, n63 of 1977 – [Pretoria: Dept of Health 1978] [mf ed Pretoria, RSA: State Library [199-]] – 51p on 1r with other items – 5 – mf#op 06480 r23 – us CRL [344]

Guide to the index of early southern artists and artisans / Museum of Early Southern Decorative Arts (MESDA) [comp] – Clearwater Publ Co – 36mf (24:1) – 9 – $315.00 – us UPA [740]

Guide to the klondike and the yukon gold fields in alaska and northwest territories : containing history of the discovery, routes of travel, necessary outfit, general and useful information, large map, corrected up to date from latest official surveys – Seattle, WA: Lowman & Hanford, 1897 [mf ed 1981] – 2mf – 9 – mf#15074 – cn Canadiana [622]

Guide to the knowledge of god : a study of the chief theodicies = De la connaissance de dieu / Gratry, Auguste – Boston: Roberts Bros, 1892 [mf ed 1991] – xi/469p on 2mf – 9 – 0-7905-9942-2 – (trans fr french into english by abby langdon alger. int by william rounseville alger) – mf#1989-1667 – us ATLA [210]

A guide to the knowledge of pottery, porcelain, and other objects of vertu : comprising an illustrated catalogue of the bernal collection of works of art / Bohn, Henry George – London. H G Bohn, 1857 – (= ser 19th c art & architecture) – 7mf – 9 – mf#4.1.55 – uk Chadwyck [730]

A guide to the lakes, in cumberland, westmorland, and lancashire / Robinson, John – London 1819 [mf ed Hildesheim 1995-98] – (= ser Fbc) – 1v on 3mf – 9 – €90.00 – 3-487-27355-1 – gw Olms [914]

Guide to the laws of england affecting roman catholics / Anstey, Thomas Chisholm – London, England. 1842 – 1r – us UF Libraries [241]

Guide to the legal profession / Maxwell, Maurice W – 4th ed. London: Sweet and Maxwell, 1994 – 2mf – 9 – $3.00 – mf#LLMC 91-068 – us LLMC [340]

Guide to the lord's table / Belfrage, Henry – Edinburgh, Scotland. 1825 – 1r – us UF Libraries [240]

A guide to the manuscript materials for the history of the united states to 1783 : in the british museum, minor london archives and the libraries of oxford and cambridge / Andrews, Charles & Davenport, Frances G – 1r – 1 – mf#2297 – uk Microform Academic [975]

Guide to the manuscript materials for the history of the united states to 1783 : in the british museum, in minor london archives, and in the libraries of oxford and cambridge / Andrews, Charles McLean – Washington, DC: Carnegie Institution of Washington, 1908 – 2mf – 9 – mf#1988-0369 – us ATLA [975]

Guide to the manuscripts and printed books exhibited in celebration of the tercentenary of the authorized version – [s.l]: printed by order of the trustees 1911 (Oxford: Horace Hart) [mf ed 1989] – 1mf – 9 – 0-7905-1911-9 – mf#1987-1911 – us ATLA [012]

Guide to the materials for american history / Perez, Luis Marino – Washington, DC. 1907 – 1r – us UF Libraries [972]

A guide to the materials for american history to 1783 in the public record office, london / Andrews, Charles – Washington, DC. 2v. 1912 – 1r – 1 – mf#2298 – uk Microform Academic [975]

Guide to the materials for american history, to 1783, in the public record office of great britain / Andrews, Charles McLean – Washington, D.C.: Carnegie Institution of Washington, 1912-1914 – (= ser Carnegie Institution Of Washington Publication) – 2mf – 9 – 0-7905-5443-7 – mf#1988-1443 – us ATLA [975]

A guide to the materials in london archives for the history of the united states since 1783 / Paullin, Charles O & Paxon, Frederick L – Washington, 1914 – 1r – 1 – mf#2337 – uk Microform Academic [025]

Guide to the mushrooms / Cole, Emma L Taylor – Toronto: Musson, c1910 – 3mf – 9 – 0-665-88058-8 – (ill by a w cole. incl glos) – mf#88058 – cn Canadiana [580]

A guide to the old observatories at delhi, jaipur, ujjain, benares / Kaye, George Rusby – Calcutta: Supt Govt Print, India, 1920 – (= ser Samp: indian books) – us CRL [520]

A guide to the principal gold and silver coins of the ancients / Head, Barclay Vincent – 4th ed. London 1895 – (= ser 19th c art & architecture) – 3mf – 9 – mf#4.2.1528 – uk Chadwyck [730]

A guide to the proper regulation of buildings in towns / Hosking, William – London 1848 – (= ser 19th c art & architecture) – 4mf – 9 – mf#4.2.1746 – uk Chadwyck [710]

A guide to the public records of southern rhodesia, under the regime of the british south africa company, 1890-1923 – Cape Town, The Archives in association with Longmans, Green, 1956 – us CRL [960]

Guide to the records of the german navy, 1850-1945 / Germany. German Navy – (= ser National archives coll of foreign records seized, 1941) – 1r – 1 – mf#M1743 – us Nat Archives [355]

Guide to the records of the guntur district, 1795-1835 / Madras. (India: State). Record Office – Madras: Printed by the Superintendent, Govt. Press, 1934 – 1 – us Wisconsin U Libr [954]

Guide to the records of the masulipatam district, 1682 to 1835 / Madras. (India: State). Record Office – Madras: Printed by the Superintendent, Govt. Press, 1935 – 1 – us Wisconsin U Libr [954]

Guide to the records of the nellore district, 1801 to 1835 / Madras. (India: State). Record Office – Madras: Printed by the Superintendent, Govt. Press, 1934 – 1 – us Wisconsin U Libr [954]

Guide to the records of the tinnevelly district, 1796 to 1835 / Madras. (India: State). Record Office – Madras: Printed by the Superintendent, Govt. Press, 1934 – 1 – us Wisconsin U Libr [954]

A guide to the religions of america : the famous look magazine series on religion, plus facts, figures, tables, charts, articles... / Look; ed by Rosten, Leo Calvin – [New York]: Simon & Schuster, 1955 [mf ed 1987] – xiii/282p – 1 – (contains complete series of articles on religion publ in look from 1952-55) – mf#8167 – us Wisconsin U Libr [243]

A guide to the royal architectural museum / Scott, George Gilbert – London [1877] – (= ser 19th c art & architecture) – 1mf – 9 – mf#4.2.949 – uk Chadwyck [720]

Guide to the ruins of kilwa / Chittick, H Neville – Dar es Salaam, Tanzania. 1965 – 1r – us UF Libraries [960]

A guide to the sculptures in the indian museum : early indian schools / Majumdar, Nani Gopal – Delhi: Archaeological Survey of India, 1937- – (= ser Samp: indian books) – us CRL [730]

Guide to the sources in the netherlands concerning the history of asia and oceania : 1796-1949 / ed by Jaquet, F G P – Leiden, 1973-1976 – 229mf – 9 – mf#SE-12103 – ne IDC [950]

Guide to the soviet union / Mandel, William M – New York, NY. 1946 – 1r – us UF Libraries [025]

A guide to the study of church history / McGlothlin, William Joseph – [new rev ed.] New York: Hodder & Stoughton: G H Doran, c1914 [mf ed 1990] – 1mf – 9 – 0-7905-5068-7 – mf#1988-1068 – us ATLA [240]

Guide to the study of dragonflies of jamaica / Whitehouse, Francis Cecil – Kingston, Jamaica. 1943 – 1r – us UF Libraries [590]

A guide to the study of the christian religion / Faunce, William Herbert Perry et al; ed by Smith, Gerald Birney – Chicago, IL: University of Chicago Press, c1916 [mf ed 1991] – 2mf – 9 – 0-7905-8648-7 – (incl bibl ref) – mf#1989-1873 – us ATLA [220]

A guide to the study of the new testament *see* Hsin-yueh yen chiu chih nan (ccm96)

A guide to the study of the old testament *see* Chiu yueh yen chiu chih nan [ccm216]

A guide to the study of theology : adapted more especially to the oxford honour school / Woods, Francis Henry – Oxford: James Thornton, 1880 [mf ed 1985] – 1mf – 9 – 0-8370-5902-X – (incl bibl) – mf#1985-3902 – us ATLA [240]

A guide to the tablets in a temple of confucius / Watters, Thomas – Shanghai, China: American Presbyterian Mission Press, 1879 [mf ed 1992] – 1mf – 9 – 0-524-02726-9 – (incl bibl ref) – mf#1990-3129 – us ATLA [290]

A guide to the textual criticism of the new testament / Miller, Edward – London: George Bell, 1886 [mf ed 1989] – 1mf – 9 – 0-7905-1246-7 – (incl bibl ref & ind) – mf#1987-1246 – us ATLA [225]

Guide to the university of illinois archives / Brichford, Maynard & Maher, William – 3mf + 10p – 9 – $10.00 – us Univ Ill Libr [020]

Guide to the west indies and bermudas / Ober, Frederick Albion – New York, NY. 1908 – 1r – us UF Libraries [972]

Guide to the zimbabwe ruins / Jones, Neville – s.l, s.l? 1951 – 1r – us UF Libraries [960]

Guide to wenli styles and chinese ideals : essays, edicts, proclamations, memorials, letters, documents, inscriptions, commercial papers / Morgan, Evan – Shanghai: Christian Literature Society for China; London: Probsthain, 1912 [mf ed 1995] – (= ser Yale coll) – vi/414p – 1 – 0-524-09459-4 – (in chinese) – mf#1995-0459 – us ATLA [480]

Guidebook for days and deeds / Gray, William Scott – Chicago, IL. 1943 – 1r – us UF Libraries [025]

A guide-book in the administration of the discipline of the methodist episcopal church / Baker, Osmon Cleander – rev ed. New York: Carlton & Porter, 1862, c1855 [mf ed 1991] – (= ser Methodist coll) – 1mf – 9 – 0-524-00962-7 – mf#1990-4020 – us ATLA [242]

A guide-book to the poetic and dramatic works of robert browning / Cooke, George Willis – Boston: Houghton Mifflin, c1919 [mf ed – xvi/451p – 1 – (incl ind) – mf#1290 – us Wisconsin U Libr [420]

Guideline sentencing : fjc in-court educational program on guideline sentencing orientation for u.s. district and circuit judges, u.s. magistrates, u.s. probation officers, supporting staff and federal public defenders – Washington: FJC, Oct 1987 – 6mf – 9 – $9.00 – mf#LLMC 95-389 – us LLMC [347]

Guideline sentencing : an outline of appellate case law on selected issues / Wood, Jefri & Sheehey, Diane – Washington: FJC, Sept 1994 – 3mf – 9 – $4.50 – mf#LLMC 95-390 – us LLMC [340]

Guide-lines for day care / South Africa. Department of National Health and Population Development [Departement van Nasionale Gesondheid en Bevolkingsontwikkeling – Pretoria: Dept of National Health & Population Development 1993 [mf ed Pretoria, RSA: State Library [199-]] – 1r with other items – 5 – (incl bibl ref) – mf#op 11276 r25 – us CRL [362]

Guidelines for dealing with learners with adhd behaviour / Holz, Tania Jacqueline – Uni of South Africa 2001 [mf ed Johannesburg 2001] – 4mf – 9 – (incl bibl ref) – mf#mfm14983 – sa Unisa [618]

Guidelines for docket clerks – 1979 – 1mf – 9 – $1.50 – mf#llmc99-011 – us LLMC [347]

Guidelines for improving juror utilization in u.s. district courts – Washington: FJC, Oct 1972 – 1mf – 9 – $1.50 – mf#LLMC 95-820 – us LLMC [347]

Guidelines for parents, teachers and professionals in the handling of rebellious children / Mathye, Lethabo Violet – Uni of South Africa 2000 [mf ed Johannesburg 2000] – 4mf – 9 – (incl bibl ref) – mf#mfm14749 – sa Unisa [150]

Guidelines for prescribing upper body exercise following open heart surgery / Huenerbein, Heidi A – University of Wisconsin-La Crosse, 1995 – 1mf – 9 – mf#PH 1497 – us Kinesology [617]

Guidelines for pupil services – v7 n3, v9 n1-v17 n3 [1969 may, 1971 mar-1980 spring] – 1r – 1 – mf#295969 – us WHS [370]

Guidelines for the safe transport of radioactive material / South Africa. Directorate Electromedical Devices and Radiological Health [comp] – rev ed. [Pretoria]: Dept of National Health & Population Development 1994 [mf ed Pretoria, RSA: State Library [199-]] – 6p on 1r with other items – 5 – (incl bibl ref) – mf#op 12117 r26 – us CRL [360]

Guides and guards in character-building / Payne, C H – New York: Phillips & Hunt; Cincinnati: Walden & Stowe, 1883 – 1mf – 9 – 0-8370-4680-7 – mf#1985-2680 – us ATLA [220]

Guides to german records microfilmed at alexandria / National Archives Trust Fund Board. Washington, DC – (= ser Records of the national archives and records administration) – 4r – 1 – mf#T733 – us Nat Archives [324]

Guides to records of the italian armed forces / U.S. National Archives and Records Service – (= ser Records of the national archives and records administration) – 1r – 1 – mf#T94 – us Nat Archives [355]

Guidi, I *see*
– Proverbi, strofe e racconti abissini
– Storia della letteratura etiopica

Guidi, Ignazio *see* Vocabulario amarico-italiano

Guidi, Jean *see* Lettres contenant le journal d'un voyage fait a rome en 1773

[Guidiccioni, L] *see* Breve racconto della trasportatione del corpo di papa paolo v dalla basilica di s pietro a'quella di s maria maggiore...
Guiding light / Church of the Lord Jesus Christ of the Apostolic Faith – v1 n9 [1949 feb] – 1r – 1 – mf#4023576 – us WHS [240]
Guido and julius : or, sin and the propitiator = Lehre von der suende und vom versoehner / Tholuck, August – Boston: Gould & Lincoln 1854 [mf ed 1993] – 1mf – 9 – 0-524-06194-7 – (trans by jonathan edwards ryland. int pref by john pye smith) – mf#1991-2450 – us ATLA [240]
Guido and lita : a tale of the riviera / Argyll, John Douglas Sutherland Campbell, Duke of – Toronto: J Campbell, 1878 – 2mf – 9 – mf#26271 – cn Canadiana [830]
Guido fischer (1877-1959) : materialien zu einer biographie / Wolf, Manfred – (mf ed 1995) – 2mf – 9 – €40.00 – 3-8267-2237-X – mf#DHS 2237 – gw Frankfurter [610]
Guido fridolin verbeek [i e verbeck] / Adriani, J H – [Rotterdam: J M Bredee, 1908] [mf ed 1995] – (= ser Yale coll; Lichtstralen op den akker der wereld [14. jaarg 1908] 5-6) – 33p (ill) – 1 – 0-524-09661-9 – (in dutch) – mf#1995-0661 – us ATLA [920]
Guido verbeck, fodt 1830, dod 1898 : et blad af det moderne japans historie / Rafn, Holger – Kobenhavn: Danske missionsselskab, 1916 [mf ed 1995] – (= ser Yale coll) – 45p – 1 – 0-524-09701-1 – (in danish) – mf#1995-0701 – us ATLA [950]
Guidon / United States Army Training Center [Fort Leonard Wood, MO] – Fort Leonard Wood MO. v15 n30-v18 n25 [1981 jan 22-dec 15] – 1r – 1 – (cont: fort leonard wood guidon) – mf#714699 – us WHS [355]
Guidon / United States Army Training Center [Fort Leonard Wood, MO] – Fort Leonard Wood MO. v21 n7-26 [1986 jul 31-dec 18] – 1r – 1 – (cont: fort leonard wood guidon [1984]; cont by: fort leonard wood guidon [1987]) – mf#1345495 – us WHS [355]
Gueysse, Paul *see* Rituel funeraire egyptien
Guignard, Ph *see*
– Monuments primitifs de la regle cistercienne
– Les monuments primitifs de la regle cistercienne
Guignard, Rene *see* Un poete romantique allemand
Guignebert, Charles *see*
– L'evolution des dogmes
– The jewish world in the time of jesus
– Modernisme et tradition catholique en france
Guignes, C L J de *see*
– Planisphere celeste, chinois...
– Voyages...peking, manille et l'ele de france
Guignes, Chretien L de *see* Voyages a peking, manille et l'ile de france, faits dans l'intervalle des annees 1784 a 1801
Guignes, Chretien-Louis-Joseph de *see* Dictionnaire chinois, francais et latin
Guignol et la revolution dans l'eglise romaine : m. veuillot et son parti condamnes par les archevaeques et evaeques de paris, tours, viviers, orleans, marseille, verdun, chartres, moulins, etc / Michaud, Eugene – Paris: Sandoz et Fischbacher, 1872 – 1mf – 9 – 0-8370-8772-4 – mf#1986-2772 – us ATLA [944]
Guihot, Julien-Marie *see* Discours prononce par l'abbe jul guihot pretre de st sulpice a l'occasion du cinquantenaire des oblats a montreal le 8 decembre 1891
El guijo, belalcazar y capilla : nuevas inscripciones romanas / Fita, Fidel – Madrid: Fortanet, 1912 – 1 – sp Bibl Santa Ana [946]
Guilbaud *see* Etapes de la guadeloupe religieuse
Guilbault, Germaine *see* Bibliographie analytique de l'oeuvre de son excellence rev'me mgr napoleon-alexandre labrie...
Guilbault, Renee *see* Bibliographie analytique de l'oeuvre de bertrand vac
Guild bulletin / North Jersey Newspaper Guild – v21 n6-v25 n12 [1964 jun-1968 dec, 1969 jan-1976 oct 25] – 1r – 1 – mf#646908 – us WHS [071]
Guild Cooperative Fellowship *see* Second city
Guild, Curtis *see* Discours prononce a vienne, france, le 12 aoaut 1909
Guild forum / Washington-Baltimore Newspaper Guild – v3 n4-v6 n1 [1979 apr 16-1982 jan] – 1r – 1 – mf#614873 – us WHS [071]
Guild journal / Central California Newspaper Guild, Local 92 – 1975 feb-1985 jul – 1r – 1 – mf#1477294 – us WHS [071]
Guild news : the voice of the metropolitan guild of pharmacists / Metropolitan Guild of Pharmacists – v2 n3-v3 n1 [1975 may-1977 mar, 1977 sep-1980 nov] – 1r – 1 – mf#679148 – us WHS [615]
Guild newsletter / Civil Service Technical Guild – 1981 mar-1992 dec – 1r – 1 – (cont: cstg press; cont by: tech guild news) – mf#1110799 – us WHS [350]
Guild notes / National Lawyers Guild – 1976 feb-1984 dec – 1r – 1 – mf#772777 – us WHS [340]
Guild notes / National Lawyers Guild. National Office – New York NY 1989+ – 1,5,9 – ISSN: 0148-0588 – mf#17452 – us UMI ProQuest [340]
Guild of Shaker Crafts *see* World of shaker
Guild practitioner – v1-58. 1940-2001 – 1,5,6 – $633.00 set – (cont: national lawyers guild quarterly. price incl national lawyers guild quaterly v1-3 1937-40. title varies: v1-20 1940-60 as lawyers guild review, v21-23 1961-63 as law in transition) – ISSN: 0017-5390 – mf#103011 – us Hein [340]
Guild reporter / American Newspaper Guild – 1933 nov 23-1938, 1939-64, 1965 jan 8-1967 dec 22, 1968 jan 12-1970 dec 11, 1971-73, 1974 jan 11-1978 dec 29, 1979-1984 oct – 11r – 1 – mf#3421149 – us WHS [071]
Guild reporter – Washington DC 1973+ – 1,5,9 – ISSN: 0017-5404 – mf#9969 – us UMI ProQuest [331]
Guild, Reuben A *see* Chaplain smith and the baptists
Guilden morden 1598-1950 – (= ser Cambridgeshire parish register transcript) – 7mf – 9 – £8.75 – uk CambsFHS [929]
Guilden sutton, st john – (= ser Cheshire monumental inscriptions) – 1mf – 9 – £2.50 – mf#17 – uk CheshireFHS [929]
Guildhall Library. London *see*
– Catalogue of the guildhall library's major archive and manuscript holdings
– Illuminated manuscripts at the guildhall library, london
– Lloyd's captains registers
Guildpaper : journal of the new york newspaper guild / Newspaper Guild of New York – v4 n6 [1946 aug] – 1r – 1 – (cont: new york guildpaper; cont by: frontpage [new york, ny]) – mf#684543 – us WHS [071]
Guilford county genealogist – v1 n1-v3 [1974 nov-1976 spring] – 1r – 1 – (continued by: guilford genealogist) – mf#1023447 – us WHS [929]
Guilford genealogist – 1976 fall-1984 summer, 1984 fall-1989 sum – 2r – 1 – (cont: guilford county genealogist) – mf#1023448 – us WHS [929]
Guilford, linda thayer, papers, ms 484 – 1855-1906 – 1r – 1 – (writings, speeches, notes, letters, clippings, and a scrapbook concerning miss guilford's activities as a teacher at the cleveland female seminary and cleveland academy, and in the young ladies temperance league) – us Western Res [920]
[Guilford-] ringing world – UK. 1982-1987 – 6r – 1 – $300.00 – mf#R63601 – us Library Micro [072]
Guilhelmus ludovicus comes nassovius... / Emmius, [U] – Groningae, 1621 – 4mf – 9 – mf#PBA-175 – ne IDC [240]
Guilhermus Parisiensis *see* Tractatus de sacramentis et de universo
Guilielmi gesenii, philosophiae et theologiae doctoris . . . thesaurus philologicus criticus linguae hebraeae et chaldaeae veteris testamenti = Thesaurus philologicus criticus linguae hebraeae et chaldaeae veteris testamenti / Gesenius, Wilhelm – Editio altera. Lipsiae: Sumtibus typisque FCG Vogelii, 1829-1858 – 18mf – 9 – 0-8370-1968-0 – (incl bibl ref) – mf#1987-6355 – us ATLA [040]
Guilielmi hesii antverpiensis e societate iesu emblemata sacra de fide, spe, charitate / Hesius, G – Antverpiae: Ex officina Plantiniana Balthasaris Moreti, 1636 – 3mf – 9 – mf#O-631 – ne IDC [090]
Guillain, Charles *see* Voyage a la cote orientale d'afrique
Guillain, M *see* Documents sur l'histoire, la geographie et le commerce de l'afrique orientale
Guillamas, Manuel de *see* De las ordenes militares de calatrava, alcantara y montesa
Guilland, Antoine *see* Modern germany and her historians
Guillard, Nicolas Francois *see* Dardanus
Guillaume *see* Branche des royaux lignages
Guillaume, Alfred *see* Prophecy and divination among the hebrews and other semites
Guillaume d'auvergne, evaeque de paris (1228-1249) : sa vie et ses ouvrages / Valois, Noel – Paris: A Picard, 1880 – 1mf – 9 – 0-7905-6902-7 – (incl bibl ref) – mf#1988-2902 – us ATLA [240]
Guillaume de Vaudoncourt, Frederic *see* Memoirs on the ionian islands
Guillaume farel, 1489-1565 / Comite, Farel – Neuchaatel: Delachaux & Niestle, 1930 – 1r – 1 – 0-8370-1117-5 – mf#1984-T022 – us ATLA [920]
Guillaume, James [comp] *see* Internatsional
Guillaume tell / Le Mierre, Antoine-Marin – Paris, France. 1808 – 1r – us UF Libraries [440]
Guillaumet, Edouard *see*
– Le soudan en 1894
– Le soudan en 1894: la verite sur tombouctou; l'esclavage au soudan
Guillaumin, Emile *see* The life of a simple man
Guillelmus a Sancto Theodorico *see*
– Expositio in epistolam ad romanos
– Expositio super cantica canticorum
Guillemard, William Henry *see* Hebraisms in the greek testament
Guillemin, Amedee *see* The world of comets
Guillemin, Henri *see* Histoire des catholiques francais au 19e siecle (1815-1905)
Guillemin, J A *see* Enumeration des plantes decouvertes
Guillemot, Eugene *see* Affaires de la plata
Guillen, Flavio *see* Fraile procer y una fabula poema
Guillen, Julio F *see*
– El primer viaje de cristobal colon
– Los tenientes de navio jorge juan y santacilia y antonio de ulloa y de la torre g y la medicion del meridiano...
Guillen Martinez, Fernando *see*
– Raiz y futuro de la revolucion
– Secreto y la imagen
Guillen, Nicolas *see*
– Antologia mayor
– Cantos para soldados y sones para turistas
– Claudio jose domingo brindis de salas
– Elegia a jesus menendez
– Espana
– Poemas de amor
– Poesias
– Prosa de prisa
– Puedes
– Son entero
Guillen, Pedro *see* Guatemala
Guillen y Flores, Agustin *see*
– Breves-geografia astronomica
– Discurso...apertura de curso sobre la importancia del catolicismo
Guillet, Didace *see* Cinquantenaire des oblats de marie immaculee en canada
Guillet, Joseph E *see* Propalladia and other works of...
Guilleux, Charles *see* Journal de route d'un caporal de tirailleurs de la mission saharienne, 1898-1900
Guillon, Aime *see*
– Lyon
– Memoires pour servir a l'histoire de la ville de lyon pendant la revolution
Guilloreau, L *see* Les memoires du r p dom bernard audebert (afm11)
Guillot De Saix, Leon *see* Poulet
Guilloteaux, Erique *see* Madagascar et la cote des somalis sainte-marie et les seychelles
La guillotine – Paris: Bonaventure et Ducessois, mar 1848 – 1 – us CRL [074]
Guilmant, A *see* Archives des maitres de l'orgue des 16e, 17e, et 18e siecles
The guilt of slavery and the crime of slaveholding / Cheever, George Barrell – Boston: JP Jewett, 1860 – 2mf – 9 – 0-7905-4618-3 – mf#1988-0618 – us ATLA [976]
Guilty land / Van Rensburg, Patrick – Harmondsworth, England. 1962 – 1r – us UF Libraries [960]
Guilty land / Van Rensburg, Patrick – New York, NY. 1962 – 1r – us UF Libraries [960]
Guilty, or not guilty? / Cox, John – London, England. 18– – 1r – us UF Libraries [240]
Guimaraes, Bernardo *see*
– Escrava Isaura
– Garimpeiro
– Lendas e romances
– Mauricio
– Rosaura, a enjeitada
Guimaraes, Jorge Maia De Oliveira *see* Invasao de mato grosso
Guimaraes, Maria A De Alencastro *see* Outline of brazilian history
Guimaraes, Osias *see* Amor a terra
Guimera, Angel *see* La aranya
Guimmaraens Filho, Alphonsus De *see* Antologia da poesia mineira, fase modernista
Guinan, Diane M *see* Predictive relationships among perceived stress, possible selves, and physical activity in elderly individuals with knee osteoarthritis
Guinard, Jean Louis *see* 'Histoires de mission, pour enfants'
Guindon, Arthur *see* Les trois combats du long-sault
La guinea espanola / Santa Isabel Misioneros Hijos del Immaculado – Corazon de Maria. v1-65 n1-1620. apr 1903-1967; Index v1-37 1903-40 – 1 – us CRL [073]
Guinea. French *see* Journal officiel
Guinea. French. Laws, Statutes, etc *see* Code penal
Guinea, Gerardo *see* Armas para ganar una nueva batalla
Guinea times – Accra: Star Pub Co, apr 3-jun 13 1958 – us CRL [079]
Guinea-Bissau (formerly Portuguese Guinea). Reparticao Provincial dos Servicos de Economia e Estatistica Geral *see* Anuario estatistico 1947-1958
La guinee / Rouget, Fernand – Corbeil, France: Impr typ E Crete, 1906 – 1 – us CRL [960]
La guinee francaise : races, religions, coutumes, production, commerce / Arcin, Andre – Paris: A Challamel, 1907 – 1 – us CRL [960]
La guinee francaise, conakry et rivieres du sud : etude economique et commerciale suivie de notes notes sur la guinee portugaise / Aspe-Fleurimont, Lucien Auguste – Paris: A Challamel, 1900 – 1 – us CRL [916]
Guinee, perspectives nouvelles – Conakry. n30-46. aug 1973-may 1975 – us CRL [980]
La guinee superieure et ses missions : etude geographique, sociale et religieuse des contrees evangelisees par les missionaires de la societe des missions africaines de lyon / Teilhard de Chardin, J – 3. ed. Keerlez-Maastricht, [1888] – 1 – us CRL [960]
Guiness, Lucy Evangeline *see*
– In the far east
– South america, the neglected continent
Guiney, Louise Imogen *see* Hurrell froude
Guiney, Louise Imogene *see* Hurrell froude
Guinnes, H Grattan *see* The new world of central africa
Guinness, G *see* In the far east
Guinness, Geraldine M *see* The story of the china inland mission
Guinness, H *see* Andrea del sarto
Guinness, H Grattan *see*
– Christ pre-eminent
– Green tree and the dry...
– The new world of central africa
– Popular view of the substance of mr h grattan guinness' book
Guinness, H. Grattan *see* The wide world and our work in it
Guinness, Henry Grattan *see*
– Creation centred in christ
– Sermons
Guinness, L E *see* Across india at the dawn of the 20th century
Guinot, Eugene *see* Ogresse
Guinto, Mariano *see* [Ang kapanimalusan. libretto]
Guion de conferencias pronunciadas en el aula de cultura de la caja de ahorros y monte de piedad de plasencia. abril-mayo 1971 / Caja de Ahorros y Monte de Piedad – Caceres: Caja de Ahorros de Plasencia, 1972 – 1 – sp Bibl Santa Ana [946]
Guion de ferias y fiestas. agosto 1959 / Jaraiz de la Vera. Ayuntamiento – Jaraiz de la era: Imp. La erata, 1959 – 1 – sp Bibl Santa Ana [390]
Guion de festejos de cuacos de la vera. septiembre de 1954 – Imprenta la Verata, 1954 – 1 – sp Bibl Santa Ana [390]
Guion, Willie K *see* Familial patterns of vo(2max) and physical activity levels
Guiral, Leon *see* Le congo francais du gabon a brazzaville
Guiral Moreno, Mario *see* En pos de la felicidad
Guiral, Paul *see*
– Immigration reglemente aux antilles francaises
Guiraud, Jean *see*
– Cartulaire de notre dame de prouille
– L'eglise romaine et les origines de la renaissance
– L'etat pontifical apres le grand schisme
– Questions d'histoire et d'archeologie chretienne
– Saint dominic
Guiraud, Pierre Marie Theresa Alexandre *see* Comte julien
Guire, Juliette de *see* Bibliographie analytique de l'oeuvre de monsieur charles-marie boissonnault
Guirior, Maria Ventura *see* Expresion festiva a la funcion que la excellentissima senora maria ventura guirior, virreyna de los reynes del peru
Guiron, J J *see* I missa yosapo
Guisain, Jacques *see* Les sages entretiens d'une ame qui desire sincerement son salut
Guisborough exchange – Guisborough, England. -w. 23 Feb, 25 Oct 1872; 23 April 1874-27 July 1878. 3 reels – 1 – uk British Libr Newspaper [072]
Guiscafre, Rosario *see*
– Hablando a tu corazon
– Oleaje intimo
Guise, Jacques de *see* Chroniques des comtes de hainault
Guiser, Moises David *see* Lider un lender
Guitar player – Manhasset NY 1967+ – 1,5,9 – ISSN: 0017-5463 – mf#6052 – us UMI ProQuest [780]
Guitarra espanola, y vandola, en dos maneras de guitarra, castellana y valenciana de cinco ordenes : la qual ensena de templar... / Amat, Joan Carles – Valencia: impr de la viuda de Agustin Laborda, vive en la Bolseria [176-?] [mf ed 19–] – 1mf – 9 – mf#fiche327 – us Sibley [780]
Guitarrero / Scribe, Eugene – Paris, France. 1841? – 1r – us UF Libraries [440]
Guitbertus Abbas Novigenti *see* Dei gesta per francos
Guiteras, Juan *see*
– Bubonic plague in cuba
– Free cuba
Guiteras, Pedro Jose *see* Historia de la conquista de la habana
Guiton, J-Ph *see* Le developpement intellectuel de l'enfant de dieu
Guizan, M *see* Traite sur les terres noye's de la guyane

GUNZO

Guizot demasque : refutation de ses derniers ecrits, sa reputation usurpee et sa profession de foi / Stephanopoli-Comnene, M N – Paris [1848?] – us CRL [920]
Guizot, Francois *see*
– Dictionnaire universel des synonymes de la langue francaise
– Meditations sur l'etat actuel de la religion chretienne
– Saint louis and the thirteenth century
Guizot, Francois Pierre Guillaume *see*
– De la democratie en france (janvier 1849)
– Des moyens de gouvernement et d'opposition dans l'etat actuel de la france
Guizot, M Francois *see*
– Meditations on the actual state of christianity
– Meditations sur la religion chretienne dans ses rapports avec l'etat actuel des societes et des esprits
– Meditations sur l'essence de la religion chretienne
– A popular history of france
– Saint louis and calvin
Gujarat – Ahmedabad, India. 1963-65 – 2r – 1 – us L of C Photodup [079]
Gujarat and the gujaratis : pictures of men and manners taken from life / Malabari, B M – Bombay, 1884 – 5mf – 9 – mf#HT-86 – ne IDC [915]
Gujarat and the gujaratis : pictures of men and manners taken from life / Malabari, Behramji Merwanji – Bombay: Ford Print Press, 1889 – (= ser Samp: indian books) – us CRL [954]
The gujarat government gazette / Gujarat. India – Admedabad. 5 May 1960-66. For earlier file see: Bombay. Bombay Government Gazette – 1 – us NY Public [954]
Gujarat. India *see* The gujarat government gazette
Gujarat samachar – Bombay, India. 1958 – 2r – 1 – us L of C Photodup [079]
Gujarata and its literature : a survey from the earliest times / Munshi, Kanaiyalal Maneklal – Calcutta: Longmans, Green & Co, 1935 – (= ser Samp: indian books) – (foreword by mahatma gandhi) – us CRL [490]
Gujarata samacara – Ahmedabad, India. Mar 1966-1992 – 143r – 1 – us L of C Photodup [079]
Gujarati bhasha ane sahitya : [lectures delivered in wilson philological lectures] = Gujarati language and literature / Divatia, Narsinhrao Bholanath – Mumbai: Sri Pharbasa Gujarati Sabha 1936- [mf ed 1997] – (= ser Samp: indian books 11918) – 1r – 1 – ([mumbai yunivarsitine asraye narasimharava bholanatha divetiya apelam vilasana bhashasastravishayaka vyakhyano, bhashantarkara ramaprasada premasankara bakshi]; filmed with other titles; in gujarati) – mf#mf-10449 reel 153 – us CRL [490]
Gujarati painting in the fifteenth century : a further essay on vasanta vilasa / Mehta, Nanalal Chamanlal – London: India Society, 1931 – (= ser Samp: indian books) – us CRL [750]
Gujastak abalish : relation d'une conference theologique presidee par le calife maamoun / Matigan-i gudshastak Abalish – Paris: F Vieweg, 1887 – (= ser Bibliotheque de l'ecole des hautes etudes) – 1mf – 9 – 0-524-01794-8 – mf#1990-2642 – us ATLA [280]
Guk kam kiles / Ken Van Sak – Bhnam Ben: Ron Bumb Deb 1963 [mf ed 1990] – 1r with other items – 1 – (in khmer; title on added t.p: prison de la passion: poemes) – mf#mf-10289 seam reel 110/4 [$] – us CRL [810]
Gulami, Abduelkadir *see* The divan project
Gulberlet, C *see* Philosophisches jahrbuch
Het gulden cabinet van de edele vrye schilderconst ontsloten door den lanck ghewenschten vrede tusschen de twee machtighe croonen van spaignien en vranckryk / Bie, C de – Antwerpen, 1662 – 11mf – 9 – mf#O-146 – ne IDC [700]
Den gulden sonnen wyser oft horologie van de passie ons heeren iesu christi / T'sogart, Aigidius – Brussel, 1626 – 9mf – 8 – €32.00 – ne Slangenburg [240]
Gulden spiegel : otte opwecking tot christelijcke deughden / Mayvogel, J C – Amsterdam: J. Bouman, 1659 – 6mf – 9 – mf#O-3120 – ne IDC [090]
Den gulden winckel der konstlievende nederlanders gestoffeert... / [Vondel, J van den] – Amstelredam: Dirck Pietersz, [1613] – 2mf – 9 – mf#O-3195 – ne IDC [090]
Gulenkov, S *see* Protiv religii za sotsializm vo vtori pyatiletke
Gulf coast baptist – Houston, TX. 1952-53 – 1 – us ABHS [071]
Gulf coast breeze – Crawfordville, FL. 1897-1915 – 2r – us UF Libraries [071]
Gulf coast news digest – 1968 sep 18-1974 dec 26, 1975 jan 1-1977 jun 1 – 2r – 1 – mf#369068 – us WHS [071]
Gulf, Colorado and Santa Fe Railway Company *see* Records
Gulf county / Lewis, N D – s.l, s.l? 193-? – 1r – us UF Libraries [978]
Gulf county breeze – Wewahitchka, FL. 1946 nov 22-1993 – 30r – (gaps) – us UF Libraries [071]
Gulf courier – v14 n46, 48, 128, 132, 144-145 [1986 dec 17, 1987 jan 7, 1988 sep 11, 1989 jan 7, apr 18-25] – 1r – 1 – mf#1714625 – us WHS [071]
Gulf defender – 1981 may 1-1982 dec 10, 1983-85, 1986 jan 11-jun, 1986 jun 27-1987 sep 18 – 4r – 1 – mf#643898 – us WHS [071]
Gulf intracoastal waterway – s.l, s.l? 193-? – 1r – us UF Libraries [978]
Gulf news – Waiheke island, NZ. 1976-dec 1989 – 44r – 1 – mf#11.31 – nz Nat Libr [079]
Gulf of mexico air quality study : final report / Systems Applications International – [New Orleans LA]: US Dept of the Interior, Minerals Management Service...[1995] [mf ed 1996] – 16mf – 9 – (incl bibl ref) – us GPO [333]
Gulf ridge groves inc... – Lakeland, FL. 1921 – 1r – us UF Libraries [630]
Gulgong advertiser – Gulgong, jan 1898-dec 1904; jun 1918-jul 1919 – 4r – A$257.71 vesicular A$279.71 silver – at Pascoe [079]
Gul-i zard – Tehran. sal-i 1, shumarah-i 1-24 27 sha'ban 1336-10 shavval 1337 [7 may 1918-july 1919]; sal-i 2, shumarah-i 1-25 24 shavval 1337-1 rabi' al-sani 1339 [23 jul 1919-13 dec 1920]; sal-i 3, shumarah-i 1-44 1 zu'l qa'dah-25 shavval; sal-i 4, shumarah-i 1-31 3-19 zu'l qa'dah – 1r – 1 – $140.00 – (missing: sal-i 3, shumarah-i 28) – us MEDOC [079]
Gulick, Luther Halsey *see* Physical education by muscular exercise
Gulick, Orramel Hinckley *see* The pilgrims of hawaii
Gulick, Sidney Lewis *see*
– Christian crusade for a warless world
– John hyde deforest
Gulick, Sydney Lewis *see* Evolution of the japanese
Gulielmus, Archbishop of Tyre *see* Histoire de la gverre saincte, dite proprement, la franciade orientale...
Gulistan : ou, le hulla de samarcande: opera en trois actes / Dalayrac, Nicolas – Paris: Chez Melles. Erard [1805?] [mf ed 1990] – 1r – 1 – (imprint date estimated fr date of first performance; libretto by lachabeaussiere & c g etienne) – mf#pres. film 87 – us Sibley [780]
Gullander, Paul *see* Tre ar i afrika jaemte minnen fran sverige och det heliga landet samt amerikas foerenta stater
Gulledge, Tracey P *see* The reproducibility of low testosterone in endurance trained males
Gullen, J A *see* De wagner-ziekte en de hedendaagsche muziekcrisis
Gullick, Thomas John *see*
– A descriptive handbook for the national pictures in the westminster palace
– The royal academy
Gulliver, Philip Hugh *see*
– The karamajong cluster
– Kinship and property among the jie and turkana
– A preliminary survey of the turkana
Gulliver's travels / Swift, Jonathan – New York, NY. 1933 – 1r – us UF Libraries [025]
Gum branch church. gum branch, south carolina : church records – 1796-Mar 1963 – 1 – us Southern Baptist [242]
Guma, Enoch S *see* Nomalizo okanya izinto zalomhlaba ngamajingiqiwu
Guma, Samson Mbizo *see*
– Form, content and technique of traditional literature
– Likoma
– Morena mohlomi, mora monyane
Gumbart, A S *see* Making the most of oneself
Gumbel, Hermann *see* Deutsche sonderrenaissance in deutscher prosa
Gumbinner kreisblatt – Gumbinnen (Gussew RUS), 1908-14, 1925-30 [gaps] – 4r – 1 – gw Misc Inst [077]
Gumbo, Mishack Thiza *see* Indigenous technologies
Gumilla, Jose *see* El orinoco...ilustrado
Gumilla, Joseph *see*
– Orinoco ilustrado
Gumlich, Gotthold Albertus *see* Christian creeds and confessions
Gummere, Amelia Mott *see*
– The quaker in the forum
– Witchcraft and quakerism
Der gummiwerker – Schoenebeck DE, 1955 aug-1973 sep, 1974-1990 13 feb – 5r – 1 – (with gaps. title varies: n7 1966-68: begutex) – gw Misc Inst [670]
Gummi-zeitung – Berlin, Dresden DE, 1892 oct-1932 sep – 35r – 1 – (with suppl: die celluloid-industrie 1912 22 mar-1914 24 jul) – uk British Libr Newspaper [670]
Gummi-zeitung, berlin *see* Die celluloid-industrie
Gumpach, Johannes von *see*
– Alttestamentliche studien
– Die zeitrechnung der babylonier und assyrer
Gumpeltzhaimer, Adam *see*
– Compendium musicae latino-germanicum
– Sacrorvm concentvvm octonis vocibvs modvlandorvm...liber primus – [secundus]
Gumpelzhaimer, Adam *see* Compendium musicae latino-germanicum
Gumppenberg, Hanns, Freiherr von *see* Der messias
Gumregah : puspita susastra djawa / Organisasi Pengarang Sastra Djawa Komisariat Djawa Tengah – Sala, 1967 – 2mf – 9 – mf#SE-1494 – ne IDC [950]
Gun and camera in southern africa / Bryden, Henry Anderson – London, England. 1893 – 1r – us UF Libraries [960]
The gun and the gospel : early kansas and chaplain fisher / Fisher, Hugh Dunn – Chicago: Kenwood Press, 1896 – 1mf – 9 – 0-524-06260-9 – mf#1991-2451 – us ATLA [240]
Gun world – Capistrano Beach CA 1972-95 – 1,5,9 – ISSN: 0017-5641 – mf#7121 – us UMI ProQuest [790]
Gunby *see* Gunby's circuit court of appeals reports
Gunby's circuit court of appeals reports / Gunby – 1v. 1885 – (= ser Louisiana Supreme Court Reports) – 2mf – 9 – $3.00 – (a pre-nrs title) – mf#LLMC 84-144 – us LLMC [347]
Gundackers von judenburg christi hort : [eine biblische dichtung] / ed by Jaksche, J – Berlin: Weidmann, 1910 [mf ed 1993] – (= ser Deutsche texte des mittelalters 18) – xviii/91p/1pl – 1 – (incl bibl ref and ind) – mf#8623 reel 4 – us Wisconsin U Libr [810]
Gundagai independent – Gundagai, sep 1898-dec 1968 – 30r – A$2281.27 vesicular A$2446.27 silver – at Pascoe [079]
Gundagai independent – Gundagi, jan 1969-dec 1988 – 20r – 9 – at Pascoe [079]
Gundagai times – Gundagai, jan 1868-dec 1931 – 24r – A$1330.12 vesicular A$1462.12 silver – at Pascoe [079]
Gundel vom koenigsee : erzaehlende dichtung aus dem bayrischen hochland in sieben gesaengen / Grosse, Julius – Berlin: F Lipperheide, 1872 [mf ed 1993] – (= ser Erzaehlende dichtungen 1) – 4/112p/2pl (ill) – 1 – mf#8701 – us Wisconsin U Libr [390]
Gundelfinger, Friedrich *see* Goethe im gespraech
Gundert, Hermann *see* Die evangelische mission
Gundissalinus' de divisione philosophiae / Baur, L – Muenster, 1903 – (= ser Bgphma 4/2-3) – 7mf – 9 – €15.00 – ne Slangenburg [100]
Gundissalinus, Dominicus *see* De divisione philosophiae
Gundogan, Nese *see* Marketing effectiveness and promotional strategies in national collegiate athletic association division 1 basketball programs
Gundolf, Ernst *see* Nietzsche als richter unsrer zeit
Gundolf, Friedrich *see*
– Ein gelegenheitsgedicht von brockes
– George
– Goethe
– Heinrich von kleist
– Martin opitz
– Rede zu goethes hundertstem todestag
– Stefan george in unsrer zeit
Gundry, Richard Simpson *see* China, present and past
Gune, Panduranga Damodara *see* An introduction to comparative philology
Gune, Vithal Trimbak *see* The judicial system of the marathas
Gunes – Baku, Azerbaijan 1910-11 [mf ed Norman Ross] – 1r – 1 – (cont: haqiqat) – mf#nrp-228 – us UMI ProQuest [077]
Gunethics, or, the ethical status of woman / Brown, W Kennedy – New York: Funk & Wagnalls, 1887. El Segundo, Ca: Micro Publication Systems, 1981 (1mf); Evanston: American Theol Lib Assoc, 1984 (1mf) – (= ser Women & the church in america) – 9 – 0-8370-1430-1 – mf#1984-2148 – us ATLA [240]
Gunfighter – Mountain Home ID. v1 n1-8,11-12,14-30,32-49 [1988 jan 21-mar 10, 31-apr 7,21-aug 11, 25-dec 22], v1 n50 [1989 jan 12], v2 n1-44,46-49 [1989 jan 19-nov 16, 30-dec 21] – 1r – 1 – (cont: wing spread) – mf#1702800 – us WHS [071]
Gunhild die reiterin / Wustmann, Erich – Reutlingen: Ensslin & Laiblin, [1939] – 1r – 1 – (incl bibl ref) – us Wisconsin U Libr [890]
Gunkel, Hermann *see*
– Ausgewaehlte psalmen
– Elias, jahve, und baal
– Esther
– Genesis
– Israel and babylon
– Reden und aufsaetze
– Schoepfung und chaos in urzeit und endzeit
– Zum religionsgeschichtlichen verstaendnis des neuen testaments
Gunkel, Richard *see*
– Georg buechner und der dandysmus
Gunn, B *see* Studies in egyptian syntax
Gunn, Charles A *see* The presbyterian church and the filipino
Gunn, Harriette Bronson *see* In a far country
Gunn, Harry C *see* Napa county
Gunn, J M *see* History of the state of california and biographical record of coast counties
Gunn, John *see* Historical enquiry respecting the performance on the harp in the highlands of scotland
Gunn, Marcus *see* An address to the public introducing a letter to the rev mr pollard
Gunn, W *see* The gospel in futuna
Gunne, Alzina Evelyn *see* The silver trail
Gunnedah advertiser – Gunnedah, jan 1898-dec 1907 – 4r – A$264.75 vesicular A$286.75 silver – at Pascoe [079]
Gunnedah courier – Gunnedah, jun-dec 1901 – 1r – 9 – A$32.82 vesicular A$38.32 silver – at Pascoe [079]
Gunnedah independent advertiser – Gunnedah, 1925-27; 1929-30; 1935; 1938-1964 – at Pascoe [079]
Gunning, J H *see*
– De godspraken van amos
– Het protestantsche nederland onzer dagen
– Van babel naar jeruzalem
Gunning, Johannes Hermanus *see* Onze eeredienst
Gunning, Mary J *see* Maximal oxygen consumption and body composition characteristics of trained male and female runners
Gunning, Mary Jo *see* Survey of aquatic programs and aquatic facility accessibility features available to and utilized by physically handicapped students at four-year pennsylvania colleges and universities
Gunnison county miscellaneous newspapers – Denver, CO (mf ed 1991) – 1r – 1 – (marston wizard (may 31 1907); crested butte citizen (jan 29 1904-feb 19, 1904); crystal river current (oct 31 1890); silver lance (dec 31 1897); pick & drill (dec 21-28 1893, feb 1 1894); daily people's champion (may 10, 18, 1898); gunnison daily review (oct 11, nov 9 1881); the echo (jun 7 1934); the free press (dec 3 1881); the marble age (jun 19 1909); marble booser (mar 18 1911; mar 30 1912); marble city times (feb 24 1911; mar 15 1912); marble times & crystal silver lance (feb 19 1904); pitkin miner (aug 16 1890, dec 28 1917); vulcan times (jan 28 1904)) – mf#MF Z99 G957 – us Colorado Hist [071]
Gunnison county miscellaneous newspapers – Denver, CO (mf ed 1991) – 1r – 1 – (the marsten wizard (may 31 1907); crested butte citizen (jun 8 1906); weekly citizen (jan 29, feb 19 1904); crystal river current (oct 31 1890); the silver lance (dec 31 1897); the pick & drill (dec 21-28 1893, feb 1 1894); the daily people's champion (may 10, 18 1898); gunnison daily review (oct 11, nov 9 1881); the echo (jun 7 1934); the free press (dec 3 1881); the marble age (jun 19 1909); the marble booster (mar 18 1911, mar 30 1912); marble city times (feb 24 1911; mar 15 1912); marble times (feb 19 1904); pitkin miner (dec 28 1917); the vulcan times (jan 28 1904)) – mf#MF Z99 G957 – us Colorado Hist [071]
Gunnison daily review *see*
– Gunnison county miscellaneous newspapers
Gunnison, John Williams *see* The mormons, or, latter-day saints in the valley of the great salt lake
Gunpowder baptist church. middletown, maryland : church records – 1806-Sep 1966 – 1 – us Southern Baptist [242]
Guns and action – 1985 jun-1986 feb – 1r – 1 – (cont: survive [boulder, co]) – mf#1102901 – us WHS [071]
Guns and ammo – Los Angeles CA 1968+ – 1,5,9 – ISSN: 0017-5684 – mf#3059 – us UMI ProQuest [790]
Gunsaulus, Frank Wakeley et al *see* Addresses at the annual meeting of the new west education commission
Gunsei keizirei kaisetu / Java. Sihobu – (Djakarta): Djawa Gunseikanbu Sihobu, 2604 – 63p 1mf – 9 – mf#SE-2002 mf87 – ne IDC [959]
Gunther, Ernst A W *see* Die deutsche heldensage des mittelalters
Gunther, John *see*
– Inside latin america
– The spanish civil war
Gunther, John T *see* Transcript of interviews with hank nelson
Gunther, Max *see* The weekenders
Guntli, Lucie *see* Goethezeit und katholizismus im werk ida hahn-hahns
Gunton's magazine – La Salle IL 1891-1904 – 1 – mf#5561 – us UMI ProQuest [320]
Gunung emoeng / Udjana – Djakarta, 1964 – 4mf – 9 – mf#SE-976 – ne IDC [950]
Gunzberg, Mordecai Aaron *see* Kiryat sefer
Gunzo, epistola ad augienses und anselm von besate, rhetorimachia (mgh quellen..: 2.bd) – 1958 – (= ser Monumenta germaniae historica. quellen zur geistesgeschichte des mittelalters (mgh quellen...)) – €11.00 – ne Slangenburg [931]

Der guote gerhart / Ems, Rudolf von; ed by Asher, John A – Tuebingen: M Niemeyer, 1989 [mf ed 1993] – (= ser Altdeutsche textbibliothek n56) – 232p – 1 – (middle high german text. int in german. incl bibl ref) – mf#8193 reel 5 – us Wisconsin U Libr [810]
Gup, Marc L *see* Conquering anxiety in grade school aged swimmers through the use of imaginative play
Guppy, Nicholas *see* Wai-wai
Gupta, B Sen *see* Mahatma gandhi and india's struggle for swaraj
Gupta, Dilip K *see* Best stories of modern bengal
The gupta empire / Mookerji, Radha Kumud – 3rd ed. rev. Bombay: Hind Kitabs, 1959.174p. illus – 1 – us Wisconsin U Libr [954]
Gupta, Hari Ram *see*
- History of the sikhs
- Life and work of mohan lal kashmiri, 1812-1877

Gupta, Jnanendra Nath *see* Life and work of romesh chunder dutt, cie
Gupta, Nagendranatha *see* Gandhi and gandhism
Gupta, Pratul C *see* Shah alam 2 and his court
Gupta, Pratulacandra *see* The last peshwa and the english commissioners, 1818-1851
The gupta temple at deogarh / Vats, Madho Sarup – [Delhi]: Manager of Publications, Delhi 1952 – (= ser Samp: indian books) – us CRL [720]
Gupte, B A *see* Hindu holidays and ceremonials
Gupte, B A [comp] *see* Selections from the historical records of the hereditary minister of baroda
Gupte, Balkrishna Atmaram *see* Hindu holidays and ceremonials
Gurashabadaratanakara : mahana kosha = Encyclopaedia of the sikh literature / Singha, Kanha – [Patiala: Darbar Patiala 1930] [mf ed 1999] – (= ser Samp: indian books 34198) – 4v on 1r [ill] – 1 – (in panjabi; v1 [reel 021], v2 [reel 023], v3 [reel 024], v4 [reel 031]; filmed with other items) – mf#mf-12110 – us CRL [490]
Gurban zuil-un uge qadamal ujekui-dur kilbar bolgagsan bicig *see* San he bian lan
Gurdon, Philip Richard Thornhagh *see*
- The khasis
- The Khasis

Gurenne-sprache in nordghana / Rapp, Eugen Ludwig – Leipzig, Germany. 1966 – 1r – us UF Libraries [470]
Gur'ev, A *see*
- Ocherk razvitiia kreditnykh uchrezhdenii v rossii
- Reforma denezhnogo obrashcheniia
- Zapiska o promyshlennykh bankakh

Gur'ev, A N *see*
- K reforme gosudarstvennogo banka
- K reforme krest'ianskogo banka

Gurevich, Aleksandr V *see* Staryi fol'klor pribaikal'ia
Gurevich, D *see* The jewish population of jerusalem
Gurevich, D et al *see* The jewish population of palestine
Gurgel Do Amaral, Luis *see* Meu velho itamarati (de amanuense a secretario de...
Gurian, W *see* De strijd om de kerk in het derde rijk
Gurian, Waldemar *see* El bolchevismo...
Guriel, Joseph *see* Elementa linguae chaldaicae
Gurjustan – Tbilisi, USSR. Nov 15-Dec 27 1990-Jan 5-Dec 21 1991 – 1r – 1 – us L of C Photodup [077]
Gurley, Ralph R *see* Life of jehudi ashmun with an appendix containing a brief sketch of the life of rev. lott cary
Gurmukh Nihal Singh *see* Landmarks in indian constitutional and national development
Gurney, Edmund et al *see* Phantasms of the living
Gurney, Jane Tritton *see* A journey to canada
Gurney, Joseph John *see*
- A letter to the followers of elias hicks
- Memoirs of joseph john gurney
- Observations on the distinguishing views and practices of the society of friends
- Puseyism traced to its root

The gurneys of earlham / Hare, Augustus John Cuthbert – New York: Dodd, Mead, 1895 – 2mf – 9 – 0-524-05877-6 – mf#1990-5171 – us ATLA [240]
Gurney-Salter, Emma *see* Franciscan legends in italian art: pictures in italian churches and galleries
Gurnhill, James *see* English retraced
Gurr, Paul *see* Bilder aus der berliner mission in lukhang-suedchina
Guru Dutt, K *see* Principles of hindu astrology: or, the book of fate
Gurudev tagore / Tan, Yun Shan et al; ed by Narasimhan, R – Bombay: Hind Kitabs, 1946 – (= ser Samp: indian books) – us CRL [920]
Gurumurti, D *see* Saptapadarthi of sivaditya
Gury, Jean Pierre *see* Compendium theologiae moralis
Gus hill's national theatrical directory *see* American theatre periodicals of the nineteenth and early twentieth centuries
Gus the bus and evelyn the exquisite checker / Lait, Jack – Toronto: T Langton [1917?] [mf ed 1998] – 4mf – 9 – 0-665-66439-7 – mf#66439 – cn Canadiana [830]
Gusel'ki-iarovchaty – Tambov, 1907-09 [mthly] – 7mf – 9 – us UMI ProQuest [780]
Gusev, F S *see* Svobodnoe slovo [tiumen': 1918-1919]
Gusev, M *see* Zapiski vilenskoi arkheologicheskoi komissii
Gusev-Orenburgskii, Sergei Ivanovich *see* Izbrannye razskazy
Gushiken, Thomas *see* The effects of leisure education on life satisfaction and leisure satisfaction among japanese american older adults
Gusii bridewealth law and custom / Mayer, Philip – Cape Town, Oxford UP, 1950 – us CRL [390]
Gusmao, Bartholomeu Lourenco De *see* Obras diversas de bartholomeu lourenco de gusmao
Gusmao, Carlos De *see* Boca da grota
Gusovius, Paul [comp] *see* Der landkreis samland
Gussago, Cesario *see* Sonate a quattro, sei et otto
Gussman, Boris *see* Out in the mid-day sun
Gussmann, Wilhelm *see* Quellen und forschungen zur geschichte des augsburgischen glaubensbekenntnisses. erster band, die ratschlaege der evangelischen reichsstaende zum reichstag von augsburg 1530
Gustafson, R P *see* The role of diet and exercise in weight control in obese women
Gustafson, Thomas F *see* The process of privatization of the public golf services in three major united states cities
Gustav adolf in deutschland, 1630-1632 / Egelhaaf, Gottlob – Halle: Verein fuer Reformationsgeschichte, 1901 – (= ser [Schriften des Vereins fuer Reformationsgeschichte]) – 1mf – 9 – 0-7905-4634-5 – mf#1988-0634 – us ATLA [943]
Gustav adolfs page : novelle / Meyer, Conrad Ferdinand – Leipzig: H Haessel, c1922 [mf ed 1996] – (= ser Saemtliche werke) – 74p – 1 – (int by emil ermatinger) – mf#9721 – us Wisconsin U Libr [830]
Gustav falke / Castelle, Friedrich – Leipzig: M Hesse [19–] [mf ed 1989] – (= ser Moderne lyriker 6) – 1r – 1 – (filmed with: zwischen den maechten / anna maria falkenstern) – mf#7229 – us Wisconsin U Libr [430]
Gustav falke : ein lebensbild / Spiero, Heinrich – Braunschweig: G Westermann, c1928 (mf ed 1990) – 1r – 1 – (filmed with: blut und eisen) – us Wisconsin U Libr [430]
Gustav frenssen : ein dichter unserer zeit / Alberts, Wilhelm – Berlin: G Grote, 1922 [mf ed 1989] – 287p/13pl (ill) – 1 – mf#7267 – us Wisconsin U Libr [430]
Gustav frenssen : ein dichter unserer zeit / Alberts, Wilhelm – Berlin: G Grote 1922 [mf ed 1989] – 1r [ill] – 1 – (filmed with: peter moors fahrt nach sudwest) – mf#7267 – us Wisconsin U Libr [430]
Gustav freytag : ein buch von deutschem leben und wirken / Zuchhold, Hans – Breslau: F Goerlich, [1926?] (mf ed 1990) – 1r – 1 – (filmed with: gustav freytag in seinen lustspielen) – us Wisconsin U Libr [430]
Gustav freytag / Lindau, Hans Rudolf David – Leipzig: S Hirzel, 1907 (mf ed 1990) – 1r – 1 – (filmed with: briefe an seine gattin. incl bibl ref and ind) – us Wisconsin U Libr [430]
Gustav freytag / Seiler, Friedrich – Leipzig: R Voigtlaender, 1898 (mf ed 1990) – 1r – 1 – (filmed with: briefe an seine gattin) – us Wisconsin U Libr [430]
Gustav freytag : sein leben und schaffen / Alberti, Conrad – 2. verb aufl. Leipzig: E Schloemp, 1886 [mf ed 1999] – (= ser Deutsche dichter der gegenwart 1) – 236p/1pl – 1 – mf#10198 – us Wisconsin U Libr [430]
Gustav freytag als politiker, journalist und mensch : mit unveroeffentlichten briefen / Freytag, Gustav; ed by Hofmann, Johannes – Leipzig: J J Weber, 1922 (mf ed 1990) – 1r – 1 – (filmed with: briefe an seine gattin. incl ind) – us Wisconsin U Libr [430]
Gustav freytag, ein publizist / Kern, Berthold – [S.l.: s.n.], 1933 (mf ed 1990) – 1r – 1 – (filmed with: gustav freytag und das junge deutschland) – us Wisconsin U Libr [430]
Gustav freytag in seinen lustspielen / Droescher, Georg – [S.l.: s.n.], 1919 (Weida i/Th.r: Buchdruckerei von Thomas & Hubert) (mf ed 1990) – 1r – 1 – (filmed with: gustav freytag. incl bibl ref) – us Wisconsin U Libr [430]
Gustav freytag und das junge deutschland / Mayrhofer, Otto – [S.l.: s.n.], 1907 (mf ed 1990) – 1r – 1 – (filmed with: gustav freytag, ein publizist) – us Wisconsin U Libr [430]
Gustav freytag und das junge deutschland / Mayrhofer, Otto – Marburg a.L.: N G Elwert, 1907 – 1 – (incl bibl ref) – us Wisconsin U Libr [430]
Gustav freytag und herzog ernst von coburg im briefwechsel 1853-1893 / Freytag, Gustav; ed by Tempeltey, Eduard – Leipzig: S Hirzel, 1904 (mf ed 1990) – 1r – 1 – (filmed with: briefe an seine gattin) – us Wisconsin U Libr [860]
Gustav freytag-galerie : nach den originalgemaelden und cartons der ersten meister der neuzeit – Jubilaeums-Ausgabe. Leipzig: E Schloemp, [1887] – 1 – us Wisconsin U Libr [750]
Gustav freytags briefe an albrecht von stosch / Freytag, Gustav; ed by Helmolt, Hans F – Stuttgart; Berlin: Deutsche Verlags-Anstalt, 1913 (mf ed 1990) – 1r – 1 – (filmed with: briefe an seine gattin. incl ind) – us Wisconsin U Libr [860]
Gustav freytags romantechnik / Ulrich, Paul – Marburg: N G Elwert, 1907 (mf ed 1990) – 1r – 1 – (filmed with: gustav freytag) – us Wisconsin U Libr [430]
Gustav freytags romantechnik / Ulrich, Paul – Marburg a.L.: N G Elwert, 1907 – 1r – 1 – (incl bibl ref (3rd prelim. leaf)) – us Wisconsin U Libr [430]
Gustav friedrich wilhelm grossmann : ein beitrag zur deutschen litteratur- und theatergeschichte des 18. jahrhunderts / Wolter, Joseph – Koeln: W Hoster, 1901 – 1r – 1 – (incl bibl ref) – us Wisconsin U Libr [430]
Gustav kuehne : sein lebensbild und briefwechsel mit zeitgenossen / ed by Pierson, Edgar – Dresden: E Pierson, [1889?] – 1 – us Wisconsin U Libr [860]
Gustav landauer / Zandbank, Jacob – Tel-Aviv, Israel. 1939 – 1r – us UF Libraries [939]
Gustav schwab's leben / Schwab, Christoph Theodor – Freiburg i.B.: J C B Mohr (Paul Siebeck), 1883 – 1r – 1 – us Wisconsin U Libr [920]
Gustav struve als politischer schriftsteller und revolutionaer / Peiser, Juergen – Frankfurt a.M., 1973 – 4mf – 9 – 3-89349-715-3 – gw Frankfurter [460]
Gustav theodor fechner / Lasswitz, Kurd – 2. verm Aufl. Stuttgart: F Frommann, 1902 – (= ser Frommanns Klassiker der Philosophie) – 1mf – 9 – 0-7905-9015-8 – mf#1989-2240 – us ATLA [190]
Gustav wasa / Brentano, Clemens; ed by Minor, J – Heilbronn: Henninger, 1883 [mf ed 1993] – (= ser Deutsche litteraturdenkmale des 18 und 19. jahrhunderts 15) – xiv/136p – 1 – mf#8676 reel 2 – us Wisconsin U Libr [820]
Gustave / Piron, Alexis – Paris, France. 1802 – 1r – us UF Libraries [440]
Gustave flaubert / Thibaudet, A – Paris, 1935 – €14.00 – ne Slangenburg [440]
Gustavus adolphus in germany, and other lectures on the thirty years' war / Trench, Richard Chenevix – 2nd ed. London: Macmillan, 1872 – 1mf – 9 – 0-524-01138-9 – mf#1990-0352 – us ATLA [943]
[Gustine-] gustine standard – CA. 1912- – 10r – 1 – $600.00 (subs $50y) – mf#B02283 – us Library Micro [071]
Gut – London, England 1960+ – 1,5,9 – ISSN: 0017-5749 – mf#1335 – us UMI ProQuest [616]
Gutachten ganganelli's, clemens 14. : in angelegenheit der blutbeschuldigung der juden – Berlin: Ph Deutsch, 1888 [mf ed 1985] – 1mf – 9 – 0-8370-2687-3 – mf#1985-0687 – us ATLA [240]
Gutachten von geistlichen der dioeces grimma ueber die vorgeschlagene verdraengung der vollstaendigen bibel aus unsern volksschulen / Steglich, Friedrich August William – Leipzig: J C Hinrichs, 1869 – 1mf – 9 – 0-8370-8551-9 – (incl bibl ref) – mf#1986-2551 – us ATLA [377]
Gutberlet, Constantin *see*
- Das buch der weisheit
- Der kampf um die seele
- Der mechanische monismus

Gutbier, Aegidius *see* Aegidii gutbirii lexicon syriacum
Gutch, Charles *see* Gloomy summer
Die gute dorfgeschichte : eine sammlung / ed by Wandrey, Horst – Rudolstadt [Germany]: Greifenverlag, c1960 – 1r – 1 – (incl bibl ref) – us Wisconsin U Libr [430]
Gute geister *see*
- Benrather tageblatt
- Landsberger nachrichtenblatt

Gute menschen und ihre geschichten : novellen / Sacher-Masoch, Leopold, Ritter von – Leipzig: E J Guenther, 1874 – 1r – 1 – us Wisconsin U Libr [830]
Das gute recht / Edschmid, Kasimir – Muenchen: K Desch c1946 [mf ed 1989] – 1r – 1 – (filmed with: eira und der gefangene / heinrich eckmann) – mf#7202 – us Wisconsin U Libr [830]
Der gute weg : roman / Flake, Otto – Berlin: S Fischer 1925, c1924 [mf ed 1989] – 1r – 1 – (filmed with: freund aller welt) – mf#7243 – us Wisconsin U Libr [830]
Gutemala : las lineas de su mano / Cardoza Y Aragon, Luis – Mexico City? Mexico. 1955 – 1r – us UF Libraries [972]
Gutemberg : orgam imparcial – Baturite, CE. 24 dez 1893 – (= ser Ps 19) – mf#P17,01,62 – bl Biblioteca [079]
Guten abend *see* Volksblatt fuer bergisch gladbach und umgebung
Die guten christen : schauspiel in einem akt / Telmann, Fritz – Leipzig: H Seemann, 1902 – 1r – 1 – us Wisconsin U Libr [820]
Die guten frauen / Goethe, Johann Wolfgang von; ed by Seuffert, Bernhard – Heilbronn: Henninger 1885 [mf ed 1993] – (= ser Deutsche litteraturdenkmale des 18. und 19. jahrhunderts 21) – 1r [ill] – 1 – (filmed with: die kindermoerderinn / h l wagner) – mf#8676 reel 2 – us Wisconsin U Libr [830]
Gutenbaum, Kalman *see* Misteryen
Gutenberg – Berlin DE, 1848 13 may-1851 – 1 – gw Misc Inst [680]
Gutenberg's illustriertes sonntags-blatt *see* Westdeutsche gewerbe-zeitung
Gutenbergs illustriertes sonntagsblatt *see* Landsberger nachrichtenblatt
Guth, Gustav *see* Der grosse alexander
Guth, William Westley *see*
- The assurance of faith
- Revelation and its record
- Spiritual values

Guthe, Hermann *see*
- Amos
- Fragmente einer lederhandschrift enthaltend mose's letzte rede an die kinder israel mitgetheilt und geprueft
- Geschichte des volkes israel
- Jesaia
- Palaestina

Gutheil, Crystal H *see* F geminiani's violin methods
Guthlac roll : british library, harley ms. roll y6 – 1200 – 1r – 14 – mf#C596 – uk Microform Academic [240]
Guthofer, Ulrich *see* Plaqueakkumulation auf gegossenen titanoberflaechen
Guthrie, A *see* The royal academy exhibition of pictures, 1870
Guthrie, Charles John Guthrie, Lord *see*
- Autobiography of thomas guthrie...
- The history of the reformation of religion within the realm of scotland

Guthrie, David *see* Christianity and natural science
Guthrie, David Kelly *see* Autobiography of thomas guthrie...
Guthrie, James H *see* Diary
Guthrie, Jas M *see* Campfires of the afro-americans
Guthrie, Malcolm *see*
- Bantu word division
- Collected papers on bantu linguistics
- Grammaire et dictionnaire de lingala

Guthrie, Thomas *see*
- The city, its sins and sorrows
- The parables

Guthrie, Thomas et al *see* The great men of god
Guthrie, William *see* A new system of modern geography
Gutierre, Nicolas Jose *see* Homenaje al ilustre habanero nicolas jose gutierre
Gutierres, Carlos *see* Fray bartolome de las casas, sus tiempos y su aposislado
Gutierrez Anzola, Jorge Enrique *see* Delitos contra la vida y la integridad personal
Gutierrez, C *see* Madrid. archivo historico national. seccion de ordenes militares. guia de la seccion de ordenes militares
Gutierrez Carrasco, Octavio *see* La federacion interamericana de abogados; memoria de prueba para optar al grado de licenciado en la facultad de ciencias juridicas y sociales de la universidad de chile
Gutierrez Cunado, Antolin *see* A varias tintas
Gutierrez Davila, Julian *see* Memorias historicas de la congregacion de el oratorio de la ciudad de mexico
Gutierrez de Arevalo, P *see* Practica de boticarios
Gutierrez de Los Rios, G *see* Noticia general para la estimacion de las artes...
Gutierrez de Santa Clara, Pedro *see* Historia de las guerras civiles del peru (1544-1548) y de otros sucesos de las indias
Gutierrez del Arroyo, C *see* Madrid. archivo historico nacional. seccion de universidades. la seccion de universidades
Gutierrez Estrada, Jose Maria *see* Mexico en 1840 y en 1847
Gutierrez Gomez, Diego *see* Dislexias
Gutierrez Gonzalez, Gregorio *see*
- Gregorio y epifanio
- Obras completas

Gutierrez, Gonzalo *see* Apuntes para la h de pacora
Gutierrez, J L *see* Febriologiae lectiones pincianae, aprendix ad febrilogiam, doloris diagnosim...
Gutierrez, Joaquin *see* Puerto limon
Gutierrez, Jose *see*
- Rebeldia colombiana
- Revolucion contra el miedo

Gutierrez, Jose Fulgencio *see*
- Bolivar y su obra
- Galan y los comuneros

Gutierrez, Juan *see*
- Canonicarum quaestionum
- Canonicarum utrusque fori liber primus
- Consilia
- Consiliorum sive responsorum
- Decisiones s. rotae romanae. opera omnia
- Iurisconsulti praeclarissimi
- Practicarum quaestionum
- Practicarum...reagias hispaniae
- Practicarum...tractatus de babellis
- Praxis criminalis civilis y canonica
- Repetitiones et allegationes...iuris
- Repetitiones sex quatordecin...allegationes
- Tractatus de iuramento confirmatorio
- Tractatus novus de tutelis curis minorum
- Tractatus...juramento

Gutierrez Lasanta, Francisco *see* Donoso cortes, el profeta de la hispanidad

Gutierrez, M *see* Nuevas consideraciones sobre la libertad absoluta de comercio y puertos francos...

Gutierrez Macias, Valeriano *see*
- El 1 congreso nacional de brujologia en san sebastian
- Alta extremadura
- Breve ensayo sobre los nombres gentilicios usado en la alta extremadura
- El cicerone del pueblo
- Comarcas naturales de la alta extremadura. la jara cacerena
- Coplas del baile del pandero
- La egregia figura de carlos de yuste. (las postrimerias de su vida y su muerte ejemplar)
- En villanueva de la sierra tuvo su origen la "fiesta del arbol"
- Figuras ilustres
- Gerifaltes extremenos
- Ligero apunte de la localidad cacerena de pescueza
- Por la geografia cacerena
- Por la geografia cacerena. (banos de montemayor)
- Por la geografia cacerena (fiestas populares)
- Villanueva de la vera. fiestas de pero palo, 1965

Gutierrez, Miguel *see* Albores.ensayos

Gutierrez, Rafael *see* Oriente heroica

Gutierrez Utrera, Benigno *see* Copia de la protesta...congreso...eleccion de coria

Gutierrez, Valeriano G *see* Cuba y espana

Gutierrez Y Salazar, Pedro *see* Reformas de cuba

Gutierrez Y Ulloa, Antonio *see* Estado general de la provincia de san salvador

Gutierrez-Marin, Claudio *see* Pastor j jezequel visits republican spain

Gutirrrez, Solano *see* Ensayo biologico sobre hernando cortes, tomo 1

Gutjahr, F S *see* Die glaubwuerdigkeit des irenaeischen zeugnisses ueber die abfassung des vierten kanonischen evangeliums

Gutman, A IA [A Gan] *see* Russkii ekonomist

Gutman, A Yosef *see* Yisra'el ba-'adam

Gutman, Golde *see* Bersarabie in nayntsen akhtsen

Gutman, Khaim *see* Azoy lakh ikh

Gutmann, M *see* Richard wagner, der judenfresser

Gutmanns reisen / Raabe, Wilhelm Karl – Berlin: Otto Janke, 1892 – 1r – 1 – us Wisconsin U Libr [830]

Gutmundsson, Einar *see* Skotlands rimur

Gutsche, Thelma *see* Microcosm

Gutscher, H *see* Erklarung der propheten nahum und zephanja

Die gutsfrau – [1] 1912-[11] 1923 [mf ed 2005] – (= ser Hq 61) – 61mf – 9 – €420.00 – 3-89131-467-1 – gw Fischer [305]

Gutsmuths, Johann C *see* Deutsches land und deutsches volk

Gutstein, Morris Aaron *see* Story of the jews of newport

Guttandin, Friedhelm *see* Genese und kritik des subjektbegriffs

Guttmacher, Adolf *see*
- Optimism and pessimism in the old and new testaments

Guttman, Mattathias Ezekial *see*
- Rabbi yisrael ba'al shem tov
- Rabi yisrael ba'al shem tov

Guttmann, Bernhard *see* Das ende der zeit

Guttmann, J *see* Die philosophische lehren des isaak ven salomon israeli

Guttmann, Jacob *see*
- Die religionsphilosophie des saadia
- Die scholastik des dreizehnten jahrhunderts
- Das verhaeltniss des thomas von aquino zum judenthum und zur juedischen literatur

Guttmann, Jakob *see* Festschrift zum siebzigsten geburtstage jakob guttmanns

Guttmann, Theodor *see* Mashal bi-tekufat ha-tana'im

Guttzeit, Emil Johannes *see* Der kreis johannisburg

Gutzkow, Karl *see*
- Anonym
- Aus der knabenzeit
- Die beiden auswanderer
- Boerne's leben
- Die deutsche revue
- Deutschland am vorabend seines falles oder seiner groesse
- Das duell wegen ems
- Fritz ellrodt
- Der gefangene von metz
- Gesammelte werke
- Die kleine narrenwelt
- Der koenigsleutnant
- Maha guru
- Patkul
- Die ritter vom geiste
- Rueckblicke auf mein leben
- Eine shakespearefeier an der ilm
- Eine shakspearefeier an der ilm
- Die soehne pestalozzi's
- Sommerreise durch oesterreich
- Das urbild des tartueffe
- Uriel acosta

Gutzkow-funde : beitraege zur literatur- und kulturgeschichte des 19. jahrhunderts / Houben, Heinrich Hubert – Berlin, 1901 (mf ed 1993) – 4mf – 9 – €49.00 – 3-89349-205-4 – mf#DHS-AR 94 – gw Frankfurter [430]

Gutzkows dramaturgische taetigkeit am dresdener hoftheater : unter besonderer beruecksichtigung seiner buehnenbearbeitungen / Baumgard, Otto Wilhelm Gustav – Bonn: Rhenania-Verlag, 1915 – 68p – 1 – (incl bibl ref) – mf#7428 – us Wisconsin U Libr [790]

Gutzkows theorie des romans in seinem roman "hohenschwangau" / Koch, Johannes Guenther – Forst (Lausitz): Buch- und Steindruckerei E Hoene, 1936 – 1r – 1 – (incl bibl ref) – us Wisconsin U Libr [430]

Gutzkows werke : auswahl in zwoelf theilen / ed by Gensel, Reinhold – Berlin: Bong, [1910?] [mf ed 1993] – (= ser Goldene klassiker-bibliothek) – 12v in 4 – 1 – (incl bibl ref & ind) – mf#8667 – us Wisconsin U Libr [430]

Gutzlaff, K *see*
- China opened
- Journal of a residence in siam and of a voyage along the coast of china to mantchou tartary
- Journal of three voyages along the coast of china, in 1831, 1832, and 1833

Gutzlaff, Karl Friedrich August *see*
- Cheng chiao an wei
- Cheng tao chih lun
- Chiu shih chu yeh-su chih sheng hsun
- Ch'uan jen chu yueh
- Fu yin chih chen kuei
- Shih fei lueh lun
- Shu tsui chih tao chuan

Guy de maupassant a traves de la correspondencia de gustavo flaubert / Segura, Enrique – Madrid, 1950. Sep. de Cuadernos de Literatura, fasc. 19-21 – sp Bibl Santa Ana [946]

Guy laviolette (michel-henri gingras) : en religion, le reverend frere achille des freres de l'instruction chretienne, auteur de "gloires nationales": bibliographie analytique / Marie-Stella, soeur – 1954 [mf ed 1978] – (= ser Bibliographies du cours...1947-66) – 2mf – 9 – (with ind; pref by lionel allard) – mf#SEM105P4 – cn Bibl Nat [241]

Guy Lut *see* Bralyn can biar

Guy, PLO *see* New light from armageddon

Guyana chronicle – Georgetown, Guyana. 1986 may-1988 dec – 5r – us UF Libraries [079]

Guyana grafic – independence souvenir – Georgetown, Guyana. 1966 may 26 – 1r – us UF Libraries [079]

Guyana star – Georgetown. Guyana. -d. 2 Mar-23 Apr 1966. (1 reel) – 1 – uk British Libr Newspaper [072]

Guyana. Statistical Bureau *see* Statistical abstract of guyana 1970-1974

La guyane : ou histoire, moeurs, usages et costumes des habitans de cette partie de l'amerique / Denis, Jean – Paris 1823 [mf ed Hildesheim 1995-98] – 2v on 4mf – 9 – €120.00 – ISBN-10: 3-487-26892-2 – ISBN-13: 978-3-487-26892-7 – gw Olms [918]

Guyane francaise / Bouyer, Frederic – Paris, France. 1867 – 1r – us UF Libraries [972]

Guyane francaise / Mourie, J F H – Paris, France. 1874 – 1r – us UF Libraries [972]

Guyane francaise : son histoire 1604-1946 / Henry, Arthur – Cayenne, French Guiana. 1950 – 1r – us UF Libraries [972]

Guyane francaise : terre de l'espace / Resse, Alix – Paris, France. 1964 – 1r – us UF Libraries [972]

Guyane francaise en 1865 / Riviere, Leon – Cayenne, French Guiana. 1866 – 1r – us UF Libraries [972]

Guyane inconnue / Bordeaux, Albert – Paris, France. 1906 – 1r – us UF Libraries [972]

Guyane inconnue / Bordeaux, Albert – Paris, France. 1914 – 1r – us UF Libraries [972]

Guyane meconnue / Bureau, Gabriel – Paris, France. 1936 – 1r – us UF Libraries [972]

Guyant family newsletter – letter 1-20 [1985 jan-1989 oct] – 1r – 1 – mf#1551048 – us WHS [929]

Guyard, Stanislas *see* Notes de lexicographie assyrienne

Guyau, Augustin *see* La philosophie et la sociologie d'alfred fouillee

Guyau, Jean Marie *see*
- Irreligion de l'avenir
- La morale anglaise contemporaine

Guye, Pierre Louis *see* La ville de neuchatel

Guy-Grand, V J *see* Dictionnaire francais-volof

Guymond De La Touche, Claude *see* Iphigenie en tauride

Guyon, Claude Marie *see* Histoire des amazones anciennes et modernes

Guyon, Louis *see* Montferrand

Guyot, Arnold *see* Creation

Guyot, Henri *see* L'infinite divine depuis philon le juif jusqu'a plotin

Guyot, Theodore *see* The royal scottish academy illustrated catalogue

Guyot, Yves *see* Quesnay et la physiocratie

Guyra argus – Guyra, 1944-53; 1993-96 – at Pascoe [079]

Guyra argus – Guyra, jul 1902-dec 1957 – 19r – A$627.00 vesicular A$731.50 silver – at Pascoe [079]

Guyra guardian – Guyra, 1964-82 – at Pascoe [079]

Guyra shire chronicle – Guyra, 1983-88 – at Pascoe [079]

Guyra weekly news – Guyra, 1889-92 – at Pascoe [079]

Guys, Pierre *see* Voyage litteraire de la grece

Guzarish'ha-yi bastanshinasi – Tehran. sal-i 1-4. shahrivar 1329-isfand 1338 [aug/sep 1950-feb/mar 1960] – 2r – 1 – $106.00 – us MEDOC [079]

Guzman Botero, Carlos A *see* Organizacion municipal

Guzman, David Joaquin *see* Apuntameintos sobre la topografia fiscia

Guzman Gundian, Lucila *see* El divorcio, estudio de legislacion comparada

Guzman, Julia Maria *see* Realismo y naturalismo en puerto rico

Guzman, Julio Alfredo *see* Visiones indo-americanas

Guzman Noguera, Ignacio De *see* Pensamiento del libertador

Guzman, Nuno De *see* Memoria de los servicios que habia hecho nuno de g...

Guzman, V L *see* Study of the growth characteristics of plants grown in sand culture

Guzman y Martinez, Jesus *see*
- Nueva seleccion de lecturas francesas
- Paginas ortograficas

Guzolik, Gerald L *see* The effects of caffeine ingestion on heart rate, blood pressure, and physical work capacity at submaximal levels in 15 caffeine habituated non-athletic male subjects

Gvam pum : [a novel] / Dhu Vam – Ran Kun: Kan To ka le Ca tuik 1965 [mf ed 1990] – 1r with other items – 1 – (in burmese) – mf#mf-10289 seam reel 182/3 [$] – us CRL [830]

Gvardeets – (city unknown), Russia 1942-43 [mf ed Norman Ross] – 1 – mf#nrp-103 – us UMI ProQuest [934]

Gvardeiskii udar – (city unknown) 1945 [mf ed Norman Ross] – 1 – mf#nrp-104 – us UMI ProQuest [934]

Gvardeiskoe znamia – (city unknown) 1942-43 [mf ed Norman Ross] – 1 – mf#nrp-105 – us UMI ProQuest [934]

Gvardiia – (city unknown) 1942-43 [mf ed Norman Ross] – 1 – mf#nrp-106 – us UMI ProQuest [934]

La gverra fatta da christiani contra barbari per la ricvperatio... / Accolti, B – Vinegia, 1549 – 3mf – 9 – mf#H-8281 – ne IDC [956]

Gwalia – Bangor, Wales. -w. Aug 1881-1898. Lacking 1883, 1897. 14 reels – 1 – uk British Libr Newspaper [072]

Gwalia – [Wales] Ynys Mon aug 1881-rhag 1921 [mf ed 2003] – 32r – 1 – (missing: 1883) – uk Newsplan [072]

Gwalther, R *see*
- Antichristus
- De incarnatione veri et aeterni filii dei...
- Der endtchrist
- In acta apostolorum..., homiliae 579
- In d pauli apostoli epistolam ad romanos homiliae
- In d pauli...epistolam ad galatas homiliae 61
- In epistolam d pauli apostoli ad romanos... homiliarum archetypi
- In euangelium iesu christi secundum marcum homiliae 89
- In hesterae historiam homiliarum sylvae vel archetypi
- In posteriorem d pauli apostoli ad corinthios epistolam homiliae
- In priorem d pauli ad corinthios epistolam homiliae...
- Die menschwerdung...vnsers herren jesu christi...
- Oiketes, sive servus ecclesiasticvs...
- Opera d. huldrychi zvinglii, pars...
- Das vatter vnser...
- Von der heiligen gschrifft vn jrem vrsprung...

[Gwalther, R *see* Opera d h'i z'ii...partim quidem ab ipso latine conscripta...

Gwatkin, Henry Melvill *see*
- The arian controversy
- The bishop of oxford's open letter
- The church, past and present
- Early church history to a.d. 313
- The eye for spiritual things
- The knowledge of god and its historical development
- Studies of arianism

Gwaza / Chafulumira, E W – Limbe, Malawi. v1-2. 1957 – 1r – us UF Libraries [960]

Y gweithiwr – Aberdare, Wales 25 sep 1858-30 jun 1860 – 1 – (incorp with: y gwron) – uk British Libr Newspaper [072]

Y gweithiwr – [Wales] LLGC 25 medi-rhag 1858 [mf ed 2004] – 1r – 1 – (missing: 1859-60) – uk Newsplan [072]

Gwekoh, Sol H *see* Distinguished 100: the book of eminent alumni of the university of the philippines

Gwelo times – Rhodesia. -w. 12 Jul 1901-17 Dec 1915. (5 reels) – 1 – uk British Libr Newspaper [072]

GWEN *see* Minamoto yo musha woukristu

Gwendolyn bennett papers : from the holdings of the schomburg center for research in black culture, manuscripts, archives and rare books division: the new york public library, astor, lenox and tilden foundations – 1995 – 2r – 1 – $170.00 – (guide which covers all coll under "literature and the arts" sold separately for $20.00 d3305.g6) – mf#D3305P19 – Dist. us Scholarly Res – us L of C Photodup [420]

Gwiazda polarna – Stevens Point WI: J Worzalla's Sons, jan 1 1916 (wkly) [mf ed 1980] – 1r – 1 – (began publ in 1908; in polish) – us Balch [071]

Gwiazda polarna – Stevens Point WI. 1908 oct 30/1909 oct 29-1994 jul/dec [gaps] – 119r – 1 – mf#851907 – us WHS [071]

Gwiazda zachodu – Omaha, NE: Gwiazda Zachodu Pub Co. -v40 n27 (29 go czerw 1945) (wkly) [mf ed 1918-45 (gaps) filmed 1978] – 7r – 1 – (in polish and english) – us NE Hist [071]

Gwiazda zachodu = Western star – Omaha: [Gwiazda Zachodu Pub Co], [-1945]. dec 14 1917-1918 – 1r – 1 – us CRL [071]

Gwilliam, George Henry *see* Tetraeuangelium sanctum juxta simplicem syrorum versionem

Gwilliam, George Henry et al *see* Biblical and patristic relics of the palestinian syriac literature

Gwillim, John Cole *see*
- A partial bibliography of publications refering [sic] to the geology and mineral industry of alberta, british columbia and the yukon
- Report of oil survey in peace river district, 1919
- Report on the atlin mining district, british columbia
- Some ores and rocks of southern slocan division, west kootenay, british columbia

Gwilt, J *see* An encyclopaedia of architecture

Gwilt, Joseph *see*
- Elements of architectural criticism for the use of students
- Elements of architectural criticism for the use of students, amateurs, and reviewers
- An encyclopaedia of architecture
- Sciography

Gwilt, Joseph et al *see* Project for a national gallery, on the site of trafalgar square, charing cross

Gwilym, David Vaughan *see* The sacrament of preparation

Gwinner, Wilhelm von *see* Goethes faustidee nach der urspruenglichen conception

Gwinnett Historical Society *see*
- Newsletter of gwinnett historical society, inc
- Quarterly

Der g'wissenswurm : bauernkomoedie mit gesang in drei akten / Anzengruber, Ludwig – Wien: L Rosner, 1874 [mf ed 1996] – (= ser Neues wiener theater 41) – 63p – 1 – mf#9561 – us Wisconsin U Libr [790]

Der gwk-aktivist – Magdeburg DE, 1949-1951 may, 1952-1990 25 jan [gaps] – 16r – 1 – (title varies: n23 1951: aktivist, veb schwermaschinenbaukombinat ernst thaelmann) – gw Misc Inst [620]

Y gwladgarwr – [Wales] Rhondda 4 medi 1858-27 hydref 1882 & 31 rhag 1884 [mf ed 2003] – 25r – 1 – uk Newsplan [072]

Gwo, Yun-Han *see* Indigenous preaching in china with a focal critique on john sung

Y gwron cymreig a chyhoeddydd cyffredinol i cymru – [Wales] Cardiff ionawr 1838-rhagfyr 1839 [mf ed 2004] – 1r – 1 – uk Newsplan [072]

Gwydir examiner – Gwydir, 1889-1926 (misc iss) – at Pascoe [079]

Y gwyliedydd – [Wales] Ynys Mon 25 tach-25 rhag 1870 [mf ed 2003] – 1r – 1 – uk Newsplan [072]

Gwyliedydd – [Wales] LLGC 12 awst 1929-rhag 1938 (incomplete) [mf ed 2003] – 3r – 1 – uk Newsplan [072]

Y gwyliedydd (rhyl) – [Wales] LLGC 2 awrth-rhagfyr 1877-rhag 1950 [mf ed 2002] – 74r – 1 – (cont by: y gwyliedydd newydd [ion 1911-50]) – uk Newsplan [072]
Gwynn, John *see*
– On a syriac ms belonging to the collection of archbishop ussher
– On a syriac ms of the new testament
– Remnants of the later syriac versions of the bible
– Selected epistles of gregory the great, bishop of rome, books 9-14
Gwynn, John Tudor *see* Indian politics
Gwynn, Stephen Lucius *see*
– Famous cities of ireland
– Life of mary kingsley
– Memorials of an eighteenth century painter (james northcote)
Gwynne, George John *see* A commentary, critical, exegetical, and doctrinal, on st paul's epistle to the galatians
Gwynne, Walker *see*
– The christian year
Gwynne-Vaughan, Helen Charlotte Isabella Fraser *see* Structure and development of the fungi
Die gwynn'schen cajus- und hippolytus-fragmente / Harnack, Adolf von – Leipzig, 1890 – (= ser Tugal 1-6/3c) – 1mf – 9 – €3.00 – ne Slangenburg [240]
Gyan pho May, Yuvati *see* Mrok cvan desa mha mran ma prannsa mya
Gyges und sein ring : ein tragoedie in fuenf akten / Hebbel, Friedrich – Leipzig: Hesse & Becker, [19–?] – 1r – 1 – (incl bibl ref) – us Wisconsin U Libr [820]
Gylden, Hugo *see* Traite analytique des orbites absolues des huit planetes principales
Gyles, J F *see* Attempt to ascertain the meaning of a passage in the twenty-second c...
Gyles, John *see* Memoirs of odd adventures, strange deliverances, etc
Gyllensten, Lars Johan Wictor *see* Diarium spirituale. roman om en roest
O gymnasial : orgao dos alumnos do gymnasio petropolis – Petropolis, RJ: Typ Nachrichten, 01 maio-07 ago 1910 – (= ser Ps 19) – mf#DIPER – bl Biblioteca [079]
O gymnasiano : orgao do centro gymnasial – Florianopolis, SC. 20 abr 1932 – (= ser Ps 19) – mf#UFSC/BPESC – bl Biblioteca [079]
Das gymnasium marienwerder : von der domschule zur oberschule / Duehring, Hans – Wuerzburg: Holzner Verlag 1964 [mf ed 1992] – (= ser Ostdeutsche beitraege aus dem goettinger arbeitskreis 30) – 1r – 1 – (incl bibl ref & ind) – mf#3180p – us Wisconsin U Libr [373]
Gymnasium patientiae / Drexelius, H – Coloniae Agr: Apud Cornelium ab Egmond, 1632 – 3mf – 9 – mf#O-1564 – ne IDC [090]
Das gymnasium von st juergen : roman / Dreyer, Max – Leipzig: L Staackmann, 1925 [mf ed 1989] – 287p – 1 – mf#7185 – us Wisconsin U Libr [830]
Gymnosophista sive indicae philosopiae documenta : voluminis 1. fasciculus 1. isvaracrishnae sankhya – caricam tenens / Lassen, Christian – Bonnae ad Rhenum: Weber 1832 – (= ser Whsb) – 1mf – 9 – €20.00 – mf#Hu 515 – gw Fischer [100]
Gympie miner – Australia. 4 Sep 1889-23 Nov 1891. -w – 2 1/2r – 1 – uk British Libr Newspaper [622]
Gynaecologia – Basel, Switzerland 1966-69 – 1 – (cont by: gynecologic investigation) – mf#2061 – us UMI ProQuest [618]
Gynaecological endoscopy – Oxford, England 1992-96 – 1,5,9 – ISSN: 0962-1091 – mf#19296 – us UMI ProQuest [618]
Gynaekologe – Dordrecht, Netherlands 1981-82 – 1,5,9 – ISSN: 0017-5994 – mf#13172 – us UMI ProQuest [618]
Gynaekologische rundschau – Basel, Switzerland 1964-73 – 1,5,9 – ISSN: 0017-6001 – mf#5927.01 – us UMI ProQuest [618]
Gynecologic investigation – Basel, Switzerland 1970-73 – 1,5 – (cont: gynaecologia) – ISSN: 0017-5986 – mf#5195.01 – us UMI ProQuest [618]
Gynecologie et obstetrique – Paris, France 1968-71 – 1,5 – ISSN: 0017-601X – mf#3398 – us UMI ProQuest [618]
The Gypsy Lore Society *see* Journal of the gypsy lore society, 1888-1973
Gypsy waters cruises south / Waters, Don – New York, NY. 1938 – 1r – us UF Libraries [978]
Gypsy wizard – v1 n2 [1970] – 1r – 1 – mf#1583459 – us WHS [071]
Gyr, Jean-Guillaume, abbe *see* La franc-maconnerie en elle meme et dans ses rapports avec les autres societes secretes de l'europe
Das gyren rupffen : hallt inn johans schmid vicarge ze costenz, mit den buechle...ist voll schimpffs vnnd ernstes – Zuerich: Froschower, 1523 – 1mf – 9 – mf#ZWI-35 – ne IDC [240]
Gyrowetz, Adalbert *see*
– Deux sonates pour le piano forte
– Notturno, no 7
– Six quatuors concertants pour deux violons, alto et basse, 4e livre de quatuors, 1e partie
– Six quatuors pour deux violons, alto et basse, oeuvre 1er & 3e livre de quatuors
– Tre quartetti per due violini, viola e violoncello, opera 47
– Trois grands quatuors concertans pour deux violons, alto et violoncelle, op 42
– Trois quatuors pour deux violons, alto et basse, oeuvre 17
– Trois quatuors pour flute, violon, alto, et basse, op 20
– Trois quatuors, pour flute, violon, alto et violoncelle, oeuvre 37me
– Trois sonates pour piano forte avec accompagnement de violon ou flute et violoncelle, oeuvre 55
Gysi, Fritz *see* Richard wagner und die schweiz
Gzhatsk. sovet rk i kd *see* Izvestiia gzhatskogo soveta rabochikh, krest'ianskikh i krasnoarmejskikh deputatov
Gzowski, Casimir Stanislaus *see*
– Description of the international bridge
– Report of c s gzowski, esq
H a-c-nachrichten *see* Hamburgische adress-comtoir-nachrichten [1810-19], h a-c-nachrichten + hamburger abendblatt [1820-1823], h a-c-nachrichten [1824-26], hamburger neue zeitung & a-c-nachrichten [1826-31]
H and s : hearing and speech action – Silver Spring. 1975-1977 (1) 1975-1977 (5) 1976-1977 (9) – (cont: hearing and speech news. cont by: hearing and speech action) – ISSN: 0098-1001 – mf#2140,01 – us UMI ProQuest [616]
H blackstone's reports : reports of cases argued and detrmined in the courts of common pleas and exchequer chamber...1788-1791 / Blackstone, Henry – London: Whieldon & Butterworth. v1-2. 1791-96 (all publ) – 15mf – 9 – $22.50 – mf#LLMC 95-271 – us LLMC [324]
H du b reports : a foreign affairs letter – 1965-1982 sep – 1r – 1 – mf#622511 – us WHS [327]
H, E *see* Scripture proof for singing of scripture psalms, hymns, and spiritual
H haringman's vormal holl kavallerie-lieutenants tagebuch einer reise nach marokko und eines achtwoechentlichen aufenthaltes in diesem lande : im gefolge einer im jahr 1788 nach mequinez abgegangenen hollaendischen gesandtschaft / Haringman, H – Weimar 1805 [mf ed Hildesheim 1995-98] – 1v on 2mf – 9 – €60.00 – ISBN-10: 3-487-26565-6 – ISBN-13: 978-3-487-26565-0 – gw Olms [916]
H heine's 'buch der lieder' und sein verhaeltnis zum deutschen volkslied / Goetze, Robert – Halle: E Karras, 1895 – 1r – 1 – (incl bibl ref) – us Wisconsin U Libr [430]
A H Heisey and Co *see* Heisey glass newscaster
Die h j-kampfblatt der hitler-jugend – Berlin DE, 1935-1939 n12 – 1 – gw Misc Inst [943]
H L Hunt *see* Why not speak?
H m elmore's britischen schiffskapitaen's, vermischte nachrichten von verschiedenen gegenden, inseln und handelsplaetzen in asien : und vorzueglich in ostindien / Elmore, H M – Weimar 1804 [mf ed Hildesheim 1995-98] – 1v on 1mf – 9 – €40.00 – ISBN-10: 3-487-26591-5 – ISBN-13: 978-3-487-26591-9 – (trans fr english) – gw Olms [915]
H m r [harrison monthly review] – 1972 aug-sep/oct – 1r – 1 – mf#1057696 – us WHS [071]
H m s parliament : or, the lady who loved a government clerk / Fuller, William Henry – Ottawa: Citizen Print & Pub Co, 1880 – 1mf – 9 – mf#03270 – cn Canadiana [830]
H Mitchell [Firm] *see* Catalogue of field, garden and flower seeds, fruit and ornamental trees, shrubs, roses, etc for sale by h mitchell
H potter's reise durch die alten und neuen oestlichen departemente des koenigreichs holland und das herzogthum oldenburg : gethan im jahre 1808 – Weimar 1811 [mf ed Hildesheim 1995-98] – 1v on 2mf – 9 – €60.00 – ISBN-10: 3-487-26541-9 – ISBN-13: 978-3-487-26541-4 – (trans fr dutch) – gw Olms [914]
H potter's reisen durch einen grossen theil von sued-holland : in den jahren 1807 und 1808 – Weimar 1811 [mf ed Hildesheim 1995-98] – 1v on 1mf – 9 – €40.00 – ISBN-10: 3-487-26540-0 – ISBN-13: 978-3-487-26540-7 – (trans fr dutch) – gw Olms [914]
H r haldeman: notes of white house meetings, 1969-1973 *see* Papers of the nixon white house
H reuterdahls teologiska askadning : med saerskild haensyn till hans staellning till schleiermacher / Aulen, Gustaf – Uppsala: W Schultz, [1907] – 1mf – 9 – 0-7905-3524-6 – (incl bibl ref) – mf#1989-0017 – us ATLA [240]
H S Feng *see* Wo so jen shih te chi-tu (ccm313)
H s knispel collection of clippings related to the lindbergh baby kidnapping – Most of the clippings are from St. Paul, MN newspapers. 1r, including filmed inventory – 1 – $30.00r – us Minn Hist [360]
H st. chamberlains vorstellungen ueber die religion der semiten, spez. der israeliten / Baentsch, Bruno – Langensalza: H Beyer, 1905 – (= ser Paedagogisches Magazin) – 1mf – 9 – 0-524-01597-X – mf#1990-2536 – us ATLA [270]
H von kleists werke / Kleist, Heinrich von – Berlin: Gustav Hempel [18–] [mf ed 1995] – 5v on 1r – 1 – (incl bibl ref & ind; biogr of poet by adolf wilbrandt) – mf#3727p – us Wisconsin U Libr [800]
H w v gerstenbergs rezensionen in der hamburgischen neuen zeitung : 1767-71 / ed by Fischer, Ottokar – Berlin: B Behr, 1904 [mf ed 1993] – (= ser Deutsche litteraturdenkmale des 18. und 19. jahrhunderts 128, 3. folge n8) – xcviii/415p – 1 – mf#8676 reel 7 – us Wisconsin U Libr [430]
H zwingli : seine stellung zur musik und seine lieder / Weber, G – Zuerich, 1884 – 1mf – 9 – mf#ZWI-58 – ne IDC [242]
An ha bao *see* An ha nhu't bao
Ha de nuestra sra. de guadalupe / Malagon, Joan – Salamanca: Imp. Cossio, 1672 – 1 – sp Bibl Santa Ana [946]
Ha eshkol – Wien, Austria. 1898-1902; 1905; 1909; 1913 – 2r – 1 – uk British Libr Newspaper [072]
Ha, Huyen Chi *see* Tren canh dong may
Ha, Minh Duc *see* Tho va may van de trong tho viet nam hie n dai
Ha, Minh Tuan *see* Vao doi
Ha tinh tan van – n6-24. Hue. 1929-juil 1930 – 1 – fr ACRPP [073]
Ha tsao chi / Lao, She – Shang-hai: K'ai ming shu tien, Min kuo 29 [1940] – (= ser P-k&k period) – us CRL [480]
Haack, Ernst *see*
– Christentum und kultur
– Christus oder buddha?
Haafner, Jacob *see*
– Jakob haafner's fussreise durch die insel ceilon
– Reize naar bengalen
Haafner, Jacob Godfried *see* Onderzoek naar het nut de zendelingen en zendelings-genootschappen
Haag, Emile *see* La france protestante
Haag, Eugene *see* La france protestante
Haage, Catherine Maria *see* Tests of functional latin for secondary school use based upon...
Haage, F W *see* Bausteine aus dem or[ient] naumburg a/s
Haagensen, Andrew *see*
– Methodismens og lutheranismens
– Den norsk-danske methodismes historie paa begge sider havet
Haak *see* Bemerkungen auf einer reise durch schlesien, boehmen und einen theil von oestreich nach salzburg
Ha'am – Los Angeles, California – 1 – (v1 n1 (apr 1972)-v2 n7 (mar 1973). v2 n10 (may 1973)-v7 n5 (18 apr 1978). v8 n5 (1 may 1979). v9 n3 (17 jan 1980). v10 n2 (18 nov 1980)-v10 n6 (june 1981). v11 n3 (dec 1981). v11 n7 (apr 1982). v11 n9 (june 1982). v12 n1 (oct 1982)-feb 1985. feb 1989-mar 1989) – us AJPC [978]
Ha-'am – New York. 1908 – 1 – us AJPC [073]
Ha-am – M., 1916-18 – 16mf – 9 – mf#J-291-27 – ne IDC [077]
Haan, Ralph Leonard *see* Het millennium of het duizendjarig rijk
Haan, Wilhelm *see* Saechsisches schriftsteller-lexicon
Haandbog i den aeldste christelige kirkes dogmehistorie *see* Dr friedrich muenter's, professor der theologie an der universitaet zu kopenhagen, handbuch der aeltesten christlichen dogmen-geschichte
Haaparannan sanomat – Haparanda, Sweden. 1917 – 1r – 1 – sw Kungliga [078]
Haaparannanlehti – Haparanda, Sweden. 1882-1921 – 1 – sw Kungliga [078]
Haapsalu baptisti kogudua: eesti baptismi sunnipaik = Haapsalu baptist church: estonian baptist birthplace – Keilas, Estonia: "Kulwaja" trukk, 1934 – 1r – 1 – $6.04 – (one part of a two-part item) – us Southern Baptist [242]
Haar, Bernard ter *see* Oratio de historica religionis christianae indole
Ha'aretz – Tel Aviv, Israel. 1970-1999 (1) – mf#60151 – us UMI ProQuest [079]
Ha-aretz – Israel, 1979- – 9 – us UMI ProQuest [079]
Haarhoff, T J *see* Why not be friends?
The haarlem legend of the invention of printing by lourens janszoon coster, : critically examined = de haarlemsche costerlegende / Linde, Antonius van der – London, 1871 – (= ser 19th c publishing...) – 3mf – 9 – mf#3.1.102 – uk Chadwyck [680]
Haas, Hans *see* "Amida buddha unsere zuflucht"
Haas, J de *see* Zionism
Haas, John Augustus William *see*
– Annotations on the gospel according to st mark
– Bible literature
– Biblical criticism
– The lutheran cyclopedia
Haas, Wilhelm *see* Antlitz der zeit
Haase, F *see* Literarkritische untersuchung zur orientalisch-apokryphen evangelien literatur
Haase, Felix *see*
– Begriff und aufgabe der dogmengeschichte
– Die koptischen quellen zum konzil von nicaea
– Literarkritische untersuchungen zur orientalisch-apokryphen evangelienliteratur
– Zur bardesanischen gnosis
Haase, Henning *see* Die subjektive diagnostische valenz von intelligenztests
Ha'asidi = Watchman – Cortez CO: Navajo Gospel Crusade [1955]-1964 [mf ed 2007] – (= ser Religious periodical literature of the hispanic and indigenous peoples of the americas, 1850-1985) – 10v on 1r – 1 – (in english & navajo; cont by: adindiin) – mf#2007i-s017 – us ATLA [240]
Ha-asif / ed by Sokolov, N – Warsaw, 1884-1893. v1-6 – 79mf – 9 – mf#J-422-23 – ne IDC [077]
Haban, Teresine *see* Hymnody of the roman catholic church
Habana : biografia de una provincia / Le Riverend, Julio – Habana, Cuba. 1960 – 1r – us UF Libraries [972]
Habana / Pan American Union – Washington, DC. 1944 – 1r – us UF Libraries [972]
Habana / Roig De Leuchsenring, Emilio – Habana, Cuba. 1939 – 1r – us UF Libraries [972]
Habana a mediados del siglo 19 / Barras Y Prado, Antonio De Las – Madrid, Spain. 1925 – 1r – us UF Libraries [972]
Habana antigua / Roig De Leuchsenring, Emilio – Habana, Cuba. 1935 – 1r – us UF Libraries [972]
Habana de ayer, de hoy y de manana / Roig De Leuchsenting, Emilio – La Habana, Cuba. 1928 – 1r – us UF Libraries [972]
Habana de cecilia valdes (siglo 19) / Torriente, Lolo De La – Habana, Cuba. 1946 – 1r – us UF Libraries [972]
Habana de velazquez / Artiles Rodriguez, Jenaro – Habana, Cuba. 1946 – 1r – us UF Libraries [972]
Habana Jose Gutierrez De La Concha Y De Irigoyen *see* Memoria sobre la guerra de la isla de cuba
Habana, Jose Gutierrez De La Concha Y De Irigoyen *see* Memorias sobre el estado politico
Habana, Jose Gutierrez de la Concha y de Irigoyen *see* Memorias sobre el estado politico, gobierno y administracion de la isla de cuba
Habanero, papel politico – Habana, Cuba. 1945 – 1r – us UF Libraries [972]
Habari – (Nairobi; in English and native languages). v3-10, 1924-31 – 1 – us CRL [960]
Habari / San Francisco Black Political Caucus – 1975 nov, 1976 jan, Apr – 1r – 1 – mf#5293150 – us WHS [321]
Habari barua / Brothers and Sisters for African Unity – 1970 – 1r – 1 – (cont: habari barua newsletter) – mf#4978676 – us WHS [305]
Habari calendar / University of Illinois at Urbana-Champaign – 1996 mar, Sep – 1r – 1 – mf#5319977 – us WHS [071]
Habari gani / National African American Club – 1996 jul/sep – 1r – 1 – mf#3975912 – us WHS [366]
Habari za leo – Dar-es-Salaam, Tanzania. -w. 21 May 1948-24 Feb 1956. Imperfect. 1 reel – 1 – uk British Libr Newspaper [072]
Habas, Braha *see* Yeladim muzalim
Habeas corpus case records,1820-1863, of the united states circuit court for the district of columbia / U.S. District Court – (= ser Records of district courts of the united states) – 2r – 1 – mf#M434 – us Nat Archives [347]
Habeas corpus, the law of war, and confiscation / Nicholas, Samuel Smith – Louisville: Bradley & Gilbert, 1862. 29p. LL-1371 – 1 – us L of C Photodup [340]
Habel, J *see* Illustre abietis cum lauro connubium, quindenis symbolorum dotibus locupletatum
Habela Patino, Eugenio *see*
– Apendice a la...salida de don quixote
– El teniente apologista universal
Haben : drama in 14 bildern / Hay, Gyula – Berlin: B Henschel, 1947 – 1r – 1 – us Wisconsin U Libr [820]
Haben wir den aechten schrifttext der evangelisten und apostel? / Tischendorf, G von – Leipzig, 1873 – 1mf – 8 – €6.00 – ne Slangenburg [226]
Ha-b'er – and supplement ha-Dli. Pietrokov. v. 1-15. 1923-1938 – 1 – us NY Public [073]
Haber – Samsun. Mueduer-i Mes'ul: Emin Refik; Idare Mueduerue: Ibrahim Naci; Sermuharriri: Aziz Samih. n83. 28 tesrinievvel 1341 [1925] – (= ser O & t journals) – 1mf – 9 – $25.00 – us MEDOC [956]

Haber anasi – Trabzon. Sahib-i Imtiyaz: Mehmed Tevfik; Mueduerue: Mustafa Sirri. n6. 5 mart 1325 [1909] – (= ser O & t journals) – 1mf – 9 – $25.00 – us MEDOC [956]
Haber-i sahih / Kureysizade (Mazhar), Mehmed Fevzi Elhac – Istanbul. 5v. 1290-93 – (= ser Ottoman histories and historical sources) – 31mf – 9 – us MEDOC [956]
Ha-berit / Wajntraub, Mordka – Champery, Switzerland. 1945 – 1r – us UF Libraries [939]
Haberl, F X *see* Orlando di lasso (1532-1594)
Haberlandt, Gottlieb Friedrich Johann *see* Physiological plant anatomy
Der habermeister : ein volksbild aus den bairischen bergen / Schmid, Herman – 2.aufl. Leipzig: Keil [18–?] – (= ser Gesammelte schriften. volks- und familien-ausgabe 17) – 1 – (bound with: der bairische hiesel and sueden und norden) – mf#film mas c438 – us Harvard [880]
Haberreiner, M F *see* Leopoldinische tugend- und stammens-benambsung
Habershon, Ada Ruth *see*
– The bible and the british museum
– The study of the parables
– The study of the types
Habershon, Matthew *see*
– The ancient half-timbered houses of england
– Glance at the events of 1848
Habert, G *see* La vie du cardinal de berulle
Habertus, I Apxiepatikon *see* Liber pontificalis ecclesiae graecae
Habesci, Elias *see* Etat actuel de l'empire ottoman
Habib, Mohammad *see*
– The desecrated bones and other stories
– Hazrat amir khusrau of delhi
Habicht, Hermann *see*
– Die einwirkung des buergerlichen gesetzbuchs auf zuvor entstandene rechtsverhaeltnisse
– Internationales privatrecht nach dem einfuehrungsgesetze zum buergerlichen gesetzbuche
Habig, Jean-Marie *see* Enseignement medico-social pour coloniaux
Habimah *see* Moskver teater habima
Habit / Gough, John B – Edinburgh, Scotland. 1853 – 1r – us UF Libraries [025]
Habit and instinct / Morgan, Conwy Lloyd – London, New York: E Arnold, 1896 [mf ed 1991] – 1mf – 9 – 0-7905-8525-1 – (incl bibl ref) – mf#1989-1750 – us ATLA [150]
Un habit par la fenetre : comedie en un acte / Renard, Jules – Montreal: J G W McGown, [189-?] [mf ed 1994] – 1mf – 9 – 0-665-94577-9 – mf#94577 – cn Canadiana [820]
Habitabec – Montreal: [s.n.] v1 n1 23 avril 1976-v20 n7 2 juin 1995 (wkly) [mf ed 1988-95] – 1 – (in french and english) – mf#SEM35P326 – cn Bibl Nat [073]
Habitabec (edition de quebec) – Ste-Foy: Habitabec inc. v1 n1 21 sep 1984- (wkly) [mf ed 1988] – 1 – mf#SEM35P327 – cn Bibl Nat [073]
Habitabec (edition d'ottawa) – Ottawa: Habitabec. v1 n1 jan 31st 1986-v5 n37 oct 19th 1990 (wkly) [mf ed 1988] – 1 – (in french and english) – mf#SEM35P328 – cn Bibl Nat [073]
Habitans des landes / Sewrin, M – Paris, France. 1811 – 1r – us UF Libraries [440]
The habitant : his origin and history / DeCelles, Alfred Duclos – Toronto: Glasgow, Brook & Co, 1914 – 2mf – 9 – 0-665-72893-X – mf#72893 – cn Canadiana [971]
Habitat / Harlem Environmental Impact Project [NY] – 1996 may/jun – 1r – 1 – (cont: harlem habitat) – mf#3603617 – us WHS [333]
Habitat for Humanity, Inc (US) *see* Habitat world
Habitat international – Oxford. 1976+ (1,5,9) – ISSN: 0197-3975 – mf#49262 – us UMI ProQuest [333]
Habitat occurrence of florida's native amphibians and reptiles / Enge, Kevin M – Tallahassee, FL. 1997 – 1r – us UF Libraries [500]
Habitat world : a quarterly publication of habitat for humanity, inc / Habitat for Humanity, Inc (US) – 1984 mar-1988 sep – 1r – 1 – mf#1336454 – us WHS [333]
Habitation of god / Smith, J – Harrow, England. 1861 – 1r – us UF Libraries [240]
The habitations of man in all ages / Viollet-Le-Duc, Eugene Emmanuel – London 1876 – (= ser 19th c art & architecture) – 5mf – 9 – mf#4.2.1755 – uk Chadwyck [720]
Habiti antichi, et moderni di tutto il mondo / Vecelli, C – Venetia, 1589 – 12mf – 9 – mf#H-8403 – ne IDC [956]
Habito de esperanza / Florit, Eugenio – Madrid, Spain. 1965 – 1r – us UF Libraries [972]
Habits, vieux galons / Sewrin, M – Paris, France. 1808 – 1r – us UF Libraries [440]
Habl al-matin – Calcutta. sal-i 8, shumarah-i 1-24. 13 jumada al-sani 1318-4 zu al-hijjah 1318 [8 oct 1900-25 mar 1901] – 1r – 1 – $325.00 – us MEDOC [079]
Habla el caudillo / Franco, Francisco – Burgos, 1938. Fiche W 891. (Blodgett Collection of Spanish Civil War Pamphlets) – 9 – us Harvard [946]
Habla y cultura popular en antioquia / Florez, Luis – Bogota, Colombia. 1957 – 1r – us UF Libraries [972]
Hablando a tu corazon / Guiscafre, Rosario – San Juan, Puerto Rico. 1955 – 1r – us UF Libraries [972]
Hablaron para la direccion de propagada de guerra : (entrevistas sobre el conflicto belico) / Rodríguez Zaldivar, Rodolfo – La Habana: P Fernandez, 1943 (mf ed 19–) – 96p – mf#ZZ-14584 – us NY Public [972]
Hablitzel, Joh Bapt *see* Hrabanus maurus
Hablo de tierra conocida / Galindo Lena, Carlos – Habana, Cuba. 1964 – 1r – us UF Libraries [972]
Habname-'i *see* The divan project
[La habra-] daily star progress – CA. 1927- – 281r – 1 – $16,860.00 (subs $840y) – mf#R02331 – us Library Micro [071]
[La habra-] la habra star – CA. 1916-94 – 15r – 1 – $900.00 – mf#R02332 – us Library Micro [071]
Das habsburgisch-oesterreichische urbarbuch / ed by Pfeiffer, Franz – Stuttgart: Literarischer Verein, 1850 [mf ed 1993] – (= ser Blvs 19) – xxviii/404p – 1 – mf#8470 reel 5 – us Wisconsin U Libr [943]
Hacettepe bulletin of social science and humanities – Hacettepe. 1972-1972 (1) 1972-1972 (5) (9) – ISSN: 0441-6058 – mf#6957 – us UMI ProQuest [300]
Hache, Patricia *see* Bio-bibliographie analytique de me marie-louis beaulieu
Hacia alla y para aca... / Bayle, Constantino – Madrid: Missionalia Hispanica, 1949 – 1 – sp Bibl Santa Ana [240]
Hacia alla y para aca... / Pancke, Florian – Madrid: Missionalia Hispanica, 1949 – 1 – sp Bibl Santa Ana [240]
Hacia dond, heroes / Munoz Rivera, Manuel – New York, NY. 1948 – 1r – us UF Libraries [972]
Hacia donde va chile : por el pan, la tierra, la paz y la libertad de chile / Contreras Labarca, Carlos – [Montivideo: Imp "Central", 1943?] – 1 – us CRL [335]
Hacia el sol / Toruno, Juan Felipe – San Salvador, El Salvador. 1940 – 1r – us UF Libraries [972]
Hacia la gnosis / Roso de Luna, Mario – Madrid: Editorial Pueyo, 1921 – 1 – sp Bibl Santa Ana [240]
Hacia la restauracion democratica y el cambio soci... / Lleras Restrepo, Carlos – Bogota, Colombia. v1-2. 1963 – 1r – us UF Libraries [972]
Hacia la unificacion basica institucional de centr... / Rolz Bennett, Jose – Guatemala, 1950 – 1r – us UF Libraries [972]
Hacia la union de los pueblos latinos / Becerro de Bengoa, Ricardo – Caceres: Delegacion de Ex-Combatientes de la Alta Extremadura, S.A. – 1 – sp Bibl Santa Ana [946]
Hacia las rutas nuevas / Juez Nieto, Antonio – Badajoz: Dip. Provincial, 1938 – 1 – sp Bibl Santa Ana [946]
Hacia mi distancia / Sendon Oreiro, Mercedes – Habana, Cuba. 1955 – 1r – us UF Libraries [972]
Hacia nuevos embrales / Brenes Mesen, Roberto – San Jose, Costa Rica. 1913 – 1r – us UF Libraries [972]
Hacia un heredia genuino / Estenger, Rafael – Santiago, Cuba. 1939 – 1r – us UF Libraries [972]
Hacia un mejor conocimiento historico cientifico de extremadura / Ramirez Ramirez, Enrique – Badajoz: Imp. Diputacion Provincial, 1974 – sp Bibl Santa Ana [946]
Hacia un sistema nacional de educacion / Chavarria Flores, Manuel – San Salvador, El Salvador. 1956 – 1r – us UF Libraries [972]
Hacia una espana sin ganado? / Mendoza Ruiz, Manuel – Madrid: Obra Sindical de Colonizacion, 1965 – 1 – sp Bibl Santa Ana [946]
Hacia una integracion metropolitana de san jose / Escuela Superior De Administracion Publica America – San Jose, Costa Rica. 1963 – 1r – us UF Libraries [972]
Hacia una ordenacion ganadera / Mendoza Ruiz, Manuel – Madrid: Obra Sindical de Colonizacion, 1965 – 1 – sp Bibl Santa Ana [946]
Hacia una sociedad nueva / Vallejo, Felix Angel – Bogota, Colombia. 1953 – 1r – us UF Libraries [972]
Hacienda colonial venezolana – Caracas, Venezuela. 1946 – 1r – us UF Libraries [972]
Hacienda publica en el salvador / Ehrhardt, Lucien Andre – Nueva York, NY. 1952 – 1r – us UF Libraries [972]
Hacivad – Istanbul. 1 sene n1-2. 13-15 saban 1326 [27-30 agustos 1324] [9-12 eyluel 1908S] – (= ser O & t journals) – 1mf – 9 – $25.00 – us MEDOC [956]
Hack, Mary Pryor *see*
– Christian womanhood
– Consecrated women
– Mary pryor
Hack, Wilton *see* Comments on the dharmapada
Hackebeils illustrierte zeitung – Berlin DE, 1931-1934 20 dec – 5r – 1 – (filmed by misc inst: 1935-36, 1940-41 [gaps]. title varies: 1931 n41: neue illustrierte zeitung) – gw Misc Inst [074]
Hackel, A A *see* Das altrussische heiligenbild. die ikone
Hackenberg, Fritz *see* Elise von hohenhausen
Hackenschmidt, Karl *see*
– Der christliche glaube in acht buechern
– Der roemische bischof im vierten jahrhundert
Hacker, Colleen M *see* Moral judgment and the perceived legitimacy of injurious acts among collegiate athletes
Hacker, Isaac Henry *see* Hundred years in travancore, 1806-1906
Hacker, J *see* Die messe in den deutschen dioezesan-gesang- und gebetbuechern
Hacker's creek journal / Hacker's Creek Pioneer Descendants (Organization) – v1 n1/2,3-5, v2 n1,2-3/4, v3 n1,2-3, v4 n2-4, v5 n1-v6 n4 [1982 oct, 1983 jan-jul, oct, 1984 jan-apr/jul, oct, 1985 jan-apr oct, 1986 jan-jul, 1986 fall-1988 summer) – 1r – 1 – mf#1685699 – us WHS [929]
Hacker's Creek Pioneer Descendants (Organization) *see* Hacker's creek journal
Hackett, Horatio Balch *see*
– Christian memorials of the war
– A commentary on the acts of the apostles
– A commentary on the original text of the acts of the apostles
– Dr. william smith's dictionary of the bible
Hackett, John *see* A history of the orthodox church of cyprus
Hacki, Michael *see* Sermones capitulares
Hackin, J et al *see* Studies in chinese art and some indian influences
Hacklaender, Friedrich Wihelm *see* Der augenblick des gluecks
Hacklaender, Friedrich Wilhelm *see*
– Handel und wandel
– Nullen
– Den nye don quixote
Hackmann, Hans *see* ...Die wiedergeburt der tanz- und gesangskunst aus dem geiste der natur
Hackmann, Heinrich Friedrich *see*
– Buddhism as a religion
– Der buddhismus
– A german scholar in the east
– Die zukunftserwartung des jesaia
Hackmann, Oskar *see*
– Die polyphemsage in der volksueberlieferung
– Sagor, referatsamling
Hackney and clapton magazine – [London & SE] Hackney jul 1829 [mf ed 2003] – 1r – 1 – uk Newsplan [072]
Hackney, Anthony C *see*
– Influence of diet and the menstrual cycle on lactate concentration during increasing exercise intensities
– The influence of the menstrual cycle and diet on metabolism during rest and exercise
– The relationship between blood testosterone levels and body composition in physically active, young adult men
– Reproductive endocrine response following anaerobic and aerobic exercise
Hackney echo and north london advertiser – London, UK. 1986-13 dec 1989; 1990; 8 jan-23 dec 1992 – 8r – 1 – uk British Libr Newspaper [072]
Hackney express and shoreditch observer *see* Shoreditch observer
Hackney express & shoreditch observer. (shoreditch observer) – London. 1886-1913.-w. 16 reels – 1 – uk British Libr Newspaper [074]
Hackney gazette – [London & SE] Hackney 1869-; BLNL 1986- – 1 – (cont: hackney & kingsland gazette [10 jul 1869-]) – uk Newsplan [072]
Hackney gazette *see* The clerkenwell chronicle, st luke's examiner, holborn reporter and north london observer
Hackney gazette daily emergency news bulletin – [London & SE] Hackney may 1826 [mf ed 2003] – 1r – 1 – uk Newsplan [072]
Hackney journal – [London & SE] Hackney mar-oct 1842 [mf ed 2003] – 1r – 1 – uk Newsplan [072]
Hackney magazine & parish reformer – [London & SE] Hackney nov 1833-feb 1838 [mf ed 2003] – 2r – 1 – uk Newsplan [072]
Hackney mercury – London, UK. 4 jul 1885-1905; 24 mar 1906-12 dec 1907; 4 jan 1908-23 jul 1910 – 22 1/2r – 1 – (aka: mercury; hackney mercury) – uk British Libr Newspaper [072]
Hackney monthly & stoke newington & stamford hill review – [London & SE] Hackney dec 1919-apr 1921 [mf ed 2003] – 2r – 1 – uk Newsplan [072]
Hackney standard, bethnal green & shoreditch chronicle – Hackney, England 25 jul 1885-10 may 1907 – 1 – (cont: borough of hackney standard & bethnal green & shoreditch chronicle [24 mar 1877-18 jul 1885]) – uk British Libr Newspaper [072]
Hacksma house genealogy hot-line – v1 n2-v4 n6 [1984 sep/oct-1987 nov/dec] – 1r – 1 – mf#1288392 – us WHS [929]
Hackwood, Frederick William *see*
– Christ lore
– Dragons and dragon slayers
Hackworth, Green Haywood *see*
– Digest of international law
Ha-Cohen, Mordecai *see* Gli ebrei in libia, usi e costumi
Hacquet, Balthasar *see* Hacquet's neueste physikalich-politische reisen...
Hacquet's neueste physikalich-politische reisen... : durch die dacischen und sarmatischen oder noerdlichen karpathen – Nuernberg 1790-96 [mf ed Hildesheim 1995-98] – 4v on 8mf – 9 – €160.00 – 3-487-29157-6 – gw Olms [930]
Had i a heart. irish air of gramachree [voice and piano] / Linley, Thomas – Philadelphia: B Carr & Co [179-?] [mf ed 1988] – 1r – 1 – (libretto by richard brinsley sheridan) – mf#pres. film 39 – us Sibley [780]
Hadamar, von Laber *see* Hadamar's von laber jagd
Hadamar's von laber jagd : und drei andere minnegedichte seiner zeit und weise, des minners klage, der minnenden zwist und versoehnung, der minne-falkner / ed by Schmeller, J A – Stuttgart: Literarischer Verein, 1850 [mf ed 1993] – (= ser Blvs 20) – xx/213p – 1 – (incl bibl ref and ind) – mf#8470 reel 5 – us Wisconsin U Libr [810]
Hadassah / Hadassah, the Women's Zionist Organization of America – 1981 fall-1985 fall – 1r – 1 – (continued by: kesher (milwaukee, wi)) – mf#1057510 – us WHS [270]
Hadassah bulletin – Cincinnati. Ohio. 1931-62 – 1 – us AJPC [071]
Hadassah headlines / Hadassah, the Women's Zionist Organization of America – 1941 mar-1965 jun – 1r – 1 – mf#369071 – us WHS [270]
Hadassah magazine – (New York). 1920-33. 1956-67 – 1 – us AJPC [939]
Hadassah magazine – New York. 1972+ (1) 1978+ (5) 1978+ (9) – ISSN: 0017-6516 – mf#6958 – us UMI ProQuest [360]
Hadassah newsletter / Hadassah, the Women's Zionist Organization of America – v11 n7-v46 n6 [[1930?] jul/aug-1965 feb] – 1r – 1 – (continued by: hadassah magazine) – mf#369072 – us WHS [071]
Hadassah p and b bulletin – Cincinnati. Ohio. 1947-62 – 1 – us AJPC [071]
Hadassah, the Women's Zionist Organization of America *see*
– Hadassah
– Hadassah headlines
– Hadassah newsletter
Hadd, James *see* History of fayette county, pennsylvania
Haddad, Jamil Almansur *see*
– Historia poetica do brasil
– Revolucao cubana e revolucao brasileira
Haddan, A W *see* Apostolical succession in the church of england
Haddan, Arthur West *see*
– Apostolical succession in the church of england
– Church patient in her mode of dealing with controversies
– Councils and ecclesiastical documents
– Councils and ecclesiastical documents relating to great britain and ireland
Hadden, James *see* Genealogical and personal history of fayette county pennsylvania, vols 1-3
Haddenham 1570-1950 – 16mf – 9 – £20.00 – uk CambsFHS [929]
Haddington, Thomas Hamilton. 9th earl of Haddington *see* A letter...on the present crisis relative to the fine arts in scotland
Haddingtonshire advertiser and east lothian journal – [Scotland] East Lothian, Haddington: A Aitken 1890-2 feb 1923 (wkly) [mf ed 2003] – 22r – 1 – uk Newsplan [072]
Haddingtonshire courier [haddington, scotland : 1859] – [Scotland] East Lothian, Haddington: D & J Croal 28 oct 1859-dec 1874, jan 1893-dec 1950 (wkly) [mf ed 2004] – 171v on 38r – 1 – (cont by: haddingtonshire courier, and east lothian advertiser [jan 1864-dec 1943]; haddingtonshire courier [jan 1944-dec 1950]) – uk Newsplan [072]
Haddock baptist church – Washington. 1972-1973 (1) 1972-1973 (5) 1972-1973 (9) – 1r – 1 – $97.56 – (1990-95 minutes, budgets, membership rolls, officers, teachers, committees, newsletters, calendars, weekly bulletins, a history, may 1987-92) – mf#6534 – us Southern Baptist [242]
Haddock, John A *see* A souvenir
Haddon, Alfred Cort *see*
– Evolution in art
– Magic and fetishism

Haddon, Alfred Cort et al *see*
– Sociology, magic and religion of the eastern islanders
– Sociology, magic and religion of the western islanders
Haddon, Eileen *see* Collection of southern rhodesia archives, manuscripts and documents
Haddon, Ernest B *see* Swahili lessons
Haddon, T C *see* Church of england's commission to her priests considered
Ha-deborah – (New York). 1911-12 – 1 – us AJPC [939]
Hadelner zeitung *see* Neuhaus-ostener-nachrichten
Haden, Francis Seymour *see*
– About etching
– The etched work of rembrandt...the unauthentic character of certain of those etchings
Hader, Berta Hoerner *see* Jamaica johnny
Ha-deror : (the swallow) – (New York). 1911 – 1 – us AJPC [939]
Haderslebener folkebladet *see* Schleswigsche grenzpost
Hades : the grave in hades or the catacombs of the bible and of egypt / Pells, Samuel Frederick – London: Skeffington, 1904 – 1mf – 9 – 0-524-07061-X – mf#1992-1024 – us ATLA [220]
Hadewijch ([Blessed] Hadewych (Hadewig, Hedwig)) *see*
– Strophische gedichten
– Die werke der hadewijch aus dem altflaemischen
Hadewijch's strophische gedichten : een studie van de minne in het kader der 12de en 13de eeuwse mystiek en profane minnelyriek / Paepe, N de – Gent, 1967 – 8mf – 8 – €17.00 – ne Slangenburg [241]
Hadfield, Hamlet *see* The tailor's preceptor
Hadfield, Wililam *see* Brazil and the river plate in 1868
Hadikat uel-su'ada / Fuzuli – [Istanbul]: Izzet Efendinin Matbaasi, 1289 [1872] – (= ser Ottoman literature, writers and the arts) – 5mf – 9 – $85.00 – us MEDOC [470]
Hadleigh weekly news and south suffolk mercury – Ipswich, Suffolk, 30 sep 1966-7 oct 1983 – 1 – (between jun 1976-18 nov 1977 merged with: stowmarket chronicle using hadleigh weekly news title; cont by: suffolk mercury extra) – uk Newsplan [072]
Hadley, Henry Kimball *see*
– [Symphony no 1, in f (youth and life). op 25. for orchestra. score]
– Symphony [no 2]. the four seasons. op 30. [for orchestra. score]
Hadley, James *see*
– A greek grammar for schools and colleges
– Introduction to roman law, in twelve academical lectures
Hadley, Samuel Hopkins *see* Down in water street
Hadley-Allen Family *see* Papers
Hadlich, Heinrich *see* Die idee des gesetzes in der praktischen vernunft
Ha-doar – (New York). 1921-50 – 1 – us AJPC [939]
Ha-doar – New York. Daily v. 1-3. Nov 1921-Oct 19 1923 – 1 – us NY Public [071]
Hadorn, W *see* Die apostelgeschichte und ihr geschichtlicher wert
Hadorn, Wilhelm *see*
– Calvins bedeutung fuer die geschichte und das leben der protestantischen kirche
– Das evangelium in der apostelgeschichte
– Geschichte des pietismus in den schweizerischen reformierten kirchen
– Jean jaques rousseau und des biblische evangelium
– Kirchengeschichte der reformierten schweiz
– Das tausendjaehrige reich
Hadriani beverlandi de peccato originali... dissertatio : psalmographus ps. 58. commate 4... / Beverland, Adriaan – [Lugduni in Batavis]: Ex typographeio [Danielis 'a Gaesbeeck], 1679. Chicago: Dep of Photodup, U of Chicago Lib, 1973 (1r); Evanston: American Theol Lib Assoc, 1984 (1r) – 1 – 0-8370-0052-1 – mf#1984-6011 – us ATLA [220]
Hadschi Chalfa, M B A *see* Rumeli und bosna geographisch beschrieben
Hae hawaii – Honolulu Hl. 1856 mar 5 – 1r – 1 – mf#850536 – us WHS [071]
Haeberle, Alfred *see* Der junge schleiermacher
Haeberle, Steven H *see* Journal of gay and lesbian politics
Haeberlin, Carl Friedr *see* Materialien und beytraege zur geschichte, den rechten und deren litteratur
Haebler, Konrad *see*
– Der deutsche kolumbus-brief
– Die religion des mittleren amerika
Haec sunt acta capituli generalis bononiae : celebrati in conventu sancti dominici in festo sanctis, pentecostes anno domini milesimo quingentesimo sexagesimoquarto, die vigesimo maij – Mexici: Apud Petrum Ocharte typographum, anno domini 1567 – (= ser Books on religion...1543/44-c1800: ordenes, etc: dominicos) – 1mf – 9 – mf#crl-184 – ne IDC [241]
Haeckel, E [H P A] *see* Das system der medusen
Haeckel, Ernst Heinrich Philipp August *see*
– The evolution of man
– The history of creation
– Last words on evolution
– Monism as connecting religion and science
– Monism as connecting religion and science; the confession of faith of a man of science
– The riddle of the universe at the close of the nineteenth century
Haeckel, Manfred *see* Warum ich dafuer bin
Haeckel, Manfred [comp] *see*
– Fuer polens freiheit
– Der wahre jacob
Haeckel's monism false : an examination of the riddle of the universe, the wonders of life, the confession of faith of a man of science by professor haeckel... / Ballard, Frank – London: Charles H. Kelly, [1905?] – 2mf – us ATLA [110]
Haeckel's monism false : an examination of the riddle of the universe, the wonders of life, the confession of faith of a man of science by professor haeckel, together with haeckel's critics answered by joseph mccabe / Ballard, Frank – London: Charles H Kelly, [1905?] – 2mf – 9 – 0-7905-3531-9 – (incl bibl ref) – mf#1989-0024 – us ATLA [140]
Haecker, Theodor *see* Soeren kierkegaard und die philosophie der innerlichkeit
Haefele, H F *see* Notkeri balbuli gesta karoli magni imp (mgh6:12.bd)
Haeften, B van *see*
– De heyr-baene des cruys
– Regia via crucis
– Schola cordis sive aversi
Haeften, Benedictus van *see* Benedictus illustratus sive disquisitionum monasticarum libri 12
Haege, Christine *see* Sozialpolitische loesungskonzepte zur absicherung des risikos der pflegebeduerftigkeit
Haegele, Sigurd *see* Grundfragen der motologie
Haegringar : reise durch schweden, lappland, norwegen und daenemark im jahre 1850 / Pancritius, Albrecht – Koenigsberg 1852 [mf ed Hildesheim 1995-98] – 3mf – 9 – €90.00 – 3-487-28923-7 – gw Olms [914]
Haehn, Johann Friedrich *see* Agenda scholastica
Haehnel, K *see* Die behandlung von goethes "faust" in den oberen klassen hoeherer schulen
Haehnelt, Wilhelm *see* Der thurmbau zu babel
H'ael S'umpha *see*
– Cor luac prak
– Vacananukram cpap camboh pad lmoes qajna
Haelsinglands folkblad – Gaevle, Sweden. 1915-49 – 96r – 1 – sw Kungliga [078]
Ha-emet : (the truth) – New York. 1894 – 1 – us AJPC [939]
Ha-emeth – New York. v. 1-2 no. 10. July 1894-June 1895 – 1 – us NY Public [071]
Die haemmer droehnen : werdestimmen / Diederich, Franz – Dresden: Kaden, [1905?] [mf ed 1989] – 110p – 1 – mf#7176 – us Wisconsin U Libr [810]
Haemodynamische und metabolische reaktionen im tennissport : unter beruecksichtigung des hoeheren lebensalters / Masuhr, Andreas – (mf ed 1994) – 3mf – 9 – €49.00 – 3-8267-2079-2 – mf#DHS 2079 – gw Frankfurter [790]
Haemstede, A C *see* De gheschiedenisse ende den doodt der vromer martelaren...
Ha-emunah ha-ramah microform / Ibn Daud, Abraham ben David, Halevi – Yerushalayim, c1966. 104p. Hebrew trans. by Solomon Ibn Labi. 104p – 1 – us Wisconsin U Libr [100]
Haendler, G H *see* Erekhe ha-noutariukin
Haenel, Johannes *see* Die aussermasorethischen uebereinstimmungen zwischen der septuaginta und der peschittha in der genesis
Haenel, Karl *see* Niggeri, am nil der schwarzen
Haenicke, Diether H *see* The challenge of german literature
Haensel, Anette *see* Aktives zuhoeren und behalten
Haensel, Frank *see* Der einfluss von angstneigung und falscher physiologischer rueckmeldung auf die kontingente negative variation
Haensel-Hohenhausen, Markus *see*
– Die deutschsprachige oscar-wilde-rezeption
– Die deutschsprachigen freimaurer-zeitschriften des 18. und 19. jahrhunderts
– Elise von hohenhausen, geb von ochs
– Frankfurt im biedermeier
– Saemtliche aufsaetze und miszellen
Haentjens, Antonie Hendrik *see* Remonstrantsche en calvinistische dogmatiek
Haering, Theodor *see*
– Aufsaetze und vortraege
– The christian faith
– The ethics of the christian life
– Die lebensfrage der systematischen theologie die lebensfrage des christlichen glaubens
– Ueber das bleibende im glauben an christus
– Zur versoehnungslehre
Haerjedalens tidning – Oestersund, Sweden. 1908-23 – 31r – 1 – sw Kungliga [078]
Haermaa nabai : the ethiopic version of pastor hermae / Schodde, George Henry – Leipzig: FA Brockhaus, 1876 – 1mf – 9 – 0-7905-6879-9 – (incl bibl ref) – mf#1988-2879 – us ATLA [240]
Haernosandsposten – Harnosand, Sweden. 1842-1951 – 108r – 1 – sw Kungliga [078]
Die haessliche herzogin margarete maultasch : roman / Feuchtwanger, Lion – Berlin: G Kiepenheuer, 1930, c1926 – 1r – 1 – us Wisconsin U Libr [830]
Ha'etgar – Berkeley, Calif. v1, no. 1; v1, no. 3-v 2, no. 3; v3, no. 2; v4, no. 2; v6, no. 1; v6, no. 3; v7, no. 1; v7, no. 3; v8, no. 1-v8, no. 2. [Nov. 1978-Winter 1986] – us AJPC [270]
Ha-'eth – Lemberg. no. 1-20. Jan 31-Mar 17 1907 – 1 – us NY Public [070]
Haettig, Christof *see* Entwicklung, implementierung und anwendung einer korrelationsmethode fuer frequenzabhaengige polarisierbarkeit
Haetzler, Klara *see* Liederbuch der clara haetzlerin
Die haeuser von ohlenhof : der roman eines dorfes / Loens, Hermann; ed by Appelt, Ewald Paul – New York: H Holt c1930 [mf ed 1995] – 1r [ill] – 1 – (german text, int & notes in english. filmed with: bert brecht / willy haas) – mf#3941p – us Wisconsin U Libr [830]
Haeusliche altenpflege : erfahrungen mit den leistungen fuer schwerpflegebeduerftige nach dem gesundheitsreformgesetz 1988 / Mayer, Renate – (mf ed 1994) – 2mf – 9 – €40.00 – 3-8267-2062-8 – mf#DHS 2062 – gw Frankfurter [362]
Haeutle, Christian *see*
– Des bamberger fuerstbischofs johann gottfried von aschhausen gesandtschafts-reise
– Des bamberger fuerstbischofs johann gottfried von aschhausen gesandtschafts-reise nach italien und rom 1612 und 1613
Haevernick, Heinrich Andreas Christoph *see*
– A general historico-critical introduction to the old testament
– Historico-critical introduction to the pentateuch
– Vorlesungen ueber die theologie des alten testaments
Hafen, Hans *see* Studien zur geschichte der deutschen prosa im 18. jahrhundert
Haffner, Karl *see* Bekannte und unbekannte grossen
Hafiz *see* Gacelas de hafiz
Hafner, Gotthilf *see* Deutsche dichtung
Hafner, Philipp *see* Philipp hafners gesammelte werke
Hafta : edebiyat ve fuenun ve sanayie dair mecmuadir – Istanbul, 1881-82. Sahibi: Mihran; Muharriri: Semseddin Sami. n1-20. 22 ramadan 1298-21 sefer 1299 [18 aug 1881-12 jan 1882] – (= ser O & t journals) – 3mf – 9 – $55.00 – us MEDOC [956]
Haftalik mecmua – Istanbul: Vatan Matbaasi, 1925-28. Mueduerue: Kemal Salih. n1-185 (30 Temmuz 1341 [1925]-28 Kanunisani 1928) – (= ser O & t journals) – 41mf – 9 – $670.00 – us MEDOC [079]
Die haftung bei wechselfaelschungen / Reissig, Helmut – Leipzig, 1936 (mf ed 1994) – 1mf – 9 – €24.00 – 3-8267-3006-2 – mf#DHS 3006 – gw Frankfurter [346]
Die haftung des herrschenden unternehmens fur verbindlichkeiten der abhangigen gesellschaft bei einem multinationalen unternehmen / Langen, Albrecht – Bonn o.J – 1 – gw Mikropress [943]
Die haftung des kommanditisten im vergleich mit der haftung des komplementars auf grundlage des franzosischen, schweizerischen und deutschen handelsrechtes / Furrer, Reinhold – Luzern: Burkhardt, 1902. 256p. LL-4072 – 1 – us L of C Photodup [346]
Hag ha herut – Tel-Aviv, Israel. 1949 or 1950 – 1r – us UF Libraries [939]
Die hagada aus aegypten : israels bedrueckung in aegypten nach den dortigen zeitgenoessischen inschriften in kurzer populaerer form / Jampel, Sigmund – Frankfurt a. M: J Kauffmann, 1911 – 1mf – 9 – 0-7905-2114-8 – mf#1987-2114 – us ATLA [939]
Hagaga – Nass River, BC: J B McCullagh, [1893-1910?] – 9 – ISSN: 1190-707X – mf#P04500 – cn Canadiana [242]
Ha-gan / ed by Rabinovich, L – Spb., 1897 – 2mf – 9 – mf#J-422-26 – ne IDC [077]
Haganah speaks – New York, NY – (v2 n1 30 jan 1948-v2 n24 21 jan 1949). cont: americans for haganah; cont by: israel speaks) – us AJPC [270]
Hagar, George Jotham *see* What the world believes
Ha-gat / ed by Rabinovich, L – Spb., 1897 – 2mf – 9 – mf#J-422-27 – ne IDC [077]
Hagborn – v1 n1-v2 n2 1979/80 winter-summer – 1r – 1 – mf#668865 – us WHS [071]
Hagedorn, Friedrich von *see* Versuch einiger gedichte
Hagedorn, Hermann *see* Roosevelt, theodore. works
Hagedorn und die erzaehlung in reimversen / Eigenbrodt, Wolrad – Berlin: Weidmann, 1884 – 1r – 1 – (incl bibl ref) – us Wisconsin U Libr [430]
Hagelganss, J H *see* Christlicher hochtheurer helden tugend-lauff
Hageman, Samuel Miller *see* St paul
Hagemann, H *see* Die roemische kirche und ihr einfluss auf disciplin und dogma in den ersten drei jahrhunderten
Hagemann, Theodor et al *see* Archiv fuer die theoretische und praktische rechtsgelehrsamkeit
Hagen, Ernst August *see*
– Norica
Hagen, Friedrich H von der *see* Briefe in die heimat aus deutschland, der schweiz und italien
Hagen, Heather L *see* A physiological comparison of chair aerobics and cycle ergometry in older females
Hagen, Martin *see*
– Atlas biblicus
– Lexicon biblicum
Hagen, Paul *see* Zwei urschriften der 'imitatio christi'
Hagen, Rosa *see* Emmendingen als schauplatz von goethes hermann und dorothea
Hagenauer anzeiger – Hagenau (Haguenau F), 1889-1892 24 sep [gaps] – 1 – fr ACRPP [074]
Hagenbach, J J *see* Symbola faunae insectorum helvetiae
Hagenbach, K R *see*
– Johann oekolampad und oswald myconius, die reformatoren basels
– Kritische geschichte der...ersten baslerkonfession...
Hagenbach, Karl Rudolf *see*
– German rationalism
– History of christian doctrines
– History of the church in the eighteenth and nineteenth centuries
– History of the reformation in germany and switzerland chiefly
– Johann oekolampad und oswald myconius, die reformatoren basels
– Leitfaden zum christlichen religionsunterrichte an hoehern gymnasien und bildungsanstalten
– Wilhelm martin leberecht de wette
Hagener freie presse – Hagen, Westf DE, 1898-1899 31 may – 1 – gw Misc Inst [074]
Hagener kreisblatt und maerkischer hausfreund fuer stadt und land *see* Der hausfreund
Hagener neues tageblatt – Hagen, Westf DE, 1952 11 oct-1954 – 6r – 1 – gw Mikrofilm [074]
Hagener volkszeitung – Hagen, Westf DE, 1874 22 sep-25 dec [gaps] – 1 – (title varies: 11 mar 1881: westfaelische post; 10 mar 1894: westfaelisches tageblatt. filmed by other misc inst: 1888-93, 1894 2 jul-1921, 1923 2 jul-1928 30 jun, 1929 1 oct-1930 31 mar, 1931 1 apr-30 jun, 1932 2 jan-30 sep, 1933, 1934 3 apr-27 jun) – gw Misc Inst [074]
Hagener zeitung – Hagenau (Haguenau F), 1890-1918 [gaps] – 1 – fr ACRPP [074]
Hagenguth, Edith *see* Hartmanns iwein
Hagenmeyer, H *see* Epistulae et chartae ad historiam primi belli sacri spectantes...
Hagenow und sohn : drama in vier akten / Gaulke, Johannes – Berlin: S Cronbach, 1901 (mf ed 1990) – 1r – 1 – (filmed with: das grosse jagen) – us Wisconsin U Libr [820]
Hagenower echo – Hagenow DE, 1963 31 oct-1968 27 mar – 1r – 1 – (publ in rostock) – gw Misc Inst [074]
Hager, Alice Rogers *see* Frontier by air
Hager, Joh Georg *see* Geographischer buechersaal zum nutzen und vergnuegen eroeffnet
Hager, Werner et al *see* Wissenschaft als dialog
Hagerman, Christopher Alexander *see* Speech of c a hagerman, esq
Die haggadischen elemente im erzaehlenden teil des korans / Schapiro, Israel – Leipzig: G Fock, 1907 – (= ser Schriften (gesellschaft zur foerderung der wissenschaft des judentums (germany))) – 1mf – 9 – 0-524-02663-7 – (incl bibl ref) – mf#1990-3093 – us ATLA [260]
Haggai and zechariah : with notes and introduction / Barnes, William Emery – Cambridge: University Press, 1917 – 1mf – 9 – 0-8370-6085-0 – (incl bibl ref and indexes) – mf#1986-0085 – us ATLA [221]
Haggai and zechariah : with notes and introduction / Perowne, T T – Cambridge: University Press, 1888 – (= ser The Cambridge Bible For Schools And Colleges) – 1mf – 9 – 0-8370-3029-3 – (incl ind) – mf#1985-1029 – us ATLA [221]
Haggard A M *see* Michael fairless, her life and writings
Haggard, Fred Porter *see* The judson centennial, 1814-1914
Haggard, Henry Rider *see* Umbuso ka shaka
Haggard, John *see* Reports of cases...in the consistory court of london

HAITI

Haggenmacher, G A *see* Reise im somali-lande, 1874
Haggenmacher, Otto *see* Zur frage nach dem ursprung der religion
Haggitt, Francis *see* Sermon preached in the cathedral church of durham
Haggitt, John *see* Two letters to a [f s a] on...gothic architecture
Hagglund, B C *see* The rebel poet
Hagin, Fred Eugene *see*
– After seventeen years
– The cross in japan
Hagioglypta sive picturae et sculpturae sacrae antiquiores... / Heureux, J – Lutetiae Parisiorum, 1856 – 5mf – 8 – mf#H-616 – ne IDC [956]
Hagiographa and apocrypha / Palfrey, John Gorham – Boston: Crosby, Nichols; New York: Charles S Francis, 1852 [mf ed 1989] – (= ser Academical lectures on the jewish scriptures and antiquities 4) – 2mf – 9 – 0-7905-2259-4 – mf#1987-2259 – us ATLA [221]
Hagner, Alexander Burton *see* Address on the life and character of william cranch, delivered january 10th, 1907, by alexander b. hagner at the request of the bar association of the district of columbia
Hagner, Hartmut *see* Reichsministerium fuer die besetzten ostgebiete (bestand r 6) bd 26
Hagood, Lewis Marshall *see* The colored man in the methodist episcopal church
Ha-goren – Berdichev, Berlin, 1897-1928 – 24mf – 9 – mf#J-291-19 – ne IDC [077]
Hags news letter / Hayward Area Genealogical Society [CA] – v10 n5-v11 n10 [1986 may-1987 oct], v12 n2-4 [1988 feb-apr] – 1r – 1 – (cont: news letter [hayward area genealogical society [ca]]) – mf#1537653 – us WHS [929]
Hague, Dyson *see*
– The protestantism of the prayer book
– Ways to win
Hague, George *see*
– Banking and commerce
– Letters to my sons from madeira, algiers, egypt, the holy land and other places
– Modern business
– The position of canada in relation to annexation, secession or independence and imperial federation
– Some practical considerations on the subject of capital and labour
– Some practical studies in the history and biography of the old testament
Hague, John *see* "Canada for the canadians"
The Hague. National Archives of the Netherlands *see*
– Images of east and west: maps, plans, views and drawings from dutch colonial archives, 1583-1950
– Images of east and west: maps, plans, views and drawings from dutch colonial archives, 1583-1963
Hague, William *see*
– Eight views of baptism
– Life notes
Ha-hinukh be-erets yisrael / Scharfstein, Zevi – New York, NY. 1928 – 1r – us UF Libraries [939]
Hahlo, H R *see* Union of south africa
Hahn, Aaron *see* History of the arguments for the existence of god
Hahn, August *see*
– Bardesanes gnosticus syrorum primus hymnologus
– Bibliothek der symbole und glaubensregeln der alten kirche
– Das evangelium marcions in seiner urspruenglichen gestalt
– Key to the massoretic notes, titles, and index generally found in the margin of the hebrew bible
Hahn, C von *see*
– Aus dem kaukasus
– Bilder aus dem kaukasus
– Kaukasische reisen und studien
– Neue kaukasische reisen und studien
Hahn, Carl Hugo Linsingen *see*
– Native tribes of south west africa
Hahn, Ferdinand *see* Blicke in die geisteswelt der heidnischen kols
Hahn, Friedrich von *see* Die materielle uebereinstimmung der roemischen und germanischen rechtsprincipien
Hahn, Georg Ludwig *see* Die lehre von den sakramenten in ihrer geschichtlichen entwickelung innerhalb der abendlaendischen kirche bis zum concil vom trient
Hahn, Heike *see* Primaerer hyperparathyreoidismus
Hahn, Heinrich *see* Die hoffnungen der katholischen kirche in china
Hahn, Heinrich August *see* Vorlesungen ueber die theologie des alten testaments
Hahn, J Ph A *see* Pommersches archiv der wissenschaften und des geschmacks
Hahn, Johann Georg von *see* Mythologische parallelen
Hahn, K A *see* Gedichte des 12. und 13. jahrhunderts
Hahn, Karl August *see*
– Kleinere gedichte
– Otte mit dem barte
Hahn, Ludwig Ernst [comp] *see* Fuerst bismarck
Hahn, Michael E *see* Locating the shoulder joint in relation to the humeral epicondyles
Hahn, S *see* Thomas bradwardinus und seine lehre von der menschlichen willensfreiheit
Hahn, Sebastian *see* Thomas bradwardinus und seine lehre von der menschlichen willensfreiheit
Hahn, Traugott *see*
– Die bibelkritik im religionsunterricht
– Ist die forderung eines modernen christentums und einer modernen theologie berechtigt?
Hahn, Werner *see* Geschichte der poetischen litteratur der deutschen
Hahn, Wilhelm *see* Der bergarbeiterstreik vom mai 1889 im rheinisch-westfalischen industriegebiet unter besonderer beruecksichtigung der stellung kaiser wilhelm 2. und furst
Hahn-Hahn, Ida, Graefin *see* Die gloecknerstochter
Hahn-Hahn, Ida M *see* Aus jerusalem
Hahn-Hahn, Ida von *see* Orientalische briefe
Hai / Ko, Hsien-ning – Shang-hai: Pei hsin shu chu, 1933 – (= ser P-k&k period) – us CRL [810]
Hai chun t'ung chi / China Hai chun pu – [China: Hai chun pu, Min kuo 21 [1932]] – (= ser P-k&k period) – us CRL [315]
Hai hsing tsa chi / Pa, Chin – Shang-hai: K'ai ming shu tien, Min kuo 30 [1941] – (= ser P-k&k period) – us CRL [480]
Hai kang yu k'ai kang chi hua / Hsia, K'ai-ju – Ch'ung-ch'ing: Ch'ing nien shu tien, Min kuo 29 [1940] – (= ser P-k&k period) – us CRL [380]
Hai kuan chin k'ou hsin shui tse chih yen chiu – Shang-hai: Chung-kuo yin hang tsung kuan li ch'u ching chi yen chiu shih, Min kuo 23 [1934] – (= ser P-k&k period) – us CRL [336]
Hai kuan fa kuei hui pien / Kuan, wu shu – [China]: Hai kuan tsung shui wu ssu kung shu t'ung chi k'o, Min kuo 26 [1937] – (= ser P-k&k period) – us CRL [336]
Hai kuang ching chi lun wen chi / Hai kuang yueh k'an she – Shang-hai: Ssu she, Min kuo 23- [1934-] – (= ser P-k&k period) – us CRL [330]
Hai kuang ching chi lun wen chi / Hai kuang yueh k'an she – Shang-hai: Ssu she, Min kuo 23- [1934-] – (= ser P-k&k period) – us CRL [330]
Hai kuang yueh k'an she *see*
– Hai kuang ching chi lun wen chi
Hai kuo ying hsiung, i ming, cheng ch'eng-kung : ssu mu li shih chu / A-ying – Shang-hai: Kuo min shu tien, Min kuo 30 [1941] – (= ser P-k&k period) – us CRL [820]
Hai nei wai : tu mu chu / Wu, T'ien – Shang-hai: Chu lin shu tien, 1941 – (= ser P-k&k period) – us CRL [820]
Hai pin ku jen / Lu, Yin – Shang-hai: Shang wu yin shu kuan, Min kuo24 [1935] – (= ser P-k&k period) – us CRL [830]
Hai shang yin / Ch'en, Hui – Shang-hai: Min chung shu chu, 1934 – (= ser P-k&k period) – us CRL [810]
Hai shih chi / Ch'ien, Hsing-ts'un – Shang-hai: Pei hsin shu chu, 1936 – (= ser P-k&k period) – us CRL [480]
Hai shih chi / Ting, Ti – Shang-hai: Shih chieh shu chu, Min kuo 30 [1941] – (= ser P-k&k period) – us CRL [480]
Hai Son *see* Khi em hai muoi
Hai ti meng / Pa, Chin – Shang-hai: K'ai ming shu tien, Min kuo 27 [1938] – (= ser P-k&k period) – us CRL [830]
Hai ti meng / Pa, Chin – Shang-hai: K'ai ming shu tien, Min kuo 28 [1939] – (= ser P-k&k period) – us CRL [830]
Hai t'ien chi / T'ang, T'ao – Shang-hai: Hsin chung shu chu, [Min kuo 25 [1936]] – (= ser P-k&k period) – us CRL [840]
Hai treis leitourgiai kata tous en athenais koodikas / Trempela, Pan N – Athenai, 1935 – €12.00 – ne Slangenburg [243]
Hai tzu men / Li, Hui-ying – [China]: Hai hui ch'u pan she, 1940 – (= ser P-k&k period) – us CRL [830]
Hai wai tai piao t'uan hsi-pei k'ao ch'a jih chi – [Nan-ching]: Chung-kuo kuo min tang chung yang shih hsing wei yuan hui hsuan ch'an wei yuan hui, min kuo 22 [1933] – (= ser P-k&k period) – us CRL [915]
Ha-ibri – New York. 1916-21 – 1 – us AJPC [071]
Haidar ali / Sinha, Narendra Krishna – Calcutta: Narendra Krishna Sinha: Agents, SC Sarkar & Sons, 1941- – (= ser Samp: indian books) – us CRL [920]
Haidbilder : neue folge von mein braunes buch / Loens, Hermann – Hannover: A Sponholtz c1913 [mf ed 1995] – 1r – 1 – (filmed with: bert brecht / willy haas) – mf#3941p – us Wisconsin U Libr [880]
Das haideroeslein von sesenheim / Gensichen, Otto Franz – Berlin: Paetel, 1896 [mf ed 1993] – 318p – 1 – mf#8653 – us Wisconsin U Libr [920]
Haifa : or life in modern palestine / Oliphant, L – Edinburgh, London, 1887 – 5mf – 9 – mf#HT-101 – ne IDC [915]
Haifa : or, life in modern palestine / Oliphant, Laurence; ed by Dana, Charles Anderson – New York: Harper, 1887 – 1mf – 9 – 0-524-05685-4 – mf#1992-0535 – us ATLA [915]
Haig, T W *see* Historic landmarks of the deccan
Haigh, Arthur Elam *see* The attic theatre
Haigh, Henry *see*
– Leading ideas of hinduism
– Some leading ideas of hinduism
Haigh, Samuel *see* Sketches of buenos ayres and chile
Haight ashbury maverick / Haight Publishing Company (San Francisco, CA) et al – San Francisco CA. v1 n2-4 – 1r – 1 – (continued by: haight & ashbury maverick) – mf#1057732 – us WHS [071]
Haight ashbury tribune – v1 n5-8 [1967], v2 n1,4,6 [1968] – 1r – 1 – mf#767321 – us WHS [071]
Haight, Canniff *see*
– Before the coming of the loyalists
– Coming of the loyalists
– Here and there in the home land
– Life in canada fifty years ago
– A united empire loyalist in great britain
Haight Publishing Company (San Francisco, CA) et al *see* Haight ashbury maverick
Haigler news – Haigler, NE: Wm J Snider, 1892-95// (wkly) [mf ed v1 n6. dec 10 1892 filmed [1973]] – 1r – 1 – (absorbed by: benkelman bee) – us NE Hist [071]
The haigler news – Haigler, NE: J P Wilson, 1907-55// (wkly) [mf ed 1909-10,1918-54 (gaps)] – 11r – 1 – (suspended with july 8 1910; resumed with dec 2 1910. suspended with mar 3 1944; resumed with nov 16 1945. some irregularities in numbering) – us NE Hist [071]
The haihayas of tripuri and their monuments / Banerji, Rakhal Das – Calcutta: Govt of India, Central Publication Branch, 1931 – (= ser Samp: indian books) – us CRL [720]
Hail, A D *see* Japan and its rescue
Hail storm / Muskingum Co. New Concord – v1 n1. aug-oct 1848 [wkly] – 1r – 1 – mf#B6739 – us Ohio Hist [071]
Haile, Martin *see*
– James francis edward, the old chevalier
– Life and letters of john lingard, 1771-1851
Haile Sellassie University. Institute of Ethiopian Studies *see* Ethnological society bulletin
Haile-selassie's government – London: Longmans, 1968 – us CRL [960]
Hailey, O L *see*
– History of the baptists of tennessee
– J r graves
– The three prophetic days of matthew 12:40 or jesus and jonas
Hailey, William M H *see* Republic of south africa and the high commission territories
Die haimonskinder in deutscher uebersetzung des 16. jahrhunderts / ed by Bachmann, Albert – Stuttgart: Litterarischer Verein, 1895 (Tuebingen: H Laupp, Jr) [mf ed 1993] – (= ser Blvs 206) – xxiii/310p – 1 – (incl bibl ref and ind) – mf#8470 reel 43 – us Wisconsin U Libr [830]
Haimowitz, Morris L *see* Population trends in florida
Hain, LF T *see* Repertorium bibliographicum
Hai-nan li jen wen shen chih yen chiu / Liu, Hsien – Nan-ching: Chung-shan wen hua chiao yu kuan, Min kuo 25 [1936] – (= ser P-k&k period) – us CRL [390]
Hainan Mission. (Pres. Church in the USA) *see* Records, 1893-1923
Hainan newsletter – Horchow, Hainan: American Presbyterian Mission. sep 1912-christmas, 1949 (frequency varies) [all publ?] – (= ser Missionary periodicals from the china mainland) – 1r – 1 – $165.00 – us UPA [242]
Haine aux femmes / Bouilly, Jean Nicolas – Paris, France. 1808 – 1r – us UF Libraries [440]
Haine d'une femme, ou, le jeune homme a marier / Scribe, Eugene – Paris, France. 1824 – 1r – us UF Libraries [440]
Haine d'une femme, ou, le jeunne homme a marier / Scribe, Eugene – Paris, France. 1825 – 1r – us UF Libraries [440]
Haines, Charles Reginald *see*
– Christianity and islam in spain, a.d. 756-1031
– Islam as a missionary religion
Haines, Elijah Middlebrook *see*
– Laws of wisconsin concerning the organization and government of towns, and the powers and duties of town officers and boards of supervisors.
– A practical treatise on the powers and duties of justices of the peace.
Haines, Herbert *see* A manual of monumental brasses
Haines, Peter George *see* Educacion comercial en centro america
Haines record – Haines OR: C Hancock, -1931 [wkly] – 1 – (merged with: courier of baker (1931) to form: haines record-the courier (1931-32)) – us Oregon Lib [071]
Haines record *see* Courier of baker county, or
Haines record-the courier – Haines OR: C M Brinton, 1931-32 [wkly] – 1 – (merger of: haines record (-1931); courier of baker county, or; cont by: record-courier (haines, or)) – us Oregon Lib [071]
Haines record-the courier *see*
– Courier of baker county, or
– Haines record
Haines, Thomas Louis *see*
– Museum of antiquity
– The royal path of life
Ha-instinkt mahu? / Fabre, Jean-Henri – Tel Aviv: Omanut, 691 [1931] (mf ed [197-]) – 1r – 1 – mf#ZZ-16578 – us NY Public [590]
Hair, Christopher Heath *see* The effects of high volume resistance training on lipid profiles and insulin sensitivity
Hair of the pale moon flower – s.l, s.l? 193-? – 1r – us UF Libraries [978]
Hair, Paul Edward Hedley *see* Early study of nigerian languages
Hairenik – Boston, MA. -w. 18 July 1946-30 Dec 1954. 4 reels – 1 – uk British Libr Newspaper [071]
Hairston, Samuel W *see* Answers to virginia bar examinations
Haithoni armeni ordinis praemonstatensis de tartaris liber / Heyt'owm Patmich – Basileae, 1532 – 2mf – 9 – mf#H-8143 – ne IDC [915]
Haiti / Bird, Mark Baker – Edimbourg, Scotland. 1881 – 1r – us UF Libraries [972]
Haiti : the black republic / Rodman, Selden – New York, NY. 1954 – 1r – us UF Libraries [972]
Haiti / Bowler, Arthur – Paris, France. 1889 – 1r – us UF Libraries [972]
Haiti / Calixte, Demosthenes Petrus – New York, NY. 1939 – 1r – us UF Libraries [972]
Haiti : conference prononcee le 28 janvier 1934 / Mayard, Constantin – Poitiers, France. 1934 – 1r – us UF Libraries [972]
Haiti / Dalbemar, Jean Joseph – Paris, France. 1903 – 1r – us UF Libraries [972]
Haiti / Eldin, F – Toulouse, France. 1878 – 1r – us UF Libraries [972]
Haiti : its dawn of progress after years / Kuser, John Dryden – Boston, MA. 1921 – 1r – us UF Libraries [972]
Haiti / Jocelyn, Marcelin – Paris, France. 1913 – 1r – us UF Libraries [972]
Haiti : land of... / Lourie, I – Port-Au-Prince, Haiti. 1945 – 1r – us UF Libraries [972]
Haiti / Legitime, Francois Denis – Port-Au-Prince, Haiti. 1888 – 1r – us UF Libraries [972]
Haiti : une page d'histoire / Laroche, Leon – Paris, France. 18885 – 1r – us UF Libraries [972]
Haiti : poetes noirs – Paris, France. 1951 – 1r – us UF Libraries [972]
Haiti : la politique a suivre / Roche-Grellier – Paris, France. 1892 – 1r – us UF Libraries [972]
Haiti – Port-au-Prince: [s.n.], feb 19, 1916 – 1 sheet – 9 – us CRL [972]
Haiti : sa lutte pour l'emancipation / Jean-Baptiste, St Victor – Paris, France. 1957 – 1r – us UF Libraries [972]
Haiti / Sillac, Max Jarousse De – Paris, France. 1934? – 1r – us UF Libraries [972]
Haiti / St John, Spenser – Paris, France. 1886 – 1r – us UF Libraries [972]
Haiti : la terre, les hommes et les dieux / Metraux, Alfred – Neuchatel, Switzerland. 1957 – 1r – us UF Libraries [972]
Haiti *see*
– Code d'instruction criminelle et code penal
– Constitution
– Haiti a l'imposition (sic) internationale de bruxe
– Legislation usuelle des conseils communaux de la r...
– Lois
– Lois et decrets du gouvernement d'haiti
– Lois modifiant la loi no 1er du code de procedure
– Le moniteur haitien
– Le moniteur jornal oficial
Haiti a handbook / International Bureau Of The American Republics – Washington, DC. 1893 – 1r – us UF Libraries [972]
Haiti a l'exposition colombienne de chicago / Gentil, Robert – Port-Au-Prince, Haiti. 1893 – 1r – us UF Libraries [972]
Haiti a l'heure du tiers-monde / Catalogne, Gerard De – Port-Au-Prince, Haiti. 1964 – 1r – us UF Libraries [972]
Haiti a l'imposition (sic) internationale de bruxe / Haiti – s.l, s.l? 191- – 1r – us UF Libraries [972]
Haiti and les etats-unis / Rosemond, Ludovic – Port-Au-Prince, Haiti. 1945 – 1r – us UF Libraries [972]
Haiti and the united states, 1714-1938 / Montague, Ludwell Lee – Durham, North Carolina . 1940 – 1r – us UF Libraries [972]

Haiti, Armee *see* Lessons in haitian creole with some information regarding
Haiti au point de vue critique / Morpeau, Emmanuel – Port-Au-Prince, Haiti. 1915 – 1r – us UF Libraries [972]
Haiti au point de vue politique / Firmin, Antenor – Paris, France. 1892 – 1r – us UF Libraries [972]
Haiti au point de vue religieux / Herivel – Alencon, France. 1887 – 1r – us UF Libraries [972]
Haiti brujo / Rodriguez, Manuel Tomas – Habana, Cuba. 1936 – 1r – us UF Libraries [972]
Haiti commerciale, industrielle et agricole – Port-au-Prince: Frederick Morin. 1ere annee, n4-n37. 6 juil 1917-16 fevr 1918 – 3 sheet – 9 – us CRL [330]
Haiti Commission De Verification Des Titres De La... *see* Rapport au secretaire d'etat des finances
Haiti Consulat (Barcelona, Spain) *see* Republica de haiti
Haiti courrier – 1992 aug 5/18, sep 16/29, oct 14/28-dec 18/1993 jan 1, 1994 jul 6/20 – 1r – 1 – mf#4695756 – us WHS [071]
Haiti culture / Haitian Communication Center – 1984 jun-1987, 1988 jan-1991 jul – 2r – 1 – mf#4695569 – us WHS [302]
Haiti demain hebdo – 1984 may 10/16, may 31/jun 6, jun 11/20, sep 5/12, oct 3/9-10/16,24/30, dec 19/25, 1985 oct 10/17, 1986 mar 26/apr 2, 1987 jul 22/28 – 1r – 1 – mf#4695633 – us WHS [071]
Haiti democratique – Port-au-Prince: D Fignole. 1ere annee, n1-n91. 31 mai-30 dec 1953 – 7 sheets – 9 – us CRL [990]
Haiti. Departement De La Justice *see* Affaire de la consolidation...
Haiti Departement De L'education Nationale *see* Plan d'etudes et programmes d'enseignement
Haiti Departement Des Finances *see* Departement des finances and du commerce d'haiti, 18...
Haiti Departement des finances *see* 55 jours de gestion de cajuste bijou secretaire d'etat des finances & du commerce, du 5 aout au 30 septembre, 1903
Haiti. Departement Des Travaux Publics *see* Album-souvenir offert par le departement...
Haiti devant les problemes interamericains / Pompee, Arsene – Port-Au-Prince, Haiti. 1947 – 1r – us UF Libraries [972]
Haiti devant son destin... / Catalogne, Gerard De – Port-Au-Prince, Haiti. 1940? – 1r – us UF Libraries [972]
Haiti diary / Orjala, Paul – Kansas City, MO. 1953 – 1r – us UF Libraries [972]
Haiti Direction Generales Des Travaux Publics *see* Notes sur haiti et sa capitale
Haiti en 1886 / Deleage, Paul – Paris, France. 1887 – 1r – us UF Libraries [972]
Haiti en marche – Miami FL. 1993 sep 29-dec 29, 1994 jan 5/11-dec 28/jan 5, 1995 jan 4/10-dec 27/jan 2, 1996 jan 3/9-dec 31/jan 7, 1997 jan 4/8-dec 30/jan 6, 1998 jan 7/13-dec 30/jan 5, 1999 jan 6/12-dec, 2000 jan 5/11-dec 27/jan 2, 2001 jan 3/9-jun 27/jul 3 – 9r – 1 – mf#2775223 – us WHS [071]
Haiti et la guerre de l'independance americaine / Nemours, Alfred – Port-Au-Prince, Haiti. 1952 – 1r – us UF Libraries [972]
Haiti et le regime parlementaire / Pauleus-Sannon, H – Paris, France. 1898 – 1r – us UF Libraries [972]
Haiti et les problemes panamericaines / L'Assaut – Port-Au-Prince, Haiti. 1933? – 1r – us UF Libraries [972]
Haiti et l'occupation americaine (usmc) / Sejourne, Georges – Port-Au-Prince, Haiti. 1931 – 1r – us UF Libraries [972]
Haiti et son peuple / Bellegarde, Dantes – Paris, France. 1953 – 1r – us UF Libraries [972]
Haiti express magazine – 1989 aocut, 1993 10 juin – 1r – 1 – (cont: express magazine) – mf#3053083 – us WHS [071]
Haiti faces tomorrow's peace / Hudicourt, Max L – New York, NY. 1945 – 1r – us UF Libraries [972]
Haiti herald – Port-au-Prince, Haiti. 1956-1962 (1) – mf#67715 – us UMI ProQuest [079]
Haiti illustree – Port-au-Prince: J Chenet. 1ere annee n7-3eme annee n48. 28 mai 1890-3 dec 1892 – 4 sheets – 9 – us CRL [079]
Haiti independante / Jeremie – Port-Au-Prince, Haiti. 1929 – 1r – us UF Libraries [972]
Haiti info – 1992 aug 31-1996 nov, 1996 nov30-1999 oct 2 – 2r – 1 – mf#2947701 – us WHS [071]
Haiti integrale – Port-au-Prince: [s.n.]. v1 n1 11-12. aug 18, sep 18-23 1915 – 1 sheet – 9 – us CRL [079]
Haiti Laws, Etc *see* Code civil d'haiti
Haiti Laws, Statutes, Etc *see*
– Code de commerce d'haiti
– Code de commerce haitien
– Code des lois usuelles, recueil des lois et de jur...
– Code du travail
– Code fiscal haitien
– Decret-loi reglementant
– Licence d'etrangers, societes et commerce
– Recueil de legislation ouvrere
– Tarif judiciaire en vigueur devent les tribunaux d...
Haiti, le reveil d'une race / Audet, Maurice – Montreal, Quebec. 1940 – 1r – us UF Libraries [972]
Haiti, l'ile enchantee / Malouin, Reine – Montreal? Quebec. 1940 – 1r – us UF Libraries [972]
Haiti litteraire and sociale – Port-au-Prince: F Marcelin. 1ere annee n1-7e annee n142. 20 janv 1905-5 juil 1911 – 50 sheets – 9 – us CRL [079]
Haiti litteraire et scientifique – Port-au-Prince: E. LaForest, 5 janv 1912-5 juil 1913 – 16 sheets – 9 – us CRL [079]
Haiti new york magazine – 1980 dec – 1r – 1 – mf#5307060 – us WHS [071]
Haiti observateur – 1984 jul 20-27, 1985 dec 6/13-1986 nov 14/21, 1986 nov 21/28-1987, 1988-94, 1995 jan-nov, 1997 jun-dec, 1998, 1999 jan-jun, 1999 july-dec 29/jan 5, 2000 jan 5/10-dec, 2001 jan 3-jun 27 – 16r – 1 – mf#1126941 – us WHS [071]
Haiti, our neighbor / Rosemond, Henri Ch – Brooklyn, NY. 1944 – 1r – us UF Libraries [972]
Haiti, portrat eines freien landes / Bouchereau, Madeleine G Sylvain – Frankfurt am Main, Germany. 1954 – 1r – us UF Libraries [972]
Haiti, primer estado negro / Vidal Y Saura, Fulgencio – Madrid, Spain. 1953 – 1r – us UF Libraries [972]
Haiti progres – Brooklyn NY. 1992 oct 28-1993 dec 29, 1994 jan 5-dec 28, 1995, jan 4-1996 jan, 1996 jan 3/9-1997 jan 7, 1997-1998 jan 6, 1998 jan 7/13-dec 30/jan 5, 1999 jan 6-jul 6, 1999 jul 7/13-dec 29/jan 4, 2000 jan-dec – 9r – 1 – mf#2576171 – us WHS [071]
Haiti progres – Brooklyn, NY – (in french. inquire for details, current subsc and backfiles) – us UMI ProQuest [071]
Haiti report / Friends of Haiti – v1 n2-v2 n3-4 [1975/76 winter-1977 spring/summer], n9-11 [1978 sum-1980 fall] – 1r – 1 – (cont: bulletins [friends of haiti]) – mf#653753 – us WHS [071]
Haiti (Republic) *see* Recueil general des lois et acts de gouvernement d...
Haiti (Republic) Delegue A La Conference *see* Rapport adresse au gouvernemet d'haiti
Haiti (Republic) Departament Des Affaires Etrange *see* Six mois de ministere en face des etats-unis
Haiti (Republic) Departement Des Affaires Etrang... *see* Rapport de m louis borno
Haiti (Republic). Departement Des Affaires Etrang... *see* Documents diplomatiques
Haiti (Republic) Departement Du Travail *see* Seminaire de l'enfance en haiti, du 9 au 18 aout 1
Haiti (Republic) Department Du Travail *see* Guide de la legislation du travail
Haiti (Republic) Laws, Statutes, Etc *see*
– Code de commerce
– Code de procedure civile
– Code penal avec les dernieres modifications
Haiti (Republic) Service D'information, De Presse *see* Etapes d'un relevement
Haiti (Republic).Laws, Statutes, Etc *see* Code de procedure civile
Haiti Section De L'enseignement Rural *see* Porgramme de l'ecole normale rurale, 1954
Haiti son passee, son avenir / Roche-Grellier – Paris, France. 1891 – 1r – us UF Libraries [972]
Haiti, un siecle d'independance / Sejourne, Georges – Anvers, Belgium. 1903 – 1r – us UF Libraries [972]
Haiti vision – n3 [du 27 janv au 3 fevr] – 1r – 1 – mf#3542380 – us WHS [071]
Haitiade / Philanthrope – Paris, France. 1878 – 1r – us UF Libraries [972]
Haitiade : poeme epique en huit chants par... / Desquiron, Antoine Toussaint – Port-Au-Prince, Haiti. 1945 – 1r – us UF Libraries [972]
Haiti-american business journal – 1989 apr/may 20, 1994 jul 29/aug 15, 1995 oct – 1r – 1 – mf#3542406 – us WHS [338]
Haitian art newsletter – 1979 apr/jun – 1r – 1 – mf#848557 – us WHS [700]
Haitian centers council newsletter – 1987 summer – 1r – 1 – mf#4872847 – us WHS [350]
Haitian Communication Center *see* Haiti culture
Haitian directory – New York, NY. 1933 – 1r – us UF Libraries [972]
Haitian revolution, 1791 to 1804 / Steward, Theophilus Gould – New York, NY. 1914 – 1r – us UF Libraries [972]
Haitian Unity Council *see* Unite
Haitian-american anthology / Cook, Mercer – Port-Au-Prince, Haiti. 1944 – 1r – us UF Libraries [972]
L'haitien : echo de l'artibonite – Gonaives [Haiti]: Bureau de l'impr du journal [sep 23-oct 7 1876] (wkly) – 1mf – 9 – us CRL [079]
L'haitien : echo des idees liberales – Gonaives [Haiti]: Bureau de limp de journal [oct 14,25-nov 25, dec 9-30 1876] (wkly) – 1mf – 9 – us CRL [079]
L'haitien – Gonaives: Bureau de l'imp du journal, sep 1876-oct-1878 – 3 sheets – 9 – us CRL [079]
Haitien parle / Bellegarde, Dantes – Port-Au-Prince, Haiti. 1934 – 1r – us UF Libraries [972]
Haiti-lutte – 1992 jan 22-1994 feb 2 – 1r – 1 – mf#4695440 – us WHS [071]
Haiti-rencontres – Lille: [s.n.], n1-n3. oct/dec 1958-aout 1959 – 2 sheets – 9 – us CRL [079]
Ha-ivri = The hebrew – New York [NY]: Sarasohn & Son, Pub. v1 n1. apr 11 1892-(wkly) – (ceased in 1902. wkly except for the first 2 iss of v1. publ and ed by gerson rosenzweig, sep 19 1895-jan 24 1902. publ in new york and philadelphia, sep 19 1895-jul 29 1898. suspended with jul 29 1898 iss; resumed with jun 7 1901 iss. first 2 iss of v4 mistakenly called v3. in hebrew) – mf#ZZAN-21212 – us NY Public [071]
Haj-Ahmad, Jumana *see* Knowledge about menopause and attitudes toward menopause among the palestinian women living in the west bank and gaza strip
Hajam wuruk – Surabaja, 1953-1955 – 26mf – 9 – (missing: 1954, v2(1); 1955, v2(10-11)) – mf#SE-743 – ne IDC [950]
Ha-jarden – New York, NY. 1923-24 – (= ser Jarden) – 1 – us AJPC [939]
Hajdu-bihari naplo – Debrecen, Hungary. 1962-Jun 1991 – 65r – 1 – us L of C Photodup [079]
Haji Khalifah *see* The history of the maritime wars of the turks
Hajnt – Warsaw, Odessa, 1908-1937 – 58r – 1 – mf#J-92-24 – ne IDC [077]
Hak Chai Huk *see*
– Bhloen maranah
– Brah pad padum raja
– Brah pad sri suriyobarn
– Camloey dyk bhnaek
– Kramum khvar ka kamloh slajham
– Krom mlap qangar
– Kum dhvoe papqun na pan!
– Nan knun lok neh poe gman pa
– Nyk qun ja nicc!
– O! phasen maranah
– Phut qanagat ralat kti sneh
– Qun sralan tae pan muay nak
– Randah tav vedamant
Ha-karmel / ed by Fuhn, S J – Vilna, 1860-1880 – 115mf – 9 – mf#J-291-8 – ne IDC [077]
Ha-kedem – Spb., 1907-09. v1-3 – 18mf – 9 – mf#J-100-183 – ne IDC [077]
Hakedem : vierteljahrschrift fuer die kunde des alten orients und die wissenschaft des judentums / ed by Markon, I B & Sarsowsky, A – St Petersburg. v1-3. 1907-09 [complete] – (= ser German-jewish periodicals...1768-1945, pt 3) – 1r – 1 – $165.00 – mf#B89 – us UPA [939]
Ha-kehilot be-erets-yisra'el ve-taktsivehen / Weinryb, Bernard Dov – Yerushalayim: [h mo l], 700 [1940] (mf ed 197-) – 1r – 1 – mf#ZZ-16975 – us NY Public [956]
Hakenkreuzbanner [main edition] – Mannheim DE, 1943 jul-1944 2 jan – 1r – 1 – gw Misc Inst [074]
Hakewill, Arthur William *see*
– An apology for the architectural monstrosities of london
– Architecture of the seventeenth century
– Modern tombs
– Thoughts upon the style of architecture...houses of parliament
Hakim, Khalifa Abdul *see* The metaphysics of rumi
Hakimiyet-i milliye – Ankara: Vilayet Matbaasi, Hakimiyet-i Milliye Matbaasi, 1920-28. Sahib-i Imtiyaz: Receb Zuehtue; Basmuharriri: Hueseyin Regib, Falih Rifki [Atay] n43. 5 temmuz 1336 [1920], 958, 1476, 2006. 5 subat 1927 – (= ser O & t journals) – 1mf – 9 – $25.00 – us MEDOC [956]
Hakki, Isma'il *see* The divan project
Hakki, Ismail [Alisan] *see*
– Avrupa bizi nasil taniyor
– Vatan ugurunda yahut yildiz mahkemesi
Hakki, Ismail [Baltacioglu] *see* Kalbin goezu
Hakkier, L *see* Untersuchungen ueber die edessenische chronik
Hakkilinnut / Uibopuu, Valev – Porvoo, Finland. 1945 – 1r – us UF Libraries [960]
Hakluyt, R *see* The principal navigations, voyages, traffiques and discoveries of the english nation made by sea or over-land...within the compasse of these 1600 yeeres
Hakluyt, Richard [comp] *see* The discovery of america and islands adjacent, 1582
Hakluyt society – London, 1847-1899; Glasgow, 1903-1947, v1-96 – 1321mf – 9 – mf#H-771c – ne IDC [910]
Hakluyt society – S1: London, 1847-1899, S2: Glasgow. v1-96. 1903-1947 – 1321mf – 8 – mf#H-771c – ne IDC [400]
Hakluyt society publications, 1847-1954 – Microcard Editions. ser1: 1847-98 (99 titles); ser2: 1899-1954 (107 titles) – 206 titles on 1484mf (18:1) – 9 – $5735.00 – us UPA [910]
Hakluyt's collection of the early voyages, travels, and discoveries of the english nation – London: Printed for R H Evans...J Mackinlay...and R Priestly...1809-1812 – 5v on 1mf – 9 – 0-665-37672-3 – mf#37672 – cn Canadiana [910]
Ha-koach – College Park, Maryland, v1, no. 4 (Feb./Mar. 1976)-v1, no. 5 (Apr. 1976); v2, no. 2 (Nov. 1976)-v3, no. 3 (Feb. 1978 – us AJPC [270]
Hakol – Stanford, CA. 1979-80 – 1 – us AJPC [071]
Haksar, Kailas Narayan *see* Federal india
Hal mizba-h ha-mada-h / Rubakin, N A – Warsaw, Poland. 1922 – 1r – us UF Libraries [939]
Haladara, Asitakumara *see* Art and tradition
Halali! : geschichteln aus den bergrevieren / Achleitner, Arthur – 5.-10. aufl. Berlin: Hermann Seemann Nachfolger, [1896?] [mf ed 1995] – 126p – 1 – mf#8918 – us Wisconsin U Libr [830]
A halaltancok torteneti : geschichte der totentaenze / Kozaky, Istvan – Budapest: Magyar Toerteneti Muzeum. 3v. 1936-1944 (mf ed 1975) – (= ser Bibliotheca humanitatis historica) – 1r – 1 – (in german and hungarian. incl bibl ref) – mf#ZZAN-9605 – us NY Public [790]
Halamantish – Umdurman: [s.n.], jun 1-22 1988-jun 1989 (weekly suppl) – us CRL [079]
Halb maer, halb mehr : erzaehlungen, skizzen und reime / Raabe, Wilhelm Karl – 3. Aufl. Berlin: G Grote, 1921 – 1 – us Wisconsin U Libr [800]
Halban, Josef et al *see* Biologie und pathologie des weibes
Halbe, Max *see*
– Frau meseck
– Jugend
– Der strom
Ein halbes jahrhundert : erinnerungen und aufzeichnungen / Schack, Adolf Friedrich von – Stuttgart: Deutsche Verlags-Anstalt, 1888 – 1r – 1 – us Wisconsin U Libr [920]
Halbmonatsschrift fuer die interessen des kunstforschers und sammlers / Der Cicerone – Leipzig, 1909-1925. v1-17 – 233mf – 9 – mf#0-498c – ne IDC [700]
Halboffizielle – Wiener Neustadt. jan-mar 1869 – 1r – 1 – us UMI ProQuest [074]
Halcrow, William *see* Transport survey of the territories of papua and new guinea
Halcyon itinerary : and true millennium messenger – Marietta. 1807-1808 (1) – mf#3580 – us UMI ProQuest [240]
Halcyon luminary : and theological repository – New York. 1812-1813 (1) – mf#3837 – us UMI ProQuest [240]
Haldane, Alexander *see* Memoirs of the lives of robert haldane of airthrey, and of his brother, james alexander haldane
Haldane, Elizabeth Sanderson *see* The wisdom and religion of a german philosopher
Haldane, J A *see* Atonement
Haldane, J B S *see* Callinicus
Haldane, Richard Burdon Haldane *see* The pathway to reality
Haldane, Robert *see*
– Address to the public concerning political opinions, and plans lately adopted to promote religion in scotland
– The books of the old and new testaments canonical and inspired
– Duty of paying tribute enforced
– Sanctification of the sabbath
Haldane Society *see* Law reform now: a programme for the next three years.
Haldar, Hiralal *see* Hegelianism and human personality
Haldar, Sukumar *see* Hinduism
Halde, J B du *see* An account of the journey of the peres boures, fontenay, gerbillon, le comte, and vesdelou
Haldeman, I. M *see* How to study the bible, the second coming and other expositions
Haldeman, Isaac Massey *see*
– The coming of christ
– Theosophy or christianity, which?
Haldeman-Julius, Emanuel *see*
– Agnostic looks at life
– Free speech and free thought in america
Haldeman-julius weekly – 1922 nov11-1925 feb 14, 1925 feb 21-1927 dec 3, 1927 dec 10-1929 apr 6 – 3r – 1 – (cont: appeal to reason [girard, ks]; cont by: american freeman [girard, ks]) – mf#964090 – us WHS [071]
Haldenslebener rundschau – Haldensleben DE, 1963 29 jan-1967 21 mar – 1r – 1 – gw Misc Inst [074]
Halderman-Julius, M *see* Clarence darrow's two great trials
The haldimand deanery magazine – [Dunnville, Ont?: s.n. 1899] – 9 – mf#P04379 – cn Canadiana [242]

The haldimand deanery magazine – Dunnville, Ont: [s.n. 1901-1902] [mf ed v2 n1 jan 1901-v3 n12 dec 1902] – 9 – mf#P04381 – cn Canadiana [242]
Haldimand, Frederick *see* Unpublished papers and correspondence, 1758-84
Hale, Brendon S *see* Effect of mental imagery of a motor task on the hoffmann reflex
Hale, Bruce D *see* The effects of internal and external imagery on muscular and ocular concomitants
Hale, Charles Reuben *see*
– A list of all the sees and bishops of the holy orthodox church of the east
– A list of sees and bishops in the holy eastern church
Hale, Donna C *see* Women and criminal justice
Hale, Edward *see*
– The psychological elements of religious faith
– Theism and the christian faith
Hale, Edward Everett *see*
– Christianity is a life / unitarianism and original congregationalism in new england / the unitarians
– A family flight over egypt and syria
– James freeman clarke
– The pilgrim covenant of 1602
– Works
Hale, Edward Everett, Jr *see* The life and letters of edward everett hale
Hale, Frederick *see* History of norwegian missionaries and immigrants in south africa
Hale, George Ellery et al *see* Annual report of the director of the mount wilson observatory
Hale, Matthew Blagden *see*
– The aborigines of australia
– The transportation question
Hale, Richard *see* Letter to the countess of harewood
Hale, Salma *see* History of the united states
Hale, Sarah *see* Diaries of miss sarah hale
Hale, Susan *see* A family flight over egypt and syria
Hale, Trevor A *see* Changes in learned motor behavior
Hale, William Benjamin *see* Handbook on the law of damages
Hale, William Gardner *see* Art of reading latin
Hale, William Hale *see*
– Charge delivered to the clergy of the archdeaconry of st alban's
– Duties of the deacons and priests in the church of england compared
– Essay on the supposed existence of a quadripartite and tripartite d...
– Method of preparation for confirmation
– Observations on clerical funds
Hale, William Pillsbury *see* Christ versus christianity
Halem, Gerhard A von *see* Erinnerungs-blaetter von einer reise nach paris im sommer 1811
Halem, Gerhard Anton von *see* Paris en 1790
Halensis, Alexander *see* In 12 aristotelis metaphycam
Halep – (= ser Vilayet salnames) – 9 – (1288 [1871] def'a 5 2mf $325; 1291 [1874] def'a 8 2mf $95; 1302 [1885] 4mf $60; 1305 [1888] def'a 16 3mf $65; 1307 [1890] def'a 18 3mf $75; 1309 [1892] def'a 20 3mf $95; 1310 [1893] def'a 21 4mf $75; 1313 [1895] def'a 23 4mf $80; 1314 [1896] def'a 24 4mf $90; 1315 [1897] def'a 25 4mf $80; 1316 [1898] def'a 26 4mf $80; 1317 [1899] def'a 27 6mf $90; 1318 [1900] def'a 28 5mf $90; 1319 [1901] def'a 29 5mf $90; 1320 [1902] def'a 30 7mf $110; 1321 [1903] def'a 31 8mf $130; 1322 [1904] def'a 32 6mf $100; 1323 [1905] def'a 33 7mf $120; 1324 [1906] def'a 34 8mf $130; 1326 [1908] def'a 34[35] 7mf $120) – us MEDOC [956]
Hales corners sunrise edition – Hales Corners WI. v2 n14 [1990 apr 2] – 1r – 1 – mf#1684763 – us WHS [071]
Hales, William *see* Methodism inspected
Halesworth times and east sufolk advertiser – Ipswich, Suffolk, 17 jul 1855-7 oct 1983 – 1 – (cont by: suffolk mercury extra) – uk Newsplan [072]
Halesworth times, southwold and general advertiser – [East Midlands] Suffolk jan 1936-dec 1950 [mf ed 2002] – 11r – 1 – uk Newsplan [072]
Halevy, Fabian S *see* Caracter de la literatura hebrea
Halevy, J *see* La guerre de sarsa-dengel contre les falachas
[Halevy, J] *see* Teezaza sanbat
Halevy, Joseph *see* Melanges d'epigraphie et d'archeologie semitiques
Halevy, Leon *see*
– Espion
– Indiana
– Sauveur
Halevy, M A *see* Moise dans l'histoire et dans la legende
Haley, Jesse James *see* Makers and molders of the reformation movement
Haley, John W *see*
– An examination of the alleged discrepancies of the bible
– The hereafter of sin
The haley record – Haley, ND: E A Hobbs, may 26 1911 (wkly) – 1 – mf#11453 – us North Dakota [071]
Half a century in china : recollections and observations / Moule, Arthur Evans – London, New York: Hodder & Stoughton [1911?] [mf ed 1990] – 1mf – 9 – 0-7905-5851-3 – mf#1988-1851 – us ATLA [915]
Half a century of sound lutheranism in america : a brief sketch of the history of the missouri synod / Graebner, Augustus Lawrence – St Louis: Concordia [1897?] [mf ed 1992] – 1mf – 9 – 0-524-02750-1 – mf#1990-4425 – us ATLA [242]
Half a dozen hints on picturesque domestic architecture / Hunt, Thomas Frederick – 2nd ed. London 1826 – (= ser 19th c art & architecture) – 1mf – 9 – mf#4.2.1130 – uk Chadwyck [720]
The half century : or, a history of the changes that have taken place and events that have transpired, chiefly in the united states, between 1800 and 1850 / Davis, Emerson – Boston: Tappan & Whittemore, 1851 [mf ed 1993] – 2mf – 9 – 0-524-08224-3 – (int by mark hopkins) – mf#1993-1009 – us ATLA [975]
A half century among the siamese and the l... : an autobiography / Mcgilvary, D – New York, Chicago, Toronto, London, Edinburgh, 1912 – 6mf – 9 – mf#HT-85 – ne IDC [915]
A half century among the siamese and the lao : an autobiography / McGilvary, Daniel; ed by Bradley, Cornelius Beach – New York: FH Revell, c1912 [mf ed 1990] – 2mf – 9 – 0-7905-7252-4 – mf#1988-3252 – us ATLA [242]
Half century discourse : the first church in buffalo. delivered on the evening of feb. 3d, 1862 / Clarke, Walter – Buffalo, NY: T Butler, 1862 – 1mf – 9 – 0-524-08673-7 – mf#1993-3198 – us ATLA [240]
A half century in burma : a memorial sketch of edward abiel stevens / Stevens, S W – Philadelphia, 1897 – 1mf – 9 – mf#HTM-186 – ne IDC [915]
Half century messages to pastors and people / Huntington, DeWitt Clinton – Cincinnati: Jennings and Graham, [c1905] Beltsville, Md: NCR Corp, 1977 (3mf); Evanston: American Theol Lib Assoc, 1984 (3mf) – 9 – 0-8370-0144-7 – mf#1984-0030 – us ATLA [240]
Half eeuwfeest van de eerste christelijke gereformeerde gemeente te pella, iowa, 1866-1916 – Pella, Iowa: Weekblad Drukkerij, [1916?] – 1mf – 9 – 0-524-06621-3 – mf#1991-2676 – us ATLA [240]
A half hour with robert elsmere / Armstrong, George Dodd – Norfolk, VA: TO Wise, 1889 [mf ed 1985] – 1mf – 9 – 0-8370-2510-9 – mf#1985-0510 – us ATLA [240]
Half hours with muhammad : being a popular account of the prophet of arabia and of his more immediate followers, together with a short synopsis of the religion he founded / Wollaston, Arthur Naylor – London: WH Allen, 1886 – 1mf – 9 – 0-524-01325-X – mf#1990-2361 – us ATLA [260]
[Half moon bay-] half moon bay news – CA. 1922-1934 – 5r – 1 – $300.00 – mf#C03235 – us Library Micro [071]
[Half moon bay-] half moon bay review and pescadero pebble – CA. 1939- – 56r – 1 – $3360.00 (subs $150y) – mf#C02284 – us Library Micro [071]
Half our future : report of the central advisory council for england (newsom report), 1963 / Great Britain. Central Advisory Committee for England – 4mf – 9 – mf#86965 – uk Microform Academic [324]
Half truths and the truth : lectures on the origin and development of prevailing forms of unbelief / Manning, Jacob Merrill – Boston: Lee & Shepard; New York: Lee, Shepard & Dillingham, 1872, c1871 [mf ed 1985] – 1mf – 9 – 0-8370-4269-0 – (incl bibl ref) – mf#1985-2269 – us ATLA [210]
Half-century magazine – Chicago. v1-18 n1. 1916-25 [all publ] – (= ser Black journals, series 1) – 1r – 1 – $200.00 – us UPA [305]
Half-century magazine – New York. 1916-1925 (1) – mf#7444 – us UMI ProQuest [305]
Half-century of science / Huxley, Thomas Henry – New York?: J Fitzgerald?, 1888? – (= ser Humboldt library of science) – 1mf – 9 – (1: the advance of science in the last half century by t h huxley; 2: the progress of science from 1836-1886 by grant allen) – mf#08974 – cn Canadiana [500]
A half-century of the unitarian controversy : with particular reference to its origin, its course, and its prominent subjects among the congregationalists of massachusetts / Ellis, George Edward – Boston: Crosby, Nichols, 1857 [mf ed 1990] – 2mf – 9 – 0-7905-5388-0 – (with app) – mf#1988-1388 – us ATLA [243]
Half-hour studies at the cross / Garrison, James Harvey – St Louis: Christian Pub Co, 1895 – 1mf – 9 – 0-524-02467-7 – mf#1990-4326 – us ATLA [240]
Half-hours with the minor prophets and the lamentations / Wiles, Joseph Pitt – London:Morgan and Scott, [1908?] – 1mf – 9 – 0-8370-5848-1 – mf#1985-3848 – us ATLA [221]
Halfmann, Jost *see* Paradigmenwechsel in der theorie der wissenschaft
Halfpenny press – Dublin, Ireland. 8 nov-22 dec 1873 – 1/2r – 1 – uk British Libr Newspaper [072]
Halfpenny weekly – [NW England] Liverpool 21 nov 1885-8 mar 1890 – 1 – uk MLA; uk Newsplan [072]
Halftermeyer, Gratus *see*
– Carrona
– Historia de managua
– Managua a traves de la historia, 1846-1946
Half-timbered houses and carved oak furniture / Sanders, William Bliss – London 1894 – (= ser 19th c art & architecture) – 5mf – 9 – mf#4.2.582 – uk Chadwyck [720]
Half-way house to infidelity / Street, James C – London, England. 1870 – 1r – us UF Libraries [240]
Halfway to heaven / Hersey, Jean – New York, NY. 1947 – 1r – us UF Libraries [972]
Halfyard, Samuel Follet *see*
– Cardinal truths of the gospel
– The spiritual basis of man and nature
A half-yearly general meeting of the british indian association...tuesday the 31st july, 1866... / Calcutta. British Indian Association – [Calcutta], [1866] – (= ser 19th c books on british colonization) – 1mf – 9 – mf#1.1.4653 – uk Chadwyck [366]
Halhed, Nathaniel Brassey *see* Revealed knowledge of the prophecies and times
Hali / Desani, Govindas Vishnoodas – London: Saturn Press, 1950 – (= ser Samp: indian books) – (foreword by t s eliot and e m forster) – us CRL [490]
Haliburton, Robert Grant *see*
– American protection and canadian reciprocity
– The coal trade of the new dominion
– Dwarf survivals
– The dwarfs of mount atlas
– Explorations in the pictou coal field
– How a race of pygmies was found in north africa and spain
– Influence of american legislation on the decline of the united states as a maritime power
– The land of the north
– The men of the north and their place in history
– New materials for the history of man, no 1
– New materials for the history of man, no 2
– New materials for the history of man, nos 1 and 2
– On the influence of american legislation on the decline of the united states as a maritime power
– The past and the future of nova scotia
– A review of british diplomacy and its fruits
– A sketch of the life and times of judge haliburton
– Survivals of dwarf races in the new world
Haliburton, the man and the writer : a study / Crofton, Francis Blake – Windsor, NS: Printed for the Haliburton by J J Anslow, 1889 – 1mf – 9 – mf#02129 – cn Canadiana [420]
Haliburton, Thomas C *see* An historical and statistical account of nova-scotia
Haliburton, Thomas Chandler *see*
– The attache
– The clockmaker
– Nature and human nature
Halich, Wasyl *see* Ukrainians in the united states
Halid *see* The divan project
Halid, Halil *see* Rodos fethinde sultan sueleyman'in tedabir-i siyasiyesi
Haliday, Charles *see* An inquiry into the influence of the excessive use of spirituous liquors
Halifax 1703-1902 – Oxford, MA (mf ed 1992) – (= ser Massachusetts vital records) – 56mf – 9 – 0-87623-152-0 – (mf 1-7: town records 1734-98. mf 4: publishments 1773-97. mf 5: births & marrs 1707-1810. mf 6: vital records 1703-1811. mf 8-14: town records 1798-1826. mf 11-12: marrs & intentions 1798-1828. mf 13: births by family 1775-1842. mf 15-21: town records 1827-55. mf 21-22: intentions & marrs 1828-55. mf 23-24: town records 1828-55. mf 25-35: town records 1853-75. mf 36-48: town records 1875-1908. mf 49-50: birth index 1842-1957. mf 50-51: marr index 1841-1963. mf 52: death index 1841-1963. mf 53: deaths 1841-99. mf 54: marriages 1841-1904. mf 55-56: births 1842-1902) – us Archive [978]
Halifax Academy *see* The academy annual
Halifax and dominion real estate register – Halifax, NS: J Naylor, [1878-18–?] – 9 – (cont: halifax and provincial real estate register) – mf#P04712 – cn Canadiana [333]
Halifax and provincial real estate register – Halifax, NS: J Naylor, [1877?-1878?] – 9 – (cont: subscribers real estate directory; cont by: halifax and dominion real estate register) – mf#P04711 – cn Canadiana [333]
Halifax baptist church. vermont : church records – 1784-1791 – 1 – 5.00 – us Southern Baptist [242]
Halifax, Charles Lindley Wood *see* Lord halifax's ghost book; a collection of stories of haunted houses, apparitions and supernatural occurrences
Halifax, Charles Lindley Wood, Viscount *see* Leo 13 and anglican orders
Halifax citizen – Canada. The Citizen The Citizen and Evening Chronicle. 20 Dec 1864; Feb 1871-Jan 1888. 36 1 2 reels – 1 – uk British Libr Newspaper [072]
Halifax citizen – Halifax, NS. 1863-77 – 16r – 1 – ISSN: 0839-3958 – cn Library Assoc [071]
Halifax comet – England, 1893-aug 1904 – 14r – 1 – uk British Libr Newspaper [072]
Halifax daily courier and guardian – England.1897-9 Jun 1898; 1899-1967. -d. 276 reels – 1 – uk British Libr Newspaper [072]
Halifax evening reporter – Halifax, NS. 1860-79 – 27r – 1 – cn Library Assoc [071]
Halifax fishery commission : closing argument of mr. doutre on behalf of her britannic majesty – S.l: s.n, 1877? – 1mf – 9 – mf#57770 – cn Canadiana [639]
Halifax gazette – Halifax, NS. 1752-1800 – 10r – 1 – ISSN: 0830-5676 – cn Library Assoc [071]
Halifax gazette – South Boston, VA. 1923-1963 (1) – mf#66865 – us UMI ProQuest [071]
Halifax guardian – Halifax, England 23 jan 1847-30 apr 1921 (wkly) [mf jan-dec 1912] – 1 – (incorp with: halifax evening courier; wanting: jan-dec 1851, jan 1897-dec 1898, 1911; cont: halifax guardian & huddersfield & bradford advertiser [1 dec 1832-16 jan 1847]) – uk British Libr Newspaper [072]
The halifax guardian *see* Our local portfolio, 1856-69
Halifax herald – Nova Scotia, Canada. Mar-jun 1941; sep 1941-sep 1942; dec 1942-may 1946; dec 1946-feb 1953 – 133r – 1 – uk British Libr Newspaper [071]
Halifax historical herald – v6 n1/2-n4/v7 n3/4 [1978 winter/spring-1979], v7 n3-4/v8 n1 [1979/1980], 1981 aug – 1r – 1 – mf#676429 – us WHS [978]
Halifax & huddersfield express – England.Feb 1831-1840.-w. 4 reels – 1 – uk British Libr Newspaper [072]
Halifax local opinion *see* Halifax comet
Halifax mercury – England.24 May 1890-5 Jan 1895. -w. 4 1/s reels – 1 – uk British Libr Newspaper [072]
The halifax monthly magazine – [Halifax, NS?: J S Cunnabell, 1830-1833] – 9 – mf#P04924 – cn Canadiana [971]
Halifax (NS). City Council *see* Annual report of the several departments of the city government of halifax, nova scotia
Halifax observer – England.9 Aug 1884-1887. -w. 3 reels – 1 – uk British Libr Newspaper [072]
The halifax philatelic magazine – Halifax, NS: Muirhead and Van Malder, [1897?] – 9 – ISSN: 1190-6553 – mf#P04575 – cn Canadiana [760]
The halifax philatelist – Halifax, NS: Halifax Philatelic Co, [1887?-1889] – 9 – mf#P04772 – cn Canadiana [760]
Halifax railway and public works = Chemin de fer de halifax et de Quebec et travaux publics / Canada (Province). Gouverneur general – [s.l.]: printed by Lovell & Gibson, 1849 [mf ed 1982] – 1mf – 9 – mf#SEM105P122 – cn Bibl Nat [380]
Halifax reporter – Daytona Beach, FL. 1974-1981 aug 28 – 8r – (gaps) – us UF Libraries [071]
Halifax times – England.16 Aug-25 Jul sic 1873; 5 Sept 1873; Jun 1889-Jan 1895. -w. 5 1/2 reels – 1 – uk British Libr Newspaper [072]
Halifax to the saskatchewan : "our boys" in the riel rebellion: a musical and dramatique burlesque / Dixon, L – Halifax, NS?: Holloway, 1886 – 1mf – 9 – (songs by r blackmore, c munro, and s h romans) – mf#30229 – cn Canadiana [790]
Halife, Mehmet *see* Tarih-i gilmani
Halim, K *see* Palawidja
Halim, Karim *see* Sandiwara chusingura
Halimbauang buhay nang turing na mag-asauang si guiadoro at si amapura at ang bilang anac na si negaderio at si rectorino... / Ignacio, Cleto R – 2nd ed. Manila: Colegio de Santo Tomas 1904 [mf ed Bloomington IN: Indiana Uni Lib, Preservation Dept 1984] – (= ser Coll...in the tagalog language 2) – 1r – 1 – us Indiana Preservation [490]
Hali's poetry : a study / Jamil, M Tahir – Bombay: DB Taraporevala Sons & Co, 1938 – (= ser Samp: indian books) – (foreword by e g hart) – us CRL [490]
Halit, Refik [Karay] *see* Bir avuc sacma
Haliwa Indian Tribe *see* Indian voice

Halk – Adana. Sahibi ve Sermuharriri: Mehmed Rasim. n107. 11 kanunisani 1340 [1924] – (= ser O & t journals) – 1mf – 9 – $25.00 – us MEDOC [956]
Halk bilgisi mecmuasi – Ankara: Iktisat Matbaasi, 1928. Nesreden: Halk Bilgisi Dernegi Umumi Merkezi; Mueduerue: Ziyaeddin Fahri Findikoglu. n1 1928 – (= ser O & t journals) – 3mf – 9 – $75.00 – us MEDOC [956]
Halk ovozi – Dushanbe, USSR. Sept 10 1991-Dec 30 1992 – 1r – 1 – us L of C Photodup [077]
Halkett, George Roland *see*
– Notes to the royal society of artist's autumn exhibition, birmingham
– Notes...glasgow institute of the fine arts
– The royal scottish academy notes
– The walker art gallery notes
Halkett, John *see*
– Historical notes respecting the indians of north america
– Precis touchant la colonie du lord selkirk sur la riviere rouge
– Statement respecting the earl of selkirk's settlement upon the red river, in north america; its destruction in 1815 and 1816; and the massacre of governor semple and his party
Halkett, Samuel *see* Dictionary of anonymous and pseudonymous english literature
Halkin, J *see* Recueil des chartes de l'abbaye de stavelot-malmedy
Halkin, Joseph *see* Les ababua (congo belge)
Hall, A C A *see* The relations of faith and life
Hall, Alexander Wilford *see* The design and importance of christian baptism
Hall, Archibald *see*
– A biographical sketch of the late a f holmes...
– The british american journal
Hall, Arthur Crawshay Alliston *see*
– Christ's temptation and ours
– Confirmation
– Considerations concerning the sacrament of our lord's body and blood
– The doctrine of the church
– The example of our lord
– A letter about the mission to be held at the church of st john the evangelist, montreal
– The use of holy scripture in the public worship of the church
– The words from and to the cross
Hall, Arthur D *see* A life of the pope (leo the thirteenth)
Hall, B M *see* The fugitive slave law
Hall, Barnes M *see* The life of rev. john clark
Hall, Basil *see*
– Entdeckungsreise nach der westkueste von korea und der grossen lutschu-insel
– Extracts from a journal
– Fragments of voyages and travels
– Travels in india, ceylon, and borneo
– Travels in north america, in the years 1827 and 1828
– Voyage dans les etats-unis de l'amerique du nord et dans le haut et bas-canada
Hall, C F *see* Narrative of the second arctic expedition
Hall, C J *see* Light from the east
Hall, Charles Cuthbert *see*
– Christ and the eastern soul
– Christ and the human race, or, the attitude of jesus christ toward foreign races and religions
– Christian belief interpreted by christian experience
– The gospel of the divine sacrifice
– Into his marvellous light
– Progress in religious and moral education
– Qualifications for ministerial power
– The redeemed life after death
– The silver cup
– Spiritual experience and theological science
– Twenty-four lessons to illustrate christian belief and christian experience by means of christian hymns
– The universal elements of the christian religion
Hall, Charles Cuthbert et al *see* Christian worship
Hall, Charles Francis *see* Life with the esquimaux
Hall, Charles Henry *see* The valley of the shadow
Hall, Charles Winslow *see* Legends of the gulf
Hall, Christopher K *see* The effects of exercise on blood volume during dialysis
Hall city, glades county, florida / Huss, Veronica E – s.l, s.l? 193-? – 1r – us UF Libraries [978]
Hall, Clayton Colman *see* Narratives of early maryland, 1633-1684
Hall county free press – Grand Island, NE: Free Press Pub Co. v1 n1. aug 18 1882 [mf ed [1996]] – 1r – 1 – us NE Hist [071]
[The hall county times] – Grand Island, NE: Albee Pub Co. v1 n1. may 6 1953- (wkly) [mf ed -jun 10 1953] – 1r – 1 – (v1 n1 lacks title) – us NE Hist [071]
Hall, Courtney D *see* Ankle strength and rate of force development
Hall, Cyrus *see* Diary
Hall, David A *see* Clarke and hall's cases in contested elections in congress, 1789-1834
Hall, David W *see* The effects of pilates-based training on balance and gait in an elderly population
Hall, Edward Hagaman *see* Alaska, the eldorado of the midnight sun
Hall, Edward Henry *see*
– Lessons on the life of st paul
– Papias and his contemporaries
– Ten lectures on orthodoxy and heresy in the christian church
Hall, Edward Hepple *see*
– Ho! for the west
– Ho! for the west!!!
– The home colony
– Lands of plenty in the new north-west
Hall, Elisa *see*
– Mostaza
– Semilla de mostaza
Hall, Elizabeth R *see* Moral development levels of athletes in sport specific and general social situations
Hall, Florence Marion Howe *see* Good form for all occasions; a manual of manners, dress and entertainment for both men and women
Hall, Francis *see*
– Colombia
– Letters written from colombia
– Travels in canada, and the united states, in 1816 and 1817
– Travels in france, in 1818 [eighteen hundred and eighteen]
Hall, Francis Joseph *see*
– Authority, ecclesiastical and biblical
– The being and attributes of god
– Creation and man
– The doctrine of god
– The doctrine of man and of the god-man
– The doctrine of the church and of last things
– Evolution and the fall
– The historical position of the episcopal church
– A history of the diocese of chicago
– The incarnation
– Introduction to dogmatic theology
– The trinity
Hall, Frederic *see* Laws of alaska pertaining to civil government, mines, and land
Hall, Frederic Thomas *see* The pedigree of the devil
Hall, Frederick Lee *see* Legal research in the south pacific
Hall, G *see* Anecdotes of the bombay mission for the conversion of the hindoos...
Hall, G Stanley (Granville Stanley) *see* Morale
Hall, Gayle C *see* Legal relationship of student teachers to public institutions of higher education and public schools
Hall, George Franklin *see* Some american evils and their remedies
Hall, H *see* Red book of the exchequer (rs99)
Hall, H R *see* 11th dynasty temple at deir el-bahari
Hall, Harry Reginald *see*
– The ancient history of the near east
– Handbook for egypt and the sudan
Hall, Herbert Byng *see* The adventures of a bric-a-brac hunter
Hall, Isaac Hollister *see* American greek testaments
Hall, J *see*
– A common apologie of the church of england
– New year's tract...
Hall, J Eugene *see* Making a will, by whom, when and how; kinds of wills, why and why not, and if not; specimens
Hall, Jacob Henry *see* Biography of gospel song and hymn writers
Hall, James *see*
– Letters from the west
– Tour through ireland, particularly the interior and least known parts
– Travels in scotland, by an unusual route
Hall, James. baronet *see* Essay on the origin, history, and principles, of gothic architecture
Hall, James Hugh Blair *see* The history of the cumberland presbyterian church in alabama prior to 1826
Hall, Janie Pauline *see* Study of florida wild flowers adapted for children of the fifth...
Hall, John *see*
– The american evangelists, d.l. moody and ira d. sankey
– Forty years' familiar letters of james w alexander
– Forty years' familiar letters of james w. alexander, d.d
– Paradoxes
– Questions of the day
– Sermons
Hall, John Smythe *see*
– Discours prononce a la salle windsor, montreal, le 16 fevrier 1892
– Discours sur le budget prononce a l'assemblee legislative de quebec vendredi, le 20 mai 1892
– A scathing exposure of the mismanagement of the provincial finances
– Speech delivered at the windsor hall, montreal
Hall, Joseph *see*
– Christ mystical
– Christian moderation
– Memorials of wesleyan methodist ministers
– The new testament in scots
– Prayer, the universal remedy
Hall, K L *see* T'ung tzu shih yeh ch'ien yen (ccm138)
Hall, Livingston *see* The livingston hall papers
Hall, Maxwell *see* Meteorology of jamaica
Hall, Newman *see*
– Atonement
– Christians
– Divine brotherhood
– The land of the forum and the vatican
– The lord's prayer
– A parting word
– Rooted in love
Hall of fame – v1 n1-v6 n4 [1977 jan-1982 oct] – 1r – 1 – mf#655280 – us WHS [071]
Hall, Peter *see*
– Fragmenta liturgica
– Reliquiae liturgicae
Hall, Prescott Farnsworth *see*
– The massachusetts law of landlord and tenant. including the cases in vol. 170 of the reports, and the legislation of 1898
– Reference list of wills construed by the supreme judicial court of massachusetts (including the cases in quincy and from vol. 1 to vol. 165 of the massachusetts reports)
Hall, Richard *see* General account of the first settlement
Hall, Richard Nicklin *see* Great zimbabwe, mashonaland, rhodesia
Hall, Richard Seymour *see* Zambia
Hall, Robert *see*
– Apology for the freedom of the press and for general liberty
– Modern infidelity considered with respect to its influence on socie...
– One day's courtship
– Sentiments proper to the present crisis
– Sermon occasioned by the death of her late royal highness the princess charlotte
– Sermon occasioned by the death of the rev john ryland
Hall, Samuel Carter *see*
– The baronial halls, and picturesque edifices of england
– The gallery of modern sculpture
– The royal gallery of art, ancient and modern
– The vernon gallery of british art
Hall, Teri-Christine R *see* Training amenorrhea in college athletes
Hall, Thomas Bond *see* The infringement of patents for inventions, not designs, with sole reference to the opinions of the supreme court of the united states
Hall, Thomas C *see* Historical setting of the early gospel
Hall, Thomas Cuming *see*
– History of ethics within organized christianity
– John hall, pastor and preacher
– Religion and life
– The social meaning of modern religious movements in england
– Social solutions in the light of christian ethics
Hallaendingen – Halmstad, Sweden. 1903-59 – 140r – 1 – sw Kungliga [078]
Hallam, Frank *see* The breath of god
Hallam, Henry *see* View of the state of europe during the middle ages
Hallam, John *see* The days of advance
The hallam progress – Hallam, NE: H L Gardner. v1 n1. oct 25 1905 [ie 1907]-08// [mf ed with gaps] – 1r – 1 – us NE Hist [071]
Hallam, Robert Alexander *see* Moses
Halland – Halmstad, Sweden. 1882-1901 – 26r – 1 – sw Kungliga [078]
Halland – Malmoe, Halmstad, Sweden. 1876-1954 – 261r – 1 – sw Kungliga [078]
Hallanding *see* Hallandingen
Hallandingen – Varberg, Sweden. 1887-89 – 1r – 1 – (aka: hallanding) – sw Kungliga [078]
Hallands folkblad – Halmstad, Sweden. 1916-43 – 108r – 1 – sw Kungliga [078]
Hallands lans tidning – Halmstad, Sweden. 1837-51 – 3r – 1 – sw Kungliga [078]
Hallands nyheter – Falkenberg, Sweden. 1919-78, 1979- – 1 – sw Kungliga [078]
Hallands nyheter – Falkenberg, Sweden. 1979- – 1 – sw Kungliga [078]
Hallands tidning – Falkenberg, Sweden. 1889-96 – 2r – 1 – sw Kungliga [078]
Hallandsposten – Halmstad, Sweden. 1857- – 136r – 1 – sw Kungliga [078]
Hallazgo de la necropolis judaica de la ciudad de teruel / Floriano Cumbreno, Antonio C – Madrid: Tip. Arch. y Biblioteca y Museos, 1926 – 1 – sp Bibl Santa Ana [946]
Hallazgo de veintiocho canciones populares de extremadura, recogidas en los anos 1884-85 / Gil Garcia, Bonifacio – Badajoz: Diputacion Provincial, 1946 – 1 – sp Bibl Santa Ana [946]
Hallazgo y descripcion de una autobiografia del primer obispo de mainas, don fray hipolito sanchez rangel / Quecedo, Francisco – Madrid: Archivo Ibero Americano, 1932 – 1 – sp Bibl Santa Ana [240]
Hallberg, C W *see* The suez canal
Hallberg, L Eugene *see* Wieland
Hallberg-Broich, Karl T von *see*
– Reise durch italien
– Reise durch skandinavien
Halle, Adam de la *see*
– Le jeu de robin et marion
– Oeuvres completes du trouvere adam de la halle
Halle aux cuirs – Paris, France. 2 jan-25 dec 1898; 1899-29 dec 1912 – 15r – 1 – uk British Libr Newspaper [072]
Halle, Fannina W *see* Su-lien hsin nu hsing
Halle, Louis Joseph *see* Transcaribbean
Halleck, Reuben Post *see* Education of the central nervous system
Hallelujahs from portsmouth : or, a report of portsmouth campmeeting held at portsmouth r i july 31 to august 17 1896 / Woodward, William D – Springfield MA: Christian Unity Pub Co 1896 [mf ed 2005] – 1 – 0-524-10549-9 – mf#b00747 – us ATLA [366]
Hallenbeck, Edwin Forrest *see* The passion for men
Hallenberg, Jonas *see* A swedish group solar heating plant with seasonal storage
Haller, Albrecht von *see*
– Albrecht hallers tagebuecher seiner reisen nach deutschland, holland und england
– Die alpen
Haller als philosoph : ein versuch / Jenny, Heinrich Ernst – Basel: Basler Druck- und Verlags-Anstalt, 1902 – 1r – 1 – (incl bibl ref) – us Wisconsin U Libr [430]
Haller, Benedictus *see* Tractatus de spiritu dei
Haller, Johannes *see*
– Deutschland und russland
– Papsttum und kirchenreform
– Die quellen zur geschichte der entstehung des kirchenstaates
– Warum und wofuer wir kaempfen
Haller, K L *see* Geschichte der kirchlichen revolution oder protestantischen reform des kantons bern und umliegender gegenden
Haller kreisblatt – Halle Westf DE, 1960 1 sep-1972 – 51r – 1 – (filmed by misc inst: 1958-1960 31 aug; 1949 2 nov-1957 (small gaps) [20r]) – gw Mikrofilm; gw Misc Inst [074]
Haller, Max *see*
– Der ausgang der prophetie
– Religion, recht und sitte in den genesissagen
Haller merkur fuer das oberamt gaildorf *see* Hallisches wochenblatt
Haller nachrichten – Schwaebisch Hall DE, 1948 28 aug-4 dec – 1 – gw Misc Inst [074]
Haller, Paul *see* 'S Juramareili
Haller, Rudolf *see* Der wilde alexander
Haller tagblatt *see* Hallisches wochenblatt
Haller, W *see* Iovianus
Haller, Wilhelm *see* Iovinianus
Hallermann, Josef *see* Freiligraths einfluss auf die lyriker der muenchener dichterschule
Hallesby, Ole *see* Johannes volkelts erkenntnistheorie
Hallesches tageblatt *see* Hallisches tageblatt
Hallet, Jean Pierre *see* Congo kitabu
Hallett, L *see* A history of henderson and macfarlane ltd
Hallett, Thomas George Palmer *see* The tenant-right question ireland
Halleux, Jean *see* La philosophie condamnee
Halley, Robert *see* The sacraments
Hallez, D-G *see* Plans d'instructions sur les sacrements d'apres le catechisme du concile de trente
Hallgarten, Charles L *see* Zum gedachtnis des herrn charles l hallgarten
Halliday, Andrew *see* The west indies
Halliday, G Y *see* Islam and christianity
Halliday, M A K (Michael Alexander Kirkwood) *see* Linguistic sicences and language teaching
Halliday, Nancy *see* The effects of contextual interference and three levels of difficulty on the acquisition, retention, and transfer of hockey striking skills by second grade children
Halliday, Samuel Byram *see*
– The church in america and its baptisms of fire
– The lost and found
Halliday, William Reginald *see*
– Greek divination
– The pagan background of early christianity
Hallier, Ludwig *see* Untersuchungen ueber die edessenische chronik
Hallische bzw deutsche jahrbuecher – Leipzig DE, 1838-42 – 3r – 1 – mf#2517 – gw Mikropress; gw Misc Inst [074]
Der hallische courier *see* Der kurier
Hallische gelehrte zeitungen *see* Neue hallische gelehrte zeitungen
Hallische kino-zeitung – Halle S DE, 1919 25 jul-1923 7 dec – 3r – 1 – gw Mikrofilm [790]
Hallische nachrichten *see* General-anzeiger
Hallische woechentliche relation... *see* Woechentliche relation
Hallisches tageblatt – Halle S DE, 1856-88 – 52r – 1 – (title varies: 2 jan 1872: hallesches tageblatt) – uk British Libr Newspaper [074]

Hallisches wochenblatt – Schwaebisch Hall DE, 1788 1 jul-1945 16 apr, 1945 1 aug-1946 30 mar [gaps], 1946 28 sep-1979 – 122r until 1939 – 1 – (filmed by other misc inst: 1978- [ca 7r/yr]. succeeding title: haller merkur fuer das oberamt gaildorf. title varies: 1842: schwaebischer hausfreund; 1848: haller tagblatt; 1 aug 1946: wuerttembergisches zeit-echo; 1 jul 1949: haller tagblatt) – gw Misc Inst [074]

Halliwell-Phillipps, James Orchard *see*
- An historical sketch of the provincial dialects of england
- Historical sketch of the provincial dialects of england

Hallman, H S *see* The law of faith

Hallmann, Georg *see* Das individualitaetsproblem bei friedrich hebbel

Hallmann, Jayne E *see* The effect of the education of third world women on family health

Hallo welt! : sechzehn erzaehlungen / Edschmid, Kasimir – Berlin: P Zsolnay, 1930 (mf ed 1990) – 1r – 1 – (filmed with: lord byron) – us Wisconsin U Libr [830]

Hallock, Leavitt Homan *see* Fifty years of plymouth church, minneapolis, minnesota

Hallock, Norman Everett *see* Study of three piano sonatas by norman dello joio

Hallowed songs / Phillips, Philip – Rev. ed. 1871 – 1 – us Southern Baptist [242]

Hallowell, Anna Davis *see* James and lucretia mott

Hallowell, Benjamin *see*
- Autobiography of benjamin hallowell
- The young friend's manual

Hallowell free press – Picton, ON. 1830-34 – 2r – 1 – cn Library Assoc [071]

Hallowell, Richard Price *see* The quaker invasion of massachusetts

The hallowing of criticism : nine sermons on elijah preached in rochester cathedral, with an essay read at the church congress, manchester, october 2nd, 1888 / Cheyne, Thomas Kelly – London: Hodder and Stoughton, 1888 – 1mf – 9 – 0-8370-2646-6 – mf#1985-0646 – us ATLA [920]

The hallowing of work : addresses / Paget, Francis – London; New York: Longmans, Green, 1913 – 1mf – 9 – 0-7905-8714-9 – mf#1989-1939 – us ATLA [240]

Halm, C *see*
- Salviani presbyteri massiliensis libri qui supersunt
- Victoris vitensis historia persecutionis africanae provinciae (mgh1:3.bd 1.-2.teil)

Halm, Friedrich *see*
- Der fechter von ravenna
- Griseldis
- Der sohn der wildniss

Halmel, A *see* Die palaestinischen maertyrer des eusebius von caesarea in ihrer zweifachen form

Halmel, Anton *see*
- Die palaestinischen martyrer des eusebius von caesarea in ihrer zweifachen form
- Ueber roemisches recht im galaterbrief
- Der zweite korintherbrief des apostels paulus

Halmhuber, A *see* Japan und die christliche mission

Halmrich, Elsie Winifred *see* History of the chorus in the german drama

Halmstads tidning – Halmstad, 1996-97 – 2r – 1 – sw Kungliga [078]

Halmstadsbladet – Halmstad, Sweden. 1853-82 – 13r – 1 – sw Kungliga [078]

Halo magazine / Spirit of Praise Outreach Ministry – 1997 apr-may – 1r – 1 – mf#4023568 – us WHS [243]

Haloean politik islam / Kartosoewirjo, S M – [Garoet, 1946] Dewan Penerangan Masjoemi. 42p – 1mf – 8 – mf#SE-1608 – ne IDC [260]

Halperin, Shim'on *see* Daber el ha-'am

Halpern, Boris *see* Pinkes fun yidishn bank

Halpern, I *see* Schleiermachers dialektik

Halpern, Jehiel *see* Revolt fun a goles-folk

Halphen, Louis *see*
- Etude sur les chroniques des comtes d'anjou et des seigneurs d'ambois
- Etudes sur l'administration de rome au moyen age

Halpin, Patrick Albert *see* Christian pedagogy

Halsbury's laws of england – Charlottesville, Michie Company. – 9 – $1773.00 set – (1st series v1-31 1907-1917 $525. 2nd series v1-37 1931-40 $650. 3rd series v1-43 $795. superseded volumes to 1st, 2nd, and 3rd series) – mf#408600 – us Hein [340]

Halsey, A W et al *see* The beloved

Halsey, Abram Woodruff *see* Go and tell john

Halsey enterprise (halsey, or: 1917) – Halsey OR: W A Priaulx, -1924 [wkly] [mf ed 1963] – 2r – 1 – (cont by: rural enterprise) – us Oregon Lib [071]

Halsey enterprise (halsey, or: 1927) – Halsey OR: H F & A A Lake, 1927-29 [wkly] [mf ed 1963] – 2r – 1 – (cont: rural enterprise (1924-27); cont by: greater oregon (1929-78)) – us Oregon Lib [071]

Halsey in the west indies / Fuller, Halsey Oakley – New York, NY. 1928 – 1r – us UF Libraries [972]

Halsey journal – Halsey OR: W C Pelham, 1932-38 [wkly] [mf ed 1963] – 1r – 1 – (cont by: halsey review (1938-63)) – us Oregon Lib [071]

Halsey, Leroy J *see*
- The beauty of immanuel
- The literary attractions of the bible
- Living christianity
- Science and the sages of the bible

Halsey, Leroy Jones *see*
- A history of the mccormick theological seminary of the presbyterian church
- Scotland's influence on civilization
- The works of philip lindsley

Halsey review – Halsey OR: C V Averill & Son, 1938-63 [wkly] [mf ed 1960-63] – 6r – 1 – (cont: halsey journal (1932-38). absorbed by: harrisburg bulletin (1925-85)) – us Oregon Lib [071]

Halsingekuriren – Soderhamn, Sweden. 1981- – 1 – sw Kungliga [078]

Halsingekuriren – Soederhamn, 1949-60 – 9 – sw Kungliga [078]

Halsingekuriren fredag – Soederhamn, 1993-94 – 9 – sw Kungliga [078]

Halsinglands tidning – Hudiksvall, Sweden. 1969 – 8r – 1 – sw Kungliga [078]

Halsinglands tidning *see*
- Hudiksvalls tidning
- Hudiksvallstidningen

Halstead, Murat *see* World on fire

Halstead, William Riley *see*
- Civil and religious forces
- A cosmic view of religion
- Future religious policy of america

Halsted, Caroline Amelia *see* Life of margaret beaufort...mother of king henry the seventh

Halt, Marie Malezieux *see*
- L'enfance de suzette
- Le menage de mme sylvain

Haltaus, Carl *see*
- Liederbuch der clara haetzlerin
- Theuerdank

Halter, Eduard *see* Die strassburger litterarische "besegard"

Die haltlosigkeit der "modernen wissenschaft" : eine kritik der kant'schen vernunftkritik fuer weitere kreise / Pesch, Tilmann – Freiburg i B: Herder, 1877 – (= ser Ergaenzungshefte zu den "Stimmen aus Maria-Laach") – 1mf – 9 – 0-524-08554-4 – mf#1993-2079 – us ATLA [190]

Halumy – New York. N.Y. The nationlist. 1888-89 – 1 – us AJPC [071]

Halunkenpostille : rumpelkammerromanzen, hafenballaden, spelunkensongs, neu: zinkenklavier / Grasshof, Fritz – rev enl ed. Duisburg: C Lange Verlag, 1959 – 89p (ill) – 1 – us Wisconsin U Libr [390]

Halversche zeitung – Halver DE, 1932 jul-dec – 1r – 1 – gw Misc Inst [074]

Halvorsen, Halvor *see* Festskrift til den norske synodes jubilaeum, 1853-1903

Halvorson, Lynnette *see* Lieder of ludwig senfl

Halvveckotidningen svenska folket – Stockholm, 1910-11 – 9 – sw Kungliga [078]

Ham : aout 1829-novembre 1832 / Mazas, Alexandre – Paris 1833 [mf ed Hildesheim 1995-98] – 1v on 3mf – 9 – €90.00 – ISBN-10: 3-487-26134-0 – ISBN-13: 978-3-487-26134-8 – gw Olms [944]

Ham and high *see* Hampstead and highgate express

Ham, Frederik Jacob van den *see* Disputatio pro religione mohammedanorum adversus christianos

Ham, George Henry *see* Our western heritage

Ham, J G van *see* Eerste boekjaar der indische partij 1912

Ham radio – Greenville. 1968-1990 (1) 1971-1990 (5) 1974-1990 (9) – ISSN: 0148-5989 – mf#3076 – us UMI ProQuest [380]

Ham radio horizons – Greenville. 1977-1980 (1,5,9) – mf#11238 – us UMI ProQuest [790]

Ha-maggid – Lyck et al, 1856-1903. – 512mf – 9 – mf#J-291-2 – ne IDC [077]

Hamakkor – Middletown, CT. May 1980-May 1984 – 1 – us AJPC [071]

Haman kloper – London, UK. Purim 1930 – 1 – uk British Libr Newspaper [072]

Hamandishe, Nicholas *see* Mashiripiti engozi

Hamann, Hermann Emil *see* Wielands bildungsideal

Hamann, Johann Georg *see*
- Schriften und briefe
- Sibyllinische blaetter des magus

Hambden, Renn Dickson *see* Introduction to the second edition of the bampton lectures of the y...

Hamberger, Georg Ernst et al *see* Jenaische gelehrte zeitungen

Hamberger, Julius *see*
- Die biblische wahrheit in ihrer harmonie mit natur und geschichte
- Die cardinalpunkte der franz baader'schen philosophie
- Christenthum und moderne cultur
- Gott und seine offenbarungen in natur und geschichte
- Physica sacra

Hamberger, Wolfgang *see* Motive und wirkungen des kommunalwahlsystems in baden-wuerttemberg

Hambloch, Ernest *see* His majesty, the president of brazil

Hamburg *see*
- Jahresberichte der verwaitungsbehosden
- Verhandlungen zwischen senat und burgerschaft

[Hamburg-] die zeit – DE. 1975-84 – 20r – 1 – $1000.00 – mf#R63612 – us Library Micro [074]

The hamburg item – Hamburg, PA. -w 1973 – 13 – $25.00r – us IMR [071]

Hamburg. Laws, statutes, etc *see* Hamburgisches gesetz-und verordnungsblatt

Hamburg und die antillen : ein seeroman / Smidt, Heinrich – Hamburg: Hammerich & Lesser, [1944] – 1r – 1 – us Wisconsin U Libr [830]

Hamburg-altonaer illustrirte zeitung – Hamburg DE, 1864-69 [gaps] – 1r – 1 – gw Misc Inst [074]

Hamburg-altonaer volksblatt – Hamburg DE, 1887 4 oct-1888 30 jun, 1892-1925 sep, 1926-1933 3 mar, 1946 3 apr-1949 – 100r – 1 – (title varies: 10 nov 1878: gerichtszeitung; 17 apr 1881: buergerzeitung; 2 oct 1887: hamburger echo; 1 jan 1964: hamburger echo am abend; 1 oct 1964: hamburger abendecho; 21 jul 1966: abendecho. filmed by misc inst: 1948 9 dec-31 dec [1r]; 1953 jul-1966. with suppls; filmed by bnl: 1946 3 apr-1951 19 jul [12r]) – gw Mikropress; uk British Libr Newspaper [074]

Hamburg.Burgerschaft *see*
- Protokolle und ausschuss-berichte
- Stenographische berichte ueber die sitzungen der burgerschaft zu hamburg

Hamburger abendblatt – Hamburg DE, 1958 22 mar-31 mar – 1 – (filmed by misc inst: 1952 27 nov-1957 sep, 1957 nov-dec, 1960 17 sep-26 sep; 1963-75; 1962 1 nov-31 dec, 1970- [ca 9r/yr]; 1948 14 oct-1958 21 mar; 1958 1 apr-1969, 1970 12 feb-10 jun) – gw Mikrofilm; gw Misc Inst [074]

Hamburger abendblatt – Hamburg: Axel Springer & Soba, 1953-1955; 1956-aug 1977; sep 1977-1980 – 1 – us CRL [074]

Hamburger abendecho – Hamburg, Germany. 1946-Jun 1953; 1962-66 – (= ser Hamburger Echo Am Abend) – 39r – 1 – (also known as: hamburger echo am abend) – us L of C Photodup [074]

Hamburger abendecho *see* Hamburg-altonaer volksblatt

Hamburger allgemeine zeitung – Hamburg, Germany 2 apr 1946-24 mar 1950 (imperfect) – 1 – (incorp with: hamburger freie presse) – uk British Libr Newspaper [074]

Hamburger anzeiger – Hamburg. Nov. 2 1932; July 10 1937-Aug 31 1939. Incomplete – 1 – us NY Public [943]

Hamburger anzeiger *see* General-anzeiger fuer hamburg-altona

Der hamburger beobachter und das archiv wissenschaft und kuenste – Hamburg DE, 1822-41 [gaps], 1849 25 jul-1852 29 dec – 1 – (title varies: 31 mar 1852: morgenzeitung; 14 apr 1852: der hamburger beobachter; filmed by other misc inst: 1822-41 [gaps], 1849 25 jul-1852 29 dec) – gw Misc Inst [074]

Hamburger echo – Germany. -d. 3 Apr 1946-19 Jul 1951. (12 reels) – 1 – uk British Libr Newspaper [072]

Hamburger echo – Hamburg, 1887-88; 1892-1925; 1926-33 – 97r – 1 – gw Mikropress [074]

Hamburger echo – Hamburg, 1946-49 – 3r – 1 – mf#6568 – gw Mikropress [074]

Hamburger echo *see* Hamburg-altonaer volksblatt

Hamburger echo am abend *see* Hamburg-altonaer volksblatt

Hamburger, Franz *see* Lehrer zwischen kaiser und fuehrer

Hamburger freie presse 1946 – Hamburg DE, 1950-1952 12 sep – 7r – 1 – gw Mikrofilm [074]

Hamburger fremden-blatt – Hamburg: Gustav Diedrich and Co, apr-jun 1932 – us CRL [074]

Hamburger fremdenblatt – Hamburg DE, 1872 3 jan-31 mar, 1882 1 feb-30 apr, 1884 2 nov-1936 30 nov [gaps], 1940-1944 31 aug [gaps], 1954 1 sep-31 oct – 297r – 1 – (filmed by other misc inst: 1913 15 apr-30 apr [1r]; 1914-19 [gaps], 1954 1 sep-31 oct) – gw Misc Inst [074]

Hamburger grundeigentuemer-zeitung – Hamburg DE, 1929-30 – 1r – 1 – gw Misc Inst [074]

Hamburger illustrierte zeitung – Hamburg DE, 1925-1930 2 aug, 1931 & 1932 [single iss], 1933-1935 24 jun, 1939 14 aug-1940 7 sep, 1941-42, 1944 16 jan-2 sep – 9r – 1 – (filmed by misc inst: 1919-44 [26r]) – gw Mikrofilm; gw Misc Inst [074]

Hamburger, Michael *see* From prophecy to exorcism

Hamburger mittag – Hamburg DE, 1954 26 jan-1957 31 mar – 1 – (bezirksausgabe von norddeutsche nachrichten, hamburg) – gw Misc Inst [074]

Hamburger morgenpost – Hamburg DE, 1949 16 sep-1971 31 jul, 1976-79 – 106r – 1 – (filmed by misc inst: 1976- [ca 10r/yr]) – mf#9837 – gw Mikropress; gw Misc Inst [074]

Hamburger nachrichten – Hamburg, Germany. Dec 1899-Jun 1910; 1912-Jun 1913; Aug 1917-1919 – 115r – 1 – us L of C Photodup [074]

Hamburger nachrichten *see* Privilegierte woechentliche gemeinnuetzige nachrichten

Hamburger nachrichten-blatt *see* Hamburger nachrichten-blatt der militaerregierung

Hamburger nachrichten-blatt der militaerregierung – Hamburg DE, 1945 9 may-1946 28 mar – 1r – 1 – (title varies: 30 nov 1945: hamburger nachrichten-blatt) – mf#6399 – gw Mikropress [074]

Hamburger neue zeitung & a-c-nachrichten *see* Hamburgische adress-comtoir-nachrichten [1810-19], h a-c-nachrichten + hamburger abendblatt [1820-1823], h a-c-nachrichten [1824-26], hamburger neue zeitung & a-c-nachrichten [1826-31]

Hamburger rundschau – Hamburg DE, 1982 26 aug-28 sep; 1983 13 oct-1992 23 dec – 16r – 1 – (incl suppls) – gw Mikrofilm [074]

Hamburger Sternwarte *see*
- Astronomische abhandlungen der hamburger sternwarte
- Mitteilungen der hamburger sternwarte in bergedorf

Hamburger Stiftung fuer Sozialgeschichte des 20. Jahrhunderts *see* Sozialstrategien der deutschen arbeitsfront

Hamburger tageblatt [main edition] – Hamburg DE, 10 mar 1939-30 may 1940 – 7r – 1 – (filmed by misc inst: regional ed: hannover-niederelbe 1933 2 apr-1937 31 mar [21r] title varies: 1 sep 1933: niederelbisches tageblatt) – uk British Libr Newspaper; gw Misc Inst [074]

Hamburger universitaetszeitung – Hamburg DE, 1919/20-1935 n2 [gaps] – 1 – gw Misc Inst [378]

Hamburger, Victor *see* Sealsfield-postl

Hamburger volks-zeitung *see* Die rote fahne

Hamburger zeitung – Hamburg DE, 1943 25 jul-17 aug, 1944 1 sep-1945 2 may – 1 – gw Misc Inst [074]

Hamburgische abend-zeitung *see* Priviligirte liste der boersenhalle

Hamburgische address-comtoir-nachrichten *see* Kayserlich-privilegirte hamburgische neue zeitung

Hamburgische adress-comtoir-nachrichten [1810-19], h a-c-nachrichten + hamburger abendblatt [1820-1823], h a-c-nachrichten [1824-26], hamburger neue zeitung & a-c-nachrichten [1826-31] / ed by Leisching, P A – Hamburg – (= ser Dz) – 22jge on 760mf – 9 – €4560.00 – mf#k/n2718 – gw Olms [342]

Hamburgische beitraege zu den werken des witzes und der sittenlehre / ed by Leyding, Johann Dietrich et al – Hamburg 1753-55 – (= ser Dz) – 2v[zu je 3st] on 10mf – 9 – €100.00 – mf#k/n4944 – gw Olms [870]

hamburgische boersen-halle *see* Priviligirte liste der boersenhalle

Hamburgische neue zeitung *see* Kayserlich-privilegirte hamburgische neue zeitung

Hamburgischer correspondent *see* Staats- und gelehrte zeitung des hollsteinischen unpartheyischen correspondenten

Hamburgischer unpartheyischer correspondent : staats- und gelehrte zeitungen des hamburgischen unpartheyischen correspondenten, durch europa und andere theile der welt... – Hamburg 1781-1800 [mf ed Hildesheim 1977-88] – (= ser Deutsche zeitungen von den anfaengen bis zur mitte des 19. jahrhunderts) – 355mf – 9 – diazo €1800.00 silver €2000.00 – gw Olms [074]

Hamburgischer unpartheyischer correspondent : staats- und gelehrte zeitungen des hollsteinischen (hamburgischen) unpartheyischen correspondenten, durch europa und andere theile der welt...teil 1 – Schiffbeck 1721-30, Hamburg 1731-1780 [mf ed Hildesheim 1977-97] – (= ser Deutsche zeitungen von den anfaengen bis zur mitte des 19. jahrhunderts) – 648mf – 9 – diazo €2900.00 silver €3400.00 – gw Olms [074]

Hamburgisches gesetz-und verordnungsblatt / Hamburg. Laws, statutes, etc – Hamburg. 1961-1968 – 1 – us NY Public [348]

Hamburgisches magazin : oder gesammlete schriften, zum unterricht und vergnuegen, aus der naturforschung und den angenehmen wissenschaften ueberhaupt / ed by Kaestner, Abraham Gotthelf et al – Hamburg, Leipzig 1747-67 – (= ser Dz. abteilung naturwissenschaft) – 26v+ind vol on 109mf – 9 – €654.00 – mf#k/n3197 – gw Olms [500]

Hamburgisches unterhaltungsblatt *see* Gemeinnuetzige unterhaltungs-blaetter

Hamdan ibn Uthman Khawajah *see* Apercu historique et statistique sur la regence d'alger

Ha-measseph – (New York). 1881 – 1 – us AJPC [939]

Ha-mediniyut ha-miktso'it veha-kalkalit shel ha-histadrut / Becker, Aaron – ha-Histadrut ha-kelalit shel ha-'ovdim ha-'Ivrim be-Erets-Yisra'el, ha-Va'ad ha-po'el, ha-Merkaz le-tarbut ule-hinukh, [1960] 720 (mf ed 197-) – 1r – 1 – mf#ZZ-16578 – us NY Public [939]
Hamel, Andre *see* Bibliographie sur le scoutisme catholique dans la province de quebec
Hamel, Anton Gerard van *see* Proeve eener kritiek van de leer der goddelijke voorzienigheid
Hamel, Hendrik *see* Corea, without and within
Hamel, Hubert *see* Memoire concernant les greves de sault-au-matelot de la chatellenie de coulnge
Hamel, J M *see* The effects of self-talk on batting performance
Hamel, Joseph *see* Lectures sur les pecheries donnees a la chambre de lecture de saint-roch
Hamel, Philippe *see* Le trust de l'electricite menace pour la securite sociale
Hamel, Thomas Etienne *see* Cours d'eloquence parlee d'apres delsarte
Hamelberg, Hendrik Anthony Lodewijk *see* Dagboek van h a l hamelberg, 1855-1871
Ha-meliz – Odessa, St Petersburg. v. 1-44 no. 18. 1860-Jan 30 1904. Incomplete – 1 – us NY Public [305]
Ha-menahel – (New York). 1920 – 1 – us AJPC [800]
Hamer Advertising and Marketing Concepts *see* Tcb
Hamer creek baptist church. montgomery association. north carolina : church records – 1862-82 – 1 – 6.39 – us Southern Baptist [242]
Hamer, Fannie Lou *see* The papers of fannie lou hamer, 1917-1977
Hamer, Hayo E *see* Mission und politik
Hamerling, Robert *see*
– Ahasver in rom
– Amor und psyche
– Danton und robespierre
– Hamerlings werke in vier baenden
– Homunculus
– Der koenig von sion
– Lehrjahre der liebe
– Ein schwanenlied der romantik
– Teut
Hamerlings werke in vier baenden / ed by Rabenlechner, Michael Maria – 3. aufl. Leipzig: M Hesse, [1900?] [mf ed 2001] – 4v – 1 – (foreword by peter rosegger) – mf#10547 – us Wisconsin U Libr [802]
Hamer's guide to taking care of business – 1986 1st qtr – 1r – 1 – (cont: seagram's guide to taking care of business) – mf#2698522 – us WHS [650]
Hamerton, Philip Gilbert *see*
– Contemporary french painters
– The etcher's handbook
– Etching and etchers
– Examples of modern etching
– The graphic arts
– Imagination in landscape painting
– Man in art
– A painter's camp in the highlands
– Painting in france after the decline of classicism
– The present state of the fine arts in france
Hamesh megiloth : shir ha-shirim, ruth, kinoth, koheleth, esther = quinque volumina: canticum canticorum, ruth, threni, ecclesiastes, esther: textum masoreticum accuratissime expressit, e fontibus masorae varie illustravit, notis criticis confirmavit – Lipsiae [Leipzig]: Bernhardi Tauchnitz, 1886 – 1mf – 9 – 0-7905-8331-3 – mf#1987-6430 – us ATLA [220]
Hamesse, J *see* The saurus librorum sententiarum petri lombardi
Hamet, Ismael *see* Chroniques de la mauritanie senegalaise
Hamevaker – Los Angeles, CA.1982-84 – 1 – us AJPC [071]
Ha-mevaser *see* Literarisze monatszriftn
Hamid, 'Abduelhak *see* The divan project
Hamid, Abduelhak *see*
– Esber
– Tayijlar gecidi
Hamid, Ismail *see* Histoire du maghreb
Hamid, V Ahmet *see* Kurun-i cedide ve navolyon'un sukuntuna kadar asr-i hazir mebadisi
Hamidoun, Mokhtar ould *see* Catalogue provisoire des manuscripts mauritaniens en langue arabe preserves en mauritanie
Ha-milhamah ha-sifrutit ben ha-haredim veha-maskilim / Katzenelson, Gide'on – Tel-Aviv, Israel. 1954 – 1r – us UF Libraries [939]
Hamill, Howard Melancthon *see* The bible and its books
Hamilton : the electric city : history, government and prosperity of the birmingham of canada... – [Hamilton, Ont]: Industrial Recorder Co, c1901 – 1mf – 9 – 0-665-78084-2 – mf#78084 – cn Canadiana [971]
Hamilton / Veitch, John – Edinburgh: W Blackwood, 1882 – (= ser Philosophical Classics for English Readers) – 1mf – 9 – 0-7905-8953-2 – mf#1989-2178 – us ATLA [190]
Hamilton 1714-1897 – Oxford, MA (mf ed 1989) – (= ser Massachusetts vital records) – 29mf – 9 – 0-87623-097-4 – (mf 1: birth index 1781-1875. mf 2: marriage index 1781-1875. mf 3-4: intentions index 1781-1861. mf 5: death index 1781-1875. mf 6: births, deaths, intentions 1750-1832. mf 7-8: b,m,d 1844-60. mf 9-10: deaths 1859-96. mf 11: marriages 1860-96. mf 12-13: births 1859-97. mf 14: rebellion records 1861-65. mf 15-21: death index 1771-1941. mf 17-21: deaths 1771-1941. mf 22-25: marriages & index 1714-1941. mf 26-29: births & index 1755-1941) – us Archive [978]
Hamilton 1750-1849 – Oxford, MA (mf ed 1996) – (= ser Massachusetts vital record transcripts to 1850) – 4mf – 9 – 0-87623-260-8 – (mf 1t: intentions 1795-1812; deaths 1784-1832; births 1750-1812; intentions & marriages 1812-38. mf 2t: intentions & marriages 1838-49; out-of-town marriages 1794-99; births & deaths 1812-48. mf 3t: deaths 1815-20; births 1844-49. mf 4t: marriages & deaths 1844-49) – us Archive [978]
Hamilton, Adelbert *see* The interstate commerce law
Hamilton advertiser – Scotland. 1862-75; 1898-99.-w. 6 reels – 1 – uk British Libr Newspaper [072]
Hamilton advertiser (1993) – Hamilton: Scottish & Universal Newspapers Ltd 1993- (wkly) [mf ed 6 jan 1995-] – 1 – (cont: hamilton advertiser and county of lanark news) – ISSN: 1353-4351 – uk Scotland NatLib [072]
Hamilton, Alexander *see*
– The federalist and other constitutional papers
– Papers
Hamilton, Alexander et al *see* The federalist
Hamilton and Gore Mechanics' Institute *see* The laws and regulations of the hamilton and gore mechanics' institute
Hamilton and gore mechanics' institute *see* Exhibition of fine arts, manufactures, machines... etc
Hamilton, Andrew *see*
– Statutory revision of the laws of new york affecting insurance companies
– Statutory revision of the laws of new york affecting miscellaneous corporations, enacted in 1892.
Hamilton Art Exposition *see* Catalogue, picture gallery, 1888
Hamilton, Augustus *see* The art workmanship of the maori race in new zealand
Hamilton Bridge Works Co *see* Some information about bridges and structural steel
The hamilton bridge works company limited : engineers, designers and contractors for railway bridges, railway turntables... – Hamilton [ON]: The Co, 1909 [mf ed 1991] – 2mf – 9 – 0-665-99501-6 – mf#99501 – cn Canadiana [624]
Hamilton, Bruce *see* Barbados and the confederation question, 1871-1885
Hamilton, Carlos Depassier *see* Nuevo lenguaje poetico de silva a neruda
Hamilton central school vocalist – [Hamilton, Ont?: s.n, 186-?] [mf ed 1994] – 1mf – 9 – 0-665-94615-5 – mf#94615 – cn Canadiana [780]
Hamilton, Charles *see* Sketches of life and sport in south-eastern africa
Hamilton, clydesdale & avondale journal and middle and upper ward advertiser – [Scotland] South Lanarkshire 23,30 jul, 6 aug 1859 (wkly) [mf ed 2004] – 3v on 1r – 1 – uk Newsplan [072]
Hamilton & clydesdale monthly advertiser – [Scotland] South Lanarkshire, Hamilton: R Allardice 1 nov 1845-3 jan 1846 (mthly) [mf ed 2004] – 1r – 1 – uk Newsplan [072]
Hamilton Co. *see* Deutsch amerikaner
Hamilton Co. Cheviot *see*
– Western hills press
Hamilton Co. Cincinnati *see*
– Advertiser and ohio phoenix
– American
– Anzeiger
– Brotherhood of rr signalman
– Christian standard
– Chronicle
– Chronicle series
– Cincinnati abend=post
– Cincinnati american
– Cincinnati chronicle
– Cincinnati chronicle and literary gazette
– Cincinnati daily chronicle
– Cincinnati daily columbian
– Cincinnati daily nonpareil
– Cincinnati emporium
– Cincinnati evening chronicle
– Cincinnati herald
– Cincinnati kurier
– Cincinnati morgan=post
– Cincinnati morning herald
– Cincinnati news journal
– Cincinnati tageblatt
– Cincinnati tagliche morgan=post
– Cincinnati taglicher abend=post
– Cincinnati telegram
– Cincinnati weekly chronicle
– Cincinnati weekly enquirer
– Cincinnati weekly gazette
– Cincinnati weekly news
– Cincinnati weekly times
– Clermont county review/w
– Commoner
– Community journal press (northern edition)
– Community journal press (southern edition)
– Community journal series (southern edition)
– Daily advertiser and journal
– Daily atlas
– Daily chronicle
– Daily chronicle and atlas
– Daily cincinnati atlas
– Daily cincinnati chronicle
– Daily cincinnati enquirer
– Daily cincinnati gazette
– Daily cincinnati republican
– Daily columbian
– Daily commercial
– Daily dispatch
– Daily enquirer
– Daily gazette
– Daily press
– Daily star series
– Deutsch-amerikanisch illustrierte zeitung
– Dollar weekly times
– Eastern hills journal series
– Enquirer
– Enquirer and message
– Evening news
– Evening telegram series
– Forest hills journal series
– Freeman's journal and miscellaneous
– Giraffe
– Great west series
– Hilltop news series
– Journal series
– Labor advocate
– Liberty hall
– Liberty hall and cincinnati gazette
– Liberty hall series
– Literary cadet and cheap city advertiser
– Mercantile daily advertiser
– Message series
– Millcreek valley news
– Mount washington press
– National crisis
– National crisis and cincinnati emporium
– National republican
– National republican and cincinnati daily mercantile advertiser
– News series
– Northeast suburban life-press
– Ohio organ temperance reform
– Organ of temperance reform
– Orton weekly bulletin
– Penny press
– People's voice
– Philanthropist
– Price current
– Price hill news
– Price hill press
– Railway clerk
– School friend
– Semi-weekly gazette
– Sentinel
– Spirit of the west
– Star
– Sun
– Taglicher cincinnati courier
– Tagliches cincinnati volksblatt
– Tribune
– Tri-weekly cincinnati gazette
– Union
– Volksblatt
– Wahrheits-freund
– Weekly cincinnati times
– Weekly times
– West and south
– Western fountain
– Western hills press/w
– Westliche blatter
– World
Hamilton Co. Cleves *see* Valley journal
Hamilton Co. Elmwood Place *see* Blade
Hamilton Co. Harrison *see*
– Press
– Record
Hamilton Co. Mariemont *see* Messenger
Hamilton Co. Price Hill *see* News (cincinnati area)
Hamilton Co. Wyoming *see* Adiramled
Hamilton conveyor / American Hamilton – 1976 dec-1988 apr – 1r – 1 – (cont: hamiltonian [two rivers, wi]) – mf#1026330 – us WHS [071]
Hamilton Co-Operative Association (Ont) *see* Constitution and by-laws of the hamilton co-operative association
The hamilton county advocate – Aurora, NE: F J Sharp. 6v. v1 n1. dec 19 1911-v6 n3. dec 26 1916 (wkly) – 1r – 1 – us Bell [071]
The hamilton county advocate – Aurora, NE: F J Sharp. 6v. v1 n1. dec 19 1911-v6 n3. dec 26 1916 (wkly) [mf ed with gaps filmed 2000] – 4r – 1 – us NE Hist [071]
Hamilton county atlas, 1869 : by titus – 1r – 1 – mf#B30575 – us Ohio Hist [978]
The hamilton county democrat – Aurora, NE: A M Glover. v1 n1. sep 17 1895- (wkly) – 1r – 1 – us Bell [071]
Hamilton County Historical Society (IL) *see* Goshen trails hamilton county historical society news
Hamilton county ledger – Noblesville, IN. 1889-1909 (1) – mf#62921 – us UMI ProQuest [071]
Hamilton county news – Hamilton, NE: C P Whitesides, -aug 1885// (wkly) – 2r – 1 – (cont by: aurora news. publ in aurora mar 15 1879-) – us Bell [071]
Hamilton county register – Aurora, NE: Register Pub Co. 38v. v1 n1. dec 6 1890-v38 n12. mar 22 1929 (wkly) – 28r – 1 – (merged with: aurora republican to form: hamilton county republican-register. v2 n1. dec 5 1891-v38 n12. mar 22 1929 called also whole n53-2013) – us Bell [071]
Hamilton county register – Noblesville, IN. 1869-1871 (1) – mf#62922 – us UMI ProQuest [071]
Hamilton county republican-register – Aurora, NE: Aurora Print Co. 4v. v56 n45. mar 29 1929-v59 n35. jan 29 1932=v38 n13-v41 n2 (wkly) – 3r – 1 – (formed by the union of: aurora republican and: hamilton county register (aurora, ne). cont by: republican-register (aurora ne)) – us Bell [071]
Hamilton county times – Noblesville, IN. 1907-1909 (1) – mf#62923 – us UMI ProQuest [071]
Hamilton daily democrat – Hamilton, OH, 11&16 aug 1887 – 1r – 1 – (daily democratic newspaper) – us Western Res [071]
Hamilton daily herald – Hamilton, OH, 28 oct 1885 – 1r – 1 – (daily democratic newspaper) – us Western Res [071]
[Hamilton-] daily inland empire – NV. 1869-70 [daily] – 2r – 1 – $120.00 – mf#U04573 – us Library Micro [071]
Hamilton daily news – Hamilton, OH, apr 27 1880-may 17 1893, apr 18 1921 (scattered) – 1r – 1 – (daily republican newspapers) – us Western Res [071]
Hamilton daily news – Ontario, Canada. 16 nov 1912-15 may 1916; 18 oct 1919-6 nov 1920 – 34r – 1 – uk British Libr Newspaper [071]
Hamilton daily republican – Hamilton, OH, may 17 1892 – 1r – 1 – (daily republican newspaper) – us Western Res [071]
Hamilton, Daniel Lee *see* Contos do brasil
Hamilton, Edith *see* Greek way
Hamilton, Edward *see* A catalogue raisonne... works of sir joshua reynolds
Hamilton, Elizabeth Charlotte *see* Pedagogy of the french horn
Hamilton evening sun – Hamilton, OH, may 28,29 1903 – 1r – 1 – (daily democratic newspaper) – us Western Res [071]
Hamilton extra – Motherwell: Archant 2003 (wkly) – 1 – (cont: hamilton people; cont by: extra (hamilton ed)) – uk Scotland NatLib [072]
Hamilton, F J *see* The syriac chronicle
Hamilton, Frederick Alexander Pollock *see* The law relating to charities in ireland
Hamilton, Gail *see*
– Divine guidance
– Sermons to the clergy
– A washington bible-class
Hamilton gazette – Hamilton, ON. 1852-55 – 3r – 1 – cn Library Assoc [071]
Hamilton, George *see*
– Heathen ceremonies adopted by the church of rome
– On extreme unction
– Protestant religion no novelty
– Protestant's reasons for not worshipping saints and images
– Sacrament of the lord's supper compared with the sacrafice of the mass
– Second letter to the most rev dr murray
Hamilton, George Alexander *see* Clergy of the church in ireland weighed in the balance
Hamilton, George E *see* Designs for rural churches
Hamilton, Graham *see* Patrol reports and related papers, milne bay and new britain, papua new guinea
Hamilton, H G *see*
– Cost of handling citrus fruit from the tree to the car in florida
– Farm management studies of truck and citrus farms in florida
– Farmers' cooperative associations in florida
– Farmer's cooperative associations in florida
– Farmers' cooperative associations in florida citrus cooperative
– Study of the cost of handling citrus fruit from the tree to the car in florida
Hamilton, Harold Francis *see* The people of god
Hamilton herald – Ontario, Canada. 1889; 16 nov 1912-27 jan 1915; 2 mar 1915; 10 may 1915-25 jan 1919; 14 may 1919-jun 1922 – 90 1/2r – 1 – uk British Libr Newspaper [071]
Hamilton herald and lanarkshire weekly news – Scotland, UK. 3 Mar 1888-1900. -w. 12 reels – 1 – uk British Libr Newspaper [072]

Hamilton herald & lanarkshire weekly news – England.3 Mar 1888-Jun 1905. -w. 17 reels – 1 – uk British Libr Newspaper [072]
Hamilton herald & lanarkshire weekly news – [Scotland] South Lanarkshire, Hamilton: T Stothers 3 mar 1888-2 oct 1930 (wkly) [mf ed 2004] – 57r – 1 – (missing: 1923; cont by: lanarkshire [mar 1905-oct 1930]) – uk Newsplan [072]
Hamilton Horticultural Society *see* Constitution and by-laws
[Hamilton-] inland empire – NV. 1869-1870 [daily] – 1r – 1 – $60.00 – mf#U04574 – us Library Micro [071]
Hamilton intelligencer – Hamilton, OH, jan 1 1857-jan 9 1862 – 2r – 1 – $230.00 – (weekly republican newspaper) – us Western Res [071]
Hamilton, J A *see* Life of daniel o'connell
Hamilton, J Taylor *see* Twenty years of pioneer missions in nyasaland
Hamilton, James *see*
– Dew of hermon
– Harp on the willows
– Historical development of christianity in the political and social li...
– Looking to christ
– A memoir of richard williams, surgeon
– Moses, the man of god
– The pearl of parables
– Psalter and hymn book
– The royal preacher
– Seeming antagonism in the real harmony of truth
Hamilton, James Cleland *see*
– The african in canada / the maroons of jamaica and nova scotia
– The development of personal liberty in great britain, france and their colonies
– Famous algonquins
– The georgian bay
– The great centre
– Osgoode hall
– The panis
– The prairie province
– Slavery in canada
Hamilton, John *see*
– Reise durch die inneren provinzen von columbien
– Reise durch die innern provinzen von columbien
– Sixty years' experience as an irish landlord
– Travels through the interior provinces of columbia
Hamilton, John Taylor *see*
– The beginnings of the moravian mission in alaska
– A history of the church known as the moravian church
– A history of the missions of the moravian church during the 18th and 19th centuries
– Twenty years of pioneer missions in nyasaland
Hamilton journal – Pickering, Ont (CDN), 1972-76, 1978-1980 5 sep – 1 – (cont: kanada-kurier, winnipeg) – gw Misc Inst [071]
Hamilton labour party records, 1918-1951 – (= ser Labour party in britain, origins and development at local level. series 1) – 2r – 1 – (int by w hamish fraser) – mf#97171 – uk Microform Academic [325]
Hamilton, Laurentine *see* The future state and free discussion
Hamilton Literary Society *see* Constitution and laws of the...
Hamilton, Louis *see* Handbuch der englischen und deutschen umgangssprache
Hamilton, Mary *see* Incubation
Hamilton Mercantile Library Association and General News Room *see* Constitution and by-laws of the...
Hamilton, N E S A *see* De gestis pontificum anglorum libri quinque [rs52]
Hamilton National Genealogical Society *see* Connector of the...
Hamilton, Patrick *see* Angel street
Hamilton people – Motherwell: Community Media Ltd [1989]-2003 (wkly) [mf ed 6 jan 1995-] – 1 – (not publ: n700; cont: lanarkshire people (hamilton ed); cont by: hamilton extra) – uk Scotland NatLib [072]
Hamilton physiog – Hamilton [Ont: s.n. 1858-18–?] – 9 – mf#P04452 – cn Canadiana [870]
Hamilton, Pierce Stevens *see*
– British american union
– The feast of saint anne and other poems
– Letter to his grace the duke of newcastle
– Nova-scotia considered as a field for emigration
– Observations upon a union of the colonies of british north america
– The repeal agitation, and what is to come of it?
– Union of the colonies of british north america
– A union of the colonies of british north america
Hamilton. Presbytery (Pres. Ch. in the U.S.A. New School) *see* Minutes, 1846-1870
Hamilton press – 1975-dec 1987 – 22r – 1 – mf#15.27 – nz Nat Libr [079]
Hamilton Public Library (Ont) *see* Quarterly bulletin
Hamilton, Pura De *see* Psiquis sin velos
Hamilton, Richard Winter *see* The revealed doctrine of rewards and punishments
Hamilton, S D *see* New zealand english language periodicals of literary interest
Hamilton Society for the Prevention of Cruelty to Animals *see* Manual of the constitution, by-laws, etc...
Hamilton spectator : summer carnival edition – [Hamilton: Spectator, 1889?] [mf ed 1982] – 1mf – 9 – 0-665-38550-1 – mf#38550 – cn Canadiana [971]
Hamilton spectator *see* Daily spectator
Hamilton sun – Hamilton, OH. 1904-1906 (1) – mf#65522 – us UMI ProQuest [071]
Hamilton, T F *see* American negligence cases
Hamilton telegraph – Hamilton, OH, aug 11 1853-dec 31 1857 – 3r – 1 – (weekly democratic newspaper) – us Western Res [071]
Hamilton telegraph – Hamilton, OH, nov 18 1847-may 2 1850 – 2r – 1 – (weekly democratic newspaper) – us Western Res [071]
Hamilton telegraph – Hamilton, OH, jun 13-oct 21 1861; jun 16 1864 – 1r – 1 – (weekly republican newspaper) – us Western Res [071]
Hamilton, the birmingham of canada – Hamilton: The Times Printing Co, 1893 – 2mf – 9 – mf#12914 – cn Canadiana [917]
Hamilton, Theodore Frank *see*
– Hamilton's cyclopedia of negligence cases
– Hamilton's new york negligence cases classified
Hamilton, Thomas *see*
– History of the irish presbyterian church
– Men and manners in america
Hamilton Thompson, A *see* Military architecture in england during the middle ages
Hamilton times – Hamilton, ON. 1858-68 – 16r – 1 – ISSN: 1181-5280 – cn Library Assoc [071]
Hamilton true telegraph / Butler Co. Hamilton – oct 27 1864-oct 18 1866 – 1r – 1 – mf#B35456 – us Ohio Hist [071]
Hamilton true telegraph – Hamilton, OH, oct 30 1862-oct 20 1864 – 1r – 1 – (weekly democratic newspaper) – us Western Res [071]
Hamilton, W J *see* Researches in asia minor, pontus, and armenia
Hamilton, Wallace *see* Christopher and gay
Hamilton, Walter *see* A geographical, statistical, and historical description of hindostan
Hamilton, Walter Kerr *see* A charge to the clergy and churchwardens of the diocese of salisbury
Hamilton Water Works (Ont) *see*
– Specifications and estimates of the three successful competitors
– Water rates, rules and regulations
Hamilton weekly journal and lanarkshire advertiser – [Scotland] South Lanarkshire, Hamilton: W Lowe & Co 15 jul 1841 (wkly) [mf ed 2004] – 1r – 1 – uk Newsplan [072]
Hamilton weekly sun – Hamilton, OH, july 18 1902 – 1r – 1 – (weekly democratic newspaper) – us Western Res [071]
Hamilton weekly telegraph – Hamilton, OH, sep 25 1851-aug 4 1853 – 1r – 1 – (weekly democratic newspaper) – us Western Res [071]
[Hamilton-] white pine news – NV. 1870-76 [wkly] – 1r – 1 – $60.00 – mf#U04575 – us Library Micro [071]
Hamilton, William *see*
– Aegyptiaca oder beschreibung des zustandes des alten und neuen aegypten
– Be not schismatics, be not martyrs, by mistake
– Details historiques des tremblemens de terre arrives en italie depuis le 5 fevrier jusqu'en mai 1783
– Diaries and pearling logs
– The metaphysics of sir william hamilton
– Philosophy of sir william hamilton, bart
Hamilton, William R *see* Aegyptiaca
Hamilton, William Richard *see*
– Letter...on the new houses of parliament
– Second letter...on the propriety of adopting the greek style
– Third letter...on the propriety of adopting the greek style
Hamilton, William Thomas *see*
– The friend of moses
– The importance of a liberal education for women
– Prejudice and its antidote
Hamilton, William Wistar *see* Sane evangelism
Hamilton world – Hamilton: Scottish & Universal Newspapers -2005 [mf ed 3 jan 1997-] – 1 – (cont by: lanarkshire world. hamilton) – uk Scotland NatLib [072]
Hamilton-Hoare, Henry William *see* The evolution of the english bible
Hamilton's cyclopedia of negligence cases / Hamilton, Theodore Frank – New York, Baker, Voorhis, 1904. 1083 p. LL-763 – 1 – us L of C Photodup [340]
Hamilton's new york negligence cases classified / Hamilton, Theodore Frank – New York, Remick, Schilling, 1898. 470 p. LL-864 – 1 – (suppl albany, bender, 1899 88p ll-864. 1899 annual albany, bender, 1900 179p ll-864) – us L of C Photodup [340]
Ha-mishnah : asher aliyah nosad ha-talmud ha-yerushalmi me-roshitah ve-ad sofah... / Lowe, William Henry – Cambridge: University Press, 1883 – 6mf – 9 – 0-524-08194-8 – mf#1991-0307 – us ATLA [270]
Ha-mishnah / Lipschutz, Eliezer Meir – Jaffa, Israel. 1913/14 – 1r – us UF Libraries [939]
Ha-mistorin be-yisrael / Horodezky, Samuel A – Tel-Aviv, Israel. v1-4. 1930/31 – 1r – us UF Libraries [939]
Ha-mitzpa – (New York). 1910-11 – 1 – us AJPC [071]
Hamiudullah, Zeb-un-Nisa *see* Indian bouquet
Hamlet / Ducis, Jean-Francois – Paris, France. 1815 – 1r – us UF Libraries [025]
Hamlet / Shakespeare, William – New York, NY. 1909 – 1r – us UF Libraries [025]
Hamlet, an ideal prince and other essays in Shakespearean interpretration : hamlet, merchant of venice, othello, king lear / Crawford, Alexander Wellington – Boston: R G Badger; Toronto: Copp Clark, c1916 – 4mf – 9 – 0-665-72015-7 – (incl ind) – mf#72015 – cn Canadiana [420]
Hamlet news – Rockingham, NC. 1975-1981 (1) – mf#65332 – us UMI ProQuest [071]
Hamlin, Charles *see*
– Diary
– The insolvent law, of maine
Hamlin, Charles S *see*
– The acts to regulate commerce
– Icc acts indexed and digested
Hamlin, Cyrus *see* My life and times
Hamlin, H *see* The youth's scripture question book on the new testament
Hamline journal of public law *see* Hamline journal of public law and policy
Hamline journal of public law and policy – v1-22. 1980-2001 – 5,6,9 – $347.00 set – (v1-5 1980-84 on reel $83. v6-22 1985-2001 on mf $264. title varies: v1-2 1980-81 as journal of minnesota public law. v3-6 1982-85 as hamline journal of public law) – ISSN: 0736-1065 – mf#108651 – us Hein [342]
Hamline law review – v1-24. 1978-2001 – 5,6,9 – $369.00 set – (v1-7 1978-84 on reel $83. v8-24 1985-2001 on mf $286) – ISSN: 0198-7364 – mf#103021 – us Hein [340]
Hamlyn, Raul *see* Cortes and the conquest of mexico
Hamm, Katherine *see* Tulare county school
Hamm, Tracy M *see* Marathon performance time in relation to age, physical characteristics, previous running experience, and various training indices of female distance runners
Hamm, Wilhelm von *see* Suedoestliche steppen und staedte
Hammaker, Charles A, Jr *see* The impact of shifting our strategic base from okinawa to micronesia
Hammann, Otto *see* Der neue kurs
Hammar, Harald E *see* Some chemical and physical properties of the florida everglades soils
Hammarskoeld, G *see* Past and present relations between the anglican communion and the church of sweden
Hammasammish gazette – v1 n1-3 [1983 jun 30-jul 21] – 1r – 1 – mf#718752 – us WHS [071]
Hammel, Patricia A *see* Changes in clinical students' perceptions of developmental physical education and effective teaching
Hammelburger conversations-lexikon : ankuendigung und erstes probeheft / Lang, Karl H von – Hammelburg 1819 [mf ed Hildesheim 1995-98] – 1mf – 9 – €40.00 – 3-487-29595-4 – gw Olms [430]
Hammelburger reise : erste bis eilfte fahrt; nebst conversationslexicon / Lang, Karl H von – Nuernberg 1818-1833 [mf ed Hildesheim 1995-98] – v3-11 on 9mf – 9 – €180.00 – 3-487-29596-2 – gw Olms [914]
Hammer – Leipzig DE, 1902-05, 1908-22 – 1 – (filmed by other misc inst: 1902-05, 1908-40 [gaps]) – gw Misc Inst [074]
Hammer and pen / Church Association for the Advancement of the Interests of Labor – 1898-1908 apr, nov, 1911 jun, 1912 jun – 1r – 1 – mf#2812097 – us WHS [366]
Hammer and steel newsletter – Boston. 1962-1973 (1) 1971-1973 (5) – ISSN: 0017-7105 – mf#3259 – us UMI ProQuest [320]
Hammer and the rock / Tayler, Charles B – London, England. 18– – 1r – us UF Libraries [240]
Hammer and tongs / Socialist Party (US) et al – (1940 01,03-09/10], [1941 02/03,08], [1942 04,10-12] [1943 02], [1944 04-05,09], [1946 01,05/06], [1953 01,03-04], [1954 05,08], [1955 05], [1956 01,05], [winter 1957/58-spring 1958], [1964 2-4, 1965 1-3, 1967, 1968 1-2], [1971 11,18], [n4, n6-7, n9-may 1977] – 2r – 1 – mf#370642 – us WHS [335]
Hammer, Anton *see* Die erkenntnistheoretische bedeutung des gefuehlsmaessigen erfassens bei schleichermeier
Hammer, Bonaventure *see*
– Die katholische kirche in den vereinigten staaten nordamerikas
– Mary, help of christians
Hammer, Darrell P *see* Ussr
Hammer, Franz *see* Hermann stehr und das junge deutschland
Hammer, Friedrich *see* Die idee der persoenlichkeit bei paul heyse
Hammer, Heinrich *see* Traktat vom samaritanermessias
Hammer, Julius *see* Auf stillen wegen
Hammer, Klaus *see* Beitraege zur literaturgeschichte und – methodologie
Hammer newsletter – 1979 summer-1989 fall – 1r – 1 – mf#1057791 – us WHS [071]
Hammer und feder : deutsche schriftsteller aus ihrem leben und schaffen / ed by Gruenberg, Karl et al – Berlin: Verlag Tribuene, 1955 [mf ed 1993] – 595p – 1 – mf#8155 – us Wisconsin U Libr [430]
Hammerich, Angelika *see* Die anwendung von asteraceae (korbbluetler) in der zahnheilkunde von der antike bis heute
Hammerich, Frederik *see* St birgitta
Hammer-Purgstall, Josef von *see* Fundgruben des orients
Hammer-Purgstall, Joseph von *see* Topographische ansichten gesammelt auf einer reise in die levante
Hammerschlaege : eine auswahl / Lersch, Heinrich; ed by Fronemann, Wilhelm – Koeln: H Schaffstein, 1933 – 1r – 1 – us Wisconsin U Libr [800]
Hammerschmidt, Andreas *see*
– Musicalischer andacht, erster theil
– Musicalischer andachten, ander theill
– Musicalischer andachten, dritter theill
Hammerschmidt, Anette C *see* Fremdverstehen
Hammerschmidt, Ferdinand *see* Goethe und der katholizismus
Hammersmith and chiswick leader – [London & SE] Hammersmith & Fulham Archives 5 jun 1987-10 jun 1988 – 1 – (cont as: hammersmith and fulham independent [17 jun 1988-20 oct 1989]) – uk Newsplan [072]
Hammersmith and fulham guardian *see* Fulham and hammersmith guardian
Hammersmith and fulham independent – [London & SE] Hammersmith & Fulham Archives 17 jun 1988, 20 oct 1989 – 1 – (merged with: hammersmith & fulham times [27 oct 1989]) – uk Newsplan [072]
Hammersmith and fulham times – mar 1927-20 nov 1987; 29 jan-5 feb; 8 apr-23 dec 1988; 1989-21 dec 1990; 1991; 1992; 8 jan 1993-1996; 10 jan 1997-jun 1998 – 17 1/2r – 1 – (aka: fulham times; hammersmith fulham and chiswick times) – uk British Libr Newspaper [072]
Hammersmith and shepherds bush gazette *see* Shepherds bush gazette and west london post
Hammersmith fulham and chiswick guardian *see* Fulham and hammersmith guardian
Hammersmith fulham and chiswick times *see* Hammersmith and fulham times
Hammersmith news and fulham post – London, UK. 1986-92 – 13r – 1 – (aka: hammersmith news and post; hammersmith post) – uk British Libr Newspaper [072]
Hammersmith post *see* Hammersmith news and fulham post
The hammersmith protestant discussion : being an authenticated report of the controversial discussion between the rev. john cumming, d.d. of the scottish national church...and daniel french, esq., barrister-at-law... / Cumming, John – New ed. London: Arthur Hall, 1852 – 2mf – 9 – 0-8370-8092-4 – mf#1986-2092 – us ATLA [242]
Hammersmith searchlight – [London & SE] Hammersmith mar 1900 [mf ed 2003] – 1r – 1 – uk Newsplan [072]
Hammersmith & shepherds bush times – [London & SE] Hammersmith 3 oct 1908 [mf ed 2003] – 1r – 1 – uk Newsplan [072]
Hammersmith socialist record – London, UK. Oct 1891-jun 1893. -irr. 1r – 1 – uk British Libr Newspaper [072]
Hammersmith times – [London & SE] Hammersmith 9 feb 1899 [mf ed 2003] – 1r – 1 – uk Newsplan [072]
Hammerstein, Ludwig von *see*
– Das christentum
– Edgar, oder, vom atheismus zur vollen wahrheit
– Erinnerungen eines alten lutheraners
– Die gegner "edgars"
– Sincerus
Hammock, James W *see* Evaluating and reshaping a model of church renewal at the first baptist church of longwood, florida
Hammon, John Kohlsaat *see* American unitarian interest in the study of non-christian religions
Hammon, Louis Lougee *see*
– The general principles of the law of contract
– A treatise on chattel mortgages for illinois
Hammon, Ulrich *see* Die wirtschaftsdemokratie im sinne der freien gewerkschaften
Hammond, Barbara *see* The village labourer, 1760-1832
Hammond, C E *see* Liturgies, eastern and western

Hammond, Charles Edward *see*
- Liturgies, eastern and western
- Outlines of textual criticism applied to the new testament

Hammond, Edward Payson *see*
- The conversion of children

Hammond first southern baptist church. hammond, indiana : church records – 1934-37 – 1 – 5.00 – us Southern Baptist [242]
Hammond, H *see* The whole duty of man
Hammond independent – Hammond WI. 1875 dec 3-1877 oct 19 – 1r – 1 – mf#921925 – us WHS [071]
Hammond independent – Hammond WI. 1875 jul 23-dec 3 – 1r – 1 – mf#921923 – us WHS [071]
Hammond, James H *see* Papers
Hammond, James Henry *see* Cotton is king, and pro-slavery arguments
Hammond, Jessica S *see* Perceived legitimacy of aggressive acts and behavioral intentions to act aggressively among beginning and experienced collegiate women rugby players
Hammond, John *see*
- Devout and moral reflections on the pious life and happy death of t...
- Physiology of reproduction in the cow

Hammond, John Lawrence *see* The village labourer, 1760-1832
Hammond, Joseph *see* The mistakes of modern nonconformity
Hammond news – Hammond WI. 1894 aug 17-1899, 1900-26, 1927-1930 jun 5, 1930 jun 12-1932, 1933-64, 1965 jan 7-1967 nov 2, 1967 nov 9-1970 jan 1 – 19r – 1 – mf#948060 – us WHS [071]
Hammond, OT *see* Diary
Hammond, S T *see* The teetotaler's companion
Hammond, S T [comp] *see* A collection of temperance dialogues
Hammond, William Andrew *see* The definitions of faith and canons of discipline of the six oecumenical councils, with the remaining canons of the code of the universal church
Hammond, William Gardiner *see*
- Synopsis of lectures delivered in the law department of iowa state university, on equity jurisprudence, as administered in the same courts with common law
- Synopsis of lectures delivered in the law department of iowa state university, on the law of real property

Hammond-Tooke, W D *see* Bhaca society
Hammonton news – Hammonton, NJ. 1987-2000 (1) – mf#61599 – us UMI ProQuest [071]
Hammurabi code / Hammurabi, King Of Babylonia – London, England. 1921 – 1r – us UF Libraries [939]
Hammurabi, King Of Babylonia *see* Hammurabi code
Hammurabi's gesetz / ed by Kohler, F J et al – Leipzig. v1-6. 1904-1923 – 31mf – 8 – €60.00 – ne Slangenburg [930]
Hammurabi's gesetz / ed by Kohler, J et al – Leipzig. v1-6. 1904-1923 – 8 – €60.00 – ne Slangenburg [340]
Hammurabi's gesetz / Kohler, Josef et al – Leipzig: Eduard Pfeiffer, 1904-1911 – 5mf – 9 – 0-8370-7541-6 – mf#1986-1541 – us ATLA [340]
Ha-modia – Poltava/Ukraine, 1910-1914 – 23mf – 9 – mf#J-291-18 – ne IDC [077]
Hamodia – [Israel], 1965- – 9 – us UMI ProQuest [079]
Hamon, Avas B *see* Soft scale insects of florida
The hamond naval papers, 1766-1825 – 3r – 1 – $255.00 – (with printed guide. originally published by the university of virginia library) – mf#D3181 – us Virginia U Pr [355]
Hampa afro-cubana : los negros brujos (apuntes para un estudio de etnologia criminal) / Ortiz, Fernando – Madrid: Editorial-America, 1917?. 406p. illus – 1 – us Wisconsin U Libr [360]
Hampa afro-cubana / Ortiz, Fernando – Madrid, Spain. 1917 – 1r – us UF Libraries [972]
Hampden 1878-1895 – Oxford, MA (mf ed 1987) – (= ser Massachusetts vital records) – 9mf – 9 – 0-87623-042-7 – (mf 1-2: birth index 1878-1986. mf 3-4: marriage index 1878-1986. mf 5-6: death index 1878-1986. mf 7: births 1878-95. mf 8: marriages 1878-95. mf 9: deaths 1878-95) – us Archive [978]
Hampden, Maine. First Baptist Church *see* Records
Hampden, R D *see* The scholastic philosophy considered in its relation to christian theology
Hampden, Renn Dickson *see*
- The fathers of greek philosophy
- Inaugural lecture read before the university of oxford in the divin...
- Lord our righteousness
- The scholastic philosophy considered in its relation to christian theology

Hampden-Cook, Ernest *see* The christ has come
Hampe, Roland *see* Fruehe griechische sagenbilder in boeotien
Hampe, T *see* Nuernberger ratsverlaesse ueber kunst und kuenstler im zeitalter der spaetgotik und renaissance [1449] 1474-1618 [1633]
Hampe, Theodor *see* Gedichte vom hausrat aus dem 15. und 16. jahrhundert
Hampel, Beate *see* Prevosts "manon lescaut" in deutschen uebersetzungen des 18., 19. und 20. jahrhunderts
Hampi ruins / Archaeological Survey of India – 3rd ed. Delhi: [Govt of India] 1933 [mf ed 1987] – 1r [ill] – 1 – (descr & ill by a h longhurst. filmed with: sketches of life and sport in southeastern africa / hamilton, o) – mf#1844 – us Wisconsin U Libr [720]
Hampi ruins / Longhurst, Albert Henry – Madras: Printed by the Supt, Govt Press, 1917 – (= ser Samp: indian books) – us CRL [930]
Hampshire advertiser. (hampshire advertiser & independent) – Southampton, England. Jan 1901-Nov 1940.-w. 54 reels – 1 – uk British Libr Newspaper [072]
Hampshire business gazette – Basingstoke, England mar 1987-mar 1996 – 1 – (cont: basingstoke & hampshire business gazette [aug 1985-feb 1987]) – uk British Libr Newspaper [650]
Hampshire chronicle – Southampton, Winchester, England. 1772-1983; 1986- – 229+ r – 1 – uk British Libr Newspaper [072]
Hampshire herald and aldershot army review – Alton, England 9 dec 1887-13 feb 1897 – 1 – (cont: hampshire herald & counties advertiser [2 jan 1886-2 dec 1887] cont by: hampshire herald & alton gazette [20 feb 1897-4 aug 1966]) – uk British Libr Newspaper [355]
Hampshire herald and alton gazette – [SW England] Alton, Hampshire 20 feb 1897-4 aug 1966 – 1 – (cont: hampshire herald & aldershot army review [jan 1887-13 feb 1897]; cont by: alton gazette, etc [11 aug 1966-]) – uk Newsplan [072]
Hampshire herald and counties advertiser – Alton, England 2 jan 1886-2 dec 1887 – 1 – (cont: north & east hants herald & general county advertiser [29 aug-26 dec 1885]; cont by: hampshire herald & aldershot army review [9 dec 1887-13 feb 1897]) – uk British Libr Newspaper [072]
Hampshire herald, winchester, basingstoke, alresford, alton, & bishop's waltham advertiser – Winchester, England 12 jul 1873-6 dec 1879 [1869-72, 1877] – 1 – (cont: winchester herald, basingstoke, alresford, & alton advertiser [4 jan-5 jul 1873; discontinued) – uk British Libr Newspaper [072]
Hampshire independent – Southampton, England. jan 1853-jul 1923 – 68r – 1 – uk British Libr Newspaper [072]
Hampshire observer and basingstoke news – Basingstoke, England 28 feb 1903-11 mar 1916 [mf 1911] – 1 – (discontinued) – uk British Libr Newspaper [072]
Hampshire record society – v1-14. 1889-1897 – 94mf – 8 – (missing: v2) – mf#H-825 – ne IDC [400]
Hampshire record society – v1-12. 1889-97 – (= ser Publications of the english record societies, 1835-1972) – 56mf – 9 – uk Chadwyck [941]
Hampshire review – Romney, WV. 1884+ (1) – mf#67462 – us UMI ProQuest [071]
Hampshire telegraph and naval chronicle – Portsmouth, England. 1911 – 1r – 1 – uk British Libr Newspaper [072]
Hampshire telegraph and sussex chronicle – Portsmouth, England. 1874; 1897-98. -w. 3 reels – 1 – uk British Libr Newspaper [072]
Hampstead advertiser – London UK, 1986-21 dec 1989; jan-20 dec 1990; jan-19 dec 1991; 1992 – 14r – 1 – (aka: hampstead local advertiser) – uk British Libr Newspaper [072]
Hampstead and highgate express – 1874, 1888, 1892, 1900-97 – 176r – 1 – (aka: ham and high) – uk British Libr Newspaper [072]
Hampstead and highgate express – London. -w. 1950-81. (62 reels) – 1 – uk British Libr Newspaper [072]
Hampstead local advertiser *see* Hampstead advertiser
Hampstead record – London UK, 1951 – 1r – 1 – (aka: hampstead and highgate record and chronicle; camden and hampstead and highgate record and chronicle) – uk British Libr Newspaper [072]
Hampton and Sons *see* Illustrated designs of cabinet furniture engraved from photographs
Hampton county guardian – Hampton, SC. 1982-1994 (1) – mf#66457 – us UMI ProQuest [071]
Hampton first baptist church. hampton, south carolina : church records – 1908-10, 1915-20, 1922-31, 1939-41, 1955-67. Board of Deacons. Church records. 1948-72 – 1 – us Southern Baptist [242]
The hampton herald – Hampton, NE: H L Hellen. v1 n1. feb 21 1884-86 (gaps) filmed 1979] – 1r – 1 – us NE Hist [071]
Hampton, Iowa. First Baptist Church *see* Minutes
Hampton journal – Hampton, NE: C F Holden, 1882-v2 no last jun 22 1883 [mf ed filmed [1979] – 1r – 1 – us NE Hist [071]
The hampton ledger – Hampton, NE: L C Huston. v1 n1. jun 10 1937-v1 n32. feb 3 1938 (gaps) (wkly) [mf ed filmed 1979] – 1r – 1 – us NE Hist [071]
Hampton park baptist church. charleston county. south carolina : church records – 1915-May 1942; Dec 1949-Apr 1970. Deacons' Minutes, May 1942-Apr 1950. Membership rolls and historical sketch. 934p – 1 – us Southern Baptist [242]
The hampton record – Hampton, NE: Geo B Pickett. v1 n1. sep 6 1901- (wkly) [mf ed -nov 15 1901 filmed [1979]] – 1r – 1 – us NE Hist [071]
Hampton reporter – Hampton, NE: Gellatly & Gray. v1 n1. aug 26 1892- (wkly) [mf ed filmed 1999] – 1r – 1 – us NE Hist [071]
The hampton star – Hampton, NE: L M Skinner, 1898 (wkly) [mf ed with gaps filmed [1979]] – 1r – 1 – us NE Hist [071]
The hampton times – Hampton, NE: Times Pub Co, 1891 (wkly) [mf ed 1891,1895-97 (gaps) filmed 1979] – 1r – 1 – us NE Hist [071]
Hampton university newspaper clipping file : 55,000 clippings from nearly 100 black newspapers – early 20th c [mf ed Chadwyck-Healey] – 790mf – 9 – (with ind) – uk Chadwyck [305]
Hamy, Ernest Theodore *see* Coup d'oeil sur l'anthropologie du cambodge
Han, Chen-yeh *see*
- Ssu wai
- Yu hsi

Han, Chen-yeh pien chi *see* Fan men
Han ch'ieh / Kuan, P'ing – Shang-hai: Kuang i shu chu, Min kuo 21 [1932] – (= ser P-k&k period) – us CRL [810]
Han chien ch'ou shih / Cheng, Ch'en-chih – [China: sn, 1940] – (= ser P-k&k period) – us CRL [951]
Han chien ti hsia ch'ang / Ho, Chu-ch'i – [Sl]: Ch'ing nien ch'u pan she, Min kuo 30 [1941] – (= ser P-k&k period) – us CRL [920]
Han de estar y estaran(cuentos y leyendas de gu... / Barnoya Galvez, Francisco – Santiago, Chile. 1938 – 1r – us UF Libraries [972]
Han, Dong H *see* Cocaine and excercise
Han fen lou ku chin wen ch'ao chien pien : [40 chuan] / Wu, Tseng-ch'i – Shang-hai: Shang wu yin shu kuan, Min kuo 24 [1935] – (= ser P-k&k period) – us CRL [840]
Han feng chi / Ch'en, Kung-po – Shang-hai: Ti fang hsing cheng she, Min kuo 34 [1945] – (= ser P-k&k period) – us CRL [951]
Han, Hsing *see* Hsueh yu
Han, Hui *see* Shang-hai hsia ch'ih shih ch'ang chih nan
Han i ku lan ching ti i chang hsiang chieh / Tao-yin – Hsiang-kang: Chung-kuo Hui chiao hsueh hui, Min kuo 30 [1941] – (= ser P-k&k period) – us CRL [260]
Han, Jen *see* Tse jen kuan nien yu hsien tai kuo min
Han liu wen yen chiu fa / Lin, Shu – Shang-hai: Shang wu yin shu kuan, Min kuo 22 [1933] – (= ser P-k&k period) – us CRL [480]
Han mien ying ta tju tien [ssu chiao hao ma] = Chinese-burmese-english dictionary / Kyin Win, U – Ran Kun: Tui tak re pum nhip tuik 1974 [mf ed 1990] – 1r with other items – 1 – mf#mf-10289 seam reel 151/2 [§] – us CRL [040]
Han, Shang-i *see* Mu i shih chiang
Han, Shih-heng *see* Wen hsueh p'ing lun chi
Han sido nombrados...en plasencia d. vicente paredes... / Fita, Fidel – Madrid: Est. Tip. Fortanet, 1897. B.R.A.H. 31, 1897, pp. 352 – sp Bibl Santa Ana [946]
Han tsai, chou leng-ch'ieh chu / Chou, Leng-ch'ieh – Shang-hai, Chung-hua shu chu, Min kuo 24 [1935] – (= ser P-k&k period) – us CRL [480]
Han tsang fo chiao kuan hsi shih liao chi = Documenta buddhismi sino-tibetici / Lu, Ch'eng – Ch'eng-tu: Hua hsi hsieh ho ta hsueh Chung-kuo wen hua yen chiu so, Min kuo 31 [1942] – (= ser P-k&k period) – us CRL [280]
Han, Tsu-te *see* Ying fu kuo nan ying yu chih ching chi cheng ts'e
Han tzu kai ko / Wang, Li – Ch'ang-sha: Shang wu yin shu kuan, Min kuo 29 [1940] – (= ser P-k&k period) – us CRL [480]
Han wei liang chin nan pei ch'ao fo chiao shih / T'ang, Yung-t'ung – [Chang-sha]: Shang wu yin shu kuan, Min kuo 27 [1938] – (= ser P-k&k period) – us CRL [280]
Han wei liu ch'ao shih yen chiu / Ch'en, Chia-ch'ing – [China]: An-hui ta hsueh ch'u pan tsu, Min kuo 23 [1934] – (= ser P-k&k period) – us CRL [810]
Han wei liu ch'ao wen hsueh / Ch'en, Chung-fan – Shang-hai: Shang wu yin shu kuan, 1932 – (= ser P-k&k period) – us CRL [480]
Han wen hsueh shih kang yao / Lu, Hsun – [China]: Lu Hsun hsien sheng chi nien wei yuan hui pien yin, Min kuo 30 [1941] – (= ser P-k&k period) – us CRL [480]
Han ya chi / Liu, Ta-chieh – Shanghai: C'i chih shu chu, min kuo 23 [1934] – (= ser P-k&k period) – us CRL [480]
Han yeh chi / Ho, Chia-huai – Shang-hai: Pei hsin shu chu, 1937 – (= ser P-k&k period) – us CRL [480]
Han yu sheng niu pien chuan chih ting lu / Fu, Tung-hua – Shang-hai: Hsueh lin she: K'ai ming shu tien, Min kuo 30 [1940] – (= ser P-k&k period) – us CRL [480]
Han, Yu-t'ung *see* Tsou hsiang min chu: hsien fa yu hsien cheng
Hana, H J *see* Kampioenen des christendoms
Hana newsletter / Harlem Neighborhoods Association, Inc – 1963 jun, nov, 1964 jan, aug, oct, 1965 apr, 1968 apr – 1r – 1 – mf#4877865 – us WHS [366]
Hanaford, Phebe Ann *see* Daughters of america, or, women of the century
Hanau herald – Hanau, Hessen. 1980 dec-1986 feb – 1r – 1 – mf#1044809 – us WHS [071]
Hanauer departements-blatt *see* Hanauer privilegirte wochen-nachricht
Hanauer, James Edward *see*
- Baptism, jewish and christian
- Walks about jerusalem

Hanauer kreisblatt *see* Hanauer privilegirte wochen-nachricht
Hanauer neue europaeische zeitung *see* Europaeische zeitung
Hanauer neue zeitung *see* Europaeische zeitung
Hanauer privilegirte wochen-nachricht – Hanau DE, 1848-49 – 2r – 1 – (title varies: fr 1 mar 1811-30 dec 1813: hanauer departements-blatt; 3 jan 1822: wochenblatt fuer die provinz hanau; fr 5 apr 1849-10 jul 1851: wochenblatt fuer den verwaltungsbezirk hanau; 8 nov 1866: wochenblatt fuer den regierungsbezirk hanau; 17 oct1867: wochenblatt fuer den vorhinnigen regierungsbezirk hanau; 6 jan 1869: hanauer wochenblatt; 1 jan 1870: hanauer kreisblatt; 1 may 1872: hanauer anzeiger. filmed by other misc inst: 1976- [ca 8r/yr]; 1804 5 jan-1806 4 nov, 1813 7 jan-1891 30 jun, 1892 2 jan-24 jun, 1893 2 jan-30 jun, 1894-1923 30 jun, 1924-1941 31 may [221r]. incl suppls) – gw Misc Inst [074]
Hanauer rundschau *see* Rundschau fuer hanau stadt und land
Hanauer zeitung *see* Europaeische zeitung
Hanauer zeitung 1943 – Hanau DE, 1943 1 jul-31 dec – 1r – 1 – gw Misc Inst [074]
Hanauisches magazin – Hanau 1778-85 – (= ser Dz) – 8v on 25mf – 9 – €250.00 – mf#k/n336 – gw Olms [943]
Hanauisches magazin : monatsblaetter fuer heimatkunde / ed by Waisenhaus-Buchdruckerei [Hanauer Anzeiger] et al – Hanau: [s.n.] v4-18[1925-39] (mthly) [mf ed 1978] – 1r – 1 – (ceased in 1939; cont by: neues magazin fuer hanauische geschichte) – mf#film mas c336 – us Harvard [943]
Hanawa, Tokinosuke *see* [Japanese commentaries on the "four shoo" or the books of the four philosophers]
Hanbury, Benjamin *see* Historical memorials relating to the independents, or congregationalists
Hanbury, D T *see* Through the barren ground of northeastern canada to the arctic coast
Hance, Gertrude Rachel *see*
- The zulu yesterday and to-day
- Zulu yesterday and to-day

Hanchen und die kuechlein / Eberhard, Christian August Gottlob – 9. Aufl. Leipzig: Rengersche Verlags-Buchhandlung, [185-?] (mf ed 1990) – 1r – 1 – (filmed with: bozena) – us Wisconsin U Libr [430]
Hancher, Heidi L *see* The influence of spousal exercise patterns and perceived social support on the quality of life and health status in regular exercisers
Hancock 1767-1896 – Oxford, MA (mf ed 1990) – (= ser Massachusetts vital records) – 29mf – 9 – 0-87623-112-1 – (mf 1-2: vital records 1767-1832. mf 3-4: vital records 1832-44. mf 5-7: town records 1776-1801. mf 8-14: town records 1801-35. mf 15-24: town records 1835-73. mf 25-26: vital records 1843-77. mf 27: births 1878-96. mf 28: marriages 1878-97. mf 29: deaths 1872-1900) – us Archive [978]
Hancock Co. Findlay *see* Daily star
Hancock Co. VanBuren *see* Bulletin
Hancock county courier – New Cumberland, WV. 1915+ (1) – mf#67396 – us UMI ProQuest [071]
Hancock eagle – Nauvoo, Ill. v. 1 no. 2-21. Apr 10-Aug 28 1846. Incomplete – 1 – us NY Public [071]
Hancock, Edward *see* Candid warning to public men in a series of letters
Hancock, Elizabeth A *see* Frances b hogan-professional educator, coach and director
The Hancock Family *see* The hancock family papers, 1728-1830
The hancock family papers, 1728-1830 – [mf ed 1977] – 2r – 1 – (with p/g. coll provides insight into one of america's best-known revolutionary-era families) – us MA Hist [975]

Hancock field mini – Syracuse NY. v11 n47-v13 n45 [1981 may 1-1983 nov 11] – 1r – 1 – mf#1520839 – us WHS [071]

Hancock news – Coloma, Hancock WI. 1897/1899 jan 20-1942/1943 mar 11 [gaps] – 38r – 1 – (continued by: hancock-coloma news) – mf#947583 – us WHS [071]

Hancock news – Hancock WI. 1894 sep 27-1895 sep 19 – r – 1 – mf#875727 – us WHS [071]

Hancock, Ralph *see* Puerto rico

Hancock, Thomas *see*
– The act of uniformity
– The peculium

Hancock, William Keith *see* Are there south africans?

Hancock, William Neilson *see* Impediments to the prosperity of ireland

Hancock-Coloma news – Coloma, Hancock WI. 1943 mar 18-1945, 1946-58, 1959-1961 feb 28 – 7r – 1 – (cont: hancock news; cont by: sun [plainfield, wi]; sun-news press) – mf#947484 – us WHS [071]

Hand – Edinburgh. 1982-1983 (1) 1982-1983 (5) 1982-1983 (9) – (cont by: journal of hand surgery) – ISSN: 0072-968X – mf#13428 – us UMI ProQuest [617]

The hand : its mechanism and vital endowments as evincing design / Bell, Charles – [2nd ed] London, 1833 – (= ser 19th c evolution & creation) – 4mf – 9 – mf#1.1.7692 – uk Chadwyck [611]

The hand : a survey of facts, legends, and beliefs pertaining to manual ceremonies, covenants, and symbols / Burdick, Lewis Dayton – Oxford, NY: Irving Co, 1905 – 1mf – 9 – 0-524-01685-2 – (incl bibl ref) – mf#1990-2587 – us ATLA [390]

A hand book for riflemen : containing the first principles of military discipline, founded on rational method... / Duane, William – 3rd ed. Philadelphia: printed for the aut, 1813 [mf ed 1983] – 2mf – 9 – 0-665-44207-6 – mf#44207 – cn Canadiana [355]

Hand book of methodist missions / John, I G – Nashville, Tenn: Pub House of the ME Church, South, 1893 – 2mf – 9 – 0-524-06547-0 – mf#1991-2631 – us ATLA [242]

Hand book of soil conservation for ccc camp scs-5, sligo, pa – Sligo, PA: [s.n.], 1935-40 – us CRL [630]

Hand book of the american republics / International Bureau Of The American Republics – Washington? DC. 1891 – 1r – us UF Libraries [972]

Hand book of the brethren mission in china, 1915 – Hankow: Central China Religious Tract Society, [1915?] – 1mf – 9 – 0-524-06868-2 – mf#1990-5287 – us ATLA [242]

Hand book of the ceylon national congress, 1919-1928 / ed by Bandaranaike, S W R D – Colombo: H W Cave & Co, 1928 – 1 – us CRL [959]

A hand book of the vedant philosophy and religion / Khedkar, Raghunath Vithal – Kolhapur: Mission Press, 1911 [mf ed 1991] – (= ser Shri shankaracharya series) – 1v on 1mf – 9 – 0-524-01611-9 – mf#1990-2550 – us ATLA [280]

Hand book to the canada tariff : with alterations and amendments to 1st august, 1879: together with exchange tables for sterling, franc... / Sargant, Robert H – [Toronto?: s.n.], 1879 (Toronto: Dudley & Burns) – 2mf – 9 – 0-665-89778-2 – mf#89778 – cn Canadiana [348]

Hand clinics – Philadelphia. 1985+ (1,5,9) – ISSN: 0749-0712 – mf#14732 – us UMI ProQuest [617]

Hand, F et al *see* Neue jenaische allgemeine literatur-zeitschrift

Hand family courier – v1 n1-v2 n4 [1985 summer-1987 sep 1] – 1r – 1 – mf#1336373 – us WHS [071]

Hand full of diamonds / Norwood, Victor George Charles – London, England. 1960 – 1r – us UF Libraries [972]

Hand, George R *see* D b ray's text book on campbellism exposed

Die hand gottes / Winnig, August – Berlin: M Warneck, 1940 – 1 – us Wisconsin U Libr [890]

Hand, J E *see* Ideals of science and faith

Hand, Julia *see* Diary

The hand of god : and other posthumous essays: together with some reprinted papers / Allen, Grant – London: Watts, 1909 – 2mf – 9 – 0-665-98005-1 – mf#98005 – cn Canadiana [210]

Hand of god acknowledged in the loss of endeared relatives / Bowden, James – London, England. 1793 – 1r – us UF Libraries [240]

The hand of god in american history : a study of national politics / Thompson, Robert Ellis – New York: T.Y. Crowell, 1902 – 1mf – 9 – 0-7905-6126-3 – mf#1988-2126 – us ATLA [975]

Hand of god in the disruption and the vital importance of free chur... / Begg, James – Edinburgh, Scotland. 1865 – 1r – us UF Libraries [240]

Hand of man / Jaquin, Noel – London, England. 1933 – 1r – us UF Libraries [025]

Hand und herz : trauerspiel in vier akten / Anzengruber, Ludwig – Wien: L Rosner, 1875 [mf ed 1993] – (= ser Neues wiener theater 45) – 58p – 1 – mf#8459 – us Wisconsin U Libr [820]

Handball – Skokie. 1971+ (1) 1974+ (5) 1975+ (9) – ISSN: 0046-6778 – mf#9816 – us UMI ProQuest [790]

Handboek der middelnederlandse geographie / Bergh, L Ph van den – Den Haag, 1949 – €12.00 – ne Slangenburg [949]

Handboek voor cultuur- en handelsondernemingen in nederlandsch-indie – Amsterdam. v1. 1888 – 7mf – 8 – mf#SE-1587 – ne IDC [959]

Handbook / Congregational Christian Churches – Milwaukee WI: National Assoc Off, 1962-1971/72 [annual] [mf ed 2002] – 9v on 1r – 1 – mf0914 – us ATLA [030]

Handbook and incidents of foreign missions of the presbyterian church, usa / Rankin, William – Newark, NJ: W H Shorts, 1893 – 1mf – 9 – 0-8370-6603-4 – mf#1986-0603 – us ATLA [242]

Hand-book and map to the gold region of frazer's and thompson's rivers : with table of distances; to which is appended chinook jargon-language used, etc, etc / Anderson, Alexander Caulfield – San Francisco: J J LeCount, c1858 – 1mf – 9 – mf#27870 – cn Canadiana [917]

Handbook and proceedings of the annual convention / National American Woman Suffrage Association – 16th, 25th-48th, 50th-51st. 1884, 1893-1916, 1919-20 – 1 – 59.00 – us L of C Photodup [977]

Hand-book for attendants at the asylum for the insane – [Toronto?: C B Robinson], 1881 – 1mf – 9 – 0-665-92339-2 – mf#92339 – cn Canadiana [360]

Handbook for canoeing councillors / Deming, Eleanor – 1926. 20p. illus – 1 – us Wisconsin U Libr [790]

Handbook for delegates to the ninth international... / Pan American Union – Washington, DC. 1947 – 1r – us UF Libraries [972]

Handbook for egypt and the sudan / ed by Hall, Harry Reginald – 11th rev enl ed. London: E Stanford, 1910 – 2mf – 9 – 0-524-05806-7 – (incl bibl ref) – mf#1992-0633 – us ATLA [916]

Handbook for elders and deacons : the nature and the duties of the offices according to the principles of reformed church polity / Heyns, William – Grand Rapids, Mich: Eerdmans, c1928 – 1mf – 9 – 0-524-06019-3 – (incl bibl ref) – mf#1991-2379 – us ATLA [240]

Handbook for federal judges' secretaries – Washington: FJC, rev Dec 1983 – 2mf – 9 – $3.00 – mf#LLMC 95-807 – us LLMC [340]

Handbook for federal judges' secretaries – Washington: FJC, rev Sept 1985 – 2mf – 9 – $3.00 – mf#LLMC 95-808 – us LLMC [340]

A handbook for magistrates, in relation to summary convictions and orders and indictable offences / McGuire, Thomas Horace – Toronto, Carswell, 1890. 92 p. LL-2321 – 1 – us L of C Photodup [340]

A handbook for notaries public and commissioners of deeds of new york. / Skinner, Joseph Osmun – 2d ed. Albany: Bender, 1927. 388p. LL-442 – 1 – us L of C Photodup [348]

A handbook for painters and art students : on the character and use of colours, their permanent or fugitive qualities, and the vehicles proper to employ / Muckley, William J – London: Bailliere, Tindall, & Cox, 1880 – (= ser 19th c art & architecture) – 2mf – 9 – mf#4.1.92 – uk Chadwyck [750]

Handbook for speakers / Capricorn Africa Society – Salisbury: [s.n.], 1955 – (filmed with: capricorn africa society papers) – us CRL [960]

A hand-book for the architecture, sculptures, tombs, and decorations of westminster abbey : with fifty-six embellishments on wood, engraved by ladies; and four etchings by david cox, jun. / Cole, Henry [pseud. Felix Summerly] – London: George Bell, 1842 – (= ser 19th c art & architecture) – 2mf – 9 – mf#4.1.129 – uk Chadwyck [720]

A hand-book for the architecture, tapestries, paintings...and grounds of hampton court / Cole, Henry Hardy – [2nd ed] London 1843 – (= ser 19th c art & architecture) – 2mf – 9 – mf#4.2.1642 – uk Chadwyck [700]

Hand-book for the city of montreal and its environs : with a plan of the city and a geological map of the surrounding country / Dawson, Samuel Edward – Montreal: Dawson, 1884 – 3mf – 9 – mf#26479 – cn Canadiana [917]

Handbook for the diplomatic history of europe, asia and africa / Anderson, Frank Maloy – Washington, DC. 1918 – 1r – us UF Libraries [025]

Handbook for the dominion of canada : prepared for the meeting of the british association for the advancement of science at montreal, 1884 / Dawson, Samuel Edward – 2nd ed. Montreal: Dawson, 1888 [mf ed 1982] – 5mf – 9 – 0-665-29122-1 – mf#29122 – cn Canadiana [917]

Handbook for the dominion of canada : prepared for the meeting of the british association for the advancement of science at montreal, 1884 / Dawson, Samuel Edward – Montreal: Dawson, 1888 – 5mf – 9 – mf#29122 – cn Canadiana [917]

Handbook for the study of egyptian topographical lists / Simons, J – 1937 – 9 – $10.00 – us IRC [930]

Handbook for the use of members and visitors : giving the rules of the society, its history, and a historical sketch of montreal with places of interest in its vicinity: 27th may, 1891 / Royal Society of Canada – Montreal: s.n, 1891 – 2mf – 9 – mf#12740 – cn Canadiana [360]

A handbook for the use of the members and friends of the protestant episcopal church / Peterkin, George William – [s.l: s.n, 1911?] [mf ed 1992] – 1mf – 9 – 0-524-04716-2 – mf#1990-5068 – us ATLA [242]

Handbook for travellers in asia minor, transcaucasia, persia, etc / ed by Wilson, Charles William – London: J Murray, 1895 – 2mf – 9 – 0-524-06214-5 – mf#1992-0852 – us ATLA [915]

Handbook for workers in evangelistic campaigns in india : revised edition of 'suggestions to workers' / ed by Popley, H A – Madras, Allaha: Christian Literature Society for India, 1917 [mf ed 1995] – (= ser Yale coll) – 68p – 1 – 0-524-09349-0 – mf#1995-0349 – us ATLA [240]

Hand-book of alabama / Berney, Saffold – Birmingham, AL. 1892 – 1r – us UF Libraries [025]

The handbook of architectural ornament... : employed in the great exhibition / Gibbs, William – London 1851 – (= ser 19th c art & architecture) – 1mf – 9 – mf#4.2.875 – uk Chadwyck [720]

A handbook of art industries in pottery and the precious metals / Wheatley, Henry Benjamin – London 1886 – (= ser 19th c art & architecture) – 3mf – 9 – mf#4.2.925 – uk Chadwyck [730]

A handbook of art smithing for the use of practical smiths, designers of ironwork technical and art schools, architects, etc / Meyer, Franz Sales – 2nd enl ed. London: B T Batsford, 1896 – (= ser 19th c art & architecture) – 3mf – 9 – mf#4.1.11 – uk Chadwyck [730]

Hand-book of bengal missions in connexion with the church of england : together with an account of general educational efforts in north india / Long, James – London: J.F. Shaw, 1848 – 1r – 1 – 0-8370-1549-9 – (incl ind) – mf#1984-B466 – us ATLA [241]

A handbook of biblical difficulties : or, reasonable solutions of perplexing things in sacred scripture / Tuck, Robert – London: Elliot Stock, 1889 [mf ed 1989] – 2mf – 9 – 0-7905-1139-8 – (incl ind) – mf#1987-1139 – us ATLA [220]

Handbook of black organizations – Durban, Black Community Programmes, 1973 – us CRL [360]

Handbook of british east africa : including zanzibar, uganda, and the territory of the imperial british east africa company – London: Printed for H M Stationery Office by Harrison, 1893 – 1 – us CRL [960]

Hand-book of canadian methodism : being an alphabetical arrangement of all the ministers and preachers whose names have appeared in connection with canadian methodism / Cornish, George Henry – Toronto: Wesleyan Printing Establishment, 1867 – 1mf – 9 – 0-7905-4787-2 – mf#1988-0787 – us ATLA [242]

Handbook of canadian methodism : being an alphabetical arrangement of all the ministers and preachers whose names have appeared in connection with canadian methodism... / Cornish, George Henry – Toronto: Wesleyan Print Establishment, 1867 [mf ed 1980] – 3mf – 9 – 0-665-00243-2 – (incl ind) – mf#00243 – cn Canadiana [242]

A handbook of chemical engineering / Davis, George Edward – Manchester, 1901-02 [mf ed 1986] – 2v in 1 (ill) – 1 – mf#8500 – us Wisconsin U Libr [660]

A hand-book of chikaranga : or the language of mashonaland / Springer, Helen Emily (Chapman) – Cincinnati: Printed by Jennings & Graham for the Methodist Episcopal Mission, Rhodesia, [1905] – 1 – (filmed with the author's transl fr the bible and her hymn) – us CRL [490]

Hand-book of chinese buddhism : being a sanskrit-chinese dictionary with vocabularies of buddhist terms in pali, singhalese, siamese, burmese, tibetan, mongolian and japanese / Eitel, Ernest John – 2nd rev enl ed. Tokyo: Sanshusha, 1904 [mf ed 1995] – (= ser Yale coll) – 324p – 1 – 0-524-09083-1 – (with chinese ind by k takakuwa) – mf#1995-0083 – us ATLA [280]

A handbook of christian apologetics / Garvie, Alfred Ernest – New York: Scribner 1915 [mf ed 1990] – (= ser Studies in theology) – 1mf – 9 – 0-7905-3937-3 – (incl bibl ref) – mf#1989-0430 – us ATLA [240]

A handbook of christian doctrine / Townsend, William John – London: G Burroughs [1897?] [mf ed 1991] – 1mf – 9 – 0-7905-8606-1 – mf#1989-1831 – us ATLA [240]

Handbook of christian doctrine / Graves, Henry Clinton – Philadelphia: American Baptist Publication Society, 1903 – 1mf – 9 – 0-7905-9943-0 – mf#1989-1668 – us ATLA [240]

Hand-book of christian evidences / Davis, Jerome Dean – Kyoto: [s.n.] 1889 [mf ed 1991] – 1mf – 9 – 0-7905-9179-0 – (incl bibl ref) – mf#1989-2404 – us ATLA [240]

Handbook of christian evidences / Stewart, Alexander – new rev enl ed. New York: Anson DF Randolph; London: Adam & Charles Black, 1895 [mf ed 1985] – (= ser The guild text books) – 1mf – 9 – 0-8370-5415-X – (incl bibl ref & ind) – mf#1985-3415 – us ATLA [240]

A handbook of church history : from the apostolic era to the dawn of the reformation / Green, Samuel Gosnell – London: Religious Tract Society [1904?] [mf ed 1991] – 2mf – 9 – 0-524-01521-X – mf#1990-0427 – us ATLA [240]

Handbook of commercial treaties etc, between great britain and foreign powers / ed by Bernhardt, G de – London, 1912 – 13mf – 9 – mf#ILM-2725 – ne IDC [380]

Handbook of commercial union : a collection of papers read before the commercial union club, toronto... / ed by Adam, Graeme Mercer – Toronto: Hunter, Rose, 1888 – 4mf – 9 – mf#00725 – cn Canadiana [337]

Handbook of common freshwater fish in florida lakes / Hoyer, Mark V – Gainesville, FL. 1994 – 1r – us UF Libraries [500]

A handbook of comparative religion / Kellogg, Samuel Henry – Philadelphia: Westminster Press, 1915 [mf ed 1991] – (= ser Westminster handbooks) – 1mf – 9 – 0-524-00913-9 – mf#1990-2136 – us ATLA [230]

A handbook of congregationalism / Dexter, Henry Martyn – Boston: Congregational Publ Society, c1880 [mf ed 1989] – 1mf – 9 – 0-7905-4343-5 – (incl bibl ref) – mf#1988-0343 – us ATLA [243]

The hand-book of dress-making : including correct rules for the pursuit of the above art, and concisely illustrating the mode of fitting at sight / Howell, Mary J – London: Simpkin, Marshall & Co, 1845 – (= ser 19th c art & architecture) – 2mf – 9 – mf#4.1.36 – uk Chadwyck [640]

Handbook of egyptian religion = Die aegyptische religion / Erman, Adolf – London: Archibald Constable, 1907 [mf ed 1991] – 1mf – 9 – 0-524-01280-6 – (english trans fr german by a s griffith) – mf#1990-2316 – us ATLA [290]

Handbook of embroidery / Higgin, Louis – London 1880 – (= ser 19th c art & architecture) – 2mf – 9 – mf#4.2.329 – uk Chadwyck [740]

Handbook of equity / McClintock, Henry Lacey – St. Paul, West, 1936. 421 p. LL-359 – 1 – us L of C Photodup [342]

The handbook of folklore – new rev enl ed. London: publ for the Folk-lore Society by Sidgwick & Jackson 1914 [mf ed 1991] – (= ser Publications of the folk-lore society 73) – 1mf – 9 – 0-524-01416-7 – (incl bibl ref) – mf#1990-2411 – us ATLA [390]

A handbook of foreign missions : containing an account of the principal protestant missionary societies in great britain – London: Religious Tract Society, 1888 [mf ed 1986] – 1mf – 9 – 0-8370-6511-9 – (incl ind) – mf#1986-0511 – us ATLA [242]

Handbook of forms and practice in bankruptcy / Menin, Abraham Isaac – New York: Industries Publishing Co., 1930. 370p. LL-1562 – 1 – us L of C Photodup [346]

A handbook of greek religion / Fairbanks, Arthur – New York: American Book Co, c1910 [mf ed 1991] – (= ser Greek series for colleges and schools) – 1mf – 9 – 0-524-01054-4 – mf#1990-2202 – us ATLA [250]

Handbook of greek vase painting / Herford, Mary Antonie Beatrice – Manchester, England. 1919 – 1r – us UF Libraries [720]

A handbook of hebrew antiquities : for the use of schools and students / Browne, Henry – London: Francis & John Rivington, 1852 [mf ed 1989] – 1mf – 9 – 0-7905-0618-1 – mf#1987-0618 – us ATLA [220]

Handbook of hindu names / Shanta – Calcutta: ARNICA International, 1969 – us CRL [490]

Handbook of home rule : being articles on the irish question / ed by Bryce, James Bruce, viscount – London, 1887 – (= ser 19th c ireland) – 4mf – 9 – mf#1.1.8621 – uk Chadwyck [941]

Handbook of homeric study / Browne, Henry – London, England. 2nd ed. 1925 – 1r – us UF Libraries [450]

A handbook of illustration / Hinton, Alfred Horsley – London: Dawbarn & Ward Ltd [1895] – (= ser 19th c art & architecture) – 2mf – 9 – mf#4.1.91 – uk Chadwyck [740]

Hand-book of india and british burmah / Robbins, W E – Cincinnati: Walden & Stowe, 1883 [mf ed 1995] – (= ser Yale coll) – 285p (ill) – 1 – 0-524-09111-0 – mf#1995-0111 – us ATLA [915]

A handbook of indian art / Havell, Ernest Binfield – London: John Murray, 1920 – (= ser Samp: indian books) – us CRL [700]

A handbook of information : touching the proposed correction of the present official title of the protestant episcopal church in the united states of america – Milwaukee: Young Churchman, 1903 [mf ed 1992] – (= ser Anglican/episcopal coll) – 1mf – 9 – 0-524-03548-2 – mf#1990-4743 – us ATLA [242]

Handbook of information of the general conference of the mennonite church of n.a. – 1947-48 [complete] – (= ser Mennonite serials coll) – 1r – 1 – mf#ATLA 1993-S007 – us ATLA [242]

Handbook of international sociometry – Beacon. 1973-1973 (1) – ISSN: 0160-4635 – mf#6906 – us UMI ProQuest [301]

Hand-book of lutheranism / Roth, J D – 2nd ed. Utica, NY: Young Lutheran Co, 1891 – 2mf – 9 – 0-8370-8858-5 – (incl ind) – mf#1986-2858 – us ATLA [242]

Handbook of marks on pottery and porcelain / Burton, William – London, England. 1928 – 1r – us UF Libraries [720]

Handbook of military hygiene, 1940 / South Africa. Dept of Defence – Pretoria: Govt Printer 1942 [mf ed Pretoria, RSA: State Library [199-]] – 126p [ill] on 1r with other items – 5 – mf#op 12814 r26 – us CRL [355]

The hand-book of millinery : comprised in a series of lessons for the formation of bonnets, capotes, turbans, caps, bows, etc / Howell, Mary J – London: Simpkin, Marshall & Co, 1847 – (= ser 19th c art & architecture) – 2mf – 9 – mf#4.1.33 – uk Chadwyck [680]

Hand-book of missions / McLean, Archibald – Cleveland, Ohio: Bethany CE Co, c1897 – (= ser Bethany c e hand-book series) – 1mf – 9 – 0-524-04264-0 – (incl bibl ref) – mf#1991-2048 – us ATLA [240]

Handbook of missions / Brethren in Christ Church – 1918-70 [mf ed 2001] – (= ser Christianity's encounter with world religions, 1850-1950) – 3r – 1 – (reports of church's missionary work in africa, india and japan; mission expanded to cuba fr 1953-60. also incl home missions) – mf#2001-s188 – us ATLA [242]

Handbook of modern japan / Clement, Ernest Wilson – 9th rev enl ed. Chicago: A C McClurg, 1913 [mf ed 1995] – (= ser Yale coll) – xvi/436p (ill) – 1 – 0-524-09808-5 – (with additional chapters on the russo-japanese war and greater japan) – mf#1995-0808 – us ATLA [950]

Handbook of moral philosophy / Calderwood, Henry – 14th ed. London; New York: Macmillan, 1895 – 1mf – 9 – 0-7905-8768-8 – (incl bibl ref) – mf#1989-1993 – us ATLA [170]

Hand-book of musical evangelism / Stephen, L I & Popley, Herbert Arthur – Madras: Methodist Publ House, 1914 [mf ed 1995] – (= ser Yale coll) – 214p – 1 – 0-524-09320-2 – (in hindi) – mf#1995-0320 – us ATLA [780]

Handbook of natural resources of british guiana / British Guiana Interior Development Committee – Georgetown, Guyana. 1946 – 1r – us UF Libraries [972]

Handbook of painting : the german, flemish, and dutch schools / Kugler, Franz Theodor – London 1860 – (= ser 19th c art & architecture) – 8mf – 9 – mf#4.2.1330 – uk Chadwyck [750]

Handbook of painting : the italian schools / Kugler, Franz Theodor – London 1887 – (= ser 19th c art & architecture) – 12mf – 9 – mf#4.2.1138 – uk Chadwyck [750]

Handbook of pictures in the international exhibition of 1862 / Taylor, Tom – London 1862 – (= ser 19th c art & architecture) – 3mf – 9 – mf#4.2.886 – uk Chadwyck [700]

Handbook of psychological literature / Louttit, Chauncey McKinley – Bloomington, IN: The Principia Press, 1932. viii,273p – 1 – us Wisconsin U Libr [150]

A handbook of psychology / Murray, John Clark – London: A Gardner, 1892 – 5mf – 9 – 0-665-91003-7 – (incl bibl ref. incl ind) – mf#91003 – cn Canadiana [150]

Handbook of revivals : for the use of winners of souls / Fish, Henry Clay – Boston: James H. Earle, 1874. Beltsville, Md: NCR Corp, 1978 (5mf); Evanston: American Theol Lib Assoc, 1984 (5mf) – 9 – 0-8370-0954-5 – mf#1984-4321 – us ATLA [240]

Handbook of revivals : for the use of winners of souls / Fish, Henry Clay – Boston: James H. Earle, 1879, c1874 – 1mf – 9 – 0-8370-6183-0 – (incl ind) – mf#1986-0183 – us ATLA [240]

A handbook of rome and its environs / Pentland, Joseph B – London 1869 [mf ed Hildesheim 1995-98] – (= ser Fbc) – 4mf – 9 – €120.00 – 3-487-29231-9 – gw Olms [914]

Handbook of sanskrit literature / Small, George – London, England. 1866 – 1r – us UF Libraries [490]

Handbook of social economy : or, the worker's abc / About, Edmond – London: Strahan; Toronto: Adam, Stevenson, 1872 [mf ed 1985] – 4mf – 9 – 0-665-05097-6 – mf#05097 – cn Canadiana [330]

Handbook of soviet lunar and planetary exploration – 1979 – 9 – $25.00 – us Univelt [550]

A handbook of suggestions on the teaching of geography : a unesco educational studies publication / Scarfe, N V – 4mf – 7 – mf#4789 – uk Microform Academic [910]

Handbook of suggestions to workers for union evangelistic movements to reach special classes : city evangelism, province-wide work, women's work / Taylor, W E – Shanghai: Assoc Press of China...YMCA of China, 1916 [mf ed 1995] – (= ser Yale coll) – 47p – 1 – 0-524-09738-0 – mf#1995-0738 – us ATLA [240]

Handbook of the american republics / International Bureau Of The American Republics – Washington, DC. 1893 – 1r – us UF Libraries [972]

A handbook of the art of illumination : as practised during the middle ages. with a description of the metals, pigments, and processes employed by the artists at different periods / Shaw, Henry – London: Bell & Daldy, 1866 – (= ser 19th c art & architecture) – 2mf – 9 – mf#4.1.194 – uk Chadwyck [740]

Handbook of the arts of the middle ages and renaissance / Labarte, Jules – London 1855 – (= ser 19th c art & architecture) – 11mf – 9 – mf#4.2.295 – uk Chadwyck [700]

Hand-book of the arya samaj / Sharma, Vishnun Lal – [2nd ed] Allahabad: Tract Dept of the Arya Pratinidhi Sabha, United Provinces, 1912 – 1mf – 9 – 0-524-02548-7 – mf#1990-3043 – us ATLA [280]

Handbook of the china mission – London: London Missionary Society, 1914 [mf ed 1995] – (= ser Yale coll) – 134p – 1 – 0-524-09137-4 – mf#1995-0137 – us ATLA [240]

A handbook of the church of scotland / Rankin, James – 4th rev enl ed. Edinburgh: W Blackwood, 1888 [mf ed 1993] – (= ser Presbyterian coll) – 5mf – 9 – 0-524-07363-5 – mf#1990-5400 – us ATLA [242]

Handbook of the churches / Federal Council of the Churches of Christ in America – 9th [1926/1927] – 1r – 1 – (cont: yearbook of the churches covering the year...; cont by: new handbook of the churches) – mf#3640622 – us WHS [243]

Handbook of the convocations, or, provincial synods of the church of england / Joyce, James Wayland – London: Rivingtons, 1887 – 1mf – 9 – 0-7905-5163-2 – (incl bibl ref) – mf#1988-1163 – us ATLA [241]

Handbook of the divine liturgy : a brief study of the historical development of the mass / Clarke, Charles Cowley – London: Kegan Paul, Trench, Truebner, 1910 – 1mf – 9 – 0-7905-5865-3 – mf#1988-1865 – us ATLA [240]

Handbook of the english presbyterian mission in south formosa : minutes of the tainan mission council jan 10 1877-jan 26 1910 / Campbell, William – Hastings: F J Parsons, 1910 [mf ed 1995] – (= ser Yale coll) – xxx/405p – 1 – 0-524-09753-4 – mf#1995-0753 – us ATLA [242]

A hand-book of the english versions of the bible : with copious examples illustrating the ancestry and relationship of the several versions, and comparative tables / Mombert, Jacob Isidor – New York: A D F Randolph, c1883 [mf ed 1990] – 2mf – 9 – 0-8370-1840-4 – mf#1987-6228 – us ATLA [220]

Handbook of the ferns of british india, ceylon and malaya peninsular / Beddome, R H – 1892 – 15mf – 7 – (with suppl) – mf#371 – uk Microform Academic [580]

A handbook of the general convention of the protestant episcopal church : giving its history and constitution, 1785-1880 / Perry, William Stevens – New York: T Whittaker, 1881 [mf ed 1990] – 1mf – 9 – 0-7905-6612-5 – mf#1988-2612 – us ATLA [242]

Hand-book of the general missionary and tract committee of the german baptist brethren church / ed by Eby, Enoch et al – Elgin, Ill: General Missionary and Tract Committee, 1899 – 1mf – 9 – 0-524-02824-9 – mf#1990-4445 – us ATLA [242]

A hand-book of the history, organization, and methods of work of the young men's christian associations / ed by Ninde, Henry Summerfield et al – New York: International Committee of YMCAs, 1892 [mf ed 1992] – 2mf – 9 – 0-524-03356-0 – mf#1990-0937 – us ATLA [240]

Handbook of the ila language (commonly called the seshukulumbwe) / Smith, Edwin William – London, England. 1907 – 1r – us UF Libraries [470]

Handbook of the irideae / Baker, John Gilbert – London: George Bell 1892 [mf ed Zug: IDC [19–]] – 3mf – 9 – ne IDC [580]

Handbook of the law of evidence / McKelvey, John Jay – St. Paul, West, 1898. 468 p. LL-1482 – 1 – (2nd ed. st. paul, west, 1907. 540p. ll-1277) – us L of C Photodup [347]

A handbook of the law of prote compiled from decisions of american and english courts / Williamson, Edward Hand – Philadelphia, 1889. 173 p. LL-1553 – 1 – us L of C Photodup [347]

Handbook of the law of trusts / Bogert, George Gleason – 2d ed. St. Paul, West, 1942. 738 p. LL-298 – 1 – us L of C Photodup [340]

Handbook of the leeward islands / Watkins, Frederick Henry – London, England. 1924 – 1r – us UF Libraries [972]

A handbook of the life of the apostle paul : an outline for class room and private study / Burton, Ernest DeWitt – 5th ed. Chicago: University of Chicago Press, 1909, c1899 [mf ed 1989] – 1mf – 9 – 0-7905-2403-1 – mf#1987-2403 – us ATLA [920]

Hand-book of the manufactures and arts of the punjab / Baden-Powell, Baden Henry – Lahore, 1872 – (= ser 19th c art & architecture) – 7mf – 9 – mf#4.2.1664 – uk Chadwyck [700]

A handbook of the mende language / Sumner, A T – Freetown: Govt Printing Office, 1917 – 1 – us CRL [490]

Hand-book of the national council of women of canada : containing constitution, standing orders, and other information... / National Council of Women of Canada – [Toronto]: The Council, 1901 – 1mf – 9 – 0-665-98990-3 – mf#98990 – cn Canadiana [360]

Handbook of the new thought / Dresser, Horatio Willis – New York: GP Putnam, 1917 – 1mf – 9 – 0-524-01957-6 – mf#1990-2748 – us ATLA [240]

Handbook of the new york state reformatory at elmira : includes...an abstract of laws relating to the reformatory / Allen, Fred C – Elmira: The Summary Press, 1916 – 4mf – 9 – $6.00 – mf#LLMC 92-141 – us LLMC [340]

Handbook of the northern sotho language / Ziervogel, D – Pretoria, South Africa. 1969 – 1r – us UF Libraries [470]

A handbook of the parochial ecclesiastical law of scotland / Black, William George – Edinburgh: William Green, 1888 [mf ed 1992] – 1mf – 9 – 0-524-03217-3 – mf#1990-0845 – us ATLA [240]

Hand-book of the presbyterian church in canada, 1883 / ed by Kemp, A F et al – Ottawa: J Durie, 1883 – 5mf – 9 – (incl ind) – mf#12156 – cn Canadiana [242]

A handbook of the romish controversy : being a refutation in detail of the creed of pope pius the fourth, on the grounds of scripture and reason / Stanford, Charles Stuart – Dublin: [s.n.], 1852 [mf ed 1986] – 1mf – 9 – 0-8370-8473-3 – mf#1986-2473 – us ATLA [241]

A handbook of the sherbo language / Sumner, A T – London: Publ by the Crown Agents for the Colonies, for the Govt of Sierra Leone, 1921 – 1 – us CRL [490]

Handbook of the speech sounds and sound changes of the bantu – Pretoria, South Africa. 1967 – 1r – us UF Libraries [470]

Handbook of the swahili language / Steere, Edward – London, England. 1884 – 1r – us UF Libraries [470]

Handbook of the swahili language / Steere, Edward – London, England. 1943 – 1r – us UF Libraries [470]

Handbook of the swahili language, as spoken at zanzibar / Steere, Edward – London, England. 1917 – 1r – us UF Libraries [470]

A handbook of the temne language / Sumner, A T – Freetown: Govt Printing Office, 1922 – 1 – us CRL [490]

A handbook of the united brethren in christ / Shuey, Edwin Longstreet – Dayton, OH: United Brethren Pub House, 1885 [mf ed 1991] – (= ser Methodist coll) – 1mf – 9 – 0-524-01335-7 – mf#1990-4084 – us ATLA [242]

Handbook of the venda language / Ziervogel, D – Pretoria, South Africa. 1961 – 1r – us UF Libraries [470]

A handbook of theology : a homiletical manual of christian doctrine / Harries, John – 2nd rev ed. London: CH Kelly, 1903 [mf ed 1991] – 1mf – 9 – 0-7905-9956-2 – (incl bibl ref) – mf#1989-1681 – us ATLA [240]

Handbook of trinidad and tobago – Port-of-Spain, Trinidad and Tobago. 1924 – 1r – us UF Libraries [972]

Handbook of tswana law and custom... / Schapera, Isaac – London, England. 1955 – 1r – us UF Libraries [340]

The handbook of uganda / Wallis, Henry Richard – 2nd ed. London: Publ for the Govt of the Uganda Protectorate by the Crown Agents for the Colonies, 1920 – 1 – us CRL [960]

Handbook of university extension : with an introduction by e.j. james / James, George Francis – 2nd ed. enl. Philadelphia: American Society for the Extension of University Teaching, 1893 – 19+425p – 1 – us Wisconsin U Libr [378]

The hand-book of water-colours / Winsor and Newton, Ltd – London [1843?] – (= ser 19th c art & architecture) – 1mf – 9 – mf#4.1.258 – uk Chadwyck [750]

Handbook of work with student enquirers in india : a symposium / ed by Walter, H A – Calcutta: Assoc Press...YMCA [1915] [mf ed 1995] – (= ser Yale coll) – 75p – 1 – 0-524-09218-4 – mf#1995-0218 – us ATLA [240]

Handbook of zoology : with examples from canadian species, recent and fossil / Dawson, John William – Montreal: Dawson, 1886 – 4mf – 9 – mf#27030 – cn Canadiana [590]

Handbook on american mining law / Costigan, George Purcell – St. Paul, West, 1908. 765 p. LL-238 – 1 – us L of C Photodup [343]

Handbook on american yacht racing rules : an explanation of their meaning and application / Parsons, H de B et al – 2nd ed. New York: Yachting Inc 1923 – 3mf – 9 – $4.50 – mf#llmc92-230 – us LLMC [790]

Handbook on baptism : or, testimonies of learned pedobaptists on the action and subjects of christian baptism. and of both baptists and pedobaptists on the design thereof / ed by Shepherd, James Walton – Nashville, TN: Gospel Advocate Pub Co, 1894 – 2mf – 9 – 0-524-02572-X – (incl bibl ref) – mf#1990-4384 – us ATLA [242]

Handbook on cyrenaica : pt 2: prehistory of libya / Myers, O H – Np, nd – 1mf – 8 – mf#A-632 – ne IDC [956]

Hand-book on hand puppets / Lynch, Dorothea Thomas – Jacksonville, FL. 193-? – 1r – us UF Libraries [978]

Handbook on jury use in the federal courts / George, Jody et al – Washington: GPO, 1989 – 2mf – 9 – $3.00 – mf#LLMC 95-345 – us LLMC [347]

A handbook on old high german literature / Bostock, John Knight – Oxford: Clarendon Press, 1955 [mf ed 1993] – viii/257p – 1 – (incl bibl ref) – mf#8166 – us Wisconsin U Libr [430]

Handbook on pastoral theology *see* Mu fan hsueh (ccm244)

Handbook on the construction and interpretation of the laws. / Black, Henry Campbell – St Paul, West, 1896. 499 p. LL-853 – 1 – us L of C Photodup [340]

Handbook on the law of damages / Hale, William Benjamin – 2d ed. St. Paul, West, 1912. 632 p. LL-390 – 1 – us L of C Photodup [346]

Handbook on the law of damages / McCormick, Charles Tilford – St. Paul, West, 1935. 811 p. LL-345 – 1 – us L of C Photodup [346]

Handbook on the law of partnership, including limited partnerships / Gilmore, Eugene Allen – St. Paul: West, 1911. 721p. LL-575 – 1 – us L of C Photodup [346]

Handbook on the law of persons and domestic relations. 2d ed / Tiffany, Walter Checkley – St. Paul, West, 1909. 656 p. LL-1248 – 1 – us L of C Photodup [340]

Handbook on the organisations dealing with rural development in the bombay state – [s.l: s.n.], 1954 – 1 – (filmed with: pakistan press yearbook) – us CRL [954]

Handbook to a collection of the minerals of the british isles:...from the ludlam collection.. / London. Museum of Practical Geology – By F.R. Rudler. London: Wyman & Sons, 1905. 241p. Includes index – 1 – us Wisconsin U Libr [550]

A handbook to agra and the taj, sikandra, fatehpur-sikri, and the neighbourhood / Havell, Ernest Binfield – London, New York: Longmans, Green and Co, 1904 – (= ser Samp: indian books) – (with ill and plans) – us CRL [915]

Hand-book to british columbia – Victoria, BC: Begg & Hoare, [1894-189- or 19–] [mf ed aug/sep 1894] – 9 – mf#P04496 – cn Canadiana [917]

A hand-book to coffee planting in southern india / Shortt, John – Madras 1864 – (= ser 19th c british colonization) – 3mf – 9 – mf#1.1.5018 – uk Chadwyck [630]

A handbook to kant's critique of pure reason / Das, Rashvihari – Bombay: Hind Kitabs, 1949 – (= ser Samp: indian books) – us CRL [140]

Handbook to king solomon's temple : containing an explanatory key and an account of the building of the model now on exhibition in this city – New York: C A Alvord, 1860 – 1mf – 9 – 0-8370-3462-0 – mf#1985-1462 – us ATLA [720]

A handbook to old testament hebrew : containing an elementary grammar of the language / ed by Green, Samuel Gosnell – New York: Fleming H Revell; London: Religious Tract Society [1901] [mf ed 1986] – 1mf – 9 – 0-8370-9148-9 – (incl ind) – mf#1986-3148 – us ATLA [470]

A handbook to the bible : being a guide to the study of the holy scriptures / Conder, Francis Roubiliac & Conder, Claude Reignier – 5th ed. London: Longmans, Green, 1890 [mf ed 1985] – 2mf – 9 – 0-8370-2720-9 – (incl ind) – mf#1985-0720 – us ATLA [220]

Handbook to the cathedrals of england / King, Richard John – London 1861 – (= ser 19th c art & architecture) – 29mf – 9 – mf#4.1.417 – uk Chadwyck [720]

Handbook to the cathedrals of wales / King, Richard John – London 1873 – (= ser 19th c art & architecture) – 5mf – 9 – mf#4.2.1559 – uk Chadwyck [720]

Handbook to the controversy with rome = Handbuch der protestantischen polemik gegen die roemisch-katholische kirche / Hase, Karl von; ed by Streane, Annesley William – London: Religious Tract Society, 1906 – 4mf – 9 – 0-8370-8114-9 – (incl bibl ref and indexes. in english) – mf#1986-2114 – us ATLA [240]

A hand-book to the courts of modern sculpture / Jameson, Anna Brownell [Murphy] – London 1854 – (= ser 19th c art & architecture) – 1mf – 9 – mf#4.2.478 – uk Chadwyck [730]

Handbook to the department of prints and drawings in the british museum / Fagan, Louis Alexander – London 1876 – (= ser 19th c art & architecture) – 3mf – 9 – mf#4.1.349 – uk Chadwyck [700]

Handbook to the 'grammar of luvale' / Horton, A E – s.l, s.l? 19–? – 1r – us UF Libraries [400]

Handbook to the grammar of the greek testament : together with a complete vocabulary, and an examination of the chief new testament synonyms / Green, Samuel Gosnell – Rev impr ed. [London]: Religious Tract Society, [ca 1885] – 2mf – 9 – 0-7905-0034-5 – (incl bibl ref and indexes) – mf#1987-0034 – us ATLA [450]

Handbook to the jeypore museum / Hendley, Thomas Holbein – Calcutta 1895 – (= ser 19th c art & architecture) – 2mf – 9 – mf#4.2.1313 – uk Chadwyck [060]

A handbook to the museum of ornamental art / Waring, John Burley – London 1857 – (= ser 19th c art & architecture) – 1mf – 9 – mf#4.2.218 – uk Chadwyck [740]

A handbook to the public galleries of art in and near london / Jameson, Anna Brownell [Murphy] – London 1842 – (= ser 19th c art & architecture) – 8mf – 9 – mf#4.2.106 – uk Chadwyck [700]

Hand-book to the public records / Great Britain. Public Record Office – By F.S. Thomas, Secretary of the Public Record Office. London, 1853. lxii, 482p. fold. col. pl – 1 – us Wisconsin U Libr [324]

Handbook to the sculptures in the museum of the bangiya sahitya parishad / Ganguly, Manomohan – Calcutta: Bangiya Sahitya Parishad, 1922 – (= ser Samp: indian books) – us CRL [730]

Handbook to the textual criticism of the new testament / Kenyon, Frederic George, Sir – 2nd ed. London: Macmillan, 1912 – 2mf – 9 – 0-8370-9394-5 – mf#1986-3394 – us ATLA [225]

A handbook to the works of william shakespeare / Luce, Morton – London: G Bell & Sons, 1924, c1906 [mf ed 1985] – x/463p – 1 – (incl bibl) – mf#8827 – us Wisconsin U Libr [420]

Handbuch bey dem generalbasse und der composition : mit zwo- drey-vier- fuenf- sechs- sieben- acht und mehrern stimmen: fuer anfaenger und geuebtere / Marpurg, Friedrich Wilhelm – Berlin: Verlegts Gottlieb August Lange 1757-62 [mf ed 19–] – 4pt on 9mf – 9 – (in 4 pt: pt [1]: 2. verb verm aufl [1762]; pt 2: 1757. pt 3: 3. & letzter theil, nebst einem hauptregister ueber alle drey theile [1758]; pt 4: anhang zum handbuche...1760) – mf#fiche 427 – us Sibley [780]

Handbuch der altorientalischen geisteskultur / Jeremias, Alfred – Leipzig: J C Hinrichs, 1913 – 1mf – 9 – 0-7905-1769-8 – (incl bibl ref and indexes) – mf#1987-1769 – us ATLA [930]

Handbuch der alttestamentlichen theologie / Dillmann, August – Leipzig: S Hirzel, 1895 – 2mf – 9 – 0-8370-2917-1 – (incl indes) – mf#1985-0917 – us ATLA [220]

Handbuch der architektur : stadtebau / ed by Durm, Josef et al – Darmstadt: A Bergstraesser 1887-1939 [4:1-4:10, 2:1-2:5 mf ed 1979] – 10r [ill] – 1 – mf#film mas c690 – us Harvard [720]

Handbuch der babylonischen astronomie / Weidner, E F – Leipzig, 1915 – (= ser Assyriologische bibliothek) – 3mf – 9 – (assyriologische bibliothek v23) – mf#NE-411 – ne IDC [956]

Handbuch der biblischen geschichte und literatur : nach den ergebnissen der heutigen wissenschaft / Langhans, Eduard – Bern: I Dalp, 1875-1880 – 2mf – 9 – 0-524-07337-6 – mf#1992-1068 – us ATLA [220]

Handbuch der buerglichen kunstaltertuemer / Bergner, H – Leipzig. v1-2. 1906 – €32.00 – ne Slangenburg [720]

Handbuch der christ-katholischen religion : zum selbstunterricht / Weninger, Francis Xavier – Cincinnati, O[hio]: Kreuzburg und Nurre, 1858 – 1mf – 9 – 0-8370-6858-4 – mf#1986-0858 – us ATLA [241]

Handbuch der christlichen archaeologie / Kaufmann, Carl Maria – Paderborn: F. Schoeningh, 1905 – (= ser Wissenschaftliche Handbibliothek) – 2mf – 9 – 0-7905-8110-8 – mf#1988-6072 – us ATLA [930]

Handbuch der christlichen kirchen- und dogmen-geschichte fuer prediger und studirende / Ebrard, Johannes Heinrich August – Erlangen: A. Deichert, 1865-67. Chicago: Dep of Photodup, U of Chicago Lib, 1978 (1r); Evanston: American Theol Lib Assoc, 1984 (1r) – 1 – 0-8370-0759-3 – (incl bibl ref and index) – mf#1984-T113 – us ATLA [240]

Handbuch der deutschen freimaurerei / Ewald, Franz [comp] – Muenchen: Rudolf Abt 1899- – v1-2 on 2mf – 9 – (filmed: t1-t2) – mf#vrl-188 – ne IDC [366]

Handbuch der deutschen literaturgeschichte / Bernt, Alois – Reichenberg in Boehmen: Gebrueder Stiepel GmbH, 1928 [mf ed 1993] – viii/816p (ill) – 1 – (incl ind) – mf#8072 – us Wisconsin U Libr [430]

Handbuch der deutschen mythologie : mit einschluss der nordischen / Simrock, Karl Joseph – 6. durchgesehene Aufl. Bonn: A Marcus, 1887 – 2mf – 9 – 0-524-02323-9 – mf#1990-2946 – us ATLA [290]

Handbuch der dogmatik der evangelisch-lutherischen kirche : oder, versuch einer beurtheilenden darstellung [sic] der grundsaetze, welche diese kirche in ihren symbolischen schriften ueber die christliche glaubenslehre ausgesprochen hat / Bretschneider, Karl Gottlieb – 3. verb und verm Aufl. Leipzig: JA Barth, 1828 – 4mf – 9 – 0-524-05646-3 – mf#1991-2315 – us ATLA [242]

Handbuch der dogmengeschichte / Bertholdt, Leonhard – Erlangen: Palm und Enke, 1822-1823 – 2mf – 9 – 0-524-00005-0 – mf#1989-2705 – us ATLA [240]

Handbuch der duala-sprache / Christaller, Th. – Basel: Verlag der Missionsbuchh, 1892 – us CRL [410]

Handbuch der ebraeischen mythologie : sage und glaube der alten ebraer in ihrem zusammenhang mit den religioesen anschauungen anderer semiten... / Schultze, Martin – Nordhausen: Ferd Foerstemann, 1876 – 1mf – 9 – 0-8370-5165-7 – (incl bibl ref & index) – mf#1985-3165 – us ATLA [270]

Handbuch der englischen und deutschen umgangssprache / Hamilton, Louis – Berlin, Germany. 1935 – 1r – us UF Libraries [400]

Handbuch der evangelisch-lutherischen synode von ohio und anderen staaten = manual of the evangelical lutheran joint synod of ohio and other states / Boehme, E A – Columbus, Ohio: Lutheran Book Concern, 1910 – 1mf – 9 – 0-524-00967-8 – mf#1990-4025 – us ATLA [242]

Handbuch der frauenbewegung / ed by Lange, Helene & Baeumer, Gertrud – Berlin 1901-06 [mf ed 1996] – (= ser Hq 26) – 5pt on 24mf – 9 – €120.00 – 3-89131-138-9 – (pt1: die geschichte der frauenbewegung in den kulturlaendern [berlin 1901]; pt2: frauenbewegung und soziale frauenthaetigkeit in deutschland nach einzelgebieten [berlin 1901]; pt3: der stand der frauenbildung in den kulturlaendern [berlin 1902]; pt4: die deutsche frau im beruf [berlin 1902]; pt5: die deutsche frau im beruf. praktische ratschlaege zur berufswahl [berlin 1906]) – gw Fischer [305]

Handbuch der geographie und statistik fuer die gebildeten staende / Stein, Christian G [Begr] – Leipzig 1855-68 [mf ed Hildesheim 1995-98] – 10v on 55mf – 9 – €550.00 – 3-487-29901-1 – gw Olms [900]

Handbuch der geschichte des franziskanerordens / Holzapfel, Heribert – Freiburg im Breisgau; St. Louis, Mo.: Herder, 1909 – 2mf – 9 – 0-7905-8106-X – (incl bibl ref) – mf#1988-6068 – us ATLA [241]

Handbuch der historisch-kritischen einleitung in das alte testament *see* Historico-critical introduction to the pentateuch

Handbuch der katholischen dogmatik *see* A manual of catholic theology

Handbuch der katholischen liturgik / Thalhofer, Valentin – Freiburg i B: Herder, 1883-[1893?] [mf ed 1992] – (= ser Theologische bibliothek) – 2v on 3mf – 9 – 0-524-03943-7 – (incl bibl ref) – mf#1990-4937 – us ATLA [241]

Handbuch der kirchengeschichte / Schmid, Heinrich – Erlangen: A. Deichert, 1880 – 2mf – 9 – 0-7905-8072-1 – (incl bibl ref) – mf#1988-6053 – us ATLA [240]

Handbuch der kirchengeschichte / Tischhauser, Christian – Basel: C Detloff, 1887 – 2mf – 9 – 0-7905-8159-0 – mf#1988-6106 – us ATLA [240]

Handbuch der kirchengeschichte *see* Manual of church history

Handbuch der land-bau-kunst : vorzueglich in ruecksicht auf die construction der wohn- und wirtschaftgebaeude / Gilly, D – Berlin. 3v. 1797-1811 – 15mf – 9 – mf#OA-104 – ne IDC [710]

Handbuch der missionsgeschichte und missionsgeographie / Blumhardt, Johann Christoph – 3., ganz neue Ausg. Calw: Vereinsbuchh; Stuttgart: J F Steinkopf. 2v. 1863 – 4mf – 9 – 0-8370-7287-5 – (incl indes) – mf#1986-1287 – us ATLA [240]

Handbuch der musikalischen literatur : oder allegemeines systematisch geordnetes verzeichnis etc / ed by Whistling, Carl Friedrich & Hofmeister, A – repr Leipzig. 27v. 1817-1900 – 11 – $395.00 set – (the handbuch was issued in 3 eds with component suppl issues. provides info about first eds, instrumentation and arrangements of the music of a great many composers) – us Univ Music [780]

Handbuch der neuesten geographie des oesterreichischen kaiserstaates / Liechtenstern, Joseph M von – Wien 1817-18 [mf ed Hildesheim 1995-98] – 3v on 12mf – 9 – €120.00 – 3-487-29453-2 – gw Olms [914]

Handbuch der pfingstbewegung / Hollenweger, Walter J – Genf, 1965-1967. Chicago, Dept. of Photodup, U of Chicago Lib, 1968 (3r); Evanston: American Theol Lib Assoc, 1984 (3r) – 1 – 0-8370-0659-7 – (includes bibliographies) – mf#1984-S143 – us ATLA [240]

Handbuch der religionsgeschichte / Wurm, Paul – 2. verm und verb Aufl. Calw: Verlag der Vereinsbuchh, 1908 – 2mf – 9 – 0-524-01995-9 – mf#1990-2786 – us ATLA [200]

Handbuch der theologischen literatur : hauptsaechlich der protestantischen / Winer, Georg Benedikt – 3. sehr erw Aufl. Leipzig: CH Reclam, 1838-1842 – 3mf – 9 – 0-524-08853-5 – (incl bibl ref and ind) – mf#1993-2138 – us ATLA [012]

Handbuch der wichtigsten grundzeichen der chinesischen und sinokoreanischen schrift / Schmidt, Wolfgang G A – (mf ed 1997) – 3mf – 9 – €49.00 – 3-8267-2467-4 – mf#DHS 2467 – gw Frankfurter [480]

Handbuch des deutschen privatrechts / Stobbe, Otto – 2. Aufl. Berlin. v1-5. 1882-85 – (= ser Civil law 3 coll) – 31mf – 9 – (vols 4-5; 1. und 2. aufl. incl bibl ref and index) – mf#LLMC 96-551 – us LLMC [346]

Handbuch des franzoesischen civilrechts / Zachariae von Lingenthal, Karl Salomo; ed by Crome, Carl – 8., verm u verb Aufl. Freiburg (Breisgau): E Mohr. v1-4. 1894-95 – (= ser Civil law 3 coll) – 33mf – 9 – mf#LLMC 96-277 – us LLMC [346]

Handbuch des organisten / Schneider, Friedrich – Halberstadt. 1830. 4v – 1 – us L of C Photodup [780]

Handbuch des personalen gelegenheitsschrifttums in europaeischen bibliotheken und archiven : die personalschriften der ehemaligen breslauer stadtbibliothek / ed by Garber, Klaus – [mf ed Hildesheim 2002] – ca 5000mf – diazo €9900.00 – (nach institutionen gegliedert: breslau: die personalschriften der ehemaligen breslauer stadtbibliothek in der biblioteka uniwersytecka w wroclawiu [mf ed 2000-03 ca 950mf €3980. akademie-bibliothek danzig ca 475mf. universitaetsbibliothek koenigsberg ca 50mf. pommersche bibliothek stettin, wojewodschafts-archiv stettin zusammen ca 330mf. universitaetsbibliothek greifswald, stadtarchiv und landesarchiv greifswald zusammen ca 475mf. nationalbibliothek warschau ca 600mf. universitaetsbibliothek thorn, wojewodschafts-bibliothek thorn zusammen [mf ed 2002] 514mf [€1480]. akademiebibliothek riga, nationalbibliothek riga, historisches staatsarchiv riga zusammen 230mf. akademiebibliothek tallinn, nationalbibliothek, geschichtsmuseum und stadtarchiv tallinn zusammen ca 100mf. universitaetsbibliothek tartu, estnisches literaturmuseum, estnisches historisches archiv zusammen [mf ed 2003] 48mf. akademiebibliothek vilnius, nationalbibliothek vilnius, universitaetsbibliothek vilnius zusammen ca 500mf. akademiebibliothek st petersburg, nationalbibliothek st petersburg zusammen ca 520mf) – gw Olms [430]

Handbuch fuer buecherfreunde und bibliothekare (ael1/4) / Lawaetz, Heinrich Wilhelm – Halle 1788-94 [mf ed 1992] – (= ser Archiv der europaeischen lexikographie, abt 1: enzyklopaedien) – 11v on 42mf – 9 – €240.00 – 3-89131-056-0 – gw Fischer [020]

Handbuch fuer das deutsche reiche auf das jahr... / ed by Germany. Reichskanzler-Amt – Berlin: C Heymann. 46v (annual) [1877-1936 mf ed 1980] – 9r – 1 – (began in 1874; ceased in 1936; iss for 1874-94 lack numerical designation) – mf#film mas c729 – us Harvard [943]

Handbuch fuer gebildete reisende durch suedfrankreich, die schweiz, italien und griechenland bis corfuin : zwei theilen – Stuttgart 1839 [mf ed Hildesheim 1995-98] – 2v on 5mf – 9 – €100.00 – 3-487-27756-5 – gw Olms [914]

Handbuch fuer harzreisende / Niemann, F – Halberstadt 1824 [mf ed Hildesheim 1995-98] – 2mf – 9 – €60.00 – 3-487-29514-8 – gw Olms [914]

Handbuch fuer reisende am rhein von seinen quellen bis holland, in die schoensten anliegenden gegenden und an die dortigen heilquellen / Schreiber, Aloys – Heidelberg [1831] [mf ed Hildesheim 1995-98] – 4mf – 9 – €120.00 – 3-487-29546-6 – gw Olms [914]

Handbuch fuer reisende im koenigreiche daenemark und in den herzogthuemern schleswig, holstein, lauenburg / Tregder, Eiler – Kopenhagen 1824 [mf ed Hildesheim 1995-98] – 4mf – 9 – €120.00 – 3-487-28950-4 – gw Olms [914]

Handbuch fuer reisende in der schweiz / Glutz-Blotzheim, Robert – Zuerich 1818 [mf ed Hildesheim 1995-98] – 4mf – 9 – €120.00 – 3-487-29393-5 – gw Olms [914]

Handbuch fuer reisende in italien / Foerster, Ernst – Muenchen 1866 [mf ed Hildesheim 1995-98] – 2v on 8mf – 9 – €160.00 – 3-487-29252-1 – gw Olms [914]

Handbuch fur die bundesrepublik deutschland – Koln, Germany. v1-2. 1953 – 1r – us UF Libraries [025]

Handbuch zum alten testament / Erbt, Wilhelm – Osterwieck (Harz): A W Zickfeldt, 1909 – 1mf – 9 – 0-7905-1381-1 – (incl bibl ref) – mf#1987-1381 – us ATLA [221]

Handbuch zur deutschen literarturgeschichte / Petry, Karl – Koeln: B Pick, 1949 – 1r – 1 – (incl bibl ref and index) – us Wisconsin U Libr [430]

Handbuecher und lexika zur militaergeschichte : militaerhistorische nachschlagwerke aus dem 19. und fruehen 20. jahrhundert – [mf ed 1997] – 198mf (1:24) – 9 – diazo €2150.00 (silver €2409 isbn: 978-3-598-33871-7) – ISBN-10: 3-598-33870-8 – ISBN-13: 978-3-598-33870-0 – gw Saur [355]

Handcock, Percy Stuart Peache *see* The latest light on bible lands

Hand-commentar zum neuen testament / Holtzmann, Heinrich Julius et al – Freiburg i B: J C B Mohr, 1889-1891 – 5mf – 9 – 0-8370-1671-1 – (incl bibl ref) – mf#1987-6101 – us ATLA [225]

Hande – Istanbul: Ahmed Saki Bey Matbaasi, 1910. Mueduerue: Hueseyin Nazmi; Sermuharriri: Cevdet Ma'suki. n3. 5 nisan 1326 [1910] – (= ser O & t journals) – 1mf – 9 – $25.00 – us MEDOC [956]

Hande – Istanbul: Karabet Matbaasi, 1916-17. Cikaran: Feridun Fahri, Sahib-i Imtiyaz: Sedat Simavi, Mueduerue: Yakup Aziz. n2,3,14. 28 temmuz-20 tesrinievvel 1332 [1916] – (= ser O & t journals) – 1mf – 9 – $25.00 – us MEDOC [956]

Handel, George Frederic *see*
- New edition of six concertos, for the harpsichord or organ
- Original manuscripts and sketches
- Samson
- Second set of six concertos
- Six concertos
- Te deum [for choir, soloists, and orchestra]
- Vocal works of handel
- Werke

Handel og samfaerdsel i oldtiden / ed by Brondsted, J – Stockholm: A Bonnier 1943 [mf ed Bloomington IN: Indiana Uni Lib, Preservation Dept 1984] – 360p on 1r [ill] – 1 – us Indiana Preservation [390]

Handel und wandel / Hacklaender, Friedrich Wilhelm – Philadelphia: Philadelphia Demokrat Publishing, [18–?] [mf ed 1993] – vi/468p (ill) – 1 – mf#8668 – us Wisconsin U Libr [430]

Handel zagraniczny – Warsaw. 1971-1980 (1) 1973-1980 (5) 1976-1980 (9) – ISSN: 0017-7245 – mf#6254 – us UMI ProQuest [337]

De handelingen der apostelen / Manen, Willem Christiaan van – Leiden: E J Brill, 1890 [mf ed 1989] – 204p on 1mf – 9 – 0-7905-1456-7 – (in dutch and greek. incl ind) – mf#1987-1456 – us ATLA [226]

Handelingen over de reglementen op het... / Bordewijk, Hugo Willem Constantijn – s'-Gravenhage, Netherlands. 1914 – 1r – us UF Libraries [972]

Handelingen van de kerkeraad der nederl gemeente te keulen 1571-1591 (de werken...1/3) / ed by Janssen, H Q & Toorenbergen, J J van – Utrecht, 1881 – (= ser De werken der marnix-vereeniging) – €15.00 – ne Slangenburg [242]

Handelmann, Heinrich *see*
- Geschichte der insel hayti
- Historia do brasil

Handels- und Gewerbekammer Ober-Oesterreichs [comp] *see* Statistische daten betreffend die volkswirthschaftlichen zustaende ober-oesterreichs

Handels- und gewerbezeitung – Sombor (YU), 1941 5 jan-12 oct – 1r – 1 – gw Misc Inst; us L of C Photodup [380]

Het handelsblad van antwerpen – Antwerp Belgium, 11 feb-7 aug 1919; 29 oct 1944-30 jul 1945 – 2r – 1 – uk British Libr Newspaper [074]

Handelsblatt – Duesseldorf DE, 1946 16 may-1966 – 62r – 1 – (1967-78 [70r]; 1979-84 [45r]; 1985-91 [63r]; 1992-94 [30r]; 1995-97 [30r]; 1998 subs) – mf#1324 – gw Mikropress [380]

Handelsblatt – Duesseldorf, 1946- – 10r per y – 1 – enquire for prices – us UMI ProQuest [074]

Handelsblatt – Duesseldorf, Germany. -d. 16 May 1946-15 July 1949; 4 Jan-30 Dec 1950; 22 Jan 1951-31 Dec 1957; 1 April 1964-31 Dec 1968. 92 reels – 1 – uk British Libr Newspaper [072]

Handelsblatt *see* Industriekurier

Handelsblatt-magazin – Duesseldorf, 22 oct 1982-25 jun 1993 – 7r – 1 – mf#12403 – gw Mikropress [380]

Das handelsmuseum / Oesterreichisches Handelsmuseum – Vienna. v20-40 1905-1925. Incomplete – 1 – us NY Public [060]

Die handelspolitik polens seit erlangung der selbstaendigkeit bis zum ablauf der genfer konvention am 15. juni 1925 / Braeutigam, Harald – Berlin, 1926 (mf ed 1993) – 2mf – 9 – €31.00 – 3-89349-328-X – mf#DHS-AR 182 – gw Frankfurter [327]

Handelspolitischen beziehungen zwischen england und deutschland / Marcus, Jacob Rader – Berlin, Germany. 1925 – 1r – us UF Libraries [025]

Handelswoche – Berlin DE, 1958-1990 23 jul – 20r – 1 – gw Misc Inst [380]

Handford, T W *see* Conversion

Handforth, st chad: index to burials/ marriages – (= ser Cheshire church registers) – 1mf – 9 – £2.50 – mf#383 – uk CheshireFHS [929]

Das handgranatenkoepfl : bilder und geschichten von der eismeerfront / Weinberger, Andreas – Muenchen: F Eher 1944 [mf ed 1991] – (= ser Soldaten-kameraden 67) – 1r – 1 – (filmed with: sieben vor verdun / josef magnus wehner) – mf#2981p – us Wisconsin U Libr [830]

Handgun body count – 1981 mar-1982 jan – 1r – 1 – mf#657348 – us WHS [360]

Handing over notes / Wilson, J B – Swaziland: [s.n.], 1965 – us CRL [960]

Handl (Gallus), Jacob *see*
- Moralia 5, 6 et 8 vocibus
- Quartus tomus musici operis
- Secundus tomus musici operis
- Tertius tomus musici operis
- Tomus primus operis musici...

Handl, Willi *see* Hermann bahr

Handledning vid undervisningen uti augustana-synodens foersamlingsskolor / ed by Rabenius, Karl Nathanael – Rock Island IL: Lutheran Augustana Book Concern c1899 [mf ed 1992] – 1mf – 9 – 0-524-05265-4 – mf#1991-2257 – us ATLA [242]

Handleiding der patrologie / Benvenutus van Venraai – 's-Hertogenbosch, 1912 – 8mf – 8 – €17.00 – ne Slangenburg [240]

Handleiding tot de kerkgeschiedvorsching en kerkgeschiedschrijving / Acquoy, Johannes Gerhardus Rijk – Tweede herziene en vermeerderde druk. 's-Gravenhage: Nijhoff, 1910 – 1mf – 9 – 0-7905-5442-9 – (incl bibl ref) – mf#1988-1442 – us ATLA [240]

Handleiding voor de maleische taal / Dutch East Indies. Koninklijk Nederlandsch-Indisch Leger – South Melbourne: Netherlands Indies Govt Print Works 1945 [mf ed 1990] – 1 – (in indonesian & dutch) – mf#mf-10289 seam reel 208/5 [§] – us CRL [490]

Handleiding voor de oudchristelijke letterkunde / Manen, Willem Christiaan van – Leiden: L van Nifterik, 1900 – 1mf – 9 – 0-524-06846-1 – mf#1992-0988 – us ATLA [240]

Handleitung zur variation : wie man den general-bass, und darueber gesetzte zahlen variiren, artige inventiones machen und aus einen schlechten, general-bass praeludia, ciaconen...dergleichen leichtlich verfertigen koenne samt andern noetigen instructionen / Niedt, Friederich Erhardt – Hamburg: B Schillern 1706 [mf ed 19–] – 1r / 4mf – 9 – mf#film 1229 / fiche 630 – us Sibley [780]

Handler, Milton *see* Cases and materials on the law of vendor and purchase

Handler, Rudolf *see* Jubilaris emlekmu

Handley, Hubert *see*
- A declaration on biblical criticism by 1725 clergy of the anglican communion
- The fatal opulence of bishops

Handley, Hubert et al *see* Anglican liberalism

Handling and shipping – Cleveland. 1959-1978 (1) 1971-1978 (5) 1977-1978 (9) – (cont by: handling and shipping management) – ISSN: 0017-7385 – mf#1401 – us UMI ProQuest [380]

Handling and shipping management – Cleveland. 1978-1987 (1) 1978-1987 (5) 1978-1987 (9) – (cont by: transportation and distribution. cont: handling and shipping) – ISSN: 0194-603X – mf#1401,01 – us UMI ProQuest [380]

The handling of prisoners of war during the korean war / U.S. Army. Army, Pacific – 1960 – 1 – us L of C Photodup [951]

Handlist of the dance collection / Hoop, Loes de & Pliester, Freek [comp] – 129 titles on 343mf – 9 – (coll incl works on dance history & aesthetics, dance methods & dance music in printed and mss form. dance forms incl french & italian renaissance dance [before 1630], the english country dance, the baroque dance [after 1630] and the contredance [before 1820]) – ne IDC [790]

Handlung oder acta gehaltner disputatio vnd gespraech zu zoffingen... / ed by Bullinger, H – Zuerich, 1532 – 4mf – 9 – mf#PBU-692 – ne IDC [240]

Handlung und dichtung der buehnenwerke richard wagners nach ihren grundlagen in sage und geschichte / Pfordten, Hermann Ludwig, Freiherr von der# – Berlin: Trowitzsch 1893 – 5mf – 9 – mf#wa-127 – ne IDC [790]

Handlungen und abhandlungen / Borchardt, Rudolf – Berlin-Grunewald: Horen-Verlag, 1928 [mf ed 1989] – 279p – 1 – mf#7052 – us Wisconsin U Libr [430]

Handlungsbibliothek / ed by Buesch, Joh Georg et al – Hamburg 1784-97 – (= ser Dz) – 3v[=4pt] on 13mf – 9 – €130.00 – mf#k/n2737 – gw Olms [330]

Handlungskontrolle und selbstkonzept(e) von hochleistungssportlern im roll- und eiskunstlauf in trainings- und wettkampfsituationen / Barkhoff, Harald – (mf ed 2000) – 2mf – 9 – €40.00 – 3-8267-2712-6 – mf#DHS 2712 – gw Frankfurter [790]

Die handlungstheoretische begruendung der identitaet : eine studie zur identitaetstheorie von georg herbert mead, erving goffman, thomas luckmann und juergen habermas / Vuletic, Andelko – (mf ed 2000) – 4mf – 9 – €56.00 – 3-8267-2749-5 – mf#DHS 2749 – gw Frankfurter [140]

Handlungs-zeitung : oder woechentliche nachrichten von dem handel, dem manufakturwesen und der oekonomie [sp:...von handel, manufakturwesen, kuensten und neuen erfindungen] / ed by Hildt, Johann Adolph [jg 7ff] – Gotha 1784-99 – (= ser Dz) – 49mf – 9 – €490.00 – mf#k/n2738 – gw Olms [380]

The handmaiden of the lord : or, wayside sketches / Cooke, Sarah A – Chicago: Arnold, 1896. El Segundo, CA: Micro Publication Systems, 1981 (2mf); Evanston: American Theol Lib Assoc, 1984 (1mf) – (= ser Women & the church in america) – 9 – 0-8370-1411-5 – mf#1984-2149 – us ATLA [240]

Handmann, R *see* Das hebraer-evangelium

Handmann, Richard *see*
- Die evangelisch-lutherische tamulen-mission in der zeit ihrer neubegruendung
- Ueberblick ueber das gebiet der ev.-luther. mission im tamulenlande
- Umschau auf dem gebiete der evangelisch-lutherischen mission in ostindien

Handmann, Rudolf *see* Das hebraeer-evangelium

Handrail assisted versus nonhandrail assisted stairmaster gauntlet ergometry / Gerhards, Marty D & Butts, Nancy Kay – 1991 – 1mf – $4.00 – us Kinesology [613]

Handrail support versus free arm wing treadmill fitness test / Dremsa, Catherine J & Hunter, Gary – 1986 – 1mf – 9 – $4.00 – us Kinesology [612]

Hands to save the soil – Washington, DC: USGPO, [1938?] – us CRL [630]

Hands, William *see* A practical treatise on fines and recoveries in the court of common pleas; with precedents

Handschin, Charles Hart *see*
- Ekkehard
- Die steinklopfer

Handschrift h 4 : gedateerd 1588 – bibl st adelbertsabdij, egmond – (= ser Werken van hendrik mande, jan van ruusbroec, meister eckhart, etc) – 18mf – 8 – €35.00 – ne Slangenburg [230]

Die handschriften, ausgaben und uebersetzungen von iamblichos de mysteriis : eine kritisch-historische studie / Sicherl, M – Berlin, 1957 – (= ser Tugal 5-62) – 5mf – 9 – €12.00 – ne Slangenburg [240]

Die handschriften der universitaetsbibliothek muenchen [hum] : gesamtedition der deutschen mittelalterlichen, der lateinischen mittelalterlichen und der musikhandschriften = Manuscripts of the munich university library – [mf ed 1995] – 3 sects on 2830mf – 9 – €24,550 diazo €29,460 silver – 3-89131-200-8 – (indiviual sects also listed separately) – gw Fischer [090]

Handschriftenproben des sechzehnten jahrhunderts nach strassburger originalen / ed by Ficker, J & Winckelmann, O – Strassburg, 1902-05. 2v – 10mf – 9 – mf#PPE-131 – ne IDC [240]

Handschriftliche briefe / Radek, Karl – 1 – gw Mikropress [400]

Die handschriftliche ueberlieferung der sogenannten historia tripartita des epiphanius-cassidor / Jacob, W – Berlin, 1954 – (= ser Tugal 5-59) – 4mf – 9 – €11.00 – ne Slangenburg [240]

Die handschriftliche ueberlieferung der zacharias- und johannes-apokryphen / Berendts, Alexander – Leipzig, 1904 – (= ser Tugal 2-26/3a) – 2mf – 9 – €5.00 – ne Slangenburg [221]

Die handschriftliche ueberlieferung des epiphanius / Holl, Karl – Leipzig, 1910 – (= ser Tugal 3-36/2) – 2mf – 9 – €5.00 – ne Slangenburg [240]

Die handschriftliche ueberlieferung des epiphanius (ancoratus und panarion) / Holl, Karl – Leipzig: J C Hinrichs, 1910 – (= ser Tugal) – 1mf – 9 – 0-7905-1897-X – mf#1987-1897 – us ATLA [240]

Handsomest and best bred trotting stallion in canada, bookmaker (standard) : a t r no 4392 – Woodstock, Ont?: s.n, 1888? – 1mf – 9 – mf#56840 – cn Canadiana [636]

Hands-on electronics – New York. 1984-1989 (1,5,9) – (cont by: popular electronics) – ISSN: 0743-2968 – mf#14798,01 – us UMI ProQuest [621]

Handsworth herald and north birmingham news – Birmingham, England 28 may 1892-19 jan 1918, 1 jan1921-28 jan 1922 (wkly) [mf 1894,1898,1902] – 1 – (cont: birmingham news [6 feb-21 may 1892]; cont by: handsworth herald [4 feb 1922-25 jan 1930]) – uk British Libr Newspaper [072]

Handsworth news – England.13 Oct 1888-1895; 11 Apr-12 Dec 1896; 1897-1899. -w. 11 reels – 1 – uk British Libr Newspaper [072]

Handt-boecxken der christelijcke gedichten, sinne-beelden ende liedekens, tot troost ende vermaeck der gelooviger zielen / [Biens, C P] – Hoorn: Marten Gerbrantz, 1635 – 3mf – 9 – mf#O-3030 – ne IDC [090]

Handtbuechlein von zweyerley nuetzlichem gebrauch vnd vbung des catechismi / Musaeus, S – [Magdeburg, 1578] – 1mf – 9 – mf#TH-1 mf 1201 – ne IDC [242]

Handweaver and craftsman – New York. 1950-1975 (1) 1971-1975 (5) – ISSN: 0017-7407 – mf#1427 – us UMI ProQuest [740]

Handwerck, Hugo *see* Gellerts aelteste fabeln

Handwerk und gewerbe – Berlin DE, 1915 jan, may, aug-oct, 1916 jan-mar, jun-sep, nov, 1925 oct, 1930 jun, 1933 apr, 1934 oct, 1935 jun, 1936 feb-1937, 1938 sep – 1r – 1 – (title varies: oct 1925: der juedische handwerker) – gw Misc Inst [939]

Der handwerker – Budweis (Ceske Budejovice CZ), 1933-35 – 1 – gw Misc Inst [077]

Handwerkszeitung – Hagen, Westf DE, 1958 3 jan-1961 1 sep, 1963 11 may-1966 – 1 – gw Misc Inst [640]

Handwoerterbuch des biblischen altertums fuer gebildete bibelleser / ed by Riehm, Eduard et al – 2. Aufl. Bielefeld: Velhagen & Klasing, 1893-1894 – 19mf – 9 – 0-524-02785-4 – (incl bibl ref) – mf#1987-6479 – us ATLA [052]

Handworterbuch des deutschen maerchens – Berlin: W de Gruyter 1930-33, 34-40 [mf ed Bloomington IN: Indiana Uni Lib, Preservation Dept 1984] – 2v on 1r – 1 – us Indiana Preservation [390]

Handwritten diaries / Taylor, George Baxton – 2400p – 1 – 84.00 – us Southern Baptist [242]

Handy andy : a tale of irish life / Lover, Samuel – Philadelphia, PA. 19-? – 1r – us UF Libraries [025]

Handy book on the dominion franchise act / Liberal-Conservative Union of Ontario – 2nd ed. [Toronto?: s.n, 1886?] [mf ed 1993] – 1mf – 9 – 0-665-91423-7 – mf#91423 – cn Canadiana [325]

A handy concordance of the septuagint : giving various readings from codices vaticanus, alexandrinus, sinaiticus, and ephraemi... – London: S Bagster [1887?] [mf ed 1990] – 3mf – 9 – 0-8370-1516-2 – mf#1987-6067 – us ATLA [221]

Haneberg, Daniel Bonifacius von *see*
- Geschichte der biblischen offenbarung
- Die religioesen alterthuemer der bibel

Hanekom, M *see* Leierskapontwikkeling as didaktiese opgawe in die primere skool

Haner, Janet A S *see* Determining types and typal profiles of adult amateur theatre participants through q-technique

Hanes, Frederick Marion *see* Lutheran church usages

Hanet-Clery, Jean *see* Journal de ce qui s'est passe a la tour du temple

[Hanford-] california social science review – CA. 1962-1968 – 1r – 1 – $60.00 – mf#R03236 – us Library Micro [071]

Hanford Engineering Works Employees' Association *see* Sage sentinel

[Hanford-] hanford journal – CA. 1891-1907; 1908-15; 1917-55 [daily, wkly, biwkly] – 173r – 1 – $10,380.00 – (aka: morning journal) – mf#BC02285 – us Library Micro [071]

[Hanford-] hanford sentinel – CA. 1901- – 334r – 1 – $20,040.00 (subs $400y) – mf#BC02286 – us Library Micro [071]

Hang chiang t'ieh lu kung ch'eng chi lueh – Hang-chou: Hang Chiang t'ieh lu chu, Min kuo 22 [1933] – (= ser P-k&k period) – us CRL [380]

Hang, Chin-fu *see* Ting ning chi

Hang k'ung ching chi cheng ts'e lun / Yu, Chi – Shang-hai: Shang wu yin shu kuan, Min kuo 23 [1934] – (= ser P-k&k period) – us CRL [380]

Hang, Li-wu *see* Fang ying chien pi

Hang yeh cheng ts'e / Wang, Kuang – Nan-ching: Chiao t'ung tsa chih she, Min kuo 23 [1934] – (= ser P-k&k period) – us CRL [380]

Hang yeh tsu ho yu chin tai ssu ch'ao / Liu, Wen-tao – Ch'ung-ch'ing: Shang wu yin shu kuan, 1943 – (= ser P-k&k period) – us CRL [190]

Hang-chou lun hsien chih ch'ien hou / Ts'ai, Ching-p'ing – Ch'ung-ch'ing: Ts'ai Ching-p'ing, [1941] – (= ser P-k&k period) – us CRL [951]

Hangchow, the "city of heaven" : with a brief historical sketch of soochow / Cloud, Frederick D – [Shanghai: Presbyterian Mission Press, 1906] [mf ed 1995] – (= ser Yale coll) – ix/110p (ill) – 1 – 0-524-09306-7 – mf#1995-0306 – us ATLA [951]

Hanging loose – Brooklyn. 1972-1995 (1) 1973-1995 (5) 1973-1995 (9) – ISSN: 0440-2316 – mf#7624 – us UMI ProQuest [810]

Hangman – iss n2-8. win 1942-fall 1943 – ea set of 4mf – 15 – (incl special comics n1) – mf#001MLJ-002MLJ – us MicroColour [740]

Hangsa Yonta *see* Histoire de hang-yon [hangsa yonta]

Hangula, Rusta J K *see* Effect of soil types, soil pasteurisation and inoculation on the growth of colophospermum mopane families

Hangyore sinmun – Seoul, Korea. May 15 1988-Dec 1990 – 10r – 1 – us L of C Photodup [079]

Han-i araha duin hacin-i hergen kamciha manju gisun-i buleku bithe = Qagan-u bicigsen gurban zuil-un usug-iyer qabsurugsan manju ugen-u toli bicig – [China: s.n, 177-] [mf ed 1966] – (= ser Tenri coll of manchu-books in manchu-characters. series 1, linguistics 21-26; Mango bunkenshu. 1, gogaku hen) – 36v on 6r – 1 – (in manchu, mongolian, tibetan and chinese) – ja Yushodo [480]

Han-i araha manju gisun-i buleku bithe = Man meng ho pi ching wen chien – [China: s.n, 1717] [mf ed 1966] – (= ser Tenri coll of manchu-books in manchu-characters. series 1, dictionaries, grammars, readers 4-9; Mango bunkenshu. 1, gogaku hen) – 29v on 6r – 1 – (in manchu and mongolian. title also in chinese: man-chou meng-ku ho pi ching wen chien) – ja Yushodo [480]

Han-i araha manju gisun-i buleku bithe = Yu zhi qing wen jian – [China: s.n, 1708] [mf ed 1966] – (= ser Tenri coll of manchu-books in manchu-characters. series 1, linguistics 1-3; Mango bunkenshu. 1, gogaku hen) – 10v on 3r – 1 – (in manchu) – ja Yushodo [480]

Han-i araha manju monggo nikan hergen ilan hachin-i mudan acaha buleku bithe = Qagan-u bicigsen mancu monggol kitad usug gurban zuil-un ayalgu neilegsen toli bicig – [China: s.n, 1780] [mf ed 1966] – (= ser Tenri coll of manchu-books in manchu-characters. series 1, linguistics 16-20; Mango bunkenshu. 1, gogaku hen) – 32v on 5r – 1 – (in manchu, mongolian and chinese) – ja Yushodo [480]

Han-i araha nonggime toktobuha manju gisun-i buleku bithe = Yu jin [i e zhi] zeng ding qing wen jian – [China: s.n, 1771] [mf ed 1966] – (= ser Tenri coll of manchu-books in manchu-characters. series 1, linguistics 10-15; Mango bunkenshu. 1, gogaku hen) – 50v on 6r – 1 – (in manchu and chinese) – ja Yushodo [480]
Hanim, Leyla *see* The divan project
Hanim, Nigar *see* Efsus
Hanim, Seref *see* The divan project
Haninge allehanda – Vasterhaninge, Sweden. 1977-81 – 1 – sw Kungliga [078]
Hanisch Espindola, W *see* Itinerario y pensamiento de los jesuitas expulsos de chile (1767-1815)
[Hank, Arthur et al] [comp] *see* Baptist general association of west virginia, 1865-1915, woman's baptist missionary society of west virginia, ministers' fraternal union
Hanka, Venceslav *see* Dalimils chronik von boehmen
Hankamer, Paul *see*
– Deutsche gegenreformation und deutsches barock
– Jakob boehme
– Spiel der maechte
– Die sprache, ihr begriff und ihre deutung im sechzehnten und siebzehnten jahrhundert
Hanke, Lewis *see*
– Cuerpo de documentos del siglo 16 sobre los derechos de espana en las indias y las filipinas descubiertos y contados por...
– Cuerpo de documentos del siglo 17 sobre los derechos de espana en las indias y las filipinas, descubiertos y anotados por lewis hanke...mexico, 1943
– The first social experiments in america. a study in the development of spanish indian policy...cambridge (usa), 1935
– The spanish struggle for justicie in their conquest of america...
Hankenstein, Johann Aloys Hanke von *see* Bibliothek der maehrischen staatskunde
Hankin, E H *see* The drawing of geometric patterns in saracenic art
Han-k'ou shih cheng kai k'uang : chung-hua min kuo erh shih erh nien chi erh shih san nien / China. Han-k'ou – [China: Han-k'ou shih, 1933-1934] – (= ser P-k&k period) – us CRL [350]
Hankukilbo *see* Korea times
Hanley, Miles L *see* Index to rimes in american and english poetry, 1500-1900
Hanley, Thomas O'Brien *see* The charles carroll papers
Hanlin papers : second series. essays on the history, philosophy, and religion of the chinese / Martin, William Alexander Parsons – Shanghai: Kelly & Walsh, 1894 [mf ed 1995] – (= ser Yale coll) – xii/427p – 1 – 0-524-09034-3 – mf#1995-0034 – us ATLA [180]
Hanmer, Lee Franklin *see* Recreation legislation
Han-min *see* Tang tai chung-kuo jen wu chih
Han-mu-la-pi fa tien / Edwards, C – Ch'ang-sha: Shang wu yin shu kuan, 1938 – (= ser P-k&k period) – us CRL [340]
Hanna, A J (Alexander John) *see* Story of the rhodesias and nyasaland
Hanna, Alexander John *see* Beginnings of nyasaland and north-eastern rhodesia, 1859-95
Hanna, Charles Augustus *see* The scotch-irish
Hanna, Hazel E B *see* Hazel e.b. hanna papers
Hanna, Henry Bathurst *see* Indian problems
Hanna, Kathryn Abbey *see* Florida
Hanna, William *see*
– The forty days after our lord's resurrection
– The last day of our lord's passion
– No man liveth to himself
– The passion week
– The patriarchs
– Selection from the correspondence of the late thomas chalmers
– A selection from the correspondence of the late thomas chalmers...
– Selection from the correspondence of the late thomas chalmers
– The wars of the huguenots
– Wycliffe and the huguenots
Hannaford, E P *see* Report on the st lawrence bridge and manufacturing scheme (shearer scheme)
Hannaford, Ebenezer *see* Map and history of cuba from the latest and best a...
Hannah corcoran : an authentic narrative of her conversion from romanism, her abduction from charlestown, and the treatment she received during her absence / Caldicott, Thomas Ford – Boston: Gould and Lincoln, 1853 – 1mf – 9 – 0-8370-8324-9 – mf#1986-2324 – us ATLA [240]
Hannah, Ian Campbell *see* Eastern asia
Hannah, J *see* Letter to the right rev the primus of the scottish episcopal churc...
Hannah, J M *see* The use of oxygen consumption and blood lactate measures in training for peak championship swimming performance
Hannah, John *see*
– Introductory lectures on the study of christian theology
– The relation between the divine and human elements in holy scripture
Hannah, John, archdeacon of Lewes *see* A plea for theology as the completion of science
Hannam, E P *see*
– Hospital manual
– Invalid's help to prayer and meditation
Hannam, Susan E *see* Smokeless tobacco use among big ten wrestlers and factors associated with use
Hannan, M *see* Standard shona dictionary
Hannas, Ruth *see* Evolution of harmonic consciousness
Hannay, James *see*
– Ballads of acadia
– The brothers d'amours
– The heroine of acadia
– The history of acadia
– History of the war of 1812 between great britain and the united states of america
– The maiden's sacrifice
– New brunswick
– Wilmot and tilley
Hannay, James O *see*
– The spirit and origin of christian monasticism
– The wisdom of the desert
Hanne, Johann Wilhelm *see*
– Der christliche glaube im kampfe mit dem modernen aufklaerungschriftenthum und der widerspruch des letztern mit der vernunft
– Friedrich schleiermacher als religioeser genius deutschlands
– Der geist des christenthums
– Die idee der absoluten persoenlichkeit, oder, gott und sein verhaeltniss zur welt, insonderheit zur menschlichen persoenlichkeit
– Das wunder des christenthums
Hannele : a dream poem / Hauptmann, Gerhart – New York: Doubleday, Page, 1908 – 1r – 1 – us Wisconsin U Libr [810]
Hannes, Ludwig *see* Des averroes abhandlung
Le hanneton : (illustre, satirique et litteraire) – Paris. nov 1862-juil 1868 – 1 – fr ACRPP [073]
Le hanneton : journal des toques – Paris. premiere annee, n1. aout 1876 – 1 – fr ACRPP [073]
Hannington, J *see* The last journals of bishop hannington
Hannoeverische politische nachrichten – Hannover DE, 1793-99 – 6r – 1 – gw Misc Inst [074]
Hannon, John *see* The devil's parables
Hannoverische beytraege zum nutzen und vergnuegen / ed by Wuellen, A C von – Hannover 1759-63 – (= ser Dz) – 4pt on 20mf – 9 – €200.00 – mf#k/n219 – gw Olms [430]
Hannoverische gelehrte anzeigen / ed by Wuellen, A C von – Hannover 1752-55 – (= ser Dz) – 4v on 24mf – 9 – €240.00 – mf#k/n168 – gw Olms [500]
Hannoverisches magazin – Hannover DE, 1775-77, 1786-1800 – more than 19r – 1 – (title varies: 1791: neues hannoverisches magazin; 1793: neues hannoeverisches magazin; later: hannoversches magazin) – gw Misc Inst [943]
Hannoverisches magazin fuer das jahr 1763 [-90] – Hannover 1764-91 – (= ser Dz) – jg1-28 on 159mf – 9 – €954.00 – mf#k/n243 – gw Olms [074]
Die hannoversche agende im auszug / ed by Meyer, Johannes – Bonn: A Marcus und E Weber, 1913 – (= ser Kleine texte fuer vorlesungen und uebungen) – 1mf – 9 – 0-524-06870-4 – mf#1990-5289 – us ATLA [240]
Hannoversche agende im auszug / ed by Meyer, J – Bonn. KIT 125, 1913 – €3.00 – ne Slangenburg [240]
Hannoversche allgemeine zeitung – Hannover DE, 1960 17 mar-1963 [gaps] – 1 – (filmed by bnl: 25 aug-7 oct 1949 (wanting n19)) – gw Misc Inst [074]
Hannoversche geschichtsblaetter : [zeitschrift des historischen vereins fuer niedersachsen] – Hannover: Historischer Verein fuer Niedersachsen 1898- (qrtly) [v1-32 (gaps) mf ed 1978] – 7r [ill] – 1 – (wkly 1898; publ suspended 1940-51; suppl accompany some iss; iss55 [2001] called also beiheft 2; some iss have also a distinctive title) – mf#film mas c436 – us Harvard [943]
Hannoversche landesblaetter – Hannover DE, 1845 27 nov-1846, 1847 5 jan & 31 mar, 1848 18 jan – 1r – 1 – gw Misc Inst [074]
Hannoversche neueste nachrichten : ausgabe suedhannover – Goettingen DE, 1946 3 jul-1949 23 aug – 2r – 1 – gw Mikrofilm [074]
Hannoversche presse – Hannover DE, 1946 19 jul-1973 – 163r – 1 – (fr 1946 19 jul-1973: neue hannoversche presse) – mf#5656 – gw Mikropress [074]
Hannoversche presse – Hannover DE, 1958-1961 6 feb, 1961 7 mar-28 apr, 1961 18 may-1963 16 aug, 1963 16 sep-1964 7 sep, 1965 11 feb-1967 31 jan, 1975-1978 17 aug, 1978 18 oct-1983 21 aug – 1 – (title varies: 22 apr 1971: neue hannoversche presse; 2 jun 1981: neue presse. filmed by other misc inst: 1973-1975 mar, 1979- [ca9r/yr]. regional ed: goettingen 1946 19 jul-1949 30 aug [2r]) – gw Misc Inst [074]
Hannoversche rundschau *see* Norddeutsche zeitung [hannover]
Hannoversche volksstimme : ausgabe suedhannover – Goettingen DE, 1946 16 aug-1947 14 oct, 1948 16 jan-1949 10 nov – 2r – 1 – (title varies: 3 dec 1948: niedersaechsische volksstimme) – gw Mikrofilm [074]
Hannoversche volksstimme – Hannover DE, 1946 16 aug-1947 14 oct, 1949 12 nov-1956 17 aug – 14r – 1 – (title varies: 19 aug 1947: niedersaechsische volksstimme; 12 nov 1949: die wahrheit; 1 feb 1956: neue niedersaechsische volksstimme) – gw Mikrofilm [074]
Hannoversche volks-zeitung – Hannover DE, 1906 7 jan-1908, 1910 17 feb & 1 jun-1911, 1913, 1915-1917 31 mar, 1925 1 jul-25 sep, 1926-1928 jun, 1928 oct-1933 30 jun – 32r – 1 – (title varies: 1925?: hannoversche volkszeitung; katholisch; publ in hildesheim. incl suppls) – gw Mikrofilm [074]
Hannoversche volkszeitung *see* Hannoversche volks-zeitung
Hannoversche zeitung *see* Niedersaechsische beobachter
Hannoversche zeitung 1832 – Hannover DE, 1832 – 1 – (title varies: 1 jan 1858: neue hannoversche zeitung. filmed by misc inst: 1848-49 [4r]) – gw Mikrofilm; gw Misc Inst [074]
Hannoverscher anzeiger – Hannover DE, 1893 1 jul-1914 25 jun; 1915-1943 27 feb – 191r – 1 – gw Mikrofilm [074]
Hannoverscher kurier – Hanover, Germany. -d. 1 Sept 1916-12 Aug 1919. Imperfect. 12 1 2 reels – 1 – uk British Libr Newspaper [072]
Hannoverscher volks-bote *see* Wochenblatt fuer die amtsgerichtsbezirke bremervoerde, beverstedt und zeven
Hannoversches magazin : worin kleine abhandlungen, einzelne gedanken, nachrichten, vorschlaege und erfahrungen...gesammelt und aufgewahret sind – Hannover 1814-50 – (= ser Dz) – 37jge on 193mf – 9 – €1158.00 – mf#k/n6206 – gw Olms [370]
Hannoversches tageblatt – Hannover DE, 1915 1 oct-31 dec – 1r – 1 – gw Misc Inst [074]
Hannoversches wochenblatt fuer handel und gewerbe – Hannover DE, 1869 2 jan-1876 23 dec – 2r – 1 – gw Mikrofilm [380]
Hanoi. Institute d'Archeologie. Comite des Sciences Sociales du Viet-Nam *see* Khao co hoc
Hanoi moi – [Ha-noi: s.n, jan 31 1968-1982; jul-dec 1986] – 1 – us CRL [079]
[Hanoi-] vietnam courier – VM. 1966-72 – 1r – 1 – $50.00 – mf#R042233 – us Library Micro [079]
Hanoka'a High School [Hanoka'a HI] et al *see* Laulima
Hanover 1712-1895 – Oxford, MA (mf ed 1992) – (= ser Massachusetts vital records) – 47mf – 9 – 0-87623-153-9 – (mf 1-2: vital records 1725-1813. mf 2-4: town records 1757-1837. mf 4-5: vital records 1741-1815. mf 5-6: town records 1746-1811. mf 7-13: town records 1727-1802. mf 10,13: vital records 1712-84. mf 14-24: town meetings 1802-57. mf 25-26: births 1727-1857. mf 26: marriages 1728-1857. mf 27: deaths & intentions 1728-1857. mf 28: vitals index 1727-1857. mf 29: births 1769-1846. mf 30: marrs & deaths 1814-44. mf 31: vital records 1844-57. mf 32: rebellion records 1861-65. mf 33-35: militia 1840-1907. mf 36-39: intentions 1815-89. mf 40: intentions 1897-1905. mf 41-43: births 1843-95. mf 43-45: marriages 1727-1895. mf 45-47: deaths 1857-95) – us Archive [978]
Hanrahan, Susan N *see* Tampon labeling and its effect on female adolescents' knowledge of tampon absorbency, knowledge of toxic shock syndrome and tampon usage patterns
Hans august vowinckel, der dichter und soldat : ein ehrenbuch / Vowinckel, Hans August; ed by Vowinckel, Renate – Stuttgart: J Engelhorn, 1942 – 1r – 1 – us Wisconsin U Libr [430]
Hans, Bruder *see* Bruder hansens marienlieder
Hans carossa : der heilkundige dichter / Schaeder, Grete – Hameln: F Seifert, 1947 – 1r – 1 – us Wisconsin U Libr [430]
Hans carossa : seine geistige haltung und sein glaubensgut / Klatt, Fritz – Wismar: H Rhein, [1937] – 1r – 1 – us Wisconsin U Libr [170]
Hans clawerts werckliche historien / Krueger, Bartholomaeus; ed by Raehse, Theobald – Halle: M Niemeyer, 1882 – 11r – 1 – (incl bibl ref) – us Wisconsin U Libr [430]
Hans david billman family bulletin – v2 n1-v5 n3 [n3-18] [1974 oct-1979 mar] – 1r – 1 – mf#640940 – us WHS [929]
Hans egede : missionary to greenland / Nieritz, Gustav – Philadelphia: Lutheran Board of Publ, 1873, c1872 – 1mf – 9 – 0-8370-7181-X – mf#1986-1181 – us ATLA [920]
Hans folz : auswahl / Spriewald, Ingeborg [comp] – Berlin: Akademie-Verlag, 1960 [mf ed 1993] – (= ser Studienausgaben zur neueren deutschen literatur 4) – 273p (ill) – 1 – (incl bibl ref) – mf#8044 – us Wisconsin U Libr [430]
Hans georg ernstingers raisbuch / ed by Walther, A F – Stuttgart: Litterarischer Verein, 1877 (Tuebingen: L F Fues) – us Wisconsin U Libr [910]
Hans georg ernstingers raisbuch / ed by Walther, A F – Stuttgart: Litterarischer Verein, 1877 (Tuebingen: L F Fues) [mf ed 1993] – (= ser Blvs 135) – 312p – 1 – mf#8470 reel 28 – us Wisconsin U Libr [880]
Hans heiner roselieb's ewiger sonntag / Schotte, Heinrich – Kempten: J Koesel & F Pustet, 1921 – 1r – 1 – us Wisconsin U Libr [830]
Hans huckebein, der ungluecksrabe : das pusterohr; das bad am samstag abend / Busch, Wilhelm – 8th ed. Stuttgart: Deutsche Verlags-Anstalt [19–?] [mf ed 1989] – 1r – 1 – (filmed with: es reiten die wilden jager / paul burre) – mf#7096 – us Wisconsin U Libr [830]
Hans jakob breunings von buchenbach relation ueber seine sendung nach england im jahr 1595 / ed by Schlossberger, August – Stuttgart: Litterarischer Verein, 1865 – 1 – (incl bibl ref) – us Wisconsin U Libr [430]
Hans jakob breunings von buchenbach relation ueber seine sendung nach england im jahr 1595 / Schlossberger, August – Stuttgart: Litterarischer Verein, 1865 [mf ed 1993] – (= ser Blvs 81) – 92p – 1 – (incl bibl ref) – mf#8470 reel 17 – us Wisconsin U Libr [880]
Hans nielsen hauge og hans samtid : et tidsbillede fra omkring aar 1800 / Bang, A Chr – Tredie oplag med billeder og facsimiler Kristiania: Gyldendalske Boghandel Nordisk Forlag, 1910 – 2mf – 9 – 0-7905-4373-7 – (incl bibl ref) – mf#1988-0373 – us ATLA [920]
Hans peter menges family newsletter – v1 n10-v4 n3 [i e v3 n4] [1974 mar-1976 dec] – 1r – 1 – (continued by: menges family association in america : [newsletter]) – mf#1277627 – us WHS [929]
Hans pfriem, oder, meister kecks : komoedie / Hayneccius, Martin; ed by Raehse, Theobald – Halle: M Niemeyer, 1882 – 11r – 1 – (incl bibl ref) – us Wisconsin U Libr [430]
Hans rothfels : ein historiker zwischen kaiserreich und nationalsozialismus / Petters, Karl Olaf – (mf ed 1995) – 2mf – 9 – €40.00 – 3-8267-2198-5 – mf#DHS 2198 – gw Frankfurter [943]
Hans Sachs : works / ed by Keller, Adelbert von & Goetze, E – Stuttgart: Litterarischer Verein, 1870-1908 (Tuebingen: H Laupp) [mf ed 1993] – (= ser Blvs 102-106, 110 etc) – 26v – 1 – (incl bibl ref and ind) – mf#8470 reel 21 etc – us Wisconsin U Libr [802]
Hans sachs : dramatisches gedicht in vier acten / Deinhardstein, Johann Ludwig – Wien: C Armbruster, 1829 [mf ed 1989] – xvi/140p – 1 – mf#7174 – us Wisconsin U Libr [820]
Hans sachs / ed by Keller, Adelbert von – Stuttgart: Litterarischer Verein, 1870-1908 (Tuebingen: H Laupp) [mf ed 1993] – (= ser Blvs 102-106, 110 etc) – 26v – 1 – (incl bibl ref and ind) – mf#8470 reel 21 etc – us Wisconsin U Libr [430]
Hans sachs and goethe : a study in meter / Burchinal, Mary Cacy – Goettingen: Vandenhoeck & Ruprecht; Baltimore: Johns Hopkins Press 1912 [mf ed 1990] – (= ser Hesperia. schriften zur germanischen philologie 2) – 1r – 1 – (incl bibl ref; filmed with: goethe und pestalozzi / gottfried bohnenblust) – mf#7384 – us Wisconsin U Libr [430]
Hans sachs und die reformation / Kawerau, Waldemar – Halle: Verein fuer Reformationsgeschichte, 1889 – (= ser [Schriften des Vereins fuer Reformationsgeschichte]) – 1mf – 9 – 0-7905-4696-5 – (incl bibl ref) – mf#1988-0696 – us ATLA [430]
Hans sachs und goethe : 1. [und] 2. teil / Wahl, Georg – Coblenz: H L Scheid, 1892 – 1r – 1 – (incl bibl ref) – us Wisconsin U Libr [430]
Hans sachsens ausgewaehlte werke / ed by Merker, Paul – Leipzig: Insel-Verlag, 1911 [mf ed 1993] – 2v (ill) – 1 – (v1: gedichte v2: dramen. biogr aft by ed) – mf#8455 – us Wisconsin U Libr [802]
Hans salat : ein schweizerischer chronist und dichter aus der ersten haelfte des 16 jahrhunderts. sein leben und seine schriften / Baechtold, J – Basel, 1876 – 4mf – 9 – mf#ZWI-15 – ne IDC [920]

Hans schiltbergers reisebuch / ed by Langmantel, Valentin – Stuttgart: Litterarischer Verein, 1885 (Tuebingen: H Laupp) [mf ed 1993] – (= ser Blvs 172) – 1 – mf#8470 reel 35 – us Wisconsin U Libr [910]

Das hans thoma-buch : freundesgabe zu des meisters 80. geburtstage / Friedrich, Karl Josef [comp] – Leipzig: E Seemann 1919 [mf ed 1991] – 1r [ill] – 1 – (filmed with: the devil's shadow / frank thiess, trans fr german by h t lowe-porter) – mf#2911p – us Wisconsin U Libr [760]

Hansa – Hamburg DE, 1916 16 sep-1919 9 aug – 5r – 1 – uk British Libr Newspaper [074]

Hansard knollys society . . . baptist writers – London: J. Haddon, 1846-1854 – 2r – 1 – 0-8370-1689-4 – mf#1984-B476 – us ATLA [242]

Hansard parliamentary debates : house of commons and house of lords – 1984/85- [mf ed Chadwyck-Healey] – (= ser House of commons parliamentary papers, 1901-1974/1975) – 9 – (final version. available for every sess fr 1984/85) – uk Chadwyck [324]

Hansa-theater : artistische nachrichten – Hamburg DE, 1896 aug-1897 sep, 1898 feb-mar, 1899 aug-1914 oct, 1915 jan – 2r – 1 – (with suppl) – gw Mikrofilm [790]

Hansbrough, John H *see* Hydrogen ion concentration of citrus leaves and its relation to certain fungus diseases

Hansconian – 1981 apr 30 [v23 n17]-1983 mar, 1983 apr-1984, 1985 dec-1987 sep, 1985 feb-nov, 1985 dec-1987 sep – 4r – 1 – mf#660864 – us WHS [071]

Hansell, Edward Halifax *see* Novum testamentum graece

Hansen, Adolfph *see* Goethes leipziger krankheit und "don sassafras"

Hansen, Adolph *see* Goethes morphologie

Hansen, Christopher A *see* The effects of ergogenic aid supplementation on the sprint capacity of male cyclists

Hansen, David E *see* "Fair play everyday"

Hansen, Deirdre Doris *see* Life and work of benjamin tyamzashe

Hansen, Emil *see* Aditi; indisk-orientalsk ballet i to akter (anden akt i to afdelinger). musiken af fr. rung. dekorationerne af w. guellich. kostumerne tegnede af pietro krohn. opfort forste gang i marts 1880

Hansen, G C *see* Kirchengeschichte (gcsej19)

Hansen, Gary F *see*
- Perceptions of agencies that market collegiate emblematic merchandise toward selected factors related to royalty income
- A study of attitudes toward high school academics reported by current football players enrolling at selected big ten conference universities from 1985 through 1989

Hansen, Joseph *see*
- Quellen und untersuchungen zur geschichte des hexenwahns und der hexenverfolgung im mittelalter
- Zauberwahn inquisition und hexenprozess im mittelalter und die entstehung der grossen hexenverfolgung

Hansen, Maurice G *see* The reformed church in the netherlands

Hansen, Peter *see* Noter til dr. g. brandes' "soeren kierkegaard"

Hansen, Peter S *see* Liber secundus mutetarum by domenico phinot

Hansen, Rolf *see* Quantitative entwicklungen und strukturelle veraenderungen der schule in der brd

Hansen, Wilhelm *see* Goethe

Hanserd Knollys Society for the Publication of the Works of Early English and other Baptist Writers *see*
- Publications

Hanserd knollys society materials / Baptist Missionary Society. Archives. London – Minutes, Journal and Correspondence, 1844-58. 540p – 1 – $18.90 – us Southern Baptist [242]

Hanserecesse, 2. abt / ed by Ropp, Goswin von der – Leipzig 1876-92 [mf ed Hildesheim 1991] – 7v on 53mf – 9 – diazo €234.00 silver €258.00 – gw Olms [943]

Hans-georg gadamer : een filosofie van het interpreteren / Vandenbuclke, J – Brugge, 1973 – 5mf – 8 – €12.00 – ne Slangenburg [140]

Hansmeier, Thomas *see* Zugangsregulation und soziale integration in der rehabilitation

Hanson 1779-1849 – Oxford, MA (mf ed 1996) – (= ser Massachusetts vital record transcripts to 1850) – 9 – 0-87623-261-6 – (mf 1t: marriage intentions 1820-21; births 1779-1847. mf 2t: deaths 1810-42; marriages 1820-36; births 1818-20. mf 3t: births 1843-49, 1827-47. mf 4t: marriages & deaths 1843-49) – us Archive [978]

Hanson 1820-1900 – Oxford, MA (mf ed 1992) – (= ser Massachusetts vital records) – 18mf – 9 – 0-87623-146-6 – (mf 1-3: index to births 1820-1970. mf 4-6: index to marriages 1820-1970. mf 7-9: index to deaths 1820-1970. mf 10-11: marriages 1852-91. mf 11-13: births 1852-91. mf 13-15: deaths 1852-91. mf 16: births 1892-1900. mf 17: marriages 1892-1900. mf 18: deaths 1892-1901) – us Archive [978]

Hanson, Earl Parker *see*
- Journey to manaos
- Puerto rico
- Transformation

Hanson, Elizabeth *see* An account of the captivity of elizabeth hanson, late of kachecky in new-england

Hanson, Felix Valentine *see* Studies in genesis

Hanson, John Wesley *see*
- Aion-aionios
- The bible hell
- Bible proofs of universal salvation
- Bible threatenings explained
- A cloud of witnesses
- The life and works of the world's greatest evangelist, dwight l. moody
- A pocket cyclopaedia
- Universalism, the prevailing doctrine of the christian church during its first five hundred years
- The world's congress of religions

Hanson, Kenneth C *see* The hymnology and the hymnals of the restoration movement

Hanson, Margaret L *see* Perceived occupational stress levels of ncaa directors of athletics

Hanson, Peter *see* Winterreise durch einen theil norwegens und schwedens nach kopenhagen, im jahre 1807

Hanson Place SDA Church [New York, NY] *see* Fort greene samaritan

Hanson, Richard Davis *see* The jesus of history

Hanson, Stanley *see* History of lee county, florida

Hanson, W Stanley *see*
- Annual events
- Artists
- Buckingham : lee county
- Crime and punishment
- Historical buildings
- History of fort myers, florida
- Indians and indian life
- Iona
- Owanita
- Parks and playgrounds
- Saint james city
- Upco hall

Hanson's latin american letter – Ithaca. 1975-1980 (1) 1979-1980 (5) 1979-1980 (9) – ISSN: 0017-7539 – mf#9319 – us UMI ProQuest [337]

Hanssens, J M *see* Amalarii episcopi opera

Hansteen, Christopher *see* Reise-erinnerungen aus sibirien

Hanstein, Penelope *see* Discipline-based dance education

Han-sung *see* Tu che hsin hsiang wai chi ti erh chi

Hantbukh fun der velt-literatur / Goldin, L – Warszawa, Poland. 1931 – 1r – us UF Libraries [939]

Hanthaler, C *see* Quinquagena symbolorum heroica

Hanthaler, Chrysostomus *see* Quinquagena symbolorum heroica in...sanctae regulae benedicti

Hanthawaddy – Rangoon, Burma. 9 Apr 1950-1951; 1959-Apr 1968 – 37r – 1 – us L of C Photodup [079]

Hantke, Friedrich *see* Biologische verfahren des pflanzenschutzes im zierpflanzenbau

Hantke's Brewers' School and Laboratories *see* Letters on brewing

Hants and berks gazette and basingstoke journal – Basingstoke, England 5 jan-17 aug 1878 – 1 – (iss twice wkly fr 1981 to aug; cont by: hants & berks gazette & middlesex & surrey journal [24 aug 1878-31 dec 1965]; hants & berks gazette and basingstoke journal [7 jan 1966-26 sep 1969]; basingstoke gazette [3 oct 1969-9 may 1975]) – uk British Libr Newspaper [072]

Hants and surrey times and aldershot advertiser – Aldershot, England 24 jul 1880-20 jul 1895 – 1 – uk British Libr Newspaper [072]

Hantu : the new england regional conference [for black studies] newsletter / New England Regional Conference for Black Studies – 1984 winter, autumn, 1985 spring, 1986 fall – 1r – 1 – mf#2698826 – us WHS [305]

Hanwell gazette and brentford observer – Ealing, England 5 nov 1898-29 sep 1923 [mf 1911] – 1 – (incorp with: west middlesex gazette) – uk British Libr Newspaper [072]

Hanzas, Barbara *see* Index to otero county newspapers 1886-1900

Hao chu pen / I, Ch'iao – [Shang-hai?]: Chu i ch'u pan she, Min kuo 29- [1940-] – (= ser P-k&k period) – us CRL [820]

Ha-olam – Cologne, Vilna, Odessa, London, Berlin, Jerusalem, 1907-14, 1919-50 – 828mf – 9 – mf#J-91-25 – ne IDC [077]

Ha-olam, 1907-1950 : from the harvard college library – 25r – 1 – (central hebrew-language publ of the world zionist organization) – us Primary [270]

Ha-'olam le-lo shomer : shirim / Locker, Malka – Tel-Aviv, Israel. 1945 – 1r – us UF Libraries [939]

Ha-or – New York, N.Y., 1981; v12, no. 5 (27 Oct. 1982); v12, no. 9 (9 Feb. 1983)-v12, no. 10 (23 Feb. 1983); v12, no. 12 (23 Mar. 1983); v13, no. 7 (14 Dec. 1983)-v13, no. 8 (7 Feb. 1984); v13, no. 10-v13, no. 11; v13, no. 13 (1 May 1984) – us AJPC [270]

Ha-or – New York, N.Y., v1, no. 2 (28 Mar. 1972); v1, no. 4 (9 May 1972); v2, no. 2 (30 Nov. 1972)-v2., no. 3 (19 Dec. 1972); v2, no. 5 (7 Mar. 1973)-v8, no. 1 (7 Aug. 1978); v8, no. 10 (28 Feb. 1979); v9, no. 2 (4 Oct. 1979)-v9, no. 3 (18 Oct. 1973); v9, no. 8 (14 Feb. 1980)-v9, no. 9 (28 Feb. 1980); v9, no. 11 (17 Apr. 1980)-v10, no. 8 (25 Feb. 1981); v11, no. 1 (3 Aug. 1981); v11, no. 3 (22 Oct. 1981)-v11, no. 5 (18 Nov.) – us AJPC [270]

Haparanda nyheter – Haparanda, 1917 – 1r – 1 – sw Kungliga [078]

Haparandabladet – Goteborg, Sweden. 1882-1955; 1975- – 1 – sw Kungliga [078]

Hapde, Jean Baptiste Augustin *see* Expedition et naufrage de la perouse

Hapde, Jean-Baptiste-Augustin *see* Fetes d'eleusis

Ha-peles – Poltava, Berlin, 1900-1904. v1-5 – 73mf – 9 – mf#J-291-17 – ne IDC [077]

Hapgood, George *see* Solitaire and patience; seventy games to test the card player's skill and make a lonely hour pass quickly

Hapgood, Powers *see* In non-union mines: the diary of a coal digger in central pennsylvania, august-september 1921

Hapisgoh – Chicago. The Summit. 1888-99 – 1 – us AJPC [071]

Ha-pisgoh – The Summit. (New York), 1888-99 – 1 – us AJPC [071]

Hapke, Ralf *see* Bad ems – struktur- und funktionswandel der baederstadt an der unterlahn

Happ, Carol K *see* Hardiness levels and coping strategies of female head women basketball coaches in the national collegiate athletic association

Happel, Eberhard G *see* Groesste denkwuerdigkeiten der welt

Happel, Julius *see*
- Die altchinesische reichsreligion
- Die anlage des menschen zur religion
- Das christentum und die heutige vergleichende religionsgeschichte
- Der eid im alten testament

Happel, Otto *see* Das buch des propheten habackuk

Happening – Bell Creek, MT. 1969-1974 (1) – mf#64241 – us UMI ProQuest [071]

Happer, Andrew P *see* A visit to peking

Happy christmas stories *see* K'uai le sheng tan ku shih (ccm142)

Happy days : a book of toasts / Madison, George Neser [comp] – Toronto: Copp, Clark, c1913 – 1mf – 9 – 0-665-98614-9 – mf#98614 – cn Canadiana [390]

Happy days / Civilian Conservation Corps (US) – 1933 may 20-1934 sep 29, 1934 oct 6-1935 oct 26, 1935 nov 2-1936 sep 19, 1936 sep 26-1937 aug 28, 1937 sep 4-1938 sep 24, 1938 oct 1-1939 oct 28, 1939 nov 4-1940 dec 28, 1941 jan 4-1942 aug 8 – 8r – 1 – mf#804263 – us WHS [071]

Happy days – Toronto: Methodist Book and Pub. House, [1886?-1906] – 9 – (cont by: playmate) – mf#P04385 – cn Canadiana [240]

Happy days ccc directory – Washington, DC: Happy Days Pub Co, 1940 – us CRL [030]

Happy death-bed / Knill, Richard – London, England. 18– – 1r – us UF Libraries [240]

Happy deaths / Stock, John – London, England. 18– – 1r – us UF Libraries [240]

Happy home and parlor magazine – Boston. 1855-1859 (1) – mf#4821 – us UMI ProQuest [640]

Happy jack – London, England. 18– – 1r – us UF Libraries [240]

Happy message to northerners from the land of year-long spring – Moore Haven, FL. 1918? – 1r – us UF Libraries [630]

Happy news : the gospel spreading church of god official publication – 1992 jan-1998 dec – 1r – 1 – mf#2963466 – us WHS [243]

Happy stories *see* K'uai le te ku shih (ccm143)

Happy times – Madison WI. 1983 aug 31-sep 28 – 1r – 1 – mf#718682 – us WHS [071]

The happy valley : our new "mission garden" in uva, ceylon / Langdon, Samuel – London: Charles H Kelly, 1890 [mf ed 1995] – (= ser Yale coll) – 137p (ill) – 1 – 0-524-09213-3 – mf#1995-0213 – us ATLA [954]

Happy world / Follower's of Jesus [Mineola, NY] – 1978 jan-1983 jul – 1r – 1 – (cont: happy worker) – mf#1208563 – us WHS [243]

Haqayiq – Baku, 1907- . sal-i 1, shumarah-'i 1-6. 7 safar-1 sha'ban 1325 [22 mar-sep 1907] – 1r – 1 – $53.00 – us MEDOC [079]

Haqiqat – Baku, 1909-10 – 2r – 1 – (cont as: guneoe) – us UMI ProQuest [077]

Har s kierkegaard fremstillet de christelige idealer – er dette sandhed? / Mynster, Christian Ludvig Nicolai; ed by Paulli, Jakob – Kobenhavn: C A Reitzel 1884 [mf ed 1991] – 1mf – 9 – 0-524-00451-X – mf#1989-3151 – us ATLA [240]

Harabony – New York. N.Y. 1912 – 1 – us AJPC [071]

Harada, Kumao *see* The saionji-harada memoirs, 1931-40

Harada, Tasuku *see* The faith of japan

Haragan : novela / Salvador, Tomas – Madrid, Spain. 1956 – 1r – us UF Libraries [972]

Harahap, P *see* Indonesie sekarang

[Harahap, P] *see* Nippon di masa perang!

Harakah – 1997-2001 – 3r per yr – 1 – us UMI ProQuest [074]

Harambee / Duke University – 1969 feb 5 – 1r – 1 – mf#4867400 – us WHS [378]

Harambee / Lincoln University (Jefferson City, MO) – 1976 sep, 1979 jun – 1r – 1 – mf#5266129 – us WHS [071]

Harambee : a newspaper for young readers that focuses on the african-american experience – 1990 feb-1996 feb/mar – 1r – 1 – mf#2690107 – us WHS [071]

Harambee Afrikan Cultural Organization *see* Harambee flame

Harambee flame / Harambee Afrikan Cultural Organization – 1980 dec-1994 dec, 1995 jan-sep – 2r – 1 – mf#1070787 – us WHS [305]

Harambee Ombudsman Project *see*
- Annual report
- Harambee speaks

Harambee speaks : news from the harambee ombudsman project / Harambee Ombudsman Project – 1977 jun-1987 apr/may – 1r – 1 – mf#1057016 – us WHS [350]

Harapan masa / Madjallah Tracee baru – Djakarta, 1967. v1(1-4) – 5mf – 9 – mf#SE-1797 – ne IDC [950]

Harar : forschungsreise nach den somal- und galla-laendern ost-afrikas, ausgefuehrt von dr kammel von hardegger und prof dr paulitschke... – Leipzig: Brockhaus, 1888 – 1 – (nebst beitraegen von gunther ritter von beck, l ganglbauer und heinrich wichmann) – us CRL [916]

Harar : forschungsreise nach den somal- und galla-laendern ost-afrikas, ausgefuehrt von dr kammel von hardegger und prof dr paulitschke... / Paulitschke, P – Leipzig, 1888 – 7mf – 9 – mf#NE-20209 – ne IDC [916]

Harb doenuesue / Cahid, Burhan [Morkaya] – Istanbul: Burhan Cahid ve Suerekasi, 1928 – (= ser Ottoman literature, writers and the arts) – 5mf – 9 – $75.00 – us MEDOC [470]

Harb mecmu'asi – v1-3. n1-27. 1331-34 [1915-18] – (= ser O & t journals) – 13mf – 9 – $210.00 – us MEDOC [956]

Harband, B M *see* The pen of brahminpeeps into hindu hearts and homes

Harbaugh, H *see* Christological theology

Harbaugh, Henry *see*
- The fathers of the german reformed church in europe and america
- The heavenly home
- The life of rev. michael schlatter

Harbaugh, Linn *see* Life of the rev. henry harbaugh, d. d

Harben, H *see* A dictionary of london

Harbin, Robert Maxwell *see* Paradoxical pain

Harbin sevk ve idaresi / Von Der Goltz Pasa – Istanbul: Sirket-i Mertebiye Matbaasi, 1315 [1897] – (= ser Ottoman histories and historical sources) – 5mf – 9 – $95.00 – us MEDOC [956]

Harbinger : devoted to social and political progress – New York. 1845-1849 (1) – mf#4376 – us UMI ProQuest [320]

Harbinger – Sebastopol CA. v1 n5 [1969 oct 6] – 1r – 1 – mf#1583463 – us WHS [071]

Harbinger – v1 n1 [1]-v3 n10 [33] [1968 may-1970 nov] – 1r – 1 – mf#1107179 – us WHS [071]

The harbinger : conducted by a committee of gentlemen – Montreal: Printed for the Committee by J Lovell, [1842-1843] – 9 – mf#P04455 – cn Canadiana [242]

The harbinger : devoted to social and political progress – v1-8. 1845-49 – 1 – us AMS Press [320]

The harbinger of peace – v1-3. 1828-31 – (= ser The library of world peace studies) – 11mf – 9 – $105.00 – us UPA [320]

Harbinger of the mississippi valley – Frankfort. 1832-1832 (1) – mf#4582 – us UMI ProQuest [240]

Harbiye ve ihtiyat zabit mektebleri talimatt – Istanbul: Harbiye Mektebi Matbaasi, 1928 – (= ser Ottoman histories and historical sources) – 2mf – 1 – $40.00 – us MEDOC [956]

Harbor grace standard – Harbour Grace Newfoundland, Canada. 14 jul 1885-1888; 22 may 1889-27 dec 1918; 10 jan-21 nov 1919 (wanting jan-may 1889) (imperfect) – 15r – 1 – uk British Libr Newspaper [071]
Harbor journal series / Ashtabula Co. Ashtabula – v1 n1. (sep 1981-jul 1988) [wkly] – 3r – 1 – mf#B32789-32791 – us Ohio Hist [071]
Harbor junior high school for the performing arts : [newsletter] – 1977 mar – 1r – 1 – mf#5306602 – us WHS [373]
Harbor of fernandina / Yulee, David Levy – Fernandina, FL. 1880 – 1r – us UF Libraries [978]
Harbor springs republican – Harbor Springs MI. 1883 may 23 – 1r – 1 – mf#1240259 – us WHS [071]
Harbor system of el salvador / Ortiz, Ricardo M – New York, NY. 1954 – 1r – us UF Libraries [972]
Harborne news and west birmingham news – Birmingham, England (ns) 26 may 1906-19 jan 1918; 2 mar 1935-9 sep 1939 [mf 1911, 1912] – 1 – (incorp with: birmingham news; cont: harborne news [28 apr-19 may 1906]) – uk British Libr Newspaper [072]
Harborough mail *see* Market harborough advertiser
Harbottle dorr collection of annotated massachusetts newspapers, 1765-1776 – [mf ed 1966] – 4r – 1 – (a unique look at pre-revolutionary new england with insight into the thinking of citizen dorr on the controversies and topics of the times) – us MA Hist [071]
Harbour grace standard – Harbour Grace, NF. 1863-73 – 2r – 1 – ISSN: 0041-5064 – cn Library Assoc [079]
Harbour, harbor, harber, and witt, whitt, whit family association bulletin – v3 n1-v9 n4 [1980 sum-1987 fall] – 1r – 1 – (cont: harbour, harbor, harber family association bulletin) – mf#1712350 – us WHS [929]
Harbour, harbor, harber family association bulletin – v1 n1-v2 n4 [1978 jun 1-1980 spring] – 1r – 1 – (continued by: harbour, harbor, harber, and witt, whitt, whit family association bulletin) – mf#1712349 – us WHS [929]
Harbour, SK *see* Heart rate responses of collegiate female volleyball players during competition
Harbour views – Coffs harbour – 1r – at Pascoe [079]
Harbrecht, Hugo *see*
– Philipp von zesen als sprachreiniger
Harburger anzeigen und nachrichten *see* Harburger anzeiger
Harburger anzeiger – Hamburg DE, 1848-49 – 1r – 1 – (title varies: 21 sep 1949: harburger anzeigen und nachrichten. filmed by misc inst: 1976- [ca 6r/yr]) – gw Misc Inst [074]
Harcourt, Eugene d' *see* Quelques remarques sur l'execution du tannhaeuser de richard wagner a l'opera de paris, mai 1895
Harcourt, Francis V *see* Hints to young officers on the principles of military law
Harcourt, Guy M *see* Banking and commercial guide
Hard core / Columbia Students for a Democratic Society – 1968 oct 15/29, 1969 mar 18, 1968 oct 15/29-1969 mar 18 – 2r – 1 – mf#1583798 – us WHS [320]
Hard hat : voice of riggers and machinery erectors local / Riggers and Machinery Erectors Local 575 – v10 n1-v18 n8 [1974 jan-1982 aug] – 1r – 1 – mf#647730 – us WHS [331]
Hard places in the way of faith / Simpson, Albert B – South Nyack, NY: Christian Alliance, c1899 [mf ed 1992] – (= ser Christian & missionary alliance coll) – 1mf – 9 – 0-524-02270-4 – mf#1990-4277 – us ATLA [240]
Hard problems of scripture / Torrey, Reuben Archer – Chicago, IL: Rams Horn, c1906 – 1mf – 9 – 0-524-04115-6 – mf#1992-0073 – us ATLA [220]
Hard sayings : a selection of meditations and studies / Tyrrell, George – London, New York: Longmans, Green, 1910 [mf ed 1986] – xix/469p on 2mf – 9 – 0-8370-8626-4 – mf#1986-2626 – us ATLA [230]
Hard times – n1-47. 1968-69 – 1 – (formerly: mayday) – us AMS Press [073]
Hard times – v2 n15-16 [1972 jan 14-28] – 1r – 1 – mf#1583804 – us WHS [071]
Hard to beat *see* From bad to worse / hard to beat / and, a terrible christmas
Hard travelin' times – v1 n4 [1972 sep 27] – 1r – 1 – mf#1583818 – us WHS [071]
The hard wheat belt – Brandon, Man: Western Publ Co, [1898-189- or 19–] – 9 – mf#P05121 – cn Canadiana [630]
Hardcastle, Mrs [Mary Scarlett Campbell] *see* Life of john, lord campbell, lord high chancellor of great britain
Hardee County (Fla) County Commission *see* Hardee county, in the heart of south florida
Hardee county herald – Wauchula, FL. 1940-1955 aug – 13r – (gaps) – us UF Libraries [071]
Hardee county, in the heart of south florida / Hardee County (Fla) County Commission – Wauchula, FL. 1926? – 1r – us UF Libraries [978]
Hardeland, Otto *see* Geschichte der lutherischen mission
Harden express – Harden, 1946; 1952-55 – 10r – A$611.67 vesicular A$666.67 silver – at Pascoe [079]
Harden Family Association *see* Harden newsletter
Harden, J M *see* The anaphoras of the ethiopic liturgy
Harden, Maximilian *see*
– Stinnes
– Die zukunft
Harden murrumburrah express – jan 1962-dec 1992 – 21r – 9 – at Pascoe [079]
Harden newsletter / Harden Family Association – v1 n2-3 [1983 dec-1984 mar], v3 n1-v6 n1 (1986 jan-1989 mat) – 1r – 1 – (continued by: harden-in-ing newsletter) – mf#1856645 – us WHS [366]
Hardenberg, Friedrich von *see* Ein kleinstaatlicher minister des achtzehnten jahrhunderts
Hardenberg, Friedrich von (Novalis) *see*
– Die analogie von natur und geist als stilprinzip in novalis' dichtung
– Der dichter vor der geschichte
– Das erlebnis und die dichtung
– Goethes einfluss auf novalis heinrich von ofterdingen
– Herder, novalis und kleist
– Hymns and thoughts on religion
– Inni alla notte e canti spirituali
– Der magische idealismus in novalis' maerchentheorie und maerchendichtung
– Mystik und lyrik bei novalis
– Novalis
– Novalis als naturphilosoph
– Novalis als philosoph
– Novalis devant la critique
– Novalis (friedrich von hardenberg)
– Novalis "heinrich von ofterdingen" und der "guido" des grafen von loeben
– Novalis' hymnen an die nacht und geistliche lieder
– Novalis' lyrik
– Novalis schriften
– Novalis und der pietismus
– Novalis und die franzoesischen symbolisten
– Novalis und die welt des ostens
– Novalis und sophie von kuehn
– Tres ensayos alemanes
– Weltschau deutscher dichter
Hardenberg, Friedrich von [Novalis] *see* Der hueter der schwelle
Hardenburgh, William Andrew *see* Operation of sewage-treatment plants
Harder, Ernst *see* Der einfluss portugals bei der wahl pius 6
Harder, Meghan *see* The effects of stage-matched intervention on the stages of change and exercise self-efficacy
Harder, Paul O *see* Orchestral style of brahms as based upon an analysis of the third symphony
[Hardesheim, C] *see*
– Consensvs orthodoxvs sacrae scriptvrae et veteris ecclesiae
– De confessione avgvstana
– Fvndamenta lvtheranae doctrinae de ubiquitate
– Historia des augspurgischen confession
– Refvtatio dogmatis de fictitiae carnis christi omnipraesentia
– Theses et sentententiae qvibvs veri corporis christi vera et realis communicatio in dominica coena breuiter explicatur...
Hardesty name index to "military history of ohio," 1885 – 1r – 1 – mf#B33715 – us Ohio Hist [355]
Hardijzer, Carol Hugo *see* The relationship between cognitive styles and personality types
Hardin county atlas, 1879 – 1r – 1 – mf#B7070 – us Ohio Hist [978]
Hardin county republican / Harding Co. Kenton – (1893-94,01,03-6/08,09-7/1911) [wkly] – 6r – 1 – mf#B9422-9427 – us Ohio Hist [071]
Hardin, John Huffman *see*
– The bible school to-day
– The sunday school helper
Hardiness levels and coping strategies of female head women basketball coaches in the national collegiate athletic association / Happ, Carol K – 1998 – 1mf – 9 – $4.00 – mf#PSY 2040 – us Kinesology [790]
Harding, AKW *see* Factors influencing family use of health care services in tamil nadu (india) villages
Harding, Bertita (Leonarz) *see* Southern empire
Harding, Bertita Leonarz *see* Amazon throne
Harding, Burcham *see* Brotherhood, nature's law
Harding, Charles Irvin *see* Evaluation of a static technique for estimating atmospheric...
Harding Co. Ada *see*
– Herald
– Record
– University herald
Harding Co. Kenton *see*
– Daily democrat
– Graphic news republican
– Graphic-news
– Hardin county republican
– News-republican
– Republican
– Republican series
Harding, Henry Alfred *see* Musical ornaments
Harding, James Duffield *see*
– Elementary art
– Lessons on trees
– The principles and practice of art
Harding, T *see*
– The decades
Harding, Thomas *see* God's predestination the confidence of his saints
Harding, Vanessa *see* Historical gazetteer of london before the great fire
Harding, W *see* Clergyman's remonstrance with a dissenting minister
Harding, Warren G *see* Speeches and addresses of warren g harding...
Harding, William *see* Remarks upon the recent conduct of mr temple
Hardings dublin impartial news letter – Dublin, Ireland. 21 jul 1724; 25 jul-1 aug 1724; 3, 6, 9 mar 1725 – 1/2r – 1 – uk British Libr Newspaper [072]
Harding's dublin impartial newsletter – Ireland, 21,25 Jul-1 Aug 1724 – 6ft – 1 – uk British Libr Newspaper [072]
Hardings weekly impartial news letter – Dublin, Ireland. 23, 30 mar; 4, 7 may; 1, 4, 8 jun; 6 jul; 6 aug 1723 – 1/4r – 1 – uk British Libr Newspaper [072]
Harding's weekly impartial newsletter – Ireland, 23,30 Mar, 4,7 May, 1,4,8 Jun, 6 Jul, 6 Aug 1723 – 8ft – 1 – uk British Libr Newspaper [072]
Hardman and Co, John *see*
– [Catalogue of designs for ecclesiastical metalwork etc]
– [Trade catalogue of sacred vessels]
Hardman, William *see* Lights and shadows of church history
Hardman's baptist church. dekalb county. georgia : church records – 1825-55 – 1 – 9.00 – us Southern Baptist [242]
Hardness reducers in drilling : a physico-chemical method of facilitating the mechanical destruction of rocks during drilling – Melbourne: Council for Scientific & Industrial Research, 1948 – 1 – us CRL [600]
Hardscrabble : or, the fall of chicago: a tale of indian warefare / Richardson, John – New-York: Pollard & Moss, 1888 [mf ed 1982] – 2mf – 9 – mf#SEM105P79 – cn Bibl Nat [830]
Hardscrabble : or, the fall of chicago: a tale of indian warefare a tale of indian warfare / Richardson, John – New York: Pollard & Moss, 1888 [mf ed 1976] – 1r – 5 – mf#SEM16P261 – cn Bibl Nat [830]
Hardt, Ernst *see* Koenig salomo
Hardt, H von der *see* Rerum concilii oecumenici constantiensis...
Hardt, Hermann von der *see* Magnum oecuminicum constantiense concilium
Hardt, Roland *see*
– Schrittmachertherapie in der geriatrie
– Vasodilatorische efferenzen der hundezunge
Harduinus, Johannes *see* Acta conciliorum
Hardware – Toronto: J B McLean, [1888?-189-?] [mf ed v1 n11 may 31 1889] – 9 – mf#P04580 – cn Canadiana [680]
Hardware age – Radnor. 1981-1983 (1) 1981-1983 (5) 1981-1983 (9) – (cont by: chilton's hardware age. cont: chilton's hardware age) – mf#946,01 – us UMI ProQuest [680]
Hardware age home improvement market – Radnor. 1995-1998 (1) 1995-1998 (5) 1995-1998 (9) – (cont: hardware age home improvement marketplace) – ISSN: 1088-6168 – mf#946,04 – us UMI ProQuest [680]
Hardware age home improvement marketplace – Radnor. 1994-1995 (1) 1994-1995 (5) 1994-1995 (9) – (cont: chilton's hardware age. cont by: hardware age home improvement market) – mf#946,03 – us UMI ProQuest [680]
Hardware world – Radnor. 1956-1963 (1) – mf#943 – us UMI ProQuest [680]
Hardwick 1564-1950 – (= ser Cambridgeshire parish register transcript) – 3mf – 9 – £3.75 – uk CambsFHS [929]
Hardwick 1670-1849 – Oxford, MA (mf ed 1996) – (= ser Massachusetts vital record transcripts to 1850) – 12mf – 9 – 0-87623-262-4 – (mf 1t-6t: births 1717-1840. mf 1t-2t,4t: publishments 1743-90. mf 1t-4t: marriages 1746-88. mf 2t-3t,6t: deaths 1746-1833. mf 6t-7t,9t: marriages 1788-1844. mf 7t-10t: publishments 1790-1849. mf 10t: deaths 1834-43; births 1800-51; out-of-town marriages 1670-1799. mf 11t: births 1843-49. mf 12t: marriages & deaths 1844-49) – us Archive [978]
Hardwick 1735-1895 – Oxford, MA (mf ed 1987) – (= ser Massachusetts vital records) – 52mf – 9 – 0-87623-038-9 – (mf 1-4: town & vital records 1730-50. mf 5-15: town & vital records 1734-1802. mf 16-24: town & vital records 1789-1833. mf 25-31: vital records 1832-51. mf 32-41: miscellaneous records 1851-88. mf 42-43: b,m,d 1843-56. mf 44-52: b,m,d 1852-95) – us Archive [978]
Hardwick, C *see* Historia monasterii s augustini cantuariensis by thomas of elmham (rs8)
Hardwick, Charles *see*
– Christ and other masters
– A history of the articles of religion
– A history of the christian church during the reformation
– A history of the christian church, middle age
Hardwicke, Herbert Junius *see* Evolution and creation
Hardy, A S *see* Life and letters of joseph hardy neesima
Hardy, Alfred *see* Life and adventure in the 'land of mud'
Hardy, Arthur Sherburne *see* Life and letters of joseph hardy neesima
Hardy county news – Moorefield, WV. 1897-1942 (1) – mf#67374 – us UMI ProQuest [071]
Hardy E L C P *see* Recueil des croniques et anchiennes istories de la grant bretagne (rs39)
Hardy, E L C P *see* A collection of chronicles and ancient histories of great britain (rs40)
Hardy, Edmund *see*
– Die allgemeine vergleichende religionswissenschaft
– Buddha
– Der buddhismus nach aeltern paali-werken
– Indische religionsgeschichte
– Die vedisch-brahmanische periode der religion des alten indiens
Hardy, Edward John *see*
– Doubt and faith
– John chinaman at home
– Mr thomas atkins
Hardy, Edwin Noah *see* The churches and educated men
Hardy, Ernest George *see*
– Christianity and the roman government
– Studies in roman history
Hardy, H *see* Recueil des croniques et anchiennes istories de la grant bretagne (rs39)
The hardy herald – Hardy, NE: R K Hill, 1880-v76 n52. sep 12 1957 (wkly) [mf ed 1882-1957 (gaps) filmed 1974-82] – 1 – (not publ jan 3 and aug 22 1890, and jan 2-9 1891. suspended with aug 31 resumed with sep 21 1944. suspended with nov 6 1947; resumed with sep 16 1948) – us NE Hist [071]
Hardy, Joseph *see* A picturesque and descriptive tour in the mountains of the high pyrenees
Hardy, M G Le *see* Etude sur la baronnie et l'abbaye d'aunay-sur-odon
Hardy, Nathaniel *see* The first general epistle of st john the apostle
Hardy, Philip Dixon *see* Ireland in 1846-7
Hardy, Robert *see* Travels in the interior of mexico, in 1825, 1826, 1827, and 1828
Hardy, Robert Spence *see*
– Christianity and buddhism compared
– Eastern monachism
– The legends and theories of the buddhists compared with history and science
– A manual of budhism in its modern development
Hardy, T D *see*
– Descriptive catalogue of materials relating to the history of great britain and ireland to the end of the reign of henry 7
– Lestorie des engles solum la translacion maistre geffrei gaimar
– Registrum palatinum dunelmense
Hardy, Thomas *see*
– The original manuscripts and papers
– The three wayfarers
Hardy, W *see* A collection of chronicles and ancient histories of great britain (rs40)
Hare, Augustus John Cuthbert *see*
– Biographical sketches
– Cities of southern italy ând sicily
– The gurneys of earlham
– Venice
Hare, Augustus William *see*
– Guesses at truth
– Letters on the religious part of the catholic question
Hare, baboon and their friends / Harmon, Roger – Cape Town, South Africa. 1967 – 1r – us UF Libraries [960]
Hare, Christopher [pseud] *see*
– The most illustrious ladies of the italian renaissance
– A queen of queens and the making of spain
Hare, Francis *see* Difficulties and discouragements which attend the study of the scriptures...
Hare, George Emlen *see* Visions and narratives of the old testament
Hare, J I C *see* Smith's leading cases
Hare, John Innes Clark *see* American leading cases

Hare, Julius Charles *see*
- Better prospects of the church
- The contest with rome
- Guesses at truth
- Sermons preached on particular occasions
- Thou shalt not bear false witness against thy neighbour
- The victory of faith
- Vindication of luther against his recent english assailants

Hare, Robert *see* Experimental investigation of the spirit manifestations

Hare, Thomas *see*
- The development of the wealth of india
- Hare's reports

Harendt, Norbert *see* Geometrische zuordnung sequentieller roentgenbilder mit hilfe drehungs- und masstabsinvarianter bildmuster-merkmale

Harengs terribles / Breffort, Alexandre – Paris, France. 1950 – 1r – us UF Libraries [440]

Hare's reports : reports of cases adjudged in the high court of chancery / Hare, Thomas – v1-11. 1841-53. London: A Maxwell & Son/W Maxwell, 1843-58 (all publ) – 86mf – 9 – $129.00 – (v11 contains a general index to the whole series and "a historical preface") – mf#LLMC 95-286 – us LLMC [324]

Harethi moallaca cum scholiis zuzenii e codicibus parisiensibus et abulolae carmina duo inedita e codice petropolitano : edidit latine vertit et commentario instruxit; typis regiis arabicis / Harit Ibn-Hilliza al- – Bonnae ad Rhenum: Habicht 1827 – (= ser Whsb) – 2mf – 9 – €30.00 – mf#Hu 212 – gw Fischer [410]

Harfield and bishopston recorder – Bristol, England. 1899-Jan 1931 – 12r – 1 – uk British Libr Newspaper [072]

Harflerimizin muedafaasi / Bey, Alican Serif – Yeni Matbaa, 1926 – (= ser Ottoman literature, writers and the arts) – 1mf – 9 – $25.00 – (transl from russian by abdullah battal) – us MEDOC [470]

Harford, Charles F *see* Pilkington of uganda

Harford, George *see* The hexateuch according to the revised version

Harford, John Scandrett *see* The life of michael angelo buonarroti

Harford, Keister *see* Woman's position in the church

Harger's times – Watertown WI. 1878 jan 5-sep 7 – 1r – 1 – mf#944235 – us WHS [071]

Hargis, Modeste *see*
- Don francisco moreno
- Escambia county history
- Escambia county place-names
- Greek study
- Information on pro-german activities of german-ame...
- Study on greeks
- Study on greeks in pensacola florida

Hargrave, Francis *see* Juridical arguments and collections

Hargrave, st peter – (= ser Cheshire monumental inscriptions) – 1mf – 9 – £2.50 – mf#18 – uk CheshireFHS [929]

Hargraves, Edward Hammond *see* Australia and its gold fields

Hargreaves, Harold *see* Excavations in baluchistan 1925, sampur mound, mastung, and sohr damb, nal

Hargreaves, Robert *see* Teaching of brass instruments in school music supervisors' courses

Hargrove, Charles *see* Reasons for retiring from the established church

Hargrove, John *see* Substance of a sermon on the leading doctrines

Hargrove, R K *see* Woman's work in the church

Hari : the jungle lad / Mukerji, Dhan Gopal – New York: EP Dutton & Co, 1937 – (= ser Samp: indian books) – (ill by morgan stinemetz) – us CRL [490]

Hari charitra : or, comparison between the ad granth and the bible / Valji Bhai – 1st ed. Lodiana: Lodiana Mission Press, 1893 – 1mf – 9 – 0-524-02945-8 – mf#1990-3157 – us ATLA [230]

Harian kami – Djakarta, Indonesia. Jul 1966-Jan 21 1974 – 12r – 1 – us L of C Photodup [079]

Harian rakjat – Jakarta, Indonesia. 1952-1965 (1) – mf#67741 – us UMI ProQuest [079]

Hariciyye nezareti salnamesi – 1302 [1885] – (= ser Ministry and special interest salnames) – 8mf – 9 – $140.00 – us MEDOC [956]

Haring, Clarence Henry *see*
- Buccaneers in the west indies
- Empire in brazil
- Spanish empire in america

Haring gangis ug haring leon : fabulas filipinas / Buyser y Aquino, Fernando – Sugbu, K P: [s n 1912?] [mf ed Bloomington IN: Indiana Uni Lib, Preservation Dept 1984] – (= ser Coll...in the bisaya language 1) – 1r – 1 – us Indiana Preservation [490]

Haringer, Michael *see*
- Theologia moralis
- Vita del beato clemente ma. hofbauer

Haringey advertiser – London, UK. jan-aug 1986; 4 sep-18 dec 1986; 1987-aug 1988; 29 sep 1988-90; 1992 – 15 1/2r – 1 – uk British Libr Newspaper [072]

Haringey independent – London, UK. 1986-25 aug 1988; jan-14 dec 1989; 1990; 1992 – 14r – 1 – (aka: haringey wood green tottenham hornsey and crouch end independent; independent (haringey wood green etc)) – uk British Libr Newspaper [072]

Haringman, H *see* H haringman's vormal holl kavallerie-lieutenants tagebuch einer reise nach marokko und eines achtwoechentlichen aufenthaltes in diesem lande

Harington, Edward Charles *see* Bull of pope pius the ninth and the ancient british church

Harit Ibn-Hilliza al- *see* Harethi moallaca cum scholiis zuzenii e codicibus parisiensibus et abulolae carmina duo inedita e codice petropolitano

Harjedalen – Sveg, Sweden. 1944-78 – 30r – 1 – sw Kungliga [078]

Harjedalen – Sveg, Sweden. 1979- – 1 – sw Kungliga [078]

Hark, Joseph Maximilian *see* The unity of the truth in christianity and evolution

Harkavy, Albert *see*
- Catalog der hebraeischen bibelhandschriften der kaiserlichen oeffentlichen bibliothek in st petersburg
- Neuaufgefundene hebraeische bibelhandschriften

Harkavy, Alexander *see* Olendorfs methode zikh grindlikh oystsulernen di englishe

Die harke – Nienburg DE, 1882 11 feb-14 dec, 1884 1 jan-23 sep, 1885 22 aug-17 dec, 1886-1941 31 may, 1950-76 – 181r – 1 – (filmed by misc inst: 1977- [ca 6r/yr]) – gw Mikrofilm; gw Misc Inst [074]

Harker, Oliver Albert *see* Three needed reforms in criminal procedure: an address before the illinois states attorneys' association at chicago, december 28th, 1915

Harkey, Simeon Walcher *see*
- The church's best state
- Justification by faith as held and taught by lutherans, together with the associated doctrines of sanctification and the union of the soul with christ

The harklean version of the epistle to the hebrews, chap. 11. 28-13. 25 / ed by Bensly, Robert Lubbock – Cambridge: University Press, 1889 – 1mf – 9 – 0-8370-1803-X – mf#1987-6191 – us ATLA [227]

Harkness, Edward S *see* Collection(mexican and peruvian documents)

Harkness, Effie *see* Excerpts from her diaries relating to her service with the methodist overseas mission in the solomon islands

Harkness, Henry *see* A description of a singular aboriginal race

Harkness, John C *see* The normal principles of education

Harkness, Margret Elise *see* Assyrian life and history

Harlan county democrat – Republican City, NE: [s.n.], 1883 [mf ed 1893,1895-1902 (gaps) filmed 1979-[93]] – 2r – 1 – (absorbed by: harlan county ranger) – us NE Hist [071]

Harlan county journal – Alma, NE: H S Wetherell, 1897 (wkly) – 17r – 1 – (absorbed: news reporter (alma ne) jun 9 1899, alma record (alma ne) jul 1 1925, orleans chronicle apr 1 1961 and: stamford star may 7 1964) – us Bell [071]

Harlan county journal – Alma, NE: H S Wetherell, 1897 (wkly) [mf ed oct 20 1899-(gaps)] – 27r – 1 – (absorbed: news reporter jun 9 1899, alma record jul 1 1925, orleans chronicle apr 1 1961 and: stamford star may 7 1964) – us NE Hist [071]

The harlan county ranger – Republican City, NE: W L Martin, jun 1902-21st yr n22. oct 12 1922 (wkly) [mf ed filmed 1973] – 6r – 1 – (absorbed: harlan county democrat. cont by: republican city ranger) – us NE Hist [071]

Harlan footprints / Genealogical Society of Harlan County, Kentucky et al – 1982 dec-1988 jan – 1r – 1 – mf#1094139 – us WHS [929]

Harlan, Henry David *see* Syllabus of the hon henry d harlan's lectures on the law of domestic relations

Harlan, Rolvix *see* John alexander dowie and the christian catholic apostolic church in zion

Harland, Marion *see* John knox

Harleian miscellany / ed by Oldys, William & Johnson, Samuel – 1744-46 – 153mf – 9 – mf#C35-23600 – us Primary [420]

The harleian miscellany – 4 reels – 1 – $175.00 – us Trans-Media [323]

Harleian Society. London *see* Publications

Harlem Arts and Cultural Consortium *see* Uptown art news

Harlem Cultural Council *see* Black arts new york

Harlem cultural review : the newsletter of the harlem cultural council – 1978 mar, [1979 dec] – 1r – 1 – (continued by: black arts new york) – mf#4879214 – us WHS [700]

Harlem daily – New York NY. 1965 sep 23, oct 12 – 1r – 1 – mf#3054997 – us WHS [071]

Harlem Environmental Impact Project [NY] *see* Habitat

Harlem Environmental Impact Project (NY) et al *see* Harlem habitat

Harlem habitat : the environmental and health newsletter of the harlem environmental impact project, a program of the international agency for minority artist affairs / Harlem Environmental Impact Project (NY) et al – v3 n2 [1992 feb/mar], iss n7-8,11-13,15-20 [1992 nov/dec-1993 feb/mar, nov/dec-1994 mar/apr, oct/nov-1996 jan/feb) – 1r – 1 – (continued by: habitat) – mf#2956791 – us WHS [362]

Harlem heights daily citizen – New York. Oct. 18, 1933-Jan. 24, 1934 – 1 – us NY Public [071]

Harlem Hospital Center *see* Outreach

Harlem Neighborhoods Association, Inc *see* Hana newsletter

Harlem news / Architects' Renewal Committee in Harlem (ARCH) – v1 n2-[n n] [1967 oct-1971 dec] – 1r – 1 – mf#1057893 – us WHS [307]

Harlem notebook / Queens College (New York, NY) – 1980 may-jun – 1r – 1 – mf#5305312 – us WHS [071]

Harlem pointer / Communist Party of the USA – 1940 aug 3 – 1r – 1 – mf#3055009 – us WHS [335]

Harlem quarterly – New York. n1-4. 1949-50 [all publ] – (= ser Black journals, series 1) – 2mf – 9 – $45.00 – us UPA [305]

Harlem School of the Arts *see*
- Newsletter
- Voices: newsletter of the harlem school of the arts

Harlem succeeds : when we support these businesses – 1989, 1991, 1993-1995, 1997/1998 winter, 1998 spring-summer, winter – 1r – 1 – mf#5308767 – us WHS [350]

Harlem tenant / West Harlem Coalition – 1988 sep-nov /dec, 1990/1991 winter – 1r – 1 – mf#5305331 – us WHS [362]

Harlem Third World Institute *see* Third world trade winds

Harlem United *see* Positively yours

Harlem urban development corporation news – 1984 nov, 1985 apr-aug – 1r – 1 – mf#5297087 – us WHS [307]

Harlem usa – 1988 jun/aug – 1r – 1 – mf#5308446 – us WHS [071]

Harlem valley news / J Raymond Jones Democratic Club – 1980 aug – 1r – 1 – mf#5297090 – us WHS [320]

Harlem youth speaks – 1970 dec – 1r – 1 – mf#5305348 – us WHS [305]

Harlem Youth Writing Workshop *see* Uptown sum...

Harlequin : a journal of the drama – London. 1829-1829 (1) – mf#5562 – us UMI ProQuest [790]

Harler, Campbell R *see* Culture and marketing of tea

Harless, Adolf von *see* Das verhaeltniss des christenthums zu cultur- und lebensfragen der gegenwart

Harless, Gottlieb Christoph Adolf von *see*
- Commentar ueber den brief pauli an die ephesier
- Jacob boehme und die alchymisten
- Staat und kirche
- System of christian ethics

Harlez, Charles de *see*
- Le livre des esprits et des immortels
- Religions de la chine

Harlington – (= ser Bedfordshire parish register series) – 1mf – 9 – £3.00 – uk BedsFHS [929]

Harlington, st mary monumental inscriptions monumental inscriptions – Arthur Weight Matthews 1907 – (= ser Bedfordshire parish register series) – 1mf – 9 – £1.25 – uk BedsFHS [929]

Harloff, A J W *see* Memorie van overgave van surakarta 1918-1922 door resident a j w harloff

The harlots and the pharisees : or, the barbary coast in a barbarous land. also, the story of a socialist mayor. letter declining mayoralty nomination / Wilson, Jackson Stitt – Berkeley, CA: J Stitt Wilson, 1913 – (= ser Social Crusade Series) – 1mf – 9 – 0-524-03437-0 – mf#1990-0991 – us ATLA [335]

Harlow, Samuel Ralph *see* The life of h. roswell bates

Harlow, Vincent Todd *see* History of barbados, 1625-1685

Harlow, William Sturtevant *see*
- Duties of sheriffs and constables, as defined by the laws, and interpreted by the supreme court, of the state of california
- Duties of sheriffs and constables particularly under the practice in california, and the pacific states and territories

Harlton 1567-1950 – (= ser Cambridgeshire parish register transcript) – 4mf – 9 – £5.00 – uk CambsFHS [929]

Harm jan huidekoper / Tiffany, Nina Moore & Tiffany, Francis – Cambridge: Riverside Press, 1904 – 2mf – 9 – 0-524-04304-3 – (incl bibl ref) – mf#1992-2024 – us ATLA [920]

Harman collection of hilton head – [S.l]: [s.n.] – 2r – 1 – (incl ind) – mf#45-352 – us South Carolina Historical [333]

Harman, Edward George *see* The countesse of pembroke's arcadia, examined and discussed

Harman, H A *see* Sounds of english speech

Harman, Henry Martyn *see* Introduction to the study of the holy scriptures

Harman, Jeanne (Perkins) *see* Virgins

Harman, Jeanne Perkins *see* Love junk

Harmar, Josiah *see* Waste books, journals, and ledger, 1788-1791

Harmer, J R *see* The apostolic fathers

Harmer, Peter A P *see* Paternalism and special olympics

Harmful algae – Amsterdam. 2002+ (1,5,9) – ISSN: 1568-9883 – mf#42888 – us UMI ProQuest [576]

Harmless people / Thomas, Elizabeth Marshall – New York, NY. 1959 – 1r – us UF Libraries [960]

Harmlose betrachtungen : gesammelt auf einer reise von hamburg nach griechenland, constantinopel und dem schwarzen meere im jahre 1822 / Dannenberg, Carl – Hamburg 1823 [mf ed Hildesheim 1995-98] – 2mf – 9 – €60.00 – 3-487-29033-2 – gw Olms [880]

Har-moad : or, the mountain of the assembly. a series of archaeological studies, chiefly from the stand-point of the cuneiform inscriptions / Miller, Orlando Dana – North Adams, Mass: Stephen M Whipple, 1892 – 2mf – 9 – 0-524-02313-1 – (incl bibl ref) – mf#1990-2936 – us ATLA [930]

The harmon home and day school : for young ladies and little girls – [Ottawa?: Harmon Home & Day School Co of Ottawa, 1902?] – 1mf – 9 – 0-665-73870-6 – mf#73870 – cn Canadiana [370]

Harmon, Roger *see* Hare, baboon and their friends

Harmonia cantionum ecclesiasticarum : kirchengesenge und geistliche lieder d lutheri und anderer frommen christen / Calvisius, Seth – [Leipzig] In Vorlegung Jacobi Apels Buchh 1597 [mf ed 19–] – 15mf – 9 – mf#fiche 240 – us Sibley [780]

Harmonia cantionum ecclesiasticarum : kirchengesenge, und geistliche lieder, d lutheri und anderer frommen christen... / Calvisius, Sethus – Leipzig: Apel 1597 – (= ser Hqab. literatur des 16. jahrh.) – 7mf – 9 – €75.00 – mf#1597a – gw Fischer [780]

Harmonia confessionum fidei orthodoxarum et reformatarum ecclesiarum / ed by Salvart, J F – Geneve, Saint-Andr – 7mf – 9 – mf#PFA-178 – ne IDC [240]

Harmonia confessionvm fidei – Genevae, 1581 – 7mf – 9 – mf#PBU-694 – ne IDC [240]

Harmonia ex tribus evangelistis composita, matthaeo, marco et luca : adiuncto seorsum johanne, quod pauca cum aliis communia habeat / Calvin, J – [Geneva]: Excudebat Robertus Stephanus, 1555 – 13mf – 9 – mf#CL-63 – ne IDC [240]

Harmoniae seu cantiones sacrae... / Berger, Andreas – 1606 – (= ser Mssa) – 4mf – 9 – €60.00 – mfchl 181 – gw Fischer [780]

Harmoniarum sacrarum continuatio... / Staden, Johann – 1621 – (= ser Mssa) – 3mf – 9 – €50.00 – mfchl 408 – gw Fischer [780]

Harmonic analysis of the requiem by giuseppe verdi / Johnson, Gordon Allen – U of Rochester 1954 [mf ed 19–] – 3mf – 9 – mf#fiche 227, 312 – us Sibley [780]

Harmonic equipment as evidenced in the work of gluck / Nakaseko, Kazu – U of Rochester 1931 [mf ed 1974] – 3mf – 9 – mf#fiche78 – us Sibley [780]

Harmonic equipment evidenced in the works of claudio monteverdi (1567-1643) / Campbell, Dana M – U of Rochester 1931 [mf ed 19–] – 1r – 1 – mf#film 994 – us Sibley [780]

Harmonic equipment of rameau / Falk, Genevieve – U of Rochester 1931 [mf ed 19–] – 2mf – 9 – mf#fiche 91, 456 – us Sibley [780]

Harmonic technique of edward elgar based upon the dream of gerontius / Dann, Mary G – U of Rochester 1937 [mf ed 19–] – 1r – 1 – mf#film 2573 – us Sibley [780]

Harmonic technique of ernest chausson as exemplified in his chamber music / Goddard, Evelyn Lily – U of Rochester 1941 [mf ed 19–] – 3mf – 9 – mf#fiche 482 – us Sibley [780]

Harmonicorvm libri 12 : in qvibvs agitvr de sonorvm natvra, cavsis, et effectibvs: de consonantiis, dissonantiis, rationibus, generibus modis... / Mersenne, Marin – ed avcta, Lvtetiae Parisiorvm, sumptibus Gvillelmi Bavdry 1648 [mf ed 19–] – 1r – 1 – mf#film 92 – us Sibley [780]

Harmonics : or the philosophy of musical sounds / Smith, Robert – 2nd augm ed, London: printed for T & J Merrill, Cambridge 1759 [mf ed 19–] – 3mf – 9 – mf#fiche 901 – us Sibley [780]

Harmonie – Recueil litteraire publiant les oeuvres des jeunes ecrivains de langue francaise. Dir. Eugene Bure. no. 1-11. oct 1891-aout 1892. devenu: Philosophie libertaire. Harmonie. Revue sociale et litteraire. no. 12-20. sept 1892-mai 1893. Marseille – 1 – fr ACRPP [800]

Die harmonie der ergebnisse der naturforschung mit den forderungen des menschlichen gemuethes, oder, die persoenliche unsterblichkeit als folge der atomistischen verfassung der natur / Drossbach, Maximilian – Leipzig: FA Brockhaus, 1858 – 1mf – 9 – 0-7905-8644-4 – mf#1989-1869 – us ATLA [110]

Die harmonie des alten und des neuen testamentes : ein beitrag zur erklaerung der biblischen geschichte / Martin, Konrad – Mainz: Franz Kirchheim, 1877 – 1mf – 9 – 0-8370-9639-1 – (incl bibl ref) – mf#1986-3639 – us ATLA [220]

Harmonie herald : the newsletter of the harmonie associates, inc – 1969 mar, 1969 sep, 1970 feb-1981 dec – 1r – 1 – mf#578615 – us WHS [071]

L'harmonie pratique : ou exemples pour le traite des accords / Roussier, Pierre Joseph, Abbe – A Paris: Chez l'editeur...A Lyon: chez M Castaud... [1775] – 3mf – 9 – mf#fiche 891 – us Sibley [780]

Harmonie super odis horatii flacci / Tritonius, Petrus – 1507 – (= ser Mssa) – 1mf – 9 – €20.00 – mfchl 419 – gw Fischer [780]

Harmonie universelle : contenant la theorie et la pratique de la musique, ou il est traite de la nature des sons & des mouvements... / Mersenne, Marin – Paris: S Cramoisy 1636-37 [mf ed 19–] – 3v in 1 on 1r / 24mf – 1,9 – mf#film 254, 234 / fiche 615 – us Sibley [780]

Harmonielehre / Swoboda, August – Wien: [F C Beck] 1828- [mf ed 19–] – 1r – 1 – mf#pres. film 120 – us Sibley [780]

Harmonielehre : [Tonalitaetslehre] / Krehl, Stephan – Berlin: Vereinigung wissenschaftlicher verleger 1922 [mf ed 1991] – (= ser Theorie der tonkunst und kompositionslehre 2.t) – 1r – 1 – (incl ind) – mf#pres. film 102 – us Sibley [780]

Harmonien / Kuehnhold, Marianne – New-York: C Schmidt, 1869 – 1r – 1 – us Wisconsin U Libr [810]

Harmonische seelen lust, musicalischer goenner und freunde : das ist, kurtze...praeludia von 2. 3. und 4 stimmen ueber die bekanntesten choral-lieder etc... / Kauffmann, Georg Friedrich – Leipzig: auf Kosten des Autoris und in Commission...bey Boeti seel, Tochter [1722] [mf ed 19–] – 4mf / 1r – 9,1 – mf#fiche 911 / pres. film 150 – us Sibley [780]

Die harmonistik im evangelientext des codex cantabrigiensis : ein beitrag zur neutestamentlichen textkritik / Vogels, Heinrich Joseph – Leipzig: J C Hinrichs, 1910 – 1mf – 9 – 0-7905-1855-4 – (incl ind) – mf#1987-1855 – us ATLA [220]

Die harmonistik im evangelientext in codex cantabrigiensis / Vogels, H J – Leipzig, 1910 – (= ser Tugal 3-36/1a) – 2mf – 9 – €5.00 – ne Slangenburg [220]

The harmonized and subject reference new testament : king james's version made into a harmonized paragraph, local, topical, textual, and subject reference edition, in modern english print – Delaware, NJ: Subject Reference Co, 1904 – 2mf – 9 – 0-524-04087-7 – mf#1992-0045 – us ATLA [225]

A harmonized exposition of the four gospels / Breen, Andrew Edward – Rochester, NY: John P Smith Print House, 1899-1904 [mf ed 1993] – 4v on 7mf – 9 – 0-524-05662-5 – mf#1992-0512 – us ATLA [226]

Harmonologia musica oder kurtze anleitung zur musicalischen composition / Werckmeister, Andreas – Franckfurth, Leipzig: T P Calvisii 1702 [mf ed 19–] – 3mf – 9 – mf#fiche 861 – us Sibley [780]

Harmony baptist church. cusseta, north carolina : church records – 1840-69 – 1 – us Southern Baptist [242]

Harmony grove baptist church : church minutes and membership rolls – Gaar's Mill, LA. 634p. 1877-1966 – 1 – mf#6966 – us Southern Baptist [242]

Harmony grove baptist church. lincoln county. missouri (extinct) : church records – 1880-1912. 288p – 1 – us Southern Baptist [242]

A harmony in greek of the gospels : with notes / Newcome, William – Andover: printed by Flagg & Gould, 1814 [mf ed 1989] – 2mf – 9 – 0-8370-1169-8 – (text in greek, notes in english & latin) – mf#1987-6005 – us ATLA [226]

The harmony of ages : a thesis on the relations between the conditions of man and the character of god / Parker, Hiram – Boston: JP Jewett, 1856 [mf ed 1993] – (= ser Baptist coll) – 1mf – 9 – 0-524-07166-7 – mf#1991-2955 – us ATLA [242]

Harmony of divine operations / Dore, James – London, England. 1804 – 1r – us UF Libraries [240]

The harmony of ethics with theology : an essay in revision: is there probation after death? is there hope for the heathen? can infants be saved? / Robins, Henry Ephraim – New York: AC Armstrong, 1891 – 1mf – 9 – 0-524-00086-7 – mf#1989-2786 – us ATLA [240]

The harmony of protestant confessions : exhibiting the faith of the churches of christ, reformed after the pure and holy doctrine of the gospel, throughout europe = Harmonia confessionum fideio orthodoxarum & reformatarum ecclesiarum / Salnar – new rev enl ed. London: J F Shaw, 1842 – 2mf – 9 – 0-7905-8146-9 – mf#1988-6093 – us ATLA [242]

A harmony of samuel, kings and chronicles / Crockett, William Day – Baker. 1957 – 9 – $12.00 – us IRC [221]

The harmony of scripture : showing the oneness between the old and new testament / ed by Fearnley, Thomas – London: W Poole, 1878 – 1mf – 9 – 0-8370-9143-8 – mf#1986-3143 – us ATLA [220]

Harmony of the acts of the apostles : and chronological arrangement of the epistles and revelation, with chronological and explanatory notes, and valuable tables / Clark, George Whitefield – new rev ed. Philadelphia: American Baptist Publ Soc, 1897 – 1mf – 9 – 0-524-05659-5 – mf#1992-0509 – us ATLA [226]

The harmony of the bible with science : or, moses and geology / Kinns, Samuel – [2nd ed] New York: Cassell, Petter, Galpin, [1882?] – 2mf – 9 – 0-524-05679-X – mf#1992-0529 – us ATLA [230]

The harmony of the collects, epistles and gospels : a devotional exposition of the continuous teaching of the church throughout the year / Scott, Melville – New York: ES Gorham, [1909?] – 1mf – 9 – 0-524-05485-1 – mf#1990-5132 – us ATLA [240]

A harmony of the four evangelists : in the words of the authorized version according to greswell's harmonia evangelica... / Mimpriss, Robert – London: Macintosh [18–?] [mf ed 1992] – 1mf – 9 – 0-524-05206-9 – mf#1992-0339 – us ATLA [225]

A harmony of the four gospels in english : according to the common version / Robinson, Edward – rev ed. Boston: Houghton, Mifflin, 1886 [mf ed 1989] – 1mf – 9 – 0-8370-1170-1 – mf#1987-6006 – us ATLA [226]

A harmony of the four gospels in greek : according to the text of tischendorf / Gardiner, Frederic – rev ed. Andover: W F Draper, 1884 [mf ed 1992] – 1mf – 9 – 0-524-05025-2 – (text in greek, notes in english. incl bibl ref and app) – mf#1992-0278 – us ATLA [226]

Harmony of the gospel narratives of the passion, resurrection, and ascension of our blessed lord from the vulgate – Dublin: M H Gill, 1879 – 1mf – 9 – 0-524-06061-4 – mf#1992-0774 – us ATLA [220]

A harmony of the gospels : being the life of jesus in the words of the four evangelists – Cincinnati: Crantson & Curts, 1894 [mf ed 1992] – 1mf – 9 – 0-524-04787-1 – (arr by william henry withrow) – mf#1992-0207 – us ATLA [226]

A harmony of the gospels : in the words of the american standard edition of the revised bible and outline of the life of christ / Kerr, John Henry – 3rd rev ed. New York: American Tract Society, c1903 [mf ed 1993] – 1mf – 9 – 0-524-08069-0 – mf#1992-1129 – us ATLA [226]

A harmony of the gospels / Robertson, A T – 1950 – 9 – $12.00 – us IRC [240]

A harmony of the gospels for historical study : an analytical synopsis of the four gospels in the version of 1881 / Stevens, William Arnold & Burton, Ernest DeWitt – Boston: Silver, Burdett, 1894, c1893 [mf ed 1986] – 1mf – 9 – 0-8370-9367-8 – mf#1986-3367 – us ATLA [226]

A harmony of the gospels in the greek of the received text : on the plan of the author's english harmony... / Strong, James – rev ed. New York: Harper, 1859, c1854 [mf ed 1991] – 4mf – 9 – 0-8370-1982-6 – mf#1987-6369 – us ATLA [226]

A harmony of the gospels in the revised version : with some new features / Broadus, John Albert – New York: A C Armstrong, 1893 [mf ed 1990] – 1mf – 9 – 0-8370-1858-7 – mf#1987-6245 – us ATLA [226]

A harmony of the life of st paul : according to the acts of the apostles and the pauline epistles / Goodwin, Frank Judson – New York: American Tract Society, c1895 [mf ed 1985] – 1mf – 9 – 0-8370-3341-1 – (incl app) – mf#1985-1341 – us ATLA [920]

The harmony of the prophetic word : a key to old testament prophecy concerning things to come / Gaebelein, Arno Clemens – New York: Francis E Fitch, 1903 – 1mf – 9 – 0-7905-1388-9 – mf#1987-1388 – us ATLA [220]

The harmony of the reformed confessions as related to the present state of evangelical theology : an essay delivered before the general presbyterian council at edinburgh, july 4, 1877 / Schaff, Philip – New York: Dodd, Mead, 1877 – 1mf – 9 – 0-8370-8786-4 – mf#1986-2786 – us ATLA [242]

The harmony society at economy, penn'a : founded by george rapp, a. d. 1805 / Williams, Aaron – Pittsburgh: printed by W. S. Haven, 1866. Chicago: Dep of Photodup, U of Chicago Lib, 1961? (1r); Evanston: American Theol Lib Assoc, 1984 (1r) – 1 – 0-8370-0460-8 – mf#1984-B007 – us ATLA [975]

Harmony society records, 1786-1951 : in the pennsylvania state archives – (mf ed 1982) – 311r – 1 – silver $9,330 diazo $6,220 – (with printed guide compiled by roland m baumann and robert m dructor (1983) $6 isbn: 0-89271-025-x) – us Penn Hist [060]

Harms, Claus *see* Briefe zu einer naehern verstaendigung ueber verschiedene meine thesen betreffende puncte

Harms, Craig A *see*
- Inadequate hyperventilation as a determinant of exercise induced hypoxemia
- Influence of body fat mass on excess post-exercise oxygen consumption

Harms, Hans *see* Neue formen und bedingungen der erwerbsarbeit

Harms, Susanne *see* Clemens brentano und die landschaft der romantik

Harms, Theodor *see* Das hohelied

Harms, Wolfgang *see*
- Der kampf mit dem freund
- Der kampf mit dem freund oder verwandten in der deutschen literatur bis um 1300

Harmssen, G W *see* Reparationen, sozialprodukt, lebensstandard; versuch einer wirtschaftsbilanz

Harn, Edith Muriel *see* Wieland's neuer amadis

Harnack, Adolf von *see*
- Die acten des karpus, des papylus und der agathonike
- Acts of the apostles
- Die adresse des epheserbriefs des paulus
- Die altercatio simonis iudaei et theophili christiani
- Die altercatio simonis judaei et theophili christiani und die acta archeali und das diatesseron tatians
- Analecta zur aeltesten geschichte des christentums in rom
- Der angebliche evangelienkommentar des theophilius von antiochien
- Antwort auf die streitschrift d cremers
- Aphrahat's des persischen weisen homilien
- Die apokryphen briefe des paulus an die laodicener und korinther
- Die apostelgeschichte
- Die apostellehre und die juedischen beiden wege
- The apostles' creed
- Das apostolische glaubensbekenntniss
- Aus wissenschaft und leben
- Bible reading in the early church
- Eine bisher nicht erkannte schrift des papstes sixtus 2
- Eine bisher nicht erkannte schrift des papstes sixtus 2. vom jahre 257/8 / zur petrusapokalypse / patristisches zu luc. 16. 19
- Eine bisher nicht erkannte schrift novatians
- Brief an die flora
- Bruchstuecke des evangeliums und der apokalypse des petrus
- Christentum und die geschichte
- The constitution and law of the church in the first two centuries
- De apellis gnosi monarchica
- Diodor van tarsus
- Diodor von tarsus
- Drei wenig beachtete cyprianische schriften und die "acta pauli"
- Das edict des antoninus pius; eine bisher nicht erkannte schrift novatian's vom jahre 249/50
- Das edict des antonius pius
- Die entstehung des neuen testaments und die wichtigsten folgen der neuen schoepfung
- Essays on the social gospel
- Das evangelienfragment von fajjum
- Festgabe von fachgenossen und freunden a von harnack
- Der gefaelschte brief des bischofs theonas an den oberkammerherrn lucian
- Geschichte der altchristkiche litteratur bis eusebius
- Die gnostischen quellen hippolyts in seiner hauptschrift gegen die haeretiker – sieben neue bruckstücke der syllogismen des apelles
- Die griechische uebersetzung des apologeticus tertullian's / medicinisches aus der aeltesten kirchengeschichte
- Grundriss der dogmengeschichte
- Die gwynn'schen cajus- und hippolytus-fragmente
- Die hypotiposen des theognost
- Ist die rede des paulus in athen ein urspruenglicher bestandteil der apostelgeschichte
- Ist die rede des paulus in athen ein urspruenglicher bestandteil der apostelgeschichte? / judentum und judenchristentum in justins dialog mit trypho
- Judentum und judenchristentum im justins dialog mit trypho
- Ein juedisch-christliches psalmbuch aus dem ersten jahrhundert
- Der ketzer-katalog des bischofs maruta von maipherkat
- Der kirchengeschichtliche ertrag der exegetischen arbeiten des origenes
- Kritik des neuen testaments von einem griechischen philosophen des 3. jahrhunderts
- Das leben cyprians von pontius
- Die lehre der zwoelf apostel
- Lukas der arzt der verfasser des dritten evangeliums und der apostelgeschichte
- Luke the physician
- Marcion
- Martin luther in seiner bedeutung fuer die geschichte der wissenschaft und der bildung
- Medicinisches aus der aeltesten kirchengeschichte
- Militia christi
- The mission and expansion of christianity in the first three centuries
- Das moenchtum, seine ideale und seine geschichte
- Monasticism
- Neue studien zu marcion
- Das neue testament um das jahr 200
- Neue untersuchungen zur apostelgeschichte
- Patristische miscellen
- Die pfaff'schen irenaeus-fragmente als faelschungen pfaffs
- Die pfaff'schen irenaeus-fragmente als faelschungen pfaffs nachgewiesen
- Prof harnack's letter to the preussische jahrbuecher
- Der pseudocyprianische tractat de aleatorbus
- Der pseudocyprianische traktat de singularitate clericorum
- Die quellen der sogenannten apostolischen kirchenordnung
- Reden und aufsaetze
- The sayings of jesus
- Der scholien-kommentar das origenes zur apokalypse johannis
- Sieben neue bruchstuecke der syllogismen des apelles
- Sources of the apostolic canons
- Die terminologie der wiedergeburt und verwandter erlebnisse in der aeltesten kirche
- Thoughts on the present position of protestantism
- Die todestage der apostel paulus und petrus
- Ueber das gnostische buch pistis sophia
- Ueber den dritten johannesbrief
- Ueber verlorene briefe und actenstuecke die sich aus der cyprianischen briefsammlung ermitteln lassen
- Die ueberlieferung der griechischen apologeten
- Die ueberlieferung der griechischen apologeten des 2. jahrhunderts in der alten kirche und im mittelalter
- Urkunden aus dem antimontanistischen kampfe des abendlandes
- Ueeber das gnostische buch pistis-sophia
- Der vorwurf des atheismus in den drei ersten jahrhunderten
- Das wesen des christentums
- What is christianity?
- Die zeit des ignatius und die chronologie der antiochenischen bischoefe bis tyrannua
- Zur abercius-inschrift
- Zur quellenkritik der geschichte des gnosticismus
- Zur ueberlieferungsgeschichte der altchristlichen litteratur

Harnack, Adolf von et al *see*
- The ologische abhandlungen
- Der vorwurf des atheismus in den drei ersten jahrhunderten – das martyrium des heiligen abo von tiflis – die frau im roemischen christenprocess

Harnack e loisy : o, le recenti polemiche intorno all'essenza del cristianesimo / Bonaccorsi, Giuseppe – Firenze: Libreria editrice Fiorentina, 1904 [mf ed 1993] – 1mf – 9 – 0-524-05972-1 – (incl bibl ref) – mf#1992-0709 – us ATLA [240]

Harnack, Otto *see*
- Der gang der handlung in goethes faust
- Zur nachgeschichte der italienischen reise

Harnack, Th *see* Ueber den kanon und die inspiration der heiligen schrift

Harnack, Theodosius *see*
- Der christliche gemeindegottesdienst im apostolischen und altkatholischen zeitalter
- Die freie lutherische volkskirche
- Die kirche, ihr amt, ihr regiment
- Luthers theologie
- Outlines of liturgics
- Praktische theologie

Harner, Nevin L *see* Hindemith

Harney county american – Burns OR: C B Cornell, 1935- [wkly] [mf ed 1967] – 3r – 1 – (cont: crane american (1916-35)) – us Oregon Lib [071]

Harney county news – Burns OR: Davey, Byrd & Davey, -1926 [wkly] [mf ed 1967-72] – 3r – 1 – (cont by: burns news (burns or)) – us Oregon Lib [071]

Harney county news (burns, or) – Burns OR: Mrs F E Wilmarth [wkly] [mf ed 1967] – 1r – 1 – us Oregon Lib [071]

Harney county tribune *see* Times-herald (burns, or)
Harney times – Harney OR: J E Roberts [mf ed 1967] – 1r – 1 – (cont by: burns times) – us Oregon Lib [071]
Harney valley items – Burns OR: H A Dillard [wkly] [mf ed 1967] – 1r – 1 – us Oregon Lib [071]
Harney, William Selby *see* Message of the president of the united states
Harnisch, Kaethe *see* Deutsche malererzaehlungen
Harnly, Henry H [Mrs] *see* A history of the harnly family
Harnoch, G A *see* Wegweiser in der kirchen- und dogmengeschichte
Harnsberger, R Scott et al *see* Popular culture in libraries
Harold and brentwood gazette – Brentwood, UK. 5 nov-dec 1992 – 1r – 1 – uk British Libr Newspaper [072]
Harold, the last of the saxon kings / Lytton, Edward Bulwer Lytton, Baron – Boston, MA. v1-2. 189- – 2r – us UF Libraries [025]
Harold, William G *see*
– Aviation
– Central florida exposition
– Eola park
– Field trials
– Flora and fauna
– Frog farms near orlando
– Golf championships
– History of orange county
– Orlando hotels
– Recreational activities
– Sunshine park
– Swimming and bathing
Harozen, Yaakov *see* Mediniyutah shel ha-tsiyonut ba-'olam
Harp / Irish Socialist Federation in America – 1908 jan-1910 jun – 1r – 1 – mf#2410842 – us WHS [335]
Harp and Thistle *see* Bourland bulletin and loving letter
Harp in the twentieth century orchestra / Nisbet, Ann – U of Rochester 1943 [mf ed 19-] – 3mf – 9 – mf#fiche 631, 823 – us Sibley [780]
The harp of canaan : or, selections from the best poets on biblical subjects / Borthwick, John Douglas – Montreal: G E Desbarats, 1871 – 3mf – 9 – mf#26673 – cn Canadiana [810]
The harp of israel : to meet the loud echo in the wilds of america / Livermore, Harriet – Philadelphia: printed for the Authoress by J. Rakestraw, 1835. El Segundo, Ca: Micro Publication Systems, 1980 (1mf); Evanston: American Theol Lib Assoc, 1984 (1mf) – (= ser Women & the church in america) – 9 – 0-8370-1389-5 – mf#1984-2124 – us ATLA [240]
The harp of prophecy / Kent, Thomas – London ON: [s.n.] 1910 [mf ed 1996] – 1mf – 9 – 0-665-80968-9 – mf#80968 – cn Canadiana [810]
Harp on the willows / Hamilton, James – London, England. 1843 – 1r – us UF Libraries [240]
Harpe, Jean F de la *see* Abrege de l'histoire generale des voyages
Harpe, Jean-Francois de la *see*
– Abrege de l'histoire generale des voyages
Harpenden advertiser – Harpenden, England. 1976- – 86+ r – 1 – uk British Libr Newspaper [072]
Harper, Alexander *see* Family papers ms 3231
Harper, Andrew *see*
– The book of deuteronomy
– The song of solomon
Harper, B *see* Post-biblical hebrew literature, an anthology
Harper, Charles George *see*
– English pen artists of to-day
– A practical handbook of drawing for modern... reproduction
Harper, Elizabeth *see* Delayed onset muscle soreness and damage in relation to electromyographic activity during concentric and eccentric contraction
Harper, Henry Andrew *see* From abraham to david
Harper, James *see* Sermons delivered on occasion of the death of the rev james peddie, d d, senior minister...
Harper, James Wilson *see*
– The books of ezra, nehemiah, and esther
– Christian ethics and social progress
Harper, John Murdoch *see*
– The annals in brief of the st andrew's society of quebec
– The annals of the war
– The battle of the plains
– Champlain
– Champlain's tomb
– The chronicles of kartdale
– Dominus domi
– The history of the irish republic
– The little sergeant
– The maritime provinces
– The montgomery siege
– Moral drill for the school room
– The origin and development of greek drama
– Sacrament sunday and bells of kartdale
– Then and now / the earliest beginnings of canada / the sillery mission
– A war-note or two
– Wolfe and montcalm
Harper, Maria Magdalena *see* Persoonlike bemagtiging van swart onderwysers in die konteks van uitkoms gebaseerde onderrig
Harper, R F et al *see* Old testament and semitic studies in memory of william, rainey, harper
Harper, Robert D *see* The church memorial
Harper, Robert Francis *see*
– Assyrian and babylonian literature
– The code of hammurabi, king of babylon about 2250 b.c
Harper, Robert Francis et al *see* Old testament and semitic studies
Harper, Robert Goodloe *see*
– Observations on the dispute between the united states and france
– Observations sur les demeles entre les etats-unis et la france
– The robert goodloe harper papers
Harper, Roland M *see* Population of florida
Harper, Samuel Northrup *see* Civic training in soviet russia
Harper, Thomas *see*
– Peace through the truth
[Harper, W R] *see* Old testament and semitic studies in memory of william, rainey, harper
Harper, William *see* Cotton is king, and pro-slavery arguments
Harper, William Edmund *see* Tests made to ascertain where conditions were most suitable for the 72-inch reflector
Harper, William Rainey *see*
– Eight books of caesar's gallic war
– Elements of hebrew syntax by an inductive method
– Hebrew vocabularies
– Introductory hebrew method and manual
– Lessons of the intermediate course
– Old testament and semitic studies
– The priestly element in the old testament
– The prophetic element in the old testament
– The prospects of the small college
– Religion and the higher life
– The stories of genesis
– The utterances of amos
– The work of the old testament sages
Harper's – New York. 1850+ (1) 1905+ (5) 1850+ (9) – ISSN: 0017-789X – mf#52 – us UMI ProQuest [073]
Harper's and queen – London. 1974-1996 (1) 1974-1996 (5) 1974-1996 (9) – ISSN: 0141-0547 – mf#9135 – us UMI ProQuest [073]
Harper's bazaar – New York. 1867+ (1) 1964+ (5) 1976+ (9) – ISSN: 0017-7873 – mf#3103 – us UMI ProQuest [740]
Harper's cases in equity / South Carolina. Supreme Court – 1v. 1824 (all publ) – (= ser Pre-nrs nominative equity reports) – 3mf – 9 – $4.50 – mf#LLMC 94-027 – us LLMC [342]
Harper's law reports / South Carolina. Supreme Court – 1v. 1823-1831 (all publ) – (= ser Pre-nrs nominative law reports) – 7mf – 9 – $10.50 – mf#LLMC 94-015 – us LLMC [340]
Harper's school geography – New York, NY. 1882, c1875 – 1r – us UF Libraries [025]
Harper's weekly – 1861 jan 5-dec 28, 1861 jan 5-dec 28 [duplicate], 1862 jan 4-dec 27, v64 3107-3120 [1975 jan 10-apr 11] – 4r – 1 – (continued by: independent (new york, ny: 1848)) – mf#1057900 – us WHS [071]
Harper's weekly : a journal of civilization – New York. v1-62. 1857-may 13 1916 and index 1857-1887 – 1 – us NY Public [073]
Harper's weekly – New York. 1857-1916 – 1,5,9 – ISSN: 0360-2397 – mf#1496 – us UMI ProQuest [073]
Harper's weekly – New York. 1975-1976 – 1 – ISSN: 0360-2397 – mf#10344 – us UMI ProQuest [073]
Harpoon / Brotherhood of Railway Postal Clerks – v1-v7 n2 [whole n1-86] [1909 jun-1917 jan] – 1r – 1 – mf#1112302 – us WHS [380]
Harpster, Mary Julia *see* Among the telugoos
L'harrach : organe politique, economique, litteraire, scientifique, industriel, agricole et financier – Alger. n1-12. oct 1921-janv 1922 – 1 – fr ACRPP [073]
Harradan, Beatrice *see* At the green dragon, a bird of passage, and the umbrella mender
Harrah, Charles Clark *see* The road
Harral, Thomas *see* Picturesque views of the severn
Harras, Philipp, Ritter von Harrasowsky *see* Der codex theresianus und seine umarbeitungen
Harrelson, Gary L *see* Predictors of success on the national athletic trainers association certification examination
Harries, John *see* A handbook of theology
Harries, Lyndon *see* Swahili poetry
Harriet taylor upton papers *see* Upton, harriet taylor, papers, ms 1746
Harriet tubman journal – 1993 oct-1995 aug – 1r – 1 – mf#2859261 – us WHS [071]
Harrigan, Anthony *see* New republic
Harrington, Bernard James *see*
– Catalogue des mineraux, roches et fossiles du canada
– On a new alkali hornblende and a titaniferous andradite
– Report on the minerals of some of the apatit-bearing veins of ottawa county, q
Harrington, Dianna J *see* Importance performance analysis of after school programs using development quality attributes
Harrington, Donald *see* Adolescents in unitarian churches
Harrington, E *see* Statistics of the manufactures and commerce of lake memphremagog
Harrington, Leicester Fitzgerald Charles Stanhope, 5th earl *see* Sketch of the history and influence of the press in british india
Harrington, Vernon Charles *see* The problem of human suffering
Harris, A C *see*
– Alaska and the klondike gold fields
Harris and gill's law reports / Maryland. Court of Appeals – v1-2. 1826-29 (all publ) – (= ser Maryland supreme court reports) – 12mf – 9 – $18.00 – (a pre-nrs title) – mf#LLMC 84-148 – us LLMC [347]
Harris and johnson's law reports / Maryland. Court of Appeals – v1-7. 1800-1826 (all publ) – (= ser Maryland supreme court reports) – 48mf – 9 – $72.00 – (a pre-nrs title) – mf#LLMC 90-301 – us LLMC [347]
Harris and mchenry's law reports / Maryland. Court of Appeals – v1-4. 1658-1799 (all publ) – (= ser Maryland supreme court reports) – 8mf – 9 – $36.00 – (a pre-nrs title) – mf#LLMC 84-146 – us LLMC [347]
Harris, Arthur Merton *see* Letters to a young lawyer
Harris, Arthur Travers *see* Bomber offensive
Harris, Beverly Dabney *see* Trade acceptance method in war financing
Harris, Carrie Jenkins *see*
– Faith and friends
– A modern evangeline
Harris, Chad *see* The influence of velocity on the metabolic and mechanical task cost of treadmill running
Harris, Charles *see*
– An investigation of some of kalidasa's views
– The position of the laity in the primitive church
– Pro fide
Harris, Charles L *see* Harris' public land guide
Harris County Genealogical Society *see* Living tree news
Harris county, georgia and her people – v1 n1-v3 n4 [1985 jan-1987/88 winter] – 1r – 1 – mf#1288349 – us WHS [071]
Harris county heritage society news – v2 n2-v3 n2 [1975 fall-1976/77 fall/winter] – 1r – 1 – (cont: heritage society news; cont by: compendium) – mf#502069 – us WHS [978]
Harris County Heritage Society [TX] *see*
– Compendium
– Heritage society news
Harris, E C *see* The genesis of the chirala station of the guntur, india, mission of the evangelical lutheran church (general synod) in the united states of america
Harris, Edward *see*
– Tractatus de benedicta incarnacione
– United empire loyalists
Harris, Errol E *see* 'White' Civilisation
Harris, George *see*
– A century's change in religion
– Doctrine of the trinity
– Inequality and progress
– Moral evolution
– A philosophical treatise on the nature and constitution of man
Harris, George Emrick *see* A treatise on the law of identification, a separate branch of the law of evidence
Harris Home for Children (Huntsville AL) *see* Homelife
Harris, J *see* Navigantium atque itinerantium bibliotheca
Harris, J Dennis *see* Summer on the borders of the caribbean sea
Harris, J Henry *see* Robert raikes
Harris, J R *see*
– The apology of aristides on behalf of the christians
– A study of codex bezae
Harris, J Rendel *see* Hermas in arcadia
Harris, Jack D *see* Fauna
Harris, James Bowmar *see* A digest of mississippi railway decisions from vol. 1 to and including vol. 71, mississippi reports
Harris, James Rendel *see*
– The ascent of olympus
– Biblical fragments from mount sinai
– Codex bezae
– The codex sangallensis
– The diatessaron of tatian
– The dioscuri in the christian legends
– The doctrine of immortality in the odes of solomon
– Four lectures on the western text of the new testament
– Fragments of the commentary of ephrem syrus upon the diatessaron
– Further researches into the history of the ferrar-group
– The gospel of the twelve apostles
– Hermas in arcadia
– The homeric centones and the acts of pilate
– Memoranda sacra
– The newly-recovered gospel of st peter
– On the origin of the ferrar-group
– The origin of the leicester codex of the new testament
– Picus who is also zeus
– A popular account of the newly-recovered gospel of peter
– The rest of the words of baruch
– Side-lights on new testament research
– Union with god
Harris, Joel C *see* Writings
Harris, Johanna L *see* The development of competency guidelines for riding instructors and equestrian coaches
Harris, John *see*
– Christian citizen
– The great teacher
– Lexicon technicum
– Patriarchy
Harris, John Andrews *see*
– The calvinistic doctrine of election and reprobation no part of st paul's teachings
– Principles of agnosticism applied to evidences of christianity
Harris, John M *see* Annexations to sierra leone
Harris, John Reese *see* Industrial entrepreneurship in nigeria
Harris, Joseph Hemington *see* Doctrine of immortality in its bearing on education
Harris, Joseph S *see* Autobiography
Harris, Josiah [comp] *see* Direct route through the north-west territories of canada to the pacific ocean
Harris, Kenneth E *see* Index to the journals of the continental congress, 1774-1789
Harris, Marjorie M *see* Basic swimming analyzed
Harris, Mattie Anstice *see* A glossary of the west saxon gospels
Harris, Merriman Colbert *see* Christianity in japan
Harris, Michael B *see* Vitamin e supplementation, delayed-onset muscular soreness, muscle tissue damage and lipid peroxidation
Harris newsletter – v1 n1-v3 n3 [1976 apr-1983 jun] – 1r – 1 – mf#379639 – us WHS [071]
Harris, P G *see* Sokoto provincial gazetteer
Harris' public land guide : a compilation of public land laws and departmental regs. as of july 1, 1911 / Harris, Charles L – Chicago: Peterson, 1912 – (= ser Land laws of the united states, 1776-1938) – 8mf – 9 – $12.00 – mf#LLMC 82-101-7 – us LLMC [343]
Harris, Reader *see* The lost tribes of israel
/Harris, Reginald V *see* The history of freemasonry in nova scotia
Harris, Robert *see* Encouragement to perseverance and holy importunity in prayer
Harris, Robert Jennings *see* The judicial power of the united states
Harris, Sam *see* Key west
Harris, Samuel *see*
– God, the creator and lord of all
– The kingdom of christ on earth
– The maxim for the times
– Pernicious fiction, or, the tendencies and results of indiscriminate novel reading
– The philosophical basis of theism
– The self-revelation of god
Harris, Shane T *see* The effect of ultrasound on temperature rise in the preheated triceps surae muscle group
Harris, Theodore *see* A preacher's and a banker's views on important subjects
Harris, Thomas *see*
– Modern entries, adapted to the american courts of justice.
– Three periods of english architecture
– Victorian architecture
Harris, Thomas Lake *see*
– Herald of light
– Truth and life in jesus
Harris, Walter D *see* Vivienda en honduras
Harris, William *see* Remarks made during a tour through the united states of america, in the years 1817, 1818, and 1819
Harris, William Cornwallis *see*
– Narrative of an expedition into southern africa
– The wild sports of southern africa
Harris, William Logan *see*
– The constitutional powers of the general conference
– The doctrines and discipline of the methodist episcopal church, 1884
– Ecclesiastical law and rules of evidence
Harris, William Richard *see*
– The catholic church in the niagara peninsula, 1626-1895
– Essays in occultism, spiritism, and demonology
Harris, William T *see* Bemerkungen auf einer reise durch die vereinten staaten von nord-amerika

Harris, William Torrey *see*
– Hegel's logic
– Introduction to the study of philosophy
– Psychologic foundations of education
– Social culture in the form of education and religion
– The spiritual sense of dante's divina commedia

Harris, William Torrey et al *see* Concord lectures on philosophy

Harris, Z S *see*
– Development of the canaanite dialects
– Grammar of the phoenecian language

Harrisburg : the monthly news – [v6 n39/40]-v10 n1 [1977 aug-1980 aug] – 1r – 1 – (cont: harrisburg independent press) – mf#570057 – us WHS [071]

Harrisburg area women's news – 1984 jan-1987 may – 1r – 1 – mf#1329884 – us WHS [305]

Harrisburg bulletin and commonwealth – Harrisburg OR: M D Morgan, -1925 [wkly] [mf ed 1975] – 3r – 1 – (merger of: harrisburg bulletin (harrisburg, or: 1901); commonwealth (harrisburg, or); cont by: harrisburg bulletin (harrisburg, or: 1925)) – us Oregon Lib [071]

Harrisburg bulletin and commonwealth *see* Commonwealth [harrisburg, or]

The harrisburg bulletin and commonwealth *see* Harrisburg bulletin (harrisburg, or: 1901)

Harrisburg bulletin (harrisburg, or: 1901) – Harrisburg OR: A P Bettersworth, Jr, 1901- [wkly] [mf ed 1975] – 1r – 1 – (ceased in 1916? merged with: commonwealth (-1916) to form: harrisburg bulletin and commonwealth (1916-25)) – us Oregon Lib [071]

Harrisburg bulletin (harrisburg, or: 1901) *see* Commonwealth [harrisburg, or]

Harrisburg bulletin (harrisburg, or: 1925) – Harrisburg OR: S P Shutt, 1925-85 [wkly] – 1 – (cont: harrisburg bulletin and commonwealth. absorbed: halsey review) – us Oregon Lib [071]

Harrisburg chronicle – Harrisburg, PA., 1836 – 13 – $25.00r – us IMR [071]

Harrisburg courier – Harrisburg, PA. 1903-42. 27 rolls – 13 – $25.00r – us IMR [071]

Harrisburg daily telegraph – Harrisburg, PA., 1866-1915 – 13 – $25.00r – us IMR [071]

Harrisburg guide – Harrisburg, PA. -w 1948 – 13 – $25.00r – us IMR [071]

[Harrisburg-] harrisburg independent press – PA. 1972-1975 – 2r – 1 – $120.00 – mf#R04991 – us Library Micro [071]

Harrisburg home news – Harrisburg, PA. -w 1948-1949 – 13 – $25.00r – us IMR [071]

Harrisburg independent press – v4 n18 [1975 feb 7/14], v5 n7 [1975 nov14/21], v6 n19-22,24-38 [1977 feb 11/18-mar 4/11, 18/25-jul 1/8] – 1r – 1 – (continued by: harrisburg) – mf#570059 – us WHS [071]

Harrisburg star independent – Harrisburg, PA., 1877-1976 – 13 – $25.00r – us IMR [071]

Harrisburg telegraph – Harrisburg, PA., 1866-1902 – 13 – $25.00r – us IMR [071]

Harrisburg weekly telegraph – Harrisburg, PA., 1944-1948 – 13 – $25.00r – us IMR [071]

Harrismith news (chronicle) – Harrismith, South Africa. 1899-1901 – 1r – 1 – sa National [079]

Harrison, Alexander James *see*
– The ascent of faith
– The church in relation to sceptics
– Problems of christianity and scepticism
– The repose of faith in view of present-day difficulties

Harrison and cadiz news / Harrison Co. Cadiz – jan 1897-apr 1917 [wkly] – 8r – 1 – mf#B7025-7032 – us Ohio Hist [071]

Harrison and Carroll Co *see* Obituaries, 1879-1914

Harrison and hodgin's upper canada municipal reports / Ontario. Canada – 1v. 1845-52 (all publ) – 9mf – 9 – $13.50 – mf#LLMC 81-060 – us LLMC [340]

Harrison, Benjamin *see* Papers

Harrison, Birge *see* Landscape painting

Harrison, Charles Edward *see* Genealogical records of the pioneers of tampa and...

Harrison Co. Cadiz *see*
– Democrat sentinel
– Democratic whig standard
– Early newspapers
– Harrison and cadiz news
– Harrison county democrat
– Liberty advocate
– Republican
– Republican series
– Sentinel and democratic sentinel

Harrison Co. Freeport *see*
– Press

Harrison Co. Scio *see*
– Herald series
– Press-herald
– Weekly herald

Harrison county atlas, 1875 : by caldwell – 1r – 1 – mf#B30575 – us Ohio Hist [978]

Harrison county democrat / Harrison Co. Cadiz – 1897-99, 1901-nov 1911 [wkly] – 6r – 1 – mf#B7016-7021 – us Ohio Hist [071]

Harrison, Cynthia *see*
– Creating the federal judicial system
– The federal appellate judiciary in the 21st century

Harrison family of amelia island – s.l, s.l? 193-? – 1r – us UF Libraries [978]

Harrison, Frederic *see*
– Autobiographic memoirs
– The creed of a layman
– John ruskin
– The positive evolution of religion
– The religion of inhumanity

Harrison, Frederic et al *see*
– The nature and reality of religion
– Theology at the dawn of the twentieth century

Harrison, Gessner *see* A treatise on the greek prepositions

Harrison gray otis papers, 1691-1870 – [mf ed 1979] – 11r – 1 – (with p/g; coll contains primarily the business, political, and personal papers of otis, a prominent federalist, lawyer etc) – us MA Hist [975]

Harrison happenings : newsletter of harrison museum of african american culture / Harrison Museum of African American Culture – 1990 summer/fall, 1991 spring, 1993 winter/spring-1994 summer/fall, 1995 summer/fall – 1r – 1 – mf#2688630 – us WHS [060]

Harrison heritage – v1 n1-v6 n4 (1981 mar-1986 dec) – 1r – 1 – mf#1140752 – us WHS [071]

Harrison, Ida Withers *see* Forty years of service

Harrison, J C *see* Glory of good deeds

Harrison, Mrs J W *see* The story of the life of mackay of uganda

Harrison, J W F *see* Historical and analytical programme of pianoforte recital

Harrison, J W [Mrs] *see* A m mackay

Harrison, James Park *see* A letter...on the fitness of gothic architecture for modern churches

Harrison, Jane Ellen *see*
– Ancient art and ritual
– Greek vase paintings
– Introductory studies in Greek art
– Myths of the odyssey in art and literature
– Prolegomena to the study of greek religion
– The religion of ancient greece
– Themis

Harrison, John *see*
– The decoration of metals chasing, repousse and saw-piercing
– On the primitive mode of making bishops
– Whose are the fathers?

Harrison, John Burchmore *see* Geology of the goldfields of british guiana

Harrison, John Smith *see* The teachers of emerson

Harrison, Maine.Harrison Baptist Church *see* Records

Harrison, Maria E G *see* The biomechanical effects of prolotherapy on traumatized achilles tendons of male rats

Harrison, Max Hunter *see* Hindu monism and pluralism

Harrison Museum of African American Culture *see* Harrison happenings

Harrison news herald – Cadiz, OH. 1991-2000 (1) – mf#65526 – us UMI ProQuest [071]

Harrison post – Indianapolis IN. 1980 sep 18-1982 jun, 1982 jul-1983, 1984 jan 12-1986 aug – 3r – 1 – mf#918540 – us WHS [071]

Harrison press-journal – Harrison, NE: Geo D Canon. v12 n2. aug 3 1899-1905// (wkly) [mf ed with gaps filmed 1975] – 3r – 1 – (formed by the union of: northwestern press and: sioux county journal) – us NE Hist [071]

Harrison, Robert *see* Outlines of german literature

Harrison, S F *see* Pipandor

Harrison, Salomay Lauderdale *see* Mexico simpatico

Harrison & Sons, London *see* Printing types

The harrison sun – Harrison, NE: Gerald Bardo, James B Griffith, Jr. v68 n23. jul 17 1969- (wkly) [mf ed 1972-] – 1 – (cont: lusk herald (1927), crawford clipper (1979) and: harrison sun and sioux county news. distributed with: the lusk herald jul 17 1969-apr 30 1981, and the crawford clipper may 7 1981- . has suppl: northwest nebraska post oct 1979-nov 1983 and: crawford clipper's northwest nebraska post dec 1983-jun/jul 1992) – us NE Hist [071]

Harrison sun and sioux county news – Chadron, NE: Gerald Bardo, James B Griffith Jr. 9v. v60 n3. may 18 1961-v68 n22. jul 10 1969 (wkly) [mf ed filmed 1972] – 6r – 1 – (formed by the union of: harrison sun and: sioux county news. cont: lusk herald (1927). cont by: harrison sun (1969). dist with: the lusk herald) – us NE Hist [071]

Harrison, Susie Frances *see*
– Crowded out!
– The forest of bourg-marie
– Ringfield

Harrison, Thomas *see* Three hundred testimonies in favor of religion and the bible

Harrison, W *see* Minutes of voyage, 1830-1

Harrison, William *see* Harrison's description of england in shakespere's youth

Harrison, William Henry *see* Papers

Harrison, William Pope *see*
– The gospel among the slaves
– The high-churchman disarmed

Harrison, William Randle *see*
– Decorative art, as applied to the ornamentation of churches
– A practical guide to decorative painting for walls, panels, screens, and terra-cotta

Harrison-bundy files relating to the development of the atomic bomb, 1942-1946 / U.S. War Dept. Office of the Chief of Engineers – (= ser Records of the office of the chief of engineers) – 9r – 1 – (with printed guide) – mf#M1108 – us Nat Archives [355]

Harrisonian / Muskingum Co. Zanesville – jan 22-oct 28 1840 [wkly] – 1r – 1 – mf#B4478 – us Ohio Hist [320]

Harrison's description of england in shakespere's youth / Harrison, William – 2nd and 3rd books of his description of Britaine and England. Ed. from first 2 eds. of Holinshed's Chronicle...1577, 1587, by Frederick J. Furnivall. London: Chatto & Windus, 1908. 427p – 1 – us Wisconsin U Libr [941]

Harrisse, Henry *see*
– Les corte-real et leurs voyages au nouveau-monde
– Histoire critique de la decouverte du mississipi sic
– Jean et sebastien cabot

Harrisson, Thomas Harnett *see* The peoples of sarawak

Harris-Sund, Valarie *see* Compliance and cardiac rehabilitation

Harrod, John J *see* Substance of a protest and arguments

Harrogate advertiser & weekly list of the visitors – [Yorkshire & Humberside] North Yorkshire 26 sep 1836-dec 1950 [mf ed 2004] – 99r – 1 – (missing: 1849, 1912, 1872, 1855, 1880) – uk Newsplan [072]

Harrogate first baptist church. harrogate, tennessee : church records – 1946-76 – 1 – us Southern Baptist [242]

Harrold – (= ser Bedfordshire parish register series) – 1mf – 9 – £3.00 – uk BedsFHS [929]

Harrop's manchester mercury – [North West] Manchester ALS 3 mar-21 apr 1752* – 1 – (title change: harrop's manchester mercury, & general advertiser [28 apr 1752-5 jul 1757]; manchester mercury, & harrop's general advertiser [12 jul 1757-1830]) – uk MLA; uk Newsplan [072]

Harrouff, George *see* Account books and miscellaneous papers

Harrow and northwood informer – London, UK. 1986-1989, 12 jan-28 dec 1990 – 12r – 1 – (aka: harrow and ruislip informer; harrow and wembley informer) – uk British Libr Newspaper [072]

Harrow and ruislip informer *see* Harrow and northwood informer

Harrow and wealdstone press *see* Harrow press and wealdstone harrow weald and watford times

Harrow and wembley observer and district reporter *see* Wealdstone harrow and wembley observer and district reporter for pinner harrow etc

Harrow and wembley recorder – London, UK. 1988-21 dec 1990; jan-jun 1991; 1992; Feb-oct 1996; 15, 29 jan, 5 feb, 2 apr, 16, 23 jul, 15, 22, 29 oct, 5 nov-30 dec 1997 – 12 1/2r – 1 – uk British Libr Newspaper [072]

Harrow gazette – London. -m. Apr 1867-Dec 1870. (40 ft) – 1 – uk British Libr Newspaper [072]

Harrow gazette and general advertiser *see* Harrow monthly gazette and general advertiser

Harrow leader – London, UK. 1986-19 dec 1991, 1992 – 20r – 1 – uk British Libr Newspaper [072]

Harrow midweek – London, UK. 1978; 20 feb 1979-19 apr 1983. -w. 10 r – 1 – uk British Libr Newspaper [072]

Harrow monthly gazette and general advertiser – London, UK. apr 1855-1 nov 1869, 1870, 1875-77, 1879, 1889 – 5 3/4r – 1 – (aka: harrow gazette and general advertiser; harrow gazette) – uk British Libr Newspaper [072]

Harrow news – London. -w. 24 apr 1925-19 feb 1926 1 r – 1 – uk British Libr Newspaper [072]

Harrow observer *see* Wealdstone harrow and wembley observer and district reporter for pinner harrow etc

Harrow observer and district reporter *see* Wealdstone harrow and wembley observer and district reporter for pinner harrow etc

Harrow observer and gazette *see* Wealdstone harrow and wembley observer and district reporter for pinner harrow etc

Harrow observer and gazette (stanmore ed) – London, UK. 1950 – 1r – 1 – uk British Libr Newspaper [072]

Harrow press and wealdstone harrow weald and watford times – London, 28 oct 1892-16 jul 1897 [wkly] – 4r – 1 – (aka: harrow & wealdstone press) – uk British Libr Newspaper [072]

Harrow & wealdstone news – London, England. -w. 1 May 1907-27 July 1910 1 r – 1 – uk British Libr Newspaper [072]

Harrowby, Dudley Ryder *see* Letter to the right honourable spencer perceval

Harrower, Charles Swartz *see* The sunday service

Harrowing of hell = Das altenglischen spiel von christi hollenfahrt / ed by Mall, E – Wuerzburg. 187-? – 1 – us CRL [820]

Harroy, Jean-Paul *see* Afrique, terre qui meurt

Harr-penn dispatch / American Postal Workers Union – 1971 nov-1974 nov – 1r – 1 – (continued by: keystone area local news and views) – mf#667619 – us WHS [331]

Harry – v1 n1-v2 n22 [1969 nov 3-1971 nov 12] – 1r – 1 – mf#701349 – us WHS [071]

Harry bridges / Meiklejohn Civil Liberties Library – 1938-52 – 1 – us AMS Press [321]

Harry bridges collection : pamphlets, clippings, letters – San Francisco Public Library – 1r – 1 – $50.00 – mf#B40301 – us Library Micro [920]

Harry, Gerard *see* Maurice maeterlinck

Harry Lundeberg School *see* Skipjack

Harry, Myriam *see*
– Radame
– Ranavalo et son amant blanc
– Routes malgaches

The harry s truman oral histories collection – (= ser Research colls in american politics) – 628mf (24:1) – 9 – $6620.00 – 1-55655-215-7 – (with p/g) – us UPA [975]

Harrydaposten – Harryda, Sweden. 1984- – 1 – sw Kungliga [078]

Hars schiltbergers reisebuch / ed by Langmantel, V – Tuebingen, 1885 – 3mf – 9 – mf#AR-1960 – ne IDC [910]

Harsavardhana, King of Thanesar and Kanauj *see*
– The dramas of shri harsha
– The nagananda of sri harsha deva
– Nagananda of sriharsa
– Priyadarsika
– Priyadarsika of sri harsha
– The priyadarsika of sri harsha-deva
– Ratnavali
– The ratnavali
– The ratnavali of sri harsha-deva
– Sri harsha's priyadarsika

Harsdoerffer, G P *see* Frauenzimmer gesprechspiele

[Harsdoerffer, G P] *see*
– Aulaea romana
– Heraclitus und democritus

Harsh, Philip Whaley *see* Studies in dramatic 'preparation' in roman comedy

Harsha / Mookerji, Radhakumud – London: Oxford University Press, 1926 – (= ser Samp: indian books) – us CRL [930]

Harsha, David Addison *see*
– The christian's present for all seasons
– The heavenly token
– Life of philip doddridge, d.d
– Life of the rev. george whitefield
– Life of the rev. james hervey
– Wanderings of a pilgrim

Harsha, William J *see* Story of iowa

Harsha, William Justin *see* Sabbath-day journeys

Harston 1599-1950 – 6mf – 9 – £7.50 – uk CambsFHS [929]

Harston, Edward *see* Grave, and the reverence due to it...

Hart, Albert Bushnell *see*
– Formation of the union, 1750-1829
– Salmon portland chase
– Slavery and abolition, 1831-1841

Hart, Alexis C et al *see* Our western congregational academies

Hart, Algerian *see* Student-athlete perceptions of the ncaa rules and regulations

Hart, Alice Marion (Rowlands) *see* Cottage industries

Hart, Amos Winfield *see* Digest of decisions of law and practice in the patent office and the united states and state courts in patents, trade-marks, copyrights, and labels

Hart and Son *see* Hart and son's illustrated catalogue of ornamental metal work

Hart and son's illustrated catalogue of ornamental metal work / Hart and Son – London [1865?] – (= ser 19th c art & architecture) – 1mf – 9 – mf#4.2.700 – uk Chadwyck [730]

Hart, Burdett *see*
– Biblical epochs
– Discourse on concluding a pastorate of thirty years

Hart, Edith *see* Chin hsing (forward march) in china

Hart, Elizabeth A *see* The effect of planned exercise as a disinhibitor of dietary restraint

Hart, Ernest Abraham *see* On the use of opium in india

Hart, Evanston Ives *see* Wake up! montreal!

Hart forum – v1 n1-v6 n1 [1984 feb-1989 feb] – 1r – 1 – mf#1109033 – us WHS [071]
Hart, Francis Russell *see*
– Disaster of darien
– Siege of havana, 1762
Hart, G H C *see* De volksraad als kristallisator van het "indische bewustzijn"
Hart, George Waldegrave *see* The church of our fathers
Hart, H A *see* Pointers being a brief digest of debt, interest, usury, mortgage and foreclosure, with comments and chapter on equity
Hart, H C *see* Col. h. c. hart's new and improved instructor for the drum
Hart, Hans-Ulrich *see* Wertsysteme von englischlehrern
Hart, Hastings Hornell *see* Juvenile court laws in the united states
Hart, Heinrich *see* Peter hille
Hart, Henry Martyn *see* Recollections and reflections
Hart, James Morgan *see*
– Faust
– German universities
Hart, James W T *see* The autobiography of judas iscariot
Hart, John Henry Arthur *see* Ecclesiasticus
Hart, Joseph Kinmont *see* A critical study of current theories of moral education
Hart, Julius *see* Sehnsucht
Hart, Lawrence Elbert *see* Approach to a practical pedagogy of piano sight-reading
Hart, Madison Ashley *see* The normal training of the child
Hart, Miss *see* Letters from the bahama islands
Hart, Robert *see* "These from the land of sinim"
Hart, Samuel *see* The book of common prayer
Hart, Solomon Alexander *see* The reminiscences of solomon alex hart
Hart, Virgil Chittenden *see*
– The temple and the sage
– Western china
Hart, W H *see*
– Cartularium monasterii de rameseia
– Historia et cartularium monsterii s petri cloucestriae
– Index expurgatorious anglicanus
Hart, William Henry *see* Everyday life in bengal
Harte, Bret *see* A waif of the plains
Harte, Frederick Edward *see* The philosophical treatment of divine personality
Das harte geschlecht : roman / Vesper, Will – Hamburg: Hanseatische Verlagsanstalt c1931 [mf ed 1991] – 1r – 1 – (filmed with: die baltische tragodie / siegfried von vegesack) – mf#2923p – us Wisconsin U Libr [830]
Das harte ja : roman / Seidl, Florian – Berlin: Volksverband der Buecherfreunde, Wegweiser-Verlag c1941 [mf ed 1996] – 1r – 1 – mf#filmed with: sterne der heimkehr / ina seidel – us Wisconsin U Libr [830]
Harten, Theodor *see* Eine hochburg der hugenotten waehrend der religionskriege
The hartford central association and the bushnell controversy : an historical address / Parker, Edwin Pond – Hartford, Conn: Case, Lockwood & Brainard, 1896 – 1mf – 9 – 0-524-02839-7 – mf#1990-4460 – us ATLA [240]
Hartford courant – Hartford, CT. 1837+ (1) – mf#60152 – us UMI ProQuest [071]
Hartford daily courant – Hartford CT. 1855 sep 3, 1867 may 17, 1871 sep 14, 1876 jul 6, 1889 oct 29, 1890 jan 31, 1900 nov 24, 1901 jun 27, 1914 oct 25 – 1r – 1 – (cont: daily courant; cont by: hartford courant) – mf#846244 – us WHS [071]
Hartford daily post – Hartford CT. 1865 jan 18, 1899 jul 12 – 1r – 1 – (continued by: evening post) – mf#872611 – us WHS [071]
Hartford daily times – Hartford CT. 1858 aug 10, 1864 aug 16 – 1r – 1 – (cont: daily times; daily times [hartford, ct]; cont by: hartford times; hartford times [hartford, ct: 1881]) – mf#854923 – us WHS [071]
Hartford inquirer – Hartford CT. 1991 apr 3-dec 25, 1992 jan 1-dec 30, 1993 jan 6-dec 29, 1994 jan 5-dec 28, 1995 jan 4-dec 27, 1996 jan 3-dec 25, 1997 jan 1-dec 31, 1998 jan 7-dec 30, 1999 jan 6-jun 30, 1999 jul 7-dec 29, 2000 jan 5-jun, 2000 jul 5-dec 27, 2001 jan-jun – 13r – 1 – mf#1886862 – us WHS [071]
Hartford pioneer – Hartford, MT. 1895-1895 (1) – mf#69182 – us UMI ProQuest [071]
Hartford press – Hartford WI. 1883 jan 10/1885 may 29-1932 jan 1/1933 oct 20 [gaps] – 31r – 1 – (cont: washington county republican; cont by: hartford times; hartford times-press) – mf#945085 – us WHS [071]
Hartford press – Hartford WI. 1867 sep 5 – 1r – 1 – mf#875507 – us WHS [071]
Hartford seminary record : issued under the auspices of the faculty of hartford theological seminary – Hartford. 1890-1913 (1) – mf#2897 – us UMI ProQuest [240]
Hartford, st john the baptist – (= ser Cheshire monumental inscriptions) – 2mf – 9 – £4.00 – mf#19 – uk CheshireFHS [929]
Hartford Steam Boiler Inspection and Insurance Co *see* Locomotive
Hartford studies in literature – West Hartford. 1969-1977 (1) 1972-1977 (5) 1974-1977 (9) – (cont by: studies in literature) – ISSN: 0196-2280 – mf#6358 – us UMI ProQuest [400]
Hartford times – Hartford WI. 1894 sep 13-1898, 1899-1901, 1902-09, 1910 jan 14-1911 sep 8, 1913 jul 25-1914 dec 25, 1915-31, 1932-1933 oct 20 – 14r – 1 – (continued by: hartford press (hartford, wi: 1883); hartford times-press) – mf#945085 – us WHS [071]
Hartford times-press – Hartford WI. 1933 oct 27/1934-1987 jul-aug 13 [gaps] – 55r – 1 – (cont: hartford press (hartford, wi: 1883); hartford times; cont by: times-press (hartford, wi: 1987)) – mf#947941 – us WHS [071]
Hartford times-press – Hartford WI. 1998 jan-dec, 1999 jan-dec, 2000 jan-dec, 2001 jan-dec – 5r – 1 – (cont: times-press (hartford, wi: 1987); cont by: times press (hartford, wi: 2000)) – us WHS [071]
Hartford weekly times – Hartford, CT: Alfred E Burr. v31 n1567. jan 2 1847-97// (semiwkly) – 1 – (cont: hartford times (1837: wkly); cont by: weekly times; suppls accompany some iss; iss fr 1869- called: hartford times suppl; daily ed: hartford daily times 1847-jul 2 1883 and: hartford times (1883) jul 3 1883-1) – us Bell [071]
Hartford Women's Center *see* Women in hartford
Hartford's other voice – n2,4 [1969 jul 14], v1 n16,19-21,23 [1970 jan 16, mar 19-apr 7, may 5], v2 n1 [1970 may 26]; v1 n12-v2 n2 [1969 nov 27-1970 jun?] – 2r – 1 – (continued by: wild raspberry) – mf#770774 – us WHS [071]
Harth, Helene *see* Dichtung und arete
Harthmuth von kronberg : eine charakterstudie aus der reformationszeit / Bogler, Wilhelm – Halle: Verein fuer Reformationsgeschichte, 1897 – (= ser Schriften des Vereins fuer Reformationsgeschichte) – 1mf – 9 – 0-7905-5262-0 – (incl bibl ref) – mf#1988-1262 – us ATLA [242]
Hartig, Georg Ludwig *see* Forst- und jagd-archiv von und fuer preussen
Hartig, Theodor *see* Jahresberichte ueber die fortschritte der forstwissenschaft und forstlichen naturkunde im jahre...
The hartington herald – Hartington, NE: lrd [ie Baird] & Watson, 1883-60th [ie 61st] yr n18. jan 27 1944 (wkly) [mf ed v5 n49. oct 5 1888,1892,1894-1905,1909-44 (gaps) filmed [1969?-73] – 18r – 1 – (absorbed: cedar county leader 1898 and: fordyce press 1915. publ as: the hartington herald and cedar county leader aug 5 1898. issues for feb 11 1943-mar 4 1943 called n20-23 but constitute 60th yr n20-23. issues for sep 30 1943-jan 27 1944 called 60th yr n1-18 but constitute 61st yr n1-18) – us NE Hist [071]
Hartknoch, Christoph *see* Preussische kirchen historia
Hartl, Eduard *see* Das benediktbeurer passionsspiel; das st galler passionsspiel
Hartland, Edwin Sidney *see*
– The legend of perseus
– Primitive paternity
– Ritual and belief
– Ritual and belief: studies in the history of religion
Hartland index – Hartland WI. 1886 oct, 1889 apr 4 – 1r – 1 – mf#876264 – us WHS [071]
Hartland news – Hartland WI. 1897 nov 13/1899, 1900-53 – 28r – 1 – (cont: news and dairyman; cont by: pewaukee post; delafield gazette; lake country reporter) – mf#945114 – us WHS [071]
Hartlaub, G F *see* Zauber des spiegels
Hartleben, Otto Erich *see*
– Die erziehung zur ehe
– Der frosch
Hartleben, Selma *see* "Mei erich"
Hartlepool gazetteer of shipping and commerce – Hartlepool, England. -w. 2 Nov 1850-1 Feb 1851. 11 ft – 1 – uk British Libr Newspaper [072]
Hartlepools daily shipping list – [NE England] Hartlepool 2 may 1904-21 dec 1916 [mf ed 2004] – 25r – 1 – uk Newsplan [380]
Hartlepools labour news – [NE England] Hartlepool 1 nov-22 dec 1922 [mf ed 2004] – 1r – 1 – uk Newsplan [331]
Hartley, Catherine Gascoigne *see* Stories of early British heroes
Hartley, Cecil B *see*
– The gentlemen's book of etiquette; and manuel of politeness.
– The three mrs judsons
Hartley, Charles Augustus *see* Notes on public works in the united states and in canada
Hartley, David *see*
– An address to the committee of the county of york
– Hartley russell papers, 1761-1783
– Letters on the american war
– Observations on man
– Prayers and religious meditations
Hartley, Edward *see*
– Extracts from the reports on the coals of pictou county, nova scotia
– Report on the coals and iron ores of pictou county, nova scotia
– Reports of sir w e logan...and edward hartley...
Hartley, G A *see* Immortality versus annihilation
Hartley, Henry Alexander Saturnin *see* "Ta tou pragma emou biou"
Hartley, Isaac Smithson *see* Oration at the dedication of the site of the fort schuyler monument
Hartley, John *see* Researches in greece and the levant
Hartley russell papers, 1761-1783 / Hartley, David – Berkshire Record Office – (= ser BRRAM series) – 1r – 1 – £67 / $134 – (int by geoffrey seed) – mf#96343 – uk Microform Academic [327]
Hartley, Thomas *see* Nine queries concerning the trinity, etc
Hartlich, C *see* Die ursprung des mythosbegriffes in der modernen bibelwissenschaft
Hartlieb, Johann *see* Johann hartliebs uebersetzung des dialogus miraculorum von caesarius von heisterbach
Hartman, Edward Randolph *see* Socialism versus christianity
Hartman, Greta B *see* The accuracy of heart rate as an indicator of metabolic rate while performing step aerobics
Hartman, H *see* The comparison between an aquatic running program versus a hard surface running program on aerobic capacity and body composition
Hartman, Levi Balmer *see* Divine penology
Hartman, Louis Oliver *see* Popular aspects of oriental religions
Hartmann, A T *see* Die hebraeerin am putztische und als braut
Hartmann, Adolf *see* Die strafrechtspflege in amerika
Hartmann, August *see* Volksschauspiele
Hartmann, Carl *see*
– Friederich carl casimir freiherr von creuz und seine dichtungen
– Geographisch-statistische beschreibung von californien
Hartmann, Christ *see* Annales heremi deiparae matris monasterii in helvetia
Hartmann, David *see* Das buch ruth in der midrasch-litteratur
Hartmann, Eduard von *see*
– Das christentum des neuen testaments
– Das judenthum in gegenwart und zukunft
– Die krisis des christenthums in der modernen theologie
– Phaenomenologie des sittlichen bewusstseins
– Philosophische fragen der gegenwart
– Philosophy of the unconscious.
– Die religion des geistes
– The religion of the future
– Schelling's positive philosophie als einheit von hegel und schopenhauer
– Die selbstzersetzung des christenthums und die religion der zukunft
– Tagesfragen
Hartmann, Ernst of Osterode *see*
– Geschichte der stadt hohenstein in ostpreussen
– Geschichte der stadt liebemuehl
– Der kreis osterode
Hartmann, Ernst Wilhelm *see* Jean jacques rousseaus einfluss auf joachim heinrich campe
Hartmann, Franz *see*
– Betrachtungen ueber die mystik in goethe's "faust"
– Capital punishment
– The life and doctrines of jacob boehme
– The life and doctrines of jacob boehme, the god-taught philosopher
– Life of a christian philosopher
– The life of jehoshua, the prophet of nazareth
– Lotusblueten
Hartmann, Georg *see* Georg leonhard hartmanns versuch einer beschreibung des bodensee's
Hartmann, Hartwig *see* Necheb und nechbet
Hartmann, Horst *see* Lucretia-dramen
Hartmann, Horst et al *see* Werkinterpretationen zur deutschen literatur
Hartmann, J *see* Uhlands tagebuch 1810-1820
Hartmann, Julius *see*
– Humanitaet und religion
– Johann brenz
– Johannes brenz
Hartmann, K et al *see* Der fraenkische republikaner
Hartmann, L *see* Das gesetz ueber die presse vom 12. mai 1851, aus der entstehungsgeschichte, der rechtslehre und der entscheidungen des koeniglichen ober-tribunals erlautert.
Hartmann, Markus *see*
– Studien zur metabolisierung und zum intrazellulaeren wirkmechanismus tumorraffiner ruthenium(iii)komplexe
– Vergleichende untersuchungen ueber stabilitaet und substitutionsverhalten tumorhemmender ruthenium(iii)komplexe in physiologischem blutpuffer
Hartmann, Martin *see*
– Die hebraeische verskunst
– Der islam
Hartmann, Max *see* Ludwig achim von arnim als dramatiker
Hartmann, Moritz *see*
– Gedichte
– Kelch und schwert
– Der krieg um den wald
– Moritz hartmann's gesammelte werke
– Zeitlosen
Hartmann, Philipp *see* Repertorium rituum
Hartmann, R *see* Reise des freiherrn von barnim durch nord-ost-afrika in den jahren 1859 und 1860
Hartmann, Richard *see* Al-kuschairis darstellung des sufitums
Hartmann, Sabine *see* Hoerschwellenbestimmungen bei neugeborenen risikokindern
Hartmann, Thekla *see* Nomenclatura botanica dos bororo
Hartmann von aue / ed by Bech, Fedor – 2. aufl. Leipzig: F A Brockhaus, 1870-73 [mf ed 1993] – (= ser Deutsche classiker des mittelalters 4-6) – 3v – 1 – (incl bibl ref and ind) – mf#8189 reel 1-2 – us Wisconsin U Libr [800]
Hartmann von aue als lyriker : eine literarhistorische untersuchung / Saran, Franz – Halle: M Niemeyer, 1889 – 1r – 1 – (incl bibl ref) – us Wisconsin U Libr [430]
Hartmanns iwein : rechtsargumentation und bildsprache / Hagenguth, Edith – Heidelberg, 1969 – 3mf – 9 – 3-89349-753-6 – gw Frankfurter [430]
Hartog, Arnold Hendrik de *see* Noodzakelijke aanvullingen tot calvijn's institutie
Hartog, Lady *see* Living india
Hartong, C *see* Danskunst
Hartranft, Chester David *see* The aims of a theological seminary
Hart's e&p – Houston. 1999+ (1,5,9) – ISSN: 1527-4063 – mf#29265 – us UMI ProQuest [550]
Hart's oil and gas world – Denver. 1994-1999 (1,5,9) – ISSN: 1075-5365 – mf#20694 – us UMI ProQuest [550]
Hart's petroleum engineer international – Dallas. 1994-1998 (1) 1994-1998 (5) 1994-1998 (9) – (cont: petroleum engineer international. cont by: petroleum engineer international) – mf#1013,01 – us UMI ProQuest [550]
Harts, William Henry *see* Diary
Hartshorne, Albert *see* Old english glasses
Hartshorne, Charles Henry *see* English medieval embroidery
Hartshorne, Francis Cope *see* The railroads and the commerce clause
Hartsinck, J J *see* Beschrijving van guiana, of de wilde kust, in zuid-america,...
Hartsville college, illinois : jefferson literary society minutes, 1863-1867 board of trustees, 2nd record book, 1865-1897 – 1r – 1 – $35.00 – mf[s]-34 – us Commission [378]
Hartsville college [illinois] account book 1860-1861 – 1r – 1 – $35.00 – mf[s]-33 – us Commission [378]
Hartsville college, illinois catalogues : 1855, 1866-1875, 1879, 1880, 1882-84, 1886, 1890 – 1r – 1 – $35.00 – mf[s]-36 – us Commission [378]
Hartsville first baptist church. hartsville, south carolina : church records – 1910-49. (Historical Report, 1954-72) – 1 – us Southern Baptist [242]
Hartte, Konstantin *see* Zum semitischen wasserkultus
Hartung, Ernst *see*
– Alles um liebe
– Das grosse leben
– Vom taetigen leben
Hartung, Fritz *see* Die zeus-kinder und die heroen
Hartung, Johann Adam *see*
– Die kronos-kinder und das reich des zeus
– Naturgeschichte der heidnischen religionen, besonders der griechischen
– Die religion der roemer
– Ungelehrte erklaerung des goethe'schen faust
– Der urwesen
– Die zeus-kinder und die heroen
Hartung, K *see* Der prophet amos
Hartungen, Hartmut von *see* Der dichter siegfried lipiner (1856-1911)
Hartungsche kriegszeitung – Koenigsberg (Kaliningrad RUS), 1914 12 aug-1918 10 jul [gaps] – 5r – 1 – (side ed of: koenigsberger hartung'sche zeitung) – gw Misc Inst [943]
Hartwell, Steven *see* Alternative dispute resolution in a bankruptcy court
Harty, Tyson H *see* The application of human motor control principles to a collective robotic arm
Hartz, Erich von *see* Odrun
Hartzel, Jonas *see*
– The baptismal controversy
– A defense of the bible against the charges of modern infidelity
Hartzler, Jonas Smucker *see* Mennonite church history

Harvard 1723-1849 – Oxford, MA (mf ed 1996) – (= ser Massachusetts vital record transcripts to 1850) – 11mf – 9 – 0-87623-263-2 – (mf 1t-3t: births & deaths 1723-66. mf 1t-4t: marriages 1732-1804. mf 1t-5t: publishments 1732-96. mf 4t-5t: births 1753-1820. mf 5t-6t: deaths a-w 1750-1819. mf 6t: out-of-town marriages 1732-99. mf 6t-10t: intentions 1796-1849. mf 7t-9t: marriages 1794-1845. mf 9t-10t: births a-y 1777-1849. mf 10t: deaths a-w 1785-1843. mf 11t: vital records 1844-49) – us Archive [978]
Harvard 1723-1900 – Oxford, MA (mf ed 1997) – (= ser Massachusetts vital records) – 119mf – 9 – 0-87623-390-6 – (mf 1-5: vitals indexed 1723-1848. mf 6-11: town records 1732-1804. mf 9-11: marriages 1732-66. mf 10-11,21,42: deaths 1732-1849. mf 9-11, 20: births 1726-1820. mf 12-21: town records 1765-1834. mf 15: marriages 1766-1805. mf 18-20: intentions 1766-95. mf 22-38: town records 1782-1843. mf 34-35: intent/marriages 1825-48. mf 37-38,40-41: births 1777-1850. mf 43-52: town records 1819-68. mf 51-52,63-64: militia 1840-90. mf 52,65: out-of-town marriages 1732-98. mf 52: perkins family 1765-1880. mf 52,64: dog licenses 1862-76. mf 53-64: mortgage/misc 1838-93. mf 66-87: church records 1733-1909. mf 88-90: military 1822-65. mf 91-104: paupers 1832-1909. mf 105: voters 1877-84. mf 106-110: intentions 1848-1906. mf 110-113: vital record indexes 1841-1900. mf 114-119: vitals 1841-1900) – us Archive [978]
Harvard advocate – Cambridge. 1968+ (1) 1970+ (5) 1975+ (9) – ISSN: 0017-8004 – mf#3379 – us UMI ProQuest [340]
Harvard african studies – v1-10. 1917-32 – 1 – us AMS Press [960]
Harvard annual legal bibliography, 1961-1981 : full cumulation / Harvard University. Law Library – 340mf – 9 – $825.00 – mf#LLMC 83-100 – us LLMC [340]
Harvard blackletter journal *see* Harvard blackletter law journal
Harvard blackletter law journal – v1-17. 1984-2001 – 9 – $185.00 set – (title varies: v1-2 1983-85 as the blackletter journal. v3-10 1986-93 as harvard blackletter journal) – ISSN: 0897-2761 – mf#111131 – us Hein [340]
Harvard business reports – Cambridge. (1) 1926-1932 (5) (9) – mf#6405 – us UMI ProQuest [338]
Harvard business review – Boston. 1922+ (1) 1967+ (5) 1960+ (9) – ISSN: 0017-8012 – mf#634 – us UMI ProQuest [338]
Harvard civil rights-civil liberties law review – Cambridge. 1966+ (1) 1975+ (5) 1976+ (9) – ISSN: 0017-8039 – mf#10519 – us UMI ProQuest [323]
Harvard civil rights-civil liberties law review – v1-36. 1966-2001 – 5,6,9 – $660.00 set – (v1-19 1966-84 on reel or mf $315. v20-36 1985-2001 on mf $345) – ISSN: 0017-8039 – mf#103031 – us Hein [321]
Harvard class reports, 1833-1975 (for the classes of 1833-1900) : one of the largest and most important biographical archives ever published – [mf ed Chadwyck-Healey] – 520+ titles on 1088mf – 9 – (with p/ind) – uk Chadwyck [378]
Harvard College. Library *see*
– The blodgett collection of spanish civil war pamphlets
– Distributable union catalog
Harvard College Observatory *see*
– Annual report of the director of the astronomical observatory of harvard college
– Bulletin of the harvard college observatory
– Circular
– Reduction tables
– Reprints [series 1]
Harvard college observatory : the anonymous gift of 1902 – Cambridge: [the Observatory] 1906 [mf ed 1998] – 1r [pl/ill] – 1 – mf#film mas 28292 – us Harvard [520]
The harvard courier – Harvard, NE: Griff. J Thomas, jan 10 1885-v91 n53. dec 30 1976 (wkly) [mf ed v8 n23. jun 4 1892,1895-1976 (gaps) -1978] – 27r – 1 – (absorbed by: clay county republican+clay county news) – us NE Hist [071]
Harvard educational review – Cambridge. 1966+ (1) 1968+ (5) 1971+ (9) – ISSN: 0017-8055 – mf#2267 – us UMI ProQuest [370]
Harvard environmental law review – v1-25. 1978-2001 – 5,6,9 – $580.00 set – (v1-8 1976-84 on reel $143. v9-25 1985-2001 on mf $437) – ISSN: 0147-8257 – mf#103041 – us Hein [344]
Harvard excavations at samaria / Reisner, George A – Harvard. 1924. 2v – 9 – $21.00 – us IRC [930]
The harvard gospels / Goodspeed, Edgar Johnson – Chicago, IL: University of Chicago Press, c1918 – (= ser Historical and linguistic studies in literature related to the new testament) – 1mf – 9 – 0-524-08177-8 – mf#1992-1163 – us ATLA [226]
Harvard graduates' magazine – Boston. 1892-1934 – 1 – mf#2898 – us UMI ProQuest [378]
Harvard human rights journal – v1-14. 1988-2001 – 9 – $218.00 set – (title varies: v1-2 1988-89 as harvard human rights yearbook) – ISSN: 1057-5057 – mf#111101 – us Hein [341]
Harvard human rights yearbook *see* Harvard human rights journal
Harvard international law club bulletin *see* Harvard international law journal
Harvard international law journal – Cambridge. 1959+ (1) 1972+ (5) 1976+ (9) – ISSN: 0017-8063 – mf#7400 – us UMI ProQuest [341]
Harvard international law journal – v1-42. 1959-2001 – 9 – $566.00 set – (title varies: v1-2 1959-60 as: bulletin of harvard international law club. v3 1961-62 as: harvard international law club bulletin. v4-7 1962-66 as: international law club journal) – ISSN: 0017-8063 – mf#101201 – us Hein [341]
Harvard journal of asiatic studies – Cambridge. 1936+ (1) 1971+ (5) 1976+ (9) – ISSN: 0073-0548 – mf#2338 – us UMI ProQuest [950]
Harvard journal of asiatic studies – Cambridge, Mass, 1936-1947. v1-10 – 86mf – 8 – mf#CH-946c – ne IDC [956]
Harvard journal of law and public policy – v1-24. 1978-2001 – 5,6,9 – $592.00 set – (v1-7 1978-84 on reel $86. v8-24 1985-2001 on mf $506) – ISSN: 0193-4872 – mf#103051 – us Hein [342]
Harvard journal of law and technology – v1-13. 1988-2000 – 9 – $219.00 set – ISSN: 0897-3393 – mf#112421 – us Hein [346]
Harvard journal on legislation – Cambridge. 1964-1975 [1]; 1971-1975 [5,9] – ISSN: 0017-808X – mf#1951 – us UMI ProQuest [340]
Harvard journal on legislation – v1-38. 1964-2001 – 1,5,6 – $823.00 set – (v1-32 1964-95 in reel $644. v33-38 1996-2001 in mf $179) – ISSN: 0017-808X – mf#103061 – us Hein [342]
Harvard lampoon – Cambridge. 1981-1996 – 1,5,9 – mf#12895 – us UMI ProQuest [870]
Harvard latino law review – v1-4. 1994, 1997-2000 – 9 – $69.00 set – mf#115461 – us Hein [340]
Harvard law record – Cambridge, MA. 1946-72 – 1 – us L of C Photodup [340]
Harvard law record – v1-111. 1946-2001 – 9 – $1276.00 set – (v1-89 1946-90 on reel $655. v90-111 1990-2001 on mf $621) – ISSN: 0017-8101 – mf#112251 – us Hein [340]
Harvard law review – Cambridge. 1887+ (1) 1968+ (5) 1976+ (9) – ISSN: 0017-811X – mf#753 – us UMI ProQuest [340]
Harvard law review – v1-39. 1889-1925/26 – 312mf – 9 – $468.00 – (more vols will be added as copyright expires) – mf#LLMC 84-477 – us LLMC [340]
Harvard law review – v1-113. 1887-2000 – 1,5,6,9 – $4184.00 set – (v1-109 1887-1996 on reel or mf $4009. v110-113 1996-2000 on mf $175) – ISSN: 0017-811X – mf#103071 – us Hein [340]
Harvard lectures on the revival of learning / Sandys, John Edwin, Sir – Cambridge: University Press, 1905 – 1mf – 9 – 0-7905-6257-X – mf#1988-2257 – us ATLA [140]
Harvard legal commentary – v1-9. 1964-72 – 34mf – 9 – $51.00 – (papers from the harvard law school 2nd-year writing program.) – mf#LLMC 84-478 – us LLMC [340]
Harvard library bulletin – Cambridge. 1947+ (1) 1975+ (5) 1976+ (9) – ISSN: 0017-8136 – mf#10303 – us UMI ProQuest [020]
Harvard lyceum – Cambridge. 1810-1811 (1) – mf#3816 – us UMI ProQuest [420]
Harvard register – Cambridge. 1827-1828 – 1 – mf#3749 – us UMI ProQuest [378]
Harvard register : an illustrated monthly – Cambridge. 1880-1881 – 1 – mf#3872 – us UMI ProQuest [378]
Harvard review – Cambridge. 1962-1968 [1]; 1966-1968 [5,9] – ISSN: 0440-3487 – mf#1680 – us UMI ProQuest [378]
The harvard theological review – 1(1908)-39(1946) – 268mf – 9 – €457.00 – ne Slangenburg [200]
Harvard union catalog – [mf ed 1993] – 2376mf (1:48) – 9 – diazo €10,400.00 – ISBN-10: 3-598-41223-1 – ISBN-13: 978-3-598-41223-3 – gw Saur [010]
Harvard University. Board of Overseers *see* Report of the committee of the overseers of harvard college appointed to visit the observatory in the year...
Harvard University Herbaria *see* Orchid herbarium of oakes ames botanical museum
Harvard University. Houghton Library *see* Inventories of the houghton manucript collection
Harvard University. Law Library *see* Harvard annual legal bibliography, 1961-1981
Harvard University. Law School Library *see* Collection of state documents
Harvard University. Library *see* Distributable union catalog
The harvard university library : a documentary history / ed by Carpenter, Kenneth E – 463mf (24:1) – 9 – $3885.00 – us UPA [020]
Harvard university. museum of comparative zoology. bulletin – v1-32. 1863-99 – 1 – $360.00 – mf#0256 – us Brook [590]
Harvard university. museum of comparative zoology. memoirs – v1-55. 1864-1940 – 9 – $690.00 – mf#0257 – us Brook [590]
Harvard University. Peabody Museum of Archaeology and Ethnology *see*
– Memoirs
– Papers
– Reports
Harvard, W M *see* A narrative of the establishment and progress of the mission to india and ceylon...
Harvard, William M *see* A narrative of the establishment and progress of the mission to ceylon and india founded by the late rev thomas coke
Harvard women's law journal – v1-24. 1978-2001 – 5,6,9 – $340.00 set – (v1-7 1978-85 on reel $83. v8-24 1985-2001 on mf $257) – ISSN: 0270-1456 – mf#103081 – us Hein [340]
Harvardiana – Cambridge. 1835-1838 (1) – mf#4377 – us UMI ProQuest [500]
Harvest – London, England. 18– – 1r – us UF Libraries [240]
Harvest bells, 1882, 1884, 1887 / Penn, WE – 1 – 86.52 – us Southern Baptist [242]
Harvest field / Macfarlan, D – Paisley, Scotland. 1843 – 1r – us UF Libraries [240]
Harvest from the desert : the life and work of sir ganga ram / Bedi, Baba Pyare Lal – Lahore: Sir Ganga Ram Trust Society, 1940 – (= ser Samp: indian books) – us CRL [920]
Harvest of justice : newsletter of the gaudette peace and justice center / Gaudette Peace and Justice Center – 1982 jan-1985 jun/jul – 1r – 1 – mf#1269508 – us WHS [340]
Harvest quarterly – n1-7 [1976 mar-1977 fall] – 1r – 1 – mf#632509 – us WHS [071]
Harvester – Exeter. 1990-1990 (1) – (cont by: harvester/aware) – ISSN: 0017-8217 – mf#16627 – us UMI ProQuest [240]
The harvester : for gathering the ripened crops on every homestead, leaving the unripe to mature – Boston: William White, 1868 – 1mf – 9 – 0-8370-3514-7 – mf#1985-1514 – us ATLA [240]
Harvester/aware – Exeter. 1990-1990 (1) – (cont: harvester. cont by: aware/harvester) – mf#16627,01 – us UMI ProQuest [240]
Harvest-home : consisting of supplementary gleanings, original dramas and poems, contributions of literary friends, and select re-publications, including sympathy, a poem... / Pratt, Samuel – London 1805 [mf ed Hildesheim 1995-98] – 3v on 12mf – 9 – €120.00 – 3-487-27966-5 – gw Olms [800]
Harvestman's feast – London, England. 1830 – 1r – us UF Libraries [240]
Harvey, Alexander *see* On the voluntary principle in relation to national responsibility
Harvey, Alfred *see* English church furniture
Harvey, Andrew Edward *see* Martin bucer in england
Harvey, Annie J *see* Chronicles of an old inn
Harvey, Arthur *see*
– Aerolites and religion
– Astronomy, in infancy, youth and maturity
– Champlain's american experiences in 1613
– Decimals and decimalisation
– The discovery of lake superior
– The distribution of aerolites in space
– The grain trade
– Harvey's guide to patents
– The manufacturing clause of the canadian patent law
– Periodicity of magnetic disturbances
– The reciprocity treaty
– The year book and almanac of canada for...
Harvey, Arthur [comp] *see*
– A statistical account of british columbia
– University question
Harvey County. Kansas. District Court *see* Records of cases
Harvey, Frederick Burn *see* Church rate opposition
Harvey, George *see* Notes of the early history of the royal scottish academy
Harvey, Hezekiah *see*
– The church
– Memoir of alfred bennett
Harvey, Jacob *see* Harvey letters, 1812-1846
Harvey, Leonise *see* Bibliographie analytique des iles-de-la-madeleine
Harvey letters, 1812-1846 / Harvey, Jacob – National Library of Ireland – (= ser BRRAM series) – 1r – 1 – £67 / $134 – (int by e r a green) – mf#r96484 – uk Microform Academic [975]
Harvey markham star tribune – Chicago Heights, IL. 1930-1992 (1) – mf#61331 – us UMI ProQuest [071]
Harvey, Moses *see*
– Across newfoundland with the governor
– The artificial propagation of marine food fishes and edible crustaceans
– The great fire in st john's newfoundland, 8 july, 1892
– Lectures, literary and biographical
– Lectures on egypt and its monuments
– Lectures on the harmony of science and revelation
– The lessons of calamity
– Newfoundland as it is in 1894
– Newfoundland as it is in 1899
– Newfoundland, the oldest british colony
– A short history of newfoundland
– The testimony of nineveh to the veracity of the bible
– This newfoundland of ours
– Thoughts on the poetry and literature of the bible
– Where are we and whither tending?
Harvey, Peter *see* Reminiscences of daniel webster
Harvey, Richard S *see*
– Rights of the minority stockholder
– Rights of the minority stockholder and of the railway security holder
Harvey, T *see* The west indies in 1837
Harvey, Thomas *see* Jamaica in 1866
Harvey, Thomas Edmund *see* The rise of the quakers
Harvey, W H *see* Algae
Harvey, William *see*
– Sketches of hayti
– Visitations of northamptonshire
Harvey, William Henry *see* Letters relating to tonga
Harvey, William Patrick *see*
– Shall woman preach?
Harvey, William Wigan *see* Sancti irenaei, episcopi lugdunensis, libros quinque adversus haereses
Harvey, Zola Emile *see*
– Problems and solutions in equity.
– Problems and solutions in personal property.
– Problems and solutions in torts.
Harvey-Jellie, Wallace *see* Le theatre classique en angleterre, dans l'age de john dryden
Harvey-Jellie, Wallace Raymond *see* Chronicles
Harvey's guide to patents / Harvey, Arthur – Ottawa: A S Woodburn, 1885 – 1mf – 9 – mf#05252 – cn Canadiana [346]
Harwich 1673-1895 – Oxford, MA (mf ed 1987) – (= ser Massachusetts vital records) – 74mf – 9 – 0-931248-95-7 – (mf 1-4: b,m,d 1673-1752. mf 5-8: vital records 1698-1790. mf 9-14: b,m,d 1731-94. mf 15-20: vital records 1731-94. mf 21-23: births & deaths 1760-1842. mf 24-36: vital records & index 1796-1845. mf 37-42: marriage intentions 1808-49. mf 43-45: births 1811-50. mf 46-51: vitals records 1697-1849. mf 52-54: births & deaths 1765-1840. mf 55-56: b,m,d 1843-49. mf 57-62: births & index 1848-95. mf 63-69: marriages & index 1848-89. mf 70-74: deaths & index 1848-95) – us Archive [978]
Harwich and manningtree standard – England.1977. -d. 2 reels – 1 – uk British Libr Newspaper [072]
Harwood, Alan *see* Witchcraft, sorcery, and social categories among the safwa
Harwood, Edwin *see* The books of the kings
Harwood, Philip *see* German anti-supernaturalism
Harwood, Thomas *see* History of new mexico spanish and english missions of the methodist episcopal church from 1850 to 1910, in decades
Haryana government gazette / Haryana. India – Chandigarh. Part 1-3. Suppl. and index. Nov. 8-Dec. 27, 1966 – 1 – us NY Public [954]
Haryana. India *see* Haryana government gazette
Der harz : seine geschichte, ruinen und sagen; zwei reisen in den jahren 1800 und 1850 / Spieker, Christian W – Berlin 1852 [mf ed Hildesheim 1995-98] – 2mf – 9 – €60.00 – 3-487-29521-0 – gw Olms [914]
Har-Zahab, Zevi *see* Leshon eboram
Harzburger zeitung – Bad Harzburg, Braunlage DE, 1988- – 9r/yr – 1 – (title varies: 24 aug 1991: goslarsche zeitung) – gw Misc Inst [074]
Harz-kurier – Herzberg i. Harz DE, 1977- – ca 7r/yr – 1 – (ed of bad lauterberger zeitung) – gw Misc Inst [074]
Harz-kurier – Wernigerode DE, 1962 20 apr-1967 27 sep [gaps] – 1r – 1 – (publ in magdeburg) – gw Misc Inst [074]
Harzreise / Heine, Heinrich; ed by Buchheim, C A – 3rd rev ed. Oxford: Clarendon Press, 1900 – 1 – us Wisconsin U Libr [830]
Harz-Verein fuer Geschichte und Alterthumskunde *see* Zeitschrift des harz-vereins fuer geschichte und alterthumskunde
Has, or is, man a soul? / Westerby, W M – London, England. 1879 – 1r – us UF Libraries [240]
Has oude been worse governed by its native princes than our indian territories by leadenhall street? / Lewin, Malcolm – London: J Ridgeway, 1857. Lucknow, 1870 – (filmed with: a historical sketch of fyzabad tehsil/p carnegy) – us CRL [950]

Has religion anything to do with our colleges? – [s.l: s.n, 1855?] [mf ed 1992] – 1mf – 9 – 0-524-03644-6 – mf#1990-1072 – us ATLA [377]
Has the church of rome any pope, priest, sacrament, or rule of fait... / Minton, Samuel – London, England. 1851 – 1r – us UF Libraries [240]
Has the english church preserved the episcopal succession? – London: SPCK [1896?] [mf ed 1992] – (= ser Church historical society (series) 1) – 1mf – 9 – 0-524-05542-4 – mf#1990-5146 – us ATLA [242]
Has the law of natural selection by survival of the fittest failed in the case of man? / Tait, Robert Lawson – Dublin, 1869 – (= ser 19th c evolution & creation) – 1mf – 9 – mf#1.1.5266 – uk Chadwyck [575]
Hasan 'Ali, B *see* Observations on the mussulmauns of india
Hasan, Mohibbul *see* History of tipu sultan
Hasan, S Badrul *see* Syed qutb shaheed
Hasan, Zafar *see* Mosque of shaikh 'abdu-n nabi
Hasan, Zafar, Khan Bahadur *see* A guide to nizamu-d din
Hasbrouck, Louise Seymour *see* Mexico, from cortes to carranza
Hasche, Johann Christian *see* Magazin der saechsischen geschichte
Hase, C *see* Historia scriptoresque alii ad res byzantinas pertinentes (cbh39)
Hase, Karl Alfred von *see*
– Die bedeutung des geschichtlichen in der religion
– Herzog albrecht von preussen und sein hofprediger
– New testament parallels in buddhistic literature
– Sebastian franck von woerd, der schwarmgeist
Hase, Karl von *see*
– Handbook to the controversy with rome
– History of the christian church
– Kirchengeschichte
Hase-Koehler, Else von *see* Ursula schreibt ins feld
Hasel, Haven Binford *see* English hymn-tunes of the eighteenth century
Haselberg, Gabriel Peter *see* Juristische bibliothek
D'haselmuus : e gschicht us em undergang vom alte baern / Tavel, Rudolf von – 3. aufl. Bern: A Francke, [1939] [mf ed 1996] – 280p – 1 – mf#9304 – us Wisconsin U Libr [830]
Haseltine, Hubert Arthur *see* Our haitian policy
Hasemann, J *see* Griechische kirche
Hasenclever, Walter *see*
– Antigone
– Jenseits
– Der juengling
– Die menschen
– Der sohn
– Tod und auferstehung
Hasenclever, Wilhelm *see* Liebe, leben, kampf
Ha-senegor – The Defender. (New York), 1890-91 – (= ser Senegor) – us AJPC [071]
Hash – 1970 jan 3/17 – 1r – 1 – (cont: warren free press) – mf#1583822 – us WHS [071]
Hash House Harriers Club *see* Only half a mind
Hash shahar – Vienna, Austria. 1870-78; 1880; 1883; 1884 – 7r – 1 – uk British Libr Newspaper [072]
Hash (wholesale to boarding houses) / Emberson, Frederick C – [Montreal: s.n, between 1885 and 1900] – 1mf – 9 – 0-665-94089-0 – mf#94089 – cn Canadiana [880]
Hashim, Talal J *see* A health knowledge test for male college freshmen in saudi arabia
Hasidishe velt / Unger, Menasheh – New York, NY. 1955 – 1r – us UF Libraries [939]
Hasidism / Buber, Martin – New York, NY. 1948 – 1r – us UF Libraries [939]
Hasidus / Zeitlin, Hillel – Warsaw, Poland. 1922 – 1r – us UF Libraries [939]
Hasimi *see* The divan project
Haskalat ha-'am ba-arets / Universitah Ha-'Ivrit Bi-Yerushalayim – Jerusalem, Israel. 1944 – 1r – us UF Libraries [939]
Haskell alumni association newsletter / Haskell Indian Junior College, Lawrence, KS – v1 n1-2 [1978 mar 28-aug] – 1r – 1 – mf#501468 – us WHS [373]
Haskell, Arnold Lionel *see* Ballet, 1945-1950
The haskell gospels / Goodspeed, Edgar Johnson – Chicago, IL: University of Chicago Press, 1918 – (= ser Historical and linguistic studies in literature related to the new testament) – 1mf – 9 – 0-524-08178-6 – mf#1992-1164 – us ATLA [226]
Haskell Indian Junior College et al *see* Indian leader
Haskell Indian Junior College, Lawrence, KS *see* Haskell alumni association newsletter
Haskell, Samuel *see* Heroes and hierarchs
Haskell, T H *see* Haskell's judgements of the honorable edward fox for the maine district and first circuit, 1866-1881
Haskell, William B *see* Two years in the klondike and alaskan gold-fields
Haskell's judgements of the honorable edward fox for the maine district and first circuit, 1866-1881 / Haskell, T H – Portland: Fessenden. v1-2. 1887 – (= ser Early federal nominative reports) – 14mf – 9 – $21.00 – (sometimes known as: fox's decisions) – mf#LLMC 81-482 – us LLMC [340]
Haskett-heskett-hiskett exchange – v1 n1-v6 n3/4 [1982 mar-1987 summer] – 1r – 1 – mf#1288298 – us WHS [071]
Haskins, Caryl Parker *see* Amazon
Haskins, Charles Hamilton *see* Galvanometer
Haskins, Dan D *see* Specialized leadership training in the tennessee tech baptist student union executive council
Haskins, Orren N *see* Collection of ancient tunes
Haskins, William C *see* Canal zone pilot
Haslam-gherai, sultan de crimee : ou voyages et souvenirs du duc de richelieu, president du conseil des ministres recueillis sur des temoignages authentiques... / Asfeld, L T d' – Paris 1827 [mf ed Hildesheim 1995-98] – (= ser Fbc) – 2mf – 9 – €60.00 – 3-487-28963-6 – gw Olms [920]
Haslbeck, Hanns *see* Ein wort zur weltanschauung richard wagners
Haslewood, Francis *see* Monumental inscriptions, in the parish of saint matthew
Haslewood, Joseph *see*
– Secret history of the green-room
Haslingden and rawtenstall express – [NW England] Haslingden Lib 17 sep 1864-22 dec 1866 – 1 – uk MLA; uk Newsplan [072]
Haslingden borough news – [NW England] Haslingden Lib 21 apr 1977-18 dec 1984 – 1 – uk MLA; uk Newsplan [072]
Haslingden chronicle and ramsbottom times – [NW England] Haslingden Lib jun 1867-dec 1871* – 1 – uk MLA; uk Newsplan [072]
Haslingden gazette – [NW England] Haslingden Lib jul 1901-dec 1912, 1914-26 – 1 – uk MLA; uk Newsplan [072]
Haslingden guardian – [NW England] Haslingden Lib nov 1891-oct 1920 – 1 – uk MLA; uk Newsplan [072]
Haslingden observer – [NW England] Haslingden Lib jan 1926-1976 – 1 – uk MLA; uk Newsplan [072]
Haslingfield 1599-1950 – 6mf – 9 – £7.50 – uk CambsFHS [929]
Haslington chronicle & ramsbottom times – Haslington, Ramsbottom, England. -w. 1 June 1867-30 Dec 1871. 1 1 2 reels – 1 – uk British Libr Newspaper [072]
Haslington, Congregational Chapel – (= ser Cheshire monumental inscriptions) – 1mf – 9 – £2.50 – mf#20 – uk CheshireFHS [929]
Haslington, primitive methodist – (= ser Cheshire monumental inscriptions) – 1mf – 9 – £2.50 – mf#95 – uk CheshireFHS [929]
Haslington, st matthew – (= ser Cheshire monumental inscriptions) – 1mf – 9 – £2.50 – mf#20 – uk CheshireFHS [929]
Hasmet *see* The divan project
Hasper volksblatt – Hagen, Westf DE, 1869 11 sep-1941 30 may, 1949 19 nov-1950 – 87r – 1 – (title varies: 1 apr 1875: hasper zeitung. filmed by other misc inst: 1951-57 [18r]). – gw Misc Inst [074]
Hasper zeitung *see* Hasper volksblatt
Haspeslagh, Louis *see*
– De godloosheyd der achthiende eeuw
– Godloosheyd der achttiende eeuw
Hass, Hans-Egon *see* Heinrich heine
Hassall, A *see* Magna carta
Hassall, Arthur *see*
– Germany in the later middle ages, 1200-1500
– Historical introductions to the rolls series
– Lectures on european history
Hassam, John Tyler *see* Bahama islands
Hassard, Albert Richard *see* Canadian constitutional history and law
Hassaurek, Friedrich *see*
– Gedichte
– The secret of the andes
Hasse, Evelyn R *see* The moravians
Hasse, Friedrich Rudolf *see* Geschichte des alten bundes
Hasse, Hermann Gustav *see*
– Grundlinien christlicher irenik
– Ueber die vereinigung der geistlichen und weltlichen obergewalt im roemischen kirchenstaate
Hasse, Joh Gottfried *see* Magazin fuer die biblisch-orientalische litteratur und gesammte philogie
Hasse, Johann Adolf *see*
– Alcide al brivo
– [Antigono]
– Attilio regolo [dramma per musica si] sig gio adolfo hasse [la poesia e del sig abbate pietro metastasio
– Hymnus ambrosianus sive te deum laudamus
– Miserere
– Solimano
Hasse, Max *see* Peter cornelius (1824-1874) musical works
Hasse, R *see* Die bergarbeiter-verhaeltnisse in grossbritannien
Hassel, Mary *see* Cap francais vu par une americaine
Hassell, John *see*
– Aqua pictura
– Beauties of antiquity
– Calcographia
– Memoirs of the life of...george morland
Hassells, C S *see* Sanctuary of god, a solemn monitor to man
Hasselquist, F *see*
– Iter palaestinum eller resa til heliga landet...
– Reise nach palaestina in den jahren von 1749 bis 1752
– Voyages dans le levant, dans les annees 1749-1752
Hasselquist, Frederik *see* Voyages dans le levant, dans les annees 1749, 50, 51 et 52
Hasselquist, Tufve Nilsson *see* Foersoek till en grundlig och dock laettfattlig foerklaring af pauli bref till eferserna
Hasselt, J L van *see* Gedenkboek van een vijf-en-twintigjarig zendelingsleven op nieuw-guinea (1862-1887)
Hasselt, Vincent B Van *see* Journal of child and adolescent substance abuse
Hassencamp, Robert *see* Neue briefe chr. mart. wielands
Hassenstein, Georg *see* Ludwig uhland
Hassert, Kurt *see* Die erforschung afrikas.
Hasskarl, Gottlieb Christopher Henry *see* Evolution, as taught in the bible
Hassler, C D *see* Evagatorium in terrae sanctae, arabiae et egypti peregrinationem
Hassler, Cunradus Dietericus *see* Fratris felicis fabri evagatorium in terrae sanctae, arabiae et egypti peregrinationem
Hassler, Hans Leo *see*
– Cantiones sacrae de festis...
– Canzonette a quatro voci libro primo
– Lustgarten neuer teutscher gesaeng...
– Madrigali a 5. 6. 7. & 8. voci
– Missae 4, 5, 6 et 8 vocibus
– Missae qvaternis, 5. 6. et 7 vocibvs
– Neuee teuetsche gesang...
– Psalmen und christliche gesaenge
– Sacri concentus 4, 5, 6, 7, 8, 9, 10 & 12 vocum
– Verbum caro factum est
Hassler, K D *see* Heinrich mynsinger von den falken, pferden und hunden
Hassler, Konrad Dietrich *see* Ott rulands handlungsbuch [1444-64]
Hasta llegar a dios / Mejias, Lola – Zaragoza: Tip. Aragonesa, 1964 – 1 – sp Bibl Santa Ana [946]
Hasta luego : poesia / Saenz Cordero, Efrain – San Jose, Costa Rica. 1958 – 1r – us UF Libraries [972]
Hastenpflug, Fritz *see* Das diminutiv in der deutschen originalliteratur des 12. und 13. jahrhunderts
Hastie, Peter A *see* Task account ability in school physical education and sports settings
Hastie, W *see*
– Hindu idolatry and english enlightenment
– Outlines of the science of jurisprudence
Hastie, William *see*
– History of german theology in the nineteenth century
– Hymns and thoughts on religion
– Kant's cosmogony
– Outlines of pastoral theology for young ministers and students
– Theology as science
– The theology of the reformed church in its fundamental principles
Hastie, William H *see* The william h hastie papers
Hastings and st leonards chronicle – [London & SE] East Sussex, Hastings Ref Lib 1849-50, 1851-52, 1858, 1863, 1884, 1892, 1896, 1898, 1899-1900 – 11r – 1 – uk Newsplan [072]
Hastings and st leonards news – [London & SE] East Sussex, Hastings Ref Lib may 1848-oct 1848, 1849-88, 1890-96, 1898-1905 – 1 – uk Newsplan [072]
Hastings, Bessie *see* Published manuscripts
Hastings Center Hastings *see* Center report
Hastings chronicle – Hastings, ON. 1861-64 – 1r – 1 – cn Library Assoc [071]
Hastings college of law : library shelf list – dec 1983 – 6r – 5 – $300.00 – mf#B50523 – us Library Micro [378]
Hastings communications and entertainment law journal (comm/ent) – Hastings College of Law. v1-22. 1977-2000 – 5,6,9 – $480.00 set – (v1-7 1977-85 on reel $124. v8-22 1985-2000 on mf $356. title varies: v1-5 1977-83 as comm/ent: a journal of communications and entertainment law. v6-10 1983-88 as comm/ent: hastings journal of communications and entertainment law) – ISSN: 1061-6578 – mf#108701 – us Hein [340]
Hastings constitutional law quarterly – v1-28. 1974-2001 – 5,6,9 – $589.00 set – (v1-12 1974-85 on reel $220. v13-28 1985-2001 on mf $369) – ISSN: 0094-5617 – mf#103091 – us Hein [323]
Hastings daily gazette – Hastings, NE: C C Babcock, 1879 (daily ex mon) [mf ed v1 n34. apr 5,8 1879 filmed [1993]] – 1r – 1 – (weekly ed: adams county gazette (juniata ne)) – us NE Hist [071]
Hastings daily nebraskan – Hastings, NE: Merritt & Creeth, 1888-95// (daily ex sun) [mf ed 1889-94 (gaps) filmed [1969?]] – 7r – 1 – (absorbed by: hastings daily republican. weekly ed: hastings weekly nebraskan) – us NE Hist [071]
Hastings daily news – Hastings, NE: Hasting News Co, 1897-97// (daily ex mon) [mf ed with gaps filmed [1974?]] – 1r – 1 – us NE Hist [071]
Hastings daily republican – Hastings, NE: Republican, 1891-v40 n110. sep 4 1915 (daily ex sun) [mf ed 1892-1915 (gaps) filmed [1967?-89] – 40r – 1 – (absorbed: hastings daily nebraskan. absorbed by: hastings daily tribune. issues for oct 2 1895-apr 10 [ie 11] 1898 also numbered v7 n38-v10 n36 cont the numbering of the hastings daily nebraskan. issues for apr 12 1898-sep 4 1915 numbered v10 n37-v40 n110. numbering very irregular) – us NE Hist [071]
Hastings daily times – Hastings, NE: Daily Times Pub Co, 1892 (daily ex sun) [mf ed n168. jun 27 1892 filmed [1974?]] – 1r – 1 – us NE Hist [071]
Hastings daily tribune – Hastings, NE: Adam Breede. 75v. v1 n1. oct 2 1905-v75 n198. may 23 1980 (daily ex sun) [mf ed with gaps [1966?-84?] – 323r – 1 – (absorbed: hastings daily republican 1915 and: morning spotlight 1942. cont by: hastings tribune) – us NE Hist [071]
The hastings democrat – Hastings, NE: Democrat Pub Co. 42nd yr n51. may 3 1923-56th yr n5. may 30 1935 (wkly) [mf ed with (gaps) filmed [1969?]] – 6r – 1 – (cont: adams county democrat; absorbed by: morning spotlight) – us NE Hist [071]
The hastings evening record – Hastings, NE: Mock Bros & Palmer. v1 n1. sep 20 1897-1900 (daily ex sun) [mf ed with gaps filmed [1972?]] – 5r – 1 – (semiweekly ed: hastings semi-weekly record. weekly ed: hastings weekly record) – us NE Hist [071]
Hastings, Frederick *see*
– The background of sacred story
– Obscure characters and minor lights of scripture
Hastings gazette – Wauchope. dec 1969-dec 1973, jan 1975-jun 1997 – at Pascoe [079]
The hastings gazette-journal – Hastings, NE: Gazette-Journal Co (wkly) [mf ed 1886-88 (gaps) filmed 1973] – 3r – 1 – (cont: hastings weekly gazette-journal. iss numbering ceased with v17 oct 5 1887; resumed with v18 n7 feb 22 1888) – us NE Hist [071]
The hastings graphic – Hastings, NE: E W Dirks and F W Kaul (wkly) [mf ed v1 n32. aug 3 1932] – 1r – 1 – us NE Hist [071]
Hastings, H L *see*
– Atheism and arithmetic
– Fourteen nuts for sceptics to crack
– Friendly hints to candid sceptics
– Is the bible inspired of god?
– Israel's greatest prophet
– Israel's messiah
– Pauline theology
– Remarks on the mistakes of moses
Hastings herald – Hastings, FL. 1918-1921 – 2r – us UF Libraries [071]
Hastings herald – Hastings, ND: Herald PublCo. v1 n1 may 11 1923-sep 10 1926?// (wkly) – 1 – (missing: 1925 oct 16) – mf#10533-10534 – us North Dakota [071]
Hastings herald – Wauchope, aug 1987-mar 1988 – 1r – A$77.92 vesicular A$83.42 silver – at Pascoe [079]
Hastings, Horace Lorenzo *see*
– Atheism and arithmetic
– The reign of christ on earth
– The signs of the times
– A square talk to young men about the inspiration of the bible
Hastings independent tribune – Hastings, NE: [Independent Tribune] 1892-v10 n22. nov 29 1895 (wkly) [mf ed with gaps filmed [1969]-75] – 2r – 1 – (cont: independent tribune. cont by: hastings tribune (1895)) – us NE Hist [071]
Hastings international and comparative law review – v1-23. 1977-2000 – 1,5,6 – $384.00 set – (v1-18 1977-95 on reel or mf $253. v19-23 1996-2000 on mf $131) – ISSN: 0149-9246 – mf#103101 – us Hein [341]
Hastings, James *see*
– The christian doctrine of prayer
– A dictionary of christ and the gospels
– A dictionary of the bible
– Sub corona
Hastings journal – Hastings, NE: A L Wigton & M K Lewis, may 1873-v8 n30. dec 2 1880 (wkly) [mf ed with gaps filmed 1973-[95]] – 4r – 1 – (merged with: adams county gazette to form: gazette-journal. daily ed: hastings journal (daily ed)) – us NE Hist [071]

The hastings journal – Hastings, NE: G E Brown (wkly) [mf ed v13 n19. jan 25, feb 8 & 22 1907 filmed 1999] – 1r – 1 – us NE Hist [071]

The hastings journal (daily edition) – Hastings, NE: Wigton Bros, 1879 (daily ex sun) [mf ed v1 n27. apr 1 1879 (gaps) filmed -[1993]] – 3r – 1 – us NE Hist [071]

Hastings law journal – San Francisco. 1949+ (1) 1982+ (5) 1982+ (9) – ISSN: 0017-8322 – mf#1588 – us UMI ProQuest [340]

Hastings law journal – v1-52. 1949-2001 – 5,6,9 – $1250.00 set – (v1-36 1949-85 on reel or mf $777. v37-52 1985-2001 on mf $473) – ISSN: 0017-8322 – mf#103111 – us Hein [340]

Hastings news – Hastings, NE: J S Williams. 3v. v3 n28. jul 15 1899-v5 n27. jun 21 1901 (wkly) [mf ed with gaps filmed [1967?]] – 2r – 1 – (cont: hastings weekly news. cont by: williams hastings news) – us NE Hist [071]

Hastings republican – Hastings, NE: Watkins Bros, 1889 (wkly) [mf ed v1 n34. aug 31 1889 filmed [1976?]] – 1r – 1 – (cont by: hastings saturday republican) – us NE Hist [071]

Hastings republican – Hastings, NE: Ed Watkins. 3v. v14 n1. aug 12 1902-v16 n78. may 11 1905 (semiwkly) [mf ed with gaps filmed 1975] – 2r – 1 – (cont: hastings weekly republican) – us NE Hist [071]

Hastings, Robert J *see*
- Published manuscripts
- Stewardship development materials

Hastings saturday republican – Hastings, NE: C L Watkins, F A Watkins (wkly) [mf ed 1895-98 (gaps)] – 2r – 1 – (cont: hastings republican. absorbed: hastings weekly nebraskan. cont by: hastings weekly republican. issues for oct 5 1895-apr 23 1898 also called v13 n52-v15 n34, cont the numbering of the hastings weekly nebraskan. weekly ed: hastings wednesday republican 1895-96. daily ed: hastings daily republican) – us NE Hist [071]

Hastings semi-weekly record – Hastings, NE: Mock Bros & Palmer, 1897 (semiwkly) [mf ed v1 n2. sep 21 1897 filmed [1974?]] – 1r – 1 – (cont by: hastings weekly record) – us NE Hist [071]

Hastings shire gazette – Wauchope, 1941-68 – at Pascoe [079]

Hastings standard – apr 1896-mar 1903; jun 1903-dec 1910; jan 1911-jan 1937// – 1 – (title changes on 12 dec 1910 to: hawkes bay tribune. jan 1911-jan 1937. ceased publ 15 jan 1937. merged with hawkes bay herald to form hawkes bay herald tribune) – mf#35.14 – nz Nat Libr [079]

Hastings, Thomas *see*
- The history of forty choirs
- The mother's hymn book
- Musica sacre
- Psalmist or choir melodies

The hastings times – Hastings, NE: H M Van Arman, jul 1873 (mthly) [mf ed aug 1873] – 1r – 1 – us NE Hist [071]

The hastings times – Hastings, Barnes Co, ND: Fred E Osborne, sep 1907; -v12 n30 jul 9 1919 (wkly) – 1 – (missing: jul 14 1915) – mf#10530-10532 – us North Dakota [071]

Hastings tribune – Hastings, NE: Hastings Tribune. v75 n199. may 24 1980- (daily ex sun and jan 1, jul 1, thanksgiving and christma) [mf ed filmed 1980-] – 1 – (cont: hastings daily tribune) – us NE Hist [071]

Hastings tribune – Hastings, NE: Adam Breed & Co. v10 n23. dec 6 1895-1917// (wkly) [mf ed with gaps filmed 1969] – 14r – 1 – (cont: hastings independent tribune. some irregularities in numbering) – us NE Hist [071]

Hastings, Truman *see* Law for the masses: for every-day use

Hastings weekly gazette-journal – Hastings, NE: Gazette-Journal Co, 1882-dec 1885// (wkly) [mf ed 1884-85 (gaps) filmed 1973] – 2r – 1 – (cont: gazette-journal. cont by: hastings gazette-journal) – us NE Hist [071]

Hastings weekly independent – Hastings, NE: Frank D Taggart. v1 n1. jul 3 1886-91// (wkly) [mf ed jul 3 1886, aug 30 1889] – 2r – 1 – (merged with: hastings tribune (1886) to form: independent tribune) – us NE Hist [071]

Hastings weekly nebraskan – Hastings, NE: A T Bratton, -1895// (wkly) [mf ed 1883-95 (gaps) filmed [1973?]-75] – 3r – 1 – (cont: central nebraskan. absorbed by: hastings saturday republican. numbering very irregular) – us NE Hist [071]

Hastings weekly news – Hastings, NE: Hastings News Co, 1897-v3 n27. jul 7 1899 (wkly) [mf ed with gaps filmed [1967?]] – 2r – 1 – (cont by: hastings news) – us NE Hist [071]

Hastings weekly republican – Hastings, NE: F A Watkins. -v14 n40. aug 8 1902 (wkly) [mf ed 1899-1902 (gaps) filmed -1975] – 3r – 1 – (cont: hastings saturday republican. cont by: hastings republican (1902). issues for feb 4 1899-aug 2 1901 also called v15 n75-v18 n39, cont the numbering of the hastings weekly nebraskan) – us NE Hist [071]

Hastings weekly republican – Hastings, NE: F A Watkins. -v14 n40. aug 8 1902 (wkly) [mf ed 1899-1902 (gaps) filmed -1975] – 3r – 1 – (cont: hastings saturday republican. cont by: hastings republican (1902). issues for feb 4 1899-aug 2 1901 also called v15 n75-v18 n39 cont the numbering of the hastings weekly nebraskan. daily ed: hastings daily republican) – us NE Hist [071]

Hastings west-northwest journal of environmental law and policy – Hastings Scholarly Publ. v1-7. 1994-2001 – 9 – $171.00 set – mf#116111 – us Hein [344]

Hastings, William F *see* Puerto rico today and tomorrow

Hastings women's law journal – v1-12. 1989-2001 – 9 – $206.00 set – ISSN: 1061-0901 – mf#113011 – us Hein [346]

Haswell, Margaret Rosary *see* Economics of agriculture in a savannah village

Haszler, K D *see*
- Die reisen des samuel kiechel
- Reisen und gefangenschaft hans ulrich kraffts

Hat der jesus der evangelien wirklich gelebt? : eine antwort an prof. dr. juelicher / Jensen, P – Frankfurt a. M.: Neuer Frankfurter Verlag, 1910 – 1mf – 9 – 0-7905-3267-0 – mf#1987-3267 – us ATLA [240]

Hat jesus das papsttum gestiftet? : eine dogmengeschichtliche untersuchung / Schnitzer, Joseph – Augsburg: Lampart, 1910 – 1mf – 9 – 0-7905-6622-2 – (incl bibl ref) – mf#1988-2622 – us ATLA [240]

Hat richard wagner eine schule hinterlassen? / Seidl, Arthur – Kiel: Lipsius und Tischer, 1892 – 1r – 1 – (incl bibl ref) – us Wisconsin U Libr [780]

Hat vlaamsche nieuws – Antwerp Belgium, 25, 26 nov 1916; 10 feb 1917-4 mar 1918 – 2r – 1 – uk British Libr Newspaper [074]

Hat worker / United Hatters, Cap and Millinery Workers International Union – 1938-45, 1946-53, 1954-59, 1960-62, 1963-65, 1966-71 – 6r – 1 – (cont: headgear worker) – mf#774334 – us WHS [680]

Hatam sofer / Reisfeder, Jacob – Varsha, Poland. 1919 – 1r – us UF Libraries [939]

Hatam-sofer ve-talmidav / Weingarten, Shmuel Hacohen – Jerusalem, Israel. 1944 – 1r – us UF Libraries [939]

Hataskori jogszabalyok es hataskori hatarozatok tara. – Budapest. On film: v1-9; 1909-17. LL-0249 – 1 – us L of C Photodup [340]

Hatch act decisions : political activities cases of the u.s. civil service commission / Irwin, James W – Washington: GPO, 1949 – 4mf – 9 – $6.00 – (pt 1-a text on principles and cases. pt 2-a casebook on commission decisions) – mf#LLMC 84-110 – us LLMC [340]

Hatch, Azel Farnsworth *see* Statutes and constitutional provisions of the states and territories of the united states and the statutes of england on libel and slander.

Hatch, E *see* Concordance to the septuagint

Hatch, Edwin *see*
- Essays in biblical greek
- Griechentum und christentum
- The growth of church institutions
- The influences of greek ideas and useages upon the christian church
- An introductory lecture on the study of ecclesiastical history
- Memorials of edwin hatch, sometime reader in ecclesiastical history in the university of oxford, and rector of purleigh
- The organization of the early christian churches

Hatch Genealogical Society *see* Genealogy and history of the hatch family

Hatch, John Charles *see* History of postwar africa

Hatcher, Eldridge Burwell *see* The bible and the monuments

Hatcher, William Eldridge *see* John jasper

Hatchet – Washington DC. 1886 jan 24 – 1r – 1 – mf#845713 – us WHS [071]

Hatchette, Wilfred Irwin *see* Youth's flight

Hatchiah – Chicago. Regeneration. 1899-1900 – 1 – us AJPC [071]

Hatem uel-enbiya / Ileri, Celal Nuri – Istanbul: Yeni Osmanli Matbaa ve Kitaphanesi, 1332 [1916] – (= ser Ottoman literature, writers and the arts) – 4mf – 9 – $75.00 – us MEDOC [470]

Hatfield, Bradley D *see* Central and autonomic nervous system activity during self-paced motor performance: a study of the activation construct in marksmen

Hatfield, Edwin Francis *see*
- The decline of popery and its doctrinal diversities
- Our ecclesiastical polity

Hatfield, James Taft *see* Gedichte

Hath god cast away his people? / Gaebelein, Arno Clemens – New York City: Gospel Publishing House; Toronto, Canada: Upper Canada Tract Society, 1905 – 1mf – 9 – 0-7905-1389-7 – (incl ind) – mf#1987-1389 – us ATLA [220]

Hathaway, Grace *see* Fate rides a tortoise; a biography of ellen spencer mussey

Hathaway, Lillie Vinal *see* German literature of the mid-nineteenth century in england and america

Hatheway, Warren Frank *see*
- Canadian nationality
- Why france lost canada

Hati emas / Phoa, Gin Hian – Soerabaia: Tan's Drukkerij, [1937] [mf ed 1998] – (= ser Penghidoepan 145) – 1r – 1 – (coll as pt of the colloquial malay collection. trans of unidentified chinese novel [salmon, claudine. literature in malay by the chinese of indonesia. paris: editions de la maison des sciences de l'homme, c1981 p290]. filmed with: pembalesan dendam hati / phoa gin hian) – mf#10003 – us Wisconsin U Libr [830]

Hatikva – Buenos Aires, Argentina. Jul 1948-oct 1958 (imperfect) – 3r – 1 – uk British Libr Newspaper [079]

Hatikvah – New York, NY, v1, no. 1(Dec. 1971)-v2, no. 1 (20 Dec. 1972); v3, no. 1 (Nov. 1973)-v3, no. 2 (Jan. 1974); v3, no. 4 (15 Apr. 1974)-v5, no. 4 (May 1976); v6, no. 2 (June 1977)-v7, no. 2 (Jan. 1978); v9, no. 2 (Mar. 1980)-v10, no. 1 (Nov. 1980); v11, no. 1 (Jan. 1982)-v15, no. 3 (Dec. 1985); v15, no. 5 (May 1986)-v16, no. 1 (Nov. 1986); v16, no. 3 (Mar. 1987) – us AJPC [270]

Hatim Tilawonu *see* Hatim's tales

Hatim's tales : kashmiri stories and songs / ed by Grierson, George A – London: John Murray for the Govt of India, 1923 – (= ser Samp: indian books) – (recorded with the assistance of govind kaul by aurel stein; with a trans, linguistic analysis, vocabulary, ind, etc; with a note on the folklore of the tales by w crooke) – us CRL [390]

Hatirat-i niyazi / Niyazi Bey, Resneli – Istanbul: Sabah Matbaasi, 1326 [1910] – (= ser Ottoman histories and historical sources) – 13mf – 9 – $210.00 – us MEDOC [956]

Hatlestad, Ole Jensen *see* Historiske meddelelser

Hatley st george 1591-1837 – 2mf – 9 – £2.50 – uk CambsFHS [929]

Hato maunten senchineru *see* Heart mountain sentinel

Hats'i hamar / Mankowni, N L – 1914 – 9 – $15.00 – us Scholars Facs [490]

Hats'i hamar / Mankowni, N L – 1962 – 9 – $15.00 – us Scholars Facs [490]

Hatsofe – New York. N.Y. 1872 – 1 – us AJPC [071]

Hatsufeh, oder, an'iberkehrenish / Shaikewitz, Nahum Meir – Vilna, Lithuania. 1889 – 1r – us UF Libraries [939]

Hatt giegen hatt : niederdeutsches bauerndrama in 3 aufzuegen / Wagenfeld, Karl – Hamburg: R Hermes, 1917 [c1916] – 1r – 1 – us Wisconsin U Libr [820]

Hatt, Paul K *see* Backgrounds of human fertility in puerto rico

Hatta, M *see* Beberapa fasal ekonomi

Hatta yoshiaki monjo : yoshiaki hatta records; vice president of south manchurian railway. in the holdings of waseda university, tokyo – 1593 items on 42r – 1 – Y630,000 – (with 168p guide. in japanese) – ja Yushodo [380]

Hattersley, Alan Frederick *see*
- British settlement of natal
- Illustrated social history of south africa

Hattersley, Allan Frederick *see* Later annals of natal.

Hattiesburg american – Hattiesburg MS. 1965 jun 19,24,26, jul 2,30, aug 11,14-16,18-19,21,25 – 1r – 1 – (cont: hattiesburg news) – mf#869132 – us WHS [071]

Hattinger zeitung *see* Maerkische blaetter

Hattingh, S C *see* Sprokiesvorsing

Hattingh, W H J et al *see* Health aspects of water supply

Hattler, Franz *see*
- Christrosen im mariengarten
- Das gnadenbild der mater ter admirabilis von ingolstadt in bayern

Hatton, Eleanor Beard *see* Follow thou me

Hatton, Joseph *see*
- By order of the czar
- Henry irving's impressions of america, vol 1
- Henry irving's impressions of america, vol 2
- Henry irving's impressions of america, vols 1 and 2
- The lyceum "faust"
- Newfoundland
- Newfoundland, the oldest british colony
- To-day in america
- Under the great seal
- The white king of manoa

Hatzfelder volksblatt – Hatzfeld (Jimbolia RO), 1924 30 nov-1932 – 2r – 1 – gw Misc Inst [077]

Hatzfelder zeitung – Hatzfeld (Jimbolia RO), 1920 10 oct-1939 10 dec – 5r – 1 – gw Misc Inst [077]

Ha-tzofe – Israel, 1979- – 9 – us UMI ProQuest [079]

Hauber, A *see* Urkundenbuch des klosters heligenkreuztal

Hauber, E D *see* Nuetzlicher discours von dem gegenwaertigen zustand der geographie besonderes in teutschland...

Haubrich, Stefan *see* Krebs durch niederfrequente magnetfelder?

Hauch, Johannes Carsten *see* Saga om thorvald vidforle eller den vidtbereiste

Hauck, A *see* Realencyclopaedie fuer protestantische theologie und kirche

Hauck, Albert *see*
- Deutschland und die paepstliche weltherrschaft
- Die entstehung der bischoeflichen fuerstenmacht
- Die entstehung des christustypus in der abendlaendischen kunst
- Der gedanke der paepstlichen weltherrschaft bis auf bonifaz 8
- Tertullian's leben und schriften
- Die trennung von kirche und staat

Hauck, Gerhard *see* Die politischen fuehrungsschichten in den neuen staaten schwarz-afrikas

Hauer, H A *see* Breevoort can ick vergeten niet

Hauff, Reinhard von *see* Nietzsches stellung zur christlichen demut

Hauff, Wilhelm *see*
- An carabhan
- Cuentos
- Hauffs maerchen
- Hauff's werke
- Phantasien im bremer ratskeller
- Wilhelm hauff's saemmtliche werkein sechs baenden
- Der wollmarkt

Hauffs maerchen / ed by Hohenstatt, Otto – Stuttgart: Union Deutsche Verlagsgesellschaft, [1949] [mf ed 1994] – 311p/11pl (ill) – 1 – mf#8749 – us Wisconsin U Libr [390]

Hauffs "memoiren des satan" : teildruck / Sommermeyer, Edwin – Berlin: E. Eberling, [1930] – 1r – 1 – (incl bibl ref) – us Wisconsin U Libr [430]

Hauff's werke / ed by Flaischlen, Caesar – Stuttgart: Deutsche Verlags-Anstalt, [pref. 1890] [mf ed 1995] – 2v (ill) – 1 – mf#8748 – us Wisconsin U Libr [800]

Haug, Balthasar *see* Schwaebisches magazin von gelehrten sachen

Haug, Karl *see* Die autoritaet der hlg schrift und die kritik

Haug, Ludwig *see* Darstellung und beurteilung der theologie ritschls

Haug, Martin *see*
- The aitareya brahmanam of the rigveda
- Brahma und die brahmanen
- Confucius, der weise china's

Haug, Rhea C *see* Development and validation of a maximal testing protocol for the nordictrack cross-country ski simulator

Hauger, Neil A *see* Physiological responses to exercise on the healthrider in males

Hauge's Norwegian Evangelical Lutheran Synod *see* Beretning om hauges synodes...kinamissions-aarsmode

Haughtelin, Jacob Diehl *see* History of coon river congregation

Haughton, Graves Chamney *see*
- Bengali selections
- A glossary, bengali and english
- Manava-sherma-sastra
- Rudiments of bengáli grammar
- Rudiments of bengali grammar

Haughton, Samuel *see* Sermon on the gospel of nature...

Hauhart, William Frederic *see* The reception of goethe's faust in england in the first half of the nineteenth century

Haulleville, Prosper Charles Alexander *see* The future of catholic peoples

Haumant, E *see* La culture francaise en russie (1700-1900)

Haunold, C *see* Institutionum theologicarum...

Haunted tower : a comic opera in three acts, as performed at the theatre-royal drury lane / Storace, Stephen – London: Longman & Broderip 1789 [mf ed 19–] – 3mf – 9 – (libretto by james cobb) – mf#fiche 754 – us Sibley [780]

Haupers, Clement B *see* Papers

Haupt, Alexander James Derbyshire *see* Historical sketch of the english evangelical lutheran synod of the northwest and of the congregations connected therewith

Haupt, C Elvin [comp] *see* Emanuel greenwald, pastor and doctor of divinity

Haupt, Erich *see*
- Die eschatologischen aussagen jesu in den synoptischen evangelien
- First epistle of st john
- The first epistle of st john
- Die gefangenschaftsbriefe
- Die paedagogische weisheit jesu in der allmaehlichen enthuellung seiner person
- Zum verstaendnis des apostolats im neuen testament

Haupt, Herman *see*
- Die deutsche bibeluebersetzung des mittelalterlichen waldenser in dem codex teplensis und der ersten gedruckten deutschen bibel nachgewiesen
- Der waldensische ursprung des codex teplensis und der vorlutherischen deutschen bibeldrucke

Haupt, M *see* Zeitschrift fuer deutsches altertum und deutsche literatur

Haupt, Marcus T von *see* Malerische wanderungen durch holland und einen theil von norddeutschland im jahr 1810
Haupt, Moritz *see* Des minnesangs fruehling
Haupt, P *see*
- Beitraege zur assyriologie und vergleichenden semitischen sprachwissenschaft
- The ship of the babylonian noah and other papers

[Haupt, P] *see* Oriental studies published in commemoration of the fortieth anniversary of paul haupt as director of the john hopkins university
Haupt, Paul *see*
- Die akkadische sprache
- Das babylonische nimrodepos
- Biblische liebeslieder
- Der keilinschriftliche sintfluthbericht
- Purim
- Sumerische studien
- Die sumerischen familiengesetze
- Ueber einen dialekt der sumerischen sprache

Haupt, Walter C *see* Die poetische form von goethes faust
Hauptergebnisse der amtlichen lohnerhebung in der schuhindustrie – Statistisches Reichsamt, 1929 – (= ser Serial publications of german trade unions) – 1 – us Wisconsin U Libr [310]
Der hauptgottesdienst der evangelisch-lutherischen kirche : zur erhaltung des liturgischen erbtheils und zur befoerderung des liturgischen studiums in der americanisch-lutherischen kirche erlaeutert und mit altkirchlichen singweisen / Lochner, Friedrich – St Louis: Concordia Pub House, 1895 – 1mf – 9 – 0-524-06875-5 – (incl bibl ref) – mf#1990-5294 – us ATLA [242]
Die hauptlehren des averroes nach seiner schrift, die widerlegung des gazali = Tahafut al-tahafut / Averroes – Bonn: A Marcus und E Weber, 1913 – 1mf – 9 – 0-524-01251-2 – (in german) – mf#1990-2287 – us ATLA [260]
Hauptling in der gesellschaft der sud... / Beukes, Wiets Taylor Heyman – Hamburg, Germany. 1931 – 1r – us UF Libraries [960]
Der hauptmann / Grote, Hans Henning, Freiherr – Oldenburg: G Stalling 1937, c1932 [mf ed 1990] – 1r – 1 – (filmed with: monteur klinkhammer / erich grisar) – mf#2693p – us Wisconsin U Libr [830]
Hauptmann, Carl *see*
- Die armseligen besenbinder
- Krieg
- Unsere wirklichkeit
- Waldleute

Hauptmann, Gerhart *see*
- Assomption de hannele mattern
- Bahnwaerter thiel
- College crampton
- The coming of peace
- Dorothea angermann
- Goethe
- Griselda
- Hannele
- Die hochzeit auf buchenhorst
- Indipohdi
- Neue gedichte
- Die ratten
- Die spitzhacke
- The sunken bell.
- Veland
- Voiturier henschel
- The weavers

Hauptmann, Moritz *see* Die lehre von der harmonik
Hauptmann-studien; untersuchungen ueber leben und schaffen gerhart hauptmanns / Voigt, Felix Alfred – Breslau: Maruschke & Berendt, 1936- – 1 – us Wisconsin U Libr [430]
Das hauptproblem der evangelienfrage und der weg zu seiner loesung : eine akademische vorlesung / Ewald, Paul – Leipzig: J C Hinrichs, 1890 – 1mf – 9 – 0-8370-3088-9 – mf#1985-1088 – us ATLA [220]
Die hauptprobleme der altisraelitischen religionsgeschichte : gegenueber den entwickelungstheoretikern / Koenig, Eduard – Leipzig: J C Hinrichs, 1884 – 1mf – 9 – 0-7905-1011-1 – (incl bibl ref) – mf#1987-1011 – us ATLA [939]
Hauptprobleme der gnosis / Bousset, Wilhelm – Goettingen: Vandenhoeck und Ruprecht, 1907 – 1mf – 9 – 0-7905-0865-6 – (incl ind) – mf#1987-0865 – us ATLA [290]
Die hauptprobleme der leben-jesu-forschung / Schmiedel, Otto – 2. verb verm Aufl. Tuebingen: J C B Mohr (Paul Siebeck), 1906 – 1mf – 9 – 0-8370-9741-X – (incl bibl ref) – mf#1986-3741 – us ATLA [240]
Die hauptprobleme der pastoralbriefe pauli / Maier, Friedrich – 1. & 2. aufl. Muenster i W: Aschendorff 1910 [mf ed 1992] – (= ser Biblische zeitfragen 3/12) – 1mf – 9 – 0-524-04105-9 – mf#1992-0063 – us ATLA [227]
Die hauptprobleme der philosophie und religion / Delff, Heinrich Karl Hugo – Leipzig: W Friedrich, 1886 – 1mf – 9 – 0-524-01270-9 – mf#1990-2306 – us ATLA [100]
Die hauptprobleme des lebens jesu : eine geschichtliche untersuchung / Barth, Fritz – 3. Aufl. Guetersloh: C Bertelsmann, 1907 – 1mf – 9 – 0-524-05966-7 – mf#1992-0703 – us ATLA [220]
Hauptstromungen der deutschen literatur, 1750-1848 : beitraege zu ihrer geschichte und kritik / Reimann, Paul – 2nd rev enl ed. Berlin: Dietz, 1963 [mf ed 1993] – 839p – 1 – (incl bibl ref and ind) – mf#8219 – us Wisconsin U Libr [430]
Die hauptwerke der deutschen literatur in zusammenhange mit ihrer gattung / Nagel, Siegfried Robert – Wien: F Deuticke, 1904 – 1r – 1 – (incl ind) – us Wisconsin U Libr [430]
Hauraki herald – Thames, NZ. mar 1979-dec 1989 – 31r – 1 – mf#16.19 – nz Nat Libr [079]
Hauraki plains gazette – Paeroa, NZ. 1976 – 5r – 1 – mf#16.2 – nz Nat Libr [079]
Haureau, A B *see*
- Provincia turonensis
- Provincia vesuntionensis
- Provincia viennensis

Haureau, B *see*
- Gallia christiana
- Histoire de la philosophie scolastique
- Les oeuvres de hugues de saint-victor
- Singularites historiques et litteraires

Haureau, Barthelemy *see*
- Bernard delicieux et l'inquisition albigeoise
- Des poemes latins attribues a saint bernard
- Histoire de la philosophie scolastique
- Singularites historiques et litteraires

Hauri, Johannes *see*
- Goethes faust
- Der islam in seinem einfluss auf das leben seiner bekenner

Haury, Jakob *see* Das eleusische fest urspruenglich identisch mit dem laubhuettenfest der juden
Das haus : erzaehlung / Faust, Philipp – Muenchen: A Langen/G Mueller 1940 [mf ed 1989] – 1r – 1 – (filmed with: gustav falke / friedrich castelle) – mf#7229 – us Wisconsin U Libr [830]
Das haus am frauenplan seit goethes tod : dokumente und stimmen von besuchern / ed by Deetjen, Werner – Weimar: Goethe-Gesellschaft, 1935 [mf ed 1993] – (= ser Schriften der goethe-gesellschaft 48) – 70p – 1 – (incl bibl ref) – mf#8657 reel 11 – us Wisconsin U Libr [060]
Haus der kunst catalogue – Munich, 1949-1975 – (= ser Art exhibition catalogues on microfiche) – 64 catalogues on 158mf – 9 – £995.00 – (individual titles not listed separately) – uk Chadwyck [700]
Das haus des dr prade : roman / Wilke, Karl – Leipzig: Koehler & Amelang c1930 [mf ed 1991] – 1r – 1 – (filmed with: armut / anton wildgans) – mf#3052p – us Wisconsin U Libr [830]
Haus, Edouard *see* Le gnosticisme et la franc-maconnerie
Haus – hof – garten – feld *see* Hoefer intelligenz-blatt
Haus-, Hof- und Staatsarchiv, Wien. Reichskanzlei *see* Akten der prinzipalkommission des immerwaehrenden reichstages zu regensburg 1663-1806
Das haus mit den drei tueren : [a novel] / Schaefer, Wilhelm – Muenchen: G Mueller c1931 [mf ed 1991] – 1r – 1 – (filmed with: heiterer guckkasten / bruno wolfgang) – mf#2865p – us Wisconsin U Libr [830]
Haus und bauernfreund = The house and farm companion – bd 19 n1 [1892 may 27]-1893 may 19; 1893 may 26-1897 may 14, 1913 jan 3-1914 mar 27, 1914 apr 3-1915 oct 22, 1915 oct 29-1917 jun 15, 1917 jun 22-1919 dec 26, 1920 aug 6-1921 dec 30, 1920 jan 2-1920 jul 30, 1922 jan 6-1922 sep 8, 1922 sep 15-1923 may 11, 1923 may 18-1924 jan 11, 1924 jan 18-sep 26, 1924 mar 10-nov – 13r – 1 – (cont: hausfreund; cont by: haus-und bauernfreund, national-farmer) – mf#370646 – us WHS [630]
Haus- und bauernfreund – Lincoln, NE (USA), 1920 3 dec-1923, 1926-1927 8 apr – 3r – 1 – gw Misc Inst [640]
Haus- und bauernfreund – Winona WI (USA), 1920 3 dec-1923 21 sep, 1926 1 jan-27 aug, 1927 7 jan-8 apr – 3r – 1 – gw Misc Inst [640]
Haus und bauernfreund der germania = The germania's agricultural and industrial weekly – v13 n5,7 [1886 jun 30, jul 14] – 1r – 1 – (cont: hausfreund; cont by: haus und bauernfreund) – mf#1003081 – us WHS [630]
Haus- und bauernfreund, national-farmer – 1924 dec 5-1925 jul 31, 1925 aug 7-1926 jul 16, 1926 jul 23-1927 apr 8 – 3r – 1 – (cont: haus und bauernfreund; national-farmer; deutsch-amerikanischer farmer) – mf#622681 – us WHS [071]
Haus und grund *see* Duesseldorfer handelszeitung fuer kapital, baugewerbe und grundstuecksmarkt
Haus- und grundbesitzer-zeitung *see* Duesseldorfer handelszeitung fuer kapital, baugewerbe und grundstuecksmarkt
Haus und siedlung im wandel der jahrtausende / Helbok, Adolf – Berlin: de Gruyter 1937 [mf ed Bloomington IN: Indiana Uni Lib, Preservation Dept 1984] – 1r – 1 – us Indiana Preservation [390]
Hausa : basic course / Foreign Service Institute (US) – Washington, DC. 1963 – 1r – us UF Libraries [960]
Hausa and fulani proverbs / Whitting, Charles Edward Jewel – Lagos, Nigeria. 1940 – 1r – us UF Libraries [960]
Hausa literature, and the hausa sound system / Abraham, Roy Clive – London, England. 1959 – 1r – us UF Libraries [960]
Hausa newspapers : assembled and filmed in university of wisconsin, memorial library – Madison: Memorial Library [for Cooperative Africana Microform Project] 1981? – us CRL [071]
Hausa superstitions and customs : an introduction to the folk-lore and the folk / ed by Tremearne, Arthur John Newman – London: J Bale, Sons & Danielsson, 1913 – 2mf – 9 – 0-524-06939-5 – (incl bibl ref) – mf#1990-3565 – us ATLA [390]
Hausa superstitions and customs / Tremearne, Arthur John Newmann – London, England. 1913 – 1r – us UF Libraries [390]
Hausaland : or fifteen hundred miles through the central soudan / Robinson, C H – London, 1900 – 4mf – 9 – mf#HT-125 – ne IDC [916]
Der hausball : eine erzaehlung, 1781 – Wien: C Konegen 1883 [mf ed 1988] – (= ser Wiener neudrucke 3) – 1r – 1 – mf#6934 n4 – us Wisconsin U Libr [880]
Haus-bote – Berlin DE, 1937 7 feb-1938 29 may – 1r – 1 – (filmed with suppl) – gw Misc Inst [640]
Hausbuch des herrn joachim von wedel auf krempzow schloss und blumberg erbgesessen / ed by Bohlendorff, Julius, Freiherr von – Stuttgart: Litterarischer Verein, 1882 (Tuebingen: L F Fues) – (incl bibl ref) – us Wisconsin U Libr [910]
Hausbuch des herrn joachim von wedel auf krempzow schloss und blumberg erbgesessen / Wedel, Joachim von; ed by Bohlendorff, Julius, Freiherr von Bohlen – Stuttgart: Litterarischer Verein, 1882 (Tuebingen: L F Fues) [mf ed 1993] – (= ser Blvs 161) – 578p/[1]pl – 1 – (incl bibl ref) – mf#8470 reel 33 – us Wisconsin U Libr [880]
Die hauschronik konrad pellikans von rufach – Strassburg, Heitz, 1892 – 2mf – 9 – mf#PBU-274 – ne IDC [240]
Hause, Benedict *see* Sechzig toaste fur alle festlichen ereignisse des israelitischen...
Hausegger, Friedrich von *see* Richard wagner und schopenhauer
Hausen, Karl Renatus *see* Historisches portefeuille zur kenntnis der vergangenen und gegenwaertigen zeit
Hauser, Henri *see* Etudes sur la reforme francaise
Hauser, Isaiah L [Mrs] *see* The orient and its people
Hauser, K *see* Die chronik des laurencius bosshart von winterthur 1485-1532
Hausfrau – 1918 oct-1919, 1920 jan-nov, 1921 jan-1922 oct, 1922 nov-1924 jun, 1924 jul-1927 mar, 1926, 1930, 1927-28, 1929 jan-1931 jun, 1931 aug-1948 nov, 1932, 1933-34, 1948 dec-1950 apr – 12r – 1 – (cont: deutsche hausfrau [1908]) – mf#568779 – us WHS [071]
Der hausfreund – Hagen, Westf DE, 1848 – 1r – 1 – (title varies: 1 dec 1841: hagener kreisblatt und maerkischer hausfreund fuer stadt und land; 1853: hagener kreisblatt; 1 jan 1864: hagener zeitung; filmed by other misc inst: 1872 n1-150, 1888 4 jan-1901 [21r], 1829 jan-jun, 1830 [gaps], 1832-34 [gaps], 1837-39, 1845-48, 1849 [l wg. belagerungszustand in hagen], 1850-1887 30 jun, 1887 27 jul-1927 10 dec, 1928-1945 13 apr) – gw Misc Inst [074]
Der hausfreund – Muelhausen / Elsass [Mulhouse F], 1905-11, 1919 14 mar-19 dec, 1920 16 apr-1921 16 dec, 1922-39 [gaps] – 1 – (title varies: 1922: d'r huesfrind) – fr ACRPP [074]
Hausfreund – v5 n35-v7 n17 [1878-80] – 1r – 1 – (continued by: haus und bauernfreund) – mf#620463 – us WHS [071]
Hausfreund *see* Die sieben tage
Haushofer, Albrecht *see* Moabiter sonette
Haushofer, Max *see* Die verbannten
Hausleutner, P W G *see* Schwaebisches archiv
Hausmann, F *see* Reichskanzlei und hofkalle unter heinrich 5 und konrad 3 (mgh schriften:14.bd)
Hausmann, J F L *see* Reise durch skandinavien in den jahren 1806 und 1807
Hausner, Karl-Heinz *see* Rwanda, burundi
Das hauspersonal – Berlin DE, 1913-14 – 1 – gw Misc Inst [640]
Hauspost – 1888 mar 10, 1890 may 31, 1891 may 31, jul 4, sep 1, oct 1, nov 1 – 1r – 1 – mf#1003091 – us WHS [071]
Hausrath, Adolf *see*
- Aleander und luther auf dem reichstage zu worms
- Geschichte der alttestamentlichen literatur in aufsaetzen
- Jesus und die neutestamentlichen schriftsteller
- Kleine schriften religionsgeschichtlichen inhalts
- Martin luthers romfahrt
- Neutestamentliche zeitgeschichte
- The time of jesus
- The time of the apostles
- Der vier-capitel-brief des paulus an die korinther

Hausrath, Adolrf *see* Jesus und die neutestamentliche schriftsteller
Haussez, Charles L d' *see* Great britain in 1833
Haussleiter, Johannes *see* Zwei apostolische zeugen fuer das johannes-evangelium
Haussmann, A *see* Voyage en chine, cochinchine, inde et malaisie...
Haussmann, Johannes *see* Untersuchungen ueber sprache und stil des jungen herder
Haussonville, Gabriel Paul Othenin De Cleron *see* Lacordaire
"Der haussradt" : ein basler gedicht vom jahre 1569: in faksimiledruck / ed by Major, E – Strassburg: J H Ed Heitz (Heitz & Muendel) 1912 [mf ed 1993] – (= ser Drucke und holzschnitte des 15. und 16. jahrhunderts in getreuer nachbildung 14) – 1r – 1 – (incl bibl ref. filmed with: kleines deutsches sagenbuch / will-erich peuckert [ed]) – mf#3367p – us Wisconsin U Libr [810]
Haustheater : sammlung kleiner lustspiele fuer gesellige kreise / Benedix, Roderich – 10. aufl. Leipzig: Weber, 1891 [mf ed 1989] – 2v – 1 – mf#7005 – us Wisconsin U Libr [820]
Der hausvater : eine oekonomische schrift / ed by Muenchhausen, Otto von – Hannover 1764-1773 – (= ser Dz) – 2v on 34mf – 9 – €340.00 – mf#k/n2879 – gw Olms [339]
Hauswedell, Ernst L [comp] *see* Dichter des deutschen barock
Hauszbuch... / Bullinger, Heinrich – Bern, Samuel Apiarius/Zuerich, Christoffel Froschower, 1558 – 12mf – 9 – mf#PBU-160 – ne IDC [240]
Haut la croix! : electeur-temperant / Hugolin, pere – [Montreal: s.n.] 1908 [mf ed 1995] – 1mf – 9 – 0-665-74646-6 – mf#74646 – cn Canadiana [170]
Haut les fourches : journal des jeunesses paysannes – n5-6. Paris. juil-aout 1936 – 1 – fr ACRPP [073]
Hautarzt – Heidelberg. 1981-1982 (1) 1981-1982 (5) 1981-1982 (9) – ISSN: 0017-8470 – mf#13173 – us UMI ProQuest [616]
Haut-Canada *see* Report of the commissioners for improvement of the navigation of the river st-lawrence
Haut-Canada. Lieutenant-gouverneur *see* Message from his excellency the lieutenant governor, of 30th january, 1836
Haut-Canada. Parliament. House of Assembly *see*
- Proceeding had in the legislature of upper canada during the years 1831-2 and 3
- Report from the select committee...appointed to report on the state of the province

Haut-Canada. Parliament. Legislative Council *see*
- The committee appointed by the honourable the legislative council and house of assembly
- Proceedings of the legislative council of upper canada
- Report from the select committee...on the state of the province

Hautcoeur, E *see*
- Cartulaire de l'abbaye de flines
- La liturgie cambresienne au 18e siecle et le projet de breviaire pour tous les dioceses des pays-bas

Haut-commissariat de France en indochine *see* Compte administratif...
La haute science : revue documentaire de la tradition esoterique et du symbolisme religieux – Paris. 1893-janv 1895 – 1 – fr ACRPP [210]
Hautecoeur, M L *see* L'aechitecture francoise
Hautefort, Charles V d' *see* Coup-d'oeil sur lisbonne et madrid en 1814
Hautekzeme bei studenten der zahnheilkunde : eine dermatologische untersuchung / Doll, Michael – (mf ed 1997) – 2mf – 9 – €40.00 – 3-8267-2451-8 – mf#DHS 2451 – gw Frankfurter [616]
La haute-saone libre – Paris, 1943-44 – 1 – (in french) – us UMI ProQuest [934]
Hautsch, E *see* Die evanglienzitate des origenes
Hautsch, Ernestus *see*
- De quattuor evangeliorum codicibus origenianis
- Die evangelienzitate des origenes

Hautsch, Ernst *see* Der lukiantext des oktateuch
Le haut-senegal et niger – Corbeil, France: E Crete, 1906 – 1 – us CRL [960]
Hauxton 1560-1950 – (= ser Cambridgeshire parish register transcript) – 4mf – 9 – £5.00 – uk CambsFHS [929]
Hava, J *see* Arab-english dictionary

Havana : see it better with mitchell / Mitchell's Tours – Miami, FL. 1930 – 1r – us UF Libraries [972]
Havana Alcalde, 1947 – Castellanos Y Rivero *see* Dos anos de labor municipal
[Havana-] casa de las americas – CU. n58-105. 1970-1977 – 8r – 1 – $400.00 – mf#R04196 – us Library Micro [440]
Havana, cinderella's city / Bradley, Hugh – Garden City, NY. 1941 – 1r – us UF Libraries [972]
Havana. Colegio De Belen *see* Album conmemorativo del quincuagesimo aniversario
[Havana-] conjunto – CU. 1964-1977 – 5r – 1 – $250.00 – mf#R04197 – us Library Micro [790]
[Havana-] cor – CU. 1967-1969 – 1r – 1 – $50.00 – mf#R04198 – us Library Micro [079]
Havana (Cuba) Oficina Del Historiador De La Ciuda... *see* Primeros monimientos revolucionarios del general n...
[Havana-] direct from cuba – CU. 1976-1978 – 3r – 1 – $150.00 – mf#R04199 – us Library Micro [079]
[Havana-] economica y desarrollo – CU. 1968-1970 – 13r – 1 – $650.00 – mf#R04200 – us Library Micro [079]
[Havana-] estudios del centro de documentacion – CU. 1965-1969 – 1r – 1 – $50.00 – mf#R04201 – us Library Micro [079]
[Havana-] granma : english edition – CU. 1966-1993 – 28r – 1 – $1400.00 – mf#R60607 – us Library Micro [320]
[Havana-] granma : spanish edition – CU. 1971-1978 – 23r – 1 – $1200.00 – mf#R04202 – us Library Micro [320]
Havana manana / Hermer, Consuelo Kamholz – New York, NY. 1941 – 1r – us UF Libraries [972]
[Havana-] obra revolucionaria – CU. 1960-1961 – 2r – 1 – $100.00 – mf#R04203 – us Library Micro [320]
[Havana-] orientador revolucionario – CU. 1967 – 1r – 1 – $50.00 – mf#R04204 – us Library Micro [071]
[Havana-] palante y palante – CU. 1969-1978 – 8r – 1 – $400.00 – mf#R04205 – us Library Micro [073]
[Havana-] pensamiento critico – CU. 1967-1971 – 5r – 1 – $250.00 – mf#R04206 – us Library Micro [073]
[Havana-] tricontinental bulletin – CU. 1966-1972 – 2r – 1 – $100.00 – mf#R04207 – us Library Micro [073]
[Havana-] tricontinental magazine – CU. 1967-1968 – 1r – 1 – $50.00 – mf#R04208 – us Library Micro [073]
[Havana-] verde olivo – CU. 1981-1986 – (= ser Latin american perspectives) – 16r – 1 – $800.00 – mf#R04209 – us Library Micro [079]
Havard, Henry *see* The dead cities of the zuyder zee: a voyage to the picturesque side of holland
Have Christ Will Travel Ministries *see* Praise and prayer letter
Have congregationalists abandoned the bible? – Boston, Mass: SD Towne, 1908 – 1mf – 9 – 0-524-04171-7 – mf#1990-4975 – us ATLA [242]
Have the sacred writers anywhere asserted that the sin or righteousness of one is imputed to another? / Stuart, Moses – [Andover: Gould & Newman 1836] [mf ed 1984] – (= ser Biblical crit – us & gb) – 1mf – 9 – 0-8370-1586-3 – (fr: the biblical repository and quarterly observer) – mf#1984-1091 – us ATLA [220]
Have we a revelation from god : being a review of professor smith's article "bible" in the "encyclopaedia britannica," ninth edition – London: Office of the "Bible Witness and Review, 1878 – 1mf – 9 – 0-8370-2823-X – mf#1985-0823 – us ATLA [220]
Have we any "word of god"? / Seeley, Robert Benton – London: S W Partridge, 1864? – 1mf – 9 – 0-8370-5214-9 – mf#1985-3214 – us ATLA [220]
Have you the spirit? / Ryle, J C – Ipswich, England. 1854 – 1r – us UF Libraries [240]
Have you understood christianity? / Carey, Walter Julius – London: Longmans, Green, 1916 – 1mf – 9 – 0-524-06807-0 – mf#1991-2794 – us ATLA [240]
Havell, Ernest Binfield *see*
– Benares, the sacred city
– Essays on indian art, industry and education
– A handbook of indian art
– A handbook to agra and the taj, sikandra, fatehpur-sikri, and the neighbourhood
– The himalayas in the indian art
– The history of aryan rule in india from the earliest times to the death of akbar
– The ideals of indian art
– Indian architecture
– Indian sculpture and painting
Havellaendisches echo – Berlin DE, 1933 jan-mar – 1r – 1 – (bereichsausgabe von spandauer zeitung; with suppl: der kleinsiedler 1933 jan-mar) – gw Misc Inst [074]
Havelock ellis: a biographical and critical survey / Goldberg, Isaac – New York: Simon & Schuster, 1926 – 1 – us Wisconsin U Libr [920]
The havelock herald – Lincoln, NE: Havelock Herald. 2v. v1 n1. jan 8 1984-v2 n6. jun 9 1985 (wkly) [mf ed 1985] – 1r – 1 – us NE Hist [071]
Havelock post – Havelock, NE: Will C Israel, 1913-13th yr n18. apr 30 1925 (wkly) [mf ed 6th yr n1. jan 17 1918-25 (gaps) filmed 1977] – 3r – 1 – (merged with: havelock times to form: havelock times-post) – us NE Hist [071]
Havelock times – Havelock, NE: Thomas S Greer, 1891-v34 n42. may 1 1925 (wkly) [mf ed jan 5 1911-1925 (gaps)] – 5r – 1 – (merged with: havelock post to form: havelock times-post) – us NE Hist [071]
Havelock times-post – Havelock, NE: J A Minder & Son. -v42 n18. mar 9 1933 (wkly) [mf ed may 7 1925-1933 (gaps)] – 3r – 1 – (formed by the union of: havelock times and: havelock post. cont by: lancaster county weekly and the havelock times-post. cont the numbering of havelock times) – us NE Hist [071]
Havel-zeitung – Berlin DE, 1927, 1929 3 may-1930 30 mar, 1930 2 jul-1932 30 apr – 9r – 1 – (covers: spandau, nauen, havelland; filmed with suppls) – gw Misc Inst [074]
The haven / Phillpotts, Eden – Toronto: Copp, Clark, 1909 – 4mf – 9 – 0-659-90455-1 – mf#9-90455 – cn Canadiana [830]
Haven, Erastus Otis *see* Autobiography of erastus o haven
Haven family history and reunions, 1896 / Goodell, Amelia – 1r – 1 – mf#B31586 – us Ohio Hist [978]
Haven, Gilbert *see* Sermons, speeches and letters on slavery and its war
Haven, Joseph *see* Studies in philosophy and theology
Haven of rest / Grace, John – Brighton, England. 1860 – 1r – us UF Libraries [240]
Haven van curacao / Lidth De Jeude, O C A Van – Hertogenbosch, Netherlands. 1910 – 1r – us UF Libraries [972]
Haver – [Istanbul: Mahmud Bey Matbaasi, 1884] Cikaran: Izmirli Ubeydullah. n1-4. 15 cemazilevvel-1 saban 1301 [12 apr-27 may 1884] – (= ser O & t journals) – 3mf – 9 – $55.00 – us MEDOC [956]
Haverford College *see* Quakeriana notes
Haverfordwest and milford haven telegraph and general weekly reporter for the county of pembroke – [Wales] Pembrokeshire feb 1854-1950 [mf ed 2003] – 76r – 1 – (missing: 1889, 1897; cont by: pembrokeshire telegraph jan 1921-jun 1934]; western telegraph and times [jul 1934-dec 1937]; western telegraph and cymric times [jan 1938-dec 1950]) – uk Newsplan [072]
Havergal, Frances Ridley *see* Lilies and shamrocks
Haverhill 1641-1849 – Oxford, MA (mf ed 1996) – (= ser Massachusetts vital record transcripts to 1850) – 19mf – 9 – 0-87623-264-0 – (mf 1t: vital records 1641-64. mf 1t-2t: family vital records 1642-1779. mf 3t-10t: vital records 1701-1849. mf 10t: deaths 1809-42. mf 10t-13t: vital records 1780-1849. mf 13t-14t: births 1844-49. mf 14t-15t: marriages 1844-49. mf 15t-16t: deaths 1844-49. mf 16t-19t: intentions 1782-1849) – us Archive [978]
Haverhill echo *see* Echo
Havering leader – Havering, UK. 1 may-dec 1992 – 2r – 1 – uk British Libr Newspaper [072]
Havering post and echo *see* Hornchurch upminster echo
Havering post and romford hornchurch express *see* Hornchurch upminster echo
Havering post news extra *see* Romford havering post
Havering yellow advertiser *see* Yellow advertiser (havering ed)
Havestadt, Bernhard *see* Chilidugu sive res chilenses
Havet, Ernest *see* La modernite des prophetes
Havich, der Kellner *see* Sankt stephans leben
Haviland, Charles Tappan *see* The general corporation law, the stock corporation law, the transportation corporations law, and the business corporation law, of the state of new york.
Haviland, Frank Wood *see* Science
Haviland, Laura Smith *see* A woman's life-work
Havret, Henri *see*
– L'ile de tsong-ming
– La mission du kiang-nan
– La stele chretienne de si-ngan-fou
Haw creek baptist church – Cumming, GA. 1392p – 1 – $125.28 – (church minutes (1841-1994). church deed; rules for decorum. history of haw creek baptist church. rules for haw creek club house; minutes of ordination service) – mf#6840 – us Southern Baptist [242]
Haw, Richard C *see* Rhodesia
Haw, William *see* Fifteen years in canada
Hawai shinpo = the hawaii shinpo – Honolulu HI. 1909 mar 28 – 1r – 1 – (continued by: shukan hawai shinpo) – mf#1219151 – us WHS [071]
Hawaian Mission Children's Society *see* Marquesas collection
Hawaii : session laws of american states and territories – 1901-2001 – 9 – $897.00 set – mf#402620 – us Hein [348]
Hawaii *see*
– Atorney general opinions
– Attorneys' ethics collection
– Reports and opinions
– Reports, pre-nrs
Hawaii afl-cio news / Hawaii State Federation of Labor, AFL-CIO – v5 n3 [1971 apr], v8 n3-v11 n5 [1974 jul-1978 nov] – 1r – 1 – (cont: hawaii state fed news; cont by: hawaii afl-cio nupepa) – mf#462821 – us WHS [331]
Hawaii appellate reports / Hawaii. Intermediate Court of Appeals – v1-10. 1980-94 – 82mf – 9 – $123.00 – (no pre-nrs vols. v9-10 ends official state set) – mf#LLMC 82-998 – us LLMC [340]
Hawaii army weekly – Fort Shafter HI. 1983 jun 9-1987 mar 26, 1987 apr 2-1989 mar 30, 1989 apr 6-nov 23 – 3r – 1 – (cont: hawaii lightning news; tropic lightning news) – mf#1220714 – us WHS [355]
Hawaii attorney general reports and opinions – 1845-2000 – 6,9 – $254.00 set – (1845-1963 on reel $70. 1961-2000 on mf $184) – mf#408200 – us Hein [340]
The hawaii baptist – Honolulu, HI. 5134p. aug 1947-99 – mf#1097 – us Southern Baptist [242]
Hawaii bar journal – Honolulu. 1959-1959 (1) – ISSN: 0438-8054 – mf#7655 – us UMI ProQuest [340]
Hawaii bar journal – Honolulu. 1963-1988 (1) 1972-1988 (5) 1973-1988 (9) – ISSN: 0440-5048 – mf#6406 – us UMI ProQuest [340]
Hawaii bar journal – v1-23. 1963-1991 (all publ) – 9 – $105.00 set – (merged with: hawaii bar news, which changed to hawaiian bar journal (ns). ceased publ with v23 n1) – ISSN: 0440-5048 – mf#401021 – us Hein [340]
Hawaii bar journal *see* Hawaii bar journal (ns)
Hawaii bar journal (ns) – v1-27. 1991-96; v1-5. 1961-2001 – 9 – $630.00 set – (titles varies: may 1961-dec 1962 as hawaii bar news; jan 1963-feb 1964 as hawaii bar journal; mar 1965-apr 1992 as hawaii bar news) – mf#116961 – us Hein [340]
Hawaii bar news – Honolulu. 1961-1962 (1) – ISSN: 0438-8062 – mf#7654 – us UMI ProQuest [340]
Hawaii bar news *see* Hawaii bar journal (ns)
Hawaii Bicentennial Commission *see* Sandwich isles gazette
Hawaii carpenter / United Brotherhood of Carpenters and Joiners of America – v2 n3 [1970 feb 14]-v15 n12 [1983 dec], 1984 jan-1990 dec, v23 n1 [1991 jan]-v26 n12 [1994 dec] – 3r – 1 – mf#1074243 – us WHS [680]
Hawaii catholic herald – 1974 nov 1-1975 dec 26, 1976 jan 2-1977 mar 25, 1977 apr 1-1978 jan 27, 1978 feb-1979 dec, 1980-1981 jun, 1981 jul-1982 sep, 1982 oct-1984 jun, 1984 jul-1985 sep, 1985 oct 4-1986 dec, 1987 jan-1988 mar, 1988 apr-1989 jun, 1989 jul-1990 aug – 12r – 1 – mf#590889 – us WHS [241]
Hawaii Education for Social Progress. People's Fund [Honolulu HI] *see* Huliau
Hawaii farm and home – Honolulu. 1950-1950 (1) – mf#389 – us UMI ProQuest [630]
Hawaii Federation of Teachers *see* Hft reporter
Hawaii filipino news – Honolulu HI. v9 n11-v11 n21 [1985 oct 11/15-1987 nov16/30] – 1r – 1 – (cont: hawaii news [honolulu, hawaii]; cont by: fil-am courier) – mf#1354158 – us WHS [071]
Hawaii filipino news – Honolulu HI. v6 n2-v8 n17 [1982 may-1985 jan 16/31] – 1r – 1 – (continued by: hawaii news (honolulu, hi)) – mf#800058 – us WHS [071]
Hawaii Foundation for American Freedoms, Inc *see* Criterion
Hawaii free people's press – v2 n5,6 [1970 nov, dec] – 1r – 1 – mf#1057987 – us WHS [071]
Hawaii Government Employees Association *see* Public employee
Hawaii. Intermediate Court of Appeals *see* Hawaii appellate reports
Hawaii jewish news / Hawaii Jewish Welfare Fund – v6 n1-v10 n14 [1983 jan-1988 may/jun] – 1r – 1 – (cont: chavurah-hillel newsletter) – mf#1724757 – us WHS [270]
Hawaii jewish news – Honolulu, HI. 1977-85 – 1 – us AJPC [071]
Hawaii Jewish Welfare Fund *see* Hawaii jewish news
Hawaii. Laws, Statutes, etc *see* Patent laws of the republic of hawaii
Hawaii Legislature *see*
– Compiled statutes
– House journals
– Senate journals
– Session laws
Hawaii marine / Kaneohe Marine Corps Air Station [Hawaii] – Kaneohe HI. 1987, 1988, 1989, 1990 – 4r – 1 – (cont: windward marine) – mf#3612595 – us WHS [355]
Hawaii medical journal – Honolulu. 1941+ (1) 1971+ (5) 1974+ (9) – ISSN: 0017-8594 – mf#332 – us UMI ProQuest [610]
Hawaii navy news – Kaneohe HI. v8 n45 [1983 nov 16]-1984 nov 29, 1985 jan 17-aug 29, 1986 feb 13-dec 18/24, 1987, 1988 jan-1989 oct 9 – 5r – 1 – mf#1034983 – us WHS [355]
Hawaii news – Honolulu HI. v8 n18-v9 n7 [1985 feb 1/16-sep 7/16] – 1r – 1 – (cont: hawaii filipino news [honolulu, hi: 1977]) – mf#1086269 – us WHS [071]
Hawaii observer – 1974 apr 2-1978 mar 14, 1977 jan 27-1978 mar 9 – 2r – 1 – mf#168599 – us WHS [071]
Hawaii revised statutes – 1905-nov 2001 update – 9 – $923.00 set – mf#402200 – us Hein [348]
Hawaii State Federation of Labor, AFL-CIO *see* Hawaii afl-cio news
Hawaii State Teachers Association *see*
– Reporter
– Teacher advocate
Hawaii supreme court reports – v1-4; v1-75. 1847-1994. Including the 4v of the U.S. – 665mf – 9 – $997.00 – (includes 4v of the us district court for the territory of hawaii. v75 ends official state set. pre-nrs run: v1-43 1847-1961 419mf $616.00) – mf#LLMC 77-101 – us LLMC [347]
Hawaii Teamsters and Allied Workers Union *see*
– Hawaiian teamster
Hawaii tribune herald – Hilo, Hawaii. 1975-2000 (1) – mf#61303 – us UMI ProQuest [071]
Hawaii Union of Socialists *see* Modern times
Hawaiian bar journal (ns) *see* Hawaii bar journal
Hawaiian bar news *see* Hawaii bar journal
Hawaiian church chronicle / Episcopal Church – 1974 oct-1987 nov/dec [v78 n8] – 1r – 1 – (cont: anglican church chronicle) – mf#1353004 – us WHS [242]
Hawaiian entomological society proceedings – Honolulu. 1904-1974 (1) – mf#8703 – us UMI ProQuest [590]
Hawaiian Evangelical Association *see*
– Friend
– Minutes of the meeting of the...
Hawaiian evangelical association. annual report : 15th-132nd, 1878-1954 – Honolulu: Missions of the Association in Hawaii 1878-1962 [mf ed 2001] – (= ser Christianity's encounter with world religions, 1850-1950) – 5r – 1 – (filmed with: hawaiian evangelical association of the congregation christian churches. annual report, 133rd-140th 1955-62) – mf#2001-s065-066 – us ATLA [240]
Hawaiian evangelical association. annual report = Hoike makahiki...o ka ahahui euanelio hawaii – Honolulu HI: Paii a ma ka hale pai o Robert Grieve 1881-1922 [mf ed 2007] – (= ser Religious periodical literature of the hispanic and indigenous peoples of the americas, 1850-1985) – 42v on 2r – 1 – (chiefly in hawaiian, lists of officers & statistical charts in english; Lacks: v27th [1890] [back cover?], v29th [1892], v41st? [1904]; cont: hawaiian evangelical association. moolelo o ka halawai makahiki o ka ahahui euanelio hawaii; cont by: hawaiian evangelical association annual report of the hawaiian evangelical association (hawaiian ed)) – mf#2007i-s029 – us ATLA [242]
Hawaiian evangelical association. annual report [hawaiian ed] – [Honolulu] HI: Honolulu Star-Bulletin 1923- (1923-31) [mf ed 2007] – (= ser Religious periodical literature of the hispanic and indigenous peoples of the americas, 1850-1985) – 1r – 1 – (chiefly in hawaiian, lists of officers & statistical charts in english; cont: hawaiian evangelical association. hoike makahiki...o ka ahahui euanelio hawaii) – mf#2007i-s030 – us ATLA [242]
Hawaiian Evangelical Association of the Congregational Christian Churches et al *see* Friend
Hawaiian falcon – Honolulu HI. 1981 apr 29-1982 mar, 1982 apr-dec, 1983 jan-jun, 1983 jul-1984 dec, 1985 jan-aug, 1985 sep 6-1986 dec 19, 1987, 1988 jan-sep, 1988 oct-1989 dec – 9r – 1 – mf#648359 – us WHS [071]
Hawaiian gazette – Honolulu HI. 1892 nov 29, 1893 mar 28 – 1r – 1 – mf#634419 – us WHS [071]
Hawaiian Mission Children's Society *see* Marquesas collection
Hawaiian mission children's society : annual report – v1-132. 1853-1991 [complete] – Inquire – 1 – mf#ATLA S0684 – us ATLA [240]

Hawaiian Mission Children's Society Library *see* Micronesian collection

Hawaiian Mission Sesquicentennial Committee *see* Sesquicentennial spectator

Hawaiian Philatelic Society *see* Po'oleka o hawaii

Hawaiian shell news – Honolulu. 1975-1979 (1) 1975-1979 (5) 1975-1979 (9) – ISSN: 0017-8624 – mf#10177 – us UMI ProQuest [590]

Hawaiian teamster / Hawaii Teamsters and Allied Workers Union – v1 n1-4 [1981 apr-nov] – 1r – 1 – (cont: hawaiian teamster [honolulu : 1978]) – mf#601821 – us WHS [331]

Hawaiian teamster / Hawaii Teamsters and Allied Workers Union – v1 n1-v2 n2 [1978 may-1979 aug] – 1r – 1 – (cont: local 5 organizer and hawaiian teamster; cont by: hawaiian teamster [hawaii : 1981]) – mf#601820 – us WHS [331]

Hawaiian teamster / Hotel and Restaurant Employees and Bartenders International Union et al – 1970 jul-1977 dec, – 1r – 1 – mf#385189 – us WHS [640]

Hawaiian/Pacific Collection of the Uni of Hawaii Library *see*
- Materials on the compact of free association for micronesia, palau and the marshalls

Hawara, biahmu, and arsinoe / Petrie, W M – London, 1889 – 3mf – 9 – mf#NE-20338 – ne IDC [956]

The hawara portfolio : paintings of the roman age / Petrie, W M – London, 1913 – 1mf – 9 – mf#NE-20363 – ne IDC [956]

Hawe hacregue : nashe nasledie, our heritage – 1988-1999 – 130mf – 9 – $1,000.00 – us UMI ProQuest [947]

Haweis, Hugh Reginald *see*
- The broad church
- The conquering cross (the church)
- Current coin
- The dead pulpit
- The key of doctrine and practice
- The light of the ages
- The picture of jesus
- Sermons
- Speech in season
- The story of the four
- "Winged words"

Haweis, Mary Eliza (Joy) *see*
- The art of beauty
- The art of decoration
- The art of dress
- Beautiful houses

Hawera and normanby star – apr 1880-jun 1888; oct 1891-jun 1897; jan 1898-dec 1924 – 1 – mf#20.5 – nz Nat Libr [079]

Hawera star – 8 sep-17 nov 1954; 2 jan-23 mar 1954; jan 1976-dec 1985 – 1 – mf#20.2 – nz Nat Libr [079]

Hawes, Charlotte E *see* New thrills in old china

Hawes, Granville Parker *see* The law relating to general voluntary assignments for the benefit of creditors, as provided for the statute of 1860, as amended

[Hawes, J] *see* Memoir of mrs mary e van lennep

Hawes, Joel *see*
- "A looking-glass for ladies"
- A tribute to the memory of the pilgrims

Hawick express and roxburghshire advertiser – Hawick: A Walker & Son Ltd 1930-64 (wkly) – 1 – (cont: hawick express & advertiser and roxburghshire gazette; cont by: hawick express; imprint varies) – uk Scotland NatLib [072]

Hawick express, and scottish border news – Hawick: Dalgleish, Rule & Co 1870-1914 (wkly) – 1 – (merged with (to form last title listed): hawick advertiser, and roxburghshire gazette and: hawick express & advertiser and roxburghshire gazette; imprint varies) – uk Scotland NatLib [072]

Hawick news and scottish border chronicle – Hawick: Vair & McNairn 1953- (wkly) [mf ed 6 jan 1995-] – 1 – (cont: hawick news and border chronicle) – uk Scotland NatLib [072]

Hawick week-end advertiser – [Scotland] Hawick: H B Little & W Simpson 15 jun 1935-17 feb 1947 [mf ed 2004] – 3r – 1 – (missing: np sep 1939, dec 1946; publ suspended between 8th sep 1939 & 9th dec 1946; cont: hawick saturday advertiser [8 jun 1935]) – uk Newsplan [072]

Hawk – Hahn. 1981 may [v28 n16]-1983 oct 28, 1984 feb 10-1986 mar 21 – 2r – 1 – (cont: hahn hawk) – mf#1044187 – us WHS [071]

Hawk killer, blind preacher, odd fellows – s.l, s.l? 1938 – 1r – us UF Libraries [978]

The hawk moths of north america / Grote, Augustus Radcliffe – Bremen: Homeyer & Meyer, 1886 – (= ser North american lepidoptera) – 1mf – 9 – mf#13494 – cn Canadiana [590]

The hawk over heron : notes on comedy and the comedy form with two special chapters on congreve's way of the world, and barrie's admirable crichton / Bhushan, V N – Bombay: Padma Publications, 1944 – (= ser Samp: indian books) – us CRL [420]

Hawken, J D *see* Upa-sastra

Hawker, George *see*
- An englishwoman's twenty-five yaers in tropical africa
- The life of george grenfell
- Open the window eastward

Hawker, Robert *see*
- Bread selling to the poor at half price
- Copy of the rules of the prayer-meetings which are established amon...
- Cottage funeral
- Death abolished, and life and immortality brought to light
- Fragments from holy scripture
- Glory of god in gathering his people to himself
- God's will and man's shall
- God's witness
- Goings forth of jehovah in his trinity of persons, in acts of perso...
- Good news from a far country
- Heirs of promise
- My birth-day
- Mystery of godliness
- Potter's house
- Royal family, a tract, proper to be put into the hands of every one

Hawke's bay herald – Napier. New Zealand. -d. Jan 1885-Dec 1886. (4 reels) – 1 – uk British Libr Newspaper [072]

Hawkes bay herald – Hastings, NZ. 24 sep 1857-25 dec 1858; 1859-65; 1867-jun 1877; jan 1878-dec 1885; jul 1886-jun 1894; jan 1895-dec 1904; may-aug 1925; 4 mar-6 may 1957; 5 sep-15 oct 1960; 26 nov 1960-17 feb 1962; 1 jul-15 nov 1966; jun 1975; dec 1976-1 may 1999 – 1 – (merged with: hawkes bay tribune on 16 jan 1937. title changes to: hawkes bay herald tribune on 4 mar 1957) – mf#35.1 – nz Nat Libr [079]

Hawkes bay herald *see* Hastings standard

Hawkes Bay herald tribune *see* Daily telegraph

Hawkes bay herald tribune *see*
- Hastings standard
- Hawkes bay herald

Hawke's bay times (napier) – jul 1861-1968 – 4r – 1 – mf#31.09 – nz Nat Libr [079]

Hawke's bay today – nov 1999-mar 2000 – 5r – 1 – mf#35.15 – nz Nat Libr [079]

Hawkes bay today *see* Daily telegraph

Hawkes bay tribune *see*
- Hastings standard
- Hawkes bay herald

Hawke's bay weekly times (napier) – 1867-68 – 1r – 1 – mf#31.10 – nz Nat Libr [079]

Hawkesbury advocate – Windsor, oct 1899-dec 1900 – 1r – 9 – A$57.24 vesicular A$62.74 silver – at Pascoe [079]

Hawkesbury chronicle / farmers advocate – Windsor, nov 1881-may 1888 – 2r – 9 – A$110.33 vesicular A$121.33 silver – (aka: farmers advocate) – at Pascoe [079]

Hawkesbury courier – Windsor, jul 1844-nov 1846 – 1r – 9 – A$38.90 vesicular A$44.40 silver – at Pascoe [079]

Hawkesbury gazette – Windsor, jan 1969-dec 1996 – at Pascoe [079]

Hawkesbury herald – Windsor, may 1902-jun 1904 – 2r – 9 – A$118.71 vesicular A$129.71 silver – at Pascoe [079]

Hawkesbury independent – Windsor – 2r – at Pascoe [079]

Hawkesbury review – Windsor – 4r – A$268.27 vesicular A$290.27 silver – at Pascoe [079]

Hawkesworth, Alan S *see* De incarnatione verbi dei

Hawkesworth, J *see* An account of the voyages by the order of his present majesty for making discoveries in the southern hemisphere...

Hawkesworth, John *see* An account of the voyages undertaken by the order of his present majesty for making discoveries in the southern hemisphere

Hawk-eye – Eugene City OR: [s.n.] [wkly] – 1 – us Oregon Lib [071]

[Adin-] hawkeye – CA. 6 sep 1978 – 1r – 1 – $60.00 – mf#B02002 – us Library Micro [071]

Hawkeye, Harry *see* Buffalo bill

Hawkeye independent : authorized newspaper, knights and women of ku klux klan, realm of iowa / Ku Klux Klan [1915-] – Des Moines IA 1923 sep 22 – 2r – 1 – mf#851250 – us WHS [366]

Hawkeye record – Mount Vernon, IA. 1924-1957 (1) – mf#63326 – us UMI ProQuest [071]

Hawkins, Chauncey Jeddie *see*
- The mind of whittier
- Samuel billings capen

Hawkins chronicle – Hawkins WI. 1922 feb 3/1923 aug 31-1957/1961 jan 13 [gaps] – 18r – 1 – mf#945233 – us WHS [071]

Hawkins, Edward *see*
- Apostolical succession
- An inquiry into the connected uses of the principal means of attaining christian truth
- Ministry of men in the economy of grace and the danger of overvalu...
- Notes on church and state

Hawkins, Ernest *see*
- Documents relative to the erection and endowment of additional bishoprics in the colonies
- Historical notices of the missions of the church of england in the north american colonies

[Hawkins, H] *see* Partheneia sacra

Hawkins, James *see* Arise, o lord

Hawkins, Jennifer C *see* Quality of life and health status perceptions of elderly participants in the purdue lifespan study

Hawkins, John *see* General history of the science and practice of music

Hawkins, John Caesar *see* Horae synopticae

Hawkins, Joshua *see* Sin and its penalty

Hawkins, Margaret C *see* Utilization of health care professionals in selected industrial settings

Hawkinsville first baptist church. hawkinsville, georgia : church records – 1839-1958 – 1 – 72.72 – us Southern Baptist [242]

Hawkridge, Emma *see* Indian gods and kings

Hawks, Francis Lister *see*
- Auricular confession in the protestant episcopal church
- Documentary history of the protestant episcopal church in the united states of america. south carolina
- A narrative of events connected with the rise and progress of the protestant epicopal church in virginia
- A narrative of events connected with the rise and progress of the protestant episcopal church in maryland

Hawk's journal – v1 n3 [1979?] – 1r – 1 – mf#5319847 – us WHS [071]

Hawles, John *see* The canadian's right the same as the englishman's

Hawley 1727-1892 – Oxford, MA (mf ed 1989) – (= ser Massachusetts vital records) – 27mf – 9 – 0-87623-095-8 – (mf 1-4: births/ deaths & index 1727-1846. mf 5: marriages 1795-1846. mf 6-12: town records 1792-1819. mf 12-13: births 1729-1826. mf 13-14: town minutes 1826-44. mf 15: civil war records 1861-65. mf 16: birth index 1844-91. mf 17-18: marriage index 1844-91. mf 19: marriage intentions index. mf 20: death index 1844-91. mf 21: b,m,d 1844-58. mf 22-23: births 1859-91. mf 24-25: marriages 1859-92. mf 26-27: deaths 1859-91) – us Archive [978]

Hawley, Bostwick *see* A treatise on the lenten season

Hawley, Charles *see*
- Early chapters of seneca history
- Jesuit missions among the cayugas

Hawley, Charles Anthony *see* Outlines, definitions, maxims, quotations, and problems in elementary law

Hawley, John Gardner *see*
- Inter-state extradition
- A treatise on the law of real property

Hawley, John S *see* Fearless bible reading

Hawley, Thomas De Riemer *see* Infallible logic, a visible and automatic system of reasoning

Haworth, Charles B *see* Congress and the courts

Hawthorn, Harry *see* A visit to babylon

[Hawthorne-] esmeralda herald – NV. 1883-84 [wkly] – 1r – 1 – $60.00 – mf#U04576 – us Library Micro [071]

[Hawthorne-] esmeralda news – NV. 1887-89 [wkly] – 1r – 1 – $60.00 – mf#U04577 – us Library Micro [071]

Hawthorne first baptist church (formerly: pleasant grove baptist church). hawthorne, florida : church records – 1853-1959 – 1 – 49.86 – us Southern Baptist [242]

Hawthorne, H W *see* Economic study of absentee ownership of citrus properties in florida

Hawthorne, James Boardman *see* Paul and the women

Hawthorne, Julian *see* One of those coincidences

[Hawthorne-] lucky boy post – NV. may-oct 1909 [wkly] – 1r – 1 – $60.00 – mf#U04578 – us Library Micro [071]

[Hawthorne-] mineral county democrat – NV. jan 1961 [wkly] – 1r – 1 – $60.00 – mf#U04579 – us Library Micro [071]

[Hawthorne-] mineral county forum – NV. feb-sep 1961 [wkly] – 1r – 1 – $60.00 – mf#U04580 – us Library Micro [071]

[Hawthorne-] mineral county independent news – NV. 1933-36 (scats); 1937- – 40r – 1 – $2400.00 (subs $60y) – (aka: mineral county and hawthorne news) – mf#UN04581 – us Library Micro [071]

Hawthorne, Nathaniel *see*
- Complete novels and selected tales of nathaniel hawthorne
- Selections from twice-told-tales

[Hawthorne-] news – NV. jan-apr 1931; 1934-35 [wkly] – 2r – 1 – $120.00 – mf#U04582 – us Library Micro [071]

[Hawthorne-] oasis – NV. 8 sep 1881 [wkly] – 1r – 1 – $60.00 – mf#U04583 – us Library Micro [071]

Hawthorne reporter – Hawthorne, FL. 1969 jan-1973 mar – 1r – (missing: 1969 feb-apr, sep; 1970 nov; 1972 jan, sep; 1973 jan-feb) – us UF Libraries [071]

[Hawthorne-] times of mineral county – MI. sep 1978-oct 1984 – 6r – 1 – $660.00 – mf#N04584 – us Library Micro [071]

[Hawthorne-] walker lake bulletin – NV. 1883-1905; 1911-12; 1914-16; 1918-19; 1922; jan-oct 1924 [wkly] – 33r – 1 – $1980.00 – mf#U04585 – us Library Micro [071]

Hawtrey, C S *see* Blessedness of dying in the lord

Haxthausen, A von *see* Transkaukasia

Haxthausen, August von *see*
- Studien ueber die innern zustaende, das volksleben und insbesondere die laendlichen einrichtungen russlands
- Transkaukasia

Haxtun herald *see* Miscellaneous newspapers of the colorado historical society

Hay, A M *see* The wisdom of the owl

Hay any work for cooper : being a reply to the "admonition to the people of england" / Marprelate, Martin – Reprinted from the black letter ed. London: John Petheram, 1845. Chicago: Dep of Photodup, U of Chicago Lib, 1975 (1r); Evanston: American Theol Lib Assoc, 1984 (1r) – (= ser Puritan Discipline Tracts) – 1 – 0-8370-0311-3 – mf#1984-B441 – us ATLA [240]

Hay, Charles Augustus *see*
- Brief notes on pastoral theology
- Memoirs of rev. jacob goering, rev. george lochman, d.d., and rev. benjamin kurtz, d.d., ll. d

Hay, David Ramsay *see*
- The laws of harmonious colouring adapted
- The natural principles of beauty

Hay, G G *see* First measures in malarial prevention for farmers and settlers

Hay, George *see* The scripture doctrine of miracles displayed

Hay, George Upham *see* Canadian history readings, vol 1

Hay, Gustavus *see* The law of railway accidents in massachusetts

Hay, Gyula *see* Haben

Hay, John *see*
- Castilian days
- Papers

Hay, John Charles Dalrymple *see* Ashanti and the gold coast

Hay levadura en las columnas / Undurraga, Antonio De – Buenos Aires, Argentina. 1960 – 1r – us UF Libraries [972]

Hay que evitar ser tan bruto como el soldado canuto, peripecias y desdichas de un mal soldado – 2nd ed. n.p. 1937? Fiche W 937. (Blodgett Collection of Spanish Civil War Pamphlets) – 9 – us Harvard [946]

Hay, Robert *see* Illustrations of cairo

Hay springs enterprise – Hay Springs, NE: F W Johansen. -v27 n26. jun 25 1915 (wkly) [mf ed 1904-15 (gaps)] – 4r – 1 – (cont: hay springs leader. merged with: hay springs news to form: hay springs news and hay springs enterprise, consolidated) – us NE Hist [071]

Hay springs leader – Hay Springs, NE: E E & N J Humphreys. v6 n5. feb 23 1894- (wkly) [mf ed -jan 24 1902 (gaps)] – 1r – 1 – (cont: sheridan county democrat. cont by: hay springs enterprise) – us NE Hist [071]

Hay springs news – Hay Springs, NE: G S Peters. 56v. v28 n1. jan 7 1916-v83 n6. mar 28 1968 (wkly) [mf ed filmed -1978] – 17r – 1 – (cont: hay springs news and hay springs enterprise, consolidated. merged with: sheridan county star to form: news-star (rushville ne)) – us NE Hist [071]

Hay springs news – Hay Springs, NE: J C Burton. 5v. 1910-v5 n41. jun 25 1915 (wkly) [mf ed v3 n4. oct 4 1912-jun 25 1915 (gaps) filmed 1999] – 2r – 1 – (merged with: hay springs enterprise to form: hay springs news and hay springs enterprise, consolidated) – us NE Hist [071]

Hay springs news and hay springs enterprise, consolidated – Hay Springs, NE: Geo S Peters. v5 n42. jul 2 1915-v27 n53. dec 31 1915 (wkly) – 1r – 1 – (formed by the union of: hay springs news and: hay springs enterprise. cont by: hay springs news (1916). issues for sep 16-dec 31 1915 called v27 n37-53 cont numbering of hay springs enterprise) – us NE Hist [071]

Hay standard – Hay, jan 1899-nov 1900 – 1r – A$60.98 vesicular A$66.48 silver – at Pascoe [079]

Hay un pais en el mundo / Mir, Pedro – Santo Domingo, Dominican Republic. 1962 – 1r – us UF Libraries [972]

Hay velas y milagros / Sanchez, Carlos Enrique, Cuban – Habana, Cuba. 1957 – 1r – us UF Libraries [972]

Hay ye hudhi – Vienna, Austria. 7 jan 1875-18 aug 1885 – 1r – 1 – uk British Libr Newspaper [072]

Haya De La Torre, Victor Raul *see* Defensa continental

Hayah aviv ba-arets / Scharfstein, Zevi – Tel-Aviv, Israel. 1952 – 1r – us UF Libraries [939]

Hayal – Istanbul: Mehmet Rauf. ed 1-5 sene n1-368. 18 tesrinievvel 1289-18 haziran 1293 [oct 1873-jun 1877] – (= ser O & t journals) – 26mf – 9 – $430.00 – us MEDOC [956]
Hayami, T *see* Church history
Hayashi, Carl T *see* Achievement motivation among anglo-american and hawaiian physical-activity participants
Hayashi, CT *see* A cross-cultural analysis of achievement motivation in anglo american and japanese marathon runners
Hayashi, Susan W *see* Understanding youth sport participation through perceived coaching behaviors, social support, anxiety and coping
Hayashida, Sandra Lynne *see* Mission and the visual expression of the gospel in the sculpture of jackson hlungwani
Hayashide kenjiro kankei monjo – The Diplomatic Record Office, Ministry of Foreign Affairs of Japan – 10r – 1 – Y150,000 – (with 94p guide. in japanese) – ja Yushodo [327]
Hayat – Adana. Sahibi: Muecavirzade Mustafa Emin; Basmuharriri: Ramazanzade Mehmed Kemal. n2. 27 mart 1334 [1918] – (= ser O & t journals) – 1mf – 9 – $25.00 – us MEDOC [956]
Hayat – Baku, 1905-06 – 3r – 1 – us UMI ProQuest [077]
Hayat – New Delhi, India. 1965 – 1r – 1 – us L of C Photodup [079]
Hayatt, Alice Nelson *see* An account of the life of dr william augustus carleton
Haycraft, John Berry *see* Darwinism and race progress
Haydee, ou, le secret / Scribe, Eugene – Paris, France. 1847 – 1r – us UF Libraries [440]
Hayden, Amos Sutton *see* Early history of the disciples in the western reserve, ohio
Hayden, Arthur *see*
– Chats on cottage and farmhouse furniture
– Chats on english eathenware
– Chats on old silver
Hayden, Chester *see* An appendix to cowen's treatise on the civil jurisdiction of justices of the peace in the state of new york.
Hayden, ferdinand v, papers, ms 3154 – 1846-65 – 1r – 1 – us Western Res [920]
Hayden, H H *see* A sketch of the geography and geology of the himalaya mountains and tibet
Hayden, Henry *see* Illustrations of astronomy
Hayden, Horace Edwin *see* An account of various silver and copper medals
Hayden, M P *see* The bible and woman
Hayden, Samuel Augustus *see* The complete conspiracy trial book
Hayden, Warren Luce *see* Centennial addresses
Haydn, Hiram Collins *see* American heroes on mission fields
Haydn, Joseph *see*
– Creation
– Dr haydn's 6 original canzonettas
– Favorite quinetetto for two violins, two tenors, and a bass
– Philemon und baucis
– Raccolta d'arie favorite ricavate di varie opere
– Second sett of dr haydn's 6 original canzonettas
– Second sett of six grand quartettos, for two violins, a tenor and violoncello obligato, opera 16
– Sept quatuors sur les sept dernieres paroles de notre seigneur j c
– Simphonie pour le clavecin ou le forte-piano
– Six sinfonies
– Trois quatuors pour deux violons, alto et violoncelle, oeuvre 71
– Trois quatuors pour deux violons, taille et violoncelle
Die haydn-drucke der hoboken-sammlung musikalischer erst- und fruehdrucke / Oesterreichische Nationalbibliothek Wien. Musiksammlung – [mf ed Hildesheim 1989] – (= ser Die europaeische musik) – 1219mf – 9 – diazo €6400.00 silver €7200.00 – gw Olms [780]
The haydock papers : a glimpse into english catholic life under the shade of persecution and in the dawn of freedom / Gillow, Joseph – London, New York: Burns & Oates, 1888 – 1mf – 9 – 0-7905-5146-2 – mf#1988-1146 – us ATLA [241]
Haydon, Benjamin Robert *see*
– Description of two pictures
– Explanation of the picture of chairing the members
– Haydon's pictures of xenophon, mock election
– Lectures on painting and design
– New churches
– Some enquiry into the causes which have obstructed...painting
– Thoughts on the relative value of fresco and oil painting
Haydon, Edwin Scott *see* Notes of selected decisions of her majesty's high court in uganda, on cases originating from the buganda courts, 1940-1958
Haydon, F S *see* Eulogium (historiarum sive temporis) (rs9)
Haydon, Roger *see* Upstate travels
Haydon's pictures of xenophon, mock election / Haydon, Benjamin Robert – London 1832 – (= ser 19th c art & architecture) – 1mf – 9 – mf#4.2.1637 – uk Chadwyck [750]
Hay-Drummond, George *see* Selection from the psalms of david
Hayduk, Alfons *see* Eichendorff-lese
Haye ha-mishnah / Wilstein, Chaim – Jerusalem, Israel. 1928 – 1r – us UF Libraries [939]
Haye yehudah / Modena, Leone – Kiev, Ukraine. 1911 – 1r – us UF Libraries [939]
Hay-Edward, C M *see* A history of clifford's inn
Ha-yekov *see* Bet ya'akov
Hayens, Kenneth Cochrane *see* Grimmelshausen
Hayes, A J *see* The source of the blue nile
Hayes, Alexander L et al *see* Formal opening of franklin and marshall college in the city of lancaster, june 7, 1853
Hayes, B *see* The via vitae of st benedict
Hayes, Bernard *see* The holy rule of st benedict
Hayes center times-republican – Hayes Center, NE: Hazel McKibbin. v76 n39. apr 13 1961- (wkly) [mf ed filmed 1969-] – 1 – (cont: times-republican) – us NE Hist [071]
Hayes center times-republican – Hayes Center, NE: Hazel McKibbin. v76 n39. apr 13 1961- (wkly) [mf ed filmed 1969-] – 1 – (cont: times-republican) – us NE Hist [071]
Hayes centre news – Hayes Centre, NE: M J Abbott. v1 n1. apr 9 1885- (wkly) [mf ed -apr 15 1886 filmed 1969?] – 1r – 1 – (cont by: hayes county herald. issues for apr 30-may 14 1885 not publ) – us NE Hist [071]
Hayes chronicle – London, UK. 1950; 1963-67; 1970; 1972-30 may 1975 – 12 1/2r – 1 – (aka: hayes uxbridge southall chronicle; hayes harlington and district chronicle) – uk British Libr Newspaper [072]
Hayes county herald – Hayes Centre, NE: Herald Co. -v5 n3. apr 25 1889 (wkly) [mf ed 1888-89 filmed 1969?] – 2r – 1 – (cont: hayes centre news. cont by: hayes county republican) – us NE Hist [071]
Hayes county republican – Hayes Centre, NE: Chas E Abbott. v5 n4. may 2 1889- (wkly) [mf ed -1902 filmed 1969?] – 5r – 1 – (cont: hayes county herald. merged with: hayes county times to form: hayes county times-republican) – us NE Hist [071]
Hayes county times – Hayes Center, NE: C L Bowman. 5v. -v5 n9. feb 26 1903 (wkly) [mf ed with gaps] – 2r – 1 – (merged with: hayes county republican to form: hayes county times-republican) – us NE Hist [071]
Hayes county times-republican – Hayes Center, NE: C A Ready. -v23 n21. may 23 1907 (wkly) [mf ed 1903-07 (lacks mar 1 1906) filmed 1969?] – 2r – 1 – (formed by the union of: hayes county times and: hayes county republican. cont by: times-republican) – us NE Hist [071]
Hayes, Doremus Almy *see*
– The gift of tongues
– The most beautiful book ever written
– Paul and his epistles
– The synoptic problem
Hayes, Everand A *see* A plain treatise on the law of marriage and divorce
Hayes harlington weekly post *see* Hayes post
Hayes, Isaac I *see* Das offene polar-meereine entdeckungsreise nach dem nordpol
Hayes, Isaac Israel *see*
– An arctic boat journey in the autumn of 1854
– The open polar sea
Hayes, James M *see* Trials and triumphs of the catholic church in america
Hayes, Jennifer M *see* Adolescent perceptions of mentoring
Hayes, Joel S *see* The foreknowledge of god
Hayes, Margaret *see* Captive of the simbas
Hayes, Marie Elizabeth *see* At work
Hayes maze – v1 n1-v7 n2 [1982 jul-1988 oct] – 1r – 1 – (continued by: hayes of america herald) – mf#1336347 – us WHS [071]
Hayes post – London, UK. 1964-73 – 18r – 1 – (aka: hillingdon district weekly post; hayes harlington weekly post; hayes weekly post; weekly post (hayes ed)) – uk British Libr Newspaper [072]
Hayes, S P *see* Some applications of behavioural research
Hayes, Thomas J *see* Journal of marketing for higher education
Hayes, Thomas Jay *see* Exterior ballistics
Hayes uxbridge southall chronicle *see* Hayes chronicle
Hayes, Watson M *see*
– Chiao hui li shih [ccm145]
– Chiao i shen hsueh [ccm147]
Hayes, Wayland Jackson *see* Visual outline of introductory sociology
Hayes, William C *see* The burial chamber of the treasurer sobk-mose
Hayford, Mark Christian *see* West africa and christianity
[Hayfork-] south trinity nugget – CA. 1984 – 1r – 1 – $60.00 – mf#B03238 – us Library Micro [071]
Haygarth, Henry W *see* Buschleben in australien
Haygood, Atticus Greene *see*
– Address of the rev atticus g haygood of the methodist episcopal church, south
– The man of galilee
– Pleas for progress
Haygood, Benjamin Iverson *see* A training program for volunteer leaders at the first baptist church, alma, georgia
Hayler, Guy *see* The master method: an enquiry into the liquor problem in america
Hayley, William *see* The life of george romney
Hayllar, Thomas *see* Scorpion-central new guinea
Haym, Rudolf *see*
– Die romantische schule
Hayman, Henry *see* The epistles of the new testament
Hayne, J C G *see* Abhandlung ueber die kriegskunst der tuerken...desgleichen derjenigen voelker...als griechen, armenier, araber, kurden...
Hayne, M H E *see* The pioneers of the klondyke
Hayne, Robert Young *see* Speeches of messrs. hayne and webster in the united states senate on the resolution of mr. foot, january, 1830
Hayne, W B *see* Qualified pastor
Hayneccius, Martin *see* Hans pfriem, oder, meister kecks
Haynel, Woldemar Claudius *see* Gellerts lustspiele
Haynes – (= ser Bedfordshire parish register series) – 3mf – 9 – £7.50 – uk BedsFHS [929]
Haynes' baptist cyclopaedia / Haynes, Thomas Wilson – 1848. v.1, Aa-Fo. 330p – 1 – $11.55 – us Southern Baptist [242]
Haynes, Dudley C *see*
– The baptist denomination
– Centennial edition of the baptist denomination
Haynes, Edmund Sidney Pollock *see*
– The belief in personal immortality
– Religious persecution
– Religious persecution; a study in political psychology
The haynes gazette – Haynes, Adams County, ND: G L Hurd. v1 n1 mar 7 1908-v2 n17 jun 26 1909 (wkly) [mf ed mar 7 1908-jun 26 1909] – 1 – (missing: 1909 mar 13) – mf#03646 – us North Dakota [071]
Haynes, H Valentine *see* Federation
Haynes, Nathaniel Smith *see* Jesus as a controversialist
Haynes register – Haynes, ND: The Register Pub Co. v1 n1 aug 6 1908-v1 n47 jun 24 1909 (wkly) – 1 – (merged with: haynes gazette to form: haynes register-gazette) – mf#03646 – us North Dakota [071]
Haynes register-gazette – Haynes, ND: The Register Pub Co. v1 n48 jul 1 1909-v12 n26 dec 9 1920 (wkly) [mf ed with gaps] – 1 – (missing: 1910 feb 17, dec 22; 1911 feb 21; 1912 feb 15, apr 18, may 23, jul 25, aug 29; 1915 may 13; 1917 jun 21. also bears numbering of: the haynes gazette v2 n18-; and the haynes register v1 n48-; formed by the union of: haynes gazette and: haynes register) – mf#03646-03650 – us North Dakota [071]
Haynes, st mary monumental inscriptions monumental inscriptions – Arthur Weight Matthews 1913 – (= ser Bedfordshire parish register series) – 1mf – 9 – £1.25 – uk BedsFHS [929]
Haynes, Thomas H *see*
– International fishery disputes
– A survey of canadian imports
Haynes, Thomas Henry *see* Legislation in western australia before responsible government
Haynes, Thomas Wilson *see* Haynes' baptist cyclopaedia
Haynes, William Casper *see* Papers
Haynovius, I *see* Via veritatis ad vitam
Haynt : ilustrirte baylage – Warsaw PL, 1924-25 – 1r – 1 – (in yiddish. with: poalej emuna isroel be polanja [lodz, poland] 1924-47) – us UMI ProQuest [939]
Haynt – New York. v. 1-52. Jan 1-Feb 21 1920 – 1 – us NY Public [071]
Ha-yom – Spb., 1886 (feb 12)-1888 (mar 12) (incomplete) – 2r – 1 – mf#J-291-5 – ne IDC [077]
Ha-yom – St Petersburg. v1-2. 1886-1887 – (= ser Yom) – 1 – us NY Public [077]
Hayrenik – Boston: Hairenik Association, Inc, nov 1916-38 – 45r – 1 – us CRL [071]
Hays, Brooks *see* Papers while president of southern baptist convention
Hays, Daniel *see* Christianity at the fountain
Hays, George Pierce *see* May women speak?
Hays, Hoffman Reynolds *see* 12 spanish american poets
Hays, J *see* The correlation of the pre-karroo succession in northern rhodesia with that of adjacent territories
Hays. Kansas. Police Court *see* Dockets
Hayslip, john, papers, ms 2944 – 1801-82 – 1r – 1 – (petitions, affidavits, and other legal papers relating to the proceedings of ohio militia and adams county common pleas courts. also included are roster, muster rolls and election books for the militia in the war of 1812) – us Western Res [976]
Haystani jayn – Paris, France. 20 jul 1922-14 jul 1923 – 1/2r – 1 – uk British Libr Newspaper [072]
Hayter, Charles *see* An introduction to perspective, drawing, and painting
Hayter, George *see*
– Descriptive catalogue of the great historical picture painted by mr george hayter
– A descriptive catalogue of the historical pictures...of the first reformed house of commons
Hayter, John *see* The landsman's log-book
Haythornthwaite, Frank *see* All the way to abenab
Hayti : alexandria black history resource center's newsletter / Alexandria Black History Resource Center – 1999 winter – 1r – 1 – mf#5076917 – us WHS [321]
Hayti / St John, Spenser Buckingham – London, England. 1884 – 1r – us UF Libraries [972]
Hayward Area Genealogical Society [CA] *see* Hags news letter
Hayward Bros and Eckstein, Ltd *see* [Hayward brothers and eckstein ltd] patent pavement lightworks
[Hayward brothers and eckstein ltd] patent pavement lightworks : trade catalogue / Hayward Bros and Eckstein, Ltd – London [1885?] – (= ser 19th c art & architecture) – 4mf – 9 – mf#4.2.1093 – uk Chadwyck [690]
Hayward, Edward Farwell *see*
– Ecce spiritus
– Lyman beecher
[Hayward-] hayward journal – CA. 1971-72 [wkly] – 1r – 1 – $60.00 – mf#B02289 – us Library Micro [071]
Hayward, John Frank *see* The conceptions of existence and essence in anthropology
Hayward journal-news – Hayward WI. 1890 may 30-dec 26, 1891 jan 2-1892 jul 1, 1892 jul 8-1893 nov 3, 1893 nov 10-1894 dec 29, 1895 jan 25-apr 26 – 5r – 1 – (cont: hayward journal; north wisconsin news) – mf#875729 – us WHS [071]
Hayward republican – Hayward WI. 1893 jul 27-1895 mar 28, 1895 apr 4-1896 dec 17, 1896 dec 24-1897 jul 15, 1897 jul 22-1898 dec 29, 1899 jan 5-jul 20, 1899 jul 27-1900 jul 19, 1900 jul 26-1901 jul 18, 1901 jul 25-1903 jul 16, 1903 jul 23-1905 jan 5, 1905 jan 12-1906 jul 26, 1909 dec 30-1910 jan 6, 27-may 19, 1910 jun 2-9, 1913 jul 24-1915 jan 7 – 11r – 1 – (continued by: sawyer county record; sawyer county record and hayward republican) – mf#945400 – us WHS [071]
Hayward, Sharman L *see* Directors of athletics' attitudes toward women
Hayward, T B *see* Musical cabinet
[Hayward-] the daily review – CA. 1895-1902; 1909-12; 1925- [daily] – 830r – 1 – $49,800.00 (subs $700y) – mf#BC02287 – us Library Micro [071]
[Hayward-] the daily review : sunrise edition – CA. 1978- [daily] – 106r – 1 – $6360.00 (subs $700y) – mf#B02288 – us Library Micro [071]
[Hayward-] the pioneer – CA: CSU Hayward, 1960-1987 – 10r – 1 – $600.00 – mf#B02290 – us Library Micro [378]
[Hayward-] the spectator – CA: Chabot College, 1974-75 – 1r – 1 – $60.00 – mf#B02291 – us Library Micro [378]
Hayward, Walter Brownell *see* Bermuda past and present
Haywood, H L *see* A story of the life and times of jacques de molay
Haywood hills baptist church. nashville, tennessee : church records – Sept 1959-1989 – 2r – 1 – $71.68 – (1,792p) – us Southern Baptist [242]
Haywood, Marshall De Lancey *see* Lives of the bishops of north carolina
Hayye yehudi shalem / Villa, Eugenio – Buenos Ayres, Argentina. 1941 – 1r – us UF Libraries [939]
Hayyuj, Judah ben David *see*
– Two treatises on verbs containing feeble and double letters
– The weak and geminative verbs in hebrew
Hazan, Lew *see* Divre yeme ha-tsiyonut
Hazan, Solomon *see* Mahalot li-shelomoh
Hazanas y La Rua, Joaquin *see* La imprenta en sevilla. noticias ineditas desde la introduccion del arte tipografico en esta ciudad hasta el siglo 29. vol 1. sevilla, 1945
Hazard, Caroline *see*
– Causation
– Freedom of mind in willing, or, every being that wills a creative first cause
– The narragansett friends' meeting in the 18 century
– Some ideals in the education of women
Hazard, John Beach *see* Attempt at the isolation of an organic toxcant in an everglade
Hazard, Marshall Custiss *see*
– Books of the bible
– A study of the life of jesus the christ

Hazard, Rowland Gibson *see*
- Causation
- Freedom of mind in willing
- Freedom of mind in willing, or, every being that wills a creative first cause

Hazardous waste news / Environmental Research Foundation (Princeton, NJ) – v7 n1-5 [1986 dec 1-29], n6-57 [1987 jan 5-dec 28], n58-98 [1988 jan 4-oct 10] – 1r – 1 – (continued by: rachel's hazardous waste news) – mf#1727749 – us WHS [360]

Hazards, pollution and legislation in the coatings field – London. 1984-1985 (1) 1984-1985 (5) 1984-1985 (9) – ISSN: 0262-7116 – mf#49474 – us UMI ProQuest [360]

Hazard's register of pennsylvania : devoted to the preservation of facts and documents, and every kind of useful information respecting pennsylvania – Philadelphia. 1828-1835 (1) – mf#3987 – us UMI ProQuest [978]

Hazard's register of pennsylvania – Philadelphia. v1-16. 1828-36 (all publ) – 80mf – 9 – $120.00 – mf#LLMC 82-926 – us LLMC [978]

The hazare de shabd of the sikhs / Macauliffe, Max – Lahore: Civil and Military Gazette Press, 1900 – 1mf – 9 – 0-524-07494-1 – mf#1991-0115 – us ATLA [280]

Hazavehei, Seyyed M M *see* The effects of 6-week and 12-week rehabilitation programs on the depression level of cardiac patients

Ha-zefirah – Warsaw, [Berlin], 1862-1931. v1-32; 36-51 (incomplete) – 39r – 1 – mf#J-291-4 – ne IDC [077]

Hazel e.b. hanna papers / Hanna, Hazel E B – 1911-19 – 1 – $100.00 – us Presbyterian [920]

Hazel grove & district postman – Hazel Grove, England. The Postman: Hazel Grove & District. -w. 24 Feb 1900-8 March 1902. 1 reel – 1 – uk British Libr Newspaper [072]

Hazel Grove, ebenezer congregational: baptisms 1878-1914 – [North Cheshire FHS] – (= ser Cheshire church registers) – 1mf – 9 – £2.50 – mf#285 – uk CheshireFHS [929]

Hazel grove, ebenezer congregational: baptisms & burials 1828-1837 – [North Cheshire FHS] – (= ser Cheshire church registers) – 1mf – 9 – £2.50 – mf#284 – uk CheshireFHS [929]

Hazel grove, methodist chapel: baptisms 1794-1879 – [North Cheshire FHS] – (= ser Cheshire church registers) – 2mf – 9 – £3.25 – mf#182 – uk CheshireFHS [929]

Hazel grove, methodist chapel: baptisms 1879-1895 – [North Cheshire FHS] – (= ser Cheshire church registers) – 1mf – 9 – £3.25 – mf#262 – uk CheshireFHS [929]

Hazel grove, methodist chapel: burials 1837-1910 – [North Cheshire FHS] – (= ser Cheshire church registers) – 1mf – 9 – £3.25 – mf#254 – uk CheshireFHS [929]

Hazel grove, refuge methodists: baptisms 1836-1886 – [North Cheshire FHS] – (= ser Cheshire church registers) – 1mf – 9 – £2.50 – mf#310 – uk CheshireFHS [929]

Hazel grove times – [NW England] Hazel Grove mar-dec 1984 – 1 – uk MLA; uk Newsplan [072]

Hazelius, Ernest L *see* History of the american lutheran church

Hazelius, Ernest Lewis *see* History of the american lutheran church

Hazelrigg, John *see* Metaphysical astrology

Hazeltine, F A *see* A year of south american travel

Hazelton, A S *see* Compiled ordinances of the city of council bluffs, iowa

Ha-zeman – St Petersburg, Wilna. 1903-July 31 1914 – 1 – us NY Public [077]

Hazen, Austin *see* Addresses delivered at richmond, vermont, june 28, 1895

Hazen, Edward Adams *see* Salvation to the uttermost

Hazim bey yahut heder / Naci, Muallim – Matbaa-i Aramiyan, 1298 [1881] – (= ser Ottoman literature, writers and the arts) – 1mf – 9 – $25.00 – us MEDOC [470]

Hazine-i evrak – Istanbul: Mihran Matbaasi, 1881-? n1. 1 mayis 1297 [1881]-20,22,26, 2 sene n1-15. 15 kanunisani 1298-30 nisan 1299 [1882-83] – (= ser O & t journals) – 6mf – 9 – $125.00 – us MEDOC [956]

Hazine-i fuenun – Istanbul: Alem Matbaasi, Ahmed Ihsan ve Suerekasi Matbaasi, Akin Matbaasi, 1892-96. Sahib-i Imtiyaz: Doktor Cerrahiyan [Cerahizade], Mueduer-i Mes'ul: Kirkor Faik. n1-52. 3 temmuz 1308-23 temmuz 1310 [1892-94] – (= ser O & t journals) – 7mf – 9 – $150.00 – us MEDOC [956]

Hazine-i fuenun – v1-4. 1311-14 [all publ] – (= ser O & t journals) – 26mf – 9 – $430.00 – us MEDOC [956]

Hazlehurst first baptist church. hazlehurst, mississippi : church records – Mar 1870-1988. 4910p – 1 – us Southern Baptist [242]

Hazlehurst first baptist church. hazlehurst, mississippi : church records – Oct 1950-1988. Lacking: 1967 – 1 – 68.27 – us Southern Baptist [242]

Hazlitt, William *see*
- Conversations of james northcote
- Life of napoleon buonaparte
- Notes of a journey through france and italy
- Sketches of the principal picture-galleries in england
- The table talk of martin luther
- Winterslow

Hazlitt, William Carew *see* The coin collector

Hazlitt, William Carew [comp] *see* British columbia and vancouver island

Hazmon (the time) – New York. N.Y. 1895-96 – 1 – us AJPC [071]

Ha-zofeh – London. v. 1-3 no. 10. Mar 2 1894-Dec 6 1895 – 1 – us NY Public [072]

Ha-zofeh – Warsaw, 1903-1905. v1-3 – 3r – 1 – mf#J-291-20 – ne IDC [077]

Ha-zofeh – Warsaw. v. 1-4 no. 11. Jan 6 1903-Feb 27 1906 – 1 – us NY Public [077]

Hazon-eropah : me'ubad u-mekutsar 'a. y. ya'akov norman; turgam bi-yede yitshak norman / Keyserling, Hermann, Graf von – Tel Aviv: defus E Strud ve-banav, [1928] (mf ed 197-) – 1r – 1 – mf#ZZ-16578 – us NY Public [940]

Hazrat amir khusrau of delhi / Habib, Mohammad – Bombay: DB Taraporevala Sons & Co, [1927] – (= ser Samp: indian books) – us CRL [954]

Hazzard and warburton's prince edward island cases / Prince Edward Island. Canada – v1-2. 1850-82 (all publ) – 12mf – 9 – $18.00 – mf#LLMC 81-061 – us LLMC [340]

Hazzledine, G D *see* The white man in nigeria

Hbw news / Northeastern University (Boston, MA) – 1992 spring – 1r – 1 – mf#3179589 – us WHS [378]

Hcgs newsletter / Howard County Genealogical Society – iss n1-2 [1977 feb 2-mar 2] – 1r – 1 – (continued by: family tree (columbia, md)) – mf#1685337 – us WHS [929]

Hco reduction tables *see* Reduction tables

Hd : the journal for healthcare design and development – Foots Cray, 2001+ [1,5,9] – mf#21408,01 – us UMI ProQuest [360]

He archaiotera gnoste morphe toon leitourgioon,... / Mooraeitis, D – Bas., Chrys., 1957 – 1mf – 8 – €3.00 – ne Slangenburg [290]

"He being dead yet speaketh" : a sermon delivered in st andrew's church, toronto, on sunday, march 14, 1847 on the occasion of the death of william campbell, esq, of olive grove, young street sic / Barclay, John – Toronto?: s.n, 1848? – 1mf – 9 – mf#41608 – cn Canadiana [240]

He being dead yet speaketh / Bucknill, George – s.l., England. 1849 – 1r – us UF Libraries [240]

He being dead yet speaketh / Tweedie, W K – Edinburgh, Scotland. 1847 – 1r – us UF Libraries [240]

He being dead yet speaketh / Winter, John S – Shenley? England. 1851? – 1r – us UF Libraries [240]

"He descended into hell" : or, an interpretation based on reason and scripture / Henderson, W – Pembroke, Ont: Observer, 1868 – 1mf – 9 – mf#05541 – cn Canadiana [240]

He is a canadian : and other verse / Kerry, Esther – [Montreal]: Regal Press, 1919 [mf ed 1998] – 1mf – 9 – 0-665-98950-4 – mf#98950 – cn Canadiana [810]

He kaine diatheke : the four gospels and acts of the apostles in greek... / ed by Spencer, Jesse Ames – New York: Harper, 1872, c1847 [mf ed 1991] – 6mf – 9 – 0-7905-8287-2 – (with english notes & ind) – mf#1987-6392 – us ATLA [225]

He kaine diatheke : the greek testament, with english notes, critical, philological, and exegetical / ed by Bloomfield, Samuel Thomas – 2nd London ed. Boston: Perkins & Marvin, 1837 [mf ed 1993] – 2v on 12mf – 9 – 0-524-08505-6 – (in greek. comm in english) – mf#1993-0030 – us ATLA [225]

He kaine diatheke : textaus stephanici a d 1550... = Novum testamentum – ed 4. Londini: G Bell; Cantabrigiae: Deighton, Bell, 1906 [mf ed 1990] – 6mf – 9 – 0-8370-1862-5 – (in greek. pref in latin) – mf#1987-6249 – us ATLA [225]

He lives – 1986 may-jul, oct, dec – 1r – 1 – mf#4027943 – us WHS [071]

He palaia diatheke kata tous ebdomekonta : secundum exemplar vaticanum romae editum accedit potior varietas codicis alexandrini = Vetus testamentum ex versione septuaginta interpretum – editio altera. Oxonii [Oxford]: E typographeo Clarendoniano, 1875 [mf ed 1991] – 3v on 20mf – 9 – 0-8370-1988-5 – mf#1987-6375 – us ATLA [221]

He palaia diatheke kata tous ebdomekonta : ex auctoritate sixti quinti pontificis maximi editum juxta exemplar originale vaticanum... = Vetus testamentum graecum juxta septuaginta interpretes – Parisiis: A Firmin-Didot, 1882 [mf ed 1991] – 8mf – 9 – 0-8370-1918-4 – mf#1987-6305 – us ATLA [221]

He palaia diatheke kata tous hebdomekonta = Vetus testamentum graece iuxta 70 interpretes / ed by Tischendorf, Constantin von – 4.ed. Lipsiae: F A Brockhaus, 1869 [mf ed 1986] – 2v on 4mf – 9 – 0-8370-9432-1 – mf#1986-3432 – us ATLA [221]

He peri ton scheseon ton autokephalon orthodoxon ekklesion kai peri allon genikon zetematon patriarchike kai synodike egkyklios tou 1902 : ai eis auten apantaseis ton hagion autokephalon orthodoxon ekklesion kai he antapantesis tou oikoumenikou patriarcheiou = En hois antepesteilan – En Konstantinoupolei: Ek tou patriarchikou typographeiou, 1904 – 1mf – 9 – 0-524-03891-0 – mf#1990-1150 – us ATLA [240]

He purposeth a crop / China Inland Mission – London: China Inland Mission [1944] (annual) [mf ed 2003] – 1v on 1r – 1 – mf#2003-s066 – us ATLA [240]

He, Qin *see* Evaluation of stretch load capacity and utilization of stored elastic energy in leg extensor muscles during vertical jumps

He tenido sujeta la palabra entre los dientes / Cayetano Rosado, Moises – Barcelona: Graficas Cromotip, 1972 – 1 – sp Bibl Santa Ana [946]

He that overcometh / Monk, Henry Wentworth – [Ottawa?: s.n, 1885?] [mf ed 1995] – 1mf – 9 – 0-665-94757-7 – (in dble clms) – mf#94757 – cn Canadiana [327]

He will do it – London, England. 18-- 1r – us UF Libraries [240]

Hea sonum ja eesti baptisti kogudused / Kaups, Richard – The Good News and the Estonian Baptist Congregations. 1974. 150p – 1 – 6.00 – us Southern Baptist [242]

Head and neck – New York. 1989+ (1,5,9) – (cont: head and neck surgery) – ISSN: 1043-3074 – mf#12710,01 – us UMI ProQuest [617]

Head and neck surgery – New York. 1979-1988 (1) 1979-1988 (5) 1979-1988 (9) – (cont by: head and neck) – ISSN: 0148-6403 – mf#12710 – us UMI ProQuest [617]

Head, Barclay Vincent *see* A guide to the principal gold and silver coins of the ancients

Head, Edith *see* Designs. sketches

Head, Francis *see* Rough notes taken during some rapid journeys across the pampas and among the andes

Head, Francis B *see* Bubbles from the brunnens of nassau

Head, Francis Bond *see*
- Copies or extracts of despatches from sir f b head
- Copy of an explanatory memorandum
- Descriptive essays contributed to the quarterly review, vol 1
- Descriptive essays contributed to the quarterly review, vol 2
- Descriptive essays contributed to the quarterly review, vols 1 and 2
- The emigrant
- A narrative
- Return to an address of the honourable the house of commons, dated 5 march 1839
- The speeches, messages, and replies of his excellency sir francis bond head...lieutenant-governor of upper canada

Head, George *see*
- Forest scenes and incidents, in the wilds of north america

Head hunters of the amazon : seven years of exploration and adventure / Up de Graff, Fritz W – Garden City NY: Garden City Publ Co Inc [1929] – 1r – 1 – us UF Libraries [918]

Head quarters – Fredericton, NB. 1844-68 – 6r – 1 – cn Library Assoc [971]

Head start newsletter / Central Wisconsin Community Action Council – 1978 oct-1982 jun – 1r – 1 – (continued by: head start of central wisconsin) – mf#601841 – us WHS [362]

Headache – Malden. 1961+ (1) 1970+ (5) 1976+ (9) – ISSN: 0017-8748 – mf#2336 – us UMI ProQuest [616]

Headdresses of the victorian era : a lecture delivered...kensington town hall, friday, may 7th, 1897, in commemoration of her majesty's record reign, g herbert thring... / Sutton, Alfred M – London: R Hovenden & Sons, 1857 – (= ser 19th c art & architecture) – 1mf – 9 – mf#4.1.59 – uk Chadwyck [390]

Headlam, Arthur Cayley *see*
- History, authority and theology
- St paul and christianity
- The teaching of the russian church

Headlam, Cecil *see* The inns of court

Headlam, Stewart Duckworth *see* The socialist's church

Headland, Edward *see*
- The epistle to the galatians
- The epistles to the thessalonians

Headley, Joel Tyler *see*
- The chaplains and clergy of the revolution
- Luther and cromwell

Headley, Phineas Camp *see* Evangelists in the church

Headley, Rowland George Allanson Allanson-Winn, Lord *see* A western awakening to islam

Headley, Russel *see* The new york criminal justice

Headlight – Flint, MI. 1941-1992 (1) – mf#63744 – us UMI ProQuest [071]

Headlight – Stromsburg, NE: I D Chamberlain, may 1885-v108 n17 [ie 27] dec 30 1993 (wkly) [mf ed with gaps filmed [1967?]] – 58r – 1 – (merged with: osceola record and: shelby sun (1904) to form: polk county news (1994)) – us NE Hist [071]

Headlight-herald – Tillamook, OR: Tillamook Pub Co. v45 n1 apr 5 1934-jul 6 1999 – 1 – (cont: tillamook headlight, and tillamook herald; aka: headlight herald, sunday headlight-herald, weekend edition headlight-herald, and high school spokesman; 1934 incl newspaper publ by tillamook high school students) – us Oregon Hist [071]

Headlight-herald – Tillamook OR: Tillamook Pub Co, 1934- [wkly] – 1 – (merger of: tillamook headlight (1888-1934); tillamook herald (1896-1934). 1934 incl newspaper pub by tillamook high school students. words sunday & weekend ed appear at head of title between feb 11 1962 and apr 16 1967) – us Oregon Lib [071]

Headlight-herald *see* Tillamook headlight

Headline series – New York. 1935+ (1) 1970+ (5) 1977+ (9) – ISSN: 0017-8780 – mf#1555 – us UMI ProQuest [327]

Headlines – 1944 sep – 1r – 1 – (continued by: headlines and pictures) – mf#5077012 – us WHS [071]

Headlines / Headlines Collective (Pasadena, CA) – v1 n2-3,7 [1970 oct 16-30, dec 23], v2 n1-2 [1971 jan 20-feb 3] – 1r – 1 – mf#927672 – us WHS [334]

Headlines and pictures – 1945 jul, oct-1946 sep – 1r – 1 – mf#4755343 – us WHS [071]

Headlines Collective (Pasadena, CA) *see* Headlines

Headquarters, 3rd regiment, florida volunteers / Shepherd, Rose – s.l, s.l? 1939 – 1r – us UF Libraries [978]

Headquarters heliogram / Council on Abandoned Military Posts – 1968 apr-1972 jan/feb, 1971 jun-1983 – 2r – 1 – mf#962299 – us WHS [355]

Headquarters records of fort cummings, new mexico, 1863-1873, 1880-1884 / New Mexico. Fort Cummings Headquarters – (= ser Records of united states army continental commands, 1821-1920) – 8r – 1 – (with printed guide) – mf#M1081 – us Nat Archives [355]

Headquarters records of fort dodge, kansas, 1866-1882 / Kansas. Fort Dodge Headquarters – (= ser Records of united states army continental commands, 1821-1920) – 25r – 1 – (with printed guide) – mf#M989 – us Nat Archives [355]

Headquarters records of fort gibson, indian territory, 1830-1857 / Indian Territory. Fort Gibson Headquarters – (= ser Records of united states army continental commands, 1821-1920) – 6r – 1 – (with printed guide) – mf#M1466 – us Nat Archives [355]

Headquarters records of fort scott, kansas, 1869-1873 / Kansas. Fort Scott Headquarters – (= ser Records of united states army continental commands, 1821-1920) – 2r – 1 – (with printed guide) – mf#M1077 – us Nat Archives [355]

Headquarters records of fort stockton, texas, 1867-1886 / Texas. Fort Stockton Headquarters – (= ser Records of united states army continental commands, 1821-1920) – 8r – 1 – (with printed guide) – mf#M1189 – us Nat Archives [355]

Headquarters records of fort sumner, new mexico, 1862-1869 / New Mexico. Fort Sumner Headquarters – (= ser Records of united states army continental commands, 1821-1920) – 5r – 1 – (with printed guide) – mf#M1512 – us Nat Archives [355]

Headquarters records of fort verde, arizona, 1886-1891 / Arizona. Fort Verde Headquarters – (= ser Records of united states army continental commands, 1821-1920) – 11r – 1 – (with printed guide) – mf#M1076 – us Nat Archives [355]

Headquarters records of the district of the pecos, 1878-1881 / District of Pecos. Headquarters – (= ser Records of united states army continental commands, 1821-1920) – 5r – 1 – (with printed guide) – mf#M1381 – us Nat Archives [355]

Headquarters star / Civilian Conservation Corps (US) – v2 n23 [1937 apr] – 1r – 1 – mf#1497302 – us WHS [355]

Heads of consideration on the case of mr ward / Keble, John – Oxford, England. 1845 – 1r – us UF Libraries [240]

Heads of english unitarian history : with appended lectures on baxter and priestley / Gordon, Alexander – London: Philip Green, 1895 – 1mf – 9 – 0-524-00735-7 – mf#1990-4004 – us ATLA [243]

Heads of hebrew grammar : containing all the principles needed by a learner / Tregelles, Samuel Prideaux – London: Samuel Bagster, [1852?] – 1mf – 9 – 0-8370-9275-2 – mf#1986-3275 – us ATLA [470]

The headship of christ and the rights of the christian people / Miller, Hugh – Edinburgh: Adam & Charles Black; London: Hamilton, Adams 1861 [mf ed 1991] – 2mf – 9 – 0-7905-9036-0 – mf#1989-2261 – us ATLA [242]

Headway – apr/may 1920-oct 1945 – (= ser The library of world peace studies) – 69mf – 9 – $470.00 – us UPA [320]

Heagle, David *see* That blessed hope

Heald, W M *see* Duties of the clergy

[Healdsburg-] healdsburg enterprise – CA. 1873-79; 1888-90; 1893-94; 1902-07; 1913-27 – 6r – 1 – $360.00 – (aka: sonoma county tribune between 1873-94) – mf#B02292a – us Library Micro [071]

[Healdsburg-] russian river flag – CA. 1880-86 – 1r – 1 – $60.00 – mf#C02293 – us Library Micro [071]

[Healdsburg-] sonoma county tribune – CA. 1888-1890; 1891-1892 – 2r – 1 – $120.00 – mf#B03239 – us Library Micro [071]

[Healdsburg-] sotoyome scimitar – CA. 1920-1930 – 4r – 1 – $240.00 – mf#B03240 – us Library Micro [071]

[Healdsburg-] the healdsburg tribune – CA. 1888-1901; 1902-27; 1945- – 90r – 1 – $5400.00 (subs $90y) – mf#BC02292 – us Library Micro [071]

Heales, Alfred Charles *see* The architecture of the churches of denmark

Healey, Joseph Graham *see* Mass media growth in kenya

Healey, Lionel Rhys *see* Documents relating to murders in telefomin, papua new guinea, 6 nov 1953

Healing : causes and effects / Phelon, William P – Chicago: Hermetic Pub Co, 1898 – 1mf – 9 – 0-524-05465-7 – mf#1990-3491 – us ATLA [150]

Healing : a poem written by w wilfrid sic campbell and decorated by franklin brownell, and issued for their friends with new year's greetings, 1898 – S.l: s.n, 1898? – 1mf – 9 – mf#60969 – cn Canadiana [810]

The healing of the nations : a treatise on medical missions: statement and appeal / Williamson, John Rutter – New York: Student Volunteer Movement for Foreign Missions, 1899 – 1mf – 9 – 0-8370-6719-7 – mf#1986-0719 – us ATLA [610]

Healing springs baptist church – Barnwell Co, SC. 464p. feb 1822-oct 1954 (scattered), feb-may 1955 – 1r – 1 – $20.88 – mf#6649 – us Southern Baptist [242]

Heals catalogues, 1844-1950 – 95mf – 9 – £700.00 – uk Matthew [380]

Health – Birmingham. 1992+ (1,5,9) – ISSN: 1059-938X – mf#18476,02 – us UMI ProQuest [610]

Health – Chicago. 1973-1978 (1) 1974-1978 (5) 1977-1978 (9) – ISSN: 0017-8853 – mf#7778 – us UMI ProQuest [613]

Health – New York. 1981-1991 (1) 1981-1991 (5) 1981-1991 (9) – (cont: family health) – ISSN: 0279-3547 – mf#10761,01 – us UMI ProQuest [613]

Health / Fuller, Russell L – s.l, s.l? 1936 – 1r – us UF Libraries [978]

Health : a home magazine devoted to physical culture and out-door life – New York. 1845-1862 [1] – mf#4583 – us UMI ProQuest [790]

**Health advocate / ** / Madison Community Health Center (WI) et al – v5 n2-v13 n2 [1977 mar/apr-1985 jul] – 1r – 1 – mf#1096695 – us WHS [362]

Health affairs – Chevy Chase. 1986+ (1,5,9) – ISSN: 0278-2715 – mf#14601 – us UMI ProQuest [360]

Health and happiness – [Scotland] Glasgow 27 nov-25 dec 1891 [mf ed 2003] – 1r – 1 – uk Newsplan [613]

Health and holiness : a study of the relations between brother ass, the body, and his rider, the soul / Thompson, Francis – 2nd ed. St Louis, MO: B Herder, [1908?] – (= ser Science of Life Series (St. Louis, Mo.)) – 1mf – 9 – 0-524-02800-1 – mf#1990-0704 – us ATLA [240]

Health and hygiene – London. 1984-1990 (1,5,9) – ISSN: 0140-2986 – mf#15528 – us UMI ProQuest [360]

Health and personal social services statistics for england : with summary tables for great britain 1973-1977 – [mf ed Chadwyck-Healey] – (= ser British government publications...1801-1977) – 13mf – 9 – uk Chadwyck [314]

Health and personal social services statistics for england and wales : with summary tables for great britain 1969-1972 – [mf ed Chadwyck-Healey] – (= ser British government publications...1801-1977) – 8mf – 9 – uk Chadwyck [314]

Health and personal social services statistics for wales 1974-1977 – [mf ed Chadwyck-Healey] – (= ser British government publications...1801-1977) – 8mf – 9 – uk Chadwyck [314]

Health and rehabilitative library services – Chicago. 1975-1975 (1) 1975-1975 (5) 1975-1975 (9) – (cont by: health and rehabilitative library services division journal) – ISSN: 0098-3462 – mf#11123 – us UMI ProQuest [020]

Health and safety bulletin / International Union of Electrical, Radio and Machine Workers – 1966 oct-1980 jun – 1r – 1 – mf#653764 – us WHS [360]

Health and safety, mines and quarries 1851-1965 – [mf ed Chadwyck-Healey] – (= ser British government publications...1801-1977) – 20r – 1 – (1852, 1853, 1919 and 1920 not publ) – uk Chadwyck [360]

Health and social care in the community – Oxford. 1993-1995 (1,5,9) – ISSN: 0966-0410 – mf#19662 – us UMI ProQuest [360]

Health and social service journal – London. 1984-1986 (1,5,9) – (cont by: health service journal) – ISSN: 0300-8347 – mf#14165,05 – us UMI ProQuest [360]

Health and social work : churches / Shepherd, Rose – s.l, s.l? 1935 – 1r – us UF Libraries [978]

Health and social work – Silver Spring. 1976+ (1,5,9) – ISSN: 0360-7283 – mf#11628 – us UMI ProQuest [360]

Health and social work / Vorhees, Lou – s.l, s.l? 1936 – 1r – us UF Libraries [978]

Health and the inner life : an analytical and historical study of spiritual healing theories / Dresser, Horatio Willis – New York: G P Putnam, 1906 – (= ser The Inner Life Series) – 1mf – 9 – 0-7905-4629-9 – mf#1988-0629 – us ATLA [130]

Health aspects of pesticides abstract bulletin – Washington. 1972-1973 (1) 1968-1973 (5) (9) – (cont by: pesticides abstracts) – ISSN: 0017-8918 – mf#6497 – us UMI ProQuest [360]

Health aspects of water supply : ten-year report / Hattingh, W H J et al – Pretoria, South Africa: National Institute for Water Research, Council for Scientific and Industrial Research 1985 [mf ed Pretoria, RSA: State Library [199-]] – 183p [ill] on 1r with other items – 5 – (incl bibl ref) – mf#A 89-1959 r26 – us CRL [360]

Health attitudes and their relation to compliance and measured cholesterol levels / Donovan, Carolyn M – 1988 – 122p 2mf – 9 – $8.00 – us Kinesology [614]

Health behaviors and attitudes of selected nigerian and american university students / Igbani, B & Maduabuchi, A – 1989 – 118p on 2mf – 9 – $8.00 – us Kinesology [613]

Health beliefs, health values, and preventive health promotion activities of african- and euro-american women : a comparative study / Herring, Rosa P & Kaplan, Leah E – 1992 – 3mf – 9 – $12.00 – us Kinesology [613]

Health book and sanitary inspectors journal of the borough of hendon, 1900-52 – 17r – 1 – mf#95794 – uk Microform Academic [610]

Health care education – New York. 1979-1981 (1) 1979-1981 (5) 1979-1981 (9) – ISSN: 0160-7006 – mf#12090,01 – us UMI ProQuest [360]

Health care financing review – Washington. 1986+ (1,5,9) – ISSN: 0195-8631 – mf#15729 – us UMI ProQuest [360]

Health care for our senior citizens = Gesondheidsorg van die bejaarde / South Africa. Department of Health [Departement van Gesondheid] [Departement van Gesondheid] – [Pretoria: Dept of Health 1977?] [mf ed Pretoria, RSA: State Library [199-]] – 9p [ill] on 1r with other items – 5 – mf#op 06709 r24 – us CRL [362]

Health care for the farmworker = Gesondheidsorg vir die plaaswerker – Pretoria: Dept of Health [1977?] [mf ed Pretoria, RSA: State Library [199-]] – 24p on 1r with other items – 5 – (in english & afrikaans) – mf#op 06680 r26 – us CRL [360]

Health care for women international – Washington. 1984+ (1,5,9) – (cont: issues in health care of women) – ISSN: 0739-9332 – mf#14242,01 – us UMI ProQuest [305]

Health care management review – Gaithersburg. 1976+ (1,5,9) – ISSN: 0361-6274 – mf#12679 – us UMI ProQuest [650]

Health care manager – Gaithersburg. 1999+ (1,5,9) – (cont: health care supervisor) – ISSN: 1525-5794 – mf#13936,01 – us UMI ProQuest [360]

Health care on the internet : a journal of methods and applications / ed by Delozier, Eric P – v1 n1. 1997- – 1,9 – $85.00 in US $119.00 outside hardcopy subsc – us Haworth [360]

Health care strategic management – Chicago. 1990+ (1,5,9) – ISSN: 0742-1478 – mf#18134 – us UMI ProQuest [360]

Health care supervisor – Gaithersburg. 1982-1998 (1) 1982-1998 (5) 1982-1998 (9) – (cont by: health care manager) – ISSN: 0731-3381 – mf#13936 – us UMI ProQuest [360]

Health care tax credits to decrease the number of uninsured : hearing...house of representatives, 107th congress, 2nd session, feb 13 2002 / United States. Congress. House. Committee on Ways and Means – Washington: US GPO 2002 [mf ed 2002] – 2mf – 9 – (incl bibl ref) – us GPO [344]

Health choices – New York, 1997-1997 – 1,5,9 – ISSN: 1087-6421 – mf#26445 – us UMI ProQuest [370]

Health communication – Mahwah. 1997+ (1,5,9) – ISSN: 1041-0236 – mf#25223 – us UMI ProQuest [360]

Health economics – Chichester. 1992+ (1,5,9) – ISSN: 1057-9230 – mf#19119 – us UMI ProQuest [332]

Health education – Bradford. 2001+ (1,5,9) – ISSN: 0965-4283 – mf#31586 – us UMI ProQuest [613]

Health education – Washington. 1975-1990 (1) 1975-1990 (5) 1977-1990 (9) – (cont: school health review. cont by: journal of health education) – ISSN: 0097-0050 – mf#7254,01 – us UMI ProQuest [360]

Health education and behavior – New York. 1997+ (1) 1997+ (5) 1997+ (9) – (cont: health education quarterly) – ISSN: 1090-1981 – mf#7933,02 – us UMI ProQuest [360]

Health education journal – London. 1943+ (1) 1971+ (5) 1975+ (9) – ISSN: 0017-8969 – mf#528 – us UMI ProQuest [360]

Health education monographs – San Francisco. 1957-1978 (1) 1972-1978 (5) 1973-1978 (9) – (cont by: health education quarterly) – ISSN: 0073-1455 – mf#7933 – us UMI ProQuest [360]

Health education quarterly – New York. 1980-1996 (1) 1980-1996 (5) 1980-1996 (9) – (cont: health education monographs. cont by: health education and behavior) – ISSN: 0195-8402 – mf#7933,01 – us UMI ProQuest [360]

Health education research – Oxford. 1986+ (1,5,9) – ISSN: 0268-1153 – mf#16451 – us UMI ProQuest [610]

Health education resources – Chicago. 1973-1974 (1) 1974-1974 (5) (9) – ISSN: 0093-5298 – mf#9679 – us UMI ProQuest [370]

Health facilities management – Chicago. 1989+ (1,5,9) – ISSN: 0899-6210 – mf#17724 – us UMI ProQuest [360]

Health forum journal – San Francisco. 1999+ (1) 1999+ (5) 1999+ (9) – (cont: healthcare forum journal) – ISSN: 1527-3547 – mf#1943,03 – us UMI ProQuest [360]

Health fruits of florida – Jacksonville, FL. 1916 – 1r – us UF Libraries [634]

Health in the household and practical recipes for the sick – Toronto: T Milburn, [1883?] [mf ed 1984] – 1mf – 9 – 0-665-01606-9 – mf#01606 – cn Canadiana [613]

Health industry today – Union. 1983-2000 (1) 1983-2000 (5) 1983-2000 (9) – (cont: surgical business) – ISSN: 0745-4678 – mf#9918,01 – us UMI ProQuest [617]

Health information and libraries journal – Oxford, 2001+ [1,5,9] – (cont: health libraries review) – ISSN: 1471-1834 – mf#15547,01 – us UMI ProQuest [020]

Health insurance in india / Agarwala, Amar Narain – Allahabad: East End Publishers, [between 1940 and 1945] – (= ser Samp: indian books) – us CRL [368]

The health journal – [Ottawa: Health Journal, 1888-1889] [mf ed v10 n7 jul/aug/sep 1888-v11 n12 dec 1889] – 9 – mf#P04587 – cn Canadiana [614]

Health knowledge competencies and essential health skills of entry level college freshmen enrolled in oregon's research universities / Beeson, Luana J & Smith, Margaret M – 1992 – 2mf – 9 – $8.00 – us Kinesology [613]

A health knowledge test for male college freshmen in saudi arabia / Hashim, Talal J – 1988 – 213p 3mf – 9 – $12.00 – us Kinesology [378]

Health laboratory science – Washington. 1964-1978 (1) 1971-1978 (5) 1977-1978 (9) – ISSN: 0017-9035 – mf#5964 – us UMI ProQuest [360]

Health lawyer (aba) – v1-12. 1982-2000 – 9 – $157.00 set – ISSN: 0736-3443 – mf#112121 – us Hein [344]

Health libraries review – Oxford. 1984-1995 (1,5,9) – ISSN: 0265-6647 – mf#15547 – us UMI ProQuest [020]

Health management technology – Nokomis. 1994+ (1,5,9) – (cont: computers in healthcare) – ISSN: 1074-4770 – mf#12663,02 – us UMI ProQuest [610]

Health marketing quarterly / ed by Winston, William J – v1- 1983- – 1, 9 ($325.00 in US $455.00 outside hardcopy subsc) – us Haworth [610]

Health matrix – Cleveland. 1988-1989 (1,5,9) – (cont by: health matrix) – ISSN: 0748-383X – mf#16158 – us UMI ProQuest [360]

Health matrix – Cleveland. 1991-1995 (1,5,9) – (cont: health matrix) – ISSN: 0748-383X – mf#18794 – us UMI ProQuest [610]

Health matrix : journal of law-medicine – Case Western Reserve University: v1-11. 1991-2001 – 9 – $211.00 set – ISSN: 0748-323X – mf#114311 – us Hein [344]

Health motivation and hiv risk behaviors among college students from urban and rural communities / Sherwood-Puzzello, Catherine M – 1998 – 2mf – 9 – $8.00 – mf#HE 616 – us Kinesology [614]

Health newsletter / Women's Health Collective (Madison, WI) et al – 1972 aug-1983 mar, jul – 1r – 1 – mf#690419 – us WHS [362]

Health physics : the radiation protection journal – v62-71. 1992-1996 – 1,5,6,9 – $106.00r – us Lippincott [613]

Health policy – Amsterdam. 1989-1995 (1,5,9) – ISSN: 0168-8510 – mf#42565,01 – us UMI ProQuest [360]

Health policy and planning – Oxford. 1986+ (1,5,9) – ISSN: 0268-1080 – mf#17355 – us UMI ProQuest [360]

Health policy council agenda – 1973 july 13-1976 sept 24, 1977 jun 24-1978 jan 13 v2, 1978 jan 12 v3-1979 dec 7, 1980 jan 17-1982 aug 20, 1983 jan 14-1984 apr 6, 1984 jun 22-1985 nov 8 – 6r – 1 – (continued by: governor's health policy council agenda, health policy council (wi)) – mf#329281 – us WHS [360]

Health policy quarterly – New York. 1981-1982 (1) 1981-1982 (5) 1981-1982 (9) – ISSN: 0163-5107 – mf#12183 – us UMI ProQuest [360]

Health Professionals for Political Activity *see* Notes on health politics

Health progress – St. Louis. 1984+ (1) 1984+ (5) 1984+ (9) – (cont: hospital progress) – ISSN: 0882-1577 – mf#2149,01 – us UMI ProQuest [360]

Health promotion – Oxford. 1986-1989 (1,5,9) – (cont by: health promotion international) – ISSN: 0268-1099 – mf#17354 – us UMI ProQuest [360]

Health promotion international – Oxford. 1990+ (1,5,9) – (cont: health promotion) – ISSN: 0957-4824 – mf#17354,01 – us UMI ProQuest [360]

Health psychology – Washington. 1989+ (1,5,9) – ISSN: 0278-6133 – mf#17604 – us UMI ProQuest [150]

Health related motivational determinants among participants in a worksite weight management intervention / Melichar, G A – 1990 – 2mf – 9 – $8.00 – us Kinesology [150]

Health rights news – Chicago. 1967-1973 (1) – ISSN: 0017-9094 – mf#7458 – us UMI ProQuest [360]

Health sciences serials / U.S. Dept of Health and Human Services – Quarterly. Citations listed represent The Nat'l. Library of Medicine's SERLINE database – 9 – $19.00y in US $23.75 outside – mf#S-N 717-012-00000-3. Sub-list ID-HSS – us GPO [610]

Health seekers', tourists' and sportsmen's guide to the sea-side, lake-side, foothill, mountain and mineral spring health and pleasure resorts of the pacific coast / Chittenden, Newton H – San Francisco?: C A Murdock, 1884 – 4mf – 9 – mf#14624 – cn Canadiana [790]

Health service journal – London. 1986+ (1,5,9) – (cont: health and social service journal) – ISSN: 0952-2271 – mf#14165,06 – us UMI ProQuest [360]

Health services management – Harlow. 1988-1992 (1) 1988-1992 (5) 1988-1992 (9) – (cont: hospital and health services review) – ISSN: 0953-8534 – mf#5135,01 – us UMI ProQuest [360]

Health services management research – London. 1988+ (1,5,9) – ISSN: 0951-4848 – mf#17171 – us UMI ProQuest [360]

Health services manager – New York. 1976-1982 (1) 1976-1982 (5) 1976-1982 (9) – (cont: hospital supervision) – ISSN: 0363-020X – mf#9115,01 – us UMI ProQuest [360]

Health services reports – Rockville. 1878-1974 [1]; 1965-1974 [5]; 1970-1974 [9] – (cont by: public health reports) – ISSN: 0090-2918 – mf#1439 – us UMI ProQuest [360]

Health services research – Chicago. 1966+ (1) 1970+ (5) 1977+ (9) – ISSN: 0017-9124 – mf#3494 – us UMI ProQuest [360]

Health situation in florida / American Public Health Association – New York, NY. 1939 – 1r – us UF Libraries [978]

Health systems review – Little Rock. 1991-1997 (1) 1991-1997 (5) 1991-1997 (9) – (cont: review – federation of american health systems) – ISSN: 1055-7466 – mf#12216,04 – us UMI ProQuest [360]

Health talks with women = Gesondheidsgeselsies met vroue / South Africa. Department of Health [Departement van Gesondheid] [Departement van Gesondheid] – Pretoria: Dept of Health 1979?] [mf ed Pretoria, RSA: State Library [199-]] – 75p on 1r with other items – 5 – (in english & afrikaans) – mf#op 06857 r26 – us CRL [362]

Health trip to the tropics / Willis, Nathaniel Parker – New York, NY. 1853 – 1r – us UF Libraries [972]
Health values – Star City. 1977-1995 (1,5,9) – (cont by: american journal of health behavior) – ISSN: 0147-0353 – mf#12296 – us UMI ProQuest [613]
Health visitor – London. 1973-1998 (1) 1973-1998 (5) 1973-1998 (9) – (cont by: community practitioner) – ISSN: 0017-9140 – mf#8320 – us UMI ProQuest [360]
Health watch news – 1995 summer/fall, 1996 winter – 1r – 1 – mf#3964601 – us WHS [360]
Healthcare executive – Chicago. 1985+ (1,5,9) – ISSN: 0883-5381 – mf#16472 – us UMI ProQuest [360]
Healthcare financial management : journal of the healthcare financial management association – Westchester. 1982+ (1,5,9) – (cont: hospital financial management) – ISSN: 0735-0732 – mf#12654,02 – us UMI ProQuest [366]
Healthcare forum – San Francisco. 1985-1987 (1) 1985-1987 (5) 1985-1987 (9) – (cont: hospital forum. cont by: healthcare forum journal) – ISSN: 0885-257X – mf#1943,01 – us UMI ProQuest [360]
Healthcare forum journal – San Francisco. 1987-1998 (1) 1987-1998 (5) 1987-1998 (9) – (cont: healthcare forum. cont by: health forum journal) – ISSN: 0899-9287 – mf#1943,02 – us UMI ProQuest [360]
Healthcare risk management – Atlanta. 1995+(1,5,9) – (cont: hospital risk management) – ISSN: 1081-6534 – mf#12281,01 – us UMI ProQuest [360]
Health-p a c bulletin – New York. 1968-1993 [1]; 1972-1993 [5]; 1976-1993 [9] – ISSN: 0017-9051 – mf#7412 – us UMI ProQuest [360]
Health-related fitness in and among youth / Lamb, Jennifer A – 1994 – 1mf – $4.00 – us Kinesology [612]
Health-related fitness levels in bahamian elementary school age children / Rowe, Daivd A & Mahar, Matthew T – 1992 – 3mf – $12.00 – us Kinesology [612]
Healthtexas – Austin. 1988-1996 (1,5,9) – (cont: texas hospitals) – ISSN: 1048-4167 – mf#12316,01 – us UMI ProQuest [360]
Healthweek – Manhasset. 1987-1992 (1,5,9) – ISSN: 0890-2259 – mf#16281 – us UMI ProQuest [613]
Healthworks – 1995 feb-1996 sep – 1r – 1 – mf#4863877 – us WHS [360]
Healthy furniture and decoration / Edis, Robert William – London 1884 – (= ser 19th c art & architecture) – 1mf – 9 – mf#4.2.147 – uk Chadwyck [740]
Healthy homes : and how to make them / Bardwell, William – London: Publ for S A Gilbert, 1854 – (= ser 19th c art & architecture) – 2mf – 9 – mf#4.1.153 – uk Chadwyck [640]
Healy, Patrick Joseph *see* The valerian persecution
Heanley, Robert Marshall *see* A memoir of edward steere
Heanor advertiser, langley mill and ripley weekly news – [East Midlands], Heanor, Derbyshire 22 mar 1890-dec 1917 [mf ed 2004] – 24r – 1 – (missing: 1897-98, 1911) – uk Newsplan [072]
Heanor and langley mill gazette – [East Midlands] Ripley, Derbyshire 20 nov 1936-8 nov 1951 [mf ed 2004] – 9r – 1 – (publ for ilkeston division labour party; previous title: the gazette [n239 20 nov 1936- n1021 8 nov 1951]) – uk Newsplan [072]
Heanor & ripley gazette – [East Midlands] Derbyshire 29 apr 1932-8 nov 1951 [mf ed 2004] – 14r – 1 – (cont by: gazette [jan 1933-dec 1936]; ripley gazette [jan 1937-nov 1951]) – uk Newsplan [072]
Heap, Charles Rogers *see* The indian famine
Hear the church / Hook, Walter Farquhar – London, England. 1838 – 1r – us UF Libraries [240]
Hear the other side / Traviss-Lockwood, J – London, England. 1885 – 1r – us UF Libraries [240]
Hear this – (v1 n1] 1946 nov 8-1978 mar 23 – 1r – 1 – mf#370645 – us WHS [071]
Hear us emerging sisters : hues – 1998 winter – 1r – 1 – mf#3357290 – us WHS [071]
Hear ye! / Milwaukee Society for the Hard of Hearing – v1 n18-v2 n14 [1939 jan-1943 feb] – 1r – 1 – mf#922955 – us WHS [366]
Heard, Albert F *see* The russian church and russian dissent
Heard, Franklin Fiske *see*
– A concise treatise on the principles of equity pleading
– A concise treatise on the principles of equity pleading.
– Curiosities of the law reporters
– Oddities of the law
– Precedents of equity pleadings
– The principles of pleading in civil actions
– A treatise adapted to the law and practice of the superior courts, and of trial justices, district, police, and municipal courts, in criminal cases
Heard, J B *see*
– Biblioatry
– The tripartite nature of man
Heard, John Bickford *see*
– Alexandrian and carthaginian theology contrasted
– The history of the extinction of paganism in the roman empire viewed in relation to the evidences of christianity
– Old and new theology
– The tripartite naaature of man, spirit, soul, and body
Heard journal and herd, hird, hurd, too – 1984 apr-1989 apr/jul – 1r – 1 – mf#1658545 – us WHS [071]
Hearer on his trial – London, England. 18– – 1r – us UF Libraries [240]
Hearing and speech action – Silver Spring. 1977-1978 (1) 1977-1978 (5) 1977-1978 (9) – (cont: h and s: hearing and speech action) – ISSN: 0162-5667 – mf#2140,02 – us UMI ProQuest [370]
Hearing and speech news – Washington. 1966-1974 (1) 1971-1974 (5) – (cont by: h and s: hearing and speech action) – ISSN: 0017-9191 – mf#2140 – us UMI ProQuest [616]
The hearing at the state house, boston, mass., before the joint committee on education, march 20th to april 25th, 1889 : upon the bill requiring that children between eight and fourteen years shall have the fundamentals of an english education... – Boston, MA: Committee of One Hundred, 1889 – 1mf – 9 – 0-8370-7907-1 – mf#1986-1907 – us ATLA [370]
Hearing instruments – New York. 1982-1995 (1,5,9) – ISSN: 0092-4466 – mf#12941,01 – us UMI ProQuest [617]
Hearing on bills to continue the civil government for the ttpi and to provide ex gratia payment to the people of bikini atoll in the marshall islands : hearing before the house committee... / U.S. Congress – mar 24 1975. Washington: GPO, 1975 (mf ed) – 1mf – 9 – $1.50 – mf#LLMC 82-100F Title 111 – us LLMC [980]
Hearing on h.r. 4689 to amend the u.c.m.j. : before the military personnel and compensation subcommittee of the committee on armed services – House of Rep. 97th Congress, 1st session, Oct 14 1981. Washington: GPO, 1981 – 2mf – 9 – $3.00 – mf#LLMC 96-088 – us LLMC [355]
Hearing on the implementation of the compact of free association act in the marshall islands and the federated states of micronesia : held before the subcommittee on insular and international affairs, 100th cong, 1st sess, nov 19, 1987 / U.S. Congress. House Committee on Interior and Insular Affairs – Washington: GPO, 1989 – (= ser Micronesia: evolution to separate political entities) – 5mf – 9 – $7.50 – mf#LLMC 82-100F, Title 108 – us LLMC [980]
Hearing on the micronesian compact : held before the subcommittee on immigration, refugees and international law, of the house committee on the judiciary, 99th cong, 1st sess, jul 18, 1985 / U.S. Congress – Washington: GPO, 1985 – (= ser Micronesia: evolution to separate political entities) – 1mf – 9 – $1.50 – mf#LLMC 82-100F, Title 107 – us LLMC [980]
Hearing professional – Livonia. 2000+ (1,5,9) – (cont: audecibel) – ISSN: 1529-1340 – mf#9865,01 – us UMI ProQuest [616]
Hearing research – Amsterdam. 1978+ (1) 1978+ (5) 1987+ (9) – ISSN: 0378-5955 – mf#42080 – us UMI ProQuest [617]
Hearings and report of the committee / ed by U.S. Congress. Joint Committee on the Investigation of the Pearl Harbor Attack – 9 – $426.00 – (hearings pursuant to joint congressional resolution 27, 79th congress) – mf#BF1007 – us Brook [324]
Hearings before and special reports made by committee on armed services of the house of representatives on subjects affecting the naval and military establishments, 1974. / U.S. Congress. House. Committee on Armed Services – Washington, Govt. Print. Off., 1975. 42p. LL-2345 – 1 – us L of C Photodup [355]
Hearings before the committee on armed services, house of representatives, 98th congress, ist session, november 15 1983 : full committee consideration of s. 974 (as amended) to amend (ucmj) to improve the quality and efficiency of the military justice system, to revise the laws concerning review of courts-martial... – Washington: GPO, 1983 – 1mf – 9 – $1.50 – mf#LLMC 96-085 – us LLMC [347]
Hearings before the committee on armed services, house of representatives, military personnel and compensation subcommittee, 98th congress, 1st session, november 9 1983 On s. 974 : to amend (ucmj) to improve the quality and efficiency of the military justice system, to revise the laws concerning review of courts-martial... – Washington: GPO, 1983 – 1mf – 9 – $1.50 – mf#LLMC 96-086 – us LLMC [347]
Hearings. foreign assistance acts, 1962-69 *see* Us congress. house. foreign affairs committee. hearings. foreign assistance acts, 1962-69
Hearings in truk, ponape and the marshall islands, jul 1973 / Joint Committee on Future Status [TTPI (U.S.)] Eastern Districts Subcommittee – n.p, nov 1973 – (= ser Micronesia: prelude to the constitutional convention) – 5mf – 9 – $7.50 – (various pagination) – mf#LLMC 82-100F, Title 49 – us LLMC [323]
Hearings in yap, palau and the marianas, jul 1973 / Joint Committee on Future Status [TTPI (U.S.)] Western Districts Subcommittee – n.p, nov 1973 – (= ser Micronesia: prelude to the constitutional convention) – 4mf – 9 – $6.00 – (various pagination) – mf#LLMC 82-100F, Title 50 – us LLMC [323]
Hearings of the general board of the u.s. navy, 1917-50 / U.S. Navy. General Board – 1983 – 15r – 1 – $1950.00 – mf#S1655 – U.S. Naval Historical Center – us Scholarly Res [355]
Hearings of the nuclear regulatory commission / U.S. Nuclear Regulatory Commission – 1975-82 – 1 – $11,160.00 coll – (basic set: jan 1975-aug 1979 40r isbn 0-89093-281-6 $6250. suppl: sep-dec 1979 6r isbn 0-89093-282-4 $935. jan-dec 1980 10r isbn 0-89093-565-3 $1575. jan-dec 1981 9r isbn 0-89093-566-1 $1395. jan-dec 1982 10r isbn 0-89093-567-X $1575. with p/g) – us UPA [350]
Hearings on national defense authorization act for fiscal year 2002 – h.r. 2586 and oversight of previously authorized programs... house of representatives, 107th congress, 1st session : military personnel subcommittee hearings on title 4 – personnel authorizations, title 5 – military personnel policy, title 6 – compensation and other personnel benefits, title 7 – health care provisions, hearings held june 21, and july 18 2001 / United States. Congress. House. Committee on Armed Services. Subcommittee on Military Personnel – Washington: US GPO 2002 [mf ed 2002] – 6mf – 9 – 0-16-068871-X – us GPO [343]
Hearings on the compact of free association (micronesia-wide) : held before the subcommittee on public lands and national parks, hse.comm on interior and insular affairs, 98th cong, 2nd sess, apr 27-dec 12, 1984 / U.S. Congress – Washington: GPO. 8pts. 1984-85 – (= ser Micronesia: evolution to separate political entities) – 29mf – 9 – $43.50 – mf#LLMC 82-100F, Title 109 – us LLMC [980]
Hearings on the compact of free association (micronesia-wide) : held before the subcommittee on public lands, hse.comm on interior and insular affairs, 99th cong, 1st sess, mar 7-may 20, 1985 / U.S. Congress – Washington: GPO. 4pts. 1986 – (= ser Micronesia: evolution to separate political entities) – 14mf – 9 – $21.00 – mf#LLMC 82-100F, Title 110 – us LLMC [980]
Hearings, prints and reports / United States. Congress – Washington, 1951-1960 [1,5,9] – mf#2578 – us UMI ProQuest [348]
Hearn Academy. Board of Trustees. (Cave Spring, Georgia) *see* Proceedings
Hearn, Edward J *see*
– Annual address of the president, mr e j hearn, barrister, etc
Hearn, Gordon Risley *see* The seven cities of delhi
Hearn, L *see* Gleanings in buddha-fields
Hearn, Lafcadio *see*
– Gleanings in buddha-fields
– Japan
– Two years in the french west indies
Hearn, William E *see* The theory of legal duties and rights
Hearne, S *see* A journey from prince of wales's fort in hudson's bay, to the northern ocean
Hearne, Samuel *see* Voyage de samuel hearne, du fort du prince de galles dans la baie de hudson, a l'ocean nord
Hearne, Thomas *see* Reliquiae hearnianae
Hearst's chicago american – Chicago IL. 1900 jul 4 – 1r – 1 – mf#1010683 – us WHS [071]
Hearst's chicago evening american – Chicago IL. 1901 sep 18, 1902 sep 2, 1904 mar 24 – 1r – 1 – mf#1010701 – us WHS [071]
Hearst's chicago morning american – Chicago IL. 1901 sep 14 – 1r – 1 – (cont: chicago evening american) – mf#992890 – us WHS [071]
Heart – London. 1996+ (1,5,9) – (cont: british heart journal) – ISSN: 1355-6037 – mf#1331,01 – us UMI ProQuest [616]
Heart – v3 [1969 jul] – 1r – 1 – mf#1583843 – us WHS [071]
Heart / Way International – 1977 nov-1988 aug/sep – 1r – 1 – mf#1046355 – us WHS [071]
Heart and lung – St. Louis. 1972+ (1) 1972+ (5) 1975+ (9) – ISSN: 0147-9563 – mf#6851 – us UMI ProQuest [611]
The "heart at work" program : factors associated with maintenance at the worksite / Frank, Laura B – 1994 – 1mf – $4.00 – us Kinesology [613]
Heart, Jonathan *see* Letterbook and orderly book
Heart messages for sabbaths at home / Simpson, Albert B – Nyack: Christian Alliance, [1899?] [mf ed 1992] – (= ser Christian & missionary alliance coll) – 1mf – 9 – 0-524-02151-1 – mf#1990-4217 – us ATLA [240]
Heart mountain sentinel – Heart Mountain, WY: Community Enterprises. v1 n1-v4 n31. oct 24 1942-jul 28 1945 – 1 – (has suppl listing rules, regulations and procedures of the relocation center: heart mountain sentinel bulletin; japanese ed: hato maunten senchineru) – us Oregon Hist [071]
Heart mountain sentinel *see* Newspapers published in internment camps
Heart of africa / Campbell, Alexander – London, England. 1954 – 1r – us UF Libraries [960]
The heart of africa : three years' travels and adventures in the unexplored regions of central africa from 1868-1871 / Schweinfurth, G – London, 1873. 2v – 21mf – 9 – mf#A-169 – ne IDC [916]
Heart of America Indian Center *see*
– Inter-tribal tribune
– Newsletter
The heart of buddhism : being an anthology of buddhist verse / ed by Saunders, Kenneth James – London: OUP 1915 [mf ed 1992] – (= ser The heritage of india) – 1mf [ill] – 9 – 0-524-02362-X – mf#1990-2973 – us ATLA [810]
The heart of christianity / Linscott, Thomas Samuel – Philadelphia, Pa: Bradley-Garretson, 1906-1907 – 2mf – 9 – 0-7905-9310-6 – mf#1989-2535 – us ATLA [240]
Heart of europe / Cram, Ralph Adams – New York: Scribner, 1915 – 1mf – 9 – 0-7905-4335-4 – mf#1988-0335 – us ATLA [914]
Heart of harlem : hh [newsletter] – 1991 summer, 1992 winter – 1r – 1 – mf#3054973 – us WHS [366]
The heart of hindusthan / Radhakrishnan, Sarvepalli – Madras: GA Natesan, [1949] – (= ser Samp: indian books) – us CRL [954]
The heart of india : sketches in the history of hindu religion and morals / Barnett, Lionel David – London: J Murray 1908 [mf ed 1991] – (= ser Wisdom of the east series (london, england)) – 1mf – 9 – 0-524-00684-9 – mf#1990-2012 – us ATLA [280]
The heart of jainism / Stevenson, Sinclair, Mrs – London: Oxford University Press, 1915 – 1mf – 9 – 0-524-01928-2 – (incl bibl ref) – mf#1990-2741 – us ATLA [280]
The heart of jesus : being addresses upon the present reality of the passion / Waggett, P N – London: SPCK; New York: E & J B Young, 1902 – 1mf – 9 – 0-7905-2396-5 – mf#1987-2396 – us ATLA [240]
The heart of john wesley's journal = Journal. Selections / Wesley, John; ed by Parker, Percy Livingstone – New York: FH Revell, [1903?] – 2mf – 9 – 0-524-08699-0 – mf#1993-3224 – us ATLA [242]
Heart of lincoln / Whipple, Wayne – Philadelphia, PA. 1915 – 1r – us UF Libraries [025]
The heart of nature : or, the quest for natural beauty / Younghusband, Francis Edward – London: John Murray, 1921 – (= ser Samp: indian books) – us CRL [900]
The heart of sz-chuan / Wallace, Edward Wilson – rev ed. Toronto: Methodist Young People's Forward Movt for Missions, [1905] [mf ed 1995] – (= ser Yale coll) – 224p (ill) – 1 – 0-524-09520-5 – mf#1995-0520 – us ATLA [242]
The heart of the antarctic / Shackleton, E H – London, 1909. 2v – 15mf – 9 – mf#H-6186 – ne IDC [590]
The heart of the bhagavad-gita / Vidyasankara Bharati – Baroda: A G Widgery 1918 [mf ed 1993] – (= ser The gaekwad studies in religion and philosophy 3) – 3mf – 9 – 0-524-07386-4 – mf#1991-0106 – us ATLA [280]
The heart of the christian message / Barton, George Aaron – new rev enl ed. New York: Macmillan, 1912 [mf ed 1991] – 1mf – 9 – 0-7905-7683-X – (1st ed publ in london, 1910) – mf#1989-0908 – us ATLA [240]
The heart of the creeds : historical religion in the light of modern thought / Eaton, Arthur Wentworth Hamilton – 3rd ed. New York: Thomas Whittaker, c1888 – 1mf – 9 – 0-8370-3649-6 – mf#1985-1649 – us ATLA [240]

The heart of the gospel : a popular exposition of the atonement / Campbell, James Mann – New York: Fleming H Revell c1907 [mf ed 1985] – 1mf – 9 – 0-8370-3140-0 – (incl bibl ref & ind) – mf#1985-1140 – us ATLA [240]
The heart of the hills / Fox, John – Toronto: McLeod & Allen, 1913 [mf ed 1995] – 5mf – 9 – 0-665-74269-X – mf#74269 – cn Canadiana [830]
Heart of the hunter / Van Der Post, Laurens – New York, NY. 1961 – 1r – us UF Libraries [960]
The heart of the jewish problem / Blackstone, William E – Chicago, IL: Chicago Hebrew Mission, [19–?] [mf ed 1992] – (= ser Christian & missionary alliance coll) – 1mf – 9 – 0-524-03687-X – mf#1990-4792 – us ATLA [270]
The heart of the old testament : a manual for christian students / Sampey, John Richard – Nashville, TN: Sunday School Board, Southern Baptist Convention, [1909?] – 1mf – 9 – 0-7905-0275-5 – mf#1987-0275 – us ATLA [221]
Heart of the stranger / Mcleod, Christian – New York, NY. 1908 – 1r – us UF Libraries [025]
Heart of the valley times-villager – Kaukauna WI. 2002 mar 7/jun-2004 nov/dec – 12r – 1 – (cont: times (appleton, wi)) – mf#5521357 – us WHS [071]
Heart rate and perceived exertion responses during climbing in beginner and recreational sport climbers / Janot, Jeffrey M – 1997 – 1mf – 9 – $4.00 – mf#PH 1553 – us Kinesology [612]
Heart rate responses of collegiate female volleyball players during competition / Harbour, SK – 1991 – 2mf – 9 – $8.00 – us Kinesology [612]
Heart rate responses to chair aerobics in healthy older women / Vande Voort, Cindy K – 1998 – 1mf – 9 – $4.00 – mf#PH 1642 – us Kinesology [612]
Heart rate responses to chair aerobics in male cardiac patients / Wenaas, Jodi L – 1998 – 1mf – 9 – $4.00 – mf#PH 1641 – us Kinesology [612]
The heart rates of elementary children during physical education classes / Burton, Catherine J – Ball State University, 1996 – 1mf – 9 – mf#PH 1489 – us Kinesology [612]
Heart searched / M'cheyne, Robert Murray – London, England. 18– – 1r – us UF Libraries [240]
Heart songs / Blewett, Jean – Toronto: G Morang, 1898 – 3mf – 9 – mf#28094 – cn Canadiana [830]
Heart stories / Blewett, Jean – Toronto: Warwick Bros & Rutter, c1919 – 1mf – 9 – 0-665-71389-4 – mf#71389 – cn Canadiana [830]
Heart that can feel for another / Smith, James – London, England. 18– – 1r – us UF Libraries [240]
Heart throbs, in prose and verse dear to the american people / Chapple, Joe Mitchell – Boston, MA. v1-2. 1905 – 1r – us UF Libraries [025]
Heart-beats / Mozoomdar, Protap Chunder – Boston: Geo H Ellis, 1894 – 1mf – 9 – 0-524-01852-9 – mf#1990-2687 – us ATLA [280]
A heart-broken coroner and other wonders / Belding, Albert Martin & Woodworth, Harry Albro' – St. John, NB?: s.n, 1895 – 1mf – 9 – mf#06158 – cn Canadiana [830]
Heart-faith – London, England. 18– – 1r – us UF Libraries [240]
Hearth and home – New York. 26 dec 1868-25 dec 1875 (wkly) – 8r – 1 – uk British Libr Newspaper [640]
Heartland / Deep West Peace Press (US) – v10-16,17-18 [1984 apr/may-1985 apr/may, summer-fall], v19-20 [1986 spring/summer-autumn] – 1r – 1 – mf#1565647 – us WHS [071]
Heartland – n7-8 [1981 summer-fall], n9,11 [1982 winter/spring, fall], n16-21 [1984 winter-1985 summer], n24 [1987 fall], n27-28 [1988 may/jun-1988 jul/aug] – 1r – 1 – mf#1712985 – us WHS [071]
The heartlander – jan-aug 1988// – 1r – 1 – (ceased publ aug 1988) – mf#45.18 – nz Nat Libr [079]
Heartline / Parents Without Partners – 1976 dec, 1977 jan-1978 nov – 1r – 1 – mf#620641 – us WHS [305]
The heartman manuscript collection : manuscripts on slavery / Xavier University Library. New Orleans – 7r or 200mf – 5,9 – £360.00 – (coverage of slave-related docs dating from 1803 to early reconstruction in new orleans. also incl 18th c materials from other areas of the us) – mf#HMN – uk World [976]
Heat and fluid flow – London. 1978-1978 (1,5,9) – (cont by: international journal of heat and fluid flow) – ISSN: 0046-7138 – mf#11218 – us UMI ProQuest [620]
Heat distribution in the lower leg from pulsed short wave diathermy and ultrasound treatments / Garrett, Candi L – 1998 – 1mf – 9 – $4.00 – mf#PE 3995 – us Kinesology [612]
Heat pumps and thermal compressors / Davies, Sydney John – London, England. 1950 – 1r – us UF Libraries [500]
Heat recovery systems and chp – Oxford. 1987-1995 (1,5,9) – (cont: journal of heat recovery systems. cont by: applied thermal engineering) – ISSN: 0890-4332 – mf#49386,01 – us UMI ProQuest [530]
Heat transfer : asian research – New York. 1999+ (1) – (cont: heat transfer: japanese research) – ISSN: 1099-2871 – mf#14354,01 – us UMI ProQuest [530]
Heat transfer : japanese research – New York. 1984-1997 (1,5,9) – (cont by: heat transfer: asian research) – ISSN: 0096-0802 – mf#14354 – us UMI ProQuest [530]
Heat transfer : soviet research – New York. 1984-1991 (1,5,9) – (cont by: heat transfer research: english ed) – ISSN: 0440-5749 – mf#14355 – us UMI ProQuest [530]
Heat transfer engineering – New York. 1979+ (1,5,9) – ISSN: 0145-7632 – mf#11991 – us UMI ProQuest [621]
Heat transfer research english ed – New York. 1992-1993 (1) 1992-1993 (5) 1992-1993 (9) – (cont: heat transfer: soviet research) – ISSN: 1064-2285 – mf#14355,01 – us UMI ProQuest [530]
Heat treating – Carol Stream. 1969-1993 (1) 1972-1993 (5) 1974-1993 (9) – (cont by: metal heat treating) – ISSN: 0017-9345 – mf#7452 – us UMI ProQuest [660]
Heat treatment for controlling the insect pests of stored corn / Grossman, Edgar F – Gainesville, FL. 1931 – 1r – us UF Libraries [630]
Heath 1737-1849 – Oxford, MA (mf ed 1996) – (= ser Massachusetts vital record transcripts to 1850) – 6mf – 9 – 0-87623-265-9 – (mf 1t: marriages 1804-40; intentions 1805-25. mf 2t: intentions 1825-37. mf 2t-5t: births & deaths 1737-1848. mf 5t-6t: marriages 1835-42, 1849. mf 5t: intentions 1838-49; deaths 1833-37. mf 6t: births 1837-49; out-of-town marriages 1790-98; deaths 1844-49) – us Archive [978]
Heath 1763-1899 – Oxford, MA (mf ed 1987) – (= ser Massachusetts vital records) – 22mf – 9 – 0-87623-049-4 – (mf 1-6: b,m,d 1763-1844. mf 7-8: births 1844-99. mf 8-9: marriages 1790-1900. mf 9-10: deaths 1844-1900. mf 11: birth index 1844-1900. mf 12: marriage intentions index 1844-1900. mf 13: marriage index 1844-1900. mf 14: death index 1844-1900. mf 15-19: church records 1785-1889. mf 20-21: church records 1805-56. mf 22: soldiers 1863; overseer of poor 1901-04) – us Archive [978]
Heath, Alan *see* Confrontation
Heath and reach, st leonard monumental inscriptions monumental inscriptions – Bedfordshire Family HS 1986 – (= ser Bedfordshire parish register series) – 1mf – 9 – £1.25 – uk BedsFHS [929]
Heath, D I *see* Fallen angels
Heath, George *see* The new bristol guide
Heath, J St George et al *see* Christ and peace
Heath, Peter S *see* An individualized self-control approach to weight reduction
Heath, Richard *see*
- Anabaptism
- Anabaptism from its rise at zwickau to its fall at munster
- The captive city of god
- Edgar quinet
- The reformation in france
- The reformation in france from the dawn of reform to the revocation of the edict of nantes

Heath, Sidney *see* Pilgrim life in the middle ages
Heath springs baptist church. heath springs, south carolina : church records – 1889-1942. Scattered records, 1932-42 – 1 – us Southern Baptist [242]
Heath, Thomas *see* The twentieth century atlas of popular astronomy
Heath, William *see* The william heath papers, 1774-1872
Heathcote, Charles William *see* The lutheran church and the civil war
Heathen ceremonies adopted by the church of rome / Hamilton, George – London, England. 18– – 1r – us UF Libraries [240]
The heathen heart : an account of the reception of the gospel among the chinese of formosa / Moody, Campbell N – Edinburgh: Oliphant, Anderson & Ferrier, 1907 [mf ed 1990] – 1mf – 9 – 0-7905-5720-7 – mf#1988-1720 – us ATLA [240]
Heathen helpers – 1882-89; The Baptist Basket. 1888 (also Southern Baptist Convention Directory, May 1880) – 1 – us Southern Baptist [242]
Heathen records to the jewish scripture history : containing all the extracts from the greek and latin writers, in which the jews and christians are named / Giles, John Allen – London: James Cornish, 1856 – 1mf – 9 – 0-7905-1048-0 – (texts in english, greek and latin; commentary in english) – mf#1987-1048 – us ATLA [939]
Heathen science monitor – Minneapolis MN. 1982 mar 4-1988 mar – 1r – 1 – mf#3230202 – us WHS [071]
Heathen scotland to the introduction of christianity / Lees, James Cameron – s.l, s.l? 18– – 1r – us UF Libraries [240]
Heathens in britain – London, England. 18– – 1r – us UF Libraries [240]
Heatherington, Alexander *see* A practical guide for tourists, miners, and investors
Heath's book of beauty – 1833-49 – (= ser English gift books and literary annuals, 1823-1857) – 68mf – 9 – uk Chadwyck [800]
Heath's french and english dictionary / Lolme, J L De – Boston, MA. 1903 – 1r – us UF Libraries [025]
Heath-Stubbs, J *see* The water wheel of love
Heating, piping, air conditioning – Cleveland. 1929-1999 (1) 1965-1999 (5) 1976-1999 (9) – (cont by: heating/piping/air conditioning engineering : hpac) – ISSN: 0017-940X – mf#281 – us UMI ProQuest [690]
Heating/piping/air conditioning engineering (hpac) – Cleveland. 1999+ (1) 1999+ (5) 1999+ (9) – (cont: heating, piping, air conditioning) – ISSN: 1527-4055 – mf#281,01 – us UMI ProQuest [690]
Heaton, Herbert *see* The letter books of joseph holroyd (cloth-factor) and sam hill (clothier)
Heaton, John Aldam *see*
- Beauty and art
- A record of work

Heaton, John Henniker *see* Australian dictionary of dates and men of the time
Heaton, Mary Margaret (Keymer) *see* The history of the life of albert duerer of nuernberg
Heaton, Mary Margaret [Keymer] et al *see* A concise history of painting
Heaton mersey, church of the upper room – (= ser Cheshire monumental inscriptions) – 1mf – 9 – £2.50 – mf#20a – uk CheshireFHS [929]
Heaton moor and heaton chapel guardian – [NW England] Stockport 1973-jun 1984 – 1 – uk MLA; uk Newsplan [072]
Heaton norris, christ church – [North Cheshire FHS] – (= ser Cheshire monumental inscriptions) – 3mf – 9 – £4.00 – mf#123 – uk CheshireFHS [929]
Heaton, Sydney Lewell *see* Ontario divorce law notebook
Heaton, William James *see*
- The bible of the reformation
- Our own english bible

Heatwole, Cornelius Jacob *see* A history of education in virginia
Heaven and its wonders and hell / Swedenborg, Emanuel – Boston, MA. 1906 – 1r – us UF Libraries [960]
Heaven on the sea / Ish-Kishor, Sulamith – New York, NY. 1924 – 1r – us UF Libraries [939]
Heaven opened : expositions of the book of revelation / Simpson, Albert B – New York: Alliance Press Co, c1899 [mf ed 1992] – (= ser Christ in the bible 24; Christian and missionary alliance coll) – 1mf – 9 – 0-524-02152-X – mf#1990-4218 – us ATLA [225]
Heavener first baptist church. heavener, oklahoma : church records – 1912-80 – 1 – 73.84 – us Southern Baptist [242]
Heavenly bridegroom's desire for his bride / Ferguson, Archibald – Aberdeen, Scotland. 1868 – 1r – us UF Libraries [240]
The heavenly home : or, the employments and enjoyments of the saints in heaven / Harbaugh, Henry – Philadelphia: Lindsay & Blakiston, 1853 – 1mf – 9 – 0-7905-0371-9 – mf#1987-0371 – us ATLA [240]
Heavenly recognition : discourses on personal immortality and identity after this life / McWhinney, Thomas Martin – New York: Fords, Howard & Hulbert, c1883 – 1mf – 9 – 0-7905-8708-4 – mf#1989-1933 – us ATLA [240]
The heavenly session of our lord : an introduction to the history of the doctrine / Tait, Arthur James – London: R Scott, 1912 – 1mf – 9 – 0-7905-8927-3 – (incl bibl ref) – mf#1989-2152 – us ATLA [240]
The heavenly token : a gift book for christians / Harsha, David Addison – New York: Dayton and Burdick, 1857. Beltsville, Md: NCR Corp, 1977 (6mf); Evanston: American Theol Lib Assoc, 1984 (6mf) – 9 – 0-8370-0140-4 – (incl bibl ref) – mf#1984-0027 – us ATLA [240]
The heavenly vision and other sermons : (1863-73) / Cochrane, William – Toronto: Adam, Stevenson, 1874 – 5mf – 9 – mf#08277 – cn Canadiana [242]
Heavenly wind = Tien feng – n83-631. 1947-63 [complete] – (= ser Chinese christian coll 39) – 5r – 1 – (in chinese) – mf#ATLA S0317 – us ATLA [240]
Heaven's antidote to the curse of labour / Quinton, John Allan – London, England. 1849 – 1r – us UF Libraries [240]
Heaven's distant lamps : poems of comfort and hope / Mack, Anna E – Boston: Lee & Shepard, 1900 [mf ed 1985] – 1mf – 9 – 0-8370-3818-9 – (incl ind) – mf#1985-1818 – us ATLA [810]
Heavy construction news – Toronto. 1975-1996 (1,5,9) – ISSN: 0017-9426 – mf#10781 – us UMI ProQuest [690]
Heavysege, Charles *see*
- The advocate
- The dark huntsman (a dream)

Hebaeische grammatik *see* Grammaire hebraique
Hebbe, Johann Gustav *see* Johann gustav hebbe's schwedischen seeoffiziers nachrichten von den azorischen inseln besonders von der insel fayal
Hebbel : das drama an der wende der zeit / Scholz, Wilhelm von – 3. Aufl. Stuttgart, Berlin: Deutsche Verlags-Anstalt, c1922 – 1r – 1 – us Wisconsin U Libr [430]
Hebbel / Wehner, Josef Magnus – Stuttgart: J G Cotta, 1938 – 1r – 1 – us Wisconsin U Libr [430]
Hebbel als dichter der frau / Engel-Mitscherlich, Hilde – Dresden: W Baensch, 1909 – 1r – 1 – us Wisconsin U Libr [430]
Hebbel als lyriker / Moeller, Hans – [S.l.: s.n.], 1908 (Cuxhaven: Gedruckt bei C Rauschenplat) – 1r – 1 – (incl bibl ref) – us Wisconsin U Libr [430]
Hebbel als novellist / Ebhardt, Rolf – Berlin: Weidmann, 1916 – 1r – 1 – (incl bibl ref) – us Wisconsin U Libr [430]
Hebbel and the dream / Schueler, Herbert – New York: [s.n.], 1941 – 1r – 1 – (incl bibl ref) – us Wisconsin U Libr [430]
Hebbel, Friedrich *see*
- Agnes bernauer
- Der diamant
- Ernst freiherrn von feuchtersleben's saemmtliche werke
- Erzaehlungen und novellen
- Friedrich hebbels demetrius
- Friedrich hebbels tagebuecher
- Gedichte
- Genoveva
- Gyges und sein ring
- Hebbel prosista
- Hebbels ausgewaehlte werke
- Hebbels werke
- Judith
- Julia
- Neue gedichte
- Die nibelungen
- Poems
- Three plays
- Ein trauerspiel in sicilien

Hebbel, Ibsen and the analytic exposition / Campbell, Thomas Moody – Heidelberg: C Winter, 1922 [mf ed 1995] – 96p – 1 – mf#8764 – us Wisconsin U Libr [410]
Hebbel in der zeitgenoessischen kritik / ed by Wuetschke, H – Berlin: B Behr, 1910 [mf ed 1993] – (= ser Deutsche litteraturdenkmale des 18. und 19. jahrhunderts 143, 3. folge n23) – vi/273p – 1 – (incl bibl ref and ind) – mf#8676 reel 9 – us Wisconsin U Libr [430]
Hebbel prosista : autobiografia, ideario, del drama / Hebbel, Friedrich; ed by Icaza, Francisco A de – Madrid: [Impr. de J. Pueyo], 1919 – 1r – 1 – us Wisconsin U Libr [830]
Hebbel und das religioese problem der gegenwart / Horneffer, Ernst – Jena: E Diederichs, 1907 [mf ed 1990] – 64p – 1 – mf#7452 – us Wisconsin U Libr [430]
Hebbel und das wiener theater seiner zeit / Kindermann, Heinz – Wien: W Frick, 1943 – 1r – 1 – us Wisconsin U Libr [430]
Hebbel und die musik / Nagler, Alois Maria – Koeln: J P Bachem, 1928 – 1r – 1 – (incl bibl ref) – us Wisconsin U Libr [430]
Hebbel und die philosophie seiner zeit / Waetzoldt, Wilhelm – [S.l.: s.n.], 1903 (Graefenhainichen: Druck von W Hecker) – 1r – (incl bibl ref) – us Wisconsin U Libr [430]
Hebbelprobleme : studien / Walzel, Oskar Franz – Leipzig: H Haessel, 1909 – 1r – 1 – us Wisconsin U Libr [430]
Hebbels ausgewaehlte werke / ed by Specht, Richard – Stuttgart: J G Cotta'sche Buchhandlung Nachfolger, [1903?] [mf ed 1995] – (= ser Cotta'sche bibliothek der weltlitteratur) – 6v – 1 – (incl bibl ref) – mf#8752 – us Wisconsin U Libr [800]
Hebbels dithmarschenfragment : anordnung / Bender, Heinrich – Bonn: P Rost, 1914 [mf ed 1990] – 111p – 1 – (incl bibl ref) – mf#7452 – us Wisconsin U Libr [430]
Hebbels frauengestalten / Kreisler, Emil – Wien: Verlag der k. k. Franz Joseph-Realschule, 1907 – 1r – 1 – us Wisconsin U Libr [430]
Hebbels herkunft und andere hebbel-fragen / Bartels, Adolf – Berlin, Leipzig: B Behr (F Feddersen), 1921 [mf ed 2001] – (= ser Hebbel-forschungen 9) – 126p – 1 – (incl ind) – mf#10627 – us Wisconsin U Libr [929]

Hebbels herodes und mariamne : vortrag / Bornstein, Paul – Hamburg: L Voss 1904 [mf ed 1990] – 1r – 1 – (filmed with: erzahlungen und novellen / friedrich hebbel) – mf#2703p – us Wisconsin U Libr [080]

Hebbels judith und maria magdalena im urteil seiner zeitgenossen / Beer, Oskar – [S.l: s.n. 19–?] (Naumburg a.S: Druck von G Paetz) [mf ed 1990] – 99p – 1 – (incl bibl ref) – mf#7449 – us Wisconsin U Libr [430]

Hebbels lyrik und epik im rahmen seines lebens : eine anregung zu hebbelstudien und hebbelstunden / Schnass, Franz – Prag: A Haase, 1921 – 1r – 1 – (incl bibl ref) – us Wisconsin U Libr [430]

Hebbels theorie und kritik poetischer muster : mit besonderer ruecksicht auf die entwicklung seiner lyrik unter uhlands einfluss / Herke, Karl – Berlin: H Lonys, 1913 – 1r – 1 – (incl bibl ref) – us Wisconsin U Libr [430]

Hebbels werke / Hebbel, Friedrich; ed by Zeiss, Karl – Leipzig: Bibliographisches Institut. 4v. [1899?] – 1 – (incl bibl ref) – us Wisconsin U Libr [800]

Hebbels werke / ed by Poppe, Theodor – Berlin: Deutsches Verlagshaus Bong [1908] [mf ed 1995] – (= ser Goldene klassiker-bibliothek) – 10v in 5 – 1 – (incl bibl ref and ind. int and ann by ed) – mf#8746 – us Wisconsin U Libr [802]

Der hebbelverein in heidelberg : die geschichte einer literarischen gesellschaft – ein rueckblick auf seine taetigkeit von 1902-1908 / Stahl, Ernst Leopold – Heidelberg: C Winter 1911 [mf ed 1990] – 1r – 1 – (filmed with: ethik und mystik in hebbels weltanschauung / ernst lahnstein) – mf#2705p – us Wisconsin U Libr [790]

Hebberd, Stephen Southric *see* The secret of christianity

Hebden, John *see* Six concertos in seven parts, for four violins, a tenor violin, a violoncello

L'hebdomadaire du temps nouveau – Lyon. dec 1940-aout 1941, sept 1944 – 1 – (dirige par les rr.pp. dominicains) – fr ACRPP [073]

Hebdomadaire du temps present – Paris, France. 22 sep 1944-6 sep 1946 – 1r – 1 – uk British Libr Newspaper [072]

Hebdomicile – Hull: F Laferriere. 29 avril 1986-(wkly) [mf ed 1988] – 1r – 1 – (ceased: avril 1988?) – mf#SEM35P296 – cn Bibl Nat [071]

Hebdo-sports – Port-au-Prince: Federation de football. v1 n1-8,10-15,17-22,24-28 jan- feb, mar 11-apr 11, apr 25-jun 6, jun 20-jul 1951; v2 n1-28 feb-aug 1952; v3 n1-7 nov 6-dec 19/26 1952; v2 n8-26 jan-may 1953 – 11r – 1 – us CRL [079]

Hebel, Johann Peter *see*
- Allemannische gedichte
- Johann peter hebel
- Johann peter hebels ausgewaehlte erzaehlungen u gedichte
- Schatzkaestlein des rheinischen hausfreundes

Hebel und kleist als meister der anekdote / Staehlin, Friedrich – Berlin: M Matthiesen, 1940 – 1 – (incl bibl ref) – us Wisconsin U Libr [430]

Heber, R *see* Narrative of a journey through the upper provinces of india

Heber, Reginald *see*
- Farewell sermon preached in the parish church of hodnet
- Narrative of a journey through the upper provinces of india
- Reise durch die obern provinzen von vorderindien
- Sermon on matthew 9, 38

Heber, Reginald, bishop of Calcutta *see* Narrative of a journey through the upper provinces of india

Heberci – Stockholm, SW. Isvicdegi Dini-Medini Tuerk-Islam Oyusmasi Karsisinda Vakitli Reviste Cigarilaridir. n1. dec 1952 – (= ser O & t journals) – 1mf – 9 – $25.00 – us MEDOC [956]

Hebert, Casimir *see* La vieille maison denis de neuville, quebec

Hebert, Charles *see*
- The lord's supper
- On clerical subscription

Hebert, Marcel *see* L'evolution de la foi catholique

Hebert-Duperron, Victor *see* Essai sur la polemique et la philosophie de saint clement d'alexandrie

Hebra en la aguja / Torres-Rioseco, Arturo – Mexico City? Mexico. 1965 – 1r – us UF Libraries [972]

Der hebraeerbrief / Bleek, Friedrich – Elberfeld: R L Friedrichs, 1868 – 2mf – 9 – 0-8370-9605-7 – mf#1986-3605 – us ATLA [227]

Der hebraeerbrief / Nikel, Johannes – 1. & 2. aufl. Muenster i W: Aschendorff 1914 [mf ed 1992] – (= ser Biblische zeitfragen 7/6) – 1mf – 9 – 0-524-03985-2 – (incl bibl ref) – mf#1992-0028 – us ATLA [225]

Der hebraeerbrief / Windisch, Hans – Tuebingen: J C B Mohr, 1913 – (= ser Handbuch zum neuen testament) – 1mf – 9 – 0-7905-2098-2 – mf#1987-2098 – us ATLA [227]

Der hebraeerbrief in zeitgeschichtlicher beleuchtung / Weiss, Bernhard – Leipzig, 1910 – (= ser Tugal 3-35/3) – 2mf – 9 – €5.00 – ne Slangenburg [227]

Der hebraeerbrief in zeitgeschichtlicher beleuchtung / Weiss, Bernhard – Leipzig: J C Hinrichs, 1910 – (= ser Tugal) – 1mf – 9 – 0-7905-1738-8 – mf#1987-1738 – us ATLA [227]

Das hebraeer-evangelium : ein beitrag zur geschichte und kritik des hebraeischen matthaeus / Handmann, Rudolf – Leipzig: J C Hinrichs, 1888 – (= ser Tugal) – 1mf – 9 – 0-7905-1761-2 – (incl bibl ref) – mf#1987-1761 – us ATLA [220]

Die hebraeerin am putztische und als braut / Hartmann, A T – Amsterdam, 1809-1810. 3v – 14mf – 9 – mf#J-412-10 – ne IDC [700]

Hebraeisch-deutsches handwoerterbuch ueber die schriften des alten testaments : mit einschluss der geographischen nahmen und der chaldaeischen woerter beym daniel und esra / Gesenius, Wilhelm – Leipzig: FCW Vogel, 1810-1812 – 14mf – 9 – 0-524-03880-5 – mf#1987-6493 – us ATLA [040]

Hebraeische archaeologie / Benzinger, I – Freiburg i. B.: Mohr, 1894 – (= ser Grundriss der theologischen wissenschaften) – 2mf – 9 – 0-7905-3304-9 – (incl bibl ref) – mf#1987-3304 – us ATLA [930]

Hebraeische bibliographie : blaetter fuer neuere und aeltere literatur des judenthums – Berlin DE, 1858-80 – 3r – 1 – us UMI ProQuest [470]

Hebraeische bibliographie : blaetter fuer neuere und aeltere literatur des judentums / ed by Steinschneider, Moritz – Berlin: A Asher & Co. v1-21. 1858-65; 1869-82 – (= ser German-jewish periodicals...1768-1945, pt 3) – 3r – 1 – $480.00 – (suspended 1866-68) – mf#B93 – us UPA [470]

Hebraeische elementargrammatik : eine zur einfuehrung in das studium der grammatischen werke ewalds und boettchers bestimmte vorschule / Grundt, Friedrich Immanuel – Leipzig: Ferdinand Hirt, 1875 – 1mf – 9 – 0-8370-9238-8 – mf#1986-3238 – us ATLA [470]

Die hebraeische elias-apokalypse : und ihre stellung in der apokalyptischen literatur des rabbinischen schrifttums und der kirche / Buttenwieser, Moses – Leipzig: Eduard Pfeiffer 1897 [mf ed 1985] – 1mf – 9 – 0-8370-2560-5 – (no more publ) – mf#1985-0560 – us ATLA [270]

Hebraeische grammatik : mit paradigmen, literatur, uebungsstuecken und woerterverzeichnis / Steuernagel, Carl – 3. & 4. verb aufl. Berlin: Reuther & Reichard; New York: Lemcke & Buechner 1909 [mf ed 1986] – (= ser Porta linguarum orientalium 1) – 1mf – 9 – 0-8370-9185-3 – (replaces ed by hermann strack) – mf#1986-3185 – us ATLA [470]

Hebraeische grammatik : mit uebungsstuecken, litteratur und vokabular / Strack, Hermann Leberecht – 2. wesentlich verm & verb aufl. Karlsruhe: H Reuther; New York: B Westermann 1885 [mf ed 1986] – (= ser Porta linguarum orientalium 1) – 1mf – 9 – 0-8370-9274-4 – mf#1986-3274 – us ATLA [470]

Hebraeische grammatik fuer den unterricht mit uebungsstuecken und woerterverzeichnissen / Koenig, Eduard – Leipzig: J C Hinrichs, 1908 – 1mf – 9 – 0-8370-9482-8 – mf#1986-3482 – us ATLA [470]

Die hebraeische praeposition lamed / Giesebrecht, Friedrich – Halle (Saale): Max Niemeyer, 1876 – 1mf – 9 – 0-8370-9147-0 – (incl bibl ref) – mf#1986-3147 – us ATLA [470]

Hebraeische schriftgestaltung in deutschland von der jahrhundertwende bis zum ausbruch des zweiten weltkrieges... / Tamari, Ittai Joseph – (mf ed 1995) – (= ser Monographien zur wissenschaft des judentums) – 7mf – 9 – €65.00 – 3-8267-2262-0 – mf#DHS 20001 – gw Frankfurter [470]

Hebraeische sprachlehre fuer anfaenger *see* Ewald's introductory hebrew grammar

Die hebraeische sprachwissenschaft vom 10. bis zum 16. jahrhundert : mit einem einleitenden abschnitte ueber die massora jahrhundert / Bacher, Wilhelm – Trier: Sigmund Mayer, 1892 – 1mf – 9 – 0-8370-2537-0 – (includes bibliographies & index of authors) – mf#1985-0537 – us ATLA [470]

Hebraeische und chaldaeische abbreviaturen : welche in dem talmudischen schriftthume und in werken der hebraeischen litteratur vorkommen / Lederer, Ph – Frankfurt a. M.: Commissions-Verlag bei J. Kauffmann, 1894 – 1mf – 9 – 0-8370-8194-7 – mf#1986-2194 – us ATLA [470]

Die hebraeische verskunst : nach dem metek s'efatajim des 'immanur'el fransis und anderen werken juedischer metriker / Hartmann, Martin – Berlin: S Calvary, 1894 – 1mf – 9 – 0-8370-3512-0 – (incl ind) – mf#1985-1512 – us ATLA [470]

Hebraeische volkskunde / Kuechler, Friedrich – Tuebingen: J C B Mohr, 1906 – (= ser Religionsgeschichtliche Volksbuecher fuer die deutsche christliche Gegenwart) – 1mf – 9 – 0-7905-3350-2 – mf#1987-3350 – us ATLA [939]

Die hebraeischen alterthuemer in briefen / Roskoff, Georg Gustav – Wien: W Braumueller, 1857 [mf ed 1989] – 1mf – 9 – 0-7905-2801-0 – mf#1987-2801 – us ATLA [939]

Die hebraeischen synonyma der zeit und ewigkeit / Orelli, Conrad von – Leipzig: A Lorentz, 1871 – 1mf – 9 – 0-7905-1016-2 – (incl bibl ref) – mf#1987-1016 – us ATLA [470]

Die hebraeischen worterklaerungen des philo und die spuren ihrer einwirkung auf die kirchenvaeter / Siegfried, Carl – Magdeburg: E Baensch, Jun.; Berolinenses [Berlin]: Prostant apud S. Calvary, 1863 – 1mf – 9 – 0-7905-2073-7 – mf#1987-2073 – us ATLA [470]

Hebraeisches familienrecht in vorprophetischer zeit / Rauh, Sigismund – Berlin: Gustav Schade, 1907 – 1mf – 9 – 0-8370-4844-3 – mf#1985-2844 – us ATLA [221]

Hebraeisches lesebuch fuer anfaenger und geuebtere : mit einem grammatischen cursus und glossarium / Brueckner, Gustav – 3. verm. und theilweise umgearb. Aufl. Leipzig: F.C.W. Vogel, 1863 – 1mf – 9 – 0-8370-9207-8 – mf#1986-3207 – us ATLA [470]

Hebraeisches schulbuch / Hollenberg, Wilhelm Adolf – 4. Aufl. Berlin: Weidmann, 1880 – 1mf – 9 – 0-8370-9248-5 – mf#1986-3248 – us ATLA [470]

Hebraeisches und chaldaeisches schul-woerterbuch ueber das alte testament / Fuerst, Julius – Leipzig: O Holtzes Nachfolger, 1894 [mf ed 1991] – 2mf – 9 – 0-7905-8301-1 – mf#1987-6406 – us ATLA [221]

Hebraeisches wurzelwoerterbuch : nebst drei anhaengen ueber die bildung der quadrilitern, erklaerung der fremdwoerter im hebraeischen, und ueber das verhaeltniss des aegyptischen sprachstammes zum semitischen / Meier, Ernst Heinrich – Mannheim: F Bassermann, 1845 – 2mf – 9 – 0-7905-2985-8 – mf#1987-2985 – us ATLA [040]

Das hebraer-evangelium / Handmann, R – Leipzig, 1888 – (= ser Tugal 1-5/3) – 3mf – 9 – €7.00 – ne Slangenburg [221]

Hebraic and yiddish catalog / U.S. Library of Congress – 39,000 entries – 1 – us L of C Photodup [010]

Hebraic literature : translations from the talmud, midrashim and kabbala – New York: M Walter Dunne, c1901 – (= ser Universal Classics Library) – 1mf – 9 – 0-524-02306-9 – mf#1990-2929 – us ATLA [470]

Hebraica – Chicago. v1-11. 1884-1895. v1-11 – 343mf – 8 – (cont as: american journal of semitic languages and literature, chicago 1895-1941 v12-58. missing: 1908 v25) – mf#H-630 – ne IDC [071]

Hebraica *see* The american journal of semitic languages and literatures

Hebraica and judaica of the tychsen collection and the rostock university library *see* Die hebraica und judaica der sammlung tychsen und der universitaetsbibliothek rostock

Die hebraica und judaica der sammlung tychsen und der universitaetsbibliothek rostock : die altjiddische (juedisch-deutsche) literatur = Hebraica and judaica of the tychsen collection and the rostock university library / Suess, Hermann & Troeger, Heike [comp]; ed by Universitaetsbibliothek Rostock – [mf ed 2001] – 600mf – 9 – €3900 diazo €4680 – 3-89131-377-2 – (with printed catalogue; titles are also listed separately) – gw Fischer [939]

Hebraicum psalterium / ed by Pellican, C – Basel, 1516 – 5mf – 9 – mf#PBU-567 – ne IDC [240]

Hebraische grammatik mit ubungsbuch / Strack, Hermann Leberecht – Berlin, Germany. 1896 – 1r – us UF Libraries [470]

Die hebraischen conditionalsaetze / Friedrich, Paul – Halle a S: Ehrhardt Karras, 1884 – 1mf – 9 – 0-524-06835-6 – (incl bibl ref) – mf#1992-0977 – us ATLA [470]

Hebraisms in the authorized version of the bible / Rosenau, William – Baltimore MD: Friedenwald 1903, c1902 [mf ed 1989] – 1mf – 9 – 0-7905-3279-4 – (incl bibl ref) – mf#1987-3279 – us ATLA [221]

Hebraisms in the greek testament : exhibited and illustrated by notes and extracts from the sacred text / Guillemard, William Henry – Cambridge: Deighton, Bell; London: George Bell, 1879 – 1mf – 9 – 0-7905-1664-0 – (includes t.p. and pref. from 1875 ed., issued under title: the greek testament, hebraistic edition; incl ind) – mf#1987-1664 – us ATLA [221]

Hebrew – (San Francisco). 1864-87 – 1 – us AJPC [071]

The hebrew – New York. N.Y. 1892-1902 – 1 – us AJPC [071]

The hebrew american – New York. N.Y. 1894-95 – 1 – us AJPC [071]

Hebrew and babylonian traditions : the haskell lectures / Jastrow, Morris – New York: Charles Scribner 1914 [mf ed 1989] – 1mf – 9 – 0-7905-2115-6 – (incl bibl ref & ind) – mf#1987-2115 – us ATLA [270]

A hebrew and english lexicon of the old testament : including the biblical chaldee = Lexicon manuale hebraicum et chaldaicum in veteris testamenti libros / Gesenius, Wilhelm; ed by Robinson, Edward – 20th rev ed. Boston: Crocker & Brewster, 1866, c1854 [mf ed 1991] – 12mf – 9 – 0-7905-8302-X – (trans into english by ed) – mf#1987-6407 – us ATLA [470]

Hebrew and judeo-arabic mss in the collections of the ussr / Katsh, A I – M, 1962 – (= ser Trudy dvadtsat'-piatogo Mezhdunarodnogo kongressa vostokovedov) – 1mf – 9 – (trudy dvadtsat'-piatogo mezhdunarodnogo kongressa vostokovedov, moskva 9-16 avgusta 1960 g. v1) – mf#R-10667 – ne IDC [956]

Hebrew and talmudical exercitations upon the gospels, the acts, some chapters of st paul's epistle to the romans, and the first epistle to the corinthians = Horae hebraicae et talmudicae / Lightfoot, John – new ed. Oxford: University Press, 1859 – 4mf – 9 – 0-8370-1365-8 – (incl bibl ref) – mf#1987-6055 – us ATLA [221]

Hebrew beginnings : old testament narratives, part 1 / Stebbins, Edna Hodgkins – Boston: American Unitarian Association, c1909 – (= ser Beacon series (boston, mass.)) – 1mf – 9 – 0-524-07065-2 – mf#1992-1028 – us ATLA [221]

The hebrew bible : revised and carefully examined by myer levi letteris – New York: J Wiley, 1889 [mf ed 2004] – 1r – 1 – 0-524-10485-9 – (with key to massoretic notes etc trans fr the latin of august hahn, with many additions & corr by a meyrowitz) – mf#b00700 – us ATLA [221]

Hebrew book review – Tel-Aviv. 1965-1974 (1) – ISSN: 0017-9469 – mf#7948 – us UMI ProQuest [470]

Hebrew books from the harvard college library / ed by Berlin, Charles – [mf ed 1990-92] – 11,448mf (1:24) – 9 – diazo €28,848.00 (silver €31,580 isbn: 978-3-598-41200-4) – ISBN-10: 3-598-41160-X – ISBN-13: 978-3-598-41160-1 – (pt 1: rabbinical works 8647mf (1:24); pt 2: secular works 2806mf; incl printed ind) – gw Saur [939]

Hebrew books of the fifteenth century – 645mf – 9 – (from incunabula: the printing revolution in europe, 1455-1500. based on the incunabula short title catalogue (istc) at the british library. a comprehensive microfiche coll of hebrew incunabula. item displayed as a full facsimile image showing original text, images and printer's typographical layout) – us Primary [090]

Hebrew Butcher Workers' Union *see* Butcher worker

Hebrew characteristics : miscellaneous papers: from the german – New York: American Jewish Publ Society, 1875 – 1mf – 9 – 0-8370-3542-2 – mf#1985-1542 – us ATLA [939]

Hebrew charts : containing the elements of the language / Irish, William Norman – Albany, NY: Weed, Parsons, 1872, c1871 – 1mf – 9 – 0-8370-9289-2 – mf#1986-3289 – us ATLA [470]

A hebrew chrestomathy – Andover: Flagg & Gould, 1829 [mf ed 1989] – 1mf – 9 – 0-7905-3235-2 – mf#1987-3235 – us ATLA [470]

A hebrew chrestomathy : or, lessons in reading and writing hebrew / Green, William Henry – New York: John Wiley, 1866, c1863 [mf ed 1986] – 1mf – 9 – 0-8370-9473-9 – (notes in english, text in hebrew) – mf#1986-3473 – us ATLA [470]

Hebrew chrestomathy *see* Course of hebrew study adapted to the use of beginners

Hebrew Christian Mission *see* Annual report of the hebrew christian mission

Hebrew fair journal – Washington, D.C. 1886-96 – 1 – us AJPC [071]

Hebrew for self-instruction : by which, with three months' study, the student will be able to read the hebrew bible and understand its grammatical structure / Wheeler, H M – London: Simpkin, Marshall, 1850 – 1mf – 9 – 0-8370-9347-3 – mf#1986-3347 – us ATLA [470]

A hebrew grammar / Lowe, William Henry – London: Hodder & Stoughton, 1887 [mf ed 1986] – (= ser The theological educator) – 1mf – 9 – 0-8370-9166-7 – mf#1986-3166 – us ATLA [470]

Hebrew grammar / Gesenius, H – 1910 – 9 – $21.00 – us IRC [470]

Hebrew Gymnasiumn (Berlin, Germany) *see* Akhsanya shel torah

Hebrew history : old testament narratives, pt 2 / Saunderson, Henry Hallam – Boston: Unitarian Sunday-School Society, c1909 – (= ser Beacon series (boston, mass.)) – 1mf – 9 – 0-524-07063-6 – mf#1992-1026 – us ATLA [221]

Hebrew history from the death of moses to the close of the scripture narrative / Cowles, Henry – New York: D Appleton, 1875, c1874 – 1mf – 9 – 0-7905-0932-6 – mf#1987-0932 – us ATLA [270]
Hebrew inscriptions, from the valleys between egypt and mount sinai : in their original characters, with translations and an alphabet / Sharpe, Samuel – London: John Russell Smith, 1875 – 1mf – 9 – 0-8370-8620-5 – (incl ind to hebrew and aramaic words) – mf#1986-2620 – us ATLA [470]
The hebrew language : its history and characteristics, including improved renderings of select passages in our authorized translation of the old testament / Craik, Henry – London: Bagster, 1860 – 1mf – 9 – 0-7905-1036-7 – mf#1987-1036 – us ATLA [470]
The hebrew language viewed in the light of assyrian research / Delitzsch, Frederic – London: Williams & Norgate, 1883 [mf ed 1989] – 1mf – 9 – 0-7905-0643-2 – (incl bibl ref and ind) – mf#1987-0643 – us ATLA [470]
Hebrew leader – New York. N.Y. 1867-70 – 1 – us AJPC [071]
Hebrew life and thought : being interpretative studies in the literature of israel / Houghton, Louise Seymour – Chicago: University of Chicago, 1906 – 1mf – 9 – 0-7905-1111-8 – (incl bibl ref and index) – mf#1987-1111 – us ATLA [939]
Hebrew literature : comprising talmudic treatises, hebrew melodies and the kabbalah unveiled – Rev. ed. New York: Colonial Press, c1901 – (= ser The World's Great Classics) – 1mf – 9 – 0-8370-3543-0 – mf#1985-1543 – us ATLA [470]
The hebrew literature of wisdom in the light of to-day : a synthesis / Genung, John Franklin – Boston: Houghton Mifflin, 1906 – 1mf – 9 – 0-8370-9780-0 – mf#1986-3780 – us ATLA [470]
The hebrew literature of wisdom in the light of to-day : a synthesis / Genung, John Franklin – Boston, New York: Houghton, Mifflin and Company, 1906. xviii,408p – 1 – us Wisconsin U Libr [939]
Hebrew men and times : from the patriarchs to the messiah / Allen, Joseph Henry – Boston: Walker, Wise; London: Chapman and Hall, 1861. Beltsville, Md: NCR Corp, 1978 (5mf); Evanston: American Theol Lib Assoc, 1984 (5mf) – 9 – 0-8370-1079-9 – (incl bibl ref and index) – mf#1984-4433 – us ATLA [270]
The hebrew messenger : an illustrated quarterly devoted to missionary work among the jews / ed by McFeeters, James Calvin – Philadelphia: Mission of the Covenant to Israel 1901-08 [qrtly] [mf ed 2005] – 8v on 1r – 1 – mf#2005c-s044 – us ATLA [242]
Hebrew observer – London. 1853-54.-w. 1men reels – 1 – uk British Libr Newspaper [290]
Hebrew observer – Cleveland, [Ohio: s.n.], 1889- – 1r – 1 – (weekly jewish newspaper. merged with: the jewish review, to form: the jewish review and observer) – us Western Res [071]
Hebrew observer *see* The jewish review and observer
The hebrew observer – Grand Rapids. Mich. 1942-55 – 1 – us AJPC [071]
The hebrew, or, the church with her surroundings : as they together appeared, the third of a century ago, to a subordinate official – New York: Holman, 1863 – 1mf – 9 – 0-8370-4512-6 – (includes appendix) – mf#1985-2512 – us ATLA [270]
The hebrew particle asher / Gaenssle, Carl – Chicago, IL: University of Chicago Press, 1915 – 1mf – 9 – 0-7905-3250-6 – (incl bibl ref) – mf#1987-3250 – us ATLA [470]
The hebrew people : or, the history and religion of the israelites from the origin of the nation to the time of christ / Smith, George – New York: Carlton & Porter, 1856 [mf ed 1992] – (= ser Sacred annals 2) – 2mf – 9 – 0-524-03994-1 – mf#1992-0037 – us ATLA [221]
The hebrew personification of wisdom : its origin, development and influence / Hesselgrave, Charles Everett – New York University, 1909 – 1mf – 9 – 0-8370-9702-9 – mf#1986-3702 – us ATLA [270]
Hebrew poetry : sunday afternoon lectures before the greensboro law school / Dick, Robert Paine – Greensboro: CF Thomas, 1883 – 1mf – 9 – 0-8370-2904-X – mf#1985-0904 – us ATLA [470]
Hebrew prophecy / Milman, Henry Hart – Oxford, England. 1865 – 1r – us UF Libraries [240]
The hebrew prophets / Ottley, Robert L – 3rd ed. New York: Edwin S Gorham, 1905 – (= ser Oxford church text books) – 1mf – 9 – 0-8370-4646-7 – (incl bibl ref and ind) – mf#1985-2646 – us ATLA [221]
The hebrew puck – (New York). 1894-96 – 1 – us AJPC [071]
Hebrew reader and grammar with exercises for translation : for the use of schools / Mannheimer, Sigmund – 2nd ed. St Louis, MO: F Roeslein, 1875 – 1mf – 9 – 0-524-05989-6 – mf#1992-0726 – us ATLA [470]
Hebrew reading lessons : consisting of the first four chapters of the book of genesis and the eighth chapter of the proverbs, with a grammatical praxis and an interlinear translation / Tregelles, Samuel Prideaux – 5th ed London: Samuel Bagster, [185-?] – 1mf – 9 – 0-8370-1859-5 – mf#1987-6246 – us ATLA [221]
Hebrew religion to the establishment of judaism under ezra / Addis, William Edward – London: Williams & Norgate; New York: G P Putnam 1906 [mf ed 1985] – (= ser Crown theological library 16) – 1mf – 9 – 0-8370-2049-2 – (incl ind) – mf#1985-0049 – us ATLA [270]
The hebrew renaissance – (New York). 1913 – 1 – us AJPC [071]
Hebrew standard – New York. N.Y. 1893-1922 – 1 – us AJPC [071]
Hebrew standard – Sydney, nov 1895-oct 1953 – 22r – A$1643.75 vesicular A$1764.75 silver – at Pascoe [079]
The hebrew student's commentary on zechariah : hebrew and 70 with excursus on syllable-dividing, metheg, initial dagesh, and siman rapheh / Lowe, William Henry – London: Macmillan, 1882 – 1mf – 9 – 0-8370-4186-4 – mf#1985-2186 – us ATLA [221]
Hebrew studies – Madison. 1985+ (1,5,9) – ISSN: 0146-4094 – mf#15247,01 – us UMI ProQuest [470]
Hebrew syntax / Davidson, Andrew Bruce – Edinburgh: T & T Clark, 1894 [mf ed 1986] – 1mf – 9 – 0-8370-9221-3 – (companion vol to: an introductory hebrew grammar. incl ind) – mf#1986-3221 – us ATLA [470]
Hebrew tenses / Driver, Samuel Rolles – 1898 – 9 – $12.00 – us IRC [470]
The hebrew text of the book of ecclesiasticus / ed by Levi, Israel – Leiden: E J Brill, 1904 – 1mf – 9 – 0-8370-9399-6 – mf#1986-3399 – us ATLA [221]
Hebrew titles : 1964-1971 – 1972 – 6r – 1 – $200.00 – us AJPC [470]
Hebrew titles : 1972-1982 – 1984 – 82mf – 9 – $360.00 – us AJPC [470]
Hebrew titles : 1983-1989 – in prep – us AJPC [470]
The hebrew tragedy / Conder, C R – Edinburgh: William Blackwood, 1900 – 1mf – 9 – 0-7905-0566-5 – mf#1987-0566 – us ATLA [939]
The hebrew twins : a vindication of god's ways with jacob and esau / Cox, Samuel – New York: Thomas Whittaker, 1894 – 1mf – 9 – 0-8370-2762-4 – mf#1985-0762 – us ATLA [920]
Hebrew union college : and other addresses / Kohler, Kaufmann – Cincinnati: Ark, 1916 [mf ed 1991] – 1mf – 9 – 0-7905-9401-3 – mf#1989-2626 – us ATLA [378]
Hebrew union college annual – (Cincinnati). 1924-66 – 1 – us AJPC [378]
Hebrew union college annual – Cincinnati. 1994-96 (1,5,9) – ISSN: 0360-9049 – mf#15991 – us UMI ProQuest [939]
Hebrew union college-jewish institute of religion – (Cincinnati, Ohio) Feb. 27, 1978-Oct. 1988 – us AJPC [270]
The hebrew utopia : a study of messianic prophecy / Adeney, Walter Frederic – London: Hodder & Stoughton, 1879 – 1mf – 9 – 0-8370-2051-4 – mf#1985-0051 – us ATLA [270]
The hebrew verb : a series of tabular studies / Carrier, Augustus Stiles – Chicago: Max Stern, 1891 – 1mf – 9 – 0-8370-9212-4 – mf#1986-3212 – us ATLA [470]
Hebrew vocabularies : lists of the most frequently occurring hebrew words / Harper, William Rainey – New York: Charles Scribner, c1890 – 1mf – 9 – 0-7905-1885-6 – mf#1987-1885 – us ATLA [470]
Hebrew watchman – Memphis, TN. 1979-84 – 1 – us AJPC [071]
Hebrew watchword – (Philadelphia). 1896-98 – 1 – us AJPC [071]
The hebrew wife : or, the law of marriage examined in relation to the lawfulness of polygamy and to the extent of the law of incest / Dwight, Sereno Edwards – New-York: Leavitt, Lord; Boston: Crocker & Brewster, 1836 – 1mf – 9 – 0-7905-0825-7 – mf#1987-0825 – us ATLA [220]
The hebrew world – New York [NY]: G L Lowenthall. v20 n48. mar 24 1905 (wkly) (mf ed [197-?]) – mf#ZZAN-21691 – us NY Public [071]
Hebrew-english vocabulary to the book of genesis / Kelso, James Anderson & Culley, David E – New York: Scribner, 1917 – 1mf – 9 – 0-524-05678-1 – mf#1992-0528 – us ATLA [221]
Hebrew-greek cairo genizah palimpsests from the taylor-schechter collection : including a fragment of the twenty-second psalm according to origen's hexapla / ed by Taylor, Charles – Cambridge: University Press, 1900 [mf ed 1993] – 3mf – 9 – 0-524-07482-8 – (in greek) – mf#1992-1114 – us ATLA [220]
Hebrews : introduction, authorized version, revised version, with notes and index / ed by Peake, Arthur Samuel – Edinburgh: T C & E C Jack, [19–] – (= ser The Century Bible) – 1mf – 9 – 0-8370-6831-2 – mf#1986-0831 – us ATLA [221]
Hebrews and the epistles general of peter, james, and jude / ed by Herkless, John – London: J M Dent; Philadelphia: J B Lippincott 1902 [mf ed 1989] – (= ser The temple bible) – 1mf – 9 – 0-7905-1825-2 – mf#1987-1825 – us ATLA [227]
Hebrews and the general epistles : with introduction and notes / Mitchell, Alexander F – New York: Fleming H Revell; London: Andrew Melrose, [1911?] – 1mf – 9 – 0-7905-1365-X – (incl ind) – mf#1987-1365 – us ATLA [227]
Hebrews, james, and 1 and 2 peter : a popular commentary upon a criticial basis, especially designed for pastors and sunday schools / Eaches, O P – Philadelphia: American Baptist Publ Society 1906 [mf ed 1989] – (= ser Clark's peoples commentary) – 1mf – 9 – 0-7905-1815-5 – (incl ind) – mf#1987-1815 – us ATLA [225]
Hebron baptist church (formerly: tinker creek baptist church). union county. south carolina : church records – 1867-87, Oct 1937 and Jan 1938. History book. 1806-1958 – 1 – 5.00 – us Southern Baptist [242]
Hebron baptist church. gaston county. north carolina : church records – 1838-1922 – 1 – 7.74 – us Southern Baptist [242]
Hebron champion – Hebron, NE: P S Mickey. 11v. v6 n43. feb 22 1901-v16 n39. jan 28 1916 (wkly) [mf ed with gaps] – 10r – 1 – (formed by the union of: people's champion and: hebron republican. merged with: hebron register to form: register-champion. v6 n43 called also v11 n13) – us NE Hist [071]
Hebron journal – Hebron, NE: E M Correll. 73v. v1 n1. feb 9 1871-v73 n52. feb 3 1944 (wkly) [mf ed with gaps filmed 1969] – 36r – 1 – (merged with: hebron register (1930) to form: hebron journal-register. v6 n27-v73 n52 called also whole n287-whole n4791) – us NE Hist [071]
Hebron journal-register – Hebron, NE: Will and Edna Long. v73 n1. feb 9 1944- (wkly) [mf ed lacks mar 12 1947] – 43r – 1 – (absorbed: davenport news (1951) and: thayer county banner-agrus. formed by the union of: hebron journal and: hebron register (hebron. v73 n1-n11 called also v59 n10-n20. issues for apr 26 1944- called v1 n12- . numbering dropped with nov 23 1977 issue. suppls accompany some issues) – us NE Hist [071]
Hebron, New Hampshire. Hebron Baptist Church *see* Records
Hebron register – Hebron, NE: Register Publishing Co, 1894-jan 28 1916// (wkly) [mf ed v11 n32. apr 13 1894-jan 21 1916 (gaps)] – 13r – 1 – (cont: hebron weekly register. merged with: hebron champion to form: register-champion) – us NE Hist [071]
Hebron register – Hebron, NE: Will Long. 12v. v48 n36. aug 28 1930-v59 n9. feb 2 1944 (wkly) [mf ed lacks jul 21 1941] – 9r – 1 – (cont: register-champion. merged with: hebron journal to form: hebron journal-register) – us NE Hist [071]
Hebron republican – Hebron, NE: L T Calkins, -feb 22 1901// (wkly) [mf ed v6 [n1] nov 15 1895-dec 28 1900 (gaps) filmed 1978] – 1r – 1 – (merged with: people's champion to form: hebron champion) – us NE Hist [071]
Hebron weekly register – Hebron, NE: H C Pershing, -1894// (wkly) [mf ed v9 n46. jul 22 1892-mar 9 1894 (gaps)] – 1r – 1 – (cont by: hebron register) – us NE Hist [071]
Hec forum – New York. 1991-1996 (1,5,9) – ISSN: 0956-2737 – mf#18621 – us UMI ProQuest [360]
Hecate – St. Lucia. 1992+ (1,5,9) – ISSN: 0311-4198 – mf#19236 – us UMI ProQuest [305]
Hecatomgraphie c'est...dire les declarations de plusieurs apophtegmes... / Corrozet, G – [Paris], n.d. – 2mf – 9 – mf#O-1544 – ne IDC [090]
Hechalutz (Organization) *see* Me'asef li-tenu'at hehaluts
Hechicero / Solorzano, Carlos – Mexico City? Mexico. 1955 – 1r – us UF Libraries [972]
Los hechos – Panama City, Panama. 7 jun-13 jul 1912 – 1r – 1 – us L of C Photodup [079]
Hechos y comentarios, nova et vetera / Rodriguez Pineres, Eduardo – Bogota, Colombia. 1956 – 1r – us UF Libraries [972]
Hecht, Emanuel *see*
– Lehrbuch der judischen geschichte und literatur
– Der pentateuch
Hecht, Georg *see* Goethes briefwechsel mit thomas carlyle
Hecht, Irene W D [coll] *see* Irene hecht collection, 1637-1950
Hecht, Wolfgang *see* Frei nach goethe
Hechtenberg, Klara *see* Das fremdwort bei grimmelshausen
Hechtle, Martha *see* Walther von der vogelweide
Heck, Fannie Exile Scudder *see* In royal service
Heck, Johann Caspar *see*
– Art of fingering
– Art of playing thorough bass with correctness according to the true principles of composition
– Complete system of harmony
– Short and fundamental instructions for learning thorough bass
Hecke, Johann *see* Reise durch die vereinigten staaten von nord-amerika
Heckel, Hans *see* Das don juan-problem in der neueren dichtung
Heckel, Karl *see*
– Die buehnenfestspiele in bayreuth
– Erlaeuterungen zu wagners tristan und isolde von karl heckel
Hecker, Friedrich *see* Reden und vorlesungen
Hecker, Isaac Thomas *see*
– Aspirations of nature
– The church and the age
– Questions of the soul
Hecker, Julius Friedrich *see* Russian sociology
Hecker, Max *see*
– Goethe
– Goethes briefwechsel mit heinrich meyer
– Jahrbuch der goethe-gesellschaft
– Wilhelm meisters wanderjahre
Heckford, Mrs *see* A lady trader in the transvaal
Heckford, Sarah *see* A lady traveler in the transvaal
Heckler, James Y *see* Ecclesianthem
Heckman, George C *see* An address on woman's work in the church
Heckrath, Goswin *see* Zur rolle der sorption beim verhalten von herbiziden im boden
Hecquard, Hyacinthe *see* Voyage sur la cote et dans l'interieur de l'afrique occidentale
Hector / Luce De Lancival, Jean-Charles-Juli – Paris, France. 1809 – 1r – us UF Libraries [440]
Hector berlioz (1803-1869) works / ed by Malherbe, Charles & Weingartner, Felix – Leipzig. 20v. 1900-07 – 11 – $225.00 set – (introductory material in french, german, english. v2 contains index to complete works) – us Univ Music [780]
Hector estudios de historia colonial venezolana... / Garcia, Huecos – Burgos: Razon y Fe, 1939 – 1 – sp Bibl Santa Ana [972]
Hed lita – Kovno. 1-2, no. 21. 1924-1925 – 1 – us NY Public [070]
Hedemann, Justus W *see* Volksgesetzbuch. grundregel und buch i
Hedemann, Justus Wilhelm *see* Die fortschritte des zivilrechts im 19. jahrhundert
Hedemora tidning – Hedemora, Sweden. 1864-80 – 5r – 1 – sw Kungliga [078]
Hedendaagsche zending in onze oost : handboek voor zendingsstudie – Hoenderloo: Stoomdrukkerij Doorgangshuis, 1909 [mf ed 1995] – (= ser Yale coll) – 267p (ill) – 1 – 0-524-09185-4 – (in dutch) – mf#1995-0185 – us ATLA [959]
De hedendaegsche lang-tonge af-gebeeld in twaelf zinne-beelden... / Pauwels, J A F – Antwerpen: J P de Cort, 1774 – 9mf – 9 – mf#O-3141 – ne IDC [090]
Hedenigg, Silvia *see* Kindheitsbegriffe japanischer strafkonzeptionen
Hederer, Edgar *see* Friedrich von hardenbergs "christenheit oder europa"
Hederich, Benjamin *see* Benjamin hederichs lateinisch-deutsche, deutsch-lateinische und griechisch-lateinische, lateinisch-griechische woerterbuecher
Hedge, Frederic Henry *see*
– Atheism in philosophy
– Hours with german classics
– Martin luther
– The primeval world of hebrew tradition
– Reason in religion
– Recent inquiries in theology, by eminent english churchmen
– Sermons
– Ways of the spirit
Hedge, Frederic Henry et al *see* Unitarian affirmations
Hedges herald – Hedgesville, MT. 1909-1917 (1) – mf#64452 – us UMI ProQuest [071]
Hedges of florida / Mowry, Harold – Gainesville, FL. 1924 – 1r – us UF Libraries [630]
Hedgpeth facts and findings – v1 n1-v4 n4 [1978 mar-1981 dec] – 1r – 1 – (cont: hedgpeth/hudspeth newsletter) – mf#636872 – us WHS [071]
Hedin, S *see* Sven hedins geologische routen-aufname durch ost persien
Hedin, Sven Anders *see* Trans-himalaya
Hedion, C *see*
– Epitome in evangelia et epistolas in usum ministrorum ecclesiae
– Radts predig
Hediye-yi sal – 1312-13 [1894-95] – (= ser Ministry and special interest salnames) – 4mf – 9 – $60.00 – us MEDOC [956]
Hedley, James *see* Canada and her commerce
Hedley, John *see* Tramps in dark mongolia

Hedley, John Cuthbert *see*
- A bishop and his flock
- The christian inheritance
- The holy eucharist
- Lex levitarum, or, preparation for the cure of souls – regula pastoralis
- Our divine saviour

Hedley, Thomas Fenwick *see* Local taxation...
Hedrich, Franz *see* Alfred meissner – franz hedrich
Heeden, Matthew *see* An analysis of athletic department operations at the dean smith center
Heer, Joseph Michael *see*
- Euangelium gatianum
- Die stammbaeume jesu nach matthaeus und lukas

Heerbrand, J *see*
- Acta des colloquij
- Compendium theologie, methodi qvaestionibvs tractatvm

Heerbrandt, Gustav *see* Gedichte in schwaebischer mundart
Heering, Hans *see* Idee und wirklichkeit bei hanns johst
Heeringen, Gustav von *see* Ein ausflug nach england
Heermann, Norbert *see* Frank duveneck
Heers, Alois *see* Das leben friedrich von matthissons
Heeter connections – v1 n1-v4 n4 [1983 jul-1987 jul] – 1r – 1 – mf#1321068 – us WHS [071]
Heever, Susanna van den *see* Strategies for large class teaching
Hefah le-toldoteha ve-yishuvah / Silman, Kadisch Jehunda – Tel-Aviv, Israel. 1931 – 1r – us UF Libraries [939]
Hefebesiedlungen im oralen bereich : eine epidemiologische studie / Svoboda, Michael – (mf ed 1996) – 1mf – 9 – €30.00 – 3-8267-2346-5 – mf#DHS 2346 – gw Frankfurter [616]
Hefele, C J *see*
- Histoire des conciles
- Histoire des conciles d'apraes les documents originaux

Hefele, Herman *see*
- Das gesetz der form
- Goethes faust

Hefele, Karl Joseph von *see*
- Beitraege zur kirchengeschichte, archaeologie und liturgik
- Geschichte der einfuehrung des christenthums im suedwestlichen deutschland
- History of the christian councils
- The life of cardinal ximenez

Heffernan b e f news / Bonus Expeditionary Force – 1932 sep 17-oct 1 – 1r – 1 – (cont: b e f news) – mf#2472742 – us WHS [071]
Heffner, Roe Merrill Secrist et al *see* The gretchen episode from goethe's faust
Heffter, Moritz Wilhelm *see* Allgemeine geographie der insel rhodos
Heflin hemlock – v1 n6-v3 n6 [1979 sep-1981 sep] – 1r – 1 – mf#634391 – us WHS [071]
Hefner, Joseph *see* Die entstehungsgeschichte des trienter rechtfertigungsdekretes
Heft meclis / Ali, Mustafa – Dersaadet: Ikdam Matbaasi, 1316 [1900] – (= ser Ottoman histories and historical sources) – 1mf – 9 – $25.00 – us MEDOC [956]
Hegarty, Joseph A *see* Journal of culinary science and technology
Hegau bote *see* Singener zeitung – hegau bote
Hegde, Sudhir S *see* Changes in clotting and fibrinolytic activity after sub-maximal exercise in males
Hege, Ruth *see* We two alone
Hegel / Caird, Edward – Edinburgh: W Blackwood, 1883 – (= ser Philosophical Classics for English Readers) – 1mf – 9 – 0-7905-7698-8 – mf#1989-0923 – us ATLA [190]
Hegel : sendschreiben an den hofrath und professor der philosophie, herrn dr. carl friedrich bachmann in jena / Rosenkranz, Karl – Koenigsberg: AW Unzer, 1834 – 1mf – 9 – 0-524-08558-7 – mf#1993-2083 – us ATLA [190]
Hegel and hegelianism / Mackintosh, Robert – Edinburgh: T & T Clark 1903 [mf ed 1991] – (= ser The world's epoch-makers) – 1mf – 9 – 0-7905-9327-0 – (incl bibl ref) – mf#1989-2552 – us ATLA [190]
Hegel, Georg Wilhelm Friedrich *see*
- Hegel's philosophy of mind
- Hegels theologische jugendschriften
- The introduction to hegel's philosophy of fine art
- Das leben jesu
- Lectures on the history of philosophy
- Lectures on the philosophy of religion
- The subjective logic of hegel
- The wisdom and religion of a german philosopher

Hegel, Georg Wilhelm Friedrich et al *see* Kritisches journal der philosophie
Hegel, Georg William Friedrich *see* Hegel's first principle
Hegel und plotin : eine kritische studie / Jong, Karel Hendrik Eduard de – Leiden: Brill, 1916 – 1mf – 9 – 0-524-00909-0 – (incl bibl ref) – mf#1990-2132 – us ATLA [100]
Hegeler, Wilhelm *see* Kleist
Hegelianism and human personality / Haldar, Hiralal – [Calcutta]: University of Calcutta, 1910 – (= ser Samp: indian books) – us CRL [140]
Hegelianism and personality / Seth Pringle-Pattison, Andrew – 2nd ed. Edinburgh: W Blackwood 1893 [mf ed 1990] – (= ser Balfour philosophical lectures) – 1mf – 9 – 0-7905-7465-9 – mf#1989-0690 – us ATLA [140]
El hegelismo juridico espanol / Elias de Tejada Spinola, Francisco – Madrid: revista de derecho privado, 1944 – sp Bibl Santa Ana [340]
El hegelismo juridico espanol : madrid, 1944 / Truyol Serra, A & Elias de Tejada, F – Madrid: Razon y Fe, 1946 – 1 – sp Bibl Santa Ana [340]
Hegel's aesthetics : a critical exposition / Kedney, John Steinfort – Chicago: SC Griggs, 1885 – (= ser German philosophical classics for english readers and students) – 1mf – 9 – 0-7905-7308-3 – mf#1989-0533 – us ATLA [140]
Hegels dialektische ontologie und die thomistische analektik / Lakebrink, B – Koeln, 1955 – 9mf – 8 – €19.00 – ne Slangenburg [110]
Hegel's first principle : an exposition of comprehension and idea (begriff und idee) = Philosophische propaedeutik. selections / Hegel, Georg William Friedrich – St Louis: Printed by G Knapp, 1869 – 1mf – 9 – 0-7905-7940-5 – (in english) – mf#1989-1165 – us ATLA [100]
Hegel's logic : a book on the genesis of the categories of the mind / Harris, William Torrey – Chicago: SC Griggs, 1890 – (= ser German philosophical classics for english readers and students) – 1mf – 9 – 0-7905-7299-0 – (incl bibl ref) – mf#1989-0524 – us ATLA [160]
Hegels offenbarungsbegriff : ein religionsphilosophischer versuch / Werner, Johannes – Leipzig: Breitkopf & Haertel, 1887 – 1mf – 9 – 0-524-00207-X – (incl bibl ref) – mf#1989-2907 – us ATLA [240]
Hegel's philosophy of mind = Philosophie des geistes / Hegel, Georg Wilhelm Friedrich – Oxford: Clarendon Press, 1894 – 1mf – 9 – 0-7905-7300-8 – (in english) – mf#1989-0525 – us ATLA [140]
Hegel's philosophy of the state and of history / Morris, George Sylvester – Chicago: Griggs, 1887. 306p – 1 – us Wisconsin U Libr [900]
Hegels theologische jugendschriften : nach den handschriften der kgl. bibliothek in berlin = Selections. 1907 / Hegel, Georg Wilhelm Friedrich; ed by Nohl, Herman – Tuebingen: JCB Mohr, 1907 – 1mf – 9 – 0-7905-9222-3 – (incl bibl ref) – mf#1989-2447 – us ATLA [200]
Hegemonius acta archelai (gcsej7) / ed by Beeson, C H – 1906 – (= ser Griechische christlichen schriftsteller der ersten jahrhunderte (gcsej)) – €11.00 – ne Slangenburg [240]
Hegendorf, Chr *see* Zwei aelteste katechismen der lutherischen reformation
Heger, Thomas *see* A tour through a part of the netherlands, france, and switzerland, in the year 1817
Hegermann, H *see* Die vorstellung vom schoepfungsmittler im hellenistischen judentum und urchristentum
Hegetschweiler, Johannes *see* Reisen in den gebirgsstock zwischen glarus und graubuenden in den jahren 1819, 1820 und 1822
Hegewisch, Dietrich Hermann et al *see* Amerikanisches magazin
Hegglin, P *see* Der benediktinische abt
Hegler, Alfred *see*
- Beitraege zur geschichte der mystik in der reformationszeit
- Geist und schrift bei sebastian franck
- Johannes brenz und die reformation in herzogtum wirtemberg [sic]
- Sebastian francks lateinische paraphrase der deutschen theologie
- Zur erinnerung an carl weizsaecker

Hehaluts Poland *see* Mir un di araber
He-halutz – Lemberg etc. v. 1-13. 1852-1889 – 1 – us NY Public [073]
Hehisch *see* Heppner gazette-times
Hehl Neiva, Artur *see* Problema imigratorio brasileiro
Hehman, Eric D *see* A survey of intercollegiate head football coaches' programs for developing racial understanding
Hehn, J *see* Siebenzahl und sabbat bei den babyloniern und im alten testament
Hehn, Johannes *see*
- Die biblische und die babylonische gottesidee
- Suende und erloesung

Hehn, Victor *see*
- Gedanken ueber goethe
- Ueber goethes gedichte
- Ueber goethes hermann und dorothea
- The wanderings of plants and animals from their first home

Hei an yu kuang ming / Lin, Tan-ch'iu – Shang-hai: Kuang ming shu chu, Min kuo 29 [1940] – (= ser P-k&k period) – us CRL [480]
Hei an yu kuang ming (ccm336) = Light and darkness / Wu, Yao-tsung – Shanghai, 1949 [mf ed 198?] – (= ser Ccm 336) – 1 – mf#1984-b500 – us ATLA [240]
Hei feng chi / Shen, Ts'ung-wen – Shang-hai: K'ai ming shu tien, Min kuo 38 [1949] – (= ser P-k&k period) – us CRL [480]
Hei mu tan / mu shih-ying teng chu / Mu, Shih-ying – Shang-hai: Liang yu t'u shu yin shua kung ssu, Min kuo 24 [1935] – (= ser P-k&k period) – us CRL [830]
Hei pai chi / Wang, Pin – Shang-hai: Hsin ti shu tien, 1940 – (= ser P-k&k period) – us CRL [480]
Hei ti yu / Ling, Ho – Shang-hai: Hsi chu shih tai ch'u pan she, 1937 – (= ser P-k&k period) – us CRL [820]
Hei t'u / Pa, Chin – Shang-hai: Wen hua sheng huo ch'u pan she, Min kuo 30 [1941] – (= ser P-k&k period) – us CRL [840]
Hei tzu erh shih pa / Ts'ao, Yu – Shang-hai: Cheng chung shu chu, Min kuo 31 [1942] – (= ser P-k&k period) – us CRL [951]
Heiberg, Knud *see* Madras, lidt om byen og missionen
Heiberg, Peter Andreas *see*
- En episode i soeren kierkegaards ungdomsliv
- Nogle bidrag til enten-eller's tilblivelseshistorie

Heichen, Walter [comp] *see* Helden der kolonien
Heidanus, A *see*
- Corpus theologiae christianae in quindecim locos digestum
- De causa dei...
- De origine erroris libri octo
- Disputationes theologicae ordinariae repetitiae
- Fasciculus disputationum theologicarum de socianismo
- Oratio de componenda inter dissidentes christianos pace et concordia
- Proeve en wederleggingbe des remostrantschen catechismi

Heidborn, A *see* Manuel de droit public et administratif de l'empire ottoman
Heide-bote – Langebrueck DE, 1927 1 aug-1941 30 may – 1r – gw Misc Inst [074]
Das heidedorf / Stifter, Adalbert – 2. aufl. Muenchen: Georg W Dietrich [19–?] [mf ed 1995] – 1r [ill] – 1 – (filmed with: abdias / adalbert stifter) – mf#3748p – us Wisconsin U Libr [830]
Heidegger, Heinrich *see* Manuel de l'etranger qui voyage en suisse
Heidegger, M *see*
- Die kategorien- und bedeutungslehre des duns scotus
- Theologie und philosophie

Heidegger, Martin *see*
- Hoelderlin und das wesen der dichtung
- Hoelderlins hymne "wie wenn am feiertage..."
- Die kategorien- und bedeutungslehre des duns scotus

Heidel, Alexander *see* The babylonian genesis, the story of creation
Heidelbach, Paul *see*
- Die neuen argonauten

Heidelberg als stoff und motiv der deutschen dichtung / Goldschmit, Rudolf K – Berlin: W de Gruyter & Co, 1929 – 2r – 1 – (incl bibl ref and index) – us Wisconsin U Libr [430]
The heidelberg catechism : historical and doctrinal studies / Richards, George Warren – Philadelphia: Publication and Sunday School Board of the Reformed Church in the United States, 1913 – (= ser Swander Lectures) – 1mf – 9 – 0-7905-6009-7 – (incl bibl ref) – mf#1988-2009 – us ATLA [240]
The heidelberg catechism : in german, latin and english – tercentenary ed. New York: Scribner, 1863 [mf ed 1991] – 1mf – 9 – 0-524-00037-9 – (hist int in english, catechism in original german with parallel latin and english trans, and a modern german transcription) – mf#1989-2737 – us ATLA [242]
The heidelberg catechism in its newest light / Good, James Isaac – Philadelphia: Publication and Sunday School Board of the Reformed Church in the United States, 1914 – 1mf – 9 – 0-7905-4909-3 – mf#1988-0909 – us ATLA [240]
Heidelberg herald-post – Heidelberg US. v8 n20-v11 n13 [1983 jul 8-1986 jan 16] – 1r – 1 – (continued by: herald-post (heidelberg, germany)) – mf#1044789 – us WHS [071]
Heidelberg Liberation Front *see* Fta with pride
Heidelberger beitraege zur mineralogie und petrographie – Heidelberg. 1947-1957 (1) 1947-1957 (5) – (cont by: beitraege zur mineralogie und petrographie) – ISSN: 0367-5769 – mf#13157 – us UMI ProQuest [550]
Heidelberger beobachter – Heidelberg DE, 1931 3 jan-1945 23 mar – 1 – (title varies: 1 mar 1932: die volksgemeinschaft) – gw Misc Inst [074]
Heidelberger general-anzeiger – Heidelberg DE, 1049 25 may-1982 – 171r – 1 – (title varies: 3 jan 1884: heidelberger tageblatt; 25 may 1949: tageblatt; 1 jun 1951: heidelberger tageblatt. filmed by bnl: 1949 4 jun-1952 nov (gaps) [20r]) – gw Mikrofilm; uk British Libr Newspaper [074]
Heidelberger journal *see* Heidelberger wochenblatt
Der heidelberger katechismus : und vier verwandte katechismen (leo jud's und micron's kleine katechismen, sowie die zwei vorarbeiten ursins) / ed by Lang, August – Leipzig: A Deichert, 1907 – (= ser Quellenschriften zur Geschichte des Protestantismus) – 4mf – 9 – 0-524-07430-5 – mf#1991-3090 – us ATLA [240]
Der heidelberger katechismus : zum 350jaehrigen gedaechtnis seiner entstehung / Lang, August – Leipzig: Verein fuer Reformationsgeschichte 1913 [mf ed 1990] – (= ser Schriften des vereins fuer reformationsgeschichte 31/1/113) – 1mf – 9 – 0-7905-4703-1 – (incl bibl ref) – mf#1988-0703 – us ATLA [240]
Der heidelberger katechismus : zum 350jaehrigen gedaechtnis seiner entstehung / Lang, August – Leipzig: Verein fuer Reformationsgeschichte, 1913. (Schriften des Vereins fuer Reformationsgeschichte; Jahrg. 31, 1. Stueck, Nr. 113) – 1mf – us ATLA [240]
Heidelberger passionsspiel = Heidelberg passion play / ed by Milchsack, Gustav – Stuttgart: Litterarischer Verein, 1880 (Tuebingen: L F Fues) [mf ed 1993] – (= ser Blvs 150) – 306p – 1 – (incl bibl ref) – mf#8470 reel 31 – us Wisconsin U Libr [790]
Heidelberger passionsspiel / ed by Milchsack, Gustav – Stuttgart: Litterarischer Verein, 1880 (Tuebingen: L F Fues) – (incl bibl ref) – us Wisconsin U Libr [430]
Heidelberger rundschau – Heidelberg DE, 1975 20 nov-1988 29 sep – 6r – 1 – (title varies: 1984: communale) – gw Mikrofilm [074]
Heidelberger tageblaetter *see* Heidelberger wochenblatt
Heidelberger tageblatt *see* Heidelberger general-anzeiger
Heidelberger wochenblaetter *see* Heidelberger wochenblatt
Heidelberger wochenblatt – Heidelberg DE, 1844-45 – 3r – 1 – (title varies: 3 jan 1831: heidelberger wochenblaetter; 1 jan 1840: heidelberger tageblaetter; 1 jul 1842: heidelberger journal. filmed by misc inst: 1842 1 jul-1844, 1846/47, 1860 [6r]; 1848-49 [3r]) – gw Mikrofilm; gw Misc Inst [074]
De heidelbergsche catechismus : in twee en vijftig leerredenen / Oosterzee, Johannes Jacobus van – Amsterdam: H De Hoogh, 1869-1870 [mf ed 1991] – 2v on 3mf – 9 – 0-7905-8538-3 – (in dutch) – mf#1989-1763 – us ATLA [240]
De heidelbergsche catechismus : toepasselijk verklaard voor de gemeente des heeren / Knap, Jan Jacob – Groningen: J B Wolters, 1912 [mf ed 1994] – 543p on 6mf – 9 – 0-524-08797-0 – (in dutch) – mf#1993-3289 – us ATLA [240]
De heidelbergsche catechismus en het boekje van de breking des broods : in het jaar 1563-64 bestreden en verdedigd / Gooszen, Maurits Albrecht – Leiden: Brill, 1892 [mf ed 1993] – viii/424p on 5mf – 9 – 0-524-07876-9 – (incl bibl ref) – mf#1991-3421 – us ATLA [242]
Heidemann *see* Ueber lessing's emilia galotti
Die heiden von kummerow : roman / Welk, Ehm – Berlin: Deutscher Verlag c1937 [mf ed 1991] – 1r – 1 – (filmed with: "und alles ist zerstoben" / werner weisbach) – mf#3038p – us Wisconsin U Libr [830]
Die heidenbekehrung im alten testament und im judentum / Sieffert, Friedrich – Berlin: Edwin Runge 1908 [mf ed 1989] – (= ser Biblische zeit- und streitfragen 4/3) – 1mf – 9 – 0-7905-1078-2 – mf#1987-1078 – us ATLA [221]
Die heidenboten friedrichs 4. von daenemark : 1. bartholomaeus ziegenbalg und seine mitarbeiter in trankebar / Brauer, Johann Hartwig – Altona: Joh Friedr Hammerich, 1837 [mf ed 1995] – (= ser Yale coll; Beitraege zur geschichte der heidenbekehrung 2) – xviii/180p – 1 – 0-524-10127-2 – (in german) – mf#1995-1127 – us ATLA [948]
Heidenheim, M *see* Die samaritanische liturgie
Heidenheimer, Heinrich *see* Petrus martyr anglerius und sein opus epistolarum
Heidenheimer zeitung – Heidenheim a. d. Brenz DE, 1975-82 – 48r – 1 – (bezirksausgabe von suedwest-presse, ulm; filmed by other misc inst: 1977- [ca 7r/yr]) – gw Misc Inst [074]
Die heidenmission nach der lehre des hl augustinus / Walter, G – Muenster, 1921 – €11.00 – ne Slangenburg [240]

HEILSLEHRE

Die heidenpredigt in indien / Hesse, Johannes – Basel: Missionsbuchh, 1883 – 1mf – 9 – 0-524-05459-2 – mf#1990-3485 – us ATLA [240]
Das heidenroeslein / Joseph, Eugen – Berlin: Gebrueder Paetel 1897 [mf ed 1990] – 1r – 1 – (incl bibl ref. filmed with: goethe / c h herford) – mf#7387 – us Wisconsin U Libr [410]
Das heidenroeslein : oder, goethe's sessenheimer lieder in ihrer veranlassung und stimmung / Baier, Adalbert – Heidelberg: G Weiss, 1877 [mf ed 1990] – 70p/xv/159p – 1 – (incl bibl ref) – mf#7320 – us Wisconsin U Libr [430]
Heidenthum und offenbarung : religionsgeschichtliche studien ueber die beruehrungspunkte der aeltesten heiligen schriften der inder, perser, babylonier, assyrer und aegypter mit der bibel, auf grund der neuesten forschungen / Fischer, Engelbert Lorenz – Mainz: F Kirchheim, 1878 – 1mf – 9 – 0-7905-7577-9 – mf#1989-0802 – us ATLA [230]
Das heidentum in der roemischen kirche : bilder aus dem religioesen und sittlichen leben sueditaliens / Trede, Theodor – Gotha: FA Perthes, 1889-1891 – 4mf – 9 – 0-524-07303-1 – mf#1991-0089 – us ATLA [241]
Das heideprinzesschen : roman / Marlitt, Eugenie [pseud] – Stuttgart, Berlin [etc]: Union Deutsche Verlagsgesellschaft [1919] [mf ed 1978] – (= ser Romane und novellen 2) – 2r – 1 – (also publ with title: das haideprinzesschen) – mf#film mas c376 – us Harvard [830]
Heider anzeiger und dithmarscher post – Heide, Holst DE, 1900-1945 9 may [gaps] – 1 – (title varies: 1903: heider anzeiger; 30 sep 1906: heider anzeiger und zeitung) – gw Misc Inst [074]
Heider anzeiger und zeitung *see* Heider anzeiger und dithmarscher post
Heider, August *see* Die aethiopische bibeluebersetzung
Heider zeitung – Heide, Holst DE, 1881-1902 – 1 – gw Misc Inst [074]
Die heidin / ed by Pretzel, Ulrich & Henschel, Erich – Leipzig: S Hirzel, 1957 [mf ed 1993] – (= ser Altdeutsche quellen heft 4) – 105/24p – 1 – (middle high german text. int in german. incl bibl ref) – mf#8377 – us Wisconsin U Libr [430]
Heidingsfelder, G *see* Albert von sachsen
Heidlauf, Felix *see* Lucidarius
Das heidnische dorf : roman / Beste, Konrad – Muenchen: A Langen/G Mueller, 1932 [mf ed 1989] – 293p – 1 – mf#7014 – us Wisconsin U Libr [830]
A heifer of the dawn – London: Medici Society, 1914 – (= ser Samp: indian books) – (trans fr original mss by f w bain) – us CRL [830]
Heigenmooser, J *see* Eremitenschule in altbayern
Heights baptist church. aiken county. south carolina : church records – Sept 1966-1983. 395p – 1 – us Southern Baptist [242]
Heighway, Osborn W Trenery *see* Leila ada, the jewish convert
Heigl, Bartholomaeus *see* Verfasser und adresse des briefes an die hebraeer
Heigl, Ferdinand *see* Die religion und kultur chinas
Heikel, I A *see* Kritische beitraege zu den constantin-schriften des eusebius
Heikel, Ivar August *see* Kritische beitraege zu den constantin-schriften des eusebius (eusebius werke band 1)
Heil, Alexander *see* Die verstaendlichkeit in der informationsgesellschaft
Heil, Daniel P *see* Body mass scaling of endurance cycling performance
Heil dir, friedrich, deutscher kaiser! / Wesendonk, Mathilde – Berlin: Haack 1888 – 1mf – 9 – mf#mw-12 – ne IDC [820]
Heil, DP *see* The effect of seat-tube angle variation on cardiorespiratory responses during submaximal bicycling
Heil, Wolfgang *see* Zur bestimmung der angiotensin i-converting enzyme-aktivitaet im serum
Heilandsflur : eine tragoedie deutscher landfahrer in drei aufzuegen / Bruees, Otto – Frankfurt/M: Verlag des Buehnenvolksbundes (Patmosverlag), 1923 [mf ed 1989] – 58p – 1 – mf#7092 – us Wisconsin U Libr [820]
Heilborn, E [comp] *see* Novalis schriften
Heilborn, Ernst *see* Das fontane-buch
Heilbronn community circle / U S Military Community Activity – Heilbronn, Germany, West. 1983 jul 4, sep 12, oct 4 – 1r – 1 – (continued by: heilbronn eagle) – mf#706827 – us WHS [355]
Heilbronner stimme – Heilbronn DE, 1946 28 mar-1967 29 apr, 1968- – ca 10r/yr – 1 – (filmed with various pp fr regional ed of eppinger zeitung & hohenloher zeitung) – gw Misc Inst [074]
Heilbrunn, Ludwig *see* Faust 2. teil als politische dichtung
Heilfron, Eduard *see* Lehrbuch des buergerlichen rechts, auf der grundlage des buergerlichen gesetzbuchs
Heilig : mit zwey choeren und einer ariette zur einleitung / Bach, Carl Philipp Emanuel – Hamburg: Im Verlag des Autors; Leipzig: aus der Breitkopfischen Buchdruckerey 1779 [mf ed 19–] – 1r – 1 – mf#film 241 – us Sibley [780]
Heilig, Otto *see* Allemannische gedichte
Heilig zaad : verhandelingen over den heiligen doop / Lonkhuijzen, Jan van – Grand Rapids, MI: Eerdmans-Sevensma, [19–?] – 1mf – 9 – 0-524-06098-3 – mf#1991-2411 – us ATLA [240]
Der heilige : novelle / Meyer, Conrad Ferdinand – 22. aufl. Leipzig: H Haessel, 1900 [mf ed 1998] – 235p – 1 – mf#9942 – us Wisconsin U Libr [830]
Der heilige alfons von liguori : der kirchenlehrer und apologet des 18. jahrhunderts / Meffert, Franz – Mainz: F Kirchheim, 1901 – (= ser Forschungen zur christlichen litteratur- und dogmengeschichte) – 1mf – 9 – 0-7905-6765-2 – mf#1988-2765 – us ATLA [240]
Heilige anliegen der kirche : vier reden / Schlatter, Adolf von – Calw: Verlag der Vereinsbuchh, 1896 – 1mf – 9 – 0-524-03470-2 – mf#1990-1013 – us ATLA [240]
Heilige augen- und gemueths-lust : vorstellend, alle sonn- fest- und feyrtaegliche nicht nur evangelien sondern auch epistelen und lectionen... / ed by Kraus, J U – Augspurg: Verfertigt und herauszgegeben von Johann Ulrich Krausen, Kupffer-Stechern, 1706 – 7mf – 9 – mf#O-19 – ne IDC [090]
Das heilige band : roman / Planner-Petelin, Rose – Hamburg: Hanseatische Verlagsanstalt c1942 [mf ed 1992] – 1r – 1 – (filmed with: der fährmann an der weichsel) – mf#3072p – us Wisconsin U Libr [830]
Der heilige berg athos : schilderung / Fallmerayer, Jakob Philipp; ed by Greinz, Rudolf – Leipzig: P Reclam [1908?] [mf ed 1992] – 1mf – 9 – 0-524-02794-3 – mf#1990-0698 – us ATLA [243]
Die heilige bischofsweihe in der katholischen kirche : nach dem roemischen pontifical lateinisch und deutsch – 2. Aufl. Eichstaett, Muenchen, 1879 (mf ed 1993) – 1mf – 9 – 3-89349-356-5 – mf#DHS-AR 356 – gw Frankfurter [241]
Der heilige born : blaetter aus dem bilderbuche des sechzehnten jahrhunderts [a novel] / Raabe, Wilhelm Karl – 2. aufl. Berlin: Otto Janke 1891 [mf ed 1995] – 1r – 1 – (filmed with: deutscher adel) – mf#3707p – us Wisconsin U Libr [830]
Heilige ceremonien, gottesdienstliche kirchenuebungen...der stadt und landschaft zuerich / Herrliberger, D – Basel, Eckenstein, 1750 – 1mf – 9 – mf#ZWI-36 – ne IDC [240]
Der heilige cyprian von karthago : bischof, kirchenvater und blutzeuge christi, in seinem leben und wirken / Peters, Johannes – Regensburg: Georg Joseph Manz, 1877 – 2mf – 9 – 0-8370-6693-X – (incl ind) – mf#1986-0693 – us ATLA [240]
De heilige dominicus / Opzoomer, Cornelis Willem – [s.l: s.n, 1859?] [mf ed 1991] – 23p on 1mf – 9 – 0-524-00295-9 – mf#1989-2995 – us ATLA [241]
Das heilige evangelium des iohannes : syrisch in harklensischer uebersetzung: mit vocalen und den puncten kuschoi und rucoch nach einer vaticanischen handschrift – Leipzig: B G Teubner, 1853 – 1mf – 9 – 0-8370-1379-8 – mf#1987-6060 – us ATLA [226]
Der heilige geist in der heilsverkuendigung des paulus : eine biblisch-theologische untersuchung / Gloel, Johannes – Halles a.S: Max Niemeyer, 1888 – 1mf – 9 – 0-8370-3314-4 – (incl ind) – mf#1985-1314 – us ATLA [220]
Der heilige georg in der griechischen ueberlieferung (abaw.pph25/3) / Krumbacher, K – Muenchen, 1911 – €19.00 – ne Slangenburg [243]
Der heilige hass : exotischer roman / Voss, Richard – Berlin: P Francke [1940?] [mf ed 1991] – 1r – 1 – (filmed with: poetische werke / johann heinrich voss) – mf#2969p – us Wisconsin U Libr [830]
Das heilige irenaeus schrift zum erweise der apostolischen verkuendigung / Ter-Mekerttschian, K – Leipzig, 1907 – (= ser Tugal 3-31/1) – 3mf – 8 – €7.00 – ne Slangenburg [240]
Das heilige land im lichte der neuesten ausgrabungen und funde / Knieschke – Berlin-Lichterfelde: E Runge 1913 [mf ed 1990] – (= ser Biblische zeit- und streitfragen 9/5) – 1mf – 9 – 0-7905-3345-6 – (incl bibl ref) – mf#1987-3345 – us ATLA [930]
Het heilige land of mededeelingen uit eene reis naar het oosten gedaan in de jaren 1849 en 1850 / Senden, G H van – Gorinchem: J Noorduyn en Zoon, 1851-1852. 2v – 9mf – 9 – mf#HT-287 – ne IDC [915]
Heilige leiden und teure blutvergissen jesu christi *see* Threnodiarvm sanctae crvcis in salutiferam passionis d n i c recordationem, continvatio historica
Die heilige mission : mit allen ihren predigten, anreden und feierlichkeiten / Weninger, Francis Xavier – Cincinnati: [s.n.], 1885 – 2mf – 9 – 0-8370-6791-X – mf#1986-0791 – us ATLA [240]
Heilige nacht : eine weihnachtslegende / Thoma, Ludwig – Muenchen; A Langen, 1930 – 1r – 1 – us Wisconsin U Libr [810]
Heilige petrus in rom und rom ohne petrum = Rome and the popes / Brandes, Karl – New York: Benziger, 1868, c1867 – 1mf – 9 – 0-8370-6651-4 – (in english) – mf#1986-0651 – us ATLA [241]
Die heilige pflicht : drei erzaehlungen / Jessen, Paul – feldpostausg. Darmstadt: L Kichler 1942 [mf ed 1990] – 1r – 1 – (filmed with: eddystone / wilhelm jensen) – mf#2743p – us Wisconsin U Libr [830]
Der heilige philippus neri : nach dem italienischen originale des cardinals capecelatro = Vita di s filippo neri / Capecelatro, Alfonso – Freiburg im Breisgau; St Louis, Mo: Herder, 1886 – 1mf – 9 – 0-8370-6725-1 – (in german) – mf#1986-0725 – us ATLA [241]
Die heilige regel fuer ein volkommenes leben : eine cisterzienserarbeit des 13. jahrhunderts / ed by Priebsch, Robert – Berlin: Weidmann, 1909 [mf ed 1993] – (= ser Deutsche texte des mittelalters 16) – xxii/104p/1pl – 1 – (incl bibl ref and ind) – mf#8623 reel 4 – us Wisconsin U Libr [241]
Die heilige sage / Gfroerer, August Friedrich – Stuttgart: C Schweizerbart, 1838 – 2mf – 9 – 0-7905-3441-X – mf#1987-3441 – us ATLA [220]
Die heilige schrift des alten testaments / ed by Kautzsch, Emil et al – 3., voellig neugearb aufl. Tuebingen: J C B Mohr, 1909-1912 – 4mf – 9 – 0-8370-1889-7 – (with int and explanations) – mf#1987-6276 – us ATLA [221]
Die heilige schrift und die negative kritik : ein beitrag zur apologetik / Johansson, Claes Elis – Leipzig: Doerffling & Franke, 1889 – 1mf – 9 – 0-524-06573-X – mf#1992-0916 – us ATLA [220]
Heilige schrift und kritik : ein beitrag zur lehre von der heiligen schrift, insonderheit alten testamentes / Volck, Wilhelm – Erlangen: A Deichert (Georg Boehme), 1897 – 1mf – 9 – 0-8370-5654-3 – mf#1985-3654 – us ATLA [220]
Das heilige schriftwerk kohelet im lichte der geschichte : neue forschung ueber ecclesiastes nebst text, uebersetzung und kommentar / Leimdoerfer, David – Hamburg: G Fritzsche, 1892 – 1mf – 9 – 0-8370-4088-4 – (text in german and hebrew introduction and commentary in german) – mf#1985-2088 – us ATLA [221]
Heilige seelenlust : heilige seelenlust oder geistliche hirtenlieder in ihren jesum verliebten psyche / Silesius, Angelus; ed by Ellinger, Georg – Halle: Max Niemeyer, 1901 – (= ser Neudrucke deutscher literaturwerke des 16. und 17. jahrhunderts n177-181) – xxxvii/312p – 1 – (incl bibl ref) – mf#8413 reel 7 – us Wisconsin U Libr [430]
Die heilige stadt und deren bewohner in ihren naturhistorischen, culturgeschichtlichen, socialen und medicinischen verhaeltnissen / Neumann, B – Hamburg, 1877 – 8mf – 9 – mf#J-27-49 – ne IDC [915]
Der heilige stuhl : eine zeitgemaesse, historisch-philosophische betrachtung / Goerres, Guido – Regensburg, 1838 (mf ed 1994) – 1mf – 9 – €24.00 – 3-8267-3080-1 – mf#DHS-AR 3080 – gw Frankfurter [240]
Der heilige stuhl und die heirat der prinzessin elisabeth von bayern mit dem kronprinzen friedrich wilhelm von preussen : nach akten des vatikanischen geheimarchivs / Bastgen, Hubert – Freiburg i.B., 1930 (mf ed 1993) – 1mf – 9 – €24.00 – 3-89349-251-8 – gw Frankfurter [241]
Heilige vnd trostliche gebaett... / Vermigli, P M – Zuerych, in der Froschouer, 1589 – 4mf – 9 – mf#PBU-651 – ne IDC [240]
Die heiligen der merowinger / Bernoulli, Carl Albrecht – Tuebingen: JCB Mohr, 1900 [mf ed 1992] – 1mf – 9 – 0-524-03512-1 – (incl bibl ref) – mf#1990-1017 – us ATLA [240]
Der heiligen kind / Clausen, Ernst Alexander – Hamburg-Grossborstel: Verlag der Deutschen Dichter...1918 [mf ed 1993] – (= ser Volksbuecher der deutschen dichter-gedaechtnis-stiftung 41) – 1r [ill] – 1 – (ill by theodor herrmann. filmed with: sein bauermaedchen / marianne fleischhack) – mf#8538 – us Wisconsin U Libr [810]
Heiligen perlen-schatzes erste [-zwoelffte] vertheilung ueber den monath januarium [-decembrem] / Lassenius, J – Ulm: Zu finden bey Matthaeo Wagnern, 1695 – 37mf – 9 – mf#O-78 – ne IDC [090]
Heiligenhauser zeitung – Heiligenhaus DE, 1998- – ca 11r/yr – 1 – (bezirksausgabe von westdeutsche allgemeine, essen) – gw Misc Inst [074]
Heiliges feuer : ein fahrtenbuch / Vershofen, Wilhelm – Leipzig: P List, c1936 – 1r – 1 – us Wisconsin U Libr [920]
Das heiligkeits-gesetz, lev 17-26 : eine historisch-kritische untersuchung / Baentsch, Bruno – Erfurt: Hugo Guether, 1893 – 1mf – 9 – 0-8370-2152-9 – mf#1985-0152 – us ATLA [221]
Heiligstedt, August *see* Praeparation zu den psalmen
Das heiligthum und die wahrheit / Gfroerer, August Friedrich – Stuttgart: C Schweizerbart, 1838 – 1mf – 9 – 0-7905-3442-8 – (incl bibl ref) – mf#1987-3442 – us ATLA [220]
Heiligtuemer des konfuzianismus in krue-fu und tschou-hien / Tschepe, Albert – Jentschoufu: Katholischen Mission, 1906 – (= ser Studien und Schilderungen aus China) – 1mf – 9 – 0-524-06974-3 – mf#1991-0048 – us ATLA [720]
Das heiligtum al-husains zu kerbelaa / Noeldeke, Arnold – Berlin: Mayer and Mueller, 1909 – (= ser Tuerkische Bibliothek) – 1mf – 9 – 0-524-01863-4 – (incl bibl ref) – mf#1990-2698 – us ATLA [260]
Heiligtum und opferstaetten in den gesetzen des pentateuch / Engelkemper, Wilhelm – Paderborn [Germany]: Ferdinand Schoeningh, 1908 – 1mf – 9 – 0-8370-3061-7 – mf#1985-1061 – us ATLA [221]
Hei-li-la / Lu, Lun – Shang-hai: Chung-kuo t'u shu ch'u pan kung ssu, Min kuo 32 [1943] – (= ser P-k&k period) – us CRL [480]
Heilman, Lee M *see* Historic sketch of the evangelical lutheran synod of northern illinois
Heilman, Paula S *see* Physiological responses obtained during exercise on the stairmaster gauntlet with and without the use of hands
Heilpaedagogische moeglichkeiten der musik in der sonderpaedagogik / Drobnitzky-Eickhoff, Barbara – (mf ed 1991) – 2mf – 9 – €49.00 – 3-89349-426-X – mf#DHS 426 – gw Frankfurter [370]
Heilperin, Eliezer Levi *see* Sefer be'urim ve-hidusshim
Die heilpflanzen der verschiedenen voelker und zeiten / Dragendorff, Georg – Stuttgart: Verlag von Ferdinand Enke, 1898. vi,884p – 1 – us Wisconsin U Libr [615]
Heilprin, Angelo *see*
– Alaska and the klondike
– Mont pelee and the tragedy of martinique
– Tower of pelee
Heilprin, Jehiel Ben Solomon *see* Seder ha-dorot
Heilprin, Michael *see* The historical poetry of the ancient hebrews
Heilsames gemisch gemasch : das ist: allerley seltsame und verwunderliche geschichten... / Abraham...Sancta Clara – Wuertzburg: Gedruckt bey Hiob Hertzen, 1704 – 6mf – 9 – mf#O-1498 – ne IDC [090]
Heilsames gemisch gemasch : das ist: allerley seltsame und verwunderliche geschichten... / Abraham...Sancta Clara – Wuertzburg: Gedruckt bey Hiob Hertzen, 1704 – 6mf – 9 – mf#O-1499 – ne IDC [090]
Heilsames gemisch gemasch : das ist: allerley seltsame und verwunderliche geschichten... / Abraham...Sancta Clara – Wuertzburg: Gedruckt bey Hiob Hertzen, 1724 – 6mf – 9 – mf#O-1497 – ne IDC [090]
Die heilsbedeutung christi bei den apostolischen vaetern / Wustmann, Georg – Guetersloh: C Bertelsmann, 1905 – (= ser Beitraege zur foerderung christlicher theologie) – 1mf – 9 – 0-524-00668-7 – (incl bibl ref) – mf#1990-0168 – us ATLA [240]
Die heilsbedeutung der taufe im neuen testamente / Althaus, Paul – Guetersloh: Bertelsmann, 1897. Chicago: Dep of Photodup, U of Chicago Lib, 1978 (1r); Evanston: American Theol Lib Assoc, 1984 (1r) – 1 – 0-8370-1120-5 – (incl bibl ref and index) – mf#1984-T133 – us ATLA [225]
Die heilsbedeutung des gesetzes : vortrag / Stange, Carl – Leipzig: Dieterich, 1904 – 1mf – 9 – 0-524-00141-3 – mf#1989-2841 – us ATLA [220]
Die heilsbedeutung des todes christi : biblisch-theologische untersuchung / Kuehl, Ernst – Berlin: Wilhelm Hertz, 1890 – 1mf – 9 – 0-8370-4415-4 – (incl ind of biblical passages cited) – mf#1985-2415 – us ATLA [220]
Der heils-bote – v3-20. 1900-17 [gaps] – (= ser Mennonite serials coll) – 1r – 1 – mf#ATLA 1994-S036 – us ATLA [242]
Die heilsgewissheit / Kaehler, Martin – Berlin: E Runge 1912 [mf ed 1990] – (= ser Biblische zeit- und streitfragen 7/9-10) – 1mf – 9 – 0-7905-3341-3 – mf#1987-3341 – us ATLA [220]
Die heilslehre des christenthums / Weisse, Christian Hermann – Leipzig: S Hirzel, 1862 – 2mf – 9 – 0-524-00802-7 – mf#1990-0234 – us ATLA [240]
Die heilslehre des hl. gregor von nyssa / Aufhauser, Johannes Baptist – Muenchen: JJ Lentner, 1910 – 3mf – 9 – 0-524-03269-6 – (incl bibl ref) – mf#1990-0880 – us ATLA [240]

Heilsthatsachen und glaubenserfahrung / Lemme, Ludwig – Heidelberg: Carl Winter, 1895 – 1mf – 9 – 0-8370-4318-2 – (incl bibl ref) – mf#1985-2318 – us ATLA [240]

Die heilung des orest in goethes iphigenie / Laehr, Hans – Berlin: G Reimer, 1902 – 1r – 1 – (incl bibl ref) – us Wisconsin U Libr [430]

Die heilung des orest in goethes iphigenie auf tauris / Primer, Paul – [S.l: s.n.], 1894; Frankfurt (Main): Enz & Rudolph [mf ed 1990] – (= ser Jahresbericht des koeniglichen kaiser-friedrichs-gymnasiums zu frankfurt a m ostern 1894) – 45p – 1 – mf#7364 – us Wisconsin U Libr [450]

Hei-lung-chiang jih-pao – Harbin, Heilungkiang, China. Heilungkiang Daily. 1956-65. 16 reels – 1 – us Chinese Res [079]

Heim, Karl *see*
– Bilden ungeloeste fragen ein hindernis fuer den glauben?
– Das gewissheitsproblem in der systematischen theologie bis zu schleiermacher
– Glaubensgewissheit
– Leitfaden der dogmatik
– Das weltbild der zukunft

Heim, Norman M *see* Use of the clarinet in published sonatas for clarinet and piano by english composers from 1880 to 1954

Heim, Richard *see* Incantamenta magica graeca latina

Heimann, Bettina *see* Aspects of sexual reproduction and their effects on population genetic structure in creeping thistle (cirsium arvense l scop)

Heimann, Ernest *see* Creative table-top photography

Heimann, Moritz *see* Faust

De heimat – Windhoek SW Africa, 1927-75 – 4r – 1 – sa National [079]

Die heimat *see*
– Der bote aus der heimat
– Der burgfried
– Verdener anzeigenblatt

Heimat – Backa Palanka, Yugoslavia. Mar-Dec 1939 – 1r – 1 – us L of C Photodup [949]

Heimat – Deutsch Palanka (Backa Palanka YU), 1940 6 jan-28 dec – 1r – 1 – gw Misc Inst [077]

Heimat : deutsches wochenblatt – Czernowitz (Cernauti RO), 1927-1931 6 sep – 2r – 1 – gw Misc Inst [077]

Heimat – Iserlohn DE, 1918 jul-1919 nov, 1921-32 – 1r – 1 – gw Mikrofilm [074]

Heimat am mittag *see* Volksblatt fuer den kreis mettmann

Heimat am mittag / bochumer tageblatt – Bochum 1931 2 jan-31 mar, 1932 1 okt-1933 31 mar, 1934 2 jan-31 mar, 1936 1 apr-30 jun & 1 okt-31 dec, 1938 3 jan-31 mar, 1940 1 okt-31 dec, 1949 15 okt-1972 29 mar [mf ed 2004] – 92r fr 1949 – 1 – (also ed bochum-hattingen or ed a; ed for bo-linden, bo-dahlhausen, bo-stiepel; 1 mai 1966: ruhr-anzeiger / bochumer tageblatt; main ed in hattingen) – gw Mikrofilm [074]

Die heimat des vierten evangeliums / Zurhellen, Otto – Tuebingen: J C B Mohr (Paul Siebeck), 1909 – 1mf – 9 – 0-8370-5980-1 – (incl bibl ref) – mf#1985-3980 – us ATLA [225]

Heimat ist arbeit : ein hausbuch deutscher geschichten / Brehm, Bruno – Karlsbad-Drahowitz: A Kraft c1934 [mf ed 1989] – 1r – 1 – (filmed with other titles) – mf#7066 – us Wisconsin U Libr [830]

Heimat ohne ende : zwei kalendergeschichten / Buerkle, Veit – Heilbronn: E Salzer, 1942 [mf ed 1989] – 75p – 1 – mf#7026 – us Wisconsin U Libr [830]

Heimat tagblatt – Saaz (Zatec CZ), 1928 23 may-1934 31 may – 3r – 1 – gw Misc Inst [077]

Heimatblaetter – Geilenkirchen DE, 1925-34 [gaps] – 1r – 1 – (suppl to geilenkirchener zeitung) – gw Misc Inst [074]

Heimatborn *see* Westfaelisches volksblatt

Heimat-bote – Chicago IL (USA), 1929, 1931 7 jan-1932 1 jun, 1937 15 sep-1939 5 oct [gaps] – 2r – 1 – gw Misc Inst [071]

Heimat-bote – Winona WI, Chicago IL (USA), 1929 2 jan-23 oct, 1931 7 jan-1932 31 may [2r] – 2r – 1 – gw Misc Inst [071]

Heimatbote – Chicago IL, Winona MN. 1926 jul 15-1927 jun 30, 1927 jul 7-1928 jun 15, 1928 jun 22-nov 21 – 3r – 1 – (continued by: america-herold und lincoln freie presse; america-herold, lincoln freie presse und heimatbote) – mf#4019760 – us WHS [071]

Heimatbote *see* Paderborner anzeiger

Der heimatdienst – Berlin DE, 1920 1 aug-1926, 1933 jan-mar – 1 – (filmed by other misc inst: 1928-32 [2r]) – gw Misc Inst [074]

Heimatruf – Prag (CZ), 1938 27 aug-31 dec – 1r – 1 – gw Misc Inst [077]

Heimatvolk und heimatflur *see* Neuss-grevenbroicher zeitung

Heimatwarte – Neustadt i. Holstein DE, 1924-40 – 1 – gw Misc Inst [074]

Heimat-zeitung *see* Kaiserswerther nachrichten

Heimat-zeitung. buedericher zeitung – Meerbusch DE, 1934-35 [gaps] – 2r – 1 – gw Misc Inst [074]

Heimat-zeitung des kreises gross-gerau – Gross-Gerau DE, 1978- – ca 8r/yr – 1 – (title varies: 14 feb 2002: gross-gerauer echo) – gw Misc Inst [074]

Heimb, Theop *see* Bernardus gutolfi monachi

Heimbucher, Max *see* Die papstwahlen unter den karolingern

Heimburg, W *see*
– Dazumal
– Kloster wendhusen; ursula
– Unter der linde

Heimgekehrt : schauspiel in drei akten / Wildermann, Ferdinand – Rheine: A Rieke, 1894 – 1r – 1 – us Wisconsin U Libr [820]

Heimg'funden! : wiener weihnachts-comoedie mit gesang in drei acten / Anzengruber, Ludwig – Wien: O F Eirich, 1885 [mf ed 1993] – 85p – 1 – (music by adolf mueller) – mf#8459 – us Wisconsin U Libr [790]

Heimkehr in die mannschaft : roman eines unvergesslichen jahres / Bloem, Walter Julius – Berlin: P Neff, c1934 [mf ed 1989] – 285p – 1 – mf#7035 – us Wisconsin U Libr [830]

Der heimkehrende gatte und sein weib in der weltlitteratur : litterarhistorische abhandlung / Splettstoesser, W – Berlin, 1899 (mf ed 1995) – 1mf – 9 – €24.00 – 3-8267-3131-X – mf#DHS-AR 3131 – gw Frankfurter [410]

Der heimkehrer – Goeppingen, Bonn DE, 1951 sep-1988 – 1 – gw Misc Inst [360]

Heimland – Moskve, Russia. n1-6. 1947-1948 – 1r – us UF Libraries [939]

Ein heimlich gespraech von der tragedia johannis hussen / Cochlaeus, Johannes; ed by Holstein, Hugo – Halle: M Niemeyer, 1900 – (incl bibl ref) – us Wisconsin U Libr [430]

Das heimliche haus : eine kleine kantate / Baumann, Hans – Muenchen: Zentralverlag der NSDAP, F Eher, [194-?] [mf ed 1989] – (= ser Reihe fahrt und feier 8) – 20p – 1 – mf#6983 – us Wisconsin U Libr [780]

Der heimliche koenig : eine dramatische dichtung in fuenf aufzuegen / Griese, Friedrich – Berlin: Theaterverlag A Langen/ G Mueller 1939 [mf ed 1990] – (= ser Buechererei der dramatischen dichtung 9) – 1r – 1 – (filmed with: christian dietrich grabbe in der nachschillerischen entwickelung / joseph gieben) – mf#2687p – us Wisconsin U Libr [810]

Der heimliche koenig : romantische komoedie in vier aufzuegen / Fulda, Ludwig – 2. aufl. Stuttgart; Berlin: J G Cotta 1906 [mf ed 1989] – 1r – 1 – (filmed with: aus der werkstatt) – mf#2595p – us Wisconsin U Libr [820]

Das heimliche leuchten : [literary sketches] / Menzel, Roderich – Berlin: R Moelich [1943?] [mf ed 1990] – 1r [ill] – 1 – (filmed with: gedichte / alfred meissner) – mf#2833p – us Wisconsin U Libr [880]

Die heimliche not : erzaehlung / Witzany, Rudolf – Jena: E Diederichs, 1941, c1939 – 1r – 1 – us Wisconsin U Libr [830]

Heimskringla : a history of the norse kings = Heimskringla / Snorri Sturluson – London: Norroena Society, 1907 – (= ser Anglo saxon classics) – 13mf – 9 – 0-524-08199-9 – (in english) – mf#1991-0312 – us ATLA [948]

Heimskringla – Canada, jan 1886-dec 1959 – 24r – 1 – (in icelandic) – cn Commonwealth Imaging [071]

Die heimstaetter : ein deutsches schicksal in kanada / Goetz, Karl – Leipzig: P Reclam, 1944 – 1r – 1 – us Wisconsin U Libr [830]

Hein, Erica J *see* The effects of an induced internal and external attentional focus upon upper body strength

Hein, Erich *see* Geheime gesellschaften in alter und neuer zeit, ihre organisation, ihre zwecke und ziele

Hein, Gustav *see* Auswahl deutscher prosa der gegenwart

Hein hoyer : [a novel] / Blunck, Hans Friedrich – Hamburg: Hanseatische Verlagsanstalt, c1940-41 [mf ed 1989] – 204p – 1 – mf#7037 – us Wisconsin U Libr [830]

Hein, Michael S *see* A swimming protocol for determination of individual anaerobic threshold

Hein wieck : eine stall- und scheunengeschichte / Kroeger, Timm – Hamburg: A Janssen, 1905 – 1r – 1 – us Wisconsin U Libr [830]

Heine : ein lesebuch fuer unsere zeit / ed by Victor, Walther – Weimar: Volksverlag, 1956 – 1 – (incl bibl ref) – us Wisconsin U Libr [430]

Heine, Bernd *see* Afrikanische verkehrssprachen

Heine, Carl *see* Der unglueckseelige todes-fall caroli 12.

Heine, Gerhard *see*
– Ernst moritz arndt
– Der mann der nach syrakus spazierenging
– Das verhaeltnis aesthetik zur ethik bei schiller

Heine, Heinrich *see*
– Buch der lieder
– Cuadros de viaje
– Daytshland
– Florentinische naechte
– Franzoesische zustaende
– Germaniia
– Harzreise
– Heinrich heine
– Heinrich heine's buch der lieder
– Heinrich heines buch der lieder
– Heinrich heines saemtliche werke
– Heinrich heine's sammtliche werke
– Letzte gedichte und gedanken
– Memorias
– Neue heine-funde
– The north sea
– Poemes et legendes
– Poems and translations
– The poems of heine, complete
– Religion and philosophy in germany
– Religion and philosophy in germany: a fragment
– Saemtliche werke in zwoelf baenden
– Selections from heine's poems

Heine, nietzsche, ibsen : essays / Berg, Leo – Berlin: Concordia Deutsche Verlagsanstalt, H Ehbock, [1908] [mf ed 1989] – 102p – 1 – mf#7008 – us Wisconsin U Libr [840]

Heine und die frau : bekenntnisse und betrachtungen des dichters / Blanck, Karl [comp] – Muenchen: G Mueller und E Rentsch, 1913 [mf ed 2001] – 195p/[4]pl – 1 – (int by comp) – mf#10601 – us Wisconsin U Libr [430]

Heine und duesseldorf : neue beitraege zu einer heine-biographie / Moos, Eugen – Duesseldorf: Schmitz & Olbertz, 1909 – 1r – 1 – (incl bibl ref) – us Wisconsin U Libr [920]

Heine, Wilhelm *see*
– Die expedition in die seen von china, japan und ochotsk unter commando von commodore colin ringgold und commodore john rodgers
– Japan und seine bewohner
– Reise um die erde nach japan an bord der expeditions-escadre unter commodore m c perry in den jahren 1853, 1854 und 1855
– Eine sommerreise nach tripolis
– Wanderbilder aus central-amerika
– Eine weltreise um die noerdliche hemisphaere in verbindung mit der ostasiatischen expedition in den jahren 1860 und 1861

Heinecke, Regina *see* Tuhfat al-wuzara

Heinecken, K H V *see* Dictionnaire des artistes, dont nous avons des estampes...

[Heinecken, K H V] *see*
– Nachrichten von kuenstlern und kunst-sachen
– Neue nachrichten von kuenstlern und kunstsachen...

Heinemann, Franz *see* Der richter und die rechtspflege in der deutschen vergangenheit

Heinemann, H *see* Shylock und nathan

Heinemann, K *see* Ausgewaehlte dichtungen

Heinemann, K [comp] *see* Goethes briefe an frau von stein

Heinemann, Karl *see*
– Die deutsche dichtung
– Goethe
– Klopstocks leben und werke / wielands leben und werke

Heinemann, Lothar *see* Modellbildung von mehrwicklungstransformatoren bei quasi-stationaeren feldstaerkeverteilungen

Heinemann, M *see* Gelasius kirchengeschichte (gcsej5)

Heinemann, O v. *see* Zur erinnerung an gotthold ephraim lessing

Heinen, Mechthild *see* Bernhard pankok

Heiner, Franz *see* Der neue syllabus pius 10

Heiner, Steven W *see* Comparison of risk factors for coronary heart disease in sedentary and physically active college students

Heines charakter und die moderne seele : eine studie mit neuen briefen und dem bisher verschollenen jugendgedicht "deutschland 1815" / Kaufmann, Max – Zuerich: A Mueller, 1902 – 1r – 1 – (incl bibl ref) – us Wisconsin U Libr [430]

Heines geburtstag / Franzos, Karl Emil – Berlin: Concordia Deutsche Verlags-Anstalt, 1900 – 1r – 1 – us Wisconsin U Libr [920]

Heines liebesleben / Kaufmann, Max – Zuerich: A Mueller, [1897] – 1r – 1 – us Wisconsin U Libr [920]

Heinichen, Johann David *see*
– Der general-bass in der composition
– Neu erfundene und gruendliche anseisung

Heinig, Kurt *see* Die finanzskandale des kaiserreichs

Heinisch, Paul *see*
– Das buch der weisheit
– Der einfluss philos auf die aelteste christliche exegese (barnabas, justin und clemens von alexandria)
– Griechentum und judentum im letzten jahrhundert vor christus
– Die griechische philosophie im buche der weisheit
– Die palaestinensischen buecher
– Personifikationen und hypostasen im alten testament und im alten orient
– Septuaginta und buch der weisheit

Heinold, William D *see* Helping responses in ambiguous and unambiguous emergencies as a function of training in first aid

Heinrich 5, der friedfertige : herzog von mecklenburg, 1503-1552 / Schnell, Heinrich – Halle: Verein fuer Reformationsgeschichte 1902 [mf ed 1990] – (= ser Schriften des vereins fuer reformationsgeschichte 19/72) – 1mf – 9 – 0-7905-5132-2 – (incl bibl ref) – mf#1988-1132 – us ATLA [242]

Heinrich, Albert *see* Musik-beilagen zu den gedichten des koenigsberger dichterkreises

Heinrich, Anthony Philipp *see* The dawning of music in kentucky, or the pleasures of harmony in the solitude of nature

Heinrich aus andernach / Unruh, Fritz von – Frankfurt am Main: Frankfurter Societaets-Druckerei, Abt. Buchverlag, 1925 – 1r – 1 – us Wisconsin U Libr [430]

Heinrich bebels facetien : drei buecher – Leipzig: K W Hiersemann, 1931 – (incl bibl ref and ind. latin text with an int in german) – us Wisconsin U Libr [430]

Heinrich bebels facetien drei buecher : historisch-kritische ausgabe / ed by Bebermeyer, Gustav – Leipzig: K W Hiersemann, 1931 [mf ed 1993] – (= ser Blvs 276) – xxix/208p – 1 – (latin text with int in german. incl bibl ref and ind) – mf#8470 reel 55 – us Wisconsin U Libr [410]

Heinrich boell: leben und werk *see* Westdeutsche prosa

Heinrich bullinger : leben und ausgewaehlte schriften / Pestalozzi, Carl – Elberfeld: RL Friderichs, 1858 – (= ser Leben und ausgewaehlte schriften der vaeter und begruender der reformirten kirche) – 2mf – 9 – 0-524-01888-X – (incl bibl ref) – mf#1990-0515 – us ATLA [242]

Heinrich bullinger : der nachfolger zwinglis / Schulthess-Rechberg, Gustav von – Halle a.d. S: Verein fuer Reformationsgeschichte 1904 [mf ed 1990] – (= ser Schriften des vereins fuer reformationsgeschichte 22/82) – 1mf – 9 – 0-7905-5315-5 – (incl bibl ref) – mf#1988-1315 – us ATLA [242]

Heinrich bullinger / Pestalozzi, C – Elberfeld, Friderichs, 1858 – 7mf – 9 – mf#PBU-431 – ne IDC [242]

Heinrich bullinger / Schulthess-Rechberg, G von – Halle, Zuerich, Max Niemeyer, Zuercher & Furrer, 1904 – (= ser Schriften des vereins fuer reformationsgeschichte 22) – 2mf – 9 – mf#PBU-442 – ne IDC [242]

Heinrich bullinger und seine gattin / Christoffel, R – Zuerich, F Schulthess, 1875 – 2mf – 9 – mf#PBU-433 – ne IDC [242]

Heinrich bullingers diarium / Bullinger, Heinrich; ed by Egli, E – Basel, Basler Buch- und Antiquariatshandlung, 1904 – 2mf – 9 – mf#PBU-273 – ne IDC [242]

Heinrich bullingers reformationsgeschichte / ed by Hottinger, J J & Voegeli, H H – Frauenfeld: C. Beyel, 1838-40. Chicago: Dep of Photodup, U of Chicago Lib, 1975 (1r); Evanston: American Theol Lib Assoc, 1984 (1r) – 1 – 0-8370-0695-3 – mf#1984-6081 – us ATLA [242]

Heinrich der loewe / Chomton, Werner – Stuttgart: K Thienemann, 1941 [mf ed 1989] – 191p (ill) – 1 – mf#7152 – us Wisconsin U Libr [830]

Heinrich ewald : orientalist and theologian, 1803-1903, a centenary appreciation / Davies, Thomas Witton – London: T Fisher Unwin, 1903 – 1mf – 9 – 0-7905-0569-X – (incl ind) – mf#1987-0569 – us ATLA [240]

Heinrich federer. seine persoenlichkeit und seine kunstform / Birnbach, Franz Bernhard – Bad Godesberg, 1935 – 1 – gw Mikropress [920]

Heinrich, Gerd H *see*
– Ichneumoninae of florida and neighboring states

Heinrich heine : aus seinem leben und aus seiner zeit / Karpeles, Gustav – Leipzig: A Titze, 1899 – 1r – 1 – us Wisconsin U Libr [920]

Heinrich heine : confessio judaica / Heine, Heinrich – Berlin, Germany. 1925 – 1r – us UF Libraries [939]

Heinrich heine / Holzamer, Wilhelm – Berlin: Schuster and Loeffler, [19–?] – 1r – 1 – us Wisconsin U Libr [430]

Heinrich heine : ein lebens- und zeitbild / Wendel, Hermann – Dresden: Kaden, [1916?] – 1 – (incl ind) – us Wisconsin U Libr [430]

Heinrich heine : sein leben, sein charatker und seine werke / Keiter, Heinrich – Koeln: J P Bachem, 1891 – 1 – (incl bibl ref) – us Wisconsin U Libr [920]

Heinrich heine : ein vortrag / Hass, Hans-Egon – Bonn: H Bouvier, 1949 – 1 – us Wisconsin U Libr [430]

Heinrich heine als dichter des judentums : ein versuch / Plotke, Georg J – Dresden: C Reissner, 1913 – 1r – 1 – us Wisconsin U Libr [430]

Heinrich heine als student / Scheuer, Oskar Franz – Bonn: A Ahn, 1922 – 1r – 1 – (incl bibl ref) – us Wisconsin U Libr [920]

Heinrich heine und das deutsche volkslied : eine kritische untersuchung nach dem stoffgebiete der heine'schen lyrik / Greinz, Rudolf – Neuwied a. Rhein: A Schupp, [1894?] – 1r – 1 – us Wisconsin U Libr [430]
Heinrich heine's an essay, read before the monday club, may 21st, 1883 / Kendig, Abby E G – Chicago, 1884. 14p – 1 – us Wisconsin U Libr [430]
Heinrich heines beziehungen zu e.t.a. hoffmann / Siebert, Wilhelm – Marburg a.L.: N G Elwert, 1908 – 1r – 1 – (incl bibl ref) – us Wisconsin U Libr [430]
Heinrich heine's biographie / Karpeles, Gustav – Hamburg: Hoffmann und Campe, 1885 – 1r – 1 – us Wisconsin U Libr [920]
Heinrich heine's buch der lieder : vervollstaendigt herausgegeben / Heine, Heinrich; ed by Lachmann, Otto F – [3. Aufl.]. Leipzig: P Reclam jun., [1839?] – 1 – us Wisconsin U Libr [810]
Heinrich heines buch der lieder : nebst einer nachlese nach den ersten drucken oder handschriften / ed by Elster, Ernst – Heilbronn: Henninger, 1887 [mf ed 1993] – (= ser Deutsche litteraturdenkmale des 18. und 19. jahrhunderts 27) – xliv/255p – 1 – (incl bibl ref. int by ed) – mf#8676 reel 3 – us Wisconsin U Libr [810]
Heinrich heine's letzte tage : erinnerungen / Selden, Camilla – Jena: H Costenoble, 1884 – 1 – us Wisconsin U Libr [920]
Heinrich heine's life told in his own words / ed by Karpeles, Gustav – New York: H Holt, 1893 – 1 – us Wisconsin U Libr [920]
Heinrich heines religioese entwicklung bis zum abschluss seiner universitaetsjahre / Puetzfeld, Carl – Berlin: G Grote, 1912 – 1r – 1 – (incl bibl ref) – us Wisconsin U Libr [430]
Heinrich heines saemtliche werke / Heine, Heinrich; ed by Walzel, Oskar – Leipzig: Insel-Verlag, 1910-1915 [mf ed 1995] – 10v on 3r – 1 – (incl bibl ref and ind) – mf#8820 – us Wisconsin U Libr [802]
Heinrich heine's sammtliche werke / Heine, Heinrich – Hamburg, Germany. v1-12. 1876 – 2r – us UF Libraries [025]
Heinrich heines verhaeltnis zur bildenden kunst / Hessel, Karl Robert Heinrich – Marburg a.L: N G Elwert, 1931 – 1r – 1 – (incl bibl ref) – us Wisconsin U Libr [700]
Heinrich heine's verhaeltnis zur religion / Kalischer, Alfred Christlieb – Dresden: F Oehlmann, 1890 – 1r – 1 – us Wisconsin U Libr [240]
Heinrich hugs villinger chronik von 1495 bis 1533 / ed by Roder, Christian – Stuttgart: Litterarischer Verein, 1883 (Tuebingen: L F Fues) [mf ed 1993] – (= ser Blvs 164) – 176p – 1 – (incl bibl ref and ind) – mf#8470 reel 34 – us Wisconsin U Libr [430]
Heinrich hugs villinger chronik von 1495-1533 / ed by Roder, Christian – Stuttgart: Litterarischer Verein, 1883 (Tuebingen: H Laupp) – (incl bibl ref and ind) – us Wisconsin U Libr [430]
Heinrich Julius, Duke of Brunswick-Wolfenbuettel *see* Die schauspiele des herzogs heinrich julius von braunschweig
Heinrich kaufringers gedichte / ed by Euling, Karl – Stuttgart: Litterarischer Verein, 1888 (Tuebingen: H Laupp) – (incl bibl ref and ind) – us Wisconsin U Libr [810]
Heinrich kaufringers gedichte / ed by Euling, Karl – Stuttgart: Litterarischer Verein, 1888 (Tuebingen: H Laupp) [mf ed 1993] – (= ser Blvs 182) – xvi/244p – 1 – (incl bibl ref and ind) – mf#8470 reel 38 – us Wisconsin U Libr [810]
Heinrich loest ueber e.t.a. hoffmann : 15. august 1823 / Loest, H W; ed by Mueller, Hans von – Koeln: P Gehly, 1922 – 1r – 1 – (incl bibl ref) – us Wisconsin U Libr [430]
Heinrich melchior muehlenberg, patriarch de lutherischen kirche nordamerikas : selbstbiographie, 1711-1743, aus den missionsarchive der franckischen stiftungen zu halle / Muhlenberg, Henry Melchior – Allentown, PA: Brobst, Diehl, 1881 – 1mf – 9 – 0-524-06819-4 – mf#1991-2806 – us ATLA [242]
Heinrich mynsinger von den falken, pferden und hunden / Albertus, Magnus, Saint; ed by Hassler, K D – Stuttgart: Litterarischer Verein, 1863 [mf ed 1993] – (= ser Blvs 71) – 98p – 1 – (trans of passages fr de animalbus. incl bibl ref) – mf#8470 reel 14 – us Wisconsin U Libr [636]
Heinrich salt's neue reise nach abyssinien in den jahren 1809 und 1810 – Weimar 1815 [mf ed Hildesheim 1995-98] – 1v on 3mf – 9 – €90.00 – 3-487-26525-7 – gw Olms [916]
Heinrich schuetz (1585-1672) : collected works / ed by Spitta, Philipp et al – Leipzig: Breitkopf & Haertel. 18v. 1885-1927 – 11 – $140.00 set – us Univ Music [780]
Heinrich seidel und der deutsche humor : eine litterarische wuerdigung / Biese, Alfred – Stuttgart: A G Liebeskind, [19-?] [mf ed 1993] – 24p – 1 – (incl bibl ref) – mf#7714 – us Wisconsin U Libr [430]
Heinrich, Ulrike *see* Reflektionsspektroskopische untersuchungen als beispiel einer nicht-invasiven messmethode zur quantitativen substanzanalyse in vivo
Heinrich und kunigunde / Erfurt, Ebernant von; ed by Bechstein, Reinhold – Quedlinburg, Leipzig: G Basse, 1860 [mf ed 1993] – (= ser Bibliothek der gesammten deutschen national-literatur von der aeltesten bis auf die neuere zeit sect1/39) – xxxii/207p – 1 – (incl ind) – mf#8438 reel 8 – us Wisconsin U Libr [810]
Heinrich v. kleist : eine studie / Eloesser, Arthur – Berlin o.J. (mf ed 1994) – 1mf – 9 – €24.00 – 3-8267-3016-X – mf#DHS-AR 3016 – gw Frankfurter [430]
Heinrich v. kleist – kant und wieland / Luther, Bernhard; ed by Wieland-Museum – Biberach-Riss: Wieland Museum, 1933 – 1r – 1 – (incl bibl ref) – us Wisconsin U Libr [430]
Heinrich von Beringen *see* Das schachgedicht
Heinrich von burgus der seele rat / ed by Rosenfeld, Hans-Friedrich – Berlin: Weidmann, 1932 [mf ed 1993] – (= ser Deutsche texte des mittelalters 37) – xlviii/146p/1pl – 1 – (incl bibl ref and ind. allegorical poem, written c1300, of which 1st pt is missing. the brixen mss apparently contains the only extant copy of the poem) – mf#8623 reel 7 – us Wisconsin U Libr [810]
Heinrich von kleist / Ayrault, Roger – Paris: Librairie Nizet et Bastard, 1934 [mf ed 1995] – 588p – 1 – (incl bibl ref) – mf#8789 – us Wisconsin U Libr [430]
Heinrich von kleist / Brahm, Otto – Berlin: Allgemeiner Verein fuer Deutsche Literatur 1884 [mf ed 1995] – 1r – 1 – (incl bibl ref; filmed with: schiller's dramas and poems... / rea, thomas & other titles) – mf#3733p – us Wisconsin U Libr [430]
Heinrich von kleist : darstellung des problems / Hellmann, Hanna – Heidelberg: C Winter, 1911 – 1r – 1 – (incl bibl ref) – us Wisconsin U Libr [430]
Heinrich von kleist : der dichter des preussentums / Fischer, Max – Stuttgart: J G Cotta, 1916 – 1r – 1 – us Wisconsin U Libr [430]
Heinrich von kleist / Graef, Hermann – Leipzig: Verlag fuer Literatur, Kunst und Musik, 1906 – 1r – 1 – us Wisconsin U Libr [920]
Heinrich von kleist / Gundolf, Friedrich – Berlin: G Bondi, 1922 – 1 – us Wisconsin U Libr [430]
Heinrich von kleist / Kiesgen, Laurenz – Leipzig: P Reclam, [1915?] – 1r – 1 – us Wisconsin U Libr [920]
Heinrich von kleist / Kuhn-Foelix, August – Murnau: U Riemerschmidt, c1948 – 1r – 1 – us Wisconsin U Libr [920]
Heinrich von kleist : lehrjahre 1799-1801 / Howe, George M – [S.l.]: Modern Language Association of America, 1926 – 1 – (incl bibl ref) – us Wisconsin U Libr [920]
Heinrich von kleist : das problem seines lebens und seiner dichtung; ein versuch / Hellmann, Hanna – Heidelberg: C Winter, 1908 – 1r – 1 – (incl bibl ref) – us Wisconsin U Libr [430]
Heinrich von kleist : eine rede / Bertram, Ernst – Bonn: F Cohen, 1925 [mf ed 1990] – 31p – 1 – mf#7514 – us Wisconsin U Libr [080]
Heinrich von kleist : "robert guiskard" – ein beitrag zur inszenierung des fragmentes / Birk, Karl – Prag: J G Calve, 1911 [mf ed 1991] – 25p – 1 – mf#7509 – us Wisconsin U Libr [790]
Heinrich von kleist / Schmidt, Erich – Leipzig, Wien: Bibliographisches Institut, [19-?] – 1r – 1 – us Wisconsin U Libr [920]
Heinrich von kleist : seine sprache und sein stil / Minde-Pouet, Georg – Weimar: E Felber, 1897 – 1 – (incl bibl ref) – us Wisconsin U Libr [430]
Heinrich von kleist : der zerbrochene krug – ein beitrag zur inszenierung des lustspieles / Birk, Karl – Prag: C Bellmann, 1910 [mf ed 1990] – 54/[1]p – 1 – (incl bibl ref) – mf#7509 – us Wisconsin U Libr [790]
Heinrich von kleist als mensch und dichter : nach neuen quellenforschungen / Rahmer, Sigismund – Berlin: G Reimer, 1909 – 1 – (incl bibl ref and index) – us Wisconsin U Libr [430]
Heinrich von kleist in seinen briefen : eine charakteristik seines lebens und schaffens / ed by Schur, Ernst – Charlottenburg: Schiller-Buchhandlung, [1911?] – 1 – us Wisconsin U Libr [920]
Heinrich von kleist in seinen briefen : ein vortrag, gehalten am 24. oktober 1898 im historisch-philosophischen verein zu heidelberg / Warkentin, Roderich – Heidelberg: C Winter, 1900 – 1r – 1 – us Wisconsin U Libr [430]
Heinrich von kleist und das deutsche theater / Kuehn, Walter – Muenchen: H Sachs-Verlag, 1912 – 1r – 1 – (incl bibl ref) – us Wisconsin U Libr [430]
Heinrich von kleist und die frauen / Kohut, Adolf – Hamburg: Verlagsgesellschaft Hamburg, 1911 – 1r – 1 – us Wisconsin U Libr [430]
Heinrich von kleist und die kantische philosophie / Cassirer, Ernst – Berlin: Reuther & Reichard 1919 [mf ed 1991] – (= ser Philosophische vortraege [kant-gesellschaft] 22) – 1r – 1 – (filmed with: kleist / wilhelm hegeler) – mf#2761p – us Wisconsin U Libr [430]
Heinrich von kleists geheimnis / Finger, Richard – Berlin: Puttkammer & Muehlbrecht, 1913 – 1r – 1 – us Wisconsin U Libr [430]
Heinrich von kleists kunst : vortrage gehalten zur feier der 150. wiederkehr von kleists geburtstag / Walzel, Oskar Franz – Bonn: L Rohrscheid, 1928 – 1r – 1 – us Wisconsin U Libr [430]
Heinrich von kleists reise nach wuerzburg / Morris, Max – Berlin: C Skopnik 1899 – 1r – 1 – (incl excerpts from kleist's letters to his sister and his fiancee) – us Wisconsin U Libr [430]
Heinrich von kleists tragischer untergang / Servaes, Franz – Berlin: E Runge, [1900?] – 1r – 1 – us Wisconsin U Libr [430]
Heinrich von ofterdingen : wartburgkrieg und verwandte dichtungen / Mess, Friedrich – Weimar: H Boehlaus, 1963 – 1r – 1 – (incl bibl ref) – us Wisconsin U Libr [430]
Heinrich von zuetphen / Iken, J Friedrich – Halle: Verein fuer Reformationsgeschichte, 1886 – (= ser [Schriften des Vereins fuer Reformationsgeschichte]) – 1mf – 9 – 0-7905-4656-6 – (incl bibl ref) – mf#1988-0656 – us ATLA [920]
Heinrich wilhelm von gerstenberg und der sturm und drang / Wagner, Albert Malte – Heidelberg: C Winter. 2v. 1920 (mf ed 1990) – 1r – 1 – (filmed with: die regulatoren in arkansas. incl bibl ref and ind) – us Wisconsin U Libr [430]
Heinrich winckel und die reformation im suedlichen niedersachsen / Jacobs, Eduard – Halle: Verein fuer Reformationsgeschichte 1896 [mf ed 1990] – (= ser Schriften des vereins fuer reformationsgeschichte 13/53) – 1mf – 9 – 0-7905-4827-5 – (incl bibl ref) – mf#1988-0827 – us ATLA [242]
Heinrich zimmermanns von wissloch in der pfalz : reise um die welt, mit capitain cook – Mannheim [Germany]: Bei C F Schwan...1781 [mf ed 1984] – 2mf – 9 – 0-665-44926-7 – mf#44926 – cn Canadiana [910]
Heinrich zschokke : ein biographischer umriss / Zschokke, Emil – 1r – 1 – us Wisconsin U Libr [943]
Heinrich zschokkes ausgewaehlte novellen / Zschokke, Heinrich – Leipzig: M Hesse [1904] [mf ed 1979] – 6v in 2 on 1r – 1 – mf#film mas 8640 – us Harvard [830]
Heinrich zschokkes jugend- und bildungsjahre (bis 1798) : ein beitrag zu seiner lebensgeschichte / Guenther, Carl – Aarau: H R Sauerlaender, 1918 – 1r – 1 – (incl bibl ref) – us Wisconsin U Libr [430]
Heinrich zschokke's novellen / Zschokke, Heinrich – Berlin: Wiener [18–] [mf ed 1979] – 10v in 5 on 1r – 1 – (v8-10 have title: humoristische novellen) – mf#film mas 8591 – us Harvard [830]
Heinrich's von freiberg tristan / ed by Bechstein, Reinhold – Leipzig: F A Brockhaus, 1877 – 1r – 1 – (incl bibl ref and ind. middle high german text with an introduction in german) – us Wisconsin U Libr [430]
Heinrich's von krolewiz uz missen vater unser / Krolewitz, Heinrich von; ed by Lisch, Ge Chr Friedrich – Quedlinburg; Leipzig: G Basse, 1839 – 10r – 1 – us Wisconsin U Libr [430]
Heinrichs von meissen des frauenlobes leiche, sprueche, streitgedichte und lieder / ed by Ettmueller, Ludwig – Quedlinburg, Leipzig: G Basse, 1843 [mf ed 1993] – (= ser Bibliothek der gesammten deutschen national-literatur von der aeltesten bis auf die neuere zeit sect1/16) – xlv/420p – 1 – mf#8438 reel 4 – us Wisconsin U Libr [430]
Heinrichs von neustadt 'apollonius von tyrland' : nach der gothaer handschrift 'gottes zukunft' und 'visio philiberti' nach der heidelberger handschrift / ed by Singer, Samuel – Berlin: Weidmann, 1906 [mf ed 1993] – (= ser Deutsche texte des mittelalters 7) – xiii/534p/6pl (ill) – 1 – (incl bibl ref and ind) – mf#8623 reel 1 – us Wisconsin U Libr [390]
Heinrichs von veldeke eneide / ed by Behaghel, Otto – Heilbronn: Gebr Henninger, 1882 [mf ed 1996] – xv/ccxxxiii/566p – 1 – (with int and notes) – mf#9737 – us Wisconsin U Libr [810]
Heinrici, Carl Friedrich Georg *see*
– Die bergpredigt (matth. 5-7, luk. 6, 20-49)
– Die bodenstaendigkeit der synoptischen ueberlieferung vom werke jesu
– Duerfen wir noch christen bleiben?
– Das erste sendschreiben des apostel paulus an die korinthier
– Die geschichtliche entwickelung der kirche im 19. jahrhundert und die ihr dadurch gestellte aufgabe – die forschungen ueber die paulinischen briefe
– Griechisch-byzantinische gespraechsbuecher und verwandtes aus sammelhandschriften
– Der leipziger papyrusfragmente der psalmen
– Der litterarische charakter der neutestamentlichen schriften
– Paulus als seelsorger
– Theologie und religionswissenschaft
– Theologische encyklopaedie
– Das urchristentum
– Der zweite brief an die korinther
– Das zweite sendschreiben des apostel paulus an die korinthier
Heinrici chronicon livoniae (mgh7:31.bd) – 1874 – (= ser Monumenta germaniae historica 7: scriptores rerum germanicarum in usum scholarum (mgh7)) – €12.00 – ne Slangenburg [240]
Hein's bar journal microfiche service – inception-2001 – 9 – $38,500.00 set (2002 subs $1560 set) – (inception-1969 $14,500 set. 1970-74 $3400 set. 1975-79 $3700 set. 1980-84 $4800 set. 1985-98 price varies per yr) – mf#400640 – us Hein [340]
Hein's early federal laws collection – Installments 1-3 – 9 – $3450.00 set – mf#402230 – us Hein [340]
Hein's federal legislative histories collection – Installments 1-4 – 9 – $7345.00 set – (incl annotated bibl and ind to officially publ sources by bernard d reams) – mf#408660 – us Hein [340]
Hein's legal theses and dissertations microfiche project – Installments 1-22 – 9 – $13,684.00 set – mf#408080 – us Hein [340]
Heins, Otto *see* Johann rist und das niederdeutsche drama des 17. jahrhunderts
Hein's state bar examinations – 1985-99 no 2 update – 9 – $1,696.00 set – 0-89941-632-2 – (updated semi-annually) – mf#401061 – us Hein [340]
Hein's united states treaties and other international agreements – 1957-98 no 50 – 9 – $6,750.00 set – (1987-96 files complete. 1957-86 additional files to follow. index provided) – mf#402210 – us Hein [327]
Heinsberger grenzpost *see* Heinsberger volkszeitung 1882
Heinsberger volkszeitung 1882 – Heinsberg DE, 1957 2 nov-1959 30 jun – 1 – (bezirksausgabe von aachener volkszeitung, aachen; title varies: 22 jan?1946: aachener volkszeitung; 31 aug 1949: heinsberger grenzpost; 15 sep 1949: aachener volkszeitung; 3 dec 1949: heinsberger volkszeitung; 4 may 1996: heinsberger zeitung. postwar ed; fr 1996 regional ed aachener zeitung, aachen) – gw Misc Inst [074]
Heinse, Gottlob *see* Reisen durch das suedliche deutschland und die schweitz in den jahren 1808 und 1809
Heinse und hoelderlin / Reuss, Theodor – Stuttgart: J F Steinkopf, 1906 – 1r – 1 – us Wisconsin U Libr [920]
Heinse, Wilhelm *see* Saemmtliche werke
Heinses stellung zur bildenden kunst und ihrer aesthetik / Jessen, Karl Detlev – Berlin: Mayer & Mueller, 1901 – 1r – 1 – (incl bibl ref) – us Wisconsin U Libr [430]
Heinsius, D *see*
– Lof-sanck van iesus christus den eenigen ende eeuwigen sone godes
– Nederduytsche poemata
[Heinsius, D] *see*
– Acta ofte handelinghen des nationalen synodi...
– Afbeeldingen van minne
– Emblemata amatoria
– Quaeris quid sit amor, quid amare...
– Quaeris quid sit amor, quid amare, cupidinis et quid castra sequi?
Heinsius, Wilhelm *see*
– Allgemeines bucher-lexicon
– Allgemeines buecher-lexikon
Heint – New York. 1, no. 1-52. Jan. 1-Feb. 21, 1920 – 1 – us NY Public [071]
Heintige nais – Kaunas, 1940 – 2r – 1 – us UMI ProQuest [077]
Heintschel-Heinegg, Bernd von *see*
– Materielles scheidungsrecht
– Das verfahren in familiensachen
Heintz, Albert *see*
– Richard wagners lohengrin
– Wegweiser durch die motivenwelt der musik zu richard wagners buehnenfestspiel der ring des nibelungen
Heintze, W *see* Der klemensroman und seine griechischen quellen
Heintze, Werner *see* Der klementsroman und seine griechischen quellen
Heintzelman, Samuel *see* Journals
Heinz, Bertram *see* Elektronenspektroskopische untersuchung von alkan- und alkanthiolfilmen auf festkoerperflaechen
Heinz, Hans-Joachim *see* Nsdap und verwaltung in der pfalz

Heinz, John *see* Quedah merchant

Heinz, Margarete *see* Ueber das politische bewusstsein von frauen in der bundesrepublik

Heinz von wolfenbuettel : ein zeitbild aus dem jahrhundert der reformation / Koldewey, Friedrich – Halle: Verein fuer Reformationsgeschichte, 1883 – (= ser [Schriften des Vereins fuer Reformationsgeschichte]) – 1mf – 9 – 0-7905-7174-9 – (incl bibl ref) – mf#1988-3174 – us ATLA [943]

Heinze, Hermann *see*
- Aufgaben aus "maria stuart"
- Aufgaben aus wallenstein

Heinze, Max *see* Die lehre vom logos in der griechischen philosophie

Heinze, Val Aug *see* Kielisches magazin vor die geschichte, staatsklugheit und staatenkunde

Heinzel, Richard *see* Geschichte der niederfraenkischen geschaeftssprache

Heinzelmann, Jacob Harold *see* The influence of the german volkslied on eichendorff's lyric

Heinzelmann, W *see* Goethes iphigenie

Heinzen, Karl Peter *see* Der teutsche editoren-kongress zu cincinnati

Heinzmann, Joh Georg *see* Patriotisches archiv fuer die schweiz angel von einer helvetischen gesellschaft

Heinzmann, Johann Georg *see* Litterarische chronik in aufsammlung zerstreuter blaetter zur gelehrsamkeit, philosophie und kritik

Heir-lines / North Oakland Genealogical Society – 1978 jan-1984 winter, 1984 spring-1988 autumn – 2r – 1 – (continued by: heir-lines (lake orion, mi: 1996)) – mf#553240 – us WHS [929]

Heirs of promise / Hawker, Robert – London, England. 1820 – 1r – us UF Libraries [240]

Heirs together of the grace of life : benjamin broomhall, amelia hudson broomhall / Broomhall, Marshall – London: Morgan & Scott; Philadelphia: China Inland Mission [1918] [mf ed 1995] – (= ser Yale coll) – xv/146p (ill) – 1 – 0-524-09630-9 – (with preface by handley c g moule) – mf#1995-0630 – us ATLA [920]

Heirs together of the grace of life / Chapel-Cure – London, England. 18– – 1r – us UF Libraries [240]

Heis, Eduard *see*
- Die feuerkugel
- Resultate der in den 43 Jahren 1833-1875 angestellten sternschnuppen-beobachtungen

Heise, Christoph Ulrich *see* Die gesetzessammlung (ulozenie) von 1649 und ihre auswirkungen auf die kirche in der aera nikons

Heise, Wolfgang *see* Bild und begriff

Heiseler, Bernt von *see* Ahnung und aussage

Heiseler, Henry von *see*
- Grischa
- Peter und alexej
- Wawas ende

Heiser, Robert F *see* The archaeology of the napa, california region

Heisey glass newscaster / A H Heisey and Co – v7 n1-v15 n4 [1977 jan/mar-1986 oct/dec] – 1r – 1 – mf#848057 – us WHS [670]

Heisey, Paul Harold *see* Psychological studies in lutheranism

Heiskell, Carrick White *see* Pioneer presbyterianism in tennessee

Heisler, Daniel Yost *see* Life pictures of the prodigal son

Heiss, Michael *see* The four gospels examined and vindicated on catholic principles

Heit, Philip *see*
- Death education and death anxiety in student nurse aides
- An hiv education needs assessment of selected teacher members of the american school health association and the american home economics association

Heitere geschichten / Sealsfield, Charles – Prag: Volk und Reich Verlag, c1944 – 1r – 1 – us Wisconsin U Libr [830]

Heitere hamsterkiste / Doering, Bruno – Leipzig, Germany. 1940 – 1r – us UF Libraries [025]

Heitere welt *see* Niederbarnimer kreisblatt

Heiterer guckkasten / Prochaska, Bruno – 4. Aufl. Berlin: C Stephenson, c1941 – 1r – 1 – us Wisconsin U Libr [830]

Heiteres darueberstehen : familienbriefe, neue folge / Fontane, Theodor; ed by Fontane, Friedrich – Berlin: G Grote, 1937 – xxiv/277p – 1 – (int by hanns martin elster. incl ind) – mf#7073 – us Wisconsin U Libr [880]

Heiteres und weiteres : kleine geschichten / Wolzogen, Ernst von – 4. aufl. Berlin: F Fontane [190-?] [mf ed 1993] – 1r – 1 – (filmed with: der kraft-mayr & other titles) – mf#7967 – us Wisconsin U Libr [830]

A der heiteri : no nes paar geschichtli / Buerki, Jakob – Langnau: Emmenthaler-Blatt, 1937 [mf ed 1989] – 187p – 1 – mf#7095 – us Wisconsin U Libr [830]

Heiterkeit des herzens : erlebnisse zwischen alltag und sonntag / Bruees, Otto – Leipzig: O Janke, [1944] [mf ed 1989] – 63p (ill) – 1 – mf#7092 – us Wisconsin U Libr [880]

Heitmann, Felix *see*
- Annette von droste-huelshoff als erzaehlerin
- Zur erzaehlungskunst der annette von droste-huelshoff

Heitmann, P *see* Beurteilung des effektes organischer loesungsmittel auf das hoervermoegen

Heitmueller, Franz Ferdinand *see* Der bookesbeutel

Heitmueller, Wilhelm *see*
- Im namen jesu
- Taufe und abendmahl bei paulus

Heitz, Friedrich C *see* Das zunftwesen in strassburg

Heitz, P *see* Strassburger holzschnitte

Heitz, Paul *see*
- Neujahrswuensche des 15. jahrhunderts
- Das wunderblut zu wilsnack

Heitzmann, Louis *see* Urinary analysis and diagnosis by microscopical and chemical...

Der heizer : ein fragment / Kafka, Franz – Leipzig: K Wolff 1913 [mf ed 1990] – 1r – 1 – (filmed with: ernst junger / wulf dieter muller) – mf#2749p – us Wisconsin U Libr [830]

Heizkostenabrechnung nach verbrauch : kommentar / Peruzzo, Guido – Frankfurt/Main: J Schweitzer Verlag, 1990 (mf ed 1996) – 2mf – 9 – €31.00 – 3-8267-9682-9 – mf#DHS 9682 – gw Frankfurter [346]

Hej – Helsingborg, 1968-69 – 1r – 1 – sw Kungliga [078]

Hekate – 1823 [mf ed 1997] – (= ser Die zeitschriften des august von kotzebue) – 9mf – 9 – €100.00 – 3-89131-236-9 – gw Fischer [074]

The hekatompathia, or passionate centurie of love / Watson, Thomas – 1582 – 9 – us Scholars Facs [810]

Heker da'at / Sivitz, Moses Simon – Jerusalem, Israel. v1-2. 1898 – 1r – us UF Libraries [939]

Heko – Dar es Salaam: Heko Publ, n3. dec 1985- – us CRL [079]

Hekserij bij de baluba van kasai / Caeneghem, E P R van – [Bruxelles, 1955] – 1 – us CRL [960]

Hektoen, Ludvig *see* An american text-book of pathology

Hela veckan – Jonkoping, Sweden. 1982-87 – 1 – sw Kungliga [078]

Helbing, Robert *see*
- Grammatik der septuaginta
- Die praepositionen bei herodot und andern historikern

Helbok, Adolf *see* Haus und siedlung im wandel der jahrtausende

Helbronner, Horace *see* Le pouvoir judiciaire aux etats-unis; son organisation et ses attributions

Held, Adolf *see* Sozialismus, sozialdemokratie und sozialpolitik

Held, Georg *see* Theorie der merkantilrechnung

Held, Hans Ludwig *see* Buddha

Held in the everglades / Spalding, Henry Stanislaus – New York, NY. 1919 – 1r – us UF Libraries [978]

Held, Johann C *see* Briefe aus paris geschrieben in den monaten sept, oct, nov 1830

Held ohne namen: ein schicksal / Hepner, Gerda – Tubingen: R. Wunderlich, 1932 – 1 – us Wisconsin U Libr [900]

Held seines landes : roman / Bloem, Walter – Leipzig: K F Koehler, 1929 [mf ed 1989] – 437p – 1 – mf#7032 – us Wisconsin U Libr [830]

Der held vom wald : schauspiel in fuenf aufzuegen / Essig, Hermann – Stuttgart: J G Cotta 1913, c1912 [mf ed 1989] – 1r – 1 – (filmed with: bozena / marie von ebner-eschenbach) – mf#7268 – us Wisconsin U Libr [820]

Helden der kolonien : der weltkrieg in unseren schutzgebieten / Heichen, Walter [comp] – Berlin: A Weichert, 1938 (mf ed 19–) – 160p/[5]pl (ill) – mf#Z-823 – us NY Public [933]

Die helden der naukluft : eine erzaehlung aus deutsch-suedwest / Bayer, Maximilian – 12. aufl. Potsdam: L Voggenreiter, 1943 [mf ed 1989] – (= ser Zeitbuecherei 26-28) – 188p (ill) – 1 – mf#7001 – us Wisconsin U Libr [880]

Helden to hus / Lau, Fritz – Hamburg: M Glogau, 1918 – 1r – 1 – us Wisconsin U Libr [830]

Helden und abenteurer / Westheim, Paul – Berlin, Germany. 1931 – 1r – us UF Libraries [720]

Die heldenbraut : ein gedicht aus dem amerikanischen befreiungs-kriege / Alpers, Wilhelm – New York: Willmer & Rogers, 1876 [mf ed 1988] – 109p – 1 – mf#6935 n7 – us Wisconsin U Libr [810]

Heldendichtung, geistlichendichtung, ritterdichtung / Schneider, Hermann – Heidelberg: C Winter, 1925 [mf ed 1993] – (= ser Geschichte der deutschen literatur v1) – xvi/532p – 1 – (incl bibl ref and ind) – mf#7848 – us Wisconsin U Libr [430]

Heldentod: studien zur vergleichenden psychologie / Spitta, Heinrich – Tuebingen: Kloeres, 1915. 32p – 1 – us Wisconsin U Libr [940]

Heldn fun der revolutsye fun noentn 'over... / Gershuni, Grigorii Andreevich – Varshe, Poland. 1938 – 1r – us UF Libraries [939]

Held's volksvertreter *see* Der volksvertreter

Helen e. moses of the christian woman's board of missions : biographical sketch, memorial tributes, missionary addresses by mrs. moses, sonnets and other verses / ed by Moses, Jasper Turney – New York: Fleming H Revell, c1909 – 1mf – 9 – 0-524-07025-3 – mf#1991-2878 – us ATLA [240]

Helena / Machado De Assis – Rio de Janeiro, Brazil. 1939 – 1r – us UF Libraries [972]

Helena / Menendez, Aldo – Habana, Cuba. 1965 – 1r – us UF Libraries [972]

Helena herald – Helena MT. [1866 nov15-1867 aug 24] – 1r – 1 – (cont: montana radiator; cont by: helena weekly herald) – mf#852129 – us WHS [071]

Helena in goethes faust / Rickert, Heinrich – Erlangen: Palm & Enke, [1925?] – 1 – us Wisconsin U Libr [430]

Helena weekly herald – Helena MT. [1867 sep 5-1869 nov 25] – 1r – 1 – (cont: helena herald [helena, mt: weekly]; cont by: helena semi-weekly herald) – mf#887775 – us WHS [071]

Helena's household : a tale of rome in the first century / De Mille, James – London, Edinburgh, New York: T Nelson, 1871 – 5mf – 9 – 0-665-90776-1 – mf#90776 – cn Canadiana [830]

Helene de la Presentation, soeur *see* Bibliographie de la croisade eucharistique

[Helene-] ferguson lode – NV. 1892-93 [wkly] – 1r – 1 – $60.00 – mf#U04586 – us Library Micro [071]

Helenes historie / Garborg, Hulda – Oslo: H Aschehoug, 1929 (mf ed 1990) – 1r – 1 – (filmed with: a passage in the night) – us Wisconsin U Libr [890]

Helensburgh advertiser – Helensburgh: C M Jeffrey Ltd 1957- (wkly) [mf ed 2 jan 1997-] – 1 – ISSN: 1356-8663 – uk Scotland NatLib [072]

Helensburgh courier : with which is incorporated the helensburgh news – [Scotland] Argyll & Bute, Helensburgh: M C Macpherson 15 nov 1935-3 dec 1937 (wkly) [mf ed 2003] – 1r – 1 – (title changes; cont in pt: west coast courier; cont by: helensburgh news and the helensburgh courier) – uk Newsplan [072]

Helensburgh news – [Scotland] Argyll & Bute, Helensburgh: R G Blair 1876, 1878, 1886, 1888-dec 1927 (wkly) [mf ed 2003] – 24r – 1 – (missing: 1889; cont as: helensburgh news and county gazette [1927]) – uk Newsplan [072]

Helferich peter sturz : nebst einer abhandlung ueber die schleswigischen literaturbriefe mit benuetzung handschriftlicher quellen / Koch, Max – Muenchen: Christian Kaiser, 1879 – 1r – 1 – (incl bibl ref and index) – us Wisconsin U Libr [430]

Helffenstein, Jacob et al *see* Addresses delivered at the inauguration of rev j w nevin

Helfferich, Adolf *see* Johann karl passavant

Helfferich, Adolph *see* Deutsche briefe aus paris

Helfferich, Karl Theodor *see* Der weltkrieg

Helga : schauspiel in fuenf akten / Hopfen, Hans – Berlin: Gebrueder Paetel 1892 [mf ed 1995] – 1r – 1 – (filmed with: fraenzchens lieder / hoffmann von fallersleben) – mf#3757p – us Wisconsin U Libr [820]

Helgans, R M *see* The role of aerobic fitness and social support in reactivity to psychological stress

Helgi und sigrun : ein episches gedicht der nordischen sage / Carus, Paul – Dresden: R von Grumbkow 1880 [mf ed 1989] – 1r – 1 – (filmed with: die poesie, ihr wesen und ihre formen / moriz carriere) – mf#7146 – us Wisconsin U Libr [810]

Helgolaender zeitung – Helgoland DE, 1921-33 – 24r – 1 – gw Misc Inst [074]

Heliand / ed by Behaghel, Otto – Halle: M Niemeyer, 1882 [mf ed 1993] – (= ser Altdeutsche textbibliothek 4) – xvi/225p – 1 – (incl bibl ref) – mf#8193 reel 1 – us Wisconsin U Libr [430]

Heliand / ed by Heyne, Moritz – 3. verb aufl. Paderborn: F Schoeningh, 1883 [mf ed 1993] – (= ser Bibliothek der aeltesten deutschen litteratur-denkmaeler 2; Altniederdeutsche denkmaeler 1) – 1 – (text in old saxon; pref material in german) – mf#8437 reel 1 – us Wisconsin U Libr [430]

Heliand / ed by Rueckert, Heinrich – Leipzig: F A Brockhaus, 1876 [mf ed 1993] – (= ser Deutsche dichtungen des mittelalters 4) – xl/308p – 1 – (old saxon text. int in german. incl bibl ref and ind) – mf#8381 – us Wisconsin U Libr [430]

Der heliand und die altsaechsische genesis / Behaghel, Otto – Giessen: J Ricker, 1902 [mf ed 1999] – 48p – 1 – mf#4678 – us Wisconsin U Libr [430]

Helianus, L *see* Ludouici heliani vercellensis chrsitanissimi franco regis senatoris...

Die helicobacter pylori-besiedlung des gesamten magens unter besonderer beruecksichtigung der fundusregion : beziehung von clo-test, serologie und historologischem befund / Langer, Doerte – 1998 – 1mf – 9 – €30.00 – 3-8267-2535-2 – mf#DHS 2535 – gw Frankfurter [616]

Helicon boemo-hercynius : in quo novem applausibus coronatur neo-rex boemiae leopoldus... – Pragae, 1656 – 1mf – 9 – mf#O-65 – ne IDC [090]

Helinadus of Froidmont *see*
- Les vers de la mort

Heliopolis, kafr ammar and shurafa / Petrie, W M – London, 1915 – 4mf – 9 – mf#NE-20367 – ne IDC [956]

Helios der titan oder rom und neapel : eine zeitschrift aus italien / Benkowitz, Carl F – Leipzig 1802-04 [mf ed Hildesheim 1995-98] – (= ser Fbc) – 3v on 8mf – 9 – €160.00 – 3-487-29314-5 – gw Olms [914]

Heliotropium seu conformatio humanae voluntatis cum divine / Drexelius, H – Coloniae Aggripp.: Sumptibus Cornelii ab Egmond et Sociorum, 1634 – 3mf – 9 – mf#O-1558 – ne IDC [090]

Helipotropium seu conformatio humanae voluntatis eum divine : editio quarta / Drexelius, H – Monachii: Apud Cornelium Leysserium, 1630 – 4mf – 9 – mf#O-1557 – ne IDC [090]

Helix – Seattle WA. 1967 apr 15-1969 may 1, 1969 may 22-1970 jun – 2r – 1 – mf#701384 – us WHS [071]

Helix herald – Helix OR: Herald Pub Co, -1907 [wkly] – 1 – us Oregon Lib [071]

Helke, Fritz *see*
- Fehde um brandenburg
- Die grosse stunde
- Preussische rebellion
- Der soldat auf dem thron

Helko shel yosef / Zismanowitz, Joseph – Kedainiai, Lithuania. 1926 – 1r – us UF Libraries [939]

Hell, Adele H de *see* Voyage dans les steppes de la mer caspienne et dans la russie meridionale

Hell, Joseph *see* The arab civilization

Hell, Theodor *see* Saengers reise

Hell upon earth : or, the town in an uproar – London, 1729. 62p – 1 – us Wisconsin U Libr [941]

Helland, Andreas *see*
- Afhandlinger og foredrag om menigheden
- Bibelske og kirkehistoriske skisser og afhandlinger
- Fra kirkens arbeidsmark
- Indledning til det gamle testamente
- Taler, afhandlinger, indberetninger ofv vedroerende augsburg seminarium og den lutherske frikirke
- Tilkomme dit rige

Hellas und rom : populaere darstellung des oeffentlichen und haeuslichen lebens der griechen und roemer / Forbiger, Albert – Leipzig: Fues 1876 [mf ed 1979] – 2v in 6 on 1r – 1 – (v1: rom in zeitalter der antonnie [1876, 72, 74]; v2: griechenland im zeitalter des perikles [1876, 78, 82]; 2nd rev enl ed by adolf winckler) – mf#film mas 9161 – us Harvard [930]

Der hellasbote – Berlin DE, 1923-1924 n8 – 1r – 1 – gw Misc Inst [074]

Hellegers, Frederick Riker *see* Die gerechtigkeit gottes im roemerbrief

Hellen, Eduard von der *see*
- Goethes briefe
- Goethes faust
- Goethes saemtliche werke
- Das journal von tiefurt
- Schillers saemtliche werke
- Ueber goethes gedichte

Hellenic free press – Canada. jan 1967-mar 1982 – 15r – 1 – (in greek) – cn Commonwealth Imaging [071]

Hellenic herald – Sydney, jan 1969-dec 1992 – (= ser Greek herald) – 97r – (aka: greek herald) – at Pascoe [079]

Hellenic herald – Sydney, nov 1926-dec 1968 – 16r – 9 – A$1069.99 vesicular A$1157.99 silver – (greek language) – at Pascoe [079]

Hellenic journal – San Francisco CA. 1975 oct 23-1976 dec 30, 1977 jan 13-dec 28, 1979 jan 11-dec 27, 1983 jul 7-1987 jul 23 – 4r – 1 – (cont: western hellenic journal) – mf#1069749 – us WHS [071]

Hellenic times – Nicosia, Cyprus. 1 mar-19 apr 1884 – 1/4r – 1 – uk British Libr Newspaper [072]

Hellenikon aima – Athens. Greece. -d. 12 Jan 1946-6 Jun 1947. (Imperfect). (2 reels) – 1 – uk British Libr Newspaper [949]

Hellenikon air base oracle – Athens, Greece. v4 n31 [1981 may 8]-v4 n46 [1982 jan 8], v4 n48 (1982 feb 12) – 1r – 1 – mf#1221023 – us WHS [355]

Hellenikos typos = Saloniki-greek press – Chicago: Greek Press Pub Co, jan 16, 1941-42; 1946-74 – us CRL [071]

HEMENWAY

Die hellenisierung des semitischen monotheismus / Deissmann, Gustav Adolf – Leipzig: B G Teubner, 1903 – 1mf – 9 – 0-7905-3327-8 – mf#1987-3327 – us ATLA [220]
Hellenism and christianity / Friedlander, Gerald – London: P Vallentine, 1912 – 1mf – 9 – 0-7905-0021-3 – (includes bibliographies and indexes) – mf#1987-0021 – us ATLA [240]
Hellenism in england : a short history of the greek people in this country from the earliest times to the present day / Dowling, Theodore Edward & Fletcher, Edwin W – London: Faith Press; Milwaukee: Young Churchman, 1915 – 1mf – 9 – 0-7905-6924-8 – (incl bibl ref) – mf#1988-2924 – us ATLA [941]
Hellenismus und christenthum, oder, die geistige reaktion des antiken heidenthums gegen das christenthum : mit besonderer ruecksicht auf die christenfeindliche literatur des klassischen alterthums so wie auch der gegenwart / Kellner, Karl Adam Heinrich – Koeln: M DuMont-Schauberg, 1866 – 2mf – 9 – 0-7905-5351-1 – (incl bibl ref) – mf#1988-1351 – us ATLA [240]
Hellenismus und judentum im neutestamentlichen zeitalter / Krueger, Paul – Leipzig: JC Hinrichs, 1908 – (= ser Schriften des Institutum Delitzschianum zu Leipzig) – 1mf – 9 – 0-8370-4003-5 – mf#1985-2003 – us ATLA [270]
Hellenistische studien / Freudenthal, Jacob – 6mf – 9 – 0-8370-1740-8 – (incl bibl ref) – mf#1987-6136 – us ATLA [930]
Hellenistische wundererzaehlungen / Reitzenstein, Richard – Leipzig: BG Teubner, 1906 – 1mf – 9 – 0-524-02101-5 – (incl bibl ref) – mf#1990-2865 – us ATLA [450]
Die hellenistischen : besonders alexandrinischen und sonst schwierigen verbalformen im griechischen neuen testamente fuer schulen und den selbstunterricht / Schirlitz, Samuel Christoph – Erfurt: Friedrich Wilhelm Otto, 1862 [mf ed 1986] – 1mf – 9 – 0-8370-9268-X – mf#1986-3268 – us ATLA [450]
Die hellenistischen mysterien-religionen : ihre grundgedanken und wirkungen / Reitzenstein, R – Leipzig, 1910 – €11.00 – ne Slangenburg [250]
Die hellenistisch-roemische kultur in ihren beziehungen zu judentum und christentum / Wendland, Paul – 2. und 3. Aufl. Tuebingen: JCB Mohr, 1912 – (= ser Handbuch zum neuen testament) – 1mf – 9 – 0-7905-2754-5 – (incl bibl ref) – mf#1987-2754 – us ATLA [230]
Heller, Hayyim *see* Untersuchung ueber die peschiattaa zur gesamten hebraeischen bibel
Heller, Otto *see* Studies in modern german literature
Das heller-blatt – Breslau [WrocLaw PL], 1834-38 – 1r – 1 – (aka: magazin zur verbreitung gemeinnuetziger kenntnisse) – gw Misc Inst [073]
Das heller-magazin – Leipzig DE, 1833 oct-1834 8 nov, 1835-42, 1844 – 1 – gw Misc Inst [073]
Helles abendlied : ausgewaehlte gedichte / Hohlbaum, Robert – Muenchen: A Langen/G Mueller, c1941 – 1r – 1 – us Wisconsin U Libr [810]
Hellier, Anna M *see* Benjamin hellier
Hellier, Benjamin *see* The universal mission of the church of christ
Hellier, Gay *see* Indian child art
Hellier, John Benjamin *see* Benjamin hellier
Helligkeits-messungen an zweihundert and acht fixsternen : angestellt mit dem steinheil'schen photometer in den jahren 1852-1860 / Seidel, Philipp Ludwig et al – Muenchen: Verlag der k Akademie, in Commission bei G Franz 1867 [mf ed 1998] – 1r – 1 – mf#film mas 28419 – us Harvard [520]
Helling, Fritz *see* Fruhgeschichte des judischen volkes
Hellinghaus, Otto *see* Friedrich leopolds grafen zu stolberg erste gattin agnes geb. von witzleben
Helliwell, Arthur Llewellyn *see* A treatise on stock and stockholders, covering watered stock, trusts, consolidations and holding companies
Hellman, S *see* Pseudo-cyprianus de 12 abusives saeculi
Hellmann, E *see* Rooiyard – sociological survey of an urban native slum yard
Hellmann, Ellen *see*
– In defence of a shared society
– Problems of urban bantu youth
– Rooiyard
– Sellgoods
Hellmann, Hanna *see*
– Heinrich von kleist
Hellmann, Othmar *see* De chronologia librorum regum
Hellmann, S *see*
– Aus den briefen der herzogin elisabeth charlotte von orleans an etienne polier de bottens
Hellmund Tello, Arturo *see* Leyendas indigenas gaujiras
Hello, Ernest *see* Studies in saintship
Hello, Henri *see* La F-M et l'ouvrier
Hellowell, S G *see* A history of cragg vale, yorkshire
Hell's angels / Thompson, Hunter S – New York, NY. 1967 – 1r – us UF Libraries [025]
Hells canyon journal – Halfway OR: Steve Backstrom [wkly] – 1 – us Oregon Lib [071]
Hellweg – Essen DE, 1921-27 – 11r – 1 – gw Misc Inst [074]
Hellweger anzeiger fuer mark und das muensterland *see* Hellweger bote
Hellweger anzeiger und bote *see* Hellweger bote
Hellweger bote – Unna DE, 1949 26 oct-1954 15 jun, 1954 21 jun-1957 – 22r – 1 – (title varies: 27 jun 1846: hellweger anzeiger fuer mark und das muensterland; 12 mar 1851: hellweger anzeiger und bote; 26 oct 1949: hellweger anzeiger. filmed by misc inst: 1958-1961 23 sep, 1962 24 apr-1963 17 jun, 1964 31 may-1967 16 apr, 1967 2 aug-1980; 1978 1 sep- [ca 7r/yr]) – gw Mikrofilm; gw Misc Inst [074]
Hellweg-maerkisches volksblatt : maerkischer anzeiger fuer dortmund-wickede,-asseln, -brackel,-husen-kurl, massen und die nachbargemeinden – Dortmund DE, 1930 1 apr-30 jun, 1930 1 oct-1931 31 mar, 1931 1 jul-30 sep, 1932 2 jan-31 mar & 1 jul-30 sep, 1933 1 apr-30 dec, 1934 1 oct-1938, 1940 2 jan-29 jun – 6mf=12df – 9 – (n229 1934: volksblatt; local ed of hoerder volksblatt) – gw Mikrofilm [074]
Hellwig, Elsa *see* Morphologischer idealismus und neue lyrikdeutung
Helm, Dagmar *see* Karl und galie
Helm, Karl *see*
– Altgermanische religionsgeschichte. erster band
– Die apokalypse
– Das buch der maccabaeer in mitteldeutscher bearbeitung
– Das buch der makkabaeer in mitteldeutscher bearbeitung
– Das evangelium nicodemi
Helman, Albert *see* Suriname aan de tweesprong
Helmann, Chayim Meir *see* Bet rabi
Helmbrecht : ein volksdrama in fuenf akten: nach wernhers von gaertners altdeutscher novelle meier helmbrecht / Ege, Ernst – Stuttgart: Strecker & Schroeder, 1906 [mf ed 1989] – 167p – 1 – mf#7204 – us Wisconsin U Libr [820]
Helme, Elizabeth *see* Instructive rambles extended in london, and the adjacent villages
Helmholtz, Hermann L *see* Vorlesungen ueber theoretische physik
Helmholtz, Hermann Ludwig Ferdinand von *see* Ueber goethe's naturwissenschaftliche arbeiten
Helmichius, W *see* Grondich bericht van de wettelijcke beroepinghe der predicanten ofte kerckendienaren...
Helmly, Ruth C *see* Utilization of visual cues by skilled and unskilled basketball players
Helmolt, Hans F *see* Gustav freytags briefe an albrecht von stosch
Helmondi presbyteri bozoviensis cronica slavorum (mgh7:32.bd) – 1909 – (= ser Monumenta germaniae historica 7: scriptores rerum germanicarum in usum scholarum (mgh7)) – €12.00 – ne Slangenburg [240]
Helms, Anton *see*
– Travels from buenos ayres, by potosi, to lima
Helms, Henrik *see* Lappland und die lapplaender
Helms, O et al *see* Dansk ornithologisk forenings tidsskrift
Helmsdorf, Konrad von *see* Der spiegel des menschlichen heils
Helmstetter, B *see* Inkcazelo yencwadi yemfundiso yobukristu
Heloise et abelard / Vailland, Roger – Paris, France. 1947 – 1r – us UF Libraries [440]
Heloise paranquet / Duratin, Armand – Paris, France. 1866 – 1r – us UF Libraries [440]
Help and guide to christian families / Burkitt, William – London, England. 1822 – 1r – us UF Libraries [240]
A help for english readers to understand mistranslated passages in our bible : with explanations and corrections / Murray, J H – London: S W Partridge, 1881 [mf ed 1990] – 1mf – 9 – 0-7905-3459-2 – mf#1987-3459 – us ATLA [220]
Help for ireland – London, 1880 – (= ser 19th c ireland) – 1mf – 9 – mf#1.1.1898 – uk Chadwyck [330]
Help in the water and the fire / Salmond, Charles A – Cults, Scotland. 1880 – 1r – us UF Libraries [240]
Help to the reading of the bible / Nicholls, Benjamin Elliott – new rev and corr ed. London: SPCK, 1892 – 2mf – 9 – 0-524-04806-1 – mf#1992-0226 – us ATLA [220]
Help yourself to better sight / Corbett, Margaret Darst – New York, NY. 1949 – 1r – us UF Libraries [025]
Helpers and hinderers – London, England. 18– – 1r – us UF Libraries [240]
Helpful amateur publishers association : mailing – n5-8 [1989 mar 1-dec 1], n11 [1990 sep] – 1r – 1 – mf#1579239 – us WHS [070]
Helpful hints on music / ed by Bixler, Marguerite Arthelda – Hartville, Ohio: MA Bixler, 1899 – 1mf – 9 – 0-524-02729-3 – mf#1990-4404 – us ATLA [780]
Helping hand / Mountain Home Air Force Base (ID) – v1 n1-v4 n1 [1971 aug-1974 oct] – 1r – 1 – mf#1002845 – us WHS [355]
Helping hand – v1-72. 1842-1914 [complete] – 5r – 1 – (title varies: v1-24 as macedonian. v25-30 as macedonian and record. v31-35 n1 as macedonian and helping hand.) – mf#ATLA R0124 – us ATLA [073]
Helping himself : or, grant thornton's ambition / Alger, Horatio – Philadelphia: Winston, c1886 [mf ed 1987] – 320p – 1 – mf#8032 – us Wisconsin U Libr [830]
Helping parents facilitate the religious education of preschool children in the home / Overman, David Gene – 1981 – 1 – 5.00 – us Southern Baptist [242]
Helping responses in ambiguous and unambiguous emergencies as a function of training in first aid / Heinold, William D – 1982 – 2mf – 9 – $8.00 – us Kinesology [610]
The helpmeet : a record of woman's work in heathen lands, in connection with the free church of scotland – Ladies Society for Female Education in India and Africa, 1891-93; Woman's Foreign Missionary Society, 1894-1900 [mf ed 2001] – (= ser Christianity's encounter with world religions, 1850-1950) – 1r – 1 – mf#2001-s045 – us ATLA [242]
Helps, Arthur, Sir *see* The life of las casas
Helps by the way – Toronto: T.J. Hamilton, [1873?-1874?] – 9 – mf#P04311 – cn Canadiana [240]
"Helps by the way" series of leaflets for letters – Toronto: [s.n, 18–] – 9 – mf#P04309 – cn Canadiana [220]
Helps, E A *see* Personal work for christ and some experiences
Helps for the profitable reading of the holy scriptures – London, England. 18– – 1r – us UF Libraries [240]
Helps from history to the true sense of the minatory clauses of the... / Dowden, John – Edinburgh, Scotland. 1897 – 1r – us UF Libraries [240]
Helps, J Sidney *see* The peach garden
Helps to bible study : with practical notes on the books of scripture designed for ministers, local preachers, s.s teachers, and all christian workers / Sims, Albert – Uxbridge, Ont: s.n, 1886 – 3mf – 9 – mf#56327 – cn Canadiana [220]
Helps to faith : a contribution to theological reconstruction / Garrison, James Harvey – St Louis: Christian Pub Co, c1903 – 1mf – 9 – 0-524-04374-4 – mf#1991-2078 – us ATLA [240]
Helps to the study of the bible : including introductions to the several books, the history and antiquities of the jews, the results of modern discoveries and the natural history of palestine... – Oxford: University Press, [1896?] [mf ed 1991] – 1v on 9mf – 9 – 0-8370-1969-9 – mf#1987-6356 – us ATLA [220]
Helps to the study of the versions of the new testament / ed by Crafts, Wilbur Fisk – teachers' ed. New York: Funk & Wagnalls, c1882 – (= ser Standard Series) – 1mf – 9 – 0-524-06121-1 – mf#1992-0788 – us ATLA [225]
Helps to the thoughtful reading of the four gospels / Stebbing, Henry – London: Virtue, Hall, and Virtue, [1856?] – 1mf – 9 – 0-524-05939-X – mf#1992-0696 – us ATLA [226]
Helrol hetre – Detroit: Magyar Hirlap], feb 17 1918-aug 1 1919 – (filmed consecutively with: magyar hirlap) – us CRL [071]
Helsey, Edouard *see* Terre d'israel
Helsingborgs dagblad – Helsingborg, Sweden. 1884-1945, 1979- – 1 – (klippans dagblad 1979-84) – sw Kungliga [078]
Helsingborgsposten skane-halland – Helsingborg, Sweden. 1900-30 – 105r – 1 – sw Kungliga [078]
Helsingen – Soederhamn, 1994- – 9 – sw Kungliga [078]
Helsingin sanomat – Helsinki. Feb-dec 1944; feb-mar 1952; may 1952-sep 1955; nov 1955-nov 1958; 1959-65 – 239r – 1 – uk British Libr Newspaper [072]
Helsinki University. Slavic Dept *see* Russian old (sic) catalog
Helston hayle and the lizard leader – Apr 9, Jun 4-Dec 24 1988; 1989-Jun 1990; Jul 7-Dec 22 1990; 1991-92; Jan 9-Jun 26, Jul 3-Dec 25 1993; Jan 8-Jun 25, Jul 2-Dec 24 1994; Jan-Dec 23 1995 – 10r – 1 – (discontinued) – uk British Libr Newspaper [072]
Helston packet – England.21 Nov 1969-25 Dec 1970. -w. 1 reel – 1 – uk British Libr Newspaper [072]
Helton, Peter *see* Instructions to juries and declarations of law
Helveg, Ludvig *see* De danske domkapitler
Helvellyn to himalaya : including an account of the first ascent of chomolhari / Chapman, Frederick Spencer – London: Chatto & Windus, 1940 – (= ser Samp: indian books) – (int by marquis of zetland) – us CRL [915]
Helvetiae gratvlatio ad galliam de henrico hvivs nominis 4 galliarum & nauarrae rege christianissimo / Stucki, J W – N p, 1591 – 3mf – 9 – mf#PBU-637 – ne IDC [240]
Helvetica physica acta – Basel. 1950-1995 (1) 1986-1995 (5) 1986-1995 (9) – ISSN: 0018-0238 – mf#564 – us UMI ProQuest [530]
Helvetische kirchen-geschichten / Hottinger, J J – Zuerich, 1698-1729. 4 v – 41mf – 9 – mf#ZWI-90 – ne IDC [242]
Helvetischer kirchen-geschichten, dritter theil / Hottinger, J J – Zuerich, Bodmerische Truckerey, 1708 – 12mf – 9 – mf#PBU-419 – ne IDC [242]
Helweg-Larsen, Sophie *see* Sollyse minder fra tropeegne, som var danske
Helwig, Werner *see* Raubfischer in hellas
Helwing, Christian Friedrich *see* Westphaelische bemuehungen zur aufnahme des geschmaks und der sitten
Helwing, Christian Friedrich et al *see* Auserlesene bibliothek der neuesten deutschen litteratur
Helwys, Thomas *see*
– A declaration of faith of the english people remaining at amsterdam in holland
– A short declaration of the mystery of iniquity
Helyot, Pierre *see* Histoire des ordres monastiques, religieux et militaires et des congregations seculieres de l'un et de l'autre sexe qui ont este etablies jusqu'a present
The hem of christ's garment, and other sermons / Mellor, Enoch – 2nd ed. London: Richard D Dickinson, 1883 – 1mf – 9 – 0-524-08483-1 – mf#1993-3128 – us ATLA [240]
Hemacandra *see* The desinamamala of hemacandra
Heman, Carl Friedrich *see* Eduard von hartmann's religion der zukunft in ihrer selbstzersetzung
Heman, Friedrich *see* Die religioese weltstellung des juedischen volkes
Hemans : the literary manuscripts of felicia hemans (1793-1835) – 5r – 1 – £475.00 – uk Matthew [810]
Hemans, Felicia *see* The breaking waves dashed high
Hematological oncology – Chichester. 1983+ (1,5,9) – ISSN: 0278-0232 – mf#12919 – us UMI ProQuest [616]
Hemdat yisra'el... – Jerusalem, Israel. 1945 – 1r – us UF Libraries [939]
Hemel hempstead advertiser, berkhamsted news, & west herts times – Hemel Hempstead, England 1 jun 1895-28 mar 1901 [mf 1897] – 1 – (wanting 1898; discontinued) – uk British Libr Newspaper [072]
Hemel hempstead and kings langley express – Hemel Hempstead, England 6 jan 1986-23 feb 1987 [mf 1986-95] – 1 – (cont: hemel hempstead, berkhamsted & tring express [4 jun 1984-30 dec 1985]; cont by: hemel hempstead, berkhamsted & tring express [2 mar 1987-4 jan 1988]; hemel hempstead & kings langley express [11 jan 1988-24 sep 1990]; hemel hempstead, berkhamsted & tring express [1 oct 1990-13 oct 1995]; discontinued) – uk British Libr Newspaper [072]
Hemel hempstead, berkhamsted & tring herald & post – Hemel Hempstead, England 21 nov 1991-3 jun 1993 – 1 – (cont: hemel hempstead herald & post [24 aug 1989-10 oct 1991]; cont by: hemel hempstead & the langleys herald & post [10 jun 1993-1 sept 1994]; herald & post [8 sep 1994-12 oct 1995]) – uk British Libr Newspaper [072]
Hemel hempstead gazette – Hemel Hempstead, England. 1980-81 – 7r – 1 – uk British Libr Newspaper [072]
Hemel hempstead gazette and west herts advertiser – 1869, 1877, 1879, 1889, 1891, feb-mar 1945; 1950, jan 4-apr 18, may 16-aug 29, sep 5-dec 19 1980, 1981-90, jul 4 1991-96 – 77 3/4r – 1 – (aka: the gazette (hemel hempstead)) – uk British Libr Newspaper [072]
Hemel Hempstead Review – Hemel Hempstead, England 15 aug 1986-19 apr 1990 – 1 – (cont by: hemel hempstead, berkhamsted, tring review [26 apr-13 sep 1990]; hemel hempstead review [26 apr-13 sep 1990]) – uk British Libr Newspaper [072]
Hemel, J B van *see* Le livre de tout le monde
Hemels-belegh... / Udemans, G C – Dordrecht, 1633 – 3mf – 9 – mf#PBA-359 – ne IDC [240]
De hemelsche morgendauw, der soete genade gods... / Lassenius, J – Amsterdam: Zacharias Romberg, 1737 – 5mf – 9 – mf#O-335 – ne IDC [090]
Hemenway, Asa *see* Story of jesus christ

Hemerken, Thomas *see*
– Alle schriften und buecher
– The authorship of the de imitatione christi
– Opera
– Prolegomena zu einer neuen ausgabe der imitatio christi
– Thomas a kempis
– Thomas a kempis and the brothers of common life
– Thomas von kempen

Hemerodromo da juventude : periodico litterario e recreativo – Rio de Janeiro, RJ: Typ de Pinheiro & C, 05 mar-25 jun 1861 – (= ser Ps 19) – bl Biblioteca [073]

[Hemet-] hemet news – CA. 1899- – 267r – 1 – $16,020.00 (subs $360y) – mf#RC02294 – us Library Micro [071]

[Hemet-] hemet week – CA. 1990-1992 – 1r – 1 – $60.00 – mf#R04031 – us Library Micro [071]

[Hemet-] ramona pagent special editions – CA. 1906-1985 – 12r – 1 – $720.00 – mf#R03240 – us Library Micro [071]

Hemet-San Jacinto Genealogical Society *see*
– Genealogy club of hemet-san jacinto
– Valley genealogist of hemet-san jacinto

Hemingford guide – Hemingford, NE: J S Paradis, -v7 n47. dec 21 1894 (wkly) – 1r – 1 – (cont by: guide (alliance ne)) – us Bell [071]

Hemingford herald – Hemingford, NE: T J O'Keefe, 1895 (wkly) – 1r – 1 – (cont by: alliance herald) – us Bell [071]

Hemingford journal – Hemingford, NE: Chas H Burleigh, 1907 (wkly) [mf ed 1908-10 (gaps)] – 1r – 1 – (cont by: journal (hemingford ne)) – us NE Hist [071]

Hemingford ledger – Hemingford, NE: Chas H Burleigh. v1 n1. oct 7 1915-v57 n32. mar 14 1963 (wkly) [mf ed with gaps] – 20r – 1 – (cont by: ledger. v30-39 not publ) – us NE Hist [071]

Hemingway and mcdonald's reports / Mississippi. Supreme Court – v1-2. 1881-87 (all publ) – (= ser Mississippi supreme court reports) – 17mf – 9 – $25.50 – (a pre-nrs title) – mf#LLMC 90-303 – us LLMC [347]

Hemingway first baptist church. hemingway, sorth carolina : church records – 1926-28, 1941-54 – 1 – 5.00 – us Southern Baptist [242]

Hemingway notes – Youngstown. 1971-1981 (1) 1971-1981 (5) 1979-1981 (9) – ISSN: 0046-7243 – mf#7427 – us UMI ProQuest [400]

Hemingway review – Moscow. 1981+ (1,5,9) – ISSN: 0276-3362 – mf#12916 – us UMI ProQuest [400]

Hemisphere : journal francais, contenant des varietes litteraires et politiques – Philadelphia. 1809-1811 (1) – mf#3817 – us UMI ProQuest [440]

Hemispherica : english edition – New York. 1971-1980 (1) 1951-1980 (5) 1975-1980 (9) – ISSN: 0018-0319 – mf#6071 – us UMI ProQuest [900]

Hemkes, Gerrit Klaas *see*
– De kinderdoop uit god
– Het rechtsbestaan der holl chr geref kerk in amerika

Hemlandet – Rock Island, IL. 1855-1914 (1) – mf#62689 – us UMI ProQuest [071]

Hemlandstoner : en haelsning fran modern svea till dotterkyrkan i amerika / Scheele, Knut Henning Gezelius von – Stockholm: PA Norstedt, [1895?] – 4mf – 9 – 0-524-07971-4 – mf#1990-5416 – us ATLA [240]

Hemlandsvaennen – Stockholm, Sweden. 1879-94 – 1 – sw Kungliga [078]

Hemmerich, Karl *see* Gerhart hauptmanns veland

Hemmes, E *see*
– Richard wagners "parsifal"
– Richard wagners parsifal

Hemmets tidning – Goteborg, Sweden. 1920-22 – sw Kungliga [078]

Hemming, Laurence Paul *see* 'No being without god'

Hemodynamics and orthostasis at rest, exercise, and recovery during -6° of head down tilt with and without a decongestant / Rosene, John M – 1996 – 3mf – 9 – $12.00 – mf#PH 1560 – us Kinesology [612]

Hemon, Louis *see* Maria chapdelaine

Hemorrhagic septicemia : the significance of pasteurella boviseptica / Sanders, D A – Gainesville, FL. 1938 – 1r – us UF Libraries [630]

Hempel, Heinrich *see* Nibelungenstudien

Hempel, Johannes *see* Die schichten des deuteronomiums

Hempel, Wilhelm *see* Ueber das apologetische element im religionsunterricht

Hemphill, Charles Robert *see* The validity and bearing of the testimony of christ and his apostles

Hemphill, Samuel *see*
– The diatessaron of tatian
– A history of the revised version of the new testament

Hempstead beacon – Hempstead, Hicksville NY. 1991 nov 1-1992 dec 25, 1993 jan 8-dec 17, 1994 jan 7-dec 24, 1995 jan 6-dec 22, 1996 jan 5-dec 27, 1997 jan 10-dec 26, 1998 jan 9-dec 25, 1999 jan 8-dec 17, 2000 – 9r – 1 – (cont: beacon [hempstead, ny]) – mf#2255626 – us WHS [071]

Hempstead, SH *see* Hempstead's reports of cases in the arkansas district and circuit courts, 1836-1856

Hempstead's reports of cases in the arkansas district and circuit courts, 1836-1856 / Hempstead, SH – Boston: Little-Brown. 1v. 1856 (all publ) – (= ser Early federal nominative reports) – 9mf – 9 – $13.50 – mf#LLMC 81-454 – us LLMC [347]

Hempstone, Smith *see*
– Katanga report
– Rebels, mercenaries, and dividends

Hemrich, Guenter *see* Entwicklungstendenzen in der landwirtschaftlichen produktion nach der einfuehrung moderner reistechnologie

Hemrick, Christina L *see* The moderating effects of humor on cognitive appraisals of stress

Hemsen, Johannes Tychsen *see* Geschichte und literatur der kirchengeschichte

Hemsley, W B *see* Report on the scientific results of the voyage of hms challenger during the years 1873-1876...botany

Hemsterhuis, Fr *see* Oeuvres philosophiques

Hen / Ch'iu-shih – Shang-hai: Ta shen shu she, 1934 – (= ser P-k&k period) – us CRL [480]

Hen hai : [ssu mu pei chu] / Sung, Yueh – Pei-ching: Wen chang shu fang, 1945 – (= ser P-k&k period) – us CRL [820]

Henao Davila, Fernando *see* Estudio de un metodo analitico para valoracion cuantitativa conjunta de los acidos organicos en vinos de tierra de barros

Henao, Jesus Maria *see* History of colombia

Henao Mejia, Gabriel *see* Juan de dios aranzazu

Henao y Munoz, Manuel *see*
– Cronica...badajoz
– El drama de la vida
– El libro del pueblo

Henatsch, Wilhelm Andreas *see* Das problem der auslaendischen wanderarbeiter

Henbury, st thomas – (= ser Cheshire monumental inscriptions) – 2mf – 9 – £4.00 – mf#21 – uk CheshireFHS [929]

Henckell, Karl *see* Hundert gedichte

Henckens, R P *see* Sainte christine l'admirable de saint-trond

Hendel, Klaus *see* Qualitative und quantitative untersuchungen der dynamik von mehrkoerpersystemen mittels stoerungsgleichungen und 1. integralen

Henderson, Alexander *see* Sermons, prayers, and pulpit addresses

Henderson, Charles Richmond *see*
– The development of doctrine in the epistles
– Introduction to the study of the dependent, defective, and delinquent classes
– Modern methods of charity
– Social duties from the christian point of view
– Social elements
– Social programmes in the west
– Social settlements

Henderson, clifford, papers, ms 4309 – 1928-39 – 14r – 1 – (correspondence, papers, press releases, clippings and memorabilia of cliff henderson, managing director of the national air races) – us Western Res [790]

Henderson, David Patterson *see* A discourse on the history, character, and design of christian baptism

Henderson, E *see* Great mystery of godliness incontrovertible

Henderson, Ebenezer *see*
– Aegidii gutbirii lexicon syriacum
– Biblical researches and travels in russia
– The book of the prophet ezekiel
– The book of the prophet isaiah
– The book of the prophet jeremiah and that of the lamentations
– The book of the twelve minor prophets
– Divine inspiration
– Iceland

Henderson, Ernest Flagg *see* Select historical documents of the middle ages

Henderson, George *see*
– The norse influence on celtic scotland
– Survivals in belief among the celts

Henderson, George A *see* Early saint john methodism and history of centenary methodist church, saint john, nb

Henderson, George E *see*
– British history notes
– Geography notes
– Geography notes for 3rd, 4th, and 5th classes

Henderson, George E et al *see*
– Exercises in composition for fourth and fifth classes
– Exercises in grammar
– Junior language lessons for first, second, and third classes

[Henderson-] henderson home news – NV. 1951-1955; 1956-1977 – 67r – 1 – $4020.00 (subs $140y) – mf#N04587 – us Library Micro [071]

[Henderson-] henderson shopping news – NV. 1947; 1948 – 2r – 1 – $120.00 – mf#U04845 – us Library Micro [071]

Henderson, Henry F *see*
– Calvin in his letters
– The dream of dante
– Erskine of linlathen
– The religious controversies of scotland

Henderson heritage – v1 n1-v5 n4 [1984 jan-1988 oct] – 1r – 1 – mf#1058081 – us WHS [071]

Henderson, J *see* Memorials of james henderson...medical missionary to china

Henderson, J B *see* Monograph of the east american scaphopod mollusks

Henderson, J Duff *see* Alvira alias orea

Henderson, J R *see*
– Effect of soil reaction on the assimilation of certain primary nutr...
– Soils of florida

Henderson, James *see*
– Forerunners of modern malawi
– Reception due to the word of god

Henderson, James Max *see*
– Questions and answers with problems and illustrative matter on conflict of laws.
– Questions and answers with problems and illustrative matter on the law of domestic relations
– Questions and answers with problems and illustrative matter on the law of equity.
– Questions and answers with problems and illustrative matter on the law of sales.

Henderson, John *see*
– Jamaica
– West indies

Henderson, John M *see* John m henderson papers, 1810-1892 [1817-1848]

Henderson, Keith *see* Palm groves and humming birds

Henderson memorial baptist church. hopkinsville, kentucky : church records – 1965-Jan 1986 – 1 – us Southern Baptist [242]

Henderson, Murdoch [Harper, John Murdoch] *see* The history of the irish republic

The henderson news – Henderson, NE: Service Press. v1 n1. nov 14 1952- (wkly) [mf ed with gaps filmed 1976-] – 1 – us NE Hist [071]

[Henderson-] post – NV. 1964 – 1r – 1 – $60.00 – mf#U04846 – us Library Micro [071]

Henderson review – Henderson, NE: H D Friesen. 1v. -v1 n20. nov 2 1937 (wkly) [mf ed lacks sep 14 filmed 1973] – 1r – 1 – (cont by: review (henderson ne)) – us NE Hist [071]

Henderson, Samuel *see* Records of samuel henderson and e b teague

Henderson, Sarah Fisher et al *see* Correspondence of the reverend ezra fisher

[Henderson-] tru-news – NV. 1964 – 1r – 1 – $60.00 – mf#U04847 – us Library Micro [071]

Henderson, W *see*
– "He descended into hell"
– Liber pontificalis chr bainbridge archiepiscopi eboracensis
– Manuale et processionale ad usum insignis ecclesiae eboracensis
– Missale ad usum insignis ecclesiae eboracensis
– Missale ad usum percelebris ecclesiae herfordensis
– Processionale ad usum insignis ac praeclarae ecclesiae sarum
– The word of god in its relation to the church

Henderson, W P M *see* Durban

Henderson, William *see*
– Examination of a pamphlet entitled "considerations on the expedienc...
– Notes on the folk-lore of the northern counties of england and the borders

Henderson, William Graham *see* A concise summary of the law of libel as it affects the press

Henderson, William James *see* The orchestra and orchestral music

Henderson, William John et al *see* The centenary volume of the baptist missionary society, 1792-1892

Henderson's british columbia gazetteer and directory and mining companies : with which is consolidated the william's b c directory for 1900-1901... – Vancouver BC: Henderson Publ Co 1901 [mf ed 1984] – 14mf – 9 – 0-665-17386-5 – mf#17386 – cn Canadiana [971]

Henderson's british columbia gazetteer and directory and mining companies, for 1898 : comprising complete alphabetical directories of the cities... – Victoria [BC], Vancouver: Henderson Pub Co [1898?] [mf ed 1984] – 10mf – 9 – 0-665-17373-3 – mf#17373 – cn Canadiana [971]

Henderson's british columbia gazetteer and directory and mining companies, for 1899-1900 : comprising complete alphabetical directories of the cities... – Victoria [BC], Vancouver: Henderson Pub Co [1899?] [mf ed 1984] – 11mf – 9 – 0-665-17374-1 – mf#17374 – cn Canadiana [971]

Henderson's british columbia gazetteer and directory and mining encyclopdia for 1897 – Victoria [BC], Vancouver: Henderson Pub Co [1898?] [mf ed 1984] – 10mf – 9 – 0-665-17387-3 – mf#17387 – cn Canadiana [971]

Henderson's city of vancouver directory for 1890 : containing a complete street directory... – Vancouver: Henderson Directory Co [1890] [mf ed 1984] – 3mf – 9 – 0-665-18074-8 – mf#18074 – cn Canadiana [971]

Hendley, J A *see* History of pasco county

Hendley, Thomas Holbein *see*
– Handbook to the jeypore museum
– Ulwar and its art treasures

Hendon advertiser – London UK – 1 – (aka: hendon and district local advertiser) – uk British Libr Newspaper [072]

Hendon advertiser – London UK, oct 1894-jan 1922 – 1 – uk British Libr Newspaper [072]

Hendon and district local advertiser – [London & SE] Barnet, Archives & Local Studies 1984-21 dec 1989 – 1 – uk Newsplan [072]

Hendon and district local advertiser *see* Hendon advertiser

Hendon And District Local Advertiser/ Hendon *see* Hendon local advertiser

Hendon And District Post *see* Hendon post

Hendon and district post – London UK, missing: 9 oct 1952-27 sep 1956 – 1 – (aka: hendon post) – uk British Libr Newspaper [072]

Hendon and finchley times *see* Hendon times, finchley, hampstead advertiser

Hendon and finchley times and guardian – London UK – 1 – (aka: hendon times and finchley and hampstead advertiser) – uk British Libr Newspaper [072]

Hendon arrow – Barnet, London 22 feb 1888-11 sep 1889 – 1/4r – 1 – (incorp with: courier) – uk British Libr Newspaper [072]

Hendon courier – London, England. 10 feb 1887-dec 1897 – 10r – 1 – (aka: the courier; courier and london & middlesex counties gazette; middlesex courier) – uk British Libr Newspaper [072]

Hendon edgware independent – London UK, 1907; 9 apr 1981-83; 19 may-22 dec 1988; 1989-13 dec 1990; 1991; 1992 – 13r – 1 – (from 4 oct 1984-12 may 1988 ed amalgamated with: the harrow – wembley independent publ as harrow – wembley – hendon – edgware independent) – uk British Libr Newspaper [072]

Hendon labour party records, 1924-1992 – (= ser Labour party in britain, origins and development at local level. series 2) – 12r – 1 – (with p/g. int by daniel weinbren) – mf#97559 – uk Microform Academic [325]

Hendon local advertiser – London Uk – 1 – (aka: hendon and district local advertiser/ hendon) – uk British Libr Newspaper [072]

Hendon local observer – London UK, 1986 – 3r – 1 – uk British Libr Newspaper [072]

Hendon post – London UK, oct 1952-sep 1956 – 1 – (aka: hendon and district post) – uk British Libr Newspaper [072]

Hendon post – London UK, 22 sep-22 dec 1988; jan-10 aug 1989 – 1 1/5r – 1 – uk British Libr Newspaper [072]

Hendon post *see* Hendon and district post

Hendon times – London UK – 1 – (aka: hendon times and finchley and hampstead advertiser) – uk British Libr Newspaper [072]

Hendon Times And Finchley And Hampstead Advertiser *see*
– Hendon and finchley times and guardian
– Hendon times

Hendon times, finchley, hampstead advertiser – London, 14 oct 1876-25 may 1878; jun 1878-1894; 1898-13 nov 1964; 1966-jun 1988; sep-dec 1998; jan-feb 1999 – 224 1/2r – 1 – uk British Libr Newspaper [072]

Hendon-edgware independent – London, UK. 9 Apr 1983 – 2r – 1 – uk British Libr Newspaper [072]

Hendren, Samuel Rivers *see* Government and religion of the virginia indians

Hendrick goltzius als maler, 1600-1617 / [Goltzius] Hirschmann, O – Haag, 1916. v9 – 2mf – 9 – mf#O-518 – ne IDC [700]

Hendrick, Kevin *see* The history of the north carolina governor's council on physical fitness and health

Hendricks county history bulletin – 1970, 1971 oct-1986 nov – 1r – 1 – mf#1528383 – us WHS [071]

Hendricks, Frank Sylvester *see* Prolepsis in afrikaans

Hendricks, Shaheed *see* Arthur nortje in port elizabeth

Hendricks, Wayne Graham *see* Die betrekkinge tussen nederland en suid-afrika, 1946-1961

Hendrickson, David K *see* Conducting job interviews

Hendrickson, Thomas L *see* The physiological responses to walking with and without power poles$_{(tm)}$ on treadmill exercise

Hendrickson, William R *see* The effects of recovery time on throwing velocity and accuracy of college baseball pitchers

Hendrik f andriessen : his life and works / Dox, Thurston J – U of Rochester 1969 [mf ed 19–] – 5mf [fiche229] 6mf [fiche534] – 9 – (incl bibl ref) – us Sibley [780]
Hendrik mande : bijdrage tot de kennis der noord-nederlandsche mystiek / Visser, G – 's-Gravenhage, 1899 – 4mf – 8 – €11.00 – ne Slangenburg [240]
Hendriks, Elma *see* 'N perspektief op die beroepsbevrediging van grondvlak maatskaplike werkers
Hendriks, Lawrence *see* The london charterhouse, its monks and its martyrs.
Hendriksen, Jorgen *see* The odor fontane og norden
Hendrix, Eugene Russell *see*
- The personality of the holy spirit
- The religion of the incarnation delivered before the vanderbilt university
- Skilled labor for the master

Hendry and glades county, florida / Huss, Veronica E – s.l, s.l? 193-? – 1r – us UF Libraries [978]
Hendry county news – Labelle, FL. 1937 oct-1972 – 20r – (gaps) – us UF Libraries [071]
Hendry county reservation : it's past and future / Sanderson, Isabelle – s.l, s.l? 1936 – 1r – us UF Libraries [978]
Henepin lawyer – v1-48. 1932-78 – 85mf – 9 – $127.00 – (lacking: v1-8. v43 no 6. v47. updates planned) – mf#LLMC 84-479 – us LLMC [340]
Henfrey, A *see* Botanical and physiological memoirs
Henfrey, Colin *see* Through indian eyes
Heng tu / Lo, Feng – Ch'ung-ch'ing: Shang wu yin shu kuan, Min kuo 32 [1943] – (= ser P-k&k period) – us CRL [480]
Hengard-Lapalice, Ovide Michel *see* Histoire de la seigneurie massue et de la paroisse de saint-aime
Heng-che san wen chi / Ch'en, Heng-che – Shang-hai: K'ai ming shu tien, Min kuo 27 [1938] – (= ser P-k&k period) – us CRL [840]
Hengel, Wessel Albertus van *see* De testamenten der twaalf patriarchen
Hengoed, cefn hengoed welsh baptist, monumental inscriptions – 4mf – 9 – £5.00 – uk Glamorgan FHS [929]
Hengst maestoso austria : liebesgeschichte zweier menschen und eines edlen pferdes / Lehmann, Arthur Heinz – Dresden: W Heyne, 1939 – 1r – 1 – us Wisconsin U Libr [830]
Hengstenberg, Ernst Wilhelm *see*
- Das buch hiob
- Christology of the old testament and a commentary on the messianic predictions
- Commentary on ecclesiastes
- Commentary on the gospel of st john
- Commentary on the psalms
- Dissertations on the genuineness of daniel and the integrity of zechariah
- Dissertations on the genuineness of the pentateuch
- Egypt and the books of moses
- Die geschichte bileams und seine weissagungen
- History of the kingdom of god under the old testament
- The lord's day
- The prophecies of the prophet ezekiel elucidated
- The revelation of st john
- Vorlesungen ueber die leidensgeschichte

Die hengstwiese : novelle / Beumelburg, Werner – Oldenburg i O: G Stalling, 1937 [mf ed 1989] – 111p – 1 – mf#7017 – us Wisconsin U Libr [830]
Henion, Doris Volz *see* Colombia
Henke, Ernst Ludwig Theodor *see*
- De epistolae quae barnabae tribuitur authentia
- Dr. e.l. th. henke's nachgelassene vorlesungen ueber liturgik und homiletik
- Dr. e.l. th. henke's neuere kirchengeschichte
- Georg calixtus und seine zeit
- Jakob friedrich fries
- Konrad von marburg, beichtvater der heiligen elisabeth und inquisitor
- The ologorum saxonicorum consensus repetitus fidei vere lutheranae
- Schleiermacher und die union

Henke, Frederick Goodrich *see* A study in the psychology of ritualism
Henke, Heinz-Werner *see* Messung der desintegrationsleistung des holmium-yag-lasers in hinblick auf den einsatz in der lithotripsie
Henke, Josef *see*
- Nachlass kurt rheindorf (bestand nl 263) bd 37
- Partei-kanzlei der nsdap (bestand ns 6) bd 23
- Persoenlicher stab reichsfuehrer ss (bestand ns 19) bd 57

Henkel, Socrates *see* History of the evangelical lutheran tennessee synod
Henkin, Yosef Eliyahu *see* Sefer perushe ivra
Henle, Fritz *see* Virgin islands
Henley and south oxfordshire standard *see* Henley free press
Henley chronicle and south oxfordshire gazette – Henley, South Oxfordshire, England. 1904-11 – 7r – 1 – uk British Libr Newspaper [072]
Henley free press – England, 21 feb 1885-17 jul 1886; jan 1889- – 106+ r – 1 – (1886, 1889, 1890 imperfect) – uk British Libr Newspaper [072]
Henley, Robert Henley Eden *see*
- A compendium of the law and practice of injunctions, and of interlocutory orders in the nature of injunctions
- Plan of church reform
- A treatise on the law of injunctions

Henley standard *see* Henley free press
Henley, William Ernest *see*
- A century of artists
- For england's sake
- Sir henry raeburn

Henlow – (= ser Bedfordshire parish register series) – 1mf – 9 – £3.00 – uk BedsFHS [929]
Henlow, st mary monumental inscriptions monumental inscriptions – Arthur Weight Matthews 1913 – (= ser Bedfordshire parish register series) – 1mf – 9 – £1.25 – uk BedsFHS [929]
Henn, Silas *see* Millennium
Henne am Rhyn, Otto *see*
- Adhuc stat!
- Anti-zarathustra
- Die deutsche volkssage im verhaeltnis zu den mythen aller zeiten und voelker
- Das jenseits

Henneberger, August *see* Briefe von johann peter uz an einen freund
Hennecke, E *see* Dei apologie des aristides ((tugal
Hennecke, Edgar *see*
- Altchristliche malerei und altkirchliche literatur
- Die apologie des aristides

Hennell, Sara S *see* On the need of dogmas in religion
Hennepin lawyer – v9-69. 1940-2000 – 9 – $513.00 set – ISSN: 0175-2000 – mf#401320 – us Hein [340]
Henner, Theodor *see* Das wesen des christentums nach thomas von aquin
Hennessy, J P I *see* A report on the first general election in basutoland, 1960
Hennessy, Joseph Patrick *see* A leading case as to hotel-keepers and guests; a summarized opinion with decision of the court of appeals of the state of new york
Hennessy, W M *see*
- Annals of loch ce
- Chronicon scotorum

Hennesthal, Rudolf *see* Deutschland unterm hakenkreuz
Hennesy, James A *see* Dictionary of grammar
Hennesy, William M *see* Annals of ulster otherwise annals of senat
Hennicke, Gayle Watts *see* Analysis of the twelve preludes and fugues of franz reizenstein
Hennig, Martin *see* Quellenbuch zur geschichte der inneren mission
Hennig, Robert *see* Zwei briefe aus amerika
Henniges, Paul Brown *see* A study of the religious social ethics of reinhold niebuhr
Hennigsdorfer lokalanzeiger – Hennigsdorf DE, 1921 jan-mar, 1922 10 may-dec, 1925 oct-dec – 1r – 1 – (bezirksausgabe von berlin-tegeler anzeiger) – gw Misc Inst [074]
Henniker, Frederick *see* Notes, during a visit to egypt, nubia, the oasis, mount sinai, and jerusalem
Henning *see* The house that albert built
Henning, Hans *see* Die deutsche literatur
Henning, James *see* The church in a workhouse
Henning, Leopold von *see* Principien der ethik in historischer entwicklung
Henning, M *see* D johannes hinrich wicherns lebenswerk in seiner bedeutung fuer das deutsche volk
Henning, Marie-Christine *see* Katalog der bibliothek ponickau
Hennings, August *see*
- Der genius der zeit
- Der genius des neunzehnten jahrhunderts
- Schleswigsches ehemals braunschweigisches journal

Hennings, August Adolph Friedrich *see* Annalen der leidenden menschheit in zwanglosen heften
Henningsen, Emanuel *see* Fra laaland: ny fortaellinger
Henri 4 et d'aubigne / Rougemont, Michel-Nicolas Balisson De – Paris, France. 1814 – 1r – us UF Libraries [440]
Henri 8 : tragedie / Chenier, Marie-Joseph – Paris, France. 1805 – 1r – us UF Libraries [440]
Henri bate de malines : speculum divinorum et quorundam naturalium / Wallerand, G – Louvain, 1931 – (= ser Philosophes belges 11) – 9mf – 8 – €18.00 – (etude critique et texte inedit) – ne Slangenburg [130]
Henri bergson : an account of his life and philosophy / Ruhe, Algot & Paul, Nancy Margaret – London: Macmillan 1914 [mf ed 1991] – 1mf – 9 – 0-7905-8573-1 – mf#1989-1798 – us ATLA [190]
Henri berneche : en religion, frere norbert de marie, novice de l'institut des freres des ecoles chretiennes, 1893-1910 – Montreal: [s.n, 1911?] – 4mf – 9 – 0-665-76907-5 – mf#76907 – cn Canadiana [241]
Henri bullinger... / Bouvier, A – Neuchatel, Paris, 1940 – 7mf – 9 – mf#PBU-684 – ne IDC [240]
Henri christophe : conference faite au lycee nation / Pierre-Paul, Antoine – Port-Au-Prince, Haiti. 1911 – 1r – us UF Libraries [972]
Henri de gand : essai sur les tendances de sa metaphysique / Paulus, J – Paris, 1938 – 11mf – 8 – €21.00 – ne Slangenburg [110]
Henri dominique lacordaire : a biographical sketch / Lear, H L Sidney – London: Rivingtons, 1882 – 1mf – 9 – 0-7905-5169-1 – mf#1988-1169 – us ATLA [920]
Henri, Ernst *see*
- Hitler over europe?
- Hitler over russia?

Henri gregoire, lami des hommes de toutes les couleurs / Grunebaum-Ballin, Paul Frederic Jean – Paris, France. 1948 – 1r – us UF Libraries [025]
Henri heine : l'homme et l'oeuvre / Bianquis, Genevieve – Paris: Boivin, c1948 [mf ed 1995] – (= ser Le livre de l'etudiant 23) – 176p – 1 – (incl bibl ref) – mf#8789 – us Wisconsin U Libr [430]
Henri Hembuche de Langestein (Henry of Langenstein (Henry of Hesse the Elder)) *see* Le miroir de l'ame
Henri le pretendant / Luchet, Auguste – Paris 1832 [mf ed Hildesheim 1995-98] – 1v on 3mf – 9 – €90.00 – ISBN-10: 3-487-26040-9 – ISBN-13: 978-3-487-26040-2 – gw Olms [944]
Henri perreyve / Gratry, Auguste – new ed. London: Rivingtons, 1880 – 1mf – 9 – 0-524-00992-9 – mf#1990-0269 – us ATLA [240]
Henri quatre : or, the days of the league / Mancur, John – London 1834 [mf ed Hildesheim 1995-98] – 11mf – 9 – €110.00 – ISBN-10: 3-487-26111-1 – ISBN-13: 978-3-487-26111-9 – gw Olms [944]
Henrich, Manuel *see* Iconografia de las editiones del quijote de miguel cervantes saavedra
Henrich, Timothy W *see* Influence of reactive hyperemia in muscle during exercise
Henrichs, Norbert *see* Briefe deutscher philosophen (1750-1850)
Henrici de Bracton *see* De legibus et consuetudines angliae libri quinque in varios tractatus distincti [rs70]
Henrici Huntenduniensis *see* Henrici huntenduniensis historia anglorum (rs74)
Henrici huntenduniensis historia anglorum (rs74) : from bc 55 to ad 1154. in eight books = The history of the english by henry, archdeacon of huntingdon / Henry of Huntingdon; ed by Arnold, T – 1879 – (= ser The rolls series (rs)) – €17.00 – ne Slangenburg [931]
Henrici Knighton *see* Chronicon henrici knighton [rs92]
Henrico county leader – Henrico, VA. 1999-2000 (1) – mf#69612 – us UMI ProQuest [071]
Henrico gazette – Richmond, VA. 1988-1994 (1) – mf#68285 – us UMI ProQuest [071]
Henrico herald – Richmond, VA. 1939-1970 (1) – mf#66823 – us UMI ProQuest [071]
Henricus de Gandovo (Henry of Ghent) *see* Summa theologica
Henrieta sold – Jerusalem, Israel. 1945 – 1r – us UF Libraries [939]
Henriette jacoby : roman / Hermann, Georg – Berlin: E Fleischel, 1912 – 1r – 1 – us Wisconsin U Libr [830]
Henrik ibsen, ein erlebnis der deutschen / Thalmann, Marianne – Marburg a.L.: N G Elwert, 1928 – 1r – 1 – (incl bibl ref) – us Wisconsin U Libr [430]
Henrik ibsens einfluss auf hermann sudermann / Juergensen, Hans – [Lausanne: s.n. 1903] – 1r – 1 – us Wisconsin U Libr [410]
Henrik steffens : ein lebensbild / Petersen, Richard – Gotha: F A Perthes, 1884 – 1r – 1 – us Wisconsin U Libr [920]
Henrik steffens romane : ein beitrag zur geschichte des historischen romans / Karsen, Fritz – Leipzig: Quelle & Meyer, 1908 [mf ed 1992] – (= ser Breslauer beitraege zur literaturgeschichte. neue folge 6) – 170p – 1 – (incl bibl ref) – mf#8014 reel 2 – us Wisconsin U Libr [430]
Henrion, Mathieu Richard Auguste, Baron *see* Histoire generale des missions catholiques depuis le 13e siecle jusqu'a nos jours
Henriot, Constant *see* Les ordres religieux au point de vue social
Henriot, Emile *see* Beautes du bresil
Henriques Castillo, Luis *see* Octava maravilla
Henriques Urena, Pedro *see* Obra critica
Henriquez Almanzar, Carmen Adolfina *see* Asistencia social
Henriquez, Chrysostomus *see*
- Fasciculus sanctorum ordinis cisterciensis
- Menologium cisterciensis
- Phoenix reviviscens, libri 2
- Quinque prudentes virgines
- Regula, constitutiones et privilegia ordinis cisterciensis

Henriquez, Constantin *see* Nos villes et nos bourgades
Henriquez, Enrique *see* Nocturnos, y otros poemas
Henriquez, Gustavo Julio *see* Contribucion de la republica dominicana
Henriquez Ureena, Max *see* Cuentos insulares
Henriquez Urena, Max *see*
- Arzobispo valera
- Breve historia del modernismo
- Conspiracion de los alcarrizos
- Continente de la esperanza
- Evocacion de jose antonio ramos
- Fosforescencias
- Garra de luz
- Independencia efimera
- Influences francaises sur la poesie hispano-americ...
- Intercambio de influencias literarias entre espana
- Liga de naciones americanas y la conferencia de bu...
- Ocaso de dogmatismo literario
- Oratoria de dos guerras
- Panorama historico de la literatura cubana
- Panorama historico de la literatura dominicana
- Programa de gramatica castellana
- Retorno de los galeones
- Retorno de los galeones y atros ensayos
- Yanquis en santo domingo

Henriquez Urena, Pedro *see*
- Antologia
- Cultura y las letras coloniales en santo domingo
- Endecasilabo castellano
- Espanol en mejico, los estados unidos, y
- Historia de la cultura en la america hispanica
- Historia de la cultura en la americana hispanica
- Literary currents in hispanic america
- Paginas escogidas
- Poesias juveniles
- Seis ensayos en busca de nuestra expresion
- Seleccion de ensayos
- Sobre el problema del andalucismo dialectal de ame
- Tablas cronologicas de la literatura espanola
- Utopia de america
- Versificacion irregular en la poesia castellana

Henriquez Y Carvajal, Federico *see*
- Cuentos
- Del amor i del dolor
- Duarte
- Generalisimo maximo gomez
- Marti, proceres heroes i martires de la independen...
- Poema de la historia
- Todo por cuba

Henry 8 / Macnalty, Arthur Salusbury – London, England. 1952 – 1r – us UF Libraries [025]
Henry 8 / Pollard, Albert Frederick – New ed. London; New York: Longmans, Green, 1905 – 2mf – 9 – 0-7905-7131-5 – (incl bibl ref) – mf#1988-3131 – us ATLA [941]
Henry 8 and the english monasteries : an attempt to illustrate the history of their suppression / Gasquet, Francis Aidan – 2nd ed. London: John Hodges, 1888-1889 – (= ser Catholic standard library) – 3mf – 9 – 0-7905-4649-3 – (incl bibl ref) – mf#1988-0649 – us ATLA [941]
The henry adams papers, 1843-1938 – [mf ed 1981] – 36r – 1 – (with p/g. the personal papers of henry adams provides another perspective on adams family history) – us MA Hist [920]
Henry and antonio : or the proselytes of the romish and evangelical churches = Heinrich und antonio / Bretschneider, Karl Gottlieb – Baltimore: Lucas & Deaver 1834 [mf ed 1993] – 1mf – 9 – 0-524-07671-5 – (trans fr german of c g bretschneider, with additional notes) – mf#1991-3256 – us ATLA [241]
Henry, Arthur *see* Guyane francaise
Henry barrow, separatist (1550?-1593), and the exiled church of amsterdam (1593-1622) / Powicke, Frederick James – London: J. Clarke, 1900 – 1mf – 9 – 0-7905-5907-2 – (incl bibl ref) – mf#1988-1907 – us ATLA [240]
Henry bazely : the oxford evangelist / Hicks, Edward Lee – London: Macmillan, 1886 – 1mf – 9 – 0-7905-5707-X – mf#1988-1707 – us ATLA [240]
Henry, Benjamin Couch *see*
- The cross and the dragon
- Ling-nam

Henry boynton smith / Stearns, Lewis French – Boston: Houghton Mifflin, 1892 – (= ser American religious leaders) – 1mf – 9 – 0-524-01020-X – mf#1990-0297 – us ATLA [240]
Henry boynton smith, his life and work / Smith, Henry Boynton; ed by Smith, Elizabeth Lee – New York: AC Armstrong, 1881, c1880 – 2mf – 9 – 0-7905-8588-X – mf#1989-1813 – us ATLA [920]
Henry bradley plant / Mclaws, Lafayette – s.l, s.l? 193-? – 1r – us UF Libraries [978]
Henry bradshaw society – London, 1891-1946. v1-81 – 453mf – 8 – mf#448c – ne IDC [240]

Henry bradshaw society (hbs) – London. v1-94. 1891-1963 – 9 – €854.00 – (vols also listed separately) – ne Slangenburg [240]
Henry Bradshaw Society, London *see* The coronation book of charles 5th of france
Henry, Bruce L *see* Black caesar
Henry, Caleb Sprague *see* The endless future of the human race
Henry, Caroline Vinton *see* Personal reminiscences of cardinal newman
Henry chronicles – v1 n1-v3 n2 [1986 mar-1988 sep/dec] – 1r – 1 – mf#1832742 – us WHS [071]
Henry clay / Schurz, Carl – Boston & NY: Houghton, Mifflin & Co. 2v. 1899 – (= ser The american statesmen series) – 11mf – 9 – $16.50 – mf#LLMC 96-026 – us LLMC [975]
The henry clay family papers – 23r – 1 – $805.00 – Dist. us Scholarly Res – us L of C Photodup [975]
Henry cloete in natal, 1843-1855 / Flanagan, Brigid – Durban, 1946 – us CRL [920]
Henry codman potter, seventh bishop of new york / Hodges, George – New York: Macmillan, 1915 – 1mf – 9 – 0-524-08386-X – mf#1993-3086 – us ATLA [240]
Henry county dial – Kewanee IL. 1862 oct 8, v9 n5 – 1r – 1 – mf#984464 – us WHS [071]
Henry county journal – Bassett, VA. 1947-1980 (1) – mf#66668 – us UMI ProQuest [071]
Henry crabb robinson diaries, travel journals and reminiscences 1790-1867 / Dr. Williams's Library – 11r – 1 – £450.00 – 1-897955-19-7 – uk Academic [090]
Henry, Dahlia *see* The effects of 15 weeks of resistive training with chromium supplementation
Henry de Bracton *see* De legibus et consuetudines angliae libri quinque in varios tractatus distincti [rs70]
The henry dispatch – Henry, NE: E P McVey. v1 n1. nov 6 1920- (wkly) [mf ed -1921 (gaps) filmed 1979] – 1r – 1 – us NE Hist [071]
Henry dreyfuss archive : drawings, designs and documents – 100mf – 9 – $730.00 – 0-907006-58-2 – (leading american industrial designer of 20th c; over 6000 reproductions; fully ind) – uk Mindata [740]
Henry drummond : a biographical sketch (with bibliography) / Lennox, Cuthbert – Toronto: W Briggs, 1901 [mf ed 1996] – 1mf – 9 – 0-665-81003-2 – mf#81003 – cn Canadiana [242]
Henry e sheffield account book *see* Sheffield, henry e, account book
Henry ford helps a child / Huss, Veronica E – s.l, s.l? 193-? – 1r – us UF Libraries [978]
Henry ford hospital medical journal – Detroit. 1953-1992 (1) 1971-1992 (5) 1977-1992 (9) – ISSN: 0018-0416 – mf#2679 – us UMI ProQuest [360]
Henry, Francoise *see* Irish art
Henry, George *see*
– Nu-gu-mo-nun o-je-boa an-oad ge-e-se-ueu-ne-gu-noo-du-be-ueng uoo muun-gou-duuz [george henry] gu-ea moo-ge-gee-seg [james evans] ge-ge-noo-ue- muu-ga-oe-ne-ne-oug
– Pastoral admonition after confirmation
Henry, George Adams *see* The probate law and practice and the laws of succession of the state of indiana.
Henry, George F *see* A layman's view of the demand for a change in the name of the church
Henry george scapbooks – New York: New York Public Library, 1978 – 3r – 1 – (george, henry) – mf#ZZ-2266 – us NY Public [330]
Henry, George W *see* Shouting, genuine and spurious
Henry griggs weston : for forty years president of crozer theological seminary – [S.l.]: Published by friends, [1909?] – 1mf – 9 – 0-524-05118-6 – mf#1992-2071 – us ATLA [240]
Henry hart milman, dd : dean of st paul's / Milman, Arthur – London: J Murray, 1900 – 1mf – 9 – 0-7905-5716-9 – mf#1988-1716 – us ATLA [240]
Henry, Henry A *see* A synopsis of jewish history
Henry, Hugh Thomas *see* Poems, charades, inscriptions of pope leo 13
Henry irving's impressions of america, vol 1 : narrated in a series of sketches, chronicles and conversations / Hatton, Joseph – London: S Low, Marston, Searle & Rivington, 1884 – v1 on 4mf – 9 – mf#29288 – cn Canadiana [970]
Henry irving's impressions of america, vol 2 : narrated in a series of sketches, chronicles and conversations / Hatton, Joseph – London: S Low, Marston, Searle & Rivington, 1884 – v2 on 4mf – 9 – mf#29289 – cn Canadiana [970]
Henry irving's impressions of america, vols 1 and 2 : narrated in a series of sketches, chronicles and conversations / Hatton, Joseph – London: S Low, Marston, Searle & Rivington, 1884 – 2v on 1mf – 9 – mf#29287 – cn Canadiana [970]
Henry, J *see* Christian pulpit
Henry j bollers fortepiano book : nazareth hall, sept 1815 – 1815 – 1r – 1 – mf#pres. film 60, 24 – us Sibley [780]
Henry james review – Baton Rouge. 1990+ (1,5,9) – ISSN: 0273-0340 – mf#18021 – us UMI ProQuest [420]
Henry, Joe, Mrs *see* Notes on pasco county and dade city, florida
Henry, Jos *see* L'ame d'un peuple africain, les bamabara, leur vie psychique, ethique, sociale, religieuse
Henry, Joseph *see* L'ame d'un peuple africain
Henry, Joseph B *see* Study of schoenberg's pierrot lunaire
The henry knox papers, 1719-1825 – New England Historic Genealogical Society [mf ed 1960] – 55r – 1 – (with p/g) – us MA Hist [355]
The henry laurens papers – 1747-92 [mf ed ProQuest] – 19r – 1 – (with p/g) – us UMI ProQuest [978]
Henry, Lord Bishop Of Exeter *see* Reply to lord john russell's letter to the remonstrance of the bish...
Henry m stanley / Brice, Arthur John Hallam Montefiore – London, England. n d – 1r – us UF Libraries [960]
Henry m stanleys reise durch den dunklen weltteil – Leipzig 1885 [mf ed Hildesheim 1995-98] – 1v on 3mf – 9 – €90.00 – 3-487-27318-7 – gw Olms [916]
The henry m wheeler collection of glass photographic plates – [mf ed 1985] – 1r – 1 – (with p/g. photographic coll of historic sites, monuments, and important buildings in massachusetts) – us MA Hist [770]
Henry, Marc *see* Beyond the rhine
Henry, Marguerite *see* West indies in story and pictures
Henry martyn / Bell, Charles Dent – New York: A C Armstrong, 1881 – 1mf – 9 – 0-8370-6564-X – mf#1986-0564 – us ATLA [920]
Henry martyn : his life and labours, cambridge-india-persia / Page, Jesse – New York: Fleming H Revell, [189-?] – 1mf – 9 – 0-8370-6292-6 – mf#1985-0292 – us ATLA [920]
Henry martyn : saint and scholar: first modern missionary to the mohammedans, 1781-1812 / Smith, George – New York: FH Revell, [1892?] – 2mf – 9 – 0-7905-7144-7 – mf#1988-3144 – us ATLA [240]
Henry, Matthew *see*
– An exposition of the shorter catechism
– Promises of god
Henry melchior muhlenberg : patriarch of the lutheran church in america / Frick, William Keller – Philadelphia: Lutheran Publ Society, c1902 – (= ser Lutheran Handbook Series) – 1mf – 9 – 0-7905-4733-3 – mf#1988-0733 – us ATLA [242]
Henry messenger – Henry, NE: J D Fugate, 1917 (wkly) [mf ed 1920 (gaps) filmed 1979] – 1r – 1 – us NE Hist [071]
Henry of Huntingdon *see* Henrici huntenduniensis historia anglorum (rs74)
Henry of Langenstein (Henry of Hesse the Elder) *see* Le miroir de l'ame
Henry of navarre and the huguenots in france / Willert, Paul Ferdinand – New York: Putnam, 1893 – (= ser Heroes of the nations) – 2mf – 9 – 0-7905-6459-9 – mf#1988-2459 – us ATLA [944]
Henry rowe schoolcraft papers – L110023 – 69r – 1 – $2,415.00 – Dist. us Scholarly Res – us L of C Photodup [305]
Henry sater / Maclay, Walker – 1690-1754. Sater Genealogy. 1897. Historical sketch of Sater Baptist Church, 1742-1917 – 1 – 8.26 – us Southern Baptist [920]
Henry, T Shuldham *see* Disembodied state
Henry the Minstrel *see* The actis and deidis of schir william wallace
Henry the third and the church : a study of his ecclesiastical policy and of the relations between england and rome / Gasquet, Francis Aidan – London: G Bell, 1905 – 2mf – 9 – 0-7905-5656-1 – mf#1988-1656 – us ATLA [240]
Henry vanderburgh papers, 1777-1808 / Vanderburgh, Henry – [mf ed 1982] – 1r – 1 – mf#ms732 – us Western Res [355]
Henry, Victor *see*
– L'agnistoma
– Le parsisme
– A short comparative grammar of english and german
Henry Vilas Park Zoological Society *see* Zoo news
Henry w chandler and his recollections of the flo... / Green, J Howard – s.l, s.l? 1936 – 1r – us UF Libraries [978]
Henry, W Ethelbert et al *see* Photo-ceramics
Henry w. longfellow : biography, anecdote, letters, criticism / Kennedy, William Sloane – Cambridge, Mass: M King, 1882 – 1mf – 9 – 0-524-04298-5 – (incl bibl ref) – mf#1992-2018 – us ATLA [920]
Henry wadsworth longfellow : seventy-fifth birthday. proceedings of the maine historical society... – Portland: Hoyt, Fogg and Donham, c1882 – 1mf – 9 – 0-524-06391-5 – mf#1991-2513 – us ATLA [975]
Henry wadsworth longfellow, sa vie, ses oeuvres litteraires, son poeme evangeline : conference donnee a moncton le 27 fevrier 1907 a l'occasion de la celebration du centenaire de naissance de longfellow / Bourgeois, Phileas Frederic – [Shediac, N-B?: s.n, 1907?] – 1mf – 9 – 0-665-71723-7 – (incl english text) – mf#71723 – cn Canadiana [420]
Henry ward beecher / Abbott, Lyman – Boston: Houghton, Mifflin, 1903 – 2mf – 9 – 0-7905-4002-9 – (incl bibl ref) – mf#1988-0002 – us ATLA [240]
Henry ward beecher : the shakespeare of the pulpit / Barrows, John Henry – New York: Funk & Wagnalls, 1893 – (= ser American reformers) – 2mf – 9 – 0-7905-6342-8 – (incl bibl ref) – mf#1988-2342 – us ATLA [240]
Henry ward beecher as his friends saw him / Abbott, Lyman et al – Boston: Pilgrim Press, c1904 – 1mf – 9 – 0-524-08245-6 – mf#1993-3000 – us ATLA [240]
Henry whitney bellows / Lion, Felix Danford – Chicago, 1938. Chicago: Dep of Photodup, U of Chicago Lib, 1971 (1r); Evanston: American Theol Lib Assoc, 1984 (1r) – 1 – 0-8370-0371-7 – (incl bibl ref) – mf#1984-B174 – us ATLA [920]
Henry wight – Edinburgh, Scotland. 1877 – 1r – us UF Libraries [240]
Henry, William J *see* Ecclesiastical law and rules of evidence
Henry, William Wirt *see* Address of the president, hon. william wirt henry, delivered at the ninth annual meeting, held at the hot springs of virginia, august 3, 4 and 5, 1897
Henry wilson : one of god's best / Wilson, Madele & Simpson, Albert B – New York: Alliance Press, c1908 [mf ed 1992] – (= ser Christian & missionary alliance coll) – 1mf – 9 – 0-524-02177-5 – mf#1990-4243 – us ATLA [240]
Henry's journal, covering adventures and experiences in the fur trade on the red river 1799-1801 : a paper read before the society, may 4, 1888 / Bell, Charles Napier – Winnipeg?: Manitoba Free Press, 1888 – 1mf – 9 – (in double clms) – mf#30244 – cn Canadiana [380]
Henschel, Erich *see*
– Frauenlist
– Die heidin
Henschenius, G *see* Acta sanctorum
Henschke, Alfred (pseud. Klabund) *see*
– Deutsche literaturgeschichte in einer stunde
– Dumpfe trommel und berauschtes gong
– Die geisha o-sen
– Kleines klabund-buch
– Laotse
– Li-tai-pe
Hensel, Sebastian *see*
– Familie mendelssohn
– Die familie mendelssohn 1729-1847
Henselt, Adolf von *see* Trio in a moll fuer pianoforte, violine und violoncello, op 24
Hensey, Andrew Fitch *see*
– A master builder on the congo
– Opals from africa
Henshall, James A *see* Camping and cruising in florida
Henshaw, Julia Wilmotte *see*
– Mountain wild flowers of canada
– The queen city of british columbia
– Vancouver
– Why not, sweetheart?
Hensley, Tammy *see* An analysis of the primary use of church sports programs in anderson, indiana
Henslow, George *see*
– Christian beliefs reconsidered in the light of modern thought
– Genesis and geology
– Present-day rationalism critically examined
– The theory of evolution of living things
– The vulgate
Henslow, George et al *see* Christian apologetics
Henslow, John Stevens *see* Syllabus of a course of lectures on botany
Hensold, Karl *see* Georg herwegh und seine deutschen vorbilder
Henson, G M *see* Christ is all
Henson, H Hensley *see*
– The liberty of prophesying
– Light and leaven
Henson, Hensley *see*
– Anglicanism and reunion
– Apostolic christianity
– Church problems
– The creed in the pulpit
– Cross-bench views of current church questions
– Dissent in england
– Ecclesiastica
– Godly union and concord
– The issue of kikuyu
– The liberty of prophesying
– Light and leaven
– Moral discipline in the christian church
– The national church
– Notes of my ministry
– Notes on popular rationalism
– Preaching to the times
– Puritanism in england
– The relation of the church of england to the other reformed churches
– Reunion and intercommunion
– The road to unity
– Robertson of brighton, 1816-1853
– Sincerity and subscription
– Studies in english religion in the 17th century
– The value of the bible and other sermons, 1902-1904
– War-time sermons
– Westminster sermons
Henson, P S *see* Manuscripts of 19 sermons delivered at first baptist church, chicago, 1891
Henson, Poindexter Smith *see* My mother's bible stories
Henssen, Gottfried *see* Sagen, maerchen und schwaenke des juelicher landes
Hentisberus, Gulielmus *see* Tractatus gulielmi hentisberi de sensu coposito et diviso
Hentschel, Ute *see* Sprache, erkenntnis und handlung
Henty, G A *see* Lion of the north
Henty, George A *see* Roving commission
Henty, George Alfred *see* A roving commission
Henty observer – Henty – 10r – 9 – A$629.79 vesicular A$684.79 silver – at Pascoe [079]
Hentz, John P *see* History of the lutheran version of the bible
Henze, Wilhelm *see*
– Eck segge man bloss
– Is duet 'ne welt!
– Sau suihste iut!
– Tau'n lustigen steebel
– Wat sei alles maket!
Henzel, J *see*
– Christelijk of heidensch
– Door de duisternis tot het licht
– In liefde vereend
– James hudson taylor
– Robert morrison
– Verloren, maar gevonden
Henzen, Wilhelm *see* Martin luther
Heolycyw, bethel independent, and rhiwceiliog, soar calvinistic methodist, monumental inscriptions – 1mf – 9 – £1.25 – uk Glamorgan FHS [929]
Heortology : a history of the christian festivals from their origin to the present day = Heortologie / Kellner, Karl Adam Heinrich – London: K. Paul, Trench, Truebner, 1908 – (= ser The International Catholic Library) – 2mf – 9 – 0-7905-4983-2 – (incl bibl ref. in english) – mf#1988-0983 – us ATLA [240]
Heortology sic : a history of the christian festivals from their origin to the present day / Kellner, Karl Adam Heinrich – Tr. from 2nd German ed. London: K. Paul, Trench, Truebner, 1908. xviii,466p – 1 – us Wisconsin U Libr [240]
Die hepatische durchblutung bei hyperthermie : untersuchungen im ueberwaermungsbad / Utsch, Sabine – Frankfurt a.M. 1984 (mf ed 1993) – 1mf – 9 – €24.00 – 3-89349-657-2 – mf#DHS-AR 657 – gw Frankfurter [616]
Hepatitis c bei haemodialysepatienten : eine einjaehrige verlaufsstudie an haemodialysepatienten, nierentransplantierten patienten und dem personal einer dialyseklinik in wuppertal / Bock, Ulrich Manfred – (mf ed 1995) – 1mf – 9 – €30.00 – 3-8267-2145-4 – mf#DHS 2145 – gw Frankfurter [616]
Hepato-gastroenterology – Stuttgart. 1980-1993 (1,5,9) – (cont: acta hepato-gastroenterologica) – ISSN: 0172-6390 – mf#10151,01 – us UMI ProQuest [616]
Hepatology – Philadelphia. 1989+ (1,5,9) – ISSN: 0270-9139 – mf#17791 – us UMI ProQuest [616]
Hepatology research – Amsterdam. 1997+ (1) – ISSN: 1386-6346 – mf#42733,01 – us UMI ProQuest [616]
Hepburn, J D *see* Twenty years in khama's country
Hepburn of japan and his wife and helpmates : a life story of toil for christ / Griffis, William Elliot – Philadelphia: Westminster Press, 1913 [mf ed 1995] – (= ser Yale coll) – ix/238p (ill) – 1 – 0-524-09795-X – mf#1995-0795 – us ATLA [920]
Hepding, Hugo *see* Attis
Hepher, Cyril *see* The fellowship of silence
Hepner, Gerda *see* Held ohne namen: ein schicksal
Hepp, Alexandre *see* Le coeurs embellis
Hepp, Michael [comp] *see* Sozialstrategien der deutschen arbeitsfront
Heppe, H *see* The odor beza
Heppe, Heinrich *see*
– Die bekenntnisschriften der altprotestantischen kirche deutschlands
– Die bekenntnisschriften der reformirten kirche deutschlands
– Die confessionelle entwicklung der altprotestantischen kirche deutschlands
– Die dogmatik der evangelisch-reformirten kirche
– Die entstehung und fortbildung des luthertums und die kirchlichen bekenntnisschriften desselben von 1548-76

– Geschichte der quietistischen mystik in der katholischen kirche
– Geschichte des deutschen volksschulwesens
– Geschichte des pietismus und der mystik in der reformierten kirche
– Die presbyteriale synodalverfassung der evangelischen kirche in norddeutschland
– The reformers of england and germany in the sixteenth century
– Das schulwesen des mittelalters und dessen reform im sechszehnten jahrhundert
– Ursprung und geschichte der bezeichnungen "reformierte" und "lutherische" kirche

Heppe, Robert A *see* The kinematic variables related to the efficiency of throwing

Heppel, Alexander *see* South africa

Hepperger, Josef von *see* Annalen der k k universitaets-sternwarte in wien [waehring]

Hepple, Alexander *see*
– Papers, 1937-1964
– South africa

Heppner, Aaron *see* Vergangenheit

Heppner gazette – Heppner OR: Patterson Pub Co, -1912 [wkly] – 1 – (began in 1892. cont: weekly heppner gazette. merged with: heppner times (-1912) to form: gazette-times (1912-25)) – us Oregon Lib [071]

Heppner gazette *see*
– Gazette-times (heppner, or)
– Heppner times

Heppner gazette-times – Heppner, OR: V and S Crawford, 1912-; v 42 n32 nov 5 1925-dec 30 1998 – 1 – (aka: heppner gazette times, and hehisch. 1932-1941 incl newspaper publ by heppner high school students) – us Oregon Hist [071]

Heppner gazette-times – Heppner OR: V & S Crawford, 1912- [wkly] – 1 – (cont: gazette-times (1912-25). 1932-41 incl newspaper pub by heppner high school students) – us Oregon Lib [071]

Heppner gazette-times *see* Gazette-times (heppner, or)

Heppner herald – Heppner OR: E G & L K Harlan, -1924 [wkly] – 1 – (absorbed: ione bulletin. absorbed by: gazette-times (1912-25)) – us Oregon Lib [071]

Heppner herald *see*
– Gazette-times (heppner, or)
– Ione bulletin

Heppner times – Heppner OR: A J Hicks [wkly] – 1 – (ceased in 1912. merged with: heppner gazette (1892-1912) to form: gazette-times (1912-25)) – us Oregon Lib [071]

Heppner times *see* Gazette-times (heppner, or)

Heppner weekly gazette – Heppner OR: J W Redington, [wkly] – 1 – (cont by: weekly heppner gazette (1890-1982)) – us Oregon Lib [071]

The heptameron; or tales and novels of marguerite, queen of navarre / Marguerite, Queen of Navarre – Trans. by Arthur Machen, with an introd.London: G. Routledge & Sons, New York: E.P. Dutton, 1911. xx,392p – 1 – us Wisconsin U Libr [830]

Her husband is dead. she must save her children – NY, n.d. Fiche W 938. (Blodgett Collection of Spanish Civil War Pamphlets) – 9 – us Harvard [946]

Her last throw : a novel / Hungerford, Margaret Wolfe – London: F V White & Co, 1890 – (= ser 19th c women writers) – 2mf – 9 – mf#5.1.101 – uk Chadwyck [830]

Her milwaukee – 1974 feb-apr – 1r – 1 – mf#843754 – us WHS [071]

Her own words : the writings of elizabeth gaskell. from the john rylands university library, manchester, england – 13r – 1 – (previous title: elizabeth gaskell and victorian literature. coll includes autograph drafts, gaskell's correspondence with patrick and charlotte bronte. also contains letters to and from w s landor, w m thackeray, matthew arnold, john ruskin, george eliot, charles dickens and other contemporaries. includes printed guide) – mf#C35-28010 – us Primary [420]

Her week's amusement / Hungerford, Margaret Wolfe (Hamilton) – London: Ward & Downey, 1884 – (= ser 19th c women writers) – 4mf – 9 – mf#5.1.89 – uk Chadwyck [420]

Heraclitus ridens : at a dialogue between jest and earnest, concerning the times – London. 1681-1682 (1) – mf#5563 – us UMI ProQuest [870]

Heraclitus ridens : a discourse between jest and earnest, concerning the times – London. 1717-1718 [1] – mf#5263 – us UMI ProQuest [870]

Heraclitus ridens in a dialogue between jest and earnest concerning the time – London. 1703-1704 (1) – mf#5564 – us UMI ProQuest [870]

Heraclitus und democritus : ...benebens 10. dreystaendigen sinnbildern von den neigungen dess gemuetes vorgestellt... / [Harsdoerffer, G P] – Nuernberg: Gedruckt bey Michael Endter, 1653 – 8mf – 9 – mf#0-01 – ne IDC [090]

Heraclius : von den farben und kuensten der roemer / Ilg, A – Wien. v4. 1873 – 3mf – 9 – mf#0-517 – ne IDC [700]

Heraeus, K G *see* Sacrae caes

Herakles : aufsatze zur griechischen religions- und sagengeschichte / Schweitzer, Bernhard – Tubingen, Germany. 1922 – 1r – us UF Libraries [930]

Herald – 1883 apr 20/1884 oct 3-1912 feb 23/mar 29 [gaps] – 1r – 1 – mf#929794 – us WHS [071]

Herald – 1897 jul-1898 dec 29, 1899 jan 5-1900 jun 14, 1900 jun 21-1901 nov 14, 1901 nov 21-1903 feb 12, 1903 feb 19-aug 27 – 1r – 1 – mf#942121 – us WHS [071]

Herald – 1899 jan 6-1900 jul 13, 1900 jul 20-1901 dec 27, 1902 jan 3-1902 jul 18 – 1r – 1 – mf#1003879 – us WHS [071]

Herald – 1939 oct, 1940 nov, 1954 dec, 1955 n1,4-5,12, 1956 n1,9-11, 1957 n1-2,4,12, 1958 n5,9, 1959 n1,3-4,6, 1960 n2-3, 1961-80 – 1r – 1 – mf#622583 – us WHS [241]

Herald – 1970 aug 13-1971 feb 25, 1971 mar 1-1972 sep 20 – 1r – 1 – mf#967263 – us WHS [071]

Herald – 1978 feb/1978 aug-1980 jul-sep – 1r – 1 – mf#1159259 – us WHS [071]

Herald – Ahoskie, NC. 1971-1977 (1) – mf#65294 – us UMI ProQuest [071]

Herald – Albany, GA. 1990-2000 (1) – mf#61294 – us UMI ProQuest [071]

Herald – Almira, WA. 1935-1971 (1) – mf#68361 – us UMI ProQuest [071]

Herald / Athens Co. Athens – v1 n1. (sep 1882-oct 1886), jan 1887-93 [wkly] – 3r – 1 – mf#B10562-10564 – us Ohio Hist [071]

Herald – Augusta, GA. 1939-1992 (1) – mf#60444 – us UMI ProQuest [071]

Herald – Aurora, IL. 1871-1886 (1) – mf#62502 – us UMI ProQuest [071]

Herald – Avondale, PA. 1896-1939 (1) – mf#68959 – us UMI ProQuest [071]

Herald – Azusa, CA. 1990-2000 (1) – mf#61222 – us UMI ProQuest [071]

Herald – Baltimore, MD. 1903-1903 (1) – mf#63587 – us UMI ProQuest [071]

Herald – Barberton, OH. 1987-2000 (1) – mf#65378 – us UMI ProQuest [071]

Herald – Bellingham, WA. 1904+ (1) – mf#61900 – us UMI ProQuest [071]

Herald / Belmont Co. Bellaire – jan 1898-jan 1899 [wkly] – 1r – 1 – mf#B12013 – us Ohio Hist [071]

Herald – Billings, MT. 1882-1885 (1) – mf#64258 – us UMI ProQuest [071]

Herald – Biloxi, MS. 1888-1898 (1) – mf#63937 – us UMI ProQuest [071]

Herald – Biloxi, MS. 1898-1985 (1) – mf#61545 – us UMI ProQuest [071]

Herald – Bossburg, WA. 1910-1910 (1) – mf#66946 – us UMI ProQuest [071]

Herald – Boston, MA. 1848-1967 (1) – mf#63637 – us UMI ProQuest [071]

Herald – Bradenton, FL. 1922-2000 (1) – mf#60433 – us UMI ProQuest [071]

Herald – Brewster, WA. 1902-1940 (1) – mf#69211 – us UMI ProQuest [071]

Herald – Bridgewater, VA. 1894-1906 (1) – mf#61167 – us UMI ProQuest [071]

Herald – Bristol, CT. 1888-1901 (1) – mf#62336 – us UMI ProQuest [071]

Herald – Brockway, MT. 1928-1930 (1) – mf#64286 – us UMI ProQuest [071]

Herald – Cairo, IL. 1917-1919 (1) – mf#62523 – us UMI ProQuest [071]

Herald – Cambridge, OH. 1882-1904 (1) – mf#65397 – us UMI ProQuest [071]

Herald – Carpenteria, CA. 1954-1975 (1) – mf#62120 – us UMI ProQuest [071]

Herald – Charlottetown, PEI: E Reilly, 1864-71 – 2r – 1 – ISSN: 0839-3265 – cn Library Assoc [071]

Herald – Cincinnati, OH. 1961-1979 (1) – mf#65414 – us UMI ProQuest [071]

Herald – Circleville, OH. 1993-2000 (1) – mf#61701 – us UMI ProQuest [071]

Herald – Clarkston, WA. 1928-1983 (1) – mf#66972 – us UMI ProQuest [071]

Herald – Clearfield, PA. 1912-1913 (1) – mf#65869 – us UMI ProQuest [071]

Herald / Clermont Co. Loveland – jan 1971-jun 1984, may 22 & 29 1986 [wkly] – 13r – 1 – mf#B34991-35003 – us Ohio Hist [071]

Herald – Clinton, IA. 1856-1865 (1) – mf#63112 – us UMI ProQuest [071]

Herald – Clinton, IA. 1867-1869 (1) – mf#63113 – us UMI ProQuest [071]

Herald – Clinton, IA. 1895-1900 (1) – mf#63114 – us UMI ProQuest [071]

Herald – Clinton, IA. 1985-2000 (1) – mf#63111 – us UMI ProQuest [071]

Herald – Clyde Park, MT. 1910-1924 (1) – mf#64329 – us UMI ProQuest [071]

Herald – Coffee Creek, MT. 1915-1923 (1) – mf#64330 – us UMI ProQuest [071]

Herald – Columbia, TN. 1994-2000 (1) – mf#68066 – us UMI ProQuest [071]

Herald – Concrete, WA. 1973-1979 (1) – mf#66980 – us UMI ProQuest [071]

Herald – Arlington, NE: Moore & Moore. v20 n36. jul 12 1902-04// (wkly) – 1r – 1 – (cont: arlington herald. merged with: arlington review to form: arlington review-herald) – us Bell [071]

Herald – Augusta, Eau Claire WI. 1872 apr 6-1873 nov 8 – 1r – 1 – (cont: augusta herald) – mf#923900 – us WHS [071]

Herald – Augusta, Eau Claire WI. 1872 apr 6-1873 nov 8 – 1r – 1 – (cont: augusta herald) – mf#923900 – us WHS [071]

Herald – Whitefish Bay WI. 1984 feb 16-jun, 1984 jul-oct, 1984 nov-dec, 1985 jan-apr, 1985 may-apr, 1985 sep-dec, 1986 jan-mar 27 – 7r – 1 – (cont: brown deer/glendale herald; cont by: herald [brown deer, wi ed : 1986]; herald [glendale, wi ed]) – mf#1159266 – us WHS [071]

Herald – Ansley, NE: Thomas Wright. -v25 n38. apr 21 1916 (wkly) – 2r – 1 – (cont: custer county herald; cont by: ansley herald) – us Bell [071]

Herald – 1945-60 – 1r – 1 – (cont: diocese of eau claire and cathedral herald) – mf#3184497 – us WHS [242]

Herald – 12 nov 2001- (mthly) [mf ed NLSA 2001-] – 1 – (cont: eastern province herald) – mf#mp1034 – sa National [079]

Herald – New Berlin WI. 1986 apr 24/dec-1995 oct-dec [small gaps] – 31r – 1 – (cont: herald [brown deer, glendale, wi ed]; cont by: glendale herald [new berlin, wi]) – mf#5490968 – us WHS [071]

Herald – Wauwatosa WI. 1986 apr 24/aug-1997 oct/dec [gaps] – 36r – 1 – (cont: herald [brown deer, glendale, wis. ed.]; cont by: brown deer herald [new berlin, wi: 1997]) – mf#5466331 – us WHS [071]

Herald – Brown Deer, Mequon, Whitefish Bay WI. 1980 jul 24-sep, 1980 oct-dec, 1981 jan-jun – 3r – 1 – (cont: herald [brown deer, wi ed: 1977]; cont by: herald [brown deer, wi ed: 1981]) – mf#1159261 – us WHS [071]

Herald – Brown Deer, Whitefish Bay WI. 1981 mar 5-jun, 1981 jul-dec, 1982 jan-jun, 1982 jul-1983 mar, 1983 apr-aug, 1983 sep-dec, 1984 jan-feb 2 – 7r – 1 – (cont: herald [brown deer-mequon, wi ed]; cont by: glendale herald; brown deer/glendale herald) – mf#1159262 – us WHS [071]

Herald – Holbrook, NE: Herald Print Co. 2v. v4 n14. nov 13 1896-v5 n21. dec 31 1897 (wkly) [mf ed with gaps filmed 1980] – 1r – 1 – (cont: holbrook herald (1895). cont by: holbrook herald (1898)) – us NE Hist [071]

Herald – Lake Geneva WI. 1883 apr 20/1884 oct 3-1912 feb 23/mar 29 [gaps] – 22r – 1 – (cont: lake geneva herald [lake geneva, wi: 1879]; cont by: lake geneva herald [lake geneva, wi: 1912]) – mf#929794 – us WHS [071]

Herald – New York: J G Bennett. v1 n1 aug 31 1835-v2 n373 may 20 1837 (daily ex sun) – 4r – 1 – (cont: morning herald (new york 1835). cont by: morning herald (new york 1837)) – Dist. us UMI ProQuest – us Eastman [071]

Herald – Brown Deer, Whitefish Bay WI. 1978 feb-1978 aug, 1978 sep-1978 dec, 1979 jan-jun, 1979 jul-dec, 1980 jan-jun, 1980 jul-sep – 6r – 1 – (cont: north shore herald (brown deer, wi ed); cont by: herald (brown deer-mequon, wi ed)) – us WHS [071]

Herald – Shorewood, Whitefish Bay WI. 1977 sep 15/dec-1997 jan – 57r – 1 – (cont: north shore herald; north shore herald; cont by: shorewood herald [west allis, wi]) – mf#1144869 – us WHS [071]

Herald – Omaha, NE: Geo L Miller, Lyman Richardson. v9 n155 apr 7 1874-v9 n177 may 2 1874 (daily ex mon) – 1r – 1 – (cont: omaha daily herald; cont by: omaha herald (daily)) – us Eastman [071]

Herald – Omaha, NE: Geo L Miller, Lyman Richardson. 1v. v9 n27 [ie 28]. apr 10 1874-v9 n31. apr 30 1874 (wkly) – 1r – 1 – (cont: omaha weekly herald (1865). cont by: omaha herald (weekly)) – us NE Hist [071]

Herald – Omaha, NE: Geo L Miller, Lyman Richardson. 1v. v9 n27 [ie n28] apr 10 1874-v9 n31. apr 30 1874 (wkly) [mf ed [S.l. : s.n.]] – 1r – 1 – (cont: omaha weekly herald (1865); cont by: omaha herald (weekly)) – us Misc Inst [071]

Herald – Rib Lake WI. 1970 aug 13-1971 feb 25, 1971 mar 1-1972 sep 20 – 2r – 1 – (cont: rib lake herald) – mf#967263 – us WHS [071]

Herald – Toronto: [s.n. 1886-1913?] – 9 – (cont: the fonetic herald) – mf#P04542 – cn Canadiana [420]

Herald – Whitefish Bay WI. 1977 sep 15/30-1997 oct/dec – 57r – 1 – (cont: whitefish bay herald; cont by: whitefish bay herald [west allis, wi]) – mf#1159065 – us WHS [071]

Herald – Oconto Falls WI. 1899 jan 6-1900 jul 13, 1900 jul 20-1901 dec 27, 1902 jan 3-1902 jul 18 – 3r – 1 – (continued by: oconto falls herald (oconto falls, wis: 1902)) – mf#1003879 – us WHS [071]

Herald – Readstown WI. 1897 jul-1898 dec 29, 1899 jan 5-1900 jun 14, 1900 jun 21-1901 nov 14, 1901 nov 21-1903 feb 12, 1903 feb 19-aug 27 – 5r – 1 – (continued by: readstown herald) – mf#942121 – us WHS [071]

Herald – Markesan WI. 1990 feb 1-dec, 1991-95, 1996 jan 4-aug 28 – 7r – 1 – (continued by: waupun leader-news; advance (randolph, wi); neighbors (beaver dam, wi)) – mf#3656469 – us WHS [071]

Herald – Cranston, RI. 1936-1992 (1) – mf#66185 – us UMI ProQuest [071]

Herald / Crawford Co. New Washington – dec 1970-may 1975 [wkly] – 3r – 1 – mf#B29836-29838 – us Ohio Hist [071]

Herald / Crawford Co. New Washington – v1. (2-12/1881,9/1885-12/70,1988-1991) [wkly] – 34r – 1 – mf#B32189-32222 – us Ohio Hist [071]

Herald / Cuyahoga Co. Cleveland – v1 n1. oct 1819-oct 1821 [wkly] – 1r – 1 – mf#B9835 – us Ohio Hist [071]

Herald / Darke Co. New Madison – dec 1907-aug 23, (aug 30-sep 1942) [wkly] – 10r – 1 – mf#B10155-10164 – us Ohio Hist [071]

Herald / Darke Co. New Madison – jan 1930-jul 1931 [wkly] – 1r – 1 – mf#B33859 – us Ohio Hist [071]

Herald – Dayton, OH. 1882-1949 (1) – mf#65462 – us UMI ProQuest [071]

Herald – Decatur, IL. 1899-1907 (1) – mf#62596 – us UMI ProQuest [071]

Herald – Decatur, IL. 1899-1931 (1) – mf#68142 – us UMI ProQuest [071]

Herald / Delaware Co. Delaware – jan 1878-nov 1885 [wkly] – 4r – 1 – mf#B8864-8867 – us Ohio Hist [071]

Herald – Denison, TX. 1993-1996 (1) – mf#61848 – us UMI ProQuest [071]

Herald – Dillon, SC. 1904-1939 (1) – mf#66484 – us UMI ProQuest [071]

Herald – Dryden, NY. 1871-1919 (1) – mf#68702 – us UMI ProQuest [071]

Herald – Duluth, MN. 1948-1982 (1) – mf#60500 – us UMI ProQuest [071]

Herald – East Providence, RI. 1941-1942 (1) – mf#66196 – us UMI ProQuest [071]

Herald – Eureka, KS. 1979-2000 (1) – mf#68156 – us UMI ProQuest [071]

Herald – Fairfax, VA. 1923-1972 (1) – mf#66704 – us UMI ProQuest [071]

Herald – Fairport, NY. 1873-1925 (1) – mf#64960 – us UMI ProQuest [071]

Herald – Fallon, MT. 1916-1920 (1) – mf#64378 – us UMI ProQuest [071]

Herald – Farmville, VA. 1915-1944 (1) – mf#66709 – us UMI ProQuest [071]

Herald : (final edition) – Miami, FL. 1969-1973 (1) – mf#62433 – us UMI ProQuest [071]

Herald – Fort Lauderdale, FL. 1919-1923 (1) – mf#62405 – us UMI ProQuest [071]

Herald – Franklin, WV. 1930-1933 (1) – mf#67291 – us UMI ProQuest [071]

Herald – Fromberg, MT. 1954-1956 (1) – mf#64392 – us UMI ProQuest [071]

Herald / Fulton Co. Archbold – oct 1893-jun 1898 (poor quality) [wkly] – 2r – 1 – mf#B558-559 – us Ohio Hist [071]

Herald – Galata, MT. 1913-1914 (1) – mf#64393 – us UMI ProQuest [071]

Herald – Geneva, IN. 1893-1971 (1) – mf#62803 – us UMI ProQuest [071]

Herald – Grand Forks, ND. 1879+ (1) – mf#61692 – us UMI ProQuest [071]

Herald – Grandview, WA. 1969-1976 (1) – mf#67005 – us UMI ProQuest [071]

Herald / Greene Co. Cedarville – jul 1890-jul 1892, nov 1899-1946 [wkly] – 12r – 1 – mf#B1228-1239 – us Ohio Hist [071]

Herald / Greene Co. Xenia – (1895-1924), sep 1928-jun 1929 [wkly] – 14r – 1 – mf#B10350-10363 – us Ohio Hist [071]

Herald – Hammondsport, NY. 1874-1931 (1) – mf#64992 – us UMI ProQuest [071]

Herald – Hardin, MT. 1922-1924 (1) – mf#64430 – us UMI ProQuest [071]

Herald / Harding Co. Ada – aug 1916-67 [wkly] – 27r – 1 – mf#B380-406 – us Ohio Hist [071]

Herald / Harding Co. Ada – jan 1968-dec 1974 (complete) [wkly] – 5r – 1 – mf#B372-376 – us Ohio Hist [071]

Herald – Harper Woods, MI. 1962-1966 (1) – mf#63767 – us UMI ProQuest [071]

Herald – Harvard, IL. 1887-1931 (1) – mf#62626 – us UMI ProQuest [071]

Herald – Hebron, IL. 1959-1967 (1) – mf#62630 – us UMI ProQuest [071]

Herald – Helena, MT. 1890-1900 (1) – mf#64459 – us UMI ProQuest [071]

Herald – Helena, MT. 1890-1900 (1) – mf#64458 – us UMI ProQuest [071]

Herald – Hudson, OH. 1929-1930 (1) – mf#65531 – us UMI ProQuest [071]

Herald – Huntington, IN. 1912-1929 (1) – mf#62826 – us UMI ProQuest [071]

Herald – Huntington, WV. 1903-1908 (1) – mf#67325 – us UMI ProQuest [071]

Herald – Hyde Park, IL. 1986-1997 (1) – mf#62554 – us UMI ProQuest [071]

Herald / Jackson Co. Jackson – (1883-2/95,2/01-19,26-1945) [wkly, semiwkly] – 36r – 1 – mf#B10038-10073 – us Ohio Hist [071]

Herald / Jackson Co. Jackson – jan 1946-may 1974 [semiwkly] – 39r – 1 – mf#B12412-12450 – us Ohio Hist [071]

Herald – Jasper, IN. 1895-1936 (1) – mf#62858 – us UMI ProQuest [071]
Herald / Jefferson Co. Steubenville – 1890-98, 1900 [wkly] – 8r – 1 – mf#B25797-25804 – us Ohio Hist [071]
Herald – Lake Geneva WI. 1941 aug 5-nov 7 – 1r – 1 – mf#932798 – us WHS [071]
Herald – Libby, MT. 1911-1914 (1) – mf#64530 – us UMI ProQuest [071]
Herald : (library edition) – Miami, FL. 1986-2000 (1) – mf#60643 – us UMI ProQuest [071]
Herald / Licking Co. Hebron – may-nov 1886 [wkly] – 1r – 1 – mf#B14122 – us Ohio Hist [071]
Herald / Licking Co. Utica – (aug 1879-dec 1993) some damaged & scattered [wkly] – 46r – 1 – mf#B34925-34970 – us Ohio Hist [071]
Herald – Little Falls, NJ. 1927-1975 (1) – mf#68990 – us UMI ProQuest [071]
Herald – Livingston, MT. 1891-1898 (1) – mf#64539 – us UMI ProQuest [071]
Herald – Lockridge, IA. 1909-1915 (1) – mf#63295 – us UMI ProQuest [071]
Herald / Logan Co. Belle Center – apr 1897-mar 1898, (1900-nov 1902) [wkly] – 1r – 1 – mf#B12917 – us Ohio Hist [071]
Herald – Louisville, OH. 1987-1998 (1) – mf#65563 – us UMI ProQuest [071]
Herald – Mansfield, OH. 1856-1883 (1) – mf#65566 – us UMI ProQuest [071]
Herald – Martinsburg, WV. 1881-1918 (1) – mf#67356 – us UMI ProQuest [071]
Herald / Medina Co. Seville – may 1918-apr 1919 (damaged iss) [wkly] – 1r – 1 – mf#B9207 – us Ohio Hist [071]
Herald / Mercer Co. Mendon – jan 1940-jul 1942,jan 1946-dec 1964 [wkly] – 7r – 1 – (suspended during ww2) – mf#B32513-32519 – us Ohio Hist [071]
Herald – Merrillville, IN. 1970-1996 (1) – mf#68450 – us UMI ProQuest [071]
Herald – Metropolis, IL. 1906-1917 (1) – mf#62650 – us UMI ProQuest [071]
Herald – Middletown, OH. 1833-1854 (1) – mf#65587 – us UMI ProQuest [071]
Herald – Missoula, MT. 1906-1911 (1) – mf#61128 – us UMI ProQuest [071]
Herald – Montague, MT. 1917-1919 (1) – mf#64583 – us UMI ProQuest [071]
Herald / Montgomery County Genealogical & Historical Society [TX] – 1978 jan-1983 win, 1984 spr-1989 win – 2r – 1 – mf#837488 – us WHS [929]
Herald – Montgomery, WV. 1941+ (1) – mf#67369 – us UMI ProQuest [071]
Herald – Mundelien, IL. 1971+ (1) – mf#62662 – us UMI ProQuest [071]
Herald / Muskingum Co. Zanesville – v1 n1. sep 1936-feb 1940 (damaged) [wkly] – 3r – 1 – mf#B32783-32785 – us Ohio Hist [071]
Herald – Nairobi, Kenya. v1 n553-718. 1998 dec 20-1999 – 3r – us UF Libraries [079]
Herald – Narragansett, RI. 1876-1898 (1) – mf#66217 – us UMI ProQuest [071]
Herald – Neihart, MT. 1891-1901 (1) – mf#64588 – us UMI ProQuest [071]
Herald – New York, NY. 1835-1919 (1) – mf#61134 – us UMI ProQuest [071]
Herald – Newport, RI. 1892-1944 (1) – mf#66222 – us UMI ProQuest [071]
Herald – Ontonagon, MI. 1973-1979 (1) – mf#63834 – us UMI ProQuest [071]
Herald – Papanui, Christchurch, NZ. 1976-87 – 10r – 1 – mf#70.17 – nz Nat Libr [079]
Herald – Pasco, WA. 1919-1947 (1) – mf#67070 – us UMI ProQuest [071]
Herald – Pascoag, RI. 1892-1918 (1) – mf#66240 – us UMI ProQuest [071]
Herald / Perry Co. New Lexington – (sep 1888-nov 1926, 1928-aug 1929) [wkly, semiwkly] – 18r – 1 – mf#B8846-8863 – us Ohio Hist [071]
Herald / Pickaway Co. Circleville – jul 1914-feb 1916 [daily] – 4r – 1 – mf#B30384-30387 – us Ohio Hist [071]
Herald / Pickaway Co. Circleville – may 1850-dec 1860,nov 1870-dec 1874 [wkly] – 6r – 1 – mf#B13138-13143 – us Ohio Hist [071]
Herald / Pickaway Co. Circleville – nov 1831-nov 1832 [wkly] – 1r – 1 – mf#B29888 – us Ohio Hist [071]
Herald – Piedmont, WV. 1888-1991 (1) – mf#67431 – us UMI ProQuest [071]
Herald – Plainview, NY. 1956-1964 (1) – mf#65170 – us UMI ProQuest [071]
Herald – Plentywood, MT. 1909-1974 (1) – mf#61035 – us UMI ProQuest [071]
Herald – Plevna, MT. 1922-1931 (1) – mf#64606 – us UMI ProQuest [071]
Herald – Port Huron, MI. 1900-1910 (1) – mf#63844 – us UMI ProQuest [071]
Herald – Portage, MI. 1958-1968 (1) – mf#63846 – us UMI ProQuest [071]
Herald / Preble Co. Camden – v1 n1. jun 1877-78, may-nov 1879 [wkly] – 1r – 1 – mf#B3948 – us Ohio Hist [071]
Herald / Preble Co. Eaton – mar 1906-mar 1915, 1916-feb 1918 [wkly] – 6r – 1 – mf#B25986-25991 – us Ohio Hist [071]
Herald / Preble Co. Eaton – may 1900-mar 1901 [wkly] – 1r – 1 – mf#B32606 – us Ohio Hist [071]
Herald – Providence, RI. 1879-1887 (1) – mf#66313 – us UMI ProQuest [071]
Herald – Pullman, WA. 1888-1988 (1) – mf#67087 – us UMI ProQuest [071]
Herald – Renton, WA. 1911-1917 (1) – mf#67094 – us UMI ProQuest [071]
Herald / Richland Co. Mansfield – 1883-86, 1889-90 [wkly] – 3r – 1 – mf#B8077-8079 – us Ohio Hist [071]
Herald – Rochester, NY. 1892-1926 (1) – mf#65193 – us UMI ProQuest [071]
Herald – Rock Hill, SC. 1880+ (1) – mf#61830 – us UMI ProQuest [071]
Herald – Rutland, VT. 1806-1907 (1) – mf#68049 – us UMI ProQuest [071]
Herald – Rutland, VT. 1861-2000 (1) – mf#60602 – us UMI ProQuest [071]
Herald – Salem, WV. 1913+ (1) – mf#67468 – us UMI ProQuest [071]
Herald – Sanford, NC. 1992-2000 (1) – mf#61691 – us UMI ProQuest [071]
Herald / Seneca Co. Green Springs – aug 23-dec 27 1890 (damaged) [semiwkly, wkly] – 1r – 1 – mf#B31708 – us Ohio Hist [071]
Herald – Sharon, PA. 1878+ (1) – mf#61805 – us UMI ProQuest [071]
Herald – Sharpsburg, PA. 1987-2000 (1) – mf#66080 – us UMI ProQuest [071]
Herald – Sidney, MT. 1908-1974 (1) – mf#64649 – us UMI ProQuest [071]
Herald – South Lyon, MI. 1929-1999 (1) – mf#68028 – us UMI ProQuest [071]
Herald – Southbridge, MA. 1902-1929 (1) – mf#63661 – us UMI ProQuest [071]
Herald – Spartanburg, SC. 1893-1982 (1) – mf#66522 – us UMI ProQuest [071]
Herald – St Joseph, MO. 1890-1899 (1) – mf#64208 – us UMI ProQuest [071]
Herald – Suffolk, VA. 1900-1926 (1) – mf#66889 – us UMI ProQuest [071]
Herald – Surry, VA. 1943-1951 (1) – mf#66890 – us UMI ProQuest [071]
Herald – Tacoma, WA. 1877-1932 (1) – mf#67149 – us UMI ProQuest [071]
Herald – Tecumseh, MI. 1863-1983 (1) – mf#63867 – us UMI ProQuest [071]
Herald – Three Forks, MT. 1908-1974 (1) – mf#64665 – us UMI ProQuest [071]
Herald – Three Rivers, MI. 1881-1883 (1) – mf#63870 – us UMI ProQuest [071]
Herald – Troy, MT. 1910-1911 (1) – mf#64671 – us UMI ProQuest [071]
Herald / Trumbull Co. Newton Falls – 6/1930-33,9/46-49,51-1983 (fire damaged thru 49) [wkly] – 23r – 1 – mf#B23135-23157 – us Ohio Hist [071]
Herald – Turner Falls, MA. 1940-1942 (1) – mf#63668 – us UMI ProQuest [071]
Herald – Tyrone, PA. 1910-1974 (1) – mf#66095 – us UMI ProQuest [071]
Herald – Utica, NY. 1866-1896 (1) – mf#65248 – us UMI ProQuest [071]
Herald – Utica, NY. 1896-1897 (1) – mf#65249 – us UMI ProQuest [071]
Herald – Vicksburg, MS. 1952-1957 (1) – mf#64141 – us UMI ProQuest [071]
Herald – Victor, NY. 1891-1952 (1) – mf#69310 – us UMI ProQuest [071]
Herald – Washington, IN. 1903-1964 (1) – mf#63003 – us UMI ProQuest [071]
Herald – Washington, OH. 1858-1862 (1) – mf#65712 – us UMI ProQuest [071]
Herald – West Union, WV. 1911-1973 (1) – mf#67505 – us UMI ProQuest [071]
Herald – Westerly, RI. 1899-1901 (1) – mf#66430 – us UMI ProQuest [071]
Herald – Wilmington, NC. 1851-1858 (1) – mf#65352 – us UMI ProQuest [071]
Herald – Wolf Point, MT. 1915-1940 (1) – mf#64698 – us UMI ProQuest [071]
Herald – Wyandotte, MI. 1880-1943 (1) – mf#63890 – us UMI ProQuest [071]
Herald – Wyandotte, MI. 1938-1962 (1) – mf#63891 – us UMI ProQuest [071]
Herald – Yakima, WA. 1889-1906 (1) – mf#67189 – us UMI ProQuest [071]
Herald – Yaounde, Cameroon n2-46. 1992 sep-1993 jul – 1r – us UF Libraries [079]
Herald – Yonkers, NY. 1863-1886 (1) – mf#65292,01 – us UMI ProQuest [071]
Herald – Yonkers, NY. 1883-1932 (1) – mf#65290 – us UMI ProQuest [071]
Herald *see*
– Montreal herald and daily commercial gazette
– Natal colonist / herald
– Richmond herald
– Welwyn garden city herald and post
The herald – Ashland, NE: Brush Bros. v1 n1. dec 21 1885- [mf ed 1996] – 1r – 1 – us NE Hist [071]
The herald – Blantyre: Midas Print & Pub, feb 16-jun 13, 29-jul 14 1994 – 1r – 1 – us CRL [071]
The herald – Eagle, NE: B C Preston, 1889-90// (wkly) [mf ed n4 jan 4 and feb 1 1890 filmed [1979] – 1r – 1 – us NE Hist [071]
The herald – Freemantle, Australia. Feb 1867-Jul 1886 – 12r – 1 – uk British Libr Newspaper [072]
The herald – Kimball, NE: Sherer & Lilly (wkly) [mf ed v5 n4. dec 1892] – 1r – 1 – us NE Hist [071]
The herald – New Castle, PA., 1920-1923 – 13 – $25.00r – us IMR [071]
The herald – Salisbury [Harare, Zimbabwe]: Rhodesian Print & Pub Co, [aug 15, 1978-] – 1 – us CRL [079]
The herald – Ulysses, NE: Thrapp & Webb [v1 n1] jun 11 1886- (wkly) – 1r – 1 – us NE Hist [071]
The herald *see*
– Daily herald
– Miscellaneous newspapers of larimer county
Herald and burrillville news gazette – Pascoag, RI. 1895-1899 (1) – mf#66241 – us UMI ProQuest [071]
Herald and gloucester farmer and columbian herald – Woodbury, NJ. 1819-1824 (1) – mf#64858 – us UMI ProQuest [071]
Herald and goshen blade – Glasgow, VA. 1890-1892 (1) – mf#66723 – us UMI ProQuest [071]
Herald and greensville register – Bayonne, NJ. 1869-1915 (1) – mf#64797 – us UMI ProQuest [071]
Herald and independent – Grant, MI. 1937-1971 (1) – mf#63764 – us UMI ProQuest [071]
Herald and independent – Harvard, IL. 1937-1944 (1) – mf#62627 – us UMI ProQuest [071]
Herald and mirror – Carlisle, PA., 1881 – 13 – $25.00r – us IMR [071]
Herald and news – Klamath Falls OR: Herald Pub Co & Klamath News Pub Co, 1942- [daily ex sun] – 1 – (merger of: evening herald (klamath falls, or); klamath news (klamath falls, or)) – us Oregon Lib [071]
Herald and news *see*
– Evening herald (klamath falls, or)
– Klamath news
Herald and predecessors – Rutland, VT. 1792-1820 (1) – mf#66659 – us UMI ProQuest [071]
Herald and reformer, and general advertiser – [Scotland] Leith: A Drummond 1868-95 (wkly) [mf ed 2004] – 1r – 1 – (formed by union of: burghs' reformer [19 mar 1859-27 jun 1863] and: leith herald and commercial advertiser [5 nov 1853-27 jun 1863]; cont by: leith herald and general advertiser [oct 1868-jul 1911]) – uk Newsplan [072]
Herald and republican series / Ottawa Co. Port Clinton – jan 1937-may 1969 [wkly] – 22r – 1 – mf#B25297-25318 – us Ohio Hist [071]
Herald and review – Decatur, IL. 1989-2000 (1) – mf#62599 – us UMI ProQuest [071]
Herald and ruralite – Sylva, NC. 1986-2000 (1) – mf#65342 – us UMI ProQuest [071]
Herald and southern democrat – St Andrews, FL. 1840-1849 – 1r – us UF Libraries [071]
Herald and torch light – Hagerstown, MD. 1865-1906 (1) – mf#63611 – us UMI ProQuest [071]
Herald and torch light – Hagerstown, MD. 1891-1895 (1) – mf#63612 – us UMI ProQuest [071]
Herald and torchlight *see* The christian herald
The herald and torchlight – Kalamazoo MI: Rev L H Trowbridge, 1873- [wkly] [mf v2-4 1874-76 filmed 1981] – 1r – 1 – (ceased in dec 1876? name changed to: the christian herald to reflect the revival of the michigan christian herald [detroit 1877]. incl: some iss of: michigan christian herald [detroit 1842] and christian herald [detroit 1877]) – mf#r0132b – us ATLA [242]
The herald and torchlight *see* Michigan christian herald
Herald and weekly free press (aberdeen) – [Scotland] Aberdeen: A Marr 18 nov 1876-nov 1922 (wkly) [mf ed 2003] – 62r – 1 – (formed by union of: aberdeen herald, and general advertiser for the counties of aberdeen, banff, & kincardine and: weekly free press; cont as: weekly free press and aberdeen herald [jan 1890-nov 1922]) – uk Newsplan [072]
Herald and western advertiser *see* Tuam herald
Herald argus – LaPorte, IN. 1994-2000 (1) – mf#69022 – us UMI ProQuest [071]
Herald (baker city, or) – Baker City OR: C W Hill, 1902-04 [daily ex sun] – 1 – (cont: baker city herald (1901-02); cont by: evening herald (1904-)) – us Oregon Lib [071]
Herald [berkhamsted] – Berkhamsted, England 14 apr 1988-17 aug 1989 – 1 – (cont by: berkhamsted herald & post [24 aug 1989-3 oct 1991]) – uk British Libr Newspaper [072]
Herald courier – Bristol, VA. 1951-2000 (1) – mf#61876 – us UMI ProQuest [071]
Y herald cymraeg – Caernarfon: Swyddfa'r 'Carnarvon & Denbigh Herald' 1855- – 1 – (lacking: 1912) – uk Wales NatLib [072]
Yr herald cymraeg – [Wales] Gwynedd 19 mai 1855-rhag 1950 [mf ed 2003] – 90r – 1 – (missing: 1856, 1859, 1861, 1874, 1888, 1893-94 & 1877; cont as: yr herald cymraeg a'r genedl [gorff 1937-rhag 1950]) – uk Newsplan [072]
Herald democrat – Denison/Sherman, TX. 1996-2000 (1) – mf#69217 – us UMI ProQuest [071]
Herald democrat *see* Miscellaneous newspapers of lake county
Herald dispatch – Decatur, IL. 1890-1895 (1) – mf#62597 – us UMI ProQuest [071]
Herald dispatch – Decatur, IL. 1895-1899 (1) – mf#62598 – us UMI ProQuest [071]
Herald dispatch – Huntington, WV. 1990-2000 (1) – mf#61917 – us UMI ProQuest [071]
Herald dispatch – Utica, NY. 1899-1922 (1) – mf#65250 – us UMI ProQuest [071]
Herald dubois county – Jasper, IN. 1946-2000 (1) – mf#61386 – us UMI ProQuest [071]
Herald examiner – Los Angeles, CA. 1937-1989 (1) – mf#60135 – us UMI ProQuest [071]
Herald express – Aurora, IL. 1887-1892 (1) – mf#62503 – us UMI ProQuest [071]
Herald express – Aurora, IL. 1894-1901 (1) – mf#62504 – us UMI ProQuest [071]
Herald express – Rutland, VT. 1983-1988 (1) – mf#68078 – us UMI ProQuest [071]
Herald express – Salem, WV. 1917-1918 (1) – mf#67469 – us UMI ProQuest [071]
Herald: gazette for the country – New York, NY. 1794-1795 (1) – mf#65078 – us UMI ProQuest [071]
The herald (gisborne) – aug 1939-apr 1941; 2-30 dec 2000 – 1 – mf#18.01 – nz Nat Libr [079]
Herald ilford barking and dagenham *see* Ilford leader
Herald independent – Winnsboro, SC. 1982-2000 (1) – mf#66526 – us UMI ProQuest [071]
Herald journal – Logan, UT. 1994-1996 [1] – mf#69084 – us UMI ProQuest [071]
Herald journal – Spartanburg, SC. 1982+ (1) – mf#61831 – us UMI ProQuest [071]
Herald journal – Syracuse, NY. 1893-2001 (1) – mf#60127 – us UMI ProQuest [071]
Herald leader – Lexington, KY. 1888-2000 (1) – mf#60479 – us UMI ProQuest [071]
Herald leader – Menominee, MI. 1987-1994 (1) – mf#61516 – us UMI ProQuest [071]
Herald mail – Fairport, NY. 1926-1992 (1) – mf#64961 – us UMI ProQuest [071]
Herald (myrtle point, or) – Myrtle Point OR: L Isenhart, 1991-97 [wkly] – 1 – (cont: myrtle point herald; cont by: myrtle point herald (myrtle point, or)) – us Oregon Lib [071]
Herald (new york, ny: 1835) – New York [NY]: James Gordon Bennett [daily ex sun] – 1 – (cont: morning herald (new york, ny: 1835); cont by: morning herald (new york, ny: 1837). related to wkly ed: weekly herald (new york, ny: 1836); evening ed: evening chronicle (new york, ny: 1837)) – us Oregon Lib [071]
Herald news – Joliet, IL. 1972-2000 (1) – mf#61335 – us UMI ProQuest [071]
Herald news – Providence, RI. 1947-1949 (1) – mf#66317 – us UMI ProQuest [071]
Herald news – Punta Gorda, FL. 1959-1971 jun – 18r – us UF Libraries [071]
Herald news – Wolf Point, MT. 1942-1974 (1) – mf#64696 – us UMI ProQuest [071]
Herald of banning, california – Banning CA. v11 n49 [1938 dec 5] – 1r – 1 – mf#918731 – us WHS [071]
Herald of co-operation – 1934 jul 26 – 1r – 1 – mf#3910486 – us WHS [071]
Herald of Freedom *see* Confidential intelligence report
Herald of freedom / Clinton Co. Wilmington – v1 n1. nov 1851-oct 1852 [wkly] – 1r – 1 – mf#B31209 – us Ohio Hist [071]
Herald of freedom – Concord. 1835-1846 (1) – mf#5302 – us UMI ProQuest [976]
Herald of freedom – 1962 feb 22-1975 dec 5, 1976-1979 feb 23 – 2r – 1 – (cont: metropolitan review; cont by: herald of freedom and metropolitan review) – mf#370641 – us WHS [071]
Herald of freedom – Hagerstown, MD. 1839-1851 (1) – mf#63613 – us UMI ProQuest [071]
Herald of freedom – Wilmington, OH, 1851-55 – 1r – 1 – (weekly abolitionist newspaper) – us Western Res [071]
Herald of freedom – Wilmington, OH. 1851-1854 (1) – mf#65725 – us UMI ProQuest [071]
Herald of freedom and metropolitan review – v35 n4-v38 n8 [1979 mar 9-1980 oct 8] – 1r – 1 – (cont: herald of freedom [staten island, ny]) – mf#667918 – us WHS [071]
Herald of freedom and torch light – Hagerstown, MD. 1851-1859 (1) – mf#63614 – us UMI ProQuest [071]
Herald of freedom and torch light *see* Daily times [baltimore, md]
Herald of gospel liberty – Portsmouth. 1808-1930 (1) – mf#4463 – us UMI ProQuest [240]

Herald of holiness / Church of the Nazarene – 1977 jul 15-aug 1, 1978 dec 15-1979 dec 15, 1980-1985 jan/dec 15, 1986, 1987-88 – 9r – 1 – (continued by: world mission; holiness today) – mf#689086 – us WHS [243]

Herald of kansas – Topeka, KS. v3 n5 feb 13 1880-v3 n21 jun 11 1880) (wkly) [mf ed 1947] – (= ser Negro newspapers on microfilm) – 1r – 1 – (cont: kansas weekly herald) – us L of C Photodup [071]

Herald of library science – Lucknow. 1962-1995 [1]; 1970-1995 [5]; 1977-1995 [9] – ISSN: 0018-0521 – mf#1928 – us UMI ProQuest [020]

Herald of life and immortality – Boston. 1819-1820 (1) – mf#3818 – us UMI ProQuest [240]

Herald of light / ed by Harris, Thomas Lake – New York: New Church Pub Assoc 1857-61 [mf ed 2005] – 6v on 2r – 1 – mf#2005C-s012 – us ATLA [243]

Herald of peace – 1868 jan 31-1869 jan 15, may 15-aug 15 – 1r – 1 – mf#1112670 – us WHS [071]

Herald of progress – Gateshead, England. -w. 16 July 1880-25 July 1884. 3 reels – 1 – uk British Libr Newspaper [072]

Herald of progress *see* New moral world, 1845

Herald of revolt – London, England. -m. Dec 1910-May 1914. 33 ft – 1 – uk British Libr Newspaper [072]

Herald of revolt : organ of the coming social revolution – v1-4 n5. 1910-14 [all publ] – (= ser Radical periodicals of great britain, 1794-1914. period 2) – 1r – 1 – $115.00 – us UPA [335]

The herald of revolt *see* The works of guy aldred

Herald of salvation / ed by Smith, Stephen R & Morse, Pitt – Philadelphia PA, 1826-27 [mf ed 2001] – (= ser Christianity's encounter with world religions, 1850-1950) – 1r – 1 – mf#2001-s128 – us ATLA [243]

Herald of salvation – Watertown. 1822-1825 – 1 – mf#3819 – us UMI ProQuest [242]

Herald of the centennial – Providence, RI. 1875-1876 (1) – mf#66315 – us UMI ProQuest [071]

Herald of the future *see* Herald of the rights of industry, 1834

The herald of the rights of industry *see* Political pamphlets... 19th c

Herald of the rights of industry, 1834 – (= ser Periodicals connected with owenite socialism and its successors in secularist, freethought and allied movements, 1934-1916) – 1r – 1 – (filmed with: bronterre's national reformer 1837; herald of the future 1839-40; oracle of reason 1841-43) – mf#97167 – uk Microform Academic [343]

Herald of the united states – Providence, RI. 1805-1807 (1) – mf#66316 – us UMI ProQuest [071]

Herald of the united states – Warren RI. 1793 sep 7 – 1r – 1 – mf#866314 – us WHS [071]

Herald of the valley – Fincastle, VA. 1820-1823 (1) – mf#66711 – us UMI ProQuest [071]

Herald of truth – Cincinnati, 1847-1848 [mnthly] – 1,5,9 – (a monthly periodical, devoted to the interests of religion, philosophy, literature, science and art.) – mf#3988 – us UMI ProQuest [073]

Herald of truth : a periodical work...the design of which is to illustrate and confirm the heavenly truths of the new jerusalem – Cincinnati, 1825-1826 [1,5,9] – mf#3820 – us UMI ProQuest [200]

The herald of truth – St John, NB: Friends of Truth, [1843-18–] – 9 – mf#P04291 – cn Canadiana [210]

The herald of truth – v1-45. 1864-1908 [complete] – (= ser Mennonite serials coll) – 12r – 1 – mf#ATLA 1991-S002 – us ATLA [242]

The herald of zion : being a series of essays, addresses, letters, etc, relating to the christian ministry / Edgar, James – Toronto?: s.n, 1856 – 1mf – 9 – mf#64755 – cn Canadiana [242]

Herald & post (edinburgh, scotland : 2001) – Edinburgh: Edinburgh Herald & Post 2001- (wkly) – 1 – (cont: edinburgh herald & post (1994)) – uk Scotland NatLib [072]

Herald & post. west lothian – Bathgate: West Lothian herald & post 2001- (wkly) – 1 – (cont: west lothian herald & post) – uk Scotland NatLib [072]

Herald press – Palestine, TX. 1995-2000 (1) – mf#61855 – us UMI ProQuest [071]

Herald reporter and bridgeport chief – Brewster, WA. 1940-1975 (1) – mf#69210 – us UMI ProQuest [071]

Herald (sat ed) *see* Montreal daily herald (saturday ed)

Herald series / Cuyahoga Co. Cleveland – jan 1872-mar 1885 (damaged) [daily] – 33r – 1 – mf#B9666-B9698 – us Ohio Hist [071]

Herald series / Harrison Co. Scio – 7/1938-41,50-59,61-1965 [wkly] – 9r – 1 – mf#B7072-7080 – us Ohio Hist [071]

Herald series / Meigs Co. Middleport – 1880-86, 1888-feb 1894 [wkly] – 6r – 1 – mf#B11595-11600 – us Ohio Hist [071]

Herald series / Morgan Co. McConnelsvill – 10/1850-53,56-58,60-62,70-74 (scattered) [wkly] – 5r – 1 – mf#B35-39 – us Ohio Hist [071]

Herald series / Morgan Co. McConnelsvill – (1875-1925, jun 1932-44) [wkly] – 33r – 1 – mf#B9428-9460 – us Ohio Hist [071]

Herald series / Pickaway Co. Circleville – jan 1875-dec 1888 [wkly] – 7r – 1 – mf#B10512-10518 – us Ohio Hist [071]

Herald statesman – Yonkers, NY. 1932-1998 (1) – mf#61661 – us UMI ProQuest [071]

Herald sun – Durham, NC. 1991-1999 (1) – mf#68604 – us UMI ProQuest [071]

Herald times – Manitowoc, WI. 1952-1954 (1) – mf#67568 – us UMI ProQuest [071]

Herald to the trades' advocate and co-operative journal – v1-36. 1820-31 [all publ] – (= ser Radical periodicals of great britain, 1794-1914. period 1) – 7mf – 9 – $125.00 – (with app) – us UPA [334]

Herald traveler – Boston, MA. 1967-1972 (1) – mf#63639 – us UMI ProQuest [071]

Herald tribune : (manatee edition) – Sarasota, FL. 1962-1964 (1) – mf#62446 – us UMI ProQuest [071]

Herald tribune – New York, NY. 1924-1966 (1) – mf#61090 – us UMI ProQuest [071]

Herald tribune – Sarasota, FL. 1925+ (1) – mf#61286 – us UMI ProQuest [071]

Herald tribune books index – New York, NY. 1924-1959 (1) – mf#65076 – us UMI ProQuest [071]

Herald weekender – Wanganui, NZ. jun 1986-dec 1987 – 3r – 1 – mf#43.10 – nz Nat Libr [079]

Herald weekly – Port Angeles, WA. 1891-1891 (1) – mf#67074 – us UMI ProQuest [071]

Herald whig – Quincy, IL. 1947+ (1) – mf#61354 – us UMI ProQuest [071]

Herald-advertiser – Portage, Tomah WI. 1888 sep 12-1890 jun 12, 1892 feb 11 – 2r – 1 – (cont: portage herald; portage advertiser) – mf#956229 – us WHS [071]

Herald-advocate – Wauchula, FL. 1937-1997 – 70r – (gaps) – us UF Libraries [071]

Herald-dispatch – Los Angeles CA. 1981 jun 19-30, 1981 jul-1982 dec, 1983-89 – 9r – 1 – (cont: weekend herald-dispatch) – mf#873871 – us WHS [071]

Herald-dispatch – Los Angeles CA. 1964 apr 25 [v14 n12]-1965 dec 25, 1966 jan 8-1967 apr 1, 1967 apr 18-1968 jul 27, 1968 aug 3-1969 sep 13, 1969 sep 18-1970 dec 3, 1970 dec 10-1971 dec 9, 1971 dec 16-1973 apr 26, 1973 may 3-1974 jun 27, 1974 jul 4-1976 jan 22, 1976 mar 18-1977 may 1, 1977 may 5-1978 dec 29 – 11r – 1 – (continued by: weekend herald-dispatch) – mf#873869 – us WHS [071]

Heraldic and genealogical manuscripts, 16th-17th centuries : armorials of english, scottish, french, german, italian, dutch, portuguese, and spanish / Lambeth Palace Library – 24 mss on 3r – 1,14 – 1-897955-54-5 – uk Academic [929]

Heraldic manuscripts ca 1300-ca 1800 / Society of Antiquaries of London – 13r – 1, 14 – £690.00 – 1-897955-64-2 – uk Academic [090]

Heraldica colombiana / Ortega Ricaurte, Enrique – Bogota, Colombia. 1952 – 1r – us UF Libraries [972]

Heraldica de extremadura / Sanchez Mateos, Dorita – Salamanca: Imp. Varona, s.a. – 1 – sp Bibl Santa Ana [946]

Heraldica general y fuentes de las armas de espana / Vicente Cascante, Ignacio – Barcelona: Salvat, 1956 – 1 – us Wisconsin U Libr [920]

Heraldica nacional / Ortega Ricaurte, Enrique – Bogota, Colombia. 1954 – 1r – us UF Libraries [972]

Herald-messenger – Crookston, NE: [Chas J Grantham] v7 n43. may 28 1920- (wkly) [mf ed -1921 (gaps) filmed [1972]] – 1r – 1 – (formed by the union of: cherry county messenger and: crookston herald) – us NE Hist [071]

O heraldo – Panjim, India. Jun 1975-1980 – 11r – 1 – us L of C Photodup [079]

Heraldo cubano – Coral Gables, FL. 1983 sep 15-oct 30 – 1r – us UF Libraries [071]

Heraldo de broward – Plantation, FL. 1974 oct 01-1995 dec 25 – 6r – (gaps) – us UF Libraries [071]

El heraldo de caceres – Caceres, 1898 – 5 – sp Bibl Santa Ana [079]

Heraldo de caceres – Caceres, 1893. 1 numero – 5 – sp Bibl Santa Ana [073]

El heraldo de la revolucion – Malalos: [s.n.], sep 29-dec 1898-jan 1899 – us CRL [079]

El heraldo de mexico – Los Angeles: C F Marburg y Cia, dec 1917-mar 1923 – 18r – 1 – us CRL [071]

Heraldo de paris – Paris. n1-69. oct 1900-mai 1904 – 1 – (lacking: n7, 53) – fr ACRPP [073]

Heraldo del exilio – Miami, FL. 1974 apr 16 – 1r – us UF Libraries [071]

Heraldo filipino – Malolos: [s.n.] jan 26-mar 23 1899 – us CRL [079]

Heraldo pinareno – Miami, FL. 1968 jul 04 – 1r – us UF Libraries [071]

Herald-observer – Belle Glade, FL. 1979-1984 – 6r – us UF Libraries [071]

Heraldos del rey – 1927. 164p – 1 – 5.74 – us Southern Baptist [242]

Herald-republic – Yakima, WA. 1969+ (1) – mf#61910 – us UMI ProQuest [071]

Heraldry in scotland, vol 1 : including a recension of 'the law and practice of heraldy in scotland' by the late george seton, advocate / Stevenson, John Horne – Glasgow: J Maclehose; Toronto: Macmillan, 1914 – 2v on 10mf – 9 – 0-665-99050-2 – (v2 99051 isbn: 0-665-99051-0) – mf#99050 – cn Canadiana [929]

The heralds of fame *see* One day's courtship

Heralds of revolt : studies in modern literature and dogma / Barry, William Francis – London: Hodder & Stoughton, 1904 [mf ed 1987] – xv/383p – 1 – mf#2030 – us Wisconsin U Libr [840]

The herald-sentinel – Shickley, NE: O L Larson. v2 n42. feb 16 1922-1925// (wkly) [mf ed with gaps] – 2r – 1 – (cont: shickley herald. union of: nebraska signal (1913)) – us NE Hist [071]

Herald-times – Bloomington, IN. 1990-2000 (1) – mf#61372 – us UMI ProQuest [071]

Herald-tribune / Meigs Co. Pomeroy – dec 1946-jun 1948 [wkly] – 1r – 1 – mf#B29910 – us Ohio Hist [071]

Herald-voice / Logan Co. Belle Center – dec 1902-may 1910, 23-28, 31-1982 [wkly] – 26r – 1 – mf#B12918-12944 – us Ohio Hist [071]

Herapath's railway and commercial journal *see*
- Herapath's railway magazine
- Railway magazine

Herapath's railway journal *see* Herapath's railway magazine

Herapath's railway journal. *see* Railway magazine

Herapath's railway magazine – London, England – 1 – uk British Libr Newspaper [380]

Herapath's railway magazine *see* Railway magazine

Heras, H *see*
- Beginnings of vijayanagara history
- The conversion policy of the jesuit in india. bombay, aug 1933

Heras, Henry *see*
- The aravidu dynasty of vijayanagara
- Studies in proto-indo-mediterranean culture

Le heraut d'armes – Paris. n1-31. avr-dec 1869 – 1 – (art, theatre, litterature. suite de: le fouet) – fr ACRPP [073]

Le heraut du grand roi jesus : ou eclaircissement de la doctrine de j. de l... / Labadie, Jean de – Amsterdam, 1667 – 5mf – 9 – mf#PPE-165 – ne IDC [240]

Heraux, Auguste Albert *see* Noveau dictionnaire des droits d'enregistrements

Heraux, Edmond *see*
- Melanges, politiques et litteraires
- Preludes

Heraux, Jules *see* Short summary of the history of haiti from 1492-19??

An herbal – 1525 – (= ser Herbal) – 9 – $10.00 – us Scholars Facs [615]

Herbal, the... : bodleian library ms. 130 – 12th c – 1r – 14 – mf#C526 – uk Microform Academic [760]

Herbals : their origin and evolution / Arber, Agnes Robertson – Cambridge, England. 1938 – 1r – us UF Libraries [500]

Herbarium : rijksherbarium / Rauwolff, L – Leiden – 38mf – 9 – mf#8303 – ne IDC [580]

Herbarium florae aegyptiacae seu collectio stirpium rariorum aegypti indigenorum / Sieber, F W – Vindobonae, 1820 – 1mf – 8 – mf#1285 – ne IDC [580]

[Herbarium p e isert [1756-1789] and p thonning [1775-1848]] – 1783-1803 – 112mf – 9 – ne IDC [580]

Herbart, Johann Friedrich *see* Herbart's abc of sense-perception

Herbart's abc of sense-perception : and minor pedagogical works / Herbart, Johann Friedrich – New York: D Appleton & Co, 1896 – (= ser International education series 36) – xxxi/288p – 1 – (trans, int, notes and comm by william j eckof) – us Wisconsin U Libr [370]

Der herbeder – Witten DE, 1993 sep-1997 – 1r – 1 – gw Misc Inst [074]

Die herberge – Freiburg Br DE, 1949 21 oct-1950 24 feb – 1 – gw Misc Inst [074]

Die herberge am tartaro / Fischer, Kurt – Muenchen: Zentralverlag der NSDAP, F Eher, [1937?] – 1r – 1 – us Wisconsin U Libr [830]

Herbermann, Charles George *see*
- The sulpicians in the united states
- Three-quarters of a century (1807 to 1882)

Herbert, A P *see* Misleading cases in the common law

Herbert, Brook Bradshaw *see* Resurrection of beauty for a postmodern church

Herbert, Charles *see* Wherefore, o god?

Herbert, Hilary A *see* Papers

Herbert j weiss collection on the belgian congo, 1947-62 – Palo Alto: Stanford University, [19–?] – 13r – 1 – us CRL [960]

Herbert, Mary E *see* Flowers by the wayside

Herbert, Samuel Asher *see* Convocation

Herbert spencer : an estimate and review / Royce, Josiah – New York: Fox, Duffield, 1904 – 1mf – 9 – 0-7905-8570-7 – mf#1989-1795 – us ATLA [100]

Herbert spencer and scientific education = Herbert spencer et l'education scientifique / Compayre, Gabriel – New York: Thomas Y Crowell, 1907 [mf ed 1986] – 1mf – 9 – 0-8370-7617-X – (in english. trans by maria e findlay) – mf#1986-1617 – us ATLA [370]

Herbert spencer's philosophy as culminated in his ethics / McCosh, James – New York: Scribner 1885 [mf ed 1991] – 1mf – 9 – 0-7905-9807-8 – mf#1989-1532 – us ATLA [190]

Herbert stanley jenkins, m.d., f.r.c.s., medical missionary, shensi, china : with some notices of the work of the baptist missionary society in that country / Glover, Richard – London: Carey Press, 1914 – 1mf – 9 – 0-524-07100-4 – mf#1991-2923 – us ATLA [242]

Herbert, T *see*
- Relation du voyage de perse et des indes orientales
- Some yeares travels into divers parts of asia and afrique

Herbert, William *see* Antiquities of the inns of court and chancery

Herbet, abbe *see* La sainte table, ou, le 4e livre de l'imitation de j.-c

Herbig, K *see* Die etruskische leinwandrolle des agramer national-museums

Herbin, John Frederic *see* Canada, and other poems

Herbordi dialogus de vita ottonis epsicopi babenbergensis (mgh7:33.bd) – 1868 – (= ser Monumenta germaniae historica 7: scriptores rerum germanicarum in usum scholarum (mgh7)) – €11.00 – ne Slangenburg [240]

Herborisations au levant...egypte, syrie et mediterranee... / Barbey, W & Barbey-Boissier, C – Lausanne, 1882 – 7mf – 9 – mf#7377 – ne IDC [956]

Herbort's von fritslar liet von troye / Fritzlar, Herbort von; ed by Frommann, Karl – Quedlinburg; Leipzig: G Basse, 1837 – (incl bibl ref and index) – us Wisconsin U Libr [430]

Herbst, Adolf *see* Ueber die von sebastian muenster und jean du tillet...evangeliums matthaei

Ein herbst in wales : land und leute, maerchen und lieder / Rodenberg, Julius – Hannover 1858 [mf ed Hildesheim 1995-98] – 3mf – 9 – €90.00 – 3-487-27972-X – gw Olms [914]

Herbst, Johann Andreas *see*
- Arte prattica et poetica
- Musica moderna prattica
- Musica practica sive instructio pro symphoniacis

Herbst, Wilhelm *see*
- Johann heinrich voss
- Wilhelm herbsts hilfsbuch fuer die deutsche litteraturgeschihte

Herbstaufbruch : gedichte / Borris, Siegfried – 3. aufl. Berlin: S Borris, 1946 [mf ed 1989] – 52p – 1 – mf#7053 – us Wisconsin U Libr [810]

Eine herbstfahrt nach spanien / Gerold, Rosa von – Wien 1880 [mf ed Hildesheim 1995-98] – 3mf – 9 – €90.00 – 3-487-29868-6 – gw Olms [914]

Herbstgesang : neue gedichte / Miegel, Agnes – Jena: E Diederichs, c1933 – 1r – 1 – us Wisconsin U Libr [810]

Herbstmonate in oberitalien : supplement zu des verfassers "eln jahr in italien" / Stahr, Adolf – Oldenburg 1860 [mf ed Hildesheim 1995-98] – 4mf – 9 – €120.00 – 3-487-29256-4 – gw Olms [914]

Herbstreise durch scandinavien / Alexis, Willibald – Berlin 1828 [mf ed Hildesheim 1995-98] – (= ser Fbc) – 2v on 5mf – 9 – €100.00 – 3-487-28932-6 – gw Olms [914]

Die herbstreise nach venedig / Raumer, Friedrich L von – Berlin 1816 [mf ed Hildesheim 1995-98] – 2v on 4mf – 9 – €120.00 – 3-487-29207-6 – gw Olms [914]

Herchner, Hans *see* Die cyropaedie in wielands werken

Herculano, Alexandre *see*
- O monge de cister
- A reaccao ultramontana em portugal

Herculano de Carvalho e Araujo, Alexandre *see* Opusculos

Hercules am scheidewege und andere antike bildstoffe / Panofsky, Erwin – Leipzig, Germany. 1930 – 1r – us UF Libraries [720]

Hercules dux ferrarie *see* Missarum josquin liber secundus

Hercules herald – 1981 may 1-1983 jun, 1983 jul 1-1986 mar 28, 1986 apr 4-1987 may 29, 1987 jun 5-1988 apr 29, 1988 may 6-nov 18 – 5r – 1 – (continued by: tiger times) – mf#651535 – us WHS [071]

Hercules prodicius seu carolus juliae... / Pakenius, J – Coloniae Agrippinae: Typis Petri Alstorff, 1679 – 7mf – 9 – mf#O-1924 – ne IDC [090]

Hercvlana, in lingva venetiana, nella vittoria dell' armata christiana contra turchi / Maganza, G B – Venetia, 1571 – 1mf – 9 – mf#H-8314 – ne IDC [956]

Hercynia : ein erinnerungsbuch fuer harzreisende – Quedlinburg [u.a.] 1823 [mf ed Hildesheim 1995-98] – (= ser Fbc) – 2mf [ill] – 9 – €60.00 – 3-487-29513-X – gw Olms [914]

Hercynia : Taschenbuch fuer reisende in den harz / ed by Hoffmann, Ludwig – Berlin 1829 [mf ed Hildesheim 1995-98] – 2mf – 9 – €60.00 – 3-487-29520-2 – gw Olms [914]

Herd, David *see* Ancient and modern scottish songs, heroic ballads, etc

Herdabref till praesterskapet och foersamlingarna i uppsala aerkestift / Soederblom, Nathan – 2. uppl. Uppsala: F.C. Askerberg, [1914?] – 1mf – 9 – 0-7905-6321-5 – mf#1988-2321 – us ATLA [240]

Herdecker zeitung – Herdecke DE, 1895 7 jul-1896 18 jun – 1 – (publ in witten-annen) – gw Misc Inst [074]

Herden, Editha Martina *see* Vom expressionismus zum "sozialistischen realismus"

Herdenking Stichting Boedi Oetomo *see* Javaansche kunstavond...gehouden door in nederland verblijvende javanen

Herdenking van zijne vijftig-jarige evangeliebediening / Hulst, Lammert J – Grand Rapids, MI: JB Hulst, 1899 – 1mf – 9 – 0-524-06626-4 – mf#1991-2681 – us ATLA [240]

Herder : ein lesebuch fuer unsere zeit / ed by Dobbek, Wilhelm – Weimar: Volksverlag, 1955 – 1 – (incl bibl ref and indexes) – us Wisconsin U Libr [430]

Herder als deutscher : ein literarhistorischer beitrag zur entwicklung der deutschen nationalidee / Goeken, Walther – Stuttgart: W Kohlhammer, 1926 – 1r – 1 – us Wisconsin U Libr [430]

Herder als faust : eine untersuchung / Jacoby, Guenther – Leipzig: F Meiner, 1911 – 1r – 1 – us Wisconsin U Libr [430]

Herder als theologe : ein beitrag zur geschichte der protestantischen theologie / Werner, August – Berlin: F Henschel, 1871 – 1mf – 9 – 0-524-00417-X – mf#1989-3117 – us ATLA [242]

Herder, Ferdinand Goffried von *see* Von und an herder

Herder, Ferdinand Gottfried von *see*
- Aus herders nachlass
- Herders reise nach italien

Herder in bueckeburg und seine bedeutung fuer die kirchengeschichte / Stephan, Horst – Tuebingen: JCB Mohr, 1905 – 1mf – 9 – 0-524-01021-8 – (incl bibl ref) – mf#1990-0298 – us ATLA [240]

Herder, Johann Gottfried *see*
- Adrastea
- Aus herders nachlass
- Auswahl
- Herder's briefwechsel mit nicolai
- Herders reise nach italien
- Herders saemmtliche werke
- Von und an herder

Herder, Johann Gottfried von *see*
- Herders cid
- Saemtliche werke

Herder, Karoline *see* Herders briefwechsel mit caroline flachsland

Herder, novalis und kleist : studien ueber die entwicklung des todesproblems in denken und dichten vom sturm und drang zur romantik / Unger, Rudolf – Frankfurt am Main: M Diesterweg, 1922 – 1 – (incl bibl ref) – us Wisconsin U Libr [430]

Herder precurseur de darwin? : histoire d'un mythe / Rouche, Max – Paris: Societe d'edition Les belles lettres, 1940 – 1r – 1 – (incl bibl ref) – us Wisconsin U Libr [430]

Herder und coleridge / Moore, Joachim Michael – Bern, 1951 (mf ed 1994) – 1mf – 9 – €24.00 – 3-8267-3032-1 – mf#DHS-AR 3032 – gw Frankfurter [140]

Herder und die deutsche kulturanschauung / Bran, Friedrich Alexander – Berlin: Junker & Duennhaupt 1932 [mf ed 1991] – 1r – 1 – (incl bibl ref; filmed with: goethe und seine eltern / herman kruger-westend) – mf#2723p – us Wisconsin U Libr [430]

Herderbuch : reisejournal; shakespeare; ossian; aus den homer (homer ein guenstling der zeit; homer u ossian): in auswahl / ed by Loeber, J – 4. Aufl. Leipzig: L Ehlermann, [between 1900 and 1915?] – 1 – (incl bibl ref) – us Wisconsin U Libr [920]

Herderbuch *see* Der freiheit eine gasse

Herders briefwechsel mit caroline flachsland / ed by Schauer, Hans – Weimar: Goethe-Gesellschaft, 1926-28 [mf ed 1993] – (= ser Schriften der goethe-gesellschaft 39, 41) – 2v – 1 – (incl bibl ref and ind) – mf#8657 reel 9 – us Wisconsin U Libr [920]

Herder's briefwechsel mit nicolai : im originaltext / Herder, Johann Gottfried; ed by Hoffmann, Otto – Berlin: Nicolai (R Stricker), 1887 – 1r – 1 – (incl bibl ref) – us Wisconsin U Libr [920]

Herder's cid : nach den besten quellen revidirte ausgabe / ed by Wollheim da Fonseca – Berlin: G Hempel, [1869?] – 1r – 1 – us Wisconsin U Libr [830]

Herders cid / Duentzer, Heinrich – 3. neubearb aufl. Leipzig: E Wartigs Verlag (E Hoppe), 1894 [mf ed 2002] – (= ser Erlaeuterungen zu den deutschen klassikern. abt 4, erlaeuterungen zu herders werken 1) – 185p – 1 – mf#10620 – us Wisconsin U Libr [430]

Herder's conception of "das volk" / Simpson, Georgiana Rose – Private ed. Chicago, IL: Distributed by the University of Chicago Libraries, 1921 – 1r – 1 – (incl bibl ref) – us Wisconsin U Libr [430]

Herders dramatische dichtungen : mit benutzung ungedruckter quellen / Treutler, Amand – Stuttgart: Metzler, 1915 [mf ed 1992] – (= ser Breslauer beitraege zur literaturgeschichte. neue folge 45) – 211p – 1 – mf#8014 reel 5 – us Wisconsin U Libr [430]

Herders "gott" / Hoffart, Elizabeth – Halle a.d.S.: M Niemeyer, 1918 – 1r – 1 – (incl bibl ref) – us Wisconsin U Libr [430]

Herders legenden / Duentzer, Heinrich – 2. neu durgesehene verm aufl. Leipzig: E Wartig's Verlag (E Hoppe), 1880 [mf ed 2002] – (= ser Erlaeuterungen zu den deutschen klassikern. abt 4, erlaeuterungen zu herders werken 2) – 127p – 1 – (comm only) – mf#10620 – us Wisconsin U Libr [430]

Herders lehre von naturschoenen im hinblick auf seinen kampf gegen die aesthetik kants / Springmeyer, Heinrich – Jena: E Diederich, 1930 [mf ed 1993] – (= ser Deutsche arbeiten der universitaet koeln 1) – 79p – 1 – (incl bibl ref) – mf#8215 reel 1 – us Wisconsin U Libr [140]

Herders reise nach italien : herders briefwechsel mit seiner gattin, vom august 1788 bis juli 1789 / ed by Duentzer, Heinrich & Herder, Ferdinand Gottfried von – Giessen: J Ricker, 1859 [mf ed 2001] – xxxii/416p – 1 – mf#10631 – us Wisconsin U Libr [860]

Herders saemmtliche werke / Herder, Johann Gottfried; ed by Suphan, Bernhard – Berlin: Weidmann. 33v. 1877-1913 – 1 – (incl bibl ref and indexes) – us Wisconsin U Libr [800]

Herders theoretische stellung zum drama / Koschmieder, Arthur – Stuttgart: J B Metzler, 1913 [mf ed 1992] – (= ser Breslauer beitraege zur literaturgeschichte. neue folge 35) – 172p – 1 – (incl bibl ref) – mf#8014 reel 4 – us Wisconsin U Libr [430]

Herder-studien : untersuchungen zu herders kritischem stil und zu seinen literaturkritischen grundeinsichten / Kohlschmidt, Werner – Berlin: Junker und Duennhaupt, 1929 – 1r – 1 – (incl bibl ref) – us Wisconsin U Libr [430]

Herdt, Ludwig *see* Immanenz und geschichte

Herdt, Ursula *see* Die verfassungstheorie karl v rottecks

Here and now : a canadian quarterly magazine of literature and art – Toronto. v1-2. dec 1947-jun 1949// – 1r – 1 – Can$86.00 – cn McLaren [073]

Here and now – oct 1949-aug 1951; oct 1953-nov 1957 – 4r – mf#ZB 20 – nz Nat Libr [079]

Here and now – Tuscon, AZ. 26 Jan 1979 – 1 – us AJPC [071]

Here and there in north india / Moore, Herbert – Westminster: The Society, 1914 [mf ed 1995] – (= ser Yale coll) – vii/99p (ill) – 1 – 0-524-09046-7 – mf#1995-0046 – us ATLA [240]

Here and there in south india / Higgens, A W B – Westminster: The Society, 1914 [mf ed 1995] – (= ser Yale coll) – vii/92p (ill) – 1 – 0-524-09040-8 – mf#1995-0040 – us ATLA [240]

Here and there in the greek new testament / Potwin, Lemuel Stoughton – Chicago: Fleming H Revell, 1898 [mf ed 1985] – 1mf – 9 – 0-8370-4788-9 – (incl bibl ref & ind) – mf#1985-2788 – us ATLA [225]

Here and there in the home land : england, scotland and ireland, as seen by a canadian / Haight, Canniff – Toronto: W Briggs, 1895 – 7mf – 9 – mf#03479 – cn Canadiana [914]

Here are a few press notices from the leading english critics on "canadian camp life"...by frances e herring, new westminster, british columbia...publ by t fisher unwin, london, england... / Herring, Frances Elizabeth – [New Westminster, BC?: s.n, 1900?] – 1mf – 9 – 0-665-94048-3 – mf#94048 – cn Canadiana [790]

Here comes joe mungin / Murray, Chalmers Swinton – 1942 [mf ed Spartanburg SC: Reprint Co, 1981?] – 7mf – 9 – mf#51-115b – us South Carolina Historical [830]

Hereafter? / Stoddard, J L – London, England. 1877 – 1r – us UF Libraries [240]

The hereafter of sin : what it will be / Haley, John W – Andover: WF Draper, 1881 – 1mf – 9 – 0-7905-9952-X – (incl bibl ref) – mf#1989-1677 – us ATLA [240]

Hereau, Joachim *see* Napoleon a sainte-helene

Heredad / Carias Reyes, Marcos – Tegucigalpa, Mexico. 1945 – 1r – us UF Libraries [972]

Heredia / Cipriano – Ciudad Trujillo, Dominican Republic. 1939 – 1r – us UF Libraries [972]

Heredia en la habana / Gonzalez Del Valle Y Ramirez, Francisco – Habana, Cuba. 1939 – 1r – us UF Libraries [972]

Heredia, Jose Felix *see* La consagracion de la republica del ecuador al sagrado corazon de jesus. quito, 1935

Heredia, Jose Maria *see*
- Antologia herediana
- Cantos partioticos
- Obras poeticas
- Poesias
- Poesias completas
- Poesias liricas
- Revisiones literarias
- Versos

Heredia, Manuel De *see* Atencion, guatemala

Hereditary genius : an inquiry into its laws and consequences / Galton, Francis – London, 1869 – (= ser 19th c evolution & creation) – 5mf – 9 – mf#1.1.10887 – uk Chadwyck [573]

Hereditas – Lund. 1920+ (1) 1970+ (5) 1977+ (9) – ISSN: 0018-0661 – mf#2707 – us UMI ProQuest [575]

Heredity : an international journal of genetics – Edinburgh. 1947+ (1) 1947+ (5) 1947+ (9) – ISSN: 0018-067X – mf#13545 – us UMI ProQuest [575]

Heredity and environment beginning with the primordial cell / Beacock, D V – S.l: s.n, 1894? – 1mf – 9 – mf#01510 – cn Canadiana [575]

Heredity, with preludes on current events / Cook, Joseph – Boston: Houghton, Mifflin, 1882, c1879 – (= ser Boston monday lectures) – 1mf – 9 – 0-7905-3775-3 – mf#1989-0268 – us ATLA [240]

Heredity, worry and intemperance as causes of insanity / Clark, Daniel – Toronto?: C B Robinson, 1880 – 1mf – 9 – mf#05726 – cn Canadiana [616]

The hereford breviary, vol 1 (hbs26) / Frere, Walter H & Brown, L E G – 1904 – (= ser Henry bradshaw society (hbs)) – 8mf – 8 – €17.00 – ne Slangenburg [241]

The hereford breviary, vol 2 (hbs40) / Frere, Walter H & Brown, L E G – 1911 – (= ser Henry bradshaw society (hbs)) – 7mf – 8 – €15.00 – ne Slangenburg [241]

The hereford breviary, vol 3 (hbs46) / Frere, Walter H & Brown, L E G – 1914 – (= ser Henry bradshaw society (hbs)) – 5mf – 8 – €12.00 – ne Slangenburg [241]

Hereford bulletin and free press – [West Midlands] Herefordshire 6 oct 1934-14 oct 1939 [mf ed 2002] – 7r – 1 – (cont as: hereford bulletin [jan 1936-oct 1939]) – uk Newsplan [072]

Hereford chronicle – [West Midlands] Herefordshire 10 apr 1858-26 may 1860 [mf ed 2002] – 1r – 1 – uk Newsplan [072]

Hereford county press – [West Midlands] Herefordshire 3 sep 1837-13 jun 1840 [mf ed 2002] – 2r – 1 – (cont as: hereford county press and shropshire mail [jan-jun 1840]) – uk Newsplan [072]

Hereford Independent / monmouth & south wales literary etc / hereford independent & ludlow etc – [West Midlands] Herefordshire 2 oct 1824-29 apr 1826; 20 oct 1827-19 jul 1828 [mf ed 2002] – 1r – 1 – uk Newsplan [072]

Hereford journal, 1770-1889 – 24r – 1 – (previously known as: british chronicle or pugh's hereford journal. lacking: 1775-82, 1787-1857, 1870-71, 1876-77) – mf#9496 – uk Microform Academic [072]

Hereford & leominster express, etc – [West Midlands] Herefordshire 20 jul 1864-15 feb 1865 [mf ed 2004] – 1r – 1 – uk Newsplan [072]

Hereford, leominster, & ross express, and advertiser for kington, tenbury, etc – [West Midlands] Herefordshire 20 jul 1864-15 feb 1865 [mf ed 2004] – 1r – 1 – uk Newsplan [072]

Hereford market express – [West Midlands] Herefordshire 2 jan-21 may 1884 [mf ed 2002] – 1r – 1 – uk Newsplan [072]

Hereford mercury – [West Midlands] Herefordshire 12 jul 1864-mar 1925 – 40r – 1 – (missing: 1870-71, 1889, 1896-97, 1912, 1922, 1924; cont as: hereford mercury and independent [jan 1872-dec 1882]; hereford mercury & independent & farmers' etc [jan 1883-dec 1888]; hereford mercury & farmers & hereford stockbreeders' journal [jan 1890-dec 1920]; hereford mercury [jan 1921-mar 1925]) – uk Newsplan [072]

Hereford weekly marvel – [West Midlands] Herefordshire 5 jun 1869-dec 1904 [mf ed 2002] – 33r – 1 – (missing: 1883, 1888) – uk Newsplan [072]

Hereford weekly news & general advertiser – [West Midlands] Herefordshire jan 1861-27 may 1863 [mf ed 2002] – 3r – 1 – (missing: 1860) – uk Newsplan [072]

Herejias y superticiones en la nueva espana. los heterodoxos en mejico / Jimenez Rueda, Julio – Madrid: Missionalia Hispanica, 1948 – 1 – sp Bibl Santa Ana [390]

Herera, Antonio de *see* Historia general de los hechos de los castellanos en las islas y tierra-firme de el mar oceano

Heres Hevia, Diego *see* Invocacion

Heresgast : eine erzaehlung aus germanischer vorzeit / Auerswald, Annmarie von – Dresden: Meinhold, 1940 [mf ed 1988] – 63p – 1 – mf#6969 – us Wisconsin U Libr [880]

L'heresie a la charite-sur-loire : et les debuts de l'inquisition monastique dans la france du nord au 13e siecle / Chenone, E – Paris, 1917 – €3.00 – ne Slangenburg [241]

Heresies of the 20th century : philosophical essays / Roy, Manabendra Nath – Moradabad: Pradeep Karyalaya, 1940 – (= ser Samp: indian books) – us CRL [100]

The heresies of the plymouth brethren / Carson, James Crawford Ledlie – London: Houlston, 1870 – 1mf – 9 – 0-524-07810-6 – mf#1991-3357 – us ATLA [242]

Heresy : its ancient wrongs and modern rights in these kingdoms / Gordon, Alexander – London: Lindsey Press, 1913 – (= ser Essex Hall Lecture) – 1mf – 9 – 0-524-00638-5 – mf#1990-0138 – us ATLA [240]

The heresy of the free spirit in the later middle ages / Lerner, R E – Los Angeles, 1972 – €15.00 – ne Slangenburg [230]

Heretic's journal – v1 n1-v4 n6 [1981 jan-1984 nov/dec] – 1r – 1 – (cont: heretics journal [1976]; cont by: heretic's journal bulletin for social and spiritual liberation) – mf#1050300 – us WHS [071]

Heretiques et revolutionnaires / Dide, Auguste – Paris: Charavay, 1886 – 1mf – 9 – 0-7905-7222-2 – mf#1988-3222 – us ATLA [240]

Herewith enclosed please find circulars relative to our new and rapidly selling family sewing machine : it not being represented in your locality as yet, we are desirous of appointing you as our agent – [Toronto?: s.n. 1867?] [mf ed 1983] – 1mf – 9 – 0-665-39772-0 – mf#39772 – cn Canadiana [090]

Herford, Brooke *see*
- The forward movement in religious thought as interpreted by unitarians
- Sermons of courage and cheer
- The small end of great problems
- The story of religion in england
- Travers madge

Herford, Charles Harold *see* Goethe

Herford, Mary Antonie Beatrice *see* Handbook of greek vase painting

Herford, Robert Travers *see*
- Christianity in talmud and midrash
- Pharisaism

Herforder kreisblatt – Herford DE, 1927 3 jan-31 mar & 30 jun-31 dec, 1929 2 jan-30 mar, 1930 2 jan-17 oct, 1930 18 dec-1931 30 jun, 1932 1 oct-31 dec, 1933 16 may-28 jun, 1935 2 jan-29 jun, 1950-57 [gaps] – 14mf=28df – 9 – (publ in bielefeld. filmed by misc inst: 1953-70 [gaps]; 1983 1 jun- [ca 7r/yr]; 1958-62 [gaps]) – gw Mikrofilm; gw Misc Inst [074]

Hergenroether, Joseph *see*
- Catholic church and christian state
- Die "irrthuemer" von mehr als vierhundert bischoefen und ihr theologischer censor
- Kritik der v. doellinger'schen erklaerung vom 28. maerz d. j
- Die lehre von der goettlichen dreieinigkeit nach dem heiligen gregor von nazianz, dem theologen
- Photius, patriarch von constantinopel

Hergenruether, Joseph *see* Anti-janus

Hergesell, H *see* Mit zeppelin nach spitzbergen

Hergesheimer, Joseph *see* San cristobal de la habana

Hergot, Hans *see* Aus dem sozialen und politischen kampf

Herguijuela. Ayuntamiento *see* Fiestas patronales en honor de san bartolome

Hericourt, Pierre *see* Arms for red spain

Hericurt, Pierre *see* Armes for red spain

Hering, E *see* Der letzte grund der dinge, oder, laesst sich das dasein gottes beweisen?

Hering, Hermann *see*
- Doktor pomeranus, johannes bugenhagen
- Die lehre von der predigt
- Die mystik luthers

Hering, J *see* Phenomenologie et philosophie religieuse
Hering, Jeanie *see* Golden days
Hering, Robert *see* Spinoza im jungen goethe
Herir en la sombra / Hurtado y Nunez de Arce, G – 1866 – 9 – sp Bibl Santa Ana [830]
Herissay, Jacques *see* Journal d'un spahi au soudan, 1897-1899
Herisson, Maurice d'Irisson d' *see* Nouveau journal d'un officier d'ordonnance
Heritage – Fern Park, FL. 1987-1996 – 10r – (gaps) – us UF Libraries [071]
Heritage – Khartoum, Sudan. Nov 5 1984-Nov 7 1988 – 1r – 1 – us L of C Photodup [079]
Heritage – Los Angeles. Calif. 1960-67 – 1 – us AJPC [071]
Heritage : a newsletter of the center for african american history and culture / Temple University – 1991 summer-1992 summer – 1r – 1 – mf#2543509 – us WHS [929]
Heritage : the quarterly bulletin of the genealogical society greater miami / Genealogical Society of Greater Miami – v6 n2-4 [1980 jun-dec], v7 n1 -v15 n4 [1981 mar-1989 oct] – 1r – 1 – (cont: genealogical society of greater miami) – mf#1685341 – us WHS [929]
Heritage / St Bernard Genealogical Society – 1978-80, 1980 index, 1981-84, 1985-1988 oct – 3r – 1 – mf#627792 – us WHS [929]
Heritage colonial en haiti / David, Placide – Madrid, Spain. 1959 – 1r – us UF Libraries [972]
Heritage conversation – v1 n1-v3 n1 [1975 winter-1977 winter] – 1r – 1 – (continued by: conversation autour du patrimoine; conversation) – mf#681429 – us WHS [929]
Heritage forum – 1992 mar/apr – 1r – 1 – mf#5307148 – us WHS [071]
Heritage news / Kentucky Heritage Commission et al – v1 n1-v6 n1 [1976 fall-1981 winter/ spring] – 1r – 1 – mf#1477149 – us WHS [929]
Un heritage notice biographique : omer rousseau, 1872-1933 / Bourk-Rousseau, Adeline – Nicolet: [s.n.], 1934 [mf ed 1996] – 1mf – 9 – mf#SEM105P2533 – cn Bibl Nat [920]
The heritage of the reformation / Pauck, Wilhelm – 2nd ed. Glencoe, 1961 – 7mf – 8 – €15.00 – ne Slangenburg [242]
The heritage of the reformation / Pauck, Wilhelm – Glencoe, 2nd ed 1961 – 7mf – 8 – €15.00 – ne Slangenburg [242]
Heritage quest : hq – 1985 sep/oct-1987 mar/ apr-1987 may/jun-1988 nov/dec – 2r – 1 – mf#1277695 – us WHS [929]
Heritage Researchers Enterprises *see* Day journal
Heritage society news / Harris County Heritage Society [TX] – v1 n1-2 [1974 spring-fall] – 1r – 1 – (continued by: harris county heritage society news) – mf#502073 – us WHS [929]
Heritage Society of Pennsylvania *see*
- Pennsylvania 1800-1850
- Pennsylvania in the civil war
- Pennsylvania in the revolution

Heritage sunday – Wyandotte, MI. 1990-1996 (1) – mf#68637 – us UMI ProQuest [071]
Heritage west / British Columbia Heritage Trust et al – v3 n1-v8 n4 [1978 winter-1984 winter] – 1r – 1 – mf#968273 – us WHS [929]
The heritagequest collection : the authoritative source for genealogical materials – 1700-present – 1 – (incl federal census records, source documents, & ind. coll is rich with demographic & genealogical detail. for more information visit: www2.heritagequest.com/ qsearch/bylocality.htm) – us UMI ProQuest [929]
Herites, Frantisek *see* Sebrane spisy
Heritiers, ou, la naufrage / Duval, Alexandre – Paris, France. 1820 – 1r – us UF Libraries [440]
Heritor And Vestryman Of The Scottish Episcopal Church *see* Letter to the members of the general assembly of the established ch...
Herivel *see* Haiti au point de vue religieux
Herizons – 1983 feb-1984, 1985-1987 mar – 2r – 1 – (cont: herizons) – mf#2437119 – us WHS [071]
Herizons – 1982 jun 12 – 1r – 1 – (cont: manitoba women's newspaper; cont by: herizons [winnipeg, manitoba : 1983]) – mf#1238822 – us WHS [071]
Herke, Karl *see* Hebbels theorie und kritik poetischer muster
Herkenne, Henr *see* De veteris latinae ecclesiastici capitibus 1-43
Herkless, John *see*
- Francis and dominic
- Hebrews and the epistles general of peter, james, and jude

Herkless, John, Sir *see*
- The early christian martyrs and their persecutions
- Richard cameron

Herkomer, Hubert von *see* Etching and mezzotint engraving
Der herkules – Kassel DE, 1925 16 may-1927 22 apr – 1r – 1 – gw Misc Inst [074]
Herkunft des christentums / Neuwinger, Rudolf – Berlin, Germany. 1941 – 1r – us UF Libraries [240]
Herlihy, Joan M *see* Papers relating to provincial and local governments in the solomon islands
Herlosssohn, C *see* Damen-conversations-lexikon
Hermae pastor graece : addita versione latina recentiore e codice palatino / Gebhardt, Oscar von – Lipsiae: J.C. Hinrichs, 1877 – (= ser Patrum Apostolicorum Opera) – 1mf – 9 – 0-8370-9624-3 – (incl bibl ref and indexes of biblical citations and greek words) – mf#1986-3624 – us ATLA [240]
The herman advertiser – Herman, NE: B A Brewster. v1 n1. oct 13 1899- (wkly) [mf ed with gaps] – 1r – 1 – us NE Hist [071]
The herman cyclone – Herman, NE: Don C VanDeusen, 1906-v2 n4. apr 7 [ie 4] 1907 (wkly) [mf ed 1979] – 1r – 1 – (absorbed by: blair courier) – us NE Hist [071]
Herman, Emily *see*
- Eucken and bergson
- The meaning and value of mysticism

Herman et verner, ou, les militaires / Favieres, Edme Guillaume Francois De – Paris, France. 1803 – 1r – us UF Libraries [440]
Herman family, 1775-1852 / Ely – 1r – 1 – (founder of elyria, ohio) – mf#B26379 – us Ohio Hist [978]
Herman melville / Arvin, Newton – New York, NY. 1950 – 1r – us UF Libraries [960]
The herman news – Herman, NE: J W Selden, 1892 (wkly) [mf ed -1895 (gaps) filmed 1979] – 1r – 1 – us NE Hist [071]
Herman, Nicolas *see*
- Die historien von der sindflut, joseph, mose, helia, elisa und der susanna, sampt etlichen historien aus den evangelisten
- Der sontagen und fuernembsten festen evangelia
- Die sontags-evangelia unnd von den fuernembsten festen uber das gantze jar

Herman record – Herman, NE: Harry L Swan. -v37 n48. nov 28 1946 (wkly) [mf ed v12 n12. jan 1 1920-nov 28 1946 (gaps) filmed 1976] – 8r – 1 – (absorbed by: pilot-tribune. vol numbering dropped with jan 1 1931; resumed with v26 n41 on oct 11 1934) – us NE Hist [071]
Herman review – Herman, NE: Chas A Robertson. 1v. jul 1896-v1 n51. jul 9 1897 (wkly) [mf ed lacks jan 15 1897 filmed 1979] – 1r – 1 – (cont by: herman weekly review) – us NE Hist [071]
Herman samuel reimarus und johann christian edelmann / Moenckeberg, Carl – (= ser Gallerie hamburgischer theologen) – 1r – 1 – 0-8370-0012-2 – mf#1984-B363 – us ATLA [240]
Herman veit simon – Berlin, Germany. 1915 – 1r – us UF Libraries [939]
The herman weekly review – Herman, NE: Geo A Byrne. v1 n52. jul 16 1897- (wkly) [mf ed -1898 (gaps) filmed 1979] – 1r – 1 – (cont: herman review) – us NE Hist [071]
Hermana matilde / Minino Gomez, Ricardo – Santo Domingo, Dominican Republic. 1964 – 1r – us UF Libraries [972]
Hermand, Jost *see*
- Das junge deutschland
- Der schein des schoenen lebens

La hermandad *see* Miscellaneous newspapers of pueblo county
Le hermandad – Pueblo, CO. 1896-1907 (1) – mf#62319 – us UMI ProQuest [071]
La hermandad de alfereces y el destino de espana / Becerro de Bengoa, Ricardo – Caceres: Imprenta Moderna, 1962 – 1 – sp Bibl Santa Ana [946]
Hermandad de Donantes de Sangre *see* Boletin informativo extraordinario
Hermandad de Trabajadores de Servicios Sociales [PR] *see* Denuncia
Hermandad del santo sepulcro. estatutos por los que ha de regirse... – Caceres: Imprenta Moderna, 1960 – sp Bibl Santa Ana [946]
Hermandad Sindical *see* Ordenanzas de la (tipo)
Hermandad Sindical de Labradores y Ganaderos *see*
- Reglamento
- Romeria de san isidro. en venta e culebrin

Hermanito menor / Chacon Y Calvo, Jose Maria – San Jose, Costa Rica. 1919 – 1r – us UF Libraries [972]
Hermann – Schwelm DE, 1950 15 mar-1957 – 21r – 1 – (bezirksausgabe von westdeutsche zeitung, wuppertal; filmed by misc inst: 1958-59; missing: jul, aug 1950. title varies: 1834?: wochenblatt fuer den land- und stadtgerichtsbezirk schwelm; 1848: der beobachter an der bergisch-maerkischen eisenbahn; 4 oct 1864: schwelmer zeitung; 11 jul 1972: wz schwelmer zeitung) – gw Mikrofilm [074]
Hermann : deutsches wochenblatt aus london – London (GB), 1859-67 – 3r – 1 – uk British Libr Newspaper [072]
Hermann – Hagen, Westf DE, 1814 [gaps] – 1 – gw Misc Inst [074]
Hermann / Wieland, Christoph Martin; ed by Muncker, Franz – Heilbronn: Henninger, 1882 [mf ed 1993] – (= ser Deutsche litteraturdenkmale des 18. und 19. jahrhunderts 6) – xxx/116p – 1 – (incl bibl ref) – mf#8676 reel 1 – us Wisconsin U Libr [890]
Hermann, A *see* Tractatus theologici in primum sententiarum librum...
Hermann and dorothea / Goethe, Johann Wolfgang von – Muenchen: F Bruckmann, [1874?] – 1r – 1 – us Wisconsin U Libr [820]
Hermann bahr / Handl, Willi – Berlin: S Fischer, 1913 [mf ed 1989] – 160/[3]/[1]pl – 1 – (incl bibl) – mf#6972 – us Wisconsin U Libr [430]
Hermann, Binger *see* The louisiana purchase and our title west of the rocky mountains
Hermann cohens juedische schriften / Cohen, Hermann; ed by Strauss, Bruno – Berlin, 1924 (mf ed 1997) – (= ser Monographien zur wissenschaft des judentums) – 9 – (band 1: ethische und religoese grundfragen 5mf €59 isbn: 3-8267-3216-2. band 2: zur juedischen zeitgeschichte 5mf €59 isbn: 3-8267-3217-0. band 3: zur juedischen religionsphilosophie und ihrer geschichte 4mf €49 isbn: 3-8267-3218-9) – gw Frankfurter [470]
Hermann, Conrad *see* Der gegensatz des classischen und des romantischen in der neueren philosophie
Hermann conradis gesammelte schriften / ed by Ssymank, Paul & Peters, Gustav Werner – Muenchen: G Mueller, 1911 [mf ed 1989] – 3v – 1 – mf#7156 – us Wisconsin U Libr [802]
Hermann flayders ausgewaehlte werke / ed by Bebermeyer, Gustav – Leipzig: K W Hiersemann, 1925 – 1 – (latin text with an introduction in german) – us Wisconsin U Libr [890]
Hermann flayders ausgewaehlte werke / Flayder, Friedrich Hermann; ed by Bebermeyer, Gustav – Leipzig: K W Hiersemann, 1925 – (latin text with an introduction in german) – us Wisconsin U Libr [450]
Hermann, Fritz H *see* Die verfassung der hessen-darmstaedtischen landstaende am ausgang des 18. jahrhunderts
Hermann, Georg *see*
- Henriette jacoby
- Jettchen gebert
- Die nacht des dr. herzfeld
- November achtzehn
- Schnee
- Eine zeit stirbt

Hermann, Georg [comp] *see* Das biedermeier im spiegel seiner zeit
Hermann goering albums / U.S. Library of Congress. Prints and Photographs Division – 47 albums (13,500 images) from 1914-18 and 1933-42. 7 reels. P&P3128 – 1 – us L of C Photodup [080]
Hermann hesse und gottfried keller : eine studie / Buehner, Karl Hans – Stuttgart: U Bonz 1927 [mf ed 1990] – 1r – 1 – (filmed with: das problem "volkstum und dichtung" bei herder / retha schmitz) – mf#2724p – us Wisconsin U Libr [430]
Hermann, Johannes *see* Die soziale predigt der propheten
Hermann, Karl Friedrich *see* Lehrbuch der griechischen antiquitaeten
Hermann, Karl Friedrich *see* Lehrbuch der griechischen privatalterthumer
Hermann kurz' saemtliche werke / Kurz, Hermann; ed by Fischer, Hermann – Leipzig: M Hesse's Verlag. 12v in 3. [19–?] – 1 – us Wisconsin U Libr [800]
Hermann lingg und seine lyrische dichtung / Knote, Walter – Wuerzburg: R Mayr, 1936 – 1 – (incl bibl ref) – us Wisconsin U Libr [430]
Hermann loens : sein leben und wirken / Deimann, Wilhelm – Dortmund: Lensing, 1922 – 1r – 1 – (incl bibl ref) – us Wisconsin U Libr [920]
Hermann loens : ein soldatisches vermaechtnis / Deimann, Wilhelm – Berlin-Dahlem: Ahnenerbe-Stiftung Verlag c1939 [mf ed 1990] – 1r [ill] – 1 – (poems and letters by loens; pref by friedhelm kaiser; biogr portrait by ernst von dombrowski. filmed with: storbonden og hans sonner) – mf#2829p – us Wisconsin U Libr [802]
Hermann loens am 20. todestage, 26. september 1934 : rede bei der loensfeier des volksbundes deutsche kriegsgraeberfuersorge am 23. sep 1934 im landeshause zu breslau / Kuehnemann, Eugen – Breslau: Trewendt & Granier [1934?] [mf ed 1996] – 1r – 1 – (filmed with: mein blaues buch / hermann loens) – mf#3942p – us Wisconsin U Libr [430]
Hermann, Rudolf *see* Christentum und geschichte bei wilhelm herrmann
Hermann schedels briefwechsel, 1452-1478 / ed by Joachimsohn, Paul – Stuttgart: Litterarischer Verein, 1893 (Tuebingen: H Laupp, Jr) [mf ed 1993] – (= ser Blvs 196) – x/218p – 1 – (incl bibl ref and ind) – mf#8470 reel 41 – us Wisconsin U Libr [860]
Hermann schedels briefwechsel, 1452-78 / ed by Joachimsohn, Paul – Stuttgart: Litterarischer Verein, 1893 (Tuebingen; H Laupp, Jr) – (incl bibl ref and ind) – us Wisconsin U Libr [860]
Hermann siebeck's religionsphilosophie dargestellt und beurteilt / Geisler, Victor – Berlin: Leopold Eber, 1908 – 1mf – 9 – 0-8370-3246-6 – mf#1985-1246 – us ATLA [100]
Hermann stark : deutsches leben / Redewitz, Oskar von – 3. Aufl. Stuttgart: J G Cotta, 1879 – 1r – 1 – us Wisconsin U Libr [920]
Hermann stehr : die geschichte eines lebens und seines werkes in 5 kapiteln / ed by Koehler, Willibald – Schweidnitz: L Heege, 1927 – 1r – 1 – us Wisconsin U Libr [430]
Hermann stehr, schlesier, deutscher, europaeer : ein gedenkbuch zum 100. geburtstag des dichters / ed by Richter, Fritz – Wuerzburg: Holzner, 1964 – 1 – (incl bibl ref) – us Wisconsin U Libr [430]
Hermann stehr und das junge deutschland : bekenntnis zum 75. geburtstag des dichters / ed by Hammer, Franz – Eisenach: E Roeth, 1939 – 1r – 1 – us Wisconsin U Libr [430]
Hermann sudermann : eine kritische studie / Kawerau, Waldemar – 2. Aufl. Leipzig: B Elischer, [1897?] – 1r – 1 – us Wisconsin U Libr [430]
Hermann, Theodor *see* Lehrbuch der symbolik
Hermann und dorothea / Goethe, Johann Wolfgang von – Muenchen: F Bruckmann, [1874?] – 1 – us Wisconsin U Libr [830]
Hermann und dorothea = Hermann und dorothea / Goethe, Johann Wolfgang von – Philadelphia: J B Lippincott, 1889 – 1r – 1 – us Wisconsin U Libr [820]
Hermann von gilm : darstellung seines dichterischen werdeganges / Sonntag, Arnulf – Muenchen: J Lindauer, 1904 (mf ed 1990) – 1r – 1 – (filmed with: meistererzaehlungen. incl bibl ref) – us Wisconsin U Libr [430]
Hermann von gilm *see* An der grenze
Hermann von gilms familien – und freundesbriefe / ed by Necker, Moritz – Wien: Literarischer Verein, 1912 – xxxii/351/16p – 1 – us Wisconsin U Libr [860]
Hermann von sachsenheim / ed by Martin, Ernst – Stuttgart: Litterarischer Verein, 1878 (Tuebingen: H Laupp) – (incl bibl ref and ind) – us Wisconsin U Libr [810]
Hermann von sachsenheim : poems / ed by Martin, Ernst – Stuttgart: Litterarischer Verein, 1878 (Tuebingen: H Laupp) [mf ed 1993] – (= ser Blvs 137) – 283p – 1 – (incl bibl ref and ind) – mf#8470 reel 29 – us Wisconsin U Libr [810]
Hermann, Wilhelm *see* Faith and morals
Das hermann-bahr-buch : zum 19. juli 1913 / ed by S Fischer Verlag Berlin – Berlin: S Fischer, [1913?] [mf ed 1989] – 318p/ [10]pl – 1 – (an anthology issued in honor of hermann bahr's 50th birthday) – mf#6972 – us Wisconsin U Libr [430]
Hermanni a kerssenbroch anabaptistici furoris : monasterium inclitam westphaliae metropolim evertentis = Anabaptistici furoris / Kerssenbrock, Hermann von; ed by Detmer, Heinrich – Muenster: Theissing, 1899-1900 – (= ser Die Geschichtsquellen des Bisthums Muenster) – 5mf – 9 – 0-8370-9002-4 – (incl bibl ref and ind) – mf#1986-3002 – us ATLA [943]
Hermanns sohn / Order of Hermann Sons in the United States – jahrg 12 n7-8 [1887 apr 1-15] – 1r – 1 – mf#1003094 – us WHS [071]
Die hermannsburger mission in indien : eine jubilaeums-gabe gewidmet seinen lieben mitarbeitern in indien und afrika, und allen lieben missionsfreunden – Hermannsburg: Missionshandlung, 1899 [mf ed 1995] – (= ser Yale coll) – vii/236p (ill) – 1 – 0-524-10192-2 – (in german) – mf#1995-1192 – us ATLA [240]
Die hermannsschlacht : ein drama in fuenf aufzuegen / Kleist, Heinrich von – Wien: K Graeser, [18–?] – 1r – 1 – (incl bibl ref) – us Wisconsin U Libr [820]
Hermannstaedter zeitung *see* Die woche
Hermannus, Contractus *see* Musica hermanni contracti
Hermannus quondam judaeus (mgh quellen..: 4.bd) : opusculum de conversione sua – 1963 – (= ser Monumenta germaniae historica. quellen zur geistesgeschichte des mittelalters (mgh quellen...)) – €7.00 – ne Slangenburg [931]
Hermanos de la salle en colombia / Rafael, Florencio – Bogota, Colombia. 1965 – 1r – us UF Libraries [972]
Los hermanos del destino (los pizarros y la conquista del peru) / Birney, Hoffman – Buenos Aires: Editorial Juventud Argentina, S.A., 1946 – sp Bibl Santa Ana [350]

Hermas in arcadia : and other essays / Harris, J Rendel – Cambridge: University Press; New York: Macmillan [distributor], 1896 – 1mf – us ATLA [240]

Hermas in arcadia : and other essays / Harris, James Rendel – Cambridge: University Press; New York: Macmillan [distributor], 1896 – 1mf – 9 – 0-7905-5469-0 – mf#1988-1469 – us ATLA [240]

Hermathena – Dublin. 1873+ (1) 1975+ (5) 1975+ (9) – ISSN: 0018-0750 – mf#10467 – us UMI ProQuest [000]

Hermbstadt, Sigismund Friedrich *see* Systematischer grundris der allgemeinen experimental-chemie zum gebrauch seiner vorlesunger entworfen

Hermbstaedt, Sigismund Friedrich *see* Museum des neuesten und wissenswuerdigsten aus dem gebiete der naturwissenschaft, der kuenste, der fabriken, der manufakturen...

Hermelink, Heinrich *see*
- Buendnis und bekenntnis 1529/1530
- Reformation und gegenreformation

Hermen : essays und etudien / Spiero, Heinrich – Leipzig: H. Finck, 1912 – 1r – 1 – (incl bibl ref) – us Wisconsin U Libr [430]

Hermeneutica biblica generalis secundum principia catholica / Szekely, Stephan – Friburgi Brisgoviae [Freiberg i B]: Herder, 1902 – 2mf – 9 – 0-524-07542-5 – (incl bibl ref) – mf#1992-1085 – us ATLA [220]

Hermeneuticae biblicae generalis principia rationalia christiana et catholica : selectis exemplis illustrata / Ranolder, Joannes – ed 2. Budae [Budapest]: C R Scient Universitatis, 1859 – 1mf – 9 – 0-8370-6934-3 – mf#1986-0934 – us ATLA [220]

Hermeneutical manual : or, introduction to the exegetical study of the scriptures of the new testament / Fairbairn, Patrick – Philadelphia: Smith, English, 1859 – 2mf – 9 – 0-8370-9541-7 – (incl bibl ref and indexes) – mf#1986-3541 – us ATLA [220]

The hermeneutical problem of yahweh war in the book of joshua 1-12 / Dobayashi, Yoichi – 1981 – 1 – 5.00 – us Southern Baptist [242]

Hermeneutics : a text book / Dungan, D R – Cincinnati, OH: Standard Publ Co, 1888 – 1mf – 9 – 0-7905-1653-5 – mf#1987-1653 – us ATLA [220]

Hermeneutics of the new testament = Hermeneutic des neuen testamentes / Immer, Albert – Andover: Warren F Draper, c1877 [mf ed 1985] – 1mf – 9 – 0-8370-3721-2 – (incl app on greek grammar and ind) – mf#1985-1721 – us ATLA [225]

Hermeneutik als allgemeine methodik der geisteswissenschaften / Betti, Emilio – Tubingen, Germany. 1962 – 1r – us UF Libraries [025]

Hermeneutik des neuen testamentes / Immer, Albert – Wittenberg: Herman Koelling, 1873 – 1mf – 9 – 0-8370-3722-0 – (incl ind) – mf#1985-1722 – us ATLA [225]

Hermeneutik und kritik : mit besonderer beziehung auf das neue testament / Schleiermacher, Friedrich [Ernst Daniel]; ed by Luecke, Friedrich – Berlin: Reimer 1838 [mf ed 1990] – 1mf – 9 – 0-7905-3984-5 – mf#1989-0477 – us ATLA [225]

Herment-Grenie *see*
- Bourdes-de-peage et pis en sont!
- Chez les civils

Hermer, Consuelo Kamholz *see* Havana manana

Hermes : zeitschrift fuer klassische philologie – Berlin. v1-61. 1866-1926 – 666mf – 8 – mf#115c – ne IDC [450]

Hermes, Georg *see*
- Christkatholische dogmatik
- Positive einleitung

Hermes mercurius trismegistus, his divine pymander, in seventeen books. together with his second book, called asclepius / Hermes Trismegistus – London, 1657 – 1 – us Wisconsin U Libr [920]

Hermes Trismegistus *see*
- Astrologica et divinatoria
- Devx livres, l'vn de la puissance & sapience de dieu, l'autre de la volonte de dieu
- Hermes mercurius trismegistus, his divine pymander, in seventeen books. together with his second book, called asclepius

Hermet, Augusto *see* Inni alla notte e canti spirituali

Hermetische philosophie und freimaurerei : ein beitrag zur vorgeschichte der freimaurerei / Hoehler, Wilhelm – Ludwigshafen am Rhein: Weiss & Hameier 1905 – 2mf – 9 – mf#vrl-205 – ne IDC [366]

Hermetischer rosenkrantz, das ist, vier schoene, auserlesene chymische tractaetlein.. – 1747. Comp. by David Herlitz. 1 reel. 1205 – 1 – us Wisconsin U Libr [540]

Hermina, Waldemar *see* The effects of different resistances on peak power during the wingate anaerobic test

L'hermine : journal de la bretagne et de la vendee – Nantes. 1837-39, 1842-48, juil 1849-6 nov 1850 – 1 – fr ACRPP [944]

Herminjard, A L *see* Correspondance des reformateurs dans les pays de langue francaise

Hermiston herald – Hermiston OR: F R Reeves, -1984 [wkly] – 1 – (cont by: hermiston herald and buyer's bonus (1984-94). 1920-25 incl newspaper by hermiston high school students) – us Oregon Lib [071]

Hermiston herald and buyer's bonus – Hermiston OR: G M Reed, 1984-94 [wkly] – 1 – (cont: hermiston herald (-1984); cont by: hermiston herald (1994-)) – us Oregon Lib [071]

Hermiston herald (hermiston or) – Hermiston OR: D Zimmerman, 1994- [wkly] – 1 – (cont: hermiston herald and buyer's bonus) – us Oregon Lib [071]

A hermit in the himalayas / Brunton, Paul – Madras: B G Paul & Co, [19–] – (= ser Samp: indian books) – us CRL [280]

Hermitage baptist church. kershaw county. south carolina : church records – 1920-23, 1943-51, 1957-72 – 1 – us Southern Baptist [242]

L'hermite a la prison des petits-carmes / Levae, Adolphe – Bruxelles 1827 [mf ed Hildesheim 1995-98] – 1v on 2mf – 9 – €60.00 – 3-487-25869-2 – gw Olms [365]

L'hermite du faubourg saint-germain : ou observations sur les moeurs et les usages francais au commencement du 19e siecle... / Colnet DuRaval, Charles – Paris 1825 [mf ed Hildesheim 1995-98] – 2v on 6mf – 9 – €120.00 – ISBN-10: 3-487-25871-4 – ISBN-13: 978-3-487-25871-3 – gw Olms [944]

L'hermite du marais : ou le rentier observateur / Lebel – Paris 1819 [mf ed Hildesheim 1995-98] – 2v on 4mf – 9 – €120.00 – 3-487-25877-3 – gw Olms [880]

L'hermite en belgique – Bruxelles 1827 [mf ed Hildesheim 1995-98] – (= ser Fbc) – 2v on 4mf – 9 – €120.00 – 3-487-29630-6 – gw Olms [914]

L'hermite en russie : ou observations sur les moeurs et les usages russes au commencement du 19e siecle; faisant suite a la collection des moeurs francaises, anglaises, italiennes, espagnoles, etc / Dupre de Saint-Maur, Emile – Paris 1829 [mf ed Hildesheim 1995-98] – 3v on 8mf – 9 – €160.00 – 3-487-29020-0 – gw Olms [910]

L'hermite en suisse, ou observations sur les moeurs et les usages suisses au commencement du 19 siecle : faisant suite a la collection des moeurs francaises, anglaises, italiennes, espagnoles, russes, etc / ed by Martin, Alexandre – Paris 1829-30 [mf ed Hildesheim 1995-98] – 3v on 8mf – 9 – €160.00 – 3-487-29337-4 – gw Olms [914]

L'hermite rodeur : ou observations sur les moeurs et usages des anglais et des francais au commencement du 19e siecle / MacDonogh, Felix – Paris 1824 [mf ed Hildesheim 1995-98] – 2v on 4mf – 9 – €120.00 – 3-487-25873-0 – gw Olms [940]

Les hermites en liberte : pour faire suite aux hermites en prison, et aux observations sur les moeurs et les usages francais au commencement du 19.e siecle / Jouy, Victor J de – Paris [mf ed Hildesheim 1995-98] – 5v on 8mf – 9 – €160.00 – ISBN-10: 3-487-25872-2 – ISBN-13: 978-3-487-25872-0 – gw Olms [944]

Les hermites en prison : pour faire suite aux observations sur les moeurs et les usages francais au commencement du 19. siecle... / Jouy, Victor J de – Paris 1823 [mf ed Hildesheim 1995-98] – 3v on 4mf – 9 – €120.00 – 3-487-25874-9 – gw Olms [944]

The hermits / Kingsley, Charles – Philadelphia: J. B. Lippincott, 1868. Chicago: Dep of Photodup, U of Chicago Lib, 1973 (1r); Evanston: American Theol Lib Assoc, 1984 (1r) – (= ser The Sunday Library For Household Reading) – 1 – 0-8370-0005-X – (incl bibl ref) – mf#1984-B385 – us ATLA [240]

The hermits and anchorites of england / Clay, Rotha Mary – London: Methuen, 1914 – (= ser The Antiquary's Books) – 1mf – 9 – 0-7905-6861-6 – mf#1988-2861 – us ATLA [941]

Hermogenes *see* Des aufrichtigen hermogensis apocalypsis spagyrica et philosophica

Hermon roots news – 1983 jan-1990 nov – 1r – 1 – mf#1091824 – us WHS [071]

[Hermosa beach-] hermosa beach review – CA. 1913-1950; 1956-1973; 1976-1979 – 34r – 1 – $2040.00 – mf#H03241 – us Library Micro [071]

Hermoso, Eugenio *see* Cometa

Hermsdorf-waidmannsluster frohnau-glienicker zeitung – Berlin DE, 1911-12, 1914-21, 1922 1 apr-30 dec, 1924-1925 30 jun, 1926-30 – 1 – (title varies: 15 jul 1927: tegel-hermsdorfer zeitung; 20 nov 1928: wittenau-borsigwalder-tegel-hermsdorfer zeitung) – gw Misc Inst [074]

Hermsen, Hugo *see* Die wiedertaeufer zu muenster in der deutschen dichtung

Hernadez, Marcial *see* Hernandez marcial

Hernan centeno. el travieso, senor del castillo de rapapelo en sierra de gata / Velo Nieto, Gervasio – Badajoz: Dip. Prov., 1958 – sp Bibl Santa Ana [946]

Hernan Cortes *see* Cartas de relacion de la conquista de mejico

Hernan cortes / Cortes, Hernando – 1868 – 9 – (1869 ed) – sp Bibl Santa Ana [946]

Hernan cortes / Justiniano Arribas, Juan – 1887 – 9 – sp Bibl Santa Ana [946]

Hernan cortes en extremadura : vision historico-literaria del preconquis tador / Reynolds, Winton A – Madrid: Castalia, 1966 – 1 – sp Bibl Santa Ana [972]

Hernan cortes, letters from mexico / Cortes, Hernando – New York: A.R. Pagden, 1971 – sp Bibl Santa Ana [350]

Hernan cortes. libertador del indio / Trueba, Alfonso – Mexico: Editorial Cempeador, 2nd ed 1954. Col.F: y episodios de la H[a] de Majico no 6 – sp Bibl Santa Ana [350]

Hernandez Acosta, Angel *see* Tierra blanca

Hernandez, Amado V *see* Isang dipang langit

Hernandez Andres, J M *see* Catalogo de una serie miscelanea procedente del convento de san antonio del prado y colegios jesuiticos...

Hernandez, Antonio Angel Delgado *see* Ernesto cardenal

Hernandez Aquino, Luis *see* Poesia puertorriquena

Hernandez B, Ernesto *see* Colombia en korea

Hernandez Briceno, Ernesto *see* Uraba heroico

Hernandez Cardenal, Garcia *see* Consideraciones sobre lo que significa...cristiano

Hernandez Cata, Alfonso *see*
- Angel de sodoma
- Corazon
- Cuentos pasionales
- Juventud de aurelio zaldivar
- Libro de amor
- Mala mujer
- Pelayo gonzalez
- Piedras preciosas
- Placer de sufrir
- Siete pecados
- Voluntad de dios

Hernandez Corujo, Enrique *see* Organizacion civil y politica de las revoluciones

Hernandez De Alba, Gregorio *see* Cuentos de la conquista

Hernandez De Alba, Guillermo *see*
- Ensayistas colombianos
- Estampas santaferenas
- Guia de bogota
- Proceso de narino a la luz de documentos ineditos

Hernandez de Soto, Sergio *see*
- Cuentos populares...extremadura
- Juegos infantiles de extremadura

Hernandez Diaz, Erasmo *see* Banos de montemayor. puerta de extremadura. apuntes 1971

Hernandez Diaz, Jose *see* Expedicion del adelantado hernando de soto a la florida

Hernandez, Felix *see* Alcazaba of merida

Hernandez Franco, Tomas Rafael *see*
- Apuntes sobre poesia popular y poesia negra en las...
- Cibao
- Juan isidro jimenez grullon
- Mas bella revolucion de america

Hernandez Gil, Antonio *see* Metodologia del derecho (ordenacion critica de las principales direcciones metodologicas). madrid, 1945

Hernandez Gil, Fernando *see* Tutela y dignificacion del trabajo

Hernandez, J Enrique *see* Revolucion es el espiritu...

Hernandez, Jesus *see*
- A los intelectuales de espana
- Atras los invasores
- El partido comunista antes, durante y despues de la crisis del gobierno largo caballero

Hernandez, Jose P H *see* Poesias

Hernandez, Juan Climaco *see* Prehistoria colombiana

Hernandez, Leopoldo *see* Pendiente

Hernandez marcial : obras completas... / Hernadez, Marcial; ed by Bayle, Constantino – Madrid: Razon y Fe, 1940 – 1 – sp Bibl Santa Ana [440]

Hernandez, Mariano *see* Ortografia espanola. colegio de san jose, villafranca (badajoz)

Hernandez Martinez, Miguel *see* Panorama de la vida

Hernandez Miyares, Enrique *see* Obras completas de enrique hernandez miyares

Hernandez Pacheco, E et al *see* El sahara espanol

Hernandez Pacheco, Eduardo *see*
- Extremadura y los extremenos
- Fisiografia del guadiana
- Pinturas prehistoricas y dolmenes de la region de alburquerque
- Las tierras negras del extremo sur de espana y sus yacimientos poleoliticos

Hernandez Pacheco, Francisco *see*
- Discurso leido en la apertura del curso academico 1943-44
- Rasgos fisiograficos y geologicos del territorio de ifni y rasgos fisiograficos y geologicos del sahara

Hernandez Poveda, Ruben *see* Desde la barra

Hernandez Rivera, Sergio Enrique *see* Compadecido bosque

Hernandez, Roberto *see* Silencio abierto

Hernandez Sanchez, Jesus *see*
- Campus
- Fidel castro

Hernandez U Urbina, Francisco *see* Hombre a traves de un libro

Hernandez Valbuena, Ramiro *see* La luz del vaticano

Hernandez y Herrero, Joaquin *see* Carta pastoral...quenta cura

Hernandez-Santana, Gilberta *see* Canto eterno, poesias

Hernandez-Santana, Gilberto *see* Semblanzas negras, poemas

Hernandez-Usera, Rafael *see* Semillas a voleo

Hernando cortes / Corona Baratech, Carlos E – Madrid: Publicaciones Espanolas, 1959 Temas Espanoles, no 57 (2nd ed.) – sp Bibl Santa Ana [920]

Hernando cortes / Corraliza, Jose V – Badajoz: Imp. de la Dip. Provincial, 1965. Sep. REE – sp Bibl Santa Ana [250]

Hernando cortes / Justiniano Arribas, Juan – 1877 – 9 – sp Bibl Santa Ana [946]

Hernando cortes / Madariaga, Salvador – Buenos Aires: Editorial Sudamericana, 1945 – 1 – sp Bibl Santa Ana [350]

Hernando cortes / Piron, Alexo – 1776 – 9 – sp Bibl Santa Ana [946]

Hernando cortes / Ramirez, Alfonso Francisco – Mexico: D.F., 1950 – sp Bibl Santa Ana [920]

Hernando cortes / Sandoval, M de – Glorias de Espana. Madrid: Laultima moda, 1898 – 1 – sp Bibl Santa Ana [920]

Hernando cortes / Torres, Luis de – 1830 – 9 – sp Bibl Santa Ana [946]

Hernando cortes and the marquesado in morelos, 1522-1547 / Riley, G Michael – Alburquerque: University of New Mexico Press, 1973 – sp Bibl Santa Ana [920]

Hernando cortes. conqueror of mexico / Madariaga, Salvador – London: Hodder-Stoughton, 1942 – 1 – sp Bibl Santa Ana [350]

Hernando cortes (conquistador de mejico) / Torres, Luis de – Madrid: Biblioteca Nueva, 2nd ed 1942 – sp Bibl Santa Ana [350]

Hernando cortes, estampas de su vida / Margarinos, Santiago – Madrid, 1947 – 1 – sp Bibl Santa Ana [910]

Hernando cortes (estudio de un caracter) / Polavieja, Marques de – Toledo, Imprenta y Libreria de la Viuda e hijos de I, Pelaez, 1909 – 1 – sp Bibl Santa Ana [920]

Hernando cortes (estudio de un caracter) por el teniente general marques de polariega / Altolaguirre, Angel de – Madrid: Fortanet, 1909. B.R.A.H. 55, pp. 506-514 – sp Bibl Santa Ana [920]

Hernando cortes, exequias, almoneda e inventario de sus bienes / Muro Orejon, Antonio – Sevilla: Escuela de Estudios Hispano-Americanos, 1958. Sep. – sp Bibl Santa Ana [946]

Hernando cortes. libertador del indio / Trueba, Alfonso – Mexico: Editorial Jus, 3rd ed 1958 – sp Bibl Santa Ana [350]

Hernando cortes o la conquista de mejico / Escofet, Jose – Barcelona: S.A. Ig. S. Barral Hermanos, 1925. Col. Los Grandes Esploradores espanoles. vol 3 – sp Bibl Santa Ana [972]

Hernando cortes y el derechos internacional en el siglo 16 / Esquivel Obregon, Toribio – Mexico: Editorial Polis, 1939 – sp Bibl Santa Ana [920]

Hernando cortes y francisco pizarro / Gonzalez Ruiz, Nicolas – Barcelona: Editorial Cervantes, 1952 – sp Bibl Santa Ana [350]

Hernando cortes y sus parientes los juarez / Ramos-Olivera, Antonio – Mexico: Cia. General de Ediciones, S.A., 1972 – sp Bibl Santa Ana [320]

Hernando de soto / Villanueva y Canedo, Luis – 1892 – 9 – sp Bibl Santa Ana [920]

Hernando de soto, paladin de florida y descubridor del missisipi / Munoz de San Pedro, Miguel – Madrid: Novelas y cuentos, 1954 – 1 – sp Bibl Santa Ana [917]

Hernandez Segui, Domingo *see* Ojeada sobre la flora medica y toxica de cuba

Herndon, Eugene Wallace *see*
- The foundation of christian hope
- A review of a lecture by eld moses e lard on future punishment!

Herndon, William *see* Exploration of the valley of the amazon

Herne bay press – England, 1883-1975 – 88r – 1 – (missing: jan-jun 1898) – uk British Libr Newspaper [072]

Herner zeitung – Herne DE, 1949 1 nov-1957 – 12mf=23df – 9 – gw Mikrofilm; gw Misc Inst [074]

Hernosandsposten halfveckoupplagen – Haernoesand, Sweden. 1909-10 – sw Kungliga [078]

The hero in hemingway: a study in development / Dahiya, Bhim S – Chandigarh: Bahri Publications, 1978. xv,225p. Includ. index and bibliog. With: Letters of two queens. Bathhurst, Hon. A.B – 1 – us Wisconsin U Libr [420]

The hero of erie (oliver hazard perry) / Barnes, James – New York: D Appleton, 1898 – (= ser Young heroes in our navy) – 2mf – 9 – mf#24310 – cn Canadiana [830]

The hero of esthonia and other studies in the romantic literature of that country / Kirby, William Forsell – Comp. from Esthonian and German sources.London: J.C. Nimmo, 1895. 2v. ill. map – 1 – us Wisconsin U Libr [460]

The hero of heroes : a life of christ for young people / Horton, Robert Forman – New York: Fleming H Revell, c1911 – 1mf – 9 – 0-524-04463-5 – mf#1992-0132 – us ATLA [240]

The hero of panama : a tale of the great canal / Brereton, Frederick Sadleir – London: Blackie; Toronto: Copp, Clark, 1912 – 5mf – 9 – 0-665-97097-8 – mf#97097 – cn Canadiana [830]

The hero of pine ridge : a story of the great prairie / Butler, William Francis – Boston: Jordon, Marsh, 188-? – 5mf – 9 – mf#01049 – cn Canadiana [390]

The hero of the monongahela : historical sketch / Beaujeu, Monongahela de – [New York?: W Post, 1913] – 1mf – 9 – 0-665-98848-6 – (trans by rev g e hawes. also available in french) – mf#98848 – cn Canadiana [920]

Heroard, Jean *see* Journal de jean heroard sur l'enfance et la jeunesse de louis 13 (1601-1628)

Herodiade / Massenet, Jules – Paris, France. 1923 – 1r – us UF Libraries [440]

Herodiade : prelude [de acte 3] pour orchestre / Massenet, Jules – Paris: Heugel & Cie [1899] [mf ed 1992] – 1r – 1 – mf#pres. film 112 – us Sibley [780]

Herodot / Pohlenz, Max – Leipzig, Germany. 1937 – 1r – us UF Libraries [025]

Herodote et la religion de l'egypte : comparaison des donnees d'herodote avec les donnees egyptiennes / Sourdille, Camille – Paris: E Leroux, 1910 – 4mf – 9 – 0-524-01874-X – (incl bibl ref) – mf#1990-2709 – us ATLA [290]

Herodotus *see*
- Historiae
- History of herodotus
- The history of herodotus. a new english version

The herods / Farrar, Frederic William – New York: E R Herrick, [1898] – 1mf – 9 – 0-8370-9697-9 – mf#1986-3697 – us ATLA [930]

El heroe serafico de san pedro de alcantara / Camberos de Yegros, Fernando – 1723 – 9 – sp Bibl Santa Ana [830]

Heroes / Kingsley, Charles – London, England. 1889 – 1r – us UF Libraries [025]

The heroes and crises of early hebrew history : from the creation to the death of moses / Kent, Charles Foster – New York: Charles Scribner, c1908 – (= ser Historical Bible) – 1mf – 9 – 0-8370-9480-1 – (incl ind) – mf#1986-3480 – us ATLA [220]

Heroes and hierarchs : or, biblical principles as held by baptists in the contention for religious liberty / Haskell, Samuel – Philadelphia: American Baptist Publ Soc, c1895 – 4mf – 9 – 0-524-07427-5 – (incl ind) – mf#1991-3087 – us ATLA [242]

Heroes and martyrs of faith / Peake, Arthur Samuel – London, New York: Hodder and Stoughton, [19–?] – (= ser The expositor's library) – 1mf – 9 – 0-8370-9645-6 – mf#1986-3645 – us ATLA [220]

Heroes and martyrs of the modern missionary enterprise : a record of their lives and labors / ed by Smith, Lucius Edwin – Hartford: P. Brockett, 1852 – 2mf – 9 – 0-7905-8091-8 – mf#1988-8027 – us ATLA [240]

Heroes de america / Servin, Felipe – Mexico City? Mexico. 1941 – 1r – us UF Libraries [972]

Heroes e bandidos / Barroso, Gustavo – Rio de Janeiro, Brazil. 1931 – 1r – us UF Libraries [972]

Heroes extremenos : alvaro de sande / Munoz Carrero, Pedro – Caceres: Tipografia Extremadura, 1923 – 1 – sp Bibl Santa Ana [946]

Heroes: narraciones para soldados / Arendt, Erich – Barcelona, 1938? Fiche W 725. (Blodgett Collection of Spanish Civil War Pamphlets) – 9 – us Harvard [946]

Heroes of bohemia : huss, jerome and zisca / Mears, John W – Philadelphia: Presbyterian Board of Publication, c1879 – 1mf – us ATLA [240]

Heroes of bohemia : huss, jerome and zisca / Mears, John William – Philadelphia: Presbyterian Board of Publ, c1879 – 1mf – 9 – 0-7905-5069-5 – mf#1988-1069 – us ATLA [943]

Heroes of faith : lectures on the eleventh chapter of the epistle to the hebrews / Vaughan, Charles John – London: Macmillan, 1876 – 1mf – 9 – 0-8370-5622-5 – mf#1985-3622 – us ATLA [220]

Heroes of israel : a teacher's manual to be used in connection with the student's textbook / Soares, Theodore Gerald – Chicago, IL: University of Chicago Press; New York: Baker & Taylor [distributor], 1910 – 1mf – 9 – 0-8370-9419-4 – (includes bibliographies) – mf#1986-3419 – us ATLA [221]

Heroes of israel : text of the hero stories with notes and questions for young students / Soares, Theodore Gerald – 2nd ed. Chicago, IL: University of Chicago Press, 1911, c1908 – (= ser Constructive Bible Studies) – 1mf – 9 – 0-7905-0232-1 – mf#1987-0232 – us ATLA [220]

The heroes of methodism : containing sketches of eminent methodist ministers, and characteristic anecdotes of their personal history / Wakeley, Joseph Beaumont – New York: Carlton & Porter, 1856 [mf ed 1991] – 2mf – 9 – 0-524-01537-6 – mf#1990-0443 – us ATLA [242]

Heroes of modern missions / Lhamon, William Jefferson – Chicago: Fleming H Revell c1899 – (= ser Bethany c.e. hand-book series; Bethany c.e. reading courses) – 1mf – 9 – 0-524-04269-1 – mf#1991-2053 – us ATLA [240]

Heroes of the cross in america / Shelton, Don Odell – New York City: Literature Dept, Presbyterian Home Missions 1904 [mf ed 1986] – (= ser Forward mission study courses) – 1mf – 9 – 0-8370-6515-1 – (incl ind) – mf#1986-0515 – us ATLA [920]

Heroes of the hour : mahatma gandhi, tilak maharaj, sir subramanya iyer – Madras: Ganesh & Co, 1918 – (= ser Samp: indian books) – us CRL [954]

Heroes of the mission field : links in the story of missionary work from the earliest ages to the close of the eighteenth century / Walsh, William Pakenham – New York: Laymen's Missionary Movement, [1879] – 1mf – 9 – 0-8370-6786-3 – mf#1986-0786 – us ATLA [920]

Heroes of the saddle bags, a history of christian denomination in the republic of texas / Smith, Jesse Guy – 1951 – 1 – 8.75 – us Southern Baptist [242]

Heroes of the south seas / Banks, Martha Burr – New York: American Tract Society, c1896 – 1mf – 9 – 0-8370-6006-0 – mf#1986-0006 – us ATLA [240]

The heroic age of india : a comparative study / Sidhanta, Normal Kumar – London: Kegan Paul, Trench, Trubner & Co; New York: Alfred A Knopf, 1929 – (= ser Samp: indian books) – us CRL [954]

An heroic bishop : the life-story of french of lahore / Stock, Eugene – London: Hodder & Stoughton, 1913 [mf ed 1990] – 1mf – 9 – 0-7905-6953-1 – mf#1988-2953 – us ATLA [240]

Heroic elegies and other pieces / Llywarc Hen – London: Owen & Williams 1792 – (= ser Whsb) – 3mf – 9 – €40.00 – (literal trans by william owen) – mf#Hu 103 – gw Fischer [800]

A heroic priest : memoir of joseph francis brophy, d d: apostle of coney island / Boyton, Paul (Mrs) – [s.l.]: George C Tilyou & Paul Boyton, 1910 [mf ed 1986] – 1mf – 9 – 0-8370-6967-X – mf#1986-0967 – us ATLA [241]

Heroic recitations of the bahima of ankole / Morris, Henry F – Oxford, England. 1964 – 1r – us UF Libraries [960]

Heroic stature : five addresses / Sheppard, Nathan – Philadelphia: American Baptist Publication Society, 1897 – 1mf – 9 – 0-524-01404-3 – mf#1990-0403 – us ATLA [240]

Heroica m. claudii paradini, belliiocensis canonici, et d. gabrielis symeonis, symbola / Paradin, Claude & Simeoni, G – Antwerpen: J Steelsius, 1563 – 4mf – 9 – mf#O-3248 – ne IDC [090]

Heroides / Ovidius – 14th c – (= ser Holkham library manuscript books 319) – 1r – 1 – mf#95969 – uk Microform Academic [450]

Heroides (cima1) : farbmikrofiche-edition der handschrift wien, oesterreichische nationalbibliothek, cod.2624 / Ovidius Naso, Publius – (mf ed 1986) – (= ser Codices illuminati medii aevi (cima) 1) – 28p on 5 color mf – 15 – €280.00 – 3-89219-001-1 – (french trans by octovien de saint-gelais. int & description by dagmar thoss) – gw Lengenfelder [090]

L'heroine de chateauguay : episode de la guerre de 1813 / Chevalier, Henri Emile – Montreal: J Lovell, 1858 [mf ed 1982] – 2mf – 9 – 0-665-33240-8 – mf#33240 – cn Canadiana [830]

Une heroine du canada : madame gamelin et ses oeuvres / Giroux, Henri – Montreal: s.n, 1885 – 1mf – 9 – mf#06420 – cn Canadiana [241]

The heroine of acadia : the romantic story of the life of frances marie jacqueline, wife of sieur de la tour, and her heroice [sic] defence of fort latour, at the mouth of the river st john in the year [1]645 / Hannay, James – St John, NB: J A Bowes, 1910 – 1mf – 9 – 0-665-73073-X – (incl bibl ref) – mf#73073 – cn Canadiana [920]

A heroine of charity and a queen by right divine / O'Meara, Kathleen – London: Burns & Oates, [1891] [mf ed 1986] – 1mf – 9 – 0-8370-6927-0 – mf#1986-0927 – us ATLA [241]

Heroine of faith – London, England. 18– – 1r – us UF Libraries [240]

Heroines of sacred history / Steele, Eliza R – 4th ed. New-York: J S Taylor, 1851 – 1mf – 9 – 0-524-06002-9 – mf#1992-0739 – us ATLA [220]

Heroines of the mission field : biographical sketches of female missionaries who have laboured in various lands among the heathen / Pitman, Emma Raymond – New York: Anson D F Randolph, [1881] Beltsville, Md: NCR Corp, 1977 (5mf); Evanston: American Theol Lib Assoc, 1984 (5mf) – 9 – 0-8370-0264-8 – mf#1984-0064 – us ATLA [240]

Heroines of the missionary enterprise : or, sketches of prominent female missionaries / Eddy, Daniel Clarke – Boston: Ticknor, Reed, and Fields, 1850. Beltsville, Md: NCR Corp, 1978 (4mf); Evanston: American Theol Lib Assoc, 1984 (4mf) – (= ser Women & the church in america) – 9 – 0-8370-1229-5 – mf#1984-2070 – us ATLA [240]

Os herois de coaro e pirapo / Bayle, Constantino & Jaeger, Luis Gonzaga – Madrid: Razon y Fe, 1943 – 1 – sp Bibl Santa Ana [946]

Herois lynenburgica : sive carminum lynenburgensium... / Mechov, G – Hagae Comitum: Apud Nicolaum Wilt, 1698 – 4mf – 9 – mf#O-1832 – ne IDC [090]

Heroisme et trahison : recits canadiens / Marmette, Joseph – Quebec?: C Darveau, 1878 – 3mf – 9 – mf#09905 – cn Canadiana [971]

Heroismes d'antan : victoires d'aujourd'hui des coureurs des bois au chemin de fer national du canada / Morin, Paul – [Montreal]: [Chemin de fer national du Canada], [1923?] (mf ed 1991) – 1mf – 9 – mf#SEM105P1472 – cn Bibl Nat [380]

Der herold – Detroit MI (USA), 1898-1918 [gaps] – 16r – 1 – gw Misc Inst [071]

Der herold – Detroit MI (USA), 1898-1918 [gaps] – 16r – 1 – gw Misc Inst [071]

Der herold – Grand Island, NE: Henry Garn & Boehl, 1880-apr 1893// [mf ed 1886-87, 1892 (gaps) filmed 1976] – 1r – 1 – (in german. merged with: grand island anzeiger to form: grand island anzeiger und herold. english ed: grand island herald) – us NE Hist [071]

Herold – Chippewa Falls, Eau Claire, Menomonie WI. 1892 sep 22-1893 dec 28, 1894 jan 4-1895 jul 25, 1895 aug 1-1898 feb 3, 1898 feb 10-1899 aug 24, 1899 aug 31-1901 mar 21, 1901 mar 28-1902 oct 9, 1902 oct 16-1904 apr 21, 1904 apr 28-1905 dec 14, 1905 dec 21-1907 feb 28, 1907 mar 7-1908 oct 29, 1908 nov 5-1910 apr 21, 1910 apr 28-1911 dec 28, 1912 jan 4-1913 oct 2, 1913 oct 9-1915 jul 8, 1915 jul 15-1916 dec 28 – 15r – 1 – (cont: eau claire und chippewa falls herold) – mf#998635 – us WHS [071]

Herold – 1866 oct 12/1867 may-1890 jan/jun [gaps] – 15r – 1 – (continued by: milwaukee herold (milwaukee, wi: 1890 : daily)) – mf#1169063 – us WHS [071]

Herold – Milwaukee WI [1863 jun 20/1865 apr 8]-1890 may/jul 4 [gaps] – 24r – 1 – (continued by: milwaukee herold (milwaukee, wi: 1890 : semiweekly)) – mf#1219080 – us WHS [071]

Herold / Franklin Co. Columbus – jan 1934-dec 1938,feb 1939-jan 1941 [wkly, semiwkly] – 6r – 1 – mf#B5624-5629 – us Ohio Hist [071]

Herold der wahrheit – Elkhart, IN. v1-45. 1912-56 [complete] – (= ser Mennonite serials coll) – 14r – 1 – ISSN: 0300-8851 – mf#ATLA 1993-S016 – us ATLA [242]

Herold der wahrheit – Chicago, IL. v1-38. 1864-1901 – (= ser Mennonite serials coll) – 10r – 1 – (lacking: v37) – mf#ATLA 1992-S002 – us ATLA [242]

Herold des glaubens – St Louis MT (USA), 1922 5 dec-1924 27 jun – 2r – 1 – gw Misc Inst [210]

Der herold fuer das deutsche volk – Berlin DE, 1845 oct-1847 aug – 1r – 1 – gw Misc Inst [074]

Herold german – Providence, RI. 1897-1898 (1) – mf#66314 – us UMI ProQuest [071]

Herold, Reinhold *see* Geschichte der reformation in der grafschaft oettingen, 1522-1569

Herold series / Defiance Co. Defiance – 4/1890-4/16,(1-12/19), jan 1920 [wkly] – 13r – 1 – (in german) – mf#B7699-7711 – us Ohio Hist [071]

Herold und volksfreund – La Crosse, Winona WI. 1898 jan 15-1900, 1901-17, 1918-1920 oct 16 – 10r – 1 – (continued by: wisconsin wochenblatt (la crosse-portage ed: 1920)) – mf#1013323 – us WHS [071]

Heron, Henry *see* Eighteenth-century english organ voluntary

Heron, James *see*
- The celtic church in ireland
- The church of the sub-apostolic age
- A short history of puritanism

Heroncio, Paulo *see* Holandeses no rio grande

Le heros de chateauguay / David, Laurent-Olivier – 2nd rev corr ed. Montreal: Cadieux & Derome, 1883 [mf ed 1982] – (= ser Bibliotheque religieuse et nationale) – 2mf – 9 – mf#24795 – cn Canadiana [355]

Le heros de st-eustache : jean olivier chenier / Frechette, Louis – Montreal: E Demers, 18–? – 1mf – 9 – mf#06540 – cn Canadiana [355]

Le heros de st-eustache, jean olivier chenier / Frechette, Louis – Montreal: E Demers, 18–? – 1mf – 9 – mf#06540 – cn Canadiana [920]

The hero's hero / Seidenman, Roger S – 1989 – 2mf – 9 – $8.00 – mf#PE 4033 – us Kinesology [306]

Herpetologica – Pittsburgh. 1936+ (1) 1936+ (5) 1936+ (9) – ISSN: 0018-0831 – mf#12792 – us UMI ProQuest [590]

Herpetology of the cayman islands / Grant, Chapman – Kingston, Jamaica. 1940 – 1r – us UF Libraries [972]

Le herpeur, m.l'oratoir de france... / ed by Bayle, Constantino – Madrid: Razon y Fe, 1926 – 1 – sp Bibl Santa Ana [944]

Herr dr cahn und der lehrerverband – Hamburg, Germany. 1905 – 1r – us UF Libraries [939]

Der herr gevatter von der strasse : genrebild in einem aufzuge / Langer, Anton – Wien: A Landvogt 1868 [mf ed 1995] – 1r – 1 – (filmed with: reichsstaedtische erzaehlugen / herman kurz) – mf#3679p – us Wisconsin U Libr [820]

Herr goldenbarg / Raboy, Isaac – NYU York, NY. 1916 – 1r – us UF Libraries [939]

Herr heinrich : die saga vom ersten deutschen reich / Vater, Fritz – 3. Aufl. Muenchen: F Eher, 1943 – 1r – 1 – us Wisconsin U Libr [943]

Herr, Johannes *see* The illustrating mirror

Herr reineke fuchs, eine unheilinge weltbibel / Reinke De Vos – Berlin, Germany. 1943 – 1r – us UF Libraries [025]

Der herr senator : novelle / Jensen, Wilhelm – Leipzig: B Elischer Nachf (B Winckler) 1890 [mf ed 1995] – 144p – 1 – mf#8795 – us Wisconsin U Libr [830]

Der herr sicher : erinnerungen aus dem leben des pfarrers j.w. ludwig / Engelhardt, E v – 3. Aufl. Basel: Missionsbuchh, 1888 – 1mf – 9 – 0-524-06614-0 – mf#1991-2669 – us ATLA [242]

Herr ulrich zwingli leerbiechlein : wie man die knaben christlich vnterweysen vnd erziehen soll... – [Augsburg], 1524 – 1mf – 9 – mf#PBU-510 – ne IDC [242]

Herr und hund : idylle / Mann, Thomas – 11.-16. Aufl. Berlin: S Fischer, 1929 – 1r – us Wisconsin U Libr [830]

Herr vetter – Medford WI. 1902 sep 11 – 1r – 1 – mf#1109175 – us WHS [071]

Herrand von Wildonie *see* Vier erzaehlungen

Herrarte, Alberto *see* Documentos de la union centroamericana

Herre, Paul *see* Dahlmann-waitz

Herren aer oefwersteprest : betraktelser oefwer ebreerbrefwets 9:de kapitel / Beskow, Gustaf Emanuel – Moline, IL: Wistrand & Thulin, [1877] – 1mf – 9 – 0-524-05247-6 – mf#1991-2239 – us ATLA [220]

Die herren der erde : [novel] / Beheim-Schwarzbach, Martin – Leipzig: Insel-Verlag, 1931 [mf ed 1989] – 309p – 1 – mf#7004 – us Wisconsin U Libr [830]

Herren ohne heer : roman des baltischen deutschtums / Vegesack, Siegfried von – Berlin: Universitas c1934 [mf ed 1991] – 1r – 1 – (filmed with: blumbergshof / siegfried von vegesack) – mf#2944p – us Wisconsin U Libr [830]

Herrera, Antonio de *see*
- Descripcion de las indias occidentales...
- Descripciones de indias occidentales
- Historia general de los hechos de los castellanos en las islas y tierra firme del mar oceano
- Historia general de los hechos de los castellanos en las islas y tierra-firme de el mar oceano
- Historia general de los hechos de los castellanos en las islas y tierra-firme del mar oceano
- Historia general de los hechos de los castellanos en las islas y tierrafirme del mar oceano
- Historia general...castellanos...oceano
- Historia...castellanos en las islas...oceano
- Historia...indias occidentales

Herrera, Benardino *see* Memorias historias... carlota joaquina y dona mariana victorias

HERRERA

Herrera, Bonifacio *see* Panegirico...santa olalla
Herrera Carrasco, F *see* Satisfaccion publica de una...calumnia sobre la constitucion pleuristico catharral...de aig
Herrera, Cesar A *see*
– Batalla de las carreras
– Poesia de salome urena en su funcion social y patr...
Herrera, Flavio *see*
– 20 (i.e. viente) rabulas en flux ensayo de picaresca
– Caos
– Poniente de sirenas
– Solera
Herrera Fritot, Rene *see*
– Caleta, joya arqueologica antillana
– Nuevo dujo taino en las colecciones
– Revision de las hachas de ceremonia
Herrera, G *see* Obra de agricultura compilada de diversos autores
Herrera Maldonado, Francisco *see* Dialogos morales de luciano
Herrera, Mariano *see* Despues de la zeta
Herrera Oria, P Enrique *see* Historia de la reconquista de espana contada a la juventud
Herrera, Philip *see* Energy
Herrera Vega, Adolfo *see* Expresion literaria de nuestra vieja raza
Herrera Velado, Francisco *see* Agua de coco
Herrera Y Reissig, Julio *see* Ciles alucinada, y otras poesias
Herrero Alvarado, Antonio *see* Huellas juveniles
Herrero, Antonio Maria *see* Carta de don...en que demuestra quan inaccesibles han sido los esfuerzos de d. bernardo arayo para defender que no que phtisis pulmonar...
Herrero, Leandro *see*
– El monge del monasterio de yuste (ultimos dias del emperador carlos 5)
– El monje del monasterio de yuste
Herrero Mediavilla, Victor [comp] *see*
– African biographical archive
– Archivo biografico de espana, portugal e iberoamerica 1960-1995 [abepi3]
– Archivo biografico de espana, portugal e iberoamerica [abepi1]
– Archivo biografico de espana, portugal e iberoamerica [abepi4]
– Archivo biografico de espana, portugal e iberoamerica. nueva serie [abepi2]
– Australasian biographical archive
– Australasian biographical archive (anzo-ba). supplement
– Deutsches biographisches archiv 1960-1999
Herrero, Miguel *see* Pedro alvarado, 4th centenario de la muerte de...1541-1941
Herrero Picado, Manuel F *see* Reemplazo del ejercito y milicias
Herreruela. Ayuntamiento *see* Grandes fiestas en honor de san juan bautista, 1974
Herrfurth, Hugo *see* Veit dietrichs predigt
Herrgott, M *see* Vetus disciplina monastica
Der herrgottsmantel : kulturbild aus dem bayrisch-boemischen waldgebirge / Schmidt, Maximilian – Berlin: Hermann Hillger [189-?] [mf ed 1995] – 1r [ill] – 1 – (ill by r a jaumann. filmed with: sueden und norden / hermann schmid) – mf#3738p – us Wisconsin U Libr [914]
Herrick, Allison Butler *see*
– Area handbook for angola
– Area handbook for mozambique
– Area handbook for tanzania
Herrick and doxsee's probate law and practice of the state of iowa. / Iowa. Laws, Statutes, etc – 2d ed. Chicago: Callaghan, 1898. 892p. LL-87 – 1 – us L of C Photodup [348]
Herrick, C Judson *see* Fatalism or freedom
Herrick, George Frederick *see* Christian and mohammedan
Herrick, Henry Martyn *see* The kingdom of god in the writings of the fathers
Herrick, Horace N *see* A history of the north indiana conference of the methodist episcopal church
Herrick, Myron Timothy *see* Myron t herrick papers, 1827-1941
Herridge, William Thomas *see*
– Anniversary sermon, 1889
– Appel aux armes
– Christianity in its relation to the state and the church
– "England's greatness"
– French and english in canada and across the sea
– In memoriam
– The ontario liquor act
– The orbit of life
– A sermon preached in st andrew's church, ottawa, on sunday morning, may 25th, 1902
Herrig, Hans *see* Luther
Herrig, L *see* Archiv fuer das studium der neueren sprachen
Herriman, Marion E *see* Survey of latin-american music for use in the junior high school
Herrin und sklave nach sacher masoch / Esau [pseud] – neu bearbeitet. Berlin [privattyposkript c1910] (mf ed 1994) – 1mf – 9 – €24.00 – 3-8267-3017-8 – mf#DHS-AR 3017 – gw Frankfurter [430]
Herring, Frances Elizabeth *see*
– Canadian camp life
– Here are a few press notices from the leading english critics on "canadian camp life"...by frances e herring, new westminster, british columbia...publ by t fisher unwin, london, england...
Herring, G *see* The people of the polar north
Herring, Hubert Clinton *see*
– Good neighbors
– Renascent mexico
– Spain, battleground of democracy
Herring, nan trammell and james alexander : correspondence. 1929-1964 = Missionaries to china – Furman University Library, Greenville, SC Baptist History Collection – 1r – 1 – $37.76 – us Southern Baptist [242]
Herring, Rosa P *see* Health beliefs, health values, and preventive health promotion activities of african- and euro-american women
Herrliberger, D *see*
– Heilige ceremonien, gottesdienstliche kirchenuebungen...der stadt und landschaft zuerich
– Kurze beschreibung der gottesdienstlichen gebraeuche
Die herrlichkeit gottes : eine biblisch-theologische untersuchung ausgedehnt ueber das alte testament, die targume, apokryphen, apokalypsen und das neue testament / Gall, August, Freiherr von – Giessen: J Ricker, 1900 – 1mf – 9 – 0-8370-3225-3 – (incl bibl ref) – mf#1985-1225 – us ATLA [220]
Herrmann, August Leberecht *see*
– Frankreichs religions- und buergerkriege im sechzehnten jahrhunderte
– Franz der erste, koenig von frankreich
Herrmann, Christian *see* Die weltanschauung gerhart hauptmanns in seinen werken
Herrmann, Emil Alfred *see*
– Der gestiefelte kater / das rotkaeppchen
– Das gottes kind
Herrmann, Eugen *see* Prolegomena zur geschichte sauls
Herrmann, F *see*
– Protestantischer schriftbeweis
– Roemischer schriftbeweis
Herrmann, Helene *see* Studien zu heines romanzero
Herrmann, Johannes *see*
– Ezechielstudien
– Die idee der suehne im alten testament
Herrmann, Klaus *see* Sturm und drang
Herrmann, Leon *see* Masques et les visages dans les bucoliques de virgile
Herrmann, Otto *see* Goethe erzaehlt sein leben
Herrmann, Peter *see* Gesellschaft und organisation
Herrmann, R *see* Erloesung
Herrmann und ulrike : ein roman / Wezel, Johann Karl; ed by Maassen, Carl Georg von – Muenchen: Georg Mueller, 1919 – 1r – 1 – (incl bibl ref) – us Wisconsin U Libr [830]
Herrmann, Walther *see*
– Der schimmelreiter
Herrmann, Wilhelm *see*
– Der begriff der offenbarung
– The communion of the christian with god
– Ethik
– Der evangelische glaube und die theologie albrecht ritschls
– Geschichte der protestantischen dogmatik von melanchthon bis schleiermacher
– Die gewissheit des glaubens und die freiheit der theologie
– Gregorii nysseni sententiae de salute adipiscenda
– Die metaphysik in der theologie
– Die religion im verhaeltniss zum welterkennen und zur sittlichkeit
– Roemische und evangelische sittlichkeit
– Die sittlichen weisungen jesu
– Die speculative theologie in ihrer entwicklung durch daub
– Warum handelt es sich in dem streit um das apostolikum?
– Die wirklichkeit gottes
Herrn figullas schaufenster : heitere geschichten / Menzel, Herybert – 2. Aufl. Hamburg: Hanseatische Verlagsanstalt, 1942 – 1r – 1 – us Wisconsin U Libr [830]
Herrn georg andreas sorgens anleitung zum generalbass und zur composition / Marpurg, Friedrich Wilhelm – Berlin: G A Lange 1760 [mf ed 19–] – 3mf – 9 – mf#fiche 666 – us Sibley [780]
Herrn peter osbeck, pastors zu hassloef und woxtorp, der koeniglichen schwedischen akademie zu stockholm und der koeniglichen gesellschaft zu upsala, mitgliedes reise nach ostindien und china : nebst d. toreens reise nach suratte und c.g. ekebergs nachricht von der landwirthschaft der chineser / Osbeck, Peter – Rostock: Johann Christian Koppe, 1765 – 7mf – 9 – mf#HT-703 – ne IDC [915]
Herrn professor gellerts geistliche oden und lieder mit melodien / Bach, Carl Philipp Emanuel – Berlin: Gedruckt...bey G L Winter 1758 [mf ed 19–] – 2mf – 9 – mf#fiche973 – us Sibley [780]
Herrn schellbogen's abenteuer : ein stuecklein aus dem alten berlin / Rodenberg, Julius – Berlin: Gebrueder Paetel, 1890 – 1r – 1 – us Wisconsin U Libr [430]
Herron, George Davis *see*
– Between caesar and jesus
– The larger christ
– The message of jesus to men of wealth
– A plea for the gospel
– Social meanings of religious experiences
Herrrera C, J Noe *see* Prensa ante el derecho
Herrschaftszeichen und staatsymbolik (mgh schriften:13.bd) : beitraege zu ihrer geschichte vom 3. bis zum 16. jahrhundert / Schramm, P E – 1954-1956 – (= ser Monumenta germaniae historica. schriften (mgh schriften)) – 3v – €56.00 – ne Slangenburg [931]
Herrscherdaemmerung und deutschlands erwachen in wagners "ring des nibelungen" *see* Beitraege zur auslegung von richard wagners "ring des nibelungen"
Hersbrucker zeitung – Hersbruck DE, 1848 7 oct-1850, 1854-59, 1860 19 may-1876, 1878-1891 27 jun, 1891 3 oct-1894, 1897-1901 5 oct, 1902-04, 1906-08, 1909 3 apr-1934 30 jun, 1935-1943 20 mar, 1949 26 aug-1962 – 52mf=102df – 9 – (incl suppls) – gw Mikrofilm [074]
Herschberger, Ruth *see* Adam's rib
Herschel, John Frederick William *see*
– Essays from the edinburgh and quarterly reviews
– Familiar lectures on scientific subjects
– A manual of scientific enquiry
– Outlines of astronomy
– A preliminary discourse on the study of natural philosopy
Herschel, John FW *see* Physical geography of the globe
Herschel, William *see* [Astronomy pamphlets]
Herse, Wilhelm *see* Die goethezeit in deutschland
Her-self – Ann Arbor. 1972-1977 (1) – mf#7031 – us UMI ProQuest [320]
Her-self – v1-v5 n6 [1972 apr-1977 dec/jan] – 1r – 1 – mf#370652 – us WHS [071]
Hersey, Jean *see* Halfway to heaven
Hersfelder anzeiger – Bad Hersfeld DE, 1854 4 jan-1864, 1866, 1868-1875 7 apr, 1876-1881 15 jun, 1881 1 oct-1923, 1924 24 jan-1927, 1929-1933 13 nov – 53r – 1 – (title varies: 1867 n78: hersfelder kreisblatt; 23 sep 1913: hersfelder tageblatt. incl suppls) – gw Misc Inst [074]
Hersfelder kreisblatt *see* Hersfelder anzeiger
Hersfelder tageblatt *see* Hersfelder anzeiger
Hersfelder zeitung – Bad Hersfeld DE, 1949 30 jul-31 dec, 1950 1 jul-1968 31 jul – 55r – 1 – (filmed by misc inst: 1969- [ca 10r/yr]) – gw Mikrofilm; gw Misc Inst [074]
Hersheleh / Dineshon, Jacob – Tel-Aviv, Israel. 1937 – 1r – us UF Libraries [939]
Hershey, Amos Shartle *see* Modern japan
Hershey citizen – Hershey, NE: Roy R Barnard. 1v. v1 n1. may 8 1941-v1 n52. apr 30 1942 (wkly) [mf ed with gaps filmed 1979] – 1r – 1 – (absorbed by: sutherland courier) – us NE Hist [071]
Hershey enterprise – Hershey, NE: Ray W and Dorothy Graham, 1946-v17 n39. jul 25 1963 (wkly) [mf ed 1947-63] – 5r – 1 – (absorbed by: sutherland courier) – us NE Hist [071]
Hershey review – Hershey, NE: Ray W and Dorothy Graham, 1946-v17 n39. jul 25 1963 (wkly) – 1 – (absorbed by: sutherland courier) – us NE Hist [071]
Hershey review – Hershey, NE: Frank M Brooks. v1 n1. may 7 1896-may 1897 (wkly) [mf ed with gaps filmed 1979] – 1r – 1 – us NE Hist [071]
Hershey, Susanne Wilcox *see* Modern japan
Hershey times – Hershey, NE: F A Rasmussen. -v27 n36. jan 13 1938 (wkly) [mf ed mar 28 1914-jan 13 1938 (gaps) filmed 1974] – 7r – 1 – (absorbed by: lincoln county tribune (1930)) – us NE Hist [071]
Hershkowitz, Leo *see* Boss tweed in court
Hershman, Shelomoh Zalman *see* Bet avot
Hershon, Paul Isaac *see* The pentateuch according to the talmud. genesis
Hershon, Paul Isaac [comp] *see* A talmudic miscellany
Herskovits, Melville J *see* Trinidad village
Herskovits, Melville Jean *see* Dahomean narrative
Herstellung und spektroskopische charakterisierung matrix-isolierter silbercluster : unter besonderer beruecksichtung der photothermischen spektroskopie / Bauer, Martin – (mf ed 1992) – 2mf – 9 – €49.00 – 3-89349-532-0 – mf#DHS 532 – gw Frankfurter [540]
Herstory – 1991 – 90r – 1 – $7,650.00 – (herstory 1, 1956-1971. supplementary set 1, update -1973. supplementary set 2, update -1974. with guide) – us National Clearing [305]
Hertefelt, Marcel D' *see* Anciens royaumes de la zone interlacustre meridionale
Hertel, Peter Ludwig *see* Ellinor
Hertener allgemeine – Herten DE, 1956 4 jun-1970 – 1 – gw Misc Inst [074]
Hertford and ware patriot *see* Ware patriot
Hertford county herald – Ahoskie, NC. 1914-1967 (1) – mf#65295 – us UMI ProQuest [071]
Hertfordshire bulletin – [London & SE] Hertfordshire oct 1926-jun 1929 [mf ed 2004] – 1r – 1 – uk Newsplan [072]
Hertfordshire east and west essex classified – [London & SE] BLNL jan-jul 1986 – 1 – (cont: classified [27 jul 1973-29 mar 1979]; bishops stortford classified [5 apr 1979-28 apr 1983]; bishops stortford news & classified [19 jun-10 jul 1986]; bishops stortford classified & news [17 jul 1986 –]) – uk Newsplan [072]
Hertfordshire express *see* Hitchin and royston express
Hertfordshire mercury – Hertford, England. Nov 1834-35; 1844-47; 1868-75; 1877-89; 1897; 1913; 1916-18; 1950; Nov 1963-Apr 1967; 1980- – 112+ r – 1 – uk British Libr Newspaper [072]
Hertfordshire mercury – Hoddesdon ed. Hertford, England. nov 1963-sep 1974 – 30r – 1 – uk British Libr Newspaper [072]
Hertfordshire news and county advertiser – [London & SE] Hertfordshire 3 dec 1919-2 may 1922 [mf ed 2003] – 3r – 1 – uk Newsplan [072]
Herting, Helga *see* Das sozialistische menschenbild in der gegenwartsliteratur
Hertlein, Eduard *see* Der daniel der roemerzeit
Hertling, G von *see* Albertus magnus
Hertling, Georg, Graf von *see*
– Albertus magnus
– Augustin
– Das princip des katholicismus und die wissenschaft
Hertling, Ludwig, Freiherr von *see* Theologiae asceticae
Hertogenbosch, Ioannes Evangelista van 'S *see*
– Het eeuwigh leven
– Het rijck godts inder zielen oft binnen u-lieden
Herts advertiser [borehamwood & radlett ed] – Radlett, England 28 mar-12 jun 1997 [mf 1981-97] – 1 – (cont, in pt of: herts advertiser (borehamwood, elstree & radlett ed); numeration irreg; cont: borehamwood & radlett advertiser [ns] 7 feb-21 mar 1997]) – uk British Libr Newspaper [072]
Herts and cambs reporter and royston crow – Royston, Cambridgeshire 29 sep 1876-to date – 1 – (jan 1855-to-date; (started out mthly titled royston crow & was free for a while c1856-58. reiss as wkly n1 29 sep 1876. took its present title on 23 nov 1877). though not a cambridgeshire title has always been very influential & circulated widely in south cambridgeshire) – uk Newsplan [072]
Herts and essex observer (saffron walden ed) *see* Saffron walden and district observer
Herts & cambs reporter & royston crow – [East Midlands] Royston, Cambridgeshire jan 1921-dec 1950 [mf ed 2002] – 21r – 1 – (started out mhly titled: royston crow & was free for a while c1856-58; reiss as wkly n1 [29 sep 1876]; took its present title on 23 nov 1877; though not a cambridgeshire title has always been very influential & circulated widely in south cambridgeshire) – uk Newsplan [072]
Herts genealogist and antiquary – S.l., S.l? v1-3. 1895-1898 – 1r – us UF Libraries [025]
Hertsel / Zitron, Samuel Leib – Vilna, Lithuania. 1921 – 1r – us UF Libraries [939]
Hertsel zal / Jewish National Fund – Jerusalem, Israel. 1929 – 1r – us UF Libraries [939]
Hertsl / Zitron, Samuel Leib – Vilna, Lithuania. 1921 – 1r – us UF Libraries [939]
Hertslet, Charles John Belcher *see* The law relating to master and servant.
Hertslet, E *see* Treaties and tariffs...between great britain and foreign nations...
Hertslet, Jessie *see* Bantu folk tales
Hertslet, Lewis E *see* Native problem
Hertslet's commercial treaties – 1827-1925. 14 reels – 1 – $850.00 – us Trans-Media [346]
Hertslet's commercial treaties – London: Butterworth, 1840-1925 [mf ed 19 1827-1925] – 9 – $858.00 – mf#0261 – us Brook [343]
Hertspieghel en andere zedeschriften... / Spieghel, H L – t'Amsterdam: Andries van Damme, 1723 – 5mf – 9 – mf#O-763 – ne IDC [090]
Hertspieghel en andere zede-schriften meest noyt voor dezen gedrukt / Spieghel, H L – Amsterdam: Hendrik Wetstein, 1694 – 4mf – 9 – mf#O-762 – ne IDC [090]
Herttell, Thomas *see* The demurrer: or, proofs of error in the decision of the supreme court of the state of new york, requiring faith in particular religious doctrines as a legal qualification of witnesses.
Hertwig, Otto Robert *see* O r hertwig's tabellen zur einleitung in die kanonischen und apokryphischen buecher des alten testaments
Hertz, Eduard *see* Voltaire und die franzosische strafrechtspflege im achtzehnten jahrhundert

Hertz, Gottfried Wilhelm *see*
– Goethes naturphilosophie im faust
– Natur und geist in goethes faust
Hertz, Simon *see* Torath s'fath eber
Hertz, Wilhelm *see* Die nibelungensage
Hertzberg, Gustav Friedrich *see*
– Geschichte des roemischen kaiserreiches
– Geschichte von hellas und rom
Hertzler, Arthur Emanuel *see* Papers
Hertzog and the south african nationalist party / Lovell, Colin Rhys – Madison 1947 – us CRL [960]
Hertzog-annale – Pretoria, Suid Afrikaanse Akademie vir Wetenskap en Kuns. v1-17, n21. jun 1952-1968 – us CRL [960]
Hertzsch, Erich *see* Karlstadts schriften aus den jahren 1523-25
Herut u.s.a – New York, NY. 1981 – 1 – us AJPC [071]
Hervas. Ayuntamiento *see*
– Ferias y fiestas, 1954
– Ferias y fiestas 1961
– Ferias y fiestas 1964
– Ferias y fiestas. hervas, 1962
– Ferias y fiestas hervas 1971
– Ferias y fietas 1963
Hervas (caceres) / Junta Provincial de Turismo – Vitoria: Tip. Fournier, s.a. – 1 – (fotos javier y grediol) – sp Bibl Santa Ana [338]
Hervas y Panduro, Lorenzo *see*
– Aritmetica delle nazioni e divisione del tempo fra l'orientali
– Catalogo delle lingue conosciute e notizia della loro affinita, e diversita
– Origine formazione, meccanismo, ed armonia degl' idiomi
– Saggio pratico delle lingue
– Vocabulario poligliotto
Hervey, A C *see*
– The genealogies of our lord and saviour jesus christ
– The inspiration of holy scripture
Hervey, Arthur Charles *see*
– The authenticity of the gospel of st luke
– The books of chronicles in relation to the pentateuch and the "higher criticism"
– The pentateuch
Hervey, George Winfred *see*
– Manual of revivals
– The story of baptist missions in foreign lands
Hervey, Hezekiah *see* Commentary on the pastoral epistles, first and second timothy and titus, and the epistle to philemon
Hervey, Maurice H *see* The trade policy of imperial federation from an economic point of view
Hervey, Walter L *see* Syllabus of a course of lessons on principles of religious teaching given at hartford theological seminary
Hervilliez, Gabriel D *see* Rente viagere
Hervormd nederland : h.n. magazine – v25-49. 1969-93 – 36r – 1 – (lacking: some iss) – ISSN: 00180939 – mf#ATLA S0274 – us ATLA [240]
De hervormde kerk in noord-amerika (1624-1664) / Eekhof, Albert – 's-Gravehage: M Nijhoff, 1913 [mf ed 1992] – 2v on 2mf – 9 – 0-524-03636-5 – (in dutch. incl bibl ref) – mf#1990-1064 – us ATLA [242]
De hervorming in spanje : in de zestiende eeuw / Lennep, Maximilian Frederik van – Haarlem: De Erven Loosjes, 1901 [mf ed 1990] – 456p on 2mf – 9 – 0-7905-4536-5 – (in dutch. incl bibl ref) – mf#1988-0536 – us ATLA [242]
Herwarth walden und die europaeische avantgarde / Berlin. Staatliche Museen – 1961 – (= ser Art exhibition catalogues on microfiche) – 9 – $9.80 – uk Chadwyck [700]
Herwegh als uebersetzer / Kilian, Werner – Stuttgart: Metzler, 1914 [mf ed 1992] – (= ser Breslauer beitraege zur literaturgeschichte. neue folge 43) – viii/112p – 1 – mf#8014 reel 5 – us Wisconsin U Libr [430]
Herwegh, Georg *see*
– Die akten ferdinand freiligrath und georg herwegh
– Der freiheit eine gasse
Herwerden, Henricus van *see* Lexicon graecum suppletorium et dialecticum
Herwig, Rachel Monika *see* Die juedische frau als mutter
Herxheimer, S *see* A key to the exercises of the new method of learning the hebrew language
Das herz befiehlt! : kleine geschichten aus krieg und alltag / Bruger, Ferdinand – Muenchen: F X Seitz, 1943 [mf ed 1989] – 126p – 1 – mf#7092 – us Wisconsin U Libr [830]
Herz in boehmen : gedichte / Hoeller, Franz – 2. Aufl. Prag: Volk und Reich Verlag, 1943 – 1r – 1 – us Wisconsin U Libr [810]
Herz jesu missionsbuch : heiliger liebes-bund: ein vollstaendiges gebet- und tugend-buch fuer alle verehrer der heiligsten herzen jesu und mariae / Weninger, Francis Xavier – 12., verb u verm Aufl. Gratz: Joseph Sirolla, 1857 – 2mf – 9 – 0-8370-7036-8 – mf#1986-1036 – us ATLA [240]
Herzberg, Wilhelm *see* Jewish family papers
Herzberger, FW *see* Pilgerklaenge
Herzberg-Fraenkel, Dr *see* Tractatus de simonia
Herz-dame – Duesseldorf DE, 1949 2 mar-1953 – 5r – 1 – gw Misc Inst [074]
Herzensergiessungen eines kunstliebenden klosterbruders / Wackenroder, Wilhelm Heinrich; ed by Jessen, Karl Detlev – Leipzig: E Diederichs, 1904 – 1r – 1 – us Wisconsin U Libr [430]
Herzensergiessungen eines kunstliebenden klosterbruders / Wackenroder, Wilhelm Heinrich – Leipzig: E Diederichs, 1904 – 1 – us Wisconsin U Libr [830]
Herzfeld, Ernst *see*
– Kushano-sasanian coins
– A new inscription of darius from hamadan
– Zoroaster and his world
Herzfeld, Hans *see* Johannes von miquel. sein anteil am ausbau des deutschen reiches bis zur jahrhundertwende
Herzfeld, Levi *see*
– Einblicke in das sprachliche der semitischen urzeit betreffend die entstehungsweise der meisten hebraeischen wortstaemme
– Geschichte des volkes jisrael von zerstoerung des ersten tempels bis zur einsetzung des mackabaeers schimon zum hohen priester und fuersten
Herzfelde, Wieland *see* Dreissig neue erzaehler des neuen deutschland
Herzig, Thomas *see* Strukturanalyse ab initio schwach streuenden kristallen mit synchrotronstrahlung
Herzkraft und koerpermasse : ballistokardiographische untersuchungen an anorektischen jungen erwachsenen und adipoesen kindern / David, Uta – (mf ed 1997) – 2mf – 9 – €40.00 – 3-8267-2441-0 – mf#DHS 2441 – gw Frankfurter [616]
Herzl, T *see* Briefe
Herzl, Theodor *see*
– Idenshtat
– Tel-aviv
Herzl-bund-blaetter – Berlin DE, 1913-18 [gaps] – 1r – 1 – gw Misc Inst [074]
Herzl-bund-blaetter – Berlin: Praesidium des Herzl-Bundes. v1-40. 1913-18 – (= ser German-jewish periodicals...1768-1945, pt 2) – 1r – 1 – $115.00 – (lacking: n20/21 sep/oct 1914, n39 1918) – mf#B98 – us UPA [939]
Die herzmaere; otto mit dem barte; der welt lohn : drei dichtungen / Wuerzburg, Konrad von – Leipzig: Reclam, [1891] [mf ed 1993] – 55p – 1 – (transposed fr middle high german by heinrich kraeger) – mf#8440 – us Wisconsin U Libr [810]
Die herzmarke : drama in zwei teilen / Langmann, Philipp – Stuttgart: J Cotta, [1902] – 1r – 1 – us Wisconsin U Libr [820]
Herzog albrecht von preussen als reformatorische persoenlichkeit / Tschackert, Paul – Halle: Verein fuer Reformationsgeschichte, 1894 [mf ed 1990] – (= ser [Schriften des vereins fuer reformationsgeschichte] 45) – 1mf – 9 – 0-7905-4715-5 – mf#1988-0715 – us ATLA [943]
Herzog albrecht von preussen und sein hofprediger : eine koenigsberger tragoedie aus dem zeitalter der reformation / Hase, Karl Alfred von – Leipzig: Breitkopf & Haertel, 1879 [mf ed 1990] – 1mf – 9 – 0-7905-6231-6 – mf#1988-2231 – us ATLA [943]
Herzog, Eduard *see*
– Leo 13. als retter der gesellschaftlichen ordnung
– Old-catholic view of confession
Herzog, Emil *see* Zsidok tortenete lipto-szt-mikloson
Herzog, Hildegard *see* Anschauungen vom wesen deutscher kunst
Herzog, Johann Jakob *see*
– Abriss der gesamten kirchengeschichte
– Die romanischen waldenser
Herzog, Johannes *see*
– Der begriff der bekehrung
– Die probleme des inneren lebens in der evangelischen verkuendigung
Herzog, Karl *see*
– Aus amerika
– Taschenbuch fuer reisende durch den thueringer wald
Herzog karl august und goethe / Wachsmuth, Wilhelm – Leipzig: Xenien-Verlag, 1911 – 1r – 1 – us Wisconsin U Libr [430]
Herzog, Peter *see* Johannes von mueller und die franzoesische literatur
Herzog, Rudolph *see* Die schlesischen musenalmanache von 1773-1823
Der herzog und das genie : friedrich schillers jugendjahre / Mueller, Ernst – Stuttgart: W Kohlhammer c1955 [mf ed 1995] – 1r – 1 – (incl ind. filmed with: schiller dem deutschen volke dargestellt / j wychgram) – mf#3736p – us Wisconsin U Libr [920]
Der herzog und sein kumpan : ein schelmenroman / Buecker, Bernd – 5. aufl. Wedel: Alster Verlag C Brauns 1943 [mf ed 1989] – 2v in 1 on 1r – 1 – (filmed with: der philister vor, in und nach geschichte / clemens brentano) – mf#7094 – us Wisconsin U Libr [830]
Herzog und vogt : roman / Finckh, Ludwig – Muenchen: Deutscher Volksverlag, [1940] (mf ed 1990) – 1r – 1 – (filmed with: der deutsche finckh) – us Wisconsin U Libr [830]
Herzog, Werner *see* Mystik und lyrik bei novalis
Herzoglich mecklenburg-schwerinsches officielles wochenblatt – Schwerin DE, 1812-1945 23 apr, 1946 25 jun-1952 4 aug – 53r – 1 – (title varies: 6 jan 1816: grossherzoglich mecklenburg-schwerinsches officielles wochenblatt; 5 jan 1850: regierungsblatt fuer das grossherzogtum mecklenburg-schwerin; 1923: regierungsblatt fuer mecklenburg-schwerin; 1934: regierungsblatt fuer mecklenburg; 25 jun 1946: amtsblatt der landesverwaltung mecklenburg-vorpommern; 12 mar 1947: regierungsblatt fuer mecklenburg [fr 4 jan 1875 with official suppls]) – gw Misc Inst [350]
Herzogs albrecht von preussen : gewesenen hochmeisters des deutschen ordens erfolgte friedrich 1., koenigs von preussen, versuchte rueckkehr zur katholischen kirche / Theiner, Augustin – Augsburg: K Kollmann, 1846 [mf ed 1992] – 1mf – 9 – 0-524-02413-8 – (in german, french, italian & latin) – mf#1990-0616 – us ATLA [241]
Das herzogspaar ferdinand und julie von anhalt-koethen, die anfaenge der katholischen pfarrei koethen und der heilige stuhl : nach den akten des vatikanischen geheimarchivs / Bastgen, Hubert – Paderborn [1937] (mf ed 1993) – 1mf – 9 – €24.00 – 3-89349-252-6 – mf#DHS-AR 106 – gw Frankfurter [241]
Hes, Else *see* Charlotte birch-pfeiffer als dramatikerin
Hesbert, R J *see*
– Antiphonale missarum sextuplex
– Corpus antiphonalium officii
Heselhaus, Clemens *see*
– Annette und levin
– Annette von droste-huelshoff
– Saemtliche werke
Heselman, George J *see* Digest of the decisions...in cases related to public lands
Heseltine, Nigel *see* Remaking africa
Hesiod / Hesiod – New York, NY. 1929 – 1r – us UF Libraries [025]
Hesiod *see* Hesiod
Hesius, G *see* Guilielmi hesii antverpiensis e societate iesu emblemata sacra de fide, spe, charitate
Hesler, Heinrich von *see*
– Die apokalypse
– Apokalypse / koenigsberger apokalypse
– Das evangelium nicodemi
Hesman, Gerrit *see*
– Christelyke aandachten of vlammende ziel-zuchten
– Cupidoos mengelwerken of minnespiegel der deugden
Hesperian : a monthly miscellany of general literature, original and select – Columbus. 1838-1839 (1) – mf#3989 – us UMI ProQuest [420]
Hesperian – Portland OR: Robert A Miller, 1883-[wkly] – 1 – us Oregon Lib [071]
Hesperides / Tolkowsky, Samuel – London, England. 1938 – 1r – us UF Libraries [630]
Hesperien : ein cicerone fuer italien, vornehmlich fuer rom und neapel / Richter, Franz – Quedlinburg [u a] 1838 [mf ed Hildesheim 1995-98] – 3mf – 9 – €90.00 – 3-487-29234-3 – gw Olms [914]
Hesperien : eine symphonie / Daeubler, Theodor – Leipzig: Insel-Verlag, 1918 [mf ed 1989] – 57p – 1 – mf#7169 – us Wisconsin U Libr [810]
Hesperos – Leipzig DE, 1881 may-1888 – 4r – 1 – (in greek) – uk British Libr Newspaper [074]
Hess, Adolf [comp] *see* Christian weises historische dramen und ihre quellen
Hess, Bernd *see* Entwicklung, implementierung und anwendung einer korrelationsmethode fuer frequenzabhaengige polarisierbarkeit
Hess, Harald *see* Kommentar zur konkursordnung
[Hess, J L von] *see* Durchfluege durch deutschland, die niederlande und frankreich
Hess, Jean *see* A l'ile du diable enquete d'un reporter aux iles du salut et a cayenne
Hess, Jonas Ludwig von *see* Journal aller journale
Hess, M *see* Rom und jerusalem – die letzte nationalitaetenfrage
Hess, Mendel *see* Ausgewahlte predigten
Hess, Michael *see* Burger- und realschule der israelitischen gemeinde zu frankfurt
Hess, Moses *see* Roym un yerusholaim
Hess, Rudolph *see* Selected documents on the flight and imprisonment of rudolph hess, 1941-1945
Hesse *see* Staats anzeiger. wiesbaden. 1959-1968
Hesse, F H *see* Die entstehung der neutestamentlichen hirtenbriefe
Hesse, Friedrich Hermann *see* Der terministische streit
Hesse, Hermann *see*
– Ausgewaehlte gedichte
– Der bluetenzweig
– Neue gedichte
– Die nuernberger reise
– Wanderung
– Der wandsbecker bote
Hesse, Johann Heinrich *see* Kurze, doch hinlaengliche anweisung zum general-basse
Hesse, Johannes *see*
– Die heidenpredigt in indien
– Lao-tsze, ein vorchristlicher wahrheitszeuge
– Vom gegensgang der bibel durch die heidenwelt
Hesse, Ludwig Friedrich *see* Konrad stolles thueringisch-erfurtische chronik
Hesse. Statistisches Landesamt *see* Hessische monatszahlen
Hessel, Franz *see* Vers und prosa (klp11)
Hessel, Frederick Adam *see* Chemistry in warfare
Hessel, Karl *see* Deutsche kolonisation in ostafrika.
Hessel, Karl Robert Heinrich *see* Heinrich heines verhaeltnis zur bildenden kunst
Hesselbarth, Hermann *see* Drei psychologische fragen zur spanischen thronkandidatur leopolds von hohenzollern, mit geheimdepeschen bismarcks
Hesselgrave, Charles Everett *see* The hebrew personification of wisdom
Hesselman, Bengt *see* Huvudlinjer i nordisk sprakhistoria
Hessen, Iulii Isidorovich *see* Yehudim be-rusiyah
Hessen, J *see* Die begruendung der erkenntnis nach dem hl augustinus
Hessen, Jozef van *see* Nikolaus lenau und das junge deutschland
Hessen, Moritz von *see* Lexicum frantzoesisch und teutsch (ael2/14)
Der hessenbote – Bad Hersfeld DE, 1837 4 nov-1845, 1850 2 jan-29 jun – 3r – 1 – gw Misc Inst [074]
Hessen-darmstaedtische privilegirte landes-zeitung – Darmstadt DE, 1848-49 – 4r – 1 – (title varies: 19 aug 1806: grossherzoglich hessische landzeitung; 2 jul 1808: grossherzoglich hessische zeitung; 22 mar 1848: darmstaedter zeitung) – gw Misc Inst [074]
Hessen-kurier – Friedberg, Hessen DE, 1948 n3 (31 jul), n6 (20 aug), n8 (30 sep) – 1 – gw Mikrofilm [074]
Hessen-schaumburgische landes-anzeigen – Rinteln DE, 1818-1821 26 dec – 1r – 1 – gw Misc Inst [074]
Hessenzeitung – Marburg DE, 1862 1 mar-1866 30 jun – 2r – 1 – gw Misc Inst [074]
Hesse-Wartegg, Ernst von *see* Mexico, land und leute
Hessey, Francis *see* Christian's thank-offering
Hessey, James Augustus *see*
– Moral difficulties connected with the bible
– Moral difficulties connected with the bible. second series
– Moral difficulties connected with the bible. third series
– Sunday
Hesshusen, T *see*
– Antidotvm contra impivm et blasphemvm dogma matthiae flacii illyrici
– Bekendtnis doctoris tilemanni heshvsii von der persoenlichen vnd in alle ewigkeit vnzertrenlichen vereinigung beyder naturen in jhesu christo
– De dvabvs natvris in christo, earvmqve vnione hypostatica tractatvs
– Examen theologicvm, complectens praecipva capita doctrinae christianae, de qvibvs interrogati
– Explicatio epistolae pavli ad galatas
– Explicatio epistolae pavli ad romanos
– Explicatio prioris epistolae pavli ad corinthios
– Explicatio psalmi 110 in qva doctrina de spirituali regno
– Explicatio secvndae epistolae pavli ad corinthios
– Postilla das ist aua legung der euangelien auff alle fest vnd apostel tage durchs gantze jar
Hessisch Oldendorf Air Station [Germany] *see* Northern scanner
Hessische allgemeine (hna) – Fritzlar DE, 1977- – ca 8r/yr – 1 – (title varies: 26 jun 1991: fritzlar-homberger allgemeine (hna). ha in kassel) – gw Misc Inst [074]
Hessische allgemeine (hna) *see* Hessische nachrichten
Hessische arbeiterzeitung – Kassel DE, 1920 31 jan-18 nov – 1r – 1 – gw Misc Inst [331]
Der hessische bauer *see* Deutsche volkswacht
Hessische beytraege zur gelehrsamkeit und kunst / ed by Forster, Georg et al – Frankfurt 1785-87 – (= ser Dz) – 8st on 10mf – 9 – €100.00 – mf#k/n374 – gw Olms [000]
Hessische blaetter – Melsungen DE, 1872 15 jun-1883 28 apr, 1883 1 aug-1921 31 mar – 17r – 1 – gw Misc Inst [074]
Hessische blaetter fuer volkskunde – Leipzig: B G Teubner 1902-25 [mf ed 1979] – 24v on 3r [ill] – 1 – (cont by neue folge with title: hessische blaetter fuer volks- und kulturforschung; cont: blaetter fuer hessische volkskunde; ceased with 1973/74 iss) – mf#film mas c642 – us Harvard [390]

Hessische dorfzeitung – (Kassel-) Wehlheiden DE, 1888-1904 20 may, 1906-1910 30 jun – 39r – 1 – (title varies: 1906: neue casseler zeitung; with suppl: wilhelmshoeher fremdenblatt (1910?: cassel-wilhelmshoeher fremdenblatt 1890 10 may-1909 25 sep, 1911 20 may-1916 16 sep) – gw Misc Inst [074]
Hessische dorfzeitung – (Wildeck-) Obersuhl DE, 1922 1 apr-1923 31 mar [gaps] – 1r – 1 – gw Misc Inst [074]
Hessische heimat – Friedberg, Hessen DE, 1950 10 may-1960 24 dec – 1 – gw Mikrofilm [943]
Hessische landeszeitung – Kassel DE, 1887 1 sep-31 dec – 1r – 1 – gw Misc Inst [074]
Hessische landeszeitung *see* Generalanzeiger fur marburg und umgebung 1887
Hessische monatszahlen / Hesse. Statistisches Landesamt – Feb 1947-Dec 1955. Jan-Jun 1948, Jul-Dec 1950 wanting – 1 – us L of C Photodup [943]
Hessische morgenzeitung – Kassel DE, 1859 10 nov-1911 [gaps] – 89r – 1 – (filmed with suppl) – gw Misc Inst [074]
Hessische nachrichten – Kassel DE, 1945 26 sep-1949 – 4r – 1 – (filmed by bnl: 26 sep 1945-30 jun 1951 [12r]) – mf#6432 – gw Mikropress; uk British Libr Newspaper [074]
Hessische nachrichten – Kassel DE, 1960 1 feb-1963 17 may [gaps] – 1 – (title varies: 28 apr 1959: hessische allgemeine [hna]) – gw Misc Inst [074]
Hessische post 1889 – Kassel DE, 1889 20 nov-1891 jun, 1892-1897 31 mar – 9r – 1 – gw Misc Inst [074]
Hessische post 1945 / ed by Die Amerikanische Armee – Kassel DE, 1945 28 apr-22 sep – 1r – 1 – (14 jul 1945: hessische post / landausgabe) – mf#6431 – gw Mikropress; gw Misc Inst [074]
Hessische post und casseler stadtanzeiger *see* Casseler stadt-anzeiger
Hessische rundschau – Kassel DE, 1906 20 feb-1907 6 jan – 1r – 1 – gw Misc Inst [074]
Hessische rundschau *see* Kirchhainer zeitung
Die hessische sonntagspost – Marburg DE, 1894 4-1897 19 sep – 4r – 1 – gw Misc Inst [074]
Hessische sonntags-post – Kassel DE, 1932 6 feb-31 dec – 1r – 1 – gw Misc Inst [074]
Der hessische volksfreund – Marburg DE, 1848 22 mar-1853 29 jun – 2r – 1 – gw Misc Inst [074]
Hessische volkswacht *see* Der sturm 1930
Hessische volkszeitung 1869 – Kassel DE, 1869 6 feb-1870 31 mar – 3r – 1 – gw Misc Inst [074]
Hessische volkszeitung 1920 – Kassel DE, 1920 27 nov-1922 31 mar – 2r – 1 – gw Misc Inst [074]
Hessische zeitung *see* Spd-mitteilungsblatt
Hessische/niedersaechsische allgemeine – Kassel – 1 – (regional ed: goettingen (nur regionalseite: suedniedersachsen, oder: kreis goettingen) 1975-92 [6r]; hofgeismar (since jun 26 1991: hofgeismarer allgemeine. hna) 1988- [ca 8r/yr]) – gw Mikrofilm [074]
Hessischer beobachter – Kassel DE, 1925 17 oct-1926 – 1r – 1 – gw Misc Inst [074]
Hessischer beobachter – Marburg DE, 1924 12 apr-28 jun – 1r – 1 – gw Misc Inst [074]
Hessischer kurier : tageszeitung fuer niederhessen, oberhessen und waldeck – Paderborn DE, 1924 1 nov-1936 1 mar – 46r – 1 – gw Misc Inst [074]
Hessischer volksbote – Kassel DE, 1896 6 jan-1899 30 sep – 1r – 1 – gw Misc Inst [074]
Hessischer vorkaempfer – Hanau DE, 1924 2 may-30 jul – 1r – 1 – (aka: deutschvolk) – gw Misc Inst [074]
Hessisches landesprivatrecht / Wolf, Paul et al – Halle (Saale): Waisenhaus, 1910 – (= ser Civil law 3 coll; Das buergerliche recht des deutschen reiches und preussens) – 7mf – 9 – (incl bibl ref and index) – mf#LLMC 96-578 – us LLMC [348]
Hessisches nachbarrecht : kommentar / Hodes, Fritz & Dehner, Walter – Muenchen: J Schweitzer Verlag, 1986 (mf ed 1996) – 3mf – 9 – €38.00 – 3-8277-9683-7 – mf#DHS 9683 – gw Frankfurter [346]
Hessisches sonntagsblatt – Kassel DE, 1888-1895 30 jun – 2r – 1 – gw Misc Inst [074]
Hessisches tageblatt – Marburg DE, 1925, 1 oct-1933 29 apr – 13r – 1 – gw Misc Inst [074]
Hessisches tageblatt *see* Hessisches wochenblatt
Hessisches volksblatt – Melsungen DE, 1890 7 sep-1911 – 5r – 1 – gw Misc Inst [074]
Hessisches volksblatt – Frankfurt/M, Darmstadt, Offenbach DE, 1857 29 mar-1863, 1865-1866 15 jul – 5r – 1 – (title varies: 1 oct 1852: volksblatt fuer rhein und main; 1 jul 1853: volksfreund fuer das mittlere deutschland. began in darmstadt, fr 2 feb 1853 in offenbach, fr 22 jun 1853 in frankfurt/m, fr 1 jul 1853 in bornheim b. frankfurt/m. filmed by misc inst: 1857 29 mar-1863, 1865-1866 15 jul) – gw Mikropress; gw Misc Inst [074]
Hessisches wochenblatt – Kassel DE, 1877 6 jan-1896 – 18r – 1 – (title varies: 16 dec 1879: hessisches tageblatt; 1881: kasseler journal. filmed with suppl) – gw Misc Inst [074]
Hessleholms tidning – Haessleholm, Sweden. 1907-08 – 1r – 1 – sw Kungliga [078]
Hessleholms tidning – Kristianstad, Sweden. 1889-1906 – 18r – 1 – sw Kungliga [078]
Hester thrale-piozzi, samuel johnson and literary society, 1755-1821 – 42r – 1 – (core of coll consists of letters, poems, translations and journals. also contains business records, sale and inventory catalogues. an appendix includes some johnson materials and the mss of david garrick, george coleman and others) – mf#C36-14800 – us Primary [420]
Hesther : erklerung vnd avsslegung ueber das buoch hesther in 47 kurtze predigen... / Lavater, L – Zuerych, Christoffel Froschower, 1583 – 6mf – 9 – mf#PBU-322 – ne IDC [240]
Hestia – Athenai, Greece, 1973-81 – (= ser Estia) – 1 – us CRL [949]
Hestia – Athens, Greece. 4 jan 1876-2 jul 1895 – 19r – 1 – uk British Libr Newspaper [074]
Hestia eikonographemene *see* Hestia
Hestia-vesta : ein cyclus religionsgeschichtlicher forschungen / Preuner, August – Tuebingen: H Laupp, 1864 – 2mf – 9 – 0-524-04864-9 – (incl bibl ref) – mf#1990-3426 – us ATLA [250]
Heswall and neston news (b) – [NW England] Heswall may 1974-dec 1983 – 1 – (title change) – uk MLA; uk Newsplan [072]
Heswall, st peter – (= ser Cheshire monumental inscriptions) – 3mf – 9 – £4.50 – mf#413 – uk CheshireFHS [929]
Heswall, st peter 1559-1729 – (= ser Cheshire church registers) – 3mf – 9 – £4.50 – mf#378 – uk CheshireFHS [929]
Het Leven *see* Algemeen geillustreerd weekblad
Heterocycles – Tokyo. 1978+ (1,5,9) – ISSN: 0385-5414 – mf#11755 – us UMI ProQuest [540]
Heterodox london : or, phases of free thought in the metropolis / Davies, Charles Maurice – London: Tinsley, 1874 – 2mf – 9 – 0-7905-5648-0 – mf#1988-1648 – us ATLA [200]
Heterodoxical voice – v1 n1-17 [1968 mar-1969 nov] – 1r – 1 – mf#769594 – us WHS [071]
Heth and moab : explorations in syria in 1881 and 1882 / Conder, Claude Reignier – London: Richard Bentley, 1883 – 2mf – 9 – 0-524-05666-8 – mf#1992-0516 – us ATLA [915]
Hetherington, Clark W *see* School program in physical education
Hetherington, W M *see* History of the church of scotland
Hetherington, William Maxwell *see*
– Coleridge and his followers
– History of the church of scotland
– History of the westminster assembly of divines
Hetherwick, Alexander *see*
– A practical handbook of the nyanja language
– Practical manual of the nyanja language
Hethitische staatsvertraege : ein beitrag zu ihrer juristischen wertung / Koresec, V – Leipzig, 1931 – 3mf – 8 – €7.00 – ne Slangenburg [341]
Heti hirek – London, UK. 1 Feb-20 Dec 1957 – 1 – uk British Libr Newspaper [072]
Hettinger, Franz *see*
– David friedrich strauss
– Die "krisis des christenthums"
– Natural religion
– Revealed religion
– Timothy
The hettinger headlight – Hettinger, Adams Co, ND: Arthur A Brundage. v1 n1 may 17 1907-v2 n47 apr 2 1909 (wkly) – 1 – (missing: 1908 feb 21, mar 13, jul 3, sep 11; 1909 feb 12; absorbed by: adams county record (hettinger, nd)) – mf#10210++ – us North Dakota [071]
The hettinger journal – Hettinger, Adams Co, ND: Journal Printing Co. dec 28 1912 -v7 n34 jul 31 1919 (wkly) [mf ed with gaps] – 1 – (official city paper, may 1916-jul 1919. missing: 1913 jan 11-18, jul 3; 1914 nov 12, dec 23 1915-nov 28 1918; 1917 apr 5) – mf#03651-03653 – us North Dakota [071]
The hettinger tribune – Hettinger, Adams Co, ND: Peter H Volbach. v1 n1 jan 14 1926-v4 n26 jun 27 1929 (wkly) – 1 – (absorbed by: adams county record (hettinger, nd)) – mf#03722-03723 – us North Dakota [071]
Die hettiter *see* The hittites
Hettner, Hermann *see*
– Geschichte der deutschen literatur im achtzehnten jahrhundert
– Das moderne drama
Hetzenauer, Michael *see*
– Epitome exegeticae biblicae
– Introductio in librum genesis
Heu, Yong Mi *see* The way of faith illustrated
Heubach, Helga *see*
– Die faserpflanze flachs/lein
– Die faserpflanze hanf
– Die fruechte ihrer haende
– Ich spinne meine aussteuer
Heuberger bote – Spaichingen DE, 1975- – 116r until 1990 – 1 – (bezirksausgabe von schwaebische zeitung, leutkirch) – gw Misc Inst [074]
Heude, W *see* A voyage up the persian gulf, and a journey overland from india to england in 1817
Heuer, Otto *see* Das werden der faustdichtung goethes
Heuermann, Adolf *see* Goethe in meinem leben
Heugh, Hugh *see*
– Civil establishments of religion unjust in their principle...
– Considerations on civil establishments of religion
– Irenicum
Heuglin, M T von *see*
– Reise in das gebiet des weissen nil und seiner westlichen zufluesse in den jahren 1862-1864
– Reise in nordost-afrika
– Reise nach abessinien, den gala-laendern, ost-sudan und chartum in den jahren 1861 und 1862
Heuglin, Theodor von *see* Reise nach abessinien, den gala-laendern, ost-sudan und chartam in den jahren 1861 und 1862
Heun, Hans Georg *see* Shakespeare in deutschen uebersetzungen
Heupel, W E *see*
– De sizilische grosshof unter kaiser friedrich 2 (mgh schriften:4.bd)
– De sizilische grosshof unter kaiser friedrich 2 (mgh schriften:4.bd)
L'heure : journal republicain du soir – Paris. janv-juin 1917 – 1 – fr ACRPP [073]
Heure – Paris, France. 1 may 1918-11 aug 1919 – 1r – 1 – uk British Libr Newspaper [072]
Heure avant / Dolley, Georges – Paris, France. 192-? – 1r – us UF Libraries [440]
Une heure d'adoration en faveur des ames du purgatoire – Montreal: Cadieux & Derome, 1883 [mf ed 1984] – 1mf – 9 – 0-665-45702-2 – mf#45702 – cn Canadiana [210]
Heure de folie / Desaugiers, Marc-Antoine – Paris, France. 1807 – 1r – us UF Libraries [440]
Heures africaines / Vandrunen, James – Bruxelles, Belgium. 1899 – 1r – us UF Libraries [025]
Heures de vie : pour apprendre a bien vivre et a bien prier dieu – 2e ed. [s.l: s.n.] 1826 [mf ed 1984] – 1mf – 9 – 0-665-45053-2 – (in french and latin) – mf#45053 – cn Canadiana [240]
Heures romaines, en gros caracteres : contenant les offices de la sainte vierge et des morts, pour l'usage des congreganistes – 2e ed. Quebec: Jean Neilson, 1812 [mf ed 1971] – 1r – 5 – mf#SEM16P30 – cn Bibl Nat [241]
Heures romaines, en gros caracteres : contenant les offices de la sainte vierge et des morts pour l'usage des congreganistes – Quebec: Jean Neilson, 1795 [mf ed 1971] – 1r – 5 – mf#SEM16P93 – cn Bibl Nat [241]
Heures solitaires : poesies / Lacasse, Arthur – Quebec: Action sociale, 1916 [mf ed 1999] – 3mf – 9 – 0-659-90932-4 – mf#9-90932 – cn Canadiana [810]
...L'heureuse ariv : e de tres-chrestien, tres grand et tres-juste monarque louys 13, roy de france et de navarre / Discours sur les arcs triomphaux dresses en la ville d'Aix – Aix: Jean Tholosan, 1624 – 2mf – 9 – mf#O-1577 – ne IDC [090]
Heureuse moisson, ou, le speculateur en defaut / Merle, Jean Toussaint – Paris, France. 1817 – 1r – us UF Libraries [440]
Heureux, J *see* Hagioglypta sive picturae et sculpturae sacrae antiquiores...
Heurnius, I *see* De legatione evangelica ad indos capessenda admonitio
Heurtevent, Raoul *see* Durand de troarn et les origines de l'heresie berengarienne
Heurtley, C A *see* Plain words about prayer
Heurtley, Charles Abel *see* De fide et symbolo
Heusch, Luc De *see* Rwanda et la civilisation interlacustre
Heuschele, Otto *see*
– Die fuerstin
– Die generalin
– Leonore
– Verse der liebe
– Die wandlung
Heuschkel, Walter *see* Untersuchungen ueber ramlers und lessings bearbeitung von sinngedichten logans
Heuser, Herman Joseph *see* Chapters of bible study
Heuser, Otto Ludwig *see* Sachen- und quellen-register zu von savigny's system des heutigen roemischen rechts
Heusler, Andreas *see*
– Institutionen des deutschen privatrechts
– Nibelungensage und nibelungenlied
Heusner, William *see* The effects of exercise training and severe caloric restriction on lean-body mass in the obese
Heuss, Theodor *see*
– Johann peter hebel
– Zwischen gestern und morgen
Heussen, J van *see* Oudheden en gestichten
Heusser, Theodor *see*
– Evangelienharmonie
– Griechische syntax zum neuen testament
Heussi, Karl *see*
– Johann lorenz mosheim
– Kompendium der kirchengeschichte
– Untersuchungen zu nilus dem asketen
– Der ursprung des moenchtums
Heustecher : roman / Burkhardt, Max – Leipzig: Verlag des Bibliographischen Instituts, 1920 [mf ed 1989] – 222p – 1 – mf#7095 – us Wisconsin U Libr [830]
Heute – Muenchen DE, 1946-1947 15 jan, 1947 1 feb-1948 1 sep, 1948 15 sep-1950 1 feb – 3r – 1 – gw Misc Inst [074]
Heute bei uns zu haus : ein anderes buch erfahrenes und erfundenes / Fallada, Hans – Stuttgart; Berlin: Rowohlt, c1943 – 1r – 1 – us Wisconsin U Libr [880]
Heute und morgen : wochenschrift fuer politik, wirtschaft und kultur – Sevres (F), 1934 sep-1936 oct – 2r – 1 – gw Misc Inst [074]
Heute und morgen/anitfaschistische revue *see* Chug kreis der buecherfreunde
Die heutige auffassung und behandlung der kirchengeschichte : fortschritte und forderungen. ein konferenz-vortrag / Schubert, Hans von – Tuebingen: JCB Mohr, 1902 – 1mf – 9 – 0-524-01534-1 – mf#1990-0440 – us ATLA [240]
Der heutige stand der roemischen rechtswissenschaft; erreichtes und erstrebtes / Wenger, Leopold – Munchen, Beck, 1927. 113 p. LL-4016 – 1 – us L of C Photodup [340]
Die heutigen auffassungen vom neuprotestantismus / Stephan, Horst – Giessen: A Toepelmann, 1911 – (= ser Vortraege der theologischen Konferenz zu Giessen) – 1mf – 9 – 0-524-02412-X – (incl bibl ref) – mf#1990-0615 – us ATLA [242]
Heuver, Gerald Dirk *see* The teachings of jesus concerning wealth
Heveningham, William *see* State papers and family documents / commonplace book
Heves ettim / Nazim, Nabizade – Istanbul: Mihran Matbaasi, 1302 [1885] – (= ser Ottoman literature, writers and the arts) – 1mf – 9 – $25.00 – us MEDOC [470]
Hevesi, Lajos *see*
– Blaue fernen
– Der zerbrochene franz
Hevesi, Ludwig *see*
– Almanaccando
– Von kalau bis saekkingen
HEW Refugee Task Force [US] *see* Doi song moi
Hewat, Matthew L *see* Bantu folk lore
Hewes, James E *see* William e borah and the image of isolation
Hewetson and Milner *see* [Hewetson and milner] illustrated furniture catalogue]
[Hewetson and milner] illustrated furniture catalogue] / Hewetson and Milner – London [1880?] – (= ser 19th c art & architecture) – 1mf – 9 – mf#4.2.1599 – uk Chadwyck [740]
Hewett, John William *see* Arrangement of parish churches considered
Hewison, James King *see*
– The covenanters
– The runic roods of ruthwell and bewcastle
Hewit, Augustine Francis *see*
– Problems of the age
– The teaching of st john the apostle to the churches of asia and the world
Hewitt, Dorothy *see* Index to otero county newspapers 1886-1900
Hewitt, James *see*
– The battle of trenton
– Three sonatas for the piano forte, opus 5
– Yankee doodle with variations
Hewitt, James Dudley Ryde *see* Creation with development or evolution
Hewitt, John *see* History and topography of the parish of wakefield and its environs
Hewitt, Theodore Brown *see* Paul gerhardt as a hymn writer
Hewitt, William *see* Papers of william hewitt, 1756-1770 (brram)
Hewlett, H G *see* Liber qui dicitur flores historiarum ab ad 1154 annoque henrici anglorum regis secundi primo (rs84)
Hewlett, Sarah Secunda *see* The well-spring of immortality
Hewlett-packard journal – Palo Alto. 1974-1997 (1) 1974-1997 (5) 1974-1997 (9) – ISSN: 0018-1153 – mf#9795 – us UMI ProQuest [621]
Hewstone Burotto, Luis *see* La fuerza mayor en el derecho mercantil.
Hexachordum seu questiones... / Raselius, Andreas – 1591 – (= ser Mssa) – lost – 9 – mfchl 66 – gw Fischer [780]

The hexaemeral literature : a study of the greek and latin commentaries on genesis / Robbins, Frank Egleston – Chicago: University of Chicago Press, c1912 – 1mf – 9 – 0-7905-3213-1 – (incl bibl ref) – mf#1987-3213 – us ATLA [221]
Der hexameter bei klopstock und voss / Linckenheld, Emil – Strassburg i.E.: C J Goeller 1906 [mf ed 1990] – 1r – 1 – (incl bibl ref. filmed with: klopstocks leben und werke / karl heinemann) – mf#2770p – us Wisconsin U Libr [430]
Hexaplarische randnoten zu isaias 1-16 : aus einer sinai-handschrift / ed by Luetkemann, Leonhard & Rahlfs, Alfred – Berlin: Weidmann, 1915 – (= ser Mitteilungen des septuaginta-unternehmens) – 1mf – 9 – 0-8370-1777-7 – mf#1987-6165 – us ATLA [221]
The hexateuch according to the revised version – Arranged by members of the Society of Historical Theology, Oxford. Ed. with introd., notes, marginal references and synoptical tables by J. Estlin Carpenter and G. Harford-Battersby. London, New York: Longmans, Green and Co., 1900. 2v – 1 – us Wisconsin U Libr [240]
The hexateuch according to the revised version / ed by Carpenter, Joseph Estlin & Harford, George – London, New York: Longmans, Green, 1900 – 7mf – 9 – 0-7905-0873-7 – (incl bibl ref) – mf#1987-0873 – us ATLA [221]
Hexem, Roger W *see* Trends in double cropping
Hexemeron dei opus / Capiton, W – Strasbourg, 1539 – 7mf – 9 – mf#PPE-105 – ne IDC [240]
Die hexen von spoek : roman / Berchtenbreiter, Maria – Werdau: O Meister, 1942 [mf ed 1989] – 159p – 1 – mf#7006 – us Wisconsin U Libr [830]
Der hexenhammer : die mittelalterliche historie von der folterung des medicus johann weyer / ed by Juhn, Kurt – New York City: F Krause 1944 [mf ed 1990] – 1r [ill] – 1 – (original-lithos by erich godal. filmed with: kunterbunt / hanns johst) – mf#2744p – us Wisconsin U Libr [830]
Die hexenprozesse und ihre gegener in tirol / Rapp, Ludwig – 2. verm Aufl. Brixen: A Weger, 1891 – 1mf – 9 – 0-524-02034-5 – mf#1990-2809 – us ATLA [240]
Hexenwahn und hexenprozess : vornehmlich im 16. jahrhundert / Paulus, Nikolaus – Freiburg i.B.: Herder, 1910 – 1mf – 9 – 0-8370-7415-0 – (incl bibl ref and index) – mf#1986-1415 – us ATLA [230]
Hexenwesen und zauberei in pommern / Jahn, Ulrich – Breslau [Wroclaw]: W Koebner, 1886 – 1mf – 9 – 0-524-02209-7 – mf#1990-2883 – us ATLA [130]
Hexham herald and northumbrian gazette – [NE England] Northumberland jan 1881-16 feb 1926 [mf ed 2004] – 40r – 1 – (missing: 1887, 1889, 1896-97, 1910) – uk Newsplan [072]
Hey, Friedrich Oskar *see* Der traumglaube der antike: ein historischer versuch
Heyck, E *see* Die kreuzzuege und das heilige land
Heyd, Guenther *see*
– Goethe als mensch und deutscher
– Ein wortweiser im faustwerk
Heydebrand, Renate von *see* Wissenschaft als dialog
Heydecker, Edward Le Moyne *see*
– Commentary on mechanic's lien law for the state of new york; chapter xlix of the general laws.
– The war revenue law of 1898
Heyden, Joseph van der *see* The louvain american college, 1857-1907
Heyden, Sebald *see*
– De arte canendi
– Der passion oder das leyden jesu christi
Heydenreich, Karl Heinrich *see*
– Kritische uebersicht der neuesten schoenen litteratur der deutschen
– Kritische uebersicht der neusten schoenen litteratur der deutschen
Heydon, John *see* The english physitians guide or a holyguide; leading the way to know all things, past, present & to come
Heydon-hayden-hyden families – 1979 jan-1984 apr – 1r – 1 – (cont: hyden families; cont by: hyden families [1997]) – mf#635158 – us WHS [929]
Heye, George Gustav *see* Exploration of a munsee cemetery near montague, new jersey
He-yehudim be-tsarfat ve-divre yemehem / Schapiro, David – Kraka, Poland. 1897 – 1r – us UF Libraries [939]
Heyer, Henri *see* L'eglise de geneve, 1555-1909
Heyes, Hermann Joseph *see* Joseph in aegypten
Heygate, Frederick William, 2nd bart *see* Ireland since 1850 and her present difficulty
Heyking, Elisabeth von *see* Briefe, die ihn nicht erreichten
Heyl, Lewis *see* Statutes of the united states relating to revenue, commerce, navigation, and the currency
Heyler, K Chr (spaeter J F Ross) *see* Archiv fuer die ausuebende erziehungskunst
Het heylich herte ver-eert aen alle godt-vrughtighe herten... / [Poirters, Adrianus] – Antwerp: C Woons, 1660 – 3mf – 9 – mf#O-3256 – ne IDC [090]
Het heylig herte ver-eert aen alle godtvrugtige herten... / Poirters, Adrianus – Antwerpen: J G J de Roveroy, n.d. – 3mf – 9 – mf#O-3149 – ne IDC [090]
Heyligh hof vanden keyser theodosius verciert... / Poirters, Adrianus – Ipere, t'Antwerpen: Joannes Baptista Moermans, Hendrick Thieullier, 1696 – 4mf – 9 – mf#O-3150 – ne IDC [090]
Heylyn, Peter *see* Historia quinqu-articularis
Heym, Georg *see* Der ewige tag
Heym, Stefan *see* Deutsches volksecho
Heym un di froy / Malitz, Charles H – NYU York, NY. 1918 – 1r – us UF Libraries [939]
Heymair, Magdalena *see*
– Das buechlein jesu sirach
– Die sonntaeglichen episteln ueber das ganze jahr
– Die sontegliche episteln vber das gantze jar: in gesangs weiss gestellt
– Die sontegliche episteln vber das gantze jar: in gesangweiss gestellt
Heymer, Juergen *see* Diagnose des technischen zustandes und des innenraumzustandes von gasleitungen unter der bedingung unvollstaendiger information
Heymland / Frisch, Daniel – New York, NY. 1947 – 1r – us UF Libraries [939]
Heymowski, Adam *see* Catalogue provisoire des manuscripts mauritaniens en langue arabe preserves en mauritanie
Heynacher, Max *see* Wie spiegelt sich die menschliche seele in goethes faust?
Heyne, Benjamin *see* Tracts, historical and statistical, on india
Heyne, Moritz *see*
– Beowulf
– Friedrich ludwig stamm's ulfilas
– Heliand
– Kleine altniederdeutsche denkmaeler
Heynen, Walter *see* Das buch deutscher briefe
Heynicke, Kurt *see* Kampf um preussen
Heyns, Maria *see* Bloemhof der doorluchtige voorbeelden
[Heyns, P] *see* Esbatiment moral, des animaux
Heyns, William *see*
– Gereformeerde geloofsleer
– Handbook for elders and deacons
– Kerkenorde der christelijke gereformeerde kerk
– Liturgiek
Heyns, Z *see*
– Emblemata
– Weg-wyser ter salicheyt
Heyse and his predecessors in the theory of the novelle / Mitchell, Robert McBurney – Frankfurt a.M.: J Baer, 1915 – 1r – 1 – us Wisconsin U Libr [430]
Heyse, Paul *see*
– Abenteuer eines blaustruempfchens
– Andrea delfin
– Aus den vorbergen
– Gesammelte werke
– Das maedchen von treppi
– Medea
– Novellen
– Der salamander
– Vetter gabriel
Heyse, Paul et al *see*
– Novellenschatz des auslandes
– A la recherche du bonheur
Heyse, Th *see* Le regime du travail au congo belge
Heyse, Theodore *see*
– Bibliographie du congo belge et du ruanda-urundi
– Index bibliographique colonial
– Le regime du travail au congo belge.
– Le regime du travail au congo belge
Heyse, Ulrich *see* Kartenspiel
Heythrop journal : a quarterly review of philosophy and theology – Oxford. 1985+ (1,5,9) – ISSN: 0018-1196 – mf#15297 – us UMI ProQuest [200]
Heyt'owm Patmich *see*
– Haithoni armeni ordinis praemonstatensis de tartaris liber
– Histoire orientale ou des tartares, de aiton, parent du roy d'armenie...
– The history of ayton
– Liber historiarvm partivm orientis...
– Parte secondo della historia del signor hayton armeno del paese
Heyward, DuBose *see* Porgy drafts
Heyward, Frank *see*
– Effect of frequent fires on chemical composition of forest soils in the longleaf pine region
– Field characteristics and partial chemical analyses off the humus layer of longleaf pine forest soil
Heywood advertiser – Heywood, England. 1865-67; 1960-74. 18 reels – 1 – uk British Libr Newspaper [072]
Heywood advertiser – [NW England] Heywood 16 jun-22 dec 1855 – 1 – uk MLA; uk Newsplan [072]
Heywood news – [NW England] Heywood 1912 – 1 – uk MLA; uk Newsplan [072]
Heywood, Oliver *see* The reverend oliver heywood, b.a. 1630-1702
Heywood, Thomas *see*
– An apology for actors (1612) by thomas heywood
– A memoir of sir benjamin heywood, baronet
Heywood, William Sweetzer *see*
– Autobiography of adin ballou, 1803-1890
– History of the hopedale community
Hezekiah and his age / Sinker, Robert – London; New York: Eyre and Spottiswoode, 1897 – 1mf – 9 – 0-7905-0335-2 – (incl bibl ref and index) – mf#1987-0335 – us ATLA [920]
Hezel, Francis X *see*
– Papers on the catholic diocese of the caroline islands
Hft reporter / Hawaii Federation of Teachers – v12 n18-19,22 [1977 jun 1-jul 11, nov 30], v12 n22,24 [1978 feb 1, jun 22] – 1r – 1 – mf#675657 – us WHS [370]
H-G, Emma *see* Syllabaire gradue et recreatif des petits enfants
Hi fi stereophonie – Stuttgart. 1977-1980 (1) 1977-1980 (5) 1977-1980 (9) – ISSN: 0018-1382 – mf#8181 – us UMI ProQuest [780]
Hi line herald – Havre, MT. 1966-1974 (1) – mf#64444 – us UMI ProQuest [071]
Hi line weekly – Hingham, MT. 1938-1952 (1) – mf#64476 – us UMI ProQuest [071]
Hi skule skrib ling *see* Seaside signal
Hialeah, florida / Warner, Lillian H – s.l, s.l? 1936 – 1r – us UF Libraries [978]
Hiapania sic – London, UK. 1912-1 Aug 1915; 1 Nov 1915-27 Jun 1916. -m. 2 reels – 1 – uk British Libr Newspaper [072]
Hiatt, Charles *see* Picture posters
Hiatt, James M *see* The voter's text book
Hiatt, Lyle S *see* Reclaimed land
Hibbard, Freeborn Garretson *see*
– The book of psalms
– Christian baptism
– Palestine
– The psalms chronologically arranged
– The religion of childhood
Hibberd, Shirley *see* Rustic adornments for homes of taste
Hibbert, Fernand *see*
– Affaire d'honneur
– Masques et visages
Hibbert journal : a quarterly review of religion, theology and philosophy – London. 1902-1967 (1) 1964-1964 (5) 1964-1964 (9) – mf#692 – us UMI ProQuest [200]
Hibbert-Ware, George *see*
– Christian missions in the telugu country
– Mass movements in india
Hibbert-Ware, Samuel *see* A description of the shetland islands
Hibernation of the cotton boll weevil under controlled temperature / Grossman, Edgar F – Gainesville, FL. 1931 – 1r – us UF Libraries [630]
Hibernia – Dublin 1937-oct 1980 – 21r – 1 – ie National [072]
Hibernia magazine and dublin monthly panorama – Dublin. 1810-1811 (1) – mf#5566 – us UMI ProQuest [420]
Hibernian advertiser *see* Clonmel gazette
Hibernian chronicle [cork mercantile chronicle] – Cork 1769-70, 1772-97, 1799-apr 1818, 1823, 1825-28 – 31r – 1 – ie National [072]
Hibernian horrors : or, the nemesis of faction / Austin, Alfred – London, 1880 – (= ser 19th c ireland) – 1mf – 9 – mf#1.1.1905 – uk Chadwyck [330]
Hibernian journal – Dublin. Ireland. 1773-76, jan-1 apr, 29 jun-dec 1778, 1780, 1781, 28 aug 1782-11 jul 1783, 7 nov-dec 1783, 1784-1801, 21 jul, 23 sep 1803, nov, dec 1804, 1805-08 – 9r – 1 – (aka: hibernian journal: or chronicle of liberty) – uk British Libr Newspaper [072]
Hibernian journal : or, chronicle of liberty – Dublin 1771-82, 1784, 1786-1806, 1809-11, 1820-21 – 35r – 1 – ie National [072]
Hibernian society minutes – 1827-1968 [mf ed 1981] – 14r – 1 – mf#45-232-244 – us South Carolina Historical [366]
Hibernicus, Thomas *see* Flores omnium doctorum illustrium...
The hiberno-argentine review : a catholic weekly – Buenos Aires: Hiberno-Argentine Review, v1 n33-v5 n216 dec 7 1906-jun 1910; v13 ns: n33 64,68-69,71,73-74,76-77,79-102 jan 9, aug 13, sep 10-17, oct 1, 15-22, nov 5-12 1920; nov 26 1920-may 6 1921; v17 ns: n218-278 aug 1923-jun 1924 – 6r – 1 – us CRL [079]
Hibiscus coaster – sep-dec 1981; 1982 – 2r – (titles changes to: the coaster) – mf#12.14 – nz Nat Libr [079]
Hibiscus coaster *see* Coaster
Hic! / Tevfik, Neyzen – Istanbul: Mahmud Bey Matbaasi, 1919 – (= ser Ottoman literature, writers and the arts) – 1mf – 9 – $25.00 – us MEDOC [470]
Hicaz – (= ser Vilayet salnames) – 9 – (1301 [1884] 4mf $60; 1305 [1888] 4mf $60; 1306 [1889] 5mf $751309 [1892] 5mf $75) – us MEDOC [956]
Hick, Barbara A *see* Freshman eligibility in intercollegiate athletics
Hickathrift, Thomas *see* The history of thomas hickathrift
Hickernell, Warren Fayette *see* Manipulation and market leadership
Hickey, Kathleen P *see* A comparison of division ia football players' grades in season and out of season
Hickey, W *see* The constitution of the united states
Hickey's bengal gazette – Calcutta, India. jul 1780-mar 1782 [wkly] – 1r – 1 – uk British Libr Newspaper [079]
The hickleton papers, 1800-1885 : from the archives of the earl of halifax, garrowby – 35r – 1 – (with ind 1r 96905) – mf#96761 – uk Microform Academic [941]
Hickman baptist church. hickman, tennessee : church records – Sept 1828-Dec 1980 – 1 – 49.14 – us Southern Baptist [242]
Hickman baptist church. tennessee : church records – Sep 1828-Dec 1980. Lacks Nov 1851-Aug 1870. 1092p – 1 – 49.14 – us Southern Baptist [242]
The hickman enterprise – Hickman, NE: E F Fassett, 1886-v67 n9. jul 27 1973 (wkly) [mf ed 1891-1952,1973- (gaps)] – 37r – 1 – (suspended with mar 28 1952; resumed with jun 22 1973. not publ jun 29 1973) – us NE Hist [071]
Hickman, James Thomas *see* Crisis on the high plains: a study of amarillo baptists, 1920 to 1940
The hickman republican – Hickman, NE: F E La Grave. v1 n1. dec 18 1896- (wkly) [mf ed -1897 (gaps) filmed [1979]] – 1r – 1 – us NE Hist [071]
Hickmann, Hans *see* La danse aux miroirs
Hickok, Dorothy Jane *see* Flute method for children
Hickok, Laurens Perseus *see*
– Creator and creation
– Empirical psychology
– Humanity immortal
– Rational cosmology
– Rational psychology
– A system of moral science
Hickory cove baptist church. rogersville, tennessee : church records – 1820-1948 – 1 – us Southern Baptist [242]
Hickory grove baptist church. lawnes county. fort deposit, alabama : church records – 1844-56 – 1 – 7.29 – us Southern Baptist [242]
Hicks, Edward *see*
– Memoirs of the life and religious labors of edward hicks, late of newtown, bucks county, pennsylvania
– Sermons
– Traces of greek philosophy and roman law in the new testament
Hicks, Edward Lee *see*
– Henry bazely
– Manual of greek historical inscriptions
Hicks, Elias *see*
– Journal of the life and religious labours of elias hicks
– Letters of elias hicks
– A series of extemporaneous discourses
– Sermons
Hicks, Frederick Charles *see* Aids to the study and use of law books: a selected list, classified and annotated, of publications relating to law literature, law study and legal ethics
Hicks, George Edgar *see* A guide to figure drawing
Hicks, George Elgar *see* Caste or christ?
Hicks, J W *see* Establishment of the church in england
Hicks, James Ernest *see* What you should know about our arms and weapons
Hicks, Lewis Ezra *see* A critique of design-arguments
Hicks newsletter – v1 iss 1-v12 iss 4 [1972 jan-1983 fall], 1984 oct-1985 jun – 1r – 1 – mf#370649 – us WHS [071]
Hicks, Robert Drew *see* Stoic and epicurean
Hicks, Wesley Jones *see* The tennessee manual of chancery practice
The hicksite quakers and their doctrines / DeGarmo, James M – New York: Christian Literature, 1897 – 1mf – 9 – 0-524-05875-X – mf#1990-5169 – us ATLA [243]
Hickson, Mary Agnes *see* Selections from old kerry records
Hicky's bengal gazette or calcutta advertiser, 1780-82 – 1r – 1 – mf#390 – uk Microform Academic [079]
Hidalgo, Carlos F *see* Estructura economica y banca central
Hidalgo de Aguero, B *see* Tesoro de la verdadera cirugia...
Hidalgo, Diego *see* Un notario espanol en rusia
Hidalgo, Dionisio *see* Diccionario general de bibliografia espanola
Hidalgo, Enrique Agustin *see* Latas y latones, poesias
Hidalgo, Juan Francisco *see* Apunte descriptivo de la serena

Hidalgo. Mexico (State) *see* Periodico oficial del gobierno del estado de hidalgo
Hidalgo, ou la grande aventure / Coradin, Jean – Port-Au-Prince, Haiti. 1945 – 1r – us UF Libraries [972]
Hidalgo y Costilla, Miguel *see* Glorias dominicanas en su esclarecido, e ilustre militar tercer orden
Hidalguia extremena / Munoz Gallardo, Juan Antonio – (Familia Morales-Arce: Condes de Torre-Arce, Condes de Casa Ayala). Badajoz: Imprenta Diputacion Provincial, 1976 – 1 – sp Bibl Santa Ana [946]
Hidatsa shrine and the beliefs respecting it / Pepper, George Hubbard – Lancaster, PA. 1908 – 1r – us UF Libraries [025]
The hidden church of the holy graal : its legends and symbolism considered in their affinity with certain mysteries of initiation... / Waite, Arthur Edward – London: Rebman, 1909 [mf ed 1993] – 2mf – 9 – 0-524-08164-6 – (with app) – mf#1991-0294 – us ATLA [390]
The hidden hand / Southworth, Emma Dorothy Eliza Nevitte – New York: Grosset & Dunlap, (1920?). 487p – 1 – us Wisconsin U Libr [830]
A hidden jewel : short sketch of the life and work of rev john alexander frey as he is known to the writer for thirty years / Kweetin, John – New York, 1920 – 1 – (1 of 5 items on a reel) – mf#6299c – us Southern Baptist [242]
Hidden life / Courtenay, C L – Eton, England. 1870 – 1r – us UF Libraries [240]
Hidden saints : life of soeur marie, the workwoman of liege / Caddell, Cecilia Mary – New York: D & J Sadlier; Boston: P H Brady, 1870 – 1mf – 9 – 0-8370-7450-9 – mf#1986-1450 – us ATLA [920]
The hidden teaching beyond yoga / Brunton, Paul – New York: EP Dutton & Co, 1941 – (= ser Samp: indian books) – us CRL [110]
Hidden valley journal / Escondido Genealogical Society – v6 n1-v11 n3 [1982 may-1987 nov] – 1r – 1 – (cont: hidden valley quarterly) – mf#1685357 – us WHS [929]
Hidden valley newsletter / Escondido Genealogical Society – v1 n1-4 [1977 feb-nov] – 1r – 1 – (continued by: hidden valley quarterly) – mf#1685353 – us WHS [929]
Hidden valley quarterly / Escondido Genealogical Society – v2 n1-v3 n4 [1978 feb-1979 nov], v4 n1-3 [1980 feb-aug], v5 n1-4 [1981 feb-nov] – 1r – 1 – (cont: hidden valley newsletter; cont by: hidden valley journal) – mf#1685355 – us WHS [929]
The hidden word : thirty devotional studies of the parables of our lord / Dover, Thomas Birkett – New York: James Pott, 1887 – 1mf – 9 – 0-8370-2959-7 – mf#1985-0959 – us ATLA [220]
Hidden words : words of wisdom and communes from the supreme pen of baha'u'llah = Kalimat al-maknunah / Bahar Allah – Chicago, IL, USA: Bahai Pub. Society, 1905 – 1mf – 9 – 0-524-01157-5 – (in english) – mf#1990-2233 – us ATLA [290]
The hidden years at nazareth / Morgan, George Campbell – New York: Fleming Revell, c1898 – 1mf – 9 – 0-524-05736-2 – mf#1992-0579 – us ATLA [220]
Hi-desert flyer – 1981 may [v10 n17]-1982, 1983 oct 14, 1984 mar 16-1985 may 31, 1985 jun 7-1986 apr 25, 1986 mar 2-1987 may – 4r – 1 – mf#651525 – us WHS [071]
Hidot ha-hagadot ha-nifla'ot / Laser, Simeon Menachem – Drohobycz, Poland. 1908 – 1r – us UF Libraries [939]
Hidrocilo paredes / Paredes Guillen, Vicente – 1891 – 9 – sp Bibl Santa Ana [946]
Hidroelectrica Espanola *see*
– Curso de instructores en higiene y seguridad del trabajo
– Curso de instructores en higiene y seguridad del trabajo. 199th curso de monitores de h.e. gabriel y galan. caceres, julio 1977
– Programa del curso...formacion para miembros de comite de seguridad...y vigilantes de seguridad de la obra del salto de cedillo
Hidroelectrica Espanola, S.A. *see* Salto de cedillo. servicio de medicina y seguridad de h.e. 122nd curso de formacion de monitores de seguridad
Hidrologia y climatologia / Santamarina, Victor – Habana, Cuba. 1937 – 1r – us UF Libraries [972]
Hidup katolik – Djakarta, 1947-1971 – 183mf – 9 – (missing: 1947-1959, v1-13(1-2, 4, 6-52); 1961, v15(1-27), 29-50, 52); 1962, v16(1-25, 28, 30-31, 34-35, 37-43, 47, 49, 51-52); 1963, v17(5, 7-52); 1964, v18(1-19, 21-31, 33-36, 38, 42, 45-48, 50-51); 1965, v19(1-16, 18-35, 40-43)) – mf#SE-368 – ne IDC [950]
Hidushe avi / Anixter, Judah Eliezer – Chicago, IL. 1903 – 1r – us UF Libraries [939]
Hidushe ha-rashba / Adret, Solomon Ben Abraham – Jerusalem, Israel. 1930 – 1r – us UF Libraries [939]
Hidushe ha-rim 'al shalosh bavot / Alter, Isaac Meir – Warsaw, Poland. 1880 – 1r – us UF Libraries [939]
Hidushe haviva / Cohen, Liber – New York, NY. 1915 – 1r – us UF Libraries [939]
Hie babel, hie bibel / Klausner, Max Albert – Berlin, Germany. 1903 – 1r – us UF Libraries [939]
Hieber, Robert *see* Nine masses of hans leo hassler
Hiecke, Katharina *see*
– Exzitonentransfer in semimagnetischen cdte/cdmnte-doppel-quantengrabenstrukturen
– Untersuchungen zum exzitonentransfer in asymmetrischen cdte/(cd,mn)te-doppelgrabenstrukturen
Hier, aujourd'hui et demain : ou origines et destinees canadiennes / Thibault, Charles – Montreal: s.n, 1880 – 1mf – 9 – mf#24706 – cn Canadiana [971]
Hier berlin und alle deutschen sender – Berlin DE, 1936 8 mar-1941 25 may – 6r – 1 – gw Mikrofilm [380]
Hier in spanien – Benisa (E), 1978 21 jul-[gaps] – 1 – gw Misc Inst [074]
Hier ist der reichssender muenchen – Muenchen DE, 1940 n1-1944 spring – 1r – 1 – gw Misc Inst [380]
Hierarchia catholica medii aevi : sive summorum pontificum, s r e cardinalium / Eubel, C – Monasterii. v1-4. 1913-35 – (= ser Ecclesiarum antistitum series) – 9 – €180.00 – ne Slangenburg [241]
Hierarchia catholica medii aevi, sive, summorum pontificum, s.r.e. cardinalium, ecclesiarum antistitum series / ed by Eubel, Konrad – Monasterii: Sumptibus et typis Librariae Regensbergianae, 1898-1910 – 5mf – 9 – 0-7905-8217-1 – mf#1988-6117 – us ATLA [241]
La hierarchie episcopale : provinces, metropolitains, primats en gaule et germanie depuis la reforme de saint boniface jusqu' a la mort d'hincmar, 742-882 / Lesne, Emile – Lille: Facultes catholiques, 1905 – (= ser Memoires et travaux) – 1mf – 9 – 0-7905-8219-8 – mf#1988-6119 – us ATLA [240]
Hierarchies : from st. paul's school, kensington, london / Dionysius of Halicarnassus – 1759. Transcript – 1r – 1 – mf#95728 – uk Microform Academic [240]
The hierarchy of the catholic church in the united states : embracing sketches of all the archbishops and bishops from the establishment of the see of baltimore to the present time / Shea, John Dawson Gilmary – New York: Office of Catholic Publications, c1886. Chicago: Dep of Photodup, U of Chicago Lib, 1969 (1r); Evanston: American Theol Lib Assoc, 1984 (1r) – 1 – 0-8370-0445-4 – mf#1984-B109 – us ATLA [241]
Hieratic papyri from kahun and gurob / ed by Griffith, F L – London, 1898 – 4mf – 9 – mf#H-240 – ne IDC [930]
Hieratische palaeographie : die aegyptische buchschrift in ihrer entwicklung von der fuenften dynastie bis zur roemischen kaiserzeit / Moeller, G – Leipzig. 3v. 1909-1912 – 12mf – 9 – mf#NE-453 – ne IDC [956]
Hieratische papyrus aus den koeniglichen museen zu berlin : generalverwaltung leipzig – 18mf – 8 – mf#H-390 – ne IDC [956]
Hieratische und hieratisch-demotische texte der sammlung aegyptischer alterthuemer des allerhoechsten kaiserhauses / Bergmann, E von – Wien, 1886 – 4mf – 9 – mf#NE-377 – ne IDC [956]
Hiereonymus, Saint *see* Rhetorica ad herennium...
Hieroclis synecdemus et nototiae graecae episcopatuum : accedunt nili doxapatrii notitia patriarchatuum et locorum nomina immutata – Berolini: In aedibus Friderici Nicolai, 1866 – 1mf – 9 – 0-8370-8605-1 – (texts in greek with latin translation. incl ind) – mf#1986-2605 – us ATLA [240]
Die hieroglyphen / Erman, A – Berlin, Leipzig, 1917 – 2mf – 9 – mf#NE-20382 – ne IDC [470]
Hieroglyphic bible : being a careful selection of the most interesting and important passages in the old and new testaments: regularly arranged from genesis to revelations – London: Leadenhall Press; New York: Scribner & Welford, 1888 – 1mf – 9 – 0-8370-7734-6 – mf#1986-1734 – us ATLA [220]
A hieroglyphic vocabulary to the theban recension of the book of the dead : with an index to all the english equivalents of the egyptian words / Budge, Ernest Alfred Wallis – new rev enl ed. London: K Paul, Trench, Truebner 1911 [mf ed 1993] – (= ser Books on egypt and chaldaea 31) – 1mf – 9 – 0-524-07992-7 – mf#1991-0214 – us ATLA [470]
Hieroglyphica : anders emblemata sacra / Groenewegen, H – 's-Gravenhage: Meyndert Uytwerf, 1693 – 7mf – 9 – mf#O-822 – ne IDC [090]
Hieroglyphica : oder denkbilder der alten voelker namentlich der aegypter, chaldaeer... / Hooghe, R de – Amsterdam: Arkstee und Markus, 1744 – 10mf – 9 – mf#O-914 – ne IDC [090]
Hieroglyphica : per bernardinum trebatium vicentinum de graecis translata... / Horapollo – [Basileae, 1518] – 1mf – 9 – mf#O-848 – ne IDC [090]
Hieroglyphica : seu de sacris aegyptiorum, aliarumque gentium literis commentarii / Valeriano Bolzani, G P – Lugduni: Apud Bartholomaeum Honoratum, 1586 – 12mf – 9 – mf#O-51 – ne IDC [090]
Hieroglyphica : sive antiqua schemata gemmarum anularium, quaesita moralia... / Licetus, F – Patavii: Typis Sebastiani Sardi, 1653 – 9mf – 9 – mf#O-25 – ne IDC [090]
Hieroglyphica : sive de sacris aegyptiorum aliarumque gentium literis, commentariorum libri 58 / Valeriano Bolzani, G P – Francofurti ad Moenum: Sumptibus Christiani Kirchneri; Typis Wendelini Moewaldi, 1678 – 22mf – 9 – mf#O-849 – ne IDC [090]
Hieroglyphica : sive de sacris aegyptiorum literis commentarii... / Valeriano Bolzani, G P – Basileae: [M. Isengrin], 1556 – 16mf – 9 – mf#O-50 – ne IDC [090]
Hieroglyphica... / Valeriano Bolzani, G P – Lugduni: Sumptibus Pauli Frelon, 1602 – 13mf – 9 – mf#O-52 – ne IDC [090]
Hieroglyphica horapollinis : a davide hoeschelio fide codicis augustani ms... / Horapollo – Augustae Vindelicorum, 1595 – 3mf – 9 – mf#O-40 – ne IDC [090]
Hieroglyphica of merkbeelden der oude volkeren / Hooghe, R de – Amsterdam: Joris van der Woude, 1735 – 12mf – 9 – mf#O-298 – ne IDC [090]
Hieroglyphica, per bernardinum trebatium vicentinum de graecis translata... / Horapollo – [Basileae, 1518] – 1mf – 9 – mf#O-848 – ne IDC [700]
The hieroglyphics of horapollo nilous / Horapollo – London: William Pickering, 1840 – 3mf – 9 – mf#O-09 – ne IDC [090]
Hieroglyphisch-demotisches woerterbuch / Brugsch, Heinrich Karl – Leipzig, 1867-1882. 7 v – 58mf – 9 – mf#NE-20001 – ne IDC [470]
Hieroglyphische urkunden der griechisch-roemischen zeit / Sethe, K – Leipzig, 1904. pt – 6mf – 8 – mf#303 – ne IDC [956]
Hierographie : archeologie et histoire religieuse / Goblet d'Alviella, Eugene, Comte – Paris: Paul Geuthner, 1911 – 1mf – 9 – 0-524-02301-8 – mf#1990-2924 – us ATLA [200]
Hierologie : questions de methode et d'origines / Goblet d'Alviella, Eugene, Comte – Paris: Paul Geuthner, 1911 – 1mf – 9 – 0-524-02205-4 – mf#1990-2879 – us ATLA [200]
Hierologus : or, the church tourists / Neale, John Mason – London: James Burns, 1843 – (= ser 19th c art & architecture) – 4mf – 9 – mf#4.1.134 – uk Chadwyck [700]
Hieron, S *see*
– A defence of the ministers reasons
– A dispute upon the question of kneeling in the acte of receiving the sacremental bread and wine, proving it to be unlawfull
– The second parte of the defence of the ministers reasons for refusal of subscription and conformitie to the book of common prayer
Hieronymi graeca in psalmos fragmenta / Waldis, Johann Joseph Klemens – Muenster i W: Aschendorff, 1908 [mf ed 1993] – (= ser Alttestamentliche abhandlungen 1/3) – 1mf – 9 – 0-524-08204-9 – mf#1992-1173 – us ATLA [221]
Hieronymi magii de tintinnabvlis liber postvmvs : franciscvs sweertivs f antuerp notis illustrabat / Maggi, Girolamo – Hanovi: Typis Wechelianis, apud C Marnium & heredes I Aubrii 1608 [mf ed 19–] – 1r – 1 – mf#film 1224 – us Sibley [780]
Hieronymous – v1 n1 [1970 jul 28] – 1r – 1 – mf#1583853 – us WHS [071]
Hieronymus *see*
– Ausgewaehlte briefe (bdk18 2.reihe)
– Ausgewaehlte briefen, 2. bd (bdk16 2.reihe)
– Ausgewaehlte historische, homiletische und dogmatische schriften, 1. bd (bdk15 1.reihe)
– Contra iohannem (sl 79a). altercatio luciferiani et orthodoxi (sl 79b)
– Contra rufinum
Hieronymus emser : ein lebensbild aus der reformationsgeschichte / Kawerau, Gustav – Halle: Verein fuer Reformationsgeschichte 1898 [mf ed 1990] – (= ser Schriften des vereins fuer reformationsgeschichte 15/61) – 1mf – 9 – 0-7905-4884-4 – (incl bibl ref) – mf#1988-0884 – us ATLA [242]
Hieronymus liber de viris inlustribus / Richardson, E C – Leipzig, 1896 – (= ser Tugal 1-14/1a) – 3mf – 9 – €5.00 – ne Slangenburg [240]
Hieronymus liber de viris inlustribus; gennadius liber de viris inlustribus – der sogenannte sophronius = De viris illustribus / Jerome, Saint; ed by Richardson, Ernest Cushing & Gebhardt, Oscar von – Leipzig: JC Hinrichs, 1896 – (= ser Tugal) – 1mf – 9 – 0-7905-1668-3 – (in greek and latin) – mf#1987-1668 – us ATLA [240]
Hieronymus, Saint *see*
– Epistolae
Hierophant : or, monthly journal of sacred symbols and prophecy – New York. 1842-1843 (1) – mf#3990 – us UMI ProQuest [240]
Hierosophie : problemes du temps present / Goblet d'Alviella, Eugene, Comte – Paris: Paul Geuthner, 1911 – 1mf – 9 – 0-524-02302-6 – (incl ind to the series) – mf#1990-2925 – us ATLA [200]
Hierro – Bilbao, Spain. -d. 11 Feb 1943-26 May 1945. Imperfect. 8 reels – 1 – uk British Libr Newspaper [074]
Hierurgia anglicana : documents and extracts illustrative of the ceremonial of the anglican church after the reformation – new rev enl ed. London: De La More Press 1902-04 [mf ed 1992] – (= ser The library of liturgiology and ecclesiology for english readers 1,3,5) – 3v on 3mf – 9 – 0-524-03705-1 – mf#1990-4810 – us ATLA [242]
Hietzinger, Carl B von *see* Statistik der militaergraenze des oesterreichischen kaiserthums
Hi-fi news – Croyden, 2000+ [1,5,9] – (cont: hi-fi news and record review) – mf#2982,01 – us UMI ProQuest [780]
Hi-fi news and record review – Croyden. 1968-2000 (1) 1971-2000 (5) 1974-2000 (9) – ISSN: 0142-6230 – mf#2982 – us UMI ProQuest [780]
Hifz al-sihhah – sal-i 1, shumarah-i 3-9. rabi' al-lani-shavval 1324 [may 1906-nov 1906] – 1r – 1 – $75.00 – us MEDOC [079]
Higdon family newsletter – n1-210 [1972 jan-1989 jun] – 1r – 1 – mf#1609805 – us WHS [929]
Higgens, A W B *see* Here and there in south india
Higgin, Louis *see*
– Art as applied to dress
– Handbook of embroidery
Higginbottom, J *see* Alcohol as a medicine
Higgin's court of civil appeals reports / Tennessee. Supreme Court – v1-8. 1911-1919 (all publ) – (= ser Tennessee Supreme Court Reports) – 64mf – 9 – $96.00 – (a pre-nrs title) – mf#LLMC 91-041 – us LLMC [347]
Higgins, George Henry *see* Procedure act of 1887, and section five of act of march 21st, 1806.
Higgins, Godfrey *see* Anacalypsis
Higgins, James F *see*
– Shop plans of useful woodworking equipment for our farms
– Survey method for determining course content of farm shop for agricultural classes
Higgins, Kathleen L *see* Validity and objectivity of a rating scale for the overhead and forearm volleyball pass
Higginson, Edward *see* Christ imitable
Higginson, T W *see*
– Poems
Higginson, Thomas Wentworth *see*
– A book of american explorers
– Book of american explorers
– Contemporaries
– Papers
– Women in christian civilization
Higginsville first baptist church. higginsville, missouri : church records – 1910-64 – 1 – us Southern Baptist [242]
Higgs, Leslie *see* Presenting nassau
High and mighty – v1-2 n2 [1970 feb 13-oct 26] – 1r – 1 – mf#1058165 – us WHS [071]
The high anglican claim and its grounds / Woods, Henry – San Francisco, Calif: Monahan, 1901 – 1mf – 9 – 0-524-04978-5 – mf#1990-1381 – us ATLA [241]
The high calling : meditations on st. paul's letter to the philippians / Jowett, John Henry – New York: Fleming H Revell, c1909 – 1mf – 9 – 0-8370-3807-3 – mf#1985-1807 – us ATLA [220]
High church episcopacy : its origin, characteristics and fruits / Annan, William – Pittsburgh: R S Davis: Presbyterian Book Store [distributor], 1874 – 1mf – 9 – 0-8370-8643-4 – (incl bibl ref) – mf#1986-2643 – us ATLA [240]
High church pretensions disproved : or, methodism and the church of england / Dewart, Edward Hartley – Toronto: Methodist Book Room, 1877 – 1mf – 9 – mf#01227 – cn Canadiana [242]
High commission fiji pamphlets / Gordon, Arthur – r1 – 1 – (available for ref) – mf#pmb1214 – at Pacific Mss [324]
High commission, fiji, pamphlets / Gordon, Arthur – 1874-1881 – 1r – 1 – mf#pmb1214 – at Pacific Mss [324]
High commission territories and the republic of south africa / Doxey, G V – London, England. 1963 – 1r – us UF Libraries [960]
High commission territories and the union of south africa – London, England. 1956 – 1r – us UF Libraries [960]
High commission territories and the union of south africa / Royal Institute Of International Affairs Information Dept – London, England. 1957 – 1r – us UF Libraries [960]

High commissioner's gazette / Great Britain. High Commissioner for Aden – Mar 1963-2 Apr 1965 – 1 – us NY Public [956]

The high commissionership as connected with the progress and prosperity of south africa / Mackenzie, John – [London] 1886 – (= ser 19th c british colonization) – 1mf – 9 – mf#1.1.4703 – uk Chadwyck [327]

High country – Gallatin, MT. 1974-1974 (1) – mf#64395 – us UMI ProQuest [071]

High country herald – Timaru, NZ. 1984-87 – 2r – 1 – mf#75.12 – nz Nat Libr [079]

High court; temporary rules of procedure in cases establishing deaths based upon absences or disappearances of persons, 1984 / Lanham, John C – apr 2 1984 – 1mf – 9 – $1.50 – mf#llmc82-100i, title 18 – us LLMC [346]

High education in india : an essay read at the bethune society on the 25th april 1878 / Chandra-Natha Vasu – Calcutta, [1878] – (= ser 19th c books on british colonization) – 1mf – 9 – mf#1.1.4890 – uk Chadwyck [378]

High education in india : a plea for the state colleges / Lethbridge, Roper – London, 1882 – (= ser 19th c british colonization) – 3mf – 9 – mf#1.1.8376 – uk Chadwyck [378]

High energy chemistry – New York. 1967-1977 (1) 1967-1977 (5) – ISSN: 0018-1439 – mf#10827 – us UMI ProQuest [540]

High fidelity – New York. 1979-1989 (1) 1979-1989 (5) 1979-1989 (9) – ISSN: 0018-1455 – mf#12027 – us UMI ProQuest [621]

High fidelity/musical america – New York. 1951-1986 (1) 1968-1986 (5) 1951-1986 (9) – ISSN: 0735-777X – mf#935 – us UMI ProQuest [621]

High gauge – Tuscaloosa AL. 1970 dec 10-1971 jan 7, feb 24 – 1r – 1 – mf#1583857 – us WHS [071]

High gear : ohio's gay journal – Cleveland, Ohio. v1-10 [i.e. 9]. sep 1974-sep 1982 (mthly) – 2r – 1 – Can$275.00 – (no more publ?) – cn McLaren [305]

High hill baptist church – Chicago. 1959-1978 (1) 1970-1978 (5) 1975-1978 (9) – 1 – $45.14 – mf#1179 – us Southern Baptist [242]

High hills of santee baptist church – Chicago. (1) 1957-1972 (5) (9) – 1r – 1 – $20.75 – (incl membership rolls, bulletins, misc historical materials 1949-1975) – mf#6477 – us Southern Baptist [242]

High intensity exercise and its effects on a ten second sprint cycle test / Smith, Jason C – 1999 – 1mf – 9 – $4.00 – mf#PE 3923 – us Kinesology [790]

High, James Lambert *see*
– A treatise on extraordinary legal remedies, embracing mandamus, quo warranto and prohibition
– A treatise on the law of injunctions, as administered in the courts of the united states and england

The high jump as performed by the 1979 united states outdoor female record holder: a biomechanical analysis / Kimura, Iris F – 1981 – 1mf – 9 – $4.00 – us Kinesology [790]

High jungle / Beebe, Charles William – New York, NY. 1949 – 1r – us UF Libraries [972]

High lane (nr stockport), st thomas – (= ser Cheshire monumental inscriptions) – 1mf – 9 – £2.50 – mf#21a – uk CheshireFHS [929]

High lights and flights in new guinea...account of the discovery and development of the morobe goldfields / Rhys, Lloyd – London: Hodder and Stoughton, 1942. 252p. illus. Maps. Plates. Bibliography – 1 – us Wisconsin U Libr [980]

High peak advertiser – [East Midlands] Derbyshire oct 1881-2 jul 1937 [mf ed 2002] – 52r – 1 – (start unknown-n3198 2 jul 1937 (localised ed of glossop advertiser)) – uk Newsplan [072]

High peak chronicle – Ashton-under-Lyne, England 3,10 sep 1998 [mf 1986-98] – 1 – (cont: high peak reporter & chronicle 12 feb-27 aug 1998) – uk British Libr Newspaper [072]

High performance – Los Angeles. 1985-1997 (1) 1985-1997 (5) 1985-1997 (9) – ISSN: 0160-9769 – mf#15661 – us UMI ProQuest [700]

High performance banking – New York. 2001+ (1,5,9) – mf#20253 – us UMI ProQuest [332]

High priesthood and sacrifice : an exposition of the epistle to the hebrews / DuBose, William Porcher – New York: Longmans, Green 1908 [mf ed 1985] – (= ser The bishop paddock lectures 1907-08) – 1mf – 9 – 0-8370-2984-8 – mf#1985-0984 – us ATLA [227]

High river times – Alberta, CN. 1905- – 1r/y – 1 – Can$93.00 – cn Commonwealth Imaging [071]

High school action / High School Youth Against War & Fascism – v1 n1-5 [1973 jun-1974 apr/may] – 1r – 1 – (cont: jailbreak) – mf#370647 – us WHS [320]

High school algebra / Crawford, John Thomas – Toronto: Macmillan, 1915 – 5mf – 9 – 0-665-71330-4 – mf#71330 – cn Canadiana [510]

High school behavioral science – New York. 1975-1977 (1) 1975-1977 (5) 1975-1977 (9) – ISSN: 0148-2211 – mf#11180,01 – us UMI ProQuest [150]

The high school book-keeping : containing illustrations of the latest and best methods of keeping accounts by single and double entry / MacLean, H S – Toronto: Copp, Clark, 1890 [mf ed 1981] – 4mf – 9 – (incl ind) – mf#09554 – cn Canadiana [650]

The high school book-keeping : containing illustrations of the latest and best methods of keeping accounts by single and double entry / MacLean, H S – Toronto: Copp, Clark, 1890 [mf ed 1986] – 4mf – 9 – 0-665-39080-7 – (incl ind) – mf#39080 – cn Canadiana [650]

The high school book-keeping : containing illustrations of the latest and best methods of keeping accounts by single and double entry / MacLean, H S – Toronto: Copp, Clark, 1890 [mf ed 1986] – 3mf – 9 – 0-665-56312-4 – (incl ind) – mf#56312 – cn Canadiana [650]

High school chemistry / Ellis, William S – Toronto: Copp, Clark, c1905 [mf ed 1998] – 3mf – 9 – 0-665-85265-7 – mf#85265 – cn Canadiana [540]

High school independent press service – n1-10 [1968 sep-dec 31] – 1r – 1 – mf#1058171 – us WHS [373]

High school journal – Chapel Hill. 1918+ (1) 1971+ (5) 1975+ (9) – ISSN: 0018-1498 – mf#2141 – us UMI ProQuest [373]

The high school journal – Orillia [Ont]: High School Literary Society, [1889?-19–] – 9 – mf#P05992 – cn Canadiana [373]

High school life – Hancock WI. v1-2 n17, v5 n1-4,7 [1917 jan 16-1918 may 21, 1920 sep 27-nov 8, 1921 feb 21] – 1r – 1 – mf#1058172 – us WHS [373]

The high school monthly – New Glasgow, NS: Students of the New Glasgow High School, [1890-189- or 19–] – 9 – mf#P04780 – cn Canadiana [373]

High School Youth Against War & Fascism *see*
– High school action
– Jailbreak

High schools and sex education / U.S. Public Health Service. Surgeon General; ed by Gruenberg, Benjamin C in collaboration with U.S Bureau – Manual GPO, 1922. vii,98p. 1236 – 1r – 1 – us Wisconsin U Libr [613]

High shoals primitive baptist church – Dayton. 1967+ (1) 1972+ (5) 1975+ (9) – 1r – 1 – $42.39 – mf#6519 – us Southern Baptist [242]

High speed ground transportation journal – Calgary. 1967-1978 (1) 1975-1978 (5) 1976-1978 (9) – (cont by: journal of advanced transportation) – ISSN: 0018-1501 – mf#9815 – us UMI ProQuest [380]

High springs herald – High Springs, FL. v1 n1-v46. 1952 may-1996 – 42r – (gaps) – us UF Libraries [071]

High springs news – High Springs, FL. 1897 jul 29- – 1r – us UF Libraries [071]

High street africa / Smith, Anthony – London, England. 1961 – 1r – us UF Libraries [960]

High technology – Boston. 1981-1987 (1) 1981-1987 (5) 1981-1987 (9) – (cont by: high technology business) – ISSN: 0277-2981 – mf#12287 – us UMI ProQuest [600]

High technology business – Boston. 1987-1989 (1) 1987-1989 (5) 1987-1989 (9) – (cont: high technology) – ISSN: 0895-8432 – mf#12287,01 – us UMI ProQuest [600]

High technology law journal – Berkeley. 1986-1995 (1,5,9) – (cont by: berkeley technology law journal) – ISSN: 0885-2715 – mf#15677 – us UMI ProQuest [346]

High technology law review *see* Berkeley technology law journal

High Technology Professionals for Peace *see* Technology and responsibility

High temperature – New York. 1963-1977 (1) 1963-1977 (5) – ISSN: 0018-151X – mf#10828 – us UMI ProQuest [530]

High tides : hyde county historical society journal, north carolina / Hyde County Historical Society (NC) – v1 n1-v9 n2 [1980 spring-1988 fall] – 1r – 1 – mf#1582080 – us WHS [978]

High volume printing: hvp – Libertyville. 1992-1996 (1) – ISSN: 0737-1020 – mf#14854 – us UMI ProQuest [680]

Higham, Charles Strachan Sanders *see* Development of the leeward islands under the resto

Higham Gobion – (= ser Bedfordshire parish register series) – 1mf – 9 – £3.00 – uk BedsFHS [929]

Higham, Robert *see* Report of the engineer on the survey of the toronto and lake huron rail-road

The high-church theory of baptism – Philadelphia: TK & PG Collins, 1853 – 1mf – 9 – 0-524-03384-6 – (incl bibl ref) – mf#1990-4696 – us ATLA [242]

The high-churchman disarmed : a defense of our methodist fathers / Harrison, William Pope – Nashville, Tenn: Southern Methodist Pub House, 1886 – 2mf – 9 – 0-524-02888-5 – (incl bibl ref) – mf#1990-4479 – us ATLA [242]

The higher aspects of greek religion : lectures...apr and may 1911 / Farnell, Lewis Richard – London: Williams & Norgate 1912 [mf ed 1990] – (= ser Hibbert lectures (london, england) 1911) – 1mf – 9 – 0-7905-7572-8 – (incl bibl ref; in english, app in greek) – mf#1989-0797 – us ATLA [250]

Higher bebington, christ church – (= ser Cheshire monumental inscriptions) – 2mf – 9 – £4.00 – mf#22 – uk CheshireFHS [929]

A higher catechism of theology / Pope, William Burt – New York: Phillips & Hunt; Cincinnati: Walden & Stowe, 1884 [mf ed 1985] – 1mf – 9 – 0-8370-5486-9 – mf#1985-3486 – us ATLA [240]

The higher christian education / Dwight, Benjamin W – New York: A S Barnes & Burr, 1859 – 1mf – 9 – 0-8370-7628-5 – (incl ind) – mf#1986-1628 – us ATLA [377]

The higher christian education of women : its mission and its method: inaugural lecture / Austin, Benjamin Fish – [St Thomas, Ont?: Journal Co], 1882 – 1mf – 9 – 0-665-91635-3 – mf#91635 – cn Canadiana [376]

The higher christian life / Boardman, William Edwin – Boston: Henry Hoyt; New York: D Appleton, 1859, c1858 – 1mf – 9 – 0-7905-9138-3 – mf#1989-2363 – us ATLA [240]

Higher criticism : some thoughts on modern theories about the old testament / Ryle, John Charles – London: Chas J Thynne, [between 1880 and 1900] – 1mf – 9 – 0-524-06683-3 – mf#1992-0936 – us ATLA [220]

Higher criticism : what is it, and where does it lead us? / Sinker, Robert – London: James Nisbet, 1899 – 1mf – 9 – 0-7905-2076-1 – (incl ind) – mf#1987-2076 – us ATLA [220]

The higher criticism : four papers / Driver, Samuel Rolles – new ed. New York: Hodder and Stoughton, 1912 – 1mf – 9 – 0-8370-9379-1 – mf#1986-3379 – us ATLA [220]

The higher criticism : an outline of modern biblical study / Rishell, Charles Wesley – rev enl ed. Cincinnati: Curts & Jennings; New York: Eaton & Mains, 1896 – 1mf – 9 – 0-8370-9981-1 – (incl bibl ref) – mf#1986-3981 – us ATLA [220]

The higher criticism and a spent bible / Dwinell, Israel Edson – New York: Funk & Wagnalls 1888 [mf ed 1985] – (= ser Essays on pentateuchal criticism by various writers 9) – 1mf – 9 – 0-8370-5785-X – mf#1985-3785 – us ATLA [221]

The higher criticism and the bible : a manual for students / Boyce, William Binnington – London: Wesleyan Conference Office 1881 [mf ed 1989] – 2mf – 9 – 0-7905-2942-4 – mf#1987-2942 – us ATLA [220]

The "higher criticism" and the verdict of the monuments / Sayce, Archibald Henry – 2nd ed. London: SPCK; New York: E & J B Young, 1894 – 2mf – 9 – 0-8370-9502-6 – (incl bibl ref and index) – mf#1986-3502 – us ATLA [220]

The higher criticism of the hexateuch / Briggs, Charles Augustus – New York: Scribner's, 1893 – 1mf – 9 – 0-8370-2452-8 – (includes subject index, index of biblical passages cited and index of hebrew words and phrases) – mf#1985-0452 – us ATLA [221]

Higher education – Amsterdam. 1986+ – 1,5,9 – ISSN: 0018-1560 – mf#16040 – us UMI ProQuest [378]

Higher education : report of the committee appointed by the prime minister under the chairmanship of lord robbins, 1962-63. command n2154 i-xiii – 52mf – 9 – mf#96938 – uk Microform Academic [324]

Higher education – Shanghai: Shanghai Mercury, 1915 – 1mf – 9 – 0-524-07798-3 – mf#1991-0175 – us ATLA [378]

Higher education – Washington. 1945-1964 – 1 – mf#1025 – us UMI ProQuest [378]

Higher education abstracts – Claremont. 1984+ (1) 1984+ (5) 1984+ (9) – (cont: college student personnel abstracts) – ISSN: 0748-4364 – mf#9170,01 – us UMI ProQuest [378]

Higher education and national affairs – Washington. 1954+ (1) 1971+ (5) 1976+ (9) – ISSN: 0018-1579 – mf#5815 – us UMI ProQuest [378]

Higher education bulletin – Lancaster. 1976-1978 – 1,5,9 – mf#10997 – us UMI ProQuest [378]

Higher education daily – Alexandria. 1974-1986 (1) 1976-1986 (5) 1976-1986 (9) – ISSN: 0194-2239 – mf#10088 – us UMI ProQuest [378]

Higher education in bengal under british rule / Ghosh, J – Calcutta: Book Co, [1926] – (= ser Samp: indian books) – us CRL [378]

Higher education in florida – s.l, s.l? 193-? – 1r – us UF Libraries [978]

Higher education in london, report on the advancement of... (selbourne commission), 1889 *see* Durham university act 1861, report on the... 1862

Higher education management – Paris. 1989+ – 1,5,9 – (cont: international journal of institutional management in higher education) – ISSN: 1013-851X – mf#17076 – us UMI ProQuest [378]

Higher education management and policy – Paris. 2002+ (1,5,9) – ISSN: 1682-3451 – mf#17076,01 – us UMI ProQuest [350]

The higher education of woman : an address delivered at the opening of queen's college, kingston, canada, session 1871-72 / Murray, John Clark – Kingston Ont: s.n, 1871 – 1mf – 9 – mf#56124 – cn Canadiana [376]

The higher education of women / Barnard, F A P – New York: [s.n.], 1882 (New York: Macgowan & Slipper) – 1mf – 9 – 0-8370-7765-6 – mf#1986-1765 – us ATLA [376]

Higher education of women in europe = Frauenbildung / Lange, Helene – New York: D Appleton, 1897, c1890 – 1mf – 9 – 0-8370-7642-0 – (in english) – mf#1986-1642 – us ATLA [376]

Higher education quarterly – Oxford. 1987+ (1) 1987+ (5) 1987+ (9) – (cont: universities quarterly : culture, education and society) – ISSN: 0951-5224 – mf#675,03 – us UMI ProQuest [378]

Higher education review : a bulletin of the association for the study of higher education – Charlottesville. 1977-1978 – 1,5,9 – (cont by: review of higher education) – ISSN: 0148-9585 – mf#12665 – us UMI ProQuest [378]

Higher education review – Croydon. 1968+ (1) 1974+ (5) 1975+ (9) – ISSN: 0018-1609 – mf#10419 – us UMI ProQuest [378]

The higher hinduism in relation to christianity : certain aspects of hindu thought from the christian standpoint / Slater, Thomas Ebenezer – 2nd and rev ed. London: E Stock, 1903 – 1mf – 9 – 0-524-01301-2 – mf#1990-2337 – us ATLA [230]

The higher individualism / Ames, Edward Scribner – Boston: Houghton Mifflin, 1915 – 1mf – 9 – 0-7905-3867-9 – mf#1989-0360 – us ATLA [240]

Higher kinnerton (flintshire), all saints : baptisms 1868-1970; marriages/burials 1894-1970 – (= ser Cheshire church registers) – 2mf – 9 – £4.00 – mf#226 – uk CheshireFHS [929]

Higher law – 1861 jan 1-apr 4 – 1r – 1 – mf#1096284 – us WHS [071]

The higher law in its relations to civil government : with particular reference to slavery and the fugitive slave law / Hosmer, William – Auburn: Derby & Miller, 1852 – 1mf – us ATLA [320]

The higher law in its relations to civil government : with particular reference to slavery and the fugitive slave law / Hosmer, William – Auburn: Derby & Miller, 1852 – 1mf – 9 – 0-7905-5408-9 – mf#1988-1408 – us ATLA [976]

The higher life : its reality, experience, and destiny / Brown, James Baldwin – 2nd ed. London: Henry S King, 1874 – 1mf – 9 – 0-7905-9161-8 – mf#1989-2386 – us ATLA [240]

The higher ministry of nature viewed in the light of modern science : and as an aid to advanced christian philosophy / Leifchild, John – London, 1872 – (= ser 19th c evolution & creation) – 6mf – 9 – mf#1.1.1517 – uk Chadwyck [210]

Higher technological education, report of the committee appointed by the minister on... (percy report), 1945 – 1mf – 9 – mf#86958 – uk Microform Academic [324]

Higher-order and symbolic computation – Boston. 1998+ (1) – (cont: lisp and symbolic computation) – ISSN: 1388-3690 – mf#16823,01 – us UMI ProQuest [000]

The highest critics vs the higher critics / Munhall, Leander Whitcomb – New York: Fleming H Revell, c1892 – 1mf – 9 – 0-8370-4541-X – mf#1985-2541 – us ATLA [220]

The highest life : a story of shortcomings and a goal: including a friendly analysis of the keswick movement / Johnson, Elias Henry – New York: Armstrong, 1901 – 1mf – 9 – 0-7905-9975-9 – mf#1989-1700 – us ATLA [240]

Highland baptist church (formerly: second street baptist). shelbyville, kentucky : church records – 1949-70 – 1 – us Southern Baptist [242]

Highland baptist church. metairie, louisiana : church records – 1953-75 – 1 – us Southern Baptist [242]

Highland bote – Highland, IL: P Voegele, [mar 31 1860-65] – 1 – us CRL [071]

Highland bote und schuetzen-zeitung – Highland, IL: T Gruaz, [jan 12 1866-mar 12 1869] – 1 – us CRL [071]

Highland Cemetery, Geary County, KS *see* Interment record no. 1-4051

Highland Co. Greenfield *see*
- Daily times
- Daily times series
- Independent
- Republican series

Highland Co. Hillsboro *see*
- Daily evening gazette
- Dispatch
- Gazette
- Highland weekly news
- Highland weekly news series
- News-herald
- Ohio news
- People's press
- Press gazette
- Press-gazette
- Saturday herald

Highland Co. Leesburg *see*
- Buckeye
- Citizen

Highland Co. Lynchburg *see*
- News

Highland county atlas, 1871 – 1r – 1 – mf#B6744 – us Ohio Hist [978]

Highland courier – Newburgh, NY. 1843-1848 (1) – mf#65117 – us UMI ProQuest [071]

Highland democrat – Peekskill, NY. 1858-1887 (1) – mf#65159 – us UMI ProQuest [071]

Highland dress, arms and ornament / Campbell, Archibald, Lord – Westminster 1899 – (= ser 19th c art & architecture) – 5mf – 9 – mf#4.2.1742 – uk Chadwyck [730]

Highland eagle – Peekskill, NY. 1851-1858 (1) – mf#65160 – us UMI ProQuest [071]

Highland echo = Guth nan gaidheal – [Scotland] Glasgow: [s.n.] 10 mar 1877-2 feb 1878 (wkly) [mf ed 2003] – 48v on 2r – 1 – uk Newsplan [072]

Highland herald – Inverness: [s.n.] 1947-70 (wkly) – 1188v – 1 – uk Scotland NatLib [072]

Highland. Kansas. Highland Presbyterian Church *see* Records

Highland news – Inverness: Highland News 1984- (wkly) [mf ed 7 jan 1995-] – 1 – (cont: inverness and highland news) – ISSN: 1354-9669 – uk Scotland NatLib [072]

[Highland park-] international socialist – CA. 1969-1975 – 2r – 1 – $120.00 – mf#R03242 – us Library Micro [071]

Highland park journal – Los Angeles, CA. 1946-1956 (1) – mf#62182 – us UMI ProQuest [071]

[Highland park-] worker's power – MI. 1970-1975 – 2r – 1 – $120.00 – mf#R04391 – us Library Micro [331]

Highland pioneer – [Scotland] Glasgow: [J Morrison sep 1875]-may 1876 (mthly) [mf ed 2003] – 1r – 1 – (cont by: pioneer (glasgow, scotland)) – uk Newsplan [072]

Highland press – Highland WI. 1943 dec 3-1947 dec 26, 1948-1949 feb 4, sep 8-1951, 1952-56 – 3r – 1 – (continued by: muscoda progressive (muscoda, wi: 1920)) – mf#945087 – us WHS [071]

Highland recorder – Monterey, VA. 1889-1892 (1) – mf#68691 – us UMI ProQuest [071]

Highland sentinel – [Scotland] Inverness: Robert Maclean & William Paterson 13 jul 1861-11 nov 1863 (wkly) [mf ed 2003] – 123v on 1r – 1 – uk Newsplan [072]

Highland sport and athletic record : devoted to the pastimes, athletics, and recreations of the north of scotland – [Scotland] Inverness: printed...for Thomas William Mackenzie 11 nov 1895-17 feb 1896 (wkly) [mf ed 2003] – 1r – 1 – uk Newsplan [790]

Highland times – [Scotland] Inverness: M Macleod & Co 8 feb 1896-29 apr 1926 (wkly) [mf ed 2003] – 20r – 1 – (missing: 1902) – uk Newsplan [072]

Highland union – Highland, Madison Co, IL: G Rutz & J S Hoerner, oct 22 1868-sep 9 1910 – 1 – us CRL [071]

Highland University *see* Records

Highland valley steelworker / Lornex Mining Corporation et al – 1980 oct-1981 jul – 1r – 1 – mf#622105 – us WHS [622]

Highland weekly news / Highland Co. Hillsboro – 1875-84, jan-mar 1886 [wkly] – 5r – 1 – mf#B9918-9922 – us Ohio Hist [071]

Highland weekly news / Highland Co. Hillsboro – jan 1885-mar 1886 [wkly] – 1r – 1 – mf#B33943 – us Ohio Hist [071]

Highland weekly news series / Highland Co. Hillsboro – jan 1852-dec 1874 [wkly] – 7r – 1 – mf#B5641-5647 – us Ohio Hist [071]

Highland weekly press – Highland WI. 1899 sep 15; 1900 may 2 – 1r – 1 – us WHS [071]

Highlander – Marble Falls, TX. 1993-2000 (1) – mf#66631 – us UMI ProQuest [071]

Highlander (inverness, scotland) – Inverness: Highlander Newspaper & Print & Publ Co 1873-81 (mthly, jul 1881-jan 1882; former frequency: wkly, 1873-may 25 1881) – 10v – 1 – uk Scotland NatLib [072]

The highlands and western isles of scotland : containing descriptions of their scenery and antiquities, with an account of the political history and ancient manners, and of the origin, language, agriculture, economy, music, present condition of the people, etc... / MacCulloch, John – London 1824 [mf ed Hildesheim 1995-98] – 4v on 12mf – 9 – €120.00 – 3-487-27857-X – gw Olms [914]

Highlands county : complete / Blanton, Kelsey – s.l, s.l? 1936 – 1r – us UF Libraries [978]

Highlands county – s.l, s.l? 193-? – 1r – us UF Libraries [978]

Highlands county news – Sebring, FL. 1927 jun 23-1957 – 19r – us UF Libraries [071]

Highlands county pilot – Avon Park, FL. 1934-1935 – 1r – us UF Libraries [071]

Highlands hammock / Blanton, Kelsey – s.l, s.l? 1936 – 1r – us UF Libraries [978]

Highlands of asiatic turkey / Percy, K – London, 1901 – 5mf – 9 – mf#AR-1986 – ne IDC [956]

Highlands post – Moss Vale – at Pascoe [079]

Highlights for children – Columbus. 1987+ – 1,5,9 – ISSN: 0018-165X – mf#16354 – us UMI ProQuest [370]

Highlights in the debates in the spanish chamber o... / Yuengling, David G – Washington, DC. 1941 – 1r – us UF Libraries [972]

High-lites / Miller Brewing Co – 1974 jun-1985 dec – 1r – 1 – (cont: new high-lites; cont by: miller time) – mf#1044337 – us WHS [640]

High-speed surface craft – Kingston-Upon-Thames. 1979-1988 (1) 1979-1988 (5) 1979-1988 (9) – (cont: hovering craft and hydrofoil. cont by: fast ferry international) – ISSN: 0144-7823 – mf#2997,01 – us UMI ProQuest [629]

High-tech news – Greencastle. 1998+ (1) – ISSN: 1092-9592 – mf#22541,01 – us UMI ProQuest [621]

[Hightstown-] book and art – NJ. 1979 – 1r – 1 – $60.00 – mf#R04975 – us Library Micro [071]

[Hightstown-] chronicle of higher education – NJ. 1966-1982 – 20r – 1 – $1200.00 – mf#R04976 – us Library Micro [378]

Highview baptist church. louisville, kentucky : church records – Sep 1947-Jul 1979 – 1 – us Southern Baptist [242]

Highway – v8-10. 1946-49 – 1r – 1 – (lacking: aug 1946. cont: kimberley and kuruman diocesan magazine) – ISSN: 0018-1684 – mf#ATLA S0727A – us ATLA [240]

Highway / Northland College [Ashland, WI] – 1964 fall, 1965 spring-winter, 1966 winter-summer, 1967 winter, sum, 1968 winter-fall, 1969 winter, 1970 mar, sum/fall-1978/79 winter, v69 n1-3 [1979 spring-fall], v70 n1-v80 n4 [1980 winter-1990 fall] – 1r – 1 – (cont: northern light [ashland, wi]; cont by: northland highway) – mf#1582737 – us WHS [378]

Highway 13 / Vietnam Veterans Against the War – v1 n4-v3 n5 [1973 apr-1975 may] – 1r – 1 – mf#370657 – us WHS [320]

Highway across the west indies / Lanks, Herbert Charles – New York, NY. 1948 – 1r – us UF Libraries [380]

Highway and heavy construction – Des Plaines. 1976-1991 (1) 1976-1991 (5) 1976-1991 (9) – (cont: roads and streets. cont by: highway and heavy construction products) – ISSN: 0362-0506 – mf#273,01 – us UMI ProQuest [690]

Highway and heavy construction products – Newton. 1992-1993 (1) 1992-1993 (5) 1992-1993 (9) – (cont: highway and heavy construction. cont by: construction products) – ISSN: 1062-5194 – mf#273,02 – us UMI ProQuest [690]

Highway engineering in australia – Klemzig. 1969-1975 (1) 1975-1975 (5) 1975-1975 (9) – ISSN: 0046-7391 – mf#8606 – us UMI ProQuest [624]

A highway in the desert : a history of southern baptist work in arizona / Maxwell, C B – 1 – $5.00 – us Southern Baptist [242]

The highway of the seas in time of war / Lord, Henry William – Cambridge [England]: Macmillan, 1862 [mf ed 1984] – 1mf – 9 – 0-665-46140-2 – mf#46140 – cn Canadiana [341]

An highway there / Scofield, William Campbell – Chicago: Fleming H Revell, c1901 [mf ed 1985] – 1mf – 9 – 0-8370-5620-9 – mf#1985-3620 – us ATLA [240]

Highway user quarterly – Washington. 1963-1976 (1) 1972-1976 (5) 1976-1976 (9) – ISSN: 0094-7393 – mf#7197 – us UMI ProQuest [625]

Highways – Croydon. 1989-1990 (1) – (cont: highways + public works) – mf#10477,04 – us UMI ProQuest [624]

Highways and byways of literary criticism in sanskrit / Kuppuswami Sastri, S – Madras: Kuppuswami Sastri Research Institute, 1945 – (= ser Samp: indian books) – (foreword by v s srinivasa sastri) – us CRL [490]

Highways and road construction – Croydon. 1974-1976 (1) 1974-1976 (5) 1974-1976 (9) – (cont by: highways and road construction international) – ISSN: 0018-1773 – mf#10477,01 – us UMI ProQuest [624]

Highways and road construction international – London. 1976-1978 (1) 1976-1978 (5) 1976-1978 (9) – (cont: highways and road construction. cont by: highways + public works) – ISSN: 0308-9533 – mf#10477,02 – us UMI ProQuest [624]

Highways of florida / Florida State Road Dept Division Of Statewide Highways – Tallahassee, FL. 1936? – 1r – us UF Libraries [500]

Highways + public works – Croydon. 1978-1980 (1) 1978-1980 (5) 1978-1980 (9) – (cont: highways and road construction international. cont by: highways) – ISSN: 0142-6168 – mf#10477,03 – us UMI ProQuest [624]

Higiene anticolerica razonada – 1885 – 9 – sp Bibl Santa Ana [610]

Higiene y profilaxis en el medio rural / Juarez, Ernesto – Caceres: Imp. Garcia Floriano, 1951 – sp Bibl Santa Ana [610]

Higuera La Real.Spain.Ayuntamiento *see* Ordenanzas municipales

Una hija de maria (maria perez de guzman y sanjuan) / Gomez Bravo, Vicente – Villafranca de los Barros; Colegio de San Jose (Imp. Bolanos), 1940 – 1 – sp Bibl Santa Ana [240]

El hijo del pueblo – Manzanillo: Imp el Comercio de la viuda de A Martin. v1 n2-10. may 24-jul 19 1885 – 1 sheet – 9 – us CRL [079]

Hijo prodigo y otros poemas / Espada Marrero, J – Puerto Rico? Puerto Rico. 19- – 1r – us UF Libraries [972]

Los hijod del mar / Diaz Maciaz, Jose – 1889 – 9 – sp Bibl Santa Ana [830]

Los hijos americanos de los pizarros de la conquista / Cuneo-Vidal, Romulo – Madrid: Tip. Rev. de Arch. Bibliot. y Museos, 1925 – 1 – sp Bibl Santa Ana [060]

Los hijos de la fortuna. novela original de costumbres espanolas / Rivera, Luis; ed by Gonzalez, J de M – Madrid, 1855 – 1 – sp Bibl Santa Ana [830]

Los hijos de lutero entre los hijos del sol / Bayle, Constantino – Madrid: Razon y Fe, 1930 – 1 – sp Bibl Santa Ana [240]

Hijos de Reus *see* Fuero de usagre (siglo 18)

Hijos del tiempo / Aparicio, Raul – Habana, Cuba. 1964 – 1r – us UF Libraries [972]

Hijos ilustres de extremadura / Velo Nieto, Gervasio – Badajoz: Imprenta Diputacion Provincial, 1971 – sp Bibl Santa Ana [946]

Hijosa del Valle, Gregorio *see* Cuaderno de lenguaje curso 2-1. otono

Hikajat pandji semirang, menoeroet naskah lama : dihiasi dengan 11 boeah gambar, tjetakan 6 – Djakarta: Balai Poestaka, 2602 (serie n48) – 157p 2mf – 9 – (at head of title: dengan idzin badan pengawasan pengoemoeman) – mf#SE-2002 mf39-40 – ne IDC [959]

Hikajat sitti rabihatoen : ditjeriterakan oleh st. p. boestami. dikeloearkan oleh balai poestaka / Bustami, Perang., sutan – Melbourne: Netherlands Indies Govt Print Works [194-?] [mf ed Ithaca NY: Cornell Uni 1991] – 1r – 1 – (..."menoeroet isi dongeng djawa jang telah dibahasa belandakan oleh...ver der pant"; filmed with other items) – mf#mf-10289 seam reel 289/7 [§] – us CRL [490]

Hikayat hang tuah : die geschichte von hang tuah; aus dem malaysichen uebersetzt von h overbeck – Muenchen: Georg Mueller 1922 [mf ed 1990] – (= ser Meisterwerke orientalischer literaturen 7,8) – 2v on 1r with other items – 1 – (trans fr malay into german) – mf#mf-10289 seam reel 266/3 [§] – us CRL [390]

Hikaye / Usakligil, Halit Ziya – Kostantiniye: Istepan Matbaasi, 1307 [188-90] – (= ser Ottoman literature, writers and the arts) – 2mf – 9 – $40.00 – us MEDOC [470]

Hikmat – Cairo. v8 n7-8 (271-272), 16-17 (280-281), 19 (283). 20 rabi al-awwal 1317-20 rajab 1317 [28 jul-24 nov 1899] – 1r – 1 – $53.00 – (r also incl: parvarish and surayya) – us MEDOC [079]

Hikmat *see*
- Parvarish
- Surayya

Hikoi, Hirotaka *see* The effects of opioid receptor antagonism on plasma catecholamines and fat metabolism during prolonged exercise above or below lactate threshold in males

Hilafet siyaseti ve tuerkluek siyaseti / Vayet – Dersaadet: Ikbal Kitaphanesi, 1331 [1915] – (= ser Ottoman histories and historical sources) – 3mf – 9 – $75.00 – us MEDOC [956]

Hilaire *see* Notre-dame de lourdes et l'immaculee-conception

Hilal – Istanbul. Turkey. -d. In French. 5 Sep 1917-27 Sep, 26 Oct 1918. (Imperfect). (2 reels) – 1 – uk British Libr Newspaper [949]

Hilarion / Mendez Ballester, Manuel – San Juan, Puerto Rico. 1943 – 1r – us UF Libraries [972]

Hilarius von poitiers : eine monographie / Reinkens, Joseph Hubert – Schaffhausen: Fr Hurter, 1864 – 1mf – 9 – 0-7905-6824-1 – (incl bibl ref) – mf#1988-2824 – us ATLA [240]

Hilarius von Poitiers (Hilary of Poitiers, Saint) *see*
- Ausgewaehlte schriften, 1. bd (bdk5 2.reihe)
- Ausgewaehlte schriften, 1. bd (bdk6 2.reihe)

Hilary, Saint, Bishop of Poitiers *see* Select works – exposition of the orthodox faith

Hilbck, A *see* Zum bergarbeiterausstand im ruhrrevier

Hilbert, Carey A *see* Comparison of resting metabolic rate and excess post-exercise oxygen consumption in normal and low calorie dieting females

Hilbert favorite – Hilbert WI. 1909 jul 29/1910 dec 29-1984 oct/1986 jul [gaps] – 50r – 1 – mf#951145 – us WHS [071]

Hilbey, Constant *see* Affreuse tentative de corruption

Hilbig, Jennifer Johnson *see* The differences between physical activity levels and percent body fat using two methods of predicting percent body fat in male senior athletes

Hilborn family journal – n1-27 [1978 nov-1986 jul] – 1r – 1 – mf#1020169 – us WHS [929]

Hilda : a story of calcutta / Cotes, Everard, mrs [Sara Jeanette Duncan] – New York: F A Stokes, c1898 – 4mf – 9 – mf#05292 – cn Canadiana [830]

Hilda wade / Allen, Grant – Toronto: Copp, Clark, 1900 – (= ser Grant richard's indian and colonial library) – 5mf – 9 – (ill by gordon browne) – mf#27570 – cn Canadiana [890]

Hildebrand, Alexander *see* Frauenlobs streitgedicht zwischen minne und welt

Hildebrand and his times / Stephens, William Richard Wood – London: Longmans, Green, 1888 – (= ser Epochs Of Church History) – 1mf – 9 – 0-8370-7910-1 – (incl ind) – mf#1986-1910 – us ATLA [240]

Hildebrand and his times / Stephens, William Richard Wood – London, New York: Longmans, Green, 1914.xvi,230p. map. Includes index – 1 – us Wisconsin U Libr [240]

Hildebrand, Karl *see* Die lieder der aelteren edda (saemundar edda)

Hildebrand, P *see* De kapucijnen in de nederlanden en het prinsbisdom luik

Hildebrand pfeiffer : ein leben aus dunkler zeit / Chezy, Wilhelm von – [Bayreuth]: Gauverlag Bayreuth, 1944 [mf ed 1989] – (= ser Bayreuther feldpostausgaben) – 127p – 1 – mf#7152 – us Wisconsin U Libr [880]

Hildebrand, the builder / Smith, Ernest Ashton – Cincinnati: Jennings and Graham; New York: Eaton and Mains, c1908 – 1mf – 9 – 0-8370-7909-8 – (incl ind) – mf#1986-1909 – us ATLA [920]

Hildebrand, Ulrich *see* Die bilanzrechtlichen beschluesse der grossen senate von rfh und bfh

Hildebrand, Wolfgang *see* Magia naturalis

Hildebrandine essays / Whitney, James Pounder – Cambridge, England: University Press, 1932 [mf ed 2003] – 1r – 1 – (incl bibl & ind) – mf#b00662 – us ATLA [240]

Hildebrandlied und waltharilied / Botticher, Gotthold – Halle a.S., Germany. 1925 – 1r – us UF Libraries [430]

Hildebrandt, August *see* Juda's verhaeltniss zu assyrien in jesaja's zeit

Hildebrandt, Friedrich *see* Karpathenbilder

Hildebrandt, Kurt *see* Wagner und nietzsche

Hildegarde / Norris, Kathleen – New York, NY. 1926 – 1r – us UF Libraries [025]

Hildegarde, H *see* Voorspellingen van de h hildegarde omtrent de belgische omwenteling

Hilde-gudrun / Panzer, Friedrich Wilhelm – Halle a.S., Germany. 1901 – 1r – us UF Libraries [960]

Hildener heimatblaetter *see* Hildener zeitung

Hildener zeitung – Hilden DE, 1953 oct-dec, 1954 apr-15 jan [gaps], 1955 jun – 1r – 1 – (with suppl: hildener heimatblaetter 1950-67 [2r]) – gw Mikrofilm [074]

Hilder, Brett *see* A research tribute by the retired officers association of papua new guinea

Hilders, J H *see* Introduction to the ateso language

Hildersham 1541-1950 – (= ser Cambridgeshire parish register transcript) – 4mf – 9 – £5.00 – uk CambsFHS [929]

Hildescheimer, Meier *see* Gedachtnisrede auf rabbiner dr meier hildescheimer

Hildesheimer allgemeine zeitung *see* Hildesheimer relations-courier

Hildesheimer, Hirsch *see* Beitrage zur geographie palastinas

Hildesheimer, L *see* Alphabetisches verzeichniss der sich in j schmidt's mondcharte befindlichen objecte

Hildesheimer relations-courier – Hildesheim DE, 1706-17, 1719, 1721, 1726-27, 1748-52, 1754-93, 1795-1801, 1809 [single iss] – over 27r – 1 – (title varies: 1751: luedemannsche zeitung; 1775: privilegirte hildesheimische zeitung; 1 oct 1949: hildesheimer allgemeine zeitung. filmed by other misc inst: 1977- [ca 6r/yr]) – gw Misc Inst [074]
Hildesheimisches magazin – Hildesheim 1786-92 – (= ser Dz) – 7jge on 20mf – 9 – €200.00 – mf#k/n5726 – gw Olms [074]
Hildreth, Richard *see*
– A letter to andrews norton on miracles as the foundation of religious faith
– Theory of politics
Hildreth telescope – Hildreth, NE: W S Ashby, 1887-jul 1 1976// (wkly) [mf ed 1892,1895-1976 (gaps) filmed -[1989]] – 16r – 1 – (suspended in 1961; resumed with special souvenir ed jul 1 1976; called v1 n1) – us NE Hist [071]
Hildrop, John *see* Husbandman's spiritual companion
Hildt, J Ad et al *see* Magazin der handels- und gewerbskunde
Hildt, Johann Adolph [jg 7ff] *see* Handlungs-zeitung
Hilf Durkh Arbet *see* Yubiley-oysgabe "hilf durkh arbet"
Die hilfe : zeitschrift fuer politik, literatur und kunst – Berlin DE, 1894 2 dec-1944 17 jun – 30r – 1 – (title varies: 1901: nationalsoziales volksblatt) – mf#6105 – gw Mikropress [335]
Hilfe, die nicht ankommt : der zielgruppenbezug bei der foerderung von entwicklungsbanken als defizitbereich deutscher entwicklungspolitik / Graef, Peter Leo – (mf ed 1995) – 2mf – 9 – €40.00 – 3-8267-2208-6 – mf#DHS 2208 – gw Frankfurter [332]
Hilfe fuer die not der kranken in china : die arbeit der aerztlichen mission des allg ev-prot missionsvereins / Witte, Johannes – Berlin-Schoeneber: Protestantischer Schriftenvertrieb [1911] [mf ed 1995] – (= ser Yale coll) – 58p (ill) – 1 – 0-524-10119-1 – (in german) – mf#1995-1119 – us ATLA [362]
Hilfskomitee fuer die evangelischen aus Danzig-Westpreussen *see* Danzig-westpreussischer kirchenbrief
Hilfsverein Der Deutschen Juden *see* Jahresbericht fur 1931
Hilfsverein Der Deutschen Juden (Germany) *see* Dreissig jahre
hilfsverein fuer die notleidende juedische bevoelkerung in galizien *see* Hillfsverein der deutschen juden
Hilfverein Deutscher Frauen *see* Weltkrieg!
Hilgardia – Berkeley. 1989-1991 (1,5,9) – ISSN: 0073-2230 – mf#17561 – us UMI ProQuest [630]
Hilgenfeld, A *see*
– Acta apostolorum
– Historisch-kritische einleitung in das neue testament
– Zeitschrift fuer wissenschaftliche theologie
Hilgenfeld, Adolf *see*
– Acta apostolorum graece et latine
– Die apostolischen vaeter
– Bardesanes
– Die clementinischen recognitionen und homilien
– Die evangelien
– Die glossolalie in der alten kirche
– Judenthum and judenchristenthum
– Die juedische apokalyptik in ihrer geschichtlichen entwickelung
– Der kanon und die kritik des neuen testaments in ihrer gescichtlichen ausbildung und gestaltung
– Die ketzergeschichte des urchristenthums
– Kritische untersuchungen ueber die evangelien justin's, der clementinischen homilien und marcion's
– Das markus-evangelium
– Der paschastreit der alten kirche
– Die propheten esra und daniel
– Das urchristenthum in den hauptwendepuncten seines entwickelungsganges
Hilger, Richard Alexander Maria *see* Raeumliche rekonstruktion embryonaler kiefergelenke des menschen
Hi-line enterprise – Curtis, NE: Harpst Pub Co. v74 n21. jun 3 1965- (wkly) [mf ed filmed 1974-] – 1 – (formed by the union of: hi-line reporter and: curtis enterprise. absorbed: farnam press) – us NE Hist [071]
Hi-Line Indian Alliance *see* Newsletter
Hi-line reporter – Curtis, NE: Harpst Pub Co. 5v. v70 n7. feb 16 1961-v74 n20. may 27 1965 (wkly) – 2r – 1 – (merged with: curtis enterprise to form: hi-line enterprise) – us NE Hist [071]
Hi-lite – [v1] n6-v2 n7 [1969 summer-1972 may/jun] – 1r – 1 – mf#1058158 – us WHS [071]
Hi-lites / American Postal Workers Union – 1956 feb-1961 nov, 1962 jan-1974 dec, 1975-81 – 3r – 1 – (cont: local 3 hi-lites) – mf#632108 – us WHS [380]
Hi-lites / Appleton Papers, Inc – 1974 nov15-1983 jun 1 – 1r – 1 – (continued by: hi-lites, appleton plant) – mf#940340 – us WHS [071]
Hi-lites / Appleton Papers, Inc – 1978 nov-1983 jun – 1r – 1 – (continued by: hi-lites, harrisburg plant) – mf#940336 – us WHS [071]
Hi-lites, appleton plant / Appleton Papers, Inc – 1983 jul-1984 may, 1985 jan-1986 may – 1r – 1 – (cont: hi-lites [appleton, wi]) – mf#940343 – us WHS [071]
Hi-lites, corporate / Appleton Papers, Inc – 1983 jul-1986 feb, may – 1r – 1 – (continued by: news break) – mf#940344 – us WHS [071]
Hi-lites, harrisburg plant / Appleton Papers, Inc – 1983 jul-1985 jun, aug-1986 may – 1r – 1 – (cont: hi-lites [harrisburg, pa]) – mf#940347 – us WHS [071]
Hi-lites, locks mill / Appleton Papers, Inc – 1985 jan-1986 apr – 1r – 1 – (cont: hi-lites [combined locks, wi]) – mf#1099391 – us WHS [071]
Hi-lites, spring mill / Appleton Papers, Inc – 1985 jan-jul, sep, dec, 1986 mar – 1r – 1 – (continued by: news break) – mf#1099392 – us WHS [071]
Hi-lites, west carrollton mill / Appleton Papers, Inc – 1985-1986 apr – 1r – 1 – mf#1099394 – us WHS [071]
Hilkene, Philipp *see* Zur entstehungsgeschichte des "goetz von berlichingen"
Hilker, Susanne *see* Graf simon iv. zur lippe und die weserrenaissance
Hilkhot de'ot : sive, canones ethici / Maimonides, Moses – Amstelodami: Apud Ioh & Cornelium Blaeu 1640 – (= ser Ethics in the early modern period) – 2mf – 9 – (in latin & hebrew) – mf#pl-312 – ne IDC [170]
Hill 303 reporter – v4 n8-v6 n7 [1986 sep-1988 sep] – 1r – 1 – mf#3462351 – us WHS [071]
Hill, A P *see* Mrs. hill's new cook book
Hill, Allan Massie *see* Sweepings frae the yarmouth curling rink
The hill bhuiyas of orissa : with comparative notes on the plains bhuiyas / Roy, Sarat Chandra, Rai Bahadur – Ranchi: Man in India Office, 1935 – (= ser Samp: indian books) – us CRL [307]
Hill, C *see* Suggestions on the teaching of history
Hill, Charles Jenkins *see* A memorial of the rev edward woolsey bacon
Hill country genealogical quarterly – Llano TX. v1 iss 3 [1974 jul], v2 iss 1-2 [1975 jan-apr] – 1r – 1 – (continued by: hill country genealogical society quarterly) – mf#1330179 – us WHS [929]
Hill country genealogical society quarterly – Llano TX. 1976 oct-1987 oct – 1r – 1 – (cont: hill country genealogical quarterly) – mf#1329637 – us WHS [929]
Hill county journal – Havre, MT. 1930-1934 (1) – mf#64445 – us UMI ProQuest [071]
Hill, David Spence *see* The education and problems of the protestant ministry
Hill, Edward Allison *see* Geological report
Hill, Edward Judson *see*
– The chancery jurisdiction and practice, according to statutes and decisions in the state of illinois, from the earliest period to 1873
– The common law jurisdiction and practice, according to statutes and decisions in the state of illinois, from the earliest period to 1872
– The probate jurisdiction and practice in the county courts
– The probate jurisdiction and practice in the courts of the state of illinois.
Hill, Edward Percy *see* Rotary converters, their principles, construction and operation
Hill Family Historical and Genealogical Society *see* Hill heritage
Hill, G F *see* The life of pophyry, bishop of gaza, by mark the deacon
Hill, Geoffry *see* English dioceses
Hill, George *see*
– Character and office of gospel-ministers
– Present happiness of great britain
Hill, George Canning *see*
– Homespun: or, five and twenty years ago
– Our parish
Hill, George S J *see* Essay on the hessian fly, wheat midge
Hill, Hamilton Andrews *see* History of the old south church (third church), boston, 1669-1884
Hill heritage / Hill Family Historical and Genealogical Society – v1 n1-4 [1979 jan-oct] – 1r – 1 – mf#501323 – us WHS [929]
Hill, J Stanley *see* The effect of the education of third world women on family health
Hill, James Hamlyn *see*
– A dissertation on the gospel commentary of s ephraem the syrian
– The earliest life of christ ever compiled from the four gospels
Hill, Jno C *see* Hints on bible reading
Hill, John *see*
– Consolation for mourners
– The family practice of physic
– A review of the works of the royal society of london
Hill, John B *see* Presbytery of kansas city, 1821-1901
Hill, John Boynton *see*
– Presbyterian home missions in missouri
– Presbyterianism in missouri
Hill, John Louis *see* As others see us, and as we are
Hill, John Ward *see* A manual of the law of fixtures
Hill, Kate Alexander *see* Correspondence
Hill, Larry B *see* The ombudsman
Hill, Larry T *see* Time motion analysis of the skating characteristics of professional ice hockey players
Hill, Maoma Frances *see* Nutritive range of copper in some typical florida soils
Hill, Merritt B *see* The laws of the united states relating to patents and trademarks.
Hill, Micaiah *see* The sabbath made for man
Hill news – Canton, NY. 1911-1994 (1) – mf#68328 – us UMI ProQuest [071]
Hill newsletter – v1-2 n3 [1976 aug-1977 oct] – 1r – 1 – mf#384138 – us WHS [071]
Hill, Nicholas S *see* New water supply system at tampa, florida
Hill of destiny / Becker, Peter – Harlow, England. 1969 – 1r – us UF Libraries [960]
Hill, Philip Carteret *see*
– Drifting away
Hill, Reuben *see*
– Family and population control
– A sermon
Hill, Richard *see* Present for your neighbour
Hill, Robert Thomas *see* Cuba and porto rico
Hill, Rowland *see* Series of letters occasioned by the late pastoral admonition of the church of scotland...
Hill, Rowley *see* The titles of our lord
Hill, S C *see* Bengal in 1756-1757
Hill, S S *see*
– The emigrant's introduction to an acquaintance with the british american colonies
– A short account of prince edward island
Hill, Thomas *see*
– Geometry and faith
– A statement of the natural sources of theology
Hill, Timothy *see* Historical outlines of the presbyterian church in missouri
Hill top times – Hill Field UT. 1981 may 1-1982 mar, 1982 apr 2-dec, 1982 dec 23-1983 jun 24, 1983 jul-1984 dec, 1985 jan 18-jun, 1985 jul-1986 mar, 1986 apr-dec, 1987 aug-1988 feb, 1987 jan-jul, 1988 mar-sep, 1989 jul 7-1990 mar 30 – 12r – 1 – mf#627587 – us WHS [071]
Hill topics : a newspaper for residents of avenue road hill district – Toronto, apr 1921-jul 7 1923//? (ill wkly) – 1r – 1 – Can$75.00 – cn McLaren [071]
The hill tribes of india : an account of the church missionary society's work among the maler, santals, gaonds, kois, bheels, hill arrians, and other tribes / Snell, C D – 2nd rev ed. London: Church Missionary House, 1899 – 1mf – 9 – 0-524-06106-8 – mf#1991-2419 – us ATLA [240]
The hill tribes of jeypore / Sahu, Lakshminarayana – [Cuttack: Orissa Mission Press, 1942] – (= ser Samp: indian books) – us CRL [307]
Hill, Victor Dwight *see* Teaching first-year latin
Hill, Walter Kent *see* Liturgical use of the organ in the sixteenth-century spanish church
Hill, William *see* History of the rise, progress, genius, and character of american presbyterianism
Hill, William Bancroft *see* The present problems of new testament study
Hillaire, J *see* Speculum heroicum principis omnium temporum poetarum homeri...
Hillard, Gustav *see* Spiel mit der wirklichkeit
Hillard, Katharine *see* On the scientific importance of dream
Hill-caves of yucatan / Mercer, Henry Chapman – Philadelphia, PA. 1896 – 1r – us UF Libraries [972]
Hillcrest baptist church. laurens county. south carolina : church records – 1955-72 – 1 – us Southern Baptist [242]
Hille, Curt *see* Die deutsche komoedie unter der einwirkung des aristophanes
Hillebiegel / Watzlik, Hans – Regensburg: G Bosse, [194-?] – 1r – 1 – us Wisconsin U Libr [430]
Hillebrandt, Alfred *see*
– Das altindische neu- und vollmondsopfer in seiner einfachsten form
– Der freiwillige feuertod in indien und die somaweihe
– Ritual-litteratur
– Varuna und mitra
Hillel? – Seattle, WA. 22 Jan 1941. Continued by: Hillel World – 1 – us AJPC [071]
Hillel banner – Bloomington, IN.Nov-Dec 1941; May 1944 – 1 – us AJPC [071]
Hillel beacon – Oxford, Ohio. 17 Nov 1941 – 1 – us AJPC [071]
Hillel herald – Evanston, IL. Nov 1946 – 1 – us AJPC [071]
Hillel herald – New Brunswick, NJ. 25 May 1946; Nov-Dec 1946 – 1 – us AJPC [071]
Hillel news – New York, NY. 8 Dec 1947; Hanukkah 1948 – 1 – us AJPC [071]
Hillel observer – New York, NY. 23 Oct-27 Nov 1946; Dec 1947 – 1 – us AJPC [071]
Hillel post – Durham, NC. June 1944; Aug-Dec 1944; Feb 1945 – 1 – us AJPC [071]
Hillel record – Detroit, MI. Feb 1948 – 1 – us AJPC [071]
Hillel review – v1 n1-v3 n13 [1925 mar 12-1927 apr 30], v5 n1-v33 n1 [1929 oct 5-1962 sep 3], 1932 oct-1940 may [incomplete] – 2r – 1 – (cont: hillel-o-grams) – mf#621235 – us WHS [071]
Hillel review – Madison, WI. 5 Oct 1929-30 Oct 1960. Many issues missing – 1 – us AJPC [071]
Hillel scribe – New York, NY. 19 Feb 1941-20 Dec 1945. Many issues missing. Continues: Hillel Scroll (New York) – 1 – us AJPC [071]
Hillel scroll – Bangor, ME. Feb 1948 – 1 – us AJPC [071]
Hillel scroll – Columbus, OH. 28 Oct 1937-27 May 1953. Many issues missing – 1 – us AJPC [071]
Hillel scroll – New York, NY. 12 Nov 1940. Continued by: Hillel Scribe – 1 – us AJPC [071]
Hillel shofar – University, AL. May 1944 – 1 – us AJPC [071]
Hillel star – Montreal, Quebec, Canada. 1982-84 – 1 – us AJPC [071]
Hillelife – Evanston, IL. Dec 1935-Jan 1936 – 1 – us AJPC [071]
Hillelites – Madison, WI. 28 Oct 1943, 10 Feb-30 Mar 1944. Continues: You Name It – 1 – us AJPC [071]
Hillel-o-gram / B'nai B'rith Hillel Foundation (University of Wisconsin) – v1 n1-v2 n19 [1927 oct 8-1929 apr 23] – 1r – 1 – mf#621276 – us WHS [270]
Hillel-o-grams / B'nai B'rith Hillel Foundation (University of Wisconsin) – v1 n1-6 [1925 feb 13-apr 24] – 1r – 1 – (continued by: hillel review) – mf#621227 – us WHS [270]
Hillen *see* Die religioesen vorstellungen im anfange der geschichte der menschheit
Hillenmayer, Dawn M *see* The effect of surface electomyography visual biofeedback on the ability to minimize mid-trapezius muscle activity during an arm flexion task in females
Hiller, Francis Hemperley *see* Juvenile court laws of the united states
Hiller, H Croft *see* Did christ claim to be son of god?
Hiller, Johann Adam *see*
– Anleitung zu der musikalischen gelahrtheit
– Lebensbeschreibungen beruehmter musikgelehrten und tonkuenstler, neuerer zeit, 1. theil
– Der lustige schuster
– Die verwandelten weiber
Hiller, Kurt *see* Kondor
Hiller, Louise *see* Gemueths-schaetze
Hiller, Oliver Prescott *see* Notes on the psalms, chiefly explanatory of their spiritual sense
Hiller, Philipp Friedrich *see*
– Neues system aller vorbilder jesu christi durch das ganze alte testament
– Die vorbilder der kirche des neuen testaments im alten testament
Hiller Von Gaertringen, Friedrich *see* Griechische epigraphik
Hillern, Wilhelmine von *see* Aus eigener kraft
Hillert, Freimut *see*
– Routenoptimierung mit greedy-algorithmen
– Tourenoptimierung im handel
Hilles, David *see* Report of the trial of friends
Hillesden housekeeping accounts 1717-21 *see* Terrars of hillesden, buckinghamshire, 1657 and 1665
Hillfielder – Hill Field UT. 1943 jul 28 – 1r – 1 – (cont: hill fielder) – mf#4364706 – us WHS [071]
Hillfoots record – [Scotland] Alva: R Cunningham 4 jul 1900-25 jan 1916 (wkly) [mf ed 2003] – 547v on 6r – 1 – (cont by: hillfoots record advertising sheet [4 apr 1911-jan 1916]) – uk Newsplan [072]
Hillfsverein der deutschen juden – Berlin DE, 1904-30 – 1r – 1 – (with: hilfsverein fuer die notleidende juedische bevoelkerung in galizien, vienna, 1906-11) – us UMI ProQuest [939]
Hilliard, Francis *see*
– American law
– The american law of real property
– The elements of law: being a comprehensive summary of american civil jurisprudence
– The law of injunctions
– The law of mortgages, of real and personal property
– The law of remedies for torts, or private wrongs
– The law of sales of personal property
– The law of torts or private wrongs
Hilliard history – 1986 dec-1989 dec – 1r – 1 – mf#1701136 – us WHS [978]
Hilliard-D'Auberteuil, Michel Rene de *see* Miss mccrea
Hillier, George *see* Narrative of the attempted escapes of king charles 1st from carisbrook castle, and of his detention in the isle of wight from november 1647, to the seizure of his person by the army at newport, in november 1648

Hilligenlei : roman / Frenssen, Gustav – Berlin: G Grote 1905 [mf ed 1989] – (= ser Grote'sche sammlung von werken zeitgenoessischer schriftsteller 86) – 1r – 1 – (filmed with: dorfpredigten) – mf#7264 – us Wisconsin U Libr [830]

Hilliger, Benno *see* Die wahl pius' 5 zum papste

Hillingdon borough recorder – London, UK. 27 sep 1990-1 jul 1992 – 5r – 1 – uk British Libr Newspaper [072]

Hillingdon district weekly post *see* Hayes post

Hillingdon leader – London, UK. 1986-19 dec 1990, 1991, 1992 – 23r – 1 – (aka: uxbridge and hillingdon leader; leader (hillingdon ed)) – uk British Libr Newspaper [072]

Hillmann, Anselm *see* Judisches genossenschaftswesen in russland

Hillquist, Morris *see* Morris hillquit papers

Hill-rosedale topics – Toronto, sep 1 1923-mar 20 1925//? (ill wkly) – 1r – 1 – Can$120.00 – (toronto social history of the early 1920's in a microcosm. articles and stylish advertisements address an upper middle-class audience) – cn McLaren [971]

Hills, Aaron Merritt *see* Holiness and power

Hill's cases in equity / South Carolina. Supreme Court – v1-2. 1833-1837 (all publ) – (= ser Pre-nrs nominative equity reports) – 14mf – 9 – $21.00 – mf#LLMC 94-030 – us LLMC [342]

Hills, Elijah Clarence *see*
- Bardos cubanos
- Some spanish-american poets

Hills family journal – v1 n1-v6 n3 [1982 jan-1987 oct], v7 n1-2 [1988 spring-fall], ns: v1 n1-v4 n2 (1982 jan-1985 may) – 1r – 1 – mf#1058200 – us WHS [929]

Hills, George *see*
- Farewell sermon
- A tour in british columbia

Hill's law reports / South Carolina. Supreme Court – v1-3. 1833-1837 (all publ) – (= ser Pre-nrs nominative law reports) – 18mf – 9 – $27.00 – mf#LLMC 94-017 – us LLMC [340]

Hills life in new south wales – Sydney, 1832-33 – 1r – 1 – A$27.50 vesicular A$33.00 silver – at Pascoe [980]

Hills, Marilla Marks Hutchins *see* Reminiscences

Hills of the boasting woman / Earl, Stephen – London, England. 1963 – 1r – us UF Libraries [972]

Hills shire times – Jul-dec 1991 – 1r – at Pascoe [079]

Hillsboro argus – Hillsboro OR: Argus Pub Co, 1895- [semiwkly] – 1 – (cont: argus (1894-95). absorbed: hillsboro independent (-1932). 1930-32 incl newspaper pub by union high school students) – us Oregon Lib [071]

Hillsboro enterprise – Hillsboro WI. 1902 mar 20-jun 5 – 1r – 1 – (continued by: hillsboro sentry; hillsboro sentry-enterprise) – mf#951050 – us WHS [071]

Hillsboro independent – Hillsboro OR: Hillsboro Pub Co [wkly] – 1 – (ceased in 1932. cont: independent (hillsboro, or). absorbed by: hillsboro argus (hillsboro, or)) – us Oregon Lib [071]

Hillsboro independent – Hillsboro, Washington County, OR: Hillsboro Pub Co. v21 n10-v59 n38. aug 4 1893-dec 25 1931 – 1 – (cont: independent; cont by: hillsboro argus; ceased in 1932) – us Oregon Hist [071]

Hillsboro sentry – Hillsboro WI. 1892 mar 12-1901 jan 11 – 1r – 1 – (cont: wisconsin sentry; cont by: hillsboro enterprise; hillsboro sentry-enterprise) – mf#951049 – us WHS [071]

Hillsboro sentry-enterprise – Hillsboro WI. 1902 jun 13/1904 dec 30-2000 jan/jun – 74r – 1 – (cont: hillsboro sentry; hillsboro enterprise) – mf#891528 – us WHS [071]

[Hillsborough-] boutique and villager – CA. 1965- [wkly] – 37r – 1 – $60.00 – mf#B02295 – us Library Micro [071]

Hillsborough county, florida – s.l, s.l? 19– – 1r – us UF Libraries [978]

Hillsborough county, florida – s.l, s.l? 193-? – 1r – us UF Libraries [978]

Hillsdale baptist church. hartsville, tennessee : church records – 1892-1953 – 1 – us Southern Baptist [242]

Hillsdale College *see* Imprimis

Hillside journal of clinical psychiatry – New York. 1989-1989 (1,5,9) – ISSN: 0193-5216 – mf#14480 – us UMI ProQuest [616]

Hillston spectator – Hillston, jan 1898-dec 1968 – 19r – A$1303.59 vesicular A$1408.09 silver – at Pascoe [079]

Hillston spectator – jan 1969-dec 1996 – 9 – at Pascoe [079]

Hilltop / Howard University – [1992 mar 6-1995 sep 8], 1996 aug, sep 6-nov 22, 1997 jan 17-apr 4 – 2r – 1 – mf#2562413 – us WHS [378]

Hilltop news series / Hamilton Co. Cincinnati – 3/1971-6/1984,6/1986-3/1991 [wkly] – 19r – 1 – mf#B35634-35652 – us Ohio Hist [071]

Hilltop record / Franklin Co. Columbus – jan 1936-dec 1963 [daily, wkly, semiwkly, wkly] – 19r – 1 – mf#B6790-6808 – us Ohio Hist [071]

Hilltop record / Franklin Co. Columbus – (mar 1914-jan 1925), feb 1925-44 [wkly] – 5r – 1 – mf#B1456-1460 – us Ohio Hist [071]

Hilltop spectator / Franklin Co. Columbus – may 1972-jul 1973 [wkly] – 2r – 1 – mf#B6752-6753 – us Ohio Hist [071]

Hill-Tout, Charles *see* Notes of the prehistoric races of british columbia and their monuments

Hillyear, Charles Wells *see* Monotheism versus priestcraft

Hillyer, Curtis *see* Code of law, practice and forms for justices' and other inferior courts in the western states

Hilmi *see* The divan project

Hilmi, Hueseyin *see* Sinop kitabeleri

Hilmi, Tueccarzade Ibrahim *see* Memalik-i osmaniye cep atlasi

Hilpert, Walter *see* Johann georg hamann als kritiker der deutschen literatur

Hilprecht anniversary volume : studies in assyriology and archaeology dedicated to hermann v. hilprecht... – Leipzig: J C Hinrichs; Chicago, IL: Open Court, 1909 – 2mf – 9 – 0-7905-1432-X – (in german, english, french etc. incl bibl ref) – mf#1987-1432 – us ATLA [470]

Hilprecht, H V *see* Old babylonian inscriptions chiefly from nippur

Hilprecht, Hermann Vollrat *see*
- Assyriaca
- The earliest version of the babylonian deluge story and the temple library of nippur
- Recent research in bible lands

Hilprecht's fragment of the babylonian deluge story : babylonian expedition of the university of pennsylvania, series d, vol 5, fasc 1 / Barton, George A – [s.l: s.n.] 1910 [mf ed 1986] – 1mf – 9 – 0-8370-7605-6 – (in english & akkadian. incl bibl ref) – mf#1986-1605 – us ATLA [470]

Hils, Hans-Peter *see* Fecht- und ringbuch / vermischtes kampfbuch (cf-lp2)

Hilsendager, Sarah A *see* A survey analysis of dance wellnessrelated curricula in american higher education

Hilsheimer, Ruth *see* Everglades flood control

Hilt, Franz *see* Des heil. gregor von nyssa lehre vom menschen

Hiltebrandt, G A *see* Neu-eroeffneter anmuthiger bilderschatz

Hiltgart von huernheim : mittelhochdeutsche prosauebersetzung des 'secretum secretorum' / ed by Moeller, Reinhold – Berlin: Akademie-Verlag, 1963 [mf ed 1993] – (= ser Deutsche texte des mittelalters 56) – cv/219p/2pl – 1 – (parallel latin and middle high german text. int in german) – mf#8623 reel 8 – us Wisconsin U Libr [430]

Hilton, James *see* Chronograms 5000 and more in number excerpted out of various authors and collected at many places

Hilton, John *see* A study of trade organisations and combinations in the united kingdom

Hilton news – sep 1974-dec 1981; feb-aug 1982 – 3r – 1 – (title changed fr hilton press in 1980) – mf#75.4 – nz Nat Libr [079]

Hilton press *see* Hilton news

Hilton, W *see* De reclusis/speculum humanae salvationis

Hilts, Joseph Henry *see* Among the forest trees

Hilula / Bension, Ariel – Jerusalem, Israel. 1927 – 1r – us UF Libraries [939]

Hilvanes y zurzidos. lo que se llama perder el tiempo. ensayos poeticos / Bachiller Cantaclaro – Madrid: Imp. Luz y Vida, 1935 – 1 – sp Bibl Santa Ana [810]

Hilversum. City and Regional Archives Gooi-en Vechtstreek, The Netherlands *see* Dudok collection of architectural plans and drawings of the city of hilversum

Himachal pradesh gazette / Himachal Pradesh. India – Simla. 1962-1966 – 1 – us NY Public [950]

Himachal Pradesh. India *see* Himachal pradesh gazette

Die himalaya-mission der bruedergemeine / Reichelt, G Th – Guetersloh: C Bertelsmann, 1896 [mf ed 1995] – (= ser Yale coll) – 87p (ill) – 1 – 0-524-09060-2 – (in german) – mf#1995-0060 – us ATLA [951]

Himalayan art / French, J C – London; New York: Oxford University Press, 1931 – (= ser Samp: indian books) – (int by laurence binyon) – us CRL [750]

Himalayan journals : or, notes of a naturalist in bengal, the sikkim and nepal himalayas, the khasia mountains... / Hooker, J D – Philadelphia. 1973+ (1) 1975+ (5) 1976+ (9) – 8mf – 9 – mf#8374 – ne IDC [915]

Himalayan Watershed Properties *see* Organ

Himalayas abode of light / Rerikh, Nikolai Konstantinovich – Bombay: Nalanda Publications; London: David Marlowe Ltd, 1947 – (= ser Samp: indian books) – us CRL [915]

The himalayas in the indian art / Havell, Ernest Binfield – London: John Murray, 1924 – (= ser Samp: indian books) – us CRL [700]

Himel un erd / Veviorka, Abram – Lemberg, Ukraine. 1909 – 1r – us UF Libraries [939]

Himelrick, David G *see* Small fruits review

Himenea. testo, introduzione e note di annamaria gallina / Torres Naharro, Bartolome – Milano: Edit. Cisalpino, (1961) – sp Bibl Santa Ana [946]

Himioben, Heinrich *see* Die schoenheit der katholischen kirche

Himle, Thorstein *see* Guds veie med et gjenstridigt folk

Das himlisch kleinot und ehrenkrentzlein : mit fuenff und zweintzig wolriechenden geruchsbluemlein, welche im krantz nicht welck werden / Mader, Johann – s.l. 1590 – (= ser Hqab. literatur des 16. jahrh.) – 1mf – 9 – €20.00 – mf#1590b – gw Fischer [780]

Himma katsina – Zaria, Nigeria: [Gaskiya Corp. n66-88. jan-oct 9 1958] – 1 – (filmed with: zaruma and other hausa newspapers) – us CRL [960]

Der himmel der enttaeuschten : novellen / Frank, Bruno – Muenchen: A Langen, c1916 [mf ed 1995] – (= ser Langens mark-buecher 12) – 1r – 1 – (filmed with: das glockenbuch / von hans franck) – mf#3842P – us Wisconsin U Libr [830]

Der himmel des christen : skizzen zu den jenseitsvorstellungen in unserer apologetischen literatur / Kropatscheck, Friedrich – Berlin-Lichterfelde: E Runge 1916 [mf ed 1992] – (= ser Biblische zeit- und streitfragen 11/1) – 1mf – 9 – 0-524-04099-0 – mf#1992-0057 – us ATLA [240]

Himmel, Friedrich Heinrich *see*
- Trois sonates pour le pianoforte
- Trois sonates pour le pianoforte avec accomp de violon et violoncelle, no 2

Himmel, Paul *see* Untersuchung ueber die entwicklung und den stand der betriebsverhaeltnisse eines schlesischen rittergutes

Der himmel und seine wunder : eine archaeologische studie nach alten juedischen mythografien / Bergel, Joseph – Leipzig: Wilhelm Friedrich, 1881 – 1mf – 9 – 0-8370-2282-7 – mf#1985-0282 – us ATLA [460]

Himmelfahrt : roman / Bahr, Hermann – Berlin: S Fischer, 1916 [mf ed 1989] – 400p – 1 – mf#830 – us Wisconsin U Libr [830]

Himmelfahrt : roman / Bahr, Hermann – Wien: H Bauer, 1946 [mf ed 1990] – 362p – 1 – mf#7078 – us Wisconsin U Libr [830]

Himmelfahrt : roman / Bahr, Hermann – Wien: H Bauer, 1946 [mf ed 1990] – 362p – 1 – mf#7078 – us Wisconsin U Libr [830]

Die himmelfahrt des mose / ed by Clemen, Carl – Bonn: A Marcus & E Weber 1904 [mf ed 1992] – (= ser Kleine texte fuer theologische vorlesungen und uebungen 10) – 1mf – 9 – 0-524-04747-2 – (text in latin, notes in german) – mf#1992-0189 – us ATLA [240]

Himmels- und weltenbild der babylonier : als grundlage der weltanschauung und mythologie aller voelker / Winckler, Hugo – 2nd rev enl ed. Leipzig: JC Hinrichs, 1903 [mf ed 1989] – (= ser Der alte orient 3/2-3) – 1mf – 9 – 0-7905-2815-0 – mf#1987-2815 – us ATLA [520]

Der himmelsbrief : ein beitrag zur allgemeinen religionsgeschichte / Stuebe, R – Tuebingen, 1918 – €5.00 – ne Slangenburg [200]

Himmelskraft : roman / Dominik, Hans – Berlin: Scherl, 1937 – 1r – 1 – us Wisconsin U Libr [830]

Himmler, Gebhard *see*
- Zur sprache des aegidius albertinus

Himmlisch, J F *see* Panthevm sive anatomia et symphonia papatvs

Himmlische landschaft / Schickele, Rene – 1.-5. Aufl. Berlin: S Fischer, 1933, [c1932] – 1r – 1 – us Wisconsin U Libr [890]

Der himmlische zecher : in sieben buechern [poems] / Mombert, Alfred – grosse ausg. [Wiesbaden]: Insel-Verlag c1951 [mf ed 1996] – 1r – 1 – (filmed with: ein volk wacht auf / walter von molo) – mf#3968p – us Wisconsin U Libr [810]

Himnos / Fernandez, Pablo Armando – Habana, Cuba. 1962 – 1r – us UF Libraries [972]

Himnos de montana / Perez, Rafael Alcides – Habana, Cuba. 1961 – 1r – us UF Libraries [972]

Himpunan mahasiswa islam – Djakarta, Jogjakarta, 1954/1955-1959. v1-6 – 23mf – 9 – (missing: 1954, v1(1-2, 5, 7, 9-12); 1955, v2(1-2, 4, 8, 10-12); 1956, v3(1-3, 7-10); 1957, v4(58)) – mf#SE-385 – ne IDC [950]

Himpunan peraturan daerah kabupaten karanganjar : karanganjar, indonesia (kabupaten) – Karanganjar, 1969 – 2mf – 9 – mf#SE-1728 – ne IDC [950]

Himpunan peraturan2 pemerintah bidang perhubungan dan perdagangan / Warta Ekonomi Maritim – Djakarta, 1969-1971. v1-10 – 20mf – 9 – (missing: 1969-1971 v2-8) – mf#SE-1990 – ne IDC [950]

Himpunan surat pernjataan : resolusi kabupaten trenggalek sekretariat dprd-gr kabupaten trenggalek, djawa timur – Trenggalek, 1968 – 1mf – 9 – mf#SE-1962 – ne IDC [950]

Himpunan surat – surat keputusan : peraturan daerah kabupaten bojolali tahun dinas dan untuk landasan pedoman kerdja dalam tahun dinas – Bojolali, 1968-1969/1970. v1-10 – 12mf – 9 – mf#SE-1361 – ne IDC [950]

Himpunan surat-surat keputusan pemerintah daerah kabupaten djombang – Djombang, 1969/1970 – 1mf – 9 – mf#SE-1457 – ne IDC [950]

Hin Pun Thun *see* Ryan jamny sneha sruk sra

Hin und zurueck : aus den papieren eines arztes / Abbott, Caroline Luxburg – 20. aufl. Halle/S: R Muehlmann, 1921 [mf ed 1989] – 328p – 1 – mf#6979 – us Wisconsin U Libr [880]

Hinchliff, P *see* The south african liturgy

Hinckley news – Leicestershire 5 oct 1861-10 dec 1892 (wkly) – 25 1/2r – 1 – (incorp with: loughborough monitor) – uk Newsplan [072]

Hinckley times – Hinckley, Leicestershire 5 jan 1889-1906, 1908-11, 1913-to date – 1 – uk Newsplan [072]

Hincks, Edward *see*
- On the assyrio-babylonian phonetic characters
- On the khorsabad inscriptions
- On the polyphony of the assyrio-babylonian cuneiform writing

Hincks, Francis *see*
- Canada and mr goldwin smith
- Expose financier de sir francis hincks, mardi, 30 avril 1872
- Religious endowments in canada
- Speech of the honorable francis hincks, inspector general

Hincks, Thomas *see*
- Living and the dead one family in christ
- Protest against the present religious agitation and a plea for the...

Hincmarus de ordine palatii (mgh leges 4:3.bd) – 1894 – (= ser Monumenta germaniae historica leges 4. fontes iuris germanici antiqui in usum scholarum separatim editi (mgh leges 4)) – €3.00 – ne Slangenburg [241]

Hind, Henry Youle *see*
- British north america
- Canadian journal
- Emigration, land and railway frauds
- Falsified departmental reports
- The ice phenomena and the tides of the bay of fundy
- North-west territory
- Reports of progress together with a preliminary and general report, on the assiniboine and saskatchewan exploring expedition
- Territoire du nord-ouest

Hind, John Russell *see* The comets

The hindee-roman orthoepigraphical ultimatum : or a systematic, discriminative view of oriental and occidental visible sounds / Gilchrist, John Borthwick – 2nd ed. London, 1820 – (= ser 19th c books on linguistics) – 4mf – 9 – mf#2.1.41 – uk Chadwyck [400]

Hindemith : third piano sonata. an analysis / Harner, Nevin L – U of Rochester 1962 [mf ed 19–] – 2mf – 9 – mf#fiche 94 – us Sibley [780]

Hindered hand / Griggs, Sutton Elbert – Nashville, TN. 1905 – 1r – us UF Libraries [025]

Hinderer, A *see* Seventeen years in the yoruba country, memorials of anna hinderer...

Hinderer, Anna Martin *see* Seventeen years in the yoruba country

Hindi first reader / Central Provinces (India). Department of Public Instruction – Bombay: Oxford UP, 1917 [mf ed 1995] – (= ser Yale coll) – 40p (ill) – 1 – 0-524-09553-1 – (in hindi) – mf#1995-0553 – us ATLA [490]

Hindi literature / Dwivedi, Ram Awadh – Banaras: Hindi Pracharak Pustakalaya, 1953 – (= ser Samp: indian books) – us CRL [490]

Hindi milap – Hyderabad, India. 7 Sept 1953-1954 – 3r – 1 – us L of C Photodup [079]

Hindi second reader / Central Provinces (India). Department of Public Instruction – Bombay: Oxford UP, 1917 [mf ed 1995] – (= ser Yale coll) – 96p (ill) – 1 – 0-524-09554-X – (in hindi) – mf#1995-0554 – us ATLA [490]

Hindi third reader / Central Provinces (India). Department of Public Instruction – Bombay: Oxford UP, 1917 [mf ed 1995] – (= ser Yale coll) – 112p (ill) – 1 – 0-524-09555-8 – (in hindi) – mf#1995-0555 – us ATLA [490]

Hindia-nederland : soerat kabar betawi – Betawi: H.M. Van Dorp & Co [1872-86] (wkly) [mf ed Jakarta: National Library of Indonesia] – 15v – 1 – (iss for jan 24 1885-jun 30 1886 have title: hindia nederland; text in indonesian; some in arabic script; iss for aug 1872-may 1877 filmed with: djawa tengah, jan 3-24 1914, feb 16-mar 1938; iss for 1883-jun 1886 filmed with: soerat chabar batawie, apr 5-jun 26 1858, and: chabar hindia-olanda, jan-apr 1888) – mf#mf-7737 seam, mf-7739 seam – us CRL [079]

HINDUISM

Hindia-olanda – Betawi: Karsseboom & Co [1890- (daily) [mf ed Jakarta: National Library of Indonesia] – reel 2 [jan 2-mar 15 1890], 8r [mar 17 1890-nov 1896] – 1 – (iss for jan 2-mar 15, 1890 filmed with: chabar hindia-olanda, apr-dec 1889) – mf#mf-7740 seam, mf-7742 seam – us CRL [079]
Hindlip, Charles Allsopp *see* British east africa
Hindoekinderen : een boek over kinderen voor kinderen / Blauenfeldt, Johanne – [Rotterdam: J M Bredee, 1909] [mf ed 1995] – (= ser Yale coll; Lichtstralen op den akker der wereld [15. jaarg 1909] 1/2) – 61p (ill) – 1 – 0-524-10064-0 – (in dutch) – mf#1995-1064 – us ATLA [306]
The hindoo traveller : comprising the geography of hindoostan with a brief view of its history, scenery etc, n1 – Manepy: American Mission Press, 1839 [mf ed 1995] – (= ser Yale coll) – 170p (ill) – 1 – 0-524-09175-7 – mf#1995-0175 – us ATLA [915]
The hindoos – London 1834-35 [mf ed Hildesheim 1995-98] – 2v on 6mf – 9 – €120.00 – 3-487-27528-7 – gw Olms [954]
Hindoostan : containing a description of the religion, manners, customs, trades, arts, sciences, literature, diversions, etc of the hindoos / ed by Shoberl, Frederick – London [1822] [mf ed Hildesheim 1995-98] – 6v on 13mf – 9 – €130.00 – 3-487-27531-7 – gw Olms [915]
Hindos, Jose de *see* Beethoven. sugestiones
Hindostan – Berlin DE, 1915 20 apr-1918 21 aug [gaps] – 1r – 1 – (for indian pows; in hindi) – gw Misc Inst [074]
L'hindoustan : ou religion, moeurs, usages, arts et metiers des hindous / Pannelier, Jean – Paris 1816 [mf ed Hildesheim 1995-98] – 6v on 12mf – 9 – €120.00 – 3-487-27524-4 – gw Olms [915]
Hinds, Allen Banks *see* A garner of saints
Hinds, Asher C *see*
– Hind's parliamentary precedents of the house of representatives
– Hind's precedents of the house of representatives
– Hinds' precedents of the house of representatives of the united states.
– Parliamentary procedures of the house of representatives of the u.s
Hinds county gazette – Raymond, MS. 1878-1883 (1) – mf#64109 – us UMI ProQuest [071]
Hind's parliamentary precedents of the house of representatives / Hinds, Asher C – Washington: GPO, 1898 (all publ) – 4mf – 9 – $14.00 – mf#llmc 84-103 – us LLMC [340]
Hind's precedents of the house of representatives / Hinds, Asher C – Washington: GPO. 8v. 1907-08 (all publ) – 29mf – 9 – $43.50 – mf#LLMC 84-104 – us LLMC [323]
Hinds' precedents of the house of representatives of the united states. / Hinds, Asher C – Washington: GPO, 1907-08. 8v. LL-1235 – 1 – us L of C Photodup [340]
Hinds, Samuel *see*
– Nature and origin of evil
– Reply to the question, "apart from supernatural revelation, what is...
Hinds, Samuel et al *see* The rise and early progress of christianity
Hinds, William Alfred *see* American communities and co-operative colonies
Hindu – Madras, India. 1951+ (1) – mf#60193 – us UMI ProQuest [079]
The hindu : illustrated sunday edition – Madras: K Gopalan, 1936-41 – us CRL [079]
The hindu – Madras, India: K Gopalan, [1942-]. mar-may 1945; jan-jul 1947; feb 28 1954; 1956-1975; 1976-1983 (weekly ed) – us CRL [079]
Hindu achievements in exact science : a study in the history of scientific development / Sarkar, Benoy Kumar – New York: Longmans, Green and Co, 1918 – (= ser Samp: indian books) – us CRL [500]
Hindu america : revealing the story of the romance of the surya vanshi hindus and depicting the imprints of hindu culture on the two americas / Chaman Lal – Bombay: New Book Co, 1948 – (= ser Samp: indian books) – us CRL [900]
Hindu astronomy / Kaye, George Rusby – Calcutta: Govt of India, Central Publication Branch, 1924 – (= ser Samp: indian books) – us CRL [520]
The hindu at home : being sketches of hindu daily life / Padfield, Joseph Edwin – 2nd ed. Madras: SPCK Depository, 1908 – 1mf – 9 – 0-524-03370-6 – mf#1990-3204 – us ATLA [280]
The hindu at home : being sketches of hindu daily life / Padfield, Joseph Edwin – Madras: SPCK, 1896 [mf ed 1995] – (= ser Yale coll) – xxiii/333p – 1 – 0-524-09742-9 – mf#1995-0742 – us ATLA [280]
Hindu castes and sects : an exposition of the origin of the hindu caste system and the bearing of the sects towards each other and towards other religious systems / Bhattacharya, Jogendra Nath – Calcutta: Thacker, Spink, 1896 – 2mf – 9 – 0-524-04505-4 – (incl bibl ref) – mf#1990-3339 – us ATLA [280]
Hindu colonies in the far east / Majumdar, Ramesh Chandra – Calcutta: General Printers & Publishers, 1944 – (= ser Samp: indian books) – us CRL [954]
The hindu colony of cambodia / Bose, Phanindra Nath – Madras: Theosophical Publ House 1927 [mf ed 1989] – 1r with other items – 1 – (with bibl) – mf#mf-10289 seam reel 001/06 [§] – us CRL [959]
The hindu conception of the deity as culminating in ramanuja / Kumarappa, Bharatan – London: Luzac & Co, 1934 – (= ser Samp: indian books) – (foreword by l d barnett) – us CRL [280]
The hindu conception of the functions of breath : a study in early hindu psycho-physics / Ewing, Arthur Henry – 1901-1903 – 1mf – 9 – 0-524-01436-1 – mf#1990-2431 – us ATLA [280]
Hindu customs and their origins / Rice, Stanley – London: George Allen & Unwin, 1937 – (= ser Samp: indian books) – (foreword by maharaja gaekwar of baroda) – us CRL [280]
The hindu (daily edition) – Madras, India: K Gopalan, [dec 5 1929-dec 1936]; mar-may 1945; jan-jul 1947; 1956-60; jan 1961-mar 1969; apr 1969-dec 1975; 1976-83 – 1 – us CRL [079]
Hindu dharma / Gandhi, Mahatma – Ahmedabad: Navajivan Pub House, 1950 – (= ser Samp: indian books) – us CRL [280]
The hindu doctrine of transmigration / Hooper, William – Madras: Christian Literature Society for India, 1916 – 1mf – 9 – 0-524-01771-9 – mf#1990-2619 – us ATLA [280]
Hindu ethics : a historical and critical essay / McKenzie, John – London; New York: Oxford University Press, 1922 – (= ser Samp: indian books) – us CRL [280]
Hindu ethics : principles of hindu religio-social regeneration / Dasa, Gobinda; ed by Jha, Mahamahopadhyaya Ganganatha – Madras: GA Natesan & Co, [1927] – (= ser Samp: indian books) – (int by bhagavan das) – us CRL [280]
Hindu exogamy / Karandikar, S V – Bombay: Advocate of India Press, 1928 – (= ser Samp: indian books) – us CRL [306]
Hindu fasts and feasts / Mukerji, Abhay Charan – [S.l.: s.n., 1916?] – 1mf – 9 – 0-524-07792-4 – mf#1991-0169 – us ATLA [280]
Hindu feasts, fasts and ceremonies / Natesa Sastri, S M – Madras: ME Pub House, 1903 – 1mf – 9 – 0-524-01801-4 – mf#1990-2649 – us ATLA [280]
Hindu holidays and ceremonials : with dissertations on origin, folklore, and symbols / Gupte, B A – Calcutta: Thacker, Spink & Co, 1919 – (= ser Samp: indian books) – us CRL [280]
Hindu holidays and ceremonials : with dissertations on origin, folklore and symbols / Gupte, Balkrishna Atmaram – 2nd rev ed. Calcutta, Simla: Thacker, Spink, 1919 [mf ed 1995] – (= ser Yale coll) – lii/285p (ill) – 1 – 0-524-09322-9 – mf#1995-0322 – us ATLA [280]
Hindu idolatry and english enlightenment / Hastie, W – Calcutta, India. 1883 – 1r – us UF Libraries [230]
Hindu infanticide : an account of the measures adopted for suppressing the practice of the systematic murder by their parents of female infants / Moor, Edward – London 1811 – (= ser 19th c british colonization) – 4mf – 9 – mf#1.1.4145 – uk Chadwyck [306]
The hindu jajmani system : a socio-economic system interrelating members of a hindu village community in services / Wiser, William Henricks – Lucknow, UP, India: Lucknow Pub House, 1936 – (= ser Samp: indian books) – us CRL [301]
Hindu kinship : an important chapter in hindu social history / Kapadia, Kanailal Motilal – Bombay: Popular Book Deopt[sic], 1947 – (= ser Samp: indian books) – us CRL [280]
Hindu law : and the methods and principles of the historical study thereof / Ketkar, Shridhar Venkatesh – Calcutta: SK Lahiri, 1914 – 1mf – 9 – 0-524-01564-3 – mf#1990-2518 – us ATLA [280]
Hindu literature : or, the ancient books of india / Reed, Elizabeth Armstrong – Chicago: SC Griggs, 1891 [mf ed 1991] – 1mf – 9 – 0-524-01868-5 – (incl bibl ref) – mf#1990-2703 – us ATLA [490]
Hindu manners, customs, and ceremonies = Moeurs, institutions et ceremonies des peuples de l'inde / Dubois, Jean Antoine; ed by Beauchamp, Henry King – 3rd ed. Oxford: Clarendon Press, c1906 [mf ed 1992] – 2mf – 9 – 0-524-05169-0 – (trans fr french into english by ed. with notes, corr and biogr) – mf#1990-3455 – us ATLA [390]
Hindu manners, customs, and ceremonies / Dubois, Jean Antoine – London; New York: Oxford University Press, [1959] – (= ser Samp: indian books) – us CRL [306]
Hindu medieval sculpture : 79 original photographs / Burnier, Raymond – Paris: Palme, 1950 – (= ser Samp: indian books) – us CRL [730]
Hindu monism and pluralism : as found in the upanishads and in the philosophies dependent upon them / Harrison, Max Hunter – London: Oxford University Press, 1932 – (= ser Samp: indian books) – us CRL [306]
Hindu music and rhythm / Shirali, Vishnudass – [Sl]: The author, 1936 – (= ser Samp: indian books) – us CRL [780]
Hindu mysticism : according to the upanishads / Sircar, Mahendranath – London: Kegan Paul, Trench, Trubner & Co, 1934 – (= ser Samp: indian books) – us CRL [280]
Hindu mysticism : six lectures / Dasgupta, Surendranath – Chicago; London: Open Court Pub Co, 1927 – (= ser Samp: indian books) – us CRL [280]
Hindu mythology : vedic and puranic / Wilkins, William Joseph – 3rd ed. Calcutta, Simla: Thacker, Spink, 1913 [mf ed 1995] – (= ser Yale coll) – xviii/517p (ill) – 1 – 0-524-09012-2 – mf#1995-0012 – us ATLA [390]
Hindu organ – Jaffna, Sri Lanka. 1899-31 Mar 1974 – (= ser Intu Catanam) – 13r – 1 – (also known as: intu catanam) – us L of C Photodup [079]
Hindu outlook – New Delhi. [v2 n44-v5 n43 jan 11 1939-jan 11 1941 1r. jan 26 1941-sep 1957 r2-5] – 5r – 1 – us CRL [954]
Hindu outlook – New Delhi. v2 n44-v5 n43. jan 1939-jan 1941 – 1r – 1 – us CRL [079]
Hindu pastors : an inquiry into the present state and probable development of the native ministry in the indian missions of the english church... / Murray, Ross – London: John Heywood, 1892 [mf ed 1995] – (= ser Yale coll) – 79p – 1 – 0-524-09685-6 – mf#1995-0685 – us ATLA [240]
Hindu philosophy / Bernard, Theos – New York: Philosophical Library, 1947 – (= ser Samp: indian books) – us CRL [180]
Hindu philosophy examined / Shaddarshana darppana – 5th ed. Allahabad: North India Christian Tract & Book Society, 1915 [mf ed 1995] – (= ser Yale coll) – 2v in 1 – 1 – 0-524-10176-0 – (in hindi) – mf#1995-1176 – us ATLA [290]
The hindu philosophy of law : the vedic and post-vedic times prior to the institutes of manu / Pal, Radhabinod – [Calcutta: sn, 19–] (Calcutta: Biswabhandar Press) – (= ser Samp: indian books) – us CRL [280]
Hindu philosophy popularly explained : the orthodox systems / Bose, Ram Chandra – New York: Funk & Wagnalls, 1884 – 1mf – 9 – 0-524-00873-6 – mf#1990-2096 – us ATLA [280]
Hindu polity : a constitutional history of india in hindu times / Jayaswal, Kashi Prasad – Calcutta: Butterworth & Co (India), 1924 – (= ser Samp: indian books) – us CRL [323]
Hindu rashtra darshan : a collection of the presidential speeches delivered from the hindu mahasabha platform / Savarkar, Vinayak Damodar – Bombay: LG Khare, 1949 – (= ser Samp: indian books) – us CRL [954]
Hindu realism : being an introduction to the metaphysics of the nyaya-vaisheshika system of philosophy / Chatterji, Jagdish Chandra – Allahabad: The Indian Press, 1912 – (= ser Samp: indian books) – us CRL [180]
Hindu religion / Ranga Rao, Venkata Swetachalapati – Madras: Addison Press, 1918 – 1mf – 9 – 0-524-08027-5 – mf#1991-0249 – us ATLA [280]
Hindu religion, customs and manners : describing the customs and manners, religious, social and domestic life, arts and sciences of the hindus / Thomas, Paul – Bombay: DB Taraporevala Sons & Co, [1948?] – (= ser Samp: indian books) – us CRL [280]
Hindu samskaras : a socio-religious study of the hindu sacraments / Pandey, Rajbali – Banaras: Vikrama Publications, 1949 – (= ser Samp: indian books) – us CRL [280]
Hindu scriptures : hymns from the rigveda, five upanishads, the bhagavadgita / ed by Macnicol, Nicol – London: JM Dent & Sons; New York: EP Dutton & Co, 1938 – (= ser Samp: indian books) – us CRL [280]
Hindu social institutions : with reference to their psychological implications / Valavalkar, Pandharinath Hari – Bombay; New York: Longmans, Green and Co, 1939 – (= ser Samp: indian books) – (foreword by s radhakrishnan) – us CRL [306]
Hindu superiority : an attempt to determine the position of the hindu race in the scale of nations / Sarda, Har Bilas – Ajmer: Vedic Yantralaya, 1922 – (= ser Samp: indian books) – us CRL [900]
Hindu superiority : an attempt to determine the position of the hindu race in the scale of nations / Sarda, Har Bilas – Ajmer: Rajputana Printing Works, [1906?] – 6mf – 9 – 0-524-07952-8 – (incl bibl ref) – mf#1991-0202 – us ATLA [322]
The hindu system of moral science : or, a few words on the sattwa, raja, and tama gunas / Sarkar, Kishori Lal – 2nd ed. Calcutta: Sarasi Lal Sarkar, 1898 [mf ed 1994] – 1mf – 9 – 0-524-08900-0 – mf#1993-4035 – us ATLA [280]
Hindu theism : a defence and exposition / Tattvabhushan, Sitanath – Calcutta: Som Brothers, 1898 – 1mf – 9 – 0-524-02616-5 – mf#1990-3066 – us ATLA [280]
Hindu thought : a short account of the religious books of india / Leonard, William Andrew – Glasgow: JS Marr, 1875 – 1mf – 9 – 0-524-01784-0 – mf#1990-2632 – us ATLA [280]
The hindu view of art / Anand, Mulk Raj – London: George Allen & Unwin Ltd, 1933 – (= ser Samp: indian books) – (int essay on art and reality by eric gill) – us CRL [700]
The hindu view of life : upton lectures delivered at manchester college, oxford, 1926 / Radhakrishnan, Sarvepalli – London: George Allen & Unwin; New York: Macmillan Co, 1949 – (= ser Samp: indian books) – us CRL [280]
Hindu weekly review – Madras, India. 10 aug 1953-aug 1968 – 11r – 1 – us L of C Photodup [280]
The hindu woman / Cormack, Margaret – New York: Bureau of Publications, Teachers College, Columbia University, 1953 – (= ser Samp: indian books) – us CRL [305]
Hindu women : with glimpses into their life and zenanas / [Lloyd], H – London, 1882 – 2mf – 9 – mf#HT-84 – ne IDC [915]
The hindu-aryan theory on evolution and involution : or, the science of raja-yoga / Rajan Iyengar, Tirumangalum Chrishna – New York: Funk & Wagnalls, 1908 [mf ed 1992] – 1mf – 9 – 0-524-02033-7 – mf#1990-2808 – us ATLA [180]
Hindubani – Bankura. v1 n1-22 oct 1947-sep 1948; v6-7 dec 1952-sep 1954; v10-15 n23 oct 1956-sep 1962; v23 n1-21 oct 1969-oct 1970 – 5r – 1 – us CRL [079]
Hinduism / Barnett, Lionel David – London: Constable, 1913 [mf ed 1991] – (= ser Religions ancient and modern) – 1mf – 9 – 0-524-00685-7 – mf#1990-2013 – us ATLA [280]
Hinduism / Monier-Williams, Monier – Calcutta: Susil Gupta, 1951 – (= ser Samp: indian books) – us CRL [280]
Hinduism / Monier-Williams, Monier – London, England. 1919 – 1r – us UF Libraries [280]
Hinduism / Monier-Williams, Monier – London: SPCK; New York: E & J B Young 1890 [mf ed 1991] – (= ser Non-christian religious systems) – 1mf [ill] – 9 – 0-524-01202-4 – mf#1990-2278 – us ATLA [280]
Hinduism : a retrospect and a prospect / Haldar, Sukumar – [S.l: s.n, s.n, 18–?] [mf ed 1995] – (= ser Yale coll) – 65p – 1 – 0-524-09896-4 – (bound with: wilford, francis: essai sur l'origine et la decadence de la religion chretienne dans l'inde, paris 1847) – mf#1995-0896 – us ATLA [280]
Hinduism / Vivekananda, Swami – Madras: Sri Ramakrishna Math, 1943 – (= ser Samp: indian books) – us CRL [280]
Hinduism ancient and modern : as taught in original sources and illustrated in practical life / Baij Nath – new rev enl ed. Meerut: Vaishya Hitkari Office, 1905 [mf ed 1991] – 1mf – 9 – 0-524-01158-3 – (1st printed 1899) – mf#1990-2234 – us ATLA [280]
Hinduism ancient and modern : viewed in the light of the incarnation / Sharrock, John A – [London]: Society for the Propagation of the Gospel in Foreign Parts, 1913 [mf ed 1992] – 1mf – 9 – 0-524-02369-7 – mf#1990-2980 – us ATLA [280]
Hinduism and buddhism / Coomaraswamy, Ananda Kentish – New York: Wisdom Library, [19–] – (= ser Samp: indian books) – us CRL [230]
Hinduism and buddhism : an historical sketch / Eliot, Charles – London: Edward Arnold & Co, 1921 – (= ser Samp: indian books) – us CRL [230]
Hinduism and christianity : a comparison and a contrast / Jones, John Peter – 1st ed. London: Christian Literature Society for India, 1898 – (= ser Papers for Thoughtful Hindus) – 1mf – 9 – 0-524-02805-2 – mf#1990-3135 – us ATLA [280]
Hinduism and christianity / Robson, John – 3rd ed. Edinburgh: Oliphant Anderson and Ferrier, 1905 – 1mf – 9 – 0-524-01295-4 – mf#1990-2331 – us ATLA [280]

Hinduism and christianity in orissa : containing a brief description of the country, religion, manners and customs of the hindus, and an account of the operations of the american freewill baptist mission in northern orissa / Bacheler, Otis Robinson – Boston: Geo C Rand & Avery, 1856 – 1mf – 9 – 0-524-07149-7 – mf#1991-2938 – us ATLA [230]

Hinduism and india : a retrospect and a prospect / Dasa, Gobinda – London: Theosophical Pub Society, 1908 – 1mf – 9 – 0-524-01689-5 – mf#1990-2591 – us ATLA [280]

Hinduism and the modern world / Panikkar, Kavalam Madhava – Allahabad: Kitabistan, 1938 – (= ser Samp: indian books) – us CRL [280]

Hinduism in europe and america / Reed, Elizabeth Armstrong – New York: GP Putnam, 1914 – 1mf – 9 – 0-524-01291-1 – (incl bibl ref) – mf#1990-2327 – us ATLA [280]

Hinduism invades america / Thomas, Wendell – New York: Beacon Press, 1930 – (= ser Samp: indian books) – us CRL [280]

Hinduism outside india / Jagadiswarananda, Swami – Rajkot: Shri Ramakrishna Ashram, 1945 – (= ser Samp: indian books) – us CRL [280]

Hinduism, the world-ideal / Maitra, Harendranath – 1st ed. London: C Palmer & Hayward, c1916 – 1mf – 9 – 0-524-01911-8 – mf#1990-2724 – us ATLA [280]

The hindu-muslim problem in india / Manshardt, Clifford – London: George Allen & Unwin Ltd, 1936 – (= ser Samp: indian books) – us CRL [954]

The hindu-muslim questions / Prasad, Beni – Allahabad: Kitabistan, 1941 – (= ser Samp: indian books) – us CRL [954]

Hindu-pad-padashahi : or, a review of the hindu empire of maharashtra / Savarkar, Vinayak Damodar – Madras: BG Paul & Co, 1925 – (= ser Samp: indian books) – us CRL [954]

Hindusaskrtipradipa / Divekara, Mahadevasastri – Miraja: Mahadevasastri Divekara, 1946 – (= ser Samp: indian books) – us CRL [900]

Hindustan – Madras, India. Mar 1947-18 Dec 1949 – 2r – 1 – us L of C Photodup [079]

Hindustan times – New Delhi, India. 1952-55; 1960-Oct 1995 – 296r – 1 – us L of C Photodup [079]

Hindustana and hindvasi – New Delhi, India. 6 Sept 1946-1953; 1961-Jul 1987 – 136r – 1 – us L of C Photodup [079]

Hindustani musalmans and musalmans of the eastern punjab / Bourne, W Fitz G [comp] – Calcutta: Superintendent Govt Printing, 1914 [mf ed 1995] – (= ser Yale coll; Handbooks for the indian army) – vi/110p – 1 – 0-524-09936-7 – mf#1995-0936 – us ATLA [260]

Hindustanu = Hindustan sindhi daily – Bombay, India. 1962-Jun 1977; 1978-93 – 79r – 1 – us L of C Photodup [072]

Hindusthan standard – Calcutta, India. Apr 1944-Jun 1982 – 150r – 1 – us L of C Photodup [079]

Hindusthani music : an outline of its physics and aesthetics / Ranade, Ganesh Hari – Sangli: GH Ranade, 1938 – (= ser Samp: indian books) – us CRL [780]

Hindvasi – Bombay, India. 1962-64 – 3r – 1 – us L of C Photodup [079]

Hine, C Vickerstaff *see* On the indian river

Hine, Charles Cole *see*
– Laws of the several states in regard to insurance companies from other states and countries
– A trip to alaska

Hine, Edward *see* Forty-seven identifications of the lost british nation and the uw with the ten lost tribes

Hine, Gerald J *see* Radiation dosimetry

Hineni Ministries *see* Jews for jesus newsletter

Hiner, Lovell David *see* Mint oils in florida

Hiner, R *see* Kentucky conference pulpit

Hines, Harvey Kimball *see* Missionary history of the pacific northwest

Hingeston, C *see* Chronicle of england (rs1)

Hingeston, F C *see*
– Liber de illustribus henricis
– Royal and historical letters during the reign of henry the fourth

Hingham 1635-1900 – Oxford, MA (mf ed 1990) – (= ser Massachusetts vital records) – 238mf – 9 – 0-87623-116-4 – (mf 1-2: vital records 1635-1780. mf 3: out-of-town marriages 1657-1799. mf 4-5: first settlers 1635-50. mf 6-11: town records 1635-1700. mf 12-18: land records 1635-77. mf 19-21: proprietors 1720-69. mf 22: valuation in 1749. mf 23: town records 1642-51. mf 24-36: selectmen books 1661-1859. mf 37-40: mortgages 1825-46. mf 41-80: town records 1635-1858. mf 81-91: town meetings 1819-66. mf 92-94: rebellion record 1861-66. mf 95-96: out-of-town deaths 1844-90. mf 97-100: intentions 1700-1823. mf 100-106: intentions 1853-97. mf 107-131: births 1635-1880. mf 132-155: marriages 1635-1880. mf 156-174: deaths 1635-1880. mf 175-188: b,m,d index 1645-1900. mf 189-208: vital records 1635-1835. mf 209-215: vital records 1790-1848. mf 216-219: vital records 1844-55. mf 220-225: vital records 1847-69. mf 226-230: vital records 1865-80. mf 231-234: deaths 1877-1900. mf 235-236: marriages 1878-1900. mf 237-238: births 1880-99) – us Archive [978]

Hingham, Massachusetts. Hingham Baptist Church *see* Records

Hingorani, Anand T *see* To the princes and their people

Hinh nhu minh quen nhau / Vo, Ha Anh – [Saigon]: Nhu Y 1974 [mf ed 1992] – on pt of 1r – 1 – mf#11052 r340 n1 – us Cornell [959]

Hinitt, Frederick W *see* Religion and education, or, "what god hath joined together, let not man put asunder!"

Hinke, William John *see*
– Minutes and letters of the coetus of the german reformed congregations in pennsylvania, 1747-1792
– A new boundary stone of nebuchadrezzar 1 from nippur
– Selected babylonian kudurru inscriptions

Hinkel, John Vincent *see* The communist network

Hinkhouse, John Frederick *see* The beloved

Hinkin, Timothy R *see* Journal of quality assurance in hospitality and tourism

Hinkley, Edward Otis *see* The law of attachments in maryland

Hinkley, Edyth *see* A struggle for a soul

Hinkmar : erzbischof von reims / Schroers, Heinrich – Freiburg im Breisgau; St Louis, MO: Herder, 1884 – 2mf – 9 – 0-8370-6946-7 – (incl bibl ref and chronological listing of hincmar's works) – mf#1986-0946 – us ATLA [240]

Hinkovic, Henrik *see* Les croates sous le joug magyar

Hinkson, Henry Albert *see* Dublin verses by members of trinity college

Hinman heritage : the hinman family association – v1 n1-n3 [1976 jan/feb-sep/dec], v2 n1-v5 n4 [1977 spring-1980 winter] – 1r – 1 – mf#625999 – us WHS [929]

Hinojois, Marques de *see* Epigrafia romana y visigotica de montemolin

Hinojosa del Valle, Gregorio *see*
– Cuaderno de lenguaje. curso no 2. pt. 2
– Lenguaje. curso segundo

Hinostroza, Rodolfo *see* Consejero del lobo

Hi-notes / American Federation of Musicians – 1946 dec-1958, 1959-1967 jul, 1967 aug-1981 jan/feb – 3r – 1 – mf#578931 – us WHS [780]

Hinrichs, Friedrich *see* Richard wagner und die neuere musik

Hinrichs' katalog – Der im deutschen Buchhandel erschienenen Buecher, Zeitschriften, Landkarten usw. 20 v. in 14. 1871-1913 – 1,9 – us AMS Press [010]

Hinschius, Paul *see* Die paepstliche unfehlbarkeit und das vatikanische koncil

Hinsdale 1784-1849 – Oxford, MA (mf ed 1996) – (= ser Massachusetts vital record transcripts to 1850) – 3mf – 9 – 0-87623-266-7 – (mf 1t: intentions & marriages 1804-43; births & deaths 1784-1851. mf 2t: births & deaths 1791-1844; intentions & marriages 1814-50. mf 3t: b,m,d 1844-49) – us Archive [978]

Hinsdale 1860-1893 – Oxford, MA (mf ed 1983) – (= ser Massachusetts vital records) – 6mf – 9 – 0-931248-43-4 – (mf 1: births 1860-79. mf 2: births 1879-93; marriages 1860-69. mf 3: marriages 1870-89. mf 4: marriages 1890-93; deaths 1860-76. mf 5: deaths 1876-90. mf 6: deaths 1890-93) – us Archive [978]

Hinsdale, Burke Aaron *see*
– Horace mann and the common school revival in the united states
– Jesus as a teacher and the making of the new testament

Hinsdale county miscellaneous newspapers – Denver, CO (mf ed 1991) – 1r – 1 – (the phonograph (may 16 1891) and san juan crescent (jul 19-26 1877)) – mf#MF Z99 H596 – us Colorado Hist [071]

Hinsdale, New Hampshire.North Hinsdale Baptist Church *see* Records

Hinshelwood, N M *see* Montreal and vicinity

Hinson, BT *see* Markers of muscle damage following prolonged swimming, cycling, and running and a triathlon competition

Hinson, William Godber *see*
– Diary
– W g hinson papers

Hint to the voluntaries in reference to the honesty and candour of... – Edinburgh, Scotland. 1838 – 1r – us UF Libraries [240]

Hintenlang, Hubert *see* Untersuchungen zu den homer-aporien des aristoteles

Hinter den mauern der senana / Rhiem, Hanna – Berlin: M Warneck, 1902 [mf ed 1995] – (= ser Yale coll) – viii/154p (ill) – 1 – 0-524-10023-3 – (in german. pref by gustav warneck) – mf#1995-1023 – us ATLA [954]

Hinter der maske : suderman und hauptmann in den dramen johannes, die drei reherfedern, schluc und jau / Gimmerthal, Armin – Berlin: C A Schwetschke, 1901 – 1 – (incl ind) – us Wisconsin U Libr [790]

Hinter pflug und schraubstock : skizzen aus dem taschenbuch eines ingenieurs / Eyth, Max – 2. aufl. Stuttgart: Deutsche Verlags-Anstalt 1899 [mf ed 1989] – 2v in 1 on 1r – 1 – (filmed with: blut und eisen) – mf#7228 – us Wisconsin U Libr [880]

Der hinterlaender bote /...// *see* Anzeige-blatt fuer den kreis biedenkopf und bezirk voehl

Hinterm gartenbusch : geschichten und skizzen / Finckh, Ludwig – Leipzig: P Reclam, [1930] (mf ed 1990) – 1r – 1 – (filmed with: der deutsche finckh) – us Wisconsin U Libr [830]

Hinton, Alfred Horsley *see* A handbook of illustration

Hinton, Charles Howard *see* Chapters on the art of thinking

Hinton, Edward Wilcox *see* Cases on the law of evidence

Hinton, James *see*
– Chapters on the art of thinking
– Life and letters of james hinton
– Man and his dwelling place

Hinton, Stephanie A *see* Contraceptive practices among division 1 collegiate women swimmers

Hintrager, Oscar *see* Geschichte von sudafrika

Hints and answers to the exercises in elements of algebra / McLellan, James Alexander – Toronto: Canada Pub Co [1886] [mf ed 1985] – 1mf – 9 – 0-665-09638-0 – mf#09638 – cn Canadiana [510]

Hints and helps in pastoral theology / Plumer, William Swan – New York: Harper & Brothers, 1874 – 1mf – 9 – 0-7905-9584-2 – mf#1989-1309 – us ATLA [240]

Hints and helps to local preachers / Hocken, J – London, England. 1845 – 1r – us UF Libraries [240]

Hints and observations on the disadvantages of emigration to british america : addressed principally to the working classes of england. by an emigrant – London, 1833 – (= ser 19th c british colonization) – 1mf – 9 – mf#1.1.8803 – uk Chadwyck [304]

Hints for an improved translataion of the new testament / Scholefield, James – Cambridge, England. 1832 – 1r – us UF Libraries [225]

Hints for finding out truth / Biggs, James – Alcester, England. 1795 – 1r – us UF Libraries [240]

Hints for picturesque improvements in ornamented cottages : and their scenery / Bartell, Edmund – London 1804 – (= ser 19th c art & architecture) – 2mf – 9 – mf#4.2.1402 – uk Chadwyck [720]

Hints for procuring employment for the labouring poor : for the better managing parish concerns; and for reducing the rates / Lovell, Thomas – Huntingdon, 1826 – (= ser 19th c economics) – 1mf – 9 – mf#1.1.288 – uk Chadwyck [331]

Hints for some improvements in the authorized version of the new testament / Scholefield, James – 4th ed. Cambridge: Deighton, Bell; London: Bell & Daldy 1857 [mf ed 1986] – 1mf – 9 – 0-8370-9269-8 – (incl ind) – mf#1986-3269 – us ATLA [225]

Hints for the formation and improvement of a church choir – London, England. 1851 – 1r – us UF Libraries [780]

Hints for the valuation of ecclesiastical and other property / Ancona, J S – London, England. 1850 – 1r – us UF Libraries [240]

Hints from a lawyer: or, legal advice to men and women. / Spencer, Edgar A – New York, Putnam, 1888. 227 p. LL-1244 – 1 – us L of C Photodup [340]

Hints on agriculture / Beckett, John Edgar – Georgetown, Guyana. 1948 – 1r – us UF Libraries [630]

Hints on bible reading : with a collection of readings from various sources / Hill, Jno C – New York: Anson D F Randolph, c1877 – 1mf – 9 – 0-524-05729-X – mf#1992-0572 – us ATLA [220]

Hints on bible study / Trumbull, Henry Clay et al – Philadelphia: John D. Wattles, 1898. Beltsville, Md: NCR Corp, 1978 (3mf); Evanston: American Theol Lib Assoc, 1984 (3mf) – 9 – 0-8370-0870-0 – mf#1984-4197 – us ATLA [220]

Hints on bible study / Trumbull, Henry Clay et al – Philadelphia: John D Wattles, 1898, c1897 – 1mf – 9 – 0-8370-5579-2 – mf#1985-3579 – us ATLA [220]

Hints on education in india : with special reference to vernacular schools / Murdoch, John [comp] – Madras 1860 – (= ser 19th c british colonization) – 2mf – 9 – mf#1.1.7842 – uk Chadwyck [370]

Hints on health / Begg, James – Edinburgh, Scotland. 1875 – 1r – us UF Libraries [613]

Hints on household taste in furniture, upholstery / Eastlake, Charles Locke – 3rd ed. London 1872 – (= ser 19th c art & architecture) – 4mf – 9 – mf#4.2.39 – uk Chadwyck [740]

Hints on how to organize new local councils of women / National Council of Women of Canada – [Toronto?: s.n, 189-?] [mf ed 1993] – 1mf – 9 – 0-665-92295-7 – mf#92295 – cn Canadiana [305]

Hints on missions to india : with notices of some proceedings of a deputation from the american board, and of reports to it from the missions / Winslow, Miron – New York: M W Dodd, 1856 [mf ed 1995] – (= ser Yale coll) – 236p – 1 – 0-524-09097-1 – mf#1995-0097 – us ATLA [954]

Hints on national education in india / Nivedita, Sister – Calcutta: Udbodhan Office, 1923 – (= ser Samp: indian books) – us CRL [370]

Hints on old testament theology / Duff, Archibald – London: Adam & Charles Black, 1908 – 1mf – 9 – 0-8370-2987-2 – (incl ind) – mf#1985-0987 – us ATLA [221]

Hints on ornamental needlework...ecclesiastical purposes – London [1843] – (= ser 19th c art & architecture) – 1mf – 9 – mf#4.2.331 – uk Chadwyck [740]

Hints on preaching *see* Ch'uan tao i yu [ccm91]

Hints on rural residences / Carlisle, Nicholas – London 1825 – (= ser 19th c art & architecture) – 2mf – 9 – mf#4.2.1021 – uk Chadwyck [720]

Hints on the cingalese and english languages : with a selection of latin and french phrases, rendered into cingalese / Callaway, John – Colombo: printed for aut at the Wesleyan Mission-press, 1821 – (= ser 19th c books on linguistics) – 1mf – 9 – mf#2.1.2 – uk Chadwyck [400]

Hints on the formation of religious opinions : addressed especially to young men and women of christian education / Palmer, Ray – New York: Anson D F Randolph, 1867 – 1mf – 9 – 0-8370-4661-0 – mf#1985-2661 – us ATLA [210]

Hints on the interpretation of prophecy / Stuart, Moses – 2nd ed. Andover: Allen, Morrill & Wardwell 1842 [mf ed 1988] – 1mf – 9 – 0-7905-0351-4 – mf#1987-0351 – us ATLA [220]

Hints relative to the construction of fire-proof buildings / Bartholomew, Alfred – London 1839 – (= ser 19th c art & architecture) – 1mf – 9 – mf#4.2.849 – uk Chadwyck [720]

Hints respecting commentaries upon the scriptures / Stuart, Moses – [Andover: Flagg, Gould & Newman 1833] [mf ed 1984] – (= ser Biblical crit – us & gb 92.1) – 1mf – 9 – 0-8370-1587-1 – (incl bibl ref) – mf#1984-6255 – us ATLA [220]

Hints to geologists : respecting the mosaic account of the creation – Belfast, Northern Ireland. 1823 – 1r – us UF Libraries [240]

Hints to the christian pilgrim / Sands, David – Newcastle upon Tyne, England. 1848 – 1r – us UF Libraries [240]

Hints to young officers on the principles of military law : and on the practice of courts-martial / Harcourt, Francis V – London: W Houghton, 1833 – 2mf – 9 – $3.00 – mf#LLMC 89-030 – us LLMC [355]

Hintze, Carl Ernst *see* Die endlose strasse

Hintze, F *see* Die berliner handschrift der sahidischen apostelgeschichte

Hinukh ha-ivri bi-tefutsot ha-golash – Jerusalem, Israel. 1948 – 1r – us UF Libraries [939]

Hinxton 1538-1950 – (= ser Cambridgeshire parish register transcript) – 5mf – 9 – £6.25 – uk CambsFHS [929]

Hiob fuer die dritte auflage nach l. hirzel und j. olshausen / Dillmann, August – 3. Aufl. Leipzig: S Hirzel, 1869 – 1mf – 9 – 0-8370-3550-3 – (incl bibl ref) – mf#1985-1550 – us ATLA [220]

Hiob, oder, die vier spiegel : gedichte / Wolfskehl, Karl – Hamburg: Claassen Verlag, vorm. Claassen & Goverts, c1950 – 1r – 1 – us Wisconsin U Libr [810]

Hip : the jazz record digest – Milwaukee, Sterling VA, McLean VA. v1-5 n5; ns v1-10 n6. sep 1962-jan 1967, mar 1967-dec 1971 [all publ] – (= ser Jazz periodicals, 1914-1977) – 1r – 1 – $230.00 – us UPA [780]

Hipi; madjalah ekonomi / Tjahaja Asia – Djakarta, 1945-1946 – 6mf – 9 – (missing: 1946 v2(1)) – mf#SE-684 – ne IDC [959]

Hipolito yrigoyen : pueblo y gobierno – 2nd ed. Buenos Aires: Editorial Raigal, 1956 – 1 – $108.00 – mf#0262 – us Brook [972]

Hipoteca / Martinez Escobar, Manuel – Habana, Cuba. 1930 – 1r – us UF Libraries [610]

Hipparchus bithynius – In Arati et Eudoxi Phaenomena Commentariorum. German. 1894. Ed. by Karl Manitius – 1 – us Wisconsin U Libr [520]

Hippeau, Edmond Gabriel *see* Parsifal et l'opera wagnerien

Hippisley, Alfred Edward *see* China

Hippocampus – New York. 1998+ (1,5,9) – ISSN: 1050-9631 – mf#19153 – us UMI ProQuest [610]

Hippocrates *see*
– Die hippokratische schrift von der siebenzahl in ihrer vierfachen ueberlieferung
– Oeuvres completes

Hippocraticas theses ex libris aphor, prognostic. y vict.... / Morera, J M – Valencia, 1745 – 1mf – 9 – sp Cultura [610]

Hippocratis opera quae feruntur omnia – Lipsiae, Germany. v1-2. 1894-1902 – 1r – us UF Libraries [610]

Hippokrates ueber aufgaben und pflichten des arztes : in einer anzahl auserlesener stellen aus dem corpus / ed by Meyer-Steineg, Theodor & Schonack, Wilhelm – Bonn: A Marcus & E Weber 1913 [mf ed 1992] – (= ser Kleine texte fuer vorlesungen und uebungen 120) – 1mf – 9 – 0-524-04699-9 – (in greek. int & notes in german) – mf#1990-3408 – us ATLA [610]

Die hippokratische schrift von der siebenzahl in ihrer vierfachen ueberlieferung = De septimanis / Hippocrates; ed by Roscher, Wilhelm Heinrich – Paderborn: F Schoeningh, 1913 – (= ser Studien zur Geschichte und Kultur des Altertums) – 1mf – 9 – 0-524-06971-9 – (polyglot) – mf#1991-0045 – us ATLA [610]

Hippolyte, Dominique *see* Baiser de l'aieul

Hippolyti, S (Hippolytus of Rome, Saint) *see* Opera graece et latine

Hippolytos' capitel gegen die magier : refut haer 4 28-42 / Ganschinietz, R – Leipzig, 1913 – (= ser Tugal 3-39/2) – 2mf – 9 – €5.00 – ne Slangenburg [240]

Hippolytos' capitel gegen die magier, refut. haer. 4 28-42 / Ganschinietz, Richard – Leipzig: J C Hinrichs, 1913 – (= ser Tugal) – 1mf – 9 – 0-7905-1758-2 – (incl bibl ref and ind) – mf#1987-1758 – us ATLA [240]

Hippolyts danielcommentar / Diobouniotis, C – Leipzig, 1911 – (= ser Tugal 3-38/1b) – 1mf – 9 – €3.00 – ne Slangenburg [240]

Hippolyts kommentar zum hohenlied : auf grund von n marrs ausgabe des grusinischen textes / Bonwetsch, G N – Leipzig, 1902 – (= ser Tugal 2-23/2c) – 2mf – 9 – €5.00 – ne Slangenburg [240]

Hippolyts schrift ueber die segnungen jakobs / Diabouniotis, C & Beis, N – Leipzig, 1911 – (= ser Tugal 3-38/1a) – 1mf – 9 – €3.00 – ne Slangenburg [240]

Hippolyts schrift ueber die segnungen jakobs – hippolyts danielcommentar in handschrift no 573 des meteoronklosters = On the blessings of jacob / Dyobouniotes, Konstantinos & Bees, Nikos A – Leipzig: J C Hinrichs, 1911 – (= ser Tugal) – 1mf – 9 – 0-7905-1768-X – (incl ind) – mf#1987-1768 – us ATLA [221]

Hippolytsstudien / Achelis, Hans – Leipzig, 1897 – (= ser Tugal 2-16/4) – 4mf – 9 – €11.00 – ne Slangenburg [240]

Hippolytstudien / Achelis, Hans – Leipzig: J C Hinrichs, 1897 – (= ser Tugal) – 1mf – 9 – 0-7905-1621-7 – (incl bibl ref and ind) – mf#1987-1621 – us ATLA [240]

Hippolytus and callistus, or, the church of rome in the first half of the third century : with special reference to the writings of bunsen, wordsworth, baur, and gieseler = Hippolytus und kallistus / Doellinger, Johann Joseph Ignaz von – Edinburgh: T. and T. Clark, 1876 – 1mf – 9 – 0-7905-4352-4 – (incl bibl ref. in english) – mf#1988-0352 – us ATLA [240]

Hippolytus and his age : or, the beginnings and prospects of christianity = Hippolytus und seine zeit / Bunsen, Christian Karl Josias, Freiherr von – London: Longman, Brown, Green, and Longmans, 1854 – 3mf – 9 – 0-7905-5023-7 – (in english) – mf#1988-1023 – us ATLA [240]

Hippolytus, Antipope *see*
- Exegetische und homiletische schriften
- Skazaniia ob antikhristie v slavianskikh perevodakh s zamiechaniiami o slavianskikh perevodakh tvorenii sv ippolita
- The statutes of the apostles, or, canones ecclesiastici

Hippolytus of Rome, Saint *see* Hippolytus werke (gcsej9)

Hippolytus und die roemischen zeitgenossen, oder, die philosophumena und die verwandten schriften nach ursprung, composition und quellen / Volkmar, Gustav – Zuerich: E Kiesling, 1855 – (= ser Quellen der Ketzergeschichte bis zum Nicaenum) – 1mf – 9 – 0-7905-6030-5 – (incl bibl ref) – mf#1988-2030 – us ATLA [240]

Hippolytus von Rom (Hippolytus of Rome, Saint) *see* Widerlegung aller haeresien (bdk40 1.reihe)

Hippolytus von rom in seiner stellung zu staat und welt : neue funde und forschungen zur geschichte von staat und kirche in der roemischen kaiserzeit / Neumann, Karl Johannes – Leipzig: Veit & Comp, 1902 [mf ed 1990] – 1mf – 9 – 0-7905-6816-0 – (incl bibl ref) – mf#1988-2816 – us ATLA [240]

Hippolytus werke (gcsej9) – (= ser Griechische christlichen schriftsteller der ersten jahrhunderte (gcsej)) – (bd1/1 ed by h achelis 1897 €18. bd1/2 ed by h achelis 1897 €15. bd3 ed by p wendland 1916 €17. bd4 ed by a bauer €23) – ne Slangenburg [240]

Hipps, Richard Sherrill *see* An investigation of the ethical dilemmas in the practice of euthanasia

Hirai, Kaichiro *see* Papers, 1924-1953

Hiral, Ange-Marie *see* Le lis refleuri

Hiram college / Portage Co. Hiram – (jun 1868-may 1982) scattered [irreg] – 9r – 1 – mf#B12629-12637 – us Ohio Hist [378]

Hiram poetry review – Hiram. 1966-1995 (1) 1972-1995 (5) 1973-1995 (9) – ISSN: 0018-2036 – mf#6407 – us UMI ProQuest [410]

Hirato, K et al *see* Sendjinkoen dan tentera soekarela

Hiren schleifer / Albertinus, A – Muenchen: Niclas Hainrich, 1618 – 8mf – 9 – mf#O-1515 – ne IDC [090]

Hiriartia, J de *see* Le cas des catholiques basques

Hiriyanna, Mysore *see*
- The essentials of indian philosophy
- Outlines of indian philosophy
- Popular essays in indian philosophy
- Prof m hiriyanna commemoration volume
- The quest after perfection

Hirlekar, K S *see* Soviet russia

Hirmondo – Budapest, Hungary. Nov. 1859-Oct 1860.-m. 20 ft – 1 – uk British Libr Newspaper [072]

Hirnschleiffer / Albertinus, A – Coellen: Bey Constantino Munich, 1664 – 6mf – 9 – mf#O-1823 – ne IDC [090]

Hiroshima – Habana, Cuba. 1962 – 1r – us UF Libraries [025]

Hiroshima journal of medical sciences – Hiroshima. 1972-1996 (1) 1976-1996 (5) 1976-1996 (9) – ISSN: 0018-2052 – mf#6954 – us UMI ProQuest [610]

Hiroyuki ichihara papers, 1942 / Ichihara, Hiroyuki – 2r – 1 – (with finding aid) – us UW Libraries [977]

Hirsch, Alan et al *see* Awarding attorneys' fees and managing fee litigation

Hirsch, Carl *see* Die gegenwart

Hirsch, Emil Gustav *see* Paul, the apostle of heathen judaism, or, christianity

Hirsch, Franz *see* Geschichte der deutschen litteratur von ihren anfaengen bis auf die neueste zeit

Hirsch, Hermann *see* Der weisse mantel faellt

Hirsch, Max *see*
- Archiv fuer frauenkunde und eugenik
- Reise in das innere von algerien durch die kabylie und sahara

Hirsch, S J S *see* Saul mozes slagter

Hirsch, Samson Raphael *see*
- Ein merkblatt
- Metav higayon
- The nineteen letters of ben uziel

Hirsch, Samuel Abraham *see*
- A book of essays
- The greek grammar of roger bacon and a fragment of his hebrew grammar

Hirschberger nachrichten – Hirschberg, Saale DE, 1936 1 jul-30 sep, 1937 2 jan-31 mar & 1 jul-31 dec, 1938 1 apr-30 jun, 1943 1 jul-31 dec – 4r – 1 – gw Misc Inst [074]

Hirschberger nachrichten – Hirschberg, Saale DE, 1936 1 jul-30 sep, 1937 2 jan-31 mar & 1 jul-31 dec, 1938 1 apr-30 jun, 1943 1 jul-31 dec – 4r – 1 – gw Misc Inst [074]

Hirsch-Davies, John Edwin de *see* A popular history of the church in wales

Hirsche, K *see* Prolegomena zu einer neuen ausgabe der imitatio christi

Hirschensohn, Chajim *see* Torat ha-hinukh ha-yisraeli

Hirschfeld, Chr Cay Lor *see* Taschenbuch fuer gartenfreunde

Hirschfeld, Georg *see* The mothers

Hirschfeld, H *see* New researches into the composition and exegesis of the qoran

Hirschfeld, Hartwig *see*
- Beitraege zur erklaerung des koraan
- New researches into the composition and exegesis of the qoran

Hirschfeld, Magnus *see*
- Geschlechtskunde
- Jahrbuch fuer sexuelle zwischenstufen unter besonderer beruecksichtigung der homosexualitaet
- Zeitschrift fuer sexualwissenschaft

Hirschfeld,Christian Cay Lorenz *see*
- Gartenkalender [auf das jahr 1782-89]
- Kleine gartenbibliothek

Hirsching, Friedrich Karl Gottlob *see* Allgemeines archiv fuer die laender- und voelkerkunde

Hirschkan, Zevi *see* Fun dervaytns

Hirschl, Andrew Jackson *see* The law of fraternities and societies...with special reference to their insurance feature

Hirschl, Samuel D *see* Business law

Hirschman, Albaert O *see* Journeys toward progress

Hirschowitz, Abraham Eber *see* Bet midrash shemu'el

Hirschstein, Hans *see* Die franzoesische revolution im deutschen drama und epos nach 1815

Hirscht, Arthur *see* Die apokalypse und ihre neueste kritik

Hirschy, Noah Calvin *see* Artaxerxes 3 ochus and his reign

Hirsh lekert – Moskve, Russia. 1922 – 1r – us UF Libraries [939]

Hirsh lekert / Pat, Jacob – Varshe, Poland. 1927 – 1r – us UF Libraries [939]

Hirst, J Crowther *see* Revivalism and revival theology

Der hirt des hermas / Bruell, Andreas – Freiburg i.B; St Louis, MO: Herder, 1882 [mf ed 1986] – 1mf – 9 – 0-8370-9607-3 – (incl bibl ref) – mf#1986-3607 – us ATLA [240]

Der hirt des hermas / Zahn, Theodor – Gotha: Friedrich Andreas Perthes, 1868 – 2mf – 9 – 0-8370-9675-8 – (incl bibl ref) – mf#1986-3675 – us ATLA [920]

Der hirte des hermas : ein beitrag zur patristik / Gaab, Ernst – Basel: Felix Schneider, 1866 – 1mf – 9 – 0-8370-9622-7 – (incl bibl ref) – mf#1986-3622 – us ATLA [920]

Hirtennovelle / Wiechert, Ernst Emil – Muenchen: K Desch, c1945 – 1r – 1 – us Wisconsin U Libr [830]

Hirtenstimmen : noch ein jahrgang epistel-predigten / Gerok, Karl – 2. Aufl. Stuttgart: Greiner & Pfeiffer, [1882?] Beltsville, Md: NCR Corp, 1978 (9mf); Evanston: American Theol Lib Assoc, 1984 (9mf) – 9 – 0-8370-1045-4 – mf#1984-4428 – us ATLA [240]

Hirth, Friedrich *see*
- The ancient history of china
- Aus friedrichs hebbels korrespondenz
- China and the roman orient

Hirwaun, glamorgan, parish church of st llewrwg : baptisms 1858-1925, burials 1901-1910, marriages 1886-1925 – [Glamorgan]: GFHS [mf ed c2003] – 1mf – 9 – £1.25 – uk Glamorgan FHS [929]

Hirzel, Heinrich *see*
- Eugenias briefe
- Eugenias briefe an ihre mutter

Hirzel, Ludwig *see*
- Albrecht hallers tagebuecher seiner reisen nach deutschland, holland und england
- Geschichte der gelehrtheit
- Salomon hirzels verzeichniss einer goethe-bibliothek
- Wieland und martin und regula kuenzli

Hirzel, R *see* Der name

Hirzel, Salomon *see* Der junge goethe

Hirzel-Escher, Hans *see* Wanderungen in weniger besuchte alpengegenden der schweiz und ihrer naechsten umgebungen

His – Chicago. 1941-1987 (1) 1972-1987 (5) 1975-1987 (9) – (cont by: u) – ISSN: 0018-2095 – mf#6693 – us UMI ProQuest [240]

His chief's wife / Anethan, Eleanora Mary (Haggard), Baronne d' – London: Chapman & Hall Ltd, 1897 – (= ser 19th c women writers) – 4mf – 9 – mf#5.1.85 – uk Chadwyck [420]

...His discours of voyages into ye easte and west indies / Linschoten, J H van – London: John Wolfe, [1598]. 4v – 11mf – 9 – mf#H-8431 – ne IDC [918]

His divine majesty, or, the living god / Humphrey, William – London: T Baker, 1897 – 2mf – 9 – 0-7905-8807-2 – mf#1989-2032 – us ATLA [210]

His excellency lord gosford, the governor-general of the canadas etc etc : will you permit me to recall your attention to a subject of all others, of a temporary kind, the most important to me... / Burroughs, Stephen – [Trois-Rivieres?: s.n, 1836?] [mf ed 1993] – 1mf – 9 – 0-665-91357-5 – mf#91357 – cn Canadiana [346]

His excellency the governor general and lady head request the honor of...company on the 4th february, to a ball – [s.l: s.n. 185-?] [mf ed 1983] – 1mf – 9 – 0-665-39763-1 – mf#39763 – cn Canadiana [090]

His footsteps : studies for edification from the life of christ / Lenski, Richard Charles Henry – Columbus, Ohio: Lutheran Book Concern, 1898 – 1mf – 9 – 0-524-06843-7 – mf#1992-0985 – us ATLA [240]

His friends : the story of the immediate disciples of jesus after his ascension and their letters to the early christians, using the text of the american standard revised bible / Soares, Theodore Gerald et al – Chicago: Hope Pub Co, c1906 – 1mf – 9 – 0-524-06036-3 – mf#1992-0749 – us ATLA [220]

His great apostle : the life and letters of paul / Strong, Sydney et al – Chicago: Hope Pub Co, c1906 – 1mf – 9 – 0-524-06195-5 – mf#1992-0833 – us ATLA [220]

His highness the aga khan : imam of the ismailis / Greenwall, Harry James – London: Cresset Press, 1952 – (= ser Samp: indian books) – (foreword on racing by hh the aga khan) – us CRL [920]

His highness the maharaja of bikaner : a biography / Panikkar, Kavalam Madhava – London; New York: Oxford University Press, 1937 – (= ser Samp: indian books) – (int by lord hardinge of penshurst) – us CRL [920]

His honor the president's speech at the opening of the present session of the legislature : the answer of both houses thereto... – [Toronto?: s.n, 1813?] [mf ed 1992 – 1mf – 9 – 0-665-91050-9 – mf#91050 – cn Canadiana [323]

His honour and a lady / Cotes, Everard, mrs [Sara Jeanette Duncan] – London; New York: Macmillan, 1896 – (= ser Macmillan's colonial library) – 4mf – 9 – mf#26975 – cn Canadiana [830]

His life : a complete story in the words of the four gospels / Barton, William Eleazar et al – Chicago: Hope Pub Co, c1905 – 1mf – 9 – 0-524-06031-2 – mf#1992-0744 – us ATLA [220]

His majesty, the president of brazil / Hambloch, Ernest – New York, NY. 1936 – 1r – us UF Libraries [972]

His majesty's theatre, montreal : mme sarah bernhardt – [Montreal]: [s.n.], [1916] (mf ed 1988) – 1mf – 9 – mf#SEM105P904 – cn Bibl Nat [790]

His name is today : report of the committee of inquiry into child mental health care services / South Africa. Committee of Inquiry into Child Mental Health Care Services – Pretoria: Dept of National Health & Population Development 1988 [mf ed Pretoria, RSA: State Library [199-]] – 181p on 1r with other items – 5 – (also available in afrikaans [op 09154 r25]; incl bibl ref) – mf#op 09155 r25 – us CRL [362]

His pilgrimes / Purchas, S – London: William Stansby, 1625-1626. 5v – 95mf – 9 – mf#HT-679 – ne IDC [910]

His presence / Ryle, J C – London, England. 1872 – 1r – us UF Libraries [240]

His sunday schools and his friends / Raikes, Robert – 336p – 1 – us Southern Baptist [242]

His version of it / Ford, Paul Leicester – Toronto: Musson, 1905, c1898 [mf ed 1994] – 2mf – 9 – 0-665-72958-8 – mf#72958 – cn Canadiana [830]

His writings / Wheelwright, John – 1 – $50.00 – (also: a memoir by charles h. bell. 1876) – us Presbyterian [240]

Hiscox, Edward Thurston *see*
- The baptist church directory
- The baptist short method with inquirers and opponents
- The standard manual for baptist churches

Hise, Charles R van *see* Wisconsin progressives

Hise, Daniel H *see* Diaries (abolitionist)

Hislop, Alexander *see*
- The two babylons

Hispalensis, Isidorus *see*
- De ecclesiasticis officiis

Hispania – University. 1918+ (1) 1969+ (5) 1975+ (9) – ISSN: 0018-2133 – mf#949 – us UMI ProQuest [370]

Hispaniae schola musica sacra / opera varia [saecal 15, 16, 17 et 18]... – Barcelona: J B Pujol 1894-98 [mf ed 19–] – 1r – 1 – (v3 has imprint: leipzig: breitkopf & haertel; contents: 1. christophorus morales; 2. francisus guerrero; 3-4. antonius a cabezon; 5. joannes ginesius perez; 6. psalmodia modulata (vulgo fabordones) a diversiis auctoribus, inter quos fr thomas a sancta maria, franciscus guerrero, thome ludovici a victoria, ceballos, aliique incenti autignotis; 7-8. antonius a cabezon) – mf#film 744 – us Sibley [780]

Hispanianarum regine. poemas guadalupenses / Corredor Garcia, Antonio – Sevilla: Editorial San Antonio, 1950 – 1 – sp Bibl Santa Ana [440]

Hispanic – Washington. 1995+ (1,5,9) – ISSN: 0898-3097 – mf#18405 – us UMI ProQuest [305]

Hispanic american historical review – Durham. 1957+ (1) 1969+ (5) 1975+ (9) – ISSN: 0018-2168 – mf#1023 – us UMI ProQuest [972]

Hispanic american report – Stanford. 1948-1964 (1) – mf#32 – us UMI ProQuest [327]

The hispanic collection – 14r – 1 – $490.00 – Dist. us Scholarly Res – us L of C Photodup [975]

Hispanic journal of behavioral sciences – Thousand Oaks. 1979+ (1,5,9) – ISSN: 0739-9863 – mf#12551 – us UMI ProQuest [150]

Hispanic law journal *see* Texas hispanic journal of law and policy

Hispanic law journal – v1-3. 1994-97 – 9 – (title varies. see: texas hispanic journal of law and policy) – mf#116652 – us Hein [322]

Hispanic link weekly report – v3, n1-51 [1985 jan 7-dec 23], v4 n46 [1986 nov 17] – 1r – 1 – mf#1238890 – us WHS [071]

Hispanic media and markets / Standard Rate & Data Service – 1988 mar 28-jun 28 – 1r – 1 – mf#1581714 – us WHS [380]

Hispanic review – Philadelphia. 1933+ (1) 1969+ (5) 1975+ (9) – ISSN: 0018-2176 – mf#1021 – us UMI ProQuest [400]

Hispaniola / Lemaire, Emmeline Carries – Habana, Cuba. 1944 – 1r – us UF Libraries [972]

Hispanismos en el guarani / Morinigo, Marcos Augusto – Buenos Aires, Argentina. 1931 – 1r – us UF Libraries [972]

Hispano – Periodicals CA. 1969 jul 1-1970 mar 10, 1976 apr 26-1978 may, 1978 jun-1980 dec, 1981-82, 1982 dec 23-1984 jan 4, 1985 nov 6-1986 dec, 1987 jan 7-jun 24, 1987 jul-1988 jun, 1988 jul-1989 jun – 9r – 1 – (continued by: hispanoamericano) – mf#384137 – us WHS [071]

El hispano-amazonense : organo de la colonia espanola en el Amazonas – Amazonas, 17 maio 1918-ago 1920; jan 1921-30 set 1922 – (= ser Ps 19) – mf#P11B,06,24 – bl Biblioteca [079]

Hispanoamericano – Sacramento CA. 5/24/68, 8/2/68/ 10/4-18/68, 12/24/68, 1/28/69, 2/18/69, 3/4-11/69, 3/25/69, 4/15/69 [v2 n11] – 1r – 1 – (continued by: hispano) – mf#691572 – us WHS [071]

Hispanofila – Chapel Hill. 1957+ (1) 1976+ (5) 1976+ (9) – ISSN: 0018-2206 – mf#3013 – us UMI ProQuest [440]

Hiss, Philip Hanson *see* Selective guide to the english literature

Hissmann, Mich [v1-6] *see* Magazin fuer die philosophie und ihre geschichte

Histadrut Ha-Kelalit Shel Ha-'Ovdim Ha-'Ivrim Be-Erets-Yisra'el *see*
- Bi-shenat ha-sheloshim
- Ve'idah ha-shev'it

Histadrut "'Ivriyah" *see* Konferentsyah ha-hagit

Histochemical journal – London. 1989-1996 (1,5,9) – ISSN: 0018-2214 – mf#14400 – us UMI ProQuest [574]

Histochemistry – Heidelberg. 1978-1991 (1) 1978-1991 (5) 1974-1991 (9) – ISSN: 0301-5564 – mf#13175,02 – us UMI ProQuest [574]

Histograms / Lawson, E W – s.l, s.l? 1938 – 1r – us UF Libraries [978]

Histoire : ou vie tiree des monuments anecdotes de l'ancienne egypte / [Terrasson, J] – Paris, 1731. 3v – 17mf – 9 – mf#VR-12.9 – ne IDC [956]

Histoire abregee de l'ancien testament – nouv rev corr ed. [Quebec?: s.n.] 1832 [mf ed 1984] – 2mf – 9 – 0-665-45466-X – mf#45466 – cn Canadiana [221]

Histoire abregee de l'ancien testament avec celle de la vie de notre seigner jesus-christ : ou sont contenues ses principales actions – nouv ed. Montreal: Chez James Brown...& James Lane...2v. 1821 [mf ed 1985] – 2v on 1mf – 9 – mf#44058 – cn Canadiana [221]

Histoire abregee de l'eglise metropolitaine d'utrecht / Dupac de Bellegarde, M G – 3e corr aug ed. Utrecht, 1852 – €26.00 – ne Slangenburg [242]

Histoire abregee du jansenisme et remarques sur l'ordonnance de m l'archeveque de paris / Louail-Fouillou-de Joncoux – Cologne, 1698 – 3mf – 8 – €7.00 – ne Slangenburg [242]

Histoire amoureuse des gaules / Bussy-Rabutin, Roger de – Paris 1829 [mf ed Hildesheim 1995-98] – 3v on 9mf – 9 – €180.00 – ISBN-10: 3-487-26091-3 – ISBN-13: 978-3-487-26091-4 – gw Olms [944]

Histoire ancienne de l'afrique du nord / Gsell, Stephane – Paris. 1920-28. 8v – 1 – us CRL [960]

Histoire ancienne de l'egypte – Paris 1830/36 [mf ed Hildesheim 1995-98] – 2v on 7mf – 9 – €140.00 – 3-487-27381-0 – gw Olms [960]

Histoire ancienne du canon du n test (etb) / Lagrange, Marie Joseph – Paris, 1933 – 4mf – 8 – €11.00 – ne Slangenburg [225]

Histoire ancienne, egypte, assyrie *see* Life in ancient egypt and assyria

Histoire anecdotique et raisonnee du theatre italien : depuis son retablissement en France jusqua l'annee 1769 / Desboulmiers, Jean-Augustin-Julien – Paris: Chez Des Ventes Deladoue 1770 [mf ed 19–] – 7v on 39mf – 9 – mf#fiche 137-143 – us Sibley [790]

Histoire anonyme de la premiere croisade. / Brehier, Louis – Paris, 1924 – 4mf – 9 – mf#H-2921 – ne IDC [931]

Histoire apologetique de la conduite des jesuites de la chine : adressee a messieurs des missions etrangeres / Daniel, Gabriel – [s.l: s.n] 1700 [mf ed 1995] – (= ser Yale coll) – 217p – 1 – 0-524-09573-6 – (in french) – mf#1995-0573 – us ATLA [241]

Histoire celeste : ou, recueil de toutes les observations astronomiques faites par ordre du roy: avec un discours preliminaire sur le progres de l'astronomie... / Monnier, Pierre-Charles le – Paris: C Briasson [1741] [mf ed 1998] – 1r [pl/ill] – 1 – (comprises the observations by jean picard fr 1666-77, & those of picard & philippe de la hire 1677 to 1685, in the form of tables) – mf#film mas 28401 – us Harvard [520]

Histoire chronologique de la nouvelle-france / Le Tac, Sixte – Paris: Fischbacher, 1888 [mf ed 1971] – 1r – 1 – mf#SEM35P71 – cn Bibl Nat [971]

Histoire chronologique des voyages vers le pole arctique : entrepris pour decouvrir un passage entre l'ocean atlantique et le grand-ocean...des scandinaves jusqu'a l'expedition faite en 1818... / Barrow, John – Paris: Libr de Gide fils... 1819 [mf ed 1985] – 2v on 1mf – 9 – 0-665-48254-X – (trans fr english) – mf#48254 – cn Canadiana [910]

Histoire comique des etats et empires de la lune et du soleil / Cyrano De Bergerac – Paris, France. 18– – 1r – us UF Libraries [025]

Histoire complete de l'idee messianique chez le peuple d'israel : ses developpements, son alteration, son rejeunissement / Lemann, Augustin – Lyon: Libr Catholique Emmanuel Vitte, 1909 [mf ed 1989] – 2mf – 9 – 0-7905-1283-1 – (incl bibl ref) – mf#1987-1283 – us ATLA [377]

Histoire complete des naufrages : evenemens es aventures de mer – Paris: [s.n.] 1836 [mf ed 1985] – 2v on 1mf – 9 – 0-665-50057-2 – mf#50058 – cn Canadiana [910]

Histoire constitutionnelle et administrative de la france depuis la mort de philippe-auguste / Capefigue, Jean Baptiste Honore Raymond – Paris 1831-1833 [mf ed Hildesheim 1995-98] – 4v on 13mf – 9 – €130.00 – ISBN-10: 3-487-26312-2 – ISBN-13: 978-3-487-26312-0 – gw Olms [944]

Histoire contemporaine / Michel, Antoine – Port-Au-Prince, Haiti. 1913 – 1r – us UF Libraries [934]

Histoire contenant vne sommaire description des genealogies, alliances : and gestes de tous les princes and grans seigneurs, dont la pluspart estoient francois, qui ont iadis commade es royaumes de hierusalem, cypre, armenie and lieux circonuoisins / Lusignan de Cypre, E de – Paris, 1579 – 2mf – 9 – mf#AR-1778 – ne IDC [956]

Histoire contenant vne sommaire description des genealogies, alliances, and gestes de tous les princes and grans seigneurs... / Lusignano, S di – Paris, 1579 – 2mf – 9 – mf#H-8353 – ne IDC [956]

Histoire critique de la creance et des coutumes des nations du levant / Moni, D de – Francfort, 1684 – 3mf – 9 – mf#AR-1678 – ne IDC [956]

Histoire critique de la decouverte du mississipi sic (1669-1673) : d'apres les documents inedits du ministere de la marine / Harrisse, Henry – Paris?: P Dupont, 1872? – 1mf – 9 – mf#34450 – cn Canadiana [917]

Histoire critique de la litterature prophetique des hebreux : depuis les origines jusqu'a la mort d'isaie / Bruston, Charles – Paris: Fischbacher, 1881 – 1mf – 9 – 0-8370-2499-4 – mf#1985-0499 – us ATLA [221]

Histoire critique de l'ecole d'alexandrie / Vacherot, Etienne – Paris: Ladrange, 1846-51 [mf ed 1991] – 3v on 15mf – 9 – 0-524-00413-7 – (incl bibl ref) – mf#1989-3113 – us ATLA [180]

Histoire critique de manichee et du manicheisme / Beausobre, M de – Amsterdam. v1-2. 1734-39 – €75.00 – ne Slangenburg [290]

Histoire critique de nicolas flamel et de perenelle sa femme; on y a joint le testament de perenelle & plusieurs autres pieces interessantes / Villain, Etienne Francois – Paris, 1761 – 1 – us Wisconsin U Libr [944]

Histoire critique des doctrines religieuses de la philosophie moderne / Bartholmess, Christian – Paris: Ch Meyrueis, 1855 – 3mf – 9 – 0-524-00003-4 – (incl bibl ref) – mf#1989-2703 – us ATLA [100]

Histoire critique des dogmes et des cultes, bons et mauvais... / Jurieu, P – Amsterdam, 1704 – 10mf – 9 – mf#PRS-156 – ne IDC [240]

Histoire critique des livres de l'ancien testament = De thora en de historische boeken des ouden verbands / Kuenen, Abraham – Paris: Michel Levy, 1866 [mf ed 1986] – 2mf – 9 – 0-8370-9801-7 – (incl bibl ref. pref by ernest renan) – mf#1986-3801 – us ATLA [221]

Histoire critique du catholicisme liberal en france : jusqu'au pontificat de leon 13: complement de toutes les histoires de l'eglise / Fevre, Justin Louis Pierre – Saint-Dizier: G Saint-Aubin et Thevenot, 1897 – 2mf – 9 – 0-8370-8571-3 – (incl bibl ref) – mf#1986-2571 – us ATLA [241]

Histoire critique du gnosticisme : et de son influence sur les sectes religieuses et philosophiques des six premiers siecles de l'ere chretienne / Matter, Jacques – 2. ed, rev et augm. Strasbourg: V Levrault, 1843-1844 – 3mf – 9 – 0-7905-9337-8 – mf#1989-2562 – us ATLA [290]

Histoire critique du vieux testament / Richard, Simon – Rotterdam, 1685 – 21mf – 8 – €41.00 – ne Slangenburg [225]

Histoire critique du vieux testament / Simon, R – Paris, 1680 – 7mf – 9 – mf#CA-150= – ne IDC [240]

L'histoire dahomeenne de la fin du 19e siecle a travers les textes – Port Novo, 195? – (= ser Etudes dahomeennes n9) – 1 – us CRL [960]

Histoire d'alger sous la domination turque / Grammont, Henri Delmas de – Paris, 1887. 16 + 420p – 1 – us Wisconsin U Libr [956]

Histoire d'angkor / Giteau, Madeleine – 1.ed. Paris: Presses universitaires de France 1974 [mf ed 1989] – (= ser Que sais -je? 1580) – 1r with other items – 1 – (with bibl) – mf#mf-10289 seam reel 022/10 [§] – us CRL [930]

Histoire de baghdad dans les temps modernes / Huart, C – Paris, 1901 – 3mf – 9 – mf#NE-20134 – ne IDC [956]

Histoire de beyrouth... / Salih Ibn Yahya; ed by Cheikho, L – Beyrouth, 1902 – 4mf – 9 – mf#NE-20168 – ne IDC [956]

Histoire de blondine : de bonne-biche et de beau-minon / Segur, Sophie, comtesse de; ed by Belisle, Louis-Alexandre – Quebec: [1942?] (mf ed 1990) – 1mf – 9 – (ill by vernier) – mf#SEM105P1269 – cn Bibl Nat [971]

Histoire de bresse et de bugey : contenant ce qui s'y est passe de memorable sous les romains... / Guichenon, Samuel – Lyon: Jean Antoine Huguetan et Marc Ant Ravaud, 1650 [mf ed 1978] – 1r – 1 – mf#SEM35P156 – cn Bibl Nat [944]

Histoire de bretagne / Daru, Pierre Antoine – Paris 1826 [mf ed Hildesheim 1995-98] – 3v on 9mf – 9 – €180.00 – ISBN-10: 3-487-25966-4 – ISBN-13: 978-3-487-25966-6 – gw Olms [944]

Histoire de christophe colomb / Roselly de Lorgues, Comte – Montreal: Librarie Saint Joseph, Cadieux & Derome, 1883 [mf ed 1982] – (= ser Bibliotheque religieuse et nationale. 1re serie in – 12) – 2mf – 9 – mf#SEM105P78 – cn Bibl Nat [910]

Histoire de france / Bignon, Louis – Paris 1829-1830 [mf ed Hildesheim 1995-98] – 6v on 18mf – 9 – €180.00 – ISBN-10: 3-487-26252-5 – ISBN-13: 978-3-487-26252-9 – gw Olms [944]

Histoire de france : depuis la mort de louis 14 jusqu'a la paix de versailles de 1783 / Fantin Desodoards, Antoine – Paris 1789 [mf ed Hildesheim 1995-98] – 8v on 40mf – 9 – €400.00 – ISBN-10: 3-487-26085-9 – ISBN-13: 978-3-487-26085-3 – gw Olms [944]

Histoire de france : pendant le dix-huitieme siecle / Lacretelle, Charles de – Paris 1812-1826 [mf ed Hildesheim 1995-98] – 15v on 41mf – 9 – €410 – ISBN-10: 3-487-26086-7 – ISBN-13: 978-3-487-26086-0 – gw Olms [944]

Histoire de france : pendant les annees 1825, 1826, 1827 et commencement de 1828... / Montgaillard, Guillaume H de – Paris [mf ed Hildesheim 1995-98] – 3v on 6mf – 9 – €120.00 – ISBN-10: 3-487-26139-1 – ISBN-13: 978-3-487-26139-3 – gw Olms [944]

Histoire de france, depuis la restauration / Lacretelle, Charles de – Paris 1829-1835 [mf ed Hildesheim 1995-98] – 4v on 13mf – 9 – €130.00 – ISBN-10: 3-487-26346-7 – ISBN-13: 978-3-487-26346-5 – gw Olms [944]

Histoire de france pendant les guerres de religion : 4 Bde / Lacretelle, Charles de – Paris 1814-1816 [mf ed Hildesheim 1995-98] – 12mf – 9 – €120.00 – ISBN-10: 3-487-26122-7 – ISBN-13: 978-3-487-26122-5 – gw Olms [944]

Histoire de france sous napoleon / Bignon, Louis – Paris 1838-1850 [mf ed Hildesheim 1995-98] – 8v on 26mf – 9 – €260 – ISBN-10: 3-487-26253-3 – ISBN-13: 978-3-487-26253-6 – gw Olms [944]

Histoire de francois premier, roi de france : dit le grand roi et le pere des lettres / Gaillard, Gabriel Henri – Paris 1766 [mf ed Hildesheim 1995-98] – 4v on 25mf – 9 – €250.00 – ISBN-10: 3-487-26215-0 – ISBN-13: 978-3-487-26215-4 – gw Olms [944]

Histoire de georges castriot, svrnomme scanderbeg, roy d'albanie : contenant ses illustres faicts, d'armes, and memorables victoires a l'encontre des turcs... / Lavardin, J de – Paris, 1598 – 11mf – 9 – mf#H-8398 – ne IDC [920]

Histoire de gregoire 7 : precedee d'un discours sur l'histoire de la papaute jusqu'au 11e siecle / Villemain, M Abel-Francois – 2e ed. Paris: Didier. 2v. 1874 – 3mf – 9 – 0-8370-7914-4 – (incl bibl ref) – mf#1986-1914 – us ATLA [920]

Histoire de guillaume 3 : roy d'angleterre, de cosse, de france, et d'irlande, prince d'orange, etc / Chevalier, N – Amsterdam, 1692 – 4mf – 9 – mf#O-1542 – ne IDC [090]

Histoire de hang-yon [hangsa yonta] : tiree de satras sur feuilles de latanier / Hangsa Yonta – 3.ed. Phmon-Penh: Editions de l'Institut bouddhique 1964 [mf ed 1990] – 1r with other items – 1 – (title & text in khmer; added t.p. in french) – mf#mf-10289 seam reel 100/02 [§] – us CRL [280]

Histoire de jerusalem / Poujoulat, M – 4e rev et corr ed. Paris: J Vermot, 1856 – 2mf – 9 – 0-524-03414-1 – mf#1990-0968 – us ATLA [956]

Histoire de kamtschatka, des isles kurilski, et des contrees voisines : publiee a petersbourg, en langue russienne, par ordre de sa majeste imperiale / Kraseninnikov, Stepan – Lyon 1767 [mf ed Hildesheim 1995-98] – 2v on 6mf – 9 – €120.00 – 3-487-28971-7 – gw Olms [910]

Histoire de kentucke, nouvelle colonie a l'ouest de la virginie : ouvrage pour servir de suite aux lettres d'un cultivateur americain / Filson, John – Paris 1785 [mf ed Hildesheim 1995-98] – 1v on 2mf – 9 – €60.00 – 3-487-27150-8 – gw Olms [978]

Histoire de la bastille : avec un appendice, contenant entr'autres choses une discussion sur le prisonnier au masque de fer / Craufurd, Quentin – [Frankfurt/Main 1798 [mf ed Hildesheim 1995-98] – 1v on 7mf – 9 – €140.00 – ISBN-10: 3-487-25923-0 – ISBN-13: 978-3-487-25923-9 – gw Olms [944]

Histoire de la bible en france : suivie de fragments relatifs a l'histoire generale de la bible et d'un apercu sur le colportage biblique en france et en indo-chine au vingtieme siecle / Lortsch, Daniel – Paris: Agence de la Societe biblique britannique et etrangere, 1910 – 2mf – 9 – 0-8370-1729-7 – mf#1987-6125 – us ATLA [220]

Histoire de la bible et de l'exegese biblique jusq'a nos jours / Wogue, L – Paris: Imprimerie nationale, 1881 – 1mf – 9 – 0-8370-7356-1 – (incl bibl ref and index) – mf#1986-1356 – us ATLA [220]

Histoire de la captivite et de la mort de toussain / Nemours, Alfred – Paris, France. 1929 – 1r – us UF Libraries [972]

Histoire de la charite / Lallemand, Leon – Paris: Alphonse Picard, 1902-1912 – 5mf – 9 – 0-524-03404-4 – (incl bibl ref) – mf#1990-0958 – us ATLA [360]

Histoire de la chute de l'empire de napoleon : ornee de huit plans ou cartes, pour servir au recit des principales batailles livrees en 1813 et 1814 / Labaume, Eugene – Paris 1820 [mf ed Hildesheim 1995-98] – 2v on 6mf – 9 – €120.00 – ISBN-10: 3-487-26395-5 – ISBN-13: 978-3-487-26395-3 – gw Olms [940]

Histoire de la chute du roi louis-philippe : de la republique de 1848 et duretablissement de l'empire (1847-1855) / Granier De Cassagnac, Adolphe – Paris 1857 [mf ed Hildesheim 1995-98] – 2v on 12mf – 9 – €120.00 – ISBN-10: 3-487-26030-1 – ISBN-13: 978-3-487-26030-3 – gw Olms [944]

Histoire de la civilisation contemporaine / Seignobos, Charles – Paris: Masson et Cie., 1890. 424p. Includes bibliographies – 1 – us Wisconsin U Libr [000]

Histoire de la clarte francaise *see* Rough justice

Histoire de la clarte francaise, ses origines, son evolution, sa valeur / Mornet, Daniel – Paris: Payot, 1929 – 1 – us Wisconsin U Libr [440]

Histoire de la classe ouvriere en france de la revolution a nos jours; la condition materielle des travailleurs, les salaires et le cout de la vie / Louis, Paul – Paris: M. Riviere, 1927.412p – 1 – us Wisconsin U Libr [944]

Histoire de la communaute de notre dame de charite du bon-pasteur de montreal : suivi d'une biographie de messire j v arraud, ss / Giroux, Henri – Montreal: Lovell, 1879 – 1mf – 9 – mf#06323 – cn Canadiana [241]

Histoire de la congregation de saint-maur (afm31) : tom 1 (1612-1630) / Martene, Edmond – (ed Charvin) 1928 – (= ser Archives de la france monastique (afm)) – €15.00 – ne Slangenburg [241]

Histoire de la congregation de saint-maur (afm32) : tom 2 (1630-1641) / Martene, Edmond – (ed Charvin) 1929 – (= ser Archives de la france monastique (afm)) – €15.00 – ne Slangenburg [241]

Histoire de la congregation de saint-maur (afm33) : tom 3 (1645-1655) / Martene, Edmond – (ed Charvin) 1929 – (= ser Archives de la france monastique (afm)) – €14.00 – ne Slangenburg [241]

Histoire de la congregation de saint-maur (afm34) : tom 4 (1656-1667) / Martene, Edmond – (ed Charvin) 1930 – (= ser Archives de la france monastique (afm)) – €12.00 – ne Slangenburg [241]

Histoire de la congregation de saint-maur (afm35) : tom 5 (1668-1680) / Martene, Edmond – (ed Charvin) 1930 – (= ser Archives de la france monastique (afm)) – €14.00 – ne Slangenburg [241]

Histoire de la congregation de saint-maur (afm42) : tom 6 (1681-1687) / Martene, Edmond – (ed Charvin) 1937 – (= ser Archives de la france monastique (afm)) – €12.00 – ne Slangenburg [241]

Histoire de la congregation de saint-maur (afm43) : tom 7 (1688-1700) / Martene, Edmond – (ed Charvin) 1937 – (= ser Archives de la france monastique (afm)) – €12.00 – ne Slangenburg [241]

Histoire de la congregation de saint-maur (afm46) : tom 8 (1701-1712) / Martene, Edmond – (ed Charvin) 1942 – (= ser Archives de la france monastique (afm)) – €12.00 – ne Slangenburg [241]

Histoire de la congregation de saint-maur (afm47) : tom 9 (1713-1747) / Martene, Edmond – (ed Charvin) 1943 – (= ser Archives de la france monastique (afm)) – €17.00 – ne Slangenburg [241]

Histoire de la congregation de saint-maur (afm48) : tables (1612-1747) / Martene, Edmond – (ed Charvin) 1954 – (= ser Archives de la france monastique (afm)) – €7.00 – ne Slangenburg [241]

Histoire de la conjuration de louis-philippe-joseph d'orleans : premier prince du sang, duc de orleans, de chartres, de nemours, de montpensier et d'etampes, comte de beaujolais... / Montjoie, Felix L de – Paris 1796 [mf ed Hildesheim 1995-98] – 3v on 7mf – 9 – €140.00 – 3-487-26286-X – gw Olms [944]

Histoire de la conquete de l'abyssinie (16e siecle) : par chihab ed-din ahmed ben 'abd el-qader surnome arab-faqih / Basset, R – Paris, 1897 2v – 11mf – 9 – mf#NE-20320 – ne IDC [956]

Histoire de la conque'te des isles moluques par les espagnols, par les portugais, & par les hollandois / Argensola, B L de – Amsterdam: Jacques Desbordes, 1706. 3v – 15mf – 9 – mf#HT-588 – ne IDC [954]

Histoire de la convention nationale / Barante, Amable G de – Bruxelles 1851-53 [mf ed Hildesheim 1995-98] – 5v on 17mf – 9 – €170.00 – 3-487-26280-0 – gw Olms [323]

Histoire de la convention nationale de france : accompagnes d'un coup-d'oeil sur les assemblees constituante et legislative... / Durdent, Rene – Paris 1818 [mf ed Hildesheim 1995-98] – 2v on 5mf – 9 – €100.00 – ISBN-10: 3-487-26283-5 – ISBN-13: 978-3-487-26283-3 – gw Olms [323]

Histoire de la decadence de la monarchie francaise : et des progres de l'autorite royale a copenhague, madrid, vienne...depuis l'epoque ou lou / Soulavie, Jean – Paris 1803 [mf ed Hildesheim 1995-98] – 3v on 7mf – 9 – €140.00 – ISBN-10: 3-487-26087-5 – ISBN-13: 978-3-487-26087-7 – gw Olms [944]

L'histoire de la decadence de l'empire grec, et establissement de celvy des turcs... / Chalcocondylas, L – Paris, 1584 – 9mf – 9 – mf#H-8362 – ne IDC [956]

Histoire de la decouverte de l'amerique / Campe, Joachim Heinrich – [Paris?: s.n.] 2v. 1836 [mf ed 1983] – 2v on 1mf – 9 – mf#43406 – cn Canadiana [917]

Histoire de la decouverte de l'amerique depuis les origines jusqu'a la mort de christophe colomb, vol 1 / Gaffarel, Paul – Paris: A Rousseau, 1892 – v1 on 6mf – 9 – mf#06184 – cn Canadiana [910]

Histoire de la decouverte de l'amerique depuis les origines jusqu'a la mort de christophe colomb, vol 2 / Gaffarel, Paul – Paris: A Rousseau, 1892 – v2 on 5 mf – 9 – mf#06185 – cn Canadiana [910]

Histoire de la decouverte de l'amerique depuis les origines jusqu'a la mort de christophe colomb, vols 1 and 2 / Gaffarel, Paul – Paris: A Rousseau, 1892 – 2v on 1mf – 9 – mf#06183 – cn Canadiana [970]

Histoire de la decouverte et de la conquete de l'amerique / Campe, Joachim Heinrich – nouv ed. Paris: Garnier, [1833?] [mf ed 1984] – 6mf – 9 – 0-665-03004-5 – mf#03004 – cn Canadiana [910]

Histoire de la derniere revolution de perse / Cerceau, J A du – Paris, Briasson, 1728. 2v – 11mf – 9 – mf#AR-1642 – ne IDC [956]

Histoire de la divination dans l'antiquite / Bouche-Leclercq, Auguste – Paris: E Leroux, 1879-1882 – 4mf – 9 – 0-524-04323-X – (incl bibl ref) – mf#1990-3307 – us ATLA [250]

Histoire de la doctrine de l'inspiration des saintes ecritures dans les pays de langue frandcaise : de la reforme a nos jours / Rabaud, Edouard – Paris: Fischbacher, 1883 – 1mf – 9 – 0-524-04474-0 – (incl bibl ref) – mf#1992-0143 – us ATLA [220]

Histoire de la double conspiration de 1800 : contre le gouvernement consulaire, et de la deportation qui eut lieu dans la deuxieme annee du consulat... / Fescourt – Paris 1819 [mf ed Hildesheim 1995-98] – 1v on 3mf – 9 – €90.00 – ISBN-10: 3-487-26404-8 – ISBN-13: 978-3-487-26404-2 – gw Olms [944]

Histoire de la esclavitud negra en puerto rico / Diaz Soler, Luis M – Rio Piedras, Puerto Rico. 1965 – 1r – us UF Libraries [972]

Histoire de la famille courtemanche, 1663-1895 / Courtemanche, Joseph Israel – Montreal: Cie d'imprimerie commerciale, 1895 – 1mf – 9 – mf#03618 – cn Canadiana [929]

Histoire de la famille et de la descendance / Nemours, Alofred – Port-Au-Prince, Haiti. 1941 – 1r – us UF Libraries [972]

Histoire de la fondation de l'eglise evangelique neuchateloise, independant de l'etat / Monvert, Charles – Neuchatel: P Attinger, 1898 [mf ed 1993] – (= ser Presbyterian coll) – 4mf – 9 – 0-524-07359-7 – (in french) – mf#1990-5396 – us ATLA [242]

Histoire de la franc-maconnerie : depuis son origine jusqu'a nos jours / Findel, Joseph Gabriel – Paris: Libr Internationale 1866 – 2v on 1mf – 9 – (trans fr german by e tandel) – mf#vrl-45 – ne IDC [366]

Histoire de la franc-maconnerie : son idee fondamental et sa constitution developpees selon l'esprit de notre siecle / Bobrik, Eduard – Lausanne: Marc Ducloux 1841 – 5mf – 9 – (trans fr german by edouard lenz) – mf#vrl-15 – ne IDC [366]

Histoire de la franc-maconnerie a liege avant 1830 / Dwelshauvers-Dery, Victor Auguste Ernest – Bruxelles: J Baertsoen 1879 – 2mf – 9 – mf#vrl-100 – ne IDC [366]

Histoire de la fronde / Sainte-Aulaire, Louis C de – Paris 1827 [mf ed Hildesheim 1995-98] – 3v on 9mf – 9 – €180.00 – ISBN-10: 3-487-26093-X – ISBN-13: 978-3-487-26093-8 – gw Olms [323]

Histoire de la guadeloupe / Blanche, Lenis – Paris, France. 1938 – 1r – us UF Libraries [972]

Histoire de la guadeloupe sous l'ancien regime / Satineau, Maurice – Paris, France. 1928 – 1r – us UF Libraries [972]

Histoire de la guerre de l'independance des etats-unis d'amerique / Botta, Carlo – Paris: J G Dentu...4v. 1812 [mf ed 1984] – 4v on 1mf – 9 – mf#48005 – cn Canadiana [975]

Histoire de la guerre des anabaptistes / Weill, Alexandre – Paris: Dentu, 1874 [mf ed 1986] – 1mf – 9 – 0-8370-8956-5 – mf#1986-2956 – us ATLA [242]

Histoire de la gverre qui c'est passee, entre les venitiens et la saincte ligue, contre les turcs, pour l'isle de cypre, es annees 1570, 1571 and 1572 / Bizari, P – Paris, 1573 – 4mf – 9 – mf#H-8338 – ne IDC [956]

Histoire de la gverre saincte, dite proprement, la franciade orientale... / Gulielmus, Archbishop of Tyre – Paris, 1573 – 14mf – 9 – mf#H-8340 – ne IDC [956]

Histoire de la justice criminelle / Allard, Alberic – Gran, 1868. 525 p. LL-4061 – 1 – us L of C Photodup [345]

Histoire de la langue et de la litterature francaise des origines a 1900 / Petit de Julleville, L – v1-8. 1896-99 – 1 – $180.00 – (in french) – mf#0447 – us Brook [440]

Histoire de la latinite de constantinople / Belin, M A – Ed 2. Paris, 1894 – 7mf – 9 – mf#NE-20124 – ne IDC [956]

Histoire de la litterature allemande / Zink, Georges et al – [Paris]: Aubier, 1970 – 1r – 1 – (incl bibl ref and indexes) – us Wisconsin U Libr [430]

Histoire de la litterature americaine de langue es... / Bazin, Robert – Paris, France. 1953 – 1r – us UF Libraries [440]

Histoire de la litterature feminine en france / Larnac, Jean – Paris: Editions Kra, (1929). 296p – 1 – us Wisconsin U Libr [440]

Histoire de la litterature haitienne / Vaval, Duracine – Port-Au-Prince, Haiti. 1933 – 1r – us UF Libraries [440]

Histoire de la litterature hindoui et hindoustani / Garcin de Tassy – 2v. 1839-47 – (= ser Royal asiatic society oriental translation fund. old series) – 1r – 1 – mf#656 – uk Microform Academic [490]

Histoire de la litterature hindoui et hindoustani / Tassy, M Garcin de – Paris: Printed under the auspices of the Oriental Translation Committee of Great Britain and Ireland, 1839-47 – us CRL [490]

Histoire de la litterature hindouie et hindoustani / Tassy, M Garcin de – 2nd rev cor aug ed. Paris: A Labitte, 1870-71 – us CRL [490]

Histoire de la milice canadienne-francaise, 1760-1897... / Sulte, Benjamin – Montreal: Desbarats, 1897 [mf ed 1976] – 1r – 1 – mf#SEM35P142 – cn Bibl Nat [355]

Histoire de la milice canadienne-francaise, 1760-1897... / Sulte, Benjamin – Montreal: Desbarats, 1897 [mf ed 1976] – 1r – 5 – mf#SEM35P142 – cn Bibl Nat [355]

Histoire de la mission du kiang-nan : jesuites de la province de france, paris 1840-99 / La Servibere, Joseph – [Zi-ka-wei]: l'Orphelinat de T'ou-se-we, [1914] [mf ed 1995] – (= ser Yale coll) – 2v – 1 – 0-524-09454-3 – (in french) – mf#1995-0454 – us ATLA [241]

Histoire de la monarchie de juillet / Thureau-Dangin, Paul – Paris 1888-1889 [mf ed Hildesheim 1995-98] – 5v on 31mf – 9 – €310.00 – ISBN-10: 3-487-26045-X – ISBN-13: 978-3-487-26045-7 – gw Olms [944]

Histoire de la musique et de ses effets : depuis son origine jusqu'a present & en quoi consiste sa beaute / Bourdelot, Pierre – Amsterdam: Chez M Charles Le Cene 1725 [mf ed 19–] – 4v on 16mf – 9 – mf#fiche370 – us Sibley [780]

Histoire de la musique, et de ses effets : depuis son origine jusqua presentet de ses effets / Bonnet, Jacques – Paris: J Cochart [etc] 1715 [mf ed 19–] – 7mf – 9 – mf#fiche369 – us Sibley [780]

Histoire de la nouvelle espagne / Diaz del Castillo, Bernal – 1878 – 9 – sp Bibl Santa Ana [972]

Histoire de la nouvelle france : contenant les navigations, decouvertes, et habitations faites... / Lescarbot, Marc – Paris: Chez Jean Milot, 1609 [mf ed 1983] – 10mf – 9 – mf#SEM105P287 – cn Bibl Nat [917]

Histoire de la nouvelle-france : contenant les navigations, decouvertes, et habitations faites par les francois es indes occidentales et nouvelle-france... / Lescarbot, Marc – A Paris: Chez Adrien Perier...1618 – 12mf – 9 – 0-665-90702-8 – mf#90702 – cn Canadiana [910]

Histoire de la papaute pendant le 14e siecle : avec des notes et des pi eces justificatives / Christophe, J-B – Paris: L Maison, 1853 – 6mf – 9 – 0-8370-8089-4 – (incl bibl ref) – mf#1986-2089 – us ATLA [240]

Histoire de la paroisse de saint-augustin (portneuf) / Bechard, Auguste – S.l: L Brousseau, 1885 – 5mf – 9 – mf#03027 – cn Canadiana [971]

Histoire de la paroisse de saint-joseph de carleton (baie des chaleurs) 1755-1906 / Chouinard, Edouard Pierre – [Rimouski, Quebec?]: impr generale de Rimouski, 1906 – 2mf – 9 – 0-665-73140-X – mf#73140 – cn Canadiana [917]

Histoire de la paroisse de saint-malachie / Kirouac, Jules-Adrien – [Quebec?: s.n.] 1909 [mf ed 1996] – 3mf – 9 – 0-665-77576-8 – (incl app) – mf#77576 – cn Canadiana [241]

Histoire de la paroisse d'yamachiche : precis historique / Caron, Napoleon – Trois-Rivieres Quebec: P V Ayotte, 1892 – 4mf – 9 – mf#00469 – cn Canadiana [929]

Histoire de la pedagogie / Guerin, Marc-Aime & Vertefeuille, Paul-Yvon – Montreal: Centre de psychologie et de pedagogie, [1959?] (mf ed 2001) – 9 – cn Bibl Nat [370]

Histoire de la philosophie en belgique / Wulf, M de – Bruxelles-Paris, 1910 – 11mf – 8 – €6.00 – ne Slangenburg [100]

Histoire de la philosophie hermetique : accompagnee d'un catalogue raisonne des ecrivains de cette science / Lenglet Du Fresnoy, Nicolas – Paris, 1744 – 1 – us Wisconsin U Libr [190]

Histoire de la philosophie scolastique / Haureau, B – Paris. v1-2. 1872-1880 – 8 – €52.00 – (v1 18mf. v2 9mf) – ne Slangenburg [100]

Histoire de la philosophie scolastique / Haureau, Barthelemy – Paris: Durand et Pedone-Lauriel, 1872-1880 – 4mf – 9 – 0-7905-3858-X – (incl bibl ref) – mf#1989-0351 – us ATLA [240]

Histoire de la philosophie scolastique dans les pays-bas et la principaute de liege / Wulf, M de – Louvain, 1893 – 7mf – 8 – €15.00 – ne Slangenburg [100]

Histoire de la ponctuation : ou, de la massore chez les syriens / Martin, Jean Pierre Paulin – Paris: Imprimerie nationale, 1875 – 1mf – 9 – 0-8370-8840-2 – (incl bibl ref) – mf#1986-2840 – us ATLA [470]

Histoire de la pragmatique sanction de bourges sous charles 7 / Valois, Noel – Paris: A Picard, 1906 – (= ser Archives de l'histoire religieuse de la france) – 2mf – 9 – 0-7905-6847-0 – (incl bibl ref) – mf#1988-2847 – us ATLA [240]

Histoire de la pragmatique sanction de bourges sous charles 7 / Valois, Noel – Paris: A. Picard, 1906 – 1 – us Wisconsin U Libr [944]

Histoire de la predication parmi les reformes de france au dix-septieme siecle / Vinet, Alexandre Rodolphe – Paris: Chez les editeurs, 1860 [mf ed 1990] – 2mf – 9 – 0-7905-9651-2 – (incl bibl ref) – mf#1989-1376 – us ATLA [242]

Histoire de la premiere mission catholique au vicariat de melanesie / Verguet, C M Leopold – Carcassonne: P Labau, 1854 [mf ed 1995] – (= ser Yale coll) – 319p (ill) – 1 – 0-524-09405-5 – (in french) – mf#1995-0405 – us ATLA [241]

Histoire de la princesse rosette / Segur, Sophie, comtesse de – Quebec: [1942?] (mf ed 1990) – 1mf – 9 – (ill by vernier) – mf#SEM105P1293 – cn Bibl Nat [920]

Histoire de la province ecclesiastique d'ottawa : et de la colonisation dans la vallee de l'ottawa / Alexis, pere – Ottawa: Impr d'Ottawa. 2v. 1897 [mf ed 1983] – 13mf – 9 – (with ind) – mf#SEM105P295 – cn Bibl Nat [241]

Histoire de la province ecclesiastique d'ottawa et de la colonisation dans la vallee de l'ottawa, vol 1 / Alexis de Barbezieux, pere – Ottawa: Impr d'Ottawa, 1897 – v1 on 7mf – 9 – mf#03592 – cn Canadiana [241]

Histoire de la province ecclesiastique d'ottawa et de la colonisation dans la vallee de l'ottawa, vol 2 / Alexis de Barbezieux, pere – Ottawa?: Impr d'Ottawa?, 1897? – v2 on 6mf – 9 – (incl ind) – mf#03593 – cn Canadiana [241]

Histoire de la province ecclesiastique d'ottawa et de la colonisation dans la vallee de l'ottawa, vols 1 and 2 / Alexis de Barbezieux, pere – S.l: s.n, 1897? – 2v on 1mf – 9 – mf#03591 – cn Canadiana [241]

Histoire de la reformation de la suisse / Ruchat, A; ed by Vulliemin, L – Paris, Lausanne, 1835-1838. 7 v – 45mf – 9 – mf#ZWI-47 – ne IDC [242]

Histoire de la reformation de la suisse *see* History of the reformation in switzerland

Histoire de la reformation et du refuge : dans le pays de neuchatel / Godet, Frederic Louis – Neuchatel: L Meyer; 1859. Beltsville, MD: NCR Corp,1977 (4mf); Evanston: American Theol Lib Assoc, 1984 (4mf) – 9 – 0-8370-0126-9 – mf#1984-0013 – us ATLA [949]

Histoire de la reformation francaise / Puaux, Francois – Paris: Michel Levy, 1859-1863 – 7mf – 9 – 0-524-03357-9 – mf#1990-0938 – us ATLA [944]

Histoire de la reforme, de la ligue, et du regne de henri 4 : 8 Bde / Capefigue, Jean Baptiste Honore Raymond – Paris 1834-1835 [mf ed Hildesheim 1995-98] – 24mf – 9 – 161=€240.00 – ISBN-10: 3-487-26129-4 – ISBN-13: 978-3-487-26129-4 – gw Olms [944]

Histoire de la refromation de la suisse : o-l'on voit tout ce qui s'est passe de plus remarquable, depuis l'an 1516 jusqu'en l'an 1556... / Ruchat, A – Geneve: Bousquet. 6v. 1727-1728 – 40mf – 9 – mf#ZWI-96 – ne IDC [242]

Histoire de la regence et de la minorite de louis 15 : jusqu'au ministere du cardinal de fleury / Lemontey, Pierre – Paris 1832 [mf ed Hildesheim 1995-98] – 2v on 6mf – 9 – €120.00 – 3-487-26213-4 – gw Olms [944]

Histoire de la religion des eglises reformees / Basnage, J – Rotterdam. 2v. 1690 – 13mf – 9 – mf#PRS-112 – ne IDC [242]

Histoire de la republique des etats-unis depuis l'etablissement des premieres colonies jusqu' a l'election du president lincoln (1620-1860) / Astie, Jean-Frederic – Paris: Grassart, 1865 – 3mf – 9 – 0-7905-5746-0 – (incl bibl ref) – mf#1988-1746 – us ATLA [240]

Histoire de la restauration / Lamartine, Alphonse de – Paris 1851-1852 [mf ed Hildesheim 1995-98] – 8v on 32mf – 9 – €320.00 – ISBN-10: 3-487-26345-9 – ISBN-13: 978-3-487-26345-8 – gw Olms [940]

Histoire de la restauration / Viel-Castel, Louis de – Paris: Michel Levy freres, 1860-78. 20v – 1 – us Wisconsin U Libr [944]

Histoire de la restauration du protestantisme en france au 18e siecle : antoine court / Hugues, Edmond – 4e ed, rev et corr. Paris: Grassart, 1875 – 3mf – 9 – 0-7905-7052-1 – (incl bibl ref) – mf#1988-3052 – us ATLA [242]

Histoire de la restauration et des causes qui ont amene la chute de la branche ainee des bourbons / Capefigue, Jean Baptiste Honore Raymond – Paris 1831-1833 [mf ed Hildesheim 1995-98] – 10v on 30mf – 9 – €300 – ISBN-10: 3-487-26080-8 – ISBN-13: 978-3-487-26080-8 – gw Olms [944]

Histoire de la revolution de 1848 / Lamartine, Alphonse de – Paris 1849 [mf ed Hildesheim 1995-98] – 2v on 6mf – 9 – €120.00 – ISBN-10: 3-487-26011-5 – ISBN-13: 978-3-487-26011-2 – gw Olms [940]

Histoire de la revolution francaise / Blanc, Louis – Paris 1847-1862 [mf ed Hildesheim 1995-98] – 12v on 69mf – 9 – €690.00 – ISBN-10: 3-487-26181-2 – ISBN-13: 978-3-487-26181-2 – gw Olms [933]

Histoire de la revolution francaise : depuis 1814 jusqu'a 1830, et annees suivantes / Dulaure, Jacques – Paris 1834 [mf ed Hildesheim 1995-98] – 1v on 3mf – 9 – €90.00 – ISBN-10: 3-487-26390-4 – ISBN-13: 978-3-487-26390-8 – gw Olms [933]

Histoire de la revolution russe – Preparee sous direction de Maxime Gorki, V. Molotov, K. Vorochilov, Serge Kirov, A Jdanov et J. Staline. Paris: Editions sociales, 1946 – 1 – us Wisconsin U Libr [947]

Histoire de la rochelle / Dupont, Edouard – La Rochelle 1830 [mf ed Hildesheim 1995-98] – 1v on 4mf – 9 – €120.00 – ISBN-10: 3-487-25908-7 – ISBN-13: 978-3-487-25908-6 – gw Olms [944]

Histoire de la saint-barthelemy d'apres les chroniques, memoires et manuscrits du 16e siecle / Audin, Jean M – Paris 1826 [mf ed Hildesheim 1995-98] – 1v on 3mf – 9 – €90.00 – 3-487-26118-9 – gw Olms [944]

Histoire de la seigneurie de st-ours / Couillard-Despres, Azarie – Impr de l'Institution des sourds-muets. 2v. 1915-1917 [mf ed 1992] – 10mf – 9 – mf#SEM105P1495 – cn Bibl Nat [929]

Histoire de la seigneurie massue et de la paroisse de saint-aime / Hengard-Lapalice, Ovide Michel – [Quebec (Province): s.n.], 1930 [mf ed 1992] – 5mf – 9 – (incl text in english) – mf#SEM105P1650 – cn Bibl Nat [971]
Histoire de la sepulture et des funerailles dans l'ancienne egypte / Amelineau, Emile – Paris: Ernest Leroux 1896 [mf ed 1992] – (= ser Annales du musee guimet 28-29) – 2v on 3mf [ill] – 9 – 0-524-04193-8 – mf#1990-3300 – us ATLA [390]
Histoire de la session de 1815 / Fievee, Joseph – Paris 1816 [mf ed Hildesheim 1995-98] – 1v on 3mf – 9 – €90.00 – ISBN-10: 3-487-26323-8 – ISBN-13: 978-3-487-26323-6 – gw Olms [944]
Histoire de la theologie chretienne au siecle apostolique *see* History of christian theology in the apostolic age
Histoire de la tolerance religieuse : evolution d'un principe social / Matagrin, Amedee – Paris: Fischbacher, 1905 – 2mf – 9 – 0-7905-6934-5 – (incl bibl ref) – mf#1988-2934 – us ATLA [200]
Histoire de la venerable mere madeleine-sophie barat : fondatrice de la societe du sacre-coeur de jesus / Baunard, Louis – Montreal: Cadieux & Derome, 1883 – 3mf – 9 – mf#03504 – cn Canadiana [241]
Histoire de la vie de m paul de chomedey : sieur de maisonneuve fondateur et premier gouverneur de villemarie / Rousseau, Pierre – Montreal: Cadieux & Derome, [1885?] [mf ed 1983] – 4mf – 9 – mf#13502 – cn Canadiana [920]
Histoire de la vie et moeurs de marie tessonniere... / La Riviere, L de – Lyon, 1650 – 7mf – 9 – mf#CA-172 – ne IDC [240]
Histoire de la vie politique, militaire et privee de napoleon bonaparte : precedee de notices biographiques sur ses fideles compagnons d'infortune... / Chennechot, L E – Paris 1825 [mf ed Hildesheim 1995-98] – 1v on 3mf – 9 – €90.00 – ISBN-10: 3-487-26230-4 – ISBN-13: 978-3-487-26230-7 – gw Olms [944]
Histoire de la vie privee des francais : depuis l'origine de la nation jusqu'a nos jours / Legrand d'Aussy, Pierre – Paris 1782 [mf ed Hildesheim 1995-98] – 3v on 9mf – 9 – €180.00 – ISBN-10: 3-487-25893-5 – ISBN-13: 978-3-487-25893-5 – gw Olms [944]
Histoire de la ville de khotan : tiree des annales de la chine et traduite du chinois... / Remusat, J P A – Paris: de Doublet, 1820 – 4mf – 9 – mf#U-616 – ne IDC [915]
Histoire de la ville et de tout le diocese de paris / Lebeuf, Jean – Paris: chez Prault pere. 15v. 1754-1758 [mf ed 1985] – 1r – 5 – mf#SEM16P347 – cn Bibl Nat [241]
Histoire de la vulgate pendant les premiers siecles du moyen age / Berger, Samuel – Paris: Hachette, 1893 – 2mf – 9 – 0-7905-0967-9 – (in french and latin. includes bibliographies and indexes) – mf#1987-0967 – us ATLA [220]
Histoire de l'abbaye de saint-germain-des-prez / Bouillard, J – Paris, 1724 – €50.00 – ne Slangenburg [241]
Histoire de l'abbaye d'ormont / Valentin, A – Reims, 1862 – €9.00 – ne Slangenburg [241]
Histoire de l'abbaye royale de jumieges / Deshayes, Charles – Rouen 1829 [mf ed Hildesheim 1995-98] – 1v on 2mf – 9 – €60.00 – ISBN-10: 3-487-25917-6 – ISBN-13: 978-3-487-25917-8 – gw Olms [241]
Histoire de l'abbaye royale de saint-denys en france / Felibien, M – Paris, 1706 – €75.00 – ne Slangenburg [241]
Histoire de l'abbaye royale et de l'ordre des chanoines reguliers de st victor de paris / Bonnard, F – Paris, 1907 – 22mf – 8 – €42.00 – ne Slangenburg [241]
Histoire de l'abbaye sainte-croix de bordeaux (afm9) / Chauliac, A – 1910 – (= ser Archives de la france monastique (afm)) – 9mf – 8 – €18.00 – ne Slangenburg [241]
Histoire de l'afrique occidentale / Niane, Djibril Tamsir – Conakry, Guinea. 1961 – 1r – us UF Libraries [960]
Histoire de l'ambassade dans le grand duche de varsovie en 1812 / Pradt, Dominique Georges Frederic de Riom de Prolhiac de – Paris 1815 [mf ed Hildesheim 1995-98] – 1v on 2mf – 9 – €60.00 – ISBN-10: 3-487-26396-3 – ISBN-13: 978-3-487-26396-0 – gw Olms [327]
Histoire de l'ancien cambodge / Aymonier, Etienne Francois – Strasbourg: Impr du Nouveau journal de Strasbourg [1920] [mf ed 1989] – 1r with other items – 1 – mf#mf-10289 seam reel 022/06 [§] – us CRL [930]
Histoire de l'art dans l'antiquite. 4, jude, sardaigne, syrie, cappadoce *see* History of art in sardinia, judaea, syria, and asia minor
Histoire de l'art monumentale dans l'antiquite et au moyen age / Batissier, L – Paris, 1845 – 8mf – 9 – mf#OA-127 – ne IDC [720]
Histoire de l'asie centrale (afghanistan. boekhara, khiva, khoquand)... / ed by Boekhary, Mir Abdoel Kerim – Paris, 1876 – 6mf – 9 – mf#U-750 – ne IDC [956]
Histoire de l'assemblee constituante / Lameth, Alexandre T de – Paris 1828-1829 [mf ed Hildesheim 1995-98] – 7mf – 9 – €140.00 – ISBN-10: 3-487-26299-1 – ISBN-13: 978-3-487-26299-4 – gw Olms [323]
Histoire de l'eau-de-vie en canada : d'apres un manuscrit recemment obtenu de france / Belmont, Francois Vachon de – [s.l: s.n, 1840?] [mf ed 1983] – 1mf – 9 – 0-665-42998-3 – mf#42998 – cn Canadiana [971]
Histoire de l'ecole d'alexandrie / Simon, Jules – Paris: Joubert, 1845 [mf ed 1993] – 2v on 3mf – 9 – 0-524-08530-7 – mf#1993-1060 – us ATLA [180]
Histoire de l'ecole d'alexandrie comparee aux principales ecoles contemporaines / Matter, Jacques – 2e ed. Paris: Hachette, 1840-48 [mf ed 1992] – 3v on 3mf – 9 – 0-524-03409-5 – mf#1990-0963 – us ATLA [370]
Histoire de l'edit de nantes / [Benoist, E] – Delft, 1693, 1695. v1-3 – 48mf – 9 – mf#PRS-115 – ne IDC [240]
Histoire de l'edition benedictine de saint augustin / Ingold, Augustin Marie Pierre – Paris: A Picard 1903 [mf ed 1990] – (= ser Documents pour servir a l'histoire religieuse des 17e et 18e siecles) – 1mf – 9 – 0-7905-6234-0 – (incl bibl ref) – mf#1988-2234 – us ATLA [240]
Histoire de l'eglise : dediee au roi / Berault-Bercastel, Antoine Henri de – Maestricht [Pays-Bas]: De l'Impr de P L Lekens. 24v. 1780 [mf ed 1984] – 24v on 1mf – 9 – mf#47854 – cn Canadiana [241]
Histoire de l'eglise d'alexandrie : depuis saint marc jusqu'a nos jours / Macaire, Georges – Le Caire [Cairo]: Imprimerie Generale, 1894 – 1mf – 9 – 0-8370-7646-3 – mf#1986-1646 – us ATLA [240]
Histoire de l'eglise d'alexandrie / Vansleb, J M; ed by Marc, S – Paris, 1677 – 5mf – 8 – €12.00 – ne Slangenburg [944]
Histoire de l'eglise de coree : precedee d'une introduction sur l'histoire, les institutions, la langue, les moeurs et coutumes coreennes / Dallet, Charles – Paris: V Palme 1874 [mf ed 1990] – 2v on 3mf – 9 – 0-7905-5691-X – mf#1988-1691 – us ATLA [241]
Histoire de l'eglise (he) : depuis les origines jusqu'a nos jours / ed by Fliche, A & Martin, V – Paris. v1-21. 1934-64 – €545.00 set – (vols also listed individually) – ne Slangenburg [941]
Histoire de l'eglise reformee d'anduze : depuis son origine jusqu'a la revolution francaise / Hugues, Jean-Pierre – 2. ed. [s.l.: s.n.] 1864 (Montpellier: Boehm & fils) [mf ed 1992] – 2mf – 9 – 0-524-03645-4 – (in french) – mf#1990-1073 – us ATLA [242]
Histoire de l'eglise vaudoise : depuis son origine et des vaudois du piemont jusqu' a nos jours / Monastier, Antoine – Lausanne: G Bridel, 1847 – 2mf – 9 – 0-7905-5492-5 – mf#1988-1492 – us ATLA [240]
Histoire de l'eglise vaudoise *see* History of the vaudois church
Histoire de l'emigration : (1789-1825) / Montrol, Francois Mongin de – Paris 1825 [mf ed Hildesheim 1995-98] – 1v on 3mf – 9 – €90.00 – ISBN-10: 3-487-26295-9 – ISBN-13: 978-3-487-26295-6 – gw Olms [320]
Histoire de l'empereur napoleon / Hugo, Abel – Stuttgart 1834 [mf ed Hildesheim 1995-98] – 1v on 3mf – 9 – €90.00 – ISBN-10: 3-487-26403-X – ISBN-13: 978-3-487-26403-5 – gw Olms [940]
Histoire de l'empire de constantinople sous les empereurs francais jusqu'a la conquete des turcs / DuCange, Charles DuFresne – Paris 1826 [mf ed Hildesheim 1995-98] – 2v on 6mf – 9 – €120.00 – ISBN-10: 3-487-26317-3 – ISBN-13: 978-3-487-26317-5 – gw Olms [931]
Histoire de l'empire de constantinople sous les empereurs francois (cbh26) / Villehardouin, Geoffroy de – Venise, 1729 – (= ser Corpus byzantinae historiae (cbh)) – €37.00 – ne Slangenburg [931]
Histoire de l'entree de la reyne mere du roy tres chrestien : dans la grande-bretaigne / Puget de la Serre, J – Londre: Par Jean Raworth, pour George Thomason & Octavian Pullen, 1639 – 3mf – 9 – mf#O-93 – ne IDC [090]
Histoire de l'entree de la reyne mere du roy tres-chrestien : dans la grande-bretaigne / Serre, de la – Londre, 1639 – 3mf – 9 – mf#O-1095 – ne IDC [700]
Histoire de l'entree de la reyne mere du roy tres-chrestien : dans les provinces unies des pays-bays / Puget de la Serre, J – Londre: J Raworth, 1639 – 4mf – 9 – mf#O-1114 – ne IDC [090]
Histoire de l'esclavage en afrique (pendant trente-quatre ans) de p j dumont, natif de paris, maintenant a l'hospice royal des incurables / Dumont, Pierre – Paris 1819 [mf ed Hildesheim 1995-98] – 1v on 1mf – 9 – €40.00 – 3-487-27345-4 – gw Olms [960]
Histoire de l'etablissement, des progres et de la decadence du christianisme dans l'empire du japon : ou l'on voit les differentes revolutions qui ont agite cette monarchie pendant plus d'un siecle / Charlevoix, Pierre Francois Xavier de – Louvain: Vanlinthout et Vandenzande, 1828-29 [mf ed 1995] – (= ser Yale coll; Bibliotheque catholique de la belgique 1828/2, 1829/1) – 2v – 1 – 0-524-09865-4 – (in french) – mf#1995-0865 – us ATLA [241]
Histoire de l'etablissement, des progres et de la decadence du christianisme dans l'empire du japon *see* Histoire du christianisme au japon
Histoire de l'etablissement des protestants francais en suede / Puaux, Frank – Paris: G Fischbacher; Stockholm: E Giron, 1891 – 1mf – 9 – 0-7905-6823-3 – mf#1988-2823 – us ATLA [242]
Histoire de l'etablissement du christianisme dans les indes orientales... : imprimee sur le manuscrit original inedit, communiquee pendant le cours de l'impression, a m sicard / Serieys, Antoine – Paris: Chez Madame Devaux, 1803 [mf ed 1995] – (= ser Yale coll) – 2v – 1 – 0-524-10242-2 – (in french) – mf#1996-1242 – us ATLA [241]
Histoire de l'ethiopie orientale...traduit en francois par...gaetan charpy / Santos Joao dos – Paris. 1684 – 1 – us CRL [960]
Histoire de l'expedition chrestienne au royaume de la chine entreprinse par les peres de la compagnie de iesus... / Ricci, M – Lille: Pierre de Rache, 1617 – 7mf – 9 – mf#HT-911 – ne IDC [915]
Histoire de l'expedition francaise en egypte / Reybaud, Louis – Paris 1830/36 [mf ed Hildesheim 1995-98] – 6v on 24mf – 9 – €240.00 – 3-487-27380-2 – gw Olms [916]
Histoire de l'harmonie au moyen age / Coussemaker, E de – Paris, 1852 – €31.00 – ne Slangenburg [931]
Histoire de l'hotel-dieu de quebec / St-Ignace, mere – A Montauban France: Chez Jerosme Legier; Paris: chez Claude-Jean-Baptiste Herissant, libraire...1751 – 7mf – 9 – mf#40272 – cn Canadiana [360]
Histoire de l'ile de la trinidad / Borde, Pierre-Gustave-Louis – Paris, France. v1-2. 1876-82 – 1r – us UF Libraries [972]
Histoire de l'ile-aux-grues et des iles voisines / Bechard, Auguste – [Arthabaskaville, Quebec?: s.n.], 1902 – 2mf – 9 – 0-665-71646-X – mf#71646 – cn Canadiana [971]
Histoire de l'insurrection du canada / Papineau, Louis Joseph – [s.l.]: [s.n.], 1839 [mf ed 1988] – 1mf – 9 – mf#SEM105P872 – cn Bibl Nat [971]
Histoire de l'isle espagnole ou de s domingue : ecrite particulierement sur des memoires manuscrits du p jean-baptiste le pers, jesuite, missionnaire a saint-domingue... / Charlevoix, Pierre-Francois-Xavier de – Amsterdam: Chez Francois L'Honore 1733 [mf ed 1985] – 4v on 1mf – 9 – 0-665-48767-3 – mf#48767 – cn Canadiana [972]
Histoire de loango et d'autres royaumes d'afrique : redigee d'apres les memoires des prefets apostoliques de la mission francaise / Proyart, Abbe – Paris, 1776 – 8mf – 9 – mf#A-135 – ne IDC [916]
Histoire de l'opera bouffon : contenant les jugemens de toutes les pieces qui ont paru depuis sa naissance jusqu'a ce jour. pour servir a l'histoire des theatres de paris / Contant d'Orville, Andre Guillaume – A Amsterdam, Paris: chez Grange' 1768 [mf ed 19–] – 2v on 7mf – 9 – mf#fiche 396 – us Sibley [780]
Histoire de l'ordre maconnique en belgique / Cordier, Adolphe – Mons: H Chevalier 1854 – 7mf – 9 – mf#vrl-33 – ne IDC [366]
Histoire de l'ouest canadien de 1822 a 1869 : epoque des troubles / Dugas, Georges – Montreal: Librairie Beauchemin, [1906?] (mf ed 1985) – 2mf – 9 – mf#SEM105P525 – cn Bibl Nat [971]
Histoire de louis 11 / Fonton, Feliks – Paris 1830 [mf ed Hildesheim 1995-98] – 2v on 6mf – 9 – €120.00 – ISBN-10: 3-487-26221-5 – ISBN-13: 978-3-487-26221-5 – gw Olms [944]
Histoire de louis 16 : avec les anecdotes de son regne / Berthre de Bourniseaux, Pierre – Paris 1829 [mf ed Hildesheim 1995-98] – 4v on 13mf – 9 – €130.00 – ISBN-10: 3-487-26199-5 – ISBN-13: 978-3-487-26199-7 – gw Olms [944]
Histoire de lyon : precedee d'une table chronologique des livres de cette histoire / Lorry, Alphonse – Lyon 1829-1845 [mf ed Hildesheim 1995-98] – 7v on 19mf – 9 – €190.00 – ISBN-10: 3-487-25916-8 – ISBN-13: 978-3-487-25916-1 – gw Olms [944]
Histoire de lyon pendant les journees des 21, 22 et 23 novembre 1831 : contenant les causes, les consequences et les suites de ces deplorables evenements / ed by Baron, Auguste – Lyon 1832 [mf ed Hildesheim 1995-98] – 1v on 2mf – 9 – €60.00 – 3-487-25913-3 – gw Olms [944]
Histoire de mme duchesne : religieuse de la societe du sacre-coeur de jesus et fondatrice des premieres maisons de cette societe en amerique / Baunard, Louis – Paris: Poussielgue, 1878 – 7mf – 9 – mf#04308 – cn Canadiana [241]
Histoire de napoleon : etudes sur les causes de son elevation et de sa chute / Bailleul, Jacques C – Paris 1829 [mf ed Hildesheim 1995-98] – 1v on 4mf – 9 – €120.00 – 3-487-26414-5 – gw Olms [944]
Histoire de napoleon / Norvins, Jacques M de – Paris 1827-1828 [mf ed Hildesheim 1995-98] – 4v on 16mf – 9 – €160.00 – ISBN-10: 3-487-26418-8 – ISBN-13: 978-3-487-26418-9 – gw Olms [920]
Histoire de napoleon : redigee d'apres les papiers d'etat, les documens officiels, les memoires et les notes secretes de ses contemporains / Tissot, Pierre – Paris 1833 [mf ed Hildesheim 1995-98] – 2v on 6mf – 9 – €120.00 – ISBN-10: 3-487-26405-6 – ISBN-13: 978-3-487-26405-9 – gw Olms [940]
Histoire de napoleon 1er / Lanfrey, Pierre – Paris 1870-1875 [mf ed Hildesheim 1995-98] – 5v on 16mf – 9 – €160.00 – ISBN-10: 3-487-26250-9 – ISBN-13: 978-3-487-26250-5 – gw Olms [944]
Histoire de napoleon buonaparte : depuis sa naissance, en 1769, jusqu'a sa translation a l'ile de sainte-helene, en 1815 – Paris 1817-18 [mf ed Hildesheim 1995-98] – 4v on 12mf – 9 – €120.00 – 3-487-26419-6 – gw Olms [944]
Histoire de napoleon d'apres lui-meme – Paris 1825 [mf ed Hildesheim 1995-98] – 1v on 4mf – 9 – €120.00 – ISBN-10: 3-487-26422-6 – ISBN-13: 978-3-487-26422-6 – gw Olms [944]
Histoire de napoleon et de la grande armee / Raisson, Horace – Paris 1830 [mf ed Hildesheim 1995-98] – 6v on 12mf – 9 – €120.00 – ISBN-10: 3-487-26406-4 – ISBN-13: 978-3-487-26406-6 – gw Olms [944]
Histoire de napoleon-le-grand / Saint-Maurice, Charles R de – Paris 1830 [mf ed Hildesheim 1995-98] – 4v on 8mf – 9 – €160.00 – ISBN-10: 3-487-26408-0 – ISBN-13: 978-3-487-26408-0 – gw Olms [940]
Histoire de notre dame de lourdes / Gros, L; ed by Bayle, Constantino – Madrid: Razon y Fe, 1928 – 9 – sp Bibl Santa Ana [240]
Histoire de paris : composee sur un plan nouveau / Touchard-Lafosse, Georges – Paris 1833-1834 [mf ed Hildesheim 1995-98] – 5v on 19mf – 9 – €190.00 – ISBN-10: 3-487-25940-0 – ISBN-13: 978-3-487-25940-6 – gw Olms [944]
Histoire de paris, et description de ses plus beaux monuments. / Poncelin de LaRoche-Tilhac, Jean – Paris 1780 [mf ed Hildesheim 1995-98] – 6mf – 9 – €120.00 – 3-487-29668-3 – gw Olms [914]
Histoire de philippe-auguste / Capefigue, Jean Baptiste Honore Raymond – Paris 1829 [mf ed Hildesheim 1995-98] – 4v on 12mf – 9 – €120.00 – ISBN-10: 3-487-26315-7 – ISBN-13: 978-3-487-26315-1 – gw Olms [931]
Histoire de photius : patriarche de constantinople / Jager, Abbe – Louvain, 1845 – 8mf – 8 – €17.00 – ne Slangenburg [243]
Histoire de pie 9 et son pontificat / Saint-Albin, Alexandre de – 2e rev et considerablement augm ed. Paris: Victor Palme. 2v. 1870 – 3mf – 9 – 0-8370-9111-X – (incl bibl ref) – mf#1986-3111 – us ATLA [920]
L'histoire de preah chinavong : d'apres les archives de l'institut bouddhique – 1.ed. Phnom-Penh: Editions de l'Institut bouddhique 1964 [mf ed 1990] – 4v on 1r with other items – 1 – (title & text in khmer; added t.p. in french) – mf#mf-10289 seam reel 099 [§] – us CRL [480]
Histoire de regne du khedive ismail. l'empire africain. tome 3, le et 2e parties / Douin, Georges – Cairo. 1936-39 – 1 – us CRL [960]
Histoire de saing selchey : tiree de satras sur feuilles de latanier – 1.ed. Phnom-Penh: Editions de l'Institut bouddhique [mf ed 1990] – 1r with other items – 1 – (title & text in khmer; added t.p. in french) – mf#mf-10289 seam reel 097/02 [§] – us CRL [790]
Histoire de saint augustin : sa vie, ses oeuvres, son siecle, influence de son genie / Poujoulat, M – 2e rev corr augm ed. Paris: A Vaton 1852 [mf ed 1990] – 2v on 3mf – 9 – 0-7905-6719-9 – mf#1988-2719 – us ATLA [240]
Histoire de saint louis, roi de france : avec un abrege de l'histoire des croisades / Bury, Richard de – Paris 1775 [mf ed Hildesheim 1995-98] – 2v on 10mf – 9 – €100.00 – ISBN-10: 3-487-26310-6 – ISBN-13: 978-3-487-26310-6 – gw Olms [944]
Histoire de saint paulin de nole / Lagrange, F – Paris: Poussielgue, 1877 – 2mf – 9 – 0-8370-6819-3 – (incl bibl ref) – mf#1986-0819 – us ATLA [240]

Histoire de saint-jacques d'embrun, russell, ontario / Forget, Jean-Urgel – Ottawa: Cie d'Impr d'Ottawa, 1910 [mf ed 1994] – 8mf – 9 – 0-665-72085-8 – mf#72085 – cn Canadiana [971]

Histoire de schisme portugais dans les indes / Bussierre, Marie Theodore Renouard, vicomte de – Paris: Jacques Lecoffre, 1854 [mf ed 1995] – (= ser Yale coll) – 363p – 1 – 0-524-09754-2 – (in french) – mf#1995-0754 – us ATLA [241]

Histoire de sratop chek : tire de satras sur feuilles de la tanier [sic] – 1.ed. Phnom-Penh: Editions de l'Institut bouddhique 1963 [mf ed 1990] – 1r with other items – 1 – (title & text in khmer; added t.p. in french) – mf#mf-10289 seam reel 098/07 [§] – us CRL [480]

Histoire de tip -sangvar : d'apres les archives de l'institut bouddhique – Phnom-Penh: Editions de l'Institut bouddhique 1963 [mf ed 1990] – 5v on 1r with other items – 1 – (title & text in khmer; added t.p. in french) – mf#mf-10289 seam reel 106/2 [§] – us CRL [480]

Histoire de touraine : depuis la conquete des gaules par les romains, jusqu'a l'annee 1790 / Chalmel, Jean – Paris 1828 [mf ed Hildesheim 1995-98] – 4v on 13mf – 9 – €130.00 – ISBN-10: 3-487-25946-X – ISBN-13: 978-3-487-25946-8 – gw Olms [930]

Histoire de toutes les villes de france / Danielo, Julien – Paris 1833 [mf ed Hildesheim 1995-98] – 1v on 4mf – 9 – €120.00 – ISBN-10: 3-487-25944-3 – ISBN-13: 978-3-487-25944-4 – gw Olms [944]

Histoire de trois ouvriers francais: richard lenoir, abraham louis breguet, michel brezin / Ernouf, Alfred Auguste – 2nd ed. Paris: Librairie Hachette, 1873. 263p. Includes bibliog. references – 1 – us Wisconsin U Libr [330]

Histoire de tum teao : texte tire de satras sur feuilles de latanier – 5.ed. Phnom-Penh: Editions de l'Institut bouddhique 1964 [mf ed 1990] – 1r with other items – 1 – (title & text in khmer; added t.p. in french) – mf#mf-10289 seam reel 098/08 [§] – us CRL [480]

Histoire des affranchis de saint-domingue, tome pr... / Lespinasse, Beauvais – Paris, France. 1882 – 1r – us UF Libraries [972]

Histoire des amazones anciennes et modernes / Guyon, Claude Marie – Paris: Villette, 1740 – (= ser Les femmes [coll]) – 2v on 6mf – 9 – mf#10175-76 – fr Bibl Nationale [305]

Histoire des arabes / Huart, Clement – Paris: P Geuthner, 1912-1913 – 3mf – 9 – 0-524-08160-3 – (incl bibl ref) – mf#1991-0290 – us ATLA [956]

Histoire des armees-bovines dans l'ancien rwanda / Kagame, Alexis – Bruxelles, Belgium. 1961 – 1r – us UF Libraries [960]

Histoire des arts... / Monier, P – Paris, 1698 – 4mf – 9 – mf#O-986 – ne IDC [720]

Histoire des bagesera, souverains du gisaka / Arianoff, A D' – Bruxelles, Belgium. 1952 – 1r – us UF Libraries [960]

Histoire des catholiques francais au 19e siecle (1815-1905) / Guillemin, Henri – Geneva, Switzerland. 1947 – 1r – us UF Libraries [025]

Histoire des chevalliers hospitaliers de s jean de jeruzalem / Vertot, R A de – Paris. v1-4. 1726 – €94.00 – ne Slangenburg [956]

Histoire des cinquante premieres annees de l'eglise evangelique libre du canton de vaud / Cart, Jacques – Lausanne: G Bridel, 1897 [mf ed 1993] – (= ser Presbyterian coll) – 5mf – 9 – 0-524-07969-2 – (in french. incl bibl ref) – mf#1990-5414 – us ATLA [242]

Histoire des colonies francaises / Besson, Maurice – Paris, France. 1931 – 1r – us UF Libraries [960]

Histoire des commandements de l'eglise *see* History of the commandments of the church

Histoire des communes de france et legislation municipale : depuis la fin du 11e siecle jusqu'a nos jours / Dufey, Pierre Joseph Spiridion – Paris 1828 [mf ed Hildesheim 1995-98] – 1v on 3mf – 9 – €90.00 – ISBN-10: 3-487-25897-8 – ISBN-13: 978-3-487-25897-3 – gw Olms [350]

Histoire des comtes de champagne et de brie / Pelletier, Robert le – Paris 1753 [mf ed Hildesheim 1995-98] – 2v on 7mf – 9 – €140.00 – ISBN-10: 3-487-25962-1 – ISBN-13: 978-3-487-25962-8 – gw Olms [944]

Histoire des comtes d'eu / Estancelin, Louis – Dieppe 1828 [mf ed Hildesheim 1995-98] – 1v on 3mf – 9 – €90.00 – ISBN-10: 3-487-25914-1 – ISBN-13: 978-3-487-25914-7 – gw Olms [944]

Histoire des conciles / ed by Hefele, C J & Leclercq, H – Paris, 1907-1911. 10v – 145mf – 9 – mf#H-2984 – ne IDC [240]

Histoire des conciles d'apraes les documents originaux / Hefele, C J & Leclercq, H – Paris. v1-11. 1907-1952 – €479.00 – ne Slangenburg [240]

Histoire des conquetes des normands : en italie, en sicile, et en grece / Gauttier d'Arc, Edouard – Paris 1830 [mf ed Hildesheim 1995-98] – 1v on 4mf – 9 – €120.00 – ISBN-10: 3-487-25953-2 – ISBN-13: 978-3-487-25953-6 – gw Olms [930]

Histoire des croisades / Michaud, J F – Paris, 1825-1829. 6v – 43mf – 9 – mf#H-2937 – ne IDC [931]

Histoire des croyances religieuses et des opinions philosophiques en chine : depuis l'origine jusqu'a nos jours / Wieger, Leon – [S.l.: s.n.], 1917 – 2mf – 9 – 0-524-05174-7 – (incl bibl ref) – mf#1990-3460 – us ATLA [290]

Histoire des decouvertes et conquestes des portugais dans le nouveau monde : avec des figures en taille-douce / Lafitau, Joseph Francois – Paris: Chez Saugrain pere...Jean-Baptiste Coignard fils...2v. 1733 [mf ed 1984] – 2v on 1mf – 9 – mf#38803 – cn Canadiana [946]

Histoire des decouvertes et conquestes des portugais dans le nouveau monde / Lafitau, J F – Paris: Saugrain Pere, Jean-Baptiste Coignard, 1733. v2 – 17mf – 9 – mf#SEP-12 – ne IDC [918]

Histoire des decouvertes et conquestes des portugais dans le nouveau monde / Lafitau, Joseph – Paris 1733 [mf ed Hildesheim 1995-98] – 2v on 17mf – 9 – €170.00 – ISBN-10: 3-487-27261-X – ISBN-13: 978-3-487-27261-0 – gw Olms [972]

Histoire des decouvertes faites par divers savans voyageurs dans plusieurs contrees de la russie et de la perse : relativement a l'histoire civile et naturelle, a l'economie rurale, au commerce, etc – Berne 1779-87 [mf ed Hildesheim 1995-98] – 6v on 36mf – 9 – €360.00 – 3-487-26724-1 – gw Olms [900]

Histoire des differens peuples du monde / Contant d'Orville, Andre Guillaume – Paris, 1770-71. v.4, Africa – 1 – us CRL [960]

Histoire des differens peuples du monde : contenant les ceremonies religieuses et civiles, l'origine des religions, leurs sectes et superstitions, et les moeurs et usages de chaque nation... / Contant Dorville, Andre Guillaume – Paris: Chez Herissant fils...J P Costard...1770-71 [mf ed 1985] – 6v on 1mf – 9 – 0-665-51044-6 – (incl bibl ref) – mf#51044 – cn Canadiana [900]

Histoire des dogmes *see* History of dogmas

Histoire des dogmes de l'eglise chretienne / Bonifas, Francois – Paris: Librairie Fischbacher, 1886 [mf ed 1991] – 2v on 3mf – 9 – 0-7905-8762-9 – (in french) – mf#1989-1987 – us ATLA [240]

Histoire des duches de lorraine et de bar : et des trois eveches (meurthe, meuse, moselle, vosges) / Begin, Emile A – Nancy 1833 [mf ed Hildesheim 1995-98] – 2v on 6mf – 9 – €120.00 – 3-487-25958-3 – gw Olms [944]

Histoire des ducs de bourgogne de la maison de valois 1364-1477 / Barante, Amable G de – Paris 1824-[1826] [mf ed Hildesheim 1995-98] – 14v on 40mf – 9 – €400.00 – 3-487-25967-2 – gw Olms [944]

Histoire des ducs d'orleans / Laurentie, Pierre – Paris 1832 [mf ed Hildesheim 1995-98] – 4v on 12mf – 9 – €120.00 – ISBN-10: 3-487-26067-0 – ISBN-13: 978-3-487-26067-9 – gw Olms [944]

Histoire des edits de pacification et des moyens que les pretendus reformez ont employe pour les obtenir... / Soulier, [P] – Paris, 1682 – 6mf – 9 – mf#CA-108 – ne IDC [240]

Histoire des eglises reformees du pays de gex / Claparede, Theodore – Geneve: Joel Cherbuliez, 1856 – 1mf – 9 – 0-524-05309-X – (incl bibl ref) – mf#1990-1427 – us ATLA [240]

Histoire des emigres francais : depuis 1789, jusqu'en 1828 / Antoine, A – Paris 1828 [mf ed Hildesheim 1995-98] – 3v on 9mf – 9 – €180.00 – 3-487-26296-7 – gw Olms [944]

Histoire des empereurs et des autres princes qui ont regne durant les six premiers siecles de l'eglise... / Lenain De Tillemont, L S – Bruxelles. v1-6. 1733 – €174.00 – ne Slangenburg [941]

Histoire des etablissements religieux britanniques fondes a douai avant la revolution francaise / Dancoisne, Louis – Douai: Lucien Crepin, 1880 – 1mf – 9 – 0-524-03517-2 – (incl bibl ref) – mf#1990-1022 – us ATLA [240]

Histoire des francais des divers etats aux cinq derniers siecles / Monteil, Amans – Paris [mf ed Hildesheim 1995-98] – 10v on 33mf – 9 – €330.00 – ISBN-10: 3-487-26308-4 – ISBN-13: 978-3-487-26308-3 – gw Olms [944]

Histoire des francs d'austrasie / Gerard, Pierre Auguste Florent – Bruxelles 1864 [mf ed Hildesheim 1995-98] – 2v on 6mf – 9 – €120.00 – ISBN-10: 3-487-26319-X – ISBN-13: 978-3-487-26319-9 – gw Olms [943]

Histoire des francs, textes de manuscrits de corbie et de bruxelles / Gregory, Saint, Bishop of Tours – Nouv. ed. par Rene Poupardin. Paris: A. Picard, 1913. xxx/501p – 1 – us Wisconsin U Libr [944]

Histoire des francs-macons : contenant les obligations & statuts de la tresvenerable confraternite de la maconnerie, conformes aux traductions les plus anciennes... / Tierce, de la – A l'Orient: Chex G de l'Etoille...1745 – 1v on 4mf – 9 – (filmed: t1 [xii, 311p]) – mf#vrl-39 – ne IDC [366]

Histoire des francs-macons / Dubreuil, J P – Bruxelles: H I G Francois 1838 – 2v on 3mf – 9 – (Filmed: t2 [212p]) – mf#vrl-139 – ne IDC [366]

Histoire des grandes familles francaises du canada : ou, apercu sur le chevalier benoist et quelques familles contemporaines / Daniel, Francois – Montreal: E Senecal, 1867 – 2v on 8mf – 9 – 0-665-90892-X – mf#90892 – cn Canadiana [929]

L'histoire des gverres faictes par les chrestiens contre les tvrcs : sovs la conduicte de godefroy de bouillon, duc de lorraine, pour le recouurement de la terre saincte / Aubert, G – Paris, 1559 – 2mf – 9 – mf#H-8293 – ne IDC [956]

Histoire des hommes illustres de la maison de medici : avec un abrege des comtes de bologne et d'auvergne / Nestor, J – Paris: Charles Perier, 1564 – 6mf – 9 – mf#O-65 – ne IDC [090]

Histoire des hommes illustres de l'ordre de saint dominique *see* Sketches of illustrious dominicans

Histoire des idees messianiques : depuis alexandre jusqu'a l'empereur hadrien / Vernes, Maurice – Paris: Sandoz et Fischbacher, 1874 – 1mf – 9 – 0-8370-6439-2 – (incl bibl ref) – mf#1986-0439 – us ATLA [270]

Histoire des idees religieuses en allemagne : depuis le 18 siecle jusqu'a nos jours / Lichtenberger, Frederic – 2e ed. Paris: Fischbacher. 3v. 1888 – 3mf – 9 – 0-8370-9078-4 – (incl bibl ref) – mf#1986-3078 – us ATLA [240]

Histoire des incas : rois du perou / Vega, Garcilaso de la; ed by Dalibard, Thomas Francois – Paris: Prault fils. 2v. 1744 [mf ed 1975] – 1r – 5 – (trans by ed) – mf#SEM16P238 – cn Bibl Nat [972]

Histoire des institutions de charite de bienfaisance et d'education du canada : depuis leur fondation jusqu'a nos jours / Drapeau, Stanislas – [Ottaoua?: s.n.], 1877 [mf ed 1980] – 2mf – 9 – 0-665-02761-3 – (incl bibl ref) – mf#02761 – cn Canadiana [360]

Histoire des israelites depuis d'edification du second temple – s.l, s.l? no date – 1r – us UF Libraries [939]

Histoire des jacobins : depuis 1789 jusqu'a ce jour, ou etat de l'europe en novembre 1820 / Lombard de Langres, Vincent – Paris 1820 [mf ed Hildesheim 1995-98] – 1v on 2mf – 9 – €60.00 – ISBN-10: 3-487-26294-0 – ISBN-13: 978-3-487-26294-9 – gw Olms [944]

Histoire des juifs en belgique / Ullmann, Salomon – Anvers, Belgium. 1932 – 1r – us UF Libraries [939]

Histoire des maitres generaux de l'ordre des freres precheurs / Mortier, D A – Paris. v1-8. 1903-20 – €229.00 – ne Slangenburg [241]

Histoire des missions de chine / Launay, Adrien – [Paris?: s.n.] 1907-08 [mf ed 1995] – (= ser Yale coll) – 3v (ill) – 1 – 0-524-10133-7 – (in french) – mf#1995-1133 – us ATLA [241]

Histoire des missions de chine : mission du kouang-tong / Launay, Adrien – Paris: Anciennes Maisons Douniol et Retaux, 1917 [mf ed 1995] – (= ser Yale coll) – vii/207p – 1 – 0-524-10124-8 – (in french) – mf#1995-1124 – us ATLA [240]

Histoire des missions de l'inde : pondichery, maissour, coimbatour / Launay, Adrien – Paris: Ancienne Maison Charles Douniol, 1898 [mf ed 1995] – (= ser Yale coll) – 5v (ill) – 1 – 0-524-10194-9 – (in french) – mf#1995-1194 – us ATLA [240]

Histoire des modes francaises : ou revolutions du costume en france depuis l'etablissement de la monarchie jusqu'a nos jours... / Mole, Guillaume – Amsterdam [u.a. 0 [mf ed Hildesheim 1995-98] – 2v on 3mf – 9 – €90.00 – ISBN-10: 3-487-25887-0 – ISBN-13: 978-3-487-25887-4 – gw Olms [740]

Histoire des monasteres de la basse-egypte : vies des saints paul, antoine, macaire, maxime et domece, jean le nain, &a [sic] / Amelineau, Emile – Paris: E Leroux 1894 [mf ed 1990] – (= ser Annales du musee guimet 25) – 2mf – 9 – 0-7905-8136-1 – mf#1988-6083 – us ATLA [243]

Histoire des naufrages : ou recueil des relations les plus interessantes des naufrages, hivernemens, delaissemens, incendies, et autres evenemens funestes arrives sur mer / Deperthes, Jean Louis Hubert Simon – Paris 1832 [mf ed Hildesheim 1995-98] – 3v on 9mf – 9 – €180.00 – 3-487-29927-5 – gw Olms [910]

Histoire des navigations aux terres australes – Paris, 1756. 2v – 12mf – 9 – mf#H-6168 – ne IDC [919]

Histoire des oracles / Fontenelle, M de; ed by Maigron, Louis – ed critique. Paris: E Cornely, 1908 [mf ed 1992] – 1mf – 9 – 0-524-02533-9 – (incl bibl ref) – mf#1990-3028 – us ATLA [250]

Histoire des ordres monastiques, religieux et militaires et des congregations seculieres de l'un et de l'autre sexe qui ont este etablies jusqu'a present / Helyot, Pierre – Paris 1833 [mf ed Hildesheim 1995-98] – 1v on 1mf – 9 – €40.00 – 3-487-27369-1 – gw Olms [240]

Histoire des ouvrages des savans, par b.. – Basnages de Beauval. Rotterdam. sept 1687-juin 1709 (1-25) – 1 – fr ACRPP [073]

Histoire des patriarches d'alexandrie depuis la mort de l'empereur anastase jusqu'a la reconciliation des eglises jacobites (518-616) / Maspero, J – Paris, 1923 – 7mf – 8 – €15.00 – ne Slangenburg [241]

Histoire des pelerinages de la sainte vierge en france / Leroy, Louis – Paris: Louis Vives. 3v. 1873-75 – 6mf – 9 – 0-8370-9077-6 – (incl bibl ref) – mf#1986-3077 – us ATLA [240]

Histoire des persecutions : et martyrs de l'eglise de paris / Chandieu, A de la Roche – Lyon, 1563 – 6mf – 9 – mf#PRS-128 – ne IDC [240]

Histoire des persecutions de l'eglise, la polemique paienne a la fin du 2e siecle / Aube, Benjamin – Paris: Didier, 1878 – 2mf – 9 – 0-7905-4016-9 – (incl bibl ref) – mf#1988-0016 – us ATLA [240]

Histoire des persecvtions de l'eglise... / Bullinger, Heinrich – N.p., 1577 – 4mf – 9 – mf#PBU-250 – ne IDC [240]

Histoire des philosophes et des theologiens musulmans (de 632 a 1258 j.-c.) : scenes de la vie religieuse en orient / Dugat, Gustave – Paris: Maisonneuve, 1878 – 1mf – 9 – 0-524-01433-7 – (incl bibl ref) – mf#1990-2428 – us ATLA [260]

Histoire des picea qui se rencontrent dans les limites du canada / Brunet, Ovide – Quebec: Aux frais de l'auteur, 1866 – 1mf – 9 – mf#11939 – cn Canadiana [580]

Histoire des polypiers corraligenes flexibles, vulgairement nommes zoophytes / Lamouroux, J F van – Caen, 1816 – 9mf – 9 – mf#8336 – ne IDC [590]

Histoire des populations de madagascar / Dandouau, Andre & Chapus, G-S – Paris: Larose, 1952 – 1 – us CRL [301]

Histoire des populations du soudan central / Urvoy, Y – Paris, 1936 – 7mf – 8 – mf#A-352 – ne IDC [956]

Histoire des populations du soudan central (colonie du niger) / Urvoy, Y – Paris: Larose, 1936 – us CRL [960]

Histoire des premiers electeurs de paris en 1789 : extraite de leur proces-verbal / Duveyrier, Honore Marie Nicolas – Paris 1828 [mf ed Hildesheim 1995-98] – 1v on 4mf – 9 – €120.00 – ISBN-10: 3-487-26298-3 – ISBN-13: 978-3-487-26298-7 – gw Olms [325]

Histoire des progres de la puissance navale de l'angleterre / Sainte-Croix, Guillaume Emmanuel Joseph Guilhem de Clermont-Lodeve, Baron de – new corr enl ed. Paris: Chez G de Bure...2v. 1786 [mf ed 1984] – 2v on 1mf – 9 – mf#47818 – cn Canadiana [350]

Histoire des progres de la puissance navale de l'angleterre : suivie d'observations sur l'acte de navigation, et de pieces justificatives / Sainte-Croix, Guillaume Emmanuel Joseph Guilhem de Clermont-Lodeve, Baron de – Yverdon [Suisse: s.n.] 2v. 1783 [mf ed 1984] – 2v on 1mf – 9 – mf#47283 – cn Canadiana [355]

Histoire des progres et de la chute de l'empire de mysore : sous les regnes d'hyder-aly et tippoo-saib... / Michaud, Joseph Fr – Paris: Giguet, 1801-09 [mf ed 1995] – (= ser Yale coll) – 2v (ill) – 1 – 0-524-09881-6 – (in french) – mf#1995-0881 – us ATLA [954]

L'histoire des quatre dernieres annees – S.l: s.n, 1891? – 1mf – 9 – mf#03675 – cn Canadiana [971]

Histoire des rapports de l'eglise et de l'etat en france de 1789 a 1870 / Debidour, Antonin – Paris: F Alcan, 1898 – (= ser Bibliotheque d'histoire contemporaine) – 2mf – 9 – 0-7905-6987-6 – (incl bibl ref) – mf#1988-2987 – us ATLA [240]

Histoire des refugies huguenots en amerique / Baird, Charles Washington – Toulouse, France: Societe des livres religieux, 1886 – 8mf – 9 – (trans by a-e meyer and de richemond. incl ind) – mf#04083 – cn Canadiana [242]

Histoire des relations internationales de toussain / Nemours, Alfred – Port-Au-Prince, Haiti. 1945 – 1r – us UF Libraries [327]

Histoire des religieuses hospitalieres de saint-joseph (france et canada) / Couanier de Launay, Etienne-Louis – Paris: Societe generale de Librairie catholique, 1887 [mf ed 1980] – 2v on 1mf – 9 – 0-665-03606-X – mf#03606 – cn Canadiana [360]

Histoire des religions et methode comparative / Foucart, George – Paris: A Picard, 1912 – (= ser Bibliotheque d'histoire religieuse) – 2mf – 9 – 0-524-04161-X – (incl bibl ref) – mf#1990-3291 – us ATLA [230]

Histoire des rois : traduction du tantaran'ny andriana du r p callet / Chapus, G S & Ratsimba, Emmanuel – Tananarive, Academie malgache, 1953-58 – us CRL [960]

Histoire des rois et des ducs de bretagne / Roujoux, Prudence G de – Paris 1828-1829 [mf ed Hildesheim 1995-98] – 4v on 12mf – 9 – €120.00 – ISBN-10: 3-487-25963-X – ISBN-13: 978-3-487-25963-5 – gw Olms [944]

Histoire des sciences mathematiques et physiques / Marie, Maximillian – v1-12. 1883-88 – 1 – $120.00 – mf#0346 – us Brook [500]

Histoire des selucides (323-64 avant j.-c.) / Bouche-Leclercq, Auguste – Paris: E Leroux, 1913-1914 – 2mf – 9 – 0-8370-1824-2 – (incl bibl ref) – mf#1987-6212 – us ATLA [930]

Histoire des societes secretes de l'armee : et des conspirations militaires qui ont eu pour objet la destruction du gouvernement de bonaparte / Nodier, Charles – Paris 1815 [mf ed Hildesheim 1995-98] – 1v on 3mf – 9 – €90.00 – ISBN-10: 3-487-26398-X – ISBN-13: 978-3-487-26398-4 – gw Olms [355]

Histoire des societes secretes et du parti republicain de 1830 a 1848 : Louis-Philippe et la revolution de fevrier, portraits, scenes de conspirations, faits inconnus / Hodde, Lucien de la – Bruxelles: Meline, Cans et Cie 1850 – 5mf – 9 – mf#vrl-133 – ne IDC [366]

Histoire des sources du droit canonique / Tardif, Adolphe – Paris: Alphonse Picard, 1887 [mf ed 1991] – 4mf – 9 – 0-524-00656-3 – (incl bibl ref) – mf#1990-0156 – us ATLA [240]

Histoire des synodes nationaux des eglises reformees de france / Felice, Guillaume de – Paris: Grassart, 1864 – 1mf – 9 – 0-7905-6289-8 – mf#1988-2289 – us ATLA [240]

Histoire des tremblemens de terre arrive's a lima, capitale du perou, et autres lieux : avec la description du perou... – La Haye 1752 [mf ed Hildesheim 1995-98] – 2v on 6mf – 9 – €120.00 – ISBN-10: 3-487-26762-4 – ISBN-13: 978-3-487-26762-3 – gw Olms [919]

Histoire des tribunaux, criminels extraordinaires, revolutionnaires et commissions militaires : crees pendant les annees 1792, 93, 94 et 1795... / Roussel, Pierre – Paris 1830 [mf ed Hildesheim 1995-98] – 2v on 4mf – 9 – €120.00 – ISBN-10: 3-487-26277-0 – ISBN-13: 978-3-487-26277-2 – gw Olms [944]

Histoire des trois premiers siecles de l'eglise chretienne *see* The religions before christ

Histoire des tuileries, du temple, et des evenemens qui y ont eu lieu pendant la revolution : contenant en outre des details secrets sur le tribunal revolutionnaire et la conciergerie – Paris 1829 [mf ed Hildesheim 1995-98] – 1v on 2mf [ill] – 9 – €60.00 – 3-487-25933-8 – gw Olms [933]

Histoire des variations et contradictions de l'eglise romaine / Ponnat, Baron de – Paris: Charpentier. 2v. 1882 – 4mf – 9 – 0-8370-9101-2 – (incl bibl ref and index) – mf#1986-3101 – us ATLA [240]

Histoire des vaudois / Comba, Emilio – Nouvelle ed. complete. Paris: Fischbacher, 1901. Chicago: Dep of Photodup, U of Chicago Lib, 1978 (1r); Evanston: American Theol Lib Assoc, 1984 (1r) – 1 – 0-8370-0763-1 – (incl bibl ref) – mf#1984-T125 – us ATLA [240]

Histoire des vaudois : refaite d'apres les plus recentes recherches / Gay, Teofilo – Florence: Claudienne, 1912 – 1mf – 9 – 0-524-04611-5 – mf#1990-1271 – us ATLA [240]

Histoire des vaudois. introduction / Comba, Emilio – nouv. ed. complete avec cartes geographiques et gravures. Paris: Fischbacher; Florence: Claudienne, 1898 – 1mf – 9 – 0-7905-5931-5 – (incl bibl ref) – mf#1988-1931 – us ATLA [240]

Histoire d'haiti / Elie, Louis E – Port-Au-Prince, Haiti. v1-2. 1944 – 1r – us UF Libraries [972]

Histoire d'haiti / Magloire, Auguste – Port-Au-Prince, Haiti. v1-5. 1909-11 – 1r – us UF Libraries [972]

Histoire d'haiti anees 1843-1846 / Madiou, Thomas – Port-Au-Prince, Haiti. 1904 – 1r – us UF Libraries [972]

Histoire documentaire de la congregation des missionnaires...otawa, 1963 / Carriere, Gaston – Madrid: Graf. Calleja, 1966 – 1 – sp Bibl Santa Ana [240]

Histoire dogmatique, liturgique et archeologique du sacrement de baptaeme / Corblet, Jules – Paris: Victor Palme, 1881-82 [mf ed 1991] – 3mf – 9 – 0-524-00254-1 – (in french. incl bibl ref) – mf#1989-2954 – us ATLA [241]

Histoire dogmatique, liturgique, et archeologique du sacrement de l'eucharistie / Corblet, Jules – Paris: Societe generale de librairie catholique, 1885-86 [mf ed 1991] – 3mf – 9 – 0-7905-9372-6 – (in french) – mf#1989-2597 – us ATLA [240]

Histoire du 19e siecle / Michelet, Jules – Paris 1872 [mf ed Hildesheim 1995-98] – 1v on 3mf – 9 – €90.00 – ISBN-10: 3-487-26260-6 – ISBN-13: 978-3-487-26260-4 – gw Olms [933]

Histoire du bienheureux jean de britto *see* Historia de la vida y martirio del beato juan de britto, de la compania de jesus

Histoire du bienheureux jean de britto de la compagnie de jesus : missionnaire du madure et martyr de la foi / Prat, Jean Marie – Paris: Societe de Saint-Victor pour la propagation des bons livres, 1853 – (= ser Yale coll) – xvi/550p (ill) – 1 – (in french) – mf#1995-0739 – us ATLA [241]

Histoire du bienheureux pierre claver de la compagnie de jesus : apotre des negres de carthagene et des indes-occidentales / Daurignac, J M S – Lyon: J B Pelagaud, 1854 – 2mf – 9 – 0-8370-6895-9 – mf#1986-0895 – us ATLA [920]

Histoire du bouddha sakya-mouni depuis sa naissance jusqu'a sa mort / Summer, Mary – Paris; New-Haven (Etats-Unis): Ernest Leroux, 1874 – 1mf – 9 – 0-524-01306-3 – (incl bibl ref) – mf#1990-2342 – us ATLA [280]

Histoire du bourbonnais et des bourbons qui l'ont possede / Coiffier de Moret, Simon – Paris 1824 [mf ed Hildesheim 1995-98] – 2v on 6mf – 9 – €120.00 – ISBN-10: 3-487-25964-8 – ISBN-13: 978-3-487-25964-2 – gw Olms [940]

Histoire du breviaire de rouen / Colette, A – Rouen, 1902 – 6mf – 8 – €14.00 – ne Slangenburg [241]

Histoire du breviaire romain = History of the roman breviary / Batiffol, Pierre – London; New York: Longmans, Green, 1898 – 1mf – 9 – 0-8370-7282-4 – (in english. includes appendix and index) – mf#1986-1282 – us ATLA [240]

Histoire du calvinisme / Maimbourg, L – Paris, 1682. 2v – 10mf – 9 – mf#CA-102 – ne IDC [242]

Histoire du calvinisme, contenant sa naissance... / Soulier, P – Paris, 1686 – 8mf – 9 – mf#CA-151 – ne IDC [242]

Histoire du cambodge / Dauphin-Meunier, Achille – 2.ed. Paris: Presses universitaires de France 1968 [mf ed 1989] – (= ser Que sais -je? 916) – 1r with other items – 1 – (with bibl) – mf#mf-10289 seam reel 003/05 [§] – us CRL [959]

Histoire du cambodge / Giteau, Madeleine – Paris: Didier [1957] [mf ed 1989] – 1r with other items – 1 – mf#mf-10289 seam reel 004/02 [§] – us CRL [959]

Histoire du cambodge depuis le 1er siecle de notre ere, d'apres les inscriptions lapidaires : les annales chinoises et annamites et les documents europeens des six derniers siecles / Leclere, Adhemard – Paris: P Geuthner 1914 [mf ed 1989] – 1r with other items – 1 – mf#mf-10289 seam reel 010/08 [§] – us CRL [959]

Histoire du cameroun / Mveng, Engelbert – Paris, France. 1963 – 1r – us UF Libraries [960]

Histoire du canada : cours elementaire / Stanislas-Joseph, frere – Montreal: [freres des ecoles chretiennes], [1883?] [mf ed 1980] – 2mf – 9 – 0-665-03256-0 – mf#03256 – cn Canadiana [971]

Histoire du canada : depuis sa decouverte jusqu'a nos jours / Garneau, Francois-Xavier – Quebec?: s.n. 4v. 1845-1852 – 1mf – 9 – (with ind) – mf#35263 – cn Canadiana [971]

Histoire du canada, 1841 a 1867 : periode comprise entre l'union legislative des provinces du haut et du bas-canada... / Royal, Joseph – Montreal: Beauchemin, 1909 [mf ed 1977] – 1r – 5 – mf#SEM16P288 – cn Bibl Nat [971]

Histoire du canada a l'usage des maisons d'education / Laverdiere, Charles-Honore – Quebec: des Presses d'Augustin Cote, 1873 [i.e. 1874] (mf ed 1989) – 3mf – 9 – mf#SEM105P1160 – cn Bibl Nat [971]

Histoire du canada, de son eglise et de ses missions : depuis la decouverte de l'amerique jusqu'a nos jours... / Brasseur de Bourbourg, abbe – Paris: Sagnier et Bray; Plancy France?: J Collin, 1852 – 2v on 1mf – 9 – mf#43018 – cn Canadiana [241]

Histoire du canada, de son eglise, et de ses missions : ecrite d'apres l'histoire du p de charlevoix, et d'autres documents... / Brasseur de Bourbourg, abbe – Paris: Putois-Crette, 1859 [mf ed 1986] – 2v on 1mf – 9 – 0-665-42828-6 – (incl bibl ref) – mf#42828 – cn Canadiana [241]

Histoire du canada depuis la confederation, 1867-1887 / David, Laurent-Olivier – Montreal: Librairie Beauchemin, [1909?] – 4mf – 9 – 0-665-72601-5 – (incl app) – mf#72601 – cn Canadiana [971]

L'histoire du canada depuis sa decouverte jusqu'a nos jours / Bourgeois, Phileas Frederic – Montreal: Librairie Beauchemin, 1913 – 3mf – 9 – 0-665-71761-X – (incl ind) – mf#71761 – cn Canadiana [971]

L'histoire du canada en 200 lecons / Bourgeois, Phileas Frederic – Montreal: Librairie Beauchemin, [1902?] – 5mf – 9 – 0-665-73731-9 – mf#73731 – cn Canadiana [971]

Histoire du canon de l'ancien testament : lecons d'ecriture sainte professees a l'ecole superieure de theologie de paris pendant l'annee 1889-1890 / Loisy, Alfred Firmin – Paris: Letouzey et Ane, 1890 – 1mf – 9 – 0-7905-3034-1 – (incl bibl ref) – mf#1987-3034 – us ATLA [221]

Histoire du canon de l'ancien testament dans l'eglise grecque et l'eglise russe / Jugie, Martin – Paris: G Beauchesne, 1909 – (= ser Etudes de Theologie Orientale) – 1mf – 9 – 0-7905-5349-X – (incl bibl ref) – mf#1988-1349 – us ATLA [221]

Histoire du canon des saintes-ecritures dans l'eglise chretienne *see* History of the canon of the holy scriptures in the christian church

Histoire du canon du nouveau testament : lecons d'ecriture sainte professees a l'ecole superieure de theologie de paris pendant l'annee 1890-1891 / Loisy, Alfred Firmin – Paris: J Maisonneuve, 1891 – 1mf – 9 – 0-7905-3083-X – (incl bibl ref) – mf#1987-3083 – us ATLA [225]

Histoire du cap-sante : depuis la fondation de cette paroisse jusqu'a 1830 / Gatien, Felix X – Quebec?: Franciscaine Missionnaire, 1899 – 4mf – 9 – (cont depuis 1830 jusqu'a 1887 par david gosselin) – mf#05763 – cn Canadiana [971]

Histoire du catholicisme liberal en france, 1828-1908 / Weill, Georges – Paris: Felix Alcan, 1909 – 1mf – 9 – 0-8370-9037-7 – (incl ind) – mf#1986-3037 – us ATLA [241]

Histoire du chevalier d'iberville, 1663-1706 / Desmazures, Adam Charles Gustave – Montreal: J M Valois, 1890 – 4mf – 9 – mf#05637 – cn Canadiana [910]

Histoire du christianisme au japon : ou l'on voit les differentes revolutions qui ont agite cette monarchie pendant plus d'un siecle / Charlevoix, Pierre Francois Xavier de – nouv ed. Liege: H Dessain, 1855 [mf ed 1995] – (= ser Yale coll) – 2v in 1 – 1 – 0-524-09867-0 – (in french. earlier eds iss under title: histoire de l'etablissement, des progres et de la decadence du christianisme dans l'empire du japon) – mf#1995-0867 – us ATLA [241]

Histoire du christianisme dans le monde paien : les missions en asie / Gindraux, Jules – Geneve: J-H Jeheber, [1908?] – 1mf – 9 – 0-8370-6498-8 – (incl ind) – mf#1986-0498 – us ATLA [240]

Histoire du christianisme d'ethiopie, et d'armenie / Veyssiere de La Croze, M – La Haye, la Veuve Le Vier and Pierre Paupie, 1739 – 5mf – 9 – mf#AR-1400 – ne IDC [956]

Histoire du concile de trente / Baguenault de Puchesse, M Fernand – Paris: Victor Palme, 1870 – 1mf – 9 – 0-8370-8320-6 – (incl bibl ref) – mf#1986-2320 – us ATLA [240]

Histoire du concile de trente / Pallavicini, S – Paris. v1-3. 1844 – €80.00 – ne Slangenburg [241]

Histoire du concile de trente / Pallavicini, S – Paris. v1-3. 1844 – 3v on 42mf – 8 – €80.00 – ne Slangenburg [241]

Histoire du congo, leopoldville / Cornevin, Robert – Paris, France. 1963 – 1r – us UF Libraries [960]

Histoire du congo pour la jeunesse / Gregoire, Herman – Bruxelles, Belgium. 1930 – 1r – us UF Libraries [960]

Histoire du couronnement, o- relation des ceremonies religieuses, politiques et militaires, qui ont eu lieu pendant les jours memorables consacres...celebrer le couronnement et le sacre de sa majeste imperiale napoleon i... – [Paris], 1805 – 7mf – 9 – mf#O-1111 – ne IDC [700]

Histoire du credo : le symbole des apaotres / Ermoni, Vincent – Paris: Bloud, [1903?] – (= ser Science et Religion) – 1mf – 9 – 0-524-03337-4 – mf#1990-0918 – us ATLA [240]

Histoire du culte de sin en babylonie et en assyrie / Combe, E – Paris, 1908 – 2mf – 9 – mf#NE-409 – ne IDC [956]

Histoire du culte des divinites d'alexandrie : serapis, isis, harpocrate et anubis, hors de l'egypte depuis les origines jusqu' a la naissance de l'ecole neo-platonicienne / Lafaye, Georges – Paris: Ernest Thorin, 1884 – (= ser Bibliotheque des ecoles francaises d'athenes et de rome) – 1mf – 9 – 0-524-04194-6 – (incl bibl ref) – mf#1990-3301 – us ATLA [250]

Histoire du cure santa cruz. : paris, 1928 / Bernoville, Gactan La Croix de Sang; ed by Bayle, Constantino – Madrid: Razon y Fe, 1928 – 9 – sp Bibl Santa Ana [946]

Histoire du dauphine / Chapuys-Montlaville, Benoit M de – Paris 1827-1828 [mf ed Hildesheim 1995-98] – 2v on 6mf – 9 – €120.00 – ISBN-10: 3-487-25961-3 – ISBN-13: 978-3-487-25961-1 – gw Olms [944]

Histoire du directoire de la republique francaise / Barante, Amable G de – Paris 1855 [mf ed Hildesheim 1995-98] – 3v on 10mf – 9 – €100.00 – 3-487-26262-2 – gw Olms [944]

Histoire du dix-huit fructidor : ou memoires contenant la verite sur les divers evenemens qui se rattachent a cette conjuration... / Rue, Isaac E de la – Paris 1821 [mf ed Hildesheim 1995-98] – 2v on 4mf – 9 – €120.00 – ISBN-10: 3-487-26257-6 – ISBN-13: 978-3-487-26257-4 – gw Olms [944]

Histoire du dogme de la divinite de jesu-christ / Reville, Albert – 2e rev augm ed. Paris: Germer Bailliere, 1876 – 1mf – 9 – 0-8370-3998-3 – (incl bibl ref) – mf#1985-1998 – us ATLA [240]

Histoire du dogme de la papaute : des origines a la fin du quatri a siecle / Turmel, Joseph – Paris: A Picard [1908?] – (= ser Bibliotheque d'histoire religieuse) – 2mf – 9 – 0-7905-6897-7 – (incl bibl ref) – mf#1988-2897 – us ATLA [240]

Histoire du dogme du peche originel / Turmel, Joseph – Macon: Protat, 1904 – 1mf – 9 – 0-7905-3565-3 – mf#1989-0058 – us ATLA [240]

Histoire du droit canadien : depuis les origines de la colonie jusqu'a nos jours / Lareau, Edmond – Montreal: A Periard, 1888-89 [mf ed 1981] – 2v on 1mf – 9 – 0-665-12248-9 – mf#12248 – cn Canadiana [340]

L'histoire du droit haitien tome premier / Jean-Jacques, Thales – Port-Au-Prince, Haiti. 1933 – 1r – us UF Libraries [972]

Histoire du gouvernement du general legitime – Paris, France. 1890 – 1r – us UF Libraries [972]

Histoire du grand royaume de la chine... / Gonzalez de Mendoza, J – Paris: Ieremenie Perier, 1588 – 8mf – 9 – mf#HT-518 – ne IDC [915]

Histoire du lutheranisme / Maimbourg, L – Paris, 1680 – 7mf – 9 – mf#CA-103 – ne IDC [242]

Histoire du maghreb : cours professe a l'institut des hautes etudes marocaines / Hamid, Ismail – Paris: E Laroux, 1923 – 1 – us CRL [960]

Histoire du montreal / Dollier de Casson, Francois – Montreal: des presses a vapeur de "La Minerve", 1868 [i.e. 1869] (mf ed 1992) – 4mf – 9 – (with ind) – mf#SEM105P1528 – cn Bibl Nat [917]

Histoire du montreal, 1640-1672 : manuscrit de paris / Dollier de Casson, Francois – Montreal: Eusebe Senecal, 1871 [i.e. 1927] (mf ed 1992) – 2mf – 9 – mf#SEM105P1529 – cn Bibl Nat [917]

Histoire du movement religieux et ecclesiastique dans le canton de vaud : pendant la premiere moitie du 19e siecle / Cart, Jacques – Lausanne: G Bridel, 1870-1880 – 7mf – 9 – 0-524-07288-4 – (incl bibl ref) – mf#1990-5375 – us ATLA [240]

Histoire du naufrage et de la captivite de m de brisson, officier de l'administration des colonies : avec la description des deserts d'afrique, depuis le senegal jusq'a maroc – Geneve 1789 [mf ed Hildesheim 1995-98] – 1v on 2mf – 9 – €60.00 – 3-487-27274-1 – gw Olms [916]

L'histoire du nouveau-monde : ou description des indes occidentales... / Laet, Jean de – Leide: Bonnaventure & Abraham Elseviers, 1640 [mf ed 1971] – 1r – 1 – mf#SEM35P69 – cn Bibl Nat [917]

Histoire du pantheisme populaire au moyen age et au seizieme siecle : suivie de pieces inedites concernant les freres du libre esprit, maitre eckhart, les libertins spirituels, etc / Jundt, Auguste – Paris: Sandoz et Fischbacher, 1875 – 1mf – 9 – 0-7905-7005-X – (incl bibl ref) – mf#1988-3005 – us ATLA [210]

Histoire du pape calixte 2 / Robert, Ulysse – Paris: Alphonse Picard; Besancon: Paul Jacquin, 1891 – 1mf – 9 – 0-8370-7980-2 – (incl bibl ref) – mf#1986-1980 – us ATLA [240]

Histoire du paraguay / Charlevoix, Pierre-Francois-Xavier de – Paris: Chez Didot... Giffart...Nyon... 1756 [mf ed 1985] – 3v on 1mf – 9 – 0-665-51970-2 – mf#51971 – cn Canadiana [972]

Histoire du parlement de paris / Voltaire, [Francois-Marie Arouet de] – Amsterdam 1769-1770 [mf ed Hildesheim 1995-98] – 4v on 4mf – 9 – €120.00 – ISBN-10: 3-487-25904-4 – ISBN-13: 978-3-487-25904-8 – gw Olms [323]

Histoire du patriarcat armenien catholique / Vernier, Donat – Lyon: Delhomme et Briguet, 1891 – 1mf – 9 – 0-8370-8073-8 – mf#1986-2073 – us ATLA [241]

Histoire du peuple haitien, 1492-1952 / Bellegarde, Dantes – Port-Au-Prince, Haiti. 1953 – 1r – us UF Libraries [972]

Histoire du psautier : des eglises reformees / Bovet, Felix – Neuchatel: Librairie General de J Sandoz, 1872 – 1mf – 9 – 0-524-04447-3 – (incl bibl ref) – mf#1992-0116 – us ATLA [220]

Histoire du regent, philippe d'orleans : divisee en quatre parties / Lapierre de Chateauneuf, Agricol H de – Paris 1829-1828 [mf ed Hildesheim 1995-98] – 2v on 4mf – 9 – €120.00 – ISBN-10: 3-487-26212-6 – ISBN-13: 978-3-487-26212-3 – gw Olms [944]

Histoire du regne de louis le grand... / Menestrier, C F – Paris: Chez Robert Pepie et JB Nolin, 1699 – 8mf – 9 – mf#O-49 – ne IDC [090]

Histoire du regne du khedive ismail : l'empire africain: tome 3, 2 i.e. parties / Douin, Georges – Caire: Imp de l'Institut francais d'archeologie orientale du Caire pour la Soc royale de geographie d'Egypte, 1936-39 – 1r – 1 – us CRL [930]

Histoire du roy louis-le-grand... / Menestrier, C F – Paris: Robert Pepie, 1693 – 3mf – 9 – mf#O-865 – ne IDC [090]

Histoire du roy louis-le-grand par les medailles, emblemes, devises, jettons, inscriptions, armoiries et autres monumens publics, recueillis et expliquez par le pere claude-francois menestrier de la compagnie de iesus / Menestrier, C F – Paris: Robert Pepie, 1693 – 3mf – 9 – mf#O-865 – ne IDC [700]

Histoire du royaume hova : ses origines jusqu'a sa fin / Malzac, V – Tananarive [Malagasy Rep]: Imprimerie Catholique, 1930 – 1 – us CRL [960]

Histoire du sacre et du couronnement des rois et reines de france : precedee d'une introduction dans laquelle l'auteur... / Noble, Alexandre le – Paris 1825 [mf ed Hildesheim 1995-98] – 1v on 4mf – 9 – €120.00 – ISBN-10: 3-487-25902-8 – ISBN-13: 978-3-487-25902-4 – gw Olms [944]

Histoire du sentiment religieux en france : depuis la fin des guerres de religion jusqu'a nos jours / Bremond, A – Paris. v1-12. 1921-1933 – 12v on 126mf – 8 – €240.00 – ne Slangenburg [240]

Histoire du sergent flavigny : ou dix annees de ma captivite sur les pontons anglais / Chomel, Auguste – Paris 1821 [mf ed Hildesheim 1995-98] – 2v on 3mf – 9 – €90.00 – 3-487-27971-1 – gw Olms [944]

Histoire du sultan djelal-eddin mankobirti, prince du kharezm / Houdas, O – (= ser Publications de l'Ecole des Langues Orientales Vivantes) – 14mf – 8 – (publications de l'ecole des langues orientales vivantes, paris 1891; 1895 s3 v9-10) – mf#U-796 – ne IDC [956]

Histoire du synode general de l'eglise reformee de france, paris, juin-juillet 1872 / ed by Bersier, Eugene – Paris: Sandoz & Fischbacher, 1872 [mf ed 1991] – 3mf – 9 – 0-524-01642-9 – (in french) – mf#1990-0463 – us ATLA [242]

Histoire du theatre de l'academie royale de musique en france : depuis son etablissement jusqu'a present / Durey de Noinville, Jacques Bernard – 2d ed. Paris: Duchesne 1757 [mf ed 19–] – 2v in 1 on 9mf – 9 – mf#fiche 719 – us Sibley [780]

Histoire du theatre de l'opera comique... / Desboulmiers, Jean-Augustin-Julien – Paris: Chez Lacombe 1769 [mf ed 19–] – 2v on 7mf / 8mf / 15mf – 9 – mf#fiche 912 / 913 / 494 – us Sibley [790]

Histoire du theatre de l'opera en france depuis l'etablissement de l'academie royale de musique, jusqu'a present / Durey de Noinville, Jacques Bernard – Paris: J Barthou 1753 [mf ed 19–] – 2v in 1 on 7mf – 9 – mf#fiche 448 – us Sibley [780]

Histoire d'un meurtre execrable : commis par un hespagnol, nomme alphonse dias, chambellan du pape, en la personne de jehan dias son frere / [Calvin, J] – [Geneva: Jean Girard], 1546 – 1mf – 9 – mf#CL-26 – ne IDC [240]

Histoire d'un voyage aux isles malouines, fait en 1763 et 1764 : avec des observations sur le detroit de magellan, et sur les patagons / Pernety, Antoine Joseph – Paris: Chez Saillant & Nyon. 2v. 1770 [mf ed 1984] – 2v on 1mf – 9 – mf#43771 – cn Canadiana [919]

Histoire d'une jeune reveuse de quarante ans – Montreal: (s.n.), 1893 [mf ed 1982] – 1mf – 9 – mf#SEM105P72 – cn Bibl Nat [971]

Histoire ecclesiastique des eglises reformees au royaume de france – ed nouvelle. Paris: Librairie Fischbacher, 1883-1889 – (= ser Les classiques de protestantisme francais) – 30mf – 9 – 0-524-07348-1 – mf#1990-5385 – us ATLA [242]

Histoire et abrege des ouvrages latins, italiens et francois : pour et contre la comedie et l'opera / Lalouette, Ambroise – Paris: Robustel 1697 [mf ed 19–] – 2mf – 9 – mf#fiche 590, 750 – us Sibley [780]

Histoire et apologie de la retraite des pasteurs... / [Benoist, E] – Francfort, 1687 – 4mf – 9 – mf#PRS-114 – ne IDC [240]

Histoire et cartulaire de l'abbaye demalbuisson / Dutilleux, A & Depoin, J – Pontoise, 1882 – €32.00 – ne Slangenburg [241]

Histoire et description des iles ioniennes : depuis les tems fabuleux et heroiques jusqu'a ce jour; avec un nouvel atlas, contenant cartes, plans, vues, costumes et medailles / Schneider, Virgile – Paris 1823 [mf ed Hildesheim 1995-98] – 3mf – 9 – €90.00 – 3-487-29050-2 – gw Olms [914]

Histoire et description generale de la nouvelle france : avec le journal historique d'un voyage...dans l'amerique septentrionnale / Charlevoix, P Fr X de – Paris, 1744 – 49mf – 9 – mf#52 – ne IDC [914]

Histoire et doctrine de la secte des cathares ou albigeois / Schmidt, Charles – Paris: J Cherbuliez, 1849 – 8mf – 9 – 0-524-08722-9 – (incl bibl ref) – mf#1993-2127 – us ATLA [240]

Histoire et geographie de madagascar / Escamps, Henri d' – Nouv. ed. Paris: Firmin-Didot, 1884 – 1 – us CRL [960]

Histoire et influence des eglises wallonnes dans les pays-bas / Poujol, David F – Paris: Librairie Fischbacher, 1902. Chicago: Dep of Photodup, U of Chicago Lib, 1902 (1r); Evanston: American Theol Lib Assoc, 1984 (1r) – 1 – 0-8370-0110-2 – (incl bibl ref and ind) – mf#1984-B054 – us ATLA [240]

Histoire et miracles de ste anne de beaupre / Giroux, Henri – Montreal: s.n, 1895 – 1mf – 9 – mf#51326 – cn Canadiana [241]

Histoire et regne de henri 2, roi de france : 2 Bde / Lambert, Claude – Paris 1755 [mf ed Hildesheim 1995-98] – 10mf – 9 – €100.00 – ISBN-10: 3-487-26123-5 – ISBN-13: 978-3-487-26123-2 – gw Olms [944]

Histoire et regne de louis 11 / Baudot de Juilly, Nicolas – Paris 1755 [mf ed Hildesheim 1995-98] – 6v on 31mf – 9 – €310.00 – 3-487-26223-1 – gw Olms [944]

Histoire et religion des nosairais / Dussaud, Rene – Paris: Emile Bouillon, 1900 – (= ser Bibliotheque de l'ecole des hautes etudes) – 1mf – 9 – 0-524-01277-6 – mf#1990-2313 – us ATLA [290]

Histoire et sagesse d'ahikar l'assyrien (fils d'anael, neveu de tobie) : traduction des versions syriaques avec les principales differences des versions arabes, armenienne, grecque, neo-syriaque, slave et roumaine / Nau, Francois – Paris: Letouzey & Ane, 1909 [mf ed 1990] – (= ser Documents pour l'etude de la bible) – 1mf – 9 – 0-8370-1882-X – (incl bibl ref) – mf#1987-6269 – us ATLA [390]

Histoire et symbolisme de la liturgie / Lerosey, A – 2. rev et corr ed. Paris: Berche et Tralin, 1912 – 2mf – 9 – 0-524-06255-2 – mf#1990-5210 – us ATLA [240]

Histoire fantastique de la revolution de juillet : en soixante-dix articles, ou recueil de varietes inseres dans la quotidienne / Nettement, Alfred – Paris 1834 [mf ed Hildesheim 1995-98] – 2v on 6mf – 9 – €120.00 – ISBN-10: 3-487-26062-X – ISBN-13: 978-3-487-26062-4 – gw Olms [944]

Histoire financiere de la france : depuis l'origine de la monarchie jusqu'a l'annee 1828... / Bresson, Jacques – Paris 1829 [mf ed Hildesheim 1995-98] – 2v on 7mf – 9 – €140.00 – ISBN-10: 3-487-25894-3 – ISBN-13: 978-3-487-25894-2 – gw Olms [332]

Histoire genealogique de la royale maison de savoye / Guichenon, Samuel – Lyon: Chez Guillaume Barbier. 2v. 1660 [mf ed 1983] – 1r – 1 – mf#SEM35P190 – cn Bibl Nat [929]

Histoire genealogique et chronologique de la maison royale de bourbon : contenant les naissances, actions memorables, alliances, et deces de tous les princes et princesses de cette illustre maison... / Achaintre, Nicolas L – Paris 1825 [mf ed Hildesheim 1995-98] – 2v on 6mf – 9 – €120.00 – 3-487-26309-2 – gw Olms [929]

Histoire genealogique et livre de famille des goyette, 1659-1959 = The goyette's family book and genealogy, 1659-1959 / Goyette, Armand – [Quebec (Province)]: [s.n.] 2v [1960?] (mf ed 1991) – 6mf – 9 – mf#SEM105P1416 – cn Bibl Nat [929]

Histoire general du jansenisme / Gerberon, Gabriel – Amsterdam: J L de Lorme. v1-5. 1701 – 35mf – 8 – €48.00 – ne Slangenburg [241]

Histoire generale, critique et philologique de la musique / Blainville, Charles Henri – Paris: Chez Pissot 1767 [mf ed 19–] – 6mf – 9 – mf#fiche 363 – us Sibley [780]

Histoire generale de france : depuis le regne de charles 9 jusqu'a la paix generale en 1815 / Dufau, Pierre Armand – Paris 1820 [mf ed Hildesheim 1995-98] – 5v on 16mf – 9 – €160.00 – ISBN-10: 3-487-26116-2 – ISBN-13: 978-3-487-26116-4 – gw Olms [944]

Histoire generale de la bastille : depuis sa fondation 1369, jusqu'a sa destruction, 1789 / Fougeret, W A – Paris 1834 [mf ed Hildesheim 1995-98] – 2v on 6mf – 9 – €120.00 – ISBN-10: 3-487-25920-6 – ISBN-13: 978-3-487-25920-8 – gw Olms [944]

Histoire generale de la franc-maconnerie en normandie : 1739 a 1875 / Loucelles, Hilaire-Pierre de – Dieppe: Emile Delevoye 1875 – 3mf – 9 – mf#vrl-99 – ne IDC [366]

Histoire generale de l'eglise : depuis la prediction des apotres jusqu'au pontificat de gregoire 16... / Berault-Bercastel, Antoine Henri de – [Paris?: s.n.] 13v. 1841 [mf ed 1985] – 13v on 1mf – 9 – mf#48900 – cn Canadiana [200]

Histoire generale de napoleon bonaparte : de sa vie privee et publique, de sa carriere politique et militaire, de son administration et de son gouvernement, tome premier / Thibaudeau, Antoine Clair – Paris 1827-1828 [mf ed Hildesheim 1995-98] – 6v on 18mf – 9 – €180.00 – ISBN-10: 3-487-26420-X – ISBN-13: 978-3-487-26420-2 – gw Olms [944]

Histoire generale de normandie : contenant les choses memorables advenues depuis les premieres courses des normands payens, tant en france quaux autres pays... / Du Moulin, Gabriel – Rouen: Chez Jean Osmont, 1631 [mf ed 1984] – 8mf – 9 – mf#SEM105P383 – cn Bibl Nat [944]

Histoire generale des auteurs sacres et ecclesiastiques / Ceillier, R – nouv ed. Paris. v1-17. 1860-1869 – 320mf – 8 – €610.00 – ne Slangenburg [240]

Histoire generale des missions catholiques depuis le 13e siecle jusqu'a nos jours / Henrion, Mathieu Richard Auguste, Baron – Paris: Gaume. 4v. 1847 [mf ed 1985] – 4v on 1mf – 9 – mf#44678 – cn Canadiana [241]

Histoire generale des roiaumes de chypre, de jerusalem, d'armenie et d'egypte, comprenant les croisades... / Jauna, D – Leide, 1747. 2v – 18mf – 9 – mf#AR-1658 – ne IDC [956]

Histoire generale des royaumes de hierusalem, cypre, armenie et lieux circonvoisins / Lusignan de Cypre, E de – Paris, 1604 – 9mf – 9 – mf#AR-1779 – ne IDC [956]

Histoire generale des voyages : ou nouvelle collection de toutes les relations de voyages par mer et par terre... / Prevost, Antoine – Paris 1746-61 [mf ed Hildesheim 1995-98] – 64v on 226mf – 9 – €1356.00 – 3-487-29951-8 – gw Olms [910]

Histoire generale des voyages : ou nouvelle collection de toutes les relations de voyages par mer et par terre... / Prevost d'Exiles, A F – Paris: Didot, 1749-1761. 64v – 376mf – 9 – mf#HT-678 – ne IDC [910]

Histoire generale des voyages : ou nouvelle collection des relations de voyages par mer et par terre / Walckenaer, Charles A – Paris 1826-1831 [mf ed Hildesheim 1995-98] – 21v on 69mf – 9 – €690 – ISBN-10: 3-487-26465-X – ISBN-13: 978-3-487-26465-3 – gw Olms [910]

Histoire generale du mouvement janseniste depuis ses origines jusqu'a nos jours / Gazier, Augustin Louis – Paris: E. Champion, 1922. 2v – 1 – us Wisconsin U Libr [240]

Histoire generale du poitou / Gueriniere, Joseph – Poitiers 1838 [mf ed Hildesheim 1995-98] – 1v on 3mf – 9 – €90.00 – ISBN-10: 3-487-25950-8 – ISBN-13: 978-3-487-25950-5 – gw Olms [944]

Histoire generale et systeme compare des langues semitiques : premiere partie, histoire generale des langues semitiques / Renan, Ernest – 3e rev augm ed. Paris: L'Imprimerie imperiale, 1863 – 2mf – 9 – 0-8370-8936-0 – (incl bibl ref) – mf#1986-2936 – us ATLA [470]

Histoire illustree des monnaies et jetons du canada : donnant l'histoire, la gravure = Illustrated history of coins and tokens relating to canada: giving illustrations with the history / Breton, Pierre Napoleon – Montreal: Breton, 1894 – 3mf – 9 – 0-665-00242-4 – mf#00242 – cn Canadiana [730]

Histoire literaire de la france / Religieux Benedictins de la Congregation de S. Maur – Paris. v1-37. 1733- – 744mf – 9 – €1418.00 – ne Slangenburg [440]

Histoire litteraire de la france – Paris: Impr Nationale. v1-32. 1733-1898 – 9 – $582.00 – (v33-41 1906-81 $150 [0265]) – mf#0264 – us Brook [440]

Histoire litteraire de l'afrique chretienne depuis les origines jusqu'a l'invasion arabe / Monceaux, P – Paris. v4-6. 1912-1923 – 27mf – 8 – €52.00 – ne Slangenburg [240]

Histoire litteraire de l'education morale et religieuse en france et dans la suisse romande / Burnier, L – Lausanne: Georges Bridel. 2v. 1864 [mf ed 1986] – 4mf – 9 – 0-8370-7609-9 – (in french. incl bibl ref and ind) – mf#1986-1609 – us ATLA [230]

Histoire litteraire de l'europe contenant l'extrait des meilleurs livres – La Haye. 1726-27 (I-VI) – 1 – fr ACRPP [410]

Histoire malacologique de la regence de tunis / Bourguignat, J R – Paris, 1868 – 2mf – 8 – mf#Z-424 – ne IDC [956]

L'histoire mariale de l'institut des soeurs de la charite de quebec : depuis son origine mil huit cent quarante-neuf jusqu'a l'annee centenaire de la definition du dogme de l'immaculee conception... / Sainte-Blanche, soeur – [Quebec]: Soeurs de la charite de Quebec, [1955?] (mf ed 1999) – 3mf – 9 – (pref by Maurice Roy) – mf#SEM105P3059 – cn Bibl Nat [366]

Histoire militaire de la guerre / Nemours, Alfred – Paris, France. v1-2. 1925- – 1r – us UF Libraries [972]

Histoire moderne de l'egypte (1801-1834) / Vaulabelle, Achille T de – Paris 1830/36 [mf ed Hildesheim 1995-98] – 2v on 6mf – 9 – €120.00 – 3-487-27379-9 – gw Olms [960]

Histoire monetaire de saint domingue et de la repu... / Lacombe, Robert – Paris, France. 1958 – 1r – us UF Libraries [972]

Histoire monetaire des colonies francaises d'apres les documents officiels / Zay, E – Paris?: s.n, 1892 – 5mf – 9 – mf#26173 – cn Canadiana [730]

Histoire nationale : ou dictionnaire geographique de toutes les communes du departement de l'aisne / Girault de Saint-Fargeau, Eusebe – Paris 1830 [mf ed Hildesheim 1995-98] – 2mf – 9 – €60.00 – 3-487-29791-4 – gw Olms [059]

Histoire naturelle a l'usage des chasseurs canadiens et des eleveurs d'animaux a fourrure / Puyjalon, Henri de – [Quebec?: s.n.], 1900 [mf ed 1981] – 5mf – 9 – mf#12292 – cn Canadiana [639]

Histoire naturelle de buffon : classee par ordres, genres et especes, d'apres le systeme de linne avec les caracteres generiques et la nomenclature linneenne / Castel, Rene-Richard – nouv ed. [Paris]: De l'impr de Crapelet a Paris, chez Deterville. an oct 1802 [mf ed 1985] – 26v on 1mf – 9 – 0-665-49305-3 – mf#49305 – cn Canadiana [500]

Histoire naturelle des animaux saus vertebres / ed by Lamarck, J B A P M de – ed 2. Paris 1835-45 – 11v [gm-320] on 115mf – 9 – mf#z-320/2 – ne IDC [590]

Histoire naturelle des animaux saus vertebres / ed by Lamarck, J B A P M de – Paris 1815-22 – 7v [gm-316] on 65mf – 9 – mf#z-319/2 – ne IDC [590]

Histoire naturelle des araign,es (aran,ides) / Simon, E – Paris, 1864 – 5mf – 9 – mf#Z-2243 – ne IDC [500]

Histoire naturelle des coquilles : contenant leur description, les moeurs des animaux qui les habitent et leurs usages... / Bosc, Louis Augustin Guillaume – [Paris]: De l'impr de Crapelet, a Paris chez Deterville... an 10 [1802] [mf ed 1985] – 5v on 1mf – 9 – 0-665-48883-1 – mf#48883 – cn Canadiana [590]

Histoire naturelle des crustaces : contenant leur description et leurs moeurs, avec figures dessinees d'apres nature / Bosc, Louis Augustin Guillaume – [Paris]: De l'impr de Guilleminet; chez Deterville... an oct [1801] [mf ed 1985] – 2v on 1mf – 9 – 0-665-51985-0 – mf#51984 – cn Canadiana [590]

Histoire naturelle des mammiferes : avec l'indication de leurs moeurs et de leurs rapports avec les arts, le commerce et l'agriculture / Gervais, Paul – Paris: L Curmer, 1854-55 [mf ed 1985] – 2v on 1mf – 9 – 0-665-51184-1 – mf#51184 – cn Canadiana [590]

Histoire naturelle des oiseaux / ed by Buffon, G L L de – Paris 1770-78 – v1-5 [1008pl] on 96mf – 9 – (one of the most extensive, important & early works on ornithology) – mf#z-368/2 – ne IDC [590]

Histoire naturelle des poissons : avec les figures dessinees d'apres nature / Bloch, Marcus Elieser – [Paris]: De l'impr de Crapelet a Paris chez Deterville. an 9 [1800] [mf ed 1985] – 10v on 1mf – 9 – 0-665-48889-0 – mf#48889 – cn Canadiana [590]

Histoire naturelle des poissons / Cuvier, G L C & Valenciennes, A – Paris 1828-49 – 22v on 204mf – 9 – mf#2122/2 – ne IDC [590]

Histoire naturelle des poissons : ou ichthyologie generale / Dumeril, A H A – Peiping. 1949-1959 (1) – 26mf – 9 – (with atlas) – mf#2613 – ne IDC [590]

Histoire naturelle des quadrupedes ovipares et des serpens / Lacepede, Bernard Germain Etienne de La Ville sur Illon, comte de – Paris: Hotel de Thou...1788-90 [mf ed 1985] – 4v on 1mf – 9 – 0-665-45480-5 – mf#45481 – cn Canadiana [590]

Histoire naturelle des reptiles et des poissons / Guichenot, A – London. 1958+ (1) 1971+ (5) 1973+ (9) – 6mf – 9 – mf#2676 – ne IDC [590]

Histoire naturelle des vers : contenant leur description et leurs moeurs, avec figures dessinees d'apres nature / Bosc, Louis Augustin Guillaume – [Paris]: De l'impr de Guilleminet; chez Deterville... an oct [1801] [mf ed 1985] – 3v on 1mf – 9 – 0-665-51987-7 – mf#51987 – cn Canadiana [590]

Histoire naturelle du senegal : coquillages / Adanson, Michel – London. 1960+ (1) 1965+ (5) 1976+ (9) – 21mf – 9 – (avec la relation abregee d'un voyage...annees 1749-1753) – mf#1335 – ne IDC [590]

Histoire naturelle et civile de la californie : contenant une description exacte de ce pays, de son sol, de ses montagnes, lacs, rivieres et mers,... / Venegas, Miguel – Paris 1767 [mf ed Hildesheim 1995-98] – 3v on 12mf – 9 – €120.00 – 3-487-27114-1 – gw Olms [917]

Histoire naturelle et civile de l'isle de minorque / Armstrong, John – Amsterdam 1769 [mf ed Hildesheim 1995-98] – 2mf [ill] – 9 – €60.00 – 3-487-29877-5 – gw Olms [914]

Histoire naturelle et morale des iles antilles de l'amerique enrichie de plusieurs belles figures des raretez les plus, considerables qui y sont d'ecrites : avec un vocabulaire caraibe / Rochefort, Cesar de – Roterdam: chez Arnould Leers, 1658 [mf ed 1997] – 6mf – 9 – mf#SEM105P2809 – cn Bibl Nat [500]

Histoire naturelle generale et particuliere / ed by Buffon, G L L de – Paris: 1749-1804. 37v – 596mf – 9 – (suppl [1774-89] 7v [gm-324]) – mf#5436/2 – ne IDC [590]

Histoire naturelle, generale et particuliere / Buffon, Georges Louis Leclerc, comte de – nouv ed. Paris: De l'Impr royale. 13v. 1769-70 [mf ed 1985] – 13v on 1mf – 9 – mf#42926 – cn Canadiana [500]

Histoire naturelle generale et particuliere des c,phalopodes ac,tabuliferes vivants et fossiles / Ferussac, A de & d'Orbigny, A – Paris, 1835-1848 – 14mf – 9 – mf#Z-2227 – ne IDC [590]

Histoire naturelle, generale et particuliere des reptiles / Daudin, F M – 1802-1805. 8v – 32mf – 9 – mf#Z-2242 – ne IDC [590]

Histoire orientale ou des tartares, de aiton, parent du roy d'armenie... / Heyt'owm Patmich – La Haye, 1735. v2 – 2mf – 9 – mf#HT-670 – ne IDC [915]

L'histoire par le theatre : 1789-1851 / Muret, Theodore – Paris 1865 [mf ed Hildesheim 1995-98] – 3v on 9mf – 9 – €180.00 – 3-487-26001-8 – gw Olms [944]

Histoire physiologique et chimique : l'air que l'on respire / Carrier, Joseph Celestin – [Canada?: s.n, 1901] – 1mf – 9 – 0-665-71924-8 – mf#71924 – cn Canadiana [612]

Histoire physiologique et chimique d'un flambeau ou bougie de cire : conference faite devant l'union catholique de montreal, le 30 novembre 1890 / Carrier, Joseph C – Montreal?: s.n, 1890? – 1mf – 9 – mf#00477 – cn Canadiana [590]

Histoire physique, civile et morale de paris : depuis les premiers temps historiques jusqu'a nos jours / Dulaure, Jacques – Paris 1821-1825 [mf ed Hildesheim 1995-98] – 8v on 49mf – 9 – €490.00 – ISBN-10: 3-487-25945-1 – ISBN-13: 978-3-487-25945-1 – gw Olms [944]

Histoire physique, civile et morale des environs de paris : depuis les premiers temps historiques jusqu'a nos jours / Dulaure, Jacques – Paris 1825-1828 [mf ed Hildesheim 1995-98] – 7v on 42mf – 9 – €420.00 – ISBN-10: 3-487-25938-9 – ISBN-13: 978-3-487-25938-3 – gw Olms [914]

Histoire physique des antilles francaises : savoir: la martinique et les îles de la guadeloupe; contenant: la geologie / Moreau de Jonnes, Alexandre – Paris 1822 [mf ed Hildesheim 1995-98] – 1v on 4mf – 9 – €120.00 – 3-487-26943-0 – gw Olms [918]

Histoire physique, naturelle et politique de madagascar / ed by Grandidier, A – Paris 1898-1905 – [tl-2/2109] on 421mf – 9 – mf#5936/2 – ne IDC [960]

Histoire physique, politique et naturelle de l'ele de cuba / ed by Sagra, R de la – New York. 1907-1974 (1) 1973-1973 (5) (9) – 113mf – 9 – mf#5707 – ne IDC [918]

Histoire pittoresque de la convention nationale et de ses principaux membres / Lamothe-Langon, Etienne L de – Paris 1833 [mf ed Hildesheim 1995-98] – 4v on 12mf – 9 – €120.00 – ISBN-10: 3-487-26281-9 – ISBN-13: 978-3-487-26281-9 – gw Olms [323]

Histoire pittoresque de la francmaconnerie et des societes secretes anciennes et modernes / Begue-Clavel, F-T – Paris: Pagnerre 1843 – 5mf – 9 – mf#vrl-96 – ne IDC [366]

Histoire pittoresque du mont-saint-michel, et de tombelene : suivie d'un fragment inedit sur tombelene... / Tellier, Charles Maurice le – [Paris 1834 [mf ed Hildesheim 1995-98] – 1v on 2mf – 9 – €60.00 – ISBN-10: 3-487-25912-5 – ISBN-13: 978-3-487-25912-3 – gw Olms [914]

Histoire pittoresque d'une famille de palestine. akkinai au pays de jesus. paris, 1933 / Miramar, Aloys – Madrid: Razon y Fe, 1934 – 1 – sp Bibl Santa Ana [920]

Histoire poetique du quinzieme siecle... / Champion, Pierre – Paris: E Champion 1923 [mf ed 1985] – 2v on 1r – 1 – mf#1261 – us Wisconsin U Libr [440]

Histoire politique de la province de quebec : premiere partie / Boissonnault, Charles-Marie – Beauceville: l'eclaireur ltee, 1936 [mf ed 1974] – 1r – 5 – mf#SEM16P108 – cn Bibl Nat [325]

Histoire politique des grandes querelles entre l'empereur charles 5, et francois 1, roi de france : avec une introduction contenant l'etat de la milice et la description de l'art de la guerre... / Goezmann, Louis V de – Paris 1777 [mf ed Hildesheim 1995-98] – 2v on 6mf – 9 – €120.00 – ISBN-10: 3-487-26132-4 – ISBN-13: 978-3-487-26132-4 – gw Olms [944]

Histoire politique du siecle : ou se voit developee la conduite de toutes les cours, d'un traite a l'autre... / Maubert de Gouvest, Jean – Londres [i.e. Lausanne 1755 [mf ed Hildesheim 1995-98] – 1v on 8mf – 9 – €160.00 – ISBN-10: 3-487-26210-X – ISBN-13: 978-3-487-26210-9 – gw Olms [900]

Histoire politique et religieuse d'abbysinie / Coulbeaux, J B – Paris, 1929. 3v – 13mf – 9 – mf#NE-20230 – ne IDC [956]

Histoire politique et religieuse de l'armenie / Tournebize, Francois – Paris: A Picard [1900?] [mf ed 1990] – 1mf – 9 – 0-7905-7153-6 – (in french) – mf#1988-3153 – us ATLA [240]

Histoire populaire de l'eglise du canada / Gosselin, David – Quebec: J A Langlais, 1887 – 3mf – 9 – mf#06333 – cn Canadiana [971]

Histoire populaire de montreal depuis son origine jusqu'a nos jours / Leblond de Brumath, Adrien – 3e rev augm ed. Montreal: Librairie Beauchemin ltee, 1926 [mf ed 1990] – (= ser Bibliotheque canadienne. coll jacques cartier 806b) – 4mf – 9 – mf#SEM105P1238 – cn Bibl Nat [971]

Histoire populaire de napoleon et de la grande armee / Raisson, Horace – Paris 1830 [mf ed Hildesheim 1995-98] – 4v on 8mf – 9 – €160.00 – ISBN-10: 3-487-26407-2 – ISBN-13: 978-3-487-26407-3 – gw Olms [944]

Histoire populaire du canada : d'apres les documents francais et americains / Baudoncourt, Jacques de – Paris: Bloud & Barral, 1888? – 6mf – 9 – mf#26288 – cn Canadiana [971]

Histoire religieuse, politique, et litteraire de la compagnie de jesus : composee sur les documents inedits et authentiques / Cretineau-Joly, Jacques – 3rd rev enl ed. Paris: Mme Ve Poussielgue-Rusand. 6v. 1851 [mf ed 1985] – 6v on 1mf – 9 – mf#46902 – cn Canadiana [241]

Histoire sainte : par demandes et par reponses, suivie d'un abrege de la vie de n s jesus-christ: a l'usage de la jeunesse – [Quebec?: s.n.] 1862 [mf ed 1985] – 1mf – 9 – 0-665-16558-7 – mf#16558 – cn Canadiana [221]

Histoire secrete du directoire / Fabre, Jean – Paris 1832 [mf ed Hildesheim 1995-98] – 4v on 12mf – 9 – €120.00 – ISBN-10: 3-487-26263-0 – ISBN-13: 978-3-487-26263-5 – gw Olms [944]

Histoire simple et veritable / Morin, Marie – Montreal: Presses de l'Universite de Montreal,1979 [mf ed 1994] – 5mf – 9 – (with ind) – mf#SEM105P2100 – cn Bibl Nat [971]

Histoire sociale des religions / Vernes, Maurice – Paris: V. Giard & E. Briere, 1911 – (= ser Etudes economiques et sociales) – 2mf – 9 – 0-7905-3491-6 – mf#1987-3491 – us ATLA [210]

Histoire socialiste – v. 1-13. 1901-08 – 1 – 93.00 – us L of C Photodup [335]

Histoire socialiste de la revolution francaise / Jaures, Jean – v1-13. 1789-1900 – 1 – $150.00 – mf#0267 – us Brook [944]

Histoire succincte de l'ile wallis / Poncet, Alexandre – 1967? – 1r – 1 – mf#pmb doc212 – at Pacific Mss [980]

Histoire universelle / Cantu, Cesare – Paris: F Didot. 1862 [mf ed 1984] – 19v on 1mf – 9 – 0-665-49285-5 – mf#49285 – cn Canadiana [930]

Histoire universelle des indes orientales et occidentales / Wytfliet, Cornelius van – Douay: Francois Fabri, 1605 [mf ed 1988] – 3mf – 9 – mf#SEM105P880 – cn Bibl Nat [900]

Histoire universelle [maille, 1616-1620] / Aubigne, T A d'; ed by Ruble, A de – Paris, 1886-1925. 10v + suppl – 47mf – 9 – mf#PRS-107 – ne IDC [240]

Histoire veritable et naturelle des moeurs et productions du pays de la nouvelle france / Boucher, Pierre, sieur de Boucherville; ed by Coffin, G – Montreal?: E Bastien, 1882 – 2mf – 9 – mf#11903 – cn Canadiana [971]

Histoire veritable et naturelle des moeurs et productions du pays de la nouvelle-france, vulgairement dite le canada / Boucher, Pierre – Paris: chez Florentin Lambert, 1664 [mf ed 1974] – 1r – 5 – mf#SEM16P6 – cn Bibl Nat [971]

Histoire vniverselle de la chine : avec l'histoire de la guerre des tartares... / Semmedo, (Semedo) A – Lyon: Hierosme Prost, 1667 – 5mf – 9 – mf#HT-551 – ne IDC [915]

Histoire...di tvtti i fatti degni di memoria nel mondo svccessi dell' anno 1524... / Guazzo, M – Vinegia, 1546 – 8mf – 9 – mf#H-8277 – ne IDC [956]

'Histoires de mission, pour enfants' / Guinard, Jean Louis – n.d. – 1r – 1 – mf#pmb452 – at Pacific Mss [241]

Histoires de vampires / Volta, Ornella – Paris, France. 1961 – 1r – us UF Libraries [130]

Histoires somalies : la malice des primitifs / Duchenet, Edouard – Paris: Larose, 1936 – 1 – us CRL [960]

Histoires vraies – [Montreal]: Les Ed Histoires vraies, [1943?]-[1957] (mf ed 1983) – 4r – 5 – mf#SEM16P332 – cn Bibl Nat [073]

Histoire...svr les choses faictes et auenues de son temps en toutes les parties du monde... / Giovio, P – Paris, 1581. 2v – 22mf – 9 – mf#H-8359 – ne IDC [910]

Histon 1599-1950 – (= ser Cambridgeshire parish register transcript) – 9mf – 9 – £11.25 – uk CambsFHS [929]

Histopathology – Oxford. 1980-1996 (1,5,9) – ISSN: 0309-0167 – mf#15549 – us UMI ProQuest [616]

Histori dess sacramentstreits / Chemnitz d A, M – np, 1591 – 8mf – 9 – mf#TH-1 mf 223-230 – ne IDC [242]

Histori dess sacramentstreits darinnen klaerlich aussgefuehret wirdt wie diese zwytracht entstanden biss auff vnsere zeit continuiret / Kirchner, T – np, 1591 – 8mf – 9 – mf#TH-1 mf 841-848 – ne IDC [242]

Historia – Lima, Peru. v. 1-3 no. 10. Mar Apr 1943-Apr July 1945 – 1 – us NY Public [900]

Historia = a magazine of local history – Norwell, Mass. v1. (1-6). Nov. 1898-Oct. 1899 – 1 – us NY Public [978]

Historia... / Diaconus, Paulus [Paul The Deacon] – 15th c – (= ser Holkham library manuscript books 459,121,451,419,647) – 1r – 1 – (filmed with: s cyprianus: epistolae. laertius diogenes: epistolae. sedulius: carmen paschale / dogale granted to niccolo bernardo) – mf#96543 – uk Microform Academic [450]

Historia abbatiae cassinensis / Gattula, Erasmus – Venetiis. v1-3. 1733 – €130.00 – ne Slangenburg [241]

Historia adminisdtrativa do brasil / Fleiuss, Max – Rio de Janeiro, Brazil. 1925 – 1r – us UF Libraries [350]

Historia administrativa, judiciaria e eclesiastica / Fortes, Amyr Borges – Porto Alegre, Brazil. 1963 – 1r – us UF Libraries [350]

Historia aethiopica : sive brevis et succincta descriptio regni habessinorum,... / Ludolfi, lobi (alias Leut-holf) – Francofurti ad Moenum (Frankfurt a. Main): prostat apud J D Zunner, 1681 – 1 – us CRL [960]

Historia aethiopica... / Ludolf, J – Francofurti ad Moenum, 1681 – 6mf – 9 – mf#SEP-14 – ne IDC [960]

Historia aliquot martyrum anglorum maxime octodecim cartusianorum : sub rege henrico octavo ob fidei confessionem et summi pontificis jura vindicanda interemptorum / Chauncy, Maurice – Monstrolii: Cartusiae S Mariae de Pratis, 1888 – 1mf – 9 – 0-524-05139-9 – mf#1990-1395 – us ATLA [240]

Historia anglicana *see* Chronia monastereii s albani 1 [rs28]

Historia anglicana (ad 449-1298) (rs16) : necnon ejusdem liber de archiepiscopis et episcopis angliae / Bartholomeus de Cotton; ed by Luard, H R – 1859 – (= ser The rolls series (rs)) – €19.00 – ne Slangenburg [242]

Historia anglorum (rs44) : sive, ut vulgo dicitur, historia minor item, ejusdem abreviatio chronicorum angliae / Paris, Matthew; ed by Madden, F – (= ser The rolls series (rs)) – (v1 1866 €19. v2 1866 €18. v3 1869 €21) – ne Slangenburg [931]

Historia animalium. libri 1 de quadrupedibus viviparis... / Gesner, C – Tiguria, 1551 – 12mf – 9 – sp Cultura [590]

Historia animalium. libri 2 de quadrupedibus viviparis... / Gesner, C – Tiguria, 1551 – 2mf – 9 – sp Cultura [590]

Historia animalium. libri 3 de de avium natura... / Gesner, C – Tiguria, 1551 – 9mf – 9 – sp Cultura [590]

Historia animalium. libri 4 de piscium et aquatilium... / Gesner, C – Tiguria, 1558 – 14mf – 9 – sp Cultura [590]

Historia animalium. libri 5 de serpentium natura... / Gesner, C – Tiguria, 1587 – 2mf – 9 – sp Cultura [590]

Historia antiga da abbadia de s paulo / Taunay, Afonso De E – Sao Paulo, Brazil. 1927 – 1r – us UF Libraries [972]

Historia antiga das minas gerais / Vasconcellos, Diego Luiz De Almeida Pereira De – Rio de Janeiro, Brazil. v1-2. 1948 – 1r – us UF Libraries [972]

Historia antigua / Fernandez Retamar, Roberto – Habana, Cuba. 1964 – 1r – us UF Libraries [972]

Historia artis grammaticae apud syros : cui accedunt severi bar sakku dialogus de grammatica, dionysii thracis grammatica syriace translata, iacobi edesseni fragmenta grammatica cum tabula photolithographica, eliae tirhanensis et duorum anonymorum de accentibus tractatus / ed by Merx, Adalbert – Leipzig: In Commission von FA Brockhaus, 1889 – (= ser Abhandlungen fuer die kunde des morgenlandes) – 1mf – 9 – 0-8370-9085-7 – (incl bibl ref & ind) – mf#1986-3085 – us ATLA [470]

Historia avgvstanae confessionis / Chytraeus, D – 8mf – 9 – mf#TH-1 mf 276-283 – ne IDC [242]

Historia belli persici, gesti inter mvrathem 3 tvrcarvm, et mehemetem hodabende, persarum regem... / Porsius, H – Francofvrti, 1583. 2pts – 3mf – 9 – mf#H-8418 – ne IDC [956]

Historia bibliografica de la medicina espanola / Fernandez Morejon, Antonio – 1852 – 9 – sp Bibl Santa Ana [610]

Historia bibliothecae romanorum pontificum : tom 1 (et unicus) / Ehrle, F – Romae, 1890 – €67.00 – ne Slangenburg [241]

Historia britonum / Nennius, Abbot of Bangor; ed by Stephenson, J – 1836 – 1r – 1 – mf#923 – uk Microform Academic [941]

Historia byzantina (cbh15) / Ducae Michaelis Nepotis; ed by Bullialdus, Ism – Parisiis, 1649 – (= ser Corpus byzantinae historiae (cbh)) – €27.00 – ne Slangenburg [243]

Historia byzantina (cshb21) / Ducae, Michaelis Ducae Nepotis – Bonnae, 1834 – (= ser Corpus scriptorum historiae byzantinae (cshb)) – €23.00 – (rec et interprete italo addito suppl imm bekkerus) – ne Slangenburg [243]

Historia byzantina duplici commentario illustrata (cbh25,1) : tom 1: familiae augustae byzantinae / ed by Cange, C du – Venetiis, 1729 – (= ser Corpus byzantinae historiae (cbh)) – €27.00 – ne Slangenburg [931]

Historia byzantina duplici commentario illustrata (cbh25,2) : tom 2: constantinopolis christiana / ed by Cange, C du – Venetiis, 1729 – (= ser Corpus byzantinae historiae (cbh)) – €27.00 – ne Slangenburg [243]

Historia can buhay, guibo asin calalagnan nin buquidnon na bertoldo, an qui bertoldino na saiyang aqui, asin an qui cacaseno niang macoapo ... : primera parte – [Nueva Caceres: Libreria Mariana 190-?] [mf ed Bloomington IN: Indiana Uni Lib, Preservation Dept 1984] – (= ser Coll...in the bikol language) – 1r – 1 – us Indiana Preservation [490]

Historia can buhay nin siete infantes na magna aqui ni busto de lara – [Nueva Caceres]: Libreria Mariana [190-?] [mf ed Bloomington IN: Indiana Uni Lib, Preservation Dept 1984] – (= ser Coll...in the bikol language) – 2v on 1r – 1 – us Indiana Preservation [490]

Historia can paghorca qui maximo bungsoan sa provinciang albay – [Nueva Caceres]: Libreria Mariana 1907 [mf ed Bloomington IN: Indiana Uni Lib, Preservation Dept 1984] – (= ser Coll...in the bikol language) – 1r – 1 – us Indiana Preservation [490]

Historia can paghorca qui valentin pavoreal na maogma asin mamondo – [Nueva Caceres?: Libreria Mariana? 190-?] [mf ed Bloomington IN: Indiana Uni Lib, Preservation Dept 1984] – (= ser Coll...in the bikol language) – 1r – 1 – us Indiana Preservation [490]

Historia (cbh13) / Nicetae Acominati; ed by Fabrotus, C A – Parisiis, 1647 – (= ser Corpus byzantinae historiae (cbh)) – €44.00 – ne Slangenburg [243]

Historia (cbh14) / Georgii Acropolitae; ed by Allatius, L – Parisiis, 1651 – (= ser Corpus byzantinae historiae (cbh)) – €37.00 – (filmed with: ioelis: chronographia compendiaria, and ioannis canani: narratio de bello cp) – ne Slangenburg [243]

Historia chronica (cbh30,3) : pars prima / Joannis Antiocheni Malalae; ed by Hodius, H – Venetiis, 1733 – (= ser Corpus byzantinae historiae (cbh)) – €17.00 – ne Slangenburg [243]

Historia chronica (cbh30,4) : pars altera. de imperatoribus christianis / Joannis Antiocheni Malalae; ed by Hodius, H – Venetiis, 1733 – (= ser Corpus byzantinae historiae (cbh)) – €18.00 – ne Slangenburg [243]

Historia comica de trujillo desde los tiempos mas remotos hasta el final del siglo 18 / Ramons Sanguino, Joaquin – Trujillo: Establecimiento Tipografico La Minerva, 1913 – 1 – sp Bibl Santa Ana [946]

Historia compostellana : formae tplila 46 – 1988 – (= ser ILL – ser a; Cccm 70) – 14mf+121p – 9 – €50.00 – 2-503-63702-7 – be Brepols [400]

Historia constitucional da republica dos estados... / Freire, Felisbello – Rio de Janeiro, Brazil. v1-3. 1894-1895 – 1r – us UF Libraries [323]

Historia constitucional de entre rios / Martinez Soler, Francisco T – Rosario, Argentina. 1922 – 1r – us UF Libraries [323]
Historia contemporanea de venezuela / Gonzalez Guinan, Francisco – Caracas, Venezuela. v1-15. 1954 – 5r – us UF Libraries [934]
Historia contemporanea de venezuela / Level De Goda, Luis – Caracas, Venezuela. 1954 – 1r – us UF Libraries [934]
Historia controversiarum de ritibus sinicis / Pray, G – Pestini Budae ac Cassoviae, 1789 – 3mf – 9 – mf#HTM-230 – ne IDC [915]
Historia critica de los sistemas filosoficos... / Nieto y Serrano, M – Madrid, 1898 – 9mf – 9 – sp Cultura [100]
Historia (cshb23) / Nicetae Choniatae; ed by Bekkeri, Imm – Bonnae, 1835 – (= ser Corpus scriptorum historiae byzantinae (cshb)) – €32.00 – ne Slangenburg [243]
Historia (cshb48) / Michaelis Attaliotae; ed by Bekkerus, Imm – Bonnae, 1853 – (= ser Corpus scriptorum historiae byzantinae (cshb)) – €14.00 – ne Slangenburg [243]
Historia da america portugueza / Rocha Pita, Sebastiao Da – Rio de Janeiro, Brazil. 1910 – 1r – us UF Libraries [972]
Historia da arte brasileira / Mattos, Anibal – Belo Horizonte, Brazil. 1937 – 1r – us UF Libraries [972]
Historia da bahia / Calmon, Pedro – Sao Paulo, Brazil. no date – 1r – us UF Libraries [972]
Historia da bahia do imperio a republica / Amaral, Braz Do – Bahia, Brazil. 1923 – 1r – us UF Libraries [972]
Historia da casa da torre / Calmon, Pedro – Rio de Janeiro, Brazil. 1939 – 1r – us UF Libraries [972]
Historia da cidade do rio de janeiro / Carvalho, Carlos Miguel Delgado De – Rio de Janeiro, Brazil. 1926 – 1r – us UF Libraries [972]
Historia da civilizacao brasileira / Calmon, Pedro – Sao Paulo, Brazil. 1940 – 1r – us UF Libraries [972]
Historia da civilizacao brasileira / Calmon, Pedro – Sao Paulo, Brazil. 1945 – 1r – us UF Libraries [972]
Historia da civilizacao brasileira / Ferreira, Tito Livio – Sao Paulo, Brazil. 1959 – 1r – us UF Libraries [972]
Historia da companhia de jesus no brasil / Leite, Serafim – Lisboa, Portugal. v1-10. 1938-1950 – 1r – us UF Libraries [972]
Historia da conjuracao mineira / Souza Silva, Joaquim Norberto De – Rio de Janeiro, Brazil. v1-2. 1948 – 1r – us UF Libraries [972]
Historia da fundacao da bahia / Calmon, Pedro – Cidade do Salvador, Brazil. 1949 – 1r – us UF Libraries [972]
Historia da guerra cisplatina / Carneiro, David – Sao Paulo, Brazil. 1946 – 1r – us UF Libraries [972]
Historia da guerra entre a triplice alianca e o pa... / Fragoso, Augusto Tasso – Rio de Janeiro, Brazil. v1-5. 1934 – 1r – us UF Libraries [972]
Historia da independencia na bahia / Amaral, Braz Do – Bahia, Brazil. 1923 – 1r – us UF Libraries [972]
Historia da literatura brasileira / Bezerra De Freitas, Jose – Porto Alegre, Brazil. 1939 – 1r – us UF Libraries [440]
Historia da literatura brasileira / Romero, Silvio – Rio de Janeiro, Brazil. t1-5. 1943 – 1r – us UF Libraries [440]
Historia da literatura brasileira / Romero, Silvio – Rio de Janeiro, Brazil. v1-5. 1960 – 1r – us UF Libraries [440]
Historia da literatura brasileira / Sodre, Nelson Werneck – Rio de Janeiro, Brazil. 1964 – 1r – us UF Libraries [440]
Historia da litteratura brazileira / Romero, Sylvio – Rio de Janeiro: B.L. Garnier, 1888. 2v – 1 – (3rd ed., aug., with preface by nelson romero. rio de janeiro: j. olympio, 1943.5v) – us Wisconsin U Libr [440]
Historia da missao dos padres capuchinhos na ilha / Claude – Sao Paulo, Brazil. 1945 – 1r – us UF Libraries [972]
Historia da policia civil de sao paulo / Vieira, Hermes – Sao Paulo, Brazil. 1955 – 1r – us UF Libraries [360]
Historia da provincia do ceara / Alencar Araripe, Tristao De – Fortaleza, Brazil. 1958 – 1r – us UF Libraries [972]
Historia da revolucao de pernambuco em 1817 / Muniz Tavares, Francisco – Recife, Brazil. 1917 – 1r – us UF Libraries [972]
Historia das expedicoes cientificas no brasil / Mello-Leitao, C De Q – Sao Paulo, Brazil. 1941 – 1r – us UF Libraries [972]
Historia das fronteiras do brasil / Vianna, Helio – Rio de Janeiro, Brazil. 1949 – 1r – us UF Libraries [972]
Historia das guerras e revolucoes do brasil de 182... / Seidler, Carl – Sao Paulo, Brazil. 1939 – 1r – us UF Libraries [972]
Historia das ideias filosoficas no brasil / Paim, Antonio – Sao Paulo, Brazil. 1967 – 1r – us UF Libraries [100]
Historia das missioes do padroado portugues de oriente / Silva Rego, Antonio da – Madrid: Missionalia Hispanica, 1949 – 1 – sp Bibl Santa Ana [915]
Historia das missoes orientais do uruguai / Porto, Aurelio – Porto Alegre, Brazil. v1-2. 1954 – 1r – us UF Libraries [972]
Historia david *see* Apokalypse / ars moriendi / biblia pauperum / antichrist / fabel vom kranken loewen / kalendarium und planetenbuecher / historia david (mxt2)
Historia de a cuerda granadina contada por algunos de sus nudos, apuntes para la misma recopilados por... / Cascales Munoz, Jose – Madrid: Tipografia de la Revista de Archivos, 1926 – 1 – sp Bibl Santa Ana [946]
Historia de abaete : (temperada com um pouco de sal...) / Oliveira, Jose Alves De – Belo Horizonte, Brazil. 1970 – 1r – us UF Libraries [972]
Historia de alburquerque / Duarte Insua, Lino – Badajoz: Tip. Libr. enc. Arqueros, 1929 – 1 – sp Bibl Santa Ana [946]
Historia de america espanola 1920-1925 / Bayle, Constantino – Madrid: Razon y Fe, 1926 – 1 – sp Bibl Santa Ana [972]
Historia de angola / Silva Correa, Elias Alexandre Da – Lisboa, Portugal. v1-2. 1937 – 1r – us UF Libraries [960]
Historia de angola, 1482-1963 / Gonzaga, Norberto – Luanda, Angola. 1969? – 1r – us UF Libraries [960]
Historia de antonio vieira / Azevedo, Joao Lucio D' – Lisboa, Portugal. v1-2. 1931 – 1r – us UF Libraries [972]
Historia de badajoz / Suarez de Figueroa, Diego – Capitulo 35-36. 1732 – 1 – sp Bibl Santa Ana [946]
Historia de belo horizonte de 1897 a 1930 / Mourao, Paulo Kruger Correa – Belo Horizonte, Brazil. 1970 – 1r – us UF Libraries [972]
Historia de caceres y su patrona / Buxoyo, Simon Benito – Caceres: Public. Dep. Prov. Seminarios FET y JONS, 1952 – 1 – sp Bibl Santa Ana [946]
Historia de cali / Arboleda, Gustavo – Cali, Colombia. v1-3. 1956 – 1r – us UF Libraries [972]
Historia de carupano / Tavera-Acosta, Bartolome – Caracas, Venezuela. 1930 – 1r – us UF Libraries [972]
Historia de castro alves / Calmon, Pedro – Rio de Janeiro, Brazil. 1947 – 1r – us UF Libraries [972]
Historia de centro-america / Alvarado Garcia, Ernesto – Tegucigalpa, Mexico. 1946 – 1r – us UF Libraries [972]
Historia de chita / Amaya Roldan, Martin – Tunja, Colombia. 1930 – 1r – us UF Libraries [972]
Historia de colombia / Gonzalez Fernandez, Hector – Bogota, Colombia. 1945 – 1r – us UF Libraries [972]
Historia de colombia / Granados, Rafael M – Bogota, Colombia. 1964 – 1r – us UF Libraries [972]
Historia de cosas del oriente primera y segunda parte : contiene vna descripcion general de los reynos de assia con las cosas mas notables dellos / Centeno, A – Cordoua, 1595 – 4mf – 9 – mf#H-8427 – ne IDC [956]
Historia de costa rica / Fernandez Guardia, Ricardo – San Jose, Costa Rica. 1924 – 1r – us UF Libraries [972]
Historia de costa rica / Fernandez Guardia, Ricardo – San Jose, Costa Rica. 1941 – 1r – us UF Libraries [972]
Historia de costa rica / Monge Alfaro, Carlos – San Jose, Costa Rica. 1958 – 1r – us UF Libraries [972]
Historia de costa rica / Monge Alfaro, Carlos – San Jose, Costa Rica. 1963 – 1r – us UF Libraries [972]
Historia de costa rica, adapta al programa oficial / Fernandez Guardia, Leon – San Jose, Costa Rica. 1939 – 1r – us UF Libraries [972]
Historia de costa rica durante / Fernandez, Leon – Madrid, Spain. 1889 – 1r – us UF Libraries [972]
Historia de cuba / Guerra, Ramiro – Habana, Cuba. v1-2. 1921- – 1r – us UF Libraries [972]
Historia de cuba : narracion humoristica / Robreno, Gustavo – Habana, Cuba. 1915 – 1r – us UF Libraries [972]
Historia de cuba / Portell Vila, Herminio – Habana, Cuba. v1-4. 1938 – 1r – us UF Libraries [972]
Historia de cuba / Portuondo Del Prado, Fernando – Habana, Cuba. 1950 – 1r – us UF Libraries [972]
Historia de el principio : y origen, progressos, venidas a mexico, y milagros de la santa ymagen de nuestra senora de los remedios, extramuros de mexico... / Cisneros, Luys de – en Mexico: Impresso...Bachiller Iuan Blanco de Alcacar, ano de 1621 – (= ser Books on religion...1543/44-c1800: milagros y culto de la virgen) – 4mf – 9 – mf#crl-77 – ne IDC [241]
Historia de espana dirigida por...tomos 5, 14 y 18, madrid, 1965-1966 / Menendez Pidal, Ramon – Madrid: Graf. Calleja, 1968 – 1 – sp Bibl Santa Ana [946]
Historia de expeditione friderici imperatoris et quidam alii rerum gestarum fontes eiusdem expeditionis (mgh6:5.bd) / ed by Chroust, A – 1928 – (= ser Monumenta germaniae historica 6: scriptores rerum germanicarum, nova series (mgh6)) – €14.00 – ne Slangenburg [240]
Historia de familias cubanas / Santa Cruz y Mallen, Francisco Xavier de – La Habana: Editorial Hercules, 1940-(86). (v1-8) – 1 – us Wisconsin U Libr [920]
Historia de francisco bilbao / Figueroa, Pedro Pablo – Santiago, Chile. 1898 – 1r – us UF Libraries [972]
Historia de gentibus septentrionalibus / Magnus Gothus, O – Romae, 1555 – 24mf – 9 – mf#N-308 – ne IDC [914]
Historia de guanabacoa / Guardia, Elpidio De La – Guanabacoa, Cuba. 1946 – 1r – us UF Libraries [972]
Historia de guatemala / Fuentes y Guzman, Francisco A – 1882-83.2v – 9 – sp Bibl Santa Ana [972]
Historia de isla de cuba / Pezuela y Lobo, Jacobo de la – Madrid. 4v. 1868-78 – 1 – us L of C Photodup [972]
Historia de la antiquisima e ilustre villa de fregenal / Martin Moreno, Rafael – Sevilla: Imp. Alvarez, 1960 – sp Bibl Santa Ana [946]
Historia de la aviacion en colombia / Forero F, Jose Ignacio – Bogota, Colombia. 1964 – 1r – us UF Libraries [972]
Historia de la aviacion en costa rica / Jimenez G, Carlos Ma – San Jose, Costa Rica. 1962 – 1r – us UF Libraries [972]
Historia de la capitania general de guatemala / Villacorta C, J Antonio – Guatemala, 1942 – 1r – us UF Libraries [972]
Historia de la casa de beneficencia de matanzas, h / Rivadulla, Julio Valdes – Cardenas, Cuba. 1928 – 1r – us UF Libraries [972]
Historia de la ciudad argentina / Razori, Amilcar – Buenos Aires, Argentina. v1-3. 1945 – 1r – us UF Libraries [972]
Historia de la ciudad de badajoz... : extractadas del doctor... / Suarez de Figueroa, Diego – Badajoz: Imprenta de Vicente Rodriguez, 1916 – 1 – sp Bibl Santa Ana [946]
Historia de la civilizacion brasilena (1) / Calmon, Pedro – Buenos Aires: [Imprenta Mercatali] 1937 (mf ed 2000) – (= ser Biblioteca de autores brasilenos traducidos al castellano) – 1r – 1 – mf#*Z-9284 – us NY Public [972]
Historia de la compania de jesus en la nueva grana / Borda, Jose Joaquin – Poissy, France. v1-2. 1872 – 1r – us UF Libraries [972]
Historia de la composicion del cuerpo humano / Valverde de Hamusco, J – Roma, 1556 – 6mf – 9 – sp Cultura [612]
Historia de la comquista de mexico...tomo 3 / Solis, Antonio de – Barcelona: Consortes Sierra, Oliver y Marti, 1789 – 1 – (tambien tomo 2 1843) – sp Bibl Santa Ana [972]
Historia de la comunicacion interoceanica / Castillero R, Ernesto J – Panama, Panama. 1941 – 1r – us UF Libraries [972]
Historia de la conquista de la habana / Guiteras, Pedro Jose – Habana, Cuba. 1932 – 1r – us UF Libraries [972]
Historia de la conquista de la provincia de itza... / Villagutierre Soto-Mayor, Juan de – Guatemala, 1933; Madrid: Razon y Fe, 1934 – 1 – sp Bibl Santa Ana [972]
Historia de la conquista de mejico / Solis Rivadeneyra, Antonio de – v1. 1843 – 9 – (v2-3 1789. v1-2 1780. v1-3 1791, 1732, 1704, 1684, 1756) – sp Bibl Santa Ana [972]
Historia de la conquista de mexico / Lopez de Gomara, Francisco – Mexico: Editorial Pedro Robleda, 1943.-v1 – sp Bibl Santa Ana [972]
Historia de la conquista de mexico / Solis, Antonio de – Barcelona: Lucas de Bezaras y Urrutia, 1756 – 1 – sp Bibl Santa Ana [972]
Historia de la conquista de mexico / Solis, Antonio de – Madrid: Antonio de Sancha, Tomo 1. 1783 – 1 – sp Bibl Santa Ana [972]
Historia de la conquista de mexico / Solis, Antonio de – Madrid: Antonio de Sancha, Tomo 2. 1784 – 1 – sp Bibl Santa Ana [972]
Historia de la conquista de mexico / Solis, Antonio de – Madrid: Bernardino Peralta, 1732 – 1 – sp Bibl Santa Ana [972]
Historia de la conquista de mexico / Solis, Antonio de – Madrid: Manuel Martin, Tomo 1. 1780 – 1 – sp Bibl Santa Ana [350]
Historia de la conquista de mexico / Solis, Antonio de – Madrid: Manuel Martin, Tomo 2, 1780 – 1 – sp Bibl Santa Ana [350]
Historia de la conquista de mexico / Solis Rivadeneira, Antonio de – Madrid: Juan de San Martin, 1756 – 1 – sp Bibl Santa Ana [972]
Historia de la conquista de mexico. tomo 2 / Solis, Antonio de – Barcelona: Consortes Sierra, Oliver y Marti, 1789 – 1 – sp Bibl Santa Ana [972]
Historia de la conquista de nueva espana / Diaz del Castillo, Bernal – 1632 – 9 – sp Bibl Santa Ana [972]
Historia de la conquista del peru...incas / Prescott, Guillermo H – 1853 – 9 – sp Bibl Santa Ana [972]
Historia de la conquista, poblacion u progresos de la merica septentrional, conocida con el nombre de nueva espana / Solis Rivadeneyra, Antonio de – 9 – sp Bibl Santa Ana [946]
Historia de la "cuerda granadina", contada por algunos de sus...madrid, 1926 / Munoz, Jose; ed by Valle, A Cascales – Madrid: Razon y Fe, 1928 – 9 – sp Bibl Santa Ana [946]
Historia de la cultura en el nuevo reino de granada / Porras Troconis, Gabriel – Sevilla, Spain. 1952 – 1r – us UF Libraries [972]
Historia de la cultura en la america hispanica / Henriquez Urena, Pedro – Mexico City? Mexico. 1947 – 1r – us UF Libraries [972]
Historia de la cultura en la americana hispanica / Henriquez Urena, Pedro – Mexico City? Mexico. 1949 – 1r – us UF Libraries [972]
Historia de la cultura en mexico / Jimenez Rueda, Julio – Mexico City? Mexico. 1957 – 1r – us UF Libraries [972]
Historia de la cultura en puerto rico / Fernandez Mendez, Eugenio – Yauco? Puerto Rico. 1964 – 1r – us UF Libraries [972]
Historia de la diocesis de siguenza y de sus obispos, por toribio minguella y arnedo / T'Serclaes, Duque de – Madrid: Fortanet, 1912. B.R.A.H. 61, pp. 154-152 – sp Bibl Santa Ana [240]
Historia de la disputa que sobre la enfermedad que quito la vida a manuel rodriguez... / Lopez de Araujo, B – Madrid, 1756 – 1mf – 9 – sp Cultura [616]
Historia de la dominacion espanola en mejico / Orozco y Berra, Manuel – Madrid: Razon y Fe, 1944 – 1 – sp Bibl Santa Ana [972]
Historia de la educacion en guatemala / Gonzalez Orellana, Carlos – Mexico City? Mexico. 1960 – 1r – us UF Libraries [370]
Historia de la enmienda platt / Roig De Leuchsenring, Emilio – Habana, Cuba. v1-2. 1935 – 1r – us UF Libraries [972]
Historia de la epidemia de calenturas benignas...en sevilla / Nieto de Pina, C – Sevilla, 1784 – 1mf – 9 – sp Cultura [614]
Historia de la epidemia...de barbastro en el ano de 1748... / Ased y Latorre, A – Zaragoza, SA – 2mf – 9 – sp Cultura [614]
Historia de la esclavitud / Saco, Jose Antonio – Habana, Cuba. v1-4. 1938-40 – 1r – us UF Libraries [972]
Historia de la esclavitud negra en puerto rico (14 / Diaz Soler, Luis M – Madrid, Spain. 1953 – 1r – us UF Libraries [972]
Historia de la filosofia y de las ciencias / Frutos Cortes, Eugenio – Zaragoza: Editora Libreria General, 1965 – sp Bibl Santa Ana [100]
Historia de la fisiologia en guatemala / Figueroa Marroquin, Horacio – Guatemala, 1958 – 1r – us UF Libraries [972]
Historia de la florida / Lasso de la Vega, Garcia – 1605 – 9 – sp Bibl Santa Ana [978]
Historia de la fundacion de granada, hoy granadilla – Caceres: Imp. Garcilasso, 1974 – 1 – sp Bibl Santa Ana [914]
Historia de la fundacion del convento de religiosas carmelitas de badajoz / Mateos, Francisco – Badajoz: Tip.Arqueros, Tomo 1. 1930 – 1 – sp Bibl Santa Ana [240]
Historia de la gobernacion de popayan seguida de l... / Arroyo, Jaime – Bogota, Colombia. v1-2. 1955 – 1r – us UF Libraries [972]
Historia de la gobernacion del tucuman (siglo 16). buenos aires, 1928 / Lizondo Borda, Manuel – Madrid: Razon y Fe, 1929 – 1 – sp Bibl Santa Ana [972]
Historia de la guerra de cuba y los estados unidos / Portell Vila, Herminio – Habana, Cuba. 1949 – 1r – us UF Libraries [972]
Historia de la guerra de los diez anos / Ponte Dominguez, Francisco J – Habana, Cuba. 1944 – 1r – us UF Libraries [972]
Historia de la historiografia espanola : tomo 1. hasta la publicacion de la cronica de ocampo. tomo 2 de ocampo a solis. madrid, 1941, 1944 / Sanchez Alonso, B – Madrid: Razon y Fe, 1946 – 1 – sp Bibl Santa Ana [946]
Historia de la imagen de nuestra senora de la fuente santa excelsa patrona de zorita / Fernandez Sanchez, Teodoro – Caceres: Tip. Extremadura, 1972 – 1 – sp Bibl Santa Ana [240]
Historia de la imaginería colonial en guatemala / Berlin-Neubart, Heinrich – Guatemala, 1952 – 1r – us UF Libraries [972]
Historia de la independencia de panama / Arrocha Graell, C – Panama, Panama. 1933 – 1r – us UF Libraries [972]
Historia de la instruccion publica en panama / Mendez Pereira, Octavio – Panama, Panama. 1916 – 1r – us UF Libraries [972]

Historia de la insurreccion de cuba (1869-1879) / Soulere, Emilio Augusto – Barcelona, Spain. v1-2. 1879-80 – 1r – us UF Libraries [972]
Historia de la inuencion de las yndias / Perez De Oliva, Fernan – Bogota, Colombia. 1965 – 1r – us UF Libraries [972]
Historia de la isla y catedral de cuba / Morell De Santa Cruz, Pedro Agustin – Habana, Cuba. 1929 – 1r – us UF Libraries [972]
Historia de la isla y catedral de cuba / Morell de Santa Cruz, Pedro Agustin – La Habana, 1929; Madrid: Razon y Fe, 1931 – 1 – sp Bibl Santa Ana [972]
Historia de la lengua y literatura castellana desde los origenes hasta carlos 5 / Cejador y Frauca, Julio – Madrid. v1-14. 1915 – 1 – $361.00 – (in spanish) – mf#0144 – us Brook [440]
Historia de la literatura americana / Sanchez, Luis Alberto – Santiago, Chile. 1937 – 1r – us UF Libraries [440]
Historia de la literatura americana / Sanchez, Luis Alberto – Santiago, Chile. 1940 – 1r – us UF Libraries [440]
Historia de la literatura americana en lengua espa... / Bazin, Robert – Buenos Aires, Argentina. 1963 – 1r – us UF Libraries [440]
Historia de la literatura colombiana / Gomez Restrepo, Antonio – Bogota, Colombia. v1-4. 1953 – 1r – us UF Libraries [440]
Historia de la literatura cubana / Bueno, Salvador – Habana, Cuba. 1954 – 1r – us UF Libraries [440]
Historia de la literatura cubana / Remos Y Rubio, Juan Nepomuceno Jose – Habana, Cuba. v1-3. 1945 – 1r – us UF Libraries [440]
Historia de la literatura cubana / Salazar Y Roig, Salvador – Habana, Cuba. 1929 – 1r – us UF Libraries [440]
Historia de la literatura de la america central / Montalban, Leonardo – EL Salvador, El Salvador. 1931 – 1r – us UF Libraries [440]
Historia de la literatura dominicana / Balaguer, Joaquin – Ciudad Trujillo, Dominican Republic. 1956 – 1r – us UF Libraries [440]
Historia de la literatura dominicana / Mejia De Fernandez, Abigail – Santiago, Dominican Republic. 1943 – 1r – us UF Libraries [440]
Historia de la literatura en nueva granada / Vergara Y Vergara, Jose Maria – Bogota, Colombia. v1-3. 1958 – 1r – us UF Libraries [440]
Historia de la literatura puertorriquena / Cabrera, Francisco Manrique – New York, NY. 1956 – 1r – us UF Libraries [440]
Historia de la medicina en mexico. mexico, 1934 / Ocaranza, Fernando – Madrid: Razon y Fe, 1935 – 1 – sp Bibl Santa Ana [610]
Historia de la milagrosa aparicion de nuestra sra de la caridad / Fonseca – Santiago, Cuba. 1935 – 1r – us UF Libraries [972]
Historia de la milagrosissima imagen de nuestra senora de occotlan : que se venera extramuros de la ciudad de tlaxcala / Loaisaga, Manuel de – Mexico: Por la Viuda de D Joseph Hogal, ano de 1750 – (= ser Books on religion...1543/44-c1800: milagros y culto de la virgen) – 2mf – 9 – mf#crl-91 – ne IDC [241]
Historia de la musica en colombia / Perdomo Escobar, Jose Ignacio – Bogota, Colombia. 1963 – 1r – us UF Libraries [780]
Historia de la musica en guatemala / Vasquez A, Rafael – Guatemala, 1950 – 1r – us UF Libraries [780]
Historia de la nueva espana / Aguilar, Francisco de – Mexico: Ediciones Botas, 1938 – sp Bibl Santa Ana [946]
Historia de la nueva espana / Zorita, Alonso de – Madrid: Libreria General de Victoriano Suarez, 1909.-v1 – 1 – sp Bibl Santa Ana [946]
Historia de la nueva granada / Restrepo, Jose Manuel – Bogota, Colombia. 1936 – 1r – us UF Libraries [972]
Historia de la orden del libertador / Planas Suarez, Simon – Caracas, Venezuela. 1955 – 1r – us UF Libraries [972]
Historia de la poesia argentina y uruguaya / Menendez Y Pelayo, Marcelino – Buenos Aires, Argentina. 1943 – 1r – us UF Libraries [972]
Historia de la provincia de la compania de jesus de nueva espana : dividida en ocho libros... / Florencia, Francisco de – en Mexico: Por Iuan Ioseph Guillena Carrascoso 1694 – (= ser Books on religion...1543/44-c1800: jesuitas) – 5mf – 9 – mf#crl-218 – ne IDC [241]
Historia de la provincia de san antonio del nuevo / Zamora, Alonso De – Bogota, Colombia. v1-4. 1945 – 1r – us UF Libraries [972]
Historia de la provincia de san antonio del nuevo reino de granada. caracas, 1930 / Zamora, Alonso de – Madrid: Razon y Fe, 1930 – 1 – sp Bibl Santa Ana [946]
Historia de la provincia de san nicolas de tolenti / Basalenque, Diego – Mexico City? Mexico. 1963 – 1r – us UF Libraries [972]
Historia de la provincia de san vicente de chiapa / Ximenez, Francisco – Guatemala, v1-3. 1920-31 – 1r – us UF Libraries [972]
Historia de la provincia de san vicente de chiapa y guatemala, de la orden de predicacdores. tomo 1. guatemala, 1929 / Ximenez, Francisco – Madrid: Razon y Fe, 1930 – 1 – sp Bibl Santa Ana [240]
Historia de la provincia de san vicente de chiapa y guatemala de la orden de predicadores...tomo 3 / Ximenez, Francisco – Guatemala, 1931; Madrid: Razon y Fe, 1932 – 1 – sp Bibl Santa Ana [972]
Historia de la publicacion monumenta historica societatis jesu / Gomez Rodeles, Cecilio – Madrid: Imprenta del Asilo de huerfanos del S. C. de Jesus, 1913 – (= ser Monumenta historica societatis jesu) – 1mf – 9 – 0-7905-6345-2 – mf#1988-2345 – us ATLA [240]
Historia de la real y general junta de comercio, moneda y minas... / Larruga y Boneta, E – Madrid, 1789 – 246mf – 9 – sp Cultura [380]
Historia de la rebelion popular de 1814 / Uslar Pietri, Juan – Caracas, Venezuela. 1962 – 1r – us UF Libraries [972]
Historia de la reconquista de espana contada a la juventud / Herrera Oria, P Enrique – Madrid. 1943 – 1 – us CRL [946]
Historia de la republica de el salvador... / Bayle, Constantino – Madrid: Razon y Fe, 1928 – 9 – sp Bibl Santa Ana [972]
Historia de la republica de guatemala, 1821-1921 / Villacorta C, J Antonio – Guatemala, 1960 – 1r – us UF Libraries [972]
Historia de la restauracion / Archambault, Pedro Maria – Paris, France. 1938 – 1r – us UF Libraries [972]
Historia de la revolucion federal en venezuela / Alvarado, Lisandro – Caracas, Venezuela. 1956 – 1r – us UF Libraries [972]
Historia de la revolucion y guerra de cuba / Gelpi Y Ferro, Gil – Habana, Cuba. v1-2. 1887-89 – 1r – us UF Libraries [972]
Historia de la serafica provincia de cataluna por el r.p... / Sanahuja, Pedro de – Madrid: Arch. Ibero Americano, 1961 – 1 – sp Bibl Santa Ana [946]
Historia de la universidad de arizona / Fitz-Gerald, John Driscoll – Ciudad Trujillo, Dominican Republic. 1942 – 1r – us UF Libraries [378]
Historia de la universidad de el salvador / Duran, Miguel Angel – San Salvador, El Salvador. 1941 – 1r – us UF Libraries [378]
Historia de la universidad de honduras / Guardiola, Esteban – Tegucigalpa, Mexico. 1955 – 1r – us UF Libraries [378]
Historia de la vida y martirio de la santa eulalia de merida / Quiros y Benavides, Felipe Bernardo – Madrid: Francisco Sanz, 1672 – 1 – sp Bibl Santa Ana [240]
Historia de la vida y martirio del beato juan de britto, de la compania de jesus : missionero del madure, muerto en odio de la fe en el reino de marava = Histoire du bienheureux jean de britto / Prat, Jean Marie – Madrid: Eusebio Aguado, 1854 [mf ed 1995] – (= ser Yale coll) – xix/420p (ill) – 1 – 0-524-09970-7 – (trans fr french into spanish) – mf#1995-0970 – us ATLA [241]
Historia de la...nueva espana / Diaz del Castillo, Bernal – 1795-96.4v – 9 – sp Bibl Santa Ana [972]
Historia de las cosas mas notables del gran reyno de china / Gonzales de Mendoza, Juan – Madrid: Missionalia Hispanica, 1945 – 1 – sp Bibl Santa Ana [951]
Historia de las guerras civiles del peru (1544-1548) y de otros sucesos de las indias / Gutierrez de Santa Clara, Pedro – Madrid: Victoriano Suarez, 1929 – 1 – sp Bibl Santa Ana [972]
Historia de las instituciones juridicas salvadoren / Rodriguez Ruiz, Napoleon – San Salvador, El Salvador. 1951 – 1r – us UF Libraries [340]
Historia de las leyes / Colombia. Laws, Statutes, etc – Bogota. On film: 1925-33. LL-062 – 1 – us L of C Photodup [340]
Historia de las leyes / Segovia, L – Cartagena, Colombia. 1953 – 1r – us UF Libraries [972]
Historia de las medidas adoptadas por la administr... / Bachiller Y Morales, Antonio – Habana, Cuba. 1860 – 1r – us UF Libraries [972]
Historia de las misiones / Unanue, H & Sobrevida, M – Madrid: Graf. Calleja, 1966 – 1 – sp Bibl Santa Ana [240]
Historia de las misiones agustinianas en china : con las licencias necesarias / Martinez, Bernardo – Madrid: Imp del Asilo de Huerfanos del S C de Jesus 1918 [mf ed 1995] – (= ser Yale coll) – 1r – 1 – 0-524-09797-6 – (in spanish. filmed with other works) – mf#1995-0797 – us ATLA [241]
Historia de las misiones franciscanas : 1619-1921 / Izaguirre, Fray Bernardino – Madrid: Razon y Fe, 1927 – 1 – sp Bibl Santa Ana [240]
Historia de las relaciones interstatuales de centr... / Moreno, Laudelino – Madrid, Spain. 1928 – 1r – us UF Libraries [972]
Historia de las revoluciones de hungaria / Brenner, D A I – Madrid. 3v. 1687-89 – 1r – 1 – mf#95873 – uk Microform Academic [943]
Historia de las vidas y milagros de nuestro beato p fr pedro de alcantara / Cogolludo, Francisco de – 1664 – 9 – sp Bibl Santa Ana [240]
Historia de las virtudes y propiedades del tabaco y de los modos de tomarse... / Castro, J – Cordoba, 1620 – 3mf – 9 – sp Cultura [630]
Historia de las yervas y plantas... / Jarava, J – Amberes, 1557 – 9mf – 9 – sp Cultura [630]
Historia de los archivos de cuba / Llaverias Y Martinez, Joaquin – Habana, Cuba. 1949 – 1r – us UF Libraries [972]
Historia de los himnos dominicanos / Ravelo, Jose De Jesus – Santo Domingo, Dominican Republic. 1934 – 1r – us UF Libraries [972]
Historia de los oraculos / Fontenelle – Merida: Imprenta de Manuel Galvan, 1868 – 1 – sp Bibl Santa Ana [946]
Historia de los partidos politicos puertorriquenos / Pagan, Bolivar – San Juan, Puerto Rico. v1-2. 1959 – 1r – us UF Libraries [972]
Historia de los pp dominicos en las islas filipinas y en sus misiones del japon, china, tung-kin y formosa : que comprende los sucesos principales de la historia general de este archipielago...hasta el ano de 1840 / Ferrando, Juan – Madrid: Imp y estereotipia de M Rivadeneyra, 1870-72 [mf ed 1995] – (= ser Yale coll) – 6v (ill) – 1 – 0-524-10055-1 – (in spanish. corr by joaquin fonseca. app by pedro payo) – mf#1995-1055 – us ATLA [241]
Historia de los reyes catolicos / Bernaldez, Andres – Sevilla: Imp. de Jose Maria Geofrin, Tomo 1. 1869 – 1 – sp Bibl Santa Ana [241]
Historia de los...don fernando y.. / Bernaldez, Andres – 1856. 2 tomas – 9 – sp Bibl Santa Ana [946]
Historia de managua / Halftermeyer, Gratus – Managua, Nicaragua. 195- – 1r – us UF Libraries [972]
Historia de mantua / Santovenia Y Echaide, Emeterio Santiago – Habana, Cuba. 1923 – 1r – us UF Libraries [972]
Historia de medio siglo / Cuadra Pasos, Carlos – Managua, Nicaragua. 1964 – 1r – us UF Libraries [972]
Historia de mejico...hernando cortes / Lopez de Gomara, Francisco – 1554 – 9 – sp Bibl Santa Ana [972]
Historia de mexico / Iturribarria, Jorge Fernando – Mexico City? Mexico. 1951 – 1r – us UF Libraries [972]
Historia de mexico, de francisco benegas galvan / Bayle, Constantino – Madrid: Razon y Fe, 1924 – 1 – sp Bibl Santa Ana [972]
Historia de minas, ademas sagad, rei de ethiopia / Esteves Pereira, F M – Lisboa, 1888 – 1mf – 9 – mf#NE-20233 – ne IDC [960]
Historia de minas, ademas sagad, rei de ethiopia / Pereira, F M E – Lisboa, 1888 – 1mf – 9 – mf#SEP-87 – ne IDC [960]
Historia de montserrat / Alboreda, A M – Monasterio de Montserrat, 1931; Madrid: Razon y Fe, 1933 – 1 – sp Bibl Santa Ana [946]
Historia de nuestra senora de guadalupe... / Rubio German, Francisco; ed by Bayle, Constantino – Madrid: Razon y Fe, 1928 – 9 – sp Bibl Santa Ana [972]
Historia de nueva espana...y notas del ilmo. d.f.a. lorenzana.. / Cortes, Hernando – 1770 – 9 – sp Bibl Santa Ana [972]
Historia de oliveira / Fonseca, Luis Gonzaga Da – Oliveira, Brazil. 1961 – 1r – us UF Libraries [972]
Historia de origine et progressu controversiae sacramentariae... / Lavater, L – Tiguri, Christoph Froschover, 1563 – 2mf – 9 – mf#PBU-307 – ne IDC [240]
Historia de pereira / Duque Gomez, Luis – Pereira, Colombia. 1963 – 1r – us UF Libraries [972]
Historia de peru / Cappa, Ricardo S J – 1885 – 9 – (1886 ed. 1887 ed) – sp Bibl Santa Ana [972]
Historia de peru...austriaca / Lorente, Sebastian – 1863 – 9 – sp Bibl Santa Ana [972]
Historia de piedra escrita / San Jose, Francisco de – 1751 – 9 – sp Bibl Santa Ana [946]
Historia de polonia / Brandenburger, C L & Laubert, M – Barcelona, 1932; Madrid: Razon y Fe, 1934 – 1 – sp Bibl Santa Ana [946]
Historia de puerto rico / Vivas Maldonado, Jose Luis – New York, NY. 1962 – 1r – us UF Libraries [972]
Historia de puerto rico / Vivas Maldonado, Jose Luis – San Juan, Puerto Rico. 1957 – 1r – us UF Libraries [972]
Historia de san martin / Mitre, Bartolome – Buenos Aires, Argentina. v1-4. 1890 – 1r – us UF Libraries [972]
Historia de santa catarina / Cabral, Oswaldo R – Rio de Janeiro, Brazil. 1970 – 1r – us UF Libraries [972]
Historia de santa maria de la victoria / Tena Fernandez, Juan; ed by Rodrigo, Sanchez – Serradilla, 1930 – 1 – sp Bibl Santa Ana [240]
Historia de santo domingo / Inchaustequi Cabral, Joaquin Marino – Mexico City? Mexico. 1958 – 1r – us UF Libraries [972]
Historia de santo domingo / Monte Y Tejada, Antonio Del – Santo Domingo, Dominican Republic. v1-3. 1890 – 1r – us UF Libraries [972]
A historia de sao paulo ensinada pela biographia dos saus vultos mais notaveis / Amaral, Tancredo do – Rio de Janeiro: Alves, 1895 [mf ed 1986] – 351p – 1 – (pref by valois de castro) – mf#8598 – us Wisconsin U Libr [972]
Historia de talavera la real / Diaz Perez, Nicolas – 1885 – 9 – (1879 ed) – sp Bibl Santa Ana [946]
Historia de talavera la real / Diaz y Perez, Nicolas – Madrid: Imp. Manuel Gines Hernandez, 2nd ed. 1879 – 1 – sp Bibl Santa Ana [946]
Historia de talavera la real... / Diaz Perez, Nicolas – Madrid: Imp. J. Antonio Garcia, 1875 – 1 – sp Bibl Santa Ana [946]
Historia de toro / Piedrahita, Diogenes – s.l, s.l? 1954 – 1r – us UF Libraries [972]
Historia de um rio (o tiete) – Sao Paulo, Brazil. 1948 – 1r – us UF Libraries [972]
Historia de un cambio de gobierno / Martinez Delgado, Luis – Bogota, Colombia. 1958 – 1r – us UF Libraries [972]
Historia de un hombre y de un pueblo / Valbuena, L Martin – Caracas, Venezuela. 1953 – 1r – us UF Libraries [972]
Historia de un homre insignificante / Acebal, Sergio – Habana, Cuba. 1929 – 1r – us UF Libraries [972]
Historia de un pepe, don bonifacio / Milla, Jose – Guatemala, 1937 – 1r – us UF Libraries [972]
Historia de un proceso / Carvallo Arvelo, Salvador – Valencia, Spain. 1943 – 1r – us UF Libraries [972]
Historia de una ciudad / Uribe Uribe, Fernando – Bogota, Colombia. 1963 – 1r – us UF Libraries [972]
Historia de una mancha de tinta (el manuscrito de longo) / Courier, Pablo Luis – Valencia: Editorial Castalia, 1948 – sp Bibl Santa Ana [946]
Historia de una monstruosa farsa / Echavarria Olozaga, Felipe – Roma, Italy. 1964 – 1r – us UF Libraries [972]
Historia de una pelea cubana contra los demonios / Ortiz, Fernando – Santa Clara, Cuba. 1959 – 1r – us UF Libraries [972]
Historia de veintiun anos / Salazar, Ramon A – Guatemala, v1-2. 1956 – 1r – us UF Libraries [972]
Historia de venezuela / Aguado, Pedro De – Madrid, Spain. v1-2. 1950 – 1r – us UF Libraries [972]
Historia de venezuela / Fuentes-Figueroa Rodriguez, Julian – Caracas, Venezuela. 1961? – 1r – us UF Libraries [972]
Historia de venezuela / Moron, Guillermo – Caracas, Venezuela. 1961 – 1r – us UF Libraries [972]
Historia de vitis pontificum romanarum *see* The lives of the popes
Historia de vitis romanorum pontificum (cbh19,2) / Anastasius Bibliothecaris; ed by Fabrotus, C A – Parisiis, 1649 – (= ser Corpus byzantinae historiae (cbh)) – €29.00 – ne Slangenburg [241]
Historia degli imperatori greci... / Nicetas, A C – Venetia, 1562 – 7mf – 9 – mf#H-8301 – ne IDC [956]
Historia del almirante don cristobal colon. tomo 1 / Colon, Hernando – Madrid, 1932 – 1 – sp Bibl Santa Ana [910]
Historia del ano de 1887 / Cruz Monclova, Lidio – Rio Piedras, Puerto Rico. 1958 – 1r – us UF Libraries [972]
Historia del arte en guatemala 1524-1962 / Chincilla Aguilar, Ernesto – Guatemala, 1963 – 1r – us UF Libraries [972]
Historia del arte hispanoamericano. tomo 1. barcelona, 1945 / Angulo, D – Madrid: Razon y Fe, 1946 – 1 – sp Bibl Santa Ana [700]
Historia del brasil / Beltran, Juan Gregorio – Buenos Aires, Argentina. 1935 – 1r – us UF Libraries [972]
Historia del cavallero cifar / ed by Michelant, Heinrich – Stuttgart: Litterarischer Verein, 1872 (Tuebingen: L F Fues) – 1 – us Wisconsin U Libr [910]
Historia del cavallero cifar / ed by Michelant, Heinrich – Stuttgart: Litterarischer Verein, 1872 (Tuebingen: L F Fues) [mf ed 1993] – (= ser Blvs 112) – 377p – 1 – mf#8470 reel 24 – us Wisconsin U Libr [440]

Historia del celebre santuario de nuestra senora de copacabana : y sus milagros e inuencion de la cruz de carabusco / Ramos Gavilan, Alonso – Lima: Por Geronymo de Cortreras, ano 1621 – (= ser Books on religion...1543/44-c1800: milagros y culto de la virgen) – 5mf – 9 – mf#crl-78 – ne IDC [241]

Historia del comercio mundial... / Schmidt, M G – Madrid: Razon y Fe, 1927 – 1 – sp Bibl Santa Ana [380]

Historia del culto y san tuario de nuestra senora de la montana patrona de caceres / Orti Belmonte, Miguel Angel – Caceres: Dip.Prov. Caceres, Tomo 1. 1949 – 1 – sp Bibl Santa Ana [240]

Historia del culto y santuario de nuestra senora de la montana / Orti Belmonte, Miguel Angel – Coleccion de Estudios Extremenos. Caceres Diput Prov. de Caceres, Tomo 2. 1950 – 1 – sp Bibl Santa Ana [240]

Historia del departamento del magdalena y del terr... / Valdeblanquez, Jose Maria – Bogota, Colombia. 1964 – 1r – us UF Libraries [972]

Historia del descubrimiento de tucuman, seguida de investigaciones historicas / Jaimes Freyre, Ricardo – Universidad de Tucuman. Buenos Aires: Impr. de Coni hermanos, 1916. 312p – 1 – us Wisconsin U Libr [972]

Historia del hombre contada por sus casas / Marti, Jose – Habana, Cuba. 1961 – 1r – us UF Libraries [972]

Historia del hombre que tuvo el mundo en la mano : johann wolfgang von goethe / Nelken, Margarita – Mexico: Ediciones de la Secretaria de Educacion Publica, 1943 – 1r – 1 – us Wisconsin U Libr [430]

Historia del hospital de san jose, 1902-1956 / Munoz, Laurentino – Bogota, Colombia. 1958 – 1r – us UF Libraries [360]

Historia del libertador don jose de san martin / Otero, Jose Pacifico – Buenos Aires, Argentina. v1-4. 1932 – 1r – us UF Libraries [972]

Historia del magnanimo, et valoroso signor georgio castrioto, detto scanderbego, dignissimo principe de gli albani / Barletius, M – Venetia, 1580 – 9mf – 9 – mf#H-8354 – ne IDC [956]

Historia del monasterio (siecle 19) / Montes, San Pedro de – Astorga – 1r – 5,6 – sp Cultura [240]

Historia del monestario de yuste / G Maria de Alboraya, Domingo de – Madrid: Suc.de Rivadeneira, 1906 – 1 – sp Bibl Santa Ana [240]

Historia del movimiento unionista / Marroquin Rojas, Clemente – Barcelona, Spain. 1929- – 1r – us UF Libraries [972]

Historia del partido liberal colombiano / Puentes, Milton – Bogota, Colombia. 1961 – 1r – us UF Libraries [972]

Historia del periodismo en colombia / Otero Munoz, Gustavo – Bogota, Colombia. 1936 – 1r – us UF Libraries [972]

Historia del periodismo en fregenal de la sierra / Real, Enrique – 1897 – 9 – sp Bibl Santa Ana [440]

Historia del peru / Lorente, Sebastian – 1861 – 9 – sp Bibl Santa Ana [972]

Historia del peru / Markham, Clements Robert – Lima, Peru. 1952 – 1r – us UF Libraries [972]

Historia del pueblo de alange / Diaz y Perez, Nicolas – Badajoz: Imp. Artes Graficas, 1930 – 1 – sp Bibl Santa Ana [946]

Historia del puerto de la santisima tinidad de sonsonate / Rubio Sanchez, Manuel – s.l, s.l? 1977 – 1r – us UF Libraries [972]

Historia del reino de badajoz durante la dominacion musulmana / Martinez Martinez, Matias Ramon – Badajoz: Tip. y Libr. de A. Arqueros, 1904 – 1 – sp Bibl Santa Ana [946]

Historia del reino de badajoz durante la dominacion musulmana. noticia / Martinez Martinez, Matias Ramon – Madrid: Fortant, 1905. B.R.A.H. XLVII, pp. 406-407 – sp Bibl Santa Ana [946]

Historia del rito mozarabe y toledano / Prado, G – Burgos, 1928 – €11.00 – ne Slangenburg [241]

Historia del santisimo cristo de la victoria que se venera en la villa de serradilla (caceres) / Cantera, Eugenio – Monachil: Tip. Santa Rita, 1922 – sp Bibl Santa Ana [240]

Historia del traslado del colegio de artilleria de badajoz / Lanuza, Francisco de – Segovia: Imp. Gabel, 1952 – 1 – sp Bibl Santa Ana [355]

Historia dela compania de jesus no brasil... / Leite, Serafin – Burgos: Razon y Fe, 1939 – 1 – sp Bibl Santa Ana [972]

Historia dela conquista de mexico / Lopez de Gomara, Francisco – Editorial Pedro Robledo, 1943 – sp Bibl Santa Ana [972]

Historia delas cosas mas notables, ritos y costvmbres... / Gonzalez de Mendoza, J – Roma: Bartholome Grassi, 1585 – 5mf – 9 – mf#H-8420 – ne IDC [910]

Historia dell' impresa di tripoli di barbaria / Ulloa, A de – Venetia, 1569 – 3mf – 9 – mf#H-8310 – ne IDC [956]

La historia dell' impresa di tripoli di barbaria... / Ulloa, A de – Venevia, 1566 – 3mf – 9 – mf#H-8305 – ne IDC [956]

Historia dell' indie orientali... / Lopes de Castanheda, F – Venetia: Apresso Giordano Ziletti, 1577-1578. 2v – 20mf – 9 – mf#H-8351 – ne IDC [915]

Historia della gverra fra tvrchi, et persiani / Minadoi, G T – Venetia, 1588 – 5mf – 9 – mf#H-8371 – ne IDC [956]

Historia der passion vnsers lieben herrn vnd heilands jesu christi / Chemnitz d A, M – np, 1590 – 8mf – 9 – mf#TH-1 mf 215-222 – ne IDC [242]

Historia des augspurgischen confession : ...item acta concordiae zwischen herren luthero und den euangelischen staetten in schweitz im jahr 38... / [Hardesheim, C] – Newstatt an der Hardt, Matthaeus Harnisch, 1580 – 5mf – 9 – mf#PBU-596 – ne IDC [242]

Historia destructionis troiae (cima3) : farbmikrofiche-edition der handschrift cologny-geneve, bibliotheca bodmeriana, cod.78 / Columnis, Guido de – (mf ed 1987) – (= ser Codices illuminati medii aevi (cima) 3) – 40p on 3 color mf – 15 – €220.00 – 3-89219-003-8 – (int by hugo buchthal) – gw Lengenfelder [930]

Historia di zighet, ispvgnata da svliman, re de' tvrchi, l'anno 1566 – Venetia, 1570 – 1mf – 9 – mf#H-8169 – ne IDC [956]

Historia diplomatica do brasil / Calmon, Pedro – Belo Horizonte, Brazil. 1941 – 1r – us UF Libraries [972]

Historia diplomatica e politica internacional / Lyra, Heitor – Rio de Janeiro, Brazil. 1941 – 1r – us UF Libraries [327]

Historia dispvtationis sev potivs colloqvii inter iacobvm colervm et mathiam flacivm illyricvm de peccato originis / Coler, J – Berlini, 1585 – 2mf – 9 – mf#TH-1 mf 336-337 – ne IDC [242]

Historia do brasil / Armitage, John – Ouro, Brazil. 1965 – 1r – us UF Libraries [972]

Historia do brasil / Calmon, Pedro – Rio de Janeiro, Brazil. v1-7. 1961 – 1r – us UF Libraries [972]

Historia do brasil / Galanti, Raphael Maria – Sao Paulo, Brazil. v1-5. 1910,1913 – 1r – us UF Libraries [972]

Historia do brasil / Handelmann, Heinrich – Rio de Janeiro, Brazil. v1-2. 1931 – 1r – us UF Libraries [972]

Historia do brasil / Veiga Cabral, Mario Vasconcellos Da – Rio de Janeiro, Brazil. 1944 – 1r – us UF Libraries [972]

Historia do brasil, 1500-1627 / Vicente Do Salvador, Father – Sao Paulo, Brazil. 1954 – 1r – us UF Libraries [972]

Historia do brasil na poesia do povo / Calmon, Pedro – Rio de Janeiro, Brazil. 1943 – 1r – us UF Libraries [972]

Historia do brazil / Rocha Pombo, Jose Francisco – Rio de Janeiro, Brazil. v1-5. 1935 – 1r – us UF Libraries [972]

Historia do brazil para o ensino secundario / Pombo, Jose Francisco Da Rocha – San Paulo, Brazil. 1918 – 1r – us UF Libraries [972]

Historia do cabo / Felipe, Israel – Recife, Brazil. 1962 – 1r – us UF Libraries [972]

Historia do discobrimento e conquista da india pelos portuguses / Lopes de Castanheda, F – Lisboa, 1833. 8v – 32mf – 9 – mf#HT-775 – ne IDC [910]

Historia do ensino no ceara / Castelo, Placido Aderaldo – Fortaleza, Brazil. 1970 – 1r – us UF Libraries [972]

Historia do espirito santo / Novaes, Maria Stella De – Vitoria, Brazil. 1968 – 1r – us UF Libraries [972]

Historia do fanatismo religioso no ceara / Montenegro, Abelardo Fernando – Fortaleza: A. Batista Fontenele, 1959. 76p. Inc. bibliog. references. 1 reel. 1188 – 1 – us Wisconsin U Libr [240]

Historia do hino nacional brasileiro / Lira, Mariza – Rio de Janeiro, Brazil. 1954 – 1r – us UF Libraries [972]

Historia do imperio / Monteiro, Tobias – Rio de Janeiro, Brazil. 1927 – 1r – us UF Libraries [972]

Historia do imperio / Monteiro, Tobias – Rio de Janeiro, Brazil. v1-2. 1939-1946 – 1r – us UF Libraries [972]

Historia do movimento politico que no anno de 1842 / Marinho, Jose Antonio – Rio de Janeiro, Brazil. 1939 – 1r – us UF Libraries [972]

Historia do parana – Curitiba, Brazil. v1-4. 1969 – 1r – us UF Libraries [972]

Historia do periodo provincial do parana / Carneiro, David – Curitiba, Brazil. 1960 – 1r – us UF Libraries [972]

Historia do povo brasileiro / Quadros, Janio – Sao Paulo, Brazil. v1-6. 1968 – 1r – us UF Libraries [972]

Historia do rio grande do norte / Lyra, Augusto Tavares De – Rio de Janeiro, Brazil. 1921 – 1r – us UF Libraries [972]

Historia doctrinae catholicae inter armenos unionisque eorum : cum ecclesia romana in concilio florentino / Balgy, Alexander – Viennae: Typis Congr Mechitharisticae, 1878 [mf ed 1986] – 1mf – 9 – 0-8370-7603-X – (discussion in latin, text in armenian) – mf#1986-1603 – us ATLA [240]

Historia documentada de la conspiracion / Garrigo, Roque E – Habana, Cuba. v1-2. 1929 – 1r – us UF Libraries [972]

Historia documentada de la conspiracion / Valle, Adrian Del – Habana, Cuba. 1930 – 1r – us UF Libraries [972]

Historia documentada de la vida y gloriosa muerte de los padres roque gonzalez de santa cruz, alonso rodriguez y juan del castillo...buenos aires, 1929 / Blanco, Jose Maria – Madrid: Razon y Fe, 1930 – 1 – sp Bibl Santa Ana [240]

Historia documentada de san cristobal / Wright, Irene Aloha – Habana, Cuba. 1930 – 1r – us UF Libraries [972]

Historia documental del canal de panama / Arosemena G, Diogenes A – Panama, Panama. 1962 – 1r – us UF Libraries [972]

Historia documental del choco / Ortega Ricaurte, Enrique – Bogota, Colombia. 1954 – 1r – us UF Libraries [972]

Historia documental do brasil / Castro, Therezinha De – Rio de Janeiro, Brazil. 1969 – 1r – us UF Libraries [972]

Historia e historiografia / Rodrigues, Jose Honorio – Petropolis, Brazil. 1970 – 1r – us UF Libraries [972]

Historia e interpretacao de 'os sertoes' / Andrade, Olimpio De Souza – Sao Paulo, Brazil. 1960 – 1r – us UF Libraries [972]

Historia ecclesiae ultrajectinae / Hoynck van Papendrecht, C P – Mechlinae, 1725 – €38.00 – ne Slangenburg [240]

Historia ecclesiastica novi testamenti, tomi 6, 8, 9 / Hottinger, J H – Tigvri, Joh Henr Hamberger, Michael Sch(a)ufelberger, 1665, 1667 – 35mf – 9 – mf#PBU-417 – ne IDC [240]

Historia ecclesiastica sive chronographia tripertita (cbh19,1) / Anastasius Bibliothecaris; ed by Fabrotus, C – Parisiis, 1649 – (= ser Corpus byzantinae historiae (cbh)) – €23.00 – ne Slangenburg [243]

Historia ecclesiasticae inclyte urbis brunsvigae / Rehtmeyer, Philipp Julius – Braunschweig: L. Schroeder, 1717-1720 – 2r – 1 – 0-8370-0769-0 – mf#1984-B502 – us ATLA [240]

Historia ecclesiatica, carmine elegiaco concinnata *see* True ecclesiastical history from moses to the time of martin luther

Historia eclesiastica de cuyo. tomo 1. milano, 1931 / Verdaguer, Jose A – Madrid: Razon y Fe, 1935 – 1 – sp Bibl Santa Ana [240]

Historia eclesiastica de la ciudad y obispado de badajoz : primera parte / Solano de Figueroa y Altamirano, Juan – Badajoz, 1929; Madrid: Razon y Fe, 1931 – 1 – sp Bibl Santa Ana [240]

Historia eclesiastica de la ciudad y obispado de badajoz. continuacion de la de solano de figueroa – Badajoz: Tip. Vda. de A. Arqueros, Tomo 2. 1945. Publ. de la Caja Rural de Badajoz – 1 – sp Bibl Santa Ana [240]

Historia eclesiastica do brasil / Camargo, Paulo Florio Da Silveira – Rio de Janeiro, Brazil. 1955 – 1r – us UF Libraries [025]

Historia eclesiastica do maranhao / Pacheco, Felipe Conduru – Sao Luis, Brazil. 1969 – 1r – us UF Libraries [972]

Historia eclesiastica indiana / Mendieta, Geronimo – Tomo I. 1870 – 9 – (tomo 2 1870) – sp Bibl Santa Ana [240]

Historia eclesiastica y civil de nueva granada / Groot, Jose Manuel – Bogota, Colombia. v1-5. 1953 – 1r – us UF Libraries [972]

Historia economica de cuba / Friedlaender, Heinrich – Havana, Cuba. v1. 1978 – 1r – us UF Libraries [972]

Historia economica de cuba / Friedlaender, Heinrich – Havana, Cuba. v2. 1978 – 1r – us UF Libraries [330]

Historia economica de cuba / Friedlander, Heinrich – Habana, Cuba. 1944 – 1r – us UF Libraries [330]

Historia economica do brasil / Prado Junior, Caio – Sao Paulo, Brazil. 1949 – 1r – us UF Libraries [330]

Historia economica do brasil / Simonsen, Roberto Cochrane – Sao Paulo, Brazil. 1969 – 1r – us UF Libraries [330]

Historia economica do brasil pesquisas e analises / Buescu, Mircea – Rio de Janeiro, Brazil. 1970 – 1r – us UF Libraries [330]

Historia elemental de cuba / Guerra, Ramiro – Habana, Cuba. 1932 – 1r – us UF Libraries [972]

Historia estadistica de cojedes (desde 1771) / Gonzalez, Eloy Guillermo – Caracas, Venezuela. 1911 – 1r – us UF Libraries [972]

Historia et cartularium monsterii s petri cloucestriae (rs33) / ed by Hart, W H – (= ser The rolls series (rs)) – (v1 1863 €18 v2 1865 €14 v3 1867 €19) – ne Slangenburg [241]

Historia fatal : asanas de la ignorancia, guerra fisica, proesas medicales sacadas a las del conosimiento por un enfermo [...] – [1690?] – (filmed with: valle y caviedes, j guerra fisica, proezas medicales, hazanas de la ignorancia) – us CRL [946]

Historia fisica, economico-politica, intelectual... / Sagra, Ramon De La – Paris, France. 1861 – 1r – us UF Libraries [330]

Historia general de chile / Barros Arana, Diego – Santiago, 1884-1902. 16 v – 1 – 80.00 – us L of C Photodup [972]

Historia general de espana : desde los tiempos primitivos hasta la muerte de fernando 7 / Lafuente y Zamlioa, Modesto – Barcelona. v1-25. 1887-91 – 1 – $300.00 – mf#0315 – us Brook [946]

Historia general de filipinas : por don jose montero y vidal. informe / Barrantes Moreno, Vicente – Madrid: Fortanet, 1887. B.R.A.H. XI, pp. 340-344 – sp Bibl Santa Ana [959]

Historia general de filipinas. navas del valle. catalogo de monumentos referentes a islas filipinas... / Pastells, Pablo – Madrid: Razon y Fe, v8. 1935 – 1 – sp Bibl Santa Ana [959]

Historia general de la yndia oriental / San Rom n d e Ribadeneyra, A – Valladolid: Luis Sanchez acosta de Diego Perez, 1603 – 15mf – 9 – mf#HT-550 – ne IDC [915]

Historia general de las cosas de nueva espana / Sahagun, Bernardino De – Mexico City? Mexico. v1-3. 1946 – 1r – us UF Libraries [972]

Historia general de las indias occidentales / Remesal, Antonio De – Guatemala. v1-2. 1932 – 1r – us UF Libraries [972]

Historia general de las indias occidentales, y particular de la gobernacion de chiapa y guatemala tomo 1 y 2 / Remesal, Antonio – Guatemala, 2nd ed 1932; Madrid: Razon y Fe, 1934 – 1 – sp Bibl Santa Ana [972]

Historia general de los hechos de los castellanos en las islas y tierra firme del mar oceano / Herrera, Antonio de – Madrid: Real Academia de la Historia, 1935.-v3 – 1 – sp Bibl Santa Ana [946]

Historia general de los hechos de los castellanos en las islas y tierra-firme de el mar oceano / Herera, Antonio de – Asuncion de Paraguay, Editorial, 1945.-v3 – 1 – sp Bibl Santa Ana [946]

Historia general de los hechos de los castellanos en las islas y tierra-firme de el mar oceano / Herrera, Antonio de – Buenos Aires: Editorial Guarania, 1945 – 1 – sp Bibl Santa Ana [946]

Historia general de los hechos de los castellanos en las islas y tierra-firme del mar oceano / Herrera, Antonio de – Buenos Aires: Talleres Graficos Continental La valle, 1944.-v1 – 1 – sp Bibl Santa Ana [946]

Historia general de los hechos de los castellanos en las islas y tierrafirme del mar oceano / Herrera, Antonio de – Madrid: Real Academia de la Historia, 1936.-v5 – 1 – sp Bibl Santa Ana [946]

Historia general de los hechos de los castellanos en las islas y tierrafirme del mar oceano / Herrera, Antonio de – Madrid: Real Acedemia de la Historia, 1936.-v4 – 1 – sp Bibl Santa Ana [946]

Historia general del derecho espanol / Chapado Garcia, Eusebio Maria – Valladolid: Montero, 1900. 971p. LL-8005 – 1 – us L of C Photodup [340]

Historia general del peru / Murua, Martin, Fray – Madrid: Imp. Gongora, Libro 2. 1964 – 1 – sp Bibl Santa Ana [972]

Historia general del peru / Murua, Martin, Fray – Madrid: Imp. Gongora, Tomo 1. 1962 – 1 – sp Bibl Santa Ana [972]

Historia general y natural de las indias / Fernandez De Oviedo Y Valdes, Gonzalo – Madrid, Spain. v1-3. 1855-61 – 1r – us UF Libraries [972]

Historia general...castellanos...oceano / Herrera, Antonio de – 1726. Decada segunda – 9 – (decada tercera 1726. decada quarta 1736. decada quinta 1728) – sp Bibl Santa Ana [972]

Historia geral das guerras angolanas, 1680 / ed by Delgado, Jose Matias – [Lisboa]: Agencia Geral das Colonias, Divisao de Publicacaoes e Bibliteca, 1940-42 – 1 – us CRL [960]

Historia geral de ethiopia a alta ou abassin / Almeida, M de – Roma, 1907-1908. 3v – 20mf – 9 – mf#SEP-58 – ne IDC [956]

Historia grafica de la republica dominicana / Estella, Jose Ramon – Trujillo, Peru. 1944 – 1r – us UF Libraries [972]

Historia hipolitin asin burac nin cabuhayan na sucat maaraan nin tauo – [Nueva Caceres?: Libreria Mariana? 1906?] [mf ed Bloomington IN: Indiana Uni Lib, Preservation Dept 1984] – (= ser Coll...in the bikol language) – 1r – 1 – us Indiana Preservation [490]

Historia hungarorum ecclesiastica : inde ab exordio novi testamenti ad nostra usque tempora ex monumentis partim editis, partim vero ineditis, fide dignis / Bod, Peter; ed by Rauwenhoff, Lodewijk Willem Ernst – Lugduni-Batavorum: E.J. Brill, 1888-1890 – 4mf – 9 – 0-7905-5571-9 – (incl bibl ref) – mf#1988-1571 – us ATLA [240]

Historia ilustrada do rio de janeiro / Mathias, Herculano Gomes – Rio de Janeiro, Brazil. 1965 – 1r – us UF Libraries [972]

Historia indiana / Federmann, Nikolaus – Madrid, Spain. 1958 – 1r – us UF Libraries [972]

Historia ivdicvm... / Wolf, J – Tigvri, Iohannes Vvolph, 1598 – 6mf – 9 – mf#PBU-663 – ne IDC [240]

Historia jacobitarum / Abudacnus, Jos – Lugd Batavorum, 1740 – €12.00 – ne Slangenburg [240]

Historia khalifatus omari 2, jazidi 2 et hischami / ed by Goeje, M J de – Lugduni Batavorum, 1865 – €5.00 – ne Slangenburg [260]

Historia liberal del juego del axedrez / Osuna Lara, Antonio J – Badajoz: Imprenta Diputacion Provincial, 1965 – sp Bibl Santa Ana [920]

Historia literaria do rio grande do sul / Pinto Da Silva, Joao – Porto Alegre, Brazil. 1930 – 1r – us UF Libraries [972]

Historia litteraria : or, an exact and early account of the most valuable books published in the several parts of europe – London. 1730-1734 (1) – mf#5568 – us UMI ProQuest [070]

Historia maior : corpus christi college, cambridge, mss 26 and 16 / Paris, Matthew – 13th c – 2v on 2r – 1 – (1 col reel [ill only] c600) – mf#96769 – uk Microform Academic [931]

Historia manichaeorvm : de fvriosae et pestiferae huius sectae origine et propagatione / Spangenberg, C – Vrsellis, 1578 – 2mf – 9 – mf#TH-1 mf 1416-1417 – ne IDC [242]

Historia media de minas gerais / Vasconcellos, Diogo Luiz De Almeida Pereira De – Rio de Janeiro, Brazil. 1948 – 1r – us UF Libraries [972]

Historia militar de cuba / Cuba Fuerzas Armadas Revolucionias Direccion Pol... – Habana, Cuba. 'folleto 3'. 19– – 1r – us UF Libraries [355]

Historia militar de el salvador / Bustamante, Gregorio – San Salvador, El Salvador. 1951 – 1r – us UF Libraries [355]

La historia militar de espana / Barado, Francisco – 1893 – 9 – sp Bibl Santa Ana [946]

Historia militar do brasil / Barroso, Gustavo – Sao Paulo, Brazil. 1938 – 1r – us UF Libraries [355]

Historia militar do brasil / Vasconcellos, Genserico De – Rio de Janeiro, Brazil. v1-2. 1941 – 1r – us UF Libraries [355]

Historia militar e politica dos portugueses em mocambique / Teixeira Botelho, Jose Justino – Lisboa, Portugal. 1936 – 1r – us UF Libraries [960]

Historia missionorum ordinis fratrum minoris 3. america septentrionalis. roma, 1968 / Barrado Manzano, Arcangel – Madrid: Graf. Calleja, 1969 – 1 – sp Bibl Santa Ana [975]

Historia missionum ordinis fratum minorum 1. asia-centro orientalis et oceania. roma, 1967 / Barrado Manzano, Arcangel – Madrid: Graf. Calleja, 1967 – 1 – (tambien africa) – sp Bibl Santa Ana [240]

Historia moderna de el salvador / Gavidia, Francisco – San Salvador, El Salvador. 1958 – 1r – us UF Libraries [972]

Historia moderna de el salvador / Gavidia, Francisco – San Salvador, El Salvador. v1 pt1-2. 1917-18 – 1r – us UF Libraries [972]

Historia monachorum und historia lausiaca / Reitzenstein, R – Goettingen, 1916 – €12.00 – ne Slangenburg [240]

Historia monasterii amorbacensis ord s benedicti / Gropp, I – Francofurti, 1736 – €40.00 – ne Slangenburg [241]

Historia monasterii s augustini cantuariensis by thomas of elmham (rs8) : formerly monk and treasurer of that foundation / Thomas of Elmham; ed by Hardwick, C – 1858 – (= ser The rolls series (rs)) – €19.00 – ne Slangenburg [241]

Historia monetaria de costa rica / Soley Guell, Tomas – San Jose, Costa Rica. 1926 – 1r – us UF Libraries [972]

Historia moschi : ad normam academiae naturae curiosorum conscripta... / Schroeck, Lucas – Augustae Vindelicorum, Impensis Theophili Goebelii, excudit Johann Jacob Schoenigkius [1682] [mf ed 1979] – 1r [pl] – 1 – mf#film mas 8913 – us Harvard [615]

Historia mvsica : nella quale si ha piena cognitione della teorica, e della pratica antica della mvsica harmonica... / Angelini Bontempi, Giovanni Andrea – Pervgia: Pe'l Costantini 1695 [mf ed 19–] – 6mf – 9 – mf#fiche 373 – us Sibley [780]

Historia natural / Rondon, Candido Mariano Da Silva – Rio de Janeiro, Brazil. 1947 – 1r – us UF Libraries [972]

Historia natural y medica del principado de asturias / Casal, G – Madrid, 1762 – 8mf – 9 – sp Cultura [610]

Historia natural y moral de las aves – SL, SA – 9mf – 9 – sp Cultura [500]

Historia naturalis = [Natural history] / Plinius [Pliny The Elder: Gaius Plinius Secundus] – Venice: Jensen, 1476 – (= ser Holkham library manuscript books 394) – 1 col r – 14 – (ill by jacometto veneziano. ed trans into italian by landino) – mf#C527 – uk Microform Academic [090]

Historia naturalis palmarum / Martius, Friedrich Philipp von – Leipzig, 1831-ca 1850 – 2r – 1 – $125.00 – us UMI ProQuest [580]

Historia ni aladino asin can princesa sa china – [Nueva Caceres: Libreria Mariana 190-?] [mf ed Bloomington IN: Indiana Uni Lib, Preservation Dept 1984] – (= ser Coll...in the bikol language) – 2v on 1r – 1 – (aka: an macagnalasgnalas na nangyari sa buhay ni aladino can pagcua nia gan lampara maravillosa asin an paga-agom nia can princesa sa china) – us Indiana Preservation [490]

Historia ni bertoldino – [Nueva Caceres: Libreria Mariana 190-?] [mf ed Bloomington IN: Indiana Uni Lib, Preservation Dept 1984] – (= ser Coll...in the bikol language) – 1r – 1 – (aka: can salyang aqui na si cacaseno) – us Indiana Preservation [490]

Historia ni princesa adelfa na napag-agom nin sarong cochero sana – [Nueva Caceres?: Libreria Mariana? 190-?] [mf ed Bloomington IN: Indiana Uni Lib, Preservation Dept 1984] – (= ser Coll...in the bikol language) – 1r [ill] – 1 – us Indiana Preservation [490]

Historia ni samuel belibet – [Mandurriao, Iloilo?: Panayana? 191-?] [mf ed Bloomington IN: Indiana Uni Lib, Preservation Dept 1984] – (= ser Coll...in the bisaya language 2) – 1r – 1 – us Indiana Preservation [490]

Historia nin bantog na d rodrigo de villas asin ni dona jimena sa cahadean sa espana – [Nueva Caceres: Libreria Mariana 190-?] [mf ed Bloomington IN: Indiana Uni Lib, Preservation Dept 1984] – (= ser Coll...in the bikol language) – 2v on 1r – 1 – us Indiana Preservation [490]

Historia nin cusganan na samson asin ni dalila traidora – Nueva Caceres: Libreria Mariana 1910 [mf ed Bloomington IN: Indiana Uni Lib, Preservation Dept 1984] – (= ser Coll...in the bikol language) – 1r – 1 – us Indiana Preservation [490]

Historia nin hadeng salomon asin ni reina saba / Ariate, Nicolas – [Nueva Caceres?: Libreria Mariana 190-?] [mf ed Bloomington IN: Indiana Uni Lib, Preservation Dept 1984] – (= ser Coll...in the bikol language) – 1r – 1 – us Indiana Preservation [490]

Historia nova : nella qvale si contengono tutti i successi della guerra turchesca, la congiura del duca de nortsolch contra la regina d'inghilterra... / Manolesso, E M – Padoua, 1572 – 3mf – 9 – mf#H-8330 – ne IDC [956]

Historia numismatica de guatemala / Prober, Kurt – Guatemala, 1957 – 1r – us UF Libraries [929]

Historia oder gschicht : von dem ursprung und fuergang der grossen zwyspaltung... / Lavater, L – Zuerych, Christoffel Froschower, 1564 – 4mf – 9 – mf#PBU-308 – ne IDC [240]

Historia palaestinorvm, tyriorvm et sidoniorvm, populorvm antiqvissimorvn... / Stucki, J W – Tigvri, Ioannes Vvolph, 1595 – 1mf – 9 – mf#PBU-641 – ne IDC [240]

Historia patria / Duarte Level, Line – Caracas, Venezuela. 1911 – 1r – us UF Libraries [972]

Historia patriarcharum alexandrinorum jacobiuarum / Renaudot, E – Parisiis, 1713 – €40.00 – ne Slangenburg [243]

Historia peregrina de un inca andaluz / ed by Bayle, Constantino – Madrid: Razon y Fe, 1927 – 1 – sp Bibl Santa Ana [910]

Historia poetica do brasil / Haddad, Jamil Almansur – Sao Paulo, Brazil. 194- – 1r – us UF Libraries [440]

Historia polemica de graecorum schismate ex ecclesiasticis monumentis concinnata / Cozza, Laurentius – Romae. v1-4. 1719-20 – €262.00 – ne Slangenburg [240]

Historia politica et patriarchica (cshb47) / ed by Bekkerus, Imm – Bonnae, 1849 – (= ser Corpus scriptorum historiae byzantinae (cshb)) – €14.00 – (incl: constantinopoleos epirotica) – ne Slangenburg [243]

Historia privada de los colombianos / Caballero Calderon, Eduardo – Bogota, Colombia. 1960 – 1r – us UF Libraries [972]

Historia quinqu-articularis : or, a declaration of the judgement of the western churches... / Heylyn, Peter – London: printed by E. C. for Thomas Johnson, 1660. Chicago: Dep of Photodup, U of Chicago Lib, 1964 (1r); Evanston: American Theol Lib Assoc, 1984 (1r) – 1 – 0-8370-1475-1 – mf#1984-B013 – us ATLA [240]

Historia reformationis ecclesiarum raeticarum / Porta, P DR – Curiae, Lindaviae, 1772-1777. 2 v – 9mf – 9 – mf#PBU-697 – ne IDC [242]

Historia rei literariae o s b / Ziegelbauer, M; ed by Legipontius, O – Augustae Vindelicorum. v1-4. 1754 – v1 31mf v2 26mf v3 30mf v4 32mf – 9 – €417.00 – ne Slangenburg [240]

Historia rerum a michaele palaeologo (cbh28,1) / Georgii Pachymeris; ed by Possinus, P – Romae, 1666 – (= ser Corpus byzantinae historiae (cbh)) – €52.00 – ne Slangenburg [243]

Historia rerum ab andronico seniore (cbh28,2) / Georgii Pachymeris; ed by Possinus, P – Romae, 1669 – (= ser Corpus byzantinae historiae (cbh)) – €56.00 – ne Slangenburg [243]

Historia rerum anglicarum, bk 5 (rs82/2) : annales furnesienses (1199-1298), a continuation of william of newburgh's history to 1298 – etienne de rouen, draco normannicus / William of Newburg – 1885 – (= ser The rolls series (rs)) – €18.00 – ne Slangenburg [242]

Historia rerum anglicarum, bks 1-4 (rs82/1) / William of Newburg – 1884 – (= ser The rolls series (rs)) – €17.00 – ne Slangenburg [242]

Historia rervm in oriente gestarvm ab exordio mvndi et orbe condito ad nostra haec vsqve tempora – Francofvrti ad Moenvm, 1587 – 12mf – 9 – mf#H-8220 – ne IDC [956]

Historia sang dalagangan nga si bernardo carpio nga anac ni d sancho diaz cag ni d a jimena sa guinharian sa espana – Mandurriao, Iloilo: Panayana 1911 [mf ed Bloomington IN: Indiana Uni Lib, Preservation Dept 1984] – (= ser Coll...in the bisaya language) – 2v on 1r – 1 – (aka: bantug nga historia ni bernardo carpio nga anac ni d a jimena sa espana) – us Indiana Preservation [490]

Historia sang hareng salomon cag ni reina saba sa guinharian sa jerusalem – Mandurriao, Iloilo: Panayana 1913 [mf ed Bloomington IN: Indiana Uni Lib, Preservation Dept 1984] – (= ser Coll...in the bisaya language 2) – 1r – 1 – us Indiana Preservation [490]

Historia sang princesa adelfa nga napangasaua sang isa ca cochero – Mandurriao: Panayana 1909 [mf ed Bloomington IN: Indiana Uni Lib, Preservation Dept 1984] – (= ser Coll...in the bisaya language 2) – 1r – 1 – us Indiana Preservation [490]

Historia scriptoresque alii ad res byzantinas pertinentes (cbh39) / Leonis Diaconi; ed by Hase, C – Parisiis, 1819 – (= ser Corpus byzantinae historiae (cbh)) – €29.00 – ne Slangenburg [243]

Historia secreta da fundacao brasil central / Telles, Carlos – Rio de Janeiro, Brazil. 1946 – 1r – us UF Libraries [972]

Historia secreta del gabinete de napoleon (anno 1811) / Goldsmith, Lewis – Santiago – 1r – 5,6 – sp Cultura [944]

Historia sive notitia episcopatus daventriensis / Lindeborn, J – Coloniae Agrippinae, 1670 – €19.00 – ne Slangenburg [240]

Historia social de chile / Amunategui Solar, Domingo – Madrid: Razon y Fe, 1934 – 1 – sp Bibl Santa Ana [972]

Historia templariorum / Guertler N – ed 2a. Amstelaedami, 1703 – €40.00 – ne Slangenburg [240]

Historia theologica-critica de vita atque doctrina sanctorum patrum / Lumper, G – Augustae Vindelicorum. v1-13. 1783-1799 – 13v on 157mf – 8 – €300.00 – ne Slangenburg [240]

Historia universal...guadalupe / San Jose, Francisco de – 1743 – 9 – sp Bibl Santa Ana [972]

Historia universitatis parisiensis / Bulaeus (du Boulay), Caesar Egassius – Parisiis. v1-6. 1665-1673 – 6v on 219mf – 8 – €418.00 – ne Slangenburg [378]

Historia utriusque belli dacici a traiacaesare gesti...quae in columna eiusdem romae visuntur... / Ciacono, A – Romae, 1616 – 10mf – 9 – mf#O-1082 – ne IDC [700]

Historia verdadera de la conquista de la neuva esp... / Diaz Del Castillo, Bernal – Mexico City? Mexico. 1950 – 1r – us UF Libraries [972]

Historia verdadera de la conquista de la neuva esp... / Diaz Del Castillo, Bernal – Mexico City? Mexico. 1961 – 1r – us UF Libraries [972]

Historia verdadera de la conquista de la nueva espana / Bayle, Constantino & Diaz del Castillo, Bernal – Madrid: Razon y Fe, 1944.-3v – 1 – sp Bibl Santa Ana [946]

Historia verdadera de la conquista de la nueva espana / Diaz del Castillo, Bernal – Madrid: Espasa-Calpe, S.A., 1928.-v1 – 1 – sp Bibl Santa Ana [946]

Historia verdadera de la conquista de la nueva espana / Diaz del Castillo, Bernal – Mexico: Editorial Pedro Robledo, 1944.-v2 – sp Bibl Santa Ana [946]

Historia verdadera de la conquista de la nueva espana / Diaz del Castillo, Bernal – Mexico: oficina Tipografica de la Secretaria de Fomento, 1904.-v1 – sp Bibl Santa Ana [946]

Historia verdadera de la conquista de la nueva espana / Diaz del Castillo, Bernal – Mexico: Porrua, 1969 – 1 – sp Bibl Santa Ana [946]

Historia verdadera de la conquista de la nueva espana : unica edicion hecha segun el codice autografo / Diaz del Castillo, Bernal – Mexico: Oficina Tip. de la Secretaria de Fomento, 1904.-v2 – sp Bibl Santa Ana [946]

Historia verdadera de la conquista de la nueva espana... / Diaz del Castillo, Bernal – Madrid: Espasa-Calpe, 1928.-v2 – sp Bibl Santa Ana [946]

Historia verdadera de la conquista de la nueva espana... / Diaz del Castillo, Bernal – Madrid: Imprenta Talleres de Silverio Aquirre, 1940 – sp Bibl Santa Ana [946]

Historia verdadera de la conquista de nueva espana / Diaz del Castillo, Bernal – Mexico: Editorial Pedro Robledo, 1944.-v1 – sp Bibl Santa Ana [946]

La historia, verdadera ensenanza en los caminos de la rura de espana / Munoz Gallardo, Juan Antonio – Badajoz: Imp. de la Diputacion Prov., 1974. Sep. Rev. Estu. Extremenos – 1 – sp Bibl Santa Ana [946]

Historia vniversale dell' origine, et imperio de' tvrchi... / Sansovino, F – Venetia, 1573 – 11mf – 9 – mf#H-8341 – ne IDC [956]

Historia von doctor johann fausten : historia d. johannis fausti des zauberers / ed by Milchsack, Gustav – Wolfenbuettel: J Zwissler, 1892 [i.e. 1892-97] (mf ed 1990) – 1 – (filmed with: fausto. issued in pts. incl bibl ref) – us Wisconsin U Libr [390]

Historia von doctor johann fausten : leben, thaten und hoellenfahrt des berufenen zauberers und schwarzkuenstlers dr. johann faust – Leipzig: O Wigand, [1842] (mf ed 1990) – 1r – 1 – (filmed with: fausto) – us Wisconsin U Libr [430]

Historia von doctor johann fausten / Saintyves, P – Paris: L'Edition d'Art, 1926 (mf ed 1990) – 1r – 1 – (filmed with: fausto) – us Wisconsin U Libr [390]

Historia von lazaro : aus dem 11. cap. des euangeli s. johannis gezogen / Suteilius, J – Schweinfurt, 1542 – 3mf – 9 – mf#PBA-425 – ne IDC [240]

Historia welcher gestalt sich die osiandrische schwermerey im lande zu preussen erhaben : vnd wie dieselbige verhandelt ist, mit allen actis, beschrieben / Moerlin, J – [Magdeburg, 1554] – 3mf – 9 – mf#TH-1 mf 1177-1179 – ne IDC [242]

Historia westfalia : opus posthumum / Schaten, N S J – Neuhussi, 1690 – €56.00 – ne Slangenburg [943]

Historia y americanidad / Congreso Nacional De Historia, 4th – Habana, Cuba. 1946 – 1r – us UF Libraries [972]

Historia y analisis del sistema contributivo de pu... / Serrano Ramirez, Francisco – Rio Piedras, Puerto Rico. 1948 – 1r – us UF Libraries [972]

Historia y antologia de la literatura costarricens / Bonilla, Abelardo – San Jose, Costa Rica. v1-2. 1957 – 1r – us UF Libraries [972]

Historia y antologia de la literatura venezolana / Diaz Seijas, Pedro – Madrid, Spain. 1955 – 1r – us UF Libraries [440]

Historia y cuadros de costumbres / Groot, Jose Manuel – Bogota, Colombia. 1951 – 1r – us UF Libraries [306]

Historia y destino / Canal Barrachina, Avelino – Habana, Cuba. 1946 – 1r – us UF Libraries [972]

Historia y fantasia / Blanchet, Emilio – Matanzas, Cuba. 1912 – 1r – us UF Libraries [972]

Historia y literatura / Armas Y Cardenas, Jose De – Habana, Cuba. 1915 – 1r – us UF Libraries [972]

Historia y patria / Congreso Nacional De Historia, 6th, Trinidad, Cuba – Habana, Cuba. 1948 – 1r – us UF Libraries [972]

Historia...castellanos en las islas...oceano / Herrera, Antonio de – 1736 – 9 – (decada primera 1601. decada segunda 1601. decada tercera 1601. decada quarta 1601. decada quinta 1728. decada sesta 1736. decada septima 1730) – sp Bibl Santa Ana [972]

Historia...comarca de la serena...cabeza del buey / Perez Jimenez, Nicolas – 1889 – 9 – sp Bibl Santa Ana [946]

Historia...cruz del casar de palomero / Martin Santivanez, Romualdo – 1870 – 9 – sp Bibl Santa Ana [946]

Historia...de los angeles / Guadelupe, Andres – 1662 – 9 – sp Bibl Santa Ana [240]

Historia...de medellin / Solano de Figueroa y Altamirano, Juan – 1650 – 9 – sp Bibl Santa Ana [946]

Historia...del castanar...bejar / Yague, Francisco – 1795 – 9 – sp Bibl Santa Ana [946]

Un historiador moderno de la tierra de la serena (d. nicolas perez jimenez) / Barrantes Moreno, Vicente – Madrid: Fortanet, 1890. B.R.A.H. 17, pp. 481-492 – sp Bibl Santa Ana [946]
Historiae / Herodotus – 15th c – (= ser Holkham library manuscript books 440) – 1r – 1 – (latin trans by laurentius valla) – mf#96613 – uk Microform Academic [900]
Historiae / Ibn-Washih; ed by Houtsma, M Th – Lugduni Batavorum, 1883 – 2pts – (pars 1: historiam ante-islamicam continens €17. pars 2: historiam islamicam continens €21) – ne Slangenburg [260]
Historiae / Orosius, Paulus – 1370 – (= ser Holkham library manuscript books 370) – 1r – 1 – mf#916 – uk Microform Academic [240]
Historiae *see*
– In orationes quasdam ciceronis...
– Livius, books 31-40/dictys...
Historiae aevi carolini *see* Scriptores rerum sangalliensium. annalium et chronicorum aevi carolini continuatio. historiae aevi carolini (mgh5:2.bd)
Historiae aevi salici (mgh5:11.bd) – 1854 – (= ser Monumenta germaniae historica 5: scriptores in folio (mgh5)) – €37.00 – ne Slangenburg [240]
Historiae aevi salici (mgh5:12.bd) – 1856 – (= ser Monumenta germaniae historica 5: scriptores in folio (mgh5)) – €48.00 – ne Slangenburg [240]
Historiae animalium... / Gessner, C – Tiguri: C Froschover, 1551-1587. 5v – 65mf – 9 – mf#Z-2262 – ne IDC [590]
Historiae coelestis britannicae.. / Flamsteed, John – Londini: Typis H. Meere, 1725 – 1 – us Wisconsin U Libr [520]
...Historiae de bello nvper venetis a selimo 2 tvrcarvm imperatore illato, liber vnvs, ex italico sermone in latinum conuersus... / Contarini, G P – Basileae, 1573 – 2mf – 9 – mf#H-8339 – ne IDC [956]
Historiae de rebus hispaniae libri trigunta : accendunt josephi emmanuelis minianae... contimeationis novae libri decem / Mariana, Juan de – Hagea-Comitum: apud Petrum de Hondt. 4v. 1733 [mf ed 1985] – 1r – 1 – (with ind) – mf#SEM35P228 – cn Bibl Nat [946]
Historiae ecclesiasticae novi testamenti / Hottinger, J H – Tiguri, 1665-1667. v6(2); v8(4) – 27mf – 9 – mf#ZWI-38 – ne IDC [240]
Historiae francorum scriptores coaetanei / Duchesne, Andre – Lutetiae Parisiorum: Sumptibus Sebastiani Cramoisy. 5v. 1636-1649 – 3r – 1 – mf#SEM35P265 – cn Bibl Nat [944]
Historiae genuensium libri 12 / Foglietta, U – Genvae, 1585 – 12mf – 9 – mf#H-8419 – ne IDC [956]
Historiae husitarum / Cochlaeus, J – S Victor, 1549 – 16mf – 8 – €31.00 – (j rokyzana: de septem sacramentis; j de przibram: de professione fidei catholicae; j cochlaeus: philippica septima...) – ne Slangenburg [241]
Historiae insectorum libellus qui est de scorpione / Gesner, C – Tiguria, 1587 – 1mf – 9 – sp Cultura [590]
Historiae libri 10 (cshb5) : et liber de velitatione bellica / Leonis Diaconi Caloensis – Bonnae, 1828 – (= ser Corpus scriptorum historiae byzantinae (cshb)) – €23.00 – (nicephori augusti e rec car ben hasii; addita ejusdem versione atque annotationibus ab ipso recognitis; accedunt theodosii acroases, de creta capta e rec fr jacobsii et luitprandi, legatio cum aliis libellis, qui nicephori phocae io tzimiscis historiam illustrant) – ne Slangenburg [243]
...historiae libri tres, ab autore innumeris locis emendati atque expoliti : in qvibvs sarracenorum, turcarum, aliarumque; genitum origines and res per annos septingentos gestae, continentur / Curio, C A – Basileae, 1568 – 6mf – 9 – mf#H-8308 – ne IDC [956]
Historiae mvsvlmanae tvrcorvm, de monvmentis ipsorvm exscriptae, libri 18 / Leunclavius, J – Francofvrti, 1591 – 10mf – 9 – mf#H-8376 – ne IDC [956]
Historiae sacramentariae pars altera / Hospinian, R – Zuerich, Johannes Wolf, 1602 – 10mf – 9 – mf#PBU-414 – ne IDC [240]
Historia...espana...badajoz / Romero Morera, Joaquin – 1878 – 9 – sp Bibl Santa Ana [946]
Historia...framontanos celtiveros / Paredes Guillen, Vicente – 1888 – 9 – sp Bibl Santa Ana [946]
Historia...guadalupe / Gabriel de Talavera, Fray – 1597 – 9 – sp Bibl Santa Ana [946]
Historia...guadalupe / Malagon, Joan – 1672 – 9 – sp Bibl Santa Ana [972]
Historia...guadalupe / Talavera, Fr. Gabriel – 1597 – 9 – sp Bibl Santa Ana [972]
Historia...indiana / Mendieta, Geronimo – 1870 – 9 – sp Bibl Santa Ana [978]
Historia...indias occidentales / Herrera, Antonio de – Tomo I. 1728 – 9 – (tomo 2 1728. tomo 4 1728) – sp Bibl Santa Ana [972]
Historia...indias...mar oceano / Fernandez de Oviedo Valdes, Gonzalo – 1851-53, 1865. 4v – 9 – sp Bibl Santa Ana [972]
Historia...indie...occidentali / Lopez de Gomara, Francisco – 1564 – 9 – sp Bibl Santa Ana [972]
Historia...japon / Orfanel, Jacinto – 1633 – 9 – sp Bibl Santa Ana [950]
Historial de cuba / Rousset, Ricardo V – Habana, Cuba. v1-3. 1918 – 1r – us UF Libraries [972]
Historial de cucuta / Ortega Ricaurte, Enrique – Bogota, Colombia. 1956 – 1r – us UF Libraries [972]
Historial de fistas y donativos, indice de caballeros y reglamento de uniformidad de la real maestranza de caballeria de sevilla, por don pedro leon y manjon / T'Serclaes, Duque de – Madrid: Fortanet, 1910. B.R.A.H. 56, 1910, pp. 437-439 – sp Bibl Santa Ana [390]
Historial genealogico del libertador / Sucre, Luis Alberto – Caracas, 2nd ed 1930; Madrid: Razon y Fe, 1932 – 1 – sp Bibl Santa Ana [920]
Historia...mejico...hernando cortes / Prescott, Guillermo H – v1-2. 1844 – 9 – (v1-3 1847 ed. v4 1850 ed) – sp Bibl Santa Ana [972]
Historia...mejico...nueva espana / Solis Rivadeneyra, Antonio de – 1766 – 9 – (1851, 1843, 1885) – sp Bibl Santa Ana [972]
Historia...merida / Moreno de Vargas, Bernabe – 1633 – 9 – (1892 ed) – sp Bibl Santa Ana [946]
Historia...montachez / Lozano Rubio, Tirso – 1894 – 9 – sp Bibl Santa Ana [900]
Historian : a journal of history – Allentown. 1938+ (1) 1970+ (5) 1976+ (9) – ISSN: 0018-2370 – mf#6056 – us UMI ProQuest [900]
Historiang totoo can buhay ni santa ana na ina ni santa maria – [Nueva Caceres?: Libreria Mariana? 190-?] [mf ed Bloomington IN: Indiana Uni Lib, Preservation Dept 1984] – (= ser Coll...in the bikol language) – 1r – 1 – us Indiana Preservation [490]
Historians in tropical africa : proceedings of the leverhulme inter-collegiate history conference, september 1930 – Salisbury, Southern Rhodesia: The College, 1962 – us CRL [960]
The historians of the church of york and its archbishops (rs71) / ed by Raine, J – (= ser The rolls series (rs)) – (v1 1879 €21. v2 1886 €19. v3 1894 €17) – ne Slangenburg [241]
Historia...peru...incas / Lasso de la Vega, Garcia – Tomos I-XII. 1800 – 9 – (tomo 13 1801) – sp Bibl Santa Ana [972]
Historiarum indicarum libri 16 / Maffei, G P – Oxford. 1952+ (1) 1982+ (5) 1982+ (9) – 16mf – 9 – mf#1372 – ne IDC [910]
Historiarum libri 4 (cbh17) / Ioannis Cantacuzeni; ed by Pontanus, J – Parisiis, 1645 – (= ser Corpus byzantinae historiae (cbh)) – €95.00 – ne Slangenburg [243]
Historiarum libri 5 (cshb1) : cum versione latina et annotationibus b vulcani / Agathiae Myrinaei – Bonnae, 1828 – (= ser Corpus scriptorum historiae byzantinae (cshb)) – €168.00 – (b g niebuhrius graeca recensuit. accedunt agathiae epigrammata) – ne Slangenburg [243]
Historiarum libri 8 (cbh2,1) / Theophylacti Simocattae; ed by Pontanus, J – Parisiis, 1648 – (= ser Corpus byzantinae historiae (cbh)) – €25.00 – ne Slangenburg [243]
Historiarum libri 8 (cshb22) / Theophylacti Simocattae; ed by Bekkerus, Imm – Bonnae, 1834 – (= ser Corpus scriptorum historiae byzantinae (cshb)) – €21.00 – (incl: genesius rec c lachmannus) – ne Slangenburg [243]
Historiarum libri decem de rebus turcicis (cbh16) / Laonici Chalcocondylae – Parisiis, 1650 – (= ser Corpus byzantinae historiae (cbh)) – €44.00 – ne Slangenburg [243]
Historiarum quae supersunt (cshb14) / Dexippi et al; ed by Bekkeri, Imm & Niebuhrii, B G – Bonnae, 1829 – (= ser Corpus scriptorum historiae byzantinae (cshb)) – €23.00 – (accedunt eclogae photii ex olympiodoro, candido, nonnoso et theophane, et procopii sophistae panegyricus, graece et latine, prisciani panegyricus, annotationes h valesii, labbei et villoisonis, et indices classeni) – ne Slangenburg [243]
Historiarum sui temporis libri 8 (cbh3,1) / Procopii Caesariensis; ed by Maltret, Cl – Parisiis, 1662 – (= ser Corpus byzantinae historiae (cbh)) – €56.00 – ne Slangenburg [243]
Historias brasileiras / Brahe, Tycho – Rio de Janeiro, Brazil. 1931 – 1r – us UF Libraries [972]
Historias brazileiras / Taunay, Alfredo D'escragnolle Taunay – Rio de Janeiro, Brazil. 1874 – 1r – us UF Libraries [972]
Historias da amazonia / Peregrino, Joao – Rio de Janeiro, Brazil. 1936 – 1r – us UF Libraries [972]
Historias da revolucoes em mato-grosso / Menodnca, Rubens De – Goiania, Brazil. 1970 – 1r – us UF Libraries [972]
Historias de merida / Madrazo, Pedro de – Madrid: Tip. Fortanet, 1895 – 1 – sp Bibl Santa Ana [946]
Historias de piratas / Perez Valenzuela, Pedro – Guatemala, 1936 – 1r – us UF Libraries [972]
Historias de tata mundo / Dobles, Fabian – San Jose, Costa Rica. 1955 – 1r – us UF Libraries [972]
Historias de venezuela / Cela, Camilo Jose – Barcelona, Spain. 1955 – 1r – us UF Libraries [972]
Las historias del origen de las indias de esta provincia de guatemala / ed by Bayle, Constantino – Madrid: Razon y Fe, 1926 – 1 – sp Bibl Santa Ana [972]
Historias infantiles – 1960. 100p – 1 – $5.00 – us Southern Baptist [242]
Las historias y los historiadores de sevilla / Perez de Guzman, Juan – 1892 – 9 – sp Bibl Santa Ana [946]
Historias y paizagens / Arinos De Melo Franco, Afonso – Rio de Janeiro, Brazil. 1921 – 1r – us UF Libraries [972]
Historias...merida / Fernandez y Perez, Gregorio – 1893 – 9 – (1857 ed) – sp Bibl Santa Ana [946]
Historic american buildings survey : a vast collection of images and documents – 2pts. 1980-88 [mf ed Chadwyck-Healey] – 4287mf – 9 – (pt1: 1933-79 [1567mf] pt2: 1980-88 [2720mf]. also available by state. coll consists of photos, text & drawings describing nearly 35,000 historically significant sites and structures) – uk Chadwyck [720]
Historic american engineering record (haer) : photographs of historically significant sites – [mf ed Chadwyck-Healey] – 870mf – 9 – (contains over 24,000 photographs and more than 20,000p documenting 1827 structures throughout america) – uk Chadwyck [620]
Historic americans / Parker, Theodore – 2nd ed. Boston: Horace B Fuller, 1871, c1870 [mf ed 1992] – 1mf – 9 – 0-524-02985-7 – (incl bibl ref) – mf#1990-0772 – us ATLA [975]
Historic and municipal documents, ireland, ad 1172-1320 (rs53) : from the archives of the city of dublin, etc / ed by Gilbert, John Th – 1870 – (= ser The rolls series (rs)) – €21.00 – ne Slangenburg [931]
Historic aspects of the priori argument concerning the being and atrributes of god : being four lectures delivered in edinburgh in nov 1884... / Cazenove, John Gibson – London: Macmillan 1886 [mf ed 1985] – 1mf – 9 – 0-8370-2614-8 – (incl ind) – mf#1985-0614 – us ATLA [210]
A historic banner : a paper read on february 8th, 1896 / FitzGibbon, Mary Agnes – Toronto: W Briggs, 1896? – 1mf – 9 – mf#07115 – cn Canadiana [355]
Historic buildings and gardens of great britain and ireland – (mf ed 1999) – 7r – 14 – £595.00 – mf#HHG – uk World [720]
Historic buildings in britain : inventories of the royal commission on ancient and historic monuments and constructions, england/ scotland/wales / Great Britain. Royal Commissions on Ancient and Historical Monuments and Constructions – pre-1714 [mf ed Chadwyck-Healey] – 333mf – 9 – (england 179mf: [buckinghamshire 14mf; cambridgeshire 10mf; city of cambridge 13mf; dorset 33mf; essex 24mf; herefordshire 18mf; hertfordshire 1910 5mf; huntingdonshire 1926 6mf; london 25mf; middlesex 1937 5mf; city of oxford 1939 6mf; westmoreland 1936 6mf; city of york 14mf]. scotland 102mf: [argyll 11mf; county of berwick 1915 (rev iss) 4mf; caithness 1911 4mf; county of dumfries 5mf; east lothian 1924 4mf; city of edinburgh 1951 6mf; fife, kinross & clackmannan 1933 7mf; galloway 10mf; midlothian & west lothian 1929 5mf; orkney & shetland 11mf; outer hebrides, skye & the small isles 1928 5mf; peeblesshire 7mf; roxburghshire 8mf; selkirkshire 1957 3mf; stirlingshire 9mf; sutherland 1911 3mf]. wales 52mf) – uk Chadwyck [720]
The historic christ in the faith of to-day / Grist, William Alexander – New York: Fleming H Revell, c1911 – 2mf – 9 – 0-7905-0259-3 – (incl bibl ref and index) – mf#1987-0259 – us ATLA [240]
The historic church : an essay on the conception of the christian church and its ministry in the sub-apostolic age / Durell, John Carlyon Vavasor – Cambridge: University Press, 1906 – 1mf – 9 – 0-524-03896-1 – (incl bibl ref) – mf#1990-1155 – us ATLA [240]
Historic churches of america / Wallington, Nellie Urner – New York: Duffield 1907 [mf ed 1990] – 1mf – 9 – 0-7905-8119-1 – (int by edward everett hale) – mf#1988-6081 – us ATLA [240]
Historic crimes and criminals / Finger, Charles Joseph – Girard, Kansas: Halderman-Julius Company, (c1922) – 1 – us Wisconsin U Libr [360]
Historic decorations at the pan-presbyterian council : a photographic souvenir of the ecclesiastical seals, symbols... used in the decorations of horticultural hall...philadelphia, a d, 1880 – Philadelphia, PA: Presbyterian Pub Co, c1880 [mf ed 1993] – (= ser Presbyterian coll) – 1mf – 9 – 0-524-07206-X – mf#1990-5364 – us ATLA [242]
Historic denver news – 1977 feb [v7 n2]-1985 dec – 1r – 1 – mf#984773 – us WHS [978]
Historic devises, badges, and war-cries / Palliser, F B – London: Sampson Low & Son & Marston, 1870 – 6mf – 9 – mf#O-814 – ne IDC [929]
Historic dress of the clergy / Tyack, George Smith – London: W Andrews, [1897?] – 1mf – 9 – 0-7905-8162-0 – mf#1988-6109 – us ATLA [240]
The historic episcopate : an essay on the four articles of church unity proposed by the american house of bishops and the lambeth conference / Shields, Charles Woodruff – New York: Scribner, 1894 – 1mf – 9 – 0-7905-6441-6 – mf#1988-2441 – us ATLA [240]
The historic episcopate / Thompson, Robert Ellis – Philadelphia: Westminster, 1910 – 1mf – 9 – 0-7905-6127-1 – mf#1988-2127 – us ATLA [240]
The historic episcopate in the columban church and in the diocese of moray : with other scottish ecclesiastical annals / Archibald, John – Edinburgh: St Giles, 1893 – 1mf – 9 – 0-524-02517-7 – mf#1990-0617 – us ATLA [240]
The historic evidence of the authorship and transmission of the books of the new testament : a lecture / Tregelles, Samuel Prideaux – 2nd ed. London:Samuel Bagster, 1881 – 1mf – 9 – 0-8370-5567-9 – mf#1985-3567 – us ATLA [225]
The historic exodus / Toffteen, Olaf Alfred – Chicago: Oriental Society of the Western Theological Seminary, 1909 – 1mf – 9 – 0-8370-5547-4 – (incl ind) – mf#1985-3547 – us ATLA [220]
The historic faith : short lectures on the apostles' creed / Westcott, Brooke Foss – 4th ed. London; New York: Macmillan, 1890 – 1mf – 9 – 0-7905-9754-3 – mf#1989-1479 – us ATLA [240]
The historic garrison at annapolis royal, n s : some of its early history / Gilmore, Andrew – Yarmouth, NS?: "Light" Office, 1898 – 1mf – 9 – mf#34953 – cn Canadiana [355]
Historic handbook of the northern tour : lakes george and champlain, niagara, montreal, quebec / Parkman, Francis – Boston: Little, Brown, 1885 – 3mf – 9 – mf#11640 – cn Canadiana [971]
Historic hawaii – 1983 feb-1987 dec – 1r – 1 – (cont: historic hawaii news) – mf#1582439 – us WHS [978]
Historic hawaii news – v4 n3-v19 n1 [1978 mar-1983 jan] – 1r – 1 – (continued by: historic hawaii) – mf#2739670 – us WHS [978]
Historic homes of the south-west mountains, virginia / Mead, Edward Campbell – Philadelphia, PA. 1899 – 1r – us UF Libraries [025]
Historic Illinois / Illinois Historic Preservation Agency – 1978 jun [v1 n1]-1985 dec – 1r – 1 – mf#970922 – us WHS [978]
Historic jamaica / Cundall, Frank – London, England. 1915 – 1r – us UF Libraries [972]
The historic jesus : being the elliott lectures / Smith, David – New York: Hodder and Stoughton, [19–?] – 1mf – 9 – 0-7905-0337-9 – (incl bibl ref and index) – mf#1987-0337 – us ATLA [240]
The historic jesus : a study of the synoptic gospels / Lester, Charles Stanley – New York: G P Putnam, 1912 – 1mf – 9 – 0-524-05618-8 – (incl bibl ref) – mf#1992-0473 – us ATLA [220]
Historic Landmarks Foundation of Indiana *see* Indiana preservationist
Historic landmarks of the deccan / Haig, T W – Allahabad: Pioneer Press, 1907 – (= ser Samp: indian books) – us CRL [915]
Historic madison newsletter – 1974 feb-1978 mar – 1r – 1 – mf#390658 – us WHS [978]
Historic manual of the reformed church in the united states / Dubbs, Joseph Henry – Lancaster, Pa.: [s.n.], 1885 (Lancaster: Inquirer) – 1mf – 9 – 0-7905-4630-2 – mf#1988-0630 – us ATLA [240]
The historic martyrs of the primitive church / Mason, Arthur James – London; New York: Longmans, Green, 1905 – 1mf – 9 – 0-7905-2176-8 – (incl ind) – mf#1987-2176 – us ATLA [240]
The historic medals of canada : a paper read before the literary and historical society of quebec, april 9, 1873 / Sandham, Alfred – Quebec?: Middleton & Dawson, 1873 – 1mf – 9 – mf#57523 – cn Canadiana [730]
Historic milwaukee news – 1982 may-1987 fall – 1r – 1 – (cont: historic walker's point news; cont by: historic milwaukee incorporated news) – mf#1278864 – us WHS [978]

Historic notes on the books of the old and new testaments / Sharpe, Samuel – 4th ed. London: Elliot Stock, 1907 – 1mf – 9 – 0-7905-0382-4 – mf#1987-0382 – us ATLA [220]

Historic origin of the bible : a handbook of principal facts from the best recent authorities, german and english / Bissell, Edwin Cone – new ed. New York: Anson D F Randolph, c1889 – 1mf – 9 – 0-8370-2352-1 – (incl app. incl subject ind and ind of biblical passages cited) – mf#1985-0352 – us ATLA [220]

Historic ornament treatise on decorative art and architectural ornament / Ward, James – London 1897 – (= ser 19th c art & architecture) – 10mf – 9 – mf#4.2.1100 – uk Chadwyck [740]

The historic policy of the united states as to annexation : a paper read before the american historical association, at chicago, july 13, 1893 / Baldwin, Simeon Eben – [New Haven, CT?: s.n.], 1893 [mf ed 1981] – 1mf – 9 – (incl bibl ref; repr fr: "yale review" for august, 1893) – mf#10191 – cn Canadiana [975]

Historic preservation – Washington. 1949-1996 (1) 1972-1996 (5) 1976-1996 (9) – (cont by: preservation) – ISSN: 0018-2419 – mf#8056 – us UMI ProQuest [900]

Historic Preservation Fund of North Carolina et al *see* North carolina preservation

Historic preservation news – Washington. 1990-1995 (1) 1990-1995 (5) 1990-1995 (9) – (cont: preservation news) – ISSN: 1065-3562 – mf#8057,01 – us UMI ProQuest [900]

Historic records of the fifth new york cavalry, first ira harris guard : its organization, marches, raids, scouts, engagements and general services, during the rebellion of 1861-1865 / Beaudry, Louis Napoleon – Albany, NY: J Munsell, 1868 – 5mf – 9 – mf#24531 – cn Canadiana [355]

Historic roll of wales, vol 17 – 1898-1904 – 3mf – 9 – £3.75 – uk Glamorgan FHS [941]

Historic sketch of the evangelical lutheran synod of northern illinois / Heilman, Lee M – Philadelphia, PA: Lutheran Publ Soc, 1892 – 1mf – 9 – 0-524-04661-1 – mf#1990-5057 – us ATLA [242]

Historic sketch of the reformed church in north carolina / ed by Clapp, Jacob Crawford et al – Philadelphia, PA: Publ Board of the Reformed Church in the United States, c1908 – 1mf – 9 – 0-524-02733-1 – mf#1990-4408 – us ATLA [242]

Historic sketches of free methodism / Kirsop, Joseph – London: Andrew Crombie, 1885 – 1mf – 9 – 0-524-06954-9 – mf#1990-5318 – us ATLA [242]

Historic society of lancashire and cheshire – v1-124. 1848-1972 – (= ser Publications of the english record societies, 1835-1972) – 423mf – 9 – uk Chadwyck [941]

Historic studies in vaud, berne, and savoy; from roman times to voltaire, rousseau, and gibbon / Read, John Meredith – With illus. London: Chatto & Windus, 1897. 2v.31 plates – 1 – us Wisconsin U Libr [900]

The historic styles of ornament / Dolmetsch, H – London 1898 – (= ser 19th c art & architecture) – 7mf – 9 – mf#4.1.443 – uk Chadwyck [740]

Historic tales of old quebec / Gale, George – [Quebec?: Telegraph Print Co], 1920 – 4mf – 9 – 0-665-71457-2 – (incl ind) – mf#71457 – cn Canadiana [971]

The historic times – Lawrence, KS. v1 n1 jul 11 1891-v1 n19 nov 14 1891 [mf ed 1947] – (= ser Negro Newspapers on Microfilm) – 1r – 1 – us L of C Photodup [071]

A historic view of the new testament : the jowett lectures delivered at the passmore edwards settlement in london, 1901 / Gardner, Percy – London: Adam & Charles Black, 1901 [mf ed 1985] – (= ser Jowett lectures 1901) – 1mf – 9 – 0-8370-3232-6 – mf#1985-1232 – us ATLA [225]

Historic walker's point news – 1974 spring-1981 may – 1r – 1 – (continued by: historic milwaukee news) – mf#601093 – us WHS [978]

Historic webster : a newsletter of the webster historical society, inc / Webster Historical Society (NC) – v1 n1-v6 n1 [1974 feb-1979/80 winter], v7 n2-v10 n2 [1981 spring-1984 summer], v10 n4-v11 n4 [1984/85 winter-1985/86 winter], v12 n1-2 [1987 spring-summer] – 1r – 1 – mf#1568626 – us WHS [978]

Historic women, 'my children shall not suffer' / Love, Phena Hudnell – s.l, s.l? 193-? – 1r – us UF Libraries [978]

Historica et critica introductio in u t libros sacros. vol 1 : introductio generalis, sive, de u t canonis, textus, interpretationis historia / Cornely, Rudolph – rev enl ed. Parisiis: P Lethielleux 1894 [mf ed 1991] – (= ser Cursus scripturae sacrae 1) – 8mf – 9 – 0-8370-1925-7 – (incl bibl ref) – mf#1987-6312 – us ATLA [221]

Historica et critica introductio in u t libros sacros. vol 2, 1 : introductio specialis in historicos veteris testamenti libros / Cornely, Rudolph – rev ed. Parisiis: P Lethielleux 1897 [mf ed 1991] – (= ser Cursus scripturae sacrae 2/1) – 5mf – 9 – 0-8370-1926-5 – (incl bibl ref) – mf#1987-6313 – us ATLA [221]

Historica et critica introductio in u t libros sacros. vol 2, 2 : introductio specialis in didacticos et propheticos vet. test. libros / Cornely, Rudolph – rev ed. Parisiis: P Lethielleux 1897 [mf ed 1991] – (= ser Cursus scripturae sacrae 2/2) – 6mf – 9 – 0-8370-1927-3 – (incl bibl ref) – mf#1987-6314 – us ATLA [221]

Historica et critica introductio in u t libros sacros. vol 3 : introductio specialis in singulos novi testamenti libros / Cornely, Rudolph – rev ed. Paris: Sumptibus P Lethielleux 1897 [mf ed 1992] – (= ser Cursus scripturae sacrae 3) – 8mf – 9 – 0-524-03878-3 – (incl bibl ref) – mf#1987-6491 – us ATLA [225]

Historica monumenta ordinis s hieronymi congregationis b petri de pisis / Sajanello, J-B – ed 2a. Venetiis. v1-3. 1758-62 – €155.00 – ne Slangenburg [240]

Historica narratio profectionis et inaugurationis serenissimorum belgii principum alberti et isabellae, austriae archiducum / Bochius, J – Antverpiae: Ex officina Plantiniana, apud Ioannem Moretum, 1602 – 14mf – 9 – mf#O-1 – ne IDC [090]

Historica narratio profectionis et inaugurationis serenissimorum belgii principum alberti et isabellae, austriae archiducum : et eorum optatissimi in belgium adventus, rerumque gestarum et memorabilium, gratulationum, apparatuum, et spectaculorum in ipsorum susceptione et inauguratione hactenus editorum accurata descriptio / Bochius, J – Antverpiae: Ex officina Plantiniana, apud Ioannem Moretum, 1602 – 14mf – 9 – mf#O-161 – ne IDC [700]

Historicae relationis continvatio / Francus, I – N p, 1593 – 2mf – 9 – mf#H-8377 – ne IDC [956]

Historical : martin county / Lyons, Isabel J – s.l, s.l? 1936 – 1r – us UF Libraries [978]

Historical : no 200 deland / Davis, Mary Irene – s.l, s.l? 1936 – 1r – us UF Libraries [978]

Historical / Shepherd, Rose – s.l, s.l? 1936 – 1r – us UF Libraries [978]

An historical account and delineation of aberdeen / Wilson, Robert – Aberdeen 1822 [mf ed Hildesheim 1995-98] – 1v on 2mf – 9 – €60.00 – 3-487-27865-0 – gw Olms [941]

A historical account of christ church, boston : an address, delivered on the one hundred and fiftieth anniversary of the opening of the church, december 29th, 1873 / Burroughs, Henry – Boston: A Williams, 1874 [mf ed 1980] – 1mf – 9 – mf#02008 – cn Canadiana [720]

An historical account of covenanting in scotland : from the first band in mearns, 1556, to the signature of the grand national covenant, 1638 / Aikman, James – Edinburgh: J Henderson, 1848 [mf ed 1992] – 2mf – 9 – 0-524-04666-2 – mf#1990-1293 – us ATLA [242]

An historical account of cumner : with some particulars of the traditions respecting the death of the countess of leicester; also en extract from ashmole's antiquities of berkshire, relative to that transaction and illustrative of the romance of keniworth / Tighe, Hugh U – Oxford 1821 [mf ed Hildesheim 1995-98] – 1v on 1mf – 9 – €40.00 – 3-487-27914-2 – gw Olms [941]

Historical account of discoveries and travels in africa / Leyden, John – Edinburgh 1817 [mf ed Hildesheim 1995-98] – 2v on 8mf – 9 – €160.00 – 3-487-27389-6 – gw Olms [916]

Historical account of discoveries and travels in asia : from the earliest ages to the present time / Murray, H – Edinburgh, London: Archibald Constable and Co. 3v. 1820 – 19mf – 9 – mf#HT-675 – ne IDC [915]

Historical account of discoveries and travels in asia : from the earliest ages to the present time / Murray, Hugh – Edinburgh 1820 [mf ed Hildesheim 1995-98] – 3v on 12mf – 9 – €120.00 – 3-487-27655-0 – gw Olms [915]

Historical account of discoveries and travels in north america : including the united states, canada, the shores of the polar sea, and the voyages in search of a north-west passage; with observations on emigration / Murray, Hugh – London 1829 [mf ed Hildesheim 1995-98] – 2v on 8mf – 9 – €160.00 – 3-487-27047-1 – gw Olms [917]

A historical account of his majesty's visit to scotland / Mudie, Robert – Edinburgh 1822 [mf ed Hildesheim 1995-98] – 1v on 3mf – 9 – €90.00 – 3-487-27854-5 – gw Olms [914]

An historical account of kenilworth castle in the county of warwick : being an historical introduction to the readers of the new novel, entitled, kenilworth, by the author of waverley, ivanhoe, etc / Nightingale, Joseph – London 1821 [mf ed Hildesheim 1995-98] – 1v on 1mf – 9 – €40.00 – 3-487-27915-0 – gw Olms [941]

Historical account of some of the more important versions and editions of the bible / Darling, Charles William – New York, 1894. 173p – 1 – us Wisconsin U Libr [240]

An historical account of the british army and of the law military... / Samuel, E – London, William Clowes, 1816 – 8mf – 9 – $12.00 – mf#LLMC 89-022 – us LLMC [355]

An historical account of the embassy to the emperor of china : undertaken by order of the king of great britain; including the manners and customs of the inhabitants / Staunton, George [comp] – London: John Stockdale, 1797 [mf ed 1995] – (= ser Yale coll) – xv/475p (ill) – 1 – 0-524-10273-2 – (abr principally fr papers of earl macartney) – mf#1996-1273 – us ATLA [915]

Historical account of the most celebrated voyages, travels, and discoveries : from the time of columbus to the present period / Mavor, William – London 1796-1801 [mf ed Hildesheim 1995-98] – 25v on 51mf – 9 – €510.00 – 3-487-29938-0 – gw Olms [910]

An historical account of the rise and development of presbyterianism in scotland / Balfour, Alexander Hugh Bruce – Cambridge: University Press; New York: G P Putnam [dist] 1911 [mf ed 1989] – (= ser The cambridge manuals of science and literature) – 1mf – 9 – 0-7905-4320-6 – (incl bibl ref) – mf#1988-0320 – us ATLA [242]

Historical account of the rise and progress of the secession / Brown, John – Edinburgh, Scotland. 1819 – 1r – us UF Libraries [240]

Historical account of the separation of victoria from new south wales / Lang, John Dunmore – Sydney, 1870 – (= ser 19th c british colonization) – 1mf – 9 – mf#1.1.3489 – uk Chadwyck [980]

Historical account of the trinidad and tobago poli... / Ottley, Carlton Robert – Port-of-Spain, Trinidad and Tobago. 1964 – 1r – us UF Libraries [972]

Historical account of the work of the american committee of revision of the authorized english version of the bible : prepared from the documents and correspondence of the committee / American Revision Committee – New York: Charles Scribner 1885 [mf ed 1985] – 1mf – 9 – 0-8370-2088-3 – mf#1985-0088 – us ATLA [220]

The Historical almanac and daily remembrancer for the year... – Ottawa: J Hope, [18–] – 9 – mf#A02448 – cn Canadiana [030]

An historical analysis of national collegiate athletic association freshman eligibility / Holzman, Lynn M – 1997 – 1mf – 9 – $4.00 – mf#PE 3759 – us Kinesology [790]

Historical analysis of religious education in malawi with a focus on specific institutions / Spencer, Stephen T – Stellenbosch: U of Stellenbosch 1998 [mf ed 1998] – 9mf – 9 – mf#mf.1275 – sa Stellenbosch [377]

Historical and analytical programme of pianoforte recital : to be given by mr o a king, pianist to h r h princess louise / Harrison, J W F – S.l: s.n, 1880? – 1mf – 9 – mf#25527 – cn Canadiana [790]

An historical and archaeological sketch of the city of goa : preceded by a short statistical account of the territory of goa / Fonseca, Jose Nicolau da – Bombay: Thacker, 1878 [mf ed 1995] – (= ser Yale coll) – xi/332p (ill) – 1 – 0-524-09821-2 – mf#1995-0821 – us ATLA [954]

An historical and architectural essay relating to redcliffe church, bristol / Britton, John – London 1813 – (= ser 19th c art & architecture) – 1mf – 9 – mf#4.2.1379 – uk Chadwyck [720]

Historical and biographical record of the cattle industry and the cattlemen of texas and adjacent territory / Cox, James – 1 – us Southern Baptist [920]

Historical and biographical sketches from borthwick's gazetteer of montreal / Borthwick, John Douglas – Montreal?: s.n, 189-? – 1mf – 9 – mf#25669 – cn Canadiana [920]

Historical and biographical sketches, kentucky, 1882-1888 – With particular coverage of Bourbon, Christian, Fayette, Harrison, Scott, Todd and Trigg Counties, with an index of many more by Bailey F. Davis. Perrin ed – 1 – us Southern Baptist [242]

Historical and biographical sketches of my mother lodge of freemasons / Willox, David – Glasgow: Andrew Holmes & Co 1922 – 2mf – 9 – mf#vrl-113 – ne IDC [366]

Historical and biographical works of john strype / Strype, John – Oxford: Clarendon Press, 1821-1840 – 7r – 1 – 0-8370-1290-2 – mf#1984-S016 – us ATLA [240]

Historical and contemporary review of bench and bar in california. march, 1926 / The Recorder. San Francisco – San Francisco, The Recorder 1926. 61p. LL-1681 – 1 – us L of C Photodup [340]

Historical and descriptive account of british india : from the most remote period to the present time / Murray, Hugh – Edinburgh 1832 [mf ed Hildesheim 1995-98] – 3v on 15mf – 9 – €150.00 – 3-487-27494-9 – gw Olms [954]

An historical and descriptive account of china : its ancient and modern history, language, literature, religion, government, industry, manners and social state... / Murray, H et al – Edinburgh, London: Oliver & Boyd, 1836. 3v – 14mf – 9 – mf#HT-543 – ne IDC [951]

An historical and descriptive account of persia : from the earliest ages to the present time, with a detailed view of its resources, government, population, natural history...including a description of afghanistan and beloochistan / Fraser, James B – Edinburgh 1834 [mf ed Hildesheim 1995-98] – 1v on 5mf – 9 – €100.00 – 3-487-27497-3 – gw Olms [956]

Historical and descriptive account of the island of cape breton : and of its memorials of the french regime / Bourinot, John George – Toronto: Copp, Clark, 1895 – 3mf – 9 – (with bibl, historical and critical notes. incl bibl ref) – mf#32552 – cn Canadiana [971]

Historical and descriptive accounts...of...english cathedrals / Britton, John – London 1836 – (= ser 19th c art & architecture) – 25mf – 9 – mf#4.2.1377 – uk Chadwyck [720]

A historical and descriptive narrative of twenty years' residence in south america : containing travels in arauco, chile, peru, and colombia; with an account of the revolution, its rise, progress, and results / Stevenson, William B – London 1825 [mf ed Hildesheim 1995-98] – 3v on 9mf – 9 – €180.00 – 3-487-26911-2 – gw Olms [918]

Historical and descriptive notes on ornament : with illustrative sketches / Glazier, Richard – [Manchester], London: John Heywood, 1887 – (= ser 19th c art & architecture) – 1mf – 9 – mf#4.1.149 – uk Chadwyck [730]

Historical and descriptive notice on the church of notre-dame of montreal / Vekeman, Gustave – Montreal: E Senecal, 1897 – 1mf – 9 – mf#16938 – cn Canadiana [720]

A historical and descriptive sketch of the county of welland in the province of ontario, in the dominion of canada : containing a succinct account of the various municipalities / Cruikshank, Ernest Alexander – Welland Ont: County Council, 1886 – 1mf – 9 – mf#03963 – cn Canadiana [971]

Historical and descriptive sketches of the maritime colonies of british america / MacGregor, John – London 1828 – (= ser 19th c british colonization) – 3mf – 9 – mf#1.1.6322 – uk Chadwyck [975]

Historical and descriptive sketches of the maritime colonies of british america / Macgregor, John – London 1828 [mf ed Hildesheim 1995-98] – 1v on 2mf – 9 – €60.00 – 3-487-27104-4 – gw Olms [975]

Historical and economic studies / ed by Karve, D G – Poona: Ferguson College, 1941 – (= ser Samp: indian books) – us CRL [330]

Historical and Genealogical Association of Mississi *see* Family trails

An historical and geographical description of formosa : an island subject to the emperor of japan. giving an account of the religion, customs, manners, etc of the inhabitants / Psalmanaazaar, George – London: Dan Brown, 1704 – 5mf – 9 – mf#HT-665 – ne IDC [951]

Historical and geographical dictionary of japan : with 300 illustrations, 18 appendixes and several maps = Dictionnaire d'histoire et de geographie au japon / Papinot, Edmond – Tokyo: Librairie Sansaisha [1910] [mf ed 1995] – (= ser Yale coll) – xiv/842p (ill) – 1 – 0-524-09969-3 – (trans fr french) – mf#1995-0969 – us ATLA [059]

An historical and geographical memoir of the north-american continent : its nations and tribes, with a summary account of his life, writings, and opinions / Gordon, James B – Dublin 1820 [mf ed Hildesheim 1995-98] – 1v on 5mf – 9 – €100.00 – 3-487-26725-X – gw Olms [975]

An historical and geographical memoir of the north-american continent, its nations and tribes / Gordon, James Bentley – Dublin: printed for John Jones, 1820 [mf ed 1971] – 1r – 1 – mf#SEM35P65 – cn Bibl Nat [970]

Historical and literary memorials of presbyterianism in ireland (1623-1731) / Witherow, Thomas – London: W Mullan, 1879 – 4mf – 9 – 0-524-08824-1 – (incl bibl ref and ind) – mf#1993-3316 – us ATLA [242]

An historical and practical guide to art illustration / Hodson, James Shirley – London 1884 – (= ser 19th c art & architecture) – 4mf – 9 – mf#4.2.131 – uk Chadwyck [740]

The historical and scientific society of manitoba : inaugural address / Bell, Charles Napier – S.l: s.n, 1889? – 1mf – 9 – (in double clms) – mf#57911 – cn Canadiana [971]

A historical and statistical account of new-brunswick, bna : with advice to emigrants / Atkinson, Christopher William – Edinburgh?: s.n, 1844 (Edinburgh: Anderson & Bryce) – 4mf – 9 – mf#28599 – cn Canadiana [917]

An historical and statistical account of nova-scotia / Haliburton, Thomas C – Halifax 1829 [mf ed Hildesheim 1995-98] – 2v on 6mf – 9 – €120.00 – 3-487-27135-4 – gw Olms [971]

A historical and statistical report of the presbyterian church in canada : in connection with the church of scotland for the year 1866 / Croil, James – Montreal?: J Lovell, 1868 – 2mf – 9 – mf#32121 – cn Canadiana [242]

Historical and statistical sketch of the schools controlled by the catholic school commission of montreal – Montreal: [s.n.], 1915 – 3mf – 9 – 0-665-71952-3 – mf#71952 – cn Canadiana [377]

A historical and topographical essay upon the islands of corfu, leucadia, cephalonia, ithaca, and zante : with remarks upon the character, manners, and customs of the ionian greeks / Goodison, William – London 1822 [mf ed Hildesheim 1995-98] – (= ser Fbc) – 2mf – 9 – €60.00 – 3-487-29056-1 – gw Olms [914]

Historical aspects of the immigration problem : select documents / Abbott, Edith – Chicago, IL: The University of Chicago Press, 1926 [mf ed 1970] – (= ser University of chicago social service series; Library of american civilization 14812) – 1mf – 9 – us Chicago U Pr [320]

Historical Association (Great Britain) *see* History

Historical atlas and chronology of the life of jesus christ : a text book and companion to a harmony of the gospels / Hodge, Richard Morse – Wytheville VA: D A St Clair Press 1899 [mf ed 1985] – 1mf – 9 – 0-8370-3609-7 – mf#1985-1609 – us ATLA [220]

Historical barometer in miami / Braman, Sidney T – s.l, s.l? 193-? – 1r – us UF Libraries [978]

The historical bases of religions : primitive, babylonian and jewish / Brown, Hiram Chellis – Boston: Herbert B Turner 1906 [mf ed 1991] – 1mf – 9 – 0-524-00698-9 – mf#1990-2026 – us ATLA [230]

The historical books / Gigot, Francis Ernest – 2nd rev ed. New York: Benziger Bros, c1901 – 1mf – 9 – 0-524-05980-2 – mf#1992-0717 – us ATLA [220]

Historical books, joshua to esther : with a brief commentary / Davey, William Harrison et al – London: SPCK, 1884 – (= ser Old testament according to the authorized version) – 3mf – 9 – 0-524-05394-4 – mf#1992-0404 – us ATLA [221]

The historical books of the holy scriptures : judges, ruth, 1 and 2... / Jamieson, Robert – Philadelphia: William S & Alfred Martien, 1860 – 1mf – 9 – 0-8370-3764-6 – (with critical and explanatory commentary) – mf#1985-1764 – us ATLA [220]

The historical books of the old testament / Kenrick, Francis Patrick – Baltimore: Kelly, Hedian & Piet, 1860 – 9mf – 9 – 0-8370-1953-2 – mf#1987-6340 – us ATLA [221]

Historical buildings : fort myers, lee county, fla / Hanson, W Stanley – s.l, s.l? 1936 – 1r – us UF Libraries [720]

Historical catalog and history / Georgetown College – 1829-1920 – 1 – 8.33 – us Southern Baptist [242]

Historical catalogue of the printed editions of holy scripture in the library of the british and foreign bible society / Darlow, Thomas Herbert & Moule, Horace Frederick – London: Bible House, 1903-1911 – 6mf – 9 – 0-8370-1825-0 – (incl bibl ref) – mf#1987-6213 – us ATLA [012]

The historical character of st john's gospel : three lectures / Robinson, Joseph Armitage – London, New York: Longmans, Green, 1908 – 1mf – 9 – 0-8370-7330-8 – mf#1986-1330 – us ATLA [226]

Historical chart of english literature for use in schools and colleges *see* Essays on german literature

An historical chart of german literature for use in schools and colleges / Greene, Nelson Lewis – 2nd ed. Chicago IL: Educational Screen c1923 [mf ed 1993] – 1r – 1 – (filmed with: saggi critici / giovanni vittorio amoretti & other titles) – mf#3160p – us Wisconsin U Libr [430]

The historical christ : or, an investigation of the views of mr j m robertson, dr a drews and prof w b smith / Conybeare, Frederick Cornwallis – Chicago: iss for the Rationalist Press Assoc Ltd, [by] Open Court, 1914 [mf ed 1989] – 1mf – 9 – 0-7905-0691-2 – (incl ind) – mf#1987-0691 – us ATLA [240]

Historical christianity : the religion of human life / Strong, Thomas Banks – London; New York: H Frowde, 1902 – 1mf – 9 – 0-7905-7473-X – mf#1989-0698 – us ATLA [240]

Historical church atlas / McClure, Edmund – London: SPCK; New York: E & J B Young 1897 [mf ed 1991] – 2mf – 9 – 0-524-01004-8 – mf#1990-0281 – us ATLA [240]

Historical (churches) : saint johns protestant epi... / Shepherd, Rose – s.l, s.l? 1936 – 1r – us UF Libraries [978]

Historical collections / Michigan State Historical Society – Lansing. v1-39. 1874-1915 – 1 – $600.00 – mf#0361 – us Brook [978]

Historical collections . . . american colonial church / Perry, William Stevens – Hartford: s.n., 1870 – 2r – 1 – 0-8370-1491-3 – mf#1984-B044 – us ATLA [240]

Historical collections of the essex institute / Essex Institute – v1 – 1r – 1 – (continued by: essex institute historical collections) – mf#2520047 – us WHS [978]

Historical commentaries on the state of christianity : during the first three hundred and twenty-five years from the christian era... = De rebus christianorum ante constantinum magnum commentarii / Mosheim, Johann Lorenz; ed by Murdock, James – New York: S Converse, 1852 [mf ed 1992] – 2v on 3mf – 9 – 0-524-03354-4 – (trans fr latin, v1 by robert studley vidal, v2 by ed) – mf#1990-0935 – us ATLA [240]

A historical commentary on st paul's epistle to the galatians / Ramsay, William Mitchell – New York: G P Putnam, 1900 [mf ed 1986] – 2mf – 9 – 0-8370-9812-2 – mf#1986-3812 – us ATLA [227]

The historical connection of the jewish people with palestine – Jerusalem, 1946 – 1mf – 9 – mf#J-28-146 – ne IDC [956]

Historical criticism and the old testament = La methode historique / Lagrange, Marie-Joseph – London: Catholic Truth Society, 1905 [mf ed 1989] – 1mf – 9 – 0-7905-2006-0 – (english by edward myers) – mf#1987-2006 – us ATLA [221]

Historical data : daytona beach, florida / Converse, Mildred – s.l, s.l? 1936 – 1r – us UF Libraries [978]

Historical data : fort george island – s.l, s.l? 193-? – 1r – us UF Libraries [978]

Historical data : old spanish mission / Dozier, H C – s.l, s.l? 193-? – 1r – us UF Libraries [978]

The historical deluge : in its relation to scientific discovery and to present questions / Dawson, John William – New York: Fleming H Revell, [1895] – 1mf – 9 – 0-8370-2852-3 – (incl app) – mf#1985-0852 – us ATLA [220]

Historical description of puerto rico / Moresi, Juana – Friend, NE. 1949 – 1r – us UF Libraries [972]

Historical development of christianity in the political and social li... / Hamilton, James – Edinburgh, Scotland. 18– – 1r – us UF Libraries [240]

The historical development of religion in china / Clennell, Walter James – New York: EP Dutton, 1917 – 1mf – 9 – 0-524-01956-8 – mf#1990-2747 – us ATLA [290]

Historical development of scoring for the wind ensemble / Gauldin, Robert – U of Rochester 1958 [mf ed 19–] – 4mf – 9 – mf#fiche 471 – us Sibley [780]

Historical development of speculative philosophy from kant to hegel = Historische entwickelung der speculativen philosophie von kant bis hegel / Chalybaeus, Heinrich Moritz – Edinburgh: T & T Clark, 1854 – 1mf – 9 – 0-7905-3930-6 – (in english) – mf#1989-0423 – us ATLA [110]

The historical development of the quran / Sell, Edward – Madras: Printed at the SPCK Press, 1898 – 1mf – 9 – 0-524-01381-0 – (incl bibl ref) – mf#1990-2393 – us ATLA [260]

Historical dictionary of guatemala by... / Moore, Richard E – Madrid: Graf. Calleja, 1968 – 1 – sp Bibl Santa Ana [972]

A historical discourse : delivered in the first reformed protestant dutch church of tarrytown, ny, may 13 1866 / Stewart, Abel T – New York: Anson DF Randolph [1866?] [mf ed 1993] – 1mf – 9 – 0-524-08689-3 – mf#1993-3214 – us ATLA [242]

An historical discourse : delivered in the central baptist meeting house, newport, ri...jan 7th, 1847 / Jackson, Henry – Newport, RI: Cranston & Norman, 1854 [mf ed 1993] – (= ser Baptist coll) – 1mf – 9 – 0-524-07202-7 – mf#1990-5360 – us ATLA [242]

An historical discourse on the 50th anniversary of the first baptist church in worcester, mass : dec 9th 1862 / Davis, Isaac – Worcester: Henry J Howland, [1863?] [mf ed 1993] – (= ser Baptist coll) – 1mf – 9 – 0-524-07195-0 – (with app) – mf#1990-5353 – us ATLA [242]

An historical discourse on the civil and religious affairs of the colony of rhode-island / Callender, John – 1838 – 1 – $9.66 – us Southern Baptist [230]

A historical documentation, an instructional manual and an annotated bibliography of selected folk dances of puerto rico / Figueroa-Cruz, Blas E & Jacobson, Phyllis C – 1990 – 4mf – $16.00 – us Kinesology [790]

Historical documents advocating christian union : epoch-making statements by leaders among the disciples of christ for the restoration of the christianity of the new testament – Chicago: Christian Century, 1904 – 1mf – 9 – 0-7905-6278-2 – mf#1988-2278 – us ATLA [240]

An historical enquiry concerning the attempt to raise a regiment of slaves by rhode island during the war of the revolution : with several tables prepared by jeremiah olney / Rider, Sidney S – Providence: S S Rider, 1880 – (filmed with: [baird, h c] washington u. jackson uber die neger als soldaten) – us CRL [976]

Historical enquiry respecting the performance on the harp in the highlands of scotland / Gunn, John – Edinburgh: A Constable 1807 [mf ed 19–] – 3mf – 9 – mf#fiche 725 – us Sibley [780]

An historical essay on architecture by the late thomas hope : illustrated from drawings made by him in italy and germany / Hope, Thomas – London: John Murray, 1835 – (= ser 19th c art & architecture) – 2v on 9mf – 9 – mf#4.1.124 – uk Chadwyck [720]

An historical essay on the magna charta of king john / Thomson, Richard – London, Major, 1829. 612 p. LL-130 – 1 – us L of C Photodup [340]

Historical essays : first published in 1902 in commemoration of the jubilee of the owens college, manchester / ed by Tout, Thomas Frederick & Tait, James – Manchester: University Press, 1907 – (= ser University of Manchester Publications) – 2mf – 9 – 0-7905-8277-5 – (incl bibl ref) – mf#1988-6155 – us ATLA [900]

Historical essays / Lightfoot, Joseph Barber – London; New York: Macmillan, 1895 – 1mf – 9 – 0-7905-5172-1 – mf#1988-1172 – us ATLA [941]

Historical essays and reviews / Creighton, Mandell; ed by Creighton, Louise – London, New York: Longmans, Green, 1902 [mf ed 1989] – 1mf – 9 – 0-7905-4452-0 – mf#1988-0452l – us ATLA [900]

Historical essays and studies / Acton (of Aldenham), John Emerich Edward Dalberg, 1st Baron; ed by Figgis, John Neville & Laurence, Reginald Vere – London, Toronto: Macmillan, 1907 – 6mf – 9 – 0-665-66726-4 – mf#66726 – cn Canadiana [900]

Historical essays on the worship of god and the ministry of the gospel of our lord and saviour : on the early christian church, a d 50-150, on the apostle paul and the gentile churches / Kimber, Thomas – New York: Taber 1889 [mf ed 1992] – 1mf – 9 – 0-524-03097-9 – (incl bibl ref) – mf#1990-0822 – us ATLA [240]

Historical evidence / George, Hereford Brooke – Oxford: Clarendon Press, 1909 – 1mf – 9 – 0-7905-5272-8 – mf#1988-1272 – us ATLA [900]

Historical evidence for the apostolical institution of episcopacy / Russell, M – Edinburgh, Scotland. 1830 – 1r – us UF Libraries [240]

Historical evidence of the new testament : an inductive study in christian evidences / Bowman, Shadrach Laycock – Cincinnati: Jennings & Pye; New York: Eaton & Mains, c1903 [mf ed 1989] – 2mf – 9 – 0-8370-1176-0 – (incl bibl ref) – mf#1987-6012 – us ATLA [225]

Historical evidences of the new testament / Maclear, George Frederick et al – New York: American Tract Society, [1895?] – 1mf – 9 – 0-8370-3594-5 – (incl bibl ref) – mf#1985-1594 – us ATLA [225]

The historical evidences of the truth of the scripture records stated anew : with special reference to the doubts and discoveries of modern times: in eight lectures / Rawlinson, George – Boston: Gould and Lincoln, 1860 – 2mf – 9 – 0-7905-0148-1 – mf#1987-0148 – us ATLA [220]

The historical evidences of the truth of the scripture records, stated anew, with special reference to the doubts and discoveries of modern times / Rawlinson, George – 1873. Eight Oxford lectures, 1859. From the London ed, 1873 – 1 – us Wisconsin U Libr [240]

A historical examination of some non-markan elements in luke / Parsons, Ernest William – Chicago, IL: University of Chicago Press, 1914 [mf ed 1989] – (= ser Historical and linguistic studies in literature related to the new testament. second series, linguistic and exegetical studies 2/6) – 1mf – 9 – 0-7905-1556-3 – (incl bibl ref) – mf#1987-1556 – us ATLA [225]

An historical exposition of the book of daniel the prophet / Rule, William Harris – London: Seeley, Jackson & Halliday, 1869 [mf ed 1985] – 1mf – 9 – 0-8370-4995-4 – mf#1985-2995 – us ATLA [221]

Historical family library : devoted to the republication of standard history – Oxford, 1835-1841 [1,5,9] – 1mf – 9 – us UMI ProQuest [900]

Historical files of the american expeditionary forces in siberia, 1918-1920 / U.S. Army – (= ser Records of united states army overseas operations and commands, 1898-1942) – 11r – 1 – (with printed guide) – mf#M917 – us Nat Archives [355]

Historical files of the american expeditionary forces, north russia, 1918-1919 / U.S. Army. American Expeditionary Forces – (= ser Records of the american expeditionary forces (world war 1), 1917-1923) – 2r – 1 – (with printed guide) – mf#M924 – us Nat Archives [355]

Historical footprints in america / Wilson, Daniel – S.l: s.n, 1864? – 1mf – 9 – (incl bibl ref) – mf#63187 – cn Canadiana [970]

Historical gazetteer of london before the great fire : pt 1: cheapside / Keene, Derek & Harding, Vanessa – 1066-1666 [mf ed Chadwyck-Healey] – 50mf – 9 – (with ind & maps. the cheapside gazetteer provides detailed histories from five parishes in london) – uk Chadwyck [914]

Historical genealogy of the lawrence family, 1635-1858 / Lawrence, Thomas – 1858 – 1 – $50.00 – us Presbyterian [920]

An historical, geographical, political and natural history of north america : and of the british and other european settlements, the united states, the general state of the laws, particularly those affecting commerce... – London, 1805 – (= ser 19th c british colonization) – 9mf – 9 – mf#1.1.7397 – uk Chadwyck [975]

The historical geography of asia minor / Ramsay, W M – London, 1890. v4 – 10mf – 9 – mf#G-155 – ne IDC [915]

The historical geography of asia minor / Ramsay, William Mitchell, Sir – London: John Murray, 1890 – (= ser Supplementary Papers (Royal Geographical Society (Great Britain))) – 2mf – 9 – 0-7905-0589-4 – (incl bibl ref and ind) – mf#1987-0589 – us ATLA [900]

Historical geography of bible lands / Calkin, John Burgess – Halifax, NS: A & W MacKinlay, 1905 – 3mf – 9 – 0-665-71564-1 – (int by by robert a falconer. incl bibl ref and ind) – mf#71564 – cn Canadiana [220]

Historical geography of south africa / Pollock, Norman Charles – London, England. 1963 – 1r – us UF Libraries [960]

Historical geography of st kitts and nevis... / Merrill, Gordon Clark – Mexico City? Mexico. 1958 – 1r – us UF Libraries [972]

An historical geography of the bible / Coleman, Lyman – new ed. Philadelphia: E H Butler, 1850, c1849 [mf ed 1989] – 2mf – 9 – 0-7905-1084-7 – (incl ind) – mf#1987-1084 – us ATLA [220]

A historical geography of the holy lands / Smith, George Adam – 1894 – 9 – $27.00 – us IRC [915]

Historical grounds of the lambeth judgment explained / Tomlinson, J T – London, England. 189- – 1r – us UF Libraries [240]

An historical guide to ancient and modern dublin / Wright, George N – London 1821 [mf ed Hildesheim 1995-98] – 1v on 3mf – 9 – €90.00 – 3-487-27896-0 – gw Olms [914]

Historical handbook and guide to the city and university of oxford / Moore, Jeames J – Oxford, England. 1871 – 1r – us UF Libraries [025]

Historical hand-book of the reformed church in the united states / Good, James Isaac – 2nd ed. Philadelphia: Heidelberg, 1901 – 1mf – 9 – 0-7905-6747-4 – mf#1988-2747 – us ATLA [242]

Historical highlights of volusia county / Fitzgerald, Thomas Edward – Daytona Beach, FL. 1939 – 1r – us UF Libraries [978]

Historical illustrations of the old testament / Rawlinson, George – London: Christian Evidence Committee of the SPCK, [1871?] – 1mf – 9 – 0-7905-0149-X – (incl bibl ref) – mf#1987-0149 – us ATLA [221]

Historical, industrial, and commercial data of mia... / Decroix, F W – St Augustine, FL. 1911? – 1r – us UF Libraries [978]

Historical information relating to military posts and other installations, ca 1700-1900 / U.S. War Dept. Adjutant General's Office – (= ser Records of the adjutant general's office, 1780's-1917) – 8r – 1 – (with printed guide) – mf#M661 – us Nat Archives [355]

An historical inquiry into the principal circumstances and events relative to the late emperor napoleonin : which are investigated the charges brought against the government and conduct of that eminent individual / Monteney, Thomas J de – London 1824 [mf ed Hildesheim 1995-98] – 1v on 4mf – 9 – €120.00 – 3-487-26229-0 – gw Olms [941]

An historical introduction to modern psychology / Murphy, Gardner – 4th ed. London: K Paul, Trench, Trubner & Co Ltd; New York: Harcourt, Brace & Co 1938 [1932] [mf ed 1986] – (= ser International library of psychology, philosophy, and scientific method) – 1r – 1 – (with suppl by heinrich kluever; first publ 1928. filmed with: essays aesthetical and philosophical / schiller, j c f) – mf#1733 – us Wisconsin U Libr [150]

An historical introduction to the marprelate tracts : a chapter in the evolution of religious and civil liberty in england / Pierce, William – London: A Constable, 1908 [mf ed 1990] – 1mf – 9 – 0-7905-5668-5 – mf#1988-1668 – us ATLA [242]

A historical introduction to the study of the books of the new testament : being an expansion of lectures / Salmon, George – 7th ed. London: John Murray, 1894 [mf ed 1988] – 2mf – 9 – 0-7905-0324-7 – (incl bibl ref and ind) – mf#1987-0324 – us ATLA [225]

Historical introductions to the rolls series / Stubbs, William; ed by Hassall, Arthur – London: Longmans, Green, 1902 – 2mf – 9 – 0-524-04974-2 – (incl bibl ref) – mf#1990-1377 – us ATLA [941]

A historical investigation into black parental involvement in the primary and secondary educational situation / Kafu, Hazel Bukiwe – Uni of South Africa 2000 [mf ed Johannesburg 2000] – 4mf [ill] – 9 – (incl bibl ref) – mf#mfm14966 – sa Unisa [370]

The historical jesus of nazareth / Schlesinger, Max – New York: Charles P Somerby, 1876 – 1mf – 9 – 0-8370-5093-6 – mf#1985-3093 – us ATLA [240]

Historical journal – Cambridge. 1958+ (1) 1971+ (5) 1976+ (9) – ISSN: 0018-246X – mf#1362 – us UMI ProQuest [900]

Historical journal of massachusetts – Westfield. 1981+ (1,5,9) – ISSN: 0276-8313 – mf#13040,01 – us UMI ProQuest [978]

An historical journal of the campaigns in north-america : for the years 1757, 1758, 1759, and 1760; containing the most remarkable occurences of that period / Knox, John – London: W Johnston & J Dodsley, 1769 [mf ed 1988] – 10mf – 9 – mf#SEM105P884 – cn Bibl Nat [970]

An historical journal of the transactions at port jackson, and norfolk island : including the journals of governors phillip and king, since the publication of phillip's voyage; with an abridged account of the new discoveries in the south seas / Hunter, John – London 1793 [mf ed Hildesheim 1995-98] – 1v on 4mf – 9 – €120.00 – 3-487-26766-7 – gw Olms [980]

Historical jurisprudence : an introduction to the systematic study of the development of law / Lee, Guy C – New York: The Macmillan Co, 1900 – 6mf – 9 – $9.00 – mf#LLMC 95-152 – us LLMC [340]

Historical labor day 1898 souvenir : official programme – Toronto: Allied Print. Trades Council, 1898? – 1mf – 9 – mf#01072 – cn Canadiana [331]

Historical lecture on teinds or tithes / Fleming, Alexander – Glasgow, Scotland. 1835 – 1r – us UF Libraries [240]

Historical lectures and addresses / Creighton, Mandell – London, New York: Longmans, Green, 1903 [mf ed 1989] – 1mf – 9 – 0-7905-4389-3 – mf#1988-0389 – us ATLA [240]

Historical lectures and essays / Kingsley, Charles – London: Macmillan, 1880 – 1mf – 9 – 0-7905-6481-5 – mf#1988-2481 – us ATLA [900]

Historical lectures on the life of our lord jesus christ : with notes, critical, historical, and explanatory / Ellicott, Charles John – Andover: Warren F Draper, 1881, c1861 – 1mf – 9 – 0-8370-3051-X – (incl indof subjects and biblical citations) – mf#1985-1051 – us ATLA [240]

Historical lectures to non-catholics / O'Connor, Joseph V – [S.l.]: Thomas P. Consedine, c1898 – 1mf – 9 – 0-8370-8284-6 – mf#1986-2284 – us ATLA [241]

Historical legal periodical series – 9 – $5,328.00 set – (incl a coll of 42 periodicals in english mostly from the early 1800's which are no longer being publ. inquire for titles) – mf#408750 – us Hein [340]

The historical, literary, theological, and miscellaneous repository *see* The nova-scotia and new-brunswick magazine

Historical magazine : and notes and queries concerning the antiquities, history, and biography of america – Boston. 1857-1874 (1) – mf#3895 – us UMI ProQuest [975]

Historical magazine of the protestant episcopal church – Austin. 1932-1986 (1) 1971-1986 (5) 1976-1986 (9) – (cont by: anglican and episcopal history) – ISSN: 0018-2486 – mf#442 – us UMI ProQuest [242]

Historical (mandarin) / Shepherd, Rose – s.l, s.l? 193-? – 1r – us UF Libraries [978]

Historical materials / Georgia Baptist Associations – 1 – 5.00 – us Southern Baptist [242]

Historical materials on baptists and other evangelicals in soviet russia and other eastern european countries – 1 – (books and booklets by evangelical, orthodox and soviet authors. 21,116p. union congresses and other union materials. 791p. historical papers. 616p. theses. 1037p. periodicals. 92,020p) – us Southern Baptist [242]

An historical memoir on the qutb, delhi / Page, James Alfred – Calcutta: Govt of India, Central Publ Branch, 1926 – (= ser Samp: indian books) – us CRL [954]

Historical memorials of westminster abbey / Stanley, Arthur Penrhyn – First American from the sixth London ed, with author's final revisions. New York: Anson D.F. Randolph, 1887. 3v. illus – 1 – us Wisconsin U Libr [941]

Historical memorials relating to the independents, or congregationalists : from their rise to the restoration of the monarchy, a.d. 1660 / Hanbury, Benjamin – London: Printed for the Congregational Union of England and Wales [by] Fisher, Son, 1839-1844 – 5mf – 9 – 0-524-03156-8 – mf#1990-4605 – us ATLA [242]

Historical methods – Washington. 1978+ (1) 1978+ (5) 1978+ (9) – (cont: historical methods newsletter) – ISSN: 0161-5440 – mf#6633,01 – us UMI ProQuest [900]

Historical methods newsletter – Pittsburgh. 1967-1977 (1) 1971-1977 (5) 1976-1977 (9) – (cont by: historical methods) – ISSN: 0018-2494 – mf#6633 – us UMI ProQuest [900]

Historical monograph : prisoner of war operations division, office of the provost marshal general / U.S. Prisoner of War Operations Division – v. 1-4. 1945-46 – 1 – 93.00 – us L of C Photodup [360]

Historical motor scrapbook – v1-3 – 1r – 1 – mf#3319699 – us WHS [620]

The historical new testament : being the literature of the new testament arranged in the order of its literary growth and according to the dates of the documents: a new translation / ed by Moffatt, James – Edinburgh: T & T Clark, 1901 – 2mf – 9 – 0-8370-9404-6 – (incl bibl ref and indexes) – mf#1986-3404 – us ATLA [225]

Historical news / Adams County Historical Society (NE) – v1 n1-v19 n1 [1968 apr-1986 jan] – 1r – 1 – mf#618173 – us WHS [978]

Historical newspapers from western ukraine – Lviv: Stefanyk Library: 1848 early 1940s [mf ed Norman Ross Publ] – 15 titles on 26r – 1 – us UMI ProQuest [077]

Historical note : woman's work in the church / Charteris, Archibald Hamilton – [New York: Scribner's; Edinburgh: T. & T. Clark, 1888] Beltsville, Md: NCR Corp, 1978 (1mf); Evanston: American Theol Lib Assoc, 1984 (1mf) – (= ser Women & the church in america) – 9 – 0-8370-0736-4 – (incl bibl ref) – mf#1984-2054 – us ATLA [240]

Historical notes / La Crosse County Historical Society – La Crosse Wl. v1 n1-v10 n2 [1974 feb-1987 may] – 1r – 1 – mf#850173 – us WHS [978]

Historical notes / DeSaussure, Wilmot Gibbes – c1885-c1897 [mf ed 1981] – 5mf – 9 – (with ind & notes) – mf#51-541 – us South Carolina Historical [976]

Historical notes on certain emirates and tribes : printed by order of his excellency, the governor / ed by Burdon, J A – London: Waterlow, 1909 – 1 – us CRL [960]

Historical notes on english catholic missions / Kelly, Bernard William – London: Kegan Paul, Trench, Truebner; St Louis, MO: Herder, 1907 – 2mf – 9 – 0-8370-7162-3 – mf#1986-1162 – us ATLA [241]

Historical notes on the archbishop's judgment (in read and others v. the lord bishop of lincoln) : particularly in reference to mr j t tomlinson's pamphlet / Wordsworth, Christopher – London: Longmans, Green, 1891 [mf ed 1993] – 1mf – 9 – 0-524-05885-7 – mf#1990-5179 – us ATLA [242]

Historical notes on the catholic mission of wairiki, taveuni / Roman Catholic Mission, Fiji – 1922 – 1r – 1 – mf#pmb440 – at Pacific Mss [241]

Historical notes on the employment of negroes in the american army of the revolution / Moore, George H – New York: C T Evans, 1862 – (filmed with: [baird, h c,] washington u. jackson uber die neger als soldaten) – us CRL [976]

Historical notes on the services of the irish officers in the french army / Dillon, Arthur – Dublin, [1890?] – (= ser 19th c ireland) – 1mf – 9 – mf#1.1.8568 – uk Chadwyck [355]

Historical notes on the tractarian movement, a.d. 1833-1845 / Oakeley, Frederick – London: Longman, Green, Longman, Roberts & Green, 1865 – 1mf – 9 – 0-7905-6606-0 – mf#1988-2606 – us ATLA [240]

Historical notes respecting the indians of north america : with remarks on the attempts made to convert and civilise them / Halkett, John – London, 1825 – (= ser 19th c books on british colonization) – 5mf – 9 – mf#1.1.9320 – uk Chadwyck [970]

Historical notes respecting the indians of north america : with remarks on the attempts made to convert and civilize them / Halkett, John – London 1825 [mf ed Hildesheim 1995-98] – 1v on 3mf – 9 – €90.00 – 3-487-27127-3 – gw Olms [307]

Historical notice of penal laws against catholics... / Madden, Richard Robert – London: Thomas Richardson & Son, 1865 – 3mf – 9 – $4.50 – mf#LLMC 91-089 – us LLMC [340]

Historical notice of penal laws against roman catholics : their operation and relaxation during the past century, of partial measures of relief in 1779, 1782, 1793, 1829, and of penal laws which remain unrepealed, or have been rendered more stringent by the latest so-called emancipation act / Madden, Richard Robert – London: T. Richardson, 1865 – 1mf – 9 – 0-7905-5061-X – mf#1988-1061 – us ATLA [241]

Historical notice of penal laws against roman catholics : their operation and relaxation during the past century, of patrial measures of relief in 1779, 1782,... / Madden, Richard Robert – London: T. Richardson, 1865 – 1mf – us ATLA [241]

Historical notices of the missions of the church of england in the north american colonies : previous to the independence of the united states / Hawkins, Ernest – London: B Fellowes, 1845 – 2mf – 9 – 0-7905-8036-5 – mf#1988-6017 – us ATLA [241]

Historical observations on grand tartary : extracted from the memoirs of the p gerbillon / Gerbillon, J F – London, 1741. v4 – 2mf – 9 – mf#HT-510 – ne IDC [915]

Historical outlines of the presbyterian church in missouri : a discourse. prepared at the request of the synod of missouri, and delivered at the annual meeting in springfield, mo... / Hill, Timothy – Kansas City, MO: Stated Clerk, 1871 – 1mf – 9 – 0-524-07199-3 – mf#1990-5357 – us ATLA [242]

Historical outlines of the rise and establishment of the papal power / Card, Henry – Margate, 1804 – (= ser 19th c ireland) – 2mf – 9 – mf#1.1.9246 – uk Chadwyck [241]

Historical overview of the national baseball library / Armitage, Maria T – 1996 – 1mf – 9 – $4.00 – mf#PE 4032 – us Kinesology [790]

Historical pageant of tallahassee / Long, Reinette Gamble – Tallahassee, FL. 192-? – 1r – us UF Libraries [978]

Historical papers and addresses of the lancaster county historical society / Lancaster County Historical Society (PA) – v27-32 [1923-1928] – 1r – 1 – (continued by: journal of the lancaster county historical society) – mf#1112894 – us WHS [978]

Historical papers and letters from the northern registers (rs61) / ed by Raine, J – 1873 – (= ser The rolls series (rs)) – €18.00 – ne Slangenburg [931]

Historical papers concerning the ashtabula baptist association for ninety years, 1817-1907 : prepared for the ninetieth anniversary, held in geneva, ohio, september 4 and 5, 1907 / ed by Leonard, George E – [S.l.]: Committee of Publication, [1907?] – 1mf – 9 – 0-524-07984-6 – mf#1990-5429 – us ATLA [240]

Historical papers on shelter island / Mallmann, Jacob E – 1899 – 1 – $50.00 – us Presbyterian [240]

Historical personal interview / Clark, John – s.l, s.l? 1936 – 1r – us UF Libraries [978]

A historical perspective on title 7 bilingual education projects in hawai'i : compendium of promising practices / Pablo, Josephine Dicsen et al – [Honolulu, Hawai'i]: Pacific Resources for Educ & Learning; [Washington DC]: US Dept of Education, Office of Educ Research & Improvement...[2000] [mf ed 2000] – 1mf – 9 – (incl bibl ref) – us GPO [350]

Historical philosophy in france and french belgium and switzerland / Flint, Robert – Edinburgh: W Blackwood, 1893 – 2mf – 9 – 0-7905-9377-7 – mf#1989-2602 – us ATLA [100]

The historical poetry of the ancient hebrews / Heilprin, Michael – New York: D Appleton. 2v. 1879-80 – 2mf – 9 – 0-7905-0132-5 – (incl bibl ref) – mf#1987-0132 – us ATLA [470]

The historical position of the episcopal church : a paper / Hall, Francis Joseph – Milwaukee, WI: Young Churchman, 1895 – 1mf – 9 – 0-8370-9870-X – mf#1986-3870 – us ATLA [241]

Historical presentation of augustinism and pelagianism : from the original sources = Versuch einer pragmatischen darstellung des augustinismus und pelagianismus / Wiggers, Gustav Friedrich – Andover: Gould, Newman & Saxton, 1840 [mf ed 1990] – 1mf – 9 – 0-7905-7264-8 – (english trans by ralph emerson; with notes & additions) – mf#1988-3264 – us ATLA [240]

Historical preservation center newsletter – South Dakota. 1978 oct-1981 apr – 1r – 1 – (cont: historical preservation newsletter) – mf#665555 – us WHS [978]

Historical prints in the british museum / British Museum. Dept of Prints & Drawings – 204mf – 9 – $1380.00 – 0-907006-49-3 – (over 10,000 prints arr in chronological order of event or scene depicted; fully captioned and with museum accession number) – uk Mindata [760]

Historical project, e-5-e-5c / U.S. Army. Signal Corps – Washington, D.C.. 1946. 4 v. illus., charts, maps, photos. By Pauline M. Oakes – 1 – us L of C Photodup [977]

Historical quarterly of the bicentennial council of the thirteen original states fund, inc / Bicentennial Council of the Thirteen Original States Fund – v2 n1-v2 n7 [1978 fall-1980 sum] – 1r – 1 – (cont: newsletter of the great american achievements program; cont by: historical quarterly of the council of the thirteen original states, inc) – mf#657158 – us WHS [975]

Historical quarterly of the council of the thirteen original states, inc / Council of the Thirteen Original States – v2 n8-v3 n2 [1981 winter-oct 1981?] – 1r – 1 – (cont: historical quarterly of the bicentennial council of the thirteen original states fund, inc; cont by: newsletter of the great american achievements program [1982]) – mf#657162 – us WHS [975]

Historical record – v1 [1886] – 2r – 1 – (continued by: historical record of wyoming valley) – mf#4339380 – us WHS [978]

An historical record of the light horse volunteers of london and westminster : with the muster rolls from the first formation of the regiment, 1779 to the relodgement of the standards in the tower 1829 / Collyer, James N – London: Wright [1843] [mf ed 1989] – 1r – 1 – (filmed with: nux elegia / wartens, s & other titles) – mf#2674 – us Wisconsin U Libr [355]

Historical record of the thirty-sixth or the herefordshire regiment of foot : containing an account of the formation of the regiment in 1701 and of its subsequent services to 1852 / Cannon, Richard – London: G E Eyre & W Spottiswoode: 1853 [mf ed 1984] – 2mf (ill) – 9 – 0-665-32313-1 – (original iss in ser: historical records of the british army) – mf#32313 – cn Canadiana [355]

Historical records 1927, [1929] / Evangelical Lutheran Joint Synod of Ohio and Other States. Women's Missionary Conference – 1r – 1 – (forms pt of: record group jso 11 women's missionary conference; contains 2 items related to the history of the women's missionary conference (wmc) of the evangelical lutheran joint synod of ohio and other states (jso); records are printed material fr 1927 and approx 1929; with finding aid) – mf#xa0115r – us ATLA [242]

Historical records and studies / United States Catholic Historical Society – Yonkers. 1899-1964 (1) – mf#1688 – us UMI ProQuest [240]

Historical records and studies thomas f mechan / Bayle, Constantino – New York, 1932; Madrid: Razon y Fe, 1933 – 1 – sp Bibl Santa Ana [900]

Historical records and studies...new york, catholic historical society. 1929 / Bayle, Constantino – Madrid: Razon y Fe, 1930 – 1 – sp Bibl Santa Ana [241]

Historical records of mare island naval shipyard vallejo, california – 1854-1961 – 1750mf – 9 – $1750.00 – (detailed ind available) – mf#B40160 – us Library Micro [978]

The historical records of the high authority of the european coal and steel community, part 1 : the records of the ecs now accessible on microfiche / Commission of the European Communities – 1951-56 [mf ed Chadwyck-Healey, 1991] – 9883mf – 9 – (with ind & inventories 1952-53) – uk Chadwyck [660]

Historical records of the new brunswick regiment, canadian artillery / Baxter, John Babington Macaulay [comp] – St John, NB: Officers of the Corps, 1896 – 4mf – 9 – (incl ind) – mf#02997 – cn Canadiana [355]

Historical records of the newport naval training station, rhode island, 1883-1948 / U.S. Post Office – (= ser Records Of Naval Districts And Shore Establishments) – 1r – 1 – mf#T1017 – us Nat Archives [355]

Historical Records Survey Florida *see*
– Preliminary list of religious bodies in florida
– Spanish land grants in florida

Historical Records Survey, Florida *see* List of municipal corporations in florida

HISTORICAL

Historical register : containing an impartial relation of all transactions, foreign and domestic – London. 1714-1738 (1) – mf#3900 – us UMI ProQuest [327]

An historical relation of ceylon : together with somewhat concerning severall remarkeable passages of my life that hath hapned since my deliverance out of my captivity / Knox, Robert – Glasgow: James MacLehose & Sons, 1911 [mf ed 1995] – (= ser Yale coll) – lxvii/459p (ill) – 1 – 0-524-09391-1 – (incl facs of t-p of original ed, london, 1681 ed by james ryan. incl "the issue for the first time of autobiogr of knox) – mf#1995-0391 – us ATLA [954]

The historical relations of christ church, philadelphia, with the province of pennsylvania : an address. delivered at the two hundreth anniversary of christ church... / Stille, Charles Janeway – Philadelphia: PC Stockhausen, 1895 – 1mf – 9 – 0-524-08590-0 – mf#1993-3175 – us ATLA [240]

Historical reports of the state acting assistant provost marshals general and district provost marshals, 1865 / U.S. Army. Provost Marshal General's Bureau – (= ser Records Of The Provost Marshal General's Bureau (Civil War)) – 5r – 1 – (with printed guide) – mf#M1163 – us Nat Archives [976]

Historical research : an outline of theory and practice / Vincent, John Martin – New York: H. Holt, 1911 – 1mf – 9 – 0-7905-6029-1 – (incl bibl ref) – mf#1988-2029 – us ATLA [900]

Historical research – Oxford. 1991-1996 (1,5,9) – ISSN: 0950-3471 – mf#17393,01 – us UMI ProQuest [900]

Historical review of the disturbance in the evangelical association / Bowman, Thomas – Cleveland, Ohio: Thomas & Mattill, 1894 – 1mf – 9 – 0-524-03257-2 – mf#1990-4660 – us ATLA [240]

An historical review of the experiences of eastern washington university african-american male athletes from the 1960s to the 1970s / Ewing, Tyrone J – 1997 – 1mf – 9 – $4.00 – mf#PE 3809 – us Kinesology [790]

Historical scenes from the old jesuit missions / Kip, William Ingraham – New York: Anson D.F. Randolph, c1875 – 1mf – 9 – 0-7905-4987-5 – mf#1988-0987 – us ATLA [241]

The historical sculptures of the vaikunthaperumal temple, kanchi / Minakshi, Cadambi – Delhi: Manager of Publications, 1941 – (= ser Samp: indian books) – us CRL [730]

Historical setting of the early gospel / Hall, Thomas C – New York: Eaton & Mains; Cincinnati: Jennings & Graham, c1912 – 1mf – 9 – 0-7905-0577-0 – (includes bibliographies) – mf#1987-0577 – us ATLA [220]

A historical sketch : or, compendious view of domestic and foreign missions in the presbyterian church of the united states of america / Green, Ashbel – Philadelphia: William S Martien, 1838 [mf ed 1993] – (= ser Presbyterian coll) – 1mf – 9 – 0-524-06538-1 – mf#1991-2622 – us ATLA [242]

Historical sketch of episcopacy in scotland : from 1688 to the present time / Drummond, David Thomas Kerr – Edinburgh: WP Kennedy, 1845 – 1mf – 9 – 0-524-05432-0 – mf#1990-1464 – us ATLA [240]

Historical sketch of evangelical reformed church, frederick, maryland / Eschbach, E R – 1894 – 1 – $50.00 – us Presbyterian [242]

Historical sketch of fyzabad tehsil *see* Bengal tenancy bill

A historical sketch of our canadian institutions for the insane : presidential address / Burgess, Thomas Joseph Workman – S.l: s.n, 1898? – 2mf – 9 – mf#16949 – cn Canadiana [366]

Historical sketch of pensacola, florida : embracing a brief retrospective / Robinson, Benjamin – Pensacola, FL. 1882 – 1r – us UF Libraries [630]

Historical sketch of presbyterianism within the bounds of the synod of central new york : the presbyterian element in our national life and history / Fowler, Philemon Halsted & Mears, John William – Utica, NY: Curtiss & Childs, 1877. Chicago: Dep of Photodup, U of Chicago Lib, 1969 (1r); Evanston: American Theol Lib Assoc, 1984 (1r) – 1 – 0-8370-0365-2 – (incl ind) – mf#1984-B118 – us ATLA [242]

Historical sketch of protestant missions in siam : 1828-1928 / ed by McFarland, George Bradley – [Bangkok?]: Printed by the Bangkok Times Press, 1928. Chicago: Dep of Photodup, U of Chicago Lib, 1971 (1r); Evanston: American Theol Lib Assoc, 1984 (1r) – 1 – 0-8370-0528-0 – mf#1984-B223 – us ATLA [242]

Historical sketch of protestant missions in siam, 1828-1928 / McFarland, George Bradley – Bankok: Bankok Times Press, 1928 – 1r – 1 – $50.00 – us Presbyterian [242]

Historical sketch of saguenay : souvenir of the ontario and quebec press excursion to chicoutimi, grand-brule and st-alphonse on the 9th august 1883 = Historique du saguenay – Quebec: Printed by Leger Brousseau, 1883 [mf ed 1984] – 1mf – 9 – mf#SEM105P375 – cn Bibl Nat [971]

Historical sketch of st luke's church, 1854-1904, montreal, canada – [Montreal?: s.n, 1904?] – 1mf – 9 – 0-665-87940-7 – mf#87940 – cn Canadiana [242]

Historical sketch of the barton lodge no.6, grc, af and am / Freemasons. Barton Lodge, No 6 (Hamilton, Ont) – Hamilton Ont: G E Mason, 1895 – 3mf – 9 – mf#03242 – cn Canadiana [360]

An historical sketch of the china mission of the protestant episcopal church in the usa : from the first appointments in 1834 to include the year 1892 – 3rd ed. New York:...Society of the Protestant Episcopal Church in USA, 1893 [mf ed 1995] – (= ser Yale coll) – 113p (ill) – 1 – 0-524-09099-8 – mf#1995-0099 – us ATLA [242]

An historical sketch of the china mission of the protestant episcopal church in the usa : from the first appointments in 1834 to include the year ending aug 31st 1884 – New York: Foreign Cttee, 1885 [mf ed 1992] – (= ser Anglican/episcopal coll) – 1mf – 9 – 0-524-03712-4 – mf#1990-4817 – us ATLA [242]

Historical sketch of the christian woman's board of missions / Dickinson, Elmira Jane – rev enl ed. Indianapolis, IN: Christian Woman's Board of Missions [1911?] [mf ed 1992] – (= ser Christian church (disciples of christ) coll) – 2mf – 9 – 0-524-04728-6 – mf#1991-2133 – us ATLA [240]

Historical sketch of the congregational society and church in bristol, conn : with the articles of faith, covenant and standing rules of the church, together with a catalogue of members since its gathering and a catalogue of members april 1st, 1852 / Peck, Tracy – Hartford: DB Moseley, 1852 – 1mf – 9 – 0-524-07292-2 – mf#1990-5379 – us ATLA [978]

Historical sketch of the english evangelical lutheran synod of the northwest and of the congregations connected therewith : illustrated with exterior and interior views of the churches and with portraits of present and former pastors / Haupt, Alexander James Derbyshire – St Paul, Minn: Press of Frank Shoop, [1902?] – 1mf – 9 – 0-524-07882-3 – mf#1991-3427 – us ATLA [242]

Historical sketch of the grande ligne mission / Ayer, Albert Azro – Grande Ligne, Quebec: s.n, 1898 – 1mf – 9 – mf#05846 – cn Canadiana [242]

Historical sketch of the hawaiian mission / Bartlett, Samuel Colcord & Hyde, Charles McEwen – Boston: American Board of Commissioners for Foreign Missions, 1900 [mf ed 1995] – (= ser Yale coll) – 46p (ill) – 1 – 0-524-09799-2 – mf#1995-0799 – us ATLA [240]

An historical sketch of the introduction of christianity into india : and its progress and present state in that and other eastern countries / Ainslie, Whitelaw – [s.l: s.n] 1835 (Edinburgh: Oliver & Boyd) [mf ed 1995] – (= ser Yale coll) – 160p – 1 – 0-524-09191-9 – (incl bibl ref) – mf#1995-0191 – us ATLA [240]

An historical sketch of the island of madeira : containing an account of its original discovery and first colonization; present produce; state of society and commerce – London 1819 [mf ed Hildesheim 1995-98] – 1mf – 9 – €40.00 – 3-487-29807-4 – gw Olms [946]

An historical sketch of the japan mission of the protestant episcopal church in the usa – 3rd ed. New York:...Society of the Protestant Episcopal Church in USA, 1891 [mf ed 1995] – (= ser Yale coll) – 42p (ill) – 1 – 0-524-09100-5 – mf#1995-0100 – us ATLA [242]

Historical sketch of the middlesex south [sic] conference of churches / Temple, Josiah Howard – [s.l: s.n. 187-?] [mf ed 1991] – 1mf – 9 – 0-524-00732-2 – mf#1990-4001 – us ATLA [240]

Historical sketch of the mission in persia under the care of the board of foreign missions of the presbyterian church / Greene, J Milton – Philadelphia: Woman's Foreign Missionary Society of the Presbyterian Church, 1881 – 1mf – 9 – 0-524-07243-4 – (incl bibl ref) – mf#1991-2984 – us ATLA [242]

Historical sketch of the mission of the general council of the evangelical lutheran church : among the telugus of india / Trabert, George Henry – Philadelphia: Jas B Rodgers, 1890 [mf ed 1986] – 1mf – 9 – 0-8370-6428-7 – mf#1986-0428 – us ATLA [242]

Historical sketch of the mission to the nestorians; and of the assyria mission / Perkins, Justin & Laurie, Thomas – New-York: John A Gray, 1862 – 1mf – 9 – 0-7905-6874-8 – mf#1988-2874 – us ATLA [240]

Historical sketch of the missions in india under the care of the board of foreign missions of the presbyterian church / Janvier, Caesar Augustus Rodney – Philadelphia: Woman's Foreign Missionary Society of the Presbyterian Church, 1903 – 1mf – 9 – 0-524-07250-7 – (incl bibl ref) – mf#1991-2991 – us ATLA [242]

Historical sketch of the missions in siam and among the laos : under the care of the board of foreign missions of the presbyterian church / Dripps, Joseph Frederick – Philadelphia: Woman's Foreign Missionary Society of the Presbyterian Church, 1881 [mf ed 1993] – 1mf – 9 – 0-524-07236-1 – mf#1991-2977 – us ATLA [242]

Historical sketch of the missions in siam under the care of the board of foreign missions of the presbyterian church in the u.s.a – 7th ed. Philadelphia: Woman's Foreign Missionary Society of the Presbyterian Church, 1915 – 1mf – 9 – 0-524-07232-9 – mf#1991-2973 – us ATLA [242]

Historical sketch of the missions of the american board among the north american indians / Bartlett, Samuel Colcord – Boston: The Board, 1880 [mf ed 1993] – 1mf – 9 – 0-524-08350-9 – mf#1993-3050 – us ATLA [240]

Historical sketch of the missions of the american board in papal lands / Worcester, Isaac Redington – Boston: The Board, 1879 – 1mf – 9 – 0-524-00663-6 – mf#1990-0163 – us ATLA [240]

Historical sketch of the missions of the american board in the sandwich islands, micronesia, and marquesas / Bartlett, Samuel Colcord – Boston: The Board, 1876 [mf ed 1995] – (= ser Yale coll) – 1 – 0-524-10262-7 – mf#1996-1262 – us ATLA [240]

Historical sketch of the montreal protestant orphan asylum : from its foundation on the 16th feb, 1822, to the present day / Montreal Protestant Orphan Asylum – Montreal?: J Lovell, 1860 – 1mf – 9 – mf#47121 – cn Canadiana [360]

Historical sketch of the origin of the secession church – the history of the rise of the relief church / Thomson, Andrew & Struthers, Gavin – Edinburgh: A Fullarton, 1848 – 1mf – 9 – 0-524-03195-9 – mf#1990-4644 – us ATLA [243]

An historical sketch of the provincial dialects of england : illustrated by numerous examples / Halliwell-Phillipps, James Orchard – Albany NY: J Munsell 1863 [mf ed 1987] – 1r – 1 – (filmed with: the grammar of science / pearson, k) – mf#1996p – us Wisconsin U Libr [420]

Historical sketch of the provincial dialects of england / Halliwell-Phillipps, James Orchard – Albany, NY. 1863 – 1r – us UF Libraries [025]

A historical sketch of the rise and progress of the unitarian christian doctrines in modern times : with a statement of the position of unitarianism in the present age in various countries and churches... – London: ET Whitfield [1876?] [mf ed 1993] – (= ser Unitarian/universalist coll) – 1mf – 9 – 0-524-07832-7 – mf#1991-3379 – us ATLA [243]

Historical sketch of the rise, progress, and decline of the reformation in poland : and of the influence which the scriptural doctrines have exercised on that country in literary, moral, and political respects / Krasinski, Valerian, Count – London: Printed for the author: Murray [distributor], 1838-1940 – 3mf – 9 – 0-7905-5849-1 – (incl bibl ref) – mf#1988-1849 – us ATLA [240]

Historical sketch of the synod of philadelphia : and biographical sketches of distinguished members of the synod of philadelphia / Patterson, Robert Mayne & Davidson, Robert – Philadelphia: Presbyterian Board of Publ, c1876 – 1mf – 9 – 0-524-01334-9 – mf#1990-4083 – us ATLA [240]

An historical sketch of the unitarian movement since the reformation / Allen, Joseph Henry – New York: Christian Literature Co, 1894 [mf ed 1986] – 1mf – 9 – 0-8370-8720-1 – (incl bibl ref & ind) – mf#1986-2720 – us ATLA [243]

Historical sketch of the young men's christian association of montreal : organized november 25th, 1851 in the st helen street baptist church – [Montreal?: s.n, 1901?] – 1mf – 9 – 0-665-76908-3 – mf#76908 – cn Canadiana [366]

Historical sketch of trinity church, 1840-1902, montreal, canada – [Montreal?: s.n, 190-?] – 1mf – 9 – 0-665-65990-3 – mf#65990 – cn Canadiana [242]

Historical sketch... synod of central new york / Fowler, Philemon H – 1877 – 1 – $50.00 – us Presbyterian [240]

Historical sketches / Newman, John Henry – London: Basil Montague Pickering, 1872-73 [mf ed 1990] – 3v on 4mf – 9 – 0-7905-7431-4 – (incl ind) – mf#1989-0656 – us ATLA [900]

Historical sketches and sidelights of miami, florida / Cohen, Isidor – Miami, FL. 1925 – 1r – us UF Libraries [978]

Historical sketches for jurisdictional and subject headings used for the letters received by the office of indian affairs, 1824-1880 / U.S. National Archives and Records Service – (= ser Records of the bureau of indian affairs) – 1r – 1 – mf#T1105 – us Nat Archives [305]

Historical sketches of ancient dekhan / Subrahmanya Aiyer, Kandadai Vaidyanatha – Madras: Modern Print Works, 1917 – (= ser Samp: indian books) – (foreword by sir s subrahmanya iyer) – us CRL [930]

Historical sketches of hymns, their writers, and their influence / Belcher, Joseph – Philadelphia: Lindsay & Blakiston; New York: Sheldon, 1859 – 1mf – 9 – 0-524-00507-9 – mf#1990-0007 – us ATLA [240]

Historical sketches of nonconformity in the county palatine of chester / ed by Urwick, William – London: Kent, 1864 – 2mf – 9 – 0-524-03666-7 – (incl bibl ref) – mf#1990-1094 – us ATLA [240]

Historical sketches of savage life in polynesia : with illustrative clan songs / Gill, W W – Wellington, 1880 – 3mf – 9 – mf#HTM-62 – ne IDC [919]

Historical sketches of the ancient native irish and their descendants / Anderson, Christopher. – Edinburgh, 1828 – (= ser 19th c ireland) – 3mf – 9 – mf#1.1.6038 – uk Chadwyck [941]

Historical sketches of the evangelical lutheran synod of south carolina : from its formation in 1824 / Schirmer, Jacob F – Charleston, SC: AJ Burke, 1875 – 1mf – 9 – 0-524-02571-1 – mf#1990-4383 – us ATLA [242]

Historical sketches of the india missions of the presbyterian church in the united states of america : known as the lodiana, the farrukhabad, and the kolhapur missions / Newton, John – Allahabad: Allahabad Mission Press, 1886 – 1mf – 9 – 0-524-04270-5 – mf#1991-2054 – us ATLA [242]

Historical sketches of the missions in japan, korea / Gosman, Abraham & Eckard, L W – 3rd rev ed. Philadelphia: Woman's Foreign Missionary Society of the Presbyterian Church, 1891 [mf ed 1995] – (= ser Yale coll) – 41p – 1 – 0-524-09765-8 – mf#1995-0765 – us ATLA [242]

Historical sketches of the missions of the united brethen / Holmes, Rev John – s.l, s.l? 1827 – 1r – us UF Libraries [025]

Historical sketches of the south of india : in an attempt to trace the history of mysoor, from the origin of the hindoo government of that state, to the extinction of the mohammedan dynasty in 1799 / Wilks, Mark – London: Longman, Hurst, Rees, and Orme, 1810-1817 – (= ser Samp: indian books) – us CRL [954]

Historical sketches of the south of india, in an attempt to trace the history of mysoor.. / Wilks, Mark – 2nd ed. Madras: Hurst, 1869. 2v. fold. map – 1 – us Wisconsin U Libr [954]

Historical sketches of woman's missionary societies in america and england / ed by Daggett, L H – Boston: Mrs. L. H. Daggett, [c.1883] Beltsville, Md: NCR Corp, 1977 (3mf); Evanston: American Theol Lib Assoc, 1984 (3mf) – 9 – 0-8370-0178-1 – mf#1984-0063 – us ATLA [240]

Historical sketches of...missions under...presbyterian church usa – Philadelphia: Women's Foreign Missionary Society, 1897 – 1mf – 9 – 0-8370-6315-9 – (includes bibliographies) – mf#1985-0315 – us ATLA [242]

Historical social organizations, no 140 / Shepherd, Rose – s.l, s.l? 1936 – 1r – us UF Libraries [978]

Historical social organizations, no 140[b] / Shepherd, Rose – s.l, s.l? 1936 – 1r – us UF Libraries [978]

Historical social organizations, no 141 / Shepherd, Rose – s.l, s.l? 1936 – 1r – us UF Libraries [978]

Historical social organizations, no 143 / Shepherd, Rose – s.l, s.l? 1936 – 1r – us UF Libraries [978]

Historical society mirror / Kansas State Historical Society – 1955 jan-1979 jan – 1r – 1 – mf#165978 – us WHS [978]

Historical Society of Carroll County [MD] *see* News letter

Historical Society of Decatur County [IN] *see* Bulletin of the historical...

Historical Society of Easton *see* Schoolhouse sentinel

Historical Society of Fort Lauderdale *see* Fort lauderdale history

Historical society of michigan newsletter – v5 n1-v11 n6 [1979 may/jun-1986 mar/apr] – 1r – 1 – (continued by: chronicle & newsletter) – mf#1126777 – us WHS [978]

Historical Society of New Mexico *see* Cronica de nuevo mexico

Historical Society of Pennsylvania *see* Pennsylvania magazine of history and biography

Historical Society of the Downs Carnegie Library *see* Osborne county researcher

The historical socrates and the platonic form of the good / Lindsay, Alexander Dunlop – Calcutta: University of Calcutta, 1932 – (= ser Samp: indian books) – us CRL [180]

Historical studies in philosophy = Etudes d'histoire de la philosophie / Boutroux, Emile – London: Macmillan, 1912 – 1mf – 9 – 0-7905-3598-X – (in english) – mf#1989-0091 – us ATLA [100]

An historical study of the developing catechesis in the emerging e u b church 1774-1946 : m a thesis, oberlin univ / Thomas, James F – 1r – 1 – $35.00 – mf^s-110 – us Commission [242]

An historical study of the terms hinayana and mahayana and the origin of mahayana buddhism / Kimura, Ryukan – [Calcutta?]: University of Calcutta, 1927 – (= ser Samp: indian books) – us CRL [280]

Historical summaries of administration measures in the several branches of public business administered in the department of revenue and agriculture, drawn up in 1896 – Calcutta: Office of the Superintendent of Govt Printing India, 1897 – 1 – us CRL [954]

Historical summary of constitutional advance in the new hebrides, 1954-1977 / Woodward, Keith – 1978 – 1r – 1 – (available for reference) – mf#pmb1151 – at Pacific Mss [323]

An historical survey of black baptist hymnody in america / Barker, George Stanley – 1981 – 1 – $5.20 – us Southern Baptist [242]

An historical survey of the ecclesiastical antiquities of france / Whittington, George Downing – London 1809 – (= ser 19th c art & architecture) – 3mf – 9 – mf#4.2.1740 – uk Chadwyck [720]

An historical survey of the first presbyterian church, caldwell, nj / Berry, Charles Treat – Newark, N.J.: Printed at the Daily Advertiser Office, 1871. Chicago: Dep of Photodup, U of Chicago Lib, 1972 (1r); Evanston: American Theol Lib Assoc, 1984 (1r) – 1 – 0-8370-0026-2 – mf#1984-B322 – us ATLA [242]

An historical text book and atlas of biblical geography / Coleman, Lyman – new rev ed. Philadelphia: E Claxton, 1881, c1854 [mf ed 1989] – 1mf – 9 – 0-7905-0979-2 – (incl ind) – mf#1987-0979 – us ATLA [220]

Historical time capsules of Monroe County / Monroe County Historical Society (WI) – v1 n1-v4 n4 [1975 aug-1984] – 1r – 1 – mf#704899 – us WHS [978]

Historical tracts and documents, state papers..., 1562-1582 – late 16th c – (= ser Holkham library family and political papers 678) – 1r – 1 – mf#2121 – uk Microform Academic [941]

The historical traveller : comprising narratives connected with the most curious epochs of european history, and with the phenomena of european countries / Gore, Catherine – London 1831 [mf ed Hildesheim 1995-98] – 2v on 4mf – 9 – €120.00 – 3-487-29919-4 – gw Olms [910]

Historical trials relevant to today's issues – 652mf (24:1) – 9 – $3145.00 coll – (transcripts of over 100 british & american trials, most from the 18th & 19th c) – us UPA [347]

The historical value of the fourth gospel / Askwith, Edward Harrison – London: Hodder & Stoughton, 1910 [mf ed 1988] – 1mf – 9 – 0-7905-0242-9 – mf#1987-0242 – us ATLA [226]

A historical view of the hindu astronomy : from the earliest dawn of that science in india, down to the present time / Bentley, John – Calcutta: Baptist Mission Press 1823 [mf ed 1998] – 1r [pl/ill] – 1 – mf#film mas 28251 – us Harvard [520]

Historical view of the languages and literature of the slavic nations : with a sketch of their popular poetry / Talvj – New-York: Putnam, 1850 – 1mf – 9 – 0-7905-8118-3 – (incl bibl ref) – mf#1988-6080 – us ATLA [460]

An historical view of the philippine islands : exhibiting their discovery, population, language, government, manners, customs, productions and commerce / Martinez de Zuniga, Joaquin – London 1814 [mf ed Hildesheim 1995-98] – 2v on 4mf – 9 – €120.00 – 3-487-27442-6 – gw Olms [915]

Historical view of the progress of discovery on the more northern coasts of america : from the earliest period to the present time / Tytler, Patrick Fraser – New York: J & J Harper, 1833 [mf ed 1983] – 5mf – 9 – 0-665-41901-5 – (original iss in ser: harper's family library. incl bibl ref.originally publ in the edinburgh cabinet library, 1832) – mf#41901 – cn Canadiana [971]

Historical view of the progress of discovery on the more northern coasts of america : from the earliest period to the present time / Wilson, James – Edinburgh 1832 [mf ed Hildesheim 1995-98] – 1v on 5mf – 9 – €100.00 – 3-487-27112-5 – gw Olms [917]

A historical vindication of the abrogation of the plan of union by the presbyterian church in the united states of america / Brown, Isaac Van Arsdale – Philadelphia: Wm S & Alfred Martien, 1855, c1854 [mf ed 1990] – 1mf – 9 – 0-7905-6462-9 – mf#1988-2462 – us ATLA [242]

Historical vindications : a discourse on the province and uses of baptist history / Cutting, Sewall Sylvester – Boston: Gould and Lincoln, 1859 – 1mf – 9 – 0-524-03316-1 – (incl bibl ref) – mf#1990-4676 – us ATLA [240]

Historical work / Screven, William – Maine Historical Society. 202p. 1629, 1663-69, 1783-87, 1820-37 – 1 – $7.07 – us Southern Baptist [978]

The historical work of master ralph de diceto, dean of london (rs68) = Radulphi de diceto decani lundoniensis, opera historica / ed by Stubbs, W – (= ser The rolls series (rs)) – (v1 1876 €19. v2 1876 €17) – ne Slangenburg [931]

Historical works (rs73) : the chronicle of the reigns of stephen, henry 2 and richard 1 / Gervase of Canterbury; ed by Stubbs, W – v1 1879 v2 1880 – (= ser The rolls series (rs)) – €23.00v – ne Slangenburg [931]

A historical-ethnographic account of a canadian woman in sport, 1920-1938 : the story of margaret (bell) gibson / Laubman, Katherine M & Schrodt, P Barbara – 1991 – 3mf – 9 – $12.00 – us Kinesology [790]

Historically speaking / Afro-American Historical Association of the Niagara Frontier – 1977 apr-1998 apr – 1r – 1 – mf#1112355 – us WHS [305]

Historici germaniae saec 12 (mgh5:21.bd) – 1869 – (= ser Monumenta germaniae historica 5: scriptores in folio (mgh5)) – €32.00 – ne Slangenburg [240]

Historici germaniae saec 12 (mgh5:22.bd) – 1872 – (= ser Monumenta germaniae historica 5: scriptores in folio (mgh5)) – €29.00 – ne Slangenburg [240]

L'historicite des trois premiers chapitres de la genese / Mechineau, Lucien – Rome: Officina Poligrafica Editrice, 1910 [mf ed 1992] – 1mf – 9 – 0-524-04107-5 – (in french) – mf#1992-0065 – us ATLA [221]

The historicity of jesus : a criticism of the contention that jesus never lived, a statement of the evidence for his existence, an estimate of his relation to christianity / Case, Shirley Jackson – Chicago, IL: University of Chicago Press, c1912 – 1mf – 9 – 0-7905-1689-6 – (incl bibl ref and indexes) – mf#1987-1689 – us ATLA [240]

Historicla memorials of a christian fellowship / Pearsall, J S – London, England. 18– – 1r – us UF Libraries [240]

Historico-critical inquiry into the origin and composition of the hexateuch (pentateuch and book of joshua) = De hexateuch / Kuenen, Abraham – London: Macmillan, 1886 [mf ed 1985] – 1mf – 9 – 0-8370-4016-7 – (trans fr dutch by philip h wicksteed; incl bibl ref & ind) – mf#1985-2016 – us ATLA [221]

Historico-critical introduction to the pentateuch = Handbuch der historisch-kritischen einleitung in das alte testament / Haevernick, Heinrich Andreas Christoph – Edinburgh: T & T Clark, 1850 [mf ed 1990] – (= ser Clark's foreign theological library 18) – 5mf – 9 – 0-7905-3449-5 – (english by alexander thomson; incl bibl ref) – mf#1987-3449 – us ATLA [221]

Historico-critico-medico-practica en que se establece el agua por remedio universal de las deolencias : historico-critico-medico-practica en que se establece el agua por remedio / Perez, V – Madrid, 1753 – 2mf – 9 – sp Cultura [615]

Historico-genealogical sketch of col. thomas lowrey, and esther fleming, his wife / Race, Henry – Flemington, NJ: H E Deats, 1892 – 1r – 1 – us Western Res [978]

A historico-geographical account of palestine in the time of christ : or, the bible student's help to a thorough knowledge of scripture / Roehr, Johann Friedrich – Edinburgh: T Clark, 1843 [mf ed 1989] – (= ser The biblical cabinet 43/1) – 1mf – 9 – 0-7905-3278-6 – (trans fr german by david esdaile) – mf#1987-3278 – us ATLA [220]

Historie de la republique centrafricaine / Serre, Jacques – Bangui, Central African Republic. 196- – 1r – us UF Libraries [960]

Le historie delle indie orientali... / Maffei, G P – Venetia, 1589 – 10mf – 9 – mf#H-8412 – ne IDC [956]

Historie der nederlantscher oorlogen... / Reyd, E v – Leeuwarden, 1650 – 9mf – 9 – mf#OA-163 – ne IDC [720]

Historie der reformatie : en andere kerkelijke geschiedenissen, in en omtrent de nederlanden...1600 / Brandt, G – Amsterdam, 1671-1704. 4v – 48mf – 9 – mf#PBA-134 – ne IDC [242]

A historie of ireland / Campion, Edmund – C.1571 – 9 – us Scholars Facs [941]

Historie of travaile into virginia, 1610-1612 / ed by Major, R H – (= ser Hakluyt society. extra series 6) – 4mf – 9 – mf#292 – uk Microform Academic [917]

Historie ofte beschrijving van 't utrechtsche bisdom – Leiden. v1-3. 1719 – €61.00 – ne Slangenburg [242]

Historie v.d. spaensche inquisitie... / Dathenus, P – n.p, 1569 – 4mf – 9 – mf#PBA-164 – ne IDC [242]

Le historie vinitiane di marco antonio sabellico... / Coccius, M A – Vinegia, 1554 – 6mf – 9 – mf#H-8287 – ne IDC [956]

Historie von groenland / Cranz, D – Barby, Leipzig, 1770. 3v – 25mf – 9 – mf#N-178 – ne IDC [917]

Historie von herzog herpin (cima17) : farbmikrofiche-edition der handschrift heidelberg, universitaetsbibliothek, cod pal germ 152 – (mf ed 1990) – (= ser Codices illuminati medii aevi (cima) 17) – 73p on 7 color mf – 15 – €360.00 – 3-89219-017-8 – (trans fr french by elisabeth von nassau-saarbruecken; int & description by ute von bloh) – gw Lengenfelder [090]

Historie von herzog herpin (cima57) : farbmikrofiche-edition der handschrift wolfenbuettel, herzog august bibliothek, cod guelf 46 novissime 2° – (mf ed 2000) – (= ser Codices illuminati medii aevi (cima) 57) – 59p on 9 color mf – 15 – €340.00 – 3-89219-057-7 – (trans fr the french by elisabeth von nassau-saarbruecken. int & description by eva wolf) – gw Lengenfelder [090]

Die "historie von vier kaufmaennern" und deren dramatische bearbeitungen in der deutschen literatur des 16. und 17. jahrhunderts / Mechel, Kurt – Halle, 1914 (mf ed 1994) – 1mf – 9 – €24.00 – 3-8267-3097-6 – mf#DHS-AR 3097 – gw Frankfurter [430]

De historie-beschouwing van den deuteronomist : met de berichten in genesis-numeri vergeleken / Kosters, Willem Hendrik – Leiden: S C van Doesburgh, 1868 [mf ed 1989] – 137p on 1mf – 9 – 0-7905-1529-6 – (in dutch and hebrew. incl bibl ref) – mf#1987-1529 – us ATLA [221]

Historie...conteneno le gverre di mahometto imperatore de turchi... / Guazzo, M – Venetia, 1545 – 1mf – 9 – mf#H-8275 – ne IDC [956]

Historien der alden e / ed by Gerhard, Wilhelm – Leipzig: K W Hiersemann, 1927 – (incl bibl ref and ind) – us Wisconsin U Libr [810]

Historien der alden e / ed by Gerhard, Wilhelm – Leipzig: K W Hiersemann, 1927 – 1 – (incl bibl ref and index) – us Wisconsin U Libr [830]

Historien von der landwirthschaft : welche sich in boehmen an verschiedenen orten zugetragen. nebst einem ewigen bauernkalender /.../ – Prag (CZ), Wien (A), 1792 – 1r – 1 – gw Misc Inst [630]

Die historien von der sindflut, joseph, mose, helia, elisa und der susanna, sampt etlichen historien aus den evangelisten : auch etliche psalmen und geistliche lieder, zu lesen und zu singen in reime gefasset: fuer christliche hausveter und ihre kinder / Herman, Nicolas – Wittemberg: [Selfisch] 1566 ([Rhau]) – (= ser Hqab. literatur des 16. jahrh.) – 3mf – 9 – €40.00 – mf#1566b – gw Fischer [780]

Historienbibel (cima6) : farbmikrofiche-edition der handschrift hamburg, staats- und universitaetsbibliothek, cod.7 in scrinio – (mf ed 1988) – (= ser Codices illuminati medii aevi (cima) 6) – 35p on 9 color mf – 15 – €290.00 – 3-89219-006-2 – (int & description by heimo reinitzer) – gw Lengenfelder [090]

Historienbibel (cima25) : farbmikrofiche-edition der handschrift heidelberg, universitaetsbibliothek, cod pal germ 60 – (mf ed 1993) – (= ser Codices illuminati medii aevi (cima) 25) – 32p on 7 color mf – 15 – €335.00 – 3-89219-025-9 – (filmed with: sankt brandans meerfahrt with int by karl a zaenker. description by ulrike bodemann.) – gw Lengenfelder [090]

Historienbibel (cima47) : farbmikrofiche-edition der handschrift hamburg, staats- und universitaetsbibliothek, cod 8 in scrinio – (mf ed 1997) – (= ser Codices illuminati medii aevi (cima) 47) – 61p on 16 color mf – 15 – €475.00 – 3-89219-047-X – (int & description by anna katharina hahn) – gw Lengenfelder [090]

Historiens grecs : recueil des historiens des croisades – Paris, 1875-1881. 2v – 74mf – 9 – mf#H-509 – ne IDC [931]

Historiens occidentaux : recueil des historiens des croisades – Paris, 1844-1895. 5v – 218mf – 9 – mf#H-510 – ne IDC [931]

Historiens orientaux : recueil des historiens des croisades – Paris, 1872-1906. 5v – 138mf – 8 – mf#H-511 – ne IDC [700]

Histories of civil war generals – v. 1-50. 1888. Duke Cigarette miniature volumes – 7 – us L of C Photodup [976]

Historiese studies – Pretoria. South Africa. 1939-49 – 2r – 1 – sa National [079]

Historiese studies – University of Pretoria. v.1-9, 1939-49 – 1 – us CRL [960]

Historikai meletai / Papadopoulos, Chrysostomos – En Ierosolumois: Ek tou typographeiou tou Hierou Koinou tou Panagiou Taphou, 1906 – 3mf – 9 – 0-7905-7128-5 – (incl bibl ref) – mf#1988-3128 – us ATLA [240]

Historikerstreit : miscellaneous materials on the debate in west germany about the holocaust and german historiography / Mosse, George [comp] – [Madison]: University of Wisconsin – Madison 1987 [mf ed 2005] – 1r – 1 – (photocopies of articles in various publ: faz, wissen, die zeit...; some of the aut: ernst nolte, renate schostack...; and many letters fr readers of the periodicals) – mf#5589p – us Wisconsin U Libr [934]

Historiografia mineira / Jose, Oiliam – Belo Horizonte, Brazil. 1959 – 1r – us UF Libraries [972]

Historique de la colonisation belge a santo-tomas... / Leysbeth, Nicolas – Bruxelles, Belgium. 1938 – 1r – us UF Libraries [972]

Historique des fonds de retraite en europe et en canada / Dorion, Eugene P – Quebec?: Hunter, Rose et Lemieux, 1862 – 2mf – 9 – mf#22947 – cn Canadiana [350]

Historique du cercle et rapport general du secretaire pour l'annee 1886-1887 / Cercle Ville-Marie (Montreal, Quebec) – Montreal: Impr de l'Etendard, 1887 [mf ed 1987] – 1mf – 9 – mf#SEM105P814 – cn Bibl Nat [971]

Historique du saguenay : souvenir de l'excursion de la presse d'ontario et de quebec a chicoutimi, au grand-brule et a st-alphonse, le 9 aout 1883 = Historical sketch of the saguenay – Quebec: Impr Leger Brousseau, 1883 [mf ed 1984] – 1mf – 9 – mf#SEM105P376 – cn Bibl Nat [971]

Historisch diplomatisches magazin fuer das vaterland und angrenzende gegenden / ed by Will, Georg Andr – Nuernberg 1780-81 – (= ser Dz. historisch-geographische abt) – 2v[zu je 4st] on 7mf – 9 – €140.00 – mf#k/n1109 – gw Olms [327]

Historisch oder mythisch? : beitraege zur beantwortung der gegenwaertigen lebensfrage der theologie / Ullmann, Karl – Hamburg: F Perthes, 1838 [mf ed 1990] – 1mf – 9 – 0-7905-3805-9 – mf#1989-0298 – us ATLA [240]

Historisch overzicht over suriname / Wolff, H J – s'-Gravenhage, Netherlands. 1934 – 1r – us UF Libraries [972]

Historisch woordenboek van zuidwestelijk celebes / Ligtvoet, A – n.p, [1880] 2v – 5mf – 8 – mf#SD-104 mf 1-5 – ne IDC [959]

Historisch-biographische studien / Ranke, Leopold von – Leipzig: Duncker & Humblot, 1877 – 2mf – 9 – 0-7905-6491-2 – mf#1988-2491 – us ATLA [940]

Historisch-biographische urkunden des mittleren reiches / Sethe, K – Leipzig, 1935 – (= ser Urkunden des aegyptischen Altertums) – 2mf – 9 – (urkunden des aegyptischen altertums, abt 7 v1) – mf#NE-399 – ne IDC [956]

Historisch-biographisches lexicon der tonkuenstler : welches nachrichten von dem leben und werken musikalischer schriftsteller, beruehmter componisten, saenger [etc]...enthaelt / Gerber, Ernst Ludwig [comp] – Leipzig: J G I Breitkopf 1790-92 [mf ed 19–] – 2v on 19mf – 9 – (imprint varies) – mf#fiche 477 – us Sibley [780]

Historisch-chronologische schwierigkeiten im zweiten makkabaeerbuche / Cigoi, Alois – Klagenfurt: Joh & Fried Leon, 1868 – 1mf – 9 – 0-7905-0311-5 – (incl bibl ref) – mf#1987-0311 – us ATLA [221]

Historisch-comparative geographie von preussen / Toeppen, Max – Gotha 1858 [mf ed Hildesheim 1995-98] – 3mf – 9 – €90.00 – 3-487-29581-4 – gw Olms [914]

Historisch-critische bijdragen : naar aanleiding van de nieuwste hypothese aangaande jezus en den paulus der vier hoofdbrieven / Scholten, Johannes Henricus – Leiden: S.C. van Doesburgh, 1882 – 1mf – 9 – 0-8370-5135-5 – (incl bibl ref) – mf#1985-3135 – us ATLA [220]

Historische arbeiten vornehmlich zur reformationzeit / Cornelius, Carl Adolf – Leipzig: Duncker & Humblot, 1899 – 2mf – 9 – 0-7905-5523-9 – mf#1988-1523 – us ATLA [943]

Historische attische inschriften / Nachmanson, Ernst – Bonn: A Marcus & E Weber, 1913 [mf ed 1992] – (= ser Kleine texte fuer vorlesungen und uebungen 110) – 1mf – 9 – 0-524-04143-1 – (in greek. notes in german. incl bibl ref & ind) – mf#1990-1213 – us ATLA [450]

Historische beschreibung der edelen sing- und klingkunst : in welcher deroselben ursprung und erfindung, fortgang...und zugleich beruehmteste ausueber von anfang der welt biss auff unsere zeit... / Printz, Wolfgang Caspar – Dresden: J C Mieth 1690 [mf ed 19–] – 4mf – 9 – mf#fiche 649, 833 – us Sibley [780]
Historische beschreibung des gantzen streits zwischen d hunnen vnd d hubern von der gnadenwahl wie derselbige entsprungen vnd biss daher zugenomen habe / Huber, S – [Oberursel], 1597 – 2mf – 9 – mf#TH-1 mf 724-725 – ne IDC [242]
Historische beschryving der stadt amsterdam: waer in de voornaemite geichiedeniffen.. / Dapper, O – Amsterdam: J. van Meurs, 1663 – 1 – us Wisconsin U Libr [949]
Historische dates sur geschichte der israelitischen gemeinde bremen – Berlin, Germany. 1926 – 1r – us UF Libraries [939]
Historische dialektwoerterbuecher aus deutschen sprachgebieten (ael2/18) – [mf ed 2001] – (= ser Archiv der europaeischen lexikographie: woerterbuecher) – 585mf – 9 – €4200.00 – 3-89131-379-9 – gw Fischer [430]
Historische einfuehrung in das achtzehngebet / Schwaab, Emil – Guetersloh: C Bertelsmann, 1913 – (= ser Beitraege zur foerderung christlicher theologie) – 1mf – 9 – 0-524-05350-2 – (incl bibl ref) – mf#1990-3471 – us ATLA [270]
Die historische entwicklung der interdependenz von atmung und herz-kreislaufsystem und der einflu_ der atmung auf die herzzeitintervalle unter besonderer beruecksichtigung der koerperposition / Geider, Stefan – (mf ed 1994) – 1mf – 9 – €30.00 – 3-8267-2030-X – mf#DHS 2030 – gw Frankfurter [611]
Historische erklaerung des zweiten teils des jesaia, capitel 40 bis capitel 66 : nach den ergebnissen aus den babylonischen keilinschriften nebst einer abhandlung, ueber die bedeutung des "knecht gottes" / Ley, Julius – Marburg: N G Elwert, 1893 – 1mf – 9 – 0-8370-4105-8 – mf#1985-2105 – us ATLA [221]
Historische Gesellschaft fuer die Provenz Posen *see* Zeitschrift der historischen gesellschaft fuer die provinz posen
Historische Gesellschaft zu Berlin *see*
– Jahresberichte der geschichtswissenschaft
– Mitteilungen aus der historischen litteratur
Historische griechische inschriften bis auf alexander den grossen – Bonn: A Marcus & E Weber, 1913 [mf ed 1992] – (= ser Kleine texte fuer vorlesungen und uebungen 121) – 1mf – 9 – 0-524-04703-0 – (in greek. int & notes in german. incl bibl ref) – mf#1990-3412 – us ATLA [450]
Der historische hans kohlhase und heinrich von kleist's michael kohlhaas / Burkhardt, Carl August Hugo – Leipzig: Vogel 1864 [mf ed 1991] – 1r – 1 – (incl bibl footnotes. filmed with: hebbels frauengestalten / emil kreisler) – mf#2710p – us Wisconsin U Libr [430]
Das historische in kants religionsphilosophie : zugleich ein beitrag zu den untersuchungen ueber kants philosophie der geschichte / Troeltsch, Ernst – Berlin: Reuter & Reichard, 1904 – 1mf – 9 – 0-524-00404-8 – mf#1989-3104 – us ATLA [200]
Der historische jesus, der mythologische christus und jesus der christ : ein kritischer gang durch die moderne jesus-forschung / Dunkmann, Karl – Leipzig: A Deichert, 1910 – 1mf – 9 – 0-7905-0490-1 – (incl bibl ref) – mf#1987-0490 – us ATLA [240]
Historische litteratur fuer das jahr 1781 [-85] / ed by Meusel, Johann Georg – Erlangen 1781-85 – (= ser Dz. historisch-geographische abt) – 5jge on 34mf – 9 – €340.00 – mf#k/n1116 – gw Olms [430]
Historische nachrichten und politische betrachtungen ueber die franzoesische revolution – Berlin DE, 1792-97, 1802-03 – 1 – gw Misc Inst [933]
Historische, politisch-geographisch-statistisch und militaerische beytraege die kgl preussischen und benachbarten staaten betreffend / ed by Fischbach, Friedrich Ludwig Joseph – Berlin 1781-85 – (= ser Dz. historisch-geographische abt) – 3v on 23mf – 9 – €230.00 – mf#k/n1112 – gw Olms [943]
Historische remarques ueber die neuesten sachen in europa – Hamburg DE, 1701-03 – 3r – 1 – gw Misc Inst [074]
Historische schets van de gemeente graafschap, mich. der chr. geref. kerk – [S.l.: s.n., 1917?] – 1mf – 9 – 0-524-07248-5 – mf#1991-2989 – us ATLA [240]
Historische syntax der griechischen comparation in der klassischen littertur / Schwab, Otto – Wuerzburg: A. Stuber, 1893-1895 – (= ser Beitraege zur historischen syntax der griechischen sprache) – 2mf – 9 – 0-8370-1599-5 – (incl bibl ref) – mf#1987-6081 – us ATLA [450]
Historische untersuchungen / ed by Meusel, Johann Georg – Nuernberg 1779-80 – (= ser Dz. historisch-geographische abt) – 1v on 6mf – 9 – €120.00 – mf#k/n1102 – gw Olms [943]
Historische zeitschrift – Munich. 1859-1973 (1) – ISSN: 0018-2613 – mf#9849 – us UMI ProQuest [900]
Die historischen volkslieder der deutschen vom 13. bis 16. jahrhundert : the historical folksongs of the germans from the 13th to the 16th century / Liliencron, Rochus von – Leipzig. 4v. 1865-69 – 11 – $80.00 set – (incl suppl) – us Univ Music [780]
Historischer bericht : von dem zu regensburgk unlangst gehaltenen colloqvio, zwischen den theologen augsburgischer confession vnd den papisten / Hunnius, A – Wittenberg, 1602 – 1mf – 9 – mf#TH-1 mf 800 – ne IDC [242]
Historischer bericht : was sich in dem grossen...koenigreich china, in verkeundigung dess h euangelij vnd fortpflantzung des catholischen glaubens, von 1604... – Augspurg: C. Dabertzhofer, 1611 [mf ed 1995] – (= ser Yale coll) – 131p – 1 – 0-524-09329-6 – (trans fr portuguese into german) – mf#1995-0329 – us ATLA [241]
Historischer bericht von dess beruemten seligen herrn philippi melanthonis meinung inn dem streit von dess herrn abendmahl / Peucer, C – Basel, 1597 – 3mf – 9 – mf#TH-1 mf 1271-1273 – ne IDC [242]
Historischer calender / ed by Westenrieder, L von – Muenchen 1790-1816 – (= ser Dz) – jg1-20 on 50mf – 9 – €500.00 – mf#k/n1237 – gw Olms [900]
Historischer Verein Bamberg *see* Bericht
Historischer verein der pfalz – Mitteilungen des Historischen Vereins der Pfalz.v1-. 1870-.-irr. 1870-1932; annual, 1953- – 1 – us Wisconsin U Libr [943]
Historischer Verein Dillingen an der Donau [Germany] *see* Jahrbuch des historischen vereins dillingen
Historischer Verein fuer das Grossherzogtum Hessen *see* Archiv fuer hessische geschichte und altertumskunde
Historischer Verein fuer die Graftschaft Ravensberg zu Bielefeld *see* Jahresbericht
Historischer Verein fuer Dortmund und die Grafschaft Mark *see* Beitraege zur geschichte dortmunds und der grafschaft mark
Historischer Verein fuer Niederbayern *see* Verhandlungen des historischen vereins fuer niederbayern
Historischer Verein fuer Oberfranken zu Bayreuth *see* Archiv fuer geschichte von oberfranken
Historischer Verein fuer Oberpfalz und Regensburg *see* Verhandlungen des historischen vereines von oberpfalz und regensburg
Historischer versuch ueber den handel und die schiffahrt auf dem schwarzen meere : oder reisen und untersuchungen um schiffahrts- und handels-verbindungen zwischen den haeven des schwarzen meeres und denen des mittellaendischen meeres zu begruenden / Anthoine de Saint-Joseph, Antoine I – Weimar 1805 [mf ed Hildesheim 1995-98] – 1v on 2mf – 9 – €60.00 – 3-487-26563-X – gw Olms [380]
Historisches geographisches lexikon von der schweiz (ael1/29) : oder vollstaendige alphabetische beschreibung aller in der ganzen schweizerischen eidgenossenschaft und den derselben zugewandten orten liegenden staedte – Ulm 1796 [mf ed 1995] – (= ser Archiv der europaeischen lexikographie, abt 1: enzyklopaedien) – 2v on 8mf – 9 – €100.00 – 3-89131-207-5 – gw Fischer [059]
Historisches jahrbuch / Goeres-gesellschaft zur pflege der wissenschaft im katholischen Deutschland – Bonn, 1880-1926. v1-46. 1914 ind v1-34 – 701mf – 8 – mf#H-628c – ne IDC [240]
Historisches jahrbuch der goerres-gesellschaft – 14(1893)-20(1899); 22(1901)-26(1905); 49(1929)-58(1938) – 314mf – 9 – €599.00 – ne Slangenburg [900]
Historisches journal / ed by Gentz, Friedrich von – Berlin 1799-1800 – (= ser Dz. historisch-politische abt) – 3v on 18mf – 9 – €180.00 – mf#k/n1315 – gw Olms [943]
Historisches journal von mitgliedern des kgl historischen instituts zu goettingen : darin: litterarischer beytrag zu dem historischen journal / ed by Gatterer, Johann Christoph – Goettingen 1772-81 – (= ser Dz. historisch-geographische abt) – 16pt on 40mf – 9 – €400.00 – mf#k/n1065 – gw Olms [943]
Historisches lesebuch der christlichen bibellehre : fuer liebhaber der wahrheit unter jungen und alten / Schoener, Johann Gottfried – 2. verm und verb aufl. Nuernberg: Raw, 1834 – 2mf – 9 – 0-524-05062-7 – mf#1992-0315 – us ATLA [220]
Historisches portefeuille zur kenntnis der vergangenen und gegenwaertigen zeit / ed by Hausen, Karl Renatus – Wien, Breslau, Berlin, Leipzig, Hamburg 1782-88 – (= ser Dz. historisch-geographische abt) – 7jge on 73mf – 9 – €730.00 – mf#k/n1135 – gw Olms [943]
Historisch-genetische darstellung von kant's verschiedenen ansichten ueber das wesen der materie / Kuttner, Otto – 1881 [mf ed 1993] – 1mf – 9 – 0-524-08453-X – mf#1993-2058 – us ATLA [190]
Historisch-geographische und genealogische anmerkungen ueber die zeitung von voriger woche – Koenigsberg (Kaliningrad RUS), 1723 [gaps] – 1 – gw Misc Inst [900]
Historisch-geographisches journal / ed by Fabri, Joh Ernst et al – Halle, Leipzig, Jena 1789-90 – (= ser Dz. historisch-geographische abt) – 6mf – 9 – €120.00 – mf#k/n1214 – gw Olms [900]
Historisch-kritische ausgabe / Deutsche National-Literatur – 163v. 1882-99 – 1 – $645.00 – us L of C Photodup [430]
Historisch-kritische beytraege zur aufnahme der musik / Marpurg, Friedrich Wilhelm – Berlin: in Verlag Joh Jacob Schutzens...1754-78 [mf ed 19–] – 5v on 1r – 1 – (imprint varies: v2-5 publ by g a lange, berlin; iss in 30pt ea with separate title pg; incl ind) – mf#film 1333 – us Sibley [780]
Historisch-kritische einleitung in das neue testament / Hilgenfeld, A – Leipzig, 1875 – 14mf – 8 – €27.00 – ne Slangenburg [225]
Historisch-kritische einleitung in den koran / Weil, Gustav – 2. verb Aufl. Bielefeld: Velhagen & Klasing, 1878 – 1mf – 9 – 0-524-02810-9 – (incl bibl ref) – mf#1990-3140 – us ATLA [260]
Historisch-kritische schriftforschung und bibelglaube : ein versuch zur theologischen wissenschaftslehre / Weber, E – 2. bedeutend erw. aufl. Guetersloh: C Bertelsmann, 1914 – 1mf – 9 – 0-7905-2205-5 – (incl bibl ref) – mf#1987-2205 – us ATLA [220]
Historisch-kritische studien zu der septuaginta : erster band, erste abtheilung, vorstudien zu der septuaginta / Frankel, Zacharias – Leipzig: Fr Chr Wilh Vogel, 1841 – 1mf – 9 – 0-7905-1660-8 – (incl bibl ref) – mf#1987-1660 – us ATLA [221]
Historisch-literarische abteilung der zeitschrift fuer mathematik und physik *see* Zeitschrift fuer mathematik und physik
Historisch-litterarisch-bibliographisches magazin / ed by Meusel, Johann Georg – Zuerich [1792ff: Chemnitz] 1788-94 – (= ser Dz) – 8st on 15mf – 9 – €150.00 – mf#k/n422 – gw Olms [074]
Historisch-litterarisches magazin / ed by Meusel, Johann Georg – Bayreuth, Leipzig 1785-86 – (= ser Dz) – 4pt on 8mf – 9 – €160.00 – mf#k/n389 – gw Olms [943]
Historisch-politische blaetter fuer das katholische deutschland – Muenchen: In Commission der Literarisch-artistischen Anstalt 1860-1923 (semimthly) [v67-171 mf ed 1978] – 36r – 1 – (v45-66 (1860-70) [mf ed 2001] 11r [film mas c4843]; cont: g phillips' und g goerres' historisch-politische blaetter fuer das katholische deutschland; cont by: gelbe hefte; v45-50 ind with v35-44 of preceding title in single vol called zweites register zu den historische-politischen blaettern; v51-130 ind in 3v called drittes-fuenftes register zu den historische-politischen blaettern) – mf#film mas c282 – us Harvard [241]
Historisch-politische blaetter (klp18) : fuer das katholische deutschland / ed by Philipps, Georg & Goerres, Guido – Muenchen 1838-1923 [mf ed 2004] – (= ser Marbacher mikrofiche-editionen (mme) 18; Kultur – literatur – politik: deutsche zeitschriften des 19./20. jahrhunderts (klp)) – 171v on ca 1650mf – 9 – €6800.00 – 3-89131-455-8 – gw Fischer [241]
Historisch-politische schriften des dietrich von nieheim (mgh10:5.bd) : 1. stueck: viridarium imperatorum et regum romanorum – 1956 – (= ser Monumenta germaniae historica scriptores 10 (mgh10): staatsschriften des spaeteren mittelalters (13. bis 15. jh)) – €11.00 – ne Slangenburg [240]
Historisch-politisches magazin, nebst litterarischen nachrichten / ed by Wittenberg, Albrecht – Hamburg 1787-95 – (= ser Dz. historisch-geographische abt) – 18v on 79mf – 9 – €790.00 – (jg1: niederelbisches h-p-litterarisches magazin) – mf#k/n1191 – gw Olms [943]
Historisch-statistische darstellung des noerdlichen englands : nebst vergleichenden bemerkungen auf einer reise durch die suedwestlichen grafschaften / Rivinus, Eduard – Leipzig 1824 [mf ed Hildesheim 1995-98] – 3mf – 9 – €90.00 – 3-487-27983-5 – gw Olms [941]
Historisk statistisk 1968 / Norway. Statistiske Sentralbyra – (= ser European official statistical serials, 1841-1984) – 7mf – 9 – uk Chadwyck [314]
Historiske meddelelser : om den norske augustana-synode samt nogle oplysninger om andre samfund i amerika / Hatlestad, Ole Jensen – Decorah IA: Decorah-Postens bogtrykkeri 1887 [mf ed 1992] – 1mf – 9 – 0-524-02472-3 – mf#1990-4331 – us ATLA [242]
Der historismus und seine ueberwindung : fuenf vortraege / Troeltsch, Ernst – Berlin: R Heise 1924 [mf ed 1987] – 1r – 1 – (int by friedrich von huegel-kensington. with: the parables of our lord / dods, m) – mf#1858 – us Wisconsin U Libr [240]
History / Comanche County. Kansas. Union Church Ediface Society – 1879 – 1 – us Kansas [978]
History / Greensburg. Kansas. Greensburg Town Company – 1884-88 – 1 – us Kansas [978]
History / Jantzen Family – 1744-1947 – 1 – us Kansas [978]
History : the journal of the historical association / Historical Association (Great Britain) – London. 1912+ (1) 1976+ (5) 1984+ (9) – ISSN: 0018-2648 – mf#10537 – us UMI ProQuest [900]
History / Nielson, Jens Christian – 1853-1915[?], In Danish – 1 – us Kansas [920]
History / Nielson, Jens Christian – 1853-[1915?], In English – 1 – us Kansas [920]
History – s.l, s.l? 193-? – 1r – us UF Libraries [978]
History – Washington. 1972+ (1) 1972+ (5) 1976+ (9) – ISSN: 0361-2759 – mf#7116 – us UMI ProQuest [900]
History (addition) : tampa / Muse, Viola B – s.l, s.l? 193-? – 1r – us UF Libraries [978]
The history and adventures of little henry : exemplified in a series of figures – 3rd ed. London: printed for S & J Fuller, 1810 [mf ed 1984] – 1mf – 9 – 0-665-45107-5 – mf#45107 – cn Canadiana [830]
History and annals of hebrew printing in the fifteenth and sixteenth centuries / Marx, Moses – 13r – 1 – $650.00 – (a list of the contents is available on request: most of the reels are available separately) – us AJPC [680]
The history and antiquities of bath abbey church / Britton, John – London 1825 – (= ser 19th c art & architecture) – 3mf – 9 – mf#4.2.1461 – uk Chadwyck [720]
History and antiquities of london, westminster, southwark and parts adjacent / Allen, Thomas – 5v. 1837. London – 1 – us L of C Photodup [941]
History and antiquities of the borough of new windsor : from the royal archives and library at windsor castle – 1811 – 1r – 1 – mf#96526 – uk Microform Academic [025]
The history and antiquities of the collegiate church of southwell / Killpack, William Bennett – London 1839 – (= ser 19th c art & architecture) – 2mf – 9 – mf#4.2.1645 – uk Chadwyck [720]
History and antiquities of the county of cumberland / Hutchinson, William – 16mf – 7 – mf#87007 – uk Microform Academic [941]
History and antiquities of the county of leicester / Nichols, John – v1-4 – 62mf – 7 – mf#87006 – uk Microform Academic [941]
The history and antiquities of the town and port of hastings / Moss, William – London 1824 [mf ed Hildesheim 1995-98] – 1v on 2mf – 9 – €60.00 – 3-487-27924-X – gw Olms [914]
History and beliefs of the major religions / Scholl, Warren – Girard, KS: Haldeman-Julius, 1924. 64p – 1 – us Wisconsin U Libr [230]
History and causes of the incorrect latitudes : as recorded in the journals of the early writers, navigators and explorers relating to the atlantic coast of north america, 1535-1740 / Slafter, Edmund Farwell – Boston?: D Clapp, 1882 – 1mf – 9 – mf#09236 – cn Canadiana [900]
History and character of american revivals of religion / Colton, Calvin – London: F Westley and AH Davis, 1832 – 1mf – 9 – 0-7905-6862-4 – mf#1988-2862 – us ATLA [240]
The history and conquests of the saracens : six lectures / Freeman, Edward Augustus – 3rd ed with new pref. London: Macmillan, 1876 – 1mf – 9 – 0-524-00720-9 – mf#1990-2048 – us ATLA [260]
History and construction of the cornett / Humfeld, Neill Hamilton – U of Rochester 1962 [mf ed 19–] – 2mf – 9 – (with app & bibl) – mf#fiche 1011 – us Sibley [780]
The history and description of africa... / Africanus, L; ed by R Brown – London, 1896. 3v – 24mf – 9 – mf#A-333 – ne IDC [916]
History and description of florida capitol at tallah / Webber, Joel Frank – s.l, s.l? 193-? – 1r – us UF Libraries [978]
A history and description of modern wine / Redding, Cyrus – 3rd corr ed. London: H G Bohn, 1851 [mf ed 1987] – (= ser Bohn's illustrated library) – viii/440p (ill) – 1 – mf#2081 – us Wisconsin U Libr [640]

History and description of mr tebbutt's observatory, windsor, new south wales / Tebbutt, John – Sydney: J Cook 1887 [mf ed 2001] – 1r – 1 – (cont by: report of mr tebbutt's observatory, the peninusula,windsor, new south wales, in 1888) – mf#film mas c4697 – us Harvard [520]

A history and description of roman political institutions / ed by Abbott, Frank F – 3rd ed. Boston: Ginn & Co, 1911 – 5mf – 9 – $7.50 – mf#LLMC 92-241 – us LLMC [340]

History and description of the styles coal mines and adjoining area of one square mile : with copies of reports, assays, etc – Halifax, NS?: s.n., 1888 (Halifax, NS: Halifax Print Co) – 1mf – 9 – mf#08418 – cn Canadiana [622]

The history and description, with graphic illustrations, of cassiobury park, hertfordshire / Britton, John – London 1837 – (= ser 19th c art & architecture) – 2mf – 9 – mf#4.2.1450 – uk Chadwyck [710]

History and development of the double bass / DeMatteo, Edward D – U of Rochester 1957 [mf ed 19–] – 3mf – 9 – mf#fiche113 – us Sibley [780]

The history and doctrines of irvingism or of the so-called catholic and apostolic church / Miller, Edward – London: C Kegan Paul, 1878 – 2mf – 9 – 0-524-04965-3 – mf#1990-1368 – us ATLA [241]

History and doctrines of the ajivikas : a vanished indian religion / Basham, Arthur Llewellyn – London: Luzac, 1951 – (= ser Samp: indian books) – (foreword by l d barnett) – us CRL [280]

History and exposition of the twenty-five articles of religion of the methodist episcopal church / Wheeler, Henry – New York: Eaton & Mains, c1908 – 1mf – 9 – 0-524-04781-2 – (incl bibl ref) – mf#1991-2167 – us ATLA [242]

History and extent of recognition of tribal law in rhodesia / Child, Harold – Salisbury, Zimbabwe. 1965 – 1r – us UF Libraries [960]

The history and fate of sacrilege / Spelman, Henry – 4th ed. London: J Hodges, 1895 – (= ser Catholic Standard Library) – 1mf – 9 – 0-524-05870-9 – (incl bibl ref) – mf#1990-3534 – us ATLA [240]

History and general description of new france / Charlevoix, Francois Xavier de – New York. v1-6. 1866-72 – 9 – $42.00 – mf#0147 – us Brook [971]

History and genius of the heidelberg catechism / Nevin, John Williamson – Chambersburg, [Pa]: German Reformed Church, 1847 – 1mf – 9 – 0-7905-9536-2 – (incl bibl ref) – mf#1989-1241 – us ATLA [240]

History and heritage – v1 n1 [1990 may/jun], v2 n1 [undated] – 1r – 1 – mf#3970389 – us WHS [975]

History and historians in the nineteenth century / Gooch, George Peabody – 2nd ed. London; New York: Longmans, Green, 1913 – 2mf – 9 – 0-7905-4685-X – (incl bibl ref) – mf#1988-0685 – us ATLA [900]

"History and historical geography of japan" / Rekishi chiri – Tokyo, 1899-1943. v1-82 – 1144mf – 9 – mf#CH-458 – ne IDC [915]

The history and life of the reverend doctor john tauler of strasbourg : with twenty-five of his sermons (temp 1340) – London: Smith, Elder, 1857 [mf ed 1990] – 2mf – 9 – 0-7905-8086-1 – (trans fr german by susanna winkworth. pref by charles kingsley. incl additional notices of tauler's life and times) – mf#1988-8022 – us ATLA [241]

The history and life of the reverend doctor john tauler of strasbourg; with twenty-five of his sermons (temp. 1340) / Tauler, Johannes – Trans. by Susanna Winkworth. Preface by Rev. Charles Kingsley. London: Smith, Elder, and comp., 1857.xl,415p – 1 – us Wisconsin U Libr [920]

The history and literature of the israelites : according to the old testament and the apocrypha / Rothschild, C de & Rothschild, A de – 2nd ed. London: Longmans, Green, 1871 [mf ed 1992] – 2v on 3mf – 9 – 0-524-04895-9 – mf#1992-0238 – us ATLA [221]

History and literature of the unitarian controversy / Gillett, Ezra Hall – Morrisania, NY: Henry B Dawson, 1871 – (= ser Historical Magazine (Boston, MA)) – 2mf – 9 – 0-524-07875-0 – (incl bibl ref) – mf#1991-3420 – us ATLA [243]

The history and mystery of methodist episcopacy : or, a glance at "the institutions of the church, as we received them from our fathers" / M'Caine, Alexander – Baltimore: Matchett 1827 – 1r – 1 – $35.00 – mf#um-15 – us Commission [242]

The history and origin of the missionary societies : containing faithful accounts of the voyages, travels, labours, and successes of the various missionaries who have been sent out, for the purpose of evangelizing the heathen... / Smith, Thomas – London 1824-39 – (= ser 19th c british colonization) – 18mf – 9 – mf#1.1.3974 – uk Chadwyck [240]

History and outline of laws relating to vessel inspection / Arzt, Frederick Karl – Washington, 1940-. 106 p. LL-392 – 1 – us L of C Photodup [340]

The history and philosophy of sport in islam / Aldousari, Badi – 2000 – 79p on 1mf – 9 – $5.00 – mf#PE 4160 – us Kinesology [260]

History and philosophy of the life sciences – London. 1991-1995 (1,5,9) – ISSN: 0391-9714 – mf#17320 – us UMI ProQuest [590]

The history and present constitution of the sheriff courts of scotland; a letter to william stirling / Robertson, Robert – Glasgow: Maclehose, 1863. 35p. LL-2287 – 1 – us L of C Photodup [347]

The history and present state of electricity, with original experiments : from warrington public library / Priestley, Joseph – J. Dodsley/ W. Eyres, 1767 – 1r – 1 – mf#96720 – uk Microform Academic [620]

The history and principles of the presbyterian church in ireland / Stewart, David – Belfast: Sabbath School Society for Ireland, 1907 – (= ser Irish Presbyterian Guild Text-Books) – 1mf – 9 – 0-524-01670-4 – mf#1990-0491 – us ATLA [242]

History and problems of moslem education in bengal / Huque, Azizul – Calcutta: Thacker, Spink & Co, 1917 – (= ser Samp: indian books) – us CRL [377]

History and problems of organized labor / Carlton, Frank T – New York, Chicago: D C Heath & Co, 1920 – 6mf – 9 – $9.00 – mf#LLMC 92-221 – us LLMC [331]

A history and record of the protestant episcopal church in the diocese of west virginia : and, before the formation of the diocese in 1878, in the territory now known as the state of west virginia / Peterkin, George William – Charleston, W VA: Tribune Co, 1902 [mf ed 1993] – (= ser Anglican/episcopal coll) – 3mf – 9 – 0-524-06476-8 – mf#1990-5250 – us ATLA [242]

The history and records of the conference : together with addresses delivered at the evening meetings / World Missionary Conference 1910 Edinburgh, Scotland – Edinburgh: Oliphant, Anderson & Ferrier; New York: Fleming H Revell, [1910?] – 1mf – 9 – 0-8370-6546-1 – (incl ind) – mf#1986-0546 – us ATLA [240]

History and repository of pulpit eloquence, deceased divines : containing the masterpieces of bossuet . . . [et al] / Fish, Henry Clay – New York: M.W. Dodd, 1856 – 3mf – 9 – 0-7905-4851-8 – mf#1988-0851 – us ATLA [240]

History and significance of the sacred tabernacle of the hebrews / Atwater, Edward Elias – New York: Dodd and Mead, 1875 – 2mf – 9 – 0-8370-9840-8 – (incl ind) – mf#1986-3840 – us ATLA [939]

History and social science teacher – Markham. 1977-1990 (1,5,9) – (cont by: canadian social studies) – ISSN: 0316-4969 – mf#11569,01 – us UMI ProQuest [300]

The history and song of deborah : judges 6 and 5 / Cooke, G A – Oxford: Horace Hart, 1892 – 1mf – 9 – 0-7905-2405-8 – mf#1987-2405 – us ATLA [221]

History and status of labor in the citrus industry of florida / Kistler, Allison Clay – s.l, s.l, s.l? 1939 – 1r – us UF Libraries [634]

The history and survey of london and its environs : from the earliest period to the present time / Lambert, B – London 1806 [mf ed Hildesheim 1995-98] – 4v on 16mf – 9 – €160.00 – 3-487-27980-0 – gw Olms [941]

The history and teaching of the plymouth brethren / Teulon, Josiah Sanders – London: Society for Promoting Christian Knowledge, [1883?] – 1mf – 9 – 0-524-08534-X – mf#1993-1064 – us ATLA [242]

The history and teachings of the early church as a basis for the re-union of christendom : lectures...1888 / Coxe, Arthur Cleveland et al – 3rd ed. New York: E & J B Young 1892, c1889 [mf ed 1992] – (= ser The church club lectures 1888) – 1mf – 9 – 0-524-02850-8 – mf#1990-0707 – us ATLA [240]

History and the mystery of good friday / Robinson, Robert – London, England. 1849 – 1r – us UF Libraries [240]

History and theology in the fourth gospel / Martyn, James Louis – New York: Harper & Row, [1968] – 1r – 1 – 0-8370-1524-3 – mf#1984-B389 – us ATLA [226]

History and theory – Middletown. 1960+ [1,5,9] – ISSN: 0018-2656 – mf#2515 – us UMI ProQuest [900]

The history and theory of vitalism = Vitalismus als geschichte und als lehre / Driesch, Hans – rev ed. London: Macmillan, 1914 – 1mf – 9 – 0-7905-7508-6 – (in english) – mf#1989-0733 – us ATLA [100]

The history and topography of bedfordshire : with biographical sketches, etc and a neat map of the county / Pinnock, William – London [ca 1819] [mf ed Hildesheim 1995-98] – 1mf – 9 – €40.00 – 3-487-28807-9 – gw Olms [941]

The history and topography of berkshire : with biographical sketches, etc and a neat map of the county / Pinnock, William – London 1819 [mf ed Hildesheim 1995-98] – 1mf – 9 – €40.00 – 3-487-28809-5 – gw Olms [941]

The history and topography of buckinghamshire : with biographical sketches, etc and a neat map of the county / Pinnock, William – London 1819 [mf ed Hildesheim 1995-98] – 1mf – 9 – €40.00 – 3-487-28808-7 – gw Olms [941]

The history and topography of cheshire : with biographical sketches, etc and a neat map of the county / Pinnock, William – London 1820 [mf ed Hildesheim 1995-98] – 1mf – 9 – €40.00 – 3-487-28806-0 – gw Olms [941]

The history and topography of devonshire : with biographical sketches, etc and a neat map of the county / Pinnock, William – London [ca 1820] [mf ed Hildesheim 1995-98] – 1v on 1mf – 9 – €40.00 – 3-487-28802-8 – gw Olms [941]

The history and topography of durham : with biographical sketches, etc and a neat map of the county / Pinnock, William – London [ca 1820] [mf ed Hildesheim 1995-98] – 1v on 1mf – 9 – €40.00 – 3-487-28803-6 – gw Olms [941]

The history and topography of gloucestershire : with biographical sketches, etc and a neat map of the county / Pinnock, William – London [ca 1820] [mf ed Hildesheim 1995-98] – 1v on 1mf – 9 – €40.00 – 3-487-28805-2 – gw Olms [941]

The history and topography of hampshire : with biographical sketches, etc and a neat map of the county / Pinnock, William – London [ca 1820] [mf ed Hildesheim 1995-98] – 1v on 1mf – 9 – €40.00 – 3-487-27998-3 – gw Olms [941]

The history and topography of herefordshire : with biographical sketches, etc and a neat map of the county / Pinnock, William – London [ca 1820] [mf ed Hildesheim 1995-98] – 1v on 1mf – 9 – €40.00 – 3-487-28801-X – gw Olms [941]

The history and topography of hertfordshire : with biographical sketches, etc and a neat map of the county / Pinnock, William – London [ca 1820] [mf ed Hildesheim 1995-98] – 1v on 1mf – 9 – €40.00 – 3-487-28800-1 – gw Olms [941]

The history and topography of lancashire : with biographical sketches, etc and a neat map of the county / Pinnock, William – London 1820 [mf ed Hildesheim 1995-98] – 1v on 1mf – 9 – €40.00 – 3-487-27997-5 – gw Olms [941]

The history and topography of leicestershire : with biographical sketches, etc and a neat map of the county / Pinnock, William – London [ca 1820] [mf ed Hildesheim 1995-98] – 1v on 1mf – 9 – €40.00 – 3-487-27996-7 – gw Olms [941]

The history and topography of london : being a correct guide to the public establishments, places of amusement, curiosities, etc in london and its immediate vicinity / Pinnock, William – London 1820 [mf ed Hildesheim 1995-98] – 1v on 1mf – 9 – €40.00 – 3-487-27995-9 – gw Olms [914]

The history and topography of norfolk : with biographical sketches, etc and a neat map of the county / Pinnock, William – London [ca 1820] [mf ed Hildesheim 1995-98] – 1v on 1mf – 9 – €40.00 – 3-487-27994-0 – gw Olms [941]

The history and topography of northamptonshire : with biographical sketches, etc and a neat map of the county / Pinnock, William – London 1820 [mf ed Hildesheim 1995-98] – 1mf – 9 – €40.00 – 3-487-27993-2 – gw Olms [941]

The history and topography of nottinghamshire : with biographical sketches, etc and a neat map of the county / Pinnock, William – London [ca 1820] [mf ed Hildesheim 1995-98] – 1mf – 9 – €40.00 – 3-487-27992-4 – gw Olms [941]

The history and topography of oxfordshire : with biographical sketches, etc and a neat map of the county / Pinnock, William – London 1819 [mf ed Hildesheim 1995-98] – 1mf – 9 – €40.00 – 3-487-27991-6 – gw Olms [941]

The history and topography of suffolk : with biographical sketches, etc and a neat map of the county / Pinnock, William – London [ca 1820] [mf ed Hildesheim 1995-98] – 1mf – 9 – €40.00 – 3-487-27990-8 – gw Olms [941]

The history and topography of surrey : with biographical sketches, etc and a neat map of the county / Pinnock, William – London [ca 1820] [mf ed Hildesheim 1995-98] – 1mf – 9 – €40.00 – 3-487-27989-4 – gw Olms [941]

The history and topography of sussex : with biographical sketches, etc and a neat map of the county / Pinnock, William – London 1820 [mf ed Hildesheim 1995-98] – 1mf – 9 – €40.00 – 3-487-27988-6 – gw Olms [941]

The history and topography of the county of essex : with biographical sketches, etc and a neat map of the county / Pinnock, William – London [ca 1820] [mf ed Hildesheim 1995-98] – 1v on 1mf – 9 – €40.00 – 3-487-28804-4 – gw Olms [941]

The history and topography of the county of kent : with biographical sketches, etc and a neat map of the county / Pinnock, William – London [ca 1820] [mf ed Hildesheim 1995-98] – 1v on 1mf – 9 – €40.00 – 3-487-27999-1 – gw Olms [941]

History and topography of the parish of wakefield and its environs / Hewitt, John – 1862 – 5mf – 9 – mf#8691 – uk Microform Academic [941]

The history and topography of wiltshire : with biographical sketches, etc and a neat map of the county / Pinnock, William – London 1820 [mf ed Hildesheim 1995-98] – 1mf – 9 – €40.00 – 3-487-27987-8 – gw Olms [941]

The history and topography of worcestershire : with biographical sketches, etc and a neat map of the county / Pinnock, William – London 1819 [mf ed Hildesheim 1995-98] – 1mf – 9 – €40.00 – 3-487-27986-X – gw Olms [941]

The history and use of hymns and hymn-tunes / Breed, David Riddle – Chicago: Fleming H Revell, 1903 – 1mf – 9 – 0-7905-4097-5 – mf#1988-0097 – us ATLA [780]

The history and work of harvard observatory, 1839 to 1927 : an outline of the origin, development, and researches of the astronomical observatory of harvard college together with brief biographies of its leading members / Bailey, Solon Irving – New York, London: McGraw-Hill Book Co 1931 [mf ed 1998] – 1r [pl/ill] – 1 – mf#film mas 28211 – us Harvard [520]

The history, art and palaeography of the manuscript styled the utrecht psalter / Birch, Walter de Gray – London: Samuel Bagster, 1876 – 1mf – 9 – 0-8370-9044-X – (incl bibl ref) – mf#1986-3044 – us ATLA [700]

History, authority and theology / Headlam, Arthur Cayley – London: J Murray 1909 [mf ed 1990] – 1mf – 9 – 0-7905-5996-X – mf#1988-1996 – us ATLA [240]

History branch office of judge advocate general with the u.s. forces european theater, 18 july, 1942-1 november, 1945 / U.S. European Theater. Office of the Judge Advocate General – v. 1-2. 1945 – 1 – us L of C Photodup [355]

History bulletin / Xavier University [New Orleans, LA] – 1946 apr – 1r – 1 – mf#4865744 – us WHS [975]

The history, civil and commercial, of the british colonies in the west indies : to which is added an historical survey of the french colony in the island of st domingo / Edwards, Bryan – London: printed for B Crosby...for Mundell & Son...1798 [mf ed 1984] – 5mf – 9 – 0-665-44108-8 – mf#44108 – cn Canadiana [972]

History department newsletter / Howard University – 1985/1986 – 1r – 1 – (continued by: biennial update, howard university. dept of history) – mf#4865829 – us WHS [378]

The history, description, and antiquities of the prebendal church of the blessed virgin mary of thame, in the county and diocese of oxford / Lee, Frederick George – London: Mitchell and Hughes, 1883 – 1mf – 9 – 0-8370-5812-0 – (incl ind) – mf#1985-3812 – us ATLA [941]

A history in diary form of civil aviation in papua and new guinea / Grabowsky, Ian – 1913-35 – 2r – 1 – mf#pmb7 – at Pacific Mss [380]

History in the making – New York, NY. 1945-1951 (1) – mf#65079 – us UMI ProQuest [071]

History, jurisdiction, and practice of the court of claims of the united states / Richardson, William Adams – Washington: Govt. Print. Off., 1882. 20p. LL-1221 – 1 – us L of C Photodup [347]

History news – Nashville. 1973+ (1) 1977+ (5) 1977+ (9) – ISSN: 0363-7492 – mf#8706 – us UMI ProQuest [900]

The history of a book / Carey, Annie – [London], Paris, New York: Cassell, Petter, & Galpin, [1873] – (= ser 19th c publishing...) – 2mf – 9 – mf#3.1.76 – uk Chadwyck [070]

The history of a colorado real estate mortgage. / Webber, Henry William – Denver, Chain & Hardy, 1895. 154 p. LL-1554 – 1 – us L of C Photodup [340]

The history of a famous court house located at carlinville, illinois / Brown, William Barrick – Carlinville, Carlinville Democrat, 1934 54 p. LL-1234 – 1 – us L of C Photodup [347]

History of a forgotten sect of baptised believers heretofore known as johnsonians / ed by Dawbarn, Robert – London: Balding & Mansell, [19–?] – 1mf – 9 – 0-524-07977-3 – mf#1990-5422 – us ATLA [242]

History of a lawsuit / Caruthers, Abraham – 3d ed. Cincinnati: Clarke, 1888. 688p. LL-739 – 1 – us L of C Photodup [340]

The history of a railroad difficulty : being an address, delivered at a public meeting of the inhabitants of port hope, in the town hall, on saturday, the 23rd apr 1859 / Fowler, John – Port Hope [Ont]: C Roger, [1859?] [mf ed 1984] – 1mf – 9 – 0-665-44672-1 – mf#44672 – cn Canadiana [380]

The history of a suit at law, according to the practice of this state / Conner, James – Charleston, S.C., Courtenay, 1857. 72 p. LL-2286 – 1 – us L of C Photodup [340]

History of a suit in equity, as prosecuted and defended in the virginia state courts. / Sands, Alexander Hamilton – 2d ed. Richmond: Randolph & English, 1882. 760, lxp. LL-1337 – 1 – us L of C Photodup [347]

The history of a title / Crocker, Uriel Haskell – Boston, The Massachusetts Title Insurance Company, 1885. 24 p. LL-540 – 1 – us L of C Photodup [340]

History of a zoological temperance convention : held in central africa in 1847 / Hitchcock, Edward – Boston: Nathaniel Noyes, 1855 – 2mf – 9 – $3.00 – mf#LLMC 91-092 – us LLMC [360]

History of abeokuta / Ajisafe, Ajayi Kolawole – [2nd ed]. Bungay, Suffolk: Clay, 1924 – 1 – us CRL [960]

The history of acadia : from its first discovery to its surrender to england by the treaty of paris / Hannay, James – [St John, NB?: J & A McMillan], 1879 – 5mf – 9 – 0-665-06689-9 – mf#06689 – cn Canadiana [971]

A history of aesthetic / Bosanquet, Bernard – London: S Sonnenschein; New York: Macmillan, 1892 [mf ed 1990] – (= ser Library of philosophy) – 2mf – 9 – 0-7905-3869-5 – (incl bibl ref) – mf#1989-0362 – us ATLA [110]

History of agriculture in the southern united states to 1860 / Gray, Lc – Washington, DC. 1933 – 1r – us UF Libraries [630]

History of alameda county / Wood, W M – Alameda Co, CA. 1964 – 1r – 1 – $50.00 – mf#B40204 – us Library Micro [978]

A history of all religions : as divided into paganism, mahometanism, judaism and christianity / Benedict, David – Providence: J Miller, printer, 1824 [mf ed 1990] – 1mf – 9 – 0-7905-6582-X – (incl bibl ref) – mf#1988-2582 – us ATLA [200]

A history of all religions : containing a statement of the origin, development, doctrines, and government of the religious denominations in the united states and europe / ed by Smucker, Samuel Mosheim – Philadelphia: Quaker City Publ House, 1859 [mf ed 1991] – 1mf – 9 – 0-524-01457-4 – (incl app) – mf#1990-2452 – us ATLA [200]

History of all the religious denominations in the united states : containing authentic accounts of the rise and progress, faith and practice, localities and statistics of the different persuasions / Cleland, W I et al – 3rd, improved and portrait ed. Harrisburg, PA: J Winebrenner, 1852 – 1mf – 9 – 0-524-05331-6 – mf#1990-1449 – us ATLA [200]

The history of allied force headquarters, 1942-1945 / U.S. Army – 1r – 1 – $130.00 – mf#S1683 – us Scholarly Res [355]

History of amelia gale – London, England. 18– – 1r – us UF Libraries [240]

The history of america : including the history of virginia to the year 1688, and new england to the year 1652 – London: publ by Richard Evans...& John Bourne...Edinburgh 1817 [mf ed 1984] – 0-665-43059-0 – mf#43059 – cn Canadiana [975]

A history of american baptist missions in asia, africa, europe and north america : under the care of the american baptist missionary union / Gammell, William – Boston: Gould & Lincoln, 1854 [mf ed 1986] – 1mf – 9 – 0-8370-6049-4 – mf#1986-0049 – us ATLA [242]

A history of american christianity / Bacon, Leonard Woolsey – New York: Christian Literature, 1897 [mf ed 1989] – (= ser The american church history series 13) – 1mf – 9 – 0-7905-4018-5 – (incl bibl ref) – mf#1988-0018 – us ATLA [240]

The history of american music / Elson, Louis Charles – New York: Macmillan, 1904 – (= ser The History Of American Art) – 1mf – 9 – 0-7905-4466-0 – (incl bibl ref) – mf#1988-0466 – us ATLA [780]

History of american painting / Isham, Samuel – New York, NY. 1927 – 1r – us UF Libraries [750]

A history of american revivals / Beardsley, Frank Grenville – 2nd rev enl ed. New York: American Tract Society, c1912 [mf ed 1990] – 1mf – 9 – 0-7905-5683-9 – mf#1988-1683 – us ATLA [240]

The history of american slavery and methodism from 1780 to 1849; and, history of the wesleyan methodist connection of america / Matlack, Lucius C – New York: [s.n.], 1849 – 1mf – 9 – 0-7905-5256-6 – mf#1988-1256 – us ATLA [242]

History of american wesleyan methodism / Jennings, Arthur T – Syracuse, NY: Wesleyan Methodist Pub Association, 1902 – 1mf – 9 – 0-524-02890-7 – mf#1990-4481 – us ATLA [242]

History of amulets, charms, and talismans : a historical investigation into their nature and origin / Rodkinson, Michael Levi – New York: [s.n.], 1893 – 1mf – 9 – 0-524-02043-4 – mf#1990-2818 – us ATLA [270]

The history of ancient art among the greeks / Winckelmann, Johann Joachim – London 1850 – (= ser 19th c art & architecture) – 4mf – 9 – mf#4.2.1102 – uk Chadwyck [930]

A history of ancient geography / Tozer, Henry Fanshawe – Cambridge: University Press, 1897 [mf ed 1990] – (= ser Cambridge geographical series) – 1mf – 9 – 0-7905-6843-8 – (incl bibl ref) – mf#1988-2843 – us ATLA [910]

History of ancient india / Tripathi, Rama Shankar – Beneres: Nand Kishore & Bros, 1942 – (= ser Samp: indian books) – us CRL [930]

History of ancient philosophy / Benn, Alfred William – New York, NY. 1912 – 1r – us UF Libraries [180]

A history of ancient sanskrit literature : so far as it illustrates the primitive religion of the brahmans / Muller, Friedrich Max – Allahabad: Bhuvaneshwari Ashrama, 1926 – (= ser Samp: indian books) – us CRL [490]

History of ancient sanskrit literature : so far as it illustrates the primitive religion of the brahmans / Meuller, Friedrich Max – Bahadurganj, Allahabad: Panini Office, Bhuvaneshwari Ashrama, [1912] [mf ed 1995] – (= ser Yale coll) – xiv/322p – 1 – 0-524-09332-6 – mf#1995-0332 – us ATLA [490]

History of ancient woodbury, connecticut : from the first indian deed in 1659 to 1854, including the present towns of washington, southbury, bethlem, roxbury, and a part of oxford and middlebury / Cothren, William – Waterbury, Conn: Bronson Bros, 1854 [mf ed 1992] – (= ser Congregational coll) – 2v on 2mf – 9 – 0-524-04043-5 – mf#1990-4951 – us ATLA [978]

History of andrew dunn : an irish catholic – London, England. 18– – 1r – us UF Libraries [241]

History of ankole / Morris, Henry Francis – Nairobi, East African Literature Bureau, 1962 – us CRL [960]

A history of anti-pedobaptism : from the rise of pedobaptism to a d 1609 / Newman, Albert Henry – Philadelphia: American Baptist Publ Society, 1897, c1896 [mf ed 1990] – 1mf – 9 – 0-7905-5615-4 – (incl bibl ref) – mf#1988-1615 – us ATLA [240]

History of anti-pedobaptism / Newman, Albert Henry – 1896. 430p – 1 – us Southern Baptist [242]

History of apartheid / Neame, Lawrence Elwin – New York, NY. 1963 – 1r – us UF Libraries [960]

A history of arabic literature = Litterature arabe / Huart, Clement – London: W Heinemann, 1903 [mf ed 1990] – (= ser Short histories of the literatures of the world 11) – 2mf – 9 – 0-7905-5409-7 – (incl bibl ref. in english) – mf#1988-1409 – us ATLA [470]

A history of architecture / Freeman, Edward Augustus – London 1849 – (= ser 19th c art & architecture) – 6mf – 9 – mf#4.2.1671 – uk Chadwyck [720]

A history of architecture for the student, craftsman, and amateur : being a comparative view of historical styles from the earliest period / Fletcher, Banister & Fletcher, Banister Flight – London: B T Batsford, 1896 – (= ser 19th c art & architecture) – 6mf – 9 – mf#4.1.154 – uk Chadwyck [720]

A history of architecture in all countries / Fergusson, James – London 1865, 1867 – (= ser 19th c art & architecture) – 17mf – 9 – mf#4.2.851 – uk Chadwyck [720]

History of architecture in london / Godfrey, Walter Hindes – London, England. 1911 – 1r – us UF Libraries [720]

History of art by its monuments : from its decline in the 4th century to its restoration / Seroux d'Agincourt, Jean Baptiste Louis Georges – London 1847 – (= ser 19th c art & architecture) – 19mf – 9 – mf#4.2.1628 – uk Chadwyck [720]

History of art in sardinia, judaea, syria, and asia minor = Histoire de l'art dans l'antiquite. 4, jude, sardaigne, syrie, cappadoce / Perrot, Georges & Chipiez, Charles; ed by Gonino, I – London: Chapman & Hall; New York: A C Armstrong, 1890 [mf ed 1990] – 2mf – 9 – 0-7905-3396-0 – (trans fr french into english by ed. incl bibl ref) – mf#1987-3396 – us ATLA [956]

The history of aryan rule in india from the earliest times to the death of akbar / Havell, Ernest Binfield – London: George G Harrap, [1918] – (= ser Samp: indian books) – us CRL [954]

A history of assam / Gait, Edward – Calcutta: Thacker, Spink & Co, 1926 – (= ser Samp: indian books) – us CRL [954]

History of assurbanipal / Ashurbanipal, King of Assyria – London: Williams and Norgate, 1871 – 1mf – 9 – 0-8370-8562-4 – (in english and akkadian) – mf#1986-2562 – us ATLA [470]

History of astronomy / Abetti, Giorgio – New York, NY. 1952 – 1r – us UF Libraries [520]

A history of auburn theological seminary, 1818-1918 / Adams, John Quincy – Auburn, NY: Auburn Seminary Press, 1918 [mf ed 1993] – 1mf – 9 – 0-524-06347-8 – (incl bibl ref) – mf#1990-1530 – us ATLA [240]

History of aurangzib mainly based on original sources / Sarkar, Jadunath – Calcutta: MC Sarkar & Sons, 1912-1924 – (= ser Samp: indian books) – us CRL [954]

A history of auricular confession and indulgences in the latin church / Lea, Henry Charles – Philadelphia: Lea Bros 1896 [mf ed 1992] – 4mf – 9 – 0-524-02891-5 – (incl bibl ref) – mf#1990-4482 – us ATLA [241]

A history of auricular confession and indulgences in the latin church / Lea, Henry Charles – London. v1-3. 1896 – 3v on 30mf – 8 – €57.00 – ne Slangenburg [241]

The history of ayton : or anthonie the armenian, of asia, and specially touching the tartar / Heyt'owm Patmich – London, 1625-1626. v3 – 1mf – 9 – mf#HT-679 – ne IDC [915]

The history of babylonia and assyria = Das alte westasien / Winckler, Hugo; ed by Craig, James Alexander – New York: Scribner, 1907 [mf ed 1989] – 1mf – 9 – 0-7905-2513-5 – (trans by ed. incl ind) – mf#1987-2513 – us ATLA [930]

History of bangor theological seminary / Clark, Calvin Montague – Boston: Pilgrim Press, c1916 – 2mf – 9 – 0-524-07515-8 – mf#1991-3145 – us ATLA [240]

The history of baptism / Robinson, Robert – reprinted from the original London edition of 1790, with introduction and notes by J.R. Graves. Nashville: Southwestern Baptist Publ. Hse., 1860. Pub. No. 6380 – 1 – $35.60 – us Southern Baptist [242]

History of baptist churches in maryland : connected with the maryland baptist union association / Adams, George F et al; ed by Weishampel, John F – Baltimore: J F Weishampel, Jr 1885 [mf ed 1992] – 1mf [ill] – 9 – 0-524-03923-2 – mf#1990-4917 – us ATLA [242]

History of baptist indian missions / McCoy, Isaac – 1840 – 1 – us Southern Baptist [242]

History of baptists and their principles : century by century, to the present time / Stokes, William – 2nd carefully rev ed. London: Elliot Stock, [1866?] – 3mf – 9 – 0-524-08816-0 – mf#1993-3308 – us ATLA [242]

History of baptists in michigan / Trowbridge, Mary Elizabeth Day – [S.l.]: Pub under the auspices of the Michigan Baptist State Convention, 1909 – 4mf – 9 – 0-524-08821-7 – mf#1993-3313 – us ATLA [242]

A history of baptists in nigeria 1849-1935 : with appropriate projections into later years / Roberson, Cecil F – 1986 – 1 – $16.56 – us Southern Baptist [242]

History of baptists in north carolina / Williams, Charles – 1901 – 1 – 9.45 – us Southern Baptist [242]

History of baptists of louisiana / Christian, John T – Shreveport: Executive Beard, Louisiana Baptist Convention, 1923 – 1 reel – 1 – $10.40 – (260p) – us Southern Baptist [242]

History of barbados, 1625-1685 / Harlow, Vincent Todd – Oxford, England. 1926 – 1r – us UF Libraries [972]

History of barnesville, ohio : newspaper scrapbook, 1883-1897 / Wilson – 1r – 1 – mf#B27433 – us Ohio Hist [978]

A history of baseball in asia : assimilating, rejecting, and remaking america's game / Reaves, Joseph A – 1998 – 3mf – 9 – $12.00 – mf#PE 4008 – us Kinesology [790]

History of beaver creek baptist church / Thompson, Mrs. J Frank – us Southern Baptist [242]

History of bengali language and literature / Sen, Dinesh Chandra – Calcutta, India. 1954 – 1r – us UF Libraries [490]

History of bengali language and literature : a series of lectures delivered as reader to the calcutta university / Sen, Dineshchandra – Calcutta: University of Calcutta, 1911 – (= ser Samp: indian books) – us CRL [490]

History of bengali literature in the nineteenth century, 1800-1825 / De, Sushil Kumar – Calcutta: University of Calcutta, 1919 – (= ser Samp: indian books) – us CRL [490]

A history of bethlehem baptist association / Kellie, E I – 62p. 1851-1896 – 1 – $5.00 – us Southern Baptist [242]

A history of bohemian literature / Luetzow, Franz Heinrich Hieronymous Valentin, Graf von – new ed. London: W Heinemann, 1907 [mf ed 1990] – (= ser Short histories of the literatures of the world 7) – 2mf – 9 – 0-7905-5184-5 – (incl bibl ref) – mf#1988-1184 – us ATLA [460]

A history of brajabuli literature : being a study of the vaisnava lyric poetry and poets of bengal / Sen, Sukumar – Calcutta: University of Calcutta, 1935 – (= ser Samp: indian books) – us CRL [490]

History of brazil / Calogeras, Joao Pandia – Chapel Hill, North Carolina. 1939 – 1r – us UF Libraries [972]

History of brazil : comprising a geographical account of that country... / Grant, Andrew – London 1809 [mf ed Hildesheim 1995-98] – 1v on 2mf – 9 – €60.00 – ISBN-10: 3-487-26879-5 – ISBN-13: 978-3-487-26879-8 – gw Olms [972]

History of british america : for the use of schools / Calkin, John Burgess – Halifax, NS: A & W Mckinlay, 1894 – (= ser Nova scotia school series) – 3mf – 9 – mf#29163 – cn Canadiana [971]

History of british columbia, 1792-1887 / Bancroft, Hubert Howe – San Francisco: History Co, 1890 – 9mf – 9 – (incl ind) – mf#14094 – cn Canadiana [971]

A history of british diplomacy at the court of the peshwas, 1786-1818 : based on english records of mahratta history / Choksey, Rustom Dinshaw – Poona: R D Choksey, 1951 [mf ed 1991] – iii/xix/399p – 1 – (with bibl) – mf#7676 – us Wisconsin U Libr [954]

History of british honduras / Donohoe, William Arlington – Montreal, Quebec. 1946 – 1r – us UF Libraries [972]

A history of british india / Hunter, William Wilson – London, New York: Longmans, Green and Co, 1899-1900 – (= ser Samp: indian books) – us CRL [954]

A history of british india / MacFarlane, Charles – London 1852 – (= ser 19th c british colonization) – 7mf – 9 – mf#1.1.5650 – uk Chadwyck [954]

The history of british india / Mill, James – London 1817 – (= ser 19th c british colonization) – 3v on 25mf – 9 – mf#1.1.3379 – uk Chadwyck [954]

History of british relations with zanzibar, 1800-86 / Lewis, O T – Cardiff, 1936 – us CRL [960]

History of broward county / Miner, Frances H – s.l, s.l? 1936 – 1r – us UF Libraries [978]

History of brown university, 1764-1914 / Bronson, Walter C – 1 – us Southern Baptist [242]

History of buddhism = Chos hbyung / Bu-ston Rin-chen-grub – Heidelberg: In Kommission bei O Harrassowitz, 1931-1932 – (= ser Samp: indian books) – (trans fr tibetan by e obermiller) – us CRL [280]

The history of buddhist thought / Thomas, Edward Joseph – London, New York: Kegan Paul, Trench, Trubner & Co, 1933 – (= ser Samp: indian books) – us CRL [280]

History of burma : including burma proper, pegu, taungu,tenasserim, and arakan / Phayre, A P – London, 1883 – 4mf – 9 – mf#SE-20180 – ne IDC [915]

History of burmese literature : Mran ma ca pe samuin / Phe Mon Tan, U – Ran Kun: Nram ca can 1987 [mf ed 1990] – 1r with other items – 1 – (in burmese; Yan Kun: Pynn Su Paccnn Kopuire rahn [1965] mf-10289 seam reel 155/5) – mf#mf-10289 seam reel 153/3 [§] – us CRL [480]

History of butte county / Mansfield, George C – Butte Co, CA. 1918 – 1r – 1 – $50.00 – mf#B40206 – us Library Micro [978]

A history of cambodia from the earliest times to the end of the french protectorate / Ghosh, Manomohan – 2nd rev ed. Calcutta [Calcutta Oriental Book Agency] 1968 [mf ed 1989] – 1r with other items – 1 – mf#mf-10289 seam reel 003/07 [§] – us CRL [959]

A history of canada : for the use of schools / Archer, Andrew – London; New York: T Nelson, 1876 – (= ser Nelson's school series) – 6mf – 9 – (incl ind) – mf#26083 – cn Canadiana [971]

History of canada : for the use of schools / Archer, Andrew – London: T Nelson; Saint John, NB: J & A McMillan, 1877 – (= ser New brunswick school series) – 6mf – 9 – (incl ind) – mf#61148 – cn Canadiana [971]

The history of canada / Kingsford, William – 10v. 1881-98 – 1 – $337.00 – mf#0313 – us Brook [971]

A history of canon law in conjunction with other branches of jurisprudence : with chapters on the royal supremacy and the report of the commission on ecclesiastical courts / Dodd, Joseph – Oxford: Parker, 1884 [mf ed 1986] – 1mf – 9 – 0-8370-9858-0 – (incl bibl ref) – mf#1986-3858 – us ATLA [242]

The history of catholic emancipation and the progress of the catholic church in the british isles : (chiefly in england) from 1771 to 1820 / Amherst, William Joseph – London: Kegan Paul, Trench, 1886 – 2mf – 9 – 0-524-03783-3 – mf#1990-4855 – us ATLA [241]

A history of catholicity in northern ohio and in the diocese of cleveland : from 1749 to december 31, 1900 / Houck, George Francis – Cleveland: J B Savage 1903 [mf ed 1992] – 1mf – 9 – 0-524-03846-5 – (incl bibl ref) – mf#1990-4893 – us ATLA [241]

A history of catholicity in northern ohio and the diocese of cleveland / Carr, Michael W – Cleveland: J B Savage 1903 [mf ed 1992] – 2mf – 9 – 0-524-04041-9 – mf#1990-4949 – us ATLA [241]

A history of cavalry from the earliest times : with lessons for the future / Denison, George Taylor – 2nd ed. London: Macmillan & Co Ltd, 1913 – xxxi/468p (ill) – 1 – mf#9876 – us Wisconsin U Libr [355]

History of central africa / Tindall, P E N – New York, NY. 1986 – 1r – us UF Libraries [960]

History of ceylon – Peradeniya: University of Ceylon Press Board. v1. [1959-] – us CRL [954]

History of ceylon, from the earliest period to the year 1815 : with characteristic details of the religion, laws, & manners of the people... / Knox, Robert – London 1817 [mf ed Hildesheim 1995-98] – 1v on 9mf – 9 – €180.00 – ISBN-10: 3-487-27259-8 – ISBN-13: 978-3-487-27259-7 – gw Olms [954]

A history of charles the great (charlemagne) / Mombert, Jacob Isidor – New York: D Appleton, 1888 [mf ed 1991] – 2mf – 9 – 0-524-00578-8 – (incl bibl ref) – mf#1990-0078 – us ATLA [940]

A history of charleston association of baptist churches in the state of south carolina / Furman, Wood – 244p – 1 – $8.54 – us Southern Baptist [242]

History of childhood quarterly – New York. 1973-1976 (1) 1973-1976 (5) (9) – (cont by: journal of psychohistory) – ISSN: 0091-4266 – mf#7465 – us UMI ProQuest [150]

History of chinese literature / Giles, Herbert Allen – New York, NY. 1901 – 1r – us UF Libraries [480]

History of christ / Fox, W J – London, England. 1823 – 1r – us UF Libraries [240]

A history of christian doctrine / Shedd, William Greenough Thayer – New York: Scribner, 1863 [mf ed 1989] – 3mf – 9 – 0-7905-4118-1 – (incl bibl ref) – mf#1988-0118 – us ATLA [240]

History of christian doctrine / Fisher, George Park – New York: Charles Scribner, 1908, c1896 – 2mf – 9 – 0-8370-6328-0 – (incl bibl ref and index) – mf#1986-0328 – us ATLA [240]

History of christian doctrine / Sheldon, Henry Clay – 2nd ed. New York: Harper, c1895 – 3mf – 9 – 0-524-08646-X – mf#1993-2106 – us ATLA [240]

History of christian doctrines = Lehrbuch der dogmengeschichte / Hagenbach, Karl Rudolf – Edinburgh: T & T Clark, New York: Scribner & Welford [dist] 1880-81 [mf ed 1990] – (= ser Clark's foreign theological library 1,3,8) – 4mf – 9 – 0-7905-4740-6 – (incl bibl ref; in english; int by e h plumptre) – mf#1988-0740 – us ATLA [240]

History of christian ethics before the reformation = Geschichte der christlichen ethik vor der reformation / Luthardt, Christoph Ernst – Edinburgh: T & T Clark 1889 [mf ed 1986] – (= ser Clark's foreign theological library. new series 40) – 1mf – 9 – 0-8370-6209-8 – (incl bibl; trans fr german by william hastie) – mf#1986-0209 – us ATLA [230]

History of christian missions / Robinson, Charles Henry – New York: Scribner, 1915 – (= ser International Theological Library) – 2mf – 9 – 0-7905-8068-3 – mf#1988-6049 – us ATLA [240]

A history of christian missions during the middle ages / Maclear, George Frederick – Cambridge: Macmillan, 1863 [mf ed 1986] – 2mf – 9 – 0-8370-6214-4 – (companion vol: history of the christian church during the middle ages by charles hardwick. incl bibl ref and ind) – mf#1986-0214 – us ATLA [240]

A history of christian missions in south africa / Plessis, Johannes du – London, New York: Longmans, Green, 1911 [mf ed 1991] – 2mf – 9 – 0-524-00632-6 – mf#1990-0132 – us ATLA [240]

History of christian missions in south africa / Du Plessis, Johannes Christiaan – London: Longmans, Green & Co 1911 – (= ser [Travel descriptions from south africa, 1711-1938]) – 6mf – 9 – mf#zah-17 – ne IDC [240]

History of christian names / Yonge, Charlotte Mary – new rev ed. London: Macmillan, 1884 – 2mf – 9 – 0-524-01143-5 – mf#1990-0357 – us ATLA [240]

The history of christian preaching / Pattison, Thomas Harwood – Philadelphia: American Baptist Publication Society, 1903 – 2mf – 9 – 0-7905-5857-2 – (incl bibl ref) – mf#1988-1857 – us ATLA [240]

History of christian theology in the apostolic age = Histoire de la theologie chretienne au siecle apostolique / Reuss, Eduard – London: Hodder & Stoughton, 1872-74 [mf ed 1988] – 2v on 4mf – 9 – 0-7905-0214-3 – (trans by annie harwood. pref & notes by robert william dale. incl ind) – mf#1987-0214 – us ATLA [225]

A history of christian thought *see* Chi-tu chiao ssu hsiang shih [ccm261]

History of christianity : comprising all that relates to the progress of the christian religion in the history of the decline and fall of the roman empire; and, a vindication of some passages in the 15th and 16th chapters / Gibbon, Edward – New York: Peter Eckler, 1891 – 3mf – 9 – 0-524-03400-1 – mf#1990-0954 – us ATLA [240]

History of christianity = Kirchengeschichte im grundriss / Sohm, Rudolf – Cincinnati: Cranston & Stowe, 1891 [mf ed 1991] – 1mf – 9 – 0-524-01823-5 – (in english; rev, notes & additions by charles w rishell) – mf#1990-0503 – us ATLA [240]

The history of christianity : consisting of the life and teachings of jesus of nazareth, the adventures of paul and the apostles and the most interesting events in the progress of christianity, from the aarliest period to the present time / Abbott, John Stevens Cabot – Cleveland, OH: American Pub., [1877?] – 2mf – 9 – 0-7905-4060-6 – mf#1988-0060 – us ATLA [240]

History of christianity in china *see*
– Chung-kuo chi-tu chiao shih kang [ccm307]

A history of christianity in japan : protestant missions / Cary, Otis – New York: Fleming H Revell, c1909 [mf ed 1986] – 1mf – 9 – 0-8370-6654-9 – (incl bibl ref and ind) – mf#1986-0654 – us ATLA [242]

A history of christianity in japan : roman catholic and greek orthodox missions / Cary, Otis – New York: Fleming H Revell, c1909 [mf ed 1986] – 1mf – 9 – 0-8370-6655-7 – (incl bibl ref and ind) – mf#1986-0655 – us ATLA [241]

A history of christian-latin poetry : from the beginnings to the close of the m a / Raby, F J E – Oxford, 1953 – €21.00 – ne Slangenburg [450]

History of cisco baptist association in texas / Brannon, J D – 1955. 480p – 1 – us Southern Baptist [242]

History of citrus in florida – s.l, s.l? 193-? – 1r – us UF Libraries [634]

A history of civilization in palestine / Macalister, Robert Alexander Stewart – Cambridge: University Press; New York: G P Putnam 1912 [mf ed 1989] – (= ser The cambridge manuals of science and literature) – 1mf – 9 – 0-7905-1428-1 – (incl ind) – mf#1987-1428 – us ATLA [956]

A history of classical scholarship / Sandys, John Edwin – Cambridge: University Press, 1903-08 [mf ed 1992] – 4mf – 9 – 0-524-03422-2 – (incl bibl) – mf#1990-0976 – us ATLA [450]

History of clear creek baptist church, kentucky / Taylor, John – 1830 – 1 – 5.00 – us Southern Baptist [242]

History of cleveland in conflict, 1876-1900 – 1r – 1 – (1951 thesis by whipple) – mf#B25857 – us Ohio Hist [978]

A history of clifford's inn : with a chapter on present owners / Hay-Edward, C M – London: T W Laurie, 1912 – 3mf – 9 – $4.50 – mf#LLMC 84-293 – us LLMC [941]

History of coconut grove / Clark, Susan – s.l, s.l? 1939 – 1r – us UF Libraries [634]

History of colgate baptist church, baltimore, maryland, 19 oct 1945-nov 1961 – 1 – 6.89 – us Southern Baptist [242]

History of collier county / Miner, Frances H – s.l, s.l? 1936 – 1r – us UF Libraries [978]

History of collier county / Russell, H – s.l, s.l? 1939 – 1r – us UF Libraries [978]

History of colombia / Henao, Jesus Maria – Chapel Hill, North Carolina. 1938 – 1r – us UF Libraries [972]

A history of colonization on the western coast of africa / Alexander, Archibald – 2nd ed. Philadelphia: William S Martien, 1849 [mf ed 1989] – 2mf – 9 – 0-7905-4367-2 – mf#1988-0367 – us ATLA [960]

History of concord association, kentucky – 1821-1906 – 1 – 5.00 – us Southern Baptist [242]

History of conferences : and other proceedings connected with the revision of the book of common prayer from the year 1558 to the year 1690 / Cardwell, Edward – 3rd ed. Oxford: University Press, 1849 [mf ed 1990] – 2mf – 9 – 0-7905-4667-1 – (cont: the two books of common prayer) – mf#1988-0667l – us ATLA [242]

A history of conferences and other proceedings connected with the revision of the book of common prayer : from the year 1558 to the year 1690 / Cardwell, Edward – 2nd ed. Oxford: University Press, 1841 [mf ed 1984] – xiii/464p – 1 – (incl bibl. sequel to...'the two books of common prayer...') – mf#8827 – us Wisconsin U Libr [242]

A history of congregational independency in scotland / Ross, James – Glasgow: J MacLehose, 1900 [mf ed 1990] – 1mf – 9 – 0-7905-6426-2 – mf#1988-2426 – us ATLA [242]

History of congregationalism : from about a d 250 to 1616 / Punchard, George – Salem: John P Jewett, 1841 [mf ed 1993] – (= ser Congregational coll) – 5mf – 9 – 0-524-07362-7 – mf#1990-5399 – us ATLA [242]

History of congregations of the presbyterian church in ireland and biographical notices of eminent presbyterian ministers and laymen / Reid, James Seaton – Belfast: J. Cleeland; Edinburgh: J. Gemmell, 1886 – 1mf – 9 – 0-7905-5378-3 – mf#1988-1378 – us ATLA [242]

History of connecticut baptist state convention, 1823-1907 / Evans, Philip Saffrey – Hartford, Conn: Smith-Linsley Co, 1909 – 1mf – 9 – 0-524-03840-6 – mf#1990-4887 – us ATLA [242]

History of coon river congregation : a history of coon river congregation of the church of the brethren, in the middle district of iowa, to march 1, 1913 / Haughtelin, Jacob Diehl – Elgin, Ill: Brethren Pub House, 1913 – 1mf – 9 – 0-524-06873-9 – mf#1990-5292 – us ATLA [242]

A history of cragg vale, yorkshire / Hellowell, S G – v1. 1959 – 1r – 1 – mf#320 – uk Microform Academic [941]

The history of creation : or the development of the earth and its inhabitants by the action of natural causes. a popular exposition of the doctrine of evolution in general, and of that of darwin, goethe, and lamarck in particular / Haeckel, Ernst Heinrich Philipp August – [London] 1876 [i.e. 1875] – (= ser 19th c evolution & creation) – 10mf – 9 – mf#1.1.4253 – uk Chadwyck [577]

A history of creeds and confessions of faith in christendom and beyond : with historical tables / Curtis, William Alexander – Edinburgh: T & T Clark; New York: Scribner [dist] 1911 [mf ed 1990] – 2mf – 9 – 0-7905-5028-8 – (incl bibl ref) – mf#1988-1028 – us ATLA [240]

History of cumberland (maryland) : from the time of the indian town, caiuctucuc, in 1728, up to the present day / Lowdermilk, William Harrison – Washington: J Anglim, 1878 – 7mf – 9 – mf#07302 – cn Canadiana [978]

History of dade county government – s.l, s.l? 193-? – 1r – us UF Libraries [978]

The history of dahomey / Dalzel, A – London, 1793 – 11mf – 9 – mf#A-301 – ne IDC [916]

History of daytona beach / Davis, Mary Irene – s.l, s.l? 193-? – 1r – us UF Libraries [978]

A history of design in painted glass / Westlake, Nat Hubert John – London 1881-94 – (= ser 19th c art & architecture) – 9mf – 9 – mf#4.2.1288 – uk Chadwyck [740]

The history of dissenters : from the revolution to the year 1808 / Bogue, David & Bennett, James – 2nd ed. London: F Westley and A H Davis, 1833 [mf ed 1986] – 2v – 1 – mf#8091 – us Wisconsin U Libr [240]

The history of dissenters during the last thirty years (from 1808 to 1838) / Bennett, James – London: Printed for Hamilton, Adams, 1839 – 7mf – 9 – 0-524-08731-8 – mf#1993-3236 – us ATLA [240]

History of dixie county / Atkinson, Dorothy – s.l, s.l? 1936 – 1r – us UF Libraries [978]

History of doctrines in the ancient church = Die dogmengeschichte der alten kirche / Seeberg, Reinhold – rev 1904. Philadelphia, PA: Lutheran Pub Soc, c1905 [mf ed 1991] – (= ser Text-book of the history of doctrines 1) – 1mf – 9 – 0-7905-9881-7 – (in english) – mf#1989-1606 – us ATLA [240]

History of doctrines in the middle and modern ages = Die dogmengeschichte der mittelalters und neuzeit / Seeberg, Reinhold – rev ed. Philadelphia, PA: Lutheran Pub Soc, c1905 [mf ed 1991] – (= ser Lehrbuch der dogmengeschichte 2) – 2mf – 9 – 0-7905-8582-0 – (english trans by charles e hay. incl bibl ref) – mf#1989-1807 – us ATLA [240]

History of dogmas = Histoire des dogmes / Tixeront, Joseph – St Louis, MO: B Herder, 1910-16 [mf ed 1991] – 3v on 4mf – 9 – 0-7905-8941-9 – (incl bibl ref. in english) – mf#1989-2166 – us ATLA [240]

History of dr rowland taylor, martyr, 1555 – London, England. 18– – 1r – us UF Libraries [240]

A history of early baptist missions among the five civilized tribes / Moffitt, James W – 1946 – 1 – $7.28 – us Southern Baptist [242]

History of early christian art / Cutts, Edward Lewes – London: S.P.C.K.; New York: E. & J. Young, 1893 – (= ser Side-Lights Of Church History) – 1mf – 9 – 0-7905-4281-1 – (incl bibl ref) – mf#1988-0281 – us ATLA [700]

History of early christian literature in the first three centuries = Geschichte der altchristlichen litteratur in den ersten drei jahrhunderten / Krueger, Gustav – New York: Macmillan, 1897 – 1mf – 9 – 0-7905-4940-9 – (incl bibl ref. in english) – mf#1988-0940 – us ATLA [240]

The history of early english literature : being the history of english poetry from its beginnings to the accession of king aelfred / Brooke, Stopford Augustus – New York: Macmillan, 1892 – 2mf – 9 – 0-7905-7920-0 – mf#1989-1145 – us ATLA [420]

History of early florida railroads and jacksonville... / Shepherd, Rose – s.l, s.l? 1937 – 1r – us UF Libraries [380]

History of eclecticism in greek philosophy = Nacharistotelische philosophie / Zeller, Eduard – London: Longmans, Green, 1883 [mf ed 1991] – 1mf – 9 – 0-7905-9774-8 – (incl bibl ref; english trans fr german by s f alleyne) – mf#1989-1499 – us ATLA [180]

History of economic thought newsletter – Loughborough. 1977-1992 (1) 1977-1981 (5) 1977-1981 (9) – ISSN: 0440-9884 – mf#10641 – us UMI ProQuest [330]

History of edmund blackett – London, England. 1808 – 1r – us UF Libraries [240]

History of education – 32508mf – 9 – (filmed from the holdings of the milbank memorial library, teachers college, columbia university. subject breakouts available) – us Primary [370]

History of education – London. 1991-1996 – 1,5,9 – ISSN: 0046-760X – mf#17298 – us UMI ProQuest [370]

History of education / Painter, Franklin Verzelius Newton – New York, NY. 1886 – 1r – us UF Libraries [370]

History of education *see* Parliamentary history of the irish land question, from 1829 to 1869

A history of education before the middle ages / Graves, Frank Pierrepont – New York: Macmillan, 1909 [mf ed 1990] – 1mf – 9 – 0-7905-5396-1 – (incl bibl ref) – mf#1988-1396 – us ATLA [370]

A history of education during the middle ages : and the transition to modern times / Graves, Frank Pierrepont – New York: Macmillan, 1910 [mf ed 1990] – 1mf – 9 – 0-7905-4736-8 – (incl bibl ref) – mf#1988-0736 – us ATLA [370]

History of education in ancient india / Mazumder, Nogendra Nath – Calcutta: Macmillan & Co, 1916 – (= ser Samp: indian books) – us CRL [370]

History of education in delaware / Powell, Lyman P – (U.S. Bureau of Education Circular of Information no. 3). 1893 – 1 – $50.00 – us Presbyterian [370]

History of education in florida / Bush, George Gary – Washington, DC. 1889 – 1r – us UF Libraries [370]

A history of education in india : during the british / Nurullah, Syed – Bombay: Macmillan & Co, 1951 – (= ser Samp: indian books) – us CRL [370]

History of education in india under the rule of the east india company / Basu, Baman Das – Calcutta: Modern Review Office, [19–] – (= ser Samp: indian books) – us CRL [370]

History of education in medieval india : rise, growth, and decay of the aryan system of education, 600-1200 ad / Patwardhan, Chintamani Nilkant – Bombay: CN Patwardhan, 1939 – (= ser Samp: indian books) – us CRL [370]

A history of education in modern times / Graves, Frank Pierrepont – New York: Macmillan, 1913 [mf ed 1990] – 1mf – 9 – 0-7905-5397-X – (incl bibl ref) – mf#1988-1397 – us ATLA [370]

A history of education in virginia / Heatwole, Cornelius Jacob – New York: Macmillan, 1916 [mf ed 1988] – (= ser Home and school series) – xviii/383p – 1 – mf#7370 – us Wisconsin U Libr [370]

History of education quarterly – Bloomington. 1961+ (1) 1975+ (5) 1975+ (9) – ISSN: 0018-2680 – mf#10131 – us UMI ProQuest [370]

History of education society bulletin – Evington. 1977-1991 (1) 1977-1980 (5) 1977-1980 (9) – ISSN: 0018-2699 – mf#11130 – us UMI ProQuest [370]

A history of egypt : from the end of the neolithic period to the death of cleopatra 7, b c 30 / Budge, Ernest Alfred Wallis – London: Kegan Paul, Trench, Truebner 1902 [mf ed 1989] – (= ser Books on egypt and chaldaea 9-16) – 8v on 5mf – 9 – 0-8370-1183-3 – mf#1987-6017 – us ATLA [930]

A history of egypt / Zaidan, Jurji – Cairo, 1889 – 1 – us NY Public [960]

A history of egypt during the 17th and 18th dynasties / Petrie, William Matthew Flinders – London: Methuen, 1896 [mf ed 1990] – (= ser A history of egypt 2) – 1mf – 9 – 0-8370-1748-3 – mf#1987-6144 – us ATLA [930]

A history of egypt from the 19th to the 30th dynasties / Petrie, William Matthew Flinders – London: Methuen, 1905 [mf ed 1989] – (= ser A history of egypt 3) – 1mf – 9 – 0-7905-3273-5 – mf#1987-3273 – us ATLA [930]

A history of egypt from the earliest times to the 16th dynasty / Petrie, William Matthew Flinders – London: Methuen, 1894 [mf ed 1990] – (= ser A history of egypt 1) – 1mf – 9 – 0-8370-1749-1 – mf#1987-6145 – us ATLA [930]

A history of egypt in the middle ages / Lane-Poole, Stanley – London: Methuen, 1901 [mf ed 1989] – (= ser A history of egypt 6) – 1mf – 9 – 0-7905-3204-2 – (incl bibl ref) – mf#1987-3204 – us ATLA [960]

History of elementary education in india / Sen, Jitendra Mohan – Calcutta: Book Co, 1933 – (= ser Samp: indian books) – us CRL [370]

History of ellis county baptist association / Brooks, A D – 1907. 200p – 1 – 7.00 – us Southern Baptist [242]

The history of emily montague / Brooke, Frances – London: J Dodsley. 4v. 1769 [mf ed 1974] – 1r – 5 – mf#SEM16P114 – cn Bibl Nat [920]

History of england : from the accession of james 1 to the outbreak of the civil war, 1603-1642 / Gardiner, Samuel R – v1-10. 1884-86 – 9 – $267.00 – mf#0229 – us Brook [941]

History of england, a.d. 1800-1815; being an introduction to the history of the peace / Martineau, Harriett – London: G. Bell and Sons, 1878. xii,548p – 1 – us Wisconsin U Libr [941]

A history of england and greater britain / Cross, Arthur Lyon – New York: Macmillan, 1914 [mf ed 1990] – 3mf – 9 – 0-7905-6407-6 – (incl bibl ref) – mf#1988-2407 – us ATLA [941]

History of england during the reigns of king william, queen anne, and king george 1 / Ralph, James – 1978 – 1r – 1 – $130.00 – mf#S1855 – us Scholarly Res [941]

The history of england from the invasion of julius caesar to the abdication of james the second, 1688 / Hume, David – New ed. with the author's last corrections and improvements. New York: Harper & Bros., 1859-64. 6v – 1 – us Wisconsin U Libr [941]

The history of england, in easy verse : from the invasion of julius caesar, to the close of the year 1809. written for the purpose of being committed to memory by young persons of both sexes / Johnson, W R – 2nd corr ed. London: Tabart & Co, 1810 – (= ser 19th c children's literature) – 2mf – 9 – mf#6.1.54 – uk Chadwyck [810]

History of england under the anglo-saxon kings / Lappenberg, Johann Martin – London, England. v1. 1881 – 1r – us UF Libraries [941]

History of english congregationalism / Dale, Robert William; ed by Dale, Alfred William Winterslow – 2nd ed. London: Hodder and Stoughton, 1907 – 2mf – 9 – 0-524-02111-2 – (incl bibl ref) – mf#1990-4177 – us ATLA [242]

The history of english democratic ideas in the seventeenth century / Gooch, George Peabody – Cambridge: University Press; New York: Macmillan [distributor], 1898 – (= ser Cambridge Historical Essays) – 1mf – 9 – 0-7905-5331-7 – (incl bibl ref) – mf#1988-1331 – us ATLA [941]

The history of english glass painting / Drake, N M – London, 1912 – 10mf – 8 – mf#H-1349 – ne IDC [700]

History of english nonconformity from wiclif to the close of the nineteenth century / Clark, Henry William – London: Chapman and Hall, 1911 – 3mf – 9 – 0-7905-5383-X – (incl bibl ref) – mf#1988-1383 – us ATLA [240]

The history of english rationalism in the nineteenth century / Benn, Alfred William – London; New York: Longmans, Green, 1906 – 3mf – 9 – 0-7905-3540-8 – (incl bibl ref) – mf#1989-0033 – us ATLA [140]

The history of english rule and policy in south africa : a lecture...on friday, the 30th may 1879, at the request of the newcastle liberal association / Watson, Robert Spence – Newcastle-upon-Tyne [1879] – (= ser 19th c british colonization) – 1mf – 9 – mf#1.1.4947 – uk Chadwyck [320]

A history of english utilitarianism / Albee, Ernest – London: Sonnenschein, 1902 [mf ed 1990] – (= ser Library of philosophy) – 1mf – 9 – 0-7905-3748-6 – (incl bibl ref) – mf#1989-0241 – us ATLA [100]

The history of esarhaddon (son of sennacherib) king of assyria, b.c. 681-668 / Budge, Ernest Alfred Wallis – Boston: J R Osgood, 1881 – 1mf – 9 – 0-8370-7694-3 – (texts in akkadian and english; discussion in english. incl ind) – mf#1986-1694 – us ATLA [930]

History of ethics within organized christianity / Hall, Thomas Cuming – New York: Charles Scribner, 1910 – 2mf – 9 – 0-8370-6062-1 – (incl bibl ref and index) – mf#1986-0062 – us ATLA [170]

History of european ideas – Oxford. 1980-1996 (1,5,9) – ISSN: 0191-6599 – mf#49375 – us UMI ProQuest [940]

History of fayette county, pennsylvania / Jordan, John W & Hadd, James – Fayette, PA. New York: Lewis Historical Publ Co. v1-3. 1912 – 1r – 1 – (genealogical and personal history of fayette county pennsylvania) – us Western Res [920]

History of fifty years : comprising the origin, establishment, progress and expansion of the methodist episcopal church in southern asia / Scott, J E – Madras: Publ by authority of the Jubilee Managing Cttee, 1906 [mf ed 1995] – (= ser Yale coll) – xvi/367p/xv (ill) – 1 – 0-524-09947-2 – (incl ind) – mf#1995-0947 – us ATLA [242]

A history of fine art in india and ceylon / Smith, Vincent Arthur – Oxford: Clarendon Press, 1930 – (= ser Samp: indian books) – us CRL [700]

History of first baptist church, leitchfield, kentucky, sesquicentennial / McBeath, William H – 1804-1954. 39p – 1 – 5.00 – us Southern Baptist [242]

History of first baptist church, william lake, british columbia, canada / Janzen, D M – 1967-72 – 1 – $5.00 – us Southern Baptist [242]

History of flagler county / Davis, Mary Irene – s.l, s.l? 1936 – 1r – us UF Libraries [978]

History of florida / Brevard, Caroline Mays – New York, NY. 1915, c1904 – 1r – us UF Libraries [978]

History of florida / Brevard, Caroline Mays – New York, NY. 1919 – 1r – us UF Libraries [978]

History of florida / Fairlie, Margaret Carrick – Kingsport, TN. 1935 – 1r – us UF Libraries [978]

History of florida from the treaty of 1763 to our... / Brevard, Caroline Mays – Deland, FL. v1-2. 1924-1925 – 1r – us UF Libraries [978]

History of fort dallas / Francis, Mabel B – s.l, s.l? 1939 – 1r – us UF Libraries [978]

History of fort myers, florida / Hanson, W Stanley – s.l, s.l? 1936 – 1r – us UF Libraries [978]

The history of forty choirs / Hastings, Thomas – New York: Mason, 1854, c1853 [mf ed 1990] – 1mf – 9 – 0-7905-6528-5 – mf#1988-2528 – us ATLA [780]

The history of france / Godwin, Parke – v1., Ancient Gaul. New York: Harper, 1860. xxiv,495p. No more publ – 1 – us Wisconsin U Libr [944]

History of franklin association (illinois) of united baptists / Throgmorton, W P – 1880 – 1 – us Southern Baptist [242]

History of franklin county / Atkinson, Dorothy – s.l, s.l? 1936 – 1r – us UF Libraries [978]

The history of freedom : and other essays / Acton (of Aldenham), John Emerich Edward Dalberg, 1st Baron; ed by Figgis, John Neville & Laurence, Reginald Vere – London, Toronto: Macmillan, 1907 – 8mf – 9 – 0-665-66727-2 – (incl some text in french, german and latin) – mf#66727 – cn Canadiana [840]

A history of freedom of thought / Bury, John Bagnell – New York: Henry Holt; London: Williams and Norgate, c1913 [mf ed 1989] – (= ser Home university library of modern knowledge 69) – 1mf – 9 – 0-7905-4445-8 – (incl bibl ref) – mf#1988-0445 – us ATLA [140]

History of freemasonry : from the year 1829 to the present time / Oliver, George – London: Richard Spencer 1841 – 2mf – 9 – mf#vrl-87 – ne IDC [366]

The history of freemasonry : its legends and traditions, its chronological history. the history of the symbolism of freemasonry: the ancient accepted scottish rite and the royal order of scotland / Mackey, Albert Gallatin & Singleton, William R – New York; London: The Masonic History Co, 1906, c1898 – us CRL [366]

The history of freemasonry in nova scotia : an outline sketch / Edwards, Joseph Plimsoll – [Londonderry, NS?: s.n, 1916?] – 1mf – 9 – 0-659-90464-0 – (incl: the masonic stone of 1606 by reginald v harris) – mf#9-90464 – cn Canadiana [366]

History of french literature in the 18th century = Histoire de la litterature francaise au 18. siecle / Vinet, Alexandre Rodolphe – Edinburgh: T & T Clark, 1854 [mf ed 1990] – 2mf – 9 – 0-7905-7670-8 – (english by james bryce) – mf#1989-0895 – us ATLA [440]

A history of french painting / Stranahan, C H [Mrs] – London 1889 – (= ser 19th c art & architecture) – 6mf – 9 – mf#4.2.184 – uk Chadwyck [750]

History of fresno county and the san joaquin valley / Winchell, Lilbourne Alsip – Fresno Co, CA. 1933 – 1r – 1 – $50.00 – mf#B40218 – us Library Micro [978]

History of fs : or, the penitent female – London, England. 18– – 1r – us UF Libraries [240]

A history of furniture / Jacquemart, Albert – London 1878 – (= ser 19th c art & architecture) – 6mf – 9 – mf#4.2.1766 – uk Chadwyck [740]

History of gadsden county / Atkinson, Dorothy – s.l, s.l? 1936 – 1r – us UF Libraries [978]

History of gallia and lawrence county histories – 1r – 1 – mf#B27275 – us Ohio Hist [978]

History of geography / Keltie, Sir John Scott & Howarth, O J R – Illus. and maps. New York, London: G.P. Putnam, 1913. vii,208p – 1 – us Wisconsin U Libr [910]

A history of german literature = Geschichte der deutschen litteratur / Scherer, Wilhelm; ed by Mueller, Max – New York: C Scribner's Sons, 1886 [mf ed 1993] – 2v on 1r – 1 – (trans fr 3rd german ed by mrs f c conybeare. incl bibl ref and ind) – mf#8134 – us Wisconsin U Libr [430]

A history of german literature / Robertson, John George – 3rd rev enl ed. Edinburgh: W Blackwood, 1959 [mf ed 1993] – 1 – (incl bibl ref & ind) – mf#7841 – us Wisconsin U Libr [430]

A history of german literature / Robertson, John George – New York: G P Putnam's Sons Ltd; Edinburgh: W Blackwood & Sons Ltd, [1931] – 1 – (incl bibl ref & ind) – mf#8059 – us Wisconsin U Libr [430]

History of german theology in the nineteenth century = Histoire des idees religieuses en allemagne depuis le milieu du dixhuitieme siecle jusqu' a nos jours. selections / Lichtenberger, Frederic; ed by Hastie, William – Edinburgh: T & T Clark, 1889 – 2mf – 9 – 0-7905-5901-3 – (incl bibl ref. in english) – mf#1988-1901 – us ATLA [240]

A history of gingee and its rulers / Srinivasachari, Chidambaram S – Annamalainagar: [Annamalai] University, 1948 – (= ser Samp: indian books) – us CRL [954]

History of glass / Corning Museum of Glass – Clearwater Publ Co – 37mf – 15 – $970.00 – 0-88354-077-0 – (contains: ancient egypt & the ancient near east 2mf $55. roman empire & the near east 2mf $55. islamic near east 1mf $30. renaissance & later venice 1mf $30. continental europe, 500 to 1980 6mf $165. great britain, 1300-1980 2mf $55. us, 1700-1985 6mf $165. steuben glass: the frederick carder era 2mf $55. steuben glass, 1933-76 2mf $55. new glass 1979 6mf $165. masterpieces from czechoslovakia 2mf $55. glass paperweights 1mf $30. cameo glass 1mf $30. brief survey of the history of glass in the corning museum 3mf $85) – us UPA [740]

A history of gloucestershire : bodleian library, oxford, ms. 1714, books 1-5, ref. top.glouc. c.2 and c.3 bpa 5564 / Wantner, A – 1r – 1 – mf#482 – uk Microform Academic [941]

A history of god's church from its origin to the present time / Pond, Enoch – Philadelphia, PA: Ziegler & McCurdy, 1871 [mf ed 1992] – 3mf – 9 – 0-524-03413-3 – (incl ind) – mf#1990-0967 – us ATLA [240]

A history of gold as a commodity and as a measure of value : its fluctuations both in ancient and modern times, with an estimate of the probable supplies from california and australia / Ward, James – London [1852] – (= ser 19th c british colonization) – 2mf – 9 – mf#1.1.5570 – uk Chadwyck [380]

A history of gothic art in england / Prior, E S – London, 1900 – 13mf – 8 – mf#H-1294 – ne IDC [700]

The history of graded exercise testing in cardiac rehabilitation / Bickum, Bonnie D – 1992 – 2mf – $8.00 – us Kinesology [615]

The history of greece from its commencement to the close of the independence of the greek nation / Holm, Adolf – Tr. from the German by Frederick Clarke.London, New York: Macmillan, 1894-98. 4v – 1 – us Wisconsin U Libr [930]

A history of greek philosophy : from the earliest period to the time of socrates = Vorsokratische philosophie / Zeller, Eduard – London: Longmans, Green, 1881 [mf ed 1991] – 2v on 3mf – 9 – 0-7905-8985-0 – (english trans fr german by s s alleyne. with int) – mf#1989-2210 – us ATLA [180]

The history of greenland : including an account of the mission carried on by the united brethren in that country; with a continuation to the present time; ...and an appendix, containing a sketch of the mission of the brethren in labrador / Cranz, David – London 1820 [mf ed Hildesheim 1995-98] – 2v on 5mf – 9 – €100.00 – 3-487-27080-3 – gw Olms [990]

A history of gujarat : including a survey of its chief architectural monuments and inscriptions / Commissariat, Manekshah Sorabshah – Bombay: Longmans, Green & Co, 1938- – (= ser Samp: indian books) – us CRL [954]

The history of gutta-percha willie: the working genius / MacDonald, George – Eight page illus. by Arthur Hughes. New ed. London: Blackie, 1901. 212p. illus – 1 – us Wisconsin U Libr [830]

History of hamilton county baptist church library organization, chattanooga, tennessee – 1951-71 – 1 – 5.00 – us Southern Baptist [242]

A history of hand-made lace : dealing with the origin of lace / Jackson, Emily Nevill – London 1900 – (= ser 19th c art & architecture) – 4mf – 9 – mf#4.2.332 – uk Chadwyck [740]

History of hanover academy / Ford, David Barnes – Boston: HM Hight, 1899 – 1mf – 9 – 0-524-03771-X – mf#1990-1118 – us ATLA [373]

History of hardee county / Plowden, Jean – Wauchula, FL. 1929 – 1r – us UF Libraries [978]

The history of harvard university / Quincy, Josiah – Boston: Crosby, Nichols, Lee, 1860 – 4mf – 9 – 0-524-07760-6 – mf#1991-3328 – us ATLA [378]

History of hebron, ohio : map and newspaper clippings c(1874-1984) – 1r – 1 – mf#B14122 – us Ohio Hist [978]

A history of henderson and macfarlane ltd / Hallett, L – 1840-1902 – 1r – 1 – mf#pmb62 – at Pacific Mss [338]

History of herodotus / Herodotus – New York, NY. v1-4. 1893 – 1r – us UF Libraries [025]

The history of herodotus. a new english version / Herodotus – By George Rawlinson assisted by Col. Sir Henry Rawlinson and Sir J.G. Wilkinson. New York: D. Appleton & Co., 1859-60. 4v.ill. plates. fold. maps. plans – 1 – us Wisconsin U Libr [930]

The history of hindostan : its arts, and its sciences, as connected with the history of the other great empires of asia, during the most ancient periods of the world, with numerous illustrative engravings / Maurice, Thomas – London: Printed by W Bulmer and Co for the author, 1795-[1799?] – 15mf – 9 – 0-524-08777-6 – mf#1993-4017 – us ATLA [950]

History of hindu civilisation : as illustrated in the vedas and their appendages / Ghosha, Ramachandra – Calcutta: Ram and Friend, 1889 – 1mf – 9 – 0-524-02018-3 – (incl bibl ref) – mf#1990-2793 – us ATLA [280]

A history of hindu political theories : from the earliest times to the end of the first quarter of seventeenth century ad / Ghoshal, Upendra Nath – London: Oxford University Press, 1923 – (= ser Samp: indian books) – us CRL [954]

History of homeopathy and its institutions in america; their founders, benefactors, faculties, officers, hospitals.. / King, William Harvey – New York, Chicago: The Lewis Publishing Company, 1905. 4v. illus, plates, ports – 1 – us Wisconsin U Libr [610]

History of homestead / Gross, Abbie Mae – s.l, s.l? 1936 – 1r – us UF Libraries [978]

History of homestead / Sanderson, Isabelle – s.l, s.l? 1936 – 1r – us UF Libraries [978]

History of hopewall church / Lathan, Robert – 1879 – 9 – $50.00 – us Presbyterian [240]

History of hungarian music / Kaldy, Gyula [Julius] – New York: Haskell House Publ 1969 [mf ed 19–] – 1r – 1 – 0-8383-0305-6 – (1st publ 1902; iss as a musical standard extra; repr fr: the millennium of hungary and its people) – mf#film 583 – us Sibley [780]

The history of hyder shah, alias, hyder ali khan bahadur : and of his son, tippoo sultan / Maistre de La Tour – London: W Thacker & Co, 1855 – (= ser Samp: indian books) – (rev and corr by gholam mohammed) – us CRL [954]

History of indi / ed by A V Williams Jackson – London: Grolier Society, 1906-1907 – (= ser Samp: indian books) – us CRL [954]

A history of india : from the earliest times to the present day / Dunbar, George – London: Nicholson & Watson, 1943 – (= ser Samp: indian books) – us CRL [954]

History of india : from the earliest times to present day / Trotter, Lionel James – London: Society for Promoting Christian Knowledge, 1917 – (= ser Samp: indian books) – (rev by w h hutton) – us CRL [954]

History of india / ed by Jackson, Abraham Valentine Williams et al – London: Grolier Society, 1906-1907 [mf ed 1995] – (= ser Yale coll) – 9v/pl (ill) – 1 – 0-524-09574-4 – (incl ind) – mf#1995-0574 – us ATLA [954]

History of india / Nilakanta Sastri, Kallidaikurichi Aiyah – Madras: S Viswanathan, 1952- – (= ser Samp: indian books) – us CRL [954]

The history of india, as told by its own historians : the muhammadan period / Elliot, Henry Miers – London: Trnbner and Co, 1867- – (= ser Samp: indian books) – us CRL [954]

The history of india from the earliest ages / Wheeler, James Talboys – London: N Truebner, 1867-81 [mf ed 1993] – 4v on 29mf – 9 – 0-524-08779-2 – (incl bibl ref) – mf#1993-4019 – us ATLA [954]

A history of india from the earliest times / Dalal, Vaman Somnarayan – Bombay: V S Dalal, 1914- – (= ser Samp: indian books) – us CRL [954]

History of india under queen victoria : from 1836 to 1880 / Trotter, Lionel James – London 1886 – (= ser 19th c british colonization) – 11mf – 9 – mf#1.1.7387 – uk Chadwyck [954]

History of indian and eastern architecture / Fergusson, James – London 1876 – (= ser 19th c art & architecture) – 8mf – 9 – mf#4.2.18 – uk Chadwyck [720]

History of indian currency and exchange / Dadachanji, Bahran Edulji – Bombay: DB Taraporevala Sons & Co, 1931 – (= ser Samp: indian books) – us CRL [332]

A history of indian literature : from vedic times to the present day / Gowen, Herbert Henry – New York: D Appleton and Co, 1931 – (= ser Samp: indian books) – us CRL [490]

A history of indian literature / Winternitz, Moriz – Calcutta: University of Calcutta, 1927- – (= ser Samp: indian books) – (trans fr original german by s ketkar and rev by aut) – us CRL [490]

The history of indian literature / Weber Albrecht – London: Kegan Paul, Trench, Truebner, [1914] [mf ed 1995] – (= ser Yale coll; Truebner's oriental series) – xxiii/360p – 1 – 0-524-09147-1 – (trans fr 2nd german ed by john mann and theodor zachariae) – mf#1995-0147 – us ATLA [490]

A history of indian missions of the pacific coast : oregon, washington, idaho / Eells, Myron – Philadelphia: American Sunday-School Union, c1882 [mf ed 1986] – 1mf – 9 – 0-8370-6181-4 – (incl bibl ref) – mf#1986-0181 – us ATLA [240]

A history of indian philosophy / Dasgupta, Surendranath – London: Cambridge University Press, 1922-1961 – (= ser Samp: indian books) – us CRL [180]

A history of indian philosophy / Sinha, Jadunath – Calcutta: Central Book Agency, 1952- – (= ser Samp: indian books) – us CRL [180]

History of indian philosophy / Belvalkar, Shripad Krishna & Ranade, S K – Poona: Bilvakunja Pub House, 1927- – (= ser Samp: indian books) – us CRL [180]

A history of indian taxation / Banerjea, Pramathanath – London: Publ for the University of Calcutta by Macmillan and Co, 1930 – (= ser Samp: indian books) – us CRL [336]

History of indians in british guiana / Nath, Dwarka – London, England. 1950 – 1r – us UF Libraries [972]

History of intellectual development on the lines of modern evolution / Crozier, John Beattie – London: Longmans, Green, 1897-1901 – 3mf – 9 – 0-524-04639-5 – mf#1990-3382 – us ATLA [120]

The history of intelligence activities under general douglas macarthur, 1942-1950 / Supreme Command for the Allied Powers – 1984 – 8r – 1 – $1040.00 – mf#S1657 – us Scholarly Res [355]

The history of intemperance / Watkins, Thomas C – [Hamilton, Ont?: s.n, 189-?] [mf ed 1994] – 9 – 0-665-94628-7 – (in dble clms. original iss in ser: prohibition series) – mf#94628 – cn Canadiana [170]

The history of intercollegiate swimming at the college of william and mary (1928-1987) / Lanchantin, Margaret M – 1989 – 107p 2mf – 9 – $8.00 – us Kinesology [790]

History of interpretation : eight lectures preached before the university of oxford in the year 1885 on the foundation of the late rev. john bampton / Farrar, Frederic William – London: Macmillan, 1886 – 2mf – 9 – 0-8370-9863-7 – (incl indes) – mf#1986-3863 – us ATLA [220]

A history of iowa baptist schools / Abernethy, Alonzo – Osage, IA: A Abernethy, 1907 [mf ed 1986] – 1mf – 9 – 0-8370-8640-X – (incl ind) – mf#1986-2640 – us ATLA [377]

The history of ireland : from its union with great britain, in jan 1801 to oct 1810 / Plowden, Francis Peter – Dublin, 1811 – (= ser 19th c ireland) – 17mf – 9 – mf#1.1.9029 – uk Chadwyck [941]

The history of israel = Geschichte des volkes israel / Ewald, Heinrich; ed by Martineau, Russell – 4th ed., thoroughly rev. and corr. London: Longmans, Green, 1878-1886 – 9mf – 9 – 0-8370-1737-8 – (in english) – mf#1987-6133 – us ATLA [939]

History of jamaica from its discovery / Gardner, William James – London, England. 1909 – 1r – us UF Libraries [972]

A history of japan : cultural and political / Mukerji, Asit – Calcutta: Susil Gupta, 1945 – (= ser Samp: indian books) – us CRL [954]

The history of java / Raffles, T S – London, 1817. 2v – 15mf – 9 – mf#SE-20158 – ne IDC [915]

History of jenny hickling – Chelsea, England. 1815 – 1r – us UF Libraries [240]

History of journalism in the philippine islands / Valenzuela, Jesus Z – With introd. by Teodoro M. Kalaw, Willard Grosvenor Bleyer, Farael Palma. Manila: J.Z. Valenzuela, 1933. xiv,217p. illus., bibliog – 1 – us Wisconsin U Libr [070]

History of kanauj to the moslem conquest / Tripathi, Rama Shankar – Benares City: Indian Book Shop, 1937 – (= ser Samp: indian books) – (foreword by l d barnett) – us CRL [954]

The history of kathiawad from the earliest times / Wilberforce-Bell, Harold – London: William Heinemann, [1916] [mf ed 1996] – (= ser Yale coll) – xix/312p (ill) – 1 – 0-524-10226-0 – (pref by c h a hill) – mf#1996-1226 – us ATLA [954]

History of katsina / Daniel, F de F – [s.l: s.n, 1940?] – 1 – (filmed with: sokoto provincial gazetteer by p g harris) – us CRL [960]

A history of king's chapel in boston : the first episcopal church in new england / Greenwood, Francis William Pitt – Boston: Carter, Hendee, 1833 [mf ed 1990] – 1mf – 9 – 0-7905-5400-3 – mf#1988-1400 – us ATLA [240]

The history of korea / Hulbert, Homer Bezaleel – Seoul: Methodist Publ House, 1905 [mf ed 1995] – (= ser Yale coll) – 2v (ill) – 1 – 0-524-09816-6 – mf#1995-0816 – us ATLA [950]

History of korean art / Eckardt, Andre – London, England. 1929 – 1r – us UF Libraries [700]

History of lafayette county / Atkinson, Dorothy – s.l, s.l? 193-? – 1r – us UF Libraries [978]

History of lake county / Allen, L – s.l, s.l? 1936 – 1r – us UF Libraries [978]

History of lakeland / Lufsey, R E – s.l, s.l? 1936 – 1r – us UF Libraries [978]

History of latin america / Webster, Hutton – Boston, MA. 1924 – 1r – us UF Libraries [972]

History of latin christianity : including that of the popes to the pontificate of nicholas 5 / Milman, Henry Hart – 8v. 1874-83 – 9 – $267.00 – mf#0364 – us Brook [240]

History of learning : giving a succinct account and narrative of the choicest new books, etc – London. 1694-1694 – 1 – mf#4261 – us UMI ProQuest [370]

History of learning – London. 1691-1692 – 1 – mf#4260 – us UMI ProQuest [370]

History of lee county, florida / Hanson, Stanley – s.l, s.l? 1936 – 1r – us UF Libraries [978]

History of leiphardt : various other spellings included – 1r – 1 – mf#B41476 – us Ohio Hist [978]

History of lexington baptist church, oglethorpe county, georgia, 1847-1974 / Brooks, Gladys C – 1 – 5.00 – us Southern Baptist [242]

History of liberty county / Atkinson, Dorothy – s.l, s.l? 1936? – 1r – us UF Libraries [978]

History of little pat : the irish chimney-sweeper – London, England. 18– – 1r – us UF Libraries [240]

A history of lloyd's from the founding of lloyd's coffee house to the present day / Wright, Charles & Fayle, C Ernest – London: Macmillan & Co, 1928 – 7mf – 9 – $10.50 – mf#LLMC 92-183 – us LLMC [366]

The history of lord seaton's regiment, the 52nd light infantry at the battle of waterloo; together with various incidents connected with that regiment.. / Leeke, William – London: Hatchard and Co., 1866. 2v – 1 – us Wisconsin U Libr [941]

A history of lutheran missions / Laury, Preston A – 2nd rev ed. Reading, PA: Pilger Publ House, c1905 [mf ed 1986] – 1mf – 9 – 0-8370-6205-5 – (incl ind) – mf#1986-0205 – us ATLA [242]

History of madagascar...the progress of the christian mission established in 1818 : and an authentic account of the...martyrdom of the native christians / Ellis, William – London: Fisher, 1838. 2v. illus. plates, map, table – 1 – us Wisconsin U Libr [240]

History of madison county / Atkinson, Dorothy – s.l, s.l? 1936 – 1r – us UF Libraries [978]

The history of magic : including a clear and precise exposition of its procedure, its rites and its mysteries = dogme et rituel de la haute magie / Levi, Eliphas – London: William Rider, 1913 – 1mf – 9 – 0-524-04163-6 – (in english) – mf#1990-3293 – us ATLA [130]

History of mahoning baptist association / Smith, M A M – 1820-30 – 1 – 8.05 – us Southern Baptist [242]

A history of maithili literature / Misra, Jayakanta – Allahabad: Tirabhukti Publ, 1949 – (= ser Samp: indian books) – us CRL [490]

History of major john montgomery who came from ireland about 1720 or '24 / Montgomery, John – 1 – $50.00 – us Presbyterian [240]

History of marie antoinette / Abbott, John Stevens Cabot – New York: Harper & Bros, 1872 [mf ed 1987] – 322p – 1 – mf#2038 – us Wisconsin U Libr [944]

History of marin county / Munro-Fraser, J P – Marin Co, CA. 1880 – 1r – 1 – $50.00 – mf#B40228 – us Library Micro [978]

The history of maritime and inland discovery / Cooley, William – London 1830-31 [mf ed Hildesheim 1995-98] – 3v on 9mf – 9 – €180.00 – 3-487-29922-4 – gw Olms [910]

The history of mary, queen of scots / Mignet, F A – 7th ed. London: R. Bentley and son, 1887. xii,466p. Trans. by Andrew Scoble – 1 – us Wisconsin U Libr [920]

History of mary white – London, England. 18– – 1r – us UF Libraries [240]

History of mason and putnam county, west virginia, histories – 1r – 1 – mf#B27276 – us Ohio Hist [978]

History of materialism and criticism of its present importance / Lange, Friedrich Albert – London, 1877-81 – (= ser 19th c evolution & creation) – 3v on 13mf – 9 – (trans by ernest chester thomas) – mf#1.1.5398 – uk Chadwyck [140]

A history of matrimonial institutions : chiefly in england and the united states / Howard, George Elliott – Chicago: University of Chicago Press, 1904 [mf ed 1993] – 3v on 4mf – 9 – 0-524-05857-1 – (incl bibl ref) – mf#1990-3521 – us ATLA [390]

A history of matrimonial institutions chiefly in england and the united states : with an introductory analysis of the literature and the theories of primitive marriage and the family / Howard, George Elliott – Chicago: University of Chicago Press, Callaghan, 1904 [mf ed 1970] – (= ser Library of american civilization 21268-69) – 3v on 2mf – 9 – (with bibl ind) – us Chicago U Pr [929]

History of mauritius or the isle of france and the neighbouring islands : from their first discovery to the present time / Grant, Charles – London 1801 [mf ed Hildesheim 1995-98] – 1v on 7mf – 9 – €140.00 – ISBN-10: 3-487-26728-4 – ISBN-13: 978-3-487-26728-9 – gw Olms [960]

History of mediaeval hindu india : being a history of india from 600 to 1200 ad / Vaidya, Chintaman Vinayak – Poona City: CV Vaidya, 1921-1926 – (= ser Samp: indian books) – us CRL [954]

History of mediaeval india / Prasad, Ishwari – Allahabad: Indian Press, 1945 – (= ser Samp: indian books) – (foreword by l f rushbrook-williams) – us CRL [954]

A history of mediaeval jewish philosophy / Husik, Isaac – New York: Macmillan, 1916 [mf ed 1991] – 2mf – 9 – 0-7905-7774-7 – (incl bibl ref) – mf#1989-0999 – us ATLA [180]

The history of medieval vaishnavism in orissa / Mukherjee, Prabhat – Calcutta: R Chatterjee, 1940 – (= ser Samp: indian books) – us CRL [954]

A history of methodism : being a volume supplemental to a history of methodism by holland n mctyeire... / Bose, Horace Mellard du – Nashville, TN: Pub House of the ME Church, South, 1916 [mf ed 1992] – 2mf – 9 – 0-524-02884-2 – mf#1990-4475 – us ATLA [242]

A history of methodism : chiefly for the use of students / Gregory, John Robinson – London: Charles H Kelly, 1911 [mf ed 1993] – 2v on 2mf – 9 – 0-524-06250-1 – (incl bibl ref) – mf#1990-5205 – us ATLA [242]

A history of methodism : comprising a view of the rise of this revival of spiritual religion in the first half of the 18th century... / McTyeire, Holland Nimmons – Nashville, TN: Pub House of the Methodist Episcopal Church, South, 1898, c1884 [mf ed 1990] – 2mf – 9 – 0-7905-4896-8 – (incl bibl ref) – mf#1988-0896 – us ATLA [242]

The history of methodism in canada : with an account of the rise and progress of the work of god among the canadian indian tribes; and occasional notices of the civil affairs of the province / Playter, George Frederick – Toronto: A Green, 1862 – 1mf – 9 – 0-7905-7182-X – mf#1988-3182 – us ATLA [242]

The history of methodism in kentucky / Redford, Albert Henry – Nashville, Tenn: Southern Methodist Pub House, 1869-1870 – 4mf – 9 – 0-524-02841-9 – (incl bibl ref) – mf#1990-4462 – us ATLA [242]

History of methodism in minnesota / Hobart, Chauncey – Red Wing: Red Wing Printing, 1887 [mf ed 1993] – (= ser Methodist coll) – 1mf – 9 – 0-524-06950-6 – mf#1990-5314 – us ATLA [242]

The history of methodism in missouri for a decade of years from 1860 to 1870 / Lewis, William Henry – Nashville, TN: Pub House of the ME Church, South, 1890 [mf ed 1993] – (= ser Methodist coll) – 2mf – 9 – 0-524-06874-7 – mf#1990-5293 – us ATLA [242]

History of methodism in north carolina / Grissom, William Lee – Nashville, Tenn: Pub House of ME Church, South, 1905 – 1mf – 9 – 0-524-06251-X – mf#1990-5206 – us ATLA [242]

History of methodism in tennessee / M'Ferrin, John Berry – Nashville, Tenn: Pub House of the ME Church, South, 1886-1895 – 4mf – 9 – 0-524-02960-1 – mf#1990-4512 – us ATLA [242]

A history of methodists in the united states / Buckley, James Monroe – New York: Christian Literature, 1896 [mf ed 1989] – (= ser The american church history series 5) – 2mf – 9 – 0-7905-4191-2 – (incl bibl ref) – mf#1988-0191 – us ATLA [242]

History of methodists in the united states / Buckley, James Monroe – New York, NY. 1907 – 1r – us UF Libraries [025]

History of mexico / Parkes, Henry Bamford – Boston, MA. 1938 – 1r – us UF Libraries [972]

History of miami valley, ohio : local history, 1921-1937 – 1r – 1 – (h burba news articles) – mf#B25909 – us Ohio Hist [978]

History of military government in newly acquired... / Thomas, David Y – New York, NY. 1904 – 1r – us UF Libraries [320]

History of military government training / U.S. Office of the Provost Marshal General – v. 1-4. 1945? – 1 – us L of C Photodup [355]

A history of missions in india / Richter, Julius – Edinburgh: Oliphant [1908] – 1 – (trans by sydney h moore.) – mf#6706 – us Wisconsin U Libr [240]

History of missions in india = Indische missionsgeschichte / Richter, Julius – Edinburgh: Oliphant Anderson & Ferrier, [1908?] [mf ed 1986] – 2mf – 9 – 0-8370-6607-7 – (in english; incl bibl ref & ind) – mf#1986-0607 – us ATLA [242]

History of missions to china – 2nd ed. Boston: Massachusetts Sabbath School Society, 1841 [mf ed 1995] – (= ser Yale coll) – xi/252p (ill) – 1 – 0-524-10245-7 – mf#1996-1245 – us ATLA [240]

History of modern italian art / Willard, Ashton Rollins – [2nd ed]. London 1900 – (= ser 19th c art & architecture) – 9mf – 9 – mf#4.2.4 – uk Chadwyck [700]

History of modern marathi literature, 1800-1938 / Bhate, Govinda Cimanaji – Mahad, Dist Kolaba: GC Bhate, 1939 – (= ser Samp: indian books) – us CRL [490]

History of modern philosophy : from nicolas of cusa to the present time = Geschichte der neueren philosophie / Falckenberg, Richard – 1st American from the 2nd German ed. New York: H Holt, 1893 – 2mf – 9 – 0-7905-7818-2 – (incl bibl ref. in english) – mf#1989-1043 – us ATLA [190]

History of modern philosophy = Grundriss der geschichte der philosophie, bd 2 / Ueberweg, Friedrich – New York: Scribner, 1873 – 6mf – 9 – 0-524-00182-0 – (incl bibl ref. in english) – mf#1989-2882 – us ATLA [100]

History of modern philosophy in france / Levy-Bruhl, Lucien – Chicago: Open Court Pub Co, 1899 – 2mf – 9 – 0-524-08541-2 – (incl bibl ref) – mf#1993-2066 – us ATLA [190]

History of monterey and santa cruz counties / Watkins, Robin G – Monterey Co, CA. 1925 – 1r – 1 – $50.00 – mf#B40235 – us Library Micro [978]

History of monterey and santa cruz counties / Watkins, Robin G – Santa Cruz Co, 1925 – 1r – 1 – $50.00 – mf#B40263 – us Library Micro [978]

The history of montgomery classis, r.c.a / Dailey, William Nelson Potter – Amsterdam, NY: Recorder Press, [1916?] – 1mf – 9 – 0-524-08675-3 – mf#1993-3200 – us ATLA [240]

History of montreal and commercial register for 1885 / Borthwick, John Douglas – Montreal: Gebhardt-Berthiaume, 1885 – 2mf – 9 – mf#10199 – cn Canadiana [971]

A history of mughal north-east frontier policy : being a study of the political relation of the mughal empire with koch bihar, kamrup, and assam / Bhattacharyya, Sudhindra Nath – Calcutta: Chuckervertty, Chatterjee & Co, 1929 – (= ser Samp: indian books) – us CRL [954]

History of music : music from the renaissance through the early classical period – 107r – 1 – (based on holdings of the houghton, loeb and widener libraries at harvard university. coll offers nearly 1,000 items of printed music along with more than 400 works of music theory. a catalogue by david a wood accompanies the coll) – mf#C14R-11200 – us Primary [780]

A history of music in new england : with biographical sketches of reformers and psalmists / Hood, George – Boston: Wilkins, Carter, 1846 [mf ed 1990] – 1mf – 9 – 0-7905-4756-2 – (incl bibl ref) – mf#1988-0756 – us ATLA [780]

History of music project – 7v. 1939-42 – (= ser History of music in san francisco series) – 1 – $54.00 – mf#0268 – us Brook [780]

History of muskingum county : history by everhart and company, 1882 / Everhart and Co – 1r – 1 – mf#B27291 – us Ohio Hist [978]

History of muslim education / Shalaby, Ahmad – Karachi: Indus Publications, 1979 – us CRL [377]

History of my religious opinions / Newman, John Henry – [2nd ed] London: Longman, Green, Longman, Roberts & Green, 1865 [mf ed 1990] – 1mf – 9 – 0-7905-7253-2 – mf#1988-3253 – us ATLA [241]

History of mysore and the yadava dynasty / Josyer, G R – [Mysore: GR Josyer, 19–] – (= ser Samp: indian books) – us CRL [954]

The history of napa and lake counties – Napa Co, 1881 – 1r – 1 – $50.00 – mf#B40238 – us Library Micro [978]

History of napa county / Wallace, W F – 1r – 1 – $50.00 – mf#B40237 – us Library Micro [978]

History of napoleon buonaparte / Lockhart, John Gibson – London 1829 [mf ed Hildesheim 1995-98] – 2v on 10mf – 9 – €100.00 – ISBN-10: 3-487-26409-9 – ISBN-13: 978-3-487-26409-7 – gw Olms [940]

The history of negro baptists in mississippi / Thompson, Patrick H. – Jackson, Miss.: R.W. Bailey Printing Co., 1898 – 1r – 1 – 0-8370-1501-4 – mf#1984-B083 – us ATLA [242]

History of neo-african literature / Jahn, Janheinz – London, England. 1968 – 1r – us UF Libraries [410]

History of nevada county – Nevada Co, CA: Thompson & West, 1880 – 1r – 1 – $50.00 – mf#B40248 – us Library Micro [978]

The history of new england from 1630 to 1649 / Winthrop, John – new ed. Boston: Little, Brown, 1853 [mf ed 1992] – (= ser Congregational coll) – 2v on 3mf – 9 – 0-524-02178-3 – mf#1990-4244 – us ATLA [978]

A history of new england theology / Boardman, George Nye – New York: ADF Randolph, 1899 [mf ed 1989] – 1mf – 9 – 0-7905-4092-4 – mf#1988-0092 – us ATLA [240]

A history of new england with particular reference to the denomination of christians called baptists / Backus, Isaac – Newton, MA: Backus Historical Society, 1871. Chicago: Dep of Photodup, U of Chicago Lib, 1963 (1r); Evanston: American Theol Lib Assoc, 1984 (1r) – 1 – 0-8370-1474-3 – mf#1984-B006 – us ATLA [240]

The history of new holland : from its first discovery in 1616 to the present time. with a particular account of its produce and inhabitants; and a description of botany bay... / Eden, W – London, 1787 – 4mf – 9 – mf#HT-41 – ne IDC [917]

The history of new horizons : 20th anniversary celebrations / Chand, Sonal – 1998 – 2mf – 9 – $8.00 – mf#HE 619 – us Kinesology [614]

History of new mexico spanish and english missions of the methodist episcopal church from 1850 to 1910, in decades / Harwood, Thomas – Albuquerque, NM: Abogado Press, 1908-1910 – 11mf – 9 – 0-524-08763-6 – mf#1993-3268 – us ATLA [242]

History of new smyrna / Sweett, Zelia Wilson – s.l, s.l? 1936 – 1r – us UF Libraries [978]

History of new south wales / O'Hara, James – London 1818 [mf ed Hildesheim 1995-98] – 1v on 3mf – 9 – €90.00 – ISBN-10: 3-487-26814-0 – ISBN-13: 978-3-487-26814-9 – gw Olms [980]

The history of new south wales : including botany bay, port jackson, parramatta, sydney and all its dependancies from the original discovery of the island... / Barrington, George – London 1810 [mf ed Hildesheim 1995-98] – 1v on 4mf [ill] – 9 – €120.00 – 3-487-26805-1 – gw Olms [980]

History of new testament criticism / Conybeare, Frederick Cornwallis – New York: G P Putnam, 1910 [mf ed 1989] – 1mf – 9 – 0-7905-0637-8 – (incl ind) – mf#1987-0637 – us ATLA [225]

A history of norfolk / Rye, Walter – London: E Stock, 1885 [mf ed 1986] – 1r – 1 – (filmed with: ideia narodovlastiia / gerle, v i) – mf#1765 – us Wisconsin U Libr [941]

History of north american pinnipeds : a monograph of the walruses, sea-lions, sea-bears and seals of north america / Allen, Joel Asaph – Washington: GPO, 1880 – 9mf – 9 – (incl ind) – mf#02388 – cn Canadiana [590]

The history of north atlantic steam navigation : with some account of early ships and shipowners / Fry, Henry – London: S Low, Marston, 1896 – 4mf – 9 – (incl ind) – mf#29966 – cn Canadiana [380]

The history of north-eastern india : extending from the foundation of the gupta empire to the rise of the pala dynasty of bengal (c320-760 ad) / Basak, Radhagovinda – Calcutta: Book Co, 1934 – (= ser Samp: indian books) – us CRL [954]

History of northern rhodesia, early days to 1953 / Gann, Lewis H – London, England. 1964 – 1r – us UF Libraries [960]

A history of norway from the earliest times / Boyesen, Hjalmar Hjorth – London: T F Unwin, c1900 [mf ed 1987] – 1r – 1 – (with a new chapter on the recent history of norway by c f keary. filmed with: fresh tracks in the belgian congo / norden, h) – mf#1840 – us Wisconsin U Libr [948]

History of norwegian missionaries and immigrants in south africa / Hale, Frederick – Pretoria: Unisa 1986 [mf ed [s.l: s.n.] 1986] – 14mf – 9 – (incl bibl) – sa Unisa [240]

History of nova scotia, cape breton, the sable islands, new brunswick, prince edward island, the bermudas, newfoundland etc / Martin, Robert Montgomery – London: Whittaker, 1837 [mf ed 1984] – 5mf – 9 – 0-665-46128-3 – (original iss in ser: the british colonial library) – mf#46128 – cn Canadiana [917]

History of old testament criticism / Duff, Archibald – London: Watts, 1910 – 1mf – 9 – 0-8370-9938-2 – (incl ind) – mf#1986-3938 – us ATLA [221]

History of orange county / Allen, L – s.l, s.l? 1936 – 1r – us UF Libraries [978]

History of orange county / Blackman, William Fremont – Deland, FL. 1927 – 1r – us UF Libraries [978]

History of orange county / Harold, William G – s.l, s.l? 1936 – 1r – us UF Libraries [978]

History of orissa : from the earliest times to the british period / Banerji, Rakhal Das – Calcutta: R Chatterjee, 1930-1931 – (= ser Samp: indian books) – us CRL [954]

History of orissa : from the earliest times to the british period / Banerji, Rakhal Das – Calcutta: R Chatterjee, 1930-31 [mf ed 1987] – 2v/pl – 1 – mf#6932 – us Wisconsin U Libr [954]

History of orlando – s.l, s.l? 1937 – 1r – us UF Libraries [978]

History of orlando, florida / Ramsdell, Nellie B – s.l, s.l? 1936 – 1r – us UF Libraries [978]

The history of orlando furioso, 1594.. / Greene, Robert – London: Printed for the Malone Society by H. Hart at the Oxford University Press, 1907. x,60p. With facsimile of the original. 1 reel. 1249 – 1 – us Wisconsin U Libr [810]

History of osceola county / Moore-Willson, Minnie – Orlando, FL. 1935 – 1r – us UF Libraries [978]

A history of our firm : some account of the firm of pollok, gilmour and co, and its off-shoots and connections / Rankin, John – Liverpool: University Press of Liverpool, 1908 [mf ed 1987] – 1r – 1 – (filmed with: harrison's description of england / harrison, w) – mf#10695 – us Wisconsin U Libr [338]

The history of our lord as exemplified in works of art / Jameson, Anna Brownell [Murphy] et al – London 1864 – (= ser 19th c art & architecture) – 11mf – 9 – mf#4.1.275 – uk Chadwyck [700]

History of paganism in caledonia : with an examination into the influence of asiatic philosophy and the gradual development of christianity in pictavia / Wise, Thomas – London: Truebner, 1884 – 1mf – 9 – 0-524-02627-0 – (incl bibl ref) – mf#1990-3077 – us ATLA [290]

The history of painting in italy : from the period of the revival of the fine arts to the end of the eighteenth century / Lanzi, L – London. 6v. 1828 – 13mf – 9 – mf#O-981 – ne IDC [700]

A history of painting in north italy... : from the 14th to the 16th century / Crowe, Joseph Archer & Cavalcaselle, Giovanni Battista – London 1871 – (= ser 19th c art & architecture) – 14mf – 9 – mf#4.2.415 – uk Chadwyck [750]

The history of painting, sculpture, architecture, graving : and of those who have excell'd in them... / Monier, P – London, 1699 – 3mf – 9 – mf#O-985 – ne IDC [700]

The history of palestine from the patriarchal age to the present time : with introductory chapters on the geography and natural history of the country, and on the customs and institutions of the hebrews / Kitto, John – Boston: American Tract Society, [1851?] – 1mf – 9 – 0-524-06211-0 – mf#1992-0849 – us ATLA [956]

A history of pali literature / Law, Bimala Churn – London: Kegan Paul, Trench, Truebner & Co 1933 [mf ed 1996] – (= ser Samp: indian books) – 1r – 1 – (foreword by wilhelm gieger; filmed with other items) – mf#mf-10881 r027 – us CRL [490]

A history of panjabi literature, 1100-1932 : a brief study of reactions between panjabi life and letters based largely on important mss and rare and select, representative published works / Uberoi, Mohan Singh – Lahore: Mohan Singh, [193-] – (= ser Samp: indian books) – us CRL [490]

History of pasco county : dedicated to the school... / Hendley, J A – s.l, s.l? no date – 1r – us UF Libraries [978]

A history of persecution for the truth's sake in louisville, ky / Evans, Silas J – Louisville, KY: printed for aut, 1858 [mf ed 1993] – 1mf – 9 – 0-524-07816-5 – mf#1991-3363 – us ATLA [242]

The history of persia, from the most early period to the present time; containing an account of the religion, government, usages, and character of the inhabitants of that kingdom / Malcolm, John – New rev. ed. London: Murray, 1829. 2v – 1 – us Wisconsin U Libr [956]

A history of philosophy / Webb, Clement Charles Julian – London: Williams & Norgate [1915?] [mf ed 1991] – (= ser Home university library of modern knowledge 102) – 1mf – 9 – 0-7905-9746-2 – (incl bibl ref) – mf#1989-1471 – us ATLA [100]

History of philosophy, eastern and western : sponsored by the ministry of education, govt of india / ed by Radhakrishnan, Sarvepalli et al – London: George Allen & Unwin, 1952-1953 – (= ser Samp: indian books) – us CRL [100]

History of philosophy in epitome = Geschichte der philosophie im umriss / Schwegler, Albert – New York: D Appleton, 1890 [mf ed 1991] – 2mf – 9 – 0-524-00120-0 – (trans fr 1st ed of german original by julius h seelyne, rev fr 9th german ed with app by benjamin e smith) – mf#1989-2820 – us ATLA [100]

The history of philosophy in islam = Geschichte der philosophie im islam / Boer, Tjitze J de – London: Luzac, 1903 – (= ser Luzac's Oriental Religions Series) – 1mf – 9 – 0-524-03363-3 – (in english) – mf#1990-3197 – us ATLA [260]

History of phoenicia / Rawlinson, George – London; New York: Longmans, Green, 1889 – 3mf – 9 – 0-8370-1596-0 – (incl bibl ref) – mf#1987-6078 – us ATLA [930]

History of photography : from the international museum of photography at the george eastman house and other sources – 489r – 1 – $51,345.00 coll – (coll of monographs and serials publ between 1830-early 1900's. periodicals: 287r c39-12010. monographs and pamphlets 202r c39-12011. printed guide available) – us Primary [770]

History of photography – London. 1991-1996 (1) – ISSN: 0308-7298 – mf#17321 – us UMI ProQuest [770]

History of physical astronomy from the earliest ages to the middle of the 19th century : comprehending a detailed account of the establishment of the theory of gravitation by newton, and its development by his successors... / Grant, Robert – London: H G Bonn [1852?] [mf ed 1998] – 1r [ill] – 1 – (incl bibl ref & ind) – mf#film mas 28415 – us Harvard [520]

History of placer and nevada counties / Lardner, W B & Brock, M J – Nevada Co, CA. 1924 – 1 – mf#B40249 – us Library Micro [978]

History of platte presbytery / Clark, Walter H – 1910 – 1 – $50.00 – us Presbyterian [242]

History of platte river / Adams, J A – 1r – 1 – $35.00 – (mss) – mf^s-92 – us Commission [242]

History of plymouth plantation 1620-1647 / Bradford, William – Boston: Published for the Massachusetts Historical Society by Houghton Mifflin, 1912 – 3mf – 9 – 0-7905-5925-0 – mf#1988-1925 – us ATLA [975]

History of point ano nuevo, san mateo county / Stanyer, F M – San Mateo Co – 1r – 1 – $50.00 – mf#B40260 – us Library Micro [978]

History of political conventions in california / Davis, Winfield J – 1893 – 1r – 1 – $50.00 – mf#B50528 – us Library Micro [978]

History of political economy – Durham. 1969+ (1) 1975+ (5) 1975+ (9) – ISSN: 0018-2702 – mf#6564 – us UMI ProQuest [330]

The history of political science : from plato to the present / Murray, Robert H – New York: D Appleton 1926 – 5mf – 9 – $7.50 – mf#llmc92-238 – us LLMC [320]

A history of political theories from luther to montesquieu / Dunning, William A – New York, London: Macmillan Co, 1919 – 5mf – 9 – $7.50 – mf#LLMC 92-139 – us LLMC [320]

History of postwar africa / Hatch, John Charles – New York, NY. 1965 – 1r – us UF Libraries [960]

A history of preaching : from the apostolic fathers to the great reformers a d 70-1572 / Dargan, Edwin Charles – New York: A C Armstrong, 1905-12 [mf ed 1990] – 2v on 3mf – 9 – 0-7905-4903-4 – (incl bibl ref) – mf#1988-0903 – us ATLA [240]

A history of pre-buddhistic indian philosophy / Barua, Beni Madhab – [Calcutta]: University of Calcutta, 1921 – (= ser Samp: indian books) – us CRL [180]

History of pre-musalman india / Rangacharya, Vijayaraghava – Madras: Huxley Press, 1929- – (= ser Samp: indian books) – us CRL [954]

History of presbyterian church, fernandina, florida – s.l, s.l? 193-? – 1r – us UF Libraries [978]

A history of presbyterian education in east tennessee : an address delivered before the alumni association of king college at the commencement of 1897 / Caldwell, John Henderson – Bristol, TE: printed at J L King's Book and Job Office, 1897 [mf ed 1986] – 1mf – 9 – 0-8370-7775-3 – mf#1986-1775 – us ATLA [242]

The history of presbyterianism in arkansas, 1828-1902 – [S.l.: s.n., 1902?] (Little Rock: Arkansas Democrat Co) – 1mf – 9 – 0-524-02483-9 – mf#1990-4342 – us ATLA [242]

A history of presbyterianism in dublin and the south and west of ireland / Irwin, Clarke Huston – London: Hodder & Stoughton, 1890 [mf ed 1992] – 1mf – 9 – 0-524-02400-6 – (incl bibl ref) – mf#1990-0603 – us ATLA [242]

A history of presbyterianism in new england : its introduction, growth, decay, revival and present mission / Blaikie, Alexander – Boston: Published for the author by A. Moore, 1882. Chicago: Dep of Photodup, U of Chicago Lib, 1970 (1r); Evanston: American Theol Lib Assoc, 1984 (1r) – 1 – 0-8370-0322-9 – (includes bibliographical refernces and index) – mf#1984-B127 – us ATLA [242]

History of presbyterianism on prince edward island / MacLeod, John – Chicago, IL: Winona, 1904 – 1mf – 9 – 0-524-01740-9 – mf#1990-4132 – us ATLA [242]

History of primitive baptists in texas, oklahoma and indian territory / Newman, J S – 1906. v.1. 338p – 1 – us Southern Baptist [242]

The history of printing / Society for Promoting Christian Knowledge, Committee of General Literature and Education – London, 1862 – (= ser 19th c publishing...) – 3mf – 9 – mf#3.1.31 – uk Chadwyck [680]

History of printing in jamaica from 1717 to 1834 / Cundall, Frank – Kingston, Jamaica. 1935 – 1r – us UF Libraries [680]

The history of protestant missions in india : from their commencement in 1706 to 1871 / Sherring, Matthew Atmore – London 1875 – (= ser 19th c british colonization) – 6mf – 9 – mf#1.1.771 – uk Chadwyck [242]

The history of protestant missions in india : from their commencement in 1706-1881 / Sherring, Matthew Atmore – new rev ed. London: Religious Tract Society, 1884 – 2mf – 9 – 0-8370-6516-X – (includes appendixes) – mf#1986-0516 – us ATLA [242]

The history of protestant missions in india from their commencement in 1706 to 1881 / Sherring, Matthew Atmore – New ed., rev. and...to date. London: Religious Tract Soc., 1884. xv,463p. tables, 4 fold. maps – 1 – us Wisconsin U Libr [242]

History of protestant missions in japan = Dreissig jahre protestantischer mission in japan / Ritter, H; ed by Christlieb, Max – Tokyo: Methodist Pub House, 1898 [mf ed 1986] – 2mf – 9 – 0-8370-6771-5 – (english trans fr german by george e albrecht; incl bibl ref & ind) – mf#1986-0771 – us ATLA [242]

History of protestant missions in the near east = Mission und evangelisation im orient / Richter, Julius – New York: Fleming H Revell, c1910 [mf ed 1986] – 1mf – 9 – 0-8370-6345-0 – (incl ind) – mf#1986-0345 – us ATLA [242]

History of protestant nonconformity in wales : from its rise in 1633 to the present time / Rees, Thomas – 2nd ed, rev and considerably enl. London: John Snow, 1883 – 2mf – 9 – 0-524-00647-4 – mf#1990-0147 – us ATLA [242]

History of protestant theology : particularly in germany: viewed according to its fundamental movement and in connection with the religious, moral, and intellectual life = Geschichte der protestantischen theologie / Dorner, Isaak August – Edinburgh: T & T Clark. 2v. 1871 – 4mf – 9 – 0-8370-8663-9 – (incl ind) – mf#1986-2663 – us ATLA [242]

History of psychology / Baldwin, James Mark – New York, NY. v1-2. 1913 – 1r – us UF Libraries [150]

History of psychology : a sketch and an interpretation / Baldwin, James Mark – New York: G P Putnam, 1913 [mf ed 1990] – (= ser A history of the sciences) – 2v on 1mf – 9 – 0-7905-3530-0 – (incl bibl ref) – mf#1989-0023 – us ATLA [150]

A history of psychology, ancient and patristic / Brett, George Sidney – London: G Allen, 1912 [mf ed 1991] – (= ser Library of philosophy) – 1mf – 9 – 0-7905-8764-5 – (incl bibl ref) – mf#1989-1989 – us ATLA [150]

History of psychology from the standpoint of a thomist / Brennan, Robert Edward – New York, NY. 1945 – 1r – us UF Libraries [150]

A history of public education in rhode island : from 1636 to 1876 / ed by Stockwell, Thomas B – Providence: Providence Press Co, 1876 [mf ed 1986] – iii/458p – 1 – mf#8064 – us Wisconsin U Libr [370]

History of public school education in florida / Cochran, Thomas Everette – Lancaster, PA. 1921 – 1r – us UF Libraries [370]

History of puerto rico / Van Middeldyk, Rudolph Adams – New York, NY. 1910 – 1r – us UF Libraries [972]

History of queen elizabeth / Abbott, Jacob – New York, NY. 1877 – 1r – us UF Libraries [025]

History of rationalism : embracing a survey of the present state of protestant theology / Hurst, John F – New York: Scribner, Armstrong, c1865 – 2mf – us ATLA [242]

History of rationalism : embracing a survey of the present state of protestant theology / Hurst, John Fletcher – 9th ed., rev. New York: Scribner, Armstrong, c1865 – 2mf – 9 – 0-7905-5340-6 – mf#1988-1340 – us ATLA [100]

History of reformatory movements resulting in a restoration of the apostolic church : with a history of the nineteen general church councils / Rowe, John Franklin – Cincinnati: GW Rice, 1884 – 1mf – 9 – 0-524-03188-6 – mf#1990-4637 – us ATLA [240]

History of religion : a sketch of primitive religious beliefs and practices, and of the origin and character of the great systems / Menzies, Allan – 4th ed. New York: Scribner, 1913 – 2mf – 9 – 0-524-00937-6 – (incl bibl ref) – mf#1990-2160 – us ATLA [200]

History of religion in the old testament / Loehr, Max Richard Hermann – New York, NY. 1936 – 1r – us UF Libraries [025]

History of religions – Chicago. 1961+ (1) 1971+ (5) 1977+ (9) – ISSN: 0018-2710 – mf#2744 – us UMI ProQuest [200]

History of religions / Hopkins, Edward Washburn – New York, NY. 1918 – 1r – us UF Libraries [200]

History of religions / Moore, George Foot – New York: Scribner, 1913-1919 – (= ser International Theological Library) – 3mf – 9 – 0-524-07299-X – (incl bibl ref) – mf#1991-0085 – us ATLA [200]

History of religions / Moore, George Foot – New York, NY. v1-2. 1929 – 1r – us UF Libraries [200]

History of religious orders : a compendious and popular sketch of the rise and progress of the principal monastic, canonical, military, mendicant, and clerical orders and congregations of the eastern and western churches / Currier, Charles Warren – New York: Murphy & McCarthy, 1895, c1894 – 2mf – 9 – 0-8370-7132-1 – (incl bibl ref and index) – mf#1986-1132 – us ATLA [240]

A history of renaissance architecture in england 1500-1800 / Blomfield, Reginald Theodore – London 1897 – (= ser 19th c art & architecture) – 8mf – 9 – mf#4.2.1108 – uk Chadwyck [720]

History of robinson crusoe / Lutie, Aunt – New York, NY. 1872 – 1r – us UF Libraries [420]

History of robinson crusoe – New York, NY. 187-? – 1r – us UF Libraries [420]

History of roman literature : with an introductory dissertation on sources and formation of the latin language / ed by Thompson, Henry – 2nd rev and enl ed. London: JJ Griffin, 1852 – (= ser Encyclopaedia Metropolitana) – 2mf – 9 – 0-524-04656-5 – mf#1990-3399 – us ATLA [450]

History of roman private law / Clark, E C – Cambridge: The University Press. pts 1-3 in 4 bks. 1906-19 – 18mf – 9 – $27.00 – mf#LLMC 95-181 – us LLMC [346]

A history of rome : amply illustrated with maps, plans, and engravings / Leighton, Robert Fowler – New York: Clark & Maynard, 1879 [mf ed 1990] – (= ser Anderson's historical series) – 1r – 1 – (filmed with: die jahre der reaktion / bernstein, a d; a history of education in virginia / heatwole, c j; the story of the irish before the conquest / ferguson, m c) – mf#7370 – us Wisconsin U Libr [930]

The history of rome : from the earliest period to the close of the empire / Corner, Julia – London: Dean & Son, 1856 – (= ser 19th c children's literature) – 4mf – 9 – mf#6.1.18 – uk Chadwyck [930]

The history of rome : in easy verse. from the earliest period to the extinction of the western empire / Johnson, W R – London, 1808 – (= ser 19th c children's literature) – 2mf – 9 – mf#6.1.57 – uk Chadwyck [810]

History of rome and the popes in the middle ages = Geschichte roms und der paepste im mittelalter / Grisar, Hartmann – London: Kegan Paul, Trench, Truebner, 1911-12 [mf ed1992] – 3v on 3mf – 9 – 0-524-02887-7 – (english trans ed by luigi cappadelta. incl bibl ref) – mf#1990-4478 – us ATLA [931]

The history of roxbury town / Ellis, Charles Mayo – Boston: Samuel G Drake, 1847 – 1mf – 9 – 0-524-02641-6 – mf#1990-0665 – us ATLA [978]

History of russell creek association of baptists in kentucky – 1954 – 1 – 5.00 – us Southern Baptist [242]

History of russia / Rambaud, Alfred – Boston, MA. v1-3. 1879 – 2r – us UF Libraries [947]

History of russia, from earliest times to the rise of commercial capitalism / Pokrovskii, Mikhail N – Trans. and ed. by J.D. Clarkson and M.R.M. Griffiths. New York: International Publishers, c1931. xvi,383p. maps – 1 – us Wisconsin U Libr [947]

History of russian nobility : (books and periodicals on microfiche) from the national library of russia, st petersburg – 14 titles 213mf – 9 – $990.00 coll – (coll covers the period from the late 18th century to the 1910s) – us UMI ProQuest [920]

History of ruth clark – London, England. 18– – 1r – us UF Libraries [240]

History of s francis of assisi = Histoire de saint francois d'assise / Le Monnier, Leon – London: Kegan Paul, Trench, Truebner 1894 [mf ed 1986] – 2mf – 9 – 0-8370-7080-5 – (trans fr french; pref by h e cardinal vaughan) – mf#1986-1080 – us ATLA [241]

History of sacerdotal celibacy in the christian church / Lea, Henry Charles – 3rd ed., rev. London: Williams and Norgate, 1907 – 3mf – 9 – 0-7905-4943-3 – (incl bibl ref) – mf#1988-0943 – us ATLA [240]

History of san mateo county / Allen, B F – San Mateo Co, CA. 1883 – 1r – 1 – $50.00 – mf#B40257 – us Library Micro [978]

History of sandy creek baptist association, north carolina / Teague, H A – 1858-1958 – 1 – 6.72 – us Southern Baptist [242]

History of sandy run baptist church, hampton, s.c – 1903-64 – 1 – 5.00 – us Southern Baptist [242]

A history of sanskrit literature : classical period / ed by Dasgupta, S N – Calcutta: University of Calcutta, 1947- – (= ser Samp: indian books) – us CRL [490]

A history of sanskrit literature / Keith, Arthur Berriedale – Oxford: Clarendon Press, 1928 – (= ser Samp: indian books) – us CRL [490]

A history of sanskrit literature / Macdonell, Arthur Anthony – [aut ed] New York: D Appleton, 1914 [mf ed 1992] – (= ser Short histories of the literatures of the world) – 2mf – 9 – 0-524-05170-4 – (incl bibl ref) – mf#1990-3456 – us ATLA [490]

History of sanskrit literature = Samskrta sahitya ka itihasa / Prasada, Mahesacandra – 1st ed. Banarasa: Hitacintaka Presa 1922 [mf ed 1998] – (= ser Samp: indian books 16784) – 1r – 1 – (filmed with other items; in hindi) – mf#mf-11256 reel 058 – us CRL [490]

History of sanskrit literature / Vaidya, Chintaman Vinayak – Poona: [sn], 1930- – (= ser Samp: indian books) – us CRL [490]

History of santa clara county / Sawyer, Eugene T – Santa Clara Co, CA. 1922 – 1r – 1 – $50.00 – mf#B40261 – us Library Micro [978]

History of science and technology, series 1 : the papers of sir hans sloane, 1660-1753 from the british library london – [mf ed Marlborough 1991] – 7pt – 1 – (pt1: science & society 1660-1773 [17r] £1600; pts2,3: mss records of voyages of discovery 1492-1750 [20r/pt] £1850/pt. pts4,5: alchemy, chemistry & magic [18r/pt] £1700/pt [mf ed 2003]; pt6/7: history of medicine, surgery and anatomy [20r] £1850/[21r] £2050; with d/g) – uk Matthew [500]

History of science and technology, series 2 : series 2: the papers of sir joseph banks, 1743-1820 – 4pts – 1 – (pt1: correspondence and papers relating to voyages of discovery 1740-1805 from the british library, london 19r $2470. pt2: papers relating to voyages of discovery 1760-1800 16r $2080. pt3: correspondence and papers relating to voyages of discovery 1743-1853 16r $2080. pt4: correspondence and papers relating to voyages of discovery 1768-1820 from the state library of new south wales 14r $1820. with guides) – uk Matthew [500]

History of science and technology, series 3 : the papers of charles babbage, 1791-1871 – 1pt – 1 – (pt1: correspondence & scientific papers fr british library, london [22r] £2050; with d/g) – uk Matthew [500]

The history of science, health, and women – (= ser History of Women) – 23r – 1 – (with printed guide) – us Primary [305]

The history of scottish song / Borthwick, John Douglas – Montreal: Murray, 1874 – 3mf – 9 – (incl ind) – mf#00184 – cn Canadiana [780]

History of seaman, ohio : "town in the making," by frank g young / Young, Frank G – 1r – 1 – mf#B26308 – us Ohio Hist [978]

A history of secular latin poetry in the middle ages / Raby, F J E – Oxford. v1-2. 1957 – €35.00 – ne Slangenburg [450]

"History of seibert and boese families" / Seibert, Grant – 1 – us Kansas [920]

History of sennacherib / Sennacherib, King of Assyria; ed by Sayce, Archibald Henry – London: Williams and Norgate, 1878 – 1mf – 9 – 0-8370-7740-0 – mf#1986-1740 – us ATLA [930]

The history of servia and the servian revolution: with a sketch of the insurrection in bosnia – the slave provinces of turkey = Serbische revolution – slaves de turquie. Selections / Ranke, Leopold von & Robert, Cyprien – 3rd ed. London: H.G. Bohn, 1853 – 2mf – 9 – 0-7905-6120-4 – (in english) – mf#1988-2120 – us ATLA [949]

History of sherbro mission, west africa : under the direction of the missionary society of the united brethren in christ / McKee, William – Dayton, Ohio: United Brethren Pub House, 1874 – 1mf – 9 – 0-524-06269-2 – mf#1991-2460 – us ATLA [242]

A history of shipwrecks, and disasters at sea : from the most authentic sources / Redding, Cyrus – London 1833 [mf ed Hildesheim 1995-98] – 2v on 5mf – 9 – €100.00 – 3-487-29921-6 – gw Olms [910]

History of shoal creek association, missouri with history of her churches and biographies of ministers / Largen, T L – 1908 – 1 – 6.93 – us Southern Baptist [242]

The history of signboards : from the earliest times to the present day / Schevichaven, Herman Diederick Johan van & Hotten, John Camden – London 1866 – (= ser 19th c art & architecture) – 7mf – 9 – mf#4.2.1743 – uk Chadwyck [740]

History of sino-japanese war / U.S. Office of the Chief of Military History – 1967 – 1 – us L of C Photodup [951]

History of sixteenth battery : ovla, history, 1861-1865 – 1r – 1 – mf#B9825 – us Ohio Hist [978]

A history of slavery in cuba, 1511 to 1868 / Aimes, Hubert Hillary Suffern – New York, London: G P Putnam's Sons, 1907 [mf ed 1986] – xi/298p – 1 – mf#1739 – us Wisconsin U Libr [306]

History of solano and napa counties, california / Gregory, Tom – Napa Co, CA. 1912 – 1r – 1 – $50.00 – (incl biographical sketches) – mf#B40239 – us Library Micro [978]

History of solano county / Munro-Fraser, J P – Solano Co, CA. 1879 – 1r – 1 – $50.00 – mf#B40273 – us Library Micro [978]

History of solano county, vols 1 and 2 – Solano Co, CA: Marguerite Hunt, 1926 – 2v on 1r – 1 – $50.00 – mf#B40272 – us Library Micro [978]

History of soldiers' and sailors' monument / Gleason, William J – Cleveland, Ohio: The Monument Commissioners, 1894 – 1 reel – 1 – us Western Res [355]

History of south africa / De Kiewiet, Cornelius William – London, England. 1966 – 1r – us UF Libraries [960]

History of south africa / Theal, George Mccall – Cape Town, South Africa. v1-11. 1964 – 3r – us UF Libraries [960]

History of south africa / Walker, Eric Anderson – London, England. 1947 – 1r – us UF Libraries [960]

History of southern baptists, 1684-1918 / Riley, B F – 1057p – 1 – us Southern Baptist [242]

The history of sprinkling : being a compilation of the best thoughts of standard authors, historians and lexicographers of ancient and modern times... / Wilson, Louis Charles – 1st ed. Oskaloosa, Iowa: Tract Pub Co, 1895 – 1mf – 9 – mf#1990-4242 – us ATLA [240]

History of sri vijaya / Nilakanta Sastri, K A – Madras, 1949 – 3mf – 9 – mf#SE-811 – ne IDC [954]

The history of st dominic : founder of the friars preachers / Drane, Augusta Theodosia – London, New York: Longmans, Green, 1891 – 2mf – 9 – 0-7905-6465-3 – (incl bibl ref) – mf#1988-2465 – us ATLA [240]

History of st edmunds college, old hall / Ward, Bernard – London: K Paul, Trench, Truebner, 1893 – 1mf – 9 – 0-7905-6851-9 – mf#1988-2851 – us ATLA [378]

History of st vincent de paul : founder of the congregation of the mission (vincentians) and of the sisters of charity = Histoire de saint-vincent de paul / Bougaud, Emile – London; New York: Longmans, Green, 1908 – 1mf – 9 – 0-8370-6965-3 – mf#1986-0965 – us ATLA [920]

History of stanislaus county – Stanislaus Co, CA: Elliot, 1880 – 1r – 1 – $50.00 – mf#B40278 – us Library Micro [978]

The history of sumatra : containing an account of the government, laws, customs, and manners of the native inhabitants, with a description of the natural productions, and a relation of the ancient political state of that island / Marsden, William – London [1811]-1811 [mf ed Hildesheim 1995-98] – 2v on 6mf – 9 – €120.00 – 3-487-27249-0 – gw Olms [959]

The history of sunday schools and of religious education from the earliest times / Pray, Lewis Glover – Boston: Crosby and Nichols, 1847 – 1mf – 9 – 0-7905-6720-2 – mf#1988-2720 – us ATLA [240]

History of szechuen riots (may-june, 1895) / Cunningham, Alfred – Shanghai: "Shanghai Mercury" Office, [1895] [mf ed 1995] – (= ser Yale coll) – 38p/xxx – 1 – 0-524-09384-9 – mf#1995-0384 – us ATLA [951]

History of tampa 1910-1920 : supplementary / Lamme, Corinne W – s.l, s.l? 193-? – 1r – us UF Libraries [978]

History of tampa and hillsborough county : 1910-19 / Lamme, Corinne W – s.l, s.l? 193-? – 1r – us UF Libraries [978]

The history of tap dance in education : 1920-1950 / Arslanian, Sharon P – 1997 – 4mf – 9 – $16.00 – mf#PE 3744 – us Kinesology [790]

The history of tasmania / West, John – Tasmania 1852 – (= ser 19th c british colonization) – 9mf – 9 – mf#1.1.5057 – uk Chadwyck [980]

A history of ten baptist churches, kentucky-virginia / Taylor, John – 1770-1818 – 1 – $10.50 – (a journal of the author's life for more than fifty years) – us Southern Baptist [242]

History of the abambo / Ayliff, John – Cape Town, South Africa. 1962 – 1r – us UF Libraries [960]

History of the abington baptist association from 1807-1857 / Bailey, Edward L – Philadelphia: JA Wagenseller, 1863 – 1mf – 9 – 0-524-03924-0 – mf#1990-4918 – us ATLA [242]

A history of the abyssinian expedition / Markham, Clements R – London: Macmillan & Co, 1869 – 1 – (with a chapter containing an account of the mission and captivity of mr rassam and his companions, by lieutenant w f prideaux) – us CRL [960]

History of the act of queen anne, 1711 : restoring church patronage / Begg, James – Edinburgh, Scotland. 1840 – 1r – us UF Libraries [240]

A history of the adult school movement / Rowntree, John Wilhelm & Binns, Henry Bryan – London: Headley Bros, 1903 [mf ed 1993] – (= ser Society of friends (quakers) coll) – 2mf – 9 – 0-524-07109-8 – mf#1991-2932 – us ATLA [374]

History of the affairs of church and state in scotland from the beginning of the reformation to the year 1568 / Keith, Robert; ed by Lawson, John Parker & Lyon, Charles Jobson – Edinburgh: Printed for the Spottiswoode Society, 1844-1850 – 5mf – 9 – 0-524-05878-4 – mf#1990-5172 – us ATLA [240]

History of the alleghany evangelical lutheran synod of pennsylvania : together with a topical handbook of the evangelical lutheran church, its ancestry, origin and development / Carney, William Harrison Bruce – Philadelphia, Pa: Printed for the Synod by the Lutheran Publication Society, c1918 – 10mf – 9 – 0-524-08740-7 – mf#1993-3245 – us ATLA [242]

The history of the almohades / Al Marrekoshi, Abdo-'L-Wahid; ed by Dozy, R – 2nd ed. Leyden, 1881 – €14.00 – ne Slangenburg [260]

History of the amandebele / Child, Harold – Salisbury, Zimbabwe. 1968 – 1r – us UF Libraries [960]

History of the american bible society : revised, and brought down to the present time / Strickland, William Peter – New York: Harper, 1856 – 2mf – 9 – 0-8370-6704-9 – mf#1986-0704 – us ATLA [220]

A history of the american church to the close of the 19th century / Coleman, Leighton – New York: E S Gorham, 1903 [mf ed 1992] – (= ser Oxford church text books; Anglican/episcopal coll) – 1mf – 9 – 0-524-05357-X – (incl bibl ref) – mf#1990-5108 – us ATLA [242]

History of the american colony in liberia, 1821-1823 / Ashmun, Jehudi – 1 – 7.21 – us Southern Baptist [960]

History of the american episcopal church / McConnell, Samuel David – 10th rev enl ed. Milwaukee: Young Churchman, 1916 – 2mf – 9 – 0-524-03621-7 – (incl bibl ref) – mf#1990-4781 – us ATLA [240]

The history of the american episcopal church, 1587-1883 / Perry, William Stevens – Boston: J.R. Osgood, 1885 – 4mf – 9 – 0-7905-8065-9 – (incl bibl ref and index) – mf#1988-6046 – us ATLA [241]

History of the american lutheran church : from its commencement in the year of our lord 1685 to the year 1842 / Hazelius, Ernest L – Zanesville, Ohio: Edwin C. Church, 1846 – 1mf – us ATLA [242]

History of the american lutheran church : from its commencement in the year of our lord 1685 to the year 1842 / Hazelius, Ernest Lewis – Zanesville, O[hio]: Edwin C. Church, 1846 – 1mf – 9 – 0-7905-4746-5 – mf#1988-0746 – us ATLA [242]

History of the american privateers, and letters-of-marque during our war with england in the years 1812, '13 and '14 : interspersed with several naval battles between american and british ships of war / Coggeshall, George – 3rd rev corr enl ed. New York: G Coggeshall, 1861 [mf ed 1984] – 6mf – 9 – 0-665-44375-7 – mf#44375 – cn Canadiana [975]

A history of the american sunday-school union / Rice, Edwin Wilbur – Philadelphia: American Sunday-School Union, 1899 [mf ed 1990] – (= ser Sunday-school missionary 24/6) – 1mf – 9 – 0-7905-7135-8 – mf#1988-3135 – us ATLA [242]

A history of the american theological library association : a master's paper prepared for librarianship 397 / Tuttle, Marcia Lee – Atlanta: [s.n.], 1961. Chicago: Dep of Photodup, U of Chicago Lib, 1963? (1r); Evanston: American Theol Lib Assoc, 1984 (1r) – 1 – 0-8370-1473-5 – mf#1984-B005 – us ATLA [020]

A history of the anabaptists in switzerland / Burrage, Henry Sweetser – Philadelphia: American Baptist Pub Society, c1882 [mf ed 1990] – 1mf – 9 – 0-7905-4543-8 – (incl bibl ref) – mf#1988-0543 – us ATLA [242]

History of the ancient and mediaeval philosophy, vol 1 = Grundriss der geschichte der philosophie, bd 1 / Ueberweg, Friedrich – New York: Scribner, 1871 [mf ed 1991] – 5mf – 9 – 0-524-00183-9 – (english by geo s morris. additions by noah porter. pref by ed of the philosophical & theological library. incl bibl ref) – mf#1989-2883 – us ATLA [180]

History of the ancient chapel of stretford / Crofton, H T – (= ser Chetham society. new series 1-3) – 1r – 1 – mf#8459 – uk Microform Academic [920]

A history of the ancient egyptians / Breasted, James Henry – New York: Scribner, c1908 [mf ed 1989] – (= ser The historical series for bible students 5) – 2mf – 9 – 0-7905-0674-2 – (incl ind) – mf#1987-0674 – us ATLA [930]

History of the ancient parish of st mary's, sandbach (1890) : includes references to goostrey and church hulme chapels – (= ser Cheshire church registers) – 7mf – 9 – £6.50 – mf#380 – uk CheshireFHS [929]

A history of the ancient world : for high schools and academies / Goodspeed, George Stephen – New York: Charles Scribner, 1904 [mf ed 1992] – 2mf – 9 – 0-524-04458-9 – (incl bibl ref) – mf#1992-0127 – us ATLA [930]

History of the andover theological seminary / Woods, Leonard – Boston: J.R. Osgood, 1885 – 2mf – 9 – 0-7905-6699-0 – mf#1988-2699 – us ATLA [240]

A history of the architecture of the abbey church of st alban : with special reference to the norman structure / Buckler, John Chessell – London: Longman, Brown, Green & Longmans 1847 – 1 – (with: energy options / league of women voters education fund) – mf#2169 – us Wisconsin U Libr [720]

History of the arguments for the existence of god / Hahn, Aaron – Cincinnati: Bloch, 1885 – 1mf – 9 – 0-8370-3448-5 – (incl bibl ref) – mf#1985-1448 – us ATLA [210]

The history of the art of cutting in england / Giles, Edward Bowyer – [London] 1887 – (= ser 19th c art & architecture) – 3mf – 9 – mf#4.2.11 – uk Chadwyck [760]

A history of the articles of religion : to which is added a series of documents, from a d 1536 to a d 1615 / Hardwick, Charles – London: George Bell, 1890 [mf ed 1986] – 2mf – 9 – 0-8370-8676-0 – (texts in english & latin, discussion in english. incl bibl ref & pref) – mf#1986-2676 – us ATLA [242]

History of the associate reformed synod of the south : to which is prefixed a history of the associate presbyterian and reformed presbyterian churches / Lathan, Robert – Harrisburg, Pa.: Published for the author, 1882 – 1mf – 9 – 0-7905-5365-1 – mf#1988-1365 – us ATLA [242]

History of the associate reformed synod of the south : to which is prefixed a history of the associate presbyterian and reformed presbyterian churches / Lathan, Robert – Harrisburg, Pa.: Published for the author, 1882 – 1mf – us ATLA [242]

History of the atlantic telegraph / Field, Henry Martyn – New York: C Scribner, 1867 – 5mf – 9 – 0-665-90780-X – mf#90780 – cn Canadiana [380]

A history of the attempts to establish the protestant reformation in ireland : and the successful resistance of that people (time, 1540-1830) / McGee, Thomas D'Arcy – Boston: Patrick Donahoe, 1853, c1852 [mf ed 1986] – 1mf – 9 – 0-8370-7168-2 – mf#1986-1168 – us ATLA [242]

History of the auglaize annual conference of the united brethren church : from 1853 to 1891 / Luttrell, John Lewis – Dayton, OH: United Brethren Pub House, 1892 [mf ed 1992] – (= ser Methodist coll) – 2mf – 9 – 0-524-03169-X – mf#1990-4618 – us ATLA [242]

The history of the augsburg confession : from its origin till the adoption of the formula of concord / Stuckenberg, John Henry Wilbrandt – rev ed. Philadelphia: Lutheran Publ Society, c1897 – 1mf – 9 – 0-7905-6685-0 – mf#1988-2685 – us ATLA [240]

History of the azores, or western islands : containing an account of the government, laws, and religion, the manners, ceremonies, and character of the inhabitants: and demonstrating the importance of these valuable islands to the british empire / Ashe, Thomas – London 1813 [mf ed Hildesheim 1995-98] – (= ser Fbc) – 1v on 4mf – 9 – €120.00 – 3-487-27944-4 – gw Olms [946]

A history of the babylonians and assyrians / Goodspeed, George Stephen – New York: Charles Scribner, 1902 [mf ed 1986] – (= ser The historical series for bible students 6) – 1mf – 9 – 0-8370-9626-X – mf#1986-3626 – us ATLA [220]

History of the bahamas house of assembly / Malcolm, Harcourt Gladstone – Nassau, Bahamas. 1921 – 1r – us UF Libraries [972]

A history of the baptist at iredell, texas / Tidwell, D D – 78p – 1 – $5.00 – us Southern Baptist [242]

History of the baptist churches composing the sturbridge association from their origin to 1843 / Committee of the Sturbridge Association – New York: J.R. Bigelow, 1844 – 1 – 5.00 – us Southern Baptist [242]

A history of the baptist churches in the united states / Newman, Albert Henry – New York: Christian Literature, 1894 [mf ed 1989] – (= ser The american church history series 2) – 2mf – 9 – 0-7905-4234-X – mf#1988-0234 – us ATLA [242]

History of the baptist mission at ishokun, oyo / Roberson, Cecil F – 1826-1925 – 1 – us Southern Baptist [960]

History of the baptist missionary association of texas / Parks, W H – n.d. 136p – 1 – 5.00 – us Southern Baptist [242]

History of the baptist missionary society: from 1792 to 1842 – a sketch of the general baptist mission / Cox, Francis Augustus & Peggs, James – London: T Ward and G & J Dyer, 1842 – 3mf – 9 – 0-524-04370-1 – (incl bibl ref) – mf#1991-2074 – us ATLA [242]

History of the baptist young people's union of america / Conley, John Wesley – Philadelphia: Griffith & Rowland, c1913 – 2mf – 9 – 0-524-07404-6 – mf#1991-3064 – us ATLA [242]

A history of the baptists : traced by their vital principles and practices from the time of our lord and saviour jesus christ to the year 1886 / Armitage, Thomas – New York: Bryan, Taylor, 1887, c1886 [mf ed 1989] – 3mf – 9 – 0-7905-4244-7 – (incl bibl ref) – mf#1988-0244 – us ATLA [242]

History of the baptists in alabama / Holcombe, Hosea – 1840 – 1 – us Southern Baptist [242]

History of the baptists in maine / Burrage, Henry Sweetser – Portland, Me: Marks Printing House, Printers, 1904 – 2mf – 9 – 0-7905-4497-0 – (incl bibl ref) – mf#1988-0497 – us ATLA [242]

A history of the baptists in missouri : embracing an account of the organization and growth of baptist churches and associations / Duncan, Robert Samuel – St Louis: Scammel, 1882 [mf ed 1990] – 3mf – 9 – 0-7905-8139-6 – mf#1988-6086 – us ATLA [242]

A history of the baptists in new england / Burrage, Henry Sweetser – Philadelphia: American Baptist Publ Soc, 1894 [mf ed 1989] – (= ser Baptist history series 1) – 1mf – 9 – 0-7905-4193-9 – (incl bibl ref) – mf#1988-0193 – us ATLA [242]

A history of the baptists in the middle states / Vedder, Henry Clay – Philadelphia: American Baptist Publ Soc, 1898 [mf ed 1990] – (= ser Baptist history series 2) – 1mf – 9 – 0-7905-6386-X – mf#1988-2386 – us ATLA [242]

A history of the baptists in the southern states east of the mississippi / Riley, Benjamin Franklin – Philadelphia: American Baptist Publ Soc, 1898 [mf ed 1990] – (= ser Baptist history series 4) – 1mf – 9 – 0-7905-6669-9 – (incl bibl ref) – mf#1988-2669 – us ATLA [242]

A history of the baptists in the western states east of the mississippi / Smith, J A – 1896 – 1 – $14.70 – us Southern Baptist [242]

A history of the baptists in the western states east of the mississippi / Smith, Justin Almerin – Philadelphia: American Baptist Publ Soc, 1896 [mf ed 1990] – (= ser Baptist history series 3) – 1mf – 9 – 0-7905-6010-0 – mf#1988-2010 – us ATLA [242]

History of the baptists in vermont / Crocker, Henry – Bellows Falls, Vt: PH Gobie Press, 1913 – 2mf – 9 – 0-524-03493-1 – mf#1990-4715 – us ATLA [242]

History of the baptists of alabama : from the time of their first occupation of alabama in 1808 until 1894 / Riley, Benjamin Franklin – Birmingham: Roberts, 1895 – 2mf – 9 – 0-524-03562-8 – (incl bibl ref) – mf#1990-4757 – us ATLA [242]

A history of the baptists of hill county, texas / Daniel, J C – 1907 – 1 – $5.00 – us Southern Baptist [242]

A history of the baptists of louisiana : from the earliest times to the present / Paxton, William Edward – St Louis: CR Barnes, 1888 [mf ed 1992] – (= ser Baptist coll) – 2mf – 9 – 0-524-04365-5 – mf#1990-5048 – us ATLA [242]

History of the baptists of south carolina, 1683-1937 – Ms. Brief unpublished history by W.C. Allen. 565p – 1 – us Southern Baptist [242]

History of the baptists of tennessee / Hailey, O L – 448p – 1 – us Southern Baptist [242]

History of the baptists of the maritime provinces / Saunders, Edward Manning – Halifax, NS: John Burgoyne, 1902 – 2mf – 9 – 0-524-04179-2 – (incl bibl ref) – mf#1990-4983 – us ATLA [242]

A history of the baushi / Chimba, Barnabas – rev ed. Cape Town, Oxford UP,1949 – us CRL [960]

The history of the bengali language / Mazumdar, Bijay Chandra – [Calcutta]: University of Calcutta, 1927 – (= ser Samp: indian books) – us CRL [490]

History of the bengali language and literature : a series of lectures delivered as reader to the calcutta university / Sen, Dineshchandra – Calcutta: The University, 1911 [mf ed 1995] – (= ser Yale coll) – xii/1030p (ill) – 1 – 0-524-09481-0 – mf#1995-0481 – us ATLA [490]

A history of the bethel baptist association in kentucky / Masters, Frank M – 1807-1944 – 1 – $23.73 – us Southern Baptist [242]

History of the bible : arranged for use as a text-book / Mutch, William James – New Haven, CT: W J Mutch, c1901 – 1mf – 9 – 0-8370-4548-7 – mf#1985-2548 – us ATLA [220]

History of the big spring presbyterian church, newville, pa., 1737-1898 / Swope, Gilbert E – 1898 – 1 – $50.00 – us Presbyterian [242]

The history of the blessed virgin mary and the history of the likeness of christ which the jews of tiberias made to mock at : the syriac texts / ed by Budge, Ernest Alfred Wallis, Sir – London: Luzac, 1899 – (= ser Luzac's Semitic Text and Translation Series) – 2mf – 9 – 0-8370-1836-6 – mf#1987-6224 – us ATLA [220]

History of the boers in south africa / Theal, George Mccall – New York, NY. 1969 – 1r – us UF Libraries [960]

The history of the book of common prayer / Pullan, Leighton – London: Longmans, Green 1900 [mf ed 1992] – (= ser The oxford library of practical theology) – 1mf – 9 – 0-524-03183-5 – mf#1990-4632 – us ATLA [242]

A history of the book of common prayer and other books of authority : with an attempt to ascertain how the rubrics and canons have been understood and observed from the reformation to the accession of george 3 / Lathbury, Thomas – 2nd ed. Oxford: John Henry & James Parker, 1859 [mf ed 1992] – (= ser Anglican/episcopal coll) – 2mf – 9 – 0-524-05361-8 – mf#1990-5112 – us ATLA [242]

The history of the book of common prayer in its bearing on present eucharistic controversies / Dimock, Nathaniel – Memorial ed. London: Longmans, Green, 1910 – 1mf – 9 – 0-524-02462-6 – mf#1990-4321 – us ATLA [240]

History of the books of the new testament / Jacquier, Eugene – London: Kegan Paul, Trench, Truebner, 1907 – (= ser The International Catholic Library) – 1mf – 9 – 0-524-06739-2 – mf#1992-0942 – us ATLA [225]

History of the borough, castle, and barony of alnwick / Tate, George – Alnwick, England. v1-2. 1866-1869 – 1r – us UF Libraries [025]

History of the boston navy yard, 1797-1874 / Preble, George Henry – (= ser Naval records coll of the office of naval records and library) – 1r – 1 – (with printed guide) – mf#M118 – us Nat Archives [355]

History of the bow : its form and technique / Stiles, Elizabeth Helen – U of Rochester 1947 [mf ed 19–] – 1r – 1 – (with bibl) – mf#film 887 – us Sibley [780]

A history of the brahma samaj : from its rise to the present day / Leonard, G S – Calcutta: W Newman, 1879 [mf ed 1991] – 1mf – 9 – 0-524-02089-2 – mf#1990-2853 – us ATLA [280]

History of the brethren in virginia / Zigler, Daniel H – [rev ed] Elgin IL: Brethren Pub House 1914 [mf ed 1992] – 1mf – 9 – 0-524-03208-4 – mf#1990-4657 – us ATLA [242]

A history of the british and foreign bible society / Canton, William – London: J Murray, 1904-10 [mf ed 1989] – 5v on 6mf – 9 – 0-8370-1189-2 – (incl bibl ref) – mf#1987-6019 – us ATLA [220]

The history of the british and foreign bible society : from its institution in 1804, to the close of its jubilee in 1854 / Browne, George Forrest – London: Bagster. 2v. 1859 – 4mf – 9 – 0-8370-6167-9 – (includes appendixes) – mf#1986-0167 – us ATLA [220]

The history of the british empire in india / Gleig, George Robert – London, 1830-1835 – (= ser 19th c books on british colonization) – 16mf – 9 – mf#1.1.8393 – uk Chadwyck [954]

History of the brooks artillery / Prince, Albert Happoldt – 1898 [mf ed Spartanburg SC: Reprint Co, 1981] – 2mf – 9 – mf#51-505 – us South Carolina Historical [355]

A history of the cambodian independence movement, 1863-1955 / Reddi, V M – Tirupati: Sri Venkateswara University [1970] [mf ed 1989] – 1r with other items – 1 – (with bibl) – mf#mf-10289 seam reel 022/07 [§] – us CRL [959]

A history of the campaigns of 1780 and 1781 in the southern provinces of north america / Tarleton, Banastre – London: printed for T Cadell...1787 [mf ed 1983] – 7mf – 9 – 0-665-41900-7 – mf#41900 – cn Canadiana [975]

History of the canon of the holy scriptures in the christian church = Histoire du canon des saintes-ecritures dans l'eglise chretienne / Reuss, Eduard – 2nd ed. Edinburgh: R W Hunter, 1891 [mf ed 1985] – 1mf – 9 – 0-8370-4882-6 – (trans fr french into english with aut's corr & rev by david hunter. incl bibl ref & ind) – mf#1985-2882 – us ATLA [220]

History of the case of professor w robertson smith : in the free church of scotland / Moncreiff, Henry Wellwood – Edinburgh: John Maclaren, [1880] Beltsville, Md: NCR Corp, 1978 (2mf); Evanston: American Theol Lib Assoc, 1984 (2mf) – (= ser Biblical crit – us & gb) – 9 – 0-8370-0741-0 – (incl bibl ref) – mf#1984-1076 – us ATLA [242]

History of the case of professor w robertson smith in the free chu... / Moncreiff, Henry Wellwood – Edinburgh, Scotland. 1880 – 1r – us UF Libraries [240]

History of the catholic church : or, christ in his church = Christus in seiner kirche / Businger, Lucas Caspar – New York: Benziger, c1881 [mf ed 1986] – 1mf – 9 – 0-8370-8008-8 – (english trans by richard brennan; with sketch of church in america by john gilmary shea) – mf#1986-2008 – us ATLA [241]

History of the catholic church : for use in seminaries and colleges = Lehrbuch fuer kirchengeschichte / Brueck, Heinrich – Einsiedeln: Benziger, 1885 – 1mf – 9 – 0-8370-8247-1 – (incl ind. in english) – mf#1986-2247 – us ATLA [241]

History of the catholic church in australasia : from authentic sources. containing many original and official documents in connection with the church in australasia, besides others from the archives of rome, westminster, and dublin / Moran, Patrick Francis, cardinal – [Sydney], [1897] – (= ser 19th c british colonization) – 14mf – 9 – mf#1.1.2445 – uk Chadwyck [241]

History of the catholic church in scotland : from the introduction of christianity to the present day = Geschichte der katholischen kirche in schottland / Bellesheim, Alphons – Edinburgh: William Blackwood, 1887-1890 – 5mf – 9 – 0-7905-5506-9 – (in english) – mf#1988-1506 – us ATLA [241]

A history of the catholic church in the dioceses of pittsburg and allegheny : from its establishment to the present time / Lambing, Andrew Arnold – New York: Benziger Bros 1880 [mf ed 1992] – 2mf – 9 – 0-524-03167-3 – mf#1990-4616 – us ATLA [241]

History of the catholic church in the nineteenth century (1789- 1908) / MacCaffrey, James – 2nd ed., rev. Dublin; M H Gill; St Louis, MO: B Herder, 1910 – 3mf – 9 – 0-7905-5111-X – (incl bibl ref) – mf#1988-1111 – us ATLA [241]

History of the catholic church in the united states : from the earliest settlement of the country to the present time = Eglise catholique dans les etats-unis / Courcy, Henri de – New York: PJ Kenedy, c1879 – 2mf – 9 – 0-524-06243-9 – (incl bibl ref. in english) – mf#1990-5198 – us ATLA [241]

History of the catholic church in the united states / Shea, John Dawson Gilmary – New York: J G Shea, 1890-92 [mf ed 1990] – 2v on 4mf – 9 – 0-7905-8076-4 – (incl bibl ref) – mf#1988-6057 – us ATLA [241]

History of the catholic missions : among the indian tribes of the united states, 1529-1854 / Shea, John Dawson Gilmary – New York: P J Kenedy, c1854 [mf ed 1986] – 2mf – 9 – 0-8370-6375-2 – (incl ind) – mf#1986-0375 – us ATLA [241]

A history of the cavalry from the earliest times : with lessons for the future / Denison, George Taylor – London: Macmillan, 1877 – 7mf – 9 – (incl ind) – mf#06213 – cn Canadiana [355]

History of the cayuga baptist association / Belden, A Russell – Auburn: Derby & Miller, 1851 – 1mf – 9 – 0-524-03927-5 – mf#1990-4921 – us ATLA [242]

History of the ceramic art : a descriptive and philosophical study / Jacquemart, Albert – London 1873 – (= ser 19th c art & architecture) – 8mf – 9 – mf#4.2.1700 – uk Chadwyck [730]

History of the chagga people of kilimanjaro / Stahl, Kathleen Mary – London, England. 1964 – 1r – us UF Libraries [960]

History of the charleston association of baptist churches in south carolina – 1683-1802 – 1 – 8.61 – us Southern Baptist [242]

The history of the charleston library society / Porcher, Elizabeth L – New York: Elizabethan Press [mf ed Spartanburg SC: Reprint Co, 1980?] – 1mf – 9 – mf#51-530 – us South Carolina Historical [020]

History of the chemung baptist association / Smiley, Thomas – 1796-1829 – 1 – us Southern Baptist [242]

A history of the cheshire county union of congregational churches / Powicke, Frederick James – Manchester: T Griffiths, 1907 [mf ed 1990] – 1mf – 9 – 0-7905-5619-7 – mf#1988-1619 – us ATLA [242]

History of the china theater / U.S. Army, China Theater – 1946 – 1 – us L of C Photodup [951]

A history of the choir and music of trinity church, new york : from its organization to the year 1897 / Messiter, Arthur Henry – New York: ES Gorham, 1906 [mf ed 1990] – 1mf – 9 – 0-7905-5436-4 – mf#1988-1436 – us ATLA [780]

History of the chorus in the german drama / Halmrich, Elsie Winifred – New York, NY. 1912 – 1r – us UF Libraries [430]

History of the christian church / Barth, Christian Gottlob, 1799-1862 – [Bombay]: American Mission, 1850 [mf ed 1995] – (= ser Yale coll) – 260p – 1 – 0-524-09444-6 – (trans into marathi) – mf#1995-0444 – us ATLA [240]

History of the christian church / Fisher, George Park – New York: Scribner, 1890, c1887 – 2mf – 9 – 0-7905-4522-5 – (incl bibl ref) – mf#1988-0522 – us ATLA [240]

History of the christian church : from its origin to the present time / Blackburn, William Maxwell – Cincinnati: Cranston & Stowe; New York: Phillips & Hunt, c1879 – 2mf – 9 – 0-8370-7286-7 – (incl bibl ref and ind) – mf#1986-1286 – us ATLA [240]

History of the christian church : from its origins to the present time / Smith, James – Nashville: Cumberland Presbyterian Church, 1835 – 1r – 1 – 0-8370-1534-0 – mf#1984-B218 – us ATLA [240]

History of the christian church : from the 4th to the 12th century / Carwithen, John Bayly Sommers & Lyall, Alfred – London: Richard Griffin, 1856 [mf ed 1991] – (= ser Encyclopaedia metropolitana. 3rd division. history and biography) – 1mf – 9 – 0-524-04307-8 – (incl bibl ref) – mf#1990-1233 – us ATLA [240]

History of the christian church : from the apostolic age to the reformation, a.d. 64-1517 / Robertson, James Craigie – new rev ed. New York: Pott, Young, 1874-1875 – 9mf – 9 – 0-524-03421-4 – mf#1990-0975 – us ATLA [240]

History of the christian church : from the earliest times to a.d. 461 / Foakes-Jackson, Frederick John – 6th ed. London: George Allen and Unwin, 1914 – 2mf – 9 – 0-524-00752-7 – (incl bibl ref) – mf#1990-0184 – us ATLA [240]

History of the christian church = Kirchengeschichte / Hase, Karl von – New York: D Appleton, 1856 [mf ed 1992] – 7mf – 9 – 0-524-03343-9 – (english trans fr 7th german ed by charles e blumenthal & conway p wing; incl bibl ref) – mf#1990-0924 – us ATLA [240]

History of the christian church / Schaff, P – New York, 1892 – 10mf – 9 – (modern christianity) – mf#ZWI-48 – ne IDC [240]

History of the christian church : from its first establishment to the present century = Short view of the history of the christian church / Reeve, Joseph – 3rd ed. Boston: Patrick Donahoe, 1857 – 2mf – 9 – 0-8370-7257-3 – (incl ind) – mf#1986-1257 – us ATLA [240]

History of the christian church, a d 1-600 / Moeller, Wilhelm – London: Swan Sonnenschein; New York: Macmillan, 1892 [mf ed 1991] – (= ser Lehrbuch der kirchengeschichte 1) – 2mf – 9 – (incl bibl ref. in english) – mf#1990-0367 – us ATLA [240]

History of the christian church, a.d. 1517-1648 : reformation and counter-reformation = Reformation und gegenreformation / Moeller, Wilhelm; ed by Kawerau, Gustav – London: Swan Sonnenschein; New York: Macmillan, 1900 – (= ser Lehrbuch der Kirchengeschichte) – 2mf – 9 – 0-524-01230-X – (incl bibl ref. in english) – mf#1990-0369 – us ATLA [240]

A history of the christian church during the first six centuries / Cheetham, S – London, New York: Macmillan, 1894 [mf ed 1989] – 2mf – 9 – 0-7905-4171-8 – (incl bibl ref) – mf#1988-0171 – us ATLA [240]

The history of the christian church during the first ten centuries : from its foundation to the full establishment of the holy roman empire and the papal power / Smith, Philip – New York: Harper, 1888 – 2mf – 9 – 0-524-03430-3 – (incl bibl ref) – mf#1990-0984 – us ATLA [240]

The history of the christian church during the middle ages : with a summary of the reformation, centuries 11 to 16 / Smith, Philip – New York: Harper & Bros, 1885 [mf ed 1991] – (= ser The student's series (new york); The student's ecclesiastical history 2) – 2mf – 9 – 0-524-01591-0 – mf#1990-0457 – us ATLA [240]

A history of the christian church during the reformation / Hardwick, Charles – new ed. London: Macmillan, 1880 [mf ed 1989] – 1mf – 9 – 0-7905-4477-6 – (incl bibl ref) – mf#1988-0477 – us ATLA [240]

History of the christian church from its establishment by christ to a.d. 1871 : including the rise of the roman heresy, all the popes, the temporal power, the abominations of popery and the reformation / Summerbell, Nicholas – 3rd ed. Cincinnati: Office of the Christian Pulpit, 1873 – 2mf – 9 – 0-524-02238-0 – mf#1990-4249 – us ATLA [240]

History of the christian church from the 13th century to the present day : including the history of the reformation / Lyall, Alfred et al – London: Richard Griffin, 1858 [mf ed 1992] – (= ser Encyclopaedia metropolitana. 3rd division. history and biography 40) – 2mf – 9 – 0-524-04138-5 – mf#1990-1208 – us ATLA [240]

History of the christian church in the apostolic times = Die kirche im apostolischen zeitalter / Thiersch, Heinrich Wilhelm Josias – 2nd ed. London: Thomas Bosworth, 1883 [mf ed 1985] – 1mf – 9 – 0-8370-5509-1 – (english trans by thomas carlyle; incl ind) – mf#1985-3509 – us ATLA [240]

History of the christian church in the middle ages = Das mittelalter / Moeller, Wilhelm – London: Swan Sonnenschein; New York: Macmillan, 1893 [mf ed 1991] – (= ser Lehrbuch der kirchengeschichte 2) – 1mf – 9 – 0-524-01229-6 – (english trans by andrew rutherfurd. incl bibl ref) – mf#1990-0368 – us ATLA [240]

History of the christian church in the second and third centuries / Jeremie, James Amiraux – London: JJ Griffin, 1852 – (= ser Encyclopaedia Metropolitana) – 1mf – 9 – 0-524-04493-7 – mf#1990-1255 – us ATLA [240]

A history of the christian church, middle age / Hardwick, Charles; ed by Stubbs, William – 3rd rev ed. London: Macmillan, 1872 [mf ed 1990] – 2mf – 9 – 0-7905-5468-2 – (incl bibl ref. 1st ed 1853) – mf#1988-1468 – us ATLA [240]

A history of the christian church since the reformation / Cheetham, S – London: Macmillan, 1907 [mf ed 1990] – 2mf – 9 – 0-7905-7162-5 – mf#1988-3162 – us ATLA [242]

History of the christian councils : from the original documents = Conciliengeschichte / Hefele, Karl Joseph von; ed by Clark, William – 2nd rev ed. Edinburgh: T & T Clark, 1872-96 [mf ed 1991] – 5v on 6mf – 9 – 0-524-00556-7 – (incl bibl ref) – mf#1990-0056 – us ATLA [240]

A history of the christian denomination in america, 1794-1911 a d / Morrill, Milo True – Dayton, OH: Christian Pub Assoc, 1912 [mf ed 1991] – (= ser Christian church (disciples of christ) coll) – 1mf – 9 – 0-524-01742-5 – (incl bibl ref) – mf#1990-4134 – us ATLA [240]

The history of the christian missions of the sixteenth, seventeenth, eighteenth, and nineteenth centuries : containing accounts of the propagation of christianity by the various missionary societies... / Brown, William – 3rd enl ed. London: Thomas Baker, 1864 – 6mf – 9 – 0-8370-6567-4 – (incl bibl ref and index) – mf#1986-0567 – us ATLA [240]

History of the christian philosophy of religion from the reformation to kant = Geschichte der christlichen religions-philosophie seit der reformation, 1. bd, bis auf kant / Puenjer, Georg Christian Bernhard – Edinburgh: T & T Clark, 1887 – 2mf – 9 – 0-524-03358-7 – (in english) – mf#1990-0939 – us ATLA [242]

A history of the church : from the edict of milan, a d 313, to the council of chalcedon, a d 451 / Bright, William – Oxford: J H & Jas Parker, 1860 [mf ed 1989] – 2mf – 9 – 0-7905-4154-8 – (incl bibl ref) – mf#1988-0154 – us ATLA [240]

History of the church : from its first establishment to our own times / Birkhaeuser, Jodocus Adolph – 13th ed [ie, 3rd ed, rev and enl] Ratisbon; New York: Frederick Pustet, [1893?] – 2mf – 9 – 0-524-00621-0 – mf#1990-0121 – us ATLA [240]

History of the church = Geschichte der christlichen kirche / Doellinger, Johann Joseph Ignaz von – London: C Dolman: T Jones, 1840-42 [mf ed 1990] – 4v on 3mf – 9 – 0-7905-4508-X – (english trans fr german by edward cox) – mf#1988-0508 – us ATLA [240]

History of the church and state in norway : from the tenth to the sixteenth century / Willson, Thomas Benjamin – Westminster: A. Constable, 1903 – 1mf – 9 – 0-7905-6217-0 – (incl bibl ref) – mf#1988-2217 – us ATLA [240]

The history of the church and state of scotland : from the accession of king charles 1 to the restoration of king charles 2 / Stevenson, Andrew – Edinburgh: [s.n.] 1753-1757 – 1r – 1 – 0-8370-0020-3 – mf#1984-B404 – us ATLA [240]

A history of the church from a d 322 to the death of theodore of mopsuestia, a d 427. and, from a d 431 to a d 594 / Theodoret, Bishop of Cyrrhus & Evagrius – London: Henry G Bohn, 1854 [mf ed 1991] – (= ser Bohn's ecclesiastical library) – 5mf – 9 – 0-524-00657-1 – (trans fr greek, rev by edward walford) – mf#1990-0157 – us ATLA [240]

The history of the church from our lord's incarnation to the year of christ / Eusebius – 1709 – 1r – 1 – mf#448 – uk Microform Academic [240]

A history of the church from the earliest ages to the reformation / Waddington, George – stereotype ed. New York: Harper, 1834 [mf ed 1992] – 2mf – 9 – 0-524-03434-6 – (incl bibl ref & ind) – mf#1990-0988 – us ATLA [242]

History of the church in eastern canada and newfoundland / Langtry, John – London: SPCK; New York: Society's agents [distributor], 1892 – (= ser Colonial church histories) – 1mf – 9 – 0-7905-5298-1 – mf#1988-1298 – us ATLA [240]

The history of the church in ireland since the scots were naturalized *see* True narrative of the rise and progress of the presbyterian church in ireland (1623-1670)

A history of the church in scotland : from the earliest times down to the present day / MacPherson, John – Paisley: Alexander Gardner, 1901 [mf ed 1990] – 2mf – 9 – 0-7905-5190-X – (incl bibl ref) – mf#1988-1190 – us ATLA [240]

History of the church in the eighteenth and nineteenth centuries = Kirchengeschichte des 18. und 19. jahrhunderts / Hagenbach, Karl Rudolf – New York: C Scribner, 1869 – 3mf – 9 – 0-7905-4593-4 – (incl bibl ref. in english) – mf#1988-0593 – us ATLA [240]

History of the church in the first seven chapters of the acts / Preston, C M – London, England. 1868 – 1r – us UF Libraries [240]

A history of the church in venezuela. chapel hill, 1933 / Watters, Mary – Madrid: Razon y Fe, 1935 – 1 – sp Bibl Santa Ana [241]

A history of the church known as the moravian church : or, the unitas fratrum, or, the unity of the brethren: during the 18th and 19th centuries / Hamilton, John Taylor – Bethlehem, PA: Times Pub Co, 1900 [mf ed 1992] – (= ser [Transactions of the moravian historical society] 61; Methodist coll) – 2mf – 9 – 0-524-03933-X – (incl bibl ref) – mf#1990-4927 – us ATLA [242]

The history of the church known as the unitas fratrum or the unity of the brethren : founded by the followers of john hus, the bohemian reformer and martyr / De Schweinitz, Edmund – Bethlehem, Pa: Moravian Publication Office, 1885 – 2mf – 9 – 0-7905-8102-7 – mf#1988-6064 – us ATLA [242]

The history of the church missionary society : its environment, its men and its work / Stock, Eugene – London: Church Missionary Society, 1899 – 5mf – 9 – 0-7905-8051-9 – mf#1988-6032 – us ATLA [240]

The history of the church missionary society, its environment, its men and its work / Stock, Eugene – London: Church Missionary Society, 1899-1916. illus., ports., 3 fold. maps. 4v – 1 – us Wisconsin U Libr [920]

The history of the church missionary society. supplementary volume, the fourth / Stock, Eugene – London: Church Missionary Society, 1916 – 2mf – 9 – 0-524-00606-7 – mf#1990-0106 – us ATLA [240]

The history of the church of christ : with a special view to the delineation of christian faith and life (from a.d. 1 to a.d. 313) / Burns, Islay – London: T Nelson, 1871 – 1mf – 9 – 0-524-03456-7 – (incl bibl ref) – mf#1990-0999 – us ATLA [240]

A history of the church of england / Patterson, Melville Watson – London: Longmans, Green, 1909 [mf ed 1992] – (= ser Anglican/episcopal coll) – 2mf – 9 – 0-524-02393-X – mf#1990-4295 – us ATLA [242]

History of the church of england / Cutts, Edward Lewes – New York: Longmans, Green, 1895 – (= ser Text-books of Religious Instruction) – 1mf – 9 – 0-524-02461-8 – mf#1990-4320 – us ATLA [241]

The history of the church of england : in the colonies and foreign dependencies of the british empire / Anderson, James Stuart Murray – London, 1845-1856 – (= ser 19th c books on british colonization) – 24mf – 9 – mf#1.1.6618 – uk Chadwyck [241]

The history of the church of england from the revolution to the last acts of convocation, a.d. 1688-1717 / Palin, William – London: Francis & John Rivington, 1851 – 2mf – 9 – 0-524-03801-5 – mf#1990-4873 – us ATLA [241]

The history of the church of england in the colonies and foreign dependencies of the british empire / Anderson, James S. M – 2nd ed. London: Rivingtons, 1856 – 5mf – 9 – 0-7905-4481-4 – (incl bibl ref) – mf#1988-0481 – us ATLA [941]

The history of the church of god during the period of revelation / Jones, Charles Colcock – New York: Scribner, 1867 – 2mf – 9 – 0-524-05615-3 – mf#1992-0470 – us ATLA [220]

The history of the church of ireland : in eight sermons. preached in westminister abbey / Wordsworth, Christopher – London: Rivingtons, 1869 – 1mf – 9 – 0-524-03207-6 – mf#1990-4656 – us ATLA [240]

History of the church of ireland, from the earliest times to the present day / ed by Phillips, Walter Alison – London: Oxford University Press, H. Milford, 1933-34. 3v. Comp. under the auspices of the General Synod of the Church of Ireland. 1 reel. 1304 – 1 – us Wisconsin U Libr [240]

The history of the church of rome : to the end of the episcopate of damasus, a.d. 384 / Shepherd, Edward John – London: Longman, Brown, Green, and Longmans, 1851 – 2mf – 9 – 0-8370-8945-X – (incl ind) – mf#1986-2945 – us ATLA [241]

A history of the church of russia / Mouravieff, Andrew Nicholaievitch – Oxford: J H Parker, 1842 [mf ed 1990] – 2mf – 9 – 0-7905-5496-8 – (trans by r w blackmore) – mf#1988-1496 – us ATLA [240]

History of the church of scotland : from the introduction of christianity to the period of the disruption, may 18, 1843 / Hetherington, William Maxwell – 7th ed. Edinburgh: J. Johnstone, 1848 – 3mf – 9 – 0-7905-5154-3 – mf#1988-1154 – us ATLA [242]

History of the church of scotland : from the introduction of christianity to the period of the disruption, may 18,1843... / Hetherington, W M – Edinburgh: J. Johnstone, 1848 – 3mf – us ATLA [242]

History of the church of scotland during the commonwealth / Beattie, James – Chicago: Dep of Photodup, U of Chicago Lib, 1974 (1r); Evanston: American Theol Lib Assoc, 1984 (1r) – 1 – 0-8370-0002-5 – mf#1984-B403 – us ATLA [242]

History of the church of st mildred the virgin / Milbourn, Thomas – London, England. 1872 – 1r – us UF Libraries [025]

A history of the church of the brethren / Eshelman, Matthew Mays – Los Angeles: District Meeting of Southern California & Arizona 1917 [mf ed 1992] – 1mf – 9 – 0-524-03701-9 – mf#1990-4806 – us ATLA [242]

A history of the church of the brethren, northeastern ohio / Moherman, Tully S – Elgin IL: Brethren Pub House 1914 [mf ed 1992] – 1mf – 9 – 0-524-02746-3 – mf#1990-4421 – us ATLA [242]

History of the church of the brethren of the eastern district of pennsylvania – Lancaster, PA: New Era Printing Company, 1915 – 2mf – 9 – 0-524-02732-3 – mf#1990-4407 – us ATLA [242]

History of the church of the brethren of the western district of pennsylvania / Blough, Jerome E – Elgin, Ill: Brethren Pub House, 1916 – 2mf – 9 – 0-524-02730-7 – mf#1990-4405 – us ATLA [242]

History of the church of the united brethren in christ / Berger, Daniel – Dayton, Ohio: United Brethren Pub House, 1897 – 2mf – 9 – 0-7905-6982-5 – mf#1988-2982 – us ATLA [242]

History of the church of the united brethren in christ / Spayth, Henry G – 1st ed. Circleville, Ohio: Conference Office of the United Brethren in Christ, 1851 – 1mf – 9 – 0-524-02966-0 – mf#1990-4518 – us ATLA [242]

The history of the church of the united brethren in christ / Lawrence, John – Dayton, OH: WJ Shuey, 1861-1868 – 2mf – 9 – 0-524-03554-7 – mf#1990-4749 – us ATLA [242]

History of the church under the roman empire, a.d. 30-476 : intended especially for the use of junior students / Crake, Augustine David – 2nd rev ed. London: Rivingtons, 1879 – 2mf – 9 – 0-524-03393-5 – mf#1990-0947 – us ATLA [240]

History of the churches and ministers connected with the presbyterian and congregational convention of wisconsin : and of the operations of the american home missionary society in the state for the past ten years / Clary, Dexter – Beloit: printed by B E Hale 1861 [mf ed 1992] – 1mf – 9 – 0-524-02816-8 – mf#1990-4437 – us ATLA [242]

History of the churches of boone's creek baptist association of kentucky, 1780-1923 / Conkwright, S J – 1 – 6.86 – us Southern Baptist [242]

The history of the cinema, 1895-1940 / ed by Short, Kenneth – [mf ed Chadwyck-Healey] – 3574mf – 9 – (with p/g & ind) – uk Chadwyck [790]

The history of the civilization of india : a sketch, with suggestions for the improvement of the country / Murdoch, John [comp] – 1st ed. London: Christian Literature Society for India, 1902 [mf ed 1995] – (= ser Yale coll) – iv/192p (ill) – 1 – 0-524-09988-X – mf#1995-0988 – us ATLA [954]

History of the clappers : the eight sons of samuel and nancy kagarice clapper / Clapper, David K – [S.l.: s.n.], 1914 – 1mf – 9 – 0-524-03930-5 – mf#1990-4924 – us ATLA [920]

History of the colonial empire of great britain / Roberts, Browne H E – London 1861 – (= ser 19th c british colonization) – 4mf – 9 – mf#1.1.3790 – uk Chadwyck [941]

A history of the colonization of africa by alien races / Johnston, Harry Hamilton – Cambridge, 1899 – (= ser 19th c british colonization) – 4mf – 9 – mf#1.1.3410 – uk Chadwyck [960]

A history of the colonization of africa by alien races / Johnston, Harry Hamilton – new rev enl ed. Cambridge: University Press; New York: G P Putnam [dist] 1913 [mf ed 1990] – (= ser Cambridge historical series 24) – 2mf – 9 – 0-7905-5346-5 – mf#1988-1346 – us ATLA [960]

History of the colony of natal, south africa : to which is added, an appendix... / Holden, William Clifford – London, 1855 – (= ser 19th c british colonization) – 6mf – 9 – mf#1.1.5557 – uk Chadwyck [960]

History of the colony of sierra leone, western africa / Crooks, John Joseph – Dublin: Browne & Nolan, 1903 – 1 – (with maps and appendices) – us CRL [960]

History of the commandments of the church = Histoire des commandements de l'eglise / Villien, Antoine – St Louis MO: B Herder, 1915 [mf ed 1991] – 1mf – 9 – 0-524-01476-0 – (incl bibl ref) – mf#1990-0425 – us ATLA [241]

History of the commerce and town of liverpool : and of the rise of the manufacturing industry in the adjoining counties / Baines, Thomas – London: Longman, Brown, Green, & Longmans, 1852 [mf ed 1990] – xvi/844p – 1 – mf#6681 – us Wisconsin U Libr [380]

History of the commonwealth and protectorate, 1649-1656 / Gardiner, Samuel Rawson – New ed. London; New York: Longmans, Green, 1903 – 4mf – 9 – 0-7905-4680-9 – (incl bibl ref) – mf#1988-0680 – us ATLA [941]

History of the commonwealth of florence, from the earliest independence of the commune to the fall of the republic in 1531 / Trollope, Thomas Adolphus – London, 1865. 4v – 1 – us Wisconsin U Libr [945]

History of the condemnation of the patriarch nicon : by a plenary council of the orthodox catholic eastern church held at moscow a.d. 1666-67 / Ligarides, Paisius – London: Truebner, 1873 – 2mf – 9 – 0-8370-7562-9 – (incl bibl ref) – mf#1986-1562 – us ATLA [241]

History of the conflict between religion and science / Draper, John William – 3rd ed. New York: D. Appleton, 1875, c1874 – (= ser International Scientific Series (New York)) – 1mf – 9 – 0-7905-4460-1 – mf#1988-0460 – us ATLA [210]

History of the conflict between religion and science / Draper, John William – 8th ed. New York: D. Appleton and Co., 1878. xxii,373p – 1 – us Wisconsin U Libr [240]

History of the congregational association of oregon and washington territory, the home missionary society of oregon and adjoining territories, and the northwestern association of congregational ministers / Eells, Myron – Portland, Oregon: Himes, 1881 – 1mf – 9 – 0-7905-6053-4 – mf#1988-2053 – us ATLA [242]

History of the congregational churches in the berks, south oxon and south bucks association : with notes on the earlier nonconformist history of the district / Summers, William Henry – London: Publ Dept; Newbury: W J Blacket 1905 [mf ed 1990] – 1mf [ill] – 9 – 0-7905-6687-7 – mf#1988-2687 – us ATLA [242]

A history of the congregational churches in the united states / Walker, Williston – New York: Christian Literature, 1894 [mf ed 1989] – (= ser The american church history series) – 2mf – 9 – 0-7905-4239-0 – (incl ind) – mf#1988-0239 – us ATLA [242]

History of the congregations of the united presbyterian church from 1733 to 1900 / Small, Robert – Edinburgh: DM Small, 1904 – 4mf – 9 – 0-524-03190-8 – mf#1990-4639 – us ATLA [242]

History of the connecticut school fund : a thesis submitted in partial fulfillment for the degree of master of art in history at trinity college, hartford, connecticut, may 3, 1939 / McCrann, Leo M – 1r – 1 – us Western Res [370]

History of the conquest of mexico : with a preliminary view of the ancient mexican civilization, and the life of the conqueror, hernando cortes / Prescott, William Hickling – 8th ed. New York: Harper, 1849 – 4mf – 9 – 0-524-03416-8 – (incl bibl ref) – mf#1990-0970 – us ATLA [972]

History of the constitutions of iowa / Shambaugh, Benjamin Franklin – Des Moines: Historical Department of Iowa, 1902. 352p. LL-1613 – 1 – us L of C Photodup [342]

A history of the convocation of the church of england : from the earliest period to the year 1742 / Lathbury, Thomas – 2nd enl ed. London: J Leslie, 1853 [mf ed 1990] – 2mf – 9 – 0-7905-5366-X – (1st ed 1842. incl bibl ref) – mf#1988-1366 – us ATLA [242]

History of the corporation of birmingham : with a sketch of the earlier government of the town / Bunce, John Thackray – Birmingham: publ... by Cornish Bros 1878-1957 [mf ed 1985] – 6v on 1r – 1 – mf#6719 – us Wisconsin U Libr [350]

A history of the council of trent / Buckley, Theodore Alois – London: George Routledge, 1852 [mf ed 1986] – 2mf – 9 – 0-8370-6250-0 – (incl bibl ref) – mf#1986-0250 – us ATLA [240]

History of the council of trent / Bungener, Laurence Louis Felix – 2nd london ed. New York: Harper 1855 [mf ed 1987] – 1r – 1 – (with summary of acts of the council, by john m'clintock; filmed with: missionary milestones / seebach, m r; & other titles) – mf#2031 – us Wisconsin U Libr [241]

History of the council of trent = Histoire du concile de trente / Bungener, Felix; ed by McClintock, John – New York: Harper, 1855 – 2mf – 9 – 0-8370-8485-7 – (incl bibl ref and ind. in english) – mf#1986-2485 – us ATLA [240]

History of the county of orange / Ruttenbur, Edward M – 1875 – 1 – $50.00 – us Presbyterian [978]

The history of the court of the king of china / Baudier, M – London: H B, 1682 – 2mf – 9 – mf#HT-501 – ne IDC [915]

The history of the creeds : (1) ante-nicene, (2) nicene and constantinopolitan, (3) the apostolic creed, (4) the quicunque, commonly called the creed of st. athanasius / Lumby, Joseph Rawson – 3rd ed. Cambridge: Deighton, Bell; London: G Bell, 1887 – 1mf – 9 – 0-524-01655-0 – mf#1990-0476 – us ATLA [240]

A history of the crusades / Runciman, St – Cambridge, 1951 – €15.00 – ne Slangenburg [241]

History of the cumberland presbyterian church / McDonnold, Benjamin Wilburn – 2nd ed. Nashville, Tenn: Board of Publication of Cumberland Presbyterian Church, 1888 – 2mf – 9 – 0-7905-4538-1 – mf#1988-0538 – us ATLA [242]

The history of the cumberland presbyterian church in alabama prior to 1826 / Hall, James Hugh Blair – Montgomery, Ala: [s.n.], 1904 – (= ser Historical Contributions) – 1mf – 9 – 0-524-02255-0 – mf#1990-4262 – us ATLA [242]

History of the cumberland presbyterian church in illinois : containing sketches of the first ministers, churches, presbyteries and synods / Logan, James B – Alton, IL: Perrin & Smith, 1878 – 3mf – 9 – 0-524-07356-2 – mf#1990-5393 – us ATLA [242]

The history of the danish mission / Eshelman, Matthew Mays – Mt Morris, IL: Western Book Exchange, 1881 – 1mf – 9 – 0-524-03149-5 – mf#1990-4598 – us ATLA [240]

History of the deaconess movement in the christian church / Golder, Christian – Cincinnati: Jennings and Pye, [c1903] El Segundo, Ca: Micro Publication Systems, 1981 (3mf); Evanston: American Theol Lib Assoc, 1984 (1mf) – (= ser Women & the church in america) – 9 – 0-8370-1451-4 – mf#1984-2155 – us ATLA [240]

History of the denton county baptist association and the sixty churches organized within its jurisdiction / Rayzor, James Newton – 1936 – 1 – 9.52 – us Southern Baptist [242]

History of the denver women's press club – [S.l: s.n., 1928] (mf ed 1977 – 1r – 1 – mf#MF Z99 Wo84h – us Colorado Hist [070]

History of the department of justice (1963-1969) / U.S. Dept of Justice – (= ser Presidential documents series) – 6r – 1 – $935.00 – 0-89093-360-X – (with p/g) – us UPA [322]

A history of the descendents of jacob and maria eva harshbarger of switzerland / Anderson, William L – [s.l: s.n, 1910?] [mf ed 1992] – 1mf – 9 – 0-524-03685-3 – mf#1990-4790 – us ATLA [929]

The history of the destruction of the colonial advocate press : by officers of the provincial government of upper canada and law students of the attorney and solicitor general... / Mackenzie, William Lyon – York [Toronto]: Printed...by W L Mackenzie...1827 – 1mf – 9 – 0-665-92892-0 – mf#92892 – cn Canadiana [070]

History of the development of the doctrine of the person of christ = Entwickelungsgeschichte von der lehre von der person christi / Dorner, Isaak August – Edinburgh: T & T Clark, 1863-78 [mf ed 1989] – (= ser Clark's foreign theological library. 3rd series 10,11,14,15,18) – 5v on 6mf – 9 – 0-7905-2947-5 – (incl bibl ref) – mf#1987-2947 – us ATLA [240]

A history of the development of the presbyterian church in north carolina : and of synodical home missions, together with evangelistic addresses by james i vance and others / Craig, David Irwin – Richmond, VA: Whittet & Shepperson, c1907 [mf ed 1992] – (= ser Presbyterian coll) – 3mf – 9 – 0-524-02249-6 – mf#1990-4256 – us ATLA [242]

The history of the devil and the idea of evil : from the earliest times to the present day / Carus, Paul – Chicago: Open Court, 1900 – 2mf – 9 – 0-524-04637-9 – (incl bibl ref) – mf#1990-3380 – us ATLA [210]

History of the diocese of central pennsylvania, 1871-1909, and the diocese of harrisburg, 1904-1909 / Miller, Jonathan Wesley – Frackville, PA: Miller, 1909 – 3mf – 9 – 0-524-06960-3 – mf#1990-5324 – us ATLA [240]

A history of the diocese of chicago : including a history of the undivided diocese of illinois from its organization in 1835 a d / Hall, Francis Joseph – Dixon, IL: De Witt C Owen [1900?] [mf ed 1993] – (= ser Anglican/episcopal coll) – 1mf – 9 – 0-524-06588-8 – (no more publ) – mf#1990-5254 – us ATLA [242]

History of the diocese of montreal, 1850-1910 / Borthwick, John Douglas – Montreal: J Lovell, 1910 – 4mf – 9 – 0-665-71872-1 – mf#71872 – cn Canadiana [242]

History of the disciples of christ in california / Ware, E B – Healdsburg, CA: [s.n.], 1916 – 1mf – 9 – 0-524-03508-3 – mf#1990-4730 – us ATLA [240]

A history of the disciples of christ in ohio / Wilcox, Alanson – Cincinnati: Standard Pub Co, c1918 [mf ed 1993] – (= ser Christian church (disciples of christ) coll) – 1mf – 9 – 0-524-07598-0 – mf#1991-3218 – us ATLA [240]

History of the disciples of christ, the society of friends, the united brethren in christ and the evangelical association / Tyler, Benjamin Bushrod et al – New York: Christian Literature, 1894 [mf ed 1989] – (= ser The american church history series 12) – 2mf – 9 – 0-7905-4238-2 – (together with: bibliography of american church history by samuel macauley jackson; incl bibl ref) – mf#1988-0238 – us ATLA [242]

History of the discipline of the methodist episcopal church / Emory, Robert – rev, and brought down to 1856. New York: Carlton and Porter, c1843 – 1mf – 9 – 0-524-01210-5 – mf#1990-4068 – us ATLA [242]

The history of the discovery and settlement, to the present time, of north and south amrica : and of the west indies / Mavor, William – London: printed for Richard Phillips... 1804 [mf ed 1983] – 5mf – 9 – 0-665-38232-4 – (also publ as v24 of 25v set by mavor entitled: universal history, ancient and modern: from the earliest records of time to the general peace of 1801) – mf#38232 – cn Canadiana [970]

History of the discovery of the northwest by john nicolet in 1634 : with a sketch of his life / Butterfield, Consul Willshire – Cincinnati: R Clarke, 1881 – 2mf – 9 – (in english, but many of the footnotes, extracts fr the jesuit relations, are in french. incl ind) – mf#06561 – cn Canadiana [910]

A history of the division of the presbyterian church in the united states of america / Presbyterian Church in the USA. (New School). Synod of New York and New Jersey – New York: MW Dodd, 1852 [mf ed 1992] – (= ser Presbyterian coll) – 1mf – 9 – 0-524-02135-X – mf#1990-4201 – us ATLA [242]

A history of the doctrine of the holy eucharist / Stone, Darwell – London, New York: Longmans, Green, 1909 [mf ed 1909] – 2v on 3mf – 9 – 0-7905-7082-3 – (incl bibl ref) – mf#1988-3082 – us ATLA [240]

A history of the doctrine of the work of christ in its ecclesiastical development / Franks, Robert Sleightholme – London: Hodder & Stoughton [1918?] [mf ed 1993] – 2v on 10mf – 9 – 0-524-08757-1 – (incl bibl ref) – mf#1993-3262 – us ATLA [240]

History of the dominican republic and... / Webster, Mamie Morris – s.l, s.l? 1940 – 1r – us UF Libraries [972]

A history of the dominion of canada / Calkin, John Burgess – Halifax, NS: A & W Mackinlay, 1898 – 6mf – 9 – (incl ind) – mf#26774 – cn Canadiana [971]

History of the donatists : with notes / Benedict, David – Memorial ed. Pawtucket, R.I.: Printed for Maria M. Benedict by Nickerson, Sibley, 1875 – 1mf – 9 – 0-7905-5626-X – mf#1988-1626 – us ATLA [240]

History of the dwelling-house and its future / Thompson, Robert Ellis – Philadelphia, PA. 1914 – 1r – us UF Libraries [720]

A history of the early dynasties of andhradesa c 200-625 ad : with a map of ancient andhradesa and daksinapatha / Krishnarao, Bhavaraju Venkata – Madras: V Ramaswami Sastrulu & Sons, 1942 – (= ser Samp: indian books) – us CRL [954]

A history of the early policy of the presbyterian church in the training of her ministry, and of the first years of the board of education / Baird, Samuel John – Philadelphia: The Board, 1865 [mf ed 1993] – (= ser Presbyterian coll) – 1mf – 9 – 0-524-07224-8 – mf#1991-2965 – us ATLA [242]

The history of the early puritans : from the reformation to the opening of the civil war in 1642 / Marsden, John Buxton – 3rd ed. London: Hamilton, Adams, 1860 – 2mf – 9 – 0-7905-6759-8 – mf#1988-2759 – us ATLA [243]

The history of the early puritans: from the reformation to the opening of the civil war in 1642 / Marsden, John Buxton – London: Hamilton, Adams, & Co., 1850. xv,426p – 1 – us Wisconsin U Libr [242]

A history of the eastern defense command / Brook, William M – New York, Raleigh. 1945. 4v. maps, charts – 1 reel – 1 – $24.00 – us L of C Photodup [977]

A history of the eastern roman empire : from the fall of irene to the accession of basil 1 (a d 802-867) / Bury, John Bagnell – London, New York: Macmillan, 1912 [mf ed 1990] – 2mf – 9 – 0-7905-4611-6 – (incl bibl ref) – mf#1988-0611 – us ATLA [931]

History of the efts summary : july 1903-july 2003 / Anderson, Mary – [s.l.]: European Federation of the Theosophical Soc, 2003 [mf ed 2004] – 1r – 1 – 0-524-10462-X – mf#2003-s124a – us ATLA [290]

History of the egyptian religion / Tiele, Cornelius Petrus – Trans. from the Dutch by James Ballingal. Boston: Houghton, Mifflin, 1882. xxiii,230p – 1 – us Wisconsin U Libr [290]

A history of the english baptists : including an investigation of the history of baptism in england from the earliest period to which it can be traced to the close of the 17th century... / Ivimey, Joseph – London: printed for aut, 1811-30 [mf ed 1993] – 4v on 6mf – 9 – 0-524-07982-X – (incl bibl ref) – mf#1990-5427 – us ATLA [242]

The history of the english bible / Brown, John – Cambridge: University Press 1911 [mf ed 1986] – (= ser The cambridge manuals of science and literature) – 1mf – 9 – 0-8370-9925-0 – (incl ind) – mf#1986-3925 – us ATLA [220]

The history of the english bible : extending from the earliest saxon translations to the present anglo-american revision / Condit, Blackford – 2nd rev enl ed. New York: A S Barnes, c1896 – 2mf – 9 – 0-7905-1810-4 – (incl bibl ref and index) – mf#1987-1810 – us ATLA [220]

The history of the english bible / Pattison, Thomas Harwood – Philadelphia: American Baptist Publ Society, 1894 – 1mf – 9 – 0-8370-4676-9 – (incl ind) – mf#1985-2676 – us ATLA [220]

History of the english church and people in south africa / Wirgman, Augustus Theodore – New York, NY. 1969 – 1r – us UF Libraries [960]

The history of the english church and people in south africa / Wirgman, Augustus Theodore – London; New York: Longmans, Green, 1895 – 1mf – 9 – 0-7905-6975-2 – mf#1988-2975 – us ATLA [240]

A history of the english church during the civil wars and under the commonwealth, 1640-1660 / Shaw, William Arthur – London, New York: Longmans, Green, 1900 [mf ed 1990] – 2v on 3mf – 9 – 0-7905-7026-2 – mf#1988-3026 – us ATLA [242]

A history of the english church in new zealand / Purchas, Henry Thomas – Christchurch: Simpson & Williams, 1914 [mf ed 1990] – 1mf – 9 – 0-7905-5912-9 – mf#1988-1912 – us ATLA [240]

A history of the english episcopacy : from the period of the long parliament to the act of uniformity... / Lathbury, Thomas – London: J W Parker, 1836 [mf ed 1990] – 1mf – 9 – 0-7905-4999-9 – mf#1988-0999 – us ATLA [242]

The history of the english revolution / Dahlmann, Friedrich Christoph – London: Longman, Brown, Green and Longmans, 1844 – 1mf – us ATLA [941]

The history of the english revolution = Geschichte der englischen revolution / Dahlmann, Friedrich Christoph – London: Longman, Brown, Green and Longmans, 1844 – 1mf – 9 – 0-7905-4394-X – (in english) – mf#1988-0394 – us ATLA [941]

History of the entertainment branch, special services division / U.S. Army. Army Service Forces. Special Services Division – 1 – us L of C Photodup [790]

History of the establishment and progress of the christian religion in the islands of the south sea : with preliminary notices of the islands and of their inhabitants – Boston: Tappan & Dennet; New York: Gould, Newman & Saxton, 1841 [mf ed 1995] – (= ser Yale coll) – xxvii/387p (ill) – 1 – 0-524-09264-8 – mf#1995-0264 – us ATLA [240]

History of the evangelical association = Geschichte der evangelischen gemeinschaft / Orwig, Wilhelm W – 1st ed. Cleveland, Ohio: C Hammer, 1858 – 1mf – 9 – 0-524-04364-7 – (in english) – mf#1990-5047 – us ATLA [242]

History of the evangelical association / Yeakel, Reuben – Cleveland, OH: Thomas & Mattill, 1894-1895 – 2mf – 9 – 0-524-06298-6 – (incl bibl ref) – mf#1990-5227 – us ATLA [242]

A history of the evangelical lutheran church in the united states / Jacobs, Henry Eyster – New York: Christian Literature, 1893 [mf ed 1989] – (= ser The american church history series 4) – 2mf – 9 – 0-7905-4233-1 – mf#1988-0233 – us ATLA [242]

History of the evangelical lutheran church in the united states *see* Geschichte der lutherischen kirche in amerika

History of the evangelical lutheran district synod of ohio : covering fifty-three years, 1857-1910 / Mechling, George Washington – [S.l.: s.n.], 1911 – 3mf – 9 – 0-524-07901-3 – mf#1991-3446 – us ATLA [242]

The history of the evangelical lutheran synod and ministerium of north carolina : in commemoration of the completion of the first century of its existence / Bernheim, Gotthardt Dellmann & Cox, George Henry – Philadelphia, PA: Pub for the Synod by the Lutheran Publication Society, 1902 – 1mf – 9 – 0-524-07553-0 – mf#1991-3173 – us ATLA [242]

A history of the evangelical lutheran synod of kansas (general synod) : together with a sketch of the augustana synod churches and a brief presentation of other lutheran bodies located in kansas / Ott, Hamilton A – [s.l]: pub by authority of Kansas Synod, 1907 [mf ed 1993] – 1mf – 9 – 0-524-08578-1 – mf#1993-3163 – us ATLA [242]

History of the evangelical lutheran tennessee synod : embracing an account of the causes, which gave rise to its organization... / Henkel, Socrates – New Market, VA: Henkel, 1890 – 1mf – 9 – 0-524-02827-3 – mf#1990-4448 – us ATLA [242]

A history of the evangelical party in the church of england / Balleine, George Reginald – London, New York: Longmans, Green, 1908 [mf ed 1990] – 1mf – 9 – 0-7905-5621-9 – (incl bibl ref) – mf#1988-1621 – us ATLA [242]

History of the expedition under the command of lewis and clark : to the sources of the missouri river, thence across the rocky mountains and down the columbia river to the pacific ocean...1804-5-6 – New York: F P Harper, 1893 – 4v on 1mf – 9 – mf#56209 – cn Canadiana [917]

The history of the extinction of paganism in the roman empire viewed in relation to the evidences of christianity / Heard, John Bickford – Cambridge: Macmillan; London: George Bell, 1852 – (= ser Hulsean Prize Essay) – 1mf – 9 – 0-7905-7240-0 – mf#1988-3240 – us ATLA [240]

The history of the factory movement : or, oastler and his times / Croft, W R – Huddersfield: George Whitehead & Sons, 1888 – (= ser 19th c economics) – 2mf – 9 – mf#1.1.477 – uk Chadwyck [331]

History of the fall of the jesuits in the eighteenth century = Histoire de la chute des jesuites au 18e siecle, 1750-1782 / Saint-Priest, Alexis, Comte de – London: John Murray, 1845 – 1mf – 9 – 0-524-02989-X – (in english) – mf#1990-0776 – us ATLA [241]

A history of the fall of the roman empire : comprising a view of the invasion and settlement of the barbarians / Sismondi, Jean Charles Leonard Simonde de – London: Longman, Rees, Orme, Brown, Green & Longman 1834 [mf ed 1988] – (= ser Lardner's cabinet cyclopaedia) – 2v – 1 – mf#2195 – us Wisconsin U Libr [930]

History of the federal trade commission (1963-1969) – (= ser Presidential documents series) – 2r – 1 – $325.00 – 0-89093-361-8 – (with p/g) – us UPA [380]

History of the fenian raid on fort erie : with an account of the battle of ridgeway / Denison, George Taylor – Toronto: Rollo & Adam; Buffalo: Breed, Butler, 1866 – (= ser The fenian raid on fort erie) – 2mf – 9 – (issued also under title: the fenian raid on fort erie) – mf#34588 – cn Canadiana [971]

History of the finances of the city and county of denver *see* University of denver theses

History of the first baptist church *see* Miscellaneous books and pamphlets

A history of the first baptist church, eldorado, texas / Hoover, L M – 1 sep 1901-31 aug 1957 – 1 – $5.00 – us Southern Baptist [242]

History of the first baptist church in providence. providence, rhode island : church records – 1639-1877 – 1 – 5.00 – us Southern Baptist [242]

The history of the first baptist church of boston (1665-1899) / Wood, Nathan Eusebius – Philadelphia: American Baptist Publication Society, 1899 – 1mf – 9 – 0-7905-8169-8 – mf#1988-6116 – us ATLA [242]

History of the first church in hartford, 1633-1883 / Walker, George Leon – Hartford: Brown & Gross, 1884 – 2mf – 9 – 0-7905-6905-1 – mf#1988-2905 – us ATLA [240]

History of the first church, oberlin, ohio : an address by the pastor rev james brand, delivered december, 1876 / Brand, James – [S.l: s.n.], 1877 [mf ed 1981] – 1mf – 9 – mf#25676 – cn Canadiana [242]

A history of the first unitarian society of chicago (1836-1933) / Newman, Herman Andrew – Chicago, 1933. Chicago: Department of Photodup, U of Chicago Lib, 1971 (1r); Evanston: American Theol Lib Assoc, 1984 (1r) – 1 – 0-8370-0288-5 – mf#1984-B162 – us ATLA [243]

History of the first west india regiment / Ellis, Alfred Burdon – London: Chapman and Hall, 1885. xii,366p. Maps, plan, col. front. With: Griechisches Erbe by F. Klatt. 1 reel. 1293 – 1 – us Wisconsin U Libr [355]

The history of the five indian nations of canada / Colden, C – London, 1747 – 9mf – 9 – mf#N-166 – ne IDC [917]

A history of the foreign missionary work of the protestant episcopal church / Denison, Samuel Dexter – New York: Foreign Committee of the Board of Missions, 1871 [mf ed 1990] – 1mf – 9 – 0-7905-6641-9 – (no more publ) – mf#1988-2641 – us ATLA [242]

A history of the formation and growth of the reformed episcopal church, 1873-1902 / Price, Annie Darling – Philadelphia: James M Armstrong, 1902 [mf ed 1992] – (= ser Anglican/episcopal coll) – 1mf – 9 – 0-524-03182-7 – (incl bibl ref) – mf#1990-4631 – us ATLA [242]

History of the formation of the constitution : of the united states of america / Bancroft, George – 3rd ed. New York: D Appleton & Co. 2v. 1883 – 12mf – 9 – $18.00 – mf#LLMC 90-365 – us LLMC [323]

History of the formation of the medical faculty, university of bishop's college, montreal / Campbell, Francis Wayland – Waterville, Quebec?: J H Osgood, 1900 – 1mf – 9 – mf#01503 – cn Canadiana [378]

History of the free baptist woman's missionary society / Davis, Mary A – Boston, Mass: Morning Star Pub House, 1900 – 1mf – 9 – 0-524-04256-X – mf#1991-2040 – us ATLA [242]

History of the free churches of england, 1688-1891 / Skeats, Herbert S & Miall, Charles S – London: Alexander & Shepheard: James Clarke [1891?] [mf ed 1991] – 2mf – 9 – 0-524-01405-1 – (from the reformation to 1851 by herbert s skeats; with continuation to 1891 by charles s miall) – mf#1990-0404 – us ATLA [240]

History of the free methodist church of north america / Hogg, Wilson Thomas – 3rd ed. Winona Lake, Ind: Free Methodist Pub House, 1938, c1915 – 3mf – 9 – 0-524-06951-4 – (incl bibl ref) – mf#1990-5315 – us ATLA [242]

The history of the freewill baptists for half a century : with an introductory chapter / Stewart, Isaac Dalton – Dover: Freewill Baptist Print Establishment, 1862, c1861 [mf ed 1990] – 2mf – 9 – 0-7905-6890-X – (no more publ) – mf#1988-2890 – us ATLA [242]

History of the french protestant refugees : from the revocation of the edict of nantes to the present time = Histoire des refugies protestants de france / Weiss, Charles – Edinburgh: William Blackwood, 1854 – 2mf – 9 – 0-524-00801-9 – (in english) – mf#1990-0233 – us ATLA [944]

The history of the gajapati kings of orissa and their successors / Mukherjee, Prabhat – Calcutta: General Trading Co, 1953 – (= ser Samp: indian books) – us CRL [954]

The history of the general conference of the mennonites of north america / Krehbiel, Henry Peter – [S.l.]: HP Krehbiel, 1898 (St Louis, Mo: A Wiebusch) – 2mf – 9 – 0-7905-6303-7 – mf#1988-2303 – us ATLA [243]

History of the general headquarters regulating system, 2 nov 1943-31 aug 1945 / U.S. Army. Forces in the Southwest Pacific – 1946 – 1 – us L of C Photodup [355]

History of the general or six principle baptists in europe and america : in two parts / Knight, Richard – Providence: Smith and Parmenter, 1827 – 4mf – 9 – 0-524-08798-9 – mf#1993-3290 – us ATLA [242]

History of the general public hospital in the city of saint john, nb / Bayard, William – S.l: s.n, 1896 – 1mf – 9 – mf#02223 – cn Canadiana [360]

History of the german baptist brethren church / Falkenstein, George N – Lancaster, PA: New Era, 1901 [mf ed 1990] – 1mf – 9 – 0-7905-5465-8 – mf#1988-1465 – us ATLA [242]

A history of the german baptist brethren in europe and america / Brumbaugh, Martin Grove – Mt Morris, IL: Brethren Pub House, 1899 [mf ed 1990] – 2mf – 9 – 0-7905-4784-8 – mf#1988-0784 – us ATLA [242]

A history of the german language / Super, Charles William – Columbus, O: Hann & Adair, 1893 [mf ed 1987] – 1r – 1 – (with: an introduction to english industrial history / allsopp, h) – mf#2029 – us Wisconsin U Libr [430]

History of the german reformed church / Mayer, Lewis – Philadelphia: Lippincott, Grambo, 1851 – 2mf – 9 – 0-524-02126-0 – (incl bibl ref) – mf#1990-4192 – us ATLA [242]

History of the gold coast and asante, based on traditions and historical facts : comprising a period of more than three centuries from about 1500 to 1860 / Reindorf, Carl Christian – Basel: Printed for the author [by] Missionbuchhandlung, 1895 – 1 – us CRL [960]

A history of the gothic revival : an attempt to show how the taste for mediaeval architecture which lingered in england during the two last centuries has since been encouraged and developed / Eastlake, Charles Locke – London: Longmans, Green, and Co, 1872 – (= ser 19th c art & architecture) – 6mf – 9 – mf#4.1.208 – uk Chadwyck [720]

A history of the gothic revival : an attempt to show how the taste for mediaeval architecture, which lingered in england during the two last centuries, has since been encouraged and developed / Eastlake, Charles Locke – London: Longmans, Green; New York: Scribner, Welford, 1872 [mf ed 1990] – 2mf – 9 – 0-7905-4675-2 – mf#1988-0675 – us ATLA [720]

A history of the grand trunk railway of canada / Brown, Thomas Storrow [comp] – Quebec: Printed for the author by Hunter, Rose, 1864 – 1mf – 9 – mf#23179 – cn Canadiana [380]

The history of the great boer trek and the origins of south african republics / Cloete, Henry; ed by Brodrick-Cloete, W – London: Murray, 1899 – 1 – us CRL [960]

History of the great civil war, 1642-1649 / Gardiner, Samuel Rawson – New ed. London: Longmans, Green, 1893 – 4mf – 9 – 0-7905-4645-0 – (incl bibl ref) – mf#1988-0645 – us ATLA [941]

The history of the great irish famine of 1847, with notices of earlier irish famines / O'Rourke, John – 3rd ed. Dublin: J. Duffy, 1902.xxiv,559p – 1 – us Wisconsin U Libr [941]

History of the great patriotic war of the soviet union, 1941-1945 – v1-3. 1960-61 – 1 – $65.00 – us L of C Photodup [947]

The history of the great patriotic war of the soviet union, 1941-1945 : official soviet history of world war 2 / ed by Pospelov, Poitr N – 1985 – 7r – 1 – $910.00 – mf#S1656 – U.S. Army Center of Military History and the Foreign Technology Division, Air Force Systems Command – us Scholarly Res [947]

History of the great reformation of the sixteenth century in germany, switzerland etc = Histoire de la reformation du seizieme siecle / Merle d'Aubigne, Jean Henri – 15th ed. New York: R Carter, 1843 – 2mf – 9 – 0-524-03410-9 – (incl bibl ref. in english) – mf#1990-0964 – us ATLA [242]

The history of the great republic : considered from a christian stand-point / Peck, Jesse Truesdell – New York: Broughton and Wyman, 1868 – 2mf – 9 – 0-524-08525-0 – mf#1993-1055 – us ATLA [975]

History of the great secession from the methodist episcopal church in the year 1845 : eventuating in the organization of the new church, entitled the "methodist episcopal church, south" / Elliott, Charles – Cincinnati: Swormstedt & Poe for the Methodist Episcopal Church, 1855, c1854 – 2mf – 9 – 0-7905-4510-1 – (incl bibl ref) – mf#1988-0510 – us ATLA [242]

History of the great secession from the methodist episcopal church in the years 1845 : eventuating in the organization of the new church, entitled the "methodist episcopal church, south" / Elliott, Charles – Cincinnati: Swormstedt & Poe for the Methodist Episcopal Church, 1855, c1854 – 2mf – us ATLA [242]

History of the greenbacks : with special reference to the economic consequences of their issue, 1862-65 / Mitchell, Wesley Clair – Chicago: University of Chicago Press, 1903 [mf ed 1970] – (= ser Decennial publications of the university of chicago 2nd ser/9; Library of american civilization 11624) – 1mf – 9 – us Chicago U Pr [332]

History of the halifax volunteer battalion and volunteer companies, 1859-1887 / Egan, Thomas J – Halifax, NS: A & W Mackinlay, 1888 – 3mf – 9 – mf#02889 – cn Canadiana [355]

A history of the harnly family : containing short biographical sketches of the harnly, hoerner, eby, hershey, sneider, and related families / Harnly, Henry H [Mrs] – Auburn IL: [s.n.] 1903 [mf ed 1992] – 1mf – 9 – 0-524-04711-1 – mf#1990-5063 – us ATLA [929]

History of the hebrew monarchy : from the administration of samuel to the babylonish captivity / Newman, Francis William – London: John Chapman 1853 [mf ed 1984] – (= ser Biblical crit – us & gb 29; Chapman's quarterly series 2) – 4mf – 9 – 0-8370-0249-4 – (incl bibl ref) – mf#1984-1029 – us ATLA [939]

The history of the hebrew nation and its literature : with an appendix on the chronology / Sharpe, Samuel – 2nd enl ed. London: John Russell Smith, 1872 – 1mf – 9 – 0-524-05238-7 – mf#1992-0371 – us ATLA [930]

A history of the hebrew people : from the division of the kingdom to the fall of jerusalem in 586 b c / Kent, Charles Foster – 6th ed. New York: Charles Scribner, 1899, c1897 [mf ed 1989] – (= ser The historical series for bible students 2) – 1mf – 9 – 0-7905-1418-4 – (incl ind) – mf#1987-1418 – us ATLA [939]

A history of the hebrew people : from the settlement in canaan to the division of the kingdom / Kent, Charles Foster – 8th ed. New York: Charles Scribner, 1903, c1896 [mf ed 1989] – (= ser The historical series for bible students 1) – 1mf – 9 – 0-7905-1419-2 – (incl ind) – mf#1987-1419 – us ATLA [939]

History of the hebrews = Geschichte der hebrer / Kittel, Rudolf – London: Williams & Norgate, 1895-96 [mf ed 1989] – 2v on 2mf – 9 – 0-7905-1212-2 – (english trans fr german by john taylor; v2 trans by hope w hogg & e b speirs; incl bibl ref & ind) – mf#1987-1212 – us ATLA [221]

History of the hebrews : their political, social and religious development and their contribution to world betterment / Sanders, Frank Knight – New York: Scribner, c1914 – 1mf – 9 – 0-524-06218-8 – (incl bibl ref) – mf#1992-0856 – us ATLA [930]

A history of the holy eucharist in great britain / Bridgett, T E – London: Burns & Oates; St Louis, MO: Herder, 1908 [mf ed 1990] – 1mf – 9 – 0-7905-6100-X – (incl bibl ref. original ed 1881 (2v)) – mf#1988-2100 – us ATLA [240]

The history of the holy, military, sovereign order of st. john of jerusalem : or knights hospitallers, knights templars, knights of rhodes, knights of malta / Taaffe, John – London: Hope, 1852 – 4mf – 9 – 0-7905-7150-1 – mf#1988-3150 – us ATLA [940]

History of the hopedale community : from its inception to its virtual submergence in the hopedale parish / Ballou, Adin; ed by Heywood, William Sweetzer – Lowell, Mass: Thompson & Hill, 1897 – 1mf – 9 – 0-524-03035-9 – mf#1990-0792 – us ATLA [978]

The history of the house of orange : william and mary, king and queen of england, scotland, france, ireland...with a sketch of the orange institution to the present day / Burton, Robert or Richard [pseud of: Nathaniel Crouch] – Toronto: Maclear; St John, NB: R A H Morrow, 18–? – 4mf – 9 – (incl publ's list) – mf#63756 – cn Canadiana [940]

History of the huguenot emigration to america, vol 1 / Baird, Charles Washington – New York: Dodd, Mead, c1885 – v1 on 5mf – 9 – mf#07409 – cn Canadiana [242]

History of the huguenot emigration to america, vol 2 / Baird, Charles Washington – New York: Dodd, Mead, c1885 – v2 on 5mf – 9 – mf#07410 – cn Canadiana [242]

History of the huguenot emigration to america, vols 1 and 2 / Baird, Charles Washington – New York: Dodd, Mead, c1885 – 2v on 1mf – 9 – mf#07408 – cn Canadiana [242]

History of the huguenots during the sixteenth century / Browning, William – London 1829 [mf ed Hildesheim 1995-98] – 2v on 6mf – 9 – €120.00 – ISBN-10: 3-487-26131-6 – ISBN-13: 978-3-487-26131-7 – gw Olms [944]

A history of the huguenots of the dispersion : at the recall of the edict of nantes / Poole, Reginald Lane – London: Macmillan, 1880 [mf ed 1990] – 1mf – 9 – 0-7905-5669-3 – (incl bibl ref) – mf#1988-1669 – us ATLA [242]

A history of the iconoclastic controversy / Martin, Edward James – London: SPCK; New York: Macmillan [1930] – (= ser Church historical society publications. new series 2) – xii/282p – 1 – mf#1343 – us Wisconsin U Libr [240]

History of the illinois river baptist association, and of its churches / Bailey, Gilbert Stephen – New York: Sheldon, Blakeman 1857 [mf ed 1993] – 1mf – 9 – 0-524-08251-0 – mf#1993-3006 – us ATLA [242]

The history of the independent or congregational church in charleston, s.c / Ramsay, David – 1815 – 1 – $50.00 – us Presbyterian [242]

History of the independent order of good templars / Parker, T[homas] F – 2nd ed. New York: Phillips & Hunt, 1887 – 1r – 1 – us Western Res [366]

History of the indian archipelago : containing an account of the manners, arts, languages, religions, institutions, and commerce of its inhabitants / Crawfurd, John – Edinburgh 1820 [mf ed Hildesheim 1995-98] – 3v on 12mf – 9 – €120.00 – 3-487-27441-8 – gw Olms [954]

History of the indian archipelago : containing an account of the manners, arts, languages, religions, institutions, and commerce of its inhabitants; with maps and engravings / Crawfurd, John – Edinburgh: Constable 1820 – (= ser Whsb) – 3v on 20mf – 9 – €140.00 – (incl: the countries, nations, and languages of the oceanic region) – mf#Hu 485 – gw Fischer [954]

History of the indian archipelago / Crawfurd, J – Edinburgh, 1820. 3v – 20mf – 9 – mf#SE-20157 – ne IDC [915]

History of the indian association, 1876-1951 / Bagala, Yogesacandra – Calcutta: The Association, [1953] – (= ser Samp: indian books) – us CRL [366]

The history of the indian national congress / Pattabhi Sitaramayya, Bhogaraju – Bombay: Padma Publications Ltd, 1946-1947 – (= ser Samp: indian books) – (int by rajendra prasad) – us CRL [323]

A history of the indian nationalist movement / Lovett, Verney – London: John Murray, 1920 – (= ser Samp: indian books) – us CRL [954]

History of the indian wars : to which is prefixed a short account of the discovery of america by columbus, and of the landing of our forefathers at plymouth... / Trumbull, Henry – new corr enl ed. Boston: Phillips & Sampson, 1846 [mf ed 1983] – 4mf – 9 – 0-665-41419-6 – mf#41419 – cn Canadiana [970]

History of the inquisition : from its establishment in the twelfth century to its extinction in the nineteenth / Rule, William Harris – London: Hamilton, Adams; New York: Scribner, Welford, 1874. 2v. illus – 1 – us Wisconsin U Libr [241]

A history of the inquisition of spain / Lea, Henry Charles – New York: Macmillan, 1906-07 [mf ed 1992] – 4v on 6mf – 9 – 0-524-03346-3 – (in english, latin, spanish. incl bibl ref) – mf#1990-0927 – us ATLA [946]

A history of the inquisition of the middle ages / Lea, Henry Charles – New York. v1-3. 1922 – 3v on 33mf – 8 – €63.00 – ne Slangenburg [241]

History of the institution of the sabbath day, its uses and abuses : with notices of the puritans, quakers, etc / Fisher, William Logan – 2nd rev enl ed. Philadelphia: TB Pugh, 1859 – 1mf – 9 – 0-524-01224-5 – mf#1990-0363 – us ATLA [243]

History of the intellectual development of europe / Draper, John William – London, 1864 – (= ser 19th c evolution & creation) – 10mf – 9 – mf#1.1.9682 – uk Chadwyck [100]

History of the intellectual development of europe / Draper, John William – Rev ed. New York: Harper, 1876 – 3mf – 9 – 0-7905-6991-4 – mf#1988-2991 – us ATLA [940]

History of the interchurch world movement in north america / Interchurch World Movement of North America – 1924. Chicago: Dep of Photodup, U of Chicago Lib, 1966 (1r); Evanston: American Theol Lib Assoc, 1984 (1r) – 1 – 0-8370-1768-8 – mf#1984-B038 – us ATLA [240]

A history of the interpretation of romans 14:1-15:13, 1845-1980 / Lamkin, Thomas Elwood – 1982 – 1 – $5.00 – us Southern Baptist [242]

History of the irish presbyterian church / Hamilton, Thomas – 2nd ed. Edinburgh: T and T Clark, [1887?] – (= ser Handbooks for bible classes and private students) – 1mf – 9 – 0-524-01652-6 – mf#1990-0473 – us ATLA [242]

A history of the irish presbyterians / Latimer, William Thomas – 2nd ed. Belfast: J Cleeland: W Mullan, 1902 [mf ed 1990] – 2mf – 9 – 0-7905-6346-0 – mf#1988-2346 – us ATLA [242]

History of the irish rebellion in 1798 / Maxwell, William Hamilton – London, England. 1866 – 1r – us UF Libraries [941]

The history of the irish republic : an extract / Henderson, Murdoch [Harper, John Murdoch] – Montreal: W Drysdale, [between 1896 and 1906] – 1mf – 9 – 0-665-93057-7 – (incl bibl ref) – mf#93057 – cn Canadiana [941]

A history of the irish settlers in north america : from the earliest period to the census of 1850 / McGee, Thomas D'Arcy – 6th ed. Boston: P Donahoe, 1855 [mf ed 1984] – 3mf – 9 – 0-665-46143-7 – mf#46143 – cn Canadiana [971]

A history of the island of madagascar : comprising a political account of the island, the religion, manners, and customs of its inhabitants, and its natural productions / Copland, Samuel – London 1822 [mf ed Hildesheim 1995-98] – 1v on 3mf – 9 – €90.00 – 3-487-27233-4 – (with app) – gw Olms [960]

A history of the island of madagascar... / Copland, S – London, 1822 – 5mf – 9 – mf#HT-35 – ne IDC [916]

A history of the island of newfoundland : containing a description of the island, the banks, the fisheries, and trade of newfoundland, and the coast of labrador / Anspach, Lewis A – London 1819 [mf ed Hildesheim 1995-98] – 1v on 4mf – 9 – €120.00 – 3-487-27139-7 – gw Olms [971]

History of the island of st helena : from its discovery by the portuguese to the year 1823 / Brooke, Thomas – London 1824 [mf ed Hildesheim 1995-98] – 1v on 3mf – 9 – €90.00 – 3-487-27282-2 – gw Olms [910]

History of the isle of man : with a comparative view of the past and present state of society and manners / Bullock, Hannah – London 1816 [mf ed Hildesheim 1995-98] – 1v on 3mf – 9 – €90.00 – 3-487-27939-8 – (with biogr anecdotes of eminent persons connected with that island) – gw Olms [941]

History of the isle of providence / Oldmixon, Mr – London, England. 1949 – 1r – us UF Libraries [972]

A history of the israelitish nation : from their origin to their dispersion at the destruction of jerusalem by the romans / Alexander, Archibald Browning Drysdale – Philadelphia: W S Martien, 1853 [mf ed 1989] – 2mf – 9 – 0-7905-0843-5 – mf#1987-0843 – us ATLA [939]

History of the japan mission of the reformed church in the united states, 1879-1904 / Noss, Christopher et al; ed by Miller, Henry K – Philadelphia: Board of Foreign Missions, Reformed Church in the United States, 1904 – 1mf – 9 – 0-524-05383-9 – mf#1991-2289 – us ATLA [242]

History of the jats : a contribution to the history of northern india / Kanunago, Kalika Ranjana – Calcutta: MC Sarkar & Sons, 1925- – (= ser Samp: indian books) – (foreword by jadunath sarkar) – us CRL [954]

History of the jesuit mission in madura south india : in the 17th and 18th centuries / Chandler, John Scudder – Madras: M E Publ House, 1909 [mf ed 1995] – (= ser Yale coll) – vii/72p (ill) – 1 – 0-524-09084-X – mf#1995-0084 – us ATLA [241]

History of the jesuits : their origin, progress, doctrines, and designs / Nicolini, Giovanni Battista – London: George Bell, 1893 – (= ser Bohn's illustrated library) – 2mf – 9 – 0-524-04314-0 – mf#1990-1240 – us ATLA [241]

The history of the jesuits in england, 1580-1773 / Taunton, Ethelred Luke – London: Methuen, 1901 – 2mf – 9 – 0-524-01900-2 – mf#1990-0527 – us ATLA [241]

The history of the jewish church / Stanley, Arthur Penrhyn – New York: Charles Scribner's, 1879. Beltsville, Md: NCR Corp, 1978 (21mf); Evanston: American Theol Lib Assoc, 1984 (21mf) – (= ser Biblical crit – us & gb) – 9 – 0-8370-0241-9 – (incl bibl ref and ind) – mf#1984-1041 – us ATLA [270]

A history of the jewish nation : from the earliest times to the present day / Palmer, Edward Henry – London: SPCK, 1874 [mf ed 1989] – 1mf – 9 – 0-7905-2263-2 – (incl ind) – mf#1987-2263 – us ATLA [939]

A history of the jewish people during the babylonian, persian, and greek periods / Kent, Charles Foster – New York: Charles Scribner, 1899 [mf ed 1989] – (= ser The historical series for bible students 3) – 1mf – 9 – 0-7905-1420-6 – (incl ind) – mf#1987-1420 – us ATLA [939]

A history of the jewish people during the maccabean and roman periods : including new testament times / Riggs, James Stevenson – New York: Charles Scribner, 1900 [mf ed 1989] – (= ser The historical series for bible students 4) – 1mf – 9 – 0-7905-2601-8 – mf#1987-2601 – us ATLA [939]

History of the jewish people in the time of jesus christ / Shurer, Emil – New York, NY. div1 v1-div2 v3. 1885-19–? – 2r – us UF Libraries [939]

History of the jews : from the war with rome to the present time / Adams, Henry Cadwallader – London, England. 1887 – 1r – us UF Libraries [939]

History of the jews / Graetz, Heinrich – Philadelphia, PA. v1-6. 1891-1898 – 2r – us UF Libraries [939]

The history of the jews : from the earliest period down to modern times / Milman, Henry Hart – 3d thoroughly rev and extended ed. London: John Murray, 1863. Beltsville, Md: NCR Corp, 1978 (17mf); Evanston: American Theol Lib Assoc, 1984 (17mf) – (= ser Biblical crit – us & gb) – 9 – 0-8370-0242-7 – (incl bibl ref and ind) – mf#1984-1028 – us ATLA [900]

A history of the jews in england / Hyamson, Albert Montefiore – London: Chatto & Windus, 1908 [mf ed 1987] – 1r – 1 – (filmed with: thoughts of the emperor.../ aurelius antoninus, m) – mf#1838 – us Wisconsin U Libr [939]

A history of the jews in england / Hyamson, Albert Montefiore – London: Methuen & Co, 1928 (mf ed 1995) – 1r – 1 – (incl bibl ref and ind) – mf#ZZ-34380 – us NY Public [939]

A history of the jews in england / Hyamson, Albert Montefiore – London: Publ for the Jewish Hist Soc of England by Chatto & Windus, 1908 (mf ed 1995) – 1r – 1 – (incl bibl ref and ind) – mf#ZZ-34380 – us NY Public [939]

History of the jews in modern times / Raisin, Max – New York, NY. 1949 – 1r – us UF Libraries [939]

History of the jews in the united states / Levinger, Lee Joseph – New York, NY. 1959 – 1r – us UF Libraries [939]

History of the karaite jews / Rule, William Harris – London: Longmans, Green, 1870 – 1mf – 9 – 0-8370-4996-2 – mf#1985-2996 – us ATLA [939]

History of the kinetograph, kinetoscope and kinetophonograph / Dickson, William Kennedy Laurie & Dickson, Antonia – [New York?]: [s.n.], [1895] (mf ed 1975) – 1r – 5 – mf#SEM16P235 – cn Bibl Nat [770]

History of the kingdom of god under the old testament = Geschichte des reiches gottes unter dem alten bunde / Hengstenberg, Ernst Wilhelm – Edinburgh: T & T Clark; New York: C Scribner [dist] 1871-72 [mf ed 1989] – (= ser Clark's foreign theological library. 4th series 32,36) – 2v on 3mf – 9 – 0-7905-2165-2 – (incl ind; trans fr german by ernst wilhelm hengstenberg) – mf#1987-2165 – us ATLA [221]

History of the kingdom of siam and of the revolutions that have caused the overthrow of the empire, up to a d 1770 / Turpin, M – Bangkok, 1908 – 3mf – 9 – mf#SE-20193 – ne IDC [915]

The history of the lake superior ring : an account of the rise and progress of the yankee combination headed by hon alexander mackenzie, premier of canada, and the browns, for the purpose of selling their interest and political power to enrich jay cooke an co and other american speculators... – [Toronto?: s.n.], 1874 [mf ed 1982] – 1mf – 9 – mf#23947 – cn Canadiana [362]

The history of the last four years – S.l: s.n, 1891? – (= ser Information for the people 3) – 1mf – 9 – mf#03691 – cn Canadiana [323]

History of the late war between great britain and the united states of america : with a retrospective view of the causes from whence it originated / Thompson, David – [Niagara-on-the-Lake, Ont: s.n.] 1832 [mf ed 1983] – 4mf – 9 – 0-665-41416-1 – (with app) – mf#41416 – cn Canadiana [975]

A history of the later roman empire : from arcadius to irene (395 a d to 800 a d) / Bury, John Bagnell – London, New York: Macmillan, 1889 [mf ed 1990] – 2v on 3mf – 9 – 0-7905-4544-6 – (incl bibl ref) – mf#1988-0544 – us ATLA [930]

History of the latin-american nations / Robertson, William Spence – New York, NY. 1922 – 1r – us UF Libraries [972]

History of the latter day saints tongan building program / Tyler, Arleah et al [comp] – 1955-1959 – 1r – 1 – mf#pmb116 – at Pacific Mss [243]

A history of the law, the courts, and the lawyers of maine, from its first colonization to the early part of the present century / Willis, William – Portland, Bailey & Noyes, 1863. 712 p. LL-1546 – 1 – us L of C Photodup [347]

History of the lemen family of illinois, virginia, and elsewhere, 1656-1898 / Lemen, Frank B – 1 – us Southern Baptist [242]

A history of the liberty baptist association : from its organization in 1832 to 1906 / Sheets, Henry – Raleigh, NC: Edwards & Broughton Print Co, 1907 [mf ed 1993] – (= ser Baptist coll) – 1mf – 9 – 0-524-07986-2 – mf#1990-5431 – us ATLA [242]

History of the library of the state historical society of colorado, 1879-1940 / Waldron, Rodney K – Denver, CO: University of Denver, 1950 (mf ed 1951) – 1r – 1 – mf#mf w148h – us Colorado Hist [020]

The history of the life of albert duerer of nuernberg : with a translation of his letters and journal, and some account of his works / Heaton, Mary Margaret (Keymer) – London: Macmillan & Co, 1870 – (= ser 19th c art & architecture) – 5mf – 9 – mf#4.1.169 – uk Chadwyck [920]

The history of the life of thomas ellwood : or an account of his birth, education, etc... / Ellwood, Thomas; ed by Crump, Charles George – New York: G P Putnam 1900 [mf ed 1992] – (= ser Putnam's library of standard literature) – 1mf – 9 – 0-524-02953-9 – (incl bibl ref) – mf#1990-4505 – us ATLA [243]

History of the life, writings, & doctrines of luther = Histoire de la vie, des ecrits, et des doctrines de martin luther / Audin, Jean Marie Vincent – London: C Dolman, 1854 – (= ser Library of Translations from Select Foreign Literature) – 3mf – 9 – 0-524-07148-9 – (incl bibl ref and ind. in english) – mf#1991-2937 – us ATLA [242]

A history of the literature of ancient israel : from the earliest times to 135 b c / Fowler, Henry Thatcher – New York: Macmillan, 1912 [mf ed 1989] – 1mf – 9 – 0-7905-0769-2 – (incl ind) – mf#1987-0769 – us ATLA [470]

The history of the litigation and legislation respecting presbyterian chapels and charities in england and ireland between 1816 and 1849 / James, Thomas Smith – London: H Adams; Birmingham: Hudson, 1867 – 3mf – 9 – 0-7905-5282-5 – mf#1988-1282 – us ATLA [340]

The history of the london missionary society, 1795-1895 / Lovett, Richard – London: H Frowde, 1899. Chicago: Dep of Photodup, U of Chicago Lib, 1969 (1r); Evanston: American Theol Lib Assoc, 1984 (1r) – 1 – 0-8370-0160-9 – (incl ind) – mf#1984-B095 – us ATLA [240]

History of the london office of the oss / U.S. Office of Strategic Services – (= ser Records Of The Office Of Strategic Services) – 10r – 1 – (with printed guide) – mf#M1623 – us Nat Archives [327]

The history of the london society for promoting christianity amongst the jews : from 1809 to 1908 / Gidney, William Thomas – London: Society for Promoting Christianity Amongst the Jews, 1908 – 2mf – 9 – 0-8370-6052-4 – (incl bibl ref and index) – mf#1986-0052 – us ATLA [230]

A history of the lutheran church in guyana / Beatty, Paul B – Chicago, 1968. Chicago: Dep of Photodup, U of Chicago Lib, 1968 (1r); Evanston: American Theol Lib Assoc, 1984 (1r) – 1 – 0-8370-0468-3 – mf#1984-B093 – us ATLA [242]

History of the lutheran church of frederick, md : a discourse / Diehl, George – Gettysburg: Printed by HC Neinstedt, 1856 – 1mf – 9 – 0-7905-8782-3 – mf#1989-2007 – us ATLA [242]

History of the lutheran version of the bible / Hentz, John P – Columbus, OH: F J Heer, 1910 – 1mf – 9 – 0-8370-9156-X – mf#1986-3156 – us ATLA [220]

A history of the mahrattas / Duff, James Grant – Calcutta: R Cambray & Co, 1912 – (= ser Samp: indian books) – us CRL [954]

A history of the maratha people / Kincaid, Charles Augustus & Parasnis, D B – London, New York: Oxford University Press, 1918-1925 – (= ser Samp: indian books) – us CRL [954]

The history of the maritime wars of the turks / Haji Khalifah – v1. 1831 – (= ser Royal asiatic society oriental translation fund. old series) – 1r – 1 – mf#95793 – uk Microform Academic [900]

A history of the mccormick theological seminary of the presbyterian church / Halsey, Leroy Jones – Chicago: The Seminary, 1893 [mf ed 1992] – (= ser Presbyterian coll) – 2mf – 9 – 0-524-02956-3 – mf#1990-4508 – us ATLA [242]

History of the mediaeval school of indian logic / Vidyabhusana, Satis Chandra – Calcutta: Calcutta University, 1909 – (= ser Samp: indian books) – us CRL [160]

The history of the melanesian mission / Armstrong, E S – London: Isbister, 1900 – 1mf – 9 – 0-7905-5380-5 – mf#1988-1380 – us ATLA [240]

History of the mennonites : historically and biographically arranged from the time of the reformation / Cassel, Daniel Kolb – Philadelphia: DK Cassel, 1888 – 2mf – 9 – 0-524-04954-8 – mf#1990-1357 – us ATLA [243]

A history of the methodist church, south, the united presbyterian church, the cumberland presbyterian church, and the presbyterian church, south, in the united states / Alexander, Gross et al – New York: Christian Literature, 1894 [mf ed 1989] – (= ser The american church history series 11) – 2mf – 9 – 0-7905-4180-7 – (incl bibl ref) – mf#1988-0180 – us ATLA [242]

History of the methodist church within the territories embraced in the late conference of eastern british america : including nova scotia, new brunswick, prince edward island, and bermuda / Smith, Thomas Watson – Halifax, NS: Methodist Book Room, 1877-[1890?] – 3mf – 9 – 0-7905-6733-4 – (incl bibl ref) – mf#1988-2733 – us ATLA [242]

A history of the methodist episcopal church / Bangs, Nathan – 3rd rev corr ed. New York: Pub...for the ME Church, 1839-42 [mf ed 1993] – (= ser Methodist coll) – 4v on 5mf – 9 – 0-524-06236-6 – mf#1990-5191 – us ATLA [242]

History of the methodist episcopal church in canada / Webster, Thomas – Hamilton: Canada Christian Advocate, 1870 – 1mf – 9 – 0-7905-7035-1 – (incl bibl ref) – mf#1988-3035 – us ATLA [242]

History of the methodist episcopal church in the united states : embracing, also, a sketch of the rise of methodism in europe, and of its origin and progress in canada / Gorrie, Peter Douglass – Philadelphia: JE Potter, [188-?] – 1mf – 9 – 0-524-03152-5 – mf#1990-4601 – us ATLA [242]

History of the methodist episcopal church in the united states of america / Stevens, Abel – New York: Carlton & Porter, 1864-1867 – 5mf – 9 – 0-7905-8154-X – mf#1988-6101 – us ATLA [242]

A history of the middle district baptist association / Moore, L W – 1 – $5.00 – us Southern Baptist [242]

History of the military intelligence division : department of the army general staff / U.S. Army – By Bruce W. Bidwell. 1959-61 – 7 – us L of C Photodup [970]

History of the military intelligence division, 7 december 1941-2 september 1945 / U.S. War Dept. General Staff. G-2 Division – 1 – us L of C Photodup [355]

History of the mission of the american board of commissioners for foreign missions to the sandwich islands / Anderson, Rufus – rev ed. Boston Congregational Pub Board, 1874 – 1mf – 9 – 0-8370-6560-7 – (incl ind) – mf#1986-0560 – us ATLA [240]

History of the mission of the secession church to nova scotia and prince edward island : from its commencement in 1765 / Robertson, James – London: J Johnstone, 1847 – 1mf – 9 – 0-524-08527-7 – mf#1993-1057 – us ATLA [240]

History of the mission of the united brethren among the indians in north america / Loskiel, G H – London, 1794 – 8mf – 9 – mf#HTM-104 – ne IDC [917]

A history of the missions in japan and paraguay / Caddell, Cecilia Mary – New York: D & J Sadlier [1856] [mf ed 1986] – 1mf – 9 – 0-8370-7049-X – mf#1986-1049 – us ATLA [241]

History of the missions of the american board of commissioners for foreign missions in india / Anderson, Rufus – Boston: Congregational Publ Society, 1875, c1874 – 2mf – 9 – 0-8370-6001-X – (incl bibl ref, list of publ and ind) – mf#1986-0001 – us ATLA [240]

History of the missions of the american board of commissioners for foreign missions to the oriental churches / Anderson, Rufus – Boston: Congregational Publ Soc 1872 [mf ed 1990] – 2v on 3mf [ill] – 9 – 0-7905-4602-7 – (incl bibl ref) – mf#1988-0602 – us ATLA [243]

History of the missions of the free church of scotland in india and africa / Hunter, Robert – London, 1873 – (= ser 19th c british colonization) – 5mf – 9 – (pref note by the rev charles j brown) – mf#1.1.1697 – uk Chadwyck [242]

History of the missions of the methodist episcopal church : from the organization of the missionary society to the present time / Strickland, William Peter – Cincinnati: L. Swormstedt & J.H. Power, 1850, c1849 – 1mf – us ATLA [242]

History of the missions of the methodist episcopal church : from the organization of the missionary society to the present time / Strickland, William Peter – Cincinnati: L. Swormstedt & J.H. Power, 1850, c1849 – 1mf – 9 – 0-7905-6572-2 – mf#1988-2572 – us ATLA [242]

A history of the missions of the moravian church during the 18th and 19th centuries / Hamilton, John Taylor – Bethlehem, PA: Times, 1901 [mf ed 1990] – 1mf – 9 – 0-7905-6175-1 – mf#1988-2175 – us ATLA [242]

History of the mit radar school in relation to army training from june 23, 1941 to june 30, 1945 / McIlroy, Malcom S & Zimmerman, Henry J – 1945. 63 p. 1 reel – 1 – us L of C Photodup [621]

History of the modern styles of architecture / Fergusson, James – 3rd ed. London 1891 – (= ser 19th c art & architecture) – 9mf – 9 – mf#4.2.1749 – uk Chadwyck [720]

History of the modern styles of architecture / Fergusson, James – London 1862 – (= ser 19th c art & architecture) – 6mf – 9 – mf#4.2.850 – uk Chadwyck [720]

A history of the modes of christian baptism : from holy scripture, the councils ecumenical and provincial, the fathers, the schoolmen, and the rubrics of the whole church east and west... / Chrystal, James – Philadelphia: Lindsay & Blakiston, 1861 [mf ed 1990] – 1mf – 9 – 0-7905-6223-5 – mf#1988-2223 – us ATLA [240]

History of the mogul dynasty in india : from its foundation by tamerlane, in the year 1399, to the accession of aurangzebe, in the year 1657 / Catrou, Fran ois – London: JM Richardson, 1826 – (= ser Samp: indian books) – (trans fr french of francois catrou) – us CRL [954]

A history of the moravian church / Hutton, Joseph Edmund – London: Moravian Publ Office, 1909 [mf ed 1986] – 1r – 1 – (filmed with: the life and letters of walter farquhar hook / stephens, w r) – mf#1701 – us Wisconsin U Libr [242]

A history of the moravian church / Hutton, Joseph Edmund – 2nd rev enl ed. London: Moravian Publ Office, 1909 [mf ed 1990] – 2mf – 9 – 0-7905-4692-2 – (incl bibl ref) – mf#1988-0692 – us ATLA [243]

History of the moravian church in philadelphia : from its foundation in 1742 to the present time / Ritter, Abraham – Philadelphia: Hayes & Zell, 1857 [mf ed 1990] – 1mf – 9 – 0-7905-8115-9 – mf#1988-6077 – us ATLA [243]

The history of the moravian mission among the indians in north america : from the commencement to the present time... – London, 1840 – 4mf – 9 – mf#HTM-86 – ne IDC [917]

The history of the morison or morrison family : with most of the "traditions of the morrisons" (clan macghillemhuire), hereditary judges of lewis, by capt f w l thomas, of scotland, and a record of the descendants of the hereditary judges to 1880... / Morrison, Leonard Allison – Boston, MA: A Williams, 1880 – 6mf – 9 – (incl ind) – mf#29776 – cn Canadiana [929]

History of the muslim world / Ahchanaulla, Khanabahadura – Calcutta: Empire Book House, [1931] – (= ser Samp: indian books) – us CRL [260]

History of the names of men, nations, and places : in their connection with the progress of civilization = Essai historique et philosophique sur les noms d'hommes, de peuples, et de lieux / Salverte, Eusebe – London: John Russell Smith, 1862 – 3mf – 9 – 0-8370-8303-6 – (incl bibl ref and index. in english) – mf#1986-2303 – us ATLA [400]

History of the national united evangelistic campaign : under the auspices of the japan continuation committee, 1914-17 – Tokyo: [s.n.]; [1917?] [mf ed 1995] – (= ser Yale coll) – 309p/64p – 1 – 0-524-10257-0 – (in japanese) – mf#1996-1257 – us ATLA [480]

History of the navy during the rebellion / Boynton, Charles Brandon – New York: Appleton 1867 [mf ed UMI 1980] – 2v on 1r [ill] – 1 – us UMI ProQuest [355]

History of the nayaks of madura / Sathianathaier, R; ed by Aiyangar, S Krishnaswami – [Madras]: Oxford University Press, 1924 – (= ser Samp: indian books) – (with int and notes) – us CRL [954]

History of the negro race in america from 1619 to 1880 / Williams, George Washington – New York, NY. v1-2. 1882 – 1r – us UF Libraries [025]

History of the negro race in america from 1619-1880 : negroes as slaves, as soldiers, and as citizens / Williams, George Washington – New York: Putnam, 1883 – 1r – 1 – 0-8370-1540-5 – mf#1984-B215 – us ATLA [975]

A history of the negro troops in the war of the rebellion, 1861-1865 109=preceded by a review of the military services of negroes in ancient and modern times / Williams, George W – New York: Harper, 1888 – (filmed with: [baird, h c] washington und jackson uber die neger als soldaten) – us CRL [976]

A history of the new school : and of the questions involved in the disruption of the presbyterian church in 1838 / Baird, Samuel John – Philadelphia: Claxton, Remsen & Haffelfinger, 1868 [mf ed 1990] – 2mf – 9 – 0-7905-4603-5 – mf#1988-0603 – us ATLA [242]

History of the new testament – Mariannhill, South Africa. 1918 – 1r – us UF Libraries [225]

The history of the new testament canon in the syrian church... / Bewer, Julius August – Chicago: University of Chicago Press, 1900. Chicago: Dep of Photodup, U of Chicago Lib, 1976 (1r); Evanston: American Theol Lib Assoc, 1984 (1r) – 1 – 0-8370-0318-0 – mf#1984-B484 – us ATLA [225]

History of the north carolina chowan baptist association, 1806-1881 / Delke, James Almerius – Raleigh: Edwards, Broughton, 1882 – 1mf – 9 – 0-524-08359-2 – mf#1993-3059 – us ATLA [242]

The history of the north carolina governor's council on physical fitness and health / Hendrick, Kevin – 2000 – 83p on 1mf – 9 – $5.00 – mf#PE4104 – us Kinesology [613]

A history of the north indiana conference of the methodist episcopal church : from its organization in 1844 to the present / Herrick, Horace N & Sweet, William Warren – Indianapolis: WK Stewart, 1917 [mf ed 1992] – (= ser Methodist coll) – 1mf – 9 – 0-524-04046-X – (incl bibl ref) – mf#1990-4954 – us ATLA [242]

History of the northwest coast – San Francisco: History Co. 2v. 1890 – 1mf – 9 – mf#14095 – cn Canadiana [978]

History of the northwest coast : vol 1: 1543-1800 / Bancroft, Hubert Howe – San Francisco: History Co, 1890 – v1 on 9mf – 9 – mf#14096 – cn Canadiana [917]

History of the northwest coast : vol 2: 1800-1846 / Bancroft, Hubert Howe – San Francisco: History Co, 1890 – v2 on 9mf – 9 – (incl ind) – mf#14097 – cn Canadiana [917]

History of the norwegian lutheran church in america = Norsk lutherske kirkes historie 1 amerika / Bergh, Johan Arndt – [1915?] [mf ed 1994] – 2v on 2mf – 9 – 0-524-08858-6 – (english trans by j a lavik) – mf#1993-3322 – us ATLA [242]

The history of the novel in england / Lovett, Robert Morss & Hughes, Helen Sard – Boston, New York: Houghton, Mifflin Co., c1932 – 1 – us Wisconsin U Libr [420]

History of the office of censorship – 3r – 1 – $490.00 – 0-89093-101-1 – (with p/g) – us UPA [350]

History of the old covenant / Kurtz, Johann Heinrich – Philadelphia: Lindsay and Blakiston, 1859. Chicago: Dep of Photodup, U of Chicago Lib, 1978 (1r); Evanston: American Theol Lib Assoc, 1984 (1r) – 1 – 0-8370-0756-9 – (includes bibliographies and index) – mf#1984-T110 – us ATLA [221]

A history of the old english letter foundries : with notes, historical and bibliographical, on the rise and progress of english typography / Reed, Talbot Baines – London, 1887 – (= ser 19th c publishing...) – 5mf – 9 – mf#3.1.57 – uk Chadwyck [670]

History of the old independent chapel, tockholes, near blackburn, lancashire : or, about two centuries and a half of nonconformity in tockholes / Nightingale, Benjamin – Manchester: John Heywood, 1886 [mf ed 1991] – 1mf – 9 – 0-524-01121-4 – mf#1990-0335 – us ATLA [240]

History of the old south church (third church), boston, 1669-1884 / Hill, Hamilton Andrews – Boston: Houghton, Mifflin, 1890 – 4mf – 9 – 0-524-02117-1 – mf#1990-4183 – us ATLA [240]

History of the organization of the methodist episcopal church, south / Redford, Albert Henry – Nashville, Tenn: Published by AH Redford, agent, for the M.E. Church, South, 1878 – 2mf – 9 – 0-524-02963-6 – (incl proceedings of the 1844 general conference of the methodist episcopal church, south and other official papers relating to the separation from the methodist episcopal church) – mf#1990-4515 – us ATLA [242]

History of the origin and development and condition of missions among the sherbro and mendi tribes in western africa / Flickinger, Daniel Kumler & McKee, William – Dayton, OH: United Brethren Pub House, 1885 – 1 – us CRL [240]

A history of the origin and development of the governing conference in methodism : and especially of the general conference of the methodist episcopal church / Neely, Thomas Benjamin – Cincinnati: Cranston & Stowe, 1892 [mf ed 1992] – (= ser Methodist coll) – 2mf – 9 – 0-524-03558-X – (incl bibl ref) – mf#1990-4753 – us ATLA [242]

History of the origin, formation and adoption of the constitution of the united states : with notices of its principal framers / Curtis, George Ticknor – New York: Harper & Bros. 2v. 1854-58 – 13mf – 9 – $19.50 – mf#LLMC 90-366 – us LLMC [323]

A history of the origin of the doctrine of the trinity in the christian church / Stannus, Hugh Hutton – London: Christian Life: Williams & Norgate, 1882 [mf ed 1985] – 1mf – 9 – 0-8370-5542-3 – (incl bibl ref) – mf#1985-3542 – us ATLA [240]

History of the origin of the free methodist church / Bowen, Elias – Rochester, N.Y.: B. T. Roberts, 1871. Beltsville, Md: NCR Corp, 1978 (4mf); Evanston: American Theol Lib Assoc, 1984 (4mf) – 9 – 0-8370-1085-3 – mf#1984-4445 – us ATLA [242]

The history of the origin, progress, and termination of the american war / Stedman, Charles – London: printed for the author and sold by J Murray, Fleet Street; J Debrett, Piccadilly; and J Kerby, corner of Wigmore-Street, Cavendish Square, 1794 – 6mf – 9 – mf#47992 – cn Canadiana [975]

The history of the origin, progress, and termination of the american war / Stedman, Charles – London: printed for the author and sold by J Murray, Fleet Street; J Debrett, Piccadilly; and J Kerby, corner of Wigmore-Street, Cavendish Square, 1794 – 1mf – 9 – mf#47990 – cn Canadiana [975]

The history of the origin, progress, and termination of the american war / Stedman, Charles – London: printed for the author and sold by J Murray, Fleet Street; J Debrett, Piccadilly; and J Kerby, corner of Wigmore-Street, Cavendish Square, 1794 – 6mf – 9 – mf#47991 – cn Canadiana [975]

A history of the original settlements on the delaware : from its discovery by hudson to the colonization under william penn / Ferris, Benjamin – Wilmington: Wilson & Heald, 1846 [mf ed 1991] – 1mf – 9 – 0-524-00989-9 – mf#1990-0266 – us ATLA [978]

History of the orthodox church in austria-hungary / Dampier, Margaret Georgiana – London: Published for the Eastern Church Association [by] Rivingtons, 1905 – 1mf – 9 – 0-7905-4286-2 – (incl bibl ref) – mf#1988-0286 – us ATLA [240]

A history of the orthodox church of cyprus : from the coming of the apostles paul and barnabas to the commencement of the british occupation (a d 45-a d 1878)... / Hackett, John – London: Methuen, 1901 [mf ed 1990] – 2mf – 9 – 0-7905-4801-1 – (incl bibl ref) – mf#1988-0801 – us ATLA [243]

History of the pacific northwest and canadian northwest – 50r – 1 – (coll emphasizes out-of-print and rare documents and consists of first-hand narratives of the events that helped shape the country. includes printed guide) – mf#C39-22900 – us Primary [971]

The history of the painters of all nations / Blanc, Charles – London 1852,53 – (= ser 19th c art & architecture) – 6mf – 9 – mf#4.1.383 – uk Chadwyck [750]

A history of the papacy : from the great schism to the sack of rome / Creighton, Mandell – new ed. London, New York: Longmans, Green, 1897 [mf ed 1986] – 6v on 7mf – 9 – 0-8370-7781-8 – (incl ind) – mf#1986-1781 – us ATLA [240]

The history of the papacy in the 19th century = Pavedommet i den nittende hundredaar / Nielsen, Fredrik – New York: E P Dutton, 1906 – 3mf – 9 – 0-8370-8131-9 – (incl bibl ref and index. in english) – mf#1986-2131 – us ATLA [240]

The history of the papacy to the period of the reformation / Riddle, Joseph Esmond – London: Richard Bentley, 1854 [mf ed 1986] – 2mf – 9 – 0-8370-9107-1 – (incl bibl ref & ind) – mf#1986-3107 – us ATLA [240]

History of the parsis : including their manners, customs, religion, and present position / Karaka, Dosabhai Framji – London: Macmillan, 1884 – 2mf – 9 – 0-524-04644-1 – mf#1990-3387 – us ATLA [280]

The history of the passion and resurrection of our lord : considered in the light of modern criticism = Die auferstehungsgeschichte des herrn / Steinmeyer, Franz Ludwig – new rev ed. Edinburgh: T & T Clark 1879 [mf ed 1985] – (= ser Clark's foreign theological library. new series 63) – 1mf – 9 – 0-8370-5390-0 – (trans fr german by thomas crerar & alexander cusin; incl bibl ref & ind) – mf#1985-3390 – us ATLA [240]

The history of the pearl fishery of the tamil coast / Arunachalam, S – Annamalai Nagar: Annamalai University, 1952 – (= ser Samp: indian books) – (foreword by r sathianathaier) – us CRL [639]

A history of the penal laws against the irish catholics / Congleton, Henry Brooke Parnell, 1st Baron – Dublin, 1808 – (= ser 19th c ireland) – 3mf – 9 – mf#1.1.403 – uk Chadwyck [241]

The history of the penal laws enacted against roman catholics : the operation and results of that system of legalized plunder, persecution, and proscription originating in rapacity and fraudulent designs, concealed under false pretences, figments of reform, and a simulated zeal for the interests of true religion / Madden, Richard Robert – London: Thomas Richardson, 1847 – 1mf – 9 – 0-7905-6869-1 – mf#1988-2869 – us ATLA [345]

History of the people of israel from the earliest times to the destruction of jerusalem by the romans = Geschichte des volkes israel von den aeltesten zeiten bis zur zerstoerung jerusalems durch die roemer / Cornill, Carl Heinrich – Chicago: Open Court; London: Kegan Paul, Trench, Truebner [distributor], 1898 – 1mf – 9 – 0-7905-0366-2 – (in english) – mf#1987-0366 – us ATLA [930]

History of the people of israel from the reign of david up to the capture of samaria / Renan, Ernest – Boston: Little, Brown, 1903 – 2mf – 9 – 0-524-05108-9 – mf#1992-0329 – us ATLA [930]

History of the people of israel from the rule of the persians to that of the greeks / Renan, Ernest – Boston: Roberts Brothers, 1895 – 1mf – 9 – 0-7905-1977-1 – mf#1987-1977 – us ATLA [930]

History of the people of israel from the time of hezekiah till the return from babylon / Renan, Ernest – London: Chapman and Hall, 1891 – 1mf – 9 – 0-7905-1978-X – mf#1987-1978 – us ATLA [930]

The history of the people of israel in pre-christian times / Sarson, Mary & Phillips, Mabel Addison – London: Longmans, Green, 1912 – 1mf – 9 – 0-524-04592-5 – (incl bibl ref) – mf#1992-0180 – us ATLA [939]

History of the people of israel till the time of king david / Renan, Ernest – Boston: Little, Brown, 1905 – 1mf – 9 – 0-524-05109-7 – mf#1992-0330 – us ATLA [930]

A history of the people of the united states / McMaster, John B – 1883-1913 – 1 – us L of C Photodup [975]

History of the people of trinidad and tobago / Willilams, Eric Eustace – Port-of-Spain, Trinidad and Tobago. 1962 – 1r – us UF Libraries [972]

History of the persecutions endured by the protestants of the south of france : and more especially of the department of the gard, during the years 1814, 1815, 1816, &c / Wilks, Mark – London 1821 [mf ed Hildesheim 1995-98] – 2v on 4mf – 9 – €120.00 – ISBN-10: 3-487-26331-9 – ISBN-13: 978-3-487-26331-1 – gw Olms [242]

History of the philippine insurrection against the u.s. 1899-1903, and documents relating to the war department project for publishing the history / U.S. War Dept. Adjutant General's Office – (= ser Records of the adjutant general's office, 1780's-1917) – 9r – 1 – (with printed guide) – mf#M719 – us Nat Archives [355]

History of the philosophy of mind : embracing the opinions of all writers on mental science from the earliest period to the present time / Blakey, Robert – London: Longman, Brown, Green and Longmans, 1850 – 6mf – 9 – 0-7905-4900-X – (incl bibl ref) – mf#1988-0900 – us ATLA [100]

History of the philosophy of mind : embracing the opinions of all writers on mental science from the earliest period to the present time / Blakey, Robert – London: Longman, Brown, Green and Longmans, 1850 – 6mf – us ATLA [100]

History of the philosophy of pedagogics : a lecture / Bennett, Charles W – New York: E Steiger, 1877 – 1mf – 9 – 0-8370-7769-9 – mf#1986-1769 – us ATLA [370]

History of the philosophy of religion from spinoza to the present day = Geschichte der religionsphilosophie von spinoza bis auf die gegenwart / Pfleiderer, Otto – London: Williams and Norgate, 1886-1887 – (= ser Religionsphilosophie auf geschichtlicher Grundlage) – 2mf – 9 – 0-7905-8720-3 – (incl bibl ref) – mf#1989-1945 – us ATLA [200]

History of the pinellas peninsula / Hunter, C M – s.l, s.l? 1936 – 1r – us UF Libraries [978]

The history of the pirates...misson, bowen, kidd, tew, halsey, white, condent, bellamy, and their several crews – Hartford: H. Benton, 1829, c1825. 283p – 1 – us Wisconsin U Libr [972]

A history of the plymouth brethren / Neatby, William Blair – 2nd ed. London: Hodder & Stoughton, 1902 [mf ed 1990] – 1mf – 9 – 0-7905-5958-7 – (incl bibl ref. original ed publ 1901) – mf#1988-1958 – us ATLA [242]

History of the popes : their church and state, and especially of their conflicts with protestantism in the 16th and 17th centuries = Geschichte der paepste / Ranke, Leopold von – London: George Bell 1884-96 [mf ed 1992] – (= ser Bohn's standard library) – 3v on 4mf – 9 – 0-524-05329-4 – (trans by e foster) – mf#1990-1447 – us ATLA [900]

The history of the popes : from the foundation of the see of rome to a.d. 1758 / Bower, Archibald – Philadelphia: Griffith & Simon, 1844-1845 – 16mf – 9 – 0-524-03380-3 – (incl bibl ref) – mf#1990-4692 – us ATLA [240]

The history of the popes during the last four centuries = Roemischen paepste in den letzten vier jahrhunderten / Ranke, Leopold von – London: G Bell and Sons, 1913 – (= ser Bohn's popular library) – 4mf – 9 – 0-524-02898-2 – (in english) – mf#1990-4489 – us ATLA [240]

History of the portuguese in bengal / Campos, Joachim Joseph A – Calcutta: Butterworth & Co (India), 1919 – (= ser Samp: indian books) – (int by f j monahan) – us CRL [954]

The history of the prayer book of the church of england / Berens, Edward – 2nd ed. Oxford: JH Parker, 1841 – 1mf – 9 – 0-524-05188-7 – (incl bibl ref) – mf#1991-2224 – us ATLA [241]

A history of the preparation of the world for christ / Breed, David Riddle – 2nd rev enl ed. New York: Fleming H Revell, c1893 [mf ed 1989] – 2mf – 9 – 0-7905-2701-4 – (incl bibl ref) – mf#1987-2701 – us ATLA [221]

History of the presbyterian and congregational churches and ministers in wisconsin : including an account of the organization of the convention and the plan of the union / Peet, Stephen – Milwaukee: S Chapman 1851 [mf ed 1992] – 1mf – 9 – 0-524-03020-0 – mf#1990-4542 – us ATLA [242]

History of the presbyterian board of publication and sabbath-school work / Rice, Willard Martin – Philadelphia: Presbyterian Board of Publication and Sabbath-School Work, [1888?] – 1mf – 9 – 0-524-01748-4 – mf#1990-4140 – us ATLA [242]

A history of the presbyterian church in america : from its origin until the year 1760, with biographical sketches of its early ministers / Webster, Richard – Philadelphia: Joseph M Wilson, 1857 [mf ed 1991] – (= ser Presbyterian coll) – 2mf – 9 – 0-524-01338-1 – mf#1990-4087 – us ATLA [242]

History of the presbyterian church in georgia / Stacy, James – 1 – $50.00 – us Presbyterian [242]

History of the presbyterian church in georgia / Stacy, James – Atlanta: Westminster [1912?] [mf ed 1992] – (= ser Presbyterian coll) – 1mf – 9 – 0-524-02169-4 – (comp after aut's death by c i stacy) – mf#1990-4235 – us ATLA [242]

History of the presbyterian church in ireland : for readers on this side of the atlantic / Cleland, William – Toronto: Hart & Co, 1890 – 4mf – 9 – mf#00681 – cn Canadiana [242]

History of the presbyterian church in south carolina / Howe, George – 1883 – 1 – $50.00 – us Presbyterian [242]

History of the presbyterian church in the dominion of canada : from the earliest times to 1834 / Gregg, William – Toronto: Presbyterian Printing and Publishing Company, 1885 – 2mf – 9 – 0-7905-4798-8 – (incl bibl ref) – mf#1988-0798 – us ATLA [242]

History of the presbyterian church in the dominion of canada : from the earliest times to 1834 : with a chronological table of events to the present time, and map / Gregg, William – Toronto: Presbyterian Printing and Publishing Company, 1885 – 2mf – us ATLA [242]

History of the presbyterian church in the state of illinois / Norton, Augustus Theodore – St Louis: WS Bryan, 1879 – 2mf – 9 – 0-524-02133-3 – mf#1990-4199 – us ATLA [242]

History of the presbyterian church in the state of kentucky : with a preliminary sketch of the churches in the valley of virginia / Davidson, Robert – New York: R Carter, 1847 – 1mf – 9 – 0-7905-7217-6 – (incl bibl ref) – mf#1988-3217 – us ATLA [242]

History of the presbyterian church in the united states of america / Gillett, Ezra Hall – Revised ed. Philadelphia: Presbyterian Board of Publication, [1873] Chicago: Dep of Photodup, U of Chicago Lib, 1975 (1r); Evanston: American Theol Lib Assoc, 1984 (1r) – 1 – 0-8370-0531-0 – (incl bibl ref and index) – mf#1984-B471 – us ATLA [242]

History of the presbyterian church in the united states of america / Gillett, Ezra Hall – Philadelphia: Presbyterian Board of Publication, 1864 – 1r – 1 – 0-8370-1545-6 – mf#1984-B470 – us ATLA [242]

History of the presbyterian church in west cameroon / Keller, Werner – Victoria, Cameroon . 1969 – 1r – us UF Libraries [242]

History of the presbyterian church of new zealand / Dickson, John – Dunedin: J Wilkie, 1899 – 2mf – 9 – 0-524-03147-9 – mf#1990-4596 – us ATLA [242]

A history of the presbyterian churches in the united states / Thompson, Robert Ellis – New York: Christian Literature, 1895 [mf ed 1989] – (= ser The american church history series 6) – 2mf – 9 – 0-7905-4236-6 – mf#1988-0236 – us ATLA [242]

History of the presbyterian churches of the world : adapted for use in the class room / Reed, Richard Clark – Philadelphia: Westminster Press 1905 [mf ed 1986] – 1mf – 9 – 0-8370-8704-X – (incl tables & ind) – mf#1986-2704 – us ATLA [242]

The history of the presbyterian controversy : with early sketches of presbyterianism / Woods, Henry – Louisville: NH White, 1843 – 1mf – 9 – 0-524-02179-1 – mf#1990-4245 – us ATLA [242]

The history of the presbyterian controversy, with early sketches of presbyterianism / Woods, Henry – Louisville: Printed by N.H. White, 1843.viii,(9)-204p – 1 – us Wisconsin U Libr [242]

History of the presbyterians in england : their rise, decline and revival / Drysdale, Alexander Hutton – London: Publication Committee of the Presbyterian Church of England, 1889 – 2mf – 9 – 0-7905-4960-3 – (incl bibl ref) – mf#1988-0960 – us ATLA [242]

History of the presbyterians in england : their rise, decline and revival / Drysdale, Alexander Hutton – London: Publication Committee of the Presbyterian Church of England, 1889 – 2mf – us ATLA [242]

History of the presbytery of erie / Eaton, Samuel J M – 1868 – 1 – $50.00 – us Presbyterian [242]

History of the presbytery of erie : embracing in its ancient boundaries the whole of northwestern pennsylvania and northeastern ohio. with biographical sketches of all its ministers and historical sketches of its churches / Eaton, Samuel John Mills – New York: Hurd and Houghton, 1868 – 2mf – 9 – 0-524-02114-7 – mf#1990-4180 – us ATLA [242]

A history of the primitive baptists in the western districts of tennessee and kentucky / Edgar, Lewis M – 1828-80 – 1 – $6.51 – us Southern Baptist [242]

A history of the primitive methodist church in the united states of america : from its origin and the landing of the first missionaries in 1829 to the present time / Acornley, John Holmes – [s.l]: NW Matthews, 1909 [mf ed 1992] – (= ser Methodist coll) – 1mf – 9 – 0-524-04354-X – mf#1990-5037 – us ATLA [242]

The history of the primitive methodist connexion : from its origin to the conference of 1860, the first jubilee year of the connexion / Petty, John – a new rev and enl ed. London: J Dickenson, 1880 – 2mf – 9 – 0-524-06879-8 – mf#1990-5298 – us ATLA [242]

History of the progress and suppression of the reformation in italy in the sixteenth century : including a sketch of the history of the reformation in the grisons / M'Crie, Thomas – Edinburgh: William Blackwood, 1856 (The works of Thomas M'Crie; v. 3a) – 1mf – us ATLA [242]

History of the progress and suppression of the reformation in italy in the sixteenth century : including a sketch of the history of the reformation in the grisons / M'Crie, Thomas; ed by M'Crie, Thomas, jr – A new ed. Edinburgh: William Blackwood, 1856 – 1mf – 9 – 0-7905-5302-3 – (incl bibl ref) – mf#1988-1302 – us ATLA [945]

History of the progress and suppression of the reformation in spain in the sixteenth century / M'Crie, Thomas; ed by M'Crie, Thomas, jr – A new ed. Edinburgh: William Blackwood, 1856 – 1mf – 9 – 0-7905-5303-1 – (incl bibl ref) – mf#1988-1303 – us ATLA [946]

History of the protestant church in hungary from the beginning of the reformation to 1850 : with special reference to transylvania = Geschichte der evangelischen kirche in ungarn vom anfange der reformation bis 1850 / Bauhofer, Janos Gyoergy – Boston: Phillips, Sampson, 1854 – 2mf – 9 – 0-524-05131-3 – (in english) – mf#1990-1387 – us ATLA [242]

History of the protestant church of the united brethren / Holmes, John – London: Printed for the author, 1825-1830 – 2mf – 9 – 0-524-03070-7 – (incl bibl ref) – mf#1990-4559 – us ATLA [242]

A history of the protestant episcopal church in america / Wilberforce, Samuel, Lord Bishop of Oxford – 2nd ed. London: James Burns, 1846 [mf ed 1993] – (= ser Anglican/episcopal coll) – 2mf – 9 – 0-524-07295-7 – mf#1990-5382 – us ATLA [242]

History of the protestant episcopal church in the county of westchester : from its foundation, a.d. 1693, to a.d. 1853 / Bolton, Robert – New-York: Stanford & Swords, 1855 – 2mf – 9 – 0-524-04203-9 – mf#1990-4994 – us ATLA [242]

A history of the protestant episcopal church in the united states / Tiffany, Charles Comfort – New York: Christian Literature, 1895 [mf ed 1989] – (= ser The american church history series 7) – 1mf – 9 – 0-7905-4237-4 – (incl bibl ref) – mf#1988-0237 – us ATLA [242]

The history of the protestant reformation : in germany and switzerland: and in england, ireland, scotland, the netherlands, france, and northern europe / Spalding, Martin John – 12th rev enl ed. Baltimore: John Murphy, c1875 – 4mf – 9 – 0-8370-6701-4 – mf#1986-0701 – us ATLA [242]

A history of the protestant reformation in england and ireland / Cobbett, William – new ed. New York: Benziger, 1896 [mf ed 1986] – 1mf – 9 – 0-8370-7374-X – (incl bibl ref) – mf#1986-1374 – us ATLA [242]

A history of the protestant reformation in england and ireland : showing how that event has impoverished the main body of the people in those countries / Cobbett, William – New York: D & J Sadlier, c1886 [mf ed 1986] – 2v on 2mf – 9 – 0-8370-7373-1 – mf#1986-1373 – us ATLA [242]

History of the protestants of france : from the commencement of the reformation to the present time / Felice, G de – London: Routledge, 1853 – 2mf – us ATLA [242]

History of the protestants of france : from the commencement of the reformation to the present time = Histoire des protestants de france / Felice, Guillaume de – London: Routledge, 1853 – 2mf – 9 – 0-7905-4565-9 – (in english) – mf#1988-0565 – us ATLA [242]

The history of the province of massachuset's bay from the first settlement thereof in 1628 : until its incorporation with the colony of plimouth, province of main etc, by the charter of king william and queen mary in 1691 / Hutchinson, Thomas – 2nd ed. London: Printed for M Richardson, 1765 – 2mf – 9 – 0-524-03716-7 – mf#1990-4821 – us ATLA [975]

The history of the province of massachusets-bay from the charter of king william and queen mary in 1691 until the year 1750 / Hutchinson, Thomas – 2nd ed. London: J Smith [printer], 1768 – 2mf – 9 – 0-524-03717-5 – mf#1990-4822 – us ATLA [975]

The history of the province of massachusetts bay from the year 1750 until june 1774 / Hutchinson, Thomas – London: J Murray, 1828 – 2mf – 9 – 0-524-03718-3 – mf#1990-4823 – us ATLA [975]

History of the province of ontario (upper canada) : containing a sketch of franco-canadian history – the bloody battles of the french and indians – the american revolution... / Canniff, William – Toronto: A H Hovey, 1872 [mf ed 1981] – 8mf – 9 – mf#26811 – cn Canadiana [971]

History of the public library in bristol : a paper read before the library association of the united kingdom, london, february, 1896 / Mathews, Edward Robert Norris – London: John Bale & Sons, 1896 – (= ser 19th c publishing...) – 1mf – 9 – mf#3.1.63 – uk Chadwyck [020]

History of the puritans in england and the pilgrim fathers / Stowell, William Hendry & Wilson, Daniel – New York: R Carter, 1849 – 1mf – 9 – 0-524-04884-3 – (incl bibl ref) – mf#1990-5082 – us ATLA [243]

The history of the puritans or protestant nonconformists : from the reformation in 1517 to the revolution in 1688... / Neal, Daniel – rev corr enl ed. New York: Harper & Bros, 1843-44 [mf ed 1993] – (= ser Congregational coll) – 2v on 3mf – 9 – 0-524-07581-6 – mf#1991-3201 – us ATLA [242]

A history of the qaraunah turks in india : based on original sources / Prasad, Ishwari – Allahabad: Indian Press, 1936 – (= ser Samp: indian books) – us CRL [956]

History of the quebec directory : since its first issue in 1844 up to the present day / Cherrier, A Benjamin – Quebec: Dawson, 1879 – 1mf – 9 – mf#02988 – cn Canadiana [917]

History of the queen's college of british guiana / Cameron, Norman Eustace – Georgetown, Guyana. 1951 – 1r – us UF Libraries [378]

The history of the rebellion and civil wars in england : begun in the year 1641 / Clarendon, Edward Hyde, Earl of; ed by Macray, William Dunn – Oxford: Clarendon Press, 1888 – 8mf – 9 – 0-524-00526-5 – (incl bibl ref) – mf#1990-0026 – us ATLA [941]

History of the recent agricultural policy of the united states / Malin, James C – undated – 1 – us Kansas [630]

A history of the reformation / Sanford, Elias Benjamin – Hartford, CT: SS Scranton, c1917 [mf ed 1992] – 1mf – 9 – 0-524-01467-1 – mf#1990-0416 – us ATLA [242]

History of the reformation : being an abridgment of burnet. together with sketches of the lives of luther, calvin, and zuingle... = History of the reformation of the church of england. selections / Burnet, Gilbert – 2nd ed. Philadelphia: E Littel, 1823 – 1mf – 9 – 0-524-05137-2 – mf#1990-1393 – us ATLA [242]

History of the reformation – London, England. 1808 – 1r – us UF Libraries [242]

History of the reformation in england / Perry, George Gresley – London; New York: Longmans, Green, 1911 – (= ser Epochs Of Church History) – 1mf – 9 – 0-7905-5547-6 – mf#1988-1547 – us ATLA [242]

History of the reformation in europe in the time of calvin = Histoire de la réformation en europe au temps de calvin / Merle d'Aubigne, Jean Henri – London: Longman, Green, Longman, Roberts, & Green, 1863-1878 – 12mf – 9 – 0-524-03653-5 – (in english) – mf#1990-1081 – us ATLA [242]

History of the reformation in germany = Deutsche geschichte im zeitalter der reformation / Ranke, Leopold von; ed by Johnson, Robert A – London: G Routledge, 1905 – 2mf – 9 – 0-524-03658-6 – (incl bibl ref. in english) – mf#1990-1086 – us ATLA [943]

History of the reformation in germany / Ranke, Leopold von; ed by Johnson, Robert A – Trans. by Sarah Austin. London: G. Routledge; New York: E.P. Dutton, 1905. xxiv,792p – 1 – us Wisconsin U Libr [242]

History of the reformation in germany and switzerland chiefly = Geschichte der reformation in deutschland und der schweiz / Hagenbach, Karl Rudolf – Edinburgh: T & T Clark 1878-79 [mf ed 1989] – (= ser Clark's foreign theological library. new series 59,62) – 2v on 2mf – 9 – 0-7905-4859-3 – (incl bibl ref; trans fr 4th rev ed german ed by evelina moore) – mf#1988-0859 – us ATLA [242]

The history of the reformation in sweden = Svenska kyrkoreformationens historia / Anjou, Lars Anton – New-York: Sheldon, 1859 – 2mf – 9 – 0-524-07287-6 – (in english) – mf#1990-5374 – us ATLA [242]

History of the reformation in switzerland = Histoire de la reformation de la suisse. premiere partie, 1516 a 1536 / Ruchat, Abraham – London: WE Painter, 1845 [mf ed 1990] – 1mf – 9 – 0-7905-6826-8 – (abr fr french by j collinson) – mf#1988-2826 – us ATLA [242]

The history of the reformation of religion within the realm of scotland = History of the reformation in scotland / Knox, John; ed by Guthrie, Charles John Guthrie, Lord – London: A. and C. Black, 1898 – 1mf – 9 – 0-7905-4299-4 – mf#1988-0299 – us ATLA [242]

History of the reformation of the sixteenth century / Peter, Philip Adam – Columbus, Ohio: Lutheran Book Concern, 1916 – 1mf – 9 – 0-524-00778-0 – mf#1990-0210 – us ATLA [242]

The history of the reformed church, dutch, the reformed church, german, and the moravian church in the united states / Corwin, Edward Tanjore et al – New York: Christian Literature, 1895, c1894 – (= ser The American Church History Series) – 2mf – 9 – 0-7905-4215-3 – (incl bibl ref) – mf#1988-0215 – us ATLA [242]

History of the reformed church in the united states, 1725-1792 / Good, James Isaac – Reading, Pa.: D. Miller, 1899 – 2mf – 9 – 0-7905-5831-9 – mf#1988-1831 – us ATLA [242]

History of the reformed church in the u.s. in the nineteenth century / Good, James Isaac – New York: Board of Publ of the Reformed Church in America, 1911 – 2mf – 9 – 0-7905-5395-3 – mf#1988-1395 – us ATLA [242]

History of the reformed church of germany, 1620-1890 / Good, James Isaac – Reading, PA: Daniel Miller, 1894 – 2mf – 9 – 0-8370-4706-4 – (incl ind) – mf#1985-2706 – us ATLA [242]

History of the reformed presbyterian church / Glasgow, W. Melanchthon – 1888 – 1 – $50.00 – us Presbyterian [242]

History of the reformed presbyterian church in america : with sketches of all her ministry, congregations, missions, institutions, publications, etc., and embellished with over fifty portraits and engravings / Glasgow, William Melanchton – Baltimore: Hill & Harvey, 1888 – 2mf – 9 – 0-7905-6995-7 – mf#1988-2995 – us ATLA [242]

History of the regular baptists / Coffey, Achilles – 1877 – 1 – 6.30 – us Southern Baptist [242]

History of the reign of queen anne : digested into annals – London. 1703-1713 (1) – mf#4262 – us UMI ProQuest [941]

The history of the reign of shah-aulum, the present emperor of hindustaun : containing the transactions of the court of delhi, and the neighbouring states, during a period of thirty-six years / Francklin, William – Allahabad: Panini Office, 1934 – (= ser Samp: indian books) – us CRL [954]

The history of the reign of tipu sultan.. / Husain Ali, Kirmani – Trans. from an original Persian ms...by Col. W. Miles. London, Printed for the Oriental Translation Fund of Great Britain and Ireland; Paris: Sold by Wm. H. Allen. 1844. xv,291p – 1 – us Wisconsin U Libr [954]

History of the reigns of louis 18 and charles 10 / Crowe, Eyre Evans – London: R Bentley 1854 – 2v – 1 – us Wisconsin U Libr [944]

The history of the religion of ancient britain : or, a succinct account of the several religious systems which have obtained in this island from the earliest times to the norman conquest / Smith, George – 3rd ed. London: Longman, Green, Longman, Roberts, and Green, 1865 – 2mf – 9 – 0-524-01349-7 – mf#1990-0395 – us ATLA [941]

The history of the religion of israel : an old testament primer / Toy, Crawford Howell – Boston: Unitarian Sunday-School Society, 1910, c1882 – 1mf – 9 – 0-8370-9321-X – (includes bibliographies and index) – mf#1986-3321 – us ATLA [270]

History of the religious house of pluscardyn : convent of the vale of saint andrew in morayshire / Macphail, Simeon Ross – Edinburgh: Oliphant, Anderson & Ferrier, 1881 – 1mf – 9 – 0-524-03824-4 – mf#1990-1140 – us ATLA [240]

History of the religious movement of the eighteenth century called methodism : considered in its different denominational forms, and its relations to british and american protestantism / Stevens, Abel – New-York: Carlton & Porter, c1858-c1861 – 4mf – 9 – 0-7905-8155-8 – (incl bibl ref) – mf#1988-6102 – us ATLA [242]

History of the religious society of friends, called by some the free quakers, in the city of philadelphia / Wetherill, Charles – [Philadelphia?]: Printed for the Society, 1894 – 1mf – 9 – 0-524-04665-4 – mf#1990-5061 – us ATLA [243]

History of the religious society of friends from its rise to the year 1828 / Janney, Samuel Macpherson – Philadelphia: TE Zell, 1867, c1859-c1867 – 5mf – 9 – 0-7905-5283-3 – mf#1988-1283 – us ATLA [240]

History of the rensselaer polytechnic institute : 1824-1894 / Ricketts, Palmer Chamberlain – New York: J. Wiley & Sons, 1895 – x/193p – 1 – us Wisconsin U Libr [378]

History of the reverend mother sacred heart of jesus : nee tezenas of montcel, second superior-general of the congregation of the sisters of st. joseph of lyons / Rivaux, Jean Joseph, Abbe – Montreal: Messenger Press, 1910 – 1mf – 9 – 0-8370-6937-8 – mf#1986-0937 – us ATLA [920]

A history of the revised version of the new testament / Hemphill, Samuel – London: Elliot Stock, 1906 [mf ed 1985] – 1mf – 9 – 0-8370-3557-0 – (incl bibl ref) – mf#1985-1557 – us ATLA [225]

History of the revisions of the discipline of the methodist episcopal church / Sherman, David – New York, Nelson & Phillips, 1874 – 1r – 1 – 0-8370-0458-6 – mf#1984-B010 – us ATLA [242]

History of the rise and influence of the spirit of rationalism in europe / Lecky, William Edward Hartpole – London, 1865 – (= ser 19th c evolution & creation) – 2v on 10mf – 9 – mf#1.1.2747 – uk Chadwyck [140]

History of the rise and progress of the alton riots, culminating in the death of rev. elijah p. lovejoy, 7 nov 1837 / Tanner, Henry – 1838 – 1 – $50.00 – us Presbyterian [920]

A history of the rise and progress of the art of printing : a lecture, delivered for the benefit of a working men's reading room / Moore, John – London: printed & publ by J Moore, 1863 – (= ser 19th c publishing...) – 1mf – 9 – mf#3.1.33 – uk Chadwyck [680]

A history of the rise and progress of the baptists in alabama : with a miniature history of the denomination from the apostolic age down to the present time... / Holcombe, Hosea – Philadelphia: King & Baird, 1840 [mf ed 1992] – (= ser Baptist coll) – 1mf – 9 – 0-524-04360-4 – mf#1990-5043 – us ATLA [242]

History of the rise and progress of the baptists in virginia / Semple, R B – 1810 – 1 – us Southern Baptist [242]

The history of the rise, increase and progress of the christian people called quakers : intermixed with several remarkable occurrences / Sewel, William – Philadelphia: Friends' Book Store, 1867 – 3mf – 9 – 0-524-06964-6 – mf#1990-5328 – us ATLA [243]

The history of the rise, increase, and progress of the christian people called quakers / Sewel, W – London, 1722 – €50.00 – ne Slangenburg [243]

A history of the rise of methodism in america : containing sketches of methodist itinerant preachers from 1736 to 1785... / Lednum, John – Philadelphia: J Lednum, 1859 [mf ed 1992] – (= ser Methodist coll) – 1mf – 9 – 0-524-04363-9 – mf#1990-5046 – us ATLA [242]

History of the rise of the huguenots of france / Baird, Henry Martyn – New York: Scribner, 1896, c1879 – 3mf – 9 – 0-7905-4483-0 – (incl bibl ref) – mf#1988-0483 – us ATLA [944]

History of the rise, progress, genius, and character of american presbyterianism : together with a review of "the constitutional history of the presbyterian church in the united states of america, by chas. hodge, d.d., professor in the theological seminary at princeton, n.j." / Hill, William – Washington City: J Gideon, Jr, 1839 – 1mf – 9 – 0-7905-6810-1 – mf#1988-2810 – us ATLA [242]

History of the ritual of the methodist episcopal church : with a commentary on its offices / Cooke, Richard Joseph – Cincinnati: Jennings & Pye, c1900 – 1mf – 9 – 0-524-02951-2 – (incl ind) – mf#1990-4503 – us ATLA [242]

History of the rock presbyterian church in cecil co., md / Johns, J H – 1872 – 1 – $50.00 – us Presbyterian [242]

History of the rocky spring church / Wylie, Samuel S et al – 1895 – 1 – $50.00 – us Presbyterian [240]

The history of the roman catholic church in lesotho, 1862-1989 / Sekoati, S M – Uni of South Africa 2001 [mf ed Johannesburg 2001] – 4mf – 9 – (incl bibl ref) – mf#mfm14845 – sa Unisa [241]

A history of the roman catholic church in the united states / O'Gorman, Thomas – New York: Christian Literature, 1895 [mf ed 1989] – (= ser The american church history series 9) – 2mf – 9 – 0-7905-4235-8 – (incl bibl ref) – mf#1988-0235 – us ATLA [241]

The history of the romeward movement in the church of england 1833-64 / Walsh, Walter – London: James Nisbet, 1900 – 1mf – 9 – 0-8370-9912-9 – (incl bibl ref and index) – mf#1986-3912 – us ATLA [241]

The history of the royal academy of arts : from its foundation in 1768 to the present time / Sandby, William – London: Longman, Green, Longman, Roberts, & Green, 1862 – (= ser 19th c art & architecture) – 2v on 10mf – 9 – mf#4.1.138 – uk Chadwyck [700]

History of the royal hospital, kilmainham, near dublin : from the original foundation as a priory of knights templar / Burton, Nathanael – Dublin, 1843 – (= ser 19th c ireland) – 3mf – 9 – mf#1.1.1448 – uk Chadwyck [360]

History of the sabbath and first day of the week / Andrews, John Nevins & Conradi, Ludwig Richard – 4th ed., rev. and enl. Washington, D.C.: Review and Herald, c1912 – 2mf – 9 – 0-7905-6400-9 – (incl bibl ref) – mf#1988-2400 – us ATLA [240]

The history of the sacramento valley / Woolridge, J W – Sacramento Co, CA. v1-3. 1931 – 1r – 1 – $50.00 – mf#B40253 – us Library Micro [978]

A history of the samskrta literature / Varadachari, Venkatadriagaram – Allahabad: Ram Narain Lal Bookseller and Publ, 1952 – (= ser Samp: indian books) – us CRL [490]

History of the san francisco theological seminary of the presbyterian church in the u.s.a. and its alumni association / Curry, James – Vacaville: Reporter Pub Co, 1907 – 3mf – 9 – 0-524-07351-1 – (incl ind) – mf#1990-5388 – us ATLA [242]

A history of the sandy creek baptist association, north carolina : from its organization, 1758-1858 / Purefoy, George W – 1 – $12.00 – us Southern Baptist [242]

The history of the saracens : comprising the lives of mohammed and his successors, to the death of abdalmelik, the eleventh caliph / Ockley, Simon [comp] – 4th rev enl ed. London: Henry G Bohn 1847 [mf ed 1992] – (= ser Bohn's standard library) – 2mf – 9 – 0-524-02224-0 – (first printed under title: the conquest of syria, persia, and aegypt, by the saracens. 1708-18 [2v]) – mf#1990-2898 – us ATLA [931]

History of the school of the reformed protestant dutch church : in the city of new-york, from 1633 to the present time / Dunshee, Henry Webb – New-York: John A Gray, 1853 – 1mf – 9 – 0-524-08571-4 – mf#1993-3156 – us ATLA [377]

History of the scottish episcopal church : from the revolution to the present time / Lawson, John Parker – Edinburgh: Gallie and Bayley, 1843 – 2mf – us ATLA [243]

History of the scottish episcopal church : from the revolution to the present time / Lawson, John Parker – Edinburgh: Gallie and Bayley, 1843 – 2mf – 9 – 0-7905-5000-8 – (incl bibl ref) – mf#1988-1000 – us ATLA [240]

History of the scottish metrical psalms : with an account of the paraphrases and hymns, and of the music of the old psalter / MacMeeken, J W – Glasgow: M'Culloch, 1872 – 1mf – 9 – 0-7905-2027-3 – mf#1987-2027 – us ATLA [220]

History of the seaboard air line railway company / Joubert, William Harry – Gainesville, FL. 1935 – 1r – us UF Libraries [380]

History of the secession church / M'Kerrow, John – rev enl ed. Edinburgh: A Fullarton, 1848 – 3mf – 9 – 0-524-07579-4 – mf#1991-3199 – us ATLA [240]

History of the separate baptist church, with a narrative of other denominations / Scott, Morgan – 1901 – 1 – us Southern Baptist [242]

History of the separation of church and state in canada / ed by Stimson, Elam Rush – Toronto: [s.n.], 1887 – 1mf – 9 – 0-524-00605-9 – mf#1990-0105 – us ATLA [240]

History of the settlement of upper canada (ontario) : with special reference to the bay quinte / Canniff, William – Toronto: Dudley & Burns, 1869 – 8mf – 9 – mf#00468 – cn Canadiana [971]

History of the seventh day baptists, 1802-1865 / Bailey, James – 1 – us Southern Baptist [242]

History of the seventh-day baptist general conference : from its origin, september, 1802, to its fifty-third session, september, 1865 / Bailey, James – Toledo, Ohio: S Bailey, 1866 – 1mf – 9 – 0-524-05354-5 – mf#1990-5105 – us ATLA [242]

History of the shaftsbury baptist association, vermont, 1781-1853 / Wright, Stevens – 1 – us Southern Baptist [242]

History of the signal corps development of u.s. army radar equipment / U.S. Army. Signal Corps – 1943, 1945. 3 v. illus., charts, maps, photos. Pts. 2 and 3 prepared by Capt. Harry M. Davis. 1 reel – 1 – us L of C Photodup [621]

History of the sikhs / Gupta, Hari Ram – Calcutta: SN Sarkar; Lahore: Sole selling agents, Minerva Book Shop, 1939-1944 – (= ser Samp: indian books) – (foreword by jadunath sarkar) – us CRL [954]

The history of the sinclair family in europe and america for eleven hundred years : giving a genealogical and biographical history of the family in normandy, france, a general record of it in scotland, england, ireland, and a full biographical and genealogical record of many branches in canada and the united states / Morrison, Leonard Allison – Boston: Damrell & Upham, 1896 – 7mf – 9 – (incl ind and bibl ref) – mf#40556 – cn Canadiana [929]

A history of the so-called jansenist church of holland : with a sketch of its earlier annals, and some account of the brothers of the common life / Neale, John Mason – Oxford: J H & J Parker, 1858 [mf ed 1990] – 1mf – 9 – 0-7905-5727-4 – mf#1988-1727 – us ATLA [241]

History of the society of dilettanti / Cust, Lionel Henry – London 1898 – (= ser 19th c art & architecture) – 5mf – 9 – mf#4.2.1166 – uk Chadwyck [700]

The history of the society of friends in america / Bowden, James – London: Charles Gilpin, 1850-1854 – 2mf – 9 – 0-524-00964-3 – mf#1990-4022 – us ATLA [240]

History of the society of jesus in north america : colonial and federal / Hughes, Thomas – London: Longmans, Green, 1907-1917 – 7mf – 9 – 0-524-03011-1 – (incl bibl ref) – mf#1990-4533 – us ATLA [241]

History of the sombansi raj and the estate of partabgarh in oudh / Tholal, Pundit Bishambhar Nath – Cawnpore, 1900 – (filmed with: carnegy, p a historical sketch of fyzabad tehsil. lucknot, 1870) – us CRL [954]

History of the south carolina college : from its incorporation, dec. 19, 1801 to dec. 19, 1865, including sketches of its presidents and professors / Laborde, Maximilian – 2nd ed. Charleston: Walker Evans & Cogswell, Printers, 1874 – xliii/596p – 1 – (with appendix and prefaced by a life of the author) – us Wisconsin U Libr [370]

History of the squares of london : topographical & historical / Chancellor, Edwin Beresford – London: K Paul, Trench, Trubner & Co 1907 [mf ed 1987] – 1r – 1 – mf#1934 – us Wisconsin U Libr [914]

History of the staffordshire potteries / Shaw, Simeon – Hanley 1829 – (= ser 19th c art & architecture) – 3mf – 9 – mf#4.2.1539 – uk Chadwyck [730]

History of the standard bank of south africa ltd, 1862-1913 / Amphlett, George Thomas – Glasgow, Scotland. 1914 – 1r – us UF Libraries [332]

History of the state of california and biographical record of coast counties / Gunn, J M – CA, 1904 – 1r – 1 – $50.00 – mf#B40203 – us Library Micro [978]

A history of the state paper office : with a view of the documents therein deposited / Thomas, Francis Sheppard – London: J Petheram, 1849 [mf ed 1983] – 1mf – 9 – 0-665-41415-3 – mf#41415 – cn Canadiana [350]

History of the struggle and progress of religious liberty in greenwich, mass – [Boston, Mass: Henry W Smith], 1886 – 1mf – 9 – 0-524-03296-3 – mf#1990-0907 – us ATLA [240]

History of the study of theology / Briggs, Charles Augustus – London: Duckworth 1916 [mf ed 1990] – (= ser Studies in theology) – 2v on 1mf – 9 – 0-7905-3551-3 – (incl bibl ref) – mf#1989-0044 – us ATLA [240]

The history of the supernatural : in all ages and nations, and in all churches, christian and pagan, demonstrating a universal faith / Howitt, William – Philadelphia: JB Lippincott, 1863 – 3mf – 9 – 0-524-02279-8 – mf#1990-0584 – us ATLA [130]

History of the suppression of infanticide in western india under the government of bombay : including notices of the provinces and tribes in which the practice has prevailed / Wilson, John – Bombay 1855 – (= ser 19th c british colonization) – 5mf – 9 – mf#1.1.4798 – uk Chadwyck [306]

History of the supreme court of the territory and state of washington. / Reinhart, Caleb S – n.p., 193?. 148p. LL-1353 – 1 – us L of C Photodup [347]

The history of the swedes = Svenska folkets historia / Geijer, Erik Gustaf – London: Whittaker, [1845?] – 4mf – 9 – 0-7905-4683-3 – (incl bibl ref. in english) – mf#1988-0683 – us ATLA [948]

History of the swedish baptists in sweden and america : being an account of the origin, progress and results of that missionary work during the last half of the nineteenth century / Schroeder, Gustavus W – Greater New York: G. Schroeder, 1898 – 1mf – us ATLA [305]

History of the swedish baptists in sweden and america : being an account of the origin, progress and results of that missionary work during the last half of the nineteenth century / Schroeder, Gustavus Wilhelm – Jubilee ed. Greater New York: G. Schroeder, 1898 – 1mf – 9 – 0-7905-6561-7 – (incl bibl ref) – mf#1988-2561 – us ATLA [242]

History of the swiss reformed church since the reformation / Good, James Isaac – Philadelphia: Publication and Sunday School Board of the Reformed Church in the United States, 1913 – 2mf – 9 – 0-7905-5832-7 – mf#1988-1832 – us ATLA [242]

History of the syrian nation and the old evangelical-apostolic church of the east : from remote antiquity to the present time / Malech, George David – Minneapolis, MN: [s.n.], c1910 – 2mf – 9 – 0-8370-8037-1 – mf#1986-2037 – us ATLA [242]

History of the tamiami trail and a brief review of... – Miami, FL. 1928 – 1r – us UF Libraries [639]

History of the teachers association of south african and the role it played in the development of education for indians in south africa / Jack, Jonathan Rajmangal – U of the Western Cape 1987 [mf ed S.l: s.n. between 1986 & 1991] – 3mf – 9 – (Summary in afrikaans; incl bibl) – sa Misc Inst [370]

History of the telugu christians / Kroot, Antonius – Trichinopoly: Printed...by St Joseph's Industrial School Press, 1910 [mf ed 1995] – (= ser Yale coll) – viii/312p/4/2/viii – 1 – 0-524-09982-0 – mf#1995-0982 – us ATLA [490]

The history of the ten "lost" tribes : anglo-israelism examined / Baron, David – 2nd ed. London: Morgan & Scott, 1915 [mf ed 1990] – 1mf – 9 – 0-7905-3180-1 – mf#1987-3180 – us ATLA [939]

A history of the textual criticism of the new testament / Vincent, Marvin Richardson – New York: Macmillan, 1899 [mf ed 1985] – 1mf – 9 – 0-8370-5641-1 – (incl ind) – mf#1985-3641 – us ATLA [225]

A history of the theology of the disciples of christ / Kirk, Hiram van – St Louis: Christian Pub Co, 1907 [mf ed 1990] – 1mf – 9 – 0-7905-6385-1 – (incl bibl ref. also publ as: the rise of the current reformation) – mf#1988-2385 – us ATLA [240]

History of the thirty years' war = Geschichte des dreissigjaehrigen krieges / Gindely, Antonin – New York: Putnam, 1884 – 3mf – 9 – 0-7905-5827-0 – (in english) – mf#1988-1827 – us ATLA [943]

History of the thirty years' war, complete: history of the revolt of the netherlands, to the confederacy of the gueux / Schiller, Friedrich von – Trans. from the German by Rev. A.J.W. Morrison. London: Bell and Daldy, 1873. vii,519p – 1 – us Wisconsin U Libr [949]

A history of the town of belfast : from the earliest times to the close of the eighteenth century / Benn, George – London, 1877-1880 – (= ser 19th c ireland) – 12mf – 9 – mf#1.1.5271 – uk Chadwyck [941]

History of the tranquebar mission : worked out from the original papers / Fenger, J Ferd – Tranquebar: Evangelical Lutheran Mission Press, 1863 – 1mf – 9 – 0-8370-7137-2 – mf#1986-1137 – us ATLA [240]

History of the tranquebar mission... / Fenger, J F – Tranquebar, 1863 – 4mf – 9 – mf#HTM-58 – ne IDC [915]

History of the translations which have been made of the scriptures / Marsh, Herbert – London, England. 1812 – 1r – us UF Libraries [220]

History of the transmission of ancient books to modern times : together with the process of historical proof, or, a concise account of the means by which the genuineness of ancient literature generally, and the authenticity of historical works especially are ascertained / Taylor, Isaac – new rev enl ed. London: Jackson & Walford, 1859 [mf ed 1988] – 1mf – 9 – 0-7905-0233-X – mf#1987-0233 – us ATLA [000]

History of the trinidad sector and base command / U.S. Army. Caribbean Defense Command. Historical Section – Lt. Robert A. Johnston and Cap't. James C. Shoultz, Jr. Port-of-Spain. 1945-47. 5v. illus., charts, maps, photos – 1 reel – 1 – $32.00 – us L of C Photodup [355]

The history of the union bank of scotland / Rait, Robert Sangster – Glasgow, J. Smith, 1930 – 1 – us Wisconsin U Libr [941]

History of the union pacific railway / White, Henry Kirke – Chicago: University of Chicago Press, 1895 [mf ed 1970] – (= ser Economic studies of the university of chicago 2; Library of american civilization 10314) – 1mf – 9 – us Chicago U Pr [380]

A history of the unitarians and the universalists in the united states / Allen, Joseph Henry & Eddy, Richard – New York: Christian Literature, 1894 [mf ed 1989] – (= ser The american church history series 10) – 2mf – 9 – 0-7905-4181-5 – mf#1988-0181 – us ATLA [243]

History of the united states : from their first settlement as colonies, to the close of the war with great britain, in 1815 / Hale, Salma – New York: C Wiley, 1825 [mf ed 1984] – 4mf – 9 – 0-665-44092-8 – mf#44092 – cn Canadiana [975]

History of the united states / Frost, John – Philadelphia, PA. 1837 – 1r – us UF Libraries [975]

History of the united states armed forces in korea, aug 1945-may 1948 / U.S. Army. Office of the Chief of Military History – v1-4. 1948 – (= ser History Of The Occupation Of Korea, Aug 1945-May 1978) – 1 – $87.00 – us L of C Photodup [977]

History of the united states in rhyme / Adams, Robert Chamblet – Boston: D Lothrop, c1884 – 1mf – 9 – mf#10188 – cn Canadiana [975]

History of the united states marine corps / U.S. Marine Corps. Historical Division – Edwin N. McClellan. Washington, D.C. 1925-1934. 2 v – 1 – us NY Public [355]

History of the united states of america / Bancroft, George – Boston, MA. v1-6. 1876 – 2r – us UF Libraries [025]

History of the united states of america, vol 1 : from the discovery of the continent / Bancroft, George – New York: D Appleton, 1883 – v1 on 7mf – 9 – mf#61405 – cn Canadiana [975]

History of the united states of america, vol 2 : from the discovery of the continent / Bancroft, George – New York: D Appleton, 1883 – v2 on 7mf – 9 – (incl bibl ref) – mf#61406 – cn Canadiana [975]

History of the united states of america, vol 3 : from the discovery of the continent / Bancroft, George – New York: D Appleton, 1883 – v3 on 6mf – 9 – (incl bibl ref) – mf#61407 – cn Canadiana [975]

History of the united states of america, vol 4 : from the discovery of the continent / Bancroft, George – New York: D Appleton, 1889 – v4 on 6mf – 9 – (incl bibl ref) – mf#61408 – cn Canadiana [975]

History of the united states of america, vol 5 : from the discovery of the continent / Bancroft, George – New York: D Appleton, 1890 – v5 on 7mf – 9 – (incl bibl ref) – mf#61409 – cn Canadiana [975]

History of the united states of america, vol 6 : from the discovery of the continent / Bancroft, George – New York: D Appleton, 1890 – v6 on 7mf – 9 – mf#61410 – cn Canadiana [975]

History of the united states of america, vols 1-6 : from the discovery of the continent / Bancroft, George – New York: D Appleton, 1883-1890 – 6v on 1mf – 9 – mf#61404 – cn Canadiana [975]

History of the universities' mission to central africa / Universities' Mission To Central Africa – London, England. v1-3. 1955-1962 – 1r – us UF Libraries [378]

The history of the university of dublin : from its foundation to the end of the eighteenth century / Stubbs, John William – Dublin: Hodges, Figgis; London: Longmans, Green, 1889 – 1mf – 9 – 0-7905-6683-4 – mf#1988-2683 – us ATLA [378]

History of the u.s. army military government in korea, sep 1945-30 jun 1946 / Korea. (Territory under U.S. Occupation, 1945-). Military Governor – v. 1-4. 1946-47 – 1 – us L of C Photodup [951]

History of the us strategic bombing survey (pacific), 1945-46 : vols 1-2 1946 / U.S. War Dept. Office of the Administrative Assistant – 1 – us L of C Photodup [950]

History of the u.s. strategic bombing survey(european). 1944-45. v. 1-2. 1946 / U.S. War Dept. Office of the Administrative Assistant – 7 – us L of C Photodup [943]

History of the use of incense in divine worship / Atchley, Edward Godfrey Cuthbert Frederic – London: Longmans, Green 1909 [mf ed 1993] – (= ser Alcuin club colls 13) – 2mf – 9 – 0-524-06942-5 – mf#1990-5306 – us ATLA [241]

History of the variations of the protestant churches = Histoire des variationes de eglises protestantes / Bossuet, Jacques Benigne – New York: D & J Sadlier 1850 [mf ed 1991] – 2v on 2mf – 9 – 0-7905-9142-1 – (trans fr french; incl bibl ref) – mf#1989-2367 – us ATLA [242]

History of the vaudois church : from its origin, and of the vaudois of piedmont to the present day = Histoire de l'eglise vaudoise / Monastier, Antoine – London: Religious Tract Society, 1848 [mf ed 1992] – 2mf – 9 – 0-524-03353-6 – (in english) – mf#1990-0934 – us ATLA [240]

A history of the vyne in hampshire / Chute, Chaloner William – Winchester 1888 – (= ser 19th c art & architecture) – 3mf – 9 – mf#4.2.1301 – uk Chadwyck [720]

History of the waddams grove church : a history of the waddams grove congregation of the church of the brethren, in stephenson and jo daviess counties, illinois, and adjoining counties of wisconsin / Lutz, Ezra – Elgin, Ill: Brethren Pub House, 1910 – 1mf – 9 – 0-524-03938-0 – mf#1990-4932 – us ATLA [242]

History of the waldenses of italy : from their origin to the reformation / Comba, Emilio – London: Truslove & Shirley, 1889 – 1mf – us ATLA [240]

History of the waldenses of italy : from their origin to the reformation = Histoire des vaudois / Comba, Emilio – London: Truslove & Shirley, 1889 – 1mf – 9 – 0-7905-4332-X – (incl bibl ref. in english) – mf#1988-0332 – us ATLA [240]

The history of the war : between the united states and great britain, which commenced in june 1812, and closed in feb 1815... – Hartford [CT]: printed & publ by B & J Russell, 1815 [mf ed 1984] – 6mf – 9 – 0-665-45105-9 – mf#45105 – cn Canadiana [975]

History of the war between france and germany *see* Story of the goths

History of the war between germany and france, with biographical sketches of the principal personages. / McCabe, James Dabney – Philadelphia, 1871. 815p. illus. fold. maps – 1 – us Wisconsin U Libr [944]

History of the war of 1812 between great britain and the united states of america / Hannay, James – [St John, NB?: J A Bowes], 1901 – 5mf – 9 – 0-665-73011-X – (incl ind) – mf#73011 – cn Canadiana [975]

History of the war of ireland from 1641 to 1653 / British Officer Of Sir John Clottworthy's Regiment – Dublin, Ireland. 1873 – 1r – us UF Libraries [025]

A history of the warfare of science with theology : in christendom / White, Andrew Dickson – New York: D Appleton, 1896 [mf ed 1990] – 2v on 3mf – 9 – 0-7905-8168-X – (incl bibl ref) – mf#1988-6115 – us ATLA [210]

History of the welsh baptists, from the year 1763 to the year 1770 / Davis, J – 1 – 7.42 – us Southern Baptist [242]

History of the wesleyan methodist church of south africa / Whiteside, Joseph – London: E. Stock, 1906. 479p. ill. Includes index – 1 – us Wisconsin U Libr [242]

History of the west africa mission / Nassau, Robert Hamill – 1919 – 1 – $100.00 – us Presbyterian [240]

History of the west indian islands of trinidad and... / Carmichael, Gertrude – London, England. 1961 – 1r – us UF Libraries [972]

History of the westminster assembly of divines / Hetherington, William Maxwell – Edinburgh: John Johnstone, 1843 – 1mf – 9 – 0-524-03820-1 – mf#1990-1136 – us ATLA [941]

A history of the wittenberg synod of the general synod of the evangelical lutheran church, 1847-1916 / Ernsberger, C S – Columbus, OH: printed...by the Lutheran Book Concern, 1917 [mf ed 1993] – 7mf – 9 – 0-524-08753-9 – mf#1993-3258 – us ATLA [242]

History of the woman's temperance crusade : a complete official history of the wonderful uprising of the christian women of the united states against the liquor traffic, which culminated in the gospel temperance movement / Wittenmyer, Annie – Boston: James H Earle [c1882] [mf ed 1984] – (= ser Women & the church in america 104) – 9mf – 9 – 0-8370-1266-X – mf#1984-2104 – us ATLA [305]

The history of the women's club movement in america / Croly, Jane Cunningham – New York: H.G. Allen & Co., c1898. illus – 1 – us Wisconsin U Libr [305]

A history of the works of sir joshua reynolds / Graves, Algernon – London 1899-1901 – (= ser 19th c art & architecture) – 27mf – 9 – mf#4.1.306 – uk Chadwyck [750]

History of the works of the learned – London. 1737-1743 (1) – mf#4263 – us UMI ProQuest [070]

History of the works of the learned : or an impartial account of books lately printed in all parts of europe – London. 1699-1712 (1) – mf#4264 – us UMI ProQuest [070]

History of the wyandott mission at upper sandusky, ohio : under the direction of the methodist episcopal church / Finley, James Bradley – Cincinnati: Pub by JF Wright and L Swormstedt for the ME Church, 1840 – 5mf – 9 – 0-524-07416-X – mf#1991-3076 – us ATLA [242]

The history of the year : a review of the events of 1891 all around the world, with special reference to canadian affairs / ed by Morrison, Charles – Toronto: W J Dyas, 1892 – 4mf – 9 – (incl ind) – mf#08015 – cn Canadiana [933]

A history of the year 1893, canadian affairs : dominion and provincial politics – S.l: s.n, 1894 – 1mf – 9 – mf#02734 – cn Canadiana [971]

A history of the year 1894 : with especial reference to canadian affairs – Toronto?: s.n, 1894 – 3mf – 9 – (incl ind) – mf#08489 – cn Canadiana [971]

History of the young men's christian association. volume 1, the founding of the association, 1844-1855 / Doggett, Laurence Locke – New York: International Committee of Young Men's Christian Associations, 1896 – 1mf – 9 – 0-7905-4729-5 – (incl bibl ref) – mf#1988-0729 – us ATLA [240]

History of the zulu war and its origin / Colenso, Frances Ellen – Westport, CT. 1970 – 1r – us UF Libraries [960]

The history of thomas hickathrift / Hickathrift, Thomas – London: The Villon Society 1885 [mf ed Bloomington IN: Indiana Uni Lib, Preservation Dept 1984] – xix 36p on 1r – 1 – us Indiana Preservation [390]

History of tipu sultan / Hasan, Mohibbul – Calcutta: Bibliophile Ltd, 1951 – (= ser Samp: indian books) – us CRL [954]

A history of tirupati / Krishnaswami Aiyangar, Sakkottai – Madras: Tirumalai-Tirupati Devastanam Committee, 1941- – (= ser Samp: indian books) – us CRL [954]

History of trial by jury / Forsyth, William – 2d ed. New York: Cockcroft, 1878. 388p. LL-112 – 1 – us L of C Photodup [340]

History of trinidad 1838 – 1838 – 1r – 1 – (in trinidad almanac and pocket register for 1840) – cn Library Assoc [972]

History of trinity church, saint john, new brunswick, 1791-1891 / Brigstocke, Frederick Hervey John [comp] – Saint John, NB: J & A McMillan, 1892 – 3mf – 9 – mf#00274 – cn Canadiana [240]

History of tugalo baptist association, georgia / Goode, J F – 1924 – 1 – 7.98 – us Southern Baptist [242]

History of turtle mound and the indian river from... / Connor, Jeanette Thurber – s.l, s.l? 193-? – 1r – us UF Libraries [978]

History of twenty-five years' four-fold gospel work in troy *see* [Autobiographical pamphlets]

The history of tyre / Fleming, Wallace Bruce – New York: Columbia University Press, 1915 – (= ser Columbia university oriental studies) – 1mf – 9 – 0-524-01057-9 – (incl bibl ref) – mf#1990-2205 – us ATLA [956]

The history of tythes / Selden, J – 1618 – 1 – us Southern Baptist [242]

A history of uganda land and surveys and of the uganda land and survey department / Thomas, Harold Beken & Spencer, A E – Entebbe: Uganda Govt Press, 1938 – 1 – us CRL [960]

History of union church / Bondfield, G H et al – [Hongkong], 1903 [mf ed 1995] – (= ser Yale coll) – 46p – 1 – 0-524-10143-4 – mf#1995-1143 – us ATLA [242]

A history of unitarianism / Graves, Charles – Boston, MA: American Unitarian Assoc, 1917 [mf ed 1993] – 1mf – 9 – 0-524-08368-1 – (incl bibl ref) – mf#1993-3068 – us ATLA [243]

A history of upper canada college : 1829-1892, with contributions by old upper canada college boys... / Dickson, George & Adam, Graeme Mercer [comp] – Toronto: Rowsell & Hutchison, 1893 – 5mf – 9 – mf#02659 – cn Canadiana [378]

A history of urdu literature / Bailey, Thomas Grahame – Calcutta: Association Press (YMCA), 1932 – (= ser Samp: indian books) – us CRL [490]

A history of urdu literature / Saksena, Ram Babu – Allahabad: Ram Narain Lal, 1927 – (= ser Samp: indian books) – us CRL [490]

History of vagrants and vagrancy, and beggars and begging / Ribton-Turner, Charles James – London: Chapman & Hall Ltd, 1887 [mf ed 1992] – (= ser 19th c: general coll: jurisprudence) – 9mf – 9 – uk Chadwyck [344]

History of venezuela / Moron, Guillermo – London, England. 1964 – 1r – us UF Libraries [972]

The history of volleyball in the united states / Flanagan, Lance – University of Colorado, 1960 – 3mf – 9 – $13.00 – us Kinesology [790]

History of volusia county, florida / Gold, Pleasant Daniel – Deland, FL. 1927 – 1r – us UF Libraries [978]

History of wesleyan methodism / Smith, George – London: Longman, Brown, Green, Longmans, and Roberts, 1857-1861 – 6mf – 9 – 0-524-06477-6 – mf#1990-5251 – us ATLA [242]

A history of western tibet : one of the unknown empires / Francke, August Hermann – London: Partridge, 1907 [mf ed 1982] – 1r – 1 – (incl bibl) – mf#237 – us Wisconsin U Libr [951]

History of william rogers – London, England. 1830 – 1r – us UF Libraries [240]

The history of witchcraft and demonology / Summers, Montague – New York: A.A. Knopf, 1926. xv,353p. (Half-title: The History of Civilization) – 1 – us Wisconsin U Libr [130]

A history of witchcraft in england from 1558 to 1718 / Notestein, Wallace – Washington: American Historical Assoc, 1911 [mf ed 1990] – (= ser Prize essays of the american historical association 1909) – 2mf – 9 – 0-7905-6661-3 – (incl bibl ref) – mf#1988-2661 – us ATLA [130]

The history of women : from the sophia smith collection at smith college, the schlesinger library at radcliffe college and other sources – 1248r coll – 1 – (coll of pre-1920 literature about the role of women throughout history. arranged chronologically. accompanied by a printed guide wh includes a reel guide summary for monographs, pamphlets, and photographs; an alphabetical main-entry list; an alphabetical periodicals title list; a subject/ added entry index; and a name index to photographs. coll available on a per-unit basis: units 1-19 monographs 934r c36-28801. unit 20: pamphlets, photographs and mss 61r c36-28802. units 21-25: periodicals 253r c36-28803) – mf#C36-28800 – us Primary [305]

The history of yucatan : from its discovery to the close of the seventeenth century / Fancourt, Charles Saint John – London: J. Murray, 1854.xvi,340p. fold. map – 1 – us Wisconsin U Libr [972]

History of zanzibar : from the middle ages to 1856 / Gray, John Milner – London, England. 1962 – 1r – us UF Libraries [960]

History on the luapula / Cunnison, Ias George – Cape Town, South Africa. 1951 – 1r – us UF Libraries [960]

History, principle, and fact : in relation to the irish question / Hutton, Henry Dix – London, 1870 – (= ser 19th c ireland) – 1mf – 9 – mf#1.1.1874 – uk Chadwyck [941]

The history, principles and practice of symbolism in christian art / Hulme, F E – London, 1908 – 3mf – 9 – mf#0-1248 – ne IDC [700]

The history, principles and practice of symbolism in christian art / Hulme, Frederick Edward – London: George Allen & Unwin, [1910?] – 1mf – 9 – 0-524-04839-8 – mf#1990-1331 – us ATLA [700]

History, prophecy and the monuments, or, israel and the nations / McCurdy, James Frederick – 3rd rev ed. New York: Macmillan. 3v. 1897-1901, c1894-1901 – 6mf – 9 – 0-7905-0102-3 – (incl bibl ref and index) – mf#1987-0102 – us ATLA [939]

History, structure, and statistics of plank roads in the united states and canada / Kingsford, William – Philadelphia: A Hart, late Carey & Hart, 1852 [mf ed 1984] – 1mf – 9 – 0-665-45231-4 – mf#45231 – cn Canadiana [380]

History teacher – Long Beach. 1967+ (1) 1971+ (5) 1975+ (9) – ISSN: 0018-2745 – mf#3331 – us UMI ProQuest [370]

History through the times; a collection of leading articles on important events, 1800-1937.. / The Times. London – Selected by Sir James Marchant, K.B.E., with introd. by Geoffrey Dawson. London: Cassell and Co. Ltd., 1937 xi/619p – 1 – us Wisconsin U Libr [900]

History today / Outagamie County Historical Society – 1976 jan 10-1985 dec – 1r – 1 – mf#501883 – us WHS [978]

History trails / Baltimore County Historical Society – 1966 sep-1982 summer – 1r – 1 – mf#615799 – us WHS [978]

History vindicated in the case of the wigtown martyrs / Stewart, Archibald – Edinburgh, Scotland. 1867 – 1r – us UF Libraries [240]

History workshop – Oxford. 1990-1994 (1,5,9) – (cont by: history workshop journal: hwj) – ISSN: 0309-2984 – mf#18496 – us UMI ProQuest [900]

History workshop journal: hwj – Oxford. 1995+ (1,5,9) – (cont: history workshop) – mf#18496,01 – us UMI ProQuest [900]

Historycke listy – Czech Republic, 1999- – 1r per y standing order – 1 – (1992-98 1r per 2yrs) – us UMI ProQuest [077]

The history...hernando de soto / Shipp, Barnard – 1881 – 9 – sp Bibl Santa Ana [910]

Hit of the week – Blantyre: [s.n.] [may 17/23, nov 1993] – 1r – 1 – us CRL [079]

Hitavada – Bhopal, India. Jul-Sept 1966 – 1r – 1 – us L of C Photodup [079]

Hitavada – Nagpur, India. Apr 1944-Jul 1994 – 214r – 1 – us L of C Photodup [079]

Hitch, Marcus *see* Goethes faust

Hitchcock, Albert Wellman *see* The psychology of jesus

Hitchcock county central – Trenton, NE: Risley & Suiter, 1885-v1 n23. dec 19 1885 (wkly) [mf ed filmed 1978] – 1r – 1 – (cont by: trenton torpedo) – us NE Hist [071]

Hitchcock county herald – Culbertson, NE: J C L Wisely. 2v. v1 n1. jan 23 1903-v2 n2. jan 26 1904 (wkly) – 1r – 1 – (absorbed in pt by: trenton leader and: culbertson era) – us NE Hist [071]

Hitchcock county news – Trenton, NE: Lester and Dorothy Power. 79th yr n47. oct 7 1965- (wkly) [mf ed with gaps filmed 1976-] – 1 – (cont: trenton register. issue numbering dropped jul 2-nov 12 1970) – us NE Hist [071]

Hitchcock county republican – Culbertson, NE: F Bert Risley, 1889 (wkly) [mf ed 1890-95 (gaps)] – 1r – 1 – (cont: hitchcock county reveille. suspended aug 1891-jan 1892) – us NE Hist [071]

[Hitchcock county reveille] – Culbertson, NE: [Reveille Pub Co] -1889// (wkly) [mf ed v4 n2. jan 6-dec 21 1888 (gaps)] – 1r – 1 – (cont by: hitchcock county republican) – us NE Hist [071]

Hitchcock, Edward *see*
- History of a zoological temperance convention
- The power of christian benevolence
- The religion of geology and its connected sciences
- Religious lectures on peculiar phenomena in the four seasons
- Religious truth

Hitchcock, Francis Ryan Montgomery *see*
- The atonement and modern thought
- Christ and his critics
- A fresh study of the fourth gospel
- Irenaeus of lugdunum

Hitchcock, Henry Russell *see* American architectural books

Hitchcock, Loranus Eaton *see* Powers and duties of sheriffs, constables, tax collectors, and other officers in the new england states

Hitchcock, Mary *see* The first soprano

Hitchcock, Mary Evelyn *see* Two women in the klondike

Hitchcock, Roswell Dwight *see*
- Hitchcock's new and complete analysis of the holy bible
- Teaching of the twelve apostles

Hitchcock, Roswell Dwight et al *see* Proceedings at the wycliffe semi-millennial celebration by the american bible society

Hitchcock's new and complete analysis of the holy bible : or, the whole of the old and new testaments arranged according to subjects in twenty-seven books / ed by Hitchcock, Roswell Dwight – New York: A J Johnson, 1870 [mf ed 1990] – 3mf – 9 – 0-8370-1742-4 – mf#1987-6138 – us ATLA [220]

Hitchens, J Hiles *see* Ritualism, and our duty in relation to it

Hitchin and royston express – mar 12-oct 8 1859; oct 15 1859-dec 27 1862; jan 3 1863-dec 30 1865; 1866-mar 26 1870; may 2 1874; may 1875-77; 1879-80; 1950; 1980-87; jan 8 1988-92; jan 8-jun 25, jul 2-dec 24 1993; jul-dec 23 1994; 1995-96 – 67 3/4r – 1 – (aka: hertfordshire express; hitchin gazette; the gazette (hitchin)) – uk British Libr Newspaper [072]

Hitchin comet – Luton, England. 1980- – 65+ r – 1 – uk British Libr Newspaper [072]

Hitchin gazette – England, 1980-81 – 8r – 1 – uk British Libr Newspaper [072]

Hitler : a menace to world peace – New York, NY. 1937 – 1r – us UF Libraries [934]

Hitler, Adolf *see*
- Speech delivered in the reichstag, february 20th, 1938..
- Speech delivered in the reichstag on september 1, 1939

Hitler, Adolph *see* Mein kampf

Hitler over europe? / Henri, Ernst – London, England. 1934 – 1r – us UF Libraries [934]

Hitler over russia? : the coming fight between the fascist and socialist armies / Henri, Ernst – Trans. by Michael Davidson. New York: Simon and Schuster, 1936. x,340p – 1r – 1 – us Wisconsin U Libr [940]

Hitler was my friend / Hoffmann, Heinrich – London, England. 1955 – 1r – us UF Libraries [025]

Der hitlerjunge quex : roman / Schenzinger, Karl Aloys – Berlin: Zeitgeschichte-Verlag, W Andermann c1932 [mf ed 1991] – 1r – 1 – (filmed with: die erzaehlungstechnik viktor scheffels / walter grebe [comp]) – mf#2870p – us Wisconsin U Libr [830]

Hitler's first foes; a study in religion and politics / Mason, John Brown – Minneapolis: Burgess Publishing Co., c1936. v,118p – 1 – us Wisconsin U Libr [943]

Hitler's war / Irving, David – 9r – 1 – (selected documents, diaries, interviews, etc. index) – mf#97125 – uk Microform Academic [943]

Hitomi, I *see* Dai-nippon

Hitopadesa / Narayana Bhatta – Poona City, India. 1933 – 1r – us UF Libraries [950]

Hitopadesas id est institutio salutaris : textum codd mss collatis recensuerunt interpretationem latinam et annotationes criticas / Narayana Pandita – Bonnae ad Rhenum: Weber 1829-30 – (= ser Whsb) – 5mf – 9 – €60.00 – (int in latin; text sanskrit in devanagari script; missing: app "the hitopadesa in the sanskrita language [london 1810]"; pt1: textum sanscritum tenens; pt2: commentarium criticum tenens) – mf#Hu 248 – gw Fischer [490]

Hitos de la raza : (cuentos tradicionales y folklori...) / Cadilla De Martinez, Maria – San Juan, Puerto Rico. 1945 – 1r – us UF Libraries [972]

Hittell, John Shertzer *see* The spirit of the papacy

Hittiter und armenier / Jensen, P – Strassburg: Karl J Truebner, 1898 – 1mf – 9 – 0-7905-2116-4 – (incl ind) – mf#1987-2116 – us ATLA [470]

The hittites = Die hettiter / Messerschmidt, Leopold – London: David Nutt, 1903 [mf ed 1989] – (= ser Der alte orient 6) – 1mf – 9 – 0-7905-1244-0 – (trans by jane hutchison) – mf#1987-1244 – us ATLA [956]

The hittites : the story of a forgotten empire / Sayce, Archibald Henry – 2nd ed. London: Religious Tract Soc 1890 [mf ed 1991] – (= ser By-paths of bible knowledge 12) – 1mf – 9 – 0-7905-8314-3 – mf#1987-6419 – us ATLA [930]

Hitzenauer, Christoph *see* Brevis et expedita ratio...

Hitzeroth, Carl *see* Johann heermann (1585-1647)

Hitzig, Ferdinand *see*
- Das buch hiob uebersetzt und ausgelegt
- Geschichte des volkes israel
- Die grabschrift des darius zu nakschi rustam
- Das hohe lied
- Der prophet ezechiel
- Der prophet jeremia
- Die sprueche salamo's – der prediger salomo's
- Urgeschichte und mythologie der philistaeer
- Vorlesungen ueber biblische theologie und messianische weissagungen des alten testaments
- Die zwoelf kleinen propheten

Hitzig, Julius Eduard *see*
- Adelbert chamisso's werke
- Leben und briefe

An hiv education needs assessment of selected teacher members of the american school health association and the american home economics association / Kerr, Dianne L & Heit, Philip – 1992 – 2mf – 9 – $8.00 – us Kinesology [613]

Hiv/aids ministries network focus paper / United Methodist Church (US) – n23 [1994 mar] – 1r – 1 – mf#5296704 – us WHS [362]

Hivale, Shamrao *see* The pardhans of the upper narbada valley

Hive – Lancaster. 1810-1810 – 1 – mf#4464 – us UMI ProQuest [370]

The hive : or weekly entertaining register (london) – aug 1822-oct 1824 – (= ser 19th c british periodicals) – reel – 1 – us Primary [073]

Un hiver au cambodge : chasses au tigre, a l'elephant et au buffle sauvage; souvenirs d'und mission officielle remplie en 1880-1881 / Boulangler, Edgar – Tours: A Mame et fils 1887 [mf ed 1989] – 1r with other items – 1 – mf#mf-10289 seam reel 001/04 [§] – us CRL [639]

Hiver caraibe : documentaire / Morand, Paul – Paris, France. 1929 – 1r – us UF Libraries [972]

Hives, Frank *see* Momo and i

Hiv-infizierte monozyten/makrophagen : produktion und sekretion immunregulatorischer proteine / Esser, Ruth Christa – (mf ed 1996) – 3mf – 9 – €49.00 – 3-8267-2365-1 – mf#DHS 2365 – gw Frankfurter [540]

Hiwale, Anand S *see* America's present opportunity in india

Hixson, Karen A *see* An epidemiologic investigation of the relationship between religiosity, selected health behaviors, and blood pressure

Hiyaban – Izmir: Kesisyan Matbaasi, 1914. Sahib-i Mecmua: Muestecabizade Ismet. n4-5. 3-18 temmuz 1330 [1914] – (= ser O & t journals) – 1mf – 9 – $25.00 – us MEDOC [956]

Hiziragazade arif aga'nin mahdumu sait bey divani / Bey, Sait – (= ser Ottoman literature, writers and the arts) – 1mf – 9 – $25.00 – us MEDOC [470]

Hizmet – Izmir, 1925-28. Mueduer-i Mes'ul: Kemal Talat; Basmuharriri: Z Besim. n710. 6 mayis 1927, 1059,1079,1107. 17 agustos 1928 – (= ser O & t journals) – 1mf – 9 – $25.00 – us MEDOC [956]

Hjaelp for bibellaesere : en praktisk bibelordbog – Chicago, IL: Norsk-Danske Boghandel, 1902 – 1mf – 9 – 0-524-06422-9 – mf#1991-2544 – us ATLA [220]

Hje nach volgend vier neuwe klaegliche vnd zu got rueffende gesang : oder lieder wider den blutdurstigen erbfeind...den tuergken... – Augspurg: Stayner 1542 – (= ser Hqab. literatur des 16. jahrh.) – 1mf – 9 – €20.00 – mf#1542c – gw Fischer [780]

Hjelt, Arthur *see*
- Die altsyrische evangelienuebersetzung und tatians diatessaron, besonders in ihrem gegenseitigen verhaeltnis untersucht...
- De johanneiska smabrefvens ursprung

Hjemmets ven – v1 n3-7 [1912 aug-dec], v2 n1,8,14 [1913 aug, 1914 apr, jun – 1r – 1 – mf#1330462 – us WHS [071]

Hjo tidning – Hjo, Sweden. 1847-60 – 1 – (aka: lordagsposten 1849-50; hjo nya tidning 1851-60) – sw Kungliga [078]

Hjo tidning – Hjo, Sweden. 1979- – 1 – sw Kungliga [078]

Hjo weckotidning – Hjo, Sweden. 1859-66 – 2r – 1 – sw Kungliga [078]

Hjokuriren – Falkoeping, 1909 – 1r – 1 – sw Kungliga [078]

Hjoposten – Falkoeping, 1913-17 – 1r – 1 – sw Kungliga [078]

Der hl theodor von studion (kgs5, 3) : sein leben und wirken. ein beitrag zur byzantinischen moenchsgeschichte / Schneider, G A – Muenster i. W, 1900 – €5.00 – ne Slangenburg [240]

Der hl thomas von aquin ueber das unfehlbare lehramt des papstes / Leitner, Franz Xaver – Freiburg i.B: Herder, 1872 [mf ed 1986] – 1mf – 9 – 0-8370-8442-3 – (incl bibl ref) – mf#1986-2442 – us ATLA [241]

Hla journal – Honolulu. 1944-1984 (1) 1971-1984 (5) 1975-1984 (9) – ISSN: 0017-8586 – mf#2192 – us UMI ProQuest [020]

Hla Phaw Zan, U *see* Vegetable gardening in burma

Hladky, James Robert *see* Twelve etudes in thumb position for solo violoncello

Hladny, Ernst *see* Hugo von hofmannsthal's griechenstuecke

Hlas = Katolicky tydennik / Cesky literarni spolek v St Louis, MO & Lincoln. University of Nebraska. Libraries University Archives Special Collections Dept – St Louis, MO: Bohemian Literary Society of St Louis, MO. roc12 cis533. 17 pros 1884 (wkly) [mf ed dec 17 1917-nov 25 1919 (gaps) filmed 1986 – 2r – 1 – us NE Hist [071]

Hlas – Cleveland OH, 1912-13 – 1r – 1 – (slovak newspaper) – us IHRC [071]

Hlas = Voice – Cleveland, OH: John Pankuch Co, jan 5 1925-nov 1927 – 1 – us CRL [071]

Hlas domova – Melbourne, Australia. 18 Feb 1957-1 Sep 1969 (imperfect) – 3r – 1 – uk British Libr Newspaper [072]

Hlas l'udu – Bratislava, Czechoslovakia. 1956-70 – 12r – 1 – us L of C Photodup [077]

Hlas l'udu – Presov, Czechoslovakia. May-Oct 1946 – 1r – 1 – us L of C Photodup [077]

Hlas naroda – Voice of the nation / Czech-American Heritage Center, Inc & Lincoln. University of Nebraska. Libraries University Archives Special Collections Dept – Chicago, IL: Velehrad. roc1 cis1. 3 led 1976- (biwkly) [mf ed jan 3 1976-oct 17 1992 (gaps) filmed 1982] – 17v in 2r – 1 – (in czech and english. issues for jan 16 1993?- publ in cicero il by czech-american heritage center, inc) – us NE Hist [071]

Hlas nitrianskeho kraja – Nitra, Czechoslovakia. 1956-Mar 1960 – 3r – 1 – us L of C Photodup [077]

Hlas (weekly edition) = Voice – [Cleveland, OH]: John Pankuch Co, dec 1917-sep 1921 – 1 – us CRL [071]

Hlasatel – Chicago: [s.n.], jan 6-apr 23 1920; 1945-jul 1975 – us CRL [071]

Hlavaek, I *see* Das urkunden- und kanzleiwesen des boehmischen und roemischen koenigs wenzel (4.) 1376-1419 (mgh schriften:23.bd)

Hlawitschka, E *see* Lotharingien und das reich an der schwelle der deutschen geschichte (mgh schriften:21.bd)

Hlm : the howard league magazine – London. 1998+ (1) – (cont: criminal justice) – ISSN: 1463-435X – mf#13457,01 – us UMI ProQuest [360]

Hlokomela lesea la hao pele le belehwa = Be good to your baby before it is born / South Africa. Department of Health [Departement van Gesondheid] [Departement van Gesondheid] – [Pretoria: The Dept 1977?] [mf ed Pretoria, RSA: State Library [199-]] – 11p [ill] on 1r with other items – 5 – (in sotho; also available in afrikaans, english, northern sotho, tsonga, venda, xhosa & zulu; incl bibl ref) – mf#op 09991 r24 – us CRL [618]

Hmb newsletter / National Baptist Convention of the United States of America – v3 [1994 aug] – 1r – 1 – mf#4024189 – us WHS [242]

Hmda. msa 1160, bridgeport-milford, ct : aggregate report / Federal Financial Institutions Examination Council (US) – Washington DC, 1990- (annual) [mf ed 1992?-] – 9 – mf#0443-E-07 (mf) – us FFIEC [332]

Hmo practice – Amherst. 1989-1998 (1,5,9) – ISSN: 0891-6624 – mf#16611 – us UMI ProQuest [615]

Hms northumberland logbook – 1 Jun 1815-11 Aug 1815 – 1r – 1 – (from the d.z. norton napoleonic collection, this vol is the logbook of the royal navy warship h.m.s. northumberland. at this time the ship transported napoleon bonaparte to exile on st. helena) – us Western Res [355]

Hno – Berlin. 1981-1981 (1,5,9) – ISSN: 0017-6192 – mf#13176 – us UMI ProQuest [617]

Hnyn hoey than suarg pan!... / Thu Dhan – Kracah: Pannagar Ghun Thai San 2508 [1965] [mf ed 1990] – 2v on 1r with other items – 1 – (in khmer) – mf#mf-10289 seam reel 111/1 [§] – us CRL [959]

–Hnyn mak khnum hoey! : pralom lok manosancetana / Thu Dhan – Kracah: Pannagar Ghun Thai San 2508 [1965] [mf ed 1990] – 1r with other items – 1 – (in khmer) – mf#mf-10289 seam reel 111/2 [§] – us CRL [959]

Ho, Alice *see* Pension plans, 1936-1954

Ho, Bieu Chanh *see* No doi

Ho, Chi Minh *see* Tho bac ho

Ho, Chia-huai *see*
- Han yeh chi
- Mao yen chi

Ho, Ch'i-fang *see*
- Huan hsiang chi
- Yeh ko
- Yu yen

Ho, Chih-hao *see* Chih-hao shih chi

Ho, Ch'ing-Ju *see* Ju ta hsueh che hsu chih

Ho, Ch'ing-ju *see* Ching-kuo ch'ing-nien chih-yeh wen-t'i [ccm149]

Ho, Chu-ch'i *see* Han chien ti hsia ch'ang

Ho, Chun *see* Tu hui ti i chiao

Ho, Chung-ying *see* Hsun ku hsueh yin lun

Ho, David Chi-hsing *see* Ai te sheng li (ccc148)

Ho, Feng-shan *see* Ou mei feng kuang

Ho! for the west : the traveller and emigrants' hand-book to canada and the north-west of the american union, comprising the states of illinois, wisconsin, and iowa and the territories of minnesota and kansas / Hall, Edward Hepple – London: Algar & Street : Tweedie, Strand, 1858 – 1mf – 9 – mf#22743 – cn Canadiana [917]

Ho! for the west!!! : the traveller and emmigrant's handbook to canada and the north-west states of america, for general circulation... / Hall, Edward Hepple – London: Algar & Street, 1856 – 1mf – 9 – mf#52966 – cn Canadiana [917]

Ho, Hsiao-i *see* Tung-pei ti chin jung

Ho, Hsin *see* Ti san t'iao lu

Ho, I *see* Hsiao ts'ao

Ho, Jo-lan *see* Ti kuo chu i yu shih chieh ta chan

Ho, Kung-hsing *see* Shang-hai chih hsiao kung yeh

Ho, Li-sheng *see* Huan shu chi

Ho liu ti ti ts'eng : ch'ang p'ien hsiao shuo / Wang, Ch'iu-ying – Ta-lien: Shih yeh yang hang ch'u pan pu, 1942 – (= ser P-k&k period) – us CRL [830]

Ho, Mei-shan *see* Tun lu shih hsuan

Ho pi hsi hsiang, i ming, mei hua meng : [t'an tz'u hsiao shuo] / Hsin-t'ieh-tao-jen – Shang-hai: Chiao ching shan fang shu chu, Min kuo 22 [1933] – (= ser P-k&k period) – us CRL [830]

Ho pien / Lu, Yen – Shang-hai: Liang yu t'u shu yin shua kung ssu, 1937 – (= ser P-k&k period) – us CRL [480]

Ho p'ing yu chan cheng : i chiu san i chih i chiu ssu i nien chien mei-kuo chih wai chiao cheng ts'e / United States. Dept of state – Ch'ung-ch'ing: Chung wai ch'u pan she, Min kuo 32 [1943] – (= ser P-k&k period) – us CRL [337]

Ho, Po-yen *see* Jen shih hsing cheng chih li lun yu shih chi

Ho, Shang *see* Tui liu ch'uan tai hui chueh i chih san ta ching chi cheng ts'e ti shang ch'ueh

Ho, Shih-chieh *see* Pu shao ch'ih hou ch'uan ling lien lo ping ch'in wu

Ho, Shih-ming *see*
- Chi-tu chiao shih erh chiang [ccm156]
- Chi-tu te hsin yang yu sheng huo [ccm150]
- Jen sheng kai lun
- Jen te chiao yu
- Ku pei li te tien kuo
- Shih chien ti hsin yang
- Ts'ung chi-tu chiao kan chung-kuo hsiao tao

Ho, Shou-liang *see*
- Chiao yu wen ta [ccm158]
- Wei li kung hui chiao yu wen ta

Ho, Ssu-yuan *see* Kuo chi ching chi cheng ts'e, yu ming, chung-kuo tui wai ching chi cheng ts'e chih yen chiu

Ho, Te-ming *see* Hsing fu ti ai ko

Ho ti kuang lin : [san mu chu] / Hung, Mo – Shang-hai: Shih chieh shu chu, 1944 (1946 printing) – (= ser P-k&k period) – us CRL [820]

Ho tso chiang hsi hui chiang i chi – [China]: Chung-kuo hua yang i chen chiu tsai tsung hui, Min kuo23 [1934] – (= ser P-k&k period) – us CRL [334]

Ho tso chin jung yao i / Chang, Tse-yao – Fu-chien Ch'ung-an: Chung-kuo ho tso ching chi yen chiu she, Min kuo 33 [1944] – (= ser P-k&k period) – us CRL [334]

Ho tso fa kuei – Ssu-ch'uan: Ssu-ch'uan sheng nung ts'un ho tso chih tao jen yuan hsun lien so, Min kuo 26 [1937] – (= ser P-k&k period) – us CRL [630]

Ho tso kai lun / T'ung, Yu-min – Shang-hai: Chung-hua shu chu, Min kuo 25 [1936] – (= ser P-k&k period) – us CRL [334]

Ho tso kuei chang hsin pien – K'un-ming: Chung-kuo ho tso shih yeh hsieh hui Yun-nan sheng fen hui yen chiu tsu, Min kuo 30 [1941] – (= ser P-k&k period) – us CRL [334]

Ho tso she / Niu, Ch'ang-yao – Shang-hai: Shang wu yin shu kuan, Min kuo 26 [1937] – (= ser P-k&k period) – us CRL [334]

Ho tso shih yeh / Chiao, Yu-t'ing & Liu, Kuang-yen – Shang-hai: Shih chieh shu chu, Min kuo 22 [1933] – (= ser P-k&k period) – us CRL [334]

Ho tsung chang ying-ch'in chiang k'ang chan ti liu nien chih chun shih / Ho, Ying-ch'in – Ch'ung-ch'ing: Meng Tsang wei yuan hui pien i shih, 1942 – (= ser P-k&k period) – us CRL [951]

Ho, Tzu-heng *see* Jih mei wen t'i

Ho, Winnie WY *see* Factorial validity of a teacher effectiveness scale for the teacher preparation program in hong kong

Ho, Ying-ch'in *see* Ho tsung chang ying-ch'in chiang k'ang chan ti liu nien chih chun shih

Ho, Yueh-seng *see* Tsung li tsung ts'ai lun li ssu hsiang chih yen chiu

Ho, Yung-chi *see* Wei chung-kuo mou cheng chih kai chin

Ho, Yu-po *see* Hsien tai chung-kuo tso chia lun ti erh chuan

Hoa lua : tieu thuyet / Dao Vu – [Ha-noi]: Phu Nu [1974] [mf ed 1992] – on pt of 1r – 1 – mf#11052 r19 n2 – us Cornell [959]

Hoadly, George *see* The constitutional guarantees of the right of property as affected by recent decisions.

Hoagland, D R *see* Growing plants without soil by the water-culture method

Hoam – New York. N.Y. 1916 – 1 – us AJPC [071]

Hoan chan tan bao – Hue. n1-7. janv-juil 1930 – 1 – (mq no. 2, 5) – fr ACRPP [073]

Hoang Ha *see*
- Bi mat phim truong
- Buon lau
- Di tim la-ng quen

Hoang, Hai Thuy *see* Ngoai cua thien duong

Hoang, Xuan Nhi *see*
- Tim hieu duong loi van nghe cua dang va su phat trien cua van hoc cach mang viet-nam hien dai
- Tim hieu tho ho chu tich

Hoar, George Frisbie *see* Church going

A hoard of silver punch marked coins from purnea / Bhattacharyya, P N – Delhi: Manager of Publ, 1940 – (= ser Samp: indian books) – us CRL [730]

Hoard's dairyman – 1907 feb-1908 jan, 1908 feb-1909 jan, 1909 feb-1910 jan, 1944 jan 10-1945 apr 25 – 4r – 1 – mf#769948 – us WHS [630]

Hoard's dairyman – Fort Atkinson. 1950+ (1) 1979+ (5) 1979+ (9) – ISSN: 0018-2885 – mf#732 – us UMI ProQuest [630]

Hoard's dairyman – Fort Atkinson, WI. 1889-1949 (1) – mf#67557 – us UMI ProQuest [071]

Hoare, Charles James *see* Kingdom of god not in word but in power

Hoare, Edward Hatch *see*
- Baptism according to scripture
- The scripture ground of justification

Hoare, Prince *see*
- Academic annals of painting, sculpture, and architecture, published by authority of the royal academy of arts, 1805-1806, 1807, 1808-1809
- Academic correspondence, 1803...
- Epochs of the arts
- Extracts from a correspondence with the academies of vienna and st petersburg
- An inquiry into the requisite cultivation

Hob ikh mir a lidele – Varshe, Poland. 1922 – 1r – us UF Libraries [939]

Hobard index and commenwealth – Gary, IN. 1934-1938 (1) – mf#62792 – us UMI ProQuest [071]

Hobart, Alfred Walters *see* William channing gannett

Hobart, Alvah Sabin *see*
- Corner stones of a baptist church
- Our silent partner

Hobart, Chauncey *see* History of methodism in minnesota

Hobart, George Vere *see*
- Idle moments in florida

Hobart, John Henry *see* Correspondence of john henry hobart

Hobart mercury – Hobart, jul 1854- – at Pascoe [079]

Hobart town courier – Hobart, 1827-43 – 3r – 1 – A$115.50 vesicular A$132.00 silver – at Pascoe [079]

Hobarton guardian – Hobart, jan-jul 1854 – 1r – A$27.50 vesicular A$33.00 silver – at Pascoe [079]

Hobbes / Robertson, George Croom – Edinburgh: William Blackwood, 1886 – (= ser Philosophical classics for english readers) – 1mf – 9 – 0-7905-9087-5 – mf#1989-2312 – us ATLA [100]

Hobbes, Thomas *see*
- The metaphysical system of hobbes as contained in twelve chapters from his elements of philosophy concerning body...and human nature...and leviathan
- Tripos in three discourses
- True ecclesiastical history from moses to the time of martin luther

Hobbies – Chicago. 1931-1985 (1) 1969-1985 (5) 1970-1985 (9) – (cont by: antiques and collecting hobbies) – ISSN: 0018-2907 – mf#2541 – us UMI ProQuest [790]

Hobbs, Alvin Ingals *see* Theological discussion

Hobbs, new mexico. first baptist church – Church Records, 1936-1988. 1,806p – 1 – $81.27 – us Southern Baptist [242]

Hobby horse – London. 1886-1894 (1) – mf#5569 – us UMI ProQuest [790]

Hobby prevue – Forest Hills. 1963-1966 (1) – mf#5800 – us UMI ProQuest [790]

Hobby-bandwagon / Circus Historical Society – v2 n7-10,12 [1947 aug-nov, 1948 jan], v3 n1-12 [1948 feb-1949 jan], v4 n1,4-v6 n6 [1949 feb, may-1951 jul] – 1r – 1 – (cont: bandwagon [farmington, mi]; hobby-swapper; cont by: c h s bandwagon) – mf#930624 – us WHS [978]

Hobby-swapper : a monthly magazine devoted to hobby – v2 n4,6 [1947 may, jul] – 1r – 1 – (continued by: bandwagon (farmington, mi); hobby bandwagon) – mf#635294 – us WHS [790]

Hobe sound, florida / Chapin, George M – Jacksonville, FL. 1913? – 1r – us UF Libraries [978]

Hoben, Allan *see* The virgin birth

Hoberg, Gottfried *see* Die genesis

Hoberg, Marcus *see* Beitrag der markophyten zu den schwebstoffen der tide-llbe

Hoberg Paper Mills *see* Tissue topics

Hobhouse, John *see* Substance of some letters written from paris during the last reign of the emperor napoleon

Hobhouse, Leonard Trelawney *see*
- Development and purpose
- Mind in evolution
- Morals in evolution
- Social evolution and political theory
- The theory of knowledge

Hobhouse letters, the... 1722-55 : from bristol central library and bristol record office / Isaac Hobhouse and Co – (= ser British records relating to america in microform) – 1r – 1 – (int by w e minchinton) – mf#3744 – uk Microform Academic [025]

Hobhouse, LT *see* Morals in evolutions

Hobhouse, Walter *see*
- The church and the world in idea and in history
- A short sketch of the first four lambeth conferences 1867-1897

The hobo : the sociology of the homeless man: a study prepared for the chicago council of social agencies under the direction of the committee on homeless men / Anderson, Nels – Chicago, IL: University of Chicago Press, 1923 [mf ed 1970] – (= ser Library of american civilization 15761) – xv/302p on 1mf – (with bibl) – us Chicago U Pr [360]

Hobo-quebec : journal d'ecritures et images – Montreal: [s.n.] v1 n1-n46/47 automne/hiver 1981 (irreg) [mf ed 1986] – 1r – 1 – mf#SEM35P261 – cn Bibl Nat [410]

Hobson, Alphonzo Augustus *see* The diatessaron of tatian and the synoptic problem

Hobson, John Atkinson *see*
- Richard cobden
- The science of wealth

Hobson, Samuel *see*
- Dialogues between a protestant and a roman catholic
- Manual for the sick

Hobson's advertiser and monthly miscellany – [NW England] Ashton-under-Lyne, Stalybridge Lib jul-aug 1854 – 1 – uk MLA; uk Newsplan [072]

Hobson's mid-monthly advertising journal for ashton-under-lyne, stalybridge and dukinfield – [NW England] Ashton-under-Lyne, Stalybridge Lib sep 1854 – 1 – uk MLA; uk Newsplan [072]

Hobstetter, Robert Goodwin *see* Study of the ensemble devices in the trio for piano, violin, and cello

Hoburg, C *see*
- Emblemata sacra
- Levendige herts-theologie

Hoby, J *see* The baptists in america

Hoby, James *see* The baptists in america

Du hoc bao : bulletin bimensuel puis mensuel de la societe d'encouragement aux etudes occidentales – Hue. 1er sept 1927-sept 1935 – 1 – fr ACRPP [073]

Hoc est novum testamentum aegyptium vulgo copticum – Oxonii: Theatro Sheldoniano Typis et Sumptibus Academiae 1716 – (= ser Whsb) – 8mf – 9 – €80.00 – (text in coptic & latin) – mf#Hu 383 – gw Fischer [470]

...Hoc est via sancta...sive biblia sacra eleganti et maivscvla characterum forma... : [derek ha-qodesh siue biblia sacra] – Hamburgi, lohannem Saxonem, 1587. 2v – 28mf – 9 – mf#H-8433 – ne IDC [956]
Hoc mien ng theo phng phap moi / Vanna, Pa et al – [Phnom-Penh: Rasmey-Moni [196-?] [mf ed 1990] – 1r with other items – 1 – (title also in khmer language) – mf#mf-10289 seam reel 124/3 [$] – us CRL [480]
Hoc sinh – Hanoi. 1936, 1939-41 – 1 – fr ACRPP [073]
Hoc tap : tap chi ly luan va chinh tri cua dang lao dong viet-nam – Hanoi: [s.n. v9 n90-v22 n252. jul 1963-1976] – 17r – 1 – us CRL [079]
Hoca nasreddin – Istanbul. n1. 1908 – (= ser O & t journals) – 1mf – 9 – $25.00 – us MEDOC [956]
Hoc-bao – Hanoi. sept 1922-aout 1923; sept 1925-juin 1929; sept 1936-juil/aout 1937, 1938-39 – 1 – fr ACRPP [073]
Hocedez, E *see* Richard de middleton
Hoch, Alexander *see*
– Lehre des johannes cassianus von natur und gnade
– Papst pius 10
Hoch, E *see* Bemba grammar notes for beginners
Hoch, Horace Lind *see* Shakespeare's influence upon grabbe
Hoch, Mark *see*
– Die aufgaben der missionspredigt in indien
– Die taufbewerber in der indischen mission, ihre beweggruende und ihre behandlung
Hoch oesterreich : patriotisches liederspiel / Purschke, Marie Sidonie – Wien: Verlag des katholischen Waisen-Hilfsvereins, 1885 – 1r – 1 – us Wisconsin U Libr [780]
Hochaufloesende absorptionsspektroskopie im weichen roentgenbereich an ausgewaehlten atomen und molekuelen / Remmers, Guido – (mf ed 1992) – 2mf – 9 – €49.00 – 3-89349-604-1 – mf#DHS 604 – gw Frankfurter [530]
Hochaufloesende interferometrische messungen in der astronomie mit ccd-detektoren : die beobachtung der objekte eta carinae, red rectangle und beteigeuze / Weghorn, Hans – (mf ed 1994) – 2mf – 9 – €40.00 – 3-89349-869-9 – mf#DHS 869 – gw Frankfurter [520]
Hochaufloesende photoelektronenspektroskopische untersuchungen des supraleitenden zustands von bi2sr2cacu208+d(delta) / Simmons, C Thomas – (mf ed 1993) – 1mf – 9 – €30.00 – 3-89349-736-6 – (in englischer sprache) – mf#DHS 736 – gw Frankfurter [540]
Das hoch-beehrte augspurg : oder wahrgruendliche vorstellung der hochwichtigen handlung- und verrichtungen... – Augsburg: Koppmayer, 1690 – 4mf – 9 – mf#O-64 – ne IDC [090]
Das hochbeehrte augspurg : wie solches nicht allein mit beeder kayserl. als auch der ungaris koenigl. majest.,...hoechsterfreulichster ankunfft... – Augspurg: Koppmayer, 1690 – 4mf – 9 – mf#O-05 – ne IDC [090]
Eine hochburg der hugenotten waehrend der religionskriege / Harten, Theodor – Halle a S: Verein fuer Reformationsgeschichte, 1898 – (= ser [Schriften fuer das deutsche Volk]) – 1mf – 9 – 0-524-01948-7 – mf#1990-0537 – us ATLA [242]
Hochdoerfer, Richard *see* Introductory studies in german literature
Hochdorf, Max *see* Gottfried keller im europaeischen gedanken
Hocheimer, Lewis *see*
– A manual of criminal law, as established in the state of maryland
– A treatise on the law relating to the custody of infants, including practice and forms
Ho-che-k'o-la tsu / Pu, Te – Fu-chien Yung-an: Kai chin ch'u pan she, 1942 – (= ser P-k&k period) – us CRL [830]
Hoch-fuerstlich brandenburg-onoltzbachischer address- und schreib-calender *see* Die amtskalender der fraenkischen fuerstentuemer ansbach und bayreuth [1737-1801]
Hochland (klp16) : monatsschrift fuer alle gebiete des wissens, der literatur und kunst / ed by Muth, Carl – 1903-1940/41 [mf ed 2004] – (= ser Marbacher mikrofiche-editionen (mme) 16; Kultur – literatur – politik: deutsche zeitschriften des 19./20. jahrhunderts (klp)) – 38v on 604mf – 9 – €3300.00 – 3-89131-452-3 – gw Fischer [074]
Hochmuth, Arno *see*
– Literatur im blickpunkt
– Literatur und dekadenz
Hochmuth und hoffaertigkeit : ein bild der hoffaertigkeit und dessen seelenmoerderischen wirkung / Weber, J, bishop – [Kitchener, Ont?: s.n, 1885?] [mf ed 1993] – 1mf – 9 – 0-665-92954-4 – mf#92954 – cn Canadiana [240]
Hochschild, Ernst *see* Kc blaetter
Das hochschulwesen – 1953-1989 – 604mf – 1 – gw Mikropress [378]
Hochstetter, Chr *see* Die geschichte der evangelisch-lutherischen missouri-synode in nord-amerika
Hochstetter, E *see* Studien zur metaphysik und erkenntnislehre wilhelms von ockham
Hochstetter, Gustav *see* Das buch der liebe
Ho-chunk wo-lduk / Wisconsin Winnebago Business Committee – Tomah WI. 1986 nov-1990 oct; 1991-93, 1994 early jan-1996 dec – 3r – 1 – (cont: wisconsin winnebago business committee; cont by: hocak worak) – mf#1054660 – us WHS [338]
Der hochverratsprozess gegen sinclai : ein beitrag zum leben hoelderlins / Kirchner, Werner – Marburg/Lahn: Simons 1949 [mf ed 1991] – (= ser Literatur und leben. neue folge 2) – 1r – 1 – (incl bibl ref. filmed with: freiheit und recht / johann gottfried seume) – mf#2943p – us Wisconsin U Libr [430]
Der hochwachter – Cincinnati: Friedrich Hassaurek, jul 1845-jul 1846 – 1r – 1 – us CRL [071]
Der hochwaechter – Stuttgart DE, 1831 1 apr-1876, 1877 4 jul-1920 30 sep – 42r – 1 – (filmed with other misc inst: 1849 1 jun-31 dec [1r]; title varies: 1 jan 1833: der beobachter) – gw Misc Inst [074]
Hochwaechter auf dem schwarzwald – Titisee-Neustadt DE, 1869 3 jan-21 feb – 1 – gw Misc Inst [074]
Hochwasser : novellen / Seidel, Ina – 3. Aufl. Berlin: E Fleischel, 1920 – 1r – 1 – us Wisconsin U Libr [830]
Die hochzeit auf buchenhorst : erzaehlung / Hauptmann, Gerhart – Berlin: S Fischer, 1932 – 1 – us Wisconsin U Libr [830]
Die hochzeit der feinde : roman / Andres, Stefan Paul – Muenchen: R Piper, 1953, c1952 [mf ed 1995] – 449p – 1 – mf#8919 – us Wisconsin U Libr [830]
Die hochzeit der sobeide : dramatisches gedicht / Hofmannsthal, Hugo von – 2. ausg. Berlin: S Fischer 1909 [mf ed 1990] – 1r – 1 – (filmed with: gestern) – mf#2729p – us Wisconsin U Libr [810]
Die hochzeit des moenchs : novelle / Meyer, Conrad Ferdinand – Leipzig: H Haessel, 1884 [mf ed 1996] – 165p – 1 – mf#9721 – us Wisconsin U Libr [830]
Hochzeitpredigten vom ehestand vnd hausswesen / Mathesius, J – Nuernberg, 1584 – 4mf – 9 – mf#TH-1 mf 999-1002 – ne IDC [242]
Der hochzeitsschmaus : und andere ergoetzlichkeiten [in verse] / Huggenberger, Alfred – Leipzig: L Staackmann 1921 [mf ed 1995] – 1r – 1 – (ill by hans witzig) – mf#3884p – us Wisconsin U Libr [810]
Hock, A *see* Argiculture au katanga
Hock, Eric *see*
– Motivgleiche gedichte
Hock, Erich *see* "Dort drueben in westfalen"
Hock, Stefan *see*
– Anton auerspergs (anastasius gruens) politische reden und schriften
– Eduard von bauernfelds gesammelte aufsaetze
Hock, Theobald *see* Schoenes blumenfeld
Hocken, J *see* Hints and helps to local preachers
Hocken, T M *see* Settlement of otago
Hockey digest – Evanston. 1972+ (1) 1972+ (5) 1973+ (9) – ISSN: 0046-7693 – mf#6277 – us UMI ProQuest [790]
Hockey news – Montreal. 1988+ (1,5,9) – ISSN: 0018-3016 – mf#16051 – us UMI ProQuest [790]
Hockin, Frederick *see* John wesley and modern wesleyanism
Hockin, John *see*
– J p hockin's bericht von den neuesten reisen nach den pelew-inseln
– Statement regarding the gold-fields of eastern canada
Hocking Co. Logan *see*
– Democrat-sentinel
– Hocking county star
– Hocking republican
– Hocking sentinel
– Hocking val republican
– Hocking valley gazette
– Hocking valley journal and gazette
– Hocking valley republican
– Morning star
– Ohio democrat
– Republican
– Republican gazette
Hocking county star / Hocking Co. Logan – jan-apr 1853 [wkly] – 1r – 1 – mf#B125 – us Ohio Hist [071]
Hocking, George Macdonald *see* Comparative study of cultural, morphological, and histological characteristics of species
Hocking, Joseph *see* Shall rome reconquer england?
Hocking republican / Hocking Co. Logan – jul 1903-mar 1909 [wkly] – 3r – 1 – mf#B10811-10813 – us Ohio Hist [071]
Hocking sentinel / Hocking Co. Logan – jan 1884-mar 1906 [wkly] – 10r – 1 – mf#B8539-8548 – us Ohio Hist [071]
Hocking sentinel – Logan, OH. 1845-1883 (1) – mf#65559 – us UMI ProQuest [071]
Hocking val republican / Hocking Co. Logan – v1 n1. sep 1855-sep 1856 [wkly] – 1r – 1 – mf#B125 – us Ohio Hist [071]
Hocking valley gazette / Hocking Co. Logan – dec 1877-jun 1883 [wkly] – 2r – 1 – mf#B11107-11108 – us Ohio Hist [071]
Hocking valley journal and gazette / Hocking Co. Logan – jan 1892-dec 1910 [wkly] – 8r – 1 – mf#B11112-11119 – us Ohio Hist [071]
Hocking valley press / Athens Co. Nelsonville – nov 1971-oct 1972 [wkly] – 1r – 1 – mf#B33876 – us Ohio Hist [071]
Hocking valley republican / Hocking Co. Logan – may 1849-mar 1850 [wkly] – 1r – 1 – mf#B125 – us Ohio Hist [071]
Hocking, William Ernest *see* The meaning of god in human experience
Hockliffe – (= ser Bedfordshire parish register series) – 1mf – 9 – £3.00 – uk BedsFHS [929]
Hockliffe, st nicholas monumental inscriptions monumental inscriptions – Arthur Weight Matthews 1912 – (= ser Bedfordshire parish register series) – 1mf – 9 – £1.25 – uk BedsFHS [929]
Hodde, Jason P *see* Remodeling characteristics of the rabbit achilles tendon complex following repair with small intestinal submucosa
Hodde, Lucien de la *see* Histoire des societes secretes et du parti republicain de 1830 a 1848
Hodder, Edwin *see*
– Conquests of the cross
– The life and work of the seventh earl of shaftesbury, k g
– Simon peter
Hodder-Williams, John Ernest *see* The life of sir george williams
Hoddesdon broxbourne mercury *see* Hoddesdon mercury
Hoddesdon journal – [London & SE] Hertfordshire apr 1935-dec 1949 [mf ed 2004] – 7r – 1 – uk Newsplan [072]
Hoddesdon mercury – England, 16 sep 1983- – 68+ r – 1 – (hoddesdon broxbourne mercury, 1986-) – uk British Libr Newspaper [072]
Hoddesdon mercury and journal *see* Hertfordshire mercury
Hodell, Charles W *see* The old yellow book
Hodes, Fritz *see* Hessisches nachbarrecht
Der hodesh – v1, n1-3. jan-mar 1921 – 1 – us NY Public [073]
Hodey, Le *see* Journal des etats generaux
Hodgdon, Norris C *see* A denominational offering from the literature of universalism
Hodge, Archibald Alexander *see*
– The atonement
– The church and its polity
– A commentary on the confession of faith
– The life of charles hodge, d.d.
– Outlines of theology
– Popular lectures on theological themes
– Questions on the text of the systematic theology of dr. charles hodge
– The system of theology contained in the westminster shorter catechism
Hodge, Caspar Wistar *see*
– Gospel history
– New testament criticism
– Syllabus of lectures on apostolic history and literature
Hodge, Charles *see*
– The church and its polity
– A commentary on the epistle to the ephesians
– Commentary on the epistle to the romans
– Conference papers, or, analyses of discourses, doctrinal and practical
– The constitutional history of the presbyterian church in the united states of america
– Cotton is king, and pro-slavery arguments
– Discussions in church polity
– Essays and reviews
– An exposition of the first epistle to the corinthians
– An exposition of the second epistle to the corinthians
– Index to systematic theology
– The reunion of the old and new-school presbyterian churches
– Sermons
– Systematic theology
– The way of life
– What is darwinism?
– What is presbyterianism?
Hodge, Frederick Arthur *see* The plea and the pioneers in virginia
Hodge, John Aspinwall *see*
– Recognition after death
– What is presbyterian law as defined by the church courts?
Hodge, John Zimmerman *see* Caste or christ?
Hodge, Richard Morse *see*
– Historical atlas and chronology of the life of jesus christ
– A syllabus of religious education
Hodge, William L *see* The law of attachment in maryland
Hodgeman County. Kansas. St. Michael's Evangelical Lutheran Church *see* Records
Hodgenville first baptist church. hodgenville, kentucky : church records – 1838-57, 1878-1966 – 1 – 82.17 – us Southern Baptist [242]
Hodges and smiths estates circular – Dublin, Ireland. aug 1854-jul 1863, oct-dec 1863, jan-jun, oct-dec 1864, jan-jun, oct-dec 1865, jan-nov 1866, jan-jul, oct-dec 1867, feb-jul, oct-dec 1868, mar-jul 1869 – 2r – 1 – (aka: hodges smith and cos estates circular) – uk British Libr Newspaper [072]
Hodges, Ann Mary *see* Songs
Hodges Brothers. Olathe, Kansas *see* Ledgers, day books
Hodges, George *see*
– Classbook of old testament history
– The confirmation rubric and christian fellowship
– The early church from ignatius to augustine
– The episcopal church
– Henry codman potter, seventh bishop of new york
– The human nature of the saints
– Three hundred years of the episcopal church in america
– When the king came
Hodges, Henry G *see* The doctrine of intervention
Hodges, James *see* Construction of the great victoria bridge in canada
Hodges, Nicola J *see* The role of instructions and demonstrations in learning a coordination skill
Hodges smith and cos estates circular *see* Hodges and smiths estates circular
Hodges, W *see* Travels in india during the years 1780-1783
Hodgin's canada election cases / Ontario. Canada – 1v. 1871-78 (all publ) – 9mf – 9 – $13.50 – mf#LLMC 81-046 – us LLMC [340]
Hodgins, Frank Egerton *see* Life insurance contracts in canada
Hodgins, J George *see* The acts relating to common schools and also separate schools in ontario
Hodgins, John George *see*
– Brief notices of "the history and legislation of separate schools in upper canada"
– Easy lessons in general geography, with maps and illustrations
– "The story of my life"
Hodgins, Thomas *see* British and american diplomacy affecting canada, 1782-1899
Hodgkin, Adrian Eliot *see* Archer's craft
Hodgkin, Henry Theodore *see*
– Friends beyond seas
– The message and mission of quakerism
– The missionary spirit and the present opportunity
Hodgkin, Howard *see* Irish land legislation and the royal commissions
Hodgkin, Thomas *see*
– The dynasty of theodosius
– George fox
– Human progress and the inward light
– The odoric the goth
– The trial of our faith, and other papers
Hodgkin, Thomas et al *see* The fellowship of silence
Hodgkins, B *see* Emma
Hodgkins, Louise Manning *see* Via christi
Hodgman, Stephen Alexander *see* Moses and the philosophers
Hodgskin, Thomas *see* Travels in the north of germany
Hodgson, Adam *see*
– Letters from north america
– Letters from north america, vol 1
– Letters from north america, vol 2
– Letters from north america, vols 1-2
Hodgson and Co, Auctioneers *see* A catalogue of...paintings
Hodgson, Brian Houghton *see* Essays on the languages, literature, and religion of nepal and tibet
Hodgson, Frederick Thomas *see* Practical bungalows and cottages for town and country
Hodgson, J E *see* Aeronautical and miscellaneous notebooks, c1799-1826
Hodgson, James Muscutt *see*
– Thelogia (sic) pectoris
– Theologia pectoris
Hodgson, John Evan *see* Fifty years of british art
Hodgson London *see* Catalogue of a valuable assemblage of paintings
Hodgson, MGS *see* The order of assassins
Hodgson, Ph *see* Cloud of unknowing and book of privy counselling
Hodgson, Richard *see* Human personality and its survival of bodily death
Hodgson, Shadworth Hollway *see*
– The metaphysic of experience
– The theory of practice
– Time and space
Hodgson, W Earl *see* Annals of my life, 1847-1856
Hodgson, William *see* Select historical memoirs of the religious society of friends, commonly called quakers
Hodius, H *see*
– Historia chronica

Hodivala, Shahpurshah Hormasji *see* Studies in indo-muslim history

Der hodscha nasreddin / ed by Wesselski, Albert – Weimar: Alexander Duncker Verlag 1911 [mf ed Bloomington IN: Indiana Uni Lib, Preservation Dept 1984] – 2v on 1r – 1 – (humorous anecdotes about nasreddin hoca, a legendary character) – us Indiana Preservation [390]

Hodson, A W *see* Seven years in southern abyssinia

Hodson, Arnold Wienholt *see* An elementary and practical grammar of the galla or oromo language

Hodson, George *see* St paul's estimate of the gospel

Hodson, James Shirley *see* An historical and practical guide to art illustration

Hodson, Thomas Callan *see* The naga tribes of manipur

Hodson, William Stephen Raikes *see* Twelve years of a soldier's life in india: being extracts from the letters of the late major w.s.r. hodson.

Hoe hana – v1-v3 n7 [1973 feb-1975 aug/sep] – 1r – 1 – mf#370655 – us WHS [071]

Hoe von Hoenegg, M *see*
- Calvinistarum vera, viva et genuina descriptio
- Gruendliche ableinung fuenffzig statlicher ausserlesener vnd in alle ewige ewigkeit unerweisslicher calvinischer ertz- vnd hauptluegen
- Gruendliche, summarische, apostolische ausfuehrung, der gantzen reinen catholischen, evangelischen lehre in funffzig predigten verfasset
- Gruendlicher bericht

Hoeber, Eduard *see* Eichendorffs jugenddichtungen

Hoechheimer, Elijah Ben Hayyim *see* Shevile de-raki'a

Die hoechste lehrgewalt des papstes = De la monarchie pontificale a propos du livre de mgr l'eveque de sura / Gueranger, Prosper – Mainz: Franz Kirchheim, 1870 [mf ed 1986] – 1mf – 9 – 0-8370-8515-2 – (german trans fr french. incl bibl ref) – mf#1986-2515 – us ATLA [241]

Hoechster kreisblatt *see* Wochenblatt fuer den kreis hoechst

Hoecker, Gustav *see* Hoffart und demut

Hoecker, R *see* Das lehrgedicht des karel van mander

Hoedemaker, Philippus Jacobus *see*
- De mozaeische oorsprong van de wetten in de boeken exodus, leviticus en numeri
- Mozaische oorsprong van de wetten in de boeken exodus, leviticus en numeri

Hoedemaker, Phillipus Jacobus *see*
- Gisberti voetii tractatus selecti de politica ecclesiastica. series prima
- Gisberti voetii tractatus selecti de politica ecclesiastica. series secunda

Hoefer, Edmund *see* Goethe und charlotte v. stein

Hoefer, F de *see* Nouvelle biographie generale depuis le temps les plus recules jusqu'a nos jours

Hoefer intelligenz-blatt – Hof DE, 1968-1984 30 apr – 1 – (filmed by other misc inst: 2000- [ca 7r/yr]; 1802 & 1807, 1809-10, 1813-1943 28 feb, 1949 1 sep-2000; 1848-49 [1r]. with suppls: fuer die jugend 1931 [1r]; haus – hof – garten – feld 1912-13, 1915-17; 1919-22, 1924-28, 1931-33, 1935 & 1938 [4r]; der erzaehler an der saale 1877-79, 1881, 1883-93, 1896 & 1904; 1915, 1918, 1920, 1923, 1925, 1928, 1931-33, 1935-55 [war gaps!], 1957-66 [b/r fr 1940: erzaehler an der saale, fr 1961: erzaehler]) – gw Mikrofilm; gw Misc Inst [074]

Hoefer, Karl-Heinz *see*
- Deutungen und bekenntnisse
- Kleine literarturfibel

Hoefer, Karlheinz et al *see* Erbe und gegenwart

Hoeffding, Harald *see* Soeren kierkegaard, som filosof

Hoeffel, Ernest *see* De la condition juridique des etrangers au cambodge

Hoeffner, Johannes *see*
- Frau rat
- Goethe und das weimarer hoftheater

Hoefig, Willi *see* The dissident press of revolutionary iran

Hoefische spuren im protestantischen schuldrama um 1600 : caspar bruelow, ein pommerscher gelehrter in strassburg (1585-1627) / Schaefer, Hildegard – [S.l.: s.n.], 1935 (Oelde in Westf: Druck, E Holterdorf) – 1 – us Wisconsin U Libr [430]

Der hoefisch-galante roman des 17. jahrhunderts bei eberhard werner happel / Lock, Gerhard – Wuerzburg: K Triltsch 1939 [mf ed 1990] – 1r – 1 – (incl bibl ref. filmed with: novalis heinrich von ofterdingen... & other titles) – mf#2697p – us Wisconsin U Libr [430]

Hoefler, Constantin *see* Albert von beham und regesten papst innocenz 4

Hoefler, Karl Adolf Constantin, Ritter von *see*
- Die avignonesischen paepste, ihre machtfuelle und ihr untergang
- Bonifatius, der apostel der deutschen, und die slavenapostel, konstantinos (cyrillus) und methodios
- Die deutschen paepste
- Magister johannes hus und der abzug der deutschen professoren und studenten aus prag, 1409
- Papst adrian 6

Hoefling, Johann Wilhelm Friedrich *see*
- Grundsaetze evangelisch-lutherischer kirchenverfassung
- Liturgisches urkundenbuch
- Das sakrament der taufe

Hoeglandet nu – Naessjoe, Sweden. 2006- – 1 – sw Kungliga [078]

Die hoehe des gefuehls : ein akt / Brod, Max – Leipzig: K Wolff [1919?] [mf ed 1989] – (= ser Buecherei "der juengste tag" 57) – 1r – 1 – (filmed with: ein gelegenheitsgedicht von brockes / friedrich gundolf) – mf#7089 – us Wisconsin U Libr [820]

Der hoehencultus : asiatischer und europaeischer voelker / Andrian-Werburg, Ferdinand, Freiherr von – Wien: Carl Konegen, 1891 – 1mf – 9 – 0-524-01153-2 – (incl bibl ref) – mf#1990-2229 – us ATLA [390]

Hoehensonne : lustspiel in drei akten / Fulda, Ludwig – Stuttgart; Berlin: J G Cotta, 1927 (mf ed 1990) – 1r – 1 – (filmed with: aus der werkstatt) – us Wisconsin U Libr [820]

Die hoehere maedchenschule *see* Die maedchenschule

Hoehlbaum, K *see* Das buch weinsberg, bd 2 1552-1577 (pgrg4)

Die hoehle von beauregard : erlebnis der westfront 1917 [a novel] / Grote, Hans Henning, Freiherr – Berlin: Brunnen-Verlag: K Winckler c1930 [mf ed 2001] – 1r – 1 – (filmed with: die beiden aeltesten drucke von grimmelshausens "simplicissimus" sprachlich verglichen / g einar toernvall) – us Wisconsin U Libr [830]

Hoehler, Wilhelm *see* Hermetische philosophie und freimaurerei

Hoehn, M *see* Fuenfzig jahre im predigtamte

Hoehn, Theodore S *see* Rare and imperiled fish species of florida

Hoehne, Frederico Carlos *see* Botanica e agricultura no brasil no seculo 16

Hoei-lan-ki : ou l'histoire du cercle de craie: drame en prose et en vers / Li, Xingdao – London 1832 [mf ed Hildesheim 1995-98] – 1v on 2mf – 9 – €60.00 – 3-487-27598-8 – gw Olms [820]

Hoek, Martin *see* De kometen van de jaren 1556, 1264, en 975

Hoekoemannja larangan ka 7 / Poei, Soey Hok – Soerabaia: Tan's Drukkery, [1934] [mf ed 1998] – (= ser Penghidoepan 117) – 1r – 1 – (varying form of title: hoekoemannja larangan ka toejoeh. coll as pt of the colloquial malay collection. filmed with: poetri satrija dewi, atawa, resia madjapait / h s t) – mf#10001 – us Wisconsin U Libr [830]

Hoekoemannja larangan ka toejoeh *see* Hoekoemannja larangan ka 7

Hoekstra, Sytze *see*
- Bronnen en grondslagen van het godsdienstig geloof
- Grondslag, wezen en openbaring van het godsdienstig geloof

Hoelderlin : choix de textes / Hoelderlin, Friedrich; ed by Leonhard, Rudolf & Rovini, Robert – [Paris]: P Seghers, [1953] – 1 – (incl bibl ref) – us Wisconsin U Libr [800]

Hoelderlin : feldauswahl / ed by Beissner, Friedrich – [Stuttgart]: Cotta, [1943] – 1r – 1 – us Wisconsin U Libr [430]

Hoelderlin / Przywara, Erich – Nuernberg: Glock und Lutz, 1949 – 1r – 1 – us Wisconsin U Libr [430]

Hoelderlin, Friedrich *see*
- Ausgewaehlte werke
- Briefe
- Gesammelte werke
- Gesang des deutschen
- Hoelderlin
- Hoelderlins gesammelte dichtungen
- Poems
- Der tod des empedokles
- Vom heiligen reich der deutschen
- Werke

Hoelderlin, friedrich : beitrage zu seinem 200. geburtstag / ed by Bundessekretariat des Deutschen Kulturbundes, Sektor Publikationen – Berlin: Deutscher Kulturbund, 1970 – 1 – (incl bibl ref) – us Wisconsin U Libr [430]

Hoelderlin und christus / Winklhofer, Alois – Nuernberg: Glock und Lutz, 1946 – 1r – 1 – us Wisconsin U Libr [430]

Hoelderlin und das deutsche theater / Kindermann, Heinz – Wien: W Frick, 1943 – 1 – us Wisconsin U Libr [430]

Hoelderlin und das wesen der dichtung / Heidegger, Martin – Muenchen: A Langen/G Mueller, c1937 – 1r – 1 – us Wisconsin U Libr [430]

Hoelderlin und die philosophie / Hoffmeister, Johannes – 2. durchgesehene Aufl. Leipzig: F Meiner, 1944 – 1r – 1 – us Wisconsin U Libr [100]

Hoelderlin und die philosophie / Hoffmeister, Johannes – 1. Aufl. Leipzig: F Meiner, 1944 – 1r – 1 – (incl bibl ref) – us Wisconsin U Libr [430]

Hoelderlin und die philosophie / Hoffmeister, Johannes – Leipzig, 1942 (mf ed 1994) – 2mf – 9 – €31.00 – 3-8267-3014-3 – mf#DHS-AR 3014 – gw Frankfurter [110]

Hoelderlin und die schweiz / ed by Boehm, Wilhelm – Frauenfeld/Leipzig: Huber, c1935 – 1 – (incl bibl ref) – us Wisconsin U Libr [920]

Hoelderlins begegnung mit goethe und schiller / Fahrner, Rudolf – Marburg a.L.: N G Elwert, 1925 – 1r – 1 – (incl bibl ref) – us Wisconsin U Libr [430]

Hoelderlins christliches erbe / Wocke, Helmut – Muenchen: Leibniz, 1949 – 1 – (incl bibl ref) – us Wisconsin U Libr [430]

Hoelderlins deutung des "oedipus" und der "antigone" : die "anmerkungen" im rahmen der klassischen und romantischen deutungen des antik-tragischen / Schrader, Hans – Bonn a. Rh.: L Roehrscheid, 1933 – 1r – 1 – (incl bibl ref) – us Wisconsin U Libr [450]

Hoelderlins gesammelte dichtungen : neue durchgesehene und vermehrte ausg. mit biographischer einleitung / Hoelderlin, Friedrich; ed by Litzmann, Berthold – Stuttgart: J G Cotta'sche Buchhandlung Nachfolger. 2v. [1895?] – 1 – (icnludes bibliographical references) – us Wisconsin U Libr [800]

Hoelderlins hymne "wie wenn am feiertage... " / Heidegger, Martin – Halle a.d.S.: M Niemeyer, [1941?] – 1r – 1 – us Wisconsin U Libr [780]

Hoelderlins studienjahre im tuebinger stift / Betzendoerfer, Walter – Heilbronn: E Salzer, 1922 [mf ed 1991] – 138p – 1 – (incl bibl ref) – mf#7487 – us Wisconsin U Libr [920]

Hoelemann, Hermann Gustav *see*
- Bibelstudien, 1. abtheilung
- Letzte bibelstudien
- Neue bibelstudien
- Die reden des satan in der heiligen schrift

Hoeller, Franz *see*
- Boehmisches wanderbuch
- Herz in boehmen

Die hoellische trinitaet : roman aus den jahren der vollendung des meisters mathis nithart, der faelschlich matthias gruenewald genannt wurde / Weismantel, Leo – Muenchen: K Alber, 1943 – 1 – (incl bibl ref) – us Wisconsin U Libr [830]

Hoellrigl, Franz *see* Die freifrauen

Hoelscher, Eberhard *see* Leben, meynungen und thaten von hieronimus jobs dem kandidaten, und wie er sich weiland viel ruhm erwarb auch endlich als nachtswaechter zu sulzburg starb

Hoelscher, Gustav *see*
- Die geschichte der juden in palaestina seit dem jahre 70 nach chr
- Kanonisch und apokryph
- Palaestina in der persischen und hellenistischen zeit
- Die profeten
- Die quellen des josephus fuer die zeit vom exil bis zum juedischen kriege
- Der sadduzaeismus

Hoelscher, U *see* Das grabdenkmal des koenigs chephren

Hoeltje, Georg *see* Zeitliche und begriffliche abgrenzung der spaetgotik innerhalb der architektur von deutschland, frankreich, und england

Hoen, K H *see*
- Ein christlicher bericht vo dem brot vnd weyn desz herren
- Von dem brot vnd weyn des herren

Hoenes, Christian *see* Ludwig uhland

Hoenig, Johannes *see* Ferdinand gregorovius als dichter

Hoeniger, R *see* Koelner schreinsurkunden des 12. jahrhundert (pgrg1)

Hoenigsberg, Julio *see* Ante la pena de muerte

Hoenigswald, Henry M *see* Spoken hindustani

Hoenigswald, Richard *see* Ernst haeckel, der monistische philosoph

Hoenne-zeitung – Balve DE, 8 jan 1955-56; 6 jan 1967-82 – 14r – 1 – (filmed by misc inst: jan 5 1957-jun 28 1958; jul 5 1958-60) – gw Mikrofilm; gw Misc Inst [074]

Hoennicke, Gustav *see*
- Die chronologie des lebens des apostels paulus
- Das judenchristentum im ersten und zweiten jahrhundert
- Die neutestamentliche weissagung vom ende

Hoens, Dirk Jan *see* Santi

Hoensbroech, Paul *see* Der syllabus

Hoensbroech, Paul, Graf von *see* Das papstthum in seiner sozial-kulturellen wirksamkeit

Hoepfl, Hildebrand *see* Kardinal wilhelm sirlets annotationen zum neuen testament

Hoepfner, Joh Georg Albrecht *see* Magazin fuer die naturkunde helvetiens

Hoer mit mir – neue funkstunde *see* Ostdeutsche illustrierte funkstunde

Hoer zu – Hamburg DE, 1946 15 dec-2003 3 jan – 287r – 1 – gw Mikrofilm [790]

Hoerder kreiszeitung – Dortmund DE, apr 14 1893-mar 31 1894, jan 5 1899-oct 16 1900, 1901 – 2r – 1 – (with gaps) – gw Misc Inst [074]

Hoerder tageblatt – Dortmund DE, 1902 1 jan-jun 27 – 1r – 1 – gw Misc Inst [074]

Hoerder volksblatt *see* Anzeiger und wochenblatt fuer hoerde, schwerte, aplerbeck, westhofen und umgebung

Hoeren und fernsehen *see*
- Westfunk

Hoeren und sehen *see* Westfunk

Hoerner, Herbert von *see*
- Der grosse baum
- Die letzte kugel
- Die welle

Hoernerklang der fruehe / Zerkaulen, Heinrich – 3. Aufl. Leipzig: E Huyke, [1943?] – 1 – us Wisconsin U Libr [830]

Hoerning, Reinhart *see* British museum karaite mss

Hoernle, August Friedrich Rudolf *see*
- Manuscript remains of buddhist literature found in eastern turkestan
- Uvasagadasao

Hoerschwellenbestimmungen bei neugeborenen risikokindern : eine prospektive verlaufsstudie / Hartmann, Sabine – (mf ed 1993) – 1mf – 9 – €30.00 – 3-89349-800-1 – mf#DHS 800 – gw Frankfurter [618]

Hoerst du nicht den eisenschritt : zeitgedichte / Claudius, Hermann – 3. aufl. Hamburg: A Janssen, 1915, c1914 [mf ed 1989] – 55p – 1 – mf#7152 – us Wisconsin U Libr [810]

Hoeser, Conrad *see* Briefwechsel zwischen joseph victor von scheffel und paul heyse

Hoestermann, Emilie *see* Beitraege zur technik in hebbels tagebuch

Hoet, G *see* Catalogus of naamlijst van schilderijen...

Hoetink, H *see* Patroon van de oude curacaose samenleving

Hoettges, Valerie *see* Die sage vom riesenspielzeug

Hoetzl, Petrus *see* Ist doellinger haeretiker?

Hoetzsch, Otto *see* Russland; eine einfuehrung auf grund seiner geschichte vom japanischen bis zum weltkrieg

Hoevell, Walter R van *see* Aus dem indischen leben

Hoeynck, F A *see* Geschichte der kirchlichen liturgie des bisthums augsburg

Hoeys dublin mercury *see* Dublin mercury

Der hof am brink : erzaehlung aus dem dreissigjaehrigen kriege / Strauss und Torney, Lulu von – Jena: E Diederichs 1935 [mf ed 1991] – 1r – 1 – (filmed with: auf dem schlachtfelde von custozza / william spindler) – mf#2905p – us Wisconsin U Libr [830]

Der hof des patrizierhauses : und andere erzaehlungen / Ehrler, Hans Heinrich – Stuttgart: Strecker & Schroeder 1919, c1918 [mf ed 1989] – 1r – 1 – (filmed with: fruehlings-lieder) – mf#7210 – us Wisconsin U Libr [830]

Hofberg, Herman *see* Swedish folk-lore

Hofe, Gerhard vom *see* Das elend des polyphem

Hofer anzeiger *see* Frankenpost [main edition]

Hofer, Hans *see* Weltanschauungen in vergangenheit und gegenwart

Hofer, Klara *see*
- Fruehling eines deutschen menschen
- Goethes ehe

Hofer neueste nachrichten – Hof DE, 1871 22 sep-1876 – 6r – 1 – gw Misc Inst [074]

Hofer ns-zeitung *see* Fraenkisches volk

Hofer, Paul J *see*
- The consequences of mandatory minimum prison terms
- Home confinement

Hofer post – Hof DE, 1878 3 dec-1886 25 jun, 1886 1 sep-1906 30 jun – 1 – (title varies: 1 sep 1886: hofer tageblatt) – gw Misc Inst [074]

Hofer presse – Hof DE, 1949 25 jun-1950 30 nov – 4r – 1 – gw Misc Inst [074]

Hofer tageblatt *see*
- Fraenkisches volk
- Hofer post

Hofer volksblatt – Hof DE, 1946 22 feb?-1971 30 mar – 60r – 1 – (title varies: 1 dec 1893: Oberfraenkische Volkszeitung. filmed by misc inst: 1893 22 jun-1894 28 feb, 1894 2 apr-1901 27 sep, 1903-1933 18 mar) – gw Mikropress; gw Misc Inst [074]

Hofer, Walther *see* Nationalsozialismus

Der hoffaertige schuster knobel : gestalten aus wildengrund / Sturm, Stefan – Karlsbad: A Kraft 1944 [mf ed 1991] – 1r – 1 – mf#2907p – us Wisconsin U Libr [830]

Hoffart, Elizabeth *see* Herders "gott"

Hoffart und demut : erzaehlung aus der zeit maria theresias / Hoecker, Gustav – Stuttgart: Gebrueder Kroener, [18–?] – 1 – us Wisconsin U Libr [830]

Hoffer, A *see* Nederduytsche poemata

Hoffert, Troy A *see* Peak torque reliability of biodex b-2000 isokinetic dynamometer during concentric loading of back flexors and extensors
Hoffman, C *see* Notes on the polish settlement at korona, flagler
Hoffman, Carl von *see* Jungle gods
Hoffman, David *see*
– An address to students of law in the united states
– A lecture; being the third of a series of lectures, introductory to a course of lectures now delivering in the university of maryland
– Magazin fuer die wissenschaft des judentums
– Views on the formation of a british and american land and emigration company
Hoffman, Diana M *see* The effects of visual imagery ability combined with visual mental practice techniques upon motor performance
Hoffman, E *see* Das konverseninstitut des cisterzienserordens
Hoffman, Frank Sargent *see* The sphere of religion
Hoffman, Frederick L *see* Malaria in florida, georgia and alabama
Hoffman, Frederick Lewis *see* The ohio supreme court reports reduced to questions and answers.
Hoffman, Jeffery D *see* Sport-confidence and preceptions of coaching behavior of male and female high scholl basketball players
Hoffman, John N *see* The broken platform
Hoffman, Laura A *see* Walking gait during pregnancy
Hoffman, Marilyn A *see* Comparison of two instruments for needs assessment and evaluation of an employee health promotion program
Hoffman, Mark *see* Study of german theoretical treatises of the nineteenth century
Hoffman, Mark A *see* Sensorimotor evaluation of post-operative anterior cruciate ligament reconstruction patients
Hoffman, Murray *see*
– Ecclesiastical law in the state of new york
– A treatise on the law of the protestant episcopal church in the united states
Hoffman, Ogden *see*
– Hoffman's reports of cases in the district court for northern california, 1853-1858
– Report of the land cases, june 1853-june 1858
Hoffman, Paul Everett *see* Background and development of pedro menendez's contribution
Hoffman, Publius V *see* Indiana addendum to green's pleading and practice
Hoffmann, Andreas Gottlieb *see*
– Grammatica syriaca
– The principles of syriac grammar
Hoffmann, C *see* Aussichten fuer die evangelische kirche deutschlands
Hoffmann, D *see*
– Antwort auff d christophori pezelii predigers zu bremen falsch gebrauchte gruende
– Apologia danielis hofmanni
– Avrea et vere theologica commentatio d danielis hofmanni
– Christliche predigt aus dem 48 capittel esaiae von den grossen wolthaten gottes
– Epistolae reverendi, danielis hoffmanni de libro concordiae
– Homiliarvm evangelicarvm notae breves
– Orthodoxa de exorcismo a qvibvsdam ecclesiis avgvstanam confessionem amplectentibus in baptismi administratione
– Von der moderation und messigung herrn phillipi melanthonis in dem betruebtem langwirigem streite vom heiligen abendmahl
Hoffmann, David *see*
– Der schulchan-uruch und die rabbinen ueber das verhaeltniss der juden zu andersglaeubigen
– Zur einleitung in die halachischen midraschim
Hoffmann, E T A [Ernst Theodor Amadeus] *see*
– Das fraeulein von scuderi
– Der goldne topf
– Lebens-ansichten des katers murr
– Meister martin der kuefner und seine gesellen
– Der sandmann
Hoffmann, Francois Benoit *see* Roman d'une heure, ou, la folle gageure
Hoffmann, Franz *see* Das papstthum im widerspruch mit vernunft, moral und christentum
Hoffmann, Franz et al *see* Franz von baader als begruender der philosophie der zukunf
Hoffmann, G F *see*
– Darstellung und kritik der von herder gegebenen ergaenzung und fortbildung der ansichten lessings in seinem laokoon
– Grundlagen, stile, gestalten der deutschen literatur
Hoffmann, Georg *see* Ueber einige phoenizische inschriften
Hoffmann, H *see* Gottesfriede und treuga dei (mgh schriften:20.bd)
Hoffmann, Heinrich *see*
– Die apostelgeschichte s lucae
– Hitler was my friend
– Die mondzuegler
– Sagen, maerchen und schwaenke des juelicher landes
– Die theologie semlers
– Ueber die beteuerungen in shakespeare's dramen
Hoffmann, Heinrich Anton *see* 3 quatuors pour deux violons, alto & violoncelle, oeuvre 7
Hoffmann, J *see* Le quodlibet 15 et trois questions ordinaires de godefroid de fontaines
Hoffmann, Karl F *see*
– Deutschland und seine bewohner
– Die erde und ihre bewohner
Hoffmann le fantastique / Mistler, Jean – Paris: Albin Michel, c1950 – 1 – (incl bibl ref) – us Wisconsin U Libr [430]
Hoffmann, Leopold Alois *see*
– Wiener zeitschrift
Hoffmann, Ludwig *see* Hercynia
Hoffmann, M *see* Der dialog bei den christlichen schriftstellern der ersten vier jahrhunderte
Hoffmann, Michael *see* Stil und stilzug im kommunikativen kontext
Hoffmann, Otto *see* Herder's briefwechsel mit nicolai
Hoffmann, Paul Theodor [comp] *see* Blut und rasse im deutschen dichter- und denkertum
Hoffmann, Professor Dr *see* Erlaeuterungen zu goethes egmont fuer schule und haus
Hoffmann, Rene *see* La notion de l'etre supreme chez les peuples non civilises
Hoffmann, Rudolf *see* Rotes metall
Hoffmann von fallersleben als vorkaempfer und erforscher der niederlaendisch-vlaemischen literatur / Berneisen, Ewald – Muenster i.W: Der Westfale, 1914 [mf ed 1995] – 102p – 1 – (incl bibl ref) – mf#7480 – us Wisconsin U Libr [430]
Hoffmann von Fallersleben, August Heinrich *see*
– Allemannische lieder
– Deutsche gassenlieder
– Deutsche salonlieder
– Fraenzchens lieder
– Spitzkugeln
– Streiflichter
Hoffmann von fallersleben und johanna kapp : begegnung in heidelberg / Derwein, Herbert – 2. aufl. [Fallersleben?]: Hoffmann von Fallersleben-Gesellschaft, [1956?] [mf ed 1995] – 36p (ill) – 1 – (incl bibl ref) – mf#8906 – us Wisconsin U Libr [920]
Hoffmann von fallersleben und sein deutsches vaterland / Gerstenberg, Heinrich – Berlin: F Fontane, 1890 [mf ed 1991] – 82p – 1 – (incl selected verse) – mf#7480 – us Wisconsin U Libr [430]
Hoffmann von fallerslebens "texanische lieder" / Goebel, Julius – S.l: s.n, 1919? [mf ed 1990 – 39p – 1 – mf#7480 – us Wisconsin U Libr [780]
Hoffmann, W *see* Franz xavier
Hoffmann, Wilhelm *see*
– Die erziehung des weiblichen geschlechts in indien und anderen heidenlaendern
– The prophecies of our lord and his apostles
Hoffmann-Harnisch, Wolfgang *see* Rio grande do sul
Hoffmann's catholic directory, almanac and clergy list – 1895, 1896 – 2r – 1 – (continued by: catholic directory, almanac and clergy list) – mf#1160156 – us WHS [241]
Hoffmann's catholic directory, almanac and clergy list – quarterly, for the year of our lord... – Milwaukee: Hoffmann Brothers, 1889, 1891-1896 – 7r – 1 – $40.00r – (includes suppls) – us Notre Dame [241]
Hoffmann's catholic directory, almanac, and clergy list – quarterly, for the year of our lord... – Milwaukee: M H Wiltzius and Co., 1897-1899 – 3r – 1 – $40.00r – (incl suppls) – us Notre Dame [241]
Hoffmannsthal, Hugo von *see* Wert und ehre deutscher sprache
Hoffman's chancery appeals reports / New York. (State) – 1v. 1839-40 (all publ) – 7mf – 9 – $10.50 – (a pre-nrs title) – mf#LLMC 80-002 – us LLMC [340]
Hoffmanns, Christiane *see* Kontinuitaet und bruch
Hoffmans, J *see*
– Le huitieme quodlibet de godefroid de fontaines
– Les quodlibet cinq, six et sept de godefroid de fontaines
– Les quodlibets onze et douze de godefroid de fontaines
Hoffman's reports of cases in the district court for northern california, 1853-1858 / Hoffman, Ogden – San Francisco: Hubert. 1v. 1862 (all publ) – (= ser Early federal nominative reports) – 7mf – 9 – $10.50 – mf#LLMC 81-455 – us LLMC [347]
Hoffmeister, Cuno *see*
– Katalog der bestimmungsgroessen fuer 611 bahnen groesser meteore
– Meteorstroeme
Hoffmeister, E von *see* Durch armenien
Hoffmeister, Franz Anton *see*
– [7e] concerto, pour la flute
– Quatuor pour le clavecin
– Un quatuor pour le clavecin ou piano forte [violon, viola et violoncelle] no 2
– Trios pour deux violons et basse, oeuvre 37
– Trois quatuors concertants pour flute, violon, alto et violoncelle, oeuvre 29
– Trois sonates pour le clavecin ou piano forte
– Trois trios progressive pour deux violons et violoncell [op 28] livre 1 & 2
Hoffmeister, Hermann *see* Der glaube unserer vaeter
Hoffmeister, Johannes *see*
– Hoelderlin und die philosophie
Hoffmeister, Karl *see*
– Schillers saemmtliche werke in zwoelf baenden
– Supplemente zu schillers werken aus seinem nachlass
Hoffmeister, Werner *see* Briefe aus indien
Hoffnung der gloeubigen / Bullinger, Heinrich – [Zuerich, Christoffel Froschouer, 1544] – 2mf – 9 – mf#PBU-140 – ne IDC [240]
Die hoffnung kuenftiger erloesung aus dem todeszustande bei den frommen des alten testaments / Klostermann, August – Gotha: Friedrich Andreas Perthes 1868 [mf ed 1989] – (= ser Untersuchungen zur alttestamentlichen theologie) – 1mf – 9 – 0-7905-0717-X – mf#1987-0717 – us ATLA [221]
Die hoffnungen der katholischen kirche in china / Hahn, Heinrich – Frankfurt a/M: G J Hamacher, 1869 [mf ed 1995] – (= ser Yale coll; Broschueren-verein 5. jahrg 3,4) – 47p – 1 – 0-524-10170-1 – (in german) – mf#1995-1170 – us ATLA [241]
Das hoffnungslose geschlecht : vier zeitgenoessische erzaehlungen / Borchardt, Rudolf – Berlin-Grunewald: Horen-Verlag, c1929 [mf ed 1989] – 375p – 1 – mf#7052 – us Wisconsin U Libr [880]
Hoffstad, Olaf Alfred *see* Norsk flora
Hoff-Wilson, Joan *see* Papers of the nixon white house
Hofgeismarer allgemeine hna *see* Hessische/ niedersaechsische allgemeine
Hofgeismarer zeitung – Hofgeismar DE, 1884-86, 1888-1944 – 64r – 1 – gw Misc Inst [074]
Hofhout, Johannes *see* Bataviasche historische, geographische, huishoudelyke en reis-almanach
Hofim / Bass, Samuel – Tel-Aviv, Israel. 1933 – 1r – us UF Libraries [939]
Die hofkapelle der deutschen koenige (mgh schriften:16.bd 1.teil) : grundlegung. die karolingische hofkapelle / Fleckenstein, J – 1959 – (= ser Monumenta germaniae historica. schriften (mgh schriften)) – €14.00 – ne Slangenburg [931]
Die hofkapelle im rahmen der ottonisch-salischen reichskirche (mgh schriften..:16.bd. 2.teil) / Fleckenstein, J – 1966 – (= ser Monumenta germaniae historica. schriften (mgh schriften)) – €15.00 – ne Slangenburg [931]
Hofland, Barbara *see* Africa described
Hofmann, Adelheid *see* Literatursoziologische untersuchungen zu der westsaechsischen stadt schneeberg an der wende vom 17. zum 18. jahrhundert
Hofmann, Bernd *see* Beitrag zur bestimmung der parameter und zur detektion der struktur von zweiphasenstroemungen mittels ultraschall
Hofmann, Carl *see* Praktisches handbuch der papier-fabrikation
Hofmann, Christian Samuel *see* Schlesische sammlung kleiner auserlesener schriften
Hofmann, Ernst *see* Die stellung der konstanzer bischoefe zu papst und kaiser waehrend des investiturstreits
Hofmann, Eva *see* Die effizienz eines individuellen intensiv-prophylaxe-programms bei koerperbehinderten patienten mit spastischer zerebralparese
Hofmann, Franz *see*
– Commentar zum oesterreichischen allgemeinen buergerlichen gesetzbuche
– Excurse ueber oesterreichisches buergerliches recht
Hofmann, Fritz *see* Koborgher quackbruennla
Hofmann, Georg *see* Predigten
Hofmann, Hans *see* Schillers flucht
Hofmann, Hans-Otto *see* Die vermoegensrechtlichen beziehungen der ehegatten nach dem rechte des bayerischen stammes in neuerer zeit
Hofmann, Johann Christian Konrad von *see*
– Biblische hermeneutik
– Encyclopaedie der theologie
– The ologische briefe der professoren delitzsch und v. hofmann
– Ueber die zukunft der theologischen fakultaeten
– Weissagung und erfuellung im alten und im neuen testamente
Hofmann, Johannes *see* Gustav freytag als politiker, journalist und mensch
Hofmann, Juergen *see* Struktur, umfeldbedingungen und motivationen bei schuelerhandballmannschaften (c, d-, e-jugend) des kreises bergstrasse
Hofmann, Karl *see* A practical treatise on the manufacture of paper in all its branches
Hofmann, Konrad *see*
– Lutwins adam und eva
Hofmann, Rudolph Hugo *see*
– Galilaea auf dem oelberg
– Die lehre von dem gewissen
– Symbolik
Hofmann, Thomas *see* Kindgerechte bestimmung der transepithelialen potentialdifferenz am respiratorischen epithel der nase
Hofmann von Wellenhof, Paul *see* Michael denis
Hofmann, Wilhelm *see* Der richard wagner-taumel
Hofmanns und ritschls lehren ueber die heilsbedeutung des todes jesu / Steffen, Bernhard – Guetersloh: C Bertelsmann, 1910 [mf ed 1991] – (= ser Beitraege zur foerderung christlicher theologie 14/5) – 1mf – 9 – 0-524-00148-0 – (incl bibl ref) – mf#1989-2848 – us ATLA [242]
Hofmannsthal, Hugo von *see*
– Ariadne auf naxos
– Deutsche epigramme
– Deutsche erzaehler
– Elektra
– Das gespraech ueber gedicht
– Gestern
– Die hochzeit der sobeide
– Das schrifttum als geistiger raum der nation
– Der tod des tizian
– Der weisse faecher
Hofmeister, A *see*
– Chronica mathiae de nuwenburg
– Handbuch der musikalischen literatur
Hofmeister, William F B *see* On the germination, development and fructification of the higher cryptogamia and on the fructification of the coniferae
Hofmevr / Paton, Alan – London, England. 1964 – 1r – us UF Libraries [960]
Hofmiller, Josef *see* Wege zu goethe
Hofmiller, Josef [comp] *see* Altbayerische sagen
Di hofnung – n1-41. 15 sep-10 nov 1907 – 1 – us NY Public [074]
Die hofschranzen des dichterfuersten : der goethecult und dessen tempeldiener... / Brunner, Sebastian – Wuerzburg, Wien: L Woerl 1889 [mf ed 1990] – 1r – 1 – (incl bibl ref; filmed with: goethe in vertraulichen briefen seiner zeitgenossen / wilhelm bode [comp]) – mf#2795p – us Wisconsin U Libr [430]
Hofstadt, Ulrich H E *see* Die schmerzzeichnung ("patient pain drawing") als screening-methode fuer lumbago-ischialgie-syndrom-patienten
Hofstede de Groot, C P *see* De 27ste october 1553
Hofstede de Groot, Petrus *see*
– Basildes am ausgange des apostolischen zeitalters
– Die groeninger theologen
Hofstede, Petrus *see* Brief aan den hooggeleerden heer..
Hofsteede, W M F *see* Decision-making processes in four west javanese villages diss
Hofstra environmental law digest – v1-10. 1984-96 – 9 – $80.00 set – ISSN: 0882-6765 – mf#109791 – us Hein [344]
Hofstra labor and employment law journal – v1-18. 1983-2001 – 9 – $268.00 set – (title varies: v1 1983 as hofstra labor law forum, v2-14 1984-97 as hofstra labor law journal) – ISSN: 1052-3332 – mf#109721 – us Hein [344]
Hofstra labor law forum *see* Hofstra labor and employment law journal
Hofstra labor law journal *see* Hofstra labor and employment law journal
Hofstra law and policy symposium – v1-3. 1996-98 – 9 – $37.00 set – mf#117341 – us Hein [340]
Hofstra law review – v1-29. 1973-2001 – 9 – $756.00 set – ISSN: 0091-4029 – mf#103121 – us Hein [340]
Hofstra property law journal – v1-6. 1988-93 (all publ) – 9 – $69.00 set – (cont: international property investment journal. ceased with v6 n1) – ISSN: 1050-2076 – mf#111671 – us Hein [346]
Hog farm management – Minnetonka. 1971-1992 (1) 1964-1992 (5) 1977-1992 (9) – ISSN: 0018-3180 – mf#6190 – us UMI ProQuest [636]
Hog, Roger *see* Tour on the continent in france, switzerland, and italy
Hogan, Benedict *see* Plain talk
Hogan, James *see* Music of the sanctuary
Hogan, John Baptist *see* Clerical studies
Hogan, John Sheridan *see* Canada
Hogan, Michael J *see* Baie des chaleurs railway co
Hogan, William *see*
– Auricular confession and popish nunneries
– Popery
Hoganaes tidning – Hoeganaes, 1889-1924 – 9 – (previous title: engelholms tidning) – sw Kungliga [078]
Hoganas tidning *see* Nordvastra skanes tidningar engelholms tidning
Hoganas tidning oresundsposten *see*
– Nordvastra skanes tidningar engelholms tidning
– Nordvastra skanes tidningar engelholms tidning och klippans tidning
Hoganasposten – Aengelholm, 1899-1901 – 3r – 1 – sw Kungliga [078]
El hogar cristiano – 1957 – 212p – 1 – us Southern Baptist [240]
Hogar Extremeno de Zaragoza *see* Fiestas patronales septiembre-octubre, 1977
Hogar Extremeno en Paris. Paris *see* Memoria, ano 1976
Hogarth, David George *see*
– Accidents of an antiquary's life
– The ancient east
– Authority and archaeology, sacred and profane
– The penetration of arabia

Hogarth, W *see* The analysis of beauty...
Hogarth's works : with life and anecdotal descriptions of his pictures / Ireland, John et al – London [1874] – (= ser 19th c art & architecture) – 15mf – 9 – mf#4.2.1045 – uk Chadwyck [700]
Hogben, Thomas *see*
– God's plan for soul-winning
– My witnesses
Hogberg, L E *see* Ett och annat fran kinesiska turkestan (vaestra kina)
Hoge, Dean R *see* Division in the protestant house
Hoge, J *see* Die geskiedenis van die lutherse kerk aan die kaap
Hoge, Peyton Harrison *see* Moses drury hoge
Hogendorp, Carel S van *see* Coup d'oeil sur l'ile de java et les autres possessions neerlandaises dans l'archipel des indes
Hogg, Alfred George *see*
– Christ's message of the kingdom
– Karma and redemption
Hogg, Bessie et al *see* In the king's service
Hogg, Charles Edgar *see*
– Equity procedure
– Pleading and forms...now in use in the state of west virginia
Hogg, R *see* Appeal to the christian public on the evils of theatrical amusement
Hogg, Robert *see*
– Death of nadab and abihu
– Principles on which a church constitution is founded directly oppos...
Hogg, Thomas *see* Two hundred and nine days
Hogg, Wilson Thomas *see*
– Christian science unmasked
– History of the free methodist church of north america
– A symposium on scriptural holiness
Hoglandet – Jonkoping, Sweden. 1981-86 – 1 – sw Kungliga [078]
Hoglund, A William *see* The immigrant in america
Hogue, L Lynn *see* Eight charges delivered at so many several general sessions...
Hogueras de cal : poemas / Bauza, Obdulio – San Juan, Puerto Rico. 1947 – 1r – us UF Libraries [972]
Der hohe befehl : opfergang und bekenntnis des werner voss [a novel] / Welk, Ehm – Berlin: Im Deutschen Verlag 1939 [mf ed 1991] – 1r – 1 – (filmed with: die lebensuhr des gottlieb grambauer & other titles) – mf#2983p – us Wisconsin U Libr [830]
Das hohe lied : die klaglieder / Hitzig, Ferdinand – Leipzig:S Hirzel, 1855 – 1mf – 9 – 0-8370-3568-6 – (incl bibl ref) – mf#1985-1568 – us ATLA [780]
Das hohe lied salomonis / Gessner, Theodor – Quakenbrueck: Rackhorst, 1881 – 1mf – 9 – 0-7905-0186-4 – mf#1987-0186 – us ATLA [221]
Die hohe warte : deutschland-dichtung 1933-1945 / Becher, Johannes Robert – Berlin: Aufbau-Verlag, 1946 [mf ed 1989] – 196p – 1 – mf#6994 – us Wisconsin U Libr [810]
Das hohelied : auf grund arabischer und anderer parallelen / Jacob, Georg – Berlin: Mayer & Mueller, 1902 – 1mf – 9 – 0-8370-3744-1 – (incl bibl ref) – mf#1985-1744 – us ATLA [221]
Das hohelied : aus dem hebraeischen originaltext ins deutsche uebertragen: wie auch sprachlich und sachlich erlaeutert und mit einer umfassenden einleitung / Kaempf, Saul Isaac – 3. neurev verm ausg. Prag: Heinr. Mercy, 1884 – 1mf – 9 – 0-8370-3827-8 – (incl ind) – mf#1985-1827 – us ATLA [221]
Das hohelied / Delitzsch, Franz – Leipzig: Doerffling und Franke, 1851 – 1mf – 9 – 0-8370-9375-9 – mf#1986-3375 – us ATLA [221]
Das hohelied / Feilchenfeld, W – Breslau: Wilhelm Koebner, 1893 – 1mf – 9 – 0-8370-3107-9 – mf#1985-1107 – us ATLA [221]
Das hohelied – Freiburg im Breisgau, St Louis MO: Herder 1908 [mf ed 1989] – (= ser Biblische studien 13/4) – 1mf – 9 – 0-7905-2414-7 – mf#1987-2414 – us ATLA [221]
Das hohelied : kritisch und metrisch untersucht / Zapletal, Vincenz – Freiburg (Schweiz): Universitaets-Buchh, 1907 – 1mf – 9 – 0-524-05660-9 – (incl bibl ref) – mf#1992-0510 – us ATLA [220]
Das hohelied : kurz erklaert / Harms, Theodor – Hermannsburg: Missionshausdruckerei, 1870 – 1mf – 9 – 0-8370-3476-0 – mf#1985-1476 – us ATLA [221]
Hohelied aus dem hebraischen originaltext in's deutsche ubertragen / Kaempf, Saul Isaac – Prag, Czechoslovakia. 1877 – 1r – us UF Libraries [939]
Das hohelied neu uebersetzt und aesthetisch-sittlich beurteilt / Thilo, Martin – Bonn: A Marcus & E Webers 1921 [mf ed 1979] – 1r – 1 – mf#mflm1888 – us Harvard [221]
Das hohelied salomos : eine biblische weissagung auf das moderne babel / Jaeger, Adolf – Berlin: Hermann Walther, 1903 – 1mf – 9 – 0-8370-3755-7 – mf#1985-1755 – us ATLA [221]
Das hohelied salomo's bei den juedischen erklaerern des mittelalters : nebst einem anhange, erklaerungsproben aus handschriften / Salfeld, Siegmund – Berlin: Julius Benzian, 1879 – 1mf – 9 – 0-8370-5026-X – (incl bibl ref, ind of aut) – mf#1985-3026 – us ATLA [221]
Hohenberg, F *see* Civitates orbis terrarvm
Hohenfellner, R Markus *see* Experimentelle und klinische untersuchungen zur therapie von neurogenen blasenfunktionsstoerungen
Hohenhausen, Elise F von *see* Natur, kunst und leben
Hohenlohe, Philipp *see* Geschichte des ennsthales
Hohenloher tagblatt – Gerabronn DE, 1983- – ca 8r/yr – 1 – (bezirksausgabe von suedwest-presse, ulm; filmed by other misc inst: 1980-1983 31 may [23r]) – gw Misc Inst [074]
Hohenloher zeitung – Kuenzelsau DE, 1948 3 jan-1952 – 9r – 1 – (bezirksausgabe von heilbronner stimme, heilbronn; filmed by misc inst: 1968-) – gw Mikrofilm; gw Misc Inst [074]
Hohenloher zeitung – Oehringen DE, 1968- – ca 9r/yr – 1 – (Bezirksausgabe von heilbronner stimme; since 1 jul 1974 regional ed of: heilbronner stimme) – gw Misc Inst [074]
Hohenloher zeitung *see* Heilbronner stimme
Hohenlohe-Waldenburg-Schillingsfuerst, Alexander, Fuerst von *see* Lichtblicke und erlebnisse aus der welt und dem priesterleben
Hohensalzaer zeitung – Hohensalza (Inowroclaw PL), 1940-1943 jun, 1943 oct-dec – 1 – gw Misc Inst [077]
Hohenschoenhausener lokalblatt – Berlin DE, 1937-40 – 4r – 1 – gw Misc Inst [074]
Hohenstatt, Otto *see* Hauffs maerchen
Hohensteiner tageblatt *see* Wochenblatt und anzeiger fuer hohenstein, ernstthal und umgegend
Hohenstein-ernstthaler anzeiger *see* Wochenblatt und anzeiger fuer ernstthal, hohenstein und oberlungwitz
Hohenstein-ernstthaler tageblatt und anzeiger *see* Wochenblatt und anzeiger fuer hohenstein, ernstthal und umgegend
Hohenzollerische volkszeitung – Sigmaringen DE, 1975 2 jan-1983 – 55r – 1 – (title varies: 2 jan 1934: verbo; 1 apr 1942: donau-bodensee-zeitung; 18 dec 1945: schwaebische zeitung; main ed in leutkirch. filmed by misc inst: 1977- [ca 6r/yr]) – gw Misc Inst [074]
Hohenzollerischer Geschichtverein *see* Mitteilungen
Hohenzollerischer landesbote *see* Schwarzwaelder bote [main edition]
Hohlbaum, Robert *see*
– Balladen vom geist
– Front der herzen
– Fruehlingssturm / charfreitag / der gang nach emmaus / pfingsten in weimar
– Der fruehlingswalzer
– Getrennt marschieren
– Grillparzer
– Helles abendlied
– Mein leben
– Stunde der sterne
– Symphonie in drei saetzen
– Unsterbliche
– Winterbrautnacht
Hohlenberg, Johannes *see* Goethes faust im zwanzigsten jahrhundert
Hohlenberg, Matthias Haquinus *see* De originibus et fatis ecclesiae christianae in india orientali
Hohler, Thomas Beaumont *see* Diplomatic petrel
Hohlfeld, A R *see*
– The goethe centenary at the university of wisconsin
– Zum irdischen ausgang von goethes faustdichtung
Hohlfeld, Alexander Rudolf *see* Zur textgestaltung der neueren faustausgaben
Hohlfeld, Paul *see* Zur religionsphilosophie und speculativen theologie
Hohn, Hermann *see* Vocations
Der hohnsteinsche erzaehler – Stolberg/Harz DE, 1816-1817 apr – 1r – 1 – (title varies: 9 jan 1817: hohnsteinsche interimsblaetter) – gw Misc Inst [074]
Hohnsteinsche interimsblaetter *see* Der hohnsteinsche erzaehler
Hohoff, Kaplan *see* Christenthum und socialismus
Hoi ky ve gia dinh nguyen tuong : nhat linh, hoang dao, thach lam / Nguyen, Thi The, 1909 – Saigon: Song 1974 [mf ed 1992] – on pt of 1r – 1 – mf#11052 r354 n2 – us Cornell [959]
Hoinkes, Carl *see* Christian und die kataloge
Hoja parroquial de coria-caceres – Caceres, 1951-1978 – 5 – (suplemento al boletin oficial del obispado) – sp Bibl Santa Ana [073]
Hojarasca / Garcia Marquez, Gabriel – Bogota, Colombia. 1960? – 1r – us UF Libraries [972]
Hojas de arbol caidas / Cordero, Juan Luis – Caceres: Tip. El Noticiero, 1954 – sp Bibl Santa Ana [946]
Hojskolebladet – Kolding, Denmark. 1942-may 1945 – 2r – 1 – uk British Libr Newspaper [074]
Hokah chief – Houston MN, La Crosse WI. 1939 jan 5-1942 dec 31 – 1r – 1 – (cont: houston county chief; cont by: la cresent times-hokah chief) – mf#772574 – us WHS [071]
Hokhmat shelomah – Warsaw, Poland. 645, 1885 – 1r – us UF Libraries [221]
Hoki bunrui taizen : collection of laws, regulations and ordinances of japan from meiji restoration to the meiji constitution (1868-1899) / Japan Cabinet. Board of Documents – Tokyo, 1889-94: 1st ser 69v; 2nd ser 16v – 52,380p on 30r – 1 – Y324,000 – (with 12p guide. in japanese) – ja Yushodo [348]
Hokianga star – 1935-feb 1937 – 2r – 1 – mf#12.23 – nz Nat Libr [079]
Hokitika guardian – jan 1917-aug 1926; may-oct 1928; jul-oct 1929; may-jun 1930; mar 1931-dec 1932; mar 1933-feb 1935; jul-aug 1935; jan 1936-dec 1940; may-jun 1973; may-jun 1975; sep 1975-dec 1985; may-aug1986; jan-dec 1987 – 1 – mf#60.3 – nz Nat Libr [079]
Hola, office of latin affairs / Indiana University, Bloomington – 1981/1982, iss 1-1986/1987, iss 7 – 1r – 1 – (cont: newsletter [indiana university, bloomigton. office of latino affairs]) – mf#1353199 – us WHS [378]
Holaind, Rene I *see* Natural law and legal practice
Holaind, Rene Isidore *see* The parent first
Holand, Patricia *see* The papers of elizabeth cady stanton and susan b. anthony
Holanda, Sergio Buarque De *see*
– Cobra de vidro
– Moncoes
– Raizes do brasil
Holandeses no brasil – Recife, Brazil. 1968 – 1r – us UF Libraries [972]
Holandeses no rio grande / Heroncio, Paulo – Rio de Janeiro, Brazil. 1937 – 1r – us UF Libraries [972]
Holas, Bohumil *see*
– Homme noir d'afrique
– Masques kono (haute-guinee francaise)
Holbach, Paul Henri Thiry, Baron d' *see* Superstition in all ages
[Holbein, H, the Younger] *see* Icones historiarum veteris testamenti...
Holbert herald – v1 n1-v5 n1 [1981 mar-1985 mar] – 1r – 1 – (continued by: holbert press) – mf#1046835 – us WHS [071]
Holbert press – v5 n3-4 [1985 sep-dec], v6 n1-3/4 [1986 mar-sep/dec], v7 n1-3 [1987 mar-sep], v8 n1 [1988 mar] – 1r – 1 – (cont: holbert herald) – mf#1712348 – us WHS [071]
Holborn guardian and bloomsbury chronicle – Holborn UK, 1892; 1951; 1986-15 dec 1988; 5 jan-16 mar 1989 – 8 1/4r – 1 – (aka: holborn and finsbury guardian; finsbury and holborn guardian; camden and holborn and finsbury guardian; holborn and finsbury guardian and westminster chronicle; holborn and city guardian) – uk British Libr Newspaper [072]
Holborn journal – London UK, 27 feb-18 sep 1858, 8 jan 1859-73 – 7 1/2r – 1 – (aka: holborn and bloomsbury journal; holborn s(ain)t pancras and bloomsbury journal) – uk British Libr Newspaper [072]
Holborn s(ain)t pancras and bloomsbury journal *see* Holborn journal
Holborow, Arthur *see* Evolution and scripture
Holbrook 1872-1900 – Oxford, MA (mf ed 1994) – (= ser Massachusetts vital records) – 20mf – 9 – 0-87623-188-1 – (mf 1: death index 1872-1900+ a-g. mf 2: death index 1872-1900+ g-o. mf death index 1872-1900+ o-z. mf 4: marriage index 1872-1900+ a-c. mf 5: marriage index 1872-1900+ d-h. mf 6: marriage index 1872-1900+ h-m. mf 7: marriage index 1872-1900+ m-s. mf 8: marriage index 1872-1900+ s-z. mf 9: birth index 1872-1900+ a-d. mf 10: birth index 1872-1900+ d-j. mf 11: birth index 1872-1900+ j-o. mf 12: birth index 1872-1900+ p-w. mf 13: birth index 1872-1900+ w-z. mf 14: births 1872-81. mf 15: births 1881-90. mf 16: births 1890-1900. mf 17: marriages 1872-90. mf 18: marriages 1890-1900. mf 19: deaths 1872-95. mf 20: deaths 1895-1904) – us Archive [978]
Holbrook collection – (refer to: massachusetts vital record transcripts to 1850 and: massachusetts vital records) – us Archive [978]
Holbrook courier – Holbrook, jan 1915-dec 1939 (misc iss) – 2r – A$131.08 vesicular A$142.08 silver – at Pascoe [079]
Holbrook, Edwin M *see* Manual of laws relating to private claims against the state of new york..
Holbrook herald – Holbrook, NE: Herald Print Co. -v4 n13. nov 6 1896 (wkly) [mf ed 1895-96 (gaps) filmed 1980] – 1r – 1 – (cont by: herald) – us NE Hist [071]
Holbrook herald – Holbrook, NE: Herald Print Co. v5 n22. jan 7 1898- (wkly) [mf ed -1899 (gaps) filmed 1980] – 1r – 1 – (cont: herald) – us NE Hist [071]
Holbrook, Jay Mack *see*
– Bibliography of massachusetts vital records 1620-1905
– Family structure in 17th-century windsor, connecticut
– Topical index to mayflower descendant 1620-1937
Holbrook, John Calvin *see* Our country's crisis
Holbrook, Jos. P *see* Worship in song
Holbrook, Martin Luther *see* How to strengthen the memory; or, natural and scientific methods of never forgetting
Holbrook observer – Holbrook, NE: Fred C Ayres. -v94 n37. jan 27 1994 (wkly) [mf ed apr 30 1908-jan 27 1994 (gaps)] – 35r – 1 – (absorbed by: arapahoe public mirror. publ in: arapahoe and holbrook jun 17 1948-july 5 1956 and: arapahoe jul 5 1962-jan 27 1994. issues for apr-aug 1979 and 1980 accompanied by a separately numbered suppl: laker (elmwood ne). issues for jan 7 1938-dec 28 1939 incl the initial pp of: public mirror (arapahoe ne)) – us NE Hist [071]
Holcad – New Wilmington, PA. 1914-1990 (1) – mf#68569 – us UMI ProQuest [071]
Holcomb, Bret E *see* The virtual athletic training room
Holcomb Family *see* Letters
Holcomb, Helen Harriet *see* Men of might in india missions
Holcomb, Helen Harriet Howe *see* In the heart of india, or, beginnings of missionary work in bundela land
Holcomb, James Foote *see* In the heart of india, or, beginnings of missionary work in bundela land
Holcomb, T L *see* Papers while executive secretary of the sunday school board, june 1935-may 1945
Holcombe, Arthur Norman *see* Chi hua ti min chu cheng chih
Holcombe, Chester *see* The real chinaman
Holcombe, Hosea *see*
– History of the baptists in alabama
– A history of the rise and progress of the baptists in alabama
Holcombe journal – Holcombe WI. 1906 feb 2-1907 sep 28; 1907 feb 2-1908 aug 22 – 2r – 1 – mf#923051 – us WHS [071]
Holcombe, Robert A *see* The effects of auditory biofeedback on the accuracy of the tennis volley
Holcombe, Theodore Isaac *see* An apostle of the wilderness
Hold back the dawn / Frings, Ketti – New York, NY. 1941, 1940 – 1r – us UF Libraries [025]
Hold that fast which thou hast / Shaw, George – Dublin, Ireland. 1848 – 1r – us UF Libraries [240]
Hold that line : powerline protest newsletter / Powerline Protest [Lowry MN] – 1979 jan 15/20-1986 aug – 1r – 1 – (cont: newsletter [powerline protest [lowry, mi]) – mf#1223779 – us WHS [320]
Holdcraft, Lola *see* Sidelines and reflections of paul holdcraft 1891-1971
Holden 1739-1849 – Oxford, MA (mf ed 1996) – (= ser Massachusetts vital record transcripts to 1850) – 14mf – 9 – 0-87623-267-5 – (mf 1t-5t: births 1739-1844. mf 1t-3t: marriages & deaths 1742-94. mf 2t-3t: intentions 1782-92. mf 5t-6t: publishments 1840-49. mf 6t-7t: deaths 1771-1844. mf 7t-10t: publishments 1803-40. mf 10t-11t: marriages 1792-1845. mf 11t-12t: births 1843-49. mf 12t-13t: marriages 1844-49. mf 13t: deaths 1844-49; out-of-town marriages 1743-99. mf 13t-14t: births a-w 1720-1847. mf 14t: vital records a-w 1742-1852) – us Archive [978]
Holden, Edith *see* Blyden of liberia
Holden, George Frederick *see* The holy ghost the comforter
Holden, Harrington William *see* John wesley in company with high churchmen
Holden, Henry *see* Re-opening of cranoe church, leicestershire
Holden, Hubert Ashton *see* M minucii felicis octavius
Holden, Oliver *see*
– American harmony
– Sacred dirges, hymns, and anthems, commemorative of the death of general george washington, the guardian of his country and the friend of man
Holden, William Clifford *see*
– British rule in south africa
– History of the colony of natal, south africa
– The past and future of the kaffir races
– Past and future of the kaffir races
Holder, Charles Frederick *see*
– Along the florida reef
– The quakers in great britain and america
Holder-Egger, O *see* Monumenta germaniae historica (mgh)
Holderlin, Friedrich *see* Hyperion
Holderness, Mary *see*
– New russia
– Notes relating to the manners and customs of the crim tatars

Holdich, Thomas Hungerford *see*
– The indian borderland, 1880-1900
– Tibet, the mysterious

Holding companies act of 1935 releases / U.S. Securities and Exchange Commission – n1-17865. 12 feb 1935-26 jan 1973 (all publ) – (= ser The sec release series preceding the sec docket) – 818mf – 9 – $1227.00 – mf#LLMC 84-356 – us LLMC [346]

Holding fast the form of sound words / Muir, William – Edinburgh, Scotland. 1865 – 1r – us UF Libraries [240]

Holding the ropes : missionary methods for workers at home / Brain, Belle Marvel – New York: Funk & Wagnalls, c1904 – 1mf – 9 – 0-8370-6648-4 – mf#1986-0648 – us ATLA [240]

Holding, Thomas Hiram *see*
– Coat cutting
– Comfort and economy in clothes
– The direct system of ladies cutting
– Trousers cutting

Holdings / Kansas State Historical Society, Manuscripts Dept – 1854-1985 – 1 – us Kansas [978]

Holditch, Robert *see* Observations on emigration to british america, and the united states

Holdnak, Andrew *see* The impacts of marine debris, weather conditions, and unexpected events on recreational boater satisfaction on the delaware inland bays

Holdrege citizen – Holdrege, NE: Holdrege Citizen Pub Co. -v13 n27. feb 26 1897 (wkly) [mf ed 1887-97 (gaps) filmed -1970] – 5r – 1 – (absorbed: holdrege republican and: nebraska nugget. merged with: political forum to form: holdrege citizen=forum) – us NE Hist [071]

Holdrege citizen – Holdrege, NE: F H Porter. 51v. v14 n43. jun 17 1898-v54 n154. jul 2 1938 (daily ex sun and holidays) [mf ed with gaps filmed -[1989]] – 30r – 1 – (cont: holdrege citizen=forum. cont by: holdrege daily citizen) – us NE Hist [071]

Holdrege citizen=forum – Holdrege, NE: F H Porter. 2v. v13 n28. mar 5 1897-v14 n42. jun 10 1898 (wkly) [mf ed filmed -1970] – 4r – 1 – (formed by the union of: holdrege citizen and: political forum. cont by: holdrege citizen (1898)) – us NE Hist [071]

Holdrege daily citizen – Holdrege, NE: Jim Hammond. v54 n155. jul 5 1938- (daily ex sat, sun & hols) – 36r – 1 – (cont: holdrege citizen (1898)) – us Bell [071]

Holdrege daily citizen – Holdrege, NE: Jim Hammond. v54 n155. jul 5 1938 (daily ex sat,sun & holidays) [mf ed 1938-70,1990-(gaps)] – 1 – (cont: holdrege citizen (1898)) – us NE Hist [071]

Holdrege daily citizen – Holdrege, Phelps Co, NE: Jim Hammond. v54 n155. jul 5 1938- (daily ex sat, sun & hols) – 11r – 1 – (cont: holdrege citizen (holdrege, ne: 1898)) – us Microfilm [071]

Holdrege daily nugget – Holdrege, NE: [s.n.] (daily) [mf ed v1 n14. nov 8 1892 filmed 1971] – 1r – 1 – us NE Hist [071]

Holdrege progress – Holdrege, NE: C Clinton Page. 38v. v20 n4. apr 5 1906-v57 [n9] apr 30 1943 (wkly) [mf ed with gaps filmed 1971] – 25r – 1 – (cont: weekly progress. cont by: irrigation farmer and holdrege progress) – us NE Hist [071]

Holdrege republican – Holdrege, NE: Guild Bros & Sears. v1 n1. sep 10 1884- (wkly) [mf ed -1886 (gaps) filmed 1970] – 1r – 1 – (absorbed by: holdrege citizen) – us NE Hist [071]

Holdren, John *see* Energy

Holdridge, Desmond *see*
– Escape to the tropics
– Pindorama

Holdsworth, W W *see*
– The christ of the gospels
– The life of faith

Hole, Charles *see*
– The early history of the church missionary society for africa and the east to the end of a.d. 1814
– Early missions to and within the british islands

Hole, Hugh Marshall *see*
– The making of rhodesia
– Passing of the black kings

Hole, S Reynolds *see* Our duty in danger

Hole, Samuel Reynolds *see* The memories of dean hole

Holgate baptist church. portland, oregon : church records – Feb1954-Jun 1973 – 1 – us Southern Baptist [242]

Holguin Arboleda, Julio *see* Mucho en serio y algo en broma

Holguin Y Caro, Margarita *see* Caros en colombia

Holiday – Dacca Bangladesh, 2 jan 1972-24 dec 1974 – 1r – 1 – uk British Libr Newspaper [079]

Holiday – Indianapolis. 1946-1977 (1) 1969-1977 (5) 1970-1977 (9) – ISSN: 0018-3520 – mf#128 – us UMI ProQuest [910]

The holiday advertiser – St John, NB: W A Barnes, [1880?] – 9 – ISSN: 1190-6596 – mf#P04538 – cn Canadiana [073]

Holiday coast times – Coffs harbour – 1r – A$72.53 vesicular A$78.03 silver – at Pascoe [079]

Holiday, Henry *see* Stained glass as an art

Holiday reporter / Champaign Co. Saint Paris – dec 19-31 1887 [daily] – 1r – 1 – mf#B11640 – us Ohio Hist [071]

Holiness : the birthright of all god's children / Crane, Jonathan Townley – New York: Nelson & Phillips; Cincinnati: Hitchcock & Walden, 1874 – 1mf – 9 – 0-8370-3244-X – mf#1985-1244 – us ATLA [240]

Holiness : its nature, hinderances, difficulties, and roots / Ryle, John Charles – 2nd ed. London: William Hunt, 1883 – 2mf – 9 – 0-524-01398-5 – mf#1990-4094 – us ATLA [240]

Holiness / Mills, Job Smith – Dayton, Ohio: United Brethren Pub House, 1902 – (= ser Doctrinal Series (Dayton, Ohio)) – 1mf – 9 – 0-7905-2182-2 – mf#1987-2182 – us ATLA [240]

Holiness : perfection in christ; sermon preached... thursday, june 2nd, 1887 / Pearse, Mark Guy – [St John, NB?: Day & Reid], 1887 – 1mf – 9 – 0-665-91956-5 – mf#91956 – cn Canadiana [240]

Holiness : a treatise on sanctification, as set forth in the new testament / Summers, Thomas Osmond – Richmond: J Early for the Methodist Episcopal Church, South, 1851 – 1mf – 9 – 0-524-00158-8 – mf#1989-2858 – us ATLA [240]

Holiness and power : for the church and the ministry / Hills, Aaron Merritt – Cincinnati, OH: M W Knapp, 1897 – 1mf – 9 – 0-8370-4877-X – (incl ind) – mf#1985-2877 – us ATLA [240]

The holiness of pascal / Stewart, Hugh Fraser – Cambridge: University Press, 1915 – (= ser Hulsean Lectures) – 1mf – 9 – 0-7905-9682-2 – (incl bibl ref) – mf#1989-1407 – us ATLA [240]

Holiness symbolic and real : a bible study / Beet, Joseph Agar – Cincinnati: Jennings & Graham, [pref. 1910] Beltsville, Md: NCR Corp, 1978 (3mf); Evanston: American Theol Lib Assoc, 1984 (3mf) – 9 – 0-8370-0844-1 – (incl ind) – mf#1984-4224 – us ATLA [220]

Holiness teachings : compiled from the editorial writings of the late rev. benjamin t roberts... / Roberts, Benson Howard – North Chili, N.Y.: "Earnest Christian" Publishing House, 1893. Beltsville, Md: NCR Corp, 1978 (3mf); Evanston: American Theol Lib Assoc, 1984 (3mf) – 9 – 0-8370-1129-9 – mf#1984-4444 – us ATLA [240]

Holiness union / United Holy Church of America – 1981 sep, 1982 sep, 1983 aug, sep, oct, nov/dec, 1984 jan, apr, sep, 1985 aug, 1986 jun, 1991 jun, sep, 1992 nov, dec, 1993 jan, 1994 jul, oct, nov, dec, 1995 jan, feb,sep, nov, 1996 jan, feb, apr, may/jun, jul, aug, 1997 jan, mar, apr, may, jun, aug, sep, dec, 1998 jan/feb, mar, apr, may, jun, jul, aug, sep – 1r – 1 – mf#3962147 – us WHS [242]

Holiness year book for... : a daily companion for the christian home / ed by Hughes, George – New York: Palmer & Hughes 1891-93 [mf ed 2004] – 1891-93 [complete] 3v on 1r – 1 – (cont: illustrated holiness year book for...; cont by: international holiness directory and year book for...) – mf1056 – us ATLA [240]

Holisso anumpa tosholi – An English and Choctaw Definer. 1852 – 1 – 9.10 – us Southern Baptist [490]

A holistic analysis of stress with implications for stress management as a function of pastoral counseling / Gray, Edward A – 1981 – 1 – $6.80 – us Southern Baptist [242]

Holistic medicine – Chichester. 1986-1990 (1) 1986-1990 (5) 1986-1990 (9) – ISSN: 0884-3988 – mf#16102 – us UMI ProQuest [616]

Holistic nursing practice – Frederick. 1986+ (1,5,9) – (cont: topics in clinical nursing) – ISSN: 0887-9311 – mf#16003 – us UMI ProQuest [610]

Holiwell's tourist guide to quebec / Anderson, William James – S.l: s.n, 1872? – 1mf – 9 – mf#04055 – cn Canadiana [917]

Holker, John *see* Papers

Holkham accounts, 1789-1814 / Crick, F [comp] – (= ser Holkham library post-1700 estate and household records) – 1r – 1 – (contains: general receipts for 1789-1814; household accounts for 1801-1807) – mf#96847 – uk Microform Academic [640]

Holkham audit books, 1707-1853 – (= ser Holkham library post-1700 estate and household records) – 11r – 1 – mf#96619 – uk Microform Academic [640]

Holkham bible picture book – 14th c – (= ser Holkham library manuscript books 666) – 1r – 14 – (notes by w o hassall) – mf#C551 – uk Microform Academic [220]

Holkham country accounts, 1722-1767, 1793-1800 – (= ser Holkham library post-1700 estate and household records) – 2r – 1 – mf#96617 – uk Microform Academic [640]

Holkham deeds, 1100-1459 – bundles 1,2,4,6 – (= ser Holkham library early estate records) – 1r – 1 – mf#97231 – uk Microform Academic [025]

Holkham deeds, 1460-1499, 1600-1619 – bundle 7 n139-193; bundle 12 n483-587 – (= ser Holkham library early estate records) – 1r – 1 – mf#97401 – uk Microform Academic [025]

Holkham deeds, 1500-1549 – bundle 9 – (= ser Holkham library early estate records) – 1r – 1 – mf#97401 – uk Microform Academic [025]

Holkham deeds, 1550-1599 – bundle 10 n297-458 – (= ser Holkham library early estate records) – 1r – 1 – mf#97376 – uk Microform Academic [025]

Holkham domestic accounts, 1719-1792 – (= ser Holkham library post-1700 estate and household records) – 3r – 1 – (incl holkham mss 730, 731, 737-741) – mf#96618 – uk Microform Academic [640]

Holkham estate papers, c1779-1833 – (= ser Holkham library post-1700 estate and household records) – 13v on 2r – 1 – mf#96872 – uk Microform Academic [346]

Holkham estates, farm premises and cottages, 1856 – (= ser Holkham library post-1700 estate and household records) – 1r – 1 – (report by h w keary) – mf#96834 – uk Microform Academic [640]

Holkham farm books, 1843/4-1860/1 – (= ser Holkham library post-1700 estate and household records) – 1r – 1 – mf#97389 – uk Microform Academic [640]

Holkham general estate deeds, 1579-1641 – bundle A n3,5-7,12-13 – (= ser Holkham library early estate records) – 1r – 1 – (incl: the nathaniel bacon papers) – mf#96709 – uk Microform Academic [025]

Holkham home farm accounts, 1814-1822 – (= ser Holkham library post-1700 estate and household records) – 1r – 1 – mf#96456 – uk Microform Academic [640]

Holkham household accounts 1698-1702 *see* Terrars of hillesden, buckinghamshire, 1657 and 1665

Holkham journal-book of accounts, 1718-1727 – (= ser Holkham library post-1700 estate and household records) – 1r – 1 – (kept by edward smith for thomas coke) – mf#96618 – uk Microform Academic [640]

Holkham letter books, 1816-1837 : agricultural letter books of francis blaikie – (= ser Holkham library post-1700 estate and household records) – 5r – 1 – mf#96727 – uk Microform Academic [630]

Holkham miscellaneous deeds, 1278-1688 – n995-1304 – (= ser Holkham library early estate records) – 1r – 1 – mf#97495 – uk Microform Academic [025]

Holkham miscellaneous documents (1) – (= ser Holkham library, the house, park & art colls 765) – 1r – 1 – (contains: the plans, elevations and sections of holkham in norfolk by matthew brettingham, 1773, p.v. lady leicester's jewels and furniture, mss inventory 1760. a catalogue of pictures, statues and busts, mss 1765. an account of the pictures and statues at holkham, mss n.d. [1774?]. an exact account of the furniture in holkham house, mss 1774. an inventory of the furniture at holkham, mss 1765. an inventory of furniture in the apartments of holkham house, mss 1774. holkham heirlooms (belonging to the late thomas, earl of leicester)) – mf#97109 – uk Microform Academic [025]

Holkham miscellaneous documents (2) – (= ser Holkham library, the house, park & art colls) – 1r – 1 – (contains: inventory and valuation at holkham hall, 1842. inventory of the furniture at holkham hall inventory of goods and pictures, 1760. inventory of furniture, household goods, pictures, statues etc, at holkham "belonging to thomas, earl of leicester, deceased". an account of furniture for holkham house begun in 1752. schedule of heirlooms at holkham hall, 1842) – mf#97110 – uk Microform Academic [025]

Holkham miscellaneous documents (3) – (= ser Holkham library, the house, park & art colls) – 1r – 1 – (contains: abstract by christopher bedingfield of coke estate deeds, 1708, containing "final draft" of catalogue of books belonging to thomas coke of holkham, 13th july 1727. particulars of holkham cottages, 1850's. "holkham's changing landscape" – various maps, plans and pictures. holkham terrier 1549, holkham deed 296. holkham commons deposition 1592, wighton deed 103) – mf#97111 – uk Microform Academic [025]

Holkham office cash accounts, 1808-1844 / Coke, Thomas William & Blaikie, Francis [comps] – (= ser Holkham library post-1700 estate and household records) – 3r – 1 – mf#96471 – uk Microform Academic [650]

Holkham servants books 1848-69 *see* Terrars of hillesden, buckinghamshire, 1657 and 1665

Holkham stock books, 1845/6-1860/1 and estate accounts, 1898/9-1899/1900 – (= ser Holkham library post-1700 estate and household records) – 1r – 1 – mf#97390 – uk Microform Academic [650]

Holkot, Rob *see* Praelectiones in librum sapientiae

Die holk'schen jaeger / Pueltz, Wilhelm – Muenchen: Deutscher Volksverlag, 1943 – 1r – 1 – us Wisconsin U Libr [830]

Holl, K *see* Epiphanius (gcsej3a)

Holl, Karl *see*
– Amphilochius von ikonium
– Enthusiasmus und bussgewalt beim griechischen moenchtum
– Fragmente vornicaenischer kirchenvaeter aus den sacra parallela
– Die geistlichen uebungen des ignatius von loyola
– Die handschriftliche ueberlieferung des epiphanius
– Die handschriftliche ueberlieferung des epiphanius (ancoratus und panarion
– Johannes calvin
– Luther und das landesherrliche kirchenregiment
– Der modernismus
– Die sacra parallela des johannes damascenus

Die hollaendische radikale kritik des neuen testaments *see* Radical views about the new testament

Hollaendisch-guiana : erlebnisse und erfahrungen waehrend eines 43 jaehrigen aufenthalts in der kolonie surinam / Kappler, A – Stuttgart, 1881. – 6'mf – 9 – mf#Z-2261 – ne IDC [590]

Holland 1771-1890 – Oxford, MA (mf ed 1984) – (= ser Massachusetts vital records) – 10mf – 9 – 0-931248-62-0 – (mf 1-2,4: marriages/intentions 1771-1824. mf 1-4: births & deaths 1767-1840. mf 2-4: church members 1784-1834. mf 5-6: bargains & sales 1834-40. mf 6-7: births 1837-84. mf 7-8: marriages 1844-80. mf 9: deaths 1876-90; marriages 1882-93; births 1885-90. mf 10: b,m,d 1781-1861) – us Archive [978]

Holland, A *see* Application of the credit valley railway for right of way and crossings at the city of toronto

Holland, A [comp] *see* The credit valley railway application for right of way and crossings at the city of toronto

Holland, C *see* Confirmation

Holland carries on / Netherlands Regeeringsvoorlichtingsdienst, New York – New York, NY. 1943 – 1r – us UF Libraries [972]

Holland, Fidele H *see* Sketches from life

Holland, Frederic May *see* Our clergywomen

Holland, George C *see* The credit valley railway application for right of way and crossings at the city of toronto

Holland, George C [comp] *see* Application of the credit valley railway for right of way and crossings at the city of toronto

Holland, Henry Fox, Baron *see* Candid reflections on the report (as published by authority) of the general-officers

Holland, Henry Richard Vassall *see* Letter to the rev dr shuttleworth, warden of new college, oxford

Holland, Henry Scott *see*
– The apostolic fathers
– A bundle of memories
– Creed and character
– Good friday
– Logic and life
– Old and new
– On behalf of belief
– The optimism of butler's 'analogy'
– Our neighbours
– Personal studies
– Vital values

Holland in den jahren 1831 [achtzehnhunderteinunddreissig] und 1832 [achtzehnhundertzweiunddreissig] / Wienbarg, Ludolf – Hamburg 1833 [mf ed Hildesheim 1995-98] – 2v on 4mf – 9 – €120.00 – 3-487-29616-0 – gw Olms [949]

Holland, John *see* Memorials of sir francis chantry, r a sculptor, in hallamshire

Holland, Josiah Gilbert *see*
– Gold-foil
– Lessons in life

Holland, Robert Afton *see* Is future punishment eternal?

Holland, Rupert Sargent *see* Builders of united italy

Holland, Saba Holland *see* A memoir of the reverend sydney smith

Holland Society of New York *see*
– Collections
– Year book of the holland society of new-york

Holland, Spencer L *see* National church of a democratic state

Holland, T J *see* Record of the expedition to abyssinia

Holland, Thomas E *see*
– The elements of jurisprudence

Holland Union Benevolent Association *see* Geschiedenis, constitutie en bij-wetten van de "holland union benevolent association" te grand rapids, mich

Holland, W *see* Meister altswert
Holland, W Lancelot *see*
– Miss ellen golding (a rescued nun) will visit edinburgh and deliver...
– Ritualism in scotland
Holland, Wes *see* The indian outlook
Holland, Wilhelm Ludwig *see*
– Briefe der herzogin elisabeth charlotte von orleans
– Das buch der beispiele der alten weisen
– Die schauspiele des herzogs heinrich julius von braunschweig
– Schreiben des kurfuersten karl ludwig von der pfalz und der seinen
– Uhlands gedichte und dramen
– Zu ludwig uhlands gedaechtnis
Holland, William Edward Sladen *see*
– The goal of india
Holland, William Jacab *see* To the river plate and back
[Hollanda, F de] Joaquim de Vasconcellos *see* Francesco de hollanda
Hollandale review – Blanchardville, Hollandale WI. 1952 mar 14-1953 jan 9, 1954 oct 15 – 2r – 1 – mf#1005493 – us WHS [071]
Hollander, Daniel B *see* The effects of social support on men's exercise-related cardiovascular reactivity
Hollandia – London, UK. 6 Nov 1897-30 Dec 1899 – 1 – uk British Libr Newspaper [072]
Hollandische kolonialreich in brasilien / Watjen, Hermann Julius Eduard – Haag, Netherlands. 1921 – 1r – us UF Libraries [972]
Holland's – Dallas. 1950-1953 (1) – mf#333 – us UMI ProQuest [630]
De hollandsch weekblad – Kalamazoo MI: Dalm Print Co Inc 1943 mar-1944 jun (semiwkly) – 1r – 1 – (aka: de hollandsche amerikaan) – us UF Libraries [071]
De hollandsche afrikanen en hunne republiek in zuid-afrika / Stuart, Jacobus – Amsterdam: G W Tielkemeijer, 1854 – 1 – us CRL [960]
De hollandsche amerikaan – Kalamazoo: Dalm Print Co, [-1945]. Dec 31 1928-1942 – 14r – 1 – us CRL [071]
Holle, Berthold von *see*
– Demantin
Holle, Hugo *see* Goethes lyrik in weisen deutscher tonsetzer bis zur gegenwart
Holle, Paul *see* De la senegambie francaise
Hollebeke, L van *see* Lissweghe
Holleman, J F *see*
– African interlude
– Chief, council and commissioner
– Shona customary law
Holleman, Johan Frederik *see* The pattern of hera kinship
Holleman, W *see* Mimus, de chuchubi bekijkt curacao
Hollenberg, Johannes *see* Zur methodik des biblischen unterrichts in den oberen gymnasialklassen
Hollenberg, Wilhelm Adolf *see*
– Hebraeisches schulbuch
– Huelfsbuch fuer den evangelischen religionsunterricht in gymnasien
Hollenweger, Walter J *see* Handbuch der pfingstbewegung
Holler, P *see*
– An english-telugu scientific dictionary
– A small english-telugu dictionary
– A telugu-english classical dictionary
– Vocabulary to the telugu bible
Holley, Horace *see* Bahaism, the modern social religion
Holley, Marietta *see*
– Josiah allen's wife as a p.a. and p. i. samentha at the centennial.
– Samantha among the brethren
Holley, Robert P *see* Resource sharing and information networks
Holliday, Corey L *see* An evaluation of carolina athletes coming together (act)
Holliday, Cyrus Kurtz *see*
– Letters and clippings
– Papers
Holliday, Fernandez C *see*
– A bible hand-book, theologically arranged
– Indiana methodism
Holliday, Omar *see* 700 years of hollidays
Holliday, Robert Cortes *see* Unmentionables from figleaves to scanties
Holliman, Susan C *see* Effects of the menstrual cycle phases on the energy intake and expenditure in physically active and inactive women
Hollinfare, st helen – (= ser Cheshire monumental inscriptions) – 2mf – 9 – £4.00 – mf#189 – uk CheshireFHS [929]
Hollings, G E *see*
– Notes respecting oudh
– Paper regarding the buddik dacoits in oude, written in 1839
Hollingsworth, Lawrence William *see* Zanzibar under the foreign office, 1890-1913
Hollingsworth, N *see* Recommendation of the madras system of instruction
Hollins, Cecil *see* Synthesis of nitrogen ring compounds containing a single...
Hollins critic – Hollins College. 1964+ (1) 1976+ (5) 1977+ (9) – ISSN: 0018-3644 – mf#8096 – us UMI ProQuest [378]
Hollin's liebeleben : ein roman / Arnim, Ludwig Achim, Freiherr von; ed by Minor, Jacob – Freiburg: J C B Mohr, 1883 [mf ed 1988] – xxxi/117p – 1 – (added tp of goettingen 1802 ed. int by ed) – mf#6956 – us Wisconsin U Libr [830]
Hollis, Alfred Claud *see* The masai
Hollis, George *see* The monumental effigies of great britain
Hollis, New Hampshire. Hollis Baptist Church *see* Records
Hollis, Thomas *see* The monumental effigies of great britain
[Hollister-] free lance – CA. 1886-1923, 1926-225r – 1 – $13,500.00 (subs $150y) – (aka: evening free lance) – mf#BC02297 – us Library Micro [071]
Hollister, George W *see*
– Li chieh sheng ching
– Tsui yu te chiu
– Wei shen me p'a chin hua lun
[Hollister-] hollister advance – CA. 1919-56; 1981; 1983 – 18r – 1 – $1080.00 – mf#BC02296 – us Library Micro [071]
[Hollister-] san benito advance – CA. 1894-1920 – 7r – 1 – $420.00 – mf#C03604 – us Library Micro [071]
Holliston 1720-1849 – Oxford, MA (mf ed 1996) – (= ser Massachusetts vital record transcripts to 1850) – 14mf – 9 – 0-87623-268-3 – (mf 1t-2t: births 1720-55. mf 3t: births 1776-98. mf 4t: births 1799-1824. mf 5t: births 1825-44. mf 6t-7t: marriages 1729-1843. mf 7t: intentions 1767-68. mf 7t-9t: deaths 1725-1844. mf 9t-11t: publishments 1766-1849. mf 11t-13t: births 1843-49, 1807-39. mf 13t-14t: marriages 1843-49. mf 14t: deaths 1843-49, 1842) – us Archive [978]
Hollmann, Georg *see* Die bedeutung des todes jesu
Hollmann, Michael *see* Reichskolonialamt (bestand r 1001) bd 98
Hollmann, Ricarda *see* Hydrolisierbarkeit von immunglobulinen und albumin durch die proteolytische aktivitaet lebender zellen der spirochaete treponema denticola
Holloman sunburst – Alamgordo NM. 1986 oct 3-1987 oct, 1987 nov-1988 nov, 1988 dec-1989 dec 22 – 3r – 1 – (cont: sunburst [alamogordo, nm]; cont by: sunburst [holloman air force base [nm]]) – mf#1702484 – us WHS [355]
Hollon, D Leslie *see* The meaning of sanctification in twentieth century southern baptist thought
Hollonius, Ludwig *see* Somnium vitae humanae
Holloway advertiser – London, UK. 16 dec 1882-11 may 1885; 18 may 1885-9 apr 1887 – 2r – 1 – uk British Libr Newspaper [072]
Holloway and islington journal – [London & SE] Islington CL 5 feb 1971-mar 1974 – 1 – (cont: north london press [-jan 1971]) – uk Newsplan [072]
Holloway, James Thomas *see* Conversion of the ethiopian
Holloway, Kari L *see* Psychological and physiological responses to 12 weeks of aerobic exercise on various modes
Holloway press – [London & SE] Islington CL 14 dec 1872-1923 – 1 – (title varies: north metropolitan press [1876-79]. subtitle: holloway, hornsey, muswell hill press [1908-15]. subtitle: holloway, hornsey and haringey and muswell hill press [1915-23]; cont as: islington and holloway press [10 mar 1923-25 dec 1942]) – uk Newsplan [072]
Holloways baptist church. liberty association. davidson county. north carolina : church records – 1832-1927 – 1 – us Southern Baptist [242]
Hollweg, A Bethmann *see* Fragmenta vaticana
Hollweg, Eduard *see* Von der getrosten verzweiflung
Holly, Arthur *see* Dra-po
The holly branch / Wilkins, Harriet Annie – [Hamilton, Ont?: s.n.] 1851 [mf ed 1983] – 2mf – 9 – 0-665-44952-6 – (incl ind) – mf#44952 – cn Canadiana [810]
Holly grange : a tale / De K, Emma – London: Joseph Cundall, 1844 – (= ser 19th c children's literature) – 3mf – 9 – mf#6.1.22 – uk Chadwyck [830]
Holly, James Theodore *see* Vindication of the capacity of the negro race
Holly, Jb *see* Stratigraphy and sedimentary
Holly springs baptist church. pickens county. south carolina : church records – 1839-1954. 702p – 1 – us Southern Baptist [242]
Holly springs baptist church. spartanburg county. south carolina : church records – 1834-1866, 1910-19, 1926-73; 754p – 1 – us Southern Baptist [242]
Hollywood close up / Screen Actors Guild – v1 n1-2 [1980 mar-may] – 1r – 1 – (continued by: hollywood) – mf#1021357 – us WHS [790]
[Hollywood-] daily variety – CA. 1972-84 – 35r – 1 – $2100.00 – mf#R02298 – us Library Micro [071]
Hollywood Free Paper *see* Jesus people
Hollywood free paper / Christian Prison Volunteers et al – v3 n11, 16 [1971 may, aug], 1974 sep, oct, nov/dec, 1975 jan/feb-nov /dec, 1976 jan-feb, n476-676 [1976 spring-fall], n177-477 [1977 jan-oct], n178,678 [1978 jan, jun], flyer [1973] – 1r – 1 – mf#203182 – us WHS [365]
Hollywood guide : hollywood, florida / Darsey, Barbara Berry – s.l, s.l? 193-? – 1r – us UF Libraries [978]
[Hollywood-] holly leaves – CA. 1918-1924 – 11r – 1 – $660.00 – mf#R03245 – us Library Micro [071]
[Hollywood-] hollywood citizen news – CA. 1903-1957 – 442r – 1 – $26,520.00 – mf#R03243 – us Library Micro [071]
[Hollywood-] hollywood reporter – CA. 1976-80 – 17r – 1 – $1020.00 – mf#R02299 – us Library Micro [071]
Hollywood lantern and quill / American Postal Workers Union – v1 n9 [1976 dec], v2 n1-8, 10-12 [1977 jan-aug, oct-dec], v3 n1-3,10 [1978 jan-mar, oct] – 1r – 1 – (continued by: lantern & quill) – mf#1546697 – us WHS [380]
Hollywood star – Salem OR: E M Sanders [wkly] – 1 – (began in 1939) – us Oregon Lib [071]
[Hollywood-] the staff – CA. 1971-1973 – 4r – 1 – $240.00 – mf#R03246 – us Library Micro [071]
[Hollywood-] valley times – CA. 1955-1961 – 73r – 1 – $4380.00 – mf#R04032 – us Library Micro [071]
Holm, Adolf *see*
– Geschichte siciliens im alterthum
– Griechische geschichte
– The history of greece from its commencement to the close of the independence of the greek nation
Holm, G *see* Den danske konebaads-expedition til gronlands ostkyst
Holm, Norbert *see* Soldat und kaempfer
Holman, Charles *see* Diary
Holman, Curt W *see* American social dance technique syllabus for the rumba, samba, mambo, and tango
Holman, J *see* Voyage round the world
Holman, James *see*
– The narrative of a journey, undertaken in the years 1819, 1820, and 1821, through france, italy, savoy, switzerland, parts of germany bordering on the rhine, holland, and the netherland
– Travels trough russia, siberia, poland, austria, saxony, prussia, hanover, etc
– Voyage round the world
– A voyage round the world
Holman, Russell *see* Fleet's in!
Holman, Sharon S *see* A syllabus of the american social dance silver level technique
Holme, Leonard Ralph *see* The extinction of the christian churches in north africa
Holmen courier – Holmen WI. 1996, 1997, 1998, 1999 – 4r – 1 – (cont: town & country courier) – mf#4856029 – us WHS [071]
Holmen times – Holmen WI. 1946 dec 19-1952 oct 31 – 1r – 1 – (continued by: la crosse county record; record-times (onalaska, wi: 1952)) – mf#945618 – us WHS [071]
Holmes, Calvin Pratt *see* Probate law and practice of the state of iowa..
Holmes chapel, ironbridge methodist chapel – (= ser Cheshire monumental inscriptions) – 1mf – 9 – £2.50 – mf#82 – uk CheshireFHS [929]
Holmes Co. Millersburg *see*
– Holmes county farmer
– Holmes county republican
– Holmes county whig
– Republican
Holmes county advertiser – Bonifay, FL. 1927 dec 2-1997 – 54r – (gaps) – us UF Libraries [071]
Holmes county atlas, 1875 – 1r – 1 – mf#B27424 – us Ohio Hist [978]
Holmes county farmer / Holmes Co. Millersburg – 1880-1917, 1919-jul 1926 [wkly] – 20r – 1 – mf#B8915-8934 – us Ohio Hist [071]
Holmes county farmer / Holmes Co. Millersburg – mar 1860-jan 1866 [wkly] – 1r – 1 – mf#B12060 – us Ohio Hist [071]
Holmes county farmer hub – Millersburg, OH. 1950-1974 (1) – mf#65593 – us UMI ProQuest [071]
Holmes county, ohio, cemetery records : and richardh. dickinson; holmes county, ohio, cemetery records / Dickinson, Marguerite – Millersburg, OH: s.p., 1970 – 1r – 1 – us Western Res [920]
Holmes county republican / Holmes Co. Millersburg – aug 1875-jul 1895 [wkly] – 9r – 1 – mf#B10390-10398 – us Ohio Hist [071]
Holmes county republican / Holmes Co. Millersburg – v1 n1. sep 1856-apr 1862,sep 1870-aug 1874 [wkly] – 4r – 1 – mf#B184-187 – us Ohio Hist [071]
Holmes county whig / Holmes Co. Millersburg – v1 n1. 7/1844-45, 3/46-3/47, 6/47-4/1853 [wkly] – 2r – 1 – mf#B212-213 – us Ohio Hist [071]
Holmes, Daniel B *see* Letters
Holmes, Eber *see* Commercial rose culture, under glass and outdoors
Holmes, Edmond Gore Alexander *see*
– The creed of buddha
– The problem of the soul
Holmes, Ernest *see* Science of mind by ernest holmes
Holmes, Ernest Edward *see*
– Immortality
– Ordination addresses
– Visitation charges delivered to the clergy and churchwardens of the dioceses of chester and oxford
Holmes, George *see* Sketches of some of the southern counties of ireland
Holmes, Henry S *see* Diary
Holmes, Isaac *see* An account of the united states of america
Holmes, J S *see* Holmes' reports of cases in the first circuit, 1870-1875
Holmes, Jean *see*
– Conferences de notre-dame de quebec
– Nouvel abrege de geographie moderne
[Holmes, Jean] *see* Nouvel abrege de geographie moderne
Holmes, Jesse Herman *see* The modern message of quakerism
Holmes, John *see* History of the protestant church of the united brethren
Holmes, ML *see* The analysis of enrollment patterns and student provile characteristics at a small rural new england university 1978-1988
Holmes, Oliver Wendell *see* Ralph waldo emerson
Holmes, Oliver Wendell, Jr *see*
– Collected legal papers
– Mechanism in thought and morals
– The oliver wendell holmes, jr papers
– Some table talk of mr. justice holmes and "the mrs."...with christmas greetings from richard walden hale
Holmes, Patricia A *see* Personality types of ncaa and naia male and female administrators
Holmes' reports of cases in the first circuit, 1870-1875 / Holmes, J S – Boston: Little-Brown. 1v. 1877 (all publ) – (= ser Early federal nominative reports) – 6mf – 9 – $9.00 – mf#LLMC 81-456 – us LLMC [340]
Holmes, Rev John *see* Historical sketches of the missions of the united brethen
Holmes, Richard J *see* A comparison of hemodynamic responses to arm and leg exercise of the same intensities
Holmes, Robert *see* Second annual account of the collation of the mss of the septuagint-version
Holmes, S *see* The journal of...as one of the guard on lord macartney's embassy to china and tartary, 1792-1793
Holmes, Samuel *see*
– Joshua
– Samuel holmes's vormals leibgardist...tagebuch einer reise nach sina und in die tatarei mit der brittischen gesandtschaft
Holmes, Thomas Scott *see* The origin and development of the christian church in gaul during the first six centuries of the christian era
Holmes, William Henry *see* A short history of the union jack
Holmested, George Smith *see* The mechanics' lien acts; being the revised statute of ontario, chapter 120, and 41 victoriae, chapter 17, with annotations
The holmesville courier – Holmesville, NE: Ray A Wild. v1 n1. dec 1 1899- (wkly) [mf ed -1900 (gaps)] – 1r – 1 – (publ in de witt ne, dec 29 1899-) – us NE Hist [071]
Holmoe, Thomas A *see* The selection and use of team captains in college football
Holmquist, Hjalmar *see* Luther, loyola, calvin i deras reformatoriska genesis
Holocaust and genocide studies – Oxford. 1986-1991 (1,5,9) – ISSN: 8756-6583 – mf#49485 – us UMI ProQuest [170]
Holograph, copy and printed plays. / Vega, Lope de – 1 – (formerly in the holland collection, now at melbury house, dorset. manuscripts 5r 96752; printed plays 2r 96753) – uk Microform Academic [790]
Holophote : critico e noticioso – Maceio, AL: Typ Mercantil, 08 nov 1896-09 ago 1897 – (= ser Ps 19) – bl Biblioteca [079]
O holophote – Juiz de Fora, MG. 30 set 1894 – (= ser Ps 19) – bl Biblioteca [079]
Holos ukrainy – Kyiv: Verkhovna Rada Ukrainy, 1991- – 1 – us East View [077]
Holos vostochnoi tserkvy = Voice of the eastern church – Perth Amboy, NJ: Vestal Publ Co, v1 n1-v5 n35. nov 1941-apr 1945 – 1 – us CRL [243]

Holoway, Ron *see* Music effects on the acquisition of a motor skill
Holscher, Kurt Heimart *see* Feinde des volkes
Holsinger, George Blackburn *see*
- Gospel songs and hymns. no. 1
- Practical exercises in music reading
- Psalms, hymns, and spiritual songs

Holsinger, Henry R *see* Holsinger's history of the tunkers and the brethren church
Holsinger's history of the tunkers and the brethren church : embracing the church of the brethren, the tunkers, the seventh-day german baptist church, the german baptist church, the old german baptists, and the brethren church, including their origin, doctrine, biography and literature = History of the tunkers and the brethren church / Holsinger, Henry R – Lathrop, Calif: Pacific Press, 1901 – 2mf – 9 – 0-524-02958-X – mf#1990-4510 – us ATLA [242]
Holst, Bernhart Paul *see* Practical home and school methods of study and instruction in the fundamental elements of education
Holst, Gustav *see* American rondo
Holst, Herman E von *see*
- Constitutional history of the united states
- John c calhoun

Holstein, Hugo *see*
- Dramen von ackermann und voith
- Ein heimlich gespraech von der tragedia johannis hussen
- Die reformation im spiegelbilde der dramatischen litteratur des sechzehnten jahrhunderts

The holstein nonpareil – Holstein, NE: W T Carson, 1890 (wkly) [mf ed v3 n16, jun 18 1892] – 1r – 1 – us NE Hist [071]
The holstein reporter – Holstein, NE: A R Oelschlager, 1924 (wkly) [mf ed v1 n31. feb 13 1925-nov 20 1925] – 1r – 1 – us NE Hist [071]
Holstein, Robyn E *see* The effects of music on patients in a cardiac rehabilitation program
Holstein world – Sandy Creek. 1980+ (1,5,9) – (cont: holstein-friesian world) – ISSN: 0199-4239 – mf#182,01 – us UMI ProQuest [636]
Holstein,...d' Vicomte *see* Saint-cloud et fontainebleau
Holsteiner nachrichten – Pinneberg DE, 1943 4 may-31 dec [gaps] – 1r – 1 – gw Misc Inst [074]
Holstein-friesian world – Sandy Creek. 1904-1979 (1) 1971-1979 (5) 1976-1979 (9) – (cont by: holstein world) – ISSN: 0018-3695 – mf#182 – us UMI ProQuest [636]
Holsteinischer courier 1872 – Neumuenster DE, 1874 8 jan-31 dec, 1876-1884 20 jun, 1885-1945 2 may, 1949 1 oct-1990 – ca 6r/yr – 1 – (filmed by other misc inst. title varies: 1 oct 1949: neuer holsteinischer courier; 1 may 1950: holsteinischer courier) – gw Misc Inst [074]
Holsten, C *see* Zum evangelium des paulus und des petrus
Holsten, Carl *see*
- Die drei urspruenglichen, noch ungeschriebenen evangelien
- Das evangelium des paulus
- Ist die theologie wissenschaft?
- Die synoptischen evangelien nach der form ihres inhaltes
- Zum evangelium des paulus und des petrus

Holsten-anzeiger – Glueckstadt DE, 1963 30 mar-1964 16 oct – 1r – 1 – gw Misc Inst [074]
Holstenius, L *see* Codex regularum
Holston messenger – Knoxville. 1827-1827 [1] – mf#5570 – us UMI ProQuest [978]
Holston methodism : from its origin to the present time / Price, Richard Nye – Nashville, TN: Pub House of the ME Church, South, 1906-1913 – 6mf – 9 – 0-524-06259-5 – (incl bibl ref) – mf#1990-5214 – us ATLA [242]
Holston pastfinder / Holston Territory Genealogical Society – n1-19 [1982 dec-1987 jun] – 1r – 1 – mf#1058349 – us WHS [978]
Holston Territory Genealogical Society *see* Holston pastfinder
Holt, Adoniram Judson *see* Pioneering in the southwest
Holt and gregson papers, the... 1778-1830 : from liverpool city libraries – (= ser British records relating to america in microform) – 1r – 1 – (int by p j buckland) – mf#96794 – uk Microform Academic [025]
Holt, Ardern *see*
- Fancy dress described
- Gentlemen's fancy dress

Holt, Basil Fenelon *see* Joseph williams and the pioneer mission to the southeastern bantu
Holt, Charles Macpherson *see* A treatise on the insurance law of canada
Holt county banner – O'Neill, NE: Cleveland Brennan & Co, 1882-84// (wkly) [mf ed with gaps] – 1r – 1 – (absorbed by: o'neill tribune) – us NE Hist [071]
Holt county democrat – O'Neill, NE: Ed S Eves. 2v. v1 n1. oct 25 1907-v2 n19. feb 26 1909 (wkly) [mf ed filmed 1993] – 1r – 1 – (absorbed by: holt county independent (o'neill, ne 1897)) – us NE Hist [071]
Holt county independent – O'Neill, NE: O F Biglin. 74v. v4 n3. jun 18 1897-v77 n21. may 27 1965 (wkly) [mf ed 1897-1903,1905-65 (gaps) filmed [1967]-70] – 31r – 1 – (cont: beacon light and holt county independent. absorbed: chambers sun,holt county democrat, o'neill sun and: page reporter. merged with: frontier to form: frontier and holt county independent) – us NE Hist [071]
Holt county independent – O'Neill, NE: Robert E Miles and George A Miles. 4v. v96 n23. jun 7 1984-v99 n47. nov 19 1987 (wkly) [mf ed filmed 1985-88] – 7r – 1 – (cont: frontier and holt county independent. cont by: frontier and holt county independent (1987)) – us NE Hist [071]
Holt county independent – O'Neill, NE: G W Lessinger, Judd Woods (wkly) [mf ed v1 n27. dec 2 1892 filmed [1965]] – 1r – 1 – (merged with: beacon light to form: beacon light and holt county independent) – us NE Hist [071]
The holt county republican – Atkinson, NE: T J Smith & J O Berkley. 1v. v1 n1. aug 19 1899-v1 n47. jul 6 1900 (wkly) – 1r – 1 – (merged with: atkinson plain dealer and graphic-consolidated (1899) to form: atkinson plain dealer, atkinson graphic and holt county republican-consolidated) – us Bell [071]
Holt, Edwin Bissell *see*
- The concept of consciousness
- The freudian wish and its place in ethics

Holt, Francis Ludlow *see* The law of libel: in which is contained a general history of this law in the ancient codes.
Holt, Hamilton *see* The way to disarm
Holt, James Maden *see* Jesuits
Holt, Pat M *see* Colombia today
Holtei als dramatiker / Moschner, Alfred – Breslau: F Hirt, 1911 [mf ed 1992] – (= ser Breslauer beitraege zur literaturgeschichte. neue folge 18) – 185p – 1 – (incl bibl ref) – mf#8014 reel 3 – us Wisconsin U Libr [430]
Holtei, Karl von *see*
- Erzaehlende schriften
- Goethe und sein sohn
- Schlesische gedichte

Holtei, Karl von [comp] *see* Briefe an ludwig tieck
Holth, Sverre *see* Chu tao wen mo hsiang lu [ccm162]
Holthaus, Arno *see* Die kontroverse um die transferproblematik 1924-1929
Holthouse, Henry James *see* A new law dictionary
Holthusen, Hans Egon *see*
- Ergriffenes dasein
- Die welt ohne transzendenz

Holthusius, Joannes *see* Compendium cantionum ecclesiasticarum
Holtinger, J J *see* Heinrich bullingers reformationsgeschichte
Holtmann, Martin *see* Ligand-interaction and receptor-regulation in the novel secretin family of g protein-coupled receptors
Holtschnitte des meisters ds / ed by Bock, Elfried – Berlin, 1924 – €5.00 – ne Slangenburg [740]
[Holtville-] holtville tribune – CA. 1923-26 – 2r – 1 – $120.00 – mf#R02300 – us Library Micro [071]
Holtz, Kurt *see*
- 1 [first] samuel 1-7, 1
- I samuel 1-7:1

Holtz, T *see*
- Die christologie der apokalypse des johannes
- Untersuchungen ueber die alttestamentlichen zitate vei lukas

Holtzendorff, Franz von *see*
- Republikanische lieder
- A short protestant commentary on the books of the new testament
- Zeitglossen des gesunden menschenverstandes

Holtzhauer, Helmut *see* Studien zur goethezeit
Holtzhauser, Helmut *see* Arbeiterbewegung und klassik
Holtzmann, Adolf *see*
- Beitraege zur erklaerung der persischen keilinschriften
- Germanische alterthuemer. mit text, uebersetzung und erklaerung von tacitus germania
- Untersuchungen uber das nibelungenlied

Holtzmann, Heinrich Julius *see*
- Akademische predigten
- Bibelgeschichte
- Eduard reuss' briefwechsel mit seinem schueler und freunde karl heinrich graf
- Die entstehung des neuen testaments
- Judenthum und christenthum im zeitalter der apokryphischen und neutestamentlichen literatur
- Kritik der epheser- und kolosserbriefe
- Kritik der epheser und kolosserbriefe auf grund einer analyse ihres verwandtschaftsverhaeltnisses
- Lehrbuch der historisch-kritischen einleitung in das neue testament
- Lehrbuch der neutestamentlichen theologie
- Das neue testament und der roemische staat
- Die pastoralbriefe
- Synoptische erklaerung der drei ersten evangelien
- Die synoptischen evangelien
- Thomas von aquino und die scholastik

Holtzmann, Heinrich Julius et al *see*
- Die anfaenge des christenthums
- Hand-commentar zum neuen testament
- Wissenschaftliche vortraege ueber religioese fragen

Holtzmann, Oskar *see*
- Das christusbild der geschichte und das christusbild der dogmatik
- Jesus christus und das gemeinschaftsleben der menschen
- Neutestamentliche zeitgeschichte
- Religionsgeschichtliche vortraege
- Der tosephtatraktat berakot

Holtzmann, R *see* Thietmari merseburgensis episcopi chronicon (mgh6:9.bd)
Holtzmann, W *see* Koenig heinrich 1 und die heilige lanze
Holub, E *see*
- Sieben jahre in sued-afrika
- Von der capstadt ins land der maschukulumbe

Holwell, John Zephaniah *see* India tracts
Holwell, st peter monumental inscriptions – Arthur Weight Matthews 1913 – (= ser Bedfordshire parish register series) – 1mf – 9 – £1.25 – uk BedsFHS [929]
Holy baptism / Stone, Darwell – London, New York: Longmans, Green 1899 [mf ed 1991] – (= ser The oxford library of practical theology) – 1mf – 9 – 0-7905-8594-4 – mf#1989-1819 – us ATLA [240]
Holy bible – London, England. v1-3. 1911 – 2r – us UF Libraries [025]
The holy bible : consisting of the old and new covenants – 2nd minion type, rev ed. Edinburgh: G A Young, 1871 – 2mf – 9 – 0-524-08171-9 – mf#1992-1157 – us ATLA [220]
The holy bible : containing the old and new covenant, commonly called the old and new testament – Philadelphia: Jane Aitken, 1808 – 5mf – 9 – 0-524-02786-2 – mf#1987-6480 – us ATLA [220]
The holy bible : containing the old and new testaments / ed by American Revision Committee – standard ed. New York: T Nelson, c1901 – 3mf – 9 – 0-8370-1955-9 – mf#1987-6342 – us ATLA [220]
The holy bible : containing the old and new testaments – Boston: Walker, Wise, 1861, c1860 – 4mf – 9 – 0-8370-1984-2 – mf#1987-6371 – us ATLA [220]
The holy bible : containing the old and new testaments / ed by Cheyne, Thomas Kelly et al – [2nd ed] London, New York: Eyre and Spottiswoode, [1888?] – 13mf – 9 – 0-8370-1890-0 – mf#1987-6277 – us ATLA [220]
The holy bible : containing the old and new testaments – London: G Morrish, 1884-1890 – 14mf – 9 – 0-7905-8285-6 – mf#1987-6390 – us ATLA [220]
The holy bible : containing the old and new testaments – New York: Oxford University Press, 1911 – 3mf – 9 – 0-8370-1954-0 – mf#1987-6341 – us ATLA [220]
The holy bible : containing the old and new testaments translated out of the original tongues – Oxford: University Press, 1886 – 14mf – 9 – 0-7905-8286-4 – mf#1987-6391 – us ATLA [220]
The holy bible : containing the old and new testaments, with the apocryphal books, in the earliest english versions / ed by Forshall, Josiah et al – OUP 1850 [mf ed 1991] – 4v on 4r – 1 – 0-7905-8326-7 – mf#1987-b004 – us ATLA [220]
The holy bible : new testament, vol 3, romans to philemon. according to the authorized version (a.d. 1611) / ed by Cook, Frederic Charles – New York: Scribner, 1900 – 8mf – 9 – 0-524-08606-0 – mf#1993-0041 – us ATLA [225]
Holy bible alone is not the rule of faith / Spratt, John – Dublin, Ireland. 1852 – 1r – us UF Libraries [220]
The holy bible and the sacred books of the east : four addresses, to which is added a fifth address on zenana missions / Monier-Williams, Monier – London: Seeley, 1887 – 1mf – 9 – 0-524-00939-2 – mf#1990-2162 – us ATLA [230]
The holy bible containing the old and new testaments – New York: T Nelson, 1903 – 4mf – 220 – 0-8370-1804-8 – mf#1987-6192 – us ATLA [240]
The holy bible in modern english : containing the complete sacred scriptures of the old and new testaments – 3rd ed. London: SW Partridge, c1903 – 3mf – 9 – 0-524-02762-5 – mf#1987-6456 – us ATLA [220]
The holy catechism of nicolas bulgaris / Bulgaris, Nicolas; ed by Bromage, Richard Raikes – London: J Masters; New York: J Pott, 1893 – 1mf – 9 – 0-8370-7532-7 – (incl indes) – mf#1986-1532 – us ATLA [240]
The holy catholic church : her faith, works, triumphs – London: Burns & Oates, 1905 – 1mf – 9 – 0-8370-8328-1 – mf#1986-2328 – us ATLA [241]
The holy catholic church, the communion of saints : a discourse...newcastle, july 29th 1873...being the 4th [fernley] lecture... / Gregory, Benjamin – London: Wesleyan Conference Off 1873 [mf ed 1990] – 1mf – 9 – 0-7905-7634-1 – mf#1989-0859 – us ATLA [241]
The holy catholic church, the communion of saints : a study in the apostles' creed / Swete, Henry Barclay – London: Macmillan, 1915 – 1mf – 9 – 0-7905-9700-4 – (incl bibl ref) – mf#1989-1425 – us ATLA [241]
The holy comforter : his person and his work / Thompson, Joseph Parrish – New York: ADF Randolph, 1866 – 1mf – 9 – 0-524-00174-X – mf#1989-2874 – us ATLA [240]
Holy communion / Dimock, James F – London, England. 1844 – 1r – us UF Libraries [240]
The holy communion : its philosophy, theology and practice / Dalgairns, John Bernard – Dublin: J Duffy, 1861 – 1mf – 9 – 0-7905-7384-9 – mf#1989-0609 – us ATLA [240]
The holy communion / Stone, Darwell – London: Longmans, Green 1904 [mf ed 1992] – (= ser The oxford library of practical theology) – 1mf – 9 – 0-524-03051-0 – (incl bibl ref) – mf#1990-0808 – us ATLA [240]
Holy communion at a visitation / Ford, James – London, England. 1851 – 1r – us UF Libraries [240]
The holy corporation again – Quebec: [s.n.] 1852 [mf ed 1984]] – 9 – 0-665-45118-0 – mf#45118 – cn Canadiana [346]
Holy cross : a history of the invention, preservation, and disappearance of the wood known as the true cross / Prime, William Cowper – New York: Anson DF Randolph, c1877 – 1mf – 9 – 0-524-02866-4 – (incl bibl ref) – mf#1990-0723 – us ATLA [240]
Holy cross journal of law and public policy – v1-5. 1996-2000 – 9 – $90.00 set – mf#117781 – us Hein [342]
The holy eastern church : a popular outline of its history, doctrines, liturgies, and vestments / Neale, John Mason – 2nd ed. London: J T Hayes, 1873 – 1mf – 9 – 0-8370-8041-X – (incl bibl ref) – mf#1986-2041 – us ATLA [240]
Holy eucharist / Anglicanus, Clemens – London, England. 18– – 1r – us UF Libraries [240]
The holy eucharist / Hedley, John Cuthbert – London; New York: Longmans, Green, 1907 – (= ser Westminster Library (London, England)) – 1mf – 9 – 0-8370-7387-1 – (incl ind) – mf#1986-1387 – us ATLA [240]
The holy father and the living christ / Forsyth, Peter Taylor – London: Hodder and Stoughton, 1897 – (= ser Little books on religion) – 1mf – 9 – 0-8370-4969-5 – mf#1985-2969 – us ATLA [240]
The holy ghost : who is he? where is he? what does he? how may we help him? the ministry of the holy angels / Vaniman, Daniel – Mt Morris IL: Brethren's Pub Co 1896 [mf ed 1992] – 1mf – 9 – 0-524-03865-1 – mf#1990-4912 – us ATLA [242]
The holy ghost dispensation / Clark, Dougan – 2nd ed. Chicago: Publishing Association of Friends, 1892, c1891 – 1mf – 9 – 0-7905-3819-9 – mf#1989-0312 – us ATLA [240]
Holy ghost greek catholic church : marriage records – Cleveland OH, 1909-67 – 1r – 1 – (in latin & english) – us IHRC [929]
The holy ghost the comforter / Holden, George Frederick – London; New York: Longmans, Green, 1912 – 1mf – 9 – 0-7905-3909-8 – mf#1989-0402 – us ATLA [240]
The holy gospel : a comparison of the gospel text as it is given in the protestant and roman catholic bible versions in the english language in use in america – New York: Fleming H Revell, c1911 – 2mf – 9 – 0-8370-1983-4 – mf#1987-6370 – us ATLA [226]
The holy gospel according to saint john / McIntyre, John – London: Catholic Truth Society, 1899 – (= ser St. Edmund's College Series of Scripture Handbooks) – 1mf – 9 – 0-7905-1463-X – mf#1987-1463 – us ATLA [226]
The holy gospel according to saint luke – London: Catholic Truth Society, [1915?] – 1mf – 9 – 0-524-03963-1 – mf#1992-0006 – us ATLA [226]
The holy lake of the acts of rama : an english translation of tulasi das's ramacaritamanasa – London, New York: Oxford University Press, 1952 – (= ser Samp: indian books) – (trans by w douglas p hill) – us CRL [490]
The holy land / Kelman, John & Fulleylove, John – London: A & C Black, 1902 – 2mf – 9 – 0-524-05404-5 – (incl bibl ref) – mf#1992-0414 – us ATLA [915]
The holy land : moslem-christian case against zionist aggression – London, 1922 – 1mf – 9 – mf#J-28-126 – ne IDC [956]
Holy land and the bible / Geikie, Cunningham – New York, NY. v1-2. 19– – 1r – us UF Libraries [025]

The holy land and the bible : a book of scripture illustrations gathered in palestine / Geikie, John Cunningham – New York: James Pott, 1888. Chicago: Dep of Photodup, U of Chicago Lib, 1978 (1r); Evanston: American Theol Lib Assoc, 1984 (1r) – 1 – 0-8370-0748-8 – (incl bibl ref and index) – mf#1984-T064 – us ATLA [915]

The holy land, egypt, constantinople, athens, etc, etc : a series of 48 photographs, taken by f bedford, for h r h the prince of wales during the tour of the east, in which, by command, he accompanied his royal highness / Bedford, F – London, [1886] – 3mf – 9 – mf#H-8131 – ne IDC [910]

The holy land of the hindus : with seven letters on religious problems / Lacey, Robert Lee – London: Robert Scott, 1913 – (= ser Samp: indian books) – us CRL [280]

The holy land of the hindus : with seven letters on religious problems / Lacey, Robert Lee – London: Robert Scott, 1913 [mf ed 1995] – (= ser Yale coll) – xii/246p (ill) – 1 – 0-524-09133-1 – mf#1995-0133 – us ATLA [280]

A holy life and how to live it / Macgregor, George Hogarth Carnaby – New York: Fleming H Revell, c1897 [mf ed 1986] – 1mf – 9 – 0-8370-7238-7 – mf#1986-1238 – us ATLA [240]

Holy life necessary to constitute a true christian / Morris, Thomas – Calcutta, India. 1826 – 1r – us UF Libraries [240]

The holy man of santa clara : or, life, virtues, and miracles of fr. magin catala, o.f.m / Engelhardt, Zephyrin – San Francisco, CA: James H Barry Co, 1909 – 1mf – 9 – 0-524-00988-0 – mf#1990-0265 – us ATLA [240]

Holy matrimony / Little, William John Knox – London: Longmans, Green 1913 [mf ed 1992] – (= ser The oxford library of practical theology) – 1mf – 9 – 0-524-04840-1 – mf#1990-1332 – us ATLA [230]

The holy of holies : sermons on fourteenth, fifteenth, and sixteenth chapters of the gospel of john / Maclaren, Alexander – London: Alexander & Shepheard, 1890 – 1mf – 9 – 0-7905-3354-5 – mf#1987-3354 – us ATLA [220]

Holy orders / Whitham, Arthur Richard – London, New York: Longmans, Green 1903 [mf ed 1989] – (= ser The oxford library of practical theology) – 1mf – 9 – 0-7905-2623-9 – (incl ind) – mf#1987-2623 – us ATLA [242]

Holy places of india / Law, Bimala Churn – [Calcutta]: Calcutta Geographical Society, 1940 – (= ser Samp: indian books) – us CRL [280]

The holy places of jerusalem / Lewis, Thomas Hayter – London: John Murray, 1888 – 1mf – 9 – 0-8370-9963-3 – (incl bibl ref and index) – mf#1986-3963 – us ATLA [720]

The holy roman empire / Bryce, James Bryce, Viscount – new rev enl ed. New York: Macmillan, 1904 – 2mf – 9 – 0-7905-5811-4 – mf#1988-1811 – us ATLA [930]

The holy rule of st benedict / Hayes, Bernard – London: R & T Washbourne [1908?] [mf ed 1993] – 1mf – 9 – 0-524-08381-9 – (int by j c hedley) – mf#1993-3081 – us ATLA [241]

Holy sacrifice of the mass / Mueller, Michael – New York: Benziger, 1883 – 2mf – 9 – 0-8370-7492-4 – (incl bibl ref) – mf#1986-1492 – us ATLA [240]

Holy scripture and the pope's supremacy contrasted / Barrow, Isaac – London, England. 1850 – 1r – us UF Libraries [240]

Holy scripture verified : or, the divine authority of the bible confirmed by an appeal to facts of science, history, and human consciousnesss / Redford, George – new ed. London: Jackson & Walford, 1853 [mf ed 1993] – (= ser The congregational lecture 5th ser) – 2mf – 9 – 0-524-07186-1 – mf#1992-1056 – us ATLA [220]

The holy scriptures – Plano, IL: Publ by the Church of Jesus Christ of Latter-Day Saints, 1867 – 3mf – 9 – 0-8370-1891-9 – mf#1987-6278 – us ATLA [220]

Holy scriptures analyzed / Cooper, Robert – London, England. 1854 – 1r – us UF Libraries [220]

The holy scriptures of the old covenant in a revised translation / Wellbeloved, Charles et al – London: Longman, Brown, Green, Longmans & Roberts, 1859-62 [mf ed 1989] – 3v on 4mf – 9 – 0-8370-1171-X – mf#1987-6007 – us ATLA [221]

Holy scriptures the only standard of divine truth / Fletcher, Joseph – London, England. 18– – 1r – us UF Libraries [220]

The holy see and the wandering of the nations : from st. leo 1. to st. gregory 1 / Allies, Thomas William – London: Burns & Oates; New York: Catholic Publication Society, 1888 – 1mf – 9 – 0-8370-7120-8 – (incl bibl ref and index) – mf#1986-1120 – us ATLA [240]

The holy sepulchre and the temple at jerusalem : being the substance of two lectures delivered in the royal institution, albemarle street, on the 21st february, 1862, and 3rd march, 1865 / Fergusson, James – London: John Murray, 1865 – 1mf – 9 – 0-7905-0941-5 – (incl bibl ref) – mf#1987-0941 – us ATLA [240]

The holy service : a short treatise on worship and the public service of god's house / Sheatsley, Jacob – Columbus, OH: Press of Lutheran Book Concern, 1897 – 1mf – 9 – 0-524-03085-5 – mf#1990-4574 – us ATLA [240]

Holy spirit / Montgomery, John – Belfast, Northern Ireland. 1859 – 1r – us UF Libraries [240]

The holy spirit : his personality, divinity, office, and agency,in the regeneration and sanctification of man / Dewar, Daniel – London: Ward, [1847?] Chicago: U of Chicago Lib, 1977 (1r); Evanston: American Theol Lib Assoc, 1984 (1r) – 1 – 0-8370-1551-0 – mf#1984-B493 – us ATLA [220]

The holy spirit : his personality, mission and modes of activity / Garrison, James Harvey – St Louis, MO: Christian Pub Co, 1905 – 1mf – 9 – 0-524-06716-3 – mf#1991-2746 – us ATLA [240]

The holy spirit : his work and mission: a discourse...hanley, staffordshire, july 25th 1870...being the 1st [fernley] lecture... / Osborn, George – London: Wesleyan Conference Off 1870 [mf ed 1989] – 1mf – 9 – 0-7905-2031-1 – mf#1987-2031 – us ATLA [240]

The holy spirit and christian privilege / Selby, Thomas Gunn – London: CH Kelly, 1894 – (= ser The Life Indeed Series) – 1mf – 9 – 0-7905-7464-0 – mf#1989-0689 – us ATLA [240]

The holy spirit and the human mind / Johnson, Ashley Sidney – Knoxville, Tenn: Gant-Ogden, 1903 – 1mf – 9 – 0-524-04097-4 – mf#1992-0055 – us ATLA [240]

The holy spirit in faith and experience / Humphries, Arthur Lewis – 2nd ed. London: WA Hammond, 1911 – (= ser Hartley Lecture) – 1mf – 9 – 0-7905-3918-7 – mf#1989-0411 – us ATLA [240]

The holy spirit in missions : six lectures / Gordon, Adoniram Judson – New York: Fleming H Revell, c1893 – 1mf – 9 – 0-8370-6054-0 – (incl bibl ref and index) – mf#1986-0054 – us ATLA [240]

The holy spirit in the new testament : a study of primitive christian teaching / Swete, Henry Barclay – London: Macmillan, 1909 – 1mf – 9 – 0-8370-9424-0 – (incl bibl ref and indexes) – mf#1986-3424 – us ATLA [225]

The holy spirit in the new testament scriptures / Scofield, William Campbell – Chicago:Fleming H. Revell, c1896 – 1mf – 9 – 0-8370-5188-6 – mf#1985-3188 – us ATLA [225]

The holy spirit of god / Thomas, William Henry Griffith – London; New York: Longmans, Green, 1913 – 1mf – 9 – 0-7905-0398-0 – (includes bibliographies and indexes) – mf#1987-0398 – us ATLA [240]

Holy spirit, the spirit of truth / Dore, James – London, England. 1805 – 1r – us UF Libraries [240]

The holy spirit then and now / Johnson, Elias Henry – Philadelphia: Griffith & Rowland, 1904 – 1mf – 9 – 0-7905-9976-7 – mf#1989-1701 – us ATLA [240]

Holy springs baptist church. raleigh association. wake county. north carolina : church records – 1822-1918 – 1 – 65.79 – us Southern Baptist [242]

Holy Tabernacle of the Most High *see* Truth: holy tabernacle of the most high nubian

Holy temple bible truth – 1992 summer-fall, 1993 spring, 1994 fall, 1995 spring, 1996 spring, 1997 spring – 1r – 1 – mf#4027060 – us WHS [220]

Holy Temple Church of God in Christ *see* Trumpet

Holy temple good news / Greater Holy Temple COGIC (Jacksonville, FL) – 1980 feb – 1r – 1 – mf#4114850 – us WHS [243]

The holy trinity : a study of the self-revelation of god / Mylne, Louis George – London; New York: Longmans, Green, 1916 – 1mf – 9 – 0-7905-9531-1 – mf#1989-1236 – us ATLA [240]

The holy war in tripoli / Abbott, George Frederick – London: E Arnold, 1912 – 1 – us CRL [960]

The holy women of old : seventeen lessons / Smith, Mary Ann – Edinburgh: John Anderson, 1897 – 1mf – 9 – 0-8370-5298-X – mf#1985-3298 – us ATLA [220]

The holy word in its own defence : addressed to bishop colenso and all other earnest seekers after truth / Silver, Abiel – 2nd rev ed. Boston: TH Carter, 1867 – 1mf – 9 – 0-524-05823-7 – mf#1992-0650 – us ATLA [220]

Holy writ and modern thought : a review of times and teachers / Coxe, Arthur Cleveland – New York:E.P. Dutton, 1892 – 1mf – 9 – 0-8370-2766-7 – mf#1985-0766 – us ATLA [220]

The holy writings of the sikhs / Macauliffe, Max – Allahabad: Christian Association Press, 1900 – 1mf – 9 – 0-524-07495-X – mf#1991-0116 – us ATLA [280]

The holy year of jubilee : an account of the history and ceremonial of the roman jubilee / Thurston, Herbert – London: Sands, 1900 – 1mf – 9 – 0-7905-6511-0 – (incl bibl ref) – mf#1988-2511 – us ATLA [240]

The holy year of jubilee : an account of the history and ceremonial of the roman jubilee / Thurston, Herbert – London: Sands, 1900 – mf – us ATLA [240]

Holyhead chronicle – [Wales] Isle of Anglesey jan 1909-dec 1950 [mf ed 2004] – 48r – 1 – uk Newsplan [072]

Holyhead mail & anglesey herald – [Wales] Isle of Anglesey jan 1885-dec 1950 [mf ed 2003] – 58r – 1 – (missing: 1893, 1897; cont as: holyhead mail & llandudno & colwyn bay herald [jan 1922-dec 1932]; holyhead mail [jan 1933-dec 1937]; holyhead & anglesey mail [jan 1938-dec 1950]) – uk Newsplan [072]

Holyland = Hilligenlei / Frenssen, Gustav – Boston: Page c1906 [mf ed 1989] – 1r – 1 – (exclusive authorized trans of 'hilligenlei'. filmed with: jorn uhl & other titles) – mf#7265 – us Wisconsin U Libr [830]

Holynski, Aleksander *see* La californie et les routes interoceaniques

Holyoake, G J *see* Reasoner series, 1846-72

Holyoake, George Jacob *see*
- Among the americans and a stranger in america
- Case of thomas pooley
- Logic of death
- Selected pamphlets, 1841-1904
- The trial of theism

Holyoake, Manfred *see* The conservation of pictures

The holyoake papers, 1831-1905 : from the bishopsgate institute, london – 9r – 1 – (with ind) – mf#96609 – uk Microform Academic [362]

The holyoake papers, 1835-1906 : from the co-operative union library, manchester – 2pt – 1 – (pt1: 1840-79 6r [96188]. pt2: 1835-1906 12r [96636]. ind: 1r [96656]. int & ind comp by edward royle) – uk Microform Academic [362]

Holyoake papers, the... *see* Religion, radicalism and freethought in victorian and edwardian britain

Holyoke 1850-1895 – Oxford, MA (mf ed 1986) – (= ser Massachusetts vital records) – 103mf – 9 – 0-87623-026-5 – (mf 1-5: births 1850-70. mf 6-11: births 1871-82. mf 12-16: births 1882-87. mf 17-21: births 1887-91. mf 22-26: births 1891-95. mf 27-32: birth index 1850-86. mf 33-37: birth index 1887-95. mf 38-42: marriages 1850-72. mf 43-47: marriages 1872-81. mf 48-53: marriages 1881-87. mf 54-55: marriages 1887-89. mf 56-64: marriages 1889-96. mf 65-70: marriage index 1850-91. mf 71-75: marriage index 1892-1906. mf 76-79: deaths 1850-71. mf 80-84: deaths 1871-84. mf 85-89: deaths 1884-92. mf 90-94: deaths 1892-98. mf 95-99: death index 1850-91. mf 100-103: death index 1892-1902) – us Archive [978]

Holyoke enterprise *see* Miscellaneous newspapers of the colorado historical society

Holyoke, Samuel *see*
- The instrumental assistant

Holyoke tribune *see* Miscellaneous newspapers of the colorado historical society

Holz als roh- und werkstoff – Heidelberg. 1981-1994 (1) 1981-1983 (5) 1981-1983 (9) – ISSN: 0018-3768 – mf#13177 – us UMI ProQuest [690]

Holz, Anita *see* Briefe

Holz, Arno *see*
- Die akte arno holz
- Die befreite deutsche wortkunst
- Briefe
- Deutsche buehnenspiele
- Emanuel geibel
- Die kunst, ihr wesen und ihre gesetze
- Socialaristokraten

Holz, Tania Jacqueline *see* Guidelines for dealing with learners with adhd behaviour

Holzamer, Wilhelm *see*
- Conrad ferdinand meyer
- Heinrich heine

Holzapfel, Heribert *see* Handbuch der geschichte des franziskanerordens

Holzapfel, Karl Maria *see* Einer baut einen dom

Holzapfel, Otto *see* Studien zur formelhaftigkeit der mittelalterlichen daenischen volksballade

Der holzarbeiter – Reichenberg (Liberec CZ), 1921 5 jan-1923 2 jan [gaps] – 1r – 1 – gw Misc Inst [634]

Holzarbeiter-zeitung – Hamburg, Berlin DE, 1968 jul-1993 – 9r – 1 – (title varies: 1933 n24-1935 n37 & 1940-41: der deutsche holzarbeiter. filmed by other misc inst: 1905 7 jan-1908 26 sep [2r]; 1893-1933 [13r]. with suppls: der betriebsrat in der holzindustrie 1920 aug-1923 jul, 1924-27; holzarbeiter frauenblatt 1914 nov-dec, 1919 oct-1923 aug) – gw Misc Inst [634]

Holzarbeiter-zeitung – v15, no.25. 1907. (Serial publications of German trade unions in the Memorial Library, University of Wisconsin-Madison.) – 1 – us Wisconsin U Libr [331]

Holzbauer, Ignaz *see* Cantata con istromenti

Holzbauer, Martin *see* Alessandro nell'indie

Holzenthal, Georg *see* Briefe ueber deutschland, frankreich, spanien, die balearischen inseln, das suedliche schottland und holland

Holzhalb, Hans Jacob *see* Allgemeines, helvetisches, eydgenoessisches oder schweizerisches lexicon (ael1/8)

Holzhey, Carl *see* Fuenfundsiebzig punkte zur beantwortung der frage, absolute oder relative wahrheit der hl schrift?

Holzinger, Heinrich *see*
- Das buch josua
- Einleitung in den hexateuch
- Exodus erklaert
- Genesis
- Numeri

Holzinger, Regina *see* Dux-gong

Holzman, Lynn M *see* An historical analysis of national collegiate athletic association freshman eligibility

Holzmann, M *see*
- Deutsches anonymen-lexikon, 1501-1910
- Deutsches pseudonymen-lexikon

Holzmann, Michael *see*
- Aus dem lager der goethe-gegner
- Ludwig boerne
- Ludwig borne

Holzmindisches wochenblatt – Holzminden DE, 1785 2 jul-1789 26 sep, 1790 & 1791, 1793-1794 4 jan – 2r – 1 – gw Misc Inst [074]

Holzner, Anton *see* Canticum virginis seu magnificat...

Hom, Jim *see* Journal of forensic neuropsychology

Homage of eminent persons to the book / Bailey, Samuel Wordswoth – New York: [s.n.], 1869, [Boston: Rand, Avery & Frye] – 1mf – 9 – 0-8370-2157-X – mf#1985-0157 – us ATLA [220]

Homage to tagore / Anand, Mulk Raj – Lahore: Sangam Publishers, 1946 – (= ser Samp: indian books) – us CRL [954]

Homan, William *see* The scottish rite

Homanner, Wilhelm *see* Die dauer der oeffentlichen wirksamkeit jesu

Homans, Isaac Smith *see*
- The commercial laws of the states: a summary of the laws relating to arrest assignments attachment...&c
- The national bank act.

Homberger anzeiger – Homberg, Bezirk Kassel DE, 1894 8 dec-1895 30 sep – 1r – 1 – (incl suppl) – gw Misc Inst [074]

Homberger kreisblatt *see* Kreisblatt fuer den kreis homberg

Homberger tageblatt – Homberg, Bezirk Kassel DE, 1902 2 nov-1906 20 sep – 10r – 1 – (incl suppl) – gw Misc Inst [074]

Homberger zeitung 1890 – Homberg, Bezirk Kassel DE, 1890 16 apr-28 jun – 1r – 1 – (kirchhain? incl suppls) – gw Misc Inst [074]

Homberger zeitung 1928 – Homberg, Bezirk Kassel DE, 1928 1 nov-1934 30 jun, 1935-1937 25 mar – 13r – 1 – gw Misc Inst [074]

Hombre a traves de un libro / Hernandez U Urbina, Francisco – Tegucigalpa, Mexico. 1943 – 1r – us UF Libraries [972]

El hombre ante el hombre / Caba, Pedro – Badajoz: imp de la dip provincial, 1969 – 1 – (sep de la revista de estudios extremenos) – sp Bibl Santa Ana [946]

El hombre contra la naturaleza / Caba, Pedro – sp Bibl Santa Ana [946]

Un hombre de estado / Lopez de Ayala, Adelardo – 1851 – 9 – sp Bibl Santa Ana [820]

Hombre de hierro : (novela) / Blanco-Fombona, Rufino – Madrid, Spain. 1917 – 1r – us UF Libraries [830]

Hombre de las leyes / Grillo, Max – Bogota, Colombia. 1940 – 1r – us UF Libraries [972]

Hombre de los pies de agua / Oscar, Armando – Ciudad Trujillo, Dominican Republic. 1959 – 1r – us UF Libraries [972]

Hombre de negocios puertorriqueno / Cochran, Thomas Childs – Rio Piedras, Puerto Rico. 1961 – 1r – us UF Libraries [972]

Hombre del 95 / Peraza Sarausa, Fermin – Habana, Cuba. 1950 – 1r – us UF Libraries [972]

Hombre del pueblo / Pedreira, Antonio Salvador – San Juan, Puerto Rico. 1937 – 1r – us UF Libraries [972]

Hombre frente a la violencia / Sevillano Quinones, Lino Antonio – Bogota, Colombia. 1965 – 1r – us UF Libraries [972]
Hombre. grabados de francisco mateos / Alvarez Lencero, Luis – Madrid: Trilce, 1961 – 1 – sp Bibl Santa Ana [700]
Hombre pequeno / Triff, Eduardo – Habana, Cuba. 1943 – 1r – us UF Libraries [972]
Hombre que parecia un caballo, y otros cuentos / Arevalo Martinez, Rafael – San Salvador, El Salvador. 1958 – 1r – us UF Libraries [972]
Hombre que parecia un cabillo, y las rosas de enga / Arevalo Martinez, Rafael – Guatemala, 1927 – 1r – us UF Libraries [972]
El hombre que perdio su sombra = Peter schlemihls wundersame geschichte / Chamisso, Adelbert von – 2nd ed. Santiago de Chile: empresa editora zig-zag 1966, c1945 [mf ed 1993] – 1r – 1 – (spanish trans of peter schlemihls wundersame geschichte; incl bibl ref; filmed with: gedichte / wilhelm rauschenbuch [ed]) – mf#8537 – us Wisconsin U Libr [830]
Hombre solo / Martin, Ramon – Habana, Cuba. 1941 – 1r – us UF Libraries [972]
Hombre y el maiz / Valladares, Leon A – Guatemala, 1957 – 1r – us UF Libraries [972]
Hombre y la encrucijada / Munoz Meany, Enrique – Guatemala, 1950 – 1r – us UF Libraries [972]
Hombre y su angustia / Franco Oppenheimer, Felix – Rio Piedras, Puerto Rico. 1960 – 1r – us UF Libraries [972]
Hombres / Estrada Monsalve, Joaquin – Bogota, Colombia. 1953 – 1r – us UF Libraries [972]
Hombres contra la muerte / Espino, Miguel Angel – Mexico City? Mexico. 1947 – 1r – us UF Libraries [972]
Hombres de america / Rodo, Jose Enrique – Barcelona, Spain. 1924 – 1r – us UF Libraries [972]
Hombres de colombia / Vallejo, Alejandro – Caracas, Venezuela. 1950 – 1r – us UF Libraries [972]
Hombres de mi tierra / Morales Otero, Pablo – San Juan, Puerto Rico. 1965 – 1r – us UF Libraries [972]
Hombres de pensamiento / Carias Reyes, Marcos – Tegucigalpa, Mexico. 1947 – 1r – us UF Libraries [972]
Hombres del 68 rafael morales y gonzalez / Morales Y Morales, Vidal – Habana, Cuba. 1904 – 1r – us UF Libraries [972]
Hombres del pasado / Nieto Caballero, Luis Eduardo – Bogota, Colombia. 1944 – 1r – us UF Libraries [972]
Hombres nuevos y nuevos cuadros : conferencia nacional de juventudes, enero de 1937 / Medrano, Trifon – [Valencia: s.n. 1937] – (= ser Blodgett coll) – 9 – mf#w1041 – us Harvard [946]
Hombres y ciudades / Otero Munoz, Gustavo – Bogota, Colombia. 1948 – 1r – us UF Libraries [972]
Hombres y cuentos / Agostini, Victor – Habana, Cuba. 1955 – 1r – us UF Libraries [972]
Hombron *see* Zoologie
Homburger, L (Lilias) *see* Prefixes nominaux dans les parlers peul, haoussa et bantous
Homburger, Lilias *see* Negro-african languages
Home – New York. 1981-1988 (1) 1981-1988 (5) 1981-1988 (9) – ISSN: 0278-2839 – mf#12844 – us UMI ProQuest [720]
Home advocate – Mapleton IA. 1887 jan 14 – 1r – 1 – (continued by: advocate (mapleton, ia: 1895)) – mf#851188 – us WHS [071]
Home, Amal *see* Rammohun roy, the man and his work
Home among the orange groves in crescent city, florida – Jacksonville, FL. 1876 – 1r – us UF Libraries [634]
Home and abroad – Oneonta, NY. 1869-1870 (1) – mf#69304 – us UMI ProQuest [071]
Home and country – New York. 1893-1897 (1) – mf#2929 – us UMI ProQuest [978]
Home and farm – Sydney, Australia 3 nov 1902-1 may 1908 (imperfect) – 1 – (cont: martin's home & farm: a monthly journal for australian farmers [ns] 1 dec 1893-1 oct 1902]) – uk British Libr Newspaper [630]
Home and foreign fields *see* Periodicals
Home and foreign journal *see* Periodicals
The home and foreign record / Presbyterian Church in the U.S.A. – Philadelphia, Pa. v1-11. 1850-1867 – 4r – 1 – $200.00 – us Presbyterian [240]
Home and foreign record of the canada presbyterian church – Toronto: Printed...by W C Chewett, [1861-1875] – 9 – (cont by: presbyterian record for the dominion of canada. incl ind) – mf#P06018 – cn Canadiana [242]
Home and foreign record of the presbyterian church of the lower provinces of british north america – Halifax, N.S.: J Barnes, [1861-1875] – 9 – (cont by: the presbyterian record for the dominion of canada) – mf#P04281 – cn Canadiana [242]
Home and foreign review – London. 1862-1864 – 1 – mf#4265 – us UMI ProQuest [073]
Home and garden supply merchandiser – Minnetonka. 1964-1979 (1) 1971-1979 (5) 1974-1979 (9) – (cont by: garden supply retailer) – ISSN: 0018-3954 – mf#1668 – us UMI ProQuest [640]
Home and school – Toronto: W Briggs, [1883?-189- or 19-] – 9 – ISSN: 1190-6235 – mf#P04663 – cn Canadiana [242]
The home and the world / Tagore, Rabindranath – London: Macmillan and Co, 1921 – (= ser Samp: indian books) – us CRL [490]
A home and work for every man : and an invincible british empire / Hunt, James – London, [1895] – (= ser 19th c british colonization) – 2mf – 9 – mf#1.1.9458 – uk Chadwyck [330]
Home and youth – Toronto: Home and Youth Pub. Co, [1897-19-] – 9 – (cont: our home) – mf#P04307 – cn Canadiana [640]
The home base of missions : with supplement, presentation and discussion of the report in the conference on 23rd june 1910 – Edinburgh: Publ for the World Missionary Conference by Oliphant, Anderson & Ferrier; New York: Fleming H Revell, [1910?] – 2mf – 9 – 0-8370-6476-7 – (incl indes) – mf#1986-0476 – us ATLA [240]
The home beyond : or, views of heaven and its relation to earth / ed by Fallows, Samuel – St Louis MO: W L Holloway 1886, c1884 [mf ed 1991] – 2mf [ill] – 9 – 0-7905-9196-0 – mf#1989-2421 – us ATLA [230]
Home building and beautification / Bryan, F Macdonald – s.l, s.l? 193-? – 1r – us UF Libraries [640]
Home bulletin – Hampton, VA. 1884-1891 (1) – mf#66726 – us UMI ProQuest [071]
Home bulletin – Newport News, VA. 1884-1886 (1) – mf#66770 – us UMI ProQuest [071]
Home center magazine – Lincolnshire. 1899-1989 (1) 1970-1989 (5) 1977-1989 (9) – (cont by: home improvement center) – ISSN: 0194-1321 – mf#841 – us UMI ProQuest [640]
Home channel news – New York. 2002+ (1,5,9) – mf#18749,01 – us UMI ProQuest [690]
Home circle leader – Toronto: Home Circle Print and Pub, v1 n1(oct 1889)- – 9 – mf#P04310 – cn Canadiana [360]
Home clippings – vol1-4 (jan-mar 1907); vol 17-19 (nov 1907-mar 1908) – 5r – mf#ZB 31 – nz Nat Libr [079]
The home colony : a guide for investors and settlers in newfoundland / Hall, Edward Hepple – London: E Stanford, [1882?] – 1mf – 9 – 0-665-32681-5 – mf#32681 – cn Canadiana [917]
Home confinement : an evolving sanction in the federal criminal justice system / Hofer, Paul J & Meierhoefer, Barbara S – Washington: FJC, 1987 – 1mf – 9 – $1.50 – mf#LLMC 95-356 – us LLMC [345]
Home cookin' – n2,4 [1972 sep 14, oct 12] – 1r – 1 – mf#1583861 – us WHS [640]
Home court advantage in men's and women's big ten intercollegiate basketball / Stoklosa, S M – 1991 – 1mf – 9 – $4.00 – us Kinesology [150]
Home, David Milne *see* Observations of the probable cause of the failure of the potato crop
Home defender – v1 n1-v3 n10 [1912 may-1914 oct 1] – 1r – 1 – mf#910297 – us WHS [071]
Home department magazine – 1914-46. (Better Home. 1935-46) – 1 – $309.19 – us Southern Baptist [640]
The home department of the sunday school : what it is, and what it does / Withrow, William Henry – Toronto: W Briggs, [1898?] – 1mf – 9 – 0-665-88973-9 – mf#88973 – cn Canadiana [240]
Home economics as applied to the choice and preparation of food / Peacock, Jean B – Fredericton NB: Dept of Agriculture, [1914?] [mf ed 1997] – 1mf – 9 – 0-665-85575-3 – mf#85575 – cn Canadiana [640]
Home economics research journal – Washington. 1972-1994 (1) 1975-1994 (5) 1975-1994 (9) – (cont by: family and consumer sciences research journal) – ISSN: 0046-7774 – mf#10308 – us UMI ProQuest [640]
Home evangelist *see* Home mission herald
Home evangelization : a view of the wants and prospects of our country – New York: American Tract Society, [185-?] – 1mf – 9 – 0-8370-6836-3 – mf#1986-0836 – us ATLA [240]
Home for Aged and Infirm Colored People [Chicago IL] *see* Annual report of the home for aged and infirm colored people
Home for Aged and Infirm Colored Persons (Philadelphia PA) *see*
– The 9th annual examination of the classes of the institute for colored youth will take place friday, may 3 1861
– The 14th annual commencement exercises of the institute for colored youth will take place 5th and 6th days, eleventh mo 1st and 2d (thursday and friday, november 1st and 2d)
– Annual commencement of the institute for colored youth at association hall
– Annual report of the board of managers of the home for aged and infirm colored persons
– Annual report of the board of managers of the institute for colored youth
– Constitution, by-laws and rules of the home for aged and infirm colored persons
– Objects and regulations of the institute for colored youth
– Objects of the institute for colored youth
– Proceedings of the...annual meeting of the home for aged and infirm colored persons, held...
Home for Aged Colored People [Chicago IL] *see* Annual report from...to...
L'home franc : ou journal tout noubel en patois – Toulouse. n1. fevr 1791 – 1 – fr ACRPP [073]
Home friend – London. 1852-1856 – 1 – mf#4721 – us UMI ProQuest [073]
Home garden's natural gardening magazine – New York. 1914-1973 (1) 1969-1973 (5) 1970-1972 (9) – ISSN: 0090-7650 – mf#882 – us UMI ProQuest [630]
Home health care management and practice – Frederick. 1995-1999 (1) 1995-1999 (5) 1995-1999 (9) – (cont: journal of home health care practice) – ISSN: 1084-8223 – mf#16712,01 – us UMI ProQuest [360]
Home health care services quarterly / ed by Simmons, W June – v1- 1979- – 1, 9 ($265.00 in US $371.00 outside hardcopy subsc) – us Haworth [610]
Home healthcare nurse – Philadelphia. 1986+ (1,5,9) – ISSN: 0884-741X – mf#15987 – us UMI ProQuest [610]
Home improvement center – Lincolnshire. 1989-1993 (1) 1989-1993 (5) 1989-1993 (9) – (cont: home center magazine) – ISSN: 1045-9367 – mf#841,01 – us UMI ProQuest [640]
Home in famous florida on ten years' time – Cocoa, FL. 19–? – 1r – us UF Libraries [978]
Home industries, canada's national policy, protection to native products, development of field and factory : speeches by leading members of parliament: free trade theories vs national prosperity – [Ottawa?: s.n.], 1876 [mf ed 1981] – 1mf – 9 – mf#24080 – cn Canadiana [330]
Home influence : a tale for mothers and daughters / Aguilar, Grace – London: R Groombridge & Sons. 2v. 1847 – (= ser 19th c women writers) – 8mf – 9 – mf#5.1.4 – uk Chadwyck [830]
Home intelligence reports, 1940-1944 : pro class inf1, boxes 264 and 292 – 4r – 1 – (whole range of subjects covered) – mf#C39-27660 – us Primary [941]
Home intelligencer – Mineral Point WI. 1860 jan 5-1863 dec 19, 1860 sep 13, 1862 may 31 – 2r – 1 – mf#1212009 – us WHS [071]
Home journal – Lafayette, IN. 1896-1904 (1) – mf#62867 – us UMI ProQuest [071]
The home journal – Toronto: [s.n, 1861-18– or 19–] – 9 – mf#P06001 – cn Canadiana [071]
The home journal almanac for... – St Thomas, Ont: s.n, 18– – 9 – mf#A00369 – cn Canadiana [971]
The home journal news – New York. N.Y – 1 – us NY Public [071]
Home knowledge high school learners have of the formation of tornadoes : a case study of high school learners in rural kwazulu-natal: research report / Zulu, Langelihle – Pretoria: Vista University 2001 [mf ed 2001] – 2mf [ill] – 9 – (incl bibl ref) – mf#mfm15218 – sa Unisa [373]
Home knowledge monthly review – Toronto: Belden Bros, [1890?-18– or 19–] – 9 – ISSN: 1190-7576 – mf#P04272 – cn Canadiana [073]
Home league – Hartford WI. 1860 aug 11-1864 mar 5 – 1r – 1 – mf#921919 – us WHS [071]
Home life – (Successor to Better Home). 1947-61 – 1 – $144.97 – us Southern Baptist [640]
Home life and reminiscences of alexander campbell / Campbell, Selina Huntington – St Louis: J Burns c1882 [mf ed 1993] – (= ser Christian church (disciples of christ) coll) – 1mf – 9 – 0-524-06986-7 – mf#1991-2839 – us ATLA [242]
Home life in florida / Warner, Helen Garnie – Louisville, KY. 1889 – 1r – us UF Libraries [978]
Home lyrics : a book of poems / Battersby, Hannah S – London: Ward, Lock and Tyler, [188-?] – 3mf – 9 – 0-665-90792-3 – mf#90792 – cn Canadiana [810]
Home maintenance and improvement – Lincolnshire. 1965-1966 (1) – mf#1825 – us UMI ProQuest [640]
Home Makers Organized for More Employment et al *see* Maine statewide newsletter
Home market and farm : how the agricultural and industrial prosperity of canada depend on each other, and will be hurt by reciprocity with the united states / Canadian National League – Toronto: The League, [1911?] – 1mf – 9 – 0-665-71272-3 – mf#71272 – cn Canadiana [630]
Home mechanix – New York. 1985-1996 (1) 1985-1996 (5) 1985-1996 (9) – (cont by: today's homeowner. cont: mechanix illustrated) – ISSN: 8755-0423 – mf#2418,01 – us UMI ProQuest [640]
Home mirror – Durand WI. 1862 may 17 – 1r – 1 – mf#964277 – us WHS [071]
Home mission college review – 1927 may-1930 jan – 1r – 1 – mf#3201533 – us WHS [071]
Home mission herald – Atlanta, Ga. v1-4. 1908-1911 – 1r – 1 – $50.00 – us Presbyterian [240]
Home mission herald – v1-17. 1849-66; v41-42. 1873-74 [complete] – 1r – 1 – (suspended: apr 1866-72. title varies) – mf#ATLA R0118 – us ATLA [240]
Home mission heroes : a series of sketches – New York City: Literature Dept, Presbyterian Home Missions, 1904 – 1mf – 9 – 0-8370-6782-0 – (includes bibliographies) – mf#1986-0782 – us ATLA [920]
The home mission journal – St John, NB: Cte of the Home Mission Board of New Brunswick, [1898-1904] – 9 – ISSN: 1190-7134 – mf#P04294 – cn Canadiana [242]
Home mission monthly – v1-38. 1886-1924 – 1 – $250.00 – us Presbyterian [240]
Home mission monthly, 1886-1924 – 5r – 1 – $425.00 – mf#D3328 – Scholarly Resources – us Presbyterian [240]
Home mission record *see* Home mission herald
The home mission task : its fundamental character, magnitude and present urgency / ed by Masters, Victor I – Atlanta: Home Mission Board of the Southern Baptist Convention, 1912 – 1mf – us ATLA [240]
The home mission task : its fundamental character, magnitude and present urgency / ed by Masters, Victor Irvine – Atlanta: Home Mission Board of the Southern Baptist Convention, 1912 – 1mf – 9 – 0-7905-5198-5 – mf#1988-1198 – us ATLA [240]
Home missionary / American Home Missionary Society – v44 n1-12 [1871 may-1872 apr], v63 n1-12 [1890 may-1891 apr], v69 – 3r – 1 – (cont: home missionary and pastor's journal; cont by: american missionary) – mf#1385165 – us WHS [242]
Home missionary and american pastor's journal / Congregational Home Missionary Society – v1 [1829] – 1r – 1 – (continued by: home missionary and pastor's journal) – mf#1384297 – us WHS [242]
Home missionary and unbelievers – London, England. 18– – 1r – us UF Libraries [240]
Home missions : encouragement from the past, exigencies of the present, hope for the future / Chapin, Aaron Lucius – New York: American Home Missionary Society, 1878 – 1mf – 9 – 0-524-06710-4 – mf#1991-2740 – us ATLA [240]
Home missions : a sermon in behalf of the american home missionary society / Barnes, Albert – New York: Printed for the American Home Missionary Society by W Osborn, 1849 – 1mf – 9 – 0-7905-5622-7 – mf#1988-1622 – us ATLA [240]
Home missions in action / Allen, Edith Hedden – New York: Revell c1915 [mf ed 1990] – (= ser Interdemoninational home mission study course) – 1mf – 9 – 0-7905-4241-2 – (incl bibl ref) – mf#1988-0241 – us ATLA [240]
The home missions task of the church of the united brethren in christ / Koontz, Harry R – Bonebrake Theological Seminary, undated – 1r – 1 – $35.00 – mf^{s}-44 [pt1] – us Commission [242]
The home missions task of the church of the united brethren in christ / MacCanon, G E – Bonebrake Theological Seminary, undated – 1r – 1 – $35.00 – mf^{s}-44 [pt2] – us Commission [242]
Home monthly and general and commercial advertiser – [Scotland] Perth: T M McGregor sep 1890-jul 1895 (mthly) [mf ed 2003] – 2r – 1 – (cont by: perth home monthly and commercial advertiser [jan 1894-jul 1895]) – uk Newsplan [072]
Home movie scenario book / Ryskind, Morrie – New York, NY. 1927 – 1r – us UF Libraries [790]
Home Mutual Insurance Company (Appleton, WI) *see* Home news
Home news – Bethel, CT. 1985-1994 (1) – mf#62323 – us UMI ProQuest [071]
Home news – Spring Green WI. 1983 mar 2/dec-2000 jul/dec – 29r – 1 – (cont: weekly home news of the river valley area) – mf#1043908 – us WHS [071]
Home news – East Brunswick, NJ. 1903-2000 (1) – mf#61606 – us UMI ProQuest [071]

Home news / Home Mutual Insurance Company (Appleton, WI) – 1981 mar 13, jul 10, oct 9, 1982 jul 9-1983 mar 11, may 13-1986 sep 5 – 1r – 1 – (continued by: secura flash) – mf#1209298 – us WHS [368]

Home news / Pickaway Co. Ashville – (mar 1907-nov 1915) [wkly] – 1r – 1 – mf#B12843 – us Ohio Hist [071]

Home news / Pickaway Co. Ashville – v1 n1. (feb 1904-nov 1915) [wkly] – 2r – 1 – mf#B3228-3229 – us Ohio Hist [071]

Home news – Richmond, VA. 1958-1960 (1) – mf#66824 – us UMI ProQuest [071]

Home news / Trumbull Co. Cortland – jan-dec 1972, jan 1977-jun 1978 [wkly] – 1r – 1 – mf#B31909 – us Ohio Hist [071]

Home news / Washington Co. Marietta – v1 n1. jan 1859-jun 1862,oct 1865-aug 1866 [wkly] – 1r – 1 – mf#B29287 – us Ohio Hist [071]

Home news and temperance advocate for cambridgeshire, bedfordshire, & huntingdonshire – [East Midlands] Cambridge, Cambridgeshire 29 sep 1876-28 dec 1877 [mf ed 2004] – 1r – 1 – (discontinued) – uk Newsplan [072]

The home of god's people / Gage, William Leonard – Hartford, CT: Worthington, Dustin, 1872, c1871 – 2mf – 9 – 0-7905-3372-3 – mf#1987-3372 – us ATLA [915]

Home of Hebrew Orphans *see* Record of the home of hebrew orphans

Home office computing – New York. 1988-2001 (1) 1988-2001 (5) 1988-2001 (9) – (cont: family and home office computing) – ISSN: 0899-7373 – mf#13402,02 – us UMI ProQuest [000]

Home office papers and records : order and authority in england, series 1 pro class ho 42, 1782-1820 – 9pt-coll – 198r – 1 – (provides insight into the inner mechanisms of domestic govt and its strivings to maintain peace, stability and order in a turbulent country. pt1: boxes 1-23, 1782-92 25r c39-17301. pt2: boxes 24-41, 1793-97 22r c39-17302. pt3: boxes 42-66, 1798-1802 22r c39-17303. pt4: boxes 67-99, 1803-09 17r c39-17304. pt5: boxes 100-131, 1810-12 26r c39-17305. pt6: boxes 132-147, 1813-15 16r c39-17306. pt7: boxes 148-172, 1816-17 25r c39-17307. pt8: boxes 173-193, 1818-aug 1819 21r c39-17308. pt 9: boxes 194-218 sep 1819-20 24r c39-17309) – mf#C39-17300 – us Primary [941]

Home owners forum – 1970 jan-jul – 1r – 1 – (continued by: taxpayers' forum) – mf#2526590 – us WHS [640]

Home portraiture : for amateur photographers / Salmon, Percy R [pseud: Richard Penlake] – London 1899 – (= ser 19th c art & architecture) – 2mf – 9 – mf#4.1.425 – uk Chadwyck [770]

Home prayers : with two services for public worship / Martineau, James – London; New York: Longmans, Green, 1891 – 1mf – 9 – 0-7905-8514-6 – mf#1989-1739 – us ATLA [240]

Home progress – Boston. 1912-1917 (1) – mf#5670 – us UMI ProQuest [640]

The home record – David City, NE: [Theo S Ward] dec 1901-03// (wkly) [mf ed 1902-03 (gaps) filmed [1992?] – 1r – 1 – us NE Hist [071]

The home record – David City, NE: [Theo S Ward] dec 1901-03 (wkly) [mf ed 1902-03 (gaps) filmed [1992?]] – 1r – 1 – us NE Hist [071]

Home rule – Ainsworth, NE: Geo A Miles, aug 1889-oct 26 1897// (wkly) – 1r – 1 – (cont by: ainsworth home rule) – us Bell [071]

Home rule : its meaning, its objects, and its hopes / O'Lynn, Cumee – Liverpool, 1880 – (= ser 19th c ireland) – 1mf – 9 – mf#1.1.1903 – uk Chadwyck [941]

Home rule – Madras, India. 29 Oct 1967-27 Jan 1971 – 4r – 1 – us L of C Photodup [079]

Home rule : a plan for the better regulation and government of the united kingdom of great britain and ireland / Dobbs, Archibald Edward – London, 1886 – (= ser 19th c ireland) – 1mf – 9 – mf#1.1.247 – uk Chadwyck [941]

Home rule, a speech : delivered in montreal on the 17th of may, 1893 / Davin, Nicholas Flood – Toronto: Hunter, Rose, 1893 – 1mf – 9 – mf#03650 – cn Canadiana [941]

Home rule and justice to ireland / Brodrick, George Charles – Oxford, 1886 – (= ser 19th c ireland) – 1mf – 9 – mf#1.1.243 – uk Chadwyck [941]

Home rule and state supremacy : or, nationality reconciled with empire / Seymour, William Digby – London, 1888 – (= ser 19th c ireland) – 3mf – 9 – mf#1.1.5106 – uk Chadwyck [941]

Home rule for ireland / Meredith, F W – London, 1886 – (= ser 19th c ireland) – 1mf – 9 – mf#1.1.412 – uk Chadwyck [941]

The home rule question eighteen years ago / Bridges, John Henry – London, 1886 – (= ser 19th c ireland) – 1mf – 9 – mf#1.1.242 – uk Chadwyck [941]

Home rule, rome rule, and civil war / Laverty, George – Belfast, Northern Ireland. 1892 – 1r – us UF Libraries [941]

Home rule step by step – London, 1880 – (= ser 19th c ireland) – 1mf – 9 – mf#1.1.1944 – uk Chadwyck [941]

Home, Ruth M *see* Ceramics for the potter

Home scenes and heart studies / Aguilar, Grace – London: Groombridge & Sons, 1853 [i.e. 1852] – (= ser 19th c women writers) – 5mf – 9 – mf#5.1.26 – uk Chadwyck [420]

The home sewing machine : w a white and co, principal office, 90 king st, east, toronto, ont – S.l: s.n, 186-? – 1mf – 9 – mf#39729 – cn Canadiana [680]

The home sewing machine : w a white & co, principal office, 90 king st, east, toronto, on – [s.l: s.n. 186-?] [mf ed 1983] – 1mf – 9 – 0-665-39729-1 – mf#39729 – cn Canadiana [680]

Home shop machinist – Traverse City. 1984+ (1,5,9) – ISSN: 0744-6640 – mf#14906 – us UMI ProQuest [621]

Home star – Harrisburg, PA. -w 1949-1964 – 13 – $25.00r – us IMR [071]

Home studies in nature / Treat, Mary Lua Adelia Davis – New York, NY. 1885 – 1r – us UF Libraries [500]

Home study – v1 Feb 1896-Jan 1897 – 1 – us CRL [370]

Home study leaflet – St John, N.B: T F Fotheringham, [1894-189- or 19–] – 9 – mf#P04473 – cn Canadiana [242]

Home study quarterly – [Toronto: Publ under authority of the General Assembly, 1898-19–] – 9 – (merger of: the home study quarterly for senior scholars and the home department; the home study quarterly for intermediate scholars) – mf#P04479 – cn Canadiana [242]

The home study quarterly for intermediate scholars – [S.l: s.n, 1895?-1898?] – 9 – mf#P04477 – cn Canadiana [220]

Home study quarterly for senior scholars and the home department – [s.l: s.n, 1895-1898] – 9 – (merged with: the home study quarterly for intermediate scholars to become: the home study quarterly) – mf#P04476 – cn Canadiana [220]

Home talks about the word : for mothers and children / Miller, Emily Huntington – New York: Hunt & Eaton; Cincinnati: Cranston & Curts, 1894 [mf ed 1989] – 1mf – 9 – 0-7905-2677-8 – mf#1987-2677 – us ATLA [240]

Home talks. vol. 1 / Noyes, John Humphrey & Barron, Alfred – Oneida, (N.Y.): Published by the Community, 1875. Chicago: Dep of Photodup, U of Chicago Lib, 1973 (1r); Evanston: American Theol Lib Assoc, 1984 (1r) – 1 – 0-8370-0338-5 – mf#1984-B344 – us ATLA [975]

Home textiles today – High Point. 1985-1995 (1) 1985-1995 (5) 1985-1995 (9) – ISSN: 0195-3184 – mf#15033 – us UMI ProQuest [650]

Home, the church and the school – v1 n1-[4] [1887 may-aug] – 1r – 1 – mf#1058356 – us WHS [071]

Home town news – Madison, WV. 1987-1996 (1) – mf#68386 – us UMI ProQuest [071]

Home towner / Guernsey Co. Quaker City – (aug 1929-dec 1974) [wkly] – 10r – 1 – mf#B30391-30400 – us Ohio Hist [071]

The home treasury of useful and entertaining knowledge on the art of making home happy : and an aid in self-education: the laws of etiquette and good society... – Toronto, London: J S Brockville, 1883? – (= ser Canadian home series of useful books 1) – 5mf – 9 – mf#08943 – cn Canadiana [640]

Home, [W] *see* Select views in mysore

Home with homeland – Forrest City AR. 1998 jun/jul-1999 jul – 1r – 1 – (cont: homeland [forrest city, ar]) – mf#4188049 – us WHS [071]

Home work : a paper presented to the american baptist missionary union at philadelphia, may 18, 1849 – Boston: Missionary Rooms 1849 [mf ed 1993] – (= ser Occasional publications (american baptist missionary union) 2) – 1mf – 9 – 0-524-08249-9 – mf#1993-3004 – us ATLA [242]

Home work bulletin / National Council of the Y.M.C.A – v1-5. 1926-31 (incomplete) – 1r – 1 – (cont by: national council of the y.m.c.a. bulletin) – us ATLA [073]

Homecare magazine – Malibu. 1999+ (1,5,9) – ISSN: 1529-1715 – mf#31910,02 – us UMI ProQuest [640]

Homecoming 2 – 1985 may 11-1988 nov 11 – 1r – 1 – mf#1611814 – us WHS [071]

Homecoming...st philip's – 1997 [1997 aug 3] – 1r – 1 – mf#5004750 – us WHS [071]

Homecraft – Mount Morris. 1948-1955 (1) – mf#208 – us UMI ProQuest [640]

The homefinder – (New York). 1922-33, 1941-57 – 1 – us AJPC [720]

Homefront – 1997 winter – 1r – 1 – (cont: african homefront; cont by: african homefront [savannah, ga]) – mf#3858546 – us WHS [071]

Homeland – Forrest City AR. 1991 oct 1-1998 apr 15/may 15 – 1r – 1 – (continued by: home with homeland) – mf#2682204 – us WHS [071]

Homeland security information sharing act : hearing...house of representatives, 107th congress, 2nd session, on h.r. 4598, june 4 2002 / United States. Congress. House. Committee on the Judiciary. Subcommittee on Crime, Terrorism, and Homeland Security – Washington: US GPO 2002 [mf ed 2002] – 1mf – 9 – us GPO [343]

Homelies populaires sur les evangiles de chaque dimanche de l'annee / Lobry, J-B – 5e ed. Paris: Louis Vives, 1885 – 2mf – 9 – 0-8370-7476-2 – (incl bibl ref) – mf#1986-1476 – us ATLA [240]

Homelife : a quarterly newsletter of the harris home for children / Harris Home for Children (Huntsville AL) – 1992 winter/spring – 1r – 1 – mf#4878346 – us WHS [362]

Homelinks – London, UK. 1898-1900. -irr. 24 feet – 1 – uk British Libr Newspaper [072]

Homely talks / Pearse, Mark Guy – London: Wesleyan Conference Office, 1881. Beltsville, Md: NCR Corp, 1978 (3mf); Evanston: American Theol Lib Assoc, 1984 (3mf) – 9 – 0-8370-0834-4 – mf#1984-4234 – us ATLA [240]

Homely truth for honest men : letters to the right reverend john hughes / Kirwan – Belfast: Ulster Tract, Book, and Bible Depository, 1850 – 1mf – 9 – 0-8370-8120-3 – mf#1986-2120 – us ATLA [240]

O homem do povo – Rio de Janeiro, RJ: Typ Imparcial de Brito, 26 fev-12 nov 1840 – (= ser Ps 19) – mf#P02,05,26 – bl Biblioteca [321]

Homem e a serra / Ribeiro Lamego, Alberto – Rio de Janeiro, Brazil. 1950 – 1r – us UF Libraries [972]

Homem e o brejo / Ribeiro Lamego, Alberto – Rio de Janeiro, Brazil. 1945 – 1r – us UF Libraries [972]

Home-maker / Fisher, Dorothy Canfield – New York, NY. 1924 – 1r – us UF Libraries [640]

Homen da independencia / Cintra, Francisco De Assis – Sao Paulo, Brazil. 1921 – 1r – us UF Libraries [972]

Homenaje a antonio caso / Salazar, Joaquin E – Ciudad Trujillo, Dominican Republic. 1946 – 1r – us UF Libraries [972]

Homenaje a bernarda toro de gomez / Gomez Carbonell, Maria – Habana, Cuba. 1932 – 1r – us UF Libraries [972]

Homenaje a enrique jose varona en el – Habana, Cuba. 1935 – 1r – us UF Libraries [972]

Homenaje a eugenio hermoso / Academia de Bellas Artes de San Fernando – Madrid: Blass, S.A. Tipografia, 1964 – 1 – sp Bibl Santa Ana [946]

Homenaje a francisco menendez – Ahuachapan, El Salvador. 1942 – 1r – us UF Libraries [972]

Homenaje a hernando cortes en mejico / Pereyra, Carlos – Madrid: Revista de Indias, 1941 – 1 – sp Bibl Santa Ana [920]

Homenaje a la benemerita sociedad – Habana, Cuba. 1936 – 1r – us UF Libraries [972]

Homenaje a los academicos de honor / Academia De La Historia De Cuba – Habana, Cuba. 1950 – 1r – us UF Libraries [972]

Homenaje a marti en el cincuentenario de... – Habana, Cuba. 1942 – 1r – us UF Libraries [972]

Homenaje a nuestros mayores – Caceres: Imp. Linea 21, 1979 – 1 – sp Bibl Santa Ana [946]

Homenaje a pedro henriquez urena / Ciudad Trujillo Universidad De Santo Domingo – Ciudad Trujillo, Dominican Republic. 1947 – 1r – us UF Libraries [972]

Homenaje al doctor enrique olaya herrera – Bogota, Colombia. 1935 – 1r – us UF Libraries [972]

Homenaje al doctor manuel amador guerrero / Susto, Juan Antonio – Panama, Panama. 1933 – 1r – us UF Libraries [972]

Homenaje al dr fermin peraza sarausa – Habana, Cuba. 1948 – 1r – us UF Libraries [972]

Homenaje al ilustre habanero – Habana, Cuba. 1935 – 1r – us UF Libraries [972]

Homenaje al ilustre habanero francisco gonzalez de... – Habana, Cuba. 1947 – 1r – us UF Libraries [972]

Homenaje al ilustre habanero nicolas jose gutierre – Habana, Cuba. 1941 – 1r – us UF Libraries [972]

Homenaje al maestro don..., caballero de la orden de alfonso 10th el sabio septiembre, 1958 / Cruz Rebosa, Maximo – Caceres: Tip. La Minerva, 1958 – 1 – sp Bibl Santa Ana [946]

Homenaje al profesor paul rivet / Academia Colombiana De Historia – Bogota, Colombia. 1958 – 1r – us UF Libraries [370]

Homenaje de alcuescar al academico de las reales academias de la lengua espanola y sevillana de buenos letras don rafael garcia-plata de osma, 22 de marzo de 1953 – Caceres: Tip. El Noticiero – sp Bibl Santa Ana [370]

Homenaje de gratitud a dr. ezequiel fernandez santana / Fernandez Santana, Ezequiel – Madrid: Tipografia Artistica, S.A. – 1 – sp Bibl Santa Ana [946]

Homenaje de los musicos al excelentisimo / Dominican Republic Secretaria De Educacion Y Bell... – Ciudad Trujillo, Dominican Republic. 1945 – 1r – us UF Libraries [972]

Homenaje. poesias dedicadas a la eximia poetisa eva cervantes en ocasion de sus onomasticas y otros varios motivos / Sanchez-Arjona, Vicente – Sevilla: Imp. Alvarez, 1960 – 1 – sp Bibl Santa Ana [810]

Homenaje que los medicos de merida rinden al que fue uno de sus mas ilustres companeros : jose fernandez dominguez – Badajoz: Manuel Huerta Esteve, 1968 – 1 – sp Bibl Santa Ana [946]

Homenajes al presidente lemus / El Salvador Presidencia De La Republica – San Salvador, El Salvador. 1959 – 1r – us UF Libraries [972]

Homens de minas / Rache, Pedro – Rio de Janeiro, Brazil. 1947 – 1r – us UF Libraries [972]

Homens e cousas do imperio / Taunay, Alfredo D'escragnolle Taunay – Sao Paulo, Brazil. 1924 – 1r – us UF Libraries [972]

Homens e factos de uma revolucao / Almeida, Guilherme de – Rio de Janeiro, Brazil. 1934 – 1r – us UF Libraries [972]

Homens e temas do brasil / Arinos De Melo Franco, Afonso – Rio de Janeiro, Brazil. 1944 – 1r – us UF Libraries [972]

Homeopathy – Kidlington. 2002+ (1,5,9) – ISSN: 1475-4916 – mf#42890 – us UMI ProQuest [615]

Homeowner – New York. 1983-1991 (1) 1983-1991 (5) 1983-1991 (9) – (cont: homeowners how to) – ISSN: 0747-3176 – mf#12206,01 – us UMI ProQuest [640]

Homeowners how to – New York. 1979-1983 (1,5,9) – (cont by: homeowner) – ISSN: 0195-2196 – mf#12206 – us UMI ProQuest [640]

Homer *see*
– The odyssey of homer
– Translations in verse from homer and virgil

The homer free press – Homer, NE: M A Bancroft, 1906-10// (wkly) [mf ed filmed 1958-71] – 2r – 1 – (cont: homer echo) – us NE Hist [071]

Homer herald – Homer, NE: Ream & Shepardson. v1 n1. apr 14 1897- (wkly) [mf ed -1898 (gaps)] – 1r – 1 – us NE Hist [071]

Homer in der fruehchristlichen literatur bis justinin / Glockmann, G – Berlin, 1968 – (= ser Tugal 5-105) – 4mf – 9 – €11.00 – ne Slangenburg [240]

The homer patriot – Homer, NE: Allen & Rockwell, 1895 (wkly) [mf ed 1896-97 (gaps)] – 1r – 1 – us NE Hist [071]

Homer star – Homer, NE: J R Taylor, 1910-v33 n4. jun 25 1942 (wkly) [mf ed 1911-42 (gaps) filmed 1974?-91] – 11r – 1 – (cont by: dakota county star) – us NE Hist [071]

Homer township star – Chicago Heights, IL. 1987-1989 (1) – mf#68365 – us UMI ProQuest [071]

Homer, William Bradford *see* Writings of rev. william bradford homer, late pastor of the congregational church in south berwick, me

Homere / Severyns, Albert – Bruxelles, Belgium. v1-3. 1944-1948 – 1r – us UF Libraries [450]

Homeri hymni – Lipsiae, Germany. 1886 – 1r – us UF Libraries [960]

Homeri odyssea – Lipsiae, Germany. 1908 – 1r – us UF Libraries [025]

The homeric centones and the acts of pilate / Harris, James Rendel – London: C J Clay; New York: Macmillan [distributor], 1898 – 1mf – 9 – 0-7905-2410-4 – mf#1987-2410 – us ATLA [220]

Homeric dictionary for use in schools and colleges / Autenrieth, Georg – New York, NY. 1876 – 1r – us UF Libraries [054]

Homes / Century 21 (Firm) – 1987 nov-dec, 1988 jan-jul, nov-dec, 1989 jan-feb – 1r – 1 – mf#1671200 – us WHS [640]

Homes – Hollywood, FL. 1940-1940 (1,5,9) – mf#62413 – us UMI ProQuest [071]

Homes / Multiple Listing Service of Dane County – 1971 jan/sep, 1980 oct 6/1981-1989 jun 26/dec 25 [gaps] – 17r – 1 – (cont: madison homes; cont by: harmon homes) – mf#652253 – us WHS [362]

Homes and happiness in the golden state of california... / Truman, Benjamin Cummings – 3rd ed. San Francisco: H S Crocker & Co. 8v. 1885 (mf ed 19–) – 88p (ill) – mf#ZH-IAG pv201 n14 – us NY Public [978]

Homes and homesteads in the land of plenty : a handbook of victoria, as a field for emigration / Ballantyne, James – Melbourne, 1871 – (= ser 19th c books on british colonization) – 3mf – 9 – mf#1.1.1345 – uk Chadwyck [980]
Homes for millions : the great canadian north-west, its resources fully described / ed by Davin, Nicholas Flood – Ottawa: B Chamberlin, 1891 – 2mf – 9 – (incl ind) – mf#27015 – cn Canadiana [917]
Homes for millions : the resources of the great canadian north-west: the reasons why agriculture is profitable there and why farmers are porsperous and independent / Davin, Nicholas Flood – Ottawa: Government Print Bureau, 1892 – 3mf – 9 – mf#30664 – cn Canadiana [630]
Homes, haunts, and works of rubens, vandyke, rembrandt, and cuyp...michael angelo and raffaelle / Fairholt, Frederick William – London 1871 – (= ser 19th c art & architecture) – 3mf – 9 – mf#4.2.1420 – uk Chadwyck [700]
Homes of the east *see* Church missionary society archive, section 2
Homes of the english over the sea : no 1, british columbia and vancouver island – [London: s.n, 1862-63] [mf ed 1987 – 1mf – 9 – 0-665-90046-5 – mf#90046 – cn Canadiana [971]
Homes of the pilgrim fathers in england and america (1620-1685) / Briggs, Martin Shaw – London, England. 1932 – 1r – us UF Libraries [720]
Homes on the east coast of florida / Florida East Coast Railway – St Augustine, FL. 1902 – 1r – us UF Libraries [720]
Homes, works, and shrines of english artists with specimens of their styles / Fairholt, Frederick William – London 1873 – (= ser 19th c art & architecture) – 3mf – 9 – mf#4.2.1336 – uk Chadwyck [700]
Homespun: or, five and twenty years ago / Hill, George Canning – By Thomas Lackland (pseud). New York: Hurd and Houghton, 1867. viii, (9),346p. Title vignette. Verse and prose. 1 reel. 1291 – 1 – us Wisconsin U Libr [810]
Homestead – Iron River WI. 1894 sep 29-1895 apr 6 – 1r – 1 – (cont by: iron river pioneer) – mf#1144728 – us WHS [071]
Homestead – South Wayne WI. 1905 feb 16-1907 feb 28, 1907 mar 1-1909 jul 8, 1909 jul 8-1910 sep 8 – 3r – 1 – (cont by: south wayne homestead) – mf#932008 – us WHS [071]
Homestead leader – Homestead, FL. 1962 – 1r – us UF Libraries [071]
Homestead leader enterprise – Homestead, FL. 1923 jun-1961 – 14r – us UF Libraries [071]
Homesteader – Osceola, NE: Osceola Print Co. v1 n1. aug 27 1873-1876//(wkly) [mf ed -mar 24 1875 (gaps) filmed [1968?]] – 1r – 1 – (cont by: osceola record) – us NE Hist [071]
Homesteaders' monthly coyote : the only paper published exclusively for public land entrymen – v1 n1,2,4,[12] [1914 may, jun, aug, 1915 apr] – 1r – 1 – mf#1058357 – us WHS [333]
Homestyle – New York, 2001+ [1,5,9] – ISSN: 1533-5771 – mf#22192,04 – us UMI ProQuest [640]
Homewood flossmoor star – Chicago Heights, IL. 1990-1992 (1) – mf#61322 – us UMI ProQuest [071]
Home-work of the church / Spurgeon, C H – Edinburgh, Scotland. 1866 – 1r – us UF Libraries [240]
Homilectical index : a handbook of texts, themes and authors for the use of preachers and bible scholars generally / Pettingell, John Hancock – New York: D Appleton, 1878 [mf ed 2003] – 1r – 1 – mf#b00669 – us ATLA [240]
Homiletic and pastoral review – New York. 1900+ [1]; 1971+ [5]; 1976+ [9] – ISSN: 0018-4268 – mf#1866 – us UMI ProQuest [240]
Homiletic monthly – New York: Funk & Wagnalls; Toronto: W Briggs, [1876?-1884] – 9 – (cont by: the homeletic review) – mf#P05104 – cn Canadiana [240]
Homiletic review – New York: Funk & Wagnalls; Toronto: W Briggs, [1885?-189- or 19–] – 9 – (cont: the homiletic monthly) – mf#P05105 – cn Canadiana [240]
Homiletics / Hoppin, James Mason – 4th ed. New York: Funk & Wagnalls, 1893, c1883. Chicago: Dep of Photodup, U of Chicago Lib, 1972 (1r); Evanston: American Theol Lib Assoc, 1984 (1r) – 1 – 0-8370-0395-2 – (incl ind) – mf#1984-B306 – us ATLA [240]
Homiletics : or, the theory of preaching / Vinet, Alexandre Rodolphe; ed by Skinner, Thomas H – New York: Ivison, Blakeman, Taylor, 1878, c1853 mf ed 1991] – 2mf – 9 – 0-7905-8955-9 – (english by ed) – mf#1989-2180 – us ATLA [240]
Homiletics *see* Hsuan tao hsueh (ccm243)
Homilia *see* Rhetorica ad herennium...
Homilia a...san juan crisostomo / Franco y Lozano, Francisco – 1883 – 9 – sp Bibl Santa Ana [240]
Homiliae / Caesarii Heisterbasensis; ed by Coppenstein, Ioan – Coloniae, 1615 – 21mf – 8 – €40.00 – ne Slangenburg [241]
Homiliae : opera et studio b pez o s b / Godefridus, abbas Admontensis olim Weingarttensis – Aug Vindelicorum, 1725 – €73.00 – ne Slangenburg [241]
Homiliae academicae in pericopas evangeliorvm et epistolarum quae diebus, tum dominicis, tum feriatis alijs in ecclesia solent proponi / Pappus, J – Argentorati, 1603. 3v – 19mf – 9 – mf#TH-1 mf 1240-1258 – ne IDC [242]
Homiliae (codice uncial, siecle 6-7) / Gregorio, San – Barcelona – 1r – 5,6 – sp Cultura [240]
Homiliae in evangelia (ccsl141 : formae tplila 120 / Gregorius Magnus – [mf ed 2000] – (= ser ILL – ser a; Ccsl) – 10mf+106p – 9 – €50.00 – 2-503-61412-4 – be Brepols [400]
Homiliae mathesii das ist : ausslegung der ersten und andern episteln an die corinthier / Mathesius, J – Leipzig, 1590 – 12mf – 9 – mf#TH-1 mf 1003-1014 – ne IDC [242]
Homiliae per circulum anni : formae tplila 82 / Autissiodorensis, Heiricus – 1994 – (= ser ILL – ser a; Cccm 116-116a-116b) – 26mf+148p – 9 – €120.00 – 2-503-64162-8 – be Brepols [400]
Homiliae qvi svnt sermones habiti de iis, qvae in christianis ecclesiis legvntvr / Camerarius, J – Lipsiae, [1573] – 5 – 9 – mf#TH-1 mf 189-193 – ne IDC [242]
Homiliae seu sermones / Clichtove, J – Coloniae, 1550 – 13mf – 9 – mf#CA-85 – ne IDC [241]
Homiliarium floriacense mss – Orleans, Bibl. Mun. 154 – 7mf – 8 – €15.00 – ne Slangenburg [240]
Das homiliarium karls des grossen auf seine urspruenglichen gestalt hin untersucht / Wiegand, F – Leipzig, 1897 – 2mf – 8 – €5.00 – ne Slangenburg [241]
Homiliarum sacrarum in pericopas evangeliorum dominicalium / Gerhard, J – Jenae, 1634-1640. 3v – 46mf – 9 – mf#TH-1 mf 552-597 – ne IDC [242]
Homiliarvm evangelicarvm notae breves / Hoffmann, D – Magdebvrgi, 1600 – 2mf – 9 – mf#TH-1 mf 701-702 – ne IDC [242]
Homilien ueber das evangelium des johannes : in den jahren 1825 und 1826 gesprochen / Schleiermacher, Friedrich [Ernst Daniel]; ed by Sydow, Adolph – Berlin: G Reimer 1847 [mf ed 1994] – 2mf – 9 – 0-524-08643-5 – mf#1993-2103 – us ATLA [225]
Homilien ueber das evangelium des johannes in den jahren 1823 und 1824 : gesprochen / Schleiermacher, Friedrich [Ernst Daniel]; ed by Sydow, Adolph – Berlin: G. Reimer 1837 [mf ed 1994] – 2mf – 9 – 0-524-08851-9 – mf#1993-2136 – us ATLA [225]
Die homilien und recognitionen des clemens romanus : nach ihrem ursprung und inhalt / Clement 1, Pope – Goettingen: Dieterich, 1854 – 5mf – 9 – 0-524-08724-5 – mf#1993-2129 – us ATLA [240]
Homilies considered / Jebb, John – Dublin, Ireland. 1826 – 1r – us UF Libraries [240]
The homilies of s. thomas aquinas upon the epistles and gospels for the sundays of the christian year : to which are appended the festival homilies = Sermons. Selections / Thomas, Aquinas, Saint – 2nd ed. London: JT Hayes, 1873 – 1mf – 9 – 0-7905-8603-7 – mf#1989-1828 – us ATLA [241]
Homilies on the book of tobias : or, a familiar explication of the practical duties of domestic life / Martyn, Francis – Baltimore: Fielding Lucas, Jr, [1831?] – 1mf – 9 – 0-524-07339-2 – mf#1992-1070 – us ATLA [221]
Homilies on the former part of the acts of the apostles : chap 1-10 / Alford, Henry – London: Rivingtons, 1858 [mf ed 2004] – 1r – 1 – 0-524-10491-3 – mf#b00706 – us ATLA [226]
Homily for good-friday – London, England. 18– – 1r – us UF Libraries [240]
Homin ukrainy – Canada. jan 1948-99 – 42r – 1 – (in ukrainian) – cn Commonwealth Imaging [071]
Hommage a edmond fleg – Paris, France. 1950 – 1r – us UF Libraries [939]
Hommage a kwame n'krumah – Conakry: Imprimerie Nationale "Patrice Lumumba", 1972 – us CRL [920]
Hommage a mafory bangoura / Toure, Ahmed Sekou – Conakry, R G: Parti-Etat de Guinee, 1976 – us CRL [920]
Hommage a monseigneur raphael merry del val, delegue apostolique au canada : souvenir de la visite de son excellence a valleyfield, 21, 22 et 23 avril 1897 – Valleyfield Quebec: E H Solis, 1897 – 1mf – 9 – mf#04487 – cn Canadiana [241]
Hommage a notre veneree ancienne mere et devouee assistante sr elizabeth f mcmullen : a l'occasion du cinquantieme anniversaire de sa profession religeuse, 22 fevrier, 1875 – Montreal?: Hopital-General? 1875? – 1mf – 9 – mf#08819 – cn Canadiana [810]
Hommage a pie 9 : 1: sermon de m colin, ptre, ss pour le 50e anniversaire de la premiere messe de pie 9... – Montreal: E Senecal, 1869 – 1mf – 9 – 0-665-55132-0 – mf#55132 – cn Canadiana [241]
Hommage a pie 9 – Montreal: E Senecal, 1869 – 1mf – 9 – 0-665-05890-X – mf#05890 – cn Canadiana [241]
Hommage au revd p lagace, ptre, superieur du college ste-anne : 15 juin 1863 – S.l: s.n, 1863? – 1mf – 9 – mf#61067 – cn Canadiana [810]
Hommage au reverend a pelletier, ptre, superieur du college de ste-anne : 9 juin 1864 – S.l: s.n, 1864? – 1mf – 9 – mf#61066 – cn Canadiana [810]
Hommage aux jeunes catholiques-liberaux / Segur, Louis Gaston de – Quebec: J A Langlais; Cercle catholique de Quebec, 1877? – 2mf – 9 – (incl latin text) – mf#13442 – cn Canadiana [241]
Hommage aux marins de l'arethuse et du hussard / Malo, J H – Montreal: s.n, 1892 – 1mf – 9 – mf#04648 – cn Canadiana [810]
Hommage du petit gazettier : aux abonnes du canadien, le premier jour de l'an 1835 – S.l: s.n, 1835? – 1mf – 9 – mf#61068 – cn Canadiana [780]
Hommaire de Hell, X *see*
– Les steppes de la mer caspienne, le caucase, la crimee et la russie meridionale, voyage pittoresque, historique et scientifique
– Voyage en turquie, et en perse pendant les annees 1846-1848
L'homme : journal de la democratie universelle – Saint Helier, Jersey [mf ed 30 nov 1853-28 dec 1855; 1 mar-23 aug 1856] – 1 – (fr 17 nov 1855 to 23 aug 1856 publ in london) – uk British Libr Newspaper [072]
L'homme : organe politique et quotidien de la federation universelle – Paris, France 9-15 mar 1871 – 1 – (cont by: l'homme libre [16 mar-7 apr 1871]; wanting: n4,11) – uk British Libr Newspaper [074]
Homme – St Helier, Channel Islands. 30 nov 1853-28 dec 1855; 1 mar-23 aug 1856 – 1/2r – 1 – uk British Libr Newspaper [072]
Homme a sentimens : ou, le tartuffe de moeurs / Cheron, Louis Claude – Paris, France. 1801 – 1r – us UF Libraries [440]
Homme blase / Duvert, Felix-Auguste – Paris, France. 1843? – 1r – us UF Libraries [440]
Homme Children's Home [Wittenberg, WI] *see* For gammel og ung
L'homme de couleur / Verdier, Cardinal et al – Paris: Plon, 1939 – 1 – us CRL [300]
L'homme du peuple – Lyon: Impr de Mme veuve Ayne, Aug 24, 1849 – us CRL [074]
L'homme enchaine : journal quotidien du matin – Paris, France 8 oct 1914-17 nov 1917 – 1 – uk British Libr Newspaper [074]
L'homme enchaine – Paris. Journal quot. du matin. Red. en chef, G. Clemenceau. 8 oct 1914-17 nov 1917, 21-23 sept 1939 – 1 – fr ACRPP [074]
Homme enchante – Paris, France. 18 nov 1917-12 aug 1919 – 2 1/2r – 1 – uk British Libr Newspaper [072]
Homme et la societe – Paris. 1966-1995 (1) 1971-1995 (5) 1973-1995 (9) – ISSN: 0018-4306 – mf#3448 – us UMI ProQuest [301]
L'homme et l'hygiene : conference faite devant l'association des instituteurs catholiques de Montreal, a l'ecole normale jacques-cartier, le 26 mai 1893 / Desroches, Joseph Israel – Montreal?: s.n, 1893? – 1mf – 9 – mf#06730 – cn Canadiana [613]
L'homme libre – Gonaives: Impr de "L'homme libre". 1ere annee n23-3eme annee n15. 5 sep 1878-23 fevr 1881 – 2 sheets – 9 – us CRL [079]
L'homme libre : ni dieu, ni maitre – Paris. n1-70. 21 juin-29 aout 1888 – 1 – fr ACRPP [073]
L'homme libre – Paris. 27 oct 1876-3 mai 1877 – 1 – fr ACRPP [073]
L'homme libre – Paris. 5 mai 1913-nov 1917, 1918-10 oct 1939, 10 mai 1943, 25 mai 1951-24 juil 1953 – 1 – fr ACRPP [073]
L'homme libre – Paris, France 18 nov 1917-12 aug 1919 (imperfect) – 1 – uk British Libr Newspaper [074]
L'homme libre – Paris, France 16 mar-7 apr 1871 – 1 – (wanting: n4,11; cont: homme: organe politique et quotidien de la federation universelle [9-15 mar 1871]) – uk British Libr Newspaper [074]
Homme libre – Brussels Belgium, 11 apr 1891-10 dec 1892 – 1/4r – 1 – uk British Libr Newspaper [074]
L'homme libre: bulletin d'informations ouvrieres – oct 1940, may 1941 – 1 – uk British Libr Newspaper [074]
Homme noir d'afrique / Holas, Bohumil – Dakar, Senegal. 1951 – 1r – us UF Libraries [960]
L'homme nouveau – Brazzaville, Congo, mar 1960-63 – 4r – 1 – us CRL [079]
L'homme nouveau – n1-35. 1934-avr 1937 [mnthly] – 1 – (devenu: idees et peuples. l'homme nouveau. paris. juil 1938) – fr ACRPP [073]
L'homme nouveau – Brazzaville jun 25 1960 – (Issues filmed with: Bartlett, Robert E: Collection of African newspapers) – us CRL [079]
Un homme pareil aux autres: roman / Maran, Rene – Paris: A. Michel, 1962 – 1 – us Wisconsin U Libr [830]
Homme que j'ai tue / Rostand, Maurice – Paris, France. 1930 – 1r – us UF Libraries [440]
L'homme qui rit jaune / Arnaud, Robert – Paris: A Michel, [c1926] – 1 – us CRL [960]
L'homme reel : revue mensuelle du syndicalisme et de l'humanisme – Paris. 1934-sept 1938 [mnthly] – 1 – fr ACRPP [320]
Homme sans facon / Sewrin, M – Paris, France. 1812 – 1r – us UF Libraries [440]
L'homme-dieu : conferences prechees a la metropole de besanon / Besson, Louis Francois Nicolas, monseigneur – 19e rev corr ed. Paris: Victor Retaux, 1897 [mf ed 1985] – 1mf – 9 – 0-8370-2677-6 – (incl bibl ref) – mf#1985-0677 – us ATLA [120]
Hommel, F *see* Ethnologie und geographie des alten orients
Hommel, Friedrich *see* Geistliche volkslieder
Hommel, Fritz *see*
– The ancient hebrew tradition as illustrated by the monuments
– The civilization of the east
– Geschichte babyloniens und assyriens
– Die goetternamen in den babylonischen siegelcylinder-legenden
– Die insel der seligen in mythus und sage der vorzeit
– Die semitischen voelker und sprachen
– Sumerische lesestuecke
Hommes celebres de la guadeloupe / Oriol, T – Basse-Terre, Guadeloupe. 1935 – 1r – us UF Libraries [972]
Les hommes du jour : a b routhier / DeCelles, Alfred Duclos – Montreal: Cie de moulins a papier de Montreal, 1891? – 1mf – 9 – mf#26004 – cn Canadiana [971]
Les hommes du jour : l r masson / DeCelles, Alfred Duclos – Montreal: Cie de moulins a papier de Montreal, 1892? – 1mf – 9 – mf#26022 – cn Canadiana [920]
Les hommes du jour – Paris. 1908-23 – 1 – (puis annales politiques, sociales, litteraires et artistiques) – fr ACRPP [073]
Les hommes du jour : a r angers / Chapais, Thomas – Montreal: Cie de moulins a papier de Montreal, 1892? – 1mf – 9 – mf#26015 – cn Canadiana [920]
Les hommes du jour : sir alexandre lacoste / DeCelles, Alfred Duclos – Montreal: Cie de moulins a papier de Montreal, 1892? – 1mf – 9 – mf#26017 – cn Canadiana [920]
Les hommes du jour : wilfrid laurier / Frechette, Louis – Montreal?: Cie de moulins a papier de Montreal, 1890? – 1mf – 9 – mf#26001 – cn Canadiana [920]
L'homme-sans-facon : ou lettres d'un voyageur allant de paris a spa / Jehin – [Neuwied] 1786 [mf ed Hildesheim 1995-98] – 2v on 4mf – 9 – €120.00 – 3-487-27890-1 – gw Olms [860]
Hommius, F *see*
– 70 disputationes theologicae adversus pontificios
– Specimen controversiarum belgicanum
Homo et ejus partes figuratus et symbolicus, anatomicus, rationalis... / Scarlattini, O – Augustae Vindelicorum, Dilingae: Sumptibus Joannis Caspari Bencard, 1695 2pts – 21mf – 9 – mf#O-859 – ne IDC [090]
Homo, Leon Pol *see* Roman political institutions
Homo sum / Ebers, Georg – Stuttgart: Deutsche Verlags-Anstalt, [1893-97?] [mf ed 1993] – (= ser Georg ebers gesammelte werke 6) – 349p – 1 – mf#8554 reel 2 – us Wisconsin U Libr [830]
Homo sum : a novel / Ebers, Georg – New York: Y L Burt, [1904?] [mf ed 1989] – 351p – 1 – (in english) – mf#7374 – us Wisconsin U Libr [830]
Homo sum : roman / Ebers, Georg – Stuttgart: Deutsche Verlags-Anstalt, [1893-1897?] [mf ed 1993] – (= ser Georg ebers gesammelte werke 6) – 349p – 1 – mf#8554 reel 2 – us Wisconsin U Libr [830]
Homo versus darwin : a judicial examination of statements recently published by mr darwin regarding "the descent of man" / Lyon, William Penman – 3rd ed. London [1872] – (= ser 19th c evolution & creation) – 2mf – 9 – (also: "man versus ape" a controversial correspondence, repr fr the eastern daily press) – mf#1.1.1502 – uk Chadwyck [575]
Die homocentrischen sphaeren des exodus des kallippus und des aristoteles / Schiaparelli, Giovani Virginio – 3mf – 7 – mf#337 – uk Microform Academic [180]
O homoeopatha : orgao de propaganda homoeopathica – Recife, PE: Typ do Homoeopatha, 26 mar-02 jul 1883 – (= ser Ps 19) – bl Biblioteca [615]

O homoeopathia : periodico de doutrinas medicas e sciencias accessorias – Rio de Janeiro, RJ: Typ Carioca de J I da Silva, 28 jul-04 ago 1850 – (= ser Ps 19) – mf#P15,01,47 n07 – bl Biblioteca [615]

Homolka, Walter *see*
– Ein dem untergang naher aramaeer war mein vater...
– The jewish attitude to homosexuality
– Traditionelles judentum in der moderne leben

Homonyme wurzeln im syrischen : ein beitrag zur semitischen lexicographie / Schulthess, Friedrich – Berlin: Reuther & Reichard, 1900 – 1mf – 9 – 0-8370-8866-6 – (incl bibl ref and index) – mf#1986-2866 – us ATLA [470]

Homosexual counseling journal – New York. 1974-1976 (1) 1976-1976 (5) 1976-1976 (9) – ISSN: 0092-3052 – mf#9185 – us UMI ProQuest [305]

Homunculus : modernes epos in zehn gesaengen / Hamerling, Robert – Hamburg: J F Richter, 1888 [mf ed 1993] – 319p – 1 – mf#8670 – us Wisconsin U Libr [810]

L'hon j a chapleau, retour d'europe : demonstration enthousiaste: adresse de bienvenue...montreal, 24 avril 1889 – S.l: s.n, 1889? – 1mf – 9 – mf#00615 – cn Canadiana [327]

Hon mr howe's speech on dr tupper's railway resolution / Howe, Joseph – [s.l: s.n, 1860?] [mf ed 1984] – 1mf – 9 – 0-665-32007-8 – mf#32007 – cn Canadiana [323]

Hon oan ve doi no mau / Nguoi Khan Trang – [Saigon]: Bong Dem 1975 [mf ed 1993] – on pt of 1r – 1 – mf#11052 r447 n1 – us Cornell [959]

The hon pandit madan mohan malaviya : his life and speeches – Madras: Ganesh & Co, Publishers, [19–?] – (= ser Samp: indian books) – us CRL [920]

L'hon pierre garneau / Bechard, Auguste – St-Hyacinthe, Quebec?: Courrier de St-Hyacinthe, 1884 – (= ser Galerie nationale, biographies 1) – 1mf – 9 – mf#08692 – cn Canadiana [920]

The hon r b sullivan's attacks upon sir charles metcalfe refuted by egerton ryerson : being a reply to the letters of "legion" – Toronto?: s.n, 1844 (Toronto: British Colonist) – 1mf – 9 – mf#21934 – cn Canadiana [320]

Ho-nan cheng chih shih ch'a ti i ts'e, ti ssu ts'e – [China: Ho-nan sheng cheng fu mi shu ch'u], 1936 – (= ser P-k&k period) – us CRL [951]

Ho-nan chih niu yang p'i – [Ho-nan]: Ho-nan nung kung yin hang ching chi tiao ch'a shih, Min kuo 32 [1943] – (= ser P-k&k period) – us CRL [338]

Ho-nan ch'uan sheng cheng li t'u ti chien i ch'ing chang shih nien wan ch'eng chi hua ts'ao an – [China: Ho-nan sheng ti cheng ch'ou pei ch'u], 1935 – (= ser P-k&k period) – us CRL [630]

Honan glimpses / ed by Carlberg, Gustav – Hsuchow, Honan: [s.n.] 1923-27 [mf ed 2006 – v2-6 (1923-27) [complete] on 1r – 1 – (publ by the augustana synod mission; absorbed by: augustana foreign missionary, in jun 1927; cont: glimpses from central honan) – mf#2006c-s003 – us ATLA [242]

Honan glimpses / ed by Carlberg, Gustav – Hsuchow, Honan: [s.n.] 1923-27 [mthly ex jul & aug] [mf ed 2006] – v2-6 (1923-27) [complete] on 1r – 1 – (publ by the augustana synod mission; absorbed by: augustana foreign missionary, in jun 1927; cont: glimpses from central honan) – mf#2006c-s003 – us ATLA [242]

Ho-nan hsiang-shih lu – List of successful candidates in the imperial examination in Honan province. Scattered years 1759-1903. 1 reel – 1 – us Chinese Res [951]

Ho-nan nan-yang hsien t'u ti ch'ing chang chuan k'an – [China: Ho-nan Nan-yang hsien cheng li t'ien fu wei yuan hui], Min kuo 25 [1936] – (= ser P-k&k period) – us CRL [630]

Ho-nan sheng cheng fu hsing cheng pao kao : min kuo erh shih ssu nien – [China: Ho-nan sheng cheng fu mi shu ch'u], 1935 – (= ser P-k&k period) – us CRL [951]

Ho-nan sheng cheng fu min kuo erh shih liu nien tu hsing cheng chi hua – [China: Ho-nan sheng cheng fu mi shu ch'u], Min kuo 26 [1937] – (= ser P-k&k period) – us CRL [350]

Ho-nan sheng cheng fu min kuo erh shih wu nien tu hsing cheng chi hua – [China: Ho-nan sheng cheng fu mi shu ch'u], 1936 – (= ser P-k&k period) – us CRL [951]

Ho-nan sheng cheng li t'u ti ko chung chi hua chang tse hui k'an – [China]: Ho-nan sheng cheng fu mi shu ch'u, 1933 – (= ser P-k&k period) – us CRL [630]

Ho-nan sheng li shui li kung ch'eng chuan k'o hsueh hsiao i lan – Ho-nan: Shang wu yin shua so, Min kuo 22 [1933] – (= ser P-k&k period) – us CRL [951]

Ho-nan sheng nung ts'un tiao ch'a / China. Hsing cheng yuan nung ts'un fu hsing wei yuan hui – Shang-hai: Shang wu yin shu kuan, Min kuo 23 [1934] – (= ser P-k&k period) – us CRL [307]

Ho-nan sheng p'u t'ung k'ao shih hui k'an – Ho-nan sheng: P'u t'ung k'ao shih tien shih wei yuan hui mi shu ch'u, Min kuo 22 [1933] – (= ser P-k&k period) – us CRL [350]

Honderd christelijke zinnebeelden naar georgette de montenay, door anna roemers visscher / Montenay, G de – ['s-Gravenhage], 1854 – 2mf – 9 – mf#O-3246 – ne IDC [090]

Honderd jaar java bode, 1852-1952 / Joel, H F – Djakarta, 1952 – 3mf – 9 – mf#SE-1454 – ne IDC [959]

[Hondius, H] *see* Perspective das ist die weit beruemhte kunst...

Hondsdolheid : inligting vir landdroste, distriksgeneeshere, veeartse en plaaslike owerhede / ed by Meredith, C D et al – [Pretoria: Staatsdrukker 1973?] [mf ed Pretoria, RSA: State Library [199-]] – 51p [ill] on 1r with other items – 5 – mf#op 11069 r23 – us CRL [616]

Hondsdolheid : vir inligting van magistrate, distriksgeneeshere, veeartse en plaaslike owerhede = Rabies / South Africa. Division of Veterinary Services – Pretoria: Staatsdrukker 1941 [mf ed Pretoria, RSA: State Library [199-]] – 31p [ill] on 1r with other items – 5 – (also available in english) – mf#OP(S) 15-23 r23 – us CRL [616]

Hondsdolheid : vir inligting van magistrate, distriksgeneeshere, veeartse en plaaslike owerhede = Rabies / South Africa. Division of Veterinary Services – Pretoria: Staatsdrukker [1949?] [mf ed Pretoria, RSA: State Library [199-]] – 31p [ill] on 1r with other items – 5 – (also available in english) – mf#op 00694 r23 – us CRL [616]

Honduras / Aguirre, Jose M – New York, NY. 1884 – 1r – us UF Libraries [972]

Honduras – National Archive of Honduras – 98r – 1 – (coll incl: documents from the colonial period (1606-1897); nineteenth-century documents; newspapers and government gazettes from honduras, guatemala, nicaragua and costa rica) – Pan-American Institute of Geography and History (IPGH) – us UMI ProQuest [972]

Honduras : internal affairs and foreign affairs, 1945-1959 / U.S. State Dept – (= ser Confidential u s state department special files) – 1 – $5890.00 coll – (1945-49 11r isbn 0-89093-878-4 $2135. 1950-54 11r isbn 0-89093-892-X $2135. 1955-59 10r isbn 0-89093-794-X $1935. with p/g) – us UPA [327]

Honduras / International Bureau Of The American Republics – Washington, DC. 1904 – 1r – us UF Libraries [972]

Honduras : the land of great depths / Charles, Cecil – Chicago, IL. 1890 – 1r – us UF Libraries [972]

Honduras / Marinas Otero, Luis – Madrid, Spain. 1963 – 1r – us UF Libraries [972]

Honduras / Stokes, William Sylvane – Madison, WI. 1950 – 1r – us UF Libraries [972]

Honduras *see*
– Arbitraje de limites entre honduras y guatemala
– Decretos del congreso nacional, 1946-1947
– La gaceta diario oficial
– Limites entre honduras y nicaragua

Honduras ante el turista / Honduras Oficina De Cooperacion Intelectual – Tegucigalpa, Mexico. 1950 – 1r – us UF Libraries [972]

Honduras Constitution *see*
– Constitucion de la republica
– Constitucion politica y leyes constitutivas de la...

Honduras Direccion General De Estadistica Y Censo *see*
– Republica de honduras
– Segundo censo nacional de vivienda de honduras, ab...

Honduras. Direccion General de Estadistica y Censos *see*
– Anuario estadistico 1952-1969
– Comercio exterior
– Investigacion industrial
– Poblacion y vivienda

Honduras, documentos microfotograifados / Pan American Institute Of Geography And History – s.l, s.l? reel 1-2. 1700-1900 – 2r – us UF Libraries [972]

Honduras ilustrada / Leyton Rodriguez, Ruben – Tegucigalpa, Mexico. 1951 – 1r – us UF Libraries [972]

Honduras Laws, Statutes, Etc *see*
– Ley electoral y sus reformas y concordancias
– Reglamento general de ensenanza primaria

Honduras. Laws, Statutes, etc *see* Codigo de procedimientos administrativos

Honduras literaria / Duron Y Gamero, Romulo Ernesto – Tegucigalpa, Mexico. v1 pt1-v2 pt4. 1957-1958 – 2r – us UF Libraries [972]

Honduras maya / Lunardi, Federico – San Pedro Sula, Honduras. 1946 – 1r – us UF Libraries [972]

Honduras Ministerio De Relaciones Exteriores *see*
– Limites entre honduras y nicaragua

Honduras Ministerio De Salud Publica Y Asistencia *see* Plan nacional de salud publica, 1958-1963

Honduras Oficina De Cooperacion Intelectual *see*
– Honduras ante el turista
– Obra material del gobierno del doctor galvez
– Sucesion presidencial en honduras

Honduras rotaria – Tegucigalpa. 1978-1990 (1) 1978-1979 (5) 1978-1979 (9) – mf#7719 – us UMI ProQuest [360]

Honduras Secretaria De Educacion Publica *see* Primera escuela normal rural de honduras

Honduras y sus problemas de educacion / Izaguirre, Carlos – Tegucigalpa, Mexico. 1935 – 1r – us UF Libraries [370]

Honduras-guatemala boundary arbitration – Washington, DC. 1932 – 1r – us UF Libraries [972]

Hondzinski, Jan M *see* Contributions of vision in aerial performances

Hone, Joseph Maunsell *see* Papers, 1901-1939

Hone, William *see* Sixty curious and authentic narratives and anecdotes respecting extraordinary characters

Honea path first baptist church. honea path, south carolina : church records – May 1869-Jul 1955 – 1 – 48.42 – us Southern Baptist [242]

Honea, Sion M *see* Symphonies of gaetano brunetti and their relationship to the contemporary viennese classical symphony of haydn

Honegger, Johann Jakob *see* Das deutsche lied der neuzeit

The honest grief of a tory expressed in a genuine letter from a burgess of...in wiltshire : to the author of the monitor, feb 17, 1759 – London: Printed for J Angel... 1759 – 1mf – 9 – mf#20285 – cn Canadiana [320]

The honest injun – Victoria, BC: D Falconer, [1897] – 9 – ISSN: 1190-7126 – mf#P04508 – cn Canadiana [870]

Honest Money for America *see* Federal reserve update

Honest penny is worth a silver shilling / Cameron, Mrs – London, England. 18– – 1r – us UF Libraries [240]

Honey bucket / Wisconsin Veterans Union – 1974 [feb?-jul?] – 1r – 1 – mf#370654 – us WHS [331]

Honey creek baptist church records, 1811-1844 / Bethel. Miami County. Ohio. Honey Creek Baptist Church – [mf ed 1974] – 1r – 1 – mf#ms670 – us Western Res [242]

Honey grove first baptist church. honey grove, texas : church records – 1847-73. Includes Church history. 1884-1912 – 1 – us Southern Baptist [242]

Honey in the horn / Davis, Harold Lenoir – New York, NY. 1935 – 1r – us UF Libraries [025]

Honey, John *see*
– Tableaux indiquant le nombre et denomination des timbres les plus convenables pour paiement en vertu du cap 5, 27-28 vic
– Tables shewing the number and denomination of law-stamps

Honey, Michael J *see* The impact of interscholastic athletics on academic performance

Honey peach group / Reimer, F C – Lake City, FL. 1904 – 1r – us UF Libraries [634]

Honeyman, Abraham Van Doren *see*
– Directory of the members of the bar in practice in new jersey
– A treatise on the jurisdiction of civil and criminal proceedings in the court for the trial of small causes in new jersey

Honeyman, David *see* On the geology of the gold-fields of nova scotia

Hong Kong *see* Statistical blue books 1844-1938

Hong kong : internal affairs and foreign affairs, 1960-jan 1963 / U.S. State Dept – (= ser Confidential u s state department central files) – 7r – 1 – $1355.00 – 1-55655-846-5 – (with p/g) – us UPA [951]

The hong kong daily press – Hong Kong: [H L Murrow, jul 1870-sep 30 1941] – 166r – 1 – us CRL [079]

Hong Kong. Laws, Statutes, etc *see*
– The companies ordinance of hongkong
– Ordinance no. 13 of 1873. special edition of "the hongkong code of civil procedure,".

Hong kong naturalist, the... 1930-41 – v1-10 with suppl 1-6 – 4r – 1 – mf#96134 – uk Microform Academic [574]

Hong kong newspaper clippings *see* Xianggang bao zhang jian bao

Hong kong press summaries *see* Survey of china mainland press

Hong Kong Salaries Commission *see* Interim report of the hong kong salaries commission

Hong kong standard – 1949- – 1 – (in english. yrly reel count varies) – us UMI ProQuest [079]

The hong kong times : daily advertiser and shipping gazette – Hong Kong: Printed and published by William Curtis, may 1873-apr 29, 1876 – 8r – 1 – us CRL [079]

Hong kong weekly press, 1895-1909 – 20r – 1 – mf#95823 – uk Microform Academic [079]

Hong kou (ccm18) = The great gulf / Chang, Wan-ju – Shanghai, 1935 [mf ed 198?] – (= ser Ccm 18) – 1 – mf#1984-b500 – us ATLA [230]

Hong, Seol E *see* The status of physical education career preparation in selected colleges and universities in seoul

Honga = The leader / American Indian Center of Omaha – 1979 jul [v2 n7]-1982 – 1r – 1 – (cont: american indian center newsletter) – mf#660741 – us WHS [305]

Hong-Joe, Christina M *see* Discipline-based dance education

Hongkong : late canton, register – Hongkong: J Slade, 1843 – (filmed with: canton register, 1841-jun 10 1843; jun 20-dec 26 1843) – us CRL [079]

Hongkong mercury and shipping gazette – Hongkong: G M Bain, jun 1-dec 4 1866 – 1r – 1 – us CRL [079]

The hongkong news – [Hong Kong: Hongkong News, feb-jul 1943; dec 29 1943-jan 22 1944; may 21-aug 17 1945 – us CRL [079]

The hongkong register / ed by Cairns, John – Victoria: publ by John Cairns, 1844- – (filmed consecutively with: general price current, mercantile register and shipping list jan 3-aug 22 1845. issue for mar 23 1858, nov 24 1860 filmed with: hongkong register, and daily advertiser jan 2 1844-dec 25 1855; jan 5-mar 30 1858; apr 6-dec 28 1858; mar 23 1858) – us CRL [079]

The hongkong register, and daily advertiser – Victoria: [s.n.], mar 23 1858, nov 24 1860 – (filmed with: hongkong register) – us CRL [079]

Hongkong sunday herald – Hong Kong: David Christian Wilson, 1929-jun 1941 – 17r – 1 – us CRL [079]

Hongkong telegraph – Victoria, HK: F B Franklin, 1946-51 – 1 – (issues for oct 1-14 1946 filmed with: hongkong telegraph (1881)) – us CRL [079]

The hongkong telegraph – Victoria, Hongkong: Robert Fraser-Smith, [oct 1-14 1946]. jun 16 1881-sep 1941 – (issues for aug 6-sep 30 1941 filmed with hongkong telegraph (victoria, hong kong: 1946)) – us CRL [079]

Hongkong tiger standard – Hongkong: Star Newspaper Enterprises, mar 1949-mar 1952; oct 1954-1961 – 1 – us CRL [079]

La hongrie – Paris: N Chaix, jul 1848 – us CRL [074]

La hongrie calviniste / Doumergue, Emile – Toulouse: Societe d'edition de Toulouse, [1912?] – 1mf – 9 – 0-7905-6163-8 – mf#1988-2163 – us ATLA [242]

La hongrie de l'adriatique au danube : impressions de voyage / Tissot, Victor – Paris 1883 [mf ed Hildesheim 1995-98] – 1v on 5mf – 9 – €100.00 – 3-487-27943-6 – gw Olms [914]

Honi soit – Sydney, 1986-94 – 9r – at Pascoe [079]

Honiara. Catholic Archdiocese *see* Archives

Honigsheim, Paul *see* Die staats- und sozial-lehren der franzoesischen jansenisten im 17. jahrhundert

Honka, Rita J M *see* Body therapy repatterning and the neuromotor system

Honky times – San Antonio TX. v1 n3 [1972 summer] – 1r – 1 – mf#1583874 – us WHS [071]

Honnefer volkszeitung – Bad Honnef DE, 1889 5 jan-1893, 1894 3 jul-1910 30 jun, 1911-1912 28 jun, 1913-1942 30 jun, 1949 1 oct-1977 – 95r – 1 – (filmed by misc inst: 1978-2002 30 jun [ca 2r/yr]) – gw Mikrofilm; gw Misc Inst [074]

Honnell, T C *see* Letters

Honneur est satisfait / Dumas, Alexandre – Paris, France. 1858 – 1r – us UF Libraries [440]

Honneur et indulgence : ou, le divorce par amour / Weiss, Mathias – Paris, France. 1803 – 1r – us UF Libraries [025]

Honolulu advertiser – Honolulu HI. 1921 mar 31-1936 apr 5, 1936 apr 5, 1940 may 10-1946 apr 3, 1959 jun 23 – 3r – 1 – (cont: pacific commercial advertiser [honolulu, hi: 1882]) – mf#846223 – us WHS [071]

Honolulu advertiser – Honolulu, Hawaii. 1882+ (1) – mf#60451 – us UMI ProQuest [071]

Honolulu drum : news from the american indian center / American Indian Center (Honolulu HI) – 1984 jan, Apr, 1985 feb,jun-jul, oct, 1986: apr, jun, aug-dec – 1r – 1 – (continued by: aisc indian news) – mf#1277965 – us WHS [305]

Honolulu star-bulletin – Honolulu HI. 1920 apr 12 [centenary no], 1941 dec 6-1942 jun 6, 1943 nov 19-1944 jan 4, 1944 apr 8-may 22, 1944 feb 21-apr 6, 1944 jan 5-feb 19, 1944 feb 21-apr 6, 1944 apr 8-may 22 – 6r – 1 – (cont: evening bulletin [honolulu, hi: 1904]; hawaiian star) – mf#846219 – us WHS [071]

Honolulu star-bulletin – Honolulu, Hawaii. 1946+ (1) – mf#60452 – us UMI ProQuest [071]

Honor awards for architecture – 1981-86, 1988-89, 1992-94, 1996-97 – 87r – 1 – $6,960.00 – (mf available for missing yrs 1987, 1990, 1991. microfilm indexes available for 1984-86, 1988-89. a paper index is available for 1987) – us UMI ProQuest [720]

Honor awards for interiors – 1992-97 – 17r – 1 – $1,360.00 – (also available in mf) – us UMI ProQuest [720]

Honor awards for urban design – 1992-97 – 10r – 1 – $800.00 – us UMI ProQuest [720]

Honor y patria / Leon Gutierrez, Florencio – 1900 – 9 – sp Bibl Santa Ana [810]

L'honorable a-n morin / Bechard, Auguste – Quebec?: "Courrier de Saint-Hyacinthe", 1885 – (= ser Galerie nationale. biographies 9) – 4mf – 9 – mf#26410 – cn Canadiana [320]

L'honorable j a chapleau : sa biographie, suivie de ses principaux discours, manifestations, etc... – Montreal: Eusebe Senecal et fils, 1887 [mf ed 1983] – 6mf – 9 – mf#SEM105P296 – cn Bibl Nat [971]

L'honorable joseph-g blanchet / Bechard, Auguste – Quebec?: L Brousseau, 1884 – (= ser Galerie nationale. biographies 2) – 1mf – 9 – mf#03025 – cn Canadiana [320]

L'honorable l a dessaules : i e dessaulles judiciaire des etats-pontificaux / Bibaud, Maximilien – Montreal?: s.n, 1862 – 1mf – 9 – mf#32740 – cn Canadiana [241]

L'honorable maurice tellier : bio-bibliographie / Payette, Ange-Albert – 1960 [mf ed 1978] – (= ser Bibliographies du cours...1947-66) – 1mf – 9 – (with inf) – mf#SEM105P4 – cn Bibl Nat [920]

L'honorable p-j-o chauveau / David, Laurent-Olivier – S.l: s.n, 1872 – 1mf – 9 – mf#03646 – cn Canadiana [920]

Honorat, Michel Lamartiniere *see* Danses folkloriques haitiennes

Honore de balzac, his life and writings / Sandars, Mary Frances – New York: Dodd, Mead & Co., 1905. xvii,377p – 1 – us Wisconsin U Libr [920]

Honorio, Cirilo S *see* Tagumpay ng manggagawa

Honorio hermeto no rio da prata / Souza, Jose Antonio Soares De – Sao Paulo, Brazil. 1959 – 1r – us UF Libraries [972]

Honour of the christian priesthood / Brett, Thomas – London, England. 1844 – 1r – us UF Libraries [240]

Honourable guest *see* Chia pin li chih [ccm89]

The honourable society of gray's inn : mediaeval and renaissance manuscripts – (mf ed 1997) – 8r – 1 – £520.00 – mf#GIR – uk World [090]

Honover College *see* Standard

Honpo shogyo kaigisho shiryo : historical documents of the chamber of commerce of japan / Chamber of Commerce of Japan – Tokyo, 1890-1927 – 176r – 1 – Y1743,000 – (mthly and annual reports of chambers of commerce; northeastern area: 15 chambers of commerce; southwestern area: 22 chambers; proceedings of the chamber of commerce assoc; with 20p guide; in japanese) – ja Yushodo [380]

Honra do brasil desafrontada de insultos da astrea espadaxin – Rio de Janeiro, RJ: Typ de Plancher-Seignot, 08 abr-20 ago 1828 – (= ser Ps 19) – mf#P01,04,15 – bl Biblioteca [321]

La honrada / Picon, Jacinto O – 1890 – 9 – sp Bibl Santa Ana [830]

Honradas / Carrion, Miguel De – Habana, Cuba. 1919 – 1r – us UF Libraries [972]

Hontheim, Joseph *see* Das buch job als strophisches kunstwerk nachgewiesen

Honved, A *see* Sketches of the hungarian emigration into turkey

Honyman, Robert *see* Diary

Hood, Edmund Lyman *see*
- The national council of congregational churches of the united states
- The new west education commission, 1880-93

Hood, Edwin Paxton *see*
- Christmas evans
- End of the curse
- The great revival of the eighteenth century
- Isaac watts
- Oliver cromwell
- The throne of eloquence
- Villages of the bible
- The vocation of the preacher
- The world of anecdote

Hood, George *see* A history of music in new england

Hood, James Walker *see*
- The negro in the christian pulpit
- One hundred years of the african methodist episcopal zion church, or, the centennial of african methodism

Hood, P *see* Social life of the chinese

Hood river county sun – Hood River OR: Sun Pub Co Inc, -1949 [daily ex sat & sun] – 1 – (absorbed: cascade locks chronicle and the bonneville dam chronicle; cont by: hood river daily sun) – us Oregon Lib [071]

Hood river daily sun – Hood River OR: Sun Pub Co Inc, 1949-52 [daily ex sat & sun] – 1 – (cont: hood river county sun (-1949)) – us Oregon Lib [071]

Hood river glacier – Hood River OR: Glacier Pub Co, 1889- [wkly] – 1 – (ceased in 1933?) – us Oregon Lib [071]

Hood river glacier – Hood River, OR: Glacier Pub Co. v1 n1-n9 jun 8-aug 3 1889; v9 n14,29 aug 27, dec 10 1897; v10 n35-v45 n23 jan 20 1899-nov 3 1933 – 1 – (ceased in 1933?) – us Oregon Hist [071]

Hood river news – Hood River OR: W H Walton & C P Sonnichsen, 1909 [semiwkly] – 1 – (cont: hood river news=letter) – us Oregon Lib [071]

Hood river news – Hood River, OR. v5 n6 feb 10 1909-apr 8 1998; nov 11 1998-jun 1999 – 1 – us Oregon Hist [071]

Hood river news=letter – Hood River OR: E R Bradley, 1905- [wkly] – 1 – (ceased in 1909; cont by: hood river news) – us Oregon Lib [071]

Hood's magazine – London. 1844-1849 – 1 – mf#5571 – us UMI ProQuest [073]

Hooft, Antonius Johannes van 't *see* De theologie van heinrich bullinger

[Hooft, P C] *see*
- Emblemata afbeeldinghen amatoria van minne
- Emblemata amatoria

Hooge, N C *see* Effects of rational behavior training on attitudes of rehabilitation support personnel

Hoogeweg, Dr *see*
- Die schriften des koelner domscholasters
- Die schriften des koelner domscholasters, spaeteren bischofs von paderborn und kardinal-bischofs von s sabina, oliverus

Hoogewerff, G J *see* Arnoldus buchelius "res pictoriae"

Hooghe, R de *see*
- Hieroglyphica
- Hieroglyphica of merkbeelden der oude volkeren

Hooghe, Romein de *see* Cartes marines a l'usage des armees du roy de la grande-bretagne

Hooghly : past and present / Dey, Shumbhoo Chunder – Calcutta: MM Dey & Co, [1906] – (= ser Samp: indian books) – us CRL [915]

Hoogstraten, F van *see* De schoole der wereld

[Hoogstraten, F van] *see*
- De schoole der wereld
- Het voorhof der ziele
- Zegepraal
- Zegepraal der goddelyke liefde

Hoogstraten, J van *see* Staat- en zedekundige zinnepренten, of leerzame fabelen

Hoogstraten, S van *see* Inleyding tot de hooge schoole der schilderkonst...

Hook, Alfred J *see* American negligence digest

Hook, James *see*
- The ascension: a sacred oratorio
- Concerto per il organo o cembalo
- Sketches of vocal and instrumental music

Hook, Theodore Edward *see* Facts, illustrative of the treatment of napoleon buonaparte in saint helena

Hook, Walter Farquhar *see*
- Auricular confession
- Call to union on the principles of the english reformation
- A church dictionary
- Duty of english churchmen and the progress of the church in leeds
- Farewell sermon
- Gorham v the bishop of exeter
- Hear the church
- Invocation of saints, a romish sin
- Letter to his parishioners on the use of the athanasian creed
- Nonentity of romish saints and the inanity of romish ordinances
- On confirmation
- On the baptismal offices
- Our holy land and our beautiful house
- Papal supremacy
- Self-deceit, a sermon preached in the church od st mary-the-virgin, oxford, on wednesday, march 4

Hooka / Humanitarian Order of Kosmic Awareness – 1971 jan 22/feb 4-1972 apr – 1r – 1 – (cont: hooka notes) – mf#1058383 – us WHS [130]

Hooka notes / Humanitarian Order of Kosmic Awareness – 1971 jan 8/22 – 1r – 1 – (cont: dallas notes; cont by: hooka) – mf#3397252 – us WHS [130]

Hooke, S H *see* Labyrinth

Hooke, SH *see*
- The labyrinth
- Myth and ritual

Hooker county tribune – Mullen, NE: Chas Shilling (wkly) [mf ed 1901-11,1914- (gaps)] – 1 – (absorbed: thedford banner (1973)) – us NE Hist [071]

Hooker, Edward William *see* Memoir of mrs sarah l huntington smith

Hooker, J D *see*
- The botany of the antartic voyage of hm discovery ships erebus and terror in the years 1839-1843
- Himalayan journals
- Journal of a tour in marocco and the great atlas

Hooker, John *see* The bible and woman suffrace

Hooker, Joseph Dalton *see* Journal of the linnean society

Hooker, Richard *see* The works of that learned and judicious divine, mr. richard hooker

Hooker, Thomas *see* Redemption: three sermons 1637-1656

Hooker, W J *see*
- The botany of captain beechey's voyage
- Journal of a tour in iceland in the summer of 1809

Hooker, William *see* Journal of a tour in iceland, in the summer of 1809

Hool, George Albert *see* Elements of structures

Hoole, all saints [church interior] – (= ser Cheshire monumental inscriptions) – 1mf – 9 – £2.50 – mf#17 – uk CheshireFHS [929]

Hoole, Charles H *see* The classical element in the new testament

Hoole, E *see* Madras, mysore, and the south of india

Hoole, Elijah *see*
- Madras, mysore, and the south of india, or, a personal narrative of a mission to those countries from mdccccxx to mdcccxxvii.
- Personal narrative of a mission to the south of india

Hoonacker, A van *see* Nehemie en l'an 20 d'artaxerxes 1

Hoonacker, Albin van *see*
- De rerum creatione ex nihilo
- Les douze petits prophetes
- Le lieu du culte dans la legislation rituelle des hebreux
- Nouvelles etudes sur la restauration juive apres l'exil de babylone

Hoop, Loes de *see* Handlist of the dance collection

Hoopa Valley Indian Reservation *see* Commonsense

Hooper, Alfred Gifford *see* Short stories from southern africa

Hooper, Charles *see*
- Brief authority

Hooper, J S M *see* The approach to the gospel

Hooper, John *see*
- Doctrine of the second advent
- Letters of hooper to bullinger

Hooper sentinel – Hooper, NE: E W Renkin. -v109 n20. aug 31 1994 (wkly) [mf ed 1892,1895-1994 (gaps) filmed -1995] – 35r – 1 – (merged with: scribner rustler (1927) to form: rustler-sentinel) – us NE Hist [071]

Hooper, Thomas *see* The story of english congregationalism

Hooper, William *see*
- Christian doctrine in contrast with hinduism and islam
- The hindu doctrine of transmigration

Hooper, William Hope *see* Letters

Hooper, William Hulme *see* Ten months among the tents of the tuski

Hooper, William Story *see* Fifty years as a presiding elder

Hoopes, Darlington *see* The socialist party of the united states

Hoopingarner, hopingardner, hoopengarner, hoopengardner family newsletter – 1987 jun-1989 winter – 1r – 1 – mf#1656922 – us WHS [929]

Hoops, J *see* Anglistische forschungen

Hoorn, J van *see* Resoluties, rapporten, brieven e.d. uitgewisseld tussen j van hoorn en de koning van bantam, 1708

Hoornbeek, J *see*
- De conversione indorum et gentilium libri duo
- De convincendis et convertendis judaeis et gustilibus
- Dissertatio de consociatione evangelica reformatorum et augustanae confessionis sive de colloquio cassellano...1661
- Oratio de ecclesiarum, inter se communione
- Orationes habitae in academia ultrajectina
- Socinianismus confutatus...
- Theologiae practicae
- Tractaat van catechisatie

Hoosharar / Armenian General Benevolent Union – 1978 sep 1-1979 dec 15, 1980 jan 15-1982 jun, 1982 sep-1985 dec, 1986-88 – 4r – 1 – (cont: mioutioun) – mf#519475 – us WHS [366]

Hoosier connection – v1 n1-v4 n2 [1984 jan-1987 apr], v1 n1 [1987 nov] – 1r – 1 – mf#1277655 – us WHS [071]

Hoosier folklore – Indianapolis, IN: Indiana Historical Bureau c1946-1950 [mf ed Bloomington IN: Indiana Uni Lib, Preservation Dept 1984] – 1r – 1 – (cont: hoosier folklore bulletin) – us Indiana Preservation [390]

Hoosier folklore bulletin – Bloomington, IN: [Hoosier Folklore Society] 1942-45 [mf ed Bloomington IN: Indiana Uni Lib, Preservation Dept 1984] – 1r – 1 – us Indiana Preservation [390]

Hoosier journal of ancestry – 1969 oct-1972 apr, 1977-78, 1979-84 – 2r – 1 – mf#622880 – us WHS [929]

Hoosier legionaire / American Legion – 1922 jun 2-1923 feb 23, 1922 jun 2, jul 28, aug 4, aug 18-sep 1, sep 29-nov 17, dec 1, dec 8, dec 22-feb 23, 1923 may, jun – 2r – 1 – mf#4360702 – us WHS [366]

Hoosier mosaics / Thompson, Maurice – 1875 – 9 – $15.00 – us Scholars Facs [830]

Hoosier observer – Fort Wayne IN. 1931 sep 25-1932 apr 10 – 1r – 1 – (cont: lebavard) – mf#857067 – us WHS [071]

Hoosier village / Sims, Newell Leroy – New York, NY. 1912 – 1r – us UF Libraries [960]

Hooton, Charles *see* St louis' isle, or, texiana

Hooton, st paul – (= ser Cheshire monumental inscriptions) – 3mf – 9 – £4.50 – mf#394 – uk CheshireFHS [929]

Hooton, Walter Stewart *see* The missionary campaign

Hoover, Herbert *see* Proclamations and executive orders

Hoover, L M *see* A history of the first baptist church, eldorado, texas

Hoover-huber family association newsletter – v4 n3-v5 n1 [1987 sep-1988 dec] – 1r – 1 – (continued by: hoover histories) – mf#1520926 – us WHS [929]

Hooykaas, Isaac *see* The bible for learners

Hooykaas, Isaac et al *see* Godsdienst

Hop / Myrick, Herbert – New York, NY. 1909, 1898 – 1r – us UF Libraries [500]

Hop tac xa thong bao : xuat ban duoi quyen kiem soat cua nong-pho ngan-quy tong-cuc / Office indochinois de credit agricole mutuel – Hanoi: Trung-Bac Tan-Van [1938-43] [v1 n1-v6 n59 [oct 1938-dec 1943] [mf ed [Hanoi, Vietnam: National Library of Vietnam] 1995] – 1r – 1 – (master neg held by crl) – mf#mf-11816 seam – us CRL [079]

Hope / Gavan Duffy, Thomas – Pondicherry, India: Catholic Mission. v1 n5-v5 n5 (easter 1919-1st dec 1929) [mf ed 2005] – 1r – 1 – (no more publ? lacks v2 n1,3-5, v3 n3-6; publ in: pondicherry, india, 1919-1921; tindivanum, south arcot, india, 1927-1929; incl poems & lectures by aut) – mf#2005C-S021 – us ATLA [241]

Hope : organ of fireside schools – Nashville TN: Fireside Schools, Woman's American Baptist Home Mission Society, v31 n10-v62 n12 (jun 1916-aug 1947) [mf ed 2005] – 1r – 1 – (lacks some iss) – mf#2005-s006 – us ATLA [242]

The hope – (Philadelphia) – 1 – us AJPC [071]

Hope, Anne Fulton *see*
- The first divorce of henry 8
- Franciscan martyrs in england
- The life of s thomas a becket of canterbury

Hope Chapel Christian Assembly, Inc of Jacksonville *see* Lighthouse of hope

Hope, Eva *see* Life of general gordon

Hope for south africa / Paton, Alan – New York, NY. 1958 – 1r – us UF Libraries [960]

Hope foundation story / Clinton, Iris – Gwelo, Zimbabwe. 1969 – 1r – us UF Libraries [960]

Hope, Gerhard Ewoud *see* Crossing boundaries

Hope, I *see* Britanny and the bible

Hope, John *see*
- Letter to the lord chancellor
- Papers of john and lugenia burns hope

Hope, Laura Lee *see* Bobbsey twins

Hope, Laurence *see* Songs from the garden of kama

Hope, Lugenia (Burns) *see* Papers of john and lugenia burns hope

The hope of immortality : an essay incorporating the lectures delivered before the university of cambridge... / Welldon, James Edward Cowell – 2nd ed. London: Seeley, 1898 – 1mf – 9 – 0-7905-0452-9 – (incl bibl ref) – mf#1987-0452 – us ATLA [240]

The hope of immortality : our reasons for it / Dole, Charles Fletcher – New York: Thomas Y Crowell, 1906 – (= ser The Ingersoll Lecture) – 1mf – 9 – 0-7905-9187-1 – mf#1989-2412 – us ATLA [240]

The hope of israel : a review of the argument from prophecy / Woods, F H – Edinburgh: T & T Clark, 1896 – (= ser Warburton lectures) – 1mf – 9 – 0-8370-6543-7 – (incl bibl ref and ind of biblical passages cited) – mf#1986-0543 – us ATLA [220]

The hope of the great community / Royce, Josiah – New York: Macmillan, 1916 – 1mf – 9 – 0-7905-8726-2 – mf#1989-1951 – us ATLA [975]

Hope standard – British Columbia, CN. apr 1968- – 1r/y – 1 – Can$93.00 – cn Commonwealth Imaging [071]

The hope that is in me / Wilberforce, Basil – New York: Dodd, Mead, [1909?] – 1mf – 9 – 0-7905-7492-6 – mf#1989-0717 – us ATLA [240]

Hope, Theodore Cracroft, Sir et al *see* The religious question in public education

Hope, Thomas *see*
- Costume of the ancients
- An historical essay on architecture by the late thomas hope
- Household furniture and interior decoration
- Observations on the plans...by j wyatt...for downing college

Hope, William Henry St John *see* English altars from illuminated manuscripts

Hopeful baptist church : burke county – Washington. 1972-1974 (1) 1964-1974 (5) 1964-1973 (9) – 1r – 1 – $11.88 – mf#6533 – us Southern Baptist [242]

Ho-pei ho tso : yu liang she chih shih k'uang – [China: sn], Min kuo 24 [1935] – (= ser P-k&k period) – us CRL [334]

Ho-pei jih-pao – Paoting, Hopeh. Aug 1, 1949-. Scattered issues missing. 5 reels – 1 – mf#87.50 – us Chinese Res [079]

Ho-pei sheng li nung hsueh yuan i lan – [Ho-pei: Sheng li nung hsueh yuan], Min kuo 25 [1936] – (= ser P-k&k period) – us CRL [951]

Hopes of an empire reversed / Jamieson, John – Edinburgh, Scotland. 1817 – 1r – us UF Libraries [240]

The hopes of the human race, hereafter and here / Cobbe, Frances Power – London: Williams and Norgate, 1874 – 1mf – 9 – 0-7905-8643-6 – mf#1989-1868 – us ATLA [240]

Hope-Scott, James Robert *see* Bishopric of the united church of england and ireland at jersualem

Hopewell baptist church – Anderson Co, SC. 1681p. 1868-1890, 1963-jul 1992 – 1 – $75.65 – (wmu 1906-26, sunday school 1958-60, history 1803-1981, scrapbooks 1803-1987, financial records 1951-nov 1957) – mf#6232 – us Southern Baptist [242]

Hopewell baptist church. arlington, kentucky : church records – Sept 1881 – Jan 1975. Incomplete – 1 – us Southern Baptist [242]

Hopewell baptist church. bethel association. south carolina : church records – 1813-37 – 1 – us Southern Baptist [242]

Hopewell baptist church. henry county. kentucky : church records – 1836-1961 – 1 – us Southern Baptist [242]

Hopewell baptist church. robertson county. springfield, tennessee : church records – 1846-1964 – 1 – us Southern Baptist [242]

Hopewell first baptist church. hopewell, virginia : church records – 1915-68 – 1 – 58.23 – us Southern Baptist [242]

Hopewell telephone – Hopewell, PA., 1891-1898 – 13 – $25.00r – us IMR [071]

Hopf, Walther *see* Jeremias gotthelf im kreise seiner amtsbrueder und als pfarrer

Hopfen, Hans *see*
- Gedichte
- Die geschichten des majors
- Helga
- Kleine leute
- Mein erstes abenteuer
- Robert leichtfuss
- Theater
- Uebereilte werbung / hotel koepf
- Zehn oder elf?

Hopgood, Cecil Robert *see* Practical introduction to tonga

Hopi action news – 1966-68 – (= ser American indian periodicals... 1) – 1r – 1 – $115.00 – us UPA [305]

Hopi action news – Winslow AZ. 1969 feb 13-1971 jan 14, 1971 jan 21-1972 dec 28 – 2r – 1 – mf#919793 – us WHS [071]

Hopi tribal news – v2 n21-n42 [1979 nov, 2nd ed.-1980 oct, 2nd ed] – 1r – 1 – mf#618272 – us WHS [307]

Hopi Tribe *see*
- Newsletter
- Tribal newsletter

Ho-ping ri-bao (peace daily) – Nanking, China. 26 sep 1946-15 apr 1949 – 6 1/4r – 1 – uk British Libr Newspaper [072]

Hopital du sacre-coeur : 25: cartierville – Montreal: Hopital du Sacre-Coeur, impr 1951 [Quebec (Province)] (mf ed 1993) – 1mf – 9 – mf#SEM105P1960 – cn Bibl Nat [360]

Hopkin-Jenkins, K *see* Basic bantu

Hopkins, Alphonso Alva *see* Geraldine

Hopkin's chancery appeals reports / New York. (State) – 1v. 1823-26 (all publ) – 7mf – 9 – $10.50 – (a pre-nrs title) – mf#LLMC 80-024 – us LLMC [340]

Hopkins' committee report : report to the u.s. secretary of the navy by the committee to study the naval administration of guam and american samoa – 1947 – 2mf – 9 – $3.00 – mf#LLMC 82-100C Title 10 – us LLMC [327]

Hopkins County Genealogical Society *see* Yesterdays tuckaways

Hopkins, Daniel C *see* True cause of all contention, strife, and civil war in christian communities

Hopkins, Earl Palmer *see* Problems and quiz on criminal procedure

Hopkins, Edward Washburn *see*
- Epic mythology
- Great epic of india
- History of religions
- India old and new
- The ordinances of manu
- The religions of india

Hopkins, Ellice *see* Life and letters of james hinton

Hopkins, Henry Whitmer *see*
- Atlas of the city and island of montreal
- Atlas of the town of sorel and county of richelieu, province of quebec

Hopkins, Joe R *see* The effects of hip position and angular velocity on quadriceps and hamstring eccentric peak torque

Hopkins, John Castell *see*
- Canadian hostility to annexation
- The maple leaf and the union jack

Hopkins, John Henry *see*
- Articles on romanism
- A candid examination of the question whether the pope of rome is the great antichrist of scripture
- "The end of controversy" controverted
- The importance of providing religious education for the poor
- The law of ritualism
- A spiritual, ecclesiastical and historical view of slavery
- Transactions of the new-york ecclesiological society

Hopkins, Josiah *see* The scripture doctrine of endless retribution candidly presented

Hopkins, Mark *see*
- Evidences of christianity
- The law of love and love as a law
- Lectures on moral science
- Miscellaneous essays and discourses
- Modern skepticism in its relations to young men
- An outline study of man
- The scriptural idea of man
- Teachings and counsels

Hopkins, Owen Johnston *see* Orders, minutes, correspondenc

Hopkins, Robert Elliott *see* Edition of four sonatas and two sets of variations for piano by alexander reinagle

Hopkins, Ruth A *see* Effects of age and ethanol on thermoregulatory responses of men to a cold air stress

Hopkin's school journal and literary advertiser – 1848 jan 1 – 1r – 1 – mf#597766 – us WHS [370]

Hopkins, T M *see* Spots on the sun, or, the plumb-line papers

Hopkins, Theodore Weld *see* The doctrine of inspiration

Hopkins, Thomas D *see* Federal user fees

Hopkinsian magazine – Providence. 1824-1832 (1) – mf#5572 – us UMI ProQuest [978]

Hopkinson, Alfred *see* The faculty of laws, and the idea of law

Hopkinson, Alfred. *see* Definite reform in english land law

Hopkinson, Francis *see* Francis hopkinson: his book

Hopkinson, Tom *see* South africa

Hopkinsville first baptist church. hopkinsville, kentucky : church records – 1818-1904, 1961-1971. Membership/Contribution Records 1913-1956; Deacon Minutes 1923-1979. Formerly: New Providence Baptist Church; includes 66th anniversary minutes, Bethel Bapt. Assn., 1890; 4229p – 1 – $190.31 – us Southern Baptist [242]

Hopkinton 1705-1849 – Oxford, MA (mf ed 1996) – (= ser Massachusetts vital record transcripts to 1850) – 15mf – 9 – 0-87623-269-1 – (mf 1t-3t: births 1705-96. mf 2t-3t: intentions 1737-95; marriages 1726-95. mf 3t: deaths 1728-91. mf 4t-6t: births & deaths 1752-1839. mf 6t-7t: intentions 1795-1835. mf 7t-9t: marriages 1793-1844. mf 9t-11t: intentions 1835-49. mf 11t-12t: births & deaths 1812-47. mf 13t-15t: vital records 1844-49. mf 15t: marriages 1720-99; births 1774-92) – us Archive [978]

Hopp, Ernst Otto *see*
- Bundesstaat und bundeskrieg in nordamerika
- Transatlantische stimmen

Hoppe, A *see* Ueber den unterschied des glaubens und wissens und ueber die mittel denselben im gymnasial-unterricht deutlich zu machen

Hoppe, Alfred *see* Die staatsauffassung heinrich von kleists

Hoppe, D H *see*
- Flora
- Tagebuch einer reise nach den kuesten des adriatischen meers und den gebirgen von krain, kaernten, tyrol, salzburg, baiern und boehmen

Hoppe, Ingeborg Marei *see* Die freundin

Hoppe, Karl *see* Der junge wieland

Hoppe, Willie *see* Thirty years of billiards

Hoppenbrouwers, H *see*
- De wandschilderingen van de slangenburg
- La plus ancienne version latine de la vie de st antoine par st athanase

Hopper, R P *see* Old-time primitive methodism in canada, 1829-1884

Hopper, Thomas *see*
- A letter to lord viscount melbourne
- A letter...in explanation of...building the new houses of parliament

Hoppe-Seyler, Felix *see* On the development of physiological chemistry and its significance for medicine

Hoppe-seyler's zeitschrift fuer physiologische chemie – Berlin. 1877-1984 (1) 1971-1984 (5) 1976-1984 (9) – (cont by: biological chemistry hoppe-seyler) – ISSN: 0018-4888 – mf#1143 – us UMI ProQuest [574]

Hoppin, James Mason *see* Homiletics

Hopson, Ella Lord *see* Memoirs of dr. winthrop hartly hopson

Hopwood, Charles H *see* Middle temple records

Hopwood, W *see* Antinomianism explained, exposed, and exploded

Hora – Athens, Greece. -d. 9 June 1876-31 Dec 1889. Imperfect. 26 reels – 1 – uk British Libr Newspaper [949]

Hora / Padilla, Heberto – Habana, Cuba. 1964 – 1r – us UF Libraries [972]

Hora – Santiago, Chile. 2 nov 1939; feb-13 aug 1945 – 6r – 1 – uk British Libr Newspaper [072]

Hora catechetica / Gilly, William Stephen – London, England. 1828 – 1r – us UF Libraries [240]

La hora de la unidad / Montes, Eugenio – Burgos, 1937. Fiche W 1058. (Blodgett Collection of Spanish Civil War Pamphlets) – 9 – us Harvard [946]

Hora de los vencidos / Rovinski, Samuel – San Jose, Costa Rica. 1963 – 1r – us UF Libraries [972]

Hora romana – London, England. 1824 – 1r – us UF Libraries [240]

Hora'at ha-haba'ah veha-sifrut – Jerusalem, Israel. 1942 – 1r – us UF Libraries [939]

Horace *see*
- Quinti horatii flacci opera
- Satires and epistles of horace
- Satires, epistles, and ars poetica

Horace and his influence / Showerman, Grant – Boston, MA. 1922 – 1r – us UF Libraries [930]

Horace bushnell : preacher and theologian / Munger, Theodore Thornton – Boston: Houghton, Mifflin, 1899 – 1mf – 9 – 0-7905-8245-7 – (incl bibl ref) – mf#1988-8108 – us ATLA [240]

Horace et lydie / Ponsard, Francois – Paris, France. 1850 – 1r – us UF Libraries [440]

Horace greeley, the editor / Zabriskie, Francis Nicoll – New York: Funk & Wagnalls, 1890 – 1mf – 9 – 0-524-06507-1 – mf#1991-2607 – us ATLA [070]

Horace greely papers, 1831-1873 : from the holdings of the rare books and manuscripts division center for the humanities, the ny. public library, astor, lenox, and tilden foundations – 1997 – ca 4r – 1 – ca $520.00 – (with guide) – mf#S3359 – us Scholarly Res [975]

The horace main bond papers – 4pts – 9 – $14,505.00 coll – (pt1: bond family papers, 1892-1971, and general correspondence, 1926-72 9r isbn 1-55655-081-2 $1395. pt2: subject files, 1926-71 36r isbn 1-55655-082-0 $5610. pt3: institutional files, 1919-72 38r isbn 1-55655-083-9 $5935. pt4: research files, 1910-71 and writings, 1926-72 15r isbn 1-55655-084-7 $2330. with p/g) – us UPA [305]

Horace mann and the common school revival in the united states / Hinsdale, Burke Aaron – [rev ed] New York: Scribner, 1900 – (= ser The great educators) – 1mf – 9 – 0-524-02399-9 – mf#1990-0602 – us ATLA [370]

The horace mann papers – early 19th c [mf ed 1989] – 40r – 1 – (with p/g) – us MA Hist [370]

Horace Mann-Lincoln Institute of School Experimentation *see* Research bulletin

Horace, Quintus Horatius Flaccus *see*
- Fifteenth century italian manuscripts
- Satirae

Horacio – Bogota, Colombia. 1954 – 1r – us UF Libraries [972]

Horack, Frank Edward *see* The organization and control of industrial corporations

Horacker : erzaehlung / Raabe, Wilhelm Karl – Berlin: Aufbau-Verlag, 1956 – 1r – 1 – us Wisconsin U Libr [830]

Horacker / Raabe, Wilhelm Karl – 4. Aufl. Berlin: G Grote, 1891 – 1 – us Wisconsin U Libr [830]

Horae : trudy russkogo entomologicheskogo obshchestva v sankt-peterburge – Washington. 1947+ (1) 1971+ (5) 1975+ (9) – 212mf – 9 – (missing: 1882/1883-1897. v14-31) – mf#2581 – ne IDC [077]

Horae aegyptiacae : or, the chronology of ancient egypt discovered from astronomical and hieroglyphic records upon its monuments: including many dates found in coeval inscriptions from the period of the building of the great pyramid to the times of the persians... / Poole, Reginald Stuart – London: John Murray, 1851 [mf ed 1989] – 1mf – 9 – 0-7905-1372-2 – (incl blbl ref) – mf#1987-1372 – us ATLA [930]

Horae apocalypticae : or, a commentary on the apocalypse: critical and historical... / Elliott, Edward Bishop – 5th corr enl ed. London: Seeley, Jackson, and Halliday, 1862 [mf ed 1989] – 4v on 7mf – 9 – 0-7905-3016-3 – (incl bibl ref and ind) – mf#1987-3016 – us ATLA [225]

Horae aramaicae : comprising concise notices of the aramean dialects in general, and of the versions of holy scripture extant in them / Etheridge, John Wesley – London: J W Etheridge, 1843 – 1mf – 9 – 0-7905-1086-3 – (in english. includes bibliographies) – mf#1987-1086 – us ATLA [220]

Horae biblicae : short studies in the old and new testaments / Carr, Arthur – London: Hodder and Stoughton, 1903 – 1mf – 9 – 0-8370-9848-3 – (incl bibl ref and indexes) – mf#1986-3848 – us ATLA [220]

Horae evangelicae : or, the internal evidence of the gospel history: being an inquiry into the structure and origin of the four gospels, and the characteristic design of each narrative / Birks, Thomas Rawson; ed by Birks, Herbert Alfred – London; New York: George Bell, 1892 – 1mf – 9 – 0-7905-0805-2 – mf#1987-0805 – us ATLA [220]

Horae hebraicae / Crawford, Francis J – London: Williams & Norgate, 1868 – 1mf – 9 – 0-8370-9217-5 – (in english and hebrew) – mf#1986-3217 – us ATLA [470]

Horae petrinae : or, studies in the life of st peter / Howson, John Saul – [London]: Religious Tract Society, [ca 1883] – 1mf – 9 – 0-8370-3682-8 – (incl app) – mf#1985-1682 – us ATLA [920]

Horae sinicae : translations from the popular literature of the chinese – London: Printed...by C Stower, 1812 [mf ed 1995] – (= ser Yale coll) – vi/71p – 1 – 0-524-09434-9 – (trans by robert morrison) – mf#1995-0434 – us ATLA [480]

Horae synopticae : contributions to the study of the synoptic problem / Hawkins, John Caesar – 2nd rev ed. Oxford, Clarendon Press, 1909 – 1mf – 9 – 0-8370-3539-2 – (includes appendixes) – mf#1985-1539 – us ATLA [220]

Horae syriacae : seu commentationes et anecdota res vel litteras syriacas spectantia / Wiseman, N – Romae. v1. 1828 – €12.00 – ne Slangenburg [240]

Hora-luz / Lleonart, Yolanda – Habana, Cuba. 1940 – 1r – us UF Libraries [972]

Horand und hilde : gedicht / Baumbach, Rudolf – Leipzig: Breitkopf und Haertel, 1878 [mf ed 1989] – 146p – 1 – mf#6983 – us Wisconsin U Libr [810]

Horapollinis hieroglyphica graece et latine : cum integris observationibus et notis... / Horapollo; ed by Pauw, J C de – Trajecti ad Rhenum: Apud Melchior, Leonardum Charlois, 1727 – 5mf – 9 – mf#O-47 – ne IDC [090]

Horapollinis niloi hieroglyphica / Horapollo; ed by Leemans, C – Amstelodami: J Muller et Socios, 1835 – 6mf – 9 – mf#O-42 – ne IDC [090]

Horapollo *see*
- Hieroglyphica
- Hieroglyphica horapollinis
- Hieroglyphica, per bernardinum trebatium vicentinum de graecis translata...
- The hieroglyphics of horapollo nilous
- Horapollinis hieroglyphica graece et latine
- Horapollinis niloi hieroglyphica
- Hori apollinis niliaci hieroglyphica
- Ori apollinis niliaci
- Ori apollinis niliaci hieroglyphica
- [Ori apollinis niliaci hieroglyphica]
- Ori apollonis niliaci
- Orus apollo niliacus de hieroglyphicis notis...

Horario e itinerario de las procesiones de semana santa...1962 / Junta de Cofradias de Penitencia – Badajoz: Imp. Dip. Provincial, 1962 – 1 – sp Bibl Santa Ana [946]

Horarium bmv – Doornik, ca. 1480 – 2mf – 8 – €5.00 – ne Slangenburg [240]

Horas de tregua / Gomez, Maximo – Habana, Cuba. 1916 – 1r – us UF Libraries [972]

Horatii flacci emblemata / Vaenius, O – Antverpiae: Ph. Lisaert, 1612 – 4mf – 9 – mf#O-784 – ne IDC [090]

Horbury express and district advertiser – [Yorkshire & Humberside] Wakefield 7 may 1870-dec 1876 [mf ed 2003] – 4r – 1 – (missing: 1871, 1872, 1874) – uk Newsplan [072]

Der horchfunk – Kamen DE, 1928 9 sep-1933 jun – 9r – 1 – gw Misc Inst [074]

Hordas azules / Velez Osorio, Antonio – s.l, s.l? 1957 – 1r – us UF Libraries [972]

Horder, William Garrett *see*
- Quaker worthies
- The treasury of american sacred song

Hore, Alexander Hugh *see*
- The church in england
- Eighteen centuries of the church in england
- Eighteen centuries of the orthodox greek church
- Student's history of the greek church

Hore, E C *see* Tanganyika

Hore presentes ad usum sarum / Catholic Church – Paris, France. 1498 – 1r – us UF Libraries [025]

Hore, Rafael *see* Contestacion...extremadura por el aviso

Horeb baptist church. nimrod hall, virginia : church records – 30 Apr 1876-1945 – 1 – 7.29 – us Southern Baptist [242]

Horei zensho : statutes at large of japan: official edition of the complete statutes of the japanese empire, 1868-1945 / Japan. Cabinet Secretariat [comp] – 235r – 1 – Y2,157,000 – (comp yrly for 1868-85 and mthly for 1886-1945. with general and aut and subject ind in 2v for 1868-85 and yrly ind fr 1886. in japanese) – ja Yushodo [348]

Die horen : eine monatsschrift / ed by Schiller, Friedrich von – Tuebingen 1795-97 – (= ser Dz. abt literatur) – 12v on 69mf – 9 – €690.00 – mf#k/n4604 – gw Olms [430]

Horetzky, Charles *see* Some startling facts relating to the canadian pacific railway and the north-west lands

Hori apollinis niliaci hieroglyphica : hoc est de sacris aegyptiorum literis libelli... / Horapollo – [Bologna, 1517] – 2mf – 9 – mf#O-08 – ne IDC [090]

Hori, Shinji *see* Wei wu chan cheng kuan

Horicon argus – Horicon WI. 1854 sep 7-1857 oct 30, 1857 nov 6-1860 nov 30 – 2r – 1 – (continued by: beaver dam argus) – mf#945747 – us WHS [071]

Horicon gazette – Horicon WI. 1861 jan 9-1862 jan 2 – 1r – 1 – (cont: weekly burlington gazette) – mf#922884 – us WHS [071]

Horicon reporter – Horicon WI. 1893/1898 jan 7-2003 jan/apr [gaps] – 82r – 1 – (cont: dodge county reporter) – mf#1139488 – us WHS [071]

Horicon volksfreund – Horicon WI. 1886 may 28 – 1r – 1 – mf#876267 – us WHS [071]

Horine, M C *see* Practical reflections on the book of ruth

Horitsu shimbu, 1900-1934 = Legal news – 91r – 1 – $3,185.00 in US $40.00r outside – (in japanese) – mf#L9400079 – us L of C Photodup [950]

Horizon – 1907 jan-1910 jul, 1907 jan-1910 jul – 2r – 1 – mf#1058398 – us WHS [071]

Horizon – London. v1-20. 1940-Jan 1950 – 1 – us NY Public [073]

Horizon – Montreal, Canada. Mar 9 1987-Dec 5 1988; Jan 9 1989-Dec 23 1991 – 2r – 1 – us L of C Photodup [071]

Horizon – Tuscaloosa. 1958-1989 (1) 1969-1989 (5) 1960-1989 (9) – ISSN: 0018-4977 – mf#1157 – us UMI ProQuest [073]

Horizon *see* Lathatar

Horizon 1980 : une etude sur l'evolution de l'economie du quebec de 1946 a 1968 et sur ses perspectives d'avenir / Lebel, Gilles – Quebec: Ministere de l'industrie et du commerce, 1970 [mf ed 1973] – 1r – 1 – mf#SEM35P72 – cn Bibl Nat [339]

Horizon of american missions / McCash, Isaac Newton – New York: Fleming H Revell, c1913 – 1mf – 9 – 0-524-04384-1 – (incl bibl ref) – mf#1991-2088 – us ATLA [240]

Horizons – Louisville. 1988-1996 (1,5,9) – (cont: concern magazine/newsfold) – ISSN: 1040-0087 – mf#16918 – us UMI ProQuest [305]

Horizons – Toronto. n19-28. aut 1966-win 1969// – 10mf – 9 – Can$65.00 – (cont: the marxist quarterly) – cn McLaren [335]

Horizons / Medical College of Wisconsin – 1979 oct-1986 sep – 1r – 1 – (cont: mcw remarks) – mf#1289677 – us WHS [610]

Horizons : a newsletter for appleton mills employees and their families – 1971 sep 30-1982 n3 – 1r – 1 – (cont: progress report [appleton mills]) – mf#618052 – us WHS [331]

Horizons – Orrville, OH. 1992-1994 [1] – mf#69134 – us UMI ProQuest [071]

Horizons – Villanova. 1980+ (1,5,9) – ISSN: 0360-9669 – mf#12809 – us UMI ProQuest [240]

Horizons in biblical theology – Pittsburgh. 1985+ (1,5,9) – ISSN: 0195-9085 – mf#15361 – us UMI ProQuest [220]

Horizons of the mind / Fernandez-Marina, Ramon – New York, NY. 1964 – 1r – us UF Libraries [972]

Horizont – Berlin DE, 1968 nov-1989 – 26r – 1 – gw Misc Inst [074]

O horizonte : ordem e progresso – Vitoria, ES: Typ do Horizonte, 25 jul, dez 1880; jan, nov-dez 1881; fev 1882-06 jun 1885 – (= ser Ps 19) – mf#P11B,05,06 – bl Biblioteca [079]

O horizonte : orgao litterario e noticioso – Rio de Janeiro, RJ. 11 nov 1877 – (= ser Ps 19) – mf#P19A,04,66 – bl Biblioteca [440]

Horizonte de la metafisica aristotelica / Gomez Nogales, Salvador – Madrid: Imp. del Colegio Maximus, 1955 – 9 – sp Bibl Santa Ana [180]

Hormayr, Joseph von *see* Wien, seine geschicke und seine denkwuerdigkeiten

Hormonal contraceptives : a guideline for nurses: family planning = gesinsbeplanning / South Africa. Department of Health [Departement van Gesondheid] [Departement van Gesondheid] – Pretoria: Dept of Health [1976?] [mf ed Pretoria, RSA: State Library [199-]] – 18p [ill] on 1r with other items – 5 – mf#op 06527 r24 – us CRL [362]

Hormone and metabolic research – Stuttgart. 1969+ (1) 1975+ (5) 1975+ (9) – ISSN: 0018-5043 – mf#10163 – us UMI ProQuest [616]

Hormone research – Basel. 1973-1974 (1) 1974-1974 (5) (9) – (cont: hormones) – ISSN: 0301-0163 – mf#5189,01 – us UMI ProQuest [616]

Hormones – Basel. 1970-1972 (1) 1970-1972 (5) (9) – (cont by: hormone research) – ISSN: 0367-617X – mf#5189 – us UMI ProQuest [616]

Horn book magazine – Boston. 1924+ (1) 1967+ (5) 1960+ (9) – ISSN: 0018-5078 – mf#889 – us UMI ProQuest [070]

Horn, Dorothy D *see* Study of the folk-hymns of southeastern america

Horn, Edward Traill *see* The christian year

Horn, Edward Traill *see*
- Annotations on the epistles of paul to the ephesians, philippians, colossians, thessalonians
- Annotations on the epistles to timothy, titus and the hebrews
- The evangelical pastor

Horn, Ernst *see* Oeffentliche rechenschaft ueber meine zwoelfjaehrige dienstfuehrung als zweiter arzt des koenigl. charite-krankenhauses zu berlin...

Horn, Ewald *see*
- Bibliographie der deutschen universitaeten

The horn of africa / Silberman, Leo – Chicago, 1969 – (includes bibliography and index) – us CRL [960]

Hornblow, Arthur *see* Lion and the mouse

Hornblower, William Henry *see*
- The duty of the general assembly to all the churches under its care
- The lamentations of jeremiah

Hornbook and nutshell series – St Paul: West Pub Co, 1921-sept 1993 – 9 – $1,550.00 set – 1-57588-270-1 – mf#402380 – us Hein [340]

Hornburg-Stralsund, Johannes *see* Bibel und babel

Horncastle news – Horncastle, Lincolnshire 5 sep 1885-to date – 1 – (fr 1935-jan 1969 horncastle and woodhall spa news; microfilm fr 1986 onward) – uk Newsplan [072]

Horncastle standard – Horncastle, Lincolnshire 6 jul 1912-27 feb 1915, 15 jul 1933-to date – 1 – (current title of: lincolnshire standard [horncastle ed]. earlier history believed to be: ls horncastle spilsby alford and lincolnshire standard (fr c1954 in north & south ed) jul 1912-jun 1915, jul 1928-25 aug 1956; ls spilsby horncastle and woodhall spa ed [1 sep 1956-28 mar 1958]; ls and boston guardian: spilsby horncastle and woodhall spa ed [4 apr 1958-19 oct 1962] (then splits into horncastle & spilsby ed); ls horncastle and woodhall spa ed [26 oct 1962-6 dec 1985]; horncastle standard [13 dec 1985-to date]) – uk Newsplan [072]

Hornchurch and upminster news – London, UK. 10 jul-23 dec 1936; 1937-23 dec 1938; 1939-27 sep 1940; 10 apr 1947-1955; 13 jan 1956-24 dec 1958; 1959-29 sep 1966 – 19 1/2r – 1 – (aka: hornchurch and upminster news; news (hornchurch and upminster); news and south east essex independent; hornchurch and upminster news etc; hornchurch news pictorial; hornchurch and upminster news) – uk British Libr Newspaper [072]

Hornchurch and upminster news *see* Hornchurch and upminster news

Hornchurch upminster echo – London, UK. 2 feb 1965-1980 – 32 1/2r – 1 – (aka: havering echo; havering post and echo; havering post and romford hornchurch express; havering hornchurch romford upminster rainham post) – uk British Libr Newspaper [072]

Hornchurch upminster observer – London, UK. 2 dec-23 dec 1992 – 1r – 1 – uk British Libr Newspaper [072]

Horne, Charles Silvester *see*
- David livingstone
- Nonconformity in the 19th century
- A popular history of the free churches
- The story of the l.m.s., 1795-1895

Horne, George *see* A commentary on the book of psalms

Horne, Herman Harrell *see*
- Free will and human responsibility
- The leadership of bible study groups

Horne, R et al *see* Whether it be mortall sinne to transgresse civil lawes

Horne, Thomas Hartwell *see*
- A compendious introduction to the study of the bible
- An introduction to the critical study and knowledge of the holy scriptures
- A manual of biblical bibliography
- Manual of parochial psalmody
- Outlines for the classification of a library
- Protestant memorial, for the commemoration

Horne, William *see* Reason and revelation

Hornedo, A M *see* Una familia de ingenios, los ramirez de prado

Hornedo, R M *see* Entrambasaguas, joaquin de. la biblioteca en ramirez de prado

Horneffer, August *see* Der bund der freimaurer

Horneffer, Ernst *see*
- Goethe als kuender des lebens
- Hebbel und das religioese problem der gegenwart
- Die tat

Horneman, F *see* The journal of frederick horneman's travels

Hornemann, Friedrich *see* Voyages dans l'interieur de l'afrique

Hornemann, Friedrich K *see*
- African researches
- Fr hornemanns tagebuch seiner reise von cairo nach murzuck, der hauptstadt des koenigreichs fessan in afrika

Horner, Emil *see* Vor dem untergang des alten reichs

Horner, Francis Asbury *see* Criminal forms for the state of indiana

Horner, G *see* The service for the consecration of a church and altar

Horner, George *see*
- The gospel of s john, register of fragments, etc, facsimiles
- The gospel of s luke
- The gospels of s matthew and s mark
- The statutes of the apostles, or, canones ecclesiastici

Horner, I B *see* The living thoughts of gotama the buddha

Horner, Isaline Blew *see* Women under primitive buddhism

Horner, J *see* The statues of the apostles or canones ecclesiastici

Horner, Ralph Cecil *see* Voice production

Horner, Susan *see* Greek vases

Hornet *see* The paladin

Horney, Julie *see* Observation and study in the federal district courts

Hornig, Josef *see*
- Ikatekism yemfundiso yetyalike ekatolike yaseroma
- Imfundiso yetyalike ekatolike yase roma

Horningsea 1599-1950 – (= ser Cambridgeshire parish register transcript) – 5mf – 9 – £6.25 – uk CambsFHS [929]

Die hornisse : zeitung fuer hessische biedermaenner – Kassel DE, 1 aug 1848-dec 1850 – 1r – 1 – (filmed with other misc inst: 1848 1 aug-1850 13 oct [2r]) – gw Misc Inst [074]

Hornovhos. Ayuntamiento *see* Ordenanzas municipales

Horns ring : roman / Flake, Otto – Berlin: S Fischer, 1921 – 1r – 1 – us Wisconsin U Libr [830]

Hornsby advocate – Sydney, 1923-931 – 4r – 1 – (1944-95 40r) – at Pascoe [079]

Hornsby advocate – Hornsby, may 1923-dec 1931; jan-dec 1944; aug 1951-dec 1973 (misc iss) – 17r – £1148.49 vesicular A$1241.99 silver – at Pascoe [079]

Hornsby, Alton Jr *see* Papers of john and lugenia burns hope

Hornsby / upper north shore advocate – Hornsby, jan 1969-dec 1996 – (aka: upper north shore advocate) – at Pascoe [079]

Hornschuch, C F *see* Tagebuch einer reise nach den kuesten des adriatischen meers und den gebirgen von krain, kaernten, tyrol, salzburg, baiern und boehmen

Hornsey and finsbury park journal *see* Seven sisters and finsbury park journal

Hornsey and middlesex messenger – London, UK. 12 oct 1888-11 oct 1889 – 1r – 1 – (aka: middlesex messenger) – uk British Libr Newspaper [072]

Hornsey and muswell hill journal – London, UK. 8 sep 1988-aug 1991; 1992; 1993; sep-dec 1994 – 19r – 1 – (aka: hornsey muswell hill and crouch end journal) – uk British Libr Newspaper [072]

Hornsey and muswell hill journal *see* Journal (hornsey)

Hornsey muswell hill and crouch end journal *see* Hornsey and muswell hill journal

[Hornsilver-] herald – NV. may-sep 1908 [wkly] – 1r – 1 – $60.00 – mf#U04588 – us Library Micro [071]

Hornyold, John Joseph *see* Real principles of catholics

Horodezky, Samuel A *see*
- Ha-mistorin be-yisrael
- Le-korot ha-rabanut
- 'Ole Tsiyon

Horontchik, Simon *see* In geroysh fun mashinen

L'horoscope a l'usage de tout le monde – [Quebec?: s.n.] 1862 [mf ed 1985] – 1mf – 9 – 0-665-16567-6 – mf#16567 – cn Canadiana [130]

Horovicz, Jonathen Benjamin *see* Gegen die blutbeschuldigung

Horovitz, Jakob *see* Babel und bibel

Horovitz, Saul *see* Einfluss der grieschischen skepsis auf die entwicklung

Horowhenua daily chronicle – jan-jun 1915; jan-dec 1916; 20 jan 1917-dec 1921; jan 1923-dec 1939; mar 1973-feb 1976; mar 1976-aug 1977; oct 1977-apr 1994; jun 1994-oct 1998 – 1 – (aka: horowhenua-kapiti chronicle (levin); horowhenua daily chronicle; title changes to: levin daily chronicle on 20 jan 1917; title changes to: the chronicle mar 1976) – mf#46.1 – nz Nat Libr [079]

Horowhenua daily chronicle *see*
- The chronicle
- Horowhenua daily chronicle

Horowhenua-kapiti chronicle *see* Horowhenua daily chronicle

Horowitz, Chaim Meir *see* Agudat agadot

Horowitz, Irving Louis *see* Revolution in brazil

Horozco y Covarrubias, J de *see*
- Emblemas morales de don juan de horozco y covarruvias arcediano de cuellar en la santa yglesia de segovia

Horrabin, James Francis *see* Atlas of africa

Horrego Estuch, Leopoldo *see*
- Placido
- Sentido revolucionario del 68

Horrell, Georgina Ann *see* White skin under an african sun

Horrell, Muriel *see*
- African education
- Days of crisis in rhodesia
- Group areas act
- Legislation and race relations
- Outline of the systems of government and the political status
- Racialism and the trade unions
- Reserves and reservations
- Rights of african women, some suggested reforms
- South africa and the olympic games
- South africa's non-white workers
- South-west africa
- Terrorism in southern africa

Horrer, Georg Adam et al *see* Almanach fuer prediger, die lesen, forschen und denken

Horribilicribrifax : scherzspiel / Gryphius, Andreas; ed by Braune, Wilhelm – Halle: Max Niemeyer, 1883 [mf ed 1993] – (= ser Neudrucke deutscher literaturwerke des 16. und 17. jahrhunderts 3) – vi/90p – 1 – (fr 1663 ed) – mf#8413 reel 1 – us Wisconsin U Libr [820]

Horridoh! : ein waidmannsleben in liedern / Bley, Fritz – 2. aufl. Berlin: E Fleischel, 1914 [mf ed 1989] – 143p – 1 – mf#7032 – us Wisconsin U Libr [780]

Horrwitz, Ernest *see* A short history of indian literature

Horrwitz, Ernest Philip *see* The indian theatre

Horry county loris sentinel – Loris, SC. 1952-1955 (1) – mf#66500 – us UMI ProQuest [071]

Horry county news – Loris, SC. 1949-1952 (1) – mf#66501 – us UMI ProQuest [071]

Horry herald – Conway, SC. 1887-1943 (1) – mf#66480 – us UMI ProQuest [071]

Horry independent – Conway, SC. 1998-2001 (1) – mf#68252 – us UMI ProQuest [071]

Horry news – Conway, SC. 1871-1876 (1) – mf#66481 – us UMI ProQuest [071]

Horsburgh, J *see* Memoirs

Horsburgh, J Heywood *see* Do not say

Horsch, John *see* Menno simons, his life, labors, and teachings

Horse and pony – Glasgow. 1971-1974 (1) – mf#8882 – us UMI ProQuest [636]

Horse cave baptist church. kentucky : church records – 1868-1968 – 1 – us Southern Baptist [242]

The horse educator / McPherson, J G – S.l: s.n, 1882? – 1mf – 9 – mf#09681 – cn Canadiana [636]

Horse lover's magazine – Temecula. 1959-1973 (1) 1970-1972 (5) – (cont by: horse lover's national magazine) – ISSN: 0018-5175 – mf#1175 – us UMI ProQuest [636]

Horse lover's national magazine – San Francisco. 1973-1980 (1) 1975-1980 (5) 1975-1980 (9) – (cont: horse lover's magazine) – ISSN: 0199-3232 – mf#1175,01 – us UMI ProQuest [636]

Horsed – Hamar [ie Mogadishu]: Society for Somali Language and Literature, n1-23 1967-1968 – us CRL [470]

Horseheath 1558-1950 – (= ser Cambridgeshire parish register transcript) – 5mf – 9 – £6.25 – uk CambsFHS [929]

The horseless age – New York: Horseless Age Co, 1895-1918. v12 n1-26 jul-dec 1903; v33 n13-25 apr-jun 24 1914 – us CRL [071]

The horseless age – New York: The Horseless Age Co, 1895-1918 – 2r – 1 – (title fr caption: the automotive trade magazine; subtitle varies. absorbed in pt by: motor age, motor world and automotive industries) – mf#MF H787 – us Colorado Hist [380]
Horseless carriage gazette – Downey. 1938-1973 (1) 1972-1972 (5) (9) – ISSN: 0018-5213 – mf#6999 – us UMI ProQuest [790]
Horseman – Navasota. 1956-1990 (1) 1972-1990 (5) 1973-1990 (9) – ISSN: 0018-5221 – mf#7499 – us UMI ProQuest [636]
Horsemen's journal – New Orleans. 1975-1992 (1) 1975-1992 (5) 1975-1992 (9) – ISSN: 0018-5256 – mf#7102 – us UMI ProQuest [636]
Horses and other animals / Royal Library. Windsor Castle – [mf ed 1986] – (= ser Leonardo drawings) – 6 colour mf – 15 – $260.00 – 0-907716-14-8 – (117 drawings, 115 details; with ind) – uk Mindata [740]
Horses nine : stories of harness and saddle / Ford, Sewell – Toronto: Copp, Clark, 1903 [mf ed 1997] – 4mf – 9 – 0-665-81896-3 – mf#81896 – cn Canadiana [830]
Horsfall, Thomas Coglan *see*
– A description of the work of the manchester art museum
– The study of beauty
Horsford, Eben Norton *see*
– The defences of norumbega and a review of the reconnaissances of col t w higginson, professor henry w haynes, dr justin winsor, dr francis parkman, and rev edmund f slafter
– Zeisberger's indian dictionary
Horsham advertiser. (west sussex times.-west sussex county times) – Horsham, England. -w. Nov 1871-1900; 1962-69; 1973-83.79 reels – 1 – uk British Libr Newspaper [072]
Horsley, C D *see* Some problems connected with the proposed scheme of church union in south india
Horst, Cornelius van der *see* Das lachen des sergeanten wassenaar
Horst e wittig, personalbibliographie : veroeffentlichungen der jahre 1950 bis 1956 / Klemm, Ulrich & Wittig, Hildja Yukino – (mf ed 1998) – 1mf – 9 – €30.00 – 3-8267-2521-2 – mf#DHS 2521 – gw Frankfurter [370]
Horst, Karl August *see* Ich und gnade
Horst, L *see* Leviticus 17-26 und hezekiel
Horsthemke, Johannes *see* Melchior von diepenbrock als uebersetzer spanischer dichtungen
Horstmann, Christina *see* Die literarhistorische gesellschaft bonn im ersten drittel des 20. jahrhunderts
Hort, A F *see* The gospel according to mark
Hort, Arthur *see* The gospel according to st mark
Hort, Fenton John Anthony *see*
– The apocalypse of st john 1-3
– The christian ecclesia
– The epistle of st james
– The epistle of st james, 1 1-4.7
– The first epistle of st peter 1. 1-2. 17
– The first epistle of st peter, 1.1-2.17
– Life and letters of fenton john anthony hort, d.d., d.c.l., ll.d.
– Memorials of the late wharton booth marriott, b.d., f.s.a
– Prolegomena to st paul's epistles to the romans and the ephesians
– Two dissertations
– Village sermons
– Village sermons in outline
– The way, the truth, the life
Hort, Fenton John Antony *see* Judaistic christianity
Hortaliza / Bobea, Joaquin Maria – San Pedro de Macoris, Dominican Republic. 1959 – 1r – us UF Libraries [972]
Horten, M *see* Das buch der ringsteine farabis
Horten, Max *see*
– Einfuehrung in die hoehere geisteskultur des islam
– Die kulturelle entwicklungsfaehigkeit des islam auf geistigem gebiete
– Die philosophischen probleme der spekulativen theologie im islam
– Die philosophischen systeme der spekulativen theologen im islam
– Die religioese gedankenwelt der gebildeten muslime im heutigen islam
Hortensius : friend of nero / Peters, Ellis – New York, NY. 1937 – 1r – us UF Libraries [025]
Hortensius the advocate : an historical essay on the office and duties of an advocate / Forsyth, William – Jersey City: N J Frederick D Linn & Co, 1882 – 5mf – 9 – $7.50 – mf#LLMC 95-159 – us LLMC [340]
Horticultural register : and gardener's magazine – Boston. 1835-1838 (1) – mf#3991 – us UMI ProQuest [634]
Horticulturist and journal of rural art and rural taste – Albany. 1846-1875 (1) – mf#4559 – us UMI ProQuest [630]
Horto simbolico che con gieroglifici di vari alberi : e diverse piante, rappresenta le virt-singulari d'alcuni santi, e molte s / Labia, Carlo – Venetia: Appresso Nicol Pezzana, 1700 – 17mf – 9 – mf#0-860 – ne IDC [090]
Horton, A E *see*
– Dictionary of luvale
– Handbook to the 'grammar of luvale'
Horton, Mary B *see* A brief exposition of gospel differences given according to the divine law of progressive instruction
Horton, R Wilmot *see* Protestant safety compatible with the remission of the civil disabi...
Horton, Robert F *see*
– The early church
– Revelation and the bible
Horton, Robert Forman *see*
– The bible
– The book of proverbs
– The cartoons of st mark
– England's danger
– Great issues
– The growth of the new testament
– The hero of heroes
– Inspiration and the bible
– My belief
– On the art of living together
– The pastoral epistles
– Reconstruction
– The reunion of english christendom
– Shall rome reconquer england?
– The springs of joy
– The teaching of jesus
– This do
– Three months in india
– The trinity
– The triumph of the cross
– Women of the old testament
Horton, Samuel Dana *see*
– The parity of moneys as regarded by adam smith, ricardo, and mill
– Silver
Horton, Walter Marshall *see* Shang ti lun (ccm163)
Horton, William *see* Memoir of the late thomas scatcherd
Hortonville shiocton community squire – Dale, Greenville, Hortonville etc WI. 1964 jan 30-nov 12 – 1r – 1 – (cont: new town and country news) – mf#923070 – us WHS [071]
Hortonville star – Dale, Greenville, Hortonville etc WI. 1972 mar 16-1973 apr 26, 1973 may 3-1973 dec 27, 1974 jan 3-1974 jun 19 – 3r – 1 – (continued by: new london press; new london press; press-star) – mf#946539 – us WHS [071]
Hortulus animae : dat is, der sielen bogaert – Antwerpen, 1606 – 9mf – 8 – €18.00 – ne Slangenburg [240]
Hortulus chelicus : das is wohlgepflantzter violinischer lust-garten darin allen kunst-begierigen musicalischen liebhabern der weeg zur volkommenheit... / Walther, Johann Jacob – Mayntz: In Verlegung Ludovici Bourgeat, Buchhaendlern 1694 [mf ed 19–] – 5mf – 9 – mf#fiche 192 – us Sibley [780]
Hortulus hermeticus flosculis philosophorum cupro incisis conformatus... / Stolcius, D – Francofurti: Impensis Lucae Jennisii, 1627 – 1mf – 9 – mf#0-1912 – ne IDC [090]
Hortus musicalis...5. 6. 7. 8. & pluribus vocibus – 1609 – (= ser Mssa) – 6mf – 9 – €80.00 – mfchl 154 – gw Fischer [780]
Horup, Ellen *see* Spain, the battlefield of capitalism
Horus : royal god of egypt / Mercer, Samuel – Grafton, MA. 1942 – 1r – us UF Libraries [025]
Horus in the pyramid texts / Allen, Thomas George – 1916 – 1mf – 9 – 0-524-00815-9 – mf#1990-2061 – us ATLA [930]
Horwich and westhoughton journal and guardian – [NW England] Horwich 1927-jan 1955 [mf ed 2002] – 1 – (title change: horwich & westhoughton journal [4 feb 1955-4 may 1979]; horwich journal [11 may 1979-31 oct 1980]) – uk Newsplan; uk MLA [072]
Horwich chronicle – [NW England] Horwich 27 oct 1888-1911, 1913-25 mar 1916 – 1 – uk MLA; uk Newsplan [072]
Horwich & west houghton supplement – [NW England] Bolton jan 1923-13 aug 1926 [mf ed 2002] – 2r – 1 – uk Newsplan [072]
Horwitz, Ludwig *see*
– Emanzipation der juden in anhalt-dessau
– Geschichte der herzoglichen franzschule in dessau, 1799-1849
Horwitz, Ralph *see*
– Expand or explode
– Political economy of south africa
Horwood, A J *see*
– Year books of the reign of king edward 1
– Year books of the reign of king edward 3
Horyzonty – Paryz; Nowy Jork: [Komitet Wspolprancy Horyzontow] 1956 [mf ed Bloomington IN: Indiana Uni Lib, Preservation Dept 1984] – 6r – 1 – us Indiana Preservation [073]
The hos of seraikella / Chatterjee, Anathnath & Tarakchandra, Das – Calcutta: University of Calcutta, 1927- – (= ser Samp: indian books) – us CRL [305]
Hosanna : hommage a sa tres gracieuse majeste victoria, reine d'angleterre et imperatrice des indes / Lemay, Pamphile – S.l: s.n, 1887? – 1mf – 9 – mf#08654 – cn Canadiana [971]
Hosea : the heart and holiness of god / Morgan, George Campbell – London: Marshall, Morgan & Scott Ltd, [19–?] [mf ed 2002] – 1r – 1 – mf#b00642 – us ATLA [221]
Hosea : with notes and introduction / Cheyne, Thomas Kelly – Cambridge: University Press, 1884 – 1mf – 9 – 0-8370-6808-8 – (incl bibl ref and indexes) – mf#1986-0808 – us ATLA [220]
Hosea ballou : a marvellous life-story / Safford, Oscar Fitzalan – 4th ed. Boston: Universalist Pub House, 1890 – 1mf – 9 – 0-524-04302-7 – mf#1992-2022 – us ATLA [240]
Hosea ballou and the gospel renaissance of the nineteenth century / Adams, John Coleman – Boston: Universalist Pub House, 1903 – 1mf – 9 – 0-524-00500-1 – mf#1990-0000 – us ATLA [220]
Hoseas illustratus chaldaica jonathanis versione and celebrium rabbinorum raschi aben-esrae et kimchi commentariis – Goettingae, 1775 – 4mf – 8 – €11.00 – ne Slangenburg [221]
Hoseas propheta commentarii illustatus / Pareus, David – Haedelbergae: Voegelinianis, [1605] – 1r – 1 – 0-8370-0970-7 – mf#1984-B511 – us ATLA [220]
Hoselitz, Berthold Frank *see* Desarrollo industrial de el salvador
Die hosen des doktors im nonnenkloster : ein weltliches lied, enthaltend das abentheyerliche fatum / Schmidt, Johann – Muenchen: Bibliographisch-artistisches Institut, [1920?] – 1r – 1 – us Wisconsin U Libr [810]
Hoshour, Samuel Klinefelter *see* Autobiography
Hosie, Alexander *see*
– Manchuria
– Three years in western china
Hosiery worker / American Federation of Hoisiery Workers – 1920 dec 31-1931, 1932-34, 1935-37, 1938-47, 1948-50, 1951-53, 1954-59, 1960-62, 1963-1965 may – 9r – 1 – mf#1058411 – us WHS [680]
Hosius, Carl *see* P vergili maronis bucolica
Hosius, Stanislaus, Cardinal *see* Opera omnia
Hoskier, H C *see*
– The complete commentary of oecumenius on the apocalypse
– Concerning the date of the bohairic version
Hoskier, Herman Charles *see*
– Concerning the genesis of the versions of the new testament
– A full account and collation of the greek cursive codex evangelium 604 (egerton 2610 in the british museum)
Hosking, William *see*
– A guide to the proper regulation of buildings in towns
– An introductory lecture...on the principles and practice of architecture
– Restoration of the church of saint mary, redcliffe, bristol
– Some observations upon the recent addition of a reading room to the british museum
– Treatises on architecture and building
The hoskins headlight – Hoskins, NE: Orin Garwood, 1905-24 (wkly) [mf ed 1908-24 (gaps) filmed 1976] – 4r – 1 – (suspended with jun 28 1923; resumed with dec 6 1923. some irregularities in numbering) – us NE Hist [071]
Hoskyns, Catherine *see* Congo since independence, january 1960-december, 1961
Hoskyns, Ed *see* The fourth gospel
Hosmer, James Kendall *see*
– The life of young sir henry vane, governor of massachusetts bay and leader of the long parliament
– A short history of german literature
– The slave power
Hosmer, William *see*
– Autobiography of rev alvin torry
– The higher law in its relations to civil government
– Slavery and the church
La hospederia real de guadalupe / Pescador del Hoyo, Maria del Carmen – Badajoz: Imp. Diputacion Provincial, 1965 – sp Bibl Santa Ana [946]
La hospederia real de guadalupe 2 / Pescador del Hoyo, Maria del Carmen – Badajoz: Imp. Diputacion Provincial, 1968 – sp Bibl Santa Ana [946]
Hospers, Gerrit Hendrik *see* Beginselen van separatie
Hospice des soeurs de la charite a quebec / Proulx, Louis – [Quebec?: s.n.] 1851 [mf ed 1984] – 1mf – 9 – 0-665-16570-6 – mf#16570 – cn Canadiana [360]
The hospice journal : the official journal of the national hospice organization / ed by Infeld, Donna Lind – v1- 1985- – 1, 9 ($200.00 in US $280.00 outside hardcopy subsc) – us Haworth [360]
Hospinian, R *see*
– Festa christianorvm...
– Historiae sacramentariae pars altera
Hospital abstracts – London. 1975-1980 (1) 1976-1980 (5) 1976-1980 (9) – ISSN: 0018-5507 – mf#7964 – us UMI ProQuest [360]
Hospital abuse / Armstrong, George E – S.l: s.n, 1898? – 1mf – 9 – mf#44794 – cn Canadiana [360]
Hospital administration – Chicago. 1956-1975 (1) 1972-1975 (5) 1975-1975 (9) – (cont by: hospital and health services administration) – ISSN: 0018-5523 – mf#6408 – us UMI ProQuest [360]
Hospital and community psychiatry : a journal of the american psychiatric association – Washington. 1950-1994 (1) 1971-1994 (5) 1971-1994 (9) – (cont by: psychiatric services) – ISSN: 0022-1597 – mf#2117 – us UMI ProQuest [616]
Hospital and health administration index – Chicago. 1995-1999 (1) 1995-1999 (5) 1995-1999 (9) – (cont: hospital literature index) – ISSN: 1077-1719 – mf#13375,03 – us UMI ProQuest [360]
Hospital and health services administration – Chicago. 1976-1997 (1,5,9) – (cont: hospital administration; cont by: journal of healthcare management) – ISSN: 8750-3735 – mf#6408,01 – us UMI ProQuest [360]
Hospital and health services review – London. 1969-1988 (1) 1971-1988 (5) 1972-1988 (9) – (cont by: health services management) – ISSN: 0308-0234 – mf#5135 – us UMI ProQuest [360]
Hospital Employees' Union, Local 180 *see* Hospital guardian
Hospital financial management – Chicago. 1968-1982 (1) 1968-1982 (5) 1968-1982 (9) – (cont by: healthcare financial management) – ISSN: 0018-5639 – mf#12654,01 – us UMI ProQuest [360]
Hospital formulary – Minneapolis. 1982-1995 (1) 1982-1995 (5) 1982-1995 (9) – (cont: hospital formulary management. cont by: formulary) – ISSN: 0098-6909 – mf#2395,01 – us UMI ProQuest [360]
Hospital formulary management – Minneapolis. 1966-1974 (1) 1970-1974 (5) – (cont by: hospital formulary) – ISSN: 0018-5655 – mf#2395 – us UMI ProQuest [360]
Hospital forum – San Francisco. 1958-1985 (1) 1970-1985 (5) 1975-1985 (9) – (cont by: healthcare forum) – ISSN: 0018-5663 – mf#1943 – us UMI ProQuest [360]
Hospital guardian / Hospital Employees' Union, Local 180 – v21 n1-2 [1983 spring-summer], v3 n3-v8 n2 [1983 nov-1990 spring] – 1r – 1 – (continued by: guardian (vancouver, bc)) – mf#698480 – us WHS [331]
Hospital infection control – Atlanta. 1979+ (1,5,9) – ISSN: 0098-180X – mf#12279 – us UMI ProQuest [614]
Hospital, Juvencio, Bishop of Cauna *see* Notas y escenas de viaje
Hospital law's regan report – Providence. 1999+ (1,5,9) – (cont: regan report on hospital law) – ISSN: 1538-8463 – mf#12805,02 – us UMI ProQuest [360]
Hospital literature index – Chicago. 1982-1994 (1) 1982-1994 (5) 1982-1994 (9) – (cont by: hospital and health administration index) – ISSN: 0018-5736 – mf#13375,02 – us UMI ProQuest [360]
Hospital management – Wilmette. 1916-1971 (1) 1970-1970 (5) – ISSN: 0018-5744 – mf#1584 – us UMI ProQuest [360]
Hospital manual / Hannam, E P – London, England. 1848 – 1r – us UF Libraries [240]
Hospital materials management – Ann Arbor. 1990+ (1,5,9) – ISSN: 0888-3068 – mf#18340,01 – us UMI ProQuest [360]
Hospital materiel management quarterly – Rockville. 1979-1999 (1) 1979-1999 (5) 1979-1999 (9) – ISSN: 0192-2262 – mf#12732 – us UMI ProQuest [360]
Hospital medical staff – Chicago. 1972-1985 (1) 1972-1985 (5) 1972-1985 (9) – (cont by: medical staff news) – ISSN: 0090-0710 – mf#7524 – us UMI ProQuest [360]
Hospital medicine – London. 1998+ (1,5,9) – (cont: british journal of hospital medicine) – mf#6627,01 – us UMI ProQuest [610]
Hospital medicine – New York. 1989-1999 (1,5,9) – ISSN: 0441-2745 – mf#17107 – us UMI ProQuest [360]
Hospital outlook – Little Rock. 1998+ (1) – ISSN: 1098-8416 – mf#27040 – us UMI ProQuest [360]
Hospital peer review – Atlanta. 1979+ (1,5,9) – ISSN: 0149-2632 – mf#12280 – us UMI ProQuest [360]
Hospital pharmacy – Saint Louis. 1966+ (1) 1971+ (5) 1974+ (9) – ISSN: 0018-5787 – mf#6889 – us UMI ProQuest [615]
Hospital physician – Wayne. 1965+ [1]; 1971+ [5]; 1976+ [9] – ISSN: 0018-5795 – mf#2010 – us UMI ProQuest [360]
Hospital practice – New York. 1966-1980 (1) 1971-1980 (5) 1975-1980 (9) – ISSN: 0018-5809 – mf#2739 – us UMI ProQuest [360]

Hospital practice : office ed – New York. 1981-2000 (1) 1981-2000 (5) 1981-2000 (9) – ISSN: 8750-2836 – mf#2739,01 – us UMI ProQuest [360]
Hospital progress – St. Louis. 1920-1984 (1) 1970-1984 (5) 1976-1984 (9) – (cont by: health progress) – ISSN: 0018-5817 – mf#2149 – us UMI ProQuest [360]
Hospital Rank and File Action Committee. Service Employee's International Union *see* Fight back!
The hospital reports of the medical missionary society in china for the year 1839 – [Canton]: Office of the Chinese Repository, 1840 – 1mf – 9 – mf#HT-1144 – ne IDC [915]
Hospital risk management – Atlanta. 1979-1994 (1,5,9) – (cont by: healthcare risk management) – ISSN: 0199-6312 – mf#12281 – us UMI ProQuest [360]
Hospital supervision – New York. 1974-1975 (1) 1975-1975 (5) 1975-1975 (9) – (cont by: health services manager) – ISSN: 0018-5841 – mf#9115 – us UMI ProQuest [360]
Hospital survey of the republic of guatemala / Kolbe, Henry W – Guatemala, 1948- – 1r – us UF Libraries [362]
Hospital topics – Sarasota. 1922+ (1) 1971+ (5) 1976+ (9) – ISSN: 0018-5868 – mf#271 – us UMI ProQuest [610]
Hospital tribune – New York. 1972-1980 (1) 1975-1980 (5) 1975-1980 (9) – ISSN: 0018-5876 – mf#6643 – us UMI ProQuest [360]
Hospital trustee – Toronto. 1980-1991 (1,5,9) – ISSN: 0704-0407 – mf#12153 – us UMI ProQuest [360]
Hospitales antiguos de la espanola / Palm, Erwin Walter – Ciudad Trujillo, Dominican Republic. 1950 – 1r – us UF Libraries [360]
Les hospitaliers en terre sainte et a chypre (1100-1810) / Delaville Le Roulx, J – Paris, 1904 – 5mf – 9 – mf#H-3080 – ne IDC [956]
Hospitality : food and lodging – Cleveland. 1976-1976 (1,5,9) – ISSN: 0015-6302 – mf#11071 – us UMI ProQuest [640]
Hospitality : lodging – Cleveland. 1976-1976 (1,5,9) – (cont by: lodging hospitality) – ISSN: 0098-3306 – mf#11072 – us UMI ProQuest [640]
Hospitality : restaurant – Chicago. 1976-1976 (1,5,9) – (cont by: restaurant hospitality) – ISSN: 0098-3292 – mf#11073,03 – us UMI ProQuest [640]
Hospitality design: hd – New York. 1992+(1,5,9) – (cont: restaurant/hotel design international) – ISSN: 1062-9254 – mf#12636,03 – us UMI ProQuest [640]
Hospitals – Chicago. 1927-1993 (1) 1970-1993 (5) 1970-1993 (9) – (cont by: hospitals and health networks) – ISSN: 0018-5973 – mf#784 – us UMI ProQuest [360]
Hospitals and health networks (h & hn) – Chicago. 1993+ (1) 1993+ (5) 1993+ (9) – (cont: hospitals) – ISSN: 1068-8838 – mf#784,01 – us UMI ProQuest [360]
Hospodar – 1928 jan 20-1939 mar 1, 1939 mar 15-1942 mar 15, 1942 apr 1-oct 1 – 3r – 1 – mf#165988 – us WHS [071]
Hospodar – Omaha, NE: Narodnitisk, 1891 (mthly) [mf ed jan 1 1962-mar 1991 (gaps) filmed 1982-91] – 17r – 1 – (absorbed: cechoslovák and westske noviny. publ west, tx: czechoslovak pub co, 1978-) – us NE Hist [071]
Hospodarske noviny – Czech Republic, 1999- – 6r per y – us UMI ProQuest [077]
Hospodarske zaznamy – London, UK. 25 Mar, 18 Aug, 22 Sept, 29 Dec 1943 – 1 – uk British Libr Newspaper [072]
Hoss, Elijah Embree *see* David morton, a biography
Hoss, Elijah Embree et al *see* The new age and its creed
Hoss, Haley A *see* Imagizationdnavisagery – solving the jumble
Hossack, Ian *see* Research and planning documents on technical education, training and manpower in papua new guinea
Hossbach, Th *see* Vorlesungen ueber die apokalypse
Host do domu – Brno: Svaz ceskoslovenskych spisovatelu, 1954- [mf ed Bloomington IN: Indiana Uni Lib, Preservation Dept 1984] – 6r – 1 – us Indiana Preservation [073]
Hostages of civilisation / Reichmann, Eva G – Boston, MA. 1951 – 1r – us UF Libraries [025]
Hostages to india : or, the life-story of anglo-indian race / Stark, Herbert Alick – [Calcutta]: Calcutta Fine Art Cottage, 1926 – (= ser Samp: indian books) – us CRL [954]
Hostelet, Georges *see* Probleme politique capital au congo et en afrique noire
Hosteling in wisconsin / American Youth Hostels, inc – 1982 mar/may-1985 winter – 1r – 1 – (continued by: wisconsin hosteler) – mf#1127376 – us WHS [640]
Hosten, Henry *see* Antiquities from san thome and mylapore
Hosterman, Charlotte O *see* Systematic giving considered under two heads
Hostetter, Karen *see* Development of a high school sports medicine/athletics training course
Hostility and coronary risk factors among native americans and caucasians / Cofrancesco, Lisa – 1993 – 2mf – $8.00 – us Kinesology [616]
Hosting of heroes / Cox, Eleanor Rogers – Dublin, Ireland. 1911 – 1r – us UF Libraries [960]
Hostos : hombre representative de america / Cestero, Tulio Manuel – Buenos Aires, Argentina. 1940 – 1r – us UF Libraries [972]
Hostos / Pedreira, Antonio Salvador – Madrid, Spain. 1932 – 1r – us UF Libraries [972]
Hostos, Adolfo De *see* Coleccion arqueologica antillana
Hostos, Eugenio Maria De *see*
– Antologia
– Essais
– Meditando
Hostos y cuba / Comision Cubana Pro Centenario De Hostos – Habana, Cuba. 1939 – 1r – us UF Libraries [972]
Hosue divided against itself / Dodsworth, William – London, England. 1850 – 1r – us UF Libraries [240]
Hot Club of Chicago *see* Jazz session
Hot coco – Peterborough. 1983-1985 (1,5,9) – ISSN: 0740-3186 – mf#13441 – us UMI ProQuest [000]
Hot line / Communications Workers of America – 1984 jan-1988 fall – 1r – 1 – mf#1886420 – us WHS [380]
Hot rod – Los Angeles. 1948+ (1) 1971+ (5) 1971+ (9) – ISSN: 0018-6031 – mf#3060 – us UMI ProQuest [380]
Hot rod annual – Los Angeles. 1986-1994 (1) 1986-1994 (5) 1985-1994 (9) – ISSN: 0735-083X – mf#15158 – us UMI ProQuest [790]
Hot rod industry news – Los Angeles. 1968-1977 (1) 1971-1977 (5) 1971-1977 (9) – ISSN: 0018-6023 – mf#3054 – us UMI ProQuest [380]
Hot rod yearbook – Los Angeles. 1961-1974 (1) 1970-1973 (5) – ISSN: 0073-3482 – mf#3140 – us UMI ProQuest [629]
Hot seat, the real taxi driver's voice / Taxi Rank and File Coalition – 1971 aug-1973, 1974 jan-1977 mar – 2r – 1 – mf#1112367 – us WHS [380]
Hot springs illustrated journal – v1 n1 [1890 sep] – 1r – 1 – mf#676157 – us WHS [071]
Hot words and hairtriggers / FitzSimons, Mabel Trott – [mf ed Spartanburg SC: Reprint Co, 1981] – 1v on 8mf – 9 – mf#51-055 – us South Carolina Historical [830]
Hotchkin, Samuel Fitch *see*
– Early clergy of pennsylvania and delaware
– First six bishops of pennsylvania
Hotchkiss, George Burton *see*
– The attention value of advertisements in a leading periodical
– Newspaper reading habits of business executives and professional men in new york
Hotchkiss, Jedediah *see* Papers
Hotchpot – [v1-2 n3] [1970 mar-1971 summer] – 1r – 1 – mf#1058417 – us WHS [071]
Hotel / New York Hotel Trades Council et al – 1956 may 7-1958, v19 n37-v20 n41 [1972 jan 3-1973 jan 29] – 2r – 1 – (continued by: hotel voice) – mf#2296591 – us WHS [640]
Hotel advertiser *see* Durban free press / hotel advertiser
Hotel and motel management – Duluth. 1967+ (1) 1970+ (5) 1976+ (9) – ISSN: 0018-6082 – mf#5651 – us UMI ProQuest [650]
Hotel and Restaurant Employees and Bartenders International Union *see*
– Cafeteria call
– Hotel voice
– Local 5 organizer
– Portland restaurant, bar, hotel news
Hotel and Restaurant Employees and Bartenders International Union et al *see*
– Hawaiian teamster
– Hotel/restaurant voice
Hotel and Restaurant Employees Union *see* Serving america
L'Hotel Drouot *see* Gazette
Hotel Employees and Restaurant Employees International Union *see* Local 11 news calendar
Hotel Employees and Restaurant employees international union *see* We're h e r e
Hotel Employees and Restaurant Employees International Union et al *see* Hotel-bar-restaurant review
Hotel Employees & Restaurant Employees International Union *see*
– Local 5
– Local 5 news
Hotel garni : ou, la lecon singuliere / Desaugiers, Marc-Antoine – Paris, France. 1842 – 1r – us UF Libraries [440]
Hotel guest registers, 1922-38; 1965-66 / Ridgecrest. North Carolina. Ridgecrest Baptist Assembly – 9746p – 1 – us Southern Baptist [242]
Hotel koepf *see* Uebereilte werbung / hotel koepf
Hotel, motel, restaurant employees and bartenders union – v1 n1-v5 n4 [1977 feb-1981 aug] – 1r – 1 – (cont: portland restaurant, bar, hotel news) – mf#637034 – us WHS [640]
Hotel register / Neodesha House. Neodesha, Kansas – 1871-74 – 1 – us Kansas [978]
Hotel, Restaurant, Institutional Employees, and Bartenders Union *see* Local 26 unity
Hotel voice / Hotel and Restaurant Employees and Bartenders International Union – 1973 feb 5-1975 dec 29, 1976 jan 5-1978 dec 25, 1979 jan-sep 24 – 3r – 1 – (cont: hotel, motel and club voice; hotel; cont by: hotel/restaurant voice) – mf#634018 – us WHS [640]
Hotel voice / Hotel and Restaurant Employees and Bartenders International Union – 1983 jan 10-1984 – 1r – 1 – (cont: hotel/restaurant voice) – mf#2296615 – us WHS [640]
Hotel-bar-restaurant review / Hotel Employees and Restaurant Employees International Union et al – v9 n1-v10 [i.e. 11] [1975 jan-1977 dec] – 1r – 1 – mf#668100 – us WHS [640]
L'hotel-dieu de quebec, 1639-1900 : notices historiques et depouillement des registres, 2e partie, 1759-1900 / Sainte-Leonie, soeur – 1964 [mf ed 1979] – (= ser Bibliographies du cours...1947-66) – 5mf – 9 – (with ind) – mf#SEM105P4 – cn Bibl Nat [360]
Hotel/restaurant voice / Hotel and Restaurant Employees and Bartenders International Union et al – 1979 oct 1-1980 n80-1889, 1981-1983 jan 10 – 2r – 1 – mf#634014 – us WHS [640]
Hotels – Des Plaines. 1989+ (1,5,9) – (cont: hotels and restaurants international) – ISSN: 1047-2975 – mf#14871,02 – us UMI ProQuest [640]
Hotels and restaurants international – Newton. 1984-1989 (1,5,9) – (cont by: hotels) – ISSN: 0744-3897 – mf#14871,01 – us UMI ProQuest [640]
Hotomanorum, F *see* Patris ac filii et clarorum virorum ad eos epistolae
Hotomanorum, J *see* Patris ac filii et clarorum virorum ad eos epistolae
"Hotsmah's kremil" fun fershidene antiken : 25 yudishe foldslider dos zenen gezungen gehoren in goldfadens yudishen theater / Tantsman, Abraham Isaac – Varsha, Poland: P. Baymritter 1891 – 1r – 1 – us UF Libraries [390]
Hotta-ke monjo : documents of the hottas, feudal load of the sakura domain (chiba pref) in edo period – 1172 items on 238r – 1 – Y2380,000 – (with 120p guide. in japanese. written in india ink) – ja Yushodo [950]
Hotten, J C *see*
– Abyssinia and its people
– Original lists of american emigrants, 1600-1700
Hotten, John Camden *see* The history of signboards
Hottenstein, Marcus S *see* The sherman anti-trust law
Hottentot hunt / Ollemans, P H – s.l, s.l? 1960? – 1r – us UF Libraries [960]
Les hotteterre : celebres jouers et facteurs de flutes, hautbois, bassons et musettes des 17e et 18e siecles; nouvelles recerces par n mauger / Mauger, Nicolas – Paris: Fischbacher 1912 [mf ed 19–] – 1r – 1 – (supplement a la brochure publice en 1894 par ernest thoinan [antoine ernest roquet]) – mf#film 1354 – us Sibley [780]
Les hotteterre et les chedevilles : celebres jouers et fracteurs de flutes, hautbois, bassoons et musettes des 17e et 18e siecles, avec portraits et fac-similes / Thoinan, Ernest – Paris: E Sagot 1894 [mf ed 19–] – 1r – 1 – mf#film 1352 – us Sibley [780]
Hotteterre, Jacques *see*
– Methode pour la musette
– Principes de la flute traversiere
Hottinger, J H *see*
– Historia ecclesiastica novi testamenti, tomi 6, 8, 9
– Historiae ecclesiasticae novi testamenti
– Schola tigurinorum carolina
Hottinger, J J *see*
– Helvetische kirchen-geschichten
– Helvetischer kirchen-geschichten, dritter theil
– The life and times of ulricus zwingli
– Reformationsgeschichte
Hottinger, Johann *see* Die schweiz in ihren ritterburgen und bergschloessern
Hottinger, Johann Jakob *see* The life and times of ulric zwingli
Hotzel, Curt *see* Der unverbluemte amor
Hou, Che-an *see* Nung yeh ts'ang k'u ching ying lun
Hou fang chi / Sha, Yen – Ch'ung-ch'ing: Chien kuo shu tien, Min kuo 31 [1942] – (= ser P-k&k period) – us CRL [480]
Hou fang hsiao hsi chu / Ch'en, Pai-ch'en – Shang-hai: Sheng hou shu tien, Min kuo 36 [1947] – (= ser P-k&k period) – us CRL [820]
Hou fang min chung ti tsung tung yuan / Hu, Sheng – Han-k'ou: Sheng huo shu tien, Min kuo 27 [1938] – (= ser P-k&k period) – us CRL [951]
Hou, Hou-p'ei *see* Jih-pen ti kuo chu i tui hua ching chi ch'in lueh
Hou lai che : san mu chu / Liu, Pei-wen – Fu-chien Nan-p'ing: Fu hsing ch'u pan she, 1945 – (= ser P-k&k period) – us CRL [820]
Hou, Tz'u-kung *see* Nung chia sheng huo
Hou, Wai-lu *see* K'ang chan chien kuo lun
Hou, Yao *see*
– Fu huo ti mei kuei
– Shan ho lei
Houben, Heinrich Hubert *see*
– Damals in weimar
– Gutzkow-funde
– Studien ueber die dramen carl gutzkows
– Tagebuecher von k a varnhagen von ense
Houbraken, A *see*
– Dichtkundige bespiegelingen op 57 gepaste in koper gebragte zinnebeelden
– De groote schouburgh der Nederlantsche konstschilders en schilderessen...
– Stichtelyke zinnebeelden gepast op deugden en ondeugden
– Stichtelyke zinnebeelden, gepast op deugden en ondeugden
[Houbraken, A] Hofstede de Groot, C *see* Arnold houbraken und seine "groote schovwburgh"
[Houbraken, A] Wurzbach, A von *see* Arnold houbraken's grosse schouburgh der niederlaendische maler und malerinnen
Houck, George Francis *see* A history of catholicity in northern ohio and in the diocese of cleveland
Houck, Louis *see* A treatise on the law of navigable rivers
Houdas, O *see* Histoire du sultan djelal-eddin mankobirti, prince du kharezm
Houde, Marguerite A *see* Bibliographie analytique de l'oeuvre de m avila bedard...
Houdeau, Serge *see* La vache et la chevre
Houdet, Antoine-Jacques *see*
– Grammaire francoise
Houdini, Harry *see* Scrapbooks
Hough, Franklin B *see* American constitutions
Hough, George H *see* English and burman vocabulary
Hough, George Henry *see* An english and burman vocabulary
Hough, Holly J *see* The effects of hormone replacement therapy and active lifestyle on immune function in postmenopausal women
Hough, James *see*
– Letters on the climate, inhabitants, productions, etc of the neilgherries
– The missionary vade mecum
– The protestant missions vindicated
– Reply to the letters of the abbe dubois on the state of christianity in india
Hough, Lynn Harold *see*
– In the valley of decision
– The quest for wonder
– The theology of a preacher
Hough, M *see* Die konstitusionele ontwikkeling van botswana
Hough, Samuel *see* Born of water and spirit
Hough, Samuel Strickler *see* Report of a visit to japan, china and the phillipine islands
Hough, Walter *see* The moki snake dance
Hough, William *see*
– Military law authorities
– The practice of courts-martial
– The practice of courts-martial and other military courts
Hough, William S *see* Specieications sic and instructions for constructing and working hough's soper improved bee-hive
Hough's vice admiralty reports / New York. (State) – 1v. 1715-88 (all publ) – 4mf – 9 – $6.00 – mf#LLMC 80-006 – us LLMC [340]
Houghton conquest – (= ser Bedfordshire parish register series) – 2mf – 9 – £5.00 – uk BedsFHS [929]
Houghton, D Hobart *see*
– Life in the ciskei
– South african economy
Houghton, J *see* Byrom hall
Houghton, Louise Seymour *see*
– Antipas, son of chuza
– From olivet to patmos
– Hebrew life and thought
– The life of the lord jesus
– Telling bible stories
Houghton, Norris *see* Great russian plays
Houghton regis – (= ser Bedfordshire parish register series) – 3mf – 9 – £7.50 – uk BedsFHS [929]
Houghton, Ross C *see*
– John the baptist
– Women of the orient
Houis, Maurice *see* Apercu sur les structures grammaticales des langues
Hould, Rejean *see* Notes historiques sur la mauricie
Houlder, J A *see* Ohabolana
Houlder, John Alden *see* Ohabolana
Houle, Alphonse *see* Bibliographie

Houle, Rosaire *see* Bio-bibliographie de feu son eminence le cardinal jean-marie-rodrigue villeneuve
Houliston, Wm *see* The coming of the great king
Houlton, John *see* Bihar
Houma ceres – Houma LA. 1860 oct 13 – 1r – 1 – mf#861617 – us WHS [071]
Houmoh – New York. N.Y. 1915 – 1 – us AJPC [071]
Hound and horn portland – Portland, Maine. v1-7. sept 1927-sept 1934. (incomplete) – 1 – us NY Public [410]
The hound of uladh : two plays in verse / Cousins, James Henry – Madras, India: Kalakshetra, 1942 – (= ser Samp: indian books) – us CRL [820]
Hounds of hell / Larteguy, Jean – New York, NY. 1966 – 1r – us UF Libraries [890]
Hounslow and brentford independent – Hounslow, England 17 feb-11 aug 1877 – 1 – (cont by: brentford & hounslow independent & west london examiner [18 aug 1877-29 may 1879]) – uk British Libr Newspaper [072]
Hounslow borough recorder *see* Hounslow recorder
Hounslow chronicle *see* Middlesex chronicle etc hounslow chronicle
Hounslow feltham and hanworth times – London, UK. 18 sep-24 dec 1987; 1988-97; 1998 – 35r – 1 – uk British Libr Newspaper [072]
Hounslow leader – London, UK. jan-aug 1986 – 1r – 1 – (aka: hounslow ed) – uk British Libr Newspaper [072]
Hounslow recorder – London, UK. Feb-dec 1986; 6 feb, 16 oct-dec 1987; 1988-18 dec 1992 – 13 1/2r – 1 – (aka: hounslow borough recorder) – uk British Libr Newspaper [072]
Der houpme lombach : berndeutsche novelle / Tavel, Rudolf von – 7. aufl. Bern: A Francke, 1931 [mf ed 1993] – 328p/1pl (ill) – 1 – mf#7743 – us Wisconsin U Libr [830]
Hour / American Council Against Nazi Propaganda – v1-153. 1939-43 [all publ] – (= ser Radical periodicals in the united states, 1881-1960. series 1) – 9mf – 9 – $115.00 – us UPA [303]
Hour : (county edition) – Norwalk, CT. 1900-1921 (1) – mf#62363 – us UMI ProQuest [071]
Hour – Norwalk, CT. 1872-1900 (1) – mf#62362 – us UMI ProQuest [071]
Hour – Norwalk, CT. 1895-2000 (1) – mf#61252 – us UMI ProQuest [071]
The hour – London. 24 Mar -Dec 1874. -f. 10mqn reels – 1 – uk British Libr Newspaper [072]
Hour after midnight / Morris, Colin M – London, England. 1961 – 1r – us UF Libraries [890]
Hour ago – London, England. 18– – 1r – us UF Libraries [240]
Hour of independence / France Ambassade (Us) Service De Presse Et D'information – New York, NY. v1-11. 1960-1961 – 1r – us UF Libraries [960]
"The hour of youth has struck" : the african national congress youth league and the struggle for a mass base, 1943-1952 / Giffard, Chris – U of Cape Town 1984 [mf ed S.l: s.n. 1984] – 2mf – 9 – (incl bibl) – sa Misc Inst [320]
Hourcade, Laurent *see* Abrege de theologie sociale d'apres les grands auteurs
Hours at home – New York. 1865-1870 (1) – mf#3892 – us UMI ProQuest [240]
Hours in the picture gallery of thirlestane house, cheltenham / Davies, Henry – new ed. Cheltenham 1846 – (= ser 19th c art & architecture) – 1mf – 9 – mf#4.1.391 – uk Chadwyck [700]
Hours of childhood and other poems / Bowman, Ariel – Montreal: Publ by A Bowman, 1820 – 2mf – 9 – mf#55147 – cn Canadiana [810]
Hours with a sceptic / Faunce, Daniel Worcester – Philadelphia: American Baptist Pub Soc [c1892] [mf ed 1985] – 1mf – 9 – 0-8370-3101-X – (incl ind) – mf#1985-1101 – us ATLA [230]
Hours with german classics / Hedge, Frederic Henry – Boston: Roberts Brothers, 1886 – 1r – 1 – us Wisconsin U Libr [430]
Hours with the bible : or, the scriptures in the light of modern knowledge: vol 1, creation to moses / Geikie, Cunningham – new rev ed. New York: James Pott 1903 [mf ed 1993] – 2mf [ill] – 9 – 0-524-08311-8 – (incl bibl ref) – mf#1993-0016 – us ATLA [221]
Hours with the bible : or, the scriptures in the light of modern discovery and knowledge: vol 2, from moses to the judges / Geikie, Cunningham – New York: James Pott 1882 [mf ed 1993] – 2mf – 9 – 0-524-08175-1 – mf#1992-1161 – us ATLA [221]
Hours with the bible : or, the scriptures in the light of modern knowledge: vol 3, from samson to solomon / Geikie, Cunningham – new rev ed. New York: James Pott 1903 [mf ed 1993] – 2mf [ill] – 9 – 0-524-08312-6 – mf#1993-0017 – us ATLA [221]
Hours with the bible : or, the scriptures in the light of modern knowledge: vol 4, from rehoboam to hezekiah, with the contemporary prophets / Geikie, Cunningham – new rev ed. New York: James Pott 1903 [mf ed 1993] – 2mf [ill] – 9 – 0-524-08313-4 – mf#1993-0018 – us ATLA [221]
Hours with the bible : or, the scriptures in the light of modern knowledge: vol 5, from manasseh to zedekiah, with the contemporary prophets / Geikie, Cunningham – new rev ed. New York: James Pott 1903 [mf ed 1993] – 2mf [ill] – 9 – 0-524-08314-2 – mf#1993-0019 – us ATLA [221]
Hours with the bible : or, the scriptures in the light of modern knowledge: vol 6, from the exile to malachi / Geikie, Cunningham – new rev ed. New York: James Pott 1903 [mf ed 1993] – 2mf – 9 – 0-524-08176-X – mf#1992-1162 – us ATLA [220]
Hours with the mystics : a contribution to the history of religious opinion / Vaughan, Robert Alfred – 8th ed. London: George Routledge, [19–?] – 2mf – 9 – 0-524-08659-1 – mf#1993-2119 – us ATLA [210]
House and garden – New York. 1901-1993 (1) 1962-1993 (5) 1960-1993 (9) – ISSN: 0018-6406 – mf#697 – us UMI ProQuest [640]
House and hearth / Spofford, Harriet Prescott – New York: Dodd, Mead and Co, 1891 – us CRL [071]
House and home – New York. 1952-1977 (1) 1969-1977 (5) 1976-1977 (9) – (cont by: housing) – ISSN: 0018-6414 – mf#1165 – us UMI ProQuest [690]
House and land agents and agriculturists gazette – Dublin, Ireland. 18 oct-22 nov 1848 – 1/4r – 1 – uk British Libr Newspaper [072]
House and senate bills and resolutions / U.S. Congress – 1 – (1st-6th. 1789-1801. $258.00. 7th-36th. 1801-61. $1,854.00. 37th-46th. 1861-81. $2,815.00. 47th-56th. 1881-1901. $9,793.00. 57th-65th. 1901-19. $12,298.00. 66th-72nd. 1919-33. $8,835.00. 92nd. 1971-72. $6,397.00) – us L of C Photodup [324]
House architecture / Stevenson, John James – London 1880 – (= ser 19th c art & architecture) – 8mf – 9 – mf#4.2.400 – uk Chadwyck [720]
House beautiful – New York. 1896+ (1) 1964+ (5) 1976+ (9) – ISSN: 0018-6422 – mf#3120 – us UMI ProQuest [640]
House beautiful's colonial homes – New York. 1979-1979 (1,5,9) – (cont by: colonial homes) – ISSN: 0164-6214 – mf#12241 – us UMI ProQuest [640]
The house decorator and painter's guide / Arrowsmith, H W & Arrowsmith, A – London 1840 – (= ser 19th c art & architecture) – 3mf – 9 – mf#4.2.40 – uk Chadwyck [640]
House in antigua / Adamic, Louis – New York, NY. 1937 – 1r – us UF Libraries [972]
House journals / Michigan Legislature – 1835-1997 – 3449mf – 9 – $3638.00 – (lacking: 1868 reg sess. 1892 spec sess. 1899 ext sess. 1900 2nd ext sess. 1944 ext sess. 1954 2nd ext sess. updates planned) – mf#LLMC 79-438H – us LLMC [340]
House journals : territorial legislature to date / Hawaii Legislature – 1901-99 – (= ser Hawaii appellate reports) – 1391mf – 9 – $1086.00 – (add vols after 1994 planned) – mf#LLMC 77-105 – us LLMC [323]
House journals 1966-79 : 2nd reg sess of the congress of micronesia to the sess of the interim congress of the federated states of micronesia (1978-79) / Congress of Micronesia – n.p, n.d. – (= ser Micronesia: interim governance) – 123mf – 9 – $184.00 – mf#LLMC 82-100F, Title 31 – us LLMC [323]
House of assembly – committee room, tuesday, 9th january, 1821 : in committee to take into consideration and examine the different accounts of the public revenue, and estimates of the civil list of this province... = Chambre d'assemblee – chambre de comite, mardi, 9e janvier, 1821... – [Quebec: Chambre d'assemblee, 1821] (mf ed 2000) – 3mf – 9 – mf#SEM105P3167 – cn Bibl Nat [336]
House of assembly, friday, 2d february 1827 : resolved, that a committee of seven members be appointed to inquire if it would be necessary to open any and what new roads... = Chambre d'assemblee, vendredi, le 2 fevrier 1827. resolu, qu'il soit nomme un comite de sept membres pour s'enquerir s'il seroit necessaire d'ouvrir quelques... – [Quebec: Chambre d'assemblee, 1827] (mf ed 1989) – 1mf – 9 – (in french and english) – mf#SEM105P1140 – cn Bibl Nat [380]
House of assembly, wednesday, 21st february 1827 : resolved, that the message from his excellency the governor in chief relating to the subdivision of parishes in this province, be referred to a committee of five members... = Chambre d'assemblee, mercredi, 21 fevrier 1827 / Bas-Canada. Parlement. Chambre d'assemblee – [Quebec: Chambre d'assemblee, 1827] (mf ed 1997) – 1mf – 9 – (in french and english) – mf#SEM105P2850 – cn Bibl Nat [323]
House of bondage / Cole, Ernest – New York, NY. 1967 – 1r – us UF Libraries [890]
House of chiefs debates – Ibadan: Govt Printer. 4th-6th sessions. 1956-1958 – 1r – 1 – (issues for 1956-1958 filmed with its western house of chiefs debates 1953) – us CRL [323]
House of commons parliamentary papers, 1901-1921 – [mf ed Chadwyck-Healey] – (= ser House of commons parliamentary papers, 1901-1974/75) – 17,944mf – 9 – uk Chadwyck [324]
House of commons parliamentary papers, 1922-1944/1945 – [mf ed Chadwyck-Healey] – (= ser House of commons parliamentary papers, 1901-1974/1975) – 5962mf – 9 – uk Chadwyck [324]
House of commons parliamentary papers, 1945/1946-1960/1961 – [mf ed Chadwyck-Healey] – (= ser House of commons parliamentary papers, 1901-1974/1975) – 5806mf – 9 – uk Chadwyck [324]
House of commons parliamentary papers, 1961/1962-1974/1975 – [mf ed Chadwyck-Healey] – (= ser House of commons parliamentary papers, 1901-1974/1975) – 6686mf – 9 – uk Chadwyck [324]
House of commons parliamentary papers, 1975/1976- : the microfiche edition of current parliamentary papers – [mf ed Chadwyck-Healey] – (= ser House of commons parliamentary papers, 1901-1974/1975) – 18,726mf – 9 – (from 1987/88, price incl free ind on computer output mf. commmand papers are available separately from 1979) – uk Chadwyck [324]
The house of dreams *see* Night / The house of dreams
House of earth / Buck, Pearl Sydenstricker – New York, NY. 1935 – 1r – us UF Libraries [830]
The house of edward winslow / Pratt, Walter E – 1949 – 1 – $5.00 – us Southern Baptist [242]
House of israel / Whitehead, E L – 1r – 1 – mf#pmb doc42 – at Pacific Mss [980]
House of lords cases (clark and finnelly) : cases on appeal and writs of error, claims of peerage, and divorces during the sessions of 1847-1866 / Clark, Charles & Finnelly, W – v1-11. 1847-66. London: Spettigue & Farance/ Butterworth, 1849-66 (all publ) – 109mf – 9 – $163.00 – mf#LLMC 95-284 – us LLMC [324]
House of lords parliamentary papers : the papers from the house of lords – 1984-88 [mf ed Chadwyck-Healey] – 9 – (1984/85-1986/87 in a single sequence, and 1987/88- in two series, bills and papers) – uk Chadwyck [324]
House of lords record office, american papers, 1621-1917 – 39r – 1 – (int by w e minchinton and peter harper) – mf#97057 – uk Microform Academic [960]
House of prayer messenger – Charles Town, West VA. 1997 fall – 1r – 1 – mf#4027635 – us WHS [240]
House of St Michael and All Angels *see*
– Annual report of the house of s michael and all angels for young colored cripples
– Annual report of the house of st michael and all angels for colored cripple children
House of shivaji : studies and documents of maratha history: royal period / Sarkar, Jadunath – Calcutta: SC Sarkar & Sons, 1948 – (= ser Samp: indian books) – us CRL [954]
The house of the lord : a study of holy sanctuaries, ancient and modern / Talmage, James Edward – Salt Lake City, UT: Deseret News, 1912 – 1mf – 9 – 0-7905-6691-5 – mf#1988-2691 – us ATLA [240]
The house of william burges – London [1885] – (= ser 19th c art & architecture) – 3mf – 9 – mf#4.2.43 – uk Chadwyck [720]
House of worth fashion designs : 1899-1929 / Victoria and Albert Museum. London – [150mf] – 9 – $920.00 – 0-907006-98-1 – (photographic documentation of over 7000 designs, grouped by type in chronological order) – uk Mindata [740]
House practice : a guide to the rules, precedents and procedures of the house / Brown, Wm Holmes – Washington: GPO 1974-94 [all publ] [mf ed 1996] – 10mf – 9 – mf#llmc97-237 – us LLMC [323]
House report approving the compact of free association with the marshall islands and the federated states of micronesia, and approving conditionally the compact of free association with palau : hse.rept n99-188, 99th cong, 1st sess, jul 1, 1985 / U.S. Congress. House Committee on Foreign Affairs – Washington: GPO. 4pts. 1985 – (= ser Micronesia: evolution to separate political entities) – 10mf – 9 – $15.00 – mf#LLMC 82-100F, Title 21 – us LLMC [980]
The house that albert built : [i.e. the crystal palace] / Henning – London [1851] – (= ser 19th c art & architecture) – 1mf – 9 – mf#4.2.1237 – uk Chadwyck [720]
The house that jack is building : and other essays / Ebey, Adam – Wawaka IN: [A Ebey] 1899 [mf ed 1992] – 1mf – 9 – 0-524-02819-2 – mf#1990-4440 – us ATLA [240]
House-flies : their life-history, destruction and prevention, and their influence on health = Huisvliee: hul lewensloop, uitroeiing en voorkoming en hul invleod op die gesondheid / South Africa. Department of Public Health – Pretoria: Dept of Public Health 1937 [mf ed Pretoria, RSA: State Library [199-]] – 14p [ill] on 1r with other items – 5 – mf#op 09743 r25 – us CRL [614]
House-flies : their life-history, destruction and prevention, and their influence on health = Huisvliee: hul lewensloop, uitroeiing en voorkoming en hul invleod op die gesondheid / South Africa. Department of Public Health – Pretoria: Dept of Public Health 1940 [mf ed Pretoria, RSA: State Library [199-]] – 14p [ill] on 1r with other items – 5 – mf#op 12818 r25 – us CRL [614]
Household accountbook, 1797-98 / Macartney, G M – 3mf – 9 – sa National [640]
Household accounts and expenses of edward 5, richard 3, henry 7 and james 1 – 1r – 1 – mf#96777 – uk Microform Academic [640]
The household companion : a monthly magazine devoted to the improvement and amusement of the family circle – Toronto: J E Bryant, [1891-189- or 19-] [mf ed v1 n1 sep 1891] – 9 – mf#P05061 – cn Canadiana [640]
Household furniture and interior decoration / Hope, Thomas – London 1807 – (= ser 19th c art & architecture) – 6mf – 9 – mf#4.2.755 – uk Chadwyck [740]
The household guide : or, domestic cyclopedia: a practical family physician, home remedies and home treatment on all diseases... / Jefferis, Benjamin Grant & Nichols, James Lawrence – Toronto: J L Nichols, [c1894] – 6mf – 9 – 0-665-91779-1 – (incl ind. also a complete cook book by mrs j l nichols) – mf#91779 – cn Canadiana [640]
The household journal : devoted to entertaining and instructive literature – Montreal: [s.n, 1878-18– or 19–] – 9 – ISSN: 1190-7150 – mf#P04271 – cn Canadiana [640]
The household life – Toronto: T H Churchill, [1884-18-?] – 9 – mf#P04251 – cn Canadiana [615]
Household magazine – West Bromwich. England 10 mar-9 jun 1888 [wkly] – 11ft – 1 – uk British Libr Newspaper [073]
Household manufacturers in the united states, 1640-1860 : a study in industrial history / Tryon, Rolla Milton – Chicago, IL: University of Chicago Press, [1917] [mf ed 1970] – (= ser Library of american civilization 16331) – xii/413p on 1mf – 9 – us Chicago U Pr [338]
Household of faith / Way, Lewis – London, England. 1823 – 1r – us UF Libraries [240]
The household of faith : portraits and essays / Russell, George William Erskine – London: Hodder and Stoughton, 1902 – 1mf – 9 – 0-524-05304-9 – (incl bibliographic references) – mf#1991-2270 – us ATLA [240]
Household words – A weekly journal conducted by Charles Dickens. v1-19. 1850-59 – 1 – us AMS Press [800]
Household words : a weekly journal conducted by charles dickens – London. 1850-1859 – 1 – mf#3899 – us UMI ProQuest [073]
Householder and the labourers / Fuller, Andrew – London, England. 18– – 1r – us UF Libraries [240]
The housekeeper's help – rev ed. [Hamilton, Ont?: s.n.] 1888 [mf ed 1994] – incl ind – 9 – 0-665-94611-2 – mf#94611 – cn Canadiana [640]
Houser hunters newsletter – v1 n1-v4 n1 [1983 jan-1986 mar] – 1r – 1 – mf#1133428 – us WHS [071]
Houses of god / Dayman, A J – London, England. 1849 – 1r – us UF Libraries [240]
Housing – New York. 1978-1982 (1) 1978-1982 (5) 1978-1982 (9) – (cont: house and home) – ISSN: 0161-0619 – mf#1165,01 – us UMI ProQuest [690]
"Housing an illegitimate aristocracy" : an urban profile of a coloured community in greenwood park from the 1950's to the 1970's / Francis, Lynette Crysta-Lee – Uni of South Africa 2001 [mf ed Johannesburg 2001] – 4mf [ill] – 9 – (incl bibl ref) – mf#mfm14985 – sa Unisa [307]
Housing and people = Habitation et les citoyens – Ottawa. 1970-1978 (1) 1972-1978 (5) 1976-1978 (9) – ISSN: 0018-6562 – mf#7169 – us UMI ProQuest [360]
Housing and transportation of the handicapped : laws, legislative histories and administrative documents / ed by Reams, Bernard D Jr – Over 250 documents – 9 – $1,750.00 set – 0-89941-247-5 – mf#400430 – us Hein [344]
Housing and urban development trends – Washington DC. v1-23. 1948-70 – 1 – $600.00 – (1971-80 $48 [0269]) – mf#0270 – us Brook [360]

Housing and urban development trends – Washington. 1974-1980 (1) 1975-1980 (5) 1975-1980 (9) – ISSN: 0018-6619 – mf#9148 – us UMI ProQuest [710]
Housing Collaborative [Madsion, WI] *see* News
Housing counseling demonstration program / United States. General Accounting Office. RCED – Washington DC: The Office [mf ed 1997?] – 1mf – 9 – us US Gen Account [360]
Housing review – London. 1975-1996 (1) 1975-1996 (5) 1975-1996 (9) – ISSN: 0018-6651 – mf#8656 – us UMI ProQuest [710]
Housing rights / Minnesota Tenants Union – 1977 apr-1978 aug – 1r – 1 – (continued by: minnesota tenants union newsletter) – mf#647642 – us WHS [362]
Housing, theory and society – Oslo. 1999+ (1) – ISSN: 1403-6096 – mf#22141,01 – us UMI ProQuest [720]
Housing wisconsin – 1971 aug-1976 winter – 1r – 1 – (continued by: dlad newsletter) – mf#329270 – us WHS [071]
Housman, John *see* John housman's reise durch die noerdlichen gegenden von england
Housman, Laurence *see* Arthur boyd houghton
Housse, Emile *see* Aves de chile
Houston bar bulletin – 1945-64 (all publ) – 7mf – 9 – $31.50 – (lacking: 1959) – mf#LLMC 84-481 – us LLMC [340]
Houston bar journal – v1 n1-10. 1930-31 (all publ) – 2mf – 9 – $3.00 – mf#LLMC 84-480 – us LLMC [340]
Houston business journal – Houston. 1975-1994 (1) 1976-1994 (5) 1976-1994 (9) – mf#10713 – us UMI ProQuest [338]
Houston chronicle – Houston, TX. 1901+ (1) – mf#60598 – us UMI ProQuest [071]
[Houston-] compass – TX. 1967-69 – 1r – 1 – $60.00 – mf#R05002 – us Library Micro [071]
Houston daily telegraph – Houston TX. 1864 may 24 – 1r – 1 – (continued by: daily telegram (houston, tx)) – mf#848654 – us WHS [071]
Houston defender – Houston TX. [1992 jan 19/25-jun 28/jul 4]-[1997 jul 6/12-dec 28/jan 3], v67 n11-v69 n46 [1998 jan 4-2000 sep 10] – 12r – 1 – mf#2337358 – us WHS [071]
Houston fire fighter : official publication of houston professional fire fighters ass[ociatio]n / Houston Professional Fire Fighters Association – 1978 aug-1986 mar, 1978 aug /sep-1987 nov, 1986 apr-1987 – 2r – 1 – (cont: professional fire fighter) – mf#1819242 – us WHS [366]
Houston informer Escene – 1998 jul, aug 21/sep 3-nov 5, nov 13/26-dec 10, 1999 jan 8/21-may 14/27, jun 4/17-dec 24/2000 jan 6, jan, feb/mar, may/jun – 1r – 1 – (continued by: escene) – mf#4194190 – us WHS [071]
[Houston-] international daily news – TX. 1989- – 30r – 1 – $1800.00 (subs $300y) – mf#H04077 – us Library Micro [071]
Houston journal of health law and policy – v1. 2001 – 9 – (filming in process) – mf#119171 – us Hein [344]
Houston journal of international law – v1-23. 1978-2001 – 9 – $386.00 set – ISSN: 0194-1879 – mf#103131 – us Hein [341]
Houston journalism review – v1 n3-v2 n9 [1972 aug-1974 apr] – 1r – 1 – mf#1058429 – us WHS [070]
Houston law review – Houston. 1972+ (1) 1972+ (5) 1972+ (9) – ISSN: 0018-6694 – mf#9543 – us UMI ProQuest [340]
Houston law review – v1-38. 1963-2001 – 9 – $859.00 – ISSN: 0018-6694 – mf#103141 – us Hein [340]
Houston newspages – Houston TX. [1992 feb 13/19-dec 23/31]-[2000 jul-dec] [gaps] – 14r – 1 – mf#2385446 – us WHS [071]
Houston post – Houston, TX. 1880-1994 (1) – mf#60599 – us UMI ProQuest [071]
Houston Professional Fire Fighters Association *see*
– Houston fire fighter
– Professional fire fighter
Houston sun – 1992 nov-dec 21, 1993 jan 11-feb 15, nov 22-dec 27, 1994 jan 24-dec 26, 1995 jan-nov, 1996 feb 10-dec 30 – 4r – 1 – mf#2604785 – us WHS [071]
Houston, Thomas *see* Christian magistrate
Houston's criminal reports – Delaware. 1v. 1856-79 (all publ) – 7mf – 9 – $10.50 – mf#LLMC 84-129 – us LLMC [345]
Houten, H R van *see* Egypte's internationaal statuut...
Houtin, Albert *see*
– L'americanisme
– La crise du clerge
– The crisis among the french clergy
– Evaeques et dioceses
– Un praetre marie
– La question biblique au 20e siecle
– La question biblique chez les catholiques de france au 19e siecle
Houtsma, M T *see* Zur geschichte der selguqen von kerman
Houtsma, M Th *see* Historiae
Houzel, Roger *see* Production et le commerce de la republique d'haiti
Hove, Masotsha Mike *see* Yesterday, today and tomorrow
Hovelacque, Abel *see*
– L'avesta, zoroastre et le mazdeisme
– The science of language
Hoveret hertsel – Buenos Aires, Argentina. 1955 – 1r – us UF Libraries [939]
Hovering craft and hydrofoil – Kingston-Upon-Thames. 1961-1979 (1) 1971-1979 (5) 1975-1979 (9) – (cont by: high-speed surface craft) – ISSN: 0018-6775 – mf#2997 – us UMI ProQuest [629]
Hovey, Alvah *see*
– Baptist pamphlets. a
– The bible – how to teach the bible
– Biblical eschatology
– The christian pastor
– Christian teaching and life
– A commentary on the acts of the apostles
– Commentary on the epistle to the galatians
– Commentary on the gospel of john
– The doctrine of the higher christian life
– God with us, or, the person and work of christ
– Manual of christian theology
– A memoir of the life and times of the rev isaac backus
– The miracles of christ
– Progress of a century
– Religion and the state
– The state of the impenitent dead
– Studies in ethics and religion, or, discourses, essays, and reviews pertaining to theism, inspiration, christian ethics, and education for the ministry
– Truth unfolded
Hovey, Alvah et al *see* The madison avenue lectures
Hovey, Richard *see*
– More songs from vagabondia
– Songs from vagabondia
How – Cincinnati. 1991-1997 (1) – ISSN: 0886-0483 – mf#16739 – us UMI ProQuest [650]
How a race of pygmies was found in north africa and spain : with comments of professors virchow, sayce and starr: and papers on other subjects / Haliburton, Robert Grant – Toronto?: Arbuthnot, 1897 – 2mf – 9 – mf#05330 – cn Canadiana [573]
How about your bible? : an argument and a plea for bible study / Neff, James Monroe – Morristown TN: Good Literature Pub Co 1902 [mf ed 1992] – 1mf – 9 – 0-524-04061-3 – mf#1990-4969 – us ATLA [220]
The how and why of the emmanuel movement : a hand-book on psycho-therapeutics / Boyd, Thomas Parker – San Francisco: Whitaker & Ray, 1909 – 1mf – 9 – 0-524-03039-1 – mf#1990-0796 – us ATLA [150]
How and why the lands were locked, with a key to unlock them : a letter on bounty immigration and the rights of labour, to the legislators and people of victoria – Melbourne, [1856] – (= ser 19th c british colonization) – 1mf – 9 – mf#1.1.7006 – uk Chadwyck [304]
How are the mighty fallen! / Medley, John – Exeter, England. 1840 – 1r – us UF Libraries [240]
How attitudes may affect the success of inclusion / Luebke, Kristine S – 1998 – 1mf – 9 – $4.00 – mf#PE 3908 – us Kinesology [790]
How best to improve and keep up the seamen of the country / Brassey, Thomas, Earl – London?: Harrison & Sons, 1876? – 1mf – 9 – mf#54873 – cn Canadiana [380]
How best to learn to speak or teach a language : better because easier – easier for being quicker / Baillairge, Charles P Florent – S.l: s.n, 1897? – 1mf – 9 – mf#17068 – cn Canadiana [400]
How can a man be born when he is old? / Cole, George – London, England. 1844 – 1r – us UF Libraries [240]
How canada is governed : a short account of its executive, legislative, judicial and municipal institutions / Bourinot, John George – Toronto: Copp, Clark, 1867? – 1mf – 9 – mf#54713 – cn Canadiana [323]
How catholics come to be misunderstood : a lecture / O'Gorman, Thomas – [St Paul]: Catholic Truth Society, [1890?] [mf ed 1986] – (= ser American series of publications 3) – 1mf – 9 – 0-8370-7971-3 – mf#1986-1971 – us ATLA [241]
How children may be brought to christ / Dayton, A C – 1859 – 1 – 5.00 – us Southern Baptist [242]
How christ said the first mass : or, the lord's last supper / Meagher, James Luke – New York: Christian Press Association Pub Co, 1906 – 2mf – 9 – 0-524-05880-6 – mf#1990-5174 – us ATLA [220]
How columbus found america : in pen and pencil / Cox, Palmer – New York?: Art Print Establishment, c1877 – 1mf – 9 – mf#29310 – cn Canadiana [810]
How continuous assessment impacts on the teaching of paragraph writing / Shabangu, Halalisani – Pretoria: Vista University 2001 [mf ed 2001] – 2mf – 9 – (incl bibl ref) – mf#mfm15320 – sa Unisa [370]
How did the satellites happen – A study of the Soveit seizure of eastern Europe, by a student of affairs. Preface by Hector McNeil. London: Batchworth Press, (1952).304p – 1 – us Wisconsin U Libr [947]
How did they get there? / Venables, George – London, England. 1862? – 1r – us UF Libraries [240]
How did we come by the reformation? / Beard, John R – London, England. 18– – 1r – us UF Libraries [242]
How do we promote democratization, poverty alleviation, and human rights to build a more secure future? : hearing...united states senate, 107th congress, 2nd session, feb 27 2002 / United States. Congress. Senate. Committee on Foreign Relations – Washington: US GPO 2002 [mf ed 2002] – 1mf – 9 – us GPO [320]
How does the death of christ save us? : or, the ethical energy of the cross / Mabie, Henry Clay – Philadelphia: American Baptist Publ Society, c1908 – 1mf – 9 – 0-8370-4283-6 – mf#1985-2283 – us ATLA [230]
How effective are different sports bra designs at attenuating forces during jumping? / Verscheure, Susan K – 1999 – 1mf – 9 – $4.00 – mf#PE 3928 – us Kinesology [612]
How effectively are state and federal agencies working together to implement the use of new dna technologies? : hearing...house of representatives, 107th congress, 1st session, june 12 2001 / United States. Congress. House. Committee on Government Reform. Subcommittee on Government Efficiency, Financial Management and Intergovernmental Relations – Washington: US GPO 2002 [mf ed 2002] – 2mf – 9 – 0-16-068410-2 – (incl bibl ref) – us GPO [362]
How england saved china / Macgowan, John – London: T Fisher Unwin [1913] [mf ed 1995] – (= ser Yale coll) – 319p (ill) – 1 – 0-524-09331-8 – mf#1995-0331 – us ATLA [951]
How europe was won for christianity : being the life-stories of the men concerned in its conquest / Stubbs, Mattie Wilma – New York: FH Revell, c1913 – 1mf – 9 – 0-524-01406-X – mf#1990-0405 – us ATLA [240]
How, Frederick Douglas *see*
– Archbishop maclagan
– William conyngham plunket
How, G E P *see* English and scottish silver spoons
How god inspired the bible : thoughts for the present disquiet / Smyth, John Paterson – Dublin: Eason; New York: James Pott, 1892 – 1mf – 9 – 0-8370-5314-5 – (incl bibl ref) – mf#1985-3314 – us ATLA [220]
How green was my father / Dodge, David – New York, NY. 1947 – 1r – us UF Libraries [972]
How half a million of the surplus revenue should be invested for the benefit of england and her colonies / Joyce, E D – London, 1851 – (= ser 19th c british colonization) – 1mf – 9 – mf#1.1.7618 – uk Chadwyck [336]
How i came out from rome : an autobiography / Trivier, C L – [London]: Religious Tract Society, [185-?] – 1mf – 9 – 0-524-03595-4 – mf#1990-1055 – us ATLA [240]
How i crossed africa / Pinto, S – London, 1881. 2v – 17mf – 9 – mf#A-173 – ne IDC [916]
How i found livingston : travels, adventures, and discoveries in central africa including four months' residence with dr livingston / Stanley, Henry Morton – Montreal: Dawson, 1872 – 10mf – 9 – (with ind) – mf#33573 – cn Canadiana [916]
How i found livingstone : travels, adventures and discoveries in central africa... / Stanley, H M – London, 1872 – 9mf – 9 – mf#H-6135 – ne IDC [916]
How i found livingstone : travels, adventures, and discoveries in central africa. including an account of four months' residence with dr. livingstone / Stanley, Henry Morton – New York: Scribner, Armstrong, 1872 – 9mf – 9 – 0-524-08786-5 – mf#1993-1094 – us ATLA [916]
How india is governed : being an account of england's work in india / Mackenzie, Alexander – London 1882 – (= ser 19th c british colonization) – 2mf – 9 – mf#1.1.8181 – uk Chadwyck [327]
How india wrought for freedom : the story of the national congress told from official records / Besant, Annie Wood – Adyar, Madras, India: Theosophical Pub House, 1915 – (= ser Samp: indian books) – us CRL [954]
How india wrought for freedom : the story of the national congress told from official records / Besant, Annie Wood – Adyar, Madras: Theosophical Publ House, 1915 [mf ed 1995] – (= ser Yale coll) – lix/709p – 1 – 0-524-09946-4 – mf#1995-0946 – us ATLA [325]
How is ireland to be governed? / Scrope, George Julius Duncombe Poulett – London, 1846 – (= ser 19th c ireland) – 1mf – 9 – mf#1.1.405 – uk Chadwyck [941]
How jesus handled holy writ / Rae, H Rose – London: Arthur H Stockwell, 1902 – 1mf – 9 – 0-7905-3162-3 – mf#1987-3162 – us ATLA [220]
How leisure beliefs relate to attitudes toward the normalization principle as perceived by service providers for people with mental retardation / Neumayer, Robert J & Lundegren, Herberta M – 1993 – 2mf – 9 – $8.00 – us Kinesology [150]
How luke was written : considerations affecting the two-document theory with special reference to the phenomena of order in the non-marcan matter common to matthew and luke / Lummis, Edward – Cambridge: University Press; New York: G P Putnam [distributor], 1915 – 1mf – 9 – 0-7905-3386-3 – mf#1987-3386 – us ATLA [226]
How much are americans at risk until congress passes terrorism insurance protection? : hearing...u.s. house of representatives, 107th congress, 2nd session, feb 27 2002 / United States. Congress. House. Committee on Financial Services. Subcommittee on Oversight and Investigations – Washington: US GPO 2002 [mf ed 2002] – 3mf – 9 – 0-16-068983-X – (incl bibl ref) – us GPO [346]
How much is left for the old doctrines? : a book for the people / Gladden, Washington – Boston: Houghton, Mifflin, 1899. Beltsville, Md: NCR Corp, 1977 (4mf); Evanston: American Theol Lib Assoc, 1984 (4mf) – 9 – 0-8370-0231-1 – (incl bibl ref) – mf#1984-0047 – us ATLA [240]
How mussolini provoked the spanish civil war – Documentary evidence. London, 1938. Fiche W948. (Blodgett Collection of Spanish Civil War Pamphlets) – 9 – us Harvard [946]
How, Samuel Blanchard *see* Slaveholding not sinful
How shall i keep the thanksgiving day? / Thompson, Henry – London, England. 1847 – 1r – us UF Libraries [240]
How shall i put thee among the children? / Cole, George – London, England. 1844 – 1r – us UF Libraries [240]
How shall we conform to the liturgy of the church of england? / Robertson, James Craigie – 3rd ed., rev. London: J. Murray, 1869 – 1mf – 9 – 0-7905-6554-4 – (incl bibl ref) – mf#1988-2554 – us ATLA [241]
How shall we know him? / Blackstone, William E – Chicago, IL: Book Store [dist] [19–?] [mf ed 1992] – 1mf – 9 – 0-524-03688-8 – mf#1990-4793 – us ATLA [240]
How shall we revise the westminster confession of faith? : a bundle of papers / Evans, Llewelyn Joan et al – New York: Charles Scribner, 1890 – 1mf – 9 – 0-8370-8735-X – mf#1986-2735 – us ATLA [240]
How shall we rightly divide the word of truth? / Salmon, George – Dublin, Ireland. 1852 – 1r – us UF Libraries [240]
How the bible was made / Wood, Ezra Morgan – Cincinnati: Walden and Stowe; New York: Phillips & Hunt, 1884 – 1mf – 9 – 0-8370-9350-3 – mf#1986-3350 – us ATLA [220]
How the boll weevil ingests poison / Grossman, Edgar F – Gainesville, FL. 1928 – 1r – us UF Libraries [630]
"How the cherokee acquired and disposed of the outlet [oklahoma]" / Chapman, Berlin B – undated – 1 – us Kansas [305]
How the church began / Rackham, Richard Belward – London, New York: Longmans, Green 1906 [mf ed 1989] – (= ser Simple guides to christian knowledge) – 1mf [ill] – 9 – 0-7905-3161-5 – mf#1987-3161 – us ATLA [226]
How the codex was found : a narrative of two visits to sinai from mrs. lewis's journals, 1892-93 / Gibson, Margaret Dunlop – Cambridge: Macmillan and Bowes, 1893 – 1mf – 9 – 0-8370-9255-8 – mf#1986-3255 – us ATLA [220]
How the disciples began and grew : a short history of the christian church / Davis, Morrison Meade – Cincinnati: Standard Pub Co, c1915 – (= ser Phillips Bible Institute Series of Efficiency Text-Books for Bible Schools and Churches) – 1mf – 9 – 0-524-02112-0 – mf#1990-4178 – us ATLA [240]
How the donkeys came to haiti : and other tales / Johnson, Gyneth – New York, NY. 1949 – 1r – us UF Libraries [390]
How the french captured fort nelson / Willson, Beckles – S.l: s.n, 1899? – 1mf – 9 – (in double columns. incl poem: the cry of the outlander by w a fraser) – mf#17840 – cn Canadiana [971]
How the german fleet shelled almeria – London, 1937? Fiche W949. (Blodgett Collection of Spanish Civil War Pamphlets) – 9 – us Harvard [946]

How the peasant lived in spain / Spain. Embajada. United States – Washington, DC, 193? Fiche W1180. (Blodgett Collection of Spanish Civil War Pamphlets) – 9 – us Harvard [946]
How the rev. dr. stone bettered his situation : an examination of the assurance of salvation, and the certainty of belief to which we are affectionately invited by his holiness the pope / Bacon, Leonard Woolsey – New York: American and Foreign Christian Union, [1870?] – 1mf – 9 – 0-8370-8002-9 – (incl bibl ref) – mf#1986-2002 – us ATLA [240]
How the spirit of god may be quenched / Burns, James C – Edinburgh, Scotland. 1859 – 1r – us UF Libraries [240]
How then can man be justified with god? / Cole, George – London, England. v1. 1844 – 1r – us UF Libraries [240]
How to answer objections to revealed religion / Whately, Elizabeth Jane – amer ed. New York: American Tract Soc [1880?] [mf ed 1985] – 1mf – 9 – 0-8370-5816-3 – (pref note by john hall) – mf#1985-3816 – us ATLA [230]
How to be a yogi / Abhedananda, Swami – New York: Vedaanta Society, 1902 – 1mf – 9 – 0-524-01146-X – mf#1990-2222 – us ATLA [280]
How to be your own lawyer – New York: Richardson, 1885. 507p. LL-1696 – 1 – us L of C Photodup [340]
How to become a child of god / Crossley, Hugh Thomas – Toronto: W Briggs; Montreal: C W Coates; Halifax: S F Huestis, 1891? – 1mf – 9 – mf#14776 – cn Canadiana [240]
How to become a good dancer / Murray, Arthur – New York, NY. 1959 – 1r – us UF Libraries [025]
How to become an efficient sunday school teacher / McKeever, William Arch – Cincinnati: Standard Pub Co, c1915 – (= ser Phillips Bible Institute Series of Efficiency Text-Books for Bible Schools and Churches) – 1mf – 9 – 0-524-06432-6 – (incl bibl ref) – mf#1991-2554 – us ATLA [240]
How to become like christ : and other papers / Dods, Marcus – London: J Clarke 1898 [mf ed 1989] – (= ser Small books on great subjects 6) – 1mf – 9 – 0-7905-1752-3 – mf#1987-1752 – us ATLA [240]
How to bring men to christ / Torrey, Reuben Archer – New York: Fleming H Revell, 1910 – 1mf – 9 – 0-8370-6426-0 – mf#1986-0426 – us ATLA [240]
How to build flying boat hulls and seaplane floats / Streeter, J – London, England. 1936 – 1r – us UF Libraries [623]
How to build up an adult bible class / Moninger, Herbert – Cincinnati, O[hio]: Standard Pub Co, c1909 – 1mf – 9 – 0-524-06648-5 – mf#1991-2703 – us ATLA [220]
How to commend christianity to the chinese *see* Hsieh chi hua jen chieh shou chi-tu chiao (ccm165)
How to compete with foreign cloth : a study of the position of hand-spinning, hand-weaving, and cotton mills in the economics of cloth production in india / Gandhi, Manmohan Purushottam – Calcutta: Book Co, 1931 – (= ser Samp: indian books) – us CRL [680]
How to conduct a meeting – Harare? Zimbabwe. 19– – 1r – us UF Libraries [650]
How to conduct a sunday school : or, twenty eight years a superintendent / Lawrance, Marion – New York: Fleming H Revell, c1905 – 1mf – 9 – 0-524-07162-4 – mf#1991-2951 – us ATLA [240]
How to deal with fenianism : and to adapt our criminal law to the times we live in – London, 1868 – (= ser 19th c ireland) – 1mf – 9 – mf#1.1.1852 – uk Chadwyck [345]
How to deal with the consumptive poor / Stone, Andrew Jackson – S.l: s.n, 1899? – 1mf – 9 – mf#17983 – cn Canadiana [360]
How to decorate our ceilings, walls, and floors / James, M E – London 1883 – (= ser 19th c art & architecture) – 2mf – 9 – mf#4.2.586 – uk Chadwyck [740]
How to do good / Patton, W J – Belfast, Northern Ireland. 1899 – 1r – us UF Libraries [240]
How to dress : a handbook for women of modest means / ed by Klickmann, Flora – London, New York, Melbourne: Ward, Lock & Co [1900] – (= ser 19th c art & architecture) – 2mf – 9 – mf#4.1.66 – uk Chadwyck [640]
How to dress on £15 a year : as a lady / Cook, Millicent Whiteside – London: Frederick Warne & Co; New York: Scribner, Welford & Armstrong [1873] – (= ser 19th c art & architecture) – 2mf – 9 – mf#4.1.65 – uk Chadwyck [640]
How to dress well on a shilling a day : a ladies' guide to home dressmaking and millinery / Sylvia [pseud] – London [1876] – (= ser 19th c art & architecture) – 2mf – 9 – mf#4.1.274 – uk Chadwyck [640]
How to employ capital in western ireland / Seymour, William Digby – London, 1851 – (= ser 19th c ireland) – 3mf – 9 – mf#1.1.5237 – uk Chadwyck [332]
How to get on / Feeney, Bernard – 4th ed. New York: Benziger, c1891 – 1mf – 9 – 0-8370-6116-4 – mf#1986-0016 – us ATLA [170]
How to get strong and how to stay so / Blaikie, William Garden – 1879 – 5mf – 9 – $15.00 – us Kinesology [790]
How to have an orange grove in florida / Porter, Charles N – Ocala, FL. 1882 – 1r – us UF Libraries [634]
How to interpret "accidents" / Calthrop, Gordon – London, England. 18– – 1r – us UF Libraries [240]
How to know the ducks, geese and swans of north america : all the species being grouped according to size and color / Cory, Charles Barney – Boston?: s.n, 1897 – 2mf – 9 – (incl ind) – mf#06201 – cn Canadiana [590]
How to know the shore birds (limicolae) of north america (south of greenland and alaska) : all the species being grouped according to size and color / Cory, Charles Barney – Boston?: s.n, 1897 – 1mf – 9 – mf#16916 – cn Canadiana [590]
How to make a saint : or, the process of canonization in the church of england / Longueville, Thomas – London: Kegan Paul, Trench, 1887 [mf ed 1986] – 1mf – 9 – 0-8370-6913-0 – mf#1986-0913 – us ATLA [242]
How to make abstracts of title and searches / Foye, Edward M – Erie, PA: Dispatch, Printing and Engraving Co., 1896. 32p. LL-280 – 1 – us L of C Photodup [340]
How to make good pictures : a book for the amateur photographer / Canadian Kodak Co – Toronto: Canadian Kodak Co, [191-?] – 2mf – 9 – 0-665-74733-0 – mf#74733 – cn Canadiana [770]
How to master the english bible : an experience, a method, a result, an illustration / Gray, James Martin – Chicago: Winona Pub Co, 1904 – 1mf – 9 – 0-524-06204-8 – mf#1992-0842 – us ATLA [220]
How to memorize / Evans, William – Chicago, IL: Moody Press, c1910 – 1mf – 9 – 0-7905-1384-6 – mf#1987-1384 – us ATLA [150]
How to organize a fundraising golf tournament / Tinkess, Jeanne S – 1997 – 2mf – 9 – $8.00 – mf#PE 3841 – us Kinesology [650]
How to organize and conduct an evening class in citrus culture / Knight, Fred Key – s.l, s.l? 1932 – 1r – us UF Libraries [634]
How to play football / ed by Camp, Walter Chauncey – New York, c1914.95p – 1 – us Wisconsin U Libr [790]
How to play golf / Vardon, Harry – Philadelphia: G.W. Jacobs, 1912?. 187p. Includes index – 1 – us Wisconsin U Libr [790]
How to prepare for confirmation : eight plain addresses with questions for candidates / Ridgeway, Charles John – 16th ed. London: Skeffington and Son, 1898 – 1mf – 9 – 0-524-05484-3 – mf#1990-5131 – us ATLA [242]
How to print and publish a book : also information about printing generally / Warren, William Thorn – Winchester, London, 1890 – (= ser 19th c publishing...) – 1mf – 9 – mf#3.1.23 – uk Chadwyck [070]
How to publish a book : being directions and hints to authors / Spon, Ernest – London, 1872 – (= ser 19th c publishing...) – 1mf – 9 – mf#3.1.28 – uk Chadwyck [070]
How to publish a book or article and how to produce a play : advice to young authors / Wagner, Leopold – London: George Redway, 1898 – (= ser 19th c publishing...) – 3mf – 9 – mf#3.1.74 – uk Chadwyck [070]
How to raise money and make an addition to your sabbath school library : or help to pay off a debt without the aid of a bazaar... – Toronto?: J Campbell, 1873? – 1mf – 9 – mf#39778 – cn Canadiana [070]
How to read josephus / Auchincloss, William Stuart – New York: D van Nostrand, 1906 – 1mf – 9 – 0-8370-2128-6 – mf#1985-0128 – us ATLA [240]
How to read the bible : hints for sunday-school teachers and other bible students / Adeney, Walter Frederic – New York: Thomas Whittaker, 1897 – 1mf – 9 – 0-8370-2052-2 – mf#1985-0052 – us ATLA [220]
How to re-construct the industrial condition of ireland / Ward, James – London, 1847 – (= ser 19th c ireland) – 1mf – 9 – mf#1.1.909 – uk Chadwyck [339]
How to remember the life of christ : an analytic arrangement of the gospel materials / Wieand, Albert Cassel – [S.l.: s.n.], c1914 – 1mf – 9 – 0-524-04244-6 – (incl wieand's an analytic diagram and outline of the life of christ) – mf#1990-5035 – us ATLA [220]
How to see montreal / Gard, Anson Albert – Montreal: the Montreal News Co Ltd, [1903?] (mf ed 1994) – 3mf – 9 – (with ind) – mf#SEM105P2148 – cn Bibl Nat [917]
How to strengthen the memory; or, natural and scientific methods of never forgetting / Holbrook, Martin Luther – New York: M.L. Holbrook & Co., (c1886). 152p – 1 – us Wisconsin U Libr [150]
How to study the bible for greatest profit : the methods and fundamental conditions of the bible study that yields the largest results / Torrey, Reuben Archer – New York: Fleming H Revell, c1896 – 1mf – 9 – 0-7905-0295-X – mf#1987-0295 – us ATLA [220]
How to study the bible, the second coming and other expositions / Haldeman, I. M – 2nd ed. New York City: Charles C. Cook, c1904 – 2mf – 9 – 0-7905-0573-8 – mf#1987-0573 – us ATLA [220]
How to study the english bible / Girdlestone, Robert Baker – New York: Fleming H Revell, [1894] – (= ser Present Day Primers) – 1mf – 9 – 0-8370-3298-9 – mf#1985-1298 – us ATLA [220]
How to study the life of christ : a handbook for sunday-school teachers and other bible students / Butler, Alford Augustus – New York: Thomas Whittaker, c1901 – 1mf – 9 – 0-8370-2555-9 – mf#1985-0555 – us ATLA [240]
How to study the new testament : the epistles (first section) / Alford, Henry – London: Strahan, 1868 – 1mf – 9 – 0-8370-2072-7 – mf#1985-0072 – us ATLA [225]
How to study the new testament : the gospels: the acts of the apostles / Alford, Henry – London, New York: Alexander Strahan, 1865 – 1mf – 9 – 0-8370-2074-3 – mf#1985-0074 – us ATLA [225]
How to study the old testament / Sanders, Frank Knight & Shermann, Henry A – New York: Scribner, c1915 – 1mf – 9 – 0-524-06157-2 – (incl bibl ref) – mf#1992-0824 – us ATLA [221]
How to succeed : a book for the young / Lister, J B – Philadelphia: American Baptist Publ Society, [18–?] – 1mf – 9 – 0-8370-9293-0 – mf#1986-3293 – us ATLA [240]
How to survive an atomic bomb / Gerstell, Richard – Washington, DC. 1950 – 1r – us UF Libraries [360]
How to teach a foreign language / Jespersen, Otto – London, England. 1912 – 1r – us UF Libraries [370]
How to teach swimming and diving / Cureton, Thomas Kirk – New York: Association Press, 1934 – 1 – (incl: tables, diagrs) – us Wisconsin U Libr [370]
How to teach the church catechism : together with a complete set of notes of lessons / Daniel, Evan – new and rev ed. London: National Society's Depository, 1890 – (= ser Religious Knowledge Manuals) – 1mf – 9 – 0-524-05372-3 – mf#1991-2278 – us ATLA [240]
How to teach the old testament / Benham, William – London: National Society's Depository, 1882 – (= ser Religious Knowledge Manuals) – 1mf – 9 – 0-524-05790-7 – mf#1992-0617 – us ATLA [221]
How to tell a caxton : with some hints where and how the same might be found / Blades, William – London, 1870 – (= ser 19th c publishing...) – 1mf – 9 – mf#3.1.22 – uk Chadwyck [680]
How to tell a story : and other essays / Twain, Mark – New York, NY. 1900 – 1r – us UF Libraries [080]
How to think about war and peace / Adler, Mortimer Jerome – New York: Simon and Schuster, c1944 [mf ed 1995] – xxiii/307p – 1 – mf#7407 – us Wisconsin U Libr [320]
How to win : or, the dignity of labor: suggestions to young men, in three lectures, for the encouragement of agriculture and the industrial arts / Newcomb, D B – Halifax, NS?: s.n, 1872 – 1mf – 9 – mf#34984 – cn Canadiana [331]
How to win souls / Willing, Jennie Fowler – Chicago: Christian Witness, 1909 – 1mf – 9 – 0-8370-6465-1 – mf#1986-0465 – us ATLA [240]
How to write a business letter : a manual for use in colleges, schools, and for private learners / Fleming, Christopher Alexander – [Owen Sound, Ont?: s.n.], 1890 [mf ed 1985] – (= ser Fleming's practical education series) – 2mf – 9 – 0-665-33726-4 – mf#33726 – cn Canadiana [650]
How to write the history of a parish : an outline guide to topographical records, manuscripts, and books / Cox, John Charles – 5th ed., rev. London: George Allen, 1909 – 1mf – 9 – 0-7905-5524-7 – (incl bibl ref) – mf#1988-1524 – us ATLA [941]
How wars arise in india : observations on mr cobden's pamphlet, entitled, "the origin of the burmese war" / Marshman, John Clark – London 1853 – (= ser 19th c british colonization) – 1mf – 9 – mf#1.1.1004 – uk Chadwyck [954]
How we find relics / Riggs, C W – Chicago, IL: W B Conkey Co, 1893 – 1r – 1 – (native american archaeology in the 19th century) – us Western Res [930]
How we got our bible / Smyth, John Paterson – New York: James Pott, c1912 – 1mf – 9 – 0-7905-3230-1 – mf#1987-3230 – us ATLA [220]
How we lived then, 1914-1918: a sketch of social and domestic life in england during the war / Peel, Dorothy Constance Bayliff – London: John Lane, 1929. xiv,235p. plates, facsims – 1 – us Wisconsin U Libr [941]
How we may best make our churches and services attractive to the pe... / Murray, J W – Dublin, Ireland. 1868 – 1r – us UF Libraries [240]
How we remember our past lives, and other essays on reincarnation / Jinarajadasa, Curuppumullage – Adyar, Madras, India: Theosophical Pub House, 1915 – 1mf – 9 – 0-524-03368-4 – mf#1990-3202 – us ATLA [280]
How we think / Dewey, John – Boston: DC Heath, [1909?] – 1mf – 9 – 0-7905-7288-5 – mf#1989-0513 – us ATLA [190]
How, William Walsham *see* The new testament of our lord and saviour jesus christ
Howald, Johann *see* Geschichte der deutschen literatur
Howard, Ar *see* Courses in agriculture for adult farmers of florida by districts
Howard County Genealogical Society *see*
– Family tree
– Hcgs newsletter
Howard county herald – St Paul, NE: M Lorkosky, Irene Lorkosky. 48v. 33rd yr n9. aug 23 1923-v81 n48. feb 24 1971 (wkly) [mf ed with gaps filmed 1971] – 17r – 1 – (cont: howard county herald and the republican. merged with: phonograph (1911) to form: phonograph-herald) – us NE Hist [071]
Howard county herald and the republican – St Paul, NE: M Lorkosky. 3v. 31st yr n4. jul 13 1922-33rd yr n8. aug 16 1923 (wkly) [mf ed filmed 1973] – 3r – 1 – (cont: republican. cont by: howard county herald) – us NE Hist [071]
Howard, Eliot *see* Studies of non-christian religions
Howard, Elizabeth Fox *see* Woman in the church and in life
Howard, Eric C *see* A survey of the desired educational preparation and employment market for high school athletic trainers in metropolitan washington, dc as perceived by high school athletic directors
Howard, Frank *see*
– The art of dress
– The st helen's crown glass company's trade book of patterns for ornamental window glass
Howard, Frederick P *see* The british columbian and victoria guide and directory for 1863
Howard, George Broadley *see*
– The christians of st thomas and their liturgies
– The christians of st. thomas and their liturgies
Howard, George Elliott *see*
– A history of matrimonial institutions
– A history of matrimonial institutions chiefly in england and the united states
Howard, Henry *see*
– A course of lectures on painting
– Remarks on the erroneous opinions entertained respecting...
– Yacht alice
Howard, James *see* The tenant farmer
Howard, John Eliot *see* Seven lectures on scripture and science
Howard, John M *see* Why? when? what?
Howard, John R *see* Bible studies
Howard, John Raymond *see* Patriotic addresses
Howard journal of criminal justice – Oxford. 1984+ (1,5,9) – ISSN: 0265-5527 – mf#13041,02 – us UMI ProQuest [360]
Howard, Laura Pratt *see* Ragtime
Howard law journal – v1-44. 1955-2001 – 5,6,9 – $828.00 set – (v1-27 1955-84 on reel $418. v28-44 1985-2001 on mf $410) – ISSN: 0018-6813 – mf#103151 – us Hein [340]
Howard medical news / Howard University. School of Medicine – Washington, D.C. v. 1, no. 7; v. 2, no. 3, 5-6, 8, 10; v. 3, no. 1-5, 7-8; v. 4, no. 1, 3-4, 6-10; v. 5, no. 5. 1925-1929 – 1 – us NY Public [610]
Howard, Overton *see* The life of the law, or, universal principles of law
Howard, Philip Eugene *see*
– The life story of henry clay trumbull
– Their call to service
Howard phillips issues and strategy bulletin : hpisb / Policy Analysis, inc – n1-293 [1984 jan 2-1989 aug 7] – 1r – 1 – mf#3048802 – us WHS [650]
Howard, Raymond Holt *see*
– Some factors affecting citrus costs, yields, and returns
– Study of the relation of grade and staple to the price of cotton gr...
Howard review – Sillery. 1923-1925 (1) – mf#7446 – us UMI ProQuest [305]

Howard, Richard A *see* Charles wright in cuba, 1856-1867
Howard, Robert Palmer *see* Circular
Howard scroll : social justice review – v1-4. 1993-2000 – 9 – $62.00 set – ISSN: 1070-3713 – mf#115431 – us Hein [344]
Howard shuman, senate service 1955-1982 : aide to senators paul douglas and william proxmire – (= ser Us senate historical office oral history coll) – 7mf – 9 – $35.00 – us Scholarly Res [323]
Howard University *see*
– Capstone
– College of medicine news
– Hilltop
– History department newsletter
– Hu newsletter
– Iah news
– Isep monitor
– Perspectives
– Stylus
Howard University Alumni Association *see* Chicago hilltop
Howard university hospital health care – 1983 feb – 1r – 1 – mf#4989629 – us WHS [360]
Howard University Medical Alumni Association *see* Medicannales
Howard University. School of Medicine *see* Howard medical news
Howard University Student Assembly *see* Omowe
Howard-Bury, Charles et al *see* Mount everest, the reconnaissance, 1921
Howard's appeal cases : unreported – New York. 1v. 1847-48 (all publ) – 9mf – 9 – $13.50 – mf#LLMC 80-010 – us LLMC [340]
Howards end / Forster, E M [Edward Morgan] – Toronto: W Briggs, 1911 [mf ed 1998] – 4mf – 9 – 0-665-66266-1 – mf#66266 – cn Canadiana [830]
Howard's practice reports / New York. (State) – 1st series: v1-67. 1844-84; ns: v1-3. 1884-86 (all publ) – 489mf – 9 – $733.00 – mf#LLMC 78-102 – us LLMC [340]
Howarth, David Armine *see* Shadow of the dam
Howarth, Henry *see* Plea for the established church
Howarth, OJ R *see* History of geography
Howarth, William *see* Modern brazil
Howat, Kenneth J *see* The effect of half-time warm-up procedures upon injuries to high school varsity football players
Howden, Jeffrey B *see* A funding plan for the renovation of the a.e. finley golf course
Howe, Clifton Durant *see* Trent watershed survey
Howe, Daniel Wait *see*
– The laws and courts of northwest and indiana territories
– The puritan republic of the massachusetts bay in new england
Howe, Eber D *see* Autobiography of a pioneer printer
Howe, Edgar Watson *see* Trip to the west indies
Howe, Elias *see*
– [First part of the] musician's companion
– Howe's new cornet instructor; containing full and complete rules, exercises, and instructions to enable the learner to play this favorite instrument, without a master
– Leviathan collection of instrumental music
– [Third part of the] musician's companion
– Young america's collection of instrumental music
Howe, Frank Clifford *see* All examination questions used for twelve years, in the regular courses in columbian university
Howe, George *see*
– A discourse on theological education
– History of the presbyterian church in south carolina
Howe, George M *see* Heinrich von kleist
Howe, John *see* Living temple
Howe, Joseph *see*
– Hon mr howe's speech on dr tupper's railway resolution
– Information for the people
– Poems and essays
– The speeches and public letters of the hon joseph howe
– To the electors of the county of cumberland
Howe, Julia Ward *see*
– Margaret fuller (marchesa ossoli)
– Trip to cuba
Howe, Laura G *see* The research and development of multimedia leisure-learning packages for the rural elderly
Howe, M A De Wolfe *see*
– Memoirs of the life and services of the rt. rev. alonzo potter, d.d., lld.
– The memory of lincoln; poems selected with an introd
Howe, Mark Antony De Wolfe *see* Memoirs of the life and services of the rt. rev. alonzo potter, d.d., ll.d
Howe, Reginald Heber *see* The creed and the year
Howe, Richard Esmond, Jr *see* Cadenza in the piano concerto
Howe, Russell Warren *see* Theirs the darkness
[Howe, William M] *see* A symposium relative to james 5:14-16

Howell, Almonte Charles, Jr *see* French organ mass in the sixteenth and seventeenth centuries
Howell, Arthur Holmes *see* Florida bird life
Howell, Charles Boynton *see* The church and the civil law
Howell, David B *see* The christology of paul's opponents in second corinthians and its relationship to their concept of apostleship
Howell, Edward Beach *see* Montana miners' code.
Howell, George *see* The selected papers of george howell, 1833-1910
Howell, Gerald Emmett *see* A study of authority from a theological perspective and its implications for buck run baptist church
Howell, Gordon P *see* Development of music in canada
Howell, Henry Spencer *see* The british union jack
Howell, Joseph Morton *see* Egypt's past, present and future
Howell, M S *see* Grammar of the classical arabic language
Howell, Mary J *see*
– The hand-book of dress-making
– The hand-book of millinery
Howell, Robert Boyte C *see*
– An address delivered before the university of nashville, 1839
– Collection, manuscripts and books
– Manuscript notes of sermons, 1838-1957
– Memorial of first baptist church, nashville, tenn., 1820-63
Howell, Robert Boyte Crawford *see*
– The covenants
– The cross
– The deaconship
– The early baptists of virginia
– The evils of infant baptism
– The terms of communion at the lord's table
– The way of salvation
Howell, Thomas Bayly *see* Observations on dr sturges's pamphlet respecting non-residence of...
[Howell], W *see* Some interesting particulars of the second voyage made by the missionary ship, the duff which was captured by the buonaparte privateer, in the year 1800
Howell, Williamson S *see* The united states and france
Howell-howel bulletin / Bodak, Shirley L – 1976 nov 1-1978 aug – 1r – 1 – mf#637873 – us WHS [071]
Howells, George *see* The soul of india
Howells journal – Howells, NE: H E Phelps, 1888-v96 n31. may 13 1981 (wkly) [mf ed 1892,1895-81 (gaps) filmed -1983] – 27r – 1 – (cont by: journal. vol numbering dropped with n42 jul 11 1963; resumed with v89 n1 oct 20 1977) – us NE Hist [071]
Howells journal – Howells, NE: Howells Journal. v99 n28. apr 8 1987- (wkly) – 1 – (cont: journal) – us NE Hist [071]
Howell's nisi prius cases – Michigan. 1v. 1868-84 (all publ) – (= ser Michigan appellate reports) – 5mf – 9 – $7.50 – mf#LLMC 81-304 – us LLMC [340]
Howell's state trials *see* Cobbett's state trials / howell's state trials
Howells, W D *see* The niagara book
Howells, William Cooper *see* Recollections of life in ohio from 1813 to 1840
Howells, William Dean *see*
– Letters of an altrurian traveller
– Prefaces to contemporaries
– Une rencontre
– Tuscan cities
Howenstine, Lydia *see* From the cradle to the grave
Howerton, Mollie W *see* Development [and] evaluation of computer-assisted instruction in smoking education for adolescents
Howe's new cornet instructor; containing full and complete rules, exercises, and instructions to enable the learner to play this favorite instrument, without a master / Howe, Elias – With a large collection of popular polkas, schottisches, waltzes, quicksteps, marches, quadrilles, &c. Boston: Elias Howe, 1860. MUSIC 1989, Item 2 – 1 – us L of C Photodup [780]
Howgrave's stamford mercury lincoln *see* Stamford mercury
Howick and pakuranga times – 1976; jan 1979-dec 1988 – 19r – 1 – mf#11.23 – nz Nat Libr [079]
Howie, John *see* The scots worthies
Howie, Robert *see*
– Reply to letter of professor blaikie, dd, lld, to rev andrew a...
– Reply to letter of professor blaikie...to rev andrew a bonar
– The state of the question in the case of rev dr marcus dods
Howison, George Holmes *see* The function of universities in religion
Howison, John *see*
– European colonies, in various parts of the world
– Foreign scenes and travelling recreations
– Sketches of upper canada, domestic, local, and characteristic

Howitt, Dr *see* An address on the formation of rifle associations for defensive purposes
Howitt, Emanuel *see* Selections from letters written during a tour through the united states, in the summer and autumn of 1819
Howitt, William *see* The history of the supernatural
Howitt's journal of literature and popular progress – London. 1847-1848 (1) – mf#2797 – us UMI ProQuest [420]
Howitzer – Greeley CO. 1866 dec 3-10 – 1r – 1 – (cont: rocky mountain howitzer) – mf#854375 – us WHS [071]
Howland, Charles R *see* "Howland's digest"
Howland, F *see* Blatchford and howland's reports of cases in the southern district court of new york, 1827-1837
Howland, John D *see* A manual for executors, administrators and guardians, embracing all the statutes in force in the state of indiana relating to the settlement of decendents' estates.
Howland, Oliver Aiken *see*
– The canadian historical exhibition, 1897
– The new empire
Howland, William *see* John howland
"Howland's digest" / Howland, Charles R – 3 sep 1862-31 jan 1912 (mf ed Washington: GPO, 1912) – 12mf – 9 – $18.00 – mf#LLMC 84-229 – us LLMC [355]
Howlett, R *see*
– Chronicles of the reigns of stephen, henry 2 and richard 1
– Monumenta franciscana
Howlett, William J *see* Life of rev. charles nerinckx
Howley, Michael Francis *see* Ecclesiastical history of newfoundland
Howling gale – New London. 1976-1978 (1) 1976-1978 (5) 1976-1978 (9) – ISSN: 0438-0185 – mf#9054 – us UMI ProQuest [370]
Howman, H Roger G *see* African local government in british east and central africa...
Howorth, Henry Hoyle *see*
– The glacial nightmare and the flood
– The mammoth and the flood
– Saint augustine of canterbury
– Saint gregory the great
Howse, Joseph *see*
– A grammar of the cree language
Howson, John Saul *see*
– Before the table
– The companions of st paul
– Deaconesses in the church of england
– The evidential value of the acts of the apostles
– Five lectures on the character of st paul
– Horae petrinae
– The life and epistles of st paul
– Meditations on the miracles of christ
– The metaphors of st paul
– Scenes from the life of saint paul and their religious lessons
Hoxton magazine : or religious & moral instructor – [London & SE] Hackney 1843 [mf ed 2003] – 1r – 1 – uk Newsplan [072]
Hoxton sausage & jerry-wags' journal – [London & SE] Hackney 1826 [mf ed 2003] – 1r – 1 – uk Newsplan [072]
Hoy – 5Jul 1941-24 Dec 1949. nos. 228-670. nos. 473, 475-478, 480-488, 567-575, 615, 627, 631-632, and 659 wanting – 1 – 579.00 – us L of C Photodup [073]
Hoy – Badajoz, 1933-1936 – 5 – sp Bibl Santa Ana [073]
Hoy – Mexico. 1950-1953 (1) – ISSN: 0018-6848 – mf#431 – us UMI ProQuest [070]
Hoy – New York, NY. 1999-2000 (1) – mf#69531 – us UMI ProQuest [071]
Hoy dia – New York. 1975-1981 (1) 1975-1981 (5) 1975-1981 (9) – ISSN: 0018-6856 – mf#8024 – us UMI ProQuest [370]
Hoya de enriquillo / Cucurullo, Oscar – Ciudad Trujillo, Dominican Republic. 1949 – 1r – us UF Libraries [972]
Hoyem, OJ *see* New eller bynes (og buvik) en bygdebeskrivelse
Hoyer, Johannes *see* Schleiermachers erkenntnistheorie in ihrem verhaeltnis zur erkenntnisstheorie kants
Hoyer, M *see* Flammulae amoris s p augustini versibus et iconibus exonatae...
Hoyer, Mark V *see* Handbook of common freshwater fish in florida lakes
Hoyer, Richard *see* Menschenschicksale
Hoyerswerdaer kreisblatt *see* Kreisblatt des hoyerswerdaer kreises
Hoyerswerdaer nachrichten *see* Kreisblatt des hoyerswerdaer kreises
Hoyerswerdaer wochenblatt – Hoyerswerda DE, 1843 8 jul-1859, 1861-81 – 12r – 1 – gw Misc Inst [074]
Hoyerswerder volksstimme – Hoyerswerda DE, 1962 30 mar-1965 29 sep – 1r – 1 – (publ in cottbus) – gw Misc Inst [074]
Hoylake, holy trinity – (= ser Cheshire monumental inscriptions) – 4mf – 9 – £5.00 – mf#371 – uk CheshireFHS [929]
Hoyland, John Somervell *see*
– The cross moves east
– Gopal krishna gokhale
– Indian crisis

Hoynck van Papendrecht, C P *see* Historia ecclesiae ultrajectinae
Hoyne, Thomas Temple *see* Speculation
Hoyo, Joseph del *see* Relacion completa, y exacta del auto publico de fe
Hoyos, F A *see*
– Barbados
– Rise of west indian democracy
Hoyos, Fabriciano Alexander *see* Story of the progressive movement
Hoyos Sainz, Luis de *see* La raza extremena
Hoyt, Edwin Palmer *see* Germans who never lost
Hoyt, James T *see* The collection laws, special, exemption, property, banking, and interest laws, of illinois, indiana, michigan, iowa, wisconsin and minnesota.
Hoyt, Wayland *see*
– Gleams from paul's prison
– The teaching of jesus concerning his own person
Hoyte, Thor A *see* ...And so we played
Hozier, H M *see*
– Der britische feldzug nach abessinien
– Record of the expedition to abyssinia
Hozumi, Nobushige *see*
– Ancestor-worship and japanese law
H.r. 1239 and h.r. 2742 : legislative hearing...u.s. house of representatives, 107th congress, 1st session, oct 17 2001 / United States. Congress. House. Committee on Resources – Washington: US GPO 2002 [mf ed 2002] – 1mf – 9 – 0-16-068850-7 – us GPO [343]
H.r. 1518, h.r. 1776, and h.r. 2114 : legislative hearing...u.s. house of representatives, 107th congress, 1st session, july 17 2001 / United States. Congress. House. Committee on Resources. Subcommittee on National Parks, Recreation, and Public Lands – Washington: US GPO 2002 [mf ed 2002] – 9 – 0-16-068854-X – us GPO [343]
H.r. 2119, "national historic forests act of 2001" : legislative hearing...u.s. house of representatives, 107th congress, 1st session, june 19 2001 / United States. Congress. House. Committee on Resources. Subcommittee on Forests and Forest Health – Washington: US GPO 2002 [mf ed 2002] – 1mf – 9 – 0-16-068374-2 – us GPO [346]
H.r. 2187, to amend title 10, u.s.c., regarding mineral receipts collected from naval oil shale reserves : legislative hearing...u.s. house of representatives, 107th congress, 1st session, june 26 2001 / United States. Congress. House. Committee on Resources. Subcommittee on Energy and Mineral Resources – Washington: US GPO 2002 [mf ed 2002] – 1mf – 9 – 0-16-068368-8 – us GPO [343]
H.r. 2436, the energy security act : legislative hearing...u.s. house of representatives, 107th congress, 1st session, july 11 2001 / United States. Congress. House. Committee on Resources – Washington: US GPO 2002 [mf ed 2002] – 9 – 0-16-068458-7 – (incl bibl ref) – us GPO [343]
H.r. 2941, brownfields redevelopment enhancement act : hearing...u.s. house of representatives, 107th congress, 2nd session, mar 6 2002 / United States. Congress. House. Committee on Financial Services. Subcommittee on Housing and Community Opportunity – Washington: US GPO 2002 [mf ed 2002] – 1mf – 9 – 0-16-068761-6 – us GPO [344]
H.r. 3951 – the financial services regulatory relief act of 2002 : hearings...u.s. house of representatives, 107th congress, 2nd session, march 14; april 25 2002 / United States. Congress. House. Committee on Financial Services. Subcommittee on Financial Institutions and Consumer Credit – Washington: US GPO 2002 [mf ed 2002] – 9 – 0-16-068746-2 – (incl bibl ref) – us GPO [346]
Hr focus – New York. 1991+ (1) 1991+ (5) 1991+ (9) – (cont: personnel) – ISSN: 1059-6038 – mf#341,01 – us UMI ProQuest [650]
Hr human resource planning – Tempe. 1978+ (1,5,9) – ISSN: 0199-8986 – mf#12852 – us UMI ProQuest [650]
Hrabanus maurus : ein beitrag zur geschichte der mittelalterlichen exegese / Hablitzel, Joh Bapt – Freiburg i Breisgau; St Louis, MO: Herder, 1906 [mf ed 1989] – (= ser Biblische studien 11/3) – 1mf – 9 – 0-7905-2163-6 – (incl bibl ref) – mf#1987-2163 – us ATLA [220]
Hraf probability sample program-part a / Human Relations Area Files – 1980. 60 cultural files constituting a worldwide sample. Available in 7 modules – 9 – (part b. 1984. 40 cultural files available in 6 modules. 2050.00 each; 9) – us HRAF [306]
Hraf topical program / Human Relations Area Files – 1984 – 9 – apply for price – (hunting and gathering societies.3990.00; 9. pastoral societies. 3325.00; hra; 9. historical societies. 3325.00; hra; 9) – us HRAF [306]
Hrdi advisory / AFL-CIO Human Resources Development Institute – 1979 oct-1994 nov/ dec – 1r – 1 – (cont: hrdi manpower advisory; cont by: advisory [washington, dc: 1995]) – mf#1056299 – us WHS [331]

Hrdlicka, Ales *see* Anthropology of florida
Hrmagazine – Alexandria. 1990+ (1) 1990+ (5) 1990+ (9) – (cont: personnel administrator) – ISSN: 1047-3149 – mf#6437,01 – us UMI ProQuest [650]
Hrn. b.h. brockes...verteutschter bethlehemitischer kinder-mord des ritters marino. / Marino, Giambattista – 3rd ed. Hamburg: J.C. Kissner, 1727. illus. Trans. of La Strage degli Innocenti with original text on facing pages – 1 – us Wisconsin U Libr [430]
Hrotsvithae opera (mgh7:34.bd) – 1902 – (= ser Monumenta germaniae historica 7: scriptores rerum germanicarum in usum scholarum (mgh7)) – €19.00 – ne Slangenburg [240]
Hrouda, Baerbel *see* Konfrontative untersuchungen zur semantischen dimension der fachlichkeit von texten
Hrovatin, Lauri A *see* The effect of different interval durations on measures of exercise intensity
Hrs society rag – New York: Hot Record Society. n1-11(?) jul 1938-mar 1941 (freq varies) [all publ] – (= ser Jazz periodicals, 1914-1977) – 1r – 1 – $115.00 – us UPA [780]
Hruschka, Alois *see* Ueber deutsche ortsnamen
Hrvatska = Croatia – Calumet: Croatian Print Co, dec 14 1917-aug 15 1924 – 3r – 1 – us CRL [079]
Hrvatska croacia – Buenos Aires. 1966-75 – 2r – 1 – uk British Libr Newspaper [072]
Hrvatska gruda – Buenos Aires, Argentina. -m. Oct 1959-Dec 1975. 2,5 reels – 1 – uk British Libr Newspaper [072]
Hrvatska rijec – Jemeppe sur Meuse, France. Sep 1953-nov 1961 – 1/2r – 1 – uk British Libr Newspaper [072]
Hrvatska smotra – Zagreb: Tiskara Merkantile 1935- [mf ed Bloomington IN: Indiana Uni Lib, Preservation Dept 1990] – 3r – 1 – us Indiana Preservation [073]
Hrvatska smotra za knjizevnost, umjetnost i drustveni zivot – Zagreb: Tiskara Merkantile 1933-34 [mf ed Bloomington IN: Indiana Uni Lib, Preservation Dept 1990] – 1r – 1 – us Indiana Preservation [073]
Hrvatska smotra za politiku, knjizevnost, znanost, umjetnost i kritiku – Zagreb: Tisak Prve hrvatske radnicke tiskare 1906- [mf ed Bloomington IN: Indiana Uni Lib, Preservation Dept 1990] – 2r – 1 – us Indiana Preservation [073]
Hrvatska straza za krscansku prosyjetu – U Krku: Tisk i nakl Kurykte 1903-17 [mf ed Bloomington IN: Indiana Uni Lib, Preservation Dept 1990] – 2r – 1 – us Indiana Preservation [073]
Hrvatska zastava – Chicago IL, aug 3 1905-oct 30 1917 – 5r – 1 – (croatian newspaper) – us IHRC [071]
Hrvatski dnevnik – Zagreb. Yugoslavia. -d. 16 Apr 1939-29 Mar 1941. (17 reels) – 1 – uk British Libr Newspaper [949]
Hrvatski glas – Winnipeg, Canada. 13 apr 1953-26 dec 1973; 9 jan-25 dec 1974; 8 jan-27 dec 1975 – 19 1/2r – 1 – uk British Libr Newspaper [072]
Hrvatski glasnik = Croatian herald – Allegheny: Hrvatsko novinarsko drustvo, [1908-]. dec 1917-sep 1 1921 – us CRL [079]
Hrvatski glasnik – Allegheny PA, dec 12 1908-sep 13 1919 – 3r – 1 – (croatian newspaper) – us IHRC [071]
Hrvatski list and danica hrvatska – New York NY, jan 20 1922-dec 29 1928; jan 2 1930-dec 30 1941 – 15r – 1 – (croatian newspaper) – us IHRC [071]
Hrvatski narod – Zagreb. Yugoslavia. -d. 2 Jan 1942-4 Jul 1944, 31 Dec 1944-22 Mar 1945. (Very imperfect). (10 reels) – 1 – uk British Libr Newspaper [949]
Hrvatski narod – Zagreb, Yugoslavia: Hrvatstski narod, oct 1941-feb 1 1945 – 12r – 1 – us CRL [079]
Hrvatski svijet – New York NY, 1914* – 1r – 1 – (croatian newspaper) – us IHRC [071]
Hrvatsko kolo – U Zagrebu: Matica hrvatska 1905-46 [mf ed Bloomington IN: Indiana Uni Lib, Preservation Dept 1989] – 2r – 1 – us Indiana Preservation [073]
Hrvatsko kolo – Zagreb: Matica hrvatska 1948-63 [mf ed Bloomington IN: Indiana Uni Lib, preservation Dept 1989] – 2r – 1 – us Indiana Preservation [073]
Hrvatsko slovo = Verbum croaticum – Zagreb: Naklada Drustva hrvatskih knjizevnika 1995- [apr 28 1995-96] [mf ed Hiawatha IA: Heritage Microfilm Inc, New York NY: available fr Norman Ross Pub 1998] – 2r – 1 – (in serbo-croatian (roman)) – mf#mf-3551 seemp – us UMI ProQuest [077]
Hryhoriiv, Nykyfor IA *see*
– Nasha pozitsiia – samostiina
– Osnovy natsioznannia
Hseih chin (ccs) – March in harmony – Cheng-tu. mar 1943-1954 [gaps] [mf ed 198?] – (= ser Chinese christian serials coll) – 1r – 1 – (incl: nanking church council bulletin) – mf0306 – us ATLA [240]
Hsi, Chin *see* Tu t'u pieh chuan
Hsi chu / Ch'ien Kung-hsia & Shih, Ying pien – Shang-hai: Ch'i ming shu chu, Min kuo 28 [1939] – (= ser P-k&k period) – us CRL [820]
Hsi chu chiang tso / Ma, Yen-hsiang – Shang-hai: Hsien tai shu chu, 1932 – (= ser P-k&k period) – us CRL [820]
Hsi chu ch'uang tso chiang hua / Ch'en, Pai-ch'en – Shang-hai: Shang-hai tsa chih kung ssu, Min kuo 29 [1940] – (= ser P-k&k period) – us CRL [790]
Hsi chu ch'un ch'iu / Hsia, Yen et al – Ch'ung-ch'ing: Wei lin ch'u pan she, 1943 – (= ser P-k&k period) – us CRL [820]
Hsi chu lun / Chang, Min – Ch'ang-sha: Shang wu yin shu kuan, Min kuo 29 [1940] – (= ser P-k&k period) – us CRL [820]
Hsi chu lun / Yu, Ta-fu – Shang-hai: Shang wu yin shu kuan, Min kuo 22 [1933] – (= ser P-k&k period) – us CRL [820]
Hsi chu yu chiao yu / Ch'en, Ming-chung – Ch'ang-sha: Shang wu yin shu kuan, Min kuo 25 [1936] – (= ser P-k&k period) – us CRL [790]
Hsi chua chi / Chang, Ya-chu – Shang-hai: Chung-kuo ying sheng she, 1933 – (= ser P-k&k period) – us CRL [810]
Hsi ch'uan chi / Yeh, Sheng-t'ao – Ch'ung-ch'ing, Wen kuang shu tien, Min kuo 34 [1945] – (= ser P-k&k period) – us CRL [840]
Hsi chun chan / Ch'en, Fei-mo – Ch'ung-ch'ing: Shang wu yin shu kuan, Min kuo 31 [1942] – (= ser P-k&k period) – us CRL [350]
Hsi hsien feng yun / Fan, Ch'ang-chiang – Shang-hai, Ta kung pao kuan, Min kuo 26 [1937] – (= ser P-k&k period) – us CRL [951]
Hsi hsien ti hsueh chan / Fan, Ch'ang-chiang – Shang-hai: Shang-hai tsa chih kung ssu, 1937 – (= ser P-k&k period) – us CRL [951]
Hsi hsing san chi / Pai, Lang – Ch'ang-sha: Shang wu yin shu kuan, Min kuo 30 [1941] – (= ser P-k&k period) – us CRL [915]
Hsi hsing shu chien / Cheng, Chen-to – Shang-hai: Shang wu yin shu kuan fa hsing, Min kuo 26 [1937] – (= ser P-k&k period) – us CRL [915]
Hsi lien / Lan, T'ien – Shang-hai: Hsin ti shu tien, Min kuo 29 [1940] – (= ser P-k&k period) – us CRL [480]
Hsi nan chiao t'ung yao lan – Ch'ung-ch'ing: Hsi nan tao pao she, 1939 – (= ser P-k&k period) – us CRL [380]
Hsi nan ching chi ti li kang yao / Chiang, Chun-chang – [Ch'ung-ch'ing]: Cheng chung shu chu, Min kuo 32 [1943] – (= ser P-k&k period) – us CRL [951]
Hsi nan ching chi tiao ch'a ho tso wei yuan *see* Ssu-ch'uan ching chi k'ao ch'a t'uan k'ao ch'a pao kao (nung lin)
Hsi nan hsing san chi / Weng, Ta-ts'ao – Ch'ung-ch'ing: Kuang t'ing ch'u pan she, 1943 – (= ser P-k&k period) – us CRL [840]
Hsi nan kung lu yeh wu kai k'uang – [China: Chiao t'ung pu kung lu tsung chu Hsi nan kung lu kung wu chu], 1944 – (= ser P-k&k period) – us CRL [625]
Hsi nan lien ho ta hsueh (K'un-ming shih, China) Hsi nan lien ta ch'u hsi fu k'an *see* Lien ta pa nien
Hsi pei chien she lun / Hsu, Hsu – Ch'ung-ch'ing: Chung hua shu chu, min kuo33 [1944] – (= ser P-k&k period) – us CRL [339]
Hsi pei chien she lun / Wang, Chao-sheng – [China]: Ch'ing nien ch'u pan she, Min kuo 32 [1943] – (= ser P-k&k period) – us CRL [951]
Hsi pei chin ying / Fan, Ch'ang-chiang, 1907- – Shang-hai: Chung-kuo chih shih she, 1937 – (= ser P-k&k period) – us CRL [951]
Hsi pei hsien / Fan, Ch'ang-chiang – Han-k'ou: Hsing sheng t'u shu kung ssu, 1937 – (= ser P-k&k period) – us CRL [951]
Hsi pei wen t'i / Chang, Ch'i-yun et al – Kuei-lin: K'o hsueh shu tien, Min kuo 32 [1943] – (= ser P-k&k period) – us CRL [915]
Hsi wang yueh k'an (ccs26) = Christian hope – Cheng-tu. v6-v9 n3. 1929-mar 1932 [gaps] [mf ed 198?] – (= ser Chinese christian serials coll 26) – 1r – 1 – (began in 1924) – mf0604 – us ATLA [240]
Hsi yu chi (ccm73) = My experience in prison / Chao, Tzu-ch'en – Shanghai, 1948 [mf ed 198?] – (= ser Ccm 73) – 1 – mf#1984-b500 – us ATLA [951]
Hsia, Cheng-nung *see* Hsien chieh tuan ti chung-kuo ssu hsiang yun tung
Hsia, Ching-kuan *see* Tz'u tiao su yuan
Hsia hsi (ccs) = News – Shanghai. dec 1930-jun 1950 [gaps] [mf ed 198?] – (= ser Chinese christian serials coll) – 1r – 1 – (began in 1928?) – mf0305 – us ATLA [240]
Hsia, Hsia *see* Kua fu yuan
Hsia hsiang chi / Hsu, Chuan-p'eng – Cha'ang-sha: Shang wu yin shu kuan, Min kuo 29 [1940] – (= ser P-k&k period) – us CRL [480]
Hsia, K'ai-ju *see*
– Hai kang yu k'ai kang chi hua
– Shih yeh chi hua t'ieh lu p'ein
Hsia, Mien-tsun *see*
– Shih nie
– Wen chang tso fa
– Wen hsin
– Yueh tu yu hsieh tso
Hsia, Mien-tsun' *see* Wen chang chiang hua
Hsia, Pang-chun *see* Jen shih kuan li chih li lun yu shih chi
Hsia, Tao-tai *see* Guide to selected legal sources of mainland china
Hsia t'ien / Lu, I-shih – Shang-hai: Shih ling t'u she, 1945 – (= ser P-k&k period) – us CRL [810]
Hsia wan-ch'un : [ssu mu li shih ko chu] / Chang, Kuang-chung – Ch'ung-ch'ing: Ch'ing nien ch'u pan she, Min kuo 31 [1942] – (= ser P-k&k period) – us CRL [780]
Hsia, Yen *see*
– Chin jih chih shang-hai
– Fu huo
– Hsin fang
– Li li ts'ao
– Pao shen kung
– Pien ku chi
– Shang-hai wu yen hsia
– Shui hsiang yin
– Tz'u shih tz'u ti chi
– Tzu yu hun
Hsia, Yen et al *see* Hsi chu ch'un ch'iu
Hsia, Yen teng *see* Tsen yang hsieh tso
Hsia, Yen-te *see* Wen i t'ung lun
Hsiang chang hsien sheng / Wang, Jen-shu – Shang-hai: Liang yu t'u shu kung ssu, 1936 – (= ser P-k&k period) – us CRL [830]
Hsiang cheng i nien – [Hu-nan: Hu-nan sheng cheng fu kung pao shih, 1940] – (= ser P-k&k period) – us CRL [951]
Hsiang chiao kung yeh pao kao shu – Shang-hai: Ch'uan kuo ching chi wei yuan hui, Min kuo 24 [1935] – (= ser P-k&k period) – us CRL [338]
Hsiang ch'ih chieh tuan chung ti hsing shih yu jen wu / Chin pu ch'u pan she – [China] Chin pu ch'u pan she, Min kuo 29 [1940] – (= ser P-k&k period) – us CRL [951]
Hsiang feng yu shih feng / Feng, Hsueh-feng – Ch'ung-ch'ing: Tso chia shu wu, 1944 – (= ser P-k&k period) – us CRL [840]
Hsiang hsi / Shen, Ts'ung-wen – Ch'ang-sha: Shang wu yin shu kuan, Min kuo 28 [1939] – (= ser P-k&k period) – us CRL [951]
Hsiang hsia hsien sheng / Huang, Mu – Shang-hai: Hsin ti shu tien, 1941 – (= ser P-k&k period) – us CRL [480]
Hsiang jih k'uei / Yuan, Shui-p'ai – Ch'ung-ch'ing: Mei hsueh ch'u pan she, Min kuo 32 [1943] – (= ser P-k&k period) – us CRL [810]
Hsiang lei – Shang-hai: T'ien ma shu tien, Min kuo 23 [1934] – (= ser P-k&k period) – us CRL [480]
Hsiang, M *see* It kie bwee / siauw eng hiong / maoe terbang tida bersajap
Hsiang nung chiao yu – Shan-tung: Hsiang ts'un chien she yen chiu yuan ch'u pan ku, Min kuo 24 [1935] – (= ser P-k&k period) – us CRL [370]
Hsiang, P'ei-liang *see* Ta shih tai ti ch'a ch'u, hsiang p'ei-liang chu
Hsiang t'ai yang / Ai, Ch'ing – Hsiang-kang: Hai yen shu tien, Min kuo 29 [1940] – (= ser P-k&k period) – us CRL [810]
Hsiang ting tso wei chi chan shu t'ung ts'ai fa chiang i lu / Nihon Rikugun Daigakko Kenkyu kai – Nan-ching: Chun yung t'u shu she, Min kuo 23 [1934] – (= ser P-k&k period) – us CRL [355]
Hsiang ts'un chiao yu / Kann, Yu-yuan – Shang-hai: Chung-hua shu chu, Min kuo 25 [1936] – (= ser P-k&k period) – us CRL [370]
Hsiang ts'un chiao yu ching yen t'an / Chang, Tsung-lin – Shang-hai: Shih chieh shu chu, Min kuo 21 [1932] – (= ser P-k&k period) – us CRL [370]
Hsiang ts'un chiao yu kai lun / Lung, Fa-chia – Shang-hai: Shang wu yin shu kuan, Min kuo 26 [1937] – (= ser P-k&k period) – us CRL [370]
Hsiang ts'un chiao yu shih tao / Li, Hsiao-nung & Li, Po-t'ang ho – Shang-hai: Li ming shu chu, Min kuo 24 [1935] – (= ser P-k&k period) – us CRL [370]
Hsiang ts'un chiao yu ts'ung chi / Chao, Shu-yu & Fang Yu-yen – Shang-hai: Erh t'ung shu chu, Min kuo 22 [1933] – (= ser P-k&k period) – us CRL [370]
Hsiang ts'un chien she li lun : i ming chung-kuo min tsu chih ch'ien t'u / Liang, Shu-ming – Ch'ung-ch'ing: Hsiang ts'un shu tien, min kuo 28 [1939] – (= ser P-k&k period) – us CRL [307]
Hsiang ts'un chien she shih yen ti erh chi : hsiang ts'un kung tso t'ao lun hui min kuo erh shih san nien shih yueh ting-hsien chi hui ko fang kung tso pao kao hui pien / Chang, Yuan-shan – Shang-hai: Chung-hua shu chu, [Min kuo 24 [1935]] – (= ser P-k&k period) – us CRL [951]
Hsiang ts'un chien she ta i / Liang, Shu-ming – Tsou-p'ing: Hsiang ts'un shu tien, Min kuo 25 [1936] – (= ser P-k&k period) – us CRL [630]
Hsiang tsun chuan tao kung tso ching yen tan (ccm350) = Christian work in rural china: a symposium of practical experiences / ed by Yu, Mu-jen – Shanghai. 2v. 1949-50 [mf ed 198?] – (= ser Ccm 350) – 1 – mf#1984-b500 – us ATLA [240]
Hsiang ts'un hsiao hsueh chiao hsueh fa / Li, Hsiao-nung – Shang-hai: Li ming shu chu, 1934 – (= ser P-k&k period) – us CRL [370]
Hsiang ts'un hsiao hsueh chiao shih hsu chih / T'ang Wen-ts'ui – Shang-hai: Erh t'ung shu chu, Min kuo 23 [1934] – (= ser P-k&k period) – us CRL [370]
Hsiang ts'un hsiao hsueh chiao ts'ai yen chiu / Chang, Tsung-lin – Shang-hai: Li ming shu chu, 1934 – (= ser P-k&k period) – us CRL [370]
Hsiang ts'un hsiao hsueh hsing cheng / Kuo, Jen-ch'uan – Shang-hai: Li ming shu chu, 1934 – (= ser P-k&k period) – us CRL [370]
Hsiang ts'un hsiao hsueh lao tso chiao yu / Fang, Ta-tsai – Shang-hai: Li ming shu chu, Min kuo 23 [1934] – (= ser P-k&k period) – us CRL [370]
Hsiang ts'un hsiao hsueh shih chi wen t'i / Chin, Ting-i – Shang-hai: Li ming shu chu, 1934 – (= ser P-k&k period) – us CRL [370]
Hsiang ts'un li pai (ccm110) = Rural worship services / Chu, Ching-i & Chang, Pei-ying – Shanghai, 1940 [mf ed 198?] – (= ser Ccm 110) – 1 – mf#1984-b500 – us ATLA [240]
Hsiang ts'un min chung chiao yu / Kan, Yu-yuan – Shang-hai: Shang wu yin shu kuan, Min kuo 23 [1934] – (= ser P-k&k period) – us CRL [370]
Hsiang ts'un she hui hsueh kang yao / T'ung, Jun-chih – [Ch'ung-ch'ing]: Cheng chung shu chu, [1941] – (= ser P-k&k period) – us CRL [301]
Hsiang ya chieh chih / Lu, Yin – Ch'ang-sha: Shang wu yin shu kuan, 1938 – (= ser P-k&k period) – us CRL [830]
Hsiang, Yu *see* Ko chu chi
Hsiang-kang chi-tu chiao hui shih (ccm228) / Liu, Yueh-sheng – Hong Kong, 1941 [mf ed 198?] – (= ser Ccm 228) – 1 – mf#1984-b500 – us ATLA [240]
Hsiang-kang hua tzu kung ch'ang tiao ch'a lu – Hsiang-kang: [Kung shang jih pao], Min kuo 23 [1934] – (= ser P-k&k period) – us CRL [030]
Hsiang-pei chih chan / Ch'en, Ho-k'un – [Sl]: Ch'ing nien ch'u pan she, 1939 – (= ser P-k&k period) – us CRL [951]
Hsiang-tsun chiao-hui (ccs) = Rural church – Cheng-tu. n1-9. 1940-47 [complete] [mf ed 198?] – (= ser Chinese christian serials coll) – 1 – mf0296j – us ATLA [240]
Hsiao, Ai *see*
– P'ing hsu chi
– Lo yeh chi
Hsiao ch'ang chi / Chou, Mu-chai – [China]: Pei she, 1940 – (= ser P-k&k period) – us CRL [840]
Hsiao ch'eng ku shih : [wu mu chu] / Chang, Chun-hsiang – Shang-hai: Wen hua sheng huo ch'u pan she, Min kuo 30 [1941] – (= ser P-k&k period) – us CRL [820]
Hsiao, Ch'eng-shen *see* Shih tao cheng ku
Hsiao chiao niang / Chang, I-p'ing – Shang-hai: Li ming shu chu, 1933 – (= ser P-k&k period) – us CRL [480]
Hsiao, Ch'ien *see*
– Hsiao shu yeh
– Hui chin
– Jih lo
– Li hsia chi
– Li tzu
Hsiao ching *see* The book of filial duty
Hsiao ching tong kao (ccm295) / Tsan, Ju-kun – Shanghai, 1937 [mf ed 198?] – (= ser Ccm 295) – 1 – mf#1984-b500 – us ATLA [480]
Hsiao, Cho-lin *see* Ch'i wang t'ien heng
Hsiao, Chueh-t'ien *see* Pa nien chan cheng shih chi
Hsiao, Chuen *see* Chiang shang
Hsiao, Chun *see*
– Lu yeh ti ku shih
– Ti san tai (ti i pu)
– Ts'e mien
– Ts'e mien ti i pu, wo liu tsai lin-fen
– Yang
Hsiao, Chung-tao *see* Shao nu shu chien
Hsiao fei ho tso / Wang, Hsiao-wen – Shang-hai: Shang wu yin shu kuan, Min kuo 22 [1933] – (= ser P-k&k period) – us CRL [334]
Hsiao feng t'u hua / Wen, hsueh chi lin she – Fu-chien Nan-p'ing: Hua chin hsin ts'un: Nan feng she wu, Min kuo 34 [1945] – (= ser P-k&k period) – us CRL [840]
Hsiao hsiao lu : cha chi hsiao shuo / Chou, Meng-tieh – Shang-hai: Ta ta t'u shu kung ying she, Min kuo 24 [1935] – (= ser P-k&k period) – us CRL [480]

Hsiao, Hsiao-jung *see*
– Erh t'ung hsin li hsueh chi ch'i ying yung
– Hsiao hsiao-jung hsiu ting mo pa liang piao
– Tsen yang ling tao

Hsiao hsiao-jung hsiu ting mo pa liang piao / Hsiao, Hsiao-jung – [China]: Shang wu yin shu kuan, [Min kuo 25 [1936]] – (= ser P-k&k period) – us CRL [000]

Hsiao hsueh chiao yu ti li lun yu shih chi – Shang-hai: Chung-hua shu chu, Min kuo 25 [1936] – (= ser P-k&k period) – us CRL [370]

Hsiao hsueh chih yeh chih tao / P'an, Wen-an – Shang-hai: Chung-hua shu chu, Min kuo 24 [1935] – (= ser P-k&k period) – us CRL [370]

Hsiao hsueh hsing cheng / Li, Ch'ing-sung – Shang-hai: Chung-hua shu chu, Min kuo 26 [1937] – (= ser P-k&k period) – us CRL [370]

Hsiao hsueh hsing cheng / Tseng, I-fu – Shang-hai: Li ming shu chu, Min kuo 24 [1935] – (= ser P-k&k period) – us CRL [370]

Hsiao hsueh hsing cheng / Yao, Wei-chun – Shang-hai: Yu chih shih fan hsueh hsiao ts'ung shu she, Min kuo 25 [1936] – (= ser P-k&k period) – us CRL [650]

Hsiao hsueh hsing cheng chi tsu chih / Jui, Chia-jui – Shang-hai: Shang wu yin shu kuan, Min kuo 23 [1934] – (= ser P-k&k period) – us CRL [370]

Hsiao hsueh hsing cheng ta kang / Tsou, Hsiang – Shang-hai: Shang wu yin shu kuan, Min kuo 25 [1936] – (= ser P-k&k period) – us CRL [370]

Hsiao hsueh sheng ch'an chiao yu ti li lun ho shih chi / Wu, Shou-ch'ien – Shang-hai: Chung-hua shu chu, 1934 – (= ser P-k&k period) – us CRL [370]

Hsiao, Hung *see* Niu ch'e shang

Hsiao jen wu k'uang hsiang ch'u / Shen, Fu – Ch'ung-ch'ing: Hsin sheng t'u shu wen chu kung ssu, Min kuo 34 [1945] – (= ser P-k&k period) – us CRL [820]

Hsiao ko erh lia / Ling, Shu-hua – Hsiang-kang: Liang yu t'u shu kung ssu, [1945] – (= ser P-k&k period) – us CRL [830]

Hsiao ko erh lia / Ling, Shu-hua – Shang-hai: Shang-hai liang yu t'u shu kung ssu, 1935 – (= ser P-k&k period) – us CRL [830]

Hsiao, K'o-mu *see* Tsou-p'ing ti ts'un hsueh hsiang hsueh

Hsiao lan hua / Ti-k'o – Ch'eng-tu: Mang yuan ch'u pan she, 1942 – (= ser P-k&k period) – us CRL [810]

Hsiao mei / Chao, Ching-shen – Shang-hai: Pei hsin shu chu, 1933 – (= ser P-k&k period) – us CRL [840]

Hsiao, Ming-hsin *see*
– Chi kuan kuan li
– Hsin hsien cheng chih kuan li
– T'u ti cheng ts'e shu yao

Hsiao niao chi / Tseng, Chin-k'o – Shang-hai: Hsin shih tai shu chu, Min kuo 22 [1933] – (= ser P-k&k period) – us CRL [840]

Hsiao p'in wen / Ch'ein, Kung-hsia & Shih Ying – Shang-hai: Ch'i ming shu chu, Min kuo 27 [1938] – (= ser P-k&k period) – us CRL [840]

Hsiao p'in wen / Ch'ein, Kung-hsia & Shih Ying – Shang-hai: Ch'i ming shu chu, Min kuo 27 [1938] – (= ser P-k&k period) – us CRL [840]

Hsiao p'in wen ho man hua / Ch'en, Wang-tao – Shang-hai: Sheng huo shu tien, Min kuo 24 [1935] – (= ser P-k&k period) – us CRL [840]

Hsiao p'in wen hsuan – Shang-hai: Shen pao yueh k'an she, Min kuo 24 [1935] – (= ser P-k&k period) – us CRL [840]

Hsiao p'in wen hsuan / T'ao, Ch'iu-ying – Shang-hai: Pei hsin shu chu, 1934 – (= ser P-k&k period) – us CRL [480]

Hsiao p'in wen yen chiu / Feng, San-mei – Shang-hai: Shih chieh shu chu, 1933 (1935 printing) – (= ser P-k&k period) – us CRL [840]

Hsiao p'in wen yen chiu – Shang-hai: Hsin Chung-kuo shu chu, 1932 – (= ser P-k&k period) – us CRL [840]

Hsiao pi-te / Chang, T'ien-i – Shang-hai: Hu feng shu chu, 1931 – (= ser P-k&k period) – us CRL [480]

Hsiao shu yeh / Hsiao, Ch'ien – Shang-hai: Shang wu yin shu kuan, Min kuo 26 [1937] – (= ser P-k&k period) – us CRL [480]

Hsiao shuo 1 / Mao, Tun et al – Shang-hai: Ch'i ming shu chu, Min kuo 35 [1946] – (= ser P-k&k period) – us CRL [480]

Hsiao shuo hsi ch'u hsin k'ao : erh chuan / Chao, Ching-shen – Shang-hai: Shih chieh shu chu, Min kuo 32 [1943] – (= ser P-k&k period) – us CRL [830]

Hsiao shuo hsien hua / Chao, Ching-shen – Shang-hai: Pei hsin shu chu, Min kuo 26 [1937] – (= ser P-k&k period) – us CRL [830]

Hsiao shuo hsien t'an / A-ying – Shang-hai: Liang yu t'u shu yin shua kung ssu, 1936 – (= ser P-k&k period) – us CRL [830]

Hsiao shuo hsuan / Lin, Hui-yin – Shang-hai: Ta kung pao kuan, 1936 – (= ser P-k&k period) – us CRL [830]

Hsiao shuo hua / Hsieh, T'ao – Shang-hai: Chung-hua shu chu, Min kuo 21 [1932] – (= ser P-k&k period) – us CRL [830]

Hsiao shuo shih tsen yang hsieh ch'eng ti / Yao, Hsueh-yin – Ch'ung-ch'ing: Shang wu yin shu kuan, Min kuo 32 [1943] – (= ser P-k&k period) – us CRL [830]

Hsiao shuo su / Pao, T'ien-hsiao et al – Shang-hai: Wen yeh shu chu, 1937 – (= ser P-k&k period) – us CRL [480]

Hsiao shuo tso fa chiang hua / Shih, Wei – Shang-hai: Kuang min shu chu, 1941 – (= ser P-k&k period) – us CRL [480]

Hsiao shuo wen hsuan – Shang-hai: Ch'i chih shu chu, Min kuo 24 [1935] – (= ser P-k&k period) – us CRL [480]

Hsiao ts'ao / Ho, I – Shang-hai: Ts'ao ya she, [Min kuo 25 [1936]] – (= ser P-k&k period) – us CRL [830]

Hsiao wen chang / Shih-heng – Shang-hai: Liang yu t'u shu yin shua kung ssu, 1934 – (= ser P-k&k period) – us CRL [840]

Hsiao, Wen-che *see*
– Hsien cheng chih tu yen chiu
– Hsing cheng hsiao lu yen chiu

Hsiao wu chi / Wu, Hsi-ju – Shang-hai: Wen hua sheng huo ch'u pan she, Min kuo 25 [1936] – (= ser P-k&k period) – us CRL [830]

Hsiao, Yang *see* San nien lai ying mei su yuan tung cheng ts'e ti t'ou shih

Hsiao yao ko sui pi chi / T'ien-lu – Shang-hai: Nu tzu shu tien, 1932 – (= ser P-k&k period) – us CRL [840]

Hsiao yao yeh t'an hsuan / T'ien-lu – Shang-hai: Kuang i shu chu, 1933 – (= ser P-k&k period) – us CRL [840]

Hsiao yeh ch'u / Tu mu chu ch'uang tso ts'ung k'an she pien – Shang-hai: Chu i ch'u pan she, 1940 (1941 printing) – (= ser P-k&k period) – us CRL [840]

Hsiao yen lun ti erh chi / T'ao-fen – Shang-hai: Sheng huo shu tien, Min kuo 22 [1933] – (= ser P-k&k period) – us CRL [840]

Hsiao yen lun ti i chi / T'ao-fen – Shang-hai: Sheng huo shu tien, Min kuo 22 [1933] – (= ser P-k&k period) – us CRL [840]

Hsiao yen lun ti san chi / T'ao-fen – Shang-hai: Sheng huo shu tien, Min kuo 23 [1934] – (= ser P-k&k period) – us CRL [840]

Hsiao-chuang chih i yeh / Fang, Yu-yen – Shang-hai: Shang-hai erh t'ung shu chu, Min kuo 23 [1934] – (= ser P-k&k period) – us CRL [370]

Hsiao-chuang i sui / Yang, Hsiao-ch'un – Shang-hai: Erh t'ung shu chu, Min kuo 24 [1935] – (= ser P-k&k period) – us CRL [370]

Hsiao-Hsiao-Sheng *see* Golden lotus

Hsiao-p'o ti sheng jih / Lao, She – Shang-hai: Sheng huo shu tien, Min kuo 26 [1937] – (= ser P-k&k period) – us CRL [830]

Hsi-ch'ang chih hsing / Lu, Ju-lin – Ch'ung-ch'ing: Shang wu yin shu kuan, Min kuo 33 [1944] – (= ser P-k&k period) – us CRL [915]

Hsi-ching shih kung yeh tiao ch'a / Shan-hsi sheng yin hang (Sian, China) Ching chi yen chiu shih – [Hsi-an: Shan-hsi sheng yin hang ching chi yen chiu shih], Min kuo 29 [1940] – (= ser P-k&k period) – us CRL [338]

Hsieh chi hua jen chieh shou chi-tu chiao (ccm165) = How to commend christianity to the chinese / Hsieh, En-kuang – Shanghai, 1917 [mf ed 198?] – (= ser Ccm 165) – 1 – mf#1984-b500 – us ATLA [240]

Hsieh, Chih-Mou *see* Leisure attitudes, motivation, participation, and satisfaction

Hsieh chin *see* Chung hua kuei chu [ccs]

Hsieh, Ching-sheng *see* Wo wei shen me tso chi-tu t'u (ccm164)

Hsieh, En-kuang *see* Hsieh chi hua jen chieh shou chi-tu chiao (ccm165)

Hsieh, Fu-ya *see*
– Chi-tu chiao ch'ing nien hui yuan li
– Chi-tu chiao yu chung-kuo
– Chi-tu chiao yu hsien tai ssu hsiang [ccm166]
– Chung-kuo san-chiao ti kung t'ung pen chih [ccm169]
– Hsin shih-tai te shin yang
– Tsung chiao che hsueh

Hsieh, Han-fu *see* Lien ho chan hsien

Hsieh hou / Nieh, Kan-nu – Shang-hai: T'ien ma shu tien, 1935 – (= ser P-k&k period) – us CRL [830]

Hsieh, Hung-lai *see*
– Cheng tao chi [ccm172]
– Chi-tu chiao yu k'o hsueh [ccm173]
– Ming mu i hui
– Sheng tao kuan k'uei

Hsieh, Jen-chao *see* Erh tz'u shih chieh ta chan chung chih mei-kuo ti wai chiao cheng ts'e

Hsieh kei ts'ing nien ti chi-tu tu (ccm315) = Talks to young christians / Wang, Ming-tao – Peiping, 1948 [mf ed 198?] – (= ser Ccm 315) – 1 – mf#1984-b500 – us ATLA [305]

Hsieh, Kuo-chen *see* Ming ch'ing chih chi tang she yun tung k'ao

Hsieh lu-yin hsien sheng chuan lueh (ccm176) = The life of h l zia / Hu, I-ku – 1st ed. Shanghai, 1917 [mf ed 198?] – (= ser Ccm 176) – 1 – mf#1984-b500 – us ATLA [920]

Hsieh, Nan-kuang *see* Jih-pen chu i ti mo lo

Hsieh ping-hsin tai piao tso / Ping-hsin – Shang-hai: San t'ung shu chu, 1941 – (= ser P-k&k period) – us CRL [480]

Hsieh, Ping-ying *see*
– Nu tso chia tzu chuan hsuan chi
– Ping-ying jih chi
– Ping-ying k'ang chan wen hsuan chi
– Tsai jih-pen yu chung
– Ts'ung chun jih chi

Hsieh p'o = Acclivity / Man-ni – Shang-hai: Hsin wen hua shu she, Min kuo 23 [1934] – (= ser P-k&k period) – us CRL [810]

Hsieh, Shou-ling *see* Sheng li te sheng huo (ccm177)

Hsieh, Sung-kao *see*
– Chiao yu chien i tu pen [ccm186]
– Chin tai k'o hsueh chia te tsung chiao kuan [ccm180]
– Chu chiao te yen chiu [ccm183]
– Hsin yueh jen wu
– Hsueh sheng men ti ku shih
– Mu-ti sheng ping
– Ping min ku shih
– A short study of education in the christian home
– Ssu-pu-chen sheng ping
– Tsung chiao chiao yu ts'ung shu
– Wang hsien sheng yu wang shih mu

Hsieh, T'ao *see* Hsiao shuo hua

Hsieh tso ching yen t'an / Su, Hsueh-lin et al – Shang-hai: Chung hsueh sheng shu chu, [1939] – (= ser P-k&k period) – us CRL [480]

Hsieh: tuan p'ien hsiao shuo chi / Mei-niang – [Pei-ching]: Wu te pao she, 1944 – (= ser P-k&k period) – us CRL [480]

Hsieh, Tung-p'ing *see* Hsin sheng lun

Hsieh, Wu-liang *see* Shih hsueh chih nan

Hsieh, Yung-yen *see* Chan huo jan shao ti mien tien

Hsien cheng chi ch'u chih shih / Ch'en, Pei-ou – Ch'ung-ch'ing: Kuo hsun shu tien, 1944 – (= ser P-k&k period) – us CRL [951]

Hsien cheng chi ch'u chih shih / Ch'en, Pei-ou – Ch'ung-ch'ing: Kuo hsun shu tien, Min kuo 33 [1944] – (= ser P-k&k period) – us CRL [951]

Hsien cheng chien she / K'ung, Ch'ung – Shang-hai: Chung-hua shu chu, min kuo 26 [1937] – (= ser P-k&k period) – us CRL [350]

Hsien cheng chih tao / Chang, Yuan-jo – [China: Cheng chung shu chu], Min kuo 34 [1945] – (= ser P-k&k period) – us CRL [323]

Hsien cheng chih tu yen chiu / Hsiao, Wen-che – Ch'ung-ch'ing: Tu li ch'u pan she, Min kuo 31 [1942] – (= ser P-k&k period) – us CRL [350]

Hsien cheng jen yuan hsun lien / Fu-chien sheng hsien cheng jen yuan hsun lien so – [China]: Fu-chien sheng hsien cheng jen yuan hsun lien so, Min kuo 28 [1939] – (= ser P-k&k period) – us CRL [350]

Hsien cheng kung tso ch'eng hsu piao chieh / Li, Ch'u-k'uang – [China]: Kuo min ch'u pan she, Min kuo 32 [1943] – (= ser P-k&k period) – us CRL [350]

Hsien cheng wen t'i ts'an k'ao tzu liao – Liu-chou: Huang t'u ch'u pan she, 1944 – (= ser P-k&k period) – us CRL [951]

Hsien cheng wen t'i tu pen / Pa, Jen – [Hsiang-kang]: Wu ming ch'u pan she, [Min kuo 29 ie 1940] – (= ser P-k&k period) – us CRL [951]

Hsien cheng yu ching chi / Wu, Ch'i-yuan – Ch'ung-ch'ing: Cheng chung shu chu, 1944 – (= ser P-k&k period) – us CRL [323]

Hsien cheng yu ti fang tzu chih / Chung yang hsuan ch'uan pu pien – Ch'ung-ch'ing: Chung-kuo wen hua fu wu she, Min kuo 28 [1939] – (= ser P-k&k period) – us CRL [323]

Hsien cheng yun tung lun wen hsuan chi / T'ao-fen – Ch'ung-ch'ing: Sheng huo shu tien, Min kuo 29 [1940] – (= ser P-k&k period) – us CRL [951]

Hsien cheng yun tung lun wen hsuan chi / T'ao-fen teng – Ch'ung-ch'ing, Sheng huo shu tien, Min kuo 29 [1940] – (= ser P-k&k period) – us CRL [951]

Hsien cheng yun tung ts'an k'ao ts'ai liao / Ch'uan min k'ang chan she pien – Ch'ung-ch'ing: Sheng huo shu tien, Min kuo 29 [1940] – (= ser P-k&k period) – us CRL [951]

Hsien cheng yun tung ts'an k'ao tzu liao ti 2 chi / Ch'uan min k'ang chan she pien – Ch'ung-ch'ing: Sheng huo shu tien, 1940 – (= ser P-k&k period) – us CRL [951]

Hsien chieh tuan ti chung-kuo chin jung / Wei, Yu-fei – Shang-hai: Ch'ien yeh yueh pao she, Min kuo 25 [1936] – (= ser P-k&k period) – us CRL [332]

Hsien chieh tuan ti chung-kuo ssu hsiang yun tung / Hsia, Cheng-nung – Shang-hai: I pan shu tien, Min kuo 26 [1937] – (= ser P-k&k period) – us CRL [180]

Hsien ch'in hsueh shuo shu lin / Kuo, Mo-jo – Fu-chien Yung-an: Tung nan ch'u pan she, 1945 – (= ser P-k&k period) – us CRL [180]

Hsien ch'in wen hsueh ta kang – Shang-hai: Hua t'ung shu chu, Min kuo 22 [1933] – (= ser P-k&k period) – us CRL [480]

Hsien ching san yueh chi / Chiang, Kung-ku – [China: Chiang Kung-ku, 1938] – (= ser P-k&k period) – us CRL [951]

Hsien, Ch'un *see* Fei hua ch'u

Hsien fa lun wen hsuan k'an – [China]: Hsien fa ts'ao an ch'i ts'ao wei yuan hui ti i, erh k'o, Min kuo 22 [1933] – (= ser P-k&k period) – us CRL [951]

Hsien hsing pao chia chih tu / Li, Tsung-huang – Shang-hai: Chung hua shu chu, Min kuo 32 [1943] – (= ser P-k&k period) – us CRL [350]

Hsien hsing shang shui / Li, Ch'uan-shih – Shang-hai: Shang wu yin shu kuan, Min kuo 22 [1933] – (= ser P-k&k period) – us CRL [336]

Hsien hsing ti fang tzu chih fa kuei shih i / Wang, Chun-an – Shang-hai: Shih chieh shu chu, Min kuo 24 [1935] – (= ser P-k&k period) – us CRL [350]

Hsien kei hsiang ts'un ti shih / Ai, Ch'ing – K'un-ming: Pei men ch'u pan she, 1945 (1947 printing) – (= ser P-k&k period) – us CRL [810]

Hsien ko chi ho tso she chang ch'eng chun tse – [China]: She hui pu ho tso shih yeh kuan li chu, Min kuo 30 [1941] – (= ser P-k&k period) – us CRL [630]

Hsien ko chi min i chi kuan / Ch'en, Nien-chung & P'an, Kung-chan – [China]: Cheng chung shu chu, Min kuo 33 [1944] – (= ser P-k&k period) – us CRL [350]

Hsien ko chi tsu chih chung chi hsiang chung yao wen t'i : chung yang hsun lien t'uan tang cheng hsun lien pan chiang yen lu / Cheng, Chen-yu – [China: Chung yang hsun lien tuan tang cheng hsun lien pan], Min kuo 31 [1942] – (= ser P-k&k period) – us CRL [350]

Hsien ko chi tsu chih kang yao chi ti fang tzu chih ts'an k'ao ts'ai liao / China – [China]: Chung yang hsun lien t'uan, 1940 – (= ser P-k&k period) – us CRL [350]

Hsien ko chi tsu chih kang yao i / Li, Tsung-huang – [Ch'ung-ch'ing]: Cheng chung shu chu, Min kuo 30 [1941] – (= ser P-k&k period) – us CRL [350]

Hsien shih yin hang shou ts'e – Kuei-lin: Kuang-hsi yin hang tsung hang, Min kuo 33 [1944] – (= ser P-k&k period) – us CRL [332]

Hsien tai chan cheng lun / Chang, Chih-ho – Shang-hai: Hsin k'en shu tien, 1936 – (= ser P-k&k period) – us CRL [303]

Hsien tai chan cheng lun – [China]: Kuo chi shih shih yen chiu hui, Min kuo 27 [1938] – (= ser P-k&k period) – us CRL [303]

Hsien tai che hsueh chih k'o hsueh chi ch'u / Fu, T'ung-hsien – Shang-hai: Shang wu yin shu kuan, 1935 – (= ser P-k&k period) – us CRL [180]

Hsien tai cheng fu chih li lun yu shih chi / Finer, Herman – Shang-hai: Shang wu yin shu kuan, Min kuo 26 [1937] – (= ser P-k&k period) – us CRL [951]

Hsien tai ch'uang tso hsiao shuo hsuan / Yao, Nai-lin – Shang-hai: Chung yang shu chu, Min kuo 24 [1935] – (= ser P-k&k period) – us CRL [480]

Hsien tai chun shih kung ch'eng hsueh / Chang, Chun – [China]: Hsi-k'ang chien she hsueh hui, Min kuo 28 [1939] – (= ser P-k&k period) – us CRL [355]

Hsien tai chung-kuo chi ch'i chiao yu, i ming, chung-kuo hsin chiao yu pei ching : shang hsia ts'e / Ku, Mei – Shang-hai: Chung-hua shu chu, Min kuo 25 [1936] – (= ser P-k&k period) – us CRL [370]

Hsien tai chung-kuo chih yeh chiao yu chih ch'an sheng yu ch'i fa chan / Chung, Tao-tsan – [China: sn] – (= ser P-k&k period) – us CRL [370]

Hsien tai chung-kuo hsi chu hsuan / Lo, Fang-chou – Shang-hai: Chung-kuo wen hua fu wu she, 1936 – (= ser P-k&k period) – us CRL [820]

Hsien tai chung-kuo nu tso chia / Ts'ao-yeh – Pei-p'ing: Jen wen shu tien, Min kuo 21 [1932] – (= ser P-k&k period) – us CRL [480]

Hsien tai chung-kuo nu tso chia ch'uang tso hsuan / Su, Fei – Shang-hai: Wen i shu chu, Min kuo 21 [1932] – (= ser P-k&k period) – us CRL [480]

Hsien tai chung-kuo shih hsuan / Sun, Wang & Ch'ang Jen-hsia – Ch'ung-ch'ing: Nan fang yin shu kuan, 1943 – (= ser P-k&k period) – us CRL [810]

Hsien tai chung-kuo shih yeh chih / Yang, Ta-chin – [Shanghai]: Shang wu yin shu kuan, min kuo 29 [1940] – (= ser P-k&k period) – us CRL [338]

Hsien tai chung-kuo tso chia lun ti erh chuan / Ho, Yu-po – Shang-hai: Kuang hua shu chu, 1932 – (= ser P-k&k period) – us CRL [480]
Hsien tai chung-kuo wen hsueh lun / Ch'ien, Hsing-ts'un – Shang-hai: Ho chung shu tien, 1933 – (= ser P-k&k period) – us CRL [480]
Hsien tai chung-kuo wen hsueh shih / Ch'ien, Chi-po – Shang-hai: Shih chieh shu chu, Min kuo 22 [1933] – (= ser P-k&k period) – us CRL [480]
Hsien tai fo hsueh (ccs63) = Modern buddhism – Pei-ching. n6. 1960 [complete] [mf ed 198?] – (= ser Chinese christian serials coll 63) – 1 – mf0296s – us ATLA [280]
Hsien tai fu nu shu hsin / Yeh, Chou – Shang-hai: Kuang ming shu chu, Min kuo 22 [1933] – (= ser P-k&k period) – us CRL [305]
Hsien tai hang cheng wen t'i / Wang, Kuang – Nan-ching: Cheng chung shu chu, Min kuo 26 [1937] – (= ser P-k&k period) – us CRL [380]
Hsien tai hsi chu hsuan / Tai, Chung-fang & Hu, Nan-hsiang – Shang-hai: Pei hsin shu chu, 1934 – (= ser P-k&k period) – us CRL [820]
Hsien tai hsiao shuo kuo yen lu / Hsu, Chieh – Fu-chien Yung-an: Li ta shu tien, 1945 – (= ser P-k&k period) – us CRL [830]
Hsien tai hsueh sheng ti ken pen wen t'i / Liu, Ch'un – Shang-hai: Hsien tai ch'u pan she, Min kuo 25 [1936] – (= ser P-k&k period) – us CRL [370]
Hsien tai huo pi yin hang chi shang yeh wen t'i / T'ang, Ch'ing-yung – Shang-hai: Shih chieh shu chu, 1935 – (= ser P-k&k period) – us CRL [332]
Hsien tai jih chi wen hsuan / Chun-sheng – Shang-hai: Fang ku shu tien, 1937 – (= ser P-k&k period) – us CRL [480]
Hsien tai kung shang ling hsiu ch'eng ming chi / Hsu, Ho-ch'un – Shang-hai: Hsin feng shu tien, Min kuo 30 [1941] – (= ser P-k&k period) – us CRL [332]
Hsien tai kuo chi fa wen t'i / Chou, Keng-sheng – Shang-hai: Shang wu yin shu kuan, Min kuo 21 [1932] – (= ser P-k&k period) – us CRL [341]
Hsien tai lao tung wen t'i lun ts'ung ti i chi / Ch'en, Chen-lu – Shang-hai: Shu pao ho tso she, Min kuo 22 [1933] – (= ser P-k&k period) – us CRL [331]
Hsien tai ming chia sui pi ts'ung hsuan / Juan, Wu-ming – Shang-hai: Nan ch'iang shu chu, Min Kuo 22 [1933] – (= ser P-k&k period) – us CRL [480]
Hsien tai ming jen ch'eng kung chih fen hsi / Morgan, John Jacob Brooke et al – Shang-hai: Shang wu yin shu kuan, 1935 – (= ser P-k&k period) – us CRL [650]
Hsien tai mo fan wen hsuan / Ta-fu – Shang-hai: Hsi wang ch'u pan she, Min kuo 25 [1936] – (= ser P-k&k period) – us CRL [840]
Hsien tai nu tso chia hsiao p'in hsuan / Chun-sheng – Shang-hai: Fang ku shu tien, 1936 – (= ser P-k&k period) – us CRL [840]
Hsien tai nu tso chia hsiao shuo hsuan ti i chi / Ping-hsin – Feng-t'ien: Sheng ching shu tien, [1942] – (= ser P-k&k period) – us CRL [830]
Hsien tai nu tso chia shih ko hsuan / Chun-sheng – Shang-hai: Fang ku shu tien, Min kuo 25 [1936] – (= ser P-k&k period) – us CRL [810]
Hsien tai nu tso chia shu hsin hsuan / Chun-sheng – Shang-hai: Fang ku shu tien, 1936 – (= ser P-k&k period) – us CRL [860]
Hsien tai nu tso chia sui pi hsuan / Chun-sheng – Shang-hai: Fang ku shu tien, Min kuo 25 [1936] – (= ser P-k&k period) – us CRL [480]
Hsien tai p'u t'ung ch'ih tu ta ch'uan / Wang, Su-ju – Shang-hai: Shang wu yin shu kuan, Min kuo 26 [1937] – (= ser P-k&k period) – us CRL [860]
Hsien tai san wen chi ti i pen / Pa, Chin – Shang-hai: Ching chi shu tien, 1936 – (= ser P-k&k period) – us CRL [840]
Hsien tai shih hsuan / Chao, Ching-shen – Shang-hai: Pei hsin shu chu, Min kuo 23 [1934] – (= ser P-k&k period) – us CRL [810]
Hsien tai shu hsin tso fa / Sun, Hsi-chen – Shang-hai: Chung-kuo wen hua fu wu she, Min kuo 25 [1936] – (= ser P-k&k period) – us CRL [860]
Hsien tai ssu hsiang chung te chi-tu chiao (ccm200) = Christianity in the light of today / Hu, I-ku – 3rd ed. Shanghai, 1926 [mf ed 198?] – (= ser Ccm 200) – 1 – mf#1984-b500 – us ATLA [240]
Hsien tan wei ho tso she tsu chih fang an ta kang – [Ho-pei: Ho-pei sheng hsien cheng chien she yen chiu yuan], Min kuo 23 [1934] – (= ser P-k&k period) – us CRL [334]
Hsien tsai shih hsing ti so te shui / K'ung, Hsiang-hsi – [China: sn], Min kuo 25 [1936] – (= ser P-k&k period) – us CRL [336]
Hsien tzu chih fa ts'ao an, hsien tzu chih fa shih hsing fa ts'ao an, shih tzu chih fa ts'ao an, shih tzu chih fa shih hsing fa ts'ao an / China – Nan-ching: Li fa yuan fa chih tzu chih fa wei yuan hui, [1934] – (= ser P-k&k period) – us CRL [350]
Hsi-erh-lieh-so *see* K'o hsueh ti shih chieh wen hsueh kuan
Hsi-k'an *see*
– Pu yao wang le
– T'ung chu ti san chia jen
Hsi-k'ang she hui chih niao k'an / K'o, Hsiang-feng – Ch'ung-ch'ing: Cheng chung shu chu, Min kuo 29 [1940] – (= ser P-k&k period) – us CRL [360]
Hsi-k'ang tsung lan / Li, I-jen et al – [Sl]: Cheng chung shu chu, Min kuo 30 [1941] – (= ser P-k&k period) – us CRL [915]
Hsi-lin tu mu chu / Ting, Hsi-lin, 1893- – Shang-hai: Hsin yueh shu tien, 1931 – (= ser P-k&k period) – us CRL [820]
Hsi-ling ti huang hun : ming chia hsiao shuo chi / Chang, T'ien-i et al – Shang-hai: Liang yu t'u shu kung ssu, 1945 – (= ser P-k&k period) – us CRL [830]
Hsi-ma-la-ya shan shang hsueh : chu pen / Lo, Yung-p'ei – Ch'ang-sha: Shang wu yin shu kuan, Min kuo 29 [1940] – (= ser P-k&k period) – us CRL [820]
Hsin che hsueh lun chi / Ai, Ssu-ch'i – [Kuei-lin]: Tu che shu fang, Min kuo 28 [1939] – (= ser P-k&k period) – us CRL [180]
Hsin che hsueh, wei wu lun – Shang-hai: Hsia she, 1939 – (= ser P-k&k period) – us CRL [306]
Hsin ch'eng-tu / Chou, Chih-ying – Ch'eng-tu: Fu hsing shu chu, Min kuo 32 [1943] – (= ser P-k&k period) – us CRL [915]
Hsin chien she = New construction – Peiping. 1949-1963 (1) – mf#2612 – us UMI ProQuest [690]
Hsin ching tao yen (ccm221) = An introduction to the creeds / Lin, Pu-chi – [Peiping, China] 1933 [mf ed 198?] – (= ser Ccm 221) – 1 – mf#1984-b500 – us ATLA [220]
Hsin ching tu pen (ccm274) = A reader on the apostles' creed – Wu-ch'ang, 1932 [mf ed 198?] – (= ser Ccm 274) – 1 – mf#1984-b500 – us ATLA [226]
Hsin, Ching-wen *see* So te shui chan hsing t'iao li shih i
Hsin chiu kung wen ch'eng shih ho shu / Wei, Wei-ch'ing – Shang-hai: Fa hsueh shu chu, Min kuo 23 [1934] – (= ser P-k&k period) – us CRL [324]
Hsin chiu yueh wen ta (ccm266) = Scripture catechism / Price, Philip Francis & Chen, Ta-san – Hankow, 1932 [mf ed 198?] – (= ser Ccm 266) – 1 – (with english pref) – mf#1984-b500 – us ATLA [220]
Hsin chu chung-kuo wen hsueh shih / Hu, Yun-i – Shang-hai: Pei hsin shu chu, 1933 – (= ser P-k&k period) – us CRL [480]
Hsin chu i tz'u tien / Sun, Chih-tseng – Shang-hai: Ta kuang shu chu, Min kuo 25 [1936] – (= ser P-k&k period) – us CRL [140]
Hsin chung-hua ching chi kai lun / Li, Ch'uan-shih – Shang-hai: Chung-hua shu chu, Min kuo 21 [1932] – (= ser P-k&k period) – us CRL [330]
Hsin chung-kuo chih hsien cheng chien she / Chin, Hui – [S l]: Kai chin ch'u pan she, min kuo 31 [1942] – (= ser P-k&k period) – us CRL [350]
Hsin chung-kuo ti hun yin wen t'i / Lu, Ssu-hung – Shang-hai: Hsin sheng t'ung hsun she, 1931 – (= ser P-k&k period) – us CRL [306]
Hsin fang : ssu mu chu / Hsia, Yen – Kuei-lin: Hsin chih shu tien, Min kuo 30 [1941] – (= ser P-k&k period) – us CRL [820]
Hsin fu / Ku, Chung-i – Shang-hai: Shih chieh shu chu, Min kuo 33 [1944] – (= ser P-k&k period) – us CRL [820]
Hsin fu mu (ccm227) = New parents / ed by Liu, Yu-chen & Sun, Hui-lan – Hong Kong. 2v. 1953 [mf ed 198?] – (= ser Ccm 227) – 1 – mf#1984-b500 – us ATLA [640]
Hsin fu nu lun / Kollontai, Aleksandra – Shang-hai: Sheng huo shu tien, Min kuo 26 [1937] – (= ser P-k&k period) – us CRL [305]
Hsin hsien cheng chih kuan li / Hsiao, Ming-hsin – [Ch'ung-ch'ing]: Cheng chung shu chu, Min kuo 32 [1943] – (= ser P-k&k period) – us CRL [350]
Hsin hsien chih chiang yen chi / Li, Tsung-huang – [np]: Hsing cheng hsien cheng chi hua wei yuan hui, Min kuo 28 [1939] – (= ser P-k&k period) – us CRL [350]
Hsin hsien chih chih li lun yu shih chi / Ch'eng, Yu-shu – Hang-chou: Cheng chung shu ch'u, Min kuo 29 [1940] – (= ser P-k&k period) – us CRL [350]
Hsin hsien chih chih li lun yu shih chi / Li, Tsung-huang – Ch'ung-ch'ing: Chung-hua, Min kuo 32 [1943] – (= ser P-k&k period) – us CRL [350]
Hsin hsien chih fa kuei hui pien / Hsing cheng yuan hsien cheng chi hua wei yuan hui (China) – Ch'ung-ch'ing: Cheng chung shu chu, Min kuo 30 [1941] – (= ser P-k&k period) – us CRL [350]
Hsin hsien chih kang yao ch'ien shuo, yu ming, ti fang tzu chih kai yao / Liu, Nai-ch'eng – Ch'ung-ch'ing: Kuo min t'u shu ch'u pan she, Min kuo 31 [1942] – (= ser P-k&k period) – us CRL [350]
Hsin hsien jih pao – Bangkok, Thailand. 1950-1974 (1) – mf#67846 – us UMI ProQuest [079]
Hsin hua pan yueh k'an = New china – Peiping. 1949-1959 [1] – mf#2613 – us UMI ProQuest [951]
Hsin hun ti meng : [san mu chu] / Hu, Yun-i – Shang-hai: Ch'i chih shu chu, Min kuo 23 [1934] – (= ser P-k&k period) – us CRL [820]
Hsin hung a tzu / Chang, Tzu-p'ing – Sang-hai: Chih hsing ch'u pan she, 1945 – (= ser P-k&k period) – us CRL [830]
Hsin i yun yun tung / Hsueh, Kuang-ch'ien – Chiang-hsi Shang-jao: Chan ti t'u shu ch'u pan she, Min kuo 29 [1940] – (= ser P-k&k period) – us CRL [380]
Hsin jen ti ku shih / I-ch'un – Shang-hai: Hsin ch'un ch'u pan she, 1947 – (= ser P-k&k period) – us CRL [480]
Hsin li chien she ti k'o hsueh chi ch'u / Sung, Shu-shih – Nan-ching: Cheng chung shu chu, Min kuo 24 [1935] – (= ser P-k&k period) – us CRL [150]
Hsin li chien she yu hsien cheng chien she / Ch'en, Kung-ch'ia – Fu-chien: Fu-chien sheng cheng fu mi shu ch'u, Min kuo 31 [1942] – (= ser P-k&k period) – us CRL [350]
Hsin li hsueh / Feng, Yu-lan – Ch'ang-sha: Shang wu yin shu kuan, Min kuo 31 [1942] – (= ser P-k&k period) – us CRL [180]
Hsin li hsueh pao = Journal of psychology – 1963-1964 [1] – mf#2614 – us UMI ProQuest [150]
Hsin miao, i ming, hsin sheng sung ti i pu, ch'ung kao ti ai / Yao, Hsueh-yin – Ch'ung-ch'ing: Hsien tai ch'u pan she, Min kuo 32 [1943] – (= ser P-k&k period) – us CRL [830]
Hsin min shih su sung fa p'ing lun / Shih, Chih-ch'uan – Pei-p'ing: Kuo li Pei-p'ing ta hsueh fa hsueh yuan ch'u pan k'o, 1932 – (= ser P-k&k period) – us CRL [340]
Hsin min shuo / Liang, Ch'i-ch'ao – Shang-hai: Chung-hua shu chu, 1941 – (= ser P-k&k period) – us CRL [306]
Hsin nien t'e k'an / Sheng Huo Pao – Djakarta, 1952-1955 – 29mf – 9 – mf#SE-961 – ne IDC [951]
Hsin nung pen chu i p'i p'an / Chou, Hsien-wen – [China]: Kuo min ch'u pan she, Min kuo 34 [1945] – (= ser P-k&k period) – us CRL [630]
Hsin pan chih chieh shui fa kuei hui pien / Fa, hsueh – Shang-hai: Hui wen t'ang Hsin chi shu chu, Min kuo 35 [1946] – (= ser P-k&k period) – us CRL [340]
Hsin pan chih chieh shui fa kuei hui pien hsu pien – Shang-hai: Fa hsueh pien i she: Hui wen t'ang Hsin chi shu chu, Min kuo 35 [1946] – (= ser P-k&k period) – us CRL [340]
Hsin pien hsi hsueh hui k'ao / Ling, Kuei-ch'ing – Shang-hai: Ta tung shu chu, Min kuo 23 [1934] – (= ser P-k&k period) – us CRL [820]
Hsin ping chih yu hsin ping fa / Chiang, Fang-chen – Ch'ung-ch'ing: Shang wu yin shu kuan, Min kuo 32 [1943] – (= ser P-k&k period) – us CRL [355]
Hsin she hui wen t'i / Ch'en, Hsi-hao – Nan-ching: Cheng chung shu chu, Min kuo 25 [1936] – (= ser P-k&k period) – us CRL [360]
Hsin sheng / Ch'eng, Lu-ting – Shang-hai: Hsin shih tai shu chu, 1932 – (= ser P-k&k period) – us CRL [810]
Hsin sheng huo yu hsiang ts'un chien she / Tsou, Shu-wen – Nan-ching: Cheng chung shu chu, Min kuo 23 [1934] – (= ser P-k&k period) – us CRL [951]
Hsin sheng huo yun tung ts'u chin tsung hui p'ei-tu hsin yun mo fan ch'u kung tso pao kao – [China: Hsin sheng huo yun tung ts'u chin tsung hui], Min kuo 31 [1942] – (= ser P-k&k period) – us CRL [390]
Hsin sheng lun / Hsieh, Tung-p'ing – Ch'ung-ch'ing: Chung-hua shu chu, Min kuo 33 [1944] – (= ser P-k&k period) – us CRL [170]
Hsin sheng tai / Ch'i, T'ung – Ch'ung-ch'ing: Sheng huo shu tien, 1940 – (= ser P-k&k period) – us CRL [830]
Hsin sheng-ming = The new life – Shanghai. v1-3. jan 1928-dec 1930 (complete) – 4r – 1 – $102.00 – us Chinese Res [073]
Hsin shih fan chiao yu shih / Wang, Ch'ih-ch'ang – Shang-hai: Chung-hua shu chu, Min kuo 21 [1932] – (= ser P-k&k period) – us CRL [370]
Hsin shih ho hsin shih jen / Feng, Shou-chu – Shang-hai: Ta tung shu chu, Min kuo 21 [1932] – (= ser P-k&k period) – us CRL [810]
Hsin shih-tai te shin yang (ccm170) = A faith for a new age / Hsieh, Fu-ya – Shanghai, 1925 [mf ed 198?] – (= ser Ccm 170) – 1 – mf#1984-b500 – us ATLA [210]
Hsin ssu-ch'uan / Chiang, Tung-pai – Ch'ung-ch'ing: Ch'ing nien shu tien, 1940 – (= ser P-k&k period) – us CRL [951]
Hsin tsung chiao kuan (ccm99) = New point of view: a compilation of articles on religion and christianity by various authors / ed by Chien, Yu-wen – 2nd ed. Shanghai, 1923 [mf ed 198?] – (= ser Ccm 99) – 1 – mf#1984-b500 – us ATLA [230]
Hsin tu hua hsu / Tuan-mu, Hung-liang – Shang-hai: Chih shih ch'u pan she, Min kuo 29 [1940] – (= ser P-k&k period) – us CRL [830]
Hsin wen fa chih lun / Shinmura, Sen'ichi – Shang-hai: Ch'un li shu tien, Min kuo 26 [1937] – (= ser P-k&k period) – us CRL [070]
Hsin wen hsueh tsung lun liu pien / Yu, Yu-lang – Ta-lien: Shih yeh Yin shu kuan, [1943] – (= ser P-k&k period) – us CRL [480]
Hsin wen i p'i p'ing t'an hua / Li, Chun-liang – Pei-p'ing: Jen wen shu tien, Min kuo 22 [1933] – (= ser P-k&k period) – us CRL [480]
Hsin wen i tz'u tien / Ku, Feng-ch'eng et al – Shang-hai: Kuang hua shu chu, 1932 – (= ser P-k&k period) – us CRL [480]
Hsin wen k'uai pao – Cholon, Vietnam. 1966-1967 (1) – mf#67667 – us UMI ProQuest [079]
Hsin wen shih yeh chien she lun – Ch'ung-ch'ing: Ch'iao sheng shu tien, Min kuo 33 [1944] – (= ser P-k&k period) – us CRL [070]
Hsin wen tzu ju men – [Shang-hai: Hsin wen tzu ch'u pan she, 1936] – (= ser P-k&k period) – us CRL [480]
Hsin wen yu hui / Tai, Kuang-te – Kuei-yang: Wen t'ung shu chu, Min kuo 31 [1942] – (= ser P-k&k period) – us CRL [070]
Hsin wu yu chi ch'i t'a / Wang, Chi-ssu – Fu-chien Nan-p'ing: Kuo min ch'u pan she, 1945 – (= ser P-k&k period) – us CRL [840]
Hsin yu : san mu chu / Yu, Ling – Ch'ung-ch'ing: Wei lin ch'u pan she, Min kuo 33 [1944] – (= ser P-k&k period) – us CRL [820]
Hsin yu chiu / Shen, Ts'ung-wen – [Hong Kong?]: Liang yu t'u shu kung ssu, min kuo 34 [1945] – (= ser P-k&k period) – us CRL [480]
Hsin yu chiu / Shen, Ts'ung-wen – Shang-hai: Liang yu t'u shu yin shua kung ssu, 1936 – (= ser P-k&k period) – us CRL [830]
Hsin yuan yang p'u : [san mu chu] / Yang, Ts'un-jen – Ch'ung-ch'ing: Nan fang yin shu kuan, 1943 – (= ser P-k&k period) – us CRL [820]
Hsin yueh cheng ching cheng li shih (ccm15) = The canon and text of the new testament / Chang, Po-huai – Hong Kong, 1954 [mf ed 198?] – (= ser Ccm 15) – 1 – mf#1984-b500 – us ATLA [225]
Hsin yueh ch'uan shu (ccm312) – Ch'ing-tao, 1933 [mf ed 198?] – (= ser Ccm 312) – 1 – mf#1984-b500 – us ATLA [225]
Hsin yueh hua pao – Bao tan viet hoa – Hanoi: Yueh-nan Hua chiao lien ho tsung hui, jul 1967-sep 2 1976 – 16r – 1 – us CRL [079]
Hsin yueh hua pao – Hanoi, Vietnam. Jan 1-31 1958; Oct 3-Dec 3 1964; Jan 1965-Oct 28 1967 – 6r – 1 – us L of C Photodup [079]
Hsin yueh jen wu (ccm181) = New testament characters / Hsieh, Sung-kao – Hong Kong, 1952 [mf ed 198?] – (= ser Ccm 181) – 1 – mf#1984-b500 – us ATLA [225]
Hsin yueh shih hsuan / Ch'en, Meng-chia – Shang-hai: Hsin yueh shu tien, 1933 – (= ser P-k&k period) – us CRL [810]
Hsin yuen hua pao – Hanoi, Vietnam. 1961-1973 (1) – mf#67808 – us UMI ProQuest [079]
Hsin yun fu nu chih tao wei yuan hui san chou nien chi nien t'e k'an – [China: Hsin yun fu nu chih tao wei yuan hui, 1941] – (= ser P-k&k period) – us CRL [951]
Hsin yun fu nu chih tao wei yuan hui ssu chou nien chi nien chuan hao – [China: Hsin yun fu nu chih tao wei yuan hui, 1942] – (= ser P-k&k period) – us CRL [951]
Hsin-chiang chih ching chi / Chang, Chih-i – [China]: Chung-hua shu chu, Min kuo 33 [1944] – (= ser P-k&k period) – us CRL [339]
Hsin-chiang chih lueh / Hsu, Ch'ung-hao – [Ch'ung-ch'ing]: Cheng chung shu chu, Min kuo 33 [1944] – (= ser P-k&k period) – us CRL [339]
Hsin-chiang ching ying lun / Chiang, Chun-chang – Nan-ching: Cheng chung shu chu, Min kuo 25 [1936] – (= ser P-k&k period) – us CRL [339]
Hsin-chiang jih-pao – Urumchi, Sinkiang, China. Sinkiang Daily. Jan 1948-Dec 1965. 26 reels – 1 – mf#632.00 – us Chinese Res [079]

Hsin-chiang nei mu / Hsu, Su-ling – [Ch'ung-ch'ing]: Ya-chou t'u shu she, Min kuo 34 [1945] – (= ser P-k&k period) – us CRL [915]

Hsin-chiang yen chiu : [2 chuan] / Li, Huan – Ch'ung-ch'ing: An ch'ing yin shu chu, min kuo 33 [1944] – (= ser P-k&k period) – us CRL [951]

Hsing : ssu mu chu / Sung, Chih-ti – Ch'ung-ch'ing: Ta tung shu chu, 1940 – (= ser P-k&k period) – us CRL [820]

Hsing / Yeh, Tzu – Shang-hai: Wen hua sheng huo ch'u pan she, Min kuo 25 [1936] – (= ser P-k&k period) – us CRL [830]

Hsing, Chao-t'ang *see* Shih shih liang mien kuan

Hsing cheng ch'uan tse hua fen lun / Liu, Tso-jen – Shao-kuan: Min tsu wen hua ch'u pan she, Min kuo 33 [1944] – (= ser P-k&k period) – us CRL [350]

Hsing cheng hsiao lu yen chiu / Hsiao, Wen-che – Ch'ung-ch'ing: Shang wu yin shu kuan, Min kuo 31 [1942] – (= ser P-k&k period) – us CRL [350]

Hsing cheng kuan li kai lun / Chang, Chin-chien – [Ch'ung-ch'ing]: Chung-kuo wen hua fu wu she, Min kuo 32 [1943] – (= ser P-k&k period) – us CRL [350]

Hsing cheng lun ts'ung ti i chi / Ts'ui, Tsung-hsuan – Yung-an: Fu-chien sheng yen chiu yuan she hui k'o hsueh yen chiu shih, Min kuo 30 [1941] – (= ser P-k&k period) – us CRL [350]

Hsing cheng san lien chih ch'ien shuo ti i chi – [China]: Kuang-tung sheng cheng fu mi shu ch'u ti erh k'o, Min kuo 31 [1942] – (= ser P-k&k period) – us CRL [350]

Hsing cheng t'ung chi / Nei cheng pu t'ung chi ch'u – [Sl]: Chung yang hsun lien wei yuan hui, Min kuo 31 [1942] – (= ser P-k&k period) – us CRL [951]

Hsing cheng yuan hsien cheng chi hua wei yuan hui (China) *see* Hsin hsien chih fa kuei hui pien

Hsing cheng yuan wen wu pao kuan wei yuan hui nien k'an / Ch'u, Min-i – [Sl: sn, 1941] – (= ser P-k&k period) – us CRL [930]

Hsing chien chai sui pi / T'ang, Hao – Shang-hai: Chung-kuo wu shu hsueh hui, Min kuo 26 [1937] – (= ser P-k&k period) – us CRL [790]

Hsing fu ti ai ko / Ho, Te-ming – Shang-hai: Pei hsin shu chu, 1933 – (= ser P-k&k period) – us CRL [810]

Hsing hsien jih pao – Bangkok, Thailand. 1960-1989 – 214r – 1 – us L of C Photodup [079]

Hsing hua *see* Hua mei chiao pao (ccs29)

Hsing hua ch'un yu chiang nan : [ssu mu chu] / Yu, Ling [pseud] – Ch'ung-ch'ing: Mei hsueh ch'u pan she, Min kuo 33 [1944] – (= ser P-k&k period) – us CRL [820]

Hsing ko lei hsing hsueh kai kuan / Juan, Ching-ch'ing – Ch'ung-ch'ing: Chung-hua shu chu, Min kuo 33 [1944] – (= ser P-k&k period) – us CRL [951]

Hsing kuo chih sheng ming / Lu, I-shih – Shang-hai: Wei ming shu wu, 1935 – (= ser P-k&k period) – us CRL [820]

Hsing lai ti shih hou / Lu, Li – Kuei-lin: Nan t'ien ch'u pan she, 1943 – (= ser P-k&k period) – us CRL [810]

Hsing ling shih – Nan-ching: Chung-hua shu chu, Min kuo 23 [1934] – (= ser P-k&k period) – us CRL [810]

Hsing nien ssu shih / Yuan, Ch'ang-ying – Shang-hai: Shang wu yin shu kuan, Min kuo 35 [1946] – (= ser P-k&k period) – us CRL [840]

Hsing shih hsiao shuo : kai chuang pen / Shih, T'ien-chi – Shang-hai: Ming te shu chu, 1933 (1941 printing) – (= ser P-k&k period) – us CRL [830]

Hsing t'ai wan pao – Bangkok, Thailand. 1950-1973 (1) – mf#67853 – us UMI ProQuest [079]

Hsing wei chih sheng li ti fen hsi / Wang, Ching-hsi – Ch'ung-ch'ing: Tu li ch'u pan she, 1944 – (= ser P-k&k period) – us CRL [150]

Hsing yun chih lien so / T'ang, Tseng-yang – Shang-hai: Hsien tai shu chu, 1931 – (= ser P-k&k period) – us CRL [840]

Hsing yun yu / Liang, Ch'iung – Kuei-lin: Wen hua kung ying she, Min kuo 30 [1941] – (= ser P-k&k period) – us CRL [810]

Hsing-an ling ti feng hsueh / Pa-lai et al – Shang-hai: Lien hua shu chu, 1937 – (= ser P-k&k period) – us CRL [480]

Hsin-hua jih-pao – Hankow and Chungking. 1938-46 – 1 – us Chinese Res [079]

Hsin-pao the sin pao – Shanghai. 1877-79; 1881-jan 1882 – 8 1/4r – 1 – uk British Libr Newspaper [072]

Hsin-t'ieh-tao-jen *see* Ho pi hsi hsiang, i ming, mei hua meng

Hsin-yueh yen chiu chih nan (ccm96) = A guide to the study of the new testament / Ch'eng, Chih-i – Hong Kong, 1954 [mf ed 198?] – (= ser Ccm 96) – 1 – mf#1984-b500 – us ATLA [225]

Hsi-pei t'e ch'u k'ang chan tung yuan chi / Shu, Ch'un – [China]: Chieh fang ch'u pan she, Min kuo 27 [1938] – (= ser P-k&k period) – us CRL [951]

Hsi-po hsien sheng / Li, Chien-wu – Shang-hai: Wen hua sheng huo ch'u pan she, 1939 – (= ser P-k&k period) – us CRL [840]

Hsi-t'ai-hou : ssu mu chu / Chou, Chien-ch'en – [China]: Chu tso hsieh she: Hsin i shu tien, 1940 – (= ser P-k&k period) – us CRL [820]

Hsi-tsang chi – Shang-hai: Shang wu yin shu kuan, 1941 – (= ser P-k&k period) – us CRL [390]

Hsi-tsang jih-pao – Lhasa, Tibet. Apr 22, 1956- . Scattered issues missing – 5r – 1 – us Chinese Res [079]

Hsiu an ying yu / Sheng, Yu-ssu – Shang-hai: K'ai ming shu tien, Min kuo 20 [1931] – (= ser P-k&k period) – us CRL [480]

Hsiu cheng lao tzu cheng i ch'u li fa : fu, hsiu cheng kung hui fa ti erh shih san t'iao t'iao wen: erh shih i nien shih erh yueh san shih jih hsiu cheng kung ch'ang fa ch'uan wen / Ku, Ping-yuan – Shang-hai: Shang-hai fa hsueh pien i she, Min kuo 22 [1933] – (= ser P-k&k period) – us CRL [340]

Hsiu cheng ping i fa chung mien huan i wen t'i / Cheng, T'ao – Ch'ung-ch'ing: Chung-hua shu chu, Min kuo 33 [1944] – (= ser P-k&k period) – us CRL [340]

Hsiu nung sheng huo yu nung ts'un chiao yu – Hu-nan: Hsiu yeh kao chi nung hsiao nung ts'un fu wu she, 1936 – (= ser P-k&k period) – us CRL [370]

Hsiu tz'u hsueh chiang hua / Chang, I-p'ing – Ch'eng-tu: Fu hsing shu chu, Min kuo 32 [1943] – (= ser P-k&k period) – us CRL [480]

Hsiu tz'u hsueh fa fan / Ch'en, Wang-tao – [Ch'ung-ch'ing]: Chung-kuo wen hua fu wu she, 1945 – (= ser P-k&k period) – us CRL [480]

Hsiu yang wei ti (ccm342) / Wu, Yung-chuan – Shanghai, 1950 [mf ed 198?] – (= ser Ccm 342) – 1 – mf#1984-b500 – us ATLA [210]

Hsiung, Fo-hsi *see*
- Fo-hsi hsi chu ti san chi, ti ssu chi
- Sai-chin-hua
- T'ieh miao

Hsiung pien shu / Jen, Pi-ming – Kuei-lin: Shih hsueh shu chu, 1943 – (= ser P-k&k period) – us CRL [400]

Hsiung, Shih-seng *see* Ch'ih tu chue chieh: wen yen tui chao

Hsiung, Ta-hui *see* Cheng li chiang-hsi kung lu ying yuen kuan li chi hua

Hsiung, Tzu-jung *see* Kung min chiao yu

Hsiung, Yu-chung *see* Tu ch'i chan cheng

Hstc hooter *see* [Arcata-] the lumberjack

Hsu, An-chen *see* Shih yeh chi hua chih li lun yu shih chien

Hsu, Cathy H C *see* Journal of teaching in travel & tourism

Hsu, Chen-chou i *see* Lun-tun ta hsueh she hui hsueh chiang yen chi

Hsu, Cheng-hsueh *see* Nung ts'un wen t'i

Hsu, Ch'en-ssu *see*
- Pa chin hsuan chi
- Ping-hsin hsuan chi

Hsu, Chen-ya *see*
- Hsueh hung lei shih
- Lang mo san chi
- Lang mo ssu chi
- Shuang huan chi

Hsu, Chia-jui *see* T'ai-wan

Hsu, Chia-Lin K *see* Motivations for participating in leisure activities between chinese and american students

Hsu, Chi-ch'ing *see* Tsui chin shang-hai chin jung shih fu k'an chih i

Hsu, Chieh *see*
- Hsien tai hsiao shuo kuo yen lu
- Mu ch'un
- Wen i, p'i p'ing yu jen sheng

Hsu, Ch'ien *see* Chi-tu chiao chiu kuo chu i k'an hsing chih san (ccm189)

Hsu, Ch'ih *see*
- Mei wen chi
- Tsui ch'iang yin

Hsu, Chih-kuei *see* Hua ch'iao kai kuan

Hsu, Chih-mo *see*
- Ai mei hsiao cha
- Meng hu chi
- Shih
- Tzu p'ou
- Yun yu

Hsu, Ch'ing-fu *see*
- Liang shih wen t'i chih yen chiu
- Wu chia wen t'i chih yen chiu

Hsu, Ching-wei *see* Fei chan kung yueh yu shih chieh ho p'ing

Hsu, Ch'in-wen *see* Wu ch'i chih lei

Hsu, Chuan-p'eng *see* Hsia hsiang chi

Hsu, Ch'ung-hao *see* Hsin-chiang chih lueh

Hsu, Chung-nien *see* Yu erh chi

Hsu, Ho-ch'un *see* Hsien tai kung shang ling hsiu ch'eng ming chi

Hsu, Hsiao-t'ien *see* K'u su

Hsu, Hsing-chih *see*
- T'ien ch'ang ti chiu
- Tsui hou ti sheng tan yeh, i ming, hsiang tao meng

Hsu, Hsing-ch'u *see* K'ang chan yu nung ts'un ching chi

Hsu, Hsu *see*
- Hsi pei chien she lun
- I chia
- Kuei lien
- Yueh liang

Hsu, Hsuan *see*
- Liang shih wen t'i
- Nung yeh ching chi hsueh

Hsu, Hsueh-yu *see* Ti fang yin hang kai lun

Hsu, Hung-Yi *see* A cinematographical and biomechanical analysis of the approach run phase for the pole vault

Hsu, Kuang-p'ing *see* Lu hsun ti ch'uang tso fang fa chi ch'i t'a

Hsu, Kung-ta et al *see* Lu nan hui chan chi

Hsu, Mao-pen *see* Mao tun

Hsu, Mao-yung *see*
- Pu ching jen chi
- Tsen yang ts'ung shih wen i hsiu yang

Hsu, Min-i *see* Nan yu shih ts'ao

Hsu, Pai-ch'i *see* Ping i

Hsu, Pao-ch'ien *see* Nung ts'un kung tso ching yen t'an

Hsu, Pao-chien *see* Nung tsun kung tso ching yen tan (ccm190)

Hsu, Pi-chang *see* Shang-ti hui kuan huai wo mo? (ccm191)

Hsu, Pi-po *see* Liu shui chi

Hsu, Ssu-t'ung *see* Tung-pei ti ch'an yeh

Hsu, Su-ling *see* Hsin-chiang nei mu

Hsu, Sung-shih *see*
- Chi-tu chiao yu chung-kuo wen hua [ccm192]
- The christian awakening of faith
- Chung-hua min tsu yen li ti yeh-su [ccm193]
- Sheng ching yu chung-kuo hsiao tao
- Yeh-su yen li ti chung-hua min tsu

Hsu, T'e-li *see* T'o p'ai tsai chung-kuo

Hsu, T'ien-t'ai *see* Fu-chien chan shih ching chi ti li

Hsu, Ti-shan *see* Tao chiao shih (ccm197)

Hsu, ti-shan *see* Lo-hua-sheng ch'uang tso hsuan

Hsu, Tseng-ming *see* Fei ch'ang shih ch'i chih ching ch'a

Hsu, Tsung-tse *see* T'ien chu san wei ih t'i lun (ccm198)

Hsu tzu yung fa / Chu, Yu-ts'ang – Shang-hai: Chu Yu-ts'ang: Hui wen t'ang hsin chi shu chu, Min kuo 28 [1939] – (= ser P-k&k period) – us CRL [480]

Hsu, Wan-ch'eng *see*
- Chan hou shang-hai chi ch'uan kuo ko ta kung ch'ang tiao ch'a lu
- Min tsu jen ko tou cheng wen hsueh chi

Hsu, Wei-nan *see*
- Pai wu shu hsin
- Shang-hai mien pu
- Shang-hai tsai t'ai p'ing t'ien kuo shih tai
- Shui mien lo hua

Hsu, Ying *see* Tang tai chung-kuo shih yeh jen wu chih

Hsu, Yin-shih *see* Yeh ts'ao chi

Hsu, Yung-p'ing *see* K'ang chan chung ti meng-ku

Hsu, Yun-ju *see* Su-o chih ou-chou kuo chi kuan hsi

Hsuan ch'uan hsueh yu hsin wen chi che / Chi, Ta – Shang-hai: Kuo li Chi nan ta hsueh wen hua shih yeh pu, 1932 – (= ser P-k&k period) – us CRL [070]

Hsuan Fo Pu *see* Guide to buddhahood

Hsuan, Hao-p'ing *see*
- Ta chung yu wen lun chan
- Ta chung yu wen lun chan erh hsu

Hsuan tao hsueh (ccm243) = Homiletics / McNeur, George Hunter – Hong Kong, 1953 [mf ed 198?] – (= ser Ccm 243) – 1 – mf#1984-b500 – us ATLA [240]

Hsuan-p'u i shu t'e chi / Shao, Yuan-ch'ung – [China]: I she, 1944 – (= ser P-k&k period) – us CRL [480]

Hsuan-Tsang *see* Si-yu-ki

Hsue, Ch'ang-lin *see* Chien pi ch'ing yeh

Hsue, Chen-ya *see* Chen-ya lang mo

Hsue, Chung-nien *see* Ch'en chi

Hsue, Kung-mei *see* Ch'i t'u

Hsue, Mao-yung *see* Chieh t'ou wen t'an

Hsue, Ying *see* Ch'ien hou fang

Hsue, Yue-no *see* Chiang lai chih hua yuean

Hsuean-tsang *see* Si-yu-ki

Hsueh / Pa, Chin – Shang-hai: Wen hua sheng huo ch'u pan she, Min kuo 25 [1936] – (= ser P-k&k period) – us CRL [830]

Hsueh : tu mu chu / Chang, Min – Han-k'ou: Hsin yen chu she, 1938 – (= ser P-k&k period) – us CRL [951]

Hsueh chan pa nien ti chiao-tung tzu ti ping – Shan-tung, Chiao-tung hsin hua shu tien, 1945 – (= ser P-k&k period) – us CRL [951]

Hsueh, Chien-wu *see* Hu-pei wu-ch'ang hsien ch'ing-shan shih yen ch'u hu k'ou yu ching chi tiao ch'a pao kao

Hsueh hsi = Study – Peiping. 1949-1958 – 1 – mf#2615 – us UMI ProQuest [370]

Hsueh hsiao sheng huo t'e chi – Shang-hai: Ti i ch'u pan she, 1934 – (= ser P-k&k period) – us CRL [370]

Hsueh hsiao tiao ch'a / Huang, Ching-ssu – [China]: Chung-hua shu chu, Min kuo 26 [1937] – (= ser P-k&k period) – us CRL [370]

Hsueh hung lei shih / Hsu, Chen-ya – Shang-hai: Ta chung shu chu, Min kuo 24 [1935] – (= ser P-k&k period) – us CRL [830]

Hsueh jen (ccm74) / Chao, Tzu-ch'en – Shanghai, 1936 [mf ed 198?] – (= ser Ccm 74) – 1 – mf#1984-b500 – us ATLA [240]

Hsueh, Kuang-ch'ien *see*
- Fu hsing pi chiao yen chiu ti 1 chi
- Hsin i yun yun tung
- I-ta-li fu hsing chih tao

Hsueh li tsuan / Ai, Ch'ing – [Shang-hai: Hsin ch'un ch'u pan she, 1944] – (= ser P-k&k period) – us CRL [810]

Hsueh mu wen chin (ccm284) = The ministry / Slattery, Charles Lewis – Hong Kong, 1924 [mf ed 198?] – (= ser Ccm 284) – 1 – (chinese trans fr the english) – mf#1984-b500 – us ATLA [240]

Hsueh, Mu-ch'iao *see* Nung ts'un ching chi ti chi pen chih shih

Hsueh, Po-k'ang *see* Jen shih hsing cheng ta kang

Hsueh sa ch'ing k'ung : fei chiang chun yen hai-wen / Yu, Ling [pseud] – Han-k'ou: Ta chung ch'u pan she, 1938 – (= ser P-k&k period) – us CRL [951]

Hsueh sha hsing ts'ao / Ch'en, Wen-chien – Ch'ung-ch'ing: Chung-kuo pien chiang hsueh hui, 1941 – (= ser P-k&k period) – us CRL [810]

Hsueh sheng hsin ch'ih tu : shang hsia ts'e ho ting pen – Shang-hai: Shih chieh shu chu, Min kuo 32 [1943] – (= ser P-k&k period) – us CRL [860]

Hsueh sheng men ti ku shih : erh chuan pa chi / Hsieh, Sung-kao – Shang-hai: Kuang hsueh hui, 1937 – (= ser P-k&k period) – us CRL [480]

Hsueh sheng ts'ung chun chi shih / China Ping i pu I cheng ssu – [Ch'ung-ch'ing]: Ping i pu i cheng ssu pien yin, Min kuo 34 [1945] – (= ser P-k&k period) – us CRL [951]

Hsueh shu ssu hsiang lun wen chi / Mu, Chi-po et al – [China]: Cheng chung shu chu, Min kuo 31 [1942] – (= ser P-k&k period) – us CRL [370]

Hsueh shu yueh k'an = Academic monthly – 1957-1959 – 1 – mf#2616 – us UMI ProQuest [370]

Hsueh shu yueh pao (ccs) = Christian school monthly – [Pei-ching?] v3 n1-10. 1899 [complete] [mf ed 198?] – (= ser Chinese christian serials coll) – 1 – mf0296k – us ATLA [240]

Hsueh t'an meng / Tseng, P'u – Shang-hai: Chen mei shan shu tien, 1931 – (= ser P-k&k period) – us CRL [820]

Hsueh, tien-tseng *see* Pao hu ch'iao min lun

Hsueh yeh hsing yu min tsu hsing / Furukawa, Takeji – Shang-hai: K'ang chien shu chu, Min kuo 25 [1936] – (= ser P-k&k period) – us CRL [612]

Hsueh yu / T'o-huang & Han, Hsing – Shang-hai: Hsin ti shu tien, 1940 – (= ser P-k&k period) – us CRL [920]

Hsueh yu ts'un chuang / Tsou, Ti-fan – Ch'eng-tu: Wen hua sheng huo ch'u pan she, 1943 – (= ser P-k&k period) – us CRL [810]

Hsun ku hsueh yin lun / Ho, Chung-ying – Shang-hai: Shang wu yin shu kuan, Min kuo 23 [1934] – (= ser P-k&k period) – us CRL [480]

Hsun yu lun / Li, Hsiang-hsu – Shang-hai: Shang wu yin shu kuan, Min kuo 24 [1935] – (= ser P-k&k period) – us CRL [370]

Hsun-t'ien hsiang-shih lu – List of successful candidates in the imperial examination in Hsun-t'ien province. Scattered years 1714-1903. 3 reels – 1 – 90.00 – us Chinese Res [951]

Htwei, Ba *see* Katvan u bha gyam nhan su tui a mran

Hu chan wen i p'ing hsuan – Shang-hai: Le hua t'u shu kung ssu, 1933 – (= ser P-k&k period) – us CRL [480]

Hu chi fa yu hu chi hsing cheng / Huang, Lun – Ch'ung-ch'ing: Chung-kuo wen hua fu wu she, 1944 – (= ser P-k&k period) – us CRL [951]

Hu, Chi-ch'en *see* Miao hsieh jen sheng tuan p'ien chih kuei yu-kuang

Hu, Ch'iu-chen *see* Nung yeh ching chi kai lun

Hu, Ch'iu-yuan *see* Min tsu wen hsueh lun

Hu ch'u / Shen, Ts'ung-wen – Shang-hai: Hsin Chung-kuo shu chu, Min kuo 21 [1932] – (= ser P-k&k period) – us CRL [480]

Hu ch'un kou tang / Wang, P'ing-ling – Ch'ung-ch'ing: Chung-kuo hsi ch'u pien k'an she, Min kuo 29 [1940] – (= ser P-k&k period) – us CRL [820]

Hu fen / Chih-hsing – Shang-hai: Tung ya t'u shu kuan, Min kuo 23 [1934] – (= ser P-k&k period) – us CRL [480]

Hu, Feng *see*
- Chi yuean ts'ao
- K'an yun jen shou chi
- Lun min tsu hsing shih wen t'i
- Mi yun ch'i feng hsi hsiao chi
- Min tsu chan cheng yu wen i hsing ko
- Min tsu hsing shih t'ao lun chi
- Wei tsu kuo erh ko
- Wen i pi t'an

Hu, Feng hsuan *see* Wo shih ch'u lai ti
Hu fu : hsin-ling-chun yu ju-chi / Kuo, Mo-jo – Ch'ung-ch'ing: Ch'un i ch'u pan she, 1942 – (= ser P-k&k period) – us CRL [820]
Hu, Han-min *see* Hu han-min hsien sheng ming chu chi hsia ts'e
Hu han-min hsien sheng ming chu chi hsia ts'e / Hu, Han-min – Shang-hai: Chun shih hsin wen she ch'u pan pu, Min kuo 25 [1936] – (= ser P-k&k period) – us CRL [951]
Hu hsiang fu shih (ccm290) = Serve one another / Surdam, T Janet – 1st ed. Hong Kong, 1958 [mf ed 198?] – (= ser Ccm 290) – 1 – mf#1984-b500 – us ATLA [230]
Hu, Hsieh-yin *see* Feng-huang shan
Hu, Huai-ch'en *see*
- Shang-hai ti hsueh i t'uan t'i
- Shih hsueh t'ao lun chi

Hu, Huan-yung *see*
- Kuo fang ti li
- Shih yeh chi hua t'ieh lu p'ein
- T'ai-wan yu liu-ch'iu

Hu, I-ku *see*
- Hsieh lu-yin hsien sheng chuan lueh
- Hsien tai ssu hsiang chung te chi-tu chiao
- Kao ching nien

Hu, Jen-k'uei *see* Yu chi ch'u ching chi wen t'i yen chiu
Hu, Kelly S *see* Cardiorespiratory responses
Hu, Kuo-hua *see* Ssu-ch'uan nung ts'un wu chia
Hu, Lan-ch'i *see* Tsai te-kuo nu lao chung
Hu, Lan-ch'i et al *see* Chan ti i nien
Hu, M *see* Effects of sensory balance training in older adults
Hu, Ming-lung *see* Fei ch'ang shih ch'i chih hsien cheng
Hu nan ch'u ti chi t'u / Shang-hai shih t'u ti – [China]: Kai chu, 1933 – (= ser P-k&k period) – us CRL [630]
Hu, Nan-hsiang *see* Hsien tai hsi chu hsuan
Hu newsletter / Howard University – 1969 may 29 – 1r – 1 – mf#4866712 – us WHS [378]
Hu p'an / Shu-wen – Ch'ung-ch'ing: Wen hua sheng huo ch'u pan she, 1941 – (= ser P-k&k period) – us CRL [480]
Hu, P'u-an *see*
- T'ang tai wen hsueh
- Ts'ung wen tzu hsueh shang k'ao chien ku tai pien se pen neng yu jan se chi shu

Hu, Shao-hsuan *see* T'ieh sha
Hu, Sheng *see* Hou fang min chung ti tsung tung yuan
Hu, Shih *see*
- The development of the logical method in ancient china
- Hu shih lun hsueh chin chu ti i chi
- Hu shih lun shuo wen hsuan
- T'ang hui shih cha chi

Hu shih chung-kuo che hsueh shih p'i p'an / Yen, Ling-feng – Chiang-hsi Kan-hsien: Chung-hua cheng ch'i ch'u pan she, 1943 – (= ser P-k&k period) – us CRL [951]
Hu shih lun hsueh chin chu ti i chi / Hu, Shih – Shang-hai: Shang wu yin shu kuan, Min kuo24 [1935] – (= ser P-k&k period) – us CRL [951]
Hu shih lun shuo wen hsuan / Hu, Shih – Shang-hai: Hsi wang ch'u pan she, Min kuo 25 [1936] – (= ser P-k&k period) – us CRL [840]
Hu, Shou-ch'uan *see* Tsui hsin ying yung chan shu chih piao chun
Hu, Su *see* Huo ti tien li
Hu, Tan-fei *see* Pa yen kuang fang yuan tien
Hu tieh pei / Chiang, Tieh-lu – Shang-hai: Kuang i shu chu, Min kuo 32 [1943] – (= ser P-k&k period) – us CRL [830]
Hu, Yeh-p'in *see*
- Shih kao
- Yeh-p'in hsiao shuo chi

Hu yu kuo ying / P'an, Tsun-hsing – Kuang-chou: Kuo li Chung-shan ta hsueh wen shih yen chiu so, Min kuo 22 [1933] – (= ser P-k&k period) – us CRL [480]
Hu, Yuan-i *see* Min fa tsung tse
Hu, Yu-chieh *see* Wo kuo ch'uang pan so te shui chih li lun yu shih shih
Hu, Yu-chih *see* Ou chan yu wo kuo wai chiao
Hu, Yun-i *see*
- Hsin chu chung-kuo wen hsueh shih
- Hsin hun ti meng

Hua, Ch'ao *see* Shih chieh ho p'ing yun tung
Hua chi shih wen chi / Yang, Ju-ch'uan – T'ien-chin: Ta kung pao she, 1933 – (= ser P-k&k period) – us CRL [480]
Hua chiao / Ch'en, Ta-tz'u – Shang-hai: Li ming shu chu, 1933 – (= ser P-k&k period) – us CRL [840]
Hua ch'iao kai kuan / Liu, Shih-mu & Hsu, Chih-kuei – Shang-hai: Chung-hua shu chu, Min kuo 24 [1935] – (= ser P-k&k period) – us CRL [951]
Hua ch'iao ko ming shih hua shang ts'e / Feng, Tzu-yu – Ch'ung-ch'ing: Hai wai ch'u pan she, Min kuo 34 [1945] – (= ser P-k&k period) – us CRL [951]
Hua ch'iao wen t'i / Ch'iu, Han-p'ing & Chuang, Tsu-t'ung – Ch'ang-sha: Shang wu yin shu kuan, Min kuo 27 [1938] – (= ser P-k&k period) – us CRL [304]
Hua chien lei : wu mu chu / Yu, Ling [pseud] – Shang-hai: Hsien tai hsi chu ch'u pan she, 1940 – (= ser P-k&k period) – us CRL [820]
Hua fa chi / Chou, Li-an – [Shang-hai]: Feng hsi shu wu: Tsung ching shou ch'u yu chou feng she, min kuo 29 [1940] – (= ser P-k&k period) – us CRL [840]
Hua, Han *see* Shen ju
Hua hsia tzu / Mao, Tun – Shang-hai: Liang yu t'u shu kung ssu, [Min kuo 23 [1934]] – (= ser P-k&k period) – us CRL [840]
Hua hsin feng : ssu mu chu / Li, Chien-wu – Shang-hai: Shih chieh shu chu, Min kuo 33 [1944] – (= ser P-k&k period) – us CRL [820]
Hua hsueh hsueh pao = Journal of the chinese chemical society – Shanghai. 1959-1964 (1) – mf#2617 – us UMI ProQuest [540]
Hua hsueh ping ch'i chih yen chiu / Kuo, Feng-kang – Shang-hai: Ta chung shu chu, Min kuo 23 [1934] – (= ser P-k&k period) – us CRL [355]
Hua hsueh t'ung pao = Bulletin of chemistry – 1962-1964 (1) – mf#2618 – us UMI ProQuest [540]
Hua hui chien wen lu / Chia, Shih-i – Shang-hai: Shang wu yin shu kuan, 1937 – (= ser P-k&k period) – us CRL [337]
Hua kung hsueh pao = Chemical industry and engineering – 1959-1960 (1) – mf#2619 – us UMI ProQuest [660]
Hua, Lin *see*
- Pa shan hsien hua
- T'i hen

Hua mei chiao pao (ccs29) = Christian advocate – Shanghai: Methodist Episcopal Church. n1-50. mar 1904-feb 1910 [gaps] [mf ed 198?] – (= ser Chinese christian serials coll 29) – 2r – 1 – (incl some iss of later title: hsing hua [chinese christian advocate] v8 n7-v24 n6 1911-27 (gaps)) – mf0307a – us ATLA [240]
Hua nien (ccs30) = Good years – Shanghai. v1-2 n25. 1932-jun 1933 [complete] [mf ed 198?] – (= ser Chinese christian serials coll 30) – 1r – 1 – mf0308 – us ATLA [240]
Hua pei ho ch'u chien she shih yeh kuan hsi ko hsien nung shih tiao ch'a pao kao shu – [Pei-ching: Hua pei cheng wu wei yuan hui chien she tsung shu shui li chu, Min kuo 31- ie 1942-] – (= ser P-k&k period) – us CRL [630]
Hua pei liu sheng k'ang jih hsueh chan shih shang chi / Fan, Ch'ang-chiang – [China]: Shih hou ch'u pan she, 1938 – (= ser P-k&k period) – us CRL [951]
Hua pei min chung shih liao ti i ko ch'u pu yen chiu / Ch'u, Chih-sheng – [China]: Ts'an mou pen pu kuo fang she chi wei yuan hui, Min kuo 23 [1934] – (= ser P-k&k period) – us CRL [630]
Hua pei nung lien t'ung hsun (ccs) = North china farmers union newsletter – T'ung-chon. n1-7. 1950 [complete] [mf ed 198?] – (= ser Chinese christian serials coll) – 1 – (primarily for christian farmers & rural churches) – mf0296l – us ATLA [240]
Hua pei nung yeh ho tso shih yeh wei yuan hui pao kao shu – [China]: Hua pei nung yeh ho tso shih yeh wei yuan hui, Min kuo 25 [1936] – (= ser P-k&k period) – us CRL [334]
Hua pei ti ch'iu / Chao, Ch'ing-ko – Shang-hai: T'ieh liu shu chu, Min kuo 26 [1937] – (= ser P-k&k period) – us CRL [480]
Lo hua shih chieh / Wen, Kuo-hsin – Pei-ching: I wen she, 1944 – (= ser P-k&k period) – us CRL [480]
Hua, Ti *see* Wen i ch'uang tso kai lun
Hua t'ing ho / Wang, T'ung-chao – Ch'ung-ch'ing: Wen hua sheng huo ch'u pan she, Min kuo 31 [1942] – (= ser P-k&k period) – us CRL [480]
Hua t'ing ho / Wang, T'ung-chao – Shang-hai: Wen hua sheng huo ch'u pan she, 1941 – (= ser P-k&k period) – us CRL [480]
Hua tung chiao hui wu nien yun tung chi hua (ccm254) = Findings on five year movement – Shanghai, 1929 [mf ed 198?] – (= ser Ccm 254) – 1 – mf#1984-b500 – us ATLA [240]
Hua tung chiao yu *see* Chia yu kung pao [ccs]
Hua yang hua chi hsin pao = The oriental – San Francisco CA. 1887 feb 19, 1888 apr 27 – 1r – 1 – mf#881855 – us WHS [071]
Hua yuan shih yeh tiao ch'a t'uan *see* Shan-hsi ch'ang-an hsien ts'ao t'an ching-yang hsien yung-lo-tien nung k'en tiao ch'a pao kao
Huachuca scout – 1980 aug 21-dec, 1981 jan-dec, 1982 jan-dec, 1983 jan-dec, 1984 jan 12-jul 26, 1984 aug 21-1986 jun 12, 1986 jul 10-1987 sep 10, 1987 oct-dec 17, 1988 jan-dec, 1989 jan-jun – 14r – 1 – (cont: scout [fort huachuca, az]; cont by: fort huachuca scout [fort huachuca, az: 1998]) – mf#570068 – us WHS [071]
Huai hsiang chi / Liu, Yu-sheng – Shang-hai: T'ai p'ing shu chu, 1944 – (= ser P-k&k period) – us CRL [840]
Huan ch'iu chung-kuo ming jen chuan lueh : shang-hai kung shang ko chieh chih pu = World chinese biographies, Shanghai commercial and professional edition / Li, Yuan-hsin – Shang-hai: Huan ch'iu ch'u pan she, min kuo 33 [1944] – (= ser P-k&k period) – us CRL [920]
Huan hsi t'uan / Li, Kuang-t'ien – Kuei-lin: Kung tso she, 1943 – (= ser P-k&k period) – us CRL [480]
Huan hsiang chi / Ho, Ch'i-fang – Kuei-lin: Kung tso she, 1943 – (= ser P-k&k period) – us CRL [840]
Huan hun ts'ao / Pa, Chin – Ch'ung-ch'ing: Wen hua sheng huo ch'u pan she, min kuo 31 [1942] – (= ser P-k&k period) – us CRL [830]
Huan shu chi / Ho, Li-sheng – [China: sn], 1932 – (= ser P-k&k period) – us CRL [810]
Huang, Ch'an-hua *see* Jo shui
Huang, Chia-te *see* Fan i lun chi
Huang, Chia-yin *see* Wo ai chiang ti ku shih
Huang, Chin-fu *see* Fang huo kai lun
Huang, ching-chai *see* K'ang chan yu chien tieh
Huang, Ching-ssu *see* Hsueh hsiao tiao ch'a
Huang, Chi-p'ing *see* Tang tai fu nu
Huang, Cho *see* Su o chi hua ching chi
Huang chun ti wei chi : k'ang chan tu mu chu chi / Yu, Ling [pseud] – Han-k'ou: Shang-hai tsa chih kung ssu, 1937 – (= ser P-k&k period) – us CRL [951]
Huang, Fen-sheng *see* Kang chan i lai chih pien chiang
Huang, Ho *see* P'an ch'iao-yun
Huang ho lou / Ch'en, Ch'uan – Ch'ung-ch'ing: Shang wu yin shu kuan, Min kuo 33 [1944] – (= ser P-k&k period) – us CRL [820]
Huang, Hsiao-fang *see* Shan-tung chiu chi-nan tao shu nung ts'un ching chi tiao ch'a
Huang, Hsu-ch'u *see* Kan pu cheng ts'e
Huang hua t'ai / Lu-fen – Shang-hai: Liang yu t'u shu yin shua kung ssu, 1937 – (= ser P-k&k period) – us CRL [840]
Huang, Hua-chieh *see* Chi-tu-chiao tao-te-kuan yu chung-kuo lun li [ccm329]
Huang hun chih hsien / Li-ni – Shanghai: Wen hua sheng huo ch'u pan she, Min kuo 24 [1935] – (= ser P-k&k period) – us CRL [840]
Huang, Jiamin *see* Creating a graduate dance curriculum model for the beijing dance academy in china
Huang, Kuo-chang *see* She hui ti ti li chi ch'u
Huang, K'u-t'ung *see* Nung ts'un tiao ch'a
Huang, Lun *see* Hu chi fa yu hu chi hsing cheng
Huang miu chi / Wang, Tsao-shih – [China]: Tzu yu yen lun she, Min kuo 24 [1935] – (= ser P-k&k period) – us CRL [951]
Huang, Mu *see* Hsiang hsia hsien sheng
Huang pai tan ch'ing : [erh mu ssu ch'ang chu] / Hung, Shen – Ch'ung-ch'ing: Wen i chiang chu chien kuan li wei yuan hui ch'u pan pu, Min kuo 31 [1942] – (= ser P-k&k period) – us CRL [951]
Huang, P'u-sheng *see* Kuang-tung liang shih wen t'i yen chiu
Huang sha / Chin, I – Shang-hai: Wen hua sheng huo ch'u pan she, Min kuo 37 [1948] – (= ser P-k&k period) – us CRL [480]
Huang, Shou-p'eng *see* K'o hsueh kuan li yu hsien tai hsing cheng
Huang, Thomas T F *see* The south-west africa question
Huang, To *see* Chen tao ch'ang shih [ccc202]
Huang ts'un / Ko, Hsien-ning – Shang-hai: Pei hsin shu chu, 1934 – (= ser P-k&k period) – us CRL [810]
Huang t'u ni / Lao, Hsiang – Shang-hai: Jen chien shu wu, Min kuo 25 [1936] – (= ser P-k&k period) – us CRL [840]
Huang, T'ung *see* T'u ti wen t'i
Huang, Yen-p'ei *see*
- Chi kuan kuan li i te
- Shu tao

Huang, Yuan-pin *see*
- Pai yin kuo yu lun
- Yin wen t'i

Huang yuan-sheng i chu fu lu / Huang, Yuan-yung – [China: sn, 1938] – (= ser P-k&k period) – us CRL [951]
Huang, Yuan-yung *see* Huang yuan-sheng i chu fu lu
Huang, Yu-shih *see* T'an hsin li wei sheng
Huang-ho chih – Shang-hai: Kuo li pien i kuan, min kuo 25-26 [1936-1937] – (= ser P-k&k period) – us CRL [915]
Huang-mei *see*
- I ko jen ti chueh hsing
- Tsai chiao t'ang ko ch'ang ti jen
- Yu yu ti ko

Huang-ming-hai *see* Komin bunkai
Huar Hen *see*
- Yoen qan

Huard, Charles *see* Nos amis les quebecquois
Huard, Victor-Alphonse *see* Monseigneur dominique racine
Huard, Victor-Amedee *see* La vie et l'ouvre de l'abbe provancher
Huart, C *see* Histoire de baghdad dans les temps modernes
Huart, Clement *see*
- Histoire des arabes
- A history of arabic literature
- Konia, la ville des derviches tourneurs

Huarte de San Juan, Juan *see* Examen de ingenios
Hua-tung cheng-fa hsueh-pao – (East China Journal of Politics and Law). Shanghai. n.1-3, Jun-Dec 1956. FA-HSUEH. (Legal Studies). Shanghai. 1957-Sep 1958. 1957: 1 missing – 1 – us Chinese Res [951]
Hua-tzu jih-pao – Hong Kong. 1895-1940 – 1 – us Chinese Res [079]
Hua-yang hsien nung ts'un kai k'uang / Yeh, Mao – [China]: Ssu-ch'uan sheng nung yeh kai chin so t'ung chi shih, Min kuo 31 [1942] – (= ser P-k&k period) – us CRL [307]
Hub – Centralia, WA. 1913-1919 (1) – mf#66959 – us UMI ProQuest [071]
Hub – Big Bend, Hales Corners, Muskego, Waterford WI. 1965 oct 7-1966 jan 27, 1966 feb 3-1967 feb 9 – 2r – 1 – (cont: hub [hales corners, wi: 1965]; cont by: tri-town news [hales corners, wi]; hub [hales corners, wi: 1965: hales corners ed]; tri-town hub [hales corners, wi: 1967: muskego ed]; tri-town hub [hales corners, wi: 1967: franklin-hales corners ed]) – mf#1125476 – us WHS [071]
Hub – Franklin, Hales Corners, Muskego WI. 1964 dec 31-1965 sep 30 – 1r – 1 – (cont: hub of hales corners & muskego; cont by: hub [hales corners, wi: 1965: muskego ed]; hub [hales corners, wi: 1965: hales corners ed]) – mf#1125344 – us WHS [071]
Hub – Stoughton WI. 1881 nov 9-1884 dec 31, 1885 jan 1-apr 23 – 2r – 1 – (continued by: stoughton hub (stoughton, wi: 1885)) – mf#939078 – us WHS [071]
Hub / Defense Depot Ogden (UT) – 1981 jul 17-1982 feb 19 – 1r – 1 – mf#599843 – us WHS [071]
Hub – Kearney, NE. 1993-2000 (1) – mf#64705 – us UMI ProQuest [071]
Hub : a publication of the 11 chapters comprising the dairyland regional council / Parents Without Partners – 1977 spring-1978 spring, 1978 summer-winter – 2r – 1 – mf#621464 – us WHS [071]
Hub / Seneca Co. Attica – jan 1971-mar 1975 [wkly] – 3r – 1 – mf#B29472-29474 – us Ohio Hist [071]
Hub of hales corners and muskego – Franklin, Hales Corners, Muskego WI. 1964 apr 9-sep 24, 1964 oct 1-dec 24 – 2r – 1 – (cont: hub of hales corners, muskego & franklin; cont by: hub [hales corners, wi: 1965]) – mf#1125343 – us WHS [071]
Hub of hales corners, muskego, and franklin – Franklin, Hales Corners, Muskego, Oak Creek WI. 1963 may 23-1964 apr 2 – 1r – 1 – (cont: suburban times; cont by: hub of hales corners & muskego) – mf#1125341 – us WHS [071]
Hub times – sunday – Hudson, OH. 1989-1999 (1) – mf#68489 – us UMI ProQuest [071]
Hub times wendesday – Hudson, OH. 1996-2000 (1) – mf#65532 – us UMI ProQuest [071]
Hubball, Harry T *see* The impact of an adult health education program on exercise self-efficacy and participation in leisure-time physical activity
Hubbard, Donald *see* Causes of the failure of the cement pipe used in sub-irrigation
Hubbard eagle series / Trumbull Co. Hubbard – jan 1995-jun 7 1996 – 1r – 1 – mf#B37489 – us Ohio Hist [071]
Hubbard enterprise – Hubbard, Marion County, OR: L C McShane, v3 n51-v21 n40. feb 23 1917-jan 4 1935 – 1 – us Oregon Hist [071]
Hubbard enterprise – Hubbard OR: L C McShane [wkly] – 1 – us Oregon Lib [071]
Hubbard enterprise : weekly independant newspaper – Hubbard, OH. 6 Feb 1913-18 Oct 1917 – 2r – 1 – us Western Res [071]
Hubbard, Ethel Daniels *see*
- Moffats
- Under marching orders

Hubbard herald – Hubbard OR: I B Muchmore [wkly] – 1 – (began 1912. ceased in 1913?) – us Oregon Lib [071]
Hubbard, J G *see* Right hon jg hubbard on tithe rent-charge
Hubbard, John Gellibrand *see* Ritual revision
Hubbard, John Waddington *see* The sobo of the niger delta

Hubbard review / Trumbull Co. Hubbard – oct 1995-dec 1996 – 1r – 1 – mf#B37488 – us Ohio Hist [071]

Hubbardston 1746-1849 – Oxford, MA (mf ed 1996) – (= ser Massachusetts vital record transcripts to 1850) – 9mf – 9 – 0-87623-270-5 – (mf 1t-4t: births & deaths 1746-1853. mf 1t-2t: marriages 1770-93. mf 4t-7t: marriages & intentions 1802-49. mf 7t-8t: births 1822-44. mf 8t: marriages 1798-1802; births 1843-49. mf 9t: births & deaths 1793-1843; marriages, deaths 1843-49) – us Archive [978]

Hubbardston 1747-1900 – Oxford, MA (mf ed 1994) – (= ser Massachusetts vital records) – 136mf – 9 – 0-87623-186-5 – (mf 1,3,7: vitals 1747-1803. mf 1-7: town records 1767-95. mf 8-11: births, deaths 1771-1853. mf 11-15: marriages 1798-1886. mf 15: births & deaths 1825-1844. mf 16-18: births, deaths 1749-1853. mf 19-21: marriages 1798-1844. mf 22-41: town records 1795-1838. mf 41: church members 1821-35. mf 42-61: town records 1838-76. mf 62-64: accounts 1829-50. mf 65-67: paupers 1820-74. mf 68-76: valuations 1836-55. mf 77-80: town orders 1831-63. mf 81-82: rebellion 1861-65. mf 83-84: veterans aid 1884-1945. mf 85-86: town farm 1868-85. mf 87-124: mortgages 1832-1905. mf 125-126: birth index 1848-1914. mf 126-127: marr index 1848-1914. mf 127-128: death index 1848-1914. mf 129-131: births 1843-1901. mf 130: marriages 1843-51. mf 130-131: deaths 1843-59. mf 132-133: marriages 1852-1904. mf 134-136: deaths 1860-1905) – us Archive [978]

Hubbell, Levi *see* Trial of impeachment of levi hubbell, judge of the second judicial circuit, by the senate of the state of wisconsin, june 1853

The hubbell standard – Hubbell, NE: H C Pershing. v1 n1. sep 29 1899-v14 n21. dec 29 1922 (wkly) [mf ed with gaps filmed 1976] – 8r – 1 – (absorbed by: belleville telescope and the belleville freeman. publ in belleville ka, apr 5 1918-dec 29 1922. suspended with jan 25 1918; resumed with apr 5 1918. numbering is irregular) – us NE Hist [071]

Hubbell times – Hubbell, NE: Jas A Harris. 7v. 1892-v7 n27. may 5 1899 (wkly) [mf ed with gaps filmed 1979] – 1r – 1 – (cont: munden times. cont by: times (hubbell ne)) – us NE Hist [071]

Hubel, Henni *see* "Belauscht!"

Huber, A *see* Indonesie

Huber, B *see* Apercu statistique de l'ile de cuba

Huber, Engelbert *see* Die personennamen in den keilschrifturkunden

Huber, Eugen *see*
- Die entwicklung des religionsbegriffs bei schleiermacher
- System und geschichte des schweizerischen privatrechtes

Huber, Fritz *see* Johann salomo semler

Huber, Johannes *see*
- Der alte und der neue glaube
- Der jesuiten-orden
- Johannes scotus erigena
- Die philosophie der kirchenvaeter

Huber, M *see* Notices generales des graveurs divises par nations et des peintres ranges par ecoles...

Huber, Michael *see* Die wanderlegende von den siebenschlaefern

Huber, S *see*
- Auslegung des 129 psalmen dauids auff gegenwertigen zustand der kirchen zu wittenberg
- Bestendige bekantnus d samuel hubers ob gott durch seinen lieben son jesum christum nur allein etlich wenig menschen order zumal alle menschen vom tode allesampt erloest habe
- Bestendige entdeckung des caluinischen geists welcher sich vnterstehe das leiden jhesu christi fuer vnsere suende zu verlaeugnen vnd auffzuheben
- Christliche predigten vber den 129 psalm dauids darinne angezeiget wird wie die caluinische schwermer der kirch zu wittemberg vnnd im gantzen churkreiss seyen mit jhrem heillosen pflug
- Clangores tubae adversvs theodorvm bezam vt priusquam in alterum exspirat secvlvm, hvc respiciat et perpendat
- Confvtatio brevis libri, sub alieno nomine editi, de controversia inter theologos vvittebergenses
- Demonstratio fallaciarum johannis calvini, in doctrina de coena domini qvibvs vsvs est in libro institutionis christianae, et ex quo suum calvinismum in omnem egurgitavit christianum orbem
- Dispvtatio secvnda contra calvinistas, et praesertim synopsin kimedoncij
- Dispvtatio tertia contra calvinistas qvod faciant devm avtorem peccati
- Drey schrifften
- Die ewige vnnd einige grundfeste auff welchem der seligmachende glaube stehen vnd verharren muss
- Historische beschreibung des gantzen streits zwischen d hunnen vnd d hubern von der gnadenwahl wie derselbige entsprungen vnd biss dahier zugenomen habe
- Kurtze anleytunge vnnd nachrichtunge wie man d egidium hunnen vnd d lucam osiandern sampt jrem anhang examinieren
- Kurtze erjnnerung von gegenwertigem zweytracht vber die lehre von der gnaden wahl
- Notwendige endeckung wie d lucas osiander in seiner predigt von der gnaden-wahl die verzweyffelte caluinische lere versteckt mit fuersatz dieselbige in die reine christliche kirchen in wuertemberg
- Protestation samuel hubers professorn der h schrifft zu wittemberg wider johan wilhelm stuck zu zuerich d johann jacob gryneum zu basel vnd johan jetzlern (welcher sich seithero hat gratianum serleyu tauffenm lassen) zu schaffhausen
- Rettung des spruchs rom 8 denn welche er zuuor versehen hat die hat er auch verordnet das sie gleich sein solten dem ebenbild seines sons
- Rettung meiner allezeit bestendigen bekantnus von der gnadenwahl darinnen auff dismal hindan gesetzt aller nebenstreitten alles vnd allein was zum hauptstreit gehoerig erkleret wird
- Sendbrieff an die burgermeister vnd rhat der loeblichen statt zuerich darjnnen sie erjnnert werden was jre kirchendiener vnter dem schein einer antwort auff d philippi nicolai buch fuer ein werck wider jesum christum
- Theses, christvm iesvm esse mortvvm pro peccatis totivs generis hvmani
- Von der caluinischen predicanten schwindelgeist vnnd dem gerechten gericht gottes vber dise sect
- Von h schrifft 2 von der christlichen kirchen 3 von der baeptischen kirchen 4 von d luthers person 5 von hans pistorij person vnd seinem schrecklichem anbljck

Huber, Siegfried *see* Pizarro et ses freres. conquerants de l'empire des incas

Huber, Thomas *see* Studien zur theorie des uebersetzens im zeitalter der deutschen aufklaerung, 1730-1770

Huber, Victor *see*
- Madrid, lisboa und die refugiados in london
- Skizzen aus ireland
- Skizzen aus Spanien

Huber, Walther *see* Gottfried keller und die frauen

Huberinus, Caspar *see* Vom[m] christlichen ritter

Hubert, F *see* Die straszburger liturgischen ordnungen im zeitalter der reformation

Hubert, Henri *see* Melanges d'histoire des religions

Hubert, Jean *see* Toussaint rwandaise et sa repression

Hubert, Jean Francois *see*
- Lettre circulaire a messieurs les cures

Hubert, Lucien *see* Une politique coloniale

Huberty, Anton *see* Neu method – messige viol d'amour stuecke aus allen thoenen

Hubl, Arthur *see* Three-colour photography

Hubley, Melissa *see* The family and alzheimer's disease

Hubli gazette – Hubli, India. 29 Oct 1944-29 Dec 1946 – 1r – 1 – us L of C Photodup [079]

Hubmaier, Balthasar *see*
- Writings

Hubo unos pinos claros / Lopez Suria, Violeta – San Juan, Puerto Rico. 1961 – 1r – us UF Libraries [972]

Hubscher, Jacob *see* Kaddisch-gebet

Huc, E R *see* Souvenirs d'un voyage dans la tartarie, le thibet, et la chine pendant les annees 1844, 1845 et 1846

Huc, Evariste *see* Souvenirs d'un voyage dans la tartarie et le thibet pendant les annees 1844, 1845 et 1846

Huc, evariste *see* L'empire chinois

Huc, Evariste R *see* Das chinesische reich

Huc, Evariste Regis *see*
- A travers les deserts de la tartarie et les neiges du thibet
- The chinese empire
- Le christianisme en chine
- Christianity in china, tartary, and thibet
- Journey through the chinese empire
- Reisherinneringen uit tartarije, thibet en china
- Travels in tartary, thibet, and china

Huc, Theophile *see* Martinique

Huch, Friedrich *see*
- Geschwister
- Traeume

Huch, Ricarda *see*
- Gedichte
- Natur und geist als die wurzeln des lebens und der kunst

Huch, Ricarda Octavia *see*
- Alte und neue gedichte
- Ausbreitung und verfall der romantik
- Bluethezeit der romantik
- Das judengrab / aus bimbos seelenwanderungen

Huch, Rudolf *see* Mein leben

Huchet, Albert *see* Chartier ancien de montmorigny

Huck, Thomas Sergej *see* Das zisterzienserkloster hardehausen in ostwestfalen von seiner gruendung im jahr 1140 bis in das 15. jahrhundert

Huckel, Oliver *see* Melody of god's love

Huckelberry herald – Belfair, WA. 1970-1971 (1) – mf#66937 – us UMI ProQuest [071]

Huck/konopacki labor cartoons – 1983 nov-1990 dec – 1r – 1 – mf#1069624 – us WHS [740]

Hucknall and bulwell dispatch – Hucknall, Nottinghamshire 30 apr 1903-to date – 1 – (hucknall torkard despatch to mar 1913 then hucknall despatch etc to june 1974; fr 1 jun 2000 hucknall dispatch; a localised ed [bulwell despatch] was publ 16 oct 1953-28 jun 1974. no copies of this ed survive) – uk Newsplan [072]

Hucknall morning star and bulwell mail – Sutton in Ashfield, Hucknall, Nottinghamshire 14 sep 1885-n1390 28 mar 1913 [mf ed 1889-95, 1898-1913] – 1 – (many title changes incl: star and rushcliffe advertiser [jan-sep 1912]; rushcliffe star and advertiser [oct 1912-21 feb 1913]) – uk Newsplan [072]

Hucknall torkard dispatch and leen valley mercury – [East Midlands] Nottinghamshire 30 apr 1903-dec 1940 [mf ed 2004] – 38r – 1 – (missing: jan-jun 1917; cont by: hucknall dispatch and leen valley mercury [jan 1914-dec 1940]) – uk Newsplan [072]

Huck's synopsis of the first three gospels – Cincinnati: Jennings and Graham, c1907 – 1mf – 9 – 0-8370-1860-9 – mf#1987-6247 – us ATLA [226]

Hud challenge / United States Dept of Housing and Urban Development – Washington. 1969-1978 (1) 1975-1978 (5) 1976-1978 (9) – (cont by: challenge) – ISSN: 0017-629X – mf#6300 – us UMI ProQuest [350]

Hud newsletter – 1970-76, 1977-82 – 2r – 1 – mf#169368 – us WHS [071]

Hud newsletter – Washington. 1974-1982 (1) 1974-1982 (5) 1975-1982 (9) – ISSN: 0017-6311 – mf#7912 – us UMI ProQuest [350]

Hudavendigar – (= ser Vilayet salnames) – 9 – (1302 [1885] 8mf $130; 1307m [1891] 7mf $110; 1310 [1892] 8mf $130; 1313 [1895] 7mf $110; 1315 [1897] 5mf $75; 1316 [1898] 6mf $90; 1318 [1900] 4mf $60; 1321 [1903] 6mf $90; 1323 [1905] 5mf $75; 1324 [1906] 12mf $195; 1325 [1907] 11mf $180; (bursa) [1927] 9mf $150) – us MEDOC [956]

Hudavendigar – Bursa: Matbaa-yi Vilayet, 1896-? n423. 11 haziran 1873, 3156 26 subat 1341 [1925] – (= ser O & t journals) – 1mf – 9 – us $25.00 – us MEDOC [956]

Hudayi, Kulliyat-i *see* The divan project

[Huddart, J] *see* The oriental navigator

Huddersfield and holmfirth examiner – England. Sep 1851-Dec 1856.-w. 5 reels – 1 – uk British Libr Newspaper [072]

Huddersfield boro' advertiser. (boro advertiser) – England. 3 Jan 1913-27 Apr 1917; 8 Dec 1922-16 Jan 1943.-w. 11 reels – 1 – uk British Libr Newspaper [072]

Huddersfield echo – England.2 Oct 1886-1887. -w. 1/2 reel – 1 – uk British Libr Newspaper [072]

Huddersfield labour party records, 1918-1951/52 – (= ser Labour party in britain, origins and development at local level. series 1) – 8r – 1 – (int by keith laybourn) – mf#97289 – uk Microform Academic [325]

Huddilston, John Homer *see* Essentials of new testament greek

Huddingeposten – Stockholm, Sweden. 1979-82 – 1 – sw Kungliga [078]

Huddleston, Trevor *see* Naught for your comfort

Huder, Karin *see* Methanreformierung mit co2 bei energieeinkopplung durch direktbestrahlung des katalysators

Hudgings, William Franklyn *see* What everybody should know about the laws of marriage and divorce...also the divorce laws of mexico, cuba, canada, england, and france

Hudicourt, Max L *see* Haiti faces tomorrow's peace

Hudicourt, Pierre L *see*
- Anexion de la republica de haiti
- Pour notre liberation economique et financiere

Hudiksvalls tidning – Hudiksvall, Sweden. 1986- – 1 – sw Kungliga [078]

Hudiksvallsposten – Hudiksvall, Soderhamn, Sweden. 1864-1954 – 1 – sw Kungliga [078]

Hudiksvallstidningen – Hudiksvall, Sweden. 1909-43, 1969-86 – 1 – (halsinglands tidning 1971-78) – sw Kungliga [078]

Hudiksvallstidningen – Hudiksvall, Sweden. 1979-86. Halsinglands Tidning – 1 – sw Kungliga [078]

Hudikswalls weckoblad – Hudiksvall, Sweden. 1845-59, 1861-64 – 1 – sw Kungliga [078]

Hudjdjatul islam : madjalah resmi pusat pimpinan persatuan islam – Bandung, 1956. v1(1) – 1mf – 9 – mf#SE-1498 – ne IDC [950]

Hudobny zivot – Bratislava, Czechoslovakia. 24 jan 1972-24 jan 1973; 21 jan-30 sep 1974; oct 1974-oct 1975; 10 nov-22 dec 1975; 19 jan 1976-18 dec 1978; 8 jan 1879-28 dec 1985 – 2 1/4r – 1 – uk British Libr Newspaper [072]

Hud's "legislative guidebook" and its potential impact on property rights and small businesses, including minority-owned businesses : hearing...house of representatives, 107th congress, 2nd session, mar 7 2002 / United States. Congress. House. Committee on the Judiciary. Subcommittee on the Constitution – Washington: US GPO 2002 [mf ed 2002] – 2mf – 9 – 0-16-068554-0 – (incl bibl ref) – us GPO [343]

Hudson 1866-1900 – Oxford, MA (mf ed 1993) – (= ser Massachusetts vital records) – 13mf – 9 – 0-87623-174-1 – (mf 1: births 1866-76. mf 2: births 1876-86. mf 3: births 1886-91. mf 4: marriages 1866-77. mf 5: marriages 1877-87. mf 6: marriages 1888-91. mf 7: deaths 1866-82. mf 8: deaths 1883-91. mf 9: births 1892-98. mf 10: births 1899-1900. mf 11: marriages 1892-1900. mf 12: deaths 1892-98. mf 13: deaths 1899-1906) – us Archive [978]

Hudson bay company : papers presented...to ascertain the legality of the powers in respect to territory, trade, taxation and government... claimed or exercised by the hudson's bay company, on the continent of north america... / Grande-Bretagne. Colonial Office – [s.l.]: the House of Commons, 1850 [mf ed 1984] – 1mf – 9 – mf#SEM105P425 – cn Bibl Nat [971]

Hudson broadcaster – Hudson WI. 1927 feb 15, mar 1 – 1r – 1 – mf#923059 – us WHS [071]

Hudson, C F *see* Tsui chin chung-kuo yu shih chieh cheng chih

Hudson, Charles Frederic *see*
- Christ our life
- Critical greek and english concordance of the new testament
- Human destiny, a discussion

Hudson chronicle – Hudson WI. 1859 may 7, 1860 mar 31 – 1r – 1 – (cont: pathfinder [hudson, wi]; cont by: hudson city times) – mf#876318 – us WHS [071]

Hudson city times – Hudson WI. 1863 jan 23, jul 10 – 1r – 1 – (cont: hudson chronicle; cont by: north star [hudson, wi]; hudson star and times) – mf#876319 – us WHS [071]

Hudson, David *see* Family papers, ms 3893

Hudson democrat – Hudson WI. 1869 apr 15, 1870 feb 11, apr 22, sep 30; 1872 jan 10, 1874 dec 16 – 2r – 1 – (cont by: true republican [hudson, wi]) – mf#876693 – us WHS [071]

Hudson dispatch – Union City, NJ. 1901-1991 (1) – mf#61143 – us UMI ProQuest [071]

Hudson Family Association [South] *see* Hudsoniana

Hudson family association, south, bulletin – Hudson WI. n1-68 [1973 jul-1988 oct 15], n9a [1974 dec, suppl], 1988 jan-1989 fall [hudsoniana] – 1r – 1 – (cont: hudsoniana bulletin; cont by: hudson family association bulletin) – mf#1687116 – us WHS [071]

Hudson family papers, 1799-1836 – [mf ed 1994] – 1r – 1 – mf#ms3893 – us Western Res [978]

Hudson, Henry *see* Descriptio ac delineatio geographica detectionis freti, sive, transitus ad occasum, sufra terras americanas, in chinam atq

The hudson highlands – New York: impr for Henry Cranston, 1883 – 1mf – 9 – mf#03623 – cn Canadiana [978]

Hudson, Hilary T *see*
- The methodist armor
- Methodist armor

Hudson journal – Hudson WI. 1853 aug 10-1854 aug 3, 1853 sep 8,15, 1854 jun 22, jul 6, jul 20 – 2r – 1 – (cont by: st croix republican) – mf#876694 – us WHS [071]

Hudson, M A *see*
- Rugwaro rwa baduku
- Rugwaro rwa vaduku

Hudson, New Hampshire. Hudson Baptist Church *see* Records

Hudson north star – Hudson WI. [1860 aug 15-1864 jul 20], 1855 jun 13-1856 jul 2 – 1r – 1 – (cont: north star [husdon, wi]; cont by: hudson city times; hudson star and times) – mf#1133656 – us WHS [071]

Hudson, ohio, taxes, ms v.f.o. – 1801 – 1r – 1 – (true copy of resident proprietors lists for the district of hudson, june 29 1801) – us Western Res [978]

Hudson review – New York. 1948+ [1]; 1968+ [5]; 1976+ [9] – ISSN: 0018-702X – mf#1424 – us UMI ProQuest [073]

Hudson, Sanford Amos *see* Law for the clergy: a compilation of the statutes of the states of illinois, indiana, iowa, michigan, minnesota, ohio, and wisconsin.

Hudson, Scott B *see* The effect of athletic participation on school discipline

HUGHES

Hudson star and times – Hudson WI. [1865 apr 19/1869 oct 13]-[1895 jun 7-1898 may 6] [gaps] – 10r – 1 – (cont: hudson north star; cont by: hudson star-times) – mf#1133845 – us WHS [071]
Hudson star-observer – Hudson WI. [1909 feb 18/1912 may 2]-[2005 nov/dec] [gaps] – 194r – 1 – (cont: hudson star-times; st croix observer) – mf#3357489 – us WHS [071]
Hudson star-times – Hudson WI. 1898 may 13-dec 30; 1899 jan 6-1902 jan 17; 1902 jan 24-1904 dec 6; 1904 dec 9-1908 dec 7; 1908 feb 14-1909 feb 12 – 5r – 1 – (cont: hudson star and times; cont by: st croix observer; hudson star-observer) – mf#1139277 – us WHS [071]
Hudson taylor and the china inland mission : the growth of a work of god / Taylor, Howard & Taylor, Howard [Mrs] – London: Morgan & Scott; Philadelphia: China Inland Mission, 1919 [mf ed 1995] – (= ser Yale coll) – xi/640p (ill) – 1 – 0-524-09669-4 – mf#1995-0669 – us ATLA [240]
Hudson, Thomson Jay *see*
– The divine pedigree of man
– The evolution of the soul
Hudson, Trevor Allan *see* Co-authoring spiritual ways of being
Hudson valley black press – Newburgh NY. [1991 jul 17/24-dec 31/jan 8 1992]-[2000 jan 5/11-2001 jan 3] – 10r – 1 – (cont by: hudson valley press) – mf#2185821 – us WHS [071]
Hudson valley news – Newburgh, NY. 1885-1992 (1) – mf#61042 – us UMI ProQuest [071]
Hudson, William *see* Anatomy of south africa
Hudson, William Henry *see* Far away and long ago
Hudson, William L *see*
– Journal of the united states exploring expedition
Hudson-Essex-Terraplane Club *see* White triangle news
Hudsoniana / Hudson Family Association [South] – n2-39 [1978 jan-1987 apr], special ed [1987 apr] – 1r – 1 – (cont by: hudson family association, south, bulletin) – mf#1344656 – us WHS [366]
Hudson's bay : or, a missionary tour in the territory of the hon. hudson's bay company / Ryerson, John – Toronto: Publ by G R Sanderson for the Missionary Soc of the Wesleyan Methodist Church, 1855 – 3mf – 9 – mf#40348 – cn Canadiana [917]
Hudson's Bay Company *see* Copy of memorial and petition from inhabitants of the red river settlement
Hudson's bay company to lord clarendon : hudson's bay house, feb 28th 1854 – [s.l: s.n, 1854?] [mf ed 1984] – 1mf – 9 – 0-665-45116-4 – mf#45116 – cn Canadiana [971]
Hudsons' Bay Company's Expedition [1837-1839] *see* Progress of north american discovery for 1838
Hue – 1953 nov-1954 oct, 1954 nov-1955 oct, 1955 nov-1956 oct, 1956 nov-1957 oct, 1957 nov-1958 oct, 1958 dec-1959 jul – 6r – 1 – mf#3745743 – us WHS [071]
Hue and cry : the story of henry and john fielding and their bow street runners / Pringle, Patrick – England: William Morrow & Co, n.d. – 3mf – 9 – $4.50 – mf#LLMC 96-052 – us LLMC [360]
Hue, Francois *see* Dernieres annees du regne et de la vie de louis 16
Hue, Otto *see* Unsere taktik beim generalstreik
Hueben und drueben *see* Argentinisches wochenblatt
Huebener, W *see* Das zertruemmerte babel, das unfehlbare gotteswort und die ewige gottesstadt
Huebler, Franz *see* Milton und klopstock
Huebner / Sanguino y Michel, Juan – Caceres: Tip. Enc. y Lib. Jimenez, 1901 – 1 – sp Bibl Santa Ana [946]
Huebner, Alfred *see*
– Rennewart
Huebner, Arthur *see* Die poetische bearbeitung des buches daniel
Huebner, E *see* Il ponte d'alcantara
Huebner, Emilio *see*
– Caceres en tiempo de los romanos
– Corpus inscriptionum latinarum
– Inscripciones romanas de merida
– Inscripciones romanas sepulcrales de ibahernando
– Situacion de la antigua norba
Huebner, Friedrich Markus *see* Die zierde der geistlichen hochzeit
Huebner, Johann *see*
– Christ-comoedia
– La geographie universelle
Huebner, Joseph A de *see* A travers l'empire britannique
Huebner, Joseph Alexander, Baron de *see* The life and times of sixtus the fifth
Huebner, Lorenz *see* Physikalisches tagebuch fuer freunde der natur
Huebner, Solomon S *see* Property insurance
Hueffer, Francis *see* Richard wagner und die musik der zukunft
Hueffer, Oliver Madox *see* The book of witches
Huegel, Friedrich, Freiherr von *see*
– Diaries, 1877-1879, 1884-1900, 1902-1924
– Eternal life
– The german soul in its attitude towards ethics and christianity, the state, and war
– The papal commission and the pentateuch
Die huegelmuehle : roman in fuenf buechern / Gjellerup, Karl – Leipzig: Quelle & Meyer, 1920 – 1r – 1 – us Wisconsin U Libr [830]
Huehnerbein-Sollmann, Christoph Matthias *see* Evaluation von assessment-centern (ac)
huei tlamahuicoltica omonexiti in ilhuicac tlatoca cihuapilli santa maria totlaconantzin guadalupe in nican huei altepenahuac mexico itocayocan tepeyacac / Lasso de la Vega, Luis – Mexico: en la Imprenta de Iuan Ruyz, ano de 1649 – (= ser Books on religion...1543/44-c1800: milagros y culto de la virgen) – 1mf – 9 – mf#crl-81 – ne IDC [241]
Huelfsbuch fuer den evangelischen religionsunterricht in gymnasien / Hollenberg, Wilhelm Adolf – 3. Aufl. Berlin: Wiegandt und Grieben, 1859 – 1mf – 9 – 0-8370-7638-2 – (incl bibl ref) – mf#1986-1638 – us ATLA [242]
La huelga / Diaz Maciaz, Jose – 1897 – 9 – sp Bibl Santa Ana [830]
Huella de tradicion / Zea Avelar, Gilberto – Guatemala, 1963 – 1r – us UF Libraries [972]
Huellas de gloria / Santovenia Y Echaide, Emeterio Santiago – La Habana, Cuba. 1944 – 1r – us UF Libraries [972]
Huellas en la arena / Vazquez Rodriguez, Benigno – Habana, Cuba. 1953 – 1r – us UF Libraries [972]
Huellas juveniles / Herrero Alvarado, Antonio – Sevilla: Editorial Catolica Espanola, S.A., 1956 – 1 – sp Bibl Santa Ana [946]
Huelsen, Charistian C F *see* Roman forum, its history and its monuments
Huelsenbeck, C *see* Der psalm 29
Huelswitt, Ignatz *see* Tagebuch einer reise nach den vereinigten staaten und der nordwestkueste von amerika
Huemer, Camillo *see* Die sage von orest in der tragischen dichtung
Huenefeld, Guenther, Freiherr von *see* Vom ewigen kampf
Hueneke, Heinrich *see* Experimentelle untersuchung zum problem des zusammenhangs zwischen erfragtem und beobachtetem verhalten
Huenerbein, Heidi A *see* Guidelines for prescribing upper body exercise following open heart surgery
Huennerkopf, Richard *see* Beitraege zur deskriptiven poetik in den mittelhochdeutschen volksepen und in der thidrekssaga
Huenziker, Rudolf *see* Wie uli der knecht gluecklich wird
Huer fikir – Izmit. Mueduer-i Mes'ul: Kilicoglu Hakki. n16. 25 nisan 1924; 85. 21 eyluel 1925 – (= ser O & t journals) – 1mf – 9 – $25.00 – us MEDOC [956]
Huer markopasa – ns: v1. n1-5. 1950 [all publ] – (= ser O & t journals) – 2mf – 9 – $40.00 – (last title in the markopasa set publ in istanbul by aziz nesin) – us MEDOC [956]
Huer markopasa – v1. n1-19 [all publ] – (= ser O & t journals) – 2mf – 9 – $40.00 – us MEDOC [956]
Huer markopasa *see* Markopasa
Huerfano cactus *see* Huerfano county miscellaneous newspapers
Huerfano county miscellaneous newspapers – Denver, CO (mf ed 1991) – 1r – 1 – (the clarion (mar 6 1945); huerfano county news (oct 20-dec 29 1951); huerfano cactus (may 10 1884, dec 19 1885); the independent (sep 8 1916-apr 24 1928); walsenburg cactus (feb 4, 25 1897); walsenburg yucca (mar 12 1903-apr 14 1904)) – mf#MF Z99 H871 – us Colorado Hist [071]
Huerfano county news *see* Huerfano county miscellaneous newspapers
Der huernen seufrid : tragoedie in sieben acten / Sachs, Hans; ed by Goetze, Edmund – Halle: M Niemeyer, 1880 [mf ed 1993] – (= ser Neudrucke deutscher literaturwerke des 16. und 17. jahrhunderts) – viii/42p – 1 – mf#8413 reel 2 – us Wisconsin U Libr [820]
Huerrem bey / Vecihi – Istanbul: Mahmud Bey Matbaasi, 1314 [1896] – (= ser Ottoman literature, writers and the arts) – 2mf – 9 – $40.00 – us MEDOC [470]
Huerriyet – London, 1895-97. Muharriri ve Nasiri: Civanpir. n71-75,77. 30 mart-1 temmuz 1897 – (= ser O & t journals) – 1mf – 9 – $25.00 – us MEDOC [956]
Huerriyet : tageszeitung fuer tuerkische arbeitnehmer in europa – Neu-Isenburg DE, 1978 1 sep- – ca 6r/yr – 1 – gw Misc Inst [074]
Huerriyet ve itilaf nasil dogdu, nasil oldu? / Nur, Riza – Dersaadet: Aksam Matbaasi, 1335 [1919] – (= ser Ottoman literature, writers and the arts) – 1mf – 9 – $25.00 – us MEDOC [470]
Huerta de Animas *see*
– Estatutos del centro juvenil nuestra senora del rosario de huerta de animas
– Fiestas patronales en honor de la santisima virgen del rosario
– Fiestas patronales en honor de la virgen del rosario. 1973
Huerta, J *see* De problemas filosoficos...
Huertas y Barrero, Francisco *see*
– Conferencias sobre la plurotonia
– La epidemia de viruela
Huerto cerrado / Sanchez Galarraga, Gustavo – Habana, Cuba. 1924 – 1r – us UF Libraries [972]
Huerto y camino / Lopez Rodriguez, Gerardo – San Juan, Puerto Rico. 1965 – 1r – us UF Libraries [972]
Hueser, Fritz *see* Wir tragen ein licht durch die nacht
Huesn ve ask / Dede, Galib – Kostantiniye: Matbaa-i Ebuezziya, 1304 [1887] – (= ser Ottoman literature, writers and the arts) – 3mf – 9 – $65.00 – us MEDOC [470]
Huet, evaeque d'avranches, ou, le scepticisme theologique / Bartholmess, Christian – Paris: Marc Ducloux, 1850 – 1mf – 9 – 0-7905-4486-5 – (incl bibl ref) – mf#1988-0486 – us ATLA [240]
Huet, Francois *see* Le regne social du christianisme
Der hueter der schwelle : von weisheit und liebe in der geisteswelt / Novalis [Friedrich von Hardenberg]; ed by Bonsels, Waldemar – 3. aufl. Muenchen: Muenchner Buchverlag, 1943 [mf ed 1990] – 141p – 1 – (int by ed) – mf#7435 – us Wisconsin U Libr [830]
Der hueter israels : kriegsnovellen aus der heimat / Pauls, Eilhard Erich – Leipzig: G Schloessmann 1915 [mf ed 1991] – 1r – 1 – (filmed with: martin opitz / friedrich gundolf) – mf#2858p – us Wisconsin U Libr [830]
Die huette – Unterwellenborn DE, 1949-1952 30 jul [gaps] – 1r – 1 – (maxhuette) – gw Misc Inst [074]
Der huettenarbeiter – Thale DE, 1949 9 jun-1960, 1962-85, 1987-1990 19 jul – 13r – 1 – (with gaps) – gw Misc Inst [622]
Huettner, Franz *see*
– Die chronik des klosters kaisheim
Huezo Cordoba De Ramirez, Transito *see* Marmoles
Huezo, Efraim *see* Prosas de efraim huezo
Huf, Stefan *see* Altern in der arbeitsgesellschaft
Hufeland, C W *see* Encyclopaedisches woerterbuch der medicinischen wissenschaften (ael3/13)
Hufeland, Christoph Wilhelm *see* Moreh darkhe ha-rafu'ah
Huff, Lonnie R *see* The press and nationalism in kenya, british east africa
Huff, O P *see* Study of essential oils occurring in the different organs and products
Huffcut, Ernest Wilson *see*
– American cases on contract.
– Cases on the law of agency
– Principles of the english law of contract and of agency
Huffcut, Wilson *see* The elements of business law
Huffman, Franklin E *see*
– Intermediate cambodian reader
– Modern spoken cambodian
Huffman, Franklin E et al *see* Cambodian system of writing and beginning reader with drills and glossary
Huffman, Franklin Eugene *see*
– Cambodian names and titles
Huffman, Jasper Abraham *see* Job, a world example
Huffmann, Scott J *see* Relationship of open chain isokenetic knee strength
Hug, Bernhard *see* Entwicklung eines kombinierten laser-elektroden-katheters zur av-knoten-koagulation bei tachykarden rhythmusstoerungen
Hug, Gall Joseph *see* Die christliche familie
Hug, Heinrich *see*
– Heinrich hugs villinger chronik von 1495 bis 1533
– Heinrich hugs villinger chronik von 1495-1533
Huge, Walter *see* Die judenbuche
Hugenholtz, Petrus Hermannus *see* Religion and liberty
Huggenberger, Alfred *see*
– Bauernerbe
– Daniel pfund
– Dem bollme si boes wuche
– Dorfgenossen
– Das ebenhoech
– Der hochzeitsschmaus
– Jakob spoendlis gluecksfall
– Jochems erste und letzte liebe
– Lebenstreue
– Von den kleinen leuten
Huggins of rhodesia / Gann, Lewis H – London, England. 1964 – 1r – us UF Libraries [960]
Huggins, William *see* Sketches in india
Hugh c thomson, of kingston : proposes to establish a candid, impartial, independent newspaper, to be entitled the upper canada herald: which will be published weekly – [Kingston ON: s.n. 1819?] [mf ed 1987] – 1mf – 9 – 0-665-64397-7 – mf#64397 – cn Canadiana [071]
Hugh latimer : a biography / Demaus, Robert – new ed. Nashville: Lamar & Barton, [1903?] – 2mf – 9 – 0-524-00747-0 – mf#1990-0179 – us ATLA [920]
Hugh mcwhirter family history research newsletter – v1 n1 [1983 sep], v2 n1 [1984 jan], v3 n1-4, [1985 mar-aug/sep], v4 n3,5 [1986 mar, may], v5 n10-11 [1986 oct-nov], v5 n1,7 [1987 jan, nov], v5 n7/v6 n6 [1988 jun 23] – 1r – 1 – mf#1058497 – us WHS [929]
Hugh of Saint-Victor *see*
– Canonici regularis scti victoris parisiensis opera omnia
– Didascalion
– Explanation of the rule of st augustine
– Le gage des divines fiancailles (de arrha animae)
– Opera
– Soliloquium de arrha animae. de vanitate mundi
– Soliloquium de arrha animae und de vanitate mundi
– Traktat ueber die hinfuehrung der kleinen zu christus
Hugh price hughes / Mantle, John Gregory – New York: Eaton & Mains, [1901?] – (= ser New century leaders) – 1mf – 9 – 0-524-08869-1 – mf#1993-3333 – us ATLA [975]
Hugh smythe papers : from the holdings of the schomburg center for research in black culture, manuscripts, archives and rare books division: the new york public library, astor, lenox and tilden foundations – 1995 – 6r – 1 – $510.00 – (guide which covers all coll under "international affairs" sold separately for $20 d3305.g5) – mf#D3305P15 – Dist. us Scholarly Res – us L of C Photodup [320]
Hughes, A *see*
– The bec missal
– The porteforium of st wulstan, vol 1-2
Hughes, Albert William *see* Outlines of indian history comprising the hindu, mahomedan and christian periods
Hughes, Arthur John Brodie *see* Kin, caste, and nation among the rhodesian ndebele
Hughes, Barry G *see*
– Rapid bimanual movement: effects of direction changes on coordination
– The representation and reproduction of two-dimensional movement patterns
Hughes, Charles Evans *see* Mr hughes' attitude toward the negro
Hughes, Donald Robert Stewart *see* Organization, implementation, and evaluation of a music program for an open secondary school
Hughes, Dorothea Price *see* The life of hugh price hughes
Hughes, Edwin Holt *see* The teaching of citizenship
Hughes, Ernest Richard *see*
– Chinese philosophy in classical times
– Chi-tu tu te lien ko wen t'i
Hughes, George *see*
– Christian holiness almanac and year book
– Fragrant memories of the tuesday meeting
– Holiness year book for...
– Illustrated holiness year book for...
– International holiness directory and year book for...
Hughes, Helen Sard *see* The history of the novel in england
Hughes, Henry *see*
– Natural morals
– Religious faith
– Supernatural morals
– A treatise on hydrophobia
Hughes, Henry Maldwyn *see* The theology of experience
Hughes heritage – v1 n1-v2 n3 [1983 jun-1985 aug/sep], v3 [1986 sep] – 1r – 1 – mf#1128187 – us WHS [929]
Hughes, Hugh Joshua *see* Life of howell harris
Hughes, Hugh Price *see*
– Essential christianity
– Ethical christianity
– The philanthropy of god
– Social christianity
Hughes, Isaac C *see* The external evidence of the bible
Hughes, J C *see* De lagardes ausgabe der arabischen sbersetzung des pentateuchs
Hughes, J E *see* Eighteen years on lake bangweulu
Hughes, James Laughlin *see*
– Canadian history
Hughes, John *see*
– Both sides of the controversy between the roman and reformed churches
– Complete works of the most rev john hughes, d d, archbishop of new york
– Controversy between rev. messrs. hughes and breckenridge

– A discussion of the question, is the roman catholic religion, in any or in all its principles or doctrines, inimical to civil or religious liberty?
– An itinerary of provence and the rhone
– Kirwan unmasked
Hughes, John Arthur *see* Garden architecture and landscape gardening
Hughes, John Caleb *see* De lagarde's ausgabe der arabischen uebersetzung des pentateuchs (cod leiden arab 377)
Hughes, Joseph *see* Attachment to life
Hughes, Kevin P *see* Influence of conceptually based physical education on student attitudes toward physical activity
Hughes' key to the revelation : paragraphed with proper headings and illustrative diagrams and an address on the apostolate of the revelation – [Holland, MI: J S Hughes], c1906 – 1mf – 9 – 0-8370-3687-9 – mf#1985-1687 – us ATLA [221]
Hughes, Langston *see*
– Poems from black africa
– Simple stakes a claim
Hughes, Laurence A *see* Truth about the movies
Hughes, Lorraine C *see* A biomechanical analysis of a sit-to-stand transfer among the elderly
Hughes, Mary *see*
– Family dialogues
– Good grandmother
Hughes, Mary Ellen *see* The rural mimeo newspaper experiment in liberia
Hughes, MaryBeth *see* The search for feminine form
Hughes, Matthew Simpson *see* Dancing and the public schools
Hughes, R M *see* The duties of judge advocates...
Hughes, R O *see* Community civics
Hughes, R W *see* Hughes' reports of cases in the fourth circuit, 1792-1883
Hughes' reports of cases in the fourth circuit, 1792-1883 / Hughes, R W – Washington/ New York: Morrison/Banks. v1-5. 1877-83 (all publ) – (= ser Early federal nominative reports) – 36mf – 9 – $54.00 – mf#LLMC 81-457 – us LLMC [340]
Hughes, Rice *see* Defence of the right reverend the lord bishop of bangor
Hughes, Rupert *see*
– Excuse me
– Music lovers' cyclopedia
Hughes, Thomas *see*
– The friendship of books, and other lectures
– History of the society of jesus in north america
– James fraser, second bishop of manchester
– Loyola and the educational system of the jesuits
– The manliness of christ
– Travels in sicily, greece and albania
Hughes, Thomas Patrick *see* Notes on muhammadanism
Hughes, Thomas Welburn *see*
– Cases on the law of evidence.
– An illustrated treatise on the law of evidence
– Sprague's illustrative cases upon the law of evidence
Hughes, William Leonard *see* Administration of health and physical education in colleges
Hughes, William Taylor *see*
– Equity
– Procedure, its theory and practice
Hughes-Games, A *see* The first and second books of chronicles
Hughes-Hallett, F *see* China looking west
Hughes's historical readers, standard 3 / Cox, George William – London: Joseph Hughes, 1882 – (= ser 19th c children's literature) – 2mf – 9 – mf#6.1.50 – uk Chadwyck [900]
Hughes's historical readers, standard 4 / Cox, George William – London: Joseph Hughes, 1882 – (= ser 19th c children's literature) – 2mf – 9 – mf#6.1.49 – uk Chadwyck [900]
Hughes's historical readers, standard 5 / Cox, George William – London: Joseph Hughes, 1883 – (= ser 19th c children's literature) – 2mf – 9 – mf#6.1.48 – uk Chadwyck [900]
Hughes's historical readers, standard 6 / Cox, George William – London: Joseph Hughes, 1884 – (= ser 19th c children's literature) – 2mf – 9 – mf#6.1.51 – uk Chadwyck [900]
Hughesville mail – Hughsville, PA. -w 1900-1912 – 13 – $25.00r – us IMR [071]
Hughey, George Washington *see*
– Baptismal remission
– Debate on the action of baptism
– Political romanism
– The scriptural mode of christian baptism
Hughson, David *see* Walks through london
Hugi, Franz Joseph *see* Ueber das wesen der gletscher: und, winterreise in das eismeer
Hugideo : eine alte geschichte / Scheffel, Joseph Viktor von – 5. Aufl. Stuttgart: A Bonz, 1887 – 1r – 1 – us Wisconsin U Libr [830]
Hugle, Richard Friedrich *see* Zur buehnentechnik adolph muellners
Hugo, Abel *see* Histoire de l'empereur napoleon
Hugo de S Caro *see* Speculum ecclesiae
Hugo, H *see*
– Affectos divinos con emblemas sagradas por el p. po de salas de la compania de jesus...
– De militia eqvestri antiqva et va ad regem philippum 4 libri qvinqve
– Goddelycke wenschen verlicht
– Gottselige begirden r.p. hermanni hugonis s. jes. verteutscht durch r.p.f. carl. stengelium ord. s. ben.
– Obsedio bredana armis phillippi
– Pia desideria emblematis elegiis et affectibus s.s. patrum...
– Pia desideria emblematis elegiis et affectibus ss. patrum...
– Pia desideria. lib. iii. ad urbanum 8
– Pia desideria tribus libris comprehensa
– Pieux desirs imites des latins du r.p. herman hugo de la compagnie de iesus
Hugo, Hermannus *see*
– L'ame amante de son dieu
– L'ame amante de son dieu...
– [L'ame amante de son dieu...]
– De godlievende ziel vertoont in zinnebeelden
– De godlievende ziel vertoont in zinnebeelden met dichtkunstige verklaringen van Jan Suderman
– Die ihren gott liebende seele
Hugo, L C *see* La vie de s norbert
Hugo, Lud *see*
– Sacrae antiquitatis monumenta
– Sacri et canonici ordinis praemonstratensis annales
Hugo, Victor *see*
– Hunchback of notre dame
– Jargal
– Marion de lorme
– Paris
– Poems and translations
– Rhine
– Ruy blas
– Selections, chiefly lyrical, from the poetical works.
Hugo von hofmannsthal : eine literarische studie / Sulger-Gebing, Emil – Leipzig: M Hesse, 1905 [mf ed 1992] – (= ser Breslauer beitraege zur literaturgeschichte 3) – 93p – 1 – (incl bibl ref. "anhang. chronologisches verzeichnis der dichtungen und aufsatz hugo von hofmannsthals") – mf#8014 reel 1 – us Wisconsin U Libr [430]
Hugo von hofmannsthal's griechenstuecke : 1 [und] 2 / Hladny, Ernst – Leoben: Im Verlag des k. k. Staatsgymnasiums in Leoben, 1910 – 1r – 1 – (incl bibl ref) – us Wisconsin U Libr [430]
Hugo von hofmannsthals mythologische oper "die aegyptische helena" / Lenz, Eva-Maria – Frankfurt a.M., 1972 – 2mf – 9 – 3-89349-743-9 – gw Frankfurter [430]
Hugo von hofmannsthals nachgelassenes lustspielfragment "die rhetorenschule" oder "timon der redner" : literarische und historische hintergruende / Fackert, Juergen – Frankfurt a.M., 1972 – 1mf – 9 – 3-89349-665-3 – gw Frankfurter [430]
Hugo von montfort / ed by Bartsch, Karl – Stuttgart: Litterarischer Verein, 1879 (Tuebingen: L F Fuess) – 1 – us Wisconsin U Libr [810]
Hugo von montfort : poems / ed by Bartsch, Karl – Stuttgart: Litterarischer Verein, 1879 (Tuebingen: L F Fues) [mf ed 1993] – (= ser Blvs 143) – 236p – 1 – mf#8470 reel 30 – us Wisconsin U Libr [810]
Hugolin, pere *see*
– Bibliographie antonienne
– Bibliographie des ouvrages concernant la temperance
– Bibliographie franciscaine
– Echos heroi-comiques du naufrage des anglais sur l'isle-aux-oeufs en 1711
– Entrez donc!
– L'etablissement des recollets a l'isle percee, 1673-1690
– L'etablissement des recollets a montreal, 1692
– Haut la croix!
– N'en buvons plus!
– Les registres paroissiaux de rimouski, des trois-pistoles et de l'ile-verte, tenus par les recollets, 1701-1769
– Saint antoine de padoue et les canadiens-francais
– Victoires et chansons
Hugolin, R P *see* Les vacances du jeune temperant
Hugolinus, B *see* Thesaurus antiquitatum sacrarum
Huguenot cemetery / Crowe, F Hilton – s.l, s.l? 1938 – 1r – us UF Libraries [978]
Huguenot records, 1578-1787 / Crottet, Alexandre Ceasar – [mf ed 1981] [Spartanburg SC: Reprint Co, dist] – 127mf – 9 – mf#51-037 – us South Carolina Historical [242]
The huguenots : or, reformed french church. their principles delineated; their character illustrated; their sufferings and successes recorded / Foote, William Henry – Richmond [Va.] Presbyterian Committee of Publication [1870]. Chicago, Dep of Photodup, U of Chicago Lib, 1975 (1r); Evanston: American Theol Lib Assoc, 1984 (1r) – 1 – 0-8370-0649-X – mf#1984-B425 – us ATLA [240]
The huguenots : their settlements, churches, and industries in england and ireland / Smiles, Samuel – New York: Harper, 1868 – 1mf – 9 – 0-524-04973-4 – (incl bibl ref) – mf#1990-1376 – us ATLA [242]
The huguenots and henry of navarre / Baird, Henry Martyn – New York: Scribner, 1886 – 3mf – 9 – 0-7905-4484-9 – (incl bibl ref) – mf#1988-0484 – us ATLA [944]
The huguenots and the revocation of the edict of nantes / Baird, Henry Martyn – New York: Scribner, 1895 – 3mf – 9 – 0-7905-4541-1 – (incl bibl ref) – mf#1988-0541 – us ATLA [944]
Les huguenots et la constitution de l'eglise reformee de france en 1559 / Castel, Elie – Paris: Grassart, 1859 [mf ed 1992] – 1mf – 9 – 0-524-02000-0 – (in french) – mf#1990-0545 – us ATLA [242]
The huguenots in the seventeenth century : including the history of the edict of nantes, from its enactment in 1598 to its revocation in 1685 / Tylor, Charles – London: Simpkin, Marshall, Hamilton, Kent, 1892 – 1mf – 9 – 0-8370-9908-0 – mf#1986-3908 – us ATLA [242]
The huguenots of la rochelle : a translation of the reformed church of la rochelle, an historical sketch = Eglise reformee de la rochelle / Delmas, Louis – New York: ADF Randolph, c1880 – 1mf – 9 – 0-524-03461-3 – (in english) – mf#1990-1004 – us ATLA [240]
The hugueonots / Punshon, William Moreley – 6th ed. London: James Nisbet, 1872. 76p – 1 – us Wisconsin U Libr [944]
Hugues, Clovis *see* Poesies choisies
Hugues, Edmond *see*
– Histoire de la restauration du protestantisme en france au 18e siecle
– Leibniz et bossuet
Hugues, Jean-Pierre *see* Histoire de l'eglise reformee d'anduze
Hui chiao ch'ien shuo / Ma, T'ien-ying – [China: Chung-kuo Hui chiao chu hsieh hui, 1940] – (= ser P-k&k period) – us CRL [260]
Hui chin / Hsiao, Ch'ien – Kuei-lin: Wen hua sheng huo ch'u pan she, 1942 – (= ser P-k&k period) – us CRL [480]
Hui hsun (ccs) = China christian educational association newsletter – Shanghai. v1-5. 1947-51 [gaps] [mf ed 198?] – (= ser Chinese christian serials coll) – 1 – mf0269m – us ATLA [240]
Hui i / Ma, Kuo-liang – Shang-hai: Liang yu fu hsing t'u shu kung ssu, 1940 – (= ser P-k&k period) – us CRL [840]
Hui k'an / Dewan Geredja2 Keristen Tionghoa di Indonesia – Surabaya, 1949-1958 – 8mf – 9 – (missing: 1949-1957(86, 89-91, 93, 95, 96); 1958(99-106=, 108-end)) – mf#SE-357 – ne IDC [959]
Hui ku lu : ti 2 ts'e – Nan-ching: Tu li ch'u pan she, Min kuo 35 [1946] – (= ser P-k&k period) – us CRL [920]
Hui ku lu / Tsou, Lu – Ch'ung-ch'ing: Tu li ch'u pan she, Min kuo 33 [1944] – (= ser P-k&k period) – us CRL [920]
Hui, Sai C *see* Comparison of acute heart rate and blood pressure responses among isometric, isotonic, and isokinetic exercise
Hui se yen ching / Chiang, Hung-chiao – Shang-hai: Ch'ang ch'eng shu chu, 1931 – (= ser P-k&k period) – us CRL [830]
Hui wu ts'ung k'an *see* Chia yu kung pao [ccs]
Huia tangata kotahi – Hastings, NZ. 1893-95 – 1r – 1 – mf#35.12 – nz Nat Libr [079]
Huidekoper, Frederic *see*
– The belief of the first three centuries concerning christ's mission to the underworld
– Indirect testimony of history to the genuineness of the gospels
– Indirect testimony to the gospels
– Judaism at rome
Huiginn, Eugene Joseph Vincent *see* The graves of myles standish and other pilgrims
Huila / Galvao, Henrique – Vila Nova de Famalicao, Italy. 1929 – 1r – us UF Libraries [972]
Huila / Vargas Motta, Gilberto – Neiva, Colombia. 1957 – 1r – us UF Libraries [972]
Hui-li *see* The life of hiuen-tsiang
Het huiselik en maa schappelik leven van de zuid-afrikaner in de eerste helft der 18de eeuw / Dominicus, Foort Cornelius – Gravenhage, 1919 – 1 – us CRL [960]
Het huiselik en maatschappelik leven van se zuid-afrikaner in de eerste helft der 18de eeuw / Dominicus, F C – s Gravenhage, M Nijhoff, 1919 – us CRL [960]
Die huisgenoot – Cape Town SA, 1 may 1916-29 dec 1950 – 123r – 1 – sa National [073]
Huish, Marcus Bourne *see*
– The art annual for 1890 birket foster his life and work by marcus b huish
– Catalogue of the collection of japanese works of art
– Greek terra-cotta statuettes
– Japan and its art
– Samplers and tapestry embroideries...also stitchery of the same
[Huisseau, I d'] *see*
– La discipline des eglises reformees de france
– La reunion du christianisme...
Huit exercices avec leur doigte : pour le piano-forte, tirees de la methode de piano / Steibelt, Daniel – 2nd ed. Offenbach: Jean Andre [c1810] [mf ed 1992] – 1r – 1 – mf#pres. film 113 – us Sibley [780]
Huit mois a madagascar / Rolland, J B – Marseille: T Samat, 1890 – 1 – us CRL [960]
Huit mois au congo / Maitrejean, Auguste – [Paris: Tolra, c1926, 1929] – 1 – us CRL [960]
Huit mois en amerique : lettres et notes de voyages 1864-1865 / Duvergier de Hauranne, Ernest – Bruxelles, Leipzig, Livourne: A Lacroix, Verboeckhoven. 2v. 1866 [mf ed 1984] – 2v on 1mf – 9 – mf#45586 – cn Canadiana [917]
Le huitieme quodlibet de godefroid de fontaines / Hoffmans, J – Louvain, 1924 – (= ser Philosophes belges 4) – 14mf – 8 – €27.00 – ne Slangenburg [110]
Huizinga, Arnold van Couthen Piccardt *see*
– The american philosophy pragmatism
– Authority
– Belief in a personal god
– Discussions on damnation
Huizinga, Henry *see* Missionary education in india
Huizinga, Wilhelmus Johannes *see* The rise of modern theology in holland
Huke ha-mizrah ha-kadmon / Korngreen, Philip – Tel-Aviv, Israel. 1944 – 1r – us UF Libraries [939]
Hukum dan masjarakat – Djakarta, 1960(1-6) – 5mf – 9 – mf#SE-646 – ne IDC [959]
Hukum dan masjarakat – Djakarta, 1947-1958 – 72mf – 9 – (missing: 1947(1, 3-5); 1948; 1949; 1950(3-5); 1951(2-4); 1954(4-5)) – mf#SE-645 – ne IDC [959]
Hukum nasional : madjalah lembaga pembinaan hukum nasional – Djakarta, 1968-1971 – 15mf – 9 – mf#SE-150-0 – ne IDC [959]
Hulasatue'l-efkar – Istanbul, 1873-74. Sahib-i Imtiyaz: Antuvan, Umur-i Tahiriyye: Luefti. n3-66. 13 haziran-25 agustos 1289 [1873] – (= ser O & t journals) – 5mf – 9 – $75.00 – us MEDOC [956]
Hulbert, Archer Butler *see* David zeisberger's history of the northern american indians
Hulbert, Eri Baker *see* The english reformation and puritanism
Hulbert, Henry Woodward *see* The church and her children
Hulbert, Homer Bezaleel *see* The history of korea
Hulbert, James *see* Complete fifer's museum
Hulcote – (= ser Bedfordshire parish register series) – 1mf – 9 – £3.00 – uk BedsFHS [929]
Hulcote, st nicholas monumental inscriptions monumental inscriptions – Bedfordshire Family HS 1982 – (= ser Bedfordshire parish register series) – 1mf – 9 – £1.25 – uk BedsFHS [929]
Hulda a rees : the pentecostal prophetess (title suggested by rev. e. i. d. pepper): or, a sketch of her life and triumph, together with seventeen of her sermons / Rees, Byron Johnson – Philadelphia: Christian Standard Co, [c1898] El Segundo, Ca: Micro Publication Systems, 1981 (1mf); Evanston: American Theol Lib Assoc, 1984 (1mf) – (= ser Women & the church in america) – 9 – 0-8370-1421-2 – mf#1984-2213 – us ATLA [240]
Hulde album / Vanderheyden, J F – Leuven, 1970 – €12.00 – ne Slangenburg [240]
Huldigingsbundel aangebied aan professor daniel pont – Kaapstad, South Africa. 1970 – 1r – us UF Libraries [960]
Die huldigung der kuenste / demetrius / marfa's monolog / der epilog zu schillers glocke / ed by Suphan, Bernhard – Weimar: Goethe-Gesellschaft, 1905 [mf ed 1993] – (= ser Schriften der goethe-gesellschaft 20) – 1 – (incl bibl ref. int by ed) – mf#8657 reel 5 – us Wisconsin U Libr [090]
Huldreich zwingli : eine darstellung seiner persoenlichkeit und seines lebenswerkes / Burckhardt, Paul – Zuerich: Rascher, 1918 – 2mf – 9 – 0-524-07855-6 – mf#1991-3400 – us ATLA [242]
Huldreich zwingli : geschichte seiner bildung zum reformator des vaterlandes / Schuler, J M – Zuerich, 1818 – 5mf – 9 – mf#ZWI-98 – ne IDC [242]
Huldreich zwingli : geschichte seiner bildung zum reformator des vaterlandes / Schuler, J M – Zuerich, Leipzig, 1819 – 5mf – 9 – mf#ZWI-99 – ne IDC [242]
Huldreich zwingli / Koehler, W – Leipzig, 1923 – (= ser Die Schweiz im deutschen Geistesleben) – 1mf – 9 – mf#ZWI-78 – ne IDC [242]
Huldreich zwingli : leben und ausgewaehlte schriften / Christoffel, Raget – Elberfeld: RL Friderichs, 1857 – 2mf – 9 – 0-524-01881-2 – mf#1990-0508 – us ATLA [242]

Huldreich zwingli : leben und ausgewaehlte schriften der vaeter und begruender der reformirten kirche / Christoffel, R – v1 – 9mf – 9 – mf#ZWI-83 – ne IDC [242]
Huldreich zwingli : the reformer of german, 1484-1531 / Jackson, S M – New York, London, 1901 – 7mf – 9 – mf#ZWI-66 – ne IDC [242]
Huldreich zwingli : the reformer of german switzerland. together with an historical survey of switzerland before the reformation. and a chapter on zwingli's theology / Jackson, Samuel Macauley et al – 2nd rev ed. New York: GP Putnam, 1903 – (= ser Heroes of the reformation) – 2mf – 9 – 0-7905-8125-6 – (incl bibl ref) – mf#1988-8042 – us ATLA [242]
Huldreich zwingli : sein leben und wirken / Staehelin, Rudolf – Basel: B. Schwabe, 1895-1897 – 3mf – 9 – 0-7905-8094-2 – (incl bibl ref) – mf#1988-8030 – us ATLA [242]
Huldreich zwingli : sein leben und wirken nach den quellen dargestellt / Staehelin, R – Basel. 2v. 1895 – 12mf – 9 – mf#ZWI-101 – ne IDC [242]
Huldreich zwingli und sein reformationswerk : zum vierhundertjaehrigen geburtstage zwinglis / Staehelin, Rudolf – Halle: Verein fuer Reformationsgeschichte, 1883 – (= ser [Schriften des Vereins fuer Reformationsgeschichte]) – 1mf – 9 – 0-7905-4714-7 – mf#1988-0714 – us ATLA [242]
Huldreich zwingli und sein reformationswerk : zum vierhundertjaehrigen geburtstage zwinglis / Staehelin, Rudolf – Halle: Verein fuer Reformationsgeschichte, 1883. ([Schriften des Vereins fuer Reformationsgeschichte; Bd. 3]) – 1mf – us ATLA [242]
Huldrichen zwinglens antwort wider hieronimum emser... / Zwingli, H – N p, 1525 – 1mf – 9 – mf#PBU-513 – ne IDC [242]
Huldrych zwinglis bibliothek / Koehler, W – Zuerich, 1921 – 1mf – 9 – mf#ZWI-77 – ne IDC [242]
Huldrych zwinglis briefe, 1512-1526 – Zuerich, 1918-1920. 2 v – 6mf – 9 – mf#ZWI-8 – ne IDC [242]
Huldrychi zuinglii epistola ad petrum gynoraeum, nunc augustae agentem, in qua nonnulla de eccio, fabro, balthazare catabaptista, comperies / Zwingli, H – N p, 1526 – 1mf – 9 – mf#ME-1217 – ne IDC [242]
Huleh – Jerusalem, Israel. 1951/52 – 1r – us UF Libraries [939]
Huli – v1 n2-v2 n8 [1971 jun 5-1973 feb] – 1r – 1 – mf#1058504 – us WHS [071]
Huliau = The turning point / Hawaii Education for Social Progress. People's Fund [Honolulu HI] – 1982 nov-1986 sep/oct – 1r – 1 – mf#1344640 – us WHS [071]
Hull 1630-1849 – Oxford, MA (mf ed 1996) – (= ser Massachusetts vital record transcripts to 1850) – 3mf – 9 – 0-87623-271-3 – (mf 1t: intentions & marriages 1686-1843; births & deaths 1630-1734. mf 2t: births 1731-1843; marriages 1808-43; deaths 1693-1844. mf 3t: vital records 1833-49) – us Archive [978]
Hull 1630-1900 – Oxford, MA (mf ed 1990) – (= ser Massachusetts vital records) – 18mf – 9 – 0-87623-117-2 – (mf 1-2: land grants 1657-1841. mf 3: b,d,m 1630-1757; intents & marriages 1696-1843. mf 4: births & deaths 1693-1749, 1771-1863; town records 1666-1734+; deaths 1691, 1749-66. mf 5-9: town records 1675-1788. mf 10-11: births 1842-92. mf 11-13: marriages 1843-92. mf 13-16: deaths 1843-1905. mf 17: marriages 1892-1900. mf 18: births 1892-1900) – us Archive [978]
Hull arrow – Hull, England 8 dec 1887-3 aug 1889 [mf 1888] – 1 – (discontinued) – uk British Libr Newspaper [072]
Hull, David William *see* The riverside preachers: models for preaching peace
Hull, Edmund C P *see* The european in india
Hull, Edward *see* Institution and abuse of ecclesiastical property
Hull, Fred H *see*
– Corn varieties and hybrids and corn improvement
– Inheritance of rest period of seeds and certain other characters in the peanut...
Hull, Hugh Munro *see* Practical hints to emigrants intending to proceed to tasmania
Hull, John *see* Observations on a petition for the revision of the liturgy of the u...
Hull news – [Yorkshire & Humberside] Hull jan 1884-17 apr 1930 [mf ed 2004] – 214r – 1 – (missing: jan-jun 1897; cont by: hull daily news etc [jan 1885-apr 1914]; daily news etc [may 1914-apr 1916]; hull daily news etc [may 1916-apr 1923]; hull evening news etc [may 1923-apr 1930]) – uk Newsplan [072]
Hull packet – England. -w. 29 Jan 1793; 7 Jan 1800-Dec 1886 (Wanting 1802, 1820-22, 1824, 1825) (1880-88 imperfect). (71 reels) – 1 – uk British Libr Newspaper [072]
Hull weekly sentinel – [Yorkshire & Humberside] Hull 14 jan 1928-dec 1950 – 6r – 1 – uk Newsplan [072]
Hull, William Winstanley *see*
– Disuse of the athanasian creed advisable in the present state of th...
– Occasional papers on church matters
Hull's crucible – Boston, MA, 1 Jan 1874-10 Nov 1877 – 1 reel – 1 – us Western Res [071]
Hulme advertiser, chorlton-upon-medlock and stretford observer – [NW England] Manchester 21 may 1870-6 may 1871 [mf ed 2003] – 1r – 1 – uk Newsplan [072]
Hulme, Edward Maslin *see* The renaissance, the protestant revolution and the catholic reformation in continental europe
Hulme, F E *see* The history, principles and practice of symbolism in christian art
Hulme, Frederick Edward *see*
– Art instruction in england
– The birth and development of ornament
– The history, principles and practice of symbolism in christian art
– Principles of ornamental art
Hulsey researcher – 1980 jan-1986 n1 – 1r – 1 – mf#1218765 – us WHS [071]
Hulshof, Abraham *see* Geschiedenis van de doopsgezinden te straatsburg van 1525 tot 1557
Hulshof, Franz *see* Alban stolz in seiner entwicklung als schriftsteller
H(ulsius), B *see* Emblemata sacra
Hulsius, B *see* Den ondergancк des roomschen arents door den noordschen leeuw
Hulsius, Levinus *see*
– Dictionaire francois allemand et allemand francois
– Dictionarium teutsch-italiaenisch und italiaenisch-teutsch
Hulst, F *see* Zamenspraak tusschen jan, pieter en hendrik
Hulst, Felix van *see* Notice sur le p hennepin d'ath
Hulst, Lammert J *see*
– Drie en zestig jaren prediker
– Herdenking van zijne vijftig-jarige evangeliebediening
– Kentering in de verbondsleer
– Open brief aan rev. n.h. dosker; een misbegrepen tekst toegelicht; en kerkje spelen
– De volmaking der gemeente
Hulstaert, G *see*
– Carte linguistique du congo belge
– Le mariage des nkundo
– Rechtspraakfabels van de nkundo
– Les sanctions coutumiere contre l'adultere chez les nkundo
– Les sanctions coutumieres contre l'adultere chez les nkundo
Hult, Ralph D *see* Ralph d hult papers 1914-1952, 1917-1928
Hultschiner zeitung – Hultschin (Hlucin CZ), 1938 30 apr-21 dec – 1r – 1 – gw Misc Inst [077]
The hum of the college – [Sackville, NS]: Mt Allison Ladies' College Rhetoric Class, [1894?-189- or 1900] [fortnightly] – 9 – mf#P05098 – The whitehouse siftings – cn Canadiana [378]
Human and experimental toxicology – Houndsmill. 1990+ (1,5,9) – (cont: human toxicology) – ISSN: 0960-3271 – mf#13460,01 – us UMI ProQuest [615]
Human behavior – Los Angeles. 1972-1979 (1) 1972-1979 (5) 1972-1979 (9) – ISSN: 0046-8134 – mf#11491 – us UMI ProQuest [150]
Human biology – Detroit. 1929+ (1) 1965+ (5) 1970+ (9) – ISSN: 0018-7143 – mf#1469 – us UMI ProQuest [574]
The human boy and the war : [novel] / Phillpotts, Eden – Toronto: S B Gundy, [191-?] – 4mf – 9 – 0-659-90453-5 – mf#9-90453 – cn Canadiana [830]
Human brain mapping – New York, 1998+ [1,5,9] – ISSN: 1065-9471 – mf#21617 – us UMI ProQuest [612]
Human communication research – Thousand Oaks. 1974+ (1,5,9) – ISSN: 0360-3989 – mf#10986 – us UMI ProQuest [150]
The human cycle / Ghose, Aurobindo – New York: Sri Aurobindo Library, Inc, c1950 – (= ser Samp: indian books) – us CRL [150]
Human depravity and moral responsibility / Mcdonald, John – Edinburgh, Scotland. 18– – 1r – us UF Libraries [240]
Human destiny, a discussion : do reason and the scriptures teach the utter extinction of an unregenerate portion of human beings, instead of the final salvation of all? / Hudson, Charles Frederic & Cobb, Sylvanus – Boston: Sylvanus Cobb, 1860 – 2mf – 9 – 0-524-08447-5 – mf#1993-2052 – us ATLA [240]
Human development – Basel. 1966+ (1) 1966+ (5) 1994+ (9) – ISSN: 0018-716X – mf#2062 – us UMI ProQuest [150]
Human development news – 1980 jun-1985 fall [inc] – 1r – 1 – mf#483895 – us WHS [301]
Human ecology – Ithaca. 2000+ (1) – (cont: human ecology forum) – ISSN: 1530-7069 – mf#18269,01 – us UMI ProQuest [574]
Human ecology – New York. 1972+ (1) 1972+ (5) 1978+ (9) – ISSN: 0300-7839 – mf#10857 – us UMI ProQuest [574]
Human ecology forum – Ithaca. 1989-1999 (1,5,9) – (cont by: human ecology) – ISSN: 0018-7178 – mf#18269 – us UMI ProQuest [574]
The human element in the gospels : a commentary on the synoptic narrative / Salmon, George – London: John Murray, 1907 – 2mf – 9 – 0-7905-0274-7 – (in english and greek. incl indes) – mf#1987-0274 – us ATLA [226]
The human element in the inspiration of the sacred scriptures / Curtis, Thomas Fenner – New York: D Appleton, 1867 – 1mf – 9 – 0-8370-2793-4 – mf#1985-0793 – us ATLA [220]
The human embryo and foetus : constitutional and other legal issues / Slabbert, Melodie Nöthling – Uni of South Africa 2000 [mf ed Johannesburg 2000] – 9mf – 9 – mf#mfm14932 – sa Unisa [342]
Human events / National Foundation for Education in American Citizenship – 1956 jun 9-1963 oct 12, 1967 feb 4-1968 nov, 1968 nov 9-1970 jun, 1970 jul-1971 dec, 1972, 1973, 1974 jan-1975 feb, 1975 mar-dec, 1976 jan-1977 feb, 1977 mar-1978 apr – 10r – 1 – (cont: clintonwatch) – mf#201266 – us WHS [322]
Human events – Washington. 1944+ [1]; 1979+ [5,9] – ISSN: 0018-7194 – mf#1987 – us UMI ProQuest [320]
Human experimentation : federal laws, legislative histories, regulations and related documents / ed by Reams, Bernard D & Gray, Carol – 6r with 1 looseleaf binder or 150mf with 1 looseleaf binder – 1,9 – (binder + current index $125.00; release 1.-9. $375.00; release 2.-9. $375.00; release 1.-1. $250.00; release 2.-1. $250.00) – us Trans-Media [348]
Human face / Picard, Max – New York, NY. 1930 – 1r – us UF Libraries [700]
Human factors – Santa Monica. 1958+ (1) 1975+ (5) 1975+ (9) – ISSN: 0018-7208 – mf#10575 – us UMI ProQuest [500]
Human factors and ergonomics in manufacturing – New York. 1997+ (1) – (cont: international journal of human factors in manufacturing) – mf#18120,01 – us UMI ProQuest [620]
Human Factors and Ergonomics Society Meeting *see* Proceedings of the human factors and ergonomics society annual meeting
Human Factors Society *see*
– Proceedings of the annual meeting...
– Proceedings of the human factors society annual meeting
Human freedom : and, a plea for philosophy: two essays / Nevin, John Williamson – Mercersburg, PA: PA Rice, 1850 [mf ed 1993] – 1mf – 9 – 0-524-08768-7 – mf#1993-3273 – us ATLA [120]
Human genetics – Heidelberg. 1981-1996 (1,5,9) – ISSN: 0340-6717 – mf#13116,01 – us UMI ProQuest [575]
Human heredity – Basel. 1967-1974 (1) 1967-1974 (5) – ISSN: 0001-5652 – mf#2044 – us UMI ProQuest [575]
Human immortality : two supposed objections to the doctrine / James, William – Boston: Houghton Mifflin, c1898 – (= ser Ingersoll Lecture) – 1mf – 9 – 0-8370-9138-1 – (incl bibl ref) – mf#1986-3138 – us ATLA [240]
Human immunology – New York. 1980-1991 (1) 1980-1991 (5) 1987-1991 (9) – ISSN: 0198-8859 – mf#42094 – us UMI ProQuest [616]
Human law / Balzac, B – Gary, IN. no date – 1r – us UF Libraries [939]
Human learning – Chichester. 1982-1986 (1,5,9) – (cont by: applied cognitive psychology) – ISSN: 0277-6707 – mf#12920 – us UMI ProQuest [150]
Human life – Boston. v. 1-13 n5. apr 1905-aug 1911 – 1 – us NY Public [073]
Human life : or, practical ethics / De Wette, Wilhelm Martin Leberecht – Boston: J, Munroe, 1842. Beltsville, Md: NCR Corp, 1978 (9mf); Evanston: American Theol Lib Assoc, 1984 (9mf) – 9 – 0-8370-0979-0 – mf#1984-4437 – us ATLA [170]
Human life review – New York. 1979+ (1,5,9) – ISSN: 0097-9783 – mf#12348 – us UMI ProQuest [301]
The human mechanism the most marvelous... : to be read by c baillairge before section 3 of the royal society of canada at its may meeting, 1901 – [Canada?: s.n, 1901?] – 1mf – 0-665-75790-5 – mf#75790 – cn Canadiana [612]
Human milk calcium, phosphorus, sodium and potassium concentrations following maximal exercise / Uhlin, Katherine L – 1996 – 1mf – 9 – $4.00 – mf#PH 1564 – us Kinesology [612]
Human molecular genetics – Oxford, 1997+ [1,5,9] – ISSN: 0964-6906 – mf#21218 – us UMI ProQuest [575]
Human mosaic – New Orleans. 1980-1989 (1) 1980-1980 (5) 1980-1980 (9) – ISSN: 0018-7240 – mf#12152 – us UMI ProQuest [300]
The human motor / Amar, Jules – 1920 – 10mf – 9 – $30.00 – us Kinesology [600]
Human movement science – Amsterdam. 1982+ (1,5,9) – ISSN: 0167-9457 – mf#42529 – us UMI ProQuest [150]
Human nature : a revelation of the divine / Robinson, Charles Henry – London: Longmans, Green, 1902 – 1mf – 9 – 0-524-05058-9 – (incl bibl ref) – mf#1992-0311 – us ATLA [220]
Human nature in politics / Wallas, Graham – New York, NY. 1921 – 1r – us UF Libraries [025]
The human nature of the saints / Hodges, George – New York: Thomas Whittaker, c1904 – 1mf – 9 – 0-524-07102-0 – mf#1991-2925 – us ATLA [220]
Human needs – Washington. 1972-1973 (1) 1972-1973 (5) (9) – mf#6918 – us UMI ProQuest [301]
Human needs and satisfactions : a global survey – 2r – 1 – $260.00 – mf#S1857 – us Scholarly Res [150]
Human neurobiology – Heidelberg. 1982-1987 (1) 1982-1987 (5) 1982-1987 (9) – ISSN: 0721-9075 – mf#13178 – us UMI ProQuest [616]
Human organization – Washington. 1941+ (1,5,9) – ISSN: 0018-7259 – mf#11750 – us UMI ProQuest [301]
Human origins / Laing, Samuel – London, 1892 – (= ser 19th c evolution & creation) – 5mf (ill) – 9 – mf#1.1.8248 – uk Chadwyck [573]
Human pathology – Philadelphia. 1970+ (1) 1973+ (5) 1976+ (9) – ISSN: 0046-8177 – mf#8735 – us UMI ProQuest [614]
Human performance – Mahwah. 1998+ (1,5,9) – ISSN: 0895-9285 – mf#25227 – us UMI ProQuest [150]
Human personality and its survival of bodily death / Myers, Frederic William Henry; ed by Hodgson, Richard & Johnson, Alice – London, New York: Longmans, Green, 1903 [mf ed 1991] – 2v on 4mf – 9 – 0-7905-8863-3 – mf#1989-2088 – us ATLA [130]
Human policy and divine truth / Mill, W H – Cambridge, England. 1850 – 1r – us UF Libraries [240]
Human potential – Philadelphia. 1967-1970 – 1 – mf#10310 – us UMI ProQuest [370]
Human progress and the inward light / Hodgkin, Thomas – London: Published for the Woodbrooke Extension Committee by Headley, 1911 – (= ser Swarthmore Lecture) – 1mf – 9 – 0-8370-8909-3 – mf#1986-2909 – us ATLA [240]
Human progress through missions / Barton, James L – New York: Fleming H. Revell, c1912 – 1mf – 9 – 0-7905-4330-3 – mf#1988-0330 – us ATLA [240]
Human psychopharmacology – Chichester. 1986-1994 (1) 1986-1994 (5) 1986-1994 (9) – ISSN: 0885-6222 – mf#16103 – us UMI ProQuest [615]
Human quest – St Petersburg. 1996+ (1,5,9) – (cont: churchman's human quest) – mf#3364,04 – us UMI ProQuest [240]
Human quest – St Petersburg. 1990-1995 (1,5,9) – (cont: churchman's human quest. cont by: churchman's human quest) – ISSN: 0897-8786 – mf#3364,02 – us UMI ProQuest [240]
The human race and other sermons / Robertson, Frederick William – 2nd ed. London: CK Paul, 1881 – 1mf – 9 – 0-7905-9617-2 – mf#1989-1342 – us ATLA [240]
Human relations – London. 1947+ (1) 1974+ (5) 1978+ (9) – ISSN: 0018-7267 – mf#993 – us UMI ProQuest [300]
Human Relations Area Files *see*
– Area studies program
– Complete archive program
– Cultural diversity program
– Cultural types program
– Hraf probability sample program-part a
– Hraf topical program
– Traditional cultures of the world program
Human Relations Area Files, Inc *see* Cross-cultural research
Human relations news of chicago / Chicago Commission on Human Relations – 1959 jun-1976 jan – 1r – 1 – mf#271312 – us WHS [302]
Human relations report – Pennsylvania, v1 n1-v7 n2 [1962 sep-1968 apr/jun] – 1r – 1 – mf#675572 – us WHS [302]
Human relations training news – Washington. 1954-1969 (1) 1970-1970 (5) (9) – ISSN: 0018-7291 – mf#5334 – us UMI ProQuest [370]
Human reproduction – Oxford. 1988+ (1,5,9) – ISSN: 0268-1161 – mf#16452 – us UMI ProQuest [618]
Human resource development quarterly – San Francisco. 1990+ (1,5,9) – ISSN: 1044-8004 – mf#17616 – us UMI ProQuest [650]
Human resource management – New York. 1961+ (1) 1971+ (5) 1975+ (9) – ISSN: 0090-4848 – mf#3070 – us UMI ProQuest [650]

Human resource management international digest – Bradford. 2001+ (1,5,9) – ISSN: 0967-0734 – mf#31617 – us UMI ProQuest [650]
Human resource management review – Greenwich. 1994+ (1,5,9) – ISSN: 1053-4822 – mf#19773 – us UMI ProQuest [650]
Human resources abstracts – Thousand Oaks. 1975+ (1) 1975+ (5) 1977+ (9) – (cont: poverty and human resources abstracts) – ISSN: 0099-2453 – mf#5051,01 – us UMI ProQuest [331]
Human rights – Chicago. 1989+ (1,5,9) – ISSN: 0046-8185 – mf#18270 – us UMI ProQuest [322]
Human rights (aba) – v1-28. 1970-2001 – 1,5,6 – $356.00 – (v1-18 1970-91 in reel or mf $220. v19-28 1992-2001 in mf $136) – ISSN: 0046-8185 – mf#100241 – us Hein [322]
Human rights and democracy in kyrgyzstan : hearing before the commission on security and cooperation in europe, 107th congress, 1st session, dec 12 2001 / United States. Congress. Commission on Security and Cooperation in Europe – Washington: US GPO 2002 [mf ed 2002] – 1mf – 9 – 0-16-068358-0 – us GPO [322]
[Human rights documents 1980-2002] – 1 title on 20,269mf – 9 – (individual titles also listed separately) – ne IDC [322]
Human rights, european politics, and the helsinki accord : the documentary evolution of the conference on security and co-operation in europe 1973-1975 / Kavass, Igor I & Granier, Jacqueline P & Dominick, Mary F – 7v – 9 – $100.00 – mf#301951 – us Hein [322]
Human rights quarterly – Baltimore. 1981+ (1,5,9) – (cont: universal human rights) – ISSN: 0275-0392 – mf#11730,01 – us UMI ProQuest [300]
Human rights quarterly – v1-19. 1979-97 – 9 – $535.00 set – (title varies: v1-2 1979-80 as: universal human rights) – ISSN: 0163-2647 – mf#108681 – us Hein [322]
Human rights quarterly *see* Universal human rights
Human rights review – Monrovia: [s.n.], jun 1993-nov/dec 1994 – 1r – 1 – us CRL [322]
Human rights, the helsinki accords and the united states : selected executive and congressional documents / ed by Kavass, Igor I & Granier, Jacquelin P – Series 1-3 in 9bks – 9 – $198.00 set – 0-89941-466-4 – mf#301961 – us Hein [324]
Human rights watch, 2000 annual update – (mf ed 2001) – 60mf – 9 – $600.00 coll – (bangladesh 1mf; bosnia and hercegovina 1mf; burundi: neglecting justice in making peace 1mf, emptying the hills 1mf; china 1mf; colombia 1mf; democratic republic of congo 1mf, general: landmine monitor report 12mf, human rights watch world report 6mf, civilian deaths in the nato air campaign 1mf; georgia 1mf; indonesia 1mf; israel 1mf, japan 3mf; kuwait 1mf; malaysia 1mf; pakistan 1mf; russia 2mf; russia/chechnya: feb 5 1mf, civilian killings...1mf, no happiness remains 1mf; rwanda 1mf; serbia & montenegro 1mf; south africa 1mf; tanzania 2mf; tunisia 1mf; turkey: small group isolation in turkish prisons 1mf, human rights and the european union accession partnership 1mf; united states: fingers to the bone 2mf, unfair advantaage 3mf, out of sight 1mf, punishment and prejudice 1mf, clinton's landmine legacy 1mf; uzbekistan: and it was hell all over again 1mf, leaving no witnesses 1mf; vietnam 1mf; federal republic of yugoslavia 1mf) – us UMI ProQuest [323]
Human rights watch publications, 1990 – 150mf 122 publ – 1 – $1,050.00 – us UMI ProQuest [322]
Human rights watch publications, 1991 – 176mf 139 publ – 9 – $1,230.00 – us UMI ProQuest [322]
Human rights watch publications, 1992 – 154mf 123 publ – 1 – $1,080.00 – us UMI ProQuest [322]
Human rights watch publications, 1993 – 149mf 111 publ – 1 – $1,040.00 – us UMI ProQuest [322]
Human rights watch publications, 1994 – 128mf 89 publ – 1 – $900.00 – us UMI ProQuest [322]
Human rights watch publications, 1995 – 94 mf 71 publ – 1 – $660.00 – us UMI ProQuest [322]
Human rights watch publications, 1996 – 94mf 65 publ – 1 – $660.00 – us UMI ProQuest [322]
Human rights watch publications, 1980-1989 – 352mf 289 publ – 9 – $2,460.00 – us UMI ProQuest [322]
Human rights watch reports, 1997 – 89mf 69 publ – 1 – $620.00 – us UMI ProQuest [322]
Human rights watch reports, 1998 – 65mf 42 publ – 1 – $450.00 – us UMI ProQuest [322]
Human rights watch reports, 1999 – 98mf 52 publ – 1 – $690.00 – us UMI ProQuest [323]
Human scale : the official publication of the center for community technology / Center for Community Technology [Madison WI] – v1 n1-v4 n1 [1980 nov-1983 feb/mar] – 1r – 1 – (cont: news & views [center for community technology [madison, wi]]) – mf#998845 – us WHS [303]
Human services in the rural environment – Cheney. 1976-1995 (1) 1976-1995 (5) 1976-1995 (9) – ISSN: 0193-9009 – mf#12178 – us UMI ProQuest [360]
Human skeletal muscle function and morphology : the effects of age and exercise / Hunter, Sandra Kay – 1998 – 5mf – 9 – $20.00 – mf#PH 1622 – us Kinesology [612]
Human society : its providential structure, relations, and offices / Huntington, Frederic Dan – New York: Robert Carter, 1860. Beltsville, Md: NCR Corp, 1978 (4mf); Evanston: American Theol Lib Assoc, 1984 (4mf) – 9 – 0-8370-0820-4 – (incl bibl ref) – mf#1984-4164 – us ATLA [301]
Human society : its providential structure, relations, and offices / Huntington, Frederic Dan – New York:Robert Carter, 1860, c1859 – 1mf – 9 – 0-8370-4509-6 – (incl bibl ref) – mf#1985-2509 – us ATLA [230]
Human studies – Dordrecht. 1989+ (1,5,9) – ISSN: 0163-8548 – mf#16790 – us UMI ProQuest [100]
Human systems management – Amsterdam. 1989-1996 (1,5,9) – ISSN: 0167-2533 – mf#19415 – us UMI ProQuest [650]
Human thinking / Fleshman, Arthur Cary – Spartanburg, SC. 1927 – 1r – us UF Libraries [025]
Human torch – iss n2-30. fall 1940-may 1948 – 15 – (n1 not publ; 2 issues of n5) – mf#017MV-021MV; 051MV – us MicroColour [740]
Human toxicology – Houndsmill. 1981-1989 (1) 1981-1989 (5) 1981-1989 (9) – (cont by: human and experimental toxicology) – ISSN: 0144-5952 – mf#13460 – us UMI ProQuest [615]
Humanae salutis monumenta b. ariae montani studio constructa et decantata / Arias Montanus, B – Antwerpiae: Ex officina Christophori Plantini, [1571] – 3mf – 9 – mf#O-127 – ne IDC [090]
Human-computer interaction – Mahwah. 1998+ (1,5,9) – ISSN: 0737-0024 – mf#25224 – us UMI ProQuest [150]
Humane policy : or justice to the aborigines of new settlements. essential to a due expenditure of british money, and to the best interests of the settlers. with suggestions how to civilise the natives by an improved administration... / Bannister, Saxe [pseud] – London, 1830 – (= ser 19th c books on british colonization) – 6mf – 9 – mf#1.1.9891 – uk Chadwyck [941]
Humaniora / ed by Usteri, Paul – Leipzig 1796-98 – (= ser Dz. abt literatur) – 10mf – 9 – €100.00 – mf#k/n4609 – gw Olms [410]
Humanism : philosophical essays / Schiller, Ferdinand Canning Scott – 2nd ed, enl. London: Macmillan, 1912 – 1mf – 9 – 0-7905-9869-8 – mf#1989-1594 – us ATLA [140]
Humanismo e o plano nacional de educacao / Lins, Ivan Monteiro De Barros – Rio de Janeiro, Brazil. 1938 – 1r – us UF Libraries [972]
Humanismo y humanitarismo / Delmonte Y Aponte, Domingo – Habana, Cuba. 1960 – 1r – us UF Libraries [972]
El humanismo y la moral de juan pablo sartre (critica) / Frutos Cortes, Eugenio – sp Bibl Santa Ana [120]
Humanist – 1851 dez-1853 okt 16 – 1r – 1 – mf#1095596 – us WHS [170]
Humanist – Buffalo. 1941+ [1]; 1971+ [5]; 1975+ [9] – ISSN: 0018-7399 – mf#5867 – us UMI ProQuest [100]
Humanist Association of Minneapolis and St Paul *see* Humanist news
Humanist educator – Falls Church. 1975-1982 (1) 1975-1982 (5) 1975-1982 (9) – (cont: student personnel association for teacher education journal. cont by: journal of humanistic education and development) – ISSN: 0362-9783 – mf#3276,01 – us UMI ProQuest [370]
Humanist in canada – Ottawa. 1964-1996 (1) 1972-1996 (5) 1974-1996 (9) – ISSN: 0018-7402 – mf#7083 – us UMI ProQuest [100]
Humanist news : newsletter of the humanist association of minneapolis and st paul / Humanist Association of Minneapolis and St Paul – 1986 oct-1990 mar – 1r – 1 – mf#1770734 – us WHS [170]
The humanist way in ancient china : essential works of confucianism / ed by Chai, Chu & Chai, Winberg – Bantam matrix eds. New York: Bantam Books, 1965 – 1mf – 9 – 0-524-08098-4 – mf#1993-9004 – us ATLA [180]
Humanistas del siglo 18 / Mendez Plancarte, Gabriel – Mexico City? Mexico. 1941 – 1r – us UF Libraries [972]
Humanistisches magazin zur gemeinnuetzlichen unterhaltung und insonderheit in beziehung auf akademische studien / ed by Wiedeburg, Friedrich August – Helmstedt, Leipzig 1787-94 – (= ser Dz) – 5v on 15mf – 9 – €150.00 – mf#k/n886 – gw Olms [000]
Humanitaet und humanismus : grundzuege einer kulturgeschichte / Weiss, Albert Maria – Freiburg im Breisgau; St Louis, MO: Herder, 1879 [mf ed 1986] – (= ser Apologie des christenthums vom standpunkte der sittenlehre 2) – 3mf – 9 – 0-8370-7352-9 – (incl bibl ref) – mf#1986-1352 – us ATLA [230]
Humanitaet und religion : eine von der gesellschaft zur vertheidigung der christlichen religion gekroente preisschrift / Hartmann, Julius – Leiden: E J Brill, 1873 – 1mf – 9 – 0-8370-3511-2 – mf#1985-1511 – us ATLA [200]
L'humanitaire : organe de la science sociale – n1-2. Paris. juil-aout 1841 – 1 – fr ACRPP [300]
Humanitarian Order of Kosmic Awareness *see*
– Hooka
– Hooka notes
Humanitas – Berlin DE, 1962 11 jan-1989 – 13r – 1 – gw Misc Inst [074]
Humanitas – Pittsburgh. 1965-1979 (1) 1975-1979 (5) 1975-1979 (9) – ISSN: 0018-7496 – mf#8215 – us UMI ProQuest [100]
Humanitat – Montpellier/Paris/Barcelona. jun 1946-mar 1952 – 1/2r – 1 – uk British Libr Newspaper [072]
La humanitat – Montpelier and Paris. France. -w. 6 Jun 1946-25 Dec 1948, 17 Feb 1949-Mar 1952. (33 ft) – 1 – uk British Libr Newspaper [072]
L'humanite : journal socialiste quotidien – Paris: Fond. J Jaures, 1904-1914 – 1 – fr ACRPP [074]
L'humanite – Paris: Imprimerie des arts et manufactures, 1949-55; 1956-69 – us CRL [074]
Humanite nouvelle – Alhiers. Feb-oct 1944 – 1/4r – 1 – uk British Libr Newspaper [072]
L'humanite-dimanche / Communist Party. France – Paris. oct 1948-1993 – 1 – fr ACRPP [335]
Humanities / National Endowment for the Humanities – 1980 jan/feb-1985 – 1r – 1 – (cont: humanities, national endowment for the humanities) – mf#455974 – us WHS [000]
Humanities : report of the national endowment for the humanities / National Endowment for the Humanities – 1969/70 winter-1978 jun – 1r – 1 – (cont by: humanities [washington, dc: 1980]) – mf#302427 – us WHS [000]
Humanities – Washington. 1980+ (1,5,9) – ISSN: 0018-7526 – mf#12328 – us UMI ProQuest [000]
Humanities collections / ed by Kinder, Robin – v1. n1. 1997- – 1,9 – $65.00 in US $91.00 outside hardcopy subsc – us Haworth [073]
Humanities journal – Tempe. 1975-1978 (1) 1975-1978 (5) 1975-1978 (9) – ISSN: 0046-8266 – mf#10694 – us UMI ProQuest [000]
Humanity : its destiny and the means to attain it = Katholische kirche und das ziel der menschheit / Denifle, Heinrich Seuse – Ratisbon; New York: Fr. Pustet, 1909 – 1mf – 9 – 0-7905-5650-2 – (in english) – mf#1988-1650 – us ATLA [241]
Humanity at the cross-roads / Randall, John Herman – New York: Dodge, c1915 – 1mf – 9 – 0-7905-8721-1 – mf#1989-1946 – us ATLA [240]
The humanity, benevolence and charity legislation of the pentateuch and the talmud : in parallel with the laws of hammurabi, the doctrines of egypt, the roman 12 tables and modern codes / Fluegel, Maurice – Baltimore: H Fluegel, 1908 – 1mf – 9 – 0-8370-3157-5 – mf#1985-1157 – us ATLA [270]
Humanity immortal : or, man tried, fallen, and redeemed / Hickok, Laurens Perseus – Boston: Lee and Shepard, 1872 – 1mf – 9 – 0-7905-3861-X – mf#1989-0354 – us ATLA [240]
Humanity's gain from unbelief / Bradlaugh, Charles – London, England. 1889 – 1r – us UF Libraries [240]
Humanizm in der elterer yidisher literatur / Stiff, Nahum – Berlin, Germany. 1922 – 1r – us UF Libraries [939]
Humarat munyati – Cairo: Muhammad Tawfiq al-Azhari & Muhammad Hilmi 'Aziz, 1898-1904; 1905-08. v1 n1-v6 n4. 1 shawwal 1315-16 jumada I 1326 [23 feb 1898-15 jun 1908] – (= ser Arabic journals and popular press) – 1r – 1 – $775.00 – (in1905 title changes to: al-mawqudhah and then back to: humarat munyati. missing: v5 n1. rl also incl: al-mawqudhah) – us MEDOC [079]
Humareda lirica / Mounier Roman, Rafael – Barcelona, Spain. 1962 – 1r – us UF Libraries [972]
Humas dprd-gr / Legislatif Jaya. Madjalah bulanan DPRD-GR DCI Djakarta – Djakarta, 1968-1971. v1-3(1-26) – 23mf – 9 – (missing: 1968, v1(10); 1970, v3(21-23)) – mf#SE-1761 – ne IDC [959]
Humayun badshah / Banerji, S K – London; New York: Oxford University Press, 1938-1941 – (= ser Samp: indian books) – (int by e denison ross) – us CRL [954]
Humayun Kabir *see*
– Mahatma and other poems
– Our heritage
– Poems
– Poetry, monads, and society
– Sarat chandra chatterjee
Humbel, F *see* Ulrich zwingli und seine reformation im spiegel der gleichzeitigen, schweizerischen volkstuemlichen literatur
Humber College of Applied Arts and Technology *see* Content
Humbert, Auguste *see*
– Frere et mari
– Les origines de la theologie moderne
Humbert, Edouard *see* Souvenirs d'etudes maconniques 1870-1882
Humbert, Pierre Hubert *see*
– Instructions chretiennes pour les jeunes gens
Humbertclaude, P *see* La doctrine ascetique de saint basile de cesaree
Humbird enterprise – Humbird WI. 1904 dec 17-1908, 1909-16, 1917-24, 1925-32, 1933-40, 1941-1949 jan 29 – 6r – 1 – mf#945613 – us WHS [071]
Humble and explanatory memorial of dr george edwards : to the honourable the commons of the united kingdom of great britain and ireland in parliament assembled... – [London, 1816?] – (= ser 19th c general coll on politics) – 1mf – 9 – mf#1.1.19 – uk Chadwyck [330]
Humble attempt to put an end to the present divisions in the church / Rose, Lewis – Glasgow, Scotland. 1840 – 1r – us UF Libraries [240]
Humble, earnest, and affectionate address to the clergy / Law, William – London, England. 1843 – 1r – us UF Libraries [240]
Humble, Henry *see* Rights of faithful laymen in the church of christ
Humble, Henry et al *see* Essays on the re-union of christendom
Humble petitions, etc : or, a prospectus of proposals for rectifying our present infinitely distressed and dangerous situation / Edwards, George – Newcastle: printed by S Hodgson, 1817 – (= ser 19th c economics) – 1mf – 9 – mf#1.1.73 – uk Chadwyck [339]
Humble way – Houston. 1945-1972 (1) – (cont by: exxon usa) – ISSN: 0018-7607 – mf#10110 – us UMI ProQuest [550]
Humboldt, Alejandro de *see* Ensayo politico sobre el reino de la nueva espana
Humboldt, Alexander Von *see* Ensayo politico sobre la isla de cuba
Humboldt, Alexander von *see*
– Briefe von alexander von humboldt an varnhagen von ense aus den jahren 1827 bis 1858
– Essai politique sur le royaume de la nouvelle-espagne
– Humboldt centennial. bound newspaper clippings concerning the proceedings in various cities in the united states
– Humboldt-perlen
– Kosmos
– Observations sur quelques phenomenes peu connus qu'offre le goitre sous les tropiques, dans les plaines et sur les plateaux des andes
– Robert hermann schomburgk's reisen in guiana und am orinoko waehrend der jahre 1835-1839
– The travels and researches of alexander von humboldt
– Voyage aux regions equinoxiales du nouveau continent, fait en 1799, 1800, 1801, 1802, 1803 et 1804
– Vues des cordilleres, et monumens des peuples indigenes de l'amerique
Humboldt bay region, 1850-1895 : a study in american colonization of california / Coy, Owen C – Eureka, CA: California Hist Assn – 1r – 1 – $50.00 – mf#B40221 – us Library Micro [978]
Humboldt centennial. bound newspaper clippings concerning the proceedings in various cities in the united states / Humboldt, Alexander von – 1869. 2v – 1 – us Wisconsin U Libr [574]
Humboldt county – 1908-33; 1992- – (= ser California telephone directory coll) – 28r – 1 – $1400.00 – mf#P00036 – us Library Micro [917]
[Humboldt county-] eureka and arcata city directories – CA. 1946-1950 – 3r – 1 – $150.00 – mf#D041 – us Library Micro [917]
[Humboldt county-] eureka census of the indians of california – CA. 1936-38; 1941-42 – 3r – 1 – $150.00 – mf#D040 – us Library Micro [978]
[Humboldt county-] humboldt county including eureka – CA. 1895-1896; 1926-1950 – 16r – 1 – $800.00 – mf#D039 – us Library Micro [978]
[Humboldt county-] humboldt, lake, marin, mendocino, napa, solano, sonoma and yolo counties – CA. 1884-1886 – 3r – 1 – $150.00 – mf#D038 – us Library Micro [978]
Humboldt daily standard *see* [Eureka-] humboldt standard

Humboldt enterprise – Humboldt, NE: Harrison Bros, jan 1888-v23 n8. jun 16 1905=whole n -1159 (wkly) [mf ed with gaps filmed 1975] – 4r – 1 – (cont: nebraska enterprise. absorbed by: falls city tribune. some irregularities in numbering) – us NE Hist [071]
Humboldt, F H A von *see* Voyage aux regions equinoxiales du nouveau continent
Humboldt, Friedrich Heinrich Alexander von *see* Views of nature
Humboldt independent news – CA. dec 13 1973-dec 19 1974 – 1r – 1 – $110.00 – mf#B03215 – us Library Micro [071]
Humboldt independent news *see* [Eureka-] north coast ripsaw; rank and file reporter; humboldt independent news
Humboldt leader – Humboldt, NE: H P and Myrtle W Marble. -v24 n2. apr 29 1920 (wkly) [mf ed 1899-1920 (gaps) filmed 1957] – 5r – 1 – (cont: standard-leader. cont by: humboldt news) – us NE Hist [071]
The humboldt news – Humboldt, NE: Simeon Beardsley. v1 n1. may 7 1920-23// (wkly) [mf ed with gaps] – 2r – 1 – (cont: humboldt leader (1899)) – us NE Hist [071]
Humboldt register – Unionville NE. 1865 mar 25 – 1r – 1 – (cont by: humboldt register and workingman's advocate) – mf#875697 – us WHS [071]
Humboldt standard – Humboldt, NE: S P Willis (wkly) [mf ed 1899-1902,1908- (gaps)] – 1 – (cont: standard-leader. absorbed: dawson herald. issues for oct 13 1899-may 17 1935 also called whole n897-2742) – us NE Hist [071]
Humboldt standard – Humboldt, NE: Geo P Monagon, 1882-v16 n5. jul 30 1897 (wkly) [mf ed 1884-97 (gaps) filmed 1958-] – 2r – 1 – (merged with: humboldt leader to form: humboldt standard and leader. issues for feb 16 1894-jul 30 1897 also called whole n599-780) – us NE Hist [071]
The humboldt standard – Humboldt, NE: Vern Gibbens. v1 n1. may 19 1921-v26 n45. nov 6 1947 (wkly) [mf ed with gaps filmed [1974?]] – 5r – 1 – (cont: verdon delphic. absorbed by: humboldt standard (1899). some irregularities in numbering) – us NE Hist [071]
Humboldt standard and leader – Humboldt, NE: H P Marble. 2v. v16 n9. aug 5 1897-v17 n10. aug 5 1898=whole n781-833 (wkly) [mf ed lacks aug 20 1897] – 1r – 1 – (formed by the union of: humboldt standard and: humboldt leader. cont by: standard-leader) – us NE Hist [071]
Humboldt star – Winnemucca, NV. 1966-1967 (1) – mf#64780 – us UMI ProQuest [071]
Humboldt star *see* [Winnemucca-] star
Humboldt, Wilhelm *see* Linguistic variability and intellectual development
Humboldt, Wilhelm, Freiherr von *see*
– Ansichten ueber aesthetik und literatur
– Sechs ungedruckte aufsaetze ueber das klassische altertum
Humboldt, Wilhelm von *see*
– Lettre a m abel-remusat
– Proben vaskischer schreibart und dichtung
– Ueber die aufgabe des geschichtsschreibers
Humboldt-perlen : ein demantkranz aus alexander von humboldt's leben und schriften – nebst einer chronologischen uebersicht seines lebens, einem verzeichnis seiner zahlreichen werke und einem portraet a. v. humboldt's nach dem von ihm selbst entworfenen spiegelbilde / Humboldt, Alexander von – Leipzig: E Wartig, 1869 – 1r – 1 – us Wisconsin U Libr [920]
Humbug – Melbourne, 1869-70 – 1r – 1 – A$27.50 vesicular A$33.00 silver – at Pascoe [079]
Hume / Huxley, Thomas Henry – New York: Harper & Brothers, 1879. vi,206p. (English Men of Letters) – 1 – us Wisconsin U Libr [920]
Hume / Knight, William Angus – Edinburgh: W Blackwood, 1886 – (= ser Philosophical Classics for English Readers) – 1mf – 9 – 0-7905-8675-4 – mf#1989-1900 – us ATLA [100]
Hume : with helps to the study of berkeley / Huxley, Thomas Henry – New York, NY. 1896 – 1r – us UF Libraries [120]
Hume, Allan O *see* The game birds of india, burmah and ceylon
Hume, David *see*
– Essay on miracles
– Essays moral, political, and literary
– The history of england from the invasion of julius caesar to the abdication of james the second, 1688
– A treatise of human nature
Hume, George Henry *see* Canada, as it is
Hume, H Harold *see*
– Anthracnose of the pomelo
– Cauliflower
– Citrus fruits and their culture
– Cultivation of citrus groves
– Diagrams for packing citrus fruits
– Japanese persimmons
– Kumquats
– Mandarin orange group
– Pecan culture
– Peen-to peach group
– Pineapple culture ii
– Pineapple culture iv
– Planting plans for florida home orchards
– Pomelos
– Potato diseases
– Some citrus troubles
– Top-working pecans
Hume, Hamilton *see* Life of edward john eyre
Hume, Harold *see* Second report on pecan culture
Hume, James Gibson *see* Socialism
Hume, Martin Andrew Sharp *see* Wives of henry the eighth
Hume, Robert Allen *see*
– An interpretation of india's religious history
– Missions from the modern view
Hume, Robert Ernest *see*
– Will jesus christ satisfy the religious needs of the world?
– World's living religions
Hume, with helps to the study of berkeley : essays / Huxley, Thomas Henry – London: Macmillan, 1908 – 1mf – 9 – 0-7905-3922-5 – mf#1989-0415 – us ATLA [190]
Hume, with helps to the study of berkeley : essays / Huxley, Thomas Henry – New York: D. Appleton and Company, 1896 – 319p. 1r – 1 – us Wisconsin U Libr [190]
Humesh-folklor / Zlotnik, Sz – Warszawa, Poland. v1-3. 1937-1938 – 1r – us UF Libraries [939]
Humfeld, Neill Hamilton *see* History and construction of the cornett
The humiliation of christ in its physical, ethical, and official aspects : the sixth series of the cunningham lectures / Bruce, Alexander Balmain – 4th ed. Edinburgh: T & T Clark, 1895 – 2mf – 9 – 0-7905-0069-8 – (incl bibl ref and index) – mf#1987-0069 – us ATLA [240]
Die hummel : wochenblatt fuer einwanderer – New York NY (USA), 1851 mar-apr – 1r – 1 – gw Misc Inst [071]
Hummelauer, Franz von *see*
– Der biblische schoepfungsbericht
– Commentarius in deuteronomium
– Commentarius in exodum et leviticum
– Commentarius in genesim
– Commentarius in libros iudicum et ruth
– Commentarius in libros samuelis
– Commentarius in librum iosue
– Commentarius in librum primum paralipomenon
– Commentarius in numeros
– Das vormosaische priesterthum in israel
Hummelstown sun – Hummelstown, PA. -w 1895-1959 – 13 – $25.00r – us IMR [071]
The hummer – Holmesville, NE: J A Church, apr 1 1902- (wkly) [mf ed -1903 (gaps) filmed [1979]] – 1r – 1 – (publ in blue springs ne, oct 24 1902- . issues for apr 7 1902- called n50-) – us NE Hist [071]
Humming bird : a monthly scientific, artistic, and industrial review – London, 1891-1894. v1-4 – 14mf – 8 – mf#Z-2020 – ne IDC [700]
Humming bird : or, herald of taste – Newfield. 1798-1798 (1) – mf#3524 – us UMI ProQuest [810]
Humo de mi pipa / Gil Fortoul, Jose – Caracas, Venezuela. 1956 – 1r – us UF Libraries [972]
Humo del tiempo / Obrador, Gina – Habana, Cuba. 1960 – 1r – us UF Libraries [972]
Humor der deutschen staemme : eine mundartensammlung / Poddel, Peter – Hamburg: Hanseatische Verlagsanstalt, c1940 – 1r – 1 – (incl bibl ref) – us Wisconsin U Libr [430]
Humor in der jungen deutschen dichtung / ed by Rockenbach, Martin – Augsburg: Orplid-Verlag, [1928?] – 1r – 1 – us Wisconsin U Libr [870]
Der humor jesu : beitrag zur skizzierung des historischen christusbildes / Wuenkhaus, Oscar W – Heidelberg: Evangelischer Verlag, 1909 – 1mf – 9 – 0-524-00220-7 – mf#1989-2920 – us ATLA [220]
Humoradas / Sanchez-Arjona, Vicente – Sevilla: Graficas Sevillanas, 1951 – 1 – sp Bibl Santa Ana [810]
Humorbuch : deutsche dichter aus fuenf jahrhunderten / ed by Riess, Richard – Muenchen: G Mueller, 1918 – 1 – us Wisconsin U Libr [870]
Humorbuch : deutsche dichter aus fuenf jahrhunderten / ed by Riess, Richard – Muenchen: G Mueller, 1918 – 1r – 1 – us Wisconsin U Libr [430]
Humoresken / Stinde, Julius – Berlin: Freund & Jeckel, 1892 – 1r – 1 – us Wisconsin U Libr [870]
Humorismo : el epigrama y la satira / Llorens, Washington – San Juan, Puerto Rico. 1960 – 1r – us UF Libraries [870]
Humorismo de machado de assis / Souza, Claudio De – Rio de Janeiro, Brazil. 1941 – 1r – us UF Libraries [972]
Humorismos y satiras / Moreno Torrado, Luis – Merida: Imprenta y Estereotipia de Corchero y Compania, 1902 – 1 – sp Bibl Santa Ana [810]
Humoriste / Dupeuty, Charles – Paris, France. 1840 – 1r – us UF Libraries [440]
Humoristische beilage *see* Dresdner nachrichten
Humoristische blaetter – Oldenburg/Oldbg DE, 1838-41 – 2r – 1 – gw Misc Inst [870]
Humoristische gedichte / Knortz, Karl – 2. verb. Aufl. Glarus: J Vogel, 1889 – 1r – 1 – us Wisconsin U Libr [810]
Humoristische reisebilder : entworfen auf einer wanderung durch berlin, dresden, die saechsische schweiz, teplitz, prag; und heimwaerts durch weimar, goettingen und hannover / Wagner, Gottlob – Meissen 1831 [mf ed Hildesheim 1995-98] – 2mf – 9 – €60.00 – 3-487-29483-4 – gw Olms [870]
Humoristisches *see* Dresdner nachrichten
Humorous phases of the law / Browne, Irving – San Francisco: Sumner Whitney & Co, 1876 – 3mf – 9 – $4.50 – mf#LLMC 95-163 – us LLMC [870]
Humour – Sao Paulo, Brazil. 1936 – 1r – us UF Libraries [870]
The humours and conversations of the town – 1693 – 9 – us Scholars Facs [390]
Humphrey democrat – Humphrey, NE: C H Swallow. v9 n32. oct 4 1895- (wkly) [mf ed with gaps filmed 1975-] – 1 – (absorbed: platte county leader 1908 and: lindsay post 1952. cont: democrat) – us NE Hist [071]
Humphrey democrat – Humphrey, NE: J W Fuchs, -1893// (wkly) [mf ed 2nd yr n3. mar 23 1888-jun 23 1893 (gaps) filmed 1975] – 1r – 1 – (cont by: democrat) – us NE Hist [071]
Humphrey, E J [Mrs] *see* Six years in india
Humphrey, George *see* Nature of learning in its relation to the living system
Humphrey, George C *see* Righteousness of god a redemptive quality: a study in biblical theology
Humphrey, George Magoffin *see* George magoffin humphrey papers, 1912-1970
Humphrey, Heman *see* Revival sketches and manual
Humphrey, Henry M *see* Dupla defesa, resposta ao pamphleto
Humphrey herald – Humphrey, NE: J T Meere, oct 1895-dec 1895// (wkly) [mf ed with gaps] – 1r – 1 – (cont by: herald) – us NE Hist [071]
Humphrey independent – Humphrey, NE: J I Robison, 1884 (wkly) [mf ed v2 n22. nov 20 1885] – 1r – 1 – us NE Hist [071]
Humphrey, James Lorenzo *see* Twenty-one years in india
Humphrey, L *see* Ioannis ivelli angli, episcopi sarisburiensis vita & mors, eiusq
Humphrey morice papers from the bank of england *see* Slave trade journals and papers
Humphrey, Simon James *see* Eshcol
Humphrey, William *see*
– Conscience and law
– Divine teacher
– His divine majesty, or, the living god
– Mary magnifying god
– Memoranda of angelical doctrine, fasciculus second
– The one mediator
Humphrey, William Ewart *see* Papers, 1903-1934
Humphreys, A E *see* The epistles to timothy and titus
Humphreys, Arthur Lee *see* Piccadilly bookmen
Humphreys, Charles *see* A compendium of the common law in force in kentucky
Humphreys, Christmas *see* Studies in zen
Humphreys' general advertiser, and railway time bill – [Wales] Caernarvon sep 1852-aug 1853 [mf ed 2004] – 1r – 1 – uk Newsplan; uk British Libr Newspaper [380]
Humphreys, Henry Noel *see*
– The art of illumination and missal painting
– The coinage of the british empire...
– The illuminated books of the middle ages
– Specimens of illuminated manuscripts of the middle ages
Humphreys, John F *see* Woman's work in the church
Humphrey's journal *see* The daguerreian journal / humphrey's journal
Humphreys, Kathryn *see* Theorizing experience
Humphreys, Mary Gay *see* Missionary explorers among the american indians
Humphreys' specific homeopathic medicines : homeopathic veterinary specifics, the mild power cures... – [Toronto?: s.n. 187-?] [mf ed 1983] – 1mf – 9 – 0-665-39747-X – mf#39747 – cn Canadiana [615]
Humphries, Arthur Lewis *see* The holy spirit in faith and experience
Humphriss, Deryck *see* Benoni
Humphry, W G *see* Intermediate state
Humphry, William Gilson *see*
– A commentary on the book of the acts of the apostles
– The doctrine of a future state
– The early progress of the gospel
Humphrys, David *see* The justice of the land league
Humpty dumpty's magazine – Indianapolis. 1979+ (1) 1979+ (5) 1979+ (9) – (cont: humpty dumpty's magazine for little children) – ISSN: 0273-7590 – mf#5901,01 – us UMI ProQuest [640]
Humpty dumpty's magazine for little children – New York. 1952-1979 (1) 1971-1979 (5) 1977-1979 (9) – (cont by: humpty dumpty's magazine) – ISSN: 0018-7666 – mf#5901 – us UMI ProQuest [640]
Humus / Caceres Lara, Victor – Comayaguela, Honduras. 1952 – 1r – us UF Libraries [972]
Hun hou : tu mu chu chi / Ch'en, Ch'uan – Ch'ung-ch'ing: Shang wu yin shu kuan, Min kuo 34 [1945] – (= ser P-k&k period) – us CRL [820]
Hun yin yu chia tsu / T'ao, Hsi-sheng – Shanghai: Shang wu, min kuo 23 [1934] – (= ser P-k&k period) – us CRL [306]
Hu-nan chih k'uang yeh / Chang, Jen-chieh – Ch'ang-sha: Hu-nan ching chi tiao ch'a so, Min kuo 23 [1934] – (= ser P-k&k period) – us CRL [550]
Hu-nan hsiang-shih lu – List of successful candidates in the imperial examination in Hunan province: 1888, 1903. 1 reel – 1 – us Chinese Res [951]
Hunan Missionary Conference (1903: Changsha, China) *see* Report of the hunan missionary conference
Hu-nan sheng cheng chih nien chien – Hunan: Sheng cheng fu mi shu ch'u, Min kuo 21 [1932] – (= ser P-k&k period) – us CRL [951]
Hu-nan sheng chin jung kai k'uang / Ch'iu, Jen-hao et al – [Lei-yang]: Hu-nan sheng yin hang ching chi yen chiu shih, Min kuo 31 [1942] – (= ser P-k&k period) – us CRL [332]
Hu-nan sheng shui tsai chiu chi tsung hui pao kao – [Hu-nan sheng: Shui tsai chiu chi tsung hui, Min kuo 25 [1936]] – (= ser P-k&k period) – us CRL [360]
Hu-nan sheng ts'ai cheng cheng li pao kao shu : min kuo erh shih ssu nien pa yueh erh shih i jih chih erh shih liu nien san yueh shih wu jih – [China: Hu-nan sheng ts'ai cheng t'ing, 1937] – (= ser P-k&k period) – us CRL [332]
Hunchback of notre dame / Hugo, Victor – Philadelphia, PA. v1-2. 1888 – 1r – us UF Libraries [025]
Hund und katz' : eine geschichte aus dem bairischen oberlande / Schmid, Herman – Leipzig: Keil [18–?] – (= ser Gesammelte schriften. volks- und familien-ausgabe 41 nf9) – 1 – (bound with: der loder) – mf#film mas c438 – us Harvard [830]
Hundert deutsche fliegerbilder aus palaestina / Dalman, Gustaf – Guetersloh, 1925 – 3mf – 9 – mf#H-2887 – ne IDC [956]
Hundert gedichte : auswahl des verfassers – mit einer selbstbiographie des dichters / Henckell, Karl – Leipzig: Hesse & Becker, [1914?] – 1r – 1 – (incl bibl ref) – us Wisconsin U Libr [810]
Hundert geistliche melodien evangelischen lieder : welche auf die fest- und andere tage, so wohl in der christlichen gemeine, als auch daheim gesungen werden... – [Berlin: Runge 1676] [mf ed 19–] – 3mf / 1r – 9,1 – mf#fiche 1038 / film 2549 – us Sibley [780]
Hundert jahre wiener stadttempel – Wien, Austria. 1926 – 1r – us UF Libraries [939]
Hundert lider / Nissenson, Aaron – NYU York, NY. 1919 – 1r – us UF Libraries [939]
Hundert mayses un mesholim / Israel Meir – Jerusalem, Israel. 1951 or 52 – 1r – us UF Libraries [939]
Hundert tage auf dem nil : reisebilder aus unter- und ober-aegypten und nubien; im anschluss an das buch desselben verfassers "von fernen ufern" / Wallner, Franz – Berlin 1873 [mf ed Hildesheim 1995-98] – 1v on 5mf – 9 – €100.00 – 3-487-27211-3 – gw Olms [916]
Hundert tage auf reisen in den oesterreichischen staaten / Kohl, Johann – Dresden [u a] 1842 [mf ed Hildesheim 1995-98] – 3v on 9mf – 9 – €180.00 – 3-487-29415-X – gw Olms [914]
Hundert vnd dreissig gemeyner fragestuecke fuer die iungen kinder ynn der deudschen meydlin schule zu eyssleben / Johannes Agricola aus Eisleben – [Wittemberg, 1528] – 1mf – 9 – mf#TH-1 mf 801 – ne IDC [242]
Die hundertjaehrige gedaechtnissfeier der kantischen kritik der reinen vernunft; johann gottlieb fichtes leben und lehre; spinozas leben und charakter / Fischer, Kuno – 2. Aufl. Heidelberg: C Winter, 1892 – (= ser Philosophische Schriften) – 1mf – 9 – 0-7905-9200-2 – mf#1989-2425 – us ATLA [100]
Hundertjahrige jubilaum der isr cultusgemeinde in wien im jahre 18... / Wolf, Gerson – Wien, Austria. 1864 – 1r – us UF Libraries [939]
Hundeshagen, C B *see* Die konflikte des zwinglianismus, luthertums und calvinismus in der bernischen landeskirche von 1532-1558
Hundeshagen, K B *see* Beitraege zur kirchenverfassungsgeschichte und kirchenpolitik
Hundeshagen, Karl Bernhard *see* Vorlesungen ueber die lehrbegriffe der kleineren protestantischen kirchenparteien

HUNDHAUSEN

Hundhausen, Ludwig Joseph *see*
- Das erste pontificalschreiben des apostelfuersten petrus
- Das zweite pontificalschreiben des apostelfuersten petrus

The hundred boston orators appointed by the municipal authorities and other public bodies, from 1770 to 1852. / Loring, James Spear – 2nd ed. enl. Boston: J.P. Jewett, 1853.viii,720p – 1 – us Wisconsin U Libr [978]
Hundred flowers – v1-3 n8 [1970 apr 17-1972 mar 24] – 1r – 1 – mf#770827 – us WHS [071]
Hundred years in travancore, 1806-1906 : a history and description of the work done by the london missionary society in travancore, south india during the past century / Hacker, Isaac Henry – London: H R Allenson, 1908 [mf ed 1995] – (= ser Yale coll) – 106p (ill) – 1 – 0-524-09074-2 – mf#1995-0074 – us ATLA [954]
A hundred years of methodism / Simpson, Matthew – New York: Nelson & Phillips; Cincinnati: Hitchcock & Walden, 1876 [mf ed 1991] – (= ser Methodist coll) – 1mf – 9 – 0-524-01397-7 – mf#1990-4093 – us ATLA [242]
A hundred years of missions : or, the story of progress since carey's beginning / Leonard Delavan Levant – New York: Funk & Wagnalls, 1895 [mf ed 1986] – 1mf – 9 – 0-8370-6678-6 – (incl ind) – mf#1986-0678 – us ATLA [240]
Hundred years of the church in india, 1814-1914 / Skipton, H P K – Westminster: Indian Church Aid Association, 1914 [mf ed 1995] – (= ser Yale coll) – 22p (ill) – 1 – 0-524-09193-5 – mf#1995-0193 – us ATLA [240]
A hundred years work for children, 1803-1903 / Groser, William H – 1 – $8.96 – (a sketch of the history and operation of sunday school union from its formation in 1803 to its centenary in 1903) – us Southern Baptist [242]
Huneker, James *see* Intimate letters of james gibbons huneker, collected and edited by josephine huneker
Hung bird : a monthly scientific, artistic, und industrial review – London 1891-94 – v1-4 on 14mf – 9 – mf#z-2020/2 – ne IDC [590]
Hung ch'i – 1958-71 – 5r – 1 – $175.00 in US $40.00r outside – (in chinese) – mf#L9300044 – Dist. us Scholarly Res – us L of C Photodup [951]
Hung, Ch'i *see* Yen
Hung chu / Chin, I – Ch'ung-ch'ing: Wen hua sheng huo ch'u pan she, 1942 – (= ser P-k&k period) – us CRL [840]
Hung chung she po *see* Chung kuoshih pao
Hung hsing ch'u ch'iang t'an tz'u / Yu, T'ing-wu & Ch'ien Yun-chan – Shang-hai: Man li shu chu, Min kuo 24 [1935] – (= ser P-k&k period) – us CRL [951]
Hung hsuan-chia : wu mu shih chu / A-ying; ed by Wie, Ju-hui – Shang-hai: Kuo min shu tien, 1941 – (= ser P-k&k period) – us CRL [480]
Hung liu / Chin, I – Ch'ung-ch'ing: Wen hua sheng huo ch'u pan she, 1941 – (= ser P-k&k period) – us CRL [480]
Hung lou meng kuang i / Ch'ing-shan-hsien-nung – Shang-hai: Chung hsi shu chu, 1932 – (= ser P-k&k period) – us CRL [951]
Hung, Mo *see* Ho ti kuang lin
Hung, Po *see* Shu fen chi
Hung, Shen *see*
- Chi ming tsao k'an t'ien
- Fei chiang chun
- Huang pai tan ch'ing
- Hung shen hsi ch'u chi
- Hung shen hsi chu lun wen chi
- I ch'ien i pai ko chi pen han tzu shih yung chiao hsueh fa
- Mi
- Nu jen nu jen
- Pao te-hsing
- T'ien ying shu yu tz'u tien
- Tsou ssu

Hung, Shen et al *see* Ti i liu
Hung shen hsi ch'u chi / Hung, Shen – Shang-hai: Hsien tai shu chu, 1933 – (= ser P-k&k period) – us CRL [820]
Hung shen hsi chu lun wen chi / Hung, Shen – Shang-hai: T'ien ma shu tien, Min kuo 23 [1934] – (= ser P-k&k period) – us CRL [820]
Hung, Su-yeh *see* Wan-nan lu hsing chi
Hung, T'ao *see* Shui ti tsui o
Hung tao er shih chou chi nien k'an (ccm139) = 20th anniversary publication of the union girls' school, hangchow – Hang-chou, 1932 [mf ed 198?] – (= ser Ccm 139) – 1 – mf#1984-b500 – us ATLA [370]
Hung teng lung : tuan p'ien hsiao shuo chi / Lo, Ai-lan – Ch'ang-sha: Shang wu yin shu kuan, 1938 – (= ser P-k&k period) – us CRL [480]
Hung tou / Yeh, Ting-lo – Hsi-an: Chien hsin shu tien, 1944 – (= ser P-k&k period) – us CRL [830]
Hung tou ti ku shih / Sun, Ling – Ch'ung-ch'ing: Feng huo she, 1940 – (= ser P-k&k period) – us CRL [480]
The hungarian and transylvanian unitarians : some account of their origin, vicissitudes, and present condition / Tagart, Margaret Lucy – London: Unitarian Christian Pub Office, 1903 – 1mf – 9 – 0-524-07836-X – mf#1991-3383 – us ATLA [243]
Hungarian heritage review / Rakoczi Foundation-International – 1986 jan-1988 dec – 1r – 1 – (cont: eighth hungarian tribe) – mf#1700807 – us WHS [929]
The hungarian peace negotiations / Hungary. Versailles Peace Conference Delegation – 1r – 1 – $175.00 – 0-89093-022-8 – us UPA [327]
Hungarian press summary – Budapest, Hungary. feb 1949-1987 – 85r – 1 – us L of C Photodup [073]
Hungarian Socialist Federation of America *see* Elore forward
Hungarian uprising, 1956 / Irving, David – 1r – 1 – mf#97283 – uk Microform Academic [943]
Hungary and its people. / Felbermann, Lajos – 2nd ed. with suppl. London: Griffith Farran, 1893. illus. map – 1 – us Wisconsin U Libr [943]
Hungary. Courts *see*
- Dontvenytar
- Uj dontvenytar

Hungary. FAO National Committee *see* Information bulletin
Hungary. Itelotablak *see* A kiralyi itelotablak felulvizsgalati tanacsainak elvi jelentosegu hatarozatai
Hungary. Kozponti Statisztikai Hivatal *see*
- Magyar orzag tiszti cim-es nevtara
- Statisztikai szemle
- Ungarisches statistisches jahrbuch

Hungary. Kuria *see*
- Felsobirosagaink elvi hatarozatai
- Maganjogi dontvenytar.
- Polgarijogi hatarozatok tara, a kir. kurianak hivatalos kiadvanya.

Hungary. Kuria. Felulvizsgalati tanacs *see* A magyar kir. curia felulvizsgalati tanacsa altal a sommas eljarasrol szolo torveny (1893: xviii. tcz.) alapjan hozott hatarozatoknak gyujtemenye.
Hungary. Laws, Statutes, etc *see*
- AZ 1948
- Az igazolo eljarasok zsebkonyve
- Az uj nepbirosagi torveny (1947: 34. t.-c.) egyseges szerkezetben a hatalyos nepbirosagi rendeletekkel
- A bunvadi perrendtartas zsebkonyve, kiegeszitve az ujabb bunvadi eljarasi szabalyokkal; irtak edvi illes karoly es vargha ferenc
- Torvenyjavaslat a kozigazgatas es az onkormanyzat rendezeserol a varmegyekben

Hungary. Nemzetgyueles *see* Iromanyok
Hungary. Orszaggyules. Felsohaz *see* Iromanyok
Hungary. Statisztikai Hivatal *see* Statisztikai e'vkoenyv
Hungary. Versailles Peace Conference Delegation *see* The hungarian peace negotiations
Hungary.Orszaggyules. Kepviselohaz *see* Naplo
Hungary.Statisztikai Evkonyv *see* Magyar statisztikai evkonyv
Hungate, Jesse Avery *see* The ordination of women to the pastorate in baptist churches
Hunger, Bertolt *see* Zum begriff der kausalitaet in der kriegsursachenforschung
Hunger, J *see* Becherwahrsagung bei den babyloniern
Hunger, Johannes *see* Becherwahrsagung bei den babyloniern
Hungerford, Margaret Wolfe *see*
- 'Airy fairy lilian'
- April's lady
- A born coquette
- The duchess
- Her last throw
- Lady branksmere
- Lady verner's flight
- A little rebel
- Molly bawn
- Mrs geoffrey
- Nor wife nor maid
- The o'connors of ballinahinch

Hungerford, Margaret Wolfe (Hamilton) *see*
- An anxious moment etc
- Her week's amusement
- A lonely girl

Hungerford's real estate journal – 1879 feb 15-1886 jul 15 – 1r – 1 – mf#1058532 – us WHS [333]
Hungering and thirsting after righteousness / Dixon, M C – Wigan, England. 1821 – 1r – us UF Libraries [240]
Hung-i-ta-shih *see* Wan-ch'ing-lao-jen chiang yen lu
Hung-i-ta-shih nien p'u / Lin, Tzu-ch'ing – Shang-hai: Chung Jih wen hua hsieh hui Shang-hai fen hui, Min kuo 33 [1944] – (= ser P-k&k period) – us CRL [951]
Hungry horse news – Columbia Falls, MT. 1949-1974 (1) – mf#64331 – us UMI ProQuest [071]
Hungry people and empty lands : an essay on population problems and international tensions / Chandrasekhar, Sripati – Baroda: Indian Institute for Population Studies, MS University of Baroda, c1952 – (= ser Samp: indian books) – us CRL [304]
Hungry stones and other stories / Tagore, Rabindranath – London: Macmillan and Co, 1916 – (= ser Samp: indian books) – (trans fr original bengali by various writers) – us CRL [830]
Huningford, George Isaac *see* Petition of the english roman catholics considered
Hunnicutt, Benjamin Harris *see*
- Brazil
- Brazil looks forward

Hunnius, A *see*
- Articvlvs de libero arbitrio, sev hvmani arbitrii viribvs, ex scriptvrae
- Articvlvs de persona christi
- Assertio sanae et orthodoxae doctrinae de persona et maiestate domini nostri iesv christi
- Calvinvs ivdaizans, hoc est
- Catechismus
- De aeterna praedestinatione filiorvm dei ad salvtem propositiones theologicae
- De praedestinatione saluandorum
- Historischer bericht
- Kurtzer vnd nuetzlicher bericht von dem heilsamen vnd christlichem buch formulae concordiae
- Methodvs concionandi praeceptis et exemplis dominicalivm qvorvndam
- Postilla
- Warhaffter gruendtlicher bericht von rechter ordenlicher wahl vnnd beruff der euangelischen prediger

Der hunnsruecken – Simmern DE, 1839 1 apr-1842 – 1 – gw Misc Inst [074]
Hunolt, Franz *see* The christian state of life
Hun's supreme court reports / New York. (State). Supreme Court – v1-92. 1874-95 (all publ) – 761mf – 9 – $1141.00 – mf#LLMC 80-007 – us LLMC [347]
Hunsruecker zeitung : ausgabe j – heimatblatt der rhein-zeitung, koblenz – Simmern DE, 1983 1 jun- – ca 7r/yr – 1 – (title varies: 2 oct 1995: rhein-hunsrueck-zeitung) – gw Misc Inst [074]
Hunt, Amy R *see* An assessment of the attitudes of college students
Hunt, Arthur Surridge *see*
- Fragment of an uncanonical gospel from oxyrhynchus
- Logia iesou
- New sayings of jesus; and, fragment of a lost gospel

Hunt, Brian R *see* Estimation of vo(2max) from a submaximal 1-mile track jog for relatively fit teenage individuals
Hunt Correctional Center *see* Hunt walk talk
Hunt, Frederick Knight *see*
- The book of art

Hunt, Galliard *see* The debates in the federal convention of 1789
Hunt, George Laird *see* Ten makers of modern protestant thought
Hunt, Horace Washington *see* Landlord and tenant laws of texas
Hunt, J E *see* Crammer's first litany 1544
Hunt, J R E *see* Lutheran home missions
Hunt, James *see* A home and work for every man
Hunt, Joel Ransom Ellis *see* Lutheran home missions
Hunt, John *see*
- Contemporary essays in theology
- An essay on pantheism
- Pantheism and christianity
- Religious thought in england, from the reformation to the end of last century
- Religious thought in england in the nineteenth century

Hunt, John Eddy *see* The acknowledgment of deeds, containing all the statutes, territorial and state, of illinois.
Hunt, L E *see* Effects of perceived quality of life between coronary artery bypass graft and heart transplantation patients with regard to cardiac rehabilitation
Hunt, Margaret (Raine) *see* Our grandmother's gowns
A hunt on snow-shoes / Ellis, Edward Sylvester – London, Toronto: Cassell, [1913?] – 5mf – 9 – 0-665-74230-4 – (ill by edwin j prittie) – mf#74230 – cn Canadiana [830]
A hunt on snow-shoes / Ellis, Edward Sylvester – London; Toronto: Cassell [1913?] [mf ed 1995] – 5mf – 9 – 0-665-74230-4 – mf#74230 – cn Canadiana [830]
Hunt, Sarah *see* Journal of the life and religious labors of sarah hunt, late of west grove, chester county, pennsylvania
Hunt, Thomas Frederick *see*
- Architettura campestre...
- Designs for parsonage houses, alms houses, etc
- Exemplars of tudor architecture
- Half a dozen hints on picturesque domestic architecture

Hunt, Thomas Sterry *see*
- The apatite deposits of canada
- Chemical and geological essays
- Esquisse geologique du canada
- A geographical, agricultural and mineralogical sketch
- The hydrometallurgy of copper
- Mineral physiology and physiography
- On the theory of types in chemistry

Hunt walk talk / Hunt Correctional Center – 1980 nov-1985 sum, 1985 fall-1986/87 win – 2r – 1 – mf#1100862 – us WHS [365]
Hunt, William *see*
- The english church from its foundation to the norman conquest
- The english church in the middles ages

Hunte, Garth S *see* Endothelial selectins and pulmonary gas exchange in female aerobic athletes
Hunter / North American Indian League – 1980 aug ed, 1981 spr ed-1983 spr ed – 1r – 1 – mf#81912 – us WHS [305]
Hunter and pickup's panoramic guide from niagara falls to quebec / Hunter, William Stewart – Montreal: Hunter & Pickup, 1865 [mf ed 1984] – 1mf – 9 – 0-665-45065-6 – mf#45065 – cn Canadiana [917]
Hunter, Andrew *see* Duties of subjects
Hunter, Archibald Macbride *see* Interpreting paul's gospel
Hunter, C M *see*
- Agriculture
- Clothes
- Education
- Florida boom
- History of the pinellas peninsula
- Largo, florida
- Modern ethnology
- Pass-a-grille beach
- Pinellas culture
- Pinellas peninsula
- Safety harbor
- St petersburg
- Tarpon fishing

Hunter College *see* Shield
Hunter College Centro de Estudios Puertorriquenos *see* Vertical file materials
Hunter, David *see* David hunter miller papers
Hunter, David Gilbert *see* Pennsylvania orphans' court common place book
Hunter education – Wisconsin, n37-68 [1982 aut-1992 win] – 1r – 1 – (cont: hunter education instructor newsletter; cont by: hunter education newsletter) – mf#2862939 – us WHS [370]
Hunter education instructor – Wisconsin, n35 [1982 win-spr] – 1r – 1 – (cont: newsletter – wisconsin hunter safety, wisconsin. dept of natural resources. bureau of law enforcement; cont by: hunter education instructor newsletter) – mf#2863130 – us WHS [370]
Hunter education instructor newsletter – Wisconsin, n35 [1982 sum] – 1r – 1 – (cont: hunter education instructor [madison, wi]; cont by: hunter education) – mf#709745 – us WHS [370]
Hunter, Elmo B *see* The judicial conference and its committee on court administration
Hunter, Fannie McDowell *see* Women preachers
Hunter, G R *see* The script of harappa and mohenjodaro
Hunter, Gary *see* Handrail support versus free arm wing treadmill fitness test
Hunter is death / Bulpin, Thomas Victor – Cape Town, South Africa. 1968 – 1r – us UF Libraries [960]
Hunter, John *see*
- Essays and observations on natural history, anatomy, physiology, psychology, and geology...
- An historical journal of the transactions at port jackson, and norfolk island
- Memoirs of a captivity among the indians of north america
- Observations and reflections on geology...

Hunter, John Dunn *see*
- Memoirs of a captivity among the indians of north america

Hunter, John Merlin *see* Emerging colombia
Hunter, L *see* Nineteenth century books on british colonisation collection
Hunter, M *see* The study of african society
Hunter, Monica *see* The study of african society
Hunter, P B *see* The effects of running with a functional knee brace on lower extremity joint moments of force in anterior cruciate ligament injured subjects
Hunter, Patrick Teaslie *see* Studies of the variations in the growth and fruiting habits of cert...
Hunter, Peter Hay *see* The story of daniel
Hunter river gazette – Maitland, dec 1841-jun 1842 – 1r – A$27.50 vesicular A$33.00 silver – at Pascoe [079]
Hunter, Robert *see* History of the missions of the free church of scotland in india and africa
Hunter safety newsletter – Wisconsin, n1-18 [1967 dec 11-1975 sum] – 1r – 1 – (cont by: newsletter – wisconsin hunter safety) – mf#389799 – us WHS [350]

Hunter, Samuel James *see* Memorial sermons and addresses
Hunter, Sandra Kay *see* Human skeletal muscle function and morphology
Hunter valley news – Muswellbrook, mar 1975-dec 1992 – 26r – at Pascoe [079]
Hunter, W A *see* Past and present of the heresy laws
Hunter, W D *see* Letters received by the bureau of entomology from w.d. hunter
Hunter, William *see* Description du pegu et de l'isle de ceylan
Hunter, William Stewart *see*
– Hunter and pickup's panoramic guide from niagara falls to quebec
– Hunter's panoramic guide from niagara falls to quebec
Hunter, William Wilson *see*
– The annals of rural bengal
– Bombay, 1885 to 1890
– Bombay 1885 to 1890 a study in indian administration
– A brief history of the indian peoples
– Brief history of the indian peoples
– A history of british india
– The indian musalmans
– Life of the earl of mayo, fourth viceroy of india
– A statistical account of bengal
Hunter-Duvar, John *see* De roberval
A hunter's adventures in the great west / Gillmore, Parker – London: Hurst and Blackett, 1871 – 4mf – 9 – mf#07137 – cn Canadiana [639]
Hunters dublin chronicle or universal journal for the year 1762 – Dublin, Ireland. 5-7 oct 1762 – 1r – 1 – uk British Libr Newspaper [072]
Hunter's hand book of the victoria bridge : illustrated with wood-cuts: a brief history of that wonderful work...1846, up to its completion in 1859 / Boxer, Frederick N – Montreal: Hunter & Pickup, 1860 – 2mf – 9 – mf#33332 – cn Canadiana [624]
Hunters of the desert land / Schoeman, Pieter Johannes – Cape Town, South Africa. 1958? – 1r – us UF Libraries [960]
Hunter's panoramic guide from niagara falls to quebec / Hunter, William Stewart – Montreal: Publ by Hunter & Pickup, 1860 [mf ed 1982] – 1mf – 9 – mf#SEM105P87 – cn Bibl Nat [917]
Hunter's torrens cases / Canada. General – 1v. 1865-93 (all publ) – 7mf – 9 – $11.50 – mf#LLMC 81-012 – us LLMC [340]
Hunter's wanderings in africa / Selous, Frederick Courteney – New York, NY. 1967 – 1r – us UF Libraries [960]
Hunterville news – nov 1972-83 – 2r – 1 – mf#44.2 – nz Nat Libr [079]
Hunting and fishing in florida / Cory, Charles B – Boston, MA. 1896 – 1r – us UF Libraries [639]
Hunting and hunted in the belgian congo / Cooper, Reginald Davey; ed by Johnston, R Keigh – London: Smith, 1914. 263p. ill – 1 – us Wisconsin U Libr [639]
Hunting for gold : reminiscences (sic) of personal experience and research in the early days of the pacific coast from alaska to panama / Downie, William – San Francisco, CA: Calif Pub Co, 1893 – 5mf – 9 – mf#02753 – cn Canadiana [920]
Hunting, Harold Bruce *see* The story of our bible
Hunting in florida in 1874 / Jenks, John Whipple Potter – Providence, RI. 1884 – 1r – us UF Libraries [639]
The hunting of the snark : an agony in eight fits / Carroll, Lewis – New York: Macmillan, 1899 – 2mf – 9 – $3.00 – mf#LLMC 91-014 – us LLMC [810]
Huntingdon gazette – Huntingdon, PA. -w 1806-1835 – 13 – $25.00r – us IMR [071]
Huntingdon literary museum : and monthly miscellany – Huntingdon. 1810-1810 (1) – mf#3992 – us UMI ProQuest [420]
Huntingdon. Presbytery *see* Minutes, 1795-1895
Huntingford, Edward *see*
– The apocalypse
– The resurrection of the body
Huntingford, George Isaac *see*
– Preparation for the holy order of deacons
– Preparation for the holy order of priests
– Sermon, preached at the anniversary of the royal humane society
Huntington 1773-1910 – Oxford, MA (mf ed 1996) – (= ser Massachusetts vital records) – 25v on 102mf – 9 – 0-87623-421-X – (mf 1-3: town meetings 1773-1796. mf 2-3: intents & marriages 1773-1796. mf 3: births & deaths 1764-1805. mf 4-13: town meetings 1797-1834. mf 13-21: town records 1834-1860. mf 22-35: town records 1860-1885. mf 36-38: vitals & index 1781-1846. mf 39-41: vital records 1844-1876. mf 41-43: intentions 1854-1904. mf 44-60: tax lists 1795-1847 61-65: chattels & index 1848-1861. mf 66: civil war soldiers 1861-1865. mf 67-68: paupers 1869-1899. mf 69: census of sick 1884-1908. mf 69-79: voters 1884-1916. mf 80-89: school records 1855-1906. mf 83-96: school census 1875-1911. mf 97-98: deaths 1865-1912. mf 99-100: marriages 1858-1911. mf 101-102: births 1876-1908) – us Archive [978]
Huntington, A K *see* Metals
Huntington, Arria Sargent *see* Memoir and letters of frederic dan huntington, first bishop of central new york
Huntington beach – 1938; 1948 – (= ser California telephone directory coll) – 2r – 1 – $100.00 – mf#P00037 – us Library Micro [917]
[Huntington beach-] huntington beach daily pilot – CA. 1966 (Scats); 1968 – 15r – 1 – $900.00 – mf#H03248 – us Library Micro [071]
[Huntington beach-] huntington beach independent – CA. 1965- – 29r – 1 – $1740.00 (subs $100y) – mf#H04043 – us Library Micro [071]
[Huntington beach-] huntington beach news – CA. 1905-1968; 1969-1982 – 48r – 1 – $2880.00 – mf#H04014 – us Library Micro [071]
Huntington, DeWitt Clinton *see*
– Half century messages to pastors and people
– Is the lord among us?
– Sin and holiness
Huntington, Ezra Abel *see* Notes on the epistle to the hebrews
Huntington, Frederic Dan *see*
– Christ in the christian year and in the life of man
– Elim
– The fitness of christianity to man
– Forty days with the master
– Human society
– Lessons on the parables of the saviour
– Massachusetts a field for church missions
– Personal religious life in the ministry and in ministering women
– The relation of the sunday school to the church
– Sermons for the people
Huntington, Frederic Dan et al *see* The new discussion of the trinity
Huntington, George *see*
– The charms of the old book
– The church's work in our large towns
– Outlines of congregational history
Huntington herald – Huntington OR: Herald Pub Co [wkly] – 1 – us Oregon Lib [071]
Huntington Historical Society *see* Quarterly of the huntington historical society
Huntington library quarterly – San Marino. 1937+ (1) 1970+ (5) 1976+ (9) – ISSN: 0018-7895 – mf#1093 – us UMI ProQuest [400]
Huntington, M C *see*
– Account of the american church mission in shanghai and the lower yangtse valley
Huntington, Mark W *see* Predictors of success on the human anatomy-related portions of the national athletic trainers' association certification examination
[Huntington park-] greater southeast bulletin – $110.00 – (see [los angeles-] south los angeles bulletin) – mf#H04015 – us Library Micro [071]
[Huntington park-] huntington park signal – CA. 1925-1983 – 451r – 1 – $27,060.00 – mf#H04016 – us Library Micro [071]
Huntington, Samuel *see* Samuel huntington correspondence, 1801-1817
Huntington, William Reed *see*
– Apocalypse
– Conditional immortality
– The curch-idea
– Four key-words of religion
– The four theories of visible church unity
– A national church
– The peace of the church
– Psyche
– A short history of the book of common prayer
– The talisman of unity
– Theology's eminent domain
Huntley beacon republican news – Marengo, IL. 1985-1986 (1) – mf#68367 – us UMI ProQuest [071]
Huntley, Ed L *see* Ed l huntley's panegyric on the jews
Huntley, Elizabeth A *see* The effects of a ten-week step aerobic training program on the body composition of college-aged women
Huntley, Henry Veel *see* Seven years' service on the savle coast of western africa
Huntley National Association *see* Annual reunion of huntley national association
Huntley news – Huntley IL. 1957 may 9 – 1r – 1 – mf#929935 – us WHS [071]
Huntly express – Huntly: A Dunbar 1863- (wkly) [mf ed 7 jan 1994-] – 1 – (title varies) – uk Scotland NatLib [072]
Huntly press – 1912-39; 1975-sep 1986 – 43r – 1 – mf#15.3 – nz Nat Libr [079]
Huntress – Washington DC. 1839 jun 29, 1847 feb 27 – 1r – 1 – (cont: paul pry [washington, dc]) – mf#874710 – us WHS [071]
Huntress – Washington. 1836-1854 (1) – mf#5573 – us UMI ProQuest [978]
Hunt's london journal – London. 1844-1844 (1) – mf#4266 – us UMI ProQuest [790]
Hunt's merchants' magazine – v1-63. 1839-70 – 1 – us AMS Press [380]
Hunt's merchants magazine yearbook – New York. 1v. 1871 – 5mf – 9 – $7.50 – mf#LLMC 84-482 – us LLMC [346]
Hunts post – Huntingdon, Cambridgeshire 18 nov 1893-to-date – 1 – (huntingdonshire post to 12 feb 1931; between 24 aug 1989-4 feb 1993 [hunts herald and post]; also a localised ed [ramsey post] & seperate property suppl [property post]) – uk Newsplan [072]
The huntsman's echo – Wood River Center, NE: J E Johnson. 1v. v1 n1. apr 19 1860-v1 n34. aug 1 1861 (wkly) [mf ed may 17 1860-aug 1 1861 (gaps) filmed 1973-[83]] – 2r – 1 – us NE Hist [071]
Huntsville first baptist church. huntsville, missouri : church records – 1837-1963 – 1 – 67.95 – us Southern Baptist [242]
Huntsville gazette – Huntsville. 1881-1894 (1) – mf#3350 – us UMI ProQuest [305]
The huntsville star – Huntsville, AL. v1 n1 jan 26 1900- (wkly) [mf ed 1947] – (= ser Negro Newspapers on Microfilm) – 1r – 1 – us L of C Photodup [071]
Huntz, Jack *see* Spotlight on spain
Hunziker, Rudolf *see*
– Erlebnisse eines schuldenbauers
– Der geldstag
– Jakobs, des handwerksgesellen wanderungen durch die schweiz
– Kalendergeschichten
– Saemtliche werke in 24 baenden
– Die wassernot im emmental / die armennot / eines schweizers wort
Hunzinger, A W *see* Das furchtmotiv in der katholischen busslehre von augustin bis petrus lombardus
Hunzinger, August Wilhelm *see*
– Das furchtproblem in der katholischen lehre von augustin bis luther
– Luthers neuplatonismus in der psalmenvorlesung von 1513-1516
– Probleme und aufgaben der gegenwaertigen systematischen theologie
– Die religioese krisis der gegenwart
– Die religionsgeschichtliche methode
– Das wunder
Huo ch'e chi / Lao, She – Ch'ung-ch'ing: Wen yu ch'u pan she, min kuo 34 [1945] – (= ser P-k&k period) – us CRL [480]
Huo hsien nei / Shen, Ch'i-yu – Shang-hai: Liang yu t'u shu kung ssu, [1932?] – (= ser P-k&k period) – us CRL [480]
Huo hua / Chin, I – Ch'ung-ch'ing: Feng huo she, Min kuo 29 [1940] – (= ser P-k&k period) – us CRL [951]
Huo pa / Ai, Ch'ing – Ch'ung-ch'ing: Feng huo she, Min kuo 30 [1941] – (= ser P-k&k period) – us CRL [810]
Huo pi yin hang yuan li / Ch'en, Chen-hua – Shang-hai: Shang wu yin shu kuan, Min kuo 24 [1935] – (= ser P-k&k period) – us CRL [332]
Huo pi yu chin jung 2 / Chou, Po-ti – Shang-hai: Chung-hua shu chu, Min kuo 24 [1935] – (= ser P-k&k period) – us CRL [332]
Huo ping / Yang, Shuo et al – K'un-ming: Hsin liu shu tien, 1940 – (= ser P-k&k period) – us CRL [830]
Huo ti chi lu / Liang, Jui-yu – Shang-hai: T'ien ma shu chu, Min kuo 25 [1936] – (= ser P-k&k period) – us CRL [070]
Huo ti tien li / Hu, Su – Kuei-yang: Wen t'ung shu chu, 1943 – (= ser P-k&k period) – us CRL [480]
Huo tsang / Lao, She – Ch'ung-ch'ing: Huang ho shu chu, Min kuo31 [1942] – (= ser P-k&k period) – us CRL [951]
Huo wu / Wang, Ya-p'ing – Ch'ung-ch'ing: Ch'un ts'ao shih she, 1945 – (= ser P-k&k period) – us CRL [810]
Huo yao / Yen, Yen-ts'un – Ch'ung-ch'ing: Shang wu yin shu kuan, 1944 (1945 printing) – (= ser P-k&k period) – us CRL [355]
Huo yen / Lu, Yin – Shang-hai: Pei hsin shu chu, Min kuo 25 [1936] – (= ser P-k&k period) – us CRL [830]
Huo yueh ti ch'ing nien chun / Lo, Shih-yang – Ch'ung-ch'ing: Ch'ing nien ch'u pan she, Min kuo 34 [1945] – (= ser P-k&k period) – us CRL [951]
Huon de Bordeaux *see* Huyge van bourdeus
Huon de bordeaux / Arnoux, Alexandre – Paris, France. 1947 – 1r – us UF Libraries [440]
Huonder, Anton *see*
– Der chinesische ritenstreit
– Der einheimische klerus in den heidenlaendern
Huot Sam Ath *see* Statement made by h e mr huot sambath
Huot Tath Vajirappano *see* L'enseignement du buddhisme des origines a nos jours
Huparikara, Balasastri *see* The problem of sanskrit teaching
Hupeh Province (China) *see* Hu-pei sheng cheng fu chiao yu t'ing hsien hsing kuei chang
Hupeh Province (China) Kung lu kuan li chu *see* Hu-pei sheng kung lu kuan li chu ch'eng li chou nien chi nien t'e k'an
Hu-pei cheng li t'u ti chi yao – [China]: Hu-pei sheng cheng fu min cheng t'ing, Min kuo 24 [1935] – (= ser P-k&k period) – us CRL [630]
Hu-pei hsiang-shih lu – List of successful candidates in the imperial examination in Hupeh province: 1879, 1885, 1900, 1901. 1 reel – 1 – us Chinese Res [951]
Hu-pei sheng cheng fu chiao yu t'ing hsien hsing kuei chang / Hupeh Province (China) – Han-k'ou: [sn], Min kuo 21 [1932] – (= ser P-k&k period) – us CRL [370]
Hu-pei sheng cheng fu san shih nien tu hsingocheng chi hua – [Hu-pei: Hu-pei sheng cheng fu mi shu ch'u], Min kuo 30 [1941] – (= ser P-k&k period) – us CRL [350]
Hu-pei sheng cheng fu yeh wu tsung chien t'ao – [Hu-pei: Hu-pei sheng cheng fu mi shu ch'u], Min kuo 29 [1940] – (= ser P-k&k period) – us CRL [350]
Hu-pei sheng kung lu kuan li chu ch'eng li chou nien chi nien t'e k'an / Hupeh Province (China) Kung lu kuan li chu – [China]: Hu-pei sheng kung lu kuan li chu, Min kuo 25 [1936] – (= ser P-k&k period) – us CRL [625]
Hu-pei sheng min cheng t'ing cheng ling chi yao ti 1 ts'e – [Hu-pei sheng: Min cheng t'ing, Min kuo 23 [1934]] – (= ser P-k&k period) – us CRL [350]
Hu-pei t'ien fu kai yao – Hu-pei: Ts'ai cheng t'ing, Min kuo 21 [1932] – (= ser P-k&k period) – us CRL [630]
Hu-pei t'u ti ts'e liang hui pien / Meng, Kuang-p'eng – Hu-pei: Sheng cheng fu min cheng t'ing, Min kuo 24 [1935] – (= ser P-k&k period) – us CRL [630]
Hu-pei wu-ch'ang hsien ch'ing-shan shih yen ch'u hu k'ou yu ching chi tiao ch'a pao kao / Wang, T'ang & Hsueh, Chien-wu – Hu-pei: Sheng li chiao yu hsueh yuan, Min kuo 25 [1936] – (= ser P-k&k period) – us CRL [339]
Hu-pei yang-lou-tung lao ch'ing ch'a chih sheng ch'an chih tsao chi yun hsiao – [China]: Chin-ling ta hsueh nung hsueh yuan nung yeh ching chi hsi, 1936 – (= ser P-k&k period) – us CRL [630]
Hupel, August Wilhelm *see*
– Ehstnische sprachlehre fuer beide hauptdialekte, den revalschen und doerptschen
– Neue nordische miscellaneen
– Nordische miscellaneen
Hupfeld, Hermann *see*
– Die psalmen
– Die quellen der genesis und die art ihrer zusammensetzung
– Ueber begriff und methode der sogenannten biblischen einleitung
Huppert, Thomas C *see* The papers of carlos montezuma, m.d
Huque, Azizul *see* History and problems of moslem education in bengal
Hur soz – Izmit: Muduru Selehattin Telser, jan 1951-sep 1953 – 3r – 1 – us CRL [079]
Hur soz – Lefkosa, Kibris: Hur soz, dec 1949-sep-dec 1954 – 3r – 1 – us CRL [079]
Huracan : su mitologia y sus simbolos / Ortiz, Fernando – Mexico City? Mexico. 1947 – 1r – us UF Libraries [972]
Hurault, Jean *see* Vie materielle des noirs refugies boni et des indi...
Hurch, Hans *see*
– Christoph von schallenberg
Hurd *see* New universal history of the religious rights, ceremonies and customs of the whole world
Hurd, Amy R *see* The influence of management styles
Hurd, John C *see* The law of freedom and bondage in the united states
Hurd, S P *see* A letter to the right honourable the earl of liverpool...
Hurdsfield, holy trinity – [Macclesfield Ferrets] – (= ser Cheshire monumental inscriptions) – 3mf – 9 – £3.50 – mf#83 – uk CheshireFHS [929]
Hurel, Eugene *see* Grammaire kinyarwanda
Hurgronje, Christiaan Snouck *see* Mohammedanism
Huria Kristen Batak Protestant *see* Almanak
Hurlbert, Jesse Beaufort *see*
– Britain and her colonies
Hurlbert, William Henry *see* Gan-eden
Hurlburt Field [FL]. United States *see* Commando
Hurlbut, William James *see* On the stairs
Hurley, Jorge *see* Amazonia cyclopica
Hurlimann, Martin *see* Picturesque india
Hurlin, William et al *see* The baptists of new hampshire
Hurly, M *see* Radeau de sauvetage baillairge-hurly
Huron and wyandot mythology : with an appendix containing earlier published records / Barbeau, Charles Marius – Ottawa: Govt Print Bureau, 1915 – (= ser Memoir (geological survey of canada); Bulletin (national museum of canada)) – 2mf – 9 – 0-524-02414-6 – mf#1990-2998 – us ATLA [290]

Huron Co. Bellevue *see*
- Bellevue central high school publications
- Evening gazette
- Gazette
- Local news
- News
- Record / evening gazette
- Rfd news
- Rfd news (firelands edition)
- Shoppers news

Huron Co. Greenwich *see* Enterprise-review
Huron Co. New London *see* Firelands farmer
Huron Co. Norwalk *see*
- Experiment and experiment news
- Huron reflector
- Reflector

Huron Co. Wakeman *see*
- Independent press
- Riverside echo

Huron county atlas, 1873 – 1r – 1 – mf#B6744 – us Ohio Hist [978]
Le huron de mont-rouge / Fournier-Verneuil – Paris 1824 [mf ed Hildesheim 1995-98] – 1v on 2mf – 9 – €60.00 – ISBN-10: 3-487-26141-3 – ISBN-13: 978-3-487-26141-6 – gw Olms [914]
Huron expositor – Seaforth, ON. 1869-1920 – 33r – 1 – ISSN: 0834-7360 – cn Library Assoc [071]
Huron news – Ontario, CN. jan 1896-dec 1911 – 16r – 1 – cn Commonwealth Imaging [071]
Huron. Presbytery (Pres. Ch. in the USA) *see* Minutes
Huron reflector / Huron Co. Norwalk – jan 1830-jan 1932,jul 1832-apr 1833 [wkly] – 1r – 1 – mf#B29833 – us Ohio Hist [071]
The huron-iroquois of canada : a typical race of american aborigines / Wilson, Daniel – Ottawa?: s.n, 1884 – 1mf – 9 – mf#29131 – cn Canadiana [572]
Hurrell froude : memoranda and comments / Guiney, Louise Imogen – London: Methuen, 1904 – 2mf – 9 – 0-7905-5403-8 – mf#1988-1403 – us ATLA [920]
Hurrell froude : memoranda and comments / Guiney, Louise Imogene – London: Methuen, 1904 – 2mf – us ATLA [240]
Hurrell, Joseph J *see* Stress among police officers
Hurricane / Roberts, Edith Kneipple – San Juan, Puerto Rico. 1929 – 1r – us UF Libraries [972]
Hurricane baptist church. clinton, south carolina : church records – 1949-62 – 1 – us Southern Baptist [242]
Hurricane baptist church. laurens county. south carolina : church records – 1893-1949, 1962-72 – 1 – us Southern Baptist [242]
Hurricane creek baptist church. stewart county. tennessee : church records – 1949-68 – 1 – us Southern Baptist [242]
Hurriyet – 1980- – 1 – (yrly reel count varies) – us UMI ProQuest [070]
Hurst, J F *see* Short history of the early church
Hurst, John F *see*
- History of rationalism
- Martyrs to the tract cause
- Our theological century

Hurst, John Fletcher *see*
- American methodism
- British methodism
- History of rationalism
- Indika
- Literature of theology
- Martyrs to the tract cause
- Our theological century
- Outline of church history
- Short history of the church in the united states, a.d. 1492-1890
- Short history of the mediaeval church
- Short history of the modern church in europe, a.d. 1558-1888
- Theological encyclopedia and methodology
- World-wide methodism

Hurst, Samuel Need *see* A complete popular encyclopedia of virginia law and forms and business guide or how-book for the businessman and citizen.
Hursthouse, Charles Flinders *see* Emigration
Hurston herald / Association to Preserve the Eatonville Community – 1991 nov – 1r – 1 – mf#4027514 – us WHS [978]
Hurston, Zora Neale *see*
- Polk county
- Tell my horse

Hurstville propeller – Hurstville, mar 1911-dec 1969 – 19r – A$627.00 vesicular A$731.50 silver – at Pascoe [079]
Hurtado Aguilar, Luis A *see* Belice es de guatemala
Hurtado, Antonio *see*
- El argumento de un drama
- Barba azul. opera
- El busto de elisa
- Una cancion de amor
- El collar de lescot
- La comedia de la vida
- Corte y cortijo
- Cosas del mundo
- En el cuarto de mi mujer
- En la sombra
- Entre el deber y el derecho
- El facedor de un entuerto...agravios
- Intriga y amor
- El matrimonio secreto
- La maya
- El medico de camara
- La nieta del zapatero
- El romancero de la princesa
- Suenos y realidades
- El toison (sic) roto
- Very well
- La virgen de la montana
- La voz del corazon

Hurtado de Mendoza, Luis *see* Cortes de la muerte
Hurtado de Mendoza, M *see* Tratado historico y fisiologico...sobre la generacion, el hombre y la mujer
Hurtado de Mendoza, P *see* Disputationes de universa philosophia
Hurtado de Mendoza, Publio *see*
- Alonso golfin. leyenda
- Amor y martirio
- La batalla de zalaca. episodio historico-extremeno
- Castillos, torres y casas fuertes de la provincia de caceres
- El cinturon de afrodita
- Extremadura en toledo. impresiones de turista
- El idolo roto (realidades de otros dias)
- Indianos cacerenos
- La parroquia de san mateos de caceres y sus agregados
- Supersticiones extremenas. prologo de urbano gonzalez serrano
- Tribunales y abogados cacerenos

Hurtado Garcia, Jose *see* Pensamiento social en la emancipacion
Hurtado Munoz de Lucas, Vicente *see* Participacion politica popular. (guiones para un cursillo monografico)
Hurtado, Oscar *see*
- Carta de un juez
- Paseo del malecon
- Seiba

Hurtado y Nunez de Arce, G *see*
- Herir en la sombra
- La jota aragonesa
- El laurel de la zubia

Hurter, H *see*
- Nomenclator literarius recentioris theologiae catholicae
- Nomenclator literarius recentioris theologiae catholicae, 1109-1899

Hurter, Hugo *see*
- Medula theologiae dogmaticae
- Theologia generalis
- Theologia specialis. pars altera
- Theologia specialis. pars prior

Hurtubise, Cheryl L *see* Comparison of waist to hip ratio measurements in relation to cardiovascular risk factors in healthy menopausal women
Hurukuro dzanyamayadenga nadzapasi – Gwelo, Zimbabwe. 1959 – 1r – us UF Libraries [960]
Hurvits, Shim'on Tsevi *see* Kol mevaser
Hurwicz, Elias *see* Die orientpolitik der 3. internationale
Hurwitz, Hayyim Dov *see* Mamon
Hurwitz, Nathan *see* Economic framework of south africa
Hurwitz, Phinehas Elijah *see* Sefer ha-berit ha-shalem
Hurwitz, Shmarya Loeb *see* Dat veha-hinukh
Hus dahlen : erzaehlung in muensterlaendischer mundart / Wibbelt, Augustin – 4. aufl. Essen: Fredebeul & Koenen, 1920 [mf ed 1992] – 289p – 1 – mf#7821 – us Wisconsin U Libr [390]
Hus, Jan *see*
- The church
- The letters of john hus

Hus, Pierre *see* Apollo e dafne
Husain, Agha Mahdi *see* The rise and fall of muhammad bin tughluq
Husain Ali, Kirmani *see* The history of the reign of tipu sultan..
Husain, Altaf *see* The complaint and the answer
Husain ibn Abdallah *see* [Al-qanun fi al-tibb]
Husain, Muhammad Ashraf *see* A record of all the quranic and non-historical epigraphs on the protected monuments in the delhi province
Husain, Wahed *see* Administration of justice during the muslim rule in india
Husband, Anna Stucki *see* Pedals of the piano
Husbandman's spiritual companion / Hildrop, John – London, England. 1819 – 1r – us UF Libraries [240]
Husbourne crawley – (= ser Bedfordshire parish register series) – 3mf – 9 – £7.50 – uk BedsFHS [929]
Husenbeth, Frederick Charles *see*
- Chain of fathers
- Discourse delivered in the catholic chapel
- Emblems of saints
- Husenbeth's defence of the catholic church
- The life of the right rev. john milner, d.d
- Notices of the english colleges and convents established on the continent

Husenbeth's defence of the catholic church : a complete refutation of the calumnies contained in a work entitled the poor man's preservative against popery by the reverend joseph blanco white... / Husenbeth, Frederick Charles – Toronto: printed for the proprietors by T Dalton, 1834 [mf ed 1984] – 2mf – 9 – 0-665-32005-1 – mf#32005 – cn Canadiana [241]
Huseyin, Husameddin *see* Amasya tarihi
Husik, Isaac *see*
- A history of mediaeval jewish philosophy
- Matter and form in aristotle

Husiti a reformacia na slovensku do zilinskej synody / Varsik, Branislav – Bratislava: [s.n.], 1932 [mf ed 1993] – (= ser Sbornik filosoficke fakulty university komenskeho v bratislave 8/62 3) – 1mf – 9 – 0-524-08147-6 – (in czech; incl bibl ref) – mf#1993-9053 – us ATLA [242]
Huske, John *see* The present state of north america
Huskvarna tidning – Joenkoeping, 1914 – 1r – 1 – sw Kungliga [078]
Huskvarnaposten – Joenkoeping, Sweden. 1911-60 – 25r – 1 – sw Kungliga [078]
Huss, Veronica E *see*
- Backwoods
- Capron trail
- Citrus center, glades county, florida
- Dwellings of the far south
- Edge, glades county, florida
- Fauna, labelle, florida, hendry county
- Felda, hendry county, florida
- Folk lore
- Fort denaud, hendry county, florida
- Fort thompson, hendry county, florida
- Hall city, glades county, florida
- Hendry and glades county, florida
- Henry ford helps a child
- Keri, hendry county, florida
- Life among the seminoles
- Live stock, game, etc
- Muse, glades county, florida
- Ortona, glades county, florida
- Palm dale, glades county, florida
- Sears, hendry county, florida
- Seminole indian reservation, hendry county, florid
- Tasmania, glades county, florida
- Timbers
- Tropical birds
- Turner's, hendry county, florida

Hussain, Iqbalunnisa *see* Purdah and polygamy
Husserl studies – The Hague. 1991-1995 (1,5,9) – ISSN: 0167-9848 – mf#16791 – us UMI ProQuest [100]
Hussey, A H *see* Divine healing in mission work
Hussey, Arthur *see* Notes on the churches in the counties of kent, sussex and surrey, mentioned in domesday book, and those of more recent date
The hussite wars / Luetzow, Franz Heinrich Hieronymous Valentin, Graf von – London: JM Dent; New York: EP Dutton, 1914 – 1mf – 9 – 0-7905-4829-1 – (incl bibl ref) – mf#1988-0829 – us ATLA [943]
Husslein, Joseph *see* The church and social problems
Husson review – Bangor. 1969-1971 – 1 – ISSN: 0018-8042 – mf#5903 – us UMI ProQuest [370]
Hust, Gerhard *see* Untersuchungen zu claudio monteverdis messkompositionen
Hust, Jerome G *see* Glass fiberboard srm for thermal resistance
Husted, Benjamin F *see* Brass ensemble
Husterer, Georg *see* Tirol im jahre 1809
Husting rolls of deeds and wills 1252-1485 : transcripts of over 22,600 deeds and wills registered in the court of husting, the principal court of medieval london / ed by Martin, G H – [mf ed Chadwyck-Healey] – 30r – 1 – (with p/g) – uk Chadwyck [929]
Hustisford journal – Hustisford WI. 1910 may 13 – 1r – 1 – mf#876274 – us WHS [071]
Hustisford news – Hustisford WI. [1840 mar 1/1843 dec 3]-[1961-1962 aug 26] [some gaps] – 15r – 1 – (cont by: independent [juneau, wi]; dodge county independent-news) – mf#946533 – us WHS [071]
Hustler – Read, NE: Carl C Crouse. v1 n1. mar 31 [1911]-14// (wkly) – 1r – 1 – (cont by: arthur enterprise. publ in calora ne, jan 2 1913-) – us Bell [071]
Hustler – Hanceville, AL. 1904-1904 (1) – mf#62020 – us UMI ProQuest [071]
Huston, Charles *see* An essay on the history and nature of original titles to land in the province and state of pennsylvania
Husumer nachrichten – Husum DE, 1977- – ca 7r/yr – 1 – gw Misc Inst [074]
Husz, Johannes (Jan Hus) *see* Opuscula
Hutchings, Pauline Linda Joan *see* Cardio-pulmonary resuscitation knowledge of registered nurses working in private hospital wards
Hutchings, Richard M *see* Trends in the marketing of florida citrus fruits
Hutchings, Samuel *see* The mode of christian baptism
Hutchings, William Henry *see* The mystery of the temptation
Hutchins, Charles Lewis *see*
- The chant and service book
- The church hymnal

Hutchins, Harry Burns *see* Illustrative cases on equity jurisprudence
Hutchins quarterly – 1984 jan-1990 oct – 1r – 1 – mf#861762 – us WHS [071]
Hutchinson, Arthur B *see* The mind of mencius
Hutchinson County Genealogical Society *see* Cenotaph
Hutchinson county herald – Stinnett, TX. 1979-1989 (1) – mf#68257 – us UMI ProQuest [071]
Hutchinson, E W *see* Adventures in siam in the 17th century
Hutchinson, John *see* A catalogue of notable middle temple templars
Hutchinson, Lester *see* The empire of the nabobs
Hutchinson, Lincoln *see* Panama canal and international trade competition
Hutchinson, Louisa *see* In tents in transvaal
Hutchinson, Margarite *see* A report of the kingdom of congo
Hutchinson, Paul *see* World revolution and religion
Hutchinson, Susan L *see* The altered self
Hutchinson, T J *see* Narrative of the niger, tsadda and binu exploration
Hutchinson, Thomas *see*
- The history of the province of massachuset's bay from the first settlement thereof in 1628
- The history of the province of massachusets-bay from the charter of king william and queen mary in 1691 until the year 1750
- The history of the province of massachusetts bay from the year 1750 until june 1774
- The witchcraft delusion of 1692

Hutchinson, William *see* History and antiquities of the county of cumberland
Hutchinson, William Francis *see* Under the southern cross
Hutchinson, Woods *see*
- The gospel according to darwin

Hutchinson's australasian encyclopaedia : comprising a description of all places in the australasian colonies, an account of the events which have taken place in australasia from its discovery to the present date / Levey, George Collins – London, 1892 – (= ser 19th c british colonization) – 5mf – 9 – mf#1.1.5059 – uk Chadwyck [919]
Hutchinson's splendour of the heavens : a popular authoritative astronomy / ed by Phillips, Theodore Evelyn Reece et al – London: Hutchinson 1923 [mf ed 1998] – 2v on 1r – 1 – (incl ind) – mf#film mas 28403 – us Harvard [520]
Hutchison, Eloise P *see* Out of the past
Hutchison, J *see* The ancient east
Hutchison, John *see*
- Lectures chiefly expository on st paul's epistle to the philippians
- Lectures chiefly expository on st paul's first and second epistles to the thessalonians
- Our lord's signs in st john's gospel

Hutchison, Matthew *see* The reformed presbyterian church in scotland
Hutchison, Thomas Dancer *see* Free-will controversy
Huth, Alfred Henry *see* The marriage of near kin
Huth, Georg *see* Geschichte des buddhismus in der mongolei
Huth, Gottfried *see* Allgemeines magazin fuer die buergerliche baukunst
Hutheesing, Krishna *see* The bride's book of beauty
Hutheesing, Krishna Nehru *see*
- Shadows on the wall
- With no regrets

Hutheesing, Raja *see* Window on china
Huther, August *see*
- Goethes goetz von berlichingen und shakespeares historische dramen
- Die verschiedenen plaene im ersten teile von goethes faust

Huther, Joh Ed *see*
- Critical and exegetical hand-book to the epistles to timothy and titus
- Critical and exegetical handbook to the general epistles of james, peter, john, and jude

Huther, Johann Eduard *see*
- Critical and exegetical handbook to the epistles of st paul to timothy and titus
- Critical and exegetical handbook to the general epistles of peter and jude

Hutschenruyter, W *see* Mahler
Hutt and petone chronicle – 1891-94; 1905-12; 1931-35; 1953-54; 1955-apr 1962; 18 apr 1964-14 dec 1966 – 10r – 1 – (title changed to: petone chronicle) – mf#49.3 – nz Nat Libr [079]
Hutt, H L *see* Planting and caring for young trees in an apple orchard
Hutt news – apr 1927-dec 1982; apr 1983-dec 1988 – 32r – 1 – mf#49.2 – nz Nat Libr [079]
Hutt sun – nov 1986-jan 1987 – 1r – 1 – mf#49.15 – nz Nat Libr [079]

Hutt valley independent – 1911-12; 1914-24; 1926-27; 1929-31; 1933 – 2r – 1 – mf#49.16 – nz Nat Libr [079]

Hutten : roman eines deutschen / Eggers, Kurt – Dortmund: Volkschaft-Verlag 1943 [mf ed 1990] – 1r – 1 – (filmed with: heiteres daruberstehen / theodor fontane & other titles) – mf#7073 – us Wisconsin U Libr [830]

Huttens letzte tage : eine dichtung / Meyer, Conrad Ferdinand – Leipzig: H Haessel, 1872 [mf ed 1996] – viii/126p – 1 – mf#9721 – us Wisconsin U Libr [810]

Huttens letzte tage *see* Angela borgia

Hutter, Leonhard *see* Compend of lutheran theology

Huttmann, Maude Aline *see* The establishment of christianity and the proscription of paganism

Huttner, J C *see*
- Nachricht von der brittischen gesandtschaftsreise durch china und einen theil der tartarei
- Voyage...la chine...

Hutton, Arthur Wollaston *see*
- The anglican ministry
- Arthur young's tour in ireland [1776-1779]
- Cardinal manning

Hutton, Caroline Amy *see* Greek terracotta statuettes

Hutton, Catherine *see* The tour of africa

Hutton, Charles *see* Calculations to determine at what point in the side of a hill its attraction will be the greatest etc

Hutton, Charles F *see* Unconscious testimony

Hutton, G H *see* An early victorian railway station

Hutton, Henry Dix *see*
- History, principle, and fact
- The prussian land-tenure reforms and a farmer-proprietary for ireland

Hutton, James *see*
- Constitution of the union of south africa
- Missionary life in the southern seas
- Theory of the earth

Hutton, John Alexander *see*
- The authority and person of our lord
- The weapons of our warfare
- The winds of god

Hutton, John Henry *see* Caste in india

Hutton, Joseph *see*
- Miracles essential to the proof of a divine commission
- Omniscience the attribute of the father only

Hutton, Joseph Edmund *see*
- A history of the moravian church

Hutton, Richard Holt *see*
- Aspects of religious and scientific thought
- Criticisms on contemporary thought and thinkers
- Essays on some of the modern guides of english thought in matters of faith
- The incarnation and principles of evidence
- Theological essays

Hutton, S K *see* Among the eskimos of labrador

Hutton, W *see* A voyage to africa

Hutton, W H *see* Brief history of the indian peoples

Hutton, William *see*
- A voyage to africa
- A voyage to africa; including a narrative of an embassy to one of the interior kingdoms.

Hutton, William Holden *see*
- The age of revolution
- The church and the barbarians
- The church of the sixth century
- The english church
- The english reformation
- The influence of christianity upon national character illustrated by the lives and legends of the english saints
- Letters of william stubbs, bishop of oxford, 1825-1901
- The reformation in europe
- Sir thomas more
- Thomas becket
- William laud

Hutzli, Walther *see*
- Jeremias gotthelf

Hu'u thanh – Hanoi. fevr 1922-sept 1924 – 1 – fr ACRPP [073]

Huvala, J M *see* Der pahlavi text

Huvudlinjer i nordisk sprakhistoria / Hesselman, Bengt – Uppsala: Almqvist & Wiksells Boktryckeri 1948-53 [mf ed Bloomington IN: Indiana Uni Lib, Preservation Dept 1984] – 2v on 1r – 1 – us Indiana Preservation [390]

Huxley, Aldous *see*
- Beyond the mexique bay
- Point counter point

Huxley, Elspeth *see* Settlers of kenya

Huxley, Francis *see* Affable savages

Huxley, Jesse *see* A reply to rev c f aked's changing creeds and social struggles

Huxley, Julian *see*
- Evolution as a process
- Religion without revelation
- We europeans

Huxley, L *see* Scott's last expedition...journals and reports...

Huxley, Leonard *see* Life and letters of thomas henry huxley

Huxley, methodist church – (= ser Cheshire monumental inscriptions) – 1mf – 9 – £2.50 – mf#23 – uk CheshireFHS [929]

Huxley, Thomas Henry *see*
- Charles darwin
- Darwiniana
- Discourses biological and geological
- Essays upon some controverted questions
- Evidence as to man's place in nature
- Evolution and ethics...
- Evolution and ethics
- Half-century of science
- Hume
- Hume, with helps to the study of berkeley
- An introduction to the classification of animals
- Lay sermons, addresses, and reviews
- Man's place in nature
- Method and results
- On our knowledge of the causes of the phenomena of organic nature
- On the educational value of the natural history sciences
- Science and christian tradition
- Science and christian tradition: essays
- Science and culture
- Science and education
- Science and hebrew tradition

Huxley, Thomas Henry et al *see* The priestley memorial at birmingham, august, 1874

Huy Can *see* Ngay hang song, ngay hang tho

Huy! und pfuy! der welt / Abraham...Sancta Clara – Wuertzburg: Gedruckt bey Martin Frantz Hertzen, 1707 – 5mf – 9 – mf#O-1493 – ne IDC [090]

Huy! und pfuy! der welt / Abraham...Sancta Clara – Wuertzburg: Gedruckt bey Martin Frantz Hertzen, 1725 – 7mf – 9 – mf#O-1494 – ne IDC [090]

Huy-fallstein-echo – Halberstadt DE, 1962 20 jan-1967 23 mar – 1r – 1 – (title varies: publ in magdeburg) – gw Misc Inst [074]

Huyge van bourdeus : ein niederlaendisches volksbuch / Huon de Bordeaux; ed by Wolf, Ferdinand – Stuttgart: Litterarischer Verein, 1860 [mf ed 1993] – (= ser Blvs 55) – 88p – 1 – mf#8470 reel 11 – us Wisconsin U Libr [390]

Huygen, Jan *see*
- De beginselen van gods koninkryk in den mensch
- Stichtelyke rymen op verscheide stoffen
- Stichtelyke rymen op verscheiden stoffen

Huygen, Pieter *see* De beginselen van gods koninkryk in den mensch

[Huygen, Pieter] *see* De beginselen van gods koninkryk in den mensch

Huysboeck : vijf decades, dat is vijftich van de vooerneemste hooftstucken der christelijcker religie / Bullinger, Heinrich – s.l, 1567 – €38.00 – ne Slangenburg [240]

Huysboeck, vijf decades : dat is... / Bullinger, Heinrich – Amsterdam, Hendrick Laurensz, 1612 – 10mf – 9 – mf#PBU-671 – ne IDC [240]

Huysboecxken... / Teelinck, W – Middelburgh, 1618 – 11mf – 9 – mf#PBA-325 – ne IDC [240]

Huyshe, Wentworth *see* Graphic history of the south african war, 1899-1900

Huysteen, Everhardus Jacobus Frick van *see* Relational selves of professional health caregivers working with hiv/aids children

Huyton, west derby and kirby reporter – [NW England] Liverpool, Kirkby Lib 9 sep 1955 – 1 – (title change: west derby & kirby reporter [16 sep 1955-28 aug 1959]; west derby reporter [4 sep 1959-1963]) – uk MLA; uk Newsplan [072]

Huzma safa / Safa, Ismail – Istanbul: Alem Matbaasi, 1308 [1891] – (= ser Ottoman literature, writers and the arts) – 3mf – 9 – $55.00 – us MEDOC [470]

Hvad christus doemmer om officiel christendom / Kierkegaard, Soeren – Kobenhavn: CA Reitzel, 1855 – 1mf – 9 – 0-7905-3792-3 – mf#1989-0285 – us ATLA [240]

Hvad har handt? – Lulea, Sweden. 1856-62 – 1 reel – 1 – sw Kungliga [078]

Hvad har handt? – Umea, Sweden. 1847 – 1 reel – 1 – sw Kungliga [078]

Hvad nytt? – Skaenninge, Sweden. 1911-12 – 1 – sw Kungliga [078]

Hverdagsliv fra syd-indien : en hindus (kupuswamis) ungdomserindringer – Kjobenhavn: Pastor Asschenfeldt Hansens Ekspedition, 1899 [mf ed 1995] – (= ser Yale coll) – 122p (ill) – 1 – 0-524-09777-1 – (trans fr english into danish) – mf#1995-0777 – us ATLA [240]

Hvetlanda tidning – Joenkoeping, Vetlanda, Sweden. 1873-1940 – 61r – 1 – sw Kungliga [078]

A hvndred sermons vpon the apocalips / Bullinger, Heinrich – [London, Iohn Day], 1561 – 8mf – 9 – mf#PBU-201 – ne IDC [240]

Hvorfor skabte gud mennesket? : fem foredrag / Kirkeberg, O L – Minneapolis, Minn: I hovedkommission hos Augsburg Pub House, 1893 – 1mf – 9 – 0-524-05255-7 – mf#1991-2247 – us ATLA [210]

Hvysboec : viif decades / Bullinger, Heinrich – [Emden, Gellius Ctematius], 1563 – 12mf – 9 – mf#PBU-161 – ne IDC [240]

Hyacinthe, Father *see* Catholic reform

Hyacinthe, pere *see* Discourses on various occasions

Hyamson, A M *see* The british consulate in jerusalem in relation to the jews of palestine

Hyamson, Albert Montefiore *see*
- A history of the jews in england

Hyannis, Massachusetts. First Baptist Church and Society *see* Records

Hyannis roundup – Hyannis, NE: R L Hamon, 1891 (wkly) [mf ed v1 n9. oct 1891 filmed [1973]] – 1r – 1 – us NE Hist [071]

Hyatt, Charles *see* Connection between ministerial character and success

Hyatt on trials: a treatise on the trial of civil and criminal cases in state and federal courts. / Hyatt, William Harvey – San Francisco: Bender-Moss Co., 1924. 2v. LL-542 – 1 – us L of C Photodup [345]

Hyatt, Thaddeus *see*
- Correspondence and papers
- Thaddeus hyatt papers

Hyatt, William Harvey *see* Hyatt on trials: a treatise on the trial of civil and criminal cases in state and federal courts.

Hybrid : the university of pennsylvania journal of law and social change – v1-5. 1993-2000 – 9 – $61.00 set – (none publ in 1995. title varies: v1-3 1993-96 as: hybrid: a journal of law and social change) – mf#115871 – us Hein [340]

Hyckel, Georg *see* Geschichte und besiedlung des ratiborer landes

Hydata – Bethesda. 1965-1977 (1) 1977-1977 (5) 1977-1977 (9) – ISSN: 0018-8115 – mf#7390 – us UMI ProQuest [333]

Hyde, Ammi Bradford *see*
- The story of methodism throughout the world
- The story of methodism throughout the world, from the beginning to the present time

Hyde, Carolyn *see* Voc emission reduction study at the hill air force base building 151 painting facility

Hyde, Charles McEwen *see* Historical sketch of the hawaiian mission

Hyde County Historical Society (NC) *see* High tides

Hyde (gee cross) unitarian: burials 1786-1877 – [North Cheshire FHS] – (= ser Cheshire church registers) – 6mf – 9 – £7.00 – mf#255 – uk CheshireFHS [929]

Hyde, John *see* The glory and divinity of the holy bible and its spiritual sense

Hyde, marple and glossop reporter – [NW England] Hyde, Stalybridge Lib – 1 – (title change: hyde reporter & telegraph [1895-18 apr 1903]; hyde reporter [25 apr 1903-21 jul 1934]) – uk MLA; uk Newsplan [072]

Hyde park 1868-1892 – Oxford, MA (mf ed 1988) – (= ser Massachusetts vital records) – 13mf – 9 – 0-931248-87-6 – (mf 1: births 1868-75. mf 2: births 1875-83. mf 3: births 1883-89. mf 4: births 1889-91. mf 5: marriages 1868-77. mf 6: marriages 1878-85. mf 7: marriages 1885-90. mf 8: marriages 1891-92. mf 9: deaths 1868-79. mf 10: deaths 1879-89. mf 11: deaths 1890-92. mf 12-13: index 1868-89) – us Archive [978]

Hyde park citizen – Chicago IL. 1991 apr 11-dec 26, 1992, 1993, 1994 jan 6/9-dec 29, 1995, 1996 jan 4-dec 26, 1997 jan 2-dec 25, 1998 jan 1-dec 31, 1999 jan 7-dec 30, 2000 jan 13-dec, 2001 jan 4-jun 28 – 11r – 1 – (cont by: citizen hyde park) – mf#1886892 – us WHS [071]

Hyde, st thomas: burials 1868-1980 – [North Cheshire FHS] – (= ser Cheshire church registers) – 4mf – 9 – £5.50 – mf#312 – uk CheshireFHS [929]

Hyde, Thomas Alexander *see* Christ the orator

Hyde, William De Witt *see*
- The five great philosophies of life
- The gospel of good will
- Outlines of social theology
- Practical ethics
- Practical idealism
- The quest of the best
- Self-measurement
- Sin and its forgiveness
- The teacher's philosophy in and out of school

Hyde, William de Witt *see* Jesus' way

Hyden families – v1 n1-10 [1978 jan-oct] – 1r – 1 – (cont by: heydon-hayden-hyden families) – mf#660032 – us WHS [929]

Hyder aly und tippo saheb : oder historisch-geographische uebersicht des mysorischen reichs nebst dessen entstehung und zertheilung / Sprengel, Matthias – Weimar 1801 [mf ed Hildesheim 1995-98] – 1v on 1mf – 9 – €40.00 – ISBN-10: 3-487-26607-5 – ISBN-13: 978-3-487-26607-7 – gw Olms [954]

Hyderabad : a guide to art and architecture – [Delhi]: Publications Division, Ministry of Information and Broadcasting, Govt of India, [1951] – (= ser Samp: indian books) – us CRL [700]

Hyderabad – Hyderabad. pt3. 1901 – (= ser Census of india) – us CRL [315]

Hyderabad. India. Dept of Statistics and Census *see* Village list of district no 9, karimnagar

Hyderabad. India. (State). Superintendent of Census Operations *see* Codes of census procedures for the hyderabad assigned districts

Hydraulic machinery / Blaine, Robert Gordon – London, England. 1913 – 1r – us UF Libraries [627]

Hydraulics and pneumatics – Cleveland. 1948+ (1) 1965+ (5) 1976+ (9) – ISSN: 0018-814X – mf#801 – us UMI ProQuest [621]

Hydrobiologia – Den Hagg. 1989-1996 (1,5,9) – ISSN: 0018-8158 – mf#16792 – us UMI ProQuest [574]

Hydrobiological journal – Silver Spring. 1984-1995 (1,5,9) – ISSN: 0018-8166 – mf#14356 – us UMI ProQuest [574]

Hydrocarbon news – Dallas. 1964-1971 (1) 1966-1971 (5) – ISSN: 0031-6326 – mf#1634 – us UMI ProQuest [540]

Hydrocarbon processing : international edition – Houston. 1922+ (1) 1965+ (5) 1975+ (9) – ISSN: 0018-8190 – mf#48 – us UMI ProQuest [660]

Hydrodynamische kenngroessen von gleitlagern mit unterbrochener ringnut und ihre anwendung bei der berechnung dynamisch beanspruchter radialgleitlager / Reischke, Gerhard – (mf ed 1994) – 2mf – 9 – €40.00 – 3-8267-2009-1 – gw Frankfurter [627]

Hydro-electric development in ontario : a history of water-power administration under the hydro-electric power commission of ontario / Biggar, Emerson Bristol – Toronto: Biggar Press, c1920 – 3mf – 9 – 0-665-73595-2 – mf#73595 – cn Canadiana [627]

Hydrogen ion concentration of citrus leaves and its relation to certain fungus diseases / Hansbrough, John H – s.l, s.l? 1925 – 1r – us UF Libraries [634]

Hydrographie : contenant la theorie et la practique de toutes les parties de la navigation / Fournier, Georges – Paris: Chez Michel Soly, 1643 [mf ed 1983] – 10mf – 9 – (with ind) – mf#SEM105P280 – cn Bibl Nat [623]

Hydrolisierbarkeit von immunglobulinen und albumin durch die proteolytische aktivitaet lebender zellen der spirochaete treponema denticola / Hollmann, Ricarda – (mf ed 1998) – 1mf – 9 – €30.00 – 3-8267-2545-X – mf#DHS 2545 – gw Frankfurter [617]

Hydrological processes – Chichester. 1986+ (1,5,9) – ISSN: 0885-6087 – mf#16104 – us UMI ProQuest [550]

Hydrological sciences bulletin = Bulletin des sciences hydrologiques – Oxford. 1980-1981 (1,5,9) – (cont by: hydrological sciences journal) – ISSN: 0303-6936 – mf#15550,03 – us UMI ProQuest [550]

Hydrological sciences journal = Journal des sciences hydrologiques – Oxford. 1982-1992 (1) 1982-1992 (5) 1982-1992 (9) – (cont: hydrological sciences bulletin) – ISSN: 0262-6667 – mf#15550,04 – us UMI ProQuest [550]

Hydrologie der deutschen kolonien in afrika / Pfalz, Richard – Berlin, Germany. 1944 – 1r – us UF Libraries [960]

Hydrometallurgy – Amsterdam. 1975+ (1) 1975+ (5) 1987+ (9) – ISSN: 0304-386X – mf#42095 – us UMI ProQuest [660]

The hydrometallurgy of copper : and its separation from the precious metals / Hunt, Thomas Sterry – S.l: s.n, 1881? – 1mf – 9 – mf#36714 – cn Canadiana [660]

Hydro-Quebec *see*
- Projet radisson – nicolet – des cantons

Hydrotechnical construction – New York. 1977-1979 (1,5,9) – ISSN: 0018-8220 – mf#11507 – us UMI ProQuest [627]

Hyffding, Harald *see* Om nogle religionfilosofiske arbejder fra den nyeste tid

Hygie – Paris. 1982-1993 [1,5,9] – (cont by: promotion and education) – ISSN: 0751-7149 – mf#14219 – us UMI ProQuest [613]

Hygie – Paris. 1982-1993 (1) 1982-1993 (5) 1982-1993 (9) – (cont: international journal of health education) – ISSN: 0751-7149 – mf#14219 – us UMI ProQuest [613]

Hygienic physiology / Steele, Joel Dorman – New York, NY. 1884 – 1r – us UF Libraries [612]

Hylan, John Perham *see* Public worship

Hylas et temire : the favorite divertissement, as performed at the king's theater, composed by mr d'egville...the music composed and arranged for the pianoforte by c bossi / Bossi, Cesare – London: Goulding, Phipps & D'Almaine [1799] [mf ed 19–] – 1r – 1 – mf#pres. film 100 – us Sibley [780]

Hym *see* Ryan brah jinavans

Hymer, Julian B *see* "As i remember kansas city from my boyhood and its townhood days."

Hymn – Boston. 1986+ (1,5,9) – ISSN: 0018-8271 – mf#16031 – us UMI ProQuest [780]

Hymn and tune book of the methodist episcopal church, south – character note ed. Nashville, Tenn: Pub House of the ME Church, South, 1889 – 6mf – 9 – 0-524-06628-0 – mf#1991-2683 – us ATLA [242]

Hymn and tune collection from library of edmond d. keith – 1 – $251.02 – (incl: the divine companion, or, david's harp new tunes [4th ed] fawcett, john; hymns...of public worship and private devotions [leeds 1782] gould, nathanial d; church music in america [boston 1852] hammond, william; psalms, hymns and spiritual songs [london 1765]) – us Southern Baptist [780]

Hymn book – (Methodist). Published by the General Conference of the Methodist Episcopal Church vca.1847 (Title page is missing) – 1 – $26.81 – us Southern Baptist [780]

Hymn book: the new casket – Charleston. 1869 – 1 – $12.32 – us Southern Baptist [780]

Hymn books – 1 – $81.48 – (includes: bradbury, william b. and sanders, charles w. the young choir... new york. 1841; bradbury, william b. bradbury's golden chain of sabbath school melodies. new york. 1861. dobwell, john. a new selection of seven hundred evangelistical hymns. morristown, nj. 1815. edson, william j. and reed, ephraim. musical monitor. ithaca, ny. 1825) – us Southern Baptist [780]

The hymn of the soul : contained in the syriac acts of st thomas / ed by Bevan, Anthony Ashley – Cambridge: University Press 1897 [mf ed 1989] – (= ser Texts and studies (cambridge, england) 5/3) – 1mf – 9 – 0-7905-1863-5 – (in english & syriac with english int) – mf#1987-1863 – us ATLA [240]

The hymn of the soul (ts5/3) : contained in the syriac acts of st thomas / ed by Bevan, Anthony Ashley – 1897 – (= ser Texts and studies (ts)) – 1mf – 9 – €3.00 – ne Slangenburg [226]

Hymn Society of America *see* Dictionary of american hymnology

The hymnal / ed by Benson, Louis FitzGerald – Philadelphia: Presbyterian Board of Publication and Sabbath-School Work, 1898, c1895 – 8mf – 9 – 0-524-07850-5 – mf#1991-3395 – us ATLA [240]

The hymnal – Philadelphia: Presbyterian Board of Publication and Sabbath-School Work, 1911 – 8mf – 9 – 0-524-08807-1 – mf#1993-3299 – us ATLA [240]

The hymnal : prepared by a union committee, for use mainly in sunday-schools – Tokyo: Kyobunkwan and Keiseisha, 1909 [mf ed 1995] – (= ser Yale coll) – ca 300p – 1 – 0-524-10006-3 – (in japanese) – mf#1995-1006 – us ATLA [780]

Hymnal of the presbyterian church – Philadelphia: Presbyterian Board of Publication, 1867 – 7mf – 9 – 0-524-06657-4 – mf#1991-2712 – us ATLA [242]

Hymnal of the presbyterian church in canada : with accompanying tunes – Toronto: C Blackett Robinson, 1881 – 5mf – 9 – 0-524-07263-9 – mf#1991-3004 – us ATLA [780]

Hymnarium quotidianum b.m.v. : ex hymnis medii aevi comparatum – Parisiis: P Lethielleux; Neo Eboraci [New York]: F Pustet, [1892?] – 2mf – 9 – 0-8370-8854-2 – (in latin and french. incl indes) – mf#1986-2854 – us ATLA [240]

Hymnarius moissiacensis : das hymnar der abtei moissac im 10. jahrhundert nach einer handschrift der rossiana / ed by Dreves, Guido Maria – Leipzig: Fues, 1888 [mf ed 1986] – (= ser Analecta hymnica medii aevi 2) – 1mf – 9 – 0-8370-7456-8 – (hymns in latin. int in german. incl ind) – mf#1986-1456 – us ATLA [450]

The hymn-book : containing a collection of the most popular catholic hymns – Enl and rev. Philadelphia: Peter F Cunningham, c1854 – 1mf – 9 – 0-8370-7189-5 – mf#1986-1189 – us ATLA [241]

The hymn-book of the modern church : brief studies of hymns and hymn-writers: the 34th fernley lecture / Gregory, Arthur Edwin – London: C H Kelly 1904 [mf ed 1990] – 1mf – 9 – 0-7905-3882-2 – (incl bibl ref) – mf#1989-0375 – us ATLA [240]

Hymne an italien / Daeubler, Theodor – Leipzig: Insel-Verlag, 1919 [mf ed 1989] – 167p – 1 – mf#7169 – us Wisconsin U Libr [810]

Hymnen / Becher, Johannes Robert – Leipzig: Insel, 1924 [mf ed 1989] – 127p – 1 – mf#6994 – us Wisconsin U Libr [780]

Hymnen an das diadem der pharaonen / Erman, A – Berlin, 1911 – (= ser Abhandlungen der Koenigl Preuss Akademie der Wissenschaften) – 1mf – 9 – mf#NE-20387 – ne IDC [960]

Hymnen gegen die irrlehrer, 2. bd (bdk61 1.reihe) / Ephraem der Syrer (Ephraem Syrus, Saint) – (= ser Bibliothek der kirchenvaeter. 1. reihe (bdk 1.reihe)) – €11.00 – ne Slangenburg [240]

Hymnen; pilgerfahrten; algabal / George, Stefan Anton – 3. Aufl. Berlin: G Bondi, 1905 (mf ed 1990) – 1r – 1 – (filmed with: gellerts lustspiele) – us Wisconsin U Libr [810]

Hymnen; pilgerfahrten; algabal / George, Stefan Anton – Godesberg. H Kuepper, 1950 (mf ed 1990) – 1r – 1 – (filmed with: gellerts lustspiele) – us Wisconsin U Libr [810]

Hymnen und gebete an nebo / Pinckert, J – Leipzig, 1920 – (= ser Leipziger semitistische Studien) – 1mf – 9 – (leipziger semitistische studien, leipzig 1920 v3 pt4) – mf#NE-20114 – ne IDC [956]

The hymnes and songs of the church, divided into two parts:...canonicall hymnes...the second part...spirituall songs.. / Wither, George – Trans. and composed by G.W. London: Printed by the assignes of George Wither, 1623. 68p. With music, by Orlando Gibbons, to some of the hymnes – 1 – us Wisconsin U Libr [240]

Hymnes latines et hymnaires / Baudot, Jules – Paris: Bloud & Gay, 1914 – (= ser Liturgie) – 1mf – 9 – 0-7905-6042-9 – (incl bibl ref) – mf#1988-2042 – us ATLA [240]

Hymni – [Paris?] [1579?] – (= ser Hqab. literatur des 16. jahrh.) – 2mf – 9 – €30.00 – mf#1579c – gw Fischer [780]

Hymni de tempore et de sanctis : in eam forma[m] qua a suis autoribus scripti sunt denuo redacti et secu[n]du[m] lege[m] carminis diligenter emendati atq[ue] interpretati – Hagenau: Rynman 1519 – (= ser Hqab. literatur des 16. jahrh.) – 2mf – 9 – €30.00 – mf#1519a – gw Fischer [780]

Hymni ecclesiae : pars 1: e breviario parisiensi, pars 2: e breviariis romano, sarisburiensi, aboracensi, et aliunde / ed by Newman, John Henry – Londini: A. Macmillan, 1865. xiv,406p – 1 – us Wisconsin U Libr [240]

Hymni ecclesiastici : praesertim qui ambrosiani dicuntur – Coloniae, 1556. Includes index – 1 – us Wisconsin U Libr [240]

Hymni et collectae, item evangelia, epistolae, introitus, gradualia et sequentiae &c : quae in diebus dominicis & festis sub precibus horariis & augustissimo missae sacrificio in ecclesia dei leguntur & canuntur &c – Coloniae: Calenius 1573 – (= ser Hqab. literatur des 16. jahrh.) – 7mf – 9 – €75.00 – mf#1573b – gw Fischer [780]

Hymni et collectae item evangelia, epistolae introitus, gradualia et sequentiae etc... – Coloniae: Calenius & Quentel 1585 – (= ser Hqab. literatur des 16. jahrh.) – 7mf – 9 – €75.00 – mf#1585a – gw Fischer [780]

Hymni et prosae ecclesiasticae vulgo sequentiae dictae – Antverpiae: Sylvius 1562 – (= ser Hqab. literatur des 16. jahrh.) – 2mf – 9 – €30.00 – mf#1562a – gw Fischer [780]

Hymni et saecula / Arias Montano, Benito – 1593 – 9 – sp Bibl Santa Ana [240]

Hymni et secula / Arias Montano, Benito – Antuerpiae: Officina. Plantiniana, 1593 – sp Bibl Santa Ana [240]

Hymni et sermones / Ephraem der Syrer (Ephraem Syrus, Saint); ed by Lamy, Thomas Josephus – Mechliniae. v1-4. 1882-19061 – (v1 1882 15mf; v2 1886 14mf; v3 1889 18mf; v4 1906 16mf) – 8 – €121.00 – ne Slangenburg [240]

Die hymni oder geistlichen lobgeseng : wie man die in der cystertienser orden durchs gantz jar singet – Nuernberg: Merckel 1555 – (= ser Hqab. literatur des 16. jahrh.) – 1mf – 9 – €20.00 – mf#1555a – gw Fischer [780]

Hymni quinque vocum / Zacharia, Cesare de – 1594 – (= ser Mssa) – 6mf – 9 – €80.00 – mfchl 450 – gw Fischer [780]

Hymni sacri in usum ludi illustris ad fontes salutares : melodiis & numeris musicis composti & collecti, a johanne nesero, musicae in eodem ludo moderatore / Neser, Johann – Curiae Variscorum (i.e. Stadt-am-Hof): ex officina Mattaei Pfeilschmidii, 1612 [mf ed 1989] – 1r – 1 – mf#pres. film 50 – us Sibley [780]

Hymni totius anni... / Palestrina (Praenestinus), Giovanni Pierluigi da – 1589 – (= ser Mssa) – 6mf – 9 – €80.00 – mfchl 362 – gw Fischer [780]

Hymnodia hispanica / Arevalo, Faustino – 1786 – 9 – sp Bibl Santa Ana [780]

Hymnody of the roman catholic church : historical survey with an analysis of musical styles / Haban, Teresine – U of Rochester 1956 [mf ed 19–] – 11mf – 9 – (with app & bibl) – mf#fiche 178, 490, 727 – us Sibley [780]

Hymnographie de l'eglise grecque : dissertation accompagnee des offices du 16 janvier, des 29 et 30 juin en l'honneur de s. pierre et des apaotres / Pitra, Jean Baptiste – Rome: impr de la Civilt a cattolica, 1867 [mf ed 1990] – 3mf – 9 – 0-7905-5906-4 – (in greek & latin. int in french) – mf#1988-1906 – us ATLA [243]

Hymnologium ecclesie : das ist, lobgesaenge der catholischen kyrchen, zur taeglichen vesperzeit... – Coeln 1545 – (= ser Hqab. literatur des 16. jahrh.) – 2mf – 9 – €30.00 – (fr: ecclesiastica liturgia / georg. wicelium) – mf#1545b – gw Fischer [780]

The hymnology and the hymnals of the restoration movement / Hanson, Kenneth C – Thesis. 1951 – 1 – us Southern Baptist [242]

Hymnorum ecclesiasticorum, ab andrea ellingero, v. cl. emendatorum, libri 3 : accessere joseph lib. 2 autore hieronymo fracastorio... – Francofurti ad Moenum: Bassaeus 1578 – (= ser Hqab. literatur des 16. jahrh.) – 7mf – 9 – €75.00 – mf#1578d – gw Fischer [780]

Hymnor[um] et threnodiarvm sanctae crvcis in devotam passionis iesu christi dei et hominis commemorationem fascicvlvs ad hebdomadam magnam... / Besler, Samuel – Wratislaviae [i.e. Breslau] ex officina typographica Georgi Bauman. Publ by aut [1611] [mf ed 19–] – 1r – 1 – mf#film 1379 – us Sibley [780]

Hymnor[um] et threnodiarvm sanctae crvcis in salvtarem passionis iesu christ, dei et hominis, memoriam pars tertia... / Besler, Samuel – Wratislaviae [i.e. Wroclaw] ex officina typographica Georgi Bauman [publ by aut] [1613] [mf ed 19–] – 1r – 1 – mf#film 1379 – us Sibley [780]

Hymnorvm liber : cvi de laudibus sacrarum scripturarum, et b ioannis euangelistae accessit oratio... / Macer, Caspar – Ingolstadii: Weissenhorn 1562 – (= ser Hqab. literatur des 16. jahrh.) – 1mf – 9 – €20.00 – mf#1562b – gw Fischer [780]

Hymns / Bowring, John – 1825 – 1 – $50.00 – us Presbyterian [780]

Hymns : their history and development in the greek and latin churches, germany and great britain / Selborne, Roundell Palmer, Earl of – London: A. and C. Black, 1892 – 1mf – 9 – 0-7905-6433-5 – mf#1988-2433 – us ATLA [240]

Hymns : their history and development in the greek and latin churches, germany and great britain / Selborne, Roundell Palmer, Earl of – London: A. and C. Black, 1892 – 1mf – us ATLA [780]

The hymns : for the use of evangelical lutheran congregations – Charleston, SC: Committee of United Synod on Common Book of Worship, 1906 – 2mf – 9 – 0-524-05245-X – mf#1992-2082 – us ATLA [242]

Hymns ancient and modern : for use in the services of the church with accompanying tunes – London, 1909 – 27mf – 8 – €52.00 – (historical ed with notes) – ne Slangenburg [780]

Hymns ancient and modern / Monk, William H – 1869 – 1 – us Southern Baptist [242]

Hymns and anthems adapted for jewish worship : selected and arranged by gustav gottheil / Gottheil, Gustav – NY: G P Putnam, 1887, c1886 – 1mf – 9 – 0-8370-3349-7 – (incl ind of first lines) – mf#1985-1349 – us ATLA [780]

Hymns and carols, old and new (annotated), for the sunday school and home : together with a short liturgy / ed by Stevens, Lorenzo Gorham – Saint John, NB: J & A McMillan, 1891 – 5mf – 9 – (incl ind) – mf#34998 – cn Canadiana [780]

The hymns and hymn writers of the church : an annotated edition of the methodist hymnal / Nutter, Charles Sumner & Tillett, Wilbur Fisk – New York: Methodist Book Concern, c1911 – 2mf – 9 – 0-524-03076-6 – mf#1990-4565 – us ATLA [242]

[Hymns and psalms] – [1804] [mf ed 19–] – 1v on 1r – 1 – mf#pres. film 33 – us Sibley [780]

Hymns and scenes of childhood : or, a sponsor's gift / Leeson, Jane Eliza – London: James Burns; Nottingham: Dearden, 1842 – (= ser 19th c children's literature) – 3mf – 9 – mf#6.1.30 – uk Chadwyck [305]

The hymns and sloks of shekh farid : contained in the granth sahib of the sikhs / Macauliffe, Max – Lahore: Caxton Printing Works, 1901 – 1mf – 9 – 0-524-07487-9 – mf#1991-0108 – us ATLA [280]

Hymns and songs of praise – Tokyo: [s.n.] 1890 [mf ed 1995] – (= ser Yale coll) – xiv/288p/16p – 1 – 0-524-09684-8 – (in japanese) – mf#1995-0684 – us ATLA [780]

Hymns and spiritual songs / Dupuy, Starke – Ed. and rev. by John M. Peck. 1843 – 1 – us Southern Baptist [242]

Hymns and spiritual songs – Newport. 1766 – 1 – 6.65 – us Southern Baptist [242]

Hymns and spiritual songs, original and selected : for the use of christians – Oshawa, CW [Ont]: Publ for the Canada Christian Conference, 1849 – 5mf – 9 – 0-665-89301-9 – (incl ind) – mf#89301 – cn Canadiana [780]

Hymns and thoughts on religion = Hymnen an die nacht / Novalis (Friedrich von Hardenberg); ed by Hastie, William – Edinburgh: T & T Clark 1888 [mf ed 1990] – 1r – 1 – 0-8370-0990-1 – (trans fr german by ed; with biogr sketch) – mf#1984-t100 – us ATLA [810]

Hymns for sunday school and church – 5th ed. Fort Wayne, Ind: Parish Press, c1915 – 2mf – 9 – 0-524-06629-9 – (incl opening service and holy communion service) – mf#1991-2684 – us ATLA [240]

Hymns for the people *see* Ming chung sheng ko chi (ccm75)

Hymns for the use of the canadian wesleyan methodist new connexion : principally from the collection of the rev john wesley, late fellow of lincoln college, oxford – 2nd Canadian ed. London [Ont]: J H Robinson, 1859 [mf ed 1985] – 3mf – 9 – 0-665-50026-2 – (incl ind) – mf#50026 – cn Canadiana [242]

Hymns from the rigveda / ed by Peterson, Peter – Poona: Bhandarkar Oriental Research Institute, 1937 – (= ser Samp: indian books) – (ed with sayana's commentary, notes and trans by peter peterson; [revised by] r d karmarkar) – us CRL [280]

Hymns in the tenni or slavi language of the indians of mackenzie river : in the north-west territory of canada – London?: SPCK, 1890? – 2mf – 9 – (in slave language) – mf#15269 – cn Canadiana [290]

The hymns of jaidev, ramanand, trilochan, pipa, bhikan, beni, parmanand, sadhna, dhanna, surdas, and mira bai : contained in the granth sahib of the sikhs / Macauliffe, Max – Lahore: Civil and Military Gazette Press, 1901 – 1mf – 9 – 0-524-07488-7 – mf#1991-0109 – us ATLA [280]

Hymns of praise / ed by Underwood, Horace Grant – Yokohama, Japan, 1894 – 1r – 1 – $50.00 – (in korean language) – us Presbyterian [240]

Hymns of salvation : selected and arranged for use in teaching the glad tidings of mercy to man, through the blood of jesus christ – Toronto?: A Lovell, 1869 – 2mf – 9 – (incl ind) – mf#08923 – cn Canadiana [240]

The hymns of the atharva-veda – Benares: E J Lazarus 1895-96 [mf ed 1993] – 2v on 3mf – 9 – 0-524-07503-4 – mf#1991-0124 – us ATLA [280]

Hymns of the atharva-veda (stbe42) – 1897 – (= ser Sacred book of the east (sbte)) – 14mf – 8 – €27.00 – (trans by m bloomfield) – ne Slangenburg [280]

The hymns of the bhagat namdev : found in the granth sahib of the sikhs / Macauliffe, Max – Lahore: Caxton Printing Works, 1901 – 1mf – 9 – 0-524-07489-5 – mf#1991-0110 – us ATLA [280]

Hymns of the church, ancient and modern : for the use of all who love to sing the praises of god in christ... / Wilson, Samuel Ramsey – Cincinnati: Robert Clarke, 1872 [mf ed 1993] – (= ser Presbyterian coll) – 1mf – 9 – 0-524-06454-7 – mf#1991-2576 – us ATLA [242]

Hymns of the faith = dhammapada : being an ancient anthology preserved in the short collection of the sacred scriptures of the buddhists – Chicago: Open Court, 1902 – 1mf – 9 – 0-524-06899-2 – mf#1991-0042 – us ATLA [280]

Hymns, psalms and spiritual songs / Graves, Absalom – 1825 – 1 – us Southern Baptist [242]

Hymns recommended for use in the reformed episcopal church – Philadelphia: James A Moore, 1875 – 1mf – 9 – 0-524-07562-X – mf#1991-3182 – us ATLA [240]

Hymns to the goddess – London: Luzac, 1913 – 1mf – 9 – 0-524-01391-8 – mf#1990-2403 – us ATLA [280]

Hymns to the goddess – Madras: Ganesh & Co, 1952 – (= ser Samp: indian books) – (trans fr sanskrit by arthur and ellen avalon) – us CRL [280]

Hymns to the holy spirit / Stratton, Joseph Buck – Richmond, VA: Presbyterian Committee of Publication, 1893 – 1mf – 9 – 0-524-06557-8 – mf#1991-2641 – us ATLA [240]

Hymns to the mystic fire : hymns to agni from the rigveda translated in their esoteric sense – Pondicherry: Sri Aurobindo Ashram, 1952 – (= ser Samp: indian books) – us CRL [280]

Hymn-tunes and their story / Lightwood, James Thomas – London: CH Kelly, [1906?] – 1mf – 9 – 0-524-07020-2 – mf#1991-2873 – us ATLA [780]

Ein hymnus abecedarius auf christus / Dold, Alban – 1959 – (= ser Texte und arbeiten. beuron (tab). beitraege zur ergruendung des aelteren lateinischen und christlichen schrifttums und gottesdienstes) – €3.00 – ne Slangenburg [240]

Hymnus ambrosianus sive te deum laudamus : [a quatro voci, 2 violini, viola, oboi, corni, trombe timpani & organo. partitura] / Hasse, Johann Adolf – [180-?] [mf ed 19–] – 1r – 1 – mf#pres. film 31 – us Sibley [780]

Hymnus jubilaeus / Weinich, Friedrich – 1617 – (= ser Mssa) – 1mf – 9 – €20.00 – mfchl 444 – gw Fischer [780]

Hymn-writers of the nineteenth century : with selections and biographical notices / Leask, George Alfred – London: E Stock 1902 [mf ed 1990] – 1mf – 9 – 0-7905-5001-6 – mf#1988-1001 – us ATLA [780]

Hymody of the 16th century anabaptists / Duerksen, Rosella R – 1956. Sacred Music thesis – 1 – us Southern Baptist [242]
Hyndman, Henry Mayers *see* The bankruptcy of india
Hyndman, P K *see* Inverness railway, cape breton
Hyner, Gerald G *see* Psychosocial factors in the development of breast cancer
Hypatia – Bloomington. 1992+ (1,5,9) – ISSN: 0887-5367 – mf#18633 – us UMI ProQuest [305]
Hyper news – Belle, WV. 1953-1967 (1) – mf#67206 – us UMI ProQuest [071]
Hyperbaric oxygen therapy in the treatment of sports injuries / Gareau, Tony – 1997 – 1mf – 9 – $4.00 – mf#PE 3844 – us Kinesology [617]
Hyperboreische briefe / ed by Wekhrlin, Wilhelm Ludwig – Nuernberg 1788-90 – (= ser Dz) – 6v on 17mf – 9 – €170.00 – mf#k/n5784 – gw Olms [860]
Hyper-evangelism / Kennedy, John – Edinburgh, Scotland. 1874 – 1r – us UF Libraries [242]
Hyperion / Holderlin, Friedrich – Berlin, Germany. 1921 – 1r – us UF Libraries [025]
Hyperius, A *see*
– Commentarii...in omnes d. pauli apostoli epistolas
– De formandis consionibus sacris...
– De theologo seu de ratione studii theologici
– Methodi theologiae
Hypermnestre : tragedie, en cinq actes et en vers / Le Mierre, Antoine-Marin – Paris, France. 1806 – 1r – us UF Libraries [820]
Hypertension – Dallas. 1979+ (1,5,9) – ISSN: 0194-911X – mf#12425 – us UMI ProQuest [616]
La hypnerotomachia di poliphilo : cioe pugna d'amore in sogno / [Colonna, F] – Venetia: [Aldus], 1545 – 9mf – 9 – mf#O-77 – ne IDC [090]
Hypnosis quarterly – New York. 1974-1982 (1) 1974-1982 (5) 1974-1982 (9) – ISSN: 0018-8344 – mf#9171 – us UMI ProQuest [615]
Hypnotism / Moll, Albert – 4th ed. rev. and enl. London, 1898. 448p. (Contemporary Science Series) – 1 – us Wisconsin U Libr [615]
Hypnotism and spiritism : a critical and medical study = Ipnotismo e spiritismo / Lapponi, Giuseppe – New York: Longmans, Green, 1907 – 1mf – 9 – 0-8370-6993-9 – (in english. incl ind) – mf#1986-0993 – us ATLA [130]
Hypnotism and suggestion in therapeutics, education, and reform / Mason, Rufus Osgood – New York: H. Holt, 1901.344p – 1 – us Wisconsin U Libr [150]
L'hypnotisme et les religions : ou, la fin du merveilleux / Skepto – 2e ed. Paris: Octave Doin; Bordeaux: Feret, 1888 [mf ed 1985] – 1mf – 9 – 0-8370-5267-X – (in french) – mf#1985-3267 – us ATLA [230]
Hypochondrische plaudereien / Amyntor, Gerhard von [pseud of: Dagobert von Gerhardt] – 2. aufl. Ebersfeld: S Lucas, [19–?] [mf ed 1990] – viii/249p – 1 – mf#7297 – us Wisconsin U Libr [880]
Der hypochondrist : eine hollsteinische wochenschrift / ed by Gerstenberg, Heinrich Wilhelm von et al – Schleswig 1762 – (= ser Dz) – 3mf – 9 – €90.00 – mf#k/n5277 – gw Olms [074]
Hypocritical priestcraft of apostolic succession / Thorn, William – London, England. 18– – 1r – us UF Libraries [240]
Les hypogees royaux de thebes / Lefebure, O – Paris, 1886 – (= ser Memoires publiees par les membres mission archaeologique francaise du caire) – 6mf – 9 – (memoires publiees par les membres mission archaeologique francaise du caire, 1882-1884 v2) – mf#NE-20009 – ne IDC [956]
Hypomnema apologeticum pro regali academia limensi in lipsianam periodum : ad limensem regium senatum: regios iudices: conscriptos senatores... / Leon Pinelo, Diego de – Limae: ex officina Iuliani de los Santos et Saldana, anno Domini 1648 – (= ser Books on religion...1543/44-c1800: teologia "culta"; derecho canonico) – 5mf – 9 – mf#crl-267 – ne IDC [241]
Die hypotiposen des theognost / Harnack, Adolf von – Leipzig, 1903 – (= ser Tugal 2-24/3b) – 1mf – 9 – €3.00 – ne Slangenburg [240]
Die hypotyposen des theognost *see* Der pseudocyprianische traktat de singularitate clericorum
Hyppolite, Michelson Paul *see*
– Litterature populaire haitienne
– Origines des variations du creole haitienne
Hysteria – v1 n1-v4 n4, v5 n1 [1980 spr-1984 spr, 1987 may] – 1r – 1 – mf#1265753 – us WHS [071]
Hyvernat, Henri *see*
– Notices sur l'histoire ancienne de l'armenie
– Petite introduction a l'etude de la massore
Hyzet *see* Einheit

I : a lecture on the immortality of the soul / Stokes, George Gabriel – London, England. 1890 – 1r – us UF Libraries [230]
I / Pa, Chin – Shang-hai: Wen hua sheng huo ch'u pan she, Min kuo 25 [1936] – (= ser P-k&k period) – us CRL [480]
I accuse / Bartlett, Vernon – London, 1937. Fiche W 748. (Blodgett Collection of Spanish Civil War Pamphlets) – 9 – us Harvard [946]
I accuse france – [High Wycombe, Buckinghamshire: s.n.] 1936 [mf ed 1977] – (= ser Blodgett coll) – 1mf – 9 – (repr fr the catholic herald) – mf#w951 – us Harvard [934]
I acta mechmeti i saracenorvm principis – [Frankfort], 1597 – 1mf – 9 – mf#H-8216 – ne IDC [956]
I am a churchman / Stowell, Hugh – Dublin, Ireland. 1842 – 1r – us UF Libraries [240]
I am afraid to do right / Smith, James – London, England. 18– – 1r – us UF Libraries [240]
I am captured *see* Wo pei tai chu le (ccm81)
I am going early – London, England. 18– – 1r – us UF Libraries [240]
I am lost! – London, England. 18– – 1r – us UF Libraries [240]
I am prepared to die / Mandela, Nelson – London, England. 1970 – 1r – us UF Libraries [920]
I am sermons / Shelton, Thomas J – Denver, Colo: Published by "Christian," [1900?] – 1mf – 9 – 0-524-02549-5 – mf#1990-3044 – us ATLA [200]
I and n reporter / U.S. Immigration and Naturalization Service – v1-25 n1. jul 1952-summer 1976 [all publ] – 25mf – 9 – $37.50 – (cont by: the i.n.s. reporter) – mf#llmc 81-505 – us LLMC [342]
I augi – [Greece], 1989- – 1 – enquire for prices – (yrly reel count varies) – us UMI ProQuest [079]
I b e w clear vision / International Brotherhood of Electrical Workers – v18 n7/8-v32 n1 [1972 jul/aug-1987 jan] – 1r – 1 – mf#1099413 – us WHS [331]
I believe in god the father almighty / Barrows, John Henry – Chicago: Fleming H Revell, c1892 [mf ed 1985] – 1mf – 9 – 0-8370-2548-6 – mf#1985-0548 – us ATLA [210]
I believe one catholic and apostolic church / Lawlor, H J – Edinburgh, Scotland. 1895 – 1r – us UF Libraries [241]
I bhava vay : [a novel] / Tan Tay – Ran Kun: Mran muir Tuik 1954 [mf ed 1990] – (= ser Mran muir tuik 20) – 1r with other items – 1 – (in burmese) – mf#mf-10289 seam reel 188/5 [$] – us CRL [830]
I c f : mensario dos estudantes do instituto commercial e orgam da mocidade do commercio de florianopolis – Florianopolis, SC. set-out 1927 – (= ser Ps 19) – mf#UFSC/BPESC – bl Biblioteca [380]
I cannot tell, god knoweth / Stock, John – London, England. 18– – 1r – us UF Libraries [240]
I ch'an shui yuan li chi shih wu / Chu, Kung-yen – Shang-hai: Chung-kuo chu tso jen ch'u pan ho tso she, Min kuo 27 [1938] – (= ser P-k&k period) – us CRL [340]
I chia / Hsu, Hsu – Shang-hai: Yeh ch'uang shu wu, Min kuo 30 [1941] – (= ser P-k&k period) – us CRL [830]
I, Ch'iao *see* Hao chu pen
I ch'ien i pai ko chi pen han tzu shih yung chiao hsueh fa / Hung, Shen – Shang-hai: Sheng huo shu tien, Min kuo 24 [1935] – (= ser P-k&k period) – us CRL [480]
I chih ti tu t'u / Tsou, Ti-fan – Kuei-lin: Nan t'ien ch'u pan she, Min kuo 31 [1942] – (= ser P-k&k period) – us CRL [810]
I chiu san erh nien chih kuo chi cheng chih ching chi / Fan, Chung-yun – Shang-hai: Hsin sheng ming shu chu, Min kuo 21 [1932] – (= ser P-k&k period) – us CRL [337]
I chiu san i, i erh, i ch'i / Yin-hung – Pei-p'ing: [sn], 1932 – (= ser P-k&k period) – us CRL [830]
I chiu san pa nien chih chung-kuo – [Sl]: Che-chiang sheng k'ang Jih tzu wei wei yuan hui chan shih chiao yu wen hua shih yeh wei yuan hui, 1939 – (= ser P-k&k period) – us CRL [951]
I chiu san san nien chih shang-hai chiao yu – Shang-hai: Shang-hai hsin wen she, Min kuo 23 [1934] – (= ser P-k&k period) – us CRL [370]
I chiu san ssu hsiao shuo nien hsuan – Shang-hai: K'ai hua shu chu, 1935 – (= ser P-k&k period) – us CRL [830]
I chiu san wu nien ti she hui tung t'ai / Lin, Meng-kung – Shang-hai: Shang wu yin shu kuan, Min kuo 25 [1936] – (= ser P-k&k period) – us CRL [360]
I chiu san wu nien ti shih chieh i shu / Lin, Feng-mien – [Shang-hai]: Shang wu yin shu kuan, Min kuo 25 [1936] – (= ser P-k&k period) – us CRL [057]
I chiu ssu erh nien ti t'ai-p'ing yang / Ch'en, Tsu-jun – Ch'ung-ch'ing: Tu li ch'u pan she, Min kuo 32 [1943] – (= ser P-k&k period) – us CRL [951]

I, Chun *see* Chien-pu-chai
I commentari i delle gverre fatto co' turchi da d giovanni d'avstria... / Carocciolo, F – Fiorenza, 1581 – 2mf – 9 – mf#H-8358 – ne IDC [956]
I commentari...dell' origine de principi turchi, and de' constumi di quella natione / Spandugino, T – Fiorenza, 1551 – 3mf – 9 – mf#H-8283 – ne IDC [956]
I commentarj della cina / Ricci, Matteo – Macerata: F Giorgetti, 1911 – 2mf – 9 – 0-524-03185-1 – (incl bibl ref) – mf#1990-4634 – us ATLA [240]
I consoli e le colonie europee nei possedimenti ottomani / Latino, A – Firenze, 1899 – 6mf – 9 – mf#AR-1841 – ne IDC [956]
I crossed the plains in the '50's / Carpenter, James C – 1 – us Kansas [978]
I d – Cincinnati. 1979-2000 (1) 1979-2000 (5) 1979-2000 (9) – ISSN: 0894-5373 – mf#1090,01 – us UMI ProQuest [720]
I dansk verstindien / Dewitz, August Karl Ludwig Von – Kobenhavn, Denmark. 1904 – 1r – us UF Libraries [972]
I dieci libri de l'architettura / Alberti, L B – Vinecia, 1546 – 5mf – 9 – mf#O-1033 – ne IDC [720]
I dieci libri dell' architettvra di m vitruvvio, tradotti e commentati da mons daniel barbaro... / Vitruvius Pollio, M – Venetia, 1584 – 6mf – 9 – mf#OA-6 – ne IDC [720]
I discorsi di m. gio. andrea palazzi sopra l'imprese / Palazzi, Giovanni Andrea – Bologna: Alessandro Benacci, 1575 – 3mf – 9 – mf#O-387 – ne IDC [090]
I doklad pravleniia pervomu ocherednomu sobraniiu aktsionerov o rabote banka za 1925-26 oper god ii izvlechenie iz bukhgalterskogo otcheta banka za 1925-26 oper god / Viatskii Gorodskoi Aktsionernyi Bank – Viatka, 1926 – 1mf – 9 – mf#REF-90 – ne IDC [332]
I dolci et harmoniosi concenti. libro secondo – 1562 – (= ser Mssa) – 2mf – 9 – €35.00 – mfchl 128 – gw Fischer [780]
I dolci et harmoniosi concenti...a cinque voci. libro primo – 1562 – (= ser Mssa) – 2mf – 9 – €35.00 – mfchl 127 – gw Fischer [780]
I f stone's bi-weekly – 1970 jan 12-1971 dec – 1r – 1 – (cont: i f stone's weekly, stone, i f [isidor f], 1907-) – mf#5295262 – us WHS [071]
I f stone's weekly – 1953 jan-1960 dec 19, 1961 jan 9-1967 dec 18, 1968 jan 8-1969 dec – 3r – 1 – (cont by: i f stone's bi-weekly, stone, i f [isidor f], 1907-) – mf#770832 – us WHS [071]
I fiori dei tre compagni...milano, 1967 / Cambell, Jacques – Madrid: Graf. Calleja, 1967 – 1 – sp Bibl Santa Ana [946]
I follow the mahatma / Munshi, Kanaiyalal Maneklal – Bombay: Allied Publishers, 1940 – (= ser Samp: indian books) – us CRL [320]
I h p maitre de dance oder tantz-meister... / Pimmer, Hans – Glueckstadt, Leipzig: 1705 (mf ed 1998) – 1mf – 9 – €30.00 – 3-8267-2560-3 – (with int and ind) – mf#DHS 2560 – gw Frankfurter [790]
I have found a christ – London, England. 18– – 1r – us UF Libraries [240]
I have not been wicked enough for that! – London, England. 18– – 1r – us UF Libraries [240]
I hay oe / hopo dzopenh *see* Vao doi
I heard the old men say / Green, Lawrence George – Cape Town, South Africa. 1964 – 1r – us UF Libraries [960]
I hope i shall go to heaven when i die – London, England. 18– – 1r – us UF Libraries [240]
I hsueh t'ao lun chi / Li, Cheng-kang – Ch'ang-sha: Shang wu yin shu kuan, Min kuo 30 [1941] – (= ser P-k&k period) – us CRL [290]
I [ie primer] simposio de professores de historia / Simposio De Professores De Historia Do Ensino Supe... – Marilia, Brazil. 1962 – 1r – us UF Libraries [972]
I ieroglifici overo commentarii delle occulte significationi de gl' egittij... / Valeriano Bolzani, G P – Venetia: Presso Gio: Battista Combi, 1625 – 24mf – 9 – mf#O-850 – ne IDC [090]
I jih i t'an / Ma, Liang – Shang-hai: Fu hsing shu chu, Min kuo 25 [1936] – (= ser P-k&k period) – us CRL [080]
I jornadas de promocion del deporte laboral / Obra Sindical de Educacion y Descanso – Caceres: T. Extremadura, 1975 – 1 – sp Bibl Santa Ana [370]
I jornadas fruticolas de extremadura – Don Benito: Graficas Sanchez Trejo, 1969 – 1 – sp Bibl Santa Ana [630]
I kathemerini – [Greece], 1987- – 1 – (yrly reel count varies) – us UMI ProQuest [079]
I kirke / Johnsen, Erik Kristian – Minneapolis, Minn: Augsburg Pub House, 1913 – 1mf – 9 – 0-524-08394-0 – mf#1993-3094 – us ATLA [240]

I ko chueh chiang ti jen / Lo, Pin-chi – Fu-chien Yung-an: Tung nan ch'u pan she, 1944 – (= ser P-k&k period) – us CRL [830]
I ko jen ti chueh hsing / Huang-mei – Ch'ung-ch'ing: Ya tien shu wu, 1945 – (= ser P-k&k period) – us CRL [830]
I ko jen ti fan nao / Yen, Wen-ching – Ch'ung-ch'ing: Chien kuo shu tien; Shang-hai: Ch'ang feng shu tien, 1946 – (= ser P-k&k period) – us CRL [830]
I ko jen ti t'an hua / Shao, Hsun-mei – Shang-hai: Ti i ch'u pan she, [1935] – (= ser P-k&k period) – us CRL [480]
I ko nu jen / Ting, Ling – Shang-hai: Chung-hua shu chu, Min kuo 21 [1932] – (= ser P-k&k period) – us CRL [830]
I ko shang-hai shang jen te kai p'ien (ccm213) = A changed exchange broker / Li, Kuan-shen – 6th rev ed. Shanghai, 1939 [mf ed 198?] – (= ser Ccm 213) – 1 – mf#1984-b500 – us ATLA [920]
I ko shih yen te hsiang tsun chiao hui (ccm109) = An experimental rural church / Chu, Ching-i – Hong Kong, 1954 [mf ed 198?] – (= ser Ccm 109) – 1 – mf#1984-b500 – us ATLA [240]
I ko t'ien ts'ai ti t'ung hsin / Shen, Ts'ung-wen – Shang-hai: Ta kuang shu chu, Min kuo 25 [1936] – (= ser P-k&k period) – us CRL [830]
I k'o wei ch'u t'ang ti ch'iang tan / Ting, Ling – Shang-hai: Chih shih ch'u pan she, Min kuo 35 [1946] – (= ser P-k&k period) – us CRL [070]
I m b o talking leaves / Indian and Metis Brotherhood Organization of Stony Mountain Institution – 1978 jun – 1r – 1 – mf#626705 – us WHS [366]
I marmi del doni : ...tre libri di lettere de doni / Doni, A F – Vinegia: Francesco Marcolino, 1552 – 7mf – 9 – mf#O-1004 – ne IDC [090]
I marmi del doni...tre libri di lettere de doni : e di termini della lingua toscana. et une quarta parte / Doni, A F – Vinegia: Francesco Marcolino, 1552 – 7mf – 9 – mf#O-1004 – ne IDC [700]
I martiri annamiti e cinesi (1798-1856) : solennemente beatificati dalla santita di papa leone 13, il 27 maggio dell'anno santo 1900 – Roma: Tipografia Vaticana, 1900 [mf ed 1995] – (= ser Yale coll) – xiii/489p (ill) – 1 – 0-524-09420-9 – (in italian) – mf#1995-0420 – us ATLA [951]
I, Men *see* Cha-pei ch'i shih san t'ien
I mensklighetens lifsfragor : populaer-filosofiska och religions-filosofiska foeredrag, uppsatser och bref / Wikner, Pontus – Stockholm: OL Lamm, [1889?] – 2mf – 9 – 0-524-00214-2 – mf#1989-2914 – us ATLA [100]
I missa yosapo / Guiron, J J – s.l, s.l? 1908 – 1r – us UF Libraries [960]
I mondi del doni / Doni, A F – Vinegia: Per Francesco Marcolini, 1552-53 – 6mf – 9 – mf#O-1551 – ne IDC [090]
I n s reporter / U.S. Immigration and Naturalization Service – v25 n2-v35 n2. sum 1976-win 1986 – 17mf – 9 – $25.50 – (cont: the i and n reporter) – mf#LLMC 81-506 – us LLMC [342]
I nien / Ting, Ling – Ch'ung-ch'ing: Sheng huo shu tien, 1939 – (= ser P-k&k period) – us CRL [951]
I nien lai chih chung-kuo kung chai : min kuo erh shih san nien fen – [Shang-hai: Che-chiang hsing yeh yin hang, Min kuo 23 [1934]] – (= ser P-k&k period) – us CRL [350]
I nien lai chih huo yun – [Ch'ung-ch'ing: Chung yang hsin t'o chu yin chih ch'u], Min kuo 32 [1943] – (= ser P-k&k period) – us CRL [380]
I nostri errori, tredici anni in eritrea : note storiche e considerazioni – Torino: F Casanova, 1898 – us CRL [945]
I nostri errori, tredici anni in eritrea – Torino. 1898 – 1 – us CRL [960]
I nostri protestanti / Comba, Emilio – Firenze: Claudiana, 1895-1897 – 3mf – 9 – 0-7905-5456-9 – (incl bibl ref) – mf#1988-1456 – us ATLA [242]
I nostri quattro evangelii : studio apologetico critico / Polidori, Eugenio – 3. ed migliorata. Roma: Civilt a Cattolica, 1913 – 1mf – 9 – 0-524-05996-9 – (incl bibl ref) – mf#1992-0733 – us ATLA [225]
I o o f proceedings of the grand lodge of the state of wisconsin, annual session / Independent Order of Odd Fellows – 49th [1895] – 1r – 1 – (cont: proceedings of the right worthy grand lodge of the state of wisconsin, independent order of odd fellows; cont by: proceedings of the grand lodge, independent order of odd fellows, jurisdiction of wisconsin [1911], independent order of odd fellows) – mf#1130649 – us WHS [366]
I obzor 1911 g 2 obzor 1912 g : ezhegodnik russkoi meditsinskoi pechati – M., 1912-1914 – 54mf – 9 – mf#R-7098 – ne IDC [077]

I p e a news : official publication of the idaho public employees association / Idaho Public Employees Association – 1974 dec-1994 nov – 1r – 1 – mf#3363566 – us WHS [366]

I pai i shih hu / Ts'ao, Yu – Kuei-lin: Chin jih wen i she, 1941 – (= ser P-k&k period) – us CRL [480]

I platonici italiani / Semprini, Giovanni – Milano: Edizioni Athena, 1926 – 119p – 1 – us Wisconsin U Libr [180]

I possedimente italiani in africa (libia, eritrea, somalia) / Stefanini, Giuseppe – Firenze. 1929 – 1 – us CRL [960]

I possedimente italiani in africa (libia, eritrea, somalia) / Stefanini, Guiseppe – Firenze: R. Bemporad and Figlio, 1929 – us CRL [960]

I primi influssi di dante, del petrarca, e del boccaccio sulla letteratura spagnuola, con appendice di documenti inediti.. / Sanvisenti, Bernardo – Milano: U. Hoepli, 1902.xvi,463p – 1 – us Wisconsin U Libr [440]

I quattro libri dell' architettura / Palladio, A – Venezia, 1570 – 7mf – 9 – mf#O-388 – ne IDC [720]

I quattro primi libri di architettura / Cataneo, P – Vinegia, [1554] – 4mf – 9 – mf#O-1008 – ne IDC [720]

I s e a news : official publicationof the idaho state employees association / Idaho State Employees Association – v9 n3 [1971 sep], v11 n2-v12 n3 [1972 nov-1974 jul], v13 n2 [1974 oct] – 1r – 1 – mf#1477178 – us WHS [366]

I sac chum : [a novel] / Kui Kui – Ran Kun: Khet Nnvan Ca pe 1963 [mf ed 1990] – (= ser Khet nnvan ca pe ca can 5) – 1r with other items – 1 – (in burmese) – mf#mf-10289 seam reel 197/4 [§] – us CRL [830]

I sacri concerti a due voci... / Fattorini, Gabriele – 1604 – (= ser Mssa) – 1mf – 9 – €20.00 – mfchl 214 – gw Fischer [780]

I samuel 1-7:1 : text- und quellen-kritisch untersucht / Holtz, Kurt – Leipzig: W. Drugulin, 1904. 1 fiche – 9 – us ATLA [221]

I sez, sez i : a series of talks to talkers on what to say and how to say it / Dezell, Robert – Allenford, Ont: R Dezell, 1911 – 3mf – 9 – 0-665-65682-3 – (ill by j e laughlin) – mf#65682 – cn Canadiana [080]

I shen tso tse / Li, Chien-wu – Shang-hai: Wen hua sheng huo ch'u pan she, Min kuo 25 [1936] – (= ser P-k&k period) – us CRL [480]

I, Shih-fang *see* Chan shih ti jen min tzu yu

I shu ch'u wei / Feng, Tzu-k'ai – Shang-hai: K'ai ming shu tien, Min kuo 37 [1948] – (= ser P-k&k period) – us CRL [057]

I shu yu sheng huo / Chou, Tso-jen – Shang-hai: Ch'un i shu she, [1930] – (= ser P-k&k period) – us CRL [480]

I simposio sobre angola / Simposio Sobre Angola (1st : 1967 : Lisbon, Portugal) – Lisboa, Portugal. 1967 – 1r – us UF Libraries [960]

I t pososhkov : zhizn i deiatelnost / Kafengauz, B B – 1961 – 4mf – 8 – mf#R-6192 – ne IDC [947]

I tai nu yu / Chang, Tzu-p'ing – Feng-t'ien: Ta tung shu chu, 1943 – (= ser P-k&k period) – us CRL [830]

I t'ien ti kung tso / Lu, Hsun – Shang-hai: Liang yu t'u shu yin shua kung ssu, 1936 – (= ser P-k&k period) – us CRL [480]

I tre primi vangeli e la critica letteraria, ossia, la questione sinottica / Bonaccorsi, Giuseppe – Monza: Artigianelli-Orfani, 1903 – 2mf – 9 – 0-524-04446-5 – (incl bibl ref) – mf#1992-0115 – us ATLA [220]

I trionfi feste, et livree fatte dalli signori conservatori, and popolo romano, and da tutte le arti di roma... / Colonna, M A – Venetia, 1571 – 1mf – 9 – mf#H-8155 – ne IDC [956]

I tuan lu ch'eng / Wang, Hsi-yen – Kuei-lin: Shih huo ch'u pan she, 1940 – (= ser P-k&k period) – us CRL [840]

I t'ung / Su, Yuan-lei – Ch'ung-ch'ing: Huang chung ch'u pan she, 1944 – (= ser P-k&k period) – us CRL [180]

I u c t journal : the official organ of the international union, commercial telegraphers / International Union of Commercial Telegraphers – v1 n1-3 [1903 jan-mar] – 1r – 1 – (cont by: journal [commercial telegrapher's union of america]) – mf#1231932 – us WHS [331]

I viaggi di messer marco polo gentil 'hvomo ventiano – Venetia, 1554-1556. v2 – 2mf – 9 – mf#HT-680 – ne IDC [910]

I wai chi / Ting, Ling – Shang-hai: Liang yu t'u shu yin shua kung ssu, 1936 – (= ser P-k&k period) – us CRL [480]

I want to feel more / Smith, James – London, England. 18– – 1r – us UF Libraries [240]

I want to feel more / Smith, James – London, England. 18– – 1r – us UF Libraries [240]

I was a franco soldier / MacKee, Seumas – London, 1938. Fiche W1011. (Bloedgett Collection of Spanish Civil War Pamphlets) – 9 – us Harvard [946]

I was defeated / Kodama, Yoshio – Transl. from the Japanese. (Tokyo?): R. Booth and T. Fukuda, (1951). 223p – 1 – us Wisconsin U Libr [950]

I was hitler's prisoner / Lorant, Stefan – London, England. 1935 – 1r – us UF Libraries [920]

I was hitler's prisoner / Lorant, Stefan – London, England. 1935 – 1r – us UF Libraries [025]

I, Wen *see* Lun chun chi

I wen ch'ein : [23 hui] / Chang, Huan-tou – Shang-hai: I hsueh shu chu, Min kuo 21 [1932] – (= ser P-k&k period) – us CRL [830]

I wen hsiao yu : wen i li lun chi i shu chieh shao p'i p'ing chi – Shang-hai: Hsin Chung-kuo pao she, [1943] – (= ser P-k&k period) – us CRL [480]

I will never leave thee, nor forsake thee / Cowan, Robert – Inverness, Scotland. 1889 – 1r – us UF Libraries [240]

I will never leave thee, nor forsake thee / Cowan, Robert – Inverness, Scotland. 1889 – 1r – us UF Libraries [240]

A i z – arbeiter-illustrierte zeitung *see* Sowjetrussland im bild

Ia nikogo ne em 365 vegetarianskikh meniu / Zelenkov, A P – St Petersburg, Russia. no date – 1r – us UF Libraries [025]

Ia nikogo ne em 365 vegetarianskikh meniu / Zelenkov, A P – St Petersburg, Russia. no date – 1r – us UF Libraries [640]

Ia – russkii / Narodnaia natsional'naia partiia – Moscow, Russia. n1(sent 1997)-n21(dek 1998) – 1 – mf#mf-12248 (reel 4) – us CRL [077]

Iablochkov, M *see* Istoriia dvorianskogo sosloviia v rossii

Iacobi arminii...disputationes...publicae et privatae / Arminius, Jacobus – Lugduni Batavorum, 1610 – 9mf – 9 – mf#PBA-126 – ne IDC [240]

Iacobi balde s societate jesu urania victrix : cum facultate superiorum / Balde, J – Monachii: Typis Joannis Wilhelmi Schell, Sumptibus Joannis Wagneri, civis ac bibliopolae Monacensis, 1663 – 4mf – 9 – mf#O-1803 – ne IDC [090]

Iacobi latomi...de confessione secreta : ioannis oecolampadii eleboron / Oecolampadius, J – Basileae, Andreas Cratander, 1525 – 2mf – 9 – mf#PBU-364 – ne IDC [240]

Iacobi philippi tomasini patavini illvstrivm virocum elogia... / Tomasini, G F – Patavii: Apud Donatum Pasquardem, & Socium, 1630 – 5mf – 9 – mf#O-1363 – ne IDC [090]

Iadrintsev, N M *see* Nauchno-literaturnoe periodicheskoe izdanie

Iagich, V *see* Codex marianus glagoliticus

Iagulli, Jonathan J *see* The importance of team chemistry

Iah news / Howard University – 1978/1979 win, 1980 win, 1982 summer – 1r – 1 – mf#4866756 – us WHS [378]

Iaia, Jim *see* Baseball and italian-americans

Iaiabc journal / International Association of Industrial Accident Boards and Commissions – 1976 spr, sep, 1977 jan, jul – 1r – 1 – (cont by: journal [international association of industrial accident boards and commissions]) – mf#463842 – us WHS [360]

Iaitskaia volia : [ezhedn polit i lit gaz] – Ural'sk: [s n] 1918 [1867 1 ianv-] – (= ser Asn 1-3) – n1-113 [1918] [gaps] item 452, on reel n88 – 1 – (suppl: iaitskie voiskovye vedomosti [asn-1 453]) – mf#asn-1 452 – ne IDC [077]

Iaitskie voiskovye vedomosti : ezhened ofits organ s pravitel'stv i voiskovymi rasporiazheniiami, prikazami i proch – Ural'sk: [s n] 1918 [1918 1 ianv-[mart]] – (= ser Asn 1-3) – n1-9 [1918] item 453, on reel n88 – 1 – (suppl of: iaitskaia volia [asn-1 452]) – mf#asn-1 453 – ne IDC [077]

Iakhinson, I *see* Mendeles epokhe

Iakhontov, I *see* Zhitiia sviatykh severnorusskikh podvizhnikov pomorskago kraia kak istoricheskii istochnik

Iakhontov, P V *see*
- Kak organizovat kreditnuiu rabotu v promyslovoi kooperatsii
- Promyslovoe kreditnoe tovarishchestvo

Iakimovich, K A *see* Termodinamicheskie svoistva gidrida litiia i ego izotopicheskikh modifikatsii v tverdoi faze

Iakinf [Bichurin, Ia] *see*
- Denkwuerdigkeiten ueber die mongolei
- Description du tubet...
- Istoriia tibeta i khukhunora s 2282 goda do r kh do 1227 goda po...
- Opisanie pekina, s prilozheniem plana sei stolitsyi, sniatogo v 1817 godu
- Opisanie tibeta v nyineshnem ego sostoianii
- Zapiski o mongolii

Iakovlev, A I *see* Za uluchshenie kooperativnykh kadrov

Iakovlev, A V *see*
- Melkii zemel'nyi kredit v rossii
- Melkii zemelnyi kredit v rssii
- Selskie ssudnye tovarichestva

Iakovlev, A.V *see* Melkii zemel'nyi kredit v rossii

Iakovlev, I *see* Russkii anarkhizm v velikoi russkoi revoliutsii

Iakovlev, I A *see*
- God sluzhby sotsialistov kapitalistam:
- K voprosu o sotsialisticheskom pereustroistve selskogo khoziaistva:

IAkovlev, R I *see*
- Golos zabaikal'ia
- Vestnik zabaikal'ia

Iakovleva, V K *see* Iazyk ioruba

Iakushkin, N V *see* Denezhnye kursy i tovarnye tseny

IAkushkin, V E *see* Iuzhnoe slovo [nikolaev: 1918]

Iakushkin, V E *see* Russkaia pechat i tsenzura v proshlom i nastoiashchem

IAkutiia – IAkutsk: IAkutiia 1991-93 (mf ed Minneapolis MN: East View Publ [199-]) – 5r – 1 – (narodnaia gazeta aug 31 1991- , ezhednevnaia respublikanskaia gazeta jan 1 1992-oct 14 1993; iss for aug 31-dec 31 1991 filmed with: sotsialisticheskaia iakutiia jul 1-aug 23 1991, iss for jul 1-oct 14 1993 filmed with: respublika sakha oct 15-dec 31 1993) – mf#mf-12134 seemp (1r) aug 31-dec 31 1991; mf-12135 seemp (4r) 1992-oct 14 1993 – us CRL [077]

Iakutskaia kooperatsiia – Iakutsk, 1921-1924(1) – 4mf – 9 – (missing:1921(2-3),1922/1923(8-9)) – mf#COR-710 – ne IDC [335]

Iakutskie eparkhial'nye vedomosti – 1887-1919 – 11r – 1 – us UMI ProQuest [243]

Iakutskii golos : organ tsentr kom iakut trudovogo soiuza federalistov / ed by Shadrin, M I – IAkutsk: [s n] 1918 [1918 3 noiab-] – (= ser Asn 1-3) – n1-4 [1918] item 454, on reel n88 – 1 – mf#asn-1 454 – ne IDC [077]

Iakutskii oblastnoi vestnik / ed by Krasnov, Iv – IAkutsk: [s n] 1918-19 [1918 3 sent-1919 [?]] – (= ser Asn 1-3) – n1-92 [1918] n66-136 [1919] [gaps] item 455, on reel n88 – 1 – mf#asn-1 455 – ne IDC [077]

IAkutskoe zemstvo : organ iakut obl zemstva / ed by Baranov, P E – IAkutsk: [s n] 1918-19 [1918 15 [2] maia-1919 21 sent] – (= ser Asn 1-3) – n33-34 [1918] n30 [1919] item 456, on reel n88 – 1 – mf#asn-1 456 – ne IDC [077]

IAlgubtsev, M P *see* Volynka

Iall journal of language learning technologies – Iowa City. 1990+ – 1,5,9 – (cont: journal of educational techniques and technologies) – ISSN: 1050-0049 – mf#12664,04 – us UMI ProQuest [370]

Ialtinskii golos / ed by Cherepnin, V A – IAlta [Tavr gub]: T-vo "IUzhnyi bereg" 1918-19 [1917 [8 marta]-[1920] [?]] – (= ser Asn 1-3) – n225, 226 [1918] n621 [1919] item 457, on reel n88 – 1 – mf#asn-1 457 – ne IDC [077]

Ialtinskii kur'er / ed by Amfiteatrov, V A – IAlta [Tavr gub]: A V Davydov 1919 [1919 [?]-1920 [?]] – (= ser Asn 1-3) – n88 [1919] item 458, on reel n88 – 1 – mf#asn-1 458 – ne IDC [077]

Ialutorovskaia zhizn' : bespartiin -demokrat gaz / ed by Mishurin, F – IAlutorovsk [Tobol gub]: Izd kollegiia 1918-19 [1918 [iiul']-1919 [?]] – (= ser Asn 1-3) – n4 [1918]-n119 [1919] [gaps] item 459, on reel n88 – 1 – mf#asn-1 459 – ne IDC [077]

Iam district 720 report / International Association of Machinists and Aerospace Workers – 1975 jun-1978 nov/dec, 1980 jan-1983 jan – 1r – 1 – mf#680032 – us WHS [366]

Iam education bulletin / International Association of Machinists and Aerospace Workers – 1967 jul-1974 jul – 1r – 1 – mf#370665 – us WHS [071]

Iam journal / International Association of Machinists and Aerospace Workers – 1995 win-2002 fall – 1r – 1 – (cont: machinist) – us WHS [629]

Iamblichus *see*
- Iamblichus on the mysteries of the egyptians, chaldeans, and assyrians
- Theurgia, or, the egyptian mysteries

Iamblichus Chalcidensis *see* De mysteriis liber

Iamblichus' exhortation to the study of philosophy / fragments of iamblichus / excerpts from the commentary of proclus on the chaldean oracles / plotinus' diverse cogitations – Osceola, MO: [s.n.] 1907 [mf ed 1992] – 1mf – 9 – 0-524-02943-1 – (in english) – mf#1990-3155 – us ATLA [180]

Iamblichus on the mysteries of the egyptians, chaldeans, and assyrians = De mysteriis / Iamblichus – 2nd ed. London: Bertram Dobell, 1895 – 1mf – 9 – 0-7905-7979-0 – (in english) – mf#1989-1264 – us ATLA [290]

Ian of the orcades : or, the armourer of girnigoe / Campbell, Wilfred – New York, Toronto: F H Revell, [1906?] – 4mf – 9 – 0-665-71562-5 – (ill by robert b m paxton) – mf#71562 – cn Canadiana [830]

IAnchevetskii, V G *see* Vpered

Ianida miani senensis ad leonem x pont max de expeditione in turcas elegeia, cu argutissimis doctissimorum uirorum epigrammatibus / Damianus, J – Basileae, 1515 – 2mf – 9 – mf#H-8227 – ne IDC [915]

Iankova, Z *see* Izmeneniia struktury sotsialnykh rolei zhenshchin v razvitom sotsialisticheskom obshchestve i model semi

Ianni, Octavio *see* Estado e capitalismo

Ianovskii, A D [comp] *see* Napoleon, sa famille et son entourage

Iapi oaye = The word carrier – 1871-1939 – (= ser American indian periodicals... 1) – 57mf – 9 – $210.00 – (in dakota language) – us UPA [305]

Iapi oaye = The word carrier – 1884-1937 – (= ser American indian periodicals... 1) – 29mf – 9 – $210.00 – (in english) – us UPA [305]

Iaponskaia pechat' i vnutrennee polozhenie v rossii 'h / Matsokin, Nikolai – 3e ed. Kharbin: Izd Ob-va Russkikh Opientalistov, 1917 [mf ed 2004] – 1r – 1 – (Filmed with: pis'mo tovarishchu emigrantu / v sukhomlinov (v1-2 1919-20)) – us Wisconsin U Libr [327]

Iardin deys mvsos provensalos : diuisat en quatre partidos / Brueys, Claude – Aix: David 1628 – (= ser Whsb) – 10mf – 9 – €90.00 – mf#Hu 071 – gw Fischer [440]

Iaroslavskaia guberniia v tsifrakh : statisticheskii spravochnik / ed by Uspenskii, VI – Iaroslavl', 1927. xii/440p – 5mf – 9 – mf#RHS-142 – ne IDC [314]

Iaroslavskie gubernskie vedomosti – Yaroslavl', 1843 – 1 – us UMI ProQuest [077]

Iaroslavskii, E *see* Anarkhizm v rossii

Iaroslavskii paterik ili zhitiia ugodnikov bozhiikh, podvizavshikhsia v nyneshnei iaroslavskoi eparkhii – Iaroslavl, 1912 – 4mf – 9 – mf#R-18294 – ne IDC [243]

Iaroslavskoe Gubernskoe Zemstvo. (Agronomicheskii otdel) *see* Melkii kredit v iaroslavskoi gubernii

IArov, IA L *see* Novosti iuga

Iasnaia poliana – n1-2. Moscou. janv-fevr 1862 – 1 – fr ACRPP [073]

Iasnopol'skii, L N *see* Bankovaia entsiklopediia

Iastrzhembskii, vA *see* O kapituliatsiiakh v ottomanskoi imperii

Iaswr buddhist sanskrit manuscripts (from nepal) / Bajracharya, M B – 1200mf – 9 – $3400.00 – (catalog of titles [1mf] $1; alphabetical title list [1mf] $1; descriptive catalogs: pt 1 [6mf] $6, pt 2 [9mf] $9, pt 3 [2mf] $2; recently comp suppl mss [314mf] $900; suppl descriptive catalog [2mf] $1) – us IASWR [090]

Iatromathematisches hausbuch *see* Regimen der gesundheit / iatromathematisches hausbuch (cima41)

Iatsevich, Andrei Grigor'evich *see* Krepostnoi peterburg pushkinskogo vremeni

Iatsunskii, Viktor Kornel'evich *see* Primechaniia k nagliadnym posobiiam po istorii narodnogo khoziaistva rossii v 8-10 vekakh vyp 8 denzhnoe obrashchenie, kredit, gosudarstvennye finansy

Iavliaetsia li nep otstupleniem / Sarab'ianov, Vladimir – Moskva: "Moskovskii rabochii," 1926 [mf ed 2004] – (= ser K itogam 14 s'ezda vkp(b) 1) – 1r – 1 – (filmed with: pis'mo tovarishchu emigrantu / v sukhomlinov (v1-2 1919-20)) – us Wisconsin U Libr [339]

IAvorskaia-Lomakina, E I *see* Sibirskii golos

IAzev, I N *see*
- Krest'ianskaia zhizn'
- Narodnoe delo [tatarsk: 1918]

IAzvitskii, V I *see* Nash put' [omsk: 1919]

Iazyk ioruba / Iakovleva, V K – Moskva: Izd-vo vostochnoi lit-ry, 1963 – us CRL [077]

Iazykov, D D *see* Novye materialy dlia istorii russkoi dukhnovoi literatury

Ibaible eli ingcwele / Bible Zulu – London, England. 1946 – 1r – us UF Libraries [960]

Ibandla lase roma lizondelwani kangaka na? / Ndlovu, Bernard – Gwelo, Zimbabwe. 1968 – 1r – us UF Libraries [960]

Ibandla lase roma lizondelwani kangaka na? / Ndlovu, Bernard – Gwelo, Zimbabwe. 1968 – 1r – us UF Libraries [960]

Ibanez, F *see* Topografia hipocratica o descripcion de la epidemia de calenturas tercianas intermitentes malignas...en la alcarria deide 1784

Ibanez, Jaime *see* Tacita doncella

Ibarguren, Carlos *see*
- Discursos a los asturianos de america
- Discursos a los vascos de america
- El paisaje y el alma argentina

Ibarra Bejarano, Georgina *see* Aquileo j echeverria

Ibarra, Cristobal Humberto *see*
- Francisco gavidia y ruben dario
- Tembladerales

Ibarra, Felipe Bartolome *see* Memorias y episodios del coronel f bartolome ibarra

Ibarra, Francisco *see* Brazilian portuguese self-taught

Ibarra Ibarra, Carlos *see* Sombra de nunez y asesinos de america

Ibarra, Jorge A *see* Apuntes de historia natural y mamiferos de guatemala
Ibarra y Berge, Javier de *see*
- De california a alaska. madrid, 1945
- El moro vizcaino (jose maria de murga)
Ibarrola, Jose *see*
- El asesinato del regato de los avellanos
- Literatura sublime e historia gloriosa y tragica. copiada la primera y relatada la segunda en articulos
Ibarruri, Dolores *see*
- Ejercito popular unido, ejercito de la victoria
- La espana franquista, satelite de hitler
- For the independence of spain, for liberty, for the republic, union of all spaniards
- No hay mas posibilidad de gobernar ni de victoria que a traves del frente popular
- Un pleno historico
- Por la independencia de espana, por la libertad, por la republica, union de todos los espanoles
Ibarzabal, Federico De *see*
- Gesta de heroes
- Problema negro
Ibarzabal, Frederico De *see* Cuentos contemporaneos
Ibbenbuerener volkszeitung – Ibbenbueren DE, 1949 1 nov-1957 – 11mf=21df – 9 – (filmed by other misc inst: 1958-1960 31 aug; 1987-[8r/yr]) – gw Mikrofilm; gw Misc Inst [074]
Ibbetson, Julius Caesar *see* An accidence, or gamut, of painting in oil
Ibbuku lya syaa-zibwene / B M And T – Mission Siding, Zambia. 1931 – 1r – us UF Libraries [960]
Ibbuku lya syaa-zibwene – Mission Siding, Zambia. 1931 – 1r – us UF Libraries [960]
Ibc's money fund report – New York. 1997+ (1,5,9) – ISSN: 1097-5019 – mf#22650,04 – us UMI ProQuest [332]
Ibel, Rudolf *see*
- Don carlos
- Der junge goethe
- Schiller, don carlos
- Weltschau deutscher dichter
Iberia first baptist church. iberia, missouri : church records – 1877-1968 – 1 – 59.31 – us Southern Baptist [242]
Iberica : english edition – New York. 1953-1966 [1,5,9] – ISSN: 0445-1708 – mf#1767 – us UMI ProQuest [320]
Iberica spanish edition – New York. 1954-1974 (1) 1971-1973 (5) – ISSN: 0019-0985 – mf#2378 – us UMI ProQuest [320]
Iberlebenishn / Lewin, Gershon – Vilna, Lithuania. 1931 – 1r – us UF Libraries [939]
Iberlebenishn / Lewin, Gershon – Vilna, Lithuania. 1931 – 1r – us UF Libraries [939]
Ibero-Amerikanisches Institut *see* Alemania y el mundo ibero-americano
D'iberville : ou le jean-bart du canada / Daniel, Francois – S.l: s.n, 1868? – 1mf – 9 – mf#27021 – cn Canadiana [917]
Ibis : a journal of general ornithology / ed by Sclater, E et al – London 1859-1946 – v1-88 on 1186mf – 9 – mf#z-244c/2 – ne IDC [590]
Ibiteekerezo: historical narratives from rwanda / Vansina, Jan – A collection of texts and translations – 1 – us CRL [960]
Ibm journal of research and development – Armonk. 1957+ (1) 1966+ (5) 1970+ (9) – ISSN: 0018-8646 – mf#1525 – us UMI ProQuest [000]
Ibm systems journal – Armonk. 1962+ (1) 1962+ (5) 1962+ (9) – ISSN: 0018-8670 – mf#3072 – us UMI ProQuest [000]
Ibn Abidin, Muhammad ibn Muhammad Amin *see* Qurrat 'uyun al-akhyar
Ibn Battuta *see* Travels in asia and africa 1325-1354
Ibn Batutah *see* Travels
Ibn Daud, Abraham ben David, Halevi *see* Ha-emunah ha-ramah microform
Ibn Ezra, Abraham Ben Meir *see*
- Buch der einheit
- Kovets hokhmat ha-ra'va'
- Margaliyot tovah
Ibn Gabirol, Solomon Ben Yehuda *see* Das weltbild gabirols
Ibn Hasan, Burhan *see* Tuzak-i-walajahi of burhan ibn hasan
Ibn khaldun's philosophy of history: a study in the philosophic foundation of the science of culture / Mahdi, Muhsin – London: G. Allen and Unwin, 1957. Bibliography.325p – 1 – us Wisconsin U Libr [180]
Ibn Malik *see* Alfiyya (quintessence de la grammaire arabe)
Ibn Qutaybah, Abd Allah ibn Muslim *see* An extract from ibn kutaiba's adab al-kaatib
Ibn Verga, Solomon *see* Shevet yehudah
Ibn Yaoish, Abu al-Baqa Yaoish ibn Ali *see* De enuntiationibus relativis semiticis. pars prior, praemisso ibn jaoi si in zamach sarii, de pronominibus relativis locum commentario, de enuntiationibus relativis arabicis agens
Ibn-al-Adim, Umar Ibn-Ahmad *see* Al-muntahab min ta'ri-h halab
Ibn-al-Balkhi *see* Descriptions of the province of fars in persia
Ibn-'Arabsah, Abu-'l-'Abbas Ahmad Ibn-Muhammad *see* Liber arabicus fakihat alhulafa' wa-mufakahat az-zurafa'
Ibn-Badroun *see*
- Comentaire...poeme...ibn abdoun
- Commentaire historique sur le poeme d'ibn-abdoen
Ibn-Washih *see* Historiae
Iboe dan anak : penghidoepan wanita di zaman baroe, tjetakan 1 / Djawa Sinbun Kai – Djakarta (2605) – 64p 1mf – 9 – mf#SE-2002 mf34 – ne IDC [305]
Ibong Adarna – Maynila: P Sayo [191-?] [mf ed Bloomington IN: Indiana Uni Lib, Preservation Dept 1984] – (= ser Coll...in the tagalog language 2) – 1r – 1 – us Indiana Preservation [490]
Ibrahim, Ahmad Hasan *see* Madinat al-aqabah
Ibrahim, Hanna et al *see* Al-aswar
Ibsen, bjoernson, nietzsche : individualismus und christentum / Weinel, Heinrich – Tuebingen: J C B Mohr (Paul Siebeck) 1908 [mf ed 1985] – (= ser Lebensfragen 20) – 1mf – 9 – 0-8370-5742-6 – mf#1985-3742 – us ATLA [230]
Ibsen, Henrik *see*
- Letters..
- Yapi ustasi solness
Ibuku lya makristo – Wankie, Zimbabwe. 1960 – 1r – us UF Libraries [960]
Ibuku lya makristo – Wankie, Zimbabwe. 1960 – 1r – us UF Libraries [960]
Ibw special report / Institute of the Black World – 1975 feb, jul, oct – 1r – 1 – mf#5319924 – us WHS [305]
Ic infection control – Thorofare. 1982-1987 (1) 1982-1987 (5) 1982-1987 (9) – (cont by: infection control and hospital epidemiology) – ISSN: 0195-9417 – mf#13522 – us UMI ProQuest [616]
Ica-kuriren – Vasteras, Sweden. 1942-78 – 55r – 1 – sw Kungliga [078]
Icaria : a chapter in the history of communism / Shaw, Albert – New York: Putnam, 1884 – 1mf – 9 – 0-7905-6952-3 – mf#1988-2952 – us ATLA [320]
Icarus : poems / Cooney, Rian – Whitethorn, CA: Holmgangers Press, c1982 (mf ed 1984) – 1mf – 9 – mf#FSN 39,874 – us NY Public [810]
Icaza, Francisco A de *see* Hebbel prosista
Icc activities : 1889-1937 / U.S. Interstate Commerce Commission – Washington: GPO. 1v. 1937 – (= ser Interstate commerce commission reports) – 3mf – 9 – $4.50 – (vol covering 1937-62 believed to exist) – mf#LLMC 81-230 – us LLMC [324]
Icc acts indexed and digested : the acts to regulate commerce / Hamlin, Charles S – Boston: Little-Brown, 1907 – 5mf – 9 – $7.50 – mf#LLMC 80-525 – us LLMC [324]
Icc practitioners' journal *see* Journal of transportation law, logistics and policy
Icc register : a daily summary of motor carrier applications and decisions and notices issued by the interstate commerce commission – [1984 apr 16/may 29]-[1995 jul-aug] – 67r – 1 – (cont by: federal highway administration register) – mf#2064004 – us WHS [380]
Ice, Charlie *see* Land tenure and power / a basis for community development planning on ponape
Ice hockey injuries : a comparative study of incidence and severity / Bullock, George E – 1997 – 2mf – 9 – $8.00 – mf#PE 3862 – us Kinesology [617]
The ice phenomena and the tides of the bay of fundy : considered in connection with the construction of the baie verte canal / Hind, Henry Youle – [S.l: s.n, 1875?] [mf ed 1982] – 1mf – 9 – 0-665-08417-X – mf#08417 – cn Canadiana [627]
Ice record – Philadelphia. v1-2. 1900-1901 (incomplete) – 1 – us NY Public [073]
Ice trade journal – Philadelphia: Thos E Cahill. r1: v1-8 oct 1877-jul 1885; r2-3: v9 n6-v23 n8 1886-mar 1900 – 1 – us CRL [380]
Iceland : or the journal of a residence in that island, during the years 1814 and 1815 containing observations on the natural phenomena... / Henderson, Ebenezer – Edinburgh 1818 [mf ed Hildesheim 1995-98] – 2v on 6mf – 9 – €120.00 – 3-487-28929-6 – (with int & app) – gw Olms [914]
Icelandic Evangelical Lutheran Synod of America *see*
- Our parish messenger
- The parish messenger
Icelandic sagas (rs88) : and other historical documents relating to the settlements and descents of the northmen on the british isles / ed by Vigfusson, G – (= ser The rolls series (rs)) – (v1,2 1887 €18v. v3,4 1894 €19v; trans by g w dasent (v3-4)) – ne Slangenburg [931]
"Ich" : liedeskunst / Wolff-Cassel, Louis – 4 enl ed. Dresden: E Pierson, 1901 [mf ed 1993] – 174p (ill) – 1 – mf#7965 – us Wisconsin U Libr [780]
Ich bin kein intellektueller : ein heiteres buch / Weiss-Ferdl – Muenchen: P Hugendubel, 1941 – 1r – 1 – us Wisconsin U Libr [870]
Ich blas auf gruenen halmen / Findeisen, Kurt Arnold; ed by Kaergel, Hans Christoph – 2. Aufl. Berlin: W Limpert, 1944 (mf ed 1990) – 1r – 1 – (filmed with: der deutsche finckh) – us Wisconsin U Libr [800]
Ich glaube! : bekenntnisse / Johst, Hanns – Muenchen: A Langen/G Mueller, 1928 – 1r – us Wisconsin U Libr [240]
Ich glaube, darum rede ich : eine kurze darlegung der lehrstellung der missouri-synode / Pieper, Franz – [St Louis?: Concordia Publishing House?, 1897?] – 1mf – 9 – 0-524-00307-6 – mf#1989-3007 – us ATLA [240]
Ich habe mich rasieren lassen : ein dramatischer scherz / Schiller, Friedrich von; ed by Kuenzel, Carl – Leipzig: Englische Kunst-Anstalt von A H Payne, [1862?] – 1r – 1 – us Wisconsin U Libr [870]
Ich moechte nach hause : roman / Bergmann, Herta – Berlin: W Limpert, 1943 [mf ed 1989] – 192p – 1 – mf#7010 – us Wisconsin U Libr [830]
Ich spinne meine aussteuer : materialien und erfahrungen / Heubach, Helga – 2000 – 2mf – 9 – 3-8267-2599-9 – mf#DHS 2599 – gw Frankfurter [300]
Ich suche land in sudbrafilien / Moeschlin, Felix – Horw-Luzern und Leipzig, Germany. 1936 – 1r – us UF Libraries [972]
Ich suche land in sudbrafilien / Moeschlin, Felix – Horw-Luzern und Leipzig, Germany. 1936 – 1r – us UF Libraries [972]
Das ich und die abwehrmechanismen *see* The ego and the mechanisms of defence
Ich und gnade : eine studie ueber friedrich schlegels bekehrung / Horst, Karl August – Freiburg: Verlag Herder, 1951 – 1r – 1 – (incl bibl ref) – us Wisconsin U Libr [190]
Ichigan christian herald *see* The christian herald
Ichihara, Hiroyuki *see* Hiroyuki ichihara papers, 1942
I-ching *see*
- Memoire compose a l'epoque de la grande dynastie t'ang sur les religieux eminents qui all erent chercher la loi dans les pays d'occident
- Record of the buddhist religion as practised in india and the malay archipelago (a d 671-695)
I-Ching, 635-713. I-Ching *see* Record of the buddhist religion as practiced in india and the malay archipelago
Ichneumoninae of florida and neighboring states / Heinrich, Gerd H – Gainesville, FL. 1977 – 1r – us UF Libraries [500]
Ichneumoninae of florida and neighboring states / Heinrich, Gerd H – Gainesville, FL. 1977 – 1r – us UF Libraries [580]
Ichthyologie : ou histoire generale et particuliere des poissons / ed by Bloch, M E – Berlin 1785-97 – 12v on 114mf – 9 – mf#2121/2 – ne IDC [590]
Ichtisar sedjarah indonesia : oentoek sekolah menengah / Siahaan, L – Bandoeng: Poestaka Ksatrian (2604) – 76p on 1mf – 9 – (disoesoen berdasarkan karangannja e f e douwes dekker, oleh I siahaan) – mf#SE-2002 mf161 – ne IDC [370]
Ichtisar tahunan / Antara – Djakarta, 1965-1969 – 74mf – 9 – mf#SE-1314 – ne IDC [959]
I-ch'un *see*
- Hsin jen ti ku shih
- Lu ch'eng chi
Ici londres – London, UK. Bulletin hebdomadaire des services europeens de la BBC. 21 Feb 1948-1 Nov 1957 – 1 – uk British Libr Newspaper [072]
The icicle – Dalhousie [N.B: s.n. 1885?-18–?] – 9 – mf#P05092 – cn Canadiana [073]
The icila dance, old style : a study in african music and dance of the lala tribe of northern rhodesia 101=jones, a m+kombe, l – Roodepoort, Longmans, Green for African Music Society, 1952 – us CRL [790]
Ickleton 1558-1950 – (= ser Cambridgeshire parish register transcript) – 7mf – 9 – £8.75 – uk CambsFHS [929]
Icmal-i netayic yahut mutira-i funun-i idadiye / Feyzi, Emin – Istanbul: Karabet Matbaasi, 1309 [1892] – (= ser Ottoman histories and historical sources) – 4mf – 9 – $75.00 – us MEDOC [956]
Icographic – Oxford. (1) 1978-1979 (5) (9) – ISSN: 0085-1698 – mf#49307 – us UMI ProQuest [680]
Icones : id est verae imagines virorum doctrina simul et pietate illustrium... / Beza, Theodor de – Genevae: Apud Ioannem Laonium, 1580 – 4mf – 9 – mf#O-145 – ne IDC [090]
Icones animalium quadrupedum viviparorum et oviparorum, quae in historia animalium c. gesneri describuntur,... / Gessner, C – Tiguri: C Froschoverus, 1553 – 2mf – 9 – mf#Z-2263 – ne IDC [590]
Icones animalium quadrupedum viviparorum et oviparorum, quae in historia animalium c. gesneri...describuntur,... / Gessner, C – Ed. 2. Tiguri: C Froschoverus, 1560 – 3mf – 9 – mf#Z-2264 – ne IDC [590]
Icones avium omnium, quae in historia avium c. gesneri describuntur,... / Gessner, C – Ed.2. Tiguri: C Froschoverus, 1560 – 3mf – 9 – mf#Z-2266 – ne IDC [590]
Icones avium omnium, quae in historia avium c. gesneri describuntur,... / Gessner, C – Tiguri: C Froschoverus, 1555 – 3mf – 9 – mf#Z-2265 – ne IDC [590]
Icones fungorum javanicorum / Penzig, Otto Albert Julius – Leiden, Netherlands. 1904 – 1r – us UF Libraries [500]
Icones fungorum javanicorum / Penzig, Otto Albert Julius – Leiden, Netherlands. 1904 – 1r – us UF Libraries [580]
Icones historiarum veteris testamenti... / [Holbein, H, the Younger] – Lugduni, 1547 – 2mf – 9 – mf#O-1035 – ne IDC [700]
Icones, id est verae imagines virorum doctrina simul et pietate illustrium...quibus adiectae sunt nonnullae picturae quas emblemata vocant / Beza, Theodor de – Genevae: Apud Ioannem Laonium, 1580 – 4mf – 9 – mf#O-145 – ne IDC [700]
Icones id est verae imagines virorvm doctrina simvl et pietate illvstrivm..., partim vera religio in variis orbis... / Beza, Theodor de – Genevae, apvd Laonivm, 1580 – 4mf – 9 – mf#ZWI-19 – ne IDC [240]
Icones livianae... / Amman, J – Francofurti ad Moenum, 1572 – 2mf – 9 – mf#O-1014 – ne IDC [700]
Icones plantarum syriae rariorum, descriptionibus et observationibus illustratae... / Labillardiere, J J H de – Lutetiae Parisiorum, 1791-1812 – 6mf – 8 – mf#5996 – ne IDC [956]
Icones rerum naturalium quas, in itinere orientali... / Forssk & l, P; ed by Niebuhr, C – Emmaus. 1971-2000 (1) 1971-2000 (5) 1975-2000 (9) – 4mf – 9 – mf#6433 – ne IDC [915]
Icones symbolicae vitae et mortis b josaphat martyris archiepiscopi polocensis expressae... / Mlodzianowski, A – Viln: Typ. Acad. Soc. Iesu, 1675 – 2mf – 9 – mf#O-58 – ne IDC [090]
Iconoclast – Dallas TX. 1971 jul 23-1972 mar 24, 1972 mar 31-dec 27, 19-74, 1974 dec 27-1975 oct 31, 1975 oct 31-1976 oct 29 – 6r – 1 – (cont: dallas news) – mf#1058625 – us WHS [071]
Iconoclast – Marcus IA. 1920 feb 15 – 1r – 1 – mf#3926404 – us WHS [071]
Iconoclast *see* Plea for atheism
Iconografia cioa disegni d'imagini de famosissimi monarchi, regi, filosofi, poeti ed oratori dell' antichit... / Canini, G A & Canini, M A – Roma, 1669 – 9mf – 9 – mf#O-1165 – ne IDC [700]
Iconografia de benito arias montano / Doetsch, Carlos – Madrid: Tip. Blass, 1927 – 1 – sp Bibl Santa Ana [946]
Iconografia de benito arias montano por carlos doetsch / Fernandez de Castro, Eduardo Felipe – Malaga: Revista Espanola de Estudios Biblicos, 1928 – 1 – sp Bibl Santa Ana [946]
Iconografia de las editiones del quijote de miguel cervantes saavedra / Henrich, Manuel – Barcelona. 1905. 3v – 1 – $23.00 – us L of C Photodup [760]
Iconografia del apostol jose marti / Cuba Secretaria De Instruccion Publica Y Bellas A... – Habana, Cuba. 1925 – 1r – us UF Libraries [740]
Iconografia del apostol jose marti / Cuba Secretaria De Instruccion Publica Y Bellas A – Habana, Cuba. 1925 – 1r – us UF Libraries [972]
Iconografia espanola de contemporaneos. coleccion de semblanzas – Madrid: Tip. Frances, 1908 – 1 – sp Bibl Santa Ana [946]
Iconographic motifs from palestine/israel and daniel 7:2-14 / Eggler, Jurg – [Stellenbosch SA: University of Stellenbosch 1998] [mf ed 1998] – 13mf – 9 – (incl bibl) – mf#mf.1339 – sa Stellenbosch [700]
Iconographie bouddique : traduit du siamois en cambodgien avec l'autorisation de l'auteur par preas maha pitou krasem. / Damrongrachanuphap, Prince, son of Mongkut, King of Siam – Phnom-Penh: A Portail 1928 [mf ed 1990] – 1r with other items – 1 – (title & text in khmer; added t.p. in french) – mf#mf-10289 SEA M reel 122/3 [§] – us CRL [280]
Iconographie calvinienne : ouvrage dedie a l'universite de geneve / Doumergue, Emile – Lausanne: G. Bridel, 1909 – 1mf – 9 – 0-7905-5984-6 – (incl bibl ref) – mf#1988-1984 – us ATLA [242]
Iconographie chretienne *see* Christian iconography
Iconographie de l'art profane au moyen-age et...la renaissance, et la decoration des demeures / Marle, R van – La Haye, 1931-1932. 2v – 23mf – 9 – mf#O-367 – ne IDC [700]
Iconographie du regne animal de g. cuvier; ou, representation d'apres nature de l'une des especes les plus remarquables et souvent non encore figurees, de chaque genre d'animaux / Guerin-Meneville, Felix Edouard – Paris, 1829-1844 – 1 – us Wisconsin U Libr [590]

Iconography of the west front of wells cathedral / Cockerell, Charles Robert – Oxford 1851 – (= ser 19th c art & architecture) – 3mf – 9 – mf#4.2.1030 – uk Chadwyck [700]

Iconologia : of beeldespraeck, of uytbeeldinge des verstands: van cesare ripa van perugien, ridder van s.s. mauritius en lazzaro / Ripa, Cesare – t'Amsteldam: Cornelis Danckerts, [1656] – 5mf – 9 – mf#O-832 – ne IDC [090]

Iconologia : of uytbeeldinghe des verstants van cesare ripa... / Ripa, Cesare – Amstelredam: Dirck Pietersz Pers, 1644 – 12mf – 9 – mf#O-734 – ne IDC [090]

Iconologia del cavaliere ceasare ripa... / Ripa, C – Perugua: Nella stamperia di Piergiovanni Costantini, 1764-1767. 5v – 43mf – 9 – mf#O-413 – ne IDC [700]

Iconologia deorum, oder abbildung der goetter... / Sandrart, J von – Nuernberg, 1680 – 6mf – 9 – mf#O-425 – ne IDC [700]

Iconologia di cesare ripa perugino cavalier di ss. mauritio et lazaro / Ripa, Cesare – Venetia: Presso Cristoforo Tomasini, 1645 – 9mf – 9 – mf#O-1264 – ne IDC [090]

Iconologia, of beeldespraeck, of uytbeeldinge des verstands : van cesare ripa van perugien, ridder van s s mauritius en lazzaro / Ripa, C – t'Amsteldam: Cornelis Danckerts, [1656] – 5mf – 9 – mf#O-832 – ne IDC [700]

Iconologia, of uytbeeldinghe des verstants van cesare ripa...waer in verscheiden afbeeldingen van deughden, ondeughden...werden verhandelt...uyt het italiaens vertaelt door d p pers / Ripa, C – Amstelredam: Dirck Pietersz Pers, 1644 – 12mf – 9 – mf#O-734 – ne IDC [700]

Iconologia overo descrittione dell'imagini universali cavate dall'antichita et da altri luoghi da cesare ripa perugino / Ripa, Cesare – Rome: Per gli heredi di Gio. Gigliotti, 1593 – 4mf – 9 – mf#O-2018 – ne IDC [090]

Iconologia overo descrittione di diverse imagini cavate dall'antichit...da cesare ripa perugino... / Ripa, Cesare – Roma: Appresso Lepido Facij, 1603 – 6mf – 9 – mf#O-87 – ne IDC [090]

Iconologia ovvero immagini di tutte le cose principali... / Pistrucci, F – Milano, 1819-1821. 2v – 26mf – 9 – mf#O-862 – ne IDC [090]

Iconologia ovvero immagini di tutte le cose principali...iconologie ou images de toutes les principales choses auxquelles le talent de l'homme a attribue un corps, bien qu'elles ne l'aient pas en realite : avec la traduction franc. par sergent marceau / Pistrucci, F – Milano, 1819-1821. 2v – 26mf – 9 – mf#O-862 – ne IDC [700]

Iconologie : oder ideen aus dem gebiethe der leidenschaften und allegorien... / Stueber, J – Wien: Im Verlag bei Rud. Sammer, [1800] – 2mf – 9 – mf#O-14 – ne IDC [090]

Iconologie : ou, explication nouvelle de plusieurs images, emblemes... / Ripa, Cesare – Paris: Mathieu Guillemot, 1644 – 11mf – 9 – mf#O-1266 – ne IDC [090]

Iconologie : ou la science des emblemes, devises etc... / Ripa, Cesare; ed by Baudoin, J – Amsterdam: Adrian Braakman, 1698 – 6mf – 9 – mf#O-1265 – ne IDC [090]

Iconologie par figures o- trait : complet des all,gories, emblemes etc / Gravelot, H Fcochin, C N – Paris: Lattr, n.d. – 9mf – 9 – mf#O-624 – ne IDC [090]

Iconologie tir : e de divers auteurs, ouvrage utile aux gens de lettres... / Boudard, J B – Vienne: Chez Jean Thomas de Trattnern, 1766. 3v – 9mf – 9 – mf#O-1234 – ne IDC [090]

Iconology : or, a collection of emblematical figures, containing four hundred and twenty-four remarkable subjects, moral and instructive; in which are displayed the beauty of virtue and deformity of vice / Richardson, G – London: G Scott, 1779. 2v – 18mf – 9 – mf#O-731 – ne IDC [700]

Iconology : or emblematic figures explained... / Pinnock, W – London: John Harris, 1830 – 5mf – 9 – mf#O-869 – ne IDC [090]

Icor, association for jewish colonization in the soviet union – Toronto, Ont., Canada. 1934-36 – 1 – us AJPC [071]

Icp business software review – Indianapolis. 1984-1985 (1,5,9) – (cont by: business software review. cont: icp software business review) – ISSN: 8750-1368 – mf#14952,01 – us UMI ProQuest [000]

Icp software business review – Indianapolis. 1982-1984 (1,5,9) – (cont by: icp business software review) – ISSN: 0744-2602 – mf#14952 – us UMI ProQuest [000]

Ictihad – Istanbul. sene 1-28 n1-358. eyluel 1904-kanunieveel 1932 – (= ser O & t journals) – 105mf – 9 – $1640.00 – (missing: 8,10-11,52,141-142) – us MEDOC [956]

Ictimaiyyat mecmuasi – Istanbul: Matbaa-i Amire, 1917. Yayimliyan: Dar'uel-Fuenun Ictimaiyat Dar'uel-Mesaisi; Mueduerue: Necmeddin Sadak. n1-3,5. nisan-agustos 1917 – (= ser O & t journals) – 4mf – 9 – $90.00 – us MEDOC [956]

Id – New York. 1993-1995 (1,5,9) – (cont: institutional distribution) – ISSN: 1080-9015 – mf#12633,01 – us UMI ProQuest [660]

Id – 1941-50* – 1r – 1 – (cont: misiones extranjeras del clero secular espanol) – mf#ATLA S0705A – us ATLA [240]

Ida graefin hahn-hahn : ein lebens- und literaturbild / Keiter, Heinrich – Wuerzburg: L Woerl, [18-?] [mf ed 1993] – 79p – 1 – (incl bibl ref) – mf#8669 – us Wisconsin U Libr [430]

Idaho : official code – Charlottesville: Michie Co, 1949-mar 99 update – 9 – $2,437.00 set – mf#401770 – us Hein [348]

Idaho : session laws of american states and territories – 1863-1999 – 9 – $1,176.50 set – mf#402630 – us Hein [348]

Idaho *see*
– Reports and opinions
– Reports, post-nrs
– Reports, pre-nrs

Idaho attorney general reports and opinions – 1893-1996 – 6,9 – $312.00 set – (1893-1978 on reel $70. 1979-96 on mf $242.00) – mf#408210 – us Hein [340]

Idaho avalanche – Silver City ID. 1876 aug 26-1878, 1879-89, 1890-1892 dec 17, 1893-1894 nov 9, 1895 aug 23-1896 aug 14, 1896 aug 21-1897 aug 12 – 7r – 1 – (cont: idaho weekly avalanche; cont by: owyhee avalanche [silver city, id]) – mf#846230 – us WHS [071]

Idaho Bicentennial Commission *see* Newsletter

Idaho daily avalanche – Silver City ID. 1875 oct-1876 apr 26, 1875 sep 6-30 – 2r – 1 – (cont: owyhee daily avalanche) – mf#1225165 – us WHS [071]

Idaho daily statesman – Boise ID. 1890 jan 1-[1921 oct/jan 2] – 36r – 1 – (cont: idaho tri-weekly statesman; cont by: idaho statesman) – mf#845990 – us WHS [071]

Idaho granger – v10 n8-v12 n9 [1940 jun-1942 jul] – 1r – 1 – mf#630007 – us WHS [071]

Idaho, her gold fields and the route to them: a handbook for emigrants / Fisk, James L – Overlander's guidebook written by the expedition leader. 99p. 1863. 1r – 1 – $5.00 – us Minn Hist [978]

Idaho labor report / Idaho State AFL-CIO [American Federation of Labor and Congress of Industrial Organizations] – v14 n3,5,12-13 [1980 feb 1,15, oct 1-nov 19], v15 n1-4 [1981 jan 1-feb 13] – 1r – 1 – (cont by: idaho labor report [1981]) – mf#676064 – us WHS [331]

Idaho law journal – Idaho School of Law. v1-3. 1931-33 – 13mf – 9 – $19.50 – mf#LLMC 84-483 – us LLMC [340]

Idaho law review – v1-37. 1964-2001+15 yr ind – 5,6,9 – $607.00 set – (v1-20 1964-84 and index on reel $185. v21-37 1985-2001 on mf $422) – ISSN: 0019-1205 – mf#103221 – us Hein [340]

Idaho leader / National Nonpartisan League – v2 n1-v3 n35/37 [1919 jan 4-1920 dec 18] – 1r – 1 – mf#918125 – us WHS [320]

Idaho librarian – Moscow. 1945-2000 (1) 1971-2000 (5) 1977-2000 (9) – ISSN: 0019-1213 – mf#5972 – us UMI ProQuest [020]

Idaho missionary baptist flag – 1946-Aug 1973 – 1 – us Southern Baptist [242]

Idaho Public Employees Association *see* I p e a news

Idaho springs advance *see* Clear creek county miscellaneous newspapers

Idaho State AFL-CIO [American Federation of Labor and Congress of Industrial Organizations] *see* Idaho labor report

Idaho. State Bar Association *see* Proceedings

Idaho state bar journal *see* Advocate (idaho state bar journal)

Idaho state bar reports of proceedings – v1-2. 1921-23. v1-44. 1925-70 (all publ) – 69mf – 9 – $103.00 – (lacking: v8,9,38) – mf#LLMC 84-484 – us LLMC [340]

Idaho State Employees Association *see* I s e a news

Idaho statesman – Boise, ID. 1864+ (1) – mf#60454 – us UMI ProQuest [071]

Idaho statesman – Boise ID. 1890 jan 1 – 1r – 1 – (cont: idaho daily statesman) – mf#1879230 – us WHS [071]

Idaho. Supreme Court *see*
– Cummins' territory reports
– Idaho supreme court reports

Idaho supreme court reports / Idaho. Supreme Court – v1-41. 1866-1925 – 72mf (1:42) 170 (1:24) – 9 – $579.00 – (pre-nrs: v1 1886-83 3mf $13.50. updates planned) – mf#LLMC 84-131 – us LLMC [347]

Idaho. (Territory). Laws, Statutes, etc *see* The compiled and revised laws of the territory of idaho.

Idaho weekly avalanche – Silver City ID. 1875 apr 24-1878 dec 28 – 1r – 1 – (cont: owyhee weekly avalanche; cont by: idaho avalanche) – mf#867963 – us WHS [071]

'Idan 'Olamim / Sossnitz, Joseph Judah Lob – Varsha, Poland. 1888 – 1r – 1 – us UF Libraries [939]

Idapo / Perera, Hilda – Miami, FL. 1971 – 1r – us UF Libraries [972]

Idapo / Perera, Hilda – Miami, FL. 1971 – 1r – us UF Libraries [972]

Idc report suvaguuq / Inuit Development Corporation – v1 n1-2 [1980 spr-1981 fall] – 1r – 1 – (cont by: nunasi report suvaguuq) – mf#921736 – us WHS [305]

Iddings' term reports / Ohio. Dayton – 1v. 1899-1900 (all publ) – (= ser Ohio appellate decisions) – 2mf – 9 – $3.00 – mf#llmc 84-185 – us LLMC [324]

Ide, George Barton *see* Instability of the pastoral relation

L'idea – Brooklyn NY, 1923 – 1r – 1 – (italian periodical) – us IHRC [073]

A idea : orgam do club dos estudantes – Curitiba, PR: Typ d'A Republica, 01 out 1888-14 jul 1889 – (= ser Ps 19) – 1,5,6 – bl Biblioteca [079]

A idea : periodico semanal – Rio de Janeiro, RJ: [s.n.] 30 dez 1880 – (= ser Ps 19) – mf#P05,04,186 – bl Biblioteca [079]

Idea : issues and ads news – Alturas CA. 1985 oct 30-1987 jan 28/feb 10 – 1r – 1 – (cont by: idea news [alturas, ca]) – mf#3366173 – us WHS [071]

Idea : journal of law and technology – v1-40. 1957-2000 – 5,6,9 – $830.00 set – (v1-25 1957-85 on reel $362. v26-40 1985-2000 on mf $468. title varies: v1-7 1957-64 as patent, trademark and copyright journal of research and educations. v8-15 1964-72 as patent, trademark and copyright journal of research education. v16-18 n1 1973-76 as idea, the ptc journal of research and education) – ISSN: 0019-1272 – mf#103231 – us Hein [346]

Idea *see* Idea

La idea – Badajoz, 1890 – 5 – sp Bibl Santa Ana [073]

The idea and reality of revelation and typical forms of christianity : two lectures / Wendt, Hans Hinrich – London: Philip Green, 1904 – 1mf – 9 – 0-8370-5786-8 – mf#1985-3786 – us ATLA [240]

Idea de la fama en la edad media castellana / Lida De Malkiel, Maria Rosa – Mexico City?, Mexico. 1952 – 1r – us UF Libraries [960]

Idea de la fama en la edad media castellana / Lida De Malkiel, Maria Rosa – Mexico City?, Mexico. 1952 – 1r – us UF Libraries [960]

L'idea de pittori, scultori, et architetti... / Zuccaro, F – Torino, 1607 – 3mf – 9 – mf#O-470 – ne IDC [700]

Idea de un perfecto prelado discurrida a la eleccion de ministro provincial... / Romera de la Torre, Francisco – 1 – sp Bibl Santa Ana [946]

Idea de un principe politico christiano : representada en cien empresas / Saavedra Faxardo, Didaco de – Amstelodami: Apud Joh. Ianssonium iuniorem, 1659 – 11mf – 9 – mf#O-3257 – ne IDC [090]

Idea de un principe politico christiano : representada en cien empresas / Saavedra Faxardo, Didaco de – Monaco: En la emprenta de Nicola Enrico, 1640 – 8mf – 9 – mf#O-1888 – ne IDC [090]

Idea de un principe politico christiano representada en cien empresas / Saavedra Faxardo, Didaco de; ed by Garcia de Diego, V – Madrid: Ediciones de la lectura, 1927-30. 4v – 14mf – 9 – mf#O-1885 – ne IDC [090]

Idea de un principe politico christiano representada en cien empresas... / Saavedra Faxardo, Didaco de – Amstelodami: Apud Ioh. Ianssonium iuniorem, 1664. 3v – 12mf – 9 – mf#O-1886 – ne IDC [090]

Idea de un principe politico christiano, representada en cien empresas... / Saavedra Faxardo, Didaco de – Amberes: En casa de Ieronymo y Ivan bapt. Verdussen, 1655 – 8mf – 9 – mf#O-1887 – ne IDC [090]

Idea de un principe politico christiano representado en cien empresas... / Saavedra Faxardo, Didaco de – Amberes: En casa de Ieronymo y Ivan Pabt. Verdussen, 1655 – 14mf – 9 – mf#O-739 – ne IDC [090]

Idea del teatro de ortega y gasset / Frutos Cortes, Eugenio – sp Bibl Santa Ana [790]

Idea del tempio della pittura... / Lomazzo, G P – Milano, [1590] – 3mf – 9 – mf#O-1002 – ne IDC [700]

L'Idea della universale architettura...divisa in 10 Libri / Scamozzi, V – Venetiis, 1615 – 15mf – 9 – mf#O-426 – ne IDC [720]

L'idea di tutte le perfezioni, introduzione al balletto de' serenissimi principi francesco, e antonio farnesi... / Lotti, L & Tosi, G – Piacenza, 1690 – 1mf – 9 – mf#O-1109 – ne IDC [790]

L'idea di un prencipe et eroe christiano in francesco i d'este... / Gamberti, D – Modona, 1659 – 15mf – 9 – mf#O-1579 – ne IDC [090]

L'idea di un principe politico christiano : di d diego saavedra fachardo / Saavedra Faxardo, Didaco de – Venetia: Per Marco Garzoni, 1648 – 5mf – 9 – mf#O-1882 – ne IDC [090]

Idea liberal – Pernambuco, 02 jan 1869 – (= ser Ps 19) – bl Biblioteca [079]

Idea locorvm commvnivm sive methodica articvlorvm praecipvorvm doctrinae christianae per thesin et antithesin tractatio / Pelargus, C – Francofurti, 1604 – 5mf – 9 – mf#TH-1 mf 1266-1270 – ne IDC [242]

L'idea nazionale – Rome, Italy 11,31 oct 1914-1 apr 1915; 28 dec 1915-27 dec 1925 – 1 – (1914 imperfect; merged with: la tribuna) – uk British Libr Newspaper [074]

L'idea nazionale – Roma: Casa Editrice Nazionale, mar 1 1911-dec 27 1925 – 1 – us CRL [074]

Idea news – Alturas CA. 1987 feb 11/24-1992 dec – 1r – 1 – (cont: idea [alturas, ca]) – mf#2577461 – us WHS [071]

The idea of a national church : an address / Creighton, Mandell – London: SPCK 1898 [mf ed 1993] – (= ser Church historical society (series) 51) – 1mf – 9 – 0-524-05498-3 – mf#1990-1493 – us ATLA [230]

The idea of a university : defined and illustrated / Newman, John Henry – 3rd ed. London: BM Pickering, 1873 – 2mf – 9 – 0-7905-7432-2 – mf#1989-0657 – us ATLA [378]

The idea of god as affected by modern knowledge / Fiske, John – London, 1885 – (= ser 19th c evolution & creation) – 2mf – 9 – mf#1.1.11256 – uk Chadwyck [110]

The idea of god in early religions / Jevons, Frank Byron – Cambridge: University Press 1910 [mf ed 1990] – (= ser The cambridge manuals of science and literature) – 1mf – 9 – 0-7905-7592-2 – mf#1989-0817 – us ATLA [200]

The idea of god in relation to theology / Read, Eliphalet Allison – Chicago: Uni of Chicago Press, 1900 – 1mf – 9 – 0-8370-5607-1 – (incl bibl ref) – mf#1985-3607 – us ATLA [240]

The idea of re-birth / Arundale, Francesca – London: Kegan Paul, Trench, Truebner, 1890 – 1mf – 9 – 0-524-07664-2 – mf#1991-014-1 – us ATLA [210]

The idea of the holy / Otto, Rudolf – 1923 – 9 – $10.00 – us IRC [210]

An idea of the perfection of painting... / Freart, R – [London], 1668 – 2mf – 9 – mf#O-1187 – ne IDC [750]

The idea of the resurrection in the ante-nicene period / Staudt, Calvin Klopp – Chicago: University of Chicago Press, 1909 [mf ed 1990] – (= ser Historical and linguistic studies in literature related to the new testament. 2nd series. linguistic and exegetical studies) – 1mf – 9 – 0-7905-3487-8 – mf#1987-3487 – us ATLA [240]

The idea of the soul / Crawley, Alfred Ernest – London: Adam and Charles Black, 1909 – 1mf – 9 – 0-524-00825-6 – mf#1990-2071 – us ATLA [240]

Idea philosophiae moralis : sive, compendiosa institutio / Burgersdijck, Franco – Editio postrema, multis in locis emendata. – Amstelodami: Apud Ioannem Ianssonium 1643 – (= ser Ethics in the early modern period) – 3mf – 9 – mf#pl-469 – ne IDC [170]

Idea politica veri christiani : sive ars oblivionis, isagogica ad artem memoriae / Luzon de Millares, A – Bruxellis: Typis Francisci Foppens, 1665 – 16mf – 9 – mf#O-679 – ne IDC [090]

Idea principis christiano-politici 101 symbolis expressa / Saavedra Faxardo, Didaco de – Amstelodami: Apud Ioh. Ianssonium Iuniorem, 1651 – 10mf – 9 – mf#O-1465 – ne IDC [090]

Idea principis christiano-politici centum symbolis... : editio noviss... / Saavedra Faxardo, Didaco de – Jenae: Sumptu Matth. Birckneri, 1686 – 7mf – 9 – mf#O-1268 – ne IDC [090]

Idea principis christiano-politici centum symbolis expressa... : editio novissima... / Saavedra Faxardo, Didaco de – Coloniae: Apud Joannem Carolum Muenich, 1669 – 9mf – 9 – (missing: 5 plates) – mf#O-1880 – ne IDC [090]

Idea principis christiano-politici centum symbolis expressa... / Saavedra Faxardo, Didaco de – Bruxellae: Excudebat Ioannes Mommartius, 1649 – 20mf – 9 – mf#O-421 – ne IDC [090]

Idea principis christiano-politici symbolis 101 expressa... / Saavedra Faxardo, Didaco de – Amstelaedami: Apud Ioannem Blaeu, 1660 – 7mf – 9 – mf#O-1881 – ne IDC [090]

Idea sacrae congregationis helveto-benedictinae : anno illius iubilaeo saeculari expressa, et orbi exposita... – S. Galli: Typis ejusdem principalis monasterij per Jacobum Mueller, 1702 – 2mf – 9 – mf#O-1847 – ne IDC [090]

Idea sapientis : id est: philosophiae morum partes tres, ethica, theo-politica, oeconomica... / [Vanossi, A] – Tyrnaviae: Typis Acad. Soc. Jesu, 1746 – 5mf – 9 – mf#O-67 – ne IDC [090]

Idea sapientis theo-politici : id est tripartita morum philosophia ethica, politica, oeconomica... / Vanossi, A – Viennae Austriae: Typis Mariae Teresiae Voigtin, 1725 – 4mf – 9 – mf#0-66 – ne IDC [090]

Idea sucinta del probabilismo : que contiene la historia abreviada de su origen, progresos y decadencia... / Lope del Rodo, Juan – en Lima: impr Real 1772 – (= ser Books on religion...1543/44-c1800: teologia "culta"; derecho canonico) – 3mf – 9 – mf#crl-274 – ne IDC [241]

La idea tradicional. destino de espana por / Becerro de Bengoa, Ricardo – Caceres: Tip. Garcia Floriano, 1945 – 1 – sp Bibl Santa Ana [946]

Ideais e lutas de um burgues progressista / Nogueira, Paulo – Sao Paulo, Brazil. v1-2. 1958 – 1r – us UF Libraries [972]

Ideais e lutas de um burgues progressista / Nogueira, Paulo – Sao Paulo, Brazil. v1-2. 1958 – 1r – us UF Libraries [972]

Ideal – Indio. 1969-1972 (1) – ISSN: 0046-8533 – mf#9036 – us UMI ProQuest [305]

Ideal and progress : essays / Ghose, Aurobindo – Calcutta: Arya Pub House: Sole agents, Indian Book Club, [between 1900 and 1946] – (= ser Samp: indian books) – us CRL [140]

Ideal and progress; essays / Ghose, Aurobindo – 2nd ed., rev. Calcutta: Arya Pub. House, 19??. 68p – 1 – us Wisconsin U Libr [280]

An ideal college for women : an address delivered before the delta sigma society of mcgill university / Dawson, John William – S.l: s.n, 1894 – 1mf – 9 – mf#03668 – cn Canadiana [376]

Ideal de familia...memoria / Soler Arques, Carlos – 1887 – 9 – sp Bibl Santa Ana [920]

Ideal de los conquistadores / Corraliza, Jose V – San Lorenzo del Escorial, 1923 – 1 – sp Bibl Santa Ana [350]

L'ideal esthetique : esquisse d'une philosophie de la beaute / Roussel-Despierres, fr – Paris: Felix Alcan, 1904 [mf ed 1986] – 1mf – 9 – 0-8370-6610-7 – (in french) – mf#1986-0610 – us ATLA [170]

Ideal husband and a woman of no importance / Wilde, Oscar – New York, NY. 19–? – 1r – us UF Libraries [025]

Ideal husband and a woman of no importance / Wilde, Oscar – New York, NY. 19–? – 1r – us UF Libraries [240]

The ideal life : and other unpublished addresses / Drummond, Henry – London: Hodder & Stoughton, 1897 [mf ed 1991] – 1mf – 9 – 0-7905-9266-5 – mf#1989-2491 – us ATLA [242]

The ideal of christian worship / Delany, Selden Peabody – Milwaukee: Young Churchman Co, 1909 – 1mf – 9 – 0-524-06244-7 – mf#1990-5199 – us ATLA [240]

The ideal of education / Abhedananda, swami – Calcutta: Ramakrishna Vedanta Math, 1945 – (= ser Samp: indian books) – us CRL [370]

The ideal of human unity / Ghose, Aurobindo – Pondicherry: Sri Aurobindo Ashram, 1950 – (= ser Samp: indian books) – us CRL [180]

The ideal of indian womanhood / Roy, Manabendra Nath – Dehradun: Indian Renaissance Association Ltd, 1941 – (= ser Samp: indian books) – us CRL [305]

Ideal of religion / Farrington, S – London, England. 1876 – 1r – us UF Libraries [240]

Ideal of religion / Farrington, S – London, England. 1876 – 1r – us UF Libraries [240]

The ideal of the karmayogin / Ghose, Aurobindo – Calcutta: Arya Pub House, 1937 – (= ser Samp: indian books) – us CRL [280]

Ideal, purpose and duties of the khmer royal socialist youth : interpretation and commentary of the statute of the k. r. s. y., by samdech norodom sihanouk = Ideal, buts et devoirs de la jeunesse socialiste royale khmere... / Norodom Sihanouk, Prince – [n. p. 196-?] [mf ed 1989] – 1r with other items – 1 – (also available in french [mf-10289 seam reel 026/08]) – mf#mf-10289 seam reel 016/16 [§] – us CRL [335]

L'ideal religieux des grecs et l'evangile / Festugiere, A – Paris, 1932 – 7mf – 8 – €15.00 – ne Slangenburg [243]

Ideal stock farm in the ocklawaha valley – Tampa, FL. 191-? – 1r – us UF Libraries [636]

Ideal stock farm in the ocklawaha valley – Tampa, FL. 191-? – 1r – us UF Libraries [636]

Ideal und leben : nach schiller und kant / Thikoetter, Julius – Bremen: M Heinsius, 1892 – 1 – us Wisconsin U Libr [430]

Die ideale der socialdemokratie und die aufgabe des zeitalters / Glogau, Gustav – Kiel: Lipsius und Tischer, 1891 – 1r – 1 – (incl bibl ref) – us Wisconsin U Libr [325]

Ideal-ehe – Berlin DE, 1927-28 – 1 – gw Misc Inst [640]

Ideal-ehe – Berlin DE, 1927-28 – 1 – gw Misc Inst [074]

Ideales culturales de la edad media : tomo 4: la vida monastica / Vedel, Waldemar – Barcelona, 1931; Madrid: Razon y Fe, 1934 – 1 – sp Bibl Santa Ana [240]

Ideales misioneros de los reyes catolicos / Bayle, Constantino – Madrid: Missionalia Hispanica, 1952 – 1 – mf#B1822 – sp Bibl Santa Ana [241]

Ideales misioneros de los reyes catolicos / Bayle, Constantino – Madrid: Missionalia Hispanica, 1952 – 1 – mf#B1823 – sp Bibl Santa Ana [241]

Idealidad / Sanchez-Arjona, Vicente – Sevilla: Imp. Carlos Acuna, 1954 – 1 – sp Bibl Santa Ana [810]

Idealism and theology : a study of presuppositions / d'Arcy, Charles Frederick – London: Hodder and Stoughton, 1899 – us ATLA [240]

Idealism and theology : a study of presuppositions / d'Arcy, Charles Frederick – London: Hodder and Stoughton, 1899 – (= ser Donnellan Lectures) – 1mf – 9 – 0-7905-3780-X – mf#1989-0273 – us ATLA [140]

Idealism as a practical creed : being the lectures on philosophy and modern life / Jones, Henry – Glasgow: J MacLehose, 1909 [mf ed 1990] – 1mf – 9 – 0-7905-7596-5 – mf#1989-0821 – us ATLA [140]

Idealism in national character : essays and addresses / Falconer, Robert – London, Toronto: Hodder & Stoughton, 1920 – 3mf – 9 – 0-665-72750-X – (incl bibl ref) – mf#72750 – cn Canadiana [370]

Idealism in theology / Ryder, H I D – London, England. 1867 – 1r – us UF Libraries [240]

Idealism in theology / Ryder, H I D – London, England. 1867 – 1r – us UF Libraries [240]

The idealism of spinoza / Murray, John Clark – Montreal: [s.n.], 1896 – (= ser Mcgill university. papers from the department of philosophy 2) – 1mf – 9 – 0-665-89781-2 – (incl bibl ref) – mf#89781 – cn Canadiana [140]

L'idealismo di georgio, berkeley... / Olgiati, Francesco – Madrid: Razon y Fe, 1927 – 1 – sp Bibl Santa Ana [120]

Idealismos de verdad y de belleza / Aizpura, Aizpuru – Panama, Panama. 1925 – 1r – us UF Libraries [972]

Idealismos de verdad y de belleza / Aizpura, Aizpuru – Panama, Panama. 1925 – 1r – us UF Libraries [972]

Der idealismus der indischen religionsphilosophie im zeitalter der opfermystik / Dahlmann, Joseph – Freiburg i B; St Louis, MO: Herder, 1901 – (= ser [Ergaenzungshefte zu den "Stimmen aus Maria-Laach"]) – 1mf – 9 – 0-524-01425-6 – (incl bibl ref) – mf#1990-2420 – us ATLA [280]

Der idealist – Vienna, Austria mar-may 1877 [mf ed Norman Ross] – 1r – 1 – mf#nrp-1943 – us UMI ProQuest [074]

An idealist view of life : being the hibbert lectures for 1929 / Radhakrishnan, Sarvepalli – London: George Allen & Unwin, 1951 – (= ser Samp: indian books) – us CRL [280]

Idealista realizador / Rezende Martins, Amelia De – Rio de Janeiro, Brazil. 1939 – 1r – us UF Libraries [972]

Idealista realizador / Rezende Martins, Amelia De – Rio de Janeiro, Brazil. 1939 – 1r – us UF Libraries [972]

The idealistic reaction against science = Reazione idealistica contro la scienza / Aliotta, Antonio – London: Macmillan, 1914 – 2mf – 9 – 0-7905-3516-5 – (incl bibl ref. in english) – mf#1989-0009 – us ATLA [140]

Idealistic thought of india / Raju, Poolla Tirupati – London: George Allen & Unwin Ltd, 1953 – (= ser Samp: indian books) – us CRL [140]

Ideals and realities : studies in education and economics / Khan, Shafa'at Ahmad – Madras: Law Printing House, 1921 – (= ser Samp: indian books) – us CRL [140]

The ideals of indian art / Havell, Ernest Binfield – London: John Murray, 1920 – (= ser Samp: indian books) – us CRL [700]

Ideals of science and faith : essays by various authors / ed by Hand, J E – New York: Longmans, Green; London: George Allen, 1904 – 1mf – 9 – 0-8370-5077-4 – (incl bibl ref and index) – mf#1985-3077 – us ATLA [210]

Ideals of the east / Baynes, Herbert – London: Swan Sonnenschein, 1898 – 1mf – 9 – 0-524-01164-8 – mf#1990-2240 – us ATLA [200]

The ideals of the prophets : sermons / Driver, Samuel Rolles – Edinburgh: T & T Clark; New York: Scribner [distributor], 1915 – 1mf – 9 – 0-7905-3248-4 – (incl bibl ref) – mf#1987-3248 – us ATLA [220]

Idealy kooperatsii / Posse, V A – 1911 – 16p 1mf – 9 – mf#COR-98 – ne IDC [335]

Ideario : ordenado / Marti, Jose – Habana, Cuba. 1930 – 1r – us UF Libraries [972]

Ideario de batista / Batista Y Zaldivar, Fulgencio – Habana, Cuba. 1940 – 1r – us UF Libraries [972]

Ideario de batista / Batista Y Zaldivar, Fulgencio – Habana, Cuba. 1940 – 1r – us UF Libraries [972]

Ideario de la colegio de egb y formacion profesional / Colegio Santiago y Santa Margarita. Sociedad Cooperativa – Caceres: Imp. Garcia Carrasco, 1979 – 1 – sp Bibl Santa Ana [060]

Ideario de un centro educativo josefino trinitario – Caceres: Imp. Moderna, 1977 – sp Bibl Santa Ana [370]

Ideario de un combatiente / Conte Aguero, Luis – Mexico City?, Mexico. 1958 – 1r – us UF Libraries [972]

Ideario de un combatiente / Conte Aguero, Luis – Mexico City?, Mexico. 1958 – 1r – us UF Libraries [972]

Ideario de varona en la filosofia social / Entralgo, Elias Jose – Habana, Cuba. 1937 – 1r – us UF Libraries [301]

Ideario de varona en la filosofia social / Entralgo, Elias Jose – Habana, Cuba. 1937 – 1r – us UF Libraries [972]

Ideario: ordenado / Marti, Jose – Habana, Cuba. 1930 – 1r – us UF Libraries [972]

Ideario politico / Briceno-Iragorry, Mario – Caracas, Venezuela. 1958 – 1r – us UF Libraries [972]

Ideario politico / Briceno-Iragorry, Mario – Caracas, Venezuela. 1958 – 1r – us UF Libraries [320]

Ideas about india / Blunt, Wilfrid Scawen – London, 1885 – (= ser 19th c books on british colonization) – 3mf – 9 – mf#1.1.7517 – uk Chadwyck [954]

Ideas actuales sobre las plagas de langosta / Moreno Marquez, Victor & Canizo Gomez, Jose del – Madrid: Direccion General de Agricultura. Secc. Plagas del campo y Fitopatologia. Servicio de lucha contra la langosta. Estacion de Fitopatologia Agricola, 1940 – 1 – sp Bibl Santa Ana [630]

Ideas and insights / Central State Hospital [WI] – v1 iss 1-scattered iss [1978-83] – 1r – 1 – (cont: grapevine) – mf#707394 – us WHS [362]

Ideas de alberto torres / Torres, Alberto – Sao Paulo, Brazil. 1932 – 1r – us UF Libraries [972]

Ideas de alberto torres / Torres, Alberto – Sao Paulo, Brazil. 1932 – 1r – us UF Libraries [972]

Ideas de alberto torres / Torres, Alberto – Sao Paulo, Brazil. 1938 – 1r – us UF Libraries [972]

Ideas de alberto torres / Torres, Alberto – Sao Paulo, Brazil. 1938 – 1r – us UF Libraries [972]

Ideas in sound – Garden City. 1972-1973 (1) – mf#8054 – us UMI ProQuest [621]

Ideas liberales / Nieto Caballero, Luis Eduardo – Bogota, Colombia. 1922 – 1r – us UF Libraries [972]

Ideas liberales / Nieto Caballero, Luis Eduardo – Bogota, Colombia. 1922 – 1r – us UF Libraries [972]

The ideas of the apostle paul / Clarke, James Freeman – Boston: James R Osgood, 1884 – 2mf – 9 – 0-7905-1033-2 – (incl ind) – mf#1987-1033 – us ATLA [225]

Ideas on liberty – Irvington-on-Hudson. 2000+ (1,5,9) – (cont: freeman) – mf#1491,01 – us UMI ProQuest [320]

Ideas politicas de angel ganivet / Elias de Tejada Spinola, Francisco – Madrid: Grafica Universal, 1939 – 1 – sp Bibl Santa Ana [320]

Ideas politicas de gabriel turbay / Turbay, Gabriel – Bogota, Colombia. 1945 – 1r – us UF Libraries [972]

Ideas politicas de gabriel turbay / Turbay, Gabriel – Bogota, Colombia. 1945 – 1r – us UF Libraries [972]

Ideas sobre educacion colombiana / Naranjo Villegas, Abel – Bogota, Colombia. 1960 – 1r – us UF Libraries [972]

Ideas sobre educacion colombiana / Naranjo Villegas, Abel – Bogota, Colombia. 1960 – 1r – us UF Libraries [972]

Ideas sociales y politicas de arevalo / Dion, Marie Berthe – Mexico City?, Mexico. 1958 – 1r – us UF Libraries [972]

Ideas sociales y politicas de arevalo / Dion, Marie Berthe – Mexico City?, Mexico. 1958 – 1r – us UF Libraries [972]

The ideas that have influenced civilization, in the original documents / ed by Thatcher, Oliver Joseph – Milwaukee, Boston: Roberts-Manchester Publ., c1901. 10v. plates – 1 – us Wisconsin U Libr [000]

Idee de dieu d'apres l'anthropologie et l'histoire *see* Lectures on the origin and growth of the conception of god

Die idee der absoluten persoenlichkeit, oder, gott und sein verhaeltniss zur welt, insonderheit zur menschlichen persoenlichkeit : eine speculativ-theologische untersuchung ueber wesen, entwicklung und ziel des christlichen theismus / Hanne, Johann Wilhelm – 2. Aufl. Hannover: C Ruempler, 1865 – 3mf – 9 – 0-7905-9385-8 – (incl bibl ref) – mf#1989-2610 – us ATLA [210]

Die idee der gottheit : eine philosophische abhandlung, als wissenschaftliche grundlegung zur philosophie der religion / Weisse, Christian Hermann – Dresden: Ch F Grimmer, 1833 – 1mf – 9 – 0-524-00357-2 – (includes bibliographic references) – mf#1989-3057 – us ATLA [210]

Die idee der persoenlichkeit bei paul heyse / Hammer, Friedrich – Dresden: Risse-Verlag, 1935 – 1r – 1 – (incl bibl ref) – us Wisconsin U Libr [430]

Die idee der seelenwanderung / Meyer, Juergen Bona – Hamburg: Meissner, 1861 – 1mf – 9 – 0-524-01847-2 – (incl bibl ref) – mf#1990-2682 – us ATLA [200]

Die idee der suehne im alten testament : eine untersuchung ueber gebrauch und bedeutung des wortes kipper / Herrmann, Johannes – Leipzig: J C Hinrichs, 1905 – 1mf – 9 – 0-8370-3571-6 – (incl ind) – mf#1985-1571 – us ATLA [221]

Die idee des gesetzes in der praktischen vernunft / Hadlich, Heinrich – Koenigsberg, 1938 (mf ed 1993) – 1mf – 9 – €24.00 – 3-89349-304-2 – mf#DHS-AR 160 – gw Frankfurter [160]

L'idee du sacrifice de la croix dans l'epitre aux hebreux / Padolskis, Vincent – 1935 [mf ed 1993] – 1mf – 9 – 0-524-08127-1 – (in french. incl bibl ref) – mf#1993-9033 – us ATLA [225]

Die idee eines goldenen zeitalters : ein geschichtsphilosophischer versuch mit besonderer beziehung auf die gegenwart / Pfleiderer, Edmund – Berlin: G Reimer, 1877 – 1mf – 9 – 0-7905-9574-5 – mf#1989-1299 – us ATLA [100]

Idee et action – n1-9 10. Paris. juin 1936-avr mai 1937 [mnthly] – 1 – (Revue mensuelle du mouvement socialiste et syndicaliste international. suite de: le combat marxiste) – fr ACRPP [325]

L'idee nouvelle *see* La revue rouge

Eine idee ueber das studium der theologie / De Wette, Wilhelm Martin Leberecht – Leipzig: TO Weigel, 1850 – 1mf – 9 – 0-7905-6864-0 – (incl bibl ref) – mf#1988-2864 – us ATLA [240]

Idee und persoenlichkeit in der kirchengeschichte / Koehler, Walther – Tuebingen: JCB Mohr, 1910 – (= ser [Sammlung gemeinverstaendlicher Vortraege und Schriften aus dem Gebiet der Theologie und Religionsgeschichte]) – 1mf – 9 – 0-524-04137-7 – (incl bibl ref) – mf#1990-1207 – us ATLA [240]

Idee und wirklichkeit bei hanns johst / Heering, Hans – Berlin: Junker und Duennhaupt, 1938 – 1 – (incl bibl ref) – us Wisconsin U Libr [430]

Ideelle kontinentalsperre / Lenard, Philipp Eduard Anton – Written in Aug 1914. Muenchen: F. Eher, 1940. 21p – 1 – us Wisconsin U Libr [940]

Ideen, Marie A *see* Changing the crosses and winning the crown

Ideen, reflexionen und betrachtungen aus schleiermachers werken / Schleiermacher, Friedrich [Ernst Daniel]; ed by Lancizolle, Ludwig von – Berlin: G Reimer 1855 [mf ed 1991] – 1mf – 9 – 0-524-00334-3 – mf#1989-3034 – us ATLA [170]

Ideen zu der organisation der teutschen kirche : ein beitrag zum kuenftigen konkordat / Kopp, Georg Ludwig Karl – Frankfurt a.M., 1814 (mf ed 1992) – 1mf – 9 – €24.00 – 3-89349-081-7 – mf#DHS-AR 54 – gw Frankfurter [240]

Der ideengehalt von richard wagners dramatischen dichtungen : im zusammenhange mit seinem leben und seiner weltanschauung, nebst einem anhang: nietzsche und wagner / Drews, Arthur – Leipzig: E Pfeiffer c1931 [mf ed 1991] – 1r – 1 – (incl bibl ref. filmed with: richard wagner's tondrama... / karl kostlin) – mf#3023p – us Wisconsin U Libr [780]

Idees – Vichy. no. 1-33. nov 1941-juil 1944 – 1 – (revue de la revolution nationale) – fr ACRPP [325]

Idees de mme aubray / Dumas, Alexandre – Paris, France. 1867 – 1r – us UF Libraries [440]

Idees de mme aubray / Dumas, Alexandre – Paris, France. 1867 – 1r – us UF Libraries [440]

Les idees des indiens algonquins relatives a la vie d'outre-tombe : Ideas of the future life held by algonkin indians / Conard, Elizabeth Laetitia Moon – Paris: E Leroux, 1901 – 1mf – 9 – 0-524-01421-3 – (incl bibl ref. in french) – mf#1990-2416 – us ATLA [290]

Idees et opinions / Rameau, Auguste – Paris, France. 1894 – 1r – us UF Libraries [972]

Idees et opinions / Rameau, Auguste – Paris, France. 1894 – 1r – us UF Libraries [972]

Idees et opinions, la reforme de l'etat / Celestin, Clement – Port-Au-Prince, Haiti. 1940 – 1r – us UF Libraries [972]

Idees et opinions, la reforme de l'etat / Celestin, Clement – Port-Au-Prince, Haiti. 1940 – 1r – us UF Libraries [972]

Idees et peuples. l'homme nouveau *see* L'homme nouveau

Idees modernes : droit international et franc-maconnerie / Nys, Ernest – Bruxelles: M Weissenbruch 1908 – 2mf – 9 – mf#vrl-157 – ne IDC [366]

Les idees morales chez les heterodoxes latins au debut du 13e siecle / Alphandery, Paul – Paris: E Leroux, 1903 [mf ed 1992] – (= ser Bibliotheque de l'ecole des hautes etudes. sciences religieuses 16/1) – 1mf – 9 – 0-524-03030-8 – (incl bibl ref) – mf#1990-0787 – us ATLA [230]

Les idees philosophiques et religieuses de philon d'alexandrie / Brehier, Emile – Paris: A. Picard, 1908. Chicago: Dep of Photodup, U of Chicago Lib, 1973 (1r); Evanston: American Theol Lib Assoc, 1984 (1r) – 1 – 0-8370-0545-0 – (incl ind) – mf#1984-B350 – us ATLA [180]

Les idees philosophiques et religieuses de philon d'alexandrie (ephm8) / Brehier, E – Paris, 1950 – €14.00 – ne Slangenburg [100]

Les idees sur dieu dans l'ancienne egypte / Amelineau, Emile – Paris: A Faivre et H Teillard, 1893 – (= ser Conferences (Ligue contre l'Atheisme)) – 1mf – 9 – 0-524-01150-8 – mf#1990-2226 – us ATLA [290]

Idees sur la meterologie / Luc, Jean Andre de – Paris, 1787 – 1 – us Wisconsin U Libr [380]

A ideia : orgao do gremio litterario maranhense – Maranhao: Typ Republicana, 01 maio-28 jul 1893 – (= ser Ps 19) – 1 – mf#DIPER – bl Biblioteca [079]

A ideia : periodico litterario e recreativo – Natal, RN: Typ Conservadora, 27 mar 1880 – (= ser Ps 19) – bl Biblioteca [079]

Ideia : revista artistica e litteraria – Rio de Janeiro, RJ: Typ e Lith de F A de Souza, 01 set-01 nov 1869 – (= ser Ps 19) – mf#P17,01,166 – bl Biblioteca [073]

Ideias do presidente getulio vargas / Vargas, Getulio – Rio de Janeiro, Brazil. 1939 – 1r – us UF Libraries [972]

Ideias do presidente getulio vargas / Vargas, Getulio – Rio de Janeiro, Brazil. 1939 – 1r – us UF Libraries [972]

Idel – Astrakhan, 1907-14 – 6r – 1 – us UMI ProQuest [077]

Idell, Albert Edward *see* Doorway in antigua

Idelsohn, Abraham Zebi *see* Sefer ha-shirim

Idel'son, A *see* Nashi sotsialisticheskiia partii

Idelson, A *see* O evreiskoi sotsial-demokratii

Ideltson, A *see* Sobranie sochinenii

Idenshtat / Herzl, Theodor – Boston, MA. 1918? – 1r – us UF Libraries [939]

Idenshtat / Herzl, Theodor – Boston, MA. 1918? – 1r – us UF Libraries [939]

Identidad y la cultura / Fernandez Mendez, Eugenio – San Juan, Puerto Rico. 1959 – 1r – us UF Libraries [972]

Identidad y la cultura / Fernandez Mendez, Eugenio – San Juan, Puerto Rico. 1959 – 1r – us UF Libraries [306]

Identification anthropometrique; instructions signaletiques / Bertillon, Alphonse – nouv ed. Melun: Impr administrative, 1893 [mf ed 1987] – lxxxiv/148p (ill) – 1 – mf#1892 – us Wisconsin U Libr [573]

Identification of athletes by athletes : at eastern washington university and the perceived media role in that identification / Krump, Jason G – 2000 – 85p on 1mf – 9 – $5.00 – mf#PSY 2132 – us Kinesology [302]

Identification of body built stereotypes in preadolescents : relationship to eating disorders / Steele, Kristy J – Purdue University, 1995 – 1mf – 9 – mf#PSY 1902 – us Kinesology [150]

The identification of employee types through q-methodology : a study of part-time and seasonal recreation and parks employees / McDade, Susan E S – 1989 – 114p 2mf – 9 – $8.00 – us Kinesology [790]

Identification of selected attributes which predict competition climbing performance / Binney, David M – 1996 – 1mf – 9 – $4.00 – mf#PE 3924 – us Kinesology [790]

The identification of the 144,000 of revelation 7 / Moulton, George Ernest – 1982 – 1 – 5.68 – us Southern Baptist [242]

An identification of the influence of various factors on athletes' cognitive-appraisal of injury / Newcomer, R Renee – 1997 – 2mf – 9 – $8.00 – mf#PSY 2004 – us Kinesology [612]

Identification of the leading citrus rootstocks by microscopical an / Deonier, Marshall T – S.l., S.l? . 1930 – 1r – us UF Libraries [634]

Identification of the leading citrus rootstocks by microscopical an... / Deonier, Marshall T – S.l., S.l? . 1930 – 1r – us UF Libraries [634]

The identification of the minimum qualifications for tennis teaching professionals to be hired at managed tennes facilities in the united states / Warrell, Theresa & Jackson, Michael W – 1992 – 2mf – 9 – $8.00 – us Kinesology [790]

Identifisering en ontleding van stresfaktore aan 'n tersiere onderwysinrigting in 'n periode van verandering / Schutte, Orgelina Fredrika – Uni of South Africa 2000 [mf ed Johannesburg 2000] – 2mf – 9 – (text in afrikaans; abstract in afrikaans and english; incl bibl ref) – mf#mfm14844 – sa Unisa [370]

Die identifisering van adolessente wat groepdruk moeilik hanteer / Fourie, Jacob Andries Cornelis – Uni of South Africa 2001 [mf ed Johannesburg 2001] – 6mf – 9 – (summary in afrikaans & english; incl bibl ref) – mf#mfm14688 – sa Unisa [305]

Die identifizierung und charakterisierung cholinerger und nitreger enterischer schaltkreise im myenterischen plexus des meerschweinchenmagens / Schaaf, Cornelia – (mf ed 1995) – 2mf – 9 – €40.00 – 3-8267-2236-1 – mf#DHS 2236 – gw Frankfurter [574]

Identifying a collective variable of locomotion : a dynamic systems analysis / Kao, Jim – 1997 – 1mf – 9 – $4.00 – mf#PE 3837 – us Kinesology [612]

Identifying challenges related to providing community-based environmental health education and promotion programmes / Witthuhn, Jacqueline – Uni of South Africa 2001 [mf ed Johannesburg 2001] – 3mf – 9 – (incl bibl ref) – mf#mfm15048 – sa Unisa [360]

Identity – Mahwah. 1998+ (1,5,9) – ISSN: 1528-3488 – mf#33088 – us UMI ProQuest [301]

Identity fraud : information on prevalence, cost, and internet impact is limited: briefing report to congressional requesters / United States. General Accounting Office – Washington DC: The Office [mf ed 1999?] – 1mf – 9 – (incl bibl ref) – us US Gen Account [364]

Identity of the free church claim from 1838 till 1875 / Moncreiff, Henry Wellwood – Edinburgh, Scotland. 1875 – 1r – us UF Libraries [240]

Identity of the free church claim from 1838 till 1875 / Moncreiff, Henry Wellwood – Edinburgh, Scotland. 1875 – 1r – us UF Libraries [240]

Ideologeme in der textsorte zeitschriftenartikel : eine untersuchung ausgewaehlter artikel der ddr-wochenillustrierten für dich / Schatz, Michael – (mf ed 2000) – (= ser Leipziger arbeiten zur fachsprachenforschung) – 1mf – 9 – €30.00 – 3-8267-2741-X – mf#DHS 2741 – gw Frankfurter [410]

L'ideologie scolaire du conseil de l'instruction publique de la province de quebec, 1927-1964 / Goyette, Gabriel – Ottawa: [G Goyette], 1970 [mf ed 2000] – 9 – cn Bibl Nat [370]

Ideology and politics of the american baptist churches in 1900-1917 / Kislov, A A – 1969. Russian. 210p – 1 – 7.35 – us Southern Baptist [242]

Ideology and power in soviet politics / Brzezinski, Zbigniew K – New York, NY. 1962 – 1r – us UF Libraries [947]

Ideology and power in soviet politics / Brzezinski, Zbigniew K – New York, NY. 1962 – 1r – us UF Libraries [025]

Ideology, technology and the historical avant garde / Knapp, Ferdinand M – (mf ed 1995) – 1mf – 9 – €30.00 – 3-8267-2209-4 – mf#DHS 2209 – gw Frankfurter [400]

Ideophones in shona / Fortune, G (George) – London, England. 1962 – 1r – us UF Libraries [470]

Ideophones in shona / Fortune, George – London, England. 1962 – 1r – us UF Libraries [470]

Ides, E I *see*
– The travels of everard isbrand ides
– Voyages d'everard isbrands ides

[Ides, E I] *see* Drie jaarige reize naar china, te lande gedaan door den moskovischen afgezant

Idh-har-haqq : ou manifestation de la verite de el-hage rahmat-ullah effendi de dehli – Paris. tom 1-2. 1880 – €35.00 – (trans fr arabic by p v carletti) – ne Slangenburg [260]

Idh-har-ul-haqq : ou, manifestation de la verite / Rahmat Allah ibn Khalil al-Rahman – Paris: E Leroux, 1880 – 3mf – 9 – 0-524-04989-0 – mf#1990-3447 – us ATLA [260]

Idilios y elegias / Moreno Torrado, Luis – 1890 – 9 – sp Bibl Santa Ana [810]

Idioma de puerto rico y el idioma escolar de puert / Fernandez Vanga, Epifanio – San Juan, Puerto Rico. 1931 – 1r – us UF Libraries [972]

Idioma de puerto rico y el idioma escolar de puert... / Fernandez Vanga, Epifanio – San Juan, Puerto Rico. 1931 – 1r – us UF Libraries [972]

Idioma nacional / Nascentes, Antenor – Rio de Janeiro, Brazil. 1960 – 1r – us UF Libraries [972]

Idioma nacional / Nascentes, Antenor – Rio de Janeiro, Brazil. 1960 – 1r – us UF Libraries [972]

Idiote / Alboise Du Pujol, Jules Edward – Paris, France. 1838? – 1r – us UF Libraries [440]

Idiote / Alboise Du Pujol, Jules Edward – Paris, France. 1838? – 1r – us UF Libraries [440]

Idioticon des christlich palaestinischen aramaeisch / Schwally, Friedrich – Giessen: J Ricker, 1893 – 1mf – 9 – 0-8370-8309-5 – (in german, aramaic, greek) – mf#1986-2309 – us ATLA [470]

Idish-amerikaner redner – New York, NY. 1922 – 1r – us UF Libraries [939]

Idish-amerikaner redner – New York, NY. 1922 – 1r – us UF Libraries [939]

Idishe biznesman firer = The jewish merchant and guide – Los Angeles, CA.1927 – 1 – us AJPC [071]

Idishe folk – London, UK. 2/9 Aug 1968- – 1 – uk British Libr Newspaper [072]

Idishe folk – London, UK. 7 Sept 1934-1 Feb 1935 – 1 – uk British Libr Newspaper [072]

Di idishe gas [evreiskaia ulitsa] – Russia, 1996- – 16mf 4 iss/yr – 9 – $80.00 standing order – (formerly: sovetish heymland 1961-91 [1340mf] $4900) – us UMI ProQuest [073]

Idishe leben (jewish life) – London, UK. 1 Mar-Aug 1923 – 1 – uk British Libr Newspaper [072]

Idishe post un ekspres (jewish post and express) – London, UK. 8 Oct 1926-7 Aug 1935 – 1 – uk British Libr Newspaper [072]

Idishe presse – Los Angeles, CA.1935-36 – 1 – us AJPC [071]

Idishe presse (the jewish press) – London, UK. 3 Feb-17 Sept 1903 – 1 – uk British Libr Newspaper [072]

Idishe shtime – New York, NY. 1941-47 – 1 – us AJPC [071]

Idishe shtime (the jewish voice) – London, UK. 19 Nov-27 Dec 1916 – 1 – uk British Libr Newspaper [072]

Idishe tribune (the jewish tribune) – London, UK. Mar 1938 – 1 – uk British Libr Newspaper [072]

Idishe un algemeine erzihung / Unterman, Isaac – Chicago, IL. 1916 – 1r – us UF Libraries [370]

Idishe un algemeine erzihung / Unterman, Isaac – Chicago, IL. 1916 – 1r – us UF Libraries [939]

Der idisher advokat – Capetown. v. 3-4, 6-9. oct 19 1906-sept 18 1908; sept 30 1910-june 5 1914 – 1 – (incomplete) – us NY Public [072]

Idisher ekspres (the jewish express) – London, UK. 6 Nov 1896-25 Dec 1924; 7 Jan-22 Sept 1926 – 1 – uk British Libr Newspaper [072]

Idisher geist – New York, NY. 1910-14 – 1 – us AJPC [071]

Idisher handels-zhurnal – London, UK. May 1920-Mar 1922 – 1 – uk British Libr Newspaper [072]

Idisher hurbn in rusland / Kossovskii, Vladimir V – New York, NY. 1915 – 1r – us UF Libraries [939]

Idisher hurbn in rusland / Kossovskii, Vladimir V – New York, NY. 1915 – 1r – us UF Libraries [939]

Idisher moment = The jewish moment – Los Angeles, CA.1941 – 1 – us AJPC [071]

Idisher monat buch = Israel's monthly magazine – New York, NY. 1900-02 – 1 – us AJPC [073]

Der idisher sotsyalist : official monthly bulletin of the jewish socialist federation of america – Chicago [IL: s.n.] [v1 n2 aug 1913]-v2 n13 jul 15 1915 (mf ed 197-?) – (semimthly: jan 30 1914-jul 15 1915; mthly: aug-nov 1913. began in jul 1913. issues for feb 15 1914-jul 15 1915 have parallel english title: the jewish socialist. no issue(s) publ dec 1913? in yiddish. cont by: naye velt) – mf#ZZAN-21803 – us NY Public [071]

Idisher treid yunionist (the jewish trade unionist) – London, UK. 15 Jan-5 Feb 1892 – 1 – uk British Libr Newspaper [072]

Idisher vechentlicher zhurnal (the jewish weekly journal) – London, UK. 28 Nov 1906-17 Apr 1907 – 1 – uk British Libr Newspaper [072]

Idisher zhurnal – New York. v. 1-7. May 26 1899-Apr 20 1906 – 1 – us NY Public [071]

Idisher zhurnal (the jewish journal) – London, UK. 19 May 1905-3 Dec 1914 – 1 – uk British Libr Newspaper [072]

Idisher zshurnal = The jewish journal – Boston, MA. 1922 – 1 – us AJPC [071]

Idishes togblat – Warsaw, 1906 (may 16)-1907 (dec 31). v1-2 (incomplete) – 2r – 1 – mf#J-92-21 – ne IDC [077]

Idisze szriftn – Warsaw. Poland. -m. Jan 1956-Jul 1968. (2 reels) – 1 – uk British Libr Newspaper [947]

Idiszes wochenblat – Warsaw PL, 1906-08 – 1r – 1 – (in yiddish) – us UMI ProQuest [939]

Idle man – New York. 1821-1822 – 1 – mf#3993 – us UMI ProQuest [073]

Idle moments in florida / Hobart, George Vere – New York, NY. 1921 – 1r – us UF Libraries [978]

Idle moments in florida / Hobart, George Vere – New York, NY. 1921 – 1r – us UF Libraries [978]

Idler : an illustrated monthly magazine – London. 1892-1911 – 1 – mf#2900 – us UMI ProQuest [073]

Idler – London. 1758-1760 – 1 – mf#4776 – us UMI ProQuest [073]

The idler in italy / Blessington, Marguerite – Paris 1839 [mf ed Hildesheim 1995-98] – 3mf – 9 – €90.00 – 3-487-29261-0 – gw Olms [914]

Idling erodible cropland : impacts on production, prices and government costs / Webb, Shwu-Eng – Washington DC: US Dept of Agriculture, Economic Research Service...1986 – (= ser Agricultural economic report 550) – 9 – us GPO [630]

Idoc : international north american edition – New York. 1970-1974 (1) 1972-1974 (5) (9) – (cont by: idoc/international documentation) – ISSN: 0018-909X – mf#7032 – us UMI ProQuest [240]

Idoc bulletin – Rome. 1977-1984 (1,5,9) – ISSN: 0254-9174 – mf#11433 – us UMI ProQuest [240]

Idoc internazionale – Rome. 1985-1996 (1,5,9) – mf#16025 – us UMI ProQuest [300]

Idoc/international documentation – New York. 1974-1976 (1) 1974-1976 (5) 1976-1976 (9) – (cont: idoc: international north american edition) – ISSN: 0160-7553 – mf#7032,01 – us UMI ProQuest [240]

Idol mis [sic] quest'alma amante : sung by sigr rubinelli in the opera of virginia / Tarchi, Angelo – London: Longman & Broderip [1786?] [mf ed 19–] – 1r – 1 – mf#pres. film 73 – us Sibley [780]

L'idolatrie huguenote figuree au patron de la vieille payenne... / Richeome, L – Lyon, 1608 – 9mf – 9 – mf#CA-146 – ne IDC [242]

Idolatries, old and new : their cause and cure / Brown, James Baldwin – London: Jackson, Walford, & Hodder, 1867 – 1mf – 9 – 0-7905-3699-4 – mf#1989-0192 – us ATLA [240]

Idolatries, old and new : their cause and cure / Brown, James Baldwin – London: Jackson, Walford & Hodder, 1867 – 1mf – us ATLA [240]

Idolatrous worship of the virgin mary in 1847 / Parretti, Giovanni Battista – London, England. 1848 – 1r – us UF Libraries [240]

Idolatrous worship of the virgin mary in 1847 / Parretti, Giovanni Battista – London, England. 1848 – 1r – us UF Libraries [240]

Idolino : erzaehlung / Penzoldt, Ernst – 1.-3. aufl. Berlin: S Fischer c1935 [mf ed 1991] – 1r – 1 – (filmed with: der mensch an der wege / rudolf paulsen) – mf#2859p – us Wisconsin U Libr [880]

Idolo / Estrella Gutierrez, Fermin – Buenos Aires, Argentina. 1928 – 1r – us UF Libraries [972]

Idolo / Estrella Gutierrez, Fermin – Buenos Aires, Argentina. 1928 – 1r – us UF Libraries [972]

El idolo roto (realidades de otros dias) / Hurtado de Mendoza, Publio – Caceres: Tip., Enc. y Lib. de Jimenez, 1904 – 1 – sp Bibl Santa Ana [946]

Idols of clay, a novel / Smythies, Harriet Maria Gordon – London: Saunders, Otley, 1867. 3v – 1 – us Wisconsin U Libr [830]

Idomenee : tragedie en musique / Campra, Andre – Paris: J B C Ballard 1731 [mf ed 1988] – 1r – 1 – mf#pres. film 37 – us Sibley [780]

Idraetsliv – Oslo, Norway. Idrettsliv. -sw. 5 April 1923-30 June 1932. 16 reels – 1 – uk British Libr Newspaper [079]

Idrisi *see* Description de l'afrique et de l'espagne

Idrott och lek – Stockholm: A Bonnier 1933 [mf ed Bloomington IN: Indiana Uni Lib, Preservation Dept 1984] – 194p on 1r – 1 – us Indiana Preservation [390]

Idrottsbladet – Stockholm, Sodertalje, Sweden. 1910-88 – 1 – sw Kungliga [790]

Idsa journal / Institute for Defence Studies and Analyses – New Delhi. 1968-1985 (1) 1974-1985 (5) 1976-1985 (9) – ISSN: 0020-2606 – mf#8642 – us UMI ProQuest [355]

Idun – Stockholm, Sweden. 1888-89 – 1 – sw Kungliga [078]

Une idylle au pays khmer : roman de moeurs cambodgiennes / Thouvenot, Maurice – Paris: Jouve & Cie [1913] [mf ed 1989] – 1r with other items – 1 – mf#mf-10289 seam reel 004/11 [§] – us CRL [306]

Idylle in bauerbach : eine schiller-novelle / Elsner, Richard – Berlin: E Sicker, [194-?] (mf ed 1990) – 1r – 1 – (filmed with: astra) – us Wisconsin U Libr [830]

Idyllia,...editio tertia / La Rue, Ch de – Parisiis: Apud Simonem Benard, 1672 – 2mf – 9 – mf#O-665 – ne IDC [090]

Idylls et caprices : fantaisies pour piano a 4 mains / Colomer, B M – Paris: Leon Grus [19-?] [mf ed 19–] – 1mf – 9 – mf#fiche 160 – us Sibley [780]

Idylls from the sanskrit / Griffith, Ralph Thomas Hotchkin – Allahabad: Panini Office, 1912 – (= ser Samp: indian books) – us CRL [490]

Idylls of our island / Boa, Myrtle J – [Montreal ?: s.n, 1923 ?] (mf ed 1992) – 1mf – 9 – mf#SEM105P1636 – cn Bibl Nat [971]

Idylls of the sea / Bullen, Frank Thomas – Toronto: Toronto News Co, 1899 – 4mf – 9 – (int by j st loe strachey) – mf#08813 – cn Canadiana [590]

[Idyllwild-] idyllwild town crier – CA. 1967-93 – 33r – 1 – $1980.00 – mf#R02301 – us Library Micro [071]

Iee proceedings : circuits, devices, and systems – London. 1994+ (1,5,9) – (cont: iee proceedings g: circuits, devices, and systems) – ISSN: 1350-2409 – mf#12533,02 – us UMI ProQuest [621]

Iee proceedings : computers and digital techniques – Stevenage. 1994+ (1,5,9) – (cont: iee proceedings e: computers and digital techniques) – ISSN: 1350-2387 – mf#12531,01 – us UMI ProQuest [621]

Iee proceedings : control theory and applications – Stevenage. 1994+ (1) 1994+ (5) 1995+ (9) – (cont: iee proceedings d: control theory and applications) – ISSN: 1350-2379 – mf#12530,01 – us UMI ProQuest [621]

Iee proceedings : electric power applications – London. 1994+ (1,5,9) – (cont: iee proceedings b: electric power applications) – ISSN: 1350-2352 – mf#12528,01 – us UMI ProQuest [621]

Iee proceedings : generation, transmission, and distribution – Stevenage. 1994+ (1,5,9) – (cont: iee proceedings c: generation, transmission, and distribution) – ISSN: 1350-2360 – mf#12529,01 – us UMI ProQuest [621]

Iee proceedings : microwaves, antennas and propagation – London. 1994+ (1,5,9) – (cont: iee proceedings h: microwaves, antennas and propagation) – ISSN: 1350-2417 – mf#12534,02 – us UMI ProQuest [621]

Iee proceedings : optoelectronics – London. 1994+ (1) 1994+ (5) 1996+ (9) – (cont: iee proceedings j: optoelectronics) – ISSN: 1350-2433 – mf#14477,01 – us UMI ProQuest [530]

Iee proceedings : radar, sonar, and navigation – Stevenage. 1994+ (1,5,9) – (cont: iee proceedings f: radar and signal processing) – ISSN: 1350-2395 – mf#12532,02 – us UMI ProQuest [621]

Iee proceedings : science, measurement and technology – London. 1994+ (1) 1994+ (5) 1996+ (9) – (cont: iee proceedings a: science, measurement and technology) – ISSN: 1350-2344 – mf#12527,02 – us UMI ProQuest [621]

Iee proceedings : software – Stevenage, 1998+ [1,5,9] – ISSN: 1462-5970 – mf#26002,01 – us UMI ProQuest [621]

Iee proceedings : vision, image and signal processing – London. 1994+ (1,5,9) – ISSN: 1350-245X – mf#20664 – us UMI ProQuest [621]

Iee proceedings 1 : communications, speech, and vision – Stevenage. 1989-1993 (1,5,9) – (cont by: iee proceedings communications) – ISSN: 0956-3776 – mf#17785 – us UMI ProQuest [621]

Iee proceedings a : physical science, measurement and instrumentation, management and education, reviews – Stevenage. 1980-1990 (1) 1980-1990 (5) 1980-1990 (9) – (cont by: iee proceedings a: science, measurement and technology) – ISSN: 0143-702X – mf#12527 – us UMI ProQuest [621]

Iee proceedings a : science, measurement and technology – London. 1991-1993 (1,5,9) – (cont: iee proceedings a: physical science, measurement and instrumentation, management and education, reviews. cont by: iee proceedings science, measurement and technology) – mf#12527,01 – us UMI ProQuest [621]

Iee proceedings b : electric power applications – Stevenage. 1980-1993 (1) 1980-1993 (5) 1980-1992 (9) – (cont by: iee proceedings: electric power applications) – ISSN: 0143-7038 – mf#12528 – us UMI ProQuest [621]

Iee proceedings c : generation, transmission, and distribution – Stevenage. 1980-1993 (1) 1980-1993 (5) 1980-1993 (9) – (cont by: iee proceedings: generation, transmission, and distribution) – ISSN: 0143-7046 – mf#12529 – us UMI ProQuest [621]

Iee proceedings communications – London. 1994+ (1,5,9) – (cont: iee proceedings 1) – ISSN: 1350-2425 – mf#17785,01 – us UMI ProQuest [380]

Iee proceedings d : control theory and applications – Stevenage. 1980-1993 (1) 1980-1993 (5) 1980-1993 (9) – (cont by: iee proceedings: control theory and applications) – ISSN: 0143-7054 – mf#12530 – us UMI ProQuest [621]

Iee proceedings e : computers and digital techniques – Stevenage. 1980-1993 (1) 1980-1993 (5) 1980-1993 (9) – (cont by: iee proceedings: computers and digital techniques) – ISSN: 0143-7062 – mf#12531 – us UMI ProQuest [621]

Iee proceedings f : communications, radar, and signal processing – Stevenage. 1980-1988 (1) 1980-1988 (5) 1980-1988 (9) – ISSN: 0143-7070 – mf#12532 – us UMI ProQuest [621]

Iee proceedings f : radar and signal processing – Stevenage. 1989-1993 (1) 1989-1993 (5) 1989-1993 (9) – (cont by: iee proceedings: radar, sonar, and navigation) – ISSN: 0956-375X – mf#12532,01 – us UMI ProQuest [621]

Iee proceedings g : circuits, devices, and systems – Stevenage. 1989-1993 (1) 1989-1993 (5) 1989-1993 (9) – (cont by: iee proceedings: circuits, devices, and systems) – ISSN: 0956-3768 – mf#12533,01 – us UMI ProQuest [621]

Iee proceedings g : electronic circuits and systems – Stevenage. 1980-1988 (1) 1980-1988 (5) 1980-1988 (9) – ISSN: 0143-7089 – mf#12533 – us UMI ProQuest [621]

Iee proceedings h : microwaves, antennas and propagation – Stevenage. 1985-1993 (1,5,9) – (cont: iee proceedings h: microwaves, optics, and antennas. cont by: iee proceedings: microwaves, antennas and propagation) – ISSN: 0950-107X – mf#12534,01 – us UMI ProQuest [621]

Iee proceedings h : microwaves, optics, and antennas – Stevenage. 1980-1985 (1,5,9) – (cont by: iee proceedings h: microwaves, antennas and propagation) – ISSN: 0143-7097 – mf#12534 – us UMI ProQuest [621]

Iee proceedings i : solid-state and electron devices – Stevenage. 1980-1988 (1) 1980-1988 (5) 1980-1988 (9) – ISSN: 0143-7100 – mf#12535 – us UMI ProQuest [621]

Iee proceedings j : optoelectronics – Stevenage. 1985-1993 (1,5,9) – (cont by: iee proceedings: optoelectronics) – ISSN: 0267-3932 – mf#14477 – us UMI ProQuest [530]

Iee review – Stevenage. 1988+ (1,5,9) – (cont: electronics and power) – ISSN: 0953-5683 – mf#10673,01 – us UMI ProQuest [621]

Ieng Sary *see* Cambodia 1972

Ienna, Tiziana M *see* The asthmatic athlete

Ier quatuor a cordes / Borowski, Felix – [1897] [mf ed 19–] – 4pt on 1r – 1 – (contains 2nd & 3rd movements only) – mf#pres. film 32 – us Sibley [780]

Ierc bulletin – 1973-78 – (= ser American indian periodicals... 1) – 7mf – 9 – $95.00 – us UPA [305]

Ieremias...propheta, expositus...concionibus 170 : brevis threnorum explicatio / Bullinger, Heinrich – Tigvri, Christoph Froschover, 1575 – 12mf – 9 – mf#PBU-202 – ne IDC [240]

Ierosolymitike bibliotheke / Papadopoulos-Kerameoos, A – St Petersburg. v1-5. 1891-1915 – 5v on 87mf – 8 – €166.00 – ne Slangenburg [240]

Ierson, Henry *see*
- Notes on the amended english bible
- Report of a visit to hungary

Iervsalem, vetvstissima illa et celeberrima totivs mvndi civitas, ex sacris literis et approbatis historicis ad unguem descripta... / Reissner, A – Francofvrti, 1563 – 13mf – 9 – mf#H-8302 – ne IDC [956]

Iesaet nassar : the story of the life of jesus the nazarene / Mamreov, Peter von Finkelstein et al – New York: Sunrise Pub Co, 1895 [mf ed 1993] – 2mf – 9 – 0-524-05620-X – mf#1992-0475 – us ATLA [830]

If stone's bi-weekly – Washington. 1953-1971 (1) 1971-1971 (5) (9) – ISSN: 0018-9758 – mf#5646 – us UMI ProQuest [320]

If this be treason / Joseph, Helen – London, England. 1963 – 1r – us UF Libraries [960]

If this be treason / Joseph, Helen – London, England. 1963 – 1r – us UF Libraries [960]

If war comes : an essay on india's military problems / Adarkar, Bhalchandra Pundlik – Allahabad: Indian Press, 1939 – (= ser Samp: indian books) – us CRL [355]

If you go to south america / Foster, Harry La Tourette – New York, NY. 1928 – 1r – us UF Libraries [972]

If you go to south america / Foster, Harry La Tourette – New York, NY. 1928 – 1r – us UF Libraries [972]

Ifac proceedings series / International Federation of Automatic Control – Oxford. 1985-1989 (1,5,9) – (cont by: ifac symposia series) – ISSN: 0742-5953 – mf#49542 – us UMI ProQuest [629]

Ifac symposia series / International Federation of Automatic Control – Oxford. 1990-1993 (1,5,9) – (cont: ifac proceedings series) – ISSN: 0962-9505 – mf#49542,01 – us UMI ProQuest [629]

Ifa-rundschau – Berlin DE, 1930-31 – 1 – gw Misc Inst [074]

Ifas administrative policy records, 1905-1962 / University Of Florida Archives Public Records Collection – Gainesville, FL. series 90a 17.1-4. 1905-1962 – 4r – us UF Libraries [025]

Ifas administrative policy records, 1905-1962 / University Of Florida Archives Public Records Collection – Gainesville, FL. series 90a 17.1-4. 1905-1962 – 4r – us UF Libraries [350]

Ifas general correspondence, 1889-1924 / University Of Florida Archives Public Records Collection – Gainesville, FL. series 87 9.1a-19a. 1889-1924 – 20r – us UF Libraries [350]

Ifas general correspondence, 1889-1924 / University Of Florida Archives Public Records Collection – Gainesville, FL. series 87 9.1a-19a. 1889-1924 – 20r – us UF Libraries [025]

Ifas records and correspondence, 1917-1971 / University Of Florida Archives Public Records Collection – Gainesville, FL. 1917-1971 – 28r – us UF Libraries [025]

Ifas records and correspondence, 1917-1971 / University Of Florida Archives Public Records Collection – Gainesville, FL. 1917-1971 – 28r – us UF Libraries [350]

Ifco news / Interreligious Foundation for Community Organization [US] – v2 n1-v5 n5 [1971 mar/apr-1974 sep/oct], v6 n1-2 [1975 jan/mar-apr/jul], v7 n1-2 [1976 spr-1977 may/jun], v8 n1-5 [1979 win-1980 oct] v8 n6 [1981 sep/oct], v9 n1 [1988 spr] – 1r – 1 – mf#405102 – us WHS [230]

[Ifda, the third system project papers] / International Foundation for Development Alternatives – 5 titles on 282mf – 9 – (with printed catalogue & ind) – ne IDC [338]

Ife, August *see* Fussreise vom brocken auf den vesuv und rueckkehr in die heimath

'Iffet, Mehmed Emin *see* The divan project

Iffland, August Wilhelm *see* Ueber meine theatralische laufbahn

Iffland in seinen schriften als kuenstler, lehrer und director der berliner buehne : zum gedaechtnis seines 100jaehrigen geburtstages am 19. april 1859 – Berlin: Duncker und Humblot, 1859 – 1r – 1 – (incl bibl ref) – us Wisconsin U Libr [790]

Ifilye ne mitekele ya muno northern rhodesia – London, England. 1948 – 1r – us UF Libraries [960]

Ifilye ne mitekele ya muno northern rhodesia – London, England. 1948 – 1r – us UF Libraries [960]

Ifriqiya – Ifriqia. n1, 3. Alger. 1919 – 1 – fr ACRPP [073]

IG Bauen – Agrar – Umwelt *see* Der grundstein 1888 bis 1933

IG Bergbau und Energie *see* Wir tragen ein licht durch die nacht

Igalaaq – v4 n3-v5 n2 [1982 jan-1983 mar/apr] – 1r – 1 – mf#669891 – us WHS [071]

Igazsag – Cluj, Romania. 1963-64; 1966-73; Jul 1975-76; 1978-80; 1982-84; 1987-89 – 25r – 1 – us L of C Photodup [949]

Igbani, B *see* Health behaviors and attitudes of selected nigerian and american university students

Igbo revision course for gce, wasc and similar examinations / Carnochan, J – London, England. 1963 – 1r – us UF Libraries [960]

Igbo revision course for gce, wasc and similar examinations / Carnochan, J – London, England. 1963 – 1r – us UF Libraries [960]

Igbo "women's war" of 1929, the... : documents relating to the aba riots in eastern nigeria – 14mf – 9 – (int by d c dorward) – mf#87277 – uk Microform Academic [960]

Igelmo Pinilia, Eulogio *see* Org. sindical-badajoz. 4 consejo economico s. prov. c. 15_a (factores humanos y sociales productividad)

Igeret r yehoshua ha-lorki = Das apologetische schreiben des josua lorki an den abtruennigen don salomon ha-lewi (paulus de santa maria) / Jeronimo de Santa fe; ed by Landau, Leo – Antwerpen: Teitelbaum & Boxenbaum, 1906 [mf ed 1985] – 1mf – 9 – 0-8370-4392-1 – (german trans and int by leo landau) – mf#1985-2392 – us ATLA [939]

Igernes schuld : ein kammerspiel in vier akten / Pulver, Max – Leipzig: Insel-Verlag, 1918 – 1r – 1 – us Wisconsin U Libr [820]

Igirama lesingisi / Bryant, D – Emugungundhlovu, South Africa. 19– – 1r – us UF Libraries [960]

Igirama lesingisi / Bryant, D – Emugungundhlovu, South Africa. 19– – 1r – us UF Libraries [960]

Iglesia : nueva frontera / Aradillas Agudo, Antonio – Madrid: Sociedad de Educacion Atenas, S.A., 1967 – 1 – sp Bibl Santa Ana [240]

Iglesia *see* Cantos liturgicos

Iglesia, Alvaro de la *see* De navidad

Iglesia ano 2000 / Aradillas Agudo, Antonio – Madrid: PPC, 1972 – 1 – sp Bibl Santa Ana [240]

La iglesia de santiago de los caballeros de caceres y el escultor alonso berruguete / Floriano Cumbreno, Antonio C – Caceres, S.L., 1918 – 1 – sp Bibl Santa Ana [240]

La iglesia de santiago de los caballeros. descripcion historico-artistica / Floriano Cumbreno, Antonio C – Caceres: Tip. de Santos Floriano y otros, 1915 – sp Bibl Santa Ana [720]

Iglesia, el subdesarrollo, y la revolucion – Mexico City?, Mexico. 1968 – 1r – us UF Libraries [972]

Iglesia, el subdesarrollo, y la revolucion – Mexico City?, Mexico. 1968 – 1r – us UF Libraries [972]

La iglesia en la independencia del uruguay / Sallaberry, Juan Faustino – Montevideo, Madrid: Razon y Fe, 1932 – 1 – sp Bibl Santa Ana [972]

Iglesia en peru y bolivia / Alonso, Isidoro – Friburgo, Switzerland. 1962 – 1r – us UF Libraries [972]

Iglesia en peru y bolivia / Alonso, Isidoro – Friburgo, Switzerland. 1962 – 1r – us UF Libraries [972]

La iglesia en toledo... / Rivera Reno, Juan Francisco – Madrid: Graf. Calleja, 1967 – 1 – sp Bibl Santa Ana [946]

Iglesia en venezuela y ecuador / Alonso, Isidoro – Friburgo, Switzerland. 1962 – 1r – us UF Libraries [972]

Iglesia en venezuela y ecuador / Alonso, Isidoro – Friburgo, Switzerland. 1962 – 1r – us UF Libraries [972]

La iglesia filipina independiente – Manila. Philippine Islands. -w. 11-26 Oct, 8 Nov 1903. (4 ft) – 1 – uk British Libr Newspaper [079]

Iglesia, N de la *see* Flores de miraflores, hieroglificos sagrados...del mysterio de la concepcion de la virgen, y madre de dios maria senora nuestra

Iglesia parroquial de san jose / Caceres. Parroquia de San Jose – Caceres: Tip. Extremadura, 1978 – sp Bibl Santa Ana [240]

Iglesia, Ramon *see* Cronista e historiadores de la conquista de mexico, fondo de cultura economica

La iglesia y la educacion popular en indias / Bayle, Constantino – Madrid: Razon y Fe, 1932 – 1 – sp Bibl Santa Ana [377]

La iglesia y la masoneria en venezuela... / Navarro, Nicolas E; ed by Bayle, Constantino – Madrid: Razon y Fe, 1928 – 9 – sp Bibl Santa Ana [290]

La iglesia y la patria. la accion catolica / Delgado Gomez, Enrique – Pamplona: Graficas Iruna, 1948 – sp Bibl Santa Ana [240]

Iglesia y mision *see* Mision

Iglesias – Mexico: Centro Nacional de Comunicacion Social y Comision Evangelica Latinoamericana de Educacion Cristiana, 1983. ns: v1 n1/2-v9 n107. 1984-1992 – 4r – 1 – us CRL [240]

Iglesias bombardeadas por los rebeles = Churches bombarded by the rebels – [s.l.]: Ministerio de Propaganda 1936 [mf ed 1977] – (= ser Blodgett coll) – 1mf – 9 – (text in english, french, german & spanish) – mf#w956 – us Harvard [946]

Las iglesias cristianas de oriente / Morillo Trivino, Santiago – Granada: publicaciones omdoc, 1946 – 1 – sp Bibl Santa Ana [240]

Iglesias, Francisco Da Assis *see* Caatingas e chapadoes

Iglesias, Luis *see* Misioneros redentoristas y la republica de el salv...

Iglesias, M *see* Memorial sobre las analogias y diferencias...entre el garrotillo...y la angina...

Iglesias mozarabes : arte espanol de los siglos 9 a 11 / Gomez-Moreno, M – Madrid, 1919 – 25mf – 8 – €48.00 – ne Slangenburg [241]

Iglesias, Santiago *see* Planificando alrededor del mundo

Ignace d'antioche, ses epitres, sa vie, sa theologie : etude critique suivie d'une traduction annotee / Bruston, Edouard – Paris: G. Fischbacher, 1897 – 1mf – 9 – 0-7905-6101-8 – (incl bibl ref) – mf#1988-2101 – us ATLA [240]

Ignacio agramonte / Marquez Sterling, Carlos – Habana, Cuba. 1936 – 1r – us UF Libraries [972]

Ignacio agramonte / Marquez Sterling, Carlos – Habana, Cuba. 1936 – 1r – us UF Libraries [972]

Ignacio agramonte y la revolucion cubana / Betancourt Agramonte, Eugenio – Habana, Cuba. 1928 – 1r – us UF Libraries [972]

Ignacio agramonte y la revolucion cubana / Betancourt Agramonte, Eugenio – Habana, Cuba. 1928 – 1r – us UF Libraries [972]

Ignacio, Cleto R *see*
- Ang cahambalhambal na pagca guho ng Troya
- Cahangahangang buhay ni santa margarita de cortona na taga toscana sa nayon ng diocesis at limang virgenes, at apat na puong soldados na pauang mga martires at ibang nadamay
- Casaysayan ng catotohanang buhay ng haring clodeveo at reyna clotilde sa reyno nang francia na tinula sa lubos na catiagaan
- Cauiliuiling dalauang pu at dalauang cuento
- Halimbauang buhay nang turing na mag-asauang si guiadoro at si amapura at ang bilang anac na si negaderio at si rectorino...
- Isang dios at tatlo sa pagca-persona na hinnago sa sagrada biblia, mula sa genesis hangang evangelio

Ignacio de azcuedo... / Costa, Manuel Goncalves da – Madrid: Missionalia Hispanica, 1949 – 1 – sp Bibl Santa Ana [240]

Ignacio, Rosendo *see* Aklat ng pagluluto

Ignat'ev, A V *see* S iu vitte – diplomat

Ignatian – Cleveland, OH, oct 8 1924-jun 15 1925 – 1r – 1 – (student newspaper of st. ignatius high school) – us Western Res [373]

The ignatian epistles entirely spurious : a reply to the right rev dr lightfoot, bishop of durham / Killen, William Dool – Edinburgh: T & T Clark, 1886 – 1mf – 9 – 0-7905-5239-6 – mf#1988-1239 – us ATLA [240]

Die ignatianischen briefe und ihr neuester kritiker : eine streitschrift gegen herrn bunsen / Baur, Ferdinand Christian – Tuebingen: Ludwig-Friedrich Fues, 1848 – 1mf – 9 – 0-8370-9600-6 – mf#1986-3600 – us ATLA [241]

Ignatienko, V *see* Bibliografiia ukrainskoi presi 1816-1916

Ignatii, Arkhim *see* Kratkiia zhizneopisaniia russkikh sviatykh

Ignatius of Loyola, Saint *see*
- Exercices spirituels d'apres saint ignace
- Exercitia spiritualia
- Die geistlichen uebungen des ignatius von loyola

Ignatius, Saint, Bishop of Antioch *see*
- Corpus ignatianum
- Epistles of st ignatius and st polycarp

Ignatius von antiochien als christ und theologe / Goltz, Eduard von der – Leipzig, 1894 – (= ser Tugal 1-12/3a) – 4mf – 9 – €11.00 – ne Slangenburg [240]

Ignatius von antiochien als christ und theologe – griechische excerpte aus homilien des origenes / Goltz, Eduard & Klostermann, Erich – Leipzig: J C Hinrichs, 1894 – (= ser Tugal) – 1mf – 9 – 0-8370-9949-8 – (incl bibl ref) – mf#1986-3949 – us ATLA [240]

Ignatius von antiochien und die paulusbriefe / Rathke, H – Berlin, 1967 – (= ser Tugal 5-99) – 2mf – 9 – €5.00 – ne Slangenburg [240]

Ignatius von loyola / Gothein, Eberhard – Halle: Verein fuer Reformationsgeschichte, 1885 – (= ser [Schriften des Vereins fuer Reformationsgeschichte]) – 1mf – 9 – 0-7905-4686-8 – mf#1988-0686 – us ATLA [241]

Ignatius von loyola und der protestantismus / Goetz, Leopold Karl – Muenchen: J.F. Lehmann, 1901 – (= ser Geschichts-Wahrheiten) – 1mf – 9 – 0-7905-5884-X – mf#1988-1884 – us ATLA [241]

Ignatius von loyola und die gegenreformation / Gothein, Eberhard – Halle: M Niemeyer, 1895 – 2mf – 9 – 0-7905-4854-2 – (incl bibl ref) – mf#1988-0854 – us ATLA [241]

Ignatov, S S *see* E T A hoffmann

Ignatovskii, I S *see* Mezhevye akty i ukrieplenie votchinnykh prav na nedvizhimyia imieniia. v dvukh chastiakh. i. s. ignatovskago

Ignatz kauffmann, 1849-1913 / Kauffmann, Ignatz – Frankfurt am Main, Germany. 1928 – 1r – us UF Libraries [920]

Ignatz kauffmann, 1849-1913 / Kauffmann, Ignatz – Frankfurt am Main, Germany. 1928 – 1r – us UF Libraries [939]

Ignaz doellingers briefe an eine junge freundin / ed by Schroers, Heinrich – Kempten: J Koesel, 1914 [mf ed 1991] – 1mf – 9 – 0-524-00987-2 – (in german & english) – mf#1990-0264 – us ATLA [860]

Ignaz von doellinger : sein leben auf grund seines schriftlichen nachlasses / Friedrich, Johann – Muenchen: Beck, 1899-1901 – 5mf – 9 – 0-7905-4642-6 – (incl bibl ref) – mf#1988-0642 – us ATLA [920]

Ignis / Albuerme Brea, P E – Hato Mayor del Rey, Dominican Republic. 1940 – 1r – us UF Libraries [972]

Ignis / Albuerme Brea, P E – Hato Mayor del Rey, Dominican Republic. 1940 – 1r – us UF Libraries [972]

Ignorance productive of atheism, faction, and superstition / Rennell, Thomas – London, England. 1798? – 1r – us UF Libraries [240]

Ignorance productive of atheism, faction, and superstition / Rennell, Thomas – London, England. 1798? – 1r – us UF Libraries [240]

La ignorancia del derecho, con un amplio estudio preliminar / Costa y Martinez, Joaquin – Buenos Aires, Editorial Partenon, 1945. 158p. LL-4093 – 1 – us L of C Photodup [340]

Ignosh, Jr, Raymond D *see* The physiological effects of cycling in three handlebar positions on trained male cyclists

Le "ignota litteratura" de jean wenck de herrenberg contre nicolas de cuse / Vansteenberghe, E – 1910 – (= ser Bgphma 8/6) – €3.00 – ne Slangenburg [100]

Ignotus *see*
- The pioneer
- Popery

Ignotus [pseud] *see* Foreshadowings

Igor stravinsky : le sacre du printemps, an harmonic analysis / Butler, William Edward – U of Rochester 1941 [mf ed 19–] – 2mf – 9 – mf#fiche383 – us Sibley [780]

Igor stravinsky's symphony of psalms : an analytical study / Mattei, Otto A – U of Rochester 1948 [mf ed 19–] – 3mf – 9 – mf#fiche 793 – us Sibley [780]

Igra = The performance – Moscow. n1-3. 1918 – 6mf – 9 – us UMI ProQuest [790]

Igreja 2001 / Aradillas Agudo, Antonio – Lisboa: Liber, 1975 – 1 – sp Bibl Santa Ana [946]

Igreja no brasil / Plaggge, Winfredo – Louvain, Belgium. 1965 – 1r – us UF Libraries [972]

Igreja no brasil / Plaggge, Winfredo – Louvain, Belgium. 1965 – 1r – us UF Libraries [972]

Igrejas de sao paulo : introducao ao estudo dos templos mais caracteristicos de sao paolo nas suas relacacoes com a cronica da cidade / Arroyo, Leonardo – Rio de Janeiro: Livraria Jose Olympio Editora, 1954 – 1 – us CRL [240]

Igrejas, desenvolvimento e participacao popular : consulta latino-americana sobre a "participacao das igrejas em programas e projetos de desenvolvimento," itaici, brasil, setembro de 1980 – Rio de Janeiro: Contro Ecumenico de Documentacao e Informacao: Tempo e Presenca Editora, 1981 – us CRL [972]

Igret erets yisrael / Yaari, Abraham – Tel-Aviv, Israel. 1942/43 – 1r – us UF Libraries [939]

Igret erets yisrael / Yaari, Abraham – Tel-Aviv, Israel. 1942/43 – 1r – us UF Libraries [939]

Igrot ba'al ha-tanya u-vene doro / Shneur Zalman – Jerusalem, Israel. 1953 – 1r – us UF Libraries [939]

Igrot ba'al ha-tanya u-vene doro / Shneur Zalman – Jerusalem, Israel. 1953 – 1r – us UF Libraries [939]

La igualdad – Madrid, Spain. -d. 2 Jan 1872-30 Dec 1874. 3 reels – 1 – uk British Libr Newspaper [072]

Iguaniona / Angulo Guridi, Javier – Trujillo, Peru. 1953 – 1r – us UF Libraries [972]

Iguaniona / Angulo Guridi, Javier – Trujillo, Peru. 1953 – 1r – us UF Libraries [972]

IH farm forum : the international harvester magazine / International Harvester Co – 1982-1985 spr – 1r – 1 – (cont: ih farm forum; cont by: case international farm forum) – mf#1006628 – us WHS [630]

Ihaza, Daniel E *see* Foreign trade of nigeria

Ihering, Rudolf von *see*
- Law as a means to an end
- The struggle for law

El-ihia : revue arabe litteraire – Alger. n1-7. fevr-mai 1907 [biwkly] – 1 – fr ACRPP [470]

Ihja ulumiddin – Djakarta, 1970/1971. v1(1-8) – 11mf – 9 – mf#SE-150-3 – ne IDC [959]

Ihl, Ralf *see* Demenz vom alzheimer typ

Ihme, Gertrud *see* Die theoretischen auffassungen vom erfolg der volkswirtschaft

Ihme, Heinrich *see* Der volksbegriff der deutschen volkskunde in seiner geschichtlichen entwicklung

Ihmels, Ludwig *see*
- Die christliche wahrheitsgewissheit
- Das evangelium von jesus christus
- Wer war jesus? was wollte jesus?

Ihne, Wilhelm *see* Roemische geschichte

Ihr aber steht im licht : eine dokumentation aus sowjetischem und sowjetzonalem gewahrsam / Pfoertner, Kurt & Natonek, Wolfgang; ed by Vereinigung der Opfer des Stalinismus – Tuebingen: F Schlichtenmayer, 1962 – 1r – 1 – (incl bibl ref and index) – us Wisconsin U Libr [947]

Ihr lebt! : hermann loens, walther flex zum gedaechtnis vortragsfolge einer totenfeier – Berlin: Verlag der Jugendlese [1917?] [mf ed 1996] – 1r – 1 – (comm by hermann bousset. filmed with: mein blaues buch / hermann loens) – mf#3942p – us Wisconsin U Libr [920]

Ihr mittel : lustspiel in vier aufzuegen / Reitler, Marzellin Adalbert – [Wien: J N Vernay, 188-?] – 1r – 1 – us Wisconsin U Libr [820]

Die ihren gott liebende seele : vorgestellt in den sinnbildern des hermanni hugonis ueber seine pia desideria, und des ottonis vaenii, ueber die liebe gottes... / Hugo, Hermannus & Vaenius, Othon – Regensburg: Verlegt von Emerich Felix Bader, 1743 – 5mf – 9 – mf#O-11 – ne IDC [090]

Ihrer vier : leben und ende einiger junger missionskaufleute / Schneider, Hermann G – Herrnhut: Verlag der Missionsbuchhandlung, [1903?] – 1mf – 9 – 0-7905-6260-X – mf#1988-2260 – us ATLA [240]

Ihsai yillik – 1928 – (= ser Ministry and special interest salnames) – 5mf – 9 – $150.00 – us MEDOC [956]

Ihsan (Harnamizade) *see* The divan project

Ihsan, Mustafa *see* Posta rehberi

Ihya' 'ulum al-din / Ghazzali – Misr, Egypt. v1-4. 1939 – 1r – us UF Libraries [956]

Ii rota overo dell'imprese. dialogo del s. scipione ammirato / Ammirato, S, the Elder – Napoli: [Apresso Gio. Maria Scotto], 1562 – 3mf – 9 – mf#O-852 – ne IDC [090]

Iie solutions – Norcross. 1995+ (1) 1995+ (5) 1995+ (9) – (cont: industrial engineering) – ISSN: 1085-1259 – mf#5936,01 – us UMI ProQuest [620]

Iie transactions / Institute of Industrial Engineers – Norcross. 1982+ (1,5,9) – (cont: aiie transactions) – ISSN: 0740-817X – mf#3180,01 – us UMI ProQuest [620]

Iii-vs review – Oxford. 1991-1991 (1,5,9) – ISSN: 0961-1290 – mf#42623,01 – us UMI ProQuest [621]

Iileka, David *see* A critical study of christian eschatology in the light of marxist thought

Iintsomi / Agar-O'connell, R M – S.I., S.I? . 19– – 1r – us UF Libraries [960]

Iintsomi / Agar-O'connell, R M – S.I., S.I? . 19– – 1r – us UF Libraries [960]

Ik beschuldig... / Mangoenkoesomo, T – Soerakarta, 1915 – 1mf – 8 – mf#SE-1295 – ne IDC [959]

Ikabala Singha *see* The ardent pilgrim

Ikarut / Levontin, Jehiel Joseph – Vilna, Lithuania. 1911 – 1r – us UF Libraries [939]

Ikarut / Levontin, Jehiel Joseph – Vilna, Lithuania. 1911 – 1r – us UF Libraries [939]

Ikatan akuntan indonesia – Djakarta, 1962-1968 – 38mf – 9 – (missing: 1962(3); 1963(1)) – mf#SE-257 – ne IDC [650]

Ikatan buruh pantjasila dpwx / Derap – Semarang, [1967]. v1-4 – 2mf – 9 – (several issues missing) – mf#SE-1399 – ne IDC [950]

Ikatan Dokter Indonesia *see* Berita

Ikatan geograf indonesia / Laporan IGI – Djakarta, 1969 – 1mf – 9 – mf#SE-150-4 – ne IDC [915]

Ikatan guru marhaenis / Suluh-pendidikan – Djakarta, 1958(1-2) – 2mf – 9 – mf#SE-431 – ne IDC [950]

Ikatan hakim indonesia / Varia peradilan – Semarang, 1961-1966. v1-6(8) – 13mf – 9 – (missing: 1961/1962, v1(1-12); 1964, v3(7-9); 1964, v4(1-3); 1965, v5(1-12)) – mf#SE-933 – ne IDC [959]

Ikatan Indonesia Untuk Perserikatan Bangsa-Bangsa *see* Good neighbourship among nations

Ikatan karyawan museum : kehidupan di museum; kumpulan karangan – Djakarta, 1967(1-5) – 2mf – 9 – mf#SE-1738 – ne IDC [950]

Ikatan karyawan museum / Manusia Indonesia. madjalah penggali budaja – Djakarta, 1967-1971. v1-5(4/6) – 28mf – 9 – mf#SE-1793 – ne IDC [060]

Ikatan kesedjahteraan keluarga hankam – Djakarta, April, 1967. v1(1-8) – 9mf – 9 – mf#SE-1848 – ne IDC [950]

Ikatan Pegawai Muda Departemen Penerangan *see* Suara berkala

Ikatan penerbit indonesia / Suara penerbit Indonesia – Djakarta, 1952-1968 – 30mf – 9 – (missing: 1952-1953, v1-2; 1955, v3; 1962-1963, v13; 1966, v16) – mf#SE-642 – ne IDC [959]

Ikatan penggemar mobil djakarta / Mobil kita – Djakarta, 1963-1964 – 7mf – 9 – (missing: 1963 v1(1, 10-12)) – mf#SE-928 – ne IDC [950]

Ikatan Sardjana Ekonomi *see* Ekonomi

Ikatekism yemfundiso yetyalike ekatolike yaseroma / Hornig, Josef – Mariannhill, South Africa. 1926 – 1r – us UF Libraries [960]

Ikatekism yemfundiso yetyalike ekatolike yaseroma / Hornig, Josef – Mariannhill, South Africa. 1926 – 1r – us UF Libraries [960]

Ikbal – Trabzon, 1908-19? Sahib-i Imtiyaz ve Mueduer-i Mes'ul: Eyuebzade A Nuri. n938. 20 haziran 1925; 961. 17 eyluel 1925 – (= ser O & t journals) – 1mf – 9 – $25.00 – us MEDOC [956]

L'ikdam : hebdomadaire de defense des interets musulmans nord-africains – El-Biar. mars 1919-avr 1923, mars 1931-janv 1935 [wkly] – 1 – (subtitle varies. suite de: L' islam) – fr ACRPP [320]

L'ikdam : organe de defense des interets des indigenes et des musulmans francais algeriens – Alger. n1-4. fevr-mars 1925 – 1 – fr ACRPP [320]

L' ikdam *see* L'islam

Ikelle-Matiba, Jean *see* Cette afrique-la!

Ikels, Marion *see* Die qualitaet von conjoint analysen

Iken, J Friedrich *see* Heinrich von zuetphen

Ikhwezi lase transkei – Umtata SA, 1966-76 – 1r – 1 – (title varies) – sa National [079]

Ikhwezi likazulu / Sikakana, J Mandlekosi – Johannesburg, South Africa. 1966 – 1r – us UF Libraries [960]

Ikhwezi likazulu / Sikakana, J Mandlekosi – Johannesburg, South Africa. 1966 – 1r – us UF Libraries [960]

Ikhwezi lomso – Queenstown: Ikhwezi Lomso, sep, nov 1958; feb, may 1959; feb 1960 – 1r – 1 – us CRL [079]

IKIP Kristen Satya Watjana *see* Warta satyawatjana

Ikonographie der christlichen kunst... / Kuenstle, K – Freiburg im Breisgau, 1926-1928. 2v – 23mf – 9 – mf#O-328 – ne IDC [700]

Ikonologisches woerterbuch... / P[rezel, L] de – Gotha, 1759 – 5mf – 9 – mf#O-1261 – ne IDC [700]

Iko-tams bulletin / Indiana, Kentucky, Ohio-Token and Medal Society – 1978 oct [v1 n1]-1985 mar – 1r – 1 – mf#941872 – us WHS [730]

Il – St Petersburg, Moscow, 1913-17 – 1r – 1 – us UMI ProQuest [077]

Il Marco Polo *see* A bulletin for the promotion of the italo-indonesian trade and cultural relations

Il Verso, Antonio *see* Di antonio il verso siciliano della citta di piazza. il primo libro de' madrigali a cinqve voci. nouamente date in luce

Ila made easy : language of the baila of northern rhodesia / Smith, Edwin William – Kasenga, NR [Zambia]: Book Room of the Baila-Batonga Mission, 1914 – 1 – us CRL [490]

Ilakecari – Chunnakam, Sri Lanka. 1932-58 – 12r – 1 – us L of C Photodup [079]

Ilan hacin-i gisun kamcibuha tuwara de ja obuha bithe *see* San he bian lan

Ilanga lase natal – Durban, South Africa. 1903-1978 – 89r – 1 – sa National [079]

Ilar (lower) hundred : parishes covered: cilcennin, henfynyw, llanbadarn trefeglwys, llanddeiniol, llanddewi aberarth, llangwyryfon, llanrhystud, llansantffraid, llanychaiarn, trefilan – (= ser Index to burials 1813-1837, 1838-1865 and 1866-1920) – 9 – (v3: 1813-37 [2mf] £3; v9: 1838-65 [2mf] £3; v15: 1866-1920 [3mf] £3) – uk Cardiganshire [941]

Ilar (upper) hundred : parishes covered: eglwys newydd, gwnnws, llanafan, llanfihangel-y-creuddyn, llanilar, lledrod, rhostie, ysbyty ystwyth, ystrad meurig – (= ser Index to burials 1813-1837, 1838-1865 and 1866-1920) – 9 – (v2: 1813-37 [1mf] £2; v8: 1838-65 [2mf] £3; v14: 1866-1920 [2mf] £2.50) – uk Cardiganshire [941]

Ilarii, Ieromonakh *see* Opisanie slavianskikh rukopisei biblioteki sviato-troitskoi sergievoi lavry

Ila-speaking peoples of northern rhodesia / Smith, Edwin William – London, England. v1-2. 1920 – 1r – us UF Libraries [306]

Ila-speaking peoples of northern rhodesia / Smith, Edwin William – London, England. v1-2. 1920 – 1r – us UF Libraries [470]

Ilaveli nuezhet uel-muenseat / Rifat, Ilaveli – [Istanbul]: Sirket Sahafiye Osmaniye Matbaasi, 1318 [1901] – (= ser Ottoman literature, writers and the arts) – 2mf – 9 – $40.00 – us MEDOC [470]

Ilberg, J *see* Neue jahrbuecher fuer das klassische altertum, geschichte und deutsche literatur und fuer paedagogik

Ilbert, Courtenay *see* Legislative methods and forms

Ilbert, Peregrine *see* Bengal tenancy bill

Ilbudet – Goeteborg, Sweden. 1892 – 1r – 1 – sw Kungliga [078]

Ildefonso pons – Minorca, Spain. v158-161. 1761-1800 – 2r – (gaps) – us UF Libraries [324]

Ildefonso pons – Minorca, Spain. v158-161. 1761-1800 – 2r – (gaps) – us UF Libraries [324]

L'ile de tsong-ming : a l'embouchure du yang-tse-kiang / Havret, Henri – 2nd ed. Chang-hai: Impr de la Mission Catholique, 1901 [mf ed 1995] – (= ser Yale coll; Varietes sinologiques 1) – 1 – 0-524-09717-8 – mf#1995-0717 – us ATLA [241]

L'ile d'orleans / Bois, Louis-Edouard – Quebec: A Cote, 1895 – 2mf – 9 – (incl ind) – mf#00167 – cn Canadiana [917]

Ile federee francaise de la martinique / Gratiant, Gilbert – Paris, France. 1961 – 1r – us UF Libraries [972]

Ile federee francaise de la martinique / Gratiant, Gilbert – Paris, France. 1961 – 1r – us UF Libraries [972]

Ile magique / Seabrook, William – Paris, France. 1932 – 1r – us UF Libraries [972]

Ile magique / Seabrook, William – Paris, France. 1932 – 1r – us UF Libraries [972]

L'ile sonnante : petite revue des lettres – Paris. n1-32. nov 1909-13 – 1 – fr ACRPP [800]

Ileach / Islay Council of Social Service – Bowmore: Islay Council of Social Service [1973]- – 1 – uk Scotland NatLib [360]

Ileri – Manastir: Neyyir-i Hakikat Matbaasi. Sahib-i Imtiyaz ve Mueduer-i Mes'ul: Matli Ziya, 1911. n3 (22) 21 temmuz 1327 [1911] – (= ser O & t journals) – 1mf – 9 – $25.00 – (cont: suengue) – us MEDOC [956]

Ileri, Celal Nuri *see*
- Hatem uel-enbiya
- Kara tehlike
- Tuercemiz

Iles, George *see* The reader's guide in economic, social and political science

Iles samoa : notes pour servir a une monographie de cet archipel / Marques, A – Lisbon: Impr Nationale, 1899 – 3mf – 9 – $4.50 – mf#LLMC 82-100C Title 27 – us LLMC [980]

Les iles samoa ou des navigateurs : le conflict entre les etats-unis et l'allemagne et la nouvelle conference de berlin / Lort-Serignam, Arthur M T (pseud Arthur de Ganniers) – Paris: Charles Bayle, 1889 – 1mf – 9 – $1.50 – mf#LLMC 82-100C Title 25 – us LLMC [980]

Les iles wallis : histoire et ethnologie / Renaud, Georges J L – 1932-33 – 1r – 1 – mf#pmb doc3 – at Pacific Mss [305]

Ilesha – and beyond! : the story of the wesley guild medical work in west africa / Souton, Arthur E – London: Cargate Press, [193-?] – 1 – us CRL [960]
Ilford and barking independent – Redbridge, England 4 oct 1984-29 nov 1990 [mf 1986-91] – 1 – (cont: ilford independent [5 may 1982-27 sep 1984]; cont by: ilford & redbridge independent [6 dec 1990-11 apr 1991]) – uk British Libr Newspaper [072]
Ilford and manor park news – London, UK. 17 mar 1900-17 aug 1901 – 1/4r – 1 – uk British Libr Newspaper [072]
Ilford and seven kings mercury – London, UK. 10 jul 1901-2 apr 1902 – 1/4r – 1 – uk British Libr Newspaper [072]
Ilford argus – London, UK. Oct 1921-31 mar 1928 – 2r – 1 – uk British Libr Newspaper [072]
Ilford independent – Redbridge, England 5 may 1982-27 sep 1984 – 1 – (cont: ilford advertiser [16 feb [4 may 1982]; cont by: ilford & barking independent [4 oct 1984-29 nov 1990]) – uk British Libr Newspaper [072]
Ilford leader – sep-oct 1994; 1995-19 dec 1997; jan-dec 1998 – 13 1/2r – 9 – (aka: herald ilford barking & dagenham; leader ilford barking dagenham) – uk British Libr Newspaper [072]
Ilford monthly – London, UK. 1925-30 – 3r – 1 – uk British Libr Newspaper [072]
Ilford observer – London, UK. 23 sep-4 nov 1919 – 1/4r – 1 – uk British Libr Newspaper [072]
Ilford recorder see Redbridge ilford recorder
Ilford recorder etc see Redbridge ilford recorder
Ilford redbridge post see Redbridge post
Ilford redbridge recorder see Redbridge ilford recorder
Ilford saturday post see Ilford saturday post and amusement guide
Ilford saturday post and amusement guide – London, UK. 1 mar-5 jul 1919 – 1/4r – 1 – (aka: ilford saturday post) – uk British Libr Newspaper [072]
Ilford yellow advertiser see
– Yellow advertiser (ilford ed)
– Yellow advertiser (redbridge edn)
Ilfracombe chronicle – England.1872-24 Jun 1876. -w.4 1/2 reels – 1 – uk British Libr Newspaper [072]
Ilfracombe gazette and observer – England.1897-98; 1901; Jan-25 Oct 1912. -w. 4 reels – 1 – uk British Libr Newspaper [072]
Ilfracombe observer and north devon review – [SW England] Devon apr 1884-27 jun 1893 [mf ed 2002] – 10r – 1 – uk Newsplan [072]
Ilfracombe times, arrival list and abc time table – Ilfracombe, England 4 jul 1905-8 jan 1907 – 1 – uk British Libr Newspaper [072]
Ilg, A see
– Beitraege zur geschichte der kunst und der kunsttechnik aus mittelhochdeutschen dichtungen
– Heraclius
– Theophilus presbyter schedula diversarun artium
Ilgen, Pedro see
– Stechaepfel
– Tiefgluth
– Unter westlichen sternen
– Welt- und gottesreichsklaenge
Ilgenstein, Heinrich see Wilhelm von polenz
Ilgwu news / International Ladies' Garment Workers' Union – [v37 n3-7,9-12 [1984 jul 13-nov 9, 1985 jan 11-apr 14]]-[v40 n5 [1987 sep 11]] [gaps] – 1r – 1 – mf#1569846 – us WHS [331]
Ilha de s thomea roca agua-ize / Sousa E Faro, Conde De – Lisboa, Portugal. 1908 – 1r – us UF Libraries [025]
Ilha de s thomea roca agua-ize / Sousa E Faro, Conde De – Lisboa, Portugal. 1908 – 1r – us UF Libraries [025]
Ilha de sao tome / Tenreiro, Francisco – Lisboa, Portugal. 1961 – 1r – us UF Libraries [960]
Ilha de sao tome / Tenreiro, Francisco – Lisboa, Portugal. 1961 – 1r – us UF Libraries [960]
Ilha grande / Lessa, Origenes – Sao Paulo, Brazil. 1933 – 1r – us UF Libraries [972]
Ilha grande / Lessa, Origenes – Sao Paulo, Brazil. 1933 – 1r – us UF Libraries [972]
The I-li : ceremonial de la chine antique – Paris: Jean Maisonneuve, 1890 – 1mf – 9 – 0-524-08002-X – mf#1991-0224 – us ATLA [390]
The I-li : or, book of etiquette and ceremonial – London: Probsthain, 1917 – (= ser Probsthain's oriental series) – 2mf – 9 – 0-524-08001-1 – mf#1991-0223 – us ATLA [390]
Iliamna volcano and its basement / Juhle, Werner – 1957 – (= ser Volcanology-archaelogy coll) – 1r – 1 – $50.00 – mf#B70064 – us Library Micro [550]
Iliffe, J H see A short guide to the exhibition of the palestine archaeological museum, illustrating the stone and bronze ages in palestine
Iliffe, John see Tanganyika under german rule, 1905-1912
Ilim fen felsefe tetebbuati mecmuasi – Ankara. Umur-i Tahririye Muediri: Hakki Baha. Umur-i Edebiye Muedirir: Ahmed Edib. n1. haziran 1338 [1922] – (= ser O & t journals) – 2mf – 9 – $40.00 – us MEDOC [956]
Ilimskii, D I see
– Kooperativnye soiuzy v sibiri 1908-1918 gg
– Ocherki po teorii kooperatsii
– Sovet vserossiiskikh kooperativnykh sezdov
Ilin, A A see Topogradiia kladov serebrianykh i zolotykh slitkov
Il'in, V see Sibirskii kazak
Il'in, Valilii see O poklonenii bogo ottstu v" vukhe i istine
Iliniwek – v1 n1-v2 n5 [1963 may/jun-1964 nov/dec], v5 n1-v6 n5 [1967 jan/feb-1968 nov/dec], 1972 sep/oct [v10 n4]-1976 mar, 1976 apr-1976 dec [v14 n3&4] – 3r – 1 – mf#370660 – us WHS [071]
Ilinskii, G A see Okhridskie glagolicheskie listki
Il'inskii, Grigorii Andreevich see O niekotorykh arkhaizmakh i novoobrazovaniiakh praslavianskago iazyka
Ilios, the city and country of the trojans : the results of researches and discoveries on the site of troy and throughout the troad in the years 1871-72-73-78-79, including an autobiography of the author = Ilios, stadt und land der trojaner / Schliemann, Heinrich – 2mf – 9 – 0-524-05061-9 – (in english) – mf#1992-0314 – us ATLA [930]
Ilk adim – Nigde, 1926-19? Sahib-i Imtiyaz: Hilmi Gueltekin; Mueduer-i Mes'ul: Kemal Turgut. n4. 15 subat 1926 – (= ser O & t journals) – 1mf – 9 – $25.00 – us MEDOC [956]
Ilkeston advertiser – Ilkeston, Derbyshire [mf ed 19 jul 1881-1911, 1914-to date] – 1 – (microfilm only 1897) – uk Newsplan [072]
Ilkeston advertiser and erewash valley weekly news – [East Midlands] Ilkeston, Derbyshire 9 jul 1881-dec 1950 [mf ed 2002] – 58r – 1 – (missing: 1897, 1912, 1913) – uk Newsplan [072]
Ilkeston gazette – [East Midlands] Ripley, Derbyshire 20 nov 1936-1 dec 1939 [mf ed 2004] – 3r – 1 – (publ for ilkeston division labour party; previous title: the gazette [n239 20 nov 1936-n397 1 dec 1939]) – uk Newsplan [072]
Ilkeston pioneer and erewash valley gazette – [East Midlands] Ilkeston, Derbyshire jan 1853-21 sep 1854, 25 apr 1856-23 mar 1967 (mthly to apr 1854] [mf ed 2002] – 78r – 1 – (not publ sep 1854-apr 1856; also a stapleford & sandiacre ed: span unknown) – uk Newsplan [072]
Ilkley gazette – England. -w. 1869-84. (5 reels) – 1 – uk British Libr Newspaper [072]
Illahun, kahun, and gurob, 1889-1890 / Petrie, W M – London, 1891 – 3mf – 9 – mf#NE-20340 – ne IDC [956]
Illawarra mercury – Wollongong, jan 1856-dec 1968 – 144r – A$4752.00 vesicular A$5544.00 silver – at Pascoe [079]
Illawarra mercury – Wollongong, jan 1969-apr 1997 – at Pascoe [079]
Illegaltitaet see Die rote fahne
Illertalbote – Dietenheim DE, 1975- – 112r [1975-90] – 1 – (bezirksausgabe von suedwest-presse, ulm) – gw Misc Inst [074]
Illgen, Christian Friedrich see Symbolarum ad vitam et doctrinam laelii socini
Ill-ia-mo searcher – v4 n1-v15 n4=n10-52 [1976 mar-1987 oct] – 1r – 1 – mf#1685662 – us WHS [071]
Illich-Svitych, V M see Opyt sravneniia nostraticheskikh iazykov
Illicit liquor problem on the witwatersrand / South African Temperance Alliance – Lovedale, South Africa. 1935 – 1r – us UF Libraries [960]
Illicit liquor problem on the witwatersrand / South African Temperance Alliance – Lovedale, South Africa. 1935 – 1r – us UF Libraries [960]
Illiger, K see Magazin der insektenkunde
Illingworth, John Richardson see
– Christian character
– Divine transcendence
– The doctrine of the trinity
– The gospel miracles
– Personality, human and divine
– Reason and revelation
– Sermons
– University and cathedral sermons
Illini times – Champaign, IL. 1982-1985 (1) – mf#62529 – us UMI ProQuest [071]
Illinois – 1977 aug/sep-oct, 1977 nov-1981, 1982-1983 may/jun – 3r – 1 – (cont: outdoor illinois; cont by: illinois magazine) – mf#396787 – us WHS [071]
Illinois : session laws of american states and territories – 1809-1997 – 9 – $3,393.00 set – mf#402640 – us Hein [348]
Illinois : smith-hurd illinois compiled statutes annotated – St Paul: West Pub Co, 1934-aug 1999 update – 9 – $6,093.00 set – mf#401161 – us Hein [348]
Illinois see
– Matthews and bangs' circuit court reports
– Reports and opinions
– Reports, pre-nrs
– State reports, post-nrs
Illinois advocate – Edwardsville: S S Brooks, feb 1831-apr 1833 – (filmed with: illinois advocate and state register (vandalia, il)) – us CRL [071]
Illinois advocate – Vandalia: J Y Sawyer, apr 15 1835-mar 16 1836 – us CRL [071]
Illinois advocate and state register – Vandalia: J Y Sawyer, apr 1833-apr 1835 – 3r – 1 – (filmed with: illinois advocate (edwardsville, il)) – us CRL [071]
Illinois agri news – LaSalle, IL. 1977-1984 (1) – mf#68085 – us UMI ProQuest [071]
Illinois antiquity : quarterly news letter of the illinois association for advancement of archeology / Illinois Association for Advancement of Archaeology – [v11 n1] 1979 jan-1981 – 1r – 1 – (cont: quarterly newsletter [illinois association for advancement of archaeology]) – mf#619797 – us WHS [930]
Illinois. Appellate Court see Illinois appellate reports
Illinois appellate reports / Illinois. Appellate Court – v1-240. 1877-1926 – 1910mf – 9 – $2865.00 – (all vols are pre-nrs. vols after v215 are still in copyright and will be filmed as they become available) – mf#LLMC 80-803 – us LLMC [340]
Illinois Association for Advancement of Archaeology see
– Illinois antiquity
– Quarterly newsletter
Illinois attorney general reports and opinions – 1872-1995 – 6,9 – $630.00 set – (1872-1978 on reel $630.00. 1979-98 on mf $140.00) – mf#408220 – us Hein [340]
Illinois banner / Prohibition Party of Illinois – v20 n50-v25 n2 [1910 dec 15-1915 jan 14] – 1r – 1 – (cont: danville banner) – mf#939667 – us WHS [325]
The illinois baptist – Springfield, IL. 41,644p. 1906-1999 – 1 – mf#0423 – us Southern Baptist [242]
Illinois baptist bulletin – Normal/Joliet, IL. Illinois Baptist State Convention. 1909-35. Incomplete. Single reels available – 1 – $121.05 – us ABHS [242]
Illinois baptist bulletin – Upper Alton, IL.Baptist General Assoc. of Illinois. 1892-98. Incomplete – 1 – $15.90 – us ABHS [242]
Illinois baptist news/baptist news – Joliet/Springfield, IL. Illinois Baptist State Convention. 1935-70. Includes Chicago ed. Single reels available – 1 – us ABHS [071]
Illinois bar journal – v1-89. 1912-2001 – 5,6,9 – $1323.00 set – (v1-72 1912-84 on reel or mf $918. v73-89 1984-2001 on mf $405. title varies: v1-20 n1 1912-32 as illinois state bar association quarterly) – ISSN: 0019-1876 – mf#103241 – us Hein [340]
Illinois Bicentennial Commission see Illinois correspondent
Illinois biological monographs – Urbana, 1914-1922 [1,5,9] – ISSN: 0073-4748 – mf#2180 – us UMI ProQuest [574]
Illinois business review – Champaign. 1944-1996 [1]; 1971-1996 [5]; 1977-1996 [9] – ISSN: 0019-1922 – mf#1816 – us UMI ProQuest [338]
Illinois. Commerce Commission see
– Illinois electric utilities: a comparative study of electric sales statistics
– Illinois gas utilities: a comparative study of gas sales statistics
– Reports
Illinois common school advocate – Springfield. 1841-1841 – 1 – mf#5020 – us UMI ProQuest [370]
Illinois conservation – v1 n1-v13 n1 [1935/36 mid-winter-1948 sum] – 1r – 1 – (cont by: outdoors in illinois) – mf#1012308 – us WHS [574]
Illinois conservator – Springfield IL. 1929 jun 20 – 1r – 1 – mf#5258951 – us WHS [071]
Illinois. Consumer Credit Division see Analysis of reports filed by consumer finance and consumer in stallment loan companies
Illinois correspondent / Illinois Bicentennial Commission – v1-4 [1972 aut-1976 sum] – 1r – 1 – mf#370658 – us WHS [978]
Illinois council of deliberation news bulletin / Scottish Rite [Masonic order] – v1 [1978 spr] – 1r – 1 – mf#4720758 – us WHS [366]
Illinois Council of State Employees see Illinois state employee
Illinois dental journal – Springfield. 1931-1995 (1) 1973-1995 (5) 1975-1995 (9) – (cont by: illinois dental news) – ISSN: 0019-1973 – mf#8386 – us UMI ProQuest [617]
Illinois dental news – Springfield. 1995-1996 (1) 1996-1996 (5) 1996-1996 (9) – (cont: illinois dental journal) – ISSN: 1084-8282 – mf#8386,01 – us UMI ProQuest [617]
Illinois. Dept of Insurance see
– Reports
– Reports by the director
Illinois. Dept of Labor see Reports
Illinois. Dept of Labor. Division of Statistics and Research see Reports on compensable work injuries in illinois
Illinois. Dept of Revenue see Reports
Illinois education – Springfield. 1913-1972 (1) 1970-1972 (5) – ISSN: 0019-2007 – mf#359 – us UMI ProQuest [370]
Illinois electric utilities: a comparative study of electric sales statistics / Illinois. Commerce Commission – 1933-79. 41 fiches. (Harvard Law School Library Collection.) – 9 – us Harvard Law [336]
Illinois Federation of Teachers see Illinois higher education
Illinois Federation of Teachers et al see Illinois union teacher
Illinois free trader – Ottawa, IL. 1840-1943 (1) – mf#62673 – us UMI ProQuest [071]
Illinois gas utilities: a comparative study of gas sales statistics / Illinois. Commerce Commission – 1931-79. 44 fiches. (Harvard Law School Library Collection.) – 9 – us Harvard Law [336]
Illinois government / University of Illinois at Urbana-Champaign – n1-37 [1959 jan-1973 sep] – 1r – 1 – (cont by: illinois government research) – mf#664184 – us WHS [350]
Illinois government research / University of Illinois at Urbana-Champaign – n38-56 [1975 jan-1983 apr] – 1r – 1 – (cont: illinois government) – mf#664182 – us WHS [350]
Illinois higher education / Illinois Federation of Teachers – v1 n1-3 [1978 oct-1979 fall] – 1r – 1 – mf#584044 – us WHS [378]
Illinois Historic Preservation Agency see Historic Illinois
Illinois historical chronicle – 1976 aug 16-1977 feb 7 – 1r – 1 – mf#1058678 – us WHS [978]
Illinois historical journal – Springfield. 1984-1998 (1) 1984-1998 (5) 1984-1998 (9) – (cont: journal of the illinois state historical society. cont by: journal of the illinois state historical society) – ISSN: 0748-8149 – mf#816,01 – us UMI ProQuest [978]
Illinois insurance – Springfield. 1979-1980 (1,5,9) – ISSN: 0094-7660 – mf#12254 – us UMI ProQuest [368]
Illinois intelligencer – Kaskaskia, Vandalia IL. 1820 jul 1-1822 mar 5 – 1r – 1 – (cont: western intelligencer [kaskaskia, il]; cont by: illinois whig; vandalia whig and illinois intelligencer) – mf#780499 – us WHS [071]
Illinois intelligencer – IL, v149 n4-v151 n28 [1967 jan 25-1969 jun 30] – 1r – 1 – mf#639201 – us WHS [071]
Illinois journal of mathematics – Urbana. 1957+ (1) 1957+ (5) 1957+ (9) – ISSN: 0019-2082 – mf#6102 – us UMI ProQuest [510]
Illinois labor – Ruskin TN. 1896 dec 12-19 – 1r – 1 – us WHS [331]
Illinois labor press – v2 n44-v3 n26 [1918 aug 30-1919 may 2, scattered iss] – 1r – 1 – mf#1058680 – us WHS [331]
The illinois law of voluntary assignments for the benefit of creditors / Candlish, William James – Chicago: Callaghan, 1896. 222p. LL-659 – 1 – us L of C Photodup [346]
Illinois law quarterly – Urbana. v1-6. 1917-24 (all publ) – (= ser Historical legal periodical series) – 1 – $50.00 – (available on reel only) – mf#103251 – us Hein [340]
Illinois law quarterly / University of Illinois College of Law, Urbana – v1-6 1917-1920/24 – 18mf – 9 – $27.00 – (v1-3 are titled: illinois law bulletin; title eventually was changed to: university of illinois law forum; add vols to be filmed) – mf#llmc95-104 – us LLMC [340]
Illinois law review – Northwestern University Law School. v1-6. 1906-12 – 42mf – 9 – $63.00 – (more vols planned as copyright expires) – mf#LLMC 84-485 – us LLMC [340]
Illinois law review see Northwestern university law review
Illinois. Laws, Statutes, etc see Private and general laws relating to the chicago, burlington and quincy railroad, 1867
Illinois libertarian / Libertarian Party of Illinois – 1975 jan-1991 dec – 1r – 1 – mf#1066732 – us WHS [325]
Illinois libraries – Springfield. 1919-2001 [1]; 1969-2001 [5]; 1977-2001 [9] – ISSN: 0019-2104 – mf#2011 – us UMI ProQuest [020]
Illinois magazine – 1983 jul/aug-1986 nov/dec – 1r – 1 – (cont: illinois) – mf#1511635 – us WHS [071]
Illinois medical journal – Chicago. 1899-1988 (1) 1967-1988 (5) 1974-1988 (9) – (cont by: illinois medicine) – ISSN: 0019-2120 – mf#3223 – us UMI ProQuest [610]
Illinois medicine – Chicago. 1989-1999 (1,5,9) – (cont: illinois medical journal) – ISSN: 1044-6400 – mf#17146 – us UMI ProQuest [610]
Illinois Mennonite Historical Society et al see Mennonite heritage
Illinois mobilizes / Illinois Veterans' Commission et al – v1 n1-v4 n4 [1942 aug-1946 feb] – 1r – 1 – (cont by: illinois mobilizes for its vets) – mf#700699 – us WHS [350]
Illinois mobilizes for its vets / Illinois Veterans' Commission – v4 n5-v7 n4 [1946 apr-1949 jan/feb], v7 n5 [1949 special report] – 1r – 1 – (cont: illinois mobilizes) – mf#700761 – us WHS [350]

Illinois monthly magazine – Vandalia; Cincinnati. 1830-1832 – 1 – mf#3994 – us UMI ProQuest [073]
Illinois. Office of Secretary of State. Securities Division *see* Reports
Illinois. Office of the Commissioner of Banks and Trust Companies *see* Reports
Illinois patriot – Jacksonville IL. 1832 feb 16 v1 n22, mar 8 v1 n25, mar 22 v1 n27, mar 29 v1 n28 – 1r – 1 – (cont: illinoian) – mf#846370 – us WHS [071]
Illinois pharmacist – Chicago. 1972-1973 (1) 1972-1972 (5) (9) – ISSN: 0019-2163 – mf#6938 – us UMI ProQuest [615]
Illinois quarterly – Normal. 1938-1982 (1) 1975-1982 (5) 1975-1982 (9) – ISSN: 0019-2295 – mf#8256 – us UMI ProQuest [300]
Illinois. Railroad and Warehouse Commission *see* Reports
Illinois school research and development – Urbana. 1976+ – 1,5,9 – ISSN: 0163-822X – mf#11627,01 – us UMI ProQuest [370]
Illinois schools journal – Chicago. 1955+ (1) 1971+ (5) 1976+ (9) – ISSN: 0019-2236 – mf#975 – us UMI ProQuest [370]
Illinois. Second Presbytery (Associate Reformed Church) *see* Minutes, 13 april 1852-13 april 1859
Illinois sentinel – Chicago IL. 1937 nov 20 – 1r – 1 – mf#5013185 – us WHS [071]
Illinois staats-herold – Chicago, apr 1932 – 1r – 1 – (bound with staats-herold) – us CRL [071]
Illinois staats-herold – Chicago IL (USA), 1929 6 dec-1931 9 oct, 1933 17 feb-1935 27 dec – 5r – 1 – gw Misc Inst [071]
Illinois staats-zeitung – Chicago, IL: [Hoffgen, 1871-87; may 1888-1900] – 1 – us CRL [071]
Illinois staats-zeitung – Chicago IL. 1915 nov 21, 1916 sep 1-24, oct 15-22, 1917 mar 4 – 1r – 1 – (cont: taegliche illinois staats-zeitung; chicagoer freie presse [chicago, il: 1882]; cont by: d a buergerzeitung; buergerzeitung und illinois staats-zeitung) – mf#2884098 – us WHS [071]
Illinois State Academy of Science *see* Transactions of the illinois state academy of science
Illinois State AFL-CIO [American Federation of Labor and Congress of Industrial Organizations] *see*
– Illinois state afl-cio laborletter
– Laborletter
– Weekly news letter
Illinois state afl-cio laborletter / Illinois State AFL-CIO [American Federation of Labor and Congress of Industrial Organizations] – 1989 oct-1993 – 1r – 1 – (cont: illinois state afl-cio news letter; cont by: laborletter) – mf#1557717 – us WHS [331]
Illinois state afl-cio news letter / Illinois State [American Federation of Labor and Congress of Industrial Organizations] – v1 n1-15 [1982 mar 13-1983 nov12], v3 n2-v6 n 5 [1984 feb 11-1987 nov 18], v7 n2-v8 n5 [1988 jan 27-1989 sep 1] – 1r – 1 – (cont: weekly news letter [illinois state afl-cio]; cont by: illinois state afl-cio laborletter) – mf#1415940 – us WHS [331]
Illinois State [American Federation of Labor and Congress of Industrial Organizations] *see* Illinois state afl-cio news letter
Illinois State Archives *see* For the record
Illinois. State Bar Association *see* Proceedings, 1877-1939
Illinois state bar association annual reports – 1st to 63rd meetings. 1877-1939 (all publ) – 262mf – 9 – $393.00 – mf#LLMC 84-486 – us LLMC [340]
Illinois state bar association quarterly *see* Illinois bar journal
Illinois state chronicle – Decatur, IL. 1855-1861 (1) – mf#62600 – us UMI ProQuest [071]
Illinois state employee / Illinois Council of State Employees – v1-v29 n1 [1945 sep-1974 dec] – 1r – 1 – mf#1058686 – us WHS [350]
Illinois State Federation of Labor *see*
– Eight hour herald
– President's report
– Proceedings...annual convention, illinois state federation of labor
– Weekly news letter
Illinois State Federation of Labor et al *see* Official labor gazette
Illinois State Grange *see*
– Journal of proceedings of the...annual session of illinois state grange, patrons of husbandry
– Proceedings of the state grange of illinois at the...annual session
Illinois State Historical Society *see*
– Journal of the illinois state historical society
– Pierre menard collection
Illinois state historical society. collections – v1-25 – 1 – $324.00 – mf#0272 – us Brook [978]
Illinois state journal : (4 star edition) – Springfield, IL. 1937-1942 (1) – mf#62692 – us UMI ProQuest [071]
Illinois state journal – Springfield, IL. 1831-1849 (1) – mf#60462 – us UMI ProQuest [071]
Illinois state journal – Springfield, IL. 1848-1974 (1) – mf#60461 – us UMI ProQuest [071]
Illinois. State Public Utilities Commission *see* Reports
Illinois state register – Springfield, IL. 1831-1851 (1) – mf#68737 – us UMI ProQuest [071]
Illinois state register – Springfield, IL. 1849-1974 (1) – mf#60463 – us UMI ProQuest [071]
Illinois. State Tax Commission *see* Reports
Illinois. Supreme Court *see* Illinois supreme court reports
Illinois supreme court reports / Illinois. Supreme Court – v1-321. 1819-1926 – 2499mf – 9 – $3748.00 – (pre-nrs: v1-112 1819 827mf $1,240.00. updates planned) – mf#LLMC 80-802 – us LLMC [347]
Illinois times – Springfield, IL. 1999-1999 (1) – mf#68190 – us UMI ProQuest [071]
Illinois tradesman / Springfield Federation of Labor [IL] – [1916 feb 2-1918 aug 28]-1954-1959 jun 25 – 16r – 1 – mf#1058690 – us WHS [331]
Illinois union teacher / Illinois Federation of Teachers et al – 1950 jan 1-1959 apr – 1r – 1 – (cont by: union news [oak brook, il]) – mf#370670 – us WHS [071]
Illinois valley community news – v1 n2-v2 n1 [1972 sep 5-1973 sep 11] – 1r – 1 – mf#1058692 – us WHS [071]
Illinois valley courier – Cave City OR: W Drews, 1935 [wkly] – 1 – (absorbed by: grants pass courier (1934-41)) – us Oregon Lib [071]
Illinois valley news – Cave City OR: Illinois Valley Pub Co, 1937- [wkly] – 1 – (absorbed: bulletin (grants pass, or: 1964)) – us Oregon Lib [071]
Illinois valley news (cave city, or) – Cave City OR: E H Weston, 1935- [wkly] – 1 – us Oregon Lib [071]
Illinois Veterans' Commission *see* Illinois mobilizes for its vets
Illinois Veterans' Commission et al *see* Illinois mobilizes
Illinois wochenblatt = Illinois weekly – Elgin, Springfield IL, Winona MN. 1920 oct 15-1921 oct 14 – 1r – 1 – (cont: elgin herold; staats-wochenblatt; zeitung [decatur, il]; volksblatt-rundschau) – mf#4137770 – us WHS [071]
Illinois wochenblatt = Illinois weekly – Winona: Westlicher Harold Pub Co, oct 1920-oct 1921 – 1r – 1 – us CRL [071]
Illinois Yearly Meeting [Society of Friends] *see* Minutes and accompanying documents of illinois yearly meeting of the society of friends
Illinoiser Volkeblatt Publishing Association *see* Neues leben
Illinois-konferensen, 1853-1903 / Scandinavian Evangelical Lutheran Augustana Synod of North America. Illinois-Konferensen – Rock Island, IL: Augustana Book Concern, [1904?] – 1mf – 9 – 0-524-08688-5 – mf#1993-3213 – us ATLA [240]
Illinois-staatszeitung – Chicago IL (USA), 1920 30 may, 14 jun, 12 sep, 1920 26 sep, 7 nov, 28 nov, 1921 27 & 28 jan – 1 – gw Misc Inst [071]
Illiustrirovannaia istoriia knigopechataniia i tipografskogo iskusstva : 1: s izobreteniia knigopechataniia po 18 vek vkliuchitelno / Bulgakov, F I – [1889] – 8mf – 8 – mf#R-4561 – ne IDC [947]
Illiustrirovannaia rossiia – Titre francais: La Russie illustree. Paris. 1924-avr 1939 – 1 – fr ACRPP [947]
Illiustrirovannoe ezhemesiachnoe izdanie : novyi zhurnal inostrannoi literatury, iskusstva i nauki – Champaign. 1944-1996 (1) 1971-1996 (5) 1977-1996 (9) – 644mf – 9 – (missing: 1901, v1(2); v2(6-8); 1907, v1(1-3), v2(6); 1908, v2(2-4), v3-4; 1909, v1(2-3), v2-4) – mf#1816 – ne IDC [077]
Illiustrirovannyi bibliograficheskii zhurnal – M., 1897-1916 – 310mf – 9 – (missing: 1905, v9) – mf#R-4313 – ne IDC [077]
Illiustrirovannyi dvukhnedelnyi vestnik sovremennoi zhizni, politiki, literatury, nauki, iskusstva i prikladnykh znanii – London. 1924-1996 (1) 1963-1996 (5) 1963-1996 (9) – 123mf – 9 – mf#1227 – ne IDC [077]
Illiustrirovannyi dvukhnedelnyi zhurnal dramy, operetki, farsa, teatra, varete, sporta i sinematografa – Spb., 1910-1912 – 57mf – 9 – (1910(9, 22-23); 1911(6, 10, 24); 1912(7, 12-24)) – mf#R-1525 – ne IDC [077]
Illiustrirovannyi dvukhnedelnyi zhurnal slovesnosti, nauki i bibliografii / Vestnik literatury – New York. 1964-1980 (1) 1971-1980 (5) 1973-1980 (9) – 12mf – 9 – mf#1969 – ne IDC [077]
Illiustrirovannyi literaturno-politicheskii shurnal – Medina. 1873-1992 (1) 1971-1992 (5) 1976-1992 (9) – 49mf – 9 – (missing: 1872(1, 14-15, 22-40); 1873(25-26)) – mf#1894 – ne IDC [077]
Illiustrirovannyi literaturnyi ezhemesiachnyi zhurnal / Beseda – Oxford. 1946+ (1) 1971+ (5) 1977+ (9) – 71mf – 8 – (preceded by: pochtal'on. spb, 1902-03. v4) – mf#1684 – ne IDC [073]
Illiustrirovannyi literaturnyi ezhemesiachnyi zhurnal – Spb., 1903, v4-1908, v1 – (= ser Pochtalon) – 71mf – 9 – (preceded by: pochtalon. spb., 1902-1903, v4. missing: 1905, v10; 1906, v1, 4-12; 1907, v1-9) – mf#1684 – ne IDC [077]
Illiustrirovannyi semeinyi zhurnal – Chicago. 1952-1995 (1) 1971-1995 (5) 1976-1995 (9) – 38mf – 9 – (missing: 1883-1885, v1-3; 1887-1888, v5-6) – mf#1842 – ne IDC [077]
Illiustrirovannyi vestnik dlia sobiratelei knig i graviur – Middletown. 1965-1974 (1) – 118mf – 9 – mf#1861 – ne IDC [077]
Illiustrirovannyi vestnik otchiznovedeniia, istorii, kultury, gosudarstvennoi, obshchestvennoi i ekonomicheskoi zhizni rossii / Zhivopisnaia Rossiia – Milano. 1976-1982 (1) 1976-1982 (5) 1976-1982 (9) – 162mf – 9 – (missing: 1905 v5) – mf#2021 – ne IDC [077]
Illiustrirovannyi voenno-obshchestvennyi zhurnal / Voennyi mir – M., 1911-1914 – 153mf – 9 – mf#R-4174 – ne IDC [077]
Illiustrirovannyi zhurnal – Lancaster. 1965+ (1) 1971+ (5) 1976+ (9) – 14mf – 9 – mf#1921 – ne IDC [077]
Illiustrirovannyi zhurnal : v mire iskusstv – Kiev, 1907-1910 – 54mf – 9 – mf#R-4199 – ne IDC [077]
Illiustrirovannyi zhurnal dlia semeinogo chteniia – Evanston. 1918+ (1) 1971+ (5) 1975+ (9) – 202mf – 9 – (missing: 1881(12); 1883(8-12); 1884-1912; 1915; 1917) – mf#1843 – ne IDC [077]
Illiustrirovannyi zhurnal iziashchnoi literatury i iskusstva – Memphis. 1940-1977 (1) 1964-1977 (5) 1970-1977 (9) – 126mf – 9 – (missing: 1892(5)) – mf#1697 – ne IDC [077]
Illiustrirovannyi zhurnal literatury, nauk i iskusstv – Houston. 1954-1988 (1) 1971-1988 (5) 1977-1988 (9) – 35mf – 9 – (missing: 1879; 1880(34); 1881-1883) – mf#1819 – ne IDC [077]
Illiustrirovannyi zhurnal literatury, politiki i sovremennoi zhizni – Spb, 1870-1919. v1-49 – 2171mf – 9 – (missing: 1870, v1(p 1-48, 145-352, 433-496, 513-576, 593-672, 737-784, 800-end); 1873, v4(p 65-80, 113-128, 561-576); 1874, v5; 1875, v6; 1881, v12; 1882, v13) – mf#1400, 1401 – ne IDC [077]
Illkirch-grafenstadener anzeiger – Illkirch, Elsass (F), 1912-1913 21 jun – 1 – fr ACRPP [074]
La illuminata de tutti i tuoni di canto fermo : con alcuni bellissimi secreti, non d'altrui piu scritti / Aiguino da Brescia, Illuminato – In Venetia: Per Antonio Gardano 1562 [mf ed 19–] – 2mf – 9 – (incl ind) – mf#fiche133 – us Sibley [780]
The illuminated book of needlework / Stone, Elizabeth – London 1847 – (= ser 19th c art & architecture) – 6mf – 9 – mf#4.2.421 – uk Chadwyck [740]
The illuminated books of the middle ages : an account of the...art of illumination / Humphreys, Henry Noel – London 1849 – (= ser 19th c art & architecture) – 10mf – 9 – mf#4.2.1729 – uk Chadwyck [740]
Illuminated byzantine gospels – 11th, 13th c – (= ser Holkham library manuscript books 3,4) – (complete mss 1r [5133]. illuminations only 1 col reel [C509]) – uk Microform Academic [090]
Illuminated byzantine gospels – 13th and 11th centuries – 1 b/w reel $85 1 col reel (27 frames) $80 – 1,14 – (notes by dr w o hassall) – us UMI ProQuest [090]
Illuminated manuscripts at the guildhall library, london / Guildhall Library. London – (mf ed 1999) – 10r – 14 – £995.00 – mfΛM – uk World [090]
Illuminating engineering (ie) – Baltimore. 1949-1971 (1) 1970-1971 (5) – ISSN: 0019-2333 – mf#95 – us UMI ProQuest [621]
Illuminating Engineering Society *see* Journal of the illuminating engineering society
Illuminating Engineering Society, London *see* Transactions of the illuminating engineering society
The illumination of joseph keeler, esq : or, on, to the land! / Bryce, Peter Henderson – Boston, MA: American journal of public health, c1915 – 2mf – 9 – 0-665-76197-X – mf#76197 – cn Canadiana [307]
Illuminations / Native American Research Institute – v1 n2-v2 n6 [1981 feb-1982 jul] – 1r – 1 – mf#654773 – us WHS [305]
Illuminator : the newsletter of the open door mission true light church / Open Door Mission True Light Church [Philadelphia PA] – 1997 may – 1r – 1 – mf#4024448 – us WHS [362]
The illuminator – Beatrice, NE: [s.n.] 1926 [mf ed v1 n2. nov 4,8 1926 filmed [1993]] – 1r – 1 – us NE Hist [071]
Illusion and delusion / Bray, Charles – London, England. 18– – 1r – us UF Libraries [240]
Illusion and delusion / Bray, Charles – London, England. 18– – 1r – us UF Libraries [240]
Illusion in religion / Abbott, Edwin Abbott – London: Francis Griffiths, [190-?] – (= ser Essays For The Times) – 1mf – 9 – 0-7905-5740-1 – mf#1988-1740 – us ATLA [240]
Illustracao academica – Pernambuco, 16 jun 1869 – (= ser Ps 19) – bl Biblioteca [370]
Illustracao anglo-brazileira – London, England. -irr. 13 Sept, 3 Nov 1870. 4 ft – 1 – uk British Libr Newspaper [072]
A illustracao brasileira – Rio de Janeiro, RJ: Off Typ da Empreza d'O Malho, 01 ago 1901-jul 1902; jun 1909-fev 1915; set 1920-dez 1930; maio 1935-fev 1958 – (= ser Ps 19) – mf#P06,03,28 – bl Biblioteca [321]
Illustracao brasileira : jornal encyclopedico – Rio de Janeiro, RJ: Typ America, 09 maio 1861 – (= ser Ps 19) – mf#P17,01,183 – bl Biblioteca [972]
Illustracao commercial do recife – Pernambuco, 03 set 1865 – (= ser Ps 19) – bl Biblioteca [380]
Illustracao mineira – Juiz de Fora, MG. Typ Pereira, 20 jul 1890 – (= ser Ps 19) – mf#P31,03,59 – bl Biblioteca [079]
Illustrated advertiser of the royal dublin society's exhibition – Dublin, Ireland. 1850 – 1/4r – 1 – uk British Libr Newspaper [072]
Illustrated american – v11 [1892], v14 [1893], v9 – 2r – 1 – (cont by: illustrated american magazine) – mf#1058694 – us WHS [071]
Illustrated archaeologist – London. 1893-1894 (1) – mf#5574 – us UMI ProQuest [930]
Illustrated atlas of the dominion of canada : containing maps of all the provinces, the north-west territories... – Toronto [ON]: H Parsell 1881 [mf ed 1987] – 18mf [ill] – 9 – 0-665-41367-X – mf#41367 – cn Canadiana [912]
Illustrated australian mail – Melbourne, dec 1861-feb 1862 – 1r – A$29.08 vesicular A$34.58 silver – at Pascoe [079]
Illustrated australian news – Melbourne, Australia Jan 1869-26 dec 1870, 24 jan-27 dec 1872 8 aug 1883-jul 1896 – 1 – uk British Libr Newspaper [079]
Illustrated berwick journal – Berwick-upon-Tweed, England 16 jun-22 dec 1855 – 1 – (cont by: berwick journal [29 dec 1855-9 feb 1928]) – uk British Libr Newspaper [072]
Illustrated bolton examiner and commercial advertiser – [NW England] Bolton Lib 30 dec 1858-6 jan 1859 – 1 – (title change: bolton examiner & commercial advertiser [13 jan-13 may 1859]; bolton examiner & south lancashire advertiser [10 may 1859-28 dec 1861]) – uk MLA; uk Newsplan [072]
Illustrated catalogue / Burlington Fine Arts Club, London – [London] 1895 – (= ser 19th c art & architecture) – 2mf – 9 – mf#4.2.535 – uk Chadwyck [700]
Illustrated catalogue / Tate Gallery, London – London 1897 – (= ser 19th c art & architecture) – 2mf – 9 – mf#4.1.284 – uk Chadwyck [700]
[Illustrated catalogue] / Goldsmiths' Alliance, Ltd – London [1870?] – (= ser 19th c art & architecture) – 2mf – 9 – mf#4.2.950 – uk Chadwyck [700]
Illustrated catalogue of general hardware / Rice, Lewis and Son Ltd – [Toronto: s.n, 1898?] [mf ed 1981] – 12mf – 9 – (incl ind) – mf#12436 – cn Canadiana [680]
An illustrated catalogue of painting and sculpture / Great Britain. Royal Commission for the Paris Exhibition, 1878 – Paris [1878] – (= ser 19th c art & architecture) – 1mf – 9 – mf#4.2.866 – uk Chadwyck [700]
Illustrated catalogue of specimens of persian and arab art : exhibited in 1885 / Burlington Fine Arts Club, London – [London?] 1885 – (= ser 19th c art & architecture) – 2mf – 9 – mf#4.2.1356 – uk Chadwyck [700]
Illustrated catalogue of stained glass windows – London [1870?] – (= ser 19th c art & architecture) – 1mf – 9 – mf#4.2.528 – uk Chadwyck [740]
The illustrated catalogue of the valuable collection of pictures...coins and medals / Brett, John Watkins – [London] 1864 – (= ser 19th c art & architecture) – 3mf – 9 – mf#4.2.1010 – uk Chadwyck [730]
Illustrated catalogue...of the permanent collection of paintings...at aston hall / Birmingham. Museum and Art Gallery – Birmingham 1899 – (= ser 19th c art & architecture) – 3mf – 9 – mf#4.2.1065 – uk Chadwyck [750]
Illustrated catechism for little children – Montreal: [s.n.], c1912 – 1mf – 9 – 0-665-77119-3 – (also available in french) – mf#77119 – cn Canadiana [241]
An illustrated commentary of the gospel according to matthew : for family use and reference, and for the great body of christian workers of all denominations / Abbott, Lyman – New York: A S Barnes, c1875 [mf ed 1985] – (= ser Abbott's commentary 1) – 1mf – 9 – 0-8370-2022-0 – (incl tabular harmony of the 4 gospels) – mf#1985-0022 – us ATLA [226]

ILLUSTRATIONS

An illustrated commentary on the acts of the apostles : for family use and reference, and for the great body of christian workers of all denominations / Abbott, Lyman – New York: A S Barnes, 1878 [mf ed 1985] – (= ser Abbott's commentary 4) – 1mf – 9 – 0-8370-2021-2 – mf#1985-0021 – us ATLA [226]

An illustrated commentary on the gospel according to st john : for family use and reference, and for the great body of christian workers of all denominations / Abbott, Lyman – New York: A S Barnes 1879 [mf ed 1985] – (= ser A popular commentary on the new testament 3) – 1mf – 9 – 0-8370-2023-9 – (incl ind) – mf#1985-0023 – us ATLA [226]

An illustrated commentary on the gospels according to mark and luke : for family use and reference, and for the great body of christian workers of all denominations / Abbott, Lyman – New York: A S Barnes, c1877 [mf ed 1985] – (= ser Abbott's commentary 2) – 1mf – 9 – 0-8370-2024-7 – mf#1985-0024 – us ATLA [226]

Illustrated designs of cabinet furniture engraved from photographs / Hampton and Sons – London [1875?] – (= ser 19th c art & architecture) – 3mf – 9 – mf#4.2.46 – uk Chadwyck [740]

Illustrated edinburgh news : the evening news weekly – [Scotland] Edinburgh: H J & J Wilson 6 feb 1897-2 jul 1898 (wkly) [mf ed 2003] – 74v on 5r – 1 – uk Newsplan [072]

An illustrated essay on the noctuid of north america : with "a colony of butterflies" / Grote, Augustus Radcliffe – London: J Van Voorst, 1882 – 1mf – 9 – (incl bibl ref) – mf#06931 – cn Canadiana [590]

Illustrated events – 1903 nov-1905 jan, v1 n1-5 [1903 nov-1905 jan] – 2r – 1 – (cont by: kilbourn weekly illustrated events[1905 may 24]) – mf#3354971 – us WHS [071]

The illustrated exhibitor : a tribute to the world's industrial jubilee / Cassell & Co Ltd – London [1851] – (= ser 19th c art & architecture) – 7mf – 9 – mf#4.2.881 – uk Chadwyck [740]

Illustrated explanation of the holy sacraments *see* Explanation of the holy sacraments

Illustrated explanation of the prayers and ceremonies of the mass / Lanslots, Ildephonse – 2nd ed. New York: Benziger Bros, 1897 – 1mf – 9 – 0-524-03618-7 – mf#1990-4778 – us ATLA [240]

Illustrated family magazine : for the diffusion of useful knowledge – Boston. 1845-1846 (1) – mf#3750 – us UMI ProQuest [640]

Illustrated gaelic-english dictionary / Dwelly, Edward – Glasgow, Scotland. 1941 – 1r – us UF Libraries [040]

Illustrated gaelic-english dictionary / Dwelly, Edward – Glasgow, Scotland. 1941 – 1r – us UF Libraries [420]

Illustrated glasgow news : and pictorial weekly newspaper for glasgow and vicinity – [Scotland] Glasgow: P Mackenzie 30 jun-17 oct 1855 (wkly) [mf ed 2004] – 1r – 1 – (preceded by 2 specimen iss dated jun 16 & june 23 1855) – uk Newsplan [072]

Illustrated guide to kensal green cemetery / Justyne, William – London [1861] – (= ser 19th c art & architecture) – 1mf – 9 – mf#4.1.426 – uk Chadwyck [730]

[Ill]ustrated guide to the world's fair and chicago and quebec – [Quebec: F Carrel, 1893] – 3mf – 9 – 0-665-91897-6 – mf#91897 – cn Canadiana [917]

The illustrated hand-book to all religions : from the earliest ages to the present time. including the rise, progress, doctrines and government of all christian denominations – Chicago, ILs: WH Harrison, c1877 – 2mf – 9 – 0-524-02872-9 – mf#1990-3145 – us ATLA [052]

Illustrated handbook to the permanent collections of industrial art objects / Birmingham. Museum and Art Gallery – [Birmingham 1895?] – (= ser 19th c art & architecture) – 4mf – 9 – mf#4.2.1064 – uk Chadwyck [740]

Illustrated heywood advertiser and general family newspaper – [NW England] Heywood 16 jun-22 dec 1855 – 1 – (title change: heywood advertiser [29 dec 1855]) – uk MLA; uk Newsplan [072]

Illustrated hindu – New Delhi, India. Oct 1942-Nov 1947 – 2r – 1 – us L of C Photodup [079]

Illustrated history of british guiana / Bennett, George Hanneman – Georgetown, Guyana. 1866 – 1r – us UF Libraries [972]

Illustrated history of british guiana / Bennett, George Hanneman – Georgetown, Guyana. 1866 – 1r – us UF Libraries [972]

The illustrated history of methodism : in great britain and america from the days of the wesleys to the present time / Daniels, William Haven – New York: Methodist Book Concern, 1880 – 2mf – 9 – 0-524-03003-0 – mf#1990-4525 – us ATLA [242]

Illustrated history of the flute : and sketch of the successive improvements made in the flute, and a statement of the principles upon which flutes... / Badger, Alfred G – 4th ed, New York: Firth, Pond & Co 1861 [mf ed 1991] – 1r – 1 – mf#pres. film 105 – us Sibley [780]

An illustrated history of the new world : containing a general history of all the various nations, states and republics of the western continent... and a complete history of the united states to the present time... / ed by Denison, John Ledyard – Norwich, CT: H Bill, 1868 [mf ed 1980] – 11mf – 9 – 0-665-05412-2 – mf#05412 – cn Canadiana [975]

Illustrated holiness year book for... : a daily companion for the christian home / ed by Hughes, George – New York: Palmer & Hughes 1889-90 [mf ed 2004] – 1889-90 [complete] 2v on 1r – 1 – (cont: christian holiness almanac and year book; cont by: holiness year book for...) – mf1055b – us ATLA [240]

Illustrated inventor – London. 31 Oct 1857-10 Apr 1858 [wkly] – 33 ft – 1 – uk British Libr Newspaper [073]

Illustrated irish weekly independent and nation *see* Irish weekly independent

Illustrated itinerary of the county of cornwall / Redding, Cyrus – London, England. 1842 – 1r – us UF Libraries [941]

Illustrated itinerary of the county of cornwall / Redding, Cyrus – London, England. 1842 – 1r – us UF Libraries [025]

Illustrated journal / Scott, Minnie – 1902-1959, Scrapbook containing clippings, certificates, photographs, souvenirs, publications, and other items relating to her training and career as a nurse. Included are items from her service in England during World War I; her job as city health nurse in Lawrence, KS; her houses in Lecompton and Topeka, KS; and a brief chronology – 1 – us Kansas [920]

Illustrated journal / Scott, Minnie – 1902-59 – 1 – us Kansas [978]

The illustrated journal of agriculture – Montreal: E Senecal, [1879-1897] – 9 – mf#P04170 – cn Canadiana [630]

Illustrated London News *see* The illustrated paris universal exhibition

Illustrated london news – London. 1842+ (1) 1978+ (5) 1978+ (9) – ISSN: 0019-2422 – mf#5407 – us UMI ProQuest [941]

The illustrated magazine of art – 1853-54 [mf ed Chadwyck-Healey] – (= ser Rare 19th Century American Art Journals) – 20mf – 9 – uk Chadwyck [700]

Illustrated montreal, the metropolis of canada : its romantic history, its beautiful scenery, its grand institutions, its present greatness, its future splendor – Montreal: J McConniff, 1890? – 2mf – 9 – mf#09889 – cn Canadiana [720]

Illustrated new era – 1878 nov – 1r – 1 – mf#4012203 – us WHS [071]

Illustrated news – Hicksville, NY. 1986-1988 (1) – mf#68173 – us UMI ProQuest [071]

Illustrated oban magazine and argyllshire advertiser – Oban: J Miller (mthly) [mf ed 1996] – 1 – (cont by: oban times, and argyllshire advertiser) – uk Scotland NatLib [072]

Illustrated oldham herald – [North West] Oldham Lib jun-aug 1855 – 1 – uk MLA; uk Newsplan [072]

Illustrated oldham journal – [North West] Oldham Lib oct 1854-feb 1855 – 1 – uk MLA; uk Newsplan [072]

Illustrated oldham telegraph – [NW England] Oldham Lib mar-nov 1859 – 1 – uk MLA; uk Newsplan [072]

Illustrated outdoor news – New York. v1-7. may 1903-sep 1906; ns: v1 n1-4. oct 1906-jan 1907 (freq varies) [all publ] – (= ser Sports periodicals, 1822-1922) – 3r – 1 – $605.00 – us UPA [790]

Illustrated oxfordshire telegraph – Bicester, England. Oxfordshire Telegraph – Oxfordshire, Buckinghamshire, and Northamptonshire Telegraph. -w. 29 Dec 1858-20 June 1894. Lacking 1872, 1874 – 11r – 1 – uk British Libr Newspaper [072]

The illustrated paris universal exhibition / Illustrated London News – [London] 1878 – (= ser 19th c art & architecture) – 7mf – 9 – mf#4.2.902 – uk Chadwyck [700]

Illustrated pattern-book of furniture, carpets, rugs...etc / Silber and Fleming Ltd, London – London [1885?] – (= ser 19th c art & architecture) – 14mf – 9 – mf#4.2.895 – uk Chadwyck [740]

Illustrated pattern-book of silver goods / Woods and Long – London [1881?] – (= ser 19th c art & architecture) – 3mf – 9 – mf#4.2.765 – uk Chadwyck [730]

Illustrated police budget – London, UK. 1899 – 1r – 1 – uk British Libr Newspaper [072]

Illustrated police news – London, UK. 1888, 1892.-w. 1 reel – 1 – uk British Libr Newspaper [072]

The illustrated police news – Parish of St Mary Le-Strand, London: George Purkess, jan 1870 – us CRL [071]

Illustrated price list : j eveleigh & co, manufacturers of trunks, valises, etc, blacksmiths' bellows and portable forges: salesroom 1753 notre dame street, montreal / J Eveleigh & Co – [Montreal?: D Bentley, c1888 – 2mf – 9 – 0-665-90955-1 – mf#90955 – cn Canadiana [680]

An illustrated quarterly art journal – New Delhi, 1948-1970. v20-40. ind v20-39 – 75mf – 9 – (missing: v39(2)) – mf#I-1079 – ne IDC [240]

An illustrated quarterly journal of oriental art : chiefly indian / ed by Ordhendra, C Gangoly – Calcutta, 1920-1930 – 45mf – 8 – mf#I-1078 – ne IDC [240]

Illustrated quebec : the story of its famous annals under french and english occupancy: being a series of pen pictures... / Adam, Graeme Mercer – Quebec: J McConniff, [1891?] [mf ed 1980] – (= ser McConniff's series) – 1mf – 9 – mf#09338 – cn Canadiana [971]

Illustrated radical rhymes : with the radical ode and the radicals pronounced and defined / Silsby, John Alfred – Shanghai: American Presbyterian Mission Press, 1912 [mf ed 1995] – (= ser Yale coll) – 36p (ill) – 1 – 0-524-10059-4 – (ill by martha layer and han ren deh) – mf#1995-1059 – us ATLA [480]

An illustrated record of the retrospective exhibition held at south kensington / Fisher, John – London 1897 – (= ser 19th c art & architecture) – 2mf – 9 – mf#4.2.164 – uk Chadwyck [700]

Illustrated review : a fortnightly journal of literature, science, and art – London. 1870-1874 – 1 – mf#4722 – us UMI ProQuest [073]

Illustrated rural industries and country produce mart – London, UK. Mar 1906. -irr – 1 – uk British Libr Newspaper [072]

Illustrated saturday reader – Montreal: R Worthington, 1866-[1867] – 9 – (cont by: saturday reader (1867)) – mf#P04514 – cn Canadiana [073]

Illustrated social history of south africa / Hattersley, Alan Frederick – Cape Town, South Africa. 1969 – 1r – us UF Libraries [960]

Illustrated social history of south africa / Hattersley, Alan Frederick – Cape Town, South Africa. 1969 – 1r – us UF Libraries [960]

Illustrated sporting and dramatic news – London. -w. 28 Feb 1874-Dec 1890. (24 reels) – 1 – uk British Libr Newspaper [700]

Illustrated sporting and theatrical news *see* Illustrated sporting news

Illustrated sporting news – London. mar 1862-mar 1870 [wkly] – 6r – 1 – uk British Libr Newspaper [790]

Illustrated story of the union in rhyme / Adams, Robert Chamblet – Boston: A M Thayer, 1891 – 2mf – 9 – mf#13576 – cn Canadiana [975]

Illustrated sutton and cheam mail – 1 – (aka: sutton & cheam mail; sutton times & cheam mail) – uk British Libr Newspaper [072]

Illustrated sydney news – Sydney, Australia. 30 jun 1855; 21 jan-23 dec 1882; 17 jan 1885-dec 1886; 16 jun 1864-16 jan 1865; 16 mar 1865-23 dec 1871; 8 jun 1872; 18 jan, 10 jun, 5 jul 1873; 7, 15 nov 1887; 26 jan-31 may 1888; 26 dec 1889; 9 jan-20 dec 1890; 3 jan 1891-6 aug 1892 – 5 1/2r – 1 – (aka: illustrated sydney news and agriculturist and grazier) – uk British Libr Newspaper [079]

Illustrated sydney news – Australia. -w. 1864-Dec 1875. Imperfect. 1 1 reels – 1 – uk British Libr Newspaper [072]

Illustrated sydney news – Australia. -w. June 1855; Jan 1882-July 1892. 4 reels – 1 – uk British Libr Newspaper [072]

Illustrated sydney news – Sydney, 1853-jun 1855; jul 1864-94 – 11r – 1 – A$423.50 vesicular A$484.00 silver – at Pascoe [079]

Illustrated sydney news and agriculturist and grazier *see* Illustrated sydney news

Illustrated tasmanian mail *see* Tasmanian mail

Illustrated times – London. 9 Jun 1855-2 Mar 1872.-w. 16mqn reels – 1 – uk British Libr Newspaper [072]

Illustrated toronto, the queen city of canada : its past, present and future, its growth, its resources, its commerce, its manufactures... – Toronto: Acme Pub and Engraving Co, [1890?] – 2mf – 9 – 0-665-92013-X – (incl ind) – mf#92013 – cn Canadiana [917]

[Illustrated trade catalogue] / Smith and Co, George. Ironfounders – London [1865] – (= ser 19th c art & architecture) – 4mf – 9 – mf#4.1.261 – uk Chadwyck [730]

An illustrated treatise on the law of evidence / Hughes, Thomas Welburn – 3d impression. Chicago: Callaghan, 1907. 678p. LL-1466 – 1 – us L of C Photodup [347]

Illustrated wrexham argus and north wales athlete – Wrexham, Wales aug 1884-dec 1916 [mf 1884-88,1892] – 1 – (discontinued; wanting: 1893, 1901) – uk British Libr Newspaper [072]

The illustrating mirror : or, a fundamental illustration of christ's sermon on the mount = Erlaeuterungs spiegel / Herr, Johannes – Lancaster, PA; E Barr, 1858 – 1mf – 9 – 0-524-05284-0 – (in english) – mf#1992-0385 – us ATLA [220]

L'illustration : journal universel – Paris. 1866-67 – 1 – fr ACRPP [073]

L'illustration du journal le moniteur acadien : organe des populations francaises des provinces maritimes... – Shediac, NB?: F Robidoux, 1892? – 1mf – 9 – mf#51510 – cn Canadiana [071]

An illustration of the architecture and sculpture of the cathedral church of lincoln / Wild, Charles – London 1819 – (= ser 19th c art & architecture) – 3mf – 9 – mf#4.2.431 – uk Chadwyck [720]

An illustration of the architecture of the cathedral church of chester / Wild, Charles – London 1813 – (= ser 19th c art & architecture) – 1mf – 9 – mf#4.2.1780 – uk Chadwyck [720]

An illustration of the architecture of the cathedral church of lichfield / Wild, Charles – London 1813 – (= ser 19th c art & architecture) – 1mf – 9 – mf#4.2.1779 – uk Chadwyck [720]

An illustration of the general evidence establishing the reality of christ's resurrection / Cook, George – Edinburgh: printed for Peter Hill, 1808 [mf ed 1993] – (= ser Presbyterian coll) – 4mf – 9 – 0-524-07406-2 – mf#1991-3066 – us ATLA [240]

Illustration of the hypothesis proposed in the dissertation on the / Marsh, Herbert – Cambridge, England. 1803 – 1r – us UF Libraries [240]

Illustration of the hypothesis proposed in the dissertation on the / Marsh, Herbert – Cambridge, England. 1803 – 1r – us UF Libraries [240]

L'illustration populaire – Montreal: [s.n.] v1 n27 7 dec 1895-v3 n109 3 juil 1897 (wkly) [mf ed 1987] – 1r – 5 – (ceased 1897?) – mf#SEM16P368 – cn Bibl Nat [073]

L'illustration populaire – Montreal: impr metropolitaine, [1895-189- ou 19–] [mf ed v1 n1 8 juin 1895-v1 n27 7 dec 1895] – 9 – mf#P06069 – cn Canadiana [073]

Illustrationes et descriptiones plantarum novarum syriae et tauri occidentalis / Fenzl, E – Stuttgart, 1843 – 9mf – 8 – mf#658 – ne IDC [956]

Illustrations 63 – Munich, 1964-73 [mf ed Chadwyck-Healey] – (= ser Art periodicals on microform) – 2r – 1 – uk Chadwyck [740]

Illustrations and descriptions of new, unfigured, or imperfectly known shells, chiefly american, in the u.s. national museum / Dall, W H – 1902. v24(p499-566, pl 27-40)) – (= ser Proceedings...U.S. National Museum Washington) – 2mf – 9 – mf#Z-2236 – ne IDC [590]

Illustrations, architectural and pictorial : of the genius of michael angelo buonarroti / Canina, Luigi et al – London 1857 – (= ser 19th c art & architecture) – 4mf – 9 – mf#4.2.1438 – uk Chadwyck [740]

Les illustrations canadiennes : premiere serie, 1494-1676 / Dupuy, Paul – Montreal: Cadieux & Derome, 1887 – 3mf – 9 – mf#02803 – cn Canadiana [971]

Illustrations from the sermons of alexander maclaren / Maclaren, Alexander; ed by Martyn, James Henry – London: Alexander & Shepheard, [1894] [mf ed 2004] – 1r – 1 – 0-524-10477-8 – (with ind) – mf#b00694 – us ATLA [242]

Illustrations of ancient buildings in kashmir / Cole, Henry Hardy – London 1869 – (= ser 19th c art & architecture) – 2mf – 9 – mf#4.2.1179 – uk Chadwyck [720]

Illustrations of architecture and ornament / Waring, John Burley – London [1865] – (= ser 19th c art & architecture) – 3mf – 9 – mf#4.2.1517 – uk Chadwyck [720]

Illustrations of art metal and woodwork / Benham and Froud – London [1878?] – (= ser 19th c art & architecture) – 1mf – 9 – mf#4.2.699 – uk Chadwyck [730]

Illustrations of astronomy / Hayden, Henry – Halifax, NS?: s.n, 1836 – 1mf – 9 – mf#64754 – cn Canadiana [520]

Illustrations of buildings near muttra and agra : showing the mixed hindu-mahomedan style / Cole, Henry Hardy – London 1873 – (= ser 19th c art & architecture) – 2mf – 9 – mf#4.2.1178 – uk Chadwyck [720]

Illustrations of cairo / Hay, Robert – London 1840 – (= ser 19th c art & architecture) – 5mf – 9 – mf#4.2.1629 – uk Chadwyck [740]

Illustrations of furniture candelabra musical instruments / Braund, John – London 1858 – (= ser 19th c art & architecture) – 3mf – 9 – mf#4.2.856 – uk Chadwyck [740]

Illustrations of her majesty's palace at brighton / Brayley, Edward Wedlake – London 1838 – (= ser 19th c art & architecture) – 5mf – 9 – mf#4.2.539 – uk Chadwyck [740]

Illustrations of indian architecture from the muhammadan conquest downwards / Kittoe, Markham – Calcutta 1838 – (= ser 19th c art & architecture) – 4mf – 9 – mf#4.1.179 – uk Chadwyck [720]

Illustrations of napa county with historical sketch / Smith and Elliott – Napa Co, CA. 1878 – 1r – 1 – $50.00 – mf#B40240 – us Library Micro [978]

The illustrations of old testament history in queen mary's psalter / Westlake, Nat Hubert John & Purdue, William – London [1865] – (= ser 19th c art & architecture) – 4mf – 9 – mf#4.2.1388 – uk Chadwyck [740]

Illustrations of the creed / Wordsworth, Elizabeth – New York: EP Dutton, 1890 – 1mf – 9 – 0-524-05390-1 – (incl bibl ref) – mf#1991-2296 – us ATLA [240]

Illustrations of the grammatical parts of the guzerattee, mahratta and english languages / Drummond, Robert – Bombay: Courier 1808 – (= ser Whsb) – 2mf – 9 – €30.00 – mf#Hu 260 – gw Fischer [410]

Illustrations of the historical works of francis parkman – Toronto: G N Morang, [190-?] – 1mf – 9 – 0-665-86120-6 – mf#86120 – cn Canadiana [740]

Illustrations of the history of mediaeval thought : in the departments of theology and ecclesiastical politics / Poole, Reginald Lane – London: Williams & Norgate, 1884 [mf ed 1990] – 1mf – 9 – 0-7905-5792-4 – (2nd ed publ 1920. incl bibl ref) – mf#1988-1792 – us ATLA [180]

Illustrations of the life of martin luther / Labouchere, Pierre Antoine – Philadelphia: Lutheran Board of Publication, 1869 – 1mf – 9 – 0-524-00567-2 – mf#1990-0067 – us ATLA [242]

The illustrations of the maqamat / Grabar, Oleg – 1984 – 10mf – 9 – $58.00f – 0-226-69057-1 – (196p accompanying text) – us Chicago U Pr [750]

Illustrations of the new palace of westminster. first series / Barry, Charles – London 1849 – (= ser 19th c art & architecture) – 3mf – 9 – mf#4.2.1519 – uk Chadwyck [720]

Illustrations of the new palace of westminster. second series / Barry, Charles – London 1865 – (= ser 19th c art & architecture) – 2mf – 9 – mf#4.2.1520 – uk Chadwyck [720]

Illustrations of the public buildings of london / Britton, John & Pugin, Augustus Charles – London 1825, 1828 – (= ser 19th c art & architecture) – 12mf – 9 – mf#4.2.819 – uk Chadwyck [740]

Illustrations of the significance of certain ancient british skull forms / Wilson, Daniel – S.l: s.n, 1863? – 1mf – 9 – (incl bibl ref) – mf#63188 – cn Canadiana [573]

Illustrations of universal progress : a series of discussions / Spencer, Herbert – New York: Appleton, 1881 – xxiv/439p – 1 – us Wisconsin U Libr [190]

Illustrations of various styles of indian architecture / Fergusson, James – London 1870 – (= ser 19th c art & architecture) – 1mf – 9 – mf#4.2.511 – uk Chadwyck [720]

Illustrations to blair's the grave : from the british museum / Blake, William – 1r – 1 – mf#96755 – uk Microform Academic [810]

Illustrations to bunyan's pilgrim's progress / Blake, William – 1 col r – 14 – mf#C604 – uk Microform Academic [760]

Illustrations to edward young's night thoughts, from british museum, dept. of prints and drawings / Blake, William – 1 col r – 14 – (a series of 537 illustrations made between 1795 and 1797) – mf#C597 – uk Microform Academic [740]

Illustrations to edward young's "night thoughts" from sir john soane's museum, london / Blake, William – 1r – 14 – mf#C97044 – uk Microform Academic [760]

Illustrative cases in equity / Pattee, William Sullivan – 2d ed. St. Paul: West, 1893. 110p. LL-1099 – 1 – (3d ed. st. paul: west, 1896. 157p. ll-1088) – us L of C Photodup [342]

Illustrative cases on equity jurisprudence / Hutchins, Harry Burns – 2d ed. St. Paul: West, 1904. 908p. LL-850 – 1 – us L of C Photodup [342]

Illustrative cases on personal rights and the domestic relations including teacher and pupil / Chadman, Charles Erehart – Chicago: American School of Law, 1907. 301p. LL-1605 – 1 – us L of C Photodup [340]

Illustrative cases upon the law of bills and notes / Johnson, Elias Finley – St. Paul, West, 1895. 219 p. LL-440 – 1 – us L of C Photodup [346]

Illustrative rules governing complaints of judicial misconduct and disability : prepared by a special committee of the chief judges of the u.s. courts of appeals – Washington: FJC, 1986 – 1mf – 9 – $1.50 – mf#LLMC 95-379 – us LLMC [347]

Illustrazione istorica del palazzo della signoria / Rastrelli, M – Firenze, 1792 – 3mf – 9 – mf#O-1050 – ne IDC [700]

Illustre abietis cum lauro connubium, quindenis symbolorum dotibus locupletatum : honori et amori...michaelis francisci ferdinandi... / Habel, J – [Pragae]: Typis Carolo Ferdinandeae, 1673 – 1mf – 9 – mf#O-1996 – ne IDC [090]

Illustreret folke-visebog – Kjobenhavn: F Woldike 1873 [mf ed Bloomington IN: Indiana Uni Lib, Preservation Dept 1984] – 1r – 1 – us Indiana Preservation [390]

Illustres disquisitiones morales / Candidus, V – Venetiis. v1-4. 1639-Romae 1643 – 4v on 76mf – 8 – €145.00 – ne Slangenburg [230]

Die illustrierte *see* Benrather tageblatt

Illustrierte berliner zeitschrift *see* Weltspiegel 1946

Das illustrierte blatt – Frankfurt/M DE, 1918-19 – 1r – 1 – (filmed by misc inst: 1938) – gw Misc Inst [074]

Illustrierte chronik der zeit – Stuttgart DE, 1876, 1879 – 9 – gw Misc Inst [900]

Illustrierte deutsche montags-zeitung *see* Deutsche montagszeitung

Illustrierte dorfzeitung des lahrer hinkenden boten – Lahr/Schwarzwald DE, 1863-73 – 2r – 1 – gw Misc Inst [074]

Illustrierte film-woche *see* Illustrierte kino-woche

Die illustrierte fuer bremen – Bremen DE, 1914 4 apr-8 aug – 1r – 1 – gw Misc Inst [074]

Die illustrierte fuer das bergische land *see* Die illustrierte fuer den industriebezirk duesseldorf

Die illustrierte fuer den industriebezirk duesseldorf – Duesseldorf DE, 1914 apr-jul – 1r – 1 – (title varies: 12 may 1914: die illustrierte fuer duesseldorf; 1914 n8: die illustrierte fuer das bergische land) – gw Misc Inst [074]

Die illustrierte fuer duesseldorf *see* Die illustrierte fuer den industriebezirk duesseldorf

Illustrierte gemeinde-zeitung : centralorgan fuer die politischen, religioesen und culturinteressen der israelitischen cultusgemeinden in oesterreich-ungarn / ed by Eibenschuetz, S – Vienna. v1-41. 1885-86 – (= ser German-jewish periodicals...1768-1945, pt 3) – 1r – 1 – $165.00 – mf#B460 – us UPA [939]

Illustrierte geschichte der deutschen literatur von den aeltesten zeiten bis zur gegenwart / Salzer, Anselm – Muenchen: Allgemeine Verlags-Gesellschaft m.b.h., [1912] – 1 – us Wisconsin U Libr [430]

Illustrierte geschichte der deutscher literatur von den aeltesten zeiten bis zur gegenwart / Salzer, Anselm – 2., neu bearb. Aufl. Regensburg: J Habbel, 1926-1932 – 1 – (incl bibl ref and index) – us Wisconsin U Libr [430]

Illustrierte geschichte des deutschen schriftthums in volksthuemlicher darstellung / Leixner-Gruenberg, Otto von – Leipzig: O Spamer, 1880-1881 – 1r – 1 – (incl bibl ref and index) – us Wisconsin U Libr [430]

Illustrierte geschichte des preussischen hofes, des adels und der diplomatie vom grossen kurfuersten bis zum tode kaiser wilhelms 1 / Vehse, Eduard – Stuttgart, 1901 (mf ed 1992) – 7mf – 9 – €74.00 – 3-89349-110-4 – mf#DHS-AR 79 – gw Frankfurter [943]

Illustrierte jugendzeitung – Leipzig DE, 1846 (gaps), 1847 – 1r – 1 – gw Misc Inst [305]

Illustrierte kino-woche – Berlin DE, 1913-19 – 3r – 1 – (title varies: 1919: illustrierte film-woche) – gw Mikrofilm [790]

Illustrierte monatshefte fuer die gesamten interessen des judentums – Vienna: A Hilberg. v1-2. 1865-66 – (= ser German-jewish periodicals...1768-1945, pt 3) – 1r – 1 – $165.00 – mf#B104 – us UPA [939]

Illustrierte republikanische zeitung – Magdeburg, Berlin DE, 1925 3 jan-1933 4 mar – 6r – 1 – (cont: illustrierte reichsbanner-zeitung) – gw Mikrofilm [074]

Illustrierte sonntagspost *see* Duesseldorfer stadtanzeiger

Illustrierte sonntags-zeitung – Duesseldorf DE, 1898-99 – 1r – 1 – gw Misc Inst [074]

Illustrierte sonntags-zeitung (gelsenkirchen) *see* Benrather zeitung

Illustrierte unterhaltungsbeilage der mecklenburgischen volks-zeitung *see* Mecklenburgische volks-zeitung

Illustrierte weltgeschichte / Mertens, O – Berlin, Germany. 196- – 1r – us UF Libraries [910]

Illustrierte weltgeschichte / Mertens, O – Berlin, Germany. 196- – 1r – us UF Libraries [025]

Illustrierte westdeutsche wochenschau – Duesseldorf, Essen DE, 1909 2 oct-1922 29 jul, 1924 29 mar-1943 13 oct – 33r – 1 – (with gaps. title varies: 24 dec 1910: die wochenschau; 29 mar 1924: westdeutsche illustrierte zeitung; 29 aug 1926: die wochenschau. fr 29 aug 1926 publ in essen) – gw Mikrofilm [074]

Illustrierte zeitung (benrath) *see* Benrather zeitung

Illustrierte zeitung fuer blechindustrie – Leipzig DE, 1886-91, 1895-96, 1899-1900, 1903, 1907-09 – 25r – 1 – uk British Libr Newspaper [670]

Illustrierte zeitung fuer buchbinderei und cartonnagen fabrikanten – Leipzig DE, 1868 apr-1874 dec, 1895-1900; 1901 jul-1903 jun – 5r – 1 – uk British Libr Newspaper [680]

Illustrierter beobachter – Muenchen DE, 1926-45 [gaps] – 1 – gw Misc Inst [074]

Illustrierter familien-freund *see* Allgemeiner anzeiger fuer die amtsgerichtsbezirke hessisch-lichtenau, grossalmerode, spangenberg und umgegend

Illustrierter familienfreund *see* Dorf-chronik 1848

Illustrierter filmkurier – Wien (A), 1919-44 – 14r – 1 – gw Mikrofilm [790]

Illustrierter juedischer kalender – Halberstadt DE, 1878-81 – 1r – 1 – us UMI ProQuest [939]

Illustrierter neuer welt-kalender – Stuttgart DE, 1883, 1888/89, 1891/92 – 1 – gw Misc Inst [900]

Illustrierter volksfreund – Kassel DE, 1912 – 1r – 1 – gw Misc Inst [074]

Illustriertes banater volksblatt *see* Banater volksblatt

Illustriertes conchylienbuch / Kobelt, W – Manchester. 1949+ (1) 1971+ (5) 1974+ (9) – 12mf – 9 – mf#6217 – ne IDC [590]

Illustriertes konversations-lexikon der frau – Berlin/Oldenburg 1900 [mf ed 1993] – (= ser HQ 12) – 2v on 20mf – 9 – €150 diazo €180 silver – 3-89131-124-9 – gw Fischer [305]

Illustriertes kreuz-blatt – Augsburg DE, 1868-70 [gaps] – 1r – 1 – gw Misc Inst [074]

Illustriertes sonntags-blatt – Stuttgart DE, 1895-1900 – 1r – 1 – gw Misc Inst [074]

Illustriertes sonntagsblatt *see* Landsberger nachrichtenblatt

Illustriertes tageblatt *see* Saechsische dorfzeitung

Illustriertes tageblatt / i *see* Saechsischer kurier

Illustriertes unterhaltungsblatt *see* Landsberger nachrichtenblatt

Illustrieter sonntag *see* Der gerade weg/ illustrierter sonntag

Illustrious chinese christians : biographical sketches / Bentley, William Preston – Cincinnati: Standard Publ, 1906 – 1mf – 9 – 0-8370-6016-8 – mf#1986-0016 – us ATLA [920]

Illustrious dames of the court of the valois kings = Vies des dames galantes / Brantome, Pierre de Bourdeille, seigneur de – New York: The Lamb Publ Co 1912 [mf ed 1987] – 1r – 1 – (trans by katharine prescott wormeley; ill with photogravures fr original paintings; with: lincoln's inn / spilsbury, w h; & other titles) – mf#9955 – us Wisconsin U Libr [944]

Illustrious irishwomen / Casey, Elizabeth – London, 1877 – (= ser 19th c ireland) – 9mf – 9 – mf#1.1.6254 – uk Chadwyck [305]

Illustrirovannyi ezhemesiachnyi istoricheskii sbornik – Washington. 1918+ (1) 1970+ (5) 1975+ (9) – 314mf – 9 – mf#1712 – ne IDC [077]

Illustrirte kreuzer-blaetter – Stuttgart DE, 1849-50 – 1 – gw Misc Inst [074]

Illustrirte kriegs-chronik – Leipzig DE, 1871 – 1r – 1 – gw Misc Inst [943]

Die illustrirte welt – Stuttgart DE, 1868 [gaps] – 1r – 1 – (filmed by other misc inst: 1853-65, 1869-70, 1896 [gaps]) – gw Misc Inst [074]

Illustrirte zeitung – Leipzig, Berlin DE, 1843 1 jul-1899 29 jun, 1899 3 aug & 24 aug, 1900 4 jan-1944 sep – 124r – 1 – (filmed by misc inst: 1907 jul-sep [1r]; 1843 jul-1897, 1899, 1900 jul-1913, 1914 [gaps], 1915-35, 1936 apr-1943 sep, 1944 jan-sep (mf nur tw. vorhanden); [1925, 1927-28, 1931, 1937-41, 1943 n5019 (with gaps)]) – gw Mikrofilm; gw Misc Inst [074]

Illustrirte zeitung fuer das katholische deutschland – Leipzig DE, 1855 – 1 – gw Misc Inst [241]

Illustrirte zeitung fuer die jugend – Leipzig DE, 1846-48, 1849 17 mar-1853 – 1 – gw Misc Inst [305]

Illustrirte zeitung (iz4) / ed by Estermann, Alfred – Leipzig/Berlin/Wien/Budapest/New York: J J Weber 1843-1944 [mf ed 2001] – (= ser Illustrierte zeitschriften (iz) 4) – 2596mf – 9 – diazo €8950 silver €11,510 – 3-89131-349-7 – gw Fischer [321]

Illustrirter dresden-prager-fuehrer : malerische beschreibung von dresden, der saechsischen schweiz mit teplitz, der dresden-prager eisenbahn und prag – Leipzig 1852 [mf ed Hildesheim 1995-98] – (= ser Fbc) – 4mf [ill] – 9 – €120.00 – 3-487-29509-1 – gw Olms [914]

Illustrirtes familien-journal – Leipzig DE, 1854-60 [gaps], 1862 (mf nur tw vorhanden) – 1 – gw Misc Inst [640]

Illustrirtes panorama – Berlin DE, 1861, 1862 [gaps], 1865, 1866 – 1r – 1 – gw Misc Inst [074]

Illustrirtes sonntags-blatt *see* Buender tageblatt

Illustrirtes unterhaltungs-blatt – Stuttgart DE, 1895-1906 – 4r – 1 – gw Misc Inst [074]

Illustrium juris tractatuum / Valencia, Melchor de – Allobrogum: Fratrum de Tournes, 1753 – 1 – sp Bibl Santa Ana [340]

Illustrograph – Dublin, Ireland. 1894-98; apr-dec 1899 – 2r – 1 – uk British Libr Newspaper [072]

Illustrowany kurier polski – Bydgoszcz, Poland. Oct 1945-Jan 1947; Jul 1950-Jun 1970 – 27r – 1 – (some issues missing) – us L of C Photodup [943]

Gli illvstri et gloriosi gesti, et vittoriose impresse, fatte contra turchi... / Francus, D – Vinegia, 1584 – 2mf – 9 – mf#H-8204 – ne IDC [956]

Illyria and dalmatia : containing a description of the manners, customs, habits, dress, and other peculiarities characteristic of their inhabitants, and those of the adjacent countries – London 1821 [mf ed Hildesheim 1995-98] – 2v on 4mf – 9 – €120.00 – 3-487-29145-2 – gw Olms [914]

Illyustrirovannaia gazeta – St Petersburg, 1868 – 1 – us UMI ProQuest [077]

Ilm va jami'ah – Alexandria, VA: Persian Journal for Science and Society, 1979- . sal-i 1, shumarah-'i 1-34. 1358-63 [1979-84] – 2r – 1 – $250.00 – us MEDOC [500]

Ilmiyye salnamesi – 1334 [1916] – (= ser Ministry and special interest salnames) – 8mf – 9 – $130.00 – us MEDOC [956]

Ilmoe toemboeh-toemboehan : oentoek dipakai di sekolah djoeroe-obat (assistant-apotheker) / Rasad, B Z – Djakarta: Ika Daigaku Yakugakubu, 2605 – 92p 2mf – 9 – mf#SE-2002 mf143-144 – ne IDC [615]

Ilmu marxis – Djakarta, 1957-1965 – 17mf – 9 – (missing: 1957, v1(2-end); 1959, v3(1); 1960, v4(2)) – mf#SE-370 – ne IDC [950]

Ilmutustraamatu seletus : An explanation to the book of revelation / Kaups, Richard – Santa Barbara, CA: Autori, 178p.1971-72 – 1 – 7.12 – us Southern Baptist [242]

El ilocano – Manila. Philippine Islands. -f. 28 Jun 1889-30 May 1890. (18 ft) – 1 – uk British Libr Newspaper [072]

Ilp news – London. -m. Apr 1897-Dec 1903. (1 reel) – 1 – uk British Libr Newspaper [331]

Ilpa reporter / International Labor Press Association – 1977 mar 3-1983 nov – 1r – 1 – (cont by: ilca reporter) – mf#705653 – us WHS [070]

Ilr report – Ithaca. 1987-1990 (1,5,9) – ISSN: 0736-6396 – mf#16273,01 – us UMI ProQuest [331]

Ilr research – Ithaca. 1954-1967 (1) – ISSN: 0536-180X – mf#5178 – us UMI ProQuest [331]

Ils sont fous ces liberaux / Barberis, Robert – Longueuil: editions R Antoine, 1974 [mf ed 1995] – 2mf – 9 – mf#SEM105P2435 – cn Bibl Nat [325]

Ilsa journal of international and comparative law – v1-7. 1995-2001 – 9 – $165.00 set – (cont: ilsa journal of international law) – ISSN: 1082-944X – mf#117041 – us Hein [341]

Ilsa journal of international law – American Society of International Law Schools. v1-16. 1977-93 (all publ) – 5,6,9 – $116.00 set – (title varies: v1-10 1977-1986 as asils international law journal; cont by: ilsa journal of international and comparative law) – ISSN: 1052-3391 – mf#100711 – us Hein [341]

Ilston, glamorgan, parish church of st illtyd : baptisms 1679-1925, burials 1683-1992, marriages 1654-1837 – [Glamorgan]: GFHS [mf ed [2001?] – 1mf – 9 – £1.25 – uk Glamorgan FHS [929]

Ilston, st illtyd, monumental inscriptions – 1mf – 9 – £1.25 – uk Glamorgan FHS [929]

Iltz, Johannes *see* De vi et usu praepositionum epi, meta, para, peri, pros, hupo apud aristophanem

Ilunga lomndeni wakho liphethwe yi-schizophrenia = Mongwe walelapa lahawo utshwere ke bolwetsi bahloho-schizophrenia – 1st ed. Pretoria: Dept of National Health & Population Development 1991 [mf ed Pretoria, RSA: State Library [199-]] – 14p [ill] on 1r with other items – 5 – (in zulu & sotho; also available in english & afrikaans with title: a member of your family suffers from schizophrenia = lid van u familie ly aan skisofrenie) – mf#op 10153 r23 – us CRL [362]

Ilustracion y valoracion / Naranjo Villegas, Abel – Bogota, Colombia. 1952 – 1r – us UF Libraries [972]

Ilustracion y valoracion / Naranjo Villegas, Abel – Bogota, Colombia. 1952 – 1r – us UF Libraries [972]

Ilustres / Briceno, Manuel – Caracas, Venezuela. 195- – 1r – us UF Libraries [972]

Ilustres / Briceno, Manuel – Caracas, Venezuela. 195- – 1r – us UF Libraries [972]

Ilustrierte folks-shtime – Warsaw PL – 1r – 1 – (in yiddish) – us UMI ProQuest [939]

Ilustrirter pojliszer manczester – Lodz PL, 1930-31 – 1r – 1 – (in yiddish) – us UMI ProQuest [939]

El ilustrisimo fray hipolito sanchez rangel, primer obispo de maynas / Bayle, Constantino & Quecedo, Francisco – Madrid: Razon y Fe, 1944 – 1 – sp Bibl Santa Ana [240]
Ilustrisimo senor doctor jose antonio ponte. caracas, 1929 / Sosa Saa, Jose Thomas – Madrid: Razon y Fe, 1930 – 1 – sp Bibl Santa Ana [946]
Ilwof, Franz *see* Der protestantismus in steiermark, kaernten und krain
Ilwu bulletin / International Longshoreman's and Warehousemen's Union – v1-v2 n21[1937 sep 17-1938 may 28] – 1r – 1 – mf#1058586 – us WHS [331]
Ilwu dispatcher / International Longshoreman's and Warehousemen's Union – 1942 dec 18-1944 feb 25 – 1r – 1 – (cont by: dispatcher [san francisco, ca: 1944]) – mf#2898165 – us WHS [331]
Ilwu reporter / International Longshoreman's and Warehousemen's Union – 1949 feb 16-1960 dec 21 – 1r – 1 – (cont by: voice of the ilwu) – mf#1058587 – us WHS [331]
Im alten deutschland : erinnerungen eines sechzigjaehrigen / Litzmann, Berthold – Berlin: G Grote, 1923 – 1r – 1 – (incl bibl ref) – us Wisconsin U Libr [943]
Im anfang liegt das ende : grillparzers epilog auf die geschichte / Schneider, Reinhold – Baden-Baden: H Buehler, 1946 – 1r – 1 – us Wisconsin U Libr [943]
Im bann der irredenta : roman / Samarow, Gregor – Stuttgart: Deutsche Verlags-Anstalt, 1889 – 1r – 1 – us Wisconsin U Libr [830]
Im banne der goetzen : schilderungen aus dem missionsleben in indien. dem aus dem englischen uebersetzten buche von amy wilson-carmichael: "tatsachen vom sudindischen missionsfelde" entnommen und frei nacherzaehlt / Bruchhaus, K – Bethel bei Bielefeld: Evangelische Missionsgesellschaft fuer Deutsch-Ostafrika, 1909 [mf ed 1995] – (= ser Yale coll) – 39p (ill) – 1 – 0-524-09974-X – (in german) – mf#1995-0974 – us ATLA [954]
Im banne des schicksals : zwei historische erzaehlungen aus der nordostmark / Wichert, Ernst – Bayreuth: Gauverlag Bayreuth, 1944 – 1r – 1 – us Wisconsin U Libr [830]
Im bannkreis babels : panbabylonistische konstruktionen und religionsgeschichtliche tatsachen / Kugler, Franz Xaver – Muenster i W: Aschendorff, 1910 – 1mf – 9 – 0-524-06890-9 – mf#1991-0033 – us ATLA [930]
Im bannkreis von gesicht und wirken / Schuemann, Kurt – Muenchen: Ner-Tamid, 1959 [mf ed 1993] – 184p – 1 – (with: vier vortragsstudien by max brod et al) – mf#8262 – us Wisconsin U Libr [430]
Im blauen hecht : roman aus dem deutschen kulturleben im anfang des sechzehnten jahrhunderts / Ebers, Georg – Stuttgart: Deutsche Verlags-Anstalt, [1893-97?] [mf ed 1993] – (= ser Georg ebers gesammelte werke 30) – 233p – 1 – mf#8554 reel 5 – us Wisconsin U Libr [830]
Im busch : australische erzaehlung / Gerstaecker, Friedrich; ed by Kaiser, Georg – Leipzig: C Grumbach, [1916?] – 1r – 1 – us Wisconsin U Libr [430]
Die im cod vat reg lat 9 vorgeheftete liste paul lesungen fuer die messfeier (tab35) / Dold, Alban – 1944 – (= ser Texte und arbeiten. beuron (tab). beitraege zur ergruendung des aelteren lateinischen und christlichen schrifttums und gottesdienstes) – €5.00 – ne Slangenburg [240]
Im deutschen reich – Berlin: Centralverein deutscher Staatsbuerger juedischen Glaubens. v1-28. 1895-1922 [complete] – (= ser German-jewish periodicals...1768-1945, pt 1) – 6r – 1 – $675.00 – mf#B105 – us UPA [939]
Im deutschen reich – Berlin DE, 1895 jul-1922 apr – 6r – 1 – (filmed by other misc inst: 1918-20 [2r]) – gw Misc Inst [943]
Im dunklen erdteil : reisen und erlebnisse stanleys, emin paschas, rohfs u. a., fur die reifere jugend / Burmann, Karl – Leipzig, O Drewitz Nachf. [189-] – us CRL [916]
Im familien-kreise – jahrg 5 n16, jahrg 9 n7, jahrg 20 [i.e.2?] n10 p73-80 [den 29 jul 1882, 27 mar 1886, 188-?] – 1r – 1 – (cont: sonntags-blatt des herold von milwaukee) – mf#1003162 – us WHS [071]
Im familien-kreise – jahrg 33 n25 [1890 jun 19] – 1r – 1 – mf#1003174 – us WHS [071]
Im familien-kreise – (Milwaukee). 1880-81 – 1 – us AJPC [071]
Im felde gegen die hereros : erlebnisse eines mitkampfers / Bulow, Franz von – Bremen: G A v Harlem, [1905?] – 1 – us CRL [960]
Im foehn : roman / Fluecht, Liselott – Berlin: R Moelich [194-?] (mf ed 1990) – 1r – 1 – (filmed with: lothar) – us Wisconsin U Libr [830]
Im foehn : roman / Fluecht, Liselott – Berlin: R Moelich, [194-?] (mf ed 1990) – 1r – 1 – (filmed with: lothar) – us Wisconsin U Libr [830]
Im garten der frau maria strom : roman / Boehlau, Helene – Stuttgart: Deutsche Verlags-Anstalt, 1922 [mf ed 1989] – 330p – 1 – mf#7042 – us Wisconsin U Libr [830]
Im grossdeutschen reiche : [essays] / Brehm, Bruno – Wien: A Luser 1940 [mf ed 1989] – 1r – 1 – (filmed with other titles) – mf#7066 – us Wisconsin U Libr [840]
Im gruenen salon : novellen vom stil in der liebe / Gleichen-Russwurm, Alexander, Freiherr von – Wien: Phaidon-Verlag, 1928 (mf ed 1990) – 1r – 1 – (filmed with: golowin) – us Wisconsin U Libr [830]
Im gruenen tann : schwarzwaldnovellen / Achleitner, Arthur – Berlin: Verein der Buecherfreunde, Schall & Grund, [1897?] [mf ed 1995] – 244p – 1 – mf#8918 – us Wisconsin U Libr [830]
'Im Hatsi Yovel / Dayan, Shmuel – Tel-Aviv, Israel. 1934/35 – 1r – 1 – us UF Libraries [939]
Im hause des kommerienrats : roman / Marlitt, Eugenie [pseud] – Stuttgart: Union Deutsche Verlagsgesellschaft [1919] [mf ed 1978] – (= ser Romane und novellen 5) – 2r – 1 – mf#film mas c376 – us Harvard [830]
Im herero- und hottentottenland / Trautmann, O – Pretoria, State Library, 1979. Orig. publ., Oldenburg: Gerhard Stalling, 1913. (Microfiche Reprint Series, State Library, no.4). 4 fiche and printed booklet – 9 – sa National [916]
Im herrgottswinkel *see* Freiburger tagespost
Im herzen der haussalaender : reise im westlichen sudan nebst bericht ueber den verlauf der deutschen niger-benue expedition... / Staudinger, Paul – Berlin: Landsberger, 1889 – 1 – us CRL [960]
Im herzen von afrika / Schweinfurth, Georg August – Leipzig, Germany. 1922 – 1r – us UF Libraries [960]
Im herzen von afrika / Schweinfurth, Georg August – Leipzig, Germany. 1922 – 1r – us UF Libraries [960]
Im hirtenhaus : eine oberfraenkische dorfgeschichte / Schaumberger, Heinrich – Leipzig: P Reclam, [1905?] – 1r – 1 – us Wisconsin U Libr [390]
Im hochland von mittel-kamerun / Thorbecke, Franz – Hamburg: L Friedrichsen, 1914-51 – 1 – us CRL [960]
Im innern afrikas : die erforschung des kassai waehrend der jahre 1883, 1884 und 1885 / Wissmann, Hermann von – Leipzig 1888 [mf ed Hildesheim 1995-98] – 1v on 4mf – 9 – €120.00 – 3-487-27316-0 – gw Olms [916]
Im innern afrikas / Wissmann, H von – Leipzig, 1891 – 10mf – 9 – mf#A-174 – ne IDC [916]
Der im irrgarten der liebe herumtaumelnde kavalier / Schnabel, Johann Gottfried – Muenchen: G Mueller, 1920 [mf ed 1992] – (= ser Die buecher der abtei thelem 26-27) – 2v – 1 – mf#7716 – us Wisconsin U Libr [830]
Im jahre 38 / Tumler, Franz – Muenchen: A. Langen, G Mueller, 1939 – 1r – 1 – us Wisconsin U Libr [943]
"Im kampf um das dritte reich" / U.S. Library of Congress. Prints and Photographs Division – sep 1936-jun 1942. 4 reels – 23v on 4r – 1 – us L of C Photodup [943]
Im kampf um die weltanschauung : bekenntnisse eines theologen / Wimmer, Richard – 10.-12. Aufl. Freiburg i.B.: J C B Mohr (Paul Siebeck), 1893 – 1mf – 9 – 0-8370-5869-4 – mf#1985-3869 – us ATLA [240]
Im kampf um gott und um das eigene ich : ernsthafte plaudereien / Koenig, Karl – Freiburg i. B: Paul Waetze, 1901 – 1mf – 9 – 0-8370-3976-2 – (incl bibl ref) – mf#1985-1976 – us ATLA [240]
Im klassenkampf : deutsche revolutionaere lieder und gedichte aus der zweiten haelfte des 19. jahrhunderts / ed by Friedrich, Wolfgang – Halle: Verlag Sprache und Literatur, 1962 [mf ed 1993] – (= ser Literarisches erbe 2) – 223p – 1 – mf#8360 – us Wisconsin U Libr [810]
Im kreuzfeuer zweier revolutionen / Boehm, Wilhelm – Munich. 1924 – 1 – us CRL [943]
Im lande der hindus : oder, kulturschilderungen aus indien, mit besonderer bereucksichtigung der evangelischen mission / Tanner, Th – St Louis: [s.n.] 1894 [mf ed 1995] – (= ser Yale coll) – 141p (ill) – 1 – 0-524-09058-0 – (in german) – mf#1995-0058 – us ATLA [954]
Im lande der verheissung : ein kolonialroman um carl peters / Buelow, Frieda, Freiin von – 3. aufl. Berlin: O Arnold 1943 [mf ed 1989] – 1r – 1 – (filmed with: der philister vor, in und nach geschichte / clemens brentano) – mf#7094 – us Wisconsin U Libr [830]
Im lande des negus / Escherich, G – Berlin, 1912 – 3mf – 9 – mf#NE-20274 – ne IDC [956]
Im letzten wagen : novelle / Frank, Leonhard – Berlin: E Rowohlt, 1925 (mf ed 1990) – 1r – 1 – (filmed with: trenck) – us Wisconsin U Libr [830]
Im morgenroth : eine muenchener geschichte aus der zeit max joseph's des dritten / Schmid, Herman – 2.aufl. Leipzig: Keil [18–?] – (= ser Gesammelte schriften. volks- und familien-ausgabe 24/25) – 2pt in 1v – 1 – mf#film mas c438 – us Harvard [830]
Im namen jesu : eine sprach- und religionsgeschichtliche untersuchung zum neuen testament, speziell zur altchristlichen taufe / Heitmueller, Wilhelm – Goettingen: Vandenhoeck & Ruprecht, 1903 – 1mf – 9 – 0-7905-1093-6 – (in german, greek, and hebrew. incl bibl ref and indexes) – mf#1987-1093 – us ATLA [220]
Im osten feuer / Czech-Jochberg, Erich – Leipzig: Grossdeutsche Buchgemeinde, 1931 [mf ed 1989] – 293p – 1 – (incl bibl) – mf#7161 – us Wisconsin U Libr [830]
Im rampenlicht – Berlin DE, 1919 n1 – 1 – gw Mikrofilm [790]
Im regiment : roman / Osten, H von – Stuttgart: Deutsche Verlags-Anstalt, 1889 – 1r – 1 – us Wisconsin U Libr [830]
I'm right enough, missus – London, England. 18– – 1r – us UF Libraries [240]
I'm right enough, missus... – London, England. 18– – 1r – us UF Libraries [240]
Im schatten der exzellenz : novelle / Moeller, Karl von – Muenchen: F Eher, [194-?] – 1r – 1 – us Wisconsin U Libr [830]
Im schillingshof : roman / Marlitt, Eugenie [pseud] – Stuttgart: Union Deutsche Verlags-Gesellschaft [1919] [mf ed 1978] – (= ser Romane und novellen 4) – 2r – 1 – mf#film mas c376 – us Harvard [830]
Im schmiedefeuer : roman aus dem alten nuernberg / Ebers, Georg – Stuttgart: Deutsche Verlags-Anstalt, [1893-97?] [mf ed 1993] – (= ser Georg ebers gesammelte werke 28-29) – 2v – 1 – mf#8554 reel 5 – us Wisconsin U Libr [830]
Im schritt der jahrhunderte : geschichtliche bilder / Molo, Walter Ritter von – 7.-10. verm. Aufl. Muenchen: A Langen, c1918 – 1r – 1 – us Wisconsin U Libr [943]
Im sommer danach / Ball, Kurt Herwarth – Halle (Saale): Mitteldeutscher Verlag, 1966 [mf ed 1995] – 262p – 1 – mf#8970 – us Wisconsin U Libr [890]
Im spiegel der form : stilkritische wege zur deutung von stefan georges maximindichtung / Aler, Jan – Amsterdam: M Hertzberger, 1947 [mf ed 1989] – 304p – 1 – (incl bibl ref) – mf#7294 – us Wisconsin U Libr [430]
Im strom : roman eines lebens / Johann, A E – Berlin: Deutscher Verlag, c1942 – 1r – 1 – us Wisconsin U Libr [830]
Im tempo der zeit *see* Der bohrkumpel
Im umkreis von vier meilen / Rothenburg, Adelheid von – Halle: Julius Fricke, 1876 – 1r – 1 – us Wisconsin U Libr [830]
Im umstrittenen gebiet : roman / Janitschek, Ellinor – Berlin: Phoenix-Verlag C Siwinna, c1921 – 1r – 1 – us Wisconsin U Libr [830]
Im urteil der dichter : die deutsche literatur von lessing bis hauptmann / Mulot, Arno – Muenchen: Bayerischer Schulbuch-Verlag, 1957 – 349p – 1 – (incl bibl ref and ind) – us Wisconsin U Libr [430]
Im urteil der dichter *see* Christian oesers geschichte der deutschen poesie in umrissen und schilderungen
Im vorderen asien : politische und andere fahrten / Rohrbach, P – Berlin, 1901 – 2mf – 9 – mf#AR-1969 – ne IDC [915]
Im weissen roessl : lustspiel in drei aufzuegen / Blumenthal, Oskar & Kadelburg, Gustav – 2. aufl. Charlottenburg: M Simson, c1898 [mf ed 1993] – 130p – 1 – mf#8520 – us Wisconsin U Libr [820]
Im, Yang Tjoe *see* Lajangan biroe
Im zwischenland : fuenf geschichten aus dem seelenleben halbwuechsiger maedchen / Andreas-Salome, Lou – 3. aufl. Stuttgart: Cotta, 1911 [mf ed 1988] – 410p – 1 – mf#6940 n6 – us Wisconsin U Libr [830]
Der imaam el-schaafiri : seine schueler und anhaenger bis zum j. 300 d. h / Wuestenfeld, Ferdinand – Goettingen: Dieterich, 1890-1891 – 1mf – 9 – 0-524-04541-0 – mf#1990-3375 – us ATLA [260]
L'image : edition: beloeil, mcmasterville, st-hilaire, otterburn park – [Longueuil]: [s.n.] v1 n1. 27 sep 1978 [mf ed 1989] – 1mf – 9 – (merged with: l'image (ed: boucherville, st-bruno, st-basile, ste-julie) to become: l'image (ed: boucheville, st-basile, st-bruno, ste-julie, beloeil, mcmasterville, st-hilaire, otterburn park); suppl to: l'image de la rive-sud) – cn Bibl Nat [073]
L'image : edition: boucherville; st-basile, st-bruno, ste-julie, beloeil, mcmasterville, st-hilaire, otterburn park – [Longueuil]: [s.n.] v1 n2 4 oct 1978- (wkly) [mf ed 1989] – 9 – (merged with: l'image (edition: boucherville, st-bruno, st-basile, ste julie) and: l'image (edition: beloeil, mcmasterville, st-hilaire, otterburn park); ceased 198-?) – cn Bibl Nat [073]
L'image : edition: boucherville, st-bruno, st-basile, ste-julie – [Longueuil]: [s.n.] v1 n1 27 sep 1978 [mf ed 1989]: – 1mf – 9 – (merged with: l'image (edition: beloeil, mcmasterville, st-hilaire, otterburn park) to become: l'image (edition: boucherville. st-basile, st-bruno, ste-julie, beloeil, mcmasterville, st-hilaire, otterburn park); suppl to: l'image de la rive-sud) – mf#SEM105P1136 – cn Bibl Nat [073]
L'image : edition: brossard, st-hubert, greenfield park, st-lambert, ville lemoyne – Longueuil: [s.n.] v1 n10 16 mai 1979- (wkly) [mf ed 1989] – 9 – (suppl to: l'image de la rive-sud) – mf#SEM105P1137 – cn Bibl Nat [073]
L'image – Paris, 1896-97 [mf ed Chadwyck-Healey] – (= ser Art periodicals on microform) – 1r + 16 col slides – 1 – uk Chadwyck [700]
L'image : revue litteraire et artistique ornee de figures sur bois – Paris. n1-17. sept 1894-nov 1899 – 1 – fr ACRPP [073]
L'image *see* L'image de la rive sud
Image – Rochester. 1952+ (1) 1972+ (5) 1975+ (9) – ISSN: 0536-5465 – mf#8194 – us UMI ProQuest [770]
Image and vision computing – Kidlington. 1983+ (1,5,9) – ISSN: 0262-8856 – mf#14189 – us UMI ProQuest [000]
L'image de la rive sud = South shore image – Longueuil: l'Image de la Rive Sud (Ville St-Laurent: Impr transcontinentale). v1 n1 8 mars 1978-v3 n27 10 sep 1980 (wkly) [mf ed 1987] – 5r – 1 – (with suppl: l'image) – mf#SEM35P273 – cn Bibl Nat [073]
L'image de la rive-sud *see*
– L'image
L'image (edition: beloeil, mcmasterville, st-hilaire, otterburn park) *see*
– L'image
L'image (edition: boucherville. st-basile, st-bruno, ste-julie, beloeil, mcmasterville, st-hilaire, otterburn park) *see* L'image
L'image (edition: boucherville, st-bruno, st-basile, ste julie) *see* L'image
L'image (edition: boucherville, st-bruno, st-basile, ste-julie) *see* L'image
L'image (edition: boucheville, st-basile, st-bruno, ste-julie, beloeil, mcmasterville, st-hilaire, otterburn park) *see* L'image
Image of a cross in pagan, christian, and anti-christian symbolism / O'connor, T Clifford – Dublin, Ireland. 1894 – 1r – us UF Libraries [240]
Image of a cross in pagan, christian, and anti-christian symbolism / O'connor, T Clifford – Dublin, Ireland. 1894 – 1r – us UF Libraries [240]
The image of god in man according to cyril of alexandria (sca14) / Burghardt, W J – Washington DC, 1957 – (= ser Studies in christian antiquity (sca)) – 4mf – 9 – €11.00 – ne Slangenburg [241]
Image technology : journal of the bksts – London. 1989-1996 (1) – (cont: bksts journal) – ISSN: 0950-2114 – mf#5327,02 – us UMI ProQuest [790]
Image technology – Washington. 1968-1973 (1) – ISSN: 0019-2651 – mf#7949 – us UMI ProQuest [621]
Image – the journal of nursing scholarship – Indianapolis. 1992-1999 (1,5,9) – (cont by: journal of nursing scholarship) – ISSN: 0743-5150 – mf#19463,01 – us UMI ProQuest [610]
Imagen de la virgen maria madre de dios de guadalupe : milagrosamente aparecida en la ciudad de mexico... / Sanchez, Miguel – en Mexico: ...Bernardo Calderon...la calle de San Agustin 1648 – (= ser Books on religion...1543/44-c1800: milagros y culto de la virgen) – 2mf – 9 – mf#crl-80 – ne IDC [241]
Imagen de varona / Ferrer Canales, Jose – Santiago, Cuba. 1964 – 1r – us UF Libraries [972]
Imagenes a la deriva / Lainez, Jose Jorge – San Salvador, El Salvador. 1962 – 1r – us UF Libraries [972]
Imagenes a la deriva / Lainez, Jose Jorge – San Salvador, El Salvador. 1962 – 1r – us UF Libraries [972]
Imagenes de chile vida y costumbres chilenas en los siglos 18 y 19 a traves de... / Slas, Mariano Picon & Cruz, Guillermo Feliu – Madrid: Razon y Fe, 1940 – sp Bibl Santa Ana [306]
Imagenes de honduras – Tegucigalpa, Mexico. 1949 – 1r – us UF Libraries [972]
Imagenes de honduras – Tegucigalpa, Mexico. 1949 – 1r – us UF Libraries [972]
Imagenes de la revolucion / Guia Civica De Guatemala – S.I., S.I? . 1951? – 1r – us UF Libraries [972]
Imagenes de la revolucion / Guia Civica De Guatemala – S.I., S.I? . 1951? – 1r – us UF Libraries [972]
Imagenes de luz y sombra / Rosas Milian, Bernardo – Habana, Cuba. 1957 – 1r – us UF Libraries [972]
Imagenes de luz y sombra / Rosas Milian, Bernardo – Habana, Cuba. 1957 – 1r – us UF Libraries [972]

Imagenes sobre el otono / Lopez Vallecillos, Italo – San Salvador, El Salvador. 1962 – 1r – us UF Libraries [972]

Imagenes sobre el otono / Lopez Vallecillos, Italo – San Salvador, El Salvador. 1962 – 1r – us UF Libraries [972]

Imagenes (versiones poeticas) rosas del tiempo antiguo. mies de logrono / Diez Canedo, Enrique – Paris: Sociedad de Ediciones Literarios y artisticos, Libreria Paul Ollendorff, S.A. – 1 – sp Bibl Santa Ana [810]

Imagens do brasil / Koseritz, Carlos Von – Sao Paulo, Brazil. 1943 – 1r – us UF Libraries [972]

Imagens do brasil / Koseritz, Carlos Von – Sao Paulo, Brazil. 1943 – 1r – us UF Libraries [972]

Images / Blue Cross Blue Shield United of Wisconsin – 1980: [v1 n1]-1983 dec – 1r – 1 – (cont: spotlight [blue cross of wisconsin]) – mf#965403 – us WHS [071]

Images – Toronto, Ontario. Mar 1980-Nov 1984. Many issues missing. Continued by: The Scribe (Jewish Students' Union-B'nai B'rith Hillel Foundation (University of Tronto) – 1 – us AJPC [071]

Images de dix congres nationaux, septembre 1955-decembre 1960 / Cambodia. Ministere de l'information – [Phnom-Penh 1961?] [mf ed 1989] – 1r with other items – 1 – (in french & cambodian) – mf#mf-10289 seam reel 029/10 [§] – us CRL [959]

Images de france : plaisir de france – Paris, France. Nov 1940-1941 – 1r – 1 – uk British Libr Newspaper [072]

Images de paris : revue libre d'art et de litterature – Paris. oct 1919-mai juil 1926 – 1 – fr ACRPP [073]

Images des antilles / Fiumi, Lionello – Paris, France. 1937 – 1r – us UF Libraries [972]

Images des antilles / Fiumi, Lionello – Paris, France. 1937 – 1r – us UF Libraries [972]

Images et symboles: essais sur le symbolisme magico-religieux / Eliade, Mircea – (Paris): Gallimard, (1952). 238p – 1 – us Wisconsin U Libr [150]

Images of dutch towns and villages in the 18th century : the historical-topographical "atlas" of andries schoemaker – [mf ed 2001] – 278 b/w mf – 9 – €2295.00 – (with p/g; also available on cd-rom) – mf#m480 – KOG, the library of the Rijksmuseum Amsterdam and other dutch institutions – ne MMF Publ [520]

Images of east and west: maps, plans, views and drawings from dutch colonial archives, 1583-1950 : pt 1: the early period, 1583-1814 / The Hague. National Archives of the Netherlands – [1840] – 1519mf – 9 – €5980.00 (€12,175.00 set) – (subsections available: atlases €4255; europe €265; africa €315; asia €1810; australia €22; north america €88; caribbean €370; south america €370; with p/g, concordances & original inventories (in repr)) – mf#m303 – ne MMF Publ [912]

Images of east and west: maps, plans, views and drawings from dutch colonial archives, 1583-1963 : pt 2: the collection of the ministry of the colonies, 1814-1963 / The Hague. National Archives of the Netherlands – [1993] – 1929mf – 9 – €7550.00 (€12,175.00 set) – (with p/g & concordance) – mf#m304 – ne MMF Publ [912]

Images of god reflected in the church / Stackpole, James Allen – Princeton: Princeton Theo. Sem., [1976] – 1r – 1 – 0-8370-1285-6 – mf#1984-T019 – us ATLA [210]

Images of the renaissance : an original classical ballet / Russell, Marsha – 1997 – 1mf – 9 – $4.00 – mf#PE 3804 – us Kinesology [790]

Les images ou tableaux de platte-peinture de philostrate lemnien sophiste grec... / Philostratus – Paris: N. Chesneau, 1578 – 13mf – 9 – mf#O-1937 – ne IDC [090]

Imaginacion de mexico / Valle, Rafael Heliodoro – Buenos Aires, Argentina. 1948 – 1r – us UF Libraries [972]

Imaginacion de mexico / Valle, Rafael Heliodoro – Buenos Aires, Argentina. 1948 – 1r – us UF Libraries [972]

Die imaginaere und die reale hexe : zwei erscheinungsformen eines sozialgeschichtlichen phaenomens in der fruehen neuzeit. untersuchungen in nord- und nordwestdeutschland vom 16. bis zum 18. jahrhundert / Gailus-Doering, Sigrid – (mf ed 1992) – 3mf – 9 – €49.00 – 3-89349-522-3 – mf#DHS 522 – gw Frankfurter [943]

Imagination / Mccosh, James – London, England. 1857? – 1r – us UF Libraries [240]

Imagination / Mccosh, James – London, England. 1857? – 1r – us UF Libraries [240]

Imagination in landscape painting / Hamerton, Philip Gilbert – London: Seeley & Co, 1887 – (= ser 19th c art & architecture) – 2mf – 9 – (with ill) – mf#4.1.9 – uk Chadwyck [750]

The imagination in spinoza and hume : a comparative study in the light of some recent contributions to psychology / Gore, Willard Clark – 1902 – (= ser University of Chicago Contributions to Philosophy) – 1mf – 9 – 0-7905-9939-2 – mf#1989-1664 – us ATLA [100]

Imagination poetique... / [Aneau, B] – Lyon: Mac, Bonhomme, 1556 – 2mf – 9 – mf#O-1813 – ne IDC [090]

Imagines deorum... / Cartari, V – Francofurti, 1687 – 4mf – 9 – mf#O-1238 – ne IDC [700]

Le imagini de i dei de gli antichi... / Cartari, V – Venetia, 1587 – 6mf – 9 – mf#O-1236 – ne IDC [700]

Imagini delli dei de gli antichi... / Cartari, V – Venetia, 1647 – 6mf – 9 – mf#O-1237 – ne IDC [700]

Imagizationdnavisagery – solving the jumble : clarification of imagery and visualization and implications for teaching dance technique / Hoss, Haley A – 1997 – 1mf – 9 – $4.00 – mf#PE 3982 – us Kinesology [370]

Imago : zeitschrift fuer psychoanalytische psychologie, ihre grenzgebiete und anwendungen / ed by Freud, Sigmund – Wien (A), 1912-37 – 1 – fr ACRPP [150]

Imago christi : the example of jesus christ / Stalker, James – London: Hodder and Stoughton, 1893 – 1mf – 9 – 0-7905-0116-3 – (incl bibl ref) – mf#1987-0116 – us ATLA [240]

Imago primi saeculi societatis jesu a provincia flandro-belgica eiusdem societatis repraesentata... – Antverpiae: Ex officina Plantiniana Balthasaris Moreti, 1640 – 17mf – 9 – mf#O-645 – ne IDC [090]

Imagracao e colonizacao no brasil / Carneiro, J Fernando – Rio de Janeiro, Brazil. 1950 – 1r – us UF Libraries [972]

Imagracao e colonizacao no brasil / Carneiro, J Fernando – Rio de Janeiro, Brazil. 1950 – 1r – us UF Libraries [972]

Imai, John Tashimichi *see* Bushido in the past and in the present

Imam-i rabbani mujaddid-i-alf-i thani shaikh ahmad sirhindi's conception of tawhid : or, the mujaddid's conception of tawhid / Faruqi, Burhan Ahmad – Lahore (India): Sh Muhammad Ashraf, 1940 – (= ser Samp: indian books) – (foreword by syed zafarul hasan) – us CRL [280]

Les imams et les derviches : pratiques, superstitions et moeurs des turcs / Osman – Paris: E Dentu, 1881 – 1mf – 9 – 0-524-01865-0 – mf#1990-2700 – us ATLA [260]

Imana et le culte des manes au rwanda / Pauwels, M – Bruxelles, Belgium. 1958 – 1r – us UF Libraries [960]

Imana et le culte des manes au rwanda / Pauwels, M – Bruxelles, Belgium. 1958 – 1r – us UF Libraries [960]

Imani / Black Allied Student Association [New York University] – 1971 feb – 1r – 1 – (cont: faith [[black allied student association [new york university]]) – mf#2530967 – us WHS [378]

Imbart de La Tour, Pierre *see*
– Les elections episcopales dans l'eglise de france du 9e au 12e siecle
– Questions d'histoire sociale et religieuse

Imbault-Huart, C *see* Le pays de hami ou khamil

Imbe-no-hironari's kogoshui : or, gleanings from ancient stories / Inbe, Hironari – Tokyo: Meiji Japan Society. ii/109p. 1924 – 1 – (transl with introd and notes by genchi kato and hikoshiro hoshino. bibl and ind in english and japanese) – us Wisconsin U Libr [950]

Imberno, Pedro Jose *see* Guia geografica y administrativa de la isla de cub...

Imbert, Jean *see*
– Enchiridon
– La practiqve lvdiciaire: tant civile qve criminelle, receve et obseruee par tout de royaume de france

Imbituba – Imbituba, SC: Typ Patria, 13 abr 1924; 18 set 1926 – (= ser Ps 19) – mf#UFSC/BPESC – bl Biblioteca [079]

Imbrie, William *see* The church of christ in japan

Imbusch, H *see* 25 jahre gewerkverein christlicher bergarbeiter

Imc journal – Silver Spring. 1967-1995 (1) 1970-1995 (5) 1970-1995 (9) – (cont by: document world) – ISSN: 0019-0012 – mf#5905 – us UMI ProQuest [020]

I-mei hsiao p'in hsu chi / Cheng, I-mei – Shang-hai: Chung fu shu chu, Min kuo 23 [1934] – (= ser P-k&k period) – us CRL [480]

I-meng tso i san ch'i tiao ch'a pao kao shu – [SI]: Meng Tsang wei yuan hui tiao ch'a shih, Min kuo 30 [1941] – (= ser P-k&k period) – us CRL [951]

I-meng yu i ssu ch'i tiao ch'a pao kao shu – [SI]: Meng Tsang wei yuan hui tiao ch'a shih, 1939 – (= ser P-k&k period) – us CRL [951]

Imes, Merle Graybill *see* "The flood-1903"

Imf reports and summary proceedings / International Monetary Fund (IMF) – (= ser Reports and summary proceedings of the imf and the world bank, 1946-1974) – 6r – 1 – $935.00 – 0-89093-008-2 – us UPA [336]

Imf staff papers / International Monetary Fund – Washington. 1999+ (1) 1999+ (5) 1999+ (9) – (cont: international monetary fund staff papers) – ISSN: 1020-7635 – mf#2212,01 – us UMI ProQuest [332]

Imf survey / International Monetary Fund – Washington. 1972+ [1,5]; 1975+ [9] – ISSN: 0047-083X – mf#6530 – us UMI ProQuest [332]

Imfinyezo yemiteto leseluleko esenzelwa ilijioni kamaria / Roggendorf, H Th – Bulawayo, Zimbabwe. 1944 – 1r – us UF Libraries [960]

Imfinyezo yemiteto leseluleko esenzelwa ilijioni kamaria / Roggendorf, H Th – Bulawayo, Zimbabwe. 1944 – 1r – us UF Libraries [960]

Imfundiso yetyalike ekatolike yase roma / Hornig, Josef – Keilands, South Africa. 1897 – 1r – us UF Libraries [960]

Imfundiso yetyalike ekatolike yase roma / Hornig, Josef – Keilands, South Africa. 1897 – 1r – us UF Libraries [960]

Imfundiso yetyalike ekatolike yase roma / Hornig, Josef – Mariannhill, South Africa. 19– – 1r – us UF Libraries [960]

Imfundiso yetyalike ekatolike yase roma / Hornig, Josef – Mariannhill, South Africa. 19– – 1r – us UF Libraries [960]

Imfundo ngesini / South Africa. Department of Health [Departement van Gesondheid] [Departement van Gesondheid] – Pretoria: Dept of Health [1979?] [mf ed Pretoria, RSA: State Library [199-]] – 22p [ill] on 1r with other items – 5 – (in xhosa) – mf#op 06754 r24 – us CRL [613]

Imhoff, Alexander Jesse *see* The life of rev. morris officer, a.m.

Imibengo / Bennie, W G – Lovedale, South Africa. 1949 – 1r – us UF Libraries [960]

Imibengo / Bennie, W G – Lovedale, South Africa. 1949 – 1r – us UF Libraries [960]

Imibono yabomdabu = Black opinion – King William's Town SA, 4 oct 1985-30 may 1986 – 1r – 1 – sa National [079]

Imikhemezelo / Nyembezi, C L Sibusiso (Cyril Lincoln Sibusiso) – Pietermaritzburg, South Africa. 1963 – 1r – us UF Libraries [960]

Imikhemezelo / Nyembezi, Cyril L S – Pietermaritzburg, South Africa. 1963 – 1r – us UF Libraries [960]

Imikhuba emihle / Preston, Hilary – Gwelo, Zimbabwe. 1970 – 1r – us UF Libraries [960]

Imikhuba emihle / Preston, Hilary – Gwelo, Zimbabwe. 1970 – 1r – us UF Libraries [960]

Imilandu ya babemba / Tanguy, F – London, England. 1948 – 1r – us UF Libraries [960]

Imilandu ya babemba / Tanguy, F – London, England. 1948 – 1r – us UF Libraries [960]

Iminshoni ya umca / Chillombo, A – Cape Town, South Africa. 1957 – 1r – us UF Libraries [960]

Iminshoni ya umca / Chillombo, A – Cape Town, South Africa. 1957 – 1r – us UF Libraries [960]

Imirce : ou la fille de la nature / DuLaurens, Henri J – Berlin: Impr du Philosophe de Sans Souci, 1765 – (= ser Les femmes [coll]) – 4mf – 9 – mf#9890 – fr Bibl Nationale [830]

Imishue, R *see* South west africa

Imitacion de una oda de m rouseau : en la que una alma se eleva al conocimiento de dios por la contemplacion de sus obras – [Lima: Se vende en la libreria Calle del Arzobispo 18-] – (= ser Books on religion... 1543/44-c1800: mistica y meditacion) – 1mf – 9 – mf#crl-286 – ne IDC [241]

L'imitateur de jesus-christ : ou la vie du venerable pere antoine yvan / Gondon, G – Paris, 1662 – 7mf – 9 – mf#CA-168 – ne IDC [240]

Imitatio christi *see* Zwei urschriften der 'imitatio christi'

Imitatio Christi Book 3 Shona 1937 *see* Chitevedzero cha kriste

Imitatio Christi Book 4 Shona 1936 *see* Chitevedzero cha kriste

Imitatio crameriana sive exercitium pietatis domesticum / Ammon, H. – Noribergae: Typis et sumptibus leremiae Duemleri, 1647 – 2mf – 9 – mf#O-540 – ne IDC [090]

L'imitation de jesus-christ : traduction nouvelle avec des reflexions a la fin de chaque chapitre par l'abbe F de Lamennais, suivie des prieres durant la sainte messe, des vepres du dimanche et du chemin de la croix – Montreal: Librairie Guay; (Turnhout (Belgique): Etabliss H Proost & co [1914?] (mf ed 1992) – 5mf – 9 – mf#SEM105P1644 – cn Bibl Nat [230]

L'imitation de jesus-christ : traduction nouvelle, avec une pratique et une priere a la fin de chaque chapitre / Gonnelieu, R P de – nouv augm ed. Quebec: Nouvelle Impr, 1813 [mf ed 1972] – 1r – 5 – mf#SEM16P21 – cn Bibl Nat [230]

Imitation de jesus-christ : ...avec une pratique et une priere a la fin de chaque chapitre, suivie de la messe et des vepres – Quebec: J P Garneau, libraire-editeur, [entre 1914 et 1920] (mf ed 1992) – 5mf – 9 – (trans by jerome de gonnelieu; incl latin text) – mf#SEM105P1646 – cn Bibl Nat [230]

The imitation of christ : new revised translation / Thomas a Kempis – San Francisco: Catholic Truth Society, 1905 – 1mf – 9 – 0-524-04189-X – mf#1990-1228 – us ATLA [240]

The imitation of srankara : being a collection of several texts bearing on the advaita / Dvivedi, Manilal Nabhubhai – Bombay: Jyestaram Mukundji, 1895 – 3mf – 9 – 0-524-07375-9 – mf#1991-0095 – us ATLA [280]

The imitation of zoroaster : quotations from zoroastrian literature – London: Cooper Pub Co, [1910?] – 1mf – 9 – 0-524-01264-4 – mf#1990-2300 – us ATLA [280]

Imkaniyat al-tanmiyah al-ziraiyah fi sina / Sadiq, Fawziyah Mahmud – al-Kuwayt: Qism al-Jughrafiya bi-Jamiat al-Kuwayt wa-al-Jamiyah al- Jughrafiyah al-Kuwaytiyah, 1983 – us CRL [956]

Imlay, Gilbert *see* The emigrants

Imlay magnet – Eden, jul 1973-jul 1993 – 45r – at Pascoe [079]

L'immacolata concezione di maria vergine e la chiesa greca ortodossa dissidente / Marini, Niccolo – Roma: Cav V Salviucci, 1908 [mf ed 1986] – 1mf – 9 – 0-8370-8200-5 – (in italian. incl bibl ref) – mf#1986-2200 – us ATLA [243]

L'immacolata davanti al razionalismo / Beninati, Giuseppe – 2a ed. Ragusa: Piccitto & Antoci, 1880 [mf ed 1986] – 1mf – 9 – 0-8370-7768-0 – (in italian. incl bibl ref) – mf#1986-1768 – us ATLA [241]

L'immacolata nel secolo 19 : panegirico detto nella cattedrale di andria il di 8 dicembre 1876 / Magno, Giuseppe – 2a ed. Roma: Tipografia della Pace, 1877 [mf ed 1986] – 1mf – 9 – 0-8370-8127-0 – (in italian) – mf#1986-2127 – us ATLA [241]

Immaculata – 1950 may-1955, 1956-61 – 2r – 1 – (cont by: immaculate) – mf#1209653 – us WHS [071]

Immaculata – 1970 may-1973, 1974-76, 1977-79, 1980-1982 dec/1983 jan, 1983 feb-1984 dec/1985 jan – 5r – 1 – (cont: immaculate) – mf#1206712 – us WHS [071]

Immaculate – 1962-65, 1966-1970 apr – 2r – 1 – (cont: immaculata [kensoha, wis.: 1950]; cont by: immaculata [kensoha, wis: 1970]) – mf#1206711 – us WHS [071]

Immaculate conception / Costa, Francesco – Glasgow, Scotland. 1855 – 1r – us UF Libraries [240]

Immaculate conception / Costa, Francesco – Glasgow, Scotland. 1855 – 1r – us UF Libraries [240]

The immaculate conception : its antecedents and concequences / Cumming, John – London: James Miller, [1861] – 1mf – 9 – 0-8370-7931-4 – mf#1986-1931 – us ATLA [240]

The immaculate conception : summary of conferences = Resume de conferences sur le dogma l'immaculate conception / Vercruysse, Bruno – Dublin: McGlashan & Gill, 1874 – 1mf – 9 – 0-8370-8395-8 – (in english) – mf#1986-2395 – us ATLA [240]

Immaculate Conception Indian Mission School *see* Wopeedah

Immaculate conception of our lord and saviour jesus christ / Walsh, Wlliam Pakenham – Dublin, Ireland. 1855 – 1r – us UF Libraries [240]

Immaculate conception of our lord and saviour jesus christ / Walsh, Wlliam Pakenham – Dublin, Ireland. 1855 – 1r – us UF Libraries [240]

The immaculate conception of the mother of god : an exposition / Ullathorne, William Bernard – London: Richardson, 1855 – 1mf – 9 – 0-8370-8071-1 – (incl bibl ref) – mf#1986-2071 – us ATLA [241]

The immaculate mary : and other poems / Gahan, James Joseph – Quebec: Barrow, 1876? – 1mf – 9 – mf#04204 – cn Canadiana [810]

L'immaculee conception : etudes sur l'origine d'un dogme / Stap, A – nouv ed. Paris: Librarie internationale; Bruxelles: Lacroix, Verboeckhoven, 1869 [mf ed 1986] – 1mf – 9 – 0-8370-8386-9 – (in french. incl bibl ref and ind) – mf#1986-2386 – us ATLA [241]

L'immaculee conception : histoire d'un dogme catholique-romain ou comment l'heresie devient un dogme / Pressense, Edmond de – 2e ed. Paris: Ch Meyrueis, 1855 [mf ed 1986] – 1mf – 9 – 0-8370-8290-0 – (in french) – mf#1986-2290 – us ATLA [241]

L'immaculee conception : poeme didactique en l'honneur de la sainte vierge / Debouge, Xavier – Bruxelles: Veuve Beugnies, 1855 [mf ed 1986] – 1mf – 9 – 0-8370-7783-4 – (in french) – mf#1986-1783 – us ATLA [810]

L'immaculee conception de la bienheureuse vierge marie consideree comme dogme de foi / Malou, J B – Bruxelles: H Goemaere, 1857 [mf ed 1986] – 4mf – 9 – 0-8370-8276-5 – (in french. incl bibl ref) – mf#1986-2276 – us ATLA [241]
L'immaculee conception de la tres-sainte vierge / Maurel, F Antoine – 3e ed. Lyon: Jules Nicolle, 1866 [mf ed 1986] – 1mf – 9 – 0-8370-8278-1 – (in french. incl bibl ref) – mf#1986-2278 – us ATLA [241]
Immanence : a book of verses / Underhill, Evelyn – London: JM Dent; New York: EP Dutton, 1912 – 1mf – 9 – 0-7905-9723-3 – mf#1989-1448 – us ATLA [420]
Immanence : essai critique sur la doctrine de m. maurice blondel / Tonquedec, Joseph de – Paris: G Beauchesne, 1913 – 1mf – 9 – 0-524-00177-4 – mf#1989-2877 – us ATLA [210]
Immanence and christian thought : implications and suggestions / Platt, Frederic – London: C H Kelly 1915 [mf ed 1992] – 2mf – 9 – 0-7905-8555-3 – (incl bibl ref) – mf#1989-1780 – us ATLA [240]
The immanence of christ in modern life / Swan, Frederick R – London: James Clarke, 1907 – 1mf – 9 – 0-8370-5551-2 – mf#1985-3551 – us ATLA [240]
The immanence of god in rabbinical literature / Abelson, J – London, 1912 – 7mf – 8 – €15.00 – ne Slangenburg [270]
Immanencia y trascendencia del ser y del conocer en heidegger / Frutos Cortes, Eugenio – Madrid, 1950. Rev. Filosofia (Tomo 9, num. 33) del Instituto Luis Vives – sp Bibl Santa Ana [100]
The immanent god, and other sermons / Jackson, Abraham Willard – Boston: Houghton, Mifflin, 1889 – 1mf – 9 – 0-524-00274-6 – mf#1989-2974 – us ATLA [240]
Immanenz und geschichte : zum begriff der kreativitaet in der metaphysik alfred n whiteheads / Herdt, Ludwig – Frankfurt a.M., 1975 – 1mf – 3-89349-851-6 – gw Frankfurter [110]
Immanuel : or, the mystery of the incarnation of the son of god / Ussher, James – London: James Nisbet, 1862 – 1mf – 9 – 0-8370-5649-7 – (repr fr the editions of 1649 and 1677) – mf#1985-3649 – us ATLA [210]
Immanuel : surat parsaoran di huria kristen batak protestant – Tarutung, [195?]1952-1955 – 2mf – 9 – mf#SE-150=5 – ne IDC [950]
Immanuel baptist church – Washington. 1972-1973 (1) 1960-1973 (5) (9) – 1 – $75.11 – (membership rolls and related records 1961-sep 1991) – mf#6311 – us Southern Baptist [242]
Immanuel baptist church. el paso, texas : church records – 1919-75. 1534p – 1 – $69.03 – us Southern Baptist [242]
Immanuel baptist church. henderson, kentucky : church records – 1914-87.6 reels – 1 – us Southern Baptist [242]
Immanuel baptist church. lexington, kentucky : church records – 1909-13 – 1 – 7.11 – us Southern Baptist [242]
Immanuel baptist church. nashville, tennessee : church records – v. 1-8. 1887-1980 – 1 – us Southern Baptist [242]
Immanuel baptist church. paducah, kentucky : church records – 1920-Oct 1964 – 1 – 6.93 – us Southern Baptist [242]
Immanuel baptist church. wichita, kansas : church records – 1914-61 – 1 – us Southern Baptist [242]
Immanuel baptist messenger/temple advocate – Chicago.Immanuel Baptist Church. 1902-03, 1905-16. Incomplete – 1 – (single reels available) – us ABHS [242]
Immanuel kant : a study and a comparison with goethe, leonardo da vinci, bruno, plato and descartes = Immanuel kant / Chamberlain, Houston Stewart – London; New York: J Lane, 1914 – 3mf – 9 – 0-7905-9258-4 – (in english) – mf#1989-2483 – us ATLA [100]
Immanuel kant in england, 1793-1838 / Wellek, Rene – Princeton: Princeton University Press, 1931 – vii/317p – 1 – us Wisconsin U Libr [190]
Immanuel kant on philosophy in general – Calcutta: University Press, 1935 – (= ser Samp: indian books) – (trans, with four introductory essays by humayun kabir) – us CRL [100]
Immanuel kant und alexander von humboldt : eine rechtfertigung kants und eine historische richtigstellung / Lind, Paul von – Erlangen: Fr Junge, 1897 – 1mf – 9 – 0-7905-9309-2 – (incl bibl ref) – mf#1989-2534 – us ATLA [520]
Immanuel kant's auferstehung aus dem grabe : die lehre des alten vom koenigsberge / Noack, Ludwig – Leipzig: Otto Wigand 1861 [mf ed 1991] – 1mf – 9 – 0-7905-8866-8 – mf#1989-2091 – us ATLA [140]
Immanuel kants auffassung von der bibel und seine auslegung derselben : ein kompendium kantscher theologie / Kuegelgen, Constantin von – Leipzig: A Deichert 1896 [mf ed 1991] – 1mf – 9 – 0-7905-8681-9 – (incl bibl ref) – mf#1989-1906 – us ATLA [220]
Immanuel kant's critique of pure reason / Kant, Immanuel – In commemoration of the centenary of its first publ. Trans. by F. Max Mueller, introd. by Ludwig Noire. London: Macmillan, 1881. 2v – 1 – us Wisconsin U Libr [190]
Immanuel kant's critique of pure reason : in commemoration of the centenary of its first publication = Kritik der reinen vernunft / Kant, Immanuel – London: Macmillan, 1881 – 4mf – 9 – 0-7905-7350-4 – (in english) – mf#1989-0575 – us ATLA [120]
Immanuel kant's kleinere schriften zur ethik und religionsphilosophie / Kant, Immanuel; ed by Kirchmann, Julius Hermann von – Berlin: L Heimann, 1870-1871 – (= ser Philosophische Bibliothek) – 1mf – 9 – 0-7905-9397-1 – mf#1989-2622 – us ATLA [170]
Immanuel kants vorlesungen ueber psychologie : mit einer einleitung, kants mystische weltanschauung = Vorlesungen ueber die metaphysik. selections / Kant, Immanuel; ed by Du Prel, Carl – Leipzig: E Guenther, 1889 – 1mf – 9 – 0-7905-7981-2 – (incl bibl ref) – mf#1989-1266 – us ATLA [150]
Immanuel, or, christian realism : a verbatim report of the riddell lectures . . . / Riddell, Newton N – Chicago: Child of Light, c1906 – 1mf – 9 – 0-7905-9610-5 – mf#1989-1335 – us ATLA [240]
Immaterialgueterrechte : familienrecht / Crome, Carl – Tuebingen: J C B Mohr, 1908 – (= ser Civil law 3 coll; System des deutschen buergerlichen rechts) – 8mf – 9 – (incl bibl ref and index) – mf#LLMC 96-549 – us LLMC [348]
The immediate cause of the indian mutiny : as set forth in the official correspondence / Crawshay, George – London, [1858] – (= ser 19th c books on british colonization) – 1mf – 9 – mf#1.1.7434 – uk Chadwyck [954]
The immediate effect of concurrent visual feedback on beginning targets archers / Trexler, James G – 1982 – 1mf – 9 – $4.00 – us Kinesology [790]
The immediate future and other lectures / Besant, Annie Wood – London: Theosophical Pub Society, 1911 – (= ser Samp: indian books) – us CRL [290]
Immel, David D *see* Physiological responses to cardio kickboxing in females
Immensee : mit 23 heliogravueren nach w. hasemann und edmund kanoldt / Storm, Theodor – 3. Aufl. Leipzig: C F Amelang, 1896 – 1r – 1 – us Wisconsin U Libr [430]
Immensee; im sonnenschein; ein gruenes blatt; abseits : novellen / Storm, Theodor – Stuttgart: Verlag Deutsche Volksbuecher, 1942 – 1r – 1 – us Wisconsin U Libr [830]
Immer, Albert *see*
– Hermeneutics of the new testament
– Hermeneutik des neuen testamentes
Immer bereit fuer die verteidigung der freiheit des volkes / Komitee der Antifaschistischen Widerstandskaempfer in der DDR – Berlin, 1956. Fiche W 957. (Blodgett Collection of Spanish Civil War Pamphlets) – 9 – us Harvard [946]
Immermann, Karl Leberecht *see*
– Der carnaval und die somnambuele
– Muenchhausen
– Der oberhof
– Das trauerspiel in tyrol
– Tristan und isolde
Immermanns alexis : eine literarhistorische untersuchung / Leffson, August – Gotha: F A Perthes, 1904 – 1r – 1 – (incl bibl ref) – us Wisconsin U Libr [430]
Immermanns "tristan und isolde" / Szymanzig, Max – Marburg a.L.: N G Elwert, 1911 – 1r – 1 – (incl bibl ref) – us Wisconsin U Libr [430]
Immersion essential to christian baptism / Broadus, John Albert – 76p – 1 – 5.00 – us Southern Baptist [242]
Immersionists against the bible : or, the babel builders confounded / Lee, Nathaniel H; ed by Summers, Thomas Osmond – Nashville TN: Pub House of the Methodist Episcopal Church, South 1870 [mf ed 1993] – 1mf – 9 – 0-524-06254-4 – mf#1990-5209 – us ATLA [242]
Immigrant – 1st issue-2nd iss [1987 may-jun] – 1r – 1 – mf#1533380 – us WHS [321]
The immigrant / Colgate, Robert – (A genealogy of the New York Colgates and some associated lines, comp. by Truman Abbe and Hubert Howson. 1703-1910 – 1 – us Southern Baptist [242]
The immigrant in america : from the new york public library, the balch institute for ethnic studies library in philadelphia and the immigration history research center at the university of minnesota / Hoglund, A William – 1789-1929 – 264r in 7 units (complete coll) – 1 – (based on the holdings of the new york public library, the balch institute for ethnic studies library in philadelphia and the immigrant history research center at the university of minnesota. derivative colls by nationality: czechs, slovaks, hungarians, ukranians 25r c39-27331. poles, slovenes, romanians, lithuanians, russians, carpatho-rusyns 35r c39-27332. norwegians, danes, finns 56r c39-27333. irish, scotch-irish, scots, english, welsh 30r c39-27334. austrians, germans, french, italians, dutch, greeks 60r c39-27335. swedish 21r c39-273336. jewish 23r c39-27337. includes printed guide) – mf#C39-27330 – us Primary [975]
Immigration : the grand desideratum for new south wales: and how to promote it effectually / Lang, John Dunmore – Sydney, 1870 – (= ser 19th c british colonization) – 1mf – 9 – mf#1.1.3492 – uk Chadwyck [320]
Immigration : select documents and case records / Abbott, Edith – Chicago: University of Chicago Press, 1924 [mf ed 1970] – (= ser University of chicago social service series 14811; Library of american civilization) – 1mf – 9 – us Chicago U Pr [342]
Immigration : the special studies series, special studies, 1969-1998 – 1 – $6635.00 coll – (1969-82 12r isbn 0-89093-596-3 $2330. 1982-85 4r isbn 0-89093-627-7 $770. 1985-88 12r isbn 1-55655-132-0 $2330. 1989-98 7r isbn 1-55655-833-3 $1355. with p/g) – us UPA [323]
Immigration and nationality : administrative decisions under the immigration and nationality laws – v1-20. 1940-95 – 190mf – 9 – $285.00 – (contains opinions of the ag, the board of immigration applications, the commissioner of immigration, and the ins) – mf#LLMC 80-900 – us LLMC [324]
Immigration and nationality acts : legislative histories and related documents / ed by Trelles, Oscar M & Bailey, James F – 15v in 16bks – 9 – $535.00 set – 0-89941-334-X – (with index) – mf#301681 – us Hein [340]
Immigration and nationality laws and regulations as of march 1 1944 / U.S. Justice Dept – Washington: GPO, 1944 – 17mf – 9 – $25.50 – (includes suppl nos 1-3 1944-46 (all publ)) – mf#LLMC 81-503 – us LLMC [348]
Immigration and naturalization service annual reports – 1955-82 – (= ser Immigration and nationality administrative decisions) – 50mf – 9 – $75.00 – mf#llmc 81-501 – us LLMC [342]
Immigration and naturalization service case files of chinese immigrants, portland, oregon, 1890-1914 / U.S. Immigration and Naturalization Service – (= ser Records Of The Immigration And Naturalization Service) – 15r – 1 – mf#M1638 – us Nat Archives [975]
Immigration and naturalization service's (ins) interior enforcement strategy : hearing... house of representatives, 107th congress, 2nd session, june 19 2002 / United States. Congress. House. Committee on the Judiciary. Subcommittee on Immigration, Border Security, and Claims – Washington: US GPO 2002 [mf ed 2002] – 1mf – 9 – us GPO [342]
Immigration and refugee services of america, 1918-1985 : from the collection of the immigration history research center, university of minnesota – [mf ed 2003] – 354r in 4pts – 1 – (foreign language information service (flis) 30r. the common council for american unity (ccau) 128r. american federation of international institutes (afii) 49r. american council for nationalities services (acns) 147r) – us Primary [360]
Immigration commission : "dillingham commission" reports / U.S. Immigration Commission – Washington: GPO. 41v. 1911 [all publ] – 320mf – 9 – $480.00 – mf#llmc 81-510 – us LLMC [342]
Immigration history newsletter – 1968 nov-1982 may – 1r – 1 – (cont by: immigration and ethnic history newsletter) – mf#619812 – us WHS [321]
Immigration newsletter / National Immigration Project [US] – v7 n1-v11 n1 [1978 jan/feb-1982 jan/feb] – 1r – 1 – mf#242628 – us WHS [321]
The immigration problem / Jenks, Jeremiah W – New York: Funk & Wagnalls, 1912 c1911.xvi,496p. incl. tables – 1 – us Wisconsin U Libr [323]
The immigration problem: a study of american immigration conditions and needs / Jenks, Jeremiah W – 3rd. ed. rev. and enl. New York, London: Funk & Wagnalls, 1913. xxiii,551p. incl. tables – 1 – us Wisconsin U Libr [323]
Immigration reglemente aux antilles francaises / Guiral, Paul – Paris, France. 1911 – 1r – us UF Libraries [972]
Immigration reglemente aux antilles francaises / Guiral, Paul – Paris, France. 1911 – 1r – us UF Libraries [972]
Imminence! – 1991 apr-aug, oct-dec, 1992 feb-mar, may-jul – 1r – 1 – mf#2562325 – us WHS [071]
Immixtio et consecratio / Andrieu, M – Paris, 1924 – 5mf – 8 – €12.00 – ne Slangenburg [240]
Immortability / Mcconnell, Samuel David – New York, NY. 1930 – 1r – us UF Libraries [960]
Immortability / Mcconnell, Samuel David – New York, NY. 1930 – 1r – us UF Libraries [960]
The immortal history of south africa : the only truthful, political, colonial...history...of the cape colony, natal, the orange free state, transvaal, and south africa / Boon, Martin James – London: W Reeves, 1885 – 1 – us CRL [960]
The immortal storm : a history of science fiction fandom / Moskowitz, Sam – [S.l]: Fantasy Commentator, 1951 (mf ed 1982) – 1r – 1 – (repr of the ed publ by atlanta science fiction organization press, atlanta) – mf#ZZ-20312 – us NY Public [830]
Immortality : the drew lecture delivered october 11, 1912 / Charles, Robert Henry – Oxford: Clarendon Press, 1912 – 1mf – 9 – 0-7905-0123-6 – mf#1987-0123 – us ATLA [270]
Immortality / Holmes, Ernest Edward – London: Longmans, Green 1908 [mf ed 1992] – (= ser The oxford library of practical theology) – 1mf – 9 – 0-524-05039-2 – (incl bibl ref) – mf#1992-0292 – us ATLA [240]
Immortality / Seabrook, William LeVin – Philadelphia: Vir Pub Co, c1905 – 1mf – 9 – 0-524-08486-6 – mf#1993-3131 – us ATLA [240]
Immortality / Woods, John Crawford – Edinburgh, Scotland. 1851 – 1r – us UF Libraries [240]
Immortality / Woods, John Crawford – Edinburgh, Scotland. 1851 – 1r – us UF Libraries [240]
Immortality, and other essays / Everett, Charles Carroll – Boston: American Unitarian Association, 1902 – 1mf – 9 – 0-7905-3734-6 – mf#1989-0227 – us ATLA [240]
Immortality and the future : the christian doctrine of eternal life / Mackintosh, Hugh Ross – New York: GH Doran, [1917?] – 1mf – 9 – 0-524-08544-7 – (incl bibl ref) – mf#1993-2069 – us ATLA [240]
Immortality newsletter – San Marcos. 1970-1973 (1) 1970-1972 (5) (9) – (cont by: theologia 21) – ISSN: 0019-2783 – mf#6565 – us UMI ProQuest [130]
Immortality of the intellect / Jellett, John Hewitt – Dublin, Ireland. 1867 – 1r – us UF Libraries [240]
Immortality of the intellect / Jellett, John Hewitt – Dublin, Ireland. 1867 – 1r – us UF Libraries [240]
Immortality of the soul / Allin, Thomas – Hanley, England. 1823 – 1r – us UF Libraries [240]
Immortality of the soul / Allin, Thomas – Hanley, England. 1823 – 1r – us UF Libraries [025]
The immortality of the soul : considered in the light of the holy scriptures, the testimony of reason and nature, and the various phenomena of life and death / Mattison, Hiram – Philadelphia: Perkinpine & Higgins, 1864. Beltsville, Md: NCR Corp, 1978 (5mf); Evanston: American Theol Lib Assoc, 1984 (5mf) – 9 – 0-8370-0778-X – (incl bibl ref and ind) – mf#1984-4146 – us ATLA [220]
The immortality of the soul : a protest / Beet, Joseph Agar – New York: Methodist Book Concern, 1901. Beltsville, Md: NCR Corp, 1978 (2mf); Evanston: American Theol Lib Assoc, 1984 (2mf) – 9 – 0-8370-0843-3 – mf#1984-4225 – us ATLA [240]
The immortality of the soul and the final condition of the wicked carefully considered / Landis, Robert Wharton – New York: Carlton & Porter, c1859 – 2mf – 9 – 0-524-04102-4 – (incl bibl ref) – mf#1992-0060 – us ATLA [240]
The immortality of the soul in the poems of tennyson and browning : a lecture / Jones, Henry – 2nd ed. London: Philip Green, 1906 – (= ser Essex Hall Lecture) – 1mf – 9 – 0-524-07824-6 – mf#1991-3371 – us ATLA [420]
Immortality versus annihilation / Hartley, G A – Saint John, NB?: s.n, 1867 (Saint John, NB: Barnes & Co) – 1mf – 9 – mf#08314 – cn Canadiana [210]
Immortellen heinrich heine's / ed by Strodtmann, Adolf – New York: S Zickel, 1872 – 1 – us Wisconsin U Libr [430]
Immun-histochemische darstellung peripherer neuraler und neuroendokriner zellelemente bei dysplasien und karzinomata in situ der harnblase / Guenther, Hans-Christian – (mf ed 1996) – 1mf – 9 – €30.00 – 3-8267-2313-9 – mf#DHS 2313 – gw Frankfurter [616]
Immunisation : a guide on the administration and storage of vaccines = Immunisering: 'n handleiding by die toediening en opberging van entstowwe / South Africa. Department of National Health and Population Development [Departement van Nasionale Gesondheid en Bevolkingsontwikkeling – 3rd ed. Pretoria: The Dept 1988 [mf ed Pretoria, RSA: State Library [199-]] – 36p on 1r with other items – 5 – (in english & afrikaans) – mf#op 08940 r26 – us CRL [615]

IMMUNISATION

Immunisation : a guide on the administration and storage of vaccines = Immunisering: 'n handleiding vir die toediening en opberging van entstowwe / South Africa. Department of National Health and Population Development [Departement van Nasionale Gesondheid en Bevolkingsontwikkeling – 2nd ed. Pretoria: Dept of National Health and Population Development: Dept van Nasionale Gesondheid en Bevolkingsontwikkeling 1986 [mf ed Pretoria, RSA: State Library [199-]] – 30p on 1r with other items – 5 – (incl ind) – mf#op 08228 r26 – us CRL [615]

Immunitaet und infektion – Baden-Baden. 1975-1979 (1) 1975-1979 (5) 1975-1979 (9) – mf#9137 – us UMI ProQuest [616]

Immunization : a guide on the administration and storage of vaccines = Immunisering: 'n handleiding by die toediening en opberging van entstowwe / South Africa. Department of Health [Departement van Gesondheid] [Departement van Gesondheid] – Pretoria: Dept of Health [1979?] [mf ed Pretoria, RSA: State Library [199-]] – 64p on 1r with other items – 5 – mf#op 07028 r26 – us CRL [615]

Immunogenetics – Heidelberg. 1974-1994 (1) 1974-1994 (5) 1974-1994 (9) – ISSN: 0093-7711 – mf#13179 – us UMI ProQuest [575]

Immunology – Oxford. 1980+ (1,5,9) – ISSN: 0019-2805 – mf#15416 – us UMI ProQuest [616]

Immunology and allergy clinics of north america – Philadelphia. 1987+ (1,5,9) – (cont: clinics in immunology and allergy) – ISSN: 0889-8561 – mf#12721,01 – us UMI ProQuest [616]

Immunology and cell biology – Adelaide. 1987+ (1,5,9) – (cont: australian journal of experimental biology and medical science) – ISSN: 0818-9641 – mf#10592,01 – us UMI ProQuest [574]

Immunology letters – Amsterdam. 1979-1992 (1) 1979-1992 (5) 1986-1992 (9) – ISSN: 0165-2478 – mf#42096 – us UMI ProQuest [616]

Immunology today – Amsterdam. 1980-1997 (1) 1980-1997 (5) 1987-1997 (9) – ISSN: 0167-5699 – mf#42262 – us UMI ProQuest [616]

Immunopharmacology – New York. 1978-1992 (1) 1978-1992 (5) 1987-1992 (9) – ISSN: 0162-3109 – mf#42097 – us UMI ProQuest [615]

The imp variety and picture hall – Stoke-on-Trent, England. -w. 26 Feb 1913-8 Dec 1915 – 2r – 1 – uk British Libr Newspaper [072]

Impact – 1968 sep-oct, 1969 jan – 1r – 1 – mf#4878416 – us WHS [071]

Impact / A O Smith Corporation – v4 n4-12 [1977 apr-dec], v5 n1-v14 n10 [1978 jan-1987 dec] – 1r – 1 – mf#1825037 – us WHS [071]

Impact / Conservative Baptist Foreign Mission Society – 1972 jul/aug [v26 n4]-1982 nov – 1r – 1 – (cont: conservative baptist impact) – mf#969532 – us WHS [242]

Impact – Kings Lynn. 1998+ (1,5,9) – (cont: assistant librarian) – mf#27916 – us UMI ProQuest [020]

Impact – Danville IL. 1969 apr n7 [1969 jul] – 1r – 1 – mf#1583876 – us WHS [071]

Impact – Manila. 1989-1995 (1,5,9) – ISSN: 0300-4155 – mf#12430 – us UMI ProQuest [300]

Impact – May 1944-71. Pub. by the Conservative Baptist Foreign Mission Society. Continues: News and Views, May 1944-46; Conservative Baptist, 1947-Aug 1965; Impact, Sept 1965-71. 2610p – 1 – 91.35 – us Southern Baptist [242]

Impact / National Office for Black Catholics – 1971 feb-1982 spr – 1r – 1 – mf#647496 – us WHS [241]

Impact – Washington, D.C.: Student National Education Association, v8, n3, jan. 1976- [-irr] – 1 – (other title. student impact) – us Wisconsin U Libr [370]

Impact / Retail Clerks Union, Local 206 – v1 n1-v3 n3 [1977 sep-1979 nov/dec] – 1r – 1 – mf#678906 – us WHS [331]

Impact and shock attenuation during landing activites from different heights on different surfaces / Yu, Yeon-Joo – 2001 – 133p on 2mf – 9 – $10.00 – mf#PE 4217 – us Kinesology [612]

The impact of an adult health education program on exercise self-efficacy and participation in leisure-time physical activity / Hubball, Harry T – 1994 – 2mf – $8.00 – us Kinesology [613]

Impact of caahep accreditation on the internship route to national certification in athletic training at ncaa 3 and naia colleges and universities / Stucky, Amy M – 1998 – 1mf – 9 – $4.00 – mf#PE 3827 – us Kinesology [790]

The impact of disrupted family life and school climate on the self-concept of the adolescent / Gasa, Velisiwe Goldencia – Uni of South Africa 2001 [mf ed Johannesburg 2001] – 3mf – 9 – (incl bibl ref) – mf#mfm14990 – sa Unisa [305]

The impact of dynamic and static flexibility programs on range of motion and athletic injury / Mann, Douglas P – 1999 – 4mf – 9 – $16.00 – mf#PE 4057 – us Kinesology [617]

The impact of extracurricular athletic participation, gender, and grade level upon elementary school students' attitudes toward physical activity / McGowan, Kathleen – 2000 – 243p on 3mf – 9 – $15.00 – mf#PE 4177 – us Kinesology [150]

The impact of foreign loans on the liberian fiscal system : an essay / Uzoaga, W Okefie – 1953 – us CRL [332]

The impact of guidance and counselling : on the attitudes and academic performance of first year undergraduate students: with reference to the university of the north / Nonyane, Dephney Leumang – Pretoria: Vista University 2002 [mf ed 2002] – 3mf – 9 – (incl bibl) – mf#mfm15192 – sa Unisa [378]

The impact of inclusive education in special and developed/mainstream schools : as perceived by parents, educators and learners in soweto schools / Mabuya, Magdeline Olivia Mmakekgathetse – Pretoria: Vista University 2003 [mf ed 2003] – 3mf – 9 – (incl bibl ref) – mf#mfm15258 – sa Unisa [370]

The impact of informal settlement environment on learning strategies : for grade seven learners in south africa / Molotsi, Nkele Julia – Pretoria: Vista University 2002 [mf ed 2003] – 2mf – 9 – (incl bibl ref) – mf#mfm15242 – sa Unisa [370]

The impact of interscholastic athletics on academic performance / Honey, Michael J – 1994 – 1mf – $4.00 – us Kinesology [370]

The impact of management styles on staff appraisal system in schools / Khumalo, Richard – Pretoria: Vista University 2002 [mf ed 2002] – 2mf – 9 – (incl bibl ref) – mf#mfm15238 – sa Unisa [370]

The impact of project adventure activities on self-perception / France, Thaddeus J & Jensen, Barbara E – 1993 – 2mf – $8.00 – us Kinesology [150]

Impact of science on society – Paris. 1950-1992 (1) 1970-1992 (5) 1975-1992 (9) – ISSN: 0019-2872 – mf#2142 – us UMI ProQuest [500]

The impact of section 21 of south african schools act no 84 of 1996 on financial school management / Lowan, Sylvia – Pretoria: Vista University 2002 [mf ed 2002] – 2mf – 9 – (incl bibl ref) – mf#mfm15326 – sa Unisa [370]

The impact of shifting our strategic base from okinawa to micronesia / Hammaker, Charles A, Jr – Carlisle Barracks PA: US Army War College 31 jan 1974 – 1mf – 9 – $1.50 – mf#llmc82-100f, title 100 – us LLMC [355]

The impact of television coverage on american "beauty" pageants / Schiller, Ginny L – Pennsylvania State University, 1986 [mf ed 1988] – 1mf – 9 – us Kinesology [790]

Impact of the acquired immunodeficiency syndrome (aids) : knowledge, attitudes, and behaviors of emergency care providers / Beaver, Kathryn L – 1989 – 54p on 1mf – 9 – $4.00 – us Kinesology [616]

The impact of the chief executive act / McDermott, John et al – 1979 – 3mf – 9 – $4.50 – mf#llmc99-012 – us LLMC [346]

The impact of the federal drug aftercare program / Eaglin, James B – Washington: FJC, 1986 – 2mf – 9 – $3.00 – mf#LLMC 95-331 – us LLMC [344]

The impact of the la crosse wellness project on the health promotion involvement of college students residing on the campus of the university of wisconsin-la crosse / Burns, Julia A & Gilmore, Gary D – 1992 – 1mf – $4.00 – us Kinesology [613]

The impact of the patent system on research / Melman, Seymour – Washington: Govt. Print. Off., 1958. 62p. LL-2308 – 1 – us L of C Photodup [346]

The impact of the principal's instructional leadership on the culture of teaching and learning in the school / Budhal, Rishichand Sookai – Uni of South Africa 2000 [mf ed Johannesburg 2000] – 4mf – 9 – (incl bibl ref) – mf#mfm15078 – sa Unisa [370]

The impact of therapeutic horseback riding on the self-concept and riding performance of children and adolescents with disabilities / Stuler, Lesley R – 1993 – 1mf – $4.00 – us Kinesology [150]

Impact of training patterns on incidence of illness and injury : during a women's basketball season / Anderson, Laura J – 2000 – 41p on 1mf – 9 – $5.00 – mf#PE 4134 – us Kinesology [617]

The impact of transformational styles of leadership on human resource management in primary schools / Lokotsch, Karl Heinz – Vista University 2000 [mf ed Johannesburg 2000] – 2mf – 9 – (incl bibl ref) – mf#mfm14758 – sa Unisa [370]

The impact of "winning weighs" weight control program on perceived body image / Kaufmann, Barbara E & Pretasky, Barbara J – 1991 – 2mf – $8.00 – us Kinesology [150]

The impact of word processing and electronic mail on us courts of appeals / Greenwood, J Michael & Farmer, Larry – Washington: FJC, Mar 1979 – 2mf – 9 – $3.00 – mf#LLMC 95-817 – us LLMC [347]

Impacto. meditaciones para militantes / Aradillas Agudo, Antonio – Madrid: Ediciones Studium, 1964 – 1 – sp Bibl Santa Ana [946]

Impacto news – Miami, FL. 1983 mar 04-1985 oct 01 – 1r – us UF Libraries [071]

Impacto news – Miami, FL. 1983 mar 4-1985 oct 0 – 1r – us UF Libraries [071]

Impacts of closing meigs field airport / United States. General Accounting Office. RCED – Washington DC: The Office [mf ed 1997?] – 1mf – 9 – us US Gen Account [380]

The impacts of marine debris, weather conditions, and unexpected events on recreational boater satisfaction on the delaware inland bays / Holdnak, Andrew & Graefe, Alan R – 1992 – 2mf – 9 – $8.00 – us Kinesology [790]

El imparcial – Madrid. Spain. -d. 1 Jan 1870-22 May 1874, 1 Jan 1875-30 May 1933. (169 reels) – 1 – uk British Libr Newspaper [074]

El imparcial – Sonora, MEXICO. 1980-2000 (1) – mf#68176 – us UMI ProQuest [079]

El imparcial – Willemstad, Netherlands Antilles. 1874-1875 (1) – mf#68602 – us UMI ProQuest [079]

Imparcial – Miami, FL. 1980 dec 04-1985 may 9 – 3r – (gaps) – us UF Libraries [071]

Imparcial – Miami, FL. 1980 dec 04-1985 may 9 – 3r – (gaps) – us UF Libraries [071]

O imparcial : diario illustrado do rio de janeiro – Rio de Janeiro, RJ. 13 ago 1912-ago 1916; jan 1917-fev 1942 – (= ser Ps 19) – mf#P11,08,43 – bl Biblioteca [321]

O imparcial : jornal politico, litterario e noticioso – Paraiba do Norte, PB: Typ de Jose Rodrigues da Costa, 13 abr 1861 – (= ser Ps 19) – mf#P11B,04,06 – bl Biblioteca [073]

O imparcial – Sacramento: Quaresma and Armas Co, sep 1917-jan 1922 – 3r – 1 – us CRL [071]

El imparcial de texas – San Antonio, TX: El Imparcial de Texas, [dec 23 1917-mar 1921] – 1 – us CRL [071]

L'Impartial – Paris: Impr Bonaventure et Ducessois, jun 1848 – us CRL [071]

L'impartial : journal de smyrne – Constantinople. mai 1848-fevr 1852, aout 1889-juin 1890 – 1 – (journal politique, commercial et litteraire) – fr ACRPP [073]

L'impartial – Tignish, PEI: F G Buote, 1893-1915 – 9r – 1 – ISSN: 0844-4080 – cn Library Assoc [071]

The impartial : a journal litteray [sic], scientific, commercial and agricultural – La Prairie [Quebec: s.n, 1834-1835?] – 9 – mf#P04159 – cn Canadiana [073]

An impartial account of the late debate at lyme in the colony of connecticut / Buckley, John – London. 204p. 1729 – 1 – $7.14 – us Southern Baptist [242]

Impartial de londres – London, UK. 21 Jun 1873 – 1 – uk British Libr Newspaper [072]

L'impartial de saone-et-loire : journal republicain hebdomadaire puis bihebdomadaire – Chalon-sur-Saone, fevr 1900-oct 1904 – 1 – fr ACRPP [073]

L'impartial francais – Paris. 1925-1er mai 1928 – 1 – (journal de critique politique, litteraire et sociale.puis quot.) – fr ACRPP [073]

Impartial herald : a periodical register of the times – Suffield CT. 1798 apr 18-may 9 – 1r – 1 – mf#871371 – us WHS [071]

Impartial observer – Providence, RI. 1800-1802 (1) – mf#66318 – us UMI ProQuest [071]

Impartial observer, and washington advertiser – Washington DC. 1795 jun 12, aug 21, sep 14 – 2r – 1 – (cont by: washington advertiser) – mf#850821 – us WHS [071]

Impartial occurrences foreign and domestic *see* Pues occurrences

Impartial occurrences, foreign and domestick – (Pue's Occurrences). Ireland. -sw. 26 Dec 1704-9 Feb 1706, 27 Jul 1714, 3 Jan-22 Dec 1719. (41 ft) – 1 – uk British Libr Newspaper [072]

Impartial reporter – Enniskillen, Ireland. 1826-29 apr 1879; 19 jun 1879-1896; 1898-1950; jan-18 dec 1986; 1987-21 dec 1989; jan-20 dec 1990; 1991-23 dec 1992; 1993-98 – 113 1/4r – 9 – (aka: impartial reporter and farmers journal) – uk British Libr Newspaper [630]

Impartial reporter *see* Enniskillen chronicle and erne packet

Impartial reporter and farmers journal *see* Impartial reporter

Impeachment: a monograph on the impeachment of the federal judiciary / Brown, Wrisley – Washington Govt. Print. Off. 1914. 18 p. LL-613 – 1 – us L of C Photodup [340]

Impeachment of christianity / Abbot, Francis Ellingwood – Ramsgate, England. 1872 – 1r – us UF Libraries [240]

Impeachment of christianity / Abbot, Francis Ellingwood – Ramsgate, England. 1872 – 1r – us UF Libraries [240]

Impedanzspektronische untersuchungen zur adsorptionsgeschwindigkeit an poly- und monokristallinen platin-elektroden / Oelgeklaus, Rainer – (mf ed 1995) – 2mf – 9 – €40.00 – 3-8267-2180-2 – mf#DHS 2180 – gw Frankfurter [540]

Impediments to the prosperity of ireland / Hancock, William Neilson – London, 1850 – (= ser 19th c ireland) – 3mf – 9 – mf#1.1.8522 – uk Chadwyck [339]

Impedimientos de misioneros / Bayle, Constantino – Madrid: Missionalia Hispanica, 1947 – 1 – sp Bibl Santa Ana [240]

The impending contact of the aryan and turanian races, with special reference to recent chinese migrations : a lecture delivered...10th february 1878 / Macfie, Matthew – London: Sunday Lecture Society, 1878 – (= ser 19th c books on china) – 1mf – 9 – mf#7.1.19 – uk Chadwyck [950]

The impending crisis of 1860 : or, the present connection of the methodist episcopal church with slavery, and our duty in regard to it / Mattison, Hiram – New York: Mason, 1859 – 1mf – 9 – 0-524-05002-3 – mf#1990-5090 – us ATLA [242]

The impending fast of mahatma gandhi : the issues explained / Rajagopalachari, Chakravarti – Delhi: Servants of Untouchables Society; Bombay: Can be had at Navajivan Karyalaya, [1933?] – (= ser Samp: indian books) – us CRL [320]

Impending judgments on the earth : or, who may abide the day of his coming / Kinnear, Beverley Oliver – New York: J Huggins, 1892 [mf ed 1991] – 1mf – 9 – 0-7905-9012-3 – mf#1989-2237 – us ATLA [220]

The impending social revolution : or, the trust problem solved / Wilson, Jackson Stitt – Berkeley: Social Crusade, [19–?] – 1mf – 9 – 0-524-04153-9 – mf#1990-1223 – us ATLA [335]

Imperador d pedro ii do brasil, proscrito em port / Martins, Francisco Jose Rocha – Porto, Portugal. 1949 – 1r – us UF Libraries [972]

Imperador d pedro ii do brasil, proscrito em port... / Martins, Francisco Jose Rocha – Porto, Portugal. 1949 – 1r – us UF Libraries [972]

Imperatorii grammatici historiarum libri seu de rebus gestis a joanne et mannuele gommensis impp (cbh11,1) / Joannis Cinnami; ed by Cange, C du – Parisiis, 1670 – (= ser Corpus byzantinae historiae (cbh)) – €42.00 – ne Slangenburg [243]

Imperatorskaia akademiia nauk (Russia) *see* Ueber die strahlenbrechung in der atmosphaere

Imperatorum symbola...quibus accedit commentarius in andreae alciati emblemata... / Vernulaeus, N – [Lovanii: Typis ac sumptibus Iudoci Coppeni, 1659] – 6mf – 9 – mf#O-1274 – ne IDC [090]

Imperatriz d leopoldina / Salgado Dos Santos, Amilcar – Sao Paulo, Brazil. 1927 – 1r – us UF Libraries [972]

Imperatriz d leopoldina / Salgado Dos Santos, Amilcar – Sao Paulo, Brazil. 1927 – 1r – us UF Libraries [972]

Imperial academy. japan, tokyo. proceedings – v1-21 1912-45 – 1 – $330.00 – mf#0273; 0274 – us Brook [500]

Imperial and asiatic quarterly review and oriental and colonial record – London. 1886-1900 (1) – mf#2853 – us UMI ProQuest [950]

The imperial and colonial institutions of the britannic empire : including indian institutions / Creasy, Edward Shepherd – London, 1872 – (= ser 19th c books on british colonization) – 5mf – 9 – mf#1.1.7420 – uk Chadwyck [954]

Imperial and commonwealth conferences *see* Empire and commonwealth

El imperial colegio de indios de la santa cruz de tlalteloco : mexico, 1934 / Ocaranza, Fernando – Madrid: Razon y Fe, 1935 – 1 – sp Bibl Santa Ana [972]

Imperial county – 1914-34; 1992-94 – (= ser California telephone directory coll) – 23r – 1 – $1150.00 – mf#P00038 – us Library Micro [917]

[Imperial county-] imperial county including imperial valley – CA. 1908; 1911-1913; 1917-1918; 1926-1949; 1952 – 14r – 1 – $700.00 – mf#D042 – us Library Micro [978]

An imperial court of appeal : or, the abolition of all overseas appeals / Ewart, John Skirving – [Ottawa?: s.n, 1919?] [mf ed 1994] – 1mf – 9 – 0-665-73197-3 – mf#73197 – cn Canadiana [347]

Imperial factor in south africa / De Kiewiet, C W (Cornelius William) – London, England. 1965 – 1r – us UF Libraries [960]

Imperial factor in south africa / De Kiewiet, Cornelius W – London, England. 1965 – 1r – us UF Libraries [960]

Imperial federation / Argyll, John Douglas Sutherland Campbell, Duke of – London: S Sonnenschein, 1885 – (= ser The imperial parliament series 1) – 2mf – 9 – (incl ind) – mf#03986 – cn Canadiana [320]

Imperial federation / Forster, William Edward – Ottawa: Government Printing Bureau, 1900 – 1mf – 9 – mf#05585 – cn Canadiana [971]

Imperial Federation League *see* Imperial federation league

Imperial federation league : the record of the past and the promise of the future / Imperial Federation League – London, Paris, New York: Cassell, 1886? – 1mf – 9 – mf#64781 – cn Canadiana [320]

Imperial Federation League. City of London Branch *see* Report of meeting of the branch held on tuesday, november 15th, 1892

Imperial Federation League Victorian branch *see* Report of public meeting...melbourne, on friday evening, 5th june, 1885, godfrey downes carter...in the chair

Imperial federation of great britain and her colonies : in letters edited by frederick young (one of the writers) – London: S W Silver, 1876 [mf ed 1984] – 3mf – 9 – 0-665-32343-3 – mf#32343 – cn Canadiana [320]

Imperial federation of great britain and her colonies in letters / ed by Young, Frederick – London 1876 – (= ser 19th c british colonization) – 3mf – 9 – mf#1.1.3718 – uk Chadwyck [320]

Imperial federation! stirring speeches by representative citizens! : his grace archbishop o'brien declares it an insult to be told that annexation is our destiny – [S.l: s.n, 1888?] [mf ed 1980] – 1mf – mf#07505 – cn Canadiana [320]

Imperial gazeteer of england and wales / Wilson, James M – Edinburgh. v1-6. 1870-72 – 9 – $349.00 – mf#0674 – us Brook [941]

The imperial gazetteer of india : district series – Milwaukee. 1954-1958 (1) – 5112mf – 9 – mf#1601 – ne IDC [915]

The imperial gazetteer of india – Northfield. 1960-1980 (1) 1971-1980 (5) 1976-1980 (9) – 218mf – 9 – mf#1608 – ne IDC [915]

Imperial history : british colonial reports, 1889-1939 – (= ser The statesman) – 60r – 1 – (incl.: the statesman (and friend of india), 1941-51) – mf#C39-21600 – us Primary [950]

[Imperial-] imperial press – CA. 1902-06 (broken issues) – 1r – 1 – $60.00 – mf#C02302 – us Library Micro [071]

Imperial intelligence department : a free press cable service around the world... / Fleming, Sandford – [Ottawa?: s.n.] 1905 [mf ed 1995] – 1mf – 9 – 0-665-74213-4 – mf#74213 – cn Canadiana [380]

The imperial legislative council manual : including the government of india act, 1915, the rules and regulations for the legislative council of the governor general, and an appendix containing the repealed indian council acts, 1861, 1892 and 1909, and government of india act, 1912 – Delhi: Supt Govt Printer, India, 1916 – 1 – us CRL [954]

Imperial loyalty "as it ought to be"... : christian philosophic on a new plan. blend and counterpoise, as remedial, for safety of the empire / O'Connor, John Hutton – London 1886 – (= ser 19th c british colonization) – 4mf – 9 – mf#1.1.6228 – uk Chadwyck [240]

Imperial magazine – London. 1819-1834 (1) – mf#3904 – us UMI ProQuest [240]

Imperial network and external dependency : the case of angola / Minter, William M – Madison 1971 – us CRL [960]

The imperial night hawk – Atlanta, Ga. v. 1-2 no. 34. Mar 28 1923-Nov 19 1924 – 1 – us NY Public [073]

The imperial night-hawk – Atlanta, GA: The Knights of the Ku Klux Klan [v1-2 (may 28 1923-nov 19 1924)] (wkly) – 1r – 1 – us CRL [071]

Imperial Observatorio do Rio de Janeiro *see*
– Annales de l'observatoire imperial de rio de janeiro
– Annuario publicado pelo imperial observatorio do rio de janeiro para o anno de...

Imperial Order of the Daughters of the Empire *see* Echoes

Imperial polk county – Lakeland, FL. 1921 – 1r – us UF Libraries [630]

Imperial polk county – Lakeland, FL. 1921 – 1r – us UF Libraries [917]

Imperial preference vis-a-vis world economy : in relation to the international trade and national economy of india / Sarkar, Benoy Kumar – Calcutta: NM Ray-Chowdhury & Co, 1934 – (= ser Samp: indian books) – us CRL [337]

The imperial record – Imperial, NE: Burton North, 1896 (wkly) [mf ed v1 n8. apr 24 1896] – 1r – 1 – us NE Hist [071]

Imperial Record Dept. India *see* An alphabetical list of the feasts and holidays of the hindus and muhammadans

Imperial republican – Imperial, NE: C M Reynolds, 1899 (wkly) [mf ed aug 11 1899-feb 10 1994 (gaps)] – 44r – 1 – (absorbed: chase county tribune and chase county enterprise, consolidated. issue for aug 25 1932 misdated aug 25 1923) – us NE Hist [071]

Imperial review : or, london, edinburgh, and dublin literary journal – London. 1804-1805 (1) – mf#5575 – us UMI ProQuest [941]

Imperial rule in india : being an examination of the principles proper to the government of dependencies / Morison, Theodore – Westminster 1899 – (= ser 19th c british colonization) – 2mf – 9 – mf#1.1.9422 – uk Chadwyck [323]

Imperial Society of Teachers of Dancing *see* Dance for people with disabilities

Imperial Society of Teachers of Dancing. Modern Theatre Dance Branch *see* Intermediate syllabus and notes, modern and tap

Imperial statutes in force in new south wales... / Bignold, H B – Sydney: The Law Book Company of Australasia. v1-3. 1913-14 – 18mf – 9 – $27.00 – (v1 contains chronological and alphabetical tables of all the imperial statutes and also of the commonwealth and nsw statutes dealing with the imperial statutes. v2-3 contain the text of all imperial statutes declared to be in force, and a selection of imperial statutes not authoritatively declared, but presumed to be in force, together with indexes and case references) – mf#LLMC 96-001 – us LLMC [323]

The imperial treasury of the indian mughuls / Aziz, Abdul – Lahore: The Author, 1942 – (= ser Samp: indian books) – us CRL [336]

Imperial unity and the dominions / Keith, Arthur Berriedale – Oxford: Clarendon Press, 1916. 626p – 1 – us Wisconsin U Libr [941]

Imperial War Museum. London *see* British 20th century war art

[L'imperiale; cantata for the paris exhibition, op 26] / Berlioz, Hector – [1855] [mf ed 19–] – 1r – 1 – (words by capitaine lafont) – mf#pres. film 73 – us Sibley [780]

Imperialism / De Thierry, C – London, 1898 – (= ser 19th c books on british colonization) – 2mf – 9 – mf#1.1.6771 – uk Chadwyck [320]

Imperialism and christ / Ottman, Ford Cyrinde – New York: CC Cook, c1912 – 1mf – 9 – 0-7905-7992-8 – mf#1989-1277 – us ATLA [240]

Imperialism and nationalism; a study of conflict in the near ea. / Page, Kirby – New York: George H. Doran Co., (c1925). vii,leaf,7-92p – 1 – us Wisconsin U Libr [321]

Imperialism in south africa / Ritchie, James Ewing – London 1879 – (= ser 19th c british colonization) – 1mf – 9 – mf#1.1.4945 – uk Chadwyck [320]

The imperialism of john marshall: a study in expediency / Bryan, George – Boston, Stratford Co., 1924. 112 p. LL-1535 – 1 – us L of C Photodup [340]

Imperialismo e angustia / Lima, Claudio De Araujo – Rio de Janeiro, Brazil. 1960 – 1r – us UF Libraries [972]

Imperialismo e angustia / Lima, Claudio De Araujo – Rio de Janeiro, Brazil. 1960 – 1r – us UF Libraries [972]

Los imperialismos de juan gines de sepulveda en su : democrates alter / Bayle, Constantino – Madrid: Missionalia Hispanica, 1948 – 1 – sp Bibl Santa Ana [320]

Imperialist – [London & SE] Hertfordshire 7 oct 1916-2 feb 1918, 9 feb-3 aug 1918 & 8 feb 1919 [mf ed 2004] – 2r – 1 – (cont as: vigilante [feb-aug 1918, feb 1919]) – uk Newsplan [072]

The imperialist / Cotes, Everard, mrs [Sara Jeanette Duncan] – Toronto: Copp, Clark, 1904 – 6mf – 9 – 0-665-77160-6 – mf#77160 – cn Canadiana [830]

O imperialista : jornal miscellaneo – Porto Alegre, RS: Typ do Imperialista, 11 jan 1840; 04 mar 1840 – (= ser Ps 19) – mf#P17,02,193 – bl Biblioteca [355]

Imperiia / Obedinenie luchshikh – Moscow, Russia. n1? [1996]-n11(1997], n13(1998]-n20[1998] – 1 – mf#mf-12248 [reel 4] – us CRL [077]

Imperio de la china : i cvltvra evangelica en sl, por los religios de la compania de iesvs / Semmedo, (Semedo) A – Madrid: Iuan Sanchez, 1642 – 5mf – 9 – mf#HT-553 – ne IDC [915]

Imperio do brazil na exposicao universal de 1876 e... / Brazil. Commissao, Exposicao Universal, Philadelph... – Rio de Janeiro, Brazil. 1875 – 1r – us UF Libraries [972]

Imperium and sacerdotium according to st basil the great (sca7) / Reilly, G F – Washington DC, 1945 – (= ser Studies in christian antiquity (sca)) – 4mf – 9 – €11.00 – ne Slangenburg [240]

Imperium orientale sive antiquitates constantinopolitani (cbh24,1) / Banduri, A – Venetiis. v1. 1729 – (= ser Corpus byzantinae historiae (cbh)) – €54.00 – ne Slangenburg [243]

Imperium orientale sive antiquitates constantinopolitani (cbh24,2) / Banduri, A – Venetiis. v2. 1729 – (= ser Corpus byzantinae historiae (cbh)) – €40.00 – ne Slangenburg [243]

Imperium romanum ferdinando secundo : ..ab augusto septem virorum sacri romani imperii senatu, votis concordibus delatum – Graecii Styriae: Ex officina typographica, Ernesti Widmanstadii, 1620 – 4mf – 9 – mf#O-2040 – ne IDC [090]

Impersonalien, eine logische untersuchung / Sigwart, Christoph – Freiburg, Germany. 1888 – 1r – us UF Libraries [160]

Impersonalien, eine logische untersuchung / Sigwart, Christoph – Freiburg, Germany. 1888 – 1r – us UF Libraries [160]

Impetu, pasion y fuga / Caba, Ruben – Madrid: Ediciones Alfaguara, 1972 – 1 – sp Bibl Santa Ana [946]

Impey, Eugene Clutterbuck *see* Delhi, agra, and rajpootana

Impietas valentini gentilis detecta, et palam traducta, qui christum non sine sacrilega blasphemia deum essentiatum esse fingit / [Calvin, J] – [Geneva: Conrad Badius], 1561 – 2mf – 9 – mf#CL-38 – ne IDC [240]

Impington 1562-1950 – (= ser Cambridgeshire parish register transcript) – 5mf – 9 – £6.25 – uk CambsFHS [929]

L'impitoyable – Paris: Blondeau, sep 1848 – us CRL [074]

Implant dentistry – v1-5. 1992-1996 – 5r – 1,5,6,9 – $65.00r – us Lippincott [617]

The implementation and assessment of a multi-media learning programme for environmental education / Jacobs, Susanne – Pretoria: Vista University 2002 [mf ed 2002] – 12mf [ill] – 9 – (abstract in afrikaans & english) – mf#mfm15230 – sa Unisa [370]

Implementation game / Bardach, Eugene – Cambridge, MA. 1977 – 1r – us UF Libraries [025]

Implementation game / Bardach, Eugene – Cambridge, MA. 1977 – 1r – us UF Libraries [025]

The implementation of a perennial program of evangelism / Jones, James Edward – 1982 – 1 – 7.28 – us Southern Baptist [242]

The implementation of obe in grade 1 at schools in mangaung / Moeca, Hamilton Thabiso – Vista University 2000 [mf ed Johannesburg 2000] – 3mf – 9 – (incl bibl ref) – mf#mfm14762 – sa Unisa [370]

Implementation of practical marketing strategies for soweto schools / Mabusela, Maria Sewela – Pretoria: Vista University 2002 [mf ed 2002] – 2mf – 9 – (incl bibl ref) – mf#mfm15317 – sa Unisa [650]

Implementation of religious symbols in a choreographic work : the revelation of john / Christensen, Karen – 1996 – 1mf – 9 – $4.00 – mf#PE 3800 – us Kinesology [790]

Implementation of the agricultural risk protection act : hearing before the subcommittee on general farm commodities and risk management of the committee on agriculture, house of representatives, 107th congress, 2nd session, feb 13 2002 / United States. Congress. House. Committee on Agriculture. Subcommittee on General Farm Commodities and Risk Management – Washington: US GPO 2002 [mf ed 2002] – 1mf – 9 – 0-16-066994-4 – us GPO [343]

Implementing and evaluating the chapter 7 filing fee waiver program / Wiggins, Elizabeth C et al – 1998 – 3mf – 9 – $4.50 – mf#llmc99-020 – us LLMC [347]

Die implikasies van pierre babin se boek "the new era in religious communication" vir 'n kontemporere jeugbedieningsmodel / Roux, Anton – Uni of South Africa 2000 [mf ed Johannesburg 2000] – 2mf [ill] – 9 – (incl bibl ref; text in afrikaans; abstract in afrikaans & english) – mf#mfm15109 – sa Unisa [302]

Import dan export dari indonesia ser 8: import indonesia : indonesia – Djakarta, 1950 – 14mf – 9 – mf#SE-174 – ne IDC [959]

The import duties enquiry 1935 *see* The report on the census of production 1907-1967

Importance of a deep and intimate knowledge of divine truth / Fuller, Andrew – London, England. 1796? – 1r – us UF Libraries [240]

Importance of a deep and intimate knowledge of divine truth / Fuller, Andrew – London, England. 1796? – 1r – us UF Libraries [240]

The importance of a liberal education for women : an address delivered january 16, 1878, before the teacher's institute (of tennessee), at jackson, tennessee / Hamilton, William Thomas – Jackson, TN: Published at request of the Institute by J G Cisco, 1878 – 1mf – 9 – 0-8370-7797-4 – mf#1986-1797 – us ATLA [376]

Importance of an early acquaintance with the scriptures / Sandys, E – Canterbury, England. 1812 – 1r – us UF Libraries [240]

Importance of an early acquaintance with the scriptures / Sandys, E – Canterbury, England. 1812 – 1r – us UF Libraries [240]

The importance of canada considered in two letters to a noble lord / Lee, Charles – London: printed for R and J Dodsley, 1761 [mf ed 1982] – 1mf – 9 – mf#SEM105P71 – cn Bibl Nat [971]

The importance of canada considered in two letters to a noble lord / Lee, Charles – London: printed for R & J Dodsley, 1761 [mf ed 1972] – 1r – 5 – mf#SEM16P55 – cn Bibl Nat [971]

Importance of civil government to society and the duty of christian / Chalmers, Thomas – Glasgow, Scotland. 1820 – 1r – us UF Libraries [240]

Importance of civil government to society and the duty of christian... / Chalmers, Thomas – Glasgow, Scotland. 1820 – 1r – us UF Libraries [240]

Importance of consideration – London, England. 18– – 1r – us UF Libraries [240]

Importance of consideration – London, England. 18– – 1r – us UF Libraries [240]

Importance of doctrinal truth in religion : and man's responsibility for his belief, a conference sermon / Clark, Davis Wasgatt – Detroit: J M Arnold, 1871. Beltsville, Md: NCR Corp, 1978 (1mf); Evanston: American Theol Lib Assoc, 1984 (1mf) – 9 – 0-8370-1027-6 – mf#1984-4390 – us ATLA [240]

The importance of historic research for the theological student of to-day : an address / Scott, Hugh Macdonald – Chicago: Jameson & Morse 1882 [mf ed 1990] – 1mf – 9 – 0-7905-6783-0 – mf#1988-2783 – us ATLA [900]

Importance of little things / Stowell, Canon – Doncaster, England. 1858 – 1r – us UF Libraries [240]

Importance of little things / Stowell, Canon – Doncaster, England. 1858 – 1r – us UF Libraries [240]

The importance of prayer meetings in promoting the revival of religion / Young, Robert – New York: Carlton & Porter, [1840?] – 1mf – 9 – 0-524-06300-1 – mf#1990-5229 – us ATLA [240]

The importance of providing religious education for the poor : connected with true principle of all christian charity: two discourses... / Hopkins, John Henry – Burlington [VT]: Smith & Harrington, 1835 [mf ed 1983] – 1mf – 9 – mf#21496 – cn Canadiana [377]

The importance of religious reserve : and the teaching of the church of england upon confession and absolution: three sermons preached in the church of st james the apostle / Norman, Richard Whitmore – [Montreal?: s.n.], 1873 [mf ed 1981] – 1mf – 9 – mf#04695 – cn Canadiana [242]

Importance of right sentiments concerning the person of christ / Belsham, Thomas – London, England. 1806 – 1r – us UF Libraries [240]

Importance of right sentiments concerning the person of christ / Belsham, Thomas – London, England. 1806 – 1r – us UF Libraries [240]

Importance of right views on baptism / Clowes, Francis – London, England. 18– – 1r – us UF Libraries [242]

Importance of right views on baptism / Clowes, Francis – London, England. 18– – 1r – us UF Libraries [242]

The importance of team chemistry : to the success of the top 25 division 3 football programs of the 1990s / Iagulli, Jonathan J – 2000 – 50p on 1mf – 9 – $5.00 – mf#PE 4112 – us Kinesology [302]

The importance of tertullian in the development of christian dogma / Morgan, James – London: K. Paul, Trench, Trubner, 1928. xviii,295p – 1 – us Wisconsin U Libr [240]

Importance of the controversy between the church of england and the / Stowell, Hugh – London, England. 1839 – 1r – us UF Libraries [241]

Importance of the controversy between the church of england and the... / Stowell, Hugh – London, England. 1839 – 1r – us UF Libraries [241]

Importance of the doctrine of the deity of christ – Northampton, England. 1829 – 1r – us UF Libraries [240]

The importance of the jews for the preservation and revival of learning during the middle ages = Die bedeutung der juden fuer erhaltung und wiederbelebung der wissenschaften im mittelalter / Schleiden, Matthias Jacob – London: Siegle, Hill, 1911 [mf ed 1990] – 1mf – 9 – 0-7905-6877-2 – (english trans by maurice kleimenhagen. int by hermann gollancz. incl bibl ref) – mf#1988-2877 – us ATLA [931]

Importance of true religion and the care of god to preserve it / Dalgliesh, William – Edinburgh, Scotland. 1808 – 1r – us UF Libraries [240]

Importance performance analysis of after school programs using development quality attributes / Harrington, Dianna J – 1997 – 2mf – 9 – $8.00 – mf#RC 512 – us Kinesology [370]

Importancia del poder naval positivo y negativo / Morales Coello, Julio – Habana, Cuba. 1950 – 1r – us UF Libraries [972]

Important decree issued by the president of the cabinet / Azana, Manuel – Barcelona, 1937. Fiche W 741. (Blodgett Collection of Spanish Civil War Pamphlets) – 9 – us Harvard [946]

Important discovery – London, England. 18– – 1r – us UF Libraries [240]

Important election information / League of Women Voters of Madison [WI] – 1959 apr 7, 1961 mar 7, 1961 apr 4 – 1r – 1 – (cont by: candidates and referenda) – mf#623409 – us WHS [325]

Important facts – Lincoln, NE: Frederick & Hamilton. v1 n(mar 28 1893)- (wkly) [mf ed [1987]] – 1r – 1 – us NE Hist [071]

Important facts about the confessional / Phayre, R – S.l., England. 1890 – 1r – us UF Libraries [240]

Important federal laws / Lapp, John A – Indianapolis: B F Bowen & Co, 1917 (all publ) – 10mf – 9 – $15.00 – mf#llmc 94-204 – us LLMC [348]

Important national information, canadian finances examined : canadian financial budget in full contrasted with the budget of the minister of finance; his budget proved nationally deluding / Griffin, George Douglas – [Parkdale, Ont?: s.n.], 1890? [mf ed 1980] – 1mf – 9 – mf#06359 – cn Canadiana [336]

Important periodicals of italian and international socialism, 1868-1917 – Clearwater Publ Co – 7r+48mf – 1,9 – $1270.00 coll – (filmed fr holdings of the feltrinelli archives in milan. individual titles listed separately) – us UPA [335]

Important religious truths / Doe, Walter P [comp] – Providence, RI: A Crawford Greene, 1883 [mf ed 1985] – 1mf – 9 – 0-8370-2942-2 – (incl biogr sketch of comp) – mf#1985-0942 – us ATLA [240]

Important speeches and writings of subhas bose : being a collection of most significant speeches, writings, and letters of subhas bose from 1927 to 1945 / ed by Bright, Jagat S – Lahore: Indian Print Works, 1947 – (= ser Samp: indian books) – us CRL [954]

Important speeches of jawaharlal nehru : being a collection of most significant speeches delivered by jawaharlal nehru, from 1922-1945 / ed by Bright, Jagat S – Lahore: Indian Print Works, 1945 – (= ser Samp: indian books) – us CRL [954]

Important subjects for consideration – Edinburgh, Scotland. 18– – 1r – us UF Libraries [240]

Important testimony to the value of tee-totalism / Jay, W – London, England. 18– – 1r – us UF Libraries [240]

Important timber trees of the united states / Ellitt, Simon Bolivar – Boston, MA. 1912 – 1r – us UF Libraries [580]

Important works and projects : new smyrna / Sweett, Zelia Wilson – S.l., S.l? . 1936 – 1r – us UF Libraries [978]

The importer's guide : a handbook of advances on sterling costs in decimal currency, from one penny to one thousand pounds, with a flannel table, from twenty to one hundred shillings per piece of forty-six yards / Campbell, Roderick & Little, John William – Montreal: printed for the authors by J Lovell, 1867 – 2mf – 9 – mf#29974 – cn Canadiana [530]

The importers' guide : a handbook of advances on sterling costs in decimal currency from one penny to one thousand pounds: with a flannel table from twenty to one hundred shillings per piece of forty-six yards / Campbell, Roderick & Little, John William – Montreal: Murray, 1869 – 2mf – 9 – mf#32015 – cn Canadiana [530]

The importers' guide : a handbook of advances on sterling costs, in decimal currency, from one penny to one thousand pounds, with a flannel table from twenty to one hundred shillings, per piece of forty-six yards / Campbell, Roderick & Little, John William – Montreal: printed for the authors by J Lovell, 1867 – 2mf – 9 – mf#04423 – cn Canadiana [530]

Imports and exports of the port of quebec for the year 1825 – Quebec: Neilson & Cowan, [1825?] [mf ed 1984] – 1mf – 9 – 0-665-32001-9 – mf#32001 – cn Canadiana [380]

Importweek – Toronto. 1979-1979 (1) 1979-1979 (5) 1979-1979 (9) – ISSN: 0702-8385 – mf#8006,01 – us UMI ProQuest [380]

L'imposition des mains et les rites connexes dans le nouveau testament et dans l'eglise ancienne / Coppens, J – Paris, 1925 – 8mf – 8 – €17.00 – ne Slangenburg [225]

Imposition du pallium a mgr l'archeveque duhamel : par son eminence le cardinal taschereau dans la basilique d'ottawa, le 29 juillet 1886 / Bruchesi, Louis Joseph Paul Napoleon – Ottawa: A Bureau, 1886 – 1mf – 9 – mf#03736 – cn Canadiana [241]

Die impossibilia des siger von brabant / Baeumker, C – Muenster, 1898 – (= ser Bgphma 2/6) – 4mf – 8 – €11.00 – ne Slangenburg [180]

Impossibility of canadian annexation / Wiman, Erastus – [New York?: E Wiman, 1891?] [mf ed 1984] – 1mf – 9 – mf#27554 – cn Canadiana [971]

The impossibility of the immaculate conception as an article of faith : in reply to several works which have appeared on that subject of late years: to which is added the author's letter to the pope = De la croyance a l'immaculee conception de la sainte vierge / Laborde, Jean Joseph, M l'abbe; ed by Coxe, A Cleveland – Philadelphia: Herman Hooker, 1855 [mf ed 1991] – 1mf – 9 – 0-524-00378-5 – (trans fr french into english) – mf#1989-3078 – us ATLA [241]

El impossible vencido : arte de la lengua bascongada / Larramendi, Manuel de – Salamanca: Alcaraz 1729 – (= ser Whsb) – 5mf – 9 – €60.00 – mf#Hu 044 – gw Fischer [440]

Impost books of the collector of customs at philadelphia, 1789-1804 / U.S. Bureau of the Customs – (= ser Records of the united states customs service) – 6r – 1 – mf#T255 – us Nat Archives [336]

The impregnable rock of holy scripture / Gladstone, William Ewart – rev enl ed. London, 1890 – (= ser 19th c evolution & creation) – 4mf – 9 – mf#1.1.11593 – uk Chadwyck [220]

The impregnable rock of holy scripture / Gladstone, William Ewart – rev enl ed. Philadelphia: John D Wattles, 1891, c1890 – 1mf – 9 – 0-8370-3306-3 – mf#1985-1306 – us ATLA [220]

A imprensa : gazeta noticiosa, litteraria e poetica – Cachoeira, BA: Typ da Imprensa, 29 dez 1884; 31 jan 1885 – (= ser Ps 19) – mf#P17,01,07 – bl Biblioteca [079]

A imprensa : periodico litterario e noticioso – 01 fev-28 mar 1880 – (= ser Ps 19) – mf#P19A,04,67 – bl Biblioteca [079]

A imprensa : periodico politico – Teresina, PI: Typ da Imprensa, 27 jul 1865-dez 1866; jan 1868-jun 1873; set 1876; abr-jun, ago-dez 1877; jan 1878-09 nov 1889 – (= ser Ps 19) – mf#P25,03,09 – bl Biblioteca [321]

Imprensa brasileira / Segismundo, Fernando – Sao Paulo, Brazil. 1962 – 1r – us UF Libraries [972]

A imprensa catharinense – Desterro, SC: Typ do Jornal do Commercio, 26 ago 1888 – (= ser Ps 19) – mf#P11A,04,02 – bl Biblioteca [079]

Imprensa da tarde *see* A imprensa

A imprensa de cuyaba : periodico politico, mercantil e litterario – Cuiaba, MT: Typ de Sousa Neves e Comp, 31 jul-ago 1859; jun-set 1860; jan-mar, dez 1861; dez 1862; jan, mar, jul-dez 1863; jan 1864-18 jun 1865 – (= ser Ps 19) – 1,5,6 – mf#P11B,01,01 – bl Biblioteca [321]

Imprensa medica : periodico de estudantes de medicina – Rio de Janeiro, RJ: Typ Cinco de Marco, 01 jul-out 1872; jun-30ago 1873 – (= ser Ps 19) – mf#P01,03,10 – bl Biblioteca [610]

A imprensa unida – Manaus, AM: Typ do Amazonas, 31 maio 1888 – (= ser Ps 19) – 1,5,6 – bl Biblioteca [079]

La imprenta en sevilla. noticias ineditas desde la introduccion del arte tipografico en esta ciudad hasta el siglo 29. vol 1. sevilla, 1945 / Hazanas y La Rua, Joaquin – Madrid: Razon y Fe, 1946 – 1 – sp Bibl Santa Ana [700]

Imprenta y los primeros periodicos de santo doming / Rodriguez Demorizi, Emilio – Ciudad Trujillo, Dominican Republic. 1941 – 1r – us UF Libraries [972]

La impresa di m. cesare trevisani amplamente da lui stesso dicchiarata... / Trevisani, C – Genova: Appresso Antonio Bellone, 1569 – 2mf – 9 – mf#O-1962 – ne IDC [090]

Impresario magazine – Ann Arbor. 1961-1976 (1) 1972-1976 (5) 1975-1976 (9) – ISSN: 0536-5813 – mf#5827 – us UMI ProQuest [790]

L'imprese della m c di d filippo d'austria 2 re di spagna / Benedetti, F – Citt...dell' Aquila: Apresso Lepido Facij, 1599 – 2mf – 9 – mf#O-1806 – ne IDC [090]

Imprese di diversi principi, duchi... / Pittoni, B & Dolce, L – [Venetia, 1562] – 3mf – 9 – mf#O-1825 – ne IDC [090]

Le imprese heroiche et morali ritrovate da m. gabriello symeoni fiorentino, al gran conestabile di francia / Simeoni, G – Lyone: Appresso Guglielmo Rovillio, 1559 – 1mf – 9 – mf#O-1915 – ne IDC [090]

Le imprese illustri del s.or ieronimo ruscelli / Ruscelli, G – Venetia: Appresso Francesco de' Franceschi Senesi, 1584 – 11mf – 9 – mf#O-420 – ne IDC [090]

Imprese illustri di diversi con discorsi di camillo camilli : et con le figure intagliate in rame di girolamo porro padovano... / Camilli, C – Venetia: Apresso Francesco Ziletti, 1586 – 6mf – 9 – mf#O-188 – ne IDC [090]

Imprese per le s.s. c.c. maest : ...dell'imperadore leopoldo e dell'imperadrice claudia – Vienna: Appresso Gio: Battista Hacque, 1674 – 1mf – 9 – mf#O-09 – ne IDC [090]

Impresiones / Arce De Vazquez, Margot – San Juan, Puerto Rico. 1950 – 1r – us UF Libraries [972]

Impresiones de un viaje por don...dedicadas a los jovenes estudiantes de las escuelas del ave maria de la ciudad de don benito / Torre Isunza de Hita, Pedro – Cabra: Tip. Manuel Cordon, 1923 – sp Bibl Santa Ana [946]

Impresiones del camino / Macau, Miguel Angel – Habana, Cuba. 1942 – 1r – us UF Libraries [972]

Impresiones intimas / Spinola de Gironza, Araceli – Madrid: Graficas Nebrija, 1966 – 1 – sp Bibl Santa Ana [946]

Impresiones martianas / Marti, Jose – Habana, Cuba. 1956 – 1r – us UF Libraries [972]

Impresiones y juicios / Aramburo Y Machado, Mariano – Habana, Cuba. 1901 – 1r – us UF Libraries [972]

Impresiones y recuerdos / Restrepo, Tomas S – Bogota, Colombia. 1922 – 1r – us UF Libraries [972]

Impresos s 18 – Minorca, Spain. no date – 1r – us UF Libraries [324]

Le imprese illustri con espositioni et discorsi del sor ieronimo ruscelli / [Ruscelli, J] – Venetia: Appresso Francesco Rampazetto, 1566 – 11mf – 9 – mf#O-738 – ne IDC [090]

Impresse nobili et ingeniose di diversi principi... / [Pittoni, B] – Venetia: Presso Francesco Ziletti, 1583 – 3mf – 9 – mf#O-853 – ne IDC [090]

Impressionism : subject collections – (= ser Art exhibition catalogues on microfiche) – 121 catalogues on 184mf – 9 – £1,160.00 – (individual titles not listed separately) – uk Chadwyck [700]

Impressionism in the arts and its influence on selected dance works / Collins, Sherry L – 1989 – 202p 3mf – 9 – $12.00 – us Kinesology [790]

Der impressionismus hofmannsthals als zeiterscheinung : eine stilkritische studie / Berendsohn, Walter Arthur – Hamburg: W Gente, 1920 [mf ed 1990] – 52p – 1 – mf#7485 – us Wisconsin U Libr [430]

Der impressionismus in der lyrik der annette von droste-huelshoff / Fruehbrodt, Gerhard – Berlin: Junker & Duennhaupt 1930 [mf ed 1989] – (= ser Neue forschung. arbeiten zur geistesgeschichte der germanischen und romanischen voelker 7) – 1r – 1 – (filmed with: annette von droste-hulshoff / clemens heselhaus & other titles) – mf#7190 – us Wisconsin U Libr [430]

Impressionist and modern paintings, drawings and sculpture – (= ser Christie's pictorial archive new york) – 122mf – 9 – $1060.00 – 0-907006-82-5 – (7300 reproductions) – uk Mindata [700]

Impressionist, modern and contemporary paintings, drawings and sculpture and modern prints – (= ser Christie's pictorial sales review) – 58mf – 9 – $450.00 – 0-907006-13-2 – (6000 images) – uk Mindata [700]

L'impressionniste : journal d'art. – no. 1-4. Paris. avr 1877 – 1 – fr ACRPP [700]

Impressionns et sourvenirs de la jamaique / La Forest, Antoine – Port-Au-Prince, Haiti. 1904 – 1r – us UF Libraries [972]

Impressions / France. Assemblee nationale – Projets de lois, propositions, rapports, etc. Sessions 1946-1985/86 – 1 – fr ACRPP [323]

Impressions / France. Chambre des Deputes – Programmes electoraux, dits Barodets. IIIe Republique, 3e-15e legislatures. 1882, 1886, 1890, 1894, 1899, 1903, 1907, 1910, 1914, 1920, 1925, 1928, 1933 – 1 – fr ACRPP [323]

Impressions / France. Senat – Projets de lois, propositions, rapports, etc. Sessions 1957/58-1985/86 – 1 – fr ACRPP [323]

[Impressions] / France. Assemblee nationale. Senat – Paris. 1907 n244-1940 n83 – 38r – 1 – us L of C Photodup [944]

Impressions and experiences of the west indies and north america in 1849 / Baird, Robert – Philadelphia: Lea & Blanchard, 1850 – 1mf – 9 – 0-7905-5749-5 – mf#1988-1749 – us ATLA [910]

Impressions de theatre / Lemaitre, Jules – Paris. 1889-1920. 11v – 1 – us L of C Photodup [790]

Impressions de voyage : [suisse] / Dumas, Alexandre (pere) – [Paris] 1833 / 1839 [mf ed Hildesheim 1995-98] – 4v on 10mf – 9 – €100.00 – 3-487-29789-2 – gw Olms [914]

Impressions d'espagne : 15 jours en espagne republicaine, 11-25 novembre 1937 / Jezequel, Jules – Paris, 1937. Fiche W971. (Blodgett Collection of Spanish Civil War Pamphlets) – 9 – us Harvard [946]

Impressions d'ethiopie / Merab, P – Paris, 1921-1929. 3v – 15mf – 9 – mf#NE-20226 – ne IDC [916]

Impressions of a careless traveler / Abbott, Lyman – New York: Outlook, 1908, c1907 – 1mf – 9 – 0-7905-5741-X – mf#1988-1741 – us ATLA [910]

Impressions of franco's spain / Rodriguez Vega, Jose – London, 1943. Fiche W1147. (Blodgett Collection of Spanish Civil War Pamphlets) – 9 – us Harvard [946]

Impressions of india / Craik, Henry – London: Macmillan and Co, 1908 – (= ser Samp: indian books) – us CRL [915]

Impressions of indian travel / Browning, Oscar – London: Hodder and Stoughton, 1903 – (= ser Samp: indian books) – us CRL [915]

Impressions of ireland and the irish / Grant, James – London, 1844 – (= ser 19th c ireland) – 8mf – 9 – mf#1.1.5720 – uk Chadwyck [941]

Impressions of japanese architecture and the allied arts / Cram, Ralph Adams – New York: Baker & Taylor, 1905 – 1mf – 9 – 0-7905-4218-8 – mf#1988-0218 – us ATLA [720]

Impressions of south africa : ...with the transvaal conventions of 1881 and 1884 / Bryce, J – Ed 3. London, 1899 – 6mf – 9 – mf#HT-14 – ne IDC [916]

Impressions of the canadian north-west / Davitt, Michael – S.l: s.n, 1892 – 1mf – 9 – mf#17974 – cn Canadiana [917]

Impressions of theophrastus such : essays and leaves from a note-book / Eliot, George – Toronto: G N Morang, 1902 – (= ser George eliot's works) – 5mf – 9 – 0-665-74183-9 – mf#74183 – cn Canadiana [840]

Impressions of theophrastus such : essays and leaves from a note-book / Eliot, George – Toronto: G N Morang, 1902 [mf ed 1995] – 5mf – 9 – 0-665-74183-9 – mf#74183 – cn Canadiana [840]

Impressions of turkey during twelve years' wanderings / Ramsay, William Mitchell – New York: G P Putnam; London: Hodder and Stoughton, 1897 – 1mf – 9 – 0-7905-0198-8 – (incl bibl ref) – mf#1987-0198 – us ATLA [949]

Impressoes da commissao rondon / Botelho De Magalhaes, Amilcar Armando – Sao Paulo, Brazil. 1942 – 1r – us UF Libraries [972]

A impressora : annunciador-commercial – Curitiba, PR: Impressora Paranaense, 01 jan, set 1899; jan, 03 maio 1900 – (= ser Ps 19) – mf#P16,02,21 – bl Biblioteca [079]

L'imprimerie : journal de la typographie, de la lithographie, etc – Paris, France. 15 jan 1900-dec 1913 – 1 – uk British Libr Newspaper [680]

Imprimerie – Paris, France. 15 jan 1900-1913 – 5r – 1 – uk British Libr Newspaper [072]

Imprimis / Hillsdale College – v7 n1-v18 n12 [1978 jan-1989 dec] – 1r – 1 – mf#1078730 – us WHS [378]

Imprint / Journal Company [Milwaukee WI] et al – 1982 oct-1983 dec – 1r – 1 – mf#703256 – us WHS [071]

The imprint – Toronto: Toronto Type Foundry, [1893-189- or 19–] – 9 – mf#P04481 – cn Canadiana [680]

Improbatio quorundam articulorum martini lutheri... / Clichtove, J – Parisiis, 1533 – 2mf – 9 – mf#CA-84 – ne IDC [240]

Impromptu du paquetage / Donnay, Maurice – Paris, France. 1916 – 1r – us UF Libraries [440]

Improve your tagalog / Aspillera, Paraluman S – 2nd ed. Manila, Philippines: [sl: sn], 1958, c1957 – us CRL [490]

Improved, Benevolent, Protective Order of Elks of the World *see* Washington eagle

Improved bridge from starvation to plenty : annexation of great britain to her colonies by means of the halifax and quebec railway, combined with ocean omnibuses – [London: s.n.], 1850 [mf ed 1984] – 9 – 0-665-45124-5 – mf#45124 – cn Canadiana [380]

The improved diaphragm ship pump and edson's diaphragm free pump : manufactured under license from jacob edson and executor estate s b loud, patented march 27, 1877, october 1, 1878, november 15, 1881 – [S.l: s.n, 1881?] [mf ed 1986] – 1mf – 9 – 0-665-61616-3 – mf#61616 – cn Canadiana [623]

Improved renderings of those passages in the english version of the new testament... / Craik, Henry – 2nd ed. London:Bagster, 1866 – 1mf – 9 – 0-8370-2768-3 – mf#1985-0768 – us ATLA [225]

Improvement era / Church of Jesus Christ of Latter-Day Saints et al – v13 n1-12 [1909 nov-1910 oct] – 1r – 1 – (cont by: ensign of the church of jesus christ of latter-day saints) – mf#166042 – us WHS [243]
Improvement of affliction : a practical sequel to a series of meditations entitled "comfort in affliction" / Buchanan, James – 2d ed. Edinburgh: John Johnstone, 1840. Beltsville, Md: NCR Corp, 1978 (3mf); Evanston: American Theol Lib Assoc, 1984 (3mf) – 9 – 0-8370-1075-6 – mf#1984-4431 – us ATLA [240]
The improvement of agriculture : and the elevation in the social scale of both husbandman and operative / Anderson, James – Montreal: printed by De Montigny & company...1858 [mf ed 1983] – 1mf – 9 – mf#SEM105P225 – cn Bibl Nat [630]
Improvement of pome fruit nursery tree quality / Steyn, Willem J – Stellenbosch: U of Stellenbosch 1998 [mf ed 1998] – 5mf – 9 – mf#mf.1274 – sa Stellenbosch [634]
The improvement of the harbor of quebec / Browne, Joseph Vincent – S.l: s.n, 1880? – 1mf – 9 – mf#04033 – cn Canadiana [627]
Improving and evaluating the child's worship at the first baptist church / Marcum, Billy Darrell – 1982 – 1 – 5.12 – us Southern Baptist [242]
Improving college and university teaching – Washington. 1953-1984 (1) 1953-1984 (5) 1953-1984 (9) – (cont by: college teaching) – ISSN: 0019-3089 – mf#6044 – us UMI ProQuest [378]
Improving exercise behavior : an application of the stages of change model in a worksite setting / Peterson, Travis – 1997 – 1mf – 9 – $4.00 – mf#PSY 2046 – us Kinesology [790]
Improving human performance quarterly – Washington. 1972-1979 (1) 1974-1979 (5) 1975-1979 (9) – ISSN: 0146-3756 – mf#9996 – us UMI ProQuest [370]
Improving the federal court library system : report and recommendations submitted to the judicial conference of the u.s. by the board of the federal judicial center – Washington: FJC, Feb 1978 – 2mf – 9 – $3.00 – mf#LLMC 95-821 – us LLMC [347]
Impuestos especiales del empresito / Abad, L V De – Habana, Cuba. 1939 – 1r – us UF Libraries [972]
Impugnacion al folleto que, con el titulo de... / Guardiola, Esteban – Tegucigalpa, Mexico. 1938 – 1r – us UF Libraries [972]
Impugnador cubano de ernesto renan / Fernandez De Castro, Jose Antonio – Habana, Cuba. 1938 – 1r – us UF Libraries [972]
Der impuls – Dessau DE, 1950 25 mar-1968, 1970-1974 sep, 1975-1990 may – 8r – 1 – (with gaps. notes: zementanlagenbau) – gw Misc Inst [621]
Impuls : bezirksdirektion deutsche post – Magdeburg DE, 1966-1968 nov, 1969-1989 1 nov – 4r – 1 – (with gaps) – gw Misc Inst [074]
Impuls – Jugenheim DE, 1959 n2-1966 n12 – 2r – 1 – (title varies: jg 2 n4: elan. incl suppl: das werdende zeitalter 1959 n2-1960 n6 [1r]) – gw Misc Inst [074]
Impulse : dance as communication – San Francisco: Impulse Publ. 18v. 1951-70 – 1r (annual) – 1 – (began publ as a student periodical of the workshop group at the halprin-lathrop dance studio, san francisco, and the first two issues, 1948 and 1949 (not in the library) were issued under the sponsorship of that group. subtitle varies. editor: 1951-70, marian van tuyl. ceased publ with 1970 issue) – mf#ZAN-MD22 – us NY Public [790]
Impulse / Syndicalist Alliance [Milwaukee WI] – Milwaukee WI. v1 n1-4 [1978 may/jun-1979 jun] – 1r – 1 – mf#665405 – us WHS [071]
Impulso inicial / Lufriu Y Alonso, Rene – Habana, Cuba. 1930 – 1r – us UF Libraries [972]
Impumelelo ngohlelo-mndeni = Successful family planning / South Africa. Department of Health [Departement van Gesondheid] [Departement van Gesondheid] – [Pretoria: Dept of Health 1978] [mf ed Pretoria, RSA: State Library [199-]] – 19p [ill] on 1r with other items – 5 – (in zulu; also available in afrikaans, english, northern sotho, sotho, tsonga, tswana & xhosa) – mf#op 06727 r24 – us CRL [360]
Impumeleo ngohlelo-mndeni = Successful family planning / South Africa. Department of National Health and Population Development [Departement van Nasionale Gesondheid en Bevolkingsontwikkeling – 3rd ed. Pretoria: Dept of National Health & Population Development 1986 [mf ed Pretoria, RSA: State Library [199-]] – 20p [ill] on 1r with other items – 5 – (in zulu; also available in afrikaans, english, northern sotho, sotho, tsonga, tswana & xhosa) – mf#op 08552 r24 – us CRL [362]
Impuras / Carrion Y Cardenas, Miguel De – Habana, Cuba. 1919 – 1r – us UF Libraries [972]
Imputation / Bates, John – London, England. 18– – 1r – us UF Libraries [240]
Imre binah / Modilevski, Isaac – Kiev, Ukraine. 1911 – 1r – us UF Libraries [939]
Imre darush / Nissenbaum, Isaac – New York, NY. 1925 or 1926 – 1r – us UF Libraries [939]
Imre haskel / Tawschunski, Jacob – Bilgoraj, Poland. 1908 – 1r – us UF Libraries [939]
Imre lev / Ennery, Jonas – New York, NY. 1910 – 1r – us UF Libraries [939]
Imre shefer / Rabinowicz, Shaga Fayvl – Vilna, Lithuania. 1929 – 1r – us UF Libraries [939]
Imre yosher / Eisenstadter, Meir – Ungvar, Ukraine. 1864 – 1r – us UF Libraries [939]
Ims – international industrial medicine and surgery – Miami. 1932-1973 (1) 1965-1973 (5) 1971-1972 (9) – (cont by: international journal of occupational health and safety) – ISSN: 0163-934X – mf#722 – us UMI ProQuest [610]
Imtiyazat ve mukavelat – Istanbul: Matbaa-i Osmaniye. v1-7. 1884-97 – (= ser O & t journals) – 70mf – 9 – $1160.00 – us MEDOC [956]
Imvaho – Kigali: Impr scolaire, nov 1984-mar 1994 – 3r – 1 – us CRL [079]
Imvaho nshya – Kigali: [s.n.], oct 19/25 1994-dec 26 1994/jan 1 1995; mar 13/19-apr 1995 – us CRL [079]
Imvo Neliso Lomsi *see* Imvo zabantsundu
Imvo neliso lomzi = Native opinion and guardian – King William's Town: J Tengo-Jabavu, jan 1895-mar 1898 – us CRL [079]
Imvo zabantsundu – King William's Town: Jabavu and Co, Ltd, mar 1912-oct 1961] – us CRL [079]
Imvo zabantsundu – King William's Town, [South Africa]: J Tengo-Jabavu, nov 3 1884-dec 19 1894 – 1 – us CRL [079]
Imvo zabantsundu – King William's Town SA, 3 nov 1884-26 dec 1936 – 31r – 1 – (title varies: imvo neliso lomsi) – sa National [079]
Imvo zabantsundu base afrika = South african native opinion – King William's Town: Jabavu and Co, feb 1903-dec 1909 – us CRL [079]
Imvo zabantsundu bomzantsi afrika = Native opinion of south africa – King William's Town: Jabavu and Co Ltd, dec 1909-feb 1912 – us CRL [079]
Imvo zabantsundu bomzantsi afrika = South african native opinion – King William's Town: Jabavu and Co, oct 1902-jan 1903] – us CRL [079]
Imvo zontsundu, neliso lomzi = Native opinion – King William's Town: Jabavu and Bokwa, apr 1898-aug 1901] – us CRL [079]
In 4 libros sententiarum commentaria / Estius, G – Duaci. v1-4. 1616 – 4v on 67mf – 8 – €128.00 – ne Slangenburg [240]
In 4 priora capita euangelij secundum matthaeum / Bugenhagen, J – Wittembergae, 1543 – 3mf – 9 – mf#TH-1 mf 168-170 – ne IDC [242]
In 12 aristotelis metaphycam / Halensis, Alexander – Venetiis: de Karera, 1572 – 43mf – 8 – €82.00 – ne Slangenburg [110]
In 12 libros metaphysicae aristotelis / Dominicus de Flandria – Agrippinae, 1621 – 55mf – 8 – €105.00 – ne Slangenburg [110]
In 12 prophetas minores explicationes svccinctae : ordinem rerum, textus sententiam, et doctrinas praecipuas strictissime indicantes / Wigand, J – Basileae, 1566 – 7mf – 9 – mf#TH-1 mf 1560-1566 – ne IDC [242]
In 31 davidis psalmos / Arias Montano, Benito – 1605. Ed. Pedro de Valencia – 9 – sp Bibl Santa Ana [240]
In 100 verrem actionis secundae libri 4, 5 / Cicero, Marcus Tullius – Lipsiae, Germany. 1949 – 1r – us UF Libraries [450]
In a far country : a story of christian heroism and achievement / Gunn, Harriette Bronson – Philadelphia: American Baptist Publ Soc, c1911 – 1mf – 9 – 0-524-07101-2 – mf#1991-2924 – us ATLA [240]
In a forshtadt / Galvez, Manuel – Buenos Ayres, Argentina. 1933 – 1r – us UF Libraries [939]
In a preacher's study / Jackson, George – London, New York: Hodder & Stoughton, 1914 [mf ed 1990] – 1mf – 9 – 0-7905-7588-4 – (incl bibl ref) – mf#1989-0813 – us ATLA [225]
In a steamer chair : and other shipboard stories / Barr, Robert – London: Chatto & Windus, 1892 – 4mf – 9 – mf#03347 – cn Canadiana [830]
In abissinia / Matteucci, P – Milano, 1880 – 4mf – 9 – mf#NE-20207 – ne IDC [916]
In acta apostolorum commentaria / Lorinus, Ioan. – Lugduni, 1605 – 45mf – 8 – €86.00 – ne Slangenburg [226]
In acta apostolorum..., homiliae 579 / Gwalther, R – Zuerich, Froschouer, 1557 – 11mf – 9 – mf#PBU-295 – ne IDC [240]
In acta apostolorvm...commentariorvm libri 6 / Bullinger, Heinrich – Tigvri, Christoph Froschouer, 1533 – 8mf – 9 – mf#PBU-118 – ne IDC [240]
In affectionate memory of the reverend doctor lewellyn pratt of norwich connecticut : who in the fullness of his years passed into the eternal light on june the fourteenth in the year nineteen hundred and thirteen – [Norwich, Ct: Norwich Free Academy, 1913?] – 1mf – 9 – 0-524-08391-6 – mf#1993-3091 – us ATLA [240]
In afric's [sic] forest and jungle : or, six years among the yorubans / Stone, Richard Henry – New York: Fleming H Revell, c1899 [mf ed 1986] – 1mf – 9 – 0-8370-6620-4 – mf#1986-0620 – us ATLA [306]
In all shades : a novel / Allen, Grant – Chicago, New York: Rand, McNally, 188-? – 4mf – mf#26239 – cn Canadiana [830]
In ambas...pauli ad corinthios epistolas commentarij / Musculus, W – Basilea, Johann Herwagen, 1559 – 5mf – 9 – mf#PBU-341 – ne IDC [240]
In amos : abdiam et ionam prophetas commentarii / Lambert, F – Strasbourg, 1525 – 4mf – 9 – mf#PPE-114 – ne IDC [240]
In anatomen corporis humani... / Vassaevs, L – Venecia, 1549 – 4mf – 9 – sp Cultura [611]
In and out : ...being a paper published from time to time by the canadian field ambulance in the field – [France] v1 n1. nov 1918// – 1r – 1 – Can$22.00 – (no more publ) – cn McLaren [071]
In and out of central america / Vincent, Frank – New York, NY. 1890 – 1r – us UF Libraries [972]
In and out of chanda : being an account of the mission of the scottish episcopal church to the city and district of chanda... / ed by Dawson, Edwin Collas – Edinburgh: Foreign Mission Board, 1906 [mf ed 1995] – (= ser Yale coll) – vi/70p (ill) – 1 – 0-524-09192-7 – (with indian folk-lore stories trans by alex wood. pref by rev the bishop of st andrews) – mf#1995-0192 – us ATLA [242]
In and out of chanda : being an account of the mission of the scottish episcopal church... and indian folklore stories / ed by Dawson, E C – Edinburgh, 1906 – 1mf – 9 – mf#HTM-48 – ne IDC [917]
In and out of the barrio: adapted from book two / Preiser, Rosa C – Manila, 1948. 316p. illus. (Philippine public school readers) – 1 – us Wisconsin U Libr [360]
In and out of the homes of india / Lee, Ada – Calcutta: Methodist press [1909] [mf ed 1995] – (= ser Yale coll) – vii/107p (ill) – 1 – 0-524-09991-X – mf#1995-0991 – us ATLA [954]
In andamans, the indian bastille / Sinha, Bejoy Kumar – Cawnpore: Profulla C Mitra, 1939 – (= ser Samp: indian books) – us CRL [920]
In aphorismo et libellum de alimento hipocratis, commentaria / Valles de Covarrubias, F – Alcala de Henares, 1561 – 12mf – 9 – sp Cultura [610]
In apocalypsim...conciones centum / Bullinger, Heinrich – Basileae, Ioannes Oporinus, 1557 – 4mf – 9 – mf#PBU-196 – ne IDC [240]
In apocalypsin / Lambert, F – Marbourg, 1528 – 8mf – 9 – mf#PPE-121 – ne IDC [240]
In apocalypsin commentarius (cima16) : farbmikrofiche-edition der handschrift manchester, the john rylands university library, latin ms 8 / Liebana, Beatus a – (mf ed 1990) – (= ser Codices illuminati medii aevi (cima) 16) – 41p on 9 color mf – 15 – €360.00 – 3-89219-016-X – (int & description by peter k klein) – gw Lengenfelder [090]
In apocalypsin johannis commentarius... / Marck, J – Trajecti ad Rhenum, 1699 – 13mf – 9 – mf#PBA-247 – ne IDC [240]
In arcane logos – v1 n1-5,5,7,10-11,14,16 [1969 apr 16-may 15, may 23, jun 5, jul 3-16, oct 24, nov 21]; v1 n1-26 [1969 apr 16-dec] – 2r – 1 – mf#770856 – us WHS [071]
In aristotelis de virtutibus librum, commentariorum libri tres, omnibus platonicae aristotelicaeq[ue] philosophae studiosis compromis utiles & necessarij... / Velsius, Justus – Coloniae: Excudebat Martinus Gymnicus, anno 1551 – (= ser Ethics in the early modern period) – 3mf – 9 – mf#pl-146 – ne IDC [170]
[In aristotelis libros quinque priores moralium ad nicomachum] : tarquinii gallutii sabini e societate iesu in aristotelis libros quinque priores moralium ad nicomachum: nova interpretatio, commentarii, quaestiones / Galluzzi, Tarquinio – Parisiis: Sumptibus Sebastiani Cramoisy...1632 – (= ser Ethics in the early modern period) – 11mf – 9 – mf#pl-337 – ne IDC [170]
In artem poeticam horatii / Sanchez de las Brozas, Francisco – 1591 – 9 – sp Bibl Santa Ana [450]
In biscayne bay / Rockwood, Caroline Washburn – New York, NY. 1891 – 1r – us UF Libraries [630]
In business – Emmaus. 1986+ (1,5,9) – ISSN: 0190-2458 – mf#15130 – us UMI ProQuest [650]
In camp and tepee : an indian mission story / Page, Elizabeth Merwin – New York: FH Revell, c1915 – 1mf – 9 – 0-7905-6940-X – mf#1988-2940 – us ATLA [240]
In canada's national park / Bell, Josiah Jones – [S.l: s.n, 1894?] [mf ed 1981] – 1mf – 9 – 0-665-14237-4 – (fr: the canadian magazine) – mf#14237 – cn Canadiana [790]
In cantica canticorum salomonis commentarii / Lambert, F – Strasbourg, 1524 – 3mf – 9 – mf#PPE-112 – ne IDC [240]
In canticum canticorum expositio : formae tplila 36 / Apponius – 1986 – (= ser ILL – ser a; Cccm 19) – 13mf+84p – 9 – €50.00 – 2-503-60192-8 – be Brepols [400]
In canticum canticorum. in librum primum regum : formae tplila 8 / Gregorius Magnus – 1982 – (= ser ILL – ser a; Cccm 144) – 18mf+108p – 9 – €40.00 – 2-503-61442-6 – be Brepols [400]
In catabaptistarvm strophas elenchus... – Tiguri: Chr. Froschouer, 1527 – 3mf – 9 – mf#ME-88 – ne IDC [242]
In catechesin religionis christianae... / Bastingius, J – [Heidelberg], 1590 – 7mf – 9 – mf#PBA-128 – ne IDC [240]
In catholicas bb iacobi : et iudae apostolorum epistolas commentarii / Lorinus, Ioan. – Moguntiae, 1622 – 21mf – 8 – €41.00 – ne Slangenburg [227]
In catholicas tres b joannis : et duas b petri epistolas commentarii / Lorinus, Ioan. – Lugduni, 1609 – 20mf – 8 – €38.00 – ne Slangenburg [226]
In cebetis thebani tabulam commentariorum libri sex, totius moralis philosophiae thesaurus : in quibus nonnulla per occasionem tum de studiorum, artium, & scientiarum abusu & corruptela, tum contra ea... / Velsius, Justus – Lugduni: [s.n.] 1551 – (= ser Ethics in the early modern period) – 6mf – 9 – mf#pl-461 – ne IDC [170]
In chordis et organo (fastes d'organiers) / mon clocher (causerie radiofusee) / Morin, Victor – Montreal: Editions des Dix, 1940 [mf ed 1987] – 1mf – 9 – mf#SEM105P783 – cn Bibl Nat [971]
In christ : or, the believer's union with his lord / Gordon, Adoniram Judson – Boston: Gould and Lincoln, 1872 – 1mf – 9 – 0-524-05212-3 – mf#1992-0345 – us ATLA [240]
In christ jesus : or, the sphere of the believer's life / Pierson, Arthur Tappan – New York: Funk & Wagnalls, 1898 – 1mf – 9 – 0-8370-4750-1 – mf#1985-2750 – us ATLA [220]
In christo, or, the monogram of st. paul / Macduff, John Ross – New York: American Tract Society, [1881?] – 1mf – 9 – 0-8370-5528-8 – mf#1985-3528 – us ATLA [240]
In common / Common Cause [US] – 1975 apr-1978 win, 1978 win-1980 win – 2r – 1 – (cont: common cause report from washington; cont by: frontline [washington, dc]; common cause [washington, dc]) – mf#646734 – us WHS [320]
In context / Context Foundation [Sequim WA] et al – n1-4 [1983 win-autumn], n6-19 [1984 summer-1988 aut] – 1r – 1 – (cont by: yes [bainbridge island, wa]) – mf#1058726 – us WHS [071]
In d apostoli pauli ad thessalonicenses... epistolas commentarii... / Bullinger, Heinrich – Tiguri, [1536] – 5mf – 9 – mf#PBU-129 – ne IDC [240]
In d apostoli pavli ad galatas, ephesios, philippen... / Bullinger, Heinrich – Tigvri, Christoph Froschouer, 1535 – 6mf – 9 – mf#PBU-125 – ne IDC [240]
In d apostoli pavli ad thessalonicenses, timotheum, titum & philemonem epistolas...commentarij / Bullinger, Heinrich – Tigvri, Christ[oph] Froschover, [1536] – 5mf – 9 – mf#PBU-128 – ne IDC [240]
In d pauli apostoli epistolam ad romanos homiliae / Gwalther, R – Zuerich, Froschouer, 1566 – 6mf – 9 – mf#PBU-297 – ne IDC [240]
In d pauli...epistolam ad galatas homiliae 61 / Gwalther, R – Zuerich, Froschouer, 1576 – 4mf – 9 – mf#PBU-302 – ne IDC [240]
In d petri apostoli epistolametranqe... commentarius / Bullinger, Heinrich – Tigvri, Christoph Frosch[auer], 1534 – 3mf – 9 – mf#PBU-119 – ne IDC [240]
In d thomae aq commentaria super libros posteriorum analyticorum aristotelis / Dominicus de Flandria – Venetiis, 1526 – 9mf – 8 – €15.00 – ne Slangenburg [180]
In danielem prophetam ioannis oecolampadij libri duo... / Oecolampadius, J – Basileae, Joannes Bebel, 1530 – 4mf – 9 – mf#PBU-384 – ne IDC [240]
In darkest africa : or the quest, rescue and retreat of emin, governor of equatoria / Stanley, H M – New York, 1890. 2v – 12mf – 9 – mf#HT-140 – ne IDC [916]
In darkest africa / Stanley, Henry Morton – New York, 1890 – 1 – us CRL [960]
In darkest cuba / Gonzales, Narciso Gener – Columbia, SC. 1922 – 1r – us UF Libraries [972]

In darkest england and the way out / Booth, William – London; New York: International Headquarters of the Salvation Army, [1890?] – 1mf – 9 – 0-7905-4432-6 – mf#1988-0432 – us ATLA [240]

In de fierabendstied : en plattduetsch geschichtenbook / Freudenthal, Friedrich – 2. Aufl. Oldenburg: G Stalling, [1889?] (mf ed 1990) – 1r – 1 – (filmed with: ein glaubensbekenntnis) – us Wisconsin U Libr [830]

In decalogum praeceptorum dei explanatio / Musculus, W – Basilea, Johan Herwagen, 1553 – 5mf – 9 – mf#PBU-337 – ne IDC [240]

In defence of a shared society / Hellmann, Ellen – Johannesburg, South Africa. 1956 – 1r – us UF Libraries [960]

In defence of the faith / Oliver, Alexander – Edinburgh: Oliphant, Anderson & Ferrier, 1886 [mf ed 1985] – 1mf – 9 – 0-8370-4618-1 – (incl bibl ref) – mf#1985-2618 – us ATLA [210]

In defence of the quebec minority / Sellar, Robert – S.l: s.n, 1894? – 1mf – 9 – mf#13264 – cn Canadiana [305]

In den alpen / Tyndall, John – Braunschweig 1872 [mf ed Hildesheim 1995-98] – 3mf – 9 – €90.00 – 3-487-29338-2 – gw Olms [914]

In den bergen, da lauert der wildschuetz : roman aus der alpenwelt / Achleitner, Arthur – Berlin: Gebrueder Paetel, [191-?] [mf ed 1995] – (= ser Paetels roman-reihe) – 254p – 1 – mf#8917 – us Wisconsin U Libr [830]

In den hochalpen : erlebnisse aus den jahren 1859-1885 / Guessfeldt, Paul – Berlin 1886 [mf ed Hildesheim 1995-98] – 3mf – 9 – €90.00 – 3-487-29468-0 – gw Olms [920]

In den pampas : eine erzaehlung aus der wilden welt / Gerstaecker, Friedrich – K"ln/Rhein: H Schaffstein, 1921 (mf ed 1990) – 1r – 1 – (filmed with: die regulatoren in arkansas) – us Wisconsin U Libr [830]

In den wohnungen des todes / Sachs, Neily – Berlin, Germany. 1947 – 1r – us UF Libraries [943]

In der badewanne zu singen / Sonnenstern, Werner – Zuerich: Sanssouci, c1969 – us Wisconsin U Libr [780]

In der brigittenau 1683 : genrebild in einem aufzuge / Langer, Anton – Wien: Wallishausser [18–?] [mf ed 1995] – 1r – 1 – (filmed with: auksines legendos / g keleris) – mf#3913p – us Wisconsin U Libr [820]

In der fremd / Mastboim, Joel – Varshe, Poland. 1920 – 1r – us UF Libraries [943]

In der grunen holle / Eichhorn, Franz – Berlin, Germany. 1937 – 1r – us UF Libraries [972]

In der heimat des konfuzius : skizzen, bilder und erlebnisse aus schantung / ed by Stenz, Georg Maria – Steyl: Missionsdruckerei, 1902 [mf ed 1995] – (= ser Yale coll) – 288p (ill) – 1 – 0-524-09333-4 – (in german) – mf#1995-0333 – us ATLA [915]

In der jodutenstrasse : roman / Gerhard, Hans Ferdinand – Berlin: G Grote, 1912 (mf ed 1990) – 1r – 1 – (filmed with: zeitgenoessische dichter) – us Wisconsin U Libr [830]

In der noth lernet man die freunde kennen : ein originallustspiel in ungebundener rede und fuenf aufzuegen / Bergobzoomer, Johann Baptist – [Wien?]: zu finden beym Logenmeister, 1777 [mf ed 1993] – [2]/98p – 1 – mf#8512 – us Wisconsin U Libr [820]

In der stille : gedanken und betrachtungen / Korn, Karl – Berlin-Schildow: E Sicker, 1944 – 1r – 1 – us Wisconsin U Libr [840]

In der veranda : eine dichterische nachlese / Gruen, Anastasius – Berlin: G Grote, 1876 – 1 – us Wisconsin U Libr [810]

In der veranda : eine dichterische nachlese / Gruen, Anastasius – Berlin: G Grote, 1876 – 1r – 1 – us Wisconsin U Libr [430]

In devteronomivm mosis enarratio / Chytraeus, D – Vvitebergae, 1575 – 9mf – 9 – mf#TH-1 mf 284-292 – ne IDC [242]

In divi lucae evangelium commentarii nunc secundo recognitae ac locupletati / Lambert, F – Strasbourg, 1525 – 6mf – 9 – mf#PPE-118 – ne IDC [240]

In divi pauli epistolas tres, ad timotheum et titum... / Grossmann, K – Basileae, [Thomas Platter et Balthasar Lasius], 1535 [1536] – 3mf – 9 – mf#PBU-609 – ne IDC [240]

In divinam ad romanos s pauli apostoli epistolam commentarius / Pareus, D – Francofurti, 1608 – 10mf – 9 – mf#PBA-3 – ne IDC [240]

In divinvm...euangelium secundum ioannem, commentariorum libri 10 / Bullinger, Heinrich – Tigvri, Christoph Froschower, 1543 – 6mf – 9 – mf#PBU-143 – ne IDC [240]

In dulci iubilo, nun singet und seid froh : ein beitrag zur geschichte der deutschen poesie / Fallersleben, Hoffmann von – Hannover, 1861 – 3mf – 8 – €7.00 – ne Slangenburg [430]

In duodecim prophetas minores scholia = Horreum mysteriorum. selections / Bar Hebraeus – Lipsiae: B G Teubneri, 1882 – 1mf – 9 – 0-8370-1783-1 – mf#1987-6171 – us ATLA [220]

In dvos libros samuelis prophetae qvi vvlgo priores libri regum... / Vermigli, P M – Tigvri, 1575 – 8mf – 9 – mf#PBU-873 – ne IDC [240]

In dvos libros samuelis...commentarij / Vermigli, P M – Tigvri, Christoph Froschouer, 1564 – 8mf – 9 – mf#PBU-284 – ne IDC [240]

In dwarf land and cannibal country : a record of travel and discovery in central africa / Lloyd, A B – London, [1900] – 5mf – 9 – mf#HT-83 – ne IDC [916]

In einem kuehlen grunde : roman / Gabele, Anton – Leipzig: P List, c1939 (mf ed 1990) – 1r – 1 – (filmed with: gustav freytag, ein publizist) – us Wisconsin U Libr [830]

In eis und schnee : die aufsuchung der jeannette-expedition und eine schlittenfahrt durch sibirien / Gilder, William – Leipzig 1884 [mf ed Hildesheim 1995-98] – 3mf – 9 – €90.00 – 3-487-27785-9 – gw Olms [910]

In- en uitvoer – Amsterdam, Netherlands. -w. 3 Oct 1917-6 Aug 1919. Imperfect. 3 reels – 1 – uk British Libr Newspaper [949]

In episolam sancti pauli ad galatas commentarius : 1531 / Luther, Martin – Chester, England. 1796 – 1r – us UF Libraries [240]

In epistolam ad hebraeos, ioannis oecolampadii, explanationes / Oecolampadius, J – Argentorati, Matthias Apiarius, 1534 – 5mf – 9 – mf#PBU-390 – ne IDC [240]

In epistolam ad romanos, pia et erudita scholia, pro rhetorica dispositione / Sarcerius, E – Francoforti, [1541] – 9mf – 9 – mf#TH-1 mf 1332-1340 – ne IDC [242]

In epistolam b pavli apost ad rhomanos, adnotationes / Oecolampadius, J – Basileae, Andreas Cratander, 1525 – 3mf – 9 – mf#PBU-363 – ne IDC [240]

In epistolam d pauli ad colossenses / Oecolampadius, J – Bern, Matthias Apiarius, 1546 – 2mf – 9 – mf#PBU-397 – ne IDC [240]

In epistolam d pauli apostoli ad romanos... homiliarum archetypi / Gwalther, R – Tigvri, officina Froschoviana, 1588 – 3mf – 9 – mf#PBU-304 – ne IDC [240]

In epistolam d pavli ad romanos scriptam commentarivs / Corner, C – Heidelbergae, 1583 – 5mf – 9 – mf#TH-1 mf 344-348 – ne IDC [242]

In epistolam ioannis apostoli catholicam primam...demegoriae, hoc est homiliae una et 20 / Oecolampadius, J – Basileae, Andreas Cratander, 1524 – 3mf – 9 – mf#PBU-358 – ne IDC [241]

In epistolam pauli ad galatas notae... / Olevianus, G – Genevae, 1578 – 2mf – 9 – mf#PBA-277 – ne IDC [240]

In epistolam pauli ad romanos notae / Olevianus, G – Genevae, 1579 – 10mf – 9 – mf#PBA-278 – ne IDC [240]

In epistolam [primam] ioannis...expositio / Bullinger, Heinrich – Tigvri, Christoph Froschover, 1532 – 2mf – 9 – mf#PBU-114 – ne IDC [240]

In epistolam s pavli ad colossenses annotationes d iohannis vvigandi / Wigand, J – Vvitebergae, 1586 – 3mf – 9 – mf#TH-1 mf 1573-1575 – ne IDC [242]

In epistolam s pavli ad romanos annotiationes / Wigand, J – Francof ad Moenvm, 1580 – 4mf – 9 – mf#TH-1 mf 1567-1570 – ne IDC [242]

In epistolam s pavli apostoli ad romanos... commentarii / Vermigli, P M – Basilea, Petrus Perna, 1558 – 8mf – 9 – mf#PBU-280 – ne IDC [240]

In epistolam...pauli ad romanos : commentarij / Musculus, W – Basileae, Sebastianus Henricpetri, 1600 – 4mf – 9 – mf#PBU-339 – ne IDC [240]

In epistolas b. pauli commentarii : the latin version / Theodore, Bishop of Mopsuestia – Cambridge: University Press, 1880-1882 – 1r – 1 – 0-8370-1038-1 – mf#1984-S031 – us ATLA [240]

In epistolas d pavli, ad galatas et ephesios, piae atque eruditae annotationes / Sarcerius, E – Francoforti, 1541 – 9mf – 9 – mf#TH-1 mf 1304-1312 – ne IDC [242]

In epistolas d pavli ad philippenses, colossenses, et thessalonicenses, pia et erudita scholia / Sarcerius, E – Francoforti, [1542] – 6mf – 9 – mf#TH-1 mf 1341-1346 – ne IDC [242]

In epistolas dominicales ac festivales expositiones / Sarcerius, E – Franc[ofortì], 1561 – 9mf – 9 – mf#TH-1 mf 1322-1330 – ne IDC [242]

In epistolas...pauli ad galatas et ephesios commentarii / Musculus, W – Basilea, Johann Herwagen, 1561 – 6mf – 9 – mf#PBU-342 – ne IDC [240]

In esaiam prophetam commentarij / Musculus, W – Basilea, Johann Herwagen, 1557 – 10mf – 9 – mf#PBU-340 – ne IDC [240]

In esaiam prophetam explicationes breves / Wigand, J – Erphordiae, 1581 – 8mf – 9 – mf#TH-1 mf 1581-1588 – ne IDC [242]

In ethicorum aristotelis interpretationem prolegomena / Caselius, Johannes – Rostochii: Excudebat Iacobus Lucius Transylvanus, anno 1575 – (= ser Ethics in the early modern period) – 35lea on 1mf – 9 – mf#pl-355 – ne IDC [170]

In euangelium iesu christi secundum marcum homiliae 89 / Gwalther, R – Zuerich, Froschouer, 1561 – 10mf – 9 – mf#PBU-296 – ne IDC [240]

In evangelistam matthaeum commentarii / Musculus, W – Basilea, Johann Herwagen, 1548 – 12mf – 9 – mf#PBU-333 – ne IDC [240]

In evangelivm s iohannis explicationes / Wigand, J – Regiomonti, 1575 – 5mf – 9 – mf#TH-1 mf 1576-1580 – ne IDC [242]

In exodvm enarratio / Chytraeus, D – Vitebergae, 1561 – 4mf – 9 – mf#TH-1 mf 293-296 – ne IDC [242]

In fact – v1-22 n2,1. 1940-50 [all publ] – (= ser Radical periodicals in the united states, 1881-1960. series 1) – 41mf – 9 – $365.00 – us UPA [335]

In farkishuftn land fun legendarn dzshugashvili / Grosman, Moisheh – Paris, France. v1-2. 1949 – 1r – us UF Libraries [939]

In florida gardens / Wilson, Millar – Jacksonville, FL. 1924 – 1r – us UF Libraries [580]

In florida's dawn / Gold, Pleasant Daniel – Jacksonville, FL. 1926 – 1r – us UF Libraries [978]

In foedvs et victoriam contra tvrcas... / Gherardius, P – Venetiis, 1572 – 6mf – 9 – mf#H-8327 – ne IDC [956]

In four continents : a sketch of the foreign missions of the presbyterian church, u.s / Williams, Henry Francis – 3rd ed. Richmond, Va: Presbyterian Committee of Publication, 1910 – 1mf – 9 – 0-524-06566-7 – mf#1991-2650 – us ATLA [242]

In frack und arbeitsbluse roman / Kretzer, Max – Dessau: C Duennhaupt 1924 [mf ed 1995] – 1r – 1 – (filmed with: berliner skizzen / max kretzer) – mf#3910p – us Wisconsin U Libr [830]

In genesim enarratio / Oecolampadius, J – Basileae, [Johann Bebel], 1536 – 4mf – 9 – mf#PBU-394 – ne IDC [240]

In genesin en arratio, tradita : vt ad lectionem textus bibliorum auditores unuitarentur / Chytraeus, D – Vitebergae, 1561 – 6mf – 9 – mf#PBA-426 – ne IDC [240]

In genesin enarratio / Chytraeus, D – Vitebergae, 1557 – 6 – 9 – mf#TH-1 mf 297-302 – ne IDC [242]

In geroysh fun mashinen / Horontchik, Simon – Warszawa, Poland. 1928 – 1r – us UF Libraries [939]

In goles bay di ukrainer / Goldelman, Salomon – Wien, Austria. 1921 – 1r – us UF Libraries [939]

In guiana wilds / Rodway, James – Boston, MA. 1899 – 1r – us UF Libraries [972]

In habakuk prophetam enarrationes / Capiton, W – Argentorati, 1526 – 2mf – 9 – mf#PPE-102 – ne IDC [240]

In health – Sausalito. 1990-1991 (1,5,9) – ISSN: 1047-0549 – mf#18476,01 – us UMI ProQuest [610]

In heavenly places / Simpson, Albert B – New York: Christian Alliance, c1892 [mf ed 1992] – (= ser Christian & missionary alliance coll) – 1mf – 9 – 0-524-02153-8 – mf#1990-4219 – us ATLA [240]

In hesterae historiam homiliarum sylvae vel archetypi / Gwalther, R – Tigvri, officina Froschoviana, 1587 – 2mf – 9 – mf#PBU-303 – ne IDC [240]

In hieremiam prophetam commentariorum libri tres ioannis oecolampadij : eivsdem in threnos hieremiae ennarationes / Oecolampadius, J – Argentinae, Matthias Apiarius, 1533 – 6mf – 9 – mf#PBU-388 – ne IDC [240]

In his name / Jinarajadasa, Curuppumullage – Chicago: Rajput Press, 1913 – 1mf – 9 – 0-524-02211-9 – mf#1990-2885 – us ATLA [280]

In his steps / Sheldon, Charles Monroe – New York, NY. no date – 1r – us UF Libraries [025]

In historiam creatonis mosaicam commentatio = Erlaeuterungen zur mosaischen schoepfungs-geschichte / Pianciani, G B – Regensburg: Friedrich Pustet, 1853 – 1mf – 9 – 0-8370-6930-0 – (in german. incl bibl ref) – mf#1986-0930 – us ATLA [220]

In historiam iudicum populi israel commentarius / Chytraeus, D – Francofurti ad Moenum, 1589 – 6mf – 9 – (missing: p463-484) – mf#TH-1 mf 303-308 – ne IDC [242]

In honorem sanctae crucis : formae tplila 100 / Rabanus Maurus – [mf ed 2000] – (= ser ILL – ser a; Cccm 100) – 6mf+80p – 9 – €40.00 – 2-503-64002-8 – be Brepols [400]

In hoseam prophetam commentarius / Capiton, W – Argentorati, 1528 – 7mf – 9 – mf#PPE-103 – ne IDC [240]

In ibin ocidi in ternarium...annotationes / Sanchez de las Brozas, Francisco – 1598 – 9 – sp Bibl Santa Ana [450]

In ieremiam prophetam commentarium / Bugenhagen, J – Vvittembergiae, 1546 – 13mf – 9 – mf#TH-1 mf 144-156 – ne IDC [242]

In iesaiam prophetam...commentariorum libri 6 / Oecolampadius, J – Basileae, Andreas Cratander, 1525 – 7mf – 9 – mf#PBU-362 – ne IDC [240]

In iesvm syrach, integra scholia in vsvm scholasticae arq; christianae iuuentutis potissimum conscripta / Sarcerius, E – Franc[ofurti], [1543] – 14mf – 9 – mf#TH-1 mf 1355-1368 – ne IDC [242]

In india : sketches of indian life and travel from letters and journals / Mitchell, Maria Hay (Flyter) – London 1876 – (= ser 19th c british colonization) – 4mf – 9 – mf#1.1.6161 – uk Chadwyck [915]

In india : sketches of indian life and travel from letters and journals / Mitchell, Maria Hay Flyter – London: T Nelson, 1876 [mf ed 1995] – (= ser Yale coll) – 319p – 1 – 0-524-09127-7 – mf#1995-0127 – us ATLA [954]

In india : sketches of indian life and travels from letters and journals / Mitchell, J Murray (Mrs) – London, 1876 – 4mf – 9 – mf#HTM-133 – ne IDC [915]

In india (the land of famine and of plague) : or, bombay the beautiful the first city of india. with incidents and experiences of pioneer mission work in western india; illustrative of the country, customs and creeds / Clutterbuck, George W – London, 1897 – (= ser 19th c books on british colonization) – 4mf – 9 – mf#1.1.7508 – uk Chadwyck [954]

In investiganda monachatus origine... / Bornemann, F W B – Goettingen, 1885 – 3mf – 8 – €7.00 – ne Slangenburg [241]

In ioannem evangelistam ivsta scholia summa diligentia / Sarcerius, E – Basileae, 1540 – 9mf – 9 – mf#TH-1 mf 1369-1377 – ne IDC [242]

In jamaica and cuba / De Lisser, Herbert George – Kingston, Jamaica. 1910 – 1r – us UF Libraries [972]

In journeyings often : glimpses of the life of bishop bompas / Soveriegn, Arthur Henry – Toronto: Church House, [1916?] – 1mf – 9 – mf#99093 – cn Canadiana [242]

In jungle depths : true stories from a missionary's diary / Carvell, Alice Maude – London: Religious Tract Society, 1919 [mf ed 1995] – (= ser Yale coll) – xiv/132p (ill) – 1 – 0-524-09167-6 – mf#1995-0167 – us ATLA [880]

In kazmerzsh / Segalowitch, Zusman – Warsaw, Poland. 1913 – 1r – us UF Libraries [939]

In lamentationes ieremiae...commentarivm / Vermigli, P M – Tigvri, Ioh Iacob Bodmer, 1629 – 2mf – 9 – mf#PBU-289 – ne IDC [240]

In leper-land : being a record of my tour of 7,000 miles among indian lepers: including some notes on missions... / Jackson, John – London: Marshall, [1901?] – 1mf – 9 – 0-8370-6128-8 – (incl ind of places) – mf#1986-0128 – us ATLA [240]

In letters from lms, london / London Missionary Society – Samoan District – 1896-1946 – 1r – 1 – mf#pmb126 – at Pacific Mss [242]

In leviticvm, complecten / Chytraeus, D – Vvitebergae, 1569 – 6mf – 9 – mf#TH-1 mf 309-314 – ne IDC [242]

In libros ethicorum aristotelis ad nicomachum, aliquot conimbricensis cursus disputationes : in quibus praecipua quaedam ethicae disciplinae capita continentur – Quarta hac in Germania editione correctiores editae. Coloniae: Impensis haered. Lazari Zetzneri 1621 – (= ser Ethics in the early modern period) – 1mf – 9 – mf#pl-63 – ne IDC [170]

In libros paralipomenon sive chronicorum...commentarius... / Lavater, L – Tiguri, Christoph Froschover, 1573 – 6mf – 9 – mf#PBU-317 – ne IDC [240]

In librum duodecim prophetarum commentarii / Ribera, Fr – Duaci, 1611 – 30mf – 8 – €58.00 – ne Slangenburg [221]

In librum iosue...homiliae / Lavater, L – Tiguri, Christoph Froschower, 1565 – 4 – 9 – mf#PBU-310 – ne IDC [240]

In librum proverbiorum...commentarii / Lavater, L – Tiguri, Christoph Froschover, 1562 – 8mf – 9 – mf#PBU-306 – ne IDC [240]

In librum psalmorum, johannis calvini commentarius / Calvin, J – [Geneva]: Robert Estienne, 1557 – 12mf – 9 – mf#CL-62 – ne IDC [242]

In librvm iob exegemata / Oecolampadius, J – Basileae, Henricus Petrus, 1532 – 5mf – 9 – mf#PBU-386 – ne IDC [240]

In librvm iudicvm...commentarij / Vermigli, P M – Tigvri, Christoph Froschouer, 1561 – 5mf – 9 – mf#PBU-283 – ne IDC [240]

In librvm solomonis qvi ecclesiastes inscribitvr ludovici lavateri...commentarius / Lavater, L – Tigvri, Christ[oph] Froschover, 1584 – 4mf – 9 – mf#PBU-605 – ne IDC [240]
In liefde vereend / Henzel, J – Rotterdam: J M Bredee, 1917 [mf ed 1995] – (= ser Yale coll; Lichtstralen op den akker der wereld [23. jaarg 1917] 6) – 34p (ill) – 1 – 0-524-09644-9 – (in dutch) – mf#1995-0644 – us ATLA [951]
In longfellows pantoffeln : und andere geschichten / Allen, Philip Schuyler – Goettingen: W F Kaestner, 1892 [mf ed 1987] – 116p – 1 – mf#6935 n5 – us Wisconsin U Libr [830]
In loving remembrance of dr eckardt : who died at unionville on, july 26th 1880, aged 47 years – [Unionville ON?: s.n. 1880?] [mf ed 1983] – 1mf – 9 – 0-665-38700-8 – mf#38700 – cn Canadiana [090]
In lower florida wilds / Simpson, Charles Torrey – New York, NY. 1920 – 1r – us UF Libraries [080]
In luculentum et sacrosanctum evengelium... secundum lucam commentariorum lib 9 / Bullinger, Heinrich – Tiguri, 1557 – 12mf – 8 – €23.00 – ne Slangenburg [240]
In luv un lee / Lau, Fritz – Hamburg: M Glogau, 1918 – 1r – 1 – us Wisconsin U Libr [830]
In lvcae evangelivm ivsta scholia, per omnes circumstantias, methodica forma conscripta / Sarcerius, E – Basileae, 1539 – 8mf – 9 – mf#TH-1 mf 1347-1354 – ne IDC [242]
In lvcvlentvm...euangeliu...secundum lucam, commentariorum lib 9 / Bullinger, Heinrich – Tigvri, Christ[oph] Froschover, 1546 – 6mf – 9 – mf#PBU-152 – ne IDC [240]
In majorem dei gloriam : ein gedaechtnissbuch aus dem 17. jahrhundert / Jensen, Wilhelm – Dresden: Carl Reissner 1905 [mf ed 1995] – 1r – 1 – (filmed with: robert leichtfuss / hans hopfen) – mf#3656p – us Wisconsin U Libr [880]
In man's own image / Roy, Ellen & Ray, Sibnarayan – Calcutta: Renaissance Publishers, 1948 – (= ser Samp: indian books) – us CRL [325]
In many keys : a book of verse / Bengough, John Wilson – Toronto: W Briggs, 1902 – 3mf – 9 – 0-665-73076-4 – mf#73076 – cn Canadiana [810]
In marcvm evangelistam ivsta scholia, iuxta perpetuam orationis seriem / Sarcerius, E – Baslieae, 1539 – 5mf – 9 – mf#TH-1 mf 1378-1382 – ne IDC [242]
In matheo : formae tplila 24 / Radbertus, Pascasius – 1984 – (= ser ILL – ser a; Cccm 56-56a-56b) – 38mf+183p – 9 – €100.00 – 2-503-63562-8 – be Brepols [400]
In matthaeum : formae tplila 140 – [mf ed 2003] – (= ser ILL – ser a; Cccm 159) – 7mf+vi/62p – 9 – €47.00 – 2-503-64592-5 – be Brepols [400]
In matthaevm evangelistam ivsta et docta scholia, per omnes rhetoricae artis circumstantias / Sarcerius, E – Basileae, 1544 – 8mf – 9 – mf#TH-1 mf 1383-1390 – ne IDC [242]
"In memoriam" : the late rev john roaf, toronto, 1863 / Clarke, William Fletcher – Toronto?: s.n, 1863? – 1mf – 9 – mf#55467 – cn Canadiana [920]
In memoriam : 8 september, 1760 / Frechette, Louis – [S.l: s.n, I884?] – 1mf – 9 – 0-665-67891-6 – (in french with english trans) – mf#67891 – cn Canadiana [810]
In memoriam – Cap-Haitien, Haiti. 1935 – 1r – us UF Libraries [972]
In memoriam : charles paschal telesphore chiniquy: docteur en theologie, l'apotre de la temperance du canada... – Montreal: [s.n.], 1899 [mf ed 1983] – 1mf – 9 – mf#04478 – cn Canadiana [242]
In memoriam : a discourse occasioned by the death of the late a w lillie, esq, and delivered in the congregational church, guelph, on sabbath evening, october 18th, 1868 / Clarke, William Fletcher – Guelph ON: s.n, 1869 – 1mf – 9 – mf#00677 – cn Canadiana [920]
In memoriam : george paul macdonell / Allen, Grant – London: P Lund, 1895 – 1mf – 9 – mf#44241 – cn Canadiana [080]
In memoriam / Nicolson, J – Dundee, Scotland. 1875 – 1r – us UF Libraries [972]
In memoriam : sermon preached by rev a b chambers...on the occasion of the death of john lovell carson...montreal, december 1885 – [Montreal?: s.n, 1886?] – 1mf – 9 – 0-665-89058-3 – mf#89058 – cn Canadiana [240]
In memoriam : a sermon preached in st andrew's church, ottawa, on sunday morning, january 27th, 1901, to commemorate the death of her most gracious majesty, queen victoria / Herridge, William Thomas – [Ottawa?]: Kirk Session, [1901?] – 1mf – 9 – 0-665-74595-8 – mf#74595 – cn Canadiana [240]
In memoriam : sir a a dorion, chevalier, juge-en-chef de la cour d'appel, ancien ministre de la justice... – Montreal: La Patrie, 1891 – 2mf – 9 – mf#04480 – cn Canadiana [340]
In memoriam : sketch of the life and thoughts upon the death of the late rev allan napier macnab... – Toronto: Church Print & Pub Co, 1872 – 1mf – 9 – mf#23786 – cn Canadiana [920]
In memoriam / Stimpson, Thomas Morrill – Essex Bar Assoc., Salem, Mass. n.p., 1899?. 23 p. LL-475 – 1 – us L of C Photodup [340]
In memoriam : william goodell, born in coventry, ny, oct 25th 1792, died in janesville, wi, feb 14th 1878 – Chicago: Guilbert & Winchell, 1879 – 1mf – 9 – 0-524-02185-6 – mf#1990-0570 – us ATLA [976]
In memoriam broadman hartwell crumpton *see* Miscellaneous books and pamphlets
In memoriam, charles joseph little : born september 21, 1840, died march 11, 1911 / Little, Charles Joseph; ed by Stuart, Charles Macaulay – Chicago: Forbes, 1912 – 1mf – 9 – 0-7905-9787-X – mf#1989-1512 – us ATLA [240]
In memoriam. constant guillou / Philadelphia. Bar – Philadelphia: Stern, 1872. 25p. LL-1198 – 1 – us L of C Photodup [340]
In memoriam, george etienne cartier / Wicksteed, Gustavus William – [S.l: s.n, 1885?] [mf ed 1981] – 1mf – 0-665-25723-6 – (french text foll english text; originally publ in the canada law journal, 1 april, 1885) – mf#25723 – cn Canadiana [920]
In memoriam jesse seligman – New York, NY. 1894 – 1r – us UF Libraries [939]
In memoriam, marshall s bidwell / Association of the Bar of the City of New York – [s.l: s.n, 1872?] [mf ed 1985] – 1mf – 9 – 0-665-01477-5 – mf#01477 – cn Canadiana [347]
In memoriam, william miller paxton, d.d., ll.d., 1824-1904 : funeral and memorial discourses with appendixes and notes / De Witt, John – New York: [s.n.], 1905 – 1mf – 9 – 0-524-06991-3 – mf#1991-2844 – us ATLA [920]
In memoriam...and other genealogical data on the field family / Field, Samuel – 1 – $50.00 – us Presbyterian [920]
In memory of the queen : an address delivered in the town hall, regina, on the 2nd of february, 1901, the day of the funeral of her late imperial majesty / Davin, Nicholas Flood – Regina: West, 1901 – 1mf – 9 – 0-665-72622-8 – mf#72622 – cn Canadiana [941]
In monsun und pori / Wenig, Richard – Berlin: Safari Verlag, c1922. 161p. plates – 1 – us Wisconsin U Libr [940]
In mosis genesim plenissimi commentarii / Musculus, W – Basilea, Johann Herwagen, 1554 – 16mf – 9 – mf#PBU-338 – ne IDC [240]
In mother's arms : for mothers of babes from birth to two years of age, including directions to pastors, churches, schools and teachers of this department / Schmauk, Theodore Emanuel – Philadelphia: General Council Publication Board, 1910 – 1mf – 9 – 0-524-07641-3 – mf#1991-3248 – us ATLA [376]
In mysticum moysi leviticum libri 20 / Radulphus Flaviacensis – Coloniae, 1563 – €27.00 – ne Slangenburg [240]
In natalitiam memoriam r patris d martini lvtheri / Cramer, D – Vvitebergae, 1595 – 1mf – 9 – mf#TH-1 mf 372 – ne IDC [242]
In nature's workshop / Allen, Grant – Toronto: W Briggs, 1901 – 3mf – 9 – 0-665-73484-0 – mf#73484 – cn Canadiana [500]
In nomine domini amen : universis & singulis praesens publicum instrumentum visuris sit notum... / Urban 8, Pope – en este convent de Santa Maria la Redonda de Mexico: Fray Joseph de San Francisco, en veinte y seite dias del mes de Junio, de mil setecientos y veinte y seis anos – (= ser Books on religion...1543/44-c1800: papas [cartas apostolicas, etc]) – 1mf – 9 – mf#crl-364 – ne IDC [241]
In nomine patris, et filii, & spiritus sancti, amen : hae sunt provisiones pro bono regmine provinciarum indiarum occidentalium ordinis fratrum praedicatorum sanctae in capitulis generalibus / Papiensis, Seraphino Sicco – Mexici: Apud Bachalaureu[m] Ioanne[m] de Alcacar, anno 1619 – (= ser Books on religion...1543/44-c1800: ordenes, etc: dominicos) – 1mf – 9 – mf#crl-187 – ne IDC [241]
In non-union mines: the diary of a coal digger in central pennsylvania, august-september 1921 / Hapgood, Powers – New York: Bureau of Industrial Research, 1922.48p – 1 – us Wisconsin U Libr [331]
In northern india : a story of mission work in zenanas, hospitals, schools and villages / Cavalier, Anthony Ramsen – London: S W Partridge; Zenana Bible and Medical Mission [1899] [mf ed 1995] – (= ser Yale coll) – xiv/174p (ill) – 1 – 0-524-09069-6 – (int by lord kinnaird) – mf#1995-0069 – us ATLA [240]
In north-western wilds : the narrative of a 2,500 mile journey of exploration in the great mackenzie river basin / Ogilvie, William – [Toronto?: s.n, 1894?] [mf ed 1981] – 1mf – 9 – 0-665-11452-4 – mf#11452 – cn Canadiana [917]
In nvmeros enarratio / Chytraeus, D – Vitebergae, 1572 – 6mf – 9 – mf#TH-1 mf 321-326 – ne IDC [242]
In old ceylon / Farrer, Reginald John – London: Edward Arnold, 1908 [mf ed 1995] – (= ser Yale coll) – ix/351p (ill) – 1 – 0-524-09103-X – mf#1995-0103 – us ATLA [954]
In omnes apostolicas epistolas, divi videlicet pavli 14. et 8 : canonicas, commentarii / Bullinger, Heinrich – Tigvri, Christoph Froschover, 1537 – 12mf – 9 – mf#PBU-131 – ne IDC [240]
In omnes apostolicas epistolas...commentarij... / Pellican, C – Tigvri, officina Froschoviana, 1539 – 1mf – 9 – mf#PBU-614 – ne IDC [240]
In omnes beati pauli et septem catholicas apostolorum epistolas commentaria / Estius, G – Parisiis, 1679 – 44mf – 8 – €84.00 – ne Slangenburg [227]
In omnes d pauli epistolas : item in catholicas commentarii / Estius, Guilielmus – Nova editio. Parisiis: Ludovico Vives. 3v. 1891 – 6mf – 9 – 0-8370-6662-X – (incl indes) – mf#1986-0662 – us ATLA [227]
In omnes divi pauli apostoli : et alias septem canonicas epistolas ... / Politus, Ambrosius Catharinus – Parisiis: Apud Bernardum Turrisanum, 1566. Chicago: Dep of Photodup, U of Chicago Lib, 1973 (1r); Evanston: American Theol Lib Assoc, 1984 (1r) – 1 – 0-8370-0008-4 – mf#1984-B382 – us ATLA [220]
In omnes pauli apostoli epistolas, atque etiam in epistolam ad hebraeos, item in canonicas petri, johannis, jacobi, et judae, quae etiam catholicae vocantur, joh. calvini commentarii / Calvin, J – [Geneva]: Robert Estienne, 1556 – 17mf – 9 – mf#CL-64 – ne IDC [240]
In omnes prophetas, scholae breves et methodicae : proposita in academia argentoratensi / Pappus, J – Francofvrti ad Moenvm, 1593 – 7mf – 9 – mf#TH-1 mf 1233-1239 – ne IDC [242]
In orationes quasdam ciceronis... / Asconius [Tiberius Catius Asconius Silius Italicus] – 14th, 15th c – (= ser Holkham library manuscript books 392,367) – 1r – 1 – (filmed with: augustinus datus: elegantiolae. eutropius et florus: historiae) – mf#96611 – uk Microform Academic [450]
In other words – Huntington Beach. 1989-1993 (1) – ISSN: 0279-3172 – mf#15324 – us UMI ProQuest [240]
In other words / Moreno Izquierdo, Juan – 1882 – 9 – sp Bibl Santa Ana [190]
In our tongues : some thoughts for readers of the english bible / Kennett, Robert Hatch – London: Edward Arnold, 1907 – 1mf – 9 – 0-8370-9958-7 – mf#1986-3958 – us ATLA [220]
In partitiones oratorias ciceronis : dialogi quator, ab ipso authore emendati et aucti / Sturm, J – Strasbourg, 1539 – 4mf – 9 – mf#PPE-139 – ne IDC [240]
In pentateuchum sive quinque libros mosis... commentarii : his accessit narratio de ortu, vita et obitu eiusdem, opera ludivici lavateri / Pellican, C – Zuerich, Froschauer, 1582 – 6mf – 9 – mf#PBU-565 – ne IDC [240]
In perils in the sea / Leifchild, J – London, England. 18– – 1r – us UF Libraries [240]
In perspective – v2 n8-9 [1980 nov 26-dec 24] – 1r – 1 – mf#635388 – us WHS [071]
In pittsburgh – Pittsburgh PA. 1989 apr 19-dec, 1989 dec 27-1990 apr 25, 1991 jul 17-1992 mar 25, 1992 apr-aug, 1992 sep-dec, 1993 jan-jun, 1993 jul-dec – 7r – 1 – mf#1070997 – us WHS [071]
In posteriorem d pauli apostoli ad corinthios epistolam homiliae / Gwalther, R – Zuerich, Froschouer, 1572 – 4mf – 9 – mf#PBU-301 – ne IDC [240]
In posteriorem d pavli ad corinthios epistolam...commentarius / Bullinger, Heinrich – Tigvri, Christoph Froscho[uer], 1535 – 3mf – 9 – mf#PBU-124 – ne IDC [240]
In praise of folly / Erasmus, Desiderius – New York, NY. 193-? – 1r – us UF Libraries [240]
In primum duodecim prophetarum, nempe oseam commentarii / Lambert, F – Strasbourg, 1525 – 5mf – 9 – mf#PPE-116 – ne IDC [240]
In primum musculi anticochlaeum replica brevis... / Cochlaeus, J – Ingolstadt, 1545 – 1mf – 9 – mf#PBU-703 – ne IDC [240]
In primum, secundum, et initium tertii libri ethicorum aristotelis ad nicomachum / Vermigli, Pietro Martire – Tiguri: Excudebat Christophorus Froschouerus lunior, mense Augusto, anno 1563 – (= ser Ethics in the early modern period) – 5mf – 9 – mf#pl-346 – ne IDC [170]
In primvm librvm mosis...commentarij / Vermigli, P M – Tigvri, Christoph Froschouer, 1569 – 4mf – 9 – mf#PBU-286 – ne IDC [240]
In primvm secvndvm et initivm tertii libri ethicorvm aristotelis ad nicomachvm... commentarius doctissimus / Vermigli, P M – Tigvri, Christoph Froschouer iunior, 1563 – 6mf – 9 – mf#PBU-650 – ne IDC [240]
In priorem d pauli ad corinthios epistolam homiliae... / Gwalther, R – Zuerich, Froschouer, 1572 – 7mf – 9 – mf#PBU-300 – ne IDC [240]
In priorem d pavli ad corinthios epistolam... commentarius / Bullinger, Heinrich – Tigvri, Christoph Froscho[uer], 1534 – 5mf – 9 – mf#PBU-120 – ne IDC [240]
In prison : being a report by kate richards o'hare to the president of the united states as to the conditions under which women federal prisoners are confined in the missouri state penitentiary... / O'Hare, Kate Richards – St Louis, MO: Publ by Frank P O'Hare, c1920 – 1 – (filmed with materials relating to the author's pardon) – us CRL [360]
In prophetam ezechielem commentarius / Oecolampadius, J – Argentorati, Matthias Apiarius, 1534 – 7mf – 9 – mf#PBU-389 – ne IDC [240]
In prophetam hoseam commentarius / Tarnow, Johannes – Rostochii: N Kilii, 1646 – 1r – 1 – 0-8370-0981-2 – mf#1984-B512 – us ATLA [221]
In psalmos 73(-77) conciones / Oecolampadius, J – Basileae, Robertus Winter, 1554 – 5mf – 9 – mf#PBU-396 – ne IDC [240]
In psalmos (siecle 7) / Augustinus, St – Autun – 1r – 5,6 – sp Cultura [220]
In psalmos v.p.d. ludolphi cartusiana enarratio clarissima : opus multo quam unquam antea accuratius postrema hac editione recognitum, et a multis mendis expurgatum = Expositio in psalterium davidis / Ludolf von Sachsen – Monsterolii [Montreuil-sur-Mer]: Typis Cartusiae Sanctae Mariae de Pratis, 1891 – 2mf – 9 – 0-524-06845-3 – mf#1992-0987 – us ATLA [220]
In psalmvm 118 praelectiones / Chytraeus, D – Rostochii, 1590 – 1mf – 9 – mf#TH-1 mf 327 – ne IDC [242]
In public service / AFL-CIO [American Federation of Labor and Congress of Industrial Organizations] – v4 n8-v9 n3 [1978 aug-1983 fall] – 1r – 1 – (cont: in the public service) – mf#801965 – us WHS [331]
In pursuit of happiness / Tolstoy, Leo – Trans. by Mrs. Aline Delano. Boston: D. Lothrop Company, c1887 – 193p 1r – 1 – us Wisconsin U Libr [460]
In quartum sententiarum petri lombardi / Richardus de Media Villa – Lugduni, 1512 – €52.00 – ne Slangenburg [240]
In quatuor sacro-sancta iesu christi evangelia..scholia / Gagneius, Ioan. – Parisiis, 1660 – 13mf – 8 – €25.00 – ne Slangenburg [240]
In quest of el dorado / Graham, Stephen – New York, NY. 1923 – 1r – us UF Libraries [972]
In quest of light / Smith, Goldwin – New York: Macmillan, 1906 – 1mf – 9 – 0-8370-6383-3 – mf#1986-0383 – us ATLA [240]
In quietness and in confidence shall be your strength / Mccheane, James H – London, England. 1866 – 1r – us UF Libraries [240]
...in qvo videtvr finis tvrcarum in praesenti eorum imperatore... / Septimus Severus – Brescia, 1596 – 2mf – 9 – mf#H-8429 – ne IDC [956]
In re corney v father evangelicus / Dropper, Eaves – London, England. 18– – 1r – us UF Libraries [240]
In re: germany : a critical bibliography of books and magazine articles on germany – New York NY (USA), 1942 feb-1944 mar – 1r – 1 – gw Misc Inst [019]
In regulam divi benedicti commentarius / Trithemius, Ioan – Valencis, 1608 – 18mf – 8 – €63.00 – ne Slangenburg [241]
In regulam s benedicti expositio / Bernardi Abbatis Casinensis – Monte Casino, 1894 – 11mf – 8 – €21.00 – ne Slangenburg [241]
In regulam sancti benedicti commentarium nunc primum editum / Petrus Boherius; ed by Allodi, L – Sublaci, 1908 – €52.00 – ne Slangenburg [241]
In regvm dvos vltimos libros, annotationes post samuelem iam primu emissae / Bugenhagen, J – Norembergae, 1526 – 3mf – 9 – mf#TH-1 mf 171-173 – ne IDC [242]
In reih' und glied : roman / Spielhagen, Friedrich – Leipzig: L Staackmann, 1890 – 1r – 1 – us Wisconsin U Libr [830]
In relief of doubt / Welsh, Robert Ethol – London: HR Allenson, 1902 [mf ed 1985] – 1mf – 9 – 0-8370-5766-3 – (int note by lord bishop of london) – mf#1985-3766 – us ATLA [210]
In remembrance-address on occasion of the death of charles greely loring / Bartol, Cyrus Augustus – Boston, The Society, 1867. 32 p. LL-470 – 1 – us L of C Photodup [340]

In review – 1967-74 – (= ser Canadian books for children) – 1r – 1 – ISSN: 0019-3259 – cn Library Assoc [420]

In rhijm ghestelt : sedighe onderwiisen der creaturen – t'Antwerpen: Iacob Mesens, 1649 – 2mf – 9 – mf#O-3036 – ne IDC [090]

In richest alaska and the gold fields of the klondike : how they were found, how worked, what fortunes have been made, the extent and richness of the gold fields, how to get there, outfit required, climate / Ingersoll, Ernest – Chicago: Dominion, c1897 [mf ed 1983] – 6mf – 9 – mf#15313 – cn Canadiana [622]

In roberti bellarmini disputationes / Pareus, D – Heidelbergae, 1612-15 – 16mf – 9 – mf#PBA-281 – ne IDC [240]

In royal service : the mission work of southern baptist women / Heck, Fannie Exile Scudder – Richmond, VA: Educational Dept, Foreign Mission Board, Southern Baptist Convention, 1913 – 1mf – 9 – 0-524-06904-2 – (incl bibl ref) – mf#1991-2817 – us ATLA [242]

In s pavli ad ephesios epistolam, annotationes d iohannis vuigandi / Wigand, J – Erphordiae, 1581 – 2mf – 9 – mf#TH-1 mf 1571-1572 – ne IDC [242]

In sacram beati johannis apostoli apocalipsin commentarii / Ribera, Fr – Antverpiae, 1594 – 11mf – 8 – €22.00 – ne Slangenburg [226]

In sacro...huor evangelia / Trejo, Gutierre – 1554 – 9 – sp Bibl Santa Ana [240]

In sacrosancta quatvor evangelia et apostolorvm acta...commentarij... / Pellican, C – Tigvri, officina Froschoviana, 1537 – 21mf – 9 – mf#PBU-612 – ne IDC [240]

In sacrosanctum davidis psalterium commentarij / Musculus, W – Basilea, Johann Herwagen, 1551. 2 v – 21mf – 9 – (incl appendices "de juramento"+"de usura") – mf#PBU-336 – ne IDC [240]

In sacrosanctum evangelium domini nostri iesu christi sec marcum commentariorum lib 6 / Bullinger, Heinrich – Tiguri, 1554 – 5mf – 8 – €12.00 – ne Slangenburg [240]

In sacrosanctum iesu christi... : evangelium secundum ioannem enarrationes / Ferus, Ioan – Antverpiae, 1556 – 19mf – 8 – €37.00 – ne Slangenburg [220]

In sacrosanctum iesu christi... : evangelium secundum matheum enarrationes / Ferus, Ioan – Lugduni, 1609 – 11mf – 8 – €21.00 – ne Slangenburg [220]

In sacrosanctum joannis evangelium commentarii / Toledo, F – Romae, 1592. 2v – 16mf – 9 – mf#CA-74 – ne IDC [240]

In sacrosanctvm euangelium...secundum marcu, commentariorum lib 6 / Bullinger, Heinrich – Tigvri, Christoph Froschover, 1545 – 2mf – 9 – mf#PBU-151 – ne IDC [240]

In sacrosanctvm...euangelium secundum matthaeum, commentariorum libri 12 / Bullinger, Heinrich – Tigvri, [Christoph] Froschover, 1542 – 7mf – 9 – mf#PBU-139 – ne IDC [240]

In search – Hilversum. 1972-1985 (1) 1972-1985 (5) 1974-1985 (9) – ISSN: 0166-4360 – mf#6748 – us UMI ProQuest [320]

In search of a relevant progressivism / Institute for the Study of Relevant Progressivism – 1973 nov 22/dec 6-1975 dec 4 – 1r – 1 – (cont by: in search of facts, ideas, challenges) – mf#772820 – us WHS [303]

In search of facts, ideas, challenges / Institute for the Study of Relevant Progressivism – v4 n21=79-v11 n22=227 [1975 dec 18/1976 jan 1-1982 dec 30/1983 jan 12] – 1r – 1 – (cont: in search of a relevant progressivism) – mf#829419 – us WHS [303]

In search of south africa / Morton, Henry Vollam – London, England. 1948 – 1r – us UF Libraries [960]

In search of south africa / Morton, Henry Vollam – New York, NY. 1948 – 1r – us UF Libraries [960]

In sententias theologicas petri lombardi commentarium libri quattuor / Durandus de S Porciano (Durandus of Saint-Pourcain) – Lugduni, 1556 – 34mf – 8 – €65.00 – ne Slangenburg [241]

In shturem fun der tseyt / Steinberg, Isaac Nachman – Varshe, Poland. 1928 – 1r – us UF Libraries [939]

In shvere teg / Barkan, H – Warsaw, Poland. 1933 – 1r – us UF Libraries [939]

In smuts's camp / Long, Basil Kellett – London, England. 1945 – 1r – us UF Libraries [960]

In somalia : note e impressioni di viaggio / Stefanini, Giuseppi – Firenze: F Le Monnier, 1922 – 1 – us CRL [960]

In south central africa / Moubray, John M – New York, NY. 1969 – 1r – us UF Libraries [960]

In southern india : a visit to some of the chief mission stations in the madras presidency / Mitchell, J Murray (Mrs) – London, 1885 – 5mf – 9 – mf#HTM-131 – ne IDC [915]

In southern india : a visit to some of the chief mission stations in the madras presidency / Mitchell, Maria Hay Flyter – [London]: Religious Tract Society, 1885 [mf ed 1995] – (= ser Yale coll) – 383p (ill) – 1 – 0-524-09128-5 – mf#1995-0128 – us ATLA [240]

In spain / Andersen, Hans Christian – London: R Bentley, 1864 [mf ed 1986] – ii/306p – 1 – (trans by mrs bushby) – mf#7375 – us Wisconsin U Libr [914]

In spain with the international brigade : a personal narrative / Bayle, Constantino – London, 1938; Burgos: Razon y Fe, 1938 – 1 – sp Bibl Santa Ana [355]

In spain with the international brigade, a personal narrative – London, 1938. Fiche W958. (Blodgett Collection of Spanish Civil War Pamphlets) – 9 – us Harvard [946]

In spiritus aufbewahrt : heitere wiener skizzen / Poetzl, Eduard – Berlin: Carl Stephenson, [1942] – 1r – 1 – us Wisconsin U Libr [830]

In subjection / Fowler, Ellen Thorneycroft – Toronto: W Briggs, 1906 [mf ed 1995] – 5mf – 9 – 0-665-76814-1 – mf#76814 – cn Canadiana [830]

In summam theologicam divi thomae aquinatis : de incarnatione p 3, qq 1.26: praelectiones – Romae: A Befani, 1888 [mf ed 1985] – 1mf – 9 – 0-8370-5261-0 – mf#1985-3261 – us ATLA [120]

In support of the raison d'etre : some aspects of the political geography of botswana / Knights, David B – Ypsilanti 1967 – us CRL [960]

In tents in transvaal / Hutchinson, Louisa – London: R Bentley, 1879 – 1 – us CRL [916]

In the andamans and nicobars / Kloss, Cecil Boden – London: John Murray, 1903 – (= ser Samp: indian books) – us CRL [915]

In the banqueting house : a series of sacramental meditations / Pearse, Mark Guy – London: Charles H. Kelly, 1902. Beltsville, Md: NCR Corp, 1978 (3mf); Evanston: American Theol Lib Assoc, 1984 (3mf) – 9 – 0-8370-0833-6 – mf#1984-4235 – us ATLA [240]

In the beginning / Sandys, Richard Hill – London, England. 1875 – 1r – us UF Libraries [240]

In the brahmans' holy land : a record of service in the mysore / Robinson, Benjamin – London: Charles H Kelly [1912] [mf ed 1995] – (= ser Yale coll) – 119p (ill) – 1 – 0-524-10137-X – (foreword by henry allyn haigh) – mf#1995-1137 – us ATLA [954]

In the bureau spotlight / Greater Madison Convention and Visitor's Bureau – 1978 may-1980 mar/apr – 1r – 1 – mf#653631 – us WHS [338]

In the case of louis riel, convicted of treason, and executed therefor : memorandum of sir alexander campbell – Ottawa?: MacLean, Roger, 1885 – 1mf – 9 – (also available in french) – mf#30086 – cn Canadiana [345]

In the circuit court of appeals for the eighth circuit : december term, 1909 no 3150-no 3163 united states appellant vs james p allen et al / U.S. Dept of Justice – Washington, Govt. Print. Off., 1909 180 p. LL-2353 – 1 – us L of C Photodup [347]

In the circuit court of appeals for the eighth circuit. december term, 1909. no.3161-3163; 3150-3160 united states, appellant v. walter p. nichols. / U.S. Dept of Justice – Washington, Govt. Print. Off., 1909 30 p. LL-2355 – 1 – us L of C Photodup [347]

In the court of error and appeal : the queen (defendant in error) vs patrick james whelan (plaintiff in error)... – [Toronto?: s.n, 1868?] [mf ed 1984] – 1mf – 9 – 0-665-32335-2 – mf#32335 – cn Canadiana [345]

In the day of the muster : sermons in time of war / Paterson, William Paterson – London; New York: Hodder and Stoughton, 1914 – 1mf – 9 – 0-7905-9434-X – mf#1989-2659 – us ATLA [240]

In the days of laggan presbytery, 1908 / Lecky, Alexander G – Also: The Laggan and its Presbyterianism, 1905; Ulster. General Synod, Records, 1691-1820, 1890, 1897, 1898 – 1 – $50.00 – us Presbyterian [242]

In the days of the company / Dewar, Douglas – Calcutta: Thacker, Spink & Co, 1920 – (= ser Samp: indian books) – us CRL [915]

In the days of the councils : a sketch of the life and times of baldassare cossa (afterward pope john the twenty-third) / Kitts, Eustace J – London: Constable, 1908 – 2mf – us ATLA [240]

In the days of the councils : a sketch of the life and times of baldassare cossa (afterward pope john the twenty-third) / Kitts, Eustace John – London: Constable, 1908 – 2mf – 9 – 0-7905-5240-X – (incl bibl ref) – mf#1988-1240 – us ATLA [240]

In the early days : the reminiscences of pioneer life on the south african diamond fields / Angove, John – Kimberley: Handel House, 1910 – 1 – us CRL [960]

In the exchequer court of canada : between francois-xavier berlinguet and marie charlotte mailloux, suppliant vs the queen, defendant... – Ottawa?: s.n, 1876? – 2mf – 9 – mf#02309 – cn Canadiana [347]

In the exchequer court of canada : petition of right: sir n f belleau et al vs the queen; henri t tashereau, attorney and counsel for suppliants – S.l: s.n, 1877? – 1mf – 9 – mf#47125 – cn Canadiana [347]

In the far east : letters from geraldine guinness in china / Taylor, Howard (Mrs); ed by Guiness, Lucy Evangeline – London: Morgan & Scott; New York: Fleming H Revell, [1889?] – 1mf – 9 – 0-8370-6154-7 – mf#1986-0154 – us ATLA [920]

In the far east : letters...edited by her sister [l e guinness] / Guinness, G – London, [1889] – 3mf – 9 – mf#HT-148 – ne IDC [915]

In the footsteps of cortes / Benitez, Fernando – New York, NY. 1952 – 1r – us UF Libraries [972]

In the footsteps of livingstone / Dolman, Alfred – London, England. 1924 – 1r – us UF Libraries [960]

In the great god's hair – London: Medici Society, 1914 – (= ser Samp: indian books) – (trans fr original mss by f w bain) – us CRL [490]

In the guatemala honduras boundary arbitration / Special Boundary Tribunal(Guatemala-Honduras Boun...) – Washington, DC. 1932 – 1r – us UF Libraries [972]

In the guiana forest / Rodway, James – London, England. 1911 – 1r – us UF Libraries [972]

In the heart of bantuland : a record of twenty-nine years' pioneering in central africa among the bantu peoples, with a description of their habits, customs, secret societies and languages / Campbell, Dugald – London: Seeley, Service, 1922 – 1 – us CRL [306]

In the heart of india : the work of the canadian presbyterian mission / Taylor, J T – Toronto: Board of Foreign Missions, Presbyterian Church in Canada, 1916 [mf ed 1995] – (= ser Yale coll) – x/225p (ill) – 1 – 0-524-09110-2 – mf#1995-0110 – us ATLA [242]

In the heart of india, or, beginnings of missionary work in bundela land : with a short chapter on the characteristics of bundelkhand and its people, and four chapters of jhansi history / Holcomb, James Foote & Holcomb, Helen Harriet Howe – Philadelphia: Westminster Press, 1905 [mf ed 1995] – (= ser Yale coll) – ix/251p (ill) – 1 – 0-524-09187-0 – mf#1995-0187 – us ATLA [954]

In the heart of the beast – n1-2 [1970 jan-nov] – 1r – 1 – mf#1583889 – us WHS [071]

In the heart of the hills : poem / Carman, Bliss – New York?: s.n, 1892 – 1mf – 9 – mf#06092 – cn Canadiana [810]

In the high court of justice : between the corporation of the city of toronto, plaintiffs, and the grand trunk railway company of canada and the canadian pacific railway company, defendants, re york street bridge – [Toronto?: s.n.], 1899 [mf ed 1983] – 2mf – 9 – mf#08928 – cn Canadiana [347]

In the high heavens / Ball, Robert Stawell – London: Isbister & Co; Philadelphia: J B Lippincott Co 1894 [mf ed 1998] – 1r [pl/ill] – 1 – (incl ind) – mf#film mas 28212 – us Harvard [520]

In the hours of meditation / Alexander, F J – Almora: Advaita Ashrama, 1944 – (= ser Samp: indian books) – us CRL [280]

In the household of faith / Smith, Charles Ernest – New York: Longmans, Green, 1896, c1895 – 1mf – 9 – 0-8370-8868-2 – mf#1986-2868 – us ATLA [230]

In the isles of the sea : the story of fifty years in melanesia / Awdry, Frances – London: Bemrose & Sons, 1902 [mf ed 1995] – (= ser Yale coll) – xiv/147p (ill) – 1 – 0-524-10081-0 – mf#1995-1081 – us ATLA [980]

In the king's german legion / Ompteda, Christian – London, England. 1894 – 1r – us UF Libraries [025]

In the king's service / Hogg, Bessie et al; ed by Watson, Charles Roger – Philadelphia, PA: Board of Foreign Missions of the United Presbyterian Church of NA, c1905 [mf ed 1990] – 1mf – 9 – 0-7905-6852-7 – mf#1988-2852 – us ATLA [240]

In the land of the afternoon / Green, Lawrence George) – Cape Town, South Africa. 1952 – 1r – us UF Libraries [960]

In the land of the blue gown / Little, Archibald [Mrs] – London, Leipsic: T Fisher Unwin, 1908 [mf ed 1995] – (= ser Yale coll) – xv/304p (ill) – 1 – 0-524-09150-1 – mf#1995-0150 – us ATLA [915]

In the land of the cherry blossom / Madden, Maude Whitmore – Cincinnati: Foreign Christian Missionary Society, c1915 – 1mf – 9 – 0-524-04382-5 – mf#1991-2086 – us ATLA [240]

In the land of the cherry blossom / Madden, Maude Whitmore – [New York]: F H Revell [for] Foreign Christian missionary Society, Cincinnati [1915] [mf ed 1995] – (= ser Yale coll) – 192p (ill) – 1 – 0-524-09617-1 – mf#1995-0617 – us ATLA [950]

In the land of the five rivers : a sketch of the work of the church of scotland in the panjab / Taylor, H F Lechmere – Edinburgh: R & R Clark; London: A & C Black, 1906 [mf ed 1995] – (= ser Yale coll) – xiv/166p (ill) – 1 – 0-524-10014-4 – (int by w mackworth young) – mf#1995-1014 – us ATLA [242]

In the land of the lamas : the story of trashilhamo, a tibetan lassie, in which are described tibetan character, life, customs, and history / Amundsen, Edward – London, Edinburgh: Marshall Bros, [1910] [mf ed 1995] – (= ser Yale coll) – xii/82p (ill) – 1 – 0-524-09962-6 – mf#1995-0962 – us ATLA [951]

In the land of the oil rivers : the story of the qua iboe mission / M'Keown, Robert L – London: Marshall Bros, 1902 – us CRL [960]

In the land of the strenuous life = Au pays de la vie intense / Klein, Felix – Chicago: A.C. McClurg, 1905 – 1mf – 9 – 0-7905-4937-9 – (in english) – mf#1988-0937 – us ATLA [910]

In the lesuto : a sketch of african mission life / Widdicombe, J – Brighton, New York, London, 1895 – 4mf – 9 – mf#HTM-213 – ne IDC [916]

In the levant / Warner, Charles Dudley – 16th ed. Boston: Houghton Mifflin, 1889 – 1mf – 9 – 0-524-04814-2 – mf#1992-0234 – us ATLA [915]

In the master's country : a geographical aid to the study of the life of christ / Tarbell, Martha – [London]: Hodder & Stoughton, c1910 – 1mf – 9 – 0-524-08512-9 – mf#1993-0037 – us ATLA [220]

In the matter of exxon corporation et al : records of the federal trade commission's case against the major oil companies / U.S. Federal Trade Commission – 23r – 1 – $3590.00 – 0-89093-200-X – (suppl 1978-81 9r isbn 0-89093-482-7 $1395. with p/g) – us UPA [380]

In the matter of the alabama and florida rr co... / Us Circuit Court – Pensacola, FL. 1869 – 1r – us UF Libraries [071]

In the matter of the provincial synod of canada : further opinion of adam crooks, esq, qc and e blake, esq – Toronto: [s.n.], 1864 [mf ed 1987] – 1mf – 9 – 0-665-63131-6 – mf#63131 – cn Canadiana [242]

In the midst of alarms : a novel / Barr, Robert – New York: F A Stokes, c1900 – 4mf – 9 – (ill by harrison fisher) – mf#32550 – cn Canadiana [830]

In the mountains / Elizabeth – Toronto: S B Gundy [1920?] [mf ed 1994] – 4mf – 9 – 0-665-72753-4 – mf#72753 – cn Canadiana [830]

In the new capital : or, the city of ottawa in 1999 / Galbraith, John – Toronto: Toronto News, 1897 – 2mf – 9 – mf#03283 – cn Canadiana [305]

In the new hebrides : reminiscences of missionary life and work, especially in the island of aneityum, from 1850 till 1877 / Inglis, J – London, 1887 – 4mf – 9 – mf#HTM-89 – ne IDC [919]

In the new hebrides : reminiscences of missionary life and work, especially on the island of aneityum, from 1850 till 1877 / Inglis, John – London, New York: T Nelson & Sons, 1887 [mf ed 1995] – (= ser Yale coll) – xvi/352p (ill) – 1 – 0-524-09512-4 – mf#1995-0512 – us ATLA [920]

In the ngombe tradition / Wolfe, Alvin William – Evanston, IL. 1961 – 1r – us UF Libraries [960]

In the nicobar islands / Whitehead, George – London: Seeley, Service & Co, 1924 – (= ser Samp: indian books) – (pref by sir richard c temple) – us CRL [915]

In the path of mahatma gandhi / Catlin, George Edward Gordon – London: Macdonald & Co, 1948 – (= ser Samp: indian books) – us CRL [920]

In the power of the spirit : or, christian experience in the light of the bible / Boardman, William Edwin – Boston: Willard Tract Repository, 1875 – 1mf – 9 – 0-8370-2733-0 – mf#1985-0733 – us ATLA [240]

In the press, and will shortly be published, railways and other ways : being reminiscences of canal and railway life during a period of sixty-five years by myles pennington... – Toronto: Williamson Book Co, 1893 – 1mf – 9 – mf#60298 – cn Canadiana [070]

In the promised land / Navarro, Mary (Anderson) de – London: Downey & Co, 1898 – (= ser 19th c women writers) – 4mf – 9 – mf#5.1.87 – uk Chadwyck [420]

In the public interest / Fund for Peace – [1975 oct 6-1977 aug] – 1r – 1 – mf#370668 – us WHS [320]

In the public interest *see* Buffalo public interest law journal

In the saddle with gomez / Carrillo, Mario – London, England. 1898 – 1r – us UF Libraries [972]

In the school of christ / McDowell, William Freser – New York: F H Revell, c1910 – (= ser Cole lectures) – 1mf – 9 – 0-7905-9508-7 – mf#1989-1213 – us ATLA [240]

In the school of christ : or, lessons from new testament characters concerning christian life and experience / Simpson, Albert B – New York: Christian Alliance Pub Co, c1890 [mf ed 1992] – (= ser Christian & missionary alliance coll) – 1mf – 9 – 0-524-03741-8 – mf#1990-4846 – us ATLA [240]

In the school of faith / Simpson, Albert B – New York: Alliance Press Co, c1907 [mf ed 1992] – (= ser Christian & missionary alliance coll) – 1mf – 9 – 0-524-03742-6 – (originally publ 1890) – mf#1990-4847 – us ATLA [240]

In the shadow of sinai : a story of travel and research from 1895 to 1897 / Lewis, Agnes Smith – Cambridge: Macmillan & Bowes, 1898 – 1mf – 9 – 0-7905-2016-8 – (cont: how the codex was found) – mf#1987-2016 – us ATLA [916]

In the shadow of the arctic / Fox, William W – [S.l: s.n, 189-?] [mf ed 1981] – 1mf – 9 – 0-665-14125-4 – (fr: the canadian magazine) – mf#14125 – cn Canadiana [990]

In the shadow of the drum tower / Garst, Laura DeLany – Cincinnati: Foreign Christian Missionary Society, c1911 – 1mf – 9 – 0-524-04259-4 – mf#1991-2043 – us ATLA [240]

In the shadow of the mahatma : a personal memoir / Birla, Ghanasyamadasa – Bombay: Orient Longmans, 1953 – (= ser Samp: indian books) – (foreword by rajendra prasad) – us CRL [920]

In the steps of the good physician : some glimpses of c e z medical work in india and china / Tiley, E S – London: Church of England Zenana missionary Society; Marshall Bros, 1913 [mf ed 1995] – (= ser Yale coll) – 60p (ill) – 1 – 0-524-09491-8 – (int chapter by c s vines) – mf#1995-0491 – us ATLA [242]

In the superior court, montreal : the reverend robert dobie, petitioner vs board for the management of the temporalities' fund of the presbyterian church of canada in connection with the church of scotland, et al, respondents: petition / Dobie, Robert – Montreal?: s.n, 1878? – 1mf – 9 – mf#12532 – cn Canadiana [242]

In the supreme court in equity, david vaughan et al, plaintiffs and james smith et al, defendants : pleadings, decree and evidence... / Vaughan, David – St John, NB?: Barnes, 1871 – 2mf – 9 – mf#27460 – cn Canadiana [346]

In the supreme court of british columbia on appeal to the divisional court : between george james findlay, john henry durham and john henry brodie, plaintiffs, and peter birrell and joseph a boscowitz, defendants – Victoria, BC?: M Miller, 1887 – 1mf – 9 – mf#14980 – cn Canadiana [336]

In the supreme court of british columbia, on appeal to the full court : between isaac j hayden, plaintiff and appellant, and the canadian pacific railway company... – [Victoria, BC?: s.n.], 1887 [mf ed 1981] – 1mf – 9 – mf#15199 – cn Canadiana [347]

In the supreme court of new brunswick, (crown side) in the matter of david s kerr, barrister : on application for an attachment against him for contempt of court: argued trinity term, june 1881 / Kerr, David Shank – St John, NB?: Daily Telegraph, 1883 – 1mf – 9 – mf#10652 – cn Canadiana [347]

In the supreme court of nova scotia, 1881 : on appeal, from the county court, district n1, insolvent act of 1875 and amending acts, in the matter of the estate of john r murray, an insolvent, and alexander mcdonald, claimant, and james g foster, assignee of said insolvent, contestant... – [Halifax, NS?: s.n.], 1881 [mf ed 1987] – 4mf – 9 – 0-665-67195-4 – mf#67195 – cn Canadiana [347]

In the supreme court of south africa (appellate division) in the matter between benjamin pogrund, appellant, and dr percy yutar, respondent – [s.l: s.n, 19–?] – 1 – us CRL [960]

In the supreme court of the northwest territories : appeal to the court in banc from the judgement of the honorable mr. justice rouleau – Calgary?: Alberta Tribune Press, 1897 – 1mf – 9 – mf#16081 – cn Canadiana [343]

In the tiger jungle : and other stories of missionary work among the telugus of india / Chamberlain, J – Edinburgh, London, 1897 – 3mf – 9 – mf#HTM-36 – ne IDC [915]

In the tiger jungle : and other stories of missionary work among the telugus of india / Chamberlain, Jacob – 3rd ed New York: Fleming H Revell, c1896 – 1mf – 9 – 0-8370-6031-1 – mf#1986-0031 – us ATLA [240]

In the time of the pharaohs = Au temps des pharaons / Moret, Alexandre – New York: GP Putnam, 1911 [mf ed 1992] – 1mf – 9 – 0-524-02093-0 – (incl bibl ref. english by madame noret) – mf#1990-2857 – us ATLA [930]

In the valley of decision / Hough, Lynn Harold – New York: Abingdon Press, c1916 – 1mf – 9 – 0-7905-7765-8 – mf#1989-0990 – us ATLA [240]

In the valley of the nile : a survey of the missionary movement in egypt / Watson, Charles Roger – 2nd ed. New York: F H Revell, c1908 – 1mf – 9 – 0-7905-6213-8 – (incl bibl ref) – mf#1988-2213 – us ATLA [240]

"In the volume of the book" : or, the profit and pleasure of bible study / Pentecost, George Frederick – [3rd ed] New York: Ward & Drummond, c1880 [mf ed 1985] – xi/200p on 1mf – 9 – 0-8370-4698-X – (incl app) – mf#1985-2698 – us ATLA [220]

In the wake of columbus / Ober, Frederick Albion – Boston, MA. 1893 – 1r – us UF Libraries [972]

In the wake of the war canoe : a stirring record of forty years' successful labour peril and adventure amongst the savage indian tribes of the pacific coast... / Collison, W H – London, 1915 – 5mf – 9 – mf#HTM-41 – ne IDC [917]

In the west indies / Van Dyke, John Charles – New York, NY. 1932 – 1r – us UF Libraries [972]

In the wind – v1 n1-6 [1980 feb-jul/aug] – 1r – 1 – mf#637043 – us WHS [071]

In the year one in the far east / Baring-Gould, Edith M E – [London]: Church missionary Society, 1914 [mf ed 1995] – (= ser Yale coll) – vi/104p (ill) – 1 – 0-524-09271-0 – (pref by eugene stock) – mf#1995-0271 – us ATLA [950]

In the ypres salient : the story of a fortnight's canadian fighting, june 2nd-16th, 1916 / Willson, Beckles – London: Simpkin, Marshall, Hamilton, Kent & Co, [1916?] – 2mf – 9 – 0-665-77885-6 – (incl french text and app) – mf#77885 – cn Canadiana [933]

In these times – Chicago. 1978+ (1) 1987+ (5) 1987+ (9) – ISSN: 0160-5992 – mf#12494 – us UMI ProQuest [320]

In these times / Institute for Policy Studies – [1976 nov 15/1977 nov 29]-[1997 jan 6-dec 28] [gaps] – 20r – 1 – mf#370667 – us WHS [320]

In tobiam. in proverbia. in cantica canticorum. in habacuc : formae tplila 15 / Beda Venerabilis – 1983 – (= ser ILL – ser a; Cccm 119b) – 13mf+112p – 9 – €40.00 – 2-503-61194-X – be Brepols [400]

In touch / Church of God in Christ of Western New York – 1992 jun – 1r – 1 – mf#4024429 – us WHS [242]

In touch *see* Canadian jewish news

In touch with canadian and b n a philately magazine – v1 n1-v2 n3 [1985 oct-1986 mar] – 1r – 1 – mf#1336427 – us WHS [730]

In touch with canadian and b n a philately stamp – v2 n4-v3 n2 [i.e. 3][1986 apr-1987 mar] – 1r – 1 – (cont: in touch with canadian and b n a philately magazine) – mf#1337433 – us WHS [730]

In transit / Amalgamated Association of Street, Electrical Railway and Motor Coach Employees of America et al – 1966 jan-1971 jul, 1971 aug-1975 aug, 1974, 1975 apr-1978 dec, 1979-1984 may – 5r – 1 – (cont: motorman, conductor, and motor coach operator) – mf#1397474 – us WHS [380]

In treue fest : geschichtlicher roman / Achleitner, Arthur – 2. durchges aufl. Leipzig: Hesse & Becker, [19–?] [mf ed 1995] – (= ser Romane der welt-literatur) – 319p – 1 – mf#8918 – us Wisconsin U Libr [830]

In unity – Australia. v14-31. 1967-84 [complete] – 1r – 1 – mf#ATLA S0809 – us ATLA [240]

"In uns ist alles" : zeugnisse vom reichtum und gesetz der deutschen seele / ed by Cerff, Karl – Berlin: W Limpert [1944?] [mf ed 1993] – 1r [ill] – 1 – (wood engravings by j leander gampp. filmed with: wegweiser zur deutschen literatur / guenther cwojdrak & other titles) – mf#3334p – us Wisconsin U Libr [430]

In veg un andere dertseylungen / Fischer, Abraham Eliezer – Buenos Aires, Argentina. 1934 – 1r – us UF Libraries [939]

In view of the end : a retrospect and a prospect / Sanday, William – Oxford: Clarendon Press, 1916 – 1mf – 9 – 0-7905-9105-7 – mf#1989-2330 – us ATLA [940]

In vitro cellular and developmental biology : animal – Columbia. 1989+ (1,5,9) – ISSN: 1071-2690 – mf#15125,01 – us UMI ProQuest [576]

In vivo insulin action on whole body and individual tissues in obese shhf/mcc-cp rats with or without acute exercise / Gao, Jiaping & Sherman, William M – 1991 – 2mf – 9 – $8.00 – us Kinesology [613]

In western india : recollections of my early missionary life / Mitchell, John Murray – Edinburgh: D Douglas, 1899 – 1mf – 9 – 0-524-03589-X – (incl bibl ref) – mf#1990-1049 – us ATLA [240]

In western india : recollections of my early missionary life / Mitchell, Murray, J – Edinburgh, 1899 – 5mf – 9 – mf#HTM-132 – ne IDC [915]

In whitest africa / Frye, William – Englewood Cliffs, NJ. 1968 – 1r – us UF Libraries [960]

In wie weit ist der bibel irrthumslosigkeit zuzuschreiben? : vortrag / Volck, Wilhelm – Dorpat: E J Karow, 1884 – 1mf – 9 – 0-524-04116-4 – mf#1992-0074 – us ATLA [220]

In yene teg – Riga, Latvia. 1937 – 1r – us UF Libraries [939]

In zululand with the british throughout the war of 1879 / Newman, Charles L Norris – London 1880 – (= ser 19th c british colonization) – 5mf – 9 – mf#1.1.7756 – uk Chadwyck [960]

In zululand with the british throughout the war of 1879 / Norris-Newman, Charles L – London: W H Allen, 1880 – 1 – us CRL [960]

Inacio de azevedo... / Costa, Manuel Goncalves da – Braga, Portugal. 1946 – 1r – us UF Libraries [972]

Inadequate hyperventilation as a determinant of exercise induced hypoxemia / Harms, Craig A – 1994 – 2mf – $8.00 – us Kinesology [612]

Inainte – Craiova, Romania. 1962-Jun 1980 – 26r – 1 – us L of C Photodup [949]

Inamura, Chikako *see* The effect of t'ai chi ch'uan upon selected fitness components of older women

Inangahua herald – 1872-1876; 1877-1900; 1902-1903; mar 1904-jun 1905; 1906-jun 1910; jan-sep 1911; 1912-1914; jul 1915-jun 1916; jul 1917-mar 1919; 1920-sep 1921; 1922-dec 1924; sep-dec 1925; mar 1926-apr 1936 – 1 – mf#50.5 – nz Nat Libr [079]

Inangahua times – jan 1877-jul 1882; nov 1882-1888; mar-oct 1889; 1890-1892; 1894-oct 1901; jan-sep 1902; jan-aug 1903; 1904; 1906-oct 1909, 1910-oct 1919;1920-1926, jan-jul 1928; 1929-1942 – 1 – (only a few issues for yrs between 1920-26) – mf#50.2 – nz Nat Libr [079]

La inapresable / Vera, Francisco – Madrid: Imp. Alrededor del Mundo, 1923 – 1 – sp Bibl Santa Ana [999]

Inauguracion de la catedra martiana / Lazo, Raimundo – Habana, Cuba. 1950 – 1r – us UF Libraries [972]

Inauguracion de la estatua ecuestre del libertador – Caracas, Venezuela. 1954 – 1r – us UF Libraries [972]

An inaugural address : "three changes in theological institutions" / Boyce, James Petigru – South Carolina, 1856 – 1 – $5.00 – us Southern Baptist [242]

Inaugural address : delivered in the convocation hall, lennoxville, at the opening of the law faculty, on the 5th october, 1880 / Ramsay, Thomas Kennedy – [Montreal?: s.n.], 1880 [mf ed 1981] – 1mf – 9 – 0-665-12360-4 – mf#12360 – cn Canadiana [378]

Inaugural address delivered before knox college metaphysical and literary society : on the evening of friday, december 1st, 1871 / Armstrong, William Dunwoodie – Toronto: The Society, 1871? – 1mf – 9 – mf#00824 – cn Canadiana [080]

Inaugural address delivered before knox college metaphysical and literary society, friday, november 28th, 1873 / McPherson, H H – Toronto: publ by the Society, [1873] – 1mf – 9 – 0-665-92306-6 – mf#92306 – cn Canadiana [080]

Inaugural address delivered by j d edgar, esq, president of the ontario literary society, february 5th, 1863 : the hon p m vankoughnet, chancellor of upper canada, in the chair / Edgar, James David – Toronto: Rollo and Adam, 1863 – 1mf – 9 – mf#23050 – cn Canadiana [080]

Inaugural address of his worship h beaugrand, esq, mayor of montreal : delivered march 9th, 1885 – S.l: s.n, 1885? – 1mf – 9 – mf#10248 – cn Canadiana [350]

The inaugural address of the rev j g binney : as president of the columbian college, dc: wed, jun 17 1855 / Binney, Joseph Getchell – Washington: RA Waters, 1857 [mf ed 1993] – 1mf – 9 – 0-524-08351-7 – mf#1993-3051 – us ATLA [378]

Inaugural address on the nature and advantages of an english and liberal education : delivered...at the opening of victoria college, june 24, 1842 / Ryerson, Egerton – Toronto: By order of the Board of Trustees and Visitors, 1842 – 1mf – 9 – mf#21866 – cn Canadiana [370]

Inaugural addresses of the presidents of the united states : from george washington to harry s truman, 1789-1949 – 82nd cong, 2nd sess. n.p., n.d. [all publ] [mf ed 1952?] – 3mf – 9 – $4.50 – mf#llmc95-236 – us LLMC [850]

Inaugural celebration held in the village of deseronto, dominion day, 1881 – Napanee [ON]: Templeton & Beeman [1881?] [mf ed 1987] – 1mf – 9 – 0-665-54711-0 – mf#54711 – cn Canadiana [390]

Inaugural discourse : delivered before the university in cambridge, august 10, 1819 / Norton, Andrews – Cambridge: Printed by Hilliard & Metcalfe at the University Press, 1819. Beltsville, Md: NCR Corp, 1978 (1mf); Evanston: American Theol Lib Assoc, 1984 (1mf) – (= ser Biblical crit – us & gb) – 9 – 0-8370-0676-7 – mf#1984-1031 – us ATLA [240]

Inaugural discourse pronounced at the first meeting of the academy / Wiseman, Nicholas Patrick – London, England. 1861 – 1r – us UF Libraries [240]

An inaugural lecture : delivered...jan 26 1903 / Bury, John Bagnell – Cambridge: University Press; New York: Macmillan [dist] 1903 [mf ed 1989] – 1mf – 9 – 0-7905-4446-6 – mf#1988-0446 – us ATLA [900]

Inaugural lecture of the department of practical science in mcgill university, montreal : delivered in the william molson hall, monday, 19th feb 1872 / Armstrong, George Frederick – [Montreal?: s.n.] 1872 [mf ed 1984] – 9 – 0-665-05102-6 – (in dble clms) – mf#05102 – cn Canadiana [620]

Inaugural lecture read before the university of oxford in the divin... / Hampden, Renn Dickson – London, England. 1836 – 1r – us UF Libraries [240]

Inaugural lectures / Tout, Thomas Frederick et al; ed by Peake, Arthur Samuel – Manchester: University Press, 1905 – (= ser University of manchester publications) – 1mf – 9 – 0-7905-0438-3 – (incl bibl ref & ind) – mf#1987-0438 – us ATLA [240]

Inaugural meeting of the local council of women of halifax : address by her excellency the countess of aberdeen, august 24th, 1894 – S.l: Morning Herald Print & Pub Co, 1894 – 1mf – 9 – mf#25532 – cn Canadiana [305]

Inaugural sermon delivered in the temple beth-el at detroit, mich... / Grossman, Louis – Cincinnati, OH. 1884? – 1r – us UF Libraries [939]

Inaugural sermon preached in christ church cathedral, montreal / Baldwin, Maurice Scollard – Montreal?: J Lovell, 1872 – 1mf – 9 – mf#01447 – cn Canadiana [242]

Inauguration du monument erige a chicoutimi a la memoire de william evan price, 24 juin 1882 – Quebec: [s.n.] 1882 [mf ed 1984] – 1mf – 9 – 0-665-04467-4 – mf#04467 – cn Canadiana [920]

Inauguration of james mccosh, d.d., ll.d., as president of the college of new jersey, princeton : october 27, 1868 – New York: Robert Carter, 1868 – 1mf – 9 – 0-7905-7975-8 – mf#1989-1260 – us ATLA [378]

Inauguration of milton valentine, d.d., as president of pennsylvania college, gettysburg, penn'a : december 21, 1868 – Gettysburg: Star & Sentinel, 1869 – 1mf – 9 – 0-7905-8950-8 – mf#1989-2175 – us ATLA [378]

Inauguration of rev. e.v. gerhart : professor of theology in the theological seminary of the german reformed church, located at tiffin, o... / Winters, D et al – Tiffin City, O[hio]: Printed at the office of the "Western Missionary", 1851 – 1mf – 9 – 0-524-08718-0 – mf#1993-1088 – us ATLA [240]

Inauguration of the european headquarters – London, England. 1890 – 1r – us UF Libraries [240]

The inauguration of the political independence of victoria : the first meeting of our parliament under the new constitution. a lecture delivered on thursday, nov 20th, 1856 / Cairns, Adam – Melbourne, 1856 – (= ser 19th c books on british colonization) – 1mf – 9 – mf#1.1.6926 – uk Chadwyck [971]

Inauguration of the rev. benjamin b. warfield, d.d., as professor of didactic and polemic theology – New York: Anson DF Randolph, 1888 – 1mf – 9 – 0-524-00383-1 – mf#1989-3083 – us ATLA [240]

Inaynem – (Chicago). 1924-25 – 1 – us AJPC [071]

Inbar, Galit *see* The post-exercise blood pressure response to acute exercise in borderline hypertensive women

Inbe, Hironari *see* Imbe-no-hironari's kogoshui

Inbreeding and outbreeding; their genetic and sociological significance / East, Edward Murray – Philadelphia, London: J.B. Lippincott, c1919. 285p. illus., plates – 1 – us Wisconsin U Libr [576]

Inc – Boston. 1979+ (1,5,9) – ISSN: 0162-8968 – mf#12288 – us UMI ProQuest [650]

Incantalupo, Patricia *see* The portrayal of women in sport advertising in two women's fitness magazines

Incantamenta magica graeca latina / ed by Heim, Richard – Lipsiae [Leipzig]: BG Teubner, 1892 – 1mf – 9 – 0-524-07071-7 – mf#1991-0053 – us ATLA [450]

Incar report : newsletter of the international committee against racism / International Committee Against Racism – v2 n1 [1974 jul], 4-5 [1974 oct/nov], 1975 dec, 1980 feb – 1r – 1 – mf#372306 – us WHS [320]

The incarnate word : being the fourth gospel elucidated by interpolation for popular use / Gill, William Hugh – Philadelphia: George W Jacobs, c 1900 [mf ed 1985] – 1mf – 9 – 0-8370-3285-7 – mf#1985-1285 – us ATLA [225]

The incarnation / Eck, Herbert Vincent Shortgrave – 2nd ed. London: Longmans, Green 1902 [mf ed 1992] – (= ser The oxford library of practical theology) – 1mf – 9 – 0-524-04890-8 – (incl bibl ref) – mf#1991-2172 – us ATLA [210]

The incarnation / Hall, Francis Joseph – New York: Longmans, Green 1915 [mf ed 1990] – (= ser Dogmatic theology 6) – 1mf – 9 – 0-7905-3891-1 – (incl bibl ref) – mf#1989-0384 – us ATLA [220]

The incarnation / Streatfeild, George Sidney – London, New York: Longmans, Green 1910 [mf ed 1989] – (= ser Anglican church handbooks) – 1mf – 9 – 0-7905-2330-2 – (incl bibl ref & ind) – mf#1987-2330 – us ATLA [240]

The incarnation : a study of philippians 2. 5-11 / Gifford, Edwin Hamilton – New York: Dodd, Mead, 1897 – 1mf – 9 – 0-8370-3802-2 – (incl bibl ref) – mf#1985-1802 – us ATLA [220]

The incarnation and modern thought / Case, Carl Delos – Chicago: University of Chicago Press, 1908 – 1mf – 9 – 0-524-07090-3 – mf#1991-2913 – us ATLA [240]

The incarnation and principles of evidence : a theological essay / Hutton, Richard Holt – New York: Pott & Amery, 1871. Chicago: Dep of Photodup, U of Chicago Lib, 1971 (1r); Evanston: American Theol Lib Assoc, 1984 (1r) – 1 – 0-8370-0293-1 – mf#1984-B213 – us ATLA [220]

The incarnation and recent criticism / Cooke, Richard Joseph – New York: Eaton & Mains; Cincinnati: Jennings & Graham, c1907 – 1mf – 9 – 0-8370-6039-7 – (incl bibl ref and index) – mf#1986-0039 – us ATLA [210]

The incarnation of the lord : a series of sermons tracing the unfolding of the doctrine of the incarnation in the new testament / Briggs, Charles Augustus – New York: Scribner's, 1902 – 1mf – 9 – 0-8370-2453-6 – (incl ind of biblical texts cited and subject index) – mf#1985-0453 – us ATLA [240]

The incarnation of the son of god : being the bampton lectures for the year 1891 / Gore, Charles – New York: Charles Scribner, 1891 – 1mf – 9 – 0-8370-4840-0 – (incl bibl ref) – mf#1985-2840 – us ATLA [240]

The incarnation of the son of god / Gore, Charles – London: J Murray, 1891 – (= ser Bampton lectures) – 1r – 1 – 0-8370-1122-1 – mf#1984-B237 – us ATLA [220]

Les incas ou la destruction de l'empire du perou / Marmontel, M – Lyon: Bruyset, Tomo 2. 1817 – 1 – sp Bibl Santa Ana [972]

Les incas...empire du perou / Marmontel, M – Tomo I. 1817 – 9 – (tomo 2 1817. tomo 3 1817) – sp Bibl Santa Ana [972]

Ince, William *see* Lord, and what shall this man do?

Incendio de conventos en espana y supresion de colegios y misiones espanolas en ultramar / Alonso Getino, G – Madrid: Razon y Fe, 1932 – 1 – sp Bibl Santa Ana [240]

El incendio de la biblioteca / Bartholino, Tomas – Valencia: editorial castalia, 1949 – sp Bibl Santa Ana [020]

Incentive – New York. 1988+ (1) 1988+ (5) 1988+ (9) – (cont: incentive marketing) – ISSN: 1042-5195 – mf#10316,01 – us UMI ProQuest [650]

Incentive marketing – New York. 1975-1988 (1) 1977-1988 (5) 1977-1988 (9) – (cont by: incentive) – ISSN: 0019-3364 – mf#10316 – us UMI ProQuest [650]

Incentive marketing – Croydon. 1965-1973 (1) 1971-1972 (5) (9) – (cont by: incentive marketing and sales promotion) – ISSN: 0019-3356 – mf#5680 – us UMI ProQuest [650]

Incentive marketing and sales promotion – Croydon. 1974-1979 (1) 1974-1979 (5) 1974-1979 (9) – (cont: incentive marketing) – ISSN: 0305-2230 – mf#5680,01 – us UMI ProQuest [650]

Incentive motivation, competitive orientation and gender in collegiate alpine skiers / Chroni, Stiliani – 1994 – 2mf – $8.00 – us Kinesology [150]

Incentive motivation differences in united states masters swimmers / Mowrey, Rebecca J – 1989 – 173p 2mf – 9 – $8.00 – us Kinesology [150]

Incentive motivation of female basketball players across three age levels / Drennan, Meredith L – Springfield College, 1994 – 2mf – 9 – $8.00 – mf#PSY1843 – us Kinesology [150]

O incentivo : periodico do collegio s jose – Bahia: Typ do Monitor, 14 maio 1878 – (= ser Ps 19) – mf#P18B,02,24 – bl Biblioteca [079]

O incentivo : semanario recreativo e de instruccao – Belem, PA. 01 fev 1851 – (= ser Ps 19) – mf#P17,02,121 – bl Biblioteca [370]

Inchambre, Diego de *see* Noticias sobre las provincias franciscanas de canarias. tenerife, 1966

Inchanted forrest : an instrumental composition expressive of the same ideas as the poem of tasso of that title / Geminiani, Francesco – London: John Johnson [1755?] [mf ed 19–] – 9pt on 3mf – 9 – mf#fiche 242 – us Sibley [780]

Inchaurre Aldape, Diego *see* Compilacion de articulos referentes a las ordenes...

Inchaustegui Cabral, Hector *see*
- En soledad de amor herido
- Insulas extranas
- Miedo en un punado de polvo
- Muerte en 'el eden'
- Rebelion vegetal
- Rumbo a la otra vigilia
- Soplo que se va y que vuelve

Inchaustegui Cabral, Joaquin Marino *see*
- Ciudad trujillo
- Cristobal colon y la isla espanola
- Geografia e historia de la republica dominicana
- Goegrafia descriptiva de la republica dominicana

Inchaustegui Cabral, Joaquin Marino *see* Historia de santo domingo

Inchiesta – Rome. 1971-1972 (1) 1971-1972 (5) 1971-1972 (9) – ISSN: 0046-8819 – mf#8331 – us UMI ProQuest [320]

Incicean, Gowkas *see* Nachrichten ueber den thrazischen bosporus oder die strasse von constantinopel

The incidence and severity of heavy episodic drinking, stages of heavy episodic drinking, stages of change of readiness and perceived normative expectations among college students / Lopez, Paulette – 1998 – 2mf – 9 – $8.00 – mf#HE 615 – us Kinesology [362]

The incidence of post-traumatic stress disorder symptoms in certified athletic trainers / Conner, Christopher P – 1997 – 1mf – 9 – $4.00 – mf#PE 3845 – us Kinesology [617]

Incidental bishop : a novel / Allen, Grant – New York: D Appleton, 1898 – (= ser Appletons' town and country library) – 4mf – 9 – (incl publ list) – mf#27426 – cn Canadiana [830]

Incidental illustrations of the economy of salvation : its doctrines and duties / Palmer, Phoebe – Boston: Degen, 1855. El Segundo, Ca: Micro Publication Systems, 1981 (1mf); Evanston: American Theol Lib Assoc, 1984 (1mf) – (= ser Women & the church in america) – 9 – 0-8370-1459-X – mf#1984-2176 – us ATLA [240]

Incidentally / Knight, John Thomas Philip – Montreal: Westmount News Press, 1913 [mf ed 1994] – 2mf – 9 – 0-665-73211-2 – mf#73211 – cn Canadiana [080]

Les incidents et les exceptions devant les tribunaux militaires en temps de paix et aux armees / Pages, E L – Nouv. ed.. Baden-Baden: Regie autonome des publications officielles 1948. 246p. LL-4114 – 1 – us L of C Photodup [355]

Incidents in the early military history of canada : with extracts from the journals of the officer commanding the queen's rangers during the war 1755 to 1763: a lecture delivered on the 12th january, 1891 / Rogers, Robert Zaccheus – [Toronto?: s.n, 1891?] – 1mf – 9 – 0-665-94338-5 – (incl app) – mf#94338 – cn Canadiana [355]

Incidents in the life of madame blavatsky / ed by Sinnett, Alfred Percy – London: George Redway, 1886 – 1mf – 9 – 0-524-01300-4 – mf#1990-2336 – us ATLA [920]

Incidents of pioneer days at guelph and the county of bruce / Kennedy, David – Toronto: [s.n.] 1903 [mf ed 1995] – 2mf – 9 – 0-665-74081-6 – mf#74081 – cn Canadiana [971]

Incidents of social life amid the european alps / Zschokke, Heinrich – New York: D Appleton & Co, 1844 – 1 – us Wisconsin U Libr [943]

Incidents of travel in central america, chiapas... / Stephens, John Lloyd – New Brunswick, NJ. v1-2. 1949 – 1r – us UF Libraries [972]

Incidents of travel in the southern states and cub... / Rogers, Carlton H – New York, NY. 1862 – 1r – us UF Libraries [972]

Incidents of western travel : in a series of letters / Pierce, George Foster; ed by Summers, Thomas Osmond – Nashville, Tenn: Southern Methodist Pub House, 1859 – 1mf – 9 – 0-524-06775-9 – mf#1991-2782 – us ATLA [240]

Incider – Peterborough. 1983-1989 (1,5,9) – (cont by: incider a+) – ISSN: 0740-0101 – mf#13348 – us UMI ProQuest [000]

Incider a+ – Peterborough. 1989-1993 (1) 1989-1993 (5) 1989-1993 (9) – (cont: incider) – ISSN: 1054-6456 – mf#13348,01 – us UMI ProQuest [000]

Incipient irish revolution : an expose of fenianism of to-day in the united kingdom and america – London, 1889 – (= ser 19th c ireland) – 1mf – 9 – mf#1.1.7086 – uk Chadwyck [941]

Incirlik Air Base *see*
- Looking glass
- Tip of the sword

Inciso, Flaviano *see*
- Comedia na dapit sa dios, o, magna cahayagan can magna misterios nin navidad, 2nd pt
- Comedia na dapit sa dios, o, magna cayayagan can magna misterios nin navidad, 3rd pt

incl bibl ref *see* Professional opportunities for the woman doctor in the republic

Inclan, M *see* Reflexiones sobre aduanas y efectos de la ley prohivitiva

[Incline village-] high sierra times – NV. 1976-1978 – 1r – 1 – $60.00 – mf#N03706 – us Library Micro [071]

[Incline village-] north lake tahoe bonanza – NV. 1981- – 50r – 1 – $3000.00 (subs $240y) – mf#N04802 – us Library Micro [071]

Die inclusen in deutschland : vornehmlich in der gegend des niederrheins um die wende des 12. und 13. jahrhunderts / Basedow, A – Heidelberg, 1895 – 1mf – 8 – €3.00 – ne Slangenburg [241]

Inclusion in the constitution of the international labour... / International Labour Office – Geneva, Switzerland. 1964 – 1r – us UF Libraries [331]

Inclusionary practices in physical education / Leaman, Nicole L – 1998 – 1mf – 9 – $4.00 – mf#PE 3907 – us Kinesology [790]

Inclusive physical education : attitudes and behaviors of students / Bulter, Rhea S – 2000 – 128p on 2mf – 9 – $10.00 – mf#PE 4167 – us Kinesology [150]

Incola e o bandeirante na historia de sao paulo / Campos, Pedro Dias De – Rio de Janeiro, Brazil. 1951 – 1r – us UF Libraries [972]

Income opportunities – Tulsa. 1973-1998 (1) 1976-1998 (5) 1976-1998 (9) – ISSN: 0019-3429 – mf#8372 – us UMI ProQuest [332]

The income tax : a study of the history, theory and practice of income taxation at home and abroad / Seligman, Edwin R A – New York: Macmillan, 1914 – 8mf – 9 – $12.00 – mf#LLMC 82-710 – us LLMC [336]

Income tax administration in the state of israel / Anderson, Wayne F – Tel Aviv, 1956. LL-4195 – 1 – us L of C Photodup [336]

The income tax and the individual : an explanation of the law as it affects very large numbers of people in canada – [Montreal]: Bank of Montreal, 1919 – 1mf – 9 – 0-665-86564-3 – mf#86564 – cn Canadiana [343]

The income tax and the individual : revised to include amendments of 1920 – [Montreal?: s.n, 1920?] – 1mf – 9 – 0-665-76880-X – mf#76880 – cn Canadiana [343]

Income tax appeal board practice : a practical treatise on procedure before the board / Fordham, Reginald Sydney Walter – Montreal: CCH Canadian, 1953. 138p. LL-2334 – 1 – us L of C Photodup [343]

The income tax fathered : as also the mode of raising the supplies, without funding / Edwards, George – [London, 1810] – (= ser 19th c economics) – 1mf – 9 – mf#1.1.217 – uk Chadwyck [336]

Income tax rulings *see* Us internal revenue service. income tax rulings. cumulative bulletin

The incoming millions / Grose, Howard Benjamin – 6th ed. New York: FH Revell, c1906 – (= ser Home Mission Study Course) – 1mf – 9 – 0-7905-4800-3 – (incl bibl ref) – mf#1988-0800 – us ATLA [240]

Incomparable india : tradition, superstition, truth / Blackham, Robert James – London: Sampson Low, Marston & Co, [19–] – (= ser Samp: indian books) – (foreword by sir william birdwood) – us CRL [954]

Incongruencias legales de las faltas contra la propiedad de corchero y compania / Carrasco Alvarez, Antonio – 1901. Imprenta. Corchero y Cia., Merida – sp Bibl Santa Ana [346]

Inconquistables / Ortega Ricaurte, Enrique – Bogota, Colombia. 1949 – 1r – us UF Libraries [972]

"Inconsistency" : and an open letter to my critics / Adams, Henry – Yarmouth, NS?: C Carey, 1888 – 1mf – 9 – mf#07219 – cn Canadiana [242]

Inconstitutionnalite de la convention americano-ha... / Morpeau, Moravia – Port-Au-Prince, Haiti. 1929 – 1r – us UF Libraries [972]

Incontrovertible facts regarding the destruction of guernica: speech / Leizaola, Jesus Maria de – Washington, DC, 193? Fiche W994. (Blodgett Collection of Spanish Civil War Pamphlets) – 9 – us Harvard [946]

La incorporacion de las masas populares a la historia : la comune primera revolucion consciente / Montseny, Frederica – [s.l.]: oficinas de propaganda [1937?] [mf ed 1977] – (= ser Blodgett coll) – 1mf – 9 – mf#w1060 – us Harvard [946]

Incorporated Society of Authors *see*
- The cost of production
- The grievances between authors and publishers

Incorporation papers 1911 / General Synod of the Evangelical Lutheran Church in the United States. Board of Foreign Missions – [mf ed 2004] – 1r – 1 – (1st sect of this doc outlines the act through wh the board as an incorporated entity was created, 2nd sect affirms that the doc contains a true & correct copy of the original papers) – mf#xa0091r – us ATLA [242]

In-court orientation programs in the federal district courts / Meierhoefer, Barbara S – Washington: FJC, Apr 1984 – 1mf – 9 – $1.50 – mf#LLMC 95-837 – us LLMC [347]

Increase and characteristics of connecticut baptists : an address. delivered at the centennial anniversary of the first baptist church, meriden, conn... / True, Benjamin Osgood – Meriden: Republican Book Dept, 1887 – 1mf – 9 – 0-524-06595-0 – mf#1990-5261 – us ATLA [242]

The increase of faith / Lee, William – 2nd ed. Edinburgh: William Blackwood, 1868 – 1mf – 9 – 0-8370-4323-9 – (includes appendix on the rule of faith) – mf#1985-2323 – us ATLA [210]

The increase of faith : some present-day aids to belief / McConnell, Francis John – New York: Eaton & Mains, c1912 – (= ser Merrick lectures) – 1mf – 9 – 0-7905-9504-4 – mf#1989-1209 – us ATLA [240]

Increase of strength / Selwyn, George Augustus – Eton, England. 1847 – 1r – us UF Libraries [240]

The increase of the israelites in egypt shewn to be probable from the statistics of modern populations : with an examination of bishop colenso's calculations on this subject / Ashpitel, Francis – Oxford: John Henry and James Parker, 1863. Beltsville, Md: NCR Corp, 1978 (1mf); Evanston: American Theol Lib Assoc, 1984 (1mf) – (= ser Biblical crit – us & gb) – 9 – 0-8370-1213-9 – (incl bibl ref) – mf#1984-1058 – us ATLA [221]

Increased communions / Pusey, E B – Aberdeen, Scotland. 18– – 1r – us UF Libraries [240]

Increased consumption of lean beef on iron status and physical performance in adolescent females / Pahnke, Thomas G – 1999 – 2mf – 9 – $8.00 – mf#PH 1680 – us Kinesology [612]

Increasing daily physical activity in postsecondary students with mental retardation / Stratton, Wendith M – 1999 – 2mf – 9 – $8.00 – mf#HE 651 – us Kinesology [613]

Increasing vertical jump : a comparison between two training programs / Timmons, Scott A – Ball State University, 1996 – 1mf – 9 – mf#PE 3675 – us Kinesology [612]

The incredibilities of part 2 of the bishop of natal's work upon the pentateuch : a lay protest / Knight, John Collyer – London: Samuel Bagster, 1863 – 1mf – 9 – 0-524-05811-3 – mf#1992-0638 – us ATLA [221]

Incubation : or, the cure of disease in pagan temples and christian churches / Hamilton, Mary – St Andrews: W C Henderson; London: Simpkin, Marshall, Hamilton, Kent 1906 [mf ed 1991] – 1mf – 9 – 0-524-00882-5 – mf#1990-2105 – us ATLA [230]

The incubi of rome and venice : or, the criminal history of the popes. and, the martyrdom of venice / Beggi, Francesco Orzzio – 2nd ed. [London?]: J Clements, 1864 – 2mf – 9 – 0-524-05133-X – mf#1990-1389 – us ATLA [240]

L'incunable – Montreal: Bibliotheque nationale du Quebec. v18 n1 mars 1984-20e annee n3 dec 1986 (qrtly) [mf ed 1984] – 3r – 5 – mf#SEM16P359 – cn Bibl Nat [073]

Incunables bogotanos, sigle 18 / Biblioteca Luis-Angel Arango – Bogota, Colombia. 1959 – 1r – us UF Libraries [972]

Incunables desconocidos : ars constructionis ordinandae / Lopez Serrano, Mathilde – Madrid: CSIC, 1947. Sep. Sup. no 1 de Rev. Bibliografica y Documental (T.1, 1947, no 2, Abril-Junio) – 1 – sp Bibl Santa Ana [946]

Incunables espanoles : obsidionis rhodie descripto. de guillermo croursin / Lopez Serrano, Mathilde – Madrid: CSIC, 1947. Sep. Sup. no 1 de Rev. Bibliografica y Documental (T.1, 1947, no 3 y 4. Julio-Diciembre) – 1 – sp Bibl Santa Ana [946]

Incunables espanoles desconocidos : "de moribus" de seneca / Lopez Serrano, Mathilde – Madrid: CSIC, 1951. Sep. Rev. Bibliografica y Documental. (T.5, 1951, Fasciculos 1,2,3 y 4, Enero-Diciembre) – 1 – sp Bibl Santa Ana [946]

Incunabula : the printing revolution in europe, 1455-1500 / Flood, John L [comp] – 47 units – 9 – (int by comp. comprehensive coll of the earliest printed books in europe. unit 1: printing in mainz to 1480 327mf. unit 2: the classics in translation 679mf. unit 3: image of the world: geography and cosmography 549mf. units 4-5: chronicles and historiography 718mf, 534mf respectively. unit 6: image of the world: travellers' tales 328mf. unit 7-10: printing in italy before 1472 377mf, 369mf, 375mf, 342mf respectively. units 11-15 & 21: medical incunabula 368mf, 365mf, 317mf, 318mf, 296mf, 370mf respectively. units 16-17: incunabula hebraica 301mf, 344mf. units 22-23: rhetoric 266mf, 269mf. units 24-25: italian humanism 322mf, 374mf. units 26-28: philosophy: ancient, medieval and renaissance 349mf, 341mf, 352mf respectively. units 29-30: grammar 261mf, 259mf. units 31-33: sermons 325mf, 286mf, 339mf respectively. units 34-39: law 387mf, 397mf, 351mf, 443mf, 355mf, 440mf respectively. units 41-44: science 283mf, 271mf, 297mf, 233mf respectively. unit 45: printing in greek 502mf. units 46-47: german vernacular literature 256mf, 192mf respectively. units 48-49: printing in england 308mf, 366mf respectively. units 50-51: liturgy 313mf, 353mf respectively. units 52, 53: current affairs 317mf, ca 320mf respectively. units 54-55: iberian printing [copubl: british library]. units 56-59: bibles and commentaries 326mf, 329mf, ca 320mf ea for units 58,59 respectively. units 60-62: academic theology 333mf, ca 320mf ea for units 61,62 respectively) – us Primary [090]

Incwadi yesisngisi nesizulu / Bryant, Alfred T – Maritzburg, South Africa. 1900 – 1r – us UF Libraries [960]

Incwadi yesixhosa yesiqibi sokuqala / Koti, Candlish – London, England. 1942 – 1r – us UF Libraries [960]

Incwadi yokuqala – Lovedale, South Africa. 19–? – 1r – us UF Libraries [960]

Inda Hernandez, Jose *see* Cantos y rumbos

Indaba – Lovedale SA, aug 1 1862-feb 28 1865 (mthly) – 1r – 1 – (one-third in english) – sa National [079]

Indagacion del choteo / Manach, Jorge – La Habana, Cuba. 1940 – 1r – us UF Libraries [972]

Indagacion y critica / Espinosa, Ciro – Habana, Cuba. 1940 – 1r – us UF Libraries [972]

Indagaciones martianas / Gonzalez, Manuel Pedro – Santa Clara, Cuba. 1961 – 1r – us UF Libraries [972]

Indagationes mathematicae – Amsterdam. 1990-1993 (1,5,9) – ISSN: 0019-3577 – mf#42572 – us UMI ProQuest [510]

Indagini su hegel, e schiarimenti filosofici / Croce, Benedetto – Bari: G Laterza, 1952 [mf ed 1990] – (= ser Saggi filosofici 14) – viii/305p – 1 – mf#7363 – us Wisconsin U Libr [190]

Indau, J *see* Wiennerisches architectur-kunst und saeulen-buch

Inday : dulang tolo kabahin / Rodriguez, Tura – [Sugbo?: Falek? 1917?] [mf ed Bloomington IN: Indiana Uni Lib, Preservation Dept 1984] – (= ser Coll...in the bisaya language 1) – 1r – 1 – us Indiana Preservation [490]

Inde ab a. 911 usque ad a. 1197 (mgh leges 2d:1.bd) – 1893 – (= ser Monumenta germaniae historica leges 2. leges in quarto. d legum sectio 4: constitutiones et acta publica imperatorum et regum (mgh leges 2)) – €37.00 – ne Slangenburg [240]

Inde ab a. 1198 usque ad a. 1272 (mgh leges 2d:2.bd) – 1896 – (= ser Monumenta germaniae historica leges 2. leges in quarto. d legum sectio 4: constitutiones et acta publica imperatorum et regum (mgh leges 2)) – €35.00 – ne Slangenburg [240]

Inde ab a. 1273 usque ad a. 1298 (mgh leges 2d:3.bd) – 1904-1906 – (= ser Monumenta germaniae historica leges 2. leges in quarto. d legum sectio 4: constitutiones et acta publica imperatorum et regum (mgh leges 2)) – €37.00 – ne Slangenburg [240]

Inde ab a. 1298 usque ad a. 1313 (mgh leges 2d:4.bd) – 1906-1911 – (= ser Monumenta germaniae historica leges 2. leges in quarto. d legum sectio 4: constitutiones et acta publica imperatorum et regum (mgh leges 2)) – €80.00 – ne Slangenburg [240]

Inde ab a. 1313 usque ad a. 1324 (mgh leges 2d:5.bd) – 1909-1913 – (= ser Monumenta germaniae historica leges 2. leges in quarto. d legum sectio 4: constitutiones et acta publica imperatorum et regum (mgh leges 2)) – €48.00 – ne Slangenburg [240]

Inde ab a. 1325 usque ad a. 1330 (mgh leges 2d:6.bd) – 1914-1927 – (= ser Monumenta germaniae historica leges 2. leges in quarto. d legum sectio 4: constitutiones et acta publica imperatorum et regum (mgh leges 2)) – €40.00 – ne Slangenburg [240]

Inde ab a. 1345 usque ad a. 1348 (mgh leges 2d:8.bd) – 1910-1926 – (= ser Monumenta germaniae historica leges 2. leges in quarto. d legum sectio 4: constitutiones et acta publica imperatorum et regum (mgh leges 2)) – €42.00 – ne Slangenburg [240]

L'inde apres le bouddha / Lamairesse – Paris: Georges Carre, 1892 [mf ed 1992] – (= ser Bibliotheque des religions comparees) – 2mf – 9 – 0-524-04861-4 – (in french) – mf#1990-3423 – us ATLA [280]

L'inde avant le bouddha / Lamairesse – Paris: Georges Carre, 1891 [mf ed 1992] – (= ser Bibliotheque des religions comparees) – 1mf – 9 – 0-524-02023-X – (in french) – mf#1990-2798 – us ATLA [280]

L'inde d'aujourd'hui : etude sociale / Metin, Albert – nouv augm ed. Paris: Armand Colin, 1918 [mf ed 1995] – (= ser Yale coll) – 362p – 1 – 0-524-09891-3 – (in french) – mf#1995-0891 – us ATLA [954]

L'inde tamoule : nos missions francaises / Suau, P – Paris, 1901 – 6mf – 9 – mf#983 – ne IDC [915]

Indecency of the marriage service of the church of england / Thorn, William – London, England. 18– – 1r – us UF Libraries [241]

Indeks biologi dan pertanian di indonesia / Departemen Pertanian, Lembaga Perpustakaan Biologi dan Pertanian "Bibliotheca Bogoriensis" – Bogor, 1969-1971 – 19mf – 9 – mf#SE-150=6 – ne IDC [959]

Indeks pers-berita djajakarta-kompas-pedoman-pos indonesia – Djakarta, 1970(4-6) – 21mf – 9 – mf#SE-150=7 – ne IDC [959]

L'indemnite des pecheries : discours prononce par m. pierre fortin depute de gaspe, dans la chambre des communes le 3 mai 1879 – [Ottawa?: s.n, 1879?] – 1mf – 9 – 0-665-92159-4 – mf#92159 – cn Canadiana [639]

Indentures of apprenticeship recorded in the orphans court, washington county, district of columbia, 1802-1811 / U.S. District Court – (= ser Records of district courts of the united states) – 1r – 1 – mf#M2011 – us Nat Archives [347]

The indepedent *see* Huerfano county miscellaneous newspapers

L'independance : chronique bimensuelle. – Paris. n1-48. mars 1911-juil 1913 – 1 – fr ACRPP [073]

L'independance – Leopoldville: Patrice Lumumba, sep 1959-feb 1960 – us CRL [079]

L'independance : organe officiel du rassemblement pour l'independance nationale – Montreal: [s.n.] v1 n1 sep 1962-v6 n20 15/30 sep 1968 (irreg) [mf ed 1969] – 1r – 1 – mf#SEM35P23 – cn Bibl Nat [320]

Independance – Charleroi Belgium, 19 oct 1944-26 jul 1945 – 1r – 1 – uk British Libr Newspaper [074]

Independance – Port-au-Prince: Imp. de "L'Action", sep 1954-dec 1956 – 24r – 1 – us CRL [079]

L'independance belge – Bruxelles. 27 fevr-30 juin 1848, juil 1850-52, juil-dec 1865 – 1 – fr ACRPP [949]

L'independance belge – Brussels, Belgium 1 jul 1843-9 may 1940 [mf 1850-1940] – 1 – (cont: l'independant [6 feb 1831-30 jun 1843]; wanting: 7 feb, 1-3 mar 1831; 6 aug – 20 oct 1914; fr 21 oct 1914-26 nov 1918 publ in london) – uk British Libr Newspaper [074]

L'independance belge : supplement diplomatique – Brussels, Belgium 1 nov 1928-28 nov 1937 – 1 – uk British Libr Newspaper [074]

Independance belge *see* Independant

L'independance belge: radio – Brussels, Belgium 9 mar 1930-22 dec 1939 – 1 – uk British Libr Newspaper [380]

Independance d'haiti devant la france / Gouraige, Ghislain – Port-Au-Prince, Haiti. 1955 – 1r – us UF Libraries [972]

L'independance economique du canada francais / Bouchette, Errol – Arthabaska: Impr d'Arthabaskaville, 1906 [mf ed 1974] – (= ser Etudes sociales et economiques sur le canada) – 1r – 5 – mf#SEM16P112 – cn Bibl Nat [330]

L'independance francaise – Paris: Impr Kugelmann, may 13-19 1871 – (Filmed as pt of: Commune de Paris newspapers) – us CRL [074]

Independance nationale d'haiti / Chancy, Emmanuel – Paris, France. 1884 – 1r – us UF Libraries [972]

L'independant : bimensuel independant d'information – Conakry: L'Independant [n10-v5 n206 (1992-1996)] (wkly) – 3r – 1 – us CRL [079]

L'independant : feuille de commerce, politique et litteraire – Rio de Janeiro, RJ: l'imprimerie Imperiale de P Plancher-Seignot, 21 abr-24 jun 1827 – (= ser Ps 19) – mf#P01,04,10 – bl Biblioteca [079]

L'independant – Port-au-Prince: Impr du Commerce. 1ere annee, n3-n6. 23 mai-13 juin 1877 – 1 sheet – 9 – us CRL [079]

Independant – Brussels Belgium, 1840, sep 1848-apr 1849, 1850-9 may 1940 – 316 1/4r – 1 – (aka: independance belge) – uk British Libr Newspaper [074]

Independant – London, UK. 8 May-19 Jun 1830 – 1 – uk British Libr Newspaper [072]

L'independant des bouches-du-rhone : journal des interets democratiques, commerciaux, artistiques et litteraires – Marseille. n2-3, 6-12. mai-juin 1848 – 1 – fr ACRPP [073]

L'independant des pyrenees – Pau. 1939-19 aout 1944 – 1 – (journal republicain quotidien) – fr ACRPP [073]

L'independant des pyrenees orientales – Perpignan. n115. . 3 fevr 1847. n222-272, fragm. 19 fevr-5 aout 1848 – 1 – (journal politique, litteraire, agricole, commercial et scientifique) – fr ACRPP [073]

Independence – Georgetown. Guyana. -w. 20 Feb 1960-22 Jul 1961. (1 reel) – 1 – uk British Libr Newspaper [072]

Independence – Maseru, Dept of Information and Broadcasting. oct 1-7, 1976 – us CRL [079]

Independence and after : a collection of the more important speeches of jawaharlal nehru from september 1946 to may 1949 – Delhi: Publications Division, Ministry of Information and Broadcasting, Govt of India, 1949 – (= ser Samp: indian books) – us CRL [954]

L'independence belge – [Brussels]: P Werll, jul 1938-may 9 1940 – 14r – 1 – us CRL [949]

Independence daily / Curative Workshop [Green Bay WI] et al – 1966 jan-1987 nov/dec – 1r – 1 – (cont by: center point) – mf#1278677 – us WHS [071]

Independence debate : official report, hansard, unrevised / Parliament of Basutoland, Senate – Maseru: The Senate, 1966 – us CRL [320]

Independence enterprise – Independence OR: Enterprise Pub Co [wkly] – 1 – (merged with: west side to form: independence enterprise and west side) – us Oregon Lib [071]

Independence enterprise and west side – Independence OR: K E Gray, -1904 [wkly] – 1 – (merger of: west side; independence enterprise (1908-69); cont by: west side enterprise (1904-08)) – us Oregon Lib [071]

Independence enterprise and west side *see* Independence enterprise

Independence enterprise (independence, or) – Independence OR: C E Hicks, 1908-69 [wkly] – 1 – (cont: west side enterprise (independence, or). merged with: monmouth herald, to form: independence enterprise monmouth herald) – us Oregon Lib [071]

Independence enterprise (independence, or) *see* Monmouth herald

Independence enterprise (independence, or: 1975) – Independence OR: F Parchman, 1975- [wkly] – 1 – (cont: independence enterprise monmouth herald. merged with: monmouth polk sun to form, sun-enterprise) – us Oregon Lib [071]

Independence enterprise monmouth herald – Independence OR: H V Irvine, 1969-75 [wkly] – 1 – (merger of: independence enterprise (independence, or); monmouth herald; cont by: independence enterprise (independence, or: 1975)) – us Oregon Lib [071]

Independence enterprise monmouth herald *see* Monmouth herald

Independence for africa / Carter, Gwendolen Margaret – New York, NY. 1960 – 1r – us UF Libraries [960]

The independence / general advertiser – Pietermaritzburg SA, 1853-54 – 1r – 1 – sa National [079]

[Independence-] inyo independent – CA. 1870-84; 1901- – 52r – 1 – $3120.00 (subs $80y) – mf#C02303 – us Library Micro [071]

Independence. Missouri. Calvary Baptist Church *see* Scrapbook

Independence news – Independence WI. [1882 mar 2-1892 feb 5] – 1r – 1 – (cont by: wave [independence, wi]; independence news wave) – mf#94864 – us WHS [071]

Independence news wave – Independence WI. [1892 jul 30/nov 19]-2001/2002 aug 8 [gaps] – 66r – 1 – (cont: independence news [independence, wi]; wave [independence, wi]) – mf#948646 – us WHS [071]

Independence of mind, the controlling element of true greatness : an oration. delivered before the calliopean society of granville college... / Tucker, Levi – Pittsburgh: Geo Parkin, 1848 – 1mf – 9 – 0-524-08601-X – mf#1993-3186 – us ATLA [100]

The independence of the holy see / Manning, Henry Edward – London: Henry S King, 1877 – 1mf – 9 – 0-7905-5114-4 – mf#1988-1114 – us ATLA [240]

Independence trade fair, limbe, jul 4-7, 1964 : natural resources exhibit – Zomba, Govt Press, 1964 – us CRL [380]

Independence weekly news – Independence WI. 1878 mar 9-1879 apr 19 – 1r – 1 – (cont by: blair bulletin; weekly news-bulletin) – mf#948636 – us WHS [071]

Independencer / Committee for an Independent Canada – v1 n1-v7 n2 [1972 feb-1978 mar/apr, 1978 sep/oct-1981 jan/feb] – 1r – 1 – mf#606289 – us WHS [320]

El independencia – Malabon: [s.n.], nov 22 1898 – us CRL [079]

Independencia – 1899 feb 28 – 1r – 1 – us WHS [071]

Independencia – v1 n12 [1898 dec 24] – 1r – 1 – mf#601397 – us WHS [071]

La independencia – Malabon: Impr del Asilo de Malabon, sep-dec 1898-jan-mar 1899 – us CRL [079]

Independencia de la costa firme justificada por th... / Paine, Thomas – Caracas, Venezuela. 1949 – 1r – us UF Libraries [972]

Independencia de las colonias hispano-americanas / Cuerto Marquez, Luis – Bogota, Colombia. v1-2. 1938 – 1r – us UF Libraries [972]

Independencia de nueva granada y venezuela / Encina, Francisco Antonio – Santiago, Chile. v1-2. 1961 – 1r – us UF Libraries [972]

Independencia de panama en 1903 / Ortega B, Ismael – Panama, Panama. 1930 – 1r – us UF Libraries [972]

Independencia de puerto rico / Geigel Polanco, Vicente – Rio Piedras, Puerto Rico. 1943 – 1r – us UF Libraries [972]

La independencia de un pueblo con un hijo ilustre fray pedro de godoy / Parron Fernandez, Felipe – Badajoz: Imprenta de la Diputacion Provincial, 1976 – sp Bibl Santa Ana [240]

Independencia de venezuela / Mitre, Bartolome – Buenos Aires, Argentina. 1902 – 1r – us UF Libraries [972]

Independencia economica do brasil / Franco, Cid – Sao Paulo, Brazil. 195? – 1r – us UF Libraries [972]

Independencia efimera / Henriquez Urena, Max – Paris, France. 1938 – 1r – us UF Libraries [972]

Independencia y otro episodios / Fernandez Guardia, Ricardo – San Jose, Costa Rica. 1928 – 1r – us UF Libraries [972]

Independency in warwickshire : a brief history of the independent or congregational churches in that county / Sibree, John & Caston, M – Coventry: G and F King, 1855 – 1mf – 9 – 0-524-05161-5 – mf#1990-1417 – us ATLA [242]

Independent – Wahoo, NE: H D Perky. 12v. v1 n1. sep 16 1875-v12 n8. nov 4 1886 (wkly) [mf ed with gaps filmed [1965]] – 4r – 1 – (absorbed: saunders county republican. cont by: wahoo wasp) – us NE Hist [071]

Independent – Anderson, SC. 1944-1973 (1) – mf#68888 – us UMI ProQuest [071]

Independent – Antelope, MT. 1922-1924 (1) – mf#64224 – us UMI ProQuest [071]

Independent – Atlanta, GA. 1904-1928 (1) – mf#62457 – us UMI ProQuest [071]

Independent – Baraboo WI. 1867 jul 16-1868 dec 9 – 1r – 1 – mf#953692 – us WHS [071]

Independent – Birmingham, AL. 1964-1968 (1) – mf#61987 – us UMI ProQuest [071]

Independent – Bronx, NY. 1990-1991 (1) – mf#68813 – us UMI ProQuest [071]

Independent – Brookport, IL. 1966-1971 (1) – mf#62520 – us UMI ProQuest [071]

Independent – Camden, SC. 1978-1981 (1) – mf#68969 – us UMI ProQuest [071]

Independent – Canton, PA. 1980-1983 (1) – mf#65855 – us UMI ProQuest [071]

Independent – Chewelah, WA. 1916-1983 (1) – mf#66970 – us UMI ProQuest [071]

Independent / Clinton Co. Wilmington – v1 n1. feb 9-nov 16, 1855 [wkly] – 1r – 1 – mf#B31406 – us Ohio Hist [071]

Independent – Colon, Panama. 1904-1914 (incomplete) – 4r – 1 – us L of C Photodup [079]

Independent / Columbiana Co. Columbiana – v1 n1. (9/1898-9/1900, 9/1901-11/1902) [wkly] – 2r – 1 – mf#B29147-29148 – us Ohio Hist [071]

Independent – Conrad, MT. 1911-1921 (1) – mf#64339 – us UMI ProQuest [071]

Independent – Baldwin WI. 1879 nov 28-1881 jan 20 – 1r – 1 – (cont: baldwin bulletin [baldwin, wi: 1873]; cont by: baldwin bulletin [baldwin, wis: 1881]) – us WHS [071]

Independent – Barnsley, England 14 oct 1980-31 jan 1996 [mf 1986-] – 1 – (cont: barnsley independent [3 jul 1973-7 oct 1980]; cont by: barnsley independent [7 feb 1996-]) – uk British Libr Newspaper [072]

Independent – Brodhead WI. 1867 mar 5-1868 apr 28, 1867 mar 5-1868 nov 17, 1868 may 5-nov 17 – 3r – 1 – (cont: brodhead independent [brodhead, wi: 1861]; cont by: brodhead weekly independent) – mf#1047282 – us WHS [071]

Independent – Brodhead WI. 1875 aug 20-1878 jul 19, 1878 jul 26-1881 sep 30 – 2r – 1 – (cont: brodhead independent [brodhead, wi: 1871]; cont by: brodhead independent [brodhead, wi: 1881]) – mf#1047287 – us WHS [071]

Independent – Clinton WI. 1875 jan 13-may 5 – 1r – 1 – (cont by: clinton independent) – mf#963574 – us WHS [071]

Independent – Gays Mills WI. 1937 may 6-1938 may 19, 1941 may 22-jul 17 – 2r – 1 – (cont by: crawford county independent) – mf#1042702 – us WHS [071]

Independent – Juneau WI. [1893 dec 13/1897]-[1959/1962 aug 30] [gaps] – 21r – 1 – (cont by: dodge county independent-news) – mf#1013660 – us WHS [071]
Independent – Cottage Grove, Deerfield WI. [1971 jul 8/1972 jun 8]-[2000 sep-dec] [gaps] – 45r – 1 – (cont: deerfield independent) – mf#938663 – us WHS [071]
Independent – Galesville WI. 1889 jul 12-aug 30 – 1r – 1 – (cont: galesville independent [galesville, wi: 1874]; cont by: galesville independent [galesville, wi: 1890]) – mf#944147 – us WHS [071]
Independent – Galesville WI. 1891 feb 27-1893 jul 28, 1893 aug 4-1894 feb 23 – 2r – 1 – (cont: galesville independent [galesville, wi: 1890]; cont by: galesville independent [galesville, wi: 1894]) – mf#944152 – us WHS [071]
Independent – Lincoln, NE: Nebraska Independent. v14 n24. nov 6 1902-20th yr. apr 25 1907 (wkly) [mf ed filmed 1962?] – 4r – 1 – (cont: nebraska independent (1896). merged with: weekly state journal and: western swine breeder to form: independent farmer and western swine breeder) – us NE Hist [630]
Independent – Durham NC. 1986 jan 17/30-1987 may 6, 1987 may-1988 sep, 1988 oct-1989 apr 20 – 3r – 1 – (cont: north carolina independent; cont by: independent weekly [durham, nc]) – mf#1321129 – us WHS [071]
Independent – Sturgeon Bay WI. 1886 mar 5-dec, 1887-1890 jun 27 – 2r – 1 – (cont: weekly expositor; cont by: republican [sturgeon bay, wi]) – mf#934413 – us WHS [071]
Independent – Corning, NY. 1874-1875 (1) – mf#64937 – us UMI ProQuest [071]
Independent / Darke Co. Hollansburg – jan 7 1904 and feb 15 1905 (only 2iss) – 1r – 1 – mf#B34657 – us Ohio Hist [071]
Independent – East Brunswick, NJ. 1988+ (1) – mf#68399 – us UMI ProQuest [071]
Independent – Evansville, Madison WI. 1878 may 2-1879 dec 11 – 1r – 1 – mf#921872 – us WHS [071]
Independent – Fenton, MI. 1931-1992 (1) – mf#63728 – us UMI ProQuest [071]
Independent – Flint, MI. 1932-1933 (1) – mf#63742 – us UMI ProQuest [071]
Independent – Forsyth, MT. 1923-1974 (1) – mf#64381 – us UMI ProQuest [071]
Independent / Geauga Co. Burton – v1 n1: jan-sep 1884 [wkly] – 1r – 1 – mf#B32792 – us Ohio Hist [071]
Independent – Harvard, IL. 1867-1913 (1) – mf#62628 – us UMI ProQuest [071]
Independent – Havre, MT. 1954-1957 (1) – mf#64446 – us UMI ProQuest [071]
Independent – Hawarden, IA. 1878-1988 (1) – mf#63245 – us UMI ProQuest [071]
Independent / Highland Co. Greenfield – v1 n1. sep 1920-dec 1922 [wkly] – 1r – 1 – mf#B12022 – us Ohio Hist [071]
Independent – Hudson, OH. 1897-1920 (1) – mf#65533 – us UMI ProQuest [071]
Independent – Huntingburg, IN. 1887-1980 (1) – mf#62820 – us UMI ProQuest [071]
Independent – Ingomer, MT. 1923-1927 (1) – mf#64489 – us UMI ProQuest [071]
Independent – Issaquah, WA. 1900-1917 (1) – mf#67013 – us UMI ProQuest [071]
Independent – Johnstown, OH. 1884-1971 (1) – mf#65542 – us UMI ProQuest [071]
Independent – Kenosha WI. 1894 jan 16 – 1r – 1 – mf#876768 – us WHS [071]
Independent – Lavina, MT. 1921-1923 (1) – mf#64521 – us UMI ProQuest [071]
Independent – Libertyville, IL. 1916-1929 (1) – mf#62642 – us UMI ProQuest [071]
Independent / Licking Co. Johnstown – dec 1978-dec 1987 [wkly] – 5r – 1 – mf#B29543-29547 – us Ohio Hist [071]
Independent / Lorain Co. Amherst – jul 1985-dec 1987 [wkly] – 1r – 1 – mf#B33430 – us Ohio Hist [071]
Independent – Manistee, MI. 1879-1880 (1) – mf#63803 – us UMI ProQuest [071]
Independent – Marinette WI. 1886 aug 14-28, sep 16-nov 4 – 1r – 1 – mf#1097870 – us WHS [071]
Independent / Marion Co. Marion – 1865-74, 1876-96 – 12r – 1 – mf#B2574-2586 – us Ohio Hist [071]
Independent – Marshall, MN. 1990+ (1) – mf#68550 – us UMI ProQuest [071]
Independent – Martinsburg, WV. 1874-1899 (1) – mf#67357 – us UMI ProQuest [071]
Independent / Meigs Co. Pomeroy – jul 1908-dec 1910 [semiwkly, wkly] – 2r – 1 – mf#B8580-8581 – us Ohio Hist [071]
Independent – Miles City, MT. 1903-1922 (1) – mf#64559 – us UMI ProQuest [071]
Independent – Moccasin, MT. 1923-1924 (1) – mf#64581 – us UMI ProQuest [071]
Independent / Montgomery Co. Dayton – (1923, 1934, 1937-apr 1953) gaps [mthly, wkly, biwkly] – 4r – 1 – mf#B5445-5448 – us Ohio Hist [071]
Independent / Montgomery Co. Englewood – v1 n1. apr 1975-mar 1981 [wkly] – 6r – 1 – mf#B33523-33528 – us Ohio Hist [071]
Independent / Montgomery Co. Germantown – v1 n1. (apr 1860-mar 1865) scattered [wkly] – 1r – 1 – mf#B5001 – us Ohio Hist [071]
Independent – Moore, MT. 1921-1930 (1) – mf#64584 – us UMI ProQuest [071]
Independent / Muskingum Co. Roseville – 5/5/1897 (1 iss only) – 1r – 1 – mf#B41472 – us Ohio Hist [071]
Independent – Nashua, MT. 1914-1933 (1) – mf#64586 – us UMI ProQuest [071]
Independent – New Cumberland, WV. 1907-1965 (1) – mf#67397 – us UMI ProQuest [071]
Independent – New York. 1848-1928 (1) – mf#4465 – us UMI ProQuest [240]
Independent – New York, NY. 1848-1928 (1) – mf#65080 – us UMI ProQuest [071]
Independent – Noblesville, IN. 1878-1888 (1) – mf#62924 – us UMI ProQuest [071]
Independent – Okanogan, WA. 1907-1975 (1) – mf#67048 – us UMI ProQuest [071]
Independent – Pittsburg, CA. 1928-1944 (1) – mf#62227 – us UMI ProQuest [071]
Independent – Port Orchard, WA. 1973-1976 (1) – mf#67077 – us UMI ProQuest [071]
Independent – Portage WI. 1855 feb 3-1857 apr 14 – 1r – 1 – mf#960222 – us WHS [071]
Independent / Preble Co. Eaton – v1 n1. may 1873-may 1874 [wkly] – 1r – 1 – mf#B32124 – us Ohio Hist [071]
Independent – Richland Center WI. 1872 aug 3-dec 13 – 1r – 1 – mf#966257 – us WHS [071]
Independent / Richland Co. Bellville – (1889-93, 94-95) scattered [wkly] – 1r – 1 – mf#B2919 – us Ohio Hist [071]
Independent – Ringling, MT. 1922-1925 (1) – mf#64630 – us UMI ProQuest [071]
Independent – Saco, MT. 1912-1971 (1) – mf#64637 – us UMI ProQuest [071]
Independent – Shelby, MT. 1901-1904 (1) – mf#64642 – us UMI ProQuest [071]
Independent – Shelton, WA. 1930-1936 (1) – mf#67124 – us UMI ProQuest [071]
Independent – Shepherdstown, WV. 1911+ (1) – mf#67470 – us UMI ProQuest [071]
Independent – Sioux Falls, SD. 1873-1876 (1) – mf#66528 – us UMI ProQuest [071]
Independent – Toronto, ON, 1849-50 – 1r – 1 – cn Library Assoc [071]
Independent – Twin Bridge, MT. 1915-1924 (1) – mf#64673 – us UMI ProQuest [071]
Independent – Vancouver, WA. 1875-1909 (1) – mf#67164 – us UMI ProQuest [071]
Independent – Wapato, WA. 1922-1983 (1) – mf#67173 – us UMI ProQuest [071]
Independent – Wausaukee WI. [1895 oct 26/1897 feb 13]-[1942 jun 26/1943 aug 21, 1946 aug 2] [gaps] – 37r – 1 – mf#948818 – us WHS [071]
Independent – Wautoma WI. 1882 jun 7, jul 28, aug 25, sep 1, oct 6,13,27, dec 8,22; 1883 jan 5, feb 2,16 – 1r – 1 – mf#950063 – us WHS [071]
Independent : weekly republican newspaper – Willoughby, OH. 11 Jan 1917-27 May 1920 – 3r – 1 – us Western Res [071]
Independent – Weston, WV. 1894+ (1) – mf#67510 – us UMI ProQuest [071]
Independent – Whitefish, MT. 1925-1931 (1) – mf#64681 – us UMI ProQuest [071]
Independent – Woonsocket, RI. 1843-1985 (1) – mf#66443 – us UMI ProQuest [071]
Independent – Wyandotte, MI. 1987-1990 (1) – mf#68292 – us UMI ProQuest [071]
Independent – Yakima, WA. 1915-1939 (1) – mf#69280 – us UMI ProQuest [071]
The independent – Freetown. Sierra Leone. -f. Dec 1874-May 1878 – 1r – 1 – uk British Libr Newspaper [079]
The independent – Harrison, NE: Independent Print. Co. v1 n1. sep 1 1892-1893// (wkly) [mf ed with gaps filmed 1975] – 1r – 1 – us NE Hist [071]
The independent – Hazard, NE: M L Whitaker, 1892 (wkly) [mf ed may 27 1892] – 1r – 1 – us NE Hist [071]
The independent – Houston, Tex. : Crawford and Osborne, 1898 [mf ed 1947] – (= ser Negro Newspapers on Microfilm) – 1r – 1 – us L of C Photodup [071]
The independent – Monrovia: [s.n.], nov 29 1954; mar 12 1955 – us CRL [071]
The independent = Nkhani mchichewa – Blantyre: [s.n, [jul 21/27 1993-dec 19 1997] (wkly) – 2r – 1 – us CRL [079]
The independent – oct 3 1859-jan 2 1860, sep 1992-1993, jul 1994-mar 2002 – 1 – mf#ZP6 – nz Nat Libr [079]
The independent – Ord, NE: R H Clayton, 1881 (wkly) [mf ed v1 n19. nov 17 1881 filmed 1973] – 1r – 1 – us NE Hist [071]
The independent – Seoul. An exponent of Korean News. v. 1-3. 7 Apr 1896-1898 – 1 – us NY Public [079]
The independent *see* The daily independent
Independent advertiser – Boston MA. 1748 jan 2-1749 oct 2, dec 5 – 1r – 1 – mf#859851 – us WHS [071]
Independent african / Shepperson, George – Edinburgh, Scotland. 1958 – 1r – us UF Libraries [960]
Independent agent – Philadelphia. 1984-1994 (1) 1984-1994 (5) 1984-1994 (9) – ISSN: 0002-7197 – mf#12495,01 – us UMI ProQuest [360]
Independent american – New Orleans LA. 1957 dec-1961, 1961 dec-1974 dec, 1975-84 – 3r – 1 – (cont: free men speak; solid south) – mf#976319 – us WHS [071]
Independent american – Littleton. 1955-1991 (1) – ISSN: 0019-3666 – mf#3263 – us UMI ProQuest [320]
Independent american / Pickaway Co. Circleville – jun 1837-apr 1838 [wkly] – 1r – 1 – mf#B8037 – us Ohio Hist [071]
Independent american – Platteville WI. 1851 sep 13-1853, 1854-1857 oct 23 – 1r – 1 – mf#960396 – us WHS [071]
Independent american and general advertiser – Platteville WI. 1845 jan 18-1849 jan 6 – 1r – 1 – (cont: american [platteville, wi]) – mf#960395 – us WHS [071]
Independent and advertiser – Huntington, WV. 1873-1876 (1) – mf#67326 – us UMI ProQuest [071]
Independent and free [for voice and piano] / Taylor, Raynor – Philadelphia: printed for aut...[c1796] [mf ed 19–] – 1mf – 9 – (words by mrs rowson) – mf#fiche 953 – us Sibley [780]
Independent and sun *see* [Rio dell-] humboldt independent
Independent (arlington, or) – Arlington OR: C E Hicks [wkly] [mf ed 1969] – 1r – 1 – us Oregon Lib [071]
The independent (auckland) – sep 1992-1993; jul 1994-sep 1998; oct-mar 2001 – 1 – mf#ZP 06 – nz Nat Libr [079]
Independent banker – Sauk Centre. 1950+ (1) 1974+ (5) 1975+ (9) – ISSN: 0019-3674 – mf#9109 – us UMI ProQuest [332]
Independent bulletin – Chicago IL. 1972 jul 20-27 – 1r – 1 – (cont: south side bulletin; cont by: southeast independent bulletin) – mf#1056116 – us WHS [071]
Independent chronicle and boston patriot – Boston MA. 1831 feb 23 – 1r – 1 – (cont: independent chronicle [boston, ma: 1801]; boston patriot and morning advertiser; cont by: boston commercial gazette [boston, ma: semiweekly]; columbian centinel [boston, ma: 1804]; new-england palladium [boston, ma: 1840]; boston semi-weekly advertiser) – mf#859986 – us WHS [071]
Independent chronicle and universal advertiser – Dublin, Ireland. 1 mar 1777 – 1/4r – 1 – uk British Libr Newspaper [072]
Independent citizen – Providence, RI. 1889-1897 (1) – mf#66319 – us UMI ProQuest [071]
The independent comment – Nairobi: [s.n.] 1959 [wkly] [mf ed [Nairobi]: Kenya National Archives Photographic Service 1970] – v14 n1-3 [jan 9 1959-jan 23 1959] on 1r – 1 – (filmed with: kenya comment [v13 n46-52 [1958 nov 14-dec 26] and: independent [nairobi, kenya] [v14 n4-n7 [1959 jan 30-feb 20]) – mf#mf-1550 camp r42 – us CRL [079]
L'independent de woonsocket – Woonsocket, RI. 1935-1942 (1) – mf#66444 – us UMI ProQuest [071]
Independent democrat – North Platte, NE: Chas Purnell. 2v. v20 n21. jun 2 1904-v21 [n26] jul 13 1905 (wkly) [mf ed lacks sep 1 1904] – 1r – 1 – (cont: independent era. absorbed by: lincoln county journal) – us NE Hist [071]
Independent democrat – Elyria, OH. 1852-1877 (1) – mf#65481 – us UMI ProQuest [071]
Independent [dundee, scotland] : or dundee periodical journal of literature and criticism / ed by Mudie, Robert – [Scotland] Dundee: printed by R S Rintoul jan-sep 1816 (qrtly) [mf ed 2004] – 3v on 1r – 1 – uk Newsplan [410]
Independent eagle – v1 n1-2 [1974 aug-campaign] – 1r – 1 – mf#372303 – us WHS [071]
Independent energy – Tulsa. 1989-1999 (1) 1989-1999 (5) 1989-1999 (9) – (cont: independent power) – ISSN: 1043-7320 – mf#9535,02 – us UMI ProQuest [333]
Independent – enterprise news – Edinboro, PA. 1943-2000 (1) – mf#61780 – us UMI ProQuest [071]
Independent era – North Platte, NE: L C Stockton. -v20 n20. may 26 1904 (wkly) [mf ed 1896-1904 (gaps)] – 3r – 1 – (absorbed: wallace herald (1895) and: north platte daily record (1898). cont by: independent democrat) – us NE Hist [071]
Independent era – South Bend IN. 1883 oct 20 – 1r – 1 – mf#857219 – us WHS [071]
An independent examination of the assuan and elephantine aramaic papyri : with eleven plates and two appendices on sundry items / Belleli, Lazare – London: Luzac [dist] 1909 [mf ed 1986] – 1mf – 9 – 0-8370-7283-2 – (discussion in english; texts in aramaic) – mf#1986-1283 – us ATLA [090]
The independent examiner – Beaver Crossing, NE: John H Waterman. 2v. v1 n1. mar 11 1905-v2 n38. nov 24 1906 (wkly) [mf ed mar 11 1905-nov 24 1906 (gaps)] – 1r – 1 – us NE Hist [071]
Independent farmer and western stock breeder – Lincoln, NE: State Journal Co. 6v. 44th yr n37. apr 20 1911-v50 n37. dec 15 1916 (semimthly) [mf ed 1975?] – 3r – 1 – (cont: independent farmer and western swine breeder. absorbed: poultry topics. cont by: nebraska ruralist) – us NE Hist [636]
Independent farmer and western swine breeder – Lincoln, NE: State Journal Co. 5v. 39th yr [n38] may 2 1907-44th yr n36. apr 13 1911 (wkly) [mf ed lacks jan 26 1911 filmed [1975?]] – 2r – 1 – (formed by the union of: independent and: weekly state journal and: western swine breeder. cont by: independent farmer and western stock breeder. 43rd yr n2-44th yr n1 not publ) – us NE Hist [636]
Independent florida alligator – Gainesville, FL. v6-93 n138. 1917 oct 10-2000 aug 10 – 149r – us UF Libraries [071]
The independent forester and forester's herald – [London, Ont?]: Independent Order of Foresters, [1880-1930] – 9 – mf#P04284 – cn Canadiana [634]
Independent (georgia edition) – Anderson, SC. 1974-1981 (1) – mf#68034 – us UMI ProQuest [071]
Independent Greek Church (Canada) *see* Khrystyianskyy katekhyzm dlia uzhytku shkilnykh ditei i molodezhy
Independent (haringey wood green etc) *see* Haringey independent
Independent herald – Bertrand, NE: L E Brown, -sep 28 1928// (wkly) [mf ed 5th yr n5. dec 3 1892-sep 21 1928 (gaps)] – 7r – 1 – (cont by: bertrand herald) – us NE Hist [071]
Independent herald – Hinton, WV. 1903-1976 (1) – mf#67318 – us UMI ProQuest [071]
Independent herald – Hinton, WV. 1909-1919 (1) – mf#67319 – us UMI ProQuest [071]
Independent herald – Johnsonville, NZ. 1974-88 – 17r – 1 – mf#41.9 – nz Nat Libr [079]
Independent herald – Pineville, WV. 1941+ (1) – mf#67433 – us UMI ProQuest [071]
Independent (hillsboro, or) – Hillsboro OR: [s,n,] [wkly] – 1 – (cont: washington county independent (hillsboro, or); cont by: hillsboro independent) – us Oregon Lib [071]
The independent hindustan – San Francisco: Hindustan Gadar Party. v1 n1-11. sep 1920-1921 – 1 – us CRL [954]
Independent idea – v1 n7 [1973 jul 25] – 1r – 1 – mf#372297 – us WHS [071]
Independent india and a new world order / Krishnamurti, Y G – Bombay: Popular Book Depot, 1943 – (= ser Samp: indian books) – (int by k m munshi; foreword by s srikantha sastri) – us CRL [954]
Independent irishman – Dublin, Ireland. 16 nov 1770, 11 jan, 8 apr 1771 – 1/2r – 1 – (aka: dublin evening post) – uk British Libr Newspaper [072]
Independent jewish press service – New York. N.Y. 1941-47 – 1 – us AJPC [071]
Independent journal – Chilton WI. 1916 dec 7-1918, 1919-21, 1922-24, 1925-27, 1928-29, 1930-1933 feb 23 – 6r – 1 – mf#960667 – us WHS [071]
Independent [kirkcaldy, scotland] – [Scotland] Fife, Kirkcaldy: J C Robertson 6 apr 1861-8 feb 1862 (wkly) [mf ed 2004] – 1r – 1 – uk Newsplan [072]
Independent Labor League of America *see*
– Revolutionary age
– Workers age
Independent Labour Party. Great Britain *see* Weekly notes for speakers, 1926-31
Independent labour party newspapers : from bradford central library – 4r – 1 – (comprising: bradford labour echo 1895-99. forward 1904-1908. west bradford gazette 1905-06. keighley labour journal 1894-1902) – mf#97017 – uk Microform Academic [325]
Independent mail – Anderson, SC. 1981-2000 (1) – mf#61822 – us UMI ProQuest [071]
Independent mechanic – New York. -w. 6 Apr 1811-26 Sep 1812. (45 ft) – 1 – uk British Libr Newspaper [071]
Independent messenger – Emporia, VA. 1988-2000 (1) – mf#66702 – us UMI ProQuest [071]
An independent monthly review *see* Fighting talk, 1954-62, johannesburg
The independent [nairobi, kenya] – Nairobi: [s.n.] 1959- [wkly] [mf ed [Nairobi]: Kenya National Archives Photographic Service 1970] – v14 n5-7 [jan 30-feb 20 1959] on 1r – 1 – (formerly comment; filmed with: kenya comment [v13 n46-52 [1958 nov 14-dec 26] and: independent commment) – mf#mf-1550 camp r42 – us CRL [079]
Independent news / Franklin Co. Gahanna – jul-aug 1973, jun 1974-oct 1975 [wkly] – 1r – 1 – mf#B29844 – us Ohio Hist [071]
Independent news – Girard KS. 1896 may 18-1897 sep 30, 1897 oct 7-1899 mar 30, 1899 apr 6-1900 jul 26, 1900 aug 2-1901 dec 26, 1902 jan 2-1903 apr 30, 1903 may 7-1904 sep 29, 1904 oct 6-1906 feb 22, 1906 mar 1-1907 jun 27, 1907 jul 4-1909 aug 14 – 9r – 1 – mf#880926 – us WHS [071]

INDEX

Independent news – Racine WI. 1932 sept 2-16, 30, oct 7-14 – 1r – 1 – mf#966620 – us WHS [071]
Independent news / Richland Co. Shelby – v1 n1. nov 1868-nov 1876 [wkly] – 3r – 1 – mf#B16054-16056 – us Ohio Hist [071]
Independent news – New Richmond, OH, jan 2 1913-jan 29 1914 (scattered) – 1r – 1 – (weekly newspaper) – us Western Res [071]
Independent news / Weirton WV – (nov 1950-mar 1980) scattered [irreg] – 1r – 1 – mf#B10349 – us Ohio Hist [331]
Independent observer – Beckley, WV. 1936-1941 (1) – mf#67197 – us UMI ProQuest [071]
Independent observer – Conrad, MT. 1923-1974 (1) – mf#64340 – us UMI ProQuest [071]
Independent observer – Scottsdale, PA. 1882-1925 (1) – mf#66072 – us UMI ProQuest [071]
Independent oil workers news – v5 n10-v10 n11 [1964 oct-1969 dec] – 1r – 1 – mf#1058742 – us WHS [622]
Independent oiler / International Union of Petroleum Workers – 1947 feb 15-1960 dec, 1961-1962 jan – 2r – 1 – (cont by: iupw views) – mf#1060551 – us WHS [622]
The independent on sunday *see* The independent / the independent on sunday
Independent or democratic church government : the divinely appointed constitution of the churches of our lord and saviour jesus christ / Slaysman, George Major – Philadelphia: SA George, c1868 – 1mf – 9 – 0-524-01469-8 – mf#1990-0418 – us ATLA [240]
Independent Order of Odd Fellows *see* International odd fellow
Independent Order of Good Templars *see*
– Proceedings of the...annual session of the grand lodge of wisconsin, i o g t
– Wisconsin good templar
– Wisconsin Good Templar
Independent Order of Good Templars. Grand Lodge of Manitoba and N.W.T. *see* The manitoba good templar
Independent Order of Good Templars of Canada. Grand Temple *see*
– Constitution of the grand and subordinate temples of the independent order of good templars of canada
– Constitution of the...good templars of canada
Independent Order of Odd Fellows *see*
– I o o f proceedings of the grand lodge of the state of wisconsin, annual session
– Journal of proceedings of the...annual communication of the sovereign grand lodge of the independent order of odd fellows
– Proceedings annual communication of the grand lodge of the united states
– Proceedings of the right worthy grand lodge of the state of wisconsin at its...annual session
– Proceedings of the...annual communication of the right worthy grand lodge of the united states of the independent order of odd fellows
– Proceedings of the...annual communication of the sovereign grand lodge of the independent order of odd fellows
– Wisconsin odd fellow
Independent Order of Odd Fellows. Eureka Lodge *see* Constitution, by-laws, rules of order, etc, no 30
Independent Order of Odd Fellows. Lynden Lodge, No 259 (Ont) *see* Constitution, by-laws, rules of order, etc...
Independent Order of Odd Fellows. Valley City Lodge, No 117 (Dundas, Ont) *see* Constitution, by-laws, rules of order etc...
Independent Order of Rechabites *see* Rechabite and family instructor
Independent Order of Vikings *see* Viking journal
Independent peoples tribunal newsletter – [v1 n1]-4 [1970 jun-1971 apr ?] – 1r – 1 – mf#1058743 – us WHS [071]
Independent power – Milaca. 1988-1989 (1,5,9) – (cont: alternative sources of energy; cont by: independent energy) – ISSN: 1042-5829 – mf#9535,01 – us UMI ProQuest [333]
Independent press – Castries, Saint Lucia. 1843-1844 (1) – mf#67957 – us UMI ProQuest [079]
Independent press – Waukesha WI. 1853 jul 20-77, 1853 oct 12-1854 feb 15 – 1r – 1 – (cont by: waukesha county democrat [waukesha, wis. : 1854]) – mf#945851 – us WHS [071]
Independent press / Huron Co. Wakeman – apr 1904-jun 1911 [wkly] – 2r – 1 – mf#B29892-29893 – us Ohio Hist [071]
Independent press – Madison WI. 1873 feb 7 – 1r – 1 – mf#916987 – us WHS [071]
Independent press / Montgomery Co. Germantown – feb 1874-feb 1876 [wkly] – 1r – 1 – mf#B5459 – us Ohio Hist [071]
Independent publisher – Traverse City. 1998+ (1,5,9) – (cont: small press) – ISSN: 1098-5735 – mf#13458,01 – us UMI ProQuest [070]
Independent reflector – 1752 nov 20-1753 nov 22 – 1r – 1 – mf#795012 – us WHS [071]
Independent reflector : or, weekly essays on sundry important subjects – New York. 1752-1753 (1) – mf#3525 – us UMI ProQuest [200]
Independent register / Columbiana Co. Columbiana – v1 n1. 4/1870-9/79,2/81-10/83,1/84-7/1896 [wkly] – 9r – 1 – mf#B7965-7973 – us Ohio Hist [071]
An independent report on the belau plebiscite of 1984 – Koror: Belau Pacific Center, nov 1984 – (= ser Republic Of Belau (Palau) – Status Negotiations With The U.S.) – 1mf – 9 – $1.50 – mf#LLMC 82-100G, Title 29 – us LLMC [323]
Independent republican : and miscellaneous magazine – Newburyport. 1805-1805 (1) – mf#3581 – us UMI ProQuest [320]
Independent republican / Belmont Co. Saint Clairsv – (mar 1856-jul 1862) fire damaged [wkly] – 2r – 1 – mf#B210-211 – us Ohio Hist [071]
Independent republican / Ross Co. Chillicothe – dec 1809-sep 1811 [wkly] – 1r – 1 – mf#B1225 – us Ohio Hist [071]
Independent school – Boston. 1976+ (1) 1976+ (5) 1976+ (9) – (cont: independent school bulletin) – ISSN: 0145-9635 – mf#7151,01 – us UMI ProQuest [370]
Independent school bulletin – Milton. 1934-1976 (1) 1972-1976 (5) (9) – (cont by: independent school) – ISSN: 0019-3755 – mf#7151 – us UMI ProQuest [370]
Independent series / Morrow Co. Cardington – 1872-feb 1875, apr 1876-88 [wkly] – 7r – 1 – mf#B9191-9197 – us Ohio Hist [071]
Independent shavian – New York. 1953+ (1) 1970+ (5) 1976+ (9) – ISSN: 0019-3763 – mf#3237 – us UMI ProQuest [400]
Independent Skilled Trades Council *see* Skilled tradesman
Independent socialist – n1-12. 1969 – (= ser International Socialist, Workers' Power) – 1 – (superseded by: international socialist, workers' power) – us AMS Press [325]
Independent (south carolina edition) – Anderson, SC. 1974-1981 (1) – mf#68035 – us UMI ProQuest [071]
Independent star / Richland Co. Bellville – may 4-jul 13 1889 [wkly] – 1r – 1 – mf#B2919 – us Ohio Hist [071]
The independent / the independent on sunday – Great Britain, 1985- mthly updates – 1 – (the independent on sunday is incl fr 1990 onwards) – us Primary [072]
Independent times – New Paltz, NY. 1868-1972 (1) – mf#65048 – us UMI ProQuest [071]
Independent times – Pardeeville WI. 1889 jan 26, mar 16,30, apr 6 – 1r – 1 – mf#1001539 – us WHS [071]
Independent tmc – Elkhorn WI. 1986 aug 11-date – 1r – 1 – mf#1107291 – us WHS [071]
Independent tribune – Hastings, NE: A H Brown & Co, 1891-92// (wkly) [mf ed v6 n50. jun 17-oct 14 1892 (gaps) filmed 1969] – 1r – 1 – (formed by the union of: hastings weekly independent and: hastings tribune (1886). cont by: hastings independent tribune) – us NE Hist [071]
Independent tribune – Hinsdale, MT. 1971-1974 (1) – mf#64478 – us UMI ProQuest [071]
Independent union / Connecticut Employees Union Independent – 1976 oct-1991 nov – 1r – 1 – (cont by: independent union news) – mf#1073109 – us WHS [331]
Independent (vernonia, or) – Vernonia OR: Dirk & Noni Anderson, 1986- [semimthly] – 1 – us Oregon Lib [071]
Independent villager – Marathon, NY. 1980-1987 (1) – mf#65021 – us UMI ProQuest [071]
Independent virginian – Chesterfield, VA. 1972-1974 (1) – mf#66686 – us UMI ProQuest [071]
Independent voice / Christian Citizens Crusade – v33 n12 [1982 aug 25], v34 n2-12 [1982 dec 25-1984 oct 20], v35 n1-12 [1985 feb 22-1987 dec] – 1r – 1 – (cont: militant truth) – mf#1799624 – us WHS [240]
Independent weekly – Durham NC. 1989 apr 27/may 3-1989 aug, 1989 sep-1990 aug, 1990 aug 29-1991 jun 26, 1992 jan 2-jun 24, 1992 jul 1/7-dec 23, 1993 jan 6-12/-jun 30, 1993 jul 7/13-dec 22 – 7r – 1 – (cont: independent [durham, nc]) – mf#5258896 – us WHS [071]
Independent whig – London. 1808-11. (1811 imperfect). -w. 1 reels – 1 – uk British Libr Newspaper [072]
Independent woman – v2 n3-4 [1987/88 win-winter interim], v2 n6 [1988 sum], v3 n1-2 [1988 summer interim-fall], v3 n3-4 [1988/89 win-winter interim], v3 n5-6 [1989 spr-spring interim], v4 n1-2 [1989 summer-summer interim], v4 n3-4 [1989 fall-winter, v4 n5-6 [1990 spr-summer]; special iss – 1r – 1 – (cont: florida womans world; cont by: radical feminist) – mf#2801796 – us WHS [305]
O independente – Hong Kong: E Ferreira, sep 1869-oct 1897* – 1r – 1 – us CRL [079]
O independente – Rio de Janeiro, RJ: Typ de Thomas B Hunt, 03 maio 1831-22 abr 1833 – (= ser Ps 19) – mf#P02,04,23-24 – bl Biblioteca [320]
Independentista – Miami, FL. 1989 may-1991 jan – 1r – us UF Libraries [071]
Independent-leader *see* [Sacramento-] sacramento tribune-progress
Independent-register – Brodhead, Juda WI [1909 jul 21/dec 29]-[1999 jul/dec] [gaps] – 86r – 1 – (cont: brodhead register; brodhead independent [brodhead, wis. : 1881]; brodhead news; juda community news) – mf#1047293 – us WHS [071]
The Independent-reporter – Skowhegan: Independent-Reporter Co, 1913-51 – 24r – 1 – us CRL [071]
Independientes de color / Portuondo Linares, Serafin – Habana, Cuba. 1950 – 1r – us UF Libraries [972]
Inder-enterprise press – 1952 feb 21-dec 31, 1953-1954 dec 29 – 2r – 1 – (cont: reminder-regional press; cudahy enterprise; cont by: reminder-enterprise [cudahy, wis. : 1955]) – mf#1001953 – us WHS [071]
Indermaur, John *see* An epitome of leading common law cases
Inderwick, Frederick Andrew *see* The interregnum (a.d. 1648-1660); studies of the commonwealth, legislative, social, and legal
Indeterminateness of unauthorized baptism / Warren, C – Cambridge, England. 1841 – 1r – us UF Libraries [242]
Index / Baptist Southern Convention – 1845-1953 – 1 – (1954-65. 5.00; 1) – us Southern Baptist [242]
Index – Petersburg, NE: C L Meyes. -v8 n10. oct 6 1898 (wkly) [mf ed 1892-98 (gaps)] – 1r – 1 – (cont by: petersburg index) – us NE Hist [071]
Index – Delafield, Dousman, Eagle etc WI [1970 jun 4/1971 apr 15]-[1992 jan-dec] [gaps] – 29r – 1 – (cont: dousman index; cont by: kettle moraine index) – mf#999844 – us WHS [071]
Index – Davison, MI. 1932+ [1] – mf#63716 – us UMI ProQuest [071]
Index – Endicott, WA. 1937-1947 (1) – mf#66990 – us UMI ProQuest [071]
Index – Fairmont, WV. 1874-1907 (1) – mf#67276 – us UMI ProQuest [071]
Index – Mitchell, NE: Bryce and Maxine Wilkins. v70 n22. oct 1 1970- (wkly) [mf ed filmed 1972-] – 1 – (formed by the union of: mitchell index and: morrill mail. absorbed: lyman nebraska leader (1971). cont the numbering of: mitchell index. issues for oct 1 1970-apr 6 1972 also called v63 n18-v67 n34. some irregularities in numbering) – us NE Hist [071]
Index – Greenwood, SC. 1897-1909 (1) – mf#66497 – us UMI ProQuest [071]
Index – Ingomer, MT. 1914-1918 (1) – mf#64490 – us UMI ProQuest [071]
Index – Mineral Wells, TX. 1977-1985 (1) – mf#66637 – us UMI ProQuest [071]
Index – Pittsburgh, PA. 1900-1929 (1) – mf#66040 – us UMI ProQuest [071]
Index – Racine WI. 1907 sep-1908 jul/aug – 1r – 1 – mf#966616 – us WHS [071]
Index – Wausaukee WI. 1904 aug 18-dec 16 – 1r – 1 – mf#948816 – us WHS [071]
Index : a weekly journal of politics, literature and news devoted to the exposition of the mutual interests, political and commercial, of great britain and the confederate states of america, 1862-1865 – (= ser Civil war journals) – 2r – 1 – £190.00 – (with d/g) – uk Matthew [976]
The index – London. -w. May 1862-Aug 1865. (2 reels) – 1 – (a weekly journal devoted to the exposition of the mutual interests of great britain and the confederate states of america) – uk British Libr Newspaper [072]
The index : a commercial and literary monthly journal / Bryant, Stratton and Odell's Business College – Toronto: [s.n, 1866-18– or 19–] – 9 – mf#P05984 – cn Canadiana [650]
The index : a weekly paper devoted to free religion – v1-18 old and new series. 1870-86 – 1 – us AMS Press [240]
Index alphabetique des noms de 3400 familles de douze enfants vivants : reconnues officiellement depuis l'origine de la loi mercier, en 1890, jusqu'a mars 1904 inclusivement / Dumais, A – Quebec: Departement des Terres, Mines et Pecheries: Departement des Terres et Forets. 2v. 1904-1906 [mf ed 1989] – 4mf – 9 – mf#SEM105P1159 – cn Bibl Nat [304]
An index and annotated bibliography of non-book materials in the john steinbeck library, salinas, c – (= ser John steinbeck coll) – 1r – 1 – $50.00 – mf#B40502 – us Library Micro [420]
Index and calendar to the papers of sir joseph paxton : from the archives of the duke of devonshire, chatsworth, derbyshire – 1r – 1 – mf#96774 – uk Microform Academic [520]
Index and catalogue to moravian archives : from muswell hill moravian church, london – 1947 – 1r – 1 – mf#96800 – uk Microform Academic [240]
Index and legislative history : uniform code of military justice, 1950 / U.S. Army Court of Military Appeals – 1950; 1985. n.p,n.d. – 34mf – 9 – $51.00 – mf#LLMC 87-320A – us LLMC [324]
Index and registers of substitute mail carriers in first- and second-class post offices, 1885-1903 / U.S. Post Office – (= ser Post Office Records) – 1r – 1 – mf#M2076 – us Nat Archives [380]
Index apologeticus : sive, clavis iustini martyris operum: aliorumque apologetarum pristinorum / Goodspeed, Edgar Johnson – Leipzig: JC Hinrichs, 1912 [mf ed 1990] – 1mf – 9 – 0-7905-5148-9 – (ind in greek. pref in latin) – mf#1988-1148 – us ATLA [240]
Index bibliographique colonial : congo belge et ruanda-urundi / ed by Heyse, Theodore – Bruxelles: Falk, [1937-40] – 1 – us CRL [960]
Index bibliorvm... / Pellican, C – Tigvri, officina Froschoviana, 1537 – 1mf – 9 – mf#PBU-613 – ne IDC [240]
Index books, 1789-1928, and minutes and bench dockets, 1789-1870 for the u.s. district court, southern district of georgia / U.S. District Court – (= ser Records of district courts of the united states) – 3r – 1 – (with printed guide) – mf#M1172 – us Nat Archives [347]
Index by district to us coast guard reports of assistance, 1917-1938 / U.S. Coast Guard – (= ser Records of the united states coast guard) – 19r – 5 – mf#T919 – us Nat Archives [360]
Index by floating unit to us coast guard reports of assistance, 1917-1935 / U.S. Coast Guard – (= ser Records of the united states coast guard) – 5r – 5 – mf#T921 – us Nat Archives [360]
Index by station to us coast guard reports of assistance, 1924-1938 / U.S. Coast Guard – (= ser Records of the united states coast guard) – 9r – 5 – mf#T920 – us Nat Archives [360]
Index canonum : the greek text, an english translation and a complete digest of the entire code of canon law of the undivided primitive church / Fulton, John – 3rd ed. New York: Thomas Whittaker, 1892, c1883 [mf ed 1989] – 1mf – 9 – 0-7905-4583-7 – mf#1988-0583 – us ATLA [240]
Index cards to bankruptcy, civil, and criminal case files of the u.s. district court for the southern district of california, southern division (san diego), 1953 to 1954 / U.S. District Court – (= ser Records of district courts of the united states) – 1r – 1 – mf#M1741 – us Nat Archives [347]
Index cards to civil and criminal case files of the us district court for the southern district of california, southern division (san diego), july 1962 to august 1966 / U.S. District Court – (= ser Records of district courts of the united states) – 2r – 1 – mf#M1736 – us Nat Archives [345]
Index cards to civil case files of the us district court for the southern district of california, southern division (san diego), january 1955 to june 1962 / U.S. District Court – (= ser Records of district courts of the united states) – 1r – 1 – mf#M1735 – us Nat Archives [347]
Index cards to overseas military petitions of the u.s. district court for the southern district of california, central division (los angeles), 1943-1945, 1954, 1955-1956 / U.S. District Court – (= ser Records of district courts of the united states) – 2r – 1 – mf#M1606 – us Nat Archives [347]
Index der antiken kunst und architektur : denkmaeler des griechisch-roemischen altertums in der photosammlung des deutschen archaeologischen instituts in rom = Index of ancient art and architecture. monuments of greek and roman cultural heritage in the photographic collection of the german archaeological institute in rome / ed by Deutsches Archaeologisches Institut. Rome – [mf ed 1988-90] – 2714mf (1:24) – 9,17 – silver €7900.00 – ISBN-10: 3-598-32070-1 – ISBN-13: 978-3-598-32070-5 – gw Saur [700]
Der index der verbotenen buecher : ein beitrag zur kirchen- und literaturgeschichte / Reusch, Franz Heinrich – Bonn: M Cohen, 1883-1885 – 5mf – 9 – 0-524-02899-0 – (incl bibl ref) – mf#1990-4490 – us ATLA [410]
Index deutschsprachiger zeitschriften : autoren-, schlagwort- und rezensionenregister zu deutschsprachigen zeitschriften 1750-1815 / Schmidt, Klaus [comp] – Hildesheim 1989 [mf ed Hildesheim 1995-98] – 28mf – 9 – diazo €498.00 – gw Olms [014]
Index du bulletin des recherches historiques / Roy, Antoine – [mf ed 1988] – 9mf – 9 – mf#SEM105P682 – cn Bibl Nat [971]
Index expurgatorious anglicanus : or a descriptive catalogue of the principal books printed or published in england, which have been suppressed or burnt by the common hangman, or censured, or for which the authors, printers, or publishers have been prosecuted / Hart, W H – London: John Russell Smith, 1872 – 4mf – 9 – $6.00 – mf#LLMC 92-105 – us LLMC [320]

Index filicum / Christiansen, Carl – v1-2. 1753-1912 – 34mf – 7 – mf#2159 – uk Microform Academic [580]

Index for assembly journal (newfoundland) – 1866-1900 – 1r – 1 – cn Library Assoc [971]

Index general des statuts de la province de quebec de 1899 a 1928 inc / Quebec. (Province). Laws, Statutes, etc – Quebec: Les Editions Themis, 1928. 222p. LL-2385 – 1 – us L of C Photodup [348]

An index in two parts to evangelical sectarians in missionerskoe obzrenie / Wardin, Albert – 1896-1916. 42p – 1 – us Southern Baptist [242]

Index journal – Greenwood, SC. 1919-2000 (1) – mf#61827 – us UMI ProQuest [071]

Index karangan-karangan dalam bidang ekonomi pertanian di indonesia dan bidang lainnja jang penting untuk servey agro ekonomi – Bogor, 1966-1972 – 11mf – 9 – (missing: 1966, v1; 1967, v2; 1968, v3(2-4); 1969, v4(1-4)) – mf#SE-150=8 – ne IDC [959]

Index kewensis plantarum phanerogamarum : supplementum – Bruxellis, Belgium. v1-10. 1901-47 – 1 – $120.00 – mf#0275 – us Brook [580]

Index, lectures et morale evangelique / Laberge, Joseph-Esdras – [Quebec: s.n.] 1914 [mf ed 1999] – 1mf – 9 – 0-665-97424-8 – mf#97424 – cn Canadiana [240]

Index library / British Record Society Ltd – v1-88. 1888-1976 – (= ser Publications of the english record societies, 1835-1972) – 387mf – 9 – uk Chadwyck [941]

Index librorum prohibitorum – [Vatican-City]: Typis Polyglottis Vaticanis, 1948 – 2mf – 9 – 0-8370-6941-6 – (bibliography in various languages; introduction in latin) – mf#1986-0941 – us ATLA [012]

Index librorum prohibitorum *see* Die indices librorum prohibitorum des sechzehnten jahrhunderts

Index librorum prohibitorum (1559) – Romae: Ex Officina Saluiana, 15 Feb 1559 – 1mf – 9 – $1.50 – (1980 reprint by the houghton library) – mf#LLMC 91-056 – us LLMC [348]

Index librorum prohibitorum (leo 13) – Romae: S C de Propaganda Fide, 1831 – 5mf – 9 – $7.50 – mf#LLMC 91-055 – us LLMC [348]

Index londinensis to illustrations of flowering plants, ferns and fern allies – 8v. 1929-31 – 1 – $180.00 – mf#0276 – us Brook [580]

Index materiarum quae in singulis... / Lopez de Tovar, Gregorio – Madrid: Juan Hefray, 1611 – 1 – sp Bibl Santa Ana [020]

Index medicus – Bethesda. 1988-1996 (9) – ISSN: 0019-3879 – mf#16307 – us UMI ProQuest [610]

Index nama penulis dalam kompas / Biro Dokumentasi Pers "Media" – Bandung, 1968 – 1mf – 9 – mf#SE-150=9 – ne IDC [950]

The index of american design (tiam) / Washington, DC. National Gallery of Art – America from settlement to 1900 [mf ed Chadwyck-Healey, 1978] – 10pts on 291mf – 15 – (coll covers every aspect of the decorative, folk & popular arts. coll divided into 10pt: pt1: textiles, costume & jewelry 57mf, with catalog on 2mf. pt2: the art & design of utopian & religious communities 40mf. pt3: architecture & naive art 21mf. pt4: tools, hardware, firearms & vehicles 16mf. pt5: domestic utensils 30mf. pt6: furniture & decorative accessories 35mf. pt7: wood carvings & weathervanes 25mf. pt8: ceramics & glass 35mf. pt9: silver, copper, pewter & toleware 11mf. pt10: toys & musical instruments 21mf. each is accompanied by its own printed catalogue. complete coll is also accompanied by a consolidated catalogue & manual on b/w mf isbn: 0-914146-94-7) – uk Chadwyck [740]

Index of births, deaths and marriages registered in the northern territory : births for 1903-1918 – deaths and marriages 1903-1913 / Northern Territory Registrar of Birth, Deaths and Marriages – 6mf – 9 – A$33.00 – at Northern [980]

Index of births, deaths and marriages registered in the northern territory for the periods 1870-1902 / Northern Territory Registrar of Birth, Deaths and Marriages – 5mf – 9 – A$55.00 – (births: surname, given names, sex, date, father, mother's maiden name, entry number, remarks [isbn: 0-949124-58-3]; deaths: surname, given name, age, sex, date, trade, residence, place of death, remarks, entry number [isbn: 0-949124-59-1]; marriage: groom, bride, date, entry number, remarks [isbn: 0-949124-60-5]) – at Northern [980]

Index of boats mentioned in northern territory newspapers : november 1873-december 1914: name of boat, date of arrival – [mf ed 1989] – 1mf – 9 – A$5.50 – 0-949124-55-9 – at Northern [980]

Index of cbi-ibt, poa, and a-p opinions / U.S. Army. Judge Advocate General. Board of Review – Washington: JAG Office, Military Justice Div, 1952 – 2mf – 9 – $3.00 – mf#LLMC 84-226 – us LLMC [348]

The index of current events / ed by Dalby, Henry – Montreal: H. Dalby, [1888-1891] – 9 – mf#P04303 – cn Canadiana [321]

The index of current events, 1889 : being an index to the dates of the principal events throughout the world which have attracted public attention during the year / ed by Dalby, Henry – Montreal: H Dalby, 1889 – 2mf – 9 – mf#06252 – cn Canadiana [321]

Index of decisions of the nlrb – classified index of the nlrb decisions and related court cases / U.S. National Labor Relations Board – (= ser National labor relations board decisions) – 262mf – 9 – $393.00 – (ind 15bks v1-177. classified ind v227-313 jan 1977-may 1994. lacking: v178-212. updates planned) – mf#LLMC 80-207 – us LLMC [331]

Index of dermatology – Washington. 1972-1979 (1) 1972-1979 (5) 1972-1979 (9) – ISSN: 0090-1245 – mf#7359 – us UMI ProQuest [616]

Index of european constitutions 1850 to 2003 *see* Constitutions of the world 1850 to the present, pt 1

Index of finnish newspapers 1771-1890 : index to microfiche – Helsinki: Helsinki University Library, 1990 – 345mf – 9 – fi Helsinki [020]

Index of flora aegyptiaco-arabica and herbarium forskilii / Christensen, C F A – Kobenhavn, 1917. MS – 4mf – 8 – mf#2205 – ne IDC [580]

Index of indonesia learned periodicals indeks madjalah ilmiah / Lembaga Ilmu Pengetahuan Indonesia – Djakarta, 1959-1968 – 10mf – 9 – (missing: 1959, v1; 1960, v2; 1961, v3; 1963, v5(1, 2, 4); 1964, v6(1)) – mf#SE-777 – ne IDC [959]

Index of materials at beginning, associational church letters at end / Falls Church. Virginia. Columbia Baptist Church – 9books, 1918-1977 – 1 – us Southern Baptist [242]

Index of microfilmed records of the german foreign ministry and the reich's chancellery covering the weimar period / Germany. Foreign Ministry – (= ser National archives coll of foreign records seized, 1941-) – 1r – 1 – mf#T407 – us Nat Archives [943]

Index of names contained in census returns of ashtabula co, ohio for 1870 : vol 1: a-h / The Western Reserve Historical Society – 1937 (mf ed 1974) – 1r – 1 – (filmed by the genealogical society of utah, 1974) – us Western Res [978]

Index of names contained in census returns of ashtabula co, ohio for 1870 : vol 2: i-z / The Western Reserve Historical Society – 1937 (mf ed 1974) – 1r – 1 – (filmed by the genealogical society of utah, 1974) – us Western Res [978]

Index of names contained on census returns of portage co, ohio for 1850 / The Western Reserve Historical Society – 1933 (mf ed 1974) – 1r – 1 – (filmed by genealogical society of utah, 1974) – us Western Res [978]

Index of noteworthy words and phrases : found in the clementine writings commonly called the homilies of clement – London, New York: Macmillan, 1893 [mf ed 1990] – 1mf – 9 – 0-7905-8081-0 – (ind in greek. pref in english) – mf#1988-6062 – us ATLA [240]

Index of people in the northern territory times and gazette : november 1873-december 1914: name, date of paper, type of article – [mf ed 1989] – 13mf – 9 – A$55.00 – 0-949124-53-2 – at Northern [920]

Index of people mentioned in the northern australian : june 1883-may 1890: name, date of paper – [mf ed 1989] – 3mf – 9 – A$16.50 – 0-949124-56-7 – at Northern [920]

Index of people mentioned in the northern territory government gazette : nov 1883-dec 1914 – [mf ed 1989] – 3mf – 9 – A$16.50 – 0-949124-57-5 – at Northern [980]

Index of presbyterian ministers : containing the names of all the ministers of the presbyterian church in the united states of america / Beecher, Willis Judson & Beecher, Mary A – Philadelphia: Presbyterian board of publication, [c 1883]. Chicago: Dep of Photodup, U of Chicago Lib, 1974 (1r); Evanston: American Theol Lib Assoc, 1984 (1r) – 1 – 0-8370-0000-9 – mf#1984-B397 – us ATLA [242]

An index of the cases overruled, reversed, denied, doubted, modified, limited, explained, and distinguished : by the courts of america, england, and ireland / Bigelow, Melville M – Boston, Little, Brown, 1873. 566 p. LL-344 – 1 – (with ind boston 1887 190p) – us L of C Photodup [347]

Index of the monthly issues of the westerners brand book – Denver, CO: The Westerners, 1946 (mf ed 1954) – 1r – 1 – mf#MF W525c – us Colorado Hist [020]

Index of title pages to pamphlet literature in the archives of the archbishop of westminster – 2r – 1 – mf#2420 – uk Microform Academic [241]

Index of veterinary specialities – Epsom. 1975-1978 (1) 1975-1978 (5) 1975-1978 (9) – ISSN: 0019-3941 – mf#10604 – us UMI ProQuest [636]

Index on censorship – [mf ed 1986] – 93mf – 9 – $650.00 – 0-907716-20-2 – (covers 63 eds fr 1972-84) – uk Mindata [360]

Index over indische onderwerpen in nederlandse couranten, 1843-1947 / Netherlands. General State Archives – 1843-1947 – (= ser Index To Colonial Subjects In Dutch Newspapers, 1843-1947) – 185mf – 9 – ne MMF Publ [324]

Index patristicus : sive, clavis patrum apostolicorum operum: ex editione minore gebhardt, harnack, zahn, lectionibus editionum minorum funk et lightfoot admissis / Goodspeed, Edgar Johnson – Leipzig: J C Hinrichs, 1907 [mf ed 1990] – 1mf – 9 – 0-7905-5220-5 – (ind in greek. pref in latin) – mf#1988-1220 – us ATLA [240]

Index photographique de l'art en france = Photographic documentation of art in france / ed by Bildarchiv Foto Marburg – Deutsches Dokumentationszentrum fuer Kunstgeschichte Philipps-Universitaet Marburg – [mf ed 1979-81] – 976mf (1:24) – 9 – silver €3660.00 – ISBN-10: 3-598-30160-X – ISBN-13: 978-3-598-30160-5 – gw Saur [700]

Index seu repertorium...septem partitarum / Lopez de Tovar, Gregorio – 1588 – 9 – sp Bibl Santa Ana [946]

Index (soundex) to naturalization petitions filed in federal, state, and local courts in new york, new york, including new york, kings, queens, and richmond counties, 1792-1906 / U.S. Circuit and District Courts – (= ser Records of district courts of the united states) – 294r – 1 – mf#M1674 – us Nat Archives [347]

Index (soundex) to passenger lists of vessels arriving at baltimore, md, 1897-1952 – (= ser Records Of The Immigrations And Naturalization Service, 1891-1957) – 43r – 5 – mf#T520 – us Nat Archives [975]

Index (soundex) to passenger lists of vessels arriving at baltimore, md, (city passenger lists), 1833-1866 – (= ser Records of the united states customs service, 1820-c1891) – 22r – 5 – (with printed guide) – mf#M326 – us Nat Archives [975]

Index (soundex) to passenger lists of vessels arriving at baltimore, md (federal passenger lists), 1820-1897 – (= ser Records of the united states customs service, 1820-c1891) – 171r – 5 – (with printed guide) – mf#M327 – us Nat Archives [975]

Index (soundex) to passenger lists of vessels arriving at new york, july 1, 1902-december 31, 1943 – (= ser Records of the immigration and naturalization service, 1891-1957) – 755r – 5 – mf#T621 – us Nat Archives [975]

Index (soundex) to passenger lists of vessels arriving at philadelphia, pa, jan 1, 1883-june 28, 1948 – (= ser Records of the immigration and naturalization service, 1891-1957) – 61r – 5 – mf#T526 – us Nat Archives [975]

Index (soundex) to passenger lists of vessels arriving at the port of new york, 1944-1948 – (= ser Records of the immigration and naturalization service, 1891-1957) – 94r – 5 – mf#M1417 – us Nat Archives [975]

Index (soundex) to the 1900 population schedules / U.S. Bureau of the Census – (= ser 1900 federal population census) – 1 – (alabama 180r t1030. alaska 15r t1031. arizona 22r t1032. arkansas 132r t1033. california 193r t1034. colorado 68r t1035. connecticut 107r t1036. delaware 21r t1037. district of columbia 42r t1038. florida 59r t1039. georgia 211r t1040. hawaii 30r t1041. idaho 19r t1042. illinois 479r t1043. indiana 252r t1044. iowa 198r t1045. kansas 147r t1046. kentucky 198r t1047. louisiana 146r t1048. maine 79r t1049. maryland 127r t1050. massachusetts 314r t1051. michigan 259r t1052. minnesota 181r t1053. mississippi 155r t1054. missouri 300r t1055. montana 40r t1056. nebraska 107r t1057. nevada 7r t1058. new hampshire 52r t1059. new jersey 203r t1060. new mexico 23r t1061. new york 766r t1062. north carolina 168r t1063. north dakota 36r t1064. ohio 395r t1065. oklahoma 43r t1066. oregon 53r t1067. pennsylvania 590r t1068. rhode island 49r t1069. south carolina 124r t1070. south dakota 44r t1071. tennessee 187r t1072. texas 286r t1073. utah 29r t1074. vermont 41r t1075. virginia 164r t1076. washington 70r t1077. west virginia 92r t1078.. wisconsin 188r t1079. wyoming 14r t1080. military and naval 32r t1081. indian territory 42r t1082. institutions 8r t1083) – us Nat Archives [317]

Index (soundex) to the 1920 federal population census schedules for [...] / U.S. Bureau of the Census – (= ser The 1920 federal poulation census) – (alabama 159r m1548. arizona 30r m1549. arkansas 131r m1550. california 327r m1551. colorado 80r m1552. connecticut 111r m1553. delaware 20r m1554. district of columbia 49r m1555. florida 74r m1556. georgia 200r m1557. idaho 33r m1558. illinois 509r m1559. indiana 230r m1560. iowa 181r m1561. kansas 129r m1562. kentucky 180r m1563. louisiana 135r m1564. maine 67r m1565. maryland 126r m1566. massachusetts 326r m1567. michigan 291r m1568. minnesota 174r m1569. mississippi 123r m1570. missouri 269r m1571. montana 46r m1572. nebraska 96r m1573. nevada 9r m1574. new hampshire 39r m1575. new jersey 253r m1576. new mexico 31r m1577. new york 885r m1578. north carolina 166r m1579. north dakota 48r m1580. ohio 476r m1581. oklahoma 155r m1582. oregon 69r m1583. pennsylvania 716r m1584. rhode island 53r m1585. south carolina 112r m1586. south dakota 48r m1587. tennessee 162r m1588. texas 373r m1589. utah 33r m1590. vermont 32r m1591. virginia 168r m1592. washington 118r m1593. west virginia 109r m1594. wisconsin 196r m1595. wyoming 17r m1596. alaska 6r m1597. hawaii 24r m1598. canal zone 3r m1599. military-naval 18r m1600. puerto rico 165r m1601. guam 1r m1602. american samoa 2r m1603. virgin islands 3r m1604. institutions 1r m1605) – us Nat Archives [317]

Index to 1851 census of bedfordshire – 24mf – 9 – £20.00 set £1.25mf – uk BedsFHS [350]

Index to annuals of southern baptist convention / Southern Baptist Convention – 1973-81, 1982-84 – 1 – 5.00 – us Southern Baptist [242]

Index to anthony lagoon police station mortuary book 1890-1949 – [mf ed 1986] – 1mf – 9 – A$5.50 – 0-949124-19-2 – (filmed with: index to newcastle waters police station mortuary book 1893-1932. katherine mortuary records 1887-1941. katherine cemetery transcriptions to july 1982. gardens road (darwin) cemetery transcriptions 1914-1980) – at Northern [929]

Index to appellate case files of the supreme court of the united states, 1792-1909 / U.S. Supreme Court – (= ser Records of the supreme court of the united states) – 20r – 5 – (with printed guide) – mf#M408 – us Nat Archives [347]

Index to archives / Baptist Missionary Society. London – 1792-1914 – 5 – 6.90 – us Southern Baptist [242]

Index to bdm articles in the northern standard [northern territory] : 1 january 1921 to 31 december 1940 – [mf ed 2004] – 1mf – 9 – $8.80 – at Northern [079]

Index to bdm articles in the northern standard [northern territory] : january 1941 to december 1953 except during world war 2 – [mf ed 2004] – 1mf – 9 – $8.80 – 0-949124-94-X – at Northern [079]

Index to b'nai brith : the early committee in yuba county – 1r – 5,9 – $50.00 – mf#B40149 – us Library Micro [978]

Index to citizens naturalized in the superior court of san diego, california, 1853-1956 / U.S. County Court – (= ser Records of county courts of the united states) – 1r – 1 – mf#M1609 – us Nat Archives [347]

Index to colonial subjects in dutch newspapers (mainly the dutch east and west indies); 1843-1947 – 185mf – 9 – €995.00 – (with ind) – mf#m109 – ne MMF Publ [079]

Index to compiled military service records of volunteer union soldiers who served in organizations from the state/territory of [...] / U.S. War Dept. – (= ser Records Of Volunteer Union Soldiers Who Served During The Civil War) – 5 – (alabama 1r m263. arizona 1r m532. arkansas 4r m383. california 7r m533. colorado 3r m534. connecticut 17r m535. dakota 1r m536. delaware 4r m537. district of columbia 3r m538. florida 1r m264. georgia 1r m385. illinois 101r m539. indiana 86r m540. iowa 29r m541. kansas 10r m542. kentucky 30r m386. louisiana 4r m387. maine 23r m543. maryland 13r m388. massachusetts 44r m544. michigan 48r m545. minnesota 10r m546. mississippi 1r m389. missouri 54r m390. nebraska 2r m547. nevada 1r m548. new hampshire 13r m549. new jersey 26r m550. new mexico 4r m242. new york 157r m551. north carolina 2r m391. ohio 122r m552. oregon 1r m553. pennsylvania 136r m554. rhode island 7r m555. tennessee 16r m392. texas 2r m393. utah 1r m556. vermont 14r m557. virginia 1r m394. washington 1r m558. west virginia 13r m507. wisconsin 33r m559) – us Nat Archives [355]

Index to compiled service records of confederate soldiers who served in organizations from the state of [...] / U.S. War Dept – (= ser Records of confederate soldiers who served during the civil war) – 5 – (alabama 49r m374. arizona territory 1r m375. arkansas 26r m376. florida 9r m225. georgia 67r m226. kentucky 14r m377. louisiana 31r m378. maryland 2r m379. mississippi 45r m232. missouri 16r m380. north carolina 43r m230. south carolina 35r m381. tennessee 48r m231. texas 41r m227. virginia 62r m382. all with printed guides) – us Nat Archives [355]

INDEX

Index to compiled service records of confederate soldiers who served in organizations raised directly by the confederate government and of confederate general and staff officers and non-regimental enlisted men / U.S. War Dept – (= ser Records of confederate soldiers who served during the civil war) – 26r – 5 – (with printed guide) – mf#M818 – us Nat Archives [355]

Index to compiled service records of revolutionary war personnel / U.S. War Dept – (= ser War department coll of revolutionary war records) – 1r – 1 – (with printed guide) – mf#M879 – us Nat Archives [355]

Index to compiled service records of revolutionary war soldiers who served with the american army in connecticut military organizations / U.S. War Dept – (= ser War department coll of revolutionary war records) – 25r – 1 – (with printed guide) – mf#M920 – us Nat Archives [355]

Index to compiled service records of revolutionary war soldiers who served with the american army in georgia military organizations / U.S. War Dept. Adjutant General's Office – (= ser War department coll of revolutionary war records) – 1r – 1 – (with printed guide) – mf#M1051 – us Nat Archives [355]

Index to compiled service records of volunteer soldiers who served during indian wars and disturbances, 1815-58 / U.S. War Dept. Adjutant General's Office – (= ser Records of volunteer soldiers who served during indian wars and disturbances) – 42r – 5 – (with printed guide) – mf#M629 – us Nat Archives [355]

Index to compiled service records of volunteer soldiers who served during the cherokee disturbances and removal in organizations from the state of [...] / U.S. War Dept. Adjutant General's Office – (= ser Records of volunteer soldiers who served during cherokee disturbances and removal) – 5 – (alabama 1r m243. georgia 1r m907. north carolina 1r m256. tennessee and the field and staff of the army of the cherokee nation 2r m908. with printed guides) – us Nat Archives [355]

Index to compiled service records of volunteer soldiers who served during the creek war in organizations from the state of alabama / U.S. War Dept. Adjutant General's Office – (= ser Records Of Volunteer Soldiers Who Served During The Creek War) – 2r – 5 – (with printed guide) – mf#M244 – us Nat Archives [355]

Index to compiled service records of volunteer soldiers who served during the florida war in organizations from the state of alabama / U.S. War Dept. Adjutant General's Office – (= ser Records of volunteer soldiers who served during the florida wars) – 1r – 5 – (with printed guide) – mf#M245 – us Nat Archives [355]

Index to compiled service records of volunteer soldiers who served during the florida war in organizations from the state of louisiana – (= ser Records of volunteer soldiers who served during the florida wars) – 1r – 5 – (with printed guide) – mf#M239 – us Nat Archives [355]

Index to compiled service records of volunteer soldiers who served during the mexican war / U.S. War Dept. Adjutant General's Office – (= ser Records of volunteer soldiers who served during the mexican war) – 41r – 5 – (with printed guide) – mf#M616 – us Nat Archives [355]

Index to compiled service records of volunteer soldiers who served during the phillippine insurrection / U.S. War Dept. Adjutant General's Office – (= ser Records of the adjutant general's office, 1780's-1917) – 24r – 1 – (with printed guide) – mf#M872 – us Nat Archives [355]

Index to compiled service records of volunteer soldiers who served during the revolutionary war in organizations from the state of north carolina / U.S. War Dept – (= ser War department coll of revolutionary war records) – 2r – 5 – (with printed guide) – mf#M257 – us Nat Archives [355]

Index to compiled service records of volunteer soldiers who served during the war of 1812 / U.S. War Dept. Adjutant General's Office – (= ser Records Of Volunteer Soldiers Who Served In The War Of 1812) – 234r – 5 – (with printed guide) – mf#M602 – us Nat Archives [355]

Index to compiled service records of volunteer soldiers who served during the war of 1812 in organizations from the state of [...] / U.S. War Dept. Adjutant General's Office – (= ser Records of volunteer soldiers who served in the war of 1812) – 5 – (louisiana 3r m229. north carolina 5r m250. south carolina 7r m652. mississippi 22r m678. with printed guides) – us Nat Archives [355]

Index to compiled service records of volunteer soldiers who served during the war of 1837-1838 in organizations from the state of louisiana / U.S. War Dept. Adjutant General's Office – (= ser Records of volunteer soldiers who served during the war of 1837-1838) – 1r – 5 – (with printed guide) – mf#M241 – us Nat Archives [355]

Index to compiled service records of volunteer soldiers who served during the war with spain in organizations from the state of [...] / U.S. War Dept. Adjutant General's Office – (= ser Records of volunteer soldiers who served during the war with spain) – 5 – (louisiana 1r m240. north carolina 2r m413. with printed guides) – us Nat Archives [355]

Index to compiled service records of volunteer soldiers who served from 1784-1811 / U.S. War Dept. Adjutant General's Office – (= ser Records Of Volunteer Soldiers Who Served From 1784 Until 1811) – 9r – 5 – (with printed guide) – mf#M694 – us Nat Archives [355]

Index to compiled service records of volunteer soldiers who served from the state of michigan during the patriot war, 1838-39 / U.S. War Dept. Adjutant General's Office – (= ser Records of volunteer soldiers who served during the patriot war, 1838-1839) – 1r – 5 – (with printed guide) – mf#M630 – us Nat Archives [355]

Index to compiled service records of volunteer soldiers who served from the state of new york during the patriot war, 1838 / U.S. War Dept. Adjutant General's Office – (= ser Records of volunteer soldiers who served during the patriot war, 1838-1839) – 1r – 5 – (with printed guide) – mf#M631 – us Nat Archives [355]

Index to compiled service records of volunteer union soldiers who served during the civil war with united states colored troops / U.S. War Dept. Adjutant General's Office – (= ser Records Of Volunteer Union Soldiers Who Served During The Civil War) – 98r – 5 – (with printed guide) – mf#M589 – us Nat Archives [355]

Index to compiled service records of volunteer union soldiers who served in organizations from the state of alabama / U.S. War Dept. Adjutant General's Office – (= ser Records Of Volunteer Union Soldiers Who Served During The Civil War) – 1r – 5 – (with printed guide) – mf#M263 – us Nat Archives [355]

Index to compiled service records of volunteer union soldiers who served in organizations from the state/territory of [...] / U.S. War Dept. Adjutant General's Office – (= ser Records Of Volunteer Union Soldiers Who Served During The Civil War) – 5 – (alabama 1r m263. arizona 1r m532. arkansas 4r m383. california 7r m533. colorado 3r m534. connecticut 17r m535. dakota 1r m536. delaware 4r m537. district of columbia 3r m538. florida 1r m264. georgia 1r m385. illinois 101r m539. indiana 86r m540. iowa 29r m541. kansas 10r m542. kentucky 30r m386. louisiana 4r m387. maine 23r m543. maryland 13r m388. massachusetts 44r m544. michigan 48r m545. minnesota 10r m546. mississippi 1r m389. missouri 54r m390. nebraska 2r m547. nevada 1r m548. new hampshire 13r m549. new jersey 26r m550. new mexico 4r m242. new york 157r m551. north carolina 2r m391. ohio 122r m552. oregon 1r m553. pennsylvania 136r m554. rhode island 7r m555. tennessee 16r m392. texas 2r m393. utah 1r m556. vermont 14r m557. virginia 1r m394. washington 1r m558. west virginia 13r m507. wisconsin 33r m559) – us Nat Archives [355]

Index to compiled service records of volunteer union soldiers who served in the veteran reserve corps / U.S. War Dept. Adjutant General's Office – (= ser Records of the adjutant general's office, 1780's-1917) – 44r – 5 – (with printed guide) – mf#M636 – us Nat Archives [355]

Index to correspondence of the office of the commander in chief, american expeditionary forces, 1917-1919 / U.S. Army. American Expeditionary Forces – (= ser Records of the american expeditionary forces (world war 1), 1917-1923) – 132r – 5 – mf#T900 – us Nat Archives [355]

Index to court exhibits in english and japanese, international prosecution section, 1945-1947 / World War 2. International Prosecution Section – (= ser Records of allied operational and occupation headquarters, world war 2) – Ir – 1 – mf#M1687 – us Nat Archives [341]

Index to declarations of intention in the superior court of san diego county, california, 1853-1956 / U.S. County Court – (= ser Records of county courts of the united states) – 1r – 1 – mf#M1612 – us Nat Archives [347]

Index to dental literature – Chicago. 1839-1999 (1) 1839-1999 (5) 1972-1999 (9) – ISSN: 0019-3992 – mf#7634 – us UMI ProQuest [616]

Index to dominion statute amendments (1907-1921) / Canada. Laws, Statutes, etc – Toronto 1922? LL-2271 – 1 – us L of C Photodup [348]

Index to dutch-language indonesian newspapers, 1810-1923 – 149mf – 9 – €890.00 – (ind newspapers: bataviasch koloniale courant, 1810-1811; java government gazette, 1812-1816; bataviasche courant, 1816-1827; javasche courant, 1828-1923; soerabaiasch handelsblad, 1866-1923; algemeen dagblad van nederlandsch indie, 1873-1923; bataviaasch nieuwsblad, 1855-1923; sumatra post, 1898-1923. ind also covers almost all other important east indies newspapers, such as: locomotief; java times; oostpost etc) – mf#m108 – ne MMF Publ [079]

Index to early tennessee baptists / Taylor, O W – Comp. by A. Stan Rescoe. 70p – 1 – 5.00 – us Southern Baptist [242]

Index to early wellington newspapers 1839-1865 – 22mf – 9 – (includes guide) – nz Nat Libr [079]

Index to east african series : pamphlet colletion – [Belfast]: Queen's Uni Dept of Photography, 1965 (mf ed) – 1 – us CRL [960]

Index to eastern provinces and dominion statute amendments to 1926. / Canada. Laws, Statutes, etc – Toronto: Garrett 1927? 122nos. LL-2313 – 1 – us L of C Photodup [348]

Index to federal bureau of investigation class 61 : treason or misprision of treason – 1921-1931 – 1992 (mf ed) – (= ser Records Of The Federal Bureau Of Investigation) – 15r – 1 – (with printed guide) – mf#M1531 – us Nat Archives [360]

Index to gazette des tribunaux – Paris, France – 172 3/4r – 1 – (incorp with: gazette du palais. aka: gazette des tribunaux) – uk British Libr Newspaper [072]

Index to general correspondence of the record and pension office, 1889-1920 / U.S. War Dept. Adjutant General's Office – (= ser Records of the adjutant general's office, 1780's-1917) – 385r – 5 – (with printed guide) – mf#M686 – us Nat Archives [360]

Index to general correspondence...1890-1917 / U.S. War Dept. Adjutant General's Office – (= ser Records of the adjutant general's office, 1780's-1917) – 1269r – 1 – (with printed guide) – mf#M698 – us Nat Archives [355]

Index to general statutes of connecticut, and public acts, from 1875 to 1882 / Connecticut. Laws, Statutes, etc – Hartford, Case, Lockwood & Brainard, 1883. 191 p. LL-1619 – 1 – us L of C Photodup [348]

Index to health information microfiche library – 1988- – 9 – apply for prices – (provides access to statistical and congressional publications on issues of public health. printed index available in combination on an annual subscription basis) – us CIS [020]

Index to history of butte county – Butte Co, CA: George C Mansfield, 1918 – 1r – 1 – $50.00 – mf#B40207 – us Library Micro [978]

Index to hoosier folklore bulletin (1942-45) and hoosier folklore (1946-50) / Posen, I Sheldon – [Bloomington: Folklore Forum Society] 1973 [mf ed Bloomington IN: Indiana Uni Lib, Preservation Dept 1984] – 1r – 1 – us Indiana Preservation [390]

Index to incorporated bodies and to private and local law... / Baudouin, Philibert – Montreal: C O Beauchemin, [1897?] – 8mf – 9 – 0-665-10524-X – (also available in french) – mf#10524 – cn Canadiana [346]

Index to indian decisions / U.S. Dept of the Interior – Office of Hearings and Appeals, 1972-95 – 18mf – 9 – $27.00 – (1st 2v entitled: "digest of indian probate law". coverage now extends to all legal matters of indians and alaska natives. updated index publ annually with 5yr cumulations. add vols planned) – mf#LLMC 87-304 – us LLMC [324]

Index to indian wars pension files, 1892-1926 / U.S. Veterans Administration – (= ser Records of the veterans administration) – 12r – 5 – mf#T318 – us Nat Archives [355]

Index to lange's commentary on the old testament : 1. hebrew. 2. topical / Pick, Bernhard – New York: Charles Scribner, c1882 [mf ed 1986] – (= ser A commentary on the holy scriptures. old testament 14/14) – 1mf – 9 – 0-8370-6150-4 – mf#1986-0150 – us ATLA [221]

Index to law school alumni publications – Littleton, Colorado: Fred B Rothman Co, 1989 – 9 – $1,475.00 set – mf#408670 – us Hein [340]

An index to legal periodical literature – v. 1-6. Boston. 1888-1924; Indianapolis. 1933-37 – 1 – 52.00 – us L of C Photodup [340]

Index to letters received by the commission to the five civilized tribes, 1897-1913 / U.S. Bureau of Indian Affairs – (= ser Records relating to census rolls and other enrollments) – 23r – 1 – mf#M1314 – us Nat Archives [350]

Index to men who worked on the overland telegraph construction 1870 to 1872 : from south australian border to palmerston in the northern territory – [mf ed 2004] – 1mf – 9 – $7.70 – 0-949124-96-6 – at Northern [380]

Index to mexican archives of monterey, vols 6-16 / Taylor, Alexander S – Monterey Co, CA. 1859 – 1r – 1 – $50.00 – mf#B03721 – us Library Micro [978]

Index to mexican war pension files, 1887-1926 / U.S. Veterans Administration – (= ser Records of the veterans administration) – 14r – 5 – mf#T317 – us Nat Archives [355]

Index to names of u.s. marshals, 1789-1960 / U.S. Dept of Justice – (= ser General records of the department of justice) – 1r – 5 – mf#T577 – us Nat Archives [340]

Index to names of witnesses and suspected war crimes perpetrators who appeared before the international military tribunal for the far east, 1945-1947 / World War 2. International Military Tribunal for the Far East – (= ser Records of allied operational and occupation headquarters, world war 2) – 1r – 1 – mf#M1695 – us Nat Archives [341]

Index to naturalization in the u(nitedstates district court for the northern district of california, 1852-ca 1989 / United.States. District Court – (= ser Records of district courts of the united states) – 165r – 1 – (with printed guide) – mf#M1744 – us Nat Archives [347]

Index to naturalization petitions and records of the u.s. district court, 1906-1966, and the u.s. circuit court, 1906-1911, for the district of massachusetts / U.S. Circuit and District Courts – (= ser Records of district courts of the united states) – 115r – 1 – mf#M1545 – us Nat Archives [347]

Index to naturalization petitions for the u.s. circuit court, 1795-1911, and district court, 1795-1928, for the district of delaware / U.S. Circuit and District Courts – (= ser Records of district courts of the united states) – 1r – 1 – mf#M1649 – us Nat Archives [347]

Index to naturalization petitions of the united states district court for the eastern district of new york, 1865-1957 / U.S. District Court – (= ser Records of district courts of the united states) – 142r – 1 – (with printed guide) – mf#M1164 – us Nat Archives [347]

Index to naturalization records of the us district court for the eastern district of tennessee at chattanooga, 1888-1955 / U.S. District Court – (= ser Records of district courts of the united states) – 1r – 1 – mf#M1611 – us Nat Archives [347]

Index to naturalization records of the u.s. district court for the southern district of california, central division, los angeles, 1887-1937 – (= ser Records of district courts of the united states) – 2r – 1 – mf#M1607 – us Nat Archives [347]

Index to naturalization records of the us [sic] supreme court for the district of columbia, 1802-1909 / U.S. District Court – (= ser Records Of U.S. District Courts) – 1r – 1 – mf#M1827 – us Nat Archives [347]

Index to neander's general history of the christian religion and church / [Torrey, Mary Cutler] – [rev ed] Boston: Houghton, Mifflin, 1881 [mf ed 1993] – 1mf – 9 – 0-524-08625-7 – mf#1993-1075 – us ATLA [240]

Index to new england naturalization records, 1791-1906 – (= ser Records of county courts of the united states) – 117r – 1 – mf#M1299 – us Nat Archives [347]

Index to northwest missions manuscripts / ed by Nute, Grace Lee – ca 1766-1926 – 5r – 5 – $150.00 $30.00r – us Minn Hist [240]

Index to office equipment and supplies – Surrey. 1967-1973 (1) – (cont by: office equipment index) – mf#1890 – us UMI ProQuest [650]

Index to officers' jackets, 1913-1925 (officers directory) / U.S. Navy. Bureau of Naval Personnel – (= ser Records relating to service in the united states navy and united states marine) – 2r – 1 – mf#T1102 – us Nat Archives [355]

Index to official and historical atlas of yuba county, california – Thompson & West, 1873 – 2r – 5,9 – $100.00 – mf#B40148 – us Library Micro [978]

Index to official published documents relating to cuba and the insular possessions of the us, 1876-1906 / U.S. Bureau of Insular Affairs – (= ser Records of the bureau of insular affairs) – 3r – 1 – (with printed guide) – mf#M24 – us Nat Archives [972]

Index to original communications in the medical journals of the united states and canada for 1877 : classified by subjects and authors / Chapin, William D – New York: s.n, 1878?] [mf ed 1984] – 2mf – 9 – 0-665-01638-7 – mf#01638 – cn Canadiana [610]

INDEX

Index to otero county newspapers 1886-1900 : with bent county papers included 1873-1879, 1898-1900 / Hewitt, Dorothy; ed by Hanzas, Barbara – La Junta, CO: Woodruff Pub Lib, 1888 (mf ed 1987) – 4r – 5 – mf#MF Ot2chs – us Colorado Hist [071]
Index to passenger arrivals at san diego, california, ca 1904-ca 1952 / U.S. Immigration and Naturalization Service – (= ser Records Of The Immigration And Naturalization Service) – 6r – 1 – mf#M1761 – us Nat Archives [975]
Index to passenger lists of vessels arriving at boston, ma, 1848-1891 – (= ser Records of the united states customs service, 1820-c1891) – 282r – 5 – mf#M265 – us Nat Archives [975]
Index to passenger lists of vessels arriving at boston, ma, jan 1, 1902-june 30, 1906 – (= ser Records of the immigration and naturalization service, 1891-1957) – 11r – 5 – mf#T521 – us Nat Archives [975]
Index to passenger lists of vessels arriving at boston, ma, jan 1, 1902-june 30, 1906 – (= ser Records of the immigration and naturalization service, 1891-1957) – 11r – 5 – mf#T617 – us Nat Archives [975]
Index to passenger lists of vessels arriving at galveston, texas, 1906-1951 – (= ser Records of the immigration and naturalization service, 1891-1957) – 7r – 5 – mf#M1358 – us Nat Archives [975]
Index to passenger lists of vessels arriving at new orleans, la, 1900-1952 – (= ser Records of the immigration and naturalization service, 1891-1957) – 22r – 5 – mf#T618 – us Nat Archives [975]
Index to passenger lists of vessels arriving at new orleans, la, before 1900 – (= ser Records of the united states customs service, 1820-c1891) – 32r – 5 – mf#T527 – us Nat Archives [975]
Index to passenger lists of vessels arriving at new york, 1820-1846 – (= ser Records of the united states customs service, 1820-c1891) – 103r – 5 – (with printed guide) – mf#M261 – us Nat Archives [975]
Index to passenger lists of vessels arriving at new york, june 16, 1897-june 30, 1902 – (= ser Records of the immigration and naturalization service, 1891-1957) – 115r – 5 – mf#T519 – us Nat Archives [975]
Index to passenger lists of vessels arriving at philadelphia, pa, 1800-1906 – (= ser Records of the united states customs service, 1820-c1891) – 151r – 5 – (with printed guide) – mf#M360 – us Nat Archives [975]
Index to passenger lists of vessels arriving at ports in alabama, florida, georgia, and south carolina, 1890-1924 – (= ser Records of the immigration and naturalization service, 1891-1957) – 26r – 5 – mf#T517 – us Nat Archives [975]
Index to passengers arriving at gulfport, ms, aug 27 1904-aug 28 1954, and at pascagoula, july 15 1903-may 21 1935 – (= ser Records of the immigration and naturalization service, 1891-1957) – 1r – 5 – mf#T523 – us Nat Archives [975]
Index to passengers arriving at new beford, ma, july 1, 1902-nov 18, 1954 – (= ser Records of the immigration and naturalization service, 1891-1957) – 2r – 5 – mf#T522 – us Nat Archives [975]
Index to passengers arriving at portland, maine, jan 29, 1893-nov 22, 1954 – (= ser Records of the immigration and naturalization service, 1891-1957) – 1r – 1 – mf#T524 – us Nat Archives [975]
Index to passengers arriving at providence, rhode island, june 18, 1911-oct 5, 1954 – (= ser Records of the immigration and naturalization service, 1891-1957) – 2r – 1 – mf#T518 – us Nat Archives [975]
Index to pension application files of remarried widows based on service in the civil war and later wars and in the regular army after the civil war / U.S. War Dept – (= ser Records of the veterans administration) – 7r – 1 – mf#M1785 – us Nat Archives [355]
Index to pension application files of remarried widows based on service in the war of 1812, indian wars, mexican war, and regular army before 1861 / U.S. War Dept – (= ser Records of the veterans administration) – 1r – 1 – mf#M1784 – us Nat Archives [355]
Index to personnel files / Great Northern Railway Company. Personnel Dept – 4r – 5 – us Minn Hist [380]
Index to personnel files / Northern Pacific Railway Company. Personnel Dep't – 3r – 5 – us Minn Hist [380]
Index to portraits in books / Royal College of Physicians – 2r – 1 – (with suppl) – mf#95854 – uk Microform Academic [920]
Index to presidential proclamations, 1789-1947 / U.S. Congress – (= ser General records of the united states government) – 2r – 1 – mf#T279 – us Nat Archives [324]
Index to private land grant cases, u.s. district court, northern district of california, 1853-1903 / U.S. District Court – (= ser Records of district courts of the united states) – 1r – 1 – mf#T1214 – us Nat Archives [346]
Index to private land grant cases, us district court, northern district of california, 1853-1903 / U.S. District Court – (= ser Records of district courts of the united states) – 1r – 1 – mf#T1216 – us Nat Archives [347]
Index to private land grant cases, u.s. district court, southern district of california / U.S. District Court – (= ser Records of district courts of the united states) – 1r – 1 – mf#T1215 – us Nat Archives [346]
Index to public notices in the northern territory news; sunday territorian & suburban newspapers january 1987 to december 2000; public notices, includes birth, death, marriage, funeral notices, and other public notices – [md ed 2004] – 8mf – 9 – $27.50 – 0-949124-93-1 – at Northern [079]
Index to records relating to war of 1812 prisoners of war / U.S. War Dept. Adjutant General's Office – (= ser Records of the adjutant general's office, 1780's-1917) – 3r – 1 – mf#M1747 – us Nat Archives [355]
Index to reference cards for work projects administration project files / U.S. Work Projects Administration – (= ser Records Of The Work Projects Administration) – 5 – (1935-37 79r t935. 1938 15r t936. 1939-42 19r t937) – us Nat Archives [324]
Index to regular and primitive baptist associations' annuals / Baptist Associations – 1 – us Southern Baptist [242]
Index to rendezvous reports, armed guard personnel, 1917-1920 / U.S. Navy. Bureau of Naval Personnel – (= ser Records relating to service in the united states navy and united states marine) – 3r – 5 – mf#T1101 – us Nat Archives [355]
Index to rendezvous reports, before and after the civil war, 1846-1861, 1865-1884 / U.S. Navy. Bureau of Naval Personnel – (= ser Records relating to service in the united states navy and united states marine) – 32r – 5 – mf#T1098 – us Nat Archives [355]
Index to rendezvous reports, civil war, 1861-1865 / U.S. Navy. Bureau of Naval Personnel – (= ser Records relating to service in the united states navy and united states marine) – 31r – 5 – mf#T1099 – us Nat Archives [355]
Index to rendezvous reports, naval auxiliary service, 1917-1918 / U.S. Navy. Bureau of Naval Personnel – (= ser Records relating to service in the united states navy and united states marine) – 1r – 5 – mf#T1100 – us Nat Archives [355]
Index to rimes in american and english poetry, 1500-1900 / Hanley, Miles L – 9 – $1864.00 – mf#0254 – us Brook [420]
Index to riyazu-s-salatin = A history of bengal / Salim, Ghulam Husain – [Calcutta: Royal Asiatic Society of Bengal, [between 1940 and 1943] – (= ser Samp: indian books) – (trans into english fr original persian with notes by abdus salam) – us CRL [954]
Index to roberson collection at jenkins memorial library, foreign missions board, sbc, richmond, va / Roberson, Cecil F – 1984 – 1 – 8.46 – us Southern Baptist [242]
Index to san francisco city licenses – San Francisco, CA. 1850-56 – 2r – 1 – $100.00 – mf#B40330 – us Library Micro [978]
Index to scripture readings – Toronto: Printed for the Dept of Education, 1888 – 1mf – 9 – mf#30483 – cn Canadiana [220]
Index to south carolina pleading and practice forms / Taylor, Esten C – Spartanburg Galloway, 1936. 202 p. LL-632 – 1 – us L of C Photodup [348]
Index to south pacific conference records – 1947-87 – 20r – 5 – mf#PMB Doc 400 – at Pacific Mss [980]
Index to st louis magazines and obituaries / S[/ain/]t Louis Public Library – 1980 jan/mar-1981 jan/dec – 1r – 1 – mf#653608 – us WHS [020]
Index to surgeons' reports in "file a and bound manuscripts," of the adjutant general's office, 1861-1865 / U.S. War Dept. Adjutant General's Office – (= ser Records of the adjutant general's office, 1780's-1917) – 1r – 1 – mf#M1828 – us Nat Archives [355]
Index to survey of china mainland press, selections from china mainland magazines and current background / U.S. Consulate General. Hong Kong – 1 Nov 1950-72; 1973- – (= ser Hong kong press summaries) – 1 – $254.00 – us L of C Photodup [951]
Index to systematic theology / Hodge, Charles – New York: Scribner, Armstrong, 1873. Beltsville, Md: NCR Corp, 1978 (1mf); Evanston: American Theol Lib Assoc, 1984 (1mf) – 9 – 0-8370-0893-X – mf#1984-4278 – us ATLA [240]
Index to telegrams collected by the office of the secretary of war (unbound), 1860-1870 / U.S. War Dept. Office of the Secretary – (= ser Records of the office of the secretary of war) – 20r – 5 – (with printed guide) – mf#M564 – us Nat Archives [324]
Index to the annexures and printed papers of the house of assembly... 1854-1897 / Cape of Good Hope. Parliament. House – Cape Town, 1899 – 1 – us CRL [960]
An index to the arkansas reports, vols 1 to 31 inclusive : and hempstead's u.s. court reports, also all the arkansas cases in woolworth's and dillon's u.s.c.c. reports / Brady, Charles B – St. Louis, Gilbert, 1878. 760 p. LL-2358 – 1 – us L of C Photodup [347]
Index to the burlington magazine 1903-1972 – [mf ed 1996] – 21mf – 9 – $135.00 – 1-900853-55-8 – uk Mindata [700]
Index to the colonial and state records of north carolina – v1-2, v3-4 – 2r – 1 – (cont: colonial records of north carolina; state records of north carolina) – mf#1013295 – us WHS [978]
Index to the criminal & penal statutes of canada, as affecting the province of quebec. / Dubreuil, Joseph Fereol – Montreal, Beauchemin & Valois, 1877. 95 p. LL-2335 – 1 – us L of C Photodup [345]
An index to the early printed books in the british museum / Proctor, Robert – 1898-1903, with Supplement for 1899-1902, 1900-03. and Registers to the four supplements, 1899-1902. 1906. 3v – 1,9 – us AMS Press [010]
Index to the eleventh census of the united states, 1890 / U.S. Bureau of the Census – (= ser 1890 census schedules) – 2r – 5 – mf#M496 – us Nat Archives [317]
Index to the federal statutes, 1874-1931 / U.S. Laws, Statutes, etc – 15mf – 9 – $22.50 – (rev of the scott/beaman ind thru 1931 by walter h. mcclenon & wilfred c. gilbert. washington: gpo, 1933?) – mf#llmc 84-100B – us LLMC [348]
Index to the general photographs of the bureau of ships, 1914-1946 – (= ser Records of the bureau of ships) – 9r – 5 – mf#M1157 – us Nat Archives [355]
Index to the great register – Tulare Co, CA. 1870's – 1r – 1 – $50.00 – mf#B40281 – us Library Micro [978]
Index to the great register – Tulare Co, CA. 1888 – 1r – 1 – $50.00 – mf#B40284 – us Library Micro [978]
Index to the journal of botany, 1863-1942 / South London Botanical Institute – London – 26mf – 9 – £185.00 – Publ in association with South London Botanical Institute – uk Chadwyck [580]
Index to the journals of the continental congress, 1774-1789 / Harris, Kenneth E & Tilley, Steven – Washington: GPO, 1976 (all publ) – (= ser Journals Of The Continental Congress) – 5mf – 9 – $7.50 – mf#LLMC 84-240 – us LLMC [020]
Index to the letters received by the confederate adjutant and inspector general and by the confederate quartermaster general, 1861-1865 / U.S. War Dept. Confederate Records – (= ser War department coll of confederate records) – 41r – 5 – (with printed guide) – mf#M410 – us Nat Archives [355]
Index to the letters received by the confederate secretary of war, 1861-1865 / U.S. War Dept. Confederate Records – (= ser War department coll of confederate records) – 34r – 5 – (with printed guide) – mf#M409 – us Nat Archives [355]
Index to the library edition of thomas jackson's life of charles wesley / Jackson, Francis M – London: Wesley Historical Society, 1899 – (= ser Publications of the Wesley Historical Society) – 1mf – 9 – 0-7905-5343-0 – mf#1988-1343 – us ATLA [941]
Index to the local and private acts of nova scotia passed by the legislature during the years 1924 to 1934 both inclusive / Nova Scotia. Laws, Statutes, etc – Halifax: Provincial Secretary, King's Printer, 1934. 1,1,xivp. LL-2337 – 1 – us L of C Photodup [342]
Index to the minutes of the uk warehousemen & clerks schools for orphan : and necessitous children warehouse administrators warehouse children – [mf ed 1999] – 1mf – 9 – A$5.50 – 0-949124-90-7 – at Northern [362]
Index to the naturalization records of the us district court for oregon, 1859-1956 / U.S. District Court – (= ser Records of district courts of the united states) – 3r – 1 – (with printed guide) – mf#M1242 – us Nat Archives [347]
Index to the new zealand listener 1939-1987 – 39mf – 9 – NZ$131.60 – (incl guide) – nz Nat Libr [079]
Index to the new zealand mail 1871-1907 – 54mf – 9 – NZ$180.00 – (with guide) – nz Nat Libr [079]
Index to the papers of the continental congress, 1774-1789 / Butler, John P – National Archives ed. Washington: GPO. v1-5. 1978 (all publ) – (= ser Journals Of The Continental Congress) – 72mf – 9 – $108.00 – mf#LLMC 84-241 – us LLMC [020]
Index to the population census schedules / Minnesota – 1860. 31r – 5 – us Minn Hist [317]
Index to the population census schedules / Minnesota – 1870 – 140r – 5 – us Minn Hist [317]
Index to the public archives (verbaal) of the ministry of the colonies, 1814-1849 – [mf ed 2003] – (= ser Finding aids for dutch colonial history from the national archives of the netherlands 1) – 2151mf – 9 – €14,430.00 – (with p/g in english & concordance) – mf#mmp103 – ne Moran [959]
Index to the public archives (verbaal) of the ministry of the colonies of the netherlands, 1850-1921 – 4509mf – 9 – €25,970.00set – (coll available in 7 chronological subsects (inquire for details); with p/g in english & concordance) – mf#m120 – ne MMF Publ [959]
Index to the published decisions of the u.s. accounting officers : 1894-1929 / U.S. Treasury Dept. General Accounting Office – Washington: GPO. 1v. 1931 – (= ser Comptroller of the treasury decisions, 1894-1921) – 9mf – 9 – $13.50 – (provides coverage for all vols of the comptroller of the treasury decisions as well as v1-8 of the comptroller general's decisions) – mf#LLMC 81-217 – us LLMC [336]
Index to the rolls of parliament, 1278-1503 / ed by Strachey, J – London, 1832 – 3r – 1 – mf#5514 – uk Microform Academic [323]
Index to the secret and cabinet archives of the ministry of the colonies, 1825-1839 – [mf ed 2003] – (= ser Finding aids for dutch colonial history from the national archives of the netherlands 2) – 143mf – 9 – €965.00 – (with p/g in english & concordance) – mf#mmp104 – ne Moran [959]
Index to the secret and cabinet archives, of the ministry of the colonies, 1901-1958 – [mf ed 2003] – (= ser Finding aids for dutch colonial history from the national archives of the netherlands 3) – 950mf – 9 – €8250.00 – (with p/g in english & concordance) – mf#mmp105 – ne Moran [959]
An index to the statutes of canada : from 3 and 4 victoria to 12 and 13 victoria, inclusive, 1840 to 1850 / Canada. Laws, Statutes, etc – Toronto: Rowsell, 1850. 72p. LL-2369 – 1 – us L of C Photodup [348]
Index to the statutes of prince edward island, in force in the year 1845 / Prince Edward Island. Laws, Statutes, etc – Charlottetown: Hazard, 1845. 96p. LL-2332 – 1 – us L of C Photodup [348]
Index to the tate gallery archive – [mf ed 1986] – 122mf – 9 – $680.00 – 0-907716-11-3 – (documentation of british & 20th c artists worldwide in 26,000 entries; with p/g) – Tate gallery archive – uk Mindata [700]
Index to the tracts for the times / Croly, David O – Oxford, England. 1842 – 1r – us UF Libraries [240]
Index to the war production board policy documentation file, 1939-1947 / U.S. War Production Board – (= ser Records of the war production board) – 86r – 1 – (with printed guide) – mf#M911 – us Nat Archives [934]
Index to the works of john henry cardinal newman / Rickaby, Joseph – London; New York: Longmans, Green, 1914 – 1mf – 9 – 0-7905-9609-1 – mf#1989-1334 – us ATLA [240]
Index to us coast guard casualty and wreck reports, 1913-1939 / U.S. Coast Guard – (= ser Records of the united states coast guard) – 7r – 5 – mf#T926 – us Nat Archives [360]
Index to war of 1812 pension application files / U.S. Veterans Administration – (= ser Records of the veterans administration) – 102r – 1 – (with printed guide) – mf#M313 – us Nat Archives [355]
Index to western provinces and dominion statute amendments to 1926. / Canada. Laws, Statutes, etc – Toronto: Garrett 1927. 101 no. LL-2327 – 1 – us L of C Photodup [348]
Index (umatilla, or) – Umatilla OR: L L McArthur [wkly] – 1 – us Oregon Lib [071]
Index van het openbaar archief van het ministerie van kolonien van nederland, 1850-1921 = Index to the public archives of the colonial ministry of the netherlands, 1850-1921 / Netherlands. General State Archives – 4509mf – 9 – (1850-60.-549mf.dfl5760.00 silver; 1861-71.-772mf.dfl8100.00 silver; 1872-82.-757mf.dfl7940.00 silver; 1883-93.-647mf.dfl6785.00 silver; 1894-1904.-647mf.dfl6785.00 silver; 1905-15.-767mf.dfl8045.00 silver; 1916-21.-370mf.dfl3885.00 silver) – ne MMF Publ [324]
Index verborum zur deutschen kaiserchronik / Tulasiewicz, Witold – Berlin: Akademie-Verlag, 1972 – (= ser Deutsche texte des mittelalters 68) – xv/387p – 1 – (int in english) – mf#8623 reel 19 – us Wisconsin U Libr [943]

Index-analysis of the federal statutes, 1789-1907 / Scott, George W & Beaman, Middleton G – Main vol, 1873-1907 – 27mf – 9 – $40.50 – (a preliminary vol covering 1789-1873 washington: gpo, 1908; 1911 by beaman and mcnamara) – mf#llmc 84-100 – us LLMC [348]

Index-catalogue of indian official publications in the library, british museum / Campbell, Francis – London: Library Supply Co 1900 – us CRL [020]

Index-digest on patent cases in the supreme court of the u.s / Lowery, Woodbury – Washington: Byrne & Co, 1897 – 5mf – 9 – $7.50 – mf#LLMC 84-334 – us LLMC [346]

Index/digest to the monographic notes in the american state reports – San Francisco: Bancroft-Whitney. 1v. 1912 (all publ) – (= ser American state reports. trinity series, pt 3) – 5mf – 9 – $7.50 – mf#LLMC 78-038C – us LLMC [348]

Index/digests of decisions of the department of the interior / U.S. Dept of the Interior – 1893-1994 – 138mf – 9 – $207.00 – (incl special vol covering impt unpubl decisions, memorandums etc for period 1943-54. coverage in later vols not so extensive. since 1991 ind covered interior decisions, and the interior boards of contract, indian and land appeals, and the decisions of the office of hearings and appeals. suppl planned) – mf#llmc 80-023 – us LLMC [340]

An indexed synopsis of the grammar of assent / Toohey, John Joseph – New York: Longmans, Green, 1906 [mf ed 1991] – 1mf – 9 – 0-7905-8605-3 – mf#1989-1830 – us ATLA [240]

Indexes : printed indexes to artists; subject index for fine arts, sculpture and decorative arts / Caisse Nationale des Monuments Historiques et des Sites. Paris – (= ser Fine and decorative arts in france) – 59mf – 9 – $485.00 – 0-907006-95-7 – uk Mindata [700]

Indexes and lists of witnesses for the defense and for the prosecution before the international military tribunal for the far east, 1946-1948 / World War 2. International Prosecution and Defense Section – (= ser Records of allied operational and occupation headquarters, world war 2) – 1r – 1 – mf#M1700 – us Nat Archives [355]

Indexes and lists to army technical and administrative publications, 1940-1979 / U.S. Govt – (= ser Publications of the united states government) – 29r – 1 – (with printed guide) – mf#M1641 – us Nat Archives [355]

Indexes and register to the correspondence of the office of the chief of naval operations and the office of the secretary of the navy, 1919-1927 / U.S. Navy. Chief of Naval Operations – (= ser General records of the department of the navy, 1798-1947) – 9r – 1 – (with printed guide) – mf#M1141 – us Nat Archives [355]

Indexes and subject cards to the secret and confidential correspondence of the secretary of the navy, mar 1917-jul 1919 / U.S. Navy. Office of the Secretary – (= ser General records of the department of the navy, 1798-1947) – 11r – 1 – (with printed guide) – mf#M1092 – us Nat Archives [355]

Indexes of exhibits of the prosecution and of the defense, introduced as evidence before the international military tribunal for the far east, 1945-1947 / World War 2. International Prosecution Section – (= ser Records of allied operational and occupation headquarters, world war 2) – 2r – 1 – mf#M1685 – us Nat Archives [355]

Indexes of marriage licences 1578 -1618 and 1791-1812 – (= ser Bedfordshire parish register series 15) – 3mf – 9 – £7.50 – uk BedsFHS [929]

Indexes of marriage licences 1747-1790 – (= ser Bedfordshire parish register series 14) – 3mf – 9 – £7.50 – uk BedsFHS [929]

Indexes of marriage licences 1813-1848 – (= ser Bedfordshire parish register series 81) – 7mf – 9 – £10.00 – uk BedsFHS [929]

Indexes of marriage licences 1848-1885 – (= ser Bedfordshire parish register series 82) – 4mf – 9 – £7.50 – uk BedsFHS [929]

Indexes of massachusetts births, marriages and deaths 1841-1895 – 9 – (birth index 54v on 417mf. death index 39v on 310mf. marriage index 42v on 327mf) – us Archive [978]

Indexes to certificates of registration and enrollment issued for merchant vessels at [...] / U.S. Bureau of Marine Inspection and Navigation – (= ser Records Of The Bureau Of Marine Inspection And Navigation) – 1 – (boston, massachusetts ca 1827-1868 1r m1866. san francisco, california, 1850-1877 1r m1867) – us Nat Archives [380]

Indexes to court documents including orders, rules of procedure, and copies of the indictment and motions of the defense, 1946-1948 / World War 2. Defense Section – (= ser Records of allied operational and occupation headquarters, world war 2) – 1r – 1 – mf#M1698 – us Nat Archives [355]

Indexes to deposit ledgers in branches of the freedmen's savings and trust company, 1865-1874 – (= ser Records of the office of the comptroller of the currency) – 5r – 1 – (with printed guide) – mf#M817 – us Nat Archives [333]

Indexes to documents presented as evidence by the defense and defense documents rejected as evidence before the international military tribunal for the far east, 1945-1947 / World War 2. Defense Section – (= ser Records of allied operational and occupation headquarters, world war 2) – 2r – 1 – mf#M1691 – us Nat Archives [355]

Indexes to eparkhial'nye vedomosti – [mf ed Norman Ross Publ] – 11 titles on 22mf – 9 – (iaroslavskiia eparkhial'nyia vedomosti 1860-92. iakutskiia eparkhial'nyia vedomosti 1908-17 (bibl ind [1897-1907] & systematic ind [1887-1897] incl). oglavleniie pribavlenii k tul'skim eparkhial'nym vedomostiam v1-20, 23-65; 1862-1903 (alphabetical ind incl). oglavlenie ufimskikh eparkhial'nykh vedomostei 1885-86, 1889, 1891, 1896, 1898, 1900, 1905. soderzhanie poltavskikh eparkhial'nykh vedomostei 1863-67 (ind of articles in the non-official pt incl [1888-1913].) oglavlenie neofitsial'nogo otdela kamchatskikh eparkhial'nykh vedomostei 1897. oglavlenie neofitsial'nogo otdela blagoveshchenskikh eparkhial'nykh vedomostei 1900. oglavlenie statei v varshavskom eparkhial'nom listke 1906-16. mogilevskie gubernskie vedomosti 1838-44. troitskie listki. khar'kovskie eparkhial'nye vedomosti 1867-82 (incl suppl vera i razum [1891-1901])) – us UMI ProQuest [020]

Indexes to files showing the receipt and distribution of defense documents and the receipt of affidavits from prisoners of war and other sources, 1946-1948 / World War 2. Defense Section – (= ser Records of allied operational and occupation headquarters, world war 2) – 2r – 1 – mf#M1696 – us Nat Archives [355]

Indexes to letters received by the secretary of war, 1861-1870 / U.S. War Dept. Office of the Secretary – (= ser Records of the office of the secretary of war) – 14r – 1 – (with printed guide) – mf#M495 – us Nat Archives [324]

Indexes to letters received...(main series), 1846, 1861-1889 / U.S. War Dept. Adjutant General's Office – (= ser Records of the adjutant general's office, 1780's-1917) – 9r – 1 – (with printed guide) – mf#M725 – us Nat Archives [355]

Indexes to letters sent by the secretary of war relating to military affairs, 1871-1889 / U.S. War Dept. Office of the Secretary – (= ser Records of the office of the secretary of war) – 12r – 1 – (with printed guide) – mf#M420 – us Nat Archives [324]

Indexes to naturalization petitions to the u.s. circuit and district courts for maryland, 1797-1951 / U.S. Circuit and District Courts – (= ser Records of district courts of the united states) – 25r – 1 – (with printed guide) – mf#M1168 – us Nat Archives [347]

Indexes to naturalization petitions to the u.s. circuit and district courts for the eastern district of pennsylvania, 1795-1951 / U.S. Circuit and District Courts – (= ser Records of district courts of the united states) – 60r – 1 – (with printed guide) – mf#M1248 – us Nat Archives [347]

Indexes to naturalization records of the king county territorial and superior courts, 1864-1889 and 1906-1928 / U.S. District Court – (= ser Records of county courts of the united states) – 1r – 1 – (with printed guide) – mf#M1233 – us Nat Archives [347]

Indexes to naturalization records of the montana territorial and federal courts, 1868-1929 / U.S. Circuit and District Courts – (= ser Records of district courts of the united states) – 1r – 1 – (with printed guide) – mf#M1236 – us Nat Archives [347]

Indexes to naturalization records of the pierce county territorial and superior courts, 1853-1923 / U.S. District Court – (= ser Records of county courts of the united states) – 2r – 1 – (with printed guide) – mf#M1238 – us Nat Archives [347]

Indexes to naturalization records of the snohomish county territorial and superior courts, 1876-1974 / U.S. Circuit and District Courts – (= ser Records of county courts of the united states) – 3r – 1 – (with printed guide) – mf#M1235 – us Nat Archives [347]

Indexes to naturalization records of the thurston county territorial and superior courts, 1850-1974 / U.S. District Court – (= ser Records of county courts of the united states) – 2r – 1 – (with printed guide) – mf#M1234 – us Nat Archives [347]

Indexes to naturalization records of the u.s. district court for western washington, northern division (seattle), 1890-1952 / U.S. District Court – (= ser Records of district courts of the united states) – 6r – 1 – (with printed guide) – mf#M1232 – us Nat Archives [347]

Indexes to naturalization records of the us district court, western district of washington, southern division (tacoma), 1890-1953 / U.S. District Court – (= ser Records of district courts of the united states) – 2r – 1 – (with printed guide) – mf#M1237 – us Nat Archives [347]

Indexes to numerical case files relating to particular incidents and suspected war criminals, international prosecution section, 1945-1947 / World War 2. International Prosecution Section – (= ser Records of allied operational and occupation headquarters, world war 2) – 4r – 1 – mf#M1682 – us Nat Archives [345]

Indexes to numerical evidentiary documents assembled by the prosecution for use as evidence before the international military tribunal for the far east, 1945-1947 / World War 2. International Prosecution Section – (= ser Records of allied operational and occupation headquarters, world war 2) – 8r – 1 – mf#M1689 – us Nat Archives [355]

Indexes to passenger lists of vessels arriving at galveston, texas, 1896-1906 – (= ser Records of the immigration and naturalization service, 1891-1957) – 3r – 5 – mf#M1357 – us Nat Archives [975]

Indexes to passenger lists of vessels arriving at san francisco, california, 1893-1934 – (= ser Records of the immigration and naturalization service, 1891-1957) – 28r – 5 – mf#M1389 – us Nat Archives [975]

Indexes to records of the presidential commission on the space shuttle challenger accident, 1986 / U.S. Temporary Committees, Commissions and Boards – (= ser Records Of Temporary Committees, Commissions, And Boards) – 30mf – 9 – mf#M1501 – us Nat Archives [324]

Indexes to records of the war college division and related general staff offices, 1903-1919 / U.S. War Dept. Adjutant General's Office – (= ser Records of the war department general and special staffs) – 49r – 1 – (with printed guide) – mf#M912 – us Nat Archives [355]

Indexes to registers and registers of declarations of intention and petitions for naturalization of the u.s. district and circuit courts for the western district of pennsylvania, 1820-1906 / U.S. District Court – (= ser Records of district courts of the united states) – 3r – 1 – (with printed guide) – mf#M1208 – us Nat Archives [347]

Indexes to rosters of railway postal clerks, ca 1883 – ca 1902 / U.S. Post Office – (= ser Post Office Records) – 1r – 1 – mf#M2099 – us Nat Archives [380]

Indexes to the ancient testamentary records of westminster / Burke, Arthur Meredyth – London, England. 1913 – 1r – us UF Libraries [941]

Indexes to the archive of the states of holland, 1524-1795 – 9 – €2085.00 set – (ind to the public resolutions of the states of holland, 1524-1795 [142mf] €865 [m101]; ind to the secret resolutions of the states of holland, 1653-1795 [22mf] €140 [m102]; ind to the resolutions of the geocommitteerde raden of the states of holland in the zuiderkwartier, 1621-1795 [191mf] €1315 [m103]) – ne MMF Publ [949]

Indexes to the naturalization records of the u.s. district court for the district and territory of alaska, 1900-1929 – (= ser Records of district courts of the united states) – 1r – 1 – mf#M1241 – us Nat Archives [347]

Indexes to vessels arriving at san francisco, ca, 1882-1957 – (= ser Records of the immigration and naturalization service, 1891-1957) – 2r – 5 – mf#M1437 – us Nat Archives [975]

India / ed by Bhandarkar, D R – Philadelphia: American Academy of Political and Social Sciences, 1929 – (= ser Samp: indian books) – us CRL [954]

India / Chirol, Valentine – London: Ernest Benn Ltd, 1926 – (= ser Samp: indian books) – (int by h a l fisher) – us CRL [954]

India / Conder, Josiah – London 1828 [mf ed Hildesheim 1995-98] – 4v on 12mf – 9 – €120.00 – 3-487-27525-2 – gw Olms [915]

India : country, people, missions / Gracey, John Talbot – Rochester: J T Gracey, 1884 [mf ed 1995] – (= ser Yale coll; Outline missionary series) – vi/207p (ill) – 1 – 0-524-09876-X – mf#1995-0876 – us ATLA [954]

India / Dodwell, Henry – [London]: Arrowsmith, 1936- – (= ser Samp: indian books) – us CRL [954]

India : from the aryan invasion to the great sepoy mutiny / Knight, Alfred Ernest – London, 1897 – (= ser 19th c british colonization) – 4mf – 9 – mf#1.1.7388 – uk Chadwyck [954]

India : internal affairs and foreign affairs, 1945-1954 / U.S. State Dept – (= ser Confidential u s state department central files) – 1 – $27,410.00 coll – (internal affairs, 1945-49: pt1: political, governmental, & national defense affairs 23r isbn 0-89093-418-5 $4455; pt2: social, economic, & industrial affairs 20r isbn 0-89093-419-3 $3885. foreign affairs, 1945-49 2r isbn 0-89093-453-3 $375. internal affairs, 1950-54 100r isbn 1-55655-435-4 $19,365. foreign affairs, 1950-54 4r isbn 1-55655-436-2 $770. with p/g) – us UPA [954]

India : its condition, religion, and missions / Bradbury, James – London: John Snow, 1884 [mf ed 1995] – (= ser Yale coll) – 253p – 1 – 0-524-09252-4 – mf#1995-0252 – us ATLA [954]

India : its life and thought / Jones, John Peter – New York: Macmillan, 1908 [mf ed 1995] – (= ser Yale coll) – xvi/448p (ill) – 1 – 0-524-09750-X – mf#1995-0750 – us ATLA [954]

India : its natives and missions / Trevor, George Herbert – London: Religious Tract Society [1859?] [mf ed 1995] – (= ser Yale coll) – xvi/344p (ill) – 1 – 0-524-09863-8 – mf#1995-0863 – us ATLA [954]

India – Madras: A F Lopes, v1-4 (1928-32) [mf ed 2005] – 1r – 1 – (lacks: v2 (1929) n2 p7-32) – mf#2005C-s003 – us ATLA [241]

India : the land and the people / Caird, James – [London], 1883 – (= ser 19th c books on british colonization) – 3mf – 9 – mf#1.1.726 – uk Chadwyck [954]

India : land of the black pagoda / Thomas, Lowell – London: Hutchinson & Co, 1931 – (= ser Samp: indian books) – (ill fr photos taken by h a chase and aut) – us CRL [915]

India – London: [British Comm of the Indian National Congress], 1890-jan 14 1921. v23-55 1905-jan 14 1921 – 1 – us CRL [954]

India – London, UK. 1890-14 Jan 1921. -w. 15 1/2 reels – 1 – uk British Libr Newspaper [072]

India : a nation. a plea for indian self-government / Besant, Annie Wood – London: T C & E C Jack; New York: Dodge Publ [1916] [mf ed 1995] – (= ser Yale coll; People's books) – 127p – 1 – 0-524-10011-X – (foreword by c p ramaswami aiyar) – mf#1995-1011 – us ATLA [954]

India : photographs and drawings of historical buildings / Griggs, William – London 1896 – (= ser 19th c art & architecture) – 5mf – 9 – mf#4.1.307 – uk Chadwyck [720]

India : a re-statement / Coupland, Reginald – London; New York: Humphrey Milford, Oxford University Press, 1945 – (= ser Samp: indian books) – us CRL [954]

India : what can it teach us? / Muller, Friedrich Max – Calcutta: Longmans Green & Co, 1934 – (= ser Samp: indian books) – (indian ed by k a nilakanta sastri) – us CRL [954]

India *see*
- Bharata ka rajapatra
- Gazette of india
- Lists and guides to official indian publications

India, a foreign view / Philip, Andre – London: Sidgwick & Jackson, Ltd, 1932 – (= ser Samp: indian books) – (int by the viscount burnham) – us CRL [301]

India a problem / Stover, Wilbur Brenner – 1st ed. Elgin IL: Brethren Pub House c1902 [mf ed 1992] – 1mf – 9 – 0-524-03504-0 – mf#1990-4726 – us ATLA [240]

India, a short cultural history / Rawlinson, Hugh George; ed by Seligman, C G – London: Cresset Press, 1937 – (= ser Samp: indian books) – us CRL [954]

India Agricultural Research Review Team *see* Report of the agricultural research review team

India, america, and world brotherhood / Sunderland, Jabez Thomas – Madras: Ganesh & Co, [1924] – (= ser Samp: indian books) – us CRL [327]

India analysed / Narain, Brij et al – London: Victor Gollangz Ltd, 1934- – (= ser Samp: indian books) – us CRL [327]

India and britain : a moral challenge / Andrews, Charles Freer – London: Student Christian Movement Press, 1935 – (= ser Samp: indian books) – us CRL [327]

India and buddhism : in translations / Aiken, Charles Francis et al – New York: Parke, Austin & Lipscomb c1917 [mf ed 1992] – (= ser The sacred books and early literature of the east 10) – 1mf – 9 – 0-524-04426-0 – (incl bibl ref) – mf#1991-0000 – us ATLA [280]

India and china : lectures delivered in china in may 1944 / Radhakrishnan, Sarvepalli – Bombay: Hind Kitabs, 1944 – (= ser Samp: indian books) – us CRL [327]

India and china : a photographic study / Nawrath, Ernst Alfred – London: Cresset Press, [1939] – (= ser Samp: indian books) – us CRL [720]

India and china : a thousand years of sino-indian cultural contact / Bagchi, Prabodh Chandra – Calcutta: China Press, 1944 – (= ser Samp: indian books) – us CRL [950]

India and christian missions / Storrow, Edward – London: John Snow, 1859 [mf ed 1995] – (= ser Yale coll) – vi/126p – 1 – 0-524-09096-3 – mf#1995-0096 – us ATLA [240]

India and christian opportunity / Beach, Harlan Page – New York: Student Volunteer Movement for Foreign Missions, 1908, c1904 – 1mf – 9 – 0-8370-6645-X – (includes appendixes) – mf#1986-0645 – us ATLA [240]

India and democracy / Schuster, George & Wint, Guy – London: Macmillan & Co, 1941 – (= ser Samp: indian books) – us CRL [954]

India and europe compared : being a popular view of the present state and future prospects of our eastern continental empire / Briggs, John – London, 1857 – (= ser 19th c books on british colonization) – 3mf – 9 – mf#1.1.4014 – uk Chadwyck [327]

India and freedom / Amery, Leopold Stennett – London: Oxford University Press, 1942 – (= ser Samp: indian books) – us CRL [954]

India and her people : a study in the social, political, educational, and religious conditions of india / Abhedananda, Swami – Calcutta: Ramakrishna Vedanta Math, 1945 – (= ser Samp: indian books) – us CRL [301]

India and imperial preference : a study in commercial policy / Madan, Bal Krishna – London; New York: Oxford University Press, 1939 – (= ser Samp: indian books) – (foreword by manohar lal) – us CRL [380]

India, and india missions : including sketches of the gigantic system of hinduism... / Duff, A – Edinburgh, 1839 – 8mf – 9 – mf#HTM-52 – ne IDC [915]

India and its faiths : a traveler's record / Pratt, James Bissett – Boston: Houghton Mifflin, 1915 – 2mf – 9 – 0-524-02659-9 – (incl bibl ref) – mf#1990-3089 – us ATLA [280]

India and its native princes : travels in central india and in the presidencies of bombay and bengal / Rousselet, L – London, 1876 – 14mf – 9 – mf#I-1133 – ne IDC [915]

India and its problems / Lilly, William Samuel – London: Sands, 1902 [mf ed 1995] – (= ser Yale coll) – xx/324p – 1 – 0-524-09817-4 – mf#1995-0817 – us ATLA [954]

India and java / Chatterjee, Bijan Raj – Calcutta: Prabasi Press, 1933 – (= ser Samp: indian books) – us CRL [327]

India and malaysia / Thoburn, James Mills – Cincinnati: Cranston & Curts; New York: Hunt & Eaton, 1892 – 2mf – 9 – 0-7905-6577-3 – mf#1988-2577 – us ATLA [240]

India and new order : an essay on human planning / Chatterjee, Sris Chandra – Calcutta: University of Calcutta, 1949 – (= ser Samp: indian books) – us CRL [710]

India and southern asia / Thoburn, James Mills – Cincinnati: Jennings & Graham; New York: Eaton & Mains [1907] [mf ed 1995] – (= ser Yale coll; Little books on missions) – 92p – 1 – 0-524-10243-0 – mf#1996-1243 – us ATLA [954]

India and the apostle thomas : an inquiry, with a critical analysis of the acta thomae / Medlycott, A E – London: David Nutt, 1905 – 1mf – us ATLA [240]

India and the apostle thomas : an inquiry, with a critical analysis of the acta thomae / Medlycott, Adolphus E – London: David Nutt, 1905 – 1mf – 9 – 0-7905-5070-9 – (incl bibl ref) – mf#1988-1070 – us ATLA [240]

India and the future / Archer, William – London: Hutchinson & Co, 1917 – (= ser Samp: indian books) – us CRL [954]

India and the gospel : or, an empire for the messiah / Clarkson, William & Archer, Thomas – 4th ed. London: John Snow, 1851 [mf ed 1995] – (= ser Yale coll) – xxiv/330p [mf ed 1995] – 1 – 0-524-10155-8 – mf#1995-1155 – us ATLA [240]

India and the hindoos : being a popular view of the geography, history, government, manners, customs, literature and religion of that ancient people / Ward, Ferdinand De Wilton – New York: Baker and Scribner, 1850 [mf ed 1995] – (= ser Yale coll) – xv/344p – 1 – 0-524-10015-2 – mf#1995-1015 – us ATLA [954]

India and the indian ocean : an essay on the influence of sea power on indian history / Panikkar, Kavalam Madhava – London: George Allen & Unwin, 1945 – (= ser Samp: indian books) – us CRL [954]

India and the pacific world / Naga, Kalidasa – Calcutta: Book Company Ltd, 1941 – (= ser Samp: indian books) – us CRL [954]

India and the simon report / Andrews, Charles Freer – London: George Allen & Unwin, 1930 – (= ser Samp: indian books) – us CRL [954]

India and the world : essays / Nehru, Jawaharlal – London: George Allen & Unwin, 1936 – (= ser Samp: indian books) – us CRL [954]

India and tibet : a history of the relations which have subsisted between the two countries from the time of warren hastings to 1910; with particular account of the mission to lhasa of 1904 / Younghusband, Francis Edward – London: John Murray, 1910 – (= ser Samp: indian books) – us CRL [327]

India and war / Roy, Manabendra Nath – Lucknow: Radical Democratic Party, 1942 – (= ser Samp: indian books) – us CRL [954]

India. Archaeological Survey *see* Annual reports

India. Archaeological Survey. Southern Circle *see* Progress report of the archaeological survey department, southern circle, for the year...

India as described in early texts of buddhism and jainism / Law, Bimala Churn – London: Luzac & Co, 1941 – (= ser Samp: indian books) – us CRL [954]

India as i knew it, 1885-1925 / O'Dwyer, Michael – London: Constable & Co, 1926 – (= ser Samp: indian books) – us CRL [954]

India as known to ancient and mediaeval europe / Ghosh, Praphullachandra – Calcutta: Hare Press, 1905 – (= ser Samp: indian books) – us CRL [915]

India as known to panini : a study of the cultural material in the ashtadhyayi / Agrawala, Vasudeva Sharana – [Lucknow: University of Lucknow, 1953] – (= ser Samp: indian books) – us CRL [954]

(India). Assam *see* List of non-confidential publications exempted from registration

India at a glance : a comprehensive reference book on india / Binani, G D & Rao, Rama – Bombay: Orient Longmans, 1954 – (= ser Samp: indian books) – us CRL [954]

India, at the death of akbar : an economic study / Moreland, William Harrison – London: Macmillan and Co, 1920 – (= ser Samp: indian books) – us CRL [330]

India awakening / Eddy, Sherwood – New York: Missionary Education Movt of the US & Canada 1912 [c1911] [mf ed 1995] – (= ser Yale coll) – 1r – 1 – 0-524-10017-9 – mf#1995-1017 – us ATLA [954]

(India). Bengal *see* List of publications (other than confidential)

(India). Bihar and Orissa Book Depot *see* List of publications (other than confidential)

(India). Bihar Book Depot *see* List of publications (other than confidential)

India, bond or free : a world problem / Besant, Annie Wood – London; New York: GP Putnam's Sons, Ltd, 1926 – (= ser Samp: indian books) – us CRL [954]

India cavalcade : some memorable yesterdays / Bhattacharya, Bhabani – Bombay: Nalanda Publications, c1948 – (= ser Samp: indian books) – us CRL [954]

India. Census Commissioner *see*
– Census commissioner's notes on census arrangements in individual provinces and states
– East india

India. Central Council of Local Self Government *see* Local self-government administration in states of india

(India). Central Provinces *see* List of official publications (other than confidential)

India. Central Provinces *see* List of chiefs and leading families

(India). Central Provinces. Government Press Book Depot *see* List of publications relating to acts, codes, rules, law books, reports, bulletins, civil list...

India Christian Mission *see* Indian christian mission

India Council of Scientific and Industrial Research 2nd Reviewing Committee *see* Report of the second reviewing committee of the council of scientific and industrial research

(India). Council of Scientific and Industrial Research et al *see* Memorandum on the programme of research for the development of national resources

India Council of Scientific and Industrial Research Reviewing Committee *see* Reviewing committee report

India cultures quarterly – v2-5. 1942-45; v15-40. 1957-85* – 4r – 1 – (cont: india's culture) – ISSN: 0019-4166 – mf#ATLA S0757 – us ATLA [954]

India. Dept. of Commercial Intelligence and Statistics *see* Accounts relating to the foreign trade and navigation of india

India. Dept of Education, Health, and Lands *see* List of publications (other than confidential)

India. Dept of Revenue and Agriculture *see* List of publications (other than confidential)

India Directorate Of National Sample Survey *see* National sample survey

India divided / Prasad, Rajendra – Bombay: Hind Kitabs, 1946 – (= ser Samp: indian books) – us CRL [954]

India dormida / Sosa, Julio Bautista – Panama, Panama. 1936 – 1r – us UF Libraries [972]

India during the raj: eyewitness accounts : diaries and related records held by the european manuscripts section in the oriental and india office collections at the british library, london – 3pt – 1 – (pt1: diaries & related records describing life in india c1750-1842 [25r] £2350 [mf ed 2004]; pt2:...c1819-59 [29r] £2750; pt3:...c1861-91 [20r] £1850; with d/g) – uk Matthew [954]

(India). Eastern Bengal and Assam *see* List of non-confidential publications exempted from registration

India. Famine Inquiry Commission *see* Mortality in bengal in 1943

India files 1880-1911 / General Council of the Evangelical Lutheran Church in North America. Board of Foreign Missions – 1r – 1 – (records consist of printed pamphlets & bklets, books & handwritten sermons; with ind; forms pt of: subgroup gc 16/2 india) – mf#xa0097r – us ATLA [242]

India for the indians / Ward, Dorothy Jane – London: Arthur Barker Ltd, 1949 – (= ser Samp: indian books) – us CRL [954]

India for the indians – and for england / Digby, William – London, 1885 – (= ser 19th c books on british colonization) – 4mf – 9 – mf#1.1.4799 – uk Chadwyck [954]

India. Ganges Canal Committee *see* Report... (n20,410c, of government of india, public works department, dated 24th feb 1866) to decide upon the propriety...

India gazette – 1782-88, 1822-43 – 46r – 1 – mf#3901 – uk Microform Academic [079]

India gazette – Calcutta, India. nov 1780-dec 1782 [wkly] – 1r – 1 – uk British Libr Newspaper [079]

India. High Commissioner in the United Kingdom *see* List of publications received in the publications branch

India. Home Dept *see* Proportions of europeans and natives in the public service

India Imperial Legislative Council *see* Speeches of the native members of the governor general's legislative council on the bengal tenancy bill

India. Imperial Record Dept *see* List of the heads of administration in india and of the india office in england

India impressions : with some notes of ceylon during a winter tour, 1906-7 / Crane, Walter – London: Methuen & Co, 1907 – (= ser Samp: indian books) – us CRL [915]

India in 1875-76 : the visit of the prince of wales. a chronicle of his royal journeyings in india, ceylon, spain, and portugal / Wheeler, George – London 1876 – (= ser 19th c british colonization) – 5mf – 9 – mf#1.1.7933 – uk Chadwyck [954]

India in kalidasa / Upadhyaya, Bhagwat Saran – Allahabad: Kitabistan, 1947 – (= ser Samp: indian books) – us CRL [490]

India in primitive christianity / Lillie, Arthur – London: K Paul, Trench, Truebner, 1909 – 1mf – 9 – 0-524-03674-8 – (incl bibl ref) – mf#1990-3252 – us ATLA [240]

India in the age of empire : the journals of michael pakenham edgeworth (1812-1881) from the bodleian library, oxford – 11r – 1 – £1050.00 – (with d/g) – uk Matthew [954]

India in the dark wood / Macnicol, Nicol – London: Edinburgh House Press, 1930 – (= ser Samp: indian books) – us CRL [280]

India in the new world order / Khanna, Radha Krishna – Lahore: Minerva Book Shop, 1942 – (= ser Samp: indian books) – us CRL [330]

India in the seventeenth century : as depicted by european travellers / Das Gupta, J N – Calcutta: University of Calcutta, 1916 – (= ser Samp: indian books) – us CRL [954]

India in world affairs, august 1947-january 1950 : a review of india's foreign relations from independence day to republic day / Karunakaran, Kotta P – London; New York: Oxford University Press, 1952 – (= ser Samp: indian books) – us CRL [327]

India in world politics : a historical analysis and appraisal / Sundaram, Lanka – Delhi: Sultan Chand & Co, 1944 – (= ser Samp: indian books) – us CRL [327]

India. Intelligence Bureau *see* The ghadr directory

India. Intelligence Bureau. Home Dept *see* Terrorism in india, 1917-1936

(India). Kodagu *see* List of non-confidential publications exempted from registration

India. Laws, Statutes, etc *see*
– The indian penal code, as modified up to the 1st august 1890
– Rules of business

India. Legislature. Legislative Assembly *see* Manual of business and procedure.

India letter links / ed by Ward, Ethel Ellen – Allahabad: Mission Press 1937-59 [irreg] [mf ed 2005] – 1r – 1 – (cont: letter links; lacks: [unknown]) – mf#2005c-s052 – us ATLA [242]

(India). Madras Public Dept *see*
– Memorandum on the indian owned english, vernacular and anglo-vernacular presses of the madras presidency and the french territories of pondicherry and karikal
– Memorandum on the indian owned english, vernacular and anglo-vernacular presses of the madras presidency, the indian states of hyderabad, mysore, coorg, travancore, cochin and pudukkottai and the french terrtories of pondicherry and karikal

India, malaysia, and the philippines : a practical study in missions / Oldham, W F – New York: Eaton & Mains; Cincinnati: Jennings & Graham, c1914 – 1mf – us ATLA [240]

India, malaysia, and the philippines : a practical study in missions / Oldham, William Fitzjames – New York: Eaton & Mains; Cincinnati: Jennings & Graham, c1914 – 1mf – 9 – 0-7905-6544-7 – mf#1988-2544 – us ATLA [240]

India. Military Finance Dept *see* List of publications (other than confidential)

India. Ministry of Education, Department of Archaeology *see* Archaeology in india

India. Ministry of Home Affairs. Office of the Registrar General *see* Census of india

The india mission of the free church of scotland : being report of the deputies to india in 1888-9; opinion of the missionaries in 1890 and minutes of the foreign missions committee...general assembly of 1891 – Edinburgh: The Committee, 1891 [mf ed 1995] – (= ser Yale coll) – vi/198p (ill) – 1 – 0-524-10135-3 – mf#1995-1135 – us ATLA [242]

India missionary bulletin / clergy monthly missionary supplement / clergy monthly supplement – Kurseong, India: St Mary's Theological College; Ranchi, India: Catholic Press, 1952-67 [mf ed 2001] – (= ser Christianity's encounter with world religions, 1850-1950) – 3r – 1 – (filmed with:) – mf#2001-s157-159 – us ATLA [240]

India missions / Macgregor, Rev Professor – Edinburgh, Scotland. 1874 – 1r – us UF Libraries [240]

India. National Archives *see*
– Confidential publications and home political files

India. National Congress *see* Years of freedom

India news, 1949-61 : from the high commission of india, london – London – 13r – 1 – mf#4899 – uk Microform Academic [072]

The india of aurangzib : topography, statistics, and roads, compared with the india of akbar / Sarkar, Jadunath – Calcutta: Bose Bros, 1901 – (= ser Samp: indian books) – us CRL [915]

India of my dreams / Gandhi, Mahatma – Bombay: Hind Kitabs, 1947 – (= ser Samp: indian books) – (comp by r k prabhu; foreword by rajendra prasad) – us CRL [954]

The india office / Seton, Malcolm Cotter Cariston – London, New York: GP Putnam's Sons, 1926 – (= ser Samp: indian books) – us CRL [350]

India. Office of the Economic Advisor *see* Guide to current official statistics.

India office records. burma materials – 111r – 1 – (ior reels 318-322: burma legislative council debates, 1923-29; ior reels 326-328: miscellaneous official publications on burma; ior reels 399-410: administrative reports, burma 1861/1862, 1864/1865-1935/1936; ior reels 450-464: burma finance and commerce proceedings, 1885-99; ior reels 466-477: burma foreign and political, 1884-99; ior reels 478-520: house proceedings, 1871-1899; ior reels 652-672: burma general proceedings) – mf$_3$r66g – us CRL [324]

India old and new : with a memorial address / Hopkins, Edward Washburn – New Haven: Yale University Press, 1913 – (= ser Yale bicentennial publications) – 1mf – 9 – 0-524-01772-7 – mf#1990-2620 – us ATLA [280]

India old and new : with a memorial address / Hopkins, Edward Washburn – New York: Charles Scribner's Sons, 1901 – (= ser Samp: indian books) – us CRL [490]

India, old and new / Chirol, Valentine – London: Macmillan and Co, 1921 – (= ser Samp: indian books) – us CRL [954]

India on the march / Nehru, Jawaharlal; ed by Bright, Jagat S – Lahore: Indian Print Works, 1946 – (= ser Samp: indian books) – us CRL [954]

India on trial : a study on present conditions / Woolacott, John Evans – London: Macmillan and Co, 1929 – (= ser Samp: indian books) – us CRL [954]

India; or facts submitted to illustrate the character and condition of the native inhabitants : with suggestions for reforming the present system of government / Rickards, Robert – London 1828-32 [mf ed Hildesheim 1995-98] – 2v on 9mf – 9 – €180.00 – 3-487-27532-5 – gw Olms [954]

India, pakistan, and the west / Spear, Thomas George Percival – London; New York: Oxford University Press, 1949 – (= ser Samp: indian books) – us CRL [954]

India, pakistan, ceylon / ed by Brown, W Norman – Ithaca, New York: Cornell University Press, 1951 – (= ser Samp: indian books) – us CRL [954]

India. Parliament *see* List of publications (periodical or ad hoc)
India, past and present : with minor essays on cognate subjects / Sasi Chandra Datt, rai bahadur – London 1880 – (= ser 19th c british colonization) – 6mf – 9 – mf#1.1.7523 – uk Chadwyck [954]
India. Public Works Dept *see* List of publications (other than confidential)
India reveals herself / Mathews, Basil Joseph – London; New York: Oxford University Press, 1937 – (= ser Samp: indian books) – us CRL [915]
India. Secretary of State *see* A list of archaeological reports published under the authority of the secretary of state, government of india, local governments, etc which are not included in the imperial series of such reports
India since cripps / Alexander, Horace Gundry – England; New York: Penguin Books, 1944 – (= ser Samp: indian books) – us CRL [915]
India speaking – Bombay: Vora & Co, Publishers, 1945 – (= ser Samp: indian books) – us CRL [954]
India steps forward : the story of the cabinet mission in india in words and pictures / Chander, Jag Parvesh – Lahore: Indian Print Works, 1946 – (= ser Samp: indian books) – us CRL [954]
India struggles for freedom : a history / Mukerjee, Hirendranath – Bombay: Kutub, 1946 – (= ser Samp: indian books) – us CRL [954]
India to-day / Dutt, Rajani Palme – Bombay: People's Pub House, 1947 – (= ser Samp: indian books) – us CRL [954]
India, today and tomorrow / Barns, Margarita – London: George Allen & Unwin, 1937 – (= ser Samp: indian books) – us CRL [301]
India tracts : containing, 1: an address to the proprietors of east-india stock; setting forth, the unavoidable necessity, and real motives, for the revolution in bengal, 1760... / Holwell, John Zephaniah – London: printed for T Becket, 1774 [mf ed 1995] – (= ser Yale coll) – vii/432p – 1 – 0-524-09181-1 – mf#1995-0181 – us ATLA [954]
India und india missions : including sketches of the gigantic system of hinduism, both in theory and practice: also, notices of some of the principal agencies employed in conducting the process of indian evangelization, &c. &c / Duff, Alexander – 2nd ed. Edinburgh: J Johnstone; London: Whittaker, 1840 – 2mf – 9 – 0-7905-4961-1 – mf#1988-0961 – us ATLA [242]
India under curzon and after / Praser, Lovat – London: William Heinemann, 1911 – (= ser Samp: indian books) – us CRL [954]
India under the british crown / Basu, Baman Das – Calcutta: R Chatterjee, 1933 – (= ser Samp: indian books) – (with the collaboration of phanindra nath bose and nagendra nath ghosh) – us CRL [954]
(India). United Provinces of Agra and Oudh *see* List of non-confidential publications exempted from registration
India weekly – London. 1977-1979 – 1 – ISSN: 0046-8959 – mf#8685 – us UMI ProQuest [954]
India, what can it teach us? : a course of lectures delivered before the university of cambridge / Meuller, Friedrich Max – London, New York: Longmans, Green, 1910 [mf ed 1995] – (= ser Yale coll) – xxii/315p – 1 – 0-524-09216-8 – mf#1995-0216 – us ATLA [954]
India, what can it teach us? : a course of lectures. delivered before the university of cambridge / Mueller, Friedrich Max – London: Longmans, Green, 1883 – 1mf – 9 – 0-524-03676-4 – mf#1990-3254 – us ATLA [470]
Indian : a paper devoted to the aborigines of north america, and especially to the indians of canada – Hagersville ON. v1 n1-24. dec 30 1885-dec 29 1886// – 1r – 1 – Can$110.00 – cn McLaren [305]
Indian / Second Division Association [US] – v1 n1-13 [1919 apr 15-jul 15] – 1r – 1 – (cont by: bulletin [second division association]) – mf#1890481 – us WHS [366]
The indian – Kuala Lumpur. Malaysia. -w. Dec 1935-Jun 1941. (6 reels) – 1 – uk British Libr Newspaper [072]
Indian Academy of Sciences *see*
– Academy proceedings in earth and planetary sciences
– Proceedings animal sciences
– Proceedings chemical sciences
– Proceedings earth and planetary sciences
– Proceedings mathematical sciences
– Proceedings plant sciences
Indian academy of sciences proceedings a – Bangalore. 1934-1979 (1) 1974-1979 (5) 1977-1979 (9) – ISSN: 0370-0089 – mf#8630 – us UMI ProQuest [500]
Indian academy of sciences proceedings b – Bangalore. 1934-1979 (1) 1975-1979 (5) 1977-1979 (9) – mf#8638 – us UMI ProQuest [500]
Indian advocate – 1846-55 – 1 – us Southern Baptist [242]
Indian affairs / American Association on Indian Affairs et al – v1 n2-3 [1933 oct-nov], v2 n1-4 [1934 apr-dec], v3 n1-3 [1935 apr-oct], v4 n1 [1936 mar], ns: v1 n2 – 1r – 1 – (cont by: news letter [american association on indian affairs]) – mf#628811 – us WHS [305]
Indian affairs / Association on American Indian Affairs – 1934-82 – (= ser American indian periodicals... 2) – 8mf – 9 – $105.00 – us UPA [305]
Indian affairs : newsletter of the american indian fund and the association of american indian affairs / American Indian Fund et al – n9 [nov 1954], n16 [may/jun 1956], n40 [feb 1961], n47 [aug 1962], n67-n105 [aug/oct 1967-1982 may] – 1r – 1 – mf#769315 – us WHS [305]
Indian affairs – Sisseton. 1949-1996 (1) 1972-1996 (5) 1975-1996 (9) – ISSN: 0046-8967 – mf#6669 – us UMI ProQuest [321]
Indian after-dinner stories / Panchapakesa Ayyar, Aiylam Subramanier – Bombay: DB Taraporevala Sons, 1927-1928 – (= ser Samp: indian books) – us CRL [390]
Indian alcohol times / California Urban Indian Health Council – v1 n1-v2 n7 [1981 jun-1983 dec] – 1r – 1 – mf#717153 – us WHS [362]
Indian and Eskimo Affairs Program [Canada] et al *see* Indian news
Indian and foreign review – New Delhi. 1963-1988 (1) 1978-1988 (5) 1978-1988 (9) – ISSN: 0019-4379 – mf#9913 – us UMI ProQuest [073]
Indian and Metis Brotherhood Organization of Stony Mountain Institution *see* I m b o talking leaves
Indian and Metis Friendship Centre *see* Masenayegan
Indian and singhalese missionary pictures – London: Baptist Missionary Society [1909] [mf ed 1995] – (= ser Yale coll) – x/220p (ill) – 1 – 0-524-09068-8 – (int by c e wilson) – mf#1995-0068 – us ATLA [242]
Indian and spanish neighbors / Johnston, Julia Harriette – New York: Fleming H Revell, c1905 [mf ed 1986] – (= ser Home mission study course 3) – 1mf – 9 – 0-8370-6582-8 – mf#1986-0582 – us ATLA [240]
Indian and white in the northwest : or, a history of catholicity in montana / Palladino, Lawrence Benedict – Baltimore: J Murphy 1894 [mf ed 1992] – 2mf – 9 – 0-524-04177-6 – (incl bibl ref) – mf#1990-4981 – us ATLA [241]
Indian architecture / Brown, Percy – Bombay: DB Taraporevala Sons & Co, 1942- – (= ser Samp: indian books) – us CRL [720]
Indian architecture / Gangoly, Ordhendra Coomar – Bombay: Kutub Publishers, 1946 – (= ser Samp: indian books) – us CRL [720]
Indian architecture : its psychology, structure, and history from the first muhammadan invasion to the present day / Havell, Ernest Binfield – London: John Murray, 1927 – (= ser Samp: indian books) – us CRL [720]
Indian archives – 1973 jan ?-1975 summer – 1r – 1 – mf#380996 – us WHS [305]
Indian archives / Antelope Indian Circle Archives – 1968-75 – (= ser American indian periodicals... 2) – 5mf – 9 – $95.00 – us UPA [305]
Indian arizona / Indian Development District of Arizona – v1 n1-6 [1978 may-oct] – 1r – 1 – (cont by: indian arizona news) – mf#396271 – us WHS [307]
Indian arizona news / Indian Development District of Arizona – v1 n7-v2 n12 [1978 nov-1980 apr] – 1r – 1 – (cont: indian arizona; cont by: arizona indian now) – mf#674418 – us WHS [307]
Indian art : essays / ed by Winstedt, Richard – London: Faber and Faber Ltd, 1947 – (= ser Samp: indian books) – us CRL [700]
Indian art of the buddhist period : with particular reference to the frescoes of ajanta / Yazdani, Ghulam – Oxford: University Press, 1937 – (= ser Samp: indian books) – us CRL [700]
Indian art series / New Mexico Association of Indian Affairs – n1-12, 1936 – (= ser American indian periodicals... 2) – 1mf – 9 – $95.00 – us UPA [700]
Indian art through the ages – [New Delhi]: Publications Division, Ministry of Information and Broadcasting, Govt of India, 1951 – (= ser Samp: indian books) – us CRL [700]
Indian baptist – Calcutta, India.v1-5, 1882-86.Baptist Mission Press – 1 – 52.20 – us ABHS [242]
The indian bazaar : 43 johnson st, victoria – S.l: s.n, 18– – 1mf – 9 – mf#15309 – cn Canadiana [390]
Indian bazaar ball : the committee of the indian bazaar, to be held at saint john, on tuesday the 7th july next – St John, NB?: s.n, 1846? – 1mf – 9 – mf#53110 – cn Canadiana [790]
Indian biographical archive (india, pakistan, bangladesh, sri lanka) (inba) = Indisches biographisches archiv (indien, pakistan, bangladesch, sri lanka) (inba) / Baillie, Laureen [comp] – [mf ed 1997-2000] – 538mf (1:24) – 9 – diazo €10,060.00 (silver €11,080 isbn: 978-3-598-34091-8) – ISBN-10: 3-598-34090-7 – ISBN-13: 978-3-598-34090-1 – (with printed ind) – gw Saur [954]
The indian borderland, 1880-1900 / Holdich, Thomas Hungerford – London: Methuen and Co, 1901 – (= ser Samp: indian books) – us CRL [355]
Indian bouquet / Hamiudullah, Zeb-un-Nisa – Calcutta: Gulistan Pub House, 1943 – (= ser Samp: indian books) – us CRL [954]
Indian Brotherhood of the NWT *see* Native press
The indian buddhist iconography : mainly based on the sadhanamala and other cognate tantric texts of rituals / Bhattacharyya, Benoytosh – London, New York: Oxford University Press, 1924 – (= ser Samp: indian books) – us CRL [700]
Indian caste / Wilson, John – Bombay: Times of India Office; Edinburgh: William Blackwood & Sons, 1877 [mf ed 1995] – (= ser Yale coll) – 2v – 1 – 0-524-09449-7 – mf#1995-0449 – us ATLA [305]
Indian census book under the act of 1928 – Fresno Co, CA – 5r – 1 – $250.00 – (a-f; f-m; m-r; r-z; suppl a-z) – mf#B06088 – us Library Micro [317]
Indian census rolls, 1885-1940 / U.S. Bureau of Indian Affairs – (= ser Records relating to census rolls and other enrollments) – 692r – 1 – (with printed guide) – mf#M595 – us Nat Archives [317]
Indian center news / Sacramento Indian Center – v1 n1-7 [1979 apr-oct] – 1r – 1 – (cont by: news [sacramento indian center]) – mf#615845 – us WHS [305]
Indian Center of San Jose [CA] *see* Newsletter
Indian Center of Topeka *see* Nish nau bah
Indian Chemical Society *see* Journal of the indian chemical society
The indian chief journeycake / Mitchell, S H – 1895 – 1 – 5.00 – us Southern Baptist [242]
The indian chief, journeycake / Mitchell, S H – Philadelphia: American Baptist Publication Society, 1895 – 1mf – 9 – 0-524-04385-X – mf#1991-2089 – us ATLA [975]
Indian child art : a handbook for teachers / Hellier, Gay – London; New York: Oxford University Press, 1951 – (= ser Samp: indian books) – us CRL [700]
Indian child welfare study, 1980-1981 / Toiyabe Indian Health Project – v1 n1-2 [1980 oct-dec], v2 n3 [1981 apr] – 1r – 1 – (cont by: family services newsletter) – mf#1336708 – us WHS [362]
Indian christian / Christian and Missionary Alliance – v10 n1-v11 n5 [1978 jan/feb-1979 sep/oct] – 1r – 1 – (cont by: indian life [rapid city, sd]; indian life [winnipeg, mb]) – mf#462820 – us WHS [240]
Indian christian mission : missionary notes / India Christian Mission – 1934-39 (complete) – 1r – 1 – mf#ATLA S0725B – us ATLA [240]
The indian christians of st. thomas otherwise called the christians of malabar : a sketch of their history, and an account of their present condition, as well as a discussion of the legend of st. thomas / Richards, William Joseph – London: Bemrose, 1908 – 1mf – 9 – 0-524-03102-9 – mf#1990-0827 – us ATLA [240]
The indian christians of st thomas: otherwise called the syrian christians of malabar.. / Richards, William Joseph – London: Bemrose, 1908. 138p. ill – 1 – us Wisconsin U Libr [240]
Indian chronicles – 32v – 9 – $234.00 – mf#0277 – us Brook [305]
The indian church during the great rebellion : an authentic narrative of the disasters that befell it... / Sherring, Matthew Atmore – 2nd ed, London: James Nisbet, 1859 [mf ed 1995] – (= ser Yale coll) – xii/355p – 1 – 0-524-09045-9 – mf#1995-0045 – us ATLA [954]
Indian church history : or, an account of the first planting of the gospel, in syria, mesopotamia, and india / Yeates, T – London, 1818 – 3mf – 9 – mf#HTM-221 – ne IDC [915]
Indian church history : or, an account of the first planting of the gospel in syria, mesopotamia, and india / Yeates, Thomas – London: printed for A Maxwell, 1818 [mf ed 1995] – (= ser Yale coll) – viii/208p – 1 – 0-524-09386-5 – mf#1995-0386 – us ATLA [240]
Indian claims commission annual reports / U.S. Dept of the Interior – 1968-77 + final report for 1946-78 [all publ] – (= ser Native american coll) – 13mf – 9 – $19.50 – (also incl in llmc's native american collection) – mf#llmc 88-004 – us LLMC [343]
Indian claims commission decisions – 1948-78. v1-43 [all publ] – (= ser Native american coll) – 485mf – 9 – $727.00 – (also included in llmc's native american collection) – mf#llmc 80-510 – us LLMC [343]
The indian colony of champa / Bose, Phanindra Nath – Madras: Theosophical Pub House, 1926 – (= ser Samp: indian books) – us CRL [930]
An indian commentary / Garrat, Geoffrey Theodore – London: Jonathan Cape, [1928] – (= ser Samp: indian books) – us CRL [954]
Indian constitutional documents, 1757-1939 / ed by Banerjee, Anil Chandra – Calcutta: A Mukherjee & Co, 1948- – (= ser Samp: indian books) – us CRL [324]
Indian constitutional reforms, government of india bill – [s.l: s.n.], 19– – 1 – (filmed with: india. legislative council: the imperial legislative council; the imperial legislative council manual 1916) – us CRL [954]
The indian contribution to english literature / Srinivasa Iyengar, K R – Bombay: Karnatak Pub House, 1945 – (= ser Samp: indian books) – us CRL [410]
Indian costumes = Bharatiya vesabhusa / Ghurye, Govind Sadashiv – Bombay: Popular Book Depot, 1951 – (= ser Samp: indian books) – us CRL [390]
Indian Country Communications [Hayward, WI] *see* News from indian country
Indian country today – Rapid City SD. 1992 jan-jun – 1r – 1 – (cont: lakota times; cont by: indian country today [rapid city, sd: northern plains ed]; indian country today [rapid city, sd: southwest ed]) – mf#2560986 – us WHS [071]
Indian crafts of guatemala and el salvador / Osborne, Lilly De Jongh – Norman, OK. 1965 – 1r – us UF Libraries [972]
The indian craftsman / Coomaraswamy, Ananda Kentish – London: Probsthain & Co., 1909. Appendices – 1 – us Wisconsin U Libr [954]
Indian creek baptist church. campbell county. jacksboro, tennessee : church records – Apr 1833-Apr 1954. Lacking: Jul 1862-Jan 1886; 1911-28. Formerly Mt. Pleasant on Indian Creek – 1 – $42.03 – us Southern Baptist [242]
Indian Creek Primitive Baptist Association *see* Minutes of the...annual session of the indian creek primitive baptist association
Indian crisis : the background / Hoyland, John Somervell – London: George Allen & Unwin, 1943 – (= ser Samp: indian books) – us CRL [954]
The indian crisis / Brockway, A Fenner – London: Victor Gollancz, 1930 – (= ser Samp: indian books) – us CRL [954]
The indian crisis : five sermons / Maurice, Frederick Denison – Cambridge: Macmillan, 1857 – 1mf – 9 – 0-524-00063-8 – mf#1989-2763 – us ATLA [954]
The indian crusader – 1969-76 – (= ser American indian periodicals... 1) – 3mf – 9 – $95.00 – us UPA [305]
Indian cultural influence in cambodia / Chatterjee, Bijan Raj – [Calcutta] University of Calcutta 1928 [mf ed 1989] – 1r with other items – 1 – (also available 2nd rev ed 1964 [mf-10289 seam reel 003/01]) – mf#mf-10289 seam reel 011/03 [§] – us CRL [930]
Indian cultural influence in cambodia / Chatterjee, Bijan Raj – Calcutta: University of Calcutta, 1928 – (= ser Samp: indian books) – us CRL [900]
Indian culture, its strands and trends : a study in contrasts / Datta, Hirendranath – Calcutta: Calcutta University, 1941 – (= ser Samp: indian books) – us CRL [900]
Indian culture through the ages / Venkateswara, Sekharipuram Vaidyanatha – London; New York: Longmans, Green, and Co, 1928-1932 – (= ser Samp: indian books) – us CRL [950]
Indian daily news – Calcutta, India. feb 1867-dec 1889 – 87r – 1 – uk British Libr Newspaper [079]
The indian deficit and the income tax / Maclean, James Mackenzie – London 1871 – (= ser 19th c british colonization) – 1mf – 9 – mf#1.1.7075 – uk Chadwyck [339]
Indian department [canada] : return to an address of the honourable the house of commons, dated 28 apr 1856...respecting alterations in the organization of the indian department in canada" – [s.l: s.n. 1856?] [mf ed 1987] – 1mf – 9 – 0-665-63353-X – mf#63353 – cn Canadiana [971]
Indian Development District of Arizona *see*
– Indian arizona
– Indian arizona news
Indian domestic economy and receipt book : comprising numerous directions for plain wholesome cookery, both oriental and english; with much miscellaneous matter answering for all general purposes of reference / Riddell, Robert Flower – Bombay 1852 – (= ser 19th c british colonization) – 7mf – 9 – mf#1.1.9621 – uk Chadwyck [640]

Indian dream lands / Mordecai, Margaret – London; New York: GP Putnams's Sons Ltd, 1925 – (= ser Samp: indian books) – us CRL [915]

The indian earthquake / Andrews, Charles Freer – London: George Allen & Unwin Ltd, 1935 – (= ser Samp: indian books) – us CRL [360]

Indian economic and social history review – New Delhi. 1963+ (1) 1976+ (5) 1976+ (9) – ISSN: 0019-4646 – mf#9323 – us UMI ProQuest [330]

Indian economic journal – Bombay. 1957+ (1) 1976+ (5) 1976+ (9) – ISSN: 0019-4662 – mf#2345 – us UMI ProQuest [338]

Indian economics : a comprehensive and critical survey of the economic problems of india / Jathar, Ganesh Bhaskar & Beri, S G – London: Oxford University Press, 1931-1932 – (= ser Samp: indian books) – us CRL [330]

Indian education / Coos County Education Service District [OR] – 1978 sep/oct – 1r – 1 – mf#639851 – us WHS [370]

Indian education / U.S. Bureau of Indian Affairs – 1936-65 – (= ser American indian periodicals... 2) – 37mf – 9 – $220.00 – us UPA [370]

Indian Education Center et al *see* Medicine bundle

Indian education newsletter – 1970-75 – (= ser American indian periodicals... 1) – 5mf – 9 – $95.00 – us UPA [370]

The indian educator – United Indians of All Tribes Foundation. sept 1981- jun-jul 1985 (irr) (mnthly) [bimnthly] – 1 – (numbering irregular) – us Wisconsin U Libr [370]

Indian election manifestos – 1967. Materials from 12 parties – 1 – us CRL [954]

Indian election results : newspapers, march 11-19 1971 – Charlottesville, NC: Alderman Library Photographic Services, 1971 – 1 – us CRL [954]

Indian election results, 1967 – Chicago: Uni of Chicago Lib, Dept of Photodup, [196-] – 3r – 1 – us CRL [954]

Indian embers / Lawrence, Lady – Oxford: George Ronald, [19–] – (= ser Samp: indian books) – us CRL [915]

The indian empire review – [London: Indian Empire Society] v1-8. nov 1931-1939 – 1 – us CRL [954]

The Indian Empire Society *see* [Pamphlets]

Indian evangelical review : a journal of missionary thought and effort – v1-29. jul 1873-1903 [complete] – 14r – 1 – mf#ATLA S0139 – us ATLA [242]

The indian exchange : shewing the enormous loss to india yearly and the remedy to be applied... / Owen, W H – Exeter 1879 – (= ser 19th c british colonization) – 1mf – 9 – mf#1.1.550 – uk Chadwyck [332]

Indian express – New Delhi, India. 1962-Jul 1995 – 194r – 1 – us L of C Photodup [079]

The indian eye on english life : or rambles of a pilgrim reformer / Malabari, Behramji Merwanji – Bombay: Apollo Print Works, 1895 – (= ser Samp: indian books) – us CRL [306]

Indian family defense / Association on American Indian Affairs – 1974-79 – (= ser American indian periodicals... 2) – 1mf – 9 – $95.00 – us UPA [305]

Indian family defense : a bulletinof the associationon american indian affairs, inc / Association on American Indian Affairs – n1-11 [1974 win-1979 feb] – 1r – 1 – mf#667625 – us WHS [366]

The indian famine : or water is the best remedy = Pani bihtar dawa hai / Heap, Charles Rogers – London, 1877 – (= ser 19th c british colonization) – 1mf – 9 – mf#1.1.2519 – uk Chadwyck [630]

Indian famines : their historical, financial, and other aspects containing remarks on their management, and some notes on preventive and mitigative measures / Blair, Charles – [Edinburgh], 1874 – (= ser 19th c books on british colonization) – 3mf – 9 – mf#1.1.7980 – uk Chadwyck [630]

Indian farming – New Delhi. 1976-1979 (1,5,9) – ISSN: 0019-4786 – mf#11012 – us UMI ProQuest [630]

Indian female evangelist *see* Church missionary society archive, section 2

The indian ferment : a traveller's tale / Alexander, Horace Gundry – London: Williams & Norgate Ltd, 1929 – (= ser Samp: indian books) – (int by c f andrews) – us CRL [915]

Indian films and film world, 1976 / ed by Jayabharathi – Madras: Jwala, 1976 – us CRL [790]

Indian finance in the days of the company / Banerjea, Pramathanath – London: Published for the University of Calcutta by Macmillan and Co, 1928 – (= ser Samp: indian books) – us CRL [332]

The indian fiscal policy / Adarkar, Bhalchandra Pundlik – Allahabad: Kitabistan, 1941 – (= ser Samp: indian books) – us CRL [332]

Indian folklore : being a collection of tales illustrating the customs and manners of the indian people / Jethabhai, Ganesh – Limbdi: Jaswatsinhji Print Press, 1903 – (= ser Samp: indian books) – us CRL [390]

Indian forerunner – v1 n2?-? [1972 feb-1974 fall] – 1r – 1 – mf#1058903 – us WHS [071]

Indian gems for the master's grown : 1. indian devotee and his disciples: 2. from bondage to freedom, or, the life of tulsi paul / Droese, Miss – London: Religious Tract Society, 1892 – 1mf – 9 – 0-8370-6734-0 – mf#1986-0734 – us ATLA [240]

Indian gods and kings : the story of a living past / Hawkridge, Emma – London: Rich & Cowan Ltd, [1935] – (= ser Samp: indian books) – us CRL [954]

Indian head / Second Indian Head Division Association – 1934 aug-1976 dec – 1r – 1 – (cont: bulletin [second division association]; cont by: indianhead [fort benning, ga]) – mf#1896077 – us WHS [366]

Indian hemp: a social menace / Johnson, Donald McIntosh – Foreword by H. Pullar-Strecker. London: C. Johnson, 1952. 112p. Bibliography – 1 – us Wisconsin U Libr [360]

Indian herald – Allahabad, India. apr 1879-mar 1882 [wkly] – 10r – 1 – uk British Libr Newspaper [079]

The indian heroes / Kincaid, Charles Augustus – London, New York: Humphrey Milford: Oxford University Press, 1915 – (= ser Samp: indian books) – us CRL [954]

Indian historical studies / Rawlinson, Hugh George – London; New York: Longmans, Green, and Co, 1913 – (= ser Samp: indian books) – us CRL [954]

Indian Home Guards. First Regiment *see* Day book

Indian home rule / Gandhi, Mahatma – Reprinted with new forward by the author. Madras: Ganesh & Co., 1919?. 136p – 1 – us Wisconsin U Libr [954]

Indian Homemakers' Association of British Columbia *see* Indian voice

Indian horizons – New Delhi. 1952-1995 (1) 1976-1995 (5) 1976-1995 (9) – ISSN: 0378-2964 – mf#7460 – us UMI ProQuest [400]

Indian idealism / Dasgupta, Surendranath – Cambridge: University Press, 1933 – (= ser Samp: indian books) – us CRL [180]

Indian ideals in education, philosophy and religion, and art / Besant, Annie Wood – Calcutta: Calcutta University Press, 1925 – (= ser Samp: indian books) – us CRL [301]

Indian idylls / Abbott, Anstice – London: Elliot Stock, 1911 [mf ed 1995] – (= ser Yale coll) – 160p (ill) – 1 – 0-524-09890-5 – (int by george smith) – mf#1995-0890 – us ATLA [954]

Indian idylls : from the sanskrit of the mahabharata / Arnold, Edwin – Boston: Little, Brown, 1907 [mf ed 1995] – (= ser Yale coll) – 318p – 1 – 0-524-09417-9 – mf#1995-0417 – us ATLA [490]

An indian in western europe / Panchapakesa Ayyar, Aiylam Subramanier – Madras: C Coomarasawmy Naidu & Sons, 1942 – (= ser Samp: indian books) – us CRL [914]

Indian industry and its problems / Soni, Hans Raj – London; New York: Longmans, Green and Co, 1932- – (= ser Samp: indian books) – us CRL [338]

Indian influences in old-balinese art / Stutterheim, Willem Frederik – London: India Society, 1935 – (= ser Samp: indian books) – (trans fr dutch by claire holt) – us CRL [700]

Indian Institute of Metals *see* Transactions of the indian institute of metals

Indian Institute of Science *see* Annual report of the director to the council

Indian Institute of Science. Bangalore *see*
– Annual report
– Annual report of the council of the indian institute of science, bangalore
– Appendix to the...annual report of the council of the indian institute of science, bangalore

Indian internal politics *see* Political pamphlets from the indian subcontinent

Indian islam : a religious history of islam in india / Titus, Murray Thurston – London; New York: Humphrey Milford: Oxford University Press, 1930 – (= ser Samp: indian books) – us CRL [260]

Indian jottings : from ten year's experience in and around poona city / Elwin, Edward Fenton – London: John Murray, 1907 [mf ed 1995] – (= ser Yale coll) – xi/314p (ill) – 1 – 0-524-09102-1 – mf#1995-0102 – us ATLA [280]

Indian jottings from ten years' experience in and around poona city / Elwin, E F – London, 1907 – 4mf – 9 – mf#HT-44 – ne IDC [915]

Indian journal – Eufaula, Muskogee OK. 1895 jul 6, 1972 dec 21 – 1r – 1 – mf#633994 – us WHS [305]

Indian journal of adult education – New Delhi. 1975-1996 (1) 1975-1996 (5) 1975-1996 (9) – ISSN: 0019-5006 – mf#10442 – us UMI ProQuest [374]

Indian journal of agricultural sciences – New Delhi. 1975-1996 (1,5,9) – ISSN: 0019-5022 – mf#10654 – us UMI ProQuest [630]

Indian journal of animal sciences – New Delhi. 1975-1995 (1,5,9) – ISSN: 0367-8318 – mf#10655 – us UMI ProQuest [636]

Indian journal of applied psychology – Madras. 1964-1982 (1) 1972-1982 (5) 1974-1982 (9) – ISSN: 0019-5073 – mf#7275 – us UMI ProQuest [150]

Indian journal of cancer – Bombay. 1975-1980 (1) 1975-1980 (5) 1975-1980 (9) – ISSN: 0019-509X – mf#10454 – us UMI ProQuest [616]

Indian journal of dermatology, venereology and leprology – Vellore. 1980-1980 (1) 1980-1980 (5) 1980-1980 (9) – ISSN: 0378-6323 – mf#688,01 – us UMI ProQuest [616]

Indian journal of experimental psychology – Madras. 1972-1972 (1) 1972-1972 (5) (9) – ISSN: 0019-5197 – mf#7274 – us UMI ProQuest [150]

Indian journal of pediatrics – New Delhi. 1972+ (1) 1972+ (5) 1974+ (9) – ISSN: 0019-5456 – mf#7252 – us UMI ProQuest [618]

Indian journal of pharmaceutical sciences – Bombay. 1978-1982 (1) 1978-1982 (5) 1978-1982 (9) – (cont: indian journal of pharmacy) – ISSN: 0250-474X – mf#7397,01 – us UMI ProQuest [615]

Indian journal of pharmacy – Bombay. 1972-1978 (1) 1972-1978 (5) 1977-1978 (9) – (cont by: indian journal of pharmaceutical sciences) – ISSN: 0019-5472 – mf#7397 – us UMI ProQuest [615]

Indian journal of political studies – Jodhpur. 1977-1978 (1,5,9) – mf#11294 – us UMI ProQuest [320]

Indian journal of power and river valley development – Calcutta. 1954-1980 (1) 1978-1980 (5) 1978-1980 (9) – ISSN: 0019-5537 – mf#7286 – us UMI ProQuest [627]

Indian journal of psychology – New Delhi. 1974-1994 (1) 1975-1994 (5) 1977-1994 (9) – ISSN: 0019-5553 – mf#9106 – us UMI ProQuest [150]

Indian journal of social research – v. 1-10. 1960-69 – 1 – us AMS Press [300]

Indian journal of social work – Bombay. 1940+ (1) 1972+ (5) 1976+ (9) – ISSN: 0019-5634 – mf#7019 – us UMI ProQuest [360]

Indian journal of technology – New Delhi. 1963-1993 (1) 1970-1993 (5) 1976-1993 (9) – ISSN: 0019-5669 – mf#3034 – us UMI ProQuest [600]

Indian key massacre – S.I., S.I? . 193-? – 1r – us UF Libraries [978]

Indian law reporter – v1-28. 1974-2001 – 5,6,9 – $1801.00 set – (v1-11 1974-84 on reel $514. v12-28 1985-2001 on mf $1287) – ISSN: 0097-1154 – mf#103301 – us Hein [340]

Indian leader / Haskell Indian Junior College et al – 1974 jan 25-1986 may 2 – 1r – 1 – mf#166068 – us WHS [373]

The indian leader – Lawrence KS: Haskell Institute. v18-77. 1914-73 – 1 – $432.00 – mf#0278 – us Brook [305]

Indian Legal Information Development Service et al *see* Legislative review

Indian legend / Buck, Gladys – S.I., S.I? . 1938 – 1r – us UF Libraries [978]

Indian liberalism : a study / Naik, Vasant Narayan – Bombay: Published for the National Liberal Federation of India by Padma Publications, 1945 – (= ser Samp: indian books) – (int by sivaswamy aiyer) – us CRL [954]

Indian librarian – Jullundur City. 1947-1982 [1]; 1976-1982 [5,9] – ISSN: 0019-5774 – mf#1942 – us UMI ProQuest [020]

Indian Library Association *see* Bulletin – indian library association

Indian life in the great north-west / Young, Egerton Ryerson – London: S W Partridge, [1900?] – 2mf – 9 – 0-665-30692-X – mf#30692 – cn Canadiana [305]

Indian literature in china and the far east / Mukherji, Probhat Kumar – Calcutta: Greater India Society [19–] [mf ed 1996] – (= ser Samp: indian books) – 1r – 1 – (filmed with other items) – mf#mf-10881 r038 – us CRL [490]

The indian literatures of today : a symposium. essays presented at Jaipur, October 20th-22nd, 1945 / ed by Bharatan Kumarappa – Bombay: Publ for the PEN All-India Centre by The International Book House, 1947 – (= ser Samp: indian books) – us CRL [490]

Indian logic and atomism : an exposition of the nyaya and vaicesika systems / Keith, Arthur Berriedale – Oxford: Clarendon Press, 1921 – (= ser Samp: indian books) – us CRL [160]

An indian looks at america / Abbas, Khwaja Ahmad – Bombay: Thacker & Co, 1943 – (= ser Samp: indian books) – us CRL [917]

The indian magazine – Ohsweken, Ont: [s.n, 1893-1897] – 9 – mf#P04020 – cn Canadiana [630]

Indian management – New Delhi. 1961-1989 (1) 1971-1971 (5) (9) – ISSN: 0019-5812 – mf#5973 – us UMI ProQuest [650]

Indian Medical Association *see* Journal of the indian medical association

Indian military of zambia, rhodesia, and malawi / Dotson, Floyd – New Haven, CT. 1968 – 1r – us UF Libraries [960]

Indian minority in south africa / Mukherji, S B – New Delhi, India. 1959 – 1r – us UF Libraries [960]

Indian mirror – Calcutta, India. -w. 1878-89. 22 reels – 1 – uk British Libr Newspaper [072]

The indian mission of the irish presbyterian church : a history of fifty years of work in kathiawar and gujarat / Jeffrey, Robert – London: Nisbet, 1890 [mf ed 1995] – (= ser Yale coll) – 279p – 1 – 0-524-09200-1 – (incl bibl ref and ind) – mf#1995-0200 – us ATLA [242]

Indian missionary – Oklahoma. 1884-91 – 1 – us Southern Baptist [242]

Indian missionary directory and memorial volume / Badley, Brenton Hamline – 3rd ed. Calcutta: Methodist Publ House; New York: Phillips & Hunt, 1886 [mf ed 1995] – (= ser Yale coll) – x/302p – 1 – 0-524-09917-0 – mf#1995-0843 – us ATLA [240]

Indian missionary manual : hints to young missionaries in india / Murdoch, John – 4th rev enl ed. London: James Nisbet, 1895 – 2mf – 9 – 0-8370-6227-6 – (incl app and ind) – mf#1986-0227 – us ATLA [240]

Indian missionary manual : hints to young missionaries in india / Murdoch, John [comp] – 3rd rev ed. London: James Nisbet, 1889 [mf ed 1995] – (= ser Yale coll) – x/613p – 1 – 0-524-09059-9 – mf#1995-0059 – us ATLA [240]

Indian missionary manual : hints to young missionaries in india: with lists of books / Murdoch, John – 2d ed., rev. London: Seeley, Jackson, and Halliday, 1870. Chicago: Dep of Photodup, U of Chicago Lib, 1971 (1r); Evanston: American Theol Lib Assoc, 1984 (1r) – 1 – 0-8370-0314-8 – mf#1984-B210 – us ATLA [240]

Indian missionary reminiscences, principally of the wyandot nation : in which is exhibited the efficacy of the gospel in elevating ignorant and savage men / Elliott, Charles – New-York: Pub by Lane & Scott for the Sunday-School Union of the ME Church, 1850 – 3mf – 9 – 0-524-07413-5 – mf#1991-3073 – us ATLA [240]

Indian missions / Frere, Bartle – 3rd ed. London: John Murray, 1874 [mf ed 1995] – (= ser Yale coll) – vi/102p – 1 – 0-524-09950-2 – (with app) – mf#1995-0950 – us ATLA [240]

Indian missions / Miller, William – Edinburgh, Scotland. 1878 – 1r – us UF Libraries [240]

Indian missions in guiana / Brett, William Henry – London, 1851 – (= ser 19th c books on british colonization) – 4mf – 9 – mf#1.1.6345 – uk Chadwyck [240]

The indian monetary policy / Adarkar, Bhalchandra Pundlik – Allahabad: Kitabistan, 1939 – (= ser Samp: indian books) – us CRL [339]

Indian money matters : the story of a famine insurance fund, and what was done with it. a speech...delivered in the house of commons, on august 27th, 1889, during the debate on the indian financial accounts / Bradlaugh, Charles – London, [1889] – (= ser 19th c books on british colonization) – 1mf – 9 – mf#1.1.4915 – uk Chadwyck [336]

Indian moral instruction and caste problems : solutions / Benton, Alexander Hay – London, New York: Longmans, Green, 1917 [mf ed 1995] – (= ser Yale coll) – xi/121p – 1 – 0-524-09089-0 – mf#1995-0089 – us ATLA [305]

Indian mounds / Coll, Aloyisus – S.I., S.I? . 193-? – 1r – us UF Libraries [978]

The indian musalmans : are they bound in conscience to rebel against the queen? / Hunter, William Wilson – London: Truebner and Co, 1871 – (= ser Samp: indian books) – us CRL [954]

The indian muse in english garb / Malabari, Behramji Merwanji – Bombay: Merwanjee Nowrojee Daboo, 1876 – (= ser Samp: indian books) – us CRL [490]

Indian music, scientific and practical : its origin, history and divisions; writers of old and modern times; description and classification of rages and ragnis, meanings and measures of tals and surs; forms and uses of musical instruments / Prasad, Nanak – Aligarh: Viddyasager 1906 [mf ed 1993] – 1r – 1 – mf#pres. film 127 – us Sibley [780]

Indian Musicological Society *see* Journal of the indian musicological society

The indian mutiny in perspective / MacMunn, George Fletcher – London: G Bell & Sons, 1931 – (= ser Samp: indian books) – us CRL [954]

Indian myth and legend / Mackenzie, Donald Alexander – London: Gresham Pub Co, 1913 – (= ser Samp: indian books) – (ill in colour by warwick goble and numerous monochrome plates) – us CRL [390]

INDIAN

Indian mythology according to the mahabharata : in outline / Fausboell, Viggo – London: Luzac, 1903 – (= ser Luzac's Oriental Religions Series) – 1mf – 9 – 0-524-01055-2 – mf#1990-2203 – us ATLA [280]
Indian nation – Patna, India. Apr 1944-Nov 1989 – 145r – 1 – us L of C Photodup [079]
Indian National Congress *see* Election manifesto
The indian national congress / Dasgupta, Hemendra Nath – [Calcutta: JK Das Gupta], 1946- – (= ser Samp: indian books) – us CRL [954]
Indian National Congress. All-India Congress Committee *see*
– Files concerning bengal, 1927-1947
– Papers, 1914-1920
Indian national evolution : a brief survey of the origin and progress of the indian national congress and the growth of indian nationalism / Mazumdar, Amvika Charan – Madras: GA Natesan & Co, 1917 – (= ser Samp: indian books) – us CRL [954]
Indian nationalism : an independent estimate / Bevan, Edwyn Robert – London: Macmillan and Co, 1913 – (= ser Samp: indian books) – us CRL [954]
Indian nationalism : its principles and personalities / Pal, Bipin Chandra – Madras, SE: SR Murthy & Co, [1918] – (= ser Samp: indian books) – us CRL [954]
Indian nationality / Gilchrist, Robert Niven – London; New York: Longmans, Green and Co, 1920 – (= ser Samp: indian books) – (int by ramsay muir) – us CRL [954]
Indian natural resources : a bulletinof the association on american indian affairs, inc / Association on American Indian Affairs – n101-6 [1977 may-1980 dec] – 1r – 1 – mf#634349 – us WHS [333]
Indian news / Indian and Eskimo Affairs Program [Canada] et al – 1966 apr-1977 dec, 1977-1982 jun – 2r – 1 – mf#662014 – us WHS [305]
Indian news / Tri-County Indian Development Council – v2 iss 4-7 [1979 jun-aug] – 1r – 1 – mf#626716 – us WHS [305]
Indian news and chronicle of eastern affairs – London, UK. 11 Jun 1840-6 Dec 1843; 1844-27 Jul 1858. -w – 1 – uk British Libr Newspaper [072]
Indian news notes – v6 n5 [1982 aug 5]-v10 n24 [1986 jun 20] – 1r – 1 – (cont: indian news clips; cont by: indian news [washington, dc]) – mf#1208386 – us WHS [305]
An indian news sheet – London: Indian Empire Society. n1-37. apr 1940-oct 1949 – 1 – us CRL [954]
Indian newspaper reports, c1868-1942 : from the british library, london – 6pt – 1 – (pt1: bengal 1874-1903 [27r] £2500; pt2: bengal 1904-16 [26r] £2450; pt3: punjab, agra, oudh, rajputana & central provinces c1868-96 [29r] £2700; pt4: north western & united provinces 1897-1937 [29r] £2700; pt5: madras 1876-1921 [32r] £3050; pt6: bombay, 1874-98 [18r] £1700; with d/g) – uk Matthew [079]
Indian notes : [especially from the field of the american presbyterian mission in western india – Poona: printed at the Orphanage Press v1-6 1893-98 [mthly] [mf ed 2005] – 6v on 1r – 1 – (printed in: poona, jan-mar 1893; bombay, by the anglo-vernacular press, apr 1893-98; lacks: v5 [1897]) – mf#2005c-s039 – us ATLA [242]
Indian notes and monographs / New York City. Museum of the American Indian, Heye Foundation – v1-12. 1919-60 – 1 – $72.00 – (misc ser: v1-59 1920-73 $162 [0408]) – mf#0407 – us Brook [305]
Indian notes and queries : including panjab notes and queries, and north indian notes and queries – Allahabad, 1883-1887 v1-4; 1891-1896 v1-5 – 35mf – 8 – mf#I-287 – ne IDC [954]
Indian observer – Calcutta, India. -w. Feb 1871-May 1872. 2 reels – 1 – uk British Libr Newspaper [072]
The indian ocean: political and strategic future: hearings. / U.S. Congress. House. Committee on Foreign Affairs. Subcommittee on National Security Policy and Scientific Developments – Washington, Govt. Print. Off., 1971. 242 p. LL-2361 – 1 – us L of C Photodup [340]
Indian opinion – Durban and Phoenix. South Africa. -w. Jun 1903-Dec 1916. (15 reels) – 1 – uk British Libr Newspaper [079]
Indian opinion – Phoenix: International Printing Press, jan 8 1960-aug 4 1961 – us CRL [975]
Indian opinion – Phoenix, Natal: International Printing Press, 1917-jun 1934; 1935; 1937-47; 1953-jun 1957 – 1 – us CRL [071]
The indian outlook : a study in the way of service / Holland, Wes – London: Church Missionary Society, 1927 – (= ser Samp: indian books) – us CRL [350]
Indian painting / Brown, Percy – Calcutta: Association Press; New York: Oxford University Press, [1918] – (= ser Samp: indian books) – us CRL [750]
Indian painting in the punjab hills : essays / Archer, William George – London: His Majesty's Stationery Office, 1952 – (= ser Samp: indian books) – us CRL [750]
Indian painting under the mughals, ad 1550 to ad 1750 / Brown, Percy – Oxford: Clarendon Press, 1924 – (= ser Samp: indian books) – us CRL [750]
The indian penal code : (act 45 of 1860) / Morgan, Walter & Macpherson, Arthur George – Calcutta 1861 – (= ser 19th c british colonization) – 6mf – 9 – (notes by w morgan & a g macpherson) – mf#1.1.5584 – uk Chadwyck [345]
The indian penal code, as modified up to the 1st august 1890 / India. Laws, Statutes, etc – Calcutta: Superintendent of Government Printing, 1890. 203p. LL-996 – 1 – us L of C Photodup [348]
Indian people in natal / Kuper, Hilda – Pietermaritzburg, South Africa. 1960 – 1r – us UF Libraries [960]
Indian philosophical review – Baroda [etc]: Indian Philosophical Assoc. [v1 n2-v3] oct 1917-1920 – 1 – us CRL [100]
Indian philosophy / Radhakrishnan, Sarvepalli – London: George Allen & Unwin Ltd, 1948 – (= ser Samp: indian books) – us CRL [180]
Indian philosophy and modern culture / Brunton, Paul – London; New York: Rider and Co, [194-] – (= ser Samp: indian books) – us CRL [180]
Indian pilgrimage / Shahani, Ranjee G – London: Michael Joseph Ltd, 1939 – (= ser Samp: indian books) – us CRL [954]
Indian pioneer – Kuala Lumpur. Malaysia. -w. Mar 1929-Aug 1930. (1 reel) – 1 – uk British Libr Newspaper [079]
Indian plastics review – Calcutta. 1971-1972 (1) 1971-1972 (5) (9) – ISSN: 0019-610X – mf#7880 – us UMI ProQuest [660]
Indian poetry : and, indian idylls / Arnold, Edwin – London: Kegan Paul, Trench, Trubner; New York: EP Dutton & Co, 1915 – (= ser Samp: indian books) – us CRL [490]
Indian poetry : containing "the indian song of songs," from the sanskrit of the gaita govinda of jayadeva, two books from "the iliad of india" (mahabharata), "proverbial wisdom" from the shlokas of the hitopadesa, and other oriental poems – London: Kegan Paul, Trench, Truebner, 1909 – (= ser Truebner's Oriental Series) – 1mf – 9 – 0-524-01155-9 – mf#1990-2231 – us ATLA [470]
Indian political science review – Delhi. 1966-1985 (1) 1975-1985 (5) 1976-1985 (9) – ISSN: 0019-6126 – mf#7474 – us UMI ProQuest [320]
Indian politics : a survey / Gwynn, John Tudor – London: Nisbet & Co, 1924 – (= ser Samp: indian books) – (int by lord meston) – us CRL [954]
Indian politics since the mutiny : being an account of the development of public life and political institutions and of prominent political personalities / Chintamani, Chirravoori Yajneswara – Waltair: Andhra University, 1937 – (= ser Samp: indian books) – us CRL [954]
Indian polity : a view of the system of administration in india / Chesney, George Tomkyns – London, 1868 – (= ser 19th c books on british colonization) – 6mf – 9 – mf#1.1.7140 – uk Chadwyck [954]
The indian press : a history of the growth of public opinion in india / Barns, Margarita – [London]: George Allen & Unwin, 1940 – (= ser Samp: indian books) – us CRL [070]
An indian priestess : the life of chundra lela / Lee, Ada – London: Morgan & Scott, [1912?] [mf ed 1990] – 1mf – 9 – 0-7905-6531-5 – (int by lord kinnaird) – mf#1988-2531 – us ATLA [920]
An indian priestess : the life of chundra lela / Lee, Ada – London: Morgan & Scott, [1902] [mf ed 1995] – (= ser Yale coll) – 121p (ill) – 1 – 0-524-10056-X – (int by lord kinnaird. also available in mf) – mf#1995-1056 – us ATLA [920]
The indian primer : or, the way of training up of our indian youth in the good knowledge of god, 1669 / Eliot, John – Edinburgh: Andrew Elliot, 1880 – 1mf – 9 – 0-524-01648-8 – mf#1990-0469 – us ATLA [490]
The indian princes in council : a record of the chancellorship of his highness, the maharaja of patiala, 1926-1931 and 1933-1936 / Panikkar, Kavalam Madhava – London: Oxford University Press: Humphrey Milford, 1936 – (= ser Samp: indian books) – (foreword by the maharaja of bikaner) – us CRL [954]
The indian princess, or la belle savage / Bray, John – Philadelphia: George E. Blake 1808. MUSIC 464 – 1 – us L of C Photodup [780]
The indian problem / Coupland, Reginald – London, New York: Humphrey Milford, Oxford University Press, 1943-1944 – (= ser Samp: indian books) – us CRL [954]
The indian problem in kenya / Maini, P L – London, 1944 – us CRL [305]
The indian problem solved : undeveloped wealth in india and state reproductive works – London, [1875] – (= ser 19th c british colonization) – 5mf – 9 – mf#1.1.4002 – uk Chadwyck [339]
Indian problems / Hanna, Henry Bathurst – [London], 1895-[1897] – (= ser 19th c books on british colonization) – 4mf – 9 – mf#1.1.6281 – uk Chadwyck [954]
Indian problems / Mitra, Siddha Mohana – London: John Murray, 1908 – (= ser Samp: indian books) – (int by sir george birdwood) – us CRL [954]
Indian problems in religion, education, politics / Whitehead, Henry – London: Constable & Co, 1924 – (= ser Samp: indian books) – us CRL [954]
Indian progress / Associated Committee of Friends on Indian Affairs – 1959-74 – (= ser American indian periodicals... 2) – 6mf – 9 – $95.00 – us UPA [305]
Indian proscribed tracts, 1907-1947 – Chicago, IL: Uni of Chicago Photodup Dept, [19–?] – 1 – us CRL [954]
Indian psychology : perception / Sinha, Jadunath – London: Kegan Paul, Trench, Trubner & Co, 1934 – (= ser Samp: indian books) – us CRL [280]
Indian public opinion – Lahore, Pakistan. -w. March 1870-Feb 1877. c49 reels – 1 – uk British Libr Newspaper [072]
Indian public opinion and punjab times – Lahore, India. mar 1870-feb 1877 – 60r – 1 – uk British Libr Newspaper [079]
Indian pulp and paper – Calcutta. 1972-1981 (1) 1975-1981 (5) 1975-1981 (9) – ISSN: 0019-6231 – mf#8172 – us UMI ProQuest [670]
The indian races of america : comprising a general view (historical and descriptive) of all the most celebrated tribes throughout the continent and adjacent islands... / Brownell, Charles De Wolf – Boston: Dayton & Wentworth, 1855 – 8mf – 9 – mf#33261 – cn Canadiana [305]
Indian realism / Sinha, Jadunath – London: Kegan Paul, Trench, Trubner & Co, 1938 – (= ser Samp: indian books) – us CRL [954]
The indian rebellion : its causes and results. in a series of letters / Duff, Alexander – New York: Robert Carter, 1858 [mf ed 1996] – (= ser Yale coll) – iv/408p – 1 – 0-524-10206-6 – mf#1996-1206 – us ATLA [954]
Indian recollections / Statham, John – London 1832 [mf ed Hildesheim 1995-98] – 1v on 3mf – 9 – €90.00 – 3-487-27483-3 – gw Olms [880]
Indian record / Oblates of Mary Immaculate – v36 n1/2-12 [1973 jan/feb-dec], 1978 jan/feb-dec 1981/1982 win-1987 dec – 2r – 1 – (cont: indian missionary record) – mf#185234 – us WHS [241]
Indian record – Winnipeg, 1964-70 – (= ser American indian periodicals... 1) – 7mf – 9 – $95.00 – us UPA [305]
Indian records : with a commercial view of the relations between the british government and the nawabs nazim of bengal, behar and orissa – London, 1870 – (= ser 19th c british colonization) – 4mf – 9 – mf#1.1.368 – uk Chadwyck [954]
Indian records of the united society for the propagation of the gospel, 1840-1861 – 4r – 1 – mf#96127 – uk Microform Academic [220]
Indian records of the united society for the propagation of the gospel, 1856-1900 – 22r – 1 – mf#96126 – uk Microform Academic [220]
Indian recreations : consisting chiefly of strictures on the domestic and rural economy of the mahomedans and hindoos / Tennant, William – 2nd rev enl corr ed. London: printed...for Longman, Hurst, Rees, and Orme, 1804 [mf ed 1995] – (= ser Yale coll) – 2v (ill) – 1 – 0-524-09752-6 – mf#1995-0752 – us ATLA [954]
Indian recreations : consisting chiefly of strictures on the domestic and rural economy of the mahommedans et hindoos / Tennant, William – Edinburgh 1803 [mf ed Hildesheim 1995-98] – 2v on 6mf – 9 – €120.00 – 3-487-27478-7 – gw Olms [954]
Indian reform bills : or legislation for india, from 1766 to 1858. also, an argument for a representative government in india... / Stokes, William – London 1858 – (= ser 19th c british colonization) – 1mf – 9 – mf#1.1.3658 – uk Chadwyck [342]
Indian relic trader – 1981 summer-1988 aut – 1r – 1 – (cont by: prehistoric antiquities and archaeological news quarterly) – mf#938155 – us WHS [930]
Indian religion and survival : a study / Davids, Caroline Augusta Foley Rhys – London: George Allen & Unwin, 1934 – (= ser Samp: indian books) – us CRL [280]
The indian religions : or, results of the mysterious buddhism: concerning that also which is to be understood in the divinity of fire / Jennings, Hargrave – London: G Redway 1890 [mf ed 1991] – 1mf – 9 – 0-524-01182-6 – mf#1990-2258 – us ATLA [230]
Indian revolt / Robberds, John – London, England. 1857 – 1r – us UF Libraries [240]
Indian Rights Association *see*
– Publications; 2nd series. no. 1-99. 1893-1915. (nos. 19 and 35 wanting)
– Reports and circulars
The indian rights association, 1885-1901 – 1972 – 26r – 1 – $3380.00 – mf#S1858 – us Scholarly Res [366]
Indian Rights Association. Executive Committee *see* Annual reports of the executive committee of the indian rights association, inc
Indian river advocate – Titusville, FL. 1889-1900 – 8r – (gaps) – us UF Libraries [071]
Indian river county / Sansbury, Walter – S.I., S.I? . 1936 – 1r – us UF Libraries [978]
Indian river news – Sebastian, FL. 1960-1965 – 7r – (gaps) – us UF Libraries [071]
The indian rural problem / Nanavati, Manilal Balabhai & Anjaria, J J – Bombay: Indian Society of Agricultural Economics, [1944] – (= ser Samp: indian books) – us CRL [630]
The indian ryot, land tax, permanent settlement, and the famine / Abhayacharana Dasa [comp] – [1st ed.] [Howrah], 1881 – (= ser 19th c books on british colonization) – 8mf – 9 – mf#1.1.10018 – uk Chadwyck [630]
Indian sadhus / Ghurye, Govind Sadashiv – Bombay: Popular Book Depot, 1953 – (= ser Samp: indian books) – (with the collaboration of l n chapekar) – us CRL [280]
The indian saint : or, buddha and buddhism: a sketch historical and critical / Mills, Charles De Berard – Northampton MA: Journal & Free Press Co 1876 [mf ed 1992] – 1mf – 9 – 0-524-01971-1 – mf#1990-2762 – us ATLA [280]
The indian scene / Spender, John Alfred – London: Methuen & Co, 1912 – (= ser Samp: indian books) – us CRL [915]
The indian scheme of life / Mukerjee, Radhakamal – Bombay: Hind Kitabs, 1951 – (= ser Samp: indian books) – us CRL [301]
Indian school journal – 1902-52 – (= ser American indian periodicals... 1) – 226mf – 9 – $1500.00 – us UPA [370]
Indian school journal / Chilocco Indian Agricultural School et al – v5 n2-10 [1905], v73 n11-v79 n11 [1974 jan 18-1980 apr 25] – 2r – 1 – (cont: chilocco farmer and stock grower) – mf#212073 – us WHS [630]
Indian school journal – Washington. 1976-1980 (1) 1976-1980 (5) 1976-1980 (9) – mf#7424 – us UMI ProQuest [305]
Indian scout – Shawnee OK. v1 n2, v2 n3, 7, 9-10, v3:n1, 8, v4 n1-7 [1915 feb, nov-1916 jun, sep-1917 apr, sep-1918 apr] – 1r – 1 – mf#3735838 – us WHS [071]
Indian sculpture / Kramrisch, Stella – Calcutta: YMCA Pub House; New York: Oxford University Press, 1933 – (= ser Samp: indian books) – us CRL [730]
Indian sculpture and painting : illustrated by typical masterpieces, with an explanation of their motives and ideals / Havell, Ernest Binfield – London: John Murray, 1908 – (= ser Samp: indian books) – us CRL [700]
Indian sculpture in bronze and stone / Singh, Madanjeet – Milan: Amilcare Pizzi Art Reproduction, [1952?] – (= ser Samp: indian books) – (int by giuseppe tucci) – us CRL [730]
Indian sentinel – 1916-62 – (= ser American indian periodicals... 1) – $570.00 – us UPA [305]
Indian sentinel – Bureau of Catholic Indian Missions. v1-40. 1902-16 – 1 – $180.00 – mf#0280 – us Brook [305]
Indian sentinel annual reports – 1902-16 – (= ser American indian periodicals... 1) – 7mf – 9 – $95.00 – us UPA [305]
Indian serpent-lore : or, the nagas in hindu legend and art / Vogel, Jean Philippe – London: Arthur Probsthain, 1926 – (= ser Samp: indian books) – us CRL [390]
Indian shadows – London: Church of England Zenana Missionary Society; Marshall Bros [1917] [mf ed 1995] – (= ser Yale coll) – 44p – 1 – 0-524-09764-X – (foreword by harrington c lees) – mf#1995-0764 – us ATLA [242]
Indian Shop [Independence KY] *see* Dig
Indian short stories / ed by Anand, Mulk Raj & Singh, Iqbal – London: New India Pub Co, 1946 – (= ser Samp: indian books) – us CRL [830]
The indian sign language : with brief explanatory notes of the gestures taught deaf-mutes in our institutions for their instruction... / Clark, William Philo – Philadelphia: L R Hamersly, 1885, c1884 – 5mf – 9 – (incl ind) – mf#14694 – cn Canadiana [410]
Indian social reformer – Madras, Bombay India. aug 1894-aug 1906 [wkly] – 5r – 1 – uk British Libr Newspaper [079]

Indian sociological bulletin – Ghaziabad. 1963-1967 (1) – (cont by: international journal of contemporary sociology) – ISSN: 0537-2550 – mf#7479 – us UMI ProQuest [300]

Indian spectator and voice of india – Bombay, India. jan 1890-dec 1900 [wkly] – 13r – 1 – uk British Libr Newspaper [079]

Indian speeches and documents on british rule, 1821-1918 / ed by Majumdar, J K – Calcutta; New York: Longmans, Green and Co, 1937 – (= ser Samp: indian books) – (foreword by ramananda chatterjee) – us CRL [954]

Indian spirituality : or, the travels and teachings of sivanarayan / Chatterjee, Mohini Mohun – London: Luzac, 1907 – 1mf – 9 – 0-524-01419-1 – mf#1990-2414 – us ATLA [180]

The indian standard – Bombay: [Printed & publ for the Proprietors at the Caxton Printing Works -1929 [mthly, wkly] [mf ed 2005] – v27-40 (1916-29) on 3r – 1 – (began in 1890; lacks: v36 [1925] n2,5,8,11-12 and v36 some pgs; damaged: v27 n3 (mar 1916) p73-74; organ of: presbyterian church in india, jan 1916-1924; united church of india (north), 1925-26; united church of northern india, 1927-nov 1929; cont by: united church review) – mf#2005c-s071 – us ATLA [242]

The indian states and princes / MacMunn, George Fletcher – London: Jarrolds Publishers, 1936 – (= ser Samp: indian books) – us CRL [954]

The indian states' problem / Gandhi, Mahatma – Ahmedabad: Navajivan Press, 1941 – (= ser Samp: indian books) – us CRL [954]

The indian story book : aining tales from the ramayana, the mahabharata, and other early stories / Wilson, Richard – London: Macmillan, 1914 – (= ser Samp: indian books) – (ill fr drawings by frank c pepe) – us CRL [390]

The indian struggle for freedom : through western eyes / ed by Kumarappa, Bharatan – Rajahmundry: Hindustan Pub Co, 1938 – (= ser Samp: indian books) – us CRL [954]

An indian study of love and death / Noble, Margaret E – London, New York: Longmans, Green, 1908 [mf ed 1991] – 1mf – 9 – 0-524-01804-9 – mf#1990-2652 – us ATLA [280]

The indian sunday school manual : specially adapted to sunday school work in india / Scott, T J – Lucknow: Methodist Episcopal Church Press, 1882 [mf ed 1995] – (= ser Yale coll) – viii/226p – 1 – 0-524-09178-1 – mf#1995-0178 – us ATLA [242]

Indian tales of love and beauty / Ransom, Josephine – Madras, India: Theosophy Office, 1912 – (= ser Samp: indian books) – us CRL [390]

The indian tariff policy : with special reference to sugar protection / Adarkar, Bhaskar Namdeo – Bombay: BN Adarkar, [1936] – (= ser Samp: indian books) – us CRL [336]

Indian tepee – v4 n4, v6 n2-v10 n3 [1922 win, 1924 summer-1928 may/jun] – 1r – 1 – mf#1058969 – us WHS [305]

Indian territories : proclamation of his royal highness the prince regent – [Toronto]: printed at Quebec, by P E Desbarats...; repr at York, in Upper Canada...by R C Horne...[1817?] [mf ed 1984] – 1mf – 9 – 0-665-44939-9 – mf#44939 – cn Canadiana [971]

Indian territory bar association reports – 1st to 5th annual meetings. 1900-04 (all publ) – 5mf – 9 – $7.50 – mf#LLMC 84-490 – us LLMC [340]

Indian Territory. Fort Gibson Headquarters *see* Headquarters records of fort gibson, indian territory, 1830-1857

Indian territory reports – v1-6. 1896-1907 (all publ) – 20mf – 9 – $90.00 – (a pre-nrs title) – mf#LLMC 84-132 – us LLMC [343]

Indian Territory. Synod (Pres. Church in the USA) *see* Minutes, 1887-1906

The indian theatre / Anand, Mulk Raj – London: Dennis Dobson Ltd, [1950?] – (= ser Samp: indian books) – (ill by usha rani) – us CRL [790]

The indian theatre : a brief survey of the sanskrit drama / Horrwitz, Ernest Philip – London: Blackie and Son, 1912 – (= ser Samp: indian books) – us CRL [490]

Indian theism from the vedic to the muhammadan period / Macnicol, Nicol – London: Oxford University Press, 1915 – (= ser The Religious Quest of India) – 1mf – 9 – 0-524-01197-4 – (incl bibl ref) – mf#1990-2273 – us ATLA [280]

Indian thought and its development / Schweitzer, Albert – London: Adams & Charles Black, 1951 – (= ser Samp: indian books) – us CRL [180]

Indian thought past and present / Frazer, Robert Watson – London: T Fisher Unwin, 1915 – 1mf – 9 – 0-524-01060-9 – (incl bibl ref) – mf#1990-2208 – us ATLA [280]

Indian time – Rooseveltown NY. 1983 jul 1 [v1 n1]-1987 mar 11, 1987 mar 25-1989 dec 21 [v7 n42] – 2r – 1 – (cont by: indian times [rooseveltown, ny]) – mf#1277696 – us WHS [071]

Indian time magazine / Native American Outreach Project [MI] et al – 1986 nov, 1987 feb,jul-aug, nov, 1988 mar, sep – 1r – 1 – mf#1574893 – us WHS [305]

Indian times / White Buffalo Council of American Indians – 1965 jan, 1970 jan-1980 may/jun[v21 n3], [ns] v1 n1-v2 n10 [1981 may 15-1982 nov 7] – 1r – 1 – mf#1001595 – us WHS [071]

Indian trader – 1987 feb-1989, 1990-91 – 2r – 1 – mf#1110918 – us WHS [071]

Indian traits : being sketches of the manners, customs, and character of the north american natives / Thatcher, Benjamin – New York, NY 1833 [mf ed Hildesheim 1995-98] – 2v on 4mf – 9 – €120.00 – 3-487-27129-x – gw Olms [306]

Indian travel newsletter / American Indian Travel Commission et al – v1 n1 [1972 may]-v9 n1 [1980 win] – 1r – 1 – mf#672498 – us WHS [910]

The indian travels of apollonius of tyana : and the indian embassies to rome from the reign of augustus to the death of justinian / Priaulx, Osmond de Beauvoir – London: Quaritch, Piccadilly, 1873 [mf ed 1993] – 1mf – 9 – 0-524-07794-0 – (incl bibl ref) – mf#1991-0171 – us ATLA [915]

Indian travels of thevenot and careri : being the third part of the travels of m de thevenot into the levant and the third part of a voyage round the world by dr john francis gemelli careri / ed by Sen, Surendranath – New Delhi: National Archives of India, 1949 – (= ser Samp: indian books) – us CRL [915]

Indian tribes of guiana / Brett, William Henry – New York, NY. 1856 – 1r – us UF Libraries [972]

The indian tribes of the united states : their history, antiquities, customs, religion, arts, language, traditions, oral legends, and myths / Schoolcraft, Henry Rowe; ed by Drake, Francis Samuel – Philadelphia: J B Lippincott, 1884, c1883 – 2v on 1mf – 9 – (individual vols also available separately) – mf#16573 – cn Canadiana [305]

Indian truth – 1924-73 – (= ser American indian periodicals... 1) – 38mf – 9 – $250.00 – us UPA [305]

Indian unrest / Chirol, Valentine – London: Macmillan and Co, 1910 – (= ser Samp: indian books) – (int by alfred lyall) – us CRL [954]

Indian Urban Affairs *see* Milwaukee indian news

Indian valley record – Greenville, CA. 1931-1974 (1) – mf#62167 – us UMI ProQuest [071]

Indian viewpoint / Motivation Through Communication [Duluth MN] – [1974 feb-1975 jun] scattered iss – 1r – 1 – mf#724676 – us WHS [302]

Indian views – Durban SA, 6 jul 1934-26 jun 1936 – 4r – 1 – sa National [079]

Indian views – (S. Africa). 5 jan 1945-1 jul 1953 – 1 – us L of C Photodup [073]

Indian village / Dube, Shyama Charan – Bombay, India. 1967 – 1r – us UF Libraries [975]

Indian village folk : their works and ways / Pandiyan, Thomas B – London: Elliot Stock, 1897 [mf ed 1995] – (= ser Yale coll) – viii/212p (ill) – 1 – 0-524-09872-7 – mf#1995-0872 – us ATLA [307]

Indian village pictures / Lester, Henry F W – London: London Missionary Society, 1910 [mf ed 1995] – (= ser Yale coll) – 212p (ill) – 1 – 0-524-10227-9 – mf#1996-1227 – us ATLA [915]

Indian voice / Canadian Indian Voice Society et al – v5 n2-v15 n7/12 [1973 feb-1983 summer/fall] – 1r – 1 – mf#601385 – us WHS [305]

Indian voice / Haliwa Indian Tribe – v1 n13-v2 n4 [1979 jan/feb-1980 jul], [v1 n1] [1981 oct]-1982 mar – 1r – 1 – mf#620714 – us WHS [305]

Indian voice / Indian Homemakers' Association of British Columbia – 1969-81 – (= ser American indian periodicals... 2) – 40mf – 9 – $250.00 – us UPA [305]

Indian voice / Small Tribes Organization of Western Washington – v6 n12-v12 n12 [1976 dec-1982 dec] – 1r – 1 – mf#657740 – us WHS [305]

Indian voices – 1963-68 – (= ser American indian periodicals... 1) – 9mf – 9 – $105.00 – us UPA [305]

The indian war of independence / Savarkar, Vinayak Damodar – Bombay: Phoenix Publications, 1947 – (= ser Samp: indian books) – us CRL [954]

The indian wars of the west and frontier army life, 1862-1898 : official histories and personal narratives – 606mf – 9 – $5810.00 – 1-55655-598-9 – (with p/g) – us UPA [355]

Indian wells valley independent – Ridgecrest, CA. 1955-1969 (1) – mf#62252 – us UMI ProQuest [071]

"Indian wigwams and northern camp-fires" : a criticism / McDougall, John – Toronto: Printed for the author by W Briggs, 1895 – 1mf – 9 – mf#30649 – cn Canadiana [390]

Indian womanhood to-day / Cousins, Margaret E – Allahabad: Kitabistan, 1941 – (= ser Samp: indian books) – us CRL [305]

Indian writers of english verse / Basu, Lotika – Calcutta: University of Calcutta, 1933 – (= ser Samp: indian books) – us CRL [420]

Indian youth / Institute for Career and Vocational Training [Culver City CA] – v3 n1-v5 n1 [1980 dec-1983 win/spring], v6 n1 [1985 win/spring] – 1r – 1 – mf#1095631 – us WHS [331]

Indiana : burns indiana statutes annotated – Charlottesville: Michie Co, 1977-aug 99 update – 9 – $4,426.00 set – mf#402260 – us Hein [348]

Indiana / Halevy, Leon – Bruxelles, Belgium. 1834 – 1r – us UF Libraries [440]

Indiana : session laws of american states and territories – 1801-1998 – 9 – $2,091.75 set – mf#402650 – us Hein [348]

Indiana *see*
– Reports and opinions
– Reports, post-nrs
– Reports, pre-nrs

Indiana addendum to green's pleading and practice / Hoffman, Publius V – St. Louis: Gilbert, 1881. 182p. LL-85 – 1 – us L of C Photodup [340]

Indiana agri news – LaSalle, IL. 1982-1984 (1) – mf#68104 – us UMI ProQuest [071]

Indiana american – Brookville, IN. 1833-1871 (1) – mf#62737 – us UMI ProQuest [071]

Indiana. Appellate Court *see* Indiana appellate reports

Indiana appellate reports / Indiana. Appellate Court – v1-83. 1890-1925 – 701mf – 9 – $1051.00 – (no pre-nrs vols. updates planned) – mf#LLMC 80-805 – us LLMC [340]

Indiana attorney general reports and opinions – 1891-1997 – 6,9 – $487.00 set – (1891-1979 on reel $385. 1979-97 $102) – mf#408230 – us Hein [340]

Indiana baptist – Georgetown. 1810-1812 (1) – 1 – mf#3537 – us Southern Baptist [242]

Indiana baptist history, 1798-1908 / Stott, William Taylor – [S.l: s.n.], c1908 – 1mf – 9 – 0-524-03388-9 – (incl bibl ref) – mf#1990-4700 – us ATLA [242]

Indiana baptist/baptist outlook – Indianapolis, IN. 1886, 1894-96, 1898-1902. Single reels available – 1 – us ABHS [242]

Indiana business review – Bloomington. 1989+ (1,5,9) – ISSN: 0019-6541 – mf#14936 – us UMI ProQuest [650]

Indiana catholic – Indianapolis, IN. 1910-1915 (1) – mf#62842 – us UMI ProQuest [071]

Indiana Conference of Teamsters *see* Official publication of teamsters joint council 69

Indiana county gazette – Indiana, PA. 1890-1912 (1) – mf#65932 – us UMI ProQuest [071]

Indiana courier – East Chicago, IN. 1914-1914 (1) – mf#62768 – us UMI ProQuest [071]

Indiana democrat – Indiana, PA. -w 1896-1912 – 13 – $25.00r – us IMR [071]

Indiana. Dept. of Insurance. Audit and Control *see* Reports of the auditor

Indiana deutsche zeitung – Indianapolis, IN. 1875-1876 (1) – mf#62839 – us UMI ProQuest [071]

Indiana deutsche zeitung – Indianapolis, IN. 1875-1877 (1) – mf#62844 – us UMI ProQuest [071]

Indiana. Employment Security Board *see* Employment security in indiana

Indiana Farmers' Alliance *see* Proceedings of the indiana farmers' alliance at its...annual meeting

Indiana folklore – Bloomington. 1968-1971 (1) – ISSN: 0019-6614 – mf#7115 – us UMI ProQuest [390]

Indiana folklore : journal of the hoosier folklore society – [Bloomington, Ind]: The Society v1, n1- 1968- [mf ed Bloomington IN: Indiana Uni Lib, Preservation Dept 1984] – 2r – 1 – (semi-annual. cont by: newsletter (hoosier folklore society 1982)) – us Indiana Preservation [390]

Indiana folklore and oral history – Indiana University, Bloomington. 1985 jan/jun-1987 jan/jun – 1r – 1 – mf#1495015 – us WHS [390]

Indiana gazette – Vincennes IN. 1804 aug 7-21, oct 23 – 1r – 1 – (cont by: western sun) – mf#850086 – us WHS [071]

Indiana gazette – Indiana, PA. 1904-2000 (1) – mf#61789 – us UMI ProQuest [071]

Indiana genealogical informer – v1 n1-v3 n12 [1979 aug-1982 jul] – 1r – 1 – mf#626150 – us WHS [929]

Indiana herald – Huntington, IN. 1848-1887 (1) – mf#68282 – us UMI ProQuest [071]

Indiana herald – Indianapolis, IN. 1953-1998 (1) – mf#62845 – us UMI ProQuest [071]

Indiana herald – Indianapolis IN [1993 jan-dec]-[1997 jan-dec] – 20r – 1 – mf#875083 – us WHS [071]

Indiana historical chronicle – 1974 nov25-1977 jan 31 – 1r – 1 – mf#1058981 – us WHS [978]

Indiana Historical Society *see* Genealogy

Indiana international and comparative law review – v1-11. 1991-2001 – 9 – $162.00 set – ISSN: 1061-4982 – mf#113221 – us Hein [341]

The indiana jewish chronicle – Indianapolis. Ind. 1945-50; 52-53. 1957-66 – 1 – us AJPC [071]

Indiana journal – Indianapolis, IN. 1831-1834 (1) – mf#62846 – us UMI ProQuest [071]

Indiana, Kentucky, Ohio-Token and Medal Society *see* Iko-tams bulletin

Indiana labor tribune – v17 n3-v22 n25 [1968 sep 6-1974 feb 8] – 1r – 1 – mf#380997 – us WHS [331]

Indiana law journal – v1. 1926 – 4mf – 9 – $6.00 – (add vols filmed) – mf#llmc97-581 – us LLMC [340]

Indiana law journal – v1-3. 1898-99 (all publ) – 10mf – 9 – $15.00 – (lacking: 1898 nos 2+4. 1899 nos 2-3) – mf#LLMC 84-487 – us LLMC [340]

Indiana law journal – v1-76. 1925-2001 – 1,5,6,9 – $1564.00 set – (v1-72 1925-97 on reel $1436. v73-76 1997-2001 on mf $128) – ISSN: 0019-6665 – mf#103311 – us Hein [340]

Indiana law magazine – 1883-85 (all publ) – 5v on 29mf – 9 – $42.50 – (incl: the corporation digest and the corporation reporter) – mf#LLMC 84-488 – us LLMC [340]

Indiana law magazine – Indianapolis. v1-5. 1883-85 (all publ) – (= ser Historical legal periodical series) – 1 – $53.00 set – mf#409010 – us Hein [340]

The indiana law reporter – v1. 1881 (all publ) – 1mf – 9 – $4.50 – mf#LLMC 84-489 – us LLMC [340]

Indiana law review – v1-34. 1967-2001 – 1,5,6 – $950.00 set – (v1-29 1967-96 on reel $792. v30-34 1997-2001 on mf $158. title varies: v1-5 1967-72 as indiana legal forum) – ISSN: 0090-4198 – mf#103331 – us Hein [340]

Indiana legal forum *see* Indiana law review

Indiana magazine of history – v1-65. 1905-1969 – 1 – us AMS Press [900]

Indiana medicine : the journal of the indiana state medical association / Indiana State Medical Association – Indianapolis. 1984-1996 (1) 1984-1996 (5) 1984-1996 (9) – (cont: journal of the indiana state medical association) – ISSN: 0746-8288 – mf#2499,01 – us UMI ProQuest [610]

Indiana methodism : being an account of the introduction, progress, and present position of methodism in the state / Holliday, Fernandez C – Cincinnati: Hitchcock and Walden, 1873. Beltsville, Md: NCR Corp, 1977 (5mf); Evanston: American Theol Lib Assoc, 1984 (5mf) – 9 – 0-8370-0121-8 – mf#1984-0008 – us ATLA [242]

Indiana military history journal – v1 n1-2 [1976 jan-jul], v2 n1 -v13 n3 [1977 jan-1988 oct] – 1r – 1 – mf#1802873 – us WHS [355]

Indiana. Morgan Raid Commission *see* Journal, ms 3506

Indiana mortgage marker's magazine – Chicago, IL. 1997+ (1) – mf#69381 – us UMI ProQuest [071]

Indiana names – 1970 spr-1974 fall – 1r – 1 – (cont by: midwestern journal of language and folklore) – mf#303010 – us WHS [978]

Indiana newspaper project – Indianapolis: Indiana Historical Society – 16,000r – 1 – $15.00r – (contains 1,000 indiana newspapers) – us IHS [071]

Indiana preservationist / Historic Landmarks Foundation of Indiana – 1975 win/spring-1980 win, 1981 1-2 – 1r – 1 – mf#667588 – us WHS [978]

Indiana progress – Indiana, PA. -w 1903-1940; 1944-1945 – 13 – $25.00r – us IMR [071]

Indiana racial study / Garrett Biblical Institute, Evanston, Ill. Bureau of Social and Religious Research – Evanston, Ill: Bureau of Social and Religious Research, [1951] Chicago: Dep of Photodup, U of Chicago Lib, 1967 (2r); Evanston: American Theol Lib Assoc, 1984 (2r) – 1 – 0-8370-0656-2 – mf#1984-6001 – us ATLA [240]

Indiana roots – v1 n1-v3 n2 [1983 aug-1985 nov] – 1r – 1 – mf#1133685 – us WHS [071]

Indiana social studies quarterly – Muncie. 1978-1986 (1,5,9) – (cont by: international journal of social education) – ISSN: 0019-6746 – mf#11381 – us UMI ProQuest [300]

Indiana socialist / Peoria Socialist Publishing Co – v1 n29-34 [1907 nov 23-dec 28] – 1r – 1 – (cont: peoria socialist [1907]) – mf#955176 – us WHS [335]

Indiana staats-herold – Hammond: Staats-Herold Pub Co, oct 1931-mar 5 1932 – 1r – 1 – us CRL [071]

INDICE

Indiana State AFL-CIO [American Federation of Labor and Congress of Industrial Organizations] *see*
- Labor news
- News and views
- News and views: indiana state afl-cio labor supplement

Indiana state afl-cio news – v1 n1-13 [1958 jan 1-1959 apr 3] – 1r – 1 – (cont by: news and views [indiana state afl-cio]) – mf#627680 – us WHS [331]

Indiana. State Bar Association *see* Proceedings

Indiana state bar association annual reports – 1st to 31st annual meetings. 1897-1927 (all publ) – 35mf – 9 – $52.50 – (lacking: 1925) – mf#LLMC 84-491 – us LLMC [340]

Indiana State Building and Construction Trades Council *see* Leader

Indiana State Building and Construction Trades Council et al *see* Labor tribune

Indiana state commercial and home advocate – Lafayette IN. 1867 sep – 1r – 1 – mf#856570 – us WHS [071]

Indiana. State Convention of Baptists *see* Executive board minutes

Indiana State Employees Association *see*
- Isea news
- Isea news letter

Indiana State Grange *see* Journal of proceedings of the...annual session

Indiana state journal – Indianapolis IN. 1846 nov 30, 1846 jun 24-1849 oct 22, 1896 jan 1-1897 may 19, 1897 may 26-1898 oct 12, 1898 oct 19-1899 dec 27 – 5r – 1 – (cont: indiana journal; cont by: indianapolis star) – mf#846284 – us WHS [071]

Indiana State Medical Association *see*
- Indiana medicine
- Journal of the indiana state medical association

Indiana state sentinel – Indianapolis IN. 1861 jul 19 – 1r – 1 – (cont: indiana democrat, and spirit of the constitution; cont by: weekly indiana state sentinel) – mf#846299 – us WHS [071]

Indiana. Superior Court *see* Wilson's superior court reports

Indiana. Supreme Court *see*
- Blackford's reports
- Indiana supreme court reports
- Smith's reports

Indiana supreme court reports / Indiana. Supreme Court – v1-197. 1848-1926 – 1505mf – 9 – $2257.00 – (pre-nrs: v1-101 1848-84 714mf $1,071.00. updates planned) – mf#LLMC 80-804 – us LLMC [347]

Indiana teacher – Indianapolis. 1869-1869 – 1 – mf#4811 – us UMI ProQuest [370]

Indiana. Territory *see* Reports, pre-nrs

The indiana times – Indiana, PA. -w 1889-1912 – 13 – $25.00r – us IMR [071]

Indiana true republican – Centreville, Richmond IN. 1864 mar 31 – 1r – 1 – (cont by: julian's indiana radical) – mf#856333 – us WHS [071]

Indiana University, Bloomington *see* Hola, office of latin affairs

Indiana university bookman – Bloomington. 1956-1979 (1) 1979-1979 (5) 1979-1979 (9) – ISSN: 0019-6800 – mf#7123 – us UMI ProQuest [400]

Indiana University. Kirkwood Observatory *see* Publications of the kirkwood observatory of indiana university

Indiana university. school of education. bulletin – v1-37. 1924-61 – 1 – $288.00 – mf#0281 – us Brook [378]

Indiana waterways – v1 n3-v4 n1 [1982 feb-1985/86 win], v5 n1-4 [1986 aug-1988 win] – 1r – 1 – (cont by: indiana canals) – mf#1058997 – us WHS [071]

Indianapolis 500-mile race history / ed by Clymer, Joseph Floyd – Deluxe ed. Los Angeles: Floyd Clymer, c1946. 320p. illus., tables, diagrs – 1 – us Wisconsin U Libr [790]

Indianapolis business journal – Indianapolis. 1990+ (1,5,9) – ISSN: 0274-4929 – mf#18466 – us UMI ProQuest [338]

Indianapolis IN *see* Freeman

Indianapolis journal – Indianapolis IN. 1871 aug 4 – 1r – 1 – (cont: indianapolis daily journal; cont by: indianapolis morning star) – mf#1147229 – us WHS [071]

Indianapolis labor news – Indianapolis IN. 1895 aug 31; 1896 feb 8 – 1r – 1 – us WHS [331]

Indianapolis ledger – Indianapolis IN. 1918 apr 13, 1922 oct 28 – 1r – 1 – mf#5258965 – us WHS [071]

Indianapolis new times – v4 n4-v5 n8 [1987 aug-1988 dec] – 1r – 1 – (cont: stepping out [indianapolis, in]; cont by: new times [indianapolis, in]) – mf#1803542 – us WHS [071]

Indianapolis news – Indianapolis, IN. 1869-1999 (1) – mf#60470 – us UMI ProQuest [071]

Indianapolis news – Indianapolis IN. [1920 mar 29-1939 jul 24] – 1r – 1 – mf#846256 – us WHS [071]

Indianapolis recorder – Indianapolis IN. [1987 aug 29/dec]-[2001 nov-dec] – 62r – 1 – (cont: recorder [indianapolis, in: 1897]) – mf#846062 – us WHS [071]

Indianapolis sentinel – Indianapolis IN. 1876 may 10-feb 5, 1877 aug 1,4,6,11,14,17,1882 jan 6, jul 1-dec 31, 1884 may 30 – 1r – 1 – (cont: indianapolis daily sentinel) – mf#850478 – us WHS [071]

Indianapolis star – Indianapolis, IN. 1903+ (1) – mf#60471 – us UMI ProQuest [071]

Indianapolis sun – Indianapolis IN. 1876 may 20 – 1r – 1 – mf#845992 – us WHS [071]

Indianapolis times – Indianapolis IN. 1884 jun 11, aug 31 – 1r – 1 – mf#845993 – us WHS [071]

Die indianer und ihr freund david zeisberger / Roemer, Hermann – Guetersloh: C Bertelsmann, 1890 – 1mf – 9 – 0-524-01764-6 – mf#1990-0498 – us ATLA [240]

Indian-eskimo association of canada bulletin – 1960-72 – (= ser American indian periodicals... 1) – $95.00 – us UPA [305]

Indian-eskimo association of canada bulletin – 1060 mar-1972 jul – 1r – 1 – (cont: canadian indians and eskimos of today; bulletin [national commission on the indian canadian]; cont by: c a s n p bulletin) – mf#470714 – us WHS [366]

Indian-Eskimo Friendship Centre *see* Newsletter

Indianhead – Camp Casey [Korea]. 1983 jun 15-1986 dec 5, 1987 jan-1989 jun – 2r – 1 – mf#1566072 – us WHS [071]

Indianhead – 1977 jan-1990 sep – 1r – 1 – (cont: indian head [washington, dc]) – mf#1896290 – us WHS [071]

Indianhead star – Clear Lake WI. 1988 nov 24-1989 feb, 1989 mar-dec, 1990 jan-dec, 1991 jan-dec, 1992 jan-dec 9 – 10r – 1 – (cont: clear lake star) – mf#2633671 – us WHS [071]

Indianian / Jeffersonville – nov 1819-may 1820 – 1r – 1 – mf#B2025 – us Ohio Hist [071]

Indianian republican – Warsaw, IN. 1882-1894 (1) – mf#62991 – us UMI ProQuest [071]

Indianische sagen von de nord-pacifischen kueste amerikas / Boas, Franz – Berlin: A Asher, 1895 – 4mf – 9 – 0-665-02419-3 – mf#02419 – cn Canadiana [390]

Indianism and its expansion / Thomas, Frederick William – Calcutta: University of Calcutta, 1942 – (= ser Samp: indian books) – us CRL [900]

Indianismo na literatura romantica brasileira / Ferreira, Maria Celeste – Rio de Janeiro, Brazil. 1949 – 1r – us UF Libraries [440]

Indianola courier – Indianola, NE: G S Bishop, 1880 (wkly) [mf ed v4 n27. jul 5 1883 filmed [1996]] – 1r – 1 – (cont by: indianola weekly courier) – us NE Hist [071]

Indianola independent – Indianola, NE: S R Smith. v8 n[31] sep 14 1900- (wkly) [mf ed -may 8 1903 (gaps) filmed 1974-] – 2r – 1 – (cont: weekly reporter. absorbed in pt by: red willow county sun nov 2 1900 and cont with new vol numbering. suspended foll oct 12 1900; resumed on nov 23 1900 with v10 n47) – us NE Hist [071]

The indianola news – Indianola, NE: C Don Harpst and Merle J Harpst, 1950 (wkly) [mf ed v1 n9. mar 16 1950- (gaps) filmed 1975] – 1r – 1 – us NE Hist [071]

Indianola reporter – Indianola, NE: E S Byfield, 1907-v54 n42. mar 6 1947 (wkly) [mf ed v16 n40. jan 3 1908-mar 6 1947 (gaps) filmed 1974] – 13r – 1 – (merged with: bartley inter-ocean to form: red willow county reporter) – us NE Hist [071]

Indianola reporter *see* Bartley inter-ocean

Indianola. Synod (Cum. Pres. Ch.Pres. Ch. in the U.S.A.) *see* Minutes, 1898-1906

Indianola weekly courier – Indianola, NE: G S Bishop. -v17 n26. jun 25 1896 (wkly) [mf ed v10 n46. nov 14 1889-96 (gaps) filmed 1972-[1996]] – 1r – 1 – (cont by: weekly courier (mccook ne)) – us NE Hist [071]

Indianos cacerenos / Hurtado de Mendoza, Publio – 1892 – 9 – sp Bibl Santa Ana [946]

Indians and indian life : food / Hanson, W Stanley – S.l., S.l? . 1936? – 1r – us UF Libraries [306]

Indians and indian life : seminoles, salient fact / Hanson, W Stanley – S.l., S.l? . 1936? – 1r – us UF Libraries [978]

Indians at work – 1933-40 – (= ser American indian periodicals... 1) – 9 – $570.00 – us UPA [305]

Indians at work : a news sheet for indians and the indian service – v1-13. 1933-45 – 1 – $120.00 – mf#0282 – us Brook [305]

The indian's friend – v. 1-52. Mar 1888-Nov 1940 – 1 – 75.00 – us L of C Photodup [360]

The indians of british columbia : a brief review of their probable origin, history and customs / MacKay, Joseph William – [S.l: s.n, 18–] [mf ed 1981] – 1mf – 9 – mf#15555 – cn Canadiana [305]

Indians of florida – S.l., S.l? . 1937 – 1r – us UF Libraries [978]

Indians of quebec = Indiens du quebec / Confederation of Indians of Quebec – v1 n1-v5 n3 [1977 jul-1982] – 1r – 1 – mf#826928 – us WHS [305]

Indiantown Gap Military Reservation [PA] *see* Libertad

Indiantown press – Indiantown, FL. v2 n1-v8 n53. 1959 nov-1966 nov – 3r – (gaps) – us UF Libraries [071]

India's armies and their costs : a century of unequal imposts for an army of occupation and a mercenary army / Sundaram, Lanka – Bombay: Avanti Prakashan, 1946 – (= ser Samp: indian books) – us CRL [355]

India's balance of indebtedness, 1898-1913 / Pandit, Yeshwant Sakharam – London: George Allen & Unwin, 1937 – (= ser Samp: indian books) – (foreword by jehangir coyajee) – us CRL [380]

India's cries to british humanity, relative to infanticide : british connection with idolatry, ghaut murders, suttee, slavery, and colonization in india; to which are added, humane hints for the melioration of the state of society in british india / Peggs, James – London 1832 – (= ser 19th c british colonization) – 6mf – 9 – mf#1.1.1397 – uk Chadwyck [306]

India's cries to british humanity, relative to the suttee, infanticide, british connection with idolatry, ghaut murders, and slavery in india / Peggs, J – London, 1830 – 6mf – 9 – mf#HT-108 – ne IDC [306]

India's cultural empire and her future / Mitra, Sisirkumar – Madras: Sri Aurobindo Library, 1947 – (= ser Samp: indian books) – us CRL [954]

India's danger and england's duty : with reference to russia's advance into the territory in dispute upon the borders of afghanistan / Russell, Richard – [London] [1885] – (= ser 19th c british colonization) – 2mf – 9 – mf#1.1.4009 – uk Chadwyck [327]

India's fighters : their mettle, history, and services to britain / Nihal Singh, Saint – London: Sampson Low, Marston & Co, 1914 – (= ser Samp: indian books) – us CRL [954]

India's hurt : and other addresses / Forrest, William Mentzel – St Louis, MO: Christian Pub Co, c1909 [mf ed 1993] – (= ser Christian church (disciples of christ) coll) – 1mf – 9 – 0-524-06406-7 – mf#1991-2528 – us ATLA [240]

India's legacy : the world's heritage / Ranganatha Punja, P R – Mangalore: Basel Mission Book Depot, 1948- – (= ser Samp: indian books) – us CRL [900]

India's mass movement / Warne, Francis Wesley – New York: Board of Foreign Missions of the Methodist Episcopal Church, 1915 [mf ed 1995] – (= ser Yale coll) – 64p (ill) – 1 – 0-524-09901-4 – mf#1995-0901 – us ATLA [242]

India's nation builders / Bannerjea, Devendra Nath – London: Headley Bros Publishers, 1919 – (= ser Samp: indian books) – us CRL [954]

India's north-east frontier in the nineteenth century / Elwin, Verrier – London: Oxford University Press, 1959 – 1 – us Wisconsin U Libr [954]

India's outlook on life : the wisdom of the vedas / Chatterji, Jagadish Chandra – New York: Kailas Press, 1931 – (= ser Samp: indian books) – (int by john dewey) – us CRL [280]

India's past : survey of her literatures, religions, languages, and antiquities / Macdonell, Arthur Anthony – Oxford: Clarendon Press, 1927 – (= ser Samp: indian books) – us CRL [954]

India's plea for men : the substance of a sermon preached in trinity church, cambridge, on sunday, nov 23 1856 / Knight, William – London, 1857 – (= ser 19th c british colonization) – 1mf – 9 – (with app) – mf#1.1.512 – uk Chadwyck [240]

India's post-war reconstruction and its international aspects / Lokanathan, Palamadai Samu – New Delhi: Indian Council of World Affairs; Bombay: Oxford University Press, 1946 – (= ser Samp: indian books) – us CRL [330]

India's problem : krishna or christ / Jones, John Peter – New York: Fleming H Revell [1903] [mf ed 1995] – (= ser Yale coll) – 369p (ill) – 1 – 0-524-10109-4 – mf#1995-1109 – us ATLA [230]

India's problem, krishna or christ / Jones, John Peter – 4th ed. New York: Laymen's Missionary Movement, c1903 – 1mf – 9 – 0-8370-6670-0 – (includes statistical tables and index) – mf#1986-0670 – us ATLA [230]

India's silent revolution / Fisher, Frederick Bohn & Williams, Gertrude Leavenworth Marvin – New York: Macmillan, 1919 [mf ed 1995] – (= ser Yale coll) – 192p (ill) – 1 – 0-524-09824-7 – (with foreword) – mf#1995-0824 – us ATLA [954]

India's social heritage / O'Malley, Lewis Sydney Steward – Oxford: Clarendon Press, 1934 – (= ser Samp: indian books) – us CRL [301]

India's struggle / Rajput, A B – Lahore: Lion Press, 1946 – (= ser Samp: indian books) – us CRL [954]

India's struggle for freedom / Chatterji, A C – Calcutta: Chuckervertty, Chatterjee & Co, 1947 – (= ser Samp: indian books) – us CRL [954]

India's teeming millions : a contribution to the study of the indian population problem / Chand, Gyan – London: George Allen & Unwin, 1939 – (= ser Samp: indian books) – us CRL [304]

India's will to freedom : writings and speeches on the present situation / Lajpat Rai, Lala – Madras: Ganesh & Co, 1921 – (= ser Samp: indian books) – us CRL [954]

India's women and china's daughters *see* Church missionary society archive, section 2

Indias y espanolas / Alfaro De Jimenez, Isabel – San Jose, Costa Rica. 1964 – 1r – us UF Libraries [972]

Indicaciones de filosofia y pedagogia / Sama, J – 1893 – 9 – sp Bibl Santa Ana [190]

Indicador da organizacao administrativa do executi... / Brazil. Departamento Administrativo do Servico Pub... – Rio de Janeiro, Brazil. 1940 – 1r – us UF Libraries [972]

L'indicateur – Montreal: G P Labat, [1895-189- ou 19–] [mf ed v1 n1 1 juin 1895] – 9 – ISSN: 1190-7797 – mf#P04094 – cn Canadiana [073]

Indicateur de dieppe : contenant le nom et la demeure de tous les magistrats, fonctionnaires publics, civils et militaires, des employes des administrations... / Charlet, Victor – Dieppe 1824 [mf ed Hildesheim 1995-98] – 1mf – 9 – €40.00 – 3-487-29720-5 – gw Olms [914]

L'indicateur de la region flamande – Hazebrouck, France 1 jan 1911-27 sep 1914 – 1 – uk British Libr Newspaper [074]

Indicateur de la region flamande – Hazebrouck, France. jan 1911-4 oct 1914 – 2r – 1 – uk British Libr Newspaper [072]

L'indicateur de quebec = The quebec indicator – Quebec: compile et publie par T L Boulanger & E Marcotte, 1889 – 9 – ISSN: 1190-7894 – mf#A00009 – cn Canadiana [917]

L'indicateur de quebec et levis = The quebec and levis directory – Quebec: compile et publie par Boulanger & Marcotte, 1890-1903 – mf#A00010 – cn Canadiana [917]

Indicateurs de francisation des entreprises : objectifs et methodes / Robert, Roger – [mf ed 1978] – 3mf – 9 – mf#SEM105P3 – cn Bibl Nat [440]

Indicator – London, UK. 2 sep 1949-1953 – 2 1/2r – 1 – (7 jan-26 aug 1949 is west london chronicle. aka: indicator and general advertiser; indicator and west london news; indicator and west london chronicle) – uk British Libr Newspaper [072]

Indicator – Providence, RI. 1883-1886 (1) – mf#66320 – us UMI ProQuest [071]

Indicator – Pueblo, CO. 1913-1948 (1) – mf#62318 – us UMI ProQuest [071]

The indicator *see* Miscellaneous newspapers of pueblo county

Indicator and general advertiser *see* Indicator

Indicator and society journal – Providence, RI. 1886-1887 (1) – mf#66321 – us UMI ProQuest [071]

The indicator (london) – sep 1829-apr 1830 – (= ser 19th c british periodicals) – r40 – 1 – (filmed with: the indicator (london), 13 oct 1819-21 mar 1821) – us Primary [073]

The indicator's digest of insurance decisions – Detroit, Leavenworth, 1899. 661 p. LL-436 – 1 – us L of C Photodup [346]

Indicators of the accountability of the financial management process of school funds in mangaung schools / Mafoyane, Mante Sarah – Pretoria: Vista University 2002 [mf ed 2002] – 2mf – 9 – (incl bibl) – mf#mfm15199 – sa Unisa [370]

Indice – Miami, FL. 1975 feb 24-oct 25 – 1r – us UF Libraries [071]

Indice alfabetico y defunciones / Cuba Ejercito Inspeccion General – Habana, Cuba. 1901 – 1r – us UF Libraries [972]

Indice de 'el repertorio colombiano – Bogota, Colombia. 1961 – 1r – us UF Libraries [972]

Indice de informes pedidos por el gobierno de s.m. y cuerpos del estado a la real academia de la historia, evacuados por esta – Madrid: Fortanet, 1900. B.R.A.H. 37, 1900, pp. 63-106 – sp Bibl Santa Ana [946]

Indice de la bibliografia hondurena / Duron, Jorge Fidel – Tegucigalpa, Mexico. 1946 – 1r – us UF Libraries [972]

Indice de la biblioteca extremena.. / Barrantes Moreno, Vicente – 1881 – 9 – sp Bibl Santa Ana [020]

Indice de la coleccion de historiadores y de documentos relativos a la independencia de chile / Villalobos R, Sergio – Santiago: Universidad de Chile, Instituto Pedagogicao, Seminario de Historia de Chile, 1956. xi,108p – 1 – us Wisconsin U Libr [972]

Indice de la coleccion salazar, tomos 20, 8-37. madrid, 1961-1966 / Cuartero, Baltasar & Vargas Zuniga, Antonio; ed by Uribe, Angel – Madrid: Graf. Calleja, 1967 – 1 – sp Bibl Santa Ana [946]
Indice de la poesia panamena contemporanea / Miro, Ricardo – Santiago, Chile. 1941 – 1r – us UF Libraries [972]
Indice de la poesia paraguaya / Buzo Gomes, Sinoforiano – Asuncion, Paraguay. 1959 – 1r – us UF Libraries [440]
Indice de la revista de occidente / Segura Covarsi, Enrique – Madrid: Instituto Miguel de Cervantes, 1952 – 1 – sp Bibl Santa Ana [946]
Indice de las leyes y glosas de las siete partidas / Lopez de Tovar, Gregorio – 1789. Tomo III – 9 – (1757) – sp Bibl Santa Ana [340]
Indice de las leyes y glosas de las siete partidas del rey don alfonso el sabio : por el licenciado gregorio lopez / Lopez de Tovar, Gregorio – Madrid: Oficina de Benito Cerro, 1789.-v4 – 1 – sp Bibl Santa Ana [946]
Indice de las pruebaj de caballeros de la orden de santiago / Vignau, V & Uragon, F de – Madrid, 1904 – 155mf – 9 – sp Cultura [025]
Indice de los documentos de la catedral – Zamora – 1r – 5,6 – sp Cultura [240]
Indice de los documentos que presento para ingresar en el real cuerpo...y...orden militar de malta... / Solar y Taboada, Antonio – Badajoz: Imprenta y Libreria La Minerva Extremena, 1927 – 1 – sp Bibl Santa Ana [355]
Indice de los libros que contiene...gaspar de molina – 1749 – 9 – sp Bibl Santa Ana [020]
Indice de los papeles de la junta central suprema gubernativa del reino y... / Carretas, J & Olavide, I – Madrid, 1904 – 2mf – 9 – sp Cultura [350]
Indice de los privilegios de la ciudad (anno 1655) / Toyuela, Nicolas Perez – Albarracin – 1r – 5,6 – sp Cultura [946]
Indice de personas nobles y otras de calidad... / Retana, WE – sp Bibl Santa Ana [920]
Indice de personas nobles y otras de calidad que han estado en filipinas (1521-1898) / Retana, WE – Madrid: Fortanet, 1920. Edit. Reus 1921, 76 p. 485-502, 77 pp. 60-67 y 245-272, y 78, pp. 68-78 y 148-161. B.R.A.H. – sp Bibl Santa Ana [920]
Indice del archivo de la ensenanza superior de gua... / Irungaray, Ezequiel C – Guatemala, 1962 – 1r – us UF Libraries [972]
Indice d'ittiologia siciliana / Rafinesque-Schmaltz, C S – Messina, 1810 – 1mf – 9 – mf#Z-2223 – ne IDC [590]
Indice do commercio – Manaus, AM. 07 ago 1890 – (= ser Ps 19) – bl Biblioteca [380]
Indice juridico colombiano / Rojas Palacio, Adelfa – Medellin, Colombia. 1965 – 1r – us UF Libraries [972]
Indice universal de inventarios / Colombia Contraloria General De La Republica – Bogota, Colombia. 1962 – 1r – us UF Libraries [972]
Die indices librorum prohibitorum des 16. jahrhunderts / ed by Reusch, Heinrich – Stuttgart: Litterarischer Verein, 1886 (Tuebingen: H Laupp) – (incl bibl ref. german and latin text with an introduction in german) – us Wisconsin U Libr [450]
Die indices librorum prohibitorum des sechzehnten jahrhunderts / ed by Reusch, Heinrich – Stuttgart: Litterarischer Verein, 1886 (Tuebingen: H Laupp) [mf ed 1993] – (= ser Blvs 176) – 598p – 1 – (incl bibl ref. german and latin text. int in german) – mf#8470 reel 36 – us Wisconsin U Libr [932]
Indices op de gewone resolutie van de staten van holland, 1524-1795 = Indexes to the public resolutions of the states of holland, 1524-1795 / Netherlands. General State Archives – 142mf – 9 – ne MMF Publ [949]
Indices op de resoluties van de gecommitteerde raden in het zuiderkwartier, 1621-1795 = Indexes to the resolutions of the gecommitteerde raden of south holland (zuiderkwartier), 1621-1795 / Netherlands. General State Archives – 191mf – 9 – ne MMF Publ [949]
Indices op de secrete resoluties van de staten van holland, 1653-1795 = Indexes to the secret resolutions of the states of holland, 1653-1795 / Netherlands. General State Archives – 22mf – 9 – ne MMF Publ [949]
Indices qvidam ionnis bvgenhagij pomerani in euangelia / Bugenhagen, J – Augsburg, 1525 – 1mf – 9 – mf#TH-1 mf 174 – ne IDC [242]
Indices to diatessarica : with a specimen of research / Abbott, Edwin Abbott – London: A. and C. Black; New York: Macmillan [distributor], 1907 – 1mf – 9 – 0-7905-3360-X – mf#1987-3360 – us ATLA [450]
An indictment of darwin / Dawson, Oswald – London, 1888 – (= ser 19th c evolution & creation) – 1mf – 9 – mf#1.1.10863 – uk Chadwyck [575]
Indie en indonesie : kranten en knipsels 1946-1947 – n.p, n.d. – 4mf – 8 – mf#SE-1619 – ne IDC [950]
Indie en indonesie kranten en knipsels 1946-1947 – Np, nd – 4mf – 9 – mf#SE-1619 – ne IDC [950]
Indie in de Nederlandsche Studentenwereld *see* Verslag van het eerste congres van het "indonesisch verbond van studerenden"
Indien : for og nu / Loventhal, Eduard – Odense: Milo'ske Boghandels Forlag, 1895 [mf ed 1995] – (= ser Yale coll) – 363p (ill) – 1 – 0-524-10029-2 – (in danish) – mf#1995-1029 – us ATLA [954]
Indien / Mantegazza, Paolo – Jena 1885 [mf ed Hildesheim 1995-98] – 1v on 3mf – 9 – €90.00 – 3-487-27520-1 – gw Olms [915]
Indien und das christentum : eine untersuchung der religionsgeschichtlichen zusammenhaenge / Garbe, Richard – Tuebingen: JCB Mohr, 1914 – 1mf – 9 – 0-524-01358-6 – (incl bibl ref) – mf#1990-2370 – us ATLA [230]
Das indienbild deutscher dichter um 1900 : dauthendey, bonsels, mauthner, gjellerup, hermann keyserling und stefan zweig: ein kapitel deutsch-indischer geistesbeziehungen im frühen 20. jahrhundert / Ganeshan, Vridhagiri – Bonn: Bouvier Verlag H Grundmann, 1975 [mf ed 1993] – (= ser Abhandlungen zur kunst-, musik- und literaturwissenschaft v187) – 425p – 1 – (Dauthendey+Bonsels+Mauthner, Fritz+Gjellerup, Karl Adolph+Keyserling, Hermann+Zweig, Stefan) – mf#8271 – us Wisconsin U Libr [430]
Indiens literatur und cultur in historischer entwicklung : ein cyclus von fuenfzig vorlesungen, zugleich als handbuch der indischen literaturgeschichte / Schroeder, Leopold von – Leipzig: H Haessel, 1887 – 1mf – 9 – 0-524-04536-4 – (incl bibl ref) – mf#1990-3370 – us ATLA [490]
De indier *see*
– Maandelijksche kronijk
Indifferent horseman / Carpenter, Maurice – London, England. 1954 – 1r – us UF Libraries [025]
Indifferentism / Maclaughlin, John – London, England. 1887 – 1r – us UF Libraries [240]
Indifferentism : or, is one religion as good as another? / MacLaughlin, John – London: Burns & Oates; New York: Benziger, 1894 – 1mf – 9 – 0-8370-7085-6 – mf#1986-1085 – us ATLA [230]
Indigena – Berkeley CA. v2 n2 [1976 fall], v3 n1-v4 n1 [1977 summer-1978 summer], 1974 summer-1977 summer – 2r – 1 – mf#637327 – us WHS [071]
Indigenas da colonia de mocambique / Cabral, Antonio Augusto Pereira – Lourenco Marques?, Mozambique. 1934? – 1r – us UF Libraries [960]
Indigene land- und selbstbestimmungsrechte in australien und kanada unter besonderer beruecksichtigung des internationalen rechts / Carstens, Margret – (mf ed 2000) – 5mf – 9 – €59.00 – 3-8267-2711-8 – mf#DHS 2711 – gw Frankfurter [323]
Les indigenes d'a o f : leur condition politique et economique / Moreau, Paul Joseph – Paris: Editions Domat-Montchrestien, 1938 – 1 – us CRL [960]
Indigenization of the ymca in china *see* Chung-hua chi-tu chiao ching nien hui shi lueh [ccm349]
The indigenous church : country church and indigenous christianity / Clark, Sidney James Wells – London, World Dominion Press [1913?] [mf ed 1995] – (= ser Yale coll; Indigenous church series) – 24p – 1 – 0-524-09787-9 – mf#1995-0787 – us ATLA [230]
Indigenous drugs inquiry : a review of the work / Chopra, Ram Nath – Simla: Liddell's Press, 1939 – 1 – us CRL [950]
Indigenous knowledge intermediate phase learners in a rural area have of rainfall : a case study of intermediate phase learners in rural kwazulu-natal: research report / Ndlovu, Ntsibeng – Pretoria: Vista University 2001 [mf ed 2001] – 2mf – 9 – (incl bibl) – mf#mfm15196 – sa Unisa [373]
Indigenous law journal – v1. 2002 – 9 – (filming in process) – ISSN: 1703-4566 – mf#119151 – us Hein [323]
Indigenous preaching in china with a focal critique on john sung / Gwo, Yun-Han – 1982 – 1 – $5.92 – us Southern Baptist [242]
Indigenous technologies : implications for a technology education curriculum / Gumbo, Mishack Thiza – Pretoria: Vista University 2003 [mf ed 2003] – 8mf – 9 – (incl bibl ref) – mf#mfm15228 – sa Unisa [370]
Indigenus flowers of the hawaiian islands : forty-four plates, painted in water-colours and described by mrs. francis sinclair, jr. / Sinclair, Francis Isabella – 2mf – 15 – $65.00 – us UMI ProQuest [580]
Indika : the country and the people of india and ceylon / Hurst, John Fletcher – New York: Harper, 1891 – 2mf – 9 – 0-7905-5341-4 – (incl bibl ref) – mf#1988-1341 – us ATLA [954]
Indio brasileiro e a revolucao francesa / Arinos De Melo Franco, Afonso – Rio de Janeiro, Brazil. 1937 – 1r – us UF Libraries [972]
[Indio-] daily news indio – CA. 1928-90 – 106r – 1 – $6360.00 – mf#RH02304 – us Library Micro [071]
[Indio-] date palm – CA. 1912-59 – 32r – 1 – $1920.00 – mf#R02305 – us Library Micro [071]
Indio e o mundo dos brancos / Oliveira, Roberto Cardoso De – Sao Paulo, Brazil. 1964 – 1r – us UF Libraries [972]
Indio en la colonia / Arboleda Llorente, Jose Maria – Bogota, Colombia. 1948 – 1r – us UF Libraries [972]
El indio liberal – Madrid, Spain. 29 apr-4 may 1820 [wkly] – 4ft – 1 – uk British Libr Newspaper [074]
Indio motilon y su historia / Reynal, Vicente – Puente Comun, Colombia. 1962 – 1r – us UF Libraries [972]
Indiographs, Inc *see* Issue
Indiologia / Costa, Angyone – Rio de Janeiro, Brazil. 1943 – 1r – us UF Libraries [972]
Indios americanos, supersticiones, hechicerias practicas / Palza S, Ernesto – Cochabamba, Bolivia. v1-2. 1946 – 1r – us UF Libraries [972]
Indios caribes / Salas, Julio C – Madrid, Spain. 1920 – 1r – us UF Libraries [972]
Indios de cuba en sus tiempos historicos / Pichardo Moya, Felipe – Habana, Cuba. 1945 – 1r – us UF Libraries [972]
Indios do brasil / Lima Figueiredo, Jose De – Sao Paulo, Brazil. 1939 – 1r – us UF Libraries [972]
Indios e a civilizacao / Ribeiro, Darcy – Rio de Janeiro, Brazil. 1970 – 1r – us UF Libraries [972]
Indios e castanheiros / Laraia, Roque De Barros – Sao Paulo, Brazil. 1967 – 1r – us UF Libraries [972]
L'indipendente : italian journal – Syracuse: [s.n.], 1917-nov 22 1918 – 1r – us CRL [073]
L'indipendente – New York NY, oct 1924-26 – 1r – 1 – (italian periodical) – us IHRC [073]
L'indipendenza italiana – Paris: Impr de Lange Levy et Comp, feb 27 1848 – us CRL [074]
Indipohdi : dramatisches gedicht / Hauptmann, Gerhart – Berlin: S Fischer, 1921 – 1r – 1 – us Wisconsin U Libr [810]
Indira and other stories / Chatterji, Bankim Chandra – Calcutta: Modern Review Office, 1925 – (= ser Samp: indian books) – (trans by j d anderson) – us CRL [390]
Indirect domestic influence : or, nova scotia as it was, as it is and as it may be... / Godfrey, L M – Boston: S M Godfrey 1855 [mf ed 1987] – 1mf – 9 – 0-665-63141-3 – mf#63141 – cn Canadiana [840]
Indirect testimony of history to the genuineness of the gospels / Huidekoper, Frederic – 5th ed. New York: David G Francis, 1886, c1879 – 1mf – 9 – 0-8370-3690-9 – (incl bibl ref, appendix & indexes) – mf#1985-1690 – us ATLA [226]
Indirect testimony to the gospels : and christ's mission to the underworld / Huidekoper, Frederic – New York: J. Miller, 1882. Beltsville, Md: NCR Corp, 1978 (5mf); Evanston: American Theol Lib Assoc, 1984 (5mf) – 9 – 0-8370-1119-1 – (incl bibl ref and ind) – mf#1984-4496 – us ATLA [226]
Indisch en nederlandsch persoverzicht / Regeerings Voorlichtings Dienst – Batavia, 1946 (jul 9-dec 18) – 7mf – 959 – mf#SE-1464 – ne IDC [959]
Indisch missietijdschrift *see* Onze missien in oost- en west-indien
Indisch tijdschrift van het recht; orgaan der nederlandsch-indische juristen-vereeniging. – Batavia. Publ. suspended from 1869 to 1874. Title varies. On film: v1-151; 1849-1940. Missing: v35-36; 1880-81. LL-0281 – 1 – us L of C Photodup [340]
Een indisch vorstenzoon / Westhoff, Johannes Peter Godfried – [Rotterdam: J M Bredee, 1903] [mf ed 1995] – (= ser Yale coll; Lichtstralen op den akker der wereld [9. jaarg 1903] 5) – 39p (ill) – 1 – 0-524-10045-4 – (in dutch) – mf#1995-1045 – us ATLA [954]
Indische (de) mercuur – Orgaan voor Handel, Landbouw, Nijverheid en Mijnwezen in Nederlansch Oost-en West-Indie. Amsterdam. Jaargang. 8-63. 1885-Apr 1940 – 1 – us NY Public [330]
Indische einfluesse auf evangelische erzaehlungen / Bergh van Eysinga, Gustaaf Adolf van den – 2. verm. Aufl. Goettingen: Vandenhoeck und Ruprecht, 1909 – (= ser Forschungen zur religion und literatur des alten und neuen testaments) – 1mf – 9 – 0-7905-0726-9 – (incl bibl ref) – mf#1987-0726 – us ATLA [220]
De indische mercuur : orgaan voor den handel op indie – Amsterdam, Netherlands [mf ed 1878-81] – 1 – uk British Libr Newspaper [074]
Indische missionsgeschichte / Richter, Julius – Guetersloh: C Bertelsmann, 1906 [mf ed 1995] – (= ser Yale coll) – iv/445p (ill) – 1 – 0-524-09203-6 – (in german) – mf#1995-0203 – us ATLA [240]
Indische missionsgeschichte *see* History of missions in india
Indische reisebriefe / Dalton, Hermann – Guetersloh: C Bertelsmann, 1899 [mf ed 1995] – (= ser Yale coll) – xii/386p – 1 – 0-524-09141-2 – (in german) – mf#1995-0141 – us ATLA [880]
Indische religionsgeschichte / Hardy, Edmund – 2. verb aufl. Leipzig: G J Goeschen 1904 [mf ed 1991] – (= ser Sammlung goeschen) – 1mf – 9 – 0-524-01550-3 – (incl bibl ref; first printed in 1898) – mf#1990-2504 – us ATLA [280]
Der indische seelenwanderungsglaube / Dilger, Wilhelm – Basel: Basler Missionsbuchh, 1910 – (= ser Basler missionsstudien) – 1mf – 9 – 0-524-01428-0 – mf#1990-2423 – us ATLA [280]
Indisch-Genootschap *see* Naamlijst der leden, 1904-1913
Indispensable james joyce / Joyce, James – New York, NY. 1949 – 1r – us UF Libraries [420]
Indispensible and absolute necessity of regeneration / Goodwin, Thomas – London, England. 1823 – 1r – us UF Libraries [240]
Inditzki, Israel Jehiel *see* Metargem
Individual – Pueblo, CO. 1896-1898 (1) – mf#62322 – us UMI ProQuest [071]
The individual : a study of life and death / Shaler, Nathaniel Southgate – New York: D Appleton, 1913, c1900 – 1mf – 9 – 0-7905-8584-7 – mf#1989-1809 – us ATLA [210]
Individual action – v1 n9 [1953 mar 31] – 1r – 1 – mf#629935 – us WHS [071]
The individual and the group : an indian study in conflict / Mallik, Basanta Kumar – London: George Allen & Unwin, 1939 – (= ser Samp: indian books) – us CRL [302]
The individual and the social gospel / Mathews, Shailer – New York: Missionary Education Movement of the United States and Canada, 1914 – 1mf – 9 – 0-7905-7996-0 – mf#1989-1201 – us ATLA [240]
Individual behaviour towards authority / Levy, Kathryn Anne – 6mf – 9 – (incl bibl ref) – mf#mfm14978 – sa Unisa [150]
Individual, corporate and firm names / McAdam, David – New York, Diossy, 1894. 84 p. LL-531 – 1 – us L of C Photodup [340]
Individual differences in variability and pattern of performance : as a consideration in the selection of a representative score from multiple trial physical performance data / Darracott, Shirley H – 1995 – 2mf – 9 – $8.00 – mf#PE 3749 – us Kinesology [370]
Individual effort – London, England. 18– – 1r – us UF Libraries [240]
Individual evangelism : christian witnessing and work: the call of christ to the laity / Beach, Charles Fisk – Philadelphia: Westminster Press, c1908 – 1mf – 9 – 0-8370-6012-5 – (incl bibl ref) – mf#1986-0012 – us ATLA [242]
Individual evangelism *see* Ko jen pu tao (ccm241)
Individual liberty : newsletter of the society for for individual liberty / Society for Individual Liberty – 1979 apr-1984 apr – 1r – 1 – (cont: society for individual liberty news) – mf#1448945 – us WHS [322]
Individual psychology – Austin. 1982-1997 (1) 1982-1997 (5) 1982-1997 (9) – (cont by: journal of individual psychology) – ISSN: 0277-7010 – mf#7717,01 – us UMI ProQuest [150]
Individual work for individuals : a record of personal experiences and convictions / Trumbull, Henry Clay – New York: International Committee of Young Men's Christian Associations, c1901 – 1mf – 9 – 0-8370-7348-0 – mf#1986-1348 – us ATLA [240]
The individualist : the journal of the personal rights association / The Personal Rights Association – 1921-69 – 1 – us AMS Press [322]
Das individualitaetsproblem bei friedrich hebbel / Hallmann, Georg – Leipzig: L Voss 1920 [mf ed 1990] – (= ser Beitraege zur aesthetik 16) – 1r – 1 – (filmed with: hebbels dithmarschenfragment / heinrich bender) – mf#2704p – us Wisconsin U Libr [430]
Individuality and immortality / Ostwald, Wilhelm – Boston: Houghton, Mifflin, 1906 – (= ser The Ingersoll Lecture) – 1mf – 9 – 0-7905-8539-1 – mf#1989-1764 – us ATLA [240]
Individuality and intimacy in pastoral marital counseling / O'Neill, James H – 1982 – 1 – 6.64 – us Southern Baptist [242]
Individualizatsiia zemlevladeniia v rossii i ee posledstviia ottisk iz "vestnik sel khoz" 1914 g / Oganovskii, N P – 1914 – 98p 2mf – 9 – mf#COR-82 – ne IDC [335]

Individualized maximal gxt is preferred over standardized bruce protocol in relatively fit college students / Spackman, Michael B – 1999 – 1mf – 9 – $4.00 – mf#PH 1687 – us Kinesology [612]

An individualized self-control approach to weight reduction / Heath, Peter S – 1982 – 1mf – 9 – $4.00 – us Kinesology [610]

Individuals for a Rational Society *see* Unbound!

Individuele : en huweliksaanpassing van die nierpasient / Bredekamp, Rosa – Uni of South Africa 2001 [mf ed Johannesburg 2001] – 8mf [ill] – 9 – (incl bibl ref; text in afrikaans) – mf#mfm14840 – sa Unisa [150]

Individuelle und gesellschaftliche repraesentationen von aids : ein vergleich zwischen bremen und rostock / Paul, Doris & Schulz, Thomas – (mf ed 1993) – 2mf – 9 – €49.00 – 3-89349-654-8 – mf#DHS 654 – gw Frankfurter [150]

Individuo / Lles Y Berdayes, Fernando – Habana, Cuba. 1934 – 1r – us UF Libraries [972]

Individuum und gemeinschaft in den romanen toni morrisons / Kielkopf, Frieder – (mf ed 1998) – 1mf – 9 – €30.00 – 3-8267-2534-4 – mf#DHS 2534 – gw Frankfurter [420]

Indledning til det gamle testamente : tilligemed en oversigt over nogle af de bibelske boeger / Sverdrup, Georg; ed by Helland, Andreas – Minneapolis, MN: Frikirkens Boghandels Forlag, 1910 [mf ed 1993] – (= ser Professor georg sverdrups samlede skrifter i udvalg 5; Lutheran coll) – 1mf – 9 – 0-524-06325-7 – mf#1991-2498 – us ATLA [220]

Indo caribbean world – 1994 dec 15-1995 dec 20, 1996 jan 10-dec 18, 1997 jan 8-dec 17, 1998 jan 7-dec 16, 1999 jan 6-dec 15, 2000 – 6r – 1 – mf#3200030 – us WHS [071]

Indo-american magazine : devoted to the great work – Chicago: Indo-American Book Co 1909 [mf ed 2005] – 1v on 1r – 1 – (cont by: life and action, the great work in america) – mf#2005C-s010 – us ATLA [290]

Indo-anglian literature / Srinivasa Iyengar, K R – Bombay: publ for PEN All-India Centre by International Book House, 1943 – (= ser Samp: indian books) – us CRL [410]

Indo-aryan polity : being a study of the economic and political condition of india as depicted in the rig veda / Basu, Praphullachandra – London: PS King & Son, Ltd, 1925 – (= ser Samp: indian books) – us CRL [954]

The indo-aryan races : a study of the origin of indo-aryan people and institutions / Chanda, Ramaprasad – Rajshahi: Published by the Varendra Research Society, 1916 – (= ser Samp: indian books) – us CRL [301]

Indo-asia – v1-12. 1959-70 – 1 – us AMS Press [073]

Indo-cambodia / Bapinidu, Maganti – Madras: Jateeya Jnana Mandir [introd 1957] [mf ed 1989] – 1r – 1 – mf#mf-10289 seam reel 001/11 [§] – us CRL [327]

Indochina : internal affairs and foreign affairs, 1945-1959 / U.S. State Dept – (= ser Confidential u s state department central files) – 1 – $19,855.00 coll – (internal affairs: 1945-49 10r isbn 0-89093-718-4 $1925. 1950-54 44r isbn 0-89093-719-2 $8515. internal & foreign affairs, 1955-59 54r isbn 1-55655-107-X $10,450. with p/g) – us UPA [959]

Indo-china and its primitive people / Baudesson, Henry – London: Hutchinson [1919?] [mf ed 1995] – (= ser Yale coll) – xii/328p (ill) – 1 – 0-524-09161-7 – (trans by e appleby holt) – mf#1995-0161 – us ATLA [950]

Indochina bulletin / Asia Information Group – n23-29 [1973 feb 23/mar 15-sep] – 1r – 1 – (cont: war bulletin [berkeley, ca]) – mf#657752 – us WHS [071]

Indochina focal point / Indochina Peace Campaign – v1-2 n14 [1973 jul-1975 nov1] – 1r – 1 – mf#381000 – us WHS [320]

Indochina. French *see*
- Bulletin administratif
- Bulletin officiel
- Bulletin periodique des actes administratifs
- Cong bao
- Guide pour l'application des lois sociales
- Journal officiel

Indochina. French. Commissariat *see*
- Bulletin officiel

Indochina. French. Direction des douanes et regies *see* Statistique mensuelle du commerce exterieur de l'indochine

Indochina. French. Laws, Statutes, etc *see*
- Recueil des reglements concernant l'organisation des regies en indochine
- Repertoire chronologique et alphabetique des lois, decrets, arretes ministeriels promulgues en indochine du 1 janvier 1926 au 1 janvier 1935

Indochina Peace Campaign *see* Indochina focal point

Indochina Resource Action Center [Washington DC] *see* Bridge

L'indochine : journal quotidien de rapprochement franco-annamite – Saigon. juin-aout 1925 [daily] – 1 – fr ACRPP [073]

L'indochine. *see* L'indochine enchainee

L'indochine enchainee – Saigon. n1-23. 1925-fevr 1926 – 1 – (ed. provisoire de: l' indochine.) – fr ACRPP [073]

L'indo-chine francaise – Hanoi: indo-chine francaise, [1896- [1896 jan 15-23, jan 25-mar 1/2,4, mar 13-apr 4, may 11-16, may 20-jun 30, aug 19-30/31, sep 10-13/14, oct 1-16, oct 23-dec 11,13-29, 1898 mar 4-17,19-26,29-31, apr 8, jun 25,28,30, jul 2-9, 1899: jul 1-aug 30, sep 7-8, 16-23, sep 28-oct 26, oct28-nov 22, dec 14-22, 1900: may 31-jun 17, jul 4,8, jul 28-oct 25, nov 6-dec 21, dec 23-28] – 1r – 1 – mf#mf-11761 seam – us CRL [079]

L'indochine vue de pekin : entretiens avec jean lacouture / Norodom Sihanouk, Prince – Paris: Editions du Seuil [1972] [mf ed 1989] – 1r with other items – 1 – mf#mf-10289 seam reel 019/05 [§] – us CRL [327]

Indo-chinese patriot – George Town and Singapore. Singapore. -w. Feb-Oct 1895, 20 Sep 1900-11 Dec 1901. (22 ft) – 1 – uk British Libr Newspaper [072]

L'indo-chinois – Hanoi: [s.n.] 1900 jul 3-7,19-21, jul 26-aug 18, 1901 jan 5-29, feb 26-jun 29, 1902 jan 4-jun 28, 1903 jan 1-dec 31, 1904 jan 2?, jan 9?-jun 27, 1905 jan 6-dec 30, 1906 jul 12-sep 1, 1907 jan 1-10, 1908 may 3-aug 8, 1910 mar 16-nov 20 – 4r – 1 – mf#mf-11748 seam – us CRL [079]

Indo-eenheids-verbond / Onze Stem – Djakarta, [1924]-1956. v1-33(2/3) – 3mf – 9 – (missing: [1924]-1953(1-7, 11-?)-1956(1)) – mf#SE-1857 – ne IDC [959]

Indogermanische eigennamen als spiegel der kulturgeschichte / Solmsen, Felix – Heidelberg, Germany. 1922 – 1r – us UF Libraries [920]

Indogermanische mythen / Meyer, Elard Hugo – Berlin: Duemmler, 1883-87 – 1 – us Wisconsin U Libr [390]

Indogermanische naturreligion / Asmus, Paul – Halle: CEM Pfeffer, 1875 – (= ser Indogermanische Religion in den Hauptpunkten ihrer Entwickelung) – 1mf – 9 – 0-524-01249-0 – mf#1990-2285 – us ATLA [200]

Indogermanische sprachwissenschaft / Krahe, Hans – Berlin: W. de Gruyter, 1943. 184p. diagrs – 1 – us Wisconsin U Libr [400]

Indogermanischer volksglaube : ein beitrag zur religionschichte der urzeit / Schwartz, Friedrich Leberecht Wilhelm – Berlin: O Seehagen, 1885 – 1mf – 9 – 0-524-03533-4 – (incl bibl ref) – mf#1990-3238 – us ATLA [290]

Indo-iranian studies...in honour of shams-ul-ullema dastur darab peshotan sanjana – London: K. Paul, Trench, Truebner & Co., 1925. viii,293p. front – 1 – us Wisconsin U Libr [490]

"Indone" djoeten – (n.p. 2603?) – 249p 3mf – 9 – (mounted label has: electrische drukkerij "tan", poerbolinggo) – mf#SE-2002 mf43-45 – ne IDC [680]

Indonesia / Anggaran dasar serikat-serikat Berita-negara RI – Djakarta – 209mf – 9 – mf#SE-218 – ne IDC [959]

Indonesia / Angkatan Darat Madjalah Angkatan Darat Menjambut pembukaan kembali AMN & nomer chusus – Djakarta, 1957 – 1mf – 9 – mf#SE-2637 – ne IDC [959]

Indonesia / Angkatan Darat Madjalah Angkatan Darat Penerangan Angkatan Darat – Djakarta, 1950-1959 – 91mf – 9 – (missing: 1951, v1(4-7); 1952, v2(p 1-641, 904-994); 1955, v5(12); 1957, v7(8, 9); 1958, v8(7, 8, 12); 1959, v9(6, 8-12)) – mf#SE-589 – ne IDC [959]

Indonesia / Biro Urusan Industrialisasi Ichtisar laporan unit2 Overheidsdienst Urusan Industrialisasi – Djakarta, 1964 – 6mf – 9 – mf#SE-1545 – ne IDC [959]

Indonesia / Biro Urusan Industrialisasi Laporan tahunan – Djakarta, 1962-1963 – 13mf – 9 – mf#SE-1546 – ne IDC [959]

Indonesia / Departemen Luar Negeri Direktorat Research Rentjana kerdja – Djakarta, 1970 – 1mf – 9 – mf#SE-155=5 – ne IDC [959]

Indonesia / Departemen Luar Negeri Direktorat Research Research brief – Djakarta, 1969-1970. v1-2(5) – 8mf – 9 – (missing: 1969 v1(1-2)) – mf#SE-155=6 – ne IDC [959]

Indonesia / Departemen Luar Negeri Direktorat Research Research diplomatik – Djakarta, 1969-1972 – 16mf – 9 – (missing: 1972(11, 14)) – mf#SE-155=7 – ne IDC [959]

Indonesia / Departemen Luar Negeri Direktorat Research Research dokumentasi – Djakarta, 1970 – 6mf – 9 – mf#SE-155-8 – ne IDC [959]

Indonesia / Departemen Luar Negeri Direktorat Research Research landasan – Djakarta, 1969-1971 – 12mf – 9 – (missing: 1970 v4) – mf#SE-1560 – ne IDC [959]

Indonesia / Departemen Luar Negeri Direktorat Research Research publikasi – Djakarta, 1969 – 3mf – 9 – mf#SE-1561 – ne IDC [959]

Indonesia / Departemen Luar Negeri Direktorat Research Research reconnaissance – Djakarta, 1969-1970 – 38mf – 9 – (missing: 1969 v1; v7) – mf#SE-1562 – ne IDC [959]

Indonesia : internal affairs and foreign affairs 1960-jan 1963 / U.S. State Dept – (= ser Confidential u s state department central files) – 20r – 1 – $3885.00 – 1-55655-836-8 – (with p/g) – us UPA [959]

Indonesia – Ithaca. 1966+ (1,5,9) – ISSN: 0019-7289 – mf#11092 – us UMI ProQuest [959]

Indonesia – Bangkok, 1949-1954 – 10mf – 9 – (missing: 1949-1953, v1-5(1-268); 1954, v6(271-275, 277-280, 288)) – mf#SE-547 – ne IDC [959]

Indonesia : news and views – Canberra, 1955-1957 – 9mf – 9 – (missing: 1955, v1(1, 11, 20); 1956/1957, v2(2-5, 7)) – mf#SE-548 – ne IDC [959]

Indonesia : republic, 1945-1949 / Kementerian Penerangan Siaran kilat – Jogjakarta, 1946. 4 v – 1mf – 9 – mf#SE-11921 – ne IDC [959]

Indonesia : republic, 1945-1949 / Kementerian Penerangan Siaran kilat – Djakarta, 1946. 45 pamphlets – 13mf – 9 – (missing: nos 2, 7, 11, 13, 15-17, 19, 20, 22, 24, 26-28, 35, 39, 41) – mf#SE-11920 – ne IDC [959]

Indonesia : republic, 1945-1949 / Voorlichtingsdienst, London, Voor Interne Circulatie – London, [1945]-1949 – 1mf – 9 – mf#SE-1519 – ne IDC [959]

Indonesia : tourism and travel gazette / Indonesian National Tourist Organization – Djakarta, 1968(1-5) – 1mf – 9 – (missing: 1967/68(1-2)) – mf#SE-1692 – ne IDC [959]

Indonesia : travel & trade / Melati Pub House – Amsterdam, 1967(1-2) – 2mf – 9 – mf#SE-1693 – ne IDC [959]

Indonesia *see*
- Badan perentjanaan pembangunan nasional bappenas
- Departemen agama agenda kementerian agama bagian publikasi dan redaksi, djawatan penerangan agama
- Departemen agama konperensi dinas
- Departemen agama laporan kementerian agama, bagian penerbitan
- Departemen anggaran negara laporan tahunan
- Departemen angkatan darat daftar singkatan-singkatan istilah resmi dalam angkatan darat
- Departemen angkatan udara lembaran keamanan penerbangan assisten direktorat keamanan terbang
- Departemen dalam negeri mimbar departemen dalam negeri bagian hubungan dan penerangan masjarakat
- Departemen dalam negeri sektor chusus irian-barat himpunan peraturan2 pemerintah tentang masalah pengurusan daerah propinsi irian barat
- Departemen dalam negeri sektor chusus irian-barat laporan pembangunan irian barat
- Departemen kesehatan pedoman dan berita
- Departemen luar negeri department of foreign affairs
- Departemen luar negeri direktorat asia timur laut dan pasifik malaysia masalah dan perkembangan selandjatnja
- Departemen luar negeri direktorat research biro research umum research kronologi dan dokumentasi
- Departemen luar negeri direktorat research pewarta dan kronologi bulanan seksi perentjanaan dan penerbitan
- Departemen pekerdjaan umum dan tenaga berita dep-pu-t
- Departemen pekerdjaan umum dan tenaga progress report; tahun kerdja 1967
- Departemen pendidikan dan kebudajaan perpustakaan sedjarah politik dan sosial press index
- Departemen pendidikan pengadjaran dan kebudajaan buku alamat sekolah landjutan dan kursus-kursus negeri subsidi dan bantuan
- Departemen pendidikan pengadjaran dan kebudajaan daftar adanja sekolah landjutan negeri, subsidi dan bantuan dimasing-masing kabupaten dan kota
- Departemen penerangan daftar harian dan madjalah seluruh indonesia
- Departemen penerangan department of information facts & figures
- Departemen penerangan department of information information bulletin
- Departemen penerangan detik peristiwa dalam negeri
- Departemen penerangan direktorat publisiteit & penerangang daerah, bagian dokumentasi madjelis permus jawaratan rakjat sementara
- Departemen penerangan ichtisar harian tanah air selama 24 djam
- Departemen penerangan ministry of informatin foreign observers on the question of west irian
- Departemen penerangan penlugri miscellany
- Departemen penerangan seri amanat
- Departemen penerangan siaran departemen penerangan melalui siaran rri pusat rtd
- Departemen penerangan siaran pemerintah
- Departemen penerangan special issue
- Departemen penerangan tanja djawab
- Departemen penerangan the fourth asian games
- Departemen penerangan upe
- Departemen penerangan uraian departemen penerangan ri melalui siaran rri pusat udp
- Departemen perburuhan laporan
- Departemen perdagangan data2 perdagangan laporan semester
- Departemen perdagangan himpunan peraturan2 dibidang perdagangan jajasan penjuluhan dan penerangan perdagangan
- Departemen perdagangan perwakilan sumatera utara laporan tahunan
- Departemen perdagangan progress report
- Departemen perhubungan bulletin perhubungan untuk dinas bagian hubungan masjarakat
- Departemen perhubungan laporan bidang organisasi dan personil departemen perhubungan biro organisasi & personil, sekretariat djenderal departemen perhubungan
- Departemen perindustrian rakjat buku laporan tahunan
- Departemen perindustrian rakjat laporan team departeman perindustrian rakjat kedaerah tahun 1960 kantor penjuluhan perindustrian, departemen perindustrian rakjat
- Departemen perindustrian rakjat laporan team departemen perindustrian rakjat kedaerah tahun 1961 kantor penjuluhan perindustrian, departemen perindustrian rakjat
- Departemen perindustrian tekstil dan keradjinan rakjat warta deptekra; bulletin bulanan bagian humas deptekra
- Departemen pertahanan keamanan pusat perlawanan dan keamanan rakjat laporan kegiatan tahun kerdja
- Departemen tenaga kerdja buku hasil raker departemen tenaga kerdja sekretariat raker
- Departemen tenaga kerdja laporan
- Departemen transmigrasi dan koperasi feasibility studies pembangunan koperasi
- Departemen transmigrasi, koperasi dan pembangunan masjarakat desa
- Departemen transmigrasi, koperasi dan pembangunan masjarakat desa biro pembangunan masjarakat desa marilah membangun masjarakat
- Departemen transmigrasi, koperasi dan pembangunan masjarakat desa biro pembukaan tanah setelah enam bulan bekerdja
- Departemen transmigrasi, koperasi dan pembangunan masjarakat desa djadikan koperasi sebagai alat untuk mentjapai masjarakat sosialis indonesia atas dasar usdek; himpunan pidato pada peringatan koperasi di istana negara 12 djuli 1960
- Departemen transmigrasi, koperasi dan pembangunan masjarakat desa koperasi dalam alam sosialisme indonesia; pidato
- Departemen transmigrasi, koperasi dan pembangunan masjarakat desa koperasi indonesia berdasarkan pantja sila dan usdek
- Departemen transmigrasi, koperasi dan pembangunan masjarakat desa pembangunan masjarakat desa dalam hubungan internasional
- Departemen transmigrasi, koperasi dan pembangunan masjarakat desa pola kerdja sama departemen transkopemada dengan departemen2 lain; himpunan keputusan2 bersama
- Departemen transmigrasi, koperasi dan pembangunan masjarakat desa pola pelaksanaan tugas departemen transkopemada th dinas 1962
- Departement van economische zaken grafieken behorende bij de economische toestand van indonesie
- Department of information republic of indonesia
- Dinas perindustrian daerah laporan kantor penjuluhan perindustrian
- Direktorat badan pimpinan umum perusahaan perkebunan dwikora laporan kerdja perusahaan
- Direktorat badan pimpinan umum perusahaan perkebunan dwikora laporan tahunan
- Direktorat djenderal kehutanan data kehutanan
- Direktorat djenderal kehutanan laporan tahun
- Direktorat djenderal kehutanan publication
- Direktorat djenderal koperasi peraturan2 tentang bimas
- Direktorat djenderal koperasi recording rapat kerdja departemen transmigrasi dan koperasi
- Direktorat djenderal padjak laporan triwulan
- Direktorat djenderal padjak musjawarah kerdja
- Direktorat djenderal pembangunan masjarakat desa lembaran pmd
- Direktorat djenderal pembangunan masjarakat desa madjalah pembangunan masjarakat desa
- Direktorat djenderal pengolahan kekajaan laut laporan tahunan
- Direktorat djenderal perguruan tinggi dan ilmu pengetahuan research journal
- Direktorat djenderal perindustrian kimia laporan kerdja
- Direktorat djenderal perindustrian kimia laporan pelaksanaan repelita
- Direktorat djenderal perindustrian ringan laporan tahunan
- Direktorat kehutanan rasionalisasi lembaga penelitian ekonomi kehutanan
- Direktorat landuse buku tahunan
- Direktorat landuse publikasi
- Direktorat pembinaan lembaga sosial desa kegiatan lsd diseluruh indonesia
- Direktorat pembinaan perusahaan2 negara industri kimia laporan tahunan direktorat djenderal perindustrian
- Direktorat pendidikan guru dan tenaga tehnis statistik pendidikan guru
- Direktorat perumahan rakjet laporan kerdja

- Djawatan kebudajaan laporan perwakilan djawatan kebudajaan nusa tenggara singaradja
- Djawatan koperasi pusat lampiran statistik pada buku tahunan
- Djawatan koperasi pusat laporan tahunan
- Djawatan pendidikan kedjuruan almanak
- Djawatan pendidikan masjarakat report of the mass education department
- Djawatan penerangan agama departemen agama
- Kementerian agama penjiaran
- Kementerian dalam negeri biro pemilihan petundjuk pemilihan daerah
- Kementerian kesehatan
- Kementerian kesehatan berita hygiene
- Kementerian keuangan nota keuangan negara
- Kementerian keuangan rantjangan anggaran
- Kementerian luar negeri
- Kementerian luar negeri direktorat 5 fakta dan dokumen2 untuk menjusun buku "indonesia memasuki gelanggang internasional"
- Kementerian penerangan bagian dokumentasi ichtisar peristiwa dalam dan luar negeri
- Kementerian penerangan dokumenta informasia
- Kementerian penerangan ichtisar indonesia sepekan
- Kementerian penerangan ichtisar parlemen
- Kementerian penerangan kepartaian dan parlementaria indonesia
- Kementerian penerangan penerbitan chusus
- Kementerian perburuhan laporan
- Kementerian perburuhan laporan kementerian perburuhan selama 2 tahun kabinet karya, april 1957-april 1959
- Kementerian perburuhan laporan singkat, 1956-1957
- Kementerian perburuhan situasi perburuhan dalam dan luar negeri
- Kementerian perhubungan djawatan pelajaran laporan masa 1950 s/d 1952
- Lembaga penjaluran perdagangan ekspor menurut negeri tudjuan dan djenis barang
- Lembaga penjaluran perdagangan impor menurut negeri asal dan djenis barang golongan ekonomi
- Lembaga penjaluran perdagangan laporan tahunan
- Lembaga perpustakaan biologi dan pertanian "bibliotheca bogoriensis" dokumentasi guntingan surat kabar mengenai biologi dan pertanian di indonesia
- Lembaga pertahanan maritim madjalah lemhanmar
- Lembaga pertahanan nasional buku peringatan
- Lembaga pertahanan nasional madjalah pertahanan nasional
- Lembaran-negara republik indonesia
- Madjelis permusjawaratan rakjat sementara bulletin
- Madjelis permusjawaratan rakjat sementara ichtisar
- Madjelis permusjawaratan rakjat sementara keputusan-keputusan republik indonesia
- Madjelis permusjawaratan rakjat sementara republik indonesia
- Madjelis permusjawaratan rakjat sementara ringkasan ketetapan republik indonesia
- Madjelis permusjawaratan sementara laporan komisi2
- Madjelis pertimbangan kesehatan dan sjara' publikasi
- Permanent mission of the republic of indonesia to the united nations pemberitaan nieuw guinea koerier mengenai persoalan irian barat
- Perseroan2 terbatas, perseroan2 firma atau komanditer dan perkumpulan2 koperasi berita-negara ri suppl 4 pertjekatan negara ri

Indonesia 1969-1971 *see* Direktorat djenderal bea dan tjukai himpunan peraturan/instruksi direktorat chusus / harga / laboratorium

Indonesia and the malay world – Oxford. 1997+ (1) – ISSN: 1363-9811 – mf#22890,01 – us UMI ProQuest [959]

Indonesia. Department of Information *see* Departemen penerangan press and broadcast releases

Indonesia. Direktorat Tata Kota dan Tata Daerah *see* Kampung improvement program

Indonesia economic bulletin / Indonesian Economic Information Foundation – Amsterdam, 1967-1970 – 20mf – 9 – (missing: 1970(141); 1971(163)) – mf#SE-1699 – ne IDC [959]

Indonesia gubernur progress report, tahun 1967 : west sumatra (province) – [Padang, 1968] – 1mf – 9 – mf#SE-6963 – ne IDC [959]

Indonesia. Kementerian pendidikan, pengadjaran dan kebudajaan *see*
- Development of education in indonesia
- Education and culture ministry of education and culture
- Perpustakaan perguruan daftar buku-buku (balai pustaka)
- Pewarta ppk
- Sekolah kita bagian naskah/madjalah, djawatan pendidikan umum, kementerian ppk
- Sekolah landjutan kita bagian naskah/madjalah, djawatan pendidikan umum, kementerian ppk
- Sekolah landjutan umum bagian naskah/ madjalah, djawatan pendidikan umum, kementerian ppk
- Sekolah rakjat kita

Indonesia Kementerian Penerangan *see* Special release on current indonesian affairs

Indonesia. Kementerian perekonomian *see*
- Berita ekonomi indonesia
- Ekonomi luar negeri
- Kronik tindakan2 ekonomi

Indonesia league of america : indonesia berkibar – New York, 1947(July) – 1mf – 9 – mf#SE-1696 – ne IDC [959]

Indonesia. madjalah kebudajaan *see* Jajasan penerbitan kebudajaan

Indonesia membangun / Departemen Pekerdjaan Umum dan Tenaga – Djakarta, 1961-1964 – 12mf – 9 – (missing: 1961 v1(1, 8, 11, 12)) – mf#SE-889 – ne IDC [959]

Indonesia merdeka / Djabatan Penerangan – Pematangsiantar, 1946-1947 – 8mf – 9 – (missing: 1946, v1(5, 10-21); 1947, v2(1-5, 7)) – mf#SE-890 – ne IDC [959]

Indonesia merdeka – Djakarta: Hokokai, Himpenan Kebaktian Rakjat, 2605 (1-4) – 1mf – 9 – (missing: 2605(1-3)) – mf#SE-2002 mf274-276 – ne IDC [959]

Indonesia. Ministry of Information *see* Indonesian review

Indonesia Neratja ringkas Bank Indonesia *see* Berita negara ri suppl 5 pertjetakan negara ri

Indonesia. Parlement *see* Penerbit

Indonesia planned parenthood association newsletter / Lembaga Keluarga Berentjana Nasional Indonesia – [Djakarta], 1968, v1(1-4); 1969-1970, v2(5-6); 1971, v3(1) – 3mf – 9 – mf#SE-1763 – ne IDC [959]

Indonesia Progresif *see* Indonesian tribune

Indonesia raya – Djakarta, Indonesia. Oct 1968-Jan 21 1974 – 14r – 1 – us L of C Photodup [079]

Indonesia raya – Jogjakarta, Indonesia. 1955-1959 (1) – mf#61194 – us UMI ProQuest [079]

Indonesia (republic, 1945-1949) / Berita Repoeblik Indonesia Departemen Penerangan – Djakarta, 1945-1946 – 9mf – 9 – (missing: 1946 v2(8, 12-14)) – mf#SE-1510 – ne IDC [959]

Indonesia (republic, 1945-1949) / Kementerian Penerangan Ichtisar pers – Jogjakarta, 1948-1949. v1-2(116) – 3mf – 9 – (missing: 1948-1949 v1-2(1-92, 95-104, 107-114)) – mf#SE-1514 – ne IDC [959]

Indonesia through foreign eyes ministry of information – Djakarta, 1955-1958 – 6mf – 9 – (missing: 1955(1-4, 6-); 1956(1-2, 6-); 1957, v2(1, 3-7)) – mf#SE-550 – ne IDC [959]

Indonesia times – Jakarta: Djamal Ali, S H, may 2 1974- – 1 – us CRL [079]

Indonesia today / Information Department, Indonesian Embassy – London, 1964-1968 – 14mf – 9 – (missing: 1968 v2(3)) – mf#SE-551 – ne IDC [959]

IndonesiaIndonesia. Kementerian pendidikan, pengadjaran dan kebudajaan *see* Djawatan pendidikan masjarakat, bahagian pemuda

Indonesian abstracts *see* Council for sciences of indonesia

Indonesian affairs ministry of information – Djakarta, 1951-1954. 4 v – 33mf – 9 – (missing: 1951 v1(6)) – mf#SE-543 – ne IDC [959]

Indonesian Chamber of Industries *see* Industrial directory of indonesia

Indonesian current affairs translation bulletin – Jakarta. 1975-1975 (1) 1975-1975 (5) (9) – (cont by: current affairs translations bulletin) – ISSN: 0046-9165 – mf#9950 – us UMI ProQuest [321]

Indonesian current affairs translation service bulletin – Kebajoran Baru, 1968-1971 – 59mf – 9 – mf#SE-1698 – ne IDC [959]

Indonesian daily news – Surabaya, Indonesia. 1964; Jul 1965-Dec 1978 – 18r – 1 – us L of C Photodup [079]

Indonesian Economic Information Foundation *see* Indonesia economic bulletin

Indonesian economic review – Djakarta, 1968(may-oct) – 7mf – 9 – mf#SE-164=0 – ne IDC [959]

Indonesian hajj : the pilgrimage to mecca from the netherlands east indies, 1872-1950. documents from the archive of the dutch consulate at jiddah, saudi arabia – [mf ed 2003] – 2807mf – 9 – €13,500.00 – (inventory in dutch; int in english) – mf#mmp106 – ne Moran [260]

Indonesian hajj: the pilgrimage to mecca from the netherlands east indies, 1872-1950 : first supplement: the archives of the dutch vice-consulate and medical officer at mecca, saudi arabia, 1937-1950 – [mf ed 2007] – 16mf – 9 – €270.00 – (with p/g & concordance; in dutch, indonesian, english & some arabic) – mf#mmp132 – ne Moran [260]

Indonesian herald – Jakarta, Indonesia. 1965-1966 (1) – mf#67742 – us UMI ProQuest [079]

Indonesian information / Information Department Indonesian Office – London, 1947-1949 – 20mf – 9 – (missing: 1947/1948, v1-2(1-17); 1948, v2(39-41, 48-49), v3(43-45); 1949, v4(2-3, 10, 14-22)) – mf#SE-1516 – ne IDC [959]

Indonesian information – New Delhi, 1948-1950 – 4mf – 9 – (missing: 1948(1-32); 35-69, 71-end); 1950, v2(1-5, 9)) – mf#SE-561 – ne IDC [959]

Indonesian information – London, 1950-1961. v1-11 – (= ser Indonesian news London) – 54mf – 9 – (several issues missing) – mf#SE-441 – ne IDC [959]

Indonesian information service – Bombay, 1947-1949 – 5mf – 9 – (missing: 1947(1); 1948-1949(20-67)) – mf#SE-560 – ne IDC [959]

Indonesian journal of natural sciences / Natuurkundig Tijdschrift voor Nederlandsch Indie... – Batavia, 1850-1940. v1-100 – 591mf – 9 – mf#SE-830 – ne IDC [500]

Indonesian language press summary – Djakarta, 1961-1965 – (= ser Djakarta Press Summary) – 38mf – 9 – (aka: djakarta press summary) – mf#SE-527 – ne IDC [959]

Indonesian Legation, Information Department *see* Indonesian news

Indonesian Mission of the Christian and Missionary Alliance *see* Pioneer

Indonesian National Scientific Documentation Center *see* Directory of special libraries in indonesia

Indonesian National Tourist Organization *see* Indonesia

Indonesian news / Indonesian Legation, Information Department – Stockholm, 1951-1968 – 71mf – 9 – (several issues missing) – mf#SE-443 – ne IDC [959]

Indonesian news and views / Information Division Embassy of Indonesia – Washington, 1967-1972(9) – 16mf – 9 – (missing: 1967(14); 1968(7-?); 1969(sep-dec); 1970; 1971(1-2, 5-12);) – mf#SE-574 – ne IDC [079]

Indonesian observer – Djakarta, Indonesia. 1959-Apr 1992 – 75r – 1 – us L of C Photodup [072]

Indonesian organization for afro-asian people's solidarity / Suara rakjat Indonesia – Peking, [1967]-1972. v1-6(11) – 46mf – 9 – (missing: [1967]-1969, v1-3(1-13); 1970, v4(16-32? dec)) – mf#SE-1939 – ne IDC [959]

Indonesian political tabloids microfilm collection – [mf ed Chicago IL: filmed by Preservation Resources, Bethlehem, PA for SEAsian MF Project, CRL 2002-03] – ca 340 titles on 41r – 1 – (reel guide incl at beginning of reel 1; pt1: tabloids (reel 1-23); pt2: newspapers (reel 24-26); pt 3: journals and newsletters (reel 27-38a, 38b-40)) – mf#mf-13314 seam – us CRL [959]

Indonesian press review / U.S. Embassy. Indonesia – Djakarta, Indonesia: US Embassy, sept 1965-30 dec 1993 – 37r – 1 – us L of C Photodup [073]

Indonesian press review / US Information Service – Djakarta, 1953-1955 – 148mf – 9 – (several issues missing) – mf#SE-578 – ne IDC [959]

Indonesian Publishing and Trade Service Coy "Alvaco" *see* National business register of indonesia

Indonesian review / Indonesia. Ministry of Information – Djakarta, 1950 – 8mf – 9 – mf#SE-894 – ne IDC [959]

Indonesian review / Jajasan Prapanca – Djakarta, 1951, v1(1-5); 1954, v2(1) – 10mf – 9 – mf#SE-164-2 – ne IDC [079]

Indonesian Socialist Party *see* Pedoman

Indonesian spectator / Nusantara Publishing Co – Djakarta, 1956/1957-1958/1959 – 46mf – 9 – (missing: 1959(23-24)) – mf#SE-376 – ne IDC [959]

Indonesian trade union news – Djakarta, Sobsi, 1957(1-8) – 1mf – 9 – (missing: 1957(1-6)) – mf#SE-164-3 – ne IDC [331]

Indonesian tribune / Indonesia Progresif – Tirana, (Albania), 1966/1967-1972. v1-6(1) – 22mf – 9 – (missing: 1966/1967, v1(1, 4/5); 1970, v4(2-4)) – mf#SE-164-4 – ne IDC [321]

Indonesie / Rutgers, S J & Huber, A – Amsterdam, 1937 – 3mf – 8 – mf#SE-1286 – ne IDC [959]

Indonesie bevrijd! *see* Manifest van de perhimpunan indonesia

Indonesie cultureel / Culturele Voorlichting RVD – Batavia, 1948-1950 – 7mf – 9 – (missing: 1948(4); 1949(1)) – mf#SE-658 – ne IDC [959]

Indonesie sekarang / Harahap, P – Djakarta, 1952. Bulan Bimbang. Tj. ke 2 jang diperb. – 3mf – 8 – mf#SE-1604 – ne IDC [959]

Indonesien / Indonesiska Legationen – Stockholm, 1956-1957 – 9mf – 9 – (missing: 1956, v1(1-6, 10); 1957, v2(2-4)) – mf#SE-552 – ne IDC [959]

Indonesie-nederland – Batavia, 1946-1947 – 5mf – 9 – mf#SE-886 – ne IDC [959]

Indonesisch bulletin – 's-Gravenhage, 1950-1957 – 100mf – 9 – (aka: indonesische documentatie; indonesische voorlichting) – mf#SE-553 – ne IDC [959]

Indonesisch Persbureau *see* Verslag van het eerste congres van het "indonesisch verbond van studerenden"

Indonesiska Legationen *see* Indonesien

The indo-sumerian seals deciphered : discovering sumerians of indus valley as phoenicians, barats, goths and famous vedic aryans 3100-2300 bc / Waddell, Laurence Austine – London: Luzac & Co, 1925 – (= ser Samp: indian books) – us CRL [490]

Indra – Balige, 1965 – 3mf – 9 – mf#SE-601 – ne IDC [959]

Indra *see* The status of women in ancient india

Induced fit aminoacyl-trna selection on the ribosome / Pape, Tillmann – (mf ed 1999) – 2mf – 9 – €40.00 – 3-8267-2642-1 – mf#DHS 2642 – gw Frankfurter [612]

Inducements to promote the fine arts in great britain / Cranch, John – Frome 1811 – (= ser 19th c art & architecture) – 1mf – 9 – mf#4.2.444 – uk Chadwyck [700]

Inductance calculations, working formulas and tables / Grover, Frederick Warren – New York, NY. 1946 – 1r – us UF Libraries [510]

Induction coils / Marshall, Percival – New York, NY. 1906 – 1r – us UF Libraries [574]

Inductive preaching: an analysis of contemporary theory and practice / Culpepper, James Edward – 1981 – 1 – 5.00 – us Southern Baptist [242]

Inductive reasoning / Bagchi, Sitansusekhar – Calcutta, India. 1953 – 1r – us UF Libraries [611]

Inductive studies in the twelve minor prophets / White, Wilbert W – Chicago: Young Men's Era, 1893 – 1mf – 9 – 0-8370-5823-6 – mf#1985-3823 – us ATLA [221]

Inductive studies in theology : including the doctrines of sin and the atonement / Burwash, Nathanael – Toronto: W Briggs; Montreal: C W Coates, 1896 – 1mf – 9 – mf#07205 – cn Canadiana [240]

Induk koperasi kopra indonesia / Madjalah kopra – Djakarta, 1959-1963 – 5mf – 9 – (missing: 1959, v1; 1960, v2; 1961, v3(1-32, 33-36); 1962, v4(1-3, 5-9)) – mf#SE-829 – ne IDC [959]

Induk koperasi perikanan indonesia / Laporan tahunan – Tjipajung, 1964-1970/71 – 17mf – 9 – (missing: 1965-1967) – mf#SE-6808 – ne IDC [959]

Die induktion mechanisch bedingter degeneration des gelenkknorpels : ein beitrag zur entwicklung eines in-vitro-modells des arthrotisch veraenderten knorpels fuer pharmakologische untersuchungen / Steinmeyer, Juergen – (mf ed 1999) – 2mf – 9 – €40.00 – 3-8267-2610-3 – mf#DHS 2610 – gw Frankfurter [615]

Indulgencias perpetuas concedidas a los congregantes de la insigne real congregacion del alumbrado y vela continua del santisimo sacramento : fundada canonicamente en la parroquia de san sebastian de esta corte... – en Mexico:...Don Felipe de Zuniga y Ontiveros, ano de 1793 – (= ser Books on religion...1543/44-c1800: real congregacion del alumbrado y vela continua del santisimo sacramento) – 1mf – 9 – mf#crl-249 – ne IDC [241]

Das indulgenz-edict des roemischen bischofs kallist / Rolffs, Ernst – Leipzig: JC Hinrichs, 1893 – (= ser Tugal) – 1mf – 9 – 0-7905-1844-9 – (incl bibl ref) – mf#1987-1844 – us ATLA [240]

Das indulgenz-edict des roemischen bischofs kallist / Rolffs, Ernst – Leipzig, 1893 – (= ser Tugal 1-11/3) – 3mf – 9 – €7.00 – ne Slangenburg [240]

The indus valley in the vedic period / Chanda, Ramaprasad – Calcutta: Govt of India, Central Publication Branch, 1926 – (= ser Samp: indian books) – us CRL [930]

Indus valley painted pottery : a comparative study of the designs on the painted wares of the harappa culture / Starr, Richard Francis Strong – Princeton: Princeton University Press, 1941 – (= ser Samp: indian books) – us CRL [730]

Industri / Madjelis Industri Indonesia – Djakarta, 1956-1962 – 72mf – 9 – (missing: 1956, v1(1-8); 1957, v2(4, 5, 8); 1958, v3(2, 4, 5); 1961, v6(5, 11, 12); 1962, v7(1-4)) – mf#SE-293 – ne IDC [959]

Industri Indonesia *see* Industrial directory of indonesia

Industria / USSR. Moscow – n1-227. 1937-40 – 1 – us L of C Photodup [947]

Industria brasileira e a amazonia – Rio de Janeiro, Brazil. 1969 – 1r – us UF Libraries [338]

Industria britanica – London, UK. Aug 1931-Jun 1965; Dec 1968- – 1 – uk British Libr Newspaper [072]

Industria quimica – Morris Plains. 1971-1973 (1) – ISSN: 0019-7726 – mf#7560 – us UMI ProQuest [540]

Industria quimica brasileira – Sao Paulo, Brazil. 1970 – 1r – us UF Libraries [338]

Industria y proteccion en colombia, 1810-1930 / Ospina Vasquez, Luis – Medellin, Colombia. 1955 – 1r – us UF Libraries [972]

Industrial accident reports / New Jersey. Division of Workmen's Compensation – 1932-55. 12 fiches. (Harvard Law School Library Collection.) – 9 – us Harvard Law [610]

Industrial accountant – Karachi. 1977-1988 (1) 1977-1988 (5) 1977-1988 (9) – ISSN: 0019-7793 – mf#10469 – us UMI ProQuest [650]

Industrial advocate / Central Labor Union of Pittston [PA] – Pittston PA. 1903 dec 26 – 1r – 1 – us WHS [331]

The industrial advocate – Halifax, [NS]: Maritime Newspaper Col, [1884-1916] – 9 – mf#P06049 – cn Canadiana [338]

Industrial age – Chicago IL. 1873 aug 20-1877 feb 24 – 1r – 1 – mf#1224218 – us WHS [338]

Industrial and agricultural advantages of miami, florida – Miami, FL. 1928 – 1r – us UF Libraries [978]

Industrial and commercial photographer – Croydon. 1962-1980 (1) 1972-1980 (5) 1977-1980 (9) – (cont by: professional photographer) – ISSN: 0019-784X – mf#1345 – us UMI ProQuest [770]

Industrial and commercial training – Guilsborough. 1975-1995 (1) 1975-1995 (5) 1975-1995 (9) – ISSN: 0019-7858 – mf#9248 – us UMI ProQuest [650]

Industrial and Commercial Workers Union of South Africa *see* Black man

Industrial and engineering chemistry – v1-62. 1909-70 – 1,5,6 – us ACS [660]

Industrial and engineering chemistry fundamentals – v1-25. 1962-86 – 1,5,6,9 – us ACS [660]

Industrial and engineering chemistry process design and development – v1-25. 1962-86 – 1,5,6,9 – us ACS [660]

Industrial and engineering chemistry product research and development – v1-25. 1962-86 – 1,5,6,9 – us ACS [660]

Industrial and engineering chemistry research – v26- 1987- – 1,5,6,9 – us ACS [660]

Industrial and labor relations forum – Cornell University. v1-15. 1969-81 (all publ) – 9 – $185.00 set – mf#105061 – us Hein [344]

Industrial and labor relations forum – Ithaca. 1964-1981 [1]; 1971-1981 [5]; 1975-1981 [9] – ISSN: 0019-7912 – mf#1814 – us UMI ProQuest [331]

Industrial and labor relations review – Ithaca. 1947+ (1) 1970+ (5) 1975+ (9) – ISSN: 0019-7939 – mf#966 – us UMI ProQuest [331]

Industrial appeal – Ottumwa IA. 1886 dec 14,28; 1887 jan 18 – 1r – 1 – us WHS [331]

Industrial art / Willms, Auguste – Birmingham 1890 – (= ser 19th c art & architecture) – 1mf – 9 – mf#4.2.712 – uk Chadwyck [740]

The industrial arts : historical sketches with numerous illustrations / Maskell, William – [London]: publ...by Chapman & Hall [1876] – (= ser 19th c art & architecture) – 4mf – 9 – mf#4.1.99 – uk Chadwyck [740]

The industrial arts of india / Birdwood, George Christopher Molesworth – London [1880] – (= ser 19th c art & architecture) – 6mf – 9 – mf#4.2.485 – uk Chadwyck [740]

The industrial arts of india / Birdwood, George Christopher Molesworth – London: Chapman and Hall, 1880 – (= ser Samp: indian books) – us CRL [740]

The industrial arts of the nineteenth century... : great exhibition of works of industry, 1851 / Wyatt, Matthew Digby – London 1851 – (= ser 19th c art & architecture) – 16mf – 9 – mf#4.1.227 – uk Chadwyck [740]

Industrial australian – Melbourne, Australia 15 oct 1932-15 feb 1933 – 1 – (cont: australian statesman & mining standard [26 nov 1914-28 dec 1916]) – uk British Libr Newspaper [622]

Industrial banner – Toronto ON. 1912 oct 18 – 1r – 1 – us WHS [338]

Industrial bulletin – New York. 1975-1979 (1) 1976-1979 (5) 1976-1979 (9) – (cont by: industrial product bulletin) – ISSN: 0019-8021 – mf#10261 – us UMI ProQuest [600]

Industrial canada – may 1966-apr 1971 – 5r – 1 – cn Commonwealth Imaging [338]

Industrial canada : a survey of canadian industries / Canadian Manufacturers' Association – [Montreal?: s.n, 1903?] – 1mf – 9 – 0-665-72266-4 – mf#72266 – cn Canadiana [338]

Industrial canada – Toronto, Canada. 1922-jul 1973 – 88 1/2r – 1 – uk British Libr Newspaper [338]

Industrial canada – Toronto: W S Johnston, [1896?-189- or 19–] – 9 – mf#P04483 – cn Canadiana [338]

Industrial Commission of Wisconsin *see*
- County by county listing of employers
- Listing by account number of employers subject to wisconsin's unemployment compensation law
- Listing by type of business and number of employees of employers subject to wisconsin's unemployment compensation law as of the first quarter of...
- Wisconsin economic indicators

Industrial defense bulletin / Industrial Workers of the World – v2 n1-v5 [i.e. 4] n5 [i.e. 6] [1975 jan-1979 feb] – 1r – 1 – (cont: industrial union newsletter) – mf#585990 – us WHS [331]

Industrial democracy / League for Industrial Democracy – v1-6 n2,6. 1932-38 – (= ser Radical periodicals in the united states, 1881-1960. series 1) – 18mf – 9 – $175.00 – us UPA [335]

Industrial democracy *see* Przemyslowa demokracja

Industrial design – Cincinnati. 1954-1978 (1) 1966-1978 (5) 1973-1978 (9) – ISSN: 0019-8110 – mf#1090 – us UMI ProQuest [740]

Industrial development – Atlanta. 1884-1984 (1) 1970-1984 (5) 1972-1984 (9) – ISSN: 0097-3033 – mf#66 – us UMI ProQuest [338]

Industrial development and site selection handbook – Atlanta. 1985-1988 (1) 1985-1988 (5) 1985-1988 (9) – (cont by: site selection and industrial development) – mf#66,01 – us UMI ProQuest [338]

Industrial development of mysore / Balakrishna, Ramachandra – Bangalore City: Bangalore Press, 1940 – (= ser Samp: indian books) – us CRL [338]

Industrial development of puerto rico and the virg... / Caribbean Commission – Port-of-Spain, Trinidad and Tobago. 1948 – 1r – us UF Libraries [338]

Industrial diamond review – London. 1958+ (1) 1971+ (5) 1975+ (9) – ISSN: 0019-8145 – mf#1237 – us UMI ProQuest [620]

Industrial directory of indonesia : madjelis industri indonesia / Industri Indonesia & Indonesian Chamber of Industries – Djakarta, 1957/1958 – 10mf – 9 – mf#SE-164-6 – ne IDC [338]

Industrial distribution – New York. 1916+ (1) 1970+ (5) 1976+ (9) – ISSN: 0019-8153 – mf#370 – us UMI ProQuest [650]

Industrial education – Southfield. 1914-1990 (1) 1968-1990 (5) 1970-1990 (9) – ISSN: 0091-8601 – mf#47 – us UMI ProQuest [370]

The industrial efficiency of india / Das, Rajani Kanta – London: PS King & Sons, 1930 – (= ser Samp: indian books) – us CRL [338]

Industrial engineering – Norcross. 1969-1995 (1) 1970-1995 (5) 1976-1995 (9) – (cont by: iie solutions) – ISSN: 0019-8234 – mf#5936 – us UMI ProQuest [620]

Industrial enterprise in india / Das, Nabagopal – London; New York: Oxford University Press, 1938 – (= ser Samp: indian books) – us CRL [338]

Industrial entrepreneurship in nigeria / Harris, John Reese – Evanston, 1967 – us CRL [338]

Industrial evolution of india / Chatterton, Alfred – Madras: Hindu Office, [1912] – (= ser Samp: indian books) – us CRL [338]

Industrial finance / ed by Shah, K T – Bombay: Vora & Co, 1948 – (= ser Samp: indian books) – us CRL [332]

Industrial finishing – Wheaton. 1924-1993 (1) 1976-1993 (5) 1976-1993 (9) – (cont by: industrial paint and powder) – ISSN: 0019-8323 – mf#315 – us UMI ProQuest [660]

Industrial finishing – London. 1952-1954 (1) – mf#599 – us UMI ProQuest [660]

Industrial gerontology – Washington. 1969-1977 (1) 1972-1977 (5) 1975-1977 (9) – ISSN: 0019-8358 – mf#6394 – us UMI ProQuest [618]

Industrial Home for Colored Girls [Peaks VA] *see* Annual report of the industrial home for colored girls

Industrial Insurance Agents Union et al *see* Debit

Industrial ireland : a practical and non-political view of "ireland for the irish" / Dennis, Robert – London, 1887 – (= ser 19th c ireland) – 3mf – 9 – mf#1.1.8100 – uk Chadwyck [330]

The industrial journal – Bangor, ME: [The Journal Pub Co]. n262-1322. 1885-sep 1918 – 1 – us CRL [338]

Industrial labor journal – Salt Lake City UT. 1902 nov 29 – 1r – 1 – mf#869135 – us WHS [331]

Industrial Labor Union Council [Portland OR] *see* Labor newsdealer

Industrial laboratory – New York. 1958-1976 (1) 1970-1976 (5) 1974-1976 (9) – ISSN: 0019-8447 – mf#1534 – us UMI ProQuest [600]

Industrial leader – Boston MA. 1890 nov 1 – 1r – 1 – us WHS [331]

Industrial leader – Dubuque IA. 1886 dec 18-1887 jan 1, jun 18-jul 2, 16-aug 6, 13-sep 3, oct 15, nov 19; 1888 feb 18 – 1r – 1 – us WHS [331]

Industrial lubrication and tribology – Droitwich. 2001+ (1,5,9) – ISSN: 0036-8792 – mf#19302,01 – us UMI ProQuest [621]

Industrial management – Des Plaines. 1977+ (1,5,9) – ISSN: 0019-8471 – mf#14433 – us UMI ProQuest [650]

Industrial management – Melbourne. 1970-1970 (1) – mf#3434 – us UMI ProQuest [650]

Industrial management – Wembley. 1976-1980 (1,5,9) – ISSN: 0007-6929 – mf#11096 – us UMI ProQuest [650]

Industrial management + data systems – Wembley. 1980-1995 (1,5,9) – ISSN: 0263-5577 – mf#11096,01 – us UMI ProQuest [650]

Industrial marketing – Chicago. 1935-1983 (1) 1965-1983 (5) 1975-1983 (9) – (cont by: business marketing) – ISSN: 0019-8498 – mf#348 – us UMI ProQuest [650]

Industrial marketing management – New York. 1972+ (1) 1972+ (5) 1983+ (9) – ISSN: 0019-8501 – mf#42098 – us UMI ProQuest [650]

Industrial mathematics – Roseville. 1950+ (1) 1970+ (5) 1977+ (9) – ISSN: 0019-8528 – mf#3045 – us UMI ProQuest [510]

Industrial mobilization in britain, 1915-1918 – 88mf – 1 – us Primary [941]

Industrial mutual news – Flint, MI. 1922-1939 (1) – mf#63743 – us UMI ProQuest [071]

Industrial news – laeger, WV. 1945-1976 (1) – mf#67329 – us UMI ProQuest [071]

Industrial opportunities in swaziland – Mbabane?, Swaziland . 196- – 1r – us UF Libraries [338]

Industrial organizer / Motor Transport and Allied Workers Industrial Union – Minneapolis MN. v1 n1-22a [1941 jul 17-1942 feb 3], 1942 may 16 – 1r – 1 – (cont: northwest organizer) – mf#594282 – us WHS [331]

Industrial paint and powder – Troy. 1993+ (1) 1993+ (5) 1993+ (9) – (cont: industrial finishing) – ISSN: 1073-4651 – mf#315,01 – us UMI ProQuest [660]

Industrial participation – London. 1975-1980 (1) 1975-1980 (5) 1975-1980 (9) – ISSN: 0950-1932 – mf#9936 – us UMI ProQuest [331]

Industrial people – Reading PA. 1876 jul 22, sep 2-9 – 1r – 1 – us WHS [071]

Industrial pioneer / Industrial Workers of the World – ser1: v1 1921-22 [all publ]. ser2: v1-4 1923-26 [all publ] – (= ser Radical periodicals in the united states, 1881-1960. series 1) – 31mf – 9 – $315.00 – us UPA [331]

Industrial press – Galena IL. 1874 feb 6-1875 jan 28, 1875 feb 4-1876 jun 29, 1876 jul 6-1877 sep 27, 1877 oct 4-1879 jan 23, 1879 jan 30-1880 sep 9, 1880 sep 16-1882 jan 26, 1882 feb 2-1882 sep 28 – 7r – 1 – (cont: galena commercial advertiser; cont by: galena industrial press) – mf#887620 – us WHS [071]

Industrial product bulletin – Pittsfield. 1979-1980 (1) 1979-1980 (5) 1979-1980 (9) – (cont: industrial bulletin) – ISSN: 0199-2074 – mf#10261,01 – us UMI ProQuest [600]

Industrial quality control – Milwaukee. 1944-1967 – 1 – ISSN: 0884-822X – mf#735 – us UMI ProQuest [310]

Industrial reform *see* Kung-yeh kai-tsao (ccs)

Industrial refrigeration – Chicago. 1949-1961 (1) – ISSN: 0096-8099 – mf#98 – us UMI ProQuest [338]

Industrial relations – Berkeley. 1961+ (1) 1975+ (5) 1977+ (9) – ISSN: 0019-8676 – mf#10384 – us UMI ProQuest [331]

Industrial relations digest *see* Industrial relations law digest

Industrial relations journal – Oxford. 1991-1994 (1) 1992-1992 (5) 1992-1992 (9) – ISSN: 0019-8692 – mf#9979 – us UMI ProQuest [331]

Industrial relations law digest – Ann Arbor. 1958-1978 (1) 1971-1978 (5) 1976-1978 (9) – ISSN: 0019-8706 – mf#6378 – us UMI ProQuest [331]

Industrial relations law digest – University of Michigan. v1-20 1958-78 (all publ) – 9 – $204.00 set – (title varies: v1-5 as industrial relations digest) – mf#103361 – us Hein [343]

Industrial relations law journal – Berkeley. 1979-1991 (1,5,9) – (cont by: berkeley journal of employment and labor law) – ISSN: 0145-188X – mf#11959 – us UMI ProQuest [344]

Industrial relations law journal *see* Berkeley journal of employment and labor law

Industrial relations review and report – London. 1976-1993 (1) 1976-1992 (5) 1976-1992 (9) – ISSN: 0309-7269 – mf#10500 – us UMI ProQuest [331]

Industrial research – 1959-1978 [1]; 1970-1978 [5]; 1975-1978 [9] – ISSN: 0019-8722 – mf#1179 – us UMI ProQuest [620]

Industrial research and development – Barrington. 1978-1983 (1) 1978-1983 (5) 1978-1983 (9) – (cont by: research and development) – ISSN: 0160-4074 – mf#1179,01 – us UMI ProQuest [620]

The industrial resources of ireland / Kane, Robert John – Dublin, 1844 – (= ser 19th c ireland) – 5mf – 9 – mf#1.1.1059 – uk Chadwyck [941]

Industrial revolution: a documentary history, series 1 : the boulton and watt archive & the matthew boulton papers from birmingham central library – [mf ed Marlborough 1993] – 13pt – 1 – (pt1: lunar society correspondence [17r] £1600; pt2: muirhead 1 – notebooks & papers of james watt & family [12r] £1125; pt3: engineering drawings – sun & planet type, c1775-1802 [8r] £750; pt4: matthew boulton correspondence (subject material: albion mill-steam engines) [23r] £2150; pt5: engineering drawings – crank, canal, dock & harbour, mint, blowing, pumping & other engines c1775-1800 [5r] £475; pt6: muirhead 2-notebooks & papers of james watt & family [33r] £3100; pt7: matthew boulton correspondence (subject material & individual correspondence incl garbett, rennie, southern & wilkinson) [20r] £1850; pt8: muirhead 3-notebooks & papers of james watt & family [28r] £2650; pt9: journal, notebooks & diaries of matthew boulton [12r] £1125 [mf ed 2000]; pt10: matthew boulton correspondence (incoming letters) [20r] £1850; pt11: engineering drawings c1801-65 [24r] £2250; pt12: boulton & watt correspondence & papers ((ms 3147/3/1-79 & 245-285) [23r] £2150 [mf ed 2003/4]. pt13:...(ms 3147/3/286-404) [27r] £2550 [mf ed 2003/4]; pt14:...(ms 3147/3/405-484) [27r] £2550; pt15:...(ms 3147/3/485-560) [14r] £1300; with d/g) – uk Matthew [941]

Industrial revolution: a documentary history, series 2 : papers of john rennie (1761), thomas telford (1757-1834) & related figures from the national library of scotland – [mf ed Marlborough 2003] – 2pt – 1 – (pt1: papers of james watt, joseph black, thomas telford & john rennie [20r] £1850; pt2: papers of john rennie, thomas telford & robert stevenson [20r] £1850; with d/g) – uk Matthew [941]

Industrial revolution: a documentary history, series 3 : the papers of james watt and his family formerly held at doldowlod house, now at birmingham central library – 3pt – 1 – (pt1: correspondence, papers & business records, 1687-1819 [20r] £1850; pt2: correspondence, papers & business records, 1736-1848 [20r] £1850; pt3: correspondence, papers & business records, 1736-1848 [25r] £2350; with d/g) – uk Matthew [941]

Industrial revolution: a documentary history, series 4 : sources from record offices in the united kingdom – 1 – (pt1: papers of boulton & watt, wedgwood and harvey & co of hayle fr cornwall record office [26r] £2450; pt2: papers of harvey & co of hayle fr cornwall record office [22r] £2050; pt3: papers of james watt (1736-1819) & james watt, jnr (1769-1848) fr james patrick muirhead coll, at glasgow university library [12r] £1125; pt4: darby family, coalbrookdale estate & the iron bridge – sources fr shropshire archives [19r] £1675; with d/g) – uk Matthew [941]

Industrial robot – Bedford. 1992-1995 (1,5,9) – ISSN: 0143-991X – mf#18831 – us UMI ProQuest [629]

Industrial safety and hygiene news – Philadelphia. 1982-1982 (1) 1982-1982 (5) 1982-1982 (9) – (cont by: chilton's industrial safety and hygiene news) – ISSN: 0278-8217 – mf#12611,02 – us UMI ProQuest [360]

Industrial School for Colored Girls of Delaware *see* Report of the industrial school for colored girls of delaware (Marshallton DE)

Industrial School for Colored Girls of Delaware (Marshallton DE) *see*
- Biennial report of the industrial school for colored girls of delaware
- Report of the board of trustees and the superintendent of the industrial school for colored girls of delaware

Industrial society – London, 1918-98+ – 41r – 1 – £1,250.00 – uk World [330]

Industrial solidarity / Industrial Workers of the World – 1921 sep 10-1922 dec 30, 1923 jan-dec, 1926 jan-dec, 1927 jan 5-dec 14, 1927 jan 5-1928 feb 1 – 10r – 1 – (cont: solidarity [chicago, il]; cont by: industrial worker [chicago, il]) – mf#981633 – us WHS [331]

Industrial solidarity / Industrial Workers of the World – Official Organ. Chicago etc. Dec 18 1909-1931. Incomplete – 1 – us NY Public [331]

Industrial supervisor – Chicago. 1975-1982 (1) 1976-1982 (5) 1976-1982 (9) – (cont by: today's supervisor) – ISSN: 0019-879X – mf#10449 – us UMI ProQuest [650]

Industrial supply and distribution in puerto rico / Economic Associates – Washington, DC. 1963 – 1r – us UF Libraries [338]

Industrial survey of ocala and marion county flori – Ocala, FL. 1928 – 1r – us UF Libraries [978]

Industrial union – Williamsport, PA., 1891 – 13 – $25.00r – us IMR [071]

Industrial union bulletin / Industrial Workers of the World – v1-2 [1907 mar 2-1909 mar 6], bulletin 4,5 [1906 dec 1, 1907 jan 10] – 1r – 1 – mf#1059237 – us WHS [331]

Industrial union bulletin / Industrial Workers of the World – v1-2 n2,31. 1907-09 [all publ] – (= ser Radical periodicals in the united states, 1881-1960. series 1) – 1r – 1 – $200.00 – us UPA [331]

The industrial union bulletin – Chicago. mar. 2, 1907-mar. 6, 1909 – 1 – us NY Public [331]

Industrial union news / Workers International Industrial Union – n1-251 [1912 jan-1924 may] – 1r – 1 – mf#1059240 – us WHS [331]

Industrial union news / Workers' International Industrial Union – n1-251. 1912-24 [all publ] – (= ser Radical periodicals in the united states, 1881-1960. series 2) – 1r – 1 – $200.00 – us UPA [331]

Industrial union newsletter / Industrial Workers of the World – v1 n1-7 [1974 jan-nov] – 1r – 1 – (cont by: industrial defense bulletin) – mf#634341 – us WHS [331]

Industrial Union of Marine and Shipbuilding Workers *see* The shipbuilder

Industrial Union of Marine and Shipbuilding Workers of America *see*
- Industrial union reporter
- Shipbuilder
- Shipyard worker

Industrial Union Party *see* Industrial unionist

Industrial Union Party [US] *see* Socialist republic

Industrial union reporter : official organ, industrial unionof marine and shipbuilding workers of america, c i o for ship, metal and railroad worker / Industrial Union of Marine and Shipbuilding Workers of America – 1948 apr 5-1950, 1948 apr 5-1951 jun, 1951 jan-jun – 3r – 1 – (cont: shipyard worker [camden, nj]; cont by: shipbuilder) – mf#1159905 – us WHS [331]

Industrial unionist / Industrial Union Party – ser1: v1-8 n2,3 1932-40 [all publ]. ser2: n1-4 1941. ser3: n1-6 1949-50 [all publ] – (= ser Radical periodicals in the united states, 1881-1960. series 1) – 1r – 1 – $200.00 – us UPA [331]

Industrial unionist / Industrial Workers of the World. Emergency Program Branches – v1-2. 1925-26 [all publ] – (= ser Radical periodicals in the united states, 1881-1960. series 1) – 1r – 1 – $200.00 – us UPA [331]

Industrial unionist – Manchester. England. -m. Mar 1908-Jun 1909. (10 ft) – 1 – uk British Libr Newspaper [072]

Industrial unionist : official bulletin of the industrial union party – v1-v8 n3 [1932 may-1940 may/jun, 1941 mar-nov] – 1r – 1 – mf#1059242 – us WHS [331]

Industrial unionist : official organ of the industrial union league – v1 n1-6 [1949 apr/may-1950 may/jun] – 1r – 1 – mf#1059244 – us WHS [331]

Industrial unionist – Portland OR: Emergency Program Branches of the IWW, 1925-26 [wkly] – 1 – us Oregon Lib [331]

Industrial unionist – Portland, OR: [Emergency Program Branches of the IWW]. v1 n1-v2 n62. apr 11 1925-jun 16 1926 – 1 – us Oregon Hist [071]

Industrial wastes – Chicago. 1976-1983 (1,5,9) – ISSN: 0046-9262 – mf#11263 – us UMI ProQuest [333]

Industrial water engineering – Littleton. 1964-1985 (1) 1972-1985 (5) 1973-1985 (9) – ISSN: 0019-8862 – mf#6986 – us UMI ProQuest [627]

Industrial welfare in india / Lokanathan, Palamadai Samu – Madras: University of Madras, 1929 – (= ser Samp: indian books) – (int by gilbert slater) – us CRL [360]

Industrial west – Atlantic IA. 1887 jan 6 – 1r – 1 – mf#851117 – us WHS [338]

Industrial west – 1887 dec, 1888 jan, apr-jun, aug – 1r – 1 – (cont: miner & manufacturer and gogebic news; reporter [milwaukee, wis. : 1887]) – mf#3557553 – us WHS [338]

Industrial worker / Industrial Workers of the World – 1919 jun 18, 1920 sep 25-1922 jun 24, 1922 jul 1-1924 jun 7, 1924 jun 11-1926 dec 18, 1926 dec 25-1930 may 17, 1930 may 24-1934 jan 30, 1934 feb 6-1937 aug 28, 1937 sep 4-1939 dec 30, 1940-43, 1944-45, 1965-1974 jul, 1975-80 – 11r – 1 – mf#1444664 – us WHS [331]

Industrial worker / Industrial Workers of the World – Spokane WA. 1909 mar 18-1910 jan 29, 1910 feb 5-1913 sep 4 – 2r – 1 – (cont by: industrial worker [seattle, wa]) – mf#1054330 – us WHS [331]

Industrial worker / Industrial Workers of the World – Chicago IL, Washington. v1 n1-v3 n14 [1916 apr 1-1918 may 25] – 1r – 1 – (cont: industrial worker [spokane, wa]; industrial solidarity) – mf#770913 – us WHS [331]

Industrial worker / Industrial Workers of the World – v1-5 n2,21. 1903-13 [all publ] – (= ser Radical periodicals in the united states, 1881-1960. series 1) – 1r – 1 – $200.00 – us UPA [331]

Industrial worker / International Union, Allied Industrial Workers of America – 1956 sep – 1r – 1 – (cont: afl auto worker [1956]; cont by: allied industrial worker) – mf#1214481 – us WHS [331]

Industrial worker – London, England. -m. Nov 1913-Aug 1914; July-Nov 1916; Oct 1917. 7 ft – 1 – uk British Libr Newspaper [072]

Industrial worker – Ypsilanti. 1968+ (1) – ISSN: 0019-8870 – mf#3230 – us UMI ProQuest [331]

The industrial worker in india / Shiva Rao, B – London: George Allen and Unwin Ltd, 1939 – (= ser Samp: indian books) – us CRL [331]

Industrial Workers of the World *see*
- Defense news bulletin
- Direct action
- Fellow worker
- General organization bulletin for ...
- Golos truzhenika
- Industrial defense bulletin
- Industrial pioneer
- Industrial solidarity
- Industrial union bulletin
- Industrial union newsletter
- Industrial worker
- Industrialen rabotnik
- Jedna velka unie
- Nuovo proletario
- Nya varlden
- One big union bulletin
- One big union monthly
- Il proletario
- Proletario
- Rebelde
- Solidaridad
- Solidarity
- Solidarity bulletin
- Solidarnosc

Industrial workers of the world *see* Department of justice investigative files

Industrial Workers of the World. Emergency Program Branches *see* Industrial unionist

Industrial world and national economist – Ottawa: Industrial World Pub Co, [1880-1882] – 9 – (cont by: canadian manufacturer and industrial world) – mf#P04362 – cn Canadiana [330]

Industrialen rabotnik / Industrial Workers of the World – v1 n1-13 [1924 jan 1-jul 1] – 1r – 1 – mf#939677 – us WHS [331]

Industrialer arbayter – Chicago, IL. 1919-20 – 1 – us AJPC [071]

Industrialisation in africa / International African Seminar – London, England. 1954 – 1r – us UF Libraries [960]

Industrialist / Allegheny County Industrial Union – East Pittsburgh PA. 1914 mar 7-28 – 1r – 1 – mf#868120 – us WHS [331]

The industrialist *see* Journals of the labour movement in trade and industry

Industrialisti – v3 n158, v4 n15, v15 n72-74, v60 n40-v62 n40 [1919 jul 8, 1920 jan 19, 1931 jan 19-mar 26, 1974 oct 8-1975 oct 21] – 1r – 1 – mf#464574 – us WHS [338]

Industrializa cao e economia natural / Paim, Gilberto – Rio de Janeiro, Brazil. 1957 – 1r – us UF Libraries [333]

Industrializacao, burguesia nacional e desenvolvim / Martins, Luciano – Rio de Janeiro, Brazil. 1968 – 1r – us UF Libraries [972]

La industrializacion de los regadios de la provincia de caceres : conferencia pronunciada...ayuntamiento de caceres... / Sanchez Torres, Clemente – Plasencia: Imprenta La Victoria, 1951 – 1 – sp Bibl Santa Ana [338]

Industrializacion y dependencia en america latina / Melazzi, Gustavo – Montevideo, Uruguay. 1969 – 1r – us UF Libraries [972]

Industrialization and balanced growth / Loeb, Gustaaf Frits – Groningen, Netherlands. 1957 – 1r – us UF Libraries [338]

Industrialization of space [aasms28] – 1978 – (= ser Aasms 1968) – 20papers on 9mf – 9 – $15.00 – 0-87703-121-5 – (suppl to v36, advances) – us Univelt [338]

Industrias carnicas / Aranguez Sanz, Bibiano – Madrid: Sociedad Veterinaria de Zootecnica, 1947. Sep. 1 Congreso Veterinario de Zootecnia. Madrid, 26 Octubre a 2 Noviembre. 1947 – 1 – sp Bibl Santa Ana [590]

Industrias paleoliticas en el tramo extremeno del tajo / Santonja Gomez, M & Queral, Maria A – Badajoz: Imp. Dipt. Provincial, 1975 – sp Bibl Santa Ana [946]

Industrias rurales / Fructuoso, Gonzalo – Caceres: Tip. El Noticiero, 1936 – 1 – sp Bibl Santa Ana [338]

Industrias santiaguinas / Martinez, Mariano – Santiago de Chile: Imprenta y encuadernacion Barcelona, 1896 – 1 – us CRL [380]

l'industrie *see* Memorial judiciaire de la loire

Die industrie am niederrhein band 2 / Thun, Alphons – Leipzig. 1879 – 1 – gw Mikropress [380]

Die industrie am niederrhein und ihre arbeiter – Bd. II, Heft 2 u. 3. Leipzig 1879 – 1 – gw Mikropress [331]

L'industrie avicole dans la province de quebec : preparation de la volaille et des oeufs pour le marche / Glebe, Jean de la – [Quebec (Province)?]: Poultry Producers Association of Eastern Canada? [1910?] [mf ed 1994] – 1mf – 9 – 0-665-73574-X – mf#73574 – cn Canadiana [636]

L'industrie du bacon dans la province de quebec : conference...des membres des societes d'agriculture, 17 mars 1903 – [Quebec (Province)?: s.n, 1903?] – 1mf – 9 – 0-665-72164-1 – mf#72164 – cn Canadiana [630]

L'industrie du sucre de betterave au canada / Musy, Alfred – Berthier, Quebec?: s.n, 1897 – 1mf – 9 – mf#11222 – cn Canadiana [635]

L'industrie electrique : revue de la science electrique et de ses applications industrielles – Paris, France 10 jan 1892-25 dec 1901; 10 jan 1906-25 dec 1909 – 1 – uk British Libr Newspaper [333]

Industrie electrique – Paris, France. 10 jan 1892-25 dec 1893; 1894-25 dec 1900; 1901; jan 1903-jun 1904; jan-oct 1905; 10 jan 1906-25 dec 1907; 10 jan 1908-25 dec 1909 – 7 1/2r – 1 – uk British Libr Newspaper [072]

L'industrie francaise : organe de la defense du travail national – Paris. mai 1877-82, 1884-89. – 1 – (mq: no.23) – fr ACRPP [073]

Le industrie, l'agricoltura, il commercio – Turin, Italy. 1874-75.-f. 1 reel – 1 – uk British Libr Newspaper [338]

L'industrie quebecoise du textile au canada = The quebec textile industry in canada / Pestieau, Caroline – Montreal: Institut de recherche C D Howe, 1978 [mf ed 1998] – 2mf – 9 – mf#SEM105P2937 – cn Bibl Nat [338]

Industrie revue hebdomadaire – Brussels Belgium, 3 oct 1897-25 dec 1898; 1899-1905; 7 jan 1906-23 jun 1907 – 9r – 1 – uk British Libr Newspaper [074]

Industrie- und Handelskammer Duesseldorf *see* Wirtschaft und verkehr

Industrie- und handels-zeitung – Berlin DE, 1922 2 jan-30 sep, 1926 9 apr-1928 jun, 1928 aug-1930 – 13r – 1 – gw Mikrofilm [380]

Die industrieansiedlung in ludwigshafen am rhein bis 1892 (chemie und metallverarbeitung) / Kube, Helga – Heidelberg, 1962 – 3mf – 9 – 3-89349-366-2 – gw Frankfurter [338]

Industriebau – Leipzig DE, 1910-12, 1919-30 – 5r – 1 – uk British Libr Newspaper [338]

Industriekurier – Duesseldorf DE, 1954 5 jan-1968 31 oct – 37r – 1 – (absorbed by: handelsblatt 1970. filmed by misc inst: 1948 30 oct-1953 [6r]; 1948 30 oct-1953, 1968 1 oct-1970 29 aug. with suppl: technik und forschung 1950-53 (2r)) – gw Misc Inst [338]

L' industriel alsacien – Mülhausen / Elsass (Mulhouse F), 1841-42, 1872-77 – 9r – 1 – (with numerous gaps) – gw Misc Inst [338]

L'industriel alsacien – Muelhausen / Elsass (Mulhouse F), 1841-42 [many gaps], 1872-77 [gaps] – 9r – 1 – gw Misc Inst [074]

L'industriel de louviers – Echo du Neubourg. Louviers. 1930-39 – 1 – fr ACRPP [073]

Industriele behuising aan die diamantvelde van griekwaland-wes met spesifieke verwysing na die oop kampongs as werkershuisvesting, 1868-1880 / Diemel, Raymond Anthony van – U of the Western Cape 1988 [mf ed S.l: s.n. between 1987 & 1991] – 2mf – 9 – (summary in english; incl bibl) – sa Misc Inst [622]

Industries : pinellas county / Phillips, Roland – S.l., S.l? . 1936 – 1r – us UF Libraries [978]

Industries et techniques – Paris. n1-246. mars 1959-73 – 1 – (le magazine de l'innovation technique puis de la productivite francaise) – fr ACRPP [073]

Industries of canada : historical and commercial sketches : kingston, prescott, brockville, belleville, trenton, picton, gananoque, sand banks, and environs... – Toronto: M G Bixby, 1887 – 2mf – 9 – (incl ind) – mf#24909 – cn Canadiana [971]

Industries of canada : historical and commercial sketches : london, guelph, berlin, brantford, paris, waterloo, chatham and environs... – Toronto: M G Bixby, 1886 – 2mf – 9 – 0-665-91596-9 – (inlc ind) – mf#91596 – cn Canadiana [971]

Industries of canada : historical and commercial sketches : london, woodstock, ingersoll, guelph, berlin, waterloo, st. thomas, windsor, and environs... – Toronto: M G Bixby, 1887 – 2mf – 9 – (incl ind) – mf#07189 – cn Canadiana [338]

Industries of canada : historical and commercial sketches, peterboro', lindsay, gravenhurst, orillia, millbrook, uxbridge, markham and environs... – Toronto: M G Bixby, 1887 – 2mf – 9 – mf#07188 – cn Canadiana [971]

Industries of canada : historical and commercial sketches, toronto, west toronto junc. and environs, its prominent places and people... – Toronto: Railway & Steamship Pub Co, 1890 [mf ed 1994] – 2mf – 9 – 0-665-94668-6 – (incl ind) – mf#94668 – cn Canadiana [917]

Industries of canada : city of montreal : historical and descriptive review, leading firms and moneyed institutions – Montreal: Historical Pub Co, 1886 – 2mf – 9 – (incl ind) – mf#07494 – cn Canadiana [338]

Industries of canada, historical and commercial sketches, hamilton and environs : its prominent places and people : representative merchants and manufacturers : its improvements, progress and enterprise – Toronto: M G Bixby, 1886 – 2mf – 9 – (incl ind) – mf#08535 – cn Canadiana [971]

The industries of philadelphia / Blodget, Lorin – 1876 – 1 – us CRL [300]

Industries sector circular ind / Commercial Advisory Foundation in Indonesia – Djakarta, 1970-1972 – 5mf – 9 – mf#SE-1390 – ne IDC [959]

Industrious men / Villamor, Ignacio – Manila: Oriental Commercial Co., 1932. xxiv,211p – 1 – us Wisconsin U Libr [920]

Industry *see* The manager, 1950-66

Industry, 1948-50 *see* Industry illustrated, 1933-1947/industry, 1948-1950

Industry and higher education – Guildford. 1987-1990 – 1,5,9 – ISSN: 0950-4222 – mf#17234 – us UMI ProQuest [378]

Industry and innovation – Sydney. 1997+ (1) – ISSN: 1366-2716 – mf#22306,01 – us UMI ProQuest [338]

Industry illustrated, 1933-1947/industry, 1948-1950 – 7r – 1 – (incorp: management review. superseded by: the manager) – mf#95570 – uk Microform Academic [338]

Industry in south africa : a survey of opportunities for industrial expansion – Cape Town: Unie-Volkspers, 1942 – 1 – us CRL [960]

Industry mart – New York. 1972-1972 (1) – ISSN: 0149-5534 – mf#8563 – us UMI ProQuest [338]

Industry, prudence, and piety – London, England. 1828 – 1r – us UF Libraries [240]

Industry reform = Kung-yeh kai-tsao – n8-10. 5 jan 1926-n18. feb 1929* – (= ser Chinese christian coll 59) – 1r – 1 – (in chinese) – mf#ATLA S02960 – us ATLA [338]

Industry week – Cleveland. 1880+ (1) 1967+ (5) 1970+ (9) – ISSN: 0039-0895 – mf#759 – us UMI ProQuest [650]

Induzierte hypervolaemie und kontrollierte volumenanpassung : zwei neue methoden zur prophylaxe und therapie der akuten tonischen kreislaufinsuffizienz / Kirchner, Erich – Marburg 1965 (mf ed 1994) – 3mf – 9 – €38.00 – 3-8267-2040-7 – mf#DHS-AR 2040 – gw Frankfurter [612]

The indwelling christ / Campbell, James Mannann – Chicago:Fleming H. Revell, c1895 – 1mf – 9 – 0-8370-3126-5 – mf#1985-1126 – us ATLA [210]

The indwelling spirit / Davison, William Theophilus – London; New York: Hodder and Stoughton, [1911] – 1mf – 9 – 0-7905-0937-7 – mf#1987-0937 – us ATLA [240]

Indy, Vincent d' *see*
- Fantaisie pour piano sur un vieil air de ronde francais[e], [op 99]
- Themes d'harmonie, 100 [2eme livre]

Indyohesha-birayi / Kagame, P Alegisi – Kabgayi, [Ruwanda]: Les Editions Royales, 1949 – 1 – us CRL [960]

Indyvelse i christendom / ed by Kierkegaard, Soeren – Andet oplag Kobenhavn: Forlagt af C A Reitzels Bo og [sic] Arvinger, 1855 – 1mf – 9 – 0-7905-3794-X – (himmelstrup) – mf#1989-0287 – us ATLA [190]

Indyvelse i christendom / ed by Kierkegaard, Soeren – Kobenhavn: C A Reitzel, 1850 – (= ser Himmelstrup) – 1mf – 9 – 0-7905-3793-1 – mf#1989-0286 – us ATLA [190]

Ineffabilis deus : the bull "ineffabilis" in four languages, or, the immaculate conception of the most blessed virgin mary defined / Pope Pius 9; ed by Bourke, Ulick Joseph – Dublin: John Mullany, 1868 – 1mf – 9 – 0-8370-8410-5 – (in english, french, irish and latin) – mf#1986-2410 – us ATLA [240]

In...epistolas ad philippenses, colossenses, thessalonicences ambas, et primam ad timotheum, commentarij / Musculus, W – Basilea, Johann Herwagen, 1565 – 5mf – 9 – mf#PBU-343 – ne IDC [240]

Inequality and progress / Harris, George – Boston: Houghton, Mifflin, 1897 – 1mf – 9 – 0-7905-7750-X – mf#1989-0975 – us ATLA [300]

Inequality in education – Cambridge. 1971-1978 (1) 1969-1978 (5) 1976-1978 (9) – ISSN: 0579-3475 – mf#6162 – us UMI ProQuest [370]

Ines de castro : tragedie / La Motte, Antoine Houdar de – 2e ed. A Paris: Chez G Dupuis et F Flahault, 1723 [mf ed 1991] – 1mf – 9 – mf#SEM105P1414 – cn Bibl Nat [790]

Ineson, Frank A *see* Forest resources of northeastern florida

Das inevitabele des honorius augustodunensis und dessen lehre : ueber das zusammenwirken von wille und gnade / Baeumker, Fr – 1914 – (= ser Bgphma 13/6) – €5.00 – ne Slangenburg [100]

Infact news / Infant Formula Action Coalition – special 1982 apr, 1983 win-1985 win, special suppl – 1r – 1 – (cont by: twin cities campaign center newsletter) – mf#1058589 – us WHS [071]

Infact update / Infant Formula Action Coalition – 1981 jul-aug, nov, 1982 jan, jun-oct, 1982 dec, 1983 jan, may-jun, oct-nov, 1984 jan-may – 1r – 1 – mf#958127 – us WHS [071]

L'Infaillibilite du pape et le syllabus : etude historique et theologique / Viollet, Paul – Besancon: Jacquin; Paris: P Lethielleux, 1904 [mf ed 1986] – 1mf – 9 – 0-8370-8396-6 – (in french. incl bibl ref and ind) – mf#1986-2396 – us ATLA [241]

Infallibilismus und katholicismus : sendschreiben an einen infallibilistisch gesinnten freund – 2., nochmals durchgesehene und in einem Nachworte vervollstaendigte Aufl. Bonn: Max Cohen, 1885 – 1mf – 9 – 0-8370-8144-0 – mf#1986-2144 – us ATLA [241]

Die infallibilitaet des oberhauptes der kirche und die zustimmungsadressen an herrn v. doellinger, namentlich die muenster'sche / Stoeckl, Albert – Muenster: Adolph Russell, 1870 – 1mf – 9 – 0-8370-8387-7 – mf#1986-2387 – us ATLA [240]

L'infaillibilite papale prise en manifeste et flagrant delit de mensonge : ou, le dogme de l'immaculee conception cite... / Durand, Louis – Bruxelles: Chretienne Evangelique, 1859 [mf ed 1986] – 2mf – 9 – 0-8370-8254-4 – (in french. incl bibl ref and ind) – mf#1986-2254 – us ATLA [241]

Infallibility : a paper / McNabb, Vincent – London; New York: Longmans, Green, 1905 – 1mf – 9 – 0-8370-8531-4 – (incl bibl ref with appendix in latin with english translation) – mf#1986-2531 – us ATLA [241]

The infallibility of the church : a course of lectures / Salmon, George – 3rd ed. London: John Murray, 1899 – 2mf – 9 – 0-7905-9865-5 – mf#1989-1590 – us ATLA [240]

Infallible logic, a visible and automatic system of reasoning / Hawley, Thomas De Riemer – Lansing: R. Smith, 1896 – xxviii/659p – 1 – us Wisconsin U Libr [160]

Infallible, the supreme, and the universal bishop / Richardson, John – London, England. 1850 – 1r – us UF Libraries [240]

A infancia – Rio de Janeiro, RJ: Typ do Magdalenense, 06 jan 1879 – (= ser Ps 19) – mf#P17,03,59 – bl Biblioteca [370]

Infancy – Mahwah. 2000+ (1,5,9) – ISSN: 1525-0008 – mf#31730 – us UMI ProQuest [305]

Infancy and manhood of christian life / Taylor, William – London: SW Partridge; New York: Nelson and Phillips, 1875 – 1mf – 9 – 0-7905-9704-7 – mf#1989-1429 – us ATLA [240]

Infancy of our lord, the... : selden supra ms. 38, sc 3426 – 1r – 14 – mf#C520 – uk Microform Academic [240]

The infancy of religion / Owen, David Cymmer – London; New York: Oxford University Press, 1914 – (= ser The S. Deiniol's Series) – 1mf – 9 – 0-524-00949-X – mf#1990-2172 – us ATLA [200]

Infant and child development – Chichester. 1999+ (1,5,9) – (cont: early development and parenting) – ISSN: 1522-7227 – mf#19118,01 – us UMI ProQuest [150]

Infant baptism : including a series of conversations on the subject and mode of baptism / Douglass, R – Philadelphia: King & Baird, printers, 1851 – 2mf – 9 – 0-524-07862-9 – mf#1991-3407 – us ATLA [242]

Infant baptism / Lumsden, James – Edinburgh, Scotland. 1856 – 1r – us UF Libraries [242]

Infant baptism : when-where-why instituted / Adams, Henry – Yarmouth, NS?: C Carey, 1888 – 1mf – 9 – mf#06150 – cn Canadiana [242]

Infant baptism a true sacrament / Gibson, John – London, England. 1851 – 1r – us UF Libraries [242]

Infant baptism and infant salvation in the calvinistic system : a review of dr. hodge's systematic theology / Krauth, Charles Porterfield – Philadelphia: Lutheran Book Store, 1874 – 1mf – 9 – 0-7905-7955-3 – (incl bibl ref) – mf#1989-1180 – us ATLA [242]

Infant baptism scriptural and reasonable / Miller, Samuel – Belfast, Northern Ireland. 1842 – 1r – us UF Libraries [242]

Infant baptism, scriptural and reasonable, and baptism by sprinkling or affusion, the most suitable and edifying mode / Miller, Samuel – Philadelphia: Presbyterian Board of Publication, 1840 – 1mf – 9 – 0-524-04223-3 – mf#1990-5014 – us ATLA [242]

Infant dedication / Tilly, Alfred – London, England. 18– – 1r – us UF Libraries [240]

Infant Formula Action Coalition *see*
– Infact news
– Infact update

Infant mental health journal – East Lansing. 1980+ (1,5,9) – ISSN: 0163-9641 – mf#12185 – us UMI ProQuest [618]

Infant projects – Leamington Spa. 1988-1988 – 1,5,9 – ISSN: 0269-9524 – mf#11790,01 – us UMI ProQuest [370]

Infant salvation in its relation to infant depravity, infant regeneration and infant baptism / Bomberger, John Henry Augustus – Philadelphia: Lindsay & Blakiston, 1859 – 1mf – 9 – 0-7905-3592-0 – mf#1989-0085 – us ATLA [240]

Infant sprinkling : weighed in the balance of the sanctuary, and found wanting in five letters addressed to the rev. george jackson, wesleyan methodist missionary... / Elder, William – Halifax [NS] Printed for the author, 1823 – 1mf – 9 – 0-665-92586-7 – mf#92586 – cn Canadiana [242]

Infantas, Don Fer(di)nando de las *see*
– Plura modulationum genera
– Sacrarum varii styli cantionum liber 3
– Sacrarum varii styli cantionum...liber 2

Infantas luntanas reinas de espana e infantas espanolas reinas de portugal / Lancastre-Laboreiro e Souza de Villalobos, Anna – Caceres: Imp. Moderna, 1931 – 1 – sp Bibl Santa Ana [946]

Infant-baptism : historically considered / McGlothlin, William Joseph – Nashville, Tenn: Sunday School Board, Southern Baptist Convention, 1916 – 1mf – 9 – 0-7905-9510-9 – mf#1989-1215 – us ATLA [242]

Infant-baptism considered / Church Of Ireland Diocese Of Dublin Archbishop – London, England. 1850 – 1r – us UF Libraries [242]

Infante, Modesto *see* Plutarco

Infanterie! : ein gedicht gewidmet dem volke in waffen / Wildgans, Anton – Wien: H Heller, 1915 – 1r – 1 – (numbered limited edition signed by the author) – us Wisconsin U Libr [810]

O infantil : orgam dos alunos do collegio camargo – Sao Paulo, SP. 31 jan 1895 – (= ser Ps 19) – mf#P17,02,218 – bl Biblioteca [079]

Infantry – Fort Benning. 1930+ (1) 1971+ (5) 1974+ (9) – ISSN: 0019-9532 – mf#3134 – us UMI ProQuest [355]

Infantry journal – Washington. 1914-1950 (1) – ISSN: 0019-9540 – mf#251 – us UMI ProQuest [355]

Infants and young children – Gaithersburg. 1988+ (1,5,9) – ISSN: 0896-3746 – mf#16711 – us UMI ProQuest [618]

Infared interactance : reliability and validity in determining body composition / Durrett, M – 1991 – 1mf – 9 – $4.00 – us Kinesology [790]

Infection and immunity – Washington. 1970+ (1) 1971+ (5) 1975+ (9) – ISSN: 0019-9567 – mf#5751 – us UMI ProQuest [616]

Infection control and hospital epidemiology – Thorofare. 1988+ (1,5,9) – (cont: ic infection control) – ISSN: 0899-823X – mf#13522,01 – us UMI ProQuest [614]

Infection, genetics and evolution – Amsterdam. 2001+ (1,5,9) – ISSN: 1567-1348 – mf#42849 – us UMI ProQuest [575]

Infection of potato tubers by alternaria solani in relation to storage conditions / Gratz, L O – Gainesville, FL. 1927 – 1r – us UF Libraries [630]

Infectious agents and disease – New York. 1993-1996 (1,5,9) – ISSN: 1056-2044 – mf#18707 – us UMI ProQuest [616]

Infectious bovine mastitis / Sanders, D A – Gainesville, FL. 1946 – 1r – us UF Libraries [636]

Infectious disease clinics of north america – Philadelphia. 1993+ (1,5,9) – ISSN: 0891-5520 – mf#20826 – us UMI ProQuest [616]

Infectious diseases in clinical practice – v1-5. 1992-1996 – 5r – 1,5,6,9 – $80.00r – us Lippincott [616]

Infeld, Donna Lind *see* The hospice journal

Inference generation in the reading of expository texts by university students / Pretorius, Elizabeth Josephine – Uni of South Africa 2000 [mf ed Johannesburg 2000] – 7mf [ill] – 9 – (incl bibl ref) – mf#mfm14944 – sa Unisa [400]

Inferno verde / Rangel, Alberto – Tours, France. 1920 – 1r – us UF Libraries [972]

Inferno verde / Rangel, Alberto – Tours, France. 1927 – 1r – us UF Libraries [972]

Infertility – New York. 1986-1990 (1) 1986-1990 (5) 1986-1990 (9) – ISSN: 0160-7626 – mf#14334 – us UMI ProQuest [618]

Infertility and reproductive medicine clinics of north america – Philadelphia. 1993+ (1,5,9) – ISSN: 1047-9422 – mf#20834 – us UMI ProQuest [618]

Infidel objections to the scriptures / Whitmore, F B – New York: Thomas Nelson, 1884 [mf ed 1985] – 1mf – 9 – 0-8370-5827-9 – (incl ind) – mf#1985-3827 – us ATLA [220]

Infidelites de lisette / Brazier, Nicholas – Paris, France. 1835 – 1r – us UF Libraries [440]

Infidelity / Sparkes, John George – London, England. 18– – 1r – us UF Libraries [240]

Infidelity among southern baptists endorsed by highest officials / Norris, J Frank – n.d – 1 – 5.84 – us Southern Baptist [242]

Infidelity disarmed / Stephens, Edward – Toronto: for sale at the Methodist Book Room, 1900 [mf ed 1985] – 1mf – 9 – 0-8370-5400-1 – mf#1985-3400 – us ATLA [230]

Infidelity dissected : the evangelical alliance prize essay on infidelity / Pearson, Thomas – Chicago: Geo MacDonald, c1874 [mf ed 1985] – 1mf – 9 – 0-8370-4690-4 – (earlier ed under title: infidelity, its aspects, causes and agencies. incl bibl ref) – mf#1985-2690 – us ATLA [140]

Infidelity in high places / Brock, William – London, England. 1864 – 1r – us UF Libraries [240]

Infidelity: its aspects, causes and agencies / Pearson, Thomas – New York: R. Carter & Bros., 1854. 620p. 'Being the prize essay of the British Organization of the Evangelical Alliance.' Bibliog. footnotes. With: Anthropologiia, by E.I. Petri – 1 – us Wisconsin U Libr [306]

Infidelity refuted by infidels / Sprecher, Samuel P – New York: Funk & Wagnalls, 1888 – 1mf – 9 – 0-8370-5348-X – mf#1985-3348 – us ATLA [240]

The infidel's text-book : being the substance of thirteen lectures on the bible / Cooper, Robert – 1st American, republ from the London ed. Boston: J P Mendum, 1876 – 1mf – 9 – 0-524-05976-4 – mf#1992-0713 – us ATLA [220]

Infierno verde : guerra del chaco / Marin Canas, Jose – Madrid, Spain. 1935 – 1r – us UF Libraries [972]

Infiesta, Ramon *see* Maximo gomez

The infinite affection / Macfarland, Charles Stedman – 2nd ed. Boston: Pilgrim Press; London: James Clarke, 1907 [mf ed 1985] – 1mf – 9 – 0-8370-4255-0 – mf#1985-2255 – us ATLA [242]

Infinite benevolence / Clarke, Joseph – London, England. 18– – 1r – us UF Libraries [240]

L'infinite divine depuis philon le juif jusqu'a plotin : avec une introduction sur le maeme sujet dans la philosophie grecque avant philon le juif / Guyot, Henri – Paris: Felix Alcan, 1906 [mf ed 1986] – 1mf – 9 – 0-8370-9700-2 – (in french) – mf#1986-3700 – us ATLA [180]

The infinitive in polybius compared with the infinitive in biblical greek / Allen, Hamilton Ford – Chicago: University of Chicago Press, 1907 – 4mf – 9 – 0-8370-9280-9 – mf#1986-3280 – us ATLA [450]

Infinity Books, Ltd *see* Cosmic landscape

Infirmiere canadienne – Ottawa. 1973-1985 (1) 1975-1985 (5) 1975-1985 (9) – ISSN: 0019-9605 – mf#9308 – us UMI ProQuest [360]

Inflammation research – Basel. 1995+ (1,5,9) – (cont: agents and actions) – ISSN: 1023-3830 – mf#5141,01 – us UMI ProQuest [650]

Inflorescencias / Valverde, Jose Antonio – San Jose, Costa Rica. 1962 – 1r – us UF Libraries [580]

L'influence de la decouverte de l'amerique sur le bonheur du genre-humain / Genty, Louis – 2nd rev corr enl ed. Orleans: De l'impr de Jacob l'Aone...1789 [mf ed 1985] – 2v on 1mf – 9 – 0-665-54496-0 – (incl bibl ref) – mf#54496 – cn Canadiana [970]

Influence des membres masculins de la famille sur l'enfant mukongo : etude descriptive et interpretative / Matota, H – Louvain, Belgium, Universite catholique, 1958 – us CRL [920]

L'influence du symbolisme francais dans le renouveau poetique de l'allemagne : les plaetter fuer die kunst de 1892 a 1900 / Duthie, Enid Lowry – Paris: Librairie ancienne H Champion, 1933 [mf ed 1993] – (= ser Bibliotheque de la revue de litterature comparee 91) – 1 – (incl bibl ref & ind) – mf#8295 – us Wisconsin U Libr [410]

Influence francaise dans l'oeuvre de ruben dario / Mapes, Erwin Kempton – Paris, France. 1925 – 1r – us UF Libraries [972]

Influence of a praying mother – London, England. 18– – 1r – us UF Libraries [240]

The influence of aerobic vs. anaerobic exercise on sex hormone-binding globulin and free testosterone concentration / Kelly, Erin W – 1997 – 2mf – 9 – $8.00 – mf#PH 1574 – us Kinesology [612]

The influence of aerobic vs. anaerobic exercise on thyroid hormone concentrations / Umscheid, Jill M – 1997 – 1mf – 9 – $4.00 – mf#PH 1578 – us Kinesology [612]

Influence of age and caffeine on resting metabolic rate, blood pressure, and mood state in younger and older individuals / Arciero, Paul J & Mahar, Matthew T – 1993 – 2mf – $8.00 – us Kinesology [612]

Influence of age on the hemodynamic adjustments to physiological stresses / Minson, Christopher T – 1997 – 150p on 2mf – 9 – $10.00 – mf#PH 1723 – us Kinesology [618]

The influence of agility on the mile run and pacer tests of aerobic endurance in fourth- and fifth-grade school children / Dinschel, Kimberly M – 1994 – 1mf – $4.00 – us Kinesology [612]

The influence of alcohol and other drugs on fatigue : the croonian lectures delivered at the royal college of physicians in 1906 / Rivers, William Halse R – London: E Arnold, 1908 – 1 – us CRL [610]

Influence of american legislation on the decline of the united states as a maritime power : an address delivered before the royal colonial institute, june 26, 1872 / Haliburton, Robert Grant – London?: European Mail, 1872 – 1mf – 9 – mf#08410 – cn Canadiana [380]

Influence of ancient hebrew music on gregorian chant / Emanuel, Nathan H – U of Rochester 1935 [mf ed 19–] – 2mf – 9 – mf#fiche98 – us Sibley [780]

The influence of animism on islam; an account of popular superstitions / Zwemer, Samuel Marinus – New York: The Macmillan Co., 1920. viii,2 leaves,246p. front., illus., plates. Bibliography p245-246 – 1 – us Wisconsin U Libr [260]

Influence of ankle orthoses on joint motion and postural stability 109=before and after exercise / Jorden, Ryan A – 2000 – 109p on 2mf – 9 – $10.00 – mf#PE 4102 – us Kinesiology [617]

The influence of aristocracies on the revolutions of nations; considered in relation to the present circumstances of the british empire / Macintyre, James J – London: Fisher, son, & Co.,1843.16p,448p – 1 – us Wisconsin U Libr [320]

Influence of bible societies on the temporal necessities of the poo... / Chalmers, Thomas – Cupar, Scotland. 1814 – 1r – us UF Libraries [240]

Influence of body fat mass on excess post-exercise oxygen consumption / Harms, Craig A & Cordain, Loren – 1990 – 1mf – 9 – $4.00 – us Kinesology [612]

Influence of caffeine on substrate utilization : during step aerobics in experienced step aerobics excersisers / Deguchi, Madoka – 2000 – 132p on 2mf – 9 – $10.00 – mf#PH 1711 – us Kinesology [612]

The influence of case discussions on physical education preservice teachers' reflection in an educational games class / Bolt, Brian R – 1996 – 3mf – 9 – $12.00 – mf#PE 3785 – us Kinesology [370]

The influence of catholicism on the sciences and on the arts / Salas y Gilavert, Andres de – London: Sands, 1900 – 1mf – 9 – 0-8370-8062-2 – (incl bibl ref) – mf#1986-2062 – us ATLA [241]

The influence of certain ocular defects in causing headache / Buller, Frank – [s.l: s.n, 1888?] [mf ed 1985] – 1mf – 9 – 0-665-01589-5 – mf#01589 – cn Canadiana [617]

The influence of christianity on war / Bethune-Baker, James Franklin – Cambridge: Macmillan and Bowes, 1888 – 1mf – 9 – 0-7905-5509-3 – (incl bibl ref) – mf#1988-1509 – us ATLA [240]

The influence of christianity upon international law / Kennedy, Charles Malcolm – Cambridge: Macmillan, 1856 – (= ser Hulsean Prize Essay) – 1mf – 9 – 0-7905-5768-1 – (incl bibl ref) – mf#1988-1768 – us ATLA [341]

The influence of christianity upon national character illustrated by the lives and legends of the english saints / Hutton, William Holden – London: W Gardner, Darton, [1903?] – (= ser Bampton lectures) – 1mf – 9 – 0-7905-6410-6 – (incl bibl ref) – mf#1988-2410 – us ATLA [240]

The influence of christianity upon social and political ideas / Carlyle, Alexander James – London: A.R. Mowbray, [1911?] – (= ser Christian Social Union Handbooks) – 1mf – 9 – 0-7905-4196-3 – (incl bibl ref) – mf#1988-0196 – us ATLA [301]

The influence of clothing on health / Treves, Frederick – London [1886] – (= ser 19th c art & architecture) – 2mf – 9 – mf#4.1.270 – uk Chadwyck [640]

Influence of commerce upon christianity / Fremantle, William Henry – Oxford, England. 1854 – 1r – us UF Libraries [240]

Influence of conceptually based physical education on student attitudes toward physical activity / Hughes, Kevin P – Springfield College, 1994 – 2mf – 9 – $8.00 – mf#PE3601 – us Kinesology [370]

The influence of context on the generalizability of children's perceptions of physical competence / Shapiro, Deborah R – 1999 – 4mf – 9 – $16.00 – mf#PSY 2123 – us Kinesology [150]

Influence of conversation, with the regulation thereof / Lucas, Richard – London, England. 1800 – 1r – us UF Libraries [240]

The influence of cryotherapy and aircast bracing on total body balance and proprioception / Rivers, Debra A – 1994 – 1mf – $4.00 – us Kinesology [617]

The influence of darwin on philosophy, and other essays in contemporary thought / Dewey, John – New York: Henry Holt, 1910 – 1mf – 9 – 0-7905-3719-2 – mf#1989-0212 – us ATLA [190]

Influence of diet and the menstrual cycle on lactate concentration during increasing exercise intensities / Berend, Julia Z & Hackney, Anthony C – 1992 – 2mf – 9 – $8.00 – us Kinesology [613]

The influence of dispositional goal orientation, perceptions of the motivational climate, and scholarship level on sport commitment in elite level athletes / Guest, Shannon M – 1998 – 215p on 3mf – 9 – $15.00 – mf#PSY 2161 – us Kinesology [150]

The influence of emerson / Mead, Edwin Doak – Boston: American Unitarian Association, 1903 – 1mf – 9 – 0-524-01087-0 – mf#1990-4052 – us ATLA [420]

The influence of english literature on urdu literature / 'Abdu'l-Latif, Sayyid – London: Forster Groom & Co, 1924 – (= ser Samp: indian books) – us CRL [410]

The influence of fitness-oriented physical activity on the physical self-perception and global self-worth of boys and girls / Fine, Deborah L & Jensen, Barbara E – 1993 – 2mf – $8.00 – us Kinesology [150]

The influence of force production and eccentric exercise on growth hormone / Kim, Junghoon – 1997 – 1mf – 9 – $4.00 – mf#PH 1575 – us Kinesology [612]

The influence of goal setting on exercise adherence of apparently healthy adults / Cobb, L E – 1991 – 2mf – 9 – $8.00 – us Kinesology [150]

The influence of goal setting on individual endurance swimming performance / LaClair, Kirsten W & Mann, Betty J – 1993 – 2mf – $8.00 – us Kinesology [150]

The influence of greek antiquity on modern german drama / Gorr, Adolph – Philadelphia: Univ of Pennsylvania, 1934 – 105p – 1 – (incl bibl ref) – us Wisconsin U Libr [430]

The influence of greek ideas and usages upon the christian church *see* Griechentum und christentum

The influence of health behavior contracting on internal locus of control / Flint, Matthew O – 1993 – 1mf – $4.00 – us Kinesology [150]

The influence of height on body image, self-confidence, and performance of female basketball players / Aardahl, Anya – 1999 – 2mf – 9 – $8.00 – mf#PSY 2064 – us Kinesology [150]

Influence of in-shoe orthotics on lower extremity function in cycling / Joganich, T G – 1991 – 2mf – 9 – $8.00 – us Kinesology [790]

Influence of international marketing upon u.s. ski resorts and japanese tour operators / Sawamura, Sachi – 1996 – 2mf – 9 – $8.00 – mf#RC 508 – us Kinesology [650]

The influence of jewish colonisation on arab development in palestine – Jerusalem, 1947 – 1mf – 9 – mf#J-28-147 – ne IDC [956]

The influence of management as a strategy for managing communication in schools : a mini-dissertation / Moemise, Punie Alfred Peter – Pretoria: Vista University 2002 [mf ed 2002] – 2mf – 9 – (incl bibl ref) – mf#mfm15175 – sa Unisa [370]

The influence of management styles : upon the use of extrinsic and intrinsic rewards in selected public park and recreational agencies / Hurd, Amy R – 1999 – 120p on 2mf – 9 – $10.00 – mf#RC 539 – us Kinesology [650]

The influence of mars – London: Grant Richards, 1900 – (= ser 19th c women writers) – 3mf – 9 – mf#5.1.119 – uk Chadwyck [420]

The influence of material on architecture / Fletcher, Banister Flight – London 1897 – (= ser 19th c art & architecture) – 1mf – 9 – mf#4.2.235 – uk Chadwyck [720]

Influence of menstrual cycle phase and oral contraceptive use on cardiovascular reactivity in women with a parental history of hypertension / Silvey, C M – 1991 – 1mf – 9 – $4.00 – us Kinesology [612]

Influence of mental imagery on tennis service accuracy of intermediate level tennis players / Choboy, Jon A & Murray, Mimi – 1992 – 1mf – $4.00 – us Kinesology [150]

The influence of mothers on differencesin role conflict and gender typing of sports for females / Mann, Lisa E – 2000 – 125p on 2mf – 9 – $10.00 – mf#PSY 2143 – us Kinesology [150]

The influence of music on motor behavior and select physiological and psychological variables / Nordvall, Michael P – 1995 – 117p on 2mf – 9 – $10.00 – mf#PSY 2157 – us Kinesology [150]

The influence of musical preference on the affective state, heart rate, and perceived exertion ratings of participants in aerobic dance/exercise classes / Patton, N W – 1991 – 2mf – 9 – $8.00 – us Kinesology [150]

The influence of oral contraceptives on running performance / Van Dyke, Alison D – 1998 – 2mf – 9 – $8.00 – mf#PH 1625 – us Kinesology [615]

The influence of participation in a sports training program on the self-concepts of the educable mentally retarded attending a one-week special olympics sports camp / Edmiston, Paula A – 1982 – 2mf – 9 – $8.00 – us Kinesology [790]

The influence of participatory development on the communication patterns of the parachute packing section of the sandf / Govender, Saravani – Uni of South Africa 2000 [mf ed Pretoria: UNISA 2000] – 2mf – 9 – (incl bibl ref) – mf#mfm14712 – sa Unisa [650]

The influence of physical conditioning and deconditioning upon cardiac structure of males and females / Al-Muhailani, Abdul-Rahman S – 1980 – 2mf – 9 – $8.00 – us Kinesology [790]

The influence of physical conditioning on the post-menopausal hot flash / Krasnoff, Joanne B – Indiana University, 1995 – 1mf – 9 – $4.00 – mf#PH1466 – us Kinesology [612]

Influence of post-exercise glucose ingestion on plasma potassium levels and ecg measurements / Reynolds, H – 1991 – 1mf – 9 – $4.00 – us Kinesology [612]

Influence of raynor taylor and benjamin carr on church music in philadelphia at the beginning of the nineteenth century / Doran, Carol – U of Rochester 1970 [19–] – 2v on 5mf – 9 – mf#fiche203, 585 – us Sibley [780]

Influence of reactive hyperemia in muscle during exercise / Henrich, Timothy W – 1988 – 152p 2mf – 9 – $8.00 – us Kinesology [612]

Influence of reinforcers on motorized bicycle on-task time of profoundly mentally retarded adolescents / Owlia, G – 1991 – 2mf – 9 – $8.00 – us Kinesology [150]

Influence of religious elites on political culture and community integration in kano, nigeria / Paden, John N – Cambridge 1968 – us CRL [305]

The influence of role status, self-efficacy and soccer performance / Mandell, Ross A – 1994 – 2mf – $8.00 – us Kinesology [150]

The influence of rules for recovery of attorneys' fees on settlement of civil cases / Shapard, J E – Washington: FJC, 1984 – 1mf – 9 – $1.50 – mf#LLMC 95-319 – us LLMC [347]

The influence of scepticism on character : being the 16th fernley lecture...london, aug 2 1886 / Watkinson, William Lonsdale – London: Charles H Kelly 1898 [mf ed 1985] – 1mf – 9 – 0-8370-5719-1 – (incl bibl ref) – mf#1985-3719 – us ATLA [170]

Influence of science on theology / Bonney, Thomas George – Cambridge, England. 1885 – 1r – us UF Libraries [240]

The influence of self-efficacy on learning : in selected secondary schools in thabong and mangaung / Segalo, Letlhoyo – Pretoria: Vista University 2002 [mf ed 2002] – 3mf – 9 – (incl bibl ref) – mf#mfm15324 – sa Unisa [150]

The influence of situation criticality on the performance of male collegiate basketball players / Roberts, N A – 1989 – 1mf – 9 – $4.00 – us Kinesology [150]

The influence of social support on athletic injury rehabilitation : the athletes' point of view / Te Selle, Lori L – 1999 – 2mf – 9 – $8.00 – mf#PSY 2093 – us Kinesology [617]

The influence of spousal exercise patterns and perceived social support on the quality of life and health status in regular exercisers / Hancher, Heidi L – 2000 – 1mf – 9 – $4.00 – mf#PSY 2113 – us Kinesology [613]

The influence of task and ego goal orientations and perceptions of competence on affect and intrinsic motivation in competitive youth tennis / Chaumeton, Nigel R – 1996 – 3mf – 9 – $12.00 – mf#PSY 1939 – us Kinesology [790]

The influence of temporal demands on continuous bimanual movements with and without a spatial component / Lantero, Dawn A – 1998 – 1mf – 9 – $4.00 – mf#PSY 2045 – us Kinesology [612]

The influence of the apostle paul on the development of christianity / Pfleiderer, Otto – New York: Charles Scribner 1885 [mf ed 1985] – (= ser Hibbert lectures (new york ny) 1885) – 1mf – 9 – 0-8370-4726-9 – mf#1985-2726 – us ATLA [227]

The influence of the coaches' expectations on the goal setting of division 1 student-athletes / Maltbey, Jamie M – 2001 – 59p on 1mf – 9 – $5.00 – mf#PSY 2173 – us Kinesology [150]

The influence of the german volkslied on eichendorff's lyric / Heinzelmann, Jacob Harold – Leipzig: G Fock, 1910 – 1r – 1 – us Wisconsin U Libr [430]

The influence of the holy spirit in conversion : a debate between asa sleeth and j.w randall: question, do the scriptures teach the direct influence of the holy spirit in conversion? / Sleeth, Asa – Cincinnati: Chase & Hall, 1876 – 1mf – 9 – 0-8370-5184-3 – mf#1985-3184 – us ATLA [220]

The influence of the home environment on the motor performance of preschool children / Botha, Marika G – 1982 – 2mf – 9 – $8.00 – us Kinesology [790]

The influence of the hoplite phalanx on the growth and change of the ancient olympic games / Ward, Paul S – 1998 – 1mf – 9 – $4.00 – mf#PE 4014 – us Kinesology [930]

The influence of the menstrual cycle and diet on metabolism during rest and exercise / Brammeier, Michele R & Hackney, Anthony C – 1992 – 1mf – 9 – $4.00 – us Kinesology [613]

The influence of the mosaic code upon subsequent legislation / Marsden, John Benjamin – London: Hamilton, Adams: Hatchard, 1862 – 1mf – 9 – 0-7905-1433-8 – mf#1987-1433 – us ATLA [221]

The influence of the netherlands in the making of the english commonwealth and the american republic : with notice of what the pilgrims learned in holland, their treatment by the government and people, and answers to criticisms made upon the proposed delfshaven memorial / Griffis, William Elliot – Boston, Mass: De Wolfe, Fiske, [1891?] – 1mf – 9 – 0-524-04014-1 – mf#1990-1186 – us ATLA [941]

Influence of the old italian school of violin playing / Schwarck, Maryls L – U of Rochester 1931 [mf ed 1974] – 1mf – 9 – (with bibl) – mf#fiche 83 – us Sibley [780]

Influence of the old masters on bach : a review of the organ works / Cossitt, Jennie – U of Rochester 1930 [mf ed 1974] – 1mf – 9 – mf#fiche75 – us Sibley [780]

The influence of the revival of classical studies on english literature during the reigns of elizabeth and james 1 : an essay which obtained the le bas prize for the year 1856 / Farrar, Frederic William – Cambridge: Macmillan, 1856 – 1mf – 9 – 0-7905-0013-2 – mf#1987-0013 – us ATLA [420]

The influence of the roman law on the law of england / Scrutton, Thomas Edward – Cambridge: University Press, 1885. 199p. LL-125 – 1 – us L of C Photodup [340]

Influence of the scottish church in christendom / Cowan, Henry – London: Adam and Charles Black, 1896 – (= ser Baird lecture) – 1mf – 9 – 0-7905-4216-1 – (incl bibl ref) – mf#1988-0216 – us ATLA [240]

The influence of the septuagint upon the pesittaa psalter / Berg, Joseph Frederic – NY: [s.n.] 1895 (Leipzig: W Drugulin) – 1mf – 9 – 0-8370-2281-9 – (includes a vita, appendixes and bibliography) – mf#1985-0281 – us ATLA [221]

The influence of the septuagint version of the old testament upon the progress of christianity / Churton, William Ralph – Cambridge: Macmillan, 1861 – 1mf – 9 – 0-524-05210-7 – (incl bibl ref) – mf#1992-0343 – us ATLA [221]

Influence of the strength shoe and three plyometric drills on the strength, velocity, and jumping ability of high school football players / Ramsey, Jill K & Kimura, Iris F – 1992 – 1mf – 9 – $4.00 – us Kinesology [612]

The influence of velocity on the metabolic and mechanical task cost of treadmill running / Harris, Chad – 1995 – 2mf – 9 – $8.00 – mf#PH 1551 – us Kinesology [612]

The influence of walter scott on the novels of theodor fontane / Shears, Lambert Armour – New York: Columbia University Press, 1922 (mf ed 1990) – 1r – 1 – (filmed with: bozena) – us Wisconsin U Libr [410]

Influence of water therapy on selected physiological variables in pregnant women / Furstenberg, Dorylee – Stellenbosch: U of Stellenbosch 1998 [mf ed 1998] – 4mf – 9 – mf#mf.1359 – sa Stellenbosch [612]

The influence of wealth in imperial rome / Davis, William Stearns – New York: Macmillan, 1910 – 1mf – 9 – 0-7905-4292-7 – mf#1988-0292 – us ATLA [930]

Influences francaises sur la poesie hispano-americ... / Henriquez Urena, Max – Paris, France. 1938 – 1r – us UF Libraries [440]

The influences of greek ideas and useages upon the christian church / Hatch, Edwin; ed by Fairbairn, Andrew Martin – 6th ed. London: Williams & Norgate 1897 [mf ed 1984] – (= ser Hibbert lectures (london, england) 1888) – 5mf – 9 – 0-8370-0196-X – (incl bibl ref & ind) – mf#1984-1071 – us ATLA [240]

The influences of indian art – London: India Society, 1925 – (= ser Samp: indian books) – us CRL [700]

Influencia africana no portugues do brasil / Mendonca, Renato – Sao Paulo, Brazil. 1935 – 1r – us UF Libraries [972]

Influencia cristiana en la emancipacion de cuba / Prio Socarras, Carlos – Habana, Cuba. 1946 – 1r – us UF Libraries [972]

Influencia de extremadura en la literatura espanola / Diaz Perez, Nicolas – 1883 – 9 – sp Bibl Santa Ana [440]

Influencia de fatores socio-culturais no inovabili... / Schneider, Joao E – Porto Alegre, Brazil. 1970 – 1r – us UF Libraries [972]

Influencia de la matematica...carreras / Leon Gutierrez, Florencio – 1898 – 9 – sp Bibl Santa Ana [510]

Influencia de la universidad de la habana / Dihigo, Juan Miguel – Habana, Cuba. 1924 – 1r – us UF Libraries [378]

Influencia del manantial de marco en el desarrolla material de caceres / Castel, Joaquin – 1895, 1896 – 9 – sp Bibl Santa Ana [000]

Influencing a broader understanding of jazz dance / Giddins, Kevin J & Black, Catherine H – 1992 – 1mf – $4.00 – us Kinesology [790]

Influential physical education books of the twentieth century / VanClief, Elizabeth W – 1982 – 6mf – 9 – $24.00 – us Kinesology [790]

Info – Berlin, Germany.West Berlin: Homosexuelle Aktion, 1972- – 1 – us Wisconsin U Libr [305]

Info – Gary, IN. 1963-1990 (1) – mf#62793 – us UMI ProQuest [071]

Info – Gary IN. 1971 feb 26 – 1r – 1 – mf#2704372 – us WHS [071]

Info – Jakarta, Indonesia. 1965-1968 (1) – mf#67744 – us UMI ProQuest [079]

Info – United States. 1980 jun, oct, dec – 1r – 1 – mf#1206616 – us WHS [071]

Info Berube / Association des familles Berube – v1 n1-10 [1986 mar-1988 sep] – 1r – 1 – (cont by: monde berrubey) – mf#1759871 – us WHS [929]

Info canada – Downsview. 1994-1994 (1,5,9) – (cont: computer data) – ISSN: 1187-7081 – mf#15027,01 – us UMI ProQuest [000]

Info journal / International Fortean Organization – College Park. 1967-1980 (1) 1976-1980 (5) 1976-1980 (9) – ISSN: 0019-0144 – mf#8257 – us UMI ProQuest [500]

Info-fneeq / Federation nationale des enseignants et des enseignantes du Quebec – v1 n1 [1984 sep], [1985 [jan], mar, may], v3 n1 [1985 undated], v4 n3,2,3 [1986 feb, oct, [dec]] – 1r – 1 – mf#1321036 – us WHS [071]

Info-fneeq / Federation nationale des enseignants et des enseignantes du Quebec – v1 n2, 3 [1983 dec, 1984 mar], v2 n10,12,15 [1984 mar, apr, jun], v3 n1,4,5 [1985 undated, apr, may], v4 n1 [1985 oct], v4 n3,2,3 [1986 feb, oct, [dec]] – 1r – 1 – mf#1321037 – us WHS [071]

Infoldings and unfoldings of the divine genius : in nature and man / Pulsford, John – London: James Clarke 1901 [mf ed 1985] – (= ser Small books on great subjects 22) – 1mf – 9 – 0-8370-4810-9 – mf#1985-2810 – us ATLA [210]

Infolink – aug 1978-may 1993 – 1r – at Pascoe [079]

Infoperspectives – New York. 1989-1994 (1) – ISSN: 0733-9305 – mf#15421 – us UMI ProQuest [000]

Infor – Ottawa. 1984+ (1,5,9) – ISSN: 0315-5986 – mf#13853,01 – us UMI ProQuest [000]

Infor-burundi : bulletin hebdomadaire d'information de l'office national de presse du burundi – Usumbura: L'Office, n1-49. jan 6-dec 10, 1962; n53-98. jan-nov 25, 1963 – 1r – 1 – us CRL [079]

Inforcongo. 1ere Direction. Presse et relations publiques *see* Bulletin de presse

Inform – WI. n1-42 [1985 sep-1995 may] – 1r – 1 – (cont: acquisition list, job service wisconsin. library; inform, jtpa resource clearinghouse) – mf#1008495 – us WHS [331]

Inform – Silver Spring. 1987-1996 (1,5,9) – (cont: journal of information and image management) – ISSN: 0892-3876 – mf#16193 – us UMI ProQuest [000]

Inform otd Shtaba Otdel'noi Vost-Sib armii *see* Izvestiia shtaba otdel'noi vostochnoi-sibirskoi armii

Informa / Costa Rica. Direccion General de Estadistica y Censo – 1897, 1908-47 – 1 – us L of C Photodup [318]

Informacao do reino do congo, 1793-95 – Lisbon, Portugal: Biblioteca Nacional de Lisboa, 1970 – us CRL [960]

Informacao sobre as minas de s paulo / Taques, Pedro – Sao Paulo, Brazil. 19-? – 1r – us UF Libraries [972]

Informacion – Santiago, Chile. Mar 1920-oct 1921; jan-nov 1922; sep 1926-1927; feb 1928-apr 1930 – 4 1/2r – 1 – uk British Libr Newspaper [072]

La informacion – Bluefields, Nicaragua. 1959-1964 (1) – mf#67661 – us UMI ProQuest [079]

La informacion – Santiago, Chile. -m. March 1920-Nov 1922; Sept 1926-April 1930. 5 reels – 1 – uk British Libr Newspaper [072]

La informacion – Santiago de los Caballeros, Dominican Republic. 1946-1950 (1) – mf#67689 – us UMI ProQuest [079]

Informacion ante el senado / Cuba Ministerio De Educacion – Habana, Cuba. 1949 – 1r – us UF Libraries [972]

Informacion consular – Mexico City. 1948-1953 (1) – mf#613 – us UMI ProQuest [338]

Informacion en derecho...juan santos cuenda...los juicios / Medinaceli, Duquesa de – 1877 – 9 – sp Bibl Santa Ana [340]

Informacion que da al publico el dr. jose gossalbes...sobre la ultima enfermedad... / Gossalbes, Jose – Valencia, 1746 – 1mf – 9 – sp Cultura [610]

Informacion sobre el linaje de hernando pizarro / Munoz de San Pedro, Miguel – Badajoz: Dip. Prov. de Badajoz, 1966. Sep. Rev. Est. Extremenos – 1 – sp Bibl Santa Ana [920]

Informacion y curacion de la peste en zaragoza y perservacion contra la peste... / Porcell, J – Zaragoza, 1565 – 5mf – 9 – sp Cultura [615]

Informaciones – Madrid, 1940-55 – 1 – us CRL [074]

Informaciones : para los inmigrantes israelitas – Quito, Guayaquil Ecuador: Asociacion de beneficencia israelita, june 1940-aug 1959 – 3r – 1 – (semi-mthly newspaper of the german jewish refugee community in ecuador. in german with some spanish) – us UMI ProQuest [079]

Informacion...marques de perales / Perez Castro, Pedro A – 1796 – 9 – sp Bibl Santa Ana [920]

Informal / International Brotherhood of Electrical Workers – [1975 nov-1978 may] – 1r – 1 – mf#665258 – us WHS [331]

Informal opinions (a-g) / Pennsylvania – 1934-56; 1951-78 (1970 never published). 15 reels – 1 – $35.00r – us Trans-Media [340]

An informal record of missionary service in the island of tongoa and the shepherd islands / Miller, J Graham et al – 1941-1947 – 1r – 1 – mf#pmb1051 – at Pacific Mss [242]

An informal record of missionary service in the island of tongoa and the shepherd islands / Miller, J Graham & Miller, Flora – 1941-1947 – 1r – 1 – mf#PMB1051 – at Pacific Mss [240]

The informal sector and its taxation system in mozambique / Alfredo, Benjamim – Uni of South Africa 2001 [mf ed Johannesburg 2001] – 2mf – 9 – (incl bibl ref) – mf#mfm15089 – sa Unisa [336]

Informant / North American Negro Department – [s.l: s.n.] (jan 1942, jul, sep, dec 1943) [mf ed 2005] – 1r – 1 – (cont by: north american informant) – mf#2005-s024 – us ATLA [242]

Informant / Southwest Wisconsin Vocational-Technical Institute – v10 n2-v11 n8 [1979 nov-1981 apr] – 1r – 1 – (cont by: visions [fennimore, wi]) – mf#709554 – us WHS [331]

Informateur – Brussels Belgium, 11 sep 1944-31 jul 1945 – 1r – 1 – uk British Libr Newspaper [074]

L'informateur haitien – Port-au-Prince: Imp Aug A Heraux, jan 15-feb 22, feb 25-apr 30, may 3-jul 9, jul 11-aug 22, 25-30 1919 – 7 sheets – 9 – us CRL [079]

Informatia bucurestiului – Bucharest, Romania. 1962-25 Aug 1990 – 40r – 1 – (cont as: libertatea as of 29 dec 1989) – us L of C Photodup [949]

Informatik – 1969-1989 – 195mf – 1 – gw Mikropress [000]

Informatik-spektrum – Heidelberg. 1981-1982 (1) 1981-1982 (5) 1981-1982 (9) – ISSN: 0170-6012 – mf#13180 – us UMI ProQuest [000]

L'Information : politique, economique, financiere – Paris. 21 oct 1899-11 juin 1940, 19 juil 1941, 30 juin 1942, 1963-6 oct 1967 – 1 – fr ACRPP [073]

L'information – Cap-Haitien: [s.n.], jan 13 1934-dec 28 1935 – 7 sheets – 9 – us CRL [079]

L'information : financiere, industrielle, miniere – Montreal: Financial Times Publ. v30 n1 24 sep 1949- (wkly) [mf ed 2000] – 1r – 1 – (cont: information financiere et industrielle; ceased 1954?) – mf#SEM35P482 – cn Bibl Nat [073]

L'information – Fort-de-France, Martinique. 1941-1962 (1) – mf#67946 – us UMI ProQuest [079]

L'information : journal de sainte-julie, de saint-amable et de la region – Sainte-Julie: Information Ste-Julie et St-Amable, v4 n1 3 janv 1978- (wkly) [mf ed 1986] – 1 – mf#SEM35P215 – cn Bibl Nat [073]

In-formation – 1965 jan – 1r – 1 – mf#4876640 – us WHS [071]

Information – Paris, France. 7 jan-dec 1917; 25 feb 1918-22 jul 1919; 21 feb 1961 – 2r – 1 – uk British Libr Newspaper [072]

Information : der vertreter des pv der spd – tyska socialdemokratiska partiets representant – Stockholm (S), 1943 mar-1947 sep – 1r – 1 – (aka: zur information between 1933/44) – gw Misc Inst [325]

Information about fernandina and nassau county in the northeast corner of florida – Fernandina, FL. 1935 – 1r – us UF Libraries [917]

Information age – Guildford. 1982-1990 (1,5,9) – (cont: information privacy. cont by: journal of strategic information systems) – ISSN: 0261-4103 – mf#13331,01 – us UMI ProQuest [000]

Information and comments : swapo of namibia – v6, n2.mar-apr 1984. London: SWAPO [mnthly] – 1 – (absorbed by: swapo information bulletin) – us Wisconsin U Libr [960]

Information and decision technologies – Amsterdam. 1988-1994 (1,5,9) – (cont: large scale systems in information and decision technologies) – ISSN: 0923-0408 – mf#42302,02 – us UMI ProQuest [000]

Information and management – Amsterdam. 1978+ (1) 1978+ (5) 1987+ (9) – ISSN: 0378-7206 – mf#42263 – us UMI ProQuest [000]

Information and organization – New York, 2001+ [1,5,9] – (cont: accounting, management and information technologies) – ISSN: 1471-7727 – mf#49616,01 – us UMI ProQuest [650]

Information and records management (irm) – Hempstead. 1966-1982 (1) 1971-1982 (5) 1976-1982 (9) – (cont by: information management) – ISSN: 0019-9966 – mf#5906 – us UMI ProQuest [650]

Information and software technology – Amsterdam. 1987+ (1) 1987+ (5) 1987+ (9) – (cont: data processing) – ISSN: 0950-5849 – mf#1323,01 – us UMI ProQuest [000]

Information bulletin – United States. v5 n4,7 [1986 apr, jul], v7 n1-11 [1988 jan-dec] – 1r – 1 – (cont by: huntsville bulletin) – mf#1520639 – us WHS [071]

Information bulletin / Fundamental Baptist Fellowship – Dec 1946-72 – 1 – 61.32 – us Southern Baptist [242]

Information bulletin / Hungary. FAO National Committee – Budapest. v. 1, no. 1-4, no. 4. 1 Mar 1947-Oct 1949 – 1 – us NY Public [330]

Information bulletin / National Association of Black and White Men Together – 1980 sep 28-1985 feb 19 – 1r – 1 – mf#4863577 – us WHS [366]

Information bulletin / Soviet Antarctic Expedition – v4 irr – 1,5,6 – $30.00v – us AGU [550]

Information bulletin / Spanish Committee in Defense of Democracy – Washington, DC, 1937. Fiche W1199. (Blodgett Collection of Spanish Civil War Pamphlets) – 9 – us Harvard [946]

Information bulletin – Toronto. 1982-1989 (1) 1982-1989 (5) 1982-1989 (9) – ISSN: 0512-3291 – mf#13065 – us UMI ProQuest [320]

Information bulletin / U.S. Library of Congress – Jan 1942-Dec 1977 – 1 – 556.00 – us L of C Photodup [324]

Information bulletin. / Partido Obrero de Unificacion Marxista. Spain – Barcelona, 1936? Fiche W1098. (Blodgett Collection of Spanish Civil War Pamphlets) – 9 – us Harvard [946]

Information bulletin of the science cooperation office for southeast asia – Djakarta. nos 1-34 – 4mf – 9 – (missing: nos 6-29) – mf#SE-977 – ne IDC [959]

Information case files, 1789-1843, and related records, 1792-1918, of the u.s. district court for the eastern district of pennsylvania / U.S. District Court – (= ser Records of district courts of the united states) – 10r – 1 – (with printed guide) – mf#M992 – us Nat Archives [324]

Information circular : division of geology / Florida Geological Survey – Tallahassee, FL. n52-60. 1968-1969 – 2r – us UF Libraries [550]

Information circular : florida geological survey / Florida Geological Survey – Tallahassee, FL. n1-51. 1949-1967 – 3r – us UF Libraries [550]

Information circular / Florida Geological Survey – Tallahassee, FL. n59-97. 1969-1992 – 3r – us UF Libraries [550]

Information circular / Florida Geological Survey – Tallahassee, FL. n98-107. 1985-1991 – 1r – us UF Libraries [550]

Information concerning cuba – New York, NY. 1904 – 1r – us UF Libraries [972]

Information concerning universities, colleges – S.l., S.l? . 193-? – 1r – us UF Libraries [378]

Information control and propaganda : records of the office of war information – 2pt – (= ser Research colls in the social history of communications) – 1 – (pt1: the director's central files, 1942-45 12r isbn 0-89093-975-6 $1865. pt2: office of policy coordination: ser a: propaganda & policy directives for overseas programs, 1942-45 15r isbn 0-89093-976-4 $2345. with p/g) – us UPA [350]

Information Department, Indonesian Embassy *see* Indonesia today

Information Department Indonesian Office *see* Indonesian information

Information Division Embassy of Indonesia *see* Indonesian news and views

L'information du vietnam – Ho Chi Minh City, Vietnam. 1963-1966 (1) – mf#67831 – us UMI ProQuest [079]

Information economics and policy – Amsterdam. 1989+ (1,5,9) – ISSN: 0167-6245 – mf#42588 – us UMI ProQuest [330]

L' information economique de cochinchine *see* Nam-ky kinh-te bao

Information executive – Park Ridge. 1997+ (1) 1997+ (5) 1997+ (9) – (cont: inside dpma) – ISSN: 1092-0374 – mf#2680,04 – us UMI ProQuest [000]

Information executive – Park Ridge. 1988-1991 (1,5,9) – ISSN: 1041-9098 – mf#17032 – us UMI ProQuest [650]

L'information financiere et economique – Montreal: Compagnie de publication de l'Information. v1 n1 4 nov 1920-v2 n5 2 dec 1921 (wkly) [mf ed 2000] – 1r – 1 – (cont by: Information financiere et industrielle) – mf#SEM35P480 – cn Bibl Nat [332]

L'information financiere et industrielle – Montreal: Compagnie de publication de l'Information. v2 n6 10 dec 1921-v29 n48 17 sep 1949 (wkly) [mf ed 2000] – 2r – 5 – (cont: information financiere et economique; cont by: information financiere, industrielle, miniere) – mf#SEM35P481 – cn Bibl Nat [332]

Information for immigrants, settlers and purchasers of public lands : with map, shewing the new[l]y surveyed townships, colonization roads, etc of canada / McDougall, William – Quebec: printed by Hunter, Rose & Lemieux, 1863 [mf ed 1983] – 1mf – 9 – mf#SEM105P238 – cn Bibl Nat [350]

Information for immigrants, settlers and purchasers of public lands : with maps, showing the newly surveyed townships, colonization roads, etc of canada / McDougall, William – Quebec: printed by Hunter, Rose & Lemieux, 1862 [mf ed 1983] – 1mf – 9 – mf#SEM105P237 – cn Bibl Nat [350]

Information for immigrants, settlers, and purchasers of public lands : with map, shewing the newly surveyed townships, colonization roads etc of canada / MacDougall, William – [Quebec?: s.n.] 1863 [mf ed 1984] – 1mf – 9 – 0-665-45792-8 – mf#45792 – cn Canadiana [304]

Information for intending emigrants of all classes to upper canada : designed principally for the small farmer, agricultural labourer, etc... / Widder, Frederick – [Toronto?: s.n.], 1850 [mf ed 1983] – 1mf – 9 – mf#22194 – cn Canadiana [304]

Information for the people – Edinburgh, Scotland. 1874 – 1r – us UF Libraries [240]

Information for the people : heads of departments: mr howe's reply to mr wilkins, feb 1846 / Howe, Joseph – [s.l: s.n, 1846?] [mf ed 1984] – 1mf – 9 – 0-665-45109-1 – (in dble clms) – mf#45109 – cn Canadiana [350]

Information for the people : the solicitor general's speech, to the people of nova scotia / Howe, Joseph – [s.l: s.n, 1841?] [mf ed 1984] – 1mf – 9 – 0-665-45110-5 – mf#45110 – cn Canadiana [320]

Information for the use of military and naval officers, proposing to settle in the british colonies – [s.l: s.n, 1834?] [mf ed 1984] – 1mf – 9 – 0-665-45097-4 – mf#45097 – cn Canadiana [333]

Information for those interested in agriculture / Zemliakoff, Alexander – St Augustine, FL. 1927 – 1r – us UF Libraries [630]

Information given regarding annexation and other matters / Johnson, George – [S.l: s.n, 1889?] [mf ed 1980] – 1mf – 9 – mf#07782 – cn Canadiana [320]

Information hotline – New York. 1976-1995 (1) 1976-1995 (5) 1976-1995 (9) – (cont: information news and sources) – ISSN: 0360-5817 – mf#10621,02 – us UMI ProQuest [380]

Information infrastructure and policy – Amsterdam. 1995-1995 (1,5,9) – (cont: informatization and the public sector) – mf#21542,01 – us UMI ProQuest [000]

Information juive – Paris. 1973-1980 (1) – ISSN: 0020-0107 – mf#8368 – us UMI ProQuest [305]

L'information litteraire – Paris, 1949-62 – 1 – fr ACRPP [400]

Information management – Woodbury. 1983-1985 (1) 1983-1985 (5) 1983-1985 (9) – (cont: information and records management) – ISSN: 0739-9049 – mf#5906,01 – us UMI ProQuest [650]

Information management and computer security – Bradford. 2001+ (1,5,9) – ISSN: 0968-5227 – mf#20056 – us UMI ProQuest [000]

Information management journal – Prairie Village. 1999+ (1,5,9) – (cont: arma records management quarterly) – mf#6778,01 – us UMI ProQuest [020]

Information management review – Frederick. 1985-1989 (1) 1985-1989 (5) 1985-1989 (9) – ISSN: 8756-1557 – mf#14923 – us UMI ProQuest [650]

L'information medicale et paramedicale – Montreal: [s.n.] (bimthly) [mf ed 1972-82] – 20r – 1 – (cont by: courrier medical) – mf#SEM35P41 – cn Bibl Nat [610]

Information news and sources – New York. 1974-1975 (1) 1974-1975 (5) 1974-1975 (9) – (cont: information part 1: news, sources, profiles) – ISSN: 0360-3148 – mf#10621,01 – us UMI ProQuest [380]

Information office – WI. 1981 nov 2-1986 dec 30 – 1r – 1 – (cont: information service, wisconsin. dept of industry, labor and human relations) – mf#698228 – us WHS [350]

Information on indonesia / Permanent Mission of the Republic of Indonesia to the United Nations – New York, 1958-1959 – 12mf – 9 – (missing: 1958(2-3, 5, 17); 1958-1959(19-94, 96, 100)) – mf#SE-545 – ne IDC [959]

Information on pro-german activities of german-ame... / Hargis, Modeste – S.l., S.l? . 191? – 1r – us UF Libraries [978]

Information outlook – Washington. 1997+ (1,5,9) – ISSN: 1091-0808 – mf#25711 – us UMI ProQuest [020]

L'information ouvriere et sociale : action, syndicale, organisation du travail evolution economique – Paris. mars 1918-35 – 1 – fr ACRPP [331]

Information papers – New York: Arab Information Center, 1955-66. n2-8 nov 1956-nov 1959; n10-13 jan-nov 1960; n15 1961; n17-19 sep-nov 1961; n22 jul 1964; n25 jan 1964 – 1r – 1 – (filmed with: league of arab states document collection) – us CRL [071]

Information pipeline : norman wells project review – Canada. 1983 jan-1985 jan – 1r – 1 – mf#1046757 – us WHS [350]

Information privacy – Guildford. 1978-1981 (1,5,9) – (cont by: information age) – ISSN: 0141-3406 – mf#13331 – us UMI ProQuest [000]

Information processing and management – Oxford. 1963+ (1,5,9) – ISSN: 0306-4573 – mf#49082 – us UMI ProQuest [020]

Information processing letters – Amsterdam. 1971+ (1) 1971+ (5) 1987+ (9) – ISSN: 0020-0190 – mf#42264 – us UMI ProQuest [000]

Information pt 1 : news, sources, profiles – New York. 1969-1973 [1] – (cont by: information news and sources) – ISSN: 0036-8776 – mf#10621 – us UMI ProQuest [020]

Information published by his majesty's commissioners for emigration : respecting the british colonies in north america – London: C Knight...[1832?] [mf ed 1984] – 1mf – 9 – 0-665-45098-2 – mf#45098 – cn Canadiana [320]

Information re: jayne's hymnal by mrs. r. t. stowe of westfield, n.j., pictures of grave and home – 9p – 1 – us Southern Baptist [242]

Information relating to municipal legislation of the liquor traffic : also of municipal franchise for women / Craig, Maria G [comp] – S.l: s.n, 1899? – 1mf – 9 – mf#03621 – cn Canadiana [344]

Information relative to the assessment and collection of taxes / Connecticut. Dept. of Revenue Services – 1907-78. 130 fiches. (Harvard Law School Library Collection.) – 9 – us Harvard Law [336]

Information resources management journal – Middletown. 1992-1996 (1,5,9) – ISSN: 1040-1628 – mf#19626 – us UMI ProQuest [000]

Information retrieval and library automation – Mt. Airy. 1965+ (1) 1976+ (5) 1976+ (9) – ISSN: 0020-0220 – mf#10706 – us UMI ProQuest [020]

L'information revolutionnaire / Toure, Ahmed Sekou – [Conakry?: s.n., 1981?] – us CRL [320]

Information sciences – New York. 1968+ (1) 1968+ (5) 1987+ (9) – ISSN: 0020-0255 – mf#42265 – us UMI ProQuest [000]

Information sciences, applications – New York. 1994-1995 (1,5,9) – ISSN: 1069-0115 – mf#42746 – us UMI ProQuest [000]

Information security technical report – Kidlington. 1998+ (1,5,9) – ISSN: 1363-4127 – mf#42786 – us UMI ProQuest [000]

Information service : catholic church. pont. consilium ad christ. unitatem fovendam – 1989-93 [complete] – Inquire – 1 – mf#ATLA S0916 – us ATLA [241]

Information service / International Committee of Coordination and Information to Aid Republican Spain – Paris, 1938. Fiche W966. (Blodgett Collection of Spanish Civil War Pamphlets) – 9 – us Harvard [946]

Information service high commissioner of indonesia for information purposes : indonesia – The Hague, 1950-1951 – 18mf – 9 – (missing: 1950(1-6)) – mf#SE-572 – ne IDC [959]

Information services – New York. 1920-1969 [1,5,9] – mf#1685 – us UMI ProQuest [240]

Information services and use – Amsterdam. 1994-1996 (1,5,9) – ISSN: 0167-5265 – mf#21538 – us UMI ProQuest [020]

Information society – New York. 1981+ (1,5,9) – ISSN: 0197-2243 – mf#12423 – us UMI ProQuest [000]

Information strategy – Pennsauken. 1984+ (1,5,9) – ISSN: 0743-8613 – mf#14374 – us UMI ProQuest [650]

Information systems – Oxford. 1975+ (1,5,9) – ISSN: 0306-4379 – mf#49084 – us UMI ProQuest [000]

Information systems journal – Oxford. 1994+ (1,5,9) – (cont: journal of information systems) – ISSN: 1350-1917 – mf#18084,01 – us UMI ProQuest [000]

Information systems management – Boston. 1991+ (1,5,9) – (cont: journal of information systems management) – ISSN: 1058-0530 – mf#14372,01 – us UMI ProQuest [020]

Information technology – Guildford. 1984-1984 (1,5,9) – (cont: information technology, research and development) – mf#13332,01 – us UMI ProQuest [000]

Information technology and libraries – Chicago. 1982+ (1,5,9) – (cont: journal of library automation) – ISSN: 0730-9295 – mf#12953 – us UMI ProQuest [020]

Information technology and people – West Linn. 2001+ (1,5,9) – ISSN: 0959-3845 – mf#15792,01 – us UMI ProQuest [650]

Information technology – essential yet vulnerable : how prepared are we for attacks?: hearing...house of representatives, 107th congress, 1st session, sep 26 2001 / United States. Congress. House. Committee on Government Reform. Subcommittee on Government Efficiency, Financial Management and Intergovernmental Relations – Washington: US GPO 2002 [mf ed 2002] – 9 – (incl bibl ref) – us GPO [343]

Information technology in childhood education annual – Charlottesville. 1999+ (1,5,9) – (cont: journal of computing in childhood education) – ISSN: 1522-8185 – mf#17105,01 – us UMI ProQuest [000]

Information technology, research and development – Guildford. 1982-1983 (1) 1982-1983 (5) 1982-1983 (9) – (cont by: information technology) – ISSN: 0144-817X – mf#13332 – us UMI ProQuest [000]

Information today – Medford. 1984+ (1,5,9) – ISSN: 8755-6286 – mf#16322 – us UMI ProQuest [020]

Information world – Arlington. 1979-1980 (1,5,9) – ISSN: 0613-0067 – mf#11859 – us UMI ProQuest [020]

Informations (criminal and quo warranto) mandamus and prohibition / Shortt, John – 1st American ed. from the English ed. 1887. Boston: Edson, 1888. 771p. LL-1628 – 1 – us L of C Photodup [345]

Informations du gouvernement militaire pour l'arrondissement de carlsuhe-ville – Karlsruhe DE, 1945 28 jun – 1 – (filmed with: military government gazette germany) – gw Misc Inst [355]

Informations indonesiennes / Service d'Information de l'Ambassade d'Indonesie en France – Paris, 1950-1960 – 63mf – 9 – (missing: 1950-1954, v1-5(1-14); 1954, v5(16); 1955, v6(1, 6, 10, 15-16); 1956, v7(16)) – mf#SE-554 – ne IDC [959]

Informations mensuelles – Kinshasa. nov 1969-jan 1970 – us CRL [079]

Informations ouvrieres / La Federation des Comites d'Alliance Ouvriere – Paris, 1968-72. Pierre Lambert, Dir – 1 – (suppl mens fevr 1964-juin; juil 1968) – fr ACRPP [331]

Les informations politiques et sociales – Paris. jan 25, 1962 – (filmed with: bartlett, robert e: collection of african newspapers) – us CRL [074]

Informations- und pressestab des bundesministeriums der verteidigung (bestand bw 1) bd 94 / ed by Loos, Werner – 2003 – 239p – €11.00 – ISBN-13: 978-3-89192-126-5 – gw Bundesarchiv [350]

Informationsblaetter – Berlin DE, 1933-38 – 1r – 1 – gw Misc Inst [074]

Informationsblaetter : im auftrage des zentral-ausschusses der deutschen juden fuer hilfe und aufbau / ed by Kreutzberger, Max et al – Berlin: Reichsvertretung der Juden in Deutschland. v1-6. 1933-38 [complete] – (= ser German-jewish periodicals...1768-1945, pt 2) – 1r – 1 – $125.00 – mf#B106 – us UPA [939]

Informationsblatt der auslandsvertretung der deutschen gewerkschaften : beilage des bulletins des internationalen gewerkschaftsbundes – Paris (F), 1937 16 nov-1939 29 aug – 1r – 1 – gw Misc Inst [331]

Informationsblatt der freien gewerkschaften – Hamburg, Schleswig-Holstein DE, 1946 19 jun-1949 10 dec – 1r – 1 – mf#3866 – gw Mikropress [331]

Informationsblatt des deutschen antifaschistischen komitees – Montevideo (ROU), 1944 1 may-1946 jul [gaps] – 1 – gw Misc Inst [320]

Informationsbrief : arbeitsgemeinschaft deutscher, oesterreicherscher und tschechoslowakischer sozialisten – Oslo (N), 1939 22 sep-1940 28 mar – 1r – 1 – gw Misc Inst [331]

Informationsbrief – Rostock, Berlin DE, 1925-38 – 1 – (publ in berlin fr 1927) – gw Misc Inst [074]

Informationsbulletin der gesellschaft fuer kulturelle verbindung der sowjetunion mit dem auslande : aus dem sowjetlande – M, 1931. v1-5 – 1mf – 8 – mf#R-8116 – ne IDC [700]

Informations-dienst – Zuerich (CH), 1945, 1946 [gaps], 1947-50 – 1 – gw Misc Inst [074]

Informationsdienst der cdu *see* Cdu-informationsdienst

Informationsdienst der cdu deutschlands *see* Cdu-informationsdienst

Informationweek – Manhasset. 1991+ (1,5,9) – ISSN: 8750-6874 – mf#19190 – us UMI ProQuest [000]

Informatization and the public sector – Amsterdam. 1994-1994 (1,5,9) – (cont by: information infrastructure and policy) – ISSN: 0925-5052 – mf#21542 – us UMI ProQuest [000]

Informe / Ecuador. Ministerio de Gobierno – 1935 36-1943 – 1 – us L of C Photodup [324]

Informe : publicaciondel comite de chicago, comite nacional pro-libertad prisoneros de guerra puertorriquenos / National Committee to Free Puerto Rican Prisoners of War – 1980 may, oct-nov, 1981 feb,apr-jun – 1r – 1 – (cont by: libertad [chicago, il]) – mf#1497906 – us WHS [320]

Informe : y parecer acerca de las razones qve ay en derecho para que los terceros de algunas de las sagradas religiones lo puedan ser juntamente de ortras qualesquier / Sanchez, Francisco – en la Puebla: impr Diego Fernandez de Leon, ano 1691 – (= ser Books on religion...1543/44-c1800: historia ecclesiastical) – 1mf – 9 – mf#crl-428 – ne IDC [241]

Informe *see* Sobre inventario...del museo de badajoz

Informe [...] / Bolivia. Ministerio de Guerra – La Paz: Impr de "El Comercio", 1886-87, 1889 – us CRL [079]

Informe [...] / Bolivia. Ministerio de Guerra – La Paz: Impr de la Union Americana, [-1881] – us CRL [079]

Informe [...] / Bolivia. Ministerio de Hacienda e Industria – La Paz: Impr de "El Nacional" de Isaac V Vila, 1891-96 – us CRL [079]

Informe [...] / Bolivia. Ministerio de Hacienda e Industria – Sucre: Impr Boliviana, 1886-88 – us CRL [079]

Informe [...] / Bolivia. Ministerio de Relaciones Exteriores y Culto – La Paz: Imp y Tip de "El Nacional", 1889-91 – us CRL [079]

Informe [...] / Bolivia. Ministerio de Relaciones Exteriores y Colonizacion – La Paz: Impr de "El Diario", 1885 – us CRL [079]

Informe.. : memoria del secretario de estado en el despacho de relaciones exteriores e instruccion publica presentado al congreso nacional de... / Costa Rica. Secretaria de Relaciones Exteriores e Instruccion Publica – San Jose, [Costa Rica]: Impr del Album [1860-1866] (annual) – 1r – 1 – us CRL [370]

Informe... / Colombia. Ministerio de Relaciones Exteriores – Bogota: Arboleda & Valencia [1915-1934] (annual) – 3r – 1 – us CRL [972]

Informe... / Colombia. Ministerio de Relaciones Exteriores – Bogota: Impr de J A Cualla [1844, 1846-47, 1849-50, 1852-55] (annual) – 2r – 1 – us CRL [972]

Informe... / Colombia. Ministerio de Relaciones Exteriores – Bogota: J J Perez [1888, 1890, 1892, 1894, 1896, 1898, 1904, 1910-1911] (annual) – 3r – 1 – us CRL [972]

Informe... / Costa Rica. Direccion General de Obras Publicas – San Jose, [Costa Rica]: Impr Nacional [1873-76, 1878-80] (annual) – 1r – 1 – us CRL [350]

Informe... / Costa Rica. Ministerio de Hacienda, Guerra, Marina i Educacion Publica – [San Jose, Costa Rica: El Ministerio [1848] (annual) – 1r – 1 – us CRL [336]

Informe... / Costa Rica. Ministerio de Hacienda y Guerra – San Jose, [Costa Rica]: Impr de la Paz [1856] (annual) – 1r – 1 – us CRL [336]

Informe... / Costa Rica. Secretaria de Hacienda y Comercio – San Jose, [Costa Rica]: Impr Nacional [1872, 1874-76, 1878, 1880] (annual) – 2r – 1 – us CRL [336]

Informe... / Costa Rica. Secretaria de Relaciones Esteriores, Instruccion Publica, Culto y Beneficencia – San Jose, [Costa Rica]: Impr Nacional [1875-1876] (annual) – 1r – 1 – us CRL [972]

Informe... / Ecuador. Ministerio de Gobiern – Quito: Talleres Tipograficos Nacionales [1935/1936, 1938, 1940-43, 1950, 1952, 1960-62, 1966] (annual) – 2r – 1 – us CRL [972]

Informe... / Ecuador. Ministerio de Gobierno y Prevision Social – Quito: [Ministerio de Gobierno y Prevision Social [1934/1935] (annual) – 1r – 1 – us CRL [360]

Informe... / Ecuador. Ministerio de lo Interior y Relaciones Exteriores – Quito: Impr del Gobierno [1886, 1892, 1894] (annual) – 1r – 1 – us CRL [972]

Informe.... / Costa Rica. Ministerio de Guerra y Marina – San Jose: Impr Nacional [1873-75, 1877-80] (annual) – 1r – 1 – us CRL [972]

Informe a la nacion / Ecuador. Ministerio de Hacienda – Quito, Ecuador: [Ministerio de Hacienda y Credito Publico, (Talleres Tipograficos Nacionales) [1934] (annual) – 1r – 1 – us CRL [336]

Informe a la nacion / Ecuador. Ministerio de Hacienda – Quito, Ecuador: Talleres Tipograficos Nacionales, [1937] (annual) – 1r – 1 – us CRL [336]

Informe a la nacion / Ecuador. Ministerio de Hacienda – Quito: El Ministerio, (Imprenta y Encuadernacion Nacionales) [1915-16] (annual) – 1r – 1 – us CRL [336]

Informe a la nacion / Ecuador. Ministerio del Tesoro – Quito, Ecuador: Talleres Graficos del Ministerio del Tesoro [1948-1962] (annual) – 2r – 1 – us CRL [336]

Informe a la nacion... : memoria anual del senor ministro de relaciones exteriores 1951- / Ecuador. Ministerio de Relaciones Exteriores – Quito: Impr del Ministerio de Gobierno, [1938/1939-1932/1943, 1944/1946, 1948-1959/1960] (annual) – 4r – 1 – us CRL [972]

Informe acerca de las medidas tomadas para dar cum... – Ginebra, Switzerland. 1939 – 1r – us UF Libraries [972]

Informe ajustado al nuncio por parte del p. general (caceres) en el pleyto con los padres diputados / Davila, Andres – s.l., s.i., s.a., 1642 – 1 – sp Bibl Santa Ana [946]

Informe anual de inmigracion de la republica argentina en el ano 1875 – Buenos Aires, 1875 – 9mf – 9 – sp Cultura [972]

Informe anual del directorio ejecutivo / International Monetary Fund – Washington. 1972-1980 [1,9]; 1964-1980 [5] – ISSN: 0250-751X – mf#6529 – us UMI ProQuest [332]

Informe anual del ministro de hacienda y credito publico / Ecuador. Ministerio de Hacienda – Quito: [Ministerio de Hacienda y Credito Publico, (Talleres Tipograficos del Ministerio de Hacienda) [1923 (anexos only)] (annual) – 1r – 1 – us CRL [336]

Informe de hacienda en... / Costa Rica. Ministerio de Hacienda, Guerra y Marina – San Jose, [Costa Rica]: Impr de la Republica [1853] (annual) – 1r – 1 – us CRL [336]

Informe de hacienda y guerra al congreso de costa-rica en... / Costa Rica. Ministerio de Hacienda y Guerra – San Jose, [Costa Rica]: Impr de la Paz [1854] (annual) – 1r – 1 – us CRL [336]

Informe de la comision de hacienda diputacion...badajoz – 1844 – 9 – sp Bibl Santa Ana [946]

Informe de la comision interamericana de mujeres a... / Inter-American Commission Of Women – Washington, DC. 1948 – 1r – us UF Libraries [972]

Informe de la comision mixta / Comision Mixta De Limites Entre Guatemala Y El Sal... – Guatemala, . 1942 – 1r – us UF Libraries [972]

Informe de la comision quinta constitucional perma / Colombia Congreso Senado Comision Quinta Consti... – Bogota, Colombia. 1946 – 1r – us UF Libraries [972]

Informe de la nacinalidad / Munoz, Laurentino – Bogota, Colombia. 1965 – 1r – us UF Libraries [972]

Informe de la obra titulada "estudio biografico de esproncedа" por jose cascales y munoz / Novo y Colson, Pedro de – Madrid: Fortanet, 1912. B.R.A.H. 60, pp. 426-428 – sp Bibl Santa Ana [946]

Informe de la primera reunion de profesores universitarios centroamerica / Reunion De Profesores Universitarios Centroamerica – San Jose, Costa Rica. 1964 – 1r – us UF Libraries [972]

Informe de labores de la secretaria de recursos hidraulicos / Mexico Secretaria de Recursos Hidraulicos – Mexico: Talleres Graficos de la Nacion, 1946/47-1959/60 – 11r – 1 – us CRL [972]

Informe de los peritos...examinar la contabilidad...ferrocarril merida-sevilla – 1897 – 9 – sp Bibl Santa Ana [314]

Informe de los resultados de la encuesta sobre la situacion del comercio, industria y servicios / Camara Oficial de Comercio e Industria de Caceres – Caceres: Imp. Maygom, 1976 – 1 – sp Bibl Santa Ana [338]

Informe del c presidente de los estados unidos mexicanos al h congreso de la union. : parte correspondiente a la secretaria de hacienda y credito publico, por el periodo de [...] – Mexico, D F: la Secretaria, 1927/28-1928/29 – us CRL [972]

Informe del departamento nacional de agricultura / Argentina Departamento Nacional de Agricultura – Buenos Aires: Impr de "La Nacion", 1872-74 – 1r – 1 – us CRL [972]

Informe del ministro de estado en el despacho de relaciones exteriores e instruccion publica de costa-rica al congreso constitucional de... / Costa Rica. Ministerio de Relaciones Exteriores e Instruccion Publica – San Jose, [Costa Rica]: Impr Nacional [1858] (annual) – 1r – 1 – us CRL [370]

Informe del ministro de hacienda... / Ecuador. Ministerio de Hacienda – Quito: [Ministerio de Hacienda (Impr del Gobierno) [1883-1912] (annual) – 4r – 1 – us CRL [336]

Informe del ministro de hacienda a la nacion / Ecuador. Ministerio de Hacienda – Quito, Ecuador: Impr del Ministerio de Hacienda [1943] (annual) – 1r – 1 – us CRL [336]

Informe del ministro de hacienda al congreso de... / Costa Rica. Ministerio de Hacienda – San Jose, [Costa Rica]: Impr de la Republica [1852] (annual) – 1r – 1 – us CRL [336]

Informe del ministro de hacienda de bolivia al congreso ordinario de [...] / Bolivia. Ministerio de Hacienda – La Paz: Impr de "El Nacional" de Isaac V Villa, 1885 – us CRL [336]

Informe del ministro de hacienda e industria a la asamblea ordinaria de [...] / Bolivia. Ministerio de Hacienda e Industria – Sucre: Impr de La Libertad, 1874 – us CRL [972]

Informe del ministro de hacienda y credito publico / Ecuador. Ministerio de Hacienda – Quito: [Ministerio de Hacienda, 1923?] (Talleres Tipograficos Nacionales) [1922] (annual) – 1r – 1 – us CRL [336]

Informe del ministro de hacienda...a la h asamblea nacional refutando el presentado por... / Ecuador. Ministerio de Hacienda – Quito: Talleres Tipograficos Nacionales [1929] (annual) – 1r – 1 – us CRL [336]

Informe del ministro de relaciones exteriores al congreso ordinario de... / Ecuador. Ministerio de Relaciones Exteriores – Quito: Impr de la Escuela de Artes y Oficios [1901-09, 1911-36] (annual) – 6r – 1 – us CRL [972]

Informe del presidente del credito publico nacional pedro agote sobre le deudo publica / Argentine Republic. Junta de Administracion del Credito Publico Nacional – Libro 1-4. 1881-87 – 1 – $46.00 – us L of C Photodup [336]

Informe del presidente honario / Dominican Republic Settlement Association, Inc – Ciudad Trujillo, Dominican Republic. 1949 – 1r – us UF Libraries [972]

Informe del secretario de estado... / Costa Rica. Ministerio de Hacienda, Guerra, Marina y Caminos – San Jose, [Costa Rica]: Impr del Album [1860-1865] (annual) – 1r – 1 – us CRL [336]

Informe del secretario de guerra de la nueva granada al congreso constitucional de... / Colombia. Ministerio de Guerra – Bogota: Impr de Jose A Cualla [1846] (annual) – 1r – 1 – us CRL [972]

Informe del secretario de relaciones esteriores de la confederacion granadina al congreso nacional de... / Colombia. Ministerio de Relaciones Exteriores – Bogota: Impr de la Nacion [1859] (annual) – 1r – 1 – us CRL [972]

Informe del secretario del interior...encargado accidentalmente de los despachos de guerra, marina y obras publicas, presenta al congreso constitucional de costa-rica en el ano de... / Costa Rica. Ministerio de lo Interior – San Jose, [Costa Rica]: Impr Nacional [1872] (annual) – 1r – 1 – us CRL [350]

Informe del secretario general, dr guillermo... / Reunion Interamericana Del Caribe – Habana, Cuba. 1940 – 1r – us UF Libraries [972]

Informe del seminario sobre organizacin y administ... – San Jose, Costa Rica. 1960 – 1r – us UF Libraries [972]

Informe del senor ministro de hacienda y credito publico al h congreso nacional / Ecuador. Ministerio de Hacienda – Quito, Ecuador: Impr del Ministerio de Hacienda [1941-1942] (annual) – 1r – 1 – us CRL [336]

Informe del subsecretario de hacienda a la convencion nacional de... / Ecuador. Ministerio de Hacienda. – Quito: Fundicion de tipos de Rivadeneira [1878] (annual) – 1r – 1 – us CRL [336]

Informe del superintendente de escuelas de cuba : con inclusion de los rendidos por los superintendentes de instruccion e inspectores pedagogicos / Cuba. Superindendencia de Escuelas – Habana: Impr de Rambla y Bouza, [1903-] – us CRL [972]

Informe detallado de la comision... / Comision Tecnica De Demarcacion De La Frontera Ent... – Washington, DC. 1937 – 1r – us UF Libraries [972]

Informe dirigido al congreso legislativo de... / Costa Rica. Ministerio de Hacienda, Guerra y Caminos – San Jose, [Costa Rica]: Impr de la Paz [1855] (annual) – 1r – 1 – us CRL [336]

Informe dirigido al honorable senor presidente de la republica [...] por el secretario de hacienda [...] sobre los trabajos realizados por el departamento desde el [...] / Cuba. Secretaria de Hacienda – Habana: Montalvo y Cardenas, 1927/28 – us CRL [972]

Informe elevado al consejo de la... / American Committee On Dependent Territories – Habana, Cuba. 1949 – 1r – us UF Libraries [972]

Informe en derecho en favor de las religiones que en esta nueva espana exercen ministerio de curas : fecha en especial para el pleyto que siguen el guarduian del conuento de san francisco del pueblo de tasuba... / Lopez de Solis, F de – [Mexico City?: s.n. 1640?] – (= ser Books on religion...1543/44-c1800: manuales de rito) – 1mf – 9 – mf#crl-71 – ne IDC [241]

Informe estadistico ganadero / Rubio Garcia, Jose – Badajoz: Junta Provincial de Fomento Pecuario, 1952 – 1 – sp Bibl Santa Ana [304]

Informe final, seminarios regionales de asuntos so... / Pan American Union Division Of Labor And Social A... – Washington, DC. 1951 – 1r – us UF Libraries [972]

Informe legal por d. francisco de ulloa y flores...con d. francisco javier escobar y torres...y dona maria luisa flores chaves... sobre la propiedad de los mayorazgos que fundaron d. juan de la hinojosa y dona teresa calderon... / Merino Ortiz, Tomas – 1 – sp Bibl Santa Ana [340]

Informe oficial / Organization Of American States Mision 105 De Asi... – Washington, DC. v1-3. 1963? – 1r – us UF Libraries [972]

Informe politico da comissao execcutiva ao comite / Partido Comunista Do Brasil Comissao Executiva – Rio de Janeiro, Brazil. 1949 – 1r – us UF Libraries [972]

Informe politico del gobierno al consejo nacional del movimiento sobre politica de vivienda, urbanismo y arquitectura / Martinez Sanchez-Arjona, Jose Maria – Madrid: Servicio Central de Publicaciones del Ministerio de la Vivienda, 1969 – sp Bibl Santa Ana [320]

Informe presentado a la junta de agricultura, industria y comercio de esta provincia acerca de las bases para la formacion de un proyecto de ensenanza agricola / Paredes Guillen, Ramon – Caceres: Tip. Bello Hermanos, Arnedo y Fernandez, s.a. 1871? – 1 – sp Bibl Santa Ana [630]

Informe presentado al congreso constitucional de la republica de costa-rica por el... / Costa Rica. Secretaria de Relaciones Esteriores, Instruccion Publica, Culto y Beneficencia – San Jose, [Costa Rica]: Impr Nacional [1872-1873] (annual) – 1r – 1 – us CRL [972]

Informe presentado al congreso ordinario de [...] / Bolivia. Ministerio de la Guerra – La Paz: Impr de La Revolucion, 1893 – us CRL [972]

Informe presentado al excelentisimo senor presidente de la republica de costa-rica por el... / Costa Rica. Secretaria de Relaciones Exteriores, Justicia, Instruccion Publica, Culto y Beneficencia – San Jose, [Costa Rica]: Impr Nacional [1877-1879] (annual) – 1r – 1 – us CRL [972]

Informe presentado por el agente financiero / Nicaragua Agente Financiero – Managua, Nicaragua. 1925 – 1r – us UF Libraries [332]

Informe presentado por el secretario de estado en el despacho de hacienda al congreso nacional de costa-rica en... / Costa Rica. Ministerio de Hacienda – [San Jose, Costa Rica: Impr Nacional [1867] (annual) – 1r – 1 – us CRL [336]

Informe presentado por el secretario de estado en el despacho de instruccion publica al congreso nacional de costa-rica en... / Costa Rica. Ministerio de Instruccion Publica – San Jose: Impr Nacional [1867] (annual) – 1r – 1 – us CRL [370]

Informe presentado por el secretario de estado en el despacho de relaciones exteriores, al congreso nacional de costa-rica en... / Costa Rica. Secretaria de Relaciones Exteriores – San Jose, [Costa Rica]: Impr Nacional [1867] (annual) – 1r – 1 – us CRL [972]

Informe presentado por el...al congreso constitucional de... / Costa Rica. Secretaria de Relaciones Exteriores, Justicia y Gracia, Culto y Beneficencia – San Jose de Costa Rica: Impr Nacional [1885] (annual) – 1r – 1 – us CRL [972]

Informe pronunciado por don...actuando como defensor en la vista seguida contra urbano calvo sanchez / Romero Carvajal, Jose Ignacio – Caceres: Tip. El Noticiero, 1949 – 1 – sp Bibl Santa Ana [946]

Informe pronunciado por...acusador privado... contra vicente sanchez perez...por asesinato, 1948 / Perez Cordoba, Luis – Caceres: Tip. El Noticiero, s.a. – 1 – sp Bibl Santa Ana [946]

Informe que el director del observatorio meteorologico central ing manuel e pastrana rinde al secretario de fomento sobre las observaciones ejecutadas durante el eclipse total de sol de 28 de mayo de 1900 / Obervatorio Meteorologico central de Mexico – Mexico: Oficina Tip de la Secretaria de Fomento 1901 [mf ed 2006] – 1r – 1 – (incl bibl ref) – mf#film mas 37490 – us Harvard [520]

Informe que el ministro de hacienda, credito publico, bancos, minas, comercio y marcas de fabrica presenta a la nacion / Ecuador. Ministerio de Hacienda – Quito: Talleres Tipograficos Nacionales [1930] (annual) – 1r – 1 – us CRL [336]

Informe que el oficial mayor encargado del ministerio de hacienda presenta a la asamblea nacional ordinaria de [...] / Bolivia. Ministerio de Hacienda – [Sucre?]: Impr del Estado a direccion de M Martinez, 1863 – us CRL [972]

Informe que haze el arzobispo de mexico al exc[ellentissi]mo senor marques de manzera virrey destos reynos : sobre la licencia que pretende para passar a aquel arzobispado... / Vega, F de la – Lima: [s.n.] 24 de Enero de 1640 anos – (= ser Books on religion... 1543/44-c1800: arzobispos: arzobispos de mexico) – 1mf – 9 – mf#crl-374 – ne IDC [241]

Informe que presenta a la legislatura ordinaria de [...] / Bolivia. Ministerio de Hacienda y Estadistica – La Paz: El Ministerio, 1940 – us CRL [972]

Informe que...ministro de hacienda, credito publico, etc, presenta a la nacion en... / Ecuador. Ministerio de Hacienda – Quito, Ecuador: [Ministerio de Hacienda, (Impr y Encuadernacion Nacionales) [1913-1914] (annual) – 2r – 1 – us CRL [336]

Informe relativo a parte de la via romana num. 25 del itinerario de antonino / Blazquez, Antonio – Madrid: Fortanet, 1912. B.R.A.H. LX, pp. 306-317 – sp Bibl Santa Ana [946]

Informe sobre : historia de la pirateria malayo-mahometana en mindanao, jolo y borneo, por d. jose montero y vidal / Barrantes Moreno, Vicente – Madrid: Fortanet, 1892. B.R.A.H. 20, pp. 155-159 – sp Bibl Santa Ana [946]

Informe sobre aranceles antillanos / Fomento Del Trabajo Nacional (Spain) – Barcelona, Spain. 1895 – 1r – us UF Libraries [972]

Informe sobre declaracion de monumento nacional de puente de alcantara / Blazquez, Antonio – Madrid: Tip. Revista de Arch. Bibliot. y Museos, 1924 – 1 – sp Bibl Santa Ana [946]

Informe sobre el censo de cuba / United States War Dept Cuban Census Office – Washington, DC. 1900 – 1r – us UF Libraries [972]

Informe sobre el censo de cuba, 1899 / U.S. War Dept. Cuban Census Office – Washington: Impr. del Gobierno, 1900. 793p. tables, maps. diagrs – 1 – us Wisconsin U Libr [318]

Informe sobre el censo de cuba de 1889 – Washington, 1900 – 17mf – 9 – sp Cultura [972]

Informe sobre el convento de san benito de alcantara y fallecimiento / Lamperez Romea, Vicente – Madrid: Ed. Reus, 1923. B.R.A.H. 82. pp. 191 y 193-194 – 1 – sp Bibl Santa Ana [946]

Informe sobre el estado actual de los trabajos de... / Costa Rica Ministerio De Educacion Publica – San Jose, Costa Rica. 1958 – 1r – us UF Libraries [972]

Informe sobre el lugar de nacimiento de hernando de soto / Munoz de San Pedro, Miguel – Badajoz: Imprenta de Diputacion Prov., 1963. Sep. Revista de Estudios Extremenos – 1 – sp Bibl Santa Ana [910]

Informe sobre el ramo de hacienda y credito publico : del cual se extracto el que rindio el c presidente de la republica al h congreso de la union el dia [...] / Mexico. Secretaria de Hacienda y Credito Publico – Mexico, D F: la Secretaria, 1934 – us CRL [972]

Informe sobre la catalogacion de la coleccion numismatica del museo de caceres / Floriano Cumbreno, Antonio C – Caceres: Imp. Santos Floriano Gonzalez, s.a. – 1 – sp Bibl Santa Ana [060]

Informe sobre la cuestion de la mosquitia... / Madriz, Jose – Managua, Nicaragua. v1-2. 1894-1895 – 1r – us UF Libraries [972]

Informe sobre la declaracion de monumento nacional de la iglesia. parroquial de santa eulalia de alamia / Monsalud, Marques de – 1 – sp Bibl Santa Ana [946]

Informe sobre la guerra civil en el pais vasco, diciembre, 1937 – Buenos Aires, 1938. Fiche W959. (Blodgett Collection of Spanish Civil War Pamphlets) – 9 – us Harvard [946]

Informe sobre la institucion de la cruz-insignia de la real maestranza de caballeria de zaragoza / Monsalud, Marques de – Madrid: Est. Tip. Fortanet, 1908. B.R.A.H.T.53, 1908, pp. 338-341 – 1 – sp Bibl Santa Ana [946]

Informe sobre la investigacion antropologico-demog... / Aloja, Ada D' – Mexico City?, Mexico. 1939 – 1r – us UF Libraries [972]

Informe sobre la reorganizacion / Puerto Rico Commission For Reorganization Of The... – San Juan, Puerto Rico. 1949 – 1r – us UF Libraries [972]

Informe (sobre) lettres intimes de j.m. alberoni... / Barrantes Moreno, Vicente – Madrid: Tip. Fortanet, 1899 – sp Bibl Santa Ana [946]

Informe sobre rebaja de las derechos que pagan... / Poey, Juan – Habana, Cuba. 1862 – 1r – us UF Libraries [972]

Informe sobreel nacimiento de francisco pizarro / Munoz de San Pedro, Miguel – Badajoz: Imp. Diputacion Prov., 1970. Separata de la Rev. de Estudios Extremenos – 1 – sp Bibl Santa Ana [910]

Un informe, una opinion y una orientacion : discurso pronunciado ante el microfono de union radio instalado en el teatro olympia, de valencia, el 31 enero 1937 / Martinez Barrio, Diego – [Valencia: s.n. 1937] [mf ed 1977] – (= ser Blodgett coll) – 1mf – 9 – mf#w1031 – us Harvard [380]

Informed Voter's League *see* National sentinel

Informe...ferrocarriles / Lorenzana y Molina, Manuel – 1869 – 9 – sp Bibl Santa Ana [380]

Informe...guadalupe...fiscal...a abadia alvaneza / Santos Calderon de la Barca, Bernardo – 1712 – 9 – sp Bibl Santa Ana [972]

Informe...presenta a la nacion / Ecuador. Ministerio de Hacienda – Quito, Ecuador: Ministerio de Hacienda [1937/1938-1940] (annual) – 1r – 1 – us CRL [336]

Informe...presenta a la nacion / Ecuador. Ministerio de Hacienda – Quito: [Ministerio de Hacienda, (Impr y Encuadernacion Nacionales) [1917-19, 1921] (annual) – 1r – 1 – us CRL [336]

Informe...presenta a la nacion / Ecuador. Ministerio de Hacienda – Quito: [Ministerio de Hacienda y Credito Publico](Talleres Tipograficos Nacionales) [jul 1931-1931/1932] (annual) – 2r – 1 – us CRL [336]

Informe...presenta a la nacion / Ecuador. Ministerio de Hacienda – Quito: Talleres Tipograficos Nacionales [1934/1935] (annual) – 1r – 1 – us CRL [336]

Informe...presenta a la nacion / Ecuador. Ministerio del Tesoro – Quito, Ecuador: El Ministerio [1946 (anexos only)] (annual) – 1r – 1 – us CRL [336]

Informe...presenta a la nacion y a sus representantes al congreso de... / Ecuador. Ministerio de Hacienda – Quito: Talleres Tipograficos del Ministerio de Hacienda [1924] (annual) – 1r – 1 – us CRL [336]

Informe...presenta el director general de estadistica referente al movimiento del ano... / Dominican Republic. Secretaria de Estado de Hacienda y Comercio – Santo Domingo: Impr "Cuna de America" [1911] (irreg) – 1r – 1 – us CRL [336]

Informe...presentado al congreso nacional de costa rica en... / Costa Rica. Secretaria de Guerra, Marina, Gobernacion, Fomento y Justicia. – [San Jose, Costa Rica]: Impr Nacional [1868] (annual) – 1r – 1 – us CRL [972]

Informe...propiedad...juan de hinojosa y.. / Ulloa y Florez, Francisco, Ma – 1797 – 9 – sp Bibl Santa Ana [946]

Informe...que al senor secretario de estado del tesoro y credito publico presenta el contralor y auditor general de la republica / Dominican Republic. Secretaria de Estado del Tesoro y Credito Publico – Ciudad Trujillo: La Secretaria [1941-1942] (annual) – 2r – 1 – us CRL [336]

Informer – 1995 may/jun-1996 nov/dec – 1r – 1 – mf#3296921 – us WHS [071]

Informer / American Federation of Government Employees – v12 n9-v16 n12 [1976 sep-1980 dec] – 1r – 1 – (cont by: local 1867 united informer) – mf#1600652 – us WHS [350]

Informer / Champaign Co. Urbana – aug 1902-oct 1920 (short roll) [mthly] – 1r – 1 – mf#B4119 – us Ohio Hist [071]

Informer / Oil Jobbers of Wisconsin – 1977 jul-1980 dec, 1981-82 – 2r – 1 – (cont: wispa informer) – mf#691842 – us WHS [331]

Informer and texas freeman – Houston TX. 1992 oct 17-dec 25, 1994 jan 1-dec 31, 1995 jan 7-dec 30, 1996 jan 6-dec 28, 1997 jan 4-dec 26, 1998 jan 2-dec 25, 1999 jan 1-2000 apr 21 – 7r – 1 – (cont: informer [houston, tx]) – mf#873101 – us WHS [071]

Informe-resumen leido en la junta general de accionistas de la sucursal del banco de espana en badajoz el...1908 – Badajoz: Tip. Lit. y Enc. de Uceda Hermanos, 1908 – sp Bibl Santa Ana [946]

Informes de gobernacion, policia y fomento correspondientes al ano de... / Costa Rica. Secretaria de Gobernacion, Policia y Fomento – San Jose, [Costa Rica]: Tip Nacional [1892/1893] (annual) – 1r – 1 – us CRL [350]

Informes de las dependencias de fomento correspondientes al ano... / Costa Rica. Secretaria de Fomento y Agricultura – San Jose, Costa Rica: Impr Nacional [1929-1930] (annual) – 2r – 1 – us CRL [350]

Informes de los consejeros legales del poder ejecutivo / Argentine Republic – Publicacion Oficial. 10v. 1890-1902 – 1 – 69.00 – us L of C Photodup [323]

Informes de relaciones exteriores, justicia y beneficencia correspondientes a los anos de... – San Jose, [Costa Rica]: Tip Nacional [1893/1894] (annual) – 1r – 1 – us CRL [972]

Informes presentados a la secretaria de fomento por el director del observatorio astronomico nacional sobre los trabajos del establecimiento desde... / Observatorio Astronomico Nacional de Tacubaya – Mexico: Officina Tip de la Secretaria de Fomento, 1o de enero de 1902 a 30 de jun de 1903 [mf ed 2006] – 1r – 1 – (began in 1902?) – mf#film mas 37491 – us Harvard [520]

Informes presentados por el secretario de estado en los despachos...al congresonacional de costa-rica en... / Costa Rica. Secretaria de Hacienda, Relaciones Exteriores, Culto e Instruccion Publica – San Jose, [Costa Rica]: Impr Nacional [1868-1869] (annual) – 1r – 1 – us CRL [336]

Informes y discursos / Colegio De Abogados De La Habana – Habana, Cuba. 1944 – 1r – us UF Libraries [972]

Les infortunes de plusieurs victimes de la tyrannie de napoleon buonaparte : ou tableau des malheurs de soixante-onze francais deportes sans jugement aux iles sechelles... / Lefranc, Jean – Paris 1816 [mf ed Hildesheim 1995-98] – 1v on 2mf – 9 – €60.00 – ISBN-10: 3-487-26402-1 – ISBN-13: 978-3-487-26402-8 – gw Olms [940]

Infosystems – Wheaton. 1959-1988 (1) 1967-1988 (5) 1975-1988 (9) – ISSN: 0364-5533 – mf#1168 – us UMI ProQuest [000]

Infotech state of the art report – Maidenhead. 1981-1981 (1) 1981-1981 (5) (9) – (cont by: state of the art report) – ISSN: 0734-8487 – mf#49555 – us UMI ProQuest [600]

Infotech update – New York. 1992 (1,5,9) – mf#19193 – us UMI ProQuest [600]

Infoworld – San Mateo. 1980+ (1,5,9) – (cont: intelligent machines journal) – ISSN: 0199-6649 – mf#12701,01 – us UMI ProQuest [000]

Infrared physics – Oxford. 1961-1993 (1,5,9) – (cont by: infrared physics and technology) – ISSN: 0020-0891 – mf#49085 – us UMI ProQuest [530]

Infrared physics and technology – Exeter. 1994-1994 (1,5,9) – (cont: infrared physics) – ISSN: 1350-4495 – mf#49085,01 – us UMI ProQuest [530]

The infringement of patents for inventions, not designs, with sole reference to the opinions of the supreme court of the united states / Hall, Thomas Bond – Cincinnati, Clarke, 1893. 275 p. LL-1508 – 1 – us L of C Photodup [346]

Infuenza = Influensa (griep) / South Africa. Department of Public Health – Pretoria: Dept of Public Health 1943 [mf ed Pretoria, RSA: State Library [199-]] – 5p on 1r with other items – 5 – mf#op 12822 r25 – us CRL [616]

Infurniture – [mthly] – 2r/yr – 1 – $200.00/yr – us Fairchild Micro [740]

Ingall, Elfric Drew *see* Summary of the mineral production of Canada

Ingalls, Daniel Henry Holmes *see* Materials for the study of navya-nyaya logic

Ingalls, Joan S *see* Cognition and athletic behavior

Ingalls, John James and Family *see* Papers

Inge, W R et al *see* Radhakrishnan

Inge, William Ralph *see*
- All saints' sermons, 1905-1907
- Christian mysticism
- The church and the age
- Faith and its psychology
- Faith and knowledge
- Personal idealism and mysticism
- The philosophy of plotinus
- The religious philosophy of plotinus and some modern philosophies of religion
- Society in rome under the caesars
- Studies of english mystics
- Truth and falsehood in religion
- Types of christian saintliness

Ingegneri, Marc Antonio *see*
- Liber sacrarum cantionum...
- Il primo libro de madrigali a quattro voci
- Il primo libro de madrigali a sei voci
- Il quarto libro de madrigali a cinque voci
- Il quinto libro de madrigali a cinque voci
- Sacrae cantiones senis vocibus decantandae liber primus
- Sacrarum cantionum cum quatuor vocibus
- Il secondo libro de madrigali a quattro voci

Ingels, Marion *see* "Baptism, in a nutshell" examined

Ingemey, Roger *see* Zweidimensionale infrarotspektroskopie an polymeren

Ingenieria britanico *see* Comercio argentino-britanico

Ingenieria de carreteras / Cuellar, Enrique – San Salvador, El Salvador. 1960 – 1r – us UF Libraries [972]

Ingeniero espanol *see* Ingeniero y ferretero espanol y sud americano

Ingeniero y ferretero espanol y sud americano – London, UK. 15 jan 1887-jul 1893 – 1 – (ingeniero espanol sept 1893-nov 1905) – uk British Libr Newspaper [072]

Ingeniero y metalista – London, UK. Jan 1932 – 1 – uk British Libr Newspaper [072]

L'ingenieur – Montreal: Association des diplomes de Polytechnique. v41 n161 (printemps 1955)-v73 n382 nov/dec 1987 (trimthly) [mf ed 1985] – 7r – 5 – (cont: la revue trimestrielle canadienne; merged with: le po, and: polytec to become: l'ingénieur (1988)) – mf#SEM16P358 – cn Bibl Nat [620]

Ingenieur – Hague. 1950-1953 (1) – ISSN: 0020-1146 – mf#645 – us UMI ProQuest [620]

Ingenieur – Montreal. 1950-1955 (1) – ISSN: 0020-1138 – mf#382 – us UMI ProQuest [620]

L'ingenieur (1988) : le journal de l'ecole polytechnique et de ses diplomes – Montreal: Association des diplomes de polytechnique. v1 n1 sep 1988- (bimthly) [mf ed 1989] – 1 – (merger of: le po and: l'ingenieur and: polytec) – mf#SEM35P332 – cn Bibl Nat [620]

L'ingenieur francais : organe mensuel corporatif, economique, social et technique – Paris.n1-24.15 avr 1926-janv fevr 1929 [mnthly] – 1 – (lacking: n7) – fr ACRPP [073]

Ingenieur industriel – London, UK. Apr 1915-Aug 1917 – 1 – uk British Libr Newspaper [072]

Ingenieur universel – Manchester, UK. 13 Sept 1878-25 Feb 1881 – 1 – uk British Libr Newspaper [072]

Ingenuas / Trigo, Felipe – Madrid, Spain. v1-2. 1917 – 1r – us UF Libraries [960]

Ingenue – New York. 1966-1972 (1) 1971-1972 (5) – ISSN: 0020-1294 – mf#2171 – us UMI ProQuest [370]

Ingersoll, Mrs. Albert Converse *see* Early laws of missouri pertaining to women

Ingersoll and moses : a reply / Curtiss, Samuel Ives – Chicago: Jansen, McClurg, 1880, c1879 – 1mf – 9 – 0-8370-9930-7 – (incl bibl ref and index) – mf#1986-3930 – us ATLA [221]

Ingersoll, Charles *see* Voyage du general lafayette aux etats-unis d'amerique, en 1824

Ingersoll, Charles Jared *see* General jackson's fine

Ingersoll chronicle – Ontario Prov., CN. aug 1854-oct 1919 [wkly] – 1 – cn Commonwealth Imaging [071]

Ingersoll, Ernest *see*
- Gold fields of the klondike and the wonders of alaska
- In richest alaska and the gold fields of the klondike

Ingersoll, LA *see* A memorial and biographical history of the coast counties of central california

Ingersoll, Robert Green *see*
- Complete lectures of col. r.g. ingersoll
- The gods and other lectures
- The philosophy of ingersoll
- A vindication of thomas paine

Ingersoll, William Halsey *see* Love and law in religion

Ingestre, Viscount *see* Meliora

Ingham, Ernest Graham *see* Sierra leone after a hundred years

Ingham, Richard *see*
- Christian baptism
- Church establishments considered

Ingigian, L *see* Description du bosphore

Inglaterra y sus pactos sobre belice / Mendoza, Jose Luis – Guatemala, . 1942 – 1r – us UF Libraries [972]

Ingle, James Addison *see* James addison ingle (yin teh-sen)

Ingle, John *see*
- Puseyites (so called) no friends of popery
- Queen's letters and state services

Ingleby, Arthur G *see* Pioneer days in darkest africa

Inglefield, E A *see* A summer search for sir john franklin

Ingler, Francis Marion *see* Quiz manual on personal property

Inglesby, John Walker *see* John w inglesby papers

Ingleses no brasil / Freyre, Gilberto – Rio de Janeiro, Brazil. 1948 – 1r – us UF Libraries [972]

[Inglewood-] daily news – CA. 1916-80 (broken series) – 203r – 1 – $12,180.00 – mf#HC02306 – us Library Micro [071]

Inglewood record and waitara age – 1918 – 1r – 1 – mf#21.6 – nz Nat Libr [079]

Inglis, Henry *see*
- Spain in 1830 [eighteen hundred and thirty]
- The tyrol

Inglis, Henry D *see* A personal narrative of a journey through norway, part of sweden

Inglis, J *see* In the new hebrides

Inglis, James *see* The bible text cyclopedia

Inglis, James William *see* The divine name in ancient china

Inglis, John *see*
- Bible illustrations from the new hebrides
- In the new hebrides
- Memory of the righteous

Ingo : the first novel of a series entitled our forefathers / Freytag, Gustav – New York: Holt & Williams, 1873 – 1r – 1 – us Wisconsin U Libr [830]

Ingold, Augustin Marie Pierre *see*
- Histoire de l'edition benedictine de saint augustin
- Le pretendu jansenisme du p. de sainte-marthe, cinqui eme superieur generale de l'oratoire

Ingpen, Arthur R *see* Master worsley's book on the history and constitution of the middle temple

Ingqumbo yeminyanya / Jordan, A C – Lovedale, South Africa. 1946 – 1r – us UF Libraries [960]

Ingraham, J H *see* Prince of the house of david

Ingraham, John Phillips Thurston *see* Why we believe the bible

Ingraham, Joseph *see* Journals, brigantine hope

Ingram, Annette *see* Argivale inligtingontsluiting en -herwinning vir die historiese navorser

Ingram, J Forsyth *see* Natalia

Ingram, James C *see* Regional payments mechanisms

Ingram, John H *see* The philosophy of handwriting, by don felix de salamanca (pseud.)... with 135 autographs

Ingram, John Henry *see* Edgar allan poe

Ingram, T Dunbar *see* England and rome

Ingram, Thomas Dunbar *see* England and rome

Ingrams, William Harold *see* Zanzibar

Ingurtha *see* Ebbe und fluth

Inhalt und auslegung des hohen liedes : vortrag gehalten in der luebeckischen schillerstiftung am 2. maerz 1892 / Leverkuehn, A – Leipzig: Akademische Buchh (W Faber) 1892 [mf ed 1985] – 1mf – 9 – 0-8370-4091-4 – mf#1985-2091 – us ATLA [221]

Der inhalt und umfang des begriffs der eigenthuemlichkeit in der philosophie schleiermacher's / Plog, Ludwig – Oldenburg: Druck von Barfuss & Isensee, 1902 – 1mf – 9 – 0-7905-9441-2 – mf#1989-2666 – us ATLA [190]

Die inheemse deliktereg van die bakwena ba mogopa van hebron in die odi 1 distrik / Merwe, Emily van der – Uni of South Africa 2000 [mf ed Johannesburg 2000] – 5mf – 9 – (Summary in English; Text in Afrikaans; incl bibl ref & ind) – mf#mfm14922 – sa Unisa [307]

Inheritance of rest period of seeds and certain other characters in the peanut... / Hull, Fred H – Gainesville, FL. 1937 – 1r – us UF Libraries [634]

Inheritance tax calculations / Wolfe, Samuel Herbert – New York, Baker, Voorhis, 1905. 300 p. LL-1505 – 1 – us L of C Photodup [343]

Inheritance taxation; a treatise on legacy succession and inheritance taxes under the laws of arkansas, california, colorado. / Ross, Peter V – San Francisco: Bancroft-Whitney, 1912. 841p. LL-1176 – 1 – us L of C Photodup [343]

Inhitat-i islam hakkinda bir tecruebe-i kalemiye / Mehmed – Istanbul: Matbaa-yi Amire, 1334 [1918] – (= ser Ottoman histories and historical sources) – 1mf – 9 – $25.00 – us MEDOC [956]

Inhlamvu zasengodlweni / Ndlovu, E M – Pietermaritzburg, South Africa. 1959? – 1r – us UF Libraries [960]

Iniciacion y desarrollo de las vias de comunicacio / Nunez, Francisco Maria – San Jose, Costa Rica. 1924 – 1r – us UF Libraries [972]

O iniciador – Pitangui, MG: Typ do Iniciador, dez 1881-set 1882 – (= ser Ps 19) – mf#P31,03,60 – bl Biblioteca [079]

Iniciadores y primeros martires / Morales Y Morales, Vidal – Habana, Cuba. v1-3. 1931 – 1r – us UF Libraries [972]

Iniguez, Dalia *see* Ofrenda al hijo sonado

Initera constantinopolitanvm et amasianvm... / Gislenius, A – Antverpiae, 1581 – 2mf – 9 – mf#H-8357 – ne IDC [956]

Initia commentariorum quaestionum et tractatuum latinorum in aristotelis libros de anima. saeculis xiii, xiv, xv editorum : bibliography / Smet, A J – Leuven: Se Wulf Mansion-Centrum, 1963 – 105 leaves – 1 – us Wisconsin U Libr [180]

Initia patrum : conlegit ac litterarum ordine disposuit / Vattaso, Marcus – Roma. v1-2. 1906-1908 – (= ser Studi e testi 16-17) – €67.00 – ne Slangenburg [240]

Initia philosophiae practicae primae acroamatice / Baumgarten, Alexander Gottlieb – Halae Magdeburgicae: Impensis Carol Herm Hemmerde 1760 – (= ser Ethics in the early modern period) – 2mf – 9 – mf#pl-11 – ne IDC [170]

Initia zwingli : beitraege zur geschichte der studien und der geistesentwicklung zwinglis... / Usteri, J M – Zuerich, 1885 – 2mf – 9 – mf#ZWI-56 – ne IDC [242]

Initiation : the perfecting of man / Besant, Annie Wood – London: Theosophical Pub Society, 1912 – (= ser Samp: indian books) – us CRL [230]

Initiation a la philosophie de la francmaconnerie / Fisch, J-C-A – Paris: L'auteur 1863 – 4mf – 9 – mf#vrl-102 – ne IDC [366]

The initiation of criminal prosecutions by indictment or information / Moley, Raymond – Ann Arbor, 1931 431p. LL-573 – 1 – us L of C Photodup [345]

Initiation pratique au buddhisme / Dhammarama, P S – [Phnom-Penh: Universite buddhique Preah Sihanouk Raj 1962] [mf ed 1989] – (= ser Culture et civilisation khmeres 2) – 1r with other items – 1 – mf#mf-10289 seam reel 002/15 [§] – us CRL [280]

Initiation rites of the bamasemola / Eiselen, Werner – Kaapstad: Nasionale Pers Beperk 1937 – (= ser [Travel descriptions from south africa, 1711-1938]) – 1mf – 9 – mf#zah-7 – ne IDC [916]

Initiative *see* Einigung

Initiative and referendum report / Free Congress Research and Education Foundation – v5 n1-v9 n10 [1984 jan-1988 dec] – 1r – 1 – (cont by: family, law & democracy report) – mf#845312 – us WHS [324]

The initiative, referendum and recall. / American Academy of Political and Social Science – Philadelphia, 1912. 352 p. LL-231 – 1 – us L of C Photodup [340]

Initiatives – Washington. 1987-2000 (1) 1987-2000 (5) 1987-2000 (9) – (cont: national association for women deans, administrators and counselors journal) – ISSN: 1042-413X – mf#2383,02 – us UMI ProQuest [378]

Injo, Bian Hien *see*
- Laijoeng
- Panah api

Injury – Kidlington. 1974+ (1,5,9) – ISSN: 0020-1383 – mf#13959 – us UMI ProQuest [617]

Injury management in professional dance companies / Smith, Tiffany J – 1997 – 2mf – 9 – $8.00 – mf#PE 3879 – us Kinesology [617]

Injury rehabilitation behavior : an investigation of stages and processes of change in the athlete-therapist relationship / Wong, Ilsa E – 1998 – 2mf – 9 – $8.00 – mf#PSY 2052 – us Kinesology [790]

Injustice within the law; a study of the case of the dorsetshire labourers / Evatt, Herbert Vere – Sydney, Law Book Co. of Australasia 1937 136 p. LL-2260 – 1 – us L of C Photodup [344]

Ink / Milwaukee Mental Health Association – 1971 apr/may, sep-1974 jul/aug, nov/dec, 1975 apr/may-1976 jan/feb, may/jul – 1r – 1 – (cont: mental health ink) – mf#1046624 – us WHS [362]

Inkanyiso yase natal – Pietermaritzburg, South Africa. 1889-96 – 3r – 1 – sa National [079]

Inkcazelo yencwadi yemfundiso yobukristu / Helmstetter, B – Umtata, South Africa. 1957 – 1r – us UF Libraries [960]

Inkilab. bittigi yerde baslar – Ankara. Cumhuriyet, Milliyet ve Demokrasi Fikrinin Koeklesmesine Calisir Tuerk Gazetesidir. Mesul Mueduerue: Ali Suereyya. n16. 22 agustos 1925 – (= ser O & t journals) – 1mf – 9 – $25.00 – us MEDOC [956]

Inkilap – Siyasi gazete. n3-6,8-9,11-13,15. 1-13 eyluel 1930 – (= ser O & t journals) – 3mf – 9 – $55.00 – us MEDOC [956]

Inkinga yomendo / Dube, B J – Pietermaritzburg, South Africa. 1961 – 1r – us UF Libraries [960]

Inkins, J *see* Baptistu zihnas lihdsekti

Inkinsela yasemgungundlovu / Nyembezi, Cyril Lincoln S – Pietermaritzburg, South Africa. 1961 – 1r – us UF Libraries [960]

Inkle and yarico : a comick opera as perfomed at the theatre royal in the hay market / Arnold, Samuel – London: Longman & Broderip [1787?] [mf ed 19–] – 2mf – 9 – (words by george coleman) – mf#fiche8 – us Sibley [790]

Inkle and yarico : an opera, in three acts... correctly given, as performed at the theatres royal / Colman, George – New York: Charles Wiley, H C Carey & I Lea; Philadelphia: M'Carty & Davis; Boston: Saml H Harper 1825 [mf ed 19–] – 1mf – 9 – (with remarks) – mf#fiche7 – us Sibley [790]

Inkoleli ya bantu – Capt Town, South Africa. nov 1940-aug 1942 – 1r – 1 – sa National [079]

Inkondlo kazulu / Vilakazi, B Wallet – Johannesburg, South Africa. 1957 – 1r – us UF Libraries [960]

Inkra : organ der bpo der sed des veb industrie- und kraftwerksrohrleitungen – Bitterfeld DE, 1964-90 – 5r – 1 – gw Misc Inst [670]

Inks : cartoon and comic art studies – Columbus. 1994-1997 (1,5,9) – ISSN: 1071-9156 – mf#20693 – us UMI ProQuest [740]

Inkululeko = Freedom – Johannesburg [n1-189] dec 1940-jun 1950 – 1 – us CRL [079]

Inkululeko (freedom), 1939-1945, johannesburg/ the passive resister, 1946-1948, johannesburg – 1r – 1 – mf#97293 – uk Microform Academic [960]

Inkundla Ya Bantu *see* The territorial magazine

Inkundla ya bantu = The bantu forum – Verulam, South Africa: Verulam Press. [n27-268] jun 1940-apr 29 1950 – 1 – us CRL [079]

Inkvizitsiyah / Rabinovitz, Alexander Siskind – Tel-Aviv, Israel. 1929 or 1930 – 1r – us UF Libraries [939]

Das inland – Muenchen DE, 1829-1831 30 jun – 6r – 1 – gw Misc Inst [074]

Inland and coastal waterways of florida / Florida Inland And Coastal Waterways Association – Washington, DC. 1929 – 1r – us UF Libraries [978]

Inland architect – Chicago. 1957-1992 [1]; 1970-1992 [5]; 1976-1992 [9] – ISSN: 0020-1472 – mf#1751 – us UMI ProQuest [720]

Inland architect and news record – Chicago. 1883-1908 [1,5,9] – mf#1745 – us UMI ProQuest [720]

Inland Boatmen's Union of the Pacific *see* Inlandboatman of the pacific

Inland empire – Hamilton NV. 1869 jun 16, sep 21 – 1r – 1 – mf#882289 – us WHS [071]

Inland empire – Moore, MT. 1905-1915 (1) – mf#64585 – us UMI ProQuest [071]

Inland empire crown – Fairchild Air Force Base, Washington [State]. v1 n1-7 [1986 apr 18-may 30] – 1r – 1 – (cont: fairchild times; cont by: fairchild crown) – mf#1153122 – us WHS [355]

Inland empire daily californian *see* [El cajon-] daily californian

Inland empire miner – Baker City, OR: Inland Empire Pub Co. v6 n34-v7 n20. may 3 1905-jan 31 1906 (mf ed 1971) – 1r – 1 – (cont: sumpter miner) – us Oregon Hist [071]

Inland empire miner – Baker City OR: Inland Empire Pub Co, 1905- [wkly] [1971] – 1r – 1 – (cont: sumpter miner (1899-1905)) – us Oregon Lib [622]

Inland empire news – Hillyard, WA. 1933-1940 (1) – mf#68651 – us UMI ProQuest [071]

Inland empire-north – 1992-94 – (= ser California telephone directory coll) – 5r – 1 – $250.00 – mf#P00039 – us Library Micro [917]

Inland empire-spanish – 1992-94 – (= ser California telephone directory coll) – 3r – 1 – $150.00 – mf#P00040 – us Library Micro [917]

Inland farmer-stockman *see* Pacific farmer-stockman

Inland herald – Spokane, WA. 1910-1911 (1) – mf#67135 – us UMI ProQuest [071]

Inland ocean – Superior WI. 1891 nov-1893 dec, 1894-96, 1897-99, 1900-1903 aug 6 – 4r – 1 – (cont: superior inter-ocean [superior, wis. : 1887]) – mf#933887 – us WHS [380]

Inland printer, american lithographer – Chicago. 1883-1978 (1) 1966-1978 (5) 1976-1978 (9) – (cont by: american printer and lithographer) – ISSN: 0020-1502 – mf#814 – us UMI ProQuest [680]

Inland register – Spokane, WA. 1942-1963 (1) – mf#69260 – us UMI ProQuest [071]

Inland Steel Co *see* Local 1010 steelworker at inland steel company

Inlandboatman : official publication of the inlandboatmen's union of the pacific [marine division of ilwu] – v9 n1 [1984 jan] – 1r – 1 – (cont: inlandboatman of the pacific) – mf#1043239 – us WHS [380]

Inlandboatman of the pacific / Inland Boatmen's Union of the Pacific – v2 n9-v8 n2 [1977 jan 31-1983 apr/jun] – 1r – 1 – (cont: pacific mariner; cont by: inlandboatmen) – mf#1043235 – us WHS [380]

Inlander news – oct 1992-mar 1993 – 1r – at Pascoe [079]

Het Inlandsch Comite tot herdenking van Neerlands honderdjarige vrijheid *see* Als ik eens nederlander was...

Inledning till psaltaren : isagogiskt-exegetisk afhandling / Nylander, K U – Upsala: E Berling, 1894 – 1mf – 9 – 0-7905-3044-9 – (incl bibl ref) – mf#1987-3044 – us ATLA [220]

Inleiding tot de muzykkunde : uit klaare, onwederspreekelyke gronden, de innerlyke geschapenheid, de oorzaaken van de zonderbaare uitwerkselen, de groote waarde, en 't regte gebruik der muzykkonst aanwyzende / Lustig, Jacob Willem – te Groningen: gedruckt voor den auteur, by H Veehnerus 1751 [mf ed 19–] – 1r – 1 – mf#film 1372 – us Sibley [780]

Inleiding tot de studie van de kongolese bantoetalen / Burssens, Amaat F S – Antwerpen, Belgium. 1954 – 1r – us UF Libraries [960]

Inleiding tot die studie van suid-sotho / Van Eeden, B I C – S.I., S.I? . 1941 – 1r – us UF Libraries [960]

Inleyding tot de hooge schoole der schilderkonst... / Hoogstraten, S van – Rotterdam, 1678 – 8mf – 9 – mf#O-301 – ne IDC [700]
Inleydinge tot de algemeene teyken-konst... / Goeree, W – Amsterdam, 1697 – 8mf – 9 – mf#O-273 – ne IDC [700]
Inman, Henry *see* Military records
The inman leader – Inman, NE: J S Jackson (wkly) – 1r – 1 – us Bell [071]
The inman news – Inman, NE: Pond & Leidy (wkly) [mf ed v2 n47. sep 4 1894-jun 25 1907 (gaps) filmed 1978] – 2r – 1 – us NE Hist [071]
Inman, Thomas *see* Ancient pagan and modern christian symbolism
Inman, W G *see* Planting and progress of the baptists' cause in tennessee
Inmed newsletter und / University of North Dakota – ed. n1-2 [1975 jun 20-jul 10] – 1r – 1 – (cont by: serpent, staff, and drum) – mf#958023 – us WHS [378]
Inmersion; el acto del bautismo cristiano / Christian, Juan T – 1907. 200p – 1 – 7.00 – us Southern Baptist [242]
Inmigracion de trabajadores espanoles / Cuba Gobierno Y Capitania General – Habana, Cuba. 1853 – 1r – us UF Libraries [972]
Inmigracion italiana y la colonizacion en cuba / Falco, Francesco Federico – Turin, Italy. 1912 – 1r – us UF Libraries [972]
Inmigracion y colonizacion en la grancolombia / Arango Cano, Jesus – Bogota, Colombia. 1953 – 1r – us UF Libraries [972]
Inmigracion y extranjeria / Estrada S, Julio – La Paz, Bolivia. 1942? – 1r – us UF Libraries [972]
Inn album / Browning, Robert – Boston, MA. 1876 – 1r – us UF Libraries [025]
The inner and middle temple : legal, literary and historic associations / Bellot, Hugh Hale L – London: Methuen & Co, 1902 – 6mf – 9 – $9.00 – mf#LLMC 84-274 – us LLMC [340]
The inner chamber and the inner life / Murray, Andrew – New York: Fleming H Revell, c1905 – 1mf – 9 – 0-8370-5998-4 – mf#1985-3998 – us ATLA [240]
Inner circle / Workmen's Circle [US] – v1-3 n3 [1974 oct-1977 jun] – 1r – 1 – mf#380995 – us WHS [331]
Inner city express – Oakland, CA. 1971-1978 (1) – mf#62198 – us UMI ProQuest [071]
Inner city news – Auckland, NZ. oct 1979-86 – 6r – 1 – mf#11.38 – nz Nat Libr [079]
Inner city voice – v1 n4-v3 n2 [[1968] jan-1971 apr] – 1r – 1 – mf#770921 – us WHS [071]
Inner history of the national convention of south africa / Walton, Edgar Harris – Westport, CT. 1970 – 1r – us UF Libraries [960]
The inner kingdom – Cambridge: John Wilson, 1870 – 1mf – 9 – 0-8370-8923-9 – mf#1986-2923 – us ATLA [240]
The inner life / Campbell, A, mrs – [Quebec?: s.n.] 1862 [mf ed 1994] – 1mf – 9 – 0-665-94649-X – mf#94649 – cn Canadiana [830]
The inner life / Jones, Rufus Matthew – New York: Macmillan, 1916 – 1mf – 9 – 0-7905-7848-4 – mf#1989-1073 – us ATLA [240]
The inner life and the tao-teh-king / Bjerregaard, Carl Henrik Andreas – New York: Theosophical Pub Co, 1912 – 1mf – 9 – 0-524-02070-1 – mf#1990-2834 – us ATLA [290]
The inner life of syria, palestine, and the holy land : from my private journal – New and cheaper ed. London: C. Kegan Paul, 1879 – 1mf – 9 – 0-7905-0562-2 – mf#1987-0562 – us ATLA [956]
The inner life of the very reverend pere lacordaire of the order of preachers = R p h-d lacordaire de l'ordre des freres praecheurs, sa vie intime et religieuse / Chocarne, pere (Bernard) – Dublin: William B Kelly; New York: Catholic Pub Society [1867?] [mf ed 1986] – 2mf – 9 – 0-8370-6809-6 – (english trans fr french by very rev father aylward. incl bibl ref) – mf#1986-0809 – us ATLA [241]
The inner mission : four addresses / Paton, John Brown – London: Wm Isbister, 1888 – 1mf – 9 – 0-7905-6818-7 – mf#1988-2818 – us ATLA [240]
The inner mission : a handbook for christian workers / Ohl, Jeremiah Franklin – Philadelphia: General Council Publication House, 1911 – 1mf – 9 – 0-7905-5777-0 – (incl bibl ref) – mf#1988-1777 – us ATLA [240]
Inner passages : a choreographic exploration of the effects of entrainment on choreography / Miller, Rebecca D – 1993 – 1mf – $4.00 – us Kinesology [790]
Inner rome : political, religious, and social / Butler, Clement Moore – Philadelphia: JB Lippincott, 1866, c1865 – 1mf – 9 – 0-7905-4111-4 – mf#1988-0111 – us ATLA [240]
Inner rome : political, religious, and social / Butler, Clement Moore – Philadelphia: J.B. Lippincott, 1866, c1865 – 1mf – us ATLA [930]
The inner sanctuary – [United States: s.n.] 5627 [i.e. 1867]-v1-5 – 8mf – 9 – (filmed: pt1 [[1870]] & pt5 [[1867]]) – mf#vrl-64 – ne IDC [366]
Inner state news – Ironwood, MI. 1890-1895 (1) – mf#63777 – us UMI ProQuest [071]
The inner teachings of the philosophies and religions of india / Ramacharaka – Chicago: Yogi Publication Society, c1909 – 1mf – 9 – 0-524-01711-5 – mf#1990-2613 – us ATLA [280]
Inner temple : masters of the bench and temple / Inner Temple. London. Library – London: Clowes & Sons, 1883 – 2mf – 9 – $3.00 – (covers: masters of the bench, 1450-1883; masters of the temple, 1540-1883) – mf#LLMC 84-298 – us LLMC [340]
Inner Temple. Library *see* Archives of the inner temple library
Inner Temple Library. London *see* Manuscripts and early printed works
Inner Temple. London. Library *see* Inner temple
Inner voice / Association of Forest Service Employees for Environmental Ethics – 1989 summer-1994 nov/dec – 1r – 1 – mf#1874542 – us WHS [366]
Innercom : a publication of nn corporation employees / NN Corporation – v10 n2, v12 n1-3, 5-v13 n12 [1977 feb, 1979 jan -apr, jul-1980 dec] – 1r – 1 – (cont by: armco insurance group news) – mf#1108877 – us WHS [331]
Die innere entwicklung des pelagianismus : beitrag zur dogmengeschichte / Klasen, Franz – Freiburg im Breisgau; St Louis, MO: Herder, 1882 [mf ed 1990] – 1mf – 9 – 0-7905-6194-8 – (in german & latin. incl bibl ref) – mf#1988-2194 – us ATLA [240]
Die innere form der romanzen vom rosenkranz von clemens brentano : erkenntnisse zum romantischen formwillen / Reichardt, Guenther – Freiburg, Schlesien: H Heiber, 1934 [mf ed 1989] – 124p – 1 – mf#7085 – us Wisconsin U Libr [430]
Die innere front – Berlin DE, 1939-40 [gaps] – 1 – gw Misc Inst [074]
Der innere gang des deutschen protestantismus / Kahnis, Karl Friedrich August – 3., erw und ueberarb Ausg. Leipzig: Doerffling und Franke, 1874 – 2mf – 9 – 0-8370-9074-1 – (incl bibl ref) – mf#1986-3074 – us ATLA [242]
Innere geschichte der entwicklung der deutschen national-litteratur : ein methodisches handbuch fuer den vortrag und zum selbststudium / Rinne, Karl-Friedrich; ed by Garber, Klaus – Leipzig 1842/43 – 2pts in 1v on 10mf – 9 – diazo €69.80 silver €84.00 – gw Olms [430]
Die innere komposition in goethe's epischer dichtung hermann und dorothea : zur ersten zentennarfeier ihrer entstehung / Neudecker, Georg – Wuerzburg: Stahel, 1896 – 1r – 1 – us Wisconsin U Libr [430]
Die innere mission in der schule : ein handbuch fuer den lehrer / Schaefer, Theodor – 3. verb. Aufl. Guetersloh: C. Bertelsmann, 1896 – 1mf – 9 – 0-8370-6408-2 – mf#1986-0408 – us ATLA [240]
Die innere motivierung in grabbes dramen / Schoettler, Wilhelm – Berlin: Junker und Duennhaupt, 1931 – 1r – 1 – (incl bibl ref) – us Wisconsin U Libr [430]
Innere rechtskraft im erbhofrecht / Roesner, Werner – Leipzig, 1938 (mf ed 1994) – 1mf – 9 – €24.00 – 3-8267-3004-6 – mf#DHS 3004 – gw Frankfurter [346]
Das innere reich – Muenchen DE, 1934 apr-1945 n1 – 5r – 1 – gw Mikrofilm [943]
Das innere reich (mme3) : zeitschrift fuer dichtung, kunst und deutsches leben – Muenchen 1934/35-1945 [mf ed 1998] – (= ser Marbacher mikrofiche-editionen (mme) 3; Kultur – literatur – politik: deutsche zeitschriften des 19./20. jahrhunderts (klp)) – 11v on 128mf – 9 – €650.00 – 3-89131-288-1 – gw Fischer [074]
Innerhalb etters / Kurz, Hermann; ed by Kurz, Isolde – Tuebingen: Rainer Wunderlich (H Leins), [1926?] – 1 – us Wisconsin U Libr [830]
Innerslavischer und slavisch-deutscher sprachvergleich / Jelitte, Herbert et al – Frankfurt am Main, New York: Lang c1995 [mf ed 1996] – 1r with other items – 1 – us Cornell [400]
Inner-view of las vegas / Eckankar of Southern Nevada – v1 n1-v2 n3 [1985 oct-1986 oct] – 1r – 1 – mf#1508364 – us WHS [071]
Innes, Alexander Taylor *see*
– Assembly of 1881 and the case of professor robertson smith
– Church and state
– Church of scotland crisis 1843 and 1874, and the duke of argyll
– Confidence of the church
– John knox
– The law of creeds in scotland
– Letters from the red beech
– Studies in scottish history
– The trial of jesus christ
Innes, Arthur Donald *see* Cranmer and the reformation in england
Innes, C A *see* Madras district gazetteers [madras manuals]
Innes, Duncan *see* Our country, our responsibility
Innes, Henry *see* Letter to the friends, in scotland, of god's ancient people, the je...
Innes, James *see* Commission of assembly
Innes review, the... 1950-58 : the journal of the scottish catholic historical committee – v1-9 – 3r – 1 – mf#4551 – uk Microform Academic [073]
Innes, W *see* Memoir of the rev levi parsons
Innes, William *see*
– Death of a christian soldier at the battle of barossa
– Liberia
Inni alla notte e canti spirituali = Hymnen an die nacht / Novalis (Friedrich von Hardenberg); ed by Hermet, Augusto – Lanciano: R Carabba, 1912 [mf ed 1993] – (= ser Cultura dell'anima) – 125p – 1 – (italian trans fr german and int by augusto hermet) – mf#8671 – us Wisconsin U Libr [810]
Innisfail free lance – Alberta. CN. 1902-08 – 1 – cn Commonwealth Imaging [071]
Innisfail monumental inscriptions – 1mf – 9 – A$24.00 – at Cairns [929]
Innisfail province – Alberta, CN. mar 1906-jan 1908; jun, aug, nov 1921; 1922-1924 – 1 – cn Commonwealth Imaging [071]
Innkeeping – Chicago. 1963-1965 (1) – mf#1586 – us UMI ProQuest [640]
Innocencia / Taunay, Alfredo D'escragnolle Taunay – Boston, MA. 1923 – 1r – us UF Libraries [972]
Innocencia / Taunay, Alfredo D'escragnolle Taunay – Sao Paulo, Brazil. 1939 – 1r – us UF Libraries [972]
Innocent 3 : la croisade des albigeois / Luchaire, Achille – Paris: Librairie Hachette, 1905 – 1mf – 9 – 0-8370-8272-2 – mf#1986-2272 – us ATLA [920]
Innocent 3 : la papaute et l'empire / Luchaire, Achille – Paris: Librairie Hachette, 1906 – 1mf – 9 – 0-8370-8273-0 – mf#1986-2273 – us ATLA [240]
Innocent 3 : la question d'orient / Luchaire, A – Paris, 1907 – 4mf – 9 – mf#H-2933 – ne IDC [956]
Innocent 3. la question d'orient / Luchaire, Achille – Paris: Librairie Hachette, 1907 – 1mf – 9 – 0-8370-8274-9 – mf#1986-2274 – us ATLA [240]
Innocent 3. les royautes vassales du saint-siege / Luchaire, Achille – Paris: Librairie Hachette, 1908 – 1mf – 9 – 0-8370-8198-X – mf#1986-2198 – us ATLA [240]
Innocent 3. rome et l'italie / Luchaire, Achille – 2e ed. Paris: Librairie Hachette, 1905 – 1mf – 9 – 0-8370-8199-8 – mf#1986-2199 – us ATLA [240]
Innocent poetry for infant minds / Elliott, Mary (Belson) et al – London: William Darton, 1823 – (= ser 19th c children's literature) – 1mf – 9 – mf#6.1.24 – uk Chadwyck [810]
Innocent the great : an essay on his life and times / Pirie-Gordon, Charles Harry Clinton – London; New York: Longmans, Green, 1907 – 1mf – 9 – 0-7905-6769-5 – (incl bibl ref) – mf#1988-2769 – us ATLA [940]
Innocent the great, an essay on his life and times / Pirie-Gordon, Charles Harry Clinton – London, New York: Longmans, Green, 1907.xxiii,273p. maps. geneal – 1 – us Wisconsin U Libr [940]
Les innocentes : ou la sagesse des femmes / Noailles, Anna Elizabeth de Brancovan – Paris: Fayard, 1923 – (= ser Les femmes [coll]) – 3mf – 9 – mf#9138 – fr Bibl Nationale [305]
Innocentia vindicata : in qua gravissimis argumentis ex s. thoma petitis ostenditur... / Sfondrati, C – Viennae Austriae: Typis Leopoldi Voigt, 1702 – 5mf – 9 – mf#O-12 – ne IDC [090]
Innocentia vindicata / Sfrondrati, C – n.p, Typis Monestarij S. Galli, 1695 – 6mf – 9 – mf#O-753 – ne IDC [090]
Innovation / UNESCO International Bureau of Education – Geneva, Switzerland. No.1- . 1975- – 1 – (cont. by: educational innovation and information) – us Wisconsin U Libr [370]
Innovation und evolutorische oekonomik : unter besonderer beruecksichtigung erkenntnistheoretischer fragestellungen / Reinhardt, Peter – (mf ed 1999) – 3mf – 9 – €49.00 – 3-8267-2668-5 – mf#DHS 2668 – gw Frankfurter [330]
Innovationen im verbaendesektor / Wiesner, Knut & Wiesner, Iris – (mf ed 2001) – 494p – 9 – €62.50 – 3-8267-2761-4 – mf#DHS 2761 – gw Frankfurter [330]
Innovations in education and training international – London. 1995-2000 – 1,5,9 – (cont: educational and training technology international: etti) – ISSN: 1355-8005 – mf#11223,02 – us UMI ProQuest [370]
Innovative accounting teaching : an action research project / Smith, Wilhelmina – Pretoria: Vista University 2000 [mf ed 2000] – 6mf [ill] – 9 – (incl bibl ref) – mf#mfm15210 – sa Unisa [650]
Innovative food science and emerging technologies – Amsterdam, 2000+ [1,5,9] – ISSN: 1466-8564 – mf#42828 – us UMI ProQuest [620]
Innovative higher education – New York. 1983+ – 1,5,9 – (cont: alternative higher education) – ISSN: 0742-5627 – mf#11172,01 – us UMI ProQuest [378]
Innovative management strategies as a tool for improving grade 12 results / Sehloho, Tsietsi Daniel – Pretoria: Vista University 2002 [mf ed 2002] – 3mf – 9 – (incl bibl ref) – mf#mfm15203 – sa Unisa [373]
Innre mission : volkserziehung und prophetenthum: drei vortraege / Zezschwitz, Gerhard von – Frankfurt a.M.: Heyder & Zimmer, 1864 – 1mf – 9 – 0-8370-6079-6 – (incl bibl ref) – mf#1986-0079 – us ATLA [240]
The inns of court / Headlam, Cecil – London: A & C Black, 1909 – 4mf – 9 – $6.00 – (painted by gordon home) – mf#LLMC 84-294 – us LLMC [347]
The inns of court : an historical description / Ringrose, Hyacinthe – Boston: Little-Brown, 1910 – 3mf – 9 – $4.50 – mf#LLMC 84-313 – us LLMC [347]
The inns of court and chancery / Loftie, William J – new ed. London: Seeley & Co, 1895 – 4mf – 9 – $6.00 – mf#LLMC 84-304 – us LLMC [347]
Inns of Court. England. Council of Legal Education *see* Council of legal education calendar, 1901-1925/26
Inns of Court Students Union. London *see* Glim
Innsbrucker nachrichten – Innsbruck, Austria. 25 feb 1942-1943; 11 jan 1944-12 apr 1945 – 4r – 1 – uk British Libr Newspaper [072]
Innumerables voces / Tejera, Nivaria – Habana, Cuba. 1964 – 1r – us UF Libraries [972]
Inocencia / Taunay, Alfredo D'escragnolle Taunay – New York, NY. 1945 – 1r – us UF Libraries [972]
Inok nikodim starodubskii : ego zhizn i literaturnaia deiatel nost / Belolikov, V Z – Kiev, 1915 – 532p 10mf – 8 – mf#R-5987 – ne IDC [243]
Inorganic and nuclear chemistry letters – Oxford. 1965-1981 (1) 1965-1981 (5) 1965-1981 (9) – ISSN: 0020-1650 – mf#49086 – us UMI ProQuest [540]
Inorganic chemistry – v1- 1962- – 1,5,6,9 – us ACS [540]
Inorganic chemistry communications – New York. 1998+ (1,5,9) – ISSN: 1387-7003 – mf#42801 – us UMI ProQuest [540]
Inorganic materials – New York. 1974-1977 (1) 1975-1977 (5) – ISSN: 0020-1685 – mf#10831 – us UMI ProQuest [540]
Inorganica chimica acta – Lausanne. 1967+ (1) 1967+ (5) 1987+ (9) – ISSN: 0020-1693 – mf#42267 – us UMI ProQuest [540]
Inorodcheskoe obozrenie – Kazan': Kazanskaia dukhovnaia akademiia 1913-17 [t1] kn 1-aia (za dek 1912 g)-t2 n4/5 (za sent/dek 1916 g) – 2v on 2r – 1 – (suppl of: pravoslavnyi sobesednik (kazan', russia: 1857)) – mf#mr-10 – ne IDC [305]
Inostrannyi kapital i russkie banki : k voprosu o finansovom kapitale v rossii / Ronin, S – M, 1926 – 3mf – 9 – mf#REF-180 – ne IDC [332]
Inoue kowashi monjo : viscount kowashi inoue records. in the holdings of kokugakuin university library, tokyo – 112,326p on 94r – 1 – Y792,000 – (with 312p guide. in japanese) – ja Yushodo [950]
Inoue, Tadashiro *see* Inoue tadashiro monjo
Inoue tadashiro monjo : tadashiro inoue records; collection of documents of the scientific and technological administration in wartime. in the holdings of kokugakuin university library, tokyo / Inoue, Tadashiro – 13,000 items on 168r – 1 – Y2,300,000 – (594p guide comp by kokugakuin university. 8p reel ind. in japanese) – ja Yushodo [025]
In...pavli ad hebraeos epistolam... commentarius / Bullinger, Heinrich – Tigvri, Christoph Frosch[auer], 1532 – 4mf – 9 – mf#PBU-115 – ne IDC [240]
In...pavli ad romanos epistolam...commentarius / Bullinger, Heinrich – Tigvri, Christoph Frosch[auer], 1533 – 4mf – 9 – mf#PBU-117 – ne IDC [240]
In-plant graphics – Philadelphia. 1996+ (1,5,9) – (cont: in-plant reproductions) – ISSN: 1043-1942 – mf#3229,06 – us UMI ProQuest [338]
In-plant printer – Northbrook. 1961-1986 (1) 1970-1986 (5) 1974-1986 (9) – (cont by: in-plant printer and electronic publisher) – ISSN: 0019-3232 – mf#1678 – us UMI ProQuest [680]
In-plant printer – Libertyville. 1993+ (1) 1993+ (5) 1993+ (9) – (cont: in-plant printer and electronic publisher) – ISSN: 1071-832X – mf#1678,02 – us UMI ProQuest [680]
In-plant printer and electronic publisher – Northbrook. 1986-1993 (1) 1986-1993 (5) 1986-1993 (9) – (cont: in-plant printer. cont by: in-plant printer) – ISSN: 0891-8996 – mf#1678,01 – us UMI ProQuest [680]

In-plant reproductions – Philadelphia. 1985-1995 (1) 1985-1995 (5) 1985-1995 (9) – (cont by: in-plant graphics) – ISSN: 1043-1942 – mf#3229,05 – us UMI ProQuest [338]

In-plant reproductions – Philadelphia. 1979-1985 (1) 1979-1985 (5) 1979-1985 (9) – (cont: reproductions review and methods) – ISSN: 0198-9065 – mf#3229,04 – us UMI ProQuest [338]

Inpress : deutsche ausgabe – Paris (F), 1936 2-18 jan [gaps] – 1 – gw Misc Inst [074]

Inqilab va azadi – Detroit, MI: Anjuman-i Azadi, 1981-83. shumarah-'i 1-9. 11 urdibihisht 1360-payiz 1362 [1 may 1981-fall 1983] – 1r – 1 – $53.00 – us MEDOC [079]

Inqilab-i islami – Tehran, 1979-80. sal-i 1, shumarah-'i 1-sal-i 2, shumarah-'i 421. 29 khurdad 1358-17 khurdad 1359 [19 jun 1979-7 jun 1980] – 1r – 1 – $350.00 – us MEDOC [079]

Inquietudes profanas / Aguilar, Manuel R – San Salvador, El Salvador. 1927 – 1r – us UF Libraries [972]

Inquietud / Liano, Manuela – Zafra: Industrias Tipograficas Extremenas, 1969 – 1 – sp Bibl Santa Ana [946]

Inquietud sosegada poetica de evaristo ribera chev... / Melendez, Concha – San Juan, Puerto Rico. 1946 – 1r – us UF Libraries [972]

L'inquietude religieuse : aubes et lendemains de conversion / Bremond, Henri – Paris: Perrin, 1909 [mf ed 1992] – 1mf – 9 – 0-524-04009-5 – (in french) – mf#1990-1181 – us ATLA [200]

Inquietudes de un ano memorable : 1944 / Fortin Magana, Romeo – San Salvador, El Salvador. 1945 – 1r – us UF Libraries [972]

Inquirer – Perth, Australia. 18, 25 dec 1844, 6 jan 1864-27 dec 1865, 1866-28 jun 1901 – 31 1/2r – 1 – (aka: inquirer and commercial news) – uk British Libr Newspaper [079]

Inquirer – Brookville, IN. 1824-1833 (1) – mf#62738 – us UMI ProQuest [071]

Inquirer – Galion, OH. 1878-1895 (1) – mf#65503 – us UMI ProQuest [071]

Inquirer – Galion, OH. 1987-2001 (1) – mf#61710 – us UMI ProQuest [071]

Inquirer – Lancaster, PA. 1799-1920 (1) – mf#61181 – us UMI ProQuest [071]

Inquirer – Owensboro, KY. 1890-1954 (1) – mf#63480 – us UMI ProQuest [071]

Inquirer / Scioto Co. Portsmouth – apr 1850-mar 53, (dec 54-jul 1855) [wkly] – 1r – 1 – mf#B29925 – us Ohio Hist [071]

The inquirer – 1842-1999+ – 74r – 1 – £3300.00 – (journal of the unitarian movement) – mf#INQ – uk World [240]

The inquirer – v1-32 n34. 17 oct 1846-27 sep 1877 – 16r – 1 – (lacking: v3 n19,29,30. title varies: christian inquirer. liberal christian) – mf#ATLA S0154 – us ATLA [240]

The inquirer – Monrovia: New Era Publications, jan 15 1991-dec 30 1994 – 5r – 1 – us CRL [071]

The inquirer – Perth, Australia. Jan 1864-Jun 1901.-w. 31 reels – 1 – uk British Libr Newspaper [079]

The inquirer – 1842- – 1 – enquire for prices – (yrly reel count varies) – us UMI ProQuest [073]

Inquirer and commercial news *see* Inquirer

Inquirer and the bible – London, England. 18– – 1r – us UF Libraries [240]

The inquirer's guide : or, mists removed from duty / Storrs, W – 12th ed. Rock Island, Ill: Haverstick, 1871 – 1mf – 9 – 0-524-02574-6 – mf#1990-4386 – us ATLA [240]

Inquiries elementary and historical in the science of law / Reddie, James – London: Longman, Orme, Brown, Green & Longmans, 1840 – 3mf – 9 – $4.50 – mf#LLMC 95-182 – us LLMC [340]

Inquiries of an emigrant : being the narrative of an english farmer, from the year 1824 to 1830 / Pickering, Joseph – new ed. London: E Wilson, 1831 [mf ed 1983] – 2mf – 9 – 0-665-39836-0 – mf#39836 – cn Canadiana [917]

Inquiries of an emigrant pickering's guide to emigrants : being the narrative of an english farmer from the year 1824 to 1830; during which period he traversed the united states of america, and the british province of canada... / Pickering, Joseph – new ed. London: E Wilson, 1831 [mf ed 1983]] – 2mf – 9 – 0-665-39836-0 – mf#39836 – cn Canadiana [917]

The inquiries of ramchandra : or, dialogues with a hindu theist on the christian religion – Calcutta: Oxford Mission Press, 1882 [mf ed 1992] – 1mf – 9 – 0-524-02668-8 – mf#1990-3098 – us ATLA [230]

Inquiry – Oslo. 1981+ (1,5,9) – ISSN: 0020-174X – mf#13026 – us UMI ProQuest [300]

Inquiry blue cross and blue shield association – Chicago. 1971+ (1) 1963+ (5) 1975+ (9) – ISSN: 0046-9580 – mf#6070 – us UMI ProQuest [610]

An inquiry concerning the relation of death to probation / Wright, George Frederick – Boston: Congregational Pub Soc, c1882 [mf ed 1994] – 1mf – 9 – 0-524-08888-8 – mf#1993-3352 – us ATLA [240]

Inquiry documents (special reports and studies), 1917-1919 / U.S. Commission to Negotiate Peace – (= ser Records of the american commission to negotiate peace) – 47r – 1 – mf#M1107 – us Nat Archives [327]

Inquiry excellus health plan – Rochester. 2002+ (1,5,9) – mf#6070,01 – us UMI ProQuest [613]

Inquiry into certain vulgar opinions concerning the catholic... / Milner, John – London, England. 1808 – 1r – us UF Libraries [241]

Inquiry into occupation and administration of hait... / United States Congress Senate – Washington, DC. 1922 – 1r – us UF Libraries [972]

An inquiry into some of the sources of channing's religious philosophy : hutcheson, ferguson, and price / Kyper, Ralph Edward – Chicago, 1941. Chicago: Dep of Photodup, U of Chicago Lib, 1971 (1r); Evanston: American Theol Lib Assoc, 1984 (1r) – 1 – 0-8370-0375-X – mf#1984-B179 – us ATLA [240]

An inquiry into some parts of christian doctrine and practice : having relation more especially to the society of friends / Ash, Edward – London: Hamilton, Adams, 1841 [mf ed 1993] – (= ser Society of friends (quakers) coll) – 1mf – 9 – 0-524-07551-4 – (with app) – mf#1991-3171 – us ATLA [243]

An inquiry into the accordancy of war with the principles of christianity : and an examination of the philosophical reasoning by which it is defended / Dymond, Jonathan & Grimke, Thomas Smith – Philadelphia: printed by I Ashmead, 1834 [mf ed 1993] – 1mf – 9 – 0-524-08537-4 – (1st publ anonymously in 1823) – mf#1993-2062 – us ATLA [240]

Inquiry into the baie des chaleurs railway matter : proceedings of the commission and depositions of witnesses / Quebec (Province). Royal Commission of Inquiry – [Quebec?: s.n.], 1891 – 12mf – 9 – 0-665-93301-0 – mf#93301 – cn Canadiana [380]

Inquiry into the baie des chaleurs railway matter : reports, proceedings of the commission and depositions of witnesses, appendices and indices / Quebec (Province). Royal Commission of Inquiry – [Quebec?: s.n.], 1892 – 3mf – 9 – 0-665-93302-9 – (incl bibl ref) – mf#93302 – cn Canadiana [380]

An inquiry into the beauties of painting : and into the merits of the most celebrated painters... / Webb, D – London, 1761 – 2mf – 9 – mf#0-1181 – ne IDC [750]

An inquiry into the character and authorship of the fourth gospel / Drummond, James – New York: publ for the Hibbert Trustees...1904 [mf ed 1986] – 2mf – 9 – 0-8370-9860-2 – (incl bibl ref & ind) – mf#1986-3860 – us ATLA [226]

Inquiry into the character of the present educational connexion bet... / Moody Stuart, A – Edinburgh, Scotland. 1848 – 1r – us UF Libraries [240]

An inquiry into the connected uses of the principal means of attaining christian truth : in eight sermons preached before the university of oxford / Hawkins, Edward – Oxford: JH Parker, 1840 [mf ed 1991] – (= ser Bampton lectures 1840) – 1mf – 9 – 0-524-00034-4 – (incl app to bampton lecture 1840) – mf#1989-2734 – us ATLA [220]

An inquiry into the difference of style...in ancient glass paintings / Winston, Charles – [2nd ed]. Oxford 1867 – (= ser 19th c art & architecture) – 7mf – 9 – mf#4.1.288 – uk Chadwyck [740]

An inquiry into the evidence relating to the charges brought by lord macaulay against william penn / Paget, John – Edinburgh: William Blackwood, 1858 [mf ed 1992] – (= ser Society of friends (quakers) coll) – 1mf – 9 – 0-524-03853-8 – mf#1990-4900 – us ATLA [941]

An inquiry into the history and theology of the ancient vallenses and albigenses : as exhibiting, agreeably to the promises, the perpetuity of the sincere church of christ / Faber, George Stanley – London: RB Seeley & W Burnside, 1838 [mf ed 1992] – 2mf – 9 – 0-524-03395-1 – mf#1990-0949 – us ATLA [240]

An inquiry into the influence of the excessive use of spirituous liquors / Haliday, Charles – Dublin, [1830] – (= ser 19th c ireland) – 2mf – 9 – mf#1.1.8489 – uk Chadwyck [345]

An inquiry into the justice and expediency of completing the publication of the authentic records of the colony of the cape of good hope relative to the aboriginal tribes / Moodie, Donald – Cape Town. 1841 – 1 – us CRL [960]

An inquiry into the laws of organized societies : as applied to the alleged decline of the society of friends / Fisher, William Logan – Philadelphia: T Ellwood Zell, 1860 [mf ed 1992] – (= ser Society of friends (quakers) coll) – 1mf – 9 – 0-524-03068-5 – mf#1990-4557 – us ATLA [243]

Inquiry into the nature, object and obligations of the religion of... / Baines, Peter Augustine – Bath, England. 1824 – 1r – us UF Libraries [240]

An inquiry into the nature of our lord's knowledge as man / Swayne, William Shuckburgh – London, New York: Longmans, Green, 1891 [mf ed 1985] – 1mf – 9 – 0-8370-5563-6 – (incl bibl ref) – mf#1985-3563 – us ATLA [210]

An inquiry into the nature, progress, and end of prophecy : in three books... / Lee, Samuel – Cambridge: University Press, 1849 [mf ed 1992] – 2mf – 9 – 0-524-06465-2 – mf#1992-0893 – us ATLA [221]

Inquiry into the obligation of religious covenants upon posterity / Paxton, George – Edinburgh, Scotland. 1801 – 1r – us UF Libraries [240]

Inquiry into the opinions of the commercial classes of great britain on the suez ship canal / Lesseps, Ferdinand Marie, vicomte de – London, 1857 – (= ser 19th c british colonization) – 2mf – 9 – mf#1.1.7617 – uk Chadwyck [330]

An inquiry into the organization and government of the apostolic church : particularly with reference to the claims of episcopacy / Barnes, Albert – Philadelphia: Presbyterian Publ Cttee; New York: Ivison & Phinney, c1855 [mf ed 1989] – 1mf – 9 – 0-7905-0854-0 – (incl bibl ref) – mf#1987-0854 – us ATLA [240]

Inquiry into the original language of st matthew's gospel : with relative discussions on the language of palestine in the time of christ and on the origin of the gospels / Roberts, Alexander – London: Samuel Bagster, [1859?] – 1mf – 9 – 0-7905-3167-4 – mf#1987-3167 – us ATLA [226]

An inquiry into the perceived ideals held by experts for regional tourism development in the state of pennsylvania for the year 2000 / Wang, P C – 1991 – 3mf – 9 – $12.00 – us Kinesology [338]

An inquiry into the principles of beauty in grecian architecture / Aberdeen, G – London, 1822 – 3mf – 9 – mf#O-1159 – ne IDC [720]

An inquiry into the principles of church-authority : or, reasons for recalling my subscription to the royal supremacy / Wilberforce, Robert Isaac – Baltimore: Hedian & O'Brien, 1855 [mf ed 1986] – 1mf – 9 – 0-8370-6876-2 – (incl bibl ref) – mf#1986-0876 – us ATLA [230]

An inquiry into the proper mode of rendering the word god in translating the sacred scriptures into the chinese language / Medhurst, Walter Henry – Shanghae: Mission Press, 1848 [mf ed 1995] – (= ser Yale coll) – 170p – 1 – 0-524-09578-7 – mf#1995-0578 – us ATLA [480]

An inquiry into the proper mode of translating ruach and pneuma : in the chinese version of the scriptures / Medhurst, Walter Henry – Shanghae: Printed at the Mission Press, 1850 – (= ser Yale coll) – 75p – 1 – 0-524-10167-1 – mf#1995-1167 – us ATLA [220]

Inquiry into the reported miraculous cure of mathew breslin / Cousins, John – Dublin, 1815 – (= ser 19th c ireland) – 1mf – 9 – mf#1.1.6669 – uk Chadwyck [230]

An inquiry into the requisite cultivation / Hoare, Prince – London 1806 – (= ser 19th c art & architecture) – 4mf – 9 – mf#4.2.140 – uk Chadwyck [700]

An inquiry into the scriptural import of the words sheol, hades, tartarus, and gehenna : translated hell in the common english version / Balfour, Walter – rev ed. Boston: Tompkins, 1863 [mf ed 1993] – (= ser Unitarian/universalist coll) – 1mf – 9 – 0-524-06381-8 – mf#1991-2503 – us ATLA [220]

An inquiry into the scriptural views of slavery / Barnes, Albert – Philadelphia: Perkins & Purves; Boston: B Perkins, 1846 [mf ed 1989] – 1mf – 9 – 0-7905-0961-X – (incl bibl ref) – mf#1987-0961 – us ATLA [220]

Inquiry into the sources of charles sealsfield's novel morton : oder, die grosse tour / Thompson, Garrett William – [s.l: s.n, 19–?] [mf ed 1993] – 1r – 1 – (incl bibl ref; filmed with: gestirn des krieges / bodo schutt) – mf#2940p – us Wisconsin U Libr [420]

An inquiry into the usage of [baptizo] and the nature of judaic baptism : as shown by jewish and patristic writings / Dale, James Wilkinson – 3rd ed. Philadelphia: Presbyterian Board of Publ & Sabbath- Social Work [c1869] [mf ed 1984] – 5mf – 9 – 0-8370-1032-2 – mf#1984-4386 – us ATLA [270]

Inquiry, whether the description of babylon, contained in the 18th... / Sharp, Granville – London, England. 1805 – 1r – us UF Libraries [240]

Inquiry whether the sentence of death pronounced at the fall of man / Buckland, William Warwick – London, England. 1839 – 1r – us UF Libraries [240]

La inquisicion en guadalupe / Fita, Fidel – Madrid: Fortanet, 1893. B.R.A.H. 23, pp 283-344 – sp Bibl Santa Ana [972]

Inquisicion. observaciones de la...avisador de badajoz – 1884 – 9 – sp Bibl Santa Ana [946]

Inquisition – v1 n5-v4 n3 [1968 nov-1969 dec 20], v4 n6 [1970 feb 19], v5 n1 [1970 mar 26] – 1r – 1 – mf#713561 – us WHS [071]

The inquisition : an essay. extracted from devivier's christian apologetics = Cours d'apologetique chretienne. Selections / Devivier, Walter; ed by Sasia, Joseph Casimir – San Francisco, Cal[if]: Catholic Truth Society, 1904 – 1mf – 9 – 0-524-03043-X – (incl bibl ref. in english) – mf#1990-0800 – us ATLA [240]

The inquisition : a critical and historical study of the coercive power of the church = Inquisition / Vacandard, Elphege – New York: Longmans, Green, 1908, c1907 – 1mf – 9 – 0-7905-6331-2 – (incl bibl ref. in english) – mf#1988-2331 – us ATLA [940]

The inquisition in the spanish dependencies : sicily, naples, sardinia, milan, the canaries, mexico, peru, new granada / Lea, Henry Charles – New York: Macmillan, 1908 – 2mf – 9 – 0-7905-5660-X – (incl bibl ref) – mf#1988-1660 – us ATLA [946]

L'inquisition protestante : les victimes de calvin / Rouquette, Jean – Paris: Bloud, 1908 [mf ed 1992] – (= ser Questions historiques) – 1mf – 9 – 0-524-02599-1 – (in french) – mf#1990-0651 – us ATLA [242]

Inquisitions : organe du groupe d'etudes pour la pheno menologie humaine – Paris. n1. juin 1936 – 1 – fr ACRPP [300]

The inquisitions : series 1: manuscripts of the spanish, portuguese and french inquisitions in the british library, london / ed by Edwards, John – [mf ed 2003] – 35r – 1 – us Primary [940]

The inquisitions : series 2: archive of the conseil des troubles, 1567-76 from les archives generales du royaume, brussels / ed by Marnef, Guido – [mf ed 2003] – ca 70r – 1 – us Primary [940]

Inquisitor – Philadelphia. 1818-1820 (1) – mf#4466 – us UMI ProQuest [420]

Inroads : a journal of opinion – Montreal. 2002+ (5,9) – ISSN: 1188-746X – mf#32053 – us UMI ProQuest [300]

In...s pavli priorem ad corinth[ios] epistolam commentarij / Vermigli, P M – Tigvri, Christ[oph] Froschouer, 1551 – 11mf – 9 – mf#PBU-279 – ne IDC [240]

Ins volle menschenleben : neue erzaehlungen / Steguweit, Heinz – Hamburg: Hanseatische Verlagsanstalt 1942 [mf ed 1991] – 1r – 1 – (filmed with: frohes leben) – mf#2897p – us Wisconsin U Libr [830]

The insane in the province of quebec : (report ot the honorable provincial secretary) / Vallee, Arthur – Quebec?: Belleau, 1890 – 1mf – 9 – mf#25326 – cn Canadiana [362]

Insaniyet – Istanbul: 1 sene n1-2. 13 saban 1328-5 agustos. 20 saban-12 agustos 1326 [1910] – (= ser O & t journals) – 42mf – 9 – $685.00 – (cont: istirak) – us MEDOC [956]

Die inschrift auf dem denkmal mesa's koenigs von moab : (9. vorchr. jahrh.): mit einem anhang betreffend die grabschrift des sid. koenigs eschmunazar / Kaempf, Saul Isaac – Prag: F Tempsky, 1870 – 1mf – 9 – 0-8370-7395-2 – mf#1986-1395 – us ATLA [930]

Die inschrift des koenigs mesa von moab : (9. jahrhundert vor christus) / Noeldeke, Theodor – Kiel: Schwers, 1870 – 1mf – 9 – 0-8370-7318-9 – (incl bibl ref) – mf#1986-1318 – us ATLA [470]

Die inschrift eschmunazars, koenigs der sidonier / Schlottmann, Konstantin – Halle: Buchh des Waisenhauses, 1868 – 1mf – 9 – 0-8370-7423-1 – (incl bibl ref) – mf#1986-1423 – us ATLA [470]

Die inschriften tiglathpileser's 1 : in transskribiertem assyrischem grundtext mit uebersetzung und kommentar / Lotz, Wilhelm – Leipzig: JC Hinrichs, 1880 – 1mf – 9 – 0-8370-7716-8 – (incl bibl ref and ind) – mf#1986-1716 – us ATLA [470]

Inschriften von cambyses, koenig von babylon (529-521 v. chr.) : von den thontafeln des britischen museums / Strassmaier, Johann Nepomuk – Leipzig: Eduard Pfeiffer, 1890 – 1mf – 9 – 0-8370-9113-6 – (texts in akkadian; preface in german. incl indes) – mf#1986-3113 – us ATLA [470]

Inschriften von cyrus, koenig von babylon (538-529 v. chr.) : von den thontafeln des britschen museums / Strassmaier, Johann Nepomuk – Leipzig: Eduard Pfeiffer, 1890 – 1mf – 9 – 0-8370-9114-4 – (texts in akkadian; preface in german. incl indes) – mf#1986-3114 – us ATLA [470]

Inschriften von nabonidus, koenig von babylon (555-538 v. chr.) : von den thontafeln des britischen museums / Strassmaier, Johann Nepomuk – Leipzig: Eduard Pfeiffer, 1889 – 2mf – 9 – 0-8370-9115-2 – (texts in akkadian; preface in german. incl indes) – mf#1986-3115 – us ATLA [470]

Inschriften von nabuchodonosor, koenig von babylon (604-561 v. chr.) – Leipzig: Eduard Pfeiffer, 1889 – 1mf – 9 – 0-8370-9187-X – (texts in akkadian; preface in german. incl indes) – mf#1986-3187 – us ATLA [470]

Inscripcion arabe en trujillo / Codera, Francisco – Madrid: Fortanet, 1914. B.R.A.H. lxiv/pp. 117-119 – 1 – sp Bibl Santa Ana [946]

Inscripcion hemisferica de santa cruz y lapida de solans de cabanas. notas a una carta de roso de luna / Fita, Fidel – Madrid: Fortanet, 1902. B.R.A.H. 40. pp. 564-566 – 1 – sp Bibl Santa Ana [946]

Inscripcion romana de la parra y de almendralejo / Fita, Fidel – Madrid: Tip. de Fortanet, 1897 – sp Bibl Santa Ana [946]

Inscripcion romana de merida / Fita, Fidel – Madrid: Tip. de Fortanet, 1898 – sp Bibl Santa Ana [946]

Inscripcion romana de riolobos / Fita, Fidel – Madrid: Tip. de Fortanet, 1896 – sp Bibl Santa Ana [946]

Inscripcion romana de titulcia / Roso de Luna, Mario – Madrid: Fortanet, 1918. B.R.A.H. 72, pp. 279-280 – sp Bibl Santa Ana [946]

Inscripcion romana de valera la vieja, junto a fregenal / Fita, Fidel – Madrid: Tip. Fortanet, 1901 – sp Bibl Santa Ana [946]

Inscripcion romana insigne de caceres / Sanguino y Michel, Juan – Madrid: Fortanet, 1913. B.R.A.H. 63, pp. 422-427 – sp Bibl Santa Ana [946]

Inscripciones / Martinez Escobar, Manuel – Habana, Cuba. v1-2. 1931 – 1r – us UF Libraries [972]

Inscripciones cacerenas ineditas / Corchon Garcia, Justo – Madrid: Imprenta y Editorial Maestre, 1955 – 1 – sp Bibl Santa Ana [946]

Inscripciones constantinianas de merida / Fita, Fidel – Madrid: Fortanet, 1913. B.R.A.H. lxii/pp. 576-580 – 1 – sp Bibl Santa Ana [946]

Inscripciones ineditas de merida, badajoz, alanje, canete de las torres y vilches / Fita, Fidel – Madrid: Fortanet, 1912. B.R.A.H. 61. pp 511-524 – 1 – sp Bibl Santa Ana [946]

Inscripciones romanas de burguillos / Martinez Martinez, Matias Ramon – Madrid: Tip. Fortanet, 1898 – sp Bibl Santa Ana [946]

Inscripciones romanas de caceres, ubeda y alcala de henares / Fita, Fidel – Madrid: Fortanet, 1885. B.R.A.H. vii/pp. 45-53 – 1 – sp Bibl Santa Ana [946]

Inscripciones romanas de merida / Huebner, Emilio – Madrid: Tip. Fortanet, 1894 – 1 – sp Bibl Santa Ana [946]

Inscripciones romanas de merida y nava de rico malillo / Fita, Fidel – Madrid: Tip. Fortanet, 1900 – sp Bibl Santa Ana [946]

Inscripciones romanas de merida y reina / Melida, Jose Ramon – Madrid: Tip. Fortanet, 1911. BRAH lviii/ pp. 187-196 – sp Bibl Santa Ana [946]

Inscripciones romanas ineditas de caceres, brandomil, naranco y lerida / Fita, Fidel – Madrid: Fortanet, 1885. B.R.A.H. vi/pp. 430-436 – 1 – sp Bibl Santa Ana [946]

Inscripciones romanas ineditas de trujillo / Fita, Fidel – Madrid: Fortanet, 1917. B.R.A.H. lxviii/p. 163-170 – 1 – sp Bibl Santa Ana [946]

Inscripciones romanas sepulcrales de ibahernando / Huebner, Emilio – Caceres: Tip. Enc. y Lib. Jimenez, 1900 – 1 – sp Bibl Santa Ana [946]

Inscripciones visigoticas : estudios hagiologicos / Fita, Fidel – Madrid: Tip. Fortanet, 1897 – sp Bibl Santa Ana [946]

Inscripcion...saturnino penitente / Salcedo, Coronel Garcia – 1890 – 9 – sp Bibl Santa Ana [440]

Inscripcion...saturnino...merida / Perez de Guzman, Juan – 1890 – 9 – sp Bibl Santa Ana [440]

Inscripcion...saturnino...merida / Salcedo, Coronel Garcia – 1890 – 9 – sp Bibl Santa Ana [440]

L'inscription de bavian : texte, traduction et commentaire philologique, avec trois appendices et un glossaire / Pognon, H – Paris: F Vieweg, 1879-80 [mf ed 1986] – (= ser Bibliotheque de l'ecole des hautes etudes, 4e section, sciences historiques et philologiques 39,42) – 2v on 2mf – 9 – 0-8370-7818-0 – (text in french and akkadian. comm in french. incl bibl ref) – mf#1986-1818 – us ATLA [470]

Une inscription fragmentaire d'augusta emerita. emerita de lusitanie a la lumiere des "histoires" de tacite / Le Roux, Patrick – 1 – sp Bibl Santa Ana [946]

Inscription historique de pinodjem 3 / Naville, E – Paris, 1883 – 1mf – 9 – mf#NE-20004 – ne IDC [956]

Inscription of tiglath pileser 1., king of assyria, b.c. 1150 – London: Royal Asiatic Society: J W Parker, 1857 – 1mf – 9 – 0-8370-7744-3 – mf#1986-1744 – us ATLA [470]

L'inscription sanscrite de han chey / Barth, Auguste – Paris: Impr Nationale 1883 [mf ed 1989] – 1r with other items – 1 – ("extrait du journal asiatique") – mf#mf-10289 seam reel 001/10 [§] – us CRL [490]

L'inscription syro-chinoise de si-ngan-fou : monument nestorien eleve en chine l'an 781 de notre ere, et decouvert en 1625 / Pauthier, Guillaume – Paris: Firmin Didot Freres, Fils, 1858 [mf ed 1995] – (= ser Yale coll; Etudes orientales 2) – xvi/96p (ill) – 1 – 0-524-09277-X – (in french) – mf#1995-0277 – us ATLA [240]

Inscriptiones graecae / ed by Kern, Otto – Bonnae, 1913. xxiiip. Bibliography – 1 – us Wisconsin U Libr [410]

Inscriptiones graecae antiqvissimae / ed by Roehl – Berolini, 1882 – 7mf – 8 – mf#125 – ne IDC [700]

Inscriptiones latinae christianae veteres / ed by Diehl, Ernst – Berolini. v1-3. 1925-1931 – 43mf – 8 – €82.00 – ne Slangenburg [240]

Inscriptiones marmori incisae centum-cellis et in ecclesia pp dominicanorum expositae : inscripciones gravadas en marmol... – [Mexico City?: s.n. 1765?] – (= ser Books on religion... 1543/44-c1800: ordenes, etc: dominicos) – 1mf – 9 – mf#crl-193 – ne IDC [241]

Inscriptions / Wisconsin Old Cemetery Society – v1 n1-v18 n5 [1972 mar-1989 sep], v19 n1-2 [1990 jan-mar] – 1r – 1 – mf#1701612 – us WHS [929]

Les inscriptions de salmanasar 2 roi d'assyrie (860-824) / Shalmaneser 2, King of Assyria; ed by Scheil, Vincent & Amiaud, Arthur – Paris: H Welter, 1890 – 1mf – 9 – 0-8370-7741-9 – (incl ind. texts in french and akkadian; commentary in french) – mf#1986-1741 – us ATLA [470]

Les inscriptions du wadi brissa et du nahr el-kelb / Langdon, Stephen – Paris: Emile Bouillon, 1905 [mf ed 1988] – 1mf – 9 – 0-7905-0044-2 – (in french and akkadian) – mf#1987-0044 – us ATLA [470]

Inscriptions from adab / Luckenbill, Daniel David; ed by Chiera, Edward – Chicago: University of Chicago Press, [1930] [mf ed 1978?] – (= ser University of chicago oriental institute publications 14; University of chicago. oriental institute. cuneiform series 2) – ix/8p/87lea on 1 sheet – 9 – (accounts...illustrating the conduct of business in the city and temple of adab during the third millenium b c) – us Chicago U Pr [470]

Inscriptions hieroglyphiques recueillies en europe et en egypte / Piehl, K Stockholm, 1884-1888 – 12mf – 8 – (publiees, traduites et commentees par karl piehl) – mf#H-391 – ne IDC [956]

Les inscriptions historiques de ninive et de babylone : aspect general de ces documents / Delattre, Alphonse J – Paris: Ernest Leroux, 1879 – 1mf – 9 – 0-8370-8566-7 – (incl bibl ref) – mf#1986-2566 – us ATLA [930]

Inscriptions in the hieratic and demotic character : from the collections of the british museum / Birch, S – London, 1868 – 2mf – 9 – mf#NE-367 – ne IDC [956]

Inscriptions juives de k'ai-fong-fou / Tobar, Jerome – Chang-hai: Impr de la Mission Catholique, 1912 [mf ed 1995] – (= ser Yale coll; Varietes sinologiques 17) – 1 – 0-524-09334-2 – (in french) – mf#1995-0334 – us ATLA [951]

Inscriptions left by early european navigators on their way / Peringuey, Louis Albert – Cape Town, South Africa. 1950 – 1r – us UF Libraries [960]

Inscriptions modernes d'angkor / Buddhasasanapandity [Institute: Phnom Penh, Cambodia] – 2.ed. Phnom-Penh 1958 [mf ed 1990] – 1r with other items – 1 – (title & text in khmer; added t.p. in french) – mf#mf-10289 seam reel 120/1 [§] – us CRL [930]

Inscriptions of kambuja / Majumdar, Ramesh Chandra – Calcutta: The Asiatic Society 1953 [mf ed 1989] – 1r with other items – 1 – mf#mf-10289 seam reel 016/01 [§] – us CRL [930]

The inscriptions of si-t and der rifeh / Griffith, F L – London, 1889 – 1mf – 9 – mf#NE-20395 – ne IDC [956]

Inscriptions of the reigns of evil-merodach (b.c. 562-559), neriglissar (b.c. 559-555) and laborosoarchod (b.c. 555) – Leipzig: Eduard Pfeiffer, 1892 – 1mf – 9 – 0-8370-9058-X – (texts in akkadian; preface in english. incl indes) – mf#1986-3058 – us ATLA [470]

Inscriptions sanscrites du cambodge / Barth, Auguste – Paris: Impr Nationale 1882 [mf ed 1989] – 1r with other items – 1 – ("extrait du journal asiatique") – mf#mf-10289 seam reel 001/09 [§] – us CRL [490]

[Inscriptions sanscrites du cambodge et de campa] / Barth, Auguste et al – [Paris: Impr nationale 1885-93] [mf ed 1989] – 2v on 1r with other items – 1 – mf#mf-10289 seam reel 012/01 [§] – us CRL [490]

Insect biochemistry – Oxford. 1971-1991 (1,5,9) – (cont by: insect biochemistry and molecular biology) – ISSN: 0020-1790 – mf#49125 – us UMI ProQuest [590]

Insect biochemistry and molecular biology – Oxford. 1992+ (1,5,9) – (cont: insect biochemistry) – ISSN: 0965-1748 – mf#49125,01 – us UMI ProQuest [590]

Insect enemies of truck and garden crops / Quaintance, A L – Lake City, FL. 1896 – 1r – us UF Libraries [634]

Insect life / United States Department of Agriculture – Washington 1888-95 – v1-7+ind on 63mf – 9 – mf#z-958/2 – ne IDC [590]

Insect science and its application – Oxford. 1980-1986 (1) 1980-1986 (5) 1982-1985 (9) – ISSN: 0191-9040 – mf#49309 – us UMI ProQuest [590]

Insect transformation / Carpenter, George Herbert – New York, NY. 1923 – 1r – us UF Libraries [590]

Insect world digest – Latham. 1975-1976 (1) – ISSN: 0090-8282 – mf#7950 – us UMI ProQuest [590]

Insecta : revue illustree d'entomologie – Rennes 1911-24 – 61mf – 9 – mf#z-959/2 – ne IDC [590]

Insectes sociaux = Social insects – Paris. 1968-1990 (1) 1971-1990 (5) 1974-1990 (9) – ISSN: 0020-1812 – mf#3414 – us UMI ProQuest [590]

Insecticides and fungicides / Gossard, H A – Lake City, FL. 1904 – 1r – us UF Libraries [630]

Insecticides and fungicides / Rolfs, P H – Lake City, FL. 1893 – 1r – us UF Libraries [630]

Insectivorous plants / Darwin, Charles Robert – London, 1875 – (= ser 19th c evolution & creation) – 5mf – 9 – (with ill) – mf#1.1.4279 – uk Chadwyck [580]

Insects and diseases of the pecan in florida / Phillips, Arthur M – Gainesville, FL. 1945 – 1r – us UF Libraries [634]

Insects and other pests of florida vegetables / Watson, J R – Gainesville, FL. 1942 – 1r – us UF Libraries [634]

Insects injurious to stored grain and cereal products / Quaintance, A L – Lake City, FL. 1896 – 1r – us UF Libraries [630]

Insects of a citrus grove / Watson, J R – Gainesville, FL. 1918 – 1r – us UF Libraries [634]

Insects of the pecan / Gossard, H A – Lake City, FL. 1905 – 1r – us UF Libraries [634]

Insecutor inscitiae menstruus : a monthly journal of entomology... – Washington 1913-26 – v1-14 on 58mf – 9 – mf#z-962/2 – ne IDC [590]

Insegnamenti del vivere del conte alberto caprara : a massimo suo nipote / Caprara, A – Bologna: Per l'Herede di Domenico Barbieri, 1672 – 3mf – 9 – mf#O-1533 – ne IDC [090]

Die insel *see* Die insel der einsamen

Die insel cypern ihrer physischen und organischen natur nach : mit ruecksicht auf ihre fruehere geschichte / Unger, Franz J – Wien 1865 [mf ed Hildesheim 1995-98] – 1v on 4mf [ill] – 9 – €120.00 – 3-487-27643-7 – gw Olms [914]

Die insel der 1000 wunder : ein utopischer roman / Daumann, Rudolf Heinrich – Berlin: Schuetzen-Verlag, 1940 [mf ed 1989] – 211p (ill) – 1 – mf#7170 – us Wisconsin U Libr [830]

Die insel der einsamen – Berlin DE, 1923-1933 n11 – 1r – 1 – (title varies: 1923 n6: die insel; 1925 n1: das freundschaftsblatt) – gw Misc Inst [074]

Insel der hoffnung : roman / Viebig, Clara – Stuttgart: Deutsche Verlags-Anstalt, 1933 – 1r – 1 – us Wisconsin U Libr [830]

Die insel der seligen in mythus und sage der vorzeit : vortrag. gehalten in der geogr. gesellschaft zu frankfurt a/m... / Hommel, Fritz – Muenchen, H Lukaschik, 1901 – 1mf – 9 – 0-524-01500-7 – mf#1990-2476 – us ATLA [230]

Die insel felsenburg : erster theil / Schnabel, Johann Gottfried; ed by Ullrich, Hermann – Berlin: B Behr (E Bock), 1902- [mf ed 1993] – (= ser Deutsche litteraturdenkmale des 18. und 19. jahrhunderts 108-120, n f n58-70) – 467p/2pl – 1 – (remaining 3v not publ in the series. repr of original ed publ under aut's pseud gisander: "wunderliche fata einiger see-fahrer, absonderlich alberti julii...) – mf#8676 reel 6 – us Wisconsin U Libr [830]

Eine insel im la plata / Fuchs, Hans – Hamburg: Hanseatische Verlagsanstalt, 1942 (mf ed 1990) – 1r – 1 – (filmed with: liebeskaempfe) – us Wisconsin U Libr [810]

Insel im seewind : schicksale vor deich und duene / Schreiner, Wilhelm – Stuttgart: J F Steinkopf, 1943 – 1r – 1 – us Wisconsin U Libr [430]

Die insel ischia : in natur-, sitten- und geschichts-bildern aus vergangenheit und gegenwart / Kaden, Woldemar – Luzern [1883] [mf ed Hildesheim 1995-98] – (= ser Fbc) – 1mf [ill] – 9 – €40.00 – 3-487-29184-3 – gw Olms [914]

Die insel mainau und der badische bodensee : mit beruecksichtigung der angrenzenden gebietstheile / Reich, Lucian – Carlsruhe 1856 [mf ed Hildesheim: 1995-98] – (= ser Fbc) – viii/289p on 2mf [ill] – 9 – €60.00 – 3-487-29488-5 – gw Olms [914]

Insel-rundschau – Stralsund DE, 1962 2 aug-1967 30 mar – 1r – 1 – (covers ruegen) – gw Misc Inst [074]

The inservice needs of south carolina public school physical educators providing instruction to handicapped students / White, C – 1989 – 2mf – 9 – $8.00 – us Kinesology [790]

Inset towards educational transformation : with reference to primary schools in the northern province / Ravhudzulo, Anniekie Ndowiseni – Uni of South Africa 2001 [mf ed Johannesburg 2001] – 6mf – 9 – (incl bibl ref) – mf#mfm15020 – sa Unisa [370]

O inseto – Rio Branco, AC. nov 1916 – (= ser Ps 19) – mf#P25,01,29 – bl Biblioteca [079]

Inshimi sha kale / Phaedrus – Lusaka, Zambia. 1956 – 1r – us UF Libraries [960]

Inside – Warwick, RI. 1974-1974 (1) – mf#66422 – us UMI ProQuest [071]

Inside america : a voyage of discovery / Nehru, Jawaharlal – New Delhi: National Book Stall, [1950] – (= ser Samp: indian books) – us CRL [920]

Inside canberra – Canberra, 1948-75 – 3r – 1 – A$115.50 vesicular A$132.00 silver – at Pascoe [079]

Inside congress / Shraddhanand, Swami – Bombay: Phoenix Publications, 1946 – (= ser Samp: indian books) – (foreword by deshbandhu gupta) – us CRL [954]

Inside coverage – iss 1-23 [1979 oct-1983 dec] – 1r – 1 – mf#962257 – us WHS [071]

Inside dpma – Park Ridge. 1988-1996 (1) 1988-1996 (5) 1988-1996 (9) – (cont by: information executive) – ISSN: 0898-171X – mf#2680,03 – us UMI ProQuest [000]

Inside education – Albany. 1915-1983 (1) 1972-1983 (5) 1976-1983 (9) – ISSN: 0020-1855 – mf#6862 – us UMI ProQuest [370]

Inside germany reports – New York NY (USA), 1939 15 apr-1944 may – 1r – 1 – (iss by the american friends of german freedom) – gw Misc Inst [943]

Inside harlem – 1989 aug 24 – 1r – 1 – mf#5307317 – us WHS [071]

Inside history of first baptist church, fort worth and temple baptist church, detroit: life story of j. f. norris / Norris, J Frank – 1 – us Southern Baptist [242]

Inside history of the white house : the complete history of the domestic and official life in washington of the nation's presidents and their families / Willets, Gilson – 1r – 1 – us WHS [920]

Inside indonesia features / Antara – Djakarta, 1965(1-5) – 3mf – 9 – mf#SE-164=7 – ne IDC [959]

Inside kashmir / Bazaz, Prem Nath – Srinagar, [Jammu and Kashmir, India]: Kashmir Pub Co, 1941 – (= ser Samp: indian books) – us CRL [954]

Inside latin america / Gunther, John – New York, NY. 1941 – 1r – us UF Libraries [972]

Inside local 1082 / United Steelworkers of America – 1973 feb-1984 jul/aug – 1r – 1 – mf#1289676 – us WHS [331]

Inside new orleans – New Orleans LA. v1 n24 [1965 may 1] – 1r – 1 – mf#4983543 – us WHS [978]

Inside sports – Evanston. 1979-1998 (1,5,9) – ISSN: 0195-3478 – mf#12395 – us UMI ProQuest [790]

Inside the bar and other occasional poems / May, John Wilder – Portland, ME: Hoyt, Fogg & Donham, 1884 – 3mf – 9 – $4.50 – mf#LLMC 91-004 – us LLMC [810]

Inside the beast / San Diego State University – v1 n7 [1973 apr 23] – 1r – 1 – mf#1583891 – us WHS [378]

Inside the cup or my 21 years in fort worth / Norris, J Frank – n.d. 218p – 1 – 7.63 – us Southern Baptist [242]

Inside the south african crucible / Du Preez, Andries Bernardus – Kaapstad, South Africa. 1959 – 1r – us UF Libraries [960]

An inside view of the vatican council : in the speech of the most reverend archbishop kenrick of st louis / ed by Bacon, Leonard Woolsey – New York: American Tract Society, [187-?] [mf ed 1986] – 1mf – 9 – 0-8370-6674-3 – mf#1986-0674 – us ATLA [241]

Inside views of methodism : or, a hand-book for inquirers and beginners / Reddy, William – New York: Carlton & Porter, c1859 [mf ed 1991] – (= ser Methodist coll) – 1mf – 9 – 0-524-01395-0 – mf#1990-4091 – us ATLA [242]

Inside your schools : the aft television magazine, show release... / American Federation of Teachers – n102-110,202,204,209-211,301,304-306,308,401-404 [1983 sep 21-1984 may 10, 1984 sep 18, nov 19, 1985 apr 10-jun 11, 1985 aug 14, dec 9-1986 feb 10, apr 15, 1986 aug 21-nov 19] – 1r – 1 – mf#1477235 – us WHS [370]

INSIDE

Inside zambia–and out / Pitch, Anthony – Cape Town, South Africa. 1967 – 1r – us UF Libraries [960]
Inside-out : taconic newsletter / Taconic Correctional Facility – 1978 jan – 1r – 1 – mf#4876630 – us WHS [365]
Insider / Loewi Financial Companies – v1 n1-2 [1979 win-spr] – 1r – 1 – (cont: employee's newsletter) – mf#626055 – us WHS [332]
Insider / Meharry Medical College – v5 n7-v6 n2 [1994 jan 7-1995 feb 10] – 1r – 1 – mf#2926717 – us WHS [610]
The insider / ed by Steel, John [d.i. Johannes Stahl] – New York, NY (USA) 1939 22 mar-14 jun [mf ed 2005] – 1 – gw Mikrofilm [327]
Insiders' chronicle – Riverside. 1978-1990 (1,5,9) – ISSN: 0162-5152 – mf#11534 – us UMI ProQuest [332]
Insieme – Montreal QC. 1983 jan-sep, 1983 oct-1984 jun, 1984 jul-1985 mar, 1985 apr-1986 feb 5, 1986 feb 12-oct 29, 1986 nov-1987 jun, 1987 jul-1988 mar, 1988 apr-dec, 1989 jan-jun, 1989 jul-dec 1, 1990 sep-1991 apr, 1991 may-dec, 1992 jan 8-aug 26 – 13r – 1 – mf#1006518 – us WHS [071]
Insight – 1986 jan 13-apr 28, 1986 may 5-aug, 1986 sep-1987 jan 5, 1987 jan-apr, 1987 may-sep 14 – 5r – 1 – mf#1311832 – us WHS [071]
Insight / AFL-CIO [American Federation of Labor and Congress of Industrial Organizations] – v12 n3-v15 n6 [1978 mar-1981 sep] – 1r – 1 – mf#615606 – us WHS [331]
Insight – Washington. 1985-1987 (1) 1985-1987 (5) 1985-1987 (9) – (cont by: insight on the news) – ISSN: 0884-9285 – mf#15662 – us UMI ProQuest [073]
Insight – Quincy. 1962-1968 (1) – ISSN: 0020-1901 – mf#8399 – us UMI ProQuest [240]
Insight / United States. Air Force – Zaragoza Air Base, Spain. 1982 feb 5-1986 apr 25 [v24 n15] – 1r – 1 – (cont by: warrior) – mf#1050894 – us WHS [071]
Insight : voice of the black student league / Black Student League [Temple University] – n1 [[1967?]] – 1r – 1 – (cont by: maji-maji [black student league [temple university]]) – mf#4877579 – us WHS [366]
Insight on the news – Washington. 1987+ – 1,5,9 – (cont: insight) – ISSN: 1051-4880 – mf#15662,01 – us UMI ProQuest [073]
Insignia del...almirante...principe de la paz – 1807 – 9 – sp Bibl Santa Ana [920]
Insinjur Indonesia *see* Persatuan insinjur indonesia
Inskip, Catherine *see* List of guide-books and handbooks dating from 1800 to the present
Insolite colombie / Dem, Marc – Paris, France. 1965 – 1r – us UF Libraries [972]
Insolvency : the other side – [Montreal?: s.n, 1879?] [mf ed 1992]] – 1mf – 9 – 0-665-94642-2 – mf#94642 – cn Canadiana [346]
The insolvent act of 1864 : with tariff, notes, forms, and a full index / Edgar, James David – Toronto: Rollo & Adam, 1864 – 2mf – 9 – mf#35040 – cn Canadiana [346]
The insolvent act of 1875 and amending acts – Toronto: R Carswell, 1877 – 6mf – 9 – (ann by samuel robinson clarke. incl ind) – mf#10555 – cn Canadiana [346]
The insolvent law, of maine / Hamlin, Charles – Portland, Loring, Short & Harmon, 1878. 162 p. LL-81 – 1 – us L of C Photodup [346]
Insomnis cura parentum / Moscherosch, Johann Michael; ed by Pariser, Ludwig – Halle: Max Niemeyer, 1893 [mf ed 1993] – (= ser Neudrucke deutscher litteraturwerke des 16. und 17. jahrhunderts 108-109) – viii/139p – 1 – (fr 1643 ed. incl bibl ref) – mf#8413 reel 5 – us Wisconsin U Libr [450]
Inspeccion Provincial de Ensenanza Prima Ria *see* Normas y cuestionarios para las clases especia les de adultos
Inspecteur grey / Gragnon, Alfred – Paris, France. 1935? – 1r – us UF Libraries [440]
Inspecteur vous demande / Priestley, John Boynton – Paris, France. 1950 – 1r – us UF Libraries [440]
Inspection et reglementation concernant la prevention des incendies / Laurin, Fernand et al – [Quebec]: Ministere des affaires municipales...1981 [mf ed 1985] – 1mf – 9 – mf#SEM105P481 – cn Bibl Nat [345]
Inspection of selected intelligence and special access program work-for-others projects / United States. Dept of Energy. Office of Inspector General – Washington DC: Office of Inspections; Oak Ridge TN: US Dept of Energy 1993 [mf ed 1994] – 1mf – 9 – us GPO [333]
Inspection reports and related records received by the inspection branch in the confederate adjutant and inspector general's office / U.S. War Dept. Confederate Records – (= ser War department coll of confederate records) – 18r – 1 – (with printed guide) – mf#M935 – us Nat Archives [324]
Inspection reports of prisoners of war camps, florence, arizona, and navajo ordnance depot, flagstaff, arizona, 1945 – Washington, DC, National Archives and Records Service, [19–] – us CRL [975]
Inspection reports of the office of the inspector general, 1814-1842 / U.S. Office of the Inspector General – (= ser Records Of The Office Of The Inspector General (Army)) – 3r – 1 – (with printed guide) – mf#M624 – us Nat Archives [355]
Inspector – Dublin, Ireland.7 Sept-19 Oct 1850. -w – 1/4r – 1 – uk British Libr Newspaper [072]
Inspector and national magazine – London. 1826-1827 – 1 – mf#4267 – us UMI ProQuest [073]
Inspector general / Gogol, Nikolai Vasilevich – New York, NY. 1931 – 1r – us UF Libraries [960]
Inspector-general sir james ranald martin / Fayrer, Joseph – London: A D Innes, 1897 – us CRL [920]
Inspector's guide to the implementation of the provisions of the foodstuffs, cosmetics and disinfectants act, 1972 (act 54 of 1972) = Gids vir inspekteurs vir die uitvoering van die bepalings van die wet op voedingsmiddels, skoonheidsmiddels en ontsmettingsmiddels, 1972 (wet 54 van 1972) / South Africa. Department of National Health and Population Development [Departement van Nasionale Gesondheid en Bevolkingsontwikkeling – rev repr. Pretoria: Dept of National Health & Population Development [1989] [mf ed Pretoria, RSA: State Library [199-]] – 8p on 1r with other items – 5 – (in english & afrikaans) – mf#op 09461 r23 – us CRL [344]
Inspeksi dinas pertanian rakjat lapuran tahunan – Medan, 1960 – 6mf – 9 – mf#SE-834 – ne IDC [950]
Inspektion des militaer-luft- und kraftfahrwesens *see* General-inspektion des militaer-verkehrswesens (bestand ph 9 5) / inspektion des militaer-luft- und kraftfahrwesens (bestand ph 9 20) bd 25
Inspiration : the infallible truth and divine authority of the holy scriptures / Bannerman, James – Edinburgh: T & T Clark, 1865. Beltsville, Md: NCR Corp, 1978 (7mf); Evanston: American Theol Lib Assoc, 1984 (7mf) – 9 – 0-8370-0985-5 – (incl bibl ref and index) – mf#1984-4331 – us ATLA [220]
Inspiration / Watson, Frederick – London: SPCK; New York: E S Gorham, 1906 – 1mf – 9 – 0-7905-2446-5 – mf#1987-2446 – us ATLA [220]
The inspiration and accuracy of the holy scriptures / Urquhart, John – London: Marshall Brothers, [1895?] – 2mf – 9 – 0-7905-2156-3 – (incl ind) – mf#1987-2156 – us ATLA [220]
The inspiration and authority of the bible / Clifford, John – 2nd rev enl ed. London: James Clarke 1895 [mf ed 1985] – 1mf – 9 – 0-8370-2688-1 – (incl ind) – mf#1985-0688 – us ATLA [220]
Inspiration and inerrancy : a history and a defense / Smith, Henry Preserved – Cincinnati: R Clarke, 1893, c1982 – 1mf – 9 – 0-7905-6786-5 – (includes charges brought against smith by the presbytery of cincinnati of the presbyterian church in the u.s.a) – mf#1988-2786 – us ATLA [240]
Inspiration and inerrancy : inaugural address; together with papers upon biblical scholarship and inspiration / Briggs, Charles Augustus et al – London: James Clarke, 1891. Chicago: Dep of Photodup, U of Chicago Lib, 1979 (1r); Evanston: American Theol Lib Assoc, 1984 (1r) – 1 – 0-8370-1319-4 – (incl bibl ref) – mf#1984-T166 – us ATLA [220]
Inspiration and interpretation : seven sermons preached before the university of oxford / Burgon, John William – Oxford: J H & Jas Parker, 1861 [mf ed 1989] – 2mf – 9 – 0-7905-0819-2 – (incl bibl ref) – mf#1987-0819 – us ATLA [240]
Inspiration and other lectures / Rooke, Thomas George – Edinburgh: T & T Clark, 1893 – 1mf – 9 – 0-8370-4961-X – (incl bibl ref) – mf#1985-2961 – us ATLA [240]
Inspiration and the bible : an inquiry / Horton, Robert Forman – [2nd ed]. New York: E P Dutton, [c1888] – 1mf – 9 – 0-8370-3667-4 – mf#1985-1667 – us ATLA [220]
Die inspiration der heiligen schrift und ihre bestreiter : eine biblisch-dogmengeschichtliche studie / Rohnert, Wilhelm – Leipzig: Georg Boehme (E Ungleich), 1889 – 1mf – 9 – 0-8370-4955-5 – (incl bibl ref) – mf#1985-2955 – us ATLA [220]
Die inspiration der helden der bibel und der schriften der bibel / Gess, Wolfgang Friedrich – Basel: R Reich 1892 [mf ed 1992] – 2mf – 9 – 0-524-04396-5 – mf#1992-0089 – us ATLA [220]
Die inspiration des neuen testamentes / Dausch, Petrus – Muenster i W: Aschendorff 1912 [mf ed 1993] – (= ser Biblische zeitfragen 5/2) – 1mf – 9 – 0-524-07117-9 – mf#1992-1033 – us ATLA [225]
Inspiration in men, books, and movements / Martin, George Currie – London: Hunter & Longhurst, [191-?] – 1mf – 9 – 0-7905-0101-5 – mf#1987-0101 – us ATLA [240]
The inspiration of holy scripture : five sermons / Hervey, A C – Cambridge: Macmillan; London: T. Hatchard, 1856 – 1mf – 9 – 0-7905-1896-1 – mf#1987-1896 – us ATLA [220]
The inspiration of holy scripture, its nature and proof : eight discourses. preached before the university of dublin / Lee, William – 2nd ed. London: Rivingtons, 1857 – 2mf – 9 – 0-524-08032-1 – mf#1992-1125 – us ATLA [220]
The inspiration of prophecy : an essay in the psychology of revelation / Joyce, Gilbert Cunningham – London; New York: Oxford University Press, 1910 – 1mf – 9 – 0-7905-0958-X – (incl bibl ref) – mf#1987-0958 – us ATLA [150]
The inspiration of responsibility, and other papers / Brent, Charles Henry – New York: Longmans, Green, 1915 – 1mf – 9 – 0-7905-3642-0 – mf#1989-0135 – us ATLA [240]
The inspiration of the holy scriptures : being the baird lecture for 1873 / Jamieson, Robert – Edinburgh: William Blackwood, 1873 – 1mf – 9 – 0-8370-3765-4 – (includes explanatory notes at end of text) – mf#1985-1765 – us ATLA [220]
The inspiration of the holy scriptures : a sermon / Smith, Henry Boynton – New-York: John A Gray, 1855 – 1mf – 9 – 0-524-00107-3 – mf#1989-2807 – us ATLA [220]
Inspiration of the holy writings of the old and new testaments : considered and improved in fourteen sermons preach'd at the merchants lecture at salters hall by edmund calamy / Calamy, Edmund – London: T Parkhurst, 1710 – 1r – 1 – 0-8370-1116-7 – mf#1984-T089 – us ATLA [220]
The inspiration of the new testament / Browne, Walter Raleigh – London: C Kegan Paul, 1880 – 1mf – 9 – 0-8370-2481-1 – mf#1985-0481 – us ATLA [225]
The inspiration of the old testament inductively considered : the seventh congregational union lecture / Cave, Alfred – 2nd ed. London: Congregational Union of England and Wales, 1888 – 2mf – 9 – 0-8370-9452-6 – (incl bibl ref) – mf#1986-3452 – us ATLA [221]
The inspiration of the scriptures / Patton, Francis Landey – Philadelphia: Presbyterian Board of Publication, c1869 – 1mf – 9 – 0-8370-4677-7 – mf#1985-2677 – us ATLA [220]
Die inspirationslehre des heiligen hieronymus : eine biblisch-geschichtliche studie / Schade, Ludwig – Freiburg i. B, St Louis MO: Herder, 1910 – (= ser [Biblische Studien]) – 1mf – 9 – 0-8370-1845-5 – mf#1987-6233 – us ATLA [220]
Inspired through suffering / Mears, David Otis – New York: F. H. Revell, [c1895] Beltsville, Md: NCR Corp, 1978 (2mf); Evanston: American Theol Lib Assoc, 1984 (2mf) – 9 – 0-8370-1007-1 – mf#1984-4363 – us ATLA [240]
Inspired through suffering / Mears, David Otis – New York: Fleming H Revell, c1895 – 1mf – 9 – 0-8370-6334-5 – mf#1986-0334 – us ATLA [240]
The inspired word : a series of papers and addresses delivered at the bible-inspiration conference, philadelphia, 1887 / ed by Pierson, Arthur Tappan – New York: Anson D F Randolph, c1888 – 1mf – 9 – 0-8370-4751-X – (incl bibl ref) – mf#1985-2751 – us ATLA [220]
Inspired word of god / Stock, John – London, England. 18– – 1r – us UF Libraries [240]
Inspiring the dance teaching/learning process through motivation / Nolan, V Lynn – 1997 – 1mf – 9 – $4.00 – mf#PE 3983 – us Kinesology [370]
Ins's march 2002 notification of approval of change of status for pilot training for terrorist hijackers mohammed atta and marwan al-shehhi : hearing...house of representatives, 107th congress, 2nd session, mar 19 2002 / United States. Congress. House. Committee on the Judiciary. Subcommittee on Immigration and Claims – Washington: US GPO 2002 [mf ed 2002] – 1mf – 9 – 0-16-068385-8 – us GPO [342]
Instabilitaeten, bifurkationen und chaos fuer einen dreidimensionalen van der pol oszillator / Suenner, Tobias – (mf ed 1995) – 2mf – 9 – €40.00 – 3-8267-2097-0 – mf#DHS 2097 – gw Frankfurter [530]
Instability of the pastoral relation / Ide, George Barton – Springfield: Samuel Bowles, 1854 – 1mf – 9 – 0-524-08390-8 – mf#1993-3090 – us ATLA [240]
L'installation des tutshokwe dans l'empire lunda 1850-1903 / N'Dua, Edouard – Leopoldville, Universite Lovanium de Kinshasa, 1971 – us CRL [960]
Instances of accessory art / Day, Lewis Foreman – London 1880 – (= ser 19th c art & architecture) – 1mf – 9 – mf#4.2.1277 – uk Chadwyck [740]
L'instant : revue franco-catalane d'art de litterature – Paris, Barcelona.v1 n1-7 8; v2, n1-4. juil 1918-sept 1919 – 1 – fr ACRPP [073]
Instante cernido, 1952-1953 / Oraa, Pedro De – Habana, Cuba. 1953 – 1r – us UF Libraries [972]
InStep – 1985 dec 12-1987 jun 10, 1987 jun 11-1988 aug 3, 1988 aug 4-1989 aug 2, 1989 aug 3-1990 jul 18, 1990 jul 19-1991 jun 19, 1991 jun 20-1992 may 6 – 6r – 1 – (cont: wisconsin step; q-voice; cont by: wisconsin in step) – mf#2437089 – us WHS [071]
Instinct and experience / Morgan, Conwy Lloyd – London: Methuen, 1912 – 1mf – 9 – 0-7905-9524-9 – mf#1989-1229 – us ATLA [150]
Institucion cultural santandereana / Arias, Juan De Dios – Bogota, Colombia. 1954 – 1r – us UF Libraries [972]
Institucion harmonica : a doctrina musical, theorica y practica, que trata del canto llano, y de organo / Roel del Rio, Antonio Ventura – Madrid: por los herederos de la viuda de J Garcia Infanzon 1748 [mf ed 19–] – 5mf – 9 – mf#fiche 135 – us Sibley [780]
Instituciones de derecho canonico / Lopez Y Lleras, Rudesindo – Bogota, Colombia. 1948 – 1r – us UF Libraries [972]
Instituciones de derecho civil patrio / Cruz, Fernando – Guatemala, v1-3. 1882-1884 – 2r – us UF Libraries [972]
Instituciones del nuevo reino de granada al tiempo / Ots Y Capdequi, Jose Maria – Madrid, Spain. 1958 – 1r – us UF Libraries [972]
Instituciones practicas de los juicios civiles / Canada, Conde de la – 1794 – 9 – sp Bibl Santa Ana [340]
Instituciones sociales de la america espanola en el periodo colonial. la plata, 1934 / Ots, Jose Maria – Madrid: Razon y Fe, 1935 – 1 – sp Bibl Santa Ana [970]
Institucion...iglesia / Nunez de Torres, Juan – 1618 – 9 – sp Bibl Santa Ana [260]
Institut agama islam negeri al-djami'ah / Al-Djami'ah – Jogjakarta, 1962-1967. v1-6(3) – 18mf – 9 – (missing: 1966 v5) – mf#SE-359 – ne IDC [950]
Institut agama islam negeri "sunan kalidjaga" dewan mahasiswa progres report dewan mahasiswa iain "sunan kalidjaga", 1385-1387, 1965-1967 / Jogjakarta, Indonesia (City) – [Jogjakarta, 1968] – 2mf – 9 – mf#SE-6491 – ne IDC [959]
Institut agama islam negeri "sunan kalidjaga" laporan tahunan / Jogjakarta, Indonesia (City) – Jogjakarta, 1951/1952-1967/1968 – 10mf – 9 – (missing: 1954/1955; 1957/1958; 1959/1960-1964/1965; 1966/1967) – mf#SE-481 – ne IDC [959]
Institut canadien de Quebec *see* Reglements du bureau de direction
Institut canadien (Montreal, Quebec) *see* Les fetes colombiennes a quebec
Institut d'afrique : le conseil superieur, par sa deliberation du 1er novembre 1842 a nomme membre president bienfaiteur de l'institut d'afrique, the honorable robert baldwin... – S.l: s.n, 1842? – 1mf – 9 – mf#47395 – cn Canadiana [360]
L'institut de recherche et d'histoire des textes (irht) : three important catalogues for the study of classical and mediaeval latin texts and their authors – [mf ed Chadwyck-Healey] – 3 catalogues on 989mf – 9 – (catalogue 1: repertoire bio-bibliographique des auteurs latins, patristiques et medievaux 492mf. catalogue 2: repertoire des fins de textes latins classiques et medievaux 224mf. catalogue 3: repertoire d'incipit de sermons latins antiquite tardive et moyen age 273mf) – uk Chadwyck [450]
Institut de recherches scientifiques au Congo *see* Bulletin
Institut Des Parcs Nationaux Du Congo Belge *see* Parcs nationaux du congo belge
L'institut des petites filles de saint-joseph : 1. methode d'oraison 2. catechisme des voeux / Pretre de Saint-Sulpice – Montreal: impr de La Salle, [1923?] (mf ed 2001) – 9 – cn Bibl Nat [241]
L'Institut d'etudes et recherches balkaniques *see* Balcania
Institut Francais d'Afrique Noire *see*
– L'agglomeration dakaroise; quelques aspects sociologiques et demographiques
– La presqu'ile du cap-vert.
L'Institut Francais d'Afrique Noire *see*
– Bulletin
Institut francais de Damas *see* Bulletin d'etudes orientales
Institut Francais d'Opinion Publique *see* Bulletin d'informations
Institut fuer Zeitgeschichte Muenchen *see*
– Akten der parteikanzlei der nsdap
– Widerstand als "hochverrat" 1933-1945
Institut genealogique Drouin *see* Une oeuvre nationale
Institut General Psychologique. Paris *see* Bulletin

Institut Haitien De Statistique see Guide economique de la republique d'haiti

Institut historique et geographique du bresil / Fleiuss, Max – Rio de Janeiro, Brazil. 1938 – 1r – us UF Libraries [972]

Institut Indochinois pour l'Etude de l'Homme. Hanoi see Bulletins et travaux compte rendu des seances

Institut istorii, filologii i filosofii SO AN SSSR see Materialy polevykh issledovanii dal'nevostochnoi arkheologicheskoi ekspeditsii

Institut Keguruan dan Ilmu Pendidikan see Gema alma mater

Institut Keguruan dan Ilmu Pendidikan, Bandung see Laporan ikip badan penerbitan institut

Institut keguruan dan ilmu pendidikan buku pedoman – Medan, 1969-1971 – 10mf – 9 – mf#SE-1779 – ne IDC [950]

Institut keguruan dan ilmu pendidikan buku tahunan – Medan, [1957]-1970 – 2mf – 9 – (several issues missing) – mf#SE-1780 – ne IDC [950]

Institut Keguruan dan Ilmu Pendidikan, Jogjakarta see Balai penelitian pendidikan bulletin bpp

Institut keguruan dan ilmu pengetahuan / Berita IKIP – Bandung, 1964-1965 – 3mf – 9 – mf#SE-457 – ne IDC [959]

Institut konkretnykh sotsialnykh issledovanii AN SSSR see
- Dinamika izmeneniia polozheniia zhenshchiny i semia

Institut konkretnykh sotsialnykh issledovanii SSSR see Dinamika izmeneniia polozheniia zhenshchiny i semia

Institut Lenina. Moscow see Zapiski

Institut mirovogo khoziaistva i mirovoi politiki (Akademiia nauk SSSR) see Mirovaia voina v tsifrakh

L'Institut Pasteur d'Algerie see Archives de l'institut pasteur d'algerie

Institut Pasteur, Paris, France see Bulletin de l'institut pasteur

Institut Pertanian Bogor see Fakultas mekanisasi dan teknologi hasil pertanian katalog

Institut Russkoi Literatury see Orevnerusskie rukopisi pushkinskogo doma

Institut selskokhoziaistvennoi i promyslovoi kooperatsii v 1922-23 v akademicheskom godu – 1923 – 31p 1mf – 9 – mf#COR-249 – ne IDC [335]

Institut Technique du Batiment et des Travaux Public see Annales

Institut teknologi madjalah proceedings – Bandung, 1961-1970 – 18mf – 9 – (missing: 1961 v1(1)) – mf#SE-451 – ne IDC [959]

Institut teknologi rentjana peladjaran – Bandung, 1951-1965 – 24mf – 9 – (missing: 1952-58; 1961) – mf#SE-452 – ne IDC [959]

Institut teoreticheskoi astronomii (Akademiia nauk SSSR) see Biulleten' instituta teoreticheskoi astronomii

Institut-canadien en 1852 / Dorion, Jean Baptiste Eric – Montreal?: W-H-Rowen, 1852 – 3mf – 9 – mf#37510 – cn Canadiana [360]

The institute / American Law Institute. Philadelphia – 1927. 107 p. LL-2342 – 1 – us L of C Photodup [340]

The institute – Vancouver: Vancouver Young People's Methodist Institute, [1889-18– or 19–] [incomplete] – 9 – mf#P05126 – cn Canadiana [242]

Institute d'estudis Catalans. Barcelona see Biblioteca de catalunya

Institute for African Studies, USSR Academy of Sciences and Institute for International Studies, University of California, Berkeley see Papers of the second soviet-american coference on sub-saharan africa, jun 26-29 1984

Institute for Career and Vocational Training [Culver City CA] see Indian youth

Institute for Colored Youth at Cheyney see Annual report

Institute for Corporate Studies [Newton MA] see Changing work

Institute for Deaf, Dumb and Blind Colored Youths of the State of Texas see
- Annual report of the board and superintendent of the institute for deaf, dumb and blind colored youths of the state of texas
- Annual report of the trustees and superintendent of the institute for deaf, dumb and blind colored youths of the state of texas
- Report of the trustees and superintendent of the institute for the deaf and dumb and blind colored youth of the state of texas
- Report of the trustees and superintendent of the institute for the deaf, dumb and blind colored youth of the state of texas

Institute for Defence Studies and Analyses see Idsa journal

Institute for Defense and Disarmament Studies [US] see Defense and disarmament news

Institute for Democratic Analysis see Democratic progress

Institute for First Amendment Studies et al see Freedom writer

Institute for Global Education see Equity

Institute for Independent Social Journalism, Inc [New York, NY] see Guardian

Institute for Islamic Involvement see Vision

Institute for Peace and Justice [US] see Ipj newsletter

Institute for Policy Studies see In these times

Institute for Rational-Emotive Therapy see Rational living

Institute for Research and Information on Multinationals see Irm multinational reports

Institute for Rubber Research and Development see Karet

Institute for Scientific Socialism see Line of march

Institute for Scientific Socialism [Oakland, CA] see Frontline

Institute for Social and Cultural Change see Zeta magazine

Institute for Social Justice see Organizer

Institute for Southern Studies see Southern exchange

Institute For The Comparative Study Of Political S... see Venezuelan elections of december 1, 1963

Institute for the Development of Indian Law see Block grants and indian tribes

Institute for the Development of Positive Relationships see Real mccoy, pages from history

Institute for the Study of Nonviolence see Journal

Institute for the Study of Relevant Progressivism see
- In search of a relevant progressivism
- In search of facts, ideas, challenges

Institute for the study of the ussr bulletin : english edition – Muenchen. 1954-1971 (1) 1970-1971 (5) – ISSN: 0020-2649 – mf#2230 – us UMI ProQuest [321]

The institute leaflet for church sunday schools – Toronto: Rowsell and Hutchison, [1881?-18– or 19–] – 9 – mf#P05078 – cn Canadiana [242]

Institute of advanced legal studies annual reports – University of London. 1st to 36th. 1947-83 – 45mf – 9 – $202.00 – mf#LLMC 84-583 – us LLMC [340]

Institute of African Studies, University of Ghana see Oral traditions of gonja

Institute of Alaska Native Arts see Newsletter of native arts

Institute of American Genealogy see Bulletin of notes and queries

Institute of American Indian Arts see Drumbeats

Institute of Certified Financial Planners (US) see Journal of financial planning

The Institute of Chartered Accountants. England and Wales see Rare books on accountancy and related subjects collection

Institute of Commonwealth Studies see Political party, trade union and pressure group materials

Institute of Contemporary Thought see Post american

Institute of economics of the communist academy, 1921-1937 : from the archive of the russian academy of sciences – (= ser The Russian Archives) – 74r – 1 – us Primary [335]

Institute of Environmental Sciences see Journal of the institute of environmental sciences

Institute of Environmental Sciences and Technology see Journal of the iest

Institute of History of the Spanish Civil War see Bibliography of the spanish civil war 1936-1939

Institute of Human Relations [American Jewish Committee] see News and views: reports from the american jewish committee

Institute of indian studies / University of South Dakota – 1956-81 – (= ser American indian periodicals... 1) – 7mf – 9 – $95.00 – us UPA [305]

Institute of Industrial Engineers see Iie transactions

Institute of International Education (New York, NY) see News bulletin of the institute of international education

Institute Of Jamaica see
- Jamaica in 1896
- Jamaica in 1897
- Jamaica in 1928

Institute of Labor and Industrial Relations see Poverty and human resources abstracts

Institute of Marine Engineers see Transactions of the institute of marine engineers

Institute of mathematical statistics bulletin – Hayward. 1974+ (1) 1974+ (5) 1977+ (9) – ISSN: 0146-3942 – mf#9320 – us UMI ProQuest [310]

Institute of Pacific Relations Conference (4th: 1931: Shanghai, China) see Tsui chin t'ai-p'ing yang wen t'i

Institute of Paper Chemistry see Bulletin of the institute of paper chemistry

Institute of Petroleum (Great Britain) see Journal of the institute of petroleum

Institute of Positive Education (Chicago IL) see Black books bulletin

Institute of Texan Cultures. Library see
- Early czech newspapers of texas
- Early texas newspapers
- Translations of statistical and census reports of texas

Institute of the Black World see
- Ibw special report
- Monthly report

Institute of Transportation Engineers see Ite journal

Institute of Welding see Transactions of the institute of welding

Institute on eminent domain (southwestern legal foundation) : proceedings – v1-9. 1959-68 – 1,5,6 – $121.00 set – mf#103401 – us Hein [340]

Institute on estate planning university of miami : annual proceedings – v1-33. 1967-99 – 1,5,6 – $1100.00 set – (v1-26 1967-92 in reel $841. v27-33 1993-99 in mf $259) – mf#103411 – us Hein [340]

Institute on federal taxation (new york university) : proceedings – v1-56. 1942-98 – 1,5,6 – $2676.00 set – (v1-50 1942-92 in reel $2375. v51-56 1993-98 in mf $301) – mf#103421 – us Hein [336]

Institute on labor southwestern legal foundation. labor law developments : annual proceedings – v11-43. 1964-97 – 1,5,6 – $471.00 set – (v11-39 1964-92 in reel $385. v40-43 1994-97 in mf $86) – mf#103431 – us Hein [344]

Institute on oil and gas law and taxation. southwestern legal foundation : annual proceedings – v1-48. 1949-97 – 1,5,6 – $856.00 set – (v1-42 1949-91 in reel $655. v43-48 1992-97 in mf $201) – mf#103451 – us Hein [343]

Institute on planning and zoning (southwestern legal foundation) : proceedings – v1-8. 1960-69 (all publ) – 1,5,6 – $83.00 set – mf#103461 – us Hein [340]

Institute on planning zoning and eminent domain (southwestern legal foundation) : proceedings – v1-22. 1971-92 – 1,5,6 – $325.00 set – (available on reel only) – mf#103471 – us Hein [340]

Institute on private investments and investors abroad : southwestern legal foundation – v1-41. 1959-98 – 1,5,6 – $595.00 set – (v1-36 1959-93 in reel $468. v37-41 1994-98 in mf $127) – mf#103481 – us Hein [073]

Institute on the Church in Urban-Industrial Society see Justice ministries

The institutes : a textbook of the history and system of roman private law = Institutionen / Sohm, Rudolf – 3rd ed. Oxford: Clarendon Press, 1907 – 2mf – 9 – 0-524-03775-2 – (incl bibl ref. in english) – mf#1990-1122 – us ATLA [340]

Institutes of common and statute law / Minor, John Barbee – 2d ed. Richmond, 1876-95. 4v. in 6. LL-1432 – 1 – (4th ed. richmond, 1891-. 2v. II-860) – us L of C Photodup [348]

The institutes of law : a treatise on the principles of jurisprudence / Lorimer, James – Edinburgh: T & T Clark, 1872 – 5mf – 9 – $7.50 – mf#LLMC 95-175 – us LLMC [340]

Institutes of masonic jurisprudence : being an exemplification of the english book of constitutions, methodically digested under appropriate heads, together with a summary view of the laws and principles of the royal arch / Oliver, George – London: R Spencer 1859 – 6mf – 9 – mf#vrl-84 – ne IDC [366]

Institutes of mussalman law : a treatise on personal law according to the hanafite school, with references to original arabic sources...1795 to 1906 / Abdur Rahman, A F M – Calcutta: Thacker, Spink, 1907 [mf ed 1987] – lxi/532p – 1 – (app contains arabic text) – mf#10695 – us Wisconsin U Libr [260]

Institutes of the christian religion / Calvin, John – 6th amer ed, rev & corr. Philadelphia: Presbyterian Board of Publ & Sabbath-School Work, 1921 [mf ed 2004] – 2v – 1 – (trans fr original latin & collated with aut's last ed in french by john allen) – mf#11046 – us Wisconsin U Libr [240]

Institutes of the christian religion / Gerhart, Emanuel Vogel – New York: AC Armstrong, 1891-1894 – 4mf – 9 – 0-7905-9276-2 – mf#1989-2501 – us ATLA [240]

The institutes of vishnu (stbe7) – 1880 – (= ser Sacred book of the east (sbte)) – 7mf – 8 – €15.00 – (trans by julius jolly) – ne Slangenburg [280]

L'instituteur des instituteurs. st jean de la salle. paris, 1929 / Laudet, Fernand – Madrid: Razon y Fe, 1930 – 1 – sp Bibl Santa Ana [060]

L'instituteur rural : revue d'education – Port-au-Prince: Publiee par le Service national de la production agricole et de l'enseignement rural, [1939-]. v1 n1,3-4. 3rd qtr 1939, 1st-2nd qtr 1940 – 1r – 1 – us CRL [370]

Institutio christianae religionis, in libros quatuor nunc primum digesta, certisque distincta capitibus, ad aptissimam methodum : aucta etiam tam magna accessione ut propemodum opus novum haberi possit / Calvin, J – Genevae: Robert Estienne, 1559 – 11mf – 9 – mf#CL-14 – ne IDC [240]

Institutio christianae religionis nunc vere demum suo titulo respondens / Calvin, J – Argentorati: Per Wendelinum Rihelium, 1539 – 9mf – 9 – mf#CL-13 – ne IDC [240]

Institutio christianae religionis nunc vere demum suo titulo respondes / Calvin, J – Argentorati: Per Wendelinum Rihelium, 1543 – 10mf – 9 – mf#CL-41 – ne IDC [240]

Institutio in musicen mensuralem – 1513 – (= ser Mssa) – 1mf – 9 – €20.00 – mfchl 459 – gw Fischer [780]

Institutio theologiae elencticae / Turrettini, F – Geneve, de Tournes, 1679-1685. 3 v – 26mf – 9 – mf#PFA-207 – ne IDC [240]

Institutio totius christianae religionis, nunc ex postrema authoris recognitione, quibusdam locis auctior, infinitis vero castigatior / Calvin, J – Genevae: Ex officina Joannis Gerardi, 1550 – 9mf – 9 – mf#CL-42 – ne IDC [240]

Institution and abuse of ecclesiastical property / Hull, Edward – London, England. 1831 – 1r – us UF Libraries [240]

Institution catholique, o- est declaree et confirmee la verite de la foy / Coton, P – Paris, 1612 – 13mf – 9 – mf#CA-122 – ne IDC [241]

Institution de la religion chrestienne / Calvin, J – Geneve: Chez Jean Crespin, 1560 – 13mf – 9 – mf#CL-16 – ne IDC [240]

Institution de la religion chrestienne / Calvin, J – Geneve: Par Jean Gerard, 1551 – 13mf – 9 – mf#CL-44 – ne IDC [240]

Institution de la religion chrestienne : en laquelle est comprinse une somme de piete, et quasi tout ce qui est necessaire a congnoistre en la doctrine de salut / Calvin, J – [Geneva: Michel Du Bois], 1541 – 10mf – 9 – mf#CL-15 – ne IDC [240]

Institution de la religion chrestienne: composee en latin par jehan calvin, et translatee en francoys par luymesme : en laquelle est comprise une somme de toute la chrestiente. avec la preface adressee au roy: par laquelle ce present livre luy est offert pour confession de foy / Calvin, J – Geneve: Jehan Girard, 1545 – 12mf – 9 – mf#CL-43 – ne IDC [242]

L'institution des sourds-muets de montreal (1848-1948) : bibliographie / Dufresne, Lise – 1964 [mf ed 1979] – (= ser Bibliographies du cours...1947-66) – 2mf – 9 – mf#SEM105P4 – cn Bibl Nat [616]

The institution of a young noble man / Cleland, James – 1607 – 9 – $20.00 – us Scholars Facs [390]

Institution of Chemical Engineers see
- Chemical engineering research and design
- Process safety and environmental protection
- Transactions of the institution of chemical engineers

Institution of Electrical Engineers see Proceedings of the institution of electrical engineers

Institution of gas engineers (london, england) journal – London. 1961-1971 (1) – (cont by: gas engineering and management) – ISSN: 0020-3432 – mf#7129 – us UMI ProQuest [550]

Institution Of Mechanical Engineers (Great Britain) see Proceedings of the general discussion on lubrication

Institution of Mechanical Engineers (Great Britain) see
- Proceedings of the institution of mechanical engineers pt 1
- Proceedings of the institution of mechanical engineers, pt a
- Proceedings of the institution of mechanical engineers pt a
- Proceedings of the institution of mechanical engineers pt b
- Proceedings of the institution of mechanical engineers pt c
- Proceedings of the institution of mechanical engineers pt d
- Proceedings of the institution of mechanical engineers pt e
- Proceedings of the institution of mechanical engineers pt f
- Proceedings of the institution of mechanical engineers pt g
- Proceedings of the institution of mechanical engineers pt h
- Proceedings of the institution of mechanical engineers pt i
- Proceedings of the institution of mechanical engineers pt j
- Proceedings of the institution of mechanical engineers pt k

Institution of mechanical engineers (great britain) proceedings – London. 1974-1982 (1) 1974-1982 (5) 1974-1982 (9) – ISSN: 0020-3483 – mf#11380 – us UMI ProQuest [621]

Institution of Nuclear Engineers see Journal of the institution of nuclear engineers

Institution of Telecommunication Engineers see Students' journal

Institution of the Rubber Industry see Journal of the iri

Institution of the rubber industry transactions and proceedings – London. 1959-1966 (1) – mf#11956,01 – us UMI ProQuest [670]

Eine institution zieht um : therapeutische nachsorge von drogenabhaengigen in laendlicher insel-idylle und szenenaher kunst- und maerchenstadt / Weil, Thomas – (mf ed 1995) – 2mf – 9 – €40.00 – 3-8267-2163-2 – mf#DHS 2163 – gw Frankfurter [360]

The institutional church : a primer in pastoral theology / Judson, Edward – New York: Lentilhon, c1899 – (= ser Hand-books for practical workers in church and philanthropy) – 1mf – 9 – 0-7905-4880-1 – mf#1988-0880 – us ATLA [240]

Institutional distribution – New York. 1984-1993 (1,5,9) – (cont by: id) – ISSN: 0020-3572 – mf#12633 – us UMI ProQuest [660]

Institutional investor : international edition – London. 1978-1999 (1,5,9) – ISSN: 0192-5660 – mf#11853 – us UMI ProQuest [338]

Institutional investor – London. 1967+ (1) 1975+ (5) 1976+ (9) – ISSN: 0020-3580 – mf#8739 – us UMI ProQuest [332]

Institutional laundry – New York. 1957-1972 (1) 1970-1972 (5) – ISSN: 0020-3599 – mf#1677 – us UMI ProQuest [660]

Institutionen des deutschen privatrechts / Heusler, Andreas – Leipzig: Duncker & Humblot. 2v in 1. 1885-86 – (= ser Civil law 3 coll; Systematisches handbuch der deutschen rechtswissenschaft) – 12mf – 9 – (incl bibl ref and index) – mf#LLMC 96-524 – us LLMC [346]

Institutiones, digestum (libri 40-50); novellae constitutiones / Justinian 1 – 12th, 14th c – (= ser Holkham library manuscript books 207,208,209) – 1r – 1 – mf#96931 – uk Microform Academic [340]

Institutiones linguae slavicae dialecti veteris : quae quum apud russos, serbos aliosque ritus graecu, tum apud dalmatas glagolitas ritus latini slavos in libris sacris obtinet / Dobrovsky, Josef – Vindobonae: Schmid 1822 – (= ser Whsb) – 9mf – 9 – €85.00 – mf#Hu 179 – gw Fischer [460]

Institutiones musicae : of korte onderwyzingen rakende de practyk van de musyk en inzonderheid van den generaalen bas... / Zumbag de Koesfelt, Coenraad – Leyden: G Potuliet 1743 [mf ed 19–] – 2mf – 9 – mf#fiche 746 – us Sibley [780]

Institutiones oratoriae / Quintilian – Lipsiae, Germany. 1886 – 1r – us UF Libraries [960]

Institutiones patrologiae quas denuo recensuit auxit / Fessler, J; ed by Jungmann, B – Oeniponte. v1-2. 1890 – €65.00 – ne Slangenburg [240]

Institutiones philosophiae moralis / Ferretti, Augustus – Romae: S C de Propaganda fide, 1893-1902 – 4mf – 9 – 0-8370-6182-2 – (incl bibl ref) – mf#1986-0182 – us ATLA [170]

Institutiones philosophicae / Palmieri, Domenico – Romae: Cuggiani, Santini, 1874-1876 – 4mf – 9 – 0-524-00297-5 – mf#1989-2997 – us ATLA [100]

Institutiones philosophicae ad usum studiosae juventutis... / Demers, Jerome – Quebeci: Ex Typis Tho Cary & Socii, 1835 [mf ed 1974] – 1r – 5 – mf#SEM16P130 – cn Bibl Nat [100]

Institutiones propaedeuticae ad sacram theologiam : 1. de christo legato divino. 2. de ecclesia christi. 3. de locis theologicis / Pesch, Christian – Friburgi Brisgoviae: Herder, 1894 – 1mf – 9 – 0-8370-4713-7 – (incl biblographical references) – mf#1985-2713 – us ATLA [240]

Institutiones que su magestad mando hacer a vis mercado para el examen de los algebristas / Mercado, L – Madrid, 1599 – 3mf – 9 – sp Cultura [500]

Institutiones theologiae dogmaticae generalis seu fundamentalis / Knoll, Albert – ed 7. Augustae Taurinorum [Turin]: Eq P Marietti, 1880 – 2mf – 9 – 0-524-07571-9 – (incl bibl ref) – mf#1991-3191 – us ATLA [240]

Institutiones theologicae antiquorum patrum / Thomasius, J M – Romae. v1-3. 1769 – 3v on 43mf – 8 – €82.00 – ne Slangenburg [240]

Institutiones theologicae seu locorum communium christianae religionis analysis / Buc, G – Geneve, J, 1625 – 10mf – 9 – mf#PFA-123 – ne IDC [240]

Institutions connected with the american church mission in china / Mosher, Gouverneur Frank – New York City: Domestic and Foreign Missionary Society, [1914?] – 1mf – 9 – 0-524-07106-3 – mf#1991-2929 – us ATLA [240]

Institutions constitutionnelles et politiques du cambodge / Gour, Claude-Gilles – Paris: Dalloz 1965 [mf ed 1989] – (= ser Systemes de droit contemporains 18) – 1r with other items – 1 – (pref by rene roblot; incl bibl ref) – mf#mf-10289 seam reel 027/03 [§] – us CRL [323]

Institutions de joliette : (details interessants) / Baillairge, Frederic-Alexandre – Joliette, PQ: Bureaux du Bon combat, du Couvent et de la Famille, 1893 – 1mf – 9 – mf#11750 – cn Canadiana [360]

Institutions ecclesiastiques de la chretiente medievale (he12) – Paris, 1959 – (= ser Histoire de l'eglise (he)) – €31.00 – ne Slangenburg [240]

The institutions of christianity : exhibited in their scriptural character and practical bearing / Jackson, Thomas – London: Wesleyan Conference Office, 1868 – 2mf – 9 – 0-524-07823-8 – mf#1991-3370 – us ATLA [240]

Les institutions politiques et administratives du pays de languedoc du 13e siecle aux guerres de religion. / Dognon, Paul – Toulouse: E. Privat, 1895. xviii/652p – 1 – us Wisconsin U Libr [944]

Les institutions politiques et l'organisation administrative du cambodge ancien, 6e-8e siecles / Sahai, Sachchidanand – Paris: Ecole francaise d'Extreme-Orient 1970 [mf ed 1989] – (= ser Publications de l'ecole francaise d'extreme-orient 75) – 1r with other items – 1 – (with bibl) – mf#mf-10289 seam reel 024/01 [§] – us CRL [959]

Institutions, religious, educational, social... – S.l., S.l? . 193-? – 1r – us UF Libraries [978]

Institutionum dialecticarum...libro octo / [Fonseca, P] – [Coloniae, 1591] – 8mf – 9 – (missing: title p) – mf#CA-15 – ne IDC [240]

Institutionum geometricarum libri quatuor... / Duerer, A – Arnhimiae, 1605 – 4mf – 9 – mf#OA-213 – ne IDC [720]

Institutionum hebraicarum, libri duo / Capiton, W – Argentorati, 1525 – 3mf – 9 – mf#PPE-101 – ne IDC [240]

Institutionum theologicarum... / Haunold, C – Ingoldstadii, 1659 – 7mf – 9 – mf#CA-53 – ne IDC [241]

Instituto Agrario Nacional (Venezuela) *see* Reforma agraria en venezuela

Instituto Brasileiro De Acao Democratica *see* Recomendacoes sobre reforma agraria

Instituto Brasileiro De Administracao Municipal *see* Municipios do brasil

Instituto Brasileiro De Estatistica *see* Nomenclatura brasileira de mercadorias

Instituto Brasileiro De Geografia E Estatistica *see*
– Brazil, 1938
– Novo paisagens do brasil
– Sinopse do censo agricola, dados gerais

Instituto Brasileiro De Geografia E Estatistica C... *see*
– Sinopse do censo comercial, dados gerais
– Sinopse do censo industrial e do censo dos servico

Instituto Brasileiro De Petroleo *see* Economia do petroleo

Instituto Brasileiro De Reforma Agraria *see* Relatorio

Instituto Caro Y Cuervo *see* Bello en colombia

Instituto Centroamericano De Administracion Public *see* Cuatro ensayos sobre administracion postal

Instituto coloniale fascista. Roma *see*
– Annuario delle colonie italiane e dei paesi vicini
– Annuario dell'impero italiano

Instituto Cubano De Estabilizacion Del Cafe *see* Segunda conferencia panamericana del cafe

Instituto de 2nd...curso 1898 a 1899 / Suarez Quintero, Valentin – Badajoz: Tip. Uceda Hermanos, 1900 – 1 – (tambien curso 1899 a 1900) – sp Bibl Santa Ana [370]

Instituto De Estudios Africanos *see* Africa en el pensamiento de donoso cortes

Instituto de Medicina Tropical de Sao Paulo *see* Revista do instituto de medicina tropical de sao paulo

Instituto de segunda ensenanza de merida. memoria del curso 1934-35 / Dominguez, Manuel – Merida: A. Rodriguez, 1936 – 1 – sp Bibl Santa Ana [946]

Instituto De Tierras Y Colonizacion *see* Estudio de la region de upala

Instituto fascista dell'Africa italiana. Roma *see* Annuario dell'africa italiana e delle isole italiane dell'egeo

Instituto general tecnico de badajoz. memoria del curso de 1904 a 1905... / Gonzalez Cuadrado, Antonio – Badajoz: Tip. La Minerva Extremena, 1906 – 1 – (tambien curso 1906-1917) – sp Bibl Santa Ana [370]

Instituto Geografico *see*
– Anuario estadistico de espana
– Censo de poblacion de espana segun el empadronamiento de...1887, tomo 1-3
– Consideraciones demograficas sobre el censo de buenos aires
– Estadistica de la emigracion e inmigracion de espana en 1882-1895
– Estadistica mortuoria de la ciudad de buenos aires
– Memoria de los trabajos realizados
– Memoria elevada al consejo de ministros
– Memoria sobre la estadistica general de espana
– Memorias, tomo 1-12
– La mortalidad infantil en buenos aires
– Movimiento de la poblacion de espana, anos 1861-1870
– Movimiento de poblacion de la plata
– Nomenclator de espana de 1888, tomo 1-5
– Nuevo nomenclator de las ciudades
– Presupuestos generales de la isla filipinas
– Resena geografica y estadistica de espana

Instituto Geografico Agustin Codazzi *see* Nivelacion geodesica

Instituto Geografico 'Agustin Codazzi' Departamen *see* Formaciones vegetales de colombia

Instituto Geologico y Minero de Espana *see* Mapa geologico de espana

Instituto GeologicoMinero de Espana Madrid *see* Mapa geologico de espana. escala 1:50000

Instituto Guatemateco De Seguridad Social *see* Reglamento sobre proteccion relativa a accidentes

Instituto historico / Feijo Bittencourt – Rio de Janeiro, Brazil. 1938 – 1r – us UF Libraries [972]

Instituto historico e geographico brasileiro... / ed by Bayle, Constantino – Madrid: Razon y Fe, 1927 – 1 – sp Bibl Santa Ana [972]

Instituto Historico e Geographico Brasileiro. Rio de Janeiro *see* Revista trimensal

Instituto Laboral Garcia de Paredes *see*
– Memoria. curso 1957-58
– Memoria del curso 1959-60
– Memoria del curso academico 1960-61, 1961-62

Instituto Laboral "Garcia de Paredes". Trujillo *see* Programa de actos que se celebraran con motivo del 10th aniversario de la implantacion de la ensenanza laboral en trujillo

Instituto Laboral General Moscardo *see* Memoria del curso academico 1956-57

Instituto Latinoamericano de Doctrina y Estudios Sociales *see* Documentacion social catolica latinoamericana docla

Instituto Latinoamericano De Mercadeo Agricola *see* Supply problems of basic agricultural products in...

Instituto militar pestolozziano de madrid, obra del extremeno manuel godoy / Guerra Guerra, Arcadio – Badajoz: Imp. Dip. Provincial, 1963 – sp Bibl Santa Ana [350]

Instituto Nacional de 2nd Ensenanza. Caceres *see*
– Memoria del curso 1935 a 1936, 1936 a 1937
– Memoria del curso de 1934 a 1935

Instituto nacional de 2nd ensenanza de badajoz...1926 – Badajoz: Tip. La Economica, 1926 – sp Bibl Santa Ana [946]

Instituto nacional de ensenanza media masculino "zurbaran". badajoz. memoria informativa – Badajoz: Imp. Comercial, 1968 – 1 – sp Bibl Santa Ana [380]

Instituto Nacional de Estadistica *see*
– Nomenclator de las ciudades, villas, lugares, aldeas y demas entidades de poblacion de espana...
– Rese na estadistica de la provincia de badajoz
– Resena estadistica de la provincia de badajoz

Instituto nacional de estadistica. nomenclator de las ciudades de espana. provincia de badajoz – Madrid: Rivadeneira, 1950 – 1 – sp Bibl Santa Ana [946]

Instituto Nacional de Industria *see*
– El plan de badajoz

Instituto Nacional de Industria. Secretaria Gestora del Plan *see* Ley y reglamento sobre el plan de obras y colonizacion, industrializacion y electrificacion de la provincia de badajoz

Instituto Nacional De Obras Sanitarias (Venezuela) *see* Normas para el diseno de los abastecimientos de ag...

Instituto Nacional De Previdencia Social Diretori *see* Atividades do inps, em 1970

El instituto nacional de prevision de laboratorio inicial y preparador del ambiente espanol para la seguridad social / Leal Ramos, Leon – Madrid: ministerio de trabajo publicaciones del instituto nacional de prevision, 1950 – sp Bibl Santa Ana [946]

Instituto Nacional De Prevision Y Reformas Sociale *see* Liga infantil de la paz

Instituto Nacional Do Negro Biblioteca *see* Relacoes de raca no brasil

L'institutrice de province / Frapie, Leon – Paris: Fasquelle, 1897 – (= ser Les femmes [coll]) – 4mf – 9 – mf#11048 – fr Bibl Nationale [370]

Instituts de chymie, ou principes elementaires de cette science / Demachy, Jacques Francois – Paris, 1766 – 1 – us Wisconsin U Libr [540]

Les instituts familiaux de notre province : ecoles de bonheur: bibliographie analytique d'une magnifique formule d'education feminine, 1937-1961 / Marie-Libermann, soeur – 1961 [mf ed 1978] – (= ser Bibliographies du cours...1947-66) – 1mf – 9 – (with ind; pref by a tessier) – mf#SEM105P4 – cn Bibl Nat [370]

Institvtio de tribvs illis religionis svmmis capitibvs : quae hodie inter euangelicas ecclesias...in controuersiam vocantur / Zepperus, W – Hanoviae, Guilielmus Antonius, 1596 – 2mf – 9 – mf#PBU-648 – ne IDC [240]

Institvtio eorvm qui...de fide examinantur... / Bullinger, Heinrich – Tigvri, Christoph Frosch[auer], 1560 – 2mf – 9 – mf#PBU-210 – ne IDC [240]

Institvtionvm grammaticarvm de lingva hebraea liber unus... / Bibliander, T – Tigvri, officina Froschoviana, 1535 – 3mf – 9 – mf#PBU-571 – ne IDC [240]

Instrucao publica no estado de sao paulo / Moacyr, Primitivo – Sao Paulo, Brazil. v1-2. 1942 – 1r – us UF Libraries [972]

Instruccao nacional : revista e pedagogia, sciencias e letras – Rio de Janeiro, RJ: Typ de Quirino F do Espirito Santo, dez 1873-jan 1874 – (= ser Ps 19) – mf#P17,01,168 – bl Biblioteca [370]

Instruccion *see* Instruccion de enfermos y modo de aplicar los remedios a todo genero de enfermedades...

Instruccion civica para las escuelas y colegios / Posada, Eduardo – Bogota, Colombia. 1913 – 1r – us UF Libraries [972]

Instruccion civica para las escuelas y colegios / Posada, Eduardo – Bogota, Colombia. 1928 – 1r – us UF Libraries [972]

Instruccion curativa de las calenturas conocidas...como tabardillo / Amar y Arguedas, J – Madrid, 1775 – 6mf – 9 – sp Cultura [616]

Instruccion curativa de las viruelas / Amar y Arguedas, J – Madrid, 1774 – 4mf – 9 – sp Cultura [615]

Instruccion curativa y preservativa de los dolores de costado y pulmones / Amar y Arguedas, J – Madrid, 1777 – 4mf – 9 – sp Cultura [610]

Instruccion de enfermos y modo de aplicar los remedios a todo genero de enfermedades... / Instruccion – Madrid, 1728 – 4mf – 9 – sp Cultura [615]

Instruccion de los barberos flebotonianos... / Munoz, A – Valencia, 1621 – 3mf – 9 – sp Cultura [615]

Instruccion de musica sobre la gvitarra espanola : y metodo de svs primeros rvdimentos, hasta tanerla con destreza / Sanz, Gaspar – En Zaragoca: por los Herederos de Diego Dormer, ano de 1697 – 3v on 2mf – 9 – (1st ed 1674) – mf#fiche 897 – us Sibley [780]

Instruccion de un passajero...guadalupe – 1697 – 9 – sp Bibl Santa Ana [918]

Instruccion general sobre la manera de redactar los documentos publicos sujetos a registro en las provincias de cuba y puerto-rico : edicion oficial – Madrid: Imprenta Nacional, 1879 – 1mf – 9 – $1.50 – mf#LLMC 92-317 – us LLMC [340]

Instruccion para defensa de los conjuntos historico-artisticos / Ministerio de Education Nacional – Caceres, Madrid: Graficas Varela, 1965 – 1 – sp Bibl Santa Ana [700]

Instruccion para la visita / Spain. Laws, Statutes, etc – 1790 – 9 – sp Bibl Santa Ana [324]

Instruccion pastoral establecidas.. / Delgado Moreno, Mateo – 1815 – 9 – sp Bibl Santa Ana [946]

Instruccion pastoral que el excmo. e iltmo. senor...dirige a los fieles de su diocesis sobre la devocion al s. corazon de jesus / Perez Munoz, Adolfo – Badajoz: Tip. Uceda hermanos, 1918 – 1 – sp Bibl Santa Ana [240]

La instruccion primaria en filipinas. desde 1596 hasta 1868 / Barrantes Moreno, Vicente – Madrid: Imp. de la Iberia, s.a. – 1 – sp Bibl Santa Ana [946]

Instruccion provisional...contribucion – 1821 – 9 – sp Bibl Santa Ana [946]

Instruccion publica / Uribe, Antonio Jose – Bogota, Colombia. 1927 – 1r – us UF Libraries [972]

Instruccion publica en alajuela – San Jose, Costa Rica. 1953 – 1r – us UF Libraries [972]

La instruccion publica en el ecuador de 1830 a 1930 / Tobar Donoso, Julio – Madrid: Razon y Fe, 1931 – 1 – sp Bibl Santa Ana [370]

Instruccion sobre cumplimiento pascual y vida religiosa en las hermandades sindicales de labradores / Rodriguez Amaya, Esteban – Badajoz: Imp. Provincial, 1946 – sp Bibl Santa Ana [240]

Instruccion sobre la peste / Mercado, M – Zaragoza, 1648 – 4mf – 9 – sp Cultura [616]

Instrucciones a los mayordomos de estancias / Rosas, Juan Manuel Jose Domingo Ortiz De – Buenos Aires, Argentina. 1951 – 1r – us UF Libraries [972]

Instrucciones importantes que...sr.d....comunica a todos... / Varela, Cipriano – Plasencia, s.i., 1828 – 1 – sp Bibl Santa Ana [946]

Instrucciones para combatir algunos parasitos del olivo. 1897 – 9 – sp Bibl Santa Ana [630]

Instrucciones para la defensa de los conjuntos historico-artisticos / Ministerio de Educacion Nacional – Caceres – 1 – sp Bibl Santa Ana [700]

Instrucciones para...la conservacion y aumento de las poblaciones / Fernandez, F – Madrid, 1769 – 3mf – 9 – sp Cultura [304]

Instrucciones sobre lapidas / Sanguino y Michel, Juan – Caceres: Tip. Enc. y Lib. Jimenez, 1905 – 1 – sp Bibl Santa Ana [946]

Instruccion...internos...colegio de humanidades de caceres – 1829 – 9 – sp Bibl Santa Ana [370]

Instruccion...plan administrativo... ayuntamientos – 1822 – 9 – sp Bibl Santa Ana [340]

Instruccion...tomar el purgante de mr. le roy – 1829 – 9 – sp Bibl Santa Ana [610]

Instruciones.. / Varela, Cipriano – 1828 – 9 – sp Bibl Santa Ana [240]

Instructio medicorum appollineam aggredientibus valde utilis / Lillo y Herrero, G – Madrid, 1679 – 9mf – 9 – sp Cultura [610]

Instructio sacerdotum locupletissima / Toledo, Francisco de – Lugduni, 1649 – 22mf – 8 – €42.00 – ne Slangenburg [240]

Instruction chrestienne en la doctrine de la loy et de l'evangile / Viret, P – Geneve, Rivery, 1564. 2 v – 28mf – 9 – mf#PFA-202 – ne IDC [240]

Instruction chrestienne et somme generale de la doctrine / Viret, P – Genvee, Badius, 1556 – 13mf – 9 – mf#PFA-199 – ne IDC [240]

Instruction generale pour la teinture des laines et manufactures de laines de toutes couleurs, & pour la culture des drogues ou ingrediens qu'on y employe / Delormois – Paris: Impr. de F. Muguet, 1671 – 1 – us Wisconsin U Libr [670]

Instruction maconnique pour le grade d'apprenti / Fisch, J-C-A – Paris: L'auteur 1863 – 2mf – 9 – mf#vrl-46 – ne IDC [366]

L'instruction obligatoire dans la province de quebec : polemique dandurand-saint-pierre – [Montreal: Ecole sociale populaire, 1912?] – 1mf – 9 – 0-659-91587-1 – mf#9-91587 – cn Canadiana [370]

Instruction of ptah-hotep and the instruction of ke'gemni / Ptah-Hetep – London, England. 1906 – 1r – us UF Libraries [930]

Instruction pastorale de monseigneur l'eveque d'orleans : sur l'immaculee conception de la tres-sainte vierge / Dupanloup, Felix – Paris: Jacques Lecoffre, 1855 [mf ed 1986] – 1mf – 9 – 0-8370-8015-0 – (incl bibl ref) – mf#1986-2015 – us ATLA [241]

Instruction pour les trois grades symboliques du rit moderne – nouv augm ed. Or[ient] de Jerusalem: Le silence 1839 – 2mf – 9 – mf#vrl-2 – ne IDC [366]

L'instruction publique au canada d'apres une publication recente / Le Roy, Alphonse – Bruxelles?: s.n, 1878 – 1mf – 9 – mf#08129 – cn Canadiana [370]

L'instruction publique dans la province de quebec : 1. bref historique 2. organisation scolaire 3. les minorites 4. communautes enseignantes et ecoles normales / Magnan, Charles-Joseph – 2e ed. Quebec: [s.n.], 1934 [mf ed 1990] – 1mf – 9 – mf#SEM105P1271 – cn Bibl Nat [370]

L'instruction publique dans la province de quebec : 1. bref historique 2. organisation scolaire 3. programmes 4. statistiques (resume) 5. appendice: les minorites 6. communautes enseignantes et ecoles normales / Magnan, Charles-Joseph – Quebec: [s.n.] 1932 [mf ed 1990] – 1mf – 9 – mf#SEM105P1270 – cn Bibl Nat [370]

L'instruction publique dans la province de quebec / Cazes, Paul de – [Quebec?: s.n.], 1905 – 1mf – 9 – 0-665-72165-X – (incl bibl ref) – mf#72165 – cn Canadiana [370]

Instruction publique en haiti / Brutus, Edner – Port-Au-Prince, Haiti. 1948 – 1r – us UF Libraries [972]

Instruction sur le cholera – Paris [1848?] – us CRL [616]

Instructional innovator – Washington. 1980-1985 (1,5,9) – (cont: audiovisual instruction with/ instructional resources. cont by: techtrends) – ISSN: 0196-6979 – mf#1475,02 – us UMI ProQuest [370]

Instructional science – Amsterdam. 1986+ (1) 1986+ (5) 1987+ (9) – ISSN: 0020-4277 – mf#16041 – us UMI ProQuest [370]

Instructions and devotions for performing the novena : or, the nine days' devotion to st francis xavier – Quebec: J E Walsh, 1890 [mf ed 1984] – 1mf – 9 – 0-665-45696-4 – mf#45696 – cn Canadiana [241]

Instructions chretiennes pour les jeunes gens : utiles a toutes sortes de personnes... / Humbert, Pierre Hubert – 14e ed. Quebec: Chez John Neilson...1802 [mf ed 1984] – 6mf – 9 – 0-665-45470-8 – (incl text in latin) – mf#45470 – cn Canadiana [240]

Instructions chretiennes pour les jeunes gens : utiles a toutes sortes de personnes: melees de plusieurs traits d'histoire et d'exemples edifians / Humbert, Pierre Hubert – Montreal: Impr & a vendre chez James Brown...1818 [mf ed 1984] – 4mf – 9 – 0-665-37936-6 – (incl latin text) – mf#37936 – cn Canadiana [240]

Instructions des h g tels qu'ils se conferent dans les chapitres de la correspondance du g o de france au rit moderne : avec les discours analogues aux receptions – Paris: Caillot 1835 – 2mf – 9 – mf#vrl-66 – ne IDC [366]

Instructions des trois degres symboliques ecossais du rit ancien et accepte – Paris: Libr Mac de Caillot [18-?] – 1mf – 9 – mf#vrl-114 – ne IDC [366]

Instructions dogmatiques sur le mariage chretien / Braun, Antoine – Montreal?: Paris?: s.n, 1873 – 3mf – 9 – mf#26694 – cn Canadiana [230]

Instructions en langue crise sur toute la doctrine catholique / Lacombe, Albert – [St-Boniface, Man?: s.n.] 1875 [mf ed 1984] – 6mf – 9 – 0-665-30289-4 – mf#30289 – cn Canadiana [241]

Instructions for children / Keach, Benjamin – 1685 – 1 – 6.26 – us Southern Baptist [242]

Instructions for the poor / Green, T – London, England. 1801 – 1r – us UF Libraries [240]

Instructions in criminal causes passed upon by the courts of missouri / Pattison, Everett Wilson – St. Louis: Gilbert, 1902. 607p. LL-1320 – 1 – us L of C Photodup [345]

Instructions maconniques des cinq premiers grades du rit ecossais – [s.l: s.n. 18–?] – 1mf – 9 – mf#vrl-115 – ne IDC [366]

Instructions of the committee to missionaries proceeding to the west africa, india, ceylon, china, and the mediterranean missions : delivered sep 28th, 1860 / Church Missionary Society – London, 1860 – (= ser 19th c books on british colonization) – 1mf – 9 – mf#1.1.514 – uk Chadwyck [240]

Instructions on the commandments and sacraments = Istruzione al popolo sovra i precetti del decalogo / Liguori, Alfonso Maria de', Saint – Dublin: J Duffy, 1869 – 1mf – 9 – 0-524-06186-6 – (in english) – mf#1991-2442 – us ATLA [240]

Instructions philosophiques sur la francmaconnerie / Fleury, Alfred – Bruxelles: Henry Kistemaeckers 1881 – 1mf – 9 – mf#vrl-104 – ne IDC [366]

Instructions populaires pour les premieres communions / Lobry, J-B – 3e ed. Paris: Louis Vives, 1883 – 1mf – 9 – 0-8370-7477-0 – mf#1986-1477 – us ATLA [240]

Instructions populaires sur la priere / Lobry, J-B – 3e ed. Paris: Louis Vives, 1883 – 1mf – 9 – 0-8370-7478-9 – (incl bibl ref) – mf#1986-1478 – us ATLA [240]

Instructions populaires sur le symbole des apaotres / Lobry, J-B – 5e ed. Paris: Louis Vives, 1885 – 2mf – 9 – 0-8370-7479-7 – (incl bibl ref) – mf#1986-1479 – us ATLA [240]

Instructions populaires sur les sacrements / Lobry, J-B – 3e ed. Paris: Louis Vives, 1883 – 1mf – 9 – 0-8370-7480-0 – (incl bibl ref) – mf#1986-1480 – us ATLA [240]

Instructions pour deux car emes et un mois de marie / Lobry, J-B – 5e ed. Paris: Louis Vives, 1885 – 1mf – 9 – 0-8370-7481-9 – (incl bibl ref) – mf#1986-1481 – us ATLA [240]

Instructions pour les trois premiers grades de la franc-maconnerie – [s.l: s.n. 18–?] – 1mf – 9 – mf#vrl-65 – ne IDC [366]

Instructions sur la navigation des indes orientales et de la chine, pour servir au neptune oriental / Apres de Mannevillette, J B – Paris: Dezauche', 1775 – 7mf – 9 – mf#HT-620 – ne IDC [915]

Instructions sur les commandements de dieu et de l'eglise / Lobry, J-B – 4e ed. Paris: Louis Vives, 1883 [mf ed 1986] – (= ser Cours d'instructions populaires) – 484p on 2mf – 9 – 0-8370-7482-7 – (in french. incl bibl ref) – mf#1986-1482 – us ATLA [230]

Instructions to agents for the publication of storm warnings issued from the meteorological office, toronto – [Toronto?: Trout & Todd], 1882 – 1mf – 9 – 0-665-92285-X – mf#92285 – cn Canadiana [550]

Instructions to architects submitting competing designs for the new city hall, quebec / Baillairge, Charles P Florent – Quebec: s.n, 1889 – 1mf – 9 – mf#06640 – cn Canadiana [720]

Instructions to catechists : in twenty chapters, with an appendix of eight chapters / Beschi, C J – Madras: American Mission Press, 1849 [mf ed 1996] – (= ser Yale coll) – 1 – 0-524-10244-9 – (in tamil) – mf#1996-1244 – us ATLA [240]

Instructions to christian converts / Clark, Dougan – Chicago: Publishing Association of Friends, 1889 – 1mf – 9 – 0-7905-3820-2 – mf#1989-0313 – us ATLA [240]

Instructions to juries and declarations of law / Helton, Peter – Springfield, MO: Tuthill, 1892. 599p. LL-890 – 1 – us L of C Photodup [340]

Instructions to juries especially adapted to the laws of texas / Goad, George Washington – Springfield, MO: Democrat, 1893. 561p. LL-669 – 1 – us L of C Photodup [340]

Instructions to lord durham for the constitution of special council – [London, England: s.n, 1838] (mf ed 1991) – 1mf – 9 – mf#SEM105P1458 – cn Bibl Nat [323]

Instructions to teachers and trustees of french-english schools / Ontario. Dept of Education – [Toronto?: s.n, 1889?] – 1mf – 9 – 0-665-89628-X – (incl bibl ref) – mf#89628 – cn Canadiana [370]

Instructions under the direction of the secretary of state for the colonial department : communicated to lieut col cockburn by the rt honble r w horton in a letter dated 26th january 1827... – S.l: s.n, 1827? – 2mf – 9 – mf#59329 – cn Canadiana [324]

Instructive rambles extended in london, and the adjacent villages : designed to amuse the mind and improve the understanding of youth / Helme, Elizabeth – London 1800 [mf ed Hildesheim 1995-98] – 2v on 4mf – 9 – €120.00 – 3-487-27951-7 – gw Olms [914]

Instructive tales / Trimmer, Mrs – London, England. 1848 – 1r – us UF Libraries [240]

Instructor – New York. 1989-1995 (1) 1989-1995 (5) 1989-1995 (9) – (cont: instructor and teacher) – ISSN: 1049-5851 – mf#252,04 – us UMI ProQuest [370]

Instructor – Cleveland. 1986-1988 (1) 1986-1988 (5) 1986-1988 (9) – (cont: instructor and teacher. cont by: instructor and teacher) – ISSN: 0892-9122 – mf#252,02 – us UMI ProQuest [370]

Instructor – [Bay Verte, NB?: s.n.], 1860- – 9 – (cont: parish school advocate and family instructor) – mf#P04012 – cn Canadiana [377]

Instructor – Dansville. 1895-1980 (1) 1968-1980 (5) 1975-1980 (9) – ISSN: 0020-4285 – mf#252 – us UMI ProQuest [370]

Instructor : intermediate ed – New York. 1996-1998 – 1.5.9 – mf#25437 – us UMI ProQuest [373]

Instructor – New York. 1755-1755 – 1 – mf#3526 – us UMI ProQuest [370]

Instructor – New York. 1998+ – 1,5,9 – ISSN: 1049-5851 – mf#28890 – us UMI ProQuest [370]

Instructor : primary ed – New York. 1996-1998 – 1,5,9 – mf#25436 – us UMI ProQuest [370]

The instructor – Montreal: J E L Miller, [1835-18–?] – 9 – ISSN: 1190-7207 – mf#P04278 – cn Canadiana [073]

Instructor and teacher – Dansville. 1980-1986 (1) 1980-1986 (5) 1980-1986 (9) – (cont by: instructor) – ISSN: 0279-3369 – mf#252,01 – us UMI ProQuest [370]

Instructor and teacher – New York. 1989-1989 (1) 1989-1989 (5) 1989-1989 (9) – (cont: instructor. cont by: instructor) – ISSN: 1048-583X – mf#252,03 – us UMI ProQuest [370]

El instruido en la corte...estremeno (sic) / Jara de Soto, Clara – 1789 – 9 – sp Bibl Santa Ana [946]

Instruktazh selskokhoziaistvennoi kooperatsii i ego osnovnye problemy / Makhov, V N – 1926 – 183p 2mf – 9 – mf#COR-489 – ne IDC [335]

Instruktsiia dlia otsenki gorodskikh imushchestv i stroenii v g. : moskve i drugikh gorodakh, prinimaemykh k zalogu moskovskim zemel'nym bankom, utverzhdennaia obshchim sobraniem gg. aktsionerov banka 24 iiunia 1879 g i g ministrom finansov 24 maia 1880 g – M, 1903 – 1mf – 9 – mf#REF-318 – ne IDC [332]

Instruktsiia kazennym palatam : utverzhdena g ministrom finansov po soglasheniiu s g gosudarstvennym kontrolerom 16 marta 1915 g / Ministerstvo Finansov – Pg, 1915 – 9mf – 9 – mf#REF-200 – ne IDC [332]

Instruktsiia Kaznacheistvam: *see* Utverzhdena g ministrom finansov po soglasheniiu s g gosudarstvennym kontrolerom, 21 iiunia 1878 g

Instruktsiia o poriadke kratko-srochnogo kreditovaniia kustarno-promyslovoi kooperatsii i operativnyi uchet / Belokurov, N G & Lapshov, I I – 1930 – 50p 1mf – 9 – mf#COR-411 – ne IDC [335]

Instrument Society of America *see* Isa transactions

Instrumenta ecclesiastica / Cambridge Camden Society – London [1850?-56] – (= ser 19th c art & architecture) – 7mf – 9 – mf#4.2.82 – uk Chadwyck [720]

The instrumental assistant / Holyoke, Samuel – Vol. I. Containing instructions for the violin, German flute, bass-viol, and hautboy...also a selection of favorite airs, marches &c. Exeter, N.H.: H. Ranlet n.d.? Bound and filmed with Vol. II. MUSIC 123, Item 13 and MUSIC 1995 – 1 – us L of C Photodup [780]

The instrumental assistant / Holyoke, Samuel – Vol. II. Containing a selection of minuets, airs, duettos, rondos and marches; with instructions for the French-horn and bassoon. Exeter, N.H.: Ranlet and Norris, 1807. MUSIC 123, Item 13 and MUSIC 1995 – 1 – us L of C Photodup [780]

Instrumental director : containing rules for all musical instruments in common use. laid down in a plain and concise manner, to which is added a variety of instrumental music... – 5th corr ed. Hallowell, Glazier, Masters & Smith [1836] [mf ed 19–] – 4mf – 9 – mf#fiche 495 – us Sibley [780]

Instrumental music / M'ewan, John – Edinburgh, Scotland. 1883 – 1r – us UF Libraries [780]

Instrumental music in christian worship : being a review of a work by m.c. kurfees entitled instrumental music in the worship / Briney, John Benton – Cincinnati: Standard Pub Co, c1914 – 1mf – 9 – 0-524-07556-5 – mf#1991-3176 – us ATLA [780]

Instrumental music in christian worship / Dick, James – Edinburgh, Scotland. 18– – 1r – us UF Libraries [240]

Instrumental music in the public worship of the church / Girardeau, John Lafayette – Richmond, Va: Whittet & Shepperson, 1888 – 1mf – 9 – 0-524-00550-8 – mf#1990-0050 – us ATLA [780]

The instrumental musician, no. 2 – Containing a large number of marches, quick-steps, waltzes, hornpipes, contra dances, cotillions &c. Arranged in three parts, for the flute, violin, clarionet, bass-viol, &c. May 1, 1843. To be issued once in two months. To be complete in six numbers. Boston: Elias Howe, Jr., 1843. "The People's Quadrille" is for five instruments and includes figures. MUSIC 1989, Item 5 – 1 – us L of C Photodup [780]

Instrumental works of georg muffat / Gore, Richard T – U of Rochester 1955 [mf ed 19–] – 1r – 1 – (with app & bibl) – mf#film 1069, 1805 – us Sibley [780]

Instrumentalmusik – Helsingfors: Svenska litteratursällskapet i Finland 1963-75 [mf ed Bloomington IN: Indiana Uni Lib, Preservation Dept 1984] – 3v on 1r – 1 – us Indiana Preservation [390]

Instrumentation and automation news (ian) – Radnor. 1993-1996 (1) 1993-1996 (5) 1993-1996 (9) – (cont: chilton's ian) – ISSN: 1072-2742 – mf#1662,01 – us UMI ProQuest [621]

Instrumentation and control systems (i&cs) – Radnor. 1992+ (1) 1992+ (5) 1992+ (9) – (cont: chilton's i and cs) – ISSN: 1074-2328 – mf#403,02 – us UMI ProQuest [621]

Die instrumentation der meistersinger von nuernberg von richard wagner : ein beitrag zur instrumentationslehre / Thomas, Eugen – Mannheim: K F Heckel 1899 – 2v on 3mf – 9 – mf#wa-108 – ne IDC [780]

Instrumentation technology – Pittsburgh. 1954-1978 (1) 1965-1978 (5) 1975-1978 (9) – (cont by: intech) – ISSN: 0020-4382 – mf#1170 – us UMI ProQuest [621]

Der instrumentator : eine orchestrationsfibel / Pimmer, Hans – [mf ed 2002] – 4mf – 9 – €56.00 – 3-8267-2781-9 – mf#DHS2781 – gw Frankfurter [780]

Instrumentos negociables / Salazar Grillo, Arturo – Bogota, Colombia. 1965 – 1r – us UF Libraries [972]

Instruments and experimental techniques – New York. 1958-1991 (1) 1966-1991 (5) 1966-1991 (9) – ISSN: 0020-4412 – mf#1532 – us UMI ProQuest [621]

Instruments and publications of the united states naval observatory / United States Naval Observatory – Washington: [s.n.] 1845-76 [mf ed 1998] – 1r [pl/ill] – 1 – (incl 6 heliotypes by james r osgood & co, boston) – mf#film mas 28292 – us Harvard [520]

Instruments-sax et fanfares civiles : etude pratique, par theodore de lajarte / Lajarte, Theodore de – Paris: Librairie des Auteurs & Compositeurs 1867 [mf ed 19–] – 1r – 1 – ("cette etude a ete faite pour etre lue dans une des reunions mensuelles de la societe des compositeurs de musique") – mf#film 1527 – us Sibley [780]

Instytut Astronomiczny U J K we Lwowie *see* Prace instytutu astronomicznego u j k we lwowie

Instytut Nastsmenshastsei (Akademiia Navuk Belaruskai Ssr) *see* Tsum fuftsentn yortog fun der oktyabr-revolyutsye

Insula – Madrid. 1946+ (1) – ISSN: 0020-4536 – mf#11570 – us UMI ProQuest [070]

Insula – Madrid. v. 1-11; 14-18; 24-26. 1946-1956; 1959-1963; 1969-1971 – 1 – us NY Public [073]

Insula : revista bibliografica de ciencias y letras – v1-. 1 enero 1946-. Madrid. [mnthly] – 1 – us Wisconsin U Libr [073]

Insula : revista bibliographica de ciencias y letras – n1-361. 1946-76 – 1 – $162.00 – mf#0283 – us Brook [010]

The insular cases / Randolph, Carman Fitz – New York? 1901 37 p. LL-1216 – 1 – us L of C Photodup [340]

The insular cases, comprising the records, briefs, and arguments of counsel in the insular cases of the october term, 1900 / U.S. Supreme Court – Washington, Govt. Print. Off., 1901. 1075 p. LL-1419 – 1 – us L of C Photodup [340]

The insular daily press – Manila: The Insular Daily Press, jul 17,19-20,22-24,27-31; aug 1-9,18,22-23,25-28,30 1899 – us CRL [079]

Insulas extranas / Inchaustegui Cabral, Hector – Mexico City?, Mexico. 1952 – 1r – us UF Libraries [972]

Insulation/circuits – Libertyville. 1979-1982 (1,5,9) – (cont by: electri-onics) – ISSN: 0020-4544 – mf#12329,02 – us UMI ProQuest [621]

Insulators – 1980 jul-1982, 1983 jan-1985 apr/may [v17 n2/3] – 2r – 1 – (cont by: crown jewels of the wire) – mf#652343 – us WHS [621]

Insunt 6 dissertationes varii argumenti / Jaeger, Gottfried et al – Lipsiae: Doerffling et Franke, 1881 – 2mf – 9 – 0-7905-9349-1 – (incl bibl ref) – mf#1989-2574 – us ATLA [220]

Insuppressible – Dublin, Ireland. 1 jan-24 jan 1891 – 1/4r – 1 – uk British Libr Newspaper [072]

Insurance advocate – Mt. Vernon. 1974+ (1) 1974+ (5) 1974+ (9) – ISSN: 0020-4587 – mf#10109 – us UMI ProQuest [368]

Insurance and finance chronicle – Montreal: R.W. Smith, [1886-1898] – 9 – (cont: insurance society; cont by: the chronicle) – mf#P04936 – cn Canadiana [368]

Insurance and technology – New York. 1990-1996 (1) 1990-1996 (5) 1990-1996 (9) – (cont: insurance software review) – ISSN: 1054-0733 – mf#14954,04 – us UMI ProQuest [368]

Insurance chronicle – New York. 2001+ (1,5,9) – mf#28804,01 – us UMI ProQuest [368]

Insurance counsel journal – Chicago. 1934-1986 (1) 1971-1986 (5) 1976-1986 (9) – (cont by: defense counsel journal) – ISSN: 0020-465X – mf#2143 – us UMI ProQuest [368]

Insurance counsel journal *see* Defense counsel journal

Insurance finance : with special reference to india / Agarwala, Amar Narain – Allahabad: Kitab-Mahal, 1939 – (= ser Samp: indian books) – us CRL [368]

Insurance forum – Ellettsville. 1980+ (1,5,9) – ISSN: 0095-2923 – mf#12405 – us UMI ProQuest [368]

Insurance gazette of ireland – Belfast Ireland, 7 aug 1879-1895 – 4 1/2r – 1 – uk British Libr Newspaper [072]

Insurance journal – San Diego. 1989-1996 (1) 1989-1989 (5) 1989-1989 (9) – ISSN: 0020-4714 – mf#12500,02 – us UMI ProQuest [368]

Insurance law journal – Chicago. 1871-1980 (1) 1954-1980 (5) 1954-1980 (9) – ISSN: 0020-4722 – mf#844 – us UMI ProQuest [346]

The insurance law journal – St Louis, New York: D T & L H Potter/C C Hine. v1-41. 1871-1912 – 501mf – 9 – $751.00 – (additional vols to be filmed) – mf#LLMC 84-492 – us LLMC [340]

The insurance law of canada : life, fire, marine, accident, guarantee, hail, burglary, employers' liability, etc, etc / Laverty, Francis Joseph – Montreal: John Lovell & Son, Ltd, 1911 [mf ed 1993] – 14mf – 9 – (incl ind) – mf#SEM105P1866 – cn Bibl Nat [348]

The insurance laws of the state of new york. / Cumming, Robert Cushing – New York, Baker, Voorhis, 1899. 653 p. LL-751 – 1 – us L of C Photodup [348]

Insurance maps / Sanborn Map Company – 1883-1950, The Sanborn fire insurance maps of Kansas – 1 – us Kansas [912]

Insurance maps of golden, colorado / Sanborn Map Company – New York: Sanborn Map Co, 1886-1887 (mf ed 1987) – 1r – 1 – mf#MF GMa5 – us Colorado Hist [368]

Insurance maps of leadville, colorado / Sanborn Map Company – New York: Sanborn Map Co, 1895 – 1r – 1 – mf#MF In8m – us Colorado Hist [368]

Insurance, mathematics and economics – Amsterdam. 1982+ (1,5,9) – ISSN: 0167-6687 – mf#42546 – us UMI ProQuest [368]

Insurance plan of the city of montreal / Underwriters Survey Bureau – Toronto; Montreal: Underwriters Survey Bureau Ltd, 1912-1935 [mf ed 1984] – 1r – 1 – (with ind) – mf#SEM35P154 – cn Bibl Nat [368]

Insurance plan of the city of montreal, quebec, canada / Goad, Charles Edward – Montreal [etc]: Chas E Goad Co, 1909-1915 [mf ed 1984] – 1r – 1 – (with ind) – mf#SEM35P153 – cn Bibl Nat [368]

Insurance record – London. 1977-1980 (1) 1977-1980 (5) 1977-1980 (9) – ISSN: 0020-479X – mf#10240 – us UMI ProQuest [368]

Insurance review – New York. 1984-1992 (1) 1984-1992 (5) 1984-1992 (9) – (cont: journal of insurance) – ISSN: 0749-8667 – mf#991,01 – us UMI ProQuest [368]

Insurance salesman – Indianapolis. 1973-1979 (1) 1973-1979 (5) 1977-1979 (9) – (cont by: is insurance sales) – ISSN: 0020-482X – mf#7745 – us UMI ProQuest [368]

Insurance society – Montreal: C E Goad, [1881-1885] – 9 – (cont by: insurance and finance chronicle) – mf#P04935 – cn Canadiana [368]

Insurance software review – Indianapolis. 1989-1990 (1,5,9) – (cont by: insurance and technology) – ISSN: 0892-8533 – mf#14954,03 – us UMI ProQuest [368]

Insurance supervision in israel / Israel. Laws, Statutes, etc – Jerusalem: Ministry of Finance, Superintendent of Insurance, 1965. 75p. LL-12040 – 1 – us L of C Photodup [346]

Insurance worker / Insurance Workers International Union – 1959 jun [v1 n1]-1964, 1965-71, 1972-82, 1983 feb-sep [v25 n6] – 1r – 1 – (cont: insurance worker [washington [dc]: 1958]) – mf#1119548 – us WHS [368]

Insurance worker / Insurance Workers of America – 3 [i.e. 4] n1-v8 n11 [1954 jan-1958 nov] – 1r – 1 – (cont: cio news. insurance edition; cont by: insurance worker [washington [dc]: 1958]) – mf#1119150 – us WHS [368]

Insurance worker / Insurance Workers of America – v1 n1-2 [1958 dec-1959 mar] – 1r – 1 – (cont: insurance worker [washington [dc]: 1954]; cont by: insurance worker [washington [dc]: 1959]) – mf#1125517 – us WHS [368]

Insurance Workers International Union *see* Insurance worker

Insurance Workers of America *see*
– Cio news
– Insurance worker

L'insurge – Paris. n1-42. janv-oct 1937 – 1 – fr ACRPP [073]

Insurgent : newsletter of the committee to fight repression / Committee to Fight Repression [US] – v1 n1-2 [1985 summer-fall], v2 n1-2,3 [1986 spr-summer, win], v3 n1-2 [1987 summer-fall], v4 n1,2,3 [1988 win, sum, win], v5 n1 [1989 spr] – 1r – 1 – mf#1772040 – us WHS [322]

Insurgent sociologist – Eugene. 1969-1987 (1) 1975-1987 (5) 1975-1987 (9) – (cont by: critical sociology) – ISSN: 0047-0384 – mf#10656 – us UMI ProQuest [301]

Insurgents admit that they lack the sympathies of the civil population – [Valencia?: s.n. 1936?] [mf ed 1977] – (= ser Blodgett coll) – 1mf – 9 – mf#w961 – us Harvard [946]

Insurreccion de los diez anos / Entralgo, Elias Jose – Habana, Cuba. 1950 – 1r – us UF Libraries [972]

Insurreccion desplomada / Vidales, Luis – Bogota, Colombia. 1948 – 1r – us UF Libraries [972]

Insurrecciones en cuba / Zaragoza, Justo – Madrid, Spain. v1-2. 1872-73 – 1r – us UF Libraries [972]

Intake / Iowa Air National Guard [US] – Des Moines IA. 1981 may [v19 n5]-1986 mar – 1r – 1 – mf#1045643 – us WHS [355]

Intech – Durham. 1979+ (1,5,9) – (cont: instrumentation technology) – ISSN: 0192-303X – mf#1170,01 – us UMI ProQuest [621]

Integer cursus philosophicus... / Oviedo, F de – Lugduni, 1640. 2v – 21mf – 9 – mf#CA-25 – ne IDC [100]

Integracion social en guatemala / Seminario De Integracion Social Guatemalteca – Guatemala, v1-2. 1956 – 1r – us UF Libraries [972]

Integralismo perante a nacao / Salgado, Plinio – Rio de Janeiro, Brazil. 1955 – 1r – us UF Libraries [972]

Integrated disease control of powdery mildew on cucurbits / Southwood, Michael J – Stellenbosch: U of Stellenbosch 1998 [mf ed 1998] – 2mf – 9 – mf#mf.1276 – sa Stellenbosch [630]

Integrated its capabilities in transit vehicles : human factors research needs – [McLean VA]: [US Dept of Transportation, Federal Highway Administration...1998 [mf ed 1999] – 1mf – 9 – us GPO [625]

Integrated management – Bangalore. 1966-1973 (1) – ISSN: 0020-4870 – mf#5725 – us UMI ProQuest [650]

Integrating dance into the study of american humanities / Jex, Amy T – 1998 – 2mf – 9 – $8.00 – mf#PE 3894 – us Kinesology [790]

Integration des umweltschutzes in die produktion hamburger industriebetriebe / Leonardi, Jaques – (mf ed 1995) – 2mf – 9 – €40.00 – 3-8267-2218-3 – mf#DHS 2218 – gw Frankfurter [660]

The integration of anatomy and physiology into fifth grade physical education / Morrison, Cary J – 1998 – 118p on 2mf – 9 – $10.00 – mf#PE 4178 – us Kinesology [370]

The integration of students with mild intellectual disabilities into regular physical education classes in victoria [australia] / Temple, Viviene A – 1995 – 3mf – 9 – $12.00 – mf#PE 3998 – us Kinesology [370]

The integration of the personality / Jung, Carl Gustav – Trans. by Stanley M. Dell. New York, Toronto: Farrar & Rinehart, (c1939). 313p, illus. PR-apply; UWM; 1 – us Wisconsin U Libr [616]

Integration testing of object-oriented software / Skelton, Gordon William – Uni of South Africa 2000 [mf ed Johannesburg 2000] – 4mf – 9 – (incl bibl ref) – mf#mfm14924 – sa Unisa [000]

Integrative and comparative biology – McLean. 2002+ (1,5,9) – mf#2201,01 – us UMI ProQuest [574]

Integrative fachtextsortenstilistik : dargestellt an historiographischen fachtexten des englischen / Baumann, Klaus-Dieter – (mf ed 1992) – 4mf – 9 – €49.00 – 3-89349-480-4 – mf#DHS 480 – gw Frankfurter [410]

Integrative physiological and behavioral science – Philadelphia. 1991+ (1) 1991+ (5) 1991+ (9) – (cont: pavlovian journal of biological science) – ISSN: 1053-881X – mf#6891,02 – us UMI ProQuest [150]

Integrierte modellkonzepte : entwurf eines klassifikationsschemas / Schoettle, Holger Hans – (mf ed 1995) – 2mf – 9 – €40.00 – 3-8267-2141-1 – mf#DHS 2141 – gw Frankfurter [330]

Integriertes curriculum fuer die faecher chemie und physik fuer den grundlegenden chemie- und physikunterricht in der sekundarstufe 1 / Potrawa, Dieter – (mf ed 1996) – 2mf – 9 – €40.00 – 3-8267-2356-2 – mf#DHS 2356 – gw Frankfurter [540]

Intellect – New York. 1915-1978 (1) 1969-1978 (5) 1960-1978 (9) – (cont by: usa today) – ISSN: 0149-0095 – mf#794 – us UMI ProQuest [370]

Intellectual and political currents in the far east / Reinsch, Paul Samuel – Boston, New York: Houghton Mifflin [1911] [mf ed 1995] – (= ser Yale coll) – viii/396p – 1 – 0-524-09365-2 – mf#1995-0365 – us ATLA [306]

The intellectual development of scotland / Macpherson, Hector – London; New York: Hodder and Stoughton, [1911?] – 1mf – 9 – 0-7905-5429-1 – mf#1988-1429 – us ATLA [941]

The intellectual ideal : three lectures on the vedanta, with an appendix illustrating the philosophy of sankaracharyya / Sen, Benoyendra Nath – Calcutta: TC Das, 1902 – 1mf – 9 – 0-524-02664-5 – mf#1990-3094 – us ATLA [280]

Intellectual liberty / Robertson, John – Ramsgate, England. 1871 – 1r – us UF Libraries [240]

Intellectual observer : review of natural history, microscopic research and recreative science – London. 1862-1868 (1) – mf#2800 – us UMI ProQuest [500]

Intellectual property and technology law journal – Clifton, 2000+ [1,5,9] – mf#24965,01 – us UMI ProQuest [346]

Intellectual property and technology law journal – v1-11. 1988-99 – 9 – $388.00 set – (title varies: v1-12 as journal of proprietary rights) – ISSN: 1041-3592 – mf#116371 – us Hein [346]

Intellectual regale : or ladies' tea tray – Philadelphia. 1814-1815 (1) – mf#3995 – us UMI ProQuest [305]

L'intellectualisme de saint thomas / Rousselot, Pierre – Paris: Felix Alcan, 1908 [mf ed 1991] – (= ser Coll historique des grands philosophes) – 1mf – 9 – 0-7905-9094-8 – (in french) – mf#1989-2319 – us ATLA [241]

Intellectuals and the spanish military rebellion – London, 1937. Fiche W962. (Blodgett Collection of Spanish Civil War Pamphlets) – 9 – us Harvard [946]

Die intellektuellen und der sozialismus / Ortner, Eugen – Berlin: Verlag Neues Vaterland, E Berger & Co., [1919] [mf ed 1987] – (= ser Flugschriften des bundes neues vaterland. [n f] 15) – 23p – 1 – mf#6929 n15 – us Wisconsin U Libr [335]

L'intelligence – Paris. sept 1837-mars 1840 – 1 – (Journal du droit commun puis Journal de la reforme sociale.) – fr ACRPP [073]

Intelligence – New York. 1998+ (1,5,9) – ISSN: 0160-2896 – mf#19216 – us UMI ProQuest [150]

Intelligence activities in the philippines during the japanese occupation / U.S. Army. Far East Command – 1948. 2 v – 1 – us L of C Photodup [959]

Intelligence and national security – London. 1990-1995 (1,5,9) – ISSN: 0268-4527 – mf#18557 – us UMI ProQuest [320]

Intelligence digest – Cheltenham. 1979-1986(1,5,9) – ISSN: 0020-4900 – mf#12396,02 – us UMI ProQuest [320]

Intelligence division, opnav, combat narratives / U.S. Navy – 1989 – 3r – 1 – $390.00 – (with printed guide) – mf#S3175 – us Scholarly Res [355]

Intelligence reports, 1941-1961 / U.S. Dept of State – (= ser General records of the department of state) – ca 9000 cards – 9 – mf#M1221 – us Nat Archives [327]

Intelligence reports on southern nigeria / Great Britain. Colonial Office. Nigeria – A collection of unpublished reports prepared by British colonial officials, 1930?-1943? – 1 – us CRL [960]

Intelligence series / U.S. Army. Far East Command – 1948-51. 10v – 1 – us L of C Photodup [950]

Intelligencer / Butler Co. Hamilton – v1 n1. aug 1828-jan 1856 [wkly] – 5r – 1 – mf#B2019-2023 – us Ohio Hist [071]

Intelligencer – Viola WI. 1891 oct 30-1894 dec 31, 1895-99, 1900-03, 1904-07, 1908-1911 oct 25 – 5r – 1 – (cont by: viola news; viola news consolidated with the intelligencer) – mf#942205 – us WHS [071]

Intelligencer – Dublin. 1728-1729 (1) – mf#5580 – us UMI ProQuest [420]

Intelligencer – Gloversville, NY. 1867-1889 (1) – mf#68688 – us UMI ProQuest [071]

Intelligencer – Lancaster, PA. 1799-1920 (1) – mf#61182 – us UMI ProQuest [071]

Intelligencer – Lexington, VA. 1823-1831 (1) – mf#66747 – us UMI ProQuest [071]

Intelligencer – Paterson, NJ. 1825-1856 (1) – mf#60221 – us UMI ProQuest [071]

Intelligencer – Petersburg, VA. 1800-1821 (1) – mf#66797 – us UMI ProQuest [071]

Intelligencer / Washington Co. Marietta – v1 n1. sep 1839-may 1862 [wkly] – 6r – 1 – mf#B12087-12092 – us Ohio Hist [071]

Intelligencer – Wheeling, WEST Virginia. 1990+ (1) – mf#68547 – us UMI ProQuest [071]

The intelligencer – n1-20. 1730 – 1 – us AMS Press [073]

Intelligencer and petersburg commercial advertiser – Petersburg, VA. 1824-1828 (1) – mf#66798 – us UMI ProQuest [071]

Intelligencer, and petersburg commercial advertiser – Richmond VA. 1792 jul 26, 1795 mar 13, sep 4,15, 1818 dec 15, 1825 mar 11 – 1r – 1 – (cont: petersburg intelligencer) – mf#881635 – us WHS [071]

Intelligencer journal – Lancaster, PA. 1864+ (1) – ISSN: 0889-4140 – mf#61792 – us UMI ProQuest [071]

Intelligencer series / Adams Co. West Union – (dec 1842-feb 49), may 51-jan 1852 [wkly] – 1r – 1 – mf#B6738 – us Ohio Hist [071]

Intelligencer series / Washington Co. Marietta – oct 1851-dec 1860 begins v1 n1 [twice wkly] – 6r – 1 – mf#B29288-29293 – us Ohio Hist [071]

Intelligent enterprise – San Mateo. 1998+ (1,5,9) – mf#28084 – us UMI ProQuest [000]

Intelligent machines journal – Woodside. 1979-1980 (1,5,9) – (cont by: infoworld) – ISSN: 0164-3878 – mf#12701 – us UMI ProQuest [000]

Intelligent man's guide to indian philosophy / Pandya, Manubhai C – Bombay: DB Taraporevala & Sons, c1935 – (= ser Samp: indian books) – us CRL [180]

Intelligente bremssysteme zur optimierung des bremsmoments / Voit, Marjan – (mf ed 1994) – 2mf – 9 – €40.00 – 3-8267-2026-1 – mf#DHS 2026 – gw Frankfurter [621]

Intelligentsia of great britain / Mirsky, D S – London, England. 1935 – 1r – us UF Libraries [941]

Intelligentsiia i revoliutsiia : sb statei / Pokrovskii, M N et al – n.d. – 181p 2mf – 9 – mf#RPP-54 – ne IDC [325]

Intelligenz nachrichten *see* Kronik der menschheit

Intelligenz und wille / Meumann, Ernst – Leipzig, Germany. 1908 – 1r – us UF Libraries [150]

Intelligenz- und wochenblatt fuer frankenberg mit sachsenburg und umgegend – Frankenberg DE, 1842-1945 4 may – 116r – 1 – (title varies: 1861: frankenberger nachrichtsblatt und bezirksanzeiger; jul 1878: frankenberger tageblatt) – gw Misc Inst [074]

Intelligenzblaetter *see* Allgemeine deutsche bibliothek [adb]

Intelligenz-blatt *see*
– Intelligenz-blatt der freyen stadt frankfurt
– Ravensburgisches gemeinnuetziges wochenblatt

Intelligenzblatt *see* Gemeinnuetziges wochenblatt

Intelligenz-blatt der freyen stadt frankfurt – Frankfurt/M DE, 1848-49, 1918-1933 31 oct – 52r – 1 – (title varies: 16 feb 1819: intelligenz-blatt; 11 oct 1910: frankfurter nachrichten und intelligenz-blatt) – gw Mikrofilm [074]

Intelligenz-blatt der herzogthuemer bremen und verden und des landes hadeln *see* Intelligenz-blatt des nord-departements

Intelligenz-blatt des nord-departements – Stade DE, 1852 – 1r – 1 – (title varies: 2 mar 1811: intelligenz-blatt; 3 apr 1813: intelligenz-blatt der herzogthuemer bremen und verden; 3 jan 1848: intelligenz-blatt der herzogthuemer bremen und verden und des landes hadeln; 1 jun 1853: anzeigen fuer die herzogthuemer bremen und verden und fuer das land hadeln; 1 jan 1862: anzeigen fuer den landdrosteibezirk stade) – gw Misc Inst [350]

Intelligenzblatt des teltower und beeskow-storkower kreises – Koenigs Wusterhausen DE, 1914 – 1r – 1 – gw Misc Inst [074]

Intelligenzblatt fuer das grossherzogtum posen *see* Posener intelligenzblatt

Intelligenz-blatt fuer den bezirk der koeniglichen regierung zu danzig *see* Danziger intelligenzblatt 1739

Intelligenzblatt fuer den kreis euskirchen und den kreis rheinbach *see* Erfa 1840

Intelligenzblatt fuer den kreis kempen und dessen umgebung – Kempen (Kepno PL), 1835-37, 1842-43, 1858-60 – 1 – (title varies: 1841?: kempener kreisblatt) – gw Misc Inst [077]

Intelligenzblatt fuer den oberamtsbezirk muensingen *see* Intelligenz-blatt fuer die oberaemter ehingen und muensingen

Intelligenzblatt fuer den oberamtsbezirk saulgau *see* Der oberlaender

Intelligenz-blatt fuer den regierungsbezirk erbach *see* Graeflich erbachisches wochen-blatt fuer den landkreis erbach

Intelligenzblatt fuer die kreise euskirchen, rheinbach und ahrweiler *see* Erfa 1840

Intelligenzblatt fuer die kreise pruem, bitburg, daun und den ehemaligen kreis st vith – Pruem DE, 1841 7 jan-1866 [gaps] – 1 – Inquire at Microfilmarchiv for details – gw Misc Inst [074]

Intelligenz-blatt fuer die kreise siegen und wittgenstein *see* Siegerlaender intelligenz-blatt

Intelligenzblatt fuer die oberaemter biberach und waldsee *see* Nuetzliches unterhaltungs- und wochenblatt fuer verschiedene leser

Intelligenz-blatt fuer die oberaemter ehingen und muensingen – Muensingen DE, 1980 2 jan-1983 31 may – 20r – 1 – (filmed by other misc inst: 1968-79 [56r], 1983- [6r/yr]. title varies: 1838: intelligenz-blatt fuer den oberamtsbezirk muensingen; 1847: amts- und intelligenzblatt fuer den oberamtsbezirk muensingen; 1849: amts-intelligenz- und politisches blatt fuer den oberamtsbezirk muensingen; 1851: amts- und intelligenzblatt fuer den oberamtsbezirk muensingen; 20 dec 1862: der alpbote; 3 jan 1863: der albbote; 1 apr 1932: albbote und rundschau; jul 1945: schwaebisches tagblatt / mr tbi [main ed in tuebingen]; 3 dec 1949: albbote [regional ed of schwaebisches tagblatt / mr tbi, tuebingen]; 1968: regional ed of suedwest-presse, ulm. with suppls) – gw Misc Inst [350]

Intelligenzblatt fuer die provinz oberhessen – Friedberg, Hessen, Darmstadt DE, 1834 4 jan-1943 28 mar, 1949 30 jul-1950 29 apr – 1 – (title varies: 1854: friedberger intelligenzblatt, 1854; 19 sep 1866: anzeiger fuer oberhessen; 5 jan 1869: oberhessischer anzeiger. filmed with suppls) – gw Mikrofilm [074]

Intelligenzblatt fuer die staedte kempen, schildberg, grabow, mixstadt und baranow – Oels (Olesnica, PL), 1860 10 jan-1866 25 sep – 1r – 1 – gw Misc Inst [077]

Intelligenzblatt fuer litthauen – Gumbinnen (Gussew RUS), 1819, 1825-1826 30 jun, 1828 4 jan-30 jun, 1830 1 jan-30 jun, 1831-1839 28 jun [gaps], 1840-48 [gaps], 1916, 1938 1 apr-30 sep, 1940-1942 30 jun [gaps] – 37r – 1 – (title varies: 1 apr 1859: preussisch-litauische zeitung; apr 1939: altpreussische volkszeitung. with suppl: verwaltungsbericht des kreises gumbinnen auf das jahr...1907-08. filmed by other misc inst: 1844 17 may-11 nov, 1912 6 jan (jubilee ed)) – gw Misc Inst [077]

Intelligenz-blatt fuer stadt und kreis bunzlau – Bunzlau (Boleslawiec PL), 1857-61 – 1r – 1 – gw Misc Inst [350]

Intendencia de extremadura: circular / Eizalde, Bernardo – 1821 – 9 – sp Bibl Santa Ana [946]

Intensive and critical care nursing – London. 1992+(1,5,9) – (cont: intensive care nursing) – ISSN: 0964-3397 – mf#15453,01 – us UMI ProQuest [610]

Intensive care medicine – Heidelberg. 1977+ (1,5,9) – (cont: european journal of intensive care medicine) – ISSN: 0342-4642 – mf#13182,01 – us UMI ProQuest [610]

Intensive care nursing – Edinburgh. 1985-1991 (1,5,9) – (cont by: intensive and critical care nursing) – ISSN: 0266-612X – mf#15453 – us UMI ProQuest [610]

Intention to use condoms for hiv/std prevention : rural southern college african-american students and the theory of planned behavior / Kanu, Andrew J – 1997 – 2mf – 9 – $8.00 – mf#HE 592 – us Kinesology [613]

Inter alia : state bar of nevada – v1-57. 1937-92 (all publ) – 9 – $479.00 set – (title varies: v1-38, n1 as nevada state bar journal; cont by: nevada lawyer) – ISSN: 0092-6086 – mf#103521 – us Hein [340]

Inter amicos : letters between james martineau and william knight, 1869-72 / Martineau, James & Knight, William Angus – London: J Murray, 1901 – 1mf – 9 – 0-7905-9792-6 – mf#1989-1517 – us ATLA [240]

Inter arma – Vien, Austria. 1918 – 1r – us UF Libraries [939]

Inter avia – 1946-62 – 1 – us L of C Photodup [629]

Inter com – NAES College. v2 n1-3,4-5 [1982 sep-dec, 1983 mar-oct], v6 n1-2,4-7 [1986 jan-mar/apr, jul-dec], v7 n1-v8 n7 [1987 jan/feb-1988 nov], v9 n8 [1989 jan] – 1r – 1 – mf#1112588 – us WHS [071]

Inter county gazette – Prairie Du Chien WI. 1880 jan 9-apr 2 – 1r – 1 – (cont: grant county gazette) – mf#930467 – us WHS [071]

Inter county gazette / Tuscarawas Co. Strasburg – nov 1944-mar 1950,jun 1950-mar 1958 [wkly] – 5r – 1 – mf#B161-165 – us Ohio Hist [071]

Inter mountain press – Manhattan, MT. 1956-1969 (1) – mf#64548 – us UMI ProQuest [071]

Inter ocean – Chicago IL. [1874 feb 15/apr 30]-[1879 sep 15/nov 11] [gaps] – 25r – 1 – (cont: chicago republican [chicago, il: 1865 : daily]; chicago republican; cont by: daily inter ocean; chicago record-herald; chicago record-herald and the inter ocean) – mf#851358 – us WHS [071]

Interaccion social y personalidad en una comunidad / Seda Bonilla, Eduardo – San Juan, Puerto Rico. 1964 – 1r – us UF Libraries [972]

Interacting with computers – Kidlington. 1989-1996 (1,5,9) – ISSN: 0953-5438 – mf#17235 – us UMI ProQuest [000]

InterAction – Boring OR: InterAct Ministries Inc 1989- [3 times/yr, qrtly] [mf ed 2006] – 1989-fall 2005 [complete] on 1r – 1 – (iss for jul-sep 2001 never publ; cont: arctic voice [2006i-s004]) – mf#2006i-s005 – us ATLA [240]

Interaction – St. Louis. 1960-1987 (1) 1976-1987 (5) 1976-1987 (9) – (cont by: teachers interaction) – ISSN: 0020-5117 – mf#7668 – us UMI ProQuest [370]

Interaction – United States. v1-v5 n8/9 [1972 oct-1977 jun] – 1r – 1 – mf#169413 – us WHS [071]

Interaction of ribosomal complexes with signal recognition particle from escherichia coli / Rauch, Gabriele – (mf ed 1999) – 2mf – 9 – €40.00 – 3-8267-2645-6 – mf#DHS 2645 – gw Frankfurter [574]

An interactional analysis of experienced and inexperienced athletic trainers' behavior in clinical instruction settings / Stemmans, Catherine L – 1998 – 1mf – 9 – $4.00 – mf#PE 3954 – us Kinesology [370]

Interactive learning international : ili – Chichester. 1984-1989 – 1,5,9 – ISSN: 0748-5743 – mf#14807 – us UMI ProQuest [370]

Inter-African Labour Institute *see* Bulletin

Interaktionshandlungen im russischunterricht / Matijaschtschuk, Evelyn – (mf ed 1992) – 2mf – 9 – €49.00 – 3-89349-538-X – mf#DHS 538 – gw Frankfurter [460]

Interaktive frueherziehung bei entwicklungsverzoegerten und entwicklungsgefaehrdeten kindern : ein beitrag zur praeventiven sondererziehung / Dietz, Gerhard – Dortmund: projekt vlg. 1992 (mf ed 1996) – 4mf – 9 – €45.00 – 3-8267-9704-3 – mf#DHS 9704 – gw Frankfurter [370]

Interamerican Children's Institute. Montevideo *see*
- Boletin
- Noticiario

Inter-American Commission Of Women *see* Informe de la comision interamericana de mujeres a...

Inter-American Conference (10th : 1954 : Caracas) *see* Final act

Inter-American Council Of Commerce And Production *see* Encuesta continental sobre el control de la inflac...

Inter-american development bank release – n1. 25 oct 1960 (all publ) – (= ser The sec release series preceding the sec docket) – 1mf – 9 – $1.50 – mf#LLMC 89-005 – us LLMC [346]

Inter-american economic affairs – Washington. 1947-1985 (1) 1975-1985 (5) 1976-1985 (9) – ISSN: 0020-4943 – mf#6164 – us UMI ProQuest [337]

Inter-american law review = Revista juridica interamericana – New Orleans. 1959-1966 (1) – ISSN: 0020-4951 – mf#8088 – us UMI ProQuest [340]

Inter-american music bulletin – Washington. 1957-1973 (1) 1971-1973 (5) (9) – ISSN: 0020-4978 – mf#6752 – us UMI ProQuest [780]

Inter-American Symposium On Linguistics And Langua *see* Simposio de cartagena

Interavia english ed – Geneva. 1946-1989 (1); 1974-1989 (5,9) – ISSN: 0020-5168 – mf#9709 – us UMI ProQuest [629]

Intercambio de influencias literarias entre espana / Henriquez Urena, Max – Habana, Cuba. 1926 – 1r – us UF Libraries [972]

Intercepted correspondence of russian revolutionaries from the special department of the police, 1906-1917 : from the state archive of the russian federation – (= ser The Russian Archives) – 175r – 1 – us Primary [947]

The intercepted correspondence of russian revolutionaries from the special departments of the police / The State Archive of the Russian Federation (GARF) – 1906-1917 – (= ser The Russian Archives) – ca 170r – 1 – us Primary [947]

Intercepted japanese messages : the documents of magic, 1938-1945 / Japan. Ministry of Foreign Affairs – 15r – 1 – $1,275.00 – (includes guide) – mf#D3254 – us L of C Photodup [950]

Interceptor / Camp New Amsterdam [Huls ter Heide, Netherlands] – Huls Ter Heide. 1981 may 1 [v2 n14]-1986 mar – 1r – 1 – mf#1043943 – us WHS [071]

Interchange / Brotherhood of Railway, Airline, and Steamship Clerks, Freight Handlers, Express, and Station Employes et al – 1985 jan-1988 dec, 1989 jan-1991 dec, 1992 jan/feb-1994 nov/dec – 3r – 1 – (cont: railway clerk interchange) – mf#1445021 – us WHS [380]

Interchange – Toronto. 1984+ (1) 1984+ (5) 1984+ (9) – (cont: interchange on education) – ISSN: 0826-4805 – mf#6802,02 – us UMI ProQuest [370]

Interchange / Minnesota Interchange Network – n10 [1980 nov], 1981 sep/oct-1982 mar/apr, 1983 spr/summer-autumn – 1r – 1 – mf#651497 – us WHS [380]

Interchange on education – Toronto. 1983-1984 (1) 1983-1984 (5) 1983-1984 (9) – (cont: interchange on educational policy. cont by: interchange) – ISSN: 0822-9856 – mf#6802,01 – us UMI ProQuest [370]

Interchange on educational policy – Toronto. 1970-1983 (1) 1972-1983 (5) 1973-1983 (9) – (cont by: interchange on education) – ISSN: 0822-9848 – mf#6802 – us UMI ProQuest [370]

Interchurch news / National Council of the Churches of Christ in the United States of America – v1-v9 n8 [1959 sep-1968 apr] – 1r – 1 – mf#1059461 – us WHS [243]

Interchurch World Movement of North America *see* History of the interchurch world movement in north america

Intercollegian – New York. 1878-1967 [1] – mf#1458 – us UMI ProQuest [230]

Intercollegiate Association of Women Studies *see* Feminine focus

Intercollegiate athletic trainer's perception of third-party reimbursement and their steps towards its implementation / McPherson, Bennetta K – 1999 – 1mf – 9 – $4.00 – mf#PE 3914 – us Kinesology [617]

Intercollegiate athletics and organizational culture / Baumgartner, Renee M – 1996 – 3mf – 9 – $12.00 – mf#PE 3922 – us Kinesology [790]

Intercollegiate law journal – New York. v1-2. 1891-93 (all publ) – 7mf – 9 – $10.50 – (cont by: university law review) – mf#LLMC 84-493 – us LLMC [340]

Intercollegiate law journal – New York. v1-2 1891-93 (all publ) – (= ser Historical legal periodical series) – 1 – $45.00 set – mf#103541 – us Hein [340]

Intercollegiate review – Wilmington. 1965+ (1) 1970+ (5) 1972+ (9) – ISSN: 0020-5249 – mf#2297 – us UMI ProQuest [378]

Intercollegiate socialist – New York. v. 1-6. Feb 1913-May 1918 – 1 – us NY Public [335]

The intercolonial journal of commerce *see* The trade review

the intercolonial journal of commerce *see* The trade review and intercolonial journal of commerce

Intercom – New York. 1959-1986 (1) 1970-1986 (5) 1977-1986 (9) – ISSN: 0020-5273 – mf#1544 – us UMI ProQuest [370]

Intercom : a publication of the journal company / Journal Communications et al – 1976 sep-1983 – 1r – 1 – (cont by: add inc link; newsprint [milwaukee, wi]; print/data; telecom [milwaukee, wi]; jci voice) – mf#703240 – us WHS [380]

Intercom : a twice monthly news bulletin of the canadian paperworkers union / Canadian Paperworkers Union – v1 n1-6 [1980 may 23-oct 2] – 1r – 1 – (cont: cpu journal; cont by: intercom [canadian paperworkers union]. french; cpu journal [1981]) – mf#666112 – us WHS [331]

Intercom – United States. 1981 may 8 [v21 n10]-1985 – 1r – 1 – mf#1289934 – us WHS [071]

Intercontinental press – New York. 1985-1986 (1) 1985-1986 (5) 1985-1986 (9) – (cont: intercontinental press combined with inprecor) – mf#6523,02 – us UMI ProQuest [070]

Intercontinental press – New York. 1963-1978 (1) 1972-1978 (5) 1976-1978 (9) – ISSN: 0020-5303 – mf#6523 – us UMI ProQuest [070]

Intercontinental press combined with inprecor – New York. 1978-1985 (1) 1978-1985 (5) 1978-1985 (9) – (cont by: intercontinental press) – ISSN: 0162-5594 – mf#6523,01 – us UMI ProQuest [070]

Inter-county leader – Centuria, Frederic WI. 1934 sep 24-1936, 1937-38, 1939-43, 1944-45, 1946-48, 1949-50, 1951-52 – 11r – 1 – (cont by: frederic star; inter-county leader and the frederic star) – mf#942370 – us WHS [071]

Inter-county leader – Frederic WI. [1954 jul 14/1954]-[1999 sep/dec] [gaps] – 62r – 1 – (cont: inter-county leader and the frederic star; burnett county leader) – mf#941883 – us WHS [071]

Inter-county leader and the frederic star – Frederic WI. 1951 aug 1-1952, 1953-1954 jul 7 – 2r – 1 – (cont: frederic star; inter-county leader [centuria, wi]; cont by: inter-county leader [frederic, wi]) – mf#942375 – us WHS [071]

Inter-county times – Lone Rock, Spring Green WI. 1877 nov 28-1878 sep 10 – 1r – 1 – (cont by: dollar times) – mf#931442 – us WHS [071]

Intercourse between india and the western world : from the earliest times to the fall of rome / Rawlinson, Hugh George – Cambridge: University Press, 1916 – (= ser Samp: indian books) – us CRL [954]

Intercourse between india and the western world : from the earliest times to the fall of rome / Rawlinson, Hugh George – Cambridge: University Press, 1916 [mf ed 1995] – (= ser Yale coll) – vi/[2]/196p/pl – 1 – 0-524-09744-5 – mf#1995-0744 – us ATLA [930]

Interdenominational conference of foreign missionary boards : 1st 1893 – 1893-1950 [mf ed 2001] – (= ser Christianity's encounter with world religions, 1850-1950) – 10r – 1 – (filmed with: conference of the officers and representatives of foreign mission board and societies in the united states and canada [2nd-10th 1894-1903]; conference of the foreign missions board in the united states and canada [11th-17th 1904-10]; foreign missions conference of north america [18th-57th 1911-50]) – mf#2001-s004-010 – us ATLA [240]

Interdenominational foreign mission association : news – 1950-83 [complete] – 2r – 1 – mf#ATLA S0449 – us ATLA [240]

The interdenominational holiness berean – Berlin [Kitchener, Ont]: J M Kerr, [1890?-19–] – 9 – mf#P05022 – cn Canadiana [230]

Interdenominational Theological Center *see* Journal of the interdenominational theological center

The interdict : its history and its operation / Krehbiel, Edward B – Washington: American Historical Association, 1909 – 1mf – 9 – 0-7905-4535-7 – (incl bibl ref) – mf#1988-0535 – us ATLA [240]

Interdisciplinary perspectives – Boston. 1976-1981 (1) 1976-1981 (5) 1976-1981 (9) – (cont: perspectives) – ISSN: 0148-1959 – mf#6698,01 – us UMI ProQuest [370]

Intereconomics – Hamburg. 1985+ (1,5,9) – ISSN: 0020-5346 – mf#15429 – us UMI ProQuest [337]

Interessante blaetter – Berlin, 1915 1 oct, 1916 11 feb, 28 jul-22 sep [gaps] – 1 – gw Mikrofilm [074]

Interessante reise-nachrichten eines suedamericanischen officiers von mainz nach london : nebst einigen, noch unbekannten notizen ueber napoleon – Eisenberg 1826 [mf ed Hildesheim 1995-98] – (= ser Fbc) – 2mf – 9 – €60.00 – 3-487-29276-9 – gw Olms [914]

The interest of america in sea power : present and future / Mahan, Alfred Thayer – London 1897 – (= ser 19th c british colonization) – 4mf – 9 – mf#1.1.9654 – uk Chadwyck [327]

Interesting case – London, England. 18– – 1r – us UF Libraries [240]

Interesting confession – Glasgow, Scotland. 18– – 1r – us UF Libraries [240]

Interesting extracts, etc on religious and moral subjects : from numerous sources, concerning the four quarters of the globe / Atkinson, Christopher William – Sheffield, England?: s.n, 1850 (Sheffield England: J Pearce) – 4mf – 9 – mf#48631 – cn Canadiana [230]

Interesting political discussion : the diplomatick sic policy of mr madison unveiled, in a series of essays containing strictures upon the late correspondence between mr smith and mr jackson / Lowell, John – S.l: s.n, 1810? – 1mf – 9 – mf#18874 – cn Canadiana [327]

An interesting trial of edward jordan and margaret his wife : who were tried at halifax, n s nov 15th, 1809, for the horrid crime of piracy and murder, committed on board the schooner three sisters... / Jordan, Edward – Boston, [1809?] [mf ed 1984] – 1mf – 9 – 0-665-45218-7 – mf#45218 – cn Canadiana [345]

L'interet general – Port-au-Prince: Imp Aug A Heraux, v1 n1-7. may-jul 1899 – 1 sheet – 9 – us CRL [079]

Interface / Black Women in Publishing, Inc – 1987 summer/fall-1988 summer/fall – 1r – 1 – mf#4877107 – us WHS [071]

Interface – Chicago. 1978+ (1,5,9) – ISSN: 0270-6717 – mf#12518 – us UMI ProQuest [020]

INTERFACE

Interface – Inglewood. 1966-1975 (1) 1974-1975 (5) 1974-1975 (9) – mf#9945 – us UMI ProQuest [000]
Interface age : computing for business – Cerritos. 1984-1985 (1,5,9) – (cont by: computing for business) – ISSN: 8756-2472 – mf#11898,01 – us UMI ProQuest [000]
Interface age – Cerritos. 1978-1984 (1,5,9) – (cont: interface age. cont by: interface age: computing for business) – ISSN: 0147-2992 – mf#11898 – us UMI ProQuest [000]
Interfaces – Linthicum. 1970+ (1,5,9) – ISSN: 0092-2102 – mf#11434 – us UMI ProQuest [650]
Interfaith Center to Reverse the Arms Race *see* Rar newsletter
Inter-faith churchman – 1941 apr 20 – 1r – 1 – mf#5258993 – us WHS [071]
Interfaith Justice and Peace Center *see* Linkages
Interfaith women's news and network – v2 n1-v10 n2 [1981 feb-1990 mar] – 1r – 1 – (cont: ecumenical women's news & network) – mf#1786762 – us WHS [305]
Interferencias / Bayle, Constantino – Madrid: Razon y Fe, 1930 – 1 – sp Bibl Santa Ana [946]
Interglacial fossils from the don valley, toronto / Coleman, Arthur Philemon – S.l: s.n, 1894? – 1mf – 9 – (incl bibl ref) – mf#61150 – cn Canadiana [560]
Intergovernmental perspective – Washington. 1979-1994 (1,5,9) – ISSN: 0362-8507 – mf#12114 – us UMI ProQuest [320]
Interhelp : a networking newsletter – v1 n1-v4 n4 [1983 nov-1987 dec] – 1r – 1 – mf#1520467 – us WHS [071]
Interieur d'un bureau / Scribe, Eugene – Paris, France. 1828 – 1r – us UF Libraries [720]
Interim – TN. 1983 jun-1991 sep/oct – 1r – 1 – (cont: bulletin [tennessee state penitentiary]) – mf#2399950 – us WHS [365]
Interim adultero-germanum : cui adjecta est, vera christianae pacificationis, et ecclesiae reformandae ratio / Calvin, J – [Geneva: Jean Girard], 1549 – 3mf – 9 – mf#CL-29 – ne IDC [240]
L'interim, c'est a dire, provision faicte sur les differens de la religion, en quelques villes et pais d'allemagne : avec la vraye facon de reformer l'eglise chrestienne, et appointer les differens qui sont en icelle / Calvin, J – [Geneva: Jean Girard], 1549 – 3mf – 9 – mf#CL-53 – ne IDC [240]
Interim colorado comprehensive outdoor recreation plan, 1974 / Colorado Division of Parks and Outdoor recreation – Denver: The Division, [1974?] – us CRL [978]
Das interim in wuerttemberg / Bossert, Gustav – Halle: Verein fuer Reformationsgeschichte, 1895 – (= ser [Schriften des Vereins fuer Reformationsgeschichte]) – 1mf – 9 – 0-7905-4661-2 – (incl bibl ref) – mf#1988-0661 – us ATLA [943]
Interim orders of government on the recommendations of the bombay economic and industrial survey committee – [s.l: s.n.], 1941 – 1 – (filmed with: pakistan press yearbook) – us CRL [954]
Interim report of the hong kong salaries commission / Hong Kong Salaries Commission – Hong Kong: S Young, Govt Printer, 1965 (mf ed 1984) – 1mf – 9 – mf#FSN 39,425 – us NY Public [350]
Interim report of the political status commission of the marshall islands – Majuro, Marshall Islands: the Commission, Apr 1976 – 1mf – 9 – $1.50 – mf#llmc82-100i, title 6 – us LLMC [324]
An interim report on the civil administration of palestine : during the period 1st july 1920-30th june 1921 – London, 1921 – 1mf – 9 – mf#J-28-191 – ne IDC [956]
Interim report on the county council of tanganyika, 1951-1956 / Tanganyika – Dar es Salaam, Tanzania. 1956 – 1r – us UF Libraries [960]
Interim report to the congress of micronesia : 2nd cong, 4th reg sess, jul 8, 1968 / Future Political Status Commission [TTPI (US)] – n.p, n.d. – (= ser Micronesia: prelude to the constitutional convention) – 3mf – 9 – $4.50 – mf#LLMC 82-100F, Title 35 – us LLMC [323]
Interim reporter / x – v1-7. aug 3 1981-apr 1996 – 34mf – 9 – $51.00 – (v4 lacks p246-309, v7 lacks p266-506; pages may not exist) – mf#llmc82-100h, title 21 – us LLMC [347]
Interim reporter / digest and updater – v1-5. 1985-92 – 5mf – 9 – $7.50 – mf#llmc82-100h, title 22 – us LLMC [347]
L'Interime...les moyenneurs, les transformateurs, les libertins... / Viret, P – Lyon, [Senneton], 1565 – 6mf – 9 – mf#PFA-203 – ne IDC [240]
Interior – Chicago IL. [1870 mar 17/1871 jun 29]-[1910 feb 24-jun 30] – 46r – 1 – (cont: western presbyterian [danville, ky]; christian hour; western presbyterian [minneapolis, mn]; cont by: westminster [philadelphia, pa]; continent [chicago, il]) – mf#1155101 – us WHS [242]
The interior castle : or, the mansions = Moradas / Teresa of Avila, St – London: Thomas Baker, 1893 – 1mf – 9 – 0-8370-7027-9 – (in english. includes appendix of spanish correspondence from saint teresa) – mf#1986-1027 – us ATLA [080]
Interior department appointment papers / U.S. Dept of the Interior. Office of the Secretary – (= ser Records of the office of the secretary of the interior) – 1 – (alaska 1871-1907 6r m1245. arizona 1857-1907 22r. california 1849-1907 29r m732. colorado 1857-1907 13r m808. florida 1849-1907 6r m1119. idaho 1862-1907 17r m693. mississippi 1849-1907 4r m849. missouri 1849-1907 9r m1058 9r. nevada 1860-1907 3r m1033. new mexico 1850-1907 18r m750. new york 1849-1906 5r m1022. north carolina 1849-92 1r m950. oregon 1849-1907 10r m814. wisconsin 1849-1907 9r m831. wyoming 1869-1907 6r m830. with printed guides) – us Nat Archives [324]
Interior department territorial papers / U.S. Dept of the Interior. Office of the Secretary – (= ser Records of the office of the secretary of the interior) – 1 – (alaska 1869-1913 17r m430. arizona 1868-1913 8r m429. colorado 1861-88 1r m431. dakota 1863-89 3r m310. hawaii 1898-1907 4r m827. idaho 1864-90 3r m191. montana, 1867-89 2r m192. new mexico 1851-1914 15r m364. oklahoma 1889-1912 5r m828. utah 1850-1902 6r m428. washington 1854-1902 4r m189. wyoming 1870-90 6r m204. with printed guides) – us Nat Archives [975]
Interior design – New York. 1933+ (1) 1977+ (5) 1977+ (9) – ISSN: 0020-5508 – mf#6670 – us UMI ProQuest [740]
Interior landscape – Chicago. 1995-1995 (1) – ISSN: 1063-1607 – mf#14245,01 – us UMI ProQuest [710]
Interiors – New York. 1917-1976 (1) 1968-1976 (5) 1975-1976 (9) – (cont by: contract interiors) – ISSN: 0020-5516 – mf#1037 – us UMI ProQuest [740]
Interiors – New York. 1978-2000 (1) 1978-2000 (5) 1978-2000 (9) – (cont: contract interiors) – ISSN: 0164-8470 – mf#1037,02 – us UMI ProQuest [740]
Interkorrelation von epidemiologischen und polysomnographischen risikofaktoren des ploetzlichen saeuglingstodes / Buschatz, Dirk – (mf ed 1999) – 1mf – 9 – €30.00 – 3-8267-2638-3 – mf#DHS 2638 – gw Frankfurter [618]
Inter-league Council of the Leagues of Women Voters of Milwaukee County *see* County view
Interlending and document supply – Bradford. 2001+ (1,5,9) – ISSN: 0264-1615 – mf#29147,02 – us UMI ProQuest [020]
The interlinear literal translation of the greek new testament : with the authorized version conveniently presented in the margins for ready reference – New York City: Arthur Hinds, [189-?] – 2mf – 9 – 0-8370-9416-X – mf#1986-3416 – us ATLA [225]
The interlinear literal translation of the hebrew old testament : with the king james version and the revised version conveniently printed in the margins for ready reference, and with explanatory textual footnotes, supplemented by tables of the hebrew verb, and the hebrew alphabet / Berry, George Ricker – New York City: Hinds & Noble, c1897 – 4mf – 9 – 0-524-05904-7 – mf#1992-0661 – us ATLA [221]
The interlineary hebrew and english psalter : in which the construction of every word is indicated, and the root of each distinguished by the use of hollow and other types / ed by Tregelles, Samuel Prideaux – London: Samuel Bagster, [1852?] – 1mf – 9 – 0-8370-1861-7 – mf#1987-6248 – us ATLA [221]
Interludio : poemas / Jordan Diaz, Alfredo Alberto – Habana, Cuba. 1958 – 1r – us UF Libraries [972]
Intermarket – Duesseldorf DE, 1956-61 – 2r – 1 – gw Mikropress [074]
Les intermedes : poesies canadiennes / Doucet, Louis-Joseph – Montreal: edition privee, 1957 [mf ed 1991] – 3mf – 9 – mf#SEM105P1417 – cn Bibl Nat [440]
Intermedia – London. 1978+ (1,5,9) – ISSN: 0309-118X – mf#11966 – us UMI ProQuest [380]
L'intermediaire des chercheurs et curieux – Paris. 1894 – 1 – fr ACRPP [073]
Intermediare des chercheurs et curieux, correspondance litteraire, historique et artistique... – Paris. v1-103. 1864-1940 – 9 – $1404.00 – mf#0284 – us Brook [440]
Intermediate and university education in ireland – Dublin, 1872 – (= ser 19th c ireland) – 4mf – 9 – mf#1.1.6023 – uk Chadwyck [378]
Intermediate cambodian reader / ed by Huffman, Franklin E – New Haven: Yale University Press 1972 [mf ed 1989] – 1r with other items – 1 – (added title in cambodian; with bibl) – mf#mf-10289 seam reel 026/01 [§] – us CRL [480]
Intermediate leader – Oct 1924-61. (Formerly: Intermediate Leader's B.Y.P.U. Quarterly. 1924-39) – 1 – us Southern Baptist [242]
Intermediate pupil, 13-16 years – 1948 – 1 – 47.88 – (intermediate teacher, 13-16 years. 1948. 48.65; 1) – us Southern Baptist [242]
Intermediate state / Humphry, W G – London, England. 1851 – 1r – us UF Libraries [240]
The intermediate state and christ among the dead : the twofold resurrection and the twofold coming of christ exhibited according to the word of god = Tod, das todtenreich und der zustand der von hier abgeschiedenen seelen / Maywahlen, Val Ulrich – London: Seeley, Jackson & Halliday, 1856 [mf ed 1992] – 1mf – 9 – 0-524-05229-8 – (trans by james frederick schoen) – mf#1992-0362 – us ATLA [240]
The intermediate state and prayers for the dead : examined in the light of scripture, and of ancient jewish and christian literature / Wright, Charles Henry Hamilton – London: James Nisbet, 1900 – 1mf – 9 – 0-7905-0474-X – (incl bibl ref and indexes) – mf#1987-0474 – us ATLA [240]
The intermediate state between death and judgment : being a sequel to after death / Luckock, Herbert Mortimer – New and cheaper ed. London; New York: Longmans, Green, 1896 – 1mf – 9 – 0-7905-9314-9 – mf#1989-2539 – us ATLA [240]
Intermediate syllabus and notes, modern and tap / Imperial Society of Teachers of Dancing. Modern Theatre Dance Branch – rev enl ed. London: The Society, 1983 (mf ed 1988) – 1mf – 9 – mf#FSN-43,162 – us NY Public [790]
Intermediate teacher, uniform series – Oct 1940-61 – 1 – us Southern Baptist [242]
Intermediate teacher's and pupil book, years 1-4 – 1928-33 – 1 – us Southern Baptist [242]
Intermediate weekly – 1929-31 – 1 – us Southern Baptist [242]
Intermedio – Bogota, Colombia. 1956-1957 (1) – mf#67681 – us UMI ProQuest [079]
Interment of the dead / Sington, A – Manchester, England. 1888 – 1r – us UF Libraries [240]
Interment record no. 1-4051 / Highland Cemetery, Geary County, KS – 1871-1930 – 1 – us Kansas [920]
Intermezzo : eine buergerliche komoedie mit sinfonischen zwischenspielen in zwei aufzuegen / Strauss, Richard – Berlin: Adolph Fuerstner, c1924 – 1r – 1 – us Wisconsin U Libr [780]
Intermission : the audelco newsletter / Audience Development Committee – 1975 jan/feb, 1980 jun, 1984 sum, sep – 2r – 1 – (cont: audelco newsletter) – mf#4852630 – us WHS [071]
Intermodal container news – Atlanta. 1991-1994 (1) 1991-1994 (5) 1991-1994 (9) – (cont: container news. cont by: intermodal shipping) – mf#8781,01 – us UMI ProQuest [380]
Intermodal shipping – Atlanta. 1994-1996 (1) 1994-1996 (5) 1994-1996 (9) – (cont: intermodal container news) – ISSN: 1076-9293 – mf#8781,02 – us UMI ProQuest [380]
Intermountain jewish news – Denver. Col. 1935-48. 1963-67 – 1 – us AJPC [071]
Intermountain liberal *see* [Reno-] nevada liberal
Intermountain observer – v21 n32-v23 n42 [1971 aug 7-1973 oct 20] – 1r – 1 – mf#1059474 – us WHS [071]
Intermountain tribune and linn county agriculturalist – Sweet Home OR: T L Dugger, 1913-14 [wkly] – 1 – (cont: lebanon tribune and linn county agriculturalist; cont by: scio tribune) – us Oregon Lib [071]
Intermural law review – New York University School of Law. v1-23. 1945-68 (all publ) – 27mf – 9 – $40.50 – mf#LLMC 84-494 – us LLMC [340]
Internacional – Tampa, FL. 1925 jan 30-1941 feb 28 [scattered] – 1r – us UF Libraries [071]
Internacional negra en colombia / Andrade, Raul – Quito, Ecuador. 1954 – 1r – us UF Libraries [972]
Internacionalismo antimperialista / Roig De Leuchsenring, Emilio – Habana, Cuba. 1935 – 1r – us UF Libraries [972]
Internacionalismo proletario / Central Organization of US Marxist-Leninists et al – 1979 marz-1980 sep, 1982 jul – 1r – 1 – mf#719372 – us WHS [335]
Internacionalizacao das colonias tropicais / Saldanha, Eduardo D'almeida – Porto, Portugal. 1932 – 1r – us UF Libraries [025]
Internal and external rotation strength values of female swimmers and water polo players / Ferry, Christopher – 1999 – 1mf – 9 – $4.00 – mf#PE 4010 – us Kinesology [611]
Internal auditing – Boston. 1987+ (1,5,9) – ISSN: 0897-0378 – mf#16346 – us UMI ProQuest [650]
Internal auditor – Altamonte Springs. 1944+ (1) 1944+ (5) 1944+ (9) – ISSN: 0020-5745 – mf#388 – us UMI ProQuest [650]
Internal bulletin – 1949 jan n2-3, 1950 jan n1-2, 1950 feb n3 – 1r – 1 – mf#669362 – us WHS [071]
Internal bulletin / Spartacist League of the US – n7-10 [1968 dec-1969 aug] – 1r – 1 – (cont: spartacist pre-conference discussion bulletin; cont by: internal information bulletin [spartacist league]) – mf#690762 – us WHS [071]
The internal christ / Wilson, Henry – New York City: Alliance Press, c1908 [mf ed 1992] – (= ser Christian & missionary alliance coll) – 1mf – 9 – 0-524-02174-0 – mf#1990-4240 – us ATLA [240]
Internal correspondence / London Missionary Society – Samoan District – 1896-1927 – 1r – 1 – mf#pmb130 – at Pacific Mss [242]
Internal discussion bulletin / Spartacist League of the US – n25,27-32 – 1r – 1 – (cont: internal information bulletin [spartacist league]) – mf#690803 – us WHS [071]
The internal evidence afforded by the historical books of the old testament... : an essay which obtained the norrisian prize for the year 1849 / Whittington, R – Cambridge: John Deighton, 1849 – 1mf – 9 – 0-7905-0531-2 – (incl bibl ref) – mf#1987-0531 – us ATLA [221]
The internal evidence of the holy bible : or, the bible proved from its own pages to be a divine revelation / Janeway, Jacob Jones – Philadelphia: Presbyterian Board of Publ, 1845 – 1mf – 9 – 0-7905-0094-9 – mf#1987-0094 – us ATLA [220]
Internal evidence of the letter "apostolicae curae" as to its own... / Collins, William Edward – London, England. 1897 – 1r – us UF Libraries [240]
The internal evidence of the letter "apostolicae curae" as to its own origin and value / Collins, William Edward – London: SPCK 1897 [mf ed 1993] – (= ser Church historical society (series) 20) – 1mf – 9 – 0-524-07194-2 – mf#1990-5352 – us ATLA [240]
Internal evidences of christianity deduced from phrenology – Edinburgh, Scotland. 1827 – 1r – us UF Libraries [240]
Internal evidences of the genuineness of the gospels / Norton, Andrews – Boston: Little, Brown, 1855. Beltsville, Md: NCR Corp, 1978 (4mf); Evanston: American Theol Lib Assoc, 1984 (4mf) – (= ser Biblical crit – us & gb) – 9 – 0-8370-0733-X – (incl bibl ref) – mf#1984-1032 – us ATLA [226]
Internal information bulletin – n11-13 [1970 apr repr], n14 [1972 jun], n15,17-20 [1972 aug, oct-nov], n21-23 [1974 jun-aug] – 2r – 1 – (cont: internal bulletin [spartacist league]; cont by: internal discussion bulletin [spartacist league of the us]) – mf#690765 – us WHS [071]
The internal management of a country bank : in a series of letters on the functions and duties of a branch manager / Rae, George – Toronto: Willing & Williamson, 1876 – 3mf – 9 – mf#11758 – cn Canadiana [332]
Internal medicine journal – Sydney, 2001+ [1,5,9] – (cont: australian and new zealand journal of medicine) – ISSN: 1444-0903 – mf#6645,01 – us UMI ProQuest [616]
Internal medicine news – Rockville. 1968-1980 (1) 1979-1980 (5) 1979-1980 (9) – (cont by: internal medicine news and cardiology news) – ISSN: 0099-152X – mf#6867 – us UMI ProQuest [616]
Internal medicine news and cardiology news – Rockville. 1980-1980 (1) 1980-1980 (5) 1980-1980 (9) – (cont: internal medicine news) – ISSN: 0274-5542 – mf#6867,01 – us UMI ProQuest [616]
The internal mission of the holy ghost / Manning, Henry Edward – 10th ed. London: Burns & Oates, [ca. 1900] – 2mf – 9 – 0-7905-9331-9 – mf#1989-2556 – us ATLA [240]
The internal parasites of the horse (entozoa) / Duncan, J T – Toronto: Presbyterian News Co, 1891 – 2mf – 9 – mf#29890 – cn Canadiana [636]
Internal revenue acts of the united states : revenue act of 1954 with legislative histories and congressional documents / ed by Reams, Bernard D Jr – v1-11 – 9 – $305.00 set – 0-89941-540-7 – mf#301981 – us Hein [340]
Internal revenue acts of the united states : revenue acts of 1953-1972 with legislative histories, laws and congressional documents / ed by Reams, Bernard D Jr – 48v. 1953-1972 – 9 – $1,735.00 set – 0-89941-624-1 – mf#201681 – us Hein [340]
Internal revenue acts of the united states, 1909-1950 : legislative histories and administrative documents – 144v – 9 – $4,795.00 set – (with guide and analytical index isbn: 0-930342-94-1) – mf#301271 – us Hein [340]

Internal revenue acts of the united states 1950-1951 : legislative histories, laws and administrative documents / U.S. Internal Revenue Service; ed by Reams, Bernard D Jr – v1-7. 1950-51 – 9 – $298.00 set – 0-89941-703-5 – mf#301971 – us Hein [340]

Internal revenue assessment lists / U.S. Internal Revenue Service – (= ser Records of the internal revenue service) – 1 – (alabama 1865-66 6r m754. arkansas 1865-66 2r m755. california 1862-66 33r m756. territory of colorado 1862-66 3r m757. connecticut 1862-66 23r m758. delaware 1862-66 8r m759. district of columbia 1862-66 8r m760. florida 1865-66 1r m761. georgia 1865-66 8r m762. territory of idaho 1865-66 1r m763. idaho 1867-74 1r t1209. illinois, 1862-66 63r m764. indiana 1862-66 42r m765. iowa 1862-66 16r m766. kansas 1862-66 3r m767. kentucky 1862-66 24r m768. louisiana 1863-66 10r m769. maine 1862-66 15r m770. maryland 1862-66 21r m771. michigan 1862-66 15r m773. minnesota, 1862-66 3r m774. mississippi 1865-66 3r m775. missouri 1862-65 22r m776. montana 1864-72 1r m777. nevada 1863-66 2r m779. new hampshire 1862-66 10r m780. territory of new mexico 1862-70, 1872-74 1r m782. new york and new jersey 1862-66 218r m603. north carolina 1864-66 2r m784. oregon district 1867-73 2r m1631. pennsylvania 1862-66 107r m787. rhode island 1862-66 10r m788. south carolina 1864-66 2r m789. texas 1865-66 2r m791. vermont 1862-66 7r m792. virginia 1862-66 6r m793. west virginia 1862-66 4r m795) – us Nat Archives [336]

Internal revenue bulletin / United States Internal Revenue Service – Washington. 1970-1976 [1]; 1973-1975 [5] – ISSN: 0020-5761 – mf#5770 – us UMI ProQuest [336]

Internal revenue cumulative bulletin : office of internal revenue. us treasury department – Washington, US. v1-1997 No2 (1919-97) – 9 – $1,850.00 set – 0-89941-234-3 – mf#400070 – us Hein [340]

Internal revenue cumulative bulletin / United States Internal Revenue Service – Washington. 1975+ (1,5,9) – ISSN: 0364-0620 – mf#9241 – us UMI ProQuest [336]

Internal revenue forms collection : beginning thru 1994 / U.S. Internal Revenue Service. Office of the Chief Counsel. Library – IRS, Office of Chief Counsel Library. Main coll – 1508mf – 9 – $650.00 – (7 annual suppl to main coll between 1987-1993/94 204mf $306.00. standing order for future suppl $45.00y. special form locator, annually updated, $15.00y. llmc 84-368) – mf#LLMC 84-368 – us LLMC [336]

The internal revenue record and customs bulletin – New York: Church. v1-40. 1865-94 (all publ) – 60mf – 9 – $270.00 – (lacking: v41 1895) – mf#LLMC 82-927 – us LLMC [330]

Internal vs external velocity : effects of strength training, protocols on velocity-specific adaptations and human skeletal muscle variables / Tricoli, Valmor – 2000 – 146p on 2mf – 9 – $10.00 – mf#PE 4130 – us Kinesology [612]

Internasjonal politikk – Oslo. 1977+ (1) 1977+ (5) 1977+ (9) – ISSN: 0020-577X – mf#7699 – us UMI ProQuest [327]

International – Johannesburg SA, 4 jan 1915-1924 – 1 – sa National [079]

International – London, UK. 7 Mar 1863-15 Nov 1871 – 1 – uk British Libr Newspaper [072]

International – London, UK. Feb-Jul 1863 – 1 – uk British Libr Newspaper [072]

International abstracts of biological sciences – Oxford. 1977-1980 (1) 1954-1980 (5) 1978-1980 (9) – (cont by: current awareness in biological sciences: cabs) – ISSN: 0020-5818 – mf#49263 – us UMI ProQuest [574]

International adoptions : problems and solutions: hearing...house of representatives, 107th congress, 2nd session, may 22 2002 / United States. Congress. House. Committee on International Relations – Washington: US GPO 2002 [mf ed 2002] – 1mf – 9 – us GPO [341]

International advertiser – New York. 1960-1972 (1) 1971-1972 (5) (9) – ISSN: 0020-5834 – mf#6167 – us UMI ProQuest [338]

International affairs – Oxford. 1944+ (1) 1944+ (5) 1944+ (9) – (cont: international affairs review supplement) – ISSN: 0020-5850 – mf#1030,04 – us UMI ProQuest [327]

International affairs : journal of the royal institute of international affairs / Royal Institute of International Affairs – London. 1931-1939 (1) 1931-1939 (5) 1931-1939 (9) – (cont: journal of the royal institute of international affairs. cont by: international affairs review supplement) – ISSN: 0020-5850 – mf#1030,02 – us UMI ProQuest [327]

International affairs review supplement – London. 1940-1943 (1) 1940-1943 (5) 1940-1943 (9) – (cont: international affairs: journal of the royal institute of international affairs. cont by: international affairs) – mf#1030,03 – us UMI ProQuest [327]

International African Institute *see*
- Practical orthography of african languages

International African Seminar *see* Industrialisation in africa

International Alliance of Theatrical Stage Employees and Moving Picture Machine Operators of the United States and Canada *see*
- Quarterly report
- Report of the general executive board

INTERNATIONAL AMERICAN CONFERENCE *see* Caso de belice ante la conciencia de america

International American Conference *see* Caso de belice ante la conciencia de america

L'international anarchiste – Marseille. n2-4. oct-nov 1886 – 1 – fr ACRPP [335]

International and comparative law quarterly – v1-19. 1952-70.8 reels – 1 – $350.00 – us Trans-Media [341]

International anesthesiology clinics – Philadelphia. 1975+(1,5,9) – ISSN: 0020-5907 – mf#10967 – us UMI ProQuest [617]

International antifascist solidarity : an appeal to the women of america / Montseny, Frederica – New York: IAS [1938] [mf ed 1977] – (= ser Blodgett coll) – 1mf – 9 – mf#w1061 – us Harvard [320]

International arbitrations : history and digest of the international arbitrations to which the united states has been a party / Moore, John Bassett – Washington: GPO. v1-6. 1898 [all publ] – 64mf – 9 – $96.00 – mf#llmc 80-912 – us LLMC [341]

International arbitrations / Moore, John Bassett – 3r – 1 – $100.00 – us Trans-Media [341]

International archives of allergy and applied immunology – Basel. 1966-1974 (1) 1966-1974 (5) 1970-1974 (9) – ISSN: 0020-5915 – mf#2063 – us UMI ProQuest [616]

International archives of occupational and environmental health – Heidelberg. 1981-1996 (1) 1981-1996 (5) 1975-1996 (9) – (cont: internationales archiv fuer arbeitsmedizin) – ISSN: 0340-0131 – mf#13118,03 – us UMI ProQuest [360]

International art market – New York. 1961-1983 (1) 1976-1983 (5) 1976-1983 (9) – ISSN: 0020-5931 – mf#9988 – us UMI ProQuest [700]

International art printer – Owen Sound [Ont]: A.M. Rutherford, [1895?-189- or 19–] – 9 – mf#P06024 – cn Canadiana [680]

International Association for Great Lakes Research *see* Conference on great lakes research proceedings

International Association for Mathematical Geology *see* Journal of the international association for mathematical geology

International association of agricultural librarians and documentalists quarterly bulletin – Beltsville. 1956-1990 (1) 1973-1990 (5) 1977-1990 (9) – (cont by: quarterly bulletin of the international association of agricultural information) – ISSN: 0020-5966 – mf#1960 – us UMI ProQuest [020]

International Association of Allied Metal Mechanics *see* Machinists and blacksmiths' monthly journal, 1870-1875 / the brass worker, 1895-1896 / official journal, 1902-1904

International Association of Black Business Educators *see* This is iabbe's newsletter

International association of factory inspectors proceedings, 1887-1914 – Columbus OH [etc] 1st-28th (all publ) – 44mf – 9 – $270.00 – us UPA [366]

International Association of Fire Fighters *see*
- Boston firefighters digest
- Fire department news
- Fire fighter
- International fire fighter
- Layout
- Local 42 pipeline
- Los angeles firefighter
- Pipeline
- Toledo firefighter
- Yonkers fire fighter

International association of industrial accident boards and commission reporter – v1-18. 1937-55 (all publ) – (= ser The abc reporter) – 17mf – 9 – $25.50 – (also known as: the abc reporter. lacking: 1938 n1,4. 1939 n3. 1954-55) – mf#LLMC 84-495 – us LLMC [331]

International Association of Industrial Accident Boards and Commissions *see* Iaiabc journal

International Association of Jewish Lawyers and Jurists *see* Bulletin

International Association of Machinists *see*
- K-t miller
- The machinist
- Machinists' monthly journal
- Monthly journal of the international association of machinists
- Transport workers bulletin

International Association of Machinists and Aerospace Workers *see*
- Cutting edge
- Educator
- Focus
- Iam district 720 report
- Iam education bulletin
- Iam journal
- Jfkapers
- Machinist
- Machinist canada
- Machinist digest
- Monitor
- Motor city news
- Newsletter
- Pioneer press
- Pride
- Pride of the international association of machinists and aerospace workers, district 87, locals 653, 1309, 2733
- Report from headquarters
- Right track
- Shoptalk
- Talking union
- Trade winds
- Turbo-lines
- Voice of '94'
- Voice of 702
- Voyager
- Way the wind blows

International Association of Officials of Bureaus of Labor, Factory Inspection and Industrial Commissions *see* Report of commissioners of the state bureaus of labor statistics on the industrial, social and economic conditions of pullman, illinois

International association of officials of bureaus of labor, factory inspection and industrial commissions proceedings, 1883-1914 – Jefferson City MO [etc] 1st-30th (all publ) – 55mf – 9 – $315.00 – us UPA [366]

International Association of Oil Field, Gas Well and Refinery Workers of America *see*
- International oil worker
- Oil worker

International Association of Pupil Personnel Workers *see* Journal of the international association of pupil personnel workers

International Association of Steam, Hot Water and Power Pipe Fitters and Helpers of America *see* Steam fitter

International bar journal – v1-10. 1970-79 (all publ) – 9 – $100.00 set – (cont by: international legal practitioner) – mf#103571 – us Hein [340]

International bar journal *see* International legal practitioners

International behaviourial scientist – Meerut. 1978-1979 (1,5,9) – ISSN: 0020-613X – mf#11840 – us UMI ProQuest [150]

International bibliography on crime and deliquency *see* Crime and delinquency abstracts

International biodeterioration and biodegradation – Barking. 1993+ (1,5,9) – ISSN: 0964-8305 – mf#42571,01 – us UMI ProQuest [574]

International black photographers newsletter – 1981 win, summer – 1r – 1 – mf#4867119 – us WHS [770]

International Black Writers' Conference *see* Black writers' news

International book trade in the 18th century : the luchtmans archive, 1697-1845 – [mf ed 2001] – 2pt – 9 – €4585.00 set – (pt1: booksellers' accounts, 1697-1803 [346mf] €2355 [m310]; pt2: private accounts and other documents, 1702-1845 [473mf] €2745 [m415]; with p/g & concordances, contemporary ind) – mf#m310/m415 – ne MMF Publ [070]

International bookbinder – 1900-55 – (= ser Labor union periodicals, pt 2: the printing trades) – 13r – 1 – $2705.00 – 1-55655-306-4 – us UPA [680]

The international brigade – Hassocks, 1939? Fiche W963. (Blodgett Collection of Spanish Civil War Pamphlets) – 9 – us Harvard [946]

International Broom and Brush Makers' Union *see* Broom maker

International Brotherhood of Blacksmiths *see* Proceedings of the...biennial convention of the international brotherhood of blacksmiths

International Brotherhood of Blacksmiths and Helpers *see*
- Blacksmiths journal
- Proceedings of the...bi-ennial convention of the international brotherhood of blacksmiths and helpers

International Brotherhood of Blacksmiths, Drop Forgers, and Helpers *see*
- Blacksmiths, drop forgers and helpers journal
- Proceedings of the...convention

International Brotherhood of Blacksmiths, Drop Forgers and Helpers bi-monthly journal – 1924 aug/sep-1929 nov/dec – 1r – 1 – (cont: international brotherhood of blacksmiths, drop forgers and helpers monthly journal) – mf#1429151 – us WHS [366]

International Brotherhood of Blacksmiths, Drop Forgers and Helpers monthly journal – 1920 aug-1922 sep, 1922 oct-1924 jul – 2r – 1 – (cont: blacksmith's journal; cont by: international brotherhood of blacksmiths, drop forgers, and helpers bi-monthly journal) – mf#1382601 – us WHS [366]

International brotherhood of boiler makers, iron ship builders and helpers of America *see* Journal

International Brotherhood of Boilermakers, Iron Ship Builders, Blacksmiths, Forgers and Helpers *see* Boilermakers-blacksmiths journal

International Brotherhood of Boilermakers, Iron Ship Builders, Blacksmiths, Forgers, and Helpers *see* Union news

International Brotherhood of Boilermakers, Iron Shipbuilders etc *see*
- Boilermaker reporter
- Boilermakers blacksmiths reporter

International Brotherhood of Bookbinders *see* Official proceedings of the convention of the international brotherhood of bookbinders of north america

International Brotherhood of Electrical Workers *see*
- Current lines
- Gang-box news
- I b e w clear vision
- Informal
- Laurel leaf
- Local 165 communicator
- Local 494 ibew relay
- Local 1484 news
- News
- Short circuit
- Stand by
- Telephone line
- Ten 49er
- Toll reporter
- Union lite review
- Union reporter
- Utility reporter
- Your union news

International Brotherhood of Electrical Workers, Local 3 *see* Electrical union world

International Brotherhood of Firemen and Oilers *see*
- Firemen and oilers' journal
- Pandora

International Brotherhood of Machinery Molders *see*
- Proceedings of the...annual convention of the international machinery molders of north america
- Report of proceedings of the...annual session of the international brotherhood of machinery molders of north america

International Brotherhood of Painters and Allied Trades *see*
- Painters' pride
- Spotlite news

International Brotherhood of Paper Makers *see*
- Paper and pulp makers' journal
- Paper maker
- Paper makers journal

International Brotherhood of Pottery & Allied Workers et al *see* Potters herald

International Brotherhood of Pulp, Sulphite, and Paper Mill Workers *see* Journal

International Brotherhood of Stationary Firemen *see* Stationary firemen's journal

International Brotherhood of Teamsters *see*
- Officers report
- Teamsters
- Voice of teamsters 391

International Brotherhood of Teamsters, Chauffeurs, Stablemen and Helpers of America *see* Official magazine

International Brotherhood of Teamsters, Chauffeurs, Warehousemen and Helpers *see*
- Local 104 reporter
- San francisco i b t local 921

International Brotherhood of Teamsters, Chauffeurs, Warehousemen and Helpers of America *see*
- Election scratch sheet
- Local 150 newsletter
- Local 150 reporter
- Local 315 teamsters
- Local union 445 dispatch
- Local vocal
- Maryland teamster
- News and report
- Northern california teamster
- Ohio teamster
- Oregon teamster
- Pegasus
- Route 142
- Voice of 743

International Brotherhood of Teamsters, Chauffeurs, Warehousemen, and Helpers of America *see*
- Joint council of teamsters no 2 news
- Local 243 mirror– mirror
- Local 243's mirror
- Midwest teamster
- Minnesota teamster
- Missouri teamster
- News service
- Team 337
- Team and wheel
- Teamster 877 news
- Teamster four-o-six news
- Teamster perspective
- Teamsters historical notes
- Teamsters Local 71 news
- Teamsters local 85 news
- Teamsters local 337 news
- Teamsters local 877 news and views
- Teamsters local 877 newsletter
- Teamsters local 1145 news

- Teamsters local union 445 dispatch
- Teamsters news release
- Teamsters news service
- Transmitter
- Washington teamster

International Brotherhood of Teamsters, Chauffeurs, Warehousemen, and Helpers of America et al *see* Midwest labor world

International Building Trades Council *see* Labor compendium

International bulletin of missionary research – New Haven. 1981+ (1,5,9) – (cont: occasional bulletin of missionary research) – ISSN: 0272-6122 – mf#12938,01 – us UMI ProQuest [240]

International Bureau Of American Republics *see* Commercial directory of latin america

International Bureau of Education *see*
- Bulletin of the international bureau of education
- Educational documentation and information

International Bureau Of The American Republics *see*
- Haiti a handbook
- Hand book of the american republics
- Handbook of the american republics
- Honduras
- Salvador
- Santo domingo a handbook
- Venezuela

International business – Rye. 1991-1997 (1,5,9) – (cont: north american international business) – ISSN: 1060-4073 – mf#18342,02 – us UMI ProQuest [337]

International business automation – Elmhurst. 1963-1968 [1,5,9] – mf#2030 – us UMI ProQuest [650]

International business equipment – New Canaan. 1964-1986 [1]; 1971-1986 [5]; 1975-1986 [9] – ISSN: 0020-6288 – mf#1775 – us UMI ProQuest [650]

International business lawyer – International Bar Association. v1-17. 1973-89 – 9 – $330.00 set – ISSN: 0309-7676 – mf#103581 – us Hein [346]

International business review – Oxford. 1993+ (1,5,9) – ISSN: 0969-5931 – mf#49632 – us UMI ProQuest [650]

International canada – Toronto. 1978-1981 (1,5,9) – ISSN: 0027-0512 – mf#11813 – us UMI ProQuest [971]

International cast metals journal – Des Plaines. 1976-1982 (1,5,9) – ISSN: 0362-1723 – mf#10795 – us UMI ProQuest [660]

International cement and lime journal – London. 2002+ (1,5,9) – ISSN: 1365-9219 – mf#32283 – us UMI ProQuest [690]

International ceramics – London. 1997+ (1,5,9) – ISSN: 1361-7605 – mf#32266 – us UMI ProQuest [730]

International chemical engineering – New York. 1961-1994 (1) 1971-1994 (5) 1977-1994 (9) – ISSN: 0020-6318 – mf#2476 – us UMI ProQuest [660]

International chemical worker – 1943 jun 25-1957, 1958-63, 1964-66, 1967-69, 1970-1973 apr, 1973 may-1976 may – 6r – 1 – (cont by: chemical worker) – mf#862468 – us WHS [331]

International Chemical Workers Union *see* Chemical worker

International child welfare review – Geneva. 1947-1985 (1) 1947-1985 (5) 1947-1985 (9) – ISSN: 0020-6342 – mf#10425 – us UMI ProQuest [640]

International Church of the Foursquare Gospel *see* Foursquare world advance

International class struggle / International Communist Opposition – v1 n1-3 [1936 summer-1937 spr] – 1r – 1 – mf#766664 – us WHS [335]

International class struggle – v1-3. 1936-37 [all publ] – (= ser Radical periodicals in the united states, 1881-1960. series 1) – 2mf – 9 – $85.00 – us UPA [335]

International Club for Collectors of Hatpins and Hatpin Holders *see*
- Pictorial journal
- Quarterly

International coden directory – 9 – us Chemical [540]

The international collection / Career Guidance Foundation – 330mf – 9 – $268.00 – (contains over 650 catalogs from 40 countries) – us Career [378]

International commerce – Washington. 1940-1970 [1]; 1969-1970 [5] – mf#1444 – us UMI ProQuest [337]

International Commission for Supervision and Control in Cambodia *see*
- Rapport de la c.i.c. sur les agressions des forces americano-sud-vietnamiennes contre le cambodge
- Verbatim proceedings of the investigation[s]

International Commission Of Jurists *see* Racial problems in the public service

International Commission Of Jurists (1952-) *see* South africa and the rule of law

International Commission on Radiological Protection *see* Annals of the icrp

International commission on the christian approach to the jews : news sheet – v2-12: 1932-34 [complete] – 1r – 1 – mf#ATLA S0664C – us ATLA [230]

International Committee Against Racism *see*
- Car-madison newsletter
- Incar report

International Committee for the Study of the Crimes of Genocide *see* Nigeria-biafra conflict

International Committee for United Front International Labor Defense *see*
- News bulletin
- News/bulletin

International Committee of Coordination and Information to Aid Republican Spain *see*
- Information service
- Les operations militaires en espagne

International communication – London, England. 1931 – 1r – us UF Libraries [960]

International communications in heat and mass transfer – New York. 1974+ (1,5,9) – ISSN: 0735-1933 – mf#49132 – us UMI ProQuest [530]

International Communist Opposition *see* International class struggle

International computer lawyer – v1-3. 1993-95 (all publ) – 9 – $55.00 set – (ceased with v3 n4. merged with: computer lawyer) – ISSN: 1067-6171 – mf#116381 – us Hein [340]

International Concatenated Order of Hoo-Hoo *see* Log and tally

International conciliation – New York. 1907-1972 (1) 1968-1972 (5) – ISSN: 0020-6407 – mf#890 – us UMI ProQuest [327]

International Conference On South West Africa (1966 : Oxford) *see* South west africa

International conference on water for peace. proceedings – Washington dc. v1-8. 23-31 may, 1967 – 9 – $300.00 – mf#0285 – us Brook [333]

International Congress for Progressive Thought. 1904: St. Louis *see* Report.

International Congress of African Historians *see* Emerging themes of african history

International Congress of Africanists (1st: 1962: Ghana) *see* [Papers presented at the congress, accra, 1962]

International Congress of European and Western Ethnology. Stockholm, Sweden. 1951 *see* Working papers

International Congress of Folklore *see* Papers

International Congress of Folklore, Sao Paulo, Brazil *see* Papers, august 16th to 22d, 1954

International Congress Of Historians Of The United... *see* New world looks at its history

International construction – Chicago. 1962-1996 (1) 1975-1996 (5) 1975-1996 (9) – ISSN: 0020-6415 – mf#10476 – us UMI ProQuest [624]

International contact lens clinic: iclc – New York. 1993-1995 (1,5,9) – ISSN: 0892-8967 – mf#17133 – us UMI ProQuest [617]

International controversy / Fabela, Isidro – S.l., S.l? . 1957 – 1r – us UF Libraries [972]

International council for exceptional children council review – Washington. 1934-1935 (1) 1934-1935 (5) 1934-1935 (9) – (cont by: journal of exceptional children) – mf#12546 – us UMI ProQuest [640]

International Council of Office Employees Unions *see* Office worker

International critical commentary – T & T Clark. 39v – 9 – $425.00 – us IRC [220]

The international critical commentary on the old and new testaments / Driver, Samuel Rolles et al – 1895-1950. 39v – 9 – $425.00 – us IRC [220]

International Defence And Aid Fund *see* Rhodesia

International development abstracts – Norwich. 1988-1990 (1,5,9) – ISSN: 0262-0855 – mf#42473 – us UMI ProQuest [337]

International development planning review (idpr) – Liverpool. 2002+ (1,5,9) – ISSN: 1474-6743 – mf#11899,01 – us UMI ProQuest [710]

International Documentation and Information Centre, Hague *see* Cambodia, problems of neutrality and independence may 1970

International dyer, textile printer, bleacher and finisher – London. 1978-1984 (1,5,9) – ISSN: 0020-658X – mf#11246 – us UMI ProQuest [670]

International economic indicators – Washington. 1979-1985 (1) 1979-1985 (5) 1979-1985 (9) – ISSN: 0149-1873 – mf#12115,01 – us UMI ProQuest [337]

International economic review – Chicago. (1) 1961-1972 (5) (9) – mf#6601 – us UMI ProQuest [330]

International economic review – Malden. 1960+ (1,5,9) – ISSN: 0020-6598 – mf#11377 – us UMI ProQuest [330]

International educational and cultural exchange – Washington. 1965-1978 (1) 1973-1978 (5) 1975-1978 (9) – ISSN: 0020-6601 – mf#6659 – us UMI ProQuest [370]

International encyclopedias of architecture from 16th to 19th century = Internationale architekturlexika des 16. bis 19. jahrhunderts / ed by Schuette, Ulrich – [mf ed 1999] – (= ser Nachschlagewerke und quellen zur kunst 2) – 233mf (1:24) + index brochure – 9 – diazo €2200.00 (silver €2600 isbn: 978-3-598-34547-0) – ISBN-10: 3-598-34546-1 – ISBN-13: 978-3-598-34546-3 – gw Saur [720]

International endodontic journal – Oxford. 1980-1996 (1,5,9) – ISSN: 0143-2885 – mf#15529,01 – us UMI ProQuest [617]

International Eugenics Congress, 2d, New York, 1921 *see* Scientific papers of the second international congress of eugenics

International examiner / Alaska Cannery Workers' Association – 1976 jan [v3 n1]-1983, 1984-87 – 2r – 1 – mf#703277 – us WHS [366]

International executive – New York. 1959-1997 (1) 1971-1997 (5) 1977-1997 (9) – (cont by: thunderbird international business review) – ISSN: 0020-6652 – mf#5167 – us UMI ProQuest [650]

The international exhibition / Wallis, George – [London] 1871 – (= ser 19th c art & architecture) – 1mf – 9 – mf#4.1.174 – uk Chadwyck [700]

International exhibition glasgow 1888 : catalogue of the fine arts section / Davison, Thomas Raffles & Walker, Robert – [Glasgow 1888] – (= ser 19th c art & architecture) – 2mf – 9 – mf#4.2.948 – uk Chadwyck [700]

International family planning perspectives – New York. 1989-1996 (1,5,9) – ISSN: 0190-3187 – mf#16862,02 – us UMI ProQuest [304]

International Federation for Documentation General Secretariat *see* Fid news bulletin

International Federation of Automatic Control *see*
- Ifac proceedings series
- Ifac symposia series

International Federation of Professional and Technical Engineers *see* Outlook

International Federation of Technical Engineers', Architects' and Draftsmen's Unions *see* Monthly outlook

International finance, global securities, and banking : special studies, 1995-2001 – 15r – 1 – $2905.00 – 1-55655-942-9 – us UPA [332]

International financial law review – London. 1994-1994 (1,5,9) – ISSN: 0262-6969 – mf#15737 – us UMI ProQuest [346]

International financial news survey – Washington. 1948-1972 (1) 1971-1972 (5) – ISSN: 0020-6717 – mf#998 – us UMI ProQuest [020]

International financial statistics / International Monetary Fund – v1-24. 1948-65 – 1 – $1296.00 – mf#0288 – us Brook [332]

International financial statistics / International Monetary Fund – Washington. 1948+ (1) 1968+ (5) 1978+ (9) – ISSN: 0020-6725 – mf#999 – us UMI ProQuest [332]

International financial statistics supplement – Washington. 1961-1973 – 1 – mf#6526 – us UMI ProQuest [310]

International Fire Buff Associates, Inc *see* Turnout

International fire fighter / International Association of Fire Fighters – v63 n1-v69 n7/8 [1980 jan-1986 jul/aug], v71 n1-v77 n4 [1988 jan/feb-1994 nov/dec] – 2r – 1 – mf#770951 – us WHS [366]

International fisherman and allied worker – 1941-51 – (= ser Labor union periodicals, pt 3: food and agricultural industries) – 1r – 1 – $210.00 – 1-55655-617-9 – us UPA [660]

International fishery disputes / Haynes, Thomas H – London, Paris: Cassell, 189-? – 1mf – 9 – mf#19015 – cn Canadiana [343]

International food and agribusiness management review – Greenwich. 1997+ (1) – ISSN: 1096-7508 – mf#26286 – us UMI ProQuest [650]

International Fortean Organization *see* Info journal

International Foundation for Development Alternatives *see* [Ifda, the third system project papers]

International Fur Workers Union of the United States and Canada *see* Report of the proceedings of the...biennial convention of the international fur workers union of the united states and canada

International Furniture Workers' Union of America *see*
- Furniture workers' journal
- General wood workers' journal

International game fish conference proceedings – Miami. 1956-1967 (1) – ISSN: 0535-0603 – mf#3332 – us UMI ProQuest [639]

International gas engineering and management – London. 1997+ (1) – (cont: gas engineering and management) – mf#7129,02 – us UMI ProQuest [550]

International genealogy consumer report – 1981 jan-1983 oct/dec – 1r – 1 – mf#696548 – us WHS [929]

International Genealogy Fellowship of Rotarians *see* Rota-gene

International geographical union igu bulletin bulletin de l'ugi – Bonn. 1972-1980 [1]; 1972-1980 [5]; 1975-1980 [9] – ISSN: 0018-9804 – mf#7393 – us UMI ProQuest [900]

International global terrorism : its links with illicit drugs as illustrated by the IRA and other groups in Colombia : hearing before the Committee on International Relations, House of Representatives, 107th congress, 2nd session, April 24, 2002. / United States. Congress. House. Committee on International Relations. – Washington: US GPO 2002 [mf ed 2002] – 2mf – 9 – us GPO [341]

International Glove Workers Union of America *see*
- Glove workers' journal
- Proceedings...annual convention

International Guards Union of America *see* Guards

International gymnast – Norman. 1977+ (1,5,9) – ISSN: 0891-6616 – mf#11561,01 – us UMI ProQuest [790]

International Harvester Co *see* IH farm forum

International herald / International Working Men's Association. British Sect – n1-81. 1872-73 [all pub] – (= ser Radical periodicals of great britain, 1794-1914. period 2) – 1r – 1 – $125.00 – us UPA [335]

International herald tribune – Paris, FRANCE. 1887+ (1) – ISSN: 0294-8052 – mf#60142 – us UMI ProQuest [074]

International herald tribune – Paris, 1985- – (yrly reel count varies) – us UMI ProQuest [074]

International history of paper and paper making : the loeber collection of the foundation for dutch paper history / Loeber, E G – 668mf – 9 – €7490.00 set – (available separately: watermarks [246mf] €2970 [m312]; drawings [82mf] €990 [m313]; photos [262mf] €3170 [m314]; dictionary [78mf] €940 [m315]; p/g to all pts with thematic ind to the watermarks (also on floppy disk); printed ind to the photos (also on floppy disk) & additional ind on mf) – mf#m309 – ne MMF Publ [680]

International holiness directory and year book for... : a daily companion for the christian home / ed by Hughes, George – New York: G Hughes; Philadelphia: International Holiness Pub House 1894- [annual] [mf ed 2004] – 1894 [complete] on 1r – 1 – (cont: holiness year book for...) – mf1057 – us ATLA [240]

International hot-spice / Metaphysical Co-op/ Motivation Producers – 1993 apr 17-1995 sep/oct – 1r – 1 – mf#2844644 – us WHS [071]

International hydrographic review – Monaco. 1973-1996 (1) 1973-1996 (5) 1973-1996 (9) – ISSN: 0020-6946 – mf#7232 – us UMI ProQuest [550]

The international illustrated commentary on the new testament / ed by Schaff, Philip – New York: C Scribner's sons 1888 [mf ed 2001] – 4v on 2r – 1 – (incl ind) – mf#2001-b018 – us ATLA [225]

International immunopharmacology – Amsterdam, 2001+ [1,5,9] – ISSN: 1567-5769 – mf#42830 – us UMI ProQuest [616]

International index to art exhibition catalogues, 1895-1991 – 867mf – 9 – €6040.00 – (incl p/g) – mf#m350 – ne MMF Publ [700]

International index to multi-media information – Pasadena. 1973-1977 (1) 1975-1977 (5) 1975-1977 (9) – (cont: film review index) – ISSN: 0094-6818 – mf#6793,01 – us UMI ProQuest [790]

International Institute for the Unification of Private Law *see* Unidroit proceedings and papers

International Institute Of Ibero-American Literatu *see* Outline history of spanish american literature

International Institute Of Ibero-American Literature *see* Anthology of spanish american literature

International Institute of Social History (IISH), Amsterdam *see* Labor issues in indonesia, 1979-1995

International institute on tax and business planning. new york university : doing business in... – v1-3. 1974-75 – 1,5,6 – $50.00 set – (available on reel only) – mf#109021 – us Hein [073]

International insurance monitor – New York. 1973-1993 (1) 1974-1993 (5) 1974-1993 (9) – ISSN: 0020-6997 – mf#9749 – us UMI ProQuest [368]

International interaction – Washington. 1976-1978 (1,5,9) – mf#11523 – us UMI ProQuest [303]

International Irrigation Congress *see* Official proceedings

International Jewelry Workers' Union *see* Jewelry workers' monthly bulletin

International Jewelry Workers' Union (founded 1916) *see* Gem

International journal – Toronto. 1946+ [1]; 1970+ [5]; 1976+ [9] – ISSN: 0020-7020 – mf#1503 – us UMI ProQuest [327]

International journal for housing science and its applications – Coral Gables. (1) 1977-1983 (5) 1977-1980 (9) – ISSN: 0146-6518 – mf#49310 – us UMI ProQuest [710]

International journal for numerical and analytical methods in geomechanics – Chichester. 1977+ (1,5,9) – ISSN: 0363-9061 – mf#11350 – us UMI ProQuest [620]

International journal for numerical methods in engineering – Chichester. 1969+ (1,5,9) – ISSN: 0029-5981 – mf#10798 – us UMI ProQuest [620]

International journal for numerical methods in fluids – Chichester. 1981+ (1,5,9) – ISSN: 0271-2091 – mf#12646 – us UMI ProQuest [627]

International journal for parasitology – Oxford. 1971-1994 (1) 1971-1994 (5) 1977-1994 (9) – ISSN: 0020-7519 – mf#49088 – us UMI ProQuest [576]

International journal for philosophy of religion – Dordrecht. 1970+ (1,5,9) – ISSN: 0020-7047 – mf#12870 – us UMI ProQuest [200]

International journal for the advancement of counselling – The Hague. 1991-1996 – 1,5,9 – ISSN: 0165-0653 – mf#16793 – us UMI ProQuest [370]

International journal for the education of the blind – Alexandria. 1951-1968 (1) – (cont by: education of the visually handicapped) – ISSN: 0538-8023 – mf#8568 – us UMI ProQuest [362]

International journal for the psychology of religion – Mahwah. 1998+ (1,5,9) – ISSN: 1050-8619 – mf#25254 – us UMI ProQuest [150]

International journal of accounting – Urbana. 1999+ (1,5,9) – mf#15738,01 – us UMI ProQuest [650]

International journal of accounting information systems – New York, 2000+ [1,5,9] – ISSN: 1467-0895 – mf#42836 – us UMI ProQuest [350]

International journal of action methods – Washington. 1997+ (1) 1997+ (5) 1997+ (9) – (cont: journal of group psychotherapy, psychodrama and sociometry) – ISSN: 1096-7680 – mf#6905,03 – us UMI ProQuest [150]

International journal of adaptive control and signal processing – Chichester. 1987+ (1,5,9) – ISSN: 0890-6327 – mf#16165 – us UMI ProQuest [000]

International journal of adhesion and adhesives – Kidlington. 1989-1996 (1,5,9) – ISSN: 0143-7496 – mf#17236 – us UMI ProQuest [660]

International journal of advertising – Eastbourne. 1992-1997 (1) 1992-1997 (5) 1992-1997 (9) – ISSN: 0265-0487 – mf#15763,01 – us UMI ProQuest [650]

International journal of american linguistics – Chicago. 1917+ (1) 1917+ (5) 1917+ (9) – ISSN: 0020-7071 – mf#10226 – us UMI ProQuest [400]

International journal of american linguistics : native american texts series – Chicago. 1976-1977 (1,5,9) – ISSN: 0361-3399 – mf#11209 – us UMI ProQuest [490]

International journal of analytical and experimental modal analysis – Bethel. 1989-1992 (1) – (cont by: modal analysis) – ISSN: 0886-9367 – mf#16027 – us UMI ProQuest [621]

International journal of andrology – Copenhagen. 1987-1994 (1) 1987-1994 (5) 1987-1994 (9) – ISSN: 0105-6263 – mf#16733 – us UMI ProQuest [616]

International journal of antimicrobial agents – Amsterdam. 1991-1996 (1,5,9) – ISSN: 0924-8579 – mf#42632 – us UMI ProQuest [576]

International journal of applied quality management – Greenwich. 1998+ (1,5,9) – ISSN: 1096-4738 – mf#26287 – us UMI ProQuest [650]

International journal of approximate reasoning – New York. 1989-1992 (1,5,9) – ISSN: 0888-613X – mf#42558 – us UMI ProQuest [000]

International journal of arts management – Montreal. 1998+ (1,5,9) – ISSN: 1480-8986 – mf#33000 – us UMI ProQuest [650]

International journal of audiology – Hamilton. 2002+ (1,5,9) – ISSN: 1499-2027 – mf#32383 – us UMI ProQuest [621]

International journal of bank marketing – Bradford. 1991-1995 (1,5,9) – ISSN: 0265-2323 – mf#15764 – us UMI ProQuest [332]

International journal of behavioral development (ijbd) – Amsterdam. 1987+ (1,5); 1987+ (9) – ISSN: 0165-0254 – mf#17095 – us UMI ProQuest [150]

International journal of behavioral development (ijbd) – Amsterdam: North-Holland. v7 n1-v8 n4. 1984-85 – us CRL [150]

International journal of biochemistry – Exeter. 1970-1994 (1) 1970-1994 (5) 1970-1994 (9) – (cont by: international journal of biochemistry and cell biology) – ISSN: 0020-711X – mf#49092 – us UMI ProQuest [574]

International journal of biochemistry and cell biology – Exeter. 1995+ (1,5,9) – (cont: international journal of biochemistry) – ISSN: 1357-2725 – mf#49092,01 – us UMI ProQuest [574]

International journal of biological macromolecules – Amsterdam. 1989-1996 (1,5,9) – ISSN: 0141-8130 – mf#17237 – us UMI ProQuest [574]

International journal of biomedical computing – Barking. 1970-1991 (1) 1970-1991 (5) 1987-1991 (9) – (cont by: international journal of medical informatics) – ISSN: 0020-7101 – mf#42271 – us UMI ProQuest [610]

International journal of cardiac imaging – Boston. 1989-1991 (1) 1989-1991 (5) 1989-1991 (9) – ISSN: 0167-9899 – mf#16794 – us UMI ProQuest [616]

International journal of cardiology – Amsterdam. 1983-1994 (1,5) 1987-1994 (9) – (cont: european journal of cardiology) – ISSN: 0167-5273 – mf#42220 – us UMI ProQuest [616]

International journal of cement composites and lightweight concrete – Harlow. 1989-1989 (1,5,9) – (cont by: cement and concrete composites) – ISSN: 0262-5075 – mf#42582,01 – us UMI ProQuest [690]

International journal of chemical kinetics – New York. 1969+ (1,5,9) – ISSN: 0538-8066 – mf#11054 – us UMI ProQuest [540]

International journal of circuit theory and applications – Chichester. 1973-1994 (1) 1973-1994 (5) 1973-1994 (9) – ISSN: 0098-9886 – mf#10797 – us UMI ProQuest [621]

International journal of climatology – Chichester. 1989+ (1,5,9) – (cont: journal of climatology) – ISSN: 0899-8418 – mf#12647,01 – us UMI ProQuest [550]

International journal of clinical and experimental hypnosis – Philadelphia. 1953+ [1]; 1971+ [5]; 1977+ [9] – ISSN: 0020-7144 – mf#1405 – us UMI ProQuest [615]

International journal of clinical monitoring and computing – Dordrecht. 1991-1995 (1,5,9) – ISSN: 0167-9945 – mf#16795 – us UMI ProQuest [617]

International journal of clinical neuropsychology – Madison. 1984-1989 (1,5,9) – (cont: clinical neuropsychology) – ISSN: 0749-8470 – mf#11955,01 – us UMI ProQuest [616]

International journal of clinical pharmacology research – Geneva. 1981-1995 (1) 1981-1995 (5) 1981-1995 (9) – ISSN: 0251-1649 – mf#12825 – us UMI ProQuest [615]

International journal of clothing science and technology – Bradford. 1991-1995 (1,5,9) – ISSN: 0955-6222 – mf#18916 – us UMI ProQuest [680]

International journal of coal geology – Amsterdam. 1980+ (1) 1980+ (5) 1987+ (9) – ISSN: 0166-5162 – mf#42272 – us UMI ProQuest [550]

International journal of cognitive ergonomics – Mahwah. 1997+ (1) – ISSN: 1088-6362 – mf#28517 – us UMI ProQuest [150]

International journal of communication systems – Chichester, 1994-1994 [1,5,9] – (cont: international journal of digital and analog communication systems) – ISSN: 1074-5351 – mf#16168,02 – us UMI ProQuest [380]

International journal of comparative and applied criminal justice – Wichita. 1977+ (1,5,9) – ISSN: 0192-4036 – mf#11831 – us UMI ProQuest [360]

International journal of comparative psychology – New York. 1987-1992 (1,5,9) – ISSN: 0889-3667 – mf#16141 – us UMI ProQuest [150]

International journal of computer algebra in mathematics education – Hemel Hempstead, 1997+ [1,5,9] – ISSN: 1362-7368 – mf#30797,01 – us UMI ProQuest [510]

International journal of computer and information sciences – New York. 1972-1977 (1) 1972-1977 (5) – (cont by: international journal of parallel programming) – ISSN: 0091-7036 – mf#10867 – us UMI ProQuest [000]

International journal of computer integrated manufacturing – London. 1988-1996 (1,5,9) – ISSN: 0951-192X – mf#17284 – us UMI ProQuest [670]

International journal of computer vision – Hingham. 1987-1996 (1,5,9) – ISSN: 0920-5691 – mf#16796 – us UMI ProQuest [000]

International journal of conflict management – Bowling Green. 2001+ (1,5,9) – ISSN: 1044-4068 – mf#23589 – us UMI ProQuest [650]

International journal of consumer studies – Oxford, 2001+ [1,5,9] – (cont: journal of consumer studies and home economics) – ISSN: 0309-3891 – mf#15558,01 – us UMI ProQuest [339]

International journal of contemporary sociology – Auburn. 1989-1996 (1) – (cont: indian sociological bulletin) – ISSN: 0019-6398 – mf#15000 – us UMI ProQuest [300]

International journal of control – London. 1989-1996 (1,5,9) – ISSN: 0020-7179 – mf#17285 – us UMI ProQuest [620]

International journal of cosmetic science – Oxford. 1979-1996 (1,5,9) – ISSN: 0142-5463 – mf#15551 – us UMI ProQuest [640]

International journal of dermatology – Philadelphia. 1975+ (1,5,9) – ISSN: 0011-9059 – mf#10390 – us UMI ProQuest [616]

International journal of developmental neuroscience : official journal of the international society for developmental neuroscience – Oxford. 1983+ (1) 1983+ (5) 1984+ (9) – ISSN: 0736-5748 – mf#49455 – us UMI ProQuest [612]

International journal of digital and analog cabled systems – Chichester. 1988-1989 (1,5,9) – (cont by: international journal of digital and analog communication systems) – ISSN: 0894-3222 – mf#16168 – us UMI ProQuest [000]

International journal of digital and analog communication systems – Chichester. 1990-1993 (1,5,9) – (cont: international journal of digital and analog cabled systems. cont by: international journal of communication systems) – ISSN: 1047-9627 – mf#16168,01 – us UMI ProQuest [000]

International journal of early childhood = Revue internationale de l'enfance prescolaire – Bakewell. 1969+ (1,5,9) – ISSN: 0020-7187 – mf#12628 – us UMI ProQuest [640]

International journal of earth sciences : geologische rundschau – Berlin. 1999+ (1) – (cont: geologische rundschau) – ISSN: 1437-3254 – mf#10146,01 – us UMI ProQuest [550]

International journal of eating disorders – New York. 1981+ (1,5,9) – ISSN: 0276-3478 – mf#13080 – us UMI ProQuest [616]

International journal of educational development – Oxford. 1981+ – 1,5,9, – ISSN: 0738-0593 – mf#49456 – us UMI ProQuest [370]

International journal of educational research – Elmsford. 1977+ – 1,5,9 – ISSN: 0883-0355 – mf#49284 – us UMI ProQuest [370]

International journal of electrical power and energy systems – Kidlington. 1979+ (1,5,9) – ISSN: 0142-0615 – mf#17238 – us UMI ProQuest [621]

International journal of energy research – Chichester. 1977+ (1,5,9) – ISSN: 0363-907X – mf#11351 – us UMI ProQuest [333]

International journal of engineering science – Oxford. 1963+ (1,5,9) – ISSN: 0020-7225 – mf#49093 – us UMI ProQuest [621]

International journal of entrepreneurial behaviour and research – Bradford. 2001+ (1,5,9) – ISSN: 1355-2554 – mf#31602 – us UMI ProQuest [338]

International journal of epidemiology – Oxford. 1972+ (1) 1972+ (5) 1972+ (9) – ISSN: 0300-5771 – mf#9854 – us UMI ProQuest [614]

International journal of estuarine and coastal law – London. 1991-1992 (1,5,9) – (cont by: international journal of marine and coastal law) – ISSN: 0268-0106 – mf#16797 – us UMI ProQuest [341]

International journal of estuarine and coastal law *see* International journal of marine and coastal law

International journal of experimental pathology – Oxford. 1990-1996 (1) 1990-1996 (5) 1990-1996 (9) – (cont: journal of experimental pathology) – ISSN: 0959-9673 – mf#2517,02 – us UMI ProQuest [619]

International journal of family counseling – New York. 1977-1978 (1) 1977-1978 (5) 1977-1978 (9) – (cont: journal of family counseling. cont by: american journal of family therapy) – ISSN: 0147-1775 – mf#8240,01 – us UMI ProQuest [150]

International journal of family therapy – New York. 1979-1985 (1) 1979-1985 (5) 1979-1985 (9) – (cont by: contemporary family therapy) – ISSN: 0148-8384 – mf#11640 – us UMI ProQuest [306]

International journal of fatigue – Kidlington. 1979-1996 (1,5,9) – ISSN: 0142-1123 – mf#17239 – us UMI ProQuest [620]

International journal of flexible manufacturing systems – Boston. 1988-1996 (1,5,9) – ISSN: 0920-6299 – mf#16798 – us UMI ProQuest [670]

International journal of food microbiology – Amsterdam. 1989-1995 (1,5,9) – ISSN: 0168-1605 – mf#42575 – us UMI ProQuest [574]

International journal of food science and technology – Oxford. 1987+ (1,5,9) – (cont: journal of food technology) – ISSN: 0950-5423 – mf#15560,01 – us UMI ProQuest [660]

International journal of food sciences and nutrition – Basingstoke. 1992+ (1,5,9) – (cont: food sciences and nutrition) – ISSN: 0963-7486 – mf#18127,04 – us UMI ProQuest [613]

International journal of forecasting – Amsterdam. 1985+ (1,5,9) – ISSN: 0169-2070 – mf#42545 – us UMI ProQuest [338]

International journal of fracture – Alpen aan den Rijn. 1989-1996 (1,5,9) – ISSN: 0376-9429 – mf#16799,01 – us UMI ProQuest [620]

International journal of geographical information science – London. 1997+ (1,5,9) – (cont: international journal of geographical information systems) – ISSN: 1365-8816 – mf#17310,01 – us UMI ProQuest [900]

International journal of geographical information systems – London. 1987-1996 (1,5,9) – (cont by: international journal of geographical information science) – ISSN: 0269-3798 – mf#17310 – us UMI ProQuest [900]

International journal of geriatric psychiatry – Chichester. 1986+ (1,5,9) – ISSN: 0885-6230 – mf#16105 – us UMI ProQuest [616]

International journal of government auditing – Washington. 1978+ (1,5,9) – ISSN: 0047-0724 – mf#11907 – us UMI ProQuest [350]

International journal of group psychotherapy – New York. 1990+ (1,5,9) – ISSN: 0020-7284 – mf#18386 – us UMI ProQuest [150]

International journal of group tensions – New York. 1971-1996 (1) 1971-1996 (5) 1971-1996 (9) – ISSN: 0047-0732 – mf#12693 – us UMI ProQuest [150]

International journal of gynecological cancer – Cambridge. 1991-1994 (1,5,9) – ISSN: 1048-891X – mf#18080 – us UMI ProQuest [616]

International journal of gynecology and obstetrics – Baltimore. 1981+ (1) 1981+ (5) 1987+ (9) – ISSN: 0020-7292 – mf#42415 – us UMI ProQuest [618]

International journal of health education – Paris. 1958-1981 (1) 1972-1981 (5) 1975-1981 (9) – (cont by: hygie) – ISSN: 0020-7306 – mf#7020 – us UMI ProQuest [360]

International journal of health planning and management – Chichester. 1985+ [1,5,9] – ISSN: 0749-6753 – mf#14808 – us UMI ProQuest [360]

International journal of heat and fluid flow – New York. 1979+ (1,5,9) – (cont: heat and fluid flow) – ISSN: 0142-727X – mf#12091 – us UMI ProQuest [621]

International journal of heat and mass transfer – Oxford. 1960+ (1) 1960+ (5) 1960+ (9) – ISSN: 0017-9310 – mf#49094 – us UMI ProQuest [621]

International journal of high performance computing applications – Thousand Oaks. 1998+ (1) – (cont: international journal of supercomputer applications and high performance computing) – ISSN: 1094-3420 – mf#16311,02 – us UMI ProQuest [000]

International journal of hospitality and tourism administration / ed by Barrows, Clayton W – ISSN: 1525-6480 – us Haworth [338]

International journal of hospitality management – Oxford. 1982+ (1,5,9) – ISSN: 0278-4319 – mf#49400 – us UMI ProQuest [650]

International journal of human factors in manufacturing – New York. 1991-1996 (1,5,9) – (cont by: human factors and ergonomics in manufacturing) – ISSN: 1045-2699 – mf#18120 – us UMI ProQuest [670]

International journal of hydrogen energy – Oxford. 1976+ (1,5,9) – ISSN: 0360-3199 – mf#49264 – us UMI ProQuest [550]

International journal of imaging systems and technology – New York. 1989-1992 (1,5,9) – ISSN: 0899-9457 – mf#18107 – us UMI ProQuest [600]

International journal of immunopharmacology – Oxford. 1979-1997 (1,5,9) – ISSN: 0192-0561 – mf#49311 – us UMI ProQuest [615]

International journal of impact engineering – Oxford. 1983-1994 (1) 1983-1994 (5) 1983-1994 (9) – ISSN: 0734-743X – mf#49457 – us UMI ProQuest [620]

International journal of industrial ergonomics – Amsterdam. 1991-1994 (1,5,9) – ISSN: 0169-8141 – mf#42507 – us UMI ProQuest [620]

International journal of industrial organization – Amsterdam. 1983+ (1,5,9) – ISSN: 0167-7187 – mf#42580 – us UMI ProQuest [331]

International journal of information management – Kidlington. 1986-1996 (1,5,9) – (cont: social science information studies: ssis) – ISSN: 0268-4012 – mf#17240,01 – us UMI ProQuest [300]

International journal of insect morphology and embryology – Oxford. 1971-1995 (1) 1971-1995 (5) 1971-1995 (9) – ISSN: 0020-7322 – mf#49095 – us UMI ProQuest [590]

International journal of institutional management in higher education – Paris. 1981-1988 (1) 1981-1988 (5) 1981-1988 (9) – (cont by: higher education management) – ISSN: 0253-0058 – mf#12896 – us UMI ProQuest [377]

International journal of instructional media – New York. 1990+ – 1,5,9 – ISSN: 0092-1815 – mf#17448 – us UMI ProQuest [370]

International journal of intelligent systems – New York. 1986+ (1,5,9) – ISSN: 0884-8173 – mf#18108 – us UMI ProQuest [000]

International journal of intensive short-term dynamic psychotherapy – New York. 1997+ (1) – (cont: international journal of short-term psychotherapy) – ISSN: 1096-7028 – mf#18146,01 – us UMI ProQuest [616]
International journal of intercultural relations: ijir – New Brunswick. 1977+ (1,5,9) – ISSN: 0147-1767 – mf#49312 – us UMI ProQuest [301]
International journal of land management – Chichester. 1997+ (1) – ISSN: 1088-4254 – mf#25596 – us UMI ProQuest [333]
International journal of law and psychiatry – New York. 1978+ (1,5,9) – ISSN: 0160-2527 – mf#49313 – us UMI ProQuest [344]
International journal of law libraries *see* International journal of legal information
International journal of legal information – v1-28. 1973-2000 – 9 – $498.00 set – (title varies: v1-9 1973-81 as international journal of law libraries) – ISSN: 0731-1265 – mf#103601 – us Hein [340]
International journal of legal medicine – Heidelberg. 1994-1996 (1) – (cont: zeitschrift fuer rechtsmedizin) – ISSN: 0937-9827 – mf#13249,02 – us UMI ProQuest [614]
International journal of legal research – Meerut. 1966-1972 (1) – ISSN: 0020-7330 – mf#7000 – us UMI ProQuest [340]
International journal of leprosy and other mycobacterial diseases – Lawrence. 1979-1996 (1,5,9) – ISSN: 0148-916X – mf#12236,01 – us UMI ProQuest [616]
International journal of lifelong education – London. 1989+ – 1,5,9 – ISSN: 0260-1370 – mf#17299 – us UMI ProQuest [370]
International journal of lighting research and technology – London. 1993-1994 (1,5,9) – (cont: lighting research and technology) – ISSN: 0024-3426 – mf#10926,01 – us UMI ProQuest [621]
International journal of machine tool design and research – Elmsford. 1961-1986 (1) 1961-1986 (5) 1961-1986 (9) – (cont by: international journal of machine tools and manufacture) – ISSN: 0020-7357 – mf#49096 – us UMI ProQuest [621]
International journal of machine tools and manufacture – Elmsford. 1987+ (1,5,9) – (cont: international journal of machine tool design and research) – ISSN: 0890-6955 – mf#49096,01 – us UMI ProQuest [621]
International journal of manpower – Bradford. 1991-1995 (1,5,9) – ISSN: 0143-7720 – mf#15768 – us UMI ProQuest [331]
International journal of marine and coastal law – London. 1993-1994 (1,5,9) – (cont: international journal of estuarine and coastal law) – ISSN: 0927-3522 – mf#16797,01 – us UMI ProQuest [341]
International journal of marine and coastal law – v1-15. 1986-2000 – 9 – $686.00 set – (title varies: v1-7 1986-92 as international journal of estuarian and coastal law) – ISSN: 0927-3522 – mf#111161 – us Hein [570]
International journal of mass spectrometry – Amsterdam. 1998+ (1,5,9) – (cont: international journal of mass spectrometry and ion processes) – ISSN: 1387-3806 – mf#42133,01 – us UMI ProQuest [530]
International journal of mass spectrometry and ion processes – Amsterdam. 1968-1997 (1) 1968-1997 (5) 1987-1997 (9) – (cont by: international journal of mass spectrometry) – ISSN: 0168-1176 – mf#42133 – us UMI ProQuest [530]
International journal of mathematical education in science and technology – London. 1991+ (1) – ISSN: 0020-739X – mf#17300 – us UMI ProQuest [510]
International journal of mechanical engineering education – Chichester. 1978+ (1,5,9) – ISSN: 0306-4190 – mf#11219 – us UMI ProQuest [621]
International journal of mechanical sciences – Oxford. 1960+ (1) 1960+ (5) 1960+ (9) – ISSN: 0020-7403 – mf#49097 – us UMI ProQuest [621]
International journal of medical informatics – Barking. 1997+ (1) – (cont: international journal of biomedical computing) – ISSN: 1386-5056 – mf#42271,01 – us UMI ProQuest [610]
International journal of medical marketing – London, 2000+ [1,5,9] – ISSN: 1469-7025 – mf#31701 – us UMI ProQuest [650]
International journal of mental health – Armonk. 1994-1996 (1,5,9) – ISSN: 0020-7411 – mf#16886 – us UMI ProQuest [150]
International journal of mental health nursing – Carlton. 2002+ (1,5,9) – ISSN: 1445-8330 – mf#21590,02 – us UMI ProQuest [610]
International journal of methods in psychiatric research – Chichester. 1991-1996 (1,5,9) – ISSN: 1049-8931 – mf#18157 – us UMI ProQuest [616]
International journal of microwave and millimeter-wave computer-aided engineering – New York. 1991-1994 (1,5,9) – (cont by: international journal of rf and microwave computer-aided engineering) – ISSN: 1050-1827 – mf#18121 – us UMI ProQuest [621]
International journal of middle east studies – Cambridge. 1970+ (1) 1976+ (5) 1976+ (9) – ISSN: 0020-7438 – mf#11034 – us UMI ProQuest [956]
International journal of mineral processing – Amsterdam. 1974+ (1) 1974+ (5) 1987+ (9) – ISSN: 0301-7516 – mf#42019 – us UMI ProQuest [660]
International journal of multiphase flow – Oxford. 1974+ (1,5,9) – ISSN: 0301-9322 – mf#49098 – us UMI ProQuest [530]
International journal of museum management and curatorship – Guildford. 1982-1989 (1,5,9) – (cont by: museum management and curatorship) – ISSN: 0260-4779 – mf#17241 – us UMI ProQuest [060]
International journal of music education – Reading. 1983+ (1,5,9) – mf#13546 – us UMI ProQuest [780]
International journal of non-linear mechanics – New York. 1966+ (1,5,9) – ISSN: 0020-7462 – mf#49099 – us UMI ProQuest [510]
International journal of nonprofit and voluntary sector marketing – London, 1999+ [1,5,9] – ISSN: 1465-4520 – mf#31710,01 – us UMI ProQuest [650]
International journal of numerical methods for heat and fluid flow – Bradford. 2001+ (1,5,9) – ISSN: 0961-5539 – mf#31587 – us UMI ProQuest [624]
International journal of numerical modelling, electronic networks devices and fields – Chichester. 1988-1994 (1,5,9) – ISSN: 0894-3370 – mf#16169 – us UMI ProQuest [000]
International journal of nursing practice – Carlton. 1998+ (1,5,9) – ISSN: 1322-7114 – mf#21959 – us UMI ProQuest [610]
International journal of nursing studies – Oxford. 1964+ (1,5,9) – ISSN: 0020-7489 – mf#49101 – us UMI ProQuest [610]
International journal of nursing terminologies and classifications – Philadelphia. 2002+ (1,5,9) – ISSN: 1541-5147 – mf#19676,01 – us UMI ProQuest [610]
International journal of obesity – Houndsmill. 1989-1991 (1) 1989-1991 (5) 1989-1991 (9) – (cont by: international journal of obesity and related metabolic disorders) – ISSN: 0307-0565 – mf#16869 – us UMI ProQuest [616]
International journal of obesity and related metabolic disorders – Houndsmill. 1992-1996 (1,5,9) – (cont: international journal of obesity) – mf#16869,01 – us UMI ProQuest [616]
International journal of occupational health and safety – Northbrook. 1974-1975 (1) 1974-1975 (5) 1975-1975 (9) – (cont by: occupational health and safety. cont: ims : international industrial medicine and surgery) – ISSN: 0093-2205 – mf#722,01 – us UMI ProQuest [610]
International journal of offender therapy and comparative criminology – London. 1957+ (1) 1974+ (5) 1974+ (9) – ISSN: 0306-624X – mf#10051 – us UMI ProQuest [360]
International journal of operations and production management – Bradford. 1992-1995 (1,5,9) – ISSN: 0144-3577 – mf#15769 – us UMI ProQuest [650]
International journal of optical computing – Chichester. 1990-1991 (1,5,9) – ISSN: 1047-8507 – mf#18145 – us UMI ProQuest [000]
International journal of optoelectronics – London. 1991-1995 (1) – ISSN: 0952-5432 – mf#17326,01 – us UMI ProQuest [530]
International journal of oral history – Westport. 1980-1989 (1) 1980-1989 (5) 1980-1989 (9) – ISSN: 0195-6787 – mf#12655 – us UMI ProQuest [390]
International journal of organization theory and behavior – Boca Raton. 1998+ (1,5,9) – ISSN: 1093-4537 – mf#27021 – us UMI ProQuest [150]
International journal of organizational analysis – Bowling Green. 1998+ (1,5,9) – ISSN: 1055-3185 – mf#23588 – us UMI ProQuest [650]
International journal of orthodontics – Milwaukee. 1962-1991 (1) 1962-1977 (5) 1962-1977 (9) – ISSN: 0020-7500 – mf#11348 – us UMI ProQuest [617]
International journal of paediatric dentistry – Oxford. 1991-1996 (1,5,9) – ISSN: 0960-7439 – mf#18302 – us UMI ProQuest [617]
International journal of parallel programming – New York. 1986+ (1,5,9) – (cont: international journal of computer and information sciences) – ISSN: 0885-7458 – mf#10867,01 – us UMI ProQuest [000]
International journal of parapsychology – v. 1-10. 1959-68 – 1 – us AMS Press [130]
International journal of pediatric otorhinolaryngology – Amsterdam. 1979+ (1) 1979+ (5) 1987+ (9) – ISSN: 0165-5876 – mf#42020 – us UMI ProQuest [617]
International journal of personal construct psychology – New York. 1988-1993 (1,5,9) – (cont by: journal of constructivist psychology) – ISSN: 0893-603X – mf#16657 – us UMI ProQuest [150]
International journal of pharmaceutics – Amsterdam. 1978+ (1) 1978+ (5) 1986+ (9) – ISSN: 0378-5173 – mf#42021 – us UMI ProQuest [615]
International journal of physical distribution and logistics management – Bradford. 1992-1993 (1) 1992-1993 (5) 1992-1993 (9) – ISSN: 0960-0035 – mf#15770,02 – us UMI ProQuest [380]
International journal of plant sciences – Chicago. 1992+ (1,5,9) – (cont: botanical gazette) – ISSN: 1058-5893 – mf#135,01 – us UMI ProQuest [580]
International journal of plant varieties and seeds – Oxford, 1997+ [1,5,9] – (cont: plant varieties and seeds) – mf#17103,01 – us UMI ProQuest [631]
International journal of plasticity – New York. 1985-1996 (1,5,9) – ISSN: 0749-6419 – mf#49486 – us UMI ProQuest [660]
International journal of political economy – Armonk. 1991+ (1) – ISSN: 0891-1916 – mf#16887,01 – us UMI ProQuest [330]
International journal of political education – Amsterdam. 1977-1983 (1) 1977-1983 (5) (9) – ISSN: 0378-5165 – mf#42092 – us UMI ProQuest [320]
International journal of politics, culture, and society – New York. 1987+ (1,5,9) – ISSN: 0891-4486 – mf#16142 – us UMI ProQuest [306]
International journal of powder metallurgy – Baltimore. 1965-1973 (1) 1971-1973 (5) – (cont by: international journal of powder metallurgy and powder technology) – ISSN: 0020-7535 – mf#2470 – us UMI ProQuest [660]
International journal of powder metallurgy – Princeton. 1986+ (1) 1986+ (5) 1986+ (9) – (cont: international journal of powder metallurgy and powder technology) – ISSN: 0888-7462 – mf#2470,02 – us UMI ProQuest [660]
International journal of powder metallurgy and powder technology – Princeton. 1974-1985 (1) 1974-1985 (5) 1976-1985 (9) – (cont: international journal of powder metallurgy. cont by: international journal of powder metallurgy) – ISSN: 0361-3488 – mf#2470,01 – us UMI ProQuest [660]
International journal of pressure vessels and piping – Barking. 1973-1994 (1) 1973-1994 (5) 1987-1994 (9) – ISSN: 0308-0161 – mf#42022 – us UMI ProQuest [621]
International journal of production economics – Amsterdam. 1991+ (1,5,9) – (cont: engineering costs and production economics) – ISSN: 0925-5273 – mf#42184,01 – us UMI ProQuest [620]
International journal of production research – London. 1988+ (1,5,9) – ISSN: 0020-7543 – mf#17286 – us UMI ProQuest [338]
International journal of project management – Kidlington. 1983+ (1,5,9) – ISSN: 0263-7863 – mf#17242 – us UMI ProQuest [650]
International journal of psychophysiology – Amsterdam. 1983+ (1) 1983+ (5) 1984+ (9) – ISSN: 0167-8760 – mf#42511 – us UMI ProQuest [150]
International journal of public opinion research – Oxford. 1989+ (1,5,9) – ISSN: 0954-2892 – mf#17500 – us UMI ProQuest [303]
International journal of purchasing and materials management – Tempe. 1991-1998 (1) 1991-1998 (5) 1991-1998 (9) – (cont: journal of purchasing and materials management. cont by: journal of supply chain management) – ISSN: 1055-6001 – mf#2757,02 – us UMI ProQuest [650]
International journal of qualitative studies in education : qse – London. 1988+ – 1,5,9 – ISSN: 0951-8398 – mf#17301 – us UMI ProQuest [370]
International journal of quality and reliability management – Bradford. 1991-1995 (1,5,9) – ISSN: 0265-671X – mf#16480 – us UMI ProQuest [650]
International journal of quantum chemistry – New York. 1967+ (1) 1967+ (5) 1967+ (9) – ISSN: 0020-7608 – mf#11055 – us UMI ProQuest [540]
International journal of quantum chemistry : quantum biology symposium – New York. 1976-1986 (1,5,9) – ISSN: 0360-8832 – mf#11776 – us UMI ProQuest [574]
International journal of quantum chemistry : quantum chemistry symposium – New York. 1967-1985 (1,5,9) – ISSN: 0161-3642 – mf#11773 – us UMI ProQuest [540]
International journal of radiation applications and instrumentation pt a : applied radiation and isotopes – Oxford. 1956-1992 (1,5,9) – (cont by: applied radiation and isotopes) – ISSN: 0883-2889 – mf#49091 – us UMI ProQuest [530]
International journal of radiation applications and instrumentation pt b : nuclear medicine and biology – Oxford. 1973-1992 (1) 1973-1992 (5) 1973-1992 (9) – (cont by: nuclear medicine and biology) – ISSN: 0883-2897 – mf#49100 – us UMI ProQuest [574]
International journal of radiation applications and instrumentation, pt c : radiation physics and chemistry – Oxford. 1969-1992 (1) 1969-1992 (5) 1969-1992 (9) – (cont by: radiation physics and chemistry) – mf#49089 – us UMI ProQuest [530]
International journal of radiation applications and instrumentation pt d : nuclear tracks and radiation measurements – Oxford. 1977-1991 (1,5,9) – (cont by: nuclear tracks and radiation measurements including thermoluminescence) – mf#49275 – us UMI ProQuest [530]
International journal of radiation applications and instrumentation pt e : nuclear geophysics – Oxford. 1987-1992 (1,5,9) – (cont by: nuclear geophysics) – ISSN: 0886-0130 – mf#49502 – us UMI ProQuest [550]
International journal of radiation oncology, biology, physics – New York. 1976+ (1,5,9) – ISSN: 0360-3016 – mf#49265 – us UMI ProQuest [530]
International journal of refrigeration = Revue internationale du froid – Kidlington. 1980-1996 (1,5,9) – ISSN: 0140-7007 – mf#17243 – us UMI ProQuest [621]
International journal of remote sensing – London. 1988+ (1,5,9) – ISSN: 0143-1161 – mf#17312 – us UMI ProQuest [621]
International journal of research in marketing – Amsterdam. 1990+ (1,5,9) – ISSN: 0167-8116 – mf#42578 – us UMI ProQuest [650]
International journal of rf and microwave computer-aided engineering – New York. 1998+ (1) – (cont: international journal of microwave and millimeter-wave computer-aided engineering) – ISSN: 1096-4290 – mf#18121,01 – us UMI ProQuest [621]
International journal of risk and safety in medicine – Amsterdam. 1990-1992 (1,5,9) – ISSN: 0924-6479 – mf#42614 – us UMI ProQuest [360]
International journal of robotics research – Thousand Oaks. 1982+ (1,5,9) – ISSN: 0278-3649 – mf#12869 – us UMI ProQuest [629]
International journal of robust and nonlinear control – Chichester. 1991-1994 (1,5,9) – ISSN: 1049-8923 – mf#18159 – us UMI ProQuest [620]
International journal of rock mechanics and mining sciences – Oxford. 1997+ (1) – mf#49636 – us UMI ProQuest [622]
International journal of rock mechanics and mining sciences and geomechanics abstracts – Oxford. 1964-1996 (1) 1964-1996 (5) 1964-1996 (9) – ISSN: 0148-9062 – mf#49102 – us UMI ProQuest [622]
International journal of satellite communications – Chichester. 1983+ (1,5,9) – ISSN: 0737-2884 – mf#14809 – us UMI ProQuest [380]
International journal of science education – London. 1991-1996 – 1,5,9 – ISSN: 0950-0693 – mf#17302,01 – us UMI ProQuest [370]
International journal of short-term psychotherapy – New York. 1986-1996 (1,5,9) – (cont by: international journal of intensive short-term dynamic psychotherapy) – ISSN: 0884-724X – mf#18146 – us UMI ProQuest [616]
International journal of slavic linguistics and poetics – The Hague. 1959-1973 (1) 1972-1973 (5) (9) – ISSN: 0538-8228 – mf#6542 – us UMI ProQuest [460]
International journal of social education – Muncie. 1989-1996 (1,5,9) – (cont: indiana social studies quarterly) – ISSN: 0889-0293 – mf#17526 – us UMI ProQuest [300]
International journal of social psychiatry – Brookmans Park. 1991+ (1,5,9) – ISSN: 0020-7640 – mf#18262 – us UMI ProQuest [616]
International journal of sociology – Armonk. 1991+ (1) – ISSN: 0020-7659 – mf#16888 – us UMI ProQuest [301]
International journal of sociology and social policy – Patrington. 1992-1993 (1) 1992-1993 (5) 1992-1993 (9) – ISSN: 0144-333X – mf#16437 – us UMI ProQuest [301]
International journal of solids and structures – New York. 1965+ (1,5,9) – ISSN: 0020-7683 – mf#49103 – us UMI ProQuest [530]
International journal of supercomputer applications – Thousand Oaks. 1991-1993 (1) – (cont by: international journal of supercomputer applications and high performance computing) – ISSN: 0890-2720 – mf#16311 – us UMI ProQuest [000]
International journal of supercomputer applications and high performance computing – Thousand Oaks. 1994-1995 (1) – (cont: international journal of supercomputer applications. cont by: international journal of high performance computing applications) – ISSN: 1078-3482 – mf#16311,01 – us UMI ProQuest [000]
International journal of sustainability in higher education – Bradford. 2001+ (1,5,9) – ISSN: 1467-6370 – mf#31577 – us UMI ProQuest [378]

INTERNATIONAL

International journal of systematic bacteriology – Washington. 1951+ [1]; 1972+ [5]; 1973+ [9] – ISSN: 0020-7713 – mf#6607 – us UMI ProQuest [576]
International journal of systems science – London. 1992-1996 (1) 1993-1993 (5) 1993-1993 (9) – ISSN: 0020-7721 – mf#17287 – us UMI ProQuest [620]
International journal of technology assessment in health care – Cambridge. 1989-1996 (1) – ISSN: 0266-4623 – mf#16533 – us UMI ProQuest [619]
International journal of testing – Mahwah. 2001+ (1,5,9) – ISSN: 1530-5058 – mf#31731 – us UMI ProQuest [380]
International journal of the addictions – New York. 1966-1995 (1) 1966-1995 (5) 1966-1995 (9) – (cont by: substance use and misuse) – ISSN: 0020-773X – mf#12925 – us UMI ProQuest [360]
International journal of the history of sport – London. 1990+ (1,5,9) – ISSN: 0952-3367 – mf#18559,01 – us UMI ProQuest [790]
International journal of theoretical physics – New York. 1968-1996 (1) 1968-1996 (5) 1978-1996 (9) – ISSN: 0020-7748 – mf#10859 – us UMI ProQuest [530]
International journal of tourism research – Chichester. 1999+ (1) – ISSN: 1099-2340 – mf#28549 – us UMI ProQuest [338]
International journal of training and development – Oxford. 1997+ (1) – ISSN: 1360-3736 – mf#25717 – us UMI ProQuest [650]
International journal of transport management – Amsterdam. 2002+ (1,5,9) – ISSN: 1471-4051 – mf#42883 – us UMI ProQuest [380]
International journal of trauma nursing – St. Louis. 1995-1996 (1,5,9) – ISSN: 1075-4210 – mf#21562 – us UMI ProQuest [610]
International journal of urban and regional research – London. 1989+ (1,5,9) – ISSN: 0309-1317 – mf#17635 – us UMI ProQuest [710]
International journal of water resources development – Guildford. 1989-1992 (1,5,9) – ISSN: 0790-0627 – mf#17244 – us UMI ProQuest [333]
International journal of women's studies – Montreal. 1978-1985 (1,5,9) – ISSN: 0703-8240 – mf#12298 – us UMI ProQuest [305]
International journal on tissue reactions – Geneva. 1985-1992 (1,5,9) – ISSN: 0250-0868 – mf#12154 – us UMI ProQuest [574]
International journal on world peace – New York. 1993+ (1,5,9) – ISSN: 0742-3640 – mf#19214 – us UMI ProQuest [327]
International juridical association bulletin – v1-10. 1932-42 (all publ) – 18mf – 9 – $27.00 – mf#LLMC 84-496 – us LLMC [340]
International labor defense / Meiklejohn Civil Liberties Library – 1933-45 – 1 – us AMS Press [321]
International Labor Press Association *see* Ilpa reporter
International Labour Defense *see* It's happening in spain
International Labour Office *see*
– Inclusion in the constitution of the international labour...
– Proposed declaration concerning the policy of 'apartheid'
International Labour Organisation *see* Legislacion social de america latina.
International labour organisation : reports and records of proceedings of the international labour conference – 1919-98+ – 128r – 1 – £6,000.00 – mf#ILO – uk World [344]
International Labour Organisation (ILO) *see* Reports and records of proceedings of the international labour conference
International labour review – Geneva. 1921+ (1) 1968+ (5) 1976+ (9) – ISSN: 0020-7780 – mf#1012 – us UMI ProQuest [331]
International Ladies Garment Workers' Union *see*
– Gerechtigkeit
– Giustizia
– Justice
– Der yunion arbeiter
International Ladies' Garment Workers' Union *see*
– Garment worker
– Giustizia
– Ilgwu news
– Justicia
– Knitgoods workers voice
– Ladies' garment worker
– Ohio sew biz
– Our aim
International Ladies' Garment Workers' Union et al *see* Local 23-25 news
International laundry worker / Laundry, Dry Cleaning and Dye House Workers' International Union – 1942 feb-1961 jan, 1961 feb/mar-1967 dec, 1968 jul-1971 apr – 2r – 1 – mf#855255 – us WHS [331]
International law bulletin *see* Columbia journal of transnational law
International law club journal *see* Harvard international law journal
International law digests (american) – Coverage includes Wharton, Moore and Hackworth – 7r – 1 – $250.00 – us Trans-Media [341]
International law in ancient / Viswanatha, Sekharipuram Vaidyanatha – Bombay; New York: Longmans, Green & Co, 1925 – (= ser Samp: indian books) – us CRL [341]
International law studies / U.S. Naval War College – v1-58. 1901-66 – 1 – $378.00 – mf#0393 – us Brook [341]
International law studies – US Naval War College. v1-73 + 2 indexes. 1900-99 – 270mf – 9 – $405.00 – mf#LLMC 79-452 – us LLMC [340]
International law studies – Washington. 1978-1978 (1) 1978-1978 (5) 1978-1978 (9) – mf#5771 – us UMI ProQuest [341]
International lawyer (aba) – v1-34. 1966-2000 – 1,5,6 – $885.00 set – (cont: american bar association, section of international and comparative law) – ISSN: 0020-7810 – mf#103641 – us Hein [341]
International legal materials – American Society of International Law. v1-40. 1962-2001 – 9 – $2500.00 set – ISSN: 0029-7829 – mf#103651 – us Hein [341]
International legal perspectives – Northwestern School of Law of Lewis and Clark. v1-10. 1987-98 – 9 – $161.00 set – mf#113031 – us Hein [341]
International legal practitioners – v5-14. 1980-89 – 9 – $55.00 set – (v1-4 1976-79 is contained in international bar journal 1976-79) – ISSN: 0029-7829 – mf#400820 – us Hein [340]
The international lesson system : the history of its origin and development / Sampey, John Richard – New York: Fleming H Revell, c1911 – 1mf – 9 – 0-7905-0276-3 – (incl bibl ref and index) – mf#1987-0276 – us ATLA [240]
International Longshoreman's and Warehousemen's Union *see*
– Ilwu bulletin
– Ilwu dispatcher
– Ilwu reporter
International Longshoremen, Marine and Transportworkers' Association *see* Directory of locals
International Longshoremen's and Warehousemen's Union *see*
– Dispatcher
– Voice of the ilwu
International Longshoremen's Association *see*
– Brooklyn longshoreman
– Directory
International Machinists and Blacksmiths of North America *see* Machinists and blacksmiths' monthly journal, 1870-1875 / the brass worker, 1895-1896 / official journal, 1902-1904
International mailer *see* Palette and graver, 1919-1923 / international mailer, 1946-1955
International management – London. 1946-1982 (1) 1956-1982 (5) 1956-1982 (9) – ISSN: 0020-7888 – mf#386 – us UMI ProQuest [650]
International management – London. 1986-1994 (1) 1986-1994 (5) 1986-1994 (9) – ISSN: 0020-7888 – mf#386,02 – us UMI ProQuest [650]
International management america : latina edition – New York. 1971-1985 (1) 1974-1985 (5) 1975-1985 (9) – ISSN: 0020-7888 – mf#10129 – us UMI ProQuest [338]
International management europe – London. 1983-1985 (1) 1983-1985 (5) 1983-1985 (9) – ISSN: 0020-7888 – mf#386,01 – us UMI ProQuest [650]
International Map Seminar (1979 : Pretoria, South Africa) *see* Proceedings on microfiche
International marketing review – London. 1992-1993 (1) 1992-1993 (5) 1992-1993 (9) – ISSN: 0265-1335 – mf#15773 – us UMI ProQuest [650]
International meat worker *see* Meat of it, 1945-1966 / international meat worker, 1946-1948
International mental health research newsletter – New York. 1975-1975 (1) 1975-1975 (5) 1975-1975 (9) – (cont by: transnational mental health research newsletter) – ISSN: 0020-7969 – mf#9898 – us UMI ProQuest [610]
International metal worker / United Metal Workers International Union of America – v1-v3 n12 [1902 dec-1905 nov] – 1r – 1 – mf#1059738 – us WHS [331]
International metal worker, 1902-1905 / united weldors' news, 1941-1945 / United Metal Workers' International Union of America & United Brotherhood of Weldors, Cutters and Helpers of America – (= ser Labor union periodicals, pt 1: the metal trades) – 1r – 1 – $210.00 – 1-55655-237-8 – us UPA [680]
The international microform journal of aesthetic-plastic surgery : transactions of the second congress of the international society of aesthetic plastic surgery jerusalem, june 1973 – us Striker [617]
International microform journal of legal medicine – Ann Arbor. 1965-1978 (1) (5) 1965-1978 (9) – (cont by: international microform journal of legal medicine and forensic sciences) – mf#3459 – us UMI ProQuest [614]
International microform journal of legal medicine and forensic sciences – Ann Arbor. 1979-1985 (1) 1979-1985 (5) 1979-1985 (9) – (cont: international microform journal of legal medicine) – mf#3459,01 – us UMI ProQuest [614]
International migration review (imr) – New York. 1964+ (1) 1971+ (5) 1975+ (9) – ISSN: 0197-9183 – mf#2218 – us UMI ProQuest [304]
International Military Tribunal *see*
– Prosecution exhibits submitted to the international military tribunal
– Trial of major war criminals
International Military Tribunal for the Far East *see*
– Court papers, journal, exhibits, and judgments of the international military tribunal for the far east, 1900-1948
– Exhibits
– General summation
– Narrative summary and transcripts of court proceedings for cases tried before the international military tribunal for the far east, 1946-1948
– Narrative summary of record
– Proceeding. judgement
– Proceedings in chamber
– Prosecution and defense summations for cases tried before international military tribunal for the far east, 1948
– Summary of evidence
– Transcript of proceedings
– Transcripts of proceedings in chambers for cases tried before the international military tribunal for the far east, 1946-1948
– Trial of major war criminals
International military tribunal for the far east : (tokyo war crimes trials) – Washington. 1946-1948 (1) – mf#2585 – us UMI ProQuest [341]
International Missionary Alliance *see*
– Annual report of the...1892-
– Report (in part) of the...year of the international missionary alliance
International Missionary Council *see*
– Cuban church in a sugar economy
– Yen pien chung te tung-ya chi-tu hua chia t'ing sheng huo
International Missionary Council [4th: 1938: Madras, India] *see* Ma-te-la-ssu ta hui yin hsiang chi (ccm337)
International Missionary Council Dept Of Social... *see* Church in puerto ricos dilemma
International Missionary Council Dept Of Social And Economic *see* Modern industry and the african
International Molders and Allied Workers Union *see* Journal
International molders and allied workers union journal – 1972 apr 21-jun 16 – 1r – 1 – mf#3277205 – us WHS [331]
International molders and foundry workers journal – 1864-1955 – (= ser Labor union periodicals, pt 1: the metal trades) – 32r – 1 – $6685.00 – 1-55655-238-6 – us UPA [680]
International Molders' Union of North America et al *see* Ironmolders' journal
International Monetary Fund *see*
– Balance of payments. yearbook
– Imf staff papers
– Imf survey
– Informe anual del directorio ejecutivo
– International financial statistics
– International monetary fund. annual report of the executive board for the financial year ended april 30
– International monetary fund. annual report on exchange arrangements and exchange restrictions
– Jahresbericht der executivdirektoren fuer das am. abgelaufene geschaeftsjahr
– Rapport annuel du conseil d'administration
International monetary fund. annual report – 1946-72 – 1 – $120.00 – mf#0286 – us Brook [332]
International monetary fund. annual report of the executive board for the financial year ended april 30 / International Monetary Fund – Washington DC 1947+ – 1,5,9 – ISSN: 0250-7498 – mf#6540 – us UMI ProQuest [332]
International monetary fund. annual report on exchange arrangements and exchange restrictions / International Monetary Fund – Washington DC 1979+ – 1,5,9 – (cont: annual report on exchange restrictions) – ISSN: 0250-7366 – mf#6525.01 – us UMI ProQuest [332]
International monetary fund. annual report on exchange restrictions *see* Annual report on exchange restrictions
International Monetary Fund (IMF) *see* Imf reports and summary proceedings
International monetary fund publications : backfile thru release no 6 – 9 – $6,125.00 – 0-89941-663-2 – mf#401640 – us Hein [332]
International monetary fund staff papers – Washington. 1950-1998 (1) 1971-1998 (5) 1975-1998 (9) – ISSN: 0020-8027 – mf#2212 – us UMI ProQuest [332]
International Monetary Fund Summary *see* Proceedings of the annual meeting of the board of governors
International monetary fund. summary proceedings of the annual meeting of the board of governors – 1st-25th. 1946-70 – 1 – $162.00 – mf#0289 – us Brook [332]
International monthly magazine of literature, science, and art – New York. 1850-1852 – 1 – mf#3869 – us UMI ProQuest [073]
International motion picture almanac – 1929-2003 – 1 – $1960.00 – mf#0290 – us Brook [790]
International Museum of Photography at George Eastman House *see* British masters of the albumen print
International Musical Society *see* Sammelbaende der internationalen musikgesellschaft
International musician / American Federation of Musicians – New York. 1909-76. 20 reels – 1 – us L of C Photodup [780]
International musician – New York. 1912+ (1) 1979+ (5) 1979+ (9) – ISSN: 0020-8051 – mf#6672 – us UMI ProQuest [780]
International musician : official journal of the american federation of musicians / American Federation of Musicians of the United States and Canada – [1901 jul/1906 jun]-[1971 jan/1974 jun] – 12r – 1 – mf#166124 – us WHS [780]
International nationalism / Day, John – London, England. 1967 – 1r – us UF Libraries [960]
International new york herald tribune – Frankfurt/M DE, Paris (F), 1959 29 mar-1969 14 aug, 1969 28 nov-1971 20 jul, 1971 2 aug-1979 4 oct, 1979 21 dec-1991 30 apr, 1991 2 aug-23 aug, 1991 1 nov-1996 26 may [gaps], 1997 8 apr-28 oct [gaps], 1998 8 apr-31 jul, 1999 2 jan-30 jun, 2000 3 jan-2002 1 nov [gaps] – 1 – gw Misc Inst [074]
International news – 1997 aug-nov/dec – 1r – 1 – mf#4024545 – us WHS [321]
International news : provisional international contact commission for the new communist – v1-12 n4. 1939-50 [all publ] – (= ser Radical periodicals in the united states, 1881-1960. series 2) – 1r – 1 – $200.00 – us UPA [335]
International news on fats, oils and related materials: inform – Champaign. 1990+ (1,5,9) – ISSN: 0897-8026 – mf#17634 – us UMI ProQuest [660]
International news : theoretical organof International Contact Commission / Revolutionary Workers League of the US et al – v8 n9-10 [1947 oct-nov], v9 n1-5 [1948 jun-dec], v10 n1 [1949 jan/feb], v11 n1-2 [1949 mar/apr-may/jun], v11 n4-6 [1949 sep/oct-dec], v12 n1-4 [1950 jan-nov] – 1r – 1 – (cont: marxist [chicago, il]) – mf#5182256 – us WHS [335]
International Non-operating Railway Unions in Canada *see* Canadian railwayman
International nursing index – Philadelphia. 1966+ (1) 1970+ (5) 1975+ (9) – ISSN: 0020-8124 – mf#3234 – us UMI ProQuest [610]
International nursing review – Oxford. 1954+ (1) 1965+ (5) 1970+ (9) – ISSN: 0020-8132 – mf#1367 – us UMI ProQuest [610]
International odd fellow / Independent Order of Odd Fellows – v14 n3 [1977 jun/jul], v16-v21 n3 [1978 aug-1984 may/jun] – 1r – 1 – mf#970989 – us WHS [366]
International oil worker : official organ, international association of oil field, gas well and refinery workers / International Association of Oil Field, Gas Well and Refinery Workers of America – v1 n1-5 [1925 july 23-dec 3] – 1r – 1 – (cont by: oil worker [long beach, ca: 1928]) – mf#1008290 – us WHS [331]
International oil worker : official publication of the international association of oil field, gas well and refinery workers of america / International Association of Oil Field, Gas Well and Refinery Workers of America – v5 n8-v6 n6 [1934 sep 12-1935 dec 13] – 1r – 1 – (cont: oil worker [long beach, ca: 1928]; cont by: international oil worker [1937]) – mf#1008294 – us WHS [331]
International oil worker / Oil Workers International Union – 1945 mar [v1 n1]-1950 dec, 1951 jan-1955 feb [v10 n24] – 2r – 1 – (cont by: union chemical workers; oil, chemical and atomic union news) – mf#837082 – us WHS [331]
International oil worker / Oil Workers International Union – v1 n2-5 [1937 dec 27-1938 apr 7] – 1r – 1 – (cont: international oil worker [1934]; cont by: cio news [oil workers ed]) – mf#1008295 – us WHS [331]
International ophthalmology – The Hague. 1991-1994 (1) 1991-1994 (5) 1991-1994 (9) – ISSN: 0165-5701 – mf#16801 – us UMI ProQuest [617]

International ophthalmology clinics – Philadelphia. 1961+ (1) 1975+ (5) 1975+ (9) – ISSN: 0020-8167 – mf#10968 – us UMI ProQuest [617]

International organization – Cambridge. 1947+ (1) 1968+ (5) 1975+ (9) – ISSN: 0020-8183 – mf#1063 – us UMI ProQuest [327]

International organization / Owen, Floyd William – Lansing, MI. 1931? – 1r – us UF Libraries [341]

International Organization of Masters, Mates and Pilots *see* Master, mate and pilot

International orthopaedics – Heidelberg. 1981-1996 (1,5,9) – ISSN: 0341-2695 – mf#13183 – us UMI ProQuest [617]

International outlook / Church of God in Christ [Los Angeles CA] – 1973 jan-may – 1r – 1 – mf#4114446 – us WHS [243]

International outlook : official publication of the seafarers international union of north america / Seafarers International Union of North America – v1 n1-v4 n2 [1980 jul-1983 mar/apr] – 1r – 1 – mf#922798 – us WHS [331]

International Pacific Salmon Fisheries Commission *see* Records, 1937-1963

International peace relations pamphlet material – 1 – (world peace foundation v1-7 1911-17; a league of nations v1-6 no 2 1917-1923; world peace foundation v6 no 3-v12 no 6 1923-30; carnegie endowment for international peace nos 1-56 1914-1937) – us AMS Press [327]

International peace research newsletter – Tampere. 1972-1987 (1) 1974-1987 (5) 1974-1987 (9) – (cont by: ipra newsletter) – ISSN: 0020-8213 – mf#8169 – us UMI ProQuest [320]

International peace research newsletter – Boulder. 1989+ (1) 1989+ (5) 1989+ (9) – (cont: ipra newsletter) – mf#8169,04 – us UMI ProQuest [320]

International Penal and Prison Commission *see* Bulletin de la commission internationale penale et penitentiaire

International perspectives on common fiscal issues : hearing...house of representatives, 107th congress, 2nd session...washington dc, june 5 2002 / United States. Congress. House. Committee on the Budget – Washington: US GPO 2002 [mf ed 2002] – 1mf – 9 – 0-16-068719-5 – us GPO [343]

International petroleum abstracts – Kingston-upon-Thames. 1982-1990 (1,5,9) – (cont by: international petroleum abstracts incorporating offshore abstracts) – ISSN: 0309-4944 – mf#13308 – us UMI ProQuest [550]

International petroleum abstracts incorporating offshore abstracts – Chichester. 1991-1992 (1,5,9) – (cont: international petroleum abstracts) – ISSN: 1052-9292 – mf#13308,01 – us UMI ProQuest [550]

International petroleum times – London. 1978-1981 (1) 1978-1981 (5) 1978-1981 (9) – (cont by: petroleum times) – ISSN: 0141-4437 – mf#3162,01 – us UMI ProQuest [550]

International pharmaceutical abstracts – Bethesda. 1964+ [1,5]; 1975+ [9] – ISSN: 0020-8264 – mf#1924 – us UMI ProQuest [615]

International pharmacopsychiatry – Basel. 1968-1974 (1) 1970-1974 (5) 1974-1974 (9) – ISSN: 0020-8272 – mf#5192 – us UMI ProQuest [616]

The international philatelist : a monthly for stamp collectors – Toronto: W S Weatherston, [1892-1893] – 9 – (absorbed: the canadian journal of philately. absorbed: the philatelic fraud reporter. absorbed: one dime) – mf#P04549 – cn Canadiana [760]

International philosophical quarterly – Bronx. 1983+ (1,5,9) – (cont: ipq) – ISSN: 0019-0365 – mf#11438,01 – us UMI ProQuest [100]

International Phonetic Association *see* Journal of the international phonetic association

International Photoengravers Union of North America *see*
- Palette and graver, 1919-1923 / international mailer, 1946-1955
- Plate makers' criterion, 1907-1909 / american photo-engraver, 1908-1955

International pipe line industry – Houston. 1989-1991 (1) 1989-1991 (5) 1989-1991 (9) – (cont: pipe line industry. cont by: pipe line industry) – ISSN: 0032-0145 – mf#1819,01 – us UMI ProQuest [550]

International Plate Printers', Die Stampers' and Engravers' Union of North America *see* Plate printer, 1902-1932

International political science review – v1-5. 1980-1984 (all publ) – 5,6 – $66.00 set – mf#400580 – us Hein [320]

International political science review: ipsr = Revue internationale de science politique: risp – London. 1983-1995 (1) 1983-1995 (5) 1983-1995 (9) – ISSN: 0192-5121 – mf#14012 – us UMI ProQuest [327]

International politics – The Hague. 1996-1996 (1,5,9) – (cont: co-existence) – ISSN: 1384-5748 – mf#16774,01 – us UMI ProQuest [300]

International population : census publications – ca 5206r – 1 – (segment 1: 1945-1967, 752r. segment 2: pre-1945, 566r. segment 3: post-1967, ca 1188r) – us Primary [010]

International population census publications : segment 1, 1945-1967 – 752r coll – 1 – (africa 73r; asia 243r; europe 209r; latin america and the caribbean 128r; north america 73r (includes 1950-60 us decennial census); oceania 26r. with guide) – us Primary [310]

International population census publications : segment 3, post-1967 – in process units 1-26 – 1188r – 1 – (africa 139r; asia 283r; europe 302r; latin america and the caribbean 260r; oceania 30r; north america (includes 1970-80 us decennial census) 174r) – us Primary [310]

International population census publications, segment 2 : pre-1945 – 566r coll – 1 – (africa 18r; asia 118r; europe 254r; latin america and the caribbean 75r; north america (includes 1950-60 us decennial census) 78r; oceania 23r. with guide) – us Primary [310]

International president's report to the...convention of the amalgamated associationof street and electric railway employes of america / Amalgamated Association of Street and Electric Railway Employees of America – 9th [1905] – 1r – 1 – (cont: international president's report to the...convention of the amalgamated association of street railway employees of america, amalgamated association of street railway employees of america; cont by: report of the international president to the...convention, amalgamated association of street, electric railway and motor coach employees of america) – mf#3397095 – us WHS [331]

International president's report to the...convention of the amalgamated associationof street railway employees of america / Amalgamated Association of Street and Electric Railway Employees of America – 7th [1901] – 1r – 1 – (cont by: international president's report to the...convention of the amalgamated association of street and electric railway employes of america, amalgamated association of street and electric railway employees of america) – mf#3397108 – us WHS [331]

International Printing and Graphic Communications Union *see* News and views

International Printing Pressman and Assistants' Union of North America *see* American pressman, 1890-1955

International Printing Pressmen and Assistants' Union of North America *see*
- Financial report of the secretary-treasurer
- News and views
- Reports of officers to the...annual convention

International problems – Tel-Aviv. 1973-1994 (1) 1976-1994 (5) 1976-1994 (9) – ISSN: 0020-840X – mf#8041 – us UMI ProQuest [337]

International projectionist – Long Island City. 1950-1955 (1) – mf#723 – us UMI ProQuest [790]

International prosecution section documents relating to witnesses for the prosecution and the defense, 1946-1947 / World War 2. International Prosecution Section – (= ser Records of allied operational and occupation headquarters, world war 2) – 21r – 1 – mf#M1684 – us Nat Archives [341]

International prosecution section staff : historical files relating to cases tried before the international military tribunal for the far east, 1945-1948 – (= ser Records of allied operational and occupation headquarters, world war 2) – 66r – 1 – mf#M1663 – us Nat Archives [355]

International public management journal (ipmj) – Stamford, 2000+ [1,5,9] – ISSN: 1096-7494 – mf#42838 – us UMI ProQuest [350]

International quarterly – Burlington. 1900-1906 (1) – mf#2903 – us UMI ProQuest [934]

The international railway and steam navigation guide – Montreal: C R Chisholm, [1869-1909] – 9 – (publ with: the dominion gazetteer) – mf#P06093 – cn Canadiana [380]

International Railway Company *see* A plan for collective bargaining and co-operative benefits

The international railway guide – Montreal: Montreal Print and Pub Co, [1866-1869] – 9 – (cont by: international railway and steam navigation guide) – mf#P04941 – cn Canadiana [380]

International railway journal – Bristol. 2001+ (1,5,9) – mf#7600,01 – us UMI ProQuest [380]

International railway journal and rapid transit review (irj) – Bristol. 1960+ (1) 1973+ (5) 1973+ (9) – ISSN: 0744-5326 – mf#7600 – us UMI ProQuest [380]

International railway journal and rapid transit review (irj) – Bristol. 1960+ [1]; 1973+ [5,9] – ISSN: 0744-5326 – mf#7600 – us UMI ProQuest [380]

International Reading Association *see* Proceedings of the annual convention

International record of medicine – New York. 1950-1961 (1) – ISSN: 0096-0632 – mf#501 – us UMI ProQuest [610]

International Red Aid. Spain. (Socorro Rojo de Espana) *see* Seis meses de solidaridad antifascista

International Reform Federation *see* Progress

International Refugee Organization *see* Occupational skills of refugees

International regional/zone communicator / Parents Without Partners – v1 n1 [1975/1976 win] – 1r – 1 – mf#621480 – us WHS [305]

International relations : university teaching of social sciences. a unesco educational studies publication – 4mf – 7 – mf#4791 – uk Microform Academic [327]

International review – Steubenville. 1989-1989 (1) – (cont: international review of natural family planning) – ISSN: 1054-0679 – mf#11811,01 – us UMI ProQuest [304]

International review – London. n1-3. 1889 [all publ] – (= ser Radical periodicals of great britain, 1794-1914. period 2) – 1r – 1 – $115.00 – us UPA [335]

International review – New York. 1874-1883 (1) – mf#4617 – us UMI ProQuest [420]

International review – New York. v1-4 n2,1. 1936-39 [all publ] – (= ser Radical periodicals in the united states, 1881-1960. series 1) – 5mf – 9 – $95.00 – us UPA [303]

International review of criminal policy – New York. 1952-1981 (1) 1977-1981 (5) 1977-1981 (9) – ISSN: 0074-7688 – mf#6409 – us UMI ProQuest [345]

International review of economics and finance – Greenwich. 1993-1996 (1,5,9) – ISSN: 1059-0560 – mf#19782 – us UMI ProQuest [332]

International review of education = Internationale zeitschrift fuer erziehungswissenschaft – Den Haag. 1988+ – 1,5,9 – ISSN: 0020-8566 – mf#16802 – us UMI ProQuest [370]

International review of history and political science – Meerut. 1964-1989 (1) 1972-1989 (5) 1976-1989 (9) – ISSN: 0020-8574 – mf#6410 – us UMI ProQuest [320]

International review of law and economics – New York. 1981+ (1,5,9) – ISSN: 0144-8188 – mf#14190 – us UMI ProQuest [340]

International review of mission – Geneva. 1912+ (1) 1975+ (5) 1976+ (9) – ISSN: 0020-8582 – mf#10210 – us UMI ProQuest [240]

International review of natural family planning – Steubenville. 1980-1980 (1,5,9) – (cont by: international review) – ISSN: 0146-1745 – mf#11811 – us UMI ProQuest [304]

International review of psychiatry – Abingdon. 1997+ (1) – ISSN: 0954-0261 – mf#20948 – us UMI ProQuest [616]

International review of social history – Assen. 1956+ (1) 1956+ (5) 1956+ (9) – ISSN: 0020-8590 – mf#13576 – us UMI ProQuest [300]

International review of strategic management – Chichester. 1990-1995 (1,5,9) – ISSN: 1047-7918 – mf#18148 – us UMI ProQuest [650]

International reviews in physical chemistry – Sevenoaks. 1991-1996 (1) – ISSN: 0144-235X – mf#17327 – us UMI ProQuest [540]

International Seaman's Union of America *see* Isu pilot

International Seamen's Union of America *see*
- Proceedings of the annual convention of the international seamen's union of america
- Seamen's journal

International securities finance – London. 2000+ (1) – mf#19530,01 – us UMI ProQuest [332]

International security – Cambridge. 1976+ (1,5,9) – ISSN: 0162-2889 – mf#12322 – us UMI ProQuest [327]

International Seminar on the Role of Women in a Developing Society *see* Lectures and summary reports

International short stories : the best from twentythree countries – New Delhi: Hindustan Times, 1952 – (= ser Samp: indian books) – us CRL [830]

International social science journal : english edition – Paris. 1949+ (1) 1971+ (5) 1976+ (9) – ISSN: 0020-8701 – mf#2448 – us UMI ProQuest [300]

International social security review – Geneva. 1989-1996 (1,5,9) – ISSN: 0020-871X – mf#17728 – us UMI ProQuest [360]

International social work – London. 1975+(1,5,9) – ISSN: 0020-8728 – mf#10540 – us UMI ProQuest [360]

International Socialist Organization *see* Socialist worker

International socialist review – Chicago, IL. july 1900-june 1907 [mnthly] – 4r – 1 – uk British Libr Newspaper [325]

International socialist review – Chicago. v1-18 n2,8. 1900-18 [all publ] – (= ser Radical periodicals in the united states, 1881-1960. series 1) – 6r – 1 – $1120.00 – us UPA [335]

International socialist review – Chicago. v1-18 n8. 1900-18 – 1 – $354.00 – mf#0292 – us Brook [335]

International socialist review – London. -m. Jul 1900-Jun 1906. (3 reels) – 1 – uk British Libr Newspaper [072]

International socialist review – New York. 1940-1975 (1) 1972-1972 (5) (9) – ISSN: 0020-8744 – mf#7103 – us UMI ProQuest [335]

International socialist review / Socialist Workers Party – v1-24. 1940-63 – (= ser Radical periodicals in the united states, 1881-1960. series 1) – 65mf – 9 – $455.00 – us UPA [335]

International Socialists [US] *see* Workers' power

International society of barristers quarterly – v1-35. 1966-2000 – 9 – $455.00 set – ISSN: 0020-8752 – mf#109391 – us Hein [340]

International Society of Christian Endeavor *see* Christian endeavor world

International society of surgery bulletin = Bulletin de la societe internationale de chirurgie – Brussels. 1971-1973 (1) 1972-1972 (5) (9) – ISSN: 0037-945X – mf#5110 – us UMI ProQuest [617]

International space safety and rescue symposia [aasms23] : first, second and third – 1975 – (= ser Aasms 1968) – 35 papers on 11mf – 9 – $20.00 – 0-87703-239-4 – us Univelt [629]

International space safety and rescue symposia [aasms24] : fourth, fifth, and sixth – 1975 – (= ser Aasms 1968) – 35papers on 11mf – 9 – $20.00 – 0-87703-240-8 – us Univelt [629]

International space safety and rescue symposia [aasms40] : ninth, tenth, and eleventh, 1976-78 – 1982 – (= ser Aasms 1968) – 33papers on 6mf – 9 – $15.00 – 0-87703-223-8 – (suppl to v54, science and technology) – us Univelt [629]

International space safety and rescue symposia [aasms41] : thirteenth and fourteenth, 1980-81 – 1982 – (= ser Aasms 1968) – 25papers on 5mf – 9 – $15.00 – 0-87703-224-6 – (suppl to v54, science and technology) – us Univelt [629]

International space safety and rescue symposium [aasms39] : twelfth, 1979 – 1982 – (= ser Aasms 1968) – 11papers on 5mf – 9 – $12.00 – 0-87703-222-X – (suppl to v54, space and technology) – us Univelt [629]

International sports journal – West Haven. 1997+ (1,5,9) – ISSN: 1094-0480 – mf#31747 – us UMI ProQuest [790]

International standard bible encyclopaedia – Chicago, IL. v1-5. 1925 – 2r – us UF Libraries [240]

International statistics microfiche library – 1983- – 9 – apply for price – (statistical publications on international intergovernmental organizations. printed index available) – us CIS [310]

International stereotyper and electrotypers union journal – 1906-55 – (= ser Labor union periodicals, pt 2: the printing trades) – 18r – 1 – $3755.00 – 1-55655-307-2 – us UPA [680]

International Stereotypers and Electrotypers Union *see* Wage scales of subordinate unions

International studies of management and organization – White Plains. 1988+ (1,5,9) – ISSN: 0020-8825 – mf#15774 – us UMI ProQuest [650]

International studies quarterly – Beverly Hills. 1957+ (1) 1971+ (5) 1976+ (9) – ISSN: 0020-8833 – mf#1410 – us UMI ProQuest [327]

International studies review – Malden, 1999+ [1,5,9] – ISSN: 1521-9488 – mf#29310 – us UMI ProQuest [327]

International Sunday-School Convention of the United States and British American Provinces *see* The development of the sunday-school, 1780-1905

International surgery – Torino. 1938+ (1) 1971+ (5) 1976+ (9) – ISSN: 0020-8868 – mf#2248 – us UMI ProQuest [617]

International symposium on earth gravity models and related problems / Rapp, R R – 1972 – 1,5,6,9 – $10.00 – us AGU [550]

International Symposium on Sierra Leone (1987: Freetown, Sierra Leone) *see* Bicentenary of the founding of the colony of sierra leone, 1787-1987

International symposium on the ecological effects of arctic airborne contaminants : hotel saga, reykjavik, iceland, oct 4-8 1993: abstracts / ed by Christie, S J & Martin, J – [Hanover NH?]: USACRREL [1993] [mf ed 1994] – 2mf – 9 – (with ind) – us GPO [574]

International tax and business lawyer *see* Berkeley journal of international law

International tax journal – Greenvale. 1993-1996 (1,5,9) – ISSN: 0097-7314 – mf#15775 – us UMI ProQuest [336]

International tax journal – v1-26. 1974-2000 – 9 – $623.00 set – ISSN: 0097-7314 – mf#103681 – us Hein [343]

International teamster – Washington. 1912-1992 (1) 1971-1992 (5) 1972-1992 (9) – (cont by: new teamster) – ISSN: 0020-8892 – mf#6192 – us UMI ProQuest [331]
International television almanac – 1938-2003 – 1 – $950.00 – mf#0293 – us Brook [790]
International textiles – Amsterdam. 1975-1980 (1) 1977-1980 (5) 1977-1980 (9) – ISSN: 0020-8914 – mf#9512 – us UMI ProQuest [680]
International Tin Council *see* Monthly statistical bulletin – international tin council
International tourism quarterly – 1971-93 – 12r – 1 – £450.00 – uk World [338]
International tourist guide for 1874 : containing information for travelers going east or west – [Chicago?: s.n. 1874?] [mf ed 1987] – 1mf [ill] – 9 – 0-665-68114-3 – mf#68114 – cn Canadiana [917]
International Tract and Missionary Society *see* Signs of the times
International trade : special studies, 1971-1988 – 1 – (1971-81 12r isbn 0-89093-506-8 $2330. 1982-85 9r isbn 0-89093-551-3. $1740. 1985-88 11r isbn 1-55655-146-0 $2135. 1989-98 15r isbn 1-55655-849-X $2915. with p/g) – us UPA [380]
International trade *see* Business and financial papers, 1780-1939
International trade forum – Geneva. 1975+ (1,5,9) – ISSN: 0020-8957 – mf#10236 – us UMI ProQuest [337]
International trade from the 17th century amsterdam : the burlamacchi archive – 1009mf – 9 – $9015.00 – us UPA [380]
International trade journal – Laredo. 1994-1996 (1,5,9) – ISSN: 0885-3908 – mf#18682 – us UMI ProQuest [337]
International trade law journal *see* Maryland journal of international law and trade
International Trade Union Committee of Negro Workers *see* Negro worker
International Training Institute Library *see* Course and syllabus materials, publications on education in papua new guinea and other rare publications relating to png
International transactions in operational research – Oxford. 1994-1994 (1,5,9) – ISSN: 0969-6016 – mf#49633 – us UMI ProQuest [000]
International Typographical Union *see*
– Bulletin
– Itu review
– Labor's daily
– Mailers n18 reports
– Official monthly bulletin of columbia typographical union, no 101
– Printers' union
– Progress
– Typographical journal
International typographical union bulletin – 1912-55 – (= ser Labor union periodicals, pt 2: the printing trades) – 7r – 1 – $1445.00 – 1-55655-305-6 – us UPA [680]
International Typographical Union of North America *see* United graphic arts forum
International typographical union review – v17 n26-v19 n23 [1974 jul 18-1976 dec 23] – 1r – 1 – (cont: itu review; itu review; cont by: itu review [1977]) – mf#1002573 – us WHS [331]
International understanding at school : english edition – Paris. 1978-1994 (1) 1978-1994 (5) 1978-1994 (9) – ISSN: 0047-1240 – mf#9199 – us UMI ProQuest [370]
International Union, Allied Industrial Workers of America *see*
– Industrial worker
– Local 232 reporter
– Local 865 newsletter
International Union, Allied Industrial Workers of america *see* Spotlight
International Union, Aluminum Workers of America [CIO] *see* Cio news
International Union for Co-operation in Solar Research. Conference *see* Transactions of the international union for co-operation in solar research
International Union of Brewery, Flour, Cereal, Soft Drink and Distillery Workers of America *see* Brewery worker
International Union of Bricklayers and Allied Craftsmen *see*
– Journal
– Line on–
International Union of Commercial Telegraphers *see* I u c t journal
International Union of District 50, Allied and Technical Workers of the United States and Canada *see* District fifty news
International Union of Electrical Radio and Machine Workers *see* Reporting to you
International Union of Electrical, Radio and Machine Workers *see*
– Health and safety bulletin
– Iue-cio news
– Local 301 news
International Union of Electrical, Radio, and Machine Workers *see* Iue local 201 electrical union news
International Union of Electrical, Radio, and Machine Workers et al *see* Iue news
International Union of Electronic, Electrical, Technical, Salaried & Machine Workers *see* Union news
International Union of Flour and Cereal Mill Employees *see*
– Eight hour miller
International Union of Life Insurance Agents *see* Our voice
International Union of Machinists and Blacksmiths of the USA *see* Machinists and blacksmiths international journal
International Union of Mine, Mill and Smelter Workers *see*
– Cio news
– Mine-mill union
International Union of Mine, Mill, and Smelter Workers *see* Cio news
International Union of Mine, Mill and Smelter Workers et al *see* Miners' magazine
International Union of Operating Engineers *see*
– Buckeye engineer
– Dredgeman
– Engineers news
– Local 30 recorder
– Local 138 news
– Local 139 wisconsin news
– Local 148 gauge
– Local 150 engineer
– News report
– News-record
– Public employee news
– Scoop
International Union of Petroleum and Industrial Workers *see* Iupiw-views
International Union of Petroleum Workers *see* Independent oiler
International Union of Petroleum Workers et al *see* Iupw views
International Union of Playthings, Jewelry and Novelty Workers *see* Union voice
International Union of the Journeymen Horseshoers of the U S and Canada *see* Report of the proceedings of the convention of the international unionof journeymenhorseshoers of the united states and canada
International Union of the Journeymen Horseshoers of the US and Canada *see* Proceedings of the...annual convention of the journeymen horseshoers' international union of the united states of america and canada
International Union of United Brewery, Flour, Cereal, Soft Drink and Distillery Workers of America *see* Brewery worker
International Union Shipwrights, Joiners and Caulkers of America *see*
– Official magazine
– Report of the proceedings of the...annual convention of the international union shipwrights, joiners and caulkers of america
International Union, United Autombile, Aerospace, and Agricultural Implement Workers of America *see* Transmitter
International Union, United Automobile, Aerospace and Agricultural Implement Workers of America *see*
– Delco antenna
– Delco sparks
– Labor beacon
– Local 261 hot-points
– Local 387 news
– Newsletter
– Propeller
– Right angle
– Solidarity
– Spicer news
– Uaw arbitration services news notes
International Union, United Automobile, Aerospace, and Agricultural Implement Workers of America *see*
– Champ
– Conveyor
– Eye opener
– Local 211 report
– Local 245 beacon
– Local 467 reporter
– Local 865 news
– News from the uaw
– News from uaw
– Quarter panel
– Rocket
– Union progress
– Union voice
International Union, United Automobile, Aerospace and Agricultural Implement Workers of America et al *see* La crosse union herald
International Union, United Automobile, Aicraft, and Agricultural Implement Workers of America et al *see* Kohlerian
International Union, United Automobile, Aircraft and Agricultural Implement Workers of America *see*
– Competitive shop organizer
– Exhaust
– Lampmaker
– Union picket
– Voice of local 212
International Union, United Automobile, Aircraft, and Agricultural Implement Workers of America *see*
– Cio news
– Cupola
– Daily strike bulletin
– Ford facts
– Kohler strike and boycott bulletin
– Solidarity
– Uaw facts
– Uaw local 833 reporter and kohlerian
– Uaw propeller
– Union daily
– Wisconsin cio news
International Union, United Automobile, Aircraft and Agricultural Workers of America *see* United automobile worker
International Union, United Automobile Workers *see* Eighth region auto worker labor news
International Union, United Automobile Workers of America *see* Searchlight
International Union, United Automobile Workers of America [CIO] *see*
– Organizer
– United automobile worker
International Union, United Automotile Workers of America [AFL] *see* West side conveyor
International voice : a quarterly newsletter published by oic international / OIC International, Inc – 1978 win – 1r – 1 – mf#5078440 – us WHS [071]
International water power and dam construction – London. 1975-1991 (1) 1975-1991 (5) 1975-1991 (9) – (cont: water power) – ISSN: 0306-400X – mf#3143,01 – us UMI ProQuest [627]
International wealth success newsletter – Merrick. 1971-1979 (1) 1971-1979 (5) 1971-1979 (9) – ISSN: 0047-1275 – mf#7968 – us UMI ProQuest [337]
International wildlife – Vienna. 1971+ (1) 1974+ (5) 1971+ (9) – ISSN: 0020-9112 – mf#10056 – us UMI ProQuest [639]
International Woman Suffrage Alliance *see* Jus suffragii
International women's suffrage – 1pt – 1 – (pt1: suffrage correspondence of rose scott (1847-1925) fr state library of new south wales [3r] £285; with d/g) – uk Matthew [322]
International wood worker / Amalgamated Woodworkers' International Union of America et al – 1895 nov-1898 jun – 1r – 1 – (cont: american wood worker) – mf#3256939 – us WHS [331]
International woodcarver – 1900 aug [v1]-1915, 1916-1953 apr [v49 n2] – 2r – 1 – mf#1059869 – us WHS [730]
International woodworker – Portland, Seattle OR. 1942 jan 21 [v8 n1]-1942 dec, 1943-69, 1970-1973 mar 28, 1973 apr 11-1977 dec 21, 1978-85 – 14r – 1 – (cont: timber worker; cont by: woodworker [gladstone, or]) – mf#1428634 – us WHS [634]
International Woodworkers of America *see*
– B c lumber worker
– B c lumber worker iwa bulletin
– B c lumber worker union bulletin
– Western canadian lumber worker
International Woodworkers of America-US *see* Woodworker
International worker – v1-v3 n6 [1974 sep 7-1976 mar 20] – 1r – 1 – mf#381002 – us WHS [335]
International Workers Aid. Committee *see* Captives of capitalism
International Workers in the Amalgamated Food Industries *see* Free voice
International Workers Party *see* Critical practice
International Working Men's Association *see* Papers of the international workingmen's association, 1868-1877
International Working Men's Association. British Sect *see* International herald
International Working People's Association *see* Freedom
L'Internationale / L'Union Communiste – puis Revue mensuelle no. 1-40. Paris. nov 1933-38 – 1 – fr ACRPP [335]
L'internationale : communiste – Paris. n3-5, 7-11, 13, 15-29, 35. mars-dec 1919 – 1 – fr ACRPP [355]
Die internationale : anarchosyndikalistisches organ / ed by Sekretariat der Internationalen Arbeiter-Assoziation – Amsterdam (NL), 1934 aug-1935 apr – 1r – 1 – (anarchosyndikalistisch) – gw Misc Inst [074]
Die internationale – Berlin, Germany 10 nov 1918 – 1 – (cont: norddeutsche allgemeine zeitung [6 dec 1913-9 nov 1918]; cont by: deutsche allgemeine zeitung [12 nov 1918-19 apr 1945]) – uk British Libr Newspaper [074]
Die internationale : zeitschrift fuer die revolutionare arbeiterbewegung.... – v1-4. 1927-31 [mnthly] – (= ser Serial publications of german trade unions) – 1 – us Wisconsin U Libr [325]
Die internationale : zeitschrift fuer praxis und theorie des marxismus – Berlin, Leipzig. v1 n1-v15 n9-10 apr 1915-sep/oct 1932 – (= ser Communist international periodicals from the feltrinelli archives) – 117mf – 9 – $720.00 – us UPA [335]
Internationale bibliographie zur deutschen klassik 1750-1850 : folge 1. 1959 bis folge 41. 1994 / ed by Stiftung Weimarer Klassik – [mf ed 1997] – 91mf (1:24) + 3 ind vol – 9 – diazo €2000.00 (silver €2300 isbn: 978-3-598-32884-8) – ISBN-10: 3-598-32883-4 – ISBN-13: 978-3-598-32883-1 – gw Saur [430]
L'internationale communiste : organe bimensuel du comite executif de l'internationale communiste – Petrograd, Paris, mai 1919-aout 1939 – 1 – fr ACRPP [335]
L'internationale communiste : organe du comite executif de l'internationale communiste / Organe du Comite Executif de l'Internationale Communiste – Moscow-Petrograd, 1919-22; Petrograd (actually Paris) 1922-24; Paris, 1925-39 – (= ser Communist international periodicals from the feltrinelli archives) – 298mf – 9 – $1725.00 – (french ed of: kommunistische internationale) – us UPA [335]
Internationale film- und kinematographen-industrie – Berlin DE, 1909-1910 30 mar – 2r – 1 – (title varies: 16 mar 1910: projektion. filmed with: mechanische musikwerke & die wissenschaftliche projektion) – gw Mikrofilm [790]
Internationale freiwirtschafts-liga-ifl – Schleiden DE, 1933 n1 – 1 – gw Misc Inst [337]
Internationale freizuegigkeit des kapitals und unterentwickelte laender / Gaebler, Joachim – Heidelberg, 1963 – 3mf – 3-89349-982-2 – gw Frankfurter [337]
Internationale kirchliche zeitschrift – v1-60. 1893-1970 [complete] – 15r – 1 – (cont: revue internationale de theologie) – ISSN: 0020-9252 – mf#ATLA S0176 – us ATLA [240]
Der internationale klassenkampf – Wolfisheim (F), 1936 feb-1939 apr – 1r – 1 – gw Misc Inst [335]
Der internationale klassenkampf kpo *see* Gegen den strom
Internationale korrespondenz : ik ueber arbeiterbewegung, sozialismus und auswaertige politik. ausgabe w – Berlin DE, 1916 apr-1917 mar – 1r – 1 – mf#3713 – gw Mikropress [335]
Internationale literatur : zentralorgan der internationalen vereinigung revolutionaerer schriftsteller – Moskau (RUS), 1933-45 – 13r – 1 – (title varies: 1937 iss1- : internationale literatur/deutsche blaetter) – gw Misc Inst [077]
Internationale literatur/deutsche blaetter *see* Internationale literatur
Internationale literatur-und musikberichte – v.1-13, 1894-1908 – 1 – 92.00 – us L of C Photodup [780]
Internationale musik-sachlexika vom 17. bis zum fruehen 19. jahrhundert = International dictionaries of musical terms from the 17th to early 19th century – [mf ed 1998-99] – (= ser Nachschlagwerke zur musik) – 160mf (1:24) in 2 installments – 9 – diazo €1790.00 (silver €2200 isbn: 978-3-598-33863-2) – ISBN-10: 3-598-33862-7 – ISBN-13: 978-3-598-33862-5 – (incl guide) – gw Saur [780]
Die internationale polarforschung 1882-1883 : die deutschen expeditionen und ihre ergebnisse. v2: beschreibende naturwissenschaften / ed by Neumayer, G – Berlin: A Asher & Co, 1890 – 12mf – 9 – mf#76 – ne IDC [919]
Internationale presse-korrespondenz : deutsche ausgabe – Berlin, 1921-23; Wien, 1923-26; Berlin, 1926-33 – (= ser Communist international periodicals from the feltrinelli archives) – 356mf – 9 – $1490.00 – us UPA [335]
Internationale situationniste / L'Internationale Situationniste – Paris. n1-12. juin 1958-sept 1969 – 1 – (bulletin central edite par les sections) – fr ACRPP [303]
L'internationale syndicale rouge : bulletin / Communist International. Comite Executif – Moscou, Paris. oct 1921-oct 1933, aout 1934. – 1 – (absorbe par: correspondance syndicale internationale) – fr ACRPP [335]
L'internationale syndicale rouge *see* La lutte de classe
Internationale tendenzen in der tiergesundheitsueberwachung und daraus abgeleitete schlussfolgerungen fuer die anpassung des nationalen tierseuchenberichtssystems / Kroschewski, Klaus – (mf ed 1993) – 2mf – 9 – €49.00 – 3-89349-714-5 – mf#DHS 714 – gw Frankfurter [636]
Internationale Vereinigung fuer Germanische Sprach- und Literaturwissenschaft *see* Bericht ueber den ersten kongress
Internationale Vereinigung fuer Germanische Sprach- und Literaturwissenschaft. Kongress *see* Ivg, bericht ueber den ersten kongress, rom, 5.-10. september 1955

Internationale zeitschrift elektrische ausstellung in wien 1883 – Vienna, Austria. 15 jul-23 dec 1883 – 1/2r – 1 – uk British Libr Newspaper [072]

Internationale zeitschrift fuer allgemeine sprachwissenschaft – Leipzig. 5v. 1884-1890+suppl – 67mf – 8 – mf#H-1381 – ne IDC [400]

Internationale zeitschrift fuer individualpsychologie – Wien (A), Leipzig DE, 1923/24-1937 n6 – 9r – 1 – (since 1927 publ in leipzig) – gw Misc Inst [150]

Internationale zeitschrift fuer psychoanalyse : offizielles organ der internationalen psychoanalytischen vereinigung – Wien (A), 1913-37, 1939-41 – 1 – (title varies: 1938/39: internationale zeitschrift und imago, paris) – fr ACRPP [150]

Internationale zeitschrift und imago *see* Internationale zeitschrift fuer psychoanalyse

Internationalen – Stockholm, Sweden. 1979- – 1 – sw Kungliga [078]

Internationalen: huvudorgan for kommunistiska arbetarforbundet : (svensk sektion av fjarde internationalen) – n4 (1974)-. Stockholm: Kommunistiska arbetarforbundet, 1974- – 1 – us Wisconsin U Libr [325]

Internationaler Germanisten-Kongress *see*
– Dichtung, sprache, gesellschaft
– Spaetzeiten und spaetzeitlichkeit
– Tradition und ursprueglichkeit

Internationales aerztliches bulletin – Paris (F), 1939 feb-jun – 1 – gw Misc Inst [610]

Internationales archiv fuer arbeitsmedizin = International archives of occupational health – Heidelberg. (1) 1970-1975 (5) 1970-1975 (9) – (cont: internationales archiv fuer gewerbepathologie und gewerbehygiene. cont by: international archives of occupational and environmental health) – ISSN: 0020-5923 – mf#13118,02 – us UMI ProQuest [360]

Internationales archiv fuer gewerbepathologie und gewerbehygiene – Heidelberg 1962-1969 (5,9) – (cont: archiv fuer gewerbepathologie und gewerbehygiene; cont by: internationales archiv fuer arbeitsmedizin international archives of occupational health) – mf#13118,01 – us UMI ProQuest [616]

Internationales jahrbuch fuer politik und arbeiterbewegung – Berlin DE, 1912-15 – 2r – 1 – gw Mikropress [331]

Internationales privatrecht nach dem einfuehrungsgesetze zum buergerlichen gesetzbuche / Habicht, Hermann; ed by Greiff, Max – Berlin: J Guttentag, 1907 – (= ser Civil law 3 coll) – 3mf – 9 – (incl bibl ref and index) – mf#LLMC 96-519 – us LLMC [346]

Internationales zentralblatt fuer bau keramik *see* Central blatt fuer die gesamte etc

Internationales zentralblatt fuer baukeramik und glasindustrie *see* Central blatt fuer glas industrie und keramik

Internationalist – v3 n11-v6 n1 [1916 mar-1918 may] – 1r – 1 – mf#1059908 – us WHS [335]

Internationalist [dublin, ireland] *see* The irish theosophist

Internationalist worker : official organ of the revolutionary communist league [internationalist] / Revolutionary Communist League – v1 n1-v10 n5/6 [1974 jun-1980 feb,aug-1985 sum] – 1r – 1 – (cont: internationalist news letter) – mf#381001 – us WHS [335]

Internationalist worker bulletin : official organ of the revolutionary communist league [internationalist] / Revolutionary Communist League – v1 n1-v2 n1 [1979 jul-1980 feb] – 1r – 1 – mf#1048440 – us WHS [335]

Internationalist Workers Party *see* Working class opposition

Internationnal Bureau Of The American Republics *see* Nicaragua a handbook

Internatsional : organ nizhegorodskogo okr kom rsdrp – Nizhny-Novgorod, Russia, 1917 – 3r – 1 – us UMI ProQuest [077]

Internatsional : vospominaniia i materialy 1864-1878 g g / Guillaume, James [comp]; ed by Lebedev, Nikolai Konstantinovich – Peterburg: Izd-vo "Golos truda", 1922- [mf ed 2002]2. – 1r – 1 – (filmed with: partiia i oppozitsionnyi blok / a i rykov & n i bukharin (1926) & other titles) – mf#5265 – us Wisconsin U Libr [331]

Internatsional molodezhi – Moscow. 14v. 1929-41 – 1 – us L of C Photodup [335]

Internatsionalnyi teatr – Moscow, jan 1932-dec 1933 – 6mf – 9 – us UMI ProQuest [790]

L'internazionale – Philadelphia PA, 1909* – 1r – 1 – (italian periodical) – us IHRC [073]

Internazionale – London, UK. 12 Jan-5 May 1901 – 1 – uk British Libr Newspaper [072]

Internet and higher education – Greenwich. 1998+ (1) – ISSN: 1096-7516 – mf#42815 – us UMI ProQuest [000]

Internet reference services quarterly : a journal of innovative information practice, technologies and resources / ed by Martin, Lyn Elizabeth M – v1 n1. 1997- – 1,9 – $48.00 in US $67.20 outside hardcopy subsc – us Haworth [020]

Internet research – Hackensack. 2001+ (1,5,9) – ISSN: 1066-2243 – mf#29177,01 – us UMI ProQuest [000]

Internet retailer – New York. 1999+ (1,5,9) – ISSN: 1527-7089 – mf#29027 – us UMI ProQuest [650]

Internet world – Drew DeSarle. 1998+ (1) – ISSN: 1097-8291 – mf#25805,02 – us UMI ProQuest [000]

Internetweek – Manhasset. 1998-2000 (1,5,9) – (cont: communicationsweek) – ISSN: 1096-9969 – mf#19185,01 – us UMI ProQuest [380]

Internist – Washington. 1985-1996 (1,5,9) – (cont by: today's internist) – ISSN: 0020-9546 – mf#12477 – us UMI ProQuest [610]

Internist – Heidelberg. 1981-1996 (1) 1981-1985 (5) 1981-1985 (9) – ISSN: 0020-9554 – mf#13184 – us UMI ProQuest [610]

Internistische praxis – Munich. 1973-1973 (1) – ISSN: 0020-9570 – mf#8203 – us UMI ProQuest [610]

Internskap as 'n integrale komponent van die professionele opleiding van onderwysers / Stander, Rudolph Hendrik – Stellenbosch: U van Stellenbosch [mf ed Claremont: Condor Microfilm [1985]] – 8mf [ill] – 9 – mf#af.2007-478 – sa Misc Inst [370]

Inter-orchestra bulletin / American Symphony Orchestra League – Kalamazoo. 1942-1946 (1) – mf#7983 – us UMI ProQuest [780]

Interpersonal development – Basel. 1970-1974 (1) 1970-1972 (5) (9) – ISSN: 0047-1283 – mf#5190 – us UMI ProQuest [150]

Interpersonal relationships as an essential skill in the effectiveness of high school principals / Mojapelo, Irene – Pretoria: Vista University 2000 [mf ed 2000] – 2mf – 9 – (incl bibl ref) – mf#mfm15171 – sa Unisa [302]

Interpretacao da literatura brasileira / Moog, Clodomir Vianna – Rio de Janeiro, Brazil. 1943 – 1r – us UF Libraries [440]

Interpretacao do brasil / Freyre, Gilberto – Rio de Janeiro, Brazil. 1947 – 1r – us UF Libraries [440]

Interpretacion de la poesia popular / Quinones Pardo, Octavio – Bogota, Colombia. 1947 – 1r – us UF Libraries [440]

Interpretacion del brasil / Freyre, Gilberto – Mexico City?, Mexico. 1945 – 1r – us UF Libraries [972]

Interpretacion judicial de la ley de registro publ... / Venezuela Corte Federal Y De Casacion – Caracas, Venezuela. 1961 – 1r – us UF Libraries [340]

Interpretacion pesimista de la sociologia hispanoa / Mijares, Augusto – Madrid, Spain. 1952 – 1r – us UF Libraries [972]

Interpretacion...horacio flaco / Saa Maldonado, Manuel – 1878 – 9 – sp Bibl Santa Ana [450]

Interpretation : a journal of bible and theology – v1-10. 1947-56 [complete] – 2r – 1 – ISSN: 0020-9643 – mf#ATLA S0032 – us ATLA [220]

Interpretation – Richmond. 1947+ (1) 1971+ (5) 1975+ (9) – ISSN: 0020-9643 – mf#1769 – us UMI ProQuest [200]

L'interpretation de la musique francaise (de lully a la revolution) / Borrel, Eugene – Paris: F Alcan 1934 [mf ed 19–] – (= ser Maitres de la musique. nouvelle serie) – 1r – 1 – mf#film 1353 – us Sibley [780]

Interpretation des 1-6 / Lewy, Israel – Breslau, Germany. v1-6. 1895-1914 – 1r – us UF Libraries [939]

Die interpretation des neuen testaments in der valentinianischen gnosis / Barth, C – Leipzig: J C Hinrichs, 1911 – (= ser Tugal) – 1mf – 9 – 0-7905-1685-3 – (incl bibl ref and ind) – mf#1987-1685 – us ATLA [225]

Die interpretation des neuen testaments in der valentinianischen gnosis / Barth, C – Leipzig, 1911 – (= ser Tugal 3-37/3) – 2mf – 9 – €5.00 – ne Slangenburg [225]

An interpretation of christianity *see* Chi-tu chiao chin chieh [ccc68]

An interpretation of genesis : including a translation into present-day english / Ramsay, Franklin Pierce – New York: Neale, 1911 [mf ed 1988] – 1mf – 9 – 0-7905-0147-3 – (incl ind) – mf#1987-0147 – us ATLA [221]

An interpretation of india's religious history / Hume, Robert Allen – New York: FH Revell, c1911 [mf ed 1991] – 1mf – 9 – 0-524-00899-X – mf#1990-2122 – us ATLA [280]

The interpretation of italy during the last two centuries : a contribution to the study of goethe's "italienische reise" / Klenze, Camillo von – Chicago: University of Chicago Press, 1907 – 1 – (incl bibl ref and index) – us Wisconsin U Libr [430]

The interpretation of nature / Shaler, Nathaniel Southgate – Boston: Houghton, Mifflin, 1893 – (= ser Winkley Lectures) – 1mf – 9 – 0-7905-9887-6 – mf#1989-1612 – us ATLA [210]

Interpretation of plato's republic / Murphy, Neville Richard – Oxford, England. 1951 – 1r – us UF Libraries [450]

The interpretation of religious experience / Watson, John – New York: J Maclehose, 1912 – (= ser Gifford lectures) – 2mf – 9 – 0-7905-9551-6 – mf#1989-1256 – us ATLA [200]

An interpretation of rudolf eucken's philosophy / Jones, William Tudor – London: Williams & Norgate, 1912 [mf ed 1990] – 1mf – 9 – 0-7905-7443-8 – mf#1989-0668 – us ATLA [170]

The interpretation of scripture and other essays / Jowett, Benjamin – London: George Routledge; New York: E P Dutton, [1906?] – (= ser The London Library) – 2mf – 9 – 0-7905-2121-0 – (incl ind) – mf#1987-2121 – us ATLA [220]

Interpretation of the bible : a short history / Gilbert, George Holley – New York: Macmillan, 1908 – 1mf – 9 – 0-7905-1662-4 – (incl bibl ref and indexes) – mf#1987-1662 – us ATLA [220]

The interpretation of the character of christ to non-christian races : an apology for christian missions / Robinson, Charles Henry – London; New York: Longmans, Green, 1910 – 1mf – 9 – 0-7905-0260-7 – (incl bibl ref and index) – mf#1987-0260 – us ATLA [230]

Interpretation of the english ordinal... / Lacey, Thomas Alexander – London, England. 1898 – 1r – us UF Libraries [240]

The interpretation of the english ordinal / Lacey, Thomas Alexander – London: SPCK 1898 [mf ed 1993] – (= ser Church historical society (series) 50) – 1mf – 9 – 0-524-05546-7 – mf#1990-5150 – us ATLA [242]

Interpretation und kritik einiger grundbegriffe der spaetphilosophie fichtes : dargestellt an den "einleitungsvorlesungen in die wissenschaftslehre" von 1813 / Lautemann, Willi – Frankfurt a.M, 1970 – 2mf – 9 – 3-89349-773-0 – gw Frankfurter [190]

Interpretations erronees et faux monuments : remarques sur quelques inscriptions recemment editees: suivies d'un sommaire analytique de l'ouvrage = An independent examination of the assuan and elephantine aramaic papyri / Belleli, Lazare – Casal Montferrat: Rossi et Lavagno, 1909 – 1mf – 9 – 0-8370-7284-0 – mf#1986-1284 – us ATLA [470]

Interpretative readings : by j h bausman, from the 'habitant' poems of dr william henry drummond of montreal, canada – S.l: s.n, 190-? – 1mf – 9 – mf#01812 – cn Canadiana [410]

L'interprete – St-Victor d'Alfred (Ont): A Lefaivre et Bertrand. 1re annee n1 20 aout 1886-(wkly) [mf ed 1992] – 1r – 1 – (ceased 189-?) – mf#SEM35P24 – cn Bibl Nat [073]

Interpreter – Evanston. 1957+ (1) 1971+ (5) 1977+ (9) – ISSN: 0020-9678 – mf#2682 – us UMI ProQuest [240]

Interpreter / United Methodist Church [US] – 1975 jul-1977 jan, 1977 feb-1978 may, 1978 jun-1979, 1979+ind – 4r – 1 – (cont: methodist story-spotlight) – mf#167560 – us WHS [242]

Interpreter / United Steelworkers of America – n3,6-8 [1975 jun, oct-dec], n10-17 [1976 feb-dec], n18-108 [1977 jan-1987 dec], n109-152 [1988 jan-1992 mar] – 2r – 1 – mf#1697259 – us WHS [331]

The interpreter / Gladden, Washington – Boston: Pilgrim Press, c1918 – 1mf – 9 – 0-524-06485-7 – mf#1991-2585 – us ATLA [240]

Interpreter and guide : cambodian-french-english / Kim-Set – Phnom-Penh: Bouth-Neang 1959 [mf ed 1989] – 1r with other items – 1 – (title also in khmer) – mf#mf-10289 seam reel 029/04 [§] – us CRL [480]

The interpreter of words and terms : used either in the common or statute laws of this realm and in tenures and jocular customs / Cowell, D – London: J Place, 1701 – 4mf – 9 – $6.00 – (with an appendix containing the ancient names of places in england) – mf#LLMC 97-107 – us LLMC [340]

Interpreter releases : an information service on immigration, naturalization and related problems – New York: American Council for Nationalities Service. v1-63. 1924-86 – 549mf – 9 – $823.00 – (includes selected decisions of the board of immigration appeals. llmc does not have permission to update after v63. missing: v1-6) – mf#LLMC 81-511 – us LLMC [340]

The interpreter with his bible / Waffle, Albert E – New York: Anson D F Randolph, c1891 – 1mf – 9 – 0-8370-5681-0 – mf#1985-3681 – us ATLA [220]

Ein interpretierendes woerterbuch der nominalabstrakta im "narrenschiff" sebastians brants von abenteuer bis zwietracht : "hie findt man der welt gantzen louff" (eine vorred in das narren schyff) / Benkartek, Dietmar – (mf ed 1995) – 6mf – 9 – €62.50 – 3-8267-2275-2 – mf#DHS 2275 – gw Frankfurter [430]

Interpreting dance : circles of perception and spheres of experience / Barnick-Ben-Ezra, Barbara – 2000 – 123p on 2mf – 9 – $10.00 – mf#PE 4149 – us Kinesology [790]

Interpreting paul's gospel / Hunter, Archibald Macbride – London: SCM Press, 1954 – 1mf – 9 – 0-524-08110-7 – mf#1993-9016 – us ATLA [226]

Interpreting the constitution: inaugural lecture of the professor of political theory and government, delivered at the college on 1 december 1955 / Rees, John Collwyn – Swansea, Wales Univ. Coll. of Swansea 1956 33 p. LL-2309 – 1 – us L of C Photodup [342]

An interpretive inquiry of preservice teachers' reflections and development during a field-based elementary physical education methods course / Sebren, Mary A & Barrett, Kate R – 192 – 3mf – 9 – $12.00 – us Kinesology [790]

An interpretive inquiry of the professional life histories of selected women dance/ physical educators / Clark, Dawn & Robinson, Sarah M – 1992 – 3mf – 9 – $12.00 – us Kinesology [790]

Interrace jr – 1989 sep – 1r – 1 – mf#4877985 – us WHS [071]

Interracial books for children – New York. 1966-1975 (1) – (cont by: interracial books for children bulletin) – ISSN: 0020-9708 – mf#12390 – us UMI ProQuest [070]

Interracial books for children bulletin – New York. 1976-1988 (1,5,9) – (cont: interracial books for children) – ISSN: 0146-5562 – mf#12390,01 – us UMI ProQuest [070]

Interracial classified – 1993 mar-1994 apr – 1r – 1 – (cont by: interracial voice) – mf#2906426 – us WHS [071]

Interracial conference reports : including background material – Chicago: General Board of Social and Economic Relations, 1955-1959. Chicago: Dep of Photodup, U of Chicago Lib, 1967 82r); Evanston: American Theol Lib Assoc, 1984 (2r) – 1 – 0-8370-0657-0 – mf#1984-6000 – us ATLA [360]

Interracial Council for Business Opportunity of New Jersey *see* Boot$trap

Interracial news service – 1951-59 – 1 – $50.00 – us Presbyterian [240]

The interregnum (a.d. 1648-1660); studies of the commonwealth, legislative, social, and legal / Inderwick, Frederick Andrew – London: S. Low, Marston, Searle & Rivington, 1891. 340p – 1 – us Wisconsin U Libr [941]

Interregnum pacificator and prospectus for "the rights of man" – Salem, OH, dec 27 1880 – 1r – 1 – (temporary national party greenback newspaper) – us Western Res [071]

Interrelationships among stress, social support, health behaviors and self-assessed health status / Colwell, Gregory B & Seffrin, John R – 1992 – 2mf – 9 – $8.00 – us Kinesology [613]

Interreligious Foundation for Community Organization [US] *see* Ifco news

Interreligious newsletter – New York, NY. May 1976-Nov 1985 – 1 – us AJPC [071]

Interrogation of japanese leaders and responses to questionnaires, 1945-1946 – (= ser National archives coll of foreign records seized, 1941-) – 9r – 1 – mf#M1654 – us Nat Archives [355]

Interrogation of japanese leaders and responses to questionnaires, 1945-1946 – 9r – 1 – mf#M1654 – us Nat Archives [355]

Interrogation records prepared for war crimes : proceedings at nuernberg, 1945-1947 / U.S. Army Commands – (= ser National archives coll of world war 2 war crimes records) – 31r – 1 – (with printed guide) – mf#M1270 – us Nat Archives [934]

Interrogationes faciendae a sacerdote ad baptismum conferendum procedente – St-Ignatii, Montanis: Typis Missionis, 1891? – 1mf – 9 – (trans by philip canestrelli) – mf#27810 – cn Canadiana [241]

Interrogatoires / Bres, G de – n.p, 1567 – 1mf – 9 – mf#PBA-437 – ne IDC [240]

Interscholastic coaching certification / Wilson, Thomas H – 1997 – 1mf – 9 – $4.00 – mf#PE 3783 – us Kinesology [370]

The inter-state commerce act; an analysis of its provisions / Dos Passos, John Randolph – New York, Putnam's, 1887. 125 p. LL-1707 – 1 – us L of C Photodup [346]

The interstate commerce act and federal anti-trust laws, including the sherman act; the act creating the bureau of corporations. / Snyder, William Lamartine – New York: Baker, Voorhis, 1906. LL-1154 – 1 – us L of C Photodup [346]

Interstate commerce acts annotated / United States Interstate Commerce Commission – Washington. 1974-1977 (1) 1974-1977 (5) 1974-1977 (9) – mf#9233 – us UMI ProQuest [380]

Interstate commerce acts annotated / U.S. Interstate Commerce Commission – Washington: GPO. v1-22. 1927-77 (all publ) – 215mf – 9 – $322.00 – mf#LLMC 79-428 – us LLMC [348]

Interstate commerce commission reports / U.S. Interstate Commerce Commission – 1st series: v1-367. 1887-1984 (all publ); 2nd series: v1-7. 1984-91 – 3756mf – 9 – $5634.00 – (updates planned) – mf#LLMC 78-220 – us LLMC [324]

The interstate commerce law / Hamilton, Adelbert – Northport, Long Island, N.Y., Thompson, 1887. 219 p. LL-171 – 1 – us L of C Photodup [346]

Interstate compact for education – Denver. 1978-1981 – 1,5,9 – (cont: compact) – ISSN: 0275-4592 – mf#11829,01 – us UMI ProQuest [370]

Inter-state extradition / Hawley, John Gardner – Detroit, 1890. 172p. LL-1269 – 1 – us L of C Photodup [340]

Inter-synodical Foreign Missionary Convention for Men *see* Men and the modern missionary enterprise

Inter-Territorial Language (Swahili) Committee To The East *see* Standard english-swahili dictionary

Intertester and intratester validity and reliability of the wisconsin wrestling minimal weight project / McHugh, Vicki L – 1999 – 1mf – 9 – $4.00 – mf#PE 3989 – us Kinesology [612]

Intertextuality in the poetry of h m l lentsoane / Makibelo, Mmakgwele Paulina – Pretoria: Vista University 2002 [mf ed 2002] – 2mf – 9 – (incl bibl ref) – mf#mfm15334 – sa Unisa [470]

Inter-tribal Council of California *see* Tribal spokesman

Inter-Tribal Council of Nevada, Inc et al *see* Native nevadan

Inter-tribal tribune / Heart of America Indian Center – 1980 may-1982 may – 1r – 1 – (cont: newsletter [heart of america indian center]; cont by: intertribal tribune [1998]) – mf#624637 – us WHS [305]

Inter-tribal voice / Native American Coalition of Tulsa – v1 n1-6 [1982 sep-1983 dec], 1984 jan/feb-aug/sep – 1r – 1 – mf#1043284 – us WHS [305]

Intervalling-effekt der betafaktoren am deutschen aktienmarkt / Sauer, Egbert – (mf ed 1993) – 2mf – 9 – €49.00 – 3-89349-666-1 – mf#DHS 666 – gw Frankfurter [332]

Inter-Varsity Christian Fellowship *see*
- Collegiate trends
- Twentyonehundred chronicle
- U

Intervencion americana / Lugo, Americo – Santo Domingo, Dominican Republic. 1916 – 1r – us UF Libraries [972]

Intervencion – conciliacion – arbitraje, en las... / Maurtua, Victor Manuel – Habana, Cuba. 1929? – 1r – us UF Libraries [972]

La intervencion del abogado en la constitucion de las sociedades mercantiles / Cuellar Grajera, Antonio – Badajoz: Dip. Provincial, 1966. Sep. REE – sp Bibl Santa Ana [946]

La intervencion federal en la provincia de buenos aires, 5 de enero a 5 de mayo de 1944 / Ojea, Julio Oscar – Buenos Aires: Goyena, 1945. 158p. LL-8018 – 1 – us L of C Photodup [340]

Intervention de monsieur thiounn prasith : chef de la delegation du front uni national du kampuchea et du gouvernement royal d'union nationale du cambodge / Thiounn Prasith – [n.p. 1970?] [mf ed 1989] – 1r with other items – 1 – (at head of title: commission internationale d'enquetes sur les crimes de guerre des etats unis d'amerique en indochine. 1. session, stockholm, 22 au 25 oct 1970) – mf#mf-10289 seam reel 016/22 [§] – us CRL [959]

Intervention in school and clinic – Austin. 1990+ (1,5,9) – (cont: academic therapy) – ISSN: 1053-4512 – mf#6344,01 – us UMI ProQuest [370]

An intervention programme to promote the choice of physical science by secondary school learners / Nukeri, Happy Jabulani – Uni of South Africa 2000 [mf ed Johannesburg 2000] – 4mf – 9 – (incl bibl ref) – mf#mfm15007 – sa Unisa [500]

Interview – New York. 1977+ (1) 1986+ (5) 1986+ (9) – (cont: andy warhol's interview) – ISSN: 0149-8932 – mf#10591,01 – us UMI ProQuest [073]

Interview transcripts / Fox, John R & Leahy, Daniel – 1973 – 1r – 1 – (available for ref) – mf#pmb1179 – at Pacific Mss [080]

Interview with mrs mattie jackson / Shepherd, Rose – S.l., S.l? . 1939 – 1r – us UF Libraries [978]

Interview with prince sihanouk / Worthy, William – Phnom Penh: Ministry of Information [1965] [mf ed 1989] – 1r with other items – 1 – ("in 'the national guardian,' jan 9 1965") – mf#mf-10289 seam reel 015/31 [§] – us CRL [080]

Interviewing japan / Kennaway, Adrienne – Allahabad: Kitabistan, 1943 – (= ser Samp: indian books) – us CRL [950]

Interviews : greeks in miami, a social ethnic stud... – S.l., S.l? . 193-? – 1r – us UF Libraries [978]

Interviews for transition in western uganda, 1891-1901 : transcripts of interviews conducted between mar 1968 and apr 1969 / Steinhart, Edward I – [Evanston, 1970?] – us CRL [960]

Interviews from villages in the njombe district, tanzania / Graham, James – [s.l.] Microsystems, Inc, 1967 – us CRL [960]

Interviews with us government officials concerning the nigeria-biafra war, 1971-72 / Robison, David – [s.l: s.n.], 1972 – us CRL [960]

Intervirology – Basel. 1973-1974 (1) 1973-1974 (5) (9) – ISSN: 0300-5526 – mf#7539 – us UMI ProQuest [610]

Interwoven gospels – New York, NY. 1889 – 1r – us UF Libraries [240]

Intestate succession in the state of new york / Remsen, Daniel Smith – New York: Baker, Voorhis, 1886. 150p. LL-1437 – 1 – us L of C Photodup [340]

Inthusathanam – Jaffna, Sri Lanka. 11 Sept 1889-31 Jan 1981 – 15r – 1 – us L of C Photodup [079]

Inti – Semarang, 1964-1967 – 4mf – 9 – (missing: 1965(6-end)-1967(1-end)) – mf#SE-897 – ne IDC [950]

L'intiero raggvaglio del svccesso di famacosta... / Martinengo, N – Np, [1571] – 1mf – 9 – mf#H-8316 – ne IDC [956]

Intikam – Geneva, 1900-02. Yayimliyan: Yeni Osmanlilar. n50. 10 mart 1902 – (= ser O & t journals) – 1mf – 9 – $25.00 – us MEDOC [956]

Intimacy / Lexington Library, Inc – 1985 aug ust-1987 dec, 1988 jan-dec, 1989, 1990 feb-dec, 1991, 1993 feb-dec, 1994 may-1995 dec, 1996 – 8r – 1 – (cont: bronze thrills) – mf#2847362 – us WHS [071]

Intimas (poesias de los tiempos idos) / Sanchez-Arjona, Vicente – Sevilla: Graficas Tirvia, Tomo 1. 1954 – 1 – sp Bibl Santa Ana [810]

Intimas (poesias de los tiempos idos) / Sanchez-Arjona, Vicente – Sevilla: Graficas Tirvia, Tomo 2. 1954 – 1 – sp Bibl Santa Ana [810]

Intimas.poesias / Real, Enrique – 1897 – 9 – sp Bibl Santa Ana [810]

Intimate Apparel Workers' Union et al *see* Our union

Intimate letters of james gibbons huneker, collected and edited by josephine huneker / Huneker, James – New York: Issued for subscribers only by Boni and Liveright, 1924. 322p – 1 – us Wisconsin U Libr [920]

Intimations of eternal life / Leighton, Caroline C – Boston: Lee and Shepard, 1891 – 1mf – 9 – 0-7905-8501-4 – mf#1989-1726 – us ATLA [240]

Intimations of immortality in the sonnets of shakespeare / Palmer, George Herbert – Boston, MA. 1912 – 1r – us UF Libraries [420]

Intiqad-i kitab – Tehran: Intisharat-i Nil, 1955-. dawrah-'i 1, shumarah-'i 1-12; dawrah-'i 2, shumarah-'i 1-8; dawrah-'i 3, shumarah-'i 1-12; dawrah-'i 4, shumarah-'i 1,3 2 day 1334-aban/azar 1347 [23 dec 1955-oct/dec 1968] – 1r – 1 – $53.00 – us MEDOC [079]

Intisari / Jajasan Intisari – Djakarta, 1963-1971 – 244mf – 9 – mf#SE-898 – ne IDC [950]

Into his marvellous light : studies in life and belief / Hall, Charles Cuthbert – Boston: Houghton Mifflin, 1892, c1891 – 1mf – 9 – 0-7905-7644-9 – mf#1989-0869 – us ATLA [240]

Into the jungles of dutch guiana : bush master / Smith, Nicol – Garden City, NY. 1943 – 1r – us UF Libraries [972]

Into tropical florida – New York, NY. 1890 – 1r – us UF Libraries [580]

L'intolerance religieuse et la politique / Bouche-Leclercq, Auguste – Paris: E Flammarion, 1911 [mf ed 1990] – (= ser Bibliotheque de philosophie scientifique) – 1mf – 9 – 0-7905-5575-1 – (in french. incl bibl ref) – mf#1988-1575 – us ATLA [240]

Intolerance the disgrace of christians / Wyvill, Christopher – London, England. 1809 – 1r – us UF Libraries [240]

Intonationes vespertinarum, primus tomus / Zacharia, Cesare de – 1594 – (= ser Mssa) – 4mf – 9 – €60.00 – mfchl 449 – gw Fischer [780]

Intorcetta, P *see* Compendiosa narratio

Intoxicants and opium in all lands and times : a twentieth-century survey of intemperance, based on a symposium of testimony from one hundred missionaries and travelers / Crafts, Wilbur Fisk et al – Washington, D.C.: International Reform Bureau, 1904, c1900 – 1mf – us ATLA [306]

Intoxicants & opium in all lands and times : a twentieth-century survey of intemperance, based on a symposium of testimony from one hundred missionaries and travelers / Crafts, Wilbur Fisk et al – Rev. 6th ed. Washington, D. C.: International Reform Bureau, 1904, c1900 – 1mf – 9 – 0-7905-4269-2 – (incl bibl ref) – mf#1988-0269 – us ATLA [360]

Intoxicating liquors; the law relating to the traffic in intoxicating liquors and drunkenness / Woollen, William Watson – Cincinnati, Anderson, 1910. 2 v. LL-1252 – 1 – us L of C Photodup [343]

Intra-abdominal pressure and rowing : the effects of inspiring versus expiring during the drive / Manning, Timothy S – 1998 – 2mf – 9 – $8.00 – mf#PH 1610 – us Kinesology [612]

Intracoastal waterway, norfolk to key west / Federal Writer's Project – Washington, DC. 1937 – 1r – us UF Libraries [978]

Intra-Community Cooperative *see*
- News and goods
- News and goods of the icc network

Intra-Community Cooperative [Madison WI] *see*
- Digester's reader
- North farm news

Intramural law review *see*
- American university law review
- Ohio northern university law review
- Saint louis university law journal

Intramuscular and subcutaneous temperature changes in the human leg due to contrast hydrotherapy / Wertz, Alice Seton – 1997 – 1mf – 9 – $4.00 – mf#PE 3821 – us Kinesology [790]

Intramuscular determinants of the vo$_2$ slow component in trained cyclists / Wadley, Glenn – 1999 – 2mf – 9 – $8.00 – mf#PH 1664 – us Kinesology [612]

L'intransigeant – Paris. 15 juil 1880-10 juin 1940, 13 mai 1947-sept 1948 – 1 – fr ACRPP [073]

Intransigeant – Paris, France. 15 jul 1880-11 jun 1940 – 211 1/2r – 1 – uk British Libr Newspaper [072]

Intransigente – Miami, FL. 1980 aug 30-1998 nov – 2r – us UF Libraries [071]

Intratester and intertester reliability when using the chatillon hand-held dynamometer to measure force production in the upper and lower extremities / Jefferson, LouAnne M – 1994 – 2mf – $8.00 – us Kinesology [574]

Intravenous therapy news – Georgetown. 1984-1985 (1,5,9) – (cont: american journal of intravenous therapy and clinical nutrition. cont by: pharmacy practice news) – ISSN: 8750-3182 – mf#14121,03 – us UMI ProQuest [616]

Intrazellulaere regulationsmechanismen der t-zell migration / Entschladen, Frank – (mf ed 1997) – 1mf – 9 – €30.00 – 3-8267-2444-5 – mf#DHS 2444 – gw Frankfurter [612]

Intrepid – nos. 1-10. 1964-68 – 1 – us AMS Press [800]

Intriga y amor / Hurtado, Antonio – 1872 – 9 – sp Bibl Santa Ana [830]

Intrigant dupe par lui-meme / Richaud Martelly, M – Paris, France. 1803 – 1r – us UF Libraries [440]

Intrigas de los rusos en espana; como y por que sali del ministerio de defensa nacional / Prieto, Indalecio – Montevideo, 1940. Fiche W1119. (Blodgett Collection of Spanish Civil War Pamphlets) – 9 – us Harvard [946]

Intrigue au bal / Bru-Thiellay, Paul – Paris, France. 1935? – 1r – us UF Libraries [440]

L'intrigue du cabinet : sous henri 4 et louis 13, terminee par la fronde / Anquetil, Louis P – Paris 1780 [mf ed Hildesheim 1995-98] – 4v on 12mf – 9 – €120.00 – 3-487-26107-3 – gw Olms [944]

Intrigues devoilees : ou louis 17, dernier roi legitime de france, decede a delft, le 10 aout 1845 / Gruau de LaBarre, Modeste – Rotterdam 1846-1848 [mf ed Hildesheim 1995-98] – 3v on 23mf – 9 – €230.00 – ISBN-10: 3-487-26182-0 – ISBN-13: 978-3-487-26182-9 – gw Olms [944]

Intrinsic, extrinsic and amotivational differences in scholarship and nonscholarship collegiate track and field athletes / Miller, Jennifer A – 2000 – 114p on 2mf – 9 – $10.00 – mf#PSY 2145 – us Kinesology [150]

Introducao a arqueologia brasileira / Costa, Angyone – Sao Paulo, Brazil. 1938 – 1r – us UF Libraries [930]

Introducao a democracia brasileira / Martins, Wilson – Porto Alegre, Brazil. 1951 – 1r – us UF Libraries [972]

Introducao a geografia das comunicacoes brasileira / Travassos, Mario – Rio de Janeiro, Brazil. 1942 – 1r – us UF Libraries [972]

Introducao a geografia das comunicacoes brasileiros / Travassos, Mario – Rio de Janeiro. 1942 – 1 – us CRL [972]

Introducao a historia da agricultura em portugal / Marques, Antonio Henrique R De Oliveira – Lisboa, Portugal. 1968 – 1r – us UF Libraries [630]

Introducao a historia das bandeiras / Cortesao, Jaime – Lisboa, Portugal. v1-2. 1964 – 1r – us UF Libraries [972]

Introducao a literatura brasileira / Lima, Alceu Amoroso – Rio de Janeiro, Brazil. 1957 – 1r – us UF Libraries [972]

Introducao a revolucao brasileira / Sodre, Nelson Werneck – Rio de Janeiro, Brazil. 1967 – 1r – us UF Libraries [972]

Introducao a sociologia das secas / Andrade, Lopes De – Rio de Janeiro, Brazil. 1948 – 1r – us UF Libraries [972]

Introducao ao estudo da amazonia brasileira / Nunes, Osorio – Rio de Janeiro, Brazil. 1949 – 1r – us UF Libraries [972]

Introducao ao estudo da amazonia brasileira / Nunes, Osorio – Rio de Janeiro, Brazil. 1950 – 1r – us UF Libraries [972]

Introducao ao estudo da nova critica no brasil / Azevedo Filho, Leodegario A De – Rio de Janeiro, Brazil. 1965 – 1r – us UF Libraries [972]

Introducao ao estudo do desenvolvimento economico / Rangel, Inacio – Salvador, Brazil. 1957 – 1r – us UF Libraries [972]

Introducao ao estudo tecnico-economico da criacao de qado bovino / Pereira, J Lima – Lisboa, Portugal. 1962 – 1r – us UF Libraries [330]

Introducao as obras do barao do rio-branco / Araujo Jorge, Arthur Guimaraes De – Rio de Janeiro, Brazil. 1945 – 1r – us UF Libraries [972]

Introducao critica a sociologia brasileira / Ramos, Alberto Guerreiro – Rio de Janeiro, Brazil. 1957 – 1r – us UF Libraries [301]

Introduccion a la ciencia del derecho / Betancur, Cayetano – Bogota, Colombia. 1953 – 1r – us UF Libraries [972]

Introduccion a la civilizacion hispanoamericana / Pattee, Richard – Boston, MA. 1948 – 1r – us UF Libraries [972]

Introduccion a la filosofia 1. introduccion y logica / Frutos Cortes, Eugenio – Zaragoza: Tip. Libreria General, 1943 – sp Bibl Santa Ana [160]

Introduccion a la filosofia 2. psicologia y etica / Frutos Cortes, Eugenio – Zaragoza: Tip. Libreria General, 1943 – sp Bibl Santa Ana [170]

Introduccion a la historia de la cultura en columb... / Lopez De Mesa, Luis – Bogota, Colombia. 1930 – 1r – us UF Libraries [972]

Introduccion a la historia eclesiastica del tucuman, 1535 a 1590. buenos aires, 1934 / Cabrera, Pablo – Madrid: Razon y Fe, 1935 – 1 – sp Bibl Santa Ana [240]

Introduccion a la poesia de la senorita armino / Coronado, Carolina – 1850 – 9 – sp Bibl Santa Ana [410]

Introduccion a la teoria constitucional guatemalte / Kestler Farnes, Maximiliano – Guatemala, 1964 – 1r – us UF Libraries [972]

Introduccion a un tratado de politica : sacado textualmente de los refranceros, romanceros y gestas de la peninsula / Costa y Martinez, Joaquin – Madrid: Impr de la Revista de Legislacion 1881 [mf ed Bloomington IN: Indiana Uni Lib, Preservation Dept 1984] – 1r – 1 – us Indiana Preservation [320]

Introduccion a una obra historica / Barrado Font, Francisco – Madrid: Imp. Sucesores de M.Minuesa de los Rios, 1901 – sp Bibl Santa Ana [946]

Introduccion al derecho, version taquigrafica de carlos argenta estable / Jimenez de Arechaga, Eduardo – Montevideo, Organizacion Taquigrafica Medina 194-? 342 p. LL-4107 – 1 – us L of C Photodup [340]

Introduccion al estudio de la filosofia de la hist... / Cuervo Marquez, Emilio – Bogota, Colombia. 1938 – 1r – us UF Libraries [100]

Introduccion al estudio del derecho notarial guate... / Rivera Toledo, Antonio – Guatemala, t1-3. 1965 – 1r – us UF Libraries [972]

Introduccion al estudio del problema immigratorio / Esguerra Camargo, Luis – Bogota, Colombia. 1940 – 1r – us UF Libraries [304]

Introduccion...que se recito...sevilla / Forner Segarra, Juan Pablo – 1796 – 9 – sp Bibl Santa Ana [840]

Introducing india – Calcutta: Royal Asiatic Society of Bengal, 1947- – (= ser Samp: indian books) – us CRL [954]

Introducing men to christ : fundamental studies / Weatherford, Willis Duke – New York: Association Press, 1911 – 1mf – 9 – 0-8370-6448-1 – mf#1986-0448 – us ATLA [240]

Introducing paris / Lucas, E V – London, England. 1928 – 1r – us UF Libraries [914]

Introductio generalis ad historiam ecclesiasticam / Smedt, Charles de – Gandavi [Ghent]: C Poelman, 1876 – 2mf – 9 – 0-8370-6839-8 – (incl bibl, appendixes on emendations to documents and chronologies and ind to works cited) – mf#1986-0839 – us ATLA [240]

Introductio in artem emblematicam... / Mueller, J J – Ienae: Sumptibus Iohannis Bielkii, 1706 – 3mf – 9 – mf#0-699 – ne IDC [090]

Introductio in chaldaicam linguae, syriacae atque armenica et dece alias linguas / Ambrosius, T – Pavia, 1539 – 5mf – 9 – mf#AR-1531 – ne IDC [470]

Introductio in historiam evangelii seculo 16 passim per europam renovati doctrinaeque reformatae / Gerdesius, D – Groningae, 1744-1752. 4 v – 31mf – 9 – mf#ZWI-34 – ne IDC [240]

Introductio in librum genesis : in qua etiam de authentia pentateuchi necnon de inspiratione et interpretatione scripturae agitur / Hetzenauer, Michael – Graecii: Styria, 1910 – 1mf – 9 – 0-524-08079-8 – (incl bibl ref) – mf#1992-1139 – us ATLA [221]

Introductio in librum psalmorum : in qua de poesi sacra hebraeorum fuse disseritur, ejusque vetustissima monumenta e libris historicis veteris testamenti collecta, necnon et cantica scripturistica breviarii romani in modum appendicis explicantur / Steenkiste, J-A van – altera ed. Brugis: Typis Modesti Delplace, 1873 – (= ser Commentarius in Librum Psalmorum) – 1mf – 9 – 0-524-07344-9 – mf#1992-1075 – us ATLA [220]

Introductio in s. theologiam dogmaticam ad mentem d. thomae aquinatis / Schaezler, Constantin, Freiherr von – Ratisbonae [Regensburg]: Typis et sumptibus GJ Manz, 1882 – 4mf – 9 – 0-524-00323-8 – mf#1989-3023 – us ATLA [241]

Introductio in sacram scripturam : ad usum scholarum pont. seminarii romani et collegii urbani de propaganda fide / Ubaldi, Ubaldo – ed tertia. Romae: Ex Typographia Polyglotta, 1886 – 6mf – 9 – 0-524-05754-0 – mf#1992-0597 – us ATLA [220]

Introductio in sacram scripturam / Lamy, Thomas Joseph – ed 6, denuo recognita. Mechliniae: H Dessain, 1901 – 2mf – 9 – 0-524-07336-8 – mf#1992-1067 – us ATLA [220]

Introduction a la critique generale de l'ancien testament. de l'origine du pentateuque : lecons provessees a l'ecole superieure de theologie de paris, en 1886-1887 / Martin, Jean Pierre Paulin – Paris: Maisonneuve: Charles Leclerc, [1889?] [mf ed 1990] – 3v on 5mf – 9 – 0-8370-1675-4 – (in french) – mf#1987-6103 – us ATLA [221]

Introduction a la critique textuelle du nouveau testament : lecons professees a l'ecole superieure de theologie de paris en 1883-1884 / Martin, Jean Pierre Paulin – Paris: Maisonneuve freres et C Leclerc [1883?-1886?] – 33mf – 9 – 0-524-08708-3 – mf#1993-0053 – us ATLA [225]

Introduction a la description de la france : et au droit public de ce royaume / Piganiol de LaForce, Jean – Paris 1752 [mf ed Hildesheim 1995-98] – 2v on 8mf – 9 – €160.00 – 3-487-29774-4 – gw Olms [914]

Introduction a la litterature orale leboue : analyse ethno-sociologique / Ndoye, Mbengue nee Mariama – 1981 – us CRL [390]

Introduction a la theologie orthodoxe / Makarii, Metropolitan of Moscow – Paris: Joel Cherbuliez, 1857 – 7mf – 9 – 0-7905-9329-7 – (incl bibl ref) – mf#1989-2554 – us ATLA [240]

Introduction a l'etude de la langue hebraique : apercu historique et philologique / Baumgartner, Antoine Jean – Paris: Libr Fischbacher [1887?] – 1mf – 9 – 0-7905-0667-X – mf#1987-0667 – us ATLA [470]

Introduction a l'etude de la theologie protestante : avec index bibliographique / Emery, Louis – Lausanne: F Rouge; Paris: Fischbacher, 1904 [mf ed 1990] – 2mf – 9 – 0-7905-7815-8 – (in french) – mf#1989-1040 – us ATLA [242]

Introduction a l'etude des idees morales dans l'egypte antique / Baillet, Jules – [Blois?]: Grande Imprimerie de Blois, 1912 – 1mf – 9 – 0-524-04155-5 – (incl bibl ref) – mf#1990-3285 – us ATLA [170]

Introduction a l'etude d'un genre satirico-laudatif : le taasu-wolof / Thiam, Mamadou Cherif – 1979 – us CRL [470]

Introduction a l'histoire de l'empire francais : ou essai sur la monarchie de napoleon / Regnault-Warin, Jean Baptiste Joseph Innocent Philadelphe – Paris 1820 [mf ed Hildesheim 1995-98] – 5mf – 9 – €100.00 – ISBN-10: 3-487-26239-8 – ISBN-13: 978-3-487-26239-0 – gw Olms [944]

Introduction a l'histoire des religions / Dussaud, Rene – Paris: Ernest Leroux, 1914 – (= ser Bibliotheque historique des religions) – 1mf – 9 – 0-524-01278-4 – mf#1990-2314 – us ATLA [200]

Introduction a l'histoire generale des religions : resume du cours public / Goblet d'Alviella, Eugene, comte – Bruxelles: C Muquardt, 1887 – 1mf – 9 – 0-524-01485-X – mf#1990-2461 – us ATLA [200]

Introduction a l'instruction economique morale et... / Alexis, Stephen – Port-Au-Prince, Haiti. 1953 – 1r – us UF Libraries [330]

Introduction a un memoire sur la propagation de l'alphabet phenicien dans l'ancien monde : couronne par l'academie des inscriptions et belles-lettres / Lenormant, Francois – Paris: A Laine et J Havard, 1866 – 1mf – 9 – 0-7905-0046-9 – (incl bibl ref) – mf#1987-0046 – us ATLA [470]

Introduction a un nouveau systeme d'harmonie / Basevi, Abramo – Florence: G G Guidi 1865 (Florence: P Tofani) [mf ed 1992] – 1r – 1 – (trans fr italian by louis delatre; with rev & corr by aut) – mf#pres. film 124 – us Sibley [780]

Introduction and history of saiva siddhanta / Cuppiramaniya Pillai, Ji – Annamalainagar: Annamalai University, 1948 – (= ser Samp: indian books) – us CRL [954]

Introduction au cambodgien / Cambefort, Gaston – Paris: Maisonneuve 1950 [mf ed 1989] – (= ser Les langues de l'orient 1. manuels, 7) – 1r with other items – 1 – mf#mf-10289 seam reel 024/14 [§] – us CRL [480]

Introduction au malgache / Faublee, Jacques – Paris: G P Maisonneuve, 1946 – 1 – us CRL [490]

Introduction au nouveau testament / Goguel, Maurice – Paris: Ernest Leroux, 1922-1926. Chicago: Dep of Photodup, U of Chicago Lib, 1971 (1r); Evanston: American Theol Lib Assoc, 1984 (1r) – (= ser Biblioteque historique des religions) – 1 – 0-8370-0495-0 – (includes bibliographies) – mf#1984-B241 – us ATLA [225]

Introduction au nouveau testament / Goguel, Maurice – Paris. v1-4/1. 1923-1925 – 8 – €63.00 – (t1: les evangelis synoptiques, paris 1923 9mf. t2: le quatrieme evangile, paris 1923 10mf. t3: le livre des actes, paris 1922 7mf. t4/1: les epitres pauliniennes. premiere partie, paris 1925 7mf) – ne Slangenburg [225]

Introduction au nouveau testament = Inledning till pauli bref / Godet, Frederic Louis – Upsala: W Schultz, [1894?] – 2mf – 9 – 0-8370-9625-1 – (in swedish. incl bibl ref) – mf#1986-3625 – us ATLA [225]

Introduction aux etudes liturgiques / Cabrol, Fernand – Paris: Bloud, 1907 – 1mf – 9 – 0-7905-6802-0 – mf#1988-2802 – us ATLA [012]

Introduction aux melodies gregoriennes / Boyer d'Agen – Paris: H Oudin, 1894 – 1mf – 9 – 0-7905-5576-X – mf#1988-1576 – us ATLA [780]

Introduction aux sources de l'histoire du culte chretien au moyen age / Vogel, C – Spoleto, 1966 – 7mf – 8 – €15.00 – ne Slangenburg [240]

L'introduction de l'imprimerie au canada : une breve histoire / Fauteux, Aegidius – Montreal: Compagnie de Papier Rolland, 1957 [mf ed 1974] – 1r – 5 – mf#SEM16P139 – cn Bibl Nat [971]

Introduction explicative concernant les statuts, regles et reglemens... / Compagnie d'assurance de Quebec contre les accidens du feu – [Quebec?: s.n.], 1827 (Quebec: P E Desbarats) – 1mf – 9 – 0-665-54967-9 – mf#54967 – cn Canadiana [360]

L'introduction frauduleuse de viande impropre sur le marche de la consommation humaine et la fraude en rapport avec la viande chevaline : rapport interimaire de l'enquete sur le crime organise / Enquete sur le crime organise (Quebec) – Ste-Foy: Commission de police du Quebec, 1975 [mf ed 1996] – 6mf – 9 – mf#SEM105P2732 – cn Bibl Nat [360]

Introduction of the art of printing into scotland / Dickson, Robert – Aberdeen: J & J P Edmond & Spark, 1885 – (= ser 19th c publishing...) – 2mf – 9 – mf#3.1.1 – uk Chadwyck [680]

Introduction of the ironclad warship / Baxter, James Phinney – Cambridge, MA. 1933 – 1r – us UF Libraries [623]

Introduction to a course of lectures on the early fathers / Blunt, J J – Cambridge, England. 1840 – 1r – us UF Libraries [240]

The introduction to a new philosophy / Bergson, Henri – Boston: J W Luce, 1912 [mf ed 1987] – 108p – 1 – (trans fr french by sidney littman) – mf#8250 – us Wisconsin U Libr [120]

Introduction to a scientific system of mythology = Prolegomena zu einer wissenschaftlichen mythologie / Mueller, Karl Otfried – London: Longman, Brown, Green, and Longmans, 1844 – 1mf – 9 – 0-524-01289-X – (in english) – mf#1990-2325 – us ATLA [250]

An introduction to adwaita philosophy : sankara school of vedanta / Bhattacharyya, Kokileswar, Pandit – Calcutta: University of Calcutta, 1924 – (= ser Samp: indian books) – us CRL [180]

Introduction to american law...11th ed / Walker, Timothy – Boston, Little, Brown, 1905. 692 p. LL-154 – 1 – us L of C Photodup [340]

Introduction to antonio soler / Carroll, Frank – U of Rochester 1960 [mf ed 19–] – 2v on 8mf – 9 – mf#fiche172 – us Sibley [780]

An introduction to astronomy : designed as a text book: for the students of yale college / Olmsted, Denison – New York: R B Collins 1850 [mf ed 1998] – 1r [pl/ill] – 1 – mf#film mas 28210 – us Harvard [520]

An introduction to astronomy : designed as a text-book for the use of students in college / Olmsted, Denison – 3rd ed. New York: Collins & Bro 1874 [mf ed 1998] – 1r [pl/ill] – 1 – (rev by ebenezer strong snell; with pref) – mf#film mas 28219 – us Harvard [520]

Introduction to bible study : the old testament / Painter, Franklin Verzelius Newton – Boston: Sibley, c1911 – 1mf – 9 – 0-7905-1553-9 – (incl bibl ref and ind) – mf#1987-1553 – us ATLA [220]

Introduction to biblical chronology : from adam to the resurrection of christ / Akers, Peter – Cincinnati: Methodist Book Concern, 1855 [mf ed 1989] – 2mf – 9 – 0-7905-0841-9 – mf#1987-0841 – us ATLA [220]

Introduction to biblical hebrew : presenting graduated instruction in the language of the old testament / Kennedy, James – London: Williams and Norgate, 1889 – 1mf – 9 – 0-8370-1564-2 – mf#1987-6073 – us ATLA [470]

An introduction to buddhist esoterism / Bhattacharyya, Benoytosh – London, New York: Oxford University Press, 1932 – (= ser Samp: indian books) – us CRL [280]

Introduction to cambodian / Jacob, Judith M – London, Bombay [etc] Oxford UP 1968 [mf ed 1989] – 1r with other items – 1 – (with bibl) – mf#mf-10289 seam reel 024/08 [§] – us CRL [480]

Introduction to christian missions / Johnson, Thomas Cary – 2nd ed. Richmond, VA: For sale by Presbyterian Comm of Publ, c1910 – 1mf – 9 – 0-7905-6111-5 – mf#1988-2111 – us ATLA [240]

Introduction to christian theology : comprising 1. a general introduction, 2. the special introduction, or, the prolegomena of systematic theology / Smith, Henry Boynton; ed by Karr, William Stevens – New York: AC Armstrong, 1883, c1882 [mf ed 1985] – 1mf – 9 – 0-8370-5290-4 – (incl bibl ref & ind) – mf#1985-3290 – us ATLA [240]

An introduction to church history *see* Chiao hui shih chi ju men [ccm271]

Introduction to classical hausa and the major dialects / Ahmed, Umaru – Zaria, Nigeria. 1970 – 1r – us UF Libraries [470]

An introduction to classical sanskrit : an introductory treatise of the history of classical sanskrit literature / Shastri, Gaurinath Bhattacharyya – Calcutta: Modern Book Agency, 1943 – (= ser Samp: indian books) – us CRL [490]

An introduction to comparative philology / Gune, Panduranga Damodara – Poona: Oriental Book-Supplying Agency, 1918 – (= ser Samp: indian books) – us CRL [490]

An introduction to dogmatic theology : based on luthardt / Weidner, Revere Franklin – 2nd rev ed. New York: Fleming H Revell c1895 [mf ed 1985] – (= ser System of dogmatic theology based on luthardt and krauth 1) – 1mf – 9 – 0-8370-5738-8 – (incl bibl ref & ind) – mf#1985-3738 – us ATLA [242]

Introduction to dogmatic theology / Hall, Francis Joseph – New York: Longmans, Green 1912 [mf ed 1990] – (= ser Dogmatic theology 1) – 1mf – 9 – 0-7905-3892-X – (first ed mar 1907) – mf#1989-0385 – us ATLA [240]

An introduction to ecclesiastes : with notes and appendices / McNeile, Alan Hugh – Cambridge: University Press; New York: Macmillan [dist] 1904 [mf ed 1986] – 1mf – 9 – 0-8370-6220-9 – (incl app & ind) – mf#1986-0220 – us ATLA [221]

An introduction to english church architecture : from the 11th to the 16th century / Bond, Francis – London, New York: Oxford UP, 1913 [mf ed 1990] – 3mf – 9 – 0-7905-6283-9 – (incl bibl ref) – mf#1988-2283 – us ATLA [720]

An introduction to english industrial history / Allsopp, Henry – London: G Bell & Sons Ltd, 1912 [mf ed 1987] – xi/1/160p – 1 – mf#8204 – us Wisconsin U Libr [338]

An introduction to ethics / Murray, John Clark – Boston: De Wolfe, Fiske, c1891 [mf ed 1986] – 1mf – 9 – 0-8370-6228-4 – (incl bibl ref & ind) – mf#1986-0228 – us ATLA [170]

An introduction to ethics / Murray, John Clark – Montreal: W F Brown, c1891 – 5mf – 9 – (incl ind) – mf#33502 – cn Canadiana [170]

Introduction to french painting / Clutton-Brock, Alan Francis – New York, NY. 1932 – 1r – us UF Libraries [750]

An introduction to greek and latin palaeography / Thompson, Edward Maunde, Sir – Oxford: Clarendon Press, 1912 [mf ed 1990] – 2mf – 9 – 0-7905-8276-7 – (incl bibl ref) – mf#1988-6154 – us ATLA [450]

Introduction to haiti / Cook, Mercer – Washington, DC. 1951 – 1r – us UF Libraries [972]

The introduction to hegel's philosophy of fine art = Vorlesungen ueber die aesthetik / Hegel, Georg Wilhelm Friedrich – London: K Paul, Trench, Truebner, 1905 – 1mf – 9 – 0-524-00268-1 – (in english) – mf#1989-2968 – us ATLA [100]

An introduction to heraldry : with nearly one thousand illustrations, including the arms of about five hundred different families / Clark, Hugh; ed by Planche, J R – 18th rev corr ed. London: Bell & Daldy 1866 [mf ed 1990] – 1r [ill] – 1 – (filmed with: roman emperor worship / sweet, l m) – mf#1770 – us Wisconsin U Libr [929]

An introduction to hindi prose composition / Dann, George James – 2nd ed. Benares: Bhagavati Prasad, 1909 [mf ed 1995] – (= ser Yale coll) – iv/110p – 1 – 0-524-09347-4 – mf#1995-0347 – us ATLA [490]

An introduction to hindu and mahommedan law : for the use of students / Markby William – Oxford: Clarendon Press, 1906 [mf ed 1995] – (= ser Yale coll) – 172p – 1 – 0-524-09206-0 – mf#1995-0206 – us ATLA [230]

An introduction to historical theology : being a sketch of doctrinal progress from the apostolic era to the reformation / Stoughton, John – London: Religious Tract Society, [1880?] [mf ed 1990] – 2mf – 9 – 0-7905-6023-2 – (incl bibl ref) – mf#1988-2023 – us ATLA [240]

An introduction to indian philosophy / Chatterjee, Satischandra & Datta, Dhirendramohan – Calcutta: University of Calcutta, 1950 – (= ser Samp: indian books) – us CRL [180]

An introduction to indonesian linguistics / Brandstetter, R – 1916 – (= ser Royal asiatic society monograph) – 1r – 1 – mf#221 – uk Microform Academic [490]

An introduction to jung's psychology / Fordham, Frieda – London, Baltimore: Penguin Books [1953] [mf ed 1986] – (= ser Penguin psychology series) – 1r – 1 – (filmed with: friends, society of / lower, t) – mf#1669 – us Wisconsin U Libr [150]

Introduction to kant's critique of pure reason / Weldon, Thomas Dewar – Oxford, England. 1945 – 1r – us UF Libraries [190]

Introduction to lutheran symbolics : a historical survey of the oecumenical and particular creeds of the lutheran church... / Neve, Juergen Ludwig – Columbus, Ohio: FJ Heer, 1917 – 1mf – 9 – 0-524-06274-9 – (incl bibl ref) – mf#1991-2465 – us ATLA [242]

An introduction to mandarin / Whitewright, John Sutherland – vl 3rd rev enl ed. Shanghai: Theodore Leslie at Christian Literature Society's Depot, [1918] [mf ed 1995] – (= ser Yale coll) – 2v – 1 – 0-524-09471-3 – (in chook. v1 in english, v2 in chinese) – mf#1995-0471 – us ATLA [480]

An introduction to metaphysics = Introduction a la metaphysique / Bergson, Henri – London: Macmillan, 1913 [mf ed 1990] – 1mf – 9 – 0-7905-7377-6 – (english trans by t e hulme) – mf#1989-0602 – us ATLA [110]

Introduction to modern brazilian poetry / Downes, Leonard Stephen – Sao Paulo, Brazil. 1954 – 1r – us UF Libraries [440]

An introduction to modern geography : with an appendix, containing an outline of astronomy and the use of the globes / Thomson, James – 3rd ed. [Belfast?: s.n.] 1831 [mf ed 1984] – 4mf – 9 – 0-665-43141-4 – mf#43141 – cn Canadiana [910]

An introduction to perspective, drawing, and painting : in a series of...dialogues / Hayter, Charles – 2nd ed. London 1815 – (= ser 19th c art & architecture) – 3mf – 9 – mf#4.2.1655 – uk Chadwyck [700]

Introduction to philosophy : an inquiry after a rational system of scientific principles in their relation to ultimate reality / Ladd, George Trumbull – New York: Scribner, 1891, c1890 – 1mf – 9 – 0-7905-7352-0 – mf#1989-0577 – us ATLA [100]

Introduction to practical astronomy : designed as a supplement to olmsted's astronomy: containing special rules for the adjustment and use of astronomical instruments... / Mason, Ebenezer Porter – New York: R Collins, 1851, c1841 [mf ed 1998] – 1r [pl/ill] – 1 – mf#film mas 28210 – us Harvard [520]

Introduction to protestant dogmatics = Essai d'une introduction a la dogmatique protestante / Lobstein, Paul – [Chicago]: A M Smith c1902 [mf ed 1985] – 1mf – 1 – 0-8370-4156-2 – (english trans by arthur maxson smith; incl bibl ref) – mf#1985-2156 – us ATLA [242]

Introduction to psychological theory / Bowne, Borden Parker – New York: Harper, 1887, c1886 – 1mf – 9 – 0-7905-3603-X – mf#1989-0096 – us ATLA [150]

Introduction to roman law, in twelve academical lectures / Hadley, James – New York: D.Appleton and Company, 1873. 332p. With: Histoire Poetique du Quinzieme Siecle by P. Champion. 1 reel. 1261 – 1 – us Wisconsin U Libr [340]

Introduction to rural sociology in india / Desai, Akshayakumar Ramanlal – Bombay: indian Society of Agricultural Economics: Sole distributors, Vora & Co, Publishers, [1953] – (= ser Samp: indian books) – us CRL [301]

Introduction to sacred philology and interpretation = Einleitung in die theologische wissenschaft. selections / Planck, G J – Edinburgh: Thomas Clark, 1834 – (= ser The biblical cabinet) – 1mf – 9 – 0-7905-1835-X – (incl bibl ref. in english) – mf#1987-1835 – us ATLA [220]

An introduction to socialism / Mukerjee, Hirendranath – Calcutta: National Book Agency, 1940 – (= ser Samp: indian books) – us CRL [335]

Introduction to spelling and reading... / Fox, Francis – London, England. 1815 – 1r – us UF Libraries [240]

An introduction to systematic philosophy / Marvin, Walter Taylor – New York: Columbia UP, 1912, c1903 [mf ed 1991] – 2mf – 9 – 0-7905-9793-4 – (incl bibl ref) – mf#1989-1518 – us ATLA [100]

Introduction to tantra shastra / Woodroffe, John George – Madras: Ganesh & Co, 1952 – (= ser Samp: indian books) – us CRL [280]

An introduction to the articles of the church of england / Maclear, George Frederick & William, Watkin Wynn – [rev ed]. London, New York: Macmillan, 1896 [mf ed 1986] – 2mf – 9 – 0-8370-8694-9 – (incl bibl ref & ind) – mf#1986-2694 – us ATLA [242]

Introduction to the ateso language / Hilders, J H – Kampala, Uganda. 1956 – 1r – us UF Libraries [470]

Introduction to the bechaunaland protectorate history / Gabatshwane, S M – Kanye, South Africa. 1957 – 1r – us UF Libraries [960]

An introduction to the bible for teachers of children : a manual for use in the sunday schools or in the home / Chamberlin, Georgia Louise – Chicago: University of Chicago Press, 1904 [mf ed 1989] – (= ser Constructive bible studies. elementary series) – 1mf – 9 – 0-7905-3188-7 – mf#1987-3188 – us ATLA [220]

Introduction to the book of genesis : with a commentary on the opening portion = Genesis / Bohlen, Peter von – London: John Chapman, 1855 – 2mf – 9 – 0-7905-1572-5 – (incl bibl ref and index. in english) – mf#1987-1572 – us ATLA [221]

Introduction to the book of isaiah : with an appendix containing the undoubted portions of the two chief prophetic writers in a translation / Cheyne, Thomas Kelly – London: Adam & Charles Black, 1895 [mf ed 1985] – 2mf – 9 – 0-8370-2648-2 – (incl ind) – mf#1985-0648 – us ATLA [221]

An introduction to the books of ezra, nehemiah, and esther / Sayce, Archibald Henry – 3rd ed. London: Religious Tract Society, 1885 [mf ed 1985] – 1mf – 1 – 0-8370-5059-6 – (incl ind) – mf#1985-3059 – us ATLA [221]

An introduction to the books of the apocrypha / Oesterley, William Oscar Emil – New York: The Macmillan Co [1946] c1935 [mf ed 1986] – 1r – 1 – ("first publ in 1935, repr in 1937 & 1946." filmed with: rozvidky mykhaila drahomanova.../ drahomaniy, m p) – mf#1631 – us Wisconsin U Libr [221]

Introduction to the books of the new testament / Allen, Willoughby Charles – Edinburgh: T & T Clark, 1913 – 1mf – 9 – 0-7905-0303-4 – (incl bibl ref and indexes) – mf#1987-0303 – us ATLA [225]

An introduction to the books of the old testament / Oesterly, William Oscar Emil & Robinson, T H – 1934 – 9 – $15.00 – us IRC [221]

Introduction to the books of the old testament : with analyses and illustrative literature / Stearns, Oakman S – Boston: Silver, Burdett, 1888 – 1mf – 9 – 0-8370-5378-1 – (includes bibliographies) – mf#1985-3378 – us ATLA [221]

Introduction to the botany of tropical crops / Cobley, Leslie S – London, England. 1956 – 1r – us UF Libraries [580]

Introduction to the catholic epistles / Gloag, Paton James – Edinburgh: T & T Clark, 1887 – 1mf – 9 – 0-8370-3310-1 – mf#1985-1310 – us ATLA [227]

An introduction to the classification of animals / Huxley, Thomas Henry – London [1869] – (= ser 19th c evolution & creation) – 2mf – 9 – mf#1.1.9543 – uk Chadwyck [590]

Introduction to the commentary on the vedas / Dayananda Sarasvati, Swami – Meerut: B Ghasi Ram, 1925 – (= ser Samp: indian books) – (trans fr original sanskrit by ghasi ram) – us CRL [280]

An introduction to the creeds / Maclear, George Frederick – London, New York: Macmillan, 1901 [mf ed 1990] – (= ser Elementary theological class-books) – 1mf – 9 – 0-7905-6415-7 – (incl bibl ref) – mf#1988-2415 – us ATLA [240]

An introduction to the creeds *see* Hsin ching tao yen (ccm221)

An introduction to the creeds and to the te deum / Burn, Andrew Eubank – London: Methuen, 1899 [mf ed 1989] – 1mf – 9 – 0-7905-4192-0 – (incl bibl ref) – mf#1988-0192 – us ATLA [240]

An introduction to the critical study and knowledge of the holy scriptures / Horne, Thomas Hartwell; ed by Ayre, John et al – 13th ed. London: Longmans, Green, 1872 [mf ed 1992] – 4v on 1mf – 9 – 0-524-02780-3 – (original ed 1818 in 2v. incl bibl ref) – mf#1987-6474 – us ATLA [220]

Introduction to the critical study of ecclesiastical history / Dowling, John Goulter – London: J G & F Rivington, 1838 [mf ed 1989] – 1mf – 9 – 0-7905-4405-9 – (with bibl ind of writers) – mf#1988-0405 – us ATLA [240]

Introduction to the devanagari script for students of sanskrit and hindi / Lambert, Hester Marjorie – London; New York: Oxford University Press, 1953 – (= ser Samp: indian books) – (foreword by j r firth) – us CRL [490]

Introduction to the devanagari script, for students of sanskrit, hindi, marathi, gujarati, and bengali / Lambert, Hester Marjorie – London; New York: Oxford University Press, 1953 – (= ser Samp: indian books) – (foreword by j r firth) – us CRL [490]

An introduction to the early history of christian doctrine : to the time of the council of chalcedon / Bethune-Baker, James Franklin – London: Methuen, 1903 [mf ed 1990] – 2mf – 9 – 0-7905-5924-2 – (incl bibl ref) – mf#1988-1924 – us ATLA [240]

An introduction to the federal probation system – Washington: FJC, 1976 – 3mf – 9 – $4.50 – mf#LLMC 95-823 – us LLMC [340]

An introduction to the fifth book of hooker's treatise of the laws of ecclesiastical polity / Paget, Francis – Oxford: Clarendon Press, 1899 [mf ed 1993] – 1mf – 9 – 0-524-08686-9 – (incl bibl footnotes) – mf#1993-3211 – us ATLA [242]

Introduction to the folk literature of mithila / Misra, Jayakanta – [Allahabad: Tirabhukti Publications, 1951?] – (= ser Samp: indian books) – us CRL [490]

The introduction to the gospel of john / Clarke, James Freeman – Boston: G H Ellis, 1890 – 1mf – 9 – 0-8370-2672-5 – mf#1985-0672 – us ATLA [226]

An introduction to the grammar of the kui or kandh language / Letchmajee, Lingum – 2nd rev ed. Calcutta: Bengal Secretariat Press, 1902 – 1 – us CRL [490]

Introduction to the hindustani language : in three parts, viz grammar, vocabulary, and reading lessons / Yates, William – 6th ed. Calcutta: Baptist Mission Press, 1855 [mf ed 1995] – (= ser Yale coll) – xiv/326p – 1 – 0-524-09477-2 – mf#1995-0477 – us ATLA [490]

Introduction to the history of architecture / Carpenter, Henry Barrett – London, England. 1936 – 1r – us UF Libraries [720]

Introduction to the history of central africa / Wills, Alfred John – London, England. 1964 – 1r – us UF Libraries [960]

Introduction to the history of central africa / Wills, Alfred John – London, England. 1967 – 1r – us UF Libraries [960]

An introduction to the history of educational theories / Browning, Oscar – Toronto: W Gage, 1886 [mf ed 1979] – 3mf – 9 – 0-665-00283-1 – (incl ind) – mf#00283 – cn Canadiana [370]

An introduction to the history of religion / Jevons, Frank Byron – London: Methuen; New York: Macmillan, 1896 [mf ed 1990] – 2mf – 9 – 0-7905-7651-1 – mf#1989-0876 – us ATLA [200]

An introduction to the history of sufism / Arberry, Arthur John – London, New York: Longmans, Green and Co, [1942] – (= ser Samp: indian books) – us CRL [260]

An introduction to the history of the assyrian church : or, the church of the sassanid persian empire, 100-640 a d / Wigram, William Ainger – London: SPCK; New York: ES Gorham, 1910 [mf ed 1986] – 1mf – 9 – 0-8370-8078-9 – (incl ind) – mf#1986-2078 – us ATLA [240]

An introduction to the history of the church of england : from the earliest times to the present day / Wakeman, Henry Offley & Ollard, Sidney Leslie – 8th ed. London: Rivingtons, 1914 [mf ed 1992] – (= ser Anglican/episcopal coll) – 2mf – 9 – 0-524-02907-5 – mf#1990-4498 – us ATLA [242]

An introduction to the history of the development of law / Morris, M F – Washington: John Byrne & Co, 1916 – 4mf – 9 – $6.00 – mf#LLMC 95-158 – us LLMC [340]

An introduction to the history of the successive revisions of the book of common prayer / Parker, James – Oxford: James Parker, 1877 [mf ed 1992] – (= ser Anglican/episcopal coll) – 2mf – 9 – 0-524-03500-8 – (incl bibl ref) – mf#1990-4722 – us ATLA [242]

Introduction to the johannine writings / Gloag, Paton James – London: James Nisbet, 1891 – 2mf – 9 – 0-8370-3311-X – (incl bibl ref & index) – mf#1985-1311 – us ATLA [227]

Introduction to the law of real property / Bigelow, Harry Augustus – 2nd ed. St. Paul, West, 1934. 95 p. LL-749 – 1 – us L of C Photodup [346]

An introduction to the life of jesus : an investigation of the historical sources / Anthony, Alfred Williams – Boston: Silver, Burdett, 1896 [mf ed 1985] – 1mf – 9 – 0-8370-2106-5 – (incl ind) – mf#1985-0106 – us ATLA [240]

An introduction to the literature of the new testament / Moffatt, James – New York: Charles Scribner, 1911 [mf ed 1986] – 2mf – 9 – 0-8370-9490-9 – (incl bibl & ind) – mf#1986-3490 – us ATLA [225]

An introduction to the literature of the old testament / Driver, Samuel Rolles – 9th rev ed. Edinburgh: T & T Clark; New York: Charles Scribner, 1913 [mf ed 1989] – (= ser International theological library (edinburgh, scotland)) – 2mf – 9 – 0-7905-0705-6 – (incl bibl ref & ind) – mf#1987-0705 – us ATLA [221]

Introduction to the massoretico-critical edition of the hebrew bible / Ginsburg, Christian David – London: Trinitarian Bible Society, 1897 – 3mf – 9 – 0-7905-1704-3 – (incl bibl ref and indexes) – mf#1987-1704 – us ATLA [221]

An introduction to the new testament / Bacon, Benjamin Wisner – New York: Macmillan, 1902, c1900 [mf ed 1986] – (= ser New testament handbooks) – 1mf – 9 – 0-8370-9524-7 – (incl bibl ref & ind) – mf#1986-3524 – us ATLA [225]

An introduction to the new testament : containing an examination of the most important questions relating to the authority, interpretation, and integrity of the canonical books... / Davidson, Samuel – London: Samuel Bagster, 1848-[51] [mf ed 1984] – (= ser Biblical crit – us & gb 66) – 3v on 19mf – 9 – 0-8370-1233-3 – (incl bibl ref & ind) – mf#1984-1066 – us ATLA [225]

An introduction to the new testament / Dods, Marcus – New York: Thomas Whittaker, 1888 [mf ed 1985] – (= ser The theological educator) – 1mf – 9 – 0-8370-2938-4 – mf#1985-0938 – us ATLA [225]

An introduction to the new testament = Einleitung in das neue testament / Bleek, Friedrich – Edinburgh: T & T Clark; New York: Scribner [dist] 1869-70 [mf ed 1989] – (= ser Clark's foreign theological library 4/24,26) – 3mf – 9 – 0-7905-1688-8 – (trans by william urwick. incl bibl ref & ind) – mf#1987-1688 – us ATLA [225]

Introduction to the new testament = Einleitung in das neue testament / Juelicher, Adolf – London: Smith, Elder, 1904 [mf ed 1989] – 2mf – 9 – 0-7905-1169-X – (english trans by janet penrose ward; pref note by mrs humphry ward; incl bibl ref & ind) – mf#1987-1169 – us ATLA [225]

Introduction to the new testament = Einleitung in das neue testament / Zahn, Theodor – Edinburgh: T & T Clark, 1909 – 4mf – 9 – 0-7905-3059-7 – (incl bibl ref and ind. in english) – mf#1987-3059 – us ATLA [225]

Introduction to the new testament : 1., the epistles of st. paul: particular introduction = Introduction au nouveau testament / Godet, Frederic Louis – Edinburgh: T & T Clark, 1894 – 2mf – 9 – 0-8370-9387-2 – (in english. incl bibl ref) – mf#1986-3387 – us ATLA [225]

Introduction to the new testament : the collection of the four gospels and the gospel of st. matthew = Introduction au nouveau testament, tome 2 / Godet, Frederic Louis – Edinburgh: T & T Clark, 1899 – 1mf – 9 – 0-8370-3320-9 – (in english) – mf#1985-1320 – us ATLA [225]

Introduction to the new testament / Michaelis, Johann David – 4th ed. London: Rivington, 1823. Beltsville, Md: NCR Corp, 1978 (30mf); Evanston: American Theol Lib Assoc, 1984 (30mf) – (= ser Biblical crit – us & gb) – 9 – 0-8370-0204-4 – (incl bibl ref) – mf#1984-1026 – us ATLA [225]

An introduction to the old testament : critical, historical, and theological... / Davidson, Samuel – Edinburgh: Williams & Norgate, 1862-63 [mf ed 1989] – 3v on 6mf – 9 – 0-7905-0642-4 – (incl bibl ref & ind) – mf#1987-0642 – us ATLA [221]

An introduction to the old testament / Wright, Charles Henry Hamilton – 2nd rev ed. New York: Thomas Whittaker [1891?] – (= ser The theological educator) – 1mf – 9 – 0-8370-5923-2 – (incl bibl) – mf#1985-3923 – us ATLA [225]

Introduction to the old testament = Einleitung in die goettlichen buecher des alten bundes / Jahn, Johann – New York: G & C Carvill, 1827 [mf ed 1989] – 2mf – 9 – 0-7905-0775-7 – (trans fr latin & german works; additonal ref & notes by samuel hulbeart turner & william rollinson whittingham; incl bibl ref & ind) – mf#1987-0775 – us ATLA [221]

An introduction to the old testament in greek / Swete, Henry Barclay – 1900 – 9 – $21.00 – us IRC [221]

An introduction to the old testament in greek / Swete, Henry Barclay – 2nd ed. Cambridge: University Press; New York: G P Putnam [dist] 1914 [mf ed 1986] – 2mf – 9 – 0-8370-9507-7 – (incl bibl & ind) – mf#1986-3507 – us ATLA [221]

Introduction to the pauline epistles / Gloag, Paton James – Edinburgh: T & T Clark, 1874 – 2mf – 9 – 0-8370-9948-X – (incl indes) – mf#1986-3948 – us ATLA [227]

An introduction to the pentateuch / Chapman, Arthur Thomas – Cambridge: University Press; New York: G P Putnam [dist] 1911 [mf ed 1986] – (= ser The cambridge bible for schools and colleges) – 1mf – 9 – 0-8370-6727-8 – (incl bibl ref & ind) – mf#1986-0727 – us ATLA [221]

Introduction to the pentateuch : an inquiry, critical and doctrinal, into the genuineness, authority, and design of the mosaic writings / Macdonald, Donald – Edinburgh: T & T Clark, 1861 – 3mf – 9 – 0-7905-1531-8 – (incl ind) – mf#1987-1531 – us ATLA [221]

An introduction to the philosophy of religion / Caird, John – new ed. New York: Macmillan, 1894 [mf ed 1985] – (= ser Croall lectures 1878-1879) – 1mf – 9 – 0-8370-2564-8 – mf#1985-0564 – us ATLA [110]

An introduction to the philosophy of sri aurobindo / Maitra, Susil Kumar – Calcutta: Culture Publ, 1941 – (= ser Samp: indian books) – us CRL [180]

Introduction to the phonology of the bantu languages / Meinhof, Carl – Berlin, Germany. 1932 – 1r – us UF Libraries [470]

Introduction to the purva mimamsa / Shastri, Pashupatinath – Calcutta: Ashoke Nath Bhattacharya, 1923 – (= ser Samp: indian books) – us CRL [490]

Introduction to the sacred scriptures : in two parts / MacDevitt, John – 2nd ed. Dublin: Sealy, Bryers & Walker; New York: Benziger, 1895 – 1mf – 9 – 0-8370-9883-1 – mf#1986-3883 – us ATLA [220]

Introduction to the science of chinese religion : a critique of max mueller and other authors / Faber, Ernest – Hong Kong: Lane, Crawford; Shanghai: Kelly & Walsh [1879] [mf ed 1995] – (= ser Yale coll) – xii/154p – 1 – 0-524-09403-9 – mf#1995-0403 – us ATLA [290]

Introduction to the science of chinese religion : a critique of max mueller and other authors / Faber, Ernst – Hongkong: Lane, Crawford & Co; Shanghai: Presbyterian Mission Press, 1879 – (= ser 19th c books on china) – 2mf – 9 – mf#7.1.21 – uk Chadwyck [290]

Introduction to the science of language / Sayce, Archibald Henry – 4th ed. London: Kegan Paul, Trench, Truebner, 1900 – 3mf – 9 – 0-7905-2374-4 – (includes bibliographies and index) – mf#1987-2374 – us ATLA [400]

Introduction to the science of law : systematic survey of the law and principles of legal study / Gareis, Karl – 3rd rev German ed. N.Y.: The Macmillan Co, 1911 (reprint 1924) – 5mf – 9 – $7.50 – mf#LLMC 95-187 – us LLMC [340]

Introduction to the science of religion : four lectures...royal institution; with two essays, on false analogies and the philosophy of mythology / Mueller, Friedrich Max – London: Longmans, Green, 1873 – 1mf – 9 – 0-524-00943-0 – mf#1990-2166 – us ATLA [200]

Introduction to the second edition of the bampton lectures of the y... / Hambden, Renn Dickson – London, England. 1837 – 1r – us UF Libraries [240]

An introduction to the singing of psalm-tunes : in a plain and easy method / Tufts, John – With a collection of tunes in three parts. 10th ed. Boston: Samuel Gerrish, 1738. MUSIC 1152 – 1 – us L of C Photodup [780]

Introduction to the skill of musick in three books / Playford, John – 10th corr enl ed, London: printed by A G[odbid] & J[ohn] P[layford the younger] for John Playford 1683 [mf ed 19–] – 1r / 3mf – 1,9 – (the first contains the grounds and rules of musick, according to the gam-ut, and other principles thereof. the second: instrvctions and lessons both for the bass-viol and treble-violin. the third, the art of descant, or composing of musick in parts, in a more plain and easie method than any heretofore published) – mf#film 240 / fiche 643 – us Sibley [780]

Introduction to the solution of the problems of the pyramid / Davie, John G – Griffin, GA, printed in U.S.A., 1934. pt.1-8 in 1v. 922p – 1 – us Wisconsin U Libr [150]

An introduction to the study and collection of ancient prints / Willshire, William Hughes – [2nd ed]. London 1877 – (= ser 19th c art & architecture) – 9mf – 9 – mf#4.2.1532 – uk Chadwyck [760]

An introduction to the study and use of the psalms / Thrupp, Joseph Francis – Cambridge: Macmillan, 1860 [mf ed 1990] – 2v on 2mf – 9 – 0-8370-1678-9 – mf#1987-6106 – us ATLA [221]

Introduction to the study of african languages = Moderne sprachforschung in afrika / Meinhof, Carl – London: JM Dent, 1915 [mf ed 1992] – 1mf – 9 – 0-524-04140-7 – (english trans by a werner) – mf#1990-1210 – us ATLA [470]

Introduction to the study of buddhism : according to material preserved in japan and china / Rozenberg, Otton Ottonovich – Tokyo: Faculty of Oriental Languages of the Imperial University of Petrograd 1916 [mf ed 1991] – 1v on 6mf – 9 – 0-524-02048-5 – (text in chinese, japanese & sanskrit; int in english & japanese; no more publ?) – mf#1990-2823 – us ATLA [052]

An introduction to the study of christian apologetics / Gray, Arthur Romeyn – Sewanee, TN: University Press at the University of the South, c1912 [mf ed 1990] – (= ser Sewanee theological library) – 1mf – 9 – 0-7905-3849-0 – (incl bibl ref) – mf#1989-0342 – us ATLA [240]

An introduction to the study of comparative religion / Jevons, Frank Byron – New York: Macmillan, 1908 [mf ed 1990] – (= ser The hartford-lamson lectures on the religions of the world) – 1mf – 9 – 0-7905-7792-5 – mf#1989-1017 – us ATLA [230]

An introduction to the study of dogmatic theology / Owen, Robert – London: Joseph Masters, 1858 [mf ed 1993] – 2mf – 9 – 0-524-08550-1 – (incl bibl ref) – mf#1993-2075 – us ATLA [240]

An introduction to the study of efforts at christian reunion / Bouquet, Alan Coates – Cambridge: W Heffer, 1914 [mf ed 1990] – 1mf – 9 – 0-7905-3545-9 – (incl bibl ref) – mf#1989-0038 – us ATLA [240]

An introduction to the study of hinduism / Sen, Guru Prosad – Calcutta: Thacker, Spink, 1893 [mf ed 1991] – 1mf – 9 – 0-524-01576-7 – mf#1990-2530 – us ATLA [280]

Introduction to the study of history = Introduction aux etudes historiques / Langlois, Charles Victor & Seignobos, Charles – New York: H. Holt, [1898?] – 1mf – 9 – 0-7905-5364-3 – (incl bibl ref. in english) – mf#1988-1364 – us ATLA [900]

An introduction to the study of indian economics / Kale, Vaman Govind – Poona: Arya Bhushan Press, 1930 – (= ser Samp: indian books) – us CRL [339]

Introduction to the study of indian languages with words, phrases, and sentences to be collected / Powell, J W – Washington: GPO. 1877 – 1 – us Kansas [490]

Introduction to the study of indian music : an attempt to reconcile modern hindustani music with ancient musical theory and to propound an accurate and comprehensive method of treatment of the subject of indian musical intonation / Clements, Ernest – London; New York: Longmans, Green, and Co, 1913 – (= ser Samp: indian books) – us CRL [780]

Introduction to the study of integral equations / Bocher, Maxime – Cambridge, England. 1914 – 1r – us UF Libraries [510]

Introduction to the study of law / Morgan, Edmund M – Chicago: Callaghan 1926 – 2mf – 9 – $3.00 – mf#llmc92-209 – us LLMC [340]

An introduction to the study of mediaeval indian sculpture / Codrington, Kenneth de Burgh – London: Edward Goldston, 1929 – (= ser Samp: indian books) – us CRL [730]

Introduction to the study of mortuary customs among north american indians / Yarrow, Henry Crecy – Washington. 1880 – 1 – us CRL [390]

An introduction to the study of new testament greek / Moulton, James Hope – London: Charles H Kelly, 1895 [mf ed 1986] – (= ser Books for bible students) – 1mf – 9 – 0-8370-9298-1 – (incl ind) – mf#1986-3298 – us ATLA [450]

An introduction to the study of obadiah / Peckham, George A – Chicago: University of Chicago Press, 1910 [mf ed 1986] – 1mf – 9 – 0-8370-9811-4 – (in english & hebrew) – mf#1986-3811 – us ATLA [221]

An introduction to the study of painted glass / Winston, Charles – Oxford 1849 – (= ser 19th c art & architecture) – 1mf – 9 – mf#4.1.254 – uk Chadwyck [740]

An introduction to the study of philosophy : with an outline treatise on logic / Gerhart, Emanuel Vogel – Philadelphia: Lindsay & Blakiston, 1858, c1857 [mf ed 1991] – 1mf – 9 – 0-524-00266-5 – mf#1989-2966 – us ATLA [100]

Introduction to the study of philosophy / Harris, William Torrey – New York: D Appleton, 1889 – 1mf – 9 – 0-7905-3943-8 – mf#1989-0436 – us ATLA [100]

Introduction to the study of philosophy / Stuckenberg, John Henry Wilbrandt – New York: AC Armstrong, 1888 – 1mf – 9 – 0-7905-8924-9 – (incl bibl ref) – mf#1989-2149 – us ATLA [100]

An introduction to the study of roman law / Cushing, Luther Stearns – Boston Little: Brown, 1854 – 3mf – 9 – $4.50 – mf#LLMC 95-198 – us LLMC [340]

An introduction to the study of the acts of the apostles / Stifler, James M – New York: Fleming H Revell, c1892 [mf ed 1985] – 1mf – 9 – 0-8370-5420-6 – mf#1985-3420 – us ATLA [226]

An introduction to the study of the books of the new testament / Kerr, John Henry – Chicago: Fleming H Revell, c1892 [mf ed 1985] – 1mf – 9 – 0-8370-3890-1 – (incl ind. int note by benjamin b warfield) – mf#1985-1890 – us ATLA [225]

An introduction to the study of the chaldee language : comprising a grammar (based upon winer's), and an analysis of the text of the chaldee portion of the book of daniel / Longfield, George – London: Whittaker; Dublin: Hodges, Smith, 1859 [mf ed 1986] – 1mf – 9 – 0-8370-9165-9 – (incl bibl & ind) – mf#1986-3165 – us ATLA [470]

Introduction to the study of the decisions of the supreme court of ohio / Rockel, William Mahlon – Norwalk, Ohio: Laning 1902. 75p. LL-618 – 1 – us L of C Photodup [347]

Introduction to the study of the dependent, defective, and delinquent classes : and of their social treatment / Henderson, Charles Richmond – 2nd ed., enl. and rewritten. Boston, U.S.A.: D.C. Heath, 1901 – 1mf – 9 – 0-7905-5896-3 – (incl bibl ref) – mf#1988-1896 – us ATLA [360]

Introduction to the study of the gospels : with historical and explanatory notes / Westcott, Brooke Foss – Boston: Gould and Lincoln, 1862 – (= ser Biblical crit – us & gb) – 9 – 0-8370-0248-6 – mf#1984-1057 – us ATLA [225]

Introduction to the study of the greek dialects : grammar, selected inscriptions, glossary / Buck, Carl Darling – Boston: Ginn, c1910 – 1mf – 9 – 0-8370-9209-4 – (incl ind) – mf#1986-3209 – us ATLA [450]

Introduction to the study of the hindu doctrines / Guenon, Rene – London: Luzac & Co, 1945 – (= ser Samp: indian books) – (trans by marco pallis) – us CRL [280]

Introduction to the study of the holy scriptures / Harman, Henry Martyn – [4th ed.] New York: Phillips & Hunt; Cincinnati: Cranston & Stowe, 1884 – (= ser Library Of Biblical And Theological Literature) – 2mf – 9 – 0-8370-1669-X – (incl bibl ref) – mf#1987-6099 – us ATLA [220]

An introduction to the study of the medieval bengali epics / Bhattacharya, Asutosh – Calcutta: Calcutta Book House, 1943 – (= ser Samp: indian books) – us CRL [490]

An introduction to the study of the middle ages (375-814) / Emerton, Ephraim – Boston: Ginn, 1888 [mf ed 1989] [mf ed 1989] – 1mf – 9 – 0-7905-4294-3 – (incl bibl ref) – mf#1988-0294 – us ATLA [931]

An introduction to the study of the new testament : critical, exegetical, and theological / Davidson, Samuel – 2d rev impr ed. London: Longmans, Green, 1882 [mf ed 1984] – (= ser Biblical crit – us & gb 67) – 2v on 13mf – 9 – 0-8370-1158-2 – (incl bibl ref & ind) – mf#1984-1067 – us ATLA [225]

Introduction to the study of the old testament = Einleitung ins alte testament. selections / Eichhorn, Johann Gottfried – [s.l]: printed for private circulation, 1888 [mf ed 1991] – 1mf – 9 – 0-7905-8300-3 – (fragment trans by george tilly gollop) – mf#1987-6405 – us ATLA [221]

Introduction to the study of the old testament : part the first / Barry, Alfred – London: John W Parker, 1856 – 1mf – 9 – 0-7905-0856-7 – (incl ind. no more published) – mf#1987-0856 – us ATLA [221]

An introduction to the study of the relations of indian states with the government of india : with illustrative documents and appendices / Panikkar, Kavalam Madhava – London: Martin Hopkinson & Co, 1927 – (= ser Samp: indian books) – us CRL [954]

An introduction to the study of the roman law / Cushing, Luther Stearns – Boston, Little, Brown, 1854. 243 p. LL-4095 – 1 – us L of C Photodup [340]

An introduction to the study of the scriptures / Carpenter, William Boyd – London: J M Dent; Philadelphia: J B Lippincott 1902 [mf ed 1984] – (= ser The temple bible) – 2v on 1mf – 9 – 0-7905-3319-7 – mf#1987-3319 – us ATLA [220]

An introduction to the study of theravada buddhism in burma : a study in indo-burmese historical and cultural relations from the earliest times to the british conquest / Ray, Niharranjan – Calcutta: University of Calcutta, 1946 – (= ser Samp: indian books) – us CRL [280]

An introduction to the study of universal history : two dissertations / Stoddart, John – 2nd ed. London: John J Griffin, 1850 [mf ed 1992] – (= ser Encyclopaedia metropolitana 3rd div. history and biography) – 1mf – 9 – 0-524-04625-5 – (incl bibl ref) – mf#1990-1285 – us ATLA [900]

Introduction to the synoptic gospels / Gloag, Paton James – Edinburgh: T & T Clark, 1895 – 1mf – 9 – 0-8370-3312-8 – (incl indes) – mf#1985-1312 – us ATLA [226]

An introduction to the textual criticism of the new testament / Warfield, Benjamin Breckenridge – Toronto: S R Briggs, 1887 – (= ser The theological educator) – 3mf – 9 – mf#26194 – cn Canadiana [225]

An introduction to the textual criticism of the new testament / Warfield, Benjamin Breckinridge – 5th ed. New York: Thomas Whittaker, [1886?] [mf ed 1985] – (= ser The theological educator) – 1mf – 9 – 0-8370-5708-6 – mf#1985-3708 – us ATLA [225]

An introduction to the theology of the church of england : in an exposition of the thirty-nine articles / Boultbee, Thomas Pownall – London: Longmans, Green, 1871 [mf ed 1986] – 1mf – 9 – 0-8370-8651-5 – (exposition in english; text in english & latin) – mf#1986-2651 – us ATLA [242]

An introduction to the thessalonian epistles : containing a vindication of the pauline authorship of both epistles and an interpretation of the eschatological section of 2 thess ii / Askwith, Edward Harrison – London, New York: Macmillan, 1902 [mf ed 1989] – 1mf – 9 – 0-7905-3002-3 – mf#1987-3002 – us ATLA [227]

Introduction to the three middle books of the pentateuch = Die buecher exodus, leviticus, numeri / Lange, Johann Peter – New York: Charles Scribner, c1876 [mf ed 1985] – (= ser A commentary on the holy scriptures. old testament 2/1) – 1mf – 9 – (english trans by howard osgood. incl bibl) – mf#1985-3036 – us ATLA [221]

Introduction to the yoruba language / Ward, Ida Caroline – Cambridge, England. 1952 – 1r – us UF Libraries [470]

An introduction to theology : its principles, its branches, its results, and its literature / Cave, Alfred – 2nd ed. Edinburgh: T & T Clark, 1896 [mf ed 1989] – 2mf – 9 – 0-7905-0918-0 – (incl bibl & ind) – mf#1987-0918 – us ATLA [240]

Introduction to theosophy / Besant, Annie Wood – London, England. 1894 – 1r – us UF Libraries [240]

An introductory address : delivered before the law class of transylvania university, on the 9th of nov 1839 / Marshall, Thomas Alexander – Lexington, Ky: Finnell & Virden, 1839. 16,3p. LL-462 – 1 – us L of C Photodup [340]

Introductory catechism *see* Chi-tu t'u chin pu wen ta [ccm344]

Introductory chemistry : suitable for use in lower schools and continuation classes / Ellis, William S – Toronto: Copp, Clark, c1904 [mf ed 1996] – 1mf – 9 – 0-665-80517-9 – mf#80517 – cn Canadiana [540]

Introductory essay to bishop horne's commentary on the book of psal... / Irving, Edward – London, England. 1859 – 1r – us UF Libraries [240]

An introductory hebrew grammar : with progressive exercises in reading and writing / Davidson, Andrew Bruce – 18th ed. Edinburgh: T & T Clark, 1909 [mf ed 1986] – 1mf – 9 – 0-8370-9222-1 – (incl english-hebrew & hebrew-english vocabularies) – mf#1986-3222 – us ATLA [470]

Introductory hebrew method and manual / Harper, William Rainey – 4th ed. New York: Charles Scribner, 1888, c1886 – 1mf – 9 – 0-8370-9152-7 – mf#1986-3152 – us ATLA [470]

Introductory hints to english readers of the old testament / Cross, John A – London: Longmans, Green, 1882 – 1mf – 9 – 0-8370-3410-8 – mf#1985-1410 – us ATLA [221]

Introductory lecture delivered at the opening of the class of moral... / Flint, Robert – Edinburgh, Scotland. 1864 – 1r – us UF Libraries [240]

Introductory lecture delivered at the opening of the second session of the medical faculty of the university of bishop's college, october 2nd, 1872 / Campbell, Francis Wayland – Montreal?: J Lovell, 1872 – 1mf – 9 – mf#05499 – cn Canadiana [378]

An introductory lecture on the study of ecclesiastical history : delivered...oxford on apr 23 1885 / Hatch, Edwin – London: Rivingtons, 1885 [mf ed 1990] – 1mf – 9 – 0-7905-6752-0 – mf#1988-2752 – us ATLA [240]

Introductory lecture upon the study of theology and of the greek te... / Dale, Thomas – London, England. 1829 – 1r – us UF Libraries [240]

Introductory lecture, winter 1853-4 / Smeaton, George – Aberdeen, Scotland. 1853 – 1r – us UF Libraries [240]

An introductory lecture...on the principles and practice of architecture / Hosking, William – London 1842 – (= ser 19th c art & architecture) – 1mf – 9 – mf#4.2.1751 – uk Chadwyck [720]

Introductory lectures delivered at the opening of the english presb... / Lorimer, Peter – London, England. 1845 – 1r – us UF Libraries [242]

Introductory lectures on the study of christian theology : with outlines of lectures on the doctrines of christianity / Hannah, John – 2nd ed. London: Wesleyan Conference Office, 1875 – 1mf – 9 – 0-8370-5072-3 – (incl bibl ref) – mf#1985-3072 – us ATLA [240]

Introductory lessons on christian evidences / Whately, Richard; ed by Tefft, Benjamin Franklin – Cincinnati: Poe & Hitchcock...1864 [mf ed 1985] – 1mf – 9 – 0-8370-5819-8 – mf#1985-3819 – us ATLA [240]

Introductory lessons on india and missions : for mission study classes – 2nd ed. London, Madras: Christian Literature Society [1909] [mf ed 1995] – (= ser Yale coll) – 109p (ill) – 1 – 0-524-09016-5 – mf#1995-0016 – us ATLA [240]

Introductory physiology and hygiene : a series of lessons in four parts, designed for use in the first four forms of the public schools / Knight, Archibald Patterson – Toronto: Copp, Clark, c1905 [mf ed 1995] – 3mf – 9 – 0-665-73598-7 – mf#73598 – cn Canadiana [613]

Introductory sketch of the bantu languages / Werner, Alice – London, England. 1919 – 1r – us UF Libraries [470]

Introductory studies in german literature / Hochdoerfer, Richard – Chautauqua, NY: The Chautauqua Press, 1904 – 1r – 1 – us Wisconsin U Libr [430]

Introductory studies in Greek art / Harrison, Jane Ellen – 2nd ed. London 1892 – (= ser 19th c art & architecture) – 4mf – 9 – mf#4.2.1397 – uk Chadwyck [700]

An introductory study of ethics / Fite, Warner – New York: Longmans, Green, 1906, c1903 [mf ed 1986] – 1mf – 9 – 0-8370-6043-5 – (incl bibl ref & ind) – mf#1986-0043 – us ATLA [170]

Introductory syriac method and manual / Wilson, Robert Dick – New York: Charles Scribner, 1891 – 1mf – 9 – 0-8370-7677-3 – mf#1986-1677 – us ATLA [470]

An introductory treatise on sanscrit hagiographa : or, the sacred literature of the hindus / Wrightson, Richard – Dublin: McGlashan & Gill, 1859 [mf ed 1992] – 1mf – 9 – 0-524-02878-8 – (incl bibl ref) – mf#1990-3151 – us ATLA [280]

Introduzione ad un nuovo sistema d'armonia / Basevi, Abramo – Firenze: Tip Tofani 1862 [mf ed 1993] – 1r – 1 – mf#pres. film 127 – us Sibley [780]

Introdvctio in lingvam arabicam... / Radtmann, B – Francofvrti, 1588 – 1mf – 9 – mf#H-8410 – ne IDC [956]

Intruduccion al estuido de las lenguas indigenas d... / Alba C, Manuel Maria – Panama, Panama. 1950 – 1r – us UF Libraries [972]

O intrujao : orgao moralisador – Rio de Janeiro, RJ. 04 dez 1883 – (= ser Ps 19) – mf#P19A,04,105 – bl Biblioteca [320]

Intruso en el jardin de academo / Llorens, Washington – San Juan, Puerto Rico. 1957 – 1r – us UF Libraries [972]

L'intuition philosophique / Bergson, H – Paris, 1927 – €5.00 – ne Slangenburg [140]

The intuitions of the mind, inductively investigated / McCosh, James – 3rd ed, rev. New York: R Carter, 1872 – 2mf – 9 – 0-7905-7531-0 – mf#1989-0756 – us ATLA [120]

A inubia : jornal de ensaios litterarios – Rio de Janeiro, RJ: Typ de Domingos Luiz dos Santos, 13 set 1871 – (= ser Ps 19) – mf#P19A,04,106 – bl Biblioteca [440]

Inuit Development Corporation *see* Idc report suvaguuq

Inuit monthly – Canada. jan 1973-dec 1976 – 3r – 1 – cn Commonwealth Imaging [971]

Inuit Tapirisat of Canada *see* ITC news

Inukshuk – 1973 mar 16-1974 mar 27, 1974 apr 3-1975 apr 30, 1975 may-1976 may, 1976 jun 18-22 – 4r – 1 – mf#1059929 – us WHS [071]

Inuktituorutit : grammaire purement esquimaude / Schneider, Lucien – Quebec: Centre de Documentation...1978 i.e. 1979 [mf ed 1988] – 2mf – 9 – (in inuit and french) – mf#SEM105P936 – cn Bibl Nat [490]

Inuktituorutit : grammaire purement esquimaude / Schneider, Lucien – Quebec: Ministere des Richesses naturelles...1972 [mf ed 1988] – 8mf – 9 – mf#SEM105P938 – cn Bibl Nat [490]

Inummarit / Inummarit Cultural Association – [v1]-v3 n1 [1972 mar-1976] – 1r – 1 – mf#281469 – us WHS [366]

Inummarit Cultural Association *see* Inummarit

Inutil combate / Martinez Sobral, Enrique – Guatemala, 1902 – 1r – us UF Libraries [972]

Inutil combate : paginas de la vida / Martinez Sobral, Enrique – Guatemala, 1957 – 1r – us UF Libraries [972]

Inuvialuit / Committee for Original Peoples Entitlement – 1981 spr-1985 jul – 1r – 1 – (cont: akana; cont by: tusaayaksat) – mf#965360 – us WHS [322]

Inuvik drum – Inuvik, Northwest Territories. 1982 jul [v17 n26]-1983 dec, 1984 jan 5-sep 27 [v19 n39] – 2r – 1 – (cont: drum [inuvik, nwt]) – mf#703266 – us WHS [071]

Der invalide : historisch-romantische bilder neuerer zeit / Spindler, Carl – Gera: C B Griesbach 1897 [mf ed 1995] – 1r – 1 – (filmed with: der jesuit) – mf#3746p – us Wisconsin U Libr [830]

Invalidez de dos poderes / Martinez Villasmil, Antonio – Caracas, Venezuela. 1964 – 1r – us UF Libraries [972]

Invalid's help to prayer and meditation / Hannam, E P – London, England. 1838 – 1r – us UF Libraries [240]

Invanga / Summers, Roger – Cambridge, England. 1958 – 1r – us UF Libraries [960]

Invasao de mato grosso / Guimaraes, Jorge Maia De Oliveira – Rio de Janeiro, Brazil. 1964 – 1r – us UF Libraries [972]

Invasion – Miami, FL. 1971 nov 13-1974 jul 03 – 1r – (1971 nov 20) – us UF Libraries [071]

Invasion du canada : collection de memoires / Verreau, Hospice Anthelme Baptiste – Montreal: E Senecal, 1873 – 5mf – 9 – (with ind) – mf#06266 – cn Canadiana [971]

Invasion inglesa de 1655 / Rodriguez Demorizi, Emilio – Ciudad Trujillo, Dominican Republic. 1957 – 1r – us UF Libraries [972]

The invasion of the crimea / Kinglake, Alexander – v1-6. 1875-88 – 1 – $120.00 – mf#0312 – us Brook [949]

Invasiones de colombia a venezuela en 1901, 1902 y / Landaeta Rosales, Manuel – Caracas, Venezuela. 1903 – 1r – us UF Libraries [972]

Invasions of india from central asia – London, 1879 – (= ser 19th c british colonization) – 4mf – 9 – mf#1.1.6028 – uk Chadwyck [954]

Inventaire chronologique / Dionne, Narcisse Eutrope – Quebec: [s.n.] 4v. 1905-1909 [mf ed 1985] – 10mf – 9 – mf#SEM105P507 – cn Bibl Nat [010]

Inventaire chronologique des livres, brochures, journaux et revues publies en diverses langues dans et hors la province de quebec / Dionne, Narcisse Eutrope – Quebec: [s.n.], 1912 [mf ed 1985] – 1mf – 9 – (with ind) – mf#SEM105P508 – cn Bibl Nat [010]

Inventaire des archives des affaires etrangeres de l'etat independant du congo et du ministere des colonies, 1885-1914 – Bruxelles, 1955 – us CRL [327]

Inventaire des documents provenant de la mission frantz cornet / Musee Royal De L'afrique Centrale – Bruxelles, Belgium. 1960 – 1r – us UF Libraries [960]

Inventaire des instruments de recherche : manuscripts occidentaux / Bibliotheque nationale departement des manuscrits. Departement des Manuscrits – [mf ed Chadwyck-Healey] – 2632mf – 9 – uk Chadwyck [090]

Les inventaires des archives nationales de paris / Paris. Archives Nationales – [mf ed Chadwyck-Healey] – 773 inventories on 8388mf – 9 – (incl suppl of 73 inventories on 2047mf. separate sects available: section ancienne [pre-1789] 3082mf. section moderne [1789-1940] et contemporaine [since 1940] 1830mf. archives privees 1429mf. la revolution francaise 1104mf. les beaux-arts 378mf. with p/g & ind) – uk Chadwyck [020]

Inventario de...museo...badajaz / Romero de Castilla, Tomas – 1896 – 9 – sp Bibl Santa Ana [900]

Inventario general de ragistros cedularios del archivo general de indias de sevilla... / Rubio Moreno, Luis – Madrid: Razon y Fe, 1929 – 1 – sp Bibl Santa Ana [350]

Inventarios e testamentos / Brazil. Sao Paulo. Departamento do Archivo do Estado – v.1-42. 1920-73 – 1 – us CRL [972]

Inventing nz surveying, science and the construction of cultural space – 1840's-1899's – 1r – 1 – mf#ZB 2 – nz Nat Libr [079]

Invention in selected sermons fo ministers opposing the election... / Walker, David Ellis – S.l., S.l? . 1961 – 1r – us UF Libraries [025]

The invention of a new religion / Chamberlain, Basil Hall – London: Watts, 1912 [mf ed 1992] – 1mf – 9 – 0-524-03303-X – mf#1990-3188 – us ATLA [290]

The invention of printing : a series of four lectures delivered in the lent term of 1897 / Middleton-Wake, Charles Henry – London, 1897 – (= ser 19th c publishing...) – 3mf – 9 – mf#3.1.53 – uk Chadwyck [680]

Inventiones mathematicae – Heidelberg. 1966+ (1,5,9) – ISSN: 0020-9910 – mf#13185 – us UMI ProQuest [510]

Invento ceres o sea metodo de proceder... propio por diez anos / Alvarez Guerra, Andres – 1827 – 9 – sp Bibl Santa Ana [999]

Inventories of the houghton manucript collection / Harvard University. Houghton Library – 331mf – 9 – £1,490.00 – uk Chadwyck [090]

Inventories of the houghton manuscript collection : medieval and renaissance manuscripts – 3000 BC-present [mf ed Chadwyck-Healey] – 331mf – 9 – uk Chadwyck [090]

Inventories of the manuscript collections / South African Library, Cape Town – Cape Town, South African Library, [19–?] – us CRL [090]

An inventory for assessment of attitudes of high school students toward health-realted physical fitness / Blackwell, E B – 1990 – 1mf – 9 – $4.00 – us Kinesology [150]

Inventory management report – New York. 2001+ (1,5,9) – mf#23049,01 – us UMI ProQuest [650]

Inventory of holdings of certain classes of materials / Midwest Inter-Library Center. Chicago – 1952. Rev. 1957 – 1 – us CRL [020]

Inventory of the church archives of tennessee / Nashville. Tennessee. Church Archives – Also includes: Nashville Baptist Assoc., 1939; Oconee Baptist Assoc., 1942; Guide to Church Vital Statistics in Tenn., 1942; Guide to Vital Statistics in the Church Records of Conn., 1942. 1392p – 1 – us Southern Baptist [242]

Inventory of the contents of holkham, 1774 – (= ser Holkham library, the house, park & art colls) – 1r – 1 – mf#775 – uk Microform Academic [025]

Inventory of uganda documentary materials – Madison: Collection Maintenance Office, Memorial Library, University of Wisconsin, 1981 – us CRL [020]

Inventory of unpublished material for american religious history in protestant church archives and other repositories / Allison, William Henry – 1910 – 1 – 9.59 – us Southern Baptist [242]

Inventory of unpublished material for american religious history in protestant church archives and other repositories / Allison, William Henry – Washington, DC: Carnegie Institution of Washington, 1910 – 1mf – 9 – 0-7905-4309-5 – mf#1988-0309 – us ATLA [975]

Invento.tercer cuaderno. inventos de ceres – 1828 – 9 – sp Bibl Santa Ana [890]

Inverell argus – Inverell, jan 1899-dec 1904 – 1r – A$93.59 vesicular A$99.09 silver – at Pascoe [079]

Inverell times – Inverell, 1938-65 – 73r – at Pascoe [079]

Inverell times – Inverell, jan 1899-dec 1937 – 36r – A$1991.84 vesicular A$2189.84 silver – at Pascoe [079]

Inverell times – Inverell, jan 1969-jun 1997 – at Pascoe [079]

Invergordon times and general advertiser – [Scotland] Invergordon: H Graham jan 1916-jan 1917 (wkly) [mf ed 2003] – 1r – 1 – uk Newsplan [072]

Inverness advertiser etc – Scotland, UK. 19 Jan 1849-1850; 1851-85. -w – 35 1/2r – 1 – uk British Libr Newspaper [072]

Inverness citizen – [Scotland] Inverness: D Grant 6 jul 1922-20 may 1926 (wkly) [mf ed 2003] – 5r – 1 – (cont: highland leader and northern weekly; absorbed by: inverness courier, and general advertiser for the counties of inverness, ross, moray, nairn, cromarty, sutherland, and caithness) – uk Newsplan [072]

Inverness courier – Inverness: R Carruthers & Sons Ltd 1990- (semiwkly) [mf ed 3 jan 1995-] – 1 – (cont: inverness courier, and general advertiser for the counties of inverness, ross, moray, nairn, cromarty, sutherland, and caithness) – ISSN: 0020-9929 – uk Scotland NatLib [072]

Inverness courier – Scotland. -w. 1870-90. Lacking 1885 – 25 1/2r – 1 – uk British Libr Newspaper [072]

Inverness football times – [Scotland] Inverness: A Bain may 1904-dec 1950 (wkly) [mf ed 2003] – 54r – 1 – (cont by: football times [jan 1908-dec 1933]; football times and highland times [jan 1934-dec 1946]; football times and highland news [jan 1947-dec 1948]; football times [jan 1949-dec 1950]) – uk Newsplan [790]

Inverness journal and northern advertiser – Scotland, UK. 3 Apr 1840; 5 Jan 1844-26 Dec 1845; 2 Jan 1856-30 Jun 1848. -w. 2 reels – 1 – uk British Libr Newspaper [072]

Inverness railway, cape breton : report to h n paint...controlling the broad cove coal mines of the exploratory and preliminary surveys from orangedale station... / Hyndman, P K – [Ottawa?: s.n.], 1890 [mf ed 1980] – 1mf – 9 – mf#07170 – cn Canadiana [380]

Invernessian : an independent journal – [Scotland] Inverness: A & W Mackenzie 30 oct 1880-aug 1881 (mthly) [mf ed 2003] – 1r – 1 – (subtitle varies; not publ: feb 1881) – uk Newsplan [072]

Invertebrate reproduction and development – Rehovot. 1989+ (1) – ISSN: 0792-4259 – mf#16922,02 – us UMI ProQuest [590]

Inverurie advertiser : a member of the advertiser series in aberdeenshire – Turiff: W Peters & Son Ltd 1979- (wkly) [mf ed 1 jul 1994-] – 1 – (cont: inverurie and district advertiser) – uk Scotland NatLib [072]

Inverurie and district advertiser – Scotland, UK. 8 May-19 Jun, 14 Aug-Dec 1959; 1960-81. -w. 23 1/2 reels – 1 – uk British Libr Newspaper [072]

Inverurie herald : serving town and country – Forfar: The Angus County Press Ltd (wkly) [mf ed 1 jul 1994-] – 1 – uk Scotland NatLib [072]

Investigacion de la naturaleza y causas de la riqueza de las naciones / Smith, Adam – Valladolid, 1794 – 31mf – 9 – sp Cultura [946]

Investigacion industrial / Honduras. Direccion General de Estadistica y Censos – 1953-58, 60, 62 – 1 – us L of C Photodup [338]

Investigaciones historicas / Davila, Vicente – Quito, Ecuador. v1-2. 1955 – 1r – us UF Libraries [972]

Investigaciones historicas. tomo 2. caracas, 1927 / Davila, Vicente – Madrid: Razon y Fe, 1930 – 1 – sp Bibl Santa Ana [946]

Investigaciones...banos de montemayor y bejar / Martinez Serrano, Francisco – 1843. Cuarta memoria – 9 – (quinta memoria 1843) – sp Bibl Santa Ana [946]

Investigating electronic commerce activities of the tourism industry in south africa / Krüger, Janine – Pretoria: Vista University 2003 [mf ed 2003] – 4mf – 9 – (incl bibl ref) – mf#mfm15260 – sa Unisa [338]

Investigating the delivery of therapeutic recreation services on the internet : a pilot study using leisure education for the prevention of alcohol abuse / Mainville, Sylvie – 1998 – 2mf – 9 – $8.00 – mf#RC 524 – us Kinesology [615]

Investigating the japanese sports travel market : a comparison of golf and ski travelers / Lee, Seonbok – 1999 – 2mf – 9 – $8.00 – mf#PE 4040 – us Kinesology [650]

Investigation and classification of non-functional harmonic music / Sherburn, Merrell L – U of Rochester 1959 [mf ed 19–] – 2v on 1r – 1 – (with bibl) – mf#film 749 – us Sibley [780]

Investigation and trial papers relating to the assasination of president lincoln / U.S. Army. Judge Advocate General – (= ser Records of the office of the judge advocate general (army)) – 16r – 1 – (with printed guide) – mf#M599 – us Nat Archives [976]

An investigation comparing the effect of different resistance levels on power production / Rash, David G – 1998 – 1mf – 9 – $4.00 – mf#PH 1690 – us Kinesology [612]

An investigation into accuracy of using rpe to monitor intensity during spinning / John, Deborah – 1998 – 1mf – 9 – $4.00 – mf#PH 1651 – us Kinesology [612]

An investigation into linkages between the formal and informal sectors in south africa : using the 1993 input-output table (1993) / Naidoo, Geevaruthnam Patrick – Pretoria: Vista University 2002 [mf ed 2002] – 4mf – 9 – (incl bibl ref) – mf#mfm15254 – sa Unisa [339]

An investigation into mentoring as a possible tool to alleviate practice shock : among primary school educators in the context of curriculum 2005 / Rapuleng, Mahlomola Abel – Pretoria: Vista University 2002 [mf ed 2002] – 3mf – 9 – (incl bibl ref) – mf#mfm15183 – sa Unisa [370]

An investigation into nurses' anxiety when dealing with hiv/aids patients / Dias, Giuliana Zorrer – Uni of South Africa 2001 [mf ed Johannesburg 2001] – 3mf – 9 – (incl bibl ref) – mf#mfm14685 – sa Unisa [610]

An investigation into pain threshold, pain tolerance and augmentation reduction levels among rugby players / Tahu, Hector – 1980 – 2mf – 9 – $8.00 – us Kinesology [790]

An investigation into the cause of the hostility of the church of rome to freemasonry : and an inquiry into freemasonry as it was, and is: with a criticism as to how far the order fulfils its functions – London: Reeves & Turner [1874] – 1mf – 9 – (by aut of "the text book of freemasonry") – mf#vrl-94 – ne IDC [366]

Investigation into the charges preferred by dr atkinson against the honorable a g blair on the eighth day of april, 1890 – S.l: s.n, 1890? – 4mf – 9 – (incl ind) – mf#05981 – cn Canadiana [325]

Investigation into the effect of race and politics on the development of south african sport (1970-1979) / Anderson, Paul Gerard – [Stellenbosch]: U of Stellenbosch 1979 [mf ed 1979] – 348 lea – 9 – (incl bibl) – mf#mf456 – sa Stellenbosch [306]

An investigation into the elastic constants of rocks : more expecially with reference to cubic compressibility / Adams, Frank Dawson & Coker, Ernest George – Washington: Carnegie Institution of Washington, 1906 – 2mf – 9 – 0-665-97211-3 – mf#97211 – cn Canadiana [550]

An investigation into the grief process and the emotional restabilization of the divorcee with some possible implications for the minister as a therapeutic agent / Arnold, Robert E – 1982 – 1 – $10.24 – us Southern Baptist [242]

An investigation into the hierarchical nature of fundamental motor skill development / O'Connor, Justen P – 2001 – 419p on 5mf – 9 – $25.00 – mf#PSY 2163 – us Kinesology [150]

Investigation into the relationship between the amount of revenue a minor league team makes and the size of the market in which it is located / Sadowsky, Mitchell – 2000 – 1mf – 9 – $4.00 – mf#PE 4064 – us Kinesology [650]

An investigation into the state of crisis management plans at national collegiate athletic association division 1-a athletic departments / Wingate, Allison M – University of North Carolina at Chapel Hill, 1995 – 2mf – 9 – $8.00 – mf#PE3624 – us Kinesology [790]

An investigation of anatomical structures associated with the site of medial tibial stress syndrome, often referred to as "shin splints" / Beck, Belinda R & Osternig, Louis R – 1991 – 2mf – 9 – $8.00 – us Kinesology [617]

An investigation of athlete satisfaction with the sport team selection process / Neu, Lois – 1993 – 2mf – $8.00 – us Kinesology [150]

Investigation of clarinet reed contour and its relation to tone quality / Willett, William Cannell – U of Rochester c1961 [mf ed 19–] – 2v on 7mf – 9 – mf#fiche 224 – us Sibley [780]

An investigation of commitment among participants in an extended day physical activity program / Schilling, Tammy A – 1999 – 3mf – 9 – $12.00 – mf#PSY 2082 – us Kinesology [150]

Investigation of communist aggression: tenth interim report of hearings...washington, dc, december 1,2,and 3, 1954 / U.S. Congress. House. Select Committee on Communist Aggression – Washington, Govt. Print. Off., 1954. 174p. LL-2312 – 1 – us L of C Photodup [340]

Investigation of concentration of economic power / U.S. Temporary National Economic Committee – 1939-41 – 9 – $570.00 – mf#0660 – us Brook [330]

Investigation of concentration of economic power. hearings / U.S. Temporary National Economic Committee – 37v. 1939-41 – 1 – us AMS Press [324]

Investigation of concentration of economic power. verbatim record of the proceedings / U.S. Temporary National Economic Committee – 14 v – 1 – us AMS Press [324]

An investigation of cost management in higher education institutions in south africa / Naidoo, Prakash – Pretoria: Vista University 2002 [mf ed 2002] – 7mf – 9 – (incl bibl ref) – mf#mfm15268 – sa Unisa [650]

An investigation of internet usage among a group of professionals in south africa : a uses and gratifications approach / Gilbert, Juliet Francis – Uni of South Africa 2001 [mf ed Johannesburg 2001] – 4mf – 9 – (incl bibl ref) – mf#mfm14984 – sa Unisa [302]

An investigation of interventions to develop entrepreneurship and new venture creation / Ngwenya, Maluleka Samson – Pretoria: Vista University 2002 [mf ed 2002] – 5mf – 9 – (incl bibl) – mf#mfm15193 – sa Unisa [338]
An investigation of medical preparation for international team travel / Suchecki,Joel D – 1998 – 219p on 3mf – 9 – $15.00 – mf#PE 4200 – us Kinesology [617]
An investigation of ministerial counseling support : problems and a proposed model / Kinchen, Thomas A – 1982 – 1 – $7.52 – us Southern Baptist [242]
An investigation of motivational climate on the perceptions of self and collective efficacy / Navarre, Michael J – 1999 – 2mf – 9 – $8.00 – mf#PSY 2105 – us Kinesology [150]
An investigation of north carolina high school football coaches : their knowledge of conditioning and strengthening the athlete / Cook, Ben T – University of North Carolina at Chapel Hill, 1995 – 1mf – 9 – $4.00 – mf#PE3586 – us Kinesology [613]
An investigation of outdoor adventure leadership and programming preparation : in physical education baccalaureate degree programs / Uhlendorf, Karen J – 1988 – 404p on 5mf – 9 – $20.00 – us Kinesology [370]
Investigation of panama canal matters / United States Senate Committee On Interoceanic... – Washington, DC. v1-4. 1907 – 2r – us UF Libraries [972]
Investigation of physical self-perceptions, fitness behavior, and program selection among fitness participants in three fitness club environments / Kiefiuk, Deborah S – Temple University, 1995 – 2mf – 9 – $8.00 – mf#PSY1847 – us Kinesology [150]
An investigation of possible selves across stages of exercise involvement with middle-aged women / Whaley, Diane E – 1998 – 2mf – 9 – $8.00 – mf#PSY 2041 – us Kinesology [790]
An investigation of self-efficacy and control theory with elite distance runners / Martin, Jeffrey J & Gill, Diane L – 1992 – 2mf – 9 – $8.00 – us Kinesology [150]
An investigation of some of kalidasa's views / Harris, Charles – Evansville, IN: Journal Co, 1884 [mf ed 1992] – 1mf – 9 – 0-524-03920-8 – mf#1990-3274 – us ATLA [280]
Investigation of some uncultivated native shrubs to determine metho... / Burgis, D S – S.l., S.l? . 1943 – 1r – us UF Libraries [630]
An investigation of static and dynamic ankle stability in a normal population of young adult females / Pascoe, Deborah A – 1998 – 2mf – 9 – $8.00 – mf#PE 3956 – us Kinesology [611]
An investigation of the career mobility patterns of national football league head coaches / Empey, Michael D – 1997 – 1mf – 9 – $4.00 – mf#PE 3831 – us Kinesology [790]
An investigation of the career mobility patterns of ncaa division 1-a head football coaches / Giles, Scott L – Brigham Young University, 1995 – 1mf – 9 – mf#PE 3646 – us Kinesology [790]
An investigation of the current shona orthography : effects of its limitations and suggested solutions / Dube, Shumirai – Uni of South Africa 2000 [mf ed Johannesburg 2000] – 3mf – 9 – (incl bibl ref) – mf#mfm14795 – sa Unisa [470]
An investigation of the effects of short-term injuries on psychological readiness for competition / Kilgore, Jennifer M – 1998 – 2mf – 9 – $8.00 – mf#PSY 2025 – us Kinesology [150]
An investigation of the ethical dilemmas in the practice of euthanasia / Hipps, Richard Sherrill – 1982 – 1 – $5.00 – us Southern Baptist [170]
An investigation of the impact of hiv/aids on small business in the vaal region / Mngomezulu, Thato David – Pretoria: Vista University 2002 [mf ed 2002] – 5mf – 9 – (incl bibl ref) – mf#mfm15179 – sa Unisa [338]
An investigation of the laws of thought : on which are founded the mathematical theories of logic and probabilities / Boole, George – London: Walton & Maberly, 1854 [mf ed 1991] – 1mf – 9 – 0-7905-9242-8 – mf#1989-2467 – us ATLA [160]
An investigation of the organizational structure and potential for intergroup conflict : between physical education and athletic departments in three secondary schools / Wyatt, T J – 1991 – 2mf – 9 – $8.00 – us Kinesology [373]
Investigation of the pathophysiological agents of nitration / Malcolm, Stuart – 1999 – 2mf – 9 – $8.00 – mf#PE 4076 – us Kinesology [612]
Investigation of the presence and change over time of water quality parameters in selected natural swimming areas in oregon / Van Ess, Erica – 1997 – 1mf – 9 – $4.00 – mf#HE 599 – us Kinesology [333]
An investigation of the process of change in the major contemporary schools of psychotherapy / Gilburth, Kenneth Riley – 1981 – 1 – $5.36 – us Southern Baptist [242]
Investigation of the properties of pliofilm / Vaughan, Paul James – S.l., S.l? . 1942 – 1r – us UF Libraries [630]
An investigation of the qualifications of contract advisors for professional athletes / Smith, Gregory K P & Jackson, Michael W – 1991 – 2mf – 9 – $8.00 – us Kinesology [790]
An investigation of the relationship between measures of kinesthesis and slected aspects of performance in beginner skiing / Solymosi, Frank – 1980 – 2mf – 9 – $8.00 – us Kinesology [790]
Investigation of the romberg test for assessing mild head injury / Riemann, Bryan L – 1997 – 2mf – 9 – $8.00 – mf#PE 3769 – us Kinesology [616]
An investigation of the trinity of plato and of philo judaeus : and of the effects which an attachment to their writings had upon the principles and reasonings of the fathers of the christian church / Morgan, Caesar – Cambridge: JW Parker, 1853 [mf ed 1990] – 1mf – 9 – 0-7905-7534-5 – (incl bibl ref) – mf#1989-0759 – us ATLA [180]
Investigation of the use of video tape recorder techniques in the identification of behavioral characteristics of music teachers / Daellenbach, C Charles – U of Rochester 1968 [mf ed 19–] – 2mf – 9 – mf#fiche1145 – us Sibley [150]
Investigational new drugs – Boston. 1989-1996 (1,5,9) – ISSN: 0167-6997 – mf#16803 – us UMI ProQuest [615]
Investigationharrison reed... / Florida Legislature – S.l., S.l? . no date – 1r – us UF Libraries [340]
Investigations into prehistoric archaeology of gujarat : being the official report of the first gujarat prehistoric expedition, 1941-42 / Sankalia, Hasmukhlal Dhirajlal – Baroda: Baroda State Press, 1946 – (= ser Samp: indian books) – us CRL [930]
Investigations on the action of certain soil constituents / Davis, A G – S.l., S.l?. 1913 – 1r – us UF Libraries [630]
Investigative and cell pathology – Chichester. 1978-1980 (1,5,9) – (cont by: diagnostic histopathology) – ISSN: 0146-7611 – mf#11772 – us UMI ProQuest [574]
Investigative case files of the bureau of investigation, 1908-1922 / U.S. Federal Bureau of Investigation – (= ser Records Of The Federal Bureau Of Investigation) – 955r – 1 – (with printed guide) – mf#M1085 – us Nat Archives [360]
Investigative ophthalmology – St Louis. 1962-1976 [1]; 1971-1976 [5,9] – (cont by: investigative ophthalmology and visual science) – ISSN: 0020-9988 – mf#1883 – us UMI ProQuest [617]
Investigative ophthalmology and visual science – Bethesda. 1977+ (1) 1977+ (5) 1977+ (9) – (cont: investigative ophthalmology) – ISSN: 0146-0404 – mf#1883,01 – us UMI ProQuest [617]
Investigative radiology – Philadelphia. 1966+ (1) 1971+ (5) 1973+ (9) – ISSN: 0020-9996 – mf#6888 – us UMI ProQuest [616]
Investigative Reporters and Editors, Inc *see* Ire journal
Investigator – Providence, RI. 1827-1830 (1) – mf#66322 – us UMI ProQuest [071]
Investigator *see* London investigator
The investigator – Omaha, NE: [Thos. H. Tibbles]. v1 n1. feb 8 1906- (wkly) [mf ed 1906-08 (gaps) filmed [1974]] – 1r – 1 – (title in publisher's box: weekly investigator) – us NE Hist [071]
The investigator – Toronto: J T White, [1872-189- or 19–] – 9 – mf#P04724 – cn Canadiana [073]
Investigator, 1843 – (= ser Periodicals connected with owenite socialism and its successors in secularist, freethought and allied movements, 1834-1916) – 1r – 1 – (filmed with: movement 1843-45; circular of the anti-persecution union 1845) – mf#97168 – uk Microform Academic [073]
Investigator and advocate of independence science, religion, literature, etc – Washington. 1845-1846 – 1 – mf#5581 – us UMI ProQuest [073]
Investigator and expositor – Troy. 1839-1840 – 1 – mf#4795 – us UMI ProQuest [073]
Investigator and general intelligencer – Providence. 1828-1828 (1) – mf#4467 – us UMI ProQuest [420]
Investment advisers act of 1940 releases / U.S. Securities and Exchange Commission – n1-358. 7 oct 1940-22 jan 1973 (all publ) – (= ser The sec release series preceding the sec docket) – 31mf – 9 – $46.50 – mf#LLMC 84-362 – us LLMC [346]
Investment company act of 1940 releases / U.S. Securities and Exchange Commission – n1-7639. 23 sept 1940-26 jan 1973 (all publ) – (= ser The sec release series preceding the sec docket) – 298mf – 9 – $447.00 – (n7143 was never released) – mf#LLMC 84-363 – us LLMC [343]
Investment dealers' digest: idd – New York. 1990+ (1,5,9) – ISSN: 0021-0080 – mf#18397 – us UMI ProQuest [332]
Investment guides – London. 1998+ (1,5,9) – mf#32376 – us UMI ProQuest [332]
Investment in union of south africa / United States Bureau Of Foreign And Domestic Commerce – Washington, DC. 1954 – 1r – us UF Libraries [960]
Investment styles and the asean stock market cycle / Frank, Ashley Gavin – Uni of South Africa 2000 [mf ed Pretoria: UNISA 2000] – 7mf – 9 – (incl bibl ref) – mf#mfm14717 – sa Unisa [332]
Investors and traders guide – 3rd ed. Jones & Baker Securities, 1920 – 1mf – 9 – $1.50 – mf#LLMC 92-204 – us LLMC [343]
Investors chronicle – London. 1972-1973 (1) 1972-1972 (5) (9) – (cont: investors chronicle and stock exchange gazette) – mf#5870,01 – us UMI ProQuest [332]
Investors chronicle – London. 1964-1967 (1) – mf#1321 – us UMI ProQuest [332]
Investors chronicle and stock exchange gazette – London. 1967-1970 [1] – (cont by: investors chronicle) – ISSN: 0021-0161 – mf#5870 – us UMI ProQuest [332]
The investors' guardian – 1863-1973 – 102r – 1 – £4,300.00 – mf#IVG – uk World [332]
Investors' monthly manual – London, UK. 15 oct 1864-jun 1930 [mnthly] – 57 1/2r – 1 – uk British Libr Newspaper [332]
Invictus : the unconquerable voice – University of Wisconsin, Milwaukee. 1986 sep 15, 1994 sep 15,23, nov 15, dec 5, 1995 feb 1 – 1r – 1 – mf#3364716 – us WHS [071]
Invisible church / Collyer, William Bengo' – London, England. 1842 – 1r – us UF Libraries [240]
Invisible fluid – London, England. 18– – 1r – us UF Libraries [240]
The invisible lodge = Unsichtbare loge / Jean Paul – New York: U S Book Co (successors to J W Lovell) c1883 – 1r – 1 – (in english) – us Wisconsin U Libr [430]
The invisible medium : the state of the art of microform and a guide to the literature / Spigai, Frances G – [Washington, American Soc for Information Science, in coop with the ASIS Special Interest Group on Reprographic Technology] 1973 – 31p – 9 – mf#FSN 23,611 – us NY Public [020]
Os invisiveis de lisboa / Lobato, Gervasio & Victor, Jayme – Lisboa. 6v. 1886-87 – 1 – $60.00 – mf#0339 – us Brook [440]
The invitation answered : a reply to dr. j. kent stone's "invitation heeded": and to his holiness, pope pius the ninth's invitation to the vatican council / Smythe, W Herbert – New York: Pliny F Smith, 1871 – 1mf – 9 – 0-8370-8789-9 – (incl bibl ref) – mf#1986-2789 – us ATLA [230]
Invitation to asia / Venkatachalam, Govindraj – Hyderabad: Chetana Prakashan, [between 1900 and 1953] – (= ser Samp: indian books) – us CRL [700]
Invitation to immortality : a one-act play / Abbas, Khwaja Ahmad – Bombay: Padma Publications, 1944 – (= ser Samp: indian books) – (ill by d d dalal) – us CRL [820]
Invitation to sinners to escape from coming wrath... / Alexander, William – Cupar, Scotland. 1834 – 1r – us UF Libraries [240]
Die invloed van bekering op pastorale psigoterapie en voorligting : 'n bibliografie / Bergh, Suzette – Stellenbosch: U van Stellenbosch 1977 – 2mf – 9 – (incl ind) – mf#mf.346 – sa Stellenbosch [016]
Die invloed van organisasiekultuur op kreatiwiteit en innovasie in 'n universiteitsbiblioteek = Influence of organisational culture on creativity and innovation in a university library / Martins, Ellen Caroline – Uni of South Africa 2000 [mf ed Pretoria: UNISA 2000] – 5mf – 9 – (incl bibl ref; abstract in afrikaans & english; text in afrikaans) – mf#mfm14884 – sa Unisa [020]
Die invloed van samelewingsveranderinge op verhoudinge binne die stedelike gesin : 'n sosio-opvoedkundige verkenning = Influence of changing societal phenomena on urban families: a socio-educational analysis / Scholtz, Renee – Uni of South Africa 2001 [mf ed Johannesburg 2001] – 5mf – 9 – (incl bibl ref) – mf#mfm14843 – sa Unisa [306]
Invocacion / Heres Hevia, Diego – Habana, Cuba. 1960 – 1r – us UF Libraries [972]
Invocacion a centroamerica / Ordonez Arguello, Alberto – San Salvador, El Salvador. 1962 – 1r – us UF Libraries [972]
Invocation and intercession of saints / Cumming, J – London, England. 1852 – 1r – us UF Libraries [240]
The invocation of saints / Percival, Henry Robert – London; New York: Longmans, Green, 1896 – 1mf – 9 – 0-524-00777-2 – mf#1990-0209 – us ATLA [240]
Invocation of saints, a romish sin / Hook, Walter Farquhar – London, England. 1847 – 1r – us UF Libraries [241]
Invocation of saints proved from the bible alone / Simpson, R – London, England. 1849 – 1r – us UF Libraries [240]
Involuntary, unmerited, perpetual, absolute, hereditary slavery examined, 1753-1819 / Barrow, David – 1 – 5.00 – us Southern Baptist [242]
Involving the laity of the north dunedin baptist church in a program of participation in the preaching event / Shaddock, Daniel Kenneth – 1 – 5.36 – us Southern Baptist [242]
The inward gospel : some familiar discourses addressed to religious who follow the rules of st. ignatius / Strappini, Walter Diver – London: Burns and Oates, 1909 – 1mf – 9 – 0-8370-7025-2 – mf#1986-1025 – us ATLA [240]
The inward light / Fielding-Hall, Harold – New York: Macmillan, 1908 [mf ed 1995] – (= ser Yale coll) – viii/228p – 1 – 0-524-09140-4 – mf#1995-0140 – us ATLA [280]
The inwardness of british annexations in india / Srinivasachari, Chidambaram S – [Madras]: University of Madras, 1951 – (= ser Samp: indian books) – us CRL [954]
Het inwendig woord : eenige bladzijden uit de geschiedenis der hervorming / Maronier, Jan Hendrick – Amsterdam: Tj van Holkema, 1890 – 1mf – 9 – 0-7905-4597-7 – (incl bibl ref) – mf#1988-0597 – us ATLA [240]
Den inwendighen christenen / Bernieres de Louvigny, Jean de – Antwerpen, 1675 – 11mf – 8 – €21.00 – ne Slangenburg [240]
Inwood, Henry William *see* The erechtheion at athens
[Inyo county-] fresno, inyo, kern, merced, san bernardino, stanislaus and tulare counties – CA. 1884-1885 – 2r – 1 – $100.00 – mf#D020 – us Library Micro [978]
Inzhenernoe delo – M., 1904-1905 – 63mf – 9 – mf#R-2351 – ne IDC [077]
Io. francisci bonomij bononiensis chiron achillis : sive navarchus humanae vitae... / Bonomi, G F – Bononiae: Typis H.H. de Duccijs, 1661 – 4mf – 9 – mf#O-4 – ne IDC [090]
Io. mercerii i.c. emblemata / Mercier, J – Bourges, 1592 – 2mf – 9 – mf#O-691 – ne IDC [090]
Io parto mio bene : sung by sigr viganoni, at the king's theater haymarket, in the opera of i due gobbi [by m a portugal], composed by sigr scismayer [sic] [piano accomp] / Suessmayr, Franz Xaver – [London: R Birchall 1796] [mf ed 1989] – 1r – 1 – mf#pres. film 53 – us Sibley [780]
Ioan bvgenhagii pomerani in hiob annotationes / Bugenhagen, J – [Zwickau, 1527] – 1mf – 9 – mf#TH-1 mf 167 – ne IDC [242]
Ioan zlatoust : margarit – Ostrog, 1595 – 19mf – 9 – mf#RHB-33 – ne IDC [460]
Ioannes ab Arnim [comp] *see* Stoicorum veterum fragmenta
Ioannes lydus (cshb31) / ed by Bekkeri, Imm – 1837 – (= ser Corpus scriptorum historiae byzantinae (cshb)) – €18.00 – ne Slangenburg [243]
Ioannes Zonarad *see* Epitomae historiarum libri 18 (cshb50)
Ioan[ni] vvolphii...nehemias sive in nehemiae de instavrata hierosolyma librum... / Wolf, J – Tigvri, Christoph Froschover, 1570 – 5mf – 9 – mf#PBU-657 – ne IDC [240]
Ioannis bisselii e societate iesu, delicae aestatis / Bissel, J – Monachii: Formis Nicolai Henrici, 1644 – 6mf – 9 – (frontispiece missing) – mf#O-91 – ne IDC [090]
Ioannis bvgenhagii pomerani commentarius : in quatuor capita prioris epistolae ad corinthios / Bugenhagen, J – Wittembergae, 1530 – 5mf – 9 – mf#TH-1 mf 159-163 – ne IDC [242]
Ioannis bvgenhagii publica : de sacramento corporis et sanguinis christi / Bugenhagen, J – [Wittembergae, 1528] – 3mf – 9 – mf#TH-1 mf 164-166 – ne IDC [242]
Ioannis bvgenhague pomerani annotationes ab ipso iam emissae in deuteronomium in samuelem propheta, id est duos libros regu / Bugenhagen, J – [Basel, 1524] – 8mf – 9 – mf#TH-1 mf 135-142 – ne IDC [242]
Ioannis calvin commentarii in epistolam pauli ad romanos / Calvin, J – Argentorati: Vuedelinum Rihelium, 1540. Chicago: Dep of Photodup, U of Chicago Lib, 1979 (1r); Evanston: American Theol Lib Assoc, 1984 (1r) – 1 – 0-8370-1335-6 – mf#1984-T182 – us ATLA [227]
Ioannis Canani *see* Historia (cbh14)
Ioannis Cantacuzeni *see* Historiarum libri 4 (cbh17)

Ioannis cantacuzeni eximperatoris historiarum libri 4 (cshb2,3,4) : graece et latine / Schopeni, Lud – Bonnae. v1-3. 1828-32 – (= ser Corpus scriptorum historiae byzantinae (cshb)) – €61.00 – ne Slangenburg [243]

Ioannis coleti enarratio in primam epistolam s pauli ad corinthios : enarratio in epistolam primam s pauli ad corinthios = An exposition of st paul's first epistle to the corinthians / Colet, John – London: George Bell, 1874 – 1mf – 9 – 0-8370-2714-4 – (incl ind. in english & latin) – mf#1985-0714 – us ATLA [227]

Ioannis ivelli angli, episcopi sarisburiensis vita & mors, eiusq : verae doctrinae defensio... / Humphrey, L – Londini: Apud Iohannem Dayum, 1573 – 4mf – 9 – mf#PW-16 – ne IDC [240]

Ioannis oecolampadii ad billibaldum pyrkaimerum de re eucharistiae responsio / Oecolampadius, J – Tiguri, Christopherus Froschouer, 1526 – 2mf – 9 – mf#PBU-369 – ne IDC [240]

Ioannis saresberiensis episcopi carnotensis policratici.. / John of Salisbury, Bishop of Chartres – Oxonnii: Typographeo Clarendoniano, 1909. 2v. C.C.J. Webb, ed – 1 – us Wisconsin U Libr [920]

Ioannis scylitzae ope ab imm bekkero suppletus et emendatus (cshb34,35) / Georgius Cedrenus – Bonnae. v1-2. 1838-1839 – (= ser Corpus scriptorum historiae byzantinae (cshb)) – €60.00 – ne Slangenburg [243]

Ioannis vvolphii...de christiana perseverantia commentationis consolatoriae... / Wolf, J – Tigvri, Christoph Froschouer, 1578 – 2mf – 9 – mf#PBU-658 – ne IDC [240]

Ioannis vvolphii...de officio praeconis euangelici oratio qua d pauli 2 timoth... / Wolf, J – Tigvri, Christoph Froschouer iunior, 1562 – 1mf – 9 – mf#PBU-655 – ne IDC [240]

Ioannis Zonarae *see* Annales (cshb42,43)

Iobvs... partim commentarijs partim paraphrasi illustratus, cui etiam additus est ecclesiastes... / Beza, Theodor de – Londini, Bishop, 1589 – 4mf – 9 – mf#PFA-114 – ne IDC [240]

IOC World Congress on Sport Sciences *see* Proceedings

Ioelis *see* Historia (cbh14)

Ioelis chronographia compendiaria *see* Breviarium historiae metricum (cshb29)

Iohannes de Caulibus *see* Meditaciones vite christi olim s bonaventurae attributae

Iohannis abbatis victoriensis liber certarum historiarum (mgh7:36.bd) – v1-2. 1909-1910 – (= ser Monumenta germaniae historica 7: scriptores rerum germanicarum in usum scholarum (mgh7)) – €21.00 – ne Slangenburg [240]

Iohannis porta de annoniaco liber de coronatione karoli 4 imperatoris (mgh7:35.bd) – 1913 – (= ser Monumenta germaniae historica 7: scriptores rerum germanicarum in usum scholarum (mgh7)) – €7.00 – ne Slangenburg [240]

Iokibe, Makoto *see* The occupation of japan

Iola herald – Iola WI. [1898 apr 28, sep 1/1902]-[2004 sep/dec] [some gaps] – 70r – 1 – mf#948068 – us WHS [071]

Iolduz – jan-jun, 1918 – 1 – (reel contains short runs of multiple titles. for complete listing of titles on a reel, please inquire) – us UMI ProQuest [077]

Ioma financial executive's news – New York. 2001+ (1,5,9) – ISSN: 1533-4929 – mf#32015 – us UMI ProQuest [332]

Ioma's dc plan investing – New York. 1997+ (1) – mf#19943,02 – us UMI ProQuest [332]

Ioma's human resource department management report – New York. 1997+ (1,5,9) – ISSN: 1092-5910 – mf#32251 – us UMI ProQuest [331]

Ioma's report on customer relationship management – New York. 2002+ (1,5,9) – ISSN: 1538-4934 – mf#32260 – us UMI ProQuest [650]

Ioma's report on financial analysis, planning and reporting – New York. 2000+ (1,5,9) – ISSN: 1532-1673 – mf#32261 – us UMI ProQuest [650]

Ioma's report on managing benefits plans – New York. 1998+ (1) – ISSN: 1098-5662 – mf#19950,01 – us UMI ProQuest [650]

Ioma's report on managing design engineering – New York. 1999+ (1) – ISSN: 1523-469X – mf#22502,02 – us UMI ProQuest [620]

Ioma's report on managing logistics – New York. 1998+ (1,5,9) – ISSN: 1097-2021 – mf#32253 – us UMI ProQuest [650]

Ioma's report on managing the general ledger – New York. 1999+ (1,5,9) – ISSN: 1523-5270 – mf#32244 – us UMI ProQuest [650]

Ioma's report on managing training and development – New York. 1999+ (1,5,9) – mf#32258 – us UMI ProQuest [650]

Ioma's safety director's report – New York. 1999+ (1,5,9) – mf#32259 – us UMI ProQuest [650]

Ioma's security director's report – New York. 1998+ (1,5,9) – ISSN: 1521-916X – mf#32250 – us UMI ProQuest [364]

Iona : lee county / Hanson, W Stanley – S.l., S.l? . 1936 – 1r – us UF Libraries [978]

Ionae vitae sanctorum columbani, vedastis, iohannis (mgh7:37.bd) – 1905 – (= ser Monumenta germaniae historica 7: scriptores rerum germanicarum in usum scholarum (mgh7)) – €15.00 – ne Slangenburg [240]

Ionas propheta / Bugenhagen, J – Wittenberge, 1550 – 9mf – 9 – mf#TH-1 mf 175-183 – ne IDC [242]

[Ione-] amador progess-news – CA. 1979 – 2r – 1 – $120.00 – (cont with: amador dispatch, jackson) – mf#B02307 – us Library Micro [071]

Ione bulletin – Ione OR: L K Harlan [wkly] – 1 – (began in 1913. ceased in 1914?. absorbed by: heppner herald) – us Oregon Lib [071]

Ione bulletin *see* Heppner herald

Ione independent – Ione OR: W E Cochran [wkly] – 1 – (began in 1916. suspended sep-oct 1917) – us Oregon Lib [071]

[Ione-] ione valley echo – CA. 1895-1898 – 1r – 1 – $60.00 – mf#C03605 – us Library Micro [071]

Ione journal – Ione OR: F W Sears, -1916 [wkly] – 1 – (began in 1914?) – us Oregon Lib [071]

[Ione-] nye county news – NV. 1865-66 [wkly] – 1r – 1 – $60.00 – mf#U04590 – us Library Micro [071]

Ione proclaimer – Ione OR: Proclaimer Pub Co [wkly] – 1 – us Oregon Lib [071]

[Ione-] the advertiser – NV. sep-oct 1964 [wkly] – 1r – 1 – $60.00 – mf#U04589 – us Library Micro [071]

Ionian Islands *see* Statistical blue books 1821-1863

The ionian islands : manners and customs; sketches of the ancient history; with anecdotes of the septinsulars / Kendrick, Tertius – London 1822 [mf ed Hildesheim 1995-98] – 2mf – 9 – €60.00 – 3-487-29057-X – gw Olms [914]

Ionisation and rhythm in the music of edgard varese / Snowden, Gloria Jean – U of Rochester 1967 [mf ed 19–] – 4mf – 9 – (with bibl) – mf#fiche 671, 939 – us Sibley [780]

Ionnis calvin commentarii in epistolam pauli ad romanos – 1540 – (= ser Commentarii In Epistolam Pauli Ad Romanos) – 1 – us ATLA [242]

Ion-selective electrode reviews – Oxford. 1979-1987 (1) 1979-1987 (5) 1979-1987 (9) – (cont by: selective electrode reviews) – ISSN: 0191-5371 – mf#49458 – us UMI ProQuest [530]

Iordanis romana et getica (mgh1:5/1) / ed by Mommsen, Theodor – 1882 – (= ser Monumenta germaniae historica 1: scriptores – auctores antiquissimi) – €15.00 – ne Slangenburg [240]

Iordanskii, N N *see* Kooperatsiia v shkole

Iorga, N *see* Breve histoire de la petite armenie

Iornada do arcebispo de goa dom frey aleixo de menezes primaz da india oriental... / Gouvea, F A de – Coimbra: Diogo Gomez Loureyro, 1606 – 4mf – 9 – mf#SEP-41 – ne IDC [915]

Ioseliani, P *see* Short history of the georgian church

Iosif, Arkhimandrit *see* Podrobnoe oglavlenie velikikh chetikh-minei vserossiiskago mitropolita makariia...

Iosif, Ieromonakh *see* Opis rukopisei perenesennykh iz biblioteki iosifova monastyria v biblioteku moskovskoi dukhovnoi akademii

Iosva : in sacram historiam iosvae...liber vnus / Wolf, J – Tigvri, Ioannes Vvolph, 1592 – 3mf – 9 – mf#PBU-662 – ne IDC [240]

Iota *see*
– The adventures of a protestant in search of a religion
– Anti-opium

The iournall of friar william de rubruquis : a french man, of the order of the minorite friars, vnto the east parts of the world, anno dom 1253 / Ruysbroek, W van – London, 1625-1626. v3 – 2mf – 9 – mf#HT-679 – ne IDC [910]

Iovianus : die fragmente seiner schriften, die quellen zu seiner geschichte, sein leben und seine lehre / Haller, W – Leipzig, 1897 – (= ser Tugal 2-17/2) – 3mf – 9 – €7.00 – ne Slangenburg [240]

Iovinianus : die fragmente seiner schriften, die quellen zu seiner geschichte, sein leben und seine lehre = Works. 1897 / Jovinian; ed by Haller, Wilhelm – Leipzig: J C Hinrichs, 1897 – (= ser Tugal) – 1mf – 9 – 0-7905-1827-9 – mf#1987-1827 – us ATLA [240]

Iovrnal de vavigation dv voyage de la coste de gvinee, isles de l'ameriqve et indes d'espagne svr le vaisseav dv roy le favcon francois arme par l'ordre de samaieste povr la royalle compagnie de la siente – [S.l: s.n., 19–?] – us CRL [910]

Iowa : code annotated – St Paul: West Pub Co, 1949-aug 99 update – 9 – $2,423.00 set – mf#401470 – us Hein [348]

Iowa : session laws of american states and territories – 1838-1998 – 9 – $907.00 set – mf#402660 – us Hein [348]

Iowa *see*
– Reports and opinions
– Reports, post-nrs
– Reports, pre-nrs

Iowa Academy of Science *see* Proceedings of the iowa academy of science

Iowa. Adjutant-Generals Office *see* Roster and records of iowa soldiers in the war of the rebellion

Iowa age – Clinton, IA. 1869-1871 (1) – mf#63115 – us UMI ProQuest [071]

Iowa agri news – LaSalle, IL. 1984-1984 (1) – mf#68086 – us UMI ProQuest [071]

Iowa Air National Guard. United States *see* On guard

Iowa Air National Guard [US] *see* Intake

Iowa American Revolution Bicentennial Commission *see* Bicen iowa

Iowa Archeological Society *see* Newsletter of the iowa archeological society

Iowa attorney general reports and opinions – 1896-1994 – 6,9 – $343.00 set – (1896-1978 on reel $245. 1979-86, 1988-94 on mf $90. 1987 not available) – mf#408240 – us Hein [340]

The iowa band / Adams, Ephraim – Boston: Congregational Publ Society, 1870 – 1mf – 9 – 0-8370-6240-3 – mf#1986-0240 – us ATLA [240]

The iowa baptist standard – Des Moines, IA. v1 n1. may 21 1897 (wkly) [mf ed 1947] – (= ser Negro Newspapers on Microfilm) – 1r – 1 – us L of C Photodup [071]

Iowa bystander – Des Moines, IA. 1967-1971 (1) – mf#63168 – us UMI ProQuest [071]

Iowa churchman / Episcopal Church – v5 n1-v20 n4 [1881 jan-1896 apr] – 1r – 1 – (cont: church missionary; cont by: iowa episcopalian) – mf#849223 – us WHS [242]

Iowa city life and times – v1 n1-6 [1979 jan 28-mar 11-18] – 1r – 1 – mf#644316 – us WHS [071]

Iowa county advocate – Dodgeville WI. 1860 aug 18 – 1r – 1 – (cont by: dodgeville chronicle) – mf#875202 – us WHS [071]

Iowa county [city directory : listing] – 1938-1939 [with lafayette county] – 1r – 1 – mf#3199940 – us WHS [917]

Iowa county democrat – Mineral Point WI. [1877 dec 14/1879 feb 7]-[1938 oct 27/dec 8] [gaps] – 30r – 1 – (cont: national democrat [mineral point, wi]; cont by: mineral point tribune [mineral point, wi: 1869]; iowa county democrat and the mineral point tribune) – mf#1131446 – us WHS [071]

Iowa county democrat and the mineral point tribune – Mineral Point WI. 1938 dec 15-1939 dec 28, 1940-57, 1958 jan-mar 27 – 9r – 1 – (cont: mineral point tribune [mineral point, wis. : 1869]; iowa county democrat; cont by: democrat-tribune [mineral point, wi]) – mf#1131447 – us WHS [071]

Iowa county republic – Dodgeville WI. 1901 apr 24-1902 feb 5, 1902 feb 12-1903 oct 29, 1903 nov 5-1904 nov 17 – 3r – 1 – (cont: semi-weekly iowa county republic; cont by: weekly republic [dodgeville, wi]) – mf#964294 – us WHS [071]

Iowa defender / University of Iowa – Iowa City IA. v9 n1-2 [1963 nov4-11], v11 n2,5,9 [1964 sep 28, oct 19, nov 16], v15 n12 [1969 jan 21], v19 n2 [1969 feb 17] – 1r – 1 – mf#933338 – us WHS [378]

Iowa Farmers Association *see*
– U[/nited/] s[/tates/] farm news
– U[/nited/] s[/tates/] farm union news

[Iowa hill-] weekly patriot – CA. 1859-1860 – 1r – 1 – $60.00 – mf#C03254 – us Library Micro [071]

Iowa history items / State Historical Society of Iowa – 1911-1940 oct – 1r – 1 – mf#2755584 – us WHS [071]

Iowa Horticultural Society *see* Fruitman and garden guest

The iowa jewish news – Des Moines. Iowa. 1932-52 – 1 – us AJPC [071]

Iowa journalist – Iowa City. 1968-1971 (1) 1971-1971 (5) – ISSN: 0021-0544 – mf#3238 – us UMI ProQuest [070]

Iowa law bulletin *see* Iowa law review

Iowa law review – v1-11. 1915-1925/26 – 48mf – 9 – $72.00 – (first 10v of series entitled: the "iowa law bulletin". add vols as copyright expires) – mf#LLMC 95-103 – us LLMC [340]

Iowa law review – v1-86. 1915-2001 – 5,6,9 – $1579.00 set – (v1-70 1915-85 on reel or mf $990. v71-86 1985-2001 on mf $589) – ISSN: 0021-0552 – mf#103721 – us Hein [340]

Iowa. Laws, Statutes, etc *see* Herrick and doxsee's probate law and practice of the state of iowa.

Iowa lawyer – v1-61. 1940-2001 – 9 – $578.00 set – (title varies: v1-50 n7 as news bulletin iowa state bar association) – mf#401340 – us Hein [340]

Iowa legal inquisitor – v1-2. 1851-53 (all publ) – 7mf – 9 – $10.50 – (lacking: 1853 no 2) – mf#LLMC 84-497 – us LLMC [340]

Iowa legionaire / American Legion – 1921 mar 4 [v1 n1]-dec 16, 1921 dec 23-1924 may 23, 1924 jun 6-1927 oct 28, 1927 nov 11-1930 feb 28 [v9 n24] – 4r – 1 – mf#955625 – us WHS [366]

Iowa letter carrier / Iowa State Association of Letter Carriers – 1982 may – 1r – 1 – (cont: iowa state letter carrier) – mf#647491 – us WHS [380]

Iowa library quarterly – Des Moines. 1901-1973 [1,5,9] – ISSN: 0021-0579 – mf#1538 – us UMI ProQuest [020]

Iowa orienting express / Iowa Refugee Service Center – v2 n1-v6 n3 [1979 jun-1983 aug] – 1r – 1 – mf#427043 – us WHS [362]

Iowa Peace Network *see* Dovetail

Iowa Peoples' Party *see* Southern iowa educator

Iowa pioneer lawmakers' association reunions – 11v. 1886, 1890-1909 (all publ) – 21mf – 9 – $31.50 – mf#LLMC 84-498 – us LLMC [340]

Iowa plain dealer – Cresco, New Oregon IA. 1884 oct [cresco plaindealer extra], 1886 nov 25, dec 16,30 – 1r – 1 – (cont: weekly new oregon plain dealer; cont by: twice-a-week plain dealer [cresco, iowa]) – mf#851149 – us WHS [071]

Iowa Press Association *see* Iowa publisher and bulletin of the iowa press association

Iowa publisher and bulletin of the iowa press association / Iowa Press Association – Iowa City. 1950-1955 [1] – mf#452 – us UMI ProQuest [070]

Iowa reform – Davenport IA (USA), 1920 2 jul-1940 7 jun – 7r – 1 – (many iss missing) – gw Misc Inst [071]

Iowa Refugee Service Center *see* Iowa orienting express

Iowa review – Iowa City. 1970+ (1) 1970+ (5) 1970+ (9) – ISSN: 0021-065X – mf#9107 – us UMI ProQuest [400]

Iowa southern baptist – Munich. 1970-1973 (1) 1972-1972 (5) (9) – 1 – mf#5864 – us Southern Baptist [242]

Iowa star – Des Moines, IA. 1849-1854 (1) – mf#63173 – us UMI ProQuest [071]

Iowa State AFL-CIO *see* Delegate

Iowa State Association of Letter Carriers *see*
– Iowa letter carrier
– Iowa State letter carrier

Iowa. State Bar Association *see* Proceedings, 1874-1968

Iowa state bar association news bulletin – v1-1940-up to copyright – 9 – mf#LLMC 84-501 – us LLMC [340]

Iowa state bar association proceedings – 1v. 1874-81 (all publ) – 3mf – 9 – $4.50 – mf#LLMC 84-499 – us LLMC [340]

Iowa state bar association proceedings – v1-103. 1895-1976 – 85mf – 9 – $127.00 – (lacking: 81st pt 1. 84th pt 2) – mf#LLMC 84-500 – us LLMC [340]

Iowa state bar association quarterly – v1-3. 1929-32 (all publ) – 5mf – 9 – $7.50 – mf#LLMC 84-502 – us LLMC [340]

Iowa state democrat – Newton, IA. 1898-1901 (1) – mf#63339 – us UMI ProQuest [071]

Iowa State Federation of Labor *see*
– Constitution and proceedings
– Tri-city labor voice

Iowa state journal of research – Ames. 1972-1988 (1) 1972-1988 (5) 1976-1988 (9) – ISSN: 0092-6345 – mf#6825 – us UMI ProQuest [500]

Iowa State letter carrier / Iowa State Association of Letter Carriers – v1 n2-10 [1980 dec-1981 oct] – 1r – 1 – (cont: new iowa letter carrier; cont by: iowa letter carrier) – mf#647489 – us WHS [380]

Iowa state tribune / National Greenback Party [IA] – 1880 sep 30 – 1r – 1 – mf#851144 – us WHS [071]

Iowa state university regulatory conferences on public utility valuation and the rate making process – 1st to 24th conferences. 1962-85 – 100mf – 9 – $150.00 – (lacking: 2nd 1963 p123-4. 4th 1965) – mf#LLMC 84-503 – us LLMC [340]

Iowa. Supreme Court *see* Iowa supreme court reports

Iowa supreme court reports / Iowa. Supreme Court – v1-201. 1855-1926 – 1952mf – 9 – $2928.00 – (pre-nrs run: v1-50 1855-79 369mf $553.00. vols after v186 to be filmed once they fall out of copyright) – mf#LLMC 80-806 – us LLMC [347]

Iowa territorial gazette and burlington advertiser – Burlington IA. 1838 jul 14-1840 jun 6 – 1r – 1 – (cont: wisconsin territorial gazette and burlington advertiser; cont by: iowa territorial gazette and advertiser) – mf#851111 – us WHS [071]

Iowa tribune – Des Moines IA. 1886 dec 1,22, 1887 jan 5, jun 8, 1889 feb 20, 1890 jan 29, 1891 jun 10 – 1r – 1 – (cont by: iowa farmers' tribune) – mf#851145 – us WHS [071]

Iowa union farmer / Educational and Cooperative Union of America et al – 1933 may 17-1936 jan 25, 1938 jan 8-1940, 1941-51, 1952-1956 apr – 3r – 1 – (cont by: u s farm union news) – mf#1330780 – us WHS [630]
Iowa unionist – Des Moines IA. 1900 feb 3-apr 28, 1901 jan 5-1903 jan 24, 1933 oct 6-1938, 1939-42 – 3r – 1 – (cont by: iowa unionist [des moines, iowa : 1945]) – mf#6024545 – us WHS [331]
Iowa weekly people – Des Moines IA. 1878 dec 26-jan 16, 1879 feb 6-13 – 1r – 1 – mf#851142 – us WHS [071]
Iowa Yearly Meeting of the Society of Friends *see* The discipline of iowa yearly meeting of the society of friends
Iowan – Des Moines. 1952+ (1) 1972-1982 (5) 1977-1982 (9) – ISSN: 0021-0722 – mf#7276 – us UMI ProQuest [073]
Iowa-posten – Des Moines IA. 1914 may 15-1915 dec 31; 1913 jan 10-1914 may 8 – 2r – 1 – (cont by: svenska tribunen-nyheter) – mf#4019763 – us WHS [071]
Ioyfull newes out of the new found world... / Monardes, N – London, 1580 – 7mf – 9 – mf#M-381 – ne IDC [917]
Ipelete mu ndebo ye tjikalanga – Vryburg, South Africa. 1935 – 1r – us UF Libraries [960]
Ipf und jagstzeitung – Ellwangen-Leutkirch DE, 1975- – 117r until 1990 – 1 – (bezirksausgabe von schwaebische zeitung) – gw Misc Inst [074]
Ipf und jagstzeitung *see* Schwaebische zeitung [main edition]
Iphigenia in tauris / Goethe, Johann Wolfgang von – London, New York: G Bell, 1901 [mf ed 1993] – (= ser Bell's modern translations) – 1 – (trans by anna swanwick) – mf#8615 – us Wisconsin U Libr [820]
Iphigenie auf tauris / Goethe, Johann Wolfgang von – Cambridge, MA: Harvard University, 1900 – (parallel german and english text with an introduction in english) – us Wisconsin U Libr [430]
Iphigenie en aulide / Racine, Jean – Paris, France. 1818 – 1r – us UF Libraries [440]
Iphigenie en tauride / Guymond De La Touche, Claude – Paris, France. 1801 – 1r – us UF Libraries [440]
Iphigenie im drama der griechen und bei goethe : eine dramaturgische studie / Vogeler, Adolf – [S.l.: s.n.], 1900; Hildesheim: Druck von Gerstenberg – 1r – 1 – (incl bibl ref) – us Wisconsin U Libr [430]
Ipj newsletter / Institute for Peace and Justice [US] – 1979 nov-1985 jul – 1r – 1 – (cont by: ipj new update) – mf#965137 – us WHS [362]
Ipnocausto, Paulo *see*
– Carta de bartolo
– La corneja sin plumas
Ipphos / Ipphos Coy Ltd – Djakarta, 1948 1959. v1-11(17) – 34mf – 9 – (missing: 1948, v1; 1949, v2(1-2, 4-end); 1950, v3; 1951, v4(1-4, 6-17, 19-end); 1952, v5(1-18, 20-21, 23-26); 1954, v6(1-10, 12-15, 17-26); 1954, v7(2-12, 14, 16-19, 23, 25-26); 1955, v8(1-2, 4-6, 8-12, 14, 16-26); 1956, v9(1-9, 12-26); 1957, v10; 1958, v11(1-16)) – mf#SE-1712 – ne IDC [950]
Ipphos Coy Ltd *see* Ipphos
IPPI Daerah Djakarta Raya *see* Pemuda masjarakat
Ippolita ed aricia [dramma in cinque atti] / Traetta, Tommaso – [n.p. 176-] [mf ed 19–] – 1r – 1 – mf#pres. film 10 – us Sibley [780]
Ipq : international philosophical quarterly – New York. 1961-1982 (1,5,9) – (cont by: international philosophical quarterly) – ISSN: 0019-0365 – mf#11438 – us UMI ProQuest [100]
Ipra newsletter – Rio de Janeiro. 1987-1987 (1) 1987-1987 (5) 1987-1987 (9) – (cont: international peace research newsletter. cont by: iprn) – mf#8169,01 – us UMI ProQuest [320]
Ipra newsletter – Rio de Janeiro. 1988-1989 (1) 1988-1989 (5) 1988-1989 (9) – (cont: iprn. cont by: international peace research newsletter) – mf#8169,03 – us UMI ProQuest [320]
Iprn – Rio de Janeiro. 1987-1987 (1) 1987-1987 (5) 1987-1987 (9) – (cont: ipra newsletter. cont by: ipra newsletter) – mf#8169,02 – us UMI ProQuest [320]
Ips papers – Beirut: Institute for Palestine Studies, n1,3-4,7-9,14. 1979-80 – 1r – 1 – us CRL [956]
Ipse, ipsa–ipse, ipsa, ipsum, which? : the latin various readings, genesis 3. 15 / Quigley, Richard F – New York: Fr Pustet, [1890?] [mf ed 1986] – 2mf – 9 – 0-8370-6933-5 – (incl app) – mf#1986-0933 – us ATLA [221]
Ipsen, Lillas F *see* Cardiovascular and body composition responses to aerobic dance training of varying frequencies and total program lengths
Ipsg newsletter : journal of indian progressive study group (england). – London, UK. oct 1971- – 1 – uk British Libr Newspaper [954]
Ipswich 1634-1892 – Oxford, MA (mf ed 1990) – (= ser Massachusetts vital records) – 123mf – 9 – 0-87623-103-2 – (mf 1-7: vital records 1664-1734. mf 8-12: b,i,m,d 1663-1733. mf 13-20: vital records 1734-83. mf 21-26: b,i,m,d 1734-83. mf 27-36: vital records 1705-1860. mf 37-43: b,i,m,d 1748-1859. mf 44-49: vital records 1935-1889. mf 50: out-of-town marriages 1648-1799. mf 51-56: town records 1634-1662. mf 57-63: town records 1634-1674. mf 64-70: town records 1674-96. mf 71-75: town records 1696-1720. mf 76-84: town & land 1634-1757. mf 85-88: death index 1850-1940. mf 89-93: marriage index 1850-1953. mf 94-97: birth index 1850-1943. mf 98-102: b,d,m 1830-66. mf 103-109: b,m,d 1867-92. mf 110-114: intentions 1860-89. mf 115-119: intentions 1890-1910. mf 120-122: deaths 1892-1921. mf 123: marriages 1892-1900) – us Archive [978]
Ipswich 1648-1849 – Oxford, MA (mf ed 1997) – (= ser Massachusetts vital record transcripts to 1850) – 28mf – 9 – 0-87623-272-1 – (mf 1t: vital records 1664-1732. mf 1t-3t: publishments 1708-34. mf 2t-5t: deaths & births 1687-1739. mf 4t,10t,12t-13t: marriages & intents 1693-1783. mf 5t-10t,18t: births & baptisms 1672-1809. mf 10t-12t: deaths & births 1734-83. mf 14t-18t,25t: intentions 1753-1849. mf 19t-20t: births 1749-1844. mf 20t-22t: marriages 1748-1849. mf 22t-25t: deaths 1775-1849. mf 25t-26t: births 1828-49. mf 26t-27t: marriages & deaths 1844-49. mf 28t: out-of-town marriages 1648-1799) – us Archive [978]
Ipswich advance – Ipswich, England 3 jul-20 nov 1885 – 1 – uk British Libr Newspaper [072]
Ipswich. Fine Arts Club *see* Constable and old suffolk artists
Ipswich journal – Suffolk, 19 nov 1720-26 jul 1902 – 1 – (lacking 1722-23, 1738, 1829, 1831-32. early yrs very incomplete; mf 1720-1800 is a composite set repr fr originals in various coll; between 1732-37 publ as: ipswich gazette; probably not publ 1737-39 then renumbered n1 17 feb 1739; 1 jul 1886-21 jul 1888 publ as daily & wkly versions: daily ipswich journal then daily journal and weekly ipswich journal then weekly journal (weekies perhaps not publ jul-dec 1886)) – uk Newsplan [072]
Ipswich mercury – Suffolk, 14 nov 1980-7 oct 1983 – 1 – (discont as paid paper) – uk Newsplan [072]
Ipw-berichte – 1972-1989 – 442mf – 1 – gw Mikropress [300]
Iqbal – Baku, 1912-15 – 7r – 1 – (cont as: yeni iqbal) – us UMI ProQuest [077]
Iqbal, Afzal *see* Select writings and speeches of maulana mohamed ali
Iqbal as a thinker – Essays by eminent scholars. (3rd ed.). Lahore: Sh. M. Ashraf, (1960). viii,304p. 1 reel. 1291 – 1 – us Wisconsin U Libr [290]
Iqbal, his art and thought / Vahid, Syed Abdul – Lahore: Shaikh Muhammad Ashraf, 1944 – (= ser Samp: indian books) – us CRL [490]
Iqbal, his poetry and message / Akbar Ali, Sheikh – Lahore: Mir Mohammad Nawab Din, 1932 – (= ser Samp: indian books) – us CRL [490]
Iqbal, Muhammad *see*
– The complaint and the answer
– Islam and ahmadism
– Six lectures on the reconstruction of religious thought in islam
Iqbal, Muhammad, Sir *see* The development of metaphysics in persia
Iqbal's educational philosophy / Saiyidain, Khwaja Ghulam – Lahore: Arafat Publications: Sole distributing agent, Sh Muhammad Ashraf, 1938 – (= ser Samp: indian books) – us CRL [180]
Iqdam – Baku, 1914-15 – 1r – 1 – (cont as: yeni iqdam) – us UMI ProQuest [077]
El-iqticad el-djazairi : l'economie algerienne. organe de l'union generale du commerce algerien – Alger. n1. oct 1956 – 1 – fr ACRPP [380]
'Ir Ha-Metim / Liwer, David – Tel-Aviv, Israel. 1945 – 1r – 1 – us UF Libraries [939]
Ir tehilah / Feinstein, Aryeh Loeb – Warsaw, Poland. 1886 – 1r – us UF Libraries [939]
Ir, venir, volver a ir / Soldevilla, Dolores – La Habana, Cuba. 1963 – 1r – us UF Libraries [972]
IRA *see* "Sakura"
Ira del cordero / Menendez, Roberto Arturo – San Salvador, El Salvador. 1959 – 1r – us UF Libraries [972]
Iracema / Alencar, Jose Martiniano de – Sao Paulo, Brazil. 1941 – 1r – us UF Libraries [972]
Irad kelley papers *see* Kelly, irad, papers, ms 485
Irade-i milliye – Sivas, 1919-22. Sahib-i Imtiyaz ve Mueduer-i Mes'ul: Selahaddin. n1. 14 eyluel 1335 [1919]-3,5,7,68,84,95,118,254. 3 kanunievvel 1922 – (= ser O & t journals) – 2mf – 9 – $40.00 – us MEDOC [956]
Iraizoz, Antonio *see* Lecturas cubanas
Iraizoz Y De Villar, Antonio *see*
– Apuntes de un turista tropical
– Critica en la literatura cubana
– Libros y autores cubanos
Iraklion Air Station [Crete] *see* Island ally
Iraklion Air Station [Crete US] *see* Cretan sun
Iral : international review of applied linguistics in language teaching = Revue internationale de linguistique appliquee enseignement des langues – Heidelberg. 1963+ (1) 1975+ (5) 1975+ (9) – ISSN: 0019-042X – mf#9776 – us UMI ProQuest [370]
Iran : internal affairs and foreign affairs, 1945-jan 1963 / U.S. State Dept – (= ser Confidential u s state department central files) – 1 – $19,295.00 coll – (1945-49 18r isbn 0-89093-676-5 $3475. 1950-54 44r isbn 0-89093-677-3 $8515. 1955-59 27r isbn 1-55655-379-X $5225. 1960-jan 1963 16r isbn 1-55655-912-7 $3100. with p/g) – us UPA [327]
Iran : the making of us policy, 1977-1980 – [mf ed Chadwyck-Healey] – (= ser National security archive, washington dc: the making of us policy) – 565mf – 9 – (with 2v p/g & ind) – uk Chadwyck [327]
Iran – Tehran, 1871-? numrah-'i 1-32,132-193. 15 muharram 1288-11 ramazan 1290 [mar 1871-nov 1873] – 1r – 1 – $60.00 – (missing: n33-131) – us MEDOC [079]
Iran – Tehran. shumarah-'i 1-216. 11 muharram 1288 – 7 rabi' al-avval [2 apr 1871-24 apr 1874] – 1r – 1 – $195.00 – us MEDOC [079]
Iran *see* Ruznamah-'i rasmi-i kishvar-i shahanshahi-i iran
Iran abad – Tehran. shumarah-'i 1-13. farvardin 1339-farvardin 1340 [mar 1960-mar 1961] – 1r – 1 – $90.00 – us MEDOC [079]
Iran al-yawm – Tihran: Wakalat al-Jumhuriyah al-Islamiyah lil-Anba', [al-'adad 1675-al-'adad 2176 (jul 17 1986-feb 9 1988)] (daily ex fri) – 2r – 1 – us CRL [079]
Iran. Laws, Statutes, etc *see* Majmu'ah salyanah
Iran tribune – Teheran. 1973-1974 (1) – ISSN: 0021-0811 – mf#9165 – us UMI ProQuest [079]
Iran va jahan – Paris: Iran Center for Documents. shumarah-'i 1-280. 10 shahrivar 1359-aban 1365 [1 sep 1980-nov 1986] – 3r – 1 – $159.00 – (missing: n221, 227-229. previously missing iss 198-202, 204-217, 219 added to 1994 holdings) – us MEDOC [079]
The iran-contra affair : the making of a scandal, 1983-1988 – [mf ed Chadwyck-Healey] – (= ser National security archive, washington dc: the making of us policy) – 664mf – 9 – (with 2v p/g & ind) – uk Chadwyck [327]
Iran-i azad – Secheron, SZ: Jibhah-'i Milli-i Iran, 1963- . sal-i 1 shumarah-'i 5-7,9-15; sal-i 2 shumarah-'i 1-5,18-19,23-25; sal-i 3 shumarah-'i 27,31-37; sal-i 4 shumarah-'i 38-41; sal-i 5 shumarah-'i 44-47; sal-i 6 shumarah-'i 51,53-55. isfand 1341-khurdad 1347 [feb/mar 1963-jun 1968] – 1r – 1 – $53.00 – us MEDOC [079]
Iran-i bastan / ed by Azad, 'Abd al-Rahman Sayf – Tihran: Kanun-i Iran-i Bastan. sal-i 1, shumarah-i 1-48; sal-i 2, shumarah-i 1-45; sal-i 3, shumarah-i 1-13. day 1311-shahrivar 1314 [jan 1933-sep 1935] – 1r – 1 – $53.00 – us MEDOC [470]
Irani, Behram S *see* Challenges to press freedom in india, 1947 to 1963
Iran-i imruz / ed by Nazirzadah – Tihran: Sal-i 1, shumarah-i 1-sal-i 4, shumarah-i 2 (Isfand 1317-Tir 1321 [march 1939-july 1942]) – $150.00 – us MEDOC [470]
Irani, K D S *see* Pahlavi texts
Iranian national census, 1976-1977 – 455mf – 9 – $6850.00 – (persian & english tables) – us MEDOC [315]
Iranian national census, 1986-1987 – 649mf – 9 – $10,000.00 – us MEDOC [315]
Iranian Student Association in the US *see* Resistance
Das iranische erloesungsmysterium. religionsgeschichtliche untersuchngen / Reitzenstein, R – Bonn, 1921 – €12.00 – ne Slangenburg [230]
Iranshahr – London: Intisharat-i Tirazh, 1978- . dawrah-'i 1, shumarah-'i 1-dawrah-'i 6, shumarah-'i 2 (shumarah musalsal-i 1-233) 28 mihr 1357-3 urdibihisht 1363 [20 oct 1978-23 apr 1984] – 2r – 1 – $200.00 – (v2-6 publ in arlington, va by intisharat-i iranshahr) – us MEDOC [079]
Iraq : internal affairs and foreign affairs, 1945-jan 1963 / U.S. State Dept – (= ser Confidential u s state department central files) – 1 – $10,820.00 – (1945-49 10r isbn 0-89093-904-7 $1935. 1950-54 18r isbn 0-89093-905-5 $3475. 1955-59 18r isbn 1-55655-380-3 $3475. 1960-jan 1963 13r isbn 1-55655-800-7 $2530. with p/g) – us UPA [327]
Iraq – v.1, 1934. Part 1 – 9 – $10.00 – (v2, 1935. part 1. $3.00. v3, 1936. part 1. $2.00. v4, 1937. $4.00. v8. 1946. $10.00. v9-10. 1947-48. $12.00v. v11-12. 1949-50. $10.00v. v13. 1951. $10.00. v14-15. 1952-53. $10.00v. v16-17. 1954-55. $12.00v. v18. 1956. $10.00) – us IRC [930]
Iraq *see*
– Alwaqai aliraqiya: official gazette of the republic of iraq
– Al-waqa'i al-iraqiyah
The iraq times – Baghdad: Times Press Ltd, 1948-50 6r; 1953-55 9r; 1956-may 1964 29r – 44r – 1 – us CRL [079]
Iraqgate : saddam hussein, us policy and the prelude to the persian gulf war, 1980-1994 – [mf ed Chadwyck-Healey] – (= ser National security archive, washington dc: the making of us policy) – 1900 docs on 331mf – 9 – (with p/g & ind) – uk Chadwyck [327]
Irazabal, Carlos *see* Venezuela esclava y feudal
Irbitskie uezdnye vedomosti / ed by Nikol'skii, V P – Irbit [Ekaterinburg gub]: Upr uezdom 1919 [1919 17 [4] apr-[iiul']] – (= ser Asn 1-3) – n1-65 [1919] item 180, on reel n37 – 1 – (cont: irbitskii vestnik) – mf#asn-1 180 – ne IDC [077]
Irbitskii vestnik / ed by Zapol'skii, A A – Irbit [Ekaterinburg gub]: Izd Tip n-kov I A Lopatkova 1918-19 [1918 [?]-1919 [31 [18] marta]] – (= ser Asn 1-3) – n59 [1918]-n189 [1919] [gaps] item 181, on reel n37, 38 – 1 – (cont by: irbitskie uezdnye vedomosti) – mf#asn-1 181 – ne IDC [077]
Irbitskij sovet rk i kd *see* Izvestiia irbitskogo soveta rabochikh, soldatskikh i krest'ianskikh deputatov
Ircd bulletin – New York. 1977-1980 – 1,5,9 – ISSN: 0536-1966 – mf#11318 – us UMI ProQuest [370]
Irdische liebe : eine alltagsgeschichte / Buelow, Frieda, Freiin von – Dresden: C Reiszner 1905 [mf ed 1993] – 1r – 1 – (filmed with: the plays of georg buechner / trans & int by geoffrey dunlop) – mf#8526 – us Wisconsin U Libr [830]
Ire journal / Investigative Reporters and Editors, Inc – Columbia. 1989+ (1,5,9) – ISSN: 0164-7016 – mf#17451 – us UMI ProQuest [070]
Ireland : as she is, as she has been, and as she ought to be / Clancy, James J – New York: T Kelly 1877 – 1r – 1 – mf#2195 – us Wisconsin U Libr [941]
Ireland / Colquhoun, John C – Glasgow, Scotland. 1836 – 1r – us UF Libraries [240]
Ireland : her landlords, her people, and their homes / French, C – Dublin, 1860 – (= ser 19th c ireland) – 2mf – 9 – mf#1.1.8485 – uk Chadwyck [941]
Ireland : its evils, and their remedies: being a refutation of the errors of the emigration committee and others / Sadler, Michael Thomas. – London, 1829 – (= ser 19th c ireland) – 6mf – 9 – mf#1.1.5445 – uk Chadwyck [941]
Ireland / Kinnear, John Boyd – London, 1880 – (= ser 19th c ireland) – 1mf – 9 – mf#1.1.2191 – uk Chadwyck [330]
Ireland : letters reprinted from the "morning post" / Munro-Butler-Johnstone, Henry Alexander – London, 1868 – (= ser 19th c ireland) – 1mf – 9 – mf#1.1.1850 – uk Chadwyck [941]
Ireland – London, 1885 – (= ser 19th c ireland) – 1mf – 9 – mf#1.1.408 – uk Chadwyck [941]
Ireland : politics and society through the press, 1760-1922 – ongoing – 5 units per yr, 40r per unit – 1 – us Primary [941]
Ireland : portions of a letter on the land question, addressed to earl grey, in 1868 / Manning, Henry Edward, Cardinal – London, 1881 – (= ser 19th c ireland) – 1mf – 9 – mf#1.1.2198 – uk Chadwyck [333]
Ireland : a word to the rt hon chichester fortescue – London, [1869?] – (= ser 19th c ireland) – 1mf – 9 – mf#1.1.1883 – uk Chadwyck [941]
Ireland *see* Reports, pre-1894
Ireland, Alleyne *see*
– The cohensive elemnts of british imperialism
– Tropical colonizations
Ireland and canada : studies in comparative constitutional law and politics / Bellot, Hugh Hale L – London: Reeves & Turner, 1893 – 1mf – 9 – mf#03567 – cn Canadiana [323]
Ireland and her churches / Godkin, James – London: Chapman and Hall, 1867. xxxv,623p – 1 – us Wisconsin U Libr [941]
Ireland and her servile war / Waveney, Robert Alexander Shafto Adair, 1st Baron – London, 1866 – (= ser 19th c ireland) – 1mf – 9 – mf#1.1.1881 – uk Chadwyck [941]
Ireland and proportional representation / De Vere, Aubrey Thomas – Dublin, 1885 – (= ser 19th c ireland) – 1mf – 9 – mf#1.1.407 – uk Chadwyck [941]
Ireland and the anglo-norman church : a history of ireland and irish christianity from the anglo-norman conquest to the dawn of the reformation / Stokes, George Thomas – London: Hodder & Stoughton, 1889 – 1mf – 9 – 0-7905-6453-X – (incl bibl ref) – mf#1988-2453 – us ATLA [240]
Ireland and the anglo-norman church : a history of ireland and irish christianity from the anglo-norman conquest to the dawn of the reformation / Stokes, George Thomas – London: Hodder & Stoughton, 1889 – 1mf – us ATLA [240]

IRISH

Ireland and the centenary of american methodism : chapters on the palatines, philip embury and mrs heck, and other irish emigrants who instrumentally laid the foundation of the methodist church in the united states of america, canada and eastern british america / Crook, William – London: Hamilton, Adams; Dublin: R Yoakley, 1866 [mf ed 1985] – 4mf – 9 – 0-665-45088-5 – (incl bibl ref and publ list) – mf#45088 – cn Canadiana [242]
Ireland and the irish – Dublin, 1881 – (= ser 19th c ireland) – 1mf – 9 – mf#1.1.2197 – uk Chadwyck [330]
Ireland by the honble. emily lawless... / Lawless, Hon Emily – London, 1887 – (= ser 19th c ireland) – 5mf – 9 – mf#1.1.4486 – uk Chadwyck [941]
Ireland exhibited to england in a political and moral survey of her population : and in a statistical and scenographic tour of certain districts... / Atkinson, A – London 1823 [mf ed Hildesheim 1995-98] – 2v on 6mf – 9 – €120.00 – 3-487-27853-7 – gw Olms [941]
Ireland for the irish : a practical, peaceable, and just solution of the irish land question / O'Neill, Charles Henry – London, 1868 – (= ser 19th c ireland) – 2mf – 9 – mf#1.1.8556 – uk Chadwyck [941]
Ireland, Gordon *see* Cursillo de derecho constitucional americano compa...
Ireland in 1868 : the battle-field for english party strife / Fitzgibbon, Gerald – London, 1868 – (= ser 19th c ireland) – 1mf – 9 – mf#1.1.1853 – uk Chadwyck [230]
Ireland in 1880 / Richardson, Ralph – London, 1881 – (= ser 19th c ireland) – 1mf – 9 – mf#1.1.2903 – uk Chadwyck [941]
Ireland in 1880 : with suggestions for the reform of her land laws / Pim, Joseph Todhunter – London, [1881] – (= ser 19th c ireland) – 1mf – 9 – mf#1.1.2201 – uk Chadwyck [339]
Ireland in 1846-7 : considered in reference to the recent rapid growth of popery / Hardy, Philip Dixon – Dublin, 1847 – (= ser 19th c ireland) – 2mf – 9 – mf#1.1.6670 – uk Chadwyck [241]
Ireland in the twentieth century / Clanchy, T J – Dublin, 1892 – (= ser 19th c ireland) – 1mf – 9 – mf#1.1.8061 – uk Chadwyck [330]
Ireland, James *see* The life of the rev. james ireland
Ireland, John *see*
– The church and modern society
– Nuptia sacra
– Papers
Ireland, John et al *see* Hogarth's works
Ireland. King's Council *see* Roll of the proceedings of the king's council in ireland (rs69)
Ireland. Laws, Statutes, etc *see* Statutes at large
Ireland of the welcomes – Dublin. 1952-1996 (1) 1970-1981 (5) 1975-1981 (9) – ISSN: 0021-0943 – mf#1521 – us UMI ProQuest [941]
Ireland. Parliament *see* The printed records of the parliament of ireland, 1613-1800
Ireland. Parliament. House of Commons *see* Transcripts of debates
Ireland, past and present : the land and the people / Wilde, William Robert Wills – Dublin, 1864 – (= ser 19th c ireland) – 1mf – 9 – mf#1.1.9686 – uk Chadwyck [941]
Ireland since '98 / Mitchel, John – Glasgow, [1871] – (= ser 19th c ireland) – 2mf – 9 – mf#1.1.4637 – uk Chadwyck [941]
Ireland since 1850 and her present difficulty / Heygate, Frederick William, 2nd bart – [London], 1880 – (= ser 19th c ireland) – 1mf – 9 – mf#1.1.1940 – uk Chadwyck [339]
Ireland, tracts and treatises, 1613-1769 – Dublin. 2v. 1860-61 – 1r – 1 – mf#96814 – uk Microform Academic [941]
Ireland versus england / Macfarlane, David Horne – London, 1880 – (= ser 19th c ireland) – 1mf – 9 – mf#1.1.1901 – uk Chadwyck [941]
Ireland violent and wilful : a plea for england's prayers / Poland, Frederick William – London, 1882 – (= ser 19th c ireland) – 1mf – 9 – mf#1.1.1952 – uk Chadwyck [941]
Ireland, William Henry *see*
– Anekdoten (zum groessten theil unbekannt) von napoleon
– France for the last seven years
– Memoirs of henry the great
– Napoleon anecdotes
Ireland's brighter prospects / Castletown, Bernard Edward Barnaby Fitzpatrick, 2nd Baron – London, 1881 – (= ser 19th c ireland) – 1mf – 9 – mf#1.1.1908 – uk Chadwyck [330]
Ireland's case for home rule considered – London, 1890 – (= ser 19th c ireland) – 1mf – 9 – mf#1.1.1947 – uk Chadwyck [941]
Ireland's case stated in reply to mr. froude / Burke, Thomas Nicholas – New York: PM Haverty, 1873, c1872 – 1mf – 9 – 0-8370-6968-8 – mf#1986-0968 – us ATLA [941]
Ireland's gazette – [Northern Ireland] Belfast jan 1891-may 1903 [mf ed 2002] – 11r – 1 – (missing: 1893) – uk Newsplan [072]
Irelands gazette – Dublin, Ireland. 1891-15 dec 1894; 1895-19 dec 1896; may 1903 – 3 3/4r – 1 – uk British Libr Newspaper [072]
Ireland's hour / Grant, Henry – London, 1850 – (= ser 19th c ireland) – 2mf – 9 – mf#1.1.261 – uk Chadwyck [330]
Ireland's hour / Grant, Henry – London: Thomas Hatchard; Dublin: Hodges & Smith, 1850 – (= ser 19th c economics) – 2mf – 9 – mf#1.1.261 – uk Chadwyck [941]
Ireland's only safety : the improvement of its waste lands / Rawstorne, Lawrence – London, 1850 – (= ser 19th c ireland) – 1mf – 9 – mf#1.1.262 – uk Chadwyck [333]
Ireland's only safety, the improvement of its waste land / Rawstorne, Lawrence – London: Longman, Brown...; Preston: H Oakey, 1850 – (= ser 19th c economics) – 1mf – 9 – mf#1.1.262 – uk Chadwyck [333]
Irelands staurday night *see* Ulster saturday night
Ireland's wrongs and how to mend them : a letter to the middle-class and operative electors / Walters, John Thomas – London, 1881 – (= ser 19th c ireland) – 1mf – 9 – mf#1.1.2203 – uk Chadwyck [339]
Iremonger, Frederic *see* Questions for the different elementary books used in the national s...
Irenaeus *see*
– Armenische irenaeusfragmente
– Des heiligen irenaeus schrift zum erweise der apostolischen verkuendigung
– Gegen den haeretiker
– Gegen die haeresien, 1. bd (bdk3 1.reihe)
– Gegen die haeresien, 2. bd (bdk4 1.reihe)
Irenaeus gegen die haeretiker : buch 4-5 / Ter-Minassiantz, E – Leipzig, 1910 – (= ser Tugal 3-35/2) – 4mf – 9 – €11.00 – ne Slangenburg [240]
Irenaeus letters : originally published in the new york observer = Correspondence. selections / Prime, Samuel Irenaeus – [New York]: New York Observer, 1881 – 1mf – 9 – 0-7905-8176-0 – mf#1988-8059 – us ATLA [240]
Irenaeus letters. second series = Correspondence. selections / Prime, Samuel Irenaeus – New York: New York Observer, c1885 – 1mf – 9 – 0-7905-8014-4 – (incl bibl ref) – mf#1988-8014 – us ATLA [240]
Irenaeus of lugdunum : a study of his teaching / Hitchcock, Francis Ryan Montgomery – Cambridge: University Press, 1914 [mf ed 2004] – 1r – 1 – 0-524-10483-2 – (incl bibl ref & ind. foreword by henry barclay swete) – mf#b00698 – us ATLA [240]
Irenaeus, Saint, Bishop of Lyon *see* Sancti irenaei, episcopi lugdunensis, libros quinque adversus haereses
Irenaeus testimony to the fourth gospel *see* Lessings fragmentenstreit
The irenaeus testimony to the fourth gospel : its extent, meaning, and value / Lewis, Frank Grant – Chicago: University of Chicago Press, 1908 – 1mf – 9 – 0-8370-9715-0 – (incl bibl ref and index) – mf#1986-3715 – us ATLA [226]
The irenaeus testimony to the fourth gospel; its extent, meaning, and value / Lewis, Frank Grant – Chicago: Univ. of Chicago Press, 1908. 64p – 1 – us Wisconsin U Libr [240]
Irenaus – London, England. 1876 – 1r – us UF Libraries [240]
Irene gallica, hoc est de pace et concordia in gallis sancita auspicijs heinrici 4...gratvlatio ad gallos / Stucki, J W – Tigvri, 1601 – 2mf – 9 – mf#PBU-645 – ne IDC [240]
Irene hecht collection, 1637-1950 / Hecht, Irene W D [coll] – [mf ed 1977] – 13r – 1 – (papers [1842-93], incl correspondence, letter books, diary & clippings, of james w taylor, govt agent, diplomat & u s consul at winnipeg, manitoba; misc diplomatic papers of u s state dept, chiefly relating to u s & canada relations [1848-70]...; with finding aid) – us UW Libraries [327]
Irene petrie : missionary to kashmir / Carus-Wilson, Ashley [Mrs] – 6th ed. London: Hodder & Stoughton, 1905 [mf ed 1991] – 1mf – 9 – 0-524-00629-6 – (amer ed publ as: a woman's life for kashmir) – mf#1990-0129 – us ATLA [242]
Irene von starenburg : roman einer brabanter frau / Gerard, Guillaume Samsoen de – Nuernberg: J L Schrag, c1943 (mf ed 1990) – 1r – 1 – (filmed with: zeitgenoessische dichter) – us Wisconsin U Libr [830]
Irenic theology : a study of some antitheses in religious thought / Mead, Charles Marsh – New York: G. P. Putnam and sons, 1905. Beltsville, Md: NCR Corp, 1978 (5mf); Evanston: American Theol Lib Assoc, 1984 (5mf) – 9 – 0-8370-1015-2 – (incl bibl ref and ind) – mf#1984-4371 – us ATLA [240]
Irenics : a series of essays showing the virtual agreement between 1. science and the bible, 2. nature and the supernatural, 3. the divine and the human in scripture, 4. the old and the new testaments, 5. calvinism and arminianism, 6. divine benevolence... / Strong, James – New York: Phillips & Hunt; Cincinnati: Walden & Stowe, 1883 [mf ed 1989] – 1mf – 9 – 0-7905-2382-5 – mf#1987-2382 – us ATLA [240]
Irenics and polemics : with sundry essays in church history / Bacon, Leonard Woolsey – New York: Christian Literature, 1895 [mf ed 1989] – 1mf – 9 – 0-7905-4063-0 – (incl bibl ref) – mf#1988-0063 – us ATLA [240]
Irenicum / Heugh, Hugh – Glasgow, Scotland. 1845 – 1r – us UF Libraries [240]
Irenicum sive de unione et synodo evangelicorum concilianda liber... / Pareus, D – Heidelbergae, 1615 – 4mf – 9 – mf#PBA-286 – ne IDC [240]
Irenicum wesleyanum : or, proposals for union with wesleyan methodists / Wordsworth, Christopher – Lincoln: J Williamson; London: Rivingtons, 1876 – 1mf – 9 – 0-7905-6916-7 – mf#1988-2916 – us ATLA [242]
Irfan – Siverek, Urfa, 1923-19? Sahib-i Imtiyaz: Siret; Mueduer-i Mes'ul: Mehmed. n56. 27 nisan 1341; 101. 16 mart 1926 – (= ser O & t journals) – 1mf – 9 – $25.00 – us MEDOC [956]
Irgun ha-yishouv ha-yehudi be-erets yisrael / Ostrovsky, Moses – Jerusalem, Israel. 1942 – 1r – us UF Libraries [939]
Irian barat : laporan kegiatan / Secretariat Koordinator Urusan Irian Barat – Djakarta, 1964. v1-2(8) – 8mf – 9 – (missing: 1964 v1-2(4-6)) – mf#SE-1713 – ne IDC [959]
Irian post / Memboramo – Ternate, 1957. v1(1-3) – 1mf – 9 – (missing: 1957 v1(1)) – mf#SE-924 – ne IDC [959]
Iriarte-A G, Joaquin *see* Vera, francisco. seneca. madrid, 1934
Iribarren Mora, Guillermo *see* Pensamientos sobre caminos
Irion, Christian *see* Malabar und die missionsstation talatscheri
Iris – 1830-31 – (= ser English gift books and literary annuals, 1823-1857) – 9mf – 9 – uk Chadwyck [800]
Iris – Badajoz.1889-90 – 9 – sp Bibl Santa Ana [074]
Iris : deutsche entomologische zeitschrift. entomologischer verein "iris" zu dresden – Dresden 1884-1926 – v1-40 on 243mf – 9 – mf#z-936c/2 – ne IDC [590]
Iris : farbenstudien und blumenstucke / Delitzsch, Franz – Leipzig: Doerffling & Franke, 1888 – 1mf – 9 – 0-524-08336-3 – mf#1993-2026 – us ATLA [240]
Iris : or literary messenger – New York. 1840-1841 (1) – mf#4378 – us UMI ProQuest [420]
Iris : ein taschenbuch fuer... / ed by Jacobi, Johann Georg – Zuerich 1803-1812 – (= ser Dz) – 23mf – 9 – €230.00 – mf#k/n6033 – gw Olms [074]
O iris : jornal litterario e instructivo – Florianopolis, SC: Liv Cysne, 21 set 1924; jan 1925; 13 jun 1926 – (= ser Ps 19) – mf#UFSC/BPESC – bl Biblioteca [073]
O iris : jornal scientifico e litterario – Sao Paulo, SP: Typ de J R de Azevedo Marques, maio, jul 1857 – (= ser Ps 19) – mf#P17,02,212 – bl Biblioteca [500]
O iris : periodico bi-mensal, dedicado ao sexo feminino – Natal, RN: Typ Conservadora, 10 nov 1875 – (= ser Ps 19) – bl Biblioteca [305]
O iris : periodico dedicado a causa do progresso – Natal, RN: Typ Conservadora, 03 mar 1876 – (= ser Ps 19) – bl Biblioteca [079]
Iris alagoense *see* O federalista alagoense
O iris da verdade : periodico religioso, literario e politico – Pernambuco, 18 mar 1865 – (= ser Ps 19) – bl Biblioteca [079]
Iris und genziane : die persianischen haeuser: [a novel] / Jensen, Wilhelm – Berlin: Deutsche Volkskultur, [18–?] [mf ed 1995] – (= ser Deutsche volkskultur in wort, bild und klang 7) – 163p – 1 – mf#8795 – us Wisconsin U Libr [830]
Iris. vierteljahrschrift fuer frauenzimmer / ed by Jacobi, Johann Georg et al – Duesseldorf [v5ff: berlin] 1774-76 – (= ser Dz) – 8v on 17mf – 9 – €170.00 – mf#k/n6478 – gw Olms [305]
Irisarri, Pantaleon *see* Reflexiones...esposa dona maria rosario mendoza y..
Das irische palimpsest-sakramentar in clm 14429 [tab53-54] / Dold, Alban – 1964 – (= ser Texte und arbeiten. beuron [tab]. beitraege zur ergruendung des aelteren lateinischen und christlichen schrifttums und gottesdienstes) – €15.00 – (incl: suppl1: manz, g: ausdrucksformen der lateinischen liturgiesprache bis ins elfte jahrhundert, 1941 €27; suppl2: fischer, b and fiala, v: colligere fragmenta: festschrift a dold, 1952 €18; suppl3: dold-gambler: das sakramentar von monza, 1957 €12; suppl4: dold-gambler: das sakramentar von salzburg, 1960 €12) – ne Slangenburg [241]
Irish advertiser farm and land list and dublin city circular *see* Irish farm list land circular and general investment reporter or real property advertiser
Irish advocate and achill missionary herald – Dublin, Ireland 1 aug 1874-1 mar 1875 – 1 – (cont by: irish advocate & missionary herald [1 apr-1 dec 1875]) – uk British Libr Newspaper [072]
Irish advocate and missionary herald – Dublin, Ireland 1 apr-1 dec 1875 – 1 – (cont: irish advocate & achill missionary herald [1 aug 1874-1 mar 1875]; cont by: irish church advocate [ns] jan 1876-1 jul 1879]) – uk British Libr Newspaper [240]
Irish advocate etc – Drogheda, Ireland. -w. 14 oct 1848-14 apr 1849 – 1/4r – 1 – uk British Libr Newspaper [072]
Irish agriculturists etc – Ireland.7 Jan-8 Dec 1849. -w. 1/4 reel – 1 – uk British Libr Newspaper [072]
Irish american – New York. aug 12 1849-feb 1915. (incomplete) [wkly] – 1 – us NY Public [073]
Irish and canadian rocks, compared / Kinahan, George Henry – London: Truebner, [1885?] [mf ed 1987] – 1mf – 9 – 0-665-64794-8 – mf#64794 – cn Canadiana [550]
Irish and scotch linen – Belfast Ireland, 1929 – 1/4r – 1 – uk British Libr Newspaper [072]
Irish art / Henry, Francoise – London, England. 1947 – 1r – us UF Libraries [700]
Irish athletic and cycling news – Dublin, Ireland. 1889, 19 aug 1890-5 apr 1892 – 4r – 1 – (aka: wheelman) – uk British Libr Newspaper [072]
Irish australian – Sydney, oct 1894-sep 1895 – 1r – A$27.50 vesicular A$33.00 silver – at Pascoe [079]
An irish beauty of the regency / Blake, Warrenne [comp] – London: John Lane; New York: John Lane Co 1911 [mf ed 1989] – 1r [ill] – 1 – (comp fr "mes souvenirs," the unpubl journals of the hon mrs calvert, 1789-1822, by mrs warrenne blake. filmed with: de gezaghebbers der oost-indische compagnie / wijnaendts van resandt, w) – mf#2763 – us Wisconsin U Libr [920]
Irish canadian – Toronto, ON. 1863-92 – 18r – 1 – cn Library Assoc [971]
Irish canadian *see* The catholic weekly review
Irish catholic – Dublin 1930 – 1r – 1 – ie National [072]
Irish catholic – Dublin, Ireland. 5 may 1888-1896; 2 jan-18 dec 1926; 3 may-20 dec 1930; 1950 – 11r – 1 – uk British Libr Newspaper [072]
Irish catholic chronicle and peoples news of the week – Dublin, Ireland. 7 sep 1867-17 jul 1869 – 2r – 1 – (aka: irish chronicle) – uk British Libr Newspaper [072]
Irish christian advocate – Belfast Ireland, 1930; 1950 – 1 1/2r – 1 – uk British Libr Newspaper [072]
Irish christian advocate *see* Christian advocate
Irish chronicle *see* Irish catholic chronicle and peoples news of the week
Irish church / Perrin, Sergeant – Edinburgh, Scotland. 1835 – 1r – us UF Libraries [241]
The irish church : a speech delivered in the house of commons on monday, march 1, 1869 / Gladstone, William Ewart – London, 1869 – (= ser 19th c ireland) – 1mf – 9 – mf#1.1.1869 – uk Chadwyck [241]
Irish church advocate – Dublin, Ireland jan 1876-1 jul 1879 – 1 – (cont: irish advocate & missionary herald [1 apr-1 dec 1875]; cont by: church advocate (ns) [aug 1879-15 nov 1891]) – uk British Libr Newspaper [072]
Irish church establishment / Gray, John – Dublin, Ireland. 1866 – 1r – us UF Libraries [241]
Irish church news – Belfast Ireland, dec 1892-1893 – 1/2r – 1 – uk British Libr Newspaper [072]
The irish church property devoted to the purchase of irish railways – London, 1869 – (= ser 19th c ireland) – 1mf – 9 – mf#1.1.1862 – uk Chadwyck [333]
Irish churchman and protestant review – Dublin, Ireland. 1889; 1890 – 2r – 1 – uk British Libr Newspaper [072]
Irish citizen – New York. oct 19 1867-oct 19 1868 (wkly) – 1 – us NY Public [073]
Irish congregational magazine and home messenger *see* Irish congregational magazine / irish congregational magazine and home messenger
Irish congregational magazine / irish congregational magazine and home messenger – Belfast: Congregational Church, Ireland; Irish Congregational Union; [mf ed 2001] – (= ser Christianity's encounter with world religions, 1850-1950) – 4r – 1 – mf#2001-s060 – us ATLA [242]
Irish constabulary records – 1816-1922 – 1 – uk National [350]

Irish convert / Bradley, Patrick – Glasgow, Scotland. 18– – 1r – us UF Libraries [241]
The irish crisis : a short speech against coercion / Manson, James Alexander – London, 1881 – (= ser 19th c ireland) – 1mf – 9 – mf#1.1.2199 – uk Chadwyck [330]
Irish daily independent – Dublin, Ireland. 18 dec 1891-1915; 5 may-dec 1916; jun-dec 1919; apr-jun 1920; 11 jan-dec 1925; 4 nov-dec 1950; 1986-jun 1987; 1988-may 1991; sep 1991-1995; jan 1996-dec 1996; 1997 [daily] – 353r – 1 – (aka: irish independent; irish daily independent and daily nation) – uk British Libr Newspaper [072]
Irish daily independent and daily nation *see* Irish daily independent
Irish daily telegraph – Londonderry, Ireland. jan-aug 1926 – 2r – 1 – (incorp with: belfast telegraph from 6 feb 1904 to 29 may 1906) – uk British Libr Newspaper [072]
Irish daily telegraph and southern reporter *see* Southern reporter and cork commercial courier
The irish deep sea fisheries / Butt, Isaac – Dublin, 1874. – (= ser 19th c ireland) – 1mf – 9 – mf#1.1.1933 – uk Chadwyck [639]
Irish democrat – England 1981-2000 – 2r – 1 – ie National [072]
Irish democrat *see* Irish freedom
Irish diamond – Dublin, Ireland. 10 mar-14 jul 1883 – 1/4r – 1 – uk British Libr Newspaper [072]
Irish diaspora and its impact on the eastern and western cape in the colonial era / Naidoo, Indren – U of Durban-Westville 1994 [mf ed S.l: s.n. 1994] – 6mf – 9 – (incl bibl) – sa Misc Inst [960]
The irish difficulty : 1: the church question; 2: the land question; 3: the education question. – London, 1868 – (= ser 19th c ireland) – 1mf – 9 – mf#1.1.1860 – uk Chadwyck [941]
The irish difficulty : and how it must be met / Scrope, George Julius Duncombe Poulett – London, 1849 – (= ser 19th c ireland) – 1mf – 9 – mf#1.1.406 – uk Chadwyck [941]
Irish digest – Dublin. 1953-1967 – 1 – mf#597 – us UMI ProQuest [073]
Irish digest, 1938-67 – v1-89 – 31r – 1 – mf#361 – uk Microform Academic [800]
Irish distress and its remedies : the land question. / Tuke, James Hack – London, 1880 – (= ser 19th c ireland) – 2mf – 9 – mf#1.1.1906 – uk Chadwyck [339]
The irish dominicans of the seventeenth century = Epilogus chronologicus exponens succinte conventus et fundationes sacri ordinis praedicatorum in regno hyberniae / O'Heyne, John – Dundalk: William Tempest 1902 [mf ed 1986] – 2mf [ill] – 9 – 0-8370-7090-2 – (in latin & english on opposite pp; incl ind; first publ at louvain in 1706) – mf#1986-1090 – us ATLA [241]
Irish draper – Dublin, Ireland. 1921 – 1r – 1 – (aka: irish draper and fashion trades journal) – uk British Libr Newspaper [740]
Irish draper and fashion trades journal *see* Irish draper
Irish druids and old irish religions / Bonwick, James – London: Griffith, Farran, 1894 – 1mf – 9 – 0-524-00695-4 – (incl bibl ref) – mf#1990-2023 – us ATLA [290]
Irish eastern counties herald – Athy. Ireland. -w. 13 feb-13 mar 1849 – 1/4r – 1 – uk British Libr Newspaper [072]
Irish ecclesiastical gazette – Dublin, Ireland. mar 1856-61, 1863-96 – 23 1/2r – 1 – (aka: irish ecclesiastical gazette or monthly repertory of miscellaneous church news; church of ireland gazette) – uk British Libr Newspaper [242]
Irish ecclesiastical journal – Dublin, Ireland. 1850-52 – 1r – 1 – uk British Libr Newspaper [072]
Irish echo – Dublin, Ireland. 6 nov 1873-1875 – 5r – 1 – uk British Libr Newspaper [072]
Irish economist – Dublin, Ireland. 29 may 1855-25 mar 1856 – 1r – 1 – uk British Libr Newspaper [072]
The irish element in mediaeval culture / Zimmer, Heinrich – New York: Putnam, c1891 – 1mf – 9 – 0-524-01826-X – mf#1990-0506 – us ATLA [940]
Irish essays : and others / Arnold, Matthew – 1st ed. London: Smith, Elder & Co, 1882 [mf ed 1985] – xiv/309p – 1 – mf#8218 – us Wisconsin U Libr [840]
Irish examiner – Dublin, Ireland. 30 sep 1848-17 feb 1879. -w – 1/2r – 1 – uk British Libr Newspaper [072]
Irish farm and land list or real property and general investment advertiser etc *see* Irish farm list land circular and general investment reporter or real property advertiser
Irish farm forest and garden – Dublin, Ireland. 1883-17 oct 1895 – 13r – 1 – (aka: farm; kennel farm poultry yard) – uk British Libr Newspaper [636]
Irish farm list land circular and general investment reporter or real property advertiser – Dublin, Ireland. mar-nov 1858 – 1/4r – 1 – (aka: irish farm and land list or real property and general investment advertiser etc; irish advertiser farm and land list and dublin city circular) – uk British Libr Newspaper [333]
Irish farmers gazette and journal of practical horticulture – Dublin, Ireland. 1861-77 – 19r – 1 – (aka: farmers gazette and journal of practical horticulture) – uk British Libr Newspaper [072]
Irish farmers gazette and journal of practical horticulture *see* Farmers gazette and journal of practical horticulture
Irish farmers journal – Dublin, Ireland. 1986 – 4r – 1 – uk British Libr Newspaper [072]
Irish farmers' journal – Dublin 1950-57, jul 1958-61, 1963-65 – 36r – 1 – ie National [630]
Irish farmers journal and weekly intelligence – 4r – 1 – uk British Libr Newspaper [072]
Irish farming world – Dublin, Ireland. 1892, 1895-96 – 3r – 1 – (incorp with: farmers gazette from dec 1920) – uk British Libr Newspaper [072]
Irish felon – Dublin, Ireland. 24 jun-22 jul 1848 – 1/4r – 1 – uk British Libr Newspaper [072]
Irish field *see* Irish sportsman and farmer
Irish field and gentlemens gazette *see* Irish sportsman and farmer
Irish figaro – Dublin, Ireland. 26 jan 1895-6 apr 1901 – 1 – (aka: irish life) – uk British Libr Newspaper [072]
Irish fireside – Dublin, Ireland. 2 jul 1883-1 oct 1887 – 4r – 1 – uk British Libr Newspaper [072]
The irish folk song society journal – London. 1904-32 – 1 – us L of C Photodup [780]
Irish free state grocery record – Dublin, Ireland. jan, feb 1924 – 1/4r – 1 – uk British Libr Newspaper [072]
Irish freedom – London, UK. 1939-80 – 4r – 1 – (irish democrat 1945-) – uk British Libr Newspaper [320]
Irish friend – Belfast Ireland, 1840 – 1/2r – 1 – uk British Libr Newspaper [072]
Irish golfer – Dublin, Ireland. 23 aug 1899-27 jun 1900 – 1 1/2r – 1 – uk British Libr Newspaper [072]
Irish grocer *see* Wine merchant and grocers review
Irish harp – Adelaide, jan 1871-dec 1872 – 1r – at Pascoe [079]
Irish history and irish character / Smith, Goldwin – Oxford, 1861 – (= ser 19th c ireland) – 3mf – 9 – mf#1.1.9683 – uk Chadwyck [941]
Irish house furnisher – Dublin, Ireland. 1924 – 1/2r – 1 – uk British Libr Newspaper [072]
Irish incumbered gazette *see* Weekly gazette
Irish independent – Dublin, 1873-74 – 1r – 1 – ie National [072]
Irish independent – Dublin: Printed & publ by the proprietors at the Offices, 1905-; 1956-oct 1977; jul-aug 1978; nov-dec 1980; nov-dec 1981 – 1 – us CRL [072]
Irish independent – 1905- – 24r per y – 1 – (sunday independent also available 8r per y $70.00r) – us UMI ProQuest [072]
Irish independent *see* Irish daily independent
Irish insurance banking and finance journal – Dublin, Ireland. nov 1882-29 may 1888, 1 jul 1888-2 jan 1890 – 2 1/2r – 1 – (cont as: finance union, london) – uk British Libr Newspaper [332]
Irish intelligence – London, England. v1-3. 1848 – 1r – us UF Libraries [420]
Irish investment journal legal and commercial advertiser assurance and railway expositor – Dublin, Ireland. 2 jan-2 nov 1855; 4 jan-15 aug 1856 – 1/4r – 1 – uk British Libr Newspaper [072]
Irish journal of education = Iris eireannach an oideachais – Dublin. 1974-1994 (1) 1974-1981 (5) 1975-1981 (9) – ISSN: 0021-1257 – mf#9974 – us UMI ProQuest [370]
Irish journal of management – Dublin, 2001+ [1,5,9] – mf#18911,01 – us UMI ProQuest [341]
Irish journal of medical science – Dublin. 1975-1996 (1) 1976-1996 (5) 1976-1996 (9) – ISSN: 0021-1265 – mf#10237 – us UMI ProQuest [610]
Irish jurist – Dublin: E Ponsonby. v1-8. 1849-66 – 1 – $378.00 – mf#0296 – us Brook [347]
Irish jurist – Dublin, Ireland. 1850-5 nov 1853; 19 nov 1853-20 oct 1855; 1 dec 1855-1 oct 1856; 1858-1 dec 1859; 1860-15 dec 1865 – 9 1/2r – 1 – uk British Libr Newspaper [072]
Irish jurist and local government review : together with reports, statutes, orders, rules and regulations – Dublin. v1-5. 1900-05 (all publ) – 1 – $80.00 set – (title varies: v1-4 as new irish jurist and local government review) – mf#103741 – us Hein [072]
Irish labour advocate – Dublin, Ireland 14,21 feb 1891 – 1 – uk British Libr Newspaper [331]
Irish lance – Dublin, 15th may 1880 – 0.25r – 1 – ie National [072]
The irish land act : will england demand it? – London, 1872 – (= ser 19th c ireland) – 2mf – 9 – mf#1.1.5938 – uk Chadwyck [348]
The irish land act, 1881 : its origin, its principles, and its working / Cook, Edward Tyas – Oxford, 1882 – (= ser 19th c ireland) – 1mf – 9 – mf#1.1.1876 – uk Chadwyck [348]
Irish land and irish rights / Montgomery, Hugh de Fellenberg – London, 1881 – (= ser 19th c ireland) – 1mf – 9 – mf#1.1.1916 – uk Chadwyck [333]
The irish land and labour question, illustrated in the history of ralahine and co-operative farming / Craig, Edward Thomas – London: Truebner & Co., 1893. xii,(3)-204p. Illus., incl. ports – 1 – us Wisconsin U Libr [331]
The irish land bill : analysis and remarks – London, 1881 – (= ser 19th c ireland) – 1mf – 9 – mf#1.1.1909 – uk Chadwyck [348]
The irish land bill – [Kingston-upon-Hull], 1881 – (= ser 19th c ireland) – 1mf – 9 – mf#1.1.1910 – uk Chadwyck [348]
Irish land legislation and the royal commissions / Hodgkin, Howard – London, 1881 – (= ser 19th c ireland) – 1mf – 9 – mf#1.1.1913 – uk Chadwyck [333]
The irish land purchase bill / Churchill, Lord Randolph Henry Spencer – London, 1890 – (= ser 19th c ireland) – 1mf – 9 – mf#1.1.1946 – uk Chadwyck [333]
The Irish land question : scheme for a peasant propriety in ireland / Moffatt, Lewis – Toronto: C Blackett Robinson, 1886 – 1mf – 9 – mf#11156 – cn Canadiana [333]
The irish land question : a problem in practical politics / Errington, George – London, 1880 – (= ser 19th c ireland) – 1mf – 9 – mf#1.1.1896 – uk Chadwyck [333]
Irish land schedule – Dublin, Ireland. may 1850-nov 1872 – 2r – 1 – (aka: alnutts irish land schedule) – uk British Libr Newspaper [333]
Irish law list – Dublin, Ireland. 9 oct-21 dec 1895; 7 jan-22 dec 1896 – 3/4r – 1 – uk British Libr Newspaper [072]
Irish law times – v1-46. 1867-1912 – 455mf – 9 – $682.00 – (updates planned) – mf#LLMC 84-504 – us LLMC [340]
Irish law times and solicitors journal – Dublin, Ireland. Feb 1867-1896 – 30r – 1 – uk British Libr Newspaper [072]
The irish liber hymnorum (hbs13-14) / Bernard, J & Atkinson, R – 1898 – (= ser Henry bradshaw society (hbs)) – 12mf – 8 – €23.00 – (vol 1: text and introduction; vol 2: translations and notes) – ne Slangenburg [780]
Irish life – Dublin, Ireland. 20 feb 1892-1895 – 4r – 1 – (aka: dublin figaro; irish figaro; figaro and irish gentlewoman) – uk British Libr Newspaper [072]
Irish life *see* Irish figaro
Irish litanies (hbs62) / Plummer, Charles – 1925 – (= ser Henry bradshaw society (hbs)) – 3mf – 8 – €7.00 – ne Slangenburg [241]
Irish literary revival / Ryan, William Patrick – London, England. 1894 – 1r – us UF Libraries [420]
Irish loyalty and english gratitude / Staples, Robert, Jr – Dublin, 1869 – (= ser 19th c ireland) – 1mf – 9 – mf#1.1.7764 – uk Chadwyck [941]
The irish magistracy – Dublin, 1885 – (= ser 19th c ireland) – 1mf – 9 – mf#1.1.7179 – uk Chadwyck [340]
Irish manufacturers journal – Dublin, Ireland. oct 1881, 16 nov, 27 dec 1882, 15 sep 1887, 1 oct-15 dec 1887, 1888-2 jul 1892 – 5 1/2r – 1 – (aka: commercial ireland) – uk British Libr Newspaper [670]
Irish manufacturers journal *see* Commercial ireland
Irish marriage question / Stoddart, John – London, 1844 – (= ser 19th c ireland) – 1mf – 9 – mf#1.1.7263 – uk Chadwyck [346]
Irish medical times – Dublin. 1974-1976 (1) – ISSN: 0047-147X – mf#8164 – us UMI ProQuest [610]
Irish medieval monasteries on the continent / Fuhrmann, J – Washington DC, 1927 – 3mf – 8 – €7.00 – ne Slangenburg [720]
Irish mercantile gazette – Dublin, Ireland. 22 sep 1877 – 1/4r – 1 – uk British Libr Newspaper [072]
Irish migration / Fitzgerald, James Edward – London, 1848 – (= ser 19th c ireland) – 1mf – 9 – mf#1.1.521 – uk Chadwyck [304]
Irish missionary record and chronicle of the reformation – Dublin, Ireland. Nov-dec 1852; jan-aug 1853; nov 1853-feb 1854 – 1/2r – 1 – uk British Libr Newspaper [072]
Irish mist and sunshine : a book of ballads / Dollard, James Bernard – Boston: R G Badger; Toronto: W E Blake, 1901 – 2mf – 9 – 0-665-88093-6 – (int by william o'brien) – mf#88093 – cn Canadiana [810]
Irish monasticism, origins and early development / Ryan, John – London, 1939 – 10mf – 8 – €19.00 – ne Slangenburg [240]
Irish monasticism. origins and early development / Ryan, John – Dublin, 1931; Madrid: Razon y Fe, 1932 – 1 – sp Bibl Santa Ana [240]
Irish nation and the peasant – Dublin, Ireland. 1909, 1 jan-3 dec 1910 [wkly] – 1 1/2r – 1 – (aka: peasant; peasant and irish ireland; irish nation and the peasant) – uk British Libr Newspaper [072]
Irish nation and the peasant *see* Irish nation and the peasant
Irish national education / Nesbitt, William – Dublin, 1864 – (= ser 19th c ireland) – 1mf – 9 – mf#1.1.928 – uk Chadwyck [370]
Irish national guard – Dublin, Ireland. 22 apr-22 jul 1848. -w – 1/4r – 1 – uk British Libr Newspaper [072]
Irish nationality in 1870 / Macdonnell, Robert – Dublin, 1870 – (= ser 19th c ireland) – 1mf – 9 – mf#1.1.1875 – uk Chadwyck [941]
Irish news – Belfast 1930 – 4r – 1 – ie National [072]
Irish news – Belfast, Ireland 15 aug 1891-27 aug 1892 – 1 – (cont by: irish news & belfast morning news [29 aug 1892-31 dec 1925, 1 jan-29 mar, 10 apr-30 sep, 11 oct 1926- (mf 1986-)]) – uk British Libr Newspaper [072]
Irish newspapers – Dublin, Ireland. 1685-1825 (1) – mf#9020 – us UMI ProQuest [072]
Irish observer – Limerick, Ireland. 1 may 1824 – 1/4r – 1 – uk British Libr Newspaper [072]
Irish packet – Dublin, Ireland. 27, 29 oct, 10 dec 1807; 5 mar 1808; 26 apr, 5 may 1810 – 1/4r – 1 – uk British Libr Newspaper [072]
Irish patriot – Dublin, Ireland. 14 sep 1878-14 feb 1880 – 1 1/2r – 1 – uk British Libr Newspaper [072]
The irish peasant : a sociological study edited from original papers – London, 1892 – (= ser 19th c ireland) – 2mf – 9 – mf#1.1.4193 – uk Chadwyck [100]
Irish peasant proprietors : facts and misrepresentations a reply to the statements of mr tuke / Sinclair, W J – [Edinburgh], 1880 – (= ser 19th c ireland) – 1mf – 9 – mf#1.1.2192 – uk Chadwyck [333]
Irish pedigrees: or, the origin and stem of the irish nation / O'Hart, John – Dublin, 1876. 2v – 1 – us Wisconsin U Libr [920]
Irish people – Dublin, Ireland. 16 sep 1899-7 nov 1903; 30 sep 1905-27 mar 1909.-w – 6r – 1 – (publ only 16 sep 1899-27 mar 1909) – uk British Libr Newspaper [072]
Irish people – Dublin, Ireland. 28 nov-dec 1863; 2 jun-19 nov 1864; 24 jun-16 sep – 1r – 1 – (publ only 28 nov 1863-16 sep 1865) – uk British Libr Newspaper [072]
Irish pictorial *see* Weekly irish times
The irish political review / Sellors, Michael – Dublin, 1832 – (= ser 19th c ireland) – 1mf – 9 – mf#1.1.1849 – uk Chadwyck [330]
The irish poor in english prisons and workhouses – London, 1866 – (= ser 19th c ireland) – 1mf – 9 – mf#1.1.101 – uk Chadwyck [345]
Irish poor law : past, present and future – London: James Ridgway, 1849 – (= ser 19th c economics) – 1mf – 9 – mf#1.1.452 – uk Chadwyck [941]
The irish poor law : how far has it failed? and why? a question addressed to the common sense of his countrymen / Scrope, George Julius Duncombe Poulett – London: James Ridgway, 1849 – (= ser 19th c economics) – 1mf – 9 – mf#1.1.451 – uk Chadwyck [941]
Irish poor law question : a letter to the rt hon lord john russell – London, 1847 – (= ser 19th c ireland) – 1mf – 9 – mf#1.1.599 – uk Chadwyck [941]
Irish post and weekly telegraph – Dublin Ireland, apr 1910-1911 – 2r – 1 – uk British Libr Newspaper [072]
Irish potato disease investigations, 1924-1925 : a preliminary report / Gratz, L O – Gainesville, FL. 1925 – 1r – us UF Libraries [630]
Irish potatoes : rye soft marl phosphate as a fertilizer – Lake City, FL. 1891 – 1r – us UF Libraries [630]
Irish potatoes in florida / Spencer, A P – Gainesville, FL. 1914 – 1r – us UF Libraries [630]
Irish potatoes in florida / Spencer, A P – Gainesville, FL. 1917 – 1r – us UF Libraries [630]
Irish presbyterian – Belfast Ireland, sep-dec 1853; jan, mar, dec 1854; 1855; jan, mar, jun, aug, sep, nov, dec 1856; 1857; jan-aug, oct-dec 1858 – 2r – 1 – uk British Libr Newspaper [242]

Irish press – 1930- – 12r per y – 1 – us UMI ProQuest [072]
Irish press – Dublin 1936 – 12r – 1 – ie National [072]
Irish press – Dublin, Ireland. 1986-25 may 1995 – 119 1/2r – 1 – uk British Libr Newspaper [072]
Irish press – Dublin, may/jun 1952 – 1r – 1 – ie National [072]
The irish press – 1760-1922 (mf ed 1999-) – 1 – (will be released in units of 40r 5 times a yr: 2 units in 1999, and 5 units per yr thereafter) – us Primary [072]
The irish problem and england's difficulty – [London], [1886] – (= ser 19th c ireland) – 4mf – 9 – mf#1.1.8483 – uk Chadwyck [941]
The irish problem and how to solve it – London, [1881] – (= ser 19th c ireland) – 5mf – 9 – mf#1.1.8482 – uk Chadwyck [941]
Irish protestant and church of ireland review – Dublin, Ireland. aug 1901-oct 1908, jan-may 1909, jan 1913, jan 1915 – 6r – 1 – (aka: irish protestant and church review; irish protestant) – uk British Libr Newspaper [242]
Irish protestant and church review *see* Irish protestant and church of ireland review
Irish quarterly review – Dublin. 1851-1859 (1) – mf#4268 – us UMI ProQuest [420]
The irish question : a speech delivered at liverpool on june 29th, 1886s / Derby, Edward Henry Smith Stanley, 15th Earl of – [London, 1886] – (= ser 19th c ireland) – 1mf – 9 – mf#1.1.251 – uk Chadwyck [941]
The irish question : union or separation? – Dublin, 1886 – (= ser 19th c ireland) – 1mf – 9 – mf#1.1.1894 – uk Chadwyck [941]
The irish question : with special reference to home rule in canada: speeches / Blake, Edward – [S.l: s.n, 1892?] – 1mf – 9 – 0-665-00175-4 – mf#00175 – cn Canadiana [941]
The irish question examined in a letter to the "new york herald" / Dunraven, Windham Thomas Wyndham-Quin, 4th Earl of – London, 1880 – (= ser 19th c ireland) – 1mf – 9 – mf#1.1.1941 – uk Chadwyck [330]
Irish racing book and sheet calendar *see* Racing calendar
Irish railway gazette etc – Ireland.4 Nov 1844-13 May 1850. -w. 3 reels – 1 – uk British Libr Newspaper [072]
Irish railway gazette mining and commercial journal etc – Dublin, Ireland. 4 nov 1844-13 may 1850 – 3r – 1 – uk British Libr Newspaper [072]
Irish railway telegraph and journal of mining banking insurance etc – Dublin, Ireland. 4 oct-27 dec 1845; 3 jan-28 feb 1846 – 1/2r – 1 – uk British Libr Newspaper [072]
Irish railways and state purchase / Findlay, George – [London], [1886] – (= ser 19th c ireland) – 1mf – 9 – mf#1.1.6933 – uk Chadwyck [380]
The irish reformation : or, the alleged conversion of the irish bishops at the accession of queen elizabeth... / Brady, William Maziere – 5th ed. London: Longmans, Green, 1867 [mf ed 1989] – 1mf – 9 – 0-7905-4489-X – mf#1988-0489 – us ATLA [241]
The irish relief measures : past and future / Scrope, George Julius Duncombe Poulett – London, 1848 – (= ser 19th c ireland) – 2mf – 9 – mf#1.1.256 – uk Chadwyck [941]
Irish reporter – Dublin, Ireland. jan-nov 1856 – 1/4r – 1 – uk British Libr Newspaper [072]
Irish saints in great britain / Moran, Patrick Francis – Dublin: M H Gill: Browne & Nolan, 1879 [mf ed 1986] – 1mf – 9 – 0-8370-6921-1 – (incl bibl ref) – mf#1986-0921 – us ATLA [240]
Irish schoolmaster – London, England. 18– – 1r – us UF Libraries [241]
Irish schoolmistress and female teachers assistant – Dublin, Ireland. 28 feb-2 may 1891 – 1/4r – 1 – uk British Libr Newspaper [072]
Irish seditions : their origin and history from 1792-1880 – London, [1883] – (= ser 19th c ireland) – 1mf – 9 – mf#1.1.1925 – uk Chadwyck [941]
Irish shield – Philadelphia. 1829-1831 (1) – mf#5582 – us UMI ProQuest [978]
Irish Socialist Federation in America *see* Harp
Irish society – Dublin, Ireland. 14 jan 1888-1896; 1917-7 oct 1922; 1923-21 jun 1924 – 26r – 1 – uk British Libr Newspaper [072]
The irish society of london – Londonderry, 1876 – (= ser 19th c ireland) – 1mf – 9 – mf#1.1.1934 – uk Chadwyck [330]
Irish sporting life – Dublin, Ireland. 14 mar-jun 1840 – 1/4r – 1 – uk British Libr Newspaper [072]
Irish sporting news – Dublin, Ireland. 18, 22 jan 1879 – 1/4r – 1 – uk British Libr Newspaper [072]
Irish sporting times – Dublin, Ireland. 14 mar-18 apr 1876 – 1/4r – 1 – uk British Libr Newspaper [072]
Irish sportsman and farmer – Dublin, Ireland. 19 feb 1870-1896, 1911, 1920, 1921, 1950, 1952, 1971 – 36 1/2r – 1 – (aka: irish field and gentlemens gazette; irish field) – uk British Libr Newspaper [790]
Irish star and catholic weekly record – Dublin, Ireland. 29 jul 1871-26 jun 1875 – 4r – 1 – uk British Libr Newspaper [072]
Irish sun – Dublin, Ireland. 12 jun-3 jul 1880 – 1/4r – 1 – uk British Libr Newspaper [072]
Irish temperance chronicle and industrial and family magazine – Dublin, Ireland. Sep, oct 1846 – 1/4r – 1 – uk British Libr Newspaper [072]
Irish temperance league journal – Belfast Ireland. feb 1863-70, 1874-96 – 8r – 1 – (imperfect; aka: everybody's monthly) – uk British Libr Newspaper [072]
Irish temperence and literary gazette – Dublin, Ireland. 12 nov 1836-29 sep 1838 – 1r – 1 – uk British Libr Newspaper [072]
Irish temperence and literary gazette – Ireland. -w. Nov 1836-29 Sept 1838. 1 reel – 1 – uk British Libr Newspaper [072]
Irish templar – Belfast Ireland, apr 1877-1882; jan-apr 1883; jul 1883-oct 1884; dec 1884-1896 – 5r – 1 – (aka: irish templar and temperance journal) – uk British Libr Newspaper [072]
Irish tenant – Dublin, Ireland. 15 jan-15 apr 1880 – 1/4r – 1 – uk British Libr Newspaper [072]
Irish tenant league – Dublin, Ireland. jun, jul, sep-dec 1851; jan, feb, apr 1852 – 1/4r – 1 – uk British Libr Newspaper [072]
The irish tenant-right question examined by a comparison of the law and practice of england with...ireland / Baxter, Robert – London, 1869 – (= ser 19th c ireland) – 1mf – 9 – mf#1.1.1882 – uk Chadwyck [346]
Irish text society. publications – London. v1-41. 1899-1941 – 9 – $850.00 – mf#0297 – us Brook [400]
Irish textile journal – Belfast Ireland, 1886-1892; 1895-1896 – 4r – 1 – uk British Libr Newspaper [072]
Irish theological quarterly – Maynooth. 1907+ (1) 1971+ (5) 1975+ (9) – ISSN: 0021-1400 – mf#5057 – us UMI ProQuest [240]
The irish theosophist – Dublin: I T Press. v1-5. 1892-97 [mthly] [mf ed 2003] – 5v on 1r – 1 – (merged with: grail to form: internationalist [dublin, ireland]) – mf#2003-s043 – us ATLA [390]
Irish times – Dublin, dec 1824-jul 1825 – 1 – (cont as: irish times and dublin journal [11 apr-18 jul 1825]) – ie National [072]
Irish times – Dublin, 2002+ [1,5,9] – mf#60261 – us UMI ProQuest [072]
Irish times – Dublin, Ireland. 15-25 oct 1823; dec 1824; 3 jan-18 jul 1825 – 1r – 1 – uk British Libr Newspaper [072]
Irish times – Dublin, Ireland. -d. 29 mar 1859-1870; 1872-20 feb 1905; 8 mar 1905-jul 1934; oct 1934-1937; 1950-1967; 1992 – 592 1/2r – 1 – uk British Libr Newspaper [072]
The irish times – Dublin: Irish Times Ltd, jul 1938-jan 1975; apr 1975-jul 1976; sep-oct 1976; feb-aug, oct-nov 1, dec 1977; jan-mar 2 1978 – 1 – us CRL [072]
Irish trades' advocate – Dublin, Ireland 13 sep-25 oct 1851 – 1 – uk British Libr Newspaper [380]
Irish tribune – Dublin, Ireland. 10 jun-8 jul 1848. -w – 1/4r – 1 – uk British Libr Newspaper [072]
Irish turf telegraph and dramatic gazette – Dublin, Ireland. 3 jul, 28 aug, 16 oct 1875 – 1/4r – 1 – uk British Libr Newspaper [072]
The irish university question / Walsh, William Joseph – Dublin, 1890 – (= ser 19th c ireland) – 2mf – 9 – mf#1.1.952 – uk Chadwyck [378]
Irish weekly – Belfast Ireland. 29 aug 1891-1926 – 36 1/4r – 1 – (aka: irish weekly and ulster examiner) – uk British Libr Newspaper [072]
Irish weekly and ulster examiner – 1 – (Belfast: irish news. running title: irish weekly, belfast) – us Wisconsin U Libr [073]
Irish weekly and ulster examiner *see* Irish weekly
Irish weekly independent – Dublin, Ireland. 8 apr 1893-96, 1898, 1916, 1 jan-26 nov 1921, 1926, 1939 – 12 1/2r – 1 – (aka: irish weekly independent and nation; illustrated irish weekly independent and nation) – uk British Libr Newspaper [072]
Irish weekly mail and sports mail *see* Warder
Irish weekly mail and warder *see* Warder
Irish, William Norman *see* Hebrew charts
Irish women writers of the romantic era : papers of mary tighe (1772-1810) and lady sydney morgan (1776-1859) from the national library of ireland – 9r – 1 – £850.00 – uk Matthew [420]
Irish worker – Dublin 1911-30th jan 1915 – 2r – 1 – ie National [072]
Irish worker – Dublin, Ireland. -m. Mar-jun 1893 – 1/4r – 1 – uk British Libr Newspaper [072]
Irish worker – Dublin, Ireland, ns: 11 nov 1911-5 dec 1914, 16 jun 1923-9 may 1925, 11 oct 1930-12 mar 1932 – 1 – (wanting: 7 jul, 8 dec 1923; 13 dec 1924; 21 feb 1925; 6,13 feb 1930; cont: irish worker & people's advocate [27 may-4 nov 1911]) – uk British Libr Newspaper [331]
Irish worker and people's advocate – Dublin, Ireland 27 may-4 nov 1911 – 1 – (cont by: irish worker [ns: 11 nov 1911-5 dec 1914, 16 jun 1923-9 may 1925, 11 oct 1930-12 mar 1932]) – uk British Libr Newspaper [331]
Irish world – New York 1876-1903, 1905-13 – 36r – 1 – ie National [071]
Irish world and american industrial liberator – New York NY. [1890 jan 11/dec 27]-[1904 apr 23/1905 apr 8] [gaps] – 12r – 1 – (cont: irish world; cont by: gaelic american; irish world and american industrial liberator and the gaelic american) – mf#854457 – us WHS [071]
The irish world and american industrial liberator – New York. nov 5 1870-dec 1950. [incomplete] – 1 – us NY Public [073]
Irish young mens journal *see* Wesleyan young mens journal
Irish-canadian representatives : their past acts, present stand, future prospects: a review of the question / Foran, Joseph Kearney – Ottawa?: Evening Journal Office, 1886 – 1mf – 9 – mf#24313 – cn Canadiana [323]
Irishman – Dublin, Ireland. 10 jun-1 jul 1848 – 1/4r – 1 – uk British Libr Newspaper [072]
Irishman – Dublin, Ireland. 1840-8 oct 1842 – 2 1/2r – 1 – uk British Libr Newspaper [072]
Irishman : or galway mayo roscommon sligo and clare chronicle – Galway, Ireland. 6 may-12 dec 1835 – 1/2r – 1 – uk British Libr Newspaper [072]
Irishman – Dublin, Ireland. 17 jul 1858-1862; 7 mar 1863-mar 1867: feb 1868-28 feb 1885 – 26r – 1 – (publ only 17 jul 1858-28 feb 1885) – uk British Libr Newspaper [072]
Irishman – Dublin, Ireland. 1849-25 may 1850 (missing 10, 17 aug 1850) – 1 1/2r – 1 – (publ only 1849-may 1850) – uk British Libr Newspaper [072]
The irishman – Belfast, Ireland. v. 1-2. Jan. 15, 1916-Feb. 10, 1917 – 1 – us NY Public [941]
The irishman – Galway. Ireland. -sw. 6 May-12 Dec 1835. (36 ft) – 1 – uk British Libr Newspaper [072]
Irj *see* International railway journal
Irka: roman / Jungfer, Victor – Karlsbad: Adam Kraft, 1945. 383p – 1 – us Wisconsin U Libr [830]
Irkutskie dni: bol'shaia ezhedn obshchestv -polit i lit gaz / ed by Sechkina, A – Irkutsk: Gruppa sotsial-demokratov 1918 [1918 16 marta-4 avg] – (= ser Asn 1-3) – n25-39 [1918] [gaps] item 183, on reel n38 – 1 – mf#asn-1 183 – ne IDC [077]
Irkutskie gubernskie vedomosti – Irkutsk, 1858 – 1 – us UMI ProQuest [077]
Irkutskie gubernskie vedomosti / ed by Porshnev, G I – Irkutsk: [s n] 1919 [1857 16 maia-] – (= ser Asn 1-3) – n6217-6277 [1919] [gaps] item 182, on reel n38 – 1 – mf#asn-1 182 – ne IDC [077]
Irkutskii gosudarstvennyi universitet : odnodn gaz / ed by Mironov, N D et al – Irkutsk: [s n] 1918 – (= ser Asn 1-3) – 1v item 184, on reel n38 – 1 – mf#asn-1 184 – ne IDC [077]
Irkutskii kooperator – Irkutsk, 1916-17 (4) – 6mf – 9 – mf#COR-598 – ne IDC [077]
Irkutskii strelok : voen -lit gaz – Krasnoufimsk [Ekaterinburg gub]: Shtab Svod korpusa gen Grivina 1919 [1919 13 marta-2 maia] – (= ser Asn 1-3) – n1-15 [1919] item 185, on reel n38 – 1 – mf#asn-1 185 – ne IDC [077]
Irkutskiia gubernskiia viedomosr / Russia – 1859, 1871-98 – 1 – $269.00 – us L of C Photodup [947]
Irkutskij gor sovet rk i kd *see* Vestnik irkutskogo soveta rabochikh deputatov
Irkutskij strelok – Krasnoufimsk, Russia, 1919 – 1r – 1 – us UMI ProQuest [077]
Irkutskoe slovo – Irkutsk, 1911-12 – 1 – us UMI ProQuest [077]
L'irlande libre : organe de la colonie irlandaise a paris – Paris. juin 1897-oct 1898, avr 1900 – 1 – fr ACRPP [073]
Irle, J *see* Deutsch-herero-worterbuch
Irm multinational reports / Institute for Research and Information on Multinationals – Chichester. 1984-1986 (1,5,9) – ISSN: 0747-6337 – mf#16106 – us UMI ProQuest [337]
Irmela : eine geschichte aus alter zeit / Steinhausen, Heinrich – 47. Aufl. Stuttgart: J F Steinkopf, [1950] – 1r – 1 – us Wisconsin U Libr [830]
Irodalmi ujsag – Budapest. Hungary. -w. 2 Nov 1950-26 Mar, 10 Sep, 3 Dec 1955, 7 Jan-20 Oct 1956. (3 reels) – 1 – uk British Libr Newspaper [079]
Irodalmi ujsag – Paris, France. 1971-75 – 1/2r – 1 – uk British Libr Newspaper [072]
Irodalmi ujsag – Paris. France. -w. Feb 1962-Dec 1965. (3 reels) – 1 – uk British Libr Newspaper [074]
Iromanyok / Hungary. Nemzetgyeules – v. 1-32. 1920-26 – 1 – 52.00 – us L of C Photodup [943]
Iromanyok / Hungary. Orszaggyules. Felsohaz – v. 1-197. 1865 68-1944. (1865-72, 1915-18 and scattered issues wanting) – 1 – us L of C Photodup [943]
Iron age – New York. 1873-1976 (1) 1965-1976 (5) 1970-1976 (9) – (cont by: chilton's iron age) – ISSN: 0021-1508 – mf#919 – us UMI ProQuest [660]
Iron age – New York. 1987-1993 (1,5,9) – (cont: iron age metals producer. cont by: iron age new steel) – ISSN: 0897-4365 – mf#16951 – us UMI ProQuest [660]
Iron age manufacturing management – Radnor. 1986-1987 (1) 1986-1987 (5) 1986-1987 (9) – (cont: chilton's iron age manufacturing management) – ISSN: 0893-2360 – mf#13869,01 – us UMI ProQuest [660]
Iron age metals producer – Radnor. 1986-1987 (1) 1986-1987 (5) 1986-1987 (9) – (cont: chilton's iron age metals producer. cont by: iron age) – ISSN: 0893-9616 – mf#13991,01 – us UMI ProQuest [660]
Iron age new steel – New York. 1993-2001 (1,5,9) – (cont: iron age) – ISSN: 1074-1690 – mf#20372 – us UMI ProQuest [660]
Iron agitator – Ishpeming MI. 1885 oct 31 – 1r – 1 – (cont: agitator [ishpeming, mi]; cont by: iron ore [ishpeming, mi]) – mf#851882 – us WHS [071]
Iron and coal trades review – Middlesbrough and London. 6 Oct 1869-dec 1886[wkly] – 30r – 1 – uk British Libr Newspaper [622]
Iron and steel : a brief historic sketch of their manufacture and use: a paper read before the hamilton association, mar 23 1882 / Freed, Augustus Toplady – [Hamilton, Ont?: s.n, 1882?] [mf ed 1993] – 1mf – 9 – 0-665-91569-1 – (in dble clms) – mf#91569 – cn Canadiana [660]
Iron and steel engineer – Pittsburgh. 1924-1999 (1) 1965-1999 (5) 1977-1999 (9) – (cont by: aise steel technology) – ISSN: 0021-1559 – mf#1473 – us UMI ProQuest [660]
Iron and steel in india : a chapter from the life of jamshedji n tata / Fraser, Lovat – Bombay: Times Press, 1919 – (= ser Samp: indian books) – us CRL [920]
Iron and Steel Institute *see* 67 special reports and bibliographies
Iron and steel institute journal – London. 1935-1973 (1) – ISSN: 0021-1567 – mf#1259 – us UMI ProQuest [660]
The iron cardinal : the romance of richelieu / McCabe, Joseph – New York: J McBride, 1909 – 1mf – 9 – 0-7905-6349-5 – mf#1988-2349 – us ATLA [944]
Iron city and pittsburgh weekly chronicle – Pittsburgh, PA. 1841-1842 (1) – mf#66041 – us UMI ProQuest [071]
Iron city socialist / Allegheny County Socialist Party – Pittsburgh PA. 1913 oct 4 – 1r – 1 – mf#868118 – us WHS [335]
Iron county citizen – Hurley WI. 1905 jan 14-21, feb 4, may 13, 1906 dec 1, 1908 feb 1, apr 18 – 1r – 1 – (cont by: montreal river miner and iron county republican; montreal river miner and iron county citizen) – mf#876270 – us WHS [071]
Iron County Historical and Museum Society [MI] *see* Mining reporter
Iron county miner – Hurley WI. [1950 apr 7/1951]-[2002] [small gaps] – 40r – 1 – (cont: montreal river miner [hurley, wis. : 1940]; iron county news) – mf#947864 – us WHS [622]
Iron county news – Hurley WI. [1913 sep 20-1915], [1929-1933 dec 29], [1933 dec 8-1939], 1908 aug 1-1913 jun 28, 1916-18, 1919 dec 27-1920, 1921-24, 1925-28, 1940-43, 1943 oct 22-1945, 1946-1950 mar 31 – 9r – 1 – (cont: our land [hurley, wi]; cont by: montreal river miner [hurley, wis. : 1940]; iron county miner) – mf#947878 – us WHS [071]
Iron county republican – Hurley WI. 1894 jun 8-1896 oct 24, 1896 oct 31-1898 may 21, 1898 may 28-1899 dec 30, 1900 jan 6-1901 aug 10, 1901 aug 17-1903 oct 31 – 5r – 1 – (cont by: montreal river miner and gogebic iron tribune; montreal river miner and iron county republican) – mf#926725 – us WHS [071]
The iron horse / Ballantyne, Robert Michael – London: Nisbet, 1871? – 5mf – 9 – mf#07472 – cn Canadiana [830]
The iron ores of pictou county, nova scotia / Gilpin, Edwin – S.l: s.n, 1885? – 1mf – 9 – mf#03478 – cn Canadiana [830]
Iron river pioneer – Iron River WI. [1895 apr 11/1897 jun 17]-[1989 jan/apr] [small gaps] – 41r – 1 – (cont: homestead [iron river, wi]) – mf#1144729 – us WHS [071]

Iron river times – Iron River WI. 1892 mar 10-1895 apr 4 – 1r – 1 – (Continued By: Times [Washburn, Wis.]) – mf#942425 – us WHS [071]

Iron valley reporter / Tuscarawas Co. Dover – jun 1872-apr 1900 poor quality [wkly] – 9r – 1 – mf#B4491-4499 – us Ohio Hist [071]

Iron valley reporter/w / Tuscarawas Co. Canal Dover – feb-dec 1876 [wkly] – 1r – 1 – mf#B34390 – us Ohio Hist [071]

Ironbridge weekly journal and borough of wenlock advertiser – England. Boro' of Wenlock Express – Wenlock & Ludlow Express. -w. 17 July 1869-4 March 1882. Lacking July-Dec 1874. 11 reels – 1 – uk British Libr Newspaper [072]

Irondequoit press – Rochester, NY. 1936-1972 (1) – mf#65194 – us UMI ProQuest [071]

Ironia y generacion / Bustamante Y Montoro, Antonio Sanchez De – Habana, Cuba. 1937 – 1r – us UF Libraries [972]

Ironias / Sanchez-Arjona, Vicente – Sevilla: Artes Graficas, Tomo 1. 1948 – 1 – sp Bibl Santa Ana [810]

Ironias / Sanchez-Arjona, Vicente – Sevilla: Graf. Tirvia, Tomo 4. 1954 – 1 – sp Bibl Santa Ana [810]

Ironias / Sanchez-Arjona, Vicente – Sevilla: Graficas Tirvia, Tomo 5. 1955 – 1 – sp Bibl Santa Ana [810]

Ironias / Sanchez-Arjona, Vicente – Sevilla: Imp. Carlos Acuna, Tomo 2. 1953 – 1 – sp Bibl Santa Ana [810]

Ironias / Sanchez-Arjona, Vicente – Sevilla: Imp. Carlos Acuna, Tomo 3. 1953 – 1 – sp Bibl Santa Ana [810]

Ironias / Sanchez-Arjona, Vicente – Sevilla: Imprenta Cuadrado, Tomo 6. 1958 – 1 – sp Bibl Santa Ana [810]

Ironias y sutilezas con honores / Sanchez-Arjona, Vicente – Sevilla: Imp. Alvarez, 1960 – 1 – sp Bibl Santa Ana [810]

Ironmaking and steelmaking – London. 1989-1996 (1,5,9) – ISSN: 0301-9233 – mf#15688 – us UMI ProQuest [660]

Ironmolders' journal / International Molders' Union of North America et al – v35 n9 [1899] – 1r – 1 – (cont: iron molders' international journal [cincinnati, oh: 1870]; cont by: international molder's journal) – mf#1428642 – us WHS [680]

Irons, David *see* A study in the psychology of ethics

Irons, Joseph *see* Beware of idolatry

Irons, William Josiah *see*
- The bible and its interpreters
- Christianity as taught by s paul
- On miracles and prophecy

Ironside, Henry Allan *see* The weeping prophet

Ironton city directories, 1893-1899, 1903 – 1r – 1 – mf#B31413 – us Ohio Hist [978]

Ironton daily news / Lawrence Co. Ironton – jan 3-jun 30 1939 – 1r – 1 – mf#B40225 – us Ohio Hist [071]

Ironton news / Lawrence Co. Ironton – jan-jun 1929; jul-dec 1931 – 2r – 1 – mf#B37486-37487 – us Ohio Hist [071]

Ironton of sweet long ago, 1872-1874 / Gilruth, James – 1r – 1 – mf#B26957 – us Ohio Hist [240]

Irontoner post / Lawrence Co. Ironton – (feb-mar 1891) [wkly] – 1r – 1 – (in german) – mf#B31208 – us Ohio Hist [071]

Ironwood times – Ironwood MI. 1915 apr 24-1917 jun 2 – 1r – 1 – mf#851888 – us WHS [071]

The iroquois beach / Coleman, Arthur Philemon – S.l: s.n, 1898? – 1mf – 9 – mf#03211 – cn Canadiana [550]

Iroquois falls enterprise – Ontario, CN. 1963- – 1r/y – 1 – Can$93.00 – cn Commonwealth Imaging [071]

Iroquois indians : a documentary history / McNickle, D'Arcy – 50r – 1 – (coll provides 8812 documents dated from the early 1600s to the 1920s wh reflect indian participation in the most important events of early american history. incl printed guide) – mf#C39-27360 – us Primary [975]

The iroquois trail : or, footprints of the six nations: in customs, traditions and history / Beauchamp, William M – Fayetteville, NY: H C Beauchamp, 1892 – 2mf – 9 – 0-665-04555-7 – (incl ind) – mf#04555 – cn Canadiana [306]

L'iroquoise du lac saint-pierre : legende / Frechette, Louis – [Quebec (Province)?: s.n, 1861?] (mf ed 1994) – 9 – cn Bibl Nat [390]

Der irre von st james : aus dem reisetagebuche eines arztes / Galen, Philipp – 7. aufl. St Louis MO: Louis Lange 1924 [mf ed 1995] – 1r – 1 – (filmed with: reichsstaedtische erzaehlugen / herman kurz) – mf#3679p – us Wisconsin U Libr [880]

Irregular periodical : newsletter of bread and roses women's health center / Bread & Roses Women's Health Center – v1 n1-v7 n1 [1979 fall-1987 spr] – 1r – 1 – mf#1406261 – us WHS [362]

Irreligion de l'avenir : a sociological study / Guyau, Jean Marie – New York: H Holt, 1897 – 2mf – 9 – 0-7905-3883-0 – (in english) – mf#1989-0376 – us ATLA [200]

Irrepressible – Winfield, WV. 1911-1913 (1) – mf#67526 – us UMI ProQuest [071]

Irrgarten der liebe : verliebte, launenhafte und moralische lieder, gedichte und sprueche aus den jahren 1885 bis 1900 / Bierbaum, Otto Julius – Leipzig: Insel-Verlag, 1901 [mf ed 1989] – xxxi/475p – 1 – mf#7020 – us Wisconsin U Libr [810]

Irrgarten gottes : oder, die komoedie des chaos / Winckler, Josef – Jena: E Diederichs, 1922 – 1r – 1 – us Wisconsin U Libr [820]

Irricab – Bet Dagan. 1982-1984 (1) 1980-1984 (5,9) – ISSN: 0376-5083 – mf#49305 – us UMI ProQuest [333]

Irrigation age – Minneapolis. 1985-1986 (1) 1985-1986 (5) 1985-1986 (9) – ISSN: 0021-1656 – mf#15039 – us UMI ProQuest [630]

Irrigation and drainage systems – Dordrecht. 1989-1995 (1,5,9) – ISSN: 0168-6291 – mf#16804 – us UMI ProQuest [630]

Irrigation and power journal – New Delhi. 1973-1973 (1) – ISSN: 0021-1664 – mf#8675 – us UMI ProQuest [630]

Irrigation by artesian wells : report of exploratory survey / McKay, E B – Victoria BC: R Wolfenden, 1888? – 1mf – 9 – mf#17903 – cn Canadiana [550]

Irrigation development / California Office Of State Engineer – Sacramento, CA. 1886 – 1r – us UF Libraries [500]

Irrigation in india : the present state of the question – Allahabad, 1869 – (= ser 19th c british colonization) – 1mf – 9 – mf#1.1.3735 – uk Chadwyck [630]

Irrigation in southern europe : being the report of a tour of inspection of the irrigation works of france, spain, and italy, undertaken in 1867-68 for the government of india / Scott-Moncrieff, Colin Campbell – London 1868 – (= ser 19th c british colonization) – 6mf – 9 – (with app) – mf#1.1.5432 – uk Chadwyck [333]

Irrigation journal – Elm Grove. 1975-1995 (1) 1976-1995 (5) 1976-1995 (9) – ISSN: 0047-1518 – mf#10597 – us UMI ProQuest [333]

Irrigation record – Leeton, feb 1915-jun 1917 – 1r – A$93.68 vesicular A$99.18 silver – at Pascoe [079]

Irrigation science – Heidelberg. 1983-1996 (1,5,9) – ISSN: 0342-7188 – mf#13186 – us UMI ProQuest [630]

Irrigation works in india and egypt / Buckley, Robert Burton – London, 1893 – (= ser 19th c books on british colonization) – 8mf – 9 – mf#1.1.6645 – uk Chadwyck [630]

Die irrlehrer der pastoralbriefe / Luetgert, Wilhelm – Guetersloh: C Bertelsmann, 1909 – (= ser Beitraege zur foerderung christlicher theologie) – 1mf – 9 – 0-8370-9637-5 – (incl bibl ref) – mf#1986-3637 – us ATLA [227]

Die irrlehrer des judas- und 2. petrusbriefes / Werdermann, Hermann – Guetersloh: C Bertelsmann, 1913 – (= ser Beitraege zur foerderung christlicher theologie) – 1mf – 9 – 0-524-05945-4 – (incl bibl ref) – mf#1992-0702 – us ATLA [227]

Die irrlehrer im ersten johannesbrief / Wurm, Alois – Freiburg i B, St Louis MO: Herder, 1903 – (= ser Biblische studien) – 1mf – 9 – 0-8370-9674-X – mf#1986-3674 – us ATLA [227]

Die irrthuemer ueber die ehe / Schneemann, G – 2., verb. und verm. Aufl. Freiburg im Breisgau: Herder, 1866 – (= ser Die Encyclica Papst Pius' 9. Vom 8. Dezember 1864) – 1mf – 9 – 0-8370-8305-2 – mf#1986-2305 – us ATLA [240]

Die "irrthuemer von mehr als vierhundert bischoefen und ihr theologischer censor" : ein beitrag zur wuerdigung der von herrn dr. von doellinger veroeffentlichten "worte ueber die unfehlbarkeitsadresse" / Hergenroether, Joseph – Freiburg im Breisgau: Herder, 1870 – 1mf – 9 – 0-8370-8433-4 – (incl bibl ref) – mf#1986-2433 – us ATLA [240]

Die irrtumslosigkeit jesu christi und der christliche glaube : vier baenden der "gottesoffenbarung in jesu christo" und zu dieser schrift / Schwartzkopff, Paul – Giessen: J Ricker, 1897 – 1mf – 9 – 0-8370-4615-7 – mf#1985-2615 – us ATLA [240]

Irs cumulative bulletin : 1922-1998 / U.S. Treasury Dept – 1413mf – 9 – $2119.00 – (updates planned) – mf#LLMC 79-416 – us LLMC [324]

Irsad – Baku, 1905-08. v2 n3-25,27-49,51-70,72-88,90-134 4 jan-29 dec 1907; v3 n1-95 1 jan-25 jun 1908 – (= ser O & t journals) – 2r – 1 – $120.00 – (in azeri) – us MEDOC [956]

Irsay, Stephen d' *see* Albrecht von haller

Irshad – Baku, 1906-08 – 4r – 1 – us UMI ProQuest [077]

Irshad al-sari fi-sharh sahih al-bukhari = Irshad al-sari / Qastallani, Ahmad ibn Muhammad – Tabah jadidah bil-ufset. al-Qahirah: al-Matbaah al-Amiriyah, 1886-1888 – 1mf – 9 – 0-524-08304-5 – (incl text of sahih al-bukhari) – mf#1993-4009 – us ATLA [470]

Irtysh : golos sibirskogo kazach'ego voiska: zhurn ezhened / ed by Bazhenov, A – Omsk [Akmol obl]: [s n] 1918-19 [1917 12 okt-[1919] [?]] – (= ser Asn 1-3) – n12-37 [1918] n2-40 [1919] [gaps] item 186, on reel n38,39 – 1 – mf#asn-1 186 – ne IDC [077]

Iruarrizaga, Jose *see* Primeros franciscanos en china

Irungaray, Ezequiel C *see* Indice del archivo de la ensenanza superior de gua...

Irurac bat – Bilbao, Spain. July 1856-Sept 16 1856 – 1r – 1 – us L of C Photodup [074]

Irvin, Samuel M *see* Diary and journal kept at ioway mission in kansas

Irvine, Andrew Alexander *see* Land of no regrets

[Irvine-] anthill – CA: UC Irvine, 1966-77 – 1r – 1 – $60.00 – (cont: the tongue) – mf#R02308 – us Library Micro [370]

[Irvine-] blade – CA. 1975-1989 – 1r – 1 – $60.00 – mf#R04034 – us Library Micro [071]

[Irvine-] east-west ties – CA. 1983-1989 – 1r – 1 – $60.00 – mf#R04033 – us Library Micro [071]

Irvine & fullarton times – [Scotland] North Ayrshire, Irvine: J S Begg 18 apr 1874-dec 1879, jan 1893-29 dec 1950 (wkly) [mf ed 2003] – 68r – 1 – (cont by: irvine times) – uk Newsplan [072]

Irvine herald and cunninghame advertiser – [Scotland] North Ayrshire, Irvine: C Murchland 6 jan 1877-28 dec 1878, jan 1893-dec 1950 (wkly) [mf ed 2003] – 42r – 1 – (missing: 1875; cont by: irvine herald and ayrshire advertiser [jan 1893-dec 1950]) – uk Newsplan [072]

Irvine, Ingram N W *see* The documents and facts in the irvine-talbot case

[Irvine-] irvine today – CA. 1979-87 – 12r – 1 – $720.00 – mf#R02313 – us Library Micro [071]

[Irvine-] irvine world news – CA. 1972-91 – 69r – 1 – $4140.00 – mf#R02315 – us Library Micro [071]

[Irvine-] la voz mestinza – CA. 1979-1989 – 1r – 1 – $60.00 – mf#R04035 – us Library Micro [071]

[Irvine-] new university – CA: UC Irvine, 1968-90 – 18r – 1 – $1080.00 – (cont: the anthill) – mf#R02310 – us Library Micro [378]

[Irvine-] outside – CA. 1983-1985 – 1r – 1 – $60.00 – mf#R04036 – us Library Micro [071]

[Irvine-] phoenix – CA. 1986-1989 – 1r – 1 – $60.00 – mf#R04037 – us Library Micro [071]

Irvine, R *see* The bereans

[Irvine-] seed – CA. 1981-1983 – 1r – 1 – $60.00 – mf#R04038 – us Library Micro [071]

[Irvine-] spectre – CA. 1966 – 1r – 1 – $60.00 – mf#R02 – us Library Micro [071]

[Irvine-] spectrum – CA. 1965-66 – 1r – 1 – $60.00 – mf#R02312 – us Library Micro [071]

[Irvine-] tapestry – CA. 1984-1986 – 1r – 1 – $60.00 – mf#R04039 – us Library Micro [071]

[Irvine-] the tongue – CA: UC Irvine, 1966 – 1r – 1 – $60.00 – (cont: anthill) – mf#R02314 – us Library Micro [378]

[Irvine-] uci journal – CA. UC Irvine: 1981-1989 – 1r – 1 – $60.00 – mf#R04157 – us Library Micro [378]

[Irvine-] university press – CA. 1979 – 1r – 1 – $60.00 – mf#R04040 – us Library Micro [071]

Irvine valley news and district advertiser – [Scotland] East Ayrshire, Newmilns: J H Greene 1 jan 1909-dec 1950 (wkly) [mf ed 2003] – 23r – 1 – (merged with: weekly supplement and advertiser for galston, newmilns, darvel, and hurlford to form: irvine valley news and galston supplement [jan 1945-dec 1950]) – uk Newsplan [072]

Irvine, W F *see* Christian ministry and its requirements

[Irvine-] women's quarterly – CA. 1987-1989 – 1r – 1 – $60.00 – mf#R04041 – us Library Micro [305]

Irving, Ann *see* Wider horizons

Irving, David *see*
- Hitler's war
- Hungarian uprising, 1956

Irving, David [comp] *see*
- The life and campaigns of field marshal rommel
- Selected documents on the flight and imprisonment of rudolph hess, 1941-1945
- Selected research documents relating to hermann goering

Irving, Edward *see*
- The church and state responsible to christ
- The collected writings of edward irving
- The coming of messiah in glory and majesty
- Farewell discourse to the congregation and parish of st john's, glasgow
- Introductory essay to bishop horne's commentary on the book of psal...

Irving gold scrapbook *see* Gold, irving, scrapbook

Irving, Theodore *see* The conquest of florida... hernando de soto

Irving, Washington *see*
- Astoria oder geschichte einer handelsexpedition jenseits der rocky mountains
- The discovery and conquest of the new world
- Notebooks containing queries, extracts from printed sources, etc. relating to astoria
- Voyages and discoveries of the companions of columbus
- Works

Irving, Washington, 1783-1859 *see* Rip van winkle

Irvingism : in its rise, progress, and present state / Baxter, Robert – 2nd ed. London: J. Nisbet, 1836 – 1mf – 9 – 0-7905-6280-4 – mf#1988-2280 – us ATLA [240]

The irvington stories / Dodge, Mary Mapes – Illus. by F.O.C. Darley.4th ed. New York: J. O'Kane, 1867. 263p – 1 – us Wisconsin U Libr [830]

Irwin, Alexander *see* Observations on the rev dr reichel's sermon

Irwin, Clarke Huston *see*
- Famous irish preachers
- A history of presbyterianism in dublin and the south and west of ireland

Irwin, E Robert *see* Analysis of the church music curriculum of selected protestant seminaries

Irwin, Eyles *see* Voyage a la mer rouge, sur les cotes de l'arabie, en egypte, et dans les deserts de la thebaide

Irwin, Frederick Chidley *see* The state and position of western australia

Irwin, Godfrey *see* American tramp and underworld slang

Irwin, James W *see* Hatch act decisions

Irwin, Melinda L *see* Development of an anthropometric regression equation to predict body density in african american women

Irwin, R L *see* Development of a collegiate licensing administrative paradigm

Irym svetem – [V Praze: Nakl Ceske graficke unie] 1924- [mf ed Bloomington IN: Indiana Uni Lib, preservation Dept 1989] – 4r – 1 – us Indiana Preservation [943]

Is a russian invasion of india feasible? / David, C – London, 1877 – (= ser 19th c books on british colonization) – 1mf – 9 – mf#1.1.2079 – uk Chadwyck [327]

I/s analyzer – Bethesda. 1987-1994 (1) 1987-1994 (5) 1987-1994 (9) – (cont: edp analyzer. cont by: i/s analyzer case studies) – ISSN: 0896-3231 – mf#1618,01 – us UMI ProQuest [000]

I/s analyzer case studies – Needham. 1994+ (1) 1994+ (5) 1994+ (9) – (cont: i/s analyzer) – ISSN: 1080-1146 – mf#1618,02 – us UMI ProQuest [000]

Is athlete burnout more than just stress? a sport commitment perspective / Raedeke, Thomas D – University of Oregon, 1995 – 3mf – 9 – $12.00 – mf#PSY 1900 – us Kinesology [150]

Is buddhism a preparation or hindrance to christianity in china? / Ball, James Dyer – Hong Kong: printed at St Paul's College 1907 [mf ed 1991] – 1mf – 9 – 0-524-01411-6 – mf#1990-2406 – us ATLA [230]

Is canada a land of sunshine or snow? : how is canada important to the british empire both from a political and domestic standpoint – S.l: s.n, 1897? – 1mf – 9 – mf#61151 – cn Canadiana [630]

Is cheap or dear bread best for the poor man? – London: James Ridgway, 1841 – (= ser 19th c economics) – 1mf – 9 – mf#1.1.220 – uk Chadwyck [339]

Is christ infallible and the bible true? / M'Intosh, Hugh – Edinburgh: T & T Clark 1901 [mf ed 1992] – 2mf – 9 – 0-524-04409-0 – mf#1992-0102 – us ATLA [220]

Is christianity a success? / Besant, Annie Wood – London, England. 1885 – 1r – us UF Libraries [240]

Is christianity from god : or, a manual of bible evidence for the people / Cumming, John – New York: MW Dodd, 1856 [mf ed 1984] – 4mf – 9 – 0-8370-0965-0 – mf#1984-4310 – us ATLA [240]

Is christianity practicable? : lectures / Brown, William Adams – New York: Scribner, 1916 – 1mf – 9 – 0-7905-7697-X – mf#1989-0922 – us ATLA [240]

Is christianity true? : answers from history, the monuments, the bible, nature, experience, and growth of christianity / Blaikie, William Garden et al – Philadelphia: Rice & Hirst, c1897 [mf ed 1985] – 1mf – 9 – 0-8370-3730-1 – (incl ind) – mf#1985-1730 – us ATLA [240]

Is conscience an emotion? : three lectures on recent ethical theories / Rashdall, Hastings – Boston: Houghton Mifflin, 1914 – (= ser Raymond Fred West Memorial Lectures at Stanford University) – 1mf – 9 – 0-7905-9599-0 – mf#1989-1324 – us ATLA [170]
Is duet 'ne welt! : schwaenke und geschichten / Henze, Wilhelm – Bad Pyrmont: F Gersbach, 1922 – 1r – 1 – us Wisconsin U Libr [830]
Is eternal punishment endless? : answered by a restatement of the original scripture doctrine / Whiton, James Morris – Boston: Lockwood, Brooks, 1876 – 1mf – 9 – 0-524-08703-2 – (incl ind) – mf#1993-3228 – us ATLA [240]
Is every statement in the bible about our heavenly father strictly... / Voysey, Charles – London, England. 1864 – 1r – us UF Libraries [240]
Is fast walking an adequate aerobic training stimulus for male and female cardiac patients? / Anthony, Ryan M – 1998 – 1mf – 9 – $4.00 – mf#PH 1644 – us Kinesology [612]
Is future punishment eternal? : a sermon. at st. george's (episcopal) church, st louis, mo... / Holland, Robert Afton – Utica, NY: Christian Leader Print, 1875 – 1mf – 9 – 0-524-06207-2 – mf#1992-0845 – us ATLA [240]
Is god able and willing to save me? / Smith, James – London, England. 18– – 1r – us UF Libraries [240]
Is god knowable? / Iverach, James – London, 1884 – (= ser 19th c evolution & creation) – 3mf – 9 – mf#1.1.11589 – uk Chadwyck [110]
Is god knowable? / Iverach, James – London: Hodder & Stoughton, 1887 [mf ed 1985] – (= ser The theological library) – 1mf – 9 – 0-8370-4608-4 – mf#1985-2608 – us ATLA [210]
Is healthful reunion impossible? : a second letter to the very rev j h newman / Pusey, Edward Bouverie – Oxford: J Parker, 1870 [mf ed 1990] – (= ser Eirenicon 3) – 1mf – 9 – 0-7905-9592-3 – (incl bibl ref) – mf#1989-1317 – us ATLA [241]
Is het woord "gereformeerd" in het vaandel der hollandsche chr ger kerk in amerika een leugen in hare rechterhand? : open brief gericht aan de afgevaardigden der e k synode der holl chr geref kerk in america / Koster, S – Holland, MI: John D Kanters, 1896 [mf ed 1993] – (= ser Reformed church coll) – 1mf – 9 – 0-524-07251-5 – (in dutch) – mf#1991-2992 – us ATLA [242]
Is immortality desirable? / Dickinson, Goldsworthy Lowes – Boston: Houghton Mifflin, 1909 – (= ser The Ingersoll Lecture) – 1mf – 9 – 0-7905-9919-8 – mf#1989-1644 – us ATLA [240]
Is india civilized? : essays on indian culture / Woodroffe, John George – Madras: Ganesh & Co, 1918 – (= ser Samp: indian books) – us CRL [954]
Is insurance sales – Indianapolis. 1980-1990 (1) 1980-1990 (5) 1980-1990 (9) – (cont: insurance salesman. cont by: life and health insurance sales) – ISSN: 0199-4581 – mf#7745,01 – us UMI ProQuest [368]
Is it mary or the lady of the jesuits? : the question as discussed in brighton, england, august 15, 1889... / Fulton, Justin Dewey – Toronto: Willard Tract Depository, 1889? – 1mf – 9 – mf#03272 – cn Canadiana [241]
Is it possible to make the best of both worlds? / Binney, Thomas – London, England. 1854 – 1r – us UF Libraries [240]
Is jesus god : an argument / Kuiper, Rienk Bouke et al – New York: American Tract Society, c1912 – 1mf – 9 – 0-7905-9777-2 – mf#1989-1502 – us ATLA [240]
Is life worth living? / Mallock, W H – New York, NY. 1879 – 1r – us UF Libraries [025]
Is life worth living? / Mallock, William Hurrell – New York: GP Putnam, 1879 – 1mf – 9 – 0-7905-8696-7 – mf#1989-1921 – us ATLA [170]
Is man an automaton? : a lecture delivered...glasgow, on 23rd feb 1875; under the auspices of the glasgow science lectures association / Carpenter, William Benjamin. – London, 1875 – (= ser 19th c evolution & creation) – 1mf – 9 – mf#1.1.7805 – uk Chadwyck [610]
Is man responsible for his belief? / Martin, William – Aberdeen, Scotland. 1849 – 1r – us UF Libraries [240]
Is mormonism true or not? – London, England. 18– – 1r – us UF Libraries [243]
Is my bible true? : where did we get it? / Leach, Charles – Chicago: Fleming H Revell, c1897 – 1mf – 9 – 0-8370-4066-3 – mf#1985-2066 – us ATLA [220]
Is "ritual" right? / Dearmer, Percy – 4th ed. London: AR Mowbray, 1911 – 1mf – 9 – 0-524-02952-0 – mf#1990-4504 – us ATLA [240]
Is romanism real christianity? / Newman, Francis William & Abbot, Francis Ellingwood – Toledo, OH: Index Association, 1872 – 1mf – 9 – 0-8370-3915-0 – mf#1985-1915 – us ATLA [240]
Is salvation conditional or unconditional? : a discussion between c.h. cayce, primitive baptist, and f.b. srygley, christian / Cayce, Claudius Hopkins & Srygley, Filo Bunyan – Nashville, TN: McQuiddy Print Co, 1912 – 1mf – 9 – 0-524-06987-5 – mf#1991-2840 – us ATLA [242]
Is sex necessary? / Thurber, James – New York, NY. 1950 – 1r – us UF Libraries [025]
Is swinton right? : or, the truth about indulgences – Boston, MA: Cttee of One Hundred, 1889 [mf ed 1992] – (= ser Committee of one hundred series 4) – 1mf – 9 – 0-524-02339-5 – mf#1990-0595 – us ATLA [900]
Is the anglo-saxon race degenerating? / Russell, James – S.l: s.n, 1900? – 1mf – 9 – mf#12787 – cn Canadiana [572]
Is the bible inspired? / Brookes, James Hall – St Louis: Gospel Book & Tract Depository, [ca 1883] – 1mf – 9 – 0-8370-2469-2 – mf#1985-0469 – us ATLA [220]
Is the bible inspired of god? / Hastings, H L – London, England. 1887 – 1r – us UF Libraries [240]
Is the bible true? : seven addresses / Brookes, James Hall – St Louis: Chas B Cox, [ca 1877] – 1mf – 9 – 0-8370-2470-6 – mf#1985-0470 – us ATLA [220]
Is the church in wales an alien institution? / Bevan, W L – London, England. 18– – 1r – us UF Libraries [240]
Is the church of england a church of christ? – London, England. 18– – 1r – us UF Libraries [241]
Is the church of rome the babylon of the book of revelation? / Wordsworth, Christopher – London, England. 1850 – 1r – us UF Libraries [241]
Is the church of scotland to stand or fall? / Williamson, Alex – Edinburgh, Scotland. 1890 – 1r – us UF Libraries [242]
Is the devil a myth? / Wimberly, Charles Franklin – New York: FH Revell, c1913 – 1mf – 9 – 0-7905-8748-3 – mf#1989-1973 – us ATLA [210]
Is the episcopal church catholic or is it protestant? : an address. delivered in the church of the saviour, philadelphia... / McKim, Randolph Harrison – Philadelphia: George W Jacobs [distributor], 1916 – 1mf – 9 – 0-524-06070-3 – mf#1990-5184 – us ATLA [240]
Is the "establishment of religion" outside of the confession? / Stuart, A Moody – Edinburgh, Scotland. 1869 – 1r – us UF Libraries [240]
Is the fascist rebellion in spain a popular movement? / Spain. Embajada. United States – Washington, DC, n.d. Fiche W1181. (Blodgett Collection of Spanish Civil War Pamphlets) – 9 – us Harvard [946]
Is the free church of scotland to continue free? / Moffat, Mr – Banff, Alberta. 1866 – 1r – us UF Libraries [242]
Is the honour or veneration given to images and relics by roman catho... / Collette, Charles Hastings – London, England. 1889? – 1r – us UF Libraries [241]
Is the judgment in the case of martin v mackonochie according to the evidence? / Layman, Thomas – London, England. 1871 – 1r – us UF Libraries [240]
Is the lord among us? / Huntington, DeWitt Clinton – Cincinnati: Jennings & Pye; New York: Eaton and Mains, c1904 [mf ed 1984] – 2mf – 9 – 0-8370-0145-5 – mf#1984-0031 – us ATLA [242]
Is the manufacturing sector an engine of growth in south africa? : an analysis of the eastern cape / Dyubhele, Noluntu Stella – Pretoria: Vista University 2000 [mf ed 2000] – 4mf – 9 – (incl bibl ref) – mf#mfm15347 – sa Unisa [650]
Is the mode of christian baptism prescribed in the new testament / Stuart, M – Andover, England. 1833 – 1r – us UF Libraries [242]
Is the new theology christian? / Egerton, Hakluyt – London: George Allen, 1907 – 1mf – 9 – 0-7905-7725-9 – mf#1989-0950 – us ATLA [240]
Is the pope independent? : or, outlines of the roman question / Prior, John – London: R and T Washburn; New York: Benziger, [190-?] – 1mf – 9 – 0-8370-7011-2 – mf#1986-1011 – us ATLA [241]
Is the reformation a blessing? / Goode, William – London, England. 1858 – 1r – us UF Libraries [242]
Is the second advent premillennial? – London, England. 18– – 1r – us UF Libraries [240]
Is the western church under anathema? : a problem for the ecumenical council of 1869 / Ffoulkes, Edmund Salusbury – London: JT Hayes, [1869?] – 1mf – 9 – 0-524-05500-9 – mf#1990-1495 – us ATLA [240]
Is theosophy anti-christian? / Besant, Annie Wood – Chicago: Rajput Press, [1904?] [mf ed 1992] – 1mf – 9 – 0-524-03301-3 – (incl bibl ref) – mf#1990-3186 – us ATLA [230]
Is there a god? : an affirmation from science / Price, Thomas Ralph – [Stirling] [1895] – (= ser 19th c evolution & creation) – 1mf – 9 – mf#1.1.10864 – uk Chadwyck [110]
Is there a god? / Bradlaugh, Charles – London, England. 1887 – 1r – us UF Libraries [240]
Is there a god? / Wieman, Henry Nelson & Macintosh, Douglas Clyde & Otto, Max Carl – Introd. by Charles Clayton Morrison. Chicago: Willet, Clark, 1932. 328p – 1 – us Wisconsin U Libr [240]
Is there a god for man to know? / Carmichael, James – Toronto: Church of England Pub Co, [1900?] – 2mf – 9 – 0-665-03842-9 – mf#03842 – cn Canadiana [210]
Is there a hell? / Bradlaugh, William Robert – London, England. 18– – 1r – us UF Libraries [240]
Is there a personal devil? / Presland, John – London, England. 187-? – 1r – us UF Libraries [240]
Is there a relationship between prenatal exercise and postpartum depression / Stephenson, Sheryl L & Bleutler, Sharon A – 1993 – 2mf – 9 – $8.00 – us Kinesology [150]
Is there salvation after death? : a treatise on the gospel in the intermediate state / Morris, Edward Dafydd – 2nd ed. New York: A C Armstrong, c1887 [mf ed 1991] – 1mf – 9 – 0-7905-9819-1 – mf#1989-1544 – us ATLA [240]
Is this peace / Radhakrishnan, Sarvepalli – Bombay: Hind Kitabs, 1945 – (= ser Samp: indian books) – us CRL [954]
Is thy heart right? / Elven, Cornelius – London, England. 18– – 1r – us UF Libraries [240]
Is winning the only thing : goal orientations and team norms predictions of legitimacy ratings of intentionally injurious sport acts / Drake, Brent M – 1997 – 2mf – 9 – $8.00 – mf#PSY 1992 – us Kinesology [150]
Is your soul in health? – London, England. 18– – 1r – us UF Libraries [240]
Isa transactions – Research Triangle Park. 1989+ (1,5,9) – (cont: isa transactions) – ISSN: 0019-0578 – mf#42743 – us UMI ProQuest [621]
Isa transactions / Instrument Society of America – Pittsburgh. 1976-1992 (1,5,9) – (cont by: isa transactions) – ISSN: 0019-0578 – mf#11114 – us UMI ProQuest [621]
Isaac and jacob : their lives and times / Rawlinson, George – New York: Anson D F Randolph, [189-?] – 1mf – 9 – 0-8370-9978-1 – (incl bibl ref) – mf#1986-3978 – us ATLA [920]
Isaac asimov's science fiction magazine – New York. 1977-1992 (1,5,9) – (cont by: asimov's science fiction) – ISSN: 1055-2146 – mf#11672 – us UMI ProQuest [420]
Isaac Ben Sheshet *see* She'elot u-teshuvot ha-ribash ha-hadashot
Isaac casaubon, 1559-1614 / Pattison, Mark; ed by Nettleship, Henry – 2nd ed Oxford: Clarendon Press, 1892 – 1mf – 9 – 0-7905-6351-7 – (incl bibl ref) – mf#1988-2351 – us ATLA [920]
Isaac, Daniel *see* Rules fo the protestant methodists brought to the test of holy scripture in a letter addressed to th
Isaac Hobhouse and Co *see* Hobhouse letters, the... 1722-55
Isaac Israeli *see* Die philosophische lehren des isaak ven salomon israeli
Isaac L Peretz *see* Three classic yiddish authors
Isaac, Max *see* Facts about bankruptcy you ought to know
Isaac mccoy: early indian missions / Wyeth, Walter N – 1895 – 1 – 8.54 – us Southern Baptist [242]
Isaac mccoy papers / McCoy, Isaac – 1808-74. In Kansas State Historical Society. Guide – 1 – us Kansas [240]
Isaac mishimens account book, 1833-1847 / Mishimens, Isaac – [mf ed 1981] – 1r – 1 – mf#ms1627 – us Western Res [630]
Isaac reid papers *see* Reid, isaac, papers, ms 4704
Isaac t. hopper : a true life / Child, Lydia Maria Francis – Boston: JP Jewett, 1853 – 2mf – 9 – 0-524-02731-5 – mf#1990-4406 – us ATLA [240]
Isaac watts : his life and writings, his homes and friends / Hood, Edwin Paxton – London: Religious Tract Society, [1875?] – 1mf – us ATLA [240]
Isaac watts : his life and writings, his homes and friends / Hood, Edwin Paxton – London: Religious Tract Society, [1875?] – 1mf – 9 – 0-7905-4817-8 – mf#1988-0817 – us ATLA [920]
Isaacs, A S *see* Step by step
Isaacs, Jorge *see* Maria (novela americana)
Isaacson, Charles S *see* The story of the later popes
Isaacson, Charles Stuteville *see* Roads from rome
Isabel 1, reina de espana y madre de america. madrid, 1943 / Gomez de Mercado y Miguel, F – Madrid: Razon y Fe, 1947 – 1 – sp Bibl Santa Ana [946]
Isabela / Carreno, Alberto Maria – Mexico City?, Mexico. 1945 – 1r – us UF Libraries [972]
Isabelle and helen mcfarland diaries *see* Mcfarland, isabelle and helen, diaries
Isaev, A A *see*
– Arteli v rossii
– Nastoiashchee i budushchee russkogo obshchestvennogo khoziaistva
Isafold – Reykjavik, Iceland. -w. 6 Jan 1894-31 Dec 1914; 21 April 1917-12 Jan 1921. 1917, 1919 imperfect. 11 reels – 1 – uk British Libr Newspaper [072]
Isagoge artis musicae ad incipientium captum maxime accommodata = Kurtze anleitung recht unnd leicht singen zu lernen / Demantius, Christoph – Norimbergae: Typis Valentini Furmanni 1607 [mf ed 19–] – 1r – 1 – mf#pres. film 231 – us Sibley [780]
Isagoge de composicione cantus / Galliculus, Johannes – 1520 – (= ser Mssa) – 2mf – 9 – €35.00 – mfchl 56 – gw Fischer [780]
Isagoge historica apologetica de las indias occide... – Guatemala, 1935 – 1r – us UF Libraries [972]
Isagoge historica apologetica de las indias occidentales y especial de la provincia de san vicente de chiapa y guatemala, de la orden de predicadores. guatemala, 1935 / Bayle, Constantino – Madrid: Razon y Fe, 1936 – 1 – sp Bibl Santa Ana [972]
Isagoge seu introductio generalis ad scripturam sacram / Rivet, A – Lugduni Batavorum, 1627 – 7mf – 9 – mf#PRS-171 – ne IDC [240]
Isagoges musicae. libri duo : tam theoricae quam practicae studiosis inservire iussi... / Snegassio, Cyriaco – Erphordiae: Typis G Baumanni 1591 [mf ed 19–] – 2mf / 1r – 9,1 – mf#fiche 156 / pres. film 231 – us Sibley [780]
Isagoges pars altera...de angelis...et de ecclesia / Daneau, Lambert – Geneve, Vignon, 1584 – 2mf – 9 – mf#PFA-128 – ne IDC [240]
Isagoges pars quinta quae est de homine / Daneau, Lambert – [Geneve], Vignon, 1588 – 6mf – 9 – mf#PFA-133 – ne IDC [240]
Isagoges...pars quarta de salutaribus dei donis erga ecclesiam... / Daneau, Lambert – Geneve, Vignon, 1586 – 8mf – 9 – mf#PFA-131 – ne IDC [240]
Isaiah / Alexander, Joseph Addison – New York: John Wiley, 1852. Beltsville, Md: NCR Corp, 1978 (10mf); Evanston: American Theol Lib Assoc, 1984 (10mf) – 9 – 0-8370-1073-X – mf#1984-4438 – us ATLA [221]
Isaiah / Barnes, William Emery – London: Methuen 1901-03 [mf ed 1989] – (= ser The churchman's bible) – 2v on 2mf – 9 – 0-7905-0664-5 – (incl ind) – mf#1987-0664 – us ATLA [221]
Isaiah : his life and times and the writings which bear his name / Driver, Samuel Rolles – New York: Anson D F Randolph, [ca 1883] – 1mf – 9 – 0-8370-2971-6 – (incl bibl ref and index of isaiah's prophecies) – mf#1985-0971 – us ATLA [920]
Isaiah / ed by Moulton, Richard Green – New York: Macmillan, 1906 – (= ser The Modern Reader's Bible) – 1mf – 9 – 0-524-08405-X – mf#1993-0020 – us ATLA [221]
Isaiah / Simpson, Albert B – 2nd rev ed. New York: Alliance Press Co, c1907 – 1mf – 9 – 0-524-02154-6 – mf#1990-4220 – us ATLA [221]
Isaiah : a study of chapters 1-12 / Mitchell, Hinckley G T – New York: Thomas Y Crowell, c1897 – 1mf – 9 – 0-8370-4455-3 – mf#1985-2455 – us ATLA [221]
Isaiah : a study of chapters 1.-12 / Mitchell, Hinckley Gilbert Thomas – New York: T Y Crowell, c1897. Chicago: Dep of Photodup, U of Chicago Lib, 1973 (1r); Evanston: American Theol Lib Assoc, 1984 (1r) – 1 – 0-8370-0412-8 – (incl ind) – mf#1984-B349 – us ATLA [221]
Isaiah : with notes, critical, explanitory [sic] and practical / Cowles, Henry – New York: D Appleton, 1869, c1868 – 2mf – 9 – 0-7905-1585-7 – mf#1987-1585 – us ATLA [221]
Isaiah 1-39 : introduction, revised version with notes, index and maps / ed by Whitehouse, Owen Charles – New York: Oxford University Press, American Branch, [19–?] – (= ser The New-Century Bible) – 1mf – 9 – 0-524-05642-0 – mf#1992-0497 – us ATLA [221]
Isaiah 40-56 : the great prophecy of israel's restoration – London: Macmillan, 1896 – 65p – 1 – us Wisconsin U Libr [221]
Isaiah 40-66 : introduction, revised version with notes, index and maps / ed by Whitehouse, Owen Charles – New York: Oxford University Press, American Branch, [19–?] – (= ser The New-Century Bible) – 1mf – 9 – 0-524-05643-9 – mf#1992-0498 – us ATLA [221]

Isaiah 40-66 : with the shorter prophecies allied to it / ed by Arnold, Matthew – London: Macmillan, 1875 – 1mf – 9 – 0-8370-2115-4 – mf#1985-0115 – us ATLA [221]
Isaiah chapters 11-55 / Smith, Sydney – 1944 – 9 – $10.00 – us IRC [221]
Isaiah, his life and times / Driver, Samuel Rolles – New York, NY. 1888 – 1r – us UF Libraries [939]
Isaiah of jerusalem in the authorised english version – London: Macmillan, 1883 – 1mf – 9 – 0-7905-0542-8 – mf#1987-0542 – us ATLA [221]
Isaiah one and his book one : an essay and an exposition / Douglas, George Cunningham Monteath – New York: Fleming H Revell, [1895?].Chicago: Dep of Photodup, U of Chicago Lib, 1971 (1r); Evanston: American Theol Lib Assoc, 1984 (1r) – 1 – 0-8370-0522-1 – mf#1984-B227 – us ATLA [221]
Isaias...expositus homilijs 190 / Bullinger, Heinrich – Tigvri, Christoph Froschover, 1567 – 8mf – 9 – mf#PBU-234 – ne IDC [240]
Isamaili/ismaili – Bombay: Ismailia Association for India. [wkly] – 1 – (in gujarati and english) – us Wisconsin U Libr [073]
Isang bansa, isang wika – Manila, Philippines: United Pub. Co., c1976. 307p. ill. map. Tagalog language-readers – 1 – us Wisconsin U Libr [490]
Isang dios at tatlo sa pagca-persona na hinnago sa sagrada biblia, mula sa genesis hangang evangelio / Ignacio, Cleto R – Maynila: P Sayo 1916 [mf ed Bloomington IN: Indiana Uni Lib, Preservation Dept 1984] – (= ser Coll...in the tagalog language 2) – 1r – 1 – us Indiana Preservation [490]
Isang dipang langit / Hernandez, Amado V – Quezon City: Tamaraw Pub Co, 1961 – us CRL [950]
Isaodaadh's kommentar zum buche hiob / Schliebitz, Johannes – Giessen: Alfred Toepelmann, 1907 – (= ser Beihefte zur zeitschrift fuer die alttestamentliche wissenschaft) – 2mf – 9 – 0-524-05986-1 – mf#1992-0723 – us ATLA [221]
Isaodaadh's stellung in der auslegungsgeschichte des alten testamentes : an seinen commentaren zu hosea, joel, jona, sacharja 9-14 und einigen angehaengten psalmen / Diettrich, Gustav – Giessen: J Ricker (Alfred Toepelmann), 1902 – (= ser Beihefte zur zeitschrift fuer die alttestamentliche wissenschaft) – 1mf – 9 – 0-7905-0700-5 – (incl bibl ref) – mf#1987-0700 – us ATLA [221]
Isar-loisach-bote – Wolfratshausen DE, 1988- – 14r/yr – 1 – (bezirksausgabe von muenchner merkur, muenchen) – gw Misc Inst [074]
Isar-post – Landshut DE, 1946 15 jan-1958 [gaps] – 24r – 1 – (filmed by misc inst: 1946 15 jan-1958 14 dec) – gw Mikrofilm; gw Misc Inst [074]
Isavnina, Z N *see* Slovo [omsk: 1918]
Isban, Samuel *see* "Umlegale" yidn shpaltn yamen
Isben, Henrik *see* Doll's house
Isbister, Alexander Kennedy *see* A proposal for a new penal settlement
Isea news / Indiana State Employees Association – v13 8-v21 2 [1977 nov-1985 may] – 1r – 1 – (cont: isea news letter; cont by: indiana state employees association sentinel) – mf#1407031 – us WHS [350]
Isea news letter / Indiana State Employees Association – v12 n11 [1976 dec] – 1r – 1 – (cont by: isea news) – mf#1409838 – us WHS [071]
Isebies : roman / Boehlau, Helene – 10. aufl. Muenchen: A Langen, 1911 [mf ed 1989] – ix/502p – 1 – mf#7042 – us Wisconsin U Libr [830]
Iseiviu draugas – [Scotland] Mossend: Rev J Norbut 1 sausio 1914-25 gruodzio 1920 (wkly) [mf ed 2003] – 7v on 4r – 1 – (imprint varies; printed & publ 8th jul 1916- in glasgow by lithuanian st joseph's society) – uk Newsplan [072]
I-se-lieh te ku shih (ccm6) / Ch'en, Shu-i – 1st ed. Hong Kong, 1955 [mf ed 198?] – (= ser Ccm 6) – 1 – mf#1984-b500 – us ATLA [939]
Iselin, Isaak *see*
– Ephemeriden der menschheit
– Filosofische und patriotische traeume eines menschenfreundes
Iselin, Jacob Christoph *see* Neu-vermehrtes historisch und geographisches allgemeines lexicon (ael1/9)
Iselin, L E *see* Eine bisher unbekannte version des ersten teiles der "apostellehre"
Isely, Christian H *see* Letters
Isenberg, C W *see*
– Dictionary of the amharic language
– Journals...detailing their proceedings in the kingdom of shoa, and journeys in other parts of abyssinia, in the years 1839, 1840, 1841, and 1842...
Isenberg, Karl *see* Der einfluss der philosophie charles bonnets auf friedrich heinrich jacobi
Isenberg, Karl Wilhelm *see* Dictionary of the amharic language
Isenhagener kreisblatt *see* Wittinger zeitung
Isep monitor / Howard University – 1979 dec, 1980 dec, 1981 jun-sep, 1982 special ed, 1983 special ed – 1r – 1 – mf#4863217 – us WHS [378]
Iserlohner anzeiger – Iserlohn DE, 1931 11 jul-28 dec [many gaps], 1932 1 apr-30 jun, 1933 2 jan-31 mar & 1 jul-30 sep, 1934 1 sep-30 sep [gaps], 1935 1 feb-29 mar – 4r – 1 – (title varies: 25 may 1906: echo der mark; 1 mar 1907: maerkisches volksblatt. with suppl) – gw Mikrofilm [074]
Iserlohner kreisanzeiger *see* Oeffentlicher anzeiger fuer die grafschaft limburg
Iserlohner kreisanzeiger und zeitung *see* Oeffentlicher anzeiger fuer die grafschaft limburg
Iserlohner zeitung *see* Westfalenpost [main edition]
Isert, Paul *see* Voyages en guinee et dans les iles caraibes en amerique
Isert, Paul Erdmann *see*
– Reise nach guinea und den caribaischen..
– Voyages en guinee et dans les iles caraibes en amerique.
Isethekeli esihle / Plummer, Gladys – Cape Town, South Africa. 1963 – 1r – us UF Libraries [960]
Isha upanishad / Ghose, Aurobindo – Calcutta: Arya Pub House, 1945 – (= ser Samp: indian books) – us CRL [280]
Isha upanishat – London: Luzac, 1918 – 1mf – 9 – 0-524-07084-9 – mf#1991-0066 – us ATLA [280]
Isham, Charles *see* The fishery question
Isham, George W *see* Two years in india
Isham, Samuel *see* History of american painting
Isham, William S *see* Journals
Ishaq al-Jundi, Khalil ibn *see* Maliki law
Ishaque, M *see* Modern persian poetry
Isherwood, Christopher *see*
– Vedanta for modern man
– Vedanta for the western world
Ishida-ke monjo : documents of the ishidas, the village squire of central japan in the edo period – 5414 items on 42r – 1 – Y450,000 – (with 224p guide. in japanese) – ja Yushodo [950]
Ishikawa ichiro monjo – 278r – 1 – Y4170,000 – (in japanese) – ja Yushodo [330]
Ishikawa, Matsutaro *see* Ohraimono bunrui shusei 2
Ishikawa, Sanshiro *see* Chi-tu chiao she hui chu i (ccm281)
Ishikawa, T *see*
– Bertanam kapas di djawa
– Nandoer kapas ing tanah djawa
Ishimaru, Tota *see* Ti erh tz'u shih chieh chan cheng
Ishimskaia step' – Petropavlovsk, Kazakhstan, 1916 – 1r – 1 – us UMI ProQuest [077]
Ishimskaia zhizn' : ezhedn bespartiin, polit i obshchestv gaz / ed by Bodrov-Poviraev, N I – Ishim [Tobol gub]: T-vo "Ishim izd" 1919 [1918 [dek]-1919 [?]] – (= ser Asn 1-3) – n20-184 [1919] [gaps] item 187, on reel n39 – 1 – mf#asn-1 187 – ne IDC [077]
Ishimskii krai : ezhedn progressiv, obshchedostup gaz / ed by Gorin, I – Ishim [Tobol gub]: T-vo "Izd delo na paiakh" 1918 [1918 [iiun']-okt – (= ser Asn 1-3) – n18-106 [1918] [gaps] item 188, on reel n39 – 1 – (suppl: telegrammy sibirskogo telegrafnogo agentstva [asn-1 411]) – mf#asn-1 188 – ne IDC [077]
Ish-Kishor, Sulamith *see* Heaven on the sea
Ishmael and the church / Cheeseman, Lewis – Philadelphia: Parry and McMillan, 1856 – 1mf – 9 – 0-7905-4201-3 – mf#1988-0201 – us ATLA [260]
Ishn – Troy, 1999+ [1,5,9] – (cont: chilton's industrial safety and hygiene news) – mf#12611,04 – us UMI ProQuest [360]
Ishq-name / Firishta, Abdulmacid ibn – [14-?] – us CRL [956]
Isibuto samavo – Newtondale SA, 1 jan 1843-31 jul 1844 – 1r – 1 – sa National [079]
Isidore de Seville, Saint *see*
– De ecclesiasticis officiis
– Opera omnia...faustino arevalo...tomus octavus et ultimus
– Opera omnia...faustino arevalo...tomus quartus...
– Opera omnia...recensente faustino arevalo...
– Opera omnia...recensente faustino arevalo, tomo 6
– Opera omnia...recensente faustino arevalo, tomus primus et secundus
– Opera omnia...recensente faustino arevalo...tomi tertius et quartus...
– Quaestiones in pentateuchum (siecle 8)
Isidore of Seville, Saint *see*
– Die altdeutschen bruchstuecke des tractats des bischof isidorus von sevilla de fide catholica contra judaeos
– Der althochdeutsche isidor
Isidorus Isolanus *see* Summae
Isigidimi sama – xosa – Lovedale SA, 1873-1884 – 4r – 1 – mf#MS00316 – sa National [079]
Isik – Giresun: Giresun Matbaasi, 1918-23. Sahib-i Imtiyaz: ve Mueduer-i Mes'ul: Cemsidzade Osman Nuri; Basmuharriri: Ussaksade Hayrunnisa Cemil, Topaloglu Osman Fikret. n21. 21 mayis 1335 [1919]; 23. 21 haziran 1335 [1919] – (= ser O & t journals) – 1mf – 9 – $25.00 – us MEDOC [956]
Isik – Izmir. Imtiyaz Sahibi: Ruscuklu Fahri. Umumi Nesriyat Mueduerue: Ibrahim. n30. 13 mayis 1933 – (= ser O & t journals) – 1mf – 9 – $25.00 – us MEDOC [956]
Isik svetlina – Sofia, Bulgaria. 5 feb 1991-30 mar 1992 – 1 – (in cyrillic) – mf#mf.688.d – uk British Libr Newspaper [077]
Isindebele – Bulawayo, Zimbabwe. 1918 – 1r – us UF Libraries [960]
Ising, Gerhard *see* Die niederdeutschen bibelfruehdrucke
Isini kofisika ebuntweni = Teenage sexuality / South Africa. Department of National Health and Population Development [Departement van Nasionale Gesondheid en Bevolkingsontwikkeling – Pretoria: Dept of National Health & Population Development 1987 [mf ed Pretoria, RSA: State Library [199-]] – 26p [ill] on 1r with other items – 5 – (in xhosa; also available in afrikaans, english, northern sotho, sotho, tsonga, tswana, venda & zulu) – mf#op 09041 r24 – us CRL [362]
Isinkwa sethu semihla ngemihla / Zama, J Mdelwa – Pietermaritzburg, South Africa. 1960 – 1r – us UF Libraries [960]
Isiqondiso ngokufa kokopha okunganqamukiyo = Guide to haemophilia / South Africa. Department of Health [Departement van Gesondheid] [Departement van Gesondheid] – [Pretoria: Dept of Health 1980?] [mf ed Pretoria, RSA: State Library [199-]] – 14p [ill] on 1r with other items – 5 – (in zulu; also available in afrikaans, english, northern sotho, sotho, tswana, venda & xhosa) – mf#op 09995 r24 – us CRL [616]
Isis : oder encyclopedische zeitung von oken – Jena, Germany. v1. 1817 – 1r – us UF Libraries [025]
Isis – Philadelphia. 1913+ (1) 1968+ (5) 1975+ (9) – ISSN: 0021-1753 – mf#707 – us UMI ProQuest [500]
Isis : tragedie mise en musique / Lully, Jean Baptiste – [167-?] [mf ed 19–] – 1r – 1 – (libretto by philippe quinault) – mf#film 608 – us Sibley [780]
Isis and asiis eastern africa's kalenjiin people and their pharaonic origin legend : a comparative study / Sambu, Kipkoeech Araap – Uni of South Africa 2000 [mf ed Johannesburg 2000] – 2v on 17mf [ill] – 9 – (incl bibl ref) – mf#mfm15059 – sa Unisa [470]
L'isis moderne : revue des sciences nouvelles – Paris. n1-6. oct 1896-mars 1897 – 1 – fr ACRPP [500]
Isis y serapis en la espana pagana : preanuncios de doctrinas y de virtudes cristianas / Tormo, Elias – Madrid, 1944 – sp Bibl Santa Ana [240]
Isitiya sompefumlo no : gokunye incwadi yemitandazo yamalungu – Mariannhill, South Africa. 1908 – 1r – us UF Libraries [960]
Isitunywa sennyanga – Mount Coke, Kingwilliamstown SA, 1 aug-31 dec 1850 – 1r – 1 – sa National [079]
Iskandar, N S *see* Tjinta tanah air
Isk-mitteilungsblatt des internationalen sozialistischen kampfbundes – Berlin, Stuttgart DE, 1926-33 – 1r – 1 – (cont as: sozialistische warte, paris [1933-40]) – mf#1609 – gw Mikropress [335]
Isk-mitteilungsblatt des internationalen sozialistischen kampfbundes – Berlin, Stuttgart DE, 1933 – 1926-33 – 1 – (cont: sozialistische warte, paris) – mf#1609 – gw Mikropress [074]
Iskodra – 1312 [1894] – (= ser Vilayet salnames) – 3mf – 9 – $55.00 – us MEDOC [956]
Iskorki : izdanie batal'onnogo komiteta i prosvetitel'skoj komissii sankt-peterburgskogo razgruzochnogo 127-go batal'ona – St Petersburg, Russia, 1917 – 1r – 1 – us UMI ProQuest [077]
Iskra : central'nyj organ rossijskoj social'demokraticeskoj rabocej partii – Muenchen, London, Geneve, 1900-05 – 1 – us NY Public [335]
Iskra – Chisinau, 1920-29 – 5r – 1 – us UMI ProQuest [077]
Iskra – Geneva, 1900-03 – 1 – us UMI ProQuest [077]
Iskra – Kazanluk, Bulgaria. 1953-Apr 1968; 1976-79 (incomplete) – 5r – 1 – us L of C Photodup [949]
Iskra – Munich DE, 1949 15 jan-1952 1 apr, 1952 15 apr-1975 – 1 – uk British Libr Newspaper [074]
Iskra – Munich, Germany. semimonthly, -irr. 1975- – 1 – us Wisconsin U Libr [949]
Iskra, Wolfgang *see* Die darstellung des sichtbaren in der dichterischen prosa um 1900
"iskra" za dva goda : sbornik statei / Akselrod, P et al – 1906 – 10mf – 9 – mf#RPP-145 – ne IDC [325]
Iskusstvo – Moscow. 1-3. 1923-1927 – 1 – us NY Public [073]
Iskusstvo farfora / Ivanov, Dmitrii Dmitrievich – Moskva: b Gos izd-vo 1924 [mf ed 2006] – (= ser Russkoe dekorativnoe iskusstvo 6) – 1r – 1 – (filmed with: oil and the germs of war / by scott nearing [c1923]) – mf#5763p – us Wisconsin U Libr [730]
Iskusstvo kino – Russia, 1999- – 3r per y standing order – 1 – (1984-95 available 3r per y) – us UMI ProQuest [077]
Iskusstvo kommuny *see* Izdanie otdela izobrazitel'nykh iskusstv komissariata narodnogo prosveshcheniia
Iskusstvo trudiashchimsia – Moscow, dec 1924-apr 1926 – 42mf – 9 – us UMI ProQuest [790]
Iskusstvo, zhivopise, grafika, khudozhestvennaia pechate – Kiev, 1911-1912 – 3mf9 – 9 – mf#R-3216 – ne IDC [077]
Isla cerrera, novela basada en la conquista de pue... / Mendez Ballester, Manuel – San Juan, Puerto Rico. 1953 – 1r – us UF Libraries [972]
Isla de aves / Mancera Galletti, Angel – Caracas, Venezuela. 1959 – 1r – us UF Libraries [972]
Isla de guijes / Barnet, Miguel – Habana, Cuba. 1964 – 1r – us UF Libraries [972]
Isla de la tortuga / Pena Batlle, Manuel Arturo – Madrid, Spain. 1951 – 1r – us UF Libraries [972]
Isla De Rodriguez, Antonia *see*
– Frente al silencio
– Poemario intimo
– Restauracion de las libertades cubanas
Isla en el tacto / Augier, Angel I – Habana, Cuba. 1965 – 1r – us UF Libraries [972]
Isla news – Miami, FL. 1983 sep-1984 jan – 1r – us UF Libraries [071]
Isla y nada / Arrivi, Francisco – San Juan, Puerto Rico. 1958 – 1r – us UF Libraries [972]
Al-islah – Biskra, 1929-30 – 6 nos – 1 – fr ACRPP [073]
Islakh : daily religious/political magazine in pashto – Kabul, 1936-71 – 15r – 1 – us UMI ProQuest [079]
L'islam : organe hebdomadaire democratique des musulmans algeriens. – Alger. IV-VI, n100-206. 1912-14 – 1 – (devenu par fusion: l' ikdam) – fr ACRPP [320]
Der islam : geschichte, glaube, recht / Hartmann, Martin – Leipzig: R Haupt, 1909 – 1mf – 9 – 0-524-02082-5 – (incl bibl ref) – mf#1990-2846 – us ATLA [260]
Der islam : zeitschrift fuer geschichte und kultur des islamischen orients – Germany: Walter de Gruyter, 1910-99 [mf ed 2001] – (= ser Christianity's encounter with world religions, 1850-1950) – 13r – 1 – (in german) – mf#2001-s013 – us ATLA [260]
Islam / Ali, Syed Ameer – London: Constable, 1909 [mf ed 1992] – (= ser Religions ancient and modern) – 1mf – 9 – 0-524-02288-7 – (incl bibl ref) – mf#1990-2911 – us ATLA [260]
Islam : its history, character, and relation to christianity / Arnold, John Muehleisen – 3rd ed. London: Longmans, Green, 1874 – 1mf – 9 – 0-524-01675-5 – mf#1990-2577 – us ATLA [260]
Islam : its rise and progress / Sell, Edward – 2nd ed. Cairo: CMS English Library, 1907 – 1mf – 9 – 0-524-02368-9 – mf#1990-2979 – us ATLA [260]
Islam : or, the religion of the turk / Wherry, Elwood Morris – New York: American Tract Society, c1886 – 1mf – 9 – 0-524-02624-6 – mf#1990-3074 – us ATLA [260]
Islam / Rabinovitz, Alexander Siskind – Tel-Aviv, Israel. 1926 – 1r – us UF Libraries [260]
L' islam *see* L'ikdam
Islam, a challenge to faith / Zwemer, Samuel Marinus – New York, 1907 – 7mf – 8 – €28.00 – ne Slangenburg [260]
L'islam algerien en l'an 1900 / Doutte, Edmond – Alger-Mustapha: Giralt, 1900 – 1 – us CRL [260]
L'islam algerien en l'an 1900 / Doutte, Edmond – Alger-Mustapha: Giralt, 1900 [mf ed 1991] – 1mf – 9 – 0-524-01275-X – (in french. incl bibl ref) – mf#1990-2311 – us ATLA [260]
Islam and ahmadism : with a reply to questions raised by pandit jawahar lal nehru / Iqbal, Muhammad – Lahore: Iqbal Academy, [between 1900 and 1944] – (= ser Samp: indian books) – us CRL [260]
Islam and christian muslim relations – Abingdon, 1998+ [1,5,9] – ISSN: 0959-6410 – mf#20951 – us UMI ProQuest [230]
Islam and christianity : or, the quran and the bible. a letter to a muslim friend / Halliday, G Y – New York: American Tract Society, c1901 – 1mf – 9 – 0-524-01705-0 – mf#1990-2607 – us ATLA [260]

Islam and christianity in india and the far east / Wherry, Elwood Morris – New York: F H Revell, c1907 – (= ser Students' Lectures On Missions) – 1mf – 9 – 0-7905-6148-4 – (incl bibl ref) – mf#1988-2148 – us ATLA [260]

Islam and its founder / Stobart, James William Hampson – London: SPCK; New York: E & J B Young [1877?] [mf ed 1991] – (= ser Non-christian religious systems) – 1mf [ill] – 9 – 0-524-01304-7 – mf#1990-2340 – us ATLA [260]

Islam and missions : being papers read at the second missionary conference on behalf of the mohammedan world at lucknow, january 23-28, 1911 / ed by Wherry, Elwood Morris et al – New York: Fleming H Revell, c1911 – 1mf – 9 – 0-524-01063-3 – mf#1990-2211 – us ATLA [260]

Islam and socialism / Kidwai, Mushir Hosain – London: Luzac, [1912?] – 1mf – 9 – 0-524-01186-9 – mf#1990-2262 – us ATLA [260]

Islam and the oriental churches : their historical relations / Shedd, William Ambrose – Philadelphia: Presbyterian Board of Publ & Sabbath-School Work 1904 [mf ed 1990] – (= ser Students' lectures on missions 1902-03) – 1mf [ill] – 9 – 0-7905-6499-8 – (incl bibl ref) – mf#1988-2499 – us ATLA [260]

Islam as a missionary religion / Haines, Charles Reginald – London: SPCK; New York: E & J B Young 1889 [mf ed 1991] – (= ser Non-christian religious systems) – 1mf [ill] – 9 – 0-524-00830-2 – mf#1990-2076 – us ATLA [260]

L'islam au service du peuple / Toure, Ahmed Sekou – 2nd ed. Conakry: Bureau de presse de la presidence de la republique, 1977 – us CRL [260]

L'islam dans l'afrique occidentale / Le Chatelier, Alfred – Paris: G Steinheil, 1899 – 1 – us CRL [260]

Islam di soematera / [Amrullah, A M K, hadji] – Medan, Badan Pembangoen Semangat Islam (2605?) – 37p 1mf – 9 – mf#SE-2002 mf6 – ne IDC [260]

L'islam en guinee : fouta-daillon / Marty, Paul – Paris: E Leroux, 1921 – 1 – us CRL [960]

L'islam en mauritanie et au senegal – Paris: E Leroux, 1915-16 – 1 – us CRL [960]

L'islam et la civilisation francaise / Ouane, Ibrahima Mamadou – Avignon: Presses universelles 1957 – us CRL [260]

L'islam et la graal : etude sur l'esoterisme du parzival de wolfram von eschenbach / Ponsoye, Pierre – Paris: Denoel, 1958, c1957 [mf ed 1993] – 231p (ill) – 1 – (incl bibl ref) – mf#8448 – us Wisconsin U Libr [410]

L'islam et la nationalite / Saba, J S – Paris, 1931 – 2mf – 9 – mf#ILM-3215 – ne IDC [956]

L'islam et la politique musulmane francaise en afrique occidentale francaise / Arnaud, Robert – Paris: Comite de l'Afrique Francaise, 1912 – 1 – us CRL [960]

L'islam et le terroir africaine / Cardaire, Michel – Koulouba, Mali, 1954 – 1 – us CRL [260]

L'islam et les tribus dans la colonie du niger / Marty, Paul – [2. ser.] Paris: P Geuthner, 1931 – 1 – us CRL [960]

Islam, her moral and spiritual value : a rational and psychological study / Leonard, Arthur Glyn – London: Luzac, 1909 – 1mf – 9 – 0-524-01617-8 – mf#1990-2556 – us ATLA [260]

Islam, her moral and spiritual value a rational and psychological study. / Leonard, Arthur Glyn – London: Luzac, 1909 – 1 – (foreword by syed ameer ali) – us Wisconsin U Libr [260]

Der islam im lichte der byzantinischen polemik / Gueterbock, Karl – Berlin: J Guttentag, 1912 – 1mf – 9 – 0-524-01365-9 – (incl bibl ref) – mf#1990-2377 – us ATLA [260]

Der islam im morgen- und abendland / Mueller, August – Berlin: G Grote 1885-87 – (= ser Allgemeine geschichte in einzeldarstellungen 2/4) – 2v [ill/pl/facs] – 1 – mf#film mas c604 – us Harvard [260]

Islam in africa / Brelvi, Mahmud – Lahore, Pakistan. 1964 – 1r – us UF Libraries [260]

Islam in africa : its effects – religious, ethical and social – upon the people of the country / Atterbury, Anson Phelps – New York: GP Putnam, 1899 – 1mf – 9 – 0-524-00681-4 – (incl bibl ref) – mf#1990-2009 – us ATLA [260]

Islam in china : a neglected problem / Broomhall, Marshall – London: Morgan and Scott; Philadelphia: China Inland Mission, 1910 – 1mf – 9 – 0-524-01046-3 – mf#1990-2194 – us ATLA [260]

Islam in india : or, the qanun-i-islam: the customs of the musalmans of india / Ja'far Sharif – London; New York: Humphrey Milford; Oxford University Press, 1921 – (= ser Samp: indian books) – (trans by g a herklots; new ed rev and rearranged with add by william crooke) – us CRL [260]

Der islam in seinem einfluss auf das leben seiner bekenner / Hauri, Johannes – Leiden: EJ Brill, [1882?] – 1mf – 9 – 0-524-02020-5 – (incl bibl ref) – mf#1990-2795 – us ATLA [260]

Islam mecmuasi – Istanbul: Tanin Matbaasi, Matbaa-i Osmaniye 1914-18. Mueduer-i Mesul: Halim Sabit. n1-63. 30 kanunisani 1330 [1914]-30 tesrinievvel 1334 [1918] – (= ser O & t journals) – 19mf – 9 – $350.00 – us MEDOC [956]

Islam mecmuasi – Istanbul: Tanin Matbaasi, Matbaa-i Osmaniye 1914-1918. Mueduer-i Mesul: Halim Sabit. n1-63 (30 Kanunisani 1330 [1914]-30 Tesrinievvel 1334 – (= ser O & t journals) – 19mf – 9 – $350.00 – us MEDOC [079]

Islam, Nehalul *see* Assessment and comparison of the stress experienced by international and american students at the university of north texas

Islam und christentum im kampf um die eroberung der animistischen heidenwelt / Simon, G – Berlin, 1914 – 8mf – 8 – €30.00 – ne Slangenburg [230]

Islam und christentum im mittelalter / Fritsch, Erdmann – Breslau, 1930 – 3mf – 8 – €7.00 – ne Slangenburg [230]

Islamic Council for Universal Curriculum *see* Views and visions

Islamic culture – Hyderabad. 1949-1954 (1) – ISSN: 0021-1834 – mf#638 – us UMI ProQuest [260]

Islamic culture – Hyderabad, Deccan, 1927-1965. v1-39 – 373mf – 8 – mf#I-116 – ne IDC [956]

The islamic mode of worship / Ahmad, Bashiruddin Mahmud – 2nd ed. Punjab, India: Qadian, [191-?] – 1mf – 9 – 0-524-01595-3 – mf#1990-2534 – us ATLA [260]

Islamic near east *see* History of glass

Islamic news letter / Dewan Dakwah Islamiyah Indonesia – Djakarta, 1970-1972 – 2mf – 9 – (missing: 1970(1-4, 7-end); 1972(jan-apr)) – mf#SE-1714 – ne IDC [260]

Islamic quarterly – London. 1989-1996 (1) – ISSN: 0021-1842 – mf#16184 – us UMI ProQuest [260]

Islamic review *see* Muslim india and islamic review / islamic review and muslim india / islamic review

Islamic review and muslim india *see* Muslim india and islamic review / islamic review and muslim india / islamic review

Islamica : a journal devoted to the study of the language, arts, and civilisations of the islamic peoples – Leipzig, Germany: Verlag der Asia Minor, 1924-38 [mf ed 2001] – (= ser Christianity's encounter with world religions, 1850-1950) – 2r – 1 – (in german and arabic) – mf#2001-s005 – us ATLA [260]

Islamism, its rise and its progress : or, the present and past condition of the turks / Neale, Fred Arthur – London: J Madden, 1854 – 2mf – 9 – 0-524-02097-3 – mf#1990-2861 – us ATLA [260]

Islamisme contre "naturisme" au soudan francais : essais de psychologie politique coloniale / Brevie, J – Paris: E Leroux, 1923 – 1 – us CRL [260]

L'islamisme et le christianisme en afrique / Bonet-Maury, Gaston – Paris: Hachette, 1906 [mf ed 1990] – vi/299p on 1mf – 9 – 0-7905-5755-X – (in french) – mf#1988-1755 – us ATLA [260]

L'islamisme et son enseignement esoterique – Paris: Publ theosophiques, 1903 [mf ed 1992] – (= ser Etudes theosophiques) – 1mf – 9 – 0-524-02535-5 – (in french. incl bibl ref) – mf#1990-3030 – us ATLA [260]

L'islamismo e la confraternita dei senussi : notizie raccolte / Bourbon del Monte Santa Maria, Giuseppe – Citta di Castello: Tip dell'Unione arti grafiche, 1912 [mf ed 1981] – 1r – 1 – mf#ZZ-18852 – us NY Public [260]

Islamyah – Medan, 1954/1955-1959/1960. v1-5(4) – 60mf – 9 – (missing: 1957 v3(1)) – mf#SE-377 – ne IDC [950]

Island – nos. 1-8. 1964-66 – 1 – us AMS Press [800]

Island : seine bewohner, landesbildung und vulcanische natur / Winkler, Gustav – Braunschweig 1861 [mf ed Hildesheim 1995-98] – 2mf – 9 – €60.00 – 3-487-28936-9 – gw Olms [914]

Island ally / Iraklion Air Station [Crete] – v1 n2-v3 n41 [1985 feb 22-1987 nov 13] – 1r – 1 – (cont: seaview [iraklion air station [crete]]; cont by: cretan sun) – mf#1611877 – us WHS [355]

Island city times – Menasha, Neenah WI. 1863 oct 22-1866 dec 24, 1867 jan 2-1870 jan 15 – 2r – 1 – (cont by: winnebago county press) – mf#1102218 – us WHS [071]

Island creek baptist church. sparta, georgia : church records – 1806-1947 – 1 reel – 1 – $57.61 – us Southern Baptist [242]

Island cricketers / Walcott, Clyde – London, England. 1958 – 1r – us UF Libraries [972]

The island empire of the east : being a short history of japan and missionary work therein... / Robinson, J Cooper – Toronto: Missionary Society of the Church of England in Canada, 1912 [mf ed 1995] – (= ser Yale coll) – 226p (ill) – 1 – 0-524-09599-X – (int by lord bishop of algoma) – mf#1995-0599 – us ATLA [242]

Island exchange of news – Washington Island WI. 1981 nov 25-1982 jan 21 – 1r – 1 – (cont by: island exchange of news & ads) – mf#1003606 – us WHS [071]

Island exchange of news and ads – Washington Island WI. 1882 feb 4-dec, 1883, 1884, 1985-89, 1990 jan-dec 20 – 9r – 1 – (cont: island exchange of news; cont by: washington island observer) – mf#1003608 – us WHS [071]

Island gazette – La Pointe WI. [1966 feb 5-1984], 1985-94 – 2r – 1 – mf#1013388 – us WHS [071]

Island guardian – Charlottetown, PEI. 1887-94 – 3r – 1 – cn Library Assoc [071]

Island life : or, the phenomena and causes of insular faunas and floras, including a revision and attempted solution of the problem of geological climates / Wallace, Alfred Russel – London, 1880 – (= ser 19th c evolution & creation) – 6mf – 9 – mf#1.1.9065 – uk Chadwyck [574]

The island of cape breton : the "long-wharf" of the dominion / Bourinot, John George – Toronto?: s.n, 1882 – 1mf – 9 – mf#05934 – cn Canadiana [917]

Island of grenada : 1650-1950 / Devas, Raymund P – St George's, GRENADA . 1964 – 1r – us UF Libraries [972]

The island of madagascar : a sketch, descriptive and historical / Phelps, John Wolcott – New York: J B Alden, 1885 – 1 – us CRL [960]

Island of the sea / Taylor, Charles Edwin – Charlotte Amalie, St Thomas. 1896 – 1r – us UF Libraries [972]

Island of tobago, the west indies / Alford, C – London, England. 1964 – 1r – us UF Libraries [972]

Island sun – Anna Maria, FL. 1990 aug-dec – 1r – (1990 aug 8,15,22; nov 28) – us UF Libraries [071]

Island sun – Holmes Beach, FL. 1991 jan-jun – 1r – us UF Libraries [071]

Island sun – Holmes Beach, FL. v3 n24-49. 1991 jul-dec – 1r – us UF Libraries [071]

The island voyage, 1876 : melanesian mission / Selwyn, John Richardson – Ludlow: Edward J Partridge, 1877 [mf ed 1995] – (= ser Yale coll) – 32p – 1 – 0-524-09104-8 – mf#1995-0104 – us ATLA [240]

Island Woolen Mill Organization *see* Wool nubs

Islander – Awali Bahrain, 19 oct 1960-22 dec 1965; 5 jan 1966-15 jan 1969 – 1 1/2r – 1 – uk British Libr Newspaper [079]

Islander – Charlottetown, Canada. 29 mar 1861-29 dec 1865; 19 jan 1866-25 dec 1868; 1869-29 sep 1871; 5 jul-12 jul 1872 – 3r – 1 – uk British Libr Newspaper [071]

Islander – Charlottetown, PEI: John Ings, 1842-71 – 12r – 1 – ISSN: 0839-2781 – cn Library Assoc [071]

Islander : linking canada and the caribbean – Toronto. v1-4. aug 15 1973-aug 11 1977// (semimthly) – 5r – 1 – Can$550.00 – cn McLaren [073]

The islander – Awali, Bahrain. Oct 1860-Jan 1969 – 2r – 1 – uk British Libr Newspaper [072]

Islands in the wind – Greenlawn, NY. 1954 – 1r – us UF Libraries [972]

The islands of the pacific : from the old to the new: a compendious sketch of missions / Alexander, James McKinney – New York: American Tract Society, c1895 – 2mf – 9 – 0-8370-6000-1 – mf#1986-0000 – us ATLA [240]

Islands of titicaca and koati / Randelier, Adolph Francis Alphonse – New York, NY. 1910 – 1r – us UF Libraries [972]

Islands to windward / Mitchell, Carleton – New York, NY. 1948 – 1r – us UF Libraries [972]

Islandske folkesagn og aeventyr / Arnason, Jon – Kjobenhavn: J H Schubothe 1877 [mf ed Bloomington IN: Indiana Uni Lib, Preservation Dept 1984] – 1r – 1 – us Indiana Preservation [390]

Islankin, F B *see*
– Kontraktatsiia, sbyt i snabzhenie v selsko-khoziaistvennykh kreditnykh tovarishchestvakh
– Kreditnye i komissionno-bankovskie operatsii v selskokhoziaistvennykh kreditnykh tovarishchestvakh

Islas desoladas / Acosta, Agustin – Habana, Cuba. 1943 – 1r – us UF Libraries [972]

Islay Council of Social Service *see* Ileach

L'Isle Andre, Yves Marie de *see* Versuch ueber das schoene da man untersucht

L'isle de cuba et la havane : ou histoire, topographie, statistique, moeurs, usages, commerce et situation politique de cette colonie, d'apres un journal ecrit sur les lieux / Masse, Etienne – Paris 1825 [mf ed Hildesheim 1995-98] – 1v on 3mf – 9 – €90.00 – ISBN-10: 3-487-26934-1 – ISBN-13: 978-3-487-26934-4 – gw Olms [918]

Isle de la megalantropogenesie / Barre, M – Paris, France. 1807 – 1r – us UF Libraries [440]

l'Isle, Guillaume de et al *see* Atlas geographique

Isle of man courier – [NW England] Douglas, Manx NHL 1974-2004 – 1 – uk MLA; uk Newsplan [072]

Isle of man courier – [NW England] Ramsey, Manx NHL 1970-74 – 1 – uk MLA; uk Newsplan [072]

Isle of man daily times (a) – [NW England] Douglas Lib, Manx NHL 1958-70 – 1 – uk MLA; uk Newsplan [072]

Isle of man daily times (b) – [NW England] Douglas, Manx NHL – 1 – (title change: isle of man daily times & womans magazine [1958-66]) – uk MLA; uk Newsplan [072]

Isle of man examiner – [NW England] Douglas, Manx NHL 1924-2004 – 1 – uk MLA; uk Newsplan [072]

Isle of man examiner and general advertiser – [NW England] Douglas, Manx Museum Lib 10 jul 1880-1887, 1894-jul 1924 – 1 – (title change: isle of man examiner [aug 1924-1975, 1982]) – uk Newsplan; uk MLA [072]

Isle of man gazette – [NW England] Douglas Lib 1985-87 – 1 – uk MLA; uk Newsplan [072]

Isle of man times and general advertiser – [NW England] Douglas Lib, Manx NHL 2 jan 1869-5 may 1900* – 1 – (title change: isle of man weekly times & general advertiser [12 may 1900-1906]) – uk MLA; uk Newsplan [072]

Isle of man times and womans magazine – [NW England] Douglas, Manx NHL 1966-70 – 1 – uk MLA; uk Newsplan [072]

Isle of man weekly gazette and general advertiser – [NW England] Douglas Lib 22 apr 1812-10 aug 1815, 1818-22 [odd copies] – 1 – uk MLA; uk Newsplan [072]

Isle of man weekly times – [NW England] Douglas, Manx NHL 1906-jan 1987 – 1 – uk MLA; uk Newsplan [072]

Isle of wight count press etc – Nov 29 1884-Dec 24 1885; 1886-88; Jan 5-Dec 28 1889; 1890-96; Jan 2-Dec 24 1897; 1898-1900; Jan 5-Dec 28 1901; 1902; Jan 3-Dec 26 1903; 1904-96 – 214 1/2r – 9 – uk British Libr Newspaper [072]

Isleham 1566-1950 – (= ser Cambridgeshire parish register transcript) – 13mf – 9 – £17.25 – uk CambsFHS [929]

Isles afar off : an illustrated handbook to the missions of the london missionary society in polynesia / Cousins, George – London: London Missionary Society, 1914 [mf ed 1995] – (= ser Yale coll; Handbooks to our mission fields) – 104p (ill) – 1 – 0-524-09687-2 – mf#1995-0687 – us ATLA [240]

Isles of disenchantment: the fletcher / jacomb correspondence: letters exchanged between r j fletcher and edward jacomb / ed by Stober, W E – 1913-1921 – 1r – 1 – mf#pmb1243 – at Pacific Mss [980]

Isles of spice and palm / Verrill, A Hyatt – New York, NY. 1915 – 1r – us UF Libraries [972]

Les isles samoa et l'arrangement anglo-allemand / Voission, Louis P – Paris: Plon-Nourrit, 1900 – 1mf – 9 – $1.50 – mf#LLMC 82-100C Title 26 – us LLMC [327]

The isles that wait : [by a lady member of the melanesian mission] – London: SPCK; New York: E S Gorham, 1912 [mf ed 1995] – (= ser Yale coll) – 128p (ill) – 1 – 0-524-09603-1 – mf#1995-0603 – us ATLA [980]

[Isleton-] delta news – CA. 1929-1944 – 7r – 1 – $420.00 – mf#C03256 – us Library Micro [071]

[Isleton-] journal – CA. 1924-41; 1960-62 [wkly] – 13r – 1 – $780.00 – mf#BC02316 – us Library Micro [071]

Islington and holloway press – [London & SE] Islington CL 10 mar 1923-25 dec 1942 – 1 – (cont as: north london press [-jan 1971]) – uk Newsplan [072]

Islington and holloway press *see* United albion circular

Islington chronicle and finsbury weekly news – London, UK. jan-19 dec 1986, 1987-93 – 15r – 1 – (aka: islington chronicle and north london advertiser) – uk British Libr Newspaper [072]

Islington chronicle and north london advertiser *see* Islington chronicle and finsbury weekly news

Islington gazette – [London & SE] Islington CL 1873- – 1 – uk Newsplan [072]

Islington guardian and north london observer – London, UK. 1951 – 1r – 1 – (aka: islington guardian north london observer and weekly news and chronicle; islington guardian and hackney news north london observer and weekly news and chronicle) – uk British Libr Newspaper [072]

Islington londoner *see* Londoner (north islington ed)
Islington news – London, UK. 1891-1892 – 2r – 1 – (aka: islington news and hornsey gazette) – uk British Libr Newspaper [072]
Islington news and hornsey gazette *see* Islington news
Islington news and hornsey press – London, UK. 8 sep-3 nov 1877 – 1/4r – 1 – uk British Libr Newspaper [072]
Ismail, Ja'kub, teungkoe *see* Teungku tjhi'di tiro
Ismail, Mirza M *see* Speeches
Ismaili tradition concerning the rise of the / Ivanow, Wladimir – London; New York: Publ for the Islamic Research Association by Oxford University Press, 1942 – (= ser Samp: indian books) – us CRL [260]
Ismailia : a narrative of the expedition to central africa for the suppression of the slave trade, organized by ismail, khedive of egypt / Baker, Samuel White – 2nd ed. London: Macmillan, 1879 [mf ed 1985] – 524p (ill) – 1 – (incl app and ind) – mf#6836 – us Wisconsin U Libr [916]
Ismay – Ismay, MT. 1908-1933 (1) – mf#64491 – us UMI ProQuest [071]
'Ismet *see* The divan project
Isms, fads and fakes : a series of sunday night discourses / Field, Jasper Newton – Indianapolis: Hollenbeck Press, 1904 – 1mf – 9 – 0-524-02854-0 – mf#1990-0711 – us ATLA [200]
Isms old and new : winter sunday evening sermon-series for 1880-81 / Lorimer, George Claude – Chicago: SC Griggs, 1881 – 1mf – 9 – 0-7905-9789-6 – mf#1989-1514 – us ATLA [240]
Isnard, Hildebert *see* Madagascar
Isn't one wife enough? / Young, Kimball – New York, NY. 1954 – 1r – us UF Libraries [025]
Iso world – Framingham. 1982-1983 (1) 1982-1983 (5) 1982-1983 (9) – (cont: computer business news. cont by: micro marketworld) – ISSN: 0745-2578 – mf#11663,01 – us UMI ProQuest [000]
Isogawa, Hiroaki *see* The economic effects of government assistance in commercial resort development
Isoglossen, isomorphen und isophonen-konzeptionelle vorstellungen und methodische ansaetze der dialektologischen forschung in deutschland und rumaenien / Maetzing, Karl-Heinrich – (mf ed 1996) – 3mf – 9 – €49.00 – 3-8267-2345-7 – mf#DHS 2345 – gw Frankfurter [410]
Isokinetic evaluation of peak torque values following closed and open kinetic chain exercise training programs / Reed, Cathy A – Springfield College, 1995 – 2mf – 9 – $8.00 – mf#PE3618 – us Kinesology [611]
Isokinetic evaluation of the knee flexors and extensors of male and female sprinters and distance runners / Kluckhohn, James C – 1997 – 1mf – 9 – $4.00 – mf#PH 1573 – us Kinesology [612]
Isokinetic evaluation of the posterior rotator cuff musculature following a strengthening program utilizing rubber tubing / O'Brian, Jeanne M & Redmond, Charles J – 1992 – 2mf – $8.00 – us Kinesology [612]
Isokinetics and exercise science – Stoneham. 1994-1995 (1,5,9) – ISSN: 0959-3020 – mf#18971 – us UMI ProQuest [613]
Isoko clans of the niger delta / Welsh, James William – 1937 – us CRL [307]
The isoko mss : a lungu view of their own history / ed by Watson, William – [SI: s.n., 195-?] – us CRL [960]
Isoko y'amajyambere / Kagame, P Alegisi – [Kabgayi, Rwanda?]: Ibitabo By'injijura Muco (Editions morales), 1949-1951 – 1 – us CRL [960]
Die isola bella im lago maggiore : ihre entstehung im seicento / Schmidt-Nechl, Barbara – (mf ed 2000) – 6mf – 9 – €62.50 – 3-8267-2701-0 – mf#DHS 2701 – gw Frankfurter [720]
Isolated anterior cruciate ligament deficiency : comparison of results following acute and chronic arthroscopic reconstruction with patellar tendon autograft / Manzour, Waleed F – 2000 – 3mf – 9 – $12.00 – mf#PE 4085 – us Kinesology [617]
Isolation in the school / Young, Ella – Chicago: University of Chicago Press, 1900 [mf ed 1970] – (= ser Library of american civilization 40014) – 57p on 1mf – 9 – us Chicago U Pr [370]
Isolde und tristan : die geschichte einer jungen liebe / Jansen-Runge, Edith – Bodenhausen: Fackeltraeger-Verlag, 1943 – 1r – 1 – us Wisconsin U Libr [390]
Isolde weisshand : ein roman aus alter zeit / Lucka, Emil – Berlin: S Fischer, 1909 – 1r – 1 – us Wisconsin U Libr [830]
Isopescul, Octavian *see* Der prophet malachias
Isore, Andre *see* La guerre et la condition privee de la femme
Isotopes and radiation technology – Washington. 1963-1972 (1) 1970-1972 (5) – ISSN: 0021-1923 – mf#2124 – us UMI ProQuest [530]
Ispolnenie dogovorov / Karavaikin, A – Moskva: Gos izd-vo Sovetskoe zakonodatel'stvo, 1934 [mf ed 2004] – 1r – 1 – (filmed with: vzaimnaia pomoshch' sredi zhivotnykh i liudei, kak dvigatel' progressa / p kropotkin (1922). incl bibl ref) – us Wisconsin U Libr [346]
Ispolnenie zhelanii / Kaverin, V – Leningrad, 1937.430p – 1 – us Wisconsin U Libr [460]
Ispolzovanie izbytochnosti v informatsionnykh sistemakh : trudy vtorogo simpoziuma, leningrad, 6-10 iiunia 1966 g / ed by Zheleznova, N A – Leningrad: Izd-vo "Nauka," Leningradskoe otd-nie, 1970 – us CRL [947]
Ispritizma – Istanbul: Tanin Matbaasi, 1909. Mueduer: Bahaeddin. n1. 1 kanunisani 1325 [1909] – (= ser 0 & t journals) – 1mf – 9 – $25.00 – us MEDOC [956]
Isprs journal of photogrammetry and remote sensing – Amsterdam. 1989+ (1,5,9) – (cont: photogrammetria) – ISSN: 0924-2716 – mf#42031,01 – us UMI ProQuest [520]
Isr : interdisciplinary science reviews – London. 1982-1996 (1,5,9) – ISSN: 0308-0188 – mf#13307 – us UMI ProQuest [500]
Isr – N7-34, 37-48, 50, 53-85. Paris. juin 1923-oct 1926 – 1 – fr ACRPP [073]
Isr journal of education personnel relations – New York. 1969-1970 – 1 – mf#7199 – us UMI ProQuest [370]
Isr newsletter / University of Michigan Institute for Social Research – Ann Arbor. 1975-1992 (1) 1976-1992 (5) 1976-1992 (9) – (cont by: profiles: the isr newsletter) – ISSN: 0020-2622 – mf#9899 – us UMI ProQuest [300]
Israel : national security files, 1963-1969 – 3r – 1 – $585.00 – 0-89093-388-X – (with p/g) – us UPA [327]
Israel : seine entwicklung im rahmen der weltgeschichte / Lehmann-Haupt, C F – Tuebingen: J C B Mohr, 1911 – 1mf – 9 – 0-7905-2013-3 – (incl ind) – mf#1987-2013 – us ATLA [956]
Israel *see*
- Kitab al-qawanin
- Majmu'at al-anzimah
- Majmu'at al-nasharat
- Mashru'at al-qawanin
- Qawanin dawlat isra'il

Israel 1931 / Paraf, Pierre – Paris: Librairie Valois, 1931 (mf ed 1995) – 1r – 1 – (incl bibl ref) – mf#ZZ-34402 – us NY Public [939]
Israel, a prince with god : the story of jacob re-told / Meyer, Frederick Brotherton – aut's ed. New York: Fleming H Revell [19–?] [mf ed 1986] – 1mf – 9 – 0-8370-9967-6 – mf#1986-3967 – us ATLA [221]
Israel among the nations : a study of the jews and antisemitism = Juifs et l'antisemitisme / Leroy-Beaulieu, Anatole – London: W Heinemann; New York: G P Putnam, 1895 [mf ed 1990] – 1mf – 9 – 0-7905-5420-8 – (incl bibl ref. english by frances hellman) – mf#1988-1420 – us ATLA [939]
Israel and babylon : the influence of babylon on the religion of israel (a reply to delitzsch) / Gunkel, Hermann – Philadelphia: John Jos McVey 1904 [mf ed 1989] – 1mf – 9 – 0-7905-1409-5 – (incl bibl) – mf#1987-1409 – us ATLA [939]
Israel and god's purpose / Wilkinson, John – London, England. 1885 – 1r – us UF Libraries [240]
Israel and human rights / Pevsner, Isaiah – Tel-Aviv: Israel Association for Human Rights, 1969. 24p. LL-12042 – 1 – us L of C Photodup [341]
Israel and palestine monthly review (i&p) – Paris. 1975-1981 (1) 1975-1981 (5) 1975-1981 (9) – (cont by: israel and palestine political report: i&p) – mf#10102 – us UMI ProQuest [956]
Israel and palestine political report (i&p) – Paris. 1981-1996 (1) 1981-1996 (5) 1981-1996 (9) – (cont: israel and palestine monthly review: i&p) – ISSN: 0294-1341 – mf#10102,01 – us UMI ProQuest [327]
Israel digest – New York. 1971-1979 (1) 1958-1979 (5) 1977-1979 (9) – ISSN: 0021-2024 – mf#6158 – us UMI ProQuest [939]
Israel economist – Jerusalem. 1972-1991 (1) 1972-1991 (5) 1972-1991 (9) – ISSN: 0021-2040 – mf#7881 – us UMI ProQuest [330]
Israel et la foi chretienne / Lubac, Henri de et al – Fribourg: Librairie de l'Universite, 1942 (mf ed 1995) – 1r – 1 – mf#ZZ-34394 – us NY Public [230]
Israel exploration journal – Jerusalem. 1950+ (1) 1970+ (5) 1975+ (9) – ISSN: 0021-2059 – mf#2514 – us UMI ProQuest [930]
Israel exploration journal (iej) – 1950-51 – 9 – $12.00 – (1952-1956 $4y) – us IRC [930]
Israel in egypt : fully illustrated by existing monuments / Osburn, William – 2nd rev ed. London: Dean, [1856?] – 1mf – 9 – 0-524-04918-1 – mf#1992-0261 – us ATLA [930]
Israel in europe / Abbott, George Frederick – London: Macmillan, 1907 – 2mf – 9 – 0-524-02195-3 – (incl bibl ref) – mf#1990-2869 – us ATLA [940]
Israel journal of medical sciences – Jerusalem. 1965-1997 (1) 1972-1997 (5) 1976-1997 (9) – ISSN: 0021-2180 – mf#6934 – us UMI ProQuest [610]
Israel journal of psychiatry and related sciences – Jerusalem. 1981-1996 (1,5,9) – ISSN: 0333-7308 – mf#12859,01 – us UMI ProQuest [616]
Israel kingdom message – 1986 mar, 1987 feb-aug, oct, dec, 1988 oct – 1r – 1 – mf#1801059 – us WHS [071]
Israel law review – v1-32. 1966-98 – 5,6,9 – $595.00 set – (v1-19 1966-84 on reel $212. v20-32 1985-98 on mf $383) – ISSN: 0021-2237 – mf#103761 – us Hein [340]
Israel. Laws, Statutes, etc *see* Insurance supervision in israel
Israel magazine – Tel-Aviv. 1972-1976 (1) 1972-1972 (5) (9) – ISSN: 0021-2245 – mf#6725 – us UMI ProQuest [939]
Israel Meir *see* Hundert mayses um mesholim
Israel my glory : or, israel's mission and missions to israel / Wilkinson, John – special ed. London: Mildmay Mission to the Jews' Book Store, 1894 – 1mf – 9 – 0-524-05337-5 – mf#1990-1455 – us ATLA [939]
The israel of the alps : a complete history of the vaudois of piedmont and their colonies = Israel des alpes / Muston, Alexis – Glasgow; New York: Blackie, 1857 – 3mf – 9 – 0-7905-8267-8 – (incl bibl ref. in english) – mf#1988-6145 – us ATLA [240]
Israel, Paul K *see* The relationship between physical fitness in university students and demographic, academic, and attitudinal factors
Israel potter, his fifty years of exile / Melville, Herman – New York, NY. 1924 – 1r – us UF Libraries [025]
Israel speaks – New York, N.Y. – (v3, no1, (4 feb. 1949)-v11, no5, (29 mar. 1957); continues: haganah speaks) – us AJPC [939]
Israel, the biblical people : israel, past, present and future. battling for mankind and civilization against anti-semitism and its real issues / Fluegel, Maurice – Baltimore, Md: H Fluegel, 1899 – 1mf – 9 – 0-7905-0886-9 – (incl bibl ref) – mf#1987-0886 – us ATLA [939]
Israel today and the jewish times *see* Jewish times
Israel und aegypten : die politischen beziehungen der koenige von israel und juda zu den pharaonen / Alt, Albrecht – Leipzig: J C Hinrichs, 1909 [mf ed 1989] – (= ser Beitraege zur wissenschaft vom alten testament 6) – 1mf – 9 – 0-7905-3000-7 – (incl bibl ref) – mf#1987-3000 – us ATLA [930]
Israel und juda : bibelkunde zum alten testamente fuer seminare und hoehere lehranstalten / Erbt, Wilhelm – Goettingen: Vandenhoeck & Ruprecht, 1903 – 1mf – 9 – 0-7905-3194-1 – mf#1987-3194 – us ATLA [221]
Israel und juda bei amos und hosea : nebst einem exkurs ueber hos 1-3 / Seesemann, Otto – Leipzig: Dieterich, 1898 [mf ed 1985] – 1mf – 9 – 0-8370-5216-5 – (incl bibl ref) – mf#1985-3216 – us ATLA [221]
Israel und juda bei amos und hosea nebst einem exkurs ueber hos 1-3 / Seesemann, Otto – Leipzig, 1898 (mf ed 1993) – 1mf – 9 – €24.00 – 3-89349-347-6 – mf#DHS-AR 200 – gw Frankfurter [221]
The israel williams papers, 1730-1785 – [mf ed 1977] – 1r – 1 – (with p/g. coll mainly covers period of french and indian wars) – us MA Hist [355]
Israeli, Samuel Michael *see* The nature of the liability of shareholders of a corporation
De israelieten te mekka van davids tijd tot in de vijfde eeuw onzer tijdrekening / Dozy, Reinhart Pieter Anne – Haarlem: A C Kruseman, 1864 [mf ed 1989] – vi/214p on 1mf – 9 – 0-7905-1878-3 – (incl bibl ref) – mf#1987-1878 – us ATLA [956]
Der israelit – Frankfurt/M DE, 1906-38 [gaps] – 1 – gw Misc Inst [074]
Israelit – Mainz. v. 1-11, 14-57, 61-79. May 15 1860-Nov 3 1938. Incomplete – 1 – us NY Public [074]
Israelite – Cincinnati, Ohio. 1854-1927; 1958-68 – 1 – us AJPC [071]
L'israelite algerien : organe des interets du judaisme en general et du judaisme d'algerie en particulier – Oran. n1-22. 1900 – 1 – fr ACRPP [270]
Israelite ostraca from samaria / Reisner, George A – 1924 – 9 – $10.00 – us IRC [930]
Die israeliten in der wueste : ein oratorium / Bach, Carl Philipp Emanuel – Hamburg: Im Verlag des Autors 1775 [mf ed 19–] – 2mf – 9 – mf#fiche238; fiche758 – us Sibley [780]
Israeliten und hyksos in aegypten : eine historisch-kritische untersuchung / Uhlemann, Max – Leipzig: Otto Wigand, 1856 – 1mf – 9 – 0-7905-2561-5 – mf#1987-2561 – us ATLA [930]
Die israeliten und ihre nachbarstaemme : alttestamentliche untersuchungen / Meyer, Eduard – Halle (Saale): Max Niemeyer, 1906 – 2mf – 9 – 0-7905-1363-3 – (incl bibl ref and indexes) – mf#1987-1363 – us ATLA [220]
Israelites algeriens de 1830 'a 1902 / Martin, Claude – Paris, France. 1936 – 1r – us UF Libraries [939]
Israelites espanoles / Bensasson, Maurice Jacques – Alicante, Spain. 1905 – 1r – us UF Libraries [939]
Israelitische annalen : ein centralblatt fuer geschichte, literatur und cultur der israeliten aller zeiten und laender – Frankfurt a.M: Isaak Markus Joest. v1-3. 1839-41 [complete] – (= ser German-jewish periodicals...1768-1945, pt 2) – 1r – 1 – $125.00 – mf#B109 – us UPA [939]
Israelitische annalen – Frankfurt/M DE, 1839-41 – 1r – 1 – gw Misc Inst [939]
Israelitische chronologie / Quandt, Ludwig; ed by Dieckmann, R – Guetersloh: C Bertelsmann, 1873 – 1mf – 9 – 0-524-05416-9 – (incl bibl ref) – mf#1992-0426 – us ATLA [221]
Israelitische friedhof in jungholz / Ginsburger, M – Gebweiler, France. 1904 – 1r – us UF Libraries [939]
Das israelitische gebetbuch fuer alle wochen-, feier- und festtage des jahres : nebst den spruechen der vaeter / Wesseln, Wolfgang – 3. verb und verm Aufl. Prag: Jakob B Brandeis, 1888 – 1mf – 9 – 0-524-08904-3 – mf#1993-4039 – us ATLA [270]
Israelitische gemeinde burgel a m / Lammertz, C – Offenbach a.M., Germany. 1924? – 1r – us UF Libraries [943]
Der israelitische lehrer – Darmstadt, Mainz DE, 1861 may-1871 – 1r – 1 – gw Misc Inst [939]
Der israelitische lehrer : wochenschrift fuer die allgemeinen angelegenheiten des judenthums und insbesondere des israelitischen lehrerstandes. organ fuer den verein "achawa" – Darmstadt, Mainz: Klingenstein. v1-12. 1861-72 – (= ser German-jewish periodicals...1768-1945, pt 1) – 1r – 1 – $125.00 – mf#B114 – us UPA [270]
Israelitische letterbode. – 1-12, n2. 1875/76-1888 – 1 – us NY Public [939]
Das israelitische pfingstfest und der plejadenkult : eine studie / Grimme, Hubert – Paderborn:Ferdinand Schoeningh, 1907 – 1mf – 9 – 0-8370-3400-0 – mf#1985-1400 – us ATLA [230]
Israelitische Religionsgemeinde (Leipgiz, Germany) *see* Gedenkblatter zur erinnerung an rabbiner dr a m goldschmidt
Israelitische rundschau *see* Berliner vereinsbote
Israelitische und judische geschichte / Wellhausen, Julius – Berlin, Germany. 1914 – 1r – us UF Libraries [939]
Israelitische und juedische geschichte : beurteilung der schrift von j wellhausen 1894 / Zahn, Adolf – Guetersloh: C Bertelsmann, 1895 [mf ed 1985] – 1mf – 9 – 0-8370-5942-9 – (with suppl. incl bibl ref) – mf#1985-3942 – us ATLA [221]
Der israelitische volksfreund – Cincinnati. Ohio. 1858-59 – 1 – us AJPC [071]
Der israelitische volksleher : eine monatsschrift erbaulichen und belehrenden inhalts, zur kenntnis des judenthums, sowie zur laeuterung und foerderung des religioesen sinnes unter den israeliten – Frankfurt a.M: Leopold Stein, Samuel Sueskind. v1-10, 1851-60 [complete] – (= ser German-jewish periodicals...1768-1945, pt 1) – 2r – 1 – $220.00 – mf#B117 – us UPA [270]
Der israelitische volkslehrer – Frankfurt/M DE, 1851 feb-1860 – 2r – 1 – gw Misc Inst [370]
Israelitische wochenschrift – Berlin DE, 1899 n12-1905 n52 – 2r – 1 – gw Misc Inst [939]
Israelitische wochenschrift – Magdeburg DE, 1874-85, 1888, 1890 – 1 – (incl suppls: das juedische literaturblatt (also: juedisches literatur-blatt, publ in magdeburg & berlin) 1872-1916 [2r]) – gw Misc Inst [939]
Israelitische wochenschrift – Magdeburg. v. 1-12, 14, 16-25. 1870-1881, 1883, 1885-1894 – 1 – us NY Public [939]
Die israelitischen vorstellungen vom zustand nach dem tode / Bertholet, Alfred – 2., gaenzlich umgearb und erw Aufl. Tuebingen: Mohr, 1914 – (= ser [Sammlung gemeinverstaendlicher Vortraege und Schriften aus dem Gebiet der Theologie und Religionsgeschichte]) – 1mf – 9 – 0-524-04563-1 – (incl bibl ref) – mf#1992-0151 – us ATLA [270]
Israelitischer jugendfreund – Berlin DE, 1898 – 1 – gw Misc Inst [939]
Israelitischer landes-lehrer-verein in boehmen : mitteilungen / ed by Freund, Max – Prague. v1-23. 1895-1917* – (= ser German-jewish periodicals...1768-1945, pt 2) – 1r – 1 – $125.00 – mf#B113 – us UPA [939]
Israelitisches familienblatt / gross-berlin – Berlin DE, 1932-34, 1936-38 [gaps] – 1 – (filmed by other misc inst: 1929-38 [7r, mpf]. main ed in hamburg. with suppl: juedische bibliothek der unterhaltung und des wissens 1898-1938 [gaps]) – gw Misc Inst [939]

Israelitisches gemeindeblatt : offizielles organ der israelitischen gemeinden mannheim und ludwigshaften – Mannheim DE, 1930-33 – 2r – 1 – us UMI ProQuest [939]
Israelitisches gemeindeblatt – Koeln DE, 1890-91, 1913-16, 1918-19 – 1 – (with gaps) – gw Misc Inst [939]
Israelitisches gemeindeblatt – Mannheim, (Ludwigshafen) DE, 1932-34, 1936-37 – 1 – (with gaps) – gw Misc Inst [270]
Israelitisches predigt-magazine : homiletische monatsschrift – Leipzig DE, 1874-94 – 3r – 1 – us UMI ProQuest [939]
Israelitisches wochenblatt fuer die schweiz – Zurich. Oct. 4, 1929-Dec. 30, 1960 – 1 – us NY Public [939]
Israel's advocate : or, the restoration of the jews contemplated and urged – New York. 1823-1827 (1) – mf#3996 – us UMI ProQuest [939]
Israels feste und gedenktage / Katz, Albert – Leipzig, Germany. 1921 – 1r – us UF Libraries [939]
Israel's future : lectures / Molyneux, Capel – 6th ed. London: James Nisbet, 1860 – 1mf – 9 – 0-7905-8710-6 – mf#1989-1935 – us ATLA [221]
Israels geschichte von alexander dem grossen bis hadrian / Schlatter, Adolf von – Calw: Vereinsbuchh., 1901 – (= ser Reiche Der Alten Welt) – 1mf – 9 – 0-7905-2036-2 – (incl bibl ref and index) – mf#1987-2036 – us ATLA [939]
Israel's greatest prophet / Hastings, H L – London, England. 188- – 1r – us UF Libraries [240]
Israels herold – (New York). 1849 – 1 – us AJPC [071]
Israel's historical and biographical narratives : from the establishment of the hebrew kingdom to the end of the maccabean struggle / Kent, Charles Foster – New York: Charles Scribner, 1905 – 2mf – 9 – 0-8370-9877-7 – mf#1986-3877 – us ATLA [956]
Israel's hope of immortality : four lectures / Burney, C F – Oxford: Clarendon Press, 1909 – 1mf – 9 – 0-7905-0561-4 – mf#1987-0561 – us ATLA [939]
Israel's ideal : or, studies in old testament theology / Adams, John – Edinburgh: T & T Clark, 1909 – 1mf – 9 – 0-8370-2041-7 – (incl ind of biblical passages cited) – mf#1985-0041 – us ATLA [221]
Israel's messiah / Hastings, H L – London, England. 188- – 1r – us UF Libraries [240]
Israel's messianic hope to the time of jesus : a study in the historical development of the foreshadowings, of the christ in the old testament and beyond / Goodspeed, George Stephen – New York: Macmillan, 1900 – 1mf – 9 – 0-8370-3339-X – (incl indes) – mf#1985-1339 – us ATLA [270]
Israel's monthly magazine – (New York). 1900-02 – 1 – us AJPC [939]
Israel's ordinances / Elizabeth, Charlotte – London, England. 1843 – 1r – us UF Libraries [240]
Israel's position on the jordan canal project : an address by ambassador abba eban before the united nations security council on oct 30, 1953 / Eban, Abba – New York: Israel Office of Information, 1953 – us CRL [956]
Israel's prophets / Petrie, George Laurens – New York: Neale, 1912 – 1mf – 9 – 0-524-05625-0 – mf#1992-0480 – us ATLA [221]
Israel's speedy restoration and conversion contemplated : or, signs of the times in familiar letters / Palmer, Phoebe – New York: J Gray 1854 [mf ed 2004] – 1r – 1 – 0-524-10505-7 – mf#b00720 – us ATLA [221]
Israel's wisdom literature / Rankin, OS – 1936 – 9 – $10.00 – us IRC [221]
Israel's world : origin and destiny of the british race, colonies, and empire; according to scripture, history, language, and signs of our times / Roe, Henry – [4th ed.] [London] [1900?] – (= ser 19th c british colonization) – 1mf – 9 – mf#1.1.7166 – uk Chadwyck [939]
Israelson, Shalom *see* Sefer divre shalom
Israil fi al-istiratijiyah al-amirikiyah fi al-thamaninat / Mansur, Kamil – Bayrut: Muassasat al-Dirasat al-Filastiniyah, 1980 – 1 – us CRL [956]
Israil wa-"mashru kartir" / Shufani, Ilyas – Bayrut: Muassasah al-Dirasat al-Filastiniyah, 1980 – 1 – us CRL [956]
Issac backus papers *see* Backus, issac, papers, ms 71
Issel, A *see*
– Catalogo dei molluschi raccolti dalla missione italiana in persia
– Viaggio nel mar rosso e tra i bogos
Issel, Ernst *see*
– Der begriff der heiligkeit im neuen testament
– Die lehre vom reiche gottes im neuen testament
– Die reformation in konstanz
Isselburg, P *see*
– Emblemata politica in aula magna curiae noribergensis depicta
Isserman, Maurice *see* The communist party usa and radical organizations, 1953-1960
Issledovanie vkladov v gosudarstvennye sberegatel'nye kassy kostromskoi gubernii za vremia s 1885 po 1911 gg vkliuchitel'no : s prilozheniem kratkogo istoricheskogo ocherka razvitiia sberegatel'nykh uchrezhdenii v... / Nazorov, I A – Kostroma, 1913. 3v – 3mf – 9 – mf#REF-455 – ne IDC [332]
Issledovanie vkladov v gosudarstvennye sberegatel'nye kassy kostromskoi gubernii za vremia s 1885 po 1911 vkliuchitel'no / Nazorov, I A – Kostroma, 1913 – 4mf – 8 – mf#RZ-175 – ne IDC [314]
Issledovanie zlatostruia po rukopisi 12 veka / Malinin, V – Kiev, 1878 – 5mf – 9 – mf#R-10160 – ne IDC [243]
Issledovaniia po russkomu iazyku *see* Izdanie otdeleniia iazyka i slovesnosti imperatorskoi akademii nauk
Issledovaniia termodinamicheskikh svoistv zhidkogo vodoroda / Pashkov, V V & Marinin, V S – Moskva: In-t vysokikh temperatur AN SSSR, 1979 – us CRL [947]
Issue / Indiographs, Inc – v1 n3 [1970 feb], v1 n7,13 [1971 apr, may 13, aug 26] – 1r – 1 – mf#1583910 – us WHS [071]
Issue – Jackson MS. 1908 apr 11,18 – 1r – 1 – mf#869168 – us WHS [071]
Issue / Socialist Party of Passaic, NJ – Passaic NJ. 1912 apr – 1r – 1 – us WHS [325]
Issue / Socialist Party of Union County [NJ] – Elizabeth NJ. 1915 feb 26 – 1r – 1 – us WHS [325]
The issue – De Witt, NE: DeWitt Times, 1890 (wkly) [mf ed with gaps filmed [1974?]] – 1r – 1 – us NE Hist [071]
Issue at stake in the alternative submitted to the presbyteries / Cousin, William – Edinburgh, Scotland. 1870 – 1r – us UF Libraries [242]
The issue of kikuyu : a sermon / Henson, Hensley – London; New York: MacMillan, 1914 – 1mf – 9 – 0-7905-6181-6 – mf#1988-2181 – us ATLA [240]
Issue update / Senate Republican Conference [US] – 1981 may-1984 oct, 1985 mar – 1r – 1 – mf#957562 – us WHS [325]
Issued by o(cuppied e(nemy) t(erritory) a(dministration) (south) to august 1919 : proclamations, ordinances and notices – [Cairo, 1920] – 2mf – 9 – mf#J-28-163 – ne IDC [956]
Issues and studies – Taipei. 1964+ (1) 1964+ (5) 1964+ (9) – ISSN: 1013-2511 – mf#9064 – us UMI ProQuest [951]
The issues before the church : letter to the clergy of the diocese of delaware / Kinsman, Frederick Joseph – 2nd ed. New York: Edwin S Gorham, 1915 – 1mf – 9 – 0-524-06632-9 – mf#1991-2687 – us ATLA [240]
Issues in accounting education – Sarasota. 1990+ (1,5,9) – ISSN: 0739-3172 – mf#18572 – us UMI ProQuest [650]
Issues in bank regulation – Park Ridge. 1980-1990 (1) 1980-1990 (5) 1980-1990 (9) – ISSN: 0164-7725 – mf#13013 – us UMI ProQuest [332]
Issues in child mental health – New York. 1977-1978 (1,5,9) – (cont: psychosocial process. cont by: family and child mental health journal) – ISSN: 0362-403X – mf#11188,01 – us UMI ProQuest [150]
Issues in comprehensive pediatric nursing – Washington. 1983+ (1,5,9) – ISSN: 0146-0862 – mf#14243 – us UMI ProQuest [610]
Issues in cooperation and power – Berkeley. 1980-1980 (1,5,9) – (cont: issues in radical therapy and cooperative power) – ISSN: 0199-8242 – mf#12488 – us UMI ProQuest [150]
Issues in criminology – Berkeley. 1965-1975 (1) 1971-1975 (5) – ISSN: 0021-2385 – mf#2545 – us UMI ProQuest [360]
Issues in engineering – New York. 1979-1982 [1]; 1979-1982 [5]; 1979-1982 [9] – (cont by: journal of professional issues in engineering) – ISSN: 0191-3271 – mf#8148,01 – us UMI ProQuest [620]
Issues in engineering – New York. 1979-1982 (1) 1979-1982 (5) 1979-1982 (9) – (cont: engineering issues) – ISSN: 0191-3271 – mf#8148,01 – us UMI ProQuest [620]
Issues in health and safety *see* Risk
Issues in health care of women – Washington. 1983-1983 (1,5,9) – (cont by: health care for women international) – ISSN: 0161-5246 – mf#14242 – us UMI ProQuest [305]
Issues in law and medicine – v1-16. 1985-2001 – 9 – $475.00set – ISSN: 8756-8160 – mf#110601 – us Hein [344]
Issues in mental health nursing – Washington. 1984+ (1,5,9) – ISSN: 0161-2840 – mf#14337 – us UMI ProQuest [610]
Issues in radical therapy – Springfield. 1982-1988 (1) 1982-1988 (5) 1982-1988 (9) – (cont by: new studies on the left) – ISSN: 0886-0629 – mf#13419 – us UMI ProQuest [616]
Issues in radical therapy and cooperative power – Berkeley. 1973-1979 (1) – (cont by: issues in cooperation and power) – mf#9499 – us UMI ProQuest [616]
Issues in reproductive and genetic engineering – Elmsford. 1990-1992 (1,5,9) – (cont: reproductive and genetic engineering) – ISSN: 0958-6415 – mf#49557,01 – us UMI ProQuest [575]
Issues in science and technology – Washington. 1989+ (1,5,9) – ISSN: 0748-5492 – mf#17554 – us UMI ProQuest [500]
Issues inreview / Greater Madison Chamber of Commerce [Madison WI] – 1984 feb-1988 sep – 1r – 1 – (cont by: issues [madison, wi]) – mf#1840766 – us WHS [071]
The issues of life / Worcester, Elwood – New York: Moffat, Yard, 1915 – 1mf – 9 – 0-7905-8981-8 – mf#1989-2206 – us ATLA [240]
Issues related to learners' discipline in schools : with reference to kwamhlanga / Kutu, Ishmael Gaster – Pretoria: Vista University 2003 [mf ed 2003] – 2mf – 9 – (incl bibl ref) – mf#mfm15236 – sa Unisa [370]
Issues surrounding the use of polygraphs : hearing...united states senate, 107th congress, 1st session, april 25 2001 / United States. Congress. Senate. Committee on the Judiciary – Washington: US GPO 2002 [mf ed 2002] – 2mf – 9 – 0-16-067030-6 – (incl bibl ref) – us GPO [360]
Issues today – Stamford. 1969-1974 (1) 1970-1972 (5) (9) – ISSN: 0021-2407 – mf#5734 – us UMI ProQuest [370]
Ist das liberale jesusbild modern? / Gruetzmacher, Richard Heinrich – Berlin: Edwin Runge 1907 [mf ed 1986] – (= ser Biblische zeit- und streitfragen 3/2) – 1mf – 9 – 0-8370-9475-5 – (incl bibl ref) – mf#1986-3475 – us ATLA [240]
Ist der sonntag heidnischen, paepstlichen oder christlichen ursprungs? / Rauschenbusch, August – Cleveland, O[hio]: H. Schulte, 1886 – 1mf – 9 – 0-7905-6313-4 – mf#1988-2313 – us ATLA [240]
Ist die forderung eines modernen christentums und einer modernen theologie berechtigt? : vortrag / Hahn, Traugott – 2. unveraenderte Aufl. Riga: Jonck & Poliewsky, 1903 – 1mf – 9 – 0-8370-3449-3 – mf#1985-1449 – us ATLA [240]
Ist die rede des paulus in athen ein urspruenglicher bestandteil der apostelgeschichte / Harnack, Adolf von – Leipzig, 1913 – (= ser Tugal 3-39/1a) – 1mf – 9 – €14.00 – ne Slangenburg [240]
Ist die rede des paulus in athen ein urspruenglicher bestandteil der apostelgeschichte? / judentum und judenchristentum in justins dialog mit trypho : nebst einer collation der pariser handschrift nr. 450 / Harnack, Adolf von – Leipzig: J C Hinrichs, 1913 – (= ser Tugal) – 1mf – 9 – 0-7905-1765-5 – (incl bibl ref and ind) – mf#1987-1765 – us ATLA [225]
Ist die theologie wissenschaft? : akademische rede zum geburtsfeste des hoechstseligen grossherzogs karl friedrich am 22. november, 1887.... / Holsten, Carl – Heidelberg:J Hoerning, 1887 – 1mf – 9 – 0-8370-3628-3 – (includes addenda) – mf#1985-1628 – us ATLA [240]
Ist doellinger haeretiker? / Hoetzl, Petrus – 2. Aufl. Muenchen: Rudolph Oldenbourg, 1870 – 1mf – 9 – 0-8370-8462-8 – (incl bibl ref) – mf#1986-2462 – us ATLA [240]
Ist duns scotus indeterminist? / Minges, P – Muenster, 1905 – (= ser Bgphma 5/4) – 3mf – 8 – €7.00 – ne Slangenburg [110]
Ist es wuenschenswerth, dass der religionsunterricht ganz in die haende von geistlichen, resp. theologen, gelegt werde? – Leipzig: Alexander Edelmann, 1902 – 1mf – 9 – 0-8370-7904-7 – (incl bibl ref) – mf#1986-1904 – us ATLA [377]
Ist gott persoenlich? : erneute untersuchung des problems der gottesfrage / Fricke, Gustav Adolf – Leipzig: Georg Wigand, 1896 – 1mf – 9 – 0-8370-4954-7 – (incl bibl ref) – mf#1985-2954 – us ATLA [210]
Istanbul ekspres – Istanbul: Tan Matbaasi, aug 30, 1951-aug 1955 – 9r – 1 – us CRL [079]
Istanbul musahabeleri / Safvet, A – Dersaadet: Saadet Kitabhanesi, 1324 [1909] – (= ser Ottoman literature, writers and the arts) – 2mf – 9 – $40.00 – us MEDOC [470]
Istanbul ticaret – Istanbul, Turkey. Feb 27 1958-Dec 1962; July 9 1965-Oct 25 1991 – 18r – 1 – us L of C Photodup [079]
Istanbul ueniversite observatoriyumu yazilari = Publications of the istanbul university observatory / Istanbul ueniversitesi Observatoryumu – Istanbul: Istanbul Universitesi 1935-42 [irreg] [mf ed 2001] – 20v on 1r – 1 – (cont by: istanbul ueniversitesi observatuari yazilari; in turkish, german, french & english; with ind) – ISSN: 0368-0657 – mf#film mas c5093 – us Harvard [520]
Istanbul ueniversite observatuari yazilari = Publications of the istanbul university observatory / Istanbul ueniversitesi Observatuari – Istanbul: Kenan Matbaasi 1943-81 [irreg] [mf ed 2001] – 1r – 1 – (cont: istanbul ueniversitesi observatoriyumu yazilari [issn 0368-0657]; cont by: publications of the istanbul university observatory; in turkish, german, french & english; imprint varies; n63-113 have title: istanbul ueniversitesi observatuari yazilari; with ind) – mf#film mas c5093 – us Harvard [520]
Istanbul ueniversitesi Observatoryumu *see* Istanbul ueniversite observatoriyumu yazilari
Istanbul ueniversitesi Observatuari *see* Istanbul ueniversite observatuari yazilari
Istanbul vilayeti meclis-i umumi mukarrerati 1329 – Istanbul: Matbaa-i Umumi, 1329 [1914] – (= ser Ottoman histories and historical sources) – 7mf – 9 – $110.00 – us MEDOC [956]
Istanbul'da guemruek muamelati hakkinda tedkik, muesahede ve muenakasalarim / Raif, Ahmet – Istanbul: Kanaat Kitabhanesi, 1926 – 4mf – 9 – $60.00 – us MEDOC [380]
Istanbul'un kara ve denizden huecm ve muedafaasi hakkinda bir kac soez / Pasa, Mahmud Mutar – Istanbul: Kitabhane-i Islam ve Askeri, 1326 [1910] – (= ser Ottoman histories and historical sources) – 1mf – 9 – $25.00 – us MEDOC [956]
Istar und saltu : ein altakkadisches lied / Zimmern, H – Leipzig, 1916 – 1mf – 9 – mf#NE-20040 – ne IDC [956]
Istel, Edgar *see* Das kunstwerk richard wagners
Isthmian diplomacy / Mcintosh, Russell Hugh – S.l., S.l? . 1941 – 1r – us UF Libraries [972]
Isthmus – [1995 jan/jun]-[2005 sep/dec] – 35r – 1 – (cont: isthmus of madison) – mf#3811123 – us WHS [071]
Isthmus of madison – [1976 apr 9/1978 jan 13]-[1994 jul/dec] [gaps] – 31r – 1 – (cont by: isthmus [madison, wi]) – mf#380998 – us WHS [071]
Isthmus of panama / Bidwell, Charles Toll – London, England. 1865 – 1r – us UF Libraries [972]
Isthmus of tehuantepec / Williams, John Jay – New York, NY. 1852 – 1r – us UF Libraries [972]
Istilah bahasa indonesia / Lembaga Bahasa Indonesia – Medan: Tokaigansyu Seityo, 2604 – 84p 1mf – 9 – mf#SE-2002 mf98 – ne IDC [490]
Istilah hoekoem / Madjallah Hoekoem – [Soerakarta, 1947] – 1mf – 9 – mf#SE-11989 – ne IDC [950]
Istilah-istilah / Lembaga Bahasa dan Kesusasteraan, Departemen PP dan K – Djakarta, [1952]-1964 – 20mf – 9 – (missing: [1952](1-5)) – mf#SE-795 – ne IDC [950]
Istimdad / Muhtar, Ahmet – Istanbul: Matbaa-yi Ceride-yi Askeri, 1304 [1886] – (= ser Ottoman literature, writers and the arts) – 1mf – 9 – $25.00 – us MEDOC [470]
Istina – Moscow, 1907 – 1 – us UMI ProQuest [077]
Istina – Paris. 1954-72 – 5 – fr ACRPP [073]
Istina – Paris. 1977+ – 1,5,9 – ISSN: 0021-2423 – mf#11279 – us UMI ProQuest [073]
Istirak etmedigimiz harekat / Nuri, Celal – [Istanbul]: Matbaa-yi Orhaniye, 1917 – (= ser Ottoman histories and historical sources) – 1mf – 9 – $40.00 – us MEDOC [956]
Istirak [ichtirak. journal socialiste] – Istanbul, 1 sene n1-20. 13 subat 1325-2 eyluel 1326. 27 feb 1909-16 sep 1910 – (= ser O & t journals) – 5mf – 9 – $150.00 – (cont by: insaniyet) – us MEDOC [956]
Istisare – Istanbul: A Asaduryan Matbaasi. Mueessisi: Suad Muhtar; Mueduer ve Sermuharriri: Mehmed Salih. n1-27. 4 eyluel-19 mart 1324 [1908-09] – (= ser O & t journals) – 18mf – 9 – $290.00 – us MEDOC [956]
Le istitutioni harmoniche / Zarlino, Gioseffo – 1561 – (= ser Mssa) – 4mf – 9 – €60.00 – mfchl 73b – gw Fischer [780]
Istitutioni harmoniche...nelle quali : oltra le materie appartenenti alla musica... / Zarlino, Gioseffo – Venetia 1558 [mf ed 19–] – 1mf – 9 – mf#fiche 963 – us Sibley [780]
Istituzione, riti e cerimonie dell'ordine de francs-macons, ossian liberi muratori : colla descrizione e disegna in rame della loro loggia... – In Venezia: Presso Leonardo Bassaglia...1785 – 2mf – 9 – mf#vrl-67 – ne IDC [366]
Istituzioni di diritto commerciale nord-americano / Lefebvre d'Ovidio, Antonio – Roma, Edizioni italiane 1945? 132 p. LL-387 – 1 – us L of C Photodup [346]
Istituzioni scolastiche in turchia / Mandalari, M – Roma, 1891 – 3mf – 9 – mf#AR-1810 – ne IDC [956]
Istitvtioni harmoniche del rev messere gioseffo zarlino...di nuono in molti luoghi migliorate, &...ampliate / Zarlino, Gioseffo – Venetia: Francesco de i Franceschi Senese 1573 [mf ed 19–] – 24mf – 9 – mf#fiche 965 – us Sibley [780]

Le istitvtioni harmoniche del reverendo gioseffo zarlino da chioggia : nelle quali; oltra le materie appartenenti alla mvsica / Zarlino, Gioseffo – Venetia: Francesco Senese 1562 [mf ed 19–] – 19mf – 9 – mf#fiche 964 – us Sibley [780]

Istochniki dlia izucheniia dvizheniia ssudnogo kapitala v rossii v kontse 19-nachale 20 v / Svishchev, M A – M, 1986 – 1mf – 9 – mf#REF-166 – ne IDC [332]

Istochniki russkoi agiografii / Barsukov, N – 1882 – 327p 9mf – 8 – mf#R-5695 – ne IDC [243]

Istochniki russkoi agiografii / Barsukov, N P – Spb, 1882 – 4mf – 9 – mf#R-18296 – ne IDC [243]

L'istoria della basilica diaconale collegiata : e parrocchiale di s maria in cosmedin di roma / Crescimbeni, G M – Roma, 1715 – 6mf – 9 – mf#0-1048 – ne IDC [700]

Istoria della compagnia di gesu il giappone : seconda parte dell'asia / Bartoli, Daniello – Napoli: Uffizio de'libri ascetici e predicabili, 1857-58 [mf ed 1995] – (= ser Yale coll; Opere del padre daniello bartoli [35]-43) – 9v in 3 – 1 – 0-524-10149-3 – (in italian) – mf#1995-1149 – us ATLA [241]

Istoria descrizione de tre regni congo, matamba et angola / Cavazzi, G A – Bologna, 1687 – 26mf – 9 – mf#A-106 – ne IDC [916]

Istoria e coltura delle piante... / Clarici, P B – Venezia, 1726 – 761p 10mf – 9 – mf#GDI-6 – ne IDC [710]

Istoria kanonizatsii sviatykh v russkoi tserkvi / Golubinskii, G G – 1903 – 600p 11mf – 8 – mf#R-4040 – ne IDC [243]

Istoria proletariata sssr – Moscow. no. 1-8. 1930-1931 – 1 – us NY Public [335]

Istoria tibeta i khukhunora s 2282 goda do r kh do 1227 goda po... / Iakinf [Bichurin, Ia] – Spb.: Pri Imperatorskoi Akademii Nauk, 1833. 2v – 3mf – 9 – (missing: v1) – mf#HT-643 – ne IDC [915]

Istoria...della sacra lega contra selim... / Foglietta, U – Genova, 1598 – 8mf – 9 – mf#H-8397 – ne IDC [956]

Istoricheskaia grammatika russkogo iazyka / Chernykh, Pavel Iakovlevich – Moskva, Russia. 1962 – 1r – us UF Libraries [460]

Istoricheskaia spravka / Gosudarev pechatnyi dvor i sinodalnaia tipografiia v Moskve – 1903 – 104p 2mf – 8 – mf#R-4567 – ne IDC [243]

Istoricheskaia zapiska : piatidesiatiletie vysochaishe utverzhdennoi kommissii po razboru i opisaniiu arkhiva sviateishego sinoda, 1865-1915 – 1915 – 7mf – 9 – mf#R-11318 – ne IDC [243]

Istoricheskii i politicheskii zhurnal – Duluth. 1933-1996 (1) 1971-1996 (5) 1976-1996 (9) – 979mf – 9 – mf#1922 – ne IDC [077]

Istoricheskii i politicheskii zhurnal – New York. 1948-1966 (1) 1966-1966 (5) 1966-1966 (9) – 37mf – 9 – (missing: 1807(1, p 61-64, 125-128, 141-144, 181-188, 197-200); 1809(1, p 35-40, 45-48, 73-76, 97-100, 129-132, 137-144, 149-150, 155-166, 179-186, 191-202, 209-210, 229-230, 237-242)) – mf#1731 – ne IDC [077]

Istoricheskii obzor uchebnikov obshchei i russkoi geografii, izdannykh so vremeni petra velikogo po 1876 god : (1710-1876 g) / Vesin, L – 1876 – 13mf – 8 – mf#R-7034 – ne IDC [947]

Istoricheskii ocherk deiatel'nosti moskovskogo gorodskogo kreditnogo obshchestva v techenii chetverti veka ego sushchestvovaniia (1863-1888) – M, 1888 – 3mf – 9 – mf#REF-364 – ne IDC [332]

Istoricheskii ocherk deiatel'nosti s-peterburgskogo gorodskogo kreditnogo obshchestva za 25 let – Spb, 1886 – 3mf – 9 – mf#REF-374 – ne IDC [332]

Istoricheskii ocherk deiatel'nosti zemskikh uchrezhdenii tverskoi gubernii, 1864-1913 gg / Veselovskii, B F – Tver', 1914 – 12mf – 8 – mf#R-3528 – ne IDC [314]

Istoricheskii ocherk dvadtsatipiatiletnei deiatel'nosti penzenskogo i rossiiskogo soiuzov obshchestv vzaimnogo ot ognia strakhovaniia, 1890-1915 / Rossiiskii Soiuz obshchestv vzaimnogo ot ognia strakhovaniia – [Pg], 1915 – 5mf – 9 – mf#REF-408 – ne IDC [332]

Istoricheskii ocherk kooperatsii v rossii / Kheisin, M L – 1918 – 182p 2mf – 9 – mf#COR-133 – ne IDC [335]

Istoricheskii ocherk piatidesiatiletnei deiatel'nosti s-peterburgskogo gorodskogo kreditnogo obshchestva, 1861 – 5 oktiabria 1911 – Spb, 1911 – 8mf – 9 – mf#REF-376 – ne IDC [332]

Istoricheskii ocherk russkoi shkoly / Grigorev, V V – 1900 – 11mf – 8 – mf#R-246 – ne IDC [947]

Istoricheskii sbornik, izdavaemyi pri obshchestve revnitelei russkogo istoricheskogo prosveshcheniia v pamiat imperatora aleksandra 3 – Montreal. 1963-1965 (1) – 162mf – 9 – mf#1908 – ne IDC [077]

Istoricheskii slovar 86 otstev i katalog ili biblioteka staroverchesko i tserkvi / Liubopytnyi, P – Saratov, 1914 – 201p 3mf – 8 – mf#R-7137 – ne IDC [243]

Istoricheskii, statisticheskii i geograficheskii zhurnal ili sovremennaia istoriia sveta – East Sussex. 1906-1980 (1) 1980-1980 (5) 1980-1980 (9) – 255mf – 9 – (missing: 1809(7-12); 1812(9-12); 1813(3); 1814(4-5); 1816(1-6); 1820(1-6); 1821(1-6); 1829(4-5)) – mf#1739 – ne IDC [077]

Istoricheskii viestnik – v. 1-150. 1880-1917 – 1 – us L of C Photodup [947]

Istoricheskii zhurnal dlia vsekh – Spb., 1908(1-12); 1909(1-5) – 31mf – 9 – mf#R-4123 – ne IDC [077]

Istoricheskiia skazaniia o zhizni sviatykh, podvizavshchikhsia v vologodskoi eparkhii / Veriuzhskii, I – Vologda, 1880 – 8mf – 9 – mf#R-18262 – ne IDC [243]

Istoricheskoe opisanie stavropigialnogo voskresenskogo, novyi ierusalem imenuemogo monastyria / Leonid, Arkhimandrit – 1876 – 768p 11mf – 9 – mf#R-11196 – ne IDC [243]

Istoricheskoe rozyskanie o russkikh povremennykh izdaniakh i sbornikakh za 1703-1802 gg / Neustroev, A N – Washington. 1910+ (1) 1967+ (5) 1970+ (9) – 17mf – 9 – mf#1155 – ne IDC [077]

Istoricheskoe znachenie nepa sbornik nauchnykh trudov / ed by Gorinov, M M et al – M, 1990 – 3mf – 9 – mf#REF-29 – ne IDC [332]

Le istorie dell' indie orientali / Maffei, Giovanni Pietro – Milano: Dalla Societa Tipografica de' Classici Italiani, 1806 [mf ed 1995] – (= ser Yale coll) – 3v (ill) – 1 – 0-524-09748-8 – (trans fr latin into italian by francesco serdonati) – mf#1995-0748 – us ATLA [241]

Istoriia bankov s drevneishikh vremen do nashikh dnei : ocherki / Malinina, E A – Spb, 1913 – 1mf – 9 – mf#REF-174 – ne IDC [332]

Istoriia biudzhetnykh issledovanii / Chaianov, A & Studenskii, G – 2nd ed. M, 1922 – 4mf – 8 – mf#RZ-185 – ne IDC [314]

Istoriia dvorianskogo sosloviia v rossii / Iablochkov, M – 1876 – 13mf – 8 – mf#R-6041 – ne IDC [947]

Istoriia evreiskoi pechati v rossii v sviazi s obshchestvennymi techeniiami / Tsinberg, S L – Pg., 1915 – 3mf – 9 – mf#R-9263 – ne IDC [077]

Istoriia finansov sssr (1917-1950 gg) / Diachenko, V P – M, 1978 – 6mf – 9 – mf#REF-13 – ne IDC [332]

Istoriia finansovykh uchrezhdenii rossii so vremeni osnovaniia gosudarstva do konchiny imperatritsy ekateriny ii / Tolstoi, D A – Spb, 1848 – 4mf – 9 – mf#R-9772 – ne IDC [332]

Istoriia gosudarstva rossiiskago / Karamzin, Nikolai M – v1-12. 1892 – 9 – $267.00 – mf#0308 – us Brook [947]

Istoriia i organizatsiia soveta vserossiiskikh kooperativnykh sezdov / Khizhniakov, V V – 1919 – 70p 1mf – 9 – mf#COR-227 – ne IDC [335]

Istoriia knigi na rusi / Bakhtiarov, A A – 1890 – 5mf – 8 – mf#R-4557 – ne IDC [947]

Istoriia knigi v rossii / Librovich, S F – 1913-1914. v1-2 – 9mf – 8 – mf#R-4579 – ne IDC [947]

Istoriia kreditnykh uchrezhdenii i sovremennoe sostoianie kreditnoi sistemy v sssr / Blium, A A – M, 1929 – 3mf – 9 – mf#REF-32 – ne IDC [332]

Istoriia meditsiny v rossii / Rikhter, V – 1814-1820. 3v – 30mf – 8 – mf#R-7814 – ne IDC [947]

Istoriia moskovskoi slaviano-greko-latinskoi akademii / Smirnov, S – 1855 – 8mf – 8 – mf#R-7879 – ne IDC [243]

Istoriia pravitelstvuiushchego senata za dvesti let : 1711-1911 gg – 1911. 5v – 106mf – 9 – mf#2316 – ne IDC [947]

Istoriia pravoslavnago monashestva v severo-vostochnoi rossii, so vremen pred sergiia radonezhskago / Kudriavtsev, Matfii – 1881. 2v – 4mf – 9 – mf#R-18304 – ne IDC [243]

Istoriia rabochei kooperatsii v rossii : ocherki po istorii rabochego kooperativnogo dvizheniia / Balabanov, M – Kiev, 1923 – 231p 3mf – 9 – mf#COR-7 – ne IDC [335]

Istoriia revoliutsionnykh dvizhenii v rossii = Geschichte der revolutionaeren bewegungen in russland / Thun, Alphons – Petrograd: Izd Petrogradskago sovieta rabochikh i krasnoarmeiskikh deputatov, 1918 [mf ed 2002] – 1r – 1 – (in russian. filmed with: internatisional: vospominaniia i materialy 1864-1878 gg / dzhems gil'om [james guillaume], (1922)) – mf#5265 – us Wisconsin U Libr [947]

Istoriia rimskogo prava / Pokrovskii, Iosif Alekseevich – Izd. 3., ispravlennoe i dopolnennoe. Petrograd, Izdanie iuridicheskogo knizhnogo sklada "Pravo", 1917. 430p. LL-4049 – 1 – us L of C Photodup [340]

Istoriia rossiiskoi sotsial-demokratii / Martov, L – 1923 – 214p 3mf – 9 – mf#RPP-150 – ne IDC [325]

Istoriia russkoi armii i flota = History of the russian army and navy / ed by Grishinskii, A S et al – Moscow: Publ Society "Obrazovanie". 15v – 57mf – 9 – $400.00 – us UMI ProQuest [355]

Istoriia russkoi literatury 18 / Blagoi, Dmitrii Dmitrievich – Moskva, Russia. 1951 – 1r – us UF Libraries [460]

Istoriia russkoi literatury 18 veka / Blagoi, D D – 1945 – 12mf – 8 – mf#R-6102 – ne IDC [947]

Istoriia russkoi literatury 18 veka *see* Bibliograficheskii ukazatel

Istoriia russkoi literatury sibiri v dvukh tomakh : akademiia nauk sssr, sibirskoe otdelenie, institut istorii, filologii i filosofii / ed by Postnov, I U S – Novosibirsk: the Institute, [1974-]. v1 1975; v2 1974; v2 pt 2 1974 – us CRL [460]

Istoriia russkoi obshchestvennoi mysli / Plekhanov, G B – 1915-1918. 3v – 17mf – 8 – mf#R-127 – ne IDC [947]

Istoriia russkoi zhurnalistiki 18 veka / Berkov, P N – M., L., 1952 – 11mf – 9 – mf#R-6098 – ne IDC [077]

Istoriia sssr / Bushchik, L P – Moskva, Russia. 1954 – 1r – us UF Libraries [947]

Istoriia teatralnogo obrazovaniia v rossii : (17-18 vv) / Vsevolodskii-Gerngross, V N – 1913. v1 – 11mf – 8 – mf#R-7965 – ne IDC [947]

Istoriia tsarstvovaniia petra velikogo / Ustrialov, N G – 1858-1863. v1-4,6 – 68mf – 8 – mf#239 – ne IDC [947]

Istoriia vtoroi russkoi revoliutsii : v1: borba burzhuaznoi i sotsialisticheskoi revoliutsii. pt1: protivorechiia revoliutsii / Miliukov, P N – Kiev, 1919 – 128p 2mf – 9 – mf#RPP-122 – ne IDC [325]

Istoriia zemstva za sorok let / Veselovskii, B – Spb, 1909-1911. v1-4 – 73mf – 8 – mf#RZ-183 – ne IDC [314]

Istorija cerkovnago razryva mezhdu gruziej i armeniej nachale 7 veka / Dzhavaxov, I A – 3mf – 8 – (bull de l'ac imp des sc de st-p, 6e serie: v2 1908 n5 p433-466; n6 p511-536) – mf#1744 mf24-26 – ne IDC [243]

Istorik i sovremennik – Berlin. v. 1-5. 1922-24 – 1 – us NY Public [335]

Istoriko-filologicheskogo fakulteta / Zapiski Novorossiiskogo universiteta – Tucson. 1959-1986 (1) 1971-1986 (5) 1977-1986 (9) – 129mf – 9 – mf#2004 – ne IDC [077]

Istoriko-iuridicheskie akty perekhodnoi epokhi 17-18 vekov / ed by Pobedonostsev, K P – 1887 – 5mf – 9 – mf#R-11135 – ne IDC [947]

Istoriko-literaturnyi i kritiko-bibliograficheskii zhurnal : v mire knig – M., 1907(2-5) – 4mf – 9 – mf#R-4305 – ne IDC [077]

Istoriko-literaturnyi i politicheskii zhurnal – Stamford. 1901-1970 (1) – 36mf – 9 – mf#1896 – ne IDC [077]

Istoriko-literaturnyi zhurnal – New York. 1933-1963 (1) – 2845mf – 9 – mf#1740 – ne IDC [077]

Istoriko-literaturnyi zhurnal / Vestnik Iugo-Zapadnoi i Zapadnoi Rossii – Des Plaines. 1963-1983 (1) 1971-1983 (5) 1976-1983 (9) – 377mf – 9 – (missing: 1862(1-6); 1868(4, 11); 1869(1-2); 1870(4, 11)) – mf#1967 – ne IDC [077]

Istoriko-literaturnyia izsliedovaniia i materialy...v perettisa : retisenziia / Zhitetiskii, P – S-Peterburg: Tip Imperatorskoi akademii nauk, 1903 [mf ed 2002] – 1r – 1 – (filmed with: zamiechaniia ob obrazovanii slov iz vyrazhenii / i i sreznevskago (1873). incl bibl ref) – mf#5235 – us Wisconsin U Libr [780]

Istoriko-politicheskoe obozrenie vestnika evropy – Spb., 1872-1874. v1-2 – 25mf – 9 – mf#R-3383 – ne IDC [077]

Istoriko-revoliutsionnyi sbornik / ed by Burtsev, V L – London, Paris, Spb., L., 1900-1933 – 338mf – 9 – (cont as: byloe, zhurnal, posviashchennyi istorii osvoboditelnogo dvizheniia) – mf#R-2343, 1427 – ne IDC [077]

Istoriko-rodoslovnoe obshchestvo – Letopis. Moscow. no. 5-12. 1906-1907 – 1 – us NY Public [920]

Istoriko-statisticheskie i ekonomicheskie tablitsy po avtonomnoi bashkirskoi ssr – Ufa, 1923. (Vserossiiskaia sel'skokhoziaistvennaia i kustarnaia vystavka 1923 g. 145p) – 2mf – 9 – mf#RHS-27 – ne IDC [314]

Istoriko-statisticheskoe obozrenie uchebnogo zavedeniia s peterburskogo uchebnogo okruga s 1715 po 1828 vkliuchitelno / Voronov, A S – 1849. v1 – 6mf – 8 – mf#R-7962 – ne IDC [947]

Istorychno-arkhivoznavchyi zhurnal / Arkhiv Radians'koi Ukrainy – Kharkiv, 1932-33. 8 nos in 5 iss + index 1958 – (= ser Radians'kyi arkhiv) – 9 – (preceded by: radians'kyi arkhiv [r-14317/1]) – mf#R-14319 – ne IDC [947]

Istorychnyi i kalendar al'manakh chervonoyi kalyny – L'viv, etc. 1928-1929, 1934-35 – 1 – us NY Public [520]

Istra – Zagreb, Yugoslavia. 20 Feb 1932; Sept 1938-1940 – 1r – 1 – us L of C Photodup [949]

Istrin, V M *see*
- Otkrovenie mefodiia patarskogo i apokrificheskiia videniia daniila v vizantiiskoi i slavianorusskoi literaturakh
- Zamechaniia o sostave tolkovoi palei

Istruzioni morali sopra la dottrina cristiana / Vicentini, Francesco – Bassano: Remondini, 1842 – 10mf – 9 – 0-524-06169-6 – mf#1991-2425 – us ATLA [240]

Isu independent news – Weirton, WV. 1950-1961 (1) – mf#67496 – us UMI ProQuest [071]

Isu pilot / International Seaman's Union of America – v1 n1-42 [1935 feb 27-dec 6], v2 n1-37,39,53 [1935 dec 13-1936 aug 28, sep 11, dec 19], v3 n1-19 [1937 jan 6-may 11] – 1r – 1 – (cont by: pilot [new york, ny]) – mf#1833702 – us WHS [380]

Isvestija – Moskau, 1966-1993ff – 57r – 1 – In zusammenarbeit mit Norman Ross Publishing Inc, New York – gw Mikropress [949]

It ain't me babe – v1-v2 n1 [1970 jan 15-1971 apr] – 1r – 1 – mf#780731 – us WHS [071]

It for industry – Toronto. 2000+ (1,5,9) – ISSN: 1498-9549 – mf#32859 – us UMI ProQuest [332]

It happened in british guiana / Oswald, Archibald – Ilfracombe, England. 1955 – 1r – us UF Libraries [972]

It has been my earnest endeavor for years to elevate the tone and character of public amusements : and to strip them of every objectionable feature / Barnum, Phineas Taylor – [s.l: s.n. 1875?] [mf ed 1983] – 1mf – 9 – 0-665-39762-3 – mf#39762 – cn Canadiana [090]

It heitelan – Leeuwarden, Netherlands. 1,8 Mar 1919; 3 Apr 1920-25 Dec 1926 – 6r – 1 – uk British Libr Newspaper [949]

It is written : a careful study of the gospels as to all the words and acts of our lord and other things contained therein touching the holy scriptures of the old testament / Bacon, Thomas Scott – New York: Wilbur B Ketcham, 1891 – 1mf – 9 – 0-8370-2145-6 – (incl ind) – mf#1985-0145 – us ATLA [220]

It kie bwee / siauw eng hiong / maoe terbang tida bersajap / Kwo, Lay Yen & Hsiang, M & Liu, Ti'enmey – Batavia: Goedang Tjerita, 1948 [mf ed 1998] – (= ser Goedang tjerita 7) – 1r – 1 – (coll as pt of the colloquial malay collection. "it kie bwee" is an indonesian, partial trans of chinese novel entitled qijian shisanxia (the seven heroes and the thirteen gallants) [salmon, claudine. literature in malay by the chinese of indonesia. paris: editions de la maison des sciences de l'homme, c1981, p329. filmed with: lajangan biroe / im yang tjoe) – mf#10005 – us Wisconsin U Libr [830]

It magazine – Toronto. 1993-1994 (1,5,9) – (cont: canadian datasystems) – ISSN: 1196-4715 – mf#10768,01 – us UMI ProQuest [000]

O itabira : jornal litterario, agricola, commercial e noticioso – Cachoeiro de Itapemirim, ES: Typ de F Carvalho & F Rios, 04 jul 1866-29 dez 1967 – (= ser Ps 19) – mf#DIPER – bl Biblioteca [073]

O itacolomy – Ouro Preto, MG: Typ do Itacolomy, 18 mar, 08 maio 1843 – (= ser Ps 19) – mf#P19B,01,15 – bl Biblioteca [321]

Itajahy : litterario e noticioso – (= ser Ps 19) – mf#UFSC/BPESC – bl Biblioteca [440]

Itajuba : organ imparcial. periodico litterario, agricola, industrial, commercial e... – Itajuba, MG: Typ de Magalhaes Baiao, 22 jul 1888 – (= ser Ps 19) – mf#P31,03,25 – bl Biblioteca [073]

Itala und vulgata / Roensch, Hermann – Marburg, 1875 – 11mf – 8 – €22.00 – ne Slangenburg [220]

Itala und vulgata : das sprachidiom der urchristlichen itala und der katholischen vulgata / Roensch, Hermann – 2., berichtigte und verm. Ausg. Marburg: N.G. Elwert, 1875 – 2mf – 9 – 0-8370-9500-X – (incl ind) – mf#1986-3500 – us ATLA [450]

I-ta-li fu hsing chih tao / Sih, Paul Kwang Tsien – Shang-hai: Shang wu yin shu kuan, Min kuo 26 [1937] – (= ser P-k&k period) – us CRL [951]

L'italia – Chicago, IL: Gentile & Durante, [1886-[1919-oct 3 1948]; jan 8 1950-jun 16 1957; 1958-59; 1961-62; 1965; Sunday issues only for 1919-jul 17 1938 – 1 – us CRL [071]

Italia : bulletin bimensuel d'informations – Paris. n1-67.avr 1929-juil 1932 – 1 – (edite par la concentration antifasciste italienne. mq no. 6, 17, 44) – fr ACRPP [320]

L'italia evangelica – Florence, Italy 1 jan 1881-27 dec 1907 – 1 – uk British Libr Newspaper [074]

Italia nostra – London, UK. 8 Sept 1928-7 Jun 1940 – 1 – uk British Libr Newspaper [072]

Italia oggi – 1986- – 4r per y – 5 – enquire for prices – us UMI ProQuest [074]

Italia pontificia : sive, repertorium privilegiorum et litterarum a romanis pontificibus ante annum 1198. vol 1, roma / Kehr, Paul Fridolin – Berolini [Berlin]: Apud Weidmannos, 1906 [mf ed 1992] – (= ser Regesta pontificum romanorum; Roman catholic coll) – 1mf – 9 – 0-524-04050-8 – mf#1990-4958 – us ATLA [241]

Italia pontificia : sive, repertorium privilegiorum et litterarum a romanis pontificibus ante annum 1198. vol 2, latium / Kehr, Paul Fridolin – Berolini [Berlin]: Apud Weidmannos, 1907 [mf ed 1992] – (= ser Regesta pontificum romanorum; Roman catholic coll) – 1mf – 9 – 0-524-04051-6 – (incl bibl ref) – mf#1990-4959 – us ATLA [241]

Italia pontificia : sive, repertorium privilegiorum et litterarum a romanis pontificibus ante annum 1198. vol 3, etruria / Kehr, Paul Fridolin – Berolini [Berlin]: Apud Weidmannos, 1908 [mf ed – (= ser Regesta pontificum romanorum; Roman catholic coll) – 2mf – 9 – 0-524-04052-4 – (incl bibl ref) – mf#1990-4960 – us ATLA [241]

Italia pontificia : sive, repertorium privilegiorum et litterarum a romanis pontificibus ante annum 1198. vol 4, umbria picenum marcia / Kehr, Paul Fridolin – Berolini [Berlin]: Apud Weidmannos, 1909 [mf ed 1992] – (= ser Regesta pontificum romanorum; Roman catholic coll) – 1mf – 9 – 0-524-04053-2 – (incl bibl ref and ind) – mf#1990-4961 – us ATLA [241]

Italia pontificia : sive, repertorium privilegiorum et litterarum a romanis pontificibus ante annum 1198. vol 5, aemilia, sive, provincia ravennas / Kehr, Paul Fridolin – Berolini [Berlin]: Apud Weidmannos, 1911 [mf ed 1992] – (= ser Regesta pontificum romanorum; Roman catholic coll) – 2mf – 9 – 0-524-04054-0 – (incl bibl ref) – mf#1990-4962 – us ATLA [241]

Italia pontificia : sive, repertorium privilegiorum et litterarum a romanis pontificibus ante annum 1198. vol 6, liguria, sive, provincia mediolanensis / Kehr, Paul Fridolin – Berolini [Berlin]: Apud Weidmannos, 1913-14 [mf ed 1992] – (= ser Regesta pontificum romanorum; Roman catholic coll) – 2v on 3mf – 9 – 0-524-04055-9 – (incl bibl ref) – mf#1990-4963 – us ATLA [241]

L'italia socialista – nuova ser. Rome, Italy 10 jun 1947-30 jan 1949 (imperfect) – 1 – uk British Libr Newspaper [335]

Italian air force in spain – London: United Editorial [1938] – (= ser Blodgett coll) – 9 – mf#w968 – us Harvard [355]

Italian Catholic Federation *see* Bollettino

The italian court in the crystal palace / Wyatt, Matthew Digby & Waring, John Burley – London 1854 – (= ser 19th c art & architecture) – 2mf – 9 – mf#4.2.479 – uk Chadwyck [720]

Italian drawings for jewelry 1700-1875 / Cooper Union Museum for the Arts of Decoration. New York – 1940 – 9 – $4.90 – uk Chadwyck [740]

Italian drawings of the 18th and 19th centuries and spanish drawings of the 17th through 19th centuries / Olsen, Sandra Haller & McCullagh, Susanne Folds – 1979 – 3 color mf – 15 – $65.00f – 0-226-68803-8 – (196p accompanying text) – us Chicago U Pr [740]

Italian duets & c – London: [various imprints] [c1790-1810] [mf ed 1992] – 1v on 1r – 1 – (binder's title: a collection of arias duets, trios etc in vocal or orchestral score, publ separately) – mf#pres. film 109 – us Sibley [780]

Italian echo – Providence, RI. 1931-1969 (1) – mf#66324 – us UMI ProQuest [071]

Italian explorers in africa / Bompiani, S – London, 1891 – 3mf – 9 – mf#NE-20184 – ne IDC [916]

Italian grammar / Vittorini, Domenico – Philadelphia, PA. 1947 – 1r – us UF Libraries [440]

Italian jewellery as worn by the peasants of italy / Castellani, Alessandro – London 1868 – (= ser 19th c art & architecture) – 1mf – 9 – mf#4.2.1212 – uk Chadwyck [730]

Italian landscape in eighteenth century england / Manwaring, Elizabeth Wheeler – New York, NY. 1925 – 1r – us UF Libraries [941]

Italian leader – Milwaukee WI. 1933 dec-1936 mar 1, 1936 dec 25-1937 apr 9 – 1r – 1 – mf#1223056 – us WHS [071]

Italian music manuscripts, c1640-c1820 : from the british library, london – 4 sects – 285r – 1 – (sect a: music mss, c1640-c1720 46r c14r-11310. sect b: music mss, c1720-c1740 79r c14r-11311. sect c: music mss, c1740-c1770 60r c14r-11312. sect d: music mss, c1770-c1820 100r c14r-11313. printed guide available for each section of the coll) – mf#C14R-11300 – us Primary [780]

Italian painters / Morelli, Giovanni – London, England. v1-2. 1893-1900 – 1r – us UF Libraries [750]

Italian painters of the renaissance / Berenson, Bernard – New York, NY. 1952 – 1r – us UF Libraries [750]

Italian parliamentary papers : 7th legislature, 1976-1979 – Clearwater Publ Co – 170r – 5 – $17,935.00 – us UPA [323]

The italian press = La stampa italiana – Kansas City: A J Tolsa. dec 25 1931-jan 3 1941 – 3r – 1 – us CRL [071]

Italian prisoners in spain / Spain. Embajada. Great Britain – N.Y., 1937. Fiche W1175. (Blodgett Collection of Spanish Civil War Pamphlets) – 9 – us Harvard [946]

Italian review – Providence, RI. 1924-1925 (1) – mf#66325 – us UMI ProQuest [071]

Italian school – (= ser Christie's pictorial archive: painting and graphic art) – 101mf – 9 – $765.00 – 0-907006-07-8 – (over 1200 artists, over 6000 reproductions) – uk Mindata [750]

The italian schools of painting : with observations on the present state of the art / James, John Thomas – London 1820 – (= ser 19th c art & architecture) – 4mf – 9 – mf#4.2.1503 – uk Chadwyck [750]

Italian sculpture of the middle ages and period of the revival of art : a descriptive catalogue of the works forming the above section of the museum / Robinson, John Charles – London: Chapman & Hall, 1862 – (= ser 19th c art & architecture) – 3mf – 9 – (with additional illus notices) – mf#4.1.162 – uk Chadwyck [730]

Italian Socialist Club of Chicago et al *see* Propaganda

Italian weekly – Rochester, NY. 1940-1949 (1) – mf#65195 – us UMI ProQuest [071]

Italian-australian bulletin of commerce – Sydney. 1973-1973 (1) – ISSN: 0047-1658 – mf#8065 – us UMI ProQuest [380]

Italianische lieder des hohenstaufischen hofes in sicilien – Stuttgart: Literarischer Verein, 1843 [mf ed 1993] – (= ser Blvs 5/2) – 67p – 1 – (italian text. int in german) – mf#8470 reel 1 – us Wisconsin U Libr [780]

Italiano – 1981 spr-1985 mar, 1985 apr-1986 nov – 2r – 1 – mf#1238963 – us WHS [071]

Italiano – London, UK. 1 May 1926-5 Dec 1928 – 1 – uk British Libr Newspaper [072]

Italiano – London, UK. 18 Sept-23 Oct 1909 – 1 – uk British Libr Newspaper [072]

L'italiano in germania – Koeln DE, 1907-11 – 1 – gw Misc Inst [074]

The italians in america / Wright, Frederick H – New York: Missionary Education Movement of the United States and Canada, 1913 – 1mf – 9 – 0-524-02997-0 – mf#1990-0784 – us ATLA [240]

Italians in spain / Friends of Democracy and Independence in Spain – London. 1937. Fiche W 900. (Blodgett Collection of Spanish Civil War Pamphlets) – 9 – us Harvard [946]

Italica – Columbus. 1924+ (1) 1971+ (5) 1977+ (9) – ISSN: 0021-3020 – mf#1088 – us UMI ProQuest [440]

L'italie confortable : manuel du touriste; appendice aux voyages historiques, litteraires et artistiques en italie / Valery, Antoine – Paris [u a] [ca 1840] [mf ed Hildesheim 1995-98] – 3mf – 9 – €90.00 – 3-487-29251-3 – gw Olms [914]

L'italie mystique : histoire de la renaissance religieuse au moyen age / Gebhart, Emile – 2eme ed. Paris: Hachette, 1893 [mf ed 1990] – 1mf – 9 – 0-7905-6591-9 – (in french. 1st publ 1890. trans as: mystics and heretics in italy 1922) – mf#1988-2591 – us ATLA [240]

Italien : erlebnisse deutscher in italien / ed by Molo, Walter Ritter von – Berlin: Wegweiser, 1921 – 1r – 1 – us Wisconsin U Libr [945]

Italien / Rehfues, Philipp J von – Berlin 1803-1804 [mf ed Hildesheim 1995-98] – 2v on 7mf – 9 – €140.00 – 3-487-29315-3 – gw Olms [914]

Italien im deutschen gedicht / ed by Riemerschmid, Werner & Bruyn, Karlheinz de – Muenchen:K Alber, 1943 – 1 – us Wisconsin U Libr [810]

Italien in eichendorffs dichtung : eine untersuchung / Bianchi, Lorenzo – Bologna: N Zanichelli, 1937 [mf ed 1989] – 139p – 1 – (with app: exkurs ueber hermann friedlaenders' ansichten von italien) – mf#7212 – us Wisconsin U Libr [410]

Italien index : bilddokumentation zur kunst in italien / ed by Bildarchiv Foto Marburg – Deutsches Dokumentationszentrum fuer Kunstgeschichte Philipps- Universitaet Marburg – [mf ed 1991-92] – 621mf (1:24) – 9 – silver €5130.00 – ISBN-10: 3-598-33130-4 – ISBN-13: 978-3-598-33130-5 – gw Saur [700]

Italien und deutschland / ed by Moritz, Karl Philipp – Berlin 1789-93 – (= ser Dz. historisch-geographische abt) – 2v on 5mf – 9 – €100.00 – mf#k/n1215 – gw Olms [940]

Italien-index. neue folge / ed by Bildarchiv Foto Marburg – Deutsches Dokumentationszentrum fuer Kunstgeschichte Philipps- Universitaet Marburg – [mf ed 2004] – 186mf (1:24) in 3 installments – 9 – silver €2040.00 – ISBN-10: 3-598-34900-9 – ISBN-13: 978-3-598-34900-3 – gw Saur [700]

Italienische analekten zur reichsgeschichte des 14. jahrhunderts (1310-1378) (mgh schriften:11.bd) / Mommsen, Theodor – 1952 – (= ser Monumenta germaniae historica. schriften (mgh schriften)) – €12.00 – ne Slangenburg [931]

Das italienische kabinet : oder merkwuerdigkeiten aus rom und neapel / Benkowitz, Carl F – Leipzig 1804 [mf ed Hildesheim 1995-98] – (= ser Fbc) – 2mf – 9 – €60.00 – 3-487-29312-9 – gw Olms [945]

Italienische reise : mit den zeichnungen goethes, seiner freunde und kunstgenossen / Goethe, Johann Wolfgang von; ed by Graevenitz, George von – Leipzig: Insel-Verlag, 1912 [mf ed 1992] – 356p/122pl (ill) – 1 – mf#7984 – us Wisconsin U Libr [914]

Italienisches abenteuer : erzaehlung / Blunck, Hans Friedrich – Muenchen: A Langen/G Mueller, c1938 [mf ed 1989] – (= ser Die kleine buecherei 92) – 55p – 1 – mf#7037 – us Wisconsin U Libr [880]

Italienisches seebuch : naturansichten und lebensbilder von den alpenseen und meereskuesten italiens / Noe, Heinrich – Stuttgart 1874 [mf ed Hildesheim 1995-98] – 3mf – 9 – €90.00 – 3-487-29250-5 – gw Olms [914]

Italimuse italic news – Fairfield. 1969-1973 (1) 1969-1973 (5) (9) – ISSN: 0021-3039 – mf#7595 – us UMI ProQuest [680]

Italische apriltage : erinnerungen aus einer confessionslosen romfahrt / Schleich, Martin – Muenchen [u a] 1880 [mf ed Hildesheim 1995-98] – 2mf – 9 – €60.00 – 3-487-29232-7 – gw Olms [914]

Italische normann in deutscher heldensage / Panzer, Friedrich Wilhelm – Frankfurt am Main: M Diesterweg, 1925 – 1r – 1 – (incl bibl ref) – us Wisconsin U Libr [410]

L'Italo australiano – Sydney, Mar 11 1905-Jan 30 1909 – 1r – 9 – A$63.67 vesicular A$69.17 silver – (italian language) – at Pascoe [079]

Italo – australiano – Sydney, mar 1905-jan 1909 – 1r – A$63.67 vesicular A$69.17 silver – at Pascoe [079]

L'italo-americano – Nuova Orleans: S Calafiore, sep 22 1917-oct 19 1918 – 1r – 1 – us CRL [071]

Italy / Balzani, Ugo – London: SPCK; New York: E & J B Young, 1883 – 340p – 1 – (publ in italian under: le cronache italiano nel medio evodescritte", milano, u hoepli, 1884) – mf#2189 – us Wisconsin U Libr [945]

Italy / Conder, Josiah – London 1831 [mf ed Hildesheim 1995-98] – 3v on 9mf – 9 – €180.00 – 3-487-29267-X – gw Olms [914]

Italy : direzione generale del demanio. pubblicazioni edite dallo stato o col suo concorso, (1861-1923); catalogo generale – Rome, Libreria dello stato, 1924 – us CRL [020]

Italy : internal affairs and foreign affairs, 1940-1954 / U.S. State Dept – (= ser Confidential u s state department central files) – 1 – $32,895.00 coll – (internal affairs, 1940-44 44r isbn 0-89093-922-5 $8515. foreign affairs, 1940-44 6r isbn 0-89093-923-3 $1155. internal affairs, 1945-49: pt1: political, governmental, & national defense affairs 29r isbn 0-89093-485-1 $5610; pt2: social, economic, & industrial affairs 34r isbn 0-89093-489-4 $6580. foreign affairs, 1945-49 9r isbn 0-89093-483-5 $1740. internal affairs, 1950-54: pt1: political, governmental, & national defense affairs 17r isbn 1-55655-038-3 $3290; pt2: social, economic, & industrial affairs 35r isbn 1-55655-039-1 $6765. foreign affairs, 1950-54 5r isbn 1-55655-040-5 $970. with p/g) – us UPA [945]

Italy : with sketches of spain and portugal / Beckford, William – London 1834 [mf ed Hildesheim 1995-98] – (= ser Fbc) – 2v on 6mf – 9 – €120.00 – 3-487-27744-1 – gw Olms [914]

Italy *see* Gazzetta ufficiale della repubblica italiana

Italy, 1847-1900 *see* The papers of queen victoria on foreign affairs

Italy, 1918-1941 – (= ser U s military intelligence reports) – 9r – 1 – $1420.00 – 0-89093-663-3 – (with p/g) – us UPA [355]

Italy and its inhabitants : an account of a tour in that country in 1816 and 1817: containing a view of characters, manners, customs...; with some remarks on the origin of rome and of the latin language / Galiffe, Jacques – London 1820 [mf ed Hildesheim 1995-98] – 2v on 6mf – 9 – €120.00 – 3-487-29299-8 – gw Olms [914]

Italy and the italians in the nineteenth century : a view of the civil, political, and moral state of that country; with a sketch of the history of italy under the french; and a treatise on modern italian literature / Vieusseux, Andre – London 1824 [mf ed Hildesheim 1995-98] – 2v on 5mf – 9 – €100.00 – 3-487-29286-6 – gw Olms [945]

Italy as it is : or narrative of an english family's residence for three years in that country / Beste, Henry – London 1828 [mf ed Hildesheim 1995-98] – 3mf – 9 – €90.00 – 3-487-29285-8 – gw Olms [914]

Italy. Commissione per l'osservazione dell' ecclisse totale di sole del 22 dicembre 1870 *see* Rapporti sulle osservazioni dell' ecclisse totale di sole del 22 dicembre 1870

Italy. Direzione Generale Della Statistica *see* Annuario statistico italiano

Italy. Dogane e Imposte Indirette. Direzione Generale delle *see* Bollettino ufficiale

Italy from dante to tasso (1300-1600), its political history.. / Cotterill, Henry Bernard – New York: Frederick A. Stokes company, 1919. xxviii,617p. illus – 1 – us Wisconsin U Libr [945]

Italy in the thirteenth century / Sedgwick, Henry Dwight – Boston: Houghton Mifflin, 1912 [mf ed 1990] – 2v on 1mf – 9 – 0-7905-7025-4 – (incl bibl ref) – mf#1988-3025 – us ATLA [931]

Italy in transition : public scenes and private opinions in the spring of 1860 / Arthur, William – New York: Harper & Bros, 1860 – 1mf – 9 – 0-524-03631-4 – (incl bibl ref) – mf#1990-1059 – us ATLA [914]

Italy, Instituto Storico Italiano *see* Fonti per la storia d'italia

Italy. Istituto Centrale di Statistica *see*
- Annali di statistica
- Annuario statistico italiano 1878-1965
- Movimento della popolazione e cause di morte
- Statistica giudiziaria civile e commerciale

Italy. Laws, Statutes, etc *see*
- Bullettino settimanale delle leggi e dei decreti del regno d'italia
- Raccolta ufficiale delle leggi e dei decreti del regno d'italia

Italy. Ministero delle Finanze *see* Bollettino di statistica e legislazione comparata

Italy. Ministero di Giustizia e dei Culti *see* Grazia e giustizia

Italy. Parlamento. Legislatura *see* Atti parlamentari

Italy through dutch eyes. dutch 17th century landscape artists in italy / Michigan. University. Museum of Art – 1964 – 9 – uk Chadwyck [700]

Itawamba Historical Society *see* Itawamba settlers

Itawamba settlers / Itawamba Historical Society – 1981 mar-1985, 1986-87 – 2r – 1 – mf#1277592 – us WHS [978]

ITC news / Inuit Tapirisat of Canada – 1976 aug/sep-1980 sep – 1r – 1 – (cont: inuit today) – mf#635797 – us WHS [305]

Ite journal / Institute of Transportation Engineers – Washington. 1978+ (1) 1978+ (5) 1978+ (9) – (cont: transportation engineering) – ISSN: 0162-8178 – mf#122,02 – us UMI ProQuest [629]

Item – O'Neill, NE: C Selah. v7 n315. jan 8 1890-v9 n6. jan 28 1892 (wkly) – 1r – 1 – (cont by: ewing item. v7 n10-v9 n6 called also whole n322-422) – us Bell [071]

Item – Hammonton, NJ. 1872-1877 (1) – mf#64821 – us UMI ProQuest [071]

Item – Huntsville, TX. 1998-1998 (1) – mf#61852 – us UMI ProQuest [071]

Item – Mobile, AL. 1912-1917 (1) – mf#62027 – us UMI ProQuest [071]

Item – New Orleans, LA. 1877-1958 (1) – mf#63513 – us UMI ProQuest [071]

Item – New Oxford, PA. 1889-1967 (1) – mf#66004 – us UMI ProQuest [071]

Item – Picayune, MS. 1980-2000 (1) – mf#61199 – us UMI ProQuest [071]

Item – Richmond, IN. 1917-1938 (1) – mf#62958 – us UMI ProQuest [071]

Item – Sumter, SC. 1960+ (1) – mf#61832 – us UMI ProQuest [071]

The item – Dallas, TX: J G Griffin & Ellis Willis, 1891 (wkly) [mf ed 1947] – (= ser Negro Newspapers on Microfilm) – 1r – 1 – us L of C Photodup [071]

The item – Williamsport, PA., 1888 – 13 – $25.00r – us IMR [071]

Itemizer-observer – Dallas OR: L Shaffer, 1990-92 [wkly] – 1 – (cont: polk county itemizer and observer (1927-90). merged with: monmouth and independence sun-enterprise (1980-92) to form: polk county itemizer observer (1992-)) – us Oregon Lib [071]

Items and issues. social science research council (US) – New York, 2000+ [1,5,9] – (cont: social science research council (u.s.) items – social science research council) – mf#30494 – us UMI ProQuest [300]

Items on pacific islands from reports of annual and semi-annual conferences / Church of Jesus Christ of Latter Day Saints – 1902-1959 – 1r – 1 – mf#pmb115 – at Pacific Mss [243]

Items on pacific islands from the minutes and reports of the annual general conferences / Reorganized Church of Jesus Christ of Latter Day Saints – 1879-1964 – 1r – 1 – mf#pmb107 – at Pacific Mss [243]

Itenarios del tropico : colombia del pacifico al at... / Diez, Jorge A – Quito, Ecuador. 1944 – 1r – us UF Libraries [972]

Itenera hierosolymitana saecvli 4-8 / Geyer, P – Wien, 1898 – 6mf – 9 – mf#H-3087 – ne IDC [956]

Das itenerarium peregrinorum (mgh schriften:18.bd) : eine zeitgenoessische englische chronik zum 3. kreuzzug in urspruenglicher gestalt / Mayer, H E – 1962 – (= ser Monumenta germaniae historica. schriften (mgh schriften)) – €18.00 – ne Slangenburg [931]

Iter ad fodinas 1733 : iter dalekarlicum 1734. iter ad experos 1735 / Linnaeus, C – 5mf – 9 – mf#168 – ne IDC [914]

Iter hispanicum, eller resa til spanska laenderna uti europa och america... / Loefling, P – Stockholm, 1758 – 7mf – 9 – mf#886 – ne IDC [910]

Iter italicum / Pflugk-Harttung, Julius von – Stuttgart: W. Kohlhammer, 1883 – 3mf – 9 – 0-8370-8139-4 – (incl ind) – mf#1986-2139 – us ATLA [940]

Iter lapponicum / Linnaeus, C – Holmiae, 1732 – 4mf – 9 – mf#169 – ne IDC [914]

Iter palaestinum eller resa til heliga landet... / Hasselquist, F – Stockholm, 1757 – 11mf – 9 – mf#2586 – ne IDC [915]

Iter per poseganam sclavoniae provinciam... / Piller, M & Mitterpacher, L – Nashville. 1962-1969 (1) 1970-1970 (5) – 6mf – 9 – mf#2457 – ne IDC [914]

Iter turcico-persicum / Nabelek, F – Brno, 1923-1929. 4 pts – 5mf – 9 – mf#11840 – ne IDC [956]

Itet. 7-37 / Russia. (1917-R.S.F.S.R.). Tsentral'nyi Irmii vspolnitel'nyi Komitet – 1 – us L of C Photodup [947]

Ithaca new times – Syracuse. 1972-1976 – 1 – mf#8770 – us UMI ProQuest [073]

I–the church of rome and recent projects for re-union / Dowden, John – Edinburgh, Scotland. 1895 – 1r – us UF Libraries [240]

I-the church of rome and recent projects for re-union / Dowden, John – Edinburgh, Scotland. 1895 – 1r – us UF Libraries [240]

Itier, J *see* Journal d'un voyage en chine en 1843, 1844, 1845, 1846

Itinera hierosolymitana et descriptiones terrae sanctae... / Molinier, A & Tobler, T – Genevae, 1879 – 5mf – 9 – mf#H-3111 – ne IDC [915]

Itinera hierosolymitana et descriptiones terrae sanctae... / ed by Tobler, T & Molinier, A – Genevae, 1879 – 6mf – 9 – mf#H-2871 – ne IDC [914]

Itinera per helvetiae alpinas regiones facta annis 1702-1707 et 1709-1711 / Scheuchzer, J J – Lugduni Batavorum, 1723. 4v – 14mf – 9 – mf#H-6126 – ne IDC [590]

Itineraire de buonaparte, de l'ile d'elbe a l'ile sainte-helene : ou memoires pour servir a l'histoire des evenemens de 1815... / Fabry, Jean – Paris 1816 [mf ed Hildesheim 1995-98] – 1v on 3mf – 9 – €90.00 – ISBN-10: 3-487-26382-3 – ISBN-13: 978-3-487-26382-3 – gw Olms [920]

Itineraire de la vallee de chamonix, d'une partie du bas-vallais et des montagnes avoisinantes / Berthoud van Berchem, Jacob – Lausanne 1790 [mf ed Hildesheim 1995-98] – 2mf – 9 – €60.00 – 3-487-29390-0 – gw Olms [914]

Itineraire de l'europe : soigneusement revu, corrige, et considerablement augmente sur la guide des voyageurs en europe de m reichard / Gandini, Francesco – Milan 1821 [mf ed Hildesheim 1995-98] – 2mf – 9 – €60.00 – 3-487-29936-4 – gw Olms [914]

Itineraire de paris a jerusalem et de jerusalem a paris : en allant par la grece, et revenant par l'egypte, la barbarie et l'espagne / Chateaubriand, Francois R de – Paris 1811 [mf ed Hildesheim 1995-98] – 3v on 9mf – 9 – €180.00 – 3-487-27689-5 – gw Olms [910]

Itineraire de paris...jerusalem et de jerusalem... paris / Chateaubriand, [F A] de – Bruxelles, 1827 – 11mf – 9 – mf#HT-276 – ne IDC [910]

Itineraire de tiflis a constantinople / Rottiers, Bernard – Bruxelles 1829 [mf ed Hildesheim 1995-98] – 1v on 3mf – 9 – €90.00 – 3-487-27659-3 – gw Olms [915]

Itineraire de tiflis...constantinople / Rottiers, B – Bruxelles: H Tarlier, 1829 – 5mf – 9 – mf#AR-1416 – ne IDC [915]

L'itineraire des francais dans la louisiane : contenant l'histoire de cette colonie francaise, sa description... / Dubroca, Louis – Paris: Chez Dubroca...oct 1802 [mf ed 1985] – 2mf – 9 – 0-665-18568-5 – mf#18568 – cn Canadiana [978]

Itineraire descriptif : ou description routiere, geographique, historique et pittoresque de la france et de l'italie / Vaysse de Villiers, Regis – Paris 1817-22 [mf ed Hildesheim 1995-98] – 7v on 16mf – 9 – €160.00 – 3-487-29696-9 – gw Olms [914]

Itineraire descriptif de la france : ou geographie complete, historique et pittoresque de ce royaume par ordre de routes / Vaysse de Villiers, Regis – Paris 1830-31 [mf ed Hildesheim 1995-98] – 6v on 13mf – 9 – €130.00 – 3-487-29697-7 – gw Olms [914]

Itineraire descriptif de l'espagne : et tableau elementaire des differentes branches de l'administration et de l'industrie de ce royaume / Laborde, Alexandre L de – Paris 1809 [mf ed Hildesheim 1995-98] – 5v+atlas on 19mf – 9 – €190.00 – 3-487-29867-8 – gw Olms [914]

Itineraire descriptif et pittoresque des hautes-pyrenees francoises : jadis territoires du bearn, du bigorre, des quatre-vallees, du comminges, et de la haute-garonne / Labouliniere, Pierre – Paris 1825 [mf ed Hildesheim 1995-98] – 3v on 9mf – 9 – €180.00 – 3-487-29750-7 – gw Olms [914]

Itineraire d'italie : contenant la description des routes les plus frequentees et des villes principales; les distances en postes et en milles, le temps qu'on doit mettre en voyage / Vallardi, Guiseppe – Milan 1810 [mf ed Hildesheim 1995-98] – 3mf – 9 – €90.00 – 3-487-29310-2 – gw Olms [914]

L'itineraire d'ou-k'ong (751-790) : note additionelle: le kipin / Levi, S & Chavannes – Paris, 1895 – v6 on 1mf – 9 – mf#U-594 – ne IDC [915]

Itineraire general topographique et hydraulique de la france : contenant les divisions militaires et judiciaires; l'itineraire administratif... / Foulon, J-b – Paris 1828 [mf ed Hildesheim 1995-98] – 2mf – 9 – €60.00 – 3-487-29795-7 – gw Olms [944]

Itineraire topographique et historique des hautes-pyrenees principalement des etablissemens thermaux de cauterets, saint-sauveur... / Abadie, A – Paris 1824 [mf ed Hildesheim 1995-98] – 2mf [ill] – 9 – €60.00 – 3-487-29757-4 – gw Olms [914]

Itineraires russes en orient / Khitrowo, B – Geneve, 1889 – €18.00 – ne Slangenburg [243]

Itineraires russes en orient : traduits pour la societe de l'orient latin / Khitrowo, B de – Geneve, 1889 – 4mf – 9 – mf#H-2931 – ne IDC [915]

Itineraires...jerusalem et descriptions de la terre sainte / Michelant, H & Raynaud, G – Geneve, 1882 – 4mf – 9 – mf#H-2935 – ne IDC [915]

An itinerant ministry : a sermon...pontiac, sep 29 1859 / Clements, S – New York: Carlton & Porter, 1860 [mf ed 1993] – (= ser Methodist coll) – 1mf – 9 – 0-524-07231-0 – mf#1991-2972 – us ATLA [242]

Itineraria romana : roemische reisewege an der hand der tabula peutingeriana / Miller, K – Stuttgart, 1916 – 11mf – 9 – mf#H-3109 – ne IDC [914]

Itinerario / Marcos Suarez, Miguel De – La Habana, Cuba. 1956 – 1r – us UF Libraries [972]

Itinerario / Rodriguez Cerna, Jose – Guatemala, 1943 – 1r – us UF Libraries [972]

Itinerario : tierras floridas, 1917-1937-1943 / Acena Duran, Ramon – Guatemala, 1964 – 1r – us UF Libraries [972]

Itinerario da historia da colonizacao da paraiba a... / Leal, Jose – Rio de Janeiro, Brazil. 1966 – 1r – us UF Libraries [972]

Itinerario de rio guadiana y todos sus afluentes / Spain. Direccion general de Obras publicas – 1883 – 9 – sp Bibl Santa Ana [914]

Itinerario de sylvio romero / Rabello, Sylvio – Rio de Janeiro, Brazil. 1944 – 1r – us UF Libraries [972]

Itinerario de un viage (sic) / Lujan, Francisco – 1837 – 9 – (2a parte 1837) – sp Bibl Santa Ana [910]

Itinerario del litigio de l : mites entre el ecuador y el peru... – Madrid, 1908 – 1mf – 9 – mf#ILM-2742 – ne IDC [918]

Itinerario historico / Colombia Junta Militar De Gobierno – Bogota, Colombia. v1-2. 1958 – 1r – us UF Libraries [972]

Itinerario sperimentale nella letteratura tedesca / Masini, Ferruccio – Parma: Studium Parmense, 1970 [mf ed 1993] – (= ser Quaderni di ricerca n1) – 414p – 1 – (incl bibl ref and ind) – mf#8257 – us Wisconsin U Libr [430]

Itinerario y pensamiento de los jesuitas expulsos de chile (1767-1815) / Hanisch Espindola, W; ed by Bello, A – Santiago, 1972 – 4mf – 9 – mf#CIDOC-1755 – ne IDC [918]

Itinerarios de el rei d. sebastiao... / Verissimo Serano, Joaquin – Madrid: Arch. Ibero Americano, 1964 – 1 – sp Bibl Santa Ana [946]

Itinerarium in terram sanctam... / Walther, P – Stuttgart, 1892 – 4mf – 9 – mf#HT-288 – ne IDC [915]

Itinerarium paradisi... / Raulin, J – Venetiis, 1585 – 4mf – 9 – mf#CA-89 – ne IDC [240]

Itinerarium septentrionale : or, a journey thro' most of the countries of scotland, and those in the north of england / Gordon, A – London, 1726. 2v – 10mf – 9 – mf#H-1133 – ne IDC [914]

Itinerary notes of plants collected in the khasyah and bootan mountains, 1837-1838, in affghanistan and neighbouring countries, 1839-1841 / Griffith, W; ed by M'Clelland, J – New York. 1968-1993 (1) 1972-1993 (5) 1976-1993 (9) – 10mf – 9 – mf#7412 – ne IDC [915]

The itinerary of greece : containing one hundred routes in attica, boeotia, phocis, locris, and thessaly / Gell, William – London 1819 [mf ed Hildesheim 1995-98] – 2mf – 9 – €60.00 – 3-487-29067-7 – gw Olms [914]

The itinerary of jacques cartier's first voyage / Ganong, William Francis – [S.l: s.n, 1890?] [mf ed 1986] – 1mf – 9 – 0-665-28317-2 – mf#28317 – cn Canadiana [910]

The itinerary of john leland in or about the years 1535-1543 : lelands itinerary in england and wales / ed by Toulmin Smith, L – London. 5v. 1907-1910 – 33mf – 9 – mf#H-1125 – ne IDC [914]

An itinerary of provence and the rhone : made during the year 1819 / Hughes, John – London 1822 [mf ed Hildesheim 1995-98] – 2mf – 9 – €60.00 – 3-487-29718-3 – gw Olms [914]

Itjeshorst, Johannes *see* De werkzaamheid van du plessis mornay in dienst van hendrik van navarre

Ito, Sayuri *see* The choreography and performance of a japanese folk tale

Ito, Takeo *see* China's challenge in manchuria

Itogi denezhnoi reformy / Sigal, B V; ed by Smushkov, VV – Khar'kov, 1925 – 2mf – 9 – mf#REF-63 – ne IDC [332]

Itogi desiatiletiia sovetskoi vlasti v tsifrakh : 1917-1927 gg – M, 1928. (Tsentral'noe statisticheskoe upravlenie sssr. xiv/514, 6p) – 6mf – 9 – mf#RHS-28 – ne IDC [314]

Itogi ekonomicheskogo issledovaniia rossii po dannym zemskoi statistiki – 2v – 24mf – 8 – mf#RZ-167 – ne IDC [314]

It's about times : abalone alliance newspaper / Abalone Alliance – 1980 nov/mid dec-1985 mar/apr – 1r – 1 – mf#1096577 – us WHS [366]

It's all in the day's work / King, Henry Churchill – New York: Macmillan, 1916 – 1mf – 9 – 0-7905-7870-0 – mf#1989-1095 – us ATLA [170]

It's happening in spain : told by the victims themselves: illustrated by scenes from the battlefront of democracy / International Labour Defense – New York City, 1937 [mf ed 1977] – (= ser Blodgett coll) – 1mf – 9 – mf#w967 – us Harvard [331]

It's up to the women / Roosevelt, Eleanor – New York: Frederick A. Stokes Company, 1933. x,263p – 1 – us Wisconsin U Libr [305]

ITT Thorp Corp *see*
– New dimensions
– Thorp insider

Ittifak – jan-jun, 1918 – 1 – (reel contains short runs of multiple titles. for complete listing of titles on a reel, please inquire) – us UMI ProQuest [077]

Ittihad – Izmir, 1908-19? Sahib-i Imtiyaz ve Mueduer-i Mes'ul: Bekir Behlul; Sermuharriri: Hueseyin Fehmi. n409 (551), n410 (552), n411 (553). 24-26 agustos 1910 – (= ser O & t journals) – 1mf – 9 – $25.00 – us MEDOC [956]

Ittihad-i buzurg – Tehran: Jibhah-'i Milli dar rah-i Ittihad-i buzurg, 1979. shumarah-'i 20-36. 7 tir 1358-17 mihr 1358 [29 jun-16 oct 1979] – 1r – 1 – $53.00 – (cont: jibhah-'i milli-i iran. missing: n33-34) – us MEDOC [079]

Ittihad-i islam ve almanya / Nuri, Celal – Istanbul: Yeni Osmanli Matbaa ve Kitabhanesi, 1333 [1917] – (= ser Ottoman histories and historical sources) – 1mf – 9 – $25.00 – us MEDOC [956]

Ittihad-i javan – Tehran, 1979. shumarah-'i 7-12,19,24-27,30-31,33-35,37-39,41-49,51-57. 23 murdad 1358-7 shahrivar 1360 [14 aug 1979-29 aug 1981] – 1r – 1 – $115.00 – us MEDOC [079]

Ittihad-i mardum – [Tehran]: Ittihad-i Dimukratik-i Mardum-i Iran, 1979- . dawrah-'i 1; shumarah-'i 1-70: 23 mihr 1358-44 isfand 1359 [15 oct 1979-23 feb 1981] – 1r – 1 – $53.00 – (missing: n11-21, 46-47, 58, 61, 63-64, 66-67. also incl: sawgand, dawrah-i 2, shumarah-i 20-22. 22 murdad 1358-16 mihr 1358 [14 aug 1979-8 oct 1979]) – us MEDOC [079]

Ittihad-i mardum *see* Sawgand

Ittihad-i osmani = La federation ottomane – Geneva: Heyet-i Muettefikiyye-i Osmaniyye, 1903-19? n1. 23 subat 1903; n3. 10 eyluel 1903 – (= ser O & t journals) – 1mf – 9 – $25.00 – us MEDOC [956]

Ittila'at : 28 hazar ruz-i tarikh-i iran va jahan – Tehran: Ittila'at, 1972?- – 1r – 1 – $53.00 – us MEDOC [079]

Ittilaat – Tihran, Iran: Muassasah-i Ittilaat, jul 10 1926- – (= ser Ettela'at) – 1 – us CRL [079]

Ittmann, Johannes *see* Grammatik des duala (kamerun)

Itu review / International Typographical Union – 1964 oct 29 [v7 n39]-1971 jun 24, 1971 jul 1-1974 jul 11 [v17 n25] – 2r – 1 – (cont by: international typographical union review) – mf#1002575 – us WHS [331]

Itu review / International Typographical Union – 1977 jan 6 [v19 [i.e. 20] n1]-1978 may 28, 1978 jun 8-1984 sep 20 [v27 n9] – 2r – 1 – (cont: international typographical union review; cont by: review [international typographical union]) – mf#1002570 – us WHS [680]

Iturbide, Agustin *see*
– Correspondencia y diario militar. 1810-1814. tomo 3
– Papers

Iturribarria, Jorge Fernando *see* Historia de mexico

Iturrios, J *see* Elias de tejada, f las doctrinas politicas de la edad media

Ityalike ekatolike yaseroma / Schweiger, Albert – Ishicelelwe, South Africa. 1931 – 1r – us UF Libraries [960]

Itzehoer nachrichten *see* Itzehoer wochenblatt

Itzehoer wochenblatt – Itzehoe DE, 1848 6 jan-1849 29 dec – 1 – (among several title changes: itzehoer nachrichten; 20 aug 1949: norddeutsche rundschau. filmed by other misc inst: 1976- [ca 7r/yr]) – gw Misc Inst [074]

Itzehoer wochenblatt – Itzehoe DE, 1961 18 feb-1965 jan [gaps] – 1r – 1 – gw Misc Inst [074]

Iubilei petra velikogo : bibliograficheskii ukazatel... / Mezhov, V I – 1881 – 3mf – 8 – mf#R-7157 – ne IDC [947]

Iucc bulletin – Cambridge. 1980-1983 (1,5,9) – (cont by: university computing: the bulletin of the iucc) – ISSN: 0142-2464 – mf#15584 – us UMI ProQuest [000]

The iud / South Africa. Department of Health [Departement van Gesondheid] [Departement van Gesondheid] – [Pretoria: Dept of Health 1976?] [mf ed Pretoria, RSA: State Library [199-]] – 6p on 1r with other items – 5 – mf#op 06603 r24 – us CRL [362]

Iud coordinated bargaining quarterly : chq / AFL-CIO [American Federation of Labor and Congress of Industrial Organizations] – v9 n1-v13 n2 [1980 jul-1985:2nd qtr] – 1r – 1 – (cont: iud spotlight on health and safety; cont by: coordinated collective bargaining quarterly) – mf#1361683 – us WHS [331]

Iud spotlight on health and safety / AFL-CIO [American Federation of Labor and Congress of Industrial Organizations] – v4 [i.e. v6] n3-v8 n2 [1977 3rd qtr-1979 2nd qtr] – 1r – 1 – mf#483100 – us WHS [331]

The iudgment of a most reverend and learned man, from beyond the seas, concerning a threefolde order of bishops... : we must needes make three bishops. 1 of god. 2 of man. 3 of the devil / Beza, Theodor de – n.p., 1580 – 1mf – 9 – mf#PW-64 – ne IDC [240]

IUdin, N A *see* Orenburgskii gubernskii vestnik

Iuditskii, A D *see* Veg tsu oktiabr

Iue local 201 electrical union news / International Union of Electrical, Radio, and Machine Workers – v34 n4-v37 n15 [1974 feb 8-1978 dec 1] – 1r – 1 – mf#673402 – us WHS [331]

Iue news / International Union of Electrical, Radio, and Machine Workers et al – 1972 jan-1978 mar, 1978 mar-1984 dec, 1985 jan-1989 dec, 1990 jan-1994 nov/dec – 4r – 1 – (cont: iue afl-cio news; cont by: iue-cwa news) – mf#3363023 – us WHS [071]

Iue news – Washington. 1975+ (1) 1980+ (5) 1980+ (9) – ISSN: 0019-0861 – mf#10443 – us UMI ProQuest [600]

Iue-cio news / International Union of Electrical, Radio and Machine Workers – 1949-55 – (= ser Labor union periodicals, pt 1: the metal trades) – 2r – 1 – $430.00 – 1-55655-239-4 – us UPA [331]

Iug rossii / ed by Fal'chenko, G I – Sevastopol' [Tavr gub]: [s n] 1920 [1920 [24 marta]-1920 [?]] – (= ser Asn 1-3) – n162 [1920] item 445, on reel n87 – 1 – mf#asn-1 445 – ne IDC [077]

Iugoslavianskiia drevnosti v izlozhenii prof l niderle : 1 ch 2 t slav drev / Lavrov, Petr Alekseevich – S-Peterburg: Tip Ministerstva Putei Soobshcheniia (T-val.N Kushnerev), 1907 [mf ed 2002] – 1r – 1 – (filmed with: smuta moskovskago gosudarstva i nizhnii-novgorod / ocherk a k kabanova (1911)) – mf#5229 – us Wisconsin U Libr [949]

Iugo-Vostochnyi Kommercheskii Aktsionernyi Bank *see* Otchet iugo-vostochnogo kommercheskogo aktsionernogo banka v rostove na donu za vtoroi operatsionnyi god. 1923-1924 god

Iulii caesaris scaligeri exoticarum exercitationum, liber quintus: de subtilitate, ad hieronymum cardanum; in extremo duo sunt indices; prior breuiusculus, continens sententias nobiliores; alter opulentissimus, pene omnia complectens / Scaliger, Julius Caesar, – Lutetia: Ex officina typographica Michaelis Vascosani, 1557 – 1 – us Wisconsin U Libr [574]

Iumoristichesko-satiricheskii zhurnal – Spb., 1859. v1-22 – 11mf – 9 – mf#R-3981 – ne IDC [077]

Iunyj tekstil'shchik – Ivanovo-Voznesensk, Russia, 1920-21 – 1r – 1 – us UMI ProQuest [077]

Iupiw-views / International Union of Petroleum and Industrial Workers – 1972 jan-dec, 1973-1981 feb/mar, 1981 apr/may-1991 jan/feb – 3r – 1 – (cont: iupw news) – mf#1058607 – us WHS [331]

Iupw views / International Union of Petroleum Workers et al – 1962-65, 1966 jan-1969 dec, 1970-72 – 3r – 1 – (cont: independent oiler; cont by: iupiw views) – mf#2504408 – us WHS [331]

Iurgens, F A *see* Vospominaniia ob e i lamanskom v sviazi s deiatel'nost'iu gosudarstvennogo banka

Iuridicheskaia gazeta – 1999- – 2r per y – 1 – $160.00 standing order – (backfile through 1998 $85r) – us UMI ProQuest [340]

Iuridicheskie zapiski : izdavaemye demidovskim iuridicheskim litseem – Iaroslavl, 1908-13. 3v – 3mf – 9 – (incomplete) – mf#R-18344 – ne IDC [077]

Iuridicheskii vestnik / ed by Kalachov, N – Spb., 1860/1861-1864 – 74mf – 9 – mf#R-957 – ne IDC [077]

Iuridicheskii vestnik – M., 1913-1916 – 80mf – 9 – mf#R-9571 – ne IDC [077]

Iuridicheskii vestnik – M., 1867-92 – 663mf – 9 – (missing: 1867-68; 1885, no 9; 1892, no 1-12) – mf#R-9570 – ne IDC [077]

Iuridicheskii zhurnal – Spb., 1860-1861. nos 1-8 – 37mf – 9 – mf#R-3202 – ne IDC [077]

Iuris allegation...enriquez de guzman con... sucesion del condado de alba de aliste – 1612 – 9 – sp Bibl Santa Ana [340]

Iuris ecclesiastici graecorum historia et monumenta / Pitra, J-B – Romae. v1-2. 1864-68 – €206.00 – ne Slangenburg [240]

Iuris ecclesiastici graecorum historia et monumenta iussu pii 9. pont. max / Pitra, Jean Baptiste – Romae: Typis Collegii Urbani, 1864-1868 – 14mf – 9 – 0-524-05160-7 – mf#1990-1416 – us ATLA [240]

Iuris processualis compendium / Roberti, Francesco – Romae: Apud custodiam Librariam Pontificii Instituti Untriusque Iuris, [19–?] – 1mf – 9 – 0-524-07713-4 – mf#1991-3298 – us ATLA [240]

Iurisconsulti praeclarissimi / Gutierrez, Juan – Libri II. 1618 – 9 – (libri 3 1618. libri 4 1611) – sp Bibl Santa Ana [240]

Iurkevich, N G *see* Trud zhenshchiny na promyshlennom predpriiatii i stabilnost braka

Iurovskii, L N *see*
- Denezhnaia politika sovetskoi vlasti
- Na putiakh k denezhnoi reforme
- Nashe denezhnoe obrashchenie

Iuvenal *see* Declaracion...sobre las satiras de..

Iuvenci carmina / Arevalo, Faustino – 1792 – 9 – sp Bibl Santa Ana [780]

Iuzhnoe slovo [khar'kov: 1918] / ed by Vladimirov-Laub, V V – Khar'kov: Izd t-vo "Pechatnoe slovo" 1918 [1918 29 [16] okt-] – (= ser Asn 1-3) – n1-7 [1918] item 448, on reel n87 – 1 – mf#asn-1 448 – ne IDC [077]

Iuzhnoe slovo [nikolaev: 1918] : ezhedn polit, ekon i obshchestv -lit gaz / ed by IAkushkin, V E – Nikolaev [Kherson gub]: V P IUritsyn 1918 [1918 [okt]-] – (= ser Asn 1-3) – n4-49 [1918] [gaps] item 446, on reel n87 – 1 – mf#asn-1 446 – ne IDC [077]

Iuzhnoe slovo [odessa: 1919] : pri blizhaishem uchastii akad i a bunina, n p kondakova / ed by Klimenko, N K – Odessa: [s n] 1919 [1919 27 avg [9 sent]-1920 12 [25] ianv – (= ser Asn 1-3) – n2-111 [1919] [gaps] item 447, on reel n87 – 1 – (cont by: rodnoe slovo) – mf#asn-1 447 – ne IDC [077]

Iuzhno-russkii literaturno-uchenyi vestnik – Washington. 1961+ (1) 1970+ (5) 1976+ (9) – 109mf – 9 – (missing: 1862(11-12)) – mf#1820 – ne IDC [077]

Iuzhnye vedomosti : ezhedn gaz tavr soiuza zhurnalistov i literatorov – Simferopol' [Tavr gub]: Tavr soiuz zhurnalistov i literatorov 1918-19 [1906 17 maia-1920 [?]] – (= ser Asn 1-3) – n1-8 [1918] n108 [1919] item 449, on reel n87 – 1 – mf#asn-1 449 – ne IDC [077]

Iuzhnyi kooperator – Odessa, 1913-1917(2) – 36mf – 9 – (missing:1913(20),1916(2,4)) – mf#COR-709 – ne IDC [335]

Iuzhnyi krai [utr vyp] / ed by Sizov, A I – Khar'kov: T-vo "IUzhnyi krai" 1918-19 [1880 1 dek-] – (= ser Asn 1-3) – n140-189 [1918] n31-133 [1919] [gaps] item 450, on reel n87,88 – 1 – mf#asn-1 450 – ne IDC [077]

Iuzhnyi krai [vech vyp] / ed by Sizov, A I – Khar'kov: T-vo "IUzhnyi krai" 1919 [1911 1 dek-] – (= ser Asn 1-3) – n13 [1919] item 451, on reel n88 – 1 – mf#asn-1 451 – ne IDC [077]

Iuzhnyi kraj – Khar'kov, 1886-98 – 1 – us UMI ProQuest [077]

Iuzhnyi muzykal'nyi vestnik – Odessa, 1915-16 [bimthly] – 8mf – 9 – us UMI ProQuest [780]

Iuzhnyi rabochii : ezhedn gaz: organ odes kom rsdrp [obed] / ed by Sukhov, A A – Odessa: Kom RSDRP v litse V M Korobkova 1918-20 [1917 16 marta-1920 [25 ianv[7 fevr]] – (= ser Asn 1-3) – n8 [1918]-n117 [1920] [gaps] item 27, on reel n2 – 1 – mf#asn-2 027 – ne IDC [077]

Iuzhnyi ural – Orenburg, 1974-88 – 4r – 1 – us UMI ProQuest [077]

IUzhnyi ural [orenburg, russia] : [obshchestvenno-politicheskaia gazeta orenburzhia] – Orenburg: Izd-vo "IUzhnyi Ural" [mf ed Minneapolis MN: East View Publ [199-]] – 10r – 1 – (crl has: mf-11782 seemp [10r] 1991-95; mf-12835 seemp [10r] 1996-2000) – us CRL [077]

Iuzio so la enfermedad que...aflige a toledo... / Vazquez, J – Toledo, 1631 – 1mf – 9 – sp Cultura [610]

Iuznyi potrebitel – Kharkov, 1914(1-5) – 22mf – 9 – (cont as:iuzhno-russkii potrebitel.kharkov,1914(1-6)-1918(4).missing:1914(6),1917(2-4, 10-12),1918(1)) – mf#COR-708 – ne IDC [335]

Iva first baptist church. iva, south carolina : church records – 1906-16, 1922-49, 1967-72. Formerly: Mizpah, 1890-1912.252p – 1 – us Southern Baptist [242]

Ivan greet's masterpiece, etc / Allen, Grant – London: Chatto & Windus, 1893 – 5mf – 9 – (with frontispiece by stanley l wood) – mf#05046 – cn Canadiana [830]

Ivanitskii, R A *see* Vozrozhdenie [tiflis: 1918-1920]

Ivanov, A *see* Izdanie olonetskogo gubernskogo statisticheskogo komiteta

Ivanov, A I *see*
- Nasha derevnia
- Sibir

Ivanov, Dmitrii Dmitrievich *see* Iskusstvo farfora

Ivanov, Lev *see* The dance of the reed pipes

Ivanov, P A *see* Obozrenie sostava i istorii reguliarnoi russkoi kavalerii ot petra velikogo i do nashikh dnei

Ivanov, S I *see* Primorskie oblastnye izvestiia

Ivanov, V I *see*
- Kapitaly kreditnogo kooperativa
- Vkladnye operatsii kreditnykh kooperativov

Ivanov, V N *see*
- Nasha gazeta
- Nasha gazeta [omsk: 1919]

Ivanov, V P *see* Orenburgskii vecher

Ivanov, V V *see* Vsemirnyi kongress" baptistov" v" londone v" 1905 godu

Ivanovich, S *see*
- Anarkhiia i anarkhisty
- Kadety i evrei

Ivanovich, V *see* Rossiiskie partii, soiuzy i ligi

Ivanovich, V G *see* Rabochii put'

Ivanov-Razumnik *see* Sobranie sochinenii v g bielinskago

Ivanovskii, V *see*
- Vedomosti tobol'skogo gubernskogo komissariata
- Vestnik tobol'skoi gubernii

Ivanow, Wladimir *see* Ismaili tradition concerning the rise of the

Ivashchenko, I S *see* Ezhegodnik russkikh kreditnykh uchrezhdenii

Ivashina, D V *see* Molodaia ukraina

Ivdicivm erasmi alberi, de spongia erasmi roterod / Alber, E – [Hagenau, 1524] – 1mf – 9 – mf#TH-1 – ne IDC [242]

"I've come a long way" : learner-identified outcomes of participation in adult literacy programs / Bingman, Mary Beth – Cambridge MA: National Center...Adult Learning & Literacy, Harvard Graduate School of Education; [Washington DC]: US Dept of Education, Office of Educ Research & Improvement...[2000] [mf ed 2000] – 2mf – 9 – us GPO [374]

I've shed my tears : a candid view of resurgent india / Karaka, Dosoo Framjee – New York: Appleton-Century Co, c1947 – (= ser Samp: indian books) – us CRL [954]

Ivens, R *see* From benguella to the territory of yacca

Ivens, W G *see* A dictionary of the language of bugotu, santa isobel island, solomon islands

Iverach, James *see*
- Descartes, spinoza and the new philosophy
- Evolution and christianity
- Is god knowable?
- The other side of greatness, and other sermons
- St paul
- Theism in the light of present science and philosophy
- Truth of christianity

Ivernois, Francis d' *see* A cursory view of the assignats and remaining resources of french finance, september 6 1795

Ives, Charles Linnaeus *see*
- The bible doctrine of the soul

Ives, J C *see* Report upon the colorado river of the west

Ives, Levi Silliman *see* The trials of a mind in its progress to catholicism

Ives, Rollin A *see* A treatise on military law

Ivey business journal – London. 1998+ (1,5,9) – (cont: ivey business quarterly) – ISSN: 1481-8248 – mf#12002,03 – us UMI ProQuest [338]

Ivey business quarterly – London. 1997-1998 (1,5,9) – (cont: business quarterly. cont by: ivey business journal) – ISSN: 1480-6746 – mf#12002,02 – us UMI ProQuest [338]

Ivg, bericht ueber den ersten kongress, rom, 5.-10. september 1955 / ed by Internationale Vereinigung fuer Germanische Sprach- und Literaturwissenschaft. Kongress – [s.l.]: Die Vereinigung, 1958 (Verona: Stamperia Valdonega) – 1 – us Wisconsin U Libr [430]

Ivimey, Joseph *see*
- A history of the english baptists
- Memoir of william fox, esq. founder of the sunday school society
- Pastoral counsels
- Reasons why the protestant dissenters lament the death of...

Ivnev, Riurik *see* Zoloto smerti

Ivo, Ledo *see*
- As aliancas
- Rio, a cidade e os dias

Ivor wilks : phyllis ferguson collection of material on ghana – Chicago, University of Chicago, Photoduplication Dept, 1974 – us CRL [960]

Ivory Coast *see*
- Journal officiel
- Journal officiel de la republique de la cote d'ivoire

Ivory Coast. Ministere du Plan *see* Annuaire statistique de la cote d'ivoire 1975

Ivory trail / Barnard, Cecil – Cape Town, South Africa. 1954 – 1r – us UF Libraries [960]

Ivresse du sage / Curel, Francois De – Paris, France. 1921 – 1r – us UF Libraries [440]

'Ivri He-Hadash – Varshah, Poland. 1912 – 1r – 1 – us UF Libraries [939]

'Ivrit Ba'ma'arav / Kressel, Getzel – Tel-Aviv, Israel. 1940 or 1941 – 1r – 1 – us UF Libraries [939]

L'ivrognerie et la loi des licences / Bedard, Joseph-Edouard – [Quebec: s.n, 1903 ?] (mf ed 1992) – 1mf – 9 – mf#SEM105P1673 – cn Bibl Nat [360]

Ivw-auflagenmeldungen – Bonn, Wiesbaden DE, 1950-55 – 5r – 1 – (publ began in wiesbaden) – gw Mikrofilm [074]

The ivy leaf, 1921-1998 – 14r – 1 – $2705.00 – 1-55655-773-6 – (filmed fr records of the national office of alpha kappa alpha sorority. with p/g) – us UPA [378]

Iwan-Mueller, Ernest Bruce *see* Lord milner and south africa

Iwanowa, Gora *see* Der nationale gedanke bei heinrich von kleist

Iweins, Henri-Marie *see* L'ordre des freres-precheurs

Iyengar, A S *see* All through the Gandhian era

Iyengar, S Kesava *see* Economists at home and abroad

Iyer, Anantha Krishna et al *see* Anthropology of the syrian christians

Iyer, L Anantha Krishna *see* The travancore tribes and castes

Iyer, T Paramasiva *see* The riks

Iyo wiliwe nta rungu / Kagame, Alexis – Kabgayi [Rwanda]: Les Editions Royales, 1949 – 1 – us CRL [960]

'Iyunim / Federbusch, Simon – Warszawa, Poland. 1929 – 1r – 1 – us UF Libraries [939]

Iz arkhiva p b akselroda : [1881-1896] – Berlin, 1924 – 255p 4mf – 9 – (materialy po istorii russkogo revoliutsionnogo dvizheniia. v2) – mf#RPP-144 – ne IDC [325]

Iz chego vyrosli kooperativy / Kablukov, N A – 1916 – 16p 1mf – 9 – mf#COR-37 – ne IDC [335]

Iz epokhi iskry : (1900-1905 gg) / Zakharova-Tsederbaum, K I & Tsederbaum, S I – 1926 – 162p 2mf – 9 – mf#RPP-143 – ne IDC [325]

Iz istorii knigi, bibliotechnogo dela i bibliografii v sibiri / Akademiia nauk SSSR. Sibirskoe otdelenie, Gosudarstvennaia publichnaia nauchno-tekhnicheskaia biblioteka; ed by Kartashov, N S – Novosibirsk: Nauka, Sibirskoe otd-nie, 1969 – us CRL [947]

Iz istorii krakha levykh eserov v turkestane / Nikishov, P P – Frunze: Kyrgyzstan, 1965 – 2mf – 9 – mf#RPP-35 – ne IDC [325]

Iz istorii mezhdunarodnykh otnoshenii nakanune i posle poltavy – 1959 – 8mf – 8 – mf#R-7780 – ne IDC [947]

Iz istorii perevoda evangeliia v iuzhnoi rossii v 16 veke : letkovskoe evangelie / Gruzinskii, A S – Kiev, 1912 – 3mf – 8 – mf#R-5930 – ne IDC [243]

Iz istorii rabochikh artelei na zapade i v rossii : ot utopistov do nashikh dnei / Pazhitnov, K A – 1924 – 260p 3mf – 9 – mf#COR-87 – ne IDC [335]

Iz istorii raskola pervoi poloviny 18 veka : po neizdannym pamiatnikam / Smirnov, P S – 1908 – 5mf – 8 – mf#R-7877 – ne IDC [243]

Iz istorii russkoi dramy : shkolnye deistva 17-18 vv i teatr iezuitov / Rezanov, V I – 1910 – 9mf – 8 – mf#R-7812 – ne IDC [947]

Iz istorii russkoi perevodnoi poviesti 18 vieka / Veselovskii, Aleksandr Nikolaevich – Sanktpeterburg: Tip Imp akademii nauk, 1887 [mf ed 2002] – 1r – 1 – (filmed with: zamiechaniia ob obrazovanii slov iz vyrazhenii / i i sreznevskago (1873)) – mf#5235 – us Wisconsin U Libr [440]

Iz materialov iskry – Geneva – 4mf – 9 – mf#R-18054 – ne IDC [077]

Iz materialov redaktsii "rabochego dela" : izdanie soiuza russkikh sotsial-demokratov... – Geneve, 1900-02 – 2mf – 9 – mf#R-18012 – ne IDC [074]

Iz materialov redaktsii "zari" : biulleten zagranichnoi ligi russkoi revoliutsionnoi sotsial-demokratii – Geneve, 1901-05 – 3mf – 9 – mf#R-18015 – ne IDC [074]

Iz moego proshlogo : vospominaniia, 1903-1919 / Kokovtsov, VN – Parizh, 1933. 2v – 18mf – 9 – mf#REF-480 – ne IDC [332]

Iz nedavnego proshlogo / Gershuni, G – 1928 – 242p 3mf – 9 – mf#RPP-225 – ne IDC [325]

Iz perezhitogo : avtobiograficheskie vospominaniia / Giliarov-Platonov, N P – 1886. 2v – 346p 13mf – 8 – mf#R-9060 – ne IDC [243]

Iz pesen starogo rabochego / Nechaev, Egor Efimovich – Moskva: Gos izd-vo, 1922 [mf ed 2004] – 1r – 1 – (filmed with: pisma proza i wierszem / mikolaj rej (1926)) – mf#5489 – us Wisconsin U Libr [810]

Iz proshlago / Ol'minskii, Mikhail – Moskva: Gos izd-vo, 1919 [mf ed 2004] – 1r – 1 – (filmed with: vzaimnaia pomoshch' sredi zhivotnykh i liudei, kak dvigatel' progressa / p kropotkin (1922). incl bibl ref) – us Wisconsin U Libr [947]

Iz proshlogo russkoi zhurnalistiki / Maksimov, V D & Evgenev-Maksimov, V – L., 1930 – 4mf – 9 – mf#R-2070 – ne IDC [077]

Iz vospominanii / Maklakov, Vasilii Alekseevich – Niu-iork, NY. 1954 – 1r – us UF Libraries [025]

Iz zapisnoi knizhki russkogo monarkhista / Cherniaev, N I – Kharkov, 1907 – 248p 3mf – 9 – mf#RPP-176 – ne IDC [325]

Iz zhizni na nerchinskoi katorge / Spiridonova, M A – 1925 – 1mf – 9 – (katorga i ssylka no1(14,p185-204);no2(15,p165-182);no3(16,p115-133)) – mf#RPP-249 – ne IDC [325]

Izaak Walton League of America *see*
- Wisconsin walton news
- Wisconsin waltonian

Izaguirre / Llado De Cosso, Jose – Tegucigalpa, Mexico. 1949 – 1r – us UF Libraries [972]

Izaguirre, Carlos *see*
- Bajo el chubasco
- Desiertos y campinas
- Honduras y sus problemas de educacion
- Nieblas
- Reflexiones y pensamientos

Izaguirre, Fray Bernardino *see* Historia de las misiones franciscanas

Izates, Esther *see* Briefe an eine christliche freundin ueber die grundwahrheiten des judenthums

Izbanda : (la victoire) – Paris. n3-6. avr-nov 1934 – 1 – fr ACRPP [073]

Izbrannya izrecheniia sviatykh ikonov i povesti iz zhizni ikh sobrannyia ep ignatiem : brianchaninovym – 1903 – 6mf – 9 – mf#R-18,235 – ne IDC [243]

Izbrannye raboty i stati / Doiarenko, A G – 1925-1926 – 6mf – 9 – mf#COR-195 – ne IDC [335]

Izbrannye razskazy = Short stories / Chirikov, Evgenii Nikolaevich – [S-Peterburg]: Izd redaktsii zhurnala "Probuzhdenie", 1913 [mf ed 2002] – 1r – 1 – (Filmed with: sokrovishche zemli / teffi (1921)) – mf#5237 – us Wisconsin U Libr [830]

Izbrannye razskazy = Short stories / Gusev-Orenburgskii, Sergei Ivanovich – [S-Peterburg]: Izd red zhurnala "Probuzhdenie", 1913 [mf ed 2002] – 1r – 1 – (filmed with: sokrovishche zemli / teffi (1921)) – mf#5237 – us Wisconsin U Libr [830]

Izbrannye sochineniia / Aksakov, Sergei Timofeevich – Moskva, Russia. 1949 – 1r – us UF Libraries [025]

Izbrannyia zhitiia sviatykh – Vladimir, 1893 – 4mf – 9 – mf#R-18240 – ne IDC [243]

Izd gazety Ekonomicheskaia Zhizn' *see* Promyshlennaia rossiia, 1923-1924 g

Izd moskovskogo narodnogo banka *see* Vestnik kooperativnogo kredita

Izd moskovskogo soiuza-potrebitelnykh obshchestv *see* Obshchee delo

Izd org biuro tsentrosoiuza et al *see* Uralskii kooperator

Izd Permskogo Gubispolkoma *see* Statisticheskii sbornik na 1923 g

Izd. russkim entomologicheskim obshchestvom = Russkoe entomologicheskoe obozrenie – Spb., L., 1900-1963. v1-42 – 390mf – 9 – mf#2579 – ne IDC [077]

Izd soiuza soiuzov promyslovoi i proizvoditelno-trudovoi koop sev r-na "arteltrudsoiuza" – 1921-1924(12) – 13mf – 9 – mf#COR-547 – ne IDC [335]

Izdaetsia armejskim komitetom : biulleten' "vestnika 4-j armii" / Armiia chetvertaia – (Russia), 1917 – 1r – 1 – us UMI ProQuest [077]

Izdaetsia gruppoi studentov-sionistov / Kodimo – Paris, 1904 (1) – 1mf – 9 – mf#R-18065 – ne IDC [077]

Izdan glavnym komitetom vseobshchego soiuza iunosheskikh khristianskikh soedinenii – Geneva, 1915-1918. nos 1-30 – 1mf – 9 – mf#R-18020 – ne IDC [077]

Izdanie anarkhistov-kommunistov – Paris-GenFve, 1905. nos 1-3 – 1mf – 9 – mf#R-3485 – ne IDC [077]

Izdanie bratstva volnykh obshchinnikov – [London], 1913. no 1 – 1mf – 9 – mf#R-18028 – ne IDC [077]

Izdanie direktsii imperatorskikh teatrov – Spb., 1892-1915 – 458mf – 9 – mf#R-2324 – ne IDC [077]

Izdanie ezhenedelenoe – Spb., 1906. v1-5 – 4mf – 9 – mf#R-3985 – ne IDC [077]

Izdanie federatsii anarkhicheskikh krasnykh krestov evropy i ameriki – London, 1910-1914. nos 1-6 – 2mf – 9 – mf#R-18018 – ne IDC [077]

Izdanie glavnogo komiteta ukrainskogo soiuza : rossiiskaia sotsial-demokraticheskaia rabochaia partiia / Izvestiia Ukrainskogo soiuza – Geneva, 1909. no 1 – 1mf – 9 – mf#R-18059 – ne IDC [077]

Izdanie glavnogo tiuremnogo upravleniia / ed by Likhachev, A – Spb., 1893-1916 – 557mf – 9 – (missing: 1893(1); 1906(1); 1907(3-7)) – mf#R-10500 – ne IDC [077]

Izdanie glavnogo upravleniia generalnogo shtaba : osnovana komitetom druzei russkogo soldata – Paris, 1916-1917. nos 4-60 – 2mf – 9 – mf#R-18026 – ne IDC [077]

Izdanie gruppy "chernoe znamia" – [Geneva], 1905 – 1mf – 9 – mf#R-18189 – ne IDC [077]

Izdanie gruppy russkikh emigrantov v londone – London, 1905. nos 1-6 – 1mf – 9 – mf#R-18120 – ne IDC [077]

Izdanie gruppy sotsialistov-revoliutsionerov / ed by Avksentev, N et al – Paris, 1912. no 1 – 1mf – 9 – mf#R-18139 – ne IDC [077]

Izdanie gruppy sotsialistov-revoliutsionerov / ed by Delevskii, I L & Agafonov, V K – London, Paris, 1908-1909. nos 1-6 – 2mf – 9 – mf#R-18160 – ne IDC [077]

Izdanie gruppy sotsialistov-revoliutsionerov – Paris, 1916-1917. nos 1-14/15 – 2mf – 9 – (missing: 1917. nos 14/15) – mf#R-18132 – ne IDC [077]

Izdanie gruppy sotsialistov-revoliutsionerov – Paris, 1905. nos 1-4 – 1mf – 9 – (missing: no 4) – mf#R-18030 – ne IDC [077]

Izdanie gruppy "vpered" : (rossiiskaia sotsial-demokraticheskaia partiia) – Paris, 1912-1914. nos 1-4 – 2mf – 9 – mf#R-18099 – ne IDC [077]

Izdanie ideinoi gruppy "vpered" / Vpered – Geneva, 1915-1917. nos 1-6 – 2mf – 9 – mf#R-3453 – ne IDC [077]

Izdanie Imperatorskago obshchestva vostokoviedieniia *see* Mir islama

Izdanie imperatorskogo obshchestva vostokovedeniia – Spb., 1912-1913. v1-2 – 30mf – 9 – mf#R-5824 – ne IDC [077]

Izdanie kruzhka sotsialistov-revoliutsionerov – Geneva, Zuerich. n1-48. 1902-03 – 6mf – 9 – mf#R-18055 – ne IDC [077]

Izdanie neperiodicheskoe – Paris, 1915. nos 1-4 – 3mf – 9 – (missing: 1915 nos 1, 3(p 1-2)) – mf#R-18048 – ne IDC [077]

Izdanie neperiodicheskoe soiuza borby za osvobozhdenie rabochego klassa – Spb., Geneva, 1897. nos 1-2 – 1mf – 9 – mf#R-18165 – ne IDC [077]

Izdanie novgorodskogo statisticheskogo komiteta / Novgorodskii sbornik; ed by Bogoslovskii, N – Novgorod, 1865-1866. 5v – 25mf – 9 – mf#RET-9 – ne IDC [314]

Izdanie obshchestva dlia sodeistviia artelnomu delu v rossii *see* Artelnoe delo

Izdanie olonetskogo gubernskogo statisticheskogo komiteta / Olonetski sbornik; ed by Ivanov, A & Blagoveshchenskii – Petrozavodsk, 1875-[1902]. 4pts – 20mf – 9 – mf#RET-10 – ne IDC [314]

Izdanie organizatsii kheirus – Berlin, 1902 – 1mf – 9 – mf#R-18075 – ne IDC [077]

Izdanie organizatsionnogo biuro pri tsentralnom komitete partii sotsialistov-revoliutsionerov – Geneva, 1908 – 1mf – 9 – mf#R-18056 – ne IDC [077]

Izdanie otdela izobrazitel'nykh iskusstv komissariata narodnogo prosveshcheniia / Iskusstvo kommuny – St Petersburg, Russia, 1918-19 – 1r – 1 – us UMI ProQuest [077]

Izdanie otdeleniia iazyka i slovesnosti imperatorskoi akademii nauk / Issledovaniia po russkomu iazyku – Spb., 1885-1895. v1 – 19mf – 9 – mf#2272 – ne IDC [077]

Izdanie partii sotsialistov-revoliutsionerov – Paris, 1911-1912. nos 1-2 – 1mf – 9 – mf#R-18152 – ne IDC [077]

Izdanie partii sotsialistov-revoliutsionerov / ed by Platonova, S – Paris, 1915. nos 1-6 – 1mf – 9 – mf#R-18116 – ne IDC [077]

Izdanie Rossiiskoi akademii nauk *see* Musul'manskii mir

Izdanie russkikh evangelskikh khristian – Stockholm, 1894-1896. v4-6(2) – 8mf – 9 – mf#R-1671 – ne IDC [077]

Izdanie russkogo bibliologicheskogo obshchestva : literaturnyi vestnik – Cincinnati. 1919-1993 (1) 1971-1993 (5) 1971-1993 (9) – 74mf – 9 – mf#1774 – ne IDC [077]

Izdanie severnogo kruzhka liubitelei iziashchnykh iskusstv / Vremennik – Vologda, 1916. v1 – 3mf – 9 – mf#R-4870 – ne IDC [077]

Izdanie smolenskoi uchenoi arkhivnoi komissii – Menlo Park. 1948-1963 (1) – 29mf – 9 – (missing: 1916. v3(1)) – mf#1493 – ne IDC [077]

Izdanie soiuza pischebumazhnykh fabrikantov v rossii – Spb., 1904-1917 – 241mf – 9 – (missing: 1904(2, 6, 11); 1905(8, 10-12); 1906(1); 1915(9)) – mf#R-2355 – ne IDC [077]

Izdanie soiuza russkikh sotsial-demokratov : listok rabotnika – Geneva, 1896-1899. v1-10 – 5mf – 9 – mf#R-3364 – ne IDC [077]

Izdanie soiuza sotsialistov-revoliutsionerov – [Geneva], 1901-04. nos 1-4 – 1mf – 9 – mf#R-18076 – ne IDC [077]

Izdanie sotsial-demokraticheskoi organizatsii "zhizn" / Listki Zhizni – London, Geneva, 1902. nos 1-12 – 3mf – 9 – mf#R-18078 – ne IDC [077]

Izdanie tsentralnogo komiteta partii sotsialistov-revoliutsionerov – Paris, 1906-1907. nos 1-7 – 2mf – 9 – mf#R-18175 – ne IDC [077]

Izdanie Verkhovnogo Soveta i Soveta Ministrov Buriatskoi SSR *see* Buriatiia

Izdanie zagranichnogo komiteta – Geneva, 1904. v1-5 – 3mf – 9 – mf#R-3462 – ne IDC [077]

Izdanie zagranichnogo komiteta bunda : rossiiskaia sotsial-demokraticheskaia rabochaia partiia... – Geneva, 1909-1911. v1-2. nos 1-5 – 3mf – 9 – mf#R-18131 – ne IDC [077]

Izdanie zagranichnogo komiteta vseobshchego evreiskogo rabochego soiuza v litve, polshe i rossii – London, Geneva, 1901-1906. nos 1-256 – 16mf – 9 – mf#R-18138 – ne IDC [077]

Izdanie zagranichnogo sekretariata organizatsionnogo komiteta rsdrp – Znrich, 1915. no 1 – 2mf – 9 – mf#R-18061 – ne IDC [077]

Izdanie Zapadnago Tsentral'nago Komiteta Samooborony Poale Tsion *see* Odesskii pogrom i samooborona

Izdanie zhenevskoi gruppy sotsialistov-revoliutsionerov – Geneva, 1916-1917. nos 1-16 – 3mf – 9 – mf#R-18101 – ne IDC [077]

Izdaniia : obshchestvo liubitelei drevnei pismennosti – London. 1922-1926 (1) 1922-1926 (5) 1922-1926 (9) – 597mf – 9 – (missing: v45; [1884], v50; v110(1, 2, p 19-end)) – mf#1030 – ne IDC [077]

Izdaniia tsentrosoiuza za 25 let (1898-1923) : sistematicheskii ukazatel / Merkulov, A V – 1924 – 144p 2mf – 9 – mf#COR-536 – ne IDC [335]

Izdaniia zemstv 34-kh gubernii po obshchei ekonomicheskoi i otsenochnoi statistike, vyshedshie za vremia s 1864 g po 1 ianvaria 1911 g – Spb, 1911 – 5mf – 8 – mf#RZ-147 – ne IDC [314]

Izdatel'stvo pechat' *see* Poslednie izvestiia

Izdavaemye istoricheskim obshchestvom pri imperatorskom moskovskom universitete – M., 1916(1-4); 1917(1-2) – 21mf – 9 – mf#R-4124 – ne IDC [077]

Izdubar-nimrod : eine altbabylonische heldensage / Jeremias, Alfred – Leipzig: BG Teubner, 1891 – 1mf – 9 – 0-8370-7071-6 – (incl bibl ref) – mf#1986-1071 – us ATLA [930]

Izett, James *see* Maori lore

Izgadda = Persian american courier – New York City: J E Werda, sep 12, 1917-jun 30, 1920 – us CRL [071]

Izgoev, A S *see*
– Obshchinnoe pravo
– Russkoe obshchestvo i revoliutsiia
– Zamechaniia k proektu obshchego kooperativnogo zakona

Izgrev – Sofia, Bulgaria. Oct 29 1944-July 1951 – 8r – 1 – (cont by: vecherni novini) – us L of C Photodup [079]

Izgrev *see* Izgruv

Izgruv – Sofia, Bulgaria. 28 oct, 25 nov 1944-25 jan 1948; 3-12 jul 1949 – 1 – (in cyrillic) – mf#mf.685.g – uk British Libr Newspaper [077]

Izhevskij sovet rabochikh, soldatskikh i krest'ianskikh deputatov Izvestiia izhevskogo soveta rabochikh, soldatskikh i krest'ianskikh deputatov

Izhevsko-votkinskaia godovshchina : [odnodn gaz, posviashch godovshchine izhev -votkin vosstaniia] / ed by Milenko, G L – Omsk [Akmol obl]: Osvedom otd Shtaba Verkhovnogo Glavnokomanduiushchego 1919 – (= ser Asn 1-3) – 1v item 170, on reel n37 – 1 – mf#asn-1 170 – ne IDC [077]

Iziashchnaia literatura – Geneva. 1947-1986 (1) 1971-1986 (5) 1976-1986 (9) – 202mf – 9 – (missing: 1885(8-12)) – mf#1741 – ne IDC [077]

Izifundo nevangeli ezecawe nentsuku ezingcwele / Schweiger, Albert – Mariannhill, South Africa. 1920 – 1r – us UF Libraries [960]

Izihlabelo zogudumisa umlimu – Lobatsi, Botswana. 1959 – 1r – us UF Libraries [960]

Izihlabelo zogudumisa umlimu / Whiteside, John – Cape Colony, South Africa. 1929 – 1r – us UF Libraries [960]

Izinakmb' eafrika – London, England. 1949 – 1r – us UF Libraries [960]

Izindaba zasencwadini engcwele – Umtata, South Africa. 1953 – 1r – us UF Libraries [960]

Izinyanga zokubula : or, divination, as existing among the amazulu – Springvale, Natal: JA Blair 1870 [i.e. 1885] [mf ed 1992] – (= ser Publications of the folk-lore society 15) – 2mf – 9 – 0-524-04508-9 – (english trans fr zulu & notes by henry callaway) – mf#1990-3342 – us ATLA [290]

Iziumov, A *see* Sel'sko-khoziaistvennyi kredit

Izler – Giresun, 1925-27. Sahibleri: Ak Engin, Nuri Ahmed, Cemil Hueseyin; Mueduerue: Nuri Ahmed. n1. 21 subat 1924 – (= ser O & t journals) – 1mf – 9 – $25.00 – us MEDOC [956]

Izlozhenie postanovlenii o tsenzure i pechati – Spb., 1865 – 3mf – 9 – mf#R-9243 – ne IDC [077]

Izmailov, N P *see*
– Tsar'-kolokol
– V moskvu!

Izmenenie polozheniia zhenshchiny i demograficheskoe razvitie semi : po materialam sotsialno-demograficheskikh obsledovanii / Volkov, A – Moskva: In-t konkretnykh sotsialnykh issledovanii AN SSSR, 1972 – (filmed with: dinamika izmeneniia polozheniia dagestanskoi zhenshchiny i semia/s gadzhieva) – us CRL [947]

Izmeneniia professionalno-kvalifikatsionnoi struktury zhenskogo truda i semia / Sonin, M – Moskva: In-t konkretnykh sotsialnykh issledovanii AN SSSR, 1981 – (filmed with: dinamika izmeneniia polozheniia dagestanskoi zhenshchiny i semia/s gadzhieva) – us CRL [947]

Izmeneniia struktury sotsialnykh rolei zhenshchin v razvitom sotsialisticheskom obshchestve i model semi / Iankova, Z – Moskva: In-t konkretnykh sotsialnykh issledovanii AN SSSR 1972 – (filmed with: dinamika izmeneniia polozheniia dagestanskoi zhenshchiny i semia/s gadzhieva) – us CRL [947]

Izmir dokuz eyluel sergisi – Izmir. Sahib-i Imtiyaz ve Mueduer-i Mes'ul: Mecdi Sadreddin. n4-7(8),9-16,18(17),18(22). 7-25 eyluel 1927 – (= ser O & t journals) – 3mf – 9 – $55.00 – us MEDOC [956]

Iznaga, Alcides *see*
– Felipe y su piel
– Patria imperecedera
– Roca y la espuma
– Valedontes

Iznaga, J M *see* Por cuba

Izoblichenie shtundistskoi bogoprotivnoi eresi na osnovanii svyashchennago pisaniya, svyashchennago predaniya i istoricheskikh" pamyatnikov" = An exposure of the impious stundist heresy on the basis of the sacred scripture, holy tradition and historic relics / Opoichenko, Iakov – Nikolaev, 1891 – 1r – 1 – $9.24 – us Southern Baptist [242]

Izotopy / Selinov, I P – Moskva: Nauka. v3. 1970 – us CRL [077]

La izquierda liberal – Badajoz, 1922. 1 numero – 5 – sp Bibl Santa Ana [073]

Izquierda republicana – Plasencia y despues Caceres, 1832 – 5 – sp Bibl Santa Ana [073]

Izquierdo, Adolfo *see* El padre la calle (primer centenario de su muerte)

Izquierdo Hernandez, Manuel *see* Godoy

Izquierdo, S *see* Praxis exercitiorum spiritualium pns ignatii

Izquierdo, Sebastian *see* Practica de los exercicios espirituales de nuestro padre san ignacio

Izraelita : organ poswiecony sprawom religii i oswiaty – Warsaw PL, 1875-1903 – 2r – 1 – us UMI ProQuest [939]

Izraelita – Warsaw. 1-25, 27-43, 45-48. 1866-90, 1892-1908, 1910-13 – 1 – us NY Public [939]

Izseljenec – Sao Paolo, Brazil, 1930* – 1r – 1 – (slovenian newspaper) – us IHRC [079]

Izuchenie raiona i postroenie godovogo operativno-khoziaistvennogo plana selskokhoziaistvennogo kreditnogo tovarishchestva / Natsentov, D I & Vasys, I M – 1929 – 112p 2mf – 9 – mf#COR-397 – ne IDC [335]

Izveshcheniia parizhskoi gruppy sotsialistov-revoliutsionerov – Paris, 1909 – 1mf – 9 – mf#R-18060 – ne IDC [077]

Izvestiia – 1999- – 3r per y – 1 – (1995-98 4r $200. backfile 1917-94 $85r. no of reels varies from yr to yr) – us UMI ProQuest [070]

Izvestiia – Moscow: Izvestiia, 1917- – 1 – us East View [321]

Izvestiia – Moskva: Izd-vo "Izvestiia", [1991-]. jan-dec 1992 – us CRL [077]

Izvestiia – Nos. 7-129. 1867-1917. Incomplete – 1 – 92.00 – us L of C Photodup [306]

Izvestiia / Orlovskaia gub ispolnitel'nyj komitet sovetov – Orlovskaya gub, Russia, 1917-19 – 10r – 1 – us UMI ProQuest [077]

Izvestiia / Kustanajskij obshchestvennyj komitet – Kustanay, Kazakhstan, 1917 – 1r – 1 – us UMI ProQuest [077]

Izvestiia / Zvenigorodskij uezdnyj ispolnitel'nyj komitet sovetov – Zvenigorod, Russia, 1917 – 1r – 1 – us UMI ProQuest [077]

Izvestiia 2-go armejskogo s"ezda 8-oj armii / Armiia 8-aia – 1917 – 1r – 1 – us UMI ProQuest [077]

Izvestiia akademii nauk sssr : otdelenie tekhnicheskikh nauk = Bulletin de l'academie des sciences de l'urss. classe des sciences techniques – Moskva: Izd-vo Akademii nauk SSSR, [-1958]. 1950-52 – us CRL [500]

Izvestiia akademii nauk sssr : seriia biologicheskaia = Bulletin de l'academie des sciences de l'ussr. serie biologique – Moskva: Izd-vo Akademii nauk SSSR, 1939-92. n2. 1950-52 – us CRL [574]

Izvestiia akademii nauk sssr : seriia fizicheskaia = Bulletin of the academy of sciences of the ussr. physical series, 1954-91 – Moskva: Izd-vo Akademii nauk SSSR, [-1992]. v14-16 n2. 1950-52 – us CRL [500]

Izvestiia akademii nauk sssr : seriia geofizicheskaia – Moskva: Izd-vo Akademii nauk SSSR, 1951-64. n2-6. 1951; n1-3 1952 – 4r – 1 – us CRL [900]

Izvestiia akademii nauk sssr : seriia geograficheskaia – Moskva: Izd-vo Akademii nauk SSSR. n2-3. 1952 – us CRL [900]

Izvestiia akademii nauk sssr : seriia geograficheskaia i geofizicheskaia – Moskva: Izd-vo Akademii nauk SSSR, [-1951]. v14 1950; v15 n1 jan/feb 1951 – 2r – 1 – us CRL [900]

Izvestiia akademii nauk sssr. otdelenie literatury i iazyka – Moskva, Izdatel Stvo: Akademii nauk SSSR, 1941-1961. 6 no. a year – 1 – us Wisconsin U Libr [460]

Izvestiia altaiskogo tsentral'nogo kreditnogo soiuza – Barnaul [Alt gub]: [s n] 1918 [1917 avg-] – (= ser Asn 1-3) – n24,26 [1918] item 171, on reel n37 – 1 – mf#asn-1 171 – ne IDC [077]

Izvestiia ariejskogo ispolnitel'nogo komiteta 5-j armii / Armiia piataia – (Russia), 1917 – 1r – 1 – us UMI ProQuest [077]

Izvestiia arkhangel'skogo gubernskogo ispolnitel'nogo komiteta sovetov rabochikh i krest'ianskikh deputatov / Arkhangel'skaia gub ispolnitel'nyj komitet sovetov – Vologda, Russia) 1917-19 – 3r – 1 – us UMI ProQuest [077]

Izvestiia arkhangel'skogo soveta rab i sold deputatov / Arkhangel'skij sovet rabochikh i krest'ianskikh deputatov – Arkhangelsk, Russia, 1917 – 1r – 1 – us UMI ProQuest [077]

Izvestiia armejskogo komiteta 7-j armii / Armiia 7-aia – (Russia), 1917 – 1r – 1 – us UMI ProQuest [077]

Izvestiia armejskogo komiteta 9-oj armii / Armiia 9-aia – 1917 – 1r – 1 – us UMI ProQuest [077]

Izvestiia bel'skogo soveta rabochikh, krest'ianskikh i armejskikh deputatov / Bel'skij sovet r ki arm deputatov – Bely, Russia, 1918 – 1r – 1 – us UMI ProQuest [077]

Izvestiia berdianskogo soveta rabochikh, soldatskikh i krest'ianskikh deputatov / Berdiansk Sovet rk i kd – Berdyansk, Ukraine, 1917-18 – 2r – 1 – us UMI ProQuest [077]

Izvestiia bezhetskogo soveta krest'ian, rabochikh i krasno-armejskikh deputatov. tversk-gub / Bezhetsk Sovet rk i kd – Bezhetsk, Russia, 1918 – 1r – 1 – us UMI ProQuest [077]

Izvestiia borisoglebskogo soveta rabochikh, soldatskikh i krest'ianskikh deputatov / Borisoglebsk Sovet rk i kd – Borisoglebsk, Russia, 1918 – 1r – 1 – us UMI ProQuest [077]

Izvestiia cheliabinskogo obshchestva potrebitelei rabochikh i sluzhashchikh – Cheliabinsk, 1909-1915(48) – 20mf – 9 – (missing:1913(32-34)) – mf#COR-596 – ne IDC [335]

Izvestiia cheliabinskogo soveta krest'ianskikh, rabochikh i soldatskikh deputatov – Chelyabinsk, Russia, 1918 – 1r – 1 – us UMI ProQuest [077]

Izvestiia dal'ne-vostochnogo kraevogo komiteta sovetov rs i kr. deputatov i khabarovskogo soveta r i s deputatov / Dal'ne-Vostochnyj kraj sovet rk i kd – Khabarovsk, Russia, 1917 – 1r – 1 – us UMI ProQuest [077]

Izvestiia ekaterinoslavskogo soveta rabochikh i soldatskikh deputatov / Ekaterinoslav. Sovet rk i kd – Dnepropetrovsk, Ukraine, 1917-18 – 3r – 1 – us UMI ProQuest [077]

Izvestiia eletskogo soveta rabochikh, soldatskikh i krest'ianskikh deputatov / Elets. sovet rk i kd – Elets, Ukraine, 1918 – 1r – 1 – us UMI ProQuest [077]

Izvestiia enakievskogo soveta rabochikh i soldatskikh deputatov / Enakievskij sovet rk i kd – Enakievo, Ukraine, 1918 – 1r – 1 – us UMI ProQuest [077]

Izvestiia georgievskogo soveta rabochikh, soldatskikh i krest'ianskikh deputatov / Georgievskij gor sovet rk i kd – Georgievsk, Russia, 1918 – 1r – 1 – us UMI ProQuest [077]

Izvestiia glavnogo komiteta vserossiiskogo zemskogo soiuza pomoshchi bolnym i ranenym voinam – M., 1914-1917. nos 1-60 – 86mf – 9 – (missing: 1915 (6/7, p 49-64; 12/13, p 47-48); 1916(37-39)) – mf#R-18346 – ne IDC [077]

Izvestiia glavnoi astronomicheskoi observatorii / Glavnaia astronomicheskaia observatoriia [Soviet Union] – St Petersburg: K Akademie der Wissenschaften 1907- (irreg) [mf ed 2000] – v1 n1-t19 n153(1905-55) on 7r – 1 – (v1-6 [1905-12] iss in 12pt; t7 [1916-17] iss in 11pt; t8 [1918-22] iss in 5pt; t9-19 [1923-55] iss in 6pt; some iss publ out of chronological sequence; v1-3 in german only; v4-6 in russian & german; v6 in russian, german & french; v11-17 in russian & french; v18- in russian only; imprint varies) – ISSN: 0367-7966 – mf#film mas c4666 – us Harvard [520]

Izvestiia gosudarstvennogo instituta opytnoi agronomii / ed by Kuznetsov, N I – Pg, L, 1923-1929 – 35mf – 9 – mf#RHS-3 – ne IDC [314]

Izvestiia gosudarstvennogo kontrolia *see* Kratkoe rukovodstvo

Izvestiia gubernskogo kaluzhskogo ispolnitel'nogo komiteta sovetov rabochikh, krest'ianskikh i krasnoarmejskikh deputatov / Kaluzhskij gub ispolnitel'nyj komitet sovetov – Kaluga, Russia, 1918 – 1r – 1 – us UMI ProQuest [077]

Izvestiia gzhatskogo soveta rabochikh, krest'ianskikh i krasnoarmejskikh deputatov / Gzhatsk. sovet rk i kd – Gagarin, Russia, 1918 – 1r – 1 – us UMI ProQuest [077]

Izvestiia iakutskogo otdela imperatorskogo russkogo geograficheskogo obshchestva – Iakutsk, 1915. v1 – 3mf – 9 – mf#R-3224 – ne IDC [077]

Izvestiia ialtinskogo soveta rabochikh i soldatskikh deputatov – Yalta, Ukraine, 1918 – 1r – 1 – us UMI ProQuest [077]

Izvestiia imperatorskogo arkheologicheskogo obshchestva – Chicago. 1883-1908 (1) – 102mf – 9 – (cont as: izvestiia imperatorskogo russkogo arkheologicheskogo obshchestva. spb., 1872-1884. v7-10) – mf#1745 – ne IDC [077]

Izvestiia imperatorskogo kazanskogo universiteta – New York. 1963+ (1) – 705mf – 9 – mf#1746 – ne IDC [077]

Izvestiia imperatorskogo nikolaevskogo universiteta – Newark. 1948+ (1) 1968+ (5) 1975+ (9) – 108mf – 9 – mf#1750 – ne IDC [077]

Izvestiia imperatorskogo russkogo geograficheskogo obshchestva – Montgomery. 1820-1886 (1) – 1023mf – 9 – mf#1743 – ne IDC [077]

Izvestiia imperatorskogo tomskogo universiteta – Galveston. 1846-1886 (1) – 1018mf – 9 – (cont as: izvestiia tomskogo universiteta [tomsk 1918-23] v67-72; izvestiia tomskogo gosudarstvennogo universiteta [tomsk 1924-29] v73-84; missing: v83) – mf#1747 – ne IDC [077]

Izvestiia imperatorskoi akademii nauk – Madison. 1951-1996 (1) 1971-1996 (5) 1975-1996 (9) – 504mf – 9 – mf#1744 – ne IDC [077]

Izvestiia imperatorskoi akademii nauk – St. Paul. 1911+ (1) 1970+ (5) 1976+ (9) – 424mf – 9 – mf#1431 – ne IDC [077]

Izvestiia imperatorskoi akademii nauk po otdeleniiu russkogo iazyka i slovesnosti – Spb., 1852-1861. v1-10 – 91mf – 9 – mf#186 – ne IDC [077]

Izvestiia imperatorskoi arkheologicheskoi komissii – London. 1979-1980 (1,5,9) – 457mf – 9 – (cont as: izvestiia arkheologicheskoi komissii [pg 1917-18] v63-65; izvestiia gosudarstvennoi rossiiskoi arkheologicheskoi komissii [pg 1918] v66) – mf#1430 – ne IDC [077]

Izvestiia irbitskogo soveta rabochikh, soldatskikh i krest'ianskikh deputatov / Irbitskij sovet rk i kd – Irbit, Russia, 1918 – 1r – 1 – us UMI ProQuest [077]

Izvestiia ispoln komitetov vladimirskogo gubernskogo i uezdnogo sovetov rabochikh, krasnoarmejskikh i krest'ianskikh deputatov / Vladimirskij gub ispolnitel'nyj komitet sovetov – Vladimir, Russia, 1918 – 3r – 1 – us UMI ProQuest [077]

Izvestiia ispolnitel'nogo komiteta / Grenaderskij korpus ispolnitel'nyj komitet – 1917 – 1r – 1 – us UMI ProQuest [077]

Izvestiia istoriko-filologicheskogo obshchestva pri institute kniazia bezborodko v nezhine – Nezhin, 1877-1918. 32v – 75mf – 9 – (missing: 1878-1897, v2-15; 1899, v17; 1914-16, v29-31) – mf#R-14706 – ne IDC [077]

Izvestiia iuga : ezhednevnaia politicheskaia gazeta khar'kovskogo soveta rabochikh i soldatskikh deputatov i oblastnogo komiteta donetskogo i krivorozhskogo bassejnov – Khar'kov, Ukraine, 1917-18 – 2r – 1 – us UMI ProQuest [077]

Izvestiia izhevskogo soveta rabochikh, soldatskikh i krest'ianskikh deputatov / Izhevskij sovet rabochikh, soldatskikh i krest'ianskikh deputatov – Izhevsk, Russia, 1917-18 – 2r – 1 – us UMI ProQuest [077]

Izvestiia kamyshlovskogo uezdnogo komiteta rossijskoj kommunisticheskoj partii (b) / VKP(b) kamyshlovskij uezd komitet – Kamyshlov, Russia, 1918 – 1r – 1 – us UMI ProQuest [077]

Izvestiia kashinskogo soveta rabochikh, krest'ianskikh i krasnoarmejskikh deputatov / Kashin. sovet rk i kd – Kashin, Russia, 1918 – 1r – 1 – us UMI ProQuest [077]

Izvestiia kavkazskogo otdela imperatorskogo russkogo geograficheskogo obshchestva – Nokomis. 1964+ (1) 1991+ (5) 1991+ (9) – 89mf – 9 – (missing: 1876, v1-4(2+); 1876, v4(4); 1877, v4(6); 1879, v6(2-6); 1880-1883, v7(1-3); 1888-1902, v10-15(1); 1902-1905, v15(3)-18(3); 1905-1906, v18(5-end); 1907-1909, v19(2)-20(1); 1910, v20(3-end); 1912, v21(3-end); 1914, v22(2); 1914, v22(5-end)) – mf#1749 – ne IDC [077]

Izvestiia kavkazskogo otdeleniia imperatorskogo moskovskogo arkheologicheskogo obshchestva – Tiflis, 1904-1915. v1-4 – 12mf – 9 – (missing: 1915, v4) – mf#R-3226 – ne IDC [077]

Izvestiia khar'kovskogo soveta i gubernskogo ispolnitel'nogo komiteta...deputatov – Khar'kov, Ukraine, 1917 – 2r – 1 – us UMI ProQuest [077]

Izvestiia kievskogo obl soiuza tekhnikov,zemlemerov, chertezhnikov, desiatnikov ok shkolu, topografov i tp / Kievskij oblastnoj soiuz tekhnikov – Kiev, Ukraine, 1917 – 1r – 1 – us UMI ProQuest [077]

Izvestiia komiteta zashchity : izd stavrop uezd kom zashchity uchred sobr – Stavropol' [Samar gub]: [s n] 1918 [1918 1 iiulia-] – (= ser Asn 1-3) – n1-10 [1918] [gaps] item 174, on reel n37 – 1 – mf#asn-1 174 – ne IDC [077]

Izvestiia kronshtadskogo soveta rabochikh, matrosskikh i krasnoarmejskikh deputatov / Kronshtad. sovet rk i kd – Kaliningrad, Russia, 1917-20 – 6r – 1 – us UMI ProQuest [077]

Izvestiia krymskoi astrofizicheskoi observatorii / Krymskaia astrofizicheskaia observatoriia – Moskva: Izd-vo Akademii nauk SSSR 1947- [irreg] [mf ed 2003] – 1r – 1 – (some vols iss in pts; in russian; imprint varies; v56-84 have title: izvestiia ordena trudovogo krasnogo znameni krymskoi astrofizicheskoi observatorii) – ISSN: 0367-8466 – mf#film mas c5530 – us Harvard [520]

Izvestiia mariupol'skogo soveta rabochikh i soldatskikh deputatov – Maryupol, Ukraine, 1917 – 1r – 1 – us UMI ProQuest [077]

Izvestiia ministerstva inostrannykh del – Cleveland. 1956-1983 (1) 1971-1983 (5) 1975-1983 (9) – 144mf – 9 – mf#1402 – ne IDC [077]

Izvestiia minskogo obshchestva liubitelei estestvoznaniia, etnografii i arkheologii – Minsk, 1914(1) – 2mf – 9 – mf#R-3230 – ne IDC [077]

Izvestiia mogilevskogo gub ispolnitel'nogo kom sovetov krest'ianskikh, soldatskikh i rabochikh deputatov / Mogilevskij gubernskij ispolnitel'nyj komitet sovetov – Mogilev, Belarus, 1918 – 1r – 1 – us UMI ProQuest [077]

Izvestiia morshanskogo soveta rabochikh, soldatskikh i krest'ianskikh deputatov / Morshanskij sovet rabochikh, soldatskikh i krest'ianskikh deputatov – Morshansk, Russia, 1918 – 1r – 1 – us UMI ProQuest [077]

Izvestiia moskovskogo gub soveta krest'ianskikh deputatov / Krest'ianskij deputat – Moscow, Russia, 1917 – 1r – 1 – us UMI ProQuest [077]

Izvestiia moskovskogo kommercheskogo instituta – M., 1913-1916. v1-4 – 32mf – 9 – mf#R-3231 – ne IDC [077]

Izvestiia moskovskogo literaturno-khudozhestvennogo kruzhka – M., 1913-1917. v1-18 – 21mf – 9 – mf#R-3326 – ne IDC [077]

Izvestiia moskovskogo narodnogo banka – M., 1916-1918 (10) – 14mf – 9 – (missing: 1917(1-6, 10-12)) – mf#COR-594 – ne IDC [077]

Izvestiia moskovskogo soiuza zashchity uchreditel'nogo sobraniia / Mosk soiuza zashchity Uchred sobr – M [Moskva: s n] 1918 [1917 22 dek-1918 [ianv]] – (= ser Asn 1-3) – n2 [1918] item 15, on reel n1 – 1 – mf#asn-2 015 – ne IDC [077]

Izvestiia moskovskogo soveta rabochikh i krasnoarmejskikh deputatov / Moskva. sovet rk i kd – Moscow, Russia, 1919 – 1r – 1 – us UMI ProQuest [077]

Izvestiia murmanskogo kraevogo soveta rabochikh i krest'ianskikh deputatov – Murmansk [Arhang gub]: [s n] 1918 [[1917]-] – (= ser Asn 1-3) – n101 [1918] item 16, on reel n1 – 1 – mf#asn-2 016 – ne IDC [077]

Izvestiia narvsogo soveta rabochikh i soldatskikh deputatov / Narvskij sovet rabochikh i soldatskikh deputatov – Narva, Russia, 1917 – 1r – 1 – us UMI ProQuest [077]

Izvestiia nikitovskogo soveta rabochikh i soldatskikh deputatov / Nikitovskij sovet rabochikh i soldatskikh deputatov – Nikitovka, Ukraine, 1917 – 2r – 1 – us UMI ProQuest [077]

Izvestiia nikolaevskogo soveta rabochikh i voennykh deputatov / Nikolaevskij sovet rabochikh i voennykh deputatov – Nikolaev, Ukraine, 1917 – 2r – 1 – us UMI ProQuest [077]

Izvestiia nizhegorodskikh sovetov rabochikh i soldatskikh deputatov / Nizhegorodskij sovet rabochikh i soldatskikh deputatov – Nizhny-Novgorod, Russia, 1917 – 2r – 1 – us UMI ProQuest [077]

Izvestiia nizhne-tagil'skogo soveta rabochikh i soldatskikh deputatov / Nizhne-tagil'skij sovet rabochikh i soldatskikh deputatov – Nizhny-Tagil, Russia, 1918 – 1r – 1 – us UMI ProQuest [077]

Izvestiia novonikolaevskogo soveta rabochikh i soldatskikh deputatov / Novonikolaevskij sovet rabochikh i solatskikh deputatov – Novosibirsk, Russia, 1917 – 1r – 1 – us UMI ProQuest [077]

Izvestiia o zaniatiiakh 4-ogo arkheologicheskogo sezda v kazani – Chicago. 1957-1992 (1) 1970-1992 (5) 1976-1992 (9) – 5mf – 9 – (missing: 1877(8)) – mf#1751 – ne IDC [077]

Izvestiia oblastnogo komiteta zagranichnoi organizatsii russkikh sotsial-revoliutsionerov – Paris, 1908-1911. nos 7-15 – 2mf – 9 – mf#R-18058 – ne IDC [077]

Izvestiia obshchestva arkheologii, istorii i etnografii pri imperatorskom kazanskom universitete – Kazane, 1878-1925. v1-33(4) – 385mf – 9 – (missing: 1922, v32(3-4)) – mf#1432 – ne IDC [077]

Izvestiia obshchestva finansovykh reform – Spb., 1910-1915. v1-12 – 19mf – 9 – mf#R-3232 – ne IDC [077]

Izvestiia obshchestva revnitelei russkogo istoricheskogo prosveshcheniia v pamiate imperatora aleksandra 3 – Spb., 1900-1904. v1-5 – 11mf – 9 – mf#R-3234 – ne IDC [077]

Izvestiia obshchestva slavianskoi kultury – M., 1912-1913 – 11mf – 9 – mf#R-3235 – ne IDC [077]

Izvestiia obshchezemskoi organizatsii – M., 1905-1906. nos 1-10 – 13mf – 9 – mf#R-4125 – ne IDC [077]

Izvestiia odesskogo bibliograficheskogo obshchestva pri imperatorskom novorossiiskom universitete – Odessa, 1912-1915. v1-4 – 24mf – 9 – (missing: 1912, v1(1-5, 9, 11-12); 1913, v2(2, 9-12); 1914, v3(7-12); 1915, v4(7-12)) – mf#R-4314 – ne IDC [077]

Izvestiia odesskogo gub ispolnitel'nogo komiteta soveta rabochikh, krest'ianskikh i krasnoarmejskikh deputatov i predstavitelej armii i flota / Odesskij gub. ispolnitel'nyj komitet soveta rabochikh, krest'ianskikh i krasnoarmejskikh deputatov – Odessa, Ukraine, 1917-20 – 4r – 1 – us UMI ProQuest [077]

Izvestiia ofitserov armii – 1917 – 1r – 1 – us UMI ProQuest [077]

Izvestiia omgubsoiuza – Omsk, 1923(1-19) – 13mf – 9 – (cont as:kooperativnaia niva omsk,1923(1-4)-1924(1-2).missing:1922,1923(8)) – mf#COR-615 – ne IDC [335]

Izvestiia orekhovo-zuevskogo soveta rabochikh deputatov / Orekhovo-zuevskij sovet rabochikh deputatov – Orekhovo-Zuevo, Russia, 1917 – 1r – 1 – us UMI ProQuest [077]

Izvestiia organizatsionnogo komiteta po sozyvu tret'ego vsesibirskogo sionistskogo sezda / ed by Evzerov, A et al – Tomsk: [s n] 1918 [1918 nenum vyp [15 noiab]] – (= ser Asn 1-3) – 1v item 175, on reel n37 – 1 – mf#asn-1 175 – ne IDC [077]

Izvestiia otdeleniia russkogo iazyka i slovesnosti imperatorskoi akademii nauk – College Park. 1949-1976 (1) 1971-1976 (5) 1976-1976 (9) – 692mf – 9 – mf#1107 – ne IDC [077]

Izvestiia pedagogicheskogo instituta imeni pavla grigorevicha shelaputina v g moskve – M., 1912-1916. v1-7 – 26mf – 9 – (missing: 1914, v4) – mf#R-4126 – ne IDC [077]

Izvestiia penzenskogo gubispolkoma i gorodskogo soveta rab i kr deputatov / Penznskaia gub ispolnitel'nyj komitet sovetov – Penza, Russia, 1917-19 – 9r – 1 – us UMI ProQuest [077]

Izvestiia peredvizhnogo biuro rosta / Krasnaia zvezda. literaturno-instruktorskij parokhod "krasnaia zvezda" – 1919 – 1r – 1 – us UMI ProQuest [077]

Izvestiia permskogo gub i permskogo uezdn ispolnitel'nykh komitetov sovetov rabochikh, krest'ianskikh i armejskikh deputatov / Permskaia gub ispolnitel'nyj komitet sovetov – Perm', Russia, 1918 – 1r – 1 – us UMI ProQuest [077]

Izvestiia petrovskoi zemledelcheskoi i lesnoi akademii – M., 1878-1889. v1-12 – 133mf – 9 – (missing: 1887. v10(1)) – mf#R-3237 – ne IDC [077]

Izvestiia po biiskomu i karakorum-altaiskomu uezdam : pril k gaz "altai" za no... – Biisk [Alt gub]: [s n] 1918 [1918-] – (= ser Asn 1-3) – n5-12 [1918] [gaps] item 176, on reel n37 – 1 – (suppl of: altai [asn-1 003]) – mf#asn-1 176 – ne IDC [077]

Izvestiia po narodnomu obrazovaniiu – Spb., 1904-1917. v1-14 – 443mf – 9 – (missing: 1904, v1(1-3); 1915, v12(25)) – mf#R-3238 – ne IDC [077]

Izvestiia polkovogo komiteta 685 pekh logishinskogo polka – 1917 – 1r – 1 – us UMI ProQuest [077]

Izvestiia pskovskogo gubsoiuza – Pskov, 1922-1923(18) – 18mf – 9 – (cont as:vestnik pskovskoi kooperatsii pskov,1923(19-24).missing:1922(1-2)) – mf#COR-565 – ne IDC [335]

Izvestiia rossiiskogo telegrafnogo agenstva – Bodaibo [Irkut gub]: vr Bod[aib] uezd zems uprava 1919 [1919 1 avg- – (= ser Asn 1-3) – n28-34 [1919] item 177, on reel n37 – 1 – mf#asn-1 177 – ne IDC [077]

Izvestiia rossiiskoi akademii – Champaign. 1964+ (1) 1970+ (5) 1977+ (9) – 38mf – 9 – mf#1752 – ne IDC [077]

Izvestiia rossijskogo telegrafnogo agenstva – Bodajbo, Russia, 1919 – 1r – 1 – us UMI ProQuest [077]

Izvestiia russkogo arkheologicheskogo instituta v konstantinopole – Odessa, Sofiia. v1-16. 1896-1912 – 90mf – 9 – (missing: 1899 v4; 1909 v14) – mf#R-17095 – ne IDC [930]

Izvestiia russkogo genealogicheskogo obshchestva – Spb. v1-4. 1900-1911 – 29mf – 9 – mf#R-3240 – ne IDC [929]

Izvestiia russkogo sobraniia – Spb., 1903, nos 1-3; 1904, nos 1-2 – 19mf – 9 – mf#R-4127 – ne IDC [077]

Izvestiia s peterburgskogo obshchestva muzykal'nykh sobranii – St Petersburg, 1896-1908 [irreg] – 22mf – 9 – (fr 1903 with a suppl: muzykal'naia bibliografiia) – us UMI ProQuest [780]

Izvestiia samarskogo obshchestva narodnykh universitetov – Samara, 1910(1-20) – 6mf – 9 – mf#R-4128 – ne IDC [077]

Izvestiia samarskogo obshchestva potrebitelei samopomoshch – Samara, 1916-1917 – 8mf – 9 – (cont as: samopomoshch. missing: 1918-19 (7); 1916 (2, 7-8); 1917 (3)-1918 (12); 1919(1-5)) – mf#COR-676 – ne IDC [077]

Izvestiia sankt-peterburgskogo lesnogo instituta – Spb., 1898-1903. v1-9 – 52mf – 9 – mf#R-3241 – ne IDC [077]

Izvestiia sankt-peterburgskogo politekhnicheskogo instituta – Spb., 1904-1916 – 93mf – 9 – (missing: 1904, v1-2, 4 (3-4)-1907, v6; 1916, v25) – mf#R-18342 – ne IDC [077]

Izvestiia semipalatinskogo soiuza kooperativov / ed by Kaplinskii, M A – Semipalatinsk: soiuz kooperativov 1918 [1918-] – (= ser Asn 1-3) – n14 [1918] item 178, on reel n37 – 1 – mf#asn-1 178 – ne IDC [077]

Izvestiia shtaba 11-oj armii : dlia doblestnykh zashchitnikov rodiny / Armiia 11-aia – 1917 – 1r – 1 – us UMI ProQuest [077]

Izvestiia shtaba otdel'noi vostochnoi-sibirskoi armii / Inform otd Shtaba Otdel'noi Vost-Sib armii – Chita [Zabaik obl]: [s n] 1919 [1919 10 apr-] – (= ser Asn 1-3) – n1-7 [1919] item 179, on reel n37 – 1 – mf#asn-1 179 – ne IDC [077]

Izvestiia shujskogo soveta – Shuya, Russia, 1918 – 3r – 1 – us UMI ProQuest [077]

Izvestiia soveta po delam strakhovaniia rabochikh – Spb., 1913-1916. v1-11 – 26mf – 9 – (missing: 1914, v3) – mf#R-9277 – ne IDC [077]

Izvestiia soveta rabochikh deputatov i predstavitelej armii i flota – Odessa, Ukraine, 1917-20 – 6r – 1 – us UMI ProQuest [077]

Izvestiia soveta rabochikh deputatov priiskovogo rajona / Bodajbo. Priiskovyj Sovet rabochikh deputatov – Bodajbo, Russia, 1917 – 1r – 1 – us UMI ProQuest [077]

Izvestiia soveta rabochikh i voennykh deputatov derbentskogo rajona / Derbentskij rajonnyj sovet rabochikh i voennykh deputatov – Derbent, Russia, 1918 – 2r – 1 – us UMI ProQuest [077]

Izvestiia soveta rabochikh, koest'ianskikh i krasnoarmejskikh deputatov g kurska i gubernii / Kurskij gor sovet rk i kd – Kursk, Russia, 1918 – 3r – 1 – us UMI ProQuest [077]

Izvestiia soveta soldatskikh deputatov elisavetpol'skogo garnizona / Elisavetpol'skij garnizonnyj sovet soldatskikh deputatov – Gyandzha, Azerbaijan, 1917 – 1r – 1 – us UMI ProQuest [077]

Izvestiia sovetov deputatov trudiashchikhsia sssr – Moskva: Prezidium Verkhovnogo Soveta SSSR, jul 1938-oct 7 1977 – 1 – us CRL [947]

Izvestiia sovetov narodnykh deputatov majkopskogo otdela / Majkopskij sovet narodnykh deputatov – Majkop, Russia, 1918 – 1r – 1 – us UMI ProQuest [077]

Izvestiia sovetov narodnykh deputatov sssr – Moskva: Prezidium Verkhovnogo Soveta SSSR, 1977-91. oct 8, 1977-jun 1991 – us CRL [077]

Izvestiia sovetov rabochikh, soldatskikh i krest'ianskikh deputatov gor moskvy i moskovskoj oblasti / Moskva. sovet rk i kd – Moscow, Russia, 1917-18 – 16r – 1 – us UMI ProQuest [077]

Izvestiia tambovskoi uchenoi arkhivnoi komissii – Washington. 1948-1994 (1) 1971-1994 (5) 1976-1994 (9) – 140mf – 9 – (missing: 1884, v1-2; 1885, v4) – mf#1489 – ne IDC [077]

Izvestiia tavricheskoi uchenoi arkhivnoi komissii – Quantico. 1916+ (1) 1971+ (5) 1975+ (9) – 66mf – 9 – (cont as: izvestiia tavricheskogo obshchestva istorii, arkheologii i etnografii. simferopol, 1927-1930[1931] 4v) – mf#1490 – ne IDC [077]

Izvestiia tiumenskogo gubernskogo i uezdnogo ispolnitel'nykh komitetov sovetov rabochikh,krest'ianskikh i krasnoarmejskikh deputatov – Tyumen', Russia, 1918 – 2r – 1 – us UMI ProQuest [077]

Izvestiia turgajskogo oblastnogo komissarmata – Orenburg, Russia, 1918 – 2r – 1 – us UMI ProQuest [077]

Izvestiia tverskogo gubsoiuza – Tver, 1920-1924(24) – 117mf – 9 – (cont as:ekho tverskoi kooperatsii tver,1925-1929(20); missing:1920(3-4,7-8),1921(14-19,21, 23-29,31,34),1922(6,13,17),1923(4)) – mf#COR-706 – ne IDC [335]

Izvestiia ufimskogo gubkoma rkp(b) i gubispolkoma soveta rabochikh, krest'ianskikh, krasnoarmejskikh deputatov / VKP(b). Ufimskij gub komitet – Ufa, Russia, 1918 – 1r – 1 – us UMI ProQuest [077]

Izvestiia Ukrainskogo soiuza *see* Izdanie glavnogo komiteta ukrainskogo soiuza

Izvestiia vladivostokskogo soveta rabochikh i soldatskikh deputatov – Vladivostok, Russia, 1917-18 – 4r – 1 – us UMI ProQuest [077]

Izvestiia vologodskogo obshchestva izucheniia severnogo kraia – Vologda, 1914-1916. v1-3 – 9mf – 9 – mf#R-3243 – ne IDC [077]

Izvestiia vologodskogo obshchestva selskogo khoziaistva – Vologda, 1910 – 74mf – 9 – (cont as: severnyi khoziain. vologda, 1911-1919 (11). missing: 1917(21, 23-24); 1918(1-10, 16, 19-21); 1919(1)) – mf#COR-680 – ne IDC [077]

Izvestiia voronezhskogo gubernskogo ispolnitel'nogo komiteta sovetov rabochikh i krest'ianskikh deputatov i gorodskogo soveta rabochikh i krasnoarmejskikh deputatov – Voronezh, Russia, 1917-19 – 5r – 1 – us UMI ProQuest [077]

Izvestiia vostochnogo instituta – Vladivostok, 1899-1916 – 687mf – 9 – (missing: 1912, v44) – mf#R-3419 – ne IDC [077]

Izvestiia vostochnogo otdeleniia imperatorskogo arkheologicheskogo obshchestva – Thorofare. 1828+ (1) 1965+ (5) 1970+ (9) – 4mf – 9 – mf#1754 – ne IDC [077]

Izvestiia vostochno-sibirskogo otdela imperatorskogo russkogo geograficheskogo obshchestva – Oak Brook. 1959-1971 (1) – 264mf – 9 – (missing: 1878-1879, v1-9; 1879, v10(3-4); 1896, v27(3-4)) – mf#1755 – ne IDC [077]

Izvestiia vremennogo komiteta samarskoi torgovo-promyshlennoi palaty / ed by Agrikov, P A – Samara: Prezidium Vrem kom Samar torgovo-prom palaty 1918 [1918 1 iiunia-[14 [1] avg]] – (= ser Asn 1-3) – nenum vyp [1] n2-5 [1918] item 172, on reel n37 – 1 – mf#asn-1 172 – ne IDC [077]

Izvestiia vremennogo revoliutsionnogo komiteta matrosov, krasnoarmeitsev i rabochikh gor kronshtadta – Kronshtadt [Petrogr gub]: [s n] 1921 [1921 [mart-mart]] – (= ser Asn 1-3) – n3-12 [1921] [gaps] item 14, on reel n1 – 1 – mf#asn-2 014 – ne IDC [077]

Izvestiia vremennogo tsentral'nogo biuro rossiiskikh musul'man – [Moscow]: Vremennoe tsentral'noe biuro Rossiiskikh musul'man [1917] n1 – 1v. item 2 on reel of 7 titles – 1 – (cont: musul'manskaia gazeta (st petersburg, russia: 1912)) – mf#mr-3 – ne IDC [077]

Izvestiia vremennogo tsentral'nogo biuro rossiiskikh musul'man – St Petersburg, 1917 – 1 – (reel contains short runs of multiple titles. for complete listing of titles on a reel, please inquire) – us UMI ProQuest [077]

Izvestiia vserossiiskogo musul'manskogo voennogo shuro – Kazan': Tipo-lit T-vo "Umid" 1917-18 n1-i (19 noiabria 1917 g)- – item 5 on reel of 7 titles – 1 – (Filmed: n1-15 (1917-18)) – mf#mr-8 – ne IDC [077]

Izvestiia vserossiiskogo natsionalenogo kluba – Spb., 1911. nos 1-3 – 6mf – 9 – mf#R-4129 – ne IDC [077]

Izvestiia vserossijskogo komiteta spaseniia rodiny i revoliutsii / Vserossijskij komitet spaseniia rodiny i revoliutsii – St Petersburg, Russia, 1917 – 1r – 1 – us UMI ProQuest [077]

Izvestiia vserossijskogo krest'ianskogo sovezta krest'ianskikh deputatov / Vserossijskij sovet krest'ianskikh deputatov – St Petersburg, Russia, 1917 – 1r – 1 – us UMI ProQuest [077]

Izvestiia vserossijskogo krest'ianskogo sovezta krest'ianskikh deputatov / Vserossijskij sovet krest'ianskikh deputatov – St Petersburg, Russia, 1917 – 1r – 1 – us UMI ProQuest [077]

Izvestiia vserossijskogo soveta krest'ianskikh deputatov – St Petersburg, Russia, 1917 – 2r – 1 – us UMI ProQuest [077]

Izvestiia v-ustiugskogo soveta rabochikh i soldatskikh deputatov / Veliko-ustiugskij sovet rabochikh i soldatskikh deputatov – Veliky Ustyug, Russia, 1917 – 1r – 1 – us UMI ProQuest [077]

Izvestiia vysochaishe uchrezhdennogo komiteta popechitelestva o russkoi ikonopisi – Spb., 1902-1903. v1-2 – 5mf – 9 – mf#R-3205 – ne IDC [077]

Izvestiia vysshikh uchebnykh zavedenii : stroitelstvo i arkhitektura – Novosibirsk: Novosibirskii inzhenerno-stroitelnyi in-t, 1958-81. n4-6 1978; n7-8 1979 – us CRL [720]

Izvestiia zabaikal'skoi oblastnoi zemskoi upravy – Chita [Zabaik obl]: [s n] 1918 [1918-] – (= ser Asn 1-3) – n11-43 [1918] [gaps] item 173, on reel n37 – 1 – mf#asn-1 173 – ne IDC [077]

Izvestiia zagranichnogo sekretariata organizatsionnogo komiteta rossiiskoi sotsial-demokraticheskoi rabochei partii – Geneva, 1915-1917. nos 1-10 – 3mf – 9 – mf#R-18057 – ne IDC [077]

Izvestiia zapadno-sibirskogo i omskogo ispolnitel'nykh komitetov sovetov krest'ianskikh... / Zapadno-sibirskaia obl ispolnitel'nyj komitet sovetov – Omsk, Russia, 1917 – 1r – 1 – us UMI ProQuest [077]

Izvestiia zapadno-sibirskogo i omskogo ispolnitel'nykh komitetov sovetov krest'ianskikh... / Zapadno-sibirskaia obl ispolnitel'nyj komitet sovetov – Omsk, Russia, 1917-18 – 3r – 1 – us UMI ProQuest [077]

Izvestiya – Leningrad, Moscow, U.S.S.R. -d. 1917-63. 1917-1932 imperfect. 113 reels – 1 – uk British Libr Newspaper [947]

Izvestiya mathematics – Providence. 1993-1995 (1,5,9) – (cont: mathematics of the ussr: izvestiya) – ISSN: 1064-5632 – mf#13416,01 – us UMI ProQuest [510]

Izvestiya narodnago komissariata po voennym delam – Moscow. May 1918-Sep 1921.-w. mqn reel – 1 – uk British Libr Newspaper [072]

Izvestiya sovetskoi meditsiny. (izvestiya narodnogo komissariata zdravookraneniya) – Moscow. 15 may-25 jul 1918; 25 aug 1918-dec 1921.-w. 28 ft – 1 – uk British Libr Newspaper [610]

Izviestiia po literaturie, naukami bibliografii / Tovarishchestvo MO Vol'f. Leningrad – 7v. 1897-1904 – 1 – us L of C Photodup [460]

Izwi labantu – East London SA, 4 jan 1901-23 dec 1902; 7 jan 1906-16 apr 1909 – 2r – 1 – (missing: 1906-09) – mf#MS00263 – sa National [079]

Izwi lama afrika – East London SA, 8 may 1931-27 feb 1932 – 1r – 1 – sa National [079]

Izwi lama swazi – Mbabane SA, 13 feb-3 jul 1934 – 1r – 1 – sa National [079]

Izwi lama swazi = The voice of the swazis – Mbabane, Swaziland: Swazi Press Co [1934]-[wkly] [mf ed Cape Town: South African Library c1985] – feb 13-jul 3 1934 on 1r – 1 – (in zulu & english) – mf#mf-14265 – sa State Libr [079]

'Izzet *see*
– The divan project

Izzi, Sueleyman *see* Tarih-i izzi

'J 3' : ou, La Nouvelle Ecole / Ferdinand, Roger – Paris, France. 1944 – 1r – 1 – us UF Libraries [440]

J A leisewitzens julius von tarent : erlaeuterung und literarhistorische wuerdigung / Kuehlhorn, Walther – Halle a.d. Saale: M Niemeyer, 1912 – 1r – 1 – (incl bibl ref) – us Wisconsin U Libr [430]

J b fraser's reise nach und in khorasan : in den jahren 1821 bis 1822; nebst nachrichten von den nordoestlich von persien gelegenen laendern... – Weimar 1828-1829 [mf ed Hildesheim 1995-98] – 2v on 8mf – 9 – €160.00 – ISBN-10: 3-487-26477-3 – ISBN-13: 978-3-487-26477-6 – gw Olms [915]

J b l durand's, vormaligen handelsdirektors am senegal, nachrichten von den senegal-laendern : in einem gedraengten auszuge, mit golberry's berichten verglichen und durch anmerkungen erlaeutert – Weimar 1803 [mf ed Hildesheim 1995-98] – 1v on 2mf – 9 – €60.00 – ISBN-10: 3-487-26597-4 – ISBN-13: 978-3-487-26597-1 – gw Olms [380]

J B Rolland & fils *see* Catalogue de la librairie de j b rolland et fils a montreal division du catalogue

J chr. k. v. hofmanns versoehnungslehre und der ueber sie gefuehrte streit : ein beitrag zur geschichte der neueren theologie / Bachmann, Philipp – Guetersloh: C Bertelsmann, 1910 – (= ser Beitraege zur foerderung christlicher theologie) – 1mf – 9 – 0-7905-9123-5 – mf#1989-2348 – us ATLA [240]

The j edgar hoover official and confidential file / ed by Theoharis, Athan – (= ser Federal bureau of investigation confidential files) – 17r – 1 – $3145.00 – 1-55655-164-9 – (with p/g) – us UPA [322]

J Eveleigh & Co *see* Illustrated price list

J f encke's astronomische abhandlungen : zusammengestellt aus den jahrgaengen 1830 bis 1862 des berliner astronomischen jahrbuches nebst drei in diesen jahrgaengen enthaltenen abhandlungen / Bessel, Friedrich Wilhelm et al – Berlin: F Duemmler 1866 [mf ed 1998] – 3v on 1r – 1 – mf#film mas 28249 – us Harvard [520]

J f j borsums reise nach constantinopel, palaestina und egypten : oder lebendiger beweis, wie gnaedig gott dem durchhilft, der seine hoffnung auf ihn setzt – Berlin 1825 [mf ed Hildesheim 1995-98] – 1v on 2mf – 9 – €60.00 – ISBN-10: 3-487-26669-5 – ISBN-13: 978-3-487-26669-5 – gw Olms [915]

J franklin little, senate service 1910-1912 : senate page – (= ser Us senate historical office oral history coll) – 1mf – 9 – $5.00 – us Scholarly Res [323]

J G herders humanitaetsidee als ausdruck seines weltbildes und seiner persoenlichkeit / Dobbek, Wilhelm – [Braunschweig]: G Westermann, 1949 – 1 – (incl bibl ref and index) – us Wisconsin U Libr [430]

J g kohl's reisen in deutschland / Kohl, Johann – Leipzig 1852 [mf ed Hildesheim 1995-98] – 2v on 6mf – 9 – €120.00 – 3-487-29600-4 – gw Olms [914]

J G schummel : leben und schaffen eines schriftstellers und reformpaedogogen: ein beitrag zur geschichte der paedagogischen literatur der aufklaerungszeit / Weigand, Georg – Frankfurt am Main: M Diesterweg, 1925 – 1 – (incl bibl ref) – us Wisconsin U Libr [430]

J gaudenz von salis-seewis / Frey, Adolf – Frauenfeld: J Huber, 1889 – 1r – 1 – (incl bibl ref and index) – us Wisconsin U Libr [920]

J griffiths, md, der koenigl gesellschaft zu edimburg und mehrerer auswaertigen gelehrten gesellschaften mitgliede, neue reise in arabien, die europaeische und asiatische tuerkey / Griffiths, John – Leipzig 1814 [mf ed Hildesheim 1995-98] – 2v on 2mf – 9 – €60.00 – 3-487-27673-9 – gw Olms [910]

J H jowett... : a character study / Morison, Frank – Boston: Pilgrim Press, 1911 – 1mf – 9 – 0-524-08577-3 – mf#1993-3162 – us ATLA [240]

J h tuckey's, esq ersten lieutenant's auf dem schiffe kalkutta, bericht von einer reise nach neu-sued-wallis : um zu port-philipp in der bass's strasse eine kolonie anzulegen, gethan in dem schiffe kalkutta in den jahren 1802, 1803 und 1804 / Tuckey, James – Weimar 1805 [mf ed Hildesheim 1995-98] – 1v on 1mf – 9 – €40.00 – ISBN-10: 3-487-26570-2 – ISBN-13: 978-3-487-26570-4 – gw Olms [919]

J haafner's landreise laengs der kueste orixa und koromandel auf der westlichen indischen halbinsel – Weimar 1809 [mf ed Hildesheim 1995-98] – 2v on 4mf – 9 – €120.00 – ISBN-10: 3-487-26547-8 – ISBN-13: 978-3-487-26547-6 – gw Olms [915]

J hudson taylor und die china-inland-mission : deutschen missionsfreunden zur glaubensstaerkung vorgefeuhrt / Stursberg, Julius – 2.halfte der 2. aufl. Neukirchen, Kreis Moers: Missionsbuchhandlung Stursberg, 1897-1909 [mf ed 1995] – (= ser Yale coll) – 148p (ill) – 1 – 0-524-09741-0 – (in german) – mf#1995-0741 – us ATLA [951]

J I Case Co *see* Drottline

J ishii and his institution : japan's chief apostle of faith, the george muller of the orient, and his unique orphanage / Pettee, James Horace – Yokohama: Yokohama Mission Press, 1892 – 1mf – 9 – 0-524-01525-2 – mf#1990-0431 – us ATLA [240]

J j wilhelm heinse und die aesthetik zur zeit der deutschen aufklaerung : eine problemgeschichtliche studie / Utitz, Emil – Halle: M Niemeyer, 1906 – 1r – 1 – (incl bibliograpical references) – us Wisconsin U Libr [430]

J k lavater und die religioesen stroemungen des achtzehnten jahrhunderts : versuch einer seelenkundlichen deutung in geistesgeschichtlichem rahmen / Forssman, Julius – Riga: Verlag der Akt.-Ges. "Ernst Plates", 1935 – 1r – 1 – (incl bibl ref) – us Wisconsin U Libr [100]

J m r lenz als zentralfigur deutschsprachiger erzaehlschriften / Schirnick, Barbara – (mf ed 1999) – 2mf – 9 – €40.00 – 3-8267-2639-1 – mf#DHS 2639 – gw Frankfurter [430]

J m r lenz und seine schriften : nachtraege zu der ausgabe von l tieck und ihren ergaenzungen / Dorer-Egloff, Edward – Baden: J Zehnder, 1857 [mf ed 1992] – 247p – 1 – mf#7586 – us Wisconsin U Libr [430]

J meerman's, herrn von dalem und vuren reise durch den norden und nordosten von europa : in den jahren 1797 bis 1800 / Meerman, Johan – Weimar 1810 [mf ed Hildesheim 1995-98] – 2v on 7mf – 9 – €140.00 – ISBN-10: 3-487-26542-7 – ISBN-13: 978-3-487-26542-1 – gw Olms [914]

J p bellaire's infanterie-hauptmann's... : beschreibung der vormals venetianischen inseln und besitzungen im jonischen meere oder der jetzigen republik der sieben vereinigten inseln... – Weimar 1806 [mf ed Hildesheim 1995-98] – 1v on 2mf – 9 – €60.00 – 3-487-26555-9 – (trans fr french) – gw Olms [914]

J p hockin's bericht von den neuesten reisen nach den pelew-inseln : besonders des kapt m'cluers und seiner gefaehrten; als nachtrag zu keate's nachricht von den pelew-inseln / Hockin, John – Weimar 1805 [mf ed Hildesheim 1995-98] – 1v on 1mf – 9 – €40.00 – ISBN-10: 3-487-26572-9 – ISBN-13: 978-3-487-26572-8 – gw Olms [919]

J p texier's, koeniglichen daenischen geheimen legations- und etats-raths und danebrog-ritters, reise durch spanien und portugal und von da nach england – Hamm 1825 [mf ed Hildesheim 1995-98] – 1v on 2mf – 9 – €60.00 – 3-487-29818-X – gw Olms [914]

J p walker and co general hardware merchants, king street, hamilton, c w / J P Walker & Co – [Hamilton, Ont?: s.n.] 1864 [mf ed 1994] – 2mf – 9 – 0-665-94674-0 – mf#94674 – cn Canadiana [680]

J P Walker & Co *see* J p walker and co general hardware merchants, king street, hamilton, c w

Der j punkt : der kleine weltlaterne zweiter schein / Bamm, Peter – Stuttgart: Deutsche Verlags-Anstalt, c1937 [mf ed 1989] – 271p (ill) – 1 – (ill by olaf gulbransson) – mf#7214 – us Wisconsin U Libr [890]

J r graves : life, times and teachings / Hailey, O L – 1909 – 1 – $5.00 – us Southern Baptist [242]

J Raymond Jones Democratic Club *see* Harlem valley news

J robert oppenheimer : fbi security file / U.S. Federal Bureau of Investigation – 1979 – 4r – 1 – $520.00 – mf#S1761 – us Scholarly Res [360]

J sarl and sons' prices of gold, silver, and plated articles / Sarl and Sons, J – London [1847?] – (= ser 19th c art & architecture) – 1mf – 9 – mf#4.2.996 – uk Chadwyck [730]

J T Arundel and Co et al *see* Correspondence files, 1892-1904

JACKSON

J T Arundel & Co and Pacific Islands Co Ltd *see* Australian office correspondence files
J v von scheffels gesammelte werke / Scheffel, Joseph Viktor von – Stuttgart: A Bonz & Co 1907 [mf ed 1979] – 6v on 1r – 1 – (with biogr int by johannes moritz proelss) – mf#film mas 8546 – us Harvard [802]
J w lazear, heroe y martir de la civilizacion am... / Portell Vila, Herminio – Habana, Cuba. 1948 – 1r – us UF Libraries [972]
J W Mansfield and Co *see* Ledger
J w v beacon – New York, NY. 1978-81 – 1 – us AJPC [071]
Ja Dan *see* Qanagles-khmaer
Ja ti man te si khyan mya / Mui Krann, Me mrui – Ran Kun: Khyui Phru ca pe 1981 [mf ed 1990] – 1r with other items – 1 – (songs, without musical notations; in burmese) – mf#mf-10289 seam reel 151/4 [§] – us CRL [780]
Jaacijfers voor nederland 1850/51-1965/66 = Statistical yearbook of the netherlands 1850/51-1965/66 / Netherlands. Centraal Bureau voor de Statistiek – (= ser European official statistical serials, 1841-1984) – 379mf – 9 – ("kolonien" sect is cont as a separate publ not incl here) – uk Chadwyck [314]
Jaacijfers voor suriname 1956-1965 / Surinam. Algemeen Bureau voor de Statistiek – (= ser Latin american & caribbean...1821-1982) – 7mf – 9 – uk Chadwyck [318]
Jaakobs traum : ein vorspiel / Beer-Hofmann, Richard – 6.-7. aufl. Berlin: S Fischer, 1919, c1918 [mf ed 1989] – 170p – 1 – mf#7004 – us Wisconsin U Libr [820]
Jaap, Walter C *see* Ecology of the south florida coral reefs
Jaapa / American Academy of Physician Assistants – Montvale. 1994+ (1,5,9) – (cont: journal of the american academy of physician assistants) – mf#16286,01 – us UMI ProQuest [610]
Jaarboek / Katholieke Missie in Nederlands Indie – Batavia: Centraal Missie Bureau 1938-42 [mf ed 2005] – 5v on 1r – 1 – (cont: katholieke missie in nederlands oost-indië. jaarboek; some pgs damaged) – mf#2005c-s048 – us ATLA [241]
Jaarboek / Katholieke Missie in Nederlands Oost-Indië – Batavia: Centraal Missie Bureau 1932-37 [mf ed 2005] – 5v on 1r – 1 – (iss for 1934/1935 combined; cont by: katholieke missie in nederlands indië. jaarboek) – mf#2005c-s047 – us ATLA [241]
Jaarboek der indologen vereeniging, 1918-1919 – Leiden, 1918 – 2mf – 8 – mf#SE-1446 – ne IDC [959]
Jaarboeken voor de israelieten in nederland – s-Gravenhage, 1835-1838. v1-4 – 20mf – 9 – mf#J-261-49 – ne IDC [270]
Jaarboeken voor wetenschappelijke theologie – Utrecht, 1(1845)-13(1856) – 136mf – 9 – €259.00 – ne Slangenburg [240]
Jaarsveld, Floris Albertus van *see*
– Afrikaner en sy geskiedenis
– Afrikaner's interpretation of south african history
Jaarverslag der deli spoorweg-maatschappij – Amsterdam, 1883/1884-1949 – 58mf – 9 – mf#SE-20125 – ne IDC [950]
Jabavu, Davidson *see* Report on the tuskegee institute, alabama, usa
Jabavu, Davidson Don Tengo *see* Black problem
Jabavu, Noni *see*
– Drawn in color
– Drawn in colour
– Ochre people
Jabez bunting : a great methodist leader / Rigg, James Harrison – London: Charles H. Kelly, [19–?] – (= ser Library Of Methodist Biography (London, England)) – 1mf – 9 – 0-7905-6551-X – mf#1988-2551 – us ATLA [242]
Jabez bunting : a great methodist leader / Rigg, James Harrison – London: Charles H. Kelly, [19–?] – 1mf – us ATLA [242]
Jabhah *see* Payam-i jibhah-'i milli
Jabhah al-Sha'biyah li-Tahrir Filastin *see* Bayan 'amaliyat raqm
Jablonski, Johann Theodor *see* Allgemeines lexicon der kuenste und wissenschaften (ael1/42)
Jablonski, Walter *see* Vom sinn der goetheschen naturforschung
Jabotinsky, V *see* State zionism
Jabotinsky, Vladimir *see* Geshikhte fun yidishen legyon
Jabuquito de haikais / Benet Y Castellon, Eduardo – Cienfuegos, Cuba. 1962 – 1r – us UF Libraries [972]
Jaca : journal of the association for communication administration / Association for Communication Administration – Annandale. 1993+ (1,5,9) – (cont: association for communication administration aca bulletin) – mf#11714,02 – us UMI ProQuest [400]
J'accuse : an address in court / Adler, Friedrich – New York: Socialist Publ Society, [191-?] (mf ed 19–) – 36p – mf#Z-BTZE pv383 n9 – us Harvard [340]
Jacimirskij, A J *see* Bibliograficeskij obzor apokrifov v juznoslavjanskoj i russkoj pismennosti
Jacinto De Palazzolo *see* Nas selvas dos vales do mucuri e do rio doce
Jack and jill – Indianapolis. 1938+ (1) 1971+ (5) 1974+ (9) – ISSN: 0021-3829 – mf#3203 – us UMI ProQuest [370]
Jack anderson's washington letter – v1 n4-v4 n12 [1984 jul-1987 jun] – 1r – 1 – (cont: jack anderson's washington letter [1977]; cont by: jack anderson confidential) – mf#1231206 – us WHS [071]
Jack, D T *see* Economic survey of sierra leone
Jack dempsey / Fleischer, Nat – New York, NY. 1949, c1939 – 1r – us UF Libraries [025]
Jack, Hendrik Cornelius *see* Die kommunikasie van die evangelie aan plattelandse kinders in die verenigende gereformeerde kerk in suider-afrika
Jack, Homer Alexander *see* Angola
Jack, J W *see*
– The date of the exodus in the light of external evidence
– Samaria in ahab's time, harvard excavations and their results
Jack, James William *see* Daybreak in livingstonia
Jack, Jonathan Rajmangal *see* History of the teachers association of south african and the role it played in the development of education for indians in south africa
Jack london newsletter – Carbondale. 1967-1988 (1) 1977-1988 (5) 1977-1988 (9) – ISSN: 0021-3837 – mf#9656 – us UMI ProQuest [400]
Jack, Robert *see*
– Discourses
– On evil speaking
Jack, Robert Logan *see* The back blocks of china
Jack tar – 1983 apr 1, may 20, 1984 jun 22, jul 20, aug 17, oct 12-19, 1986 aug 8, oct 3,17-24, 1987 may 1, 1988 feb 5, may 21 – 1r – 1 – (cont by: rota coastline) – mf#1060159 – us WHS [071]
Jack the rapper – 1992 jan 15-dec 28, 1993 jan-jun, 1993 jul 7-dec, 1994 jan-mar, 1995 feb-dec, 1996 jan-mar – 6r – 1 – (cont by: jack the rapper's mello yello) – mf#2339182 – us WHS [071]
Jack, Thomas Godfrey *see* Anti-papa
Jackie robinson and the integration of organized baseball / Cable, Dale – 1979 – 141p on 2mf – 9 – $10.00 – mf#PE 4174 – us Kinesology [790]
Jackman, William James *see* Legal features of commerce regulation
Jackovic, Terence J *see* A comparison of student development outcomes among male revenue athletes, non-revenue athletes, and club sport athletes at an ncaa divison 1 university
Jack-pine warbler – Lansing. 1972-1996 (1) 1972-1980 (5) 1976-1980 (9) – ISSN: 0021-3845 – mf#6852 – us UMI ProQuest [590]
Jackpot – iss n1-9. spr 1941-spr 1943 – 15 – (sum 1942-spr 1943 set of 4mf $29.35) – mf#007MLJ-008MLJ – us MicroColour [740]
Jacks, Lawrence Pearsall *see*
– The alchemy of thought
– Among the idolmakers
– Life and letters of stopford brooke
Jackson 2000 / Jackson State University – 1994, 1996 – 1r – 1 – mf#5026485 – us WHS [378]
Jackson, Abner *see* Discourses
Jackson, Abraham Valentine Williams *see*
– From constantinople to the home of omar khayyam
– Persia past and present
– Zoroaster
Jackson, Abraham Valentine Williams et al *see* History of india
Jackson, Abraham Willard *see*
– The immanent god, and other sermons
– James martineau
Jackson advocate – Jackson MS. [v46 n27,29-51 [1984 mar 22, apr 5/dec]]-[1999 jan 7/13-jun 24/30] [gaps] – 31r – 1 – mf#807212 – us WHS [071]
Jackson afl-cio news / Jackson County AFL-CIO Central Labor Council [MI] – v1 n4-5 [1983 jan-feb] – 1r – 1 – (cont by: jackson labor news) – mf#1477185 – us WHS [331]
[Jackson-] amador dispatch – CA. 1863-75, 1878-89, 1890-1980 – 57r – 1 – $3420.00 – (cont by: ledger and progress-news) – mf#BC02317 – us Library Micro [071]
[Jackson-] amador ledger and amador record – CA. 1900-18; 1921-1980 – 43r – 1 – $2580.00 – mf#BC02318 – us Library Micro [071]
Jackson and Sons, George *see* Part of the collection of releivo [sic] decorations as executed in papier mache and carton pierre
Jackson, Andrew *see* Papers
Jackson baptist church. butts county. georgia : church records – 1851-89 – 1 – us Southern Baptist [242]
Jackson, Blomfield *see* Twenty-five agrapha
Jackson Co. Jackson *see*
– Democratic / jackson herald
– Herald
– Journal
– Journal herald
– Journal-herald
– Oak hill press
– Standard
– Standard-journal
– Sun
– Sun-journal
Jackson Co. Jackson C.H. *see* Express / standard
Jackson Co. Oak Hill *see* Press
Jackson Co. Wellston *see*
– Daily sentinel
– Daily sentinel series
– Sentinel
– Sentry
– Telegram
– Telegram series
– Wellston telegram series
Jackson county / Woltz, Larry – S.I., S.I? . 193-? – 1r – us UF Libraries [978]
Jackson County AFL-CIO Central Labor Council [MI] *see*
– Jackson afl-cio news
– Jackson labor news
Jackson county atlas, 1875 – 1r – 1 – mf#B27425 – us Ohio Hist [978]
Jackson county banner – Black River Falls WI. 1857 sep 17-1858 oct 7, 1861 apr 11, sep 21, 1862 apr 26, jun 21, jul 5,12,19, aug 2, sep 6, 1864 apr 23, 1867 jan 5-1868 mar 28 – 3r – 1 – (cont: badger state banner) – mf#1008014 – us WHS [071]
Jackson County Citizens for a Nuclear Arms Freeze [NC] *see* Freeze
Jackson county floridan – Marianna, FL. 1873 oct-1997 aug – 145r – (gaps) – us UF Libraries [071]
Jackson county floridan – Marianna, FL. 1940 – 1r – us UF Libraries [071]
Jackson county floridan – Marianna, FL. 1948 – 1r – us UF Libraries [071]
Jackson county floridan – Marianna, FL. 1958-1975 (1) – mf#62430 – us UMI ProQuest [071]
Jackson County Genealogical Society [MO] *see* Pioneer wagon
Jackson county journal – Black River Falls WI. [1890 aug 6/1891 dec 9]-[1925 feb 25/1926 mar 3] [gaps] – 23r – 1 – mf#963581 – us WHS [071]
Jackson county miscellaneous newspapers – Denver, CO (mf ed 1991) – 1r – 1 – (pearl mining times (apr 28 1905); jackson county times (mar 23 1911-jan 8 1912); the new era (jul 6 1912-jun 27 1913); north park news (feb 3, apr 14 1899)) – mf#MF Z99 J132 – us Colorado Hist [071]
Jackson county news – Medford OR: James W Young & L B Tuttle, 1924-26 [wkly] – 1 – (cont: clarion (medford, or: 1920); cont by: daily news (medford, or: 1926)) – us Oregon Lib [071]
Jackson county news – Ravenswood, WV. 1878-1879 (1) – mf#67453 – us UMI ProQuest [071]
Jackson county times *see* Jackson county miscellaneous newspapers
The jackson criterion – Jackson, NE: Wm T Bartlett (wkly) [mf ed v10 n17. jul 23 1896-jan 23 1902 (gaps)] – 1r – 1 – us NE Hist [071]
Jackson, Emily Nevill *see* A history of hand-made lace
Jackson first baptist church. jackson, louisiana : church records – 1835-1958 – 1 – 59.76 – us Southern Baptist [242]
Jackson first baptist church. jackson, missouri : church records – 1856-1905 – 1 – us Southern Baptist [242]
Jackson Foundation for Medical Research and Education *see*
– Tracings
Jackson, Francis M *see* Index to the library edition of thomas jackson's life of charles wesley
Jackson, Frank G *see*
– Lessons on decorative design
– Theory and practice of design
Jackson, G K *see* New miscellaneous musical work
Jackson gazette – Jackson TN. 1824 may-1826 may, 1828 jun-1830 – 1r – 1 – (cont: pioneer [jackson, tn: 1822]) – mf#858259 – us WHS [071]
Jackson, George *see*
– The fact of conversion
– In a preacher's study
– The old methodism and the new
– A series of letters, on the subjects and mode of christian baptism
– Studies in the old testament
– The table-talk of jesus
– The teaching of jesus
– A young man's religion
Jackson, George Anson *see*
– The apostolic fathers and the apologists of the second century
– The fathers of the third century
– The post-nicene greek fathers
– The post-nicene latin fathers
Jackson, George H *see* The medicinal value of french brandy
Jackson, George & Sons *see* First part of the collection of detached enrichments
The jackson headlight – Jackson, TN: C A Leftwich, Joshua W Lane, Rev W H Daniel, 1900 (wkly) [mf ed 1947] – (= ser Negro Newspapers on Microfilm) – 1r – 1 – us L of C Photodup [071]
Jackson, Helen Hunt *see* A century of dishonor
Jackson, Helen Maria Fiske Hunt *see* Ramona
Jackson, Henry *see*
– An account of the churches in rhode-island
– A discourse in commemoration of the 46th anniversary of the mite society
– An historical discourse
Jackson, Henry Latimer *see*
– The eschatology of jesus
– The fourth gospel and some recent german criticism
Jackson herald – Ripley, WV. 1914+ (1) – mf#67460 – us UMI ProQuest [071]
Jackson, J A *see* National emigration
Jackson, J G *see*
– An account of timbuctoo and housa
– Designs for villas
Jackson, James *see* Remarks on mr ewing's attempt towards a statement of the doctrine...
Jackson, James G *see* An account of timbuctoo and housa
Jackson, John *see*
– In leper-land
– Lepers
– Mary reed
– Tagebuch einer im jahre 1797 unternommenen landreise aus ostindien nach europa
– A treatise on wood engraving
Jackson, Joseph Henry *see* Notes on a drum
Jackson journal – v1 n1-v3 n4 [1984 jan-1986 oct] – 1r – 1 – mf#1093105 – us WHS [071]
Jackson, Kristin M *see* Current world wide web use in park and recreation departments
Jackson labor news / Jackson County AFL-CIO Central Labor Council [MI] – v2 n1-22 [1984 jan-1985 oct] – 1r – 1 – (cont: jackson afl-cio news) – mf#1477187 – us WHS [331]
[Jackson-] ledger-dispatch : ione progress-news – CA. 1981- – 32r – 1 – $1920.00 (subs $150y) – mf#B02319 – us Library Micro [071]
Jackson, Luther Porter *see* Negro office-holders in virginia, 1865-1895
Jackson, Michael *see*
– A study of the perceived effects of the repeal of the pennsylvania interscholastic athletic association constitutional bylaw article 11, section 2
– A survey of the desired educational preparation and employment market for high school athletic trainers in metropolitan washington, dc as perceived by high school athletic directors
Jackson, Michael W *see*
– The identification of the minimum qualifications for tennis teaching professionals to be hired at managed tennes facilities in the united states
– An investigation of the qualifications of contract advisors for professional athletes
– A study of high school track and field outdoor championships based on the events endorsed by the national federation of state high school associations
– Survey of aquatic programs and aquatic facility accessibility features available to and utilized by physically handicapped students at four-year pennsylvania colleges and universities
Jackson, Miles *see*
– Communicant's remembrancer
– Constraining power of the love of christ
Jackson, Mrs *see* Moonlight scene
Jackson, Nathaniel P *see* Select texas statutes, annotations and forms and notaries' manual
Jackson, Robert Houghwout *see* The case against the nazi war criminals
Jackson, S K *see*
– Chiwororo chavakuru
– Ciwororo cavakuru
– Kuzadzwa nomweya mutsvene
– Madambudziko
– Nyaya
– Shona lessons
Jackson, S M *see*
– Huldreich zwingli
– The latin works and the correspondence of huldreich zwingli
Jackson, Samuel Macauley *see* Selected works of huldreich zwingli, 1484-1531
Jackson, Samuel Macauley et al *see*
– The concise dictionary of religious knowledge and gazetteer
– Huldreich zwingli
Jackson, Samuel Trevena *see* Lincoln's use of the bible

Jackson, Sheldon *see*
- Alaska and missions on the north pacific coast
- Collection
- Correspondence
- Cruise of the u.s. revenue marine steamer bear
- Photograph collection
- Scrapbooks

Jackson, Sheldon et al *see* Addresses at the celebration of the 250th anniversary of the westminster assembly
Jackson, Spencer *see* The land monopolists of ireland
Jackson star news – Ravenswood, WV. 1987-1991 (1) – mf#67454 – us UMI ProQuest [071]
Jackson State University *see* Jackson 2000
Jackson, Susan A *see* Elite athletes in flow
Jackson, T *see* Book of trinidad
Jackson, Thomas *see*
- The centenary of wesleyan methodism
- The duties of christianity
- Expository discourses on various scripture facts and characters
- Faithful pastor
- Fulfilment of the christian ministry
- The institutions of christianity
- The life of john goodwin
- The providence of god
- Recollections of my own life and times

[Jackson, Thomas] *see* A catalogue of books and manuscripts
Jackson, Thomas Graham, Baronet *see* Architecture
Jackson, Thomas Graham. baronet *see*
- The church of st mary the virgin oxford
- Modern gothic architecture

Jackson, Thomas Graham, Sir *see* Gothic architecture in france, england, and italy
Jackson, Thomas Jefferson *see* Researches in non-euclidian geometry and the theory of relativity
Jackson vinton journal herald – Jackson, OH. 1995-1997 (1) – mf#69167 – us UMI ProQuest [071]
Jackson, William *see*
- The doctrine of retribution
- The philosophy of natural theology

Jacksonian – Wooster, OH. 1881-1899 (1) – mf#65730 – us UMI ProQuest [071]
Jacksoniana – 1978-1987 jul – 1r – 1 – mf#2173787 – us WHS [071]
Jackson's oxford journal – Oxford. England. -w. 1756-66; 1826-37 (1756-66 imperfect). (6mqn reels) – 1 – uk British Libr Newspaper [072]
Jacksons oxford journal – may 5 1753-dec 27 1755, 1756-71, jan 4-may 2 1772, may 15 1773, jan 4 1777-nov 27 1779, 1780-82, feb 1 1783-dec 25 1786, jan 20 1787-dec 18 1790, 1791-1826, jan 6 1827-dec 19 1829, 1930-dec 29 1832, 1833-1908, jan 2-29 1909, 1910-27, jan 4-nov 28 1928, 1986-jun 1991, 1992-sep, oct 7-dec 23 1993, 1994-jun 1997 – 127 1/2r – 1 – (aka: oxford journal) – uk British Libr Newspaper [072]
Jackson-slinger herald – Addison, Allenton, Cedarburg etc WI. 1983 nov 24-dec 27, 1985-90, 1991 jan 2-may 1,15-22, jun 5,19, jul 10,24, aug 7-14, sep 25, oct 9,23, nov 13, dec 4,24, 1992 jan 22, feb 5,19, mar 11, apr 15, 1993 aug 25, sep 8,22, oct 6,27, nov 10 – 7r – 1 – mf#1012307 – us WHS [071]
Jacksonville advocate – Jacksonville FL. v9 n32-35 [1987 aug 26/sep 2-16/23] – 10r – 1 – (cont by: jacksonville advocate free press) – mf#1789844 – us WHS [071]
Jacksonville advocate – Jacksonville FL. 1992 nov 16/29, 1993 feb-jun, 1993 aug 9/15-dec 13/19, 1994 jan 10/16-dec 26/jan 1, 1995 jan 9/15-jun 5/11, 1996 jan-dec, 1997 jan 6/12-jun 30/jul 6, 1997 jul 7/13-dec 29/jan 4, 1998 jan-jun, 1999 jan-dec – 10r – 1 – mf#4034256 – us WHS [071]
Jacksonville advocate free press – Jacksonville, FL. 1989 – 1r – (1989 jun 29) – us UF Libraries [071]
Jacksonville advocate free press – Jacksonville FL. 1987 sep 23/30-1988 sep 28, 1988 sep 28-1990 jun 21/27 – 2r – 1 – (cont: jacksonville advocate; cont by: jacksonville free press) – mf#1789815 – us WHS [071]
Jacksonville advocate-free press – Jacksonville, FL. 1987 jul 29-1988 dec 22 – 2r – (gaps) – us UF Libraries [071]
Jacksonville american – Jacksonville, FL. 1950-1956 jul – 8r – (1950 feb 24; may 19) – us UF Libraries [071]
Jacksonville arlingtonian – Jacksonville, FL. 1936 jan 29-1949 sep – 2r – (gaps) – us UF Libraries [071]
Jacksonville Auxiliary Sanitary Association, Jacks *see* Report of the jacksonville auxiliary sanitary association
Jacksonville board of health / Shepherd, Rose – S.l., S.l?. 1936 – 1r – us UF Libraries [978]
Jacksonville buildings : duval county court house / Shepherd, Rose – S.l., S.l?. 1937 – 1r – us UF Libraries [978]
Jacksonville bulletin – [Jacksonville OR: Jacksonville Booster's Club] 1964- [wkly] – 1 – (ceased with v11 iss 1 (feb 18 1974)?) – us Oregon Lib [071]
Jacksonville business journal – Jacksonville. 1991-1994 (1) – (cont by: business journal) – ISSN: 0885-453X – mf#16679 – us UMI ProQuest [650]
Jacksonville Chamber Of Commerce *see* Jacksonville, florida
Jacksonville courier – Jacksonville FL. 1825 jan 29 – 1r – 1 – (cont by: jacksonville courier and southern index) – mf#850712 – us WHS [071]
Jacksonville courier – Jacksonville, FL. 1835 jan 29-1836 feb 25 – 1r – us UF Libraries [071]
Jacksonville Crime Justice Commission *see* Report
Jacksonville, early history / Shepherd, Rose – S.l., S.l?. 1937 – 1r – us UF Libraries [978]
Jacksonville, Fla City Council *see* Jacksonville, florida's dominant city
Jacksonville (Fla) Community Service *see* Florida historical pageant
Jacksonville, florida / Jacksonville Chamber Of Commerce – Jacksonville, FL. 191-? – 1r – us UF Libraries [978]
Jacksonville florida dispatch – Jacksonville, FL. 1886-1889 – 6r – (gaps) – us UF Libraries [071]
Jacksonville, florida's dominant city / Jacksonville, Fla City Council – Jacksonville, FL. 193-? – 1r – us UF Libraries [978]
Jacksonville free press – Jacksonville FL. 1990 jul-dec, 1991 jan-dec, 1992 jan-dec, 1993, 1994 jan 13-dec, 1995 jan-dec, 1996 jan-dec, 1999 jan-jun – 12r – 1 – (cont: jacksonville advocate free press) – mf#1743205 – us WHS [071]
Jacksonville free press – Jacksonville, FL. v5 n12-v11 n1. 1991-1996 – 6r – (gaps) – us UF Libraries [071]
Jacksonville goldrush gazette – Jacksonville OR: Jacksonville Lion's Club, [various dates] [mf ed 1959] – 1r – 1 – (incl on reel: jacksonville, or: miscellaneous newspapers, 1863-1963) – us Oregon Lib [071]
Jacksonville jewish news – Jacksonville, FL. v1 n1-v9 n4. 1988 aug-1996 – 2r – us UF Libraries [071]
Jacksonville journal – Jacksonville, FL. 1884 may 26-1929 aug – 3r – (gaps) – us UF Libraries [071]
Jacksonville miner – Jacksonville OR: L Hall, 1932-35 [wkly] [mf ed 1967] – 1r – 1 – (cont by: southern oregon miner) – us Oregon Lib [071]
Jacksonville nugget – Jacksonville OR: Donald W Wendt, 1977- [semiwkly] – 1 – us Oregon Lib [071]
Jacksonville, religion, richardson : sanctified ch... – S.l., S.l?. 193-? – 1r – us UF Libraries [978]
Jacksonville republican – Jacksonville, AL: J F Grant, feb 2 1841-jun 19 1862 – 2r – 1 – us CRL [071]
Jacksonville reveille – Jacksonville OR: Short & Owen [wkly] – 1 – us Oregon Lib [071]
Jacksonville seafarer – Jacksonville. 1972-1981 (1) 1976-1981 (5) 1976-1981 (9) – (cont by: seafarer) – ISSN: 0447-2462 – mf#8227 – us UMI ProQuest [380]
Jacksonville sentinel (jacksonville, or: 1903) – Jacksonville OR: Charles Meserve [wkly] – 1 – (ceased in 1906) – us Oregon Lib [071]
Jacksonville sentinel (jacksonville, or: 1961) – Jacksonville OR: R E Lowe [wkly] – 1 – us Oregon Lib [071]
Jacksonville social : the young women's christian / Shepherd, Rose – S.l., S.l?. 1937 – 1r – us UF Libraries [978]
Jacksonville, St Augustine And Indian River Railway *see* Florida
Jacksonville suburbs : glynlea school, south jacks / Shepherd, Rose – S.l., S.l?. 1937 – 1r – us UF Libraries [978]
Jacksonville utilities : municipal airport – S.l., S.l?. 193-? – 1r – us UF Libraries [978]
Jacksonville woman's club / Shepherd, Rose – S.l., S.l? . 1937 – 1r – us UF Libraries [978]
Jacl reporter / Japanese American Citizens League – v1 n4 [1945 apr] – 1r – 1 – mf#629945 – us WHS [071]
Jacob : three sermons. preached before the university of cambridge in lent, 1870 / Moorhouse, James – London: Macmillan, 1870 – 1mf – 9 – 0-524-05049-X – mf#1992-0302 – us ATLA [221]
Jacob albright and his co-laborers / Yeakel, Reuben [comp] – Cleveland, OH: Evangelical Assoc, c1883 [mf ed 1990] – 1mf – 9 – 0-7905-6337-1 – (in english) – mf#1988-2337 – us ATLA [240]
Jacob at bethel : the vision, the stone, the anointing / Palmer, Abram Smythe – London: David Nutt, 1899 – (= ser Studies in biblical subjects) – 1mf – 9 – 0-524-07341-4 – (incl bibl ref & ind) – mf#1992-1072 – us ATLA [221]
Jacob, B *see* Das erste buch der tora
Jacob behmen : an appreciation / Whyte, Alexander – Edinburgh: Oliphant Anderson & Ferrier, 1894 – 1mf – 9 – 0-524-05524-6 – mf#1990-1519 – us ATLA [240]
Jacob boehme, his life and teaching, or, studies in theosophy = Jacob boehme / Martensen, Hans – London: Hodder and Stoughton, 1885 – 1mf – 9 – 0-7905-8701-7 – (in english) – mf#1989-1926 – us ATLA [210]
Jacob boehme und die alchymisten : ein beitrag zum verstaendniss j. boehme's: nebst zwei anhaengen, j.g. gichtel's leben und irrthuemer und ueber ein rosenkreuzerisches manuscript / Harless, Gottlieb Christoph Adolf von – 2. verm Ausg. Leipzig: JC Hinrichs, 1882 – 1mf – 9 – 0-524-00368-8 – (incl bibl ref) – mf#1989-3068 – us ATLA [130]
Jacob boehmes deutsches christentum / Elert, Werner – Berlin: Edwin Runge 1914 [mf ed 1989] – (= ser Biblische zeit- und streitfragen 9/6) – 1mf – 9 – 0-7905-4293-5 – (incl bibl ref) – mf#1988-0293 – us ATLA [240]
Jacob boehme's the way to christ = Weg zu christo – 1st ed. New York, London: Harper, c1947 [mf ed 1998] – xxxix/254p (ill) on 1r – 1 – (trans by john joseph stoudt. foreword by rufus m jones. incl bibl ref) – mf#9963 – us Wisconsin U Libr [230]
Jacob, C A *see* Neuere nachrichten ueber sicilien und ueber die jetzige eintheilung dieser insel in districte oder intendenzen
Jacob, Edgar *see* The divine society
Jacob, Edwin *see* Annual discourse delivered by edwin jacob...before the fredericton atheneum, february 21, 1853
Jacob Foutz Family Association *see* Pfautz-fouts-foutz newsletter
Jacob, G A *see* Perpetuity of the sabbath law in the fourth commandment
Jacob, G A [comp] *see* Laukikanyayanjalih
Jacob, Georg *see* Das hohelied
Jacob, George Andrew *see* The ecclesiastical polity of the new testament
Jacob, H *see*
- A defence of the churches and ministery of englande
- To the right high and mightie prince, iames by the grace of god, king of great britennie, france, and irelande, defender of the faith, etc

Jacob Isaac *see* Sefer nifle'os hayehudi
Jacob, John *see* Remarks on the native troops of the indian army
Jacob, Joseph *see* Daffodils
Jacob, Judith M *see* Introduction to cambodian
Jacob, Karl *see* Die grossen kriege in der geschichte des deutschen volkes
Jacob, Kleber Georges *see*
- Contribution a l'etude de l'homme haitien
- Ethnie haitienne

Jacob, Michael P *see* College women athletes' knowledge and perceptions of title 9
Jacob, mlle [Victoire, Jeanne] *see* Aux femmes
Jacob morier's, secretaer's des englischen gesandten sir hartford jones an den persischen hof : reise durch persien, armenien und klein-asien nach constantinopel in den jahren 1808 / Morier, James – Weimar 1815 [mf ed Hildesheim 1995-98] – 1v on 2mf – 9 – €60.00 – ISBN-10: 3-487-26527-3 – ISBN-13: 978-3-487-26527-8 – (trans fr english) – gw Olms [915]
Jacob of Edessa *see* A letter by mar jacob, bishop of edessa, on syriac orthography
Jacob, P L *see*
- Marechale d'ancre
- Memoirs of cardinal dubois

Jacob ruffs adam und heva / ed by Kottinger, Hermann Marcus – Quedlinburg, Leipzig: G Basse, 1848 [mf ed 1993] – (= ser Bibliothek der gesammten deutschen national-literatur von der aeltesten bis auf die neuere zeit sect1/26) – viii/216p – 1 – mf#8438 reel 6 – us Wisconsin U Libr [430]
Jacob ruffs etter heini uss dem schwizerland : sammt einem vorspiel / ed by Kottinger, H M – Quedlinburg, Leipzig: G Basse, 1847 – (= ser Bibliothek der gesammten deutschen national-literatur von der aeltesten bis auf die neuere zeit sect1/14) – xxxviii/251p – 1 – (incl bibl ref) – mf#8438 reel 4 – us Wisconsin U Libr [430]
Jacob, Son of Aaron *see* The book of enlightenment for the instruction of the inquirer
Jacob sturm : rede. gehalten bei uebernahme des rektorats der universitaet strassburg... / Baumgarten, Hermann – Strassburg: Karl J Truebner, 1876 – 1mf – 9 – 0-524-03886-4 – mf#1990-1145 – us ATLA [943]
Jacob the wrestler / Kennedy, Henry Dawson – Toronto: W Briggs [1901?] [mf ed 1995] – 1mf – 9 – 0-665-74725-X – mf#74725 – cn Canadiana [221]
Jacob thomson : ein vergessener dichter des achtzehnten jahrhunderts / Schmeding, G – Braunschweig, 1889 (mf ed 1992) – 1mf – 9 – €24.00 – 3-89349-065-5 – mf#DHS-AR 28 – gw Frankfurter [430]
Jacob und seine zwoelf soehne : ein evangelisches schulspiel aus steyr / Brunner, Thomas; ed by Stumpfl, Robert – Halle: M Niemeyer 1928 [mf ed 1993] – (= ser Neudrucke deutscher literaturwerke des 16. und 17. jahrhunderts 258-260) – 1r – 1 – (incl bibl ref; filmed with: neudrucke deutscher literaturwerke des 16. und 17. jahrhunderts) – mf#3387p – us Wisconsin U Libr [820]
Jacob unrest (mgh6:11.bd) : oesterreichische chronik / ed by Grossmann, K – 1957 – (= ser Monumenta germaniae historica 6: scriptores rerum germanicarum, nova series (mgh6)) – €15.00 – ne Slangenburg [240]
Jacob unrest, oesterreichische chronik (mgh6:11.bd) / ed by Grossmann, K – 1957 – (= ser Monumenta germaniae historica 6: scriptores rerum germanicarum, nova series (mgh6)) – €15.00 – ne Slangenburg [240]
Jacob van reenen and the grosvenor expedition of 1790-1791 / Kirby, Percival Robson – Johannesburg, South Africa. 1958 – 1r – us UF Libraries [960]
Jacob, W *see* Die handschriftliche ueberlieferung der sogenannten historia tripartita des epiphanius-cassidor
Jacob, Walter *see* Rampenlicht
Jacob, William *see* A view of the agriculture, manufactures, statistics, and state of society, of germany
Jacobi, A *see* Ethnographische beobachtungen ueber die voelker des beringsmeeres, 1789-1791
Jacobi, Andreas Ludolph et al *see* Annalen der braunschweig-lueneburgischen churlande
Jacobi, Charles Thomas *see*
- On the making and issuing of books
- The printers' vocabulary
- Some notes on books and printing

Jacobi, Franz *see* Das thorner blutgericht 1724
Jacobi, Johann Georg *see* Iris
Jacobi, Johann Georg et al *see* Iris. vierteljahrschrift fuer frauenzimmer
Jacobi, Justus Ludwig *see*
- Lectures on the history of christian dogmas
- Wissenschaftliche abhandlungen

Jacobi, Wilhelm Heinrich *see* Die philosophie der persoenlichkeit nach friedrich heinrich jacobi
Jacobins noirs / James, C L R – Paris, France. 1949 – 1r – us UF Libraries [972]
The jacobite – nov 1919-oct 1952 – 1r – mf#ZB 30 – nz Nat Libr [079]
Jacobite's journal – London. 1747-1748 (1) – mf#4790 – us UMI ProQuest [420]
Jacobowski, Ludwig *see*
- Aus deutscher seele
- Klinger und shakespeare

Jacobs, Aletta Henriette *see* La femme et le feminisme
Jacob's band monthly : a music magazine published exclusively in the interests of the professional and amateur player of band instruments – Boston: Walter Jacobs 1916-27 – 26v on 1mf – 9 – mf#fiche 33 – us Sibley [780]
Jacobs, C H *see* Die schiffahrtsfreiheit im suezkanal
Jacob's dream : a prologue / Beer-Hofmann, Richard – Philadelphia: The Jewish Publication Society of America, 1946 [mf ed 1995] – 188p – 1 – (trans fr german by ida bension wynn) – mf#8973 – us Wisconsin U Libr [820]
Jacobs, Eduard *see* Heinrich winckel und die reformation im suedlichen niedersachsen
Jacobs, H *see* Complete aas microfiche series collection
Jacobs, Henry E *see* Annotations on the epistles of paul to the romans and 1. corinthians, chaps 1.-6
Jacobs, Henry Eyster *see*
- The book of concord, or, the symbolical books of the evangelical lutheran church
- The doctrine of the ministry as taught by the dogmaticians of the lutheran church
- Elements of religion
- The english augsburg confession of 1536
- Geschichte der lutherischen kirche in amerika
- A history of the evangelical lutheran church in the united states
- The lutheran cyclopedia
- The lutheran movement in england during the reigns of henry 8 and edward 6
- Martin luther
- A summary of the christian faith

Jacobs, Henry Eyster et al *see* Annotations on the epistles of paul to 1. corinthians 7-16, 2. corinthians and galatians
Jacobs, Joseph *see*
- As others saw him
- The earliest english version of the fables of bidpai / "the morall philosophie of doni" by sir thomas north
- Jewish ideals
- Studies in biblical archaeology

Jacobs, Peter *see* Journal of the reverend peter jacobs, indian wesleyan missionary
Jacobs, Reinhold *see* Jamaika, seine physikalisch-politische geographie
Jacobs, Susanne *see* The implementation and assessment of a multi-media learning programme for environmental education
Jacobs, Walter Darnell *see* Special study of south west africa in law and politics
Jacobs, William States *see* Presbyterianism in nashville
Jacobsen, Jerome V *see* Educational formations of the jesuits...

Jacobshagen, Burkhard *see* Die variabilitaet des langzeitspektrums der menschlichen sprechstimme

Jacobskoetter, Ludwig *see* Goethes faust im lichte der kulturphilosophie spenglers

Jacobsohn, B *see* Deutsch-israelitische gemeindebund nach ablauf

Jacobsohn, Jacob *see* Mitteilungen des gesamtarchivs der deutschen juden

Jacobson, B J *see* The relationship between sport-confidence, competitive orientation and performance on a muscular leg-endurance task

Jacobson, Harold Karan *see* Diplomats, scientists, and politicians

Jacobson, Hermann *see* Altitalische inschriften

Jacobson, Jonathan et al *see* Educational achievement and black-white inequality

Jacobson, Lynn B *see* The relapse prevention model and exercise maintenance behavior

Jacobson, Phyllis C *see* A historical documentation, an instructional manual and an annotated bibliography of selected folk dances of puerto rico

Jacobson, Sharon A *see* An examination of leisure in the lives of old lesbians from an ecological perspective

Jacobson, Wolf S *see* Worte des gedenkens an drei auf dem felde der ehre gefallene freund

Jacobstein, J Myron *see* Supreme court of the us hearings and reports on successful and unsuccessful nominations of supreme court justices by the senate judiciary committee

Jacobstown baptist church. wrightstown, new jersey : church records – 1785-c1943, 1842-1910. list of members, Original Deed, etc – 1 – us Southern Baptist [242]

Jacobstroer, Bernhard *see* Die romantechnik bei friedrich gerstaecker

Jacobus arminius : een biografie (met portret en handteekening) / Maronier, Jan Hendrick – Amsterdam: Y. Rogge, 1905 – 1mf – 9 – 0-7905-6240-5 – (incl bibl ref) – mf#1988-2240 – us ATLA [949]

Jacobus baradaeus : de stichter der syrische monophysietische kerk / Kleyn, Hendrik Gerrit – Leiden: E J Brill, 1882 – 1mf – 9 – 0-8370-7880-6 – (incl bibl ref and index) – mf#1986-1880 – us ATLA [920]

Jacobus baradaeus de stichter der syrische monophysietische ker / Kleyn, H G – Leiden, 1883 – €13.00 – ne Slangenburg [243]

Jacobus de Boragine (Blessed Jacopo de Voragine) *see*
- Sermones de tempore per totum annum
- Sermones super evangelia per quadragesimam

Jacobus Mediolanensis *see* Stimulus amoris – canticum pauperis

Jacobus, Melancthon W *see* Matthew

Jacobus, Melancthon Williams *see*
- Memorial addresses upon the late chester david hartranft
- Notes, critical and explanatory, on the acts of the apostles
- Notes, critical and explanatory, on the book of genesis
- A problem in new testament criticism

Jacobus Veritas *see* Jacobus veritas's legacy to the franchised portion of the british empire

Jacobus veritas's legacy to the franchised portion of the british empire – 2nd ed. London, 1839 – (= ser 19th c british colonization) – 1mf – 9 – mf#1.1.335 – uk Chadwyck [320]

Jacoby, Albert *see* Verband der juedischen jugendvereine deutschlands

Jacoby, Daniel *see* Richard der dritte

Jacoby, Elfriede *see* Zur geschichte des wandels von lat u zu y im galloromanischen

Jacoby, Guenther *see* Herder als faust

Jacoby, Hermann *see* Neutestamentliche ethik

Jacoby, James Calvin *see* Around the home table

Jacoby, Leopold *see*
- Ein ausflug nach comacchio
- Deutsche lieder aus italien
- Es werde licht

Jacoby, Yoram K *see* Juedisches leben in koenigsberg/pr. im 20. jahrhundert

Jacolliot, Louis *see*
- L'olympe brahmanique
- Voyage au pays des bayaderes

Jacomb correspondence *see* Isles of disenchantment: the fletcher / jacomb correspondence: letters exchanged between r j fletcher and edward jacomb

Jacomb, Edward *see* The future of the kanaka

Jacombe, Thomas *see* Sermons on the eighth chapter of the epistle to the romans (verses 1-4)

Jacopo torriti / Bertos, Rigas – Muenchen, 1963 – 1 – gw Mikropress [920]

Jacottet, Edouard *see*
- Grammar of the sesuto language
- Practical method to learn sesuto

Jacoubovitch, M-Daniel *see* Summary jury trials in the northern district of ohio

Jacquelin, Jacques Andre *see* Amour a l'anglaise

Jacqueline d'olzebourg / Bazin, Jacques Rigomer – Paris, France. 1803 – 1r – us UF Libraries [440]

Jacquemart, Albert *see*
- A history of furniture
- History of the ceramic art

Jacquemin, C *see* Lettre du pere jacquemin

Jacquemin, Charles *see* Aux iles caraibes

Jacquemin le franc-macon : aventures du dix-neuvieme siecle / Collin de Plancy, Jacques-Albin-Simon – 2e ed. Bruxelles: De Mortier Freres 1845 – 3mf – 9 – mf#vrl-169 – ne IDC [366]

Jacquemont, V *see* Voyage dans l'inde...pendant les annees 1828...1832...

Jacquemont, Victor *see*
- Correspondance de victor jacquemont avec sa famille et plusieurs de ses amis

Jacquerie / Langle, Ferdinand – Paris, France. 1839? – 1r – us UF Libraries [440]

Jacques bonhomme / Benoit-Jean & Delvau, Alfred – Paris: Ad Blondeau. n1. jul 1850 – us CRL [920]

Jacques bonhomme – London, UK. 27 Mar 1857-26 Jun 1858 – 1 – uk British Libr Newspaper [072]

Jacques bonhomme d'haiti, en sept tableaux / Thoby, Armand – Port-Au-Prince, Haiti. 1901 – 1r – us UF Libraries [972]

Jacques cartier / Dionne, Narcisse Eutrope – 2e ed. Quebec: impr Emile Robitaille, 1933 [mf ed 1985] – 2mf – 9 – mf#SEM105P499 – cn Bibl Nat [910]

Jacques cartier : questions de calendrier civil et ecclesiastique / Verreau, Hospice Anthelme Baptiste – S.l: s.n, 1890? – 1mf – 9 – mf#35798 – cn Canadiana [917]

Jacques cartier : questions de droit public, de legislation et d'usages maritimes / Verreau, Hospice Anthelme Baptiste – Ottawa?: s.n, 1891 – 1mf – 9 – mf#29150 – cn Canadiana [910]

Jacques cartier : questions de lois et coutumes maritimes / Verreau, Hospice Anthelme Baptiste – Ottawa?: s.n, 1897 (Ottawa: J Durie; Toronto: Copp-Clark) – 1mf – 9 – mf#25444 – cn Canadiana [340]

Jacques cartier 1534 / Maxine – Montreal: Editions Albert Levesque, 1933 [mf ed 1992] – 1mf – 9 – (ill by j-arthur lemay) – mf#SEM105P1657 – cn Bibl Nat [917]

Jacques cartier (1491-1557) : decouvreur du canada / Laviolette, Guy – 3e ed, 48e mille. Sherbrooke; Montreal [etc]: Apostolat de la presse, [1958?] (mf ed 1993) – 1mf – 9 – (cartier, jacques) – mf#SEM105P1965 – cn Bibl Nat [917]

Jacques cartier, decouvreur, explorateur, colonisateur du canada, 1491-1557 / Leymarie, A Leo – [Saint-Jerome, Quebec: s.n.] 1913 [mf ed 1996] – 1mf – 9 – 0-665-81346-5 – mf#81346 – cn Canadiana [910]

Jacques coeur, l'argentier du roi / Anicet-Bourgeois, Auguste – Paris, France. 1841 – 1r – us UF Libraries [440]

Jacques, Colette *see* Bio-bibliographie de me jean-charles bonenfant

Jacques, Dh *see* Florida as a permanent home

Jacques le chouan : madame en vendee / Muret, Theodore – Paris 1833 [mf ed Hildesheim 1995-98] – 1v on 3mf – 9 – €90.00 – ISBN-10: 3-487-26039-5 – ISBN-13: 978-3-487-26039-6 – gw Olms [440]

Jacques marquette et la decouverte de la vallee du mississipi sic / Brucker, Joseph – Lyon: Pitra Aine, 1880 – 1mf – 9 – mf#08696 – cn Canadiana [917]

Jacques, Marthe *see* Antoine goulet, de la societe des poetes canadiens-francais

Jacques riviere et alain fournier: correspondance, 1905-14 / Riviere, Jacques – (Paris): Gallimard, (1940, c. 1926-38). 4v. fronts – 1 – us Wisconsin U Libr [860]

Jacques rousseau : bio-bibliographie / Millo, Valentino [comp] – [Montreal]: Universite de Montreal, ecole des bibliothecaires, 1959-1960 [mf ed 2001] – 9 – cn Bibl Nat [580]

Jacquier *see*
- La credibilite des evangiles
- La resurrection de jesus-christ; les miracles evangeliques

Jacquier, Eugene *see*
- History of the books of the new testament
- Preparation, formation et definition du canon du nouveau testament
- Le texte du nouveau testament

Jacquin, Robert *see* Taparelli

Jacquinot *see* Zoologie

Jacquot / Gabriel, M – Paris, France. 1843 – 1r – us UF Libraries [440]

Jacqz, Jane W *see* Development needs in botswana and lesotho

Jad, Taha Muhammad *see* Al-khasais al-jimruflujiyah li-nahr al-sahl al-faydi

Jadassohn, Salomon *see* Die kunst zu moduliren und zu praeludiren

Jaderboeg, Elizabeth *see* Notes, correspondence and clippings relating to joseph kinchen griffis

Jadeson, Sam *see* The alien in american law

Jadin, Hyacinthe *see*
- Trois trios pour deux violons et basse, oeuvre 1er de trios
- Trois trios pour violon, alto et basse, oeuvre 2e de trios

Jadin, Louis-Emmanuel *see*
- 3me quatuor concertant pour piano, violon, alto et violoncelle
- Trois grands quatuors pour deux violons, alto et violoncelle
- Trois nocturnes en trois livraisons pour piano et violon, [3] livraison

Jadis et aujourd'hui / Sewrin, M – Paris, France. 1808 – 1r – us UF Libraries [440]

Jadran – San Francisco CA, feb 26 1908-dec 29 1910 – 2r – 1 – (croatian newspaper) – us IHRC [071]

Jae gaell, so geit's : e luschtigi gschicht us truuriger zyt / Tavel, Rudolf von – 10. aufl. Bern: A Francke, 1928 [mf ed 1993] – 220p/1pl (ill) – 1 – mf#7743 – us Wisconsin U Libr [830]

Jaeck, Joachim *see* Reise durch frankreich, england und die beiden niederlande

Jaeger, Abraham *see* Mind and heart in religion

Jaeger, Adolf *see* Das hohelied salomos

Jaeger, B *see* Reise von st petersburg in die krim und die laender des kaukasus im jahre 1825

Jaeger, Fritz *see* Beitraege zur landeskunde von suedwestafrika

Jaeger, Gottfried et al *see* Insunt 6 dissertationes varii argumenti

Jaeger, Hans *see* Clemens brentanos fruehlyrik

Jaeger, Hella *see* Naivitaet

Jaeger, Karl *see*
- Das bauernhaus in palaestina
- Johann brenz

Jaeger, Katrin *see* 'Nektar der unsterblichkeit'

Jaeger, Luis Gonzaga *see* Os herois de coaro e pirapo

Jaeger, Paul *see* Zur ueberwindung des zweifels

Jaeger, Samuel *see* Der weg zu gott unserm vater

Jaeger, Tobias Ludwig Ulrich *see* Juristisches magazin fuer die deutschen reichsstaedte

Jaeger, W *see* Two rediscovered works of ancient christian literature

Jaeger, Werner Wilhelm *see* Nemesios von emesa

Die jaegerin : [a novel] / Blunck, Hans Friedrich – Hamburg: Hanseatische Verlagsanstalt, 1940 [mf ed 1989] – 278p – 1 – mf#7037 – us Wisconsin U Libr [830]

Jaegerndorfer zeitung – Jaegerdorf (Krnov CZ), 1938 – 1r – 1 – gw Misc Inst [077]

Jaehnert, Katrin *see* Untersuchungen zur semantischen kongruenz substantivisch-verbaler lexemverbindungen im russischen und deutschen (manuskript 1984)

Jaell, Marie *see* Voix du printemps

Jaentsch, H *see* Denkschrift zum entwurf eines buergerlichen gesetzbuchs nebst drei anlagen, ergaenzt durch hinweise auf die beschluesse des reichstages sowie auf die paragraphen des buergerlichen gesetzbuchs und seiner nebengesetze

Ja'far Sharif *see* Islam in india

Jafar Sharif *see* Qanoon-e-islam

Jafari, Salih ibn al-Husayn *see* Disputatio pro religione mohammedanorum adversus christianos

Jaffe, Abraham Nissan *see* Bikure nisan

Jaffe, Ph *see*
- Bibliotheca rerum germanicarum
- Regesta pontificum romanorum ab condita ecclesia ad annum 1198

Jaffe, Philipp *see* Monumenta gregoriana

Jaffray, Robert *see* Essay on the reasons of secession from the national church of scotland

Jag journal – Alexandria. 1947-1984 (1) 1971-1984 (5) 1975-1984 (9) – (cont by: naval law review) – ISSN: 0021-3519 – mf#2090 – us UMI ProQuest [340]

Jag law review / United States Air Force – Washington. 1959-1974 (1) 1971-1973 (5,9) – (cont by: air force law review) – ISSN: 0021-3527 – mf#5750 – us UMI ProQuest [355]

Jag law review *see* Air force law review

Les jaga et les bayaka du kwango : contribution historico-ethnographique / Plancquaert, M – Bruxelles: G van Campenhout, 1932 – us CRL [306]

Jagadisa Ayyar, P V *see*
- South indian festivities
- South indian shrines

Jagadisan, T N *see*
- My master gokhale
- The other harmony
- The wisdom of a modern rishi

Jagadiswarananda, Swami *see* Hinduism outside india

Jagannatha Panditaraja *see* Bhaminivilasa of panditaraja jagannatha

Jagemann, Hans C G von *see* Goethe's dichtung und wahrheit

Jager, Abbe *see* Histoire de photius

Jager, B *see* Reise von st petersburg in die krim und die laender der kaukasus im jahre 1825...

Jager, Esme de *see* Facilitators and learners

Jaggan, Vijay Aheer Jaggan *see* The educational background of the gifted indian pre-school child

Jaggar, Thomas James *see*
- Diary, journal and letters (wesleyan mission in fiji)
- Fiji journals and letters (wesleyan mission in fiji)

Jaggard, W *see* A view of all the right honourable lord mayors of this honourable city of london, 1558-1601

Jagged sword – v8 n5-9,11 [1983 jul-oct, dec], v9 n1-7 [1984 jan-aug], v10 n8 [1985 sep], v11 n9-v15 n5 [1986 dec-1990 jul] – 1r – 1 – (cont by: seoulword) – mf#1060172 – us WHS [071]

Jagic, I V *see*
- Slovenskaia psaltyr
- Sluzhebnye minei za sentiabr, oktiabr i noiabr v tserkovnoslavianskom perevode po russkim rukopisiam 1095-1097 gg
- Zografskoe evangelie

Jagic, Vatroslav *see* Entstehungsgeschichte der kirchenslavischen sprache

Jagirdar, R V *see* Drama in sanskrit literature

Jagow, Eugen von *see* Die chauvinisten

Der jagteuffel bestendiger vnd wolgegruendter bericht : wie fern die jagten rechtmessig und zugelassen / Spangenberg, C – [Eisleben, 1560] – 3mf – 9 – mf#TH-1 mf 1439-1441 – ne IDC [242]

Jahan – [Tehran]: Danishjuyan-i Havadar-i Sazman-i Chirik'ha-yi Fida'i-i Khalq-i Iran dar kharij az Iran. sal-i 1, shumarah-'i 1-sal-i 3, shumarah-'i 26. 12 day 1360-azar 1363 [2 jan 1982-dec 1984] – 1r – 1 – $65.00 – us MEDOC [079]

Jahandiez, E *see* Catalogue des plantes du maroc (spermatophytes et pteridophytes

Jahangir and the jesuits / Guerreiro, Fernao – London: George Routledge & Sons, 1930 – (= ser Samp: indian books) – (trans by c h payne) – us CRL [241]

Jahan-i zanan / ed by Siyasifar, Farzanah – Tihran: Jahan-i Zanan. dawrah-'i jadid, sal-i duvvum, shumarah-'i 10,12 khurdad-murdad 1360 [may-aug 1981]; dawrah-'i jadid, sal-i sivvum, shumarah-'i 2,5,6,7,8,12 mihr 1360-murdad 1361 [sep 1981-jul 1982] – 9mf – 9 – $150.00 – us MEDOC [079]

Jahn, A *see* Des h eustathius beurtheiling des origenes

Jahn, Alfred *see* Studies on geology of the sudetic mountains

Jahn, August *see* Reise von mainz nach egypten, jerusalem und konstantinopel in den jahren 1826-27

Jahn, Gustav *see*
- Beitraege zur beurtheilung der septuaginta
- Die buecher esra (a und b) und nehemja
- Die elephantiner papyri und die buecher esra-nehemja
- Ueber den gottesbegriff der alten hebraeer und ihre geschichtsschreibung
- Ueber die person jesu und ueber die entstehung des christenthums und den werth desselben fuer modern gebildete

Jahn, Gustav Adolph *see* Der grosse komet von 1556 und seine bevorstehende wiederkehr

Jahn, Janheinz *see* History of neo-african literature

Jahn, Johann *see* Introduction to the old testament

Jahn, Kurt *see* Edward youngs gedanken ueber die originalwerke

Jahn, M *see* Sittlichkeit und religion

Jahn, Moritz *see* Frangula, oder, die himmlischen weiber im wald

Jahn, O W A *see* Mozart

Jahn, Otto *see*
- Gesammelte aufsaetze
- Ludwig uhland

Jahn, Ulrich *see*
- Hexenwesen und zauberei in pommern
- Schwaenke und schnurren aus bauern mund

Jahn, Walter *see* Dramatische elemente in hebbels jugendballaden, 1829-1839

Das jahr 1848 im deutschen drama und epos / Dohn, Walter – Stuttgart: J B Metzler, 1912 [mf ed 1992] – (= ser Breslauer beitraege zur literaturgeschichte. neue folge 32) – vii/294p – 1 – (incl bibl ref and ind) – mf#8014 reel 3 – us Wisconsin U Libr [430]

Ein jahr aus dem leben einer hausfrau in sued-afrika / Broome, Mary A – Wien [u a] 1878 [mf ed Hildesheim 1995-98] – 1v on 3mf [ill] – 9 – €90.00 – 3-487-27187-7 – gw Olms [920]

Ein jahr aus meinem leben : oder reise von den westlichen ufern der donau an die nara, suedlich von moskwa, und zurueck an die beresina mit der grossen armee napoleons, im jahre 1812 / Roos, Heinrich U von – St Petersburg 1832 [mf ed Hildesheim 1995-98] – (= ser Fbc) – 1v on 3mf – 9 – €90.00 – 3-487-27800-6 – gw Olms [914]

Das jahr der schoenen taeuschungen / Carossa, Hans – Leipzig: Insel-Verlag 1942 [mf ed 1989] – 1r – 1 – ("cont of: 'verwandlungen einer jugend' ") – mf#7143 – us Wisconsin U Libr [880]

Das jahr der seele / George, Stefan Anton – Godesberg: H. Kuepper vormals G Bondi 1948 [mf ed 1989] – 1r – 1 – (filmed with: gellerts lustspiele / wold. haynel) – mf#7293 – us Wisconsin U Libr [810]

Ein jahr in italien / Stahr, Adolf W – Oldenburg 1853/54 [mf ed Hildesheim 1995-98] – (= ser Fbc) – 3v on 10mf – 9 – €100.00 – 3-487-29257-2 – gw Olms [920]

Ein jahr in italien *see* Herbstmonate in oberitalien

Das jahr in vier gesaengen : ein laendliches epos / Donelaitis, Kristijonas – Koenigsberg: Hartung 1818 – (= ser Whsb) – 2mf – 9 – €30.00 – (trans into german fr lithuanian by ludwig jedemin rhesa) – mf#Hu 205 – gw Fischer [780]

Jahr- und handbuch / Deutscher Metallarbeiterverband – Stuttgart, Berlin. 1904, 1906-09, 1911-20, 1922-31. (Serial publications of German trade unions in the Memorial Library, University of Wisconsin-Madison.) – 1 – us Wisconsin U Libr [330]

Jahrbuch / Allgemeiner deutscher Gewerkschaftsbund – Berlin. 1922-27, 1929-31 – (= ser Serial publications of german trade unions) – 1 – us Wisconsin U Libr [331]

Jahrbuch / Arbeiterrat Gross-Hamburg – Hamburg. 1928-29, 1931. Some vols. issued as Bericht ueber das Jahr. (Serial publications of German trade unions in the Memorial Library, University of Wisconsin-Madison) – 1 – us Wisconsin U Libr [330]

Jahrbuch / Bund der Bau-, Maurer- und Zimmermeister zu Berlin – Berlin. v16. 1926 27. List of members in each vol. (Serial publications of German trade unions in the Memorial Library, University of Wisconsin-Madison) – 1 – us Wisconsin U Libr [330]

Jahrbuch / Deutscher Bauarbeiterverband. Hamburg – 1914, 1920-21. (Serial Publications of German trade unions in the Memorial Library, University of Wisconsin-Madison.) – 1 – us Wisconsin U Libr [330]

Jahrbuch / Deutscher Baugewerksbund – Berlin. 1926-30. Ceased publ. with 1931 issue. (Serial publications of German trade unions in the Memorial library, University of Wisconsin-Madison.) – 1 – us Wisconsin U Libr [330]

Jahrbuch / Deutscher Holzarbeiter-Verband. Berlin – 1906, 1908, 1911-19, 1921-22, 1924-26, 1930-31. 26v. illus. (Serial publications of German trade unions in the Memorial Library, University of Wisconsin-Madison.) – 1 – us Wisconsin U Libr [330]

Jahrbuch / Deutscher Lederarbeiter-Verband – Berlin. 1928-31. illus. (Serial publications of German trade unions in the Memorial Library, University of Wisconsin-Madison) – 1 – us Wisconsin U Libr [330]

Jahrbuch / Deutscher Lehrerverein – Leipzig. v46. 1920. (Serial publications of German trade unions in the Memorial Library, University of Wisconsin-Madison.) – 1 – us Wisconsin U Libr [330]

Jahrbuch / Deutscher Nahrungs- und Genussmitterlarbeiter-verband – 1922 24-26. Hamburg. (Serial publications of German trade unions in the Memorial Library, University of Wisconsin-Madison.) – 1 – us Wisconsin U Libr [330]

Jahrbuch / Deutscher Sattler-, Tapezierer- und Portefeuiller-Verband – Berlin. 1924, 1926-31. (Serial publications of German trade unions in the Memorial Library, University of Wisconsin-Madison.) – 1 – us Wisconsin U Libr [330]

Jahrbuch / Deutscher Textilarbeiter-Verband – Berlin. 1912. 1914-30. (Serial publications of German trade unions in the Memorial Library, University of Wisconsin-Madison.) – 1 – us Wisconsin U Libr [330]

Jahrbuch / Deutscher Transportarbeiter-Verband – Berlin. 1913-14. (Serial publications of German trade unions in the Memorial Library, University of Wisconsin-Madison.) – 1 – us Wisconsin U Libr [330]

Jahrbuch / Deutscher Verkehrsbund – Berlin. 1927. (Serial publications of German trade unions in the Memorial Library, University of Wisconsin-Madison.) – 1 – us Wisconsin U Libr [330]

Jahrbuch / Gesamtverband der Arbeiternehmer der offentlichen betriebe und des Personen- und verkehrs – Berlin. v1, 1930. (Serial publications of German trade unions in the Memorial Library, University of Wisconsin-Madison.) – 1 – us Wisconsin U Libr [330]

Jahrbuch / Paul Zsolnay Verlag – Berlin: Der Verlag, 1927- – 1r – 1 – us Wisconsin U Libr [430]

Jahrbuch : paul zsolnay verlag – Berlin: Paul Zsolnay Verlag, 1927- – 1 – us Wisconsin U Libr [800]

Jahrbuch / Verband der Baecker, Konditoren und verwandten Berufsgenossen Deutschlands – Hamburg. 1907, 1910-14. (Serial publications of German trade unions in the Memorial Library, University of Wisconsin-Madison) – 1 – us Wisconsin U Libr [330]

Jahrbuch / Verband der Bergarbeiter Deutschlands – Bochum. 1921, 1923, 1926. (Serial publications of German trade unions in the Memorial Library, University of Wisconsin-Madison.) – 1 – us Wisconsin U Libr [330]

Jahrbuch / Verband der Bergbauindustriearbeiter Deutschlands – Bochum. 1930-31. (Serial publications of German trade unions in the Memorial Library, University of Wisconsin-Madison.) – 1 – us Wisconsin U Libr [330]

Jahrbuch / Verband der Brauerei- und Muehlenarbeiter und verwandter Berufsgenossen – Berlin. 1912, 1914-15. (Serial publications of German trade unions in the Memorial Library, University of Wisconsin-Madison.) – 1 – us Wisconsin U Libr [330]

Jahrbuch / Verband der Fabrikarbeiter Deutschlands – 1912, 1925, 1927-28, 1930-31. Hannover. Serial publications of German trade unions in the Memorial Library, University of Wisconsin-Madison.) – 1 – us Wisconsin U Libr [330]

Jahrbuch / Verband der Maler, Lackierer, Anstreicher, Tuencher und Weissbinder Deutschlands – 1921, 1927-31. Hamburg. (Serial publications of German trade unions in the Memorial Library, University of Wisconsin-Madison) – 1 – us Wisconsin U Libr [330]

Jahrbuch / Verband der Nahrungsmittel- und Getraenkearbeiter – 1928-31. Berlin. (Serial publications of German trade unions in the Memorial Library, University of Wisconsin-Madison.) – 1 – us Wisconsin U Libr [330]

Jahrbuch / Zentralverband der Maurer Deutschlands – 1907-10. Hamburg. Continued as: Deutscher Bauarbeiterverband. At head of title 1910 Zentralverbande der Maurer und Bauhilfsarbeiter. (Serial publications of German trade unions in the Memorial Library, University of Wisconsin-Madison.) – 1 – us Wisconsin U Libr [330]

Jahrbuch / Zentralverband der Schuhmacher – 1918 19, 1921, 1923-31. Nurnberg. (Serial publications of German trade unions in the Memorial Library, University of Wisconsin-Madison.) – 1 – us Wisconsin U Libr [330]

Jahrbuch / Zentralverband deutscher Konsumgenossenschaften – v1-31. 1903-33. (Serial publications of German trade unions in the Memorial Library, University of Wisconsin-Madison.) – 1 – us Wisconsin U Libr [330]

Jahrbuch *see* Nsdap (national socialist german workers party) nazi publications

Jahrbuch.. / Allgemeiner deutscher Gewerkschaftsbund. Ortsausschuss Muenchen – Muenchen. v23 24, 28-32. 1920-29 – (= ser Serial publications of german trade unions) – 1 – (title and subtitle varies) – us Wisconsin U Libr [331]

Jahrbuch... / Gesamtverband der Christlichen Gewerkschaften Deutschlands – Berlin. 1908, 1930, 1932. Some vols. issued as Schriften; some include Bericht. (Serial publications of German trade unions in the Memorial Library, University of Wisconsis-Madison) – 1 – us Wisconsin U Libr [330]

Jahrbuch der angestelltenbewegung – Berlin. 1913. (Serial publications of German trade unions in the Memorial Library, University of Wisconsin-Madison.) – 1 – us Wisconsin U Libr [331]

Jahrbuch der deutschen schillergesellschaft – Stuttgart, 1957-59 – 1r – 1 – gw Mikropress [943]

Jahrbuch der frauenarbeit – v6-8. 1930-32. Verband der weiblichen Handels- und Buroangestellten. (Serial publications of German trade unions in the Memorial Library, University of Wisconsin-Madison.) – 1 – us Wisconsin U Libr [331]

Jahrbuch der frauenbewegung / ed by Altmann-Gottheiner, Elisabeth et al – v1-10 1912-21, v10 [11]-11 [12]-12 [13] 1921/27-1928-31 [mf ed 1998] – (= ser Hq 35) – 45mf – 9 – €200.00 – 3-89131-294-6 – (with various titles between v4 1915-v7 1918; later under title: jahrbuch des bundes deutscher frauenvereine) – gw Fischer [305]

Jahrbuch der gesellschaft fuer geschichte der juden : gesellschaft fuer geschichte der juden in der tsechoslovakischen republik / ed by Steinherz, Samuel – Prague. v1-9. 1929-38 – (= ser German-jewish periodicals...1768-1945, pt 3) – 4r – 1 – $605.00 – mf#B125 – us UPA [939]

Jahrbuch der gesellschaft fuer geschichte der juden in der cechoslovakischen republik – Prague XR, 1929-37 – 5r – 1 – us UMI ProQuest [939]

Jahrbuch der goethe-gesellschaft – Weimar: Verlag der Goethe-Gesellschaft, 1914-1935 [mf ed 1996] – 22v – 1 – (none publ for 1923. ed for 1914-22 hans gerhard graef, 1924-35 max hecker) – mf#9655 – us Wisconsin U Libr [030]

Jahrbuch der innung bund der bau-, maurer- und zimmermeister zu berlin... *see* Berichte und protokoll vom verbandstag...

Jahrbuch der kleist-gesellschaft – Berlin: Weidmann, 1921-1939 [mf ed 1994] – (= ser Schriften der kleist-gesellschaft 1-3/4, 7/8-13/14, 17-18) – 9v on 2r – 1 – (incl bibl ref. some vols are combined yrs) – mf#8707 – us Wisconsin U Libr [430]

Jahrbuch der koeniglich preussischen kunstsammlungen – Berlin, 1880-1918. v1-39 – (= ser Jahrbuch der preussischen kunstsammlungen) – 348mf – 9 – (cont as: jahrbuch der preussischen kunstsammlungen. berlin, 1919-1925. v40-46+ind 1891, v1-10; 1900, v11-20; 1910, v21-30) – mf#O-503c – ne IDC [700]

Jahrbuch der kunsthistorischen sammlungen in wien [jahrbuch der kunsthistorischen sammlungen des allerhoechsten kaiserhauses. v 1-34] – Wien, 1883-1925, v1-36; N S, 1926-1944, v1-13 – 929mf – 9 – mf#O-502c – ne IDC [700]

Jahrbuch der musikbibliothek peters / Musikbibliothek Peters – Leipzig. 1894-1939. 3 reels – 1 – 76.00 – us L of C Photodup [780]

Jahrbuch der preussischen kunstsammlungen – Berlin. v1-61. 1880-1940 – 9 – $550.00 – mf#0301 – us Brook [700]

Jahrbuch der psychoanalyse. *see* Jahrbuch fuer psychoanalytische und psychopathologische forschungen

Jahrbuch der sammlung kippenberg – Leipzig: Insel-Verlag, 1921- [mf ed 1994] – (ill) – 1 – mf#8635 – us Wisconsin U Libr [430]

Jahrbuch der schweizerfrauen = Annuaire feminin suisse / ed by Bern. Sektion des Schweizerischen Verbandes fuer Frauenstimmrecht – 1915, 1940/41 [mf ed 2001] – (= ser Hq 51) – 20v on 43mf – 9 – €190.00 – 3-89131-382-9 – gw Fischer [305]

Jahrbuch der seele / Thoma, Hans – Jena: E Diederichs, 1922 – 1r – 1 – us Wisconsin U Libr [750]

Jahrbuch der sozialdemokratischen partei – Hannover, Bonn DE, 1946 – 1 – gw Misc Inst [325]

Jahrbuch der weltpolitik / Deutsches Auslandswissenschaftliches Institut; ed by Six, Franz Alfred – Berlin: Junker & Duennhaupt 1942-44 (annual) [mf ed 1979] – 3v on 3r – 1 – (cont: jahrbuch fuer politik und auslandskunde) – mf#film mas c686 – us Harvard [320]

Jahrbuch des bundes / German Baptist Convention – 1878-1959 60 – 1 – us Southern Baptist [242]

Jahrbuch des bundes deutscher frauenvereine *see* Jahrbuch der frauenbewegung

Jahrbuch des deutschen rechtes – Berlin. On film: v1, pt.2-v39; 1903-41. LL-0245 – 1 – us L of C Photodup [340]

Jahrbuch des deutschen rechtes – Berlin. v1-38+ind (2v). 1903-40 – 9 – $966.00 – mf#0302 – us Brook [348]

Jahrbuch des historischen vereins dillingen / Historischer Verein Dillingen an der Donau [Germany] – Dillingen a/D: J Keller 1897- [v1-37 mf ed 1978] – 4r [ill] – 1 – (cont: jahresbericht des historischen vereins dillingen) – mf#film mas c385 – us Harvard [943]

Jahrbuch des kaiserlich deutschen archaeologischen instituts / Kaiserlich Deutsches Archaeologisches Institut – Berlin: Georg Reimer 1887-1917 (annual) [v1-30(1886-1915)] [mf ed 1978] – 13r – 1 – (suppls accompany v4-32, with title: archaeologischer anzeiger; cont: archaeologischer zeitung; cont by: jahrbuch des deutschen archaeologischen instituts; ceased with vol for 1917; incl bibl) – mf#film mas c387 – us Harvard [930]

Jahrbuch des kaiserlich deutschen archaeologischen instituts. ergänzungsheft / Kaiserlich Deutsches Archaeologisches Institut – Berlin: G Reimer 1888-1913 [mf ed 198-] – 1 – (cont by: jahrbuch des deutschen archaeologischen instituts. ergaenzungsheft) – mf#film mas 8163 – us Harvard [930]

Jahrbuch des vereines fuer geschichte der deutschen in boehmen / Verein fuer Geschichte der Deutschen in Boehmen [Prague, Czechoslovakia] – Prag: Im Selbstverlage des Vereines 1926-34 – 3v [ill] – 1 – mf#film mas 8957 – us Harvard [943]

Jahrbuch fuer bremische statistik / Bremen. Statistisches Landesamt – v. 1-46. 1876-1915 16. 1875-80, 1881 wanting – 1 – us L of C Photodup [943]

Jahrbuch fuer deutschnationale handlungsgehilfen – v1-7, 9-10, 12-14, 17-21. 1900-20. -ann. (Serial publications of German trade unions in the Memorial Library, University of Wisconsin-Madison.) – 1 – us Wisconsin U Libr [331]

Jahrbuch fuer die amtliche statistik des bremischen staats / Bremen. Statistisches Amt – v. 1-8. 1868-75 – 1 – 52.00 – us L of C Photodup [943]

Jahrbuch fuer die geschichte der juden und des judenthums – Leipzig DE,1860-71 – 2r – 1 – us UMI ProQuest [939]

Jahrbuch fuer die geschichte der juden und des judentums / ed by Steinherz, Samuel – Leipzig: Oskar Leiner. v1-4. 1859-62; 1968 – (= ser German-jewish periodicals...1768-1945, pt 3) – 1r – 1 – $165.00 – mf#B124 – us UPA [939]

Jahrbuch fuer die israelitischen cultus-gemeinden in ungarn und seinen ehemaligen nebenlaendern / ed by Rosenberg, Leopold – Arad: H Goldschneider. v1. 1860 – (= ser German-jewish periodicals...1768-1945, pt 3) – 1r – 1 – $165.00 – mf#B127 – us UPA [939]

Jahrbuch fuer die juedische gemeinden preussens – Berlin DE, 1856-58 – 1r – 1 – us UMI ProQuest [939]

Jahrbuch fuer die juedische Gemeinden Schleswig-Holstein – Hamburg DE, 1929-32 – 1r – 1 – us UMI ProQuest [939]

Jahrbuch fuer die juedischen gemeinden schleswig-holsteins und der hansestaedte und der landesgemeinde oldenberg... – Hamburg: Ackermann & Wulff. v1-9. 1929-37 – (= ser German-jewish periodicals...1768-1945, pt 3) – 1r – 1 – mf#B129 – us UPA [939]

Jahrbuch fuer die menschheit – Hannover DE, 1788-90 – 4r – 1 – gw Misc Inst [900]

Jahrbuch fuer die menschheit : oder beytraege zur befoerderung haeuslicher erziehung, haeuslicher glueckseligkeit und praktischer menschenkenntnis / ed by Benecken, Friedrich Burch – Hannover 1788-91 – (= ser Dz) – 4jge[zu je 12st] on 25mf – 9 – €250.00 – mf#k/n705 – gw Olms [370]

Jahrbuch fuer frauenarbeit / ed by Silbermann, J – 1924-32 [mf ed 2000] – (= ser Hq 44) – 8v on 17mf – 9 – €100.00 – 3-89131-363-2 – (successor to: archiv fuer frauenarbeit) – gw Fischer [305]

Jahrbuch fuer juedische geschichte und literatur – Berlin. bd 1-31. 1898-1937 – 1 – us NY Public [939]

Jahrbuch fuer juedische geschichte und literatur – Berlin DE, 1898-1936 – 9r – 1 – us UMI ProQuest [939]

Jahrbuch fuer liturgiewissenschaft – v1-15. 1921-35 [complete] – Inquire – 1 – mf#ATLA 1993-S500 – us ATLA [240]

Jahrbuch fuer photographie und reproduktionstechnik – Halle S DE, 1890-1920 – 7r – 1 – gw Mikrofilm [770]

Jahrbuch fuer psychoanalytische und psychopathologische forschungen – Wien (A), 1909-14 – 1 – (title varies: 1914: jahrbuch der psychoanalyse) – fr ACRPP [150]

Jahrbuch fuer sexuelle zwischenstufen *see* Jahrbuch fuer sexuelle zwischenstufen unter besonderer beruecksichtigung der homosexualitaet

Jahrbuch fuer sexuelle zwischenstufen unter besonderer beruecksichtigung der homosexualitaet / ed by Hirschfeld, Magnus – Leipzig. v1-9. 1899-1908 [mf ed 1996] – (= ser Hq 20) – 78mf – 9 – €830.00 set – 3-89131-132-X – (filmed with: vierteljahrsberichte des wissenschaftlich-humanitaeren komitees. fortsetzung der monatsberichte und des jahrbuchs fuer sexuelle zwischenstufen [v1-4 1909-12]; jahrbuch fuer sexuelle zwischenstufen...[v13 1913, v14 1914]; vierteljahrsberichte...waehrend der kriegszeit [v15-18 1915-18]; jahrbuch fuer sexuelle zwischenstufen...[v19-23 1919-23]) – gw Fischer [618]

Jahrbuch zum conversations-lexikon *see* Unsere zeit, leipzig 1857-1891

Jahrbuecher / Sozialdemokratische Partei Deutschlands – 1946-1968/69 – 1 – gw Mikropress [943]

Jahrbuecher der deutschen malakozoologischen gesellschaft – Frankfurt am Main 1874-87 – v1-14 on 107mf – 9 – mf#2802/2 – ne IDC [590]

Jahrbuecher der insektenkunde... / ed by Klug, R – Berlin 1834 – v1 on 6mf – 9 – mf#z-963/2 – ne IDC [590]

Jahrbuecher der literatur – Wien 1818-49 [mf ed 1991] – (= ser Wiener literaturzeitschriften der ersten haelfte des 19. jahrhunderts) – 357mf – 9 – €1790.00 – 3-89131-038-2 – gw Fischer [430]

Jahrbuecher des vereins von alterthumsfreunden – Bonn, 1842-1926. v1-131 – 856mf – 8 – mf#H-360c – ne IDC [700]

Jahrbuecher fuer die dogmatik des heutigen ... privatrechts – Jena: F Mauke. Annual, v1-50 only. 1857-1906 – (= ser Civil law 3 coll) – 268mf – 9 – (series continues to 1943. in 1893 title became jherings jahrbuecher fuer die dogmatik des heutigen roemischen und deutschen privatrechts. in 1897 title changed again to jherings jahrbuecher fuer die dogmatik des buergerlichen rechts. the 50 volumes of this publication which are provided by llmc are alternately numbered as: 1st series, v1-12, 1857-73; new series, v1-24, 1874-96; and second series, v1-14, 1897-1906) – mf#LLMC 96-572 – us LLMC [346]

Jahrbuecher fuer juedische geschichte und literatur / ed by Bruell, Nehemias – Frankfurt 1874-90 [mf ed Hildesheim 1989] – (= ser Bibliothek des deutschen judentums) – 10v on 23mf – 9 – diazo €118.00 silver €138.00 – gw Olms [939]

Jahrbuecher fuer juedische geschichte und literatur – Frankfurt a.M: Nehemias Bruell. 1874, 1876, 1877, 1879, 1883-85, 1887, 1889, 1890 [complete] – (= ser German-jewish periodicals...1768-1945, pt 1) – 1r – 1 – $125.00 – mf#B133 – us UPA [939]

Jahrbuecher fuer juedische geschichte und literatur – Frankfurt/M DE, 1874, 1877, 1879, 1883-85, 1887, 1889-90 – 1r – 1 – gw Misc Inst [939]

Jahrbuecher fuer kultur und geschichte der slaven – Breslau, 1924-1935. v1-11(4) – 107mf – 8 – mf#R-8064 – ne IDC [700]

Jahrbuecher fuer protestantische theologie – 1(1875)-18(1892) – 222mf – 9 – €423.00 – ne Slangenburg [242]

Jahrbuecher fuer statistik und landeskunde / Wuerttemberg – 1866-1951 52. 1876, 1938-50 wanting – 1 – us L of C Photodup [310]

Jahrbuecher fuer wissenschaftliche kritik – Stuttgart/Tuebingen 1827-33, Berlin 1833-46 [mf ed 1990] – 261mf – 9 – €940.00 – 3-89131-035-8 – gw Fischer [500]

Die jahre der reaktion : historische skizze / Bernstein, Aaron David – Berlin: M Bading, 1881 [mf ed 1990] – iv/260p – 1 – mf#7370 – us Wisconsin U Libr [943]

Jahre und zeiten : erinnerungen / Wiechert, Ernst Emil – Erlenbach-Zuerich: E Rentsch, [1948, c1949] – 1r – 1 – us Wisconsin U Libr [080]

Jahres- und rechenschaftsbericht / Zentralverband der Glasarbeiter und -Arbeiterinnen Deutschlands – Berlin. 1921 23. Title varies slightly. (Serial publications of German trade unions in the Memorial Library, University of Wisconsin-Madison.) – 1 – us Wisconsin U Libr [330]

Jahres-bericht... / Arbeiter-Sekretariat. Muenchen – v1-4, 8-11, 17-21 22. 1898-1918 19. (Serial publications of German trade unions in the Memorial Library, University of Wisconsin-Madison) – 1 – us Wisconsin U Libr [330]

Jahresbericht / Deutsche Angestelltenschaft. Gau Brandenburg-Pommern – Berlin. 1929. (Serial publications of German trade unions in the Memorial Library, University of Wisconsin-Madison.) – 1 – us Wisconsin U Libr [330]

Jahresbericht / Deutscher Forstverein – [Donaueschingen, etc] (annual) [1925-39][mf ed 2005] – 5r [ill] – 1 – (cont: deutscher forstverein. hauptversammlung. bericht ueber die...hauptversammlung des deutschen forstvereins; vol for 1925 contains: deutscher forstverein. hauptversammlung. bericht ueber die...hauptversammlung des deutschen forstvereines; some vols for 1926- contain: deutscher forstverein. mitgliederversammlung. bericht ueber die...mitgliederversammlung des deutschen forstvereins) – mf#film mas c6059 – us Harvard [634]

Jahresbericht / Deutscher Tabakarbeiter-Verband – Bremen. 1908, 1910, 1922 24, 1928-29. (Serial publications of German trade unions in the Memorial Library, University of Wisconsin-Madison.) – 1 – us Wisconsin U Libr [330]

Jahresbericht / Historischer Verein fuer die Graftschaft Ravensberg zu Bielefeld – Bielefeld: Historischer Verein fuer die Grafschaft Ravensberg 1877- [v1-56 mf ed 1978] – 3r – 1 – (1901 incl register 1877-1900; 1984/85 incl register jahrg 1-75) – mf#film mas c302 – us Harvard [943]

Jahresbericht / Verband der Bergarbeiter Deutschlands – Bochum. 1907 08. (Serial publications of German trade unions in the Memorial Library, University of Wisconsin-Madison.) – 1 – us Wisconsin U Libr [330]

Jahresbericht / Verband der Bergbauindustriearbeiter Deutschlands. Bezirk Saarbruecken – 1930-32. Saarbrucken. (Serial publications of German trade unions in the Memorial Library, University of Wisconsin-Madison.) – 1 – us Wisconsin U Libr [330]

Jahresbericht / Verband der deutschen Buchdrucker – Berlin. 1919-31. Title varies: Bericht. Includes some suppls. for regional unions. (Serial publications of German trade unions in the Memorial Library, University of Wisconsin-Madison.) – 1 – us Wisconsin U Libr [330]

Jahresbericht / Verband der Gemeinde- und Staatsarbeiter – 1919-21. Berlin. (Serial publications of German trade unions in the Memorial Library, University of Wisconsin-Madison.) – 1 – us Wisconsin U Libr [330]

Jahresbericht / Verband der Steinsetzer, Pflasterer und Berufsgenossen Deutschlands. Berlin – 1907-08, 1910 11, 1912, 1914 15. (Serial publications of German trade unions in the Memorial Library, University of Wisconsin-Madison.) – 1 – us Wisconsin U Libr [330]

Jahresbericht / Verband deutscher Berufsfeuerwehrmaenner – 1925. Berlin. (Serial publications of German trade unions in the Memorial Library, University of Wisconsin-Madison.) – 1 – us Wisconsin U Libr [330]

Jahresbericht / Verein fuer Naturkunde in OEsterreich, Ob der Enns, Linz – v1-44 1870-1918 – 1 – us Wisconsin U Libr [500]

Jahresbericht / Zentralverband der Lederarbeiter und Arbeiterinnen Deutschlands – 1912-13, 1915. Berlin. (Serial publications of German trade unions in the Memorial Library, University of Wisconsin-Madison.) – 1 – us Wisconsin U Libr [330]

Jahresbericht... / Arbeiter-Sekretariat. Bremen – v3, 1902. (Serial publications of German trade unions in the Memorial Library, University of Wisconsin-Madison) – 1 – us Wisconsin U Libr [330]

Jahresbericht... / Arbeiter-Sekretariat. Nuremberg – v20. 1914-20. (Serial publications of German trade unions in the Memorial Library, University of Wisconsin-Madison) – 1 – us Wisconsin U Libr [330]

Jahresbericht am...dem comite der nicolai-hauptsternwarte abgestattet vom director der sternwarte / Nikolaevskaia glavnaia astronomicheskaia observatoriia – St Petersburg: Buchdr der Kaiserlichen Akademie der Wissenschaften -1887 (annual) [mf ed 2006] – 1r – 1 – (trans fr russian; cont by: nikolaevskaia glavnaia astronomicheskaia observatoriia. bericht fuer der periode...den comite der nicolai-hauptsternwarte ueber deren thaetigkeit abgestattet von director der sternwarte; began in 1862? vol for 1884 covers 1882/1883-1883/1884; no separate iss publ for 1883; title varies slightly) – mf#film mas 37493 – us Harvard [520]

Jahresbericht der aktiengesellschaft reichskohlenverband – Berlin, 1920-38 – 2r – 1 – gw Mikropress [380]

Jahresbericht der deutschen gesellschaft von milwaukee – 1881-1912 – 1r – 1 – mf#1321908 – us WHS [366]

Jahresbericht der executivdirektoren fuer das am. abgelaufene geschaeftsjahr / International Monetary Fund – Washington. 1972-1974 [1]; 1964-1974 [5]; 1964-1973 [9] – ISSN: 0250-7528 – mf#6533 – us UMI ProQuest [332]

Jahresbericht der gesellschaft zur foerderung der wissenschaft des judenthums *see* Altneuland

Jahresbericht der israelitisch-theologischen lehranstalt in wien – Vienna AU, 1894-1919 – 3r – 1 – us UMI ProQuest [939]

Jahresbericht der landes-rabbinerschule in budapest – Budapest HU, 1877-1917 – 7r – 1 – us UMI ProQuest [270]

Jahresbericht der ortsverwaltung fuer.../ zentralverband.. / Zentralverband der Schuhmacher. Zahlstelle Berlin – 1923 24, 1925 27. Berlin. Title from cover. Description based on: 1923-24. (Serial publications of German trade unions in the Memorial Library, University of Wisconsin-Madison.) – 1 – us Wisconsin U Libr [330]

Jahresbericht der ostasien-mission / Allgemeiner Evangelisch-Protestantischer Missionsverein – Berlin, 1887-1939 [mf ed 2001] – (= ser Christianity's encounter with world religions, 1850-1950) – 5r – 1 – mf#2001-s500-503 – us ATLA [242]

Jahresbericht des entomologischen vereins zu stettin – Stettin 1839. v1 – 661mf – 9 – (cont as: entomologische zeitung [stettin 1840-1914] v1-75; cont as: stettiner entomologische zeitung [stettin 1915-26] v76-87) – mf#z-964c/2 – ne IDC [590]

Jahresbericht des juedisch-theologischen seminars fraenckelscher stiftung – Wroclaw PL, 1865-1934 – 6r – 1 – us UMI ProQuest [939]

Jahresbericht des physikalischen vereins zu frankfurt am main / Physikalischer Verein. Frankfurt am Main – 15v 1858-1931 – 1 – $92.00 – us L of C Photodup [530]

Jahres-bericht des rabbiner-seminars fuer das orthodoxe judenthum – Berlin DE, 1876-1927 – 3r – 1 – us UMI ProQuest [270]

Jahresbericht des vereins fuer die bergbaulichen interessen – Essen, 1861-1913; 1925-30; 1934-36 – 5r – 1 – gw Mikropress [943]

Jahresbericht des vereins fuer geschichte der stadt nuernberg / Verein fuer Geschichte der Stadt Nuernberg – Nuernberg: Im Selbstverlage des Vereins 1881-90 (annual) [v3-76 mf ed 1978] – 10v on 2r – 1 – (1st & 2nd yearly reports iss in: mitteilungen des vereins fuer geschichte der stadt nuernberg, v1 [1878] & v2 [1879] respectively) – mf#film mas c424 – us Harvard [943]

Jahresbericht fuer das geschaftsjahr / Deutscher Metallarbeiterverband. Verwaltungsstelle. Berlin – Berlin. 1918-19, 1921-22, 1924, 1926-27. (Serial publications of German trade unions in the Memorial Library, University of Wisconsin-Madison.) – 1 – us Wisconsin U Libr [330]

Jahresbericht fur 1931 / Hilfsverein Der Deutschen Juden – Berlin, Germany. 1932 – 1r – us UF Libraries [943]

Jahresbericht uber die entwickelung der schutzgebiete in afrika und der sudsee = Annual reports of the german colonies in afrika and the south seas – 1898-1908 – 1r – 1 – (this series for pacific colonies only) – mf#PMB Doc401 – at Pacific Mss [980]

Jahresbericht ueber das geschaftsjahr / Deutscher Verkehrsbund – Berlin. 1929. (Serial publications of German trade unions in the Memorial Library, University of Wisconsin-Madison.) – 1 – us Wisconsin U Libr [330]

Jahresbericht ueber die erscheinungen auf dem gebiete der germanischen philologie / ed by Gesellschaft fuer Deutsche Philologie in Berlin – Berlin: S Calvary 1880- (annual) [v1-22 1879-1900 mf ed 1979] – 3r – 1 – (merged with: jahresbericht ueber die wissenschaftlichen erscheinungen auf dem gebiete der neueren deutschen literatur, to form: jahresbericht fuer deutsche sprache und literatur; ceased in 1935) – mf#film mas c681 – us Harvard [430]

Jahresbericht ueber die fortschritte der klassischen altertumswissenschaft – Berlin: S Calvary 1874- [v1-285] [mf ed 1979] – 39r – 1 – (title varies slightly; publ: leipzig: o r reisland 1898-) – mf#film mas c675 – us Harvard [450]

Jahresbericht ueber die fortschritte der klassischen altertumswissenschaft 1899-1909 – Leipzig. v100-144. 1900-1910 – 280mf – 8 – mf#H-539 – ne IDC [450]

Jahres-bericht ueber die religionschule der synagogen-gemeinde zu hannover – Hannover DE, 1896-1915 – 1r – 1 – us UMI ProQuest [270]

Jahresbericht ueber die taetigkeit des zentralvorstundes / Deutscher Lederarbeiter-Verband – Berlin. 1922, 1925-27. illus. (Serial publications of German trade unions in the Memorial Library, University of Wisconsin-Madison.) – 1 – us Wisconsin U Libr [330]

Jahresbericht ueber die thaetigkeit der kaiserlichen nicolai-hauptsternwarte / Nikolaevskaia glavnaia astronomicheskaia observatoriia – St Petersburg: Buckdr der Kaiserlichen Akademie der Wissenschaften (annual) [mf ed 2006] – 1r – 1 – (trans fr russian; cont: nikolaevskaia glavnaia astronomicheskaia observatoriia. bericht fuer der periode...dem comite der nicolai-hauptsternwarte ueber deren thaetigkeit abgestattet von director der sternwarte) – mf#film mas 37493 – us Harvard [520]

Jahresbericht ueber die wissenschaftlichen erscheinungen auf dem gebiete der neuren deutschen literatur / ed by Berlin. Literaturarchiv-Gesellschaft – Berlin: De Gruyter. 19v. 1924-56 (annual) [nf: v1 1921] [mf ed 19–] – 1 – (merged with: jahresbericht ueber die erscheinungen auf dem gebiete der germanischen philologie, to form: jahresbericht fuer deutsche sprache und literatur; cont: jahresberichte fuer neuere deutsche literaturgeschichte; v6-7 & v16-19 iss combined; v16-19 iss by: deutsche akademie der wissenschaften zu berlin & publ: berlin: akademie-verlag) – mf#film mas 517 – us Harvard [430]

Jahresbericht und abrechnung / Verband der deutschen Hutarbeiter – Altenberg. 1919. tables. (Serial publications of German trade unions in the Memorial Library, University of Wisconsin-Madison.) – 1 – us Wisconsin U Libr [330]

Jahresberichte der fabrikinspektoren : jahresberichte der gewerbeaufsichtsbeamten und bergbehoerden berlin – Berlin DE, 1875-1937/38 – 41r – 1 – gw Mikropress [670]

Jahresberichte der geschichtswissenschaft / Historische Gesellschaft zu Berlin; ed by Abraham, Fritz et al – Berlin: E S Mittler 1880- [v1-36(1878-1913)] (annual) [mf ed 1978] – 36v on 19r – 1 – (cont by: jahresberichte der deutschen geschichte; ceased with: 36. jahrg [1913], publ in 1916) – mf#film mas c390 – us Harvard [900]

Jahresberichte der verwaitungsbehosden / Hamburg – Hamburg.1877-1927. 16 reels – 1 – us L of C Photodup [943]

Jahresberichte des literarischen zentralblattes *see* Literarisches centralblatt fuer deutschland

Jahresberichte ueber die fortschritte der forstwissenschaft und forstlichen naturkunde im jahre... : nebst original-abhandlungen aus dem gebiete dieser wissenschaften / ed by Hartig, Theodor – Berlin: A Foerstner 1837- (irreg) [1. jahrg 1.-4.heft(1836/37)] [mf ed 2005] – 1r [ill] – 1 – (ceased with: 1. jahrg 4. heft [1836 und 1837]?) – mf#film mas 36989 – us Harvard [634]

Jahresbriefe des berneuchener kreises – 1(1931)-5(1936) – 19mf – 9 – €37.00 – ne Slangenburg [242]

Jahresbuch / Verband der Maler, Lackierer, Anstreicher, Tuencher und Weissbinder Deutschlands – 1911-12. Hamburg. (Serial publications of German trade unions in the Memorial Library, University of Wisconsin-Madison.) – 1 – us Wisconsin U Libr [330]

Die jahresernte *see* Literarisches centralblatt fuer deutschland

Jahres-sammlung des vereins der deutschen presse von wisconsin / Verein der Deutschen Presse von Wisconsin – 28th-33th [1910-1915] – 1r – 1 – (cont: verhandlungen des vereins der deutschen presse von wisconsin, verein der deutschen presse von wisconsin) – mf#3254449 – us WHS [070]

Die jahre-versammlung des forstverein fuer das grossherzogthum hessen zu... / Forstverein fuer das Grossherzogthum Hessen. Jahres-Versammlung – Darmstadt: J C Herbert 1878-80 [mf ed 2005] – 2v on 1r – 1 – (cont: forstverein fuer das grossherzogthum hessen. bericht ueber die...jahre-versammlung des forstverein fuer das grossherzogthum hessen zu...; cont by: forstverein fuer das grossherzogthum hessen. bericht ueber die...jahre-versammlung des forstverein fuer das grossherzogthum hessen zu...[1883]) – mf#film mas 99999 – us Harvard [634]

Das jahrhundert des heils / Gfroerer, August Friedrich – Stuttgart: C. Schweizerbart, 1838 – 3mf – 9 – 0-7905-3443-6 – (incl bibl ref) – mf#1987-3443 – us ATLA [270]

Ein jahrhundert deutscher literaturkritik (1750-1850) : ein lesebuch und studienwerk / ed by Fambach, Oscar – Berlin: Akademie-Verlag, 1957-63 [mf ed 1993] – 1 – (v1+6 never publ. incl bibl ref) – mf#8223 – us Wisconsin U Libr [430]

Ein jahrhundert rheinische mission / Bonn, Alfred – Barmen: Verlag des Missionshauses, 1928. Chicago: Dep of Photodup, U of Chicago Lib, l967 (1r); Evanston: American Theol Lib Assoc, 1984 (1r) – 1 – 0-8370-0082-3 – (incl ind) – mf#1984-B064 – us ATLA [240]

Ein jahrtausend lateinischer hymnendichtung / Dreves, G M & -Blume, J – Leipzig. v1-2. 1909 – 2v on 18mf – 8 – €35.00 – ne Slangenburg [450]

Jahve et moloch, sive, de ratione inter deum israelitarum et molochum intercedente / Baudissin, Wolf Wilhelm – Lipsiae: Fr Guil Grunow, 1874 – 1mf – 9 – 0-7905-0180-5 – (incl bibl ref) – mf#1987-0180 – us ATLA [270]

Jai hind – Hubli, India. 1947-54 – 4r – 1 – us L of C Photodup [079]

Jai hind – Rajkot, India. Apr-Sept 1966 – 2r – 1 – us L of C Photodup [079]

Jaicoa, cuentos y leyendas / Rodriguez Escudero, Nestor A – Aguadilla, Puerto Rico. 1958 – 1r – us UF Libraries [972]

Jail administration digest – Annandale. 1978-1980 (1,5,9) – mf#11542 – us UMI ProQuest [345]

Jail calendar / Sedgwick County. Kansas. Sheriff – 1886-1907 – 1 – us Kansas [360]

Jail diary of albie sachs / Sachs, Albie – London, England. 1966 – 1r – us UF Libraries [960]

Jail diary of albie sachs / Sachs, Albie – London, England. 1969 – 1r – us UF Libraries [960]

Jail records of shawnee county, kansas / Shawnee County. Kansas. Sheriff – 1890-97, 1935-48 – 1 – us Kansas [360]

Jailbreak / High School Youth Against War & Fascism – v1 n3-v3 n5 [1970 oct 1-1972 dec] – 1r – 1 – mf#1112708 – us WHS [320]

Jailly, Hector de *see*
- Une annee
- Encore deux annees
- Le mois de henri

Jaime 1 : diezmos, testamento; conquista de valencia (siecle 13) – Barcelona – 1r – 5,6 – sp Cultura [946]

Jaime 1 y lugartenencia del infante pedro (anno 1231-1276) – Barcelona – 1r – 5,6 – sp Cultura [946]

Jaime, E *see* Diable a quatre

Jaime el Justo *see*
- The letter of james the just
- The wisdom of james the just

Jaime, G *see* De koulikoro a tombouctou a bord du "mage", 1889-1890

Jaimes Freyre, Ricardo *see* Historia del descubrimiento de tucuman, seguida de investigaciones historicas

Jaimini *see*
- Mimansa
- The sacred books of the hindus

Jain, Champat Rai *see* Nyaya, the science of thought

The jaina gazette – Lucknow: Bharat Jaina Mahamandal. [v 9 n10-v14]. oct 1913-1918 – us CRL [954]

Jaina system of education / Das Gupta, Debendra Chandra – Calcutta: Bharati Mahavidyalaya: Sole agents, Sree Bharatee Pub Co, 1942 – (= ser Samp: indian books) – (foreword by syama prasad mookerjee) – us CRL [280]

Jaini, Jagmandar Lal *see* Outlines of jainism

Jaini, Jagomandar Lal *see* Outlines of jainism

Jainism : in western garb as a solution to life's great problems / Warren, Herbert – 2nd rev enl ed. Arrah, India: Kumar Devendra Prasad, 1916 – (= ser Library of jaina literature) – 1mf – 9 – 0-524-02558-4 – mf#1990-3053 – us ATLA [280]

Jainism and karnataka culture / Sharma, Sri Ram – Dharwar: Karnatak Historical Research Society, 1940 – (= ser Samp: indian books) – (foreword by a b latthe) – us CRL [280]

Jainism in north india, 800 bc-ad 526 / Shah, Chimanlal J – London; New York: Longmans, Green, and Co, 1932 – (= ser Samp: indian books) – (foreword by h heras) – us CRL [280]

Jaire – Badajoz, 1954 y 1955 – 5 – sp Bibl Santa Ana [073]

Jais, Regina *see* Legendary germany, oberammergau and bayreuth

Jajasan akademi populer / Akademi popular – Djakarta, 1954 – 1mf – 9 – (missing: 1954(1-2); 1954(4)) – mf#SE-448 – ne IDC [959]

Jajasan badan penerbit godjah mada / Gadjah-Mada – Jogjakarta, 1950-1960 – 97mf – 9 – (missing: 1955, v5(12); 1955/1956, v6(9-12); 1956, v7(1, 2, 4); 1957, v8(5); 1958, v9(2, 5-7, 9-12); 1959, v10(2, 5, 6)) – mf#SE-881 – ne IDC [950]

Jajasan badan penerbit pekerdjaan umum / Madjalah pekerdjaan umum – Djakarta, [1964]-1972. v1-9(4) – 13mf – 9 – (missing: [1964]-1970 v1-7(1-2)) – mf#SE-1819 – ne IDC [950]

Jajasan badan penerbit "pembimbing rakjat" / Mingguan membimbing – Djakarta, 1953-1955 – 37mf – 9 – (missing: 1953 v1(1-20)) – mf#SE-927 – ne IDC [950]

Jajasan carya dharma praja mukti / Pradja – Medan, Sept, 1967-1970. v1-3(1-13) – 18mf – 9 – (missing: 1968 v1-2(mar-dec)) – mf#SE-1909 – ne IDC [950]

Jajasan dana kesedjahteraan sosial / Daja sosial – Djakarta, 1958-1960 – 16mf – 9 – (missing: 1960 v3(5-8)) – mf#SE-850 – ne IDC [950]

Jajasan Da'watul Islam *see* Pikiran islam

Jajasan dharma / Mimbar Indonesia – Djakarta, Nov 1947-1966 – 517mf – 9 – (missing: 1948, v2(14-28, 30-38, 40-45, 47-48, 50); 1950, v4(1-7, 32); 1951, v5(1-2); 1952, v6(49); 1955, v9(47); 1956, v10(11, 14, 40); 1957, v11(27); 1966, v20(5-8, 11-12)) – mf#SE-563 – ne IDC [959]

"Jajasan dharma karya" / Madjalah perbankan – Djakarta, 1967-1971. v1-4(1-25) – 18mf – 9 – (missing: 1967 v1(1, 3-11)) – mf#SE-1823 – ne IDC [959]

Jajasan "gadjah mada" : almanak dan buku tjatatan militer – Djakarta, 1954 – 8mf – 9 – mf#SE-580 – ne IDC [959]

Jajasan harapan kita : indonesia magazine – Djakarta, 1969-1971(1-16) – 34mf – 9 – (missing: 1969(1); 1970(3)) – mf#SE-1697 – ne IDC [073]

Jajasan hikmah – Djakarta, 1948-1960 – 211mf – 9 – (missing: many iss) – mf#SE-369 – ne IDC [959]

Jajasan indonesia : horison; madjalah sastra – Djakarta, 1966-1972 – 67mf – 9 – (missing: 1971 v6(11)) – mf#SE-1497 – ne IDC [959]

Jajasan "indonesia baru" / Djalan rajat – Bandung, 1952-1953 – 2mf – 9 – (missing: 1952, v1(1-3); 1953, v2(1)) – mf#SE-749 – ne IDC [959]

Jajasan Intisari *see* Intisari

Jajasan kebudajaan sulawesi selatan dan tenggara : bingkisan – Makassar, 1967-1970. v1-3 – 26mf – 9 – mf#SE-1360 – ne IDC [950]

Jajasan keluarga – Djakarta, 1953-1971 – 230mf – 9 – (missing: several iss) – mf#SE-854 – ne IDC [640]

Jajasan kesedjahteraan nasional / Manipol – Makassar, 1963-1965 – 10mf – 9 – mf#SE-922 – ne IDC [950]

Jajasan kesedjakteraan mahasiswa veteran ri, perwakilan surakarta : api marhaenisme the light of marhaenism – Sala, 1963-1964 – 4mf – 9 – (missing: 1963(2)) – mf#SE-336 – ne IDC [950]

Jajasan Kesehatan Djiwa "Dharmawangsa" *see* Djiwa

Jajasan komunikasi – Djakarta, 1969/1970-1970/1971. v1-2(1-48) – 42mf – 9 – mf#SE-1754 – ne IDC [950]

Jajasan lembaga ilmiah indonesia untuk penjelidikan sedjarah / Penelitian sedjarah – Djakarta, Sept, 1960-1965. v1-6(10) – 9mf – 9 – mf#SE-742 – ne IDC [959]

Jajasan lembaga pendidikan nasional / Pendidikan Nasional – Djakarta, 1961-1965 – 9mf – 9 – (missing: 1961(2, 4-5, 8-12); 1962(1-2, 5-6, 9-10); 1963(1-10); 1964(1-8, 11-12); 1965(1-2, 5-12)) – mf#SE-501 – ne IDC [950]

Jajasan lembaga research dan afiliasi industri, universitas negeri diponegoro / Madjalah Universitas Diponegoro – Semarang, 1962-1965 – 7mf – 9 – mf#SE-489 – ne IDC [950]

Jajasan Makara Chandradimuka *see* Tjerpen

Jajasan maritim press / Warta ekonomi maritim – Tandjung Priok, 1969-1971 – 461mf – 9 – (missing: 1967-1969 v1-3(1-180, 182, 185, 217, 260)) – mf#SE-1986 – ne IDC [959]

Jajasan maritim press / Warta ekonomi maritim – Tandjung Priok, 1968-1973. v1-6(1-499) – 304mf – 9 – (missing: 1968, v1-2(1-16), v2(18-49, 51-55, 57-87); 1969, v3(89, 92, 123); 1971, v4(285-286, 292-296), v5(319-332, 336-338, 343-344)) – mf#SE-1987 – ne IDC [959]

Jajasan maritim press / Warta ekonomi maritim review – Tandjung Priok, 1969-1972. v1-3(1-41) – 55mf – 9 – (missing: 1971, v3(23); 1972, v4(33)) – mf#SE-1988 – ne IDC [959]

Jajasan masdjid mudjahidien : almanak mudjahidien – Djakarta, 1957 – 7mf – 9 – mf#SE-330 – ne IDC [950]

Jajasan melati : api kartini – Djakarta, 1959-1964 – 26mf – 9 – (missing: 1960(12); 1961(9, 10); 1962(11, 12); 1963(7-12); 1964(1, 8, 10)) – mf#SE-335 – ne IDC [950]

Jajasan merah putih / Merah putih – Djakarta, 1953-1968 – 91mf – 9 – (missing: 1953(3, 7, 28, 29, 33, 42); 1954(78, 80, 82, 86, 90, 91, 93, 95-97, 99, 100); 1955(102, 107-109); 1955, v3(111, 112, 114-117, 121, 124, 127, 128, 130-134, 136, 147-152); 1956, v4(159-163, 166, 167, 169-173, 176-end); 1957; 1958; 1959; 1960; 1961; 1962; 1963; 1964; 1965) – mf#SE-925 – ne IDC [950]

Jajasan Museum Perdjoangan Bogor *see* Puspa merdeka

Jajasan "pantja murti" : bharata; berkala seni & budaya – Djakarta, 1967-1968. v1-2(2) – 2mf – 9 – (missing: 1968 v2(1)) – mf#SE-1359 – ne IDC [950]

Jajasan pembangunan – Djakarta, 1948-1950 – 28mf – 9 – mf#SE-780 – ne IDC [950]

Jajasan pembangunan sosial / Teratai – Djakarta, 1970-1971. v1-2(6) – 21mf – 9 – (1970 v1(12)) – mf#SE-1958 – ne IDC [950]

Jajasan Pembaruan *see* Pki dan perwakilan

Jajasan pembaruan : bintang merah – Djakarta, 1945-1965 v1-21 – 84mf – 9 – (missing: 1946-1949, v2-5; 1950, v6(1, 10); 1951, v7(4)) – mf#SE-353 – ne IDC [950]

Jajasan pembina darussalam : sinar darussalam – Banda, March, 1968-1971. v1-4(1-38) – 38mf – 9 – mf#SE-1930 – ne IDC [950]

Jajasan pembina hukum adat / Sosiografi Indonesia dan hukum adat – Jogjakarta, 1959-1963 – 10mf – 9 – (missing: 1962) – mf#SE-648 – ne IDC [950]

Jajasan Pembina Kesedjahteraan Mahasiswa Islam *see* Prima

Jajasan Pembina Ruhul Islam *see* Ruhul islam

Jajasan pemeliharaan anak-anak tjatjad : berita jpat – Solo, 1954-1959 – 11mf – 9 – (missing: 1954-1955/1956, v1-2(1-8); 1957, v3(jan-sept); 1958, v4(5-8, 10)) – mf#SE-849 – ne IDC [950]

Jajasan pemuda / Suara pemuda – Gorontalo, [1946]-1956. v1-11 – 10mf – 9 – (missing: [1946]-1949, v1-4; 1950, v5(1-3, 9-end)-1954, v9; 1955, v10(1-14, 17, 18, 46-52); 1956, v11(1-20, 32, 35-39)) – mf#SE-422 – ne IDC [950]

Jajasan pendirian tempat2 peribadatan / Laporan tahunan – Bandung, 1961 – 1mf – 9 – mf#SE-1716 – ne IDC [950]

Jajasan penerbit dan pertjetakan pribudi patria indonesia – Makassar, 1969-1970. v1-2(6) – 7mf – 9 – (1969 v1(1-9)) – mf#SE-1875 – ne IDC [959]

Jajasan penerbit pantjasila : berita fakta indonesia & internasional – Djakarta, 1966-1967 – 16mf – 9 – mf#SE-1349 – ne IDC [959]

Jajasan penerbit pesat / Almenak "Waspada" – Ngajogyakarta, 1954-1965 – 63mf – 9 – (missing: 1954-56 v1-3; 1958 v5; 1961 v8) – mf#SE-613 – ne IDC [959]

Jajasan penerbitan "djiwa baru" : bhakti; madjalah bulanan tentang pendidikan dalam keluarga – Jogjakarta, 1953-1956 – 6mf – 9 – (missing: 1953-1956 v1-4(1, 7)) – mf#SE-1358 – ne IDC [950]

Jajasan penerbitan dr gssj ratu langie : komentar nasional – Surabaja, 1966. v1(1-23) – 10mf – 9 – (missing: 1966(8)) – mf#SE-1751 – ne IDC [950]

Jajasan penerbitan karya sastra ikatan sardjana sastra indonesia / Madjalah ilmu-ilmu sastra Indonesia – Djakarta, 1963-1968 – 23mf – 9 – mf#SE-659 – ne IDC [959]

Jajasan penerbitan kebudajaan / Indonesia. madjalah kebudajaan – Djakarta, 1950-1965 – 136mf – 9 – mf#SE-656 – ne IDC [959]

Jajasan penerbitan maritim departemen perhubungan laut : dunia maritim – Djakarta, 1950-1972. v1-22(10) – 98mf – 9 – (missing: 1950, v1-2; 1963, v13(6-7); 1966, v16(14-15, 34-35); 1968, v17(7-12); 1971, v21(4-end)) – mf#SE-586 – ne IDC [950]

Jajasan Penerbitan Pantjasila (Japenpa) *see* Japenpa features

Jajasan penerbitan pembina perekonomian nasional / Warta perusahaan – Djakarta, 1963-1964 – 7mf – 9 – (missing: 1963/1964 v1(2-23, 25-26, 28)) – mf#SE-308 – ne IDC [950]

Jajasan penerbitan pesat / Pesat – Jogjakarta [1945]-1965 – 426mf – 9 – (missing: [1945]-1951 v1-7(1-21, 23-35, 29, 31-37, 39-46, 48-end)) – mf#SE-952 – ne IDC [950]

Jajasan perdjalanan hadji indonesia / Madjalah Islam Kiblat – Djakarta, 1953-1972. v1-20(5) – 233mf – 9 – (missing: 1953-1959, v1-6; 1960-1962, v7-9(1-6); 1963, v9-10(jan-oct); 1964, v10-11(7-end); 1966, v13(1); 1967, v13-14(24); 1969, v16(24); 1970, v17(17); 1970, v18(1)) – mf#SE-1827 – ne IDC [959]

Jajasan perpustakaan nasional / Mayapada – Djakarta, 1967-1971 – 126mf – 9 – (missing: 1969, v3(48, 55); 1970, v4(97-98); 1971, v5(108-109)) – mf#SE-1778 – ne IDC [950]

Jajasan perpustakaan nasional (japernas) jajasan bina sedjahtera (jbs) : almanak ekonomi – Djakarta, 1967/1968 – 6mf – 9 – mf#SE-1305 – ne IDC [950]

Jajasan pertanian nasional : hanura tani – Djakarta, 1968. v1(1) – 1mf – 9 – mf#SE-1495 – ne IDC [950]

Jajasan Prapanca *see* Indonesian review

Jajasan Psychologi *see* Psychologi

Jajasan puspa / Madjalah pekerdja – Djakarta, 1964-1965 – 11mf – 9 – (missing: 1964(1-16)) – mf#SE-382 – ne IDC [959]

Jajasan pustaka industri rakjat / Madjalah industri rakjat – Djakarta, 1962-1965 – 53mf – 9 – (missing: 1963, v2(2, 9); 1964, v3(11-12)) – mf#SE-693 – ne IDC [959]

Jajasan serba/guna / Genta massa – Djakarta, 1964-1965 – 7mf – 9 – (missing: 1965 v2(13)) – mf#SE-720 – ne IDC [950]

Jajasan Sosial Tani Membangun *see* Trubus

Jajasan suara tani / Suara tani – Jogjakarta, [1945]-1964 – 31mf – 9 – (missing: 1945-1950, v1-5(1); 1951, v6(1-8); 1952-1955, v7; 1964, v15(11, 12)) – mf#SE-429 – ne IDC [950]

Jajasan sundabudaja / Warga – Bogor, 1951-1965 – 76mf – 9 – (missing: 1951/1952, v1-2(1-48); 1953, v3(53, 57, 75); 1954, v4(109, 110); 1955, v5(121-162); 1956, v6(187-195, 197-232, 234, 235, 239); 1958/1959, v7-8(241-245); 1959, v9(251-253, 255, 256); 1960/1963, v9-13(258-262)) – mf#SE-980 – ne IDC [950]

Jajasan tjampaka / Tjampaka – Bandung, 1965-1966 – 6mf – 9 – (missing: 1965, v1(1-3); 1966, v2(5-16, 20-22, 24, 26-32)) – mf#SE-959 – ne IDC [950]

The jakarta times – Jakarta: Zein Effendi, sep 23 1972-jan 21 1974 – us CRL [079]

Jakko : der roman eines jungen / Weidenmann, Alfred – 4. Aufl. Stuttgart: Loewes (F Carl), 1941 – 1r – 1 – us Wisconsin U Libr [830]

Jakmi lesbumi : gelanggang sastera, seni dan pemikiran – Djakarta, 1966-1967 – 4mf – 9 – mf#SE-1484 – ne IDC [950]

Jakob ayrers "sidea", shakespeares "tempest" und das maerchen / Fouquet, Karl – Marburg a.L.: N G Elwert, 1929 – 1r – 1 – (incl bibl ref) – us Wisconsin U Libr [410]

Jakob boehme : gestalt und gestaltung / Hankamer, Paul – Bonn: F Cohen, 1924 [mf ed 1989] – 427p – 1 – mf#7044 – us Wisconsin U Libr [140]

Jakob boehme : gestalt und gestaltung / Hankamer, Paul – Bonn: F Cohen, 1924 [mf ed 1989] – 427p – 1 – mf#7044 – us Wisconsin U Libr [140]

Jakob boehme's saemmtliche werke / ed by Schiebler, K W – Leipzig: J A Barth, 1831-47 [mf ed 1989] – 7v in 5 – 1 – mf#7046 – us Wisconsin U Libr [802]

Jakob freys gartengesellschaft / ed by Bolte, Johannes – Stuttgart: Litterarischer Verein, 1896 [mf ed 1989] (Tuebingen: H Laupp, Jr) [mf ed 1993] – (= ser Blvs 209) – xxxiv/312p – 1 – (coll of facetiae based upon h bebel, poggio, adelphus et al. early modern german, dutch and latin text. int and comm in german) – mf#8470 reel 43 – us Wisconsin U Libr [880]

Jakob friedrich fries : aus seinem handschriftlichen nachlasse / Henke, Ernst Ludwig Theodor – Leipzig: FA Brockhaus, 1867 – 1mf – 9 – 0-7905-6808-X – (incl bibl ref) – mf#1988-2808 – us ATLA [920]

Jakob grant's koenigl grossbritannischen schiffs-lieutenant's bericht von einer entdeckungs-reise nach neu-sued-wallis : gethan in dem schiffe lady nelson, in den jahren 1800, 1801 und 1802 / Grant, James – Weimar 1807 [mf ed Hildesheim 1995-98] – 1v on 2mf – 9 – €60.00 – ISBN-10: 3-487-26557-5 – ISBN-13: 978-3-487-26557-5 – gw Olms [919]

Jakob haafner's fussreise durch die insel ceilon / Haafner, Jacob – Magdeburg 1816 [mf ed Hildesheim 1995-98] – 1v on 2mf – 9 – €60.00 – 3-487-27451-5 – gw Olms [915]

Jakob naumann's reise nach den vereinigten staaten von nordamerika, siebenjaehriger aufenthalt in denselben und rueckkehr nach deutschland : mittheilungen fuer auswanderungslustige, mit besonderer beziehung auf ackerbau, handel und gewerbe – Leipzig 1850 [mf ed Hildesheim 1995-98] – 1v on 3mf – 9 – €90.00 – 3-487-27004-8 – gw Olms [917]

Jakob spoendlis gluecksfall : erzaehlung / Huggenberger, Alfred – Feldpostausg. Stuttgart: Verlag Deutsche Volksbuecher 1942 [mf ed 1995] – (= ser Wiesbadener volksbuecher 183) – 1r – 1 – (filmed with: daniel pfund) – mf#3884p – us Wisconsin U Libr [390]

Jakob wassermann und sein werk / Wassermann-Speyer, Julie – Wien: Deutsch-Oesterreichischer Verlag, 1923 – 1r – 1 – us Wisconsin U Libr [430]

Jakob ziegler aus landau an der isar : ein gelehrtenleben aus der zeit des humanismus und der reformation / Schottenloher, Karl – Muenster i W: Aschendorff, 1910 – (= ser Reformationsgeschichtliche Studien und Texte) – 1mf – 9 – 0-524-01533-3 – (incl bibl ref) – mf#1990-0439 – us ATLA [943]

Jakob ziegler und adam reissner : eine quellenkritische untersuchung ueber eine streitschrift der reformationszeit gegen das papsttum / Schottenloher, Karl – Muenchen: C Wolf, 1908 [mf ed 1993] – 1mf – 9 – 0-524-08528-5 – (incl bibl ref) – mf#1993-1058 – us ATLA [230]

Der jakobiner-klub : ein beitrag zur geschichte der parteien und der politischen sitten im revolutions-zeitalter / Zinkeisen, Johann Wilhelm – Berlin 1852-53 [mf ed Hildesheim 1995-98] – 2v on 10mf – 9 – €100.00 – 3-487-26292-4 – gw Olms [325]

Eine jakobitische einleitung in den psalter : in verbindung mit zwei homilien aus dem grossen psalmenkommentar des daniel von salah – Giessen: J.Ricker, 1901 – (= ser Beihefte zur zeitschrift fuer die alttestamentliche wissenschaft) – 1mf – 9 – 0-7905-1867-8 – (incl bibl ref) – mf#1987-1867 – us ATLA [221]

Jakobs, des handwerksgesellen wanderungen durch die schweiz / Gotthelf, Jeremias [pseud: Albert Bitzius]; ed by Hunziker, Rudolf – Muenchen: Verlegt von Eugen Rentsch im Delphin-Verlag, 1917 [mf ed 1993] – (= ser Saemtliche werke in 24 baenden 9) – 640p – 1 – mf#8522 reel 3 – us Wisconsin U Libr [830]

Jakobsen, Marikje *see* Die gebruik van die ontwikkelingsgefasiliteerde groepmodel vir egskeidingsgetraumatiseerde adolessente

Die jakobsleiter : [a novel] / Finckh, Ludwig – Stuttgart: Deutsche Verlags-Anstalt 1923, c1920 [mf ed 1995] – 1r – 1 – (filmed with: lion feuchtwanger) – mf#3838p – us Wisconsin U Libr [830]

Der jakobusbrief und die johannisbriefe : ausgelegt fuer bibelleser / Schlatter, Adolf von – Calw: Verlag der Vereinsbuchh, 1893 – 1mf – 9 – 0-524-05420-7 – mf#1992-0430 – us ATLA [227]

Der jakobusbrief und die neuere kritik / Weiss, Bernhard – Leipzig: A Deichert (Georg Boehme), 1904 – 1mf – 9 – 0-8370-7437-1 – mf#1986-1437 – us ATLA [227]

Jaksche, J *see* Gundackers von judenburg christi hort

Jakubczyk, Karl *see* Eichendorffs weltbild

Jal, Auguste *see* Resume de l'histoire du lyonnais, (rhone)

Jalal al-Din Rumi, Maulana *see* Selected poems from the divani shamsi tabriz

Jalisco. Mexico *see*
- El estado de jalisco
- El estado del jalisco

Jallet, Jacques *see* Journal inedit de jallet

Jaloux, Edmond *see* Du reve a la realite

Jama : the journal of the american medical association / American Medical Association – Chicago. 1883+ (1) 1964+ (5) 1970+ (9) – ISSN: 0098-7484 – mf#1161 – us UMI ProQuest [610]

Jamaica / Henderson, John – London, England. 1906 – 1r – us UF Libraries [972]

Jamaica, 1683-1818 : from the public record office, london – (= ser Naval office shipping lists; British records relating to america in microform) – 7r – 1 – (with guide) – mf#97013 – uk Microform Academic [972]

Jamaica advocate – Kingston, Jamaica, sep 1898-nov 1899 [incomplete] – 1r – 1 – ie National [079]

Jamaica and the colonial office : who caused the crisis? / Price, George – London 1866 – (= ser 19th c british colonization) – 4mf – 9 – mf#1.1.3788 – uk Chadwyck [972]

Jamaica arise!, 1947-50 : the political and labour guide [official organ of the pnp] – (= ser The private coll of richard hart) – 4mf – 9 – mf#87550 – uk Microform Academic [321]

Jamaica. Assembly *see* Journals of the assembly of jamaica, 1663-1826

Jamaica at the colonial and indian exhibition – London, England. 1886 ? – 1r – us UF Libraries [972]

Jamaica, c1765-1848: the taylor & vanneck-arcedekne papers... *see* Plantation life in the caribbean

Jamaica christian chronicle – (People's Paper). Kingston. Jamaica. -w. 3 May 1888, 6 Apr 1893-10 Nov 1894. (Very imperfect). (38 ft) – 1 – uk British Libr Newspaper [072]

Jamaica churchman – Kingston, Jamaica. sept 1899-sept 1915 – 2r – 1 – (lacking jan 1903) – uk British Libr Newspaper [240]

Jamaica creole – Kingston. Jamaica. -d. 24 Sep 1878, 9, 24 Jan, 8, 24 Feb, 11, 24 Mar, 9 Apr, 10 May, 9 Jul, 9 Aug 1879. (10 ft) – 1 – uk British Libr Newspaper [079]

Jamaica daily telegraph – Kingston, Jamaica. 10, 27 Apr 1899-31 Dec 1909.-d. 58 reels – 1 – uk British Libr Newspaper [079]

Jamaica Dept of Statistics *see* 1960 census of british virgin islands

Jamaica. Dept of Statistics *see*
– Annual abstract of statistics 1947-1968
– Statistical abstract 1972-1976

Jamaica despatch and jamaica gazette – Kingston. Jamaica. -d. 1 Jan-21 Apr 1840. (1 reel) – 1 – uk British Libr Newspaper [072]

Jamaica in 1866 / Harvey, Thomas – London, England. 1867 – 1r – us UF Libraries [972]

Jamaica in 1896 / Institute Of Jamaica – Kingston, Jamaica. 1896 – 1r – us UF Libraries [972]

Jamaica in 1897 / Institute Of Jamaica – Kingston, Jamaica. 1897 – 1r – us UF Libraries [972]

Jamaica in 1905 / Cundall, Frank – Kingston, Jamaica. 1905 – 1r – us UF Libraries [972]

Jamaica in 1928 / Institute Of Jamaica – London, England. 1928 – 1r – us UF Libraries [972]

Jamaica johnny / Hader, Berta Hoerner – New York, NY. 1935 – 1r – us UF Libraries [972]

Jamaica journal – Kingston. 1968-1996 (1) 1972-1992 (5) 1974-1992 (9) – ISSN: 0021-4124 – mf#6949 – us UMI ProQuest [073]

Jamaica labour weekly, 1938-39 – (= ser The private coll of richard hart) – 3mf – 9 – mf#87548 – uk Microform Academic [331]

Jamaica mail – Kingston, Jamaica. 2 Jan 1930-13 Jun 1931.-d 9 reels – 1 – uk British Libr Newspaper [072]

The jamaica maroons : how they came to nova scotia: how they left it / Brymner, Douglas – S.l: s.n, 1894? – 1mf – 9 – mf#02111 – cn Canadiana [972]

Jamaica mercury and kingston weekly advertiser – Kingston, Jamaica 1 may 1779-25 mar 1780 (wkly) – 1 – (cont by: royal gazette [kingston]) – uk British Libr Newspaper [079]

Jamaica of today / Verrill, A Hyatt – New York, NY. 1931 – 1r – us UF Libraries [972]

Jamaica people's national party, 1938-56 : pamphlets, leaflets, etc – 12mf – 9 – (with guide) – mf#87539 – uk Microform Academic [321]

Jamaica place-names / Cundall, Frank – Kingston, Jamaica. 1909 – 1r – us UF Libraries [918]

Jamaica plantation records from the dickinson papers, 1675-1849 – Somerset & Wiltshire Record Offices – (= ser BRRAM series) – 4r – 1 – £268 / $536 – (with p/g; int by w e minchinton) – mf#r96977 – uk Microform Academic [972]

Jamaica post – Kingston, Jamaica. 11 Oct 1892-8 Apr 1899.-tw. 13 reels – 1 – uk British Libr Newspaper [072]

Jamaica standard – Montego Bay. Jamaica. -sw. 9 Jan 1839-Apr 1840. (44 ft) – 1 – uk British Libr Newspaper [072]

Jamaica, the blessed island / Olivier, Sydney Haldane – London, England. 1936 – 1r – us UF Libraries [972]

Jamaica times – Kingston, Jamaica. 25 Aug 1900-24 Jun 1922; 4 Jan 1930-30 Jul 1938; 7 Jan-30 Sep, 23 Dec 1939-14 Sep 1940; 3 Mar 1951-27 Jun 1953; 22 Jun 1957-22 Jan 1963 (1939, 40 imperfect).-w. 50 reels – 1 – uk British Libr Newspaper [072]

Jamaica times – Kingston. Jamaica. -w. 3 Mar 1951-27 Jun 1953, 22 Jun 1957-26 Jan 1963. (7 reels) – 1 – uk British Libr Newspaper [072]

Jamaica, trinidad and tobago, leeward islands / University Of The West Indies (Mona, Jamaica) – Jerusalem, Israel. 1964 – 1r – us UF Libraries [972]

Jamaica under the spaniards / Cundall, Frank – Kingston, Jamaica. 1919 – 1r – us UF Libraries [972]

Jamaica witness – Falmouth. Jamaica. -w. 15 Jan 1877-4 Oct 1878, 6 Jan-1 May 1879, 1 Jan 1883-1 Oct 1887. (Imperfect). (1 reel) – 1 – uk British Libr Newspaper [079]

The jamaican – Kingston. Jamaica. -sw. 2 Nov-28 Dec 1907. (31 ft) – 1 – uk British Libr Newspaper [079]

Jamaican journey / Brown, William John – London, England. 1948 – 1r – us UF Libraries [972]

Jamaican weekly gleaner – 1994 dec 9/15, 1995 mar 17/23, 1995 jun 9/15-1995 dec 29/1996 jan 6 – 2r – 1 – mf#3215227 – us WHS [071]

Jamaican weekly gleaner – Kingston, Jamaica. 1988 jan-1998 jun – 17r – (gaps) – us UF Libraries [079]

Jamaican weekly gleaner – Kingston, Jamaica. 1958-1987 (1) – mf#67779 – us UMI ProQuest [079]

Jamaica's part in the great war, 1914-1918 / Cundall, Frank – London, England. 1925 – 1r – us UF Libraries [972]

Jamaika, seine physikalisch-politische geographie / Jacobs, Reinhold – Chemnitz, Germany. 1916 – 1r – us UF Libraries [972]

Jambalaya / East Ascension Genealogical and Historical Society [Ascension Parish LA] – v7 n1-v8 n4 [1987 mar-1988 dec] – 1r – 1 – (cont by: journal of the east ascension genealogical and historical society) – mf#1519312 – us WHS [929]

Jambar / Trumbull Co. Youngstown – (feb 1969-nov 1970) [semiwkly] – 1r – 1 – mf#B29901 – us Ohio Hist [378]

Jambert *see* Les saints

Jam-e-jamshed – Bombay, India. 1947-Jul 1987 – 125r – 1 – us L of C Photodup [079]

James a garfield, 1853-1913 – 1r – 1 – mf#B25946 – us Ohio Hist [920]

James a garfield family papers, 1855-1938 – [mf ed 1991] – 2r – 1 – (correspondence, diaries, deeds, herbariums, receipts, architectural drawings, and probate documents of president garfield, his mother eliza, his wife lucretia, and descendants) – mf#ms4575 – us Western Res [975]

James a. garfield papers – 177r – 1 – $6,195.00 – (with guide) – Dist. us Scholarly Res – us L of C Photodup [975]

James addison ingle (yin teh-sen) : first bishop of the missionary district of hankow, china / Ingle, James Addison; ed by Jefferys, William Hamilton – New York: Domestic and Foreign Missionary Society, 1913 – 1mf – 9 – 0-524-05119-4 – mf#1992-2072 – us ATLA [240]

James and lucretia mott : life and letters / ed by Hallowell, Anna Davis – Boston: Houghton, Mifflin, 1884 [mf ed 1984] – (= ser Women & the church in america 117) – 2mf – 9 – 0-8370-1409-3 – (incl ind) – mf#1984-2117 – us ATLA [305]

James backus papers, ms 1548 / Backus, James – 1791-1833. 1 reel – 1 – us Western Res [920]

The james bay and northern quebec agreement : agreement between the government of quebec, the societe d'energie de la baie james... – Quebec: editeur officiel du Quebec, c1976 – (= ser Convention de la Baie James et du Nord quebecois (1975)) – 6mf – 9 – mf#SEM105P446 – cn Bibl Nat [333]

James bishop/allied family information interchange – v1 n1-3 [1983 jul/sep-1984 jan/jun] – 1r – 1 – mf#1096812 – us WHS [929]

James brand : twenty-six years pastor of the first congregational church, oberlin / Brand, James – Oberlin, Ohio: LD Harkness, 1899 – 1mf – 9 – 0-524-05243-3 – mf#1992-2080 – us ATLA [976]

James buchanan and harriet (lane) johnston papers – 4r – 1 – $140.00 – (with guide) – Dist. us Scholarly Res – us L of C Photodup [975]

James, C L R *see*
– Jacobins noirs
– Notes on the life of george padmore
– Party politics in the west indies

James, C Roger et al *see* Effects of fatigue on mechanical and muscular components of performance during drop landings

James calvert : or, from dark to dawn in fiji / Vernon, R (Mrs) – 2nd ed. New York: Fleming H Revell, [189-] – 1mf – 9 – 0-8370-6710-3 – mf#1986-0710 – us ATLA [920]

James, Catherine *see* Davis family newsletter

James, Charles *see* A collection of the charges, opinions and sentences of general courts-martial

James, Charles Canniff *see*
– Pitting the sugar beet
– The teaching of agriculture in our public schools
– The teaching of agriculture in the public schools

James, Charles F *see* The struggles for religious liberty in virginia

James, Charles R *see* Effects of overuse injury proneness and task difficulty on joint kinetic variability during landing

James chesnut papers, 1815-1900 – 115mf – 9 – (consist of financial and property records, correspondence, estate records, & other items) – us South Carolina Historical [978]

James, Croake *see*
– Curiosities of law and lawyers

James, Daniel *see*
– Red design for the americas
– Tacticas rojas en las americas

James, David *see* Peter without a primacy

James de mille's works, vol 1 – New York: D Appleton; Harper & Bros, 1871-73 – 2v on 15mf – 9 – 0-665-90851-2 – (v2 90852 isbn: 0-665-90852-0) – mf#90851 – cn Canadiana [802]

James duane doty papers, ms 1090 / Doty, James Duane – 1820-40. Collection includes letters, deeds, legal opinions relating in part to Indian matters and lands in Wisconsin. 1 reel – 1 – us Western Res [920]

James, E *see* Account of an expedition from pittsburgh to the rocky mountains

James, Edwin *see* Account of an expedition from pittsburgh to the rocky mountains

James, F L *see* The unknown horn of africa

James, Francis Bacon *see* The ohio law of opinion evidence, expert and non-expert

James francis edward, the old chevalier / Haile, Martin – With 11 photogravure illus. London: J.M. Dent & Co; New York: E.P. Dutton & Co., 1907. xii,479,(1)p. 11 ports. on 10 pl – 1 – us Wisconsin U Libr [920]

James, Francis Huberty *see* Tan tao pen yuan (ccm187)

James fraser, second bishop of manchester : a memoir, 1818-1885 / Hughes, Thomas – London, New York: Macmillan and Co, 1987 – 1r – 1 – us Wisconsin U Libr [920]

James freeman and king's chapel, 1782-87 : a chapter in the early history of the unitarian movement in new england / Foote, Henry Wilder – Boston: Leonard C Bowles, 1873 – 1mf – 9 – 0-524-08756-3 – mf#1993-3261 – us ATLA [243]

James freeman clarke : autobiography, diary and correspondence / ed by Hale, Edward Everett – Boston: Houghton, Mifflin, 1891 [mf ed 1989] – 1mf – 9 – 0-7905-4254-4 – mf#1988-0254 – us ATLA [920]

James g birney and his times : the genesis of the republican party with some account of abolition movements in the south before 1828 / Birney, William – New York: D Appleton, 1890, c1889 [mf ed 1989] – 2mf – 9 – 0-7905-4090-8 – (incl bibl ref) – mf#1988-0090 – us ATLA [976]

James g swan papers – Vancouver BC: UBCL 1994 – 9r – 1 – (incl: inventory of james g swan papers, 1852-1907 / jane turner) – cn UBC Preservation [971]

James, G Wharton *see* The klondyke

James, George Francis *see* Handbook of university extension

James, George Moffat *see* A model course in touch typewriting

James, George Payne Rainsford *see*
– A brief history of the united states boundary question
– Margaret graham

James, George Wharton *see* Old missions and mission indians of california

James gilchrist swan papers, 1833-1909 / Swan, James Gilchrist – 3.42 cubic ft [13 boxes, 1 vertical file] – (some material on mf; with finding aid) – us UW Libraries [305]

James gilmour : de apostel van mongolie / Marang, Gerardus Pieter – [Rotterdam: J M Bredee, 1912] [mf ed 1995] – (= ser Yale coll) – 39p (ill) – 1 – 0-524-09942-1 – (in dutch) – mf#1995-0942 – us ATLA [920]

James gilmour and john horden : the story of their lives / Bryson, Mary Isabella & Buckland, Augustus Robert – London: Sunday School Union [1—] [mf ed 1995] – (= ser Yale coll) – 144p/141p (ill) – 1 – 0-524-09850-6 – mf#1995-0850 – us ATLA [920]

James gilmour of mongolia : his diaries, letters and reports / Gilmour, J – London, 1892 – 4mf – 9 – mf#HTM-64 – ne IDC [920]

James gilmour of mongolia : his diaries, letters, and reports / Gilmour, James; ed by Lovett, Richard – London: Religious Tract Society, 1892 – 1mf – 9 – 0-8370-6053-2 – (incl ind) – mf#1986-0053 – us ATLA [240]

James, H R *see* Problems of higher education in india

James harris fairchild, or, sixty-eight years with a christian college / Swing, Albert Temple – New York: FH Revell, c1907 – 1mf – 9 – 0-7905-6840-3 – mf#1988-2840 – us ATLA [240]

James, Henry *see*
– Christianity, the logic of creation
– Lectures and miscellanies
– The literary remains of the late henry james
– Moralism and christianity
– The nature of evil
– The secret of swedenborg
– The social significance of our institutions
– Society the redeemed form of man
– William wetmore story and his friends

James, Henry James, Baron *see* The work of the irish leagues

James, Henry Rosher *see* Education and statesmanship in india

James hepburn : free church minister / Veitch, Sophie Frances Fane – Toronto: Williamson, 1888 [mf ed 1984] – 5mf – 9 – 0-665-32318-2 – mf#32318 – cn Canadiana [830]

James, Herbert A *see* School ideals

James, Herman Gerlach *see*
– Brazil after a century of independence
– Republics of latin america

James holmes and john varley / Story, Alfred Thomas – London 1894 – (= ser 19th c art & architecture) – 4mf – 9 – mf#4.2.1033 – uk Chadwyck [750]

James hudson taylor / Henzel, J – [Rotterdam: J M Bredee, 1907] [mf ed 1995] – (= ser Yale coll; Lichtstralen op den akker der wereld [13. jaarg 1907] 3-4) – 65p (ill) – 1 – 0-524-09601-5 – (in dutch) – mf#1995-0601 – us ATLA [920]

James hutchison stirling : his life and work / Stirling, Amelia Hutchison – London: TF Unwin, 1912 – 1mf – 9 – 0-7905-8917-6 – mf#1989-2142 – us ATLA [920]

James I *see* The workes of the most high and mightie prince, iames by the grace of god...

James, J A *see*
– Attraction of the cross
– Parental desire, duty, and encouragement

James, J D *see* The genuineness and authorship of the pastoral epistles

The james j hill papers / ed by White, W Thomas et al – 3pt – 1 – $7135.00 coll – (pt1: personal & private ser 1874, 1877-1916 17r $2645. pt2: pre-railroad business ser 1866-78 4r $605. pt3: railroad ser 1877-98 27r $4220. with p/g) – us UPA [330]

James, Janice *see* Qualifications of gymnastic coaches in utah

James, John Angell *see*
– The anxious inquirer after salvation, directed and encouraged
– Bearing of the american revival on the duties and hopes of british...
– Character and translation of enoch
– Christian citizen in life and in death
– The life and letters of john angell james
– Means and methods to be adopted for a successful ministry
– Ministerial duties stated and enforced
– Principles of dissent and the duties of dissenters

James, John Hall *see* Correspondence and diary of ba campaign

James, John Henry *see* Military commissions for the trial of citizens

James, John Thomas *see* The italian schools of painting

James johnstone vs the minister and trustees of st andrew's church : the ecclesiastical bearings of the case: being a review of the judgement rendered thereon by his honor, mr justice johnson, 30th december, 1873 / Campbell, Robert – Montreal?: s.n, 1873 – 1mf – 9 – mf#08531 – cn Canadiana [242]

James, Joseph H *see* The life of mrs mary d james

James joyce quarterly – Tulsa. 1963+ (1) 1976+ (5) 1976+ (9) – ISSN: 0021-4183 – mf#11347 – us UMI ProQuest [420]

James k. polk papers – 67r – 1 – $2,345.00 – (with guide) – Dist. us Scholarly Res – us L of C Photodup [975]

James knox polk and a history of his administration : embracing the annexation of texas, the difficulties with mexico, the settlement of the oregon question, and other important events / Jenkins, John Stilwell – Auburn [NY]; Buffalo: J E Beardsley, [1850?] [mf ed 1982] – 5mf – 9 – mf#36685 – cn Canadiana [975]

James, M E *see* How to decorate our ceilings, walls, and floors

James m parker daybook *see* Daybook, ms 2067

James, M R *see*
– Apocrypha anecdota
– Apocrypha anedocta
– The apocryphal new testament
– The testament of abraham

James madison / Gay, Sydney Howard – Boston & NY: Houghton, Mifflin & Co, 1899 – (= ser The american statesmen series) – 4mf – 9 – $6.00 – mf#LLMC 96-029 – us LLMC [975]

James madison papers – 28r – 1 – $980.00 – Dist. us Scholarly Res – us L of C Photodup [975]

James madison's notes of debates in the federal convention of 1789 : and their relation to a more perfect society of nations / Scott, James Brown – New York: Oxford UP, American Branch, 1918 – 2mf – 9 – $3.00 – mf#LLMC 95-078 – us LLMC [323]

James, Marquis *see* Life of andrew jackson, complete in one volume

James martineau : a biography and study / Jackson, Abraham Willard – Boston: Little, Brown, 1900 – 2mf – 9 – 0-7905-7780-1 – mf#1989-1005 – us ATLA [920]

James martineau, theologian and teacher : a study of his life and thought / Carpenter, Joseph Estlin – 2nd issue. London: Philip Green, 1905 – 2mf – 9 – 0-524-08231-6 – (incl ind) – mf#1993-2006 – us ATLA [240]

James martineaus ethik : darstellung, kritik und paedagogische konsequenzen / Wilkinson, John J – Leipzig, 1898 (mf ed 1993) – 2mf – 9 – €31.00 – 3-89349-299-2 – mf#DHS-AR 159 – gw Frankfurter [170]

JAMES

James, Maurice see Series of letters touching the church in england and ireland, addressed to the dean of hereford...
James mcgill and the origin of his university / Dawson, John William – S.l: s.n, 1870? – 1mf – 9 – mf#23636 – cn Canadiana [378]
James monroe buckley / Mains, George Preston – New York: Methodist Book Concern, c1917 – 1mf – 9 – 0-524-04004-4 – mf#1992-2004 – us ATLA [240]
James monroe papers – 11r – 1 – $385.00 – Dist. us Scholarly Res – us L of C Photodup [975]
James monroe papers – 1758-1839 (mf ed 1960) – (= ser Presidential papers microfilm) – 11r – 1 – us L of C Photodup [975]
The james monroe papers – Rare Books and Manuscript Division: The New York Public Library, Astor, Lenox and Tilden Foundations 1995 – ca 8r – 1 – ca $680.00 – (with printed guide) – mf#D3336 – us NY Public [320]
James, Montague Rhodes see
– Apocrypha anecdota
– A descriptive catalogue of the manuscripts... fitzwilliam museum
– The life and miracles of st. william of norwich
– Old testament legends
– Psalms of the pharisees, commonly called the psalms of solomon
– The sculptures in the lady chapel at ely
– The testament of abraham
James morier's zweite reise durch persien, armenien und kleinasien nach constantinopel : in den jahren 1810 bis 1816 – Weimar 1820 [mf ed Hildesheim 1995-98] – 1v on 3mf – 9 – €90.00 – ISBN-10: 3-487-26504-4 – ISBN-13: 978-3-487-26504-9 – gw Olms [915]
James Officer Descendants Family Association see Newsletter
James ormsbee murray : a memorial sermon / De Witt, John – [Princeton, NJ]: Princeton University Press, 1899 – 1mf – 9 – 0-524-08678-8 – mf#1993-3203 – us ATLA [240]
James parnell : died in colchester castle 4th may 1656, aetat 19 / Fell-Smith, Charlotte – 2nd ed. London: Headley Bros, 1907 – 2mf – 9 – 0-524-07414-3 – mf#1991-3074 – us ATLA [920]
James, Philip Gilbert see British policy in relation to the gold coast 1815-1850
James, Preston Everett see
– Latin america
James r williams' defence against the charges exhibited by his prosecutors – Baltimore: Toy 1827 – 1r – 1 – $35.00 – mf#um-15 – us Commission [242]
James, Robert see
– Dictionnaire universel de medecine
– A medicinal dictionary
James shoolbred gibbes letterbooks see Letterbooks
James shoolbred letterbooks – [mf ed 1981] [Spartanburg SC: Reprint Co, dist] – 44mf – 9 – mf#51-148 – us South Carolina Historical [025]
James skinner : a memoir / Trench, Maria – London: K. Paul, Trench, 1883 – 1mf – 9 – 0-7905-6134-4 – mf#1988-2134 – us ATLA [920]
James talbot / Savage, Sarah – London, England. 1826 – 1r – us UF Libraries [240]
James, the lord's brother / Patrick, William – Edinburgh: T & T Clark, 1906 – 1mf – 9 – 0-8370-4674-2 – (includes an appendix and indexes) – mf#1985-2674 – us ATLA [225]
James, Thomas Smith see The history of the litigation and legislation respecting presbyterian chapels and charities in england and ireland between 1816 and 1849
James, W R see Effectiveness of fundraising techniques for collegiate women's and olympic sports' facilities
James, Walter Stevens see John cooper
James weldon and stanton high school / Johnson, James Weldon – S.l., S.l? . 193-? – 1r – us UF Libraries [978]
James whitcomb riley : an essay; and some letters to him from james whitcomb riley, august 30, 1898-october 12, 1915 / Carman, Bliss – New York: Printed for G D Smith, [1918?] – 1mf – 9 – 0-665-77779-5 – mf#77779 – cn Canadiana [840]
James, William see
– Essays in radical empiricism
– Essays, philosophical and psychological
– Human immortality
– The literary remains of the late henry james
– The meaning of truth
– Memories and studies
– On vital reserves
– A pluralistic universe
– Pragmatism, a new name for some old ways of thinking
– Principles of psychology
– Psychology
– Some problems of philosophy
– Talks to teachers on psychology
– The will to believe
James, William Powell see Guesses at purpose in nature
James, Winifred Lewellin see Mulberry tree
Jameson, A B see Legends of the madonna as represented in the fine art...
Jameson, Anna see
– Characteristics of women, moral, poetical, and historical
– Memoirs of celebrated female sovereigns
Jameson, Anna Brownell (Murphy) see
– Companion to the most celebrated private galleries of art
– Legends of the monastic orders, as represented in the fine arts
– Sacred and legendary art
Jameson, Anna Brownell [Murphy] see
– A description...of the great picture by paul delaroche
– A hand-book to the courts of modern sculpture
– A handbook to the public galleries of art in and near london
– Legends of the madonna
– Memoirs of the early italian painters
Jameson, Anna Brownell [Murphy] et al see The history of our lord as exemplified in works of art
Jameson, J Franklin see Essays in the constitutional history of the united states in the formative period, 1775-1789
Jameson, Robert see
– Letters from the havana, during the year 1820
– Narrative of discovery and adventure in africa, from the earliest ages to the present time
[Jamestown-] jamestown news – NV. 1908 – 1r – 1 – $60.00 – mf#U04848 – us Library Micro [071]
[Jamestown-] mother lode magnet – CA. 1898-1938 – 12r – 1 – $720.00 – mf#C03606 – us Library Micro [071]
The jami masjid at badaun : and other buildings in the united provinces / Blakiston, J F – Calcutta: Govt of India, Central Publication Branch, 1926 – (= ser Samp: indian books) – us CRL [720]
Jamiat al-Malik Saud. Kulliyat al-Ulum see Journal of the college of science, king saud university
Jamieson, George see The silver question
Jamieson, John see
– Duty, excellency, and pleasantness, of brotherly unity
– Etymological dictionary of the scottish language
– Hopes of an empire reversed
Jamieson, Robert see
– Eastern manners illustrative of the old testament history
– The historical books of the holy scriptures
– The inspiration of the holy scriptures
– The pentateuch and the book of joshua
Jamieson, William F see The clergy a source of danger to the american republic
Jamil, M Tahir see Hali's poetry
Jamini roy : 15 coloured plates – New Delhi: Dhoomi Mal Dharam Das, [19–] – (= ser Samp: indian books) – us CRL [920]
Jamis, Fayad see
– Cuatro poemas en china
– Pedrada
– Por esta libertad
– Victoria de playa giron
Jamis, Fayed see Puentes
Jamison, Monroe Franklin see Autobiography and work of bishop m f jamison...
Jammes, Francis see Les georgiques chretiennes: poeme couronne par l'academie francaise
Jammu and kashmir / Jammu and Kashmir. India. Census Commissioner – Jammu: Ranbir Govt Press. pts1-4. 1941 – us CRL [324]
The jammu and kashmir government gazette / Kashmir – Jammu. Apr. 4, 1963-Mar. 1967 – 1 – us NY Public [324]
Jammu and Kashmir. India. Census Commissioner see Jammu and kashmir
The jamnagar experiment / Central Institute of Research in Indigenous Systems of Medicine – [Delhi: s.n.], 1956 – (filmed with: india (republic) committee appointed...to establish a research centre in the indigenous systems of medicine. report ...; and others) – us CRL [610]
Jamnuay sgal ti nyn prajajan loe libhab lok / Kan Man – [Bhnam Ben]: Vappadharm 1984 [mf ed 1990] – 1r with other items – 1 – (in khmer) – mf#mf-10289 seam reel 121/6 [§] – us CRL [900]
Jampel, Sigmund see
– Das buch esther
– Die hagada aus aegypten
– Vom kriegsschauplatze der israelitischen religionswissenschaft
– Die wiederherstellung israels unter den achaemeniden
Jamtlands allehanda – Stockholm, Sweden. 1889-95 – sw Kungliga [078]
Jamtlands allehanda see Jamtlands tidning
Jamtlands folkblad – Oestersund, 1911-14 – 3r – 1 – sw Kungliga [078]
Jamtlands folkblad – Stockholm, 1942-49 – 7r – 1 – sw Kungliga [078]
Jamtlands tidning – Ostersund, Sweden. 1895-1957, 1960 – 185r – 1 – (jamtlands allehanda, 1889-95) – sw Kungliga [078]
Jan chih yeh kuo huo cheng hsin chi – [Shang-hai: Shang-hai shih chi ch'i jan chih yeh t'ung yeh kung hui], Min kuo 27 [1938] – (= ser P-k&k period) – us CRL [338]
Jan hofmeyr / Macdonald, Tom – London, England. 1948 – 1r – us UF Libraries [960]
Jan, Jean Marie see Congregations religieuses a saint-domingue, 1681-1
Jan + khyac, nok chum 2 – Ran kun: Sa pre Phru Ca pe 1976 [mf ed 1995] – on pt of 1r – 1 – mf#11052 r1960 n2 – us Cornell [959]
Jan may vatthu to kri : [short stories] / Thvan Sin, Dhammacariya U – Ran Kun: Tan Aon ca pe phran Khyi re 1981- [mf ed 1990] – 1r with other items – 1 – (in burmese) – mf#mf-10289 seam reel 174/4 [§] – us CRL [830]
Jan pieterszoon sweelinck (1562-1621) : collected works = Werke van jan pietersen sweelinck, uitgegeven door de vereenigung voor noord-nederlands muziekgeschiedenis / ed by Seiffert, Max – Leipzig. 10v. 1894-1901 – 11 – $115.00 set – us Univ Music [780]
The jan vansina collection : ibiteekerezo: historical narratives from rwanda: a collection of texts and translations, 1957-1961 – Chicago, IL, 1973 – us CRL [960]
Jan yug – Delhi, India. 10 Aug 1952-1953 – 1r – 1 – us L of C Photodup [079]
Janakabhivamsa, Ashin see
– A na gat sasana re
– A rhan janakabhivamsa e dasama kyam ca anagat sasana re
– A rhan janakabhivamsa e nok chumchay la mrat buddha
Janarajaye gasat patraya / Sri Lanka – 1967- – 1 – us L of C Photodup [954]
Janasakti – Cuttack, India. Jul-Sept 1966 – 1r – 1 – (oriya language) – us L of C Photodup [079]
Janasakti – Patna, India. 1954 – 1r – 1 – us L of C Photodup [079]
Janasatta – Ahmedabad, India. Jul-Sept 1966 – 1r – 1 – us L of C Photodup [079]
Janauschek, P L see Originum cisterciensum
Janayuga – New Delhi, India. Sept 1973-11 May 1985 – 27r – 1 – us L of C Photodup [079]
Janayuga (janyug) – Lucknow, India. 1954; 1957-60; 1962-63 – (= ser Janyug) – 7r – 1 – us L of C Photodup [079]
Jancke, Oskar [comp] see Kunst und reichtum deutscher prosa
A jandaia : revista da classe estudantil – Fortaleza, CE: Typ Universal, 07 set 1895 – (= ser Ps 19) – mf#P17,01,48 – bl Biblioteca [440]
Jander, Konrad see Oratorum et rhetorum graecorum fragmenta nuper reperta
Jane – [mthly] – 2r/yr – 1 – $400.00/yr – us Fairchild Micro [740]
Jane clement jones / Burwash, Nathanael – S.l: s.n, 1895 – 1mf – 9 – mf#07206 – cn Canadiana [242]
Jane eyre / Bronte, Charlotte – New York, NY. 1941 – 1r – us UF Libraries [025]
Jane, Fred T see British battle fleet
Jane grey / Soumet, Alexandre – Paris, France. 1844 – 1r – us UF Libraries [440]
Jane, Lionel Cecil see Select documents illustrating the four voyages of columbus
Jane routledge : or, married misery – London, England. 18– – 1r – us UF Libraries [240]
Janelle, Christopher M see The relationship of physical self-perception to injury potential of college football athletes
Janelle, Joseph-Emile see La famille janelle
Janentzky, Christian see Johann caspar lavater
Janer, F see
– Catalogo del museo de ciencias naturales
– Madrid. museo arqueologico nacional inventario de la seccion de etnografia
Janes, Lewis George see A study of primitive christianity
Janes, Lewis George et al see Sociology
Janesville city times – Janesville WI. 1873 jan 2-1876 apr 27, 1876 may 4-1879 aug 28, 1879 sep 4-1883 feb 22, 1883 mar 1-1884 dec 25 – 4r – 1 – (cont by: janesville times) – mf#926862 – us WHS [071]
Janesville daily gazette – Janesville WI. [1901 jun 1/12]-[1904 jul 2-oct 10] – 14r – 1 – (cont: daily gazette [janesville, wi]; cont by: janesville gazette [janesville, wis. : 1969 : daily]) – mf#1145197 – us WHS [071]
Janesville daily gazette – Janesville WI. [1880 aug 19/sep 20]-1893 jun 10-1894 jan 5] [small gaps] – 25r – 1 – (cont: janesville gazette [janesville, wis. : 1860 : daily]; cont by: daily gazette [janesville, wi]) – mf#1145175 – us WHS [071]
Janesville daily gazette – Janesville WI. 1860 mar 19-jun, 1860 jul-dec, 1861 jan-dec, 1862 jan-dec, 1863 jan-dec, 1864 jan-dec, 1865 jan-mar 2 – 11r – 1 – (cont: janesville morning gazette; cont by: janesville gazette [janesville, wis. : 1865 : daily]) – mf#1145204 – us WHS [071]
Janesville daily gazette – Janesville WI. 1854 jul 10-oct 7 – 1r – 1 – mf#1145234 – us WHS [071]
Janesville daily recorder – Janesville WI. [1879 jul 21/1880 jan 21]-[1913 jul 12-1913 sep 21] [small gaps] – 20r – 1 – mf#876700 – us WHS [071]
Janesville democrat – Janesville WI. 1860 sep 7-dec 7 – 1r – 1 – (cont by: rock county republican) – mf#951945 – us WHS [071]
Janesville free press – Janesville WI. 1854 jan 17-1855 dec 25 – 2r – 1 – (cont by: janesville weekly free press) – mf#876697 – us WHS [071]
Janesville free press – Janesville WI. 1853 oct 11, 1856 apr 10-nov 28, 1856 dec 2-1857 mar 4 – 3r – 1 – (cont by: janesville weekly free press; janesville gazette [janesville, wi: 1845: weekly]; weekly gazette and free press) – mf#952437 – us WHS [071]
Janesville gazette – Janesville WI. 1847 jan 16-1852 sep 4, 1852 sep 4-1857 mar 14 – 2r – 1 – (cont by: janesville weekly free press; janesville free press [daily]; weekly gazette and free press) – mf#1145241 – us WHS [071]
Janesville gazette – Janesville WI. [1865 mar 3/jun]-[1880 apr 9/aug 18] [small gaps] – 3r – 1 – (cont: janesville daily gazette [janesville, wi: 1860]; cont by: janesville daily gazette [janesville, wi 1880]) – mf#1141085 – us WHS [071]
Janesville gazette – Janesville WI. 1865 may 25-1865 jun 29, 1866 jul 5-1866 dec 27, 1869 jan 7-1869 dec 30, 1878 may 17 – 4r – 1 – mf#1141086 – us WHS [071]
Janesville journal – 1910 may 12-19, jun 23 – 1r – 1 – mf#4735469 – us WHS [071]
Janesville morning gazette – Janesville WI. 1857 mar 9-jun 15, 1857 jun 16-dec 31, 1858 jan-jun, 1858 jul-dec, 1859 jan-jun, 1859 jul-dec, 1860 jan-mar 17 – 7r – 1 – (cont by: janesville daily gazette [janesville, wi: 1860]) – mf#1145189 – us WHS [071]
Janesville republican – Janesville WI. 1893 jul 20, 1893 sep 7 – 1r – 1 – mf#1012304 – us WHS [071]
Janesville semi-weekly gazette – Janesville WI. 1876 jan 18, 1876 oct 3, 1878 may 17 – 3r – 1 – mf#877752 – us WHS [071]
Janesville times – Janesville WI. 1885 jan 1-1886 apr 22 – 1r – 1 – (cont: janesville city times) – mf#926863 – us WHS [071]
Janesville weekly free press – Janesville WI. 1856 jan 8-1857 mar 3 – 1r – 1 – (cont: janesville free press [weekly]; cont by: janesville free press [daily]; janesville gazette [janesville, wi: 1845: daily]; weekly gazette and free press) – mf#952436 – us WHS [071]
Janesville weekly gazette – Janesville WI. 1864 jan 22-1865 may 18 – 1r – 1 – (cont: weekly gazette and free press; cont by: janesville gazette [janesville, wi: 1865: weekly]) – mf#1143962 – us WHS [071]
Janesville weekly recorder – Janesville WI. 1885 nov 6-1886 apr 16 – 1r – 1 – (cont: rock county recorder [1870]; cont by: janesville times; janesville weekly recorder and times) – mf#951935 – us WHS [071]
Janesville weekly recorder and times – Janesville WI. 1886 apr 23-1887 feb 24, 1887 mar 10-1890 dec 25, 1891 jan 1-1893 apr 20 – 3r – 1 – (cont: janesville weekly recorder; janesville times; cont by: recorder and times) – mf#951936 – us WHS [071]
Janesville weekly sun – Janesville WI. 1882 jun 17-1883 dec 2, 1884 jan 5-1885 may 16, 1885 may 23-1885 dec 26 – 3r – 1 – (cont by: saturday morning sun) – mf#927047 – us WHS [071]
Janesville weekly times – Janesville WI. 1859 feb 15-1859 dec 22 – 1r – 1 – mf#951949 – us WHS [071]
Janet, Charles see Sur la phylogenese de l'orthobionte
Janet, Paul see
– Fenelon, his life and works
– The materialism of the present day
– The materialism of the present time
– Principes de metaphysique et de psychologie
– The theory of morals
Janeway, Jacob Jones see
– Antidote to the poison of popery
– The internal evidence of the holy bible
Janeway, Thomas Leiper see Memoir of the rev. jacob j. janeway
Janeyyabhivamsa, a rhan see Buddha e vi nann vatthu to kri
Jangan, Cheddi see What happened in british guiana
Jani, Mirza see Kitab-i nuqtatu'l-kaf
Janiche : y otros cuentos / Rodriguez Ruiz, Napoleon – San Salvador, El Salvador. 1960 – 1r – us UF Libraries [972]
Janin, Jules see
– The american in paris
– The american in paris during the summer
– Le mois de mai a londres et l'exposition de 1851
Janin, Jules Gabriel see Causeries litteraires et historiques

Janitschek, Ellinor *see* Im umstrittenen gebiet

Janitschek, H *see* Leone battista alberti's kleinere kunsttheoretische schriften...

Jank, Martin *see* Eckert auf grossfahrt

Jankovich, Gina *see* Comparison of unstructured feedback to structured feedback on initial learning of cpr

Jankowski, Joachim *see* Einfluss von wasserdampf auf den ablauf der heterogen katalysierten oxidativen kupplung von methan

Janmabhoomi – Bombay, India. 1958; 1964-Aug 1966 – 10r – 1 – us L of C Photodup [079]

Janmabhoomi – Coorg, India. 1962 – 1r – 1 – us L of C Photodup [079]

Janmabhumi – Masulipatam: M Krishnarao. v1 n1-25 dec 4 1919-may 27 1929; v2 n1-v5 n50 dec 2 1920-nov 27 1924-7. Dec 1919-Nov 1926 – 3r – 1 – us CRL [079]

Jann, Adelhelm *see* Die katholischen missionen in indien, china und japan

Jannasch, Lilli *see* Schwarze schmach und schwarz-weiss-rote schande

Janneau, Gustave Jean Auguste *see* Oeuvres

Janney, Oliver Edward *see* Quakerism and its application to some modern problems

Janney, Samuel Macpherson *see*
- An examination of the causes which led to the separation of the religious society of friends in america, in 1827-28
- History of the religious society of friends from its rise to the year 1828
- The life of george fox
- Memoirs of samuel m. janney
- Peace principles exemplified in the early history of pennsylvania
- Summary of christian doctrines as held by the religious society of friends

Jannsen, Johannes *see* Geschichte des deutschen volkes seit dem ausgang des mittelalters

Janot, Jeffrey M *see* Heart rate and perceived exertion responses during climbing in beginner and recreational sport climbers

Jansen, Brigitte E S *see*
- Einige gedanken zur ausdifferenzierung von staat und recht
- Geistesgeschichtliche aspekte des genossenschaftlichen bildungsgedankens
- Zum verstaendnis von bildung und ausbildung in der soziallehre v a hubers

Jansen enikels werke (mgh8:3.bd) : 1. abt: die weltchronik 1891. 2. abt: fuerstenbuch 1900 – (= ser Monumenta germaniae historica 8: scriptores qui vernacula lingua usi sunt (mgh8)) – €46.00 – ne Slangenburg [240]

Jansen, Gottfried *see* Der streit um die praedestination im ausgehenden 16. jahrhundert

Jansen news – Jansen, NE: J J Fast, 1915-v10 n27 apr 30 1925 (wkly) [mf ed 1916-25] – 6r – 1 – (absorbed by: fairbury journal) – us NE Hist [071]

Jansen, Werner *see* Absonderliche charaktere bei wilhelm raabe

Le jansenisme au 18e siecle et joachim colbert evaeque de montpellier (1696-1738) / Durand, Valentin – Toulouse: Edouard Privat, 1907 – 1mf – 9 – 0-8370-9616-2 – mf#1986-3616 – us ATLA [920]

Le jansenisme convulsionnaire et l'affaire de la planchette : d'apres les archives de la bastille / Gagnol, abbe – Paris: Libraire generale catholique, 1911 – 1mf – 9 – 0-8370-8424-5 – mf#1986-2424 – us ATLA [240]

The jansenists : their rise, persecutions by the jesuits, and existing remnant / Tregelles, Samuel Prideaux – London: Samuel Bagster, 1851 – 1mf – 9 – 0-8370-8230-7 – mf#1986-2230 – us ATLA [241]

Jansenius, Corn. *see*
- Pentateuchus
- Tetrateuchus

Jansen-Runge, Edith *see* Isolde und tristan

Janson, Charles W *see* A view of the present condition of the states of barbary

Janson, cobb, pearson and co solicitors : archives, 1728-1928 – 44r – 1 – £2100.00 – mf#JCP – uk World [340]

Janson, Florence Edith Alfreda *see* The background of swedish immigration, 1840-1930

Janson, John M *see* Collection of booklets and pamphlets obtained in 1965-66 in the lower congo

Janson, Meredith *see* Wu wei

Janssen, Bob Ronald *see* 'N ondersoek na die gebruik van krygsgeskiedenis in die ontwikkeling van militere doktrine

Janssen, H Q *see*
- Acten van de classicale en synodale vergaderingen der verschillenden gemeenten in het land van cleef, sticht van keulen en aken 1571-1589 (de werken..2,2)
- Handelingen van de kerkeraad der nederl gemeente te keulen 1571-1591 (de werken... 1/3)

Janssen, Hendrik Quirinus *see* De synode te emden in 1571

Janssen, Johannes *see* Schiller als historiker

De janssen kwestie en nog iets / Kuiper, Barend Klaas – Grand Rapids, MI: Eerdmans-Sevensma, [1922?] [mf ed 1993] – (= ser Reformed church coll) – 62p on 1mf – 9 – 0-524-06096-7 – mf#1991-2409 – us ATLA [242]

Janssen, N A *see* Een woord over het gregoriaansch

Janssen, O *see* L'expressivite chez salvien de marseille, premiere partie

Janssen, Philip F *see* The development of a design for a total evaluation system for professional baseball umpires

Janssen, R *see* Das johannes-evangelium

Janssen, Ralph *see* Das johannes-evangelium nach der paraphrase des nonnus panopolitanus

Janssens, Laurent *see*
- Tractatus de deo creatore et de angelis
- Tractatus de deo trino
- Tractatus de deo uno
- Tractatus de deo-homine, sive, de verbo incarnato

Jansz, Pieter *see* Java's zendingveld

Jantzen Family *see* History

Jantzen, Hermann *see* Die deutsche romantik

Jantzen Hillsboro Creamery *see* Records

Jantzen, Peter *see* Records of the hillsboro creamery

Januarii-novembris : acta sanctorum / ed by Bollandus, J – Bruxelles, Paris, 1863-1931 – 2096mf – 8 – mf#13 – ne IDC [700]

Januca / Gambach, Nesim – Habana, Cuba. 1960 – 1r – us UF Libraries [025]

Janus : archives internationales pour l'histoire de la medecine et la geographie medicale – v1-22. 1897-1917 – 1 – $486.00 – mf#0303 – us Brook [610]

Janus – Cahiers of young French and American poetry. nos. 1-5. 1950-51 – 1 – us AMS Press [410]

Janus : the edinburgh literary almanac – 1826 – (= ser English gift books and literary annuals, 1823-1857) – 6mf – 9 – uk Chadwyck [800]

Janus / Ripamonte Y Toledo, Carlos P – Buenos Aires, Argentina. 1926 – 1r – us UF Libraries [720]

Janvier, Caesar Augustus Rodney *see* Historical sketch of the missions in india under the care of the board of foreign missions of the presbyterian church

Janvier, Louis Joseph *see*
- Affaires d'haiti
- Antinationaux
- Caisse d'epargne et l'ecole en haiti
- Chercheuse
- Elections legislatives de 1908
- Government civil en haiti
- Republique d'haiti et ses visiteurs

Janvier, Pierre Desire *see* Vie de m. dupont

Janvier, Thomas Allibone *see* Legends of the city of mexico

Janze, Comte de *see* Vertical land

Janzen, D M *see* History of first baptist church, william lake, british columbia, canada

Janzen, Tami M *see* Spontaneous kicking in infants

Jaoa – the journal of the american osteopathic association / American Osteopathic Association – Chicago. 1901+ (1) 1971+ (5) 1973+ (9) – ISSN: 0098-6151 – mf#2527 – us UMI ProQuest [615]

Jaocs – journal of the american oil chemists' society – Champaign. 1980+ (1,5,9) – (cont: journal of the american oil chemists' society) – ISSN: 0003-021X – mf#131,01 – us UMI ProQuest [540]

Japan : an attempt at interpretation / Hearn, Lafcadio – New York: Macmillan, 1905 [c1904] [mf ed 1995] – (= ser Yale coll) – v/549p (ill) – 1 – 0-524-10075-6 – mf#1995-1075 – us ATLA [241]

Japan : containing illustrations of the character, manners, customs, religion, dress, amusements, commerce, agriculture, etc of the people of that empire / ed by Shoberl, Frederick – London [1823] [mf ed Hildesheim 1995-98] – 1v on 2mf – 9 – €60.00 – 3-487-27543-0 – gw Olms [915]

Japan : internal affairs and foreign affairs, 1945-1966 / U.S. State Dept – (= ser Confidential u s state department central files) – 1 – $35,015.00 coll – (internal affairs, 1945-49 42r isbn 0-89093-731-1 $6250. 1950-54 62r isbn 0-89093-732-X $8725. internal affairs & foreign affairs, 1960-jan 1963 38r isbn 1-55655-701-9 $7370. feb 1963-1966 50r isbn 1-55655-702-7 $9680. subject-numeric files, 1967-69: pt1: political, governmental, & national defense affairs 18r isbn 1-55655-892-9 $3485. with p/g) – us UPA [950]

Japan : its architecture, art, and art manufactures / Dresser, Christopher – London 1882 – (= ser 19th c art & architecture) – 5mf – 9 – mf#4.2.493 – uk Chadwyck [700]

Japan / Pratt, Helen Gay – New York, NY. 1937 – 1r – us UF Libraries [025]

Japan 21st – Tokyo. 1992-1996 (1,5,9) – (cont: business japan) – ISSN: 0916-877X – mf#13095,02 – us UMI ProQuest [338]

Japan, 1914-1941 – 3pt – (= ser Confidential u s diplomatic post records) – 1 – $23,310.00 coll – (pt1: 1914-18 11r isbn 0-89093-503-3 $1915. pt2: 1919-29 50r isbn 0-89093-504-1 $8710. pt3: 1930-41 80r isbn 0-89093-505-X $13,920. with p/g) – us UPA [327]

Japan, 1918-1941 – (= ser U s military intelligence reports) – 31r – 1 – $5400.00 – 0-89093-448-7 – (with p/g) – us UPA [355]

Japan, 1947-1956 / U.S. State Dept – (= ser Confidential u s state department special files) – 39r – 1 – $6785.00 – 1-55655-197-5 – (1st suppl, 1946-66 25r isbn 1-55655-847-3 $4840. with p/g) – us UPA [327]

The japan advertiser – Tokyo: B W Fleisher, jul 1938-nov 9 1940 – (issues for nov 1-9 1940 filmed with: japan times and advertiser (morning ed), nov 10-30 1940, and japan times and advertiser (evening ed), nov 11-30 1940) – us CRL [079]

Japan and america, c1930-1955 – the pacific war and the occupation of japan, series 1 : the papers of general robert l eichelberger (1886-1961) from the william r perkins library, duke university – 4pt – 1 – (pt1: subject files on world war 2 & japan (boxes 32-53) [23r] £2150; pt2: subject files on japan & diaries (boxes 54-65 & boxes 1-4) [20r] £1850; pt3: correspondence (boxes 5-27) [27r] £2550; pt4: subject files, writings, speeches, photographs & oversize material (boxes 28-31, 66-69, 79-88 and 93-98) [17r] £1600; with d/g) – uk Matthew [934]

Japan and america, c1930-1955 – the pacific war and the occupation of japan, series 2 : the o'ryan mission to japan and occupied china, 1940 the whitney diary; correspondence and papers of dr whitney, general o'ryan and other members of the economic and trade mission – 2r – 1 – £190.00 – (with d/g) – uk Matthew [934]

Japan and india : The outgrouth of a trip to japan by a delegation from india to the convention of the world's student christian federation, held in tokyo in 1907 / Eddy, Sherwood – Calcutta: Student Volunteer Movt of India and Ceylon [1908?] [mf ed 1995] – (= ser Yale coll) – 115p (ill) – 1 – 0-524-09992-8 – mf#1995-0992 – us ATLA [240]

Japan and its art / Huish, Marcus Bourne – London 1889 – (= ser 19th c art & architecture) – 3mf – 9 – mf#4.2.494 – uk Chadwyck [700]

Japan and its occupied territories during world war 2 : 1942-1945 / U.S. Office of Strategic Services & U.S. State Dept – (= ser Oss/state department intelligence and research reports 1) – 16r – 1 – $2460.00 – 0-89093-117-8 – (with p/g) – us UPA [327]

Japan and its regeneration / Cary, Otis – rev ed. New York: Student Volunteer Movt for Foreign Missions, 1908, c1904 – 1mf – 9 – 0-8370-6094-X – (incl ind) – mf#1986-0094 – us ATLA [240]

Japan and its rescue : a brief sketch of the geography, history, religion and evangelization of japan / Hail, A D – Nashville,TN: Cumberland Presbyterian Publ House, 1898 [mf ed 1986] – 1mf – 9 – 0-8370-6061-3 – mf#1986-0061 – us ATLA [242]

Japan and the japan mission of the church missionary society / Stock, Eugene – 2nd rev ed in part re-written, and continued to date. London: Church Missionary House, 1887 – 1mf – 9 – 0-524-04560-7 – mf#1991-2124 – us ATLA [240]

Japan and the united states : diplomatic, security, and economic relations, 1960-1976 – [mf ed Chadwyck-Healey] – (= ser National security archive, washington dc: the making of us policy) – 2000+ docs on 316mf – 9 – (with p/g & ind) – uk Chadwyck [327]

Japan and the west : sources from dutch archives – 5pt – 9 – €13,845.00 set – (pt1: documents concerning the negotiation of a trade agreement with japan, 1852-1870 [124mf] €1170 [m481]; pt2: political reports on japan, 1887-1940 [84mf] €745 [m482]; pt3: the archive of the dutch legation in japan, 1870-1890 [741mf] €6370 [m483]; pt4: the archive of the dutch consulate at nagasaki 1860-1915 [406mf] €3490 [m484]; pt5: the archive of the dutch consulate at yokohama, 1860-1870 [420mf] €3610 [m485]) – mf#m481-m485 – ne MMF Publ [327]

Japan and the world economy – Amsterdam. 1990-1991 (1,5,9) – ISSN: 0922-1425 – mf#42559 – us UMI ProQuest [332]

Japan as a mission field / Worcester, Isaac Redington – Boston: ABCFM, 1878 – 1mf – 9 – 0-524-00664-4 – mf#1990-0164 – us ATLA [240]

Japan. Bureau of Customs of the Ministry of Finance *see* Nippon boeki nempyo

Japan Cabinet. Board of Documents *see* Hoki bunrui taizen

Japan. Cabinet Secretariat [comp] *see* Horei zensho

Japan chemical quarterly – Tokyo. 1965-1969 (1) – (cont by: chemical economy and engineering review: ceer) – ISSN: 0448-8571 – mf#10569 – us UMI ProQuest [540]

Japan christian quarterly – Tokyo. 1989-1991 (1) 1989-1991 (5) 1989-1991 (9) – (cont by: japan christian review) – ISSN: 0021-4361 – mf#16014 – us UMI ProQuest [240]

Japan christian review – Tokyo. 1992-1996 (1,5,9) – (cont: japan christian quarterly) – ISSN: 0918-516X – mf#16014,01 – us UMI ProQuest [240]

Japan chronicle – Kobe. Japan. -d. 1 Mar 1929-11 Apr 1937, 24 Jul 1938-25 Dec 1940. (41 reels) – 1 – uk British Libr Newspaper [072]

Japan correspondence, 1856-1905 : registers, 1856-1905 / British Foreign Office – 6r – 1 – $780.00 – (with printed guide) – mf#S0056-05 – us Scholarly Res [324]

Japan correspondence, 1856-1948 : conflict over korea, 1883-1893 / British Foreign Office – 65r – 1 – $8450.00 – (with printed guide) – mf#S0183-93 – us Scholarly Res [324]

Japan correspondence, 1856-1948 : dominance of the genro, 1906-1913 / British Foreign Office – 47r – 1 – $6110.00 – (with printed guide) – mf#S0406-13 – us Scholarly Res [324]

Japan correspondence, 1856-1948 : the early meiji period, 1868-1875 / British Foreign Office – 61r – 1 – $7930.00 – (with printed guide) – mf#S0168-75 – us Scholarly Res [324]

Japan correspondence, 1856-1948 : emergence of a military clique, 1930-1936 / British Foreign Office – 38r – 1 – $4940.00 – (with printed guide) – mf#S0430-36 – us Scholarly Res [324]

Japan correspondence, 1856-1948 : emergence of japan as a pacific power, 1914-1923 / British Foreign Office – 73r – 1 – $9490.00 – (with printed guide) – mf#S0414-23 – us Scholarly Res [324]

Japan correspondence, 1856-1948 : the end of feudal rule, 1856-1867 / British Foreign Office – 48r – 1 – $6240.00 – (with printed guides) – mf#S0156-67 – us Scholarly Res [324]

Japan correspondence, 1856-1948 : the occupation, 1946-1948 / British Foreign Office – 85r – 1 – $11,050.00 – (with printed guide) – mf#S0446-49 – us Scholarly Res [324]

Japan correspondence, 1856-1948 : period of intense westernization, 1876-1882 / British Foreign Office – 59r – 1 – $7670.00 – (with printed guides) – mf#S0176-82 – us Scholarly Res [324]

Japan correspondence, 1856-1948 : registers, 1906-1919 / British Foreign Office – 3r – 1 – $390.00 – (with printed guide) – mf#S0106-19 – us Scholarly Res [324]

Japan correspondence, 1856-1948 : rise of the kwangtung army, 1924-1929 / British Foreign Office – 15r – 1 – $1950.00 – (with printed guide) – mf#S0424-29 – us Scholarly Res [324]

Japan correspondence, 1856-1948 : the russo-japanese war, 1905 / British Foreign Office – 57r – 1 – $7410.00 – (with printed guide) – mf#S0205 – us Scholarly Res [324]

Japan correspondence, 1856-1948 : the sino-japanese war, 1937-1941 / British Foreign Office – 48r – 1 – $6240.00 – (with printed guide) – mf#S0437-41 – us Scholarly Res [324]

Japan correspondence, 1856-1948 : the sino-japanese war and expansionism, 1894-1904 / British Foreign Office – 78r – 1 – $10,140.00 – (with printed guide) – mf#S0194-04 – us Scholarly Res [324]

Japan correspondence, 1856-1948 : the war in the pacific, 1942-1945 / British Foreign Office – 28r – 1 – $3640.00 – (with printed guide) – mf#S0442-45 – us Scholarly Res [324]

Japan correspondence, 1856-1951 : the restoration of sovereignty, 1949-1951 / British Foreign Office – 1996 – 39r – 1 – $5070.00 – (guide also sold separately $20 s0449-51.g) – mf#S0449-51 – us Scholarly Res [950]

Japan Dental Association *see* Journal of the japan dental association

Japan echo – Tokyo. 1993-1996 (1,5,9) – ISSN: 0388-0435 – mf#19212 – us UMI ProQuest [950]

Japan economic journal : international weekly edition – Tokyo. 1985-1991 (1,5,9) – (cont by: nikkei weekly) – ISSN: 0021-4388 – mf#14487 – us UMI ProQuest [330]

Japan forum – Oxford. 1989+ (1,5,9) – ISSN: 0955-5803 – mf#17502 – us UMI ProQuest [950]

Japan in the year of the war : containing encouraging facts from missions in the sunrise kingdom / Pettee, James Horace – Boston: Young People's Dept, American Board of Commissioners for Foreign Missions, 1904 [mf ed 1995] – (= ser Yale coll) – 30p – 1 – 0-524-09704-6 – mf#1995-0704 – us ATLA [950]

Japan in world politics / Kawakami, Kiyoshi Karl – New York: The Macmillan Co., 1917. xxvii,300p – 1 – us Wisconsin U Libr [950]

Japan, its weakness and strength / Chattopadhyaya, Kamaladevi – Bombay: Padma Publications, 1943 – (= ser Samp: indian books) – us CRL [950]

Japan, korea and the security of asia, 1946-1976 / U.S. Central Intelligence Agency – (= ser Cia research reports) – 5r – 1 – $770.00 – 0-89093-450-9 – (with p/g) – us UPA [327]

Japan, korea, southeast asia and the far east generally : 1950-1961 supplement / U.S. Office of Strategic Services & U.S. State Dept – (= ser Oss/state department intelligence and research reports 8) – 7r – 1 – $1085.00 – 0-89093-345-6 – (with p/g) – us UPA [327]

Japan. Laws, Statutes, etc *see* Kampo

Japan mail – Yokohama, Japan. 21 jun 1873-23 dec 1893 [wkly] – 10r – 1 – (lacking: 1876, 1888, 1890, 1891; aka: japan weekly mail summary; japan mail summary) – uk British Libr Newspaper [079]

Japan mail summary *see* Japan mail

Japan. Ministry of Agriculture and Forestry *see* Tochi keizai shiryo

Japan. Ministry of Agriculture and Forestry [comp] *see* Nihon rinseishi chosa shiryo

Japan. Ministry of Foreign Affairs *see* Intercepted japanese messages

Japan. Ministry of War *see*
– Rikugun
– Sendjinkoen

Japan mission annual, 1919 : american board of commissioners for foreign missions, featuring the japan mission's semicentennial – [Ginza: Tokyo: Methodist Publ House (Kyobunkan) 1919] [mf ed 1995] – (= ser Yale coll) – 182p (ill) – 1 – 0-524-09424-1 – mf#1995-0424 – us ATLA [950]

Japan. mission of the protestant episcopal church : station, nagasaki. missionaries, rev j liggins, rev c m williams – New York: Bible House, 1859 [mf ed 1995] – (= ser Yale coll; Occasional missionary paper 20 mar 1859) – 15p (ill) – 1 – 0-524-09316-4 – mf#1995-0316 – us ATLA [242]

Japan plastics age – Tokyo. 1978-1987 (1) 1978-1987 (5) 1978-1987 (9) – ISSN: 0021-4582 – mf#11490 – us UMI ProQuest [660]

Japan plastics industry annual – Tokyo. 1980-1980 (1,5,9) – ISSN: 0448-8679 – mf#11618 – us UMI ProQuest [660]

Japan quarterly – Tokyo. 1954-2001 (1) 1954-2001 (5) 1954-2001 (9) – ISSN: 0021-4590 – mf#12745 – us UMI ProQuest [073]

Japan Society of Accounting *see* Accounting

Japan through western eyes : manuscript records of traders, travellers, missionaries and diplomats, 1853-1941 – [mf ed Marlborough 1996] – 8pt – 1 – (pt1: sources fr william r perkins library, duke university [20r] £1850; pt2: william elliot griffis coll fr rutgers university library – journals & student essays [6r] £570; pt3: william elliot griffis coll fr rutgers university library – correspondence & scrapbooks [24r] £2250; pt4: william elliot griffis coll fr rutgers university library – collected papers of brown, perry & others [21r] £1975; pt5: william elliot griffis coll fr rutgers university library – writings by griffis [12r] £1125; pt6: correspondence & papers of sir ernest satow (1843-1929) relating to japan fr public record office class pro 30/33 [21r] £1975; pt7: papers of harold s williams (1898-1987) fr national library of australia – the green subject files (folders 1-138) [21r] £1975; pt8:... – additional subject files [13r] £1250; pt9: siebold mss fr oriental manuscripts coll at the british library [c19r] £1850; with d/g) – uk Matthew [327]

Japan times – Tokyo: The Japan Times Ltd, mar 22 1897-1998 – 844r – 1 – Y6232,700 – (in english) – ja Yushodo [079]

The japan times – Tokyo: Japan Times Ltd, jul 1956-70 – us CRL [079]

The japan times and advertiser – Tokyo: The Japan Times Ltd, nov 10 1940-aug 2 1941 (Morning ed) – (filmed consecutively with: japan times and advertiser (evening ed); issues for nov 10-30 1940 filmed with: japan advertiser, nov 1-9 1940) – us CRL [072]

The japan times and advertiser – Tokyo: The Japan Times Ltd, nov 11 1940-aug 2 1941 (evening ed) – (filmed consecutively with: japan times and advertiser (morning ed); issues for nov 11-30 1940 filmed with: japan advertiser, nov 1-9 1940) – us CRL [079]

Japan times. international – Tokyo, 1999-2000 [1,5,9] – (cont: japan times. weekly international edition) – mf#18231,01 – us UMI ProQuest [079]

Japan times weekly : international edition – Tokyo, Japan. 1990-1997 (1,5,9) – ISSN: 0447-5763 – mf#18231 – us UMI ProQuest [070]

Japan today and tomorrow – (annual). Osaka. Japan. 1927-39. (2 reels) – 1 – uk British Libr Newspaper [072]

Japan und die christliche mission / Halmhuber, A – Cleveland: Verlagshaus der Evangelischen Gemeinschaft, 1896 [c1884] [mf ed 1995] – (= ser Yale coll) – x/391p (ill) – 1 – 0-524-09927-8 – (in german) – mf#1995-0927 – us ATLA [240]

Japan und seine bewohner : geschichtliche rueckblicke und ethnographische schilderungen von land und leuten / Heine, Wilhelm – Leipzig 1860 [mf ed Hildesheim 1995-98] – 1v on 3mf – 9 – €90.00 – 3-487-27544-9 – gw Olms [915]

Japan weekly mail – jan 1871-dec 1913 – 14r – A$1035.45 vesicular A$1112.45 silver – at Pascoe [079]

Japan weekly mail – Yokohama. Japan. -w. 7 Jan 1871-25 Dec 1897. (Wanting 1872-77, 1889, 1893, 1895, 1896). (19 reels) – 1 – uk British Libr Newspaper [072]

Japan weekly mail summary *see* Japan mail

Japan weekly times – (Japan Times. Weekly ed. Japan Times and Mail. Weekly ed.). Tokyo. Japan. -w. 27 Mar 1897-23 Dec 1922. (51 reels) – 1 – uk British Libr Newspaper [072]

Die japaner : wanderungen durch das geistige, soziale und religioese leben des japanischen volkes / Munzinger, Carl – Berlin: A Haack, 1898 [mf ed 1995] – (= ser Yale coll) – 417p – 1 – 0-524-09830-1 – (in german) – mf#1995-0830 – us ATLA [950]

Japanese air target analyses, objective folders, and aerial photographs, 1942-1945 – (= ser National archives coll of foreign records seized, 1941-) – 7r – 1 – mf#M1653 – us Nat Archives [355]

Japanese American Citizens League *see* Jacl reporter

Japanese American Citizens' League et al *see* Pacific citizen

Japanese art : [catalogue of the collection in the national art library] / South Kensington Museum, London – London 1893, 1898 – (= ser 19th c art & architecture) – 3mf – 9 – mf#4.1.352 – uk Chadwyck [700]

Japanese Association of North America *see* Records, 1916-1941

Japanese biographical archive (jaba) = Japanisches biographisches archiv (jaba) / Wispelwey, Berend [comp] – [mf ed 2000-03] – 425mf (1:24) in 12 installments – 9 – diazo €10,060.00 (silver €11,080 isbn: 978-3-598-34001-7) – ISBN-10: 3-598-34000-1 – ISBN-13: 978-3-598-34000-0 – (with printed ind) – gw Saur [950]

Japanese camp newspapers – 1942-45 – 1 – $40.00r outside US – (15 japanese- and english-language papers available. apply for complete listing) – Dist. us Scholarly Res – us L of C Photodup [071]

Japanese canadian research collection – Vancouver: microfilmed for UBCL... [mf ed 1996] – 11r – 1r – (incl: an inventory to the papers and records in the japanese canadian research collection / terry nabata [1975]; rev 1996 by norman amor, with new int by tsuneharu gonnami) – cn UBC Preservation [971]

Japanese cane / Scott, John M – Gainesville, FL. 1916 – 1r – us UF Libraries [630]

Japanese cane for forage / Scott, John M – Gainesville, FL. 1911 – 1r – us UF Libraries [630]

A japanese collection / Tomkinson, Michael – London 1898 – (= ser 19th c art & architecture) – 11mf – 9 – mf#4.2.491 – uk Chadwyck [700]

[Japanese commentaries on the "four shoo" or the books of the four philosophers] / Hanawa, Tokinosuke; ed by Fukai, Kanichiro – [11th ed] [Tokyo: s.n, 1900] [mf ed 1993] – 5v on 5mf – 9 – 0-524-08045-3 – (in japanese) – mf#1991-0261 – us ATLA [180]

Japanese Congregational Church [Seattle WA] *see* Records, 1907-1949

Japanese economic studies – White Plains. 1988-1993 (1) – (cont by: japanese economy) – ISSN: 0021-4841 – mf#16889 – us UMI ProQuest [330]

Japanese economy – White Plains. 1997+ (1) – (cont: japanese economic studies) – ISSN: 1097-203X – mf#16889,01 – us UMI ProQuest [330]

Japanese enamels / Bowes, James Lord – Liverpool 1884 – (= ser 19th c art & architecture) – 2mf – 9 – mf#4.2.702 – uk Chadwyck [730]

Japanese government documents and censored publications – 230r – 1 – $8,050.00 in US $40.00r outside – ((pre-1946) japanese thought control police materials 3r I9400015. naimusho keihokyoku, 1910 (japanese police intelligence reports) 3r I9400016. censored issues of gakan (my views) 3r I9400027. kokuhon, organ of kokuhonsha, established in 1924 3r I9400031. koron (public opinions) 8r I9400033. gekkan nihon (japanese monthly) 3r I9400035. kokumin hyoron (national review) 4r I9400036. kosaku mondai ni kansuru shiryo (documents pertaining to farm tenancy question) 4r I9400042. sayoku undo ni kansuru shiryo, 1922-1940 (documents concerning japanese left-wing movements) 4r I9400043. musan seito ni kansuru shiryo (materials concerning japanese proletarian political parties) 4r I9400044. rodo mondai ni kansuru shiryo (materials on japanese labor problems, 1917-1939) 9r I9400045. documents on japanese police activities, 1921-1945 4r I9400046. documents on japanese police activities, 1912-1946 15r I9400047. man ju dai nikki (great daily records of documents received concerning manchurian incident-classified) 4r I9400049. taisho 3-4 nen kaigun senshi (history of the naval warfare during the war of 1914-15) 3r I9400052. showa 6-7-nen jihen kaigun senshi (history of the naval warfare in the incident of 1931-32) 3r I9400059. pre-war (i.e., pre-1946) japanese government materials, 1913-1945 5r I9400060. foreign affairs documents, 1917-1945 6r I9400061. taisei yokusankai kyoryoku kaigi kankei shorui (documents concerning the meetings to cooperate with the imperial rule assistance association), 1941-1942 4r I9400062. gaimusho genson kiroku mokuroku (catalog of extant documents in gaimusho) 3r I9400065. japanese mongraphs compiled under auspices of scap, 1945-1954 12r I9400068. banned japanese publications, 1919-1943 57r I9400077. banned japanese publications, 1923-1944 12r I9400078. in japanese) – mf#L9400014-9400078 – Dist. us Scholarly Res – us L of C Photodup [324]

Japanese government reports to the league of nations : on the administration of the south seas islands under japanese mandate – 1921-37 – r1-3 – 1 – (available for ref) – mf#pmb doc443 – at Pacific Mss [324]

Japanese heart journal – Tokyo. 1960+ (1) 1974+ (5) 1974+ (9) – ISSN: 0021-4868 – mf#7262 – us UMI ProQuest [616]

Japanese illustration : a history of the arts of wood-cutting / Strange, Edward Fairbrother – London 1897 – (= ser 19th c art & architecture) – 4mf – 9 – mf#4.2.530 – uk Chadwyck [760]

Japanese journal of cancer research : gann – Tokyo. 1985+ (1) 1985+ (5) 1985+ (9) – (cont: gann) – ISSN: 0910-5050 – mf#7995,01 – us UMI ProQuest [616]

Japanese journal of ophthalmology – Tokyo. 1957-1989 [1]; 1972-1980 [5]; 1974-1980 [9] – ISSN: 0021-5155 – mf#7651 – us UMI ProQuest [617]

Japanese journal of parasitology = Kiseichugaku zasshi – Tsukuba Science City. 1951-1989 (1) 1972-1979 (5) 1974-1979 (9) – (cont by: parasitology international) – ISSN: 0021-5171 – mf#7223 – us UMI ProQuest [616]

Japanese journal of pharmacology – Kyoto. 1975-1996 (1,5,9) – ISSN: 0021-5198 – mf#10517 – us UMI ProQuest [615]

Japanese journal of religious studies – Nagoya. 1987+ (1,5,9) – ISSN: 0304-1042 – mf#15988 – us UMI ProQuest [200]

Japanese journal of urology – Tokyo. 1973-1996 (1) 1973-1980 (5) 1975-1980 (9) – ISSN: 0021-5287 – mf#7612 – us UMI ProQuest [616]

Japanese language program at the university of michigan / U.S. War Dept. General Staff. G-2 Division – 1943-46 – 1 – $35.00 – us L of C Photodup [480]

Japanese life in town and country / Knox, George William – New York: Putnam, 1904 – (= ser Our asiatic neighbours) – 1mf – 9 – 0-524-00917-1 – mf#1990-2140 – us ATLA [950]

Japanese Methodist Episcopal Church [Seattle WA] *see* Records, 1904-1950

Japanese ministry of foreign affairs, 1868-1945 – 2,116r – 1 – $40.00r outside US – (showa documents, 1926-1945 722r. unindexed documents 52r. special studies 185r. ppaers of the parliamentary vice minister matsumoto tadao 76r. biographical materials 6r. documents of the international military tribunal, ghq, scap 94r. treaties 13r. telegraphs 164r. meiji-taisho documents, 1867-1912 804r. checklist available. in japanese) – Dist. us Scholarly Res – us L of C Photodup [324]

Japanese monographs on the war in the pacific – 18r – 1 – $630.00 set in US $40.00r outside – (in english. guide to japanese monographs on the war in the pacific 1r I9400093. japanese monographs on the war in the pacific 14r I9400094. japanese studies on manchuria 3r I9400095) – mf#L9400093-L9400095 – Dist. us Scholarly Res – us L of C Photodup [355]

Japanese newspapers – 299r – 1 – (some titles in japanese; some in english; akahata: tokyo, july 1955-feb 1958; 1959-1960. 8 reels. I9400100; asahi evening news: tokyo, 1969-1974; 1977-1990. 62 reels. I9400101; choson sinbo: tokyo, 1962-1965; 1968. 7 reels. I9400102; hokkai taimushu: sapporo, 1940; feb-apr, 1941; july-nov 1941; feb-oct 1942. 17 reels. I9400103; hokkaidu shimbun: sapporo, nov 1942-dec 1945. 12 reels. I9400104; japan advertiser: tokyo, jan 1916-june 1938. 72 reels. I9400105; japan times (international airmail edition): tokyo, 1977-1987. 31 reels. I9400106; japan times weekly (international edition): tokyo, 1964-1969; 1977; 1988-1989. 11 reels, I9400107; nippon dokusho shimbun: tokyo, mar 1937-1960; 1962-1964. 6 reels. I9400108; nippon times: tokyo, jan 1946-1954. 14 reels. I9400109; osaka mainichi and the tokyo nichi nichi: osaka, july 1922-1925; apr 1926-may 1927; 1932-1940. 25 reels. I9400110; osaka mainichi: osaka, jan 1941-1945. 12 reels. I9400111; otaru shimbun: otaru, 1938; jan 1940-apr 1941; sept-oct 1941. 8 reels. I9400112; people's korea: tokyo, jan 1967-1989. 8 reels. I9400113; teikoku diagaku shimbun: tokyo, apr 1929-may 1944; may-june 1946. 6 reels. I9400114) – Dist. us Scholarly Res – us L of C Photodup [079]

Japanese occupation *see* War and decolonization in indonesia, 1940-1950

Japanese persimmon in florida / Camp, A F – Gainesville, FL. 1929 – 1r – us UF Libraries [634]

Japanese persimmons / Hume, H Harold – Lake City, FL. 1904 – 1r – us UF Libraries [634]

Japanese philosopher = Shundai zatsuwa / Knox, George William; ed by and other papers upon the chinese philosophy in japan – Yokohama: R Meiklejohn, 1892 [mf ed 1991] – 2mf – 9 – 0-524-02096-5 – (english trans fr japanese by muro kyuso; notes by t haga & t inoue) – mf#1990-2860 – us ATLA [180]

Japanese pottery / Bowes, James Lord – Liverpool 1890 – (= ser 19th c art & architecture) – 7mf – 9 – mf#4.2.534 – uk Chadwyck [730]

Japanese pottery / Franks, Augustus Wollaston – [London] 1880 – (= ser 19th c art & architecture) – 2mf – 9 – mf#4.2.536 – uk Chadwyck [730]

Japanese Presbyterian Church [Seattle WA] *see* Records, 1947-1955

Japanese relocation camp and assembly center newspapers – 22r – 1 – $770.00 set in US $40.00r outside – (mostly in english) – mf#L9400013 – Dist. us Scholarly Res – us L of C Photodup [978]

Japanese resources reference notebooks, 1945-1947 – (= ser Records of the united statgic bombing survey) – 6r – 1 – mf#M1199 – us Nat Archives [355]

Japanese wood engravings / Anderson, William – London 1895 – (= ser 19th c art & architecture) – 2mf – 9 – mf#4.2.717 – uk Chadwyck [730]

Japanese wood-block prints in the british museum – 139mf – 9 – $840.00 – 0-907006-33-7 – (over 6000 images) – uk Mindata [760]

Japanese-american evacuation claims act : adjudications of the attorney general; precedent decisions under japanese-american evacuation claims act, 1950-1956 / U.S. Dept of Justice – Washington, GPO: 1956 [all publ] – 5mf – 9 – $7.50 – mf#llmc 84-114 – us LLMC [342]

Japanese-american war relocation camps : u.s. department of interior war relocation authority reports. 1942-1945 evacuation of japanese-americans from west coast usa – 4r – 1 – $200.00 – mf#B63003 – us Library Micro [324]

Japanese-english dictionary / Nitobe, Inazo & Takakusu, Junjiro – Tokyo: Sanseido, [1916] [mf ed 1995] – (= ser Yale coll) – 1206p – 1 – 0-524-09412-8 – mf#1995-0412 – us ATLA [040]

Japanische mythologie : nihongi "zeitalter der goetter" nebst ergaenzungen aus andern alten quellenwerken / Florenz, Karl – Tokyo: Druck der Hobunsha, 1901 [mf ed 1996] – (= ser Yale coll) – ix/336p (ill) – 1 – 0-524-10224-4 – (in german. suppl of: 'mittheilungen' der deutschen gesellschaft fuer natur- und voelkerkunde ostasiens 4) – mf#1996-1224 – us ATLA [390]

Japan's modernization / Singh, Saint Nihal – London: Charles H Kelly [1914] [mf ed 1995] – (= ser Yale coll; Manuals for christian thinkers) – 136p – 1 – 0-524-09551-5 – mf#1995-0551 – us ATLA [950]

The japan-us semiconductor cases : documentary case studies in international trade – 382mf – 9 – $2030.00 – 0-89093-987-X – (with p/g) – us UPA [380]

Japca – Pittsburgh. 1987-1989 (1) 1987-1989 (5) 1987-1989 (9) – (cont: journal of the air pollution control association. cont by: journal of the air and waste management association) – ISSN: 0894-0630 – mf#6210,01 – us UMI ProQuest [360]

Japenpa *see* Development progress in indonesia

Japenpa features / Jajasan Penerbitan Pantjasila (Japenpa) – Djakarta, 1968-1972 – 104mf – 9 – (missing: 1970, v3(7); 1972, v5) – mf#SE-1715 – ne IDC [959]

Japenpa foreign languages publishing institute special issue : indonesia – Djakarta, 1968-1969(1-3) – 1mf – 9 – (missing: 1968(1)) – mf#SE-1691 – ne IDC [959]

Japhet, Kirilo *see* Meru land case

Japon, Cuentas del *see* Barcelona, 1933

Le japon, ou moeurs, usages et costumes des habitans de cet empire : d'apres les relations recentes de krusenstern, langsdorf, titzing, etc, et ce que les voyageurs precedens offrent de plus avere; suivi de la relation du voyage et de la captivite du capitaine russe golownin / Breton de LaMartiniere, Jean – Paris 1818 [mf ed Hildesheim 1995-98] – 4v on 8mf – 9 – €160.00 – 3-487-27540-6 – gw Olms [915]

Le japon, par un missionnaire / Launay, Adrien – Paris: Societe de Saint-Augustin; Desclee, De Brouwer [1895] [mf ed 1995] – (= ser Yale coll) – 204p (ill) – 1 – 0-524-09680-5 – (in french) – mf#1995-0680 – us ATLA [241]
El japon su evolucion, cultura, religiones / Domenzain, Moises – Madrid: Razon y Fe, 1943 – 1 – sp Bibl Santa Ana [306]
Japp, Alexander Hay *see* Master-missionaries
Japura, Miguel Maria Lisboa *see* Relacion de un viaje a venezuela, nueva granada y...
Jaquet, F G P *see* Guide to the sources in the netherlands concerning the history of asia and oceania
Jaquin, Noel *see* Hand of man
Jara de Soto, Clara *see* El instruido en la corte...estremeno (sic)
Jaragua / Rodriguez Ruiz, Napoleon – San Salvador, El Salvador. 1950 – 1r – us UF Libraries [972]
Jaraiz de la Vera. Ayuntamiento *see*
- Feria y fiestas de jaraiz de la vera 1974
- Fiestas del tabaco y del pimiento 1980
- Guion de ferias y fiestas. agosto 1959

Jaramillo Londono, Agustin *see* Testamento del paisa
Jaramillo, Luis E H *see* Kant und die idealismusfrage
Jarava, J *see*
- Historia de las yervas y plantas...
- Problemas o preguntas problematicas ansi de amor...y acerca del vino

Jardeni, M *see* Daber 'ivrit!
Jardim, Germano Goncalves *see* Rumos da organizacao estatistica brasileira
Jardim, Luis *see* Boi aru a
Jardim, Renato *see* Aventura de outubro e a invasao de s paulo
Jardin de liliana / Arguello, Agenor – Managua, Nicaragua. 1961 – 1r – us UF Libraries [972]
Jardin del alma cristiana / Diaz Tanco, Vasco – 1552 – 9 – sp Bibl Santa Ana [810]
Le jardin d'honneur : contenant plusieurs apologies... – Paris: Estienne Groulleau, 1559 – 2mf – 9 – mf#O-13 – ne IDC [090]
Jardin du paradis / Bruneau, Alfred – Paris, France. 1923 – 1r – us UF Libraries [440]
Le jardin litteraire illustre – Montreal: s.n. [1898] – 9 – mf#P04119 – cn Canadiana [440]
Jardin musical : ...chansons a quatre parties – 1556 – (= ser Mssa) – 2mf – 9 – €35.00 – mfchl 120 – gw Fischer [780]
Jardin musical : ...chansons a trois parties...le premier livre – 1555 – (= ser Mssa) – 2mf – 9 – €35.00 – mfchl 121 – gw Fischer [780]
Jardin musical : contenant plusieurs belles fleurs de chansons spirituelles – 1556 – (= ser Mssa) – 2mf – 9 – €35.00 – mfchl 119 – gw Fischer [780]
Jardin musiqual : contenant plusieurs belles fleurs de chansons – 1556 – (= ser Mssa) – 2mf – 9 – €35.00 – mfchl 118 – gw Fischer [780]
Jardine, Douglas James *see* The mad mullah of somaliland
Jardine, Robert *see* What to believe
Jarfalla nyheter – Stockholm, Sweden. 1982-87 – 1 – sw Kungliga [078]
Jarfalla nyheter *see* Vasterort
Jargal / Hugo, Victor – New York, NY. 1866 – 1r – us UF Libraries [972]
Jaridat al-Ikhwan al-muslimin *see* Al-ikhwan al-muslimun
Jaridat al-ikhwan al-muslimin – Cairo: Tantawi Jawhari, 1933-37? v1 n3,5-8,10,12,14-32,34-35; v2 n3,8-9,11,14,16-17,25-26,36,38; v3 n6-7,13,23,25,29,32,35,42; v4 n7,32-33,40,42-43. 6 rabi' I 1352-20 dhu al-Qa'dah 1355 [29 jun 1933-3 feb 1937] – (= ser Arabic journals and popular press) – 1r – 1 – $600.00 – (r also incl: al-nadhir and al-ikhwan al-muslimun) – us MEDOC [079]
Jaridat al-shacb – Cairo, 1979-1981 – 50mf – 9 – (missing: 1981(122)) – mf#NE-20325 – ne IDC [956]
Jariges, Karl F von *see* Bruchstuecke einer reise durch das suedliche frankreich, spanien und portugal [im jahr 1802]
Jarman, Thomas *see* A treatise on wills
Jarmatz, Klaus *see* Literatur im exil
Jarrah : the official journal of the australian forest league. – Vol. 1, no. 1 (May 1918)- – 1r [ill] – 1 – (qrtly [1918]; irreg 1919-aug 1920) – mf#film mas c6020 – us Harvard [634]
Jarrapellejos / Trigo, Felipe – Madrid: Renacimiento, 1914 – sp Bibl Santa Ana [946]
Jarratt, Devereux *see*
- Sermons on various and important subjects

Jarratt, Frederick *see* Our lord's sabbath-keeping
Jarrel, Willis Anselm *see*
- Baptist church perpetuity
- Baptizo-dip-only
- The gospel in water, or campbellism
- Old testament ethics vindicated

Jarrett, Harold Reginald *see* The gambia
Jarrin, Francisco *see* Moral
El jarro ritual lusitano de la coleccion calzadilla / Garcia y Bellido, Antonio – Madrid, 1957 – 1 – sp Bibl Santa Ana [946]
Jarrow chronicle and tyneside news – England. -w. 1 Apr-16 Dec 1871. (33 ft) – 1 – uk British Libr Newspaper [072]
Jarrow express – England. -w. Dec 1873-Dec 1913. 36 1 2 reels – 1 – uk British Libr Newspaper [072]
Jarrow guardian – England. -w. Jan 1872-Dec 1880. (9 reels) – 1 – uk British Libr Newspaper [072]
Jarrow labour herald – England. -w. 6 Apr 1906-15 Mar 1907. (1 reel) – 1 – uk British Libr Newspaper [072]
Jaruqueno en miami – Miami, FL. 1973 may 20-oct 10 – 1r – us UF Libraries [071]
Jarva nyheter *see* Vasterort
Jarves, James Jackson *see* Parisian sights and french principles
Jarvey – Dublin, Ireland. 1889-90 – 2r – 1 – uk British Libr Newspaper [072]
Jarvis Christian College *see* Jarvisonian
Jarvis, Jose Antonio *see*
- Brief history of the virgin islands
- Virgin islands and their people
- Virgin islands picture book

Jarvis, Lucy Cushing *see* Sketches of church life in colonial connecticut
Jarvis, Mary Eloise *see* Latin motets of hans leo hassler
Jarvis, Robert Edward Lee *see* The making of a christian
Jarvis, Samuel Peters *see*
- Correspondence relative to the accounts of the indian department in canada west
- Statement of facts relating to the trespass on the printing press in the possession of mr william lyon mackenzie, in june, 1826

Jarvisonian / Jarvis Christian College – 1987 4th qtr, fall, 1989 spr/sum, 1990 spr/sum, fall, 1990/1991 2nd qtr, 1992 spr, 1994 sum, 1996 sum – 1r – 1 – mf#3751362 – us WHS [071]
Jas – V Praze: Nakl a knihkupectvi Ceskoslovenske obce sokolske 1927 [mf ed Bloomington IN: Indiana Uni Lib, Preservation Dept 1989] – 5r – 1 – us Indiana Preservation [073]
Ja-sagen zum judentum – Berlin, Germany. 1933 – 1r – us UF Libraries [939]
Jasche, G B *see* Der pantheismus nach seinen verschiedenen hauptformen, seinem ursprung und fortgange
Jashar : fragmenta archetypa carminum hebraicorum in masorethico veteris testamenti textu passim tessellata / Donaldson, John William – Londini: Williams et Northgate, 1854 – 5mf – 9 – 0-524-07962-5 – mf#1992-1117 – us ATLA [221]
Jasiewicz, Jan *see* Neural mechanisms of chick posture control
O jasmin : orgao do atheneu dramatico esther de carvalho – Rio de Janeiro, RJ. 31 mar-21 abr 1888 – (= ser Ps 19) – mf#DIPER – bl Biblioteca [073]
Jasny, A Wolf *see* Geshikhte fun der yidisher arbeter-bavegung in lodzsh
Jason and lily iby / Darsey, Barbara Berry – s.l, s.l? 1938 – 1r – us UF Libraries [978]
Jason von kyrene : ein beitrag zu seiner wiederherstellung / Schlatter, Adolf von – Muenchen: C.H. Beck, 1891 – 1mf – 9 – 0-7905-3409-6 – mf#1987-3409 – us ATLA [240]
Jaspan, Meryn Aubrey *see* Australia and cambodia
Jasper banner – Rensselaer, IN. 1853-1858 (1) – mf#62945 – us UMI ProQuest [071]
Jasper county democrat – Rensselaer, IN. 1898-1948 (1) – mf#62946 – us UMI ProQuest [071]
Jasper county news – Rensselaer, IN. 1949-1958 (1) – mf#62947 – us UMI ProQuest [071]
Jasper news – Jasper, FL. 1948 oct-1997 – 48r – (gaps) – us UF Libraries [071]
Jasper republican – Rensselaer, IN. 1874-1876 (1) – mf#62948 – us UMI ProQuest [071]
Jaspers, Karl *see* Nietzsche und das christentum
Jaspersen, Ursula *see* Georg trakl
Jaspis, Johannes Sigmund *see* Koran und bibel
Jassin, H B *see* Sandiwara chusingura
Jast, Louis Stanley *see* Reincarnation and karma
Jastin, A T *see* A theoretical model to enhance the popularity of college soccer
Jastram, Gervais *see* Juvenilia
Jastrow, Marcus *see*
- A dictionary of the targumim, the talmud babli and yerushalmi, and the midrashic literature
- Vier jahrhunderte aus der geschichte der juden

Jastrow, Morris *see*
- Aspects of religious belief and practice in babylonia and assyria
- Babylonian-assyrian birth-omens and their cultural significance
- Bildermappe
- The civilization of babylonia and assyria
- A dictionary of the targumim
- A fragment of the babylonian "dibbarra" epic
- Hebrew and babylonian traditions
- The religion of babylonia and assyria
- Selected essays of james darmesteter
- The study of religion
- The war and the bagdad railway
- The weak and geminative verbs in hebrew

Jat kri chay bhvai / Thvan Mran, U, Dagun – Yan Kun: Nah Lum Lah 1975 [mf ed 1990] – 1r with other items – 1 – (in burmese) – mf#mf-10289 seam reel 133/7 [§] – us CRL [480]
Jataka dhammapada : morceaux modeles pour apprendre a composer dans les ecoles primaires de pali / Chhim-Soum – [Phnom-Penh] 1958 [mf ed 1990] – 1r with other items – 1 – (title & text in khmer; title also in french) – mf#mf-10289 seam reel 114/2 [§] – us CRL [490]
Jataka tales / ed by Thomas, Edward Joseph & Francis, Henry Thomas – Cambridge: University Press, 1916 [mf ed 1995] – (= ser Yale coll) – xiv/488p (ill) – 1 – 0-524-09698-8 – (int and notes by ed) – mf#1995-0698 – us ATLA [280]
Jathar, Ganesh Bhaskar *see* Indian economics
Jatho, Carl *see*
- Der ewig kommende gott
- Predigten

Jatho, Georg Friedrich *see*
- Pauli brief an die galater nach seinem inneren gedankengange
- Pauli brief an die philipper nach seinem inneren gedankengange

Jatun rijchari-h : manuel nunez butron, precursor de la medicina rural / Frisancho Pineda, David – Lima, Peru: Editorial Juan Mejia Baca, 1981 (mf ed 1999) – 1 – mf#ZZ-32202 – us NY Public [610]
Jatvaraar saga Helga / ed by Sigurasson, Jon – Kjobenhavn: Trykt hos J D Qvist 1852 [mf ed Bloomington IN: Indiana Uni Lib, Preservation Dept 1984] – 1r – 1 – (text & transl on opposite pgs) – us Indiana Preservation [941]
Jaubert de Passa, M *see* Canales de riego de cataluna y reino de valencia...
Jaubert, P A *see* Reise durch armenien und persien im jahr 1805 und 1806
Jaubert, Pierre *see*
- Reise durch armenien und persien
- Reise durch armenien und persien im jahr 1805 und 1806

Jaubert, Pierre Amedee Emilien Probe *see* Elements de la grammaire turke
Jaud, Leon *see* Vie des saints pour tous les jours de l'annee
Jaul / Jimenez, Max – Santiago, Chile. 1937 – 1r – us UF Libraries [972]
The jaulaan / Schumacher, Gottlieb – London: R. Bentley, 1888 – 1mf – 9 – 0-7905-3479-7 – mf#1987-3479 – us ATLA [956]
Jaume, Adela *see* Genesis
Jauna, D *see* Histoire generale des roiaumes de chypre, de jerusalem, d'armenie et d'egypte, comprenant les croisades...
Jaures, Jean *see* Histoire socialiste de la revolution francaise
Jaussen, Antonin *see* Coutumes des arabes au pays de moab
Java : deszelfs gedaante, bekleeding en inwendige structuur / Junghuhn, F W – Atlanta. 1948+ (1) 1973+ (5) 1976+ (9) – 34mf – 9 – mf#8384 – ne IDC [915]
Java and its challenge / Brooks, Elizabeth Harper – [Cincinnati: Jennings & Graham] 1911 [mf ed 1995] – (= ser Yale coll) – 196p (ill) – 1 – 0-524-09370-9 – mf#1995-0370 – us ATLA [240]
The java gazette / British Chamber of Commerce for the Netherlands East Indies – London, 1947-1950 – 13mf – 9 – mf#SE-705 – ne IDC [959]
Java. (Japanese Military Administration) *see*
- Boekoe petoendjoek praktek teknik bagi pemimpin seinendan
- Bunkyokyoku
- Jawa boei giyugun
- Jawa hokokai
- Kobijitsu kenkusho
- Peladjaran bahasa nippon

Java. (Japanese Military Administration). Laws, statutes, etc *see* Boekoe pengoempoelan oendang-oendang
Java. (Japanese Military Administration). Naimubu *see* Bunkyo kyoku nichi ma jiten
Java. (Japanese Military Administration). Saiko Shikikan *see* Keterangan saikoo sikikan
Java. Sihobu *see* Gunsei keizirei kaisetu
Javaansche kunstavond...gehouden door in nederland verblijvende javanen / Herdenking Stichting Boedi Oetomo – 's-Gravenhage, 1918 – 1mf – 8 – mf#SE-1431 – ne IDC [959]
Javana *see* Pum pran
Javanan-i tudah – Tehran: Hizb-i Tudah-'i Iran. sal-i 1, shumarah-'i 1-4. shahrivar-day 1359 [sep-dec 1980] – 1r – 1 – $53.00 – us MEDOC [079]
Java's zendingveld : beschouwd na de beoordeeling door s. e. harthoorn in zijn werkje: de evangelische zending in oost-java / Jansz, Pieter – Amsterdam: H De Hoogh, 1865 [mf ed 1995] – (= ser Yale coll) – 204p – 1 – 0-524-09343-1 – (in dutch) – mf#1995-0343 – us ATLA [240]
Javasche courant – 1828-1940 – 5451mf – 9 – €21,530.00 – (also available in subsets: 1828-37 [134mf] € 583 [m151]; 1838-47 [149mf] €649 [m152]; 1848-57 [210mf] €913 [m153];1858-67 [185mf] €805 [m154]; 1868-77 [247mf] €1070 [m155]; 1878-87 [325mf] €1410 [m156]; 1888-92 [259mf] €1120 [m157]; 1893-97 [257mf] €1120 [m158]; 1898-1902 [256mf] €1120 [m159]; 1903-07 [285mf] €1235 [m160]; 1908-12 [336mf] €1455 [m161]; 1913-17 [364mf] €1580 [m162]; 1918-22 [580mf] €2515 [m163]; 1923-27 [500mf] €2170 [m164]; 1928-32 [606mf] €2625 [m165]; 1933-39 [758mf] €3290 [m166]) – mf#m150 – ne MMF Publ [079]
The javelin / Seiss, Joseph Augustus – Philadelphia: Lutheran Book Store, 1871 – 1mf – 9 – 0-524-06445-8 – mf#1991-2567 – us ATLA [240]
Javens, J A *see* Effect of acupuncture tens on second degree ankle sprains
Javeri, Shanti *see* Deluge
Javierre, A *see* Madrid. archivo historico national. seccion de ordenes militares. guia de la seccion de ordenes militares
Javierre, Jose Maria et al *see* Control de natalidad. informe para expertos, llos documentos de roma
Javorskij, Ivan *see* Reise der russischen gesandtschaft in afghanistan und buchara in den jahren 1878-79
Jawa boei giyugun : djawa boei giyugun kyoren-kyotei / Java. (Japanese Military Administration) – (Djakarta? 2603) 3v – 8mf – 9 – mf#SE-2002 mf59-66 – ne IDC [959]
Jawa hokokai : peratoeranperatoeran himpoenan kebaktian rakjat (boelan 2 tahoen 2604) / Java. (Japanese Military Administration) – Djakarta: Panitia Persiapan Himpoenan Kebaktian Rakjat, 2604 – 47p 1mf – 9 – mf#SE-2002 mf191 – ne IDC [959]
Jawa nenkan – Djakarta: Jawa Shinbunsha, 2604 – 473p 6mf – 9 – mf#SE-2002 mf199-204 – ne IDC [959]
Jawaharlal nehru : an autobiography: with musings on recent events in india / Nehru, Jawaharlal – London: John Lane, 1936 – (= ser Samp: indian books) – us CRL [920]
Jawaharlal nehru : the man and his ideas / Krishnamurti, Y G – Bombay: Popular Book Depot, 1942 – (= ser Samp: indian books) – (pref by bhulabhai j desai and rameshuri nehru) – us CRL [920]
Jawaharlal nehru / Roy, Manabendra Nath – Delhi: Radical Democratic Party, 1945 – (= ser Samp: indian books) – us CRL [954]
Jawi peranakkan – Singapore, nov 7 1881 – us CRL [079]
Jax air news – Jacksonville, FL. 1952 sep-1998 apr – 44r – (gaps) – us UF Libraries [071]
Jax postal worker / American Postal Workers Union – 1977 apr-1979 jan, mar-oct, dec-1980 may, aug/sep, 1981 feb – 1r – 1 – (cont by: voice and views of the northeast florida area local) – mf#625962 – us WHS [380]
Jaxa-Roniker, Bogdan H *see* The red executioner dzierjinsi, the good heart
Jay, Allen *see* Autobiography of allen jay
Jay, Cyrus *see* The law
Jay, G F le *see*
- Bibliotheca rhetorum praecepta et exempla complectens quae ad poeticam facultatem pertinent...
- Le triomphe de la religion sous louis le grand...

Jay jamnah : su slap khluan oy tae pan samrec subhamangal / Nup Savan – Bhnam Ben: Pannagar Sar Vitti 2502 [1959] [mf ed 1990] – 1r with other items – 1 – (in khmer) – mf#mf-10289 seam reel 109/3 [§] – us CRL [959]
Jay, John *see* Papers
Jay, W *see*
- Important testimony to the value of tee-totalism
- Perfection of the heavenly state

Jay, William *see*
- Autobiography of the rev william jay
- Essay on marriage
- Lectures on female scripture characters
- Mutual duties of husbands and wives
- Paul's commission explained and applied
- Sensibility at the fall of eminence
- Sermons preached on various and particular occasions
- Value of life
- The works of the rev william jay

"Jaya kerta eka rasa" : almanak "telaga djaja" – Jogjakarta, 1962-1966 – 13mf – 9 – (missing: 1963-65) – mf#SE-608 – ne IDC [959]
Jayabharathi *see* Indian films and film world, 1976
Jayakar, M R *see* Papers, 1916-1925
Jayakar, Mukund R *see* Studies in vedanta
Jayaraman, Roop *see* The use of functional magnetic resonance imaging in the study of delayed muscle scoreness
Jayaswal, Kashi Prasad *see*
- Hindu polity
- Manu and yajnavalkya

Jayawardena, Chandra *see* Conflict and solidarity in a guianese plantation

Jaycees magazine – Greensboro. 1987+ (1,5,9) – (cont: future) – ISSN: 0893-0031 – mf#16461 – us UMI ProQuest [380]

Jayewardene, Gustavus *see* Centenary souvenir, 1851-1951

Jayne, Ebenezer *see* Jayne's hymnal

Jayne's hymnal / Jayne, Ebenezer – Misc. information. 1909 – 1 – 5.00 – us Southern Baptist [242]

Jaynes, Julian Clifford *see* [Unitarian interpretations of jesus christ]

Jayo kambuja prajadhipateayy : camrian pativatt – [Paris?: s.n. 195-?] [mf ed 1990] – 1r with other items – 1 – (in khmer) – mf#mf-10289 seam reel 122/2 [§] – us CRL [780]

Jayo khuap di 17 nai paks kummuynis kambuja? / Pol Pot – Gentilly: Comite des patriotes du Kampuchea democratique en France 1977 [mf ed 1990] – 1r with other items – 1 – (in khmer At head of title: Kambuja Prajadhipateyy.) – mf#mf-10289 seam reel 119/11 [§] – us CRL [335]

Jazz – Forest Hills NY. v1 n1-10. jun 1942-dec 1943 (irreg) [all publ] – (= ser Jazz periodicals, 1914-1977) – 1r – 1 – $105.00 – us UPA [780]

Jazz : a quarterly of american music – Berkeley. n1-5. oct 1958-winter 1960 [all publ] – (= ser Jazz periodicals, 1914-1977) – 1r – 1 – $125.00 – us UPA [780]

Jazz, 1990-1993 – Guardian/Observer newspaper group – 17mf – 9 – (with ill) – mf#87520 – uk Microform Academic [780]

Jazz digest – McLean VA. v1-3 n6. jan/feb 1972-jun 1974 (freq varies) [all publ] – (= ser Jazz periodicals, 1914-1977) – 1r – 1 – $165.00 – us UPA [780]

Jazz forum : english edition – New York. 1974-1992 (1) 1974-1992 (5) 1974-1992 (9) – ISSN: 0021-5635 – mf#8610 – us UMI ProQuest [780]

Jazz hot – Paris: Federation International des Hot Clubs Francais. n1-32 mar 1935-jul 1939; ns: n1-288 mar 1945-nov 1972 (irreg) – (= ser Jazz periodicals, 1914-1977) – 11r – 1 – $2020.00 – us UPA [780]

Jazz information – New York. v1-2 n16. sep 1939-nov 1941 (irreg) [all publ] – (= ser Jazz periodicals, 1914-1977) – 1r – 1 – $155.00 – us UPA [780]

Jazz journal – Sevenoaks. 1948-1977 (1) 1975-1977 (5) 1975-1977 (9) – (cont by: jazz journal international) – mf#6854 – us UMI ProQuest [780]

Jazz journal international – London. 1977+ (1) 1977+ (5) 1977+ (9) – (cont: jazz journal) – ISSN: 0140-2285 – mf#6854,01 – us UMI ProQuest [780]

Jazz magazine – Dir. D. Filipachi et F. Tenot. no. 1-125. Paris. dec 1954-65 – 1 – fr ACRPP [780]

Jazz magazine – Paris. 1972+ (1) 1972+ (5) 1972+ (9) – ISSN: 0021-566X – mf#8183 – us UMI ProQuest [780]

Jazz magazine – Paris. n1-148. dec 1954-nov 1967 (freq varies) – (= ser Jazz periodicals, 1914-1977) – 7r – 1 – $1445.00 – us UPA [780]

Jazz periodicals, 1914-1977 – Greenwood Press – 41r – 1 – $5970.00 coll – (coll of 22 titles; individual titles also listed separately) – us UPA [780]

Jazz quarterly – Chicago: Jazz Quarterly Society. v(?)-2 n4. spring 1942-(?) (irreg) [all publ] – (= ser Jazz periodicals, 1914-1977) – 1r – 1 – $115.00 – us UPA [780]

The jazz record – New York. n1-60. feb 1943-nov 1947 (freq varies) [all publ?] – (= ser Jazz periodicals, 1914-1977) – 1r – 1 – $165.00 – us UPA [780]

Jazz session / Hot Club of Chicago – Chicago: Hot Club of Chicago. n1-13. sep 1944-jul 1946 (irreg) [all publ?] – (= ser Jazz periodicals, 1914-1977) – 1r – 1 – $115.00 – us UPA [780]

Jazz spotlite news – 1979 jun-1980 feb/mar – 1r – 1 – mf#3055264 – us WHS [780]

Jazz times – Washington. 1985+ (1,5,9) – ISSN: 0272-572X – mf#15086 – us UMI ProQuest [780]

Jca news – Los Angeles, CA.Jewish Centers Assoc. of Los Angeles. 1976-80 – 1 – us AJPC [071]

Jci world : (jaycees international) – Coral Gables. 1972-1980 (1) 1972-1980 (5) 1975-1980 (9) – ISSN: 0021-3578 – mf#7085 – us UMI ProQuest [360]

Jcje *see* Journal of criminal justice education (jcje)

Jck – Secaucus. 2000+ (1,5,9) – ISSN: 1534-2719 – mf#948,02 – us UMI ProQuest [680]

Jct – journal of coatings technology – Blue Bell. 1976+ (1) 1976+ (5) 1976+ (9) – (cont: journal of paint technology) – ISSN: 0361-8773 – mf#3497,01 – us UMI ProQuest [660]

Je fais mes farces – Paris, France. 1817 – 1r – us UF Libraries [440]

Je lieh chu ho hsi-ha-nu-ko chin wang shih cha chien-pu-chai chieh fang chu ti chu ta cheng kung = Re lie zhu he xihanuke qin wang shi cha jianpuzhai jie fang qu di ju da cheng gong – Pei-ching: Jen min chu pan she [mf ed 1990] – 1r with other items – 1 – mf#mf-10289 seam reel 030/09 [§] – us CRL [920]

Je sais tout – Paris. 1922-juil 1939 – 1 – fr ACRPP [073]

Je suis le veritable pere duchesne, foutre – Paris. 30 no. nov-dec 1790, n1-355. janv 1791-mars 1794 – 1 – fr ACRPP [944]

Je suis partout : le grand hebdomadaire de la vie mondiale – Paris. 29 nov 1930-16 aout 1944 – 1 – (n'a pas paru du 7 juin 1940 au 7 fevr 1941) – fr ACRPP [073]

Je suis partout – Paris, France. 17 may 1940-7 jul 1941; 4 dec 1942-14 jul 1944 – 1r – 1 – uk British Libr Newspaper [072]

Je Yya *see* Mran ma ca, mran ma cit, mran ma man

Jeaffreson, John C *see*
- A book about lawyers
- Pleasantries of english courts and lawyers

Jean / Theaulon, M (Marie-Emmanuel-Guillaume-Marguerite) – Bruxelles, Belgium. 1829 – 1r – us UF Libraries [025]

Jean, Alfred *see*
- Il y a soixante ans, 1883-1943

Jean, Auguste *see*
- Le madure

Jean, B *see* Vie de sainte marie madeleine. barcelona

Jean baptiste : a story of french canada / Rossignol, James Edward le – London, Toronto: J M Dent, 1915 [mf ed 1998] – 3mf – 9 – 0-665-98947-4 – mf#98947 – cn Canadiana [830]

Le jean baptiste – Pawtucket, RI. 1897-1933 (1) – mf#66250 – us UMI ProQuest [071]

Jean calvin : 1. l'homme, 2. quelques accusations. bolsec, servet. deux conferences / Felice, Paul de – [S.l.]: Imprimerie de Nessonvaux, 1909 – 1mf – 9 – 0-524-02587-8 – mf#1990-0639 – us ATLA [242]

Jean de lasco : baron de pologne, evaeque catholique, reformateur protestant, 1499-1560 / Pascal, George – Paris: Fischbacher, 1894 – 1mf – 9 – 0-524-01236-9 – (incl bibl ref) – mf#1990-0375 – us ATLA [242]

Jean de lasco 1499-1560 : son temps, sa vie, ses oeuvres / Pascal, G – Paris, 1894 – €17.00 – ne Slangenburg [242]

Jean de passy / Martainville, A (Alphonse) – Paris, France. 1812 – 1r – us UF Libraries [440]

Jean dominique, saint dominique... – Madrid: Razon y Fe, 1927 – 1 – sp Bibl Santa Ana [240]

Jean du met, 1662...a jacques demers, 1965 / Demers, Louis-Philippe – [Sherbrooke: en vente chez l'auteur, 1965?] [mf ed 1994] – 3mf – 9 – mf#SEM105P2261 – cn Bibl Nat [920]

Jean et sebastien cabot : leur origine et leurs voyages, etude d'histoire critique... / Harrisse, Henry – Paris: Leroux, 1882 [mf ed 1980] – 5mf – 9 – 0-665-05392-4 – (incl ind and bibl ref) – mf#05392 – cn Canadiana [910]

Jean geiler de kaysersberg, predicateur a la cathedrale de strasbourg, 1478-1510 : etude sur sa vie et son temps / Dacheux, Leon – Paris: C. Delagrave; Strasbourg: Derivaux, 1876 – 2mf – 9 – 0-7905-5814-9 – (incl bibl ref) – mf#1988-1814 – us ATLA [240]

Jean guiton et le siege de la rochelle / Blanchon, Pierre – 1911 – 1 – $50.00 – us Presbyterian [944]

Jean jacques dessalines, fundador de haiti / Pattee, Richard – Habana, Cuba. 1936 – 1r – us UF Libraries [972]

Jean jacques lartigue : par la misericorde de dieu et la grace du siege apostolique premier eveque de montreal, et suffragant immediat de la sainte eglise romaine... / Catholic Church. Diocese de Montreal Eveque (1836-1840: Lartigue) – [s.l: s.n, 1838?] [mf ed 1985] – 1mf – 9 – 0-665-05100-X – mf#05100 – cn Canadiana [241]

Jean jacques lartigue, premier eveque de montreal, etc : au clerge et a tous les fideles de notre diocese, salut et benediction en notre seigneur – [Montreal?: s.n, 1837?] [mf ed 1985] – 1mf – 9 – 0-665-07992-3 – mf#07992 – cn Canadiana [241]

Jean jacques rousseau and education from nature = Jean jacques rousseau et l'education de la nature / Compayre, Gabriel – New York: Thomas Y Crowell, 1907 [mf ed 1986] – (= ser Pioneers in education) – 1mf – 9 – 0-8370-7618-8 – (trans fr french into english by r p jago) – mf#1986-1618 – us ATLA [370]

Jean jacques rousseaus einfluss auf joachim heinrich campe / Hartmann, Ernst Wilhelm – [S.l.: s.n.], 1904 (Neuenburg Wpr.: Buchdruckerei von F Nelson) – 1r – 1 – us Wisconsin U Libr [430]

Jean jaques rousseau und des biblische evangelium : ein nachwort zur rousseaufeier / Hadorn, Wilhelm – Berlin-Lichterfelde: Edwin Runge 1913 [mf ed 1989] – (= ser Biblische zeit- und streitfragen 9/1) – 1mf – 9 – 0-7905-3141-0 – mf#1987-3141 – us ATLA [220]

Jean le theologien (etb) : les grandes traditions d'israel. l'accord des ecritures d'apres le quatrieme evangile / Braun, F M – Paris, 1964 – €15.00 – ne Slangenburg [240]

Jean le theologien (etb) : sa theologie. le christ, notre seigneur / Braun, F M – Paris, 1972 – €12.00 – ne Slangenburg [240]

Jean le theologien (etb) : sa theologie. le mystere de jesus christ / Braun, F M – Paris, 1966 – €14.00 – ne Slangenburg [240]

Jean marie guyaus religionsphilosophie (l'irreligion de l'avenir) : dargestellt und kritisch untersucht / Schumm, Felix – Tuebingen, 1913 (mf ed 1994) – 2mf – 9 – €31.00 – 3-8267-3075-5 – mf#DHS-AR 3075 – gw Frankfurter [200]

Jean, Michele *see* Quebecoises du 20e siecle

Jean migault : or, the trials of a french protestant family during the period of the revocation of the edict of nantes = Journal de jean migault / Migault, Jean – Edinburgh: Johnstone and Hunter, 1852 – 1mf – 9 – 0-524-01656-9 – (in english) – mf#1990-0477 – us ATLA [944]

Jean Paul *see*
- Flower, fruit, and thorn pieces
- The invisible lodge
- Leben des quintus fixlein
- Werke

Jean paul : weltgedanken und gedankenwelt / ed by Benz, Richard – Stuttgart: Alfred Kroener, c1938 – 1 – (incl bibl ref & index) – us Wisconsin U Libr [430]

Jean paul marat : the people's friend / Bax, Ernest Belfort – 2nd ed. London: G Richards, 1901 – xvi/353p/4pl – 1 – mf#2199 – us Wisconsin U Libr [944]

Jean pauls verhaeltnis zu rousseau : nach den haupt-romanen dargestellt / Kommerell, Max – Marburg a.L.: N G Elwert, 1924 – 1 – (incl bibl ref) – us Wisconsin U Libr [410]

Jean raisin : revue joyeuse et vinicole – Paris. n1-10. 15 oct 1854-mars 1855 – 1 – fr ACRPP [073]

Jean rivard : scenes de la vie reelle / Gerin-Lajoie, Antoine – nouv ed. Montreal: J B Rolland, 1877 [mf ed 1974] – 1r – 5 – mf#SEM16P143 – cn Bibl Nat [830]

Jean rivard, economiste : pour faire suite a jean rivard le defricheur / Gerin-Lajoie, Antoine – Montreal: J B Rolland, 1876 – 3mf – 9 – mf#33004 – cn Canadiana [440]

Jean rivard, le defricheur : recit de la vie reelle / Gerin-Lajoie, Antoine – Montreal: J B Rolland, 1874 – 3mf – 9 – mf#33005 – cn Canadiana [440]

Jean sbogar : melodrame en trois actes / Cuvelier, J-G-G (Jean-Guillaume-Antoine) – Paris, France. 1818 – 1r – us UF Libraries [440]

Jean te theologien et son evangile dans l'eglise ancienne (etb) / Braun, F M – Paris, 1964 – €17.00 – ne Slangenburg [240]

Jean-adam moehler et l'ecole catholique de tubingue, 1815-1840 : etude sur la theologie romantique en wurtemberg et les origines germaniques du modernisme / Vermeil, Edmond – Paris: A Colin, 1913 [mf ed 1990] – 2mf – 9 – 0-7905-7031-9 – mf#1988-3031 – us ATLA [241]

Jean-baptiste blanchard au dahomey. : journal de la campagne par un marsouin / Badin, Adolphe – Paris: A Colin, 1895 – us CRL [960]

Jean-baptiste de la salle, fondateur des ecoles chretiennes : (poeme lyrique) / Frechette, Louis – Montreal: s.n, 1889 – 1mf – 9 – mf#29755 – cn Canadiana [810]

Jean-Baptiste, St Victor *see*
- Deux concepts d'independance a saint-domingue
- Fondateur devant l'histoire
- Haiti

Jean-daniel dumas, le heros de la monongahela : esquisse biographique / Audet, Francis-Joseph – Montreal: G Ducharme, 1920 – 2mf – 9 – 0-665-71793-8 – (incl bibl ref) – mf#71793 – cn Canadiana [920]

Jean-dominique mansi et les grandes collections conciliaires : etude d'histoire litteraire: suivie d'une correspondance inedite de baluze avec le cardinal casanate, et de lettres de pierre morin, hardouin, lupus, mabillon et montfaucon / Quentin, Henri – Paris: E Leroux, 1900 – 1mf – 9 – 0-7905-7191-9 – (incl bibl ref) – mf#1988-3191 – us ATLA [240]

Jean-francois de la rocque : seigneur de roberval, vice-roi du canada / Morel, Emile – Compiegne, France?: s.n, 1892? – 1mf – 9 – (with bibl ref) – mf#58621 – cn Canadiana [971]

Jean-francois de la roque : seigneur de roberval / Dionne, Narcisse-Eutrope – Ottawa?: s.n, 1899 – 1mf – 9 – (incl bibl ref) – mf#59521 – cn Canadiana [917]

Jean-francois de la roque : seigneur de roberval, vice-roi du canada / Morel, Emile – Paris: E Leroux, 1893 – 1mf – 9 – mf#34049 – cn Canadiana [917]

Jean-francois millet : his life and letters – London 1896 – (= ser 19th c art & architecture) – 5mf – 9 – mf#4.2.312 – uk Chadwyck [750]

Jean-francois millet : peasant and painter / Sensier, Alfred – London 1881 – (= ser 19th c art & architecture) – 3mf – 9 – mf#4.2.1242 – uk Chadwyck [750]

Jean-jacques rousseau, a ses derniers momens : trait historique: en un acte et en prose: represente pour la premiere fois, a paris, par les comediens italiens ordinaires du roi, le 31 decembre 1790 / Bouilly, Jean Nicolas – Paris: Brunet 1791 [mf ed 1980] – 1r – 1 – mf#38 – us Wisconsin U Libr [820]

Jean-Jacques, Thales *see* Histoire du droit haitien tome premier

Jeanne avec nous / Vermorel, Claude – Paris, France. 1942 – 1r – us UF Libraries [025]

Jeanne d'arc / Peguy, Charles – Paris, France. 1948 – 1r – us UF Libraries [440]

Jeanne d'arc / Puymaigre, Th De (Theodore) – Paris, France. 1843 – 1r – us UF Libraries [440]

Jeanne d'arc a rouen / Avrigni, C J L d' – Paris, France. 1819 – 1r – us UF Libraries [440]

Jeanne d'arc et l'ame francaise : conference donnee au cercle ville-marie de montreal, le 16 avril 1903 / Lemerre, A J – Montreal: Libr Granger, 1903 [mf ed 1994] – 1mf – 9 – 0-665-72098-X – mf#72098 – cn Canadiana [240]

Jeanne et jeanneton / Scribe, Eugene – Paris, France. 1845 – 1r – us UF Libraries [440]

Jeanne la fileuse : episode de l'emigration franco-canadienne aux etats-unis / Beaugrand, Honore – Montreal: La Patrie, 1888 – 4mf – 9 – mf#26490 – cn Canadiana [830]

Jeannette / Bechmann, Trude – 1. aufl. Weimar: Volksverlag, 1960 [mf ed 1995] – 225p – 1 – mf#8973 – us Wisconsin U Libr [810]

Jeannotte, Adhemar *see* Vaudreuil

Jeannotte, Hormidas *see* Communication par h jeannotte, ecr, mp, a ses electeurs du comte de l'assomption

Jean-phillipe rameau (1683-1764) : complete works / ed by Saint-Saens, Camille et al – Paris: A Durand. 18v. 1895-1913 – 11 – $350.00 set – (complete ed never finished) – us Univ Music [780]

Jean-pierre boyer bazelais et le drame de miragoan / Mars, Jean Price – Port-Au-Prince, Haiti. 1948 – 1r – us UF Libraries [972]

Jeanroy, A *see* Giousue carducci, l'homme et le poete

Jeanroy, Alfred et al *see* Lais et descorts francais du 13e siecle

Jeanson, Henri *see* Amis comme avant

Jebb, Camilla *see* Mary wollstonecraft

Jebb, John *see*
- Homilies considered
- Tract for all times

Jec : journal jeciste mensuel – Ville Saint-Laurent: [Jeunesse etudiante catholique] (mf ed 1983) – 6r – 1 – mf#SEM35P181 – cn Bibl Nat [241]

Jedburgh post – [Scotland] Jedburgh: Lunn & Mabon 10 jan 1896-6 apr 1906 (wkly) [mf ed 2004] – 530v on 5r – 1 – uk Newsplan [072]

[Jeddah-] saudi gazette – SU. 1979-85 – 25r – 1 – $1250.00 – mf#R63579 – us Library Micro [072]

"Ein jeder wird nach seinem mass gerichtet... " : richter, gerichtete und die gerechtigkeit in duerrenmatts kriminalromanen / Farago, Lydia – Uni of South Africa 2001 [mf ed Pretoria: UNISA 2000] – 3mf – 9 – (incl bibl ref) – mf#mfm14721 – sa Unisa [430]

Jedermann : geschichte eines namenlosen / Wiechert, Ernst Emil – Muenchen: A Langen/ G Mueller, 1935, c1931 – 1r – 1 – us Wisconsin U Libr [830]

Jedidja : eine religioese, moralische und paedagogische zeitschrift – Berlin DE, 1817-21 – 2r – 1 – us UMI ProQuest [939]

Jedinstvo – Canada. dec 1948-jan 1970 – 23r – 1 – (in yugoslavian) – cn Commonwealth Imaging [071]

Jedinstvo = Unity – Chicago, IL: Palandech's Pub House, nov 11 1948-oct 1 1953 – 3r – 1 – us CRL [071]

Jedinstvo = Unity – Gary, IN: Carpathian-Russian Unity. v1 n1-v2 n5,6,7. mar 1942-jul/aug/sep 1943 – us CRL [073]

Jedlicska, Johann *see*
- Der angebliche turmbau zu babel, die erlebnisse der familie abrahams und die beschneidung
- Die entstehung der welt
- Die zweite entstehung der welt, das angebliche paradies und die angebliche sintflut

Jedna velka unie = One big union / Industrial Workers of the World – Chicago IL, Cleveland OH. 1926 oct-1929 mar 16, 1929 apr 6-1931 aug [v9 n42-v11 n25] – 2r – 1 – mf#981176 – us WHS [331]

Jednosc polek / Cuyahoga Co. Cleveland – jul 1923-aug 1931 [wkly] – 2r – 1 – mf#B30366-30367 – us Ohio Hist [071]
Jednosc polek : official organ of the association of polish women of the united states of america – Cleveland, OH. v43 n1. jan 11 1966-apr 5 1988 – 4r – 1 – (semi-monthly polish language fraternal newspaper. weekly 1923-56. in polish and english) – mf#(M) 34 C9.3 145 – us Western Res [071]
Jednota – Cleveland, OH: S Furdek, 1941-1960 – 20r – 1 – us CRL [071]
Jednota : official organ of the first catholic slovak union of the united states and canada – Cleveland, OH, nov 20 1895-jun 13 1900 – 2r – 1 – (weekly catholic slovak newspaper. in english and slovak. aka: union. publ from 1911 onwards in middletown, pa.) – mf#11 D1.1 002 – us Western Res [071]
Jednota – Middletown PA, 1893-1940 – 15r – 1 – (slovak newspaper) – us IHRC [071]
Jeens, C H *see* Memorials of john mcleod campell, d.d.
Jeep, I *see* Zur ueberlieferung des philostorgios
The jeep: its development and procurement under the quartermaster corp, 1940-1942 / U.S. Army. Quartermaster Corp. General Administrative Services – 1943 – 1 – $26.00 – us L of C Photodup [629]
Jeep, Ludwig *see* Quellenuntersuchungen zu den griechischen kirchenhistorikern
Jeepney journal – United States. 1984 jan 5-1987 jun 25, 1987 jul 2-1989 nov 30, 1989 dec 7-1991 jul 18 – 3r – 1 – mf#1608175 – us WHS [071]
Jeetzel zeitung – Dannenberg DE, 1856, 1911-23, 1926-1941 31 may – 24r – 1 – gw Misc Inst [074]
Jefatura de obras publicas de la provincia de Badajoz *see* Escalofon de capataces y camineros de la misma en 31 de diciembre de 1959
Jefatura provincial Servicio Pesca Continental, Caza y Parques Nacionales *see* Catalogo exposicion de trofeos de caza mayor 1970
Jefferds, Chester Daniels *see* Select remains of rev. c. d. jefferds, pastor of the congregational church in chester, vermont
Jefferis, Benjamin Grant *see* The household guide
Jefferis, Shelly J *see* Aerobic certification
Jeffers, W *see* The cherubim / the ordering of human life
Jefferson *see* Music part books
Jefferson banner – Jefferson, Johnson Creek WI. [1864 jan 7/1886 jun 30]-[1986/1987 may] [gaps] – 72r – 1 – (cont: jefferson county republican) – mf#1032492 – us WHS [071]
Jefferson baptist church. jefferson county. south carolina : church records – 1899-1977 – 1 – us Southern Baptist [242]
Jefferson, Ceroy *see* Educational performance of athletes and nonathletes in two mississippi rural high schools
Jefferson, Charles Edward *see*
– The building of the church
– The cause of the war
– The character of jesus
– Christianity and international peace
– Congregationalism
– Doctrine and deed
– The new crusade
– Quiet hints to growing preachers in my study
– Things fundamental
– What the war is teaching
– Why we may believe in life after death
Jefferson city first baptist church. jefferson city, tennessee : church records – Apr 1834-Feb 1987 – 1 – us Southern Baptist [242]
Jefferson Co. Mount Pleasan *see* Philanthropist
Jefferson Co. Steubenville *see*
– American union
– Daily gazette
– Daily news
– Gazette series
– Germania
– Herald
– Ohio press
– True american
– Weekly gazette
Jefferson Co. Toronto *see* Tribune
Jefferson county 1939 / Federal Writers' Project (FL) – s.l, s.l? n d – 1r – us UF Libraries [978]
Jefferson county advertiser – Jefferson WI. 1994 mar 29-jun 21 – 1r – 1 – (cont: advertiser [jefferson, wi: south ed]) – mf#3103869 – us WHS [071]
Jefferson county branch naacp annual report / National Association for the Advancement of Colored People et al – 1992 aug 14 – 1r – 1 – mf#5004730 – us WHS [305]
Jefferson county bureau scope – v1-2 [1953 aug-1955 dec] – 1r – 1 – mf#1060476 – us WHS [071]
Jefferson county business news – v1 n1-v2 n7 [1981 apr-1982 jul] – 1r – 1 – mf#637826 – us WHS [338]
Jefferson county democrat – Fort Atkinson WI. 1904 mar 12-1905 aug 17, 1905 aug 24-1906 dec 6, 1906 dec 13-1908 mar 12, 1908 mar 19-1909 jul 15, 1909 jul 22-1910 oct 20, 1910 oct 27-1912 jan 25, 1912 feb 1-1913 may 1, 1913 aug 29, may 8-1914 sep 24, 1914 oct 1-1916 feb 24, 1916 mar 2-1917 jul 26, 1917 aug 2-1918 apr 18, 1919 may 1-1920 apr 22, 1921 jun 30, jul 7,14 – 11r – 1 – (cont by: fort atkinson news) – mf#1223181 – us WHS [071]
Jefferson county journal – Fairbury, NE: Hammond & Andrews. 11v. v1 n1. apr 9 1892-v11 n9. may 24 1902 (wkly) [mf ed with gaps] – 4r – 1 – (formed by the union of: sun (fairbury ne) and: fairbury world. cont by: fairbury journal) – us NE Hist [071]
Jefferson County, KS *see*
– Cemetery tombstone and obituary cards
– Obituary card files bro-cly
– Obituary card files gob-sch
– Obituary card files sch-zwy
Jefferson county miscellaneous newspapers – Denver, CO (mf ed 1991) – 1r – 1 – (colorado democrat (feb 18-mar 16 1863); golden globe (dec 14 1907); golden weekly globe (jul 24 1875); kernals of political thots n' observations (jun 15, sep 15 1962); kernals (nov 1 1962-feb 1964)) – mf#MF Z99 J356 – us Colorado Hist [071]
Jefferson county news – Fairbury, NE: Albert H Hammond, dec 1897 (wkly) [mf ed 1898-jun 30 1899 (gaps)] – 1r – 1 – (cont by: fairbury news) – us NE Hist [071]
Jefferson county, ohio, newsletter – v1 n1-4 [1980 apr-1981 jan] – 1r – 1 – mf#641206 – us WHS [978]
Jefferson county record – Endicott, NE: Frank T Pearce. 15v. v1 n[1] mar 25 1887-v15 n37 [ie 39] dec 6 1901 (wkly) [mf ed 1888-1901 (gaps) – 2r – 1 – (cont by: diller record. publ in diller ne, nov 18 1887-1901. issue for nov 11 1887 not publ. some irregularities in numbering) – us NE Hist [071]
Jefferson county record – Metolius OR: Record Pub Co, 1915- [wkly] – 1 – (cont: jefferson county searchlight (1915)) – us Oregon Lib [071]
Jefferson county republican – Fairfield, IA. 1897-1917 (1) – mf#63209 – us UMI ProQuest [071]
Jefferson county republican – Jefferson WI. [1860 aug 15-1862 nov 16] – 1r – 1 – mf#1045245 – us WHS [071]
Jefferson county searchlight – Metolius OR: B K Leach, 1915 [wkly] – 1 – (cont by: jefferson county record (metolius, or: 1915-)) – us Oregon Lib [071]
Jefferson county sentinel – Boulder, MT. 1886-1900 (1) – mf#64274 – us UMI ProQuest [071]
Jefferson county union – Fort Atkinson, Lake Mills WI. 1870 mar 24-1873, 1872 mar 29, 1874-1878 mar 8, 1939, 1940 jan 5-dec 27, 1941 jan 3-1942 feb 27, 1942 mar 6-1943 jan 1, 1943 jan 8-1944 jun 30, 1944 jul 7-1946 feb 21 – 9r – 1 – (cont by: daily jefferson county union [fort atkinson, wi: 1946]) – mf#929766 – us WHS [071]
Jefferson county union – Fort Atkinson, WI. 1870-2000 (1) – mf#61932 – us UMI ProQuest [071]
Jefferson Federation of Teachers, Local 1559 *see* Jefferson teacher
Jefferson first baptist church. jefferson, georgia : church records – Jun 1866-Sep 1973 – 1 – 63.63 – us Southern Baptist [242]
Jefferson, LouAnne M *see* Intratester and intertester reliability when using the chatillon hand-held dynamometer to measure force production in the upper and lower extremities
Jefferson, New Hampshire. Jefferson Baptist Church *see* Records
Jefferson. Ohio. Bethel Union Baptist Church *see* Church records, ms 1585
Jefferson. Ohio. First Baptist Church *see* Church records, ms 642
Jefferson parish american – New Orleans, LA. 1944-1947 (1) – mf#63514 – us UMI ProQuest [071]
Jefferson republican – Ranson, WV. 1943-1955 (1) – mf#67452 – us UMI ProQuest [071]
Jefferson review – Jefferson OR: G A Sanford, 1890- [wkly] – 1 – (began with 1890) – us Oregon Lib [071]
Jefferson star – Jefferson, PA., 1851 – 13 – $25.00r – us IMR [071]
Jefferson teacher / Jefferson Federation of Teachers, Local 1559 – v1 n1-v2 n6 [1975 jan-1976 may] – 1r – 1 – mf#642440 – us WHS [331]
Jefferson, Thomas *see*
– Papers
– A summary view of the rights of british america
– Thomas jefferson papers, 1705-1827
– Writings of thomas jefferson
– The writings of thomas jefferson
Jefferson valley news – Whitehall, MT. 1911-1974 (1) – mf#64686 – us UMI ProQuest [071]
Jefferson valley zepher – Whitehall, MT. 1894-1901 (1) – mf#64687 – us UMI ProQuest [071]
Jeffersonian – Albany. 1838-1839 (1) – mf#3784 – us UMI ProQuest [323]
Jeffersonian – Cambridge, OH. 1885-1905 (1) – mf#65398 – us UMI ProQuest [071]
Jeffersonian – Jefferson WI. 1853 may 12-1854 nov 2 – 1r – 1 – (cont by: weekly jeffersonian [jefferson, wi]) – mf#927036 – us WHS [071]
Jeffersonian – Jefferson, PA., 1854-1859 – 13 – $25.00r – us IMR [071]
Jeffersonian – Jefferson, PA., 1859-1871 – 13 – $25.00r – us IMR [071]
Jeffersonian – New Orleans LA. 1842 may 30 – 1r – 1 – mf#1044066 – us WHS [071]
Jeffersonian americana – 1999mf (18:1) – 9 – $7325.00 – (with p/g) – us UPA [975]
Jeffersonian brookville – Brookville, PA., 1834-1835 – 13 – $25.00r – us IMR [071]
Jeffersonian democrat – Brookville, PA. 1958-1971 (1) – mf#65848 – us UMI ProQuest [071]
Jeffersonian democrat – Monroe WI. 1856 aug 14-1857 mar 26 – 1r – 1 – (cont by: independent press; independent press) – mf#1125590 – us WHS [071]
Jeffersonian democrat / Geauga Co. Chardon – jan 1859-dec 1865 [wkly] – 2r – 1 – mf#B276-277 – us Ohio Hist [071]
Jeffersonian democrat – Jefferson, PA., 1886-1983 – 13 – $25.00r – us IMR [071]
Jeffersonian democrat : weekly republican newspaper – Chardon, Ohio. 1854-65 – 5r – 1 – us Western Res [071]
Jeffersonian democrat *see* Free democrat
Jeffersonian republican – Charlottesville VA. 1855 apr 19, aug 30, 1880 dec 22 – 1r – 1 – (cont by: daily progress) – mf#882370 – us WHS [071]
Jeffersonian/brookville demo – Jefferson, PA., 1871-1886 – 13 – $25.00r – us IMR [071]
Jefferson's reports / Virginia. Supreme Court. General Court – 1v. 1730-1740 and 1768-1772 (all publ) – (= ser Virginia supreme court reports) – 2mf – 9 – $3.00 – (a pre-nrs title) – mf#LLMC 91-043 – us LLMC [347]
Jeffersonville *see* Indianian
Jeffersonville baptist church – Twiggs County, GA. 272p. 1849-94 – 1 – $12.24 – mf#6524 – us Southern Baptist [242]
Jeffersonville baptist church. jeffersonville, goergia : church records – 1849-1894 – 1 reel – 1 – $12.24 – (272p) – us Southern Baptist [242]
Jeffery, A *see* The qur'an as scripture
Jeffery, Arthur *see* The foreign vocabulary of the qur'an
Jefferys, William Hamilton *see* James addison ingle (yin teh-sen)
Jeffrey, Edward Charles *see* Anatomy of woody plants
Jeffrey, Robert *see* The indian mission of the irish presbyterian church
Jeffreys, Arcelia T *see* Experiences and relations in the work of women teacher/coaches
Jeffreys, Charles *see* Van dieman's land
Jeffreys, K *see* The widowed missionary's journal
Jeffreys, Keturah *see* The widowed missionary's journal
Jeffreys, Letitia D *see* Ancient hebrew names
Jeffreys, M K *see* Kaapse plakkaatboek
Jeffreys, Mervyn David Waldegrave *see* Old calabar and notes on the ibibio language
Jeffs, Robin *see* Fast sermons to parliament: reproductions in facsimile with notes
Jegp – journal of english and germanic philology – Urbana. 1897+ (1) 1969+ (5) 1975+ (9) – ISSN: 0363-6941 – mf#1085 – us UMI ProQuest [400]
Jehan de Wavrin, seigneur du Forestel *see*
– A collection of chronicles and ancient histories of great britain
– Recueil des croniques et anchiennes istories de la grant bretagne
Jehin *see* L'homme-sans-facon
Jehle, Robert *see* Auswirkungen von ausleitungen zur wasserkraftnutzung auf die besiedlung durch makroobenthon in gewaesserstrecken des nordschwarzwaldes
Jehovah / Elisabeth, Queen – Leipzig: W Friedrich, 1882 (mf ed 1990) – 1r – 1 – (filmed with: astra) – us Wisconsin U Libr [810]
Jehovah glorified, and his church secured in christ – London, England. 1824 – 1r – us UF Libraries [240]
Jehovah our righteousness – Kelso, Scotland. 18-- – 1r – us UF Libraries [240]
Jehovah, the redeemer god : the scriptural interpretation of the divine name jehovah / Tyler, Thomas – London: Ward, 1861 – 1mf – 9 – 0-8370-5593-8 – (incl bibl ref) – mf#1985-3593 – us ATLA [221]
Jehovah-jesus : the oneness of god, the true trinity / Weeks, Robert Dodd – New York: Dodd, Mead, 1880, c1876 [mf ed 1985] – 1mf – 9 – 0-8370-5666-7 – (incl app & ind) – mf#1985-3666 – us ATLA [242]
Jehovah-jesus / Whitelaw, Thomas – New York: Scribner, 1913 – (= ser The Short Course Series) – 1mf – 9 – 0-524-05760-5 – mf#1992-0603 – us ATLA [240]
Jehovah-jireh : a treatise on providence / Plumer, William Swan – Philadelphia: JB Lippincott, 1866, c1865 – 1mf – 9 – 0-7905-9585-0 – mf#1989-1310 – us ATLA [210]
Jehovah's decree of predestination / Bleby, Henry – London, England. 1873 – 1r – us UF Libraries [240]
Jehovah's war against false gods : and other addresses / Atwater, John Milton; ed by Atwater, Anna Robison – St Louis: Christian Pub Co, c1903 [mf ed 1993] – (= ser Christian church (disciples of christ) coll) – 4mf – 9 – 0-524-07846-7 – mf#1991-3391 – us ATLA [243]
Jehowah-jovis und die drei soehne noah's : ein beitrag zur vergleichenden goetterlehre / Glaser, Eduard – Muenchen: Hermann Lukaschik, 1901 – 1mf – 9 – 0-8370-3307-1 – mf#1985-1307 – us ATLA [221]
Jehrings jahrbuecher fuer die dogmatik des burgerlichen rechts – Jena. v. 1-90. 1857-1942; Index. 1857-1906 – 1 – us L of C Photodup [943]
Jehuda halevi : zweiundneunzig hymnen und gedichte / Rosenzweig, Franz – Berlin, 1927 – 7mf – 8 – €15.00 – ne Slangenburg [270]
Jekyll, G *see* Treatise on the right use of the fathers
Jekyll, Gertrude *see*
– Wall and water gardens
– Wood and garden
Jelf, R W *see* Grounds for laying before the council of king's college, london
Jelf, Richard William *see*
– Grounds for laying before the council of king's college, london, certain statements contained in a recent publication entitled theological essays by the rev. f.d. maurice, m.a., professor of divinity in king's college
– Specific evidence of unsoundness in the volume entitled "essays and reviews"
– The thirty-nine articles of the church of england
Jelf, William Edward *see*
– Christian faith, comprehensive, not partial; definite, not uncertain
– An examination into the doctrine and practice of confession
– A grammar of the greek language
– Ritualism, romanism and the english reformation
– Supremacy of scripture
Jelitte, Herbert et al *see* Innerslavischer und slavisch-deutscher sprachvergleich
Jellett, John H *see* Retrospect of a christian's work
Jellett, John Hewitt *see*
– Church membership in the past and the future
– The efficacy of prayer
– Immortality of the intellect
Jellicoe, S *see* The septuagint and modern study
Jellinek, Adolph *see* Peninim me-derashot dr yellinek
Jellinek, M Hermann *see* Adriatische rosemund
Jellinek, Max Hermann *see* Die psalmenuebersetzung des paul schede melissus
Jellinghaus, H *see* Das buch sidrach
Jellinghaus, Hermann *see* Niederdeutsche bauernkomoedien des siebzehnten jahrhunderts
Jellinghaus, Hermann Friedrich *see* Das buch sidrach
Jelusich, Mirko *see* Sickingen und karl 5
Jem : Journal of educational measurement – Washington. 1964+ – 1,5,9 – ISSN: 0022-0655 – mf#11835 – us UMI ProQuest [370]
Jems : a journal of emergency medical services – Solana Beach. 1985+ (1) 1985+ (5) 1986+ (9) – ISSN: 0197-2510 – mf#15209 – us UMI ProQuest [610]
Jemtlands tidning – Ostersund, Sweden. 1845-89 – 16r – 1 – sw Kungliga [078]
Jen chien sui pi – Shang-hai: Liang yu t'u shu yin shua kung ssu, 1935 – (= ser P-k&k period) – us CRL [840]
Jen chien tsa chi / Ch'en, Shih – Shang-hai: Shang wu yin shu kuan, Min kuo 25 [1936] – (= ser P-k&k period) – us CRL [840]
Jen chien tz'u chi jen chien tz'u hua / Wang, Kuo-wei – Pei-p'ing: Jen wen shu tien, 1933 – (= ser P-k&k period) – us CRL [951]
Jen, Chi-kao *see* Wei wu chan cheng kuan
Jen, Cho-hsuan *see*
– Min sheng chu i chen chieh
– San min chu i ti che hsueh chi ch'u
Jen ch'uan tsai na li – [China]: Chin-men ch'u pan she, 1941 – (= ser P-k&k period) – us CRL [323]
Jen, Chun *see* Wei sheng li erh ko
Jen ho jen men / Chao, Hsiao-sung – Hsin-ching: I wen shu fang, 1943 – (= ser P-k&k period) – us CRL [830]
Jen ko chiao yu hsueh kai / Ch'ien, Ho – Shang-hai: Shih chieh shu chu, Min kuo 23 [1934] – (= ser P-k&k period) – us CRL [370]
Jen k'ou wen t'i / Ch'en, Ta – Shang-hai: Shang wu yin shu kuan, min kuo 23 [1934] – (= ser P-k&k period) – us CRL [304]

Jen li tung yuan fa kuei hui pien / China – Ch'ung-ch'ing: She hui pu lao tung chu, 1943 – (= ser P-k&k period) – us CRL [323]

Jen li tung yuan lun / Chu, Hsiao-ch'un – Ch'ung-ch'ing: Kuo min t'u shu ch'u pan she, Min kuo 32 [1943] – (= ser P-k&k period) – us CRL [331]

Jen min / Yuan, Shui-p'ai – [China]: Hsin shih she, 1940 – (= ser P-k&k period) – us CRL [810]

Jen min jih pao so yin = Index to the people's daily – 1951-1959 [1] – mf#2621 – us UMI ProQuest [072]

Jen min shui = People's taxation – 1956-1958 (1) – mf#2622 – us UMI ProQuest [336]

Jen, Pai-t'ao *see* K'ang-chan ch'i chien ti hsin wen hsuan ch'uan

Jen, Pi-ming *see* Hsiung pien shu

Jen sheng / Yin, Hsi – Ch'ang-ch'un: Wen hua she ch'u pan pu, 1942 – (= ser P-k&k period) – us CRL [830]

Jen sheng = Young son – Hong Kong. n198-267. 1959-61 [gaps] [mf ed 198?] – (= ser Chinese christian serials coll) – 1r – 1 – (began in 1951. ceased 1971?) – mf0320 – us ATLA [230]

Jen sheng che hsueh chuan shang / Li, Shih-ts'en – [Shang-hai]: Shang wu yin shu kuan, Min kuo 30 [1941] – (= ser P-k&k period) – us CRL [180]

Jen sheng fo chiao / T'ai-hsu – Ch'ung-ch'ing: Hai ch'ao yin yueh k'an she, Min kuo 34 [1945] – (= ser P-k&k period) – us CRL [280]

Jen sheng hsing ch'u / Ts'ao, Fu – Ch'ung-ch'ing: Kuang t'ing ch'u pan she, Min kuo 32 [1943] – (= ser P-k&k period) – us CRL [390]

Jen sheng kai lun (ccm154) = The key to life / Ho, Shih-ming – 1st ed. Hong Kong, 1959 [mf ed 198?] – (= ser Ccm 154) – 1 – mf#1984-b500 – us ATLA [240]

Jen sheng pei hsi chu / Ting, Ti – Shang-hai: T'ai p'ing shu chu, 1944 – (= ser P-k&k period) – us CRL [480]

Jen shih hsing cheng chih li lun yu shih chi / Ho, Po-yen – [Ch'ung-ch'ing]: Cheng chung shu chu, Min kuo 33 [1944] – (= ser P-k&k period) – us CRL [350]

Jen shih hsing cheng ta kang / Hsueh, Po-k'ang – [Ch'ung-ch'ing]: Cheng chung shu chu, Min kuo 32 [1943] – (= ser P-k&k period) – us CRL [350]

Jen shih hsing cheng yuan li yu chi shu / Chang, Chin-chien – Ch'ung-ch'ing: Shang wu yin shu kuan, Min kuo 34 [1945] – (= ser P-k&k period) – us CRL [650]

Jen shih kuan li / Wang, Shih-Hsien – Ch'ung-ch'ing: Shang wu yin shu kuan, Min kuo 32 [1943] – (= ser P-k&k period) – us CRL [650]

Jen shih kuan li / Wang, Shih-Hsien – Ch'ung-ch'ing: Shang wu yin shu kuan, Min kuo 32 [1943] – (= ser P-k&k period) – us CRL [650]

Jen shih kuan li chih li lun yu shih chi / Hsia, Pang-chun – Ch'ung-ch'ing: Kuo hsun shu tien, Min kuo 33 [1944] – (= ser P-k&k period) – us CRL [650]

Jen shih pai t'u / Chin, I – Fu-chien Nan-p'ing: Kuo min ch'u pan she, Min kuo 32 [1943] – (= ser P-k&k period) – us CRL [840]

Jen shih t'ai-wan / Ch'en, T'ing-t'ing – Fu-chien: Hua sheng t'ung hsun she, 1945 – (= ser P-k&k period) – us CRL [951]

Jen shih yu hsing tung (ccm119) – Shanghai, 1936 [mf ed 198?] – (= ser Ccm 119) – 1 – mf#1984-b500 – us ATLA [951]

Jen te chiao yu (ccm153) = Education of life / Ho, Shih-ming – 1st ed. Hong Kong, 1958 [mf ed 198?] – (= ser Ccm 153) – 1 – (missing: p46-47) – mf#1984-b500 – us ATLA [370]

Jen ti hsi wang / Ssu-ma, Wen-sen – Ch'ung-ch'ing: Lien i ch'u pan she, Min kuo 34 [1945] – (= ser P-k&k period) – us CRL [830]

Jen ti hua to / Lu, Ying – Shang-hai: Hsin hsin ch'u pan she, Min kuo 37 [1948] – (= ser P-k&k period) – us CRL [840]

Jena first baptist church (formerly salem baptist church). jena, louisiana : church records – Feb 1850-Sep 1872. 88p – 1 – 5.00 – us Southern Baptist [242]

Jena oder sedan? : roman / Beyerlein, Franz Adam – 3. aufl. Berlin: Vita, 1903 [mf ed 1989] – 737p – 1 – mf#7019 – us Wisconsin U Libr [830]

'Jena' or 'sedan'? : [novel] / Beyerlein, Franz Adam – New York: G H Doran, 1914 [mf ed 1989] – 361p – 1 – (fr german of franz adam beyerlein) – mf#7019 – us Wisconsin U Libr [830]

Jenaer volksblatt – Jena 1890 15 apr-1941 31 mai [gaps] [mf ed 2005] – 108r – 1 – (with suppl) – gw Mikrofilm [074]

Jenaische allgemeine literatur-zeitung – Jena 1804-41 [mf ed 1992] – 680mf – 9 – €2410.00 – 3-89131-049-8 – (filmed with: neue jenaische allgemeine literatur-zeitung [1842-48]; incl suppl pp, advertisers & ind) – gw Fischer [430]

Jenaische allgemeine literatur-zeitung : nebst intelligenzblatt / ed by Eichstaedt, Heinrich Karl Abrah – Jena 1804-41 – (= ser Dz) – jg1-37 on 455mf – 9 – €2730.00 – mf#k/n479 – gw Olms [410]

Jenaische gelehrte anzeigen – Jena 1787 – (= ser Dz) – st1-102 on 6mf – 9 – €120.00 – mf#k/n423 – gw Olms [500]

Jenaische gelehrte zeitungen / ed by Hamberger, Georg Ernst et al – Jena 1749-56 – (= ser Dz) – 8jge on 41mf – 9 – €410.00 – mf#k/n166 – gw Olms [500]

Jenaische monathliche auszuege aus den merkwuerdigsten neuen schriften nebst gelehrten neuigkeiten und beitraegen – Jena 1765-67 – (= ser Dz) – 4v on 15mf – 9 – €150.00 – (v3ff:...nebst der neuesten geschichte der akademie und gelehrten beitraegen) – mf#k/n246 – gw Olms [378]

Jenaische zeitungen von gelehrten sachen – Jena 1765-86 – (= ser Dz) – 126mf – 9 – €756.00 – (1782ff: jenaische gelehrte anzeigen) – mf#k/n253 – gw Olms [500]

Jenbach, Bela *see* Tzarewitch

Jendro, Frank *see* Eingriffsqualitaet und rechtliche regelung polizeilicher videoaufnahmen

Jenkens, Charles Augustus *see* Baptist doctrines

Jenkin, Fleeming *see* Papers literary, scientific etc

Jenkin lloyd jones : a free catholic / Seebode, Richard William F – Chicago, 1929. Chicago: Dep of Photodup, U of Chicago Lib, 1971 (1r); Evanston: American Theol Lib Assoc, 1984 (1r) – 1 – 0-8370-0389-X – mf#1984-B153 – us ATLA [241]

Jenkins, Barry *see* The various effects of cryogenic modalities with regards to restriction of blood volume in the male forearm

Jenkins, Burris *see* The man in the street and religion

Jenkins, Charles Francis *see*
- Autographs of the signers of the declaration
- Quaker poems
- Tortola

Jenkins, D E *see* The atonement and intercession of christ

Jenkins, Daniel Edward *see* Present-day attitude toward doctrinal theology

Jenkins Dobles, Eduardo *see*
- Otro sol de faenas
- Tierra doliente

Jenkins, Ebenezer Evans *see* Modern atheism, its position and promise

Jenkins, Edward *see*
- The colonial question
- The colonies and imperial unity or the "barrel without the hoops"

Jenkins, F *see* Tour in arracan, 1831

Jenkins, Floyd Thomas, jr *see* The design and implementation of a program of ministry to non-participating resident members of berea baptist church

Jenkins, George *see* A bibliography of nigerian history

Jenkins, Herbert George *see* The life of george borrow

Jenkins, John *see*
- Christian giving illustrated and enforced by ancient tithing
- Life of the rev alex mathieson...minister of st. andrew's church, montreal

Jenkins, John Stilwell *see*
- James knox polk and a history of his administration
- The new clerk's assistant, or book of practical forms; containing numerous precedents and forms for ordinary business transactions.

Jenkins, Max *see* Voyage of the yacht bounty

Jenkins, Robert Charles *see*
- Canterbury
- The jesuits in china and the legation of cardinal de tournon
- The last crusader
- Romanism

Jenkins, Stanley John *see* The administration of cecil john rhodes as prime minister of the cape colony, 1890-1896

Jenkins, Stuart *see* Arctic temperatures and exploration

Jenkinsburg baptist church. butts county. georgia : church records – 1912-41 – 1 – us Southern Baptist [242]

Jenkinson, A *see* Early voyages and travels to russia and persia, by [him] and other englishmen...

Jenkinson, Thomas B *see*
- Amazulu

Jenks, Jeremiah W *see*
- The immigration problem
- The immigration problem: a study of american immigration conditions and needs

Jenks, Jeremiah Whipple *see*
- The political and social significance of the life and teachings of jesus
- The testing of a nation's ideals
- Twelve studies on the making of a nation

Jenks, John Whipple Potter *see* Hunting in florida in 1874

Jenks' portland gazette – Portland, ME: Elezer A Jenks, oct 31 1803-mar 12 1805 – us CRL [071]

Jenkyn, Thomas William *see*
- The extent of the atonement
- The union of the holy spirit and the church in the conversion of the world

Jenkyn, William *see* An exposition upon the epistle of jude

Jenne, Franz *see* Jenne's reisen

Jenner, Thomas *see* The nanking monument of the beatitudes

Jenne's reisen / Jenne, Franz – Frankfurt [u.a. 1790 [mf ed Hildesheim 1995-98] – 3v on 18mf – 9 – €180.00 – ISBN-10: 3-487-26677-6 – ISBN-13: 978-3-487-26677-0 – gw Olms [910]

Jennie baxter, journalist / Barr, Robert – New York: F A Stokes, c1899 – 4mf – 9 – mf#32093 – cn Canadiana [830]

Jennings, A *see* The effect of perception of performance outcomes on mood following exercise

Jennings, A C *see* The psalms

Jennings, Abraham G *see*
- The last days of jesus christ on the earth
- The mosaic record of the creation explained

Jennings, Arthur Charles *see* The mediaeval church and the papacy

Jennings, Arthur T *see* History of american wesleyan methodism

Jennings county review – North Vernon, IN. 1921-1922 (1) – mf#62930 – us UMI ProQuest [071]

Jennings, David *see* Jewish antiquities

Jennings, Hargrave *see*
- The indian religions
- Phallicism, celestial and terrestrial, heathen and christian
- The rosicrucians

Jennings, Henry James *see* Cardinal newman

Jennings, Ivor *see* Some characteristics of the indian constitution

Jennings, J G *see* The vedantic buddhism of the buddha

Jennings, John *see* Reason or revelation

Jennings, Samuel K *see* An address, intended when written

Jennings, Walter Wilson *see* Origin and early history of the disciples of christ

Jenns, Eustace Alvanley *see* Orpheus and eurydice

Jenny / Lewald, Fanny – Berlin: O Janke, 1872 – 1r – 1 – us Wisconsin U Libr [830]

Jenny, Hans *see* Sudwestafrika

Jenny, Hans Heinrich *see* Die amerikanischen antitrust-gesetze

Jenny, Heinrich Ernst *see* Haller als philosoph

Jenny jenkins – London, England. 18– – 1r – us UF Libraries [240]

Jenofonte / Xenophon – Bogota, Colombia. 1952 – 1r – us UF Libraries [972]

Jens baggesen : en litteraer-psykologisk studie / Clausen, Julius – Kobenhavn: Brodrene Salmonsen, 1895 – (incl bibl ref) – us Wisconsin U Libr [430]

Jens baggesen : en litteraer-psykologisk studie / Clausen, Julius – Kobenhavn: Brodrene Salmonsen, 1895 – 1 – (incl bibl ref) – us Wisconsin U Libr [430]

Jens, Walter *see* Von deutscher rede

Das jenseits : kulturgeschichtliche darstellung der ansichten ueber schoepfung und weltuntergang, die andere welt und das geisterreich / Henne am Rhyn, Otto – Leipzig: O Wigand, 1881 – 1mf – 9 – 0-524-01552-X – mf#1990-2506 – us ATLA [210]

Jenseits : drama in 5 akten / Hasenclever, Walter – Berlin: E Rowohlt 1920 [mf ed 1990] – 1r – 1 – (filmed with: der frosch / otto erich hartleben) – mf#2699p – us Wisconsin U Libr [820]

Jenseits des rationalitaetsprinzips : ueber den (auto-)suggestiven charakter der vernunft / Erdmann, Stephan – (mf ed 2000) – 3mf – 9 – €49.00 – 3-8267-2724-X – mf#DHS 2724 – gw Frankfurter [140]

Das jenseits im mythos der hellenen : untersuchungen ueber antiken jenseitsglauben / Radermacher, Ludwig – Bonn: A Marcus und E Weber, 1903 – 1mf – 9 – 0-524-01513-9 – (incl bibl ref) – mf#1990-2489 – us ATLA [250]

Jenseitsmotive im deutschen volksmaerchen / Siuts, Hans – Leipzig: In Kommission bei E Avenarius 1911 [mf ed Bloomington IN: Indiana Uni Lib, Preservation Dept 1984] – 1r – 1 – us Indiana Preservation [390]

Jensen, Amy Elizabeth *see* Guatemala

Jensen, Andrew *see* Society islands mission

Jensen, Barbara E *see*
- Conquering anxiety in grade school aged swimmers through the use of imaginative play
- The impact of project adventure activities on self-perception
- The influence of fitness-oriented physical activity on the physical self-perception and global self-worth of boys and girls
- Modification and revision of the leadership scale for sport
- Testosterone and physical activity

Jensen beach mirror – Jensen Beach, FL. v1 n1-v21 n52. 1961 sep 14-1980 dec – 16r – (gaps) – us UF Libraries [071]

Jensen beach mirror (jensen beach, fla : 1985) – Jensen Beach, FL. v24 n14-26. 1985 apr 3-june – 1r – us UF Libraries [071]

Jensen, Christian *see* Soeren kierkegaards religioese udvikling

Jensen, Harald *see* Stormaend

Jensen, Herman *see* A practical tamil reading book for european beginners

Jensen, J Keith *see* The effects of two educational processes on energy, nutrient, and food group intakes of sedentary, overweight women who are consuming self-help, low-fat, ad libitum diets

Jensen, Marian *see* Comparison of risk factors for coronary heart disease in sedentary and physically active college students

Jensen, P *see*
- Hat der jesus der evangelien wirklich gelebt?
- Hittiter und armenier
- Texte zur assyrisch-babylonischen religion

Jensen, Peter *see*
- Assyrisch-babylonische mythen und epen
- Moses, jesus, paulus

Jensen, Petrus *see* De incantamentis nonnullis sumerico-assyriis

Jensen, Wilhelm *see*
- Auf der feuerstaette
- Aus den tagen der hansa
- Aus schwerer vergangenheit
- Eddystone
- Der herr senator
- In majorem dei gloriam
- Iris und genziane
- Karin von schweden
- Neue novellen
- Nirwana
- Norddeutsche erzaehler
- Runensteine
- Sanct-elmsfeuer
- Ein ton
- Um den kaiserstuhl
- Um die wende des jahrhunderts (1789-1806)

Jensma, Wopko *see* Sing for our execution

Jensson, Jens Christian *see* American lutheran biographies

Jenty, C *see* Methodo de hacer la amputacion del muslo...

Jentzsch, Franz *see* Briefe aus china

Jenyns, L *see* The zoology of the voyage of hms beagle...during the years 1832-1836

Jenyns, Soame *see* A free inquiry into the nature and origin of evil

Jeografia fisica i politica de las provincias de l... / Colombia Comision Corografica – Bogota, Colombia. 1856 – 1r – us UF Libraries [972]

Jeografia fisica i politica de las provincias de l... / Colombia Comision Corografica – Bogota, Colombia. v1-4. 1957 – 1r – us UF Libraries [972]

Jephet Inb Ali the Karaite *see* A commentary on the book of daniel

Jepheth ben Eli *see* A commentary on the book of daniel

Jephson, A J Mounteney *see* Emin pasha and the rebellion at the equator

Jepson, John James *see* The latinity of the vulgate psalter

The jeptha homer wade family papers, 1771-1957 – [mf ed 1998] – 17r – 1 – mf#ms3292 – us Western Res [380]

Jeqe, the bodyservant of king tshaka / Dube, J L – Lovedale, South Africa. 1951 – 1r – us UF Libraries [960]

Jerabek B, Carlos *see* Tikal

Jeremia / Liechtenhan, R – Tuebingen: J C B Mohr 1909 [mf ed 1989] – (= ser Religionsgeschichtliche volksbuecher fuer die deutsche christliche gegenwart 2/11) – 1mf – 9 – 0-7905-2128-8 – mf#1987-2128 – us ATLA [221]

Jeremia : vortrag / Coerper, F – Elberfeld [Wuppertal]: Buchhandlung der Evangelischen Gesellschaft, 1900 – 1mf – 9 – 0-8370-2699-7 – mf#1985-0699 – us ATLA [920]

Jeremia im fruehjudentum und urchristentum / Wolff, Chr – 1976 – (= ser Tugal 5-118) – 5mf – 9 – €12.00 – ne Slangenburg [221]

Jeremia und seine zeit : die geschichte der letzten fuenfzig jahre des vorexilischen juda / Erbt, Wilhelm – Goettingen: Vandenhoeck und Ruprecht, 1902 – 1mf – 9 – 0-8370-3064-1 – (incl ind of biblical citations) – mf#1985-1064 – us ATLA [221]

Jeremiad – Los Angeles. 1968-1973 (1) 1972-1973 (5) (9) – ISSN: 0047-1968 – mf#7987 – us UMI ProQuest [338]

The jeremiad – (Lafayette). 1900 – 1 – us AJPC [073]

Jeremiah : a drama in nine scenes / Zweig, Stefan – new ed. New York: The Viking Press, 1929 [mf ed 1992] – ix/336p – 1 – (trans fr german by eden and cedar paul. pref by aut) – mf#7982 – us Wisconsin U Libr [820]

Jeremiah : his time and his work / Welch, A C – Oxford, 1955 – €12.00 – ne Slangenburg [221]

Jeremiah : the man and his message / Gillies, James Robertson – London: Hodder & Stoughton, 1907 – 1mf – 9 – 0-8370-3289-X – (includes a chronological table) – mf#1985-1289 – us ATLA [221]

Jeremiah : priest and prophet / Meyer, Frederick Brotherton – New York: Fleming H Revell, c1894 – 1mf – 9 – 0-8370-4408-1 – mf#1985-2408 – us ATLA [221]
Jeremiah and his lamentations : with notes, critical, explanatory and practical: designed for both pastors and people / Cowles, Henry – New York: D Appleton, 1880, c1869 – 1mf – 9 – 0-8370-6098-2 – mf#1986-0098 – us ATLA [221]
Jeremiah and lamentations : introduction, revised version with notes, map, and index / ed by Peake, Arthur Samuel – Edinburgh: T C & E C Jack, [1910?-1911?] – 2mf – 9 – 0-524-05026-0 – (incl bibl ref) – mf#1992-0279 – us ATLA [221]
Jeremiah, his life and times / Cheyne, Thomas Kelly – New York: Anson D F Randolph, [1888] – 1mf – 9 – 0-8370-2649-0 – mf#1985-0649 – us ATLA [920]
Jeremiah the prophet : a study in personal religion / Calkins, R – New York, 1930 – 7mf – 8 – €15.00 – ne Slangenburg [221]
Jeremiah theus / Middleton, Margaret Simons – Uni of South Carolina Press, 1953 [mf ed Spartanburg SC: Reprint Co, 1981? – 5mf – 9 – mf#51-111 – us South Carolina Historical [700]
Jeremiah walker: georgia general baptist / Gardner, Robert G – 1976. 49p – 1 – 5.00 – us Southern Baptist [920]
Jeremias : eine dramatische dichtung in neun bildern / Zweig, Stefan – Leipzig: Insel-Verlag, 1922 [mf ed 1992] – 216p – 1 – mf#7982 – us Wisconsin U Libr [820]
Jeremias, Alfred *see*
– Das alte testament im lichte des alten orients
– Das alter der babylonischen astronomie
– The babylonian conception of heaven and hell
– Die babylonisch-assyrischen vorstellungen vom leben nach dem tode
– Babylonisches im neuen testament
– Der einfluss babyloniens auf das verstaendnis des alten testamentes
– Handbuch der altorientalischen geisteskultur
– Izdubar-nimrod
– Monotheistische stroemungen innerhalb der babylonischen religion
– The old testament in the light of the ancient east
Jeremias gotthelf / Bartels, Adolf – Leipzig: G H Meyer, 1902 [mf ed 1993] – 225p – 1 – mf#8518 – us Wisconsin U Libr [430]
Jeremias gotthelf : eine einfuehrung in seine werke / Muschg, Walter – Muenchen: Lehnen, c1954 – 1 – us Wisconsin U Libr [430]
Jeremias gotthelf : eine einfuehrung in seine werke / Muschg, Walter – Muenchen: Lehnen, c1954 – us Wisconsin U Libr [430]
Jeremias gotthelf : das kirchliche leben im spiegel seiner werke / Hutzli, Walther – Bern: B Haller, c1953 – 1 – (incl bibl ref) – us Wisconsin U Libr [430]
Jeremias gotthelf : das kirchliche leben im spiegel seiner werke / Hutzli, Walther – Bern: B Haller, c1953 – 1r – 1 – (incl bibl ref) – us Wisconsin U Libr [430]
Jeremias gotthelf : sein gottes- und menschenverstaendnis / Buess, Eduard – Zuerich: Evangelischer Verlag, 1948 [mf ed 1989] – 301p – 1 – (incl ind) – mf#7030 – us Wisconsin U Libr [920]
Jeremias gotthelf : ein staatsbuergerlicher mahner: ein vortrag / Bloesch, Hans – Erlenbach-Zuerich: E Rentsch, [1940] [mf ed 1989] – 36p – 1 – mf#7030 – us Wisconsin U Libr [080]
Jeremias gotthelf : unbekanntes und ungedrucktes ueber pestalozzi, fellenberg und die bernische schule / ed by Bloesch, Hans – Bern: H Lang, 1938 [mf ed 1989] – (= ser Schriften der literarischen gesellschaft bern. neue folge der neujahrsblaetter 1) – 79p – 1 – mf#7028 – us Wisconsin U Libr [430]
Jeremias gotthelf im kreise seiner amtsbrueder und als pfarrer / Hopf, Walther – Bern: A Francke, 1927 [mf ed 1989] – 168p – 1 – (incl bibl) – mf#7030 – us Wisconsin U Libr [920]
Jeremias gotthelf in seinen beziehungen zu deutschland / Muret, Gabriel – Muenchen: G Mueller & E Rentsch, 1913 [mf ed 1989] – 106p – 1 – (incl bibl ref) – mf#7030 – us Wisconsin U Libr [430]
Jeremias gotthelf's ausgewaehlte werke / Gotthelf, Jeremias [pseud: Albert Bitzius] – Berlin: J Springer, 1896-1901 [mf ed 1989] – 5v in 1r – 1 – mf#7027 – us Wisconsin U Libr [802]
Jeremias gotthelfs geld und geist : studien zur kuenstlerischen gestaltung / Grob, Fritz – Olten: Hauenstein-Verlag, [1948] [mf ed 1989] – 124p – 1 – mf#7028 – us Wisconsin U Libr [430]
Jeremias gotthelfs persoenlichkeit : erinnerungen von zeitgenossen / ed by Muschg, Walter – Basel: B Schwabe, c1944 [mf ed 1989] – (= ser Sammlung klosterberg. schweizerische reihe) – 205p/pl – 1 – mf#7028 – us Wisconsin U Libr [920]
Jeremias, Johannes *see* Moses und hammurabi
Jeremias metrik / Giesebrecht, Friedrich – Goettingen: Vandenhoeck & Ruprecht, 1905 – 1mf – 9 – 0-8370-3274-1 – mf#1985-1274 – us ATLA [430]
Jeremie *see*
– Effort
– Haiti independante
– Mission de l'homme dans la vie
– Paroisse sainte-anne
Jeremie gotthelf : sa vie et ses oeuvres / Muret, Gabriel – [S.l: s.n, 1912?] [mf ed 1989] – xv/496p – 1 – mf#7028 – us Wisconsin U Libr [430]
Jeremie, James Amiraux *see*
– History of the christian church in the second and third centuries
– Sermons, doctrinal and practical. second series
Jeremie, John *see* Four essays on colonial slavery
Jeremy bentham and american jurisprudence / Reeves, Jesse Siddall – Address delivered at the tenth annual meeting of the Indiana State Bar Association, July 11-12, 1906. n.p., 1906 26 p. LL-1079 – 1 – us L of C Photodup [340]
The jeremy robinson papers – 4r – 1 – $140.00 – Dist. us Scholarly Res – us L of C Photodup [355]
Jeremy taylor / Gosse, Edmund – London: Macmillan 1903 [mf ed 1990] – (= ser English men of letters (london, england)) – 1mf – 9 – 0-7905-6593-5 – mf#1988-2593 – us ATLA [378]
Jeremy taylor : a sketch of his life and times with a popular exposition of his works / Worley, George – London: Longmans, Green, 1904 – 1mf – 9 – 0-524-04945-9 – (incl bibl ref) – mf#1992-2066 – us ATLA [920]
Jerez de Los Caballeros *see* Ordenanzas. ordenanzas. ordenanzas para el gobierno de la m.m. y m.l. ciudad de xerez de los caballeros probadas por los senores del real...
Jerez de los Caballeros. Badajoz *see* Feria y fiestas, 1945
Jerez, Francisco de *see* Verdadera...conquista de mejico
Jericho / National Moratorium on Prison Construction et al – v1 n1-44 [1975 sep/oct-1987 fall] – 1r – 1 – mf#681349 – us WHS [365]
Jerichower zeitung – Jerichow DE, 1924-32 – 1 – gw Misc Inst [074]
Jerilderie coleanbally herald – Jerilderie, jan 1969-sep 1972 – 4r – 9 – at Pascoe [079]
Jerilderie herald – Jerilderie, jan 1898-dec 1968 – 15r – A$1081.12 vesicular A$1163.62 silver – at Pascoe [079]
Jernegan, Marcus Wilson *see* Laboring and dependent classes in colonial america, 1607-1783
Jerningham, Frederick William *see* Steam communication with the cape of good hope, australia, and new zealand
Jerome, A *see* La vie intellectuelle dans une abbaye loraine au 17e et 18e siecles
Jerome, Saint *see*
– Hieronymus liber de viris inlustribus; gennadius liber de viris inlustribus – der sogenannte sophronius
– Omnium operum divi eusebii hieronymi stridonesis
– Selections
– Vitae patrum
Jerome savonarola : a sketch / O'Neil, James Louis – Boston: Marlier, Callanan, 1898 – 1mf – 9 – 0-7905-5779-7 – (incl bibl ref) – mf#1988-1779 – us ATLA [240]
Die jeromin-kinder : roman / Wiechert, Ernst Emil – Muenchen: K Desch, 1945 – 1r – 1 – us Wisconsin U Libr [830]
Jeronimo de Guadalupe *see* Sanctissimi maximinge...
Jeronimo de Santa Cruz, M *see* Libro primero de la aritmetica en el cual se contienen las siete especies principales
Jeronimo de Santa fe *see* Igeret r yehoshua ha-lorki
Jeronimo zapata, natural de azuaga, en notas de bibliografia franciscana / Castro, Manuel – Madrid: Graf. Calleja, 1968 – 1 – sp Bibl Santa Ana [240]
Jeronimos *see*
– Ordinis s. hieronymi...observationis...memoriale additionale
– Ordinis s. hieronymi...observationis...summarium
Jerrmann, Eduard *see* Unpolitische bilder aus st petersburg
Jerrold, Douglas *see* Espana; impresiones y reflejos
Jerrold, William Blanchard *see*
– Jerrold's guide to the exhibition
– The life of george cruikshank in two epochs
– Life of gustave dore
– The official guide to the universal exhibition of 1855
Jerrold's guide to the exhibition / Jerrold, William Blanchard – Manchester 1857 – (= ser 19th c art & architecture) – 1mf – 9 – mf#4.2.224 – uk Chadwyck [700]
Jerry mcauley : an apostle to the lost / ed by Offord, Robert Marshall – 5th rev enl ed. New York: American Tract Society, c1907 [mf ed 1992] – 1mf – 9 – 0-524-02407-3 – mf#1990-0610 – us ATLA [240]
Jerry t verkler, senate service 1963-1974 : staff director of the senate interior and insular affairs committee – (= ser Us senate historical office oral history coll) – 2mf – 9 – $10.00 – us Scholarly Res [323]
Jersey baptist church. liberty association. davidson county. north carolina : church records – 1784-May 1964 – 1 – 63.45 – us Southern Baptist [242]
The jersey jewish voice – Bayonne. N.J. 1932 – 1 – us AJPC [071]
Jersey journal – Jersey city, NJ. 1867-2000 (1) – mf#60112 – us UMI ProQuest [071]
Jersey libertarian / New Jersey Libertarian Party – 1980 mar-oct, 1981 jan, mar-apr – 1r – 1 – (cont: nj libertarian [1975]; cont by: nj libertarian [1981]) – mf#1223567 – us WHS [325]
Jersey shore herald – Jersey Shore, PA. -w 1889-1912 – 13 – $25.00r – us IMR [071]
Jersey shore vidette – Jersey Shore, PA. -w 1890-1912 – 13 – $25.00r – us IMR [071]
Jersey times and british press – St Helier, Jersey aug 1888-16 jul 1910 (wkly) [mf jul-oct 1897, jan-jun 1899, jan-jul 1910] – 1 – (incorp with: jersey weekly post; wanting: jan-jun 1897, jan-dec 1898; cont: british press & jersey times, naval & military chronicle [jan 1870-31 jul 1888]) – uk British Libr Newspaper [072]
Jerseyanna : the official publication of the new jersey exonumia society / New Jersey Exonumia Society – iss n1-39 [1980 sep-1987 jan/feb] – 1r – 1 – mf#922033 – us WHS [366]
Jersiais – Saint Helier. 6 Jan-29 Sept 1838; 11 Apr 1840 – 1 – uk British Libr Newspaper [072]
Jerte, Ayuntamiento de *see*
– Centenario de un episodio de la guerra de la independencia ocurrido el 21 de agosto de 1809
– Ferias y fiestas de san gil abad, septiembre, 1953
– Ferias y fiestas de san gil abad, septiembre de 1952
Jerubbaal : eine zeitschrift der juedischen jugend / ed by Bernfeld, Siegfried – Berlin, Vienna: R Loewit. v1. 1918/19 [complete] – (= ser German-jewish periodicals...1768-1945, pt 3) – 1r – 1 – mf#B135 – us UPA [939]
Jerusalem : publications = Al-quds – Chicago, IL: 1993 – us CRL [071]
Jerusalem / Mendelssohn, Moses – Berlin, Germany. 1919 – 1r – us UF Libraries [939]
Jerusalem : a sketch of the city and temple from the earliest times to the siege by titus / Lewin, Thomas – London: Longman, Green, Longman & Roberts, 1861 [mf ed 1989] – 1mf – 9 – 0-7905-2049-4 – (incl bibl ref & ind) – mf#1987-2049 – us ATLA [956]
Jerusalem / Smith, George Adam – New York, NY. v1-2. 1908 – 1r – us UF Libraries [939]
Jerusalem : the topography, economics, and history from the earliest times to a d 70 / Smith, George Adam – London: Hodder & Stoughton, 1907-08 [mf ed 1989] – 2v on 3mf – 9 – 0-7905-2935-1 – (incl bibl ref) – mf#1987-2935 – us ATLA [939]
Jerusalem / Wolff, Philipp – Leipzig 1872 [mf ed Hildesheim 1995-98] – 2mf – 9 – €60.00 – 3-487-27672-0 – gw Olms [915]
Jerusalem, ancient and modern : outlines of its history and antiquities / Warren, Israel Perkins – Boston: Elliot, Blakeslee & Noyes, c1873 [mf ed 1988] – 1mf – 9 – 0-7905-0411-1 – mf#1987-0411 – us ATLA [915]
Jerusalem and east mission archive 1842-1976 – 47 titles on 2571mf – 9 – ne IDC [956]
Jerusalem and tiberias / Etheridge, John Wesley – London, England. 1856 – 1r – us UF Libraries [939]
Jerusalem antique / Vincent, Hugues – Paris: Victor Lecoffre, 1912 – 2mf – 9 – 0-7905-8350-X – (incl bibl ref) – mf#1987-6449 – us ATLA [915]
Jerusalem, bethany, and bethlehem / Porter, Josias Leslie – London: T Nelson 1887 [mf ed 1993] – 1mf [ill] – 9 – 0-524-05626-9 – mf#1992-0481 – us ATLA [915]
Jerusalem delivree / Baour-Lormian, Pierre Marie Francois Louis – Paris, France. 1813 – 1r – us UF Libraries [440]
Jerusalem, die opfer und die orgel / Gudemann, Moritz – Wien, Austria. 1871 – 1r – us UF Libraries [939]
Jerusalem in bible times / Paton, Lewis Bayles – Chicago: University of Chicago Press; London: Luzac, 1908 – 1mf – 9 – 0-7905-3093-7 – mf#1987-3093 – us ATLA [930]
[Jerusalem-] israel economist – IS. 1972-76 – 4r – 1 – $200.00 – mf#R63572 – us Library Micro [330]
[Jerusalem-] jerusalem post – IS. 1977-87 – 9r – 1 – $450.00 – mf#R63573 – us Library Micro [079]
Jerusalem, Karl Wilhelm *see* Philosophische aufsaetze
The jerusalem mission : under the direction of the american christian missionary society / Barclay, James Turner – Cincinnati: American Christian Pub Soc, 1853 [mf ed 1992] – (= ser Christian church (disciples of christ) coll) – 1mf – 9 – 0-524-04254-3 – mf#1991-2038 – us ATLA [240]
Jerusalem nouvelle. fascicule 1 et 2, aelia capitolina, le saint-sepulcre et le mont des oliviers / Vincent, Hugues & Abel, Felix-Marie – Paris: Victor Lecoffre, 1914 – 5mf – 9 – 0-7905-8351-8 – (incl bibl ref) – mf#1987-6450 – us ATLA [930]
Jerusalem. [planches] / Vincent, Hugues – [S.l.: s.n., 1912-1914?] – 2mf – 9 – 0-7905-8349-6 – mf#1987-6448 – us ATLA [915]
Jerusalem post – 1948- mf updates – (english-lang daily newspaper documents and records events in palestine, israel and the middle east) – us Primary [072]
The jerusalem post – Jerusalem, [Israeli]: Palestine Post Publ Ltd, apr 23 1950-feb 1997 – us CRL [079]
Jerusalem quarterly – Jerusalem. 1987-1990 (1,5,9) – ISSN: 0334-4800 – mf#15670 – us UMI ProQuest [956]
Jerusalem, recherches de topographie, d'archeologie et d'histoire / Vincent, H – Paris, 1912-1926. 2v – 30mf – 9 – mf#H-3132 – ne IDC [956]
Jerusalem Shaare Zedek Hospital *see* Statuten fur die stiftung allgemeines judisches krakenhaus
Jerusalem star *see* Radical leader, 1888
Jerusalem und das heilige land / Sepp, J N – Schaffhausen, 1863. 2v – 1mf9 – 9 – mf#H-2976 – ne IDC [956]
Jerusalem und sein gelaende / Dalman, Gustaf – Guetersloh, 1930 – 5mf – 9 – mf#H-2926 – ne IDC [915]
Jerusalem under the high-priests : five lectures on the period betweeen nehemiah and the new testament / Bevan, Edwyn Robert – London: E. Arnold, 1904 – 1mf – 9 – 0-7905-3244-1 – mf#1987-3244 – us ATLA [930]
Der jerusalemische talmud in seinen haggadischen bestandtheilen = Talmud yerushalmi. selections / Wuensche, August – Zuerich: Verlags-Magazin (J Schabelitz), 1880 – 1mf – 9 – 0-8370-9756-8 – mf#1986-3756 – us ATLA [270]
Jervis, H *see* Narrative of a journey to the falls of the cavery
Jervis, Humphrey *see* Narrative of a journey to the falls of the cavery
Jervis, John B *see*
– John b jervis papers, 1795-1885
– Sir john jervis on the office and duties of coroners with forms and precedents
Jervis, John Bloomfield *see* Report of messrs j b jervis and alfred w craven, esq's, civil engineers, new york
Jervis, John Jervis White *see* Brief statement of the rise, progress, and decline of the ancient c...
Jervis, William Henley *see*
– The gallican church
– The gallican church and the revolution
Jesaia / Guthe, Hermann – Tuebingen: J C B Mohr (Paul Siebeck) 1906, c1905 – (= ser Religionsgeschichtliche volksbuecher fuer die deutsche christliche gegenwart 2/10) – 1mf – 9 – 0-8370-9477-1 – mf#1986-3477 – us ATLA [221]
Jesaia und jeremia : ihr leben und wirken aus ihren schriften / Koestlin, Friedrich – Berlin: G Reimer, 1879 – 1mf – 9 – 0-8370-3979-7 – mf#1985-1979 – us ATLA [221]
Jesaias : exegetisch-kritische studien / Reich, Wilhelm – Wien: Oskar Frank, 1892 – 1mf – 9 – 0-8370-3994-0 – (incl bibl ref) – mf#1985-1994 – us ATLA [221]
Jesaja 53 : das prophetenwort vom suehnleiden des gottesknechtes / Dalman, Gustaf – 2. umgearb. aufl. Leipzig:J.C. Hinrichs, 1914 – 1mf – 9 – 0-8370-2809-4 – (incl bibl ref) – mf#1985-0809 – us ATLA [221]
Jesaja mit den uebrigen aelteren propheten / Ewald, Heinrich – 2. ausg. Goettingen: Vandenhoeck & Ruprecht, 1867 – 2mf – 9 – 0-8370-9382-1 – (incl bibl ref) – mf#1986-3382 – us ATLA [221]
Jesaja und assur : eine exegetisch-historische untersuchung zur politik des propheten jesaja / Wilke, Fritz – Leipzig: Dieterich (Theodor Weicher), 1905 – 1mf – 9 – 0-8370-5849-X – mf#1985-3849 – us ATLA [221]
Jesaja und seine zeit / Meinhold, Johannes – Freiburg i.B: J C B Mohr, 1898 – 1mf – 9 – 0-8370-6281-0 – (incl bibl ref) – mf#1986-00281 – us ATLA [220]
Jeschke, Gunnar *see* Fehler bei der messung und auswertung von festkoerper-mas-nmr-spektren
Jeschurun – Berlin. v. 1-17. 1914-1930 and Hebrew suppl. v. 1-7. 1920-1926 – 1 – us NY Public [270]
Jeschurun – Breslau (Wroclaw PL), 1868 n1-2, 1871 n1, 1871/72 n1-4, 1873/78 n1-2 – 1 – gw Misc Inst [077]

Jeschurun – Frankfurt/M DE, 1854/55-60/61, 1863/64-64/65 – 1 – gw Misc Inst [074]

Jeschurun : monatsschrift fuer lehre und leben im judentum – Berlin DE, 1914-29 – 5r – 1 – us UMI ProQuest [270]

Jeschurun : monatsschrift fuer lehre und leben im judentum – Berlin, 1(1914)-17(1930) – 197mf – 9 – €376.00 – ne Slangenburg [270]

Jeschurun : organ fuer die geistigen und sozialen interessen des judenthums – Posen: Bernhard Koenigsberger. v1-4. 1901-04 [complete] – (= ser German-jewish periodicals...1768-1945, pt 1) – 3r – 1 – $325.00 – mf#B140 – us UPA [270]

Jeschurun – Pleschen (Pleszew PL), 1901 4 jan-1904 30 jun – 3r – 1 – gw Misc Inst [077]

Jeschurun : zeitschrift fuer die wissenschaft des judenthums – Wroclaw, L'viv, various, 1866 – 1r – 1 – us UMI ProQuest [939]

Jeschurun : zeitschrift fuer die wissenschaft des judentums / ed by Kobak, Joseph – Fuerth, Bamberg, 1856-78: deutsche abt v1-9; hebraeische abt v1-5 – (= ser German-jewish periodicals...1768-1945, pt 3) – 2r – 1 – mf#B142 – us UPA [939]

Jeshurun – (New York). 1915 – 1 – us AJPC [939]

Jeske, Werner *see* Lernstoerungen und leistungshemmungen

Jespersen, Otto *see*
- How to teach a foreign language
- Language, its nature, development and origin

"Jesse chisholm" / Matthews, Warren L – 1 – us Kansas [920]

Jesse, F Tennyson (Fryniwyd Tennyson) *see* Pin to see the peep show

Jesse, John Heneage *see* Memoirs of the pretenders and their adherents

Jesse lee : a methodist apostle / Meredith, William Henry – New York: Eaton & Mains, c1909 – 1mf – 9 – 0-524-00575-3 – mf#1990-0075 – us ATLA [242]

Jesse N. Smith Family Association *see* Kinsman

Jessel, E E *see* The unknown history of the jews

Jessen, Hans *see* Briefe an freunde

Jessen, Karl Detlev *see*
- Heinses stellung zur bildenden kunst und ihrer aesthetik
- Herzensergiessungen eines kunstliebenden klosterbruders

Jessen, Paul *see*
- Die heilige pflicht
- Der krautsteig

Jessen, Wolfgang *see* Kinder zwischen arbeit und schule

Jessen's weekly – Fairbanks AK. 1954 nov 11, 1956 feb 6 – 1r – 1 – mf#867444 – us WHS [071]

Jessopp, Augustus *see*
- Before the great pillage
- The coming of the friars and other historic essays
- Emblems of saints
- John donne
- The life and miracles of st. william of norwich

Jessup, Henry Harris *see*
- Fifty-three years in syria
- The mohammedan missionary problem
- The setting of the crescent and the rising of the cross
- Syrian home life
- Syrian home-life
- The women of the arabs

Jessup, Walter Edgar *see* Law and specifications for engineers and scientists

Jestedsky obzor – Liberec, Czechoslovakia. Jun 1937-Sept 1938 – 1r – 1 – us L of C Photodup [077]

Jestem polakiem – London, UK. 4 Aug 1940-15 May/1 Jun 1941 – 1 – uk British Libr Newspaper [072]

Jestin, R *see* Le verbe sumerien

Jesu barndom og ungdom : i anledning af henning jensen's kritiske angreb / Poulsen, Alfred Sveistrup – Kobenhavn: Gyldendalske Boghandels, 1891 – 1mf – 9 – 0-524-06854-2 – (incl bibl ref) – mf#1992-0996 – us ATLA [220]

Jesu blut, ein geheimnis? / Fiebig, Paul – Tuebingen: J C B Mohr (Paul Siebeck) 1906 [mf ed 1985] – (= ser Lebensfragen 14) – 1mf – 9 – 0-8370-3124-9 – mf#1985-1124 – us ATLA [240]

Jesu evangelium : en historisk fremstilling av jesu forkyndelse / Brun, Lyder – Kristiania [Oslo]: H Aschehoug, 1917 – 2mf – 9 – 0-524-04448-1 – (incl bibl ref) – mf#1992-0117 – us ATLA [220]

Jesu gottheit und das kreuz / Schlatter, Adolf von – 2. Aufl. Guetersloh: C. Bertelsmann, 1913 – (= ser Beitraege zur foerderung christlicher theologie) – 1mf – 9 – 0-7905-3220-4 – mf#1987-3220 – us ATLA [240]

Jesu irrtumslosigkeit / Lemme, Ludwig – Berlin: Edwin Runge 1907 [mf ed 1989] – (= ser Biblische zeit- und streitfragen 3/1) – 1mf – 9 – 0-7905-0504-5 – (incl bibl ref) – mf#1987-0504 – us ATLA [240]

Jesu muttersprache : das galilaeische aramaeisch in seiner bedeutung fuer die erklaerung der reden jesu und der evangelien ueberhaupt / Meyer, Arnold – Freiburg i.B.: J C B Mohr (Paul Siebeck), 1896 – 1mf – 9 – 0-8370-7171-2 – (incl bibl ref and indexes) – mf#1986-1171 – us ATLA [470]

Jesu persoenlichkeit : eine psychologische studie / Weidel, Karl – Halle a S: Carl Marhold, 1908 – 1mf – 9 – 0-524-05758-3 – mf#1992-0601 – us ATLA [220]

Jesu pinas och uppstandelses historia / Waldenstroem, Paul – Stockholm: Pietistens Expedition, [1896] – 2mf – 9 – 0-524-06452-0 – mf#1991-2574 – us ATLA [220]

Jesu predigt in ihrem gegensatz zum judentum : ein religiongeschichtlicher vergleich / Bousset, Wilhelm – Goettingen:Vandenhoeck & Ruprecht, 1892 – 1mf – 9 – 0-8370-2420-X – mf#1985-0420 – us ATLA [240]

Jesu tid : en fremstilling af den nytestamentlige tids / Seidel, Martin – Bergen: Fr Nygaard, 1883 – 1mf – 9 – 0-524-06803-8 – mf#1992-0966 – us ATLA [220]

Jesu wissen und weisheit / Lemme, Ludwig – Berlin: Edwin Runge 1907 [mf ed 1989] – (= ser Biblische zeit- und streitfragen 3/7) – 1mf – 9 – 0-7905-1424-9 – mf#1987-1424 – us ATLA [240]

Jesucristo a traves de las edades. instruccion pastoral dirigida a sus diocesanos con motivo del 19th centenario de la redencion del genero humano... / Martinez Zarate, Jose de Jesus – Madrid: Razon y Fe, 1934 – 1 – sp Bibl Santa Ana [240]

Jesucristo redentor / Goma, Isidoro – Barcelona, 1933; Madrid: Razon y Fe, 1933 – 1 – sp Bibl Santa Ana [240]

Jesucristo redentor : programa para los circulos de estudio de accion catolica de la diocesis de coris. curso 1938-1939 / Junta Diocesana de Accion Catolica. (Coria) – Caceres: Editorial Extremadura, s.a. – 1 – sp Bibl Santa Ana [240]

Der jesuit : charaktergemaelde aus dem ersten viertel des achtzehnten jahrhunderts / Spindler, Carl – 3. aufl. Stuttgart: Hallberger [1904?] [mf ed 1995] – 3v in 1 on 1r – 1 – (filmed with: der invalide) – mf#3746p – us Wisconsin U Libr [880]

The jesuit conspiracy : the secret plan of the order = Conjuraton des jesuites / Leone, Jacopo – London: Chapman and Hall, 1848 – 1mf – 9 – 0-524-04962-9 – (in english) – mf#1990-1365 – us ATLA [241]

Jesuit education : its history and principles viewed in the light of modern educational problems / Schwickerath, Robert – 2nd ed. St Louis, MO: B Herder, 1904, c1903 – 2mf – 9 – 0-8370-7587-4 – (incl ind) – mf#1986-1587 – us ATLA [377]

The jesuit mission press in japan, 1591-1610 / Satow, Ernest Mason – [London?, 1888] – (= ser 19th c publishing...) – 1mf – 9 – mf#3.1.104 – uk Chadwyck [070]

The jesuit mission press in japan, 1591-1610 / Satow, Ernest Mason – [Tokyo]: privately printed, 1888 [mf ed 1996] – (= ser Yale coll) – 2v – 1 – 0-524-10235-X – (v2 in japanese) – mf#1996-1235 – us ATLA [241]

Jesuit missions – s.l, s.l? 193-? – 1r – us UF Libraries [978]

Jesuit missions among the cayugas : from 1656 to 1684 / Hawley, Charles – Auburn, NY: s.n, 1876 – 1mf – 9 – mf#34455 – cn Canadiana [241]

Jesuit or catholic sentinel – Boston. 1829-1834 (1) – ISSN: 0275-097X – mf#5939 – us UMI ProQuest [241]

The jesuit order : or, an infallible pope, who "being dead, speaketh" about the jesuits: a reply / Roy, Jesse J – Winnipeg?: s.n, 1889 – 1mf – 9 – mf#12881 – cn Canadiana [241]

Jesuit Relations, and Allied Documents *see* Travels and explorations of the jesuit missionaries in new france, 1610-1791

Jesuit relations and allied documents, 1610-1791 – 508mf (20:1) – 9 – $1575.00 – us UPA [241]

Un jesuita "a palos", jeronimo del portillo / Bayle, Constantino – Madrid: Missionalia Hispanica, 1945 – 1 – sp Bibl Santa Ana [241]

Los jesuitas desde sus origenes hasta nuestros dias : apuntes historicos / Rosa, Enrico – Madrid: Administracion de Razon y Fe, 1924. 477p. Trans. from the Italian – 1 – us Wisconsin U Libr [241]

Jesuitas en el mar... / Plattner, Felix Alfredo – Madrid: Missionalia Hispanica, 1955 – 1 – sp Bibl Santa Ana [241]

Los jesuitas en la provincia de quito de 1570 a 1774 / Bayle, Constantino – Madrid: Razon y Fe, 1945 – 1 – sp Bibl Santa Ana [241]

Los jesuitas germanos en la conquista espiritual de hispano america. siglos 16-17 / Sierra, Vicente D – Buenos Aires, 1944; Madrid: Missionalia Hispanica, 1946 – 1 – sp Bibl Santa Ana [241]

Jesuitas no brasil : (seculo 16) / Cabral, Luis Gonzaga – Sao Paulo, Brazil. 1925? – 1r – us UF Libraries [241]

Jesuitas no grao-para / Azevedo, Joao Lucio D' – Lisboa, Portugal. 1901 – 1r – us UF Libraries [972]

Jesuiten-fabeln : ein beitrag zur culturgeschichte / Duhr, Bernhard – 2., unveraend Aufl. Freiburg i.B.; St Louis, MO: Herder, 1892 – 2mf – 9 – 0-8370-7457-6 – (incl bibl ref and index) – mf#1986-1457 – us ATLA [241]

Der jesuiten-orden : nach seiner verfassung und doctrin, wirksamkeit und geschichte / Huber, Johannes – Berlin: C.G. Luederitz, 1873 – 2mf – us ATLA [241]

Der jesuiten-orden : nach seiner verfassung und doctrin, wirksamkeit und geschichte / Huber, Johannes – Berlin: C.G. Luederitz, 1873 – 2mf – 9 – 0-7905-4758-9 – (incl bibl ref) – mf#1988-0758 – us ATLA [241]

Der jesuitenorden und der freimaurerorden : vortrag in der oeffentlichen katholikenversammlung zu aachen am 5. november 1871 / Thissen, Eugen Theodor – Aachen: A Jacobi, [1871?] – 1mf – 9 – 0-524-05269-7 – mf#1991-2261 – us ATLA [241]

Les jesuites et l'universite / Genin, Francois – Paris: Paulin, 1844 – 2mf – 9 – 0-524-05317-0 – mf#1990-1435 – us ATLA [241]

Jesuites hors la loi / Cayla, Jean-Mamert – Paris: E Dentu, 1869 – 1mf – 9 – 0-524-05308-1 – mf#1990-1426 – us ATLA [241]

Les jesuites-martyrs du canada / Bressani, Francisco Giuseppe – Montreal: Compagnie d'Impr Canadienne, 1877 – 4mf – 9 – mf#09844 – cn Canadiana [241]

Jesuitism / White, Verner M – Liverpool, England. 1851 – 1r – us UF Libraries [241]

Jesuits / Duff, Alexander – Edinburgh, Scotland. 1845 – 1r – us UF Libraries [241]

Jesuits / Holt, James Maden – Oxford, England. 1859? – 1r – us UF Libraries [241]

Jesuits / Waller, Henry – London, England. 1852 – 1r – us UF Libraries [241]

Jesuits / Whytehead, Robert – London, England. 1848? – 1r – us UF Libraries [241]

Jesuits! = Jesuites / Feval, Paul – Baltimore: John Murphy, 1879, c1878 – 1mf – 9 – 0-8370-7459-2 – (in english) – mf#1986-1459 – us ATLA [241]

The jesuits / Michelet, Jules; ed by Lester, Charles Edwards – New York: Gates & Stedman, 1845 – 1mf – 9 – 0-524-03825-2 – mf#1990-1141 – us ATLA [241]

The jesuits : their foundation and history / Neave, Benjamin – New York: Benziger Bros, 1879 – 2mf – 9 – 0-524-07133-0 – mf#1990-5340 – us ATLA [241]

The jesuits : their origin, history, aims, principles, immoral teaching, their expulsions from catholic and protestant authorities: with the bull of pope clement 14, abolishing the society and a chapter on the jesuits estates act / Austin, Benjamin Fish – London, Ont: Advertiser Print & Pub Co, 1890 – 1mf – 9 – mf#26462 – cn Canadiana [241]

Jesuits, 1534-1921 / Campbell, Thomas Joseph – New York, NY. v1-2. 1921 – 1r – us UF Libraries [025]

The jesuits and the great mogul / Maclagan, Edward – London: Burns, Oates & Washbourne Ltd, 1932 – (= ser Samp: indian books) – us CRL [954]

The jesuits and the great mogul / Melagan, Edward – London, Madrid: Razon y Fe, 1933 – 1 – sp Bibl Santa Ana [241]

The jesuits as educators / Magevney, Eugene – 2nd ed. New York: Cathedral Library Association, 1900 – 1mf – 9 – 0-8370-7961-6 – mf#1986-1961 – us ATLA [241]

The jesuits' estates act : a speech delivered in the house of commons of canada on the 30th of april, 1890 / Davin, Nicholas Flood – Ottawa: J Durie, 1890 – 1mf – 9 – mf#59183 – cn Canadiana [348]

Jesuits exposed – London, England. 1839 – 1r – us UF Libraries [241]

The jesuits in china and the legation of cardinal de tournon : an examination of conflicting evidence and an attempt at an impartial judgment / Jenkins, Robert Charles – London: Nutt, 1894 – 1mf – 9 – 0-524-03617-9 – (incl bibl ref) – mf#1990-4777 – us ATLA [241]

The jesuits in great britain : an historical inquiry into their political influence / Walsh, Walter – London: G Routledge; New York: E P Dutton, 1903 – 1mf – 9 – 0-7905-6579-X – (incl bibl ref) – mf#1988-2579 – us ATLA [241]

The jesuits in malabar / Ferroli, Domenico – Bangalore: Bangalore Press, 1939-51 – 1r – 1 – 0-8370-1495-6 – (incl bibl ref and index) – mf#1984-B052 – us ATLA [241]

The jesuits in north america in the seventeenth century / Parkman, Francis – Boston: Little, Brown, 1867 – 2mf – 9 – 0-7905-7254-0 – mf#1988-3254 – us ATLA [241]

Jesuits. Provincia de Mexico *see* Catalogus personarum, & domiciliorum

The jesuits unmasked : being an illustration of the existing evils of popery in a protestant government... / Parker, William – London: L B Seeley & Son, 1823 – us CRL [241]

Jesus / Bousset, Wilhelm – 3 aufl. Tuebingen: J C B Mohr (Paul Siebeck) 1907 [mf ed 1985] – (= ser Religionsgeschichtliche volksbuecher fuer die deutsche christliche gegenwart 1/2-3) – 1mf – 9 – 0-8370-2422-6 – mf#1985-0422 – us ATLA [240]

Jesus / Bousset, Wilhelm; ed by Morrison, William Douglas – New York: Putnam's 1906 [mf ed 1985] – (= ser Crown theological library 14) – 1mf – 9 – 0-8370-2421-8 – (trans by janet penrose trevelyan) – mf#1985-0421 – us ATLA [240]

Jesus : a christmas sermon, preached in the unitarian church, montreal, on christmas day, 1851 / Cordner, John – [Montreal?: J C Becket], 1851 – 1mf – 9 – 0-665-93366-5 – mf#93366 – cn Canadiana [240]

Jesus : his self-introspection / Armitage, Thomas – New York: Putnam's, 1877 – 1mf – 9 – 0-8370-2112-X – mf#1985-0112 – us ATLA [920]

Jesus / Neumann, Arno – London: Adam and Charles Black; New York: Macmillan [distributor], 1906 – 1mf – 9 – 0-8370-4572-X – (transl fr german. incl bibl ref and index) – mf#1985-2572 – us ATLA [240]

Jesus : ein spiel / Avenarius, Ferdinand – Muenchen: G D W Callwey, 1921 [mf ed 1988] – 57p – 1 – mf#6970 – us Wisconsin U Libr [820]

Jesus : an unfinished portrait / Norden, Charles van – NY: Funk & Wagnalls, 1906 – 1mf – 9 – 0-8370-56195 – mf#1985-3619 – us ATLA [240]

Jesus : vier vortraege / Bornemann, Wilhelm et al – Frankfurt am Main: Moritz Diesterweg, 1910 – 1mf – 9 – 0-7905-9243-6 – (incl bibl ref) – mf#1989-2468 – us ATLA [240]

Jesus : was er uns heute ist / Koenig, Alfred – Freiburg i. B:Paul Waetzel, 1903 – 1mf – 9 – 0-8370-4423-5 – mf#1985-2423 – us ATLA [240]

Jesus *see* Yeh-su (ccm291)

Jesus according to s mark / Thompson, J M – 2nd ed. New York: E.P. Dutton, 1910 – 1mf – 9 – 0-7905-3172-0 – mf#1987-3172 – us ATLA [225]

Jesus all good / Gallerani, Alessandro – New York: P.J. Kenedy, c1908 – 1mf – 9 – 0-8370-7460-6 – mf#1986-1460 – us ATLA [240]

Jesus als charakter : eine untersuchung / Ninck, Johannes – 2. teilweise geaend ausg. Leipzig: J C Hinrichs, 1910 – 2mf – 9 – 0-8370-9974-9 – (incl ind) – mf#1986-3974 – us ATLA [220]

Jesus and modern religion / Rumball, Edwin Alfred – Chicago: Open Court, 1908 – (= ser Christianity of To-day Series) – 1mf – 9 – 0-8370-4997-0 – (incl bibl ref) – mf#1985-2997 – us ATLA [240]

Jesus and my character *see* Chi-tu yu wo ti jen ko [ccm71]

Jesus and the future : an investigation into the eschatological teaching attributed to our lord in the gospels, together with an estimate of the significance and practical value thereof for our own time / Winstanley, Edward William – Edinburgh: T & T Clark, 1913 – 1mf – 9 – 0-7905-0465-0 – (incl ind) – mf#1987-0465 – us ATLA [240]

Jesus and the greeks: or, early christianity in the tideway of hellenism / Fairweather, William – Edinburgh: T. & T. Clark, 1924. xvi,407p. With: The Royal Woman by H. Mann. 1 reel.1297 – 1 – us Wisconsin U Libr [240]

Jesus and the men about him / Dole, Charles Fletcher – Boston: G H Ellis, 1888 [mf ed 1985] – 1mf – 9 – 0-8370-2947-3 – mf#1985-0947 – us ATLA [240]

Jesus and the resurrection : thirty addresses for good friday and easter / Mortimer, Alfred Garnett – New York: Longmans, Green, 1898 – 1mf – 9 – 0-524-05411-8 – mf#1992-0421 – us ATLA [240]

Jesus and the seekers : the saviour of the world and the sages of the world / Marshall, Newton Herbert – London: J Clarke, [190-?] – 1mf – 9 – 0-7905-7973-1 – mf#1989-1198 – us ATLA [240]

Jesus as a controversialist / Haynes, Nathaniel Smith – Cincinnati, Ohio: Standard Pub Co, 1911 – 1mf – 9 – 0-524-06719-8 – mf#1991-2749 – us ATLA [240]

Jesus as a penologist : read before the national prison congress, kansas city, mo., november, 1901 / Barrows, Samuel June – Louisville, KY.: Press of the Industrial School Gem, 1902 – 1mf – 9 – 0-8370-2191-X – mf#1985-0191 – us ATLA [240]

Jesus as a teacher and the making of the new testament / Hinsdale, Burke Aaron – St Louis: Christian Pub Co, 1895 – 1mf – 9 – 0-524-06264-1 – mf#1991-2455 – us ATLA [225]

Jesus as problem, teacher, personality and force : four lectures / Bornemann, Wilhelm et al – New York: Funk & Wagnalls, 1910 – 1mf – 9 – 0-7905-0422-7 – (includes bibliographies and index) – mf#1987-0422 – us ATLA [240]
Jesus at the well, john 4. 1-42 / Taylor, William Mackergo – New York: Anson D F Randolph, c1884 – 1mf – 9 – 0-8370-5488-5 – mf#1985-3488 – us ATLA [221]
Jesus bar rabba or jesus bar abba? / Pratt, Henry – London: Williams and Norgate, 1887 – 1mf – 9 – 0-524-05934-9 – mf#1992-0691 – us ATLA [240]
Jesus castellanos / Dominguez Y Roldan, Guillermo – Habana, Cuba. 1914 – 1r – us UF Libraries [972]
Jesus Castro, Tomas de *see* Emboscada a morfeo
Jesus (ccm325) : a life of jesus taken from the records of matthew, mark and luke / ed by Willmott, Lesslie Earl – Chengtu, Szechwa, 1936 [mf ed 198?] – (= ser Ccm 325) – 1 – mf#1984-b500 – us ATLA [225]
Jesus christ : conferences delivered at notre dame in paris = Conferences de notre-dame de paris / Lacordaire, Henri-Dominique – New York: Scribner, Welford, 1870 [mf ed 1985] – 1mf – 9 – 0-8370-4029-9 – mf#1985-2029 – us ATLA [240]
Jesus christ : our saviour's person, mission, and spirit / Didon, H – Philadelphia, PA: American Catholic Historical Book and News Publ, c1891 – 1mf – 9 – 0-7905-2102-4 – (in english) – mf#1987-2102 – us ATLA [240]
Jesus christ : a pattern of religious virtue – S.l, England? . 18–? – 1r – us UF Libraries [240]
Jesus christ and the christian character : an examination of the teaching of jesus in its relation to some of the moral problems of personal life / Peabody, Francis Greenwood – New York: Macmillan, 1905 – (= ser Lyman Beecher Lectures) – 1mf – 9 – 0-7905-1727-2 – (incl bibl ref and indexes) – mf#1987-1727 – us ATLA [240]
Jesus christ and the old commandments : a study in the development of religion / Peters, John Punnett – [New York]: Edwin S Gorham, 1914 – 1mf – 9 – 0-7905-9058-1 – mf#1989-2283 – us ATLA [240]
Jesus christ and the people / Pearse, Mark Guy – Cincinnati: Jennings and Pye, [190-] Beltsville, MD: NCR Corp, 1978 (3mf); Evanston: American Theol Lib Assoc, 1984 (3mf) – 9 – 0-8370-0832-8 – mf#1984-4236 – us ATLA [240]
Jesus christ and the people / Pearse, Mark Guy – Cincinnati: Jennings and Pye; New York: Eaton and Mains, [1904] – (= ser Little Books on Devotion) – 1mf – 9 – 0-8370-5224-6 – mf#1985-3224 – us ATLA [240]
Jesus christ and the present age : being the 25th fernley lecture...plymouth, aug 2 1895 / Chapman, James – London: C H Kelly 1895 [mf ed 1990] – 1mf – 9 – 0-7905-3771-0 – (incl bibl ref) – mf#1989-0264 – us ATLA [240]
Jesus christ and the social question : an examination of the teaching of jesus in its relation to some of the problems of modern social life / Peabody, Francis Greenwood – New York: Macmillan, 1901, c1900 – 1mf – 9 – 0-7905-3096-1 – (incl bibl ref) – mf#1987-3096 – us ATLA [240]
Jesus christ d'apres mahomet : ou les notions et les doctrines musulmanes sur le christianisme / Sayous, E – Paris, 1880 – €13.00 – ne Slangenburg [230]
Jesus christ during his ministry = Jesus-christ pendant son ministere / Stapfer, Edmond – New York: Charles Scribner, 1897 – 1mf – 9 – 0-8370-5374-9 – (in english) – mf#1985-3374 – us ATLA [240]
Jesus christ et les croyances messianiques de son temps / Colani, Timothee – 2e rev augm ed. Strasbourg: Treuttel & Wurtz, 1864 – 1mf – 9 – 0-8370-2706-3 – mf#1985-0706 – us ATLA [240]
Jesus christ, his times, life and work = Jesus-christ, son temps, sa vie, son oeuvre / Pressense, Edmond de – 2nd rev ed. New York: Scribner, Welford, 1868 [mf ed 1986] – 2mf – 9 – 0-8370-9646-4 – (trans by annie harwood. incl bibl ref) – mf#1986-3646 – us ATLA [225]
Jesus christ in his homeland : lectures / Mountford, Lydia Mary von Finkelstein – Cincinnati: Jennings and Graham, c1911 – 1mf – 9 – 0-7905-2188-1 – mf#1987-2188 – us ATLA [240]
Jesus christ in human experience / Dutt, Meade Ervin – 2nd ed. Cincinnati: Standard Pub Co, c1916 – 1mf – 9 – 0-524-05947-0 – mf#1991-2347 – us ATLA [240]
Jesus christ in the talmud, midrash, zohar, and the liturgy of the synagogue : texts and translations = Jesus christus im thalmud / Laible, Heinrich – Cambridge: Deighton, Bell, 1893 – 1mf – 9 – 0-8370-2810-8 – (in english) – mf#1985-0810 – us ATLA [270]
Jesus christ our lord : an english bibliography of christology comprising over five thousand titles / Ayres, Samuel Gardiner – New York: A C Armstrong, 1906 – 2mf – 9 – 0-7905-0244-5 – (incl indes) – mf#1987-0244 – us ATLA [240]
Jesus christ the divine man : his life and times / Vallings, James Frederick – NY: Anson D F Randolph, [c1887] – 1mf – 9 – 0-8370-5614-4 – (incl bibl ref) – mf#1985-3614 – us ATLA [240]
Jesus christ, the propitiation for our sins : being the 3rd [fernely] lecture...london, july 30th 1872... / Lomas, John – London: Wesleyan Conference Off 1872 [mf ed 1989] – 1mf – 9 – 0-7905-2022-2 – mf#1987-2022 – us ATLA [240]
Jesus christ, the son of god : sermons and interpretations / Macgregor, William Malcolm – 2nd ed. Edinburgh: T & T Clark, 1907 – 1mf – 9 – 0-7905-8511-1 – mf#1989-1736 – us ATLA [240]
Jesus christus im bewusstsein und in der froemmigkeit der kirche / Bonwetsch, Gottlieb Nathanael – Berlin: Edwin Runge 1908 [mf ed 1989] – (= ser Biblische zeit- und streitfragen 4/1) – 1mf – 9 – 0-7905-0613-0 – (incl bibl ref) – mf#1987-0613 – us ATLA [240]
Jesus christus und das gemeinschaftsleben der menschen / Holtzmann, Oskar – Freiburg i. B: J C B Mohr, 1893 – 1mf – 9 – 0-7905-3264-6 – (includes biblographical references) – mf#1987-3264 – us ATLA [240]
Jesus cristo, a vida do mundo : sexta assembleia do conselho mundial de igrejas, vancouver, canada: 24 de julho a 10 agosto 1983 = World council of churches 6th assembly,1983 vancouver, BC – Rio de janeiro: Centro Ecumenico de Documentacao e Informacao, 1984 – us CRL [240]
Jesus der menschensohn : oder, das berufsbewusstsein jesu / Voelter, Daniel – Strassburg: Heitz & Muendel, 1914 – 1mf – 9 – 0-7905-3109-7 – mf#1987-3109 – us ATLA [240]
Jesus, der menschensohn / Tillmann, Fritz – 1. & 2. aufl. Muenster i W: Aschendorff 1908 [mf ed 1992] – (= ser Biblische zeitfragen 1/11) – 1mf – 9 – 0-524-05425-8 – mf#1992-0435 – us ATLA [220]
Jesus devant caiphe et pilate : ou, proces de jesus- christ / Dupin, Andre-Marie-Jean-Jacques – Paris: F-H Barba, 1864 – 1mf – 9 – 0-8370-3005-6 – mf#1985-1005 – us ATLA [220]
Jesus, die haeretiker und die christen nach den aeltesten juedischen angaben / Strack, Hermann Leberecht – Leipzig: J C Hinrichs, 1910 – (= ser Schriften des Institutum Judaicum in Berlin) – 1mf – 9 – 0-7905-2091-5 – (incl ind) – mf#1987-2091 – us ATLA [270]
Jesus, doctor / Caron, Max – Madrid: Razon y Fe, 1927 – 1 – sp Bibl. Santa Ana [240]
Jesus en de ziel : een geestelijcke spiegel voor 't gemoed / Luyken, Jan – t'Amsterdam: Pieter Arentsz, 1685 – 3mf – 9 – mf#O-3113 – ne IDC [090]
Jesus en de ziel / Luyken, Jan – Amsteldam: Kornelis van der Sys, 1722 – 3mf – 9 – mf#O-675 – ne IDC [090]
Jesus en de ziel / Luyken, Jan – Amsterdam, 1687 – 3mf – 9 – mf#O-674 – ne IDC [090]
Jesus en de ziel / Luyken, Jan – t'Amsterdam: Jan Rieuwertsz, 1680 – 2mf – 9 – mf#O-3239 – ne IDC [090]
Jesus et la tradition evangelique / Loisy, Alfred Firmin – Paris: Emile Nourry, 1910 – 1mf – 9 – 0-7905-2021-4 – mf#1987-2021 – us ATLA [220]
Jesus, Gabriel de *see*
– Dios esta aqui
– Vida grafica de santa teresa de jesus
– Vida grafica de santa teresa de jesus. la santa de la raza. volumen 1
– Vida grafica de santa teresa de jesus. tomo 4. madrid, 1935
– Vida...de santa teresa de jesus...
The jesus i know *see*
– Wo so jen shih te chi-tu
– Wo so jen shih te yeh-su
Jesus im glauben des urchristentums / Weiss, Johannes – Tuebingen: J C B Mohr (Paul Siebeck), 1910 – 1mf – 9 – 0-8370-9915-3 – mf#1986-3915 – us ATLA [240]
Jesus im neunzehnten jahrhundert / Weinel, Heinrich – Neue Bearbeitung. Tuebingen: J C B Mohr (Paul Siebeck), 1907, c1905 – 1mf – 9 – 0-8370-5743-4 – (incl ind of names) – mf#1985-3743 – us ATLA [240]
Jesus in bildern aus seinem leben / Zuendel, Friedrich – 2., neu durchgearbeitete und verm. Aufl. Zuerich: S Hoehr, 1885 – 1mf – 9 – 0-8370-5976-3 – (incl bibl ref) – mf#1985-3976 – us ATLA [240]
Jesus in his offices : thirty discourses on the offices of jesus / Comings, Albert Gallatin – Boston: Damrell and Moore, c1860 – 2mf – 9 – 0-524-08570-6 – mf#1993-3155 – us ATLA [240]
Jesus in modern life / Logan, Algernon Sydney – Philadelphia:J B Lippincott, 1888 – 1mf – 9 – 0-8370-4169-4 – mf#1985-2169 – us ATLA [240]
Jesus in the qur'an / Parrinder, Geoffrey – London, 1965 – 4mf – 8 – €11.00 – ne Slangenburg [230]
Jesus is coming / Blackstone, William E – Chicago: F H Revell: Moody Bible Institute, c1908 [mf ed 1989] – 1mf – 9 – 0-7905-2833-9 – mf#1987-2833 – us ATLA [240]
Jesus Mejia, Manuel De *see* Guia viaria de ciudad trujillo
Jesus, my physician *see* [Autobiographical pamphlets]
Jesus nazarenus und die erste christliche zeit : mit den beiden ersten erzaehlern / Volkmar, Gustav – Zuerich: Caesar Schmidt, 1882 – 1mf – 9 – 0-8370-9330-9 – mf#1986-3330 – us ATLA [240]
The jesus of history / Glover, Terrot Reaveley – New York: Association Press, 1917 – 1mf – 9 – 0-524-04457-0 – (incl bibl ref) – mf#1992-0126 – us ATLA [240]
The jesus of history / Hanson, Richard Davis – London: Williams and Norgate, 1869 – 1mf – 9 – 0-7905-1410-9 – (incl bibl ref) – mf#1987-1410 – us ATLA [240]
Jesus of history and the jesus of tradition identified, by george s... / Radcliffe, J – Jamaica, Jamaica. 1880 – 1r – us UF Libraries [240]
Jesus of nazareth : 1: his personal character 2: his ethical teachings 3: his supernatural works: three lectures... / Broadus, John Albert – 3rd ed. New York: A C Armstrong, 1890 – 1mf – 9 – 0-8370-2462-5 – mf#1985-0462 – us ATLA [240]
Jesus of nazareth : his life and teachings / Abbott, Lyman – New York: Harper, 1869, c1868 – 2mf – 9 – 0-7905-1683-7 – (incl ind) – mf#1987-1683 – us ATLA [240]
Jesus of nazareth : an historical and critical survey of his life and teaching = Jesus de nazareth / Giran, Etienne – London: Sunday School Association, 1907 – 1mf – 9 – 0-8370-3295-4 – (in english. includes an appendix) – mf#1985-1295 – us ATLA [240]
Jesus of nazareth : the life of our lord / Meek, Jessie – Kansas City, MO: Publ House of the Pentecostal Church of the Nazarene, c1914 – 1mf – 9 – 0-7905-2180-6 – mf#1987-2180 – us ATLA [240]
Jesus of nazareth / Park, Charles Edwards – [Teachers ed, with helper] Boston: Unitarian Sunday-School Society, c1909 – (= ser Beacon series (boston, mass.)) – 1mf – 9 – 0-524-04287-X – mf#1992-0079 – us ATLA [220]
Jesus of nazareth : the story of his life / Mary Loyola – 4th ed. New York: Benziger, 1906 – 1mf – 9 – 0-7905-2175-X – mf#1987-2175 – us ATLA [240]
Jesus of nazareth in the light of today / Russell, Elbert – Philadelphia: John C Winston, c1909 – 1mf – 9 – 0-8370-5002-2 – (incl bibl ref) – mf#1985-3002 – us ATLA [240]
Jesus of nazareth passeth by... / Greig, B F – Edinburgh, Scotland. 1880 – 1r – us UF Libraries [240]
The jesus of the evangelists : his historical character vindicated, or an examination of the internal evidence for our lord's divine mission with reference to modern controversy / Row, Charles Adolphus – 3rd ed. London: Williams & Norgate, 1868 [mf ed 1985] – 1mf – 1 – 0-8370-4985-7 – mf#1985-2985 – us ATLA [225]
Jesus or christ? : essays / Tyrrell, George et al – Boston: Sherman, French; London: Williams and Norgate, 1909 [mf ed 1985] – (= ser Hibbert journal supplement 1909) – 1mf – 9 – 0-8370-4721-8 – mf#1985-2721 – us ATLA [240]
Jesus our worship / Forbes, A P – Edinburgh, Scotland. 1848 – 1r – us UF Libraries [240]
Jesus people / Hollywood Free Paper – n4-[n.n.] [1973? win?-1974 spr] – 1r – 1 – mf#1060512 – us WHS [071]
Jesus People of Milwaukee *see* Street level
Jesus seen of angels / Toplady, Augustus – London, England. v1. 1827 – 1r – us UF Libraries [240]
Jesus: seven questions : chapters in reconstruction / Warschauer, Joseph – London: James Clarke, 1908 – 1mf – 9 – 0-8370-4647-5 – (incl bibl ref) – mf#1985-2647 – us ATLA [240]
Jesus, the carpenter of nazareth / Bird, Robert – 2nd rev ed. New York: Scribner, 1891 – 2mf – 9 – 0-524-04900-9 – mf#1992-0243 – us ATLA [220]
Jesus the christ, historical or mythical? : a reply to professor drews' die christusmythe / Thorburn, Thomas James – Edinburgh: T & T Clark, 1912 – 1mf – 9 – 0-524-05941-1 – mf#1992-0698 – us ATLA [240]
Jesus, the heart of christianity / Furness, William Henry – Philadelphia: JB Lippincott, 1882 [mf ed 1986] – 1mf – 9 – 0-8370-9945-5 – mf#1986-3945 – us ATLA [225]
Jesus the jew : and other addresses / Weinstock, Harris – New York: Funk & Wagnalls, 1902 – 1mf – 9 – 0-8370-5747-7 – (incl ind) – mf#1985-3747 – us ATLA [270]
Jesus the nazarene : a brief life of our savior, with a parallel harmony / Kephart, Cyrus Jeffries – Dayton, Ohio: W J Shuey, 1894 – 1mf – 9 – 0-524-04753-7 – (incl bibl ref) – mf#1992-0195 – us ATLA [220]
Jesus, the prophet of god / Street, Christopher James – Croydon: Pelling, 1890 – 1mf – 9 – 0-524-05751-6 – mf#1992-0594 – us ATLA [220]
Jesus the son of god, or, primitive christology : three essays and a discussion / Bacon, Benjamin Wisner – New Haven: Yale University Press; London: Oxford University Press, 1911 – 1mf – 9 – 0-7905-0246-1 – mf#1987-0246 – us ATLA [240]
Jesus the son of mary, or, the doctrine of the catholic church upon the incarnation of god the son : considered in its bearings upon the reverence shewn by catholics to his blessed mother / Morris, John Brande – London: James Toovey. 2v. 1851 – 4mf – 9 – 0-8370-8842-9 – (incl bibl ref and index) – mf#1986-2842 – us ATLA [241]
Jesus, the source of spiritual blessing to men / Alexander, Lindsay – London, England. 1874? – 1r – us UF Libraries [240]
Jesus the true messiah / Fuller, Andrew – London, England. 1810 – 1r – us UF Libraries [240]
Jesus the unknown / Merezhkovsky, Dmitry Sergeyevich – New York, NY. 1934 – 1r – us UF Libraries [240]
Jesus through the eyes of the chinese nation *see* Chung-hua min tsu yen li ti yeh-su [ccm193]
Jesus und das alte testament : erlaeuterungen zu thesen / Kaehler, Martin – Leipzig: A Deichert, 1896 [mf ed 1989] – 1mf – 9 – 0-7905-1335-8 – (incl bibl ref) – mf#1987-1335 – us ATLA [225]
Jesus und das alte testament : ein zweites ernstes wort an die evangelischen christen / Meinhold, Johannes – Freiburg i. B: JCB Mohr (Paul Siebeck) 1896 [mf ed 1985] – 1mf – 9 – 0-8370-4381-6 – (incl bibl ref) – mf#1985-2381 – us ATLA [221]
Jesus und das sacaeenopfer : religionsgeschichtliche streiflichter / Vollmer, Hans – Giessen: Alfred Toepelmann, 1905 – 1mf – 9 – 0-524-05755-9 – mf#1992-0598 – us ATLA [240]
Jesus und die heidenmission / Meinertz, M – Muenster i.W, 1925 – 16mf – 8 – €12.00 – ne Slangenburg [240]
Jesus und die heidenmission / Spitta, Friedrich – Giessen: Alfred Toepelmann, 1909 – 1mf – 9 – 0-8370-9582-4 – (incl bibl ref) – mf#1986-3582 – us ATLA [240]
Jesus und die modernen jesusbilder / Jordan, Hermann – Berlin: Edwin Runge 1909 [mf ed 1989] – (= ser Biblische zeit- und streitfragen 5/5-6) – 1mf – 9 – 0-7905-2719-7 – mf#1987-2719 – us ATLA [240]
Jesus und die neutestamentliche schriftsteller / Hausrath, Adolrf – Berlin. bd 1-2. 1908-1909 – €41.00 – ne Slangenburg [225]
Jesus und die neutestamentlichen schriftsteller / Hausrath, Adolf – Berlin. v1-2. 1908-1909 – 2v on 21mf – 8 – €41.00 – ne Slangenburg [225]
Jesus und die rabbinen / Kittel, Gerhard – Berlin-Lichterfelde: Edwin Runge 1914 [mf ed 1993] – (= ser Biblische zeit- und streitfragen 9/7) – 1mf – 9 – 0-524-06144-0 – (incl bibl ref) – mf#1992-0811 – us ATLA [220]
Jesus und die religionsgeschichte : vortrag auf dem ersten religionswissenschaftlichen kongress in stockholm / Larsen, H Martensen – Freiburg i. B: J C B Mohr (Paul Siebeck), 1898 – 1mf – 9 – 0-8370-4385-9 – (transl fr the danish) – mf#1985-2385 – us ATLA [240]
Jesus und paulus / Dausch, Petrus – 1. & 2. aufl. Muenster i W: Aschendorff 1910 [mf ed 1992] – (= ser Biblische zeitfragen 1mf) – 1mf – 9 – 0-524-04091-5 – (incl bibl ref) – mf#1992-0049 – us ATLA [227]
Jesus und paulus : eine freundschaftliche streitschrift gegen die religionsgeschichtlichen volksbuecher von d bousset und d wrede / Kaftan, Julius – Tuebingen: J C B Mohr 1906 [mf ed 1985] – 1mf – 9 – 0-8370-3829-4 – mf#1985-1829 – us ATLA [240]
Jesus' view of himself in mark's gospel : an inductive study / Moxom, Philip Stafford – Pittsfield, Mass: Sun Printing, 1904 – 1mf – 9 – 0-8370-4515-0 – mf#1985-2515 – us ATLA [240]
Jesus von nazareth, mythus oder geschichte? : eine auseinandersetzung mit kalthoff, drews, jensen: vortraege / Weiss, Johannes – Tuebingen: J C B Mohr (Paul Siebeck), 1910 – 1mf – 9 – 0-7905-0415-4 – (incl bibl ref) – mf#1987-0415 – us ATLA [240]
Jesus' way : an aprreciation of the teaching in the synoptic gospels / Hyde, William de Witt – Boston: Houghton, Mifflin, 1902 – 1mf – 9 – 0-8370-3714-X – mf#1985-1714 – us ATLA [240]

JESUS-CHRIST

Jesus-christ d'apres mahomet : ou les notions et les doctrines musulmanes sur le chrstianisme / Sayous, E – Paris, 1880 – 3mf – 8 – €13.00 – ne Slangenburg [230]
Jesus-christ d'apres mahomet, ou, les notions et les doctrines musulmanes sur le christianisme / Sayous, Edouard – Paris: E Leroux; Leipzig: O Schulze, 1880 – 1mf – 9 – 0-7905-9867-1 – (incl bibl ref) – mf#1989-1592 – us ATLA [230]
Jesus-christ devant les aristos – Paris: Impr de Beaule et Maignand, may 1849 – us CRL [240]
Jesus-christ par l. cl. fillion / Vie, N -S de – Madrid: Razon y Fe, 1923 – 1 – sp Bibl Santa Ana [240]
Jesus-christ pendant son ministere / Stapfer, Edmond – 2e ed. Paris: Librairie Fischbacher, 1897 – 1mf – 9 – 0-8370-5375-7 – mf#1985-3375 – us ATLA [240]
Jet – Chicago. 1951+ (1) 1968+ (5) 1976+ (9) – ISSN: 0021-5996 – mf#5404 – us UMI ProQuest [305]
Jet forty eight – Lakenheath, England. 1981 may 8 [v7 n18]-1984 dec 21, 1985 jan 18-1986 feb 28, 1986 mar 7-1987 apr 24, 1987 may-1988 apr – 4r – 1 – mf#1043951 – us WHS [071]
Jet gazette – Austin TX. 1981 apr 30-1982, 1983 jan-nov, 1984 feb 16-1985 mar, 1985 apr-1986 apr, 1986 may-1987 jun, 1987 jul-1988 jun, 1989 jul-1990 jun – 7r – 1 – (cont: bergstrom jet gazette) – mf#643831 – us WHS [071]
Jet journal / Naval Air Station [Miramar CA] – Miramar CA. v33 n35,41 [1984 aug 13, oct 12], v35 n47 [1986 dec 6], v37 n14,16,26 [1988 apr 8,22, jul 8], v38 n14-28,46 [1989 apr 7-jul 14, nov 17], v39 n35-44,46,48-50 [1990 sep 7-nov 10,21, dec 8-21] – 1r – 1 – (cont by: compass [san diego, ca]) – us WHS [355]
Jet line – Great Falls, MT. 1956-1957 (1) – mf#64417 – us UMI ProQuest [071]
Jet stone news series / Montgomery Co. Dayton – apr 1974-may 1981 [wkly] – 5r – 1 – mf#B34444-34448 – us Ohio Hist [071]
Jet stream / Beaufort Marine Corps Air Station – Beaufort SC. 1972, 1973, 1974 jan 4-dec 20, 1975, 1976 jan 9-dec 31, 1977-79, 1980, 1983, 1984, 1990 – 12r – 1 – mf#702658 – us WHS [355]
Jeter, Helen Rankin *see* The chicago juvenile court
Jeter, Henry Norval *see* Pastor henry n jeter's twenty-five years experience with the shiloh baptist church and her history
Jeter, Jeremiah Bell *see*
– Campbellism examined
– Campbellism re-examined
– The seal of heaven
Jeter, Jeremiah Bell et al *see* Baptist principles reset
Jethabhai, Ganesh *see* Indian folklore
Jeton / Ambrogi, Arturo – San Salvador, El Salvador. 1961 – 1r – us UF Libraries [972]
Jetp letters – v1- 1965- – 1,5,6 – us AIP [530]
Jett : journal of english teaching techniques – Flint. 1968-1976 (1) – ISSN: 0022-0884 – mf#10302 – us UMI ProQuest [420]
Jetta : historischer roman aus der zeit der voelkerwanderung / Taylor, George – 2. aufl. Leipzig: S Hirzel, 1884 [mf ed 1994] – 525p – 1 – mf#8749 – us Wisconsin U Libr [830]
Jettchen gebert : roman in zwei baenden / Hermann, Georg – Berlin: E Fleischel, 1912 [mf ed 1989] – 2v – 1 – mf#7051 – us Wisconsin U Libr [830]
Jetter *see* Die gemeinschaften und sekten wuerttembergs
Jetton newsletter – v1 n3-v4 n1 [1977 aug-1980 feb] – 1r – 1 – mf#642463 – us WHS [071]
Die jetzige lehre der synode von missouri von der ewigen wahl gottes : ein vortrag. gehalten in der ev. luth. immanuels-gemeinde von lebanon, wisconsin... / Allwardt, Henry August – 2. Aufl. Columbus, OH: Lutheran Book Concern, 1909 – 1mf – 9 – 0-524-05070-8 – mf#1991-2194 – us ATLA [240]
Jeu de l'amour et du hasard / Marivaux, Pierre Carlet De Chamblain De – Paris, France. 1842 – 1r – us UF Libraries [440]
Le jeu de robin et marion / Halle, Adam de la – Paris [1872] [mf ed 1988] – 1r – 1 – (notation...charles edmond henri de coussemaker; harmonise au piano par jean-baptist weckerlin) – mf#pres. film 25 – us Sibley [780]
Jeune afrique – 1993- – 5r per y – 1 – (also available 1967-1992 on 16mm) – us UMI ProQuest [079]
Le jeune annam : tribune de liberation nationale – Saigon. Mars 1926 – 1 – fr ACRPP [959]
Jeune et la vieille garde / Clairville, M – Paris, France. 1843? – 1r – us UF Libraries [440]
Jeune femme colere / Etienne, Charles Guillaume – Paris, France. 1807 – 1r – us UF Libraries [440]
Jeune force de france – Paris. n1-29. nov 1942-juil aout 1944 – 1 – (mq no. 28) – fr ACRPP [073]
La jeune garde : organe des jeunesses socialistes s.f.i.o. de la seine – Paris. juil 1936-mai 1938, fevr 1939 – 1 – fr ACRPP [325]
La jeune haiti – Port-au-Prince: Impr de la Jeunesse, 2me annee, n5-3me, n4. 14 sep 1894-mai 1896 – 2 sheets – us CRL [972]
Le jeune homme et la litterature : lecture faite au cercle ville-marie de montreal / Bedard, M H – Montreal: E Senecal, 1892 – 1mf – 9 – mf#03539 – cn Canadiana [080]
La jeune indochine – Saigon. n1-12. 10 nov 1927-2 fevr 1928 – 1 – fr ACRPP [073]
Le jeune latour : tragedie canadienne en trois actes / Gerin-Lajoie, Antoine – Montreal: [s.n, 1845?] – 9 – 0-665-92746-0 – mf#92746 – cn Canadiana [820]
Jeune, Louis le *see* Tableaux synoptiques de l'histoire de l'acadie
Jeune mari / Mazeres, M (Edouard) – Paris, France. 1826 – 1r – us UF Libraries [440]
Jeune medecin : ou, l'influence des perruques / Picard, L-B (Louis-Benoit) – Paris, France. 1807 – 1r – us UF Libraries [440]
Jeune menage / Verneuil, Louis – Paris, France. 1922 – 1r – us UF Libraries [025]
Une jeune mere dans les prisons de franco / Fidalgo Carasa, Pilar – Paris, 1939? Fiche W 884. (Blodgett Collection of Spanish Civil War Pamphlets) – 9 – us Harvard [946]
Le jeune ouvrier – Revue destinee au patronage des apprentis et des jeunes ouvriers. Angers. sept 1856-61 – 1 – fr ACRPP [073]
La jeune ouvriere *see* La jeunesse ouvriere
Jeune, Paul Le *see* Relation de ce qui s'est passe de plus remarquable aux missions des peres de la compagnie de jesus
Jeune, Paul le *see*
– Relation de ce qui s'est passe en la nouvelle france en l'annee 1640
– Relation du voyage fait a canada pour la prise de possession du fort de quebec par les francois
Jeune republique : pour une gauche unie et constructive au service de l'homme – Paris. juin 1920-juin 1940, oct 1944-juin 1972 – 1 – fr ACRPP [325]
Le jeune voyageur : en egypte et en nubie / Belzoni, Giovanni B – Paris [1826] [mf ed Hildesheim 1995-98] – 1v on 2mf [ill] – 9 – €60.00 – 3-487-27328-4 – gw Olms [916]
Les jeunes captifs : drame en trois actes... / Lebardin, abbe – 7e ed. Montreal: C O Beauchemin, 1888 [mf ed 1994] – 9 – 0-665-94690-2 – mf#94690 – cn Canadiana [820]
Jeunes coeurs soyez fidelles *see* Beaux yeux and jeunes coeurs soyez fidelles
Jeunes officiers / Berquin, M (Arnaud) – Paris, France. 18–? – 1r – us UF Libraries [440]
Les jeunes voyageurs en asie : ou description raisonnee des divers pays compris dans cette belle partie du monde; contenant des details sur le sol, les productions, les curiosites, les moeurs et coutumes des habitans... / Briand, Pierre – Paris 1829 [mf ed Hildesheim 1995-98] – 8v on 16mf – 9 – €160.00 – 3-487-27657-7 – gw Olms [915]
Les jeunes voyageurs, ou lettres sur la france : en prose et en vers / Taillard, Constant – Paris 1821 [mf ed Hildesheim 1995-98] – 6v on 12mf – 9 – €120.00 – 3-487-29803-1 – gw Olms [910]
Jeunesse / Augier, Emile – Paris, France. 1858 – 1r – us UF Libraries [440]
Jeunesse aux antilles / Delmond, Stany – Paris, France. 1937 – 1r – us UF Libraries [972]
La jeunesse de calvin / Lefranc, Abel – Paris: Fischbacher, 1888 – 1mf – 9 – 0-7905-5251-5 – (incl bibl ref) – mf#1988-1251 – us ATLA [242]
Jeunesse de charles-quint / Melesville, M – Paris, France. 1841 – 1r – us UF Libraries [440]
Jeunesse de henri v / Duval, Alexandre – Paris, France. 1812 – 1r – us UF Libraries [440]
Jeunesse du grand frederic / Boirie, Jean-Bernard-Eugene Cantiran De – Paris, France. 1817 – 1r – us UF Libraries [440]
La jeunesse ouvriere : journal jociste – Montreal. v1 n1 oct 1932- (mthly) [mf ed 1983] – 10r – 1 – (fait suite a: la jeune ouvriere; fusionne avec: le mouvement ouvrier et devient: le front ouvrier) – mf#SEM35P180 – cn Bibl Nat [073]
La jeunesse ouvriere *see*
– Le front ouvrier
Jeunesse ouvriere (1956) : journal mensuel des jeunes travailleurs – Montreal: [s.n.], 1956- [mf ed 1989] – 1r – 1 – (cont: front ouvrier; ceased 1965?) – mf#SEM35P335 – cn Bibl Nat [073]
La jeunesse socialiste : Revue mensuelle de socialisme scientifique. / Groupe des Etudiants Socialistes – no. 1-11 12. Toulouse. 1895 – 1 – fr ACRPP [335]
Jeunesses national-populaires *see* Essor
Jeuthe, Lothar *see* Friedrich de la motte fouque als erzaehler
Jeux et divertissements abyssins / Griaule, M – Paris, 1935 – 4mf – 9 – mf#NE-20241 – ne IDC [956]
Jeux populaires au cambodge : d'apres les travaux d'etudes et recherches de la commission des moeurs et coutumes avec le concours des membres correspondants. redige par chap pin [et al] illus: chap nou / Cambodia. Commission des murs et coutumes – 1.ed. Phnom-Penh: Editions de l'Institut bouddhique 1964 [mf ed 1990] – 1r with other items – 1 – (title & text in khmer; added t.p. in french) – mf#mf-10289 seam reel 122/5 [§] – us CRL [790]
Jeverische woechentliche anzeigen und nachrichten – Jever DE, 1978- – ca 6r/yr – 1 – (title varies: 1812: affiches, annonces et avis divers de jever; 5 aug 1813: woechentliche anzeigen und nachrichten von jever; 1817: jeverisches wochenblatt) – gw Misc Inst [074]
Jeverisches wochenblatt *see* Jeverische woechentliche anzeigen und nachrichten
Jevnin, Samuel *see* Nahalath 'olamim
Jevons, Frank Byron *see*
– Comparative religion
– Evolution
– The idea of god in early religions
– An introduction to the history of religion
– An introduction to the study of comparative religion
– Personality
– Religion in evolution
Jevons, Harriet A *see* Pure logic and other minor works
Jevons, Herbert Stanley *see* The future of exchange and the indian currency
Jevons, William *see* Book of common prayer examined in the light of the present age
Jevons, William Stanley *see*
– Economists' papers
– Pure logic and other minor works
Jew – New York. 1823-1825 (1) – mf#3775 – us UMI ProQuest [939]
The jew – (New York). 1823-25 – 1 – us AJPC [939]
Jew and american ideals / Spargo, John – New York, NY. 1921 – 1r – us UF Libraries [939]
Jew and gentile : being a report of a conference of israelites and christians regarding their mutual relations and welfare / Goodwin, E P et al – New York:Revell, c1890 – 1mf – 9 – 0-8370-2724-1 – mf#1985-0724 – us ATLA [230]
Jew and his daughter – Dublin, Ireland. 1823 – 1r – us UF Libraries [240]
The jew and human sacrifice : an historical and sociological inquiry = Human blood and jewish ritual / Strack, Hermann Leberecht – London: Cope and Fenwick, 1909 – 1mf – 9 – 0-8370-5443-5 – (incl ind) – mf#1985-3443 – us ATLA [270]
The jew as a patriot / Peters, Madison Clinton – New York: Baker & Taylor, 1902 – 1mf – 9 – 0-7905-6942-6 – mf#1988-2942 – us ATLA [975]
The jew exile : a pedestrian tour and residence in the most remote and untravelled districts of the highlands and islands of scotland, under persecution – London 1828 [mf ed Hildesheim 1995-98] – 2v on 4mf – 9 – €120.00 – 3-487-27880-4 – gw Olms [914]
Jewelers' circular-keystone (jck) – Radnor. 1990+ (1) 1990+ (5) 1990+ (9) – (cont: chilton's jewelers' circular/keystone) – ISSN: 1070-0242 – mf#948,01 – us UMI ProQuest [730]
Jewell County. Kansas. School District No. 106 *see* Records
Jewell, David A *see* The effects of dietary carbohydrates on resting metabolic rate
Jewell, Elizabeth A *see* Eating disorder symptomatology in a male athletic population
Jewell, Frederick Swartz *see* The claims of christian science as so styled
Jewell, Pliny *see* The jewell register, containing a list of the descendants of thomas jewell
The jewell register, containing a list of the descendants of thomas jewell / Jewell, Pliny – 1860 – 1 – $50.00 – us Presbyterian [920]
Jewelry making gems and minerals – Redlands. 1972-1986 (1) 1972-1986 (5) 1975-1986 (9) – ISSN: 0274-8193 – mf#6849 – us UMI ProQuest [740]
Jewelry workers bulletin / Amalgamated Jewelry, Diamond and Watchcase Workers Union – v44 n2-v47 n6 [1979 nov/dec-1982 jan/feb] – 1r – 1 – (cont by: local n1-j news) – mf#622503 – us WHS [331]
Jewelry workers' monthly bulletin / International Jewelry Workers' Union – 1917 jun-1924 jun – 1r – 1 – mf#1429078 – us WHS [331]
Jewelry Workers Organizing Committee *see* Links and chains
Jewett, Ann E *see*
– Curriculum development for exercise behavioral change
– Development of an inventory to assess multicultural education attitudes, competencies and knowledge of physical education professionals
– Value orientations of preservice physical education teachers
Jewett, Edwin Hurtt *see* Diabolology
Jewett, Frances Gulick *see* Luther halsey gulick
Jewett, Sarah Orne *see* The mate of the daylight, and friends ashore
Jewett's book of duets, trios, and quartets – The duets composed and arranged for two violins and two flutes; the trios for three violins and three flutes; and a beautiful selection and arrangement of quartets for four instruments. Boston: John P. Jewett & Co., 1851. Includes: "Pot Pourri for two flutes, on popular airs, comprising Hail Columbia, Oh Susanna, The Last Rose of Summer, and Sweet Home, with variations." MUSIC 1986 – 1 – us L of C Photodup [780]
Jewish action / Union of Orthodox Jewish Congregations of America – 1974 apr [v27 n4]-1976 oct, 1975 jan-1981 dec – 2r – 1 – mf#382323 – us WHS [270]
Jewish advance – Chicago. Ill. 1878 – 1 – us AJPC [071]
The jewish advance – Detroit. Mich. 1904 – 1 – us AJPC [071]
Jewish advocate – Boston. 1976-1980 (1) – mf#8014 – us UMI ProQuest [939]
The jewish advocate – Rochester, NY. 1898 – 1 – us AJPC [071]
Jewish advocate and connecticut hebrew record – Boston. Mass. 1923-36 – 1 – us AJPC [071]
Jewish advocate of south broward – Hollywood, FL. v1 n1-v6 n26. 1986 nov-1992 oct – 3r – us UF Libraries [071]
Jewish advocate springfield special – Boston. Mass. 1923-24 – 1 – us AJPC [071]
Jewish Agency For Israel *see* Some legal aspects of the jewish case
Jewish Agency For Israel. Dept For Aliyah And Absorption *see* Dape 'aliyah
Jewish Agency for Palestine *see*
– Memorandum submitted to the bermuda refugee conference, april, 1943
– Memorandum submitted to the palestine royal commission
Jewish agency statement at san fransisco conference : memorandum to united nations – Tel Aviv, [1945] – 1mf – 9 – mf#J-28-142 – ne IDC [956]
The jewish altar : an inquiry into the spirit and intent of the expiatory offerings of the mosaic ritual: with special reference to their typical character / Leighton, John – NY: Funk & Wagnalls, 1886 – 1mf – 9 – 0-8370-4086-8 – mf#1985-2086 – us ATLA [270]
Jewish american – Dallas, TX. 1938-41 – 1 – us AJPC [071]
The jewish american – Detroit, MI: S M Goldsmith. v3 n1. oct 18 1901 (wkly) [mf ed 197-?] – mf#ZZAN-17401 – us NY Public [071]
Jewish american women's magazine and gazette – New York, NY. -w. Jan 1929-Dec 1931. – 6r – 1 – (in Yiddish.) – uk British Libr Newspaper [073]
Jewish antiquities / Jennings, David – London, England. 1837 – 1r – us UF Libraries [939]
Jewish artisan life in the time of our lord : to which is appended a critical comparison between jesus and hillel = Juedisches handwerkerleben zur zeit jesu / Delitzsch, Franz – London: S Bagster, 1877 – 1mf – 9 – 0-7905-3329-4 – (in english) – mf#1987-3329 – us ATLA [220]
The jewish attitude to homosexuality / Mariner, Rodney J & Homolka, Walter – (mf ed 1999) – (= ser Monographien zur wissenschaft des judentums) – 2mf – 9 – €40.00 – 3-8267-2659-6 – mf#DHS 40002 – gw Frankfurter [270]
The jewish banner – (New York). 1905 – 1 – us AJPC [939]
Jewish bookland – New York. 1976-1977 (1) 1976-1976 (5,9) – (cont by: books in review) – mf#7665 – us UMI ProQuest [939]
Jewish bulletin – Omaha. Neb. 1919-21 – 1 – us AJPC [071]
The jewish bulletin – Omaha, NE: Isaac Konecky (wkly) [mf ed 1919-21 (lacks jan 16 1920)] – 1r – 1 – us NE Hist [071]
The jewish business record – New York. N.Y. 1916-18 – 1 – us AJPC [071]
The jewish case : before the anglo-american committee of inquiry on palestine as presented by the jewish agency for palestine. statements and memoranda – Jerusalem, 1947 – 9mf – 9 – mf#J-28-144 – ne IDC [956]
The jewish case – New York, nd – 1mf – 9 – mf#J-28-27 – ne IDC [956]
The jewish case against the palestine white paper – London, 1939 – 1mf – 9 – mf#J-28-139 – ne IDC [956]

JEWISH

Jewish charities – Baltimore. Md. 1910-21 – 1 – us AJPC [071]
Jewish chautauqua society. assembly. jewish chautauqua assembly record – 21 July 1899-1903 – 1 – us AJPC [060]
The jewish chess journal – (New York). 1906 – 1 – us AJPC [790]
The jewish children's world – (New York). 1917-18 – 1 – us AJPC [830]
Jewish christians and judaism : a study in the history of the first two centuries / Sorley, William Ritchie – Cambridge: Deighton Bell; London: George Bell, 1881 – 1mf – 9 – 0-7905-8902-8 – mf#1989-2127 – us ATLA [240]
Jewish chronicle – Baltimore. Md. 1873-75 – 1 – us AJPC [071]
Jewish chronicle – Baltimore. v. 1 no. 1-5, 8-9, 15-16, 18-22, 35-44, 47; v. 2 no. 1. Jan-Dec 1875 – 1 – us NY Public [071]
Jewish chronicle – Boston. Mass. 1891-93 – 1 – us AJPC [071]
Jewish chronicle – London, England. 1841-thrice-yrly updates – 1 – (index 1880-1988 avialble) – us Primary [072]
Jewish chronicle – Mobile. Ala. 1899-1901 – 1 – us AJPC [071]
Jewish chronicle / Montgomery Co. Dayton – mar 1962-5/1968, 3/1971-may 1972 [wkly] – 3r – 1 – mf#B5339-5342 – us Ohio Hist [071]
Jewish chronicle – Pittsburgh. Pa. 1962-68 – 1 – us AJPC [071]
Jewish chronicle – Providence, RI. 1919-1919 (1) – mf#66327 – us UMI ProQuest [071]
Jewish chronicle – (Syracuse, New York). v7 n37 (30 Mar 1951); v8 n12 (12 Oct 1951)-v8 n13 (19 Oct 1951); v8 n27 (4 Jan 1952) – us AJPC [270]
The jewish church in its relations to the jewish nation and to the "gentiles" : or, the people of the congregation in their relations to the people of the land, and to the peoples of the lands / Kerr, Samuel C – Cincinnati: William Scott, 1866 [mf ed 1985] – 1mf – 9 – 0-8370-3892-8 – (incl app) – mf#1985-1892 – us ATLA [270]
Jewish citizen – Jacksonville, FL. v1 n2-v14 n38. 1938 nov 25-1939 aug 4 – 1r – us UF Libraries [071]
The jewish civic leader – Framingham. Mass. 1961-67 – 1 – us AJPC [071]
The jewish civic leader – Worcester. Mass. 1960-67 – 1 – us AJPC [071]
Jewish claims on christian sympathy / Stowell, Hugh – London, England. 1838 – 1r – us UF Libraries [240]
Jewish comment – Baltimore. Md. 1900-18 – 1 – us AJPC [071]
Jewish community advocate of south broward – Hollywood, FL. v1 n1-v8 n17. 1992 oct 30-1994 sep 2 – 1r – us UF Libraries [071]
Jewish community blue book – 1926 – 1r – 1 – (cont: jewish community blue book of milwaukee and wisconsin) – mf#941081 – us WHS [939]
Jewish community blue book of milwaukee and wisconsin – 1924, 1925 – 2r – 1 – (cont by: jewish community blue book) – mf#2695100 – us WHS [939]
Jewish community bulletin – Peoria, Illinois – 1 – (dec. 1948-jan. 1949; apr. 1949-may 1949; nov. 1949-apr. 1950; sept. 1950-oct. 1950; feb. 1967; nov. 1967; mar. 1968-june 1968; continued by: jewish community journal) – us AJPC [939]
Jewish community bulletin – Los Angeles, CA.1954-86 – 1 – us AJPC [071]
The jewish community bulletin consolidated with emanu-el – San Francisco. Calif. 1895-49 – 1 – us AJPC [939]
Jewish Community Center of Milwaukee *see* On center
Jewish community federation of cleveland minutes, 1902-1987 – [mf ed 1959-88] – 28r – 1 – us Western Res [360]
Jewish community journal – Peoria, IL. 1968-83 – 1 – us AJPC [071]
Jewish community news – Belleville, IL. 1972-84 – 1 – us AJPC [071]
The jewish community news : organ of the san diego jewish community / ed by Dubin, M H – San Diego, CA: [s.n.] v7 n14. feb 7 1924 (wkly) [mf ed 197-?] – mf#ZZAN-21942 – us NY Public [071]
Jewish community press – Los Angeles, CA: Consolidated Pub Co. v2 n52 mar 13 1936-v4 n89 nov 25 1938 (wkly) [mf ed 197-?] – (began in 1934?) – mf#ZZAN-17397 – us NY Public [071]
Jewish connection – New York, N.Y. – 1 – (v1, n1, (mar. 1981)-v8, n1, (fall 1987); lacking: v1, n2; v5, n1) – us AJPC [939]
Jewish conservator – Chicago. Ill. 1904-05 – 1 – us AJPC [071]
Jewish criterion – Pittsburgh. Pa. 1958-62 – 1 – us AJPC [071]
Jewish current events – Elmont, NY. 1974-86 – 1 – us AJPC [071]
Jewish current events – Fall River. Mass. 1959-65 – 1 – us AJPC [071]
Jewish currents – New York. 1946+ (1) 1972+ (5) 1977+ (9) – ISSN: 0021-6399 – mf#6498 – us UMI ProQuest [305]
Jewish daily bulletin – New York: Jewish Daily Bulletin Co, 1926-30 – us CRL [071]
Jewish daily forward – 1999- – 2r per y – 1 – (available in yiddish or english. backfiles 1897- $80r) – us UMI ProQuest [072]
Jewish daily forward – 1999- – 1r per y – 1 – (in russian) – us UMI ProQuest [072]
Jewish daily forward – New York, 1897-1998 – 676r – 1 – $39,990.00 – (in yiddish) – us UMI ProQuest [071]
Jewish daily news *see* Philadelphia jewish morning journal, and the jewish daily news
Jewish daily press – Milwaukee WI. 1919 apr 22, 1919 feb 25-1920 nov 21, 1921 jan 6-oct 26 – 2r – 1 – mf#1165554 – us WHS [071]
The jewish daily press – Cleveland. Ohio. 1908-13 – 1 – us AJPC [071]
Jewish daily press and "der weg" – Detroit. Mich. 1920 – 1 – us AJPC [071]
Jewish deaf – New York. N.Y. 1915-25 – 1 – us AJPC [939]
Jewish dialog – Montreal, Quebec, Canada. 1972-83 – 1 – us AJPC [071]
Jewish displaced persons : periodicals from the collections of the yivo institute / Yivo Institute – (= ser Research colls in judaica) – 33r – 1 – $5915.00 – 1-55655-210-6 – with p/g – us UPA [939]
The jewish doctrine of mediation / Oesterley, W O E – London: Skeffington, 1910 – 1mf – 9 – 0-7905-1612-8 – (incl bibl ref and index) – mf#1987-1612 – us ATLA [270]
The jewish dramatic world – (New York). 1909 – 1 – us AJPC [790]
Jewish echo : scotland's only jewish newspaper – [Scotland] Glasgow: City Press jan 1928-dec 1950 (wkly) [mf ed 2003] – 40r – 1 – uk Newsplan [939]
Jewish education – New York. 1980-1993 (1) 1980-1993 (5) 1980-1993 (9) – (cont by: journal of jewish education) – ISSN: 0021-6429 – mf#12037 – us UMI ProQuest [939]
The jewish encyclopedia – New York, London. v1-12. 1902-1906 – 12v on 226mf – 9 – €431.00 – ne Slangenburg [270]
Jewish examiner – Brooklyn, NY. 10 Jan 1936-26 Dec 1941. Continues: Brookly Jewish Examiner. Continued by: The Examiner – 1 – us AJPC [071]
Jewish exponent – Philadelphia. Pa. 1887-1955 – 1 – us AJPC [071]
Jewish family papers / Herzberg, Wilhelm – New York, NY. 1875 – 1r – us UF Libraries [939]
The jewish farmer – (New York). 1891-92 – 1 – us AJPC [630]
The jewish festivals / Lehrman, Simon Maurice – London: Shapiro, Vallentine & Co., 1936.Illus. by Vivienne S. Lehrman. 191p. illus. plates (2 fold.) – 1 – us Wisconsin U Libr [270]
Jewish floridian – Miami, FL. v1 n1-v63 n1-26. 1928 oct 19-1990 jan-jun – 91r – (gaps) – us UF Libraries [071]
Jewish floridian – Miami, FL. v8 n1-v11 n42. 1982 jan 01-1985 – 2r – us UF Libraries [071]
Jewish floridian and shofar of greater hollywood – Hollywood, FL. v1 n1-v13 n26. 1970 nov-1983 – 5r – us UF Libraries [071]
Jewish floridian of greater fort lauderdale – Miami, FL. v3 n7-v19 n13. 1974 apr 05-1990 jul – 7r – (gaps) – us UF Libraries [071]
Jewish floridian of north broward – Ft Lauderdale, FL. v1 n1-v2 n29. 1971 oct 22-1974 mar 22 – 1r – us UF Libraries [071]
Jewish floridian of palm beach county – Miami, FL. v1 n1-v16 n13. 1975 feb 28-1990 jun 29 – 6r – (gaps) – us UF Libraries [071]
Jewish floridian of pinellas county – Miami, FL. v1 n1-v7 n13. 1980 apr-1986 jun 27 – 3r – us UF Libraries [071]
Jewish floridian of south broward – Hollywood, FL. v14 n1-v20 n13. 1984-1990 jul – 3r – (gaps) – us UF Libraries [071]
Jewish floridian of south county – Boca Raton, FL. v1 n1-v12 n13. 1979 dec-1990 jul – 5r – (gaps) – us UF Libraries [071]
Jewish floridian of tampa – Miami, FL. v1 n1-v10 n13. 1979 apr 06-1988 jul – 4r – (gaps) – us UF Libraries [071]
Jewish floridian/floridian newspaper – Miami, FL. v63 n21-26. 1990 may 25 jun 29 – 1r – us UF Libraries [071]
Jewish free press – St. Louis. Mo. 1885-87 – 1 – us AJPC [071]
Jewish frontier – New York. 1933+ (1) 1971+ (5) 1976+ (9) – ISSN: 0021-6453 – mf#3333 – us UMI ProQuest [071]
Jewish gazette – Manchester, England. 1950; 1955-81. -w. 22 reels – 1 – uk British Libr Newspaper [072]
The jewish gazette – New York. N.Y. 1876-27 – 1 – us AJPC [071]
Jewish gazette [manchester ed] – [NW England] Manchester ALS jan 1991-17 feb 1995 – 1 – uk MLA; uk Newsplan [072]
Jewish guardian – London. v. 1-12. Oct 3 1919-Aug 14 1931 – 1 – us NY Public [072]
Jewish guardian / Neturei Karta of the USA – [v1] n3-v2 n4 [1974 nov 10-1980 mar] – 1r – 1 – (cont: guardian [brooklyn, ny]) – mf#429503 – us WHS [071]
The jewish guardian – New York, NY. 1898 – 1 – us AJPC [071]
The jewish guardian – New York, NY. 1912 – 1 – us AJPC [071]
Jewish herald – Boston. Mass. 1893-94 – 1 – us AJPC [071]
Jewish herald – Houston. Texas. 1911-13 – 1 – us AJPC [071]
Jewish herald – Providence, RI. 1930-1958 (1) – mf#66328 – us UMI ProQuest [071]
Jewish herald – Sydney, jan 1902-jun 1926 – 9r – A$643.46 vesicular A$692.96 silver – at Pascoe [079]
The jewish herald – New York, NY. 1894 – 1 – us AJPC [071]
Jewish heritage – Washington. 1972-1974 (1) 1972-1974 (5) (9) – ISSN: 0021-6496 – mf#8055 – us UMI ProQuest [939]
Jewish Historical Institute, Warsaw *see* Rare serials and books from the zydowski instytut historyczny (jewish historical institute, warsaw)
Jewish historical society of greater hartford – Hartford, CT. 1974-82 – 1 – us AJPC [071]
Jewish Historical Society of Maryland *see* Generations
The jewish home prayer-book : a manual of household devotion / ed by Jewish Ministers' Association of America – New York: Publ for the Jewish Ministers' Assoc [by] Philip Cowen, 1888, c1887 – 1mf – 9 – 0-8370-3783-2 – mf#1985-1783 – us ATLA [270]
Jewish ideals : and other essays / Jacobs, Joseph – New York: Macmillan, 1896 – 1mf – 9 – 0-524-03832-5 – mf#1990-3270 – us ATLA [939]
Jewish immigration to the united states : from 1881 to 1910 / Joseph, Samuel – New York: Columbia University, 1914 – (= ser Studies in History, Economics and Public Law) – 1mf – 9 – 0-7905-4934-4 – (incl bibl ref) – mf#1988-0934 – us ATLA [975]
Jewish immigration to the united states : from 1881 to 1910 / Joseph, Samuel – New York: Columbia University, 1914. (Studies in history, economics and public law; v59, whole no. 4: no. 145) – 1mf – us ATLA [304]
Jewish independent – Cleveland OH. 1916 jan 7-1918 mar 29, 1918 apr 5-1918 oct 18 – 2r – 1 – (cont by: cleveland jewish news; jewish review and observer) – mf#852028 – us WHS [071]
Jewish independent *see* The jewish review and observer
The jewish inquirer – London (GB), 1938-1939 19 may [gaps] – 1 – gw Misc Inst [072]
The jewish institute quarterly – New York, NY. 1924-30 – 1 – us AJPC [939]
Jewish journal – Brooklyn, NY. 1979-87 – 1 – us AJPC [071]
Jewish journal – Ft Lauderdale, FL. 1986-1996 aug – 36r – (gaps) – us UF Libraries [071]
Jewish journal – New Brunswick, NJ. 1968-81 – 1 – us AJPC [071]
The jewish journal – New York. N.Y. 1889-1906 – 1 – us AJPC [071]
Jewish journal of greater l.a – Los Angeles, CA.1986-87 – 1 – us AJPC [071]
Jewish journal of raritan valley – Highland Park, NJ. 1981-85 – 1 – us AJPC [071]
Jewish journal/jewish voice – Highland Park, NJ. 1985 – 1 – us AJPC [071]
Jewish leader : a newspaper and magazine – [Scotland] Glasgow: N Lewis 14 mar-21 nov 1930 (wkly) [mf ed 2004] – 37v on 2r – 1 – uk Newsplan [939]
Jewish ledger – Hartford. Conn. 1929-37; 1958-68 – 1 – us AJPC [071]
Jewish ledger – Rochester, NY. 1937-1982 (1) – mf#65196 – us UMI ProQuest [071]
The jewish ledger – New Orleans. La. 1951-60 – 1 – us AJPC [071]
Jewish legends of the middle ages / Pascheles, Wolf et al – New York: Bloch Pub Co, [1912?] [mf ed 1992] – 1mf – 9 – 0-524-04649-2 – mf#1990-3392 – us ATLA [390]
Jewish life – New York. 1980-1981 (1,5,9) – ISSN: 0021-6577 – mf#12553 – us UMI ProQuest [939]
Jewish life in the middle ages / Abrahams, Israel – New York, London: MacMillan & Co., 1911.xxvi,452p – 1 – us Wisconsin U Libr [939]
Jewish literature and modern education / Maitland, Edward – Ramsgate, England. 1871? – 1r – us UF Libraries [240]
Jewish magic and superstition / Trachtenberg, Joshua – New York, NY. 1939 – 1r – us UF Libraries [939]
Jewish merchants in colonial america / Freund, Miriam K – New York, NY. 1939 – 1r – us UF Libraries [960]
Jewish messenger – (New York), 1857-1902 – (= ser Hebraica) – 1 – us AJPC [939]
Jewish messenger – New York. v. 1-92. 1857-1902 – 1 – us NY Public [071]
The jewish messiah : a critical histor of the messianic idea among the jews from the rise of the maccabees to the closing of the talmud / Drummond, James – London: Longmans, Green, 1877 [mf ed 1985] – 1mf – 9 – 0-8370-2978-3 – (incl bibl ref, app & ind) – mf#1985-0978 – us ATLA [270]
Jewish Ministers' Association of America *see* The jewish home prayer-book
Jewish monitor – Birmingham. Ala. 1960-65 – 1 – us AJPC [071]
The jewish monthly – Rochester, NY. 1898 – 1 – us AJPC [071]
Jewish morning times – New York. N.Y. Morgen-zeitung. 1906 – 1 – us AJPC [071]
The jewish musical world and theater magazine – (New York). 1923 – 1 – us AJPC [780]
Jewish nation – London, England. 1860 – 1r – us UF Libraries [939]
The jewish nation – New York. N.Y. 1909-10 – 1 – us AJPC [071]
The Jewish National and University Library. Jerusalem *see* The collective catalogue of hebrew manuscripts
Jewish National Fund *see*
– Bericht des hauptburos an den 17
– Hertsel zal
Jewish news – London (GB), 1942-1945 n46 – 1 – gw Misc Inst [939]
Jewish news – Richmond, VA. 1983-85 – 1 – us AJPC [071]
Jewish news – Southfield, MI. 1942-1984 (1) – mf#63857 – us UMI ProQuest [071]
The jewish news – Detroit. Mich. 1942-62 – 1 – us AJPC [071]
The jewish news – Newark. N.J. 1959-67 – 1 – us AJPC [071]
Jewish news-morris-sussex – Ledgewood, NJ. 1979-83 – 1 – us AJPC [071]
The jewish newspaper – Los Angeles, CA.28 Feb-6 Jun 1985. Missing issues – 1 – us AJPC [071]
Jewish newspapers from latvia – 7r – 1 – (inquire for complete listing) – us UMI ProQuest [077]
Jewish observer – New York. 1973+ (1) 1976+ (5,9) – ISSN: 0021-6615 – mf#8410 – us UMI ProQuest [071]
The jewish observer of the east bay – Oakland, CA. 1968-78 – 1 – us AJPC [071]
Jewish opinion – New York. N.Y. 1949-52 – 1 – us AJPC [071]
The jewish outlook : devoted to the interests of southern jewry – New Orleans, LA: Jewish Outlook Pub Co. v1 n1 mar 19 1937- (biwkly) [mf ed 197-?] – mf#ZZAN-17457 – us NY Public [071]
The jewish outlook : a weekly devoted to traditional judaism – New York, NY: The Jewish Outlook Inc. v2 n18. feb 20 1930 (wkly) [mf ed 197-?] – (began in 1928) – mf#ZZAN-22065 – us NY Public [071]
The jewish passover and the lord's supper / Beer, Joseph W – Lancaster, PA: Inquirer Print and Pub, 1874 – 1mf – 9 – 0-524-03379-X – mf#1990-4691 – us ATLA [230]
The jewish people and palestine / Weizmann, C – Jerusalem, [1936] – 1mf – 9 – mf#J-28-22 – ne IDC [956]
Jewish people from holocaust to nationhood : from the archives of the central british fund for world jewish relief – 1933-1960 – 74r in 3 units – 1 – (coll reveals activities of the central british fund (cbf), a philanthropic organisation founded in 1933 to secure refugees from nazi germany. valuable source in the study of the jewish people from 1933 to 1960) – us Primary [940]
The jewish people from holocaust to nationhood : archives of the central british fund for world jewish relief, 1933-1960 – 74r – 1 – (coll consists of minutes, records and reports fr the central british fund (cbf), an organization wh organized and coordinated activities to secure refugees from nazi germany. with printed guide and detailed reel listing) – mf#C39-27940 – us Primary [939]
Jewish pictorial leader – Pittsburgh, PA. Jan 1952; Apr 1953 – 1 – us AJPC [071]
Jewish pioneers in america, 1492-1848 / Lebeson, Anita Libman – New York, NY. 1931 – 1r – us UF Libraries [939]
The jewish plan for palestine – Jerusalem, 1947 – 7mf – 9 – mf#J-28-149 – ne IDC [956]
The jewish population of jerusalem / Gurevich, D – Jerusalem, 1940 – 3mf – 9 – mf#J-28-181 – ne IDC [956]
The jewish population of palestine / Gurevich, D et al – Jerusalem, 1944 – 7mf – 9 – mf#J-28-183 – ne IDC [956]
Jewish post *see* Vochenzaitung
Jewish press – New York. N.Y. 1961-67 – 1 – us AJPC [071]
Jewish press – Omaha. Neb. 1920-53 – 1 – us AJPC [071]

Jewish press – (The voice of the Torah Jewry). 1962-June 1970 – 1 – us NY Public [290]
The jewish press – Omaha, NE: Jewish Press Pub Co, 1920- (wkly) – 1 – (some numbering irregularities, 1923-41. not publ from mid-july to the last wk of aug each yr, 1960-68) – us NE Hist [071]
Jewish press of pinellas county – Clearwater, FL. v4 n6-v10 n11. 1989 sep 22-1995 – 4r – (gaps) – us UF Libraries [071]
The jewish problem, its solution : or, israel's present and future / Baron, David – Chicago: Fleming H Revell, c1891 – 1mf – 9 – 0-524-04783-9 – mf#1992-0203 – us ATLA [956]
The jewish progress – San Francisco. Calif. 1878-96 – 1 – us AJPC [071]
Jewish Publication Society of America *see*
– The jewish publication society of america twenty-fifth anniversary
– Jps bookmark
The jewish publication society of america twenty-fifth anniversary : april fifth and sixth, nineteen hundred and thirteen, philadelphia / Jewish Publication Society of America – Philadelphia: Jewish Publication Society of America, 1913 – 1mf – 9 – 0-524-01342-X – mf#1990-0388 – us ATLA [070]
Jewish publicatons on microfiche, rare russian – 225mf – 9 – $990.00 – (inquire for complete listing) – us UMI ProQuest [090]
Jewish quarterly – New Brunswick. 1975+ (1,5,9) – ISSN: 0449-010X – mf#9726 – us UMI ProQuest [071]
Jewish quarterly review – London. 1888-1908 (1) – mf#2904 – us UMI ProQuest [939]
The jewish quarterly review – 1(1889)-20(1908) – 262mf – 9 – €500.00 – (incl ind. ns: 1(1910)-55(1964/65) €862) – ne Slangenburg [270]
The jewish question or debates on zionism – (Philadelphia). 1900 – 1 – us AJPC [270]
Jewish record – Philadelphia. Pa. 1875-86 – 1 – us AJPC [071]
The jewish record – Chicago. Ill. 1911-22 – 1 – us AJPC [071]
Jewish record (glasgow) – [Scotland] Glasgow: M C D Ltd 30 oct 1931-3 mar 1933 (wkly) [mf ed 2004] – 2r – 1 – (cont by: jewish standard & record) – uk Newsplan [939]
The jewish recorder – New York. N.Y. 1893-95 – 1 – us AJPC [071]
Jewish refugees in shanghai / Ginsbourg, Anna – Shanghai, CHINA . 1940 – 1r – us UF Libraries [939]
Jewish religious life after the exile / Cheyne, Thomas Kelly – New York: Putnam's, 1898 [mf ed 1985] – (= ser The american lectures on the history of religions 3ser, 1897-98) – 1mf – 9 – 0-8370-2650-4 – (incl ind) – mf#1985-0650 – us ATLA [939]
The jewish repository : or, monthly communications respecting the jews, and proceedings of the london society – [London: Soc for Promoting Christianity amonf the Jews, c1813-15 [mthly] [mf ed 2003] – 3v on 1r – 1 – (in english & hebrew) – mf#2003-s078 – us ATLA [230]
Jewish review – London, England. 1910-14 [mf ed 2001] – (= ser Christianity's encounter with world religions, 1850-1950) – 1r – 1 – mf#2001-s004 – us ATLA [939]
Jewish review : weekly jewish newspaper – Cleveland, OH: The Wertheimer-Machol Pub Co. v4 n6. nov 8 1895-1899 – 2r – 1 – (weekly jewish newspaper. merged with: jewish review and observer) – mf#(M) 34 C9.3 128 – us Western Res [071]
Jewish review *see* The jewish review and observer
Jewish review and observer – 1916 jan 7 [v42 n3]-1917 mar 16, 1917 mar 23-1918 aug 2 [v44 n33] – 2r – 1 – mf#467579 – us WHS [939]
The jewish review and observer – Cleveland. Ohio. 1959-62 – 1 – us AJPC [071]
The jewish review and observer – Cleveland, OH: Dan S. Wertheimer, nov 24 1899-oct 11 1907; The Dan S. Wertheimer Co, cct 18 1907-aug 28 1964 – 32r – 1 – (the weekly newspaper was a reflection of cleveland's established jewish community. merger of: jewish review and hebrew observer. titled: jewish independent. merged: cleveland jewish news) – mf#(M) 34 C9.3 129 – us Western Res [071]
Jewish rights at the congresses of vienna (1814-1815) and aix-la-chapelle (1818) / Kohler, Max James – New York: American Jewish Committee, 1918 (mf ed 1995) – 1r – 1 – (incl bibl ref and ind) – mf#ZZ-34398 – us NY Public [939]
Jewish school / Morris, Nathan – London, England. 1937 – 1r – us UF Libraries [939]
The jewish scriptures : the books of the old testament in the light of their origin and history / Fiske, Amos Kidder – New York: Scribner's, 1896 [mf ed 1985] – 1mf – 9 – 0-8370-3146-X – mf#1985-1146 – us ATLA [221]
Jewish sheet music : folk songs and rare musical scores from russia and the ukraine – late 19th-early 20th c [mf ed Norman Ross Publ] – 2v on 8r – 1 – (with p/g) – us UMI ProQuest [780]
Jewish socialist critique – v1 n1-3 [1979 fall-1980 spr/sum] – 1r – 1 – mf#635894 – us WHS [939]
The jewish sources of the sermon on the mount / Friedlander, Gerald – London: George Routledge; New York: Bloch, 1911 – 1mf – 9 – 0-7905-0022-1 – (includes bibliographies and indexes) – mf#1987-0022 – us ATLA [220]
Jewish south – Richmond, VA. 1893-99 – 1 – us AJPC [939]
The jewish south – Atlanta. Ga. 1877-79 – 1 – us AJPC [939]
Jewish spectator – Santa Monica. 1972+ (1) 1972+ (5) 1975+ (9) – ISSN: 0021-6720 – mf#7264 – us UMI ProQuest [071]
The jewish spectator – Memphis, TN: Jewish Pub Co. v6 n2. apr 20 1888 (wkly) [mf ed 197-?] – (began in 1885. ceased in 1926? publ in new orleans la and memphis tn may 7 1897-) – mf#ZZAN-22173 – us NY Public [071]
The jewish spectator – Memphis, TN. 1885, 1902 – 1 – us AJPC [071]
The jewish spirit – Portland. Or. 1916 – 1 – us AJPC [071]
Jewish standard – London. -w. Jan 1890-Jun 1891. (1 reel) – 1 – uk British Libr Newspaper [072]
The jewish standard : from the early english newspapers collection – mar 16 1888-jun 26 1891 – 3r – 1 – us Primary [072]
The jewish star – Edison, NJ. 1985 – 1 – us AJPC [071]
Jewish struggle – London. -m. Dec 1945-Oct Nov 1946. (4 ft) – 1 – uk British Libr Newspaper [072]
Jewish student's companion / De Solla, Jacob Mendes – New York, NY. 1880 – 1r – us UF Libraries [939]
Jewish students' organisation – New York. N.Y. 1925 – 1 – us AJPC [071]
Jewish studies, part 1 : rare printed sources from the parkes collection, university of southampton – [mf ed Marlborough 1993] – 322mf – 9 – £2100.00 – (with d/g) – uk Matthew [939]
The jewish tabernacle : two lectures / Chase, Ira Joy – Cincinnati: Standard Pub, 1890 – 1mf – 9 – 0-524-03965-8 – (incl bibl ref) – mf#1992-0008 – us ATLA [220]
The jewish tabernacle and its furniture : in their typical teachings / Newton, Richard – New York: Robert Carter, 1878, c1863 [mf ed 1985] – 1mf – 9 – 0-8370-4579-7 – mf#1985-2579 – us ATLA [270]
Jewish telegraph – [NW England] Manchester ALS jan 1956-Dec 1967, jan 1970-jun 1975, jan 1991-feb 1995 – 1 – (incorp: jewish gazette [feb 1995]) – uk MLA; uk Newsplan [072]
Jewish Telegraphic Agency *see*
– Bulletin
– Community news reporter
– News
– News bulletin
Jewish Telegraphic Agency, Inc *see* Daily news bulletin
The jewish temple and the christian church : a series of discourses on the epistle to the hebrews / Dale, Robert William – 2nd ed. London:Hodder and Stoughton, 1871 – 1mf – 9 – 0-8370-2803-5 – mf#1985-0803 – us ATLA [220]
Jewish Theological Seminary of America. Library *see*
– Biblical manuscripts and books in the library of the jewish theological seminary (mostly from the sulzberger collection)
– Maimonides' mishneh torah
The jewish theosophist : devoted to the study of judaism in the light of theosophy and theosophy in the light of judaism – Seattle WA: Assoc of Hebrew Theosophists in America. v1-2. 1926-32 [qrterly] [mf ed 2003] – 2v on 1r – 1 – (none publ jul 1928-oct 1929) – mf#2003-s501 – us ATLA [290]
Jewish tidings – Rochester. N.Y. 1887-94 – 1 – us AJPC [071]
Jewish times – Baltimore. Md. 1958-68 – 1 – us AJPC [071]
Jewish times – New Orleans, LA. v1 n1-v14 n26. 1974 mar 24-1988 dec – 4r – (gaps) – us UF Libraries [071]
Jewish times – Glasgow, Scotland. 23 oct 1964-jul 1972 – 5r – 1 – (israel today and the jewish times 1968-72) – uk British Libr Newspaper [072]
Jewish times – Los Angeles. Calif. 1930-31 – 1 – us AJPC [071]
Jewish times / Mahoning Co. Youngstown – jul 1967-dec 1979 [biwkly] – 4r – 1 – mf#B3439-3442 – us Ohio Hist [072]
Jewish times – Miami Beach, Fla., v2, no. 16 (20 Apr. 1983); v2, no. 18 (4 May 1983) – us AJPC [071]
Jewish times – New York. N.Y. 1869-79 – 1 – us AJPC [071]
Jewish times – Toronto, Ontario. 9 Nov 1979-20 Dec 1985. English. Incomplete – 1 – us AJPC [071]
Jewish times / Trumbull Co. Youngstown – jul 1967-dec 1979 [biwkly] – 4r – 1 – mf#B3439-3442 – us Ohio Hist [072]
The jewish times – Brookline, MA. 1978-83 – 1 – us AJPC [071]
The jewish times – Downsview, Ont., Canada.1979-85 – 1 – us AJPC [071]
The jewish times – New Orleans, LA.14 Mar 1975-27 Mar 1981. English. Some issues missing – 1 – us AJPC [071]
The jewish times – New Orleans, LA.1894 – 1 – us AJPC [071]
Jewish tribune – San Francisco. Calif. 1936-47 – 1 – us AJPC [071]
Jewish tribune – St. Louis. Mo. 1879-84 – 1 – us AJPC [071]
Jewish tribune – Westchester, CT.sic 1978-81 – 1 – us AJPC [071]
The jewish tribune – Bombay (IND), 1936 n6-1940 n2 – 1 – gw Misc Inst [079]
Jewish veteran / Jewish War Veterans of the United States of America – 1948 aug-1957 jan, 1957 mar-1968 dec, 1969-74, 1975-81 – 4r – 1 – mf#1394503 – us WHS [939]
Jewish veteran – Washington. D.C. 1963-64 – 1 – us AJPC [071]
Jewish voice – Providence, RI. 1991-1992 (1) – mf#68909 – us UMI ProQuest [071]
Jewish voice – St. Louis. Mo. 1888-1920 – 1 – us AJPC [071]
The jewish voice – Cincinnati. Ohio. 1911-12 – 1 – us AJPC [071]
The jewish voice – Edison, NJ. 1977-85 – 1 – us AJPC [071]
The jewish voice – Houston, TX. 1937-38 – 1 – us AJPC [071]
The jewish voice – Philadelphia. Pa. 1934 – 1 – us AJPC [071]
The jewish voice far east : die juden in europa – Schanghai (VR), 1945 n51-1946 n24 [gaps] – 1 – gw Misc Inst [939]
Jewish voices – Cincinnati, Ohio. v1 n1 (Nov 1989)-v2 n2 (Mar 1991) – us AJPC [939]
Jewish War Veterans of the United States of America *see* Jewish veteran
Jewish week and the american examiner – 1973 nov 1-1974 dec 28, 1975 jan 11-1975 dec 27, 1976 sep 26-1977 may 29, 1977 jun 5-1978 jan 29, 1978 feb 5-jul 15 – 5r – 1 – (cont: american examiner; jewish week, national jewish ledger; cont by: new york jewish week and the american examiner) – mf#585904 – us WHS [071]
The jewish weekly *see* Vochenzaitung
Jewish weekly news – Springfield. Mass. 1952-68 – 1 – us AJPC [071]
Jewish western bulletin – Vancouver, BC. 9 Jun 1939; 20 May-30 Sept 1948; 23 Mar 1956; 12 Sept 1958; 14 Dec 1978 – 1 – us AJPC [071]
The jewish woman's home journal – (New York). 1922-23 – 1 – us AJPC [939]
The jewish women's journal – New York, NY. v1 n4 (Summer 1993)-v2 n4 (Dec 1994) – us AJPC [939]
The jewish workers circle minutes (1935-1952) *see* Fascist and anti-fascist archives from the hackney archives, london
Jewish world – Albany, NY. 1979-86 – 1 – us AJPC [071]
Jewish world – West Palm Beach, FL. 1987 sep 25-1989 – 7r – (gaps) – us UF Libraries [071]
Jewish world – London. Feb 14 1873-Feb 1 1934. Incomplete – 1 – us NY Public [939]
The jewish world – New York. N.Y. 1902-04 – 1 – us AJPC [071]
The jewish world – Philadelphia. Pa. 1914-42 – 1 – us AJPC [071]
The jewish world in the time of jesus / Guignebert, Charles – Trans. from the French by S.H. Hooke. New York: Dutton, 1939. xiv,288p – 1 – us Wisconsin U Libr [240]
Jewish world of long island – Commack, NY. 1976-86 – 1 – us AJPC [071]
Jewish yearbooks and calendars : from the national library of russia, st petersburg – 1901-19 [mf ed Norman Ross Publ] – 51mf – 9 – us UMI ProQuest [270]
The jewish-american people's calendar – (New York). 1894-1900 – 1 – us AJPC [939]
Jewitt, John *see* The adventures and sufferings of john r jewitt, only survivor of the ship boston, during a captivity of nearly three years among the savages of nootka sound
Jewitt, Llewellynn Frederick William *see*
– The ceramic art of great britain
– Chatsworth...illustrated by upwards of fifty engravings
– Grave-mounds and their contents
Jewitt, Llewellynn Frederick William et al *see* The stately homes of england
Jews : or, the voice of the new testament concerning them / Marsh, William – Leamington, England. 1841 – 1r – us UF Libraries [240]
Jews a blessing to the nations and christians bound to seek their c... / Scott, Thomas – London, England. 1810 – 1r – us UF Libraries [240]
Jews and arabs in palestine – London, 1936 – 1mf – 9 – mf#J-28-135 – ne IDC [956]
Jews and judaism : in the nineteenth century / Karpeles, Gustav – Philadelphia: Jewish Publ Soc of America, 1905 – (= ser Special Series (Jewish Publ Soc of America)) – 1mf – 9 – 0-524-04985-8 – mf#1990-3443 – us ATLA [270]
Jews and judaism in the united states / Levinger, Lee J – Cincinnati, OH. 1925 – 1r – us UF Libraries [939]
The jews and the israelites : their religion, philosophy, traditions and literature, in connection with their past and present condition, and their future prospects / Freshman, Charles – Toronto: A Dredge, 1870 – 6mf – 9 – mf#03259 – cn Canadiana [939]
Jews and the national question / Levy, H – New York, NY. 1958 – 1r – us UF Libraries [939]
The jews and their evangelization / Gidney, William Thomas – London: Student Volunteer Missionary Union, 1899 – 1mf – 9 – 0-8370-6119-9 – mf#1986-0119 – us ATLA [242]
Jews College (London, England) Literary Society *see* Papers read before the jews college literary society
Jews for jesus newsletter / Hineni Ministries – iss 5736:v10 [1976 jun], iss 5737:v5 [1977 feb]-1984 aug – 1r – 1 – mf#1000949 – us WHS [270]
The jews in america : a short story of their part in the building of the republic / Peters, Madison Clinton – Philadelphia: J C Winston, 1905 – 1mf – 9 – 0-7905-6352-5 – mf#1988-2352 – us ATLA [975]
The jews in babylonia in the time of ezra and nehemiah : according to babylonian inscriptions / Daiches, Samuel – London: Jews' College 1910 [mf ed 1989] – 1mf – 9 – 0-7905-1926-7 – (incl bibl ref) – mf#1987-1926 – us ATLA [939]
Jews in our time / Bentwich, Norman De Mattos – Baltimore, MD. 1960 – 1r – us UF Libraries [939]
The jews in relation to the church and the world : a course of lectures / Cairns, John – London: Hodder and Stoughton, 1877 – 1mf – 9 – 0-7905-0070-1 – mf#1987-0070 – us ATLA [270]
Jews in south africa / Saron, Gustav – Cape Town, South Africa. 1955 – 1r – us UF Libraries [939]
Jews Liturgy And Ritual Mourners' Prayers *see* Memorial prayers and meditations
The jews of eastern europe / Adeney, John Howard – London: Central Board of Missions and SPCK, New York: Macmillan, 1921 (mf ed 1995) – 1r – 1 – (incl bibl ref and ind) – mf#ZZ-34398 – us NY Public [939]
Jews of philadelphia / Morais, Henry Samuel – Philadelphia, PA. 1894 – 1r – us UF Libraries [939]
The jews of philadelphia : their history from the earliest settlements to the present time / Morais, Henry Samuel – Philadelphia: Levytype, 1894 [mf ed 1989] – 2mf – 9 – 0-7905-4117-3 – mf#1988-0117 – us ATLA [305]
The jews' who's who : israelite finance, its sinister influence – London: Judaic Pub Co, 1920 (mf ed 1995) – 1r – 1 – mf#ZZ-34380 – us NY Public [939]
Jews without money / Gold, Michael – Garden City, NY. 1946, c1930 – 1r – us UF Libraries [939]
Jewsbury : the collected writings of geraldine jewsbury (1812-1880) – [mf ed Marlborough 1994] – 6r – 1 – £570.00 – (with d/g) – uk Matthew [420]
Jewsbury, Geraldine *see* Jewsbury
Jex, Amy T *see* Integrating dance into the study of american humanities
Jeypur, land und leute : eine volksteumliche schilderung des hauptgebietes der breklumer mission / Bracker, pastor – Breklum: Missionshauses, 1902 [mf ed 1995] – (= ser Yale coll) – 179p (ill) – 1 – 0-524-09063-7 – (in german) – mf#1995-0063 – us ATLA [240]
Jeyyasena, U *see* Vimana vatthu to kri
Jezebel / Mcneile, Hugh – Dublin, Ireland. 1840 – 1r – us UF Libraries [240]
Jezequel, Jules *see* Impressions d'espagne
Jezus en de ziel / Luyken, Jan – Amsteldam: Wed P Arentsz en C van der Sys, 1704 – 3mf – 9 – mf#O-3240 – ne IDC [090]
Jezus en de ziel / Luyken, Jan – Amsteldam: Wed P Arentz, en K vander Sys, 1714 – 4mf – 9 – mf#O-347 – ne IDC [090]
Jezus in de islam / Bakker, F L – Den Haag, 1955 – 1mf – 8 – €3.00 – ne Slangenburg [230]
Jfkapers / International Association of Machinists and Aerospace Workers – 1983 nov-1986 oct – 1r – 1 – mf#1802543 – us WHS [366]

Jh H Men Jin *see* Bansavatar prades kambuja sankhep

J-H newman : essai de psychologie religieuse / Grappe, Georges – 2e ed. Paris: P-J Beduchaud, 1902 – (= ser Les grands hommes de l'eglise au 19e siecle) – 1mf – 9 – 0-7905-4795-3 – mf#1988-0795l – us ATLA [240]

Jha, Amaranatha *see* Shakespearean comedy and other studies

Jha, Ganganatha *see* The philosophical discipline

Jha, Hari Bansh *see* Buddhist economics and the modern world

Jha, Mahamahopadhyaya Ganganatha *see* Hindu ethics

Jham snehkpaer joen bhnam : pralom lok knun manosancetana tam lam nam bit / Thu Dhan – Kracah: Ghun Thai San 2508 [1965] [mf ed 1990] – 1r with other items – 1 – (at head of title: ryan; in khmer) – mf#mf-10289 seam reel 110/8 [§] – us CRL [480]

Jham thmi hur croh lap loe jralan cas / Lak Sarum – Bhnam Ben: Bhmaek Bhluk, Edition Oeil D'lvoire 2517 [1972]- [mf ed 1989] – 1r with other items – 1 – (in khmer) – mf#mf-10289 seam reel 023/09 [§] – us CRL [959]

Jhaveri, Krishnalala Mohanalala *see* Milestones in gujarati literature

Jhering, Rudolf von *see*
- Der zweck im recht
- Der zweck im recht, von rudolph von jhering.

Jherings jahrbucher fuer die dogmatik des burgerlichen rechts – Jena. On film: v1-90; 1857-1942; index 1857-1906. LL-0226 – 1 – us L of C Photodup [346]

Jhyam Duc *see* Prasna gihi patipatti

Jhym Duc *see* Vacanadhippay

Jhym Duc, Lok Qacarr *see* Prasna traigun

Jhym Krasem *see* Ryan maha bharatayuddh

Jhym Sum *see* Vijja gun katha

Ji phru son pay lay na pham lup nan / Takkatho Thet Win – Ran Kun: Ca pe bhiman 1970 – 1r with other items – 1 – (in burmese) – mf#mf-10289 seam reel 175/5 [§] – us CRL [639]

Jiang, Peixing *see* The effect of foot landing position on foot mechanics during gait

Jiao shi bao – Peking. Sep-dec 1957 – 1/4r – 1 – uk British Libr Newspaper [072]

Jiaoshi bao – Peking. 1 mar 1957-8 jul 1958 – 1/2r – 1 – uk British Libr Newspaper [072]

Jias news – Montreal, Quebec, Canada. 1966-83. Jewish Immigrant Aid Services of Canada – 1 – us AJPC [071]

Jibaro en la literatura de puerto rico / Silva, Ana Margarita – San Juan, Puerto Rico. 1957 – 1r – us UF Libraries [440]

Jibhah – Tehran, (1981-83); London (1983-87): Nashriyah-'i milliyun-i Iran. sal-i 1-2, shumarah-'i 2-97. 1 mihr 1360-21 farvardin 1366 [23 sep 1981-11 april 1987] – 2r – 1 – $106.00 – (missing nos 1, 6, 12-13, 15, 23-24, 38, 52, 70-71, 75-79, 84, 86, 90, 95. r also incl payam-i jibhah-'i milli) – us MEDOC [079]

Jibhah-'i milli-i iran – Tehran: Jibhah-'i Milli-i Iran. dawrah-'i jadid, shumarah-'i 1-19. 12 isfand 1357-31 khurdad 1358 [3 mar 1979-21 jun 1980] – 1r – 1 – $53.00 – (cont by: ittihad-i buzurg) – us MEDOC [079]

Jibhah-'i milli-i iran – [Tehran]: Asnad-i Jibhah-'i Milli-i Duvvum. 30 tir 1339-31 shahrivar 1342 [21 jul 1940-22 sep 1943] – 1r – 1 – $53.00 – us MEDOC [079]

Jibkenyan / Walpole Island Band Council – 1981 jun 19-1987 dec 18 – 1r – 1 – (cont: jikenyan) – mf#1277692 – us WHS [350]

Jicaras tristes / Espino, Alfredo – San Salvador, El Salvador. 1947? – 1r – us UF Libraries [972]

Jie fang ri bao – Yan'an: Jie fang ri bao she, n1(may 16 1941)-n2130(mar 27 1947) [mf ed 1981] – 12r – 1 – (iss for may 16 1941 called also: chuang kan hao. ind publ under title: jie fang ri bao suo yin) – located: HKUST – cc Misc Inst [951]

Jie fang ri bao = Jiefang daily – Shanghai: Jie fang ri bao she, min guo 38- may 28 1949-31 dec 1997 [daily] [mf ed 1989] – 1 – (iss for may 28, 1949 called also chuang kan hao) – located: HKUST – cc Misc Inst [951]

Jie fang ri bao suo yin *see* Jie fang ri bao

Jiefang daily *see* Jie fang ri bao

Jiefang junbao = People's liberation army daily – 1992- – 1 – (yrly reel count varies) – us UMI ProQuest [070]

Jiefang ribao *see* Jie fang ri bao

Jigs-med nam-mka *see* Geschichte des buddhismus in der mongolei

Jih chi hsuan / Liu, Chih – Shang-hai: Pei hsin shu chu, 1934 – (= ser P-k&k period) – us CRL [880]

Jih chi wen hsueh ts'ung hsuan : wen yen chuan / Juan, Wu-ming – Shang-hai: Nan ch'iang shu chu, Min kuo 22 [1933] – (= ser P-k&k period) – us CRL [480]

Jih chi wen hsueh ts'ung hsuan : yu t'i chuan – Shang-hai: Nan ch'iang shu chu, Min kuo 22 [1933] – (= ser P-k&k period) – us CRL [480]

Jih ch'u / Ts'ao, Yu – Shang-hai: Wen hua sheng huo ch'u pan she, Min kuo 25 [1936] – (= ser P-k&k period) – us CRL [820]

Jih ch'u erh tso (ccm226) = Prayers for the daily task: a collection of prayers for women / Liu, Mei-li – 1st ed. Hong Kong, 1955 [mf ed 198?] – (= ser Ccm 226) – 1 – mf#1984-b500 – us ATLA [240]

Jih hsien lang jen tsai chung-kuo ko ti fei fa hsing tung – [China]: Kuo nan she, [1937] – (= ser P-k&k period) – us CRL [951]

Jih lo : hsien tai san wen hsin chi / Hsiao, Ch'ien – Kuei-lin: Liang yu fu hsing t'u shu kung ssu, 1943 – (= ser P-k&k period) – us CRL [840]

Jih mei kuan hsi kai kuan / Kao, Tsung-wu – Shang-hai: T'ai-p'ing yang shu tien, Min kuo 22 [1933] – (= ser P-k&k period) – us CRL [327]

Jih mei wen t'i / Ho, Tzu-heng – Ch'ang-sha, Shang wu yin shu kuan, Min kuo 27 [1938] – (= ser P-k&k period) – us CRL [327]

Jih pao ch'i k'an shih / Weill, G – Shang-hai: Shang wu yin shu kuan, Min kuo 29 [1940] – (= ser P-k&k period) – us CRL [079]

Jih su kuan hsi lun / Chou, I-wu – Ch'ang-sha: Shang wu yin shu kuan, Min kuo 27 [1938] – (= ser P-k&k period) – us CRL [327]

Lo jih sung / Ts'ao, Pao-hua – Shang-hai: Hsin yueh shu tien, 1932 – (= ser P-k&k period) – us CRL [810]

Jiho = Shih pao – The Eastern Times, Shanghai, Shanghai, 1909-37 – 299 – 1 – Y2100,000 – (in chinese. lacking: 1911 nov, dec; 1921 jan-1923 jun; 1927 jan, feb. ceased publ 1939) – ja Yushodo [079]

Jihoceska pravda – Budweis, Czechoslovakia. 1956-65 – 9r – 1 – us L of C Photodup [077]

Jih-pen chan shih mao i cheng ts'e / Fu, Ts'an-yen – Ch'ang-sha: Shang wu yin shu kuan, Min kuo 27 [1938] – (= ser P-k&k period) – us CRL [380]

Jih-pen chan shih ts'ai cheng ching chi ti wei chi / Su, Hsiang-yu – Kuei-lin: Wen hua kung ying she, Min kuo 29 [1940] – (= ser P-k&k period) – us CRL [332]

Jih-pen cheng chih ti mo lu / Ou-yang, Fan – Ch'ung-ch'ing: Kuo min t'u shu ch'u pan she, Min kuo 33 [1944] – (= ser P-k&k period) – us CRL [951]

Jih-pen cheng chih yen chiu / Wang, Chi-yuan – Shang-hai: Sheng huo shu tien, Min kuo 26 [1937] – (= ser P-k&k period) – us CRL [951]

Jih-pen cheng fu / Chin, Ch'ang-yu – Shang-hai: Shang wu yin shu kuan, Min kuo 26 [1937] – (= ser P-k&k period) – us CRL [951]

Jih-pen chi lieh kuo chih lu chun – [China: Lu chun ts'an mou pen pu ti erh t'ing, 1940] – (= ser P-k&k period) – us CRL [951]

Jih-pen ch'in lueh man meng chih chi chi cheng ts'e – [China]: Chung-hua tzu chiu she, Min kuo 22 [1931] – (= ser P-k&k period) – us CRL [951]

Jih-pen ch'in lueh ti hsin chieh tuan yu chung-kuo tou cheng ti hsin shih ch'i / Wang, Ming – [China]: Ch'ing nien shu pao she, 1937 – (= ser P-k&k period) – us CRL [951]

Jih-pen ching chi kai k'uang / Chao, Lan-p'ing – Shang-hai: Li ming shu chu, 1931 – (= ser P-k&k period) – us CRL [330]

Jih-pen chu i ti mo lo / Hsieh, Nan-kuang – Ch'ung-ch'ing: Kuo min t'u shu ch'u pan she, Min kuo 33 [1944] – (= ser P-k&k period) – us CRL [951]

Jih-pen chuan mai yen chiu yu wo kuo chuan mai wen t'i / Wu, Meng-tso – [Ch'ung-ch'ing]: Cheng chung shu chu, Min kuo 32 [1943] – (= ser P-k&k period) – us CRL [350]

Jih-pen fu nu yun tung k'ao ch'a chi lueh / Ch'en, Wei – Shang-hai: Shang wu yin shu kuan, Min kuo 17 [1928] – (= ser P-k&k period) – us CRL [305]

Jih-pen hsien wu t'ai chih yao chiao / Wu, Po-ming – Ch'ung-ch'ing: Ch'ing nien shu tien, Min kuo 29 [1940] – (= ser P-k&k period) – us CRL [951]

Jih-pen jen min tui tung-pei shih chien kung lun / Shen, Shu-chih & Wu, Chueh-nung – Shang-hai: Li ming shu chu, Min kuo 21 [1932] – (= ser P-k&k period) – us CRL [951]

Jih-pen jen ti chung-kuo kuan / Yu, Chung-yao – Han-k'ou: Hua-chung t'u shu kung ssu, Min kuo 27 [1938] – (= ser P-k&k period) – us CRL [327]

Jih-pen kuo chia chi kou lueh chieh – Shang-hai: Chung hua shu chu, Min kuo 26 [1937] – (= ser P-k&k period) – us CRL [951]

Jih-pen so ts'ang chung-kuo i pen hsiao shuo shu k'ao / T'an, Cheng-pi – Shang-hai: Chih hsing pien i she, 1945 – (= ser P-k&k period) – us CRL [830]

Jih-pen ti kuo chu i tui hua ching chi ch'in lueh / Hou, Hou-p'ei – Shang-hai: Li ming shu chu, 1931 – (= ser P-k&k period) – us CRL [332]

Jih-pen tui hua mei t'ieh tzu yuan chih ch'in lueh / Wu, Shih-han – Ch'ung-ch'ing: Chung-kuo wen hua fu wu she, Min kuo 30 [1941] – (= ser P-k&k period) – us CRL [951]

Jih-pen tui hua shang yeh / Chao, Lan-p'ing – Shang-hai: Shang wu yin shu kuan, [Min kuo 23 ie 1934] – (= ser P-k&k period) – us CRL [380]

Jih-pen t'ung chih t'ai-wan ching kuo : chung yang hsun lien t'uan t'ai-wan hsing cheng kan pu hsun lien pan chiang yen lu / Ch'en, I – [T'ai-wan]: Chung yang hsun lien t'uan T'ai-wan hsing cheng kan pu hsun lien pan, Min kuo 34 [1945] – (= ser P-k&k period) – us CRL [951]

Jim kobak's kirkus reviews – New York. 1985-1991 – 1,5,9 – (cont: kirkus reviews; cont by: kirkus reviews) – mf#13371,01 – us UMI ProQuest [073]

Jim thorpe times-news – Jim Thorpe, PA. -w 1954-1967 – 13 – $25.00r – us IMR [071]

Jimboliaer zeitung – Hatzfeld (Jimbolia RO), 1938 3 apr-17 jul – 1r – 1 – gw Misc Inst [077]

Jimenez Andrades, Ildefonso *see* Recuerdos de mi campana en rusia

Jimenez, Antonio *see* Erudicion evangelica...santa oracion

Jimenez Arias, Diego *see*
- Lexicon ecclesiasticum...concilis
- Lexicon...adiciones de juan de lama escudero
- Lexicon...divorum vitis
- Sacris bibliis
- Tablas reformadas segun el calendario gregoriano

Jimenez Borja, Arturo *see* Coreografia colonial

Jimenez Canossa, Salvador *see*
- Cuentos de trapiche
- Del viento y de las nubes
- Tierra del cielo

Jimenez de Arechaga, Eduardo *see* Introduccion al derecho, version taquigrafica de carlos argenta estable

Jimenez De Barbosa, Frances *see* Verbo iluminado

Jimenez de la Espada, Marcos *see*
- Correspondencia del doctor benito arias montano con el licenciado juan de ovando
- No fue tea, fue barreno

Jimenez de la Llave, Luis *see* Lapida romana inedita del villar del pedroso

Jimenez De Quesada, Gonzalo *see* Antijovio

Jimenez de samaniego, jose, general de la orden y obispo de plasencia, en la inmaculada en la literatura franciscano-espanola / Uribe, Angel – Archivo Ibero Americano, 1955 – 1 – sp Bibl Santa Ana [240]

Jimenez de Savariego, J *see* Tratado de peste, donde se contienen las causas, preservacion y cura

Jimenez de Zalamea, Fr. Juan *see* Sermon...la limpieza de la virgen en su inmaculada

Jimenez G, Carlos Ma *see* Historia de la aviacion en costa rica

Jimenez Guillen, F *see*
- Animadversiones...acerca de la receta del unguento de mercurio
- Respuesta a los pareceres...acerca del mal...en sevilla

Jimenez, Juan *see* Vida y...d. juan de ribera... obispo de badajoz

Jimenez, Juan Ramon *see* Poesia cubana en 1936

Jimenez Lugo, Angel *see* Apuntes y pinchazos

Jimenez Malaret, Rene *see*
- Camino de sombras
- Epistolario historico
- Pandemonium
- Puntos de vista

Jimenez, Manuel De Jesus *see* Paginas escogidas

Jimenez, Max *see*
- Jaul
- Poesia

Jimenez, Miguel Angel *see* Merengue

Jimenez Navarro, E et al *see* Arqueologia de magacela

Jimenez Navarro, Ernesto *see* La coleccion de lapidas de d. claudio constanzo

Jimenez Oreamuno, Ricardo *see* Seleccion de articulos originales del procer...

Jimenez Priego, Teresa *see* Guadalupe en los siglos 17; 18

Jimenez Quilez, Manuel *see* Wheels within wheels

Jimenez, Ramon Emilio *see*
- Del lenguaje dominicano
- Oracion panegirica en memoria del academico feneci
- Patria en la cancion
- Savia dominicana

Jimenez Rodriguez, Fernando *see* Merida, roma de occidente. guion literario para un documental cinematografico

Jimenez Rodriguez, Manuel Antonio *see* Cuatro articulos y un prologo

Jimenez Rueda, Julio *see*
- Herejias y superticiones en la nueva espana. los heterodoxos en mejico
- Historia de la cultura en mexico

Jimenez Salas, Maria *see* Vida y obras de don juan pablo forner y segarra, madrid, 1944

Jimenez, Salvador *see* Elementos de derecho civil y penal de costa rica

Jimenez Samaniego, Jose *see*
- Constitutiones et acta generalium
- Prologo galeato.relacion...de agreda
- Statutorum...familia...observantiae...sancti francisci
- Synodo del obispado de plasencia
- Vida de juan duns scoto
- Vida...juan dunsio escoto

Jimenez, Sebastian *see*
- Concordantiae iuris canonici cum legibus partitarum...
- Concordiantiae...glossematibusque gregorii lopez

Jimenez Tobon, Gerardo *see* Gobernantes de caldas

Jimenez, Tomas Fidias *see* Nueva geografia de el salvador

Jimenez-Quiros, Otto *see* Arbol criollo

Jimeno Agius, Jose *see* Puerto rico

Jimenz Vasco, Felipe *see* Como nace un monasterio y muere un cesar

Jimeson, Allen Alexander *see*
- Notes on the twenty-five articles of religion, as received and taught by methodists in the united states
- The sacred literature of the lord's prayer

Jimmy Swaggart Ministries *see* Evangelist

Jin ko-niu : a brief sketch of the life of jessie m johnston, for eighteen years w m a missionary in amoy, china / Johnston, Meta L & Johnston, Lena E – London: T French Downie, 1907 [mf ed 1995] – (= ser Yale coll) – xii/203p (ill) – 1 – 0-524-09595-7 – (pref by her mother) – mf#1995-0595 – us ATLA [920]

Jinaalankaara : or, embellishments of buddha / Buddharakkhita, Mahathera; ed by Gray, James – London: Luzac, 1894 – 1mf – 9 – 0-524-07665-0 – mf#1991-0142 – us ATLA [280]

Jinacarita : or, the career of the conqueror. a pali poem / Medhamkara, Vanaratna; ed by Duroiselle, Charles – Rangoon: British Burma Press, 1906 – 1mf – 9 – 0-524-07142-X – (incl bibl ref. in english and pali) – mf#1991-0072 – us ATLA [280]

Jinaprabha Suri *see* A legend of the jaina stupa at mathura

Jinarajadasa, Curuppumullage *see*
- How we remember our past lives, and other essays on reincarnation
- In his name
- The meeting of the east and the west
- The message of the future
- Theosophy and modern thought
- What we shall teach

Jinaratnakosa: an alphabetical register of jain works and authors / Velankar, Hari Damodar – Poona: Bhandarkar Oriental Research Institute, 1944. Text and commentary in English – 1 – us Wisconsin U Libr [280]

Jinbu ribao – Tientsin, China. 12 aug-18 aug 1949; 22 mar-dec 1951; oct-dec 1952 – 4r – 1 – uk British Libr Newspaper [072]

Jinesta, Carlos *see*
- Bronces de mexico
- Mar y pensamiento
- Ruben dario en costa rica, loanza

Jingzhai *see* San he bian lan

Jinnah, Mahomed Ali *see* Some recent speeches and writings of mr jinnah

Jinruigaku zashi, 1886-1964 : journal of the anthropological society of japan – v1-72 – 24r – 1 – $840.00 set in US $40.00r outside – (in english) – mf#L9400097 – Dist. us Scholarly Res – us L of C Photodup [306]

Jirafa sagrada / Madariaga, Salvador De – Buenos Aires, Argentina. 1941 – 1r – us UF Libraries [972]

Jirak, Antonin *see* Uchebnoe dielo u slavianskikh narodov

Jirku, Anton *see*
- Die daemonen und ihre abwehr im alten testament
- Die juedische gemeinde von elephantine
- Materialien zur volksreligion israels

Jisabu, jiheng'ele, ifika ni jinongonongo, josoneke mu limoundu ni putu, kua mon'angola jakim ria matta / Cordeiro da Matta, J D – Lisboa: Typo. e Stereotypia Moderna, 1891 – (filmed with his ensaio de diccionario kimbundu-portuguez) – us CRL [470]

Jiskaur! / Botschko, R E – Montreux?, Switzerland. 1943? – 1r – us UF Libraries [939]

Jiskra – Jihlava, Czechoslovakia. 1956-Mar 1960 – 2r – 1 – us L of C Photodup [077]

Jiu yi nian xianggang bao zhang jian bao mu lu *see* Xianggang bao zhang jian bao

Jivavilya : thnak di 5 – [Bhnam Ben]: Krasuan Qaparam 1984 [mf ed 1990] – 1r with other items – 1 – (at head of title: krasuan qaparam; in khmer) – mf#mf-10289 seam reel 121/2 [§] – us CRL [500]

Jivit it nay : a novel / Suddh Pulin – Paris: Institut de l'Asia du Sud-Est 1981 [mf ed 1990] – 1r with other items – 1 – (also available in khmer [mf-10289 seam reel 105/8]) – mf#mf-10289 seam reel 100/10 [§] – us CRL [830]

Jivit it sanghym : pralom lok knun manosancetana / Ravivans Govid – Bhnam Ben: Pannagar Indradevi 1962 [mf ed 1989] – 1r with other items – 1 – (in khmer) – mf#mf-10289 seam reel 104/1 [§] – us CRL [480]

Jivit kamloh : kamran kaby manosancetana / Nay Brahm – Bhnam Ben: Pannagar Sen Nuan Huat 2509 [1966] [mf ed 1990] – 1r with other items – 1 – (in khmer) – mf#mf-10289 seam reel 111/6 [§] – us CRL [810]

Jiyumurfulujiyat al-huwwat fi al-jabal al-akhdar / Awdah, Samih Ahmad – [Kuwait]: Qism al-Jughrafiya, Jamiat al-Kuwayt: al-Jamiyah al-Jughrafiyah al-Kuwaytiyah, 1984 – us CRL [956]

J-j olier, 1608-1657 : cure de saint-sulpice et fondateur des seminaires: essai d'histoire religieuse sur le 17e siecle / Fruges, G-M, de – Paris: chez l'auteur, [1904] (mf ed 1990) – 5mf – 9 – mf#SEM105P1209 – cn Bibl Nat [241]

J-J rousseau : le protestantisme et la revolution francaise / Dide, Auguste – Paris: Ernest Flammarion, [1911?] – 1mf – 9 – 0-7905-4506-3 – mf#1988-0506 – us ATLA [100]

J-j rousseau en anggleterre au 18e siecle / Roddier, Henri – Paris, France. 1950 – 1r – us UF Libraries [440]

Jjit Khyai *see*
– Khmoc bray qasurakay
– Memay pti pram

Jl : jornal de letras, artes e ideias – Lisbao. v2-. 1982- [biwkly] – 1 – us Wisconsin U Libr [073]

Jlius [sic] pamphilius und di ambrosia / Arnim, Bettina von – Berlin: im Propylaeen-Verlag, c1920 [mf ed 1993] – (= ser Saemtliche werke v5) – 563p/pl – 1 – mf#8196 reel 2 – us Wisconsin U Libr [890]

Jlmc highlights / Overseas Federation of Teachers – 1979 dec-1981 apr – 1r – 1 – (cont: jlmc summary record; cont by: jlmc summary record [1983]) – mf#635402 – us WHS [370]

Jmj reglamento para el colegio de senoritas... sagrada familia... plasencia / Colegio de la Inmaculada Concepcion – 1871 – 9 – sp Bibl Santa Ana [241]

Jmpt : journal of manipulative and physiological therapeutics – v1-19. 1978-96 – 1,5,6,9 – $80.00r – us Lippincott [615]

Jmr – journal of marketing research – Chicago. 1964+ [1]; 1969+ [5]; 1975+ [9] – ISSN: 0022-2437 – mf#1920 – us UMI ProQuest [650]

Jne – journal of nursing education – Thorofare. 1962-1983 (1) 1972-1983 (5) 1975-1983 (9) – (cont by: journal of nursing education) – mf#6469 – us UMI ProQuest [610]

Jo, Boen Ek *see*
– Ampir ke noraka
– Boekan impian, boekan lamoenan
– Etty dan erry

Jo Gyi *see*
– Beda lam kabya mya
– Rasa cape aphvan nhan nidan
– Rhe khet pu gam kabya mya nhan a khra kabya mya

Jo Gyi et al *see* Kam ko mruin ca tan

Jo shui / Huang, Ch'an-hua – Shang-hai: Nan-ching shu tien, Min kuo 21 [1932] – (= ser P-k&k period) – us CRL [480]

Jo thvan on nhan mui sita / Ei Lwin, Sitkaing – Ran kun: Mui Jo Ca pe 1971 [mf ed 1993] – on pt of 1r – 1 – mf#11052 r639 n2 – us Cornell [959]

J.O. Wright & Co *see* Price list of the barlow library

Joachim, Abbot of Fiore *see* Vaticinia

Joachim, Harold Henry *see*
– The nature of truth
– A study of the ethics of spinoza

Joachim heinrich campe. ein lebensbild aus dem zeitalter der aufklaerung / Leyser, J – Braunschweig 1877 – 1 – gw Mikropress [920]

Joachim murat, roi des deux siciles : sa sentence, sa mort, drame historique et a sensation en un acte / Doin, Ernest – Montreal: Payette & Bourgeault, 1880 – 1mf – 9 – mf#04890 – cn Canadiana [820]

Joachim murat roi des deux-siciles : sa sentence, sa mort: drame historique et a sensation en un acte / Doin, Ernest – Montreal: C O Beauchemin & fils...[1879] (mf ed 1985) – 1mf – 9 – mf#SEM105P465 – cn Bibl Nat [830]

Joachim raucheis satyrische gedichte / ed by Drescher, Karl – Halle: M Niemeyer, 1903 – (incl bibl ref) – us Wisconsin U Libr [810]

Joachim vadian / Pressel, T – Elberfeld, R L Friderichs, 1861 – 2mf – 9 – mf#PBU-464 – ne IDC [240]

Joachim vadian : der reformator und geschichtschreiber von st gallen / Goetzinger, Ernst – Halle: Verein fuer Reformationsgeschichte 1895 [mf ed 1990] – (= ser Schriften des vereins fuer reformationsgeschichte 13/50) – 1mf – 9 – 0-7905-4793-7 – (incl bibl ref) – mf#1988-0793 – us ATLA [920]

Joachim von watt : deutsche historische schriften / Vadian, J; ed by Gaetzinger, E – St Gallen, Zollikofer'sche Buchdruckerei, 1875-1879. 3 v – 19mf – 9 – mf#PBU-406 – ne IDC [240]

Joachimi vadiani vita / Kessler, J – St Gallen, Zollikofer, 1865 – 1mf – 9 – mf#PBU-465 – ne IDC [240]

Joachimsohn, Paul *see*
– Hermann schedels briefwechsel, 1452-1478
– Hermann schedels briefwechsel, 1452-78

Joad, Cyril Edwin Mitchinson *see* The story of indian civilization

Joalland, Jules *see* Le drame de dankori

Joannes, de Janduno *see* Quaestiones in libros de coelo et mundo aristotelis stagiritae

Joannis Antiocheni Malalae *see*
– Chronographia
– Historia chronica

Joannis calvini commentarii in isaiam prophetam... / Calvin, J – Genevae: Apud Jo. Crispinum, 1559 – 11mf – 9 – mf#CL-61 – ne IDC [242]

Joannis calvini in librum josue brevis commentarius, quem paulo ante mortem absolvit : addita sunt quaedam de eiusdem morbo et obitu / Calvin, J – Genevae: Ex officina Francisci Perrini, 1564 – 4mf – 9 – mf#CL-60 – ne IDC [242]

Joannis calvini praelectiones : in librum prophetiarum jeremiae, et lamentationes / Calvin, J – Genevae: Apud Jo Crispinum, 1563 – 16mf – 9 – mf#CL-65 – ne IDC [242]

Joannis calvini praelectiones in duodecim prophetas (quos vocant) minores : ad serenissimum suetiae et gothiae regem / Calvin, J – Genevae: Apud Joannem Crispinum, 1559 – 15mf – 9 – mf#CL-67 – ne IDC [242]

Joannis calvini praelectiones in librum prophetiarum danielis, joannis budaei et caroli jonvillaei labore et industria exceptae / Calvin, J – Genevae: Excudebat Joannes Laonius, 1561 – 7mf – 9 – mf#CL-66 – ne IDC [242]

Joannis calvini responsio ad balduini convicia : ad leges de transfugis desertoribus et emansoribus, francisci balduini epistolae quaedam ad joannem calvinum pro commentariis... / Calvin, J – [Geneva: Jean Crespin], 1562 – 2mf – 9 – mf#CL-39 – ne IDC [242]

Joannis calvini, sacrarum literarum in ecclesia genevensi professoris, epistolae duae, de rebus hoc saeculo cognitu apprime necessariis : prior, de fugiendis impiorum illicitis sacris, et puritate christianae religionis observanda... / Calvin, J – Basileae: Per Balthasarem Lasium et Thomam Platterum, 1537 – 1mf – 9 – mf#CL-17 – ne IDC [242]

Joannis Cinnami *see*
– Epitome rerum ab ioanne et alexio comnenis gestarum
– Imperatorii grammatici historiarum libri seu de rebus gestis a joanne et mannuele gommensis impp

Joannis kepleri astronomi opera omnia / ed by Frisch, Christian – Frankofurti a M, Erlangae: Heyder & Zimmer 1858-71 [mf ed 1998] – 8v on 1r [ill] – 1 – (incl bibl ref & ind) – mf#film mas c3887 – us Harvard [520]

Joannis Zonarae *see* Annales (cbh22)

Joannou, Petros-Perikles *see* Die erfahrung in platons ideenlehre

Joao fernandes vieira / Mello, Jose Antonio Gonsalves De – Recife, Brazil. v1-2. 1956 – 1r – us UF Libraries [972]

Joao fernandes vieira / Mello, Jose Antonio Gonsalves De – Recife, Brazil. v1-2. 1967 – 1r – us UF Libraries [972]

Joao ramalho a nove de julho / Salgado, Cesar – Sao Paulo, Brazil. 1934 – 1r – us UF Libraries [972]

Joao semmedo / Nasser, David – Rio de Janeiro, Brazil. 1965 – 1r – us UF Libraries [972]

Job : introduction, revised version, with notes and index / ed by Peake, Arthur Samuel – Edinburgh: T C & E C Jack, 1905 – (= ser The Century Bible) – 1mf – 9 – 0-7905-3072-4 – mf#1987-3072 – us ATLA [221]

Job, a world example / Huffman, Jasper Abraham – New Carlisle, OH: Bethel, c1914 [mf ed 1989] – 1mf – 9 – 0-7905-2106-7 – (int by bud robinson) – mf#1987-2106 – us ATLA [221]

Job and his comforters : studies in the theology of the book of job / Marshall, J T – London: James Clark: Kingsgate Press, [1905?] – 1mf – 9 – 0-7905-1232-7 – (incl ind) – mf#1987-1232 – us ATLA [221]

Job and prelude for trombones / Moore, Jeff – 1982 – 1 – 5.00 – us Southern Baptist [242]

Job and solomon : or, the wisdom of the old testament / Cheyne, Thomas Kelly – New York: Thomas Whittaker, 1887 – 1mf – 9 – 0-8370-9370-8 – (includes bibliographies and index) – mf#1986-3370 – us ATLA [221]

Job and the problem of suffering / Royds, Thomas Fletcher – London: Wells Gardner, Darton, 1911 – 1mf – 9 – 0-7905-0272-0 – (incl bibl ref and index) – mf#1987-0272 – us ATLA [221]

Job corps happenings – v17-18 [1978-1982 jun] – 1r – 1 – (cont: corpsman [washington, dc]) – mf#1938644 – us WHS [071]

Job et l'egypte : le redempteur et la vie future dans les civilisations primitives / Ancessi, Victor – Paris: Ernest Leroux, 1877 – 1mf – 9 – 0-7905-0241-0 – (incl bibl ref) – mf#1987-0241 – us ATLA [240]

Job family series. no a[rmy] – United States – 1r – 1 – us WHS [331]

Job family series. no [industrial] 1- – United States. 1942 oct-1943 jan, 1943 mar-sep, 1943 nov-1944 oct – 3r – 1 – us WHS [331]

Job family series. no o[ccupational]-1-89 – United States – 1r – 1 – us WHS [331]

The job master's price-book – London: printed & publ by W H Tickle, by Cowie & Strange, [1813?] – (= ser 19th c publishing...) – 1mf – 9 – mf#3.1.26 – uk Chadwyck [680]

Job openings – United States. v5 n12-v7 n5 [1981 dec-1983 may] – 1r – 1 – (cont: occupations in demand at job service offices) – mf#2032453 – us WHS [331]

Job parsons diaries – [mf ed University of West Virginia, Morgantown] – 2r – 1 – (1r contains: 1874, 1884, 1886, 1887, 1888, 1893 & 1894; 1r contains: 1875, 1879, 1880 thru 1883) – us UMI ProQuest [880]

Job, proverbs, ecclesiastes, and solomon's song / Burr, Jonathan Kelsey et al; ed by Whedon, Daniel Denison – New York: Hunt & Eaton; Cincinnati: Cranston & Stowe c1881 [mf ed 1990] – (= ser Commentary on the old testament 6) – 2mf [ill] – 9 – 0-8370-1903-6 – mf#1987-6290 – us ATLA [221]

Job safety and health – 1983 fall-1994 fall – 1r – 1 – (cont: r i safety and health news; cont by: ricosh quarterly) – mf#1060554 – us WHS [360]

Job safety and health – Washington. 1972-1978 (1) 1972-1978 (5) 1975-1978 (9) – ISSN: 0090-4589 – mf#7360 – us UMI ProQuest [360]

Job safety and health *see* Occupational safety and health administration reports

Job satisfaction among secondary level teachers / Lambeth, Kelly K & Myers, Betty – 1991 – 1mf – 9 – $4.00 – us Kinesology [150]

Job scott, an eighteenth century friend / Wilbur, Henry Watson – Philadelphia: Friends' General Conference Advancement Committee, 1911 – 1mf – 9 – 0-524-02848-6 – mf#1990-4469 – us ATLA [240]

Job Service Wisconsin *see* Lmi review

Jobber – United States. v10 n13-v11 n18 [1983 jun 20-1984 sep 10], v13 n18-v15 n15 [1986 jul 28-1988 aug 1] – 1r – 1 – (cont by: champion [neilingen, germany]) – mf#1060556 – us WHS [071]

Jobbins, John Richard *see* An analysis of ancient domestic architecture

Jobim, Anisio *see*
– Amazonas
– Aspectos socio-geograficos do amazonas

Jobim, Jose *see* Brazil in the making

Job–problem analysis of the ten major truck crops in florida / Munoz, Vedasto Zabala – s.l, s.l? 1928 – 1r – us UF Libraries [630]

Jobs and communities action newsletter : the newsletter of the working group on economic dislocation / Working Group on Economic Dislocation [Saint Paul MN] – n1-2,3-5,6-7 [1986-1987 feb, sum-winter, 1988 fall-1989 spr], 1989 fall – 1r – 1 – (cont by: working notes) – mf#1800086 – us WHS [331]

Job's conversion : or, god the justifier – London, England. 18– – 1r – us UF Libraries [240]

Jobson, R *see* The golden trade

Jobus...eccleciastes / Beza, Theodor de – Genevae, 1589 – 3mf – 9 – mf#PFA-11 – ne IDC [240]

Joc week – New York, NY. 2000-2000 (1) – mf#60763 – us UMI ProQuest [071]

Jocelin of Furness *see* Lives of s ninian and s kentigern

Jocelyn Brook, E *see* Uyterste wille van een moeder aan haar toekomende kind

[Jocelyn Brook, E] *see* Uiterste wille van een moeder aan haar toekomende kind

Jocelyn, Marcelin *see*
– Haiti
– Reponse a m pouget, ancien ministre des finances

Jochems erste und letzte liebe : humoristicher roman / Huggenberger, Alfred – Leipzig: L Staackmann Verlag 1922 [mf ed 1995] – 1r [ill] – 1 – (ill by hans witzig. filmed with: daniel pfund) – mf#3884p – us Wisconsin U Libr [830]

Jockers, Ernst *see* Soziale polaritaet in goethes klassik

Le jockey – Paris. 1893, janv-juin 1895, 1911, janv-aout 1914 – 1 – (Sport. Sportsman, vie sportive, turf et jockey reunis) – fr ACRPP [790]

Jodl, Alfred *see* War diaries and correspondence of general alfred jodl, 1937-1945

Jodl, Friedrich *see*
– Geschichte der ethik als philosophischer wissenschaft
– Ludwig feuerbach

Jodlowski, Stanislaw *see* Slowniczek ortograficzny z zasadami pisowni

Joe louis scrapbooks, 1935-1944 : full length articles, sketches, cartoons, photographs, records,and statistics – [mf ed Chadwyck-Healey] – 304mf – 9 – (with p/g) – uk Chadwyck [790]

Joe weider's men's fitness – Woodland Hills. 1992-1996 (1) – ISSN: 0893-4460 – mf#20391,02 – us UMI ProQuest [613]

Joe weider's muscle and fitness – Woodland Hills. 1992+ (1,5,9) – ISSN: 0744-5105 – mf#19211 – us UMI ProQuest [613]

Joe weider's shape – Woodland Hills. 1992-2000 (1,5,9) – ISSN: 0744-5121 – mf#19258 – us UMI ProQuest [613]

Joecher, Christian Gottlieb *see* Allgemeines gelehrten-lexicon

Joel, H F *see* Honderd jaar java bode, 1852-1952

Joel, Manuel *see* Blicke in die religionsgeschichte zu anfang des

Joelson, Ferdinand Stephen *see*
– Eastern africa to-day
– Tanganyika territory (formerly german east africa)

Joenkoeping nu – Joenkoeping, Sweden. 2005- – 1 – sw Kungliga [078]

Joenkoepings laens tidning – Joenkoeping, Sweden. 1910-35 – 62r – 1 – sw Kungliga [078]

Joenkoepings laens tidning varannandagsupplagan – Joenkoeping, Sweden. 1925-26 – 1r – 1 – sw Kungliga [078]

Joenkoepings laens veckoblad – Joenkoeping, Sweden. 1911-13 – 2r – 1 – sw Kungliga [078]

Joerg / Beumelburg, Werner – Stuttgart: Deutsche Volksbuecher, 1943 [mf ed 1989] – (= ser Wiesbadener volksbuecher 290) – 54p – 1 – (excerpt fr novel mount royal) – mf#7017 – us Wisconsin U Libr [830]

Joerg wickrams romantechnik / Fauth, Gertrud – Strassburg, 1914 (mf ed 1994) – 1mf – 9 – €24.00 – 3-8267-3102-6 – mf#DHS-AR 3102 – gw Frankfurter [430]

Joergen, Juan *see* Don bosco

Joergensen, Adolf Ditlev *see* Den nordiske kirkes

Joergensen, Alfred Theodor *see* Soeren kierkegaard und das biblische christentum

Joerger, M J *see* Waldveilchen

Joern uhl : roman / Frenssen, Gustav – Berlin: G Grote 1903 [mf ed 1989] – (= ser Grote'sche sammlung von werken zeitgenoessischer schriftsteller 73) – 1r [ill] – 1 – (filmed with: holyland) – mf#7265 – us Wisconsin U Libr [830]

Joernaal van dirk gysbert van reenen 1803 / Reenen, Dirk Gysbert Van – Kaapstad, South Africa. 1937 – 1r – us UF Libraries [960]

Joernale van die landtogte van die edele vaandrig olof bergh [1682 en 1683] en die vaandrig isaq schrijver [1689] / Mossop, E E – Kaapstad: Die Van Riebeeck vereniging 1931 – (= ser [Travel descriptions from south africa, 1711-1938]) – 4mf – 9 – mf#zah-68 – ne IDC [916]

Joffe, J G *see* The rivonia trial

Joffre, A *see* Le mandat de la france sur la syrie...

Jofroi / Grenier, Jean-Pierre – Grenoble, Switzerland. 1943, c1942 – 1r – us UF Libraries [440]

Jog, Narayan Gopal *see*
– Judge or judas?
– Onions and opinions

Joganich, T G *see* Influence of in-shoe orthotics on lower extremity function in cycling

Jogendra Singh *see* Kamla

Jogging in a laminar flow resistance pool *see* Energy cost of walking/jogging in a laminar flow resistance pool

Jogjakarta, Indonesia *see* Universitas gadjah mada balai pembinaan bulletin

Jogjakarta, Indonesia (City) *see*
– Institut agama islam negeri "sunan kalidjaga" dewan mahasiswa progres report dewan mahasiswa iain "sunan kalidjaga", 1385-1387, 1965-1967
– Institut agama islam negeri "sunan kalidjaga" laporan tahunan

Jogjakarta, indonesia (city) / Anggaran keuangan daerah istimewa Jogjakarta – Jogjakarta, 1957. v1+suppl – 10mf – 9 – mf#SE-1717 – ne IDC [915]

Jogjakarta, indonesia (city) / Dinas Pertanian dan Perikanan Laporan tahunan – Jogjakarta, 1968. v1-8 – 11mf – 9 – mf#SE-1720 – ne IDC [915]

Jogjakarta, indonesia (city) / Djawatan Penerangan Daerah istimewa Jogjakarta – Jogjakarta, 1958-1962 – 16mf – 9 – mf#SE-1721 – ne IDC [915]

Jogjakarta, indonesia (city) / Djawatan Penerangan Siaran Kotamadya Jogjakarta – Jogjakarta, 1968 – 1mf – 9 – (missing: [19?]-1968 v1-15(1-2)) – mf#SE-1722 – ne IDC [915]

Joglar Cacho, Manuel *see*
- Canto a los angeles
- Faena intima
- Soliloquios de lazaro

Jogn nursing – Hagerstown. 1972-1984 (1) 1972-1984 (5) 1975-1984 (9) – (cont by: journal of obstetric, gynecologic, and neonatal nursing: jognn) – ISSN: 0090-0311 – mf#8778 – us UMI ProQuest [610]

Jogues, Isaac, Saint *see* Novum belgium

Jogyi *see* Mran ma nuin nam rajavat tara civan thum pon khyup, [1956-68]

Joh caspar lavater 1741-1801 : ein lebensbild / Voemel, Alexander – Elberfeld: Buchhandlung des Erziehungs-Vereins, [1923?] – 1 – (incl bibl ref) – us Wisconsin U Libr [920]

Joh Chr, Freiherr von Aretin et al *see* Aurora

Joh chr gottscheds sterbender cato *see* Rudolf von gottschall

Joh. chr. gottscheds sterbender cato / Gottsched, Johann Christoph; ed by Lachmann, Otto F – Leipzig: P Reclam [1885] – 1r – 1 – us Wisconsin U Libr [430]

Joh christophori wagenseilii de sacri rom imperii libera civitate noribergensi commentatio... / Wagenseil, Johann Christoph – Altdorfi Noricorum: typis impensisque J W Kohlesius 1697 [mf ed 19–] – 10mf – 9 – mf#fiche 960 – us Sibley [450]

Joh david koehlers p p kurtze und gruendliche anleitung zu der alten und mittlern geographie / Koehler, Johann – Nuernberg 1745 -65 [mf ed Hildesheim 1995-98] – 3v on 5mf – 9 – €100.00 – 3-487-29963-1 – gw Olms [910]

Joh georg schoch's comoedia vom studentenleben / ed by Fabricius, Wilhelm – Muenchen: Seitz & Schauer, 1892 [mf ed 1993] – (= ser Auswahl litterarischer denkmaeler des deutschen studententhums) – x/122p/1pl (ill) – 1 – mf#8456 – us Wisconsin U Libr [820]

Joh. gottfr. herder zwischen riga und bueckeburg : die aesthetik und sprachphilosophie der fruehzeit nach ihren existenziellen motiven / Kuentzel, Gerhard – Frankfurt a.M.: M Diesterweg, 1936 – 1 – (incl bibl ref) – us Wisconsin U Libr [100]

Johaentgen, Franz *see* Ueber das gesetzbuch des manu

Johan maurits van nassau en de korte bloeitijd / Molengraaff, Cornelia (Gerlings) – 'S-Gravenhage, Netherlands. 1928? – 1r – us UF Libraries [972]

Johann, A E *see* Im strom

Johann adam moehler der symboliker / Friedrich, Johann – Muenchen, 1894 – 2mf – 8 – €5.00 – ne Slangenburg [241]

Johann adam moehler, der symboliker : ein beitrag zu seinem leben und seiner lehre / Friedrich, Johann – Muenchen: C H Beck, 1894 [mf ed 1990] – 1mf – 9 – 0-7905-5876-9 – (incl bibl ref) – mf#1988-1876 – us ATLA [240]

Johann agricola von eisleben : ein beitrag zur reformationsgeschichte / Kawerau, Gustav – Berlin: W Hertz, 1881 [mf ed 1990] – 1mf – 9 – 0-7905-4885-2 – (in german and latin. incl bibl ref and ind) – mf#1988-0885 – us ATLA [242]

Johann albrecht 1. : herzog von mecklenburg / Schreiber, Heinrich – Halle: Verein fuer Reformationsgeschichte 1899 [mf ed 1990] – (= ser Schriften des vereins fuer reformationsgeschichte 16/64) – 1mf – 9 – 0-7905-5379-1 – (incl bibl ref) – mf#1988-1379 – us ATLA [943]

Johann amos comenius als theolog : ein beitrag zur comeniusliteratur / Criegern, Hermann Ferdinand von – Leipzig: C F Winter, 1881 [mf ed 1990] – 1mf – 9 – 0-7905-5459-3 – mf#1988-1459 – us ATLA [240]

Johann arndt, der verfasser des "wahren christentums" : ein christliches lebensbild / Winter, Friedrich Julius – Leipzig: Verein fuer Reformationsgeschichte, 1911 – (= ser [Schriften des Vereins fuer Reformationsgeschichte]) – 1mf – 9 – 0-7905-4719-8 – (incl bibl ref) – mf#1988-0719 – us ATLA [242]

Johann arndts vier buecher vom wahren christenthum : ...aufs neue ausgefertiget von joachim langen / Arndt, Johann – Halle: In Verlegung des Waysenhauses, 1734 – 17mf – 9 – mf#O-867 – ne IDC [090]

Johann balthasar schupp : beitraege zu seiner wuerdigung / Luehmann, Johann – Marburg a.L.: N G Elwert, 1907 – 1r – 1 – (incl bibl ref) – us Wisconsin U Libr [430]

Johann balthasar schupp, corinna / ed by Vogt, Karl – Halle: M Niemeyer, 1911 – us Wisconsin U Libr [430]

Johann balthasar schupp streitschriften / ed by Vogt, Carl – Halle: M Niemeyer. 2v. 1910-11 – (incl bibl ref) – us Wisconsin U Libr [430]

Johann barrow's, esq vormaligen privatsekretaers des grafen von macartney, jetzigen sekretaers der admiralitaet reise durch china von peking nach canton : im gefolge der grossbrittannischen gesandtschaft in den jahren 1793 – Weimar 1804 [mf ed Hildesheim 1995-98] – 2v on 6mf – 9 – €120.00 – 3-487-26587-7 – gw Olms [915]

Johann barrow's reisen durch die inneren gegenden des suedlichen africa – Weimar 1801-05 [mf ed Hildesheim 1995-98] – 2v on 6mf – 9 – €120.00 – 3-487-26604-0 – gw Olms [916]

Johann beers kurtzweilige sommer-taege / ed by Schmitt, Wolfgang – Halle: M Niemeyer, 1958 – (incl bibl ref and index) – us Wisconsin U Libr [430]

Johann brenz / Hartmann, Julius & Jaeger, Karl – Hamburg: F. Perthes, 1840-1842 – 3mf – 9 – 0-7905-4297-8 – mf#1988-0297 – us ATLA [920]

Johann calvin / Baur, August – Tuebingen: J C B Mohr 1909 [mf ed 1994] – (= ser Religionsgeschichtliche volksbuecher fuer die deutsche christliche gegenwart 4/9) – 1mf – 9 – 0-8370-9122-5 – mf#1986-3122 – us ATLA [242]

Johann calvin : rede bei der calvin-feier der universitaet giessen / Eck, Samuel – Tuebingen: JCB Mohr, 1909 – 1mf – 9 – 0-7905-7624-4 – mf#1989-0849 – us ATLA [242]

Johann calvin : seine kirche und sein staat in genf / Kampschulte, Franz Wilhelm; ed by Goetz, Walther – Leipzig: Duncker & Humblot, 1869-1899 – 3mf – 9 – 0-7905-4882-8 – (incl bibl ref) – mf#1988-0882 – us ATLA [242]

Johann calvin : seine kirche und sein staat in genf / Kampschulte, FW – Leipzig: Duncker & Humblot, 1869-99 – 3mf – us ATLA [242]

Johann calvin / Sodeur, Gottlieb – Leipzig: BG Teubner, 1909 – (= ser Aus natur und geisteswelt) – 1mf – 9 – 0-7905-9637-7 – (incl bibl ref) – mf#1989-1362 – us ATLA [242]

Johann calvins religioese entwicklung und sittliche grundrichtung : festrede / Sieffert, Friedrich – Leipzig: R Haupt, 1909 – 1mf – 9 – 0-7905-7667-8 – (incl bibl ref) – mf#1989-0892 – us ATLA [242]

Johann carl weyands reisen durch europa, asien und afrika von dem jahre 1818 bis 1821 incl / Weyand, Johann – Amberg 1822-1825 [mf ed Hildesheim 1995-98] – 3v on 6mf – 9 – €120.00 – ISBN-10: 3-487-26675-X – ISBN-13: 978-3-487-26675-6 – gw Olms [910]

Johann caspar lavater / Janentzky, Christian – Frauenfeld: Huber, 1928 – 1r – 1 – us Wisconsin U Libr [920]

Johann caspar lavater, 1741-1801 : denkschrift zur hundertsten wiederkehr seines todestages / ed by Stiftung von Schnyder von Wartensee – Zuerich: A Mueller 1902 [mf ed 1990] – 1r [ill] – 1 – (incl bibl ref. filmed with: katenlud / fritz lau) – mf#2818p – us Wisconsin U Libr [140]

Johann christian guenthers saemtliche werke : historisch-kritische gesamtausgabe / Guenther, Johann Christian; ed by Kraemer, Wilhelm – Leipzig: K W Hiersemann. 6v. 1930-36 – us Wisconsin U Libr [430]

Johann christian krueger als lustspieldichter / Wittekindt, Wilhelm – Marburg, 1898 (mf ed 1995) – 1mf – 9 – €24.00 – 3-8267-3138-7 – mf#DHS-AR 3138 – gw Frankfurter [430]

Johann daniel schoepflins brieflicher verkehr : mit goennern, freunden und schuelern / ed by Fester, Richard – Stuttgart: Litterarischer Verein, 1906 (Tuebingen: H Laupp, Jr) [mf ed 1993] – (= ser Blvs 240) – xxvii/425p – 1 – (german and french text. int and notes in german. incl bibl ref and ind) – mf#8470 reel 49 – us Wisconsin U Libr [860]

Johann daniel schoepflins brieflicher verkehr mit goennern, freunden und schuelern / Schoepflin, Johann Daniel; ed by Fester, Richard – Stuttgart: Litterarischer Verein, 1906 (Tuebingen: H Laupp, Jr) – (incl bibl ref and ind. german and french text. int and notes in german) – us Wisconsin U Libr [860]

Johann eck als junger gelehrter : eine literar- und dogmengeschichtliche untersuchung ueber seinen chrysopassus praedestinationis aus dem jahre 1514 / Greving, Joseph – Muenster i W: Aschendorff, 1906 – (= ser Reformationsgeschichtliche Studien und Texte) – 1mf – 9 – 0-524-00639-3 – (incl bibl ref) – mf#1990-0139 – us ATLA [240]

Johann ecks predigttaetigkeit an u.l. frau zu ingolstadt, 1525-1542 / Brandt, August – Muenster i W: Aschendorff, 1914 – (= ser Reformationsgeschichtliche Studien und Texte) – 1mf – 9 – 0-524-04949-1 – (incl bibl ref) – mf#1990-1352 – us ATLA [240]

Johann elias schlegels aesthetische und dramaturgische schriften / ed by Antoniewicz, Johann von – Heilbronn: Henninger, 1887 [mf ed 1993] – (= ser Deutsche litteraturdenkmale des 18. und 19. jahrhunderts 26) – clxxx/226p – 1 – (incl bibl ref) – mf#8676 reel 3 – us Wisconsin U Libr [430]

Johann, Ernst *see*
- Georg buechner in selbstzeugnissen und bilddokumenten

Johann faust : ein allegorisches drama von fuenf aufzuegen: zum erstenmahl aufgefuehrt auf der koenigl. prager schaubuehne von der von brunianischen gesellschaft, 1775 / Lessing, G C; ed by Thurn, R Payer von – Wien: Rosenbaum, 1911 – 1 – us Wisconsin U Libr [430]

Johann fischarts geschichtklitterung (gargantua) / ed by Alsleben, A – Halle: Max Niemeyer, 1891 – 11r – 1 – (includes reproduction of t.p. of 1590 edition) – us Wisconsin U Libr [430]

Johann friedrich august tischbein : leben und werk / Franke, Martin – (mf ed 1993) – 11mf – 9 – €87.50 – 3-89349-698-X – (mit umfassendem werkverzeichnis) – mf#DHS 698 – gw Frankfurter [750]

Johann friedrich freiherr cotta von cottendorf (1764-1832) : ein beitrag zur berufsgeschichte der verleger / Muench, Roger – (mf ed 1993) – 9mf – 9 – €74.00 – 3-89349-700-5 – mf#DHS 700 – gw Frankfurter [070]

Johann friedrich reichardt's vertraute briefe aus paris geschrieben in den jahren 1802 und 1803 – Hamburg 1804 [mf ed Hildesheim 1995-98] – 3v on 9mf – 9 – €180.00 – 3-487-29645-4 – gw Olms [860]

Johann gabriel seidl / Fuchs, Karl – Wien: Carl Fromme, 1904 – 1r – 1 – (incl bibl ref) – us Wisconsin U Libr [920]

Johann gabriel seidl, seine sagen und geschichten.. / Seidl, Johann Gabriel – Graz: P Cieslar 1881 [mf ed Bloomington IN: Indiana Uni Lib, Preservation Dept 1984] – 1r – 1 – us Indiana Preservation [390]

Johann georg albrechtsbergers... gruendliche anweisung zur composition : mit deutlichen und ausfuehrlichen exempeln, zum selbstunterrichte, erlaeutert / Albrechtsberger, Johann Georg – Leipzig: J G I Breitkopf 1790 [mf ed 19–] – 10mf – 9 – mf#fiche324 – us Sibley [780]

Johann georg august galletti's allgemeine weltkunde : oder encyclopaedie fuer geographie, statistik und staatengeschichte; mittelst einer geographisch-statistisch-historischen uebersicht aller laender, hinsichtlich ihrer lage, groesse, bevoelkerung, kultur, ihrer vorzueglichen staedte, ihrer verfassung... – Pest [u a] 1854 [mf ed Hildesheim 1995-98] – 6mf – 9 – €120.00 – 3-487-29989-5 – gw Olms [059]

Johann georg hamann als kritiker der deutschen literatur / Hilpert, Walter – Koenigsberg i.Pr.: P Escher, 1933 – 1r – 1 – (incl bibl ref) – us Wisconsin U Libr [430]

Johann georg hamann, der magus im norden : sein leben und mittheilungen aus seinen schriften / Poel, Gustav – Hamburg: Agentur des Rauhen Hauses, 1874-1876 – 12mf – 9 – 0-524-08783-0 – mf#1993-1091 – us ATLA [240]

Johann georg hamann in seiner bedeutung fuer die sturm- und drangperiode / Minor, Jacob – Frankfurt a/M.: Ruetten & Loening, 1881 – 1r – 1 – us Wisconsin U Libr [430]

Johann georg jacobis iris / Manthey-Zorn, Otto – [S.l.: s.n.], 1905 (Zwickau: Druck von J Herrmann) – 1 – (incl bibl ref) – us Wisconsin U Libr [430]

Johann georg zimmermann u johann gottfried herder : nach bisher ungedruckten briefen / Bonin, Daniel – Worms: [s.n.], 1910 [mf ed 1991] – 32p – 1 – mf#7472 – us Wisconsin U Libr [860]

Johann gerhard oncken, his life and work / Cooke, John H – London. 1800-79 – 1 – $6.54 – us Southern Baptist [242]

Johann Ghro *see* Dreissig neue ausserlesene padovane und galliard

Johann gottfried herder : sein leben in selbstzeugnissen, briefen und berichten / ed by Reisiger, Hans – Berlin: Im Propylaeen-Verlag, c1942 – 1 – us Wisconsin U Libr [430]

Johann gottfried herder : der weg, das werk, die zeit / Baete, Ludwig – Stuttgart: S Hirzel, 1948 [mf ed 1995] – vii/174p/1pl – 1 – (incl bibl ref and ind) – mf#8773 – us Wisconsin U Libr [430]

Johann gottfried schadow : das bluecher-denkmal in rostock unter besonderer beruecksichtigung des verwendeten materials bronze / Schmidt, Martin – (mf ed 1992) – 2mf – 9 – €49.00 – 3-89349-496-0 – mf#DHS 496 – gw Frankfurter [730]

Johann gottfried schadows auseinandersetzung mit johann wolfgang v. goethe – bezogen auf die jahre 1800 bis 1823 / Schmidt, Martin H – 1994 – 2mf – 3-8267-2055-5 – gw Frankfurter [700]

Johann gottfried seume als mensch, dichter, patriot und denker / Kohut, Adolph – Berlin: Gotthold Auerbach, [1910?] – 1r – 1 – us Wisconsin U Libr [920]

Johann gustav hebbe's schwedischen seeoffiziers nachrichten von den azorischen inseln besonders von der insel fayal / Hebbe, Johann Gustav – Weimar 1805 [mf ed Hildesheim 1995-98] – 1v on 1mf – 9 – €40.00 – ISBN-10: 3-487-26561-3 – ISBN-13: 978-3-487-26561-2 – gw Olms [914]

Johann hartliebs uebersetzung des dialogus miraculorum von caesarius von heisterbach / ed by Drescher, Karl – Berlin: Weidmann, 1929 [mf ed 1993] – (= ser Deutsche texte des mittelalters 33) – xxiii/474p/2pl – 1 – (middle high german trans fr latin. int in german. foreword by konrad burdach. incl bibl ref and ind) – mf#8623 reel 7 – us Wisconsin U Libr [430]

Johann heermann (1585-1647) : ein beitrag zur geschichte der geistlichen lyrik im siebzehnten jahrhundert / Hitzeroth, Carl – Marburg a.L.: N G Elwert, 1907 – 1r – 1 – (incl bibl ref) – us Wisconsin U Libr [430]

Johann heinrich merck : seine umgebung und zeit / Zimmermann, Georg – Frankfurt/M: J D Sauerlaender, 1871 [mf ed 1993] – viii/587p – 1 – mf#8642 – us Wisconsin U Libr [920]

Johann heinrich voss / Herbst, Wilhelm – Leipzig: B G Teubner, 1872-1876 – 1r – 1 – (incl bibl ref and index) – us Wisconsin U Libr [430]

Johann hermann schein (1586-1630) : collected works / ed by Pruefer, Arthur – Leipzig. 7v. 1901-23 – 11 – $105.00 set – us Univ Music [780]

Johann huebners curieuses natur- kunst-gewerck- und handlungs-lexicon (ael1/26) – 1712-1792 [mf ed 1995] – (= ser Archiv der europaeischen lexikographie, abt 1: enzyklopaedien) – 140mf – 9 – €1280 set €120 per ed – 3-89131-199-0 – (gesamtedition aller deutschsprachigen ausgaben: 1. 1712; 2.1714; 3.1717; 4.1722; 5.1727; 6.1731; 7.1736; 8.1739; 9.1741; 10.1746; 11.1755; 12. 1762; 13.1776; 14.1792; vols available individually) – gw Fischer [030]

Johann hus : ein lebensbild / Friedrich, Johann – Frankfurt am Main: Verlag fuer Kunst und Wissenschaft, 1864 – 1mf – 9 – 0-7905-6744-X – mf#1988-2744 – us ATLA [240]

Johann jaenicke : der evangelisch-lutherische prediger an der boehmischen- oder bethlehems-kirche zu berlin, nach seinem leben und wirken / Ledderhose, Karl Friedrich; ed by Knak, G – Berlin: Im Selbstverlage des Herausgebers: In Commission bei F. Beck, 1863 – 1mf – 9 – 0-7905-5250-7 – mf#1988-1250 – us ATLA [242]

Johann jakob bodmer : denkschrift zum cc. geburtstag (19. juli 1898) / ed by Stiftung von Schnyder von Wartensee – Zuerich: A Mueller, 1900 [mf ed 1989] – xii/418p/pl (ill) – 1 – (incl bibl) – mf#7040 – us Wisconsin U Libr [140]

Johann Jakob Bodmer und die geschichte der literatur / Wehrli, Max – Frauenfeld/Leipzig: Huber, 1936 [mf ed 1989] – (= ser Wege zur dichtung 27) – 163p – 1 – mf#7061 – us Wisconsin U Libr [430]

Johann jakob bodmer und die geschichte der literatur *see* Narciss

Johann jakob von willemer : der mensch und buerger / Mueller, Adolf – Frankfurt am Main: Englert und Schlosser, 1925 – 1r – 1 – (incl bibl ref) – us Wisconsin U Libr [943]

Johann joachim quantzens...versuch einer anweisung die floetetraversiere zu spielen : mit verfchiedenen zur befoerderung des guten geschmackes in der praktischen musik dienlichen anmerkungen / Quantz, Johann Joachim – Berlin: J F Voss 1752 [mf ed 19–] – 9mf – 9 – mf#fiche 651, 835 – us Sibley [780]

Johann joachim winckelmann : ausgewaehlte briefe / ed by Uhde-Bernays, Hermann – Leipzig: Insel-Verlag, 1925 – 1 – us Wisconsin U Libr [860]

Johann karl passavant : ein christliches charakterbild / Helfferich, Adolf – Frankfurt, a. M.: C. Winter, 1867. Chicago: Dep of Photodup, U of Chicago Lib, 1971 (1r); Evanston: American Theol Libr Assoc, 1984 (1r) – 1 – 0-8370-0464-0 – (incl bibl ref) – mf#1984-B222 – us ATLA [240]

Johann karl wezel : sein leben und seine schriften / Kreymborg, Gustav – [Vechta, Germany]: Vechtaer Druckerei und Verlag, 1913 – 1r – 1 – (incl bibl ref) – us Wisconsin U Libr [430]

Johann kaspar friedrich manso : der schlesische schulmann, dichter und historiker / Lux, Konrad – Leipzig: Quelle & Meyer, 1908 [mf ed 1992] – (= ser Breslauer beitraege zur literaturgeschichte. neue folge 4) – 244p – 1 – (incl bibl ref) – mf#8014 reel 2 – us Wisconsin U Libr [920]

Johann kaspar lavater : eine skizze seines lebens und wirkens / Muncker, Franz – Stuttgart: J G Cotta, 1883 – 1r – 1 – (incl ind) – us Wisconsin U Libr [920]

Johann kepler und die bibel : ein beitrag zur geschichte der schriftautoritaet / Deissmann, Gustav Adolf – Marburg: N G Elwert; Tuebingen: J C B Mohr [distributor], 1894 – 1mf – 9 – 0-7905-1870-8 – (incl bibl ref) – mf#1987-1870 – us ATLA [220]

Johann klaj : ein beitrag zur deutschen literaturgeschichte des 17. jahrhunderts / Franz, Albin – Marburg a.L.: N G Elwert, 1908 – 1r – 1 – (incl bibl ref) – us Wisconsin U Libr [430]

Johann knipstro : der erste generalsuperintendent von pommern-wolgast: sein leben und wirken / Bahlow, Ferdinand – Halle: Verein fuer Reformationsgeschichte 1898 [mf ed 1990] – (= ser Schriften des vereins fuer reformationsgeschichte 16/62) – 1mf – 9 – 0-7905-5260-4 – (incl bibl ref) – mf#1988-1260 – us ATLA [242]

Johann konrad dippel : der freigeist aus dem pietismus: ein beitrag zur entstehungsgeschichte der aufklaerung / Bender, Wilhelm – Bonn: E Weber 1882 [mf ed 1992] – 1mf – 9 – 0-524-03037-5 – (incl bibl ref) – mf#1990-0794 – us ATLA [242]

Johann lane buchanans, missionars der schottischen kirche, reisen durch die westlichen hebriden, waehrend der jahre 1782 bis 1790 – Berlin 1812 [mf ed Hildesheim 1995-98] – 1v on 2mf – 9 – €60.00 – 3-487-27883-9 – (trans fr english) – gw Olms [914]

Johann lorenz mosheim : ein beitrag zur kirchengeschichte des achtzehnten jahrhunderts / Heussi, Karl – Tuebingen: J C B Mohr, 1906 – 1mf – 9 – 0-7905-6529-3 – (incl bibl ref) – mf#1988-2529 – us ATLA [240]

Johann ludwig burckhardt's reisen in arabien : enthaltend eine beschreibung derjenigen gebiete in hedjaz, welche die mohammedaner fuer heilig achten – Weimar 1830 [mf ed Hildesheim 1995-98] – 1v on 5mf – 9 – €100.00 – 3-487-26472-2 – gw Olms [915]

Johann ludwig burckhardt's reisen in nubien – Weimar 1820 [mf ed Hildesheim 1995-98] – 1v on 5mf – 9 – €100.00 – ISBN-10: 3-487-26502-8 – ISBN-13: 978-3-487-26502-5 – gw Olms [916]

Johann ludwig burckhardt's reisen in syrien, palaestina und der gegend des berges sinai – Weimar 1823-1824 [mf ed Hildesheim 1995-98] – 2v on 8mf – 9 – €160.00 – ISBN-10: 3-487-26489-7 – ISBN-13: 978-3-487-26489-9 – (trans fr english) – gw Olms [915]

Johann matthesons grosse general-bass-schule : oder, der exemplarischen organisten-probe: bestehend in dreien klassen, als: in einer gruendlichen vorbereitung, in 24 leichten exempeln, in 24 schwerern prob-stuecken / Mattheson, Johann – 2. verb verm aufl, Hamburg: J C Kissner 1731 [mf ed 19–] – 11mf – 9 – mf#fiche 876, 438, 796 – us Sibley [780]

Johann oekolampad und oswald myconius, die reformatoren basels / Hagenbach, K R – Elberfeld, R L Friderichs, 1859 – 6mf – 9 – mf#PBU-462 – ne IDC [242]

Johann oekolampad und oswald myconius, die reformatoren basels : leben und ausgewaehlte schriften / Hagenbach, Karl Rudolf – Elberfeld: RL Friderichs, 1859 – (= ser Leben und ausgewaehlte Schriften der Vaeter und Begruender der reformirten Kirche) – 2mf – 9 – 0-7905-7109-9 – (incl bibl ref) – mf#1988-3109 – us ATLA [242]

Johann peter hebel / Altwegg, Wilhelm – Frauenfeld: Huber, c1935 [mf ed 2001] – (= ser Schweiz im deutschen geistesleben 22) – 296p/16lea/4pl (ill) – 1 – (incl bibl ref) – mf#10590 – us Wisconsin U Libr [430]

Johann peter hebel / Heuss, Theodor – Tuebingen und Stuttgart: Rainer Wunderlich Verlag Hermann Leins, 1952 [mf ed 1995] – 1 – mf#8764 – us Wisconsin U Libr [430]

Johann peter hebel : leben und briefe / Strauss, Emil – Muenchen: A Langen, G Mueller, c1939 [mf ed 1995] – (= ser Kleine buecherei 225) – 74p – 1 – mf#8764 – us Wisconsin U Libr [860]

Johann peter hebel / Zentner, Wilhelm – Karlsruhe: C F Mueller, 1948 [mf ed 2001] – 255p/8lea/4pl (ill) – 1 – (incl bibl ref and ind) – mf#10590 – us Wisconsin U Libr [430]

Johann peter hebels ausgewaehlte erzaehlungen u gedichte / ed by Fritz, Otto – Karlsruhe i.B: I Langs Buchhandlung, 1907 [mf ed 1994] – 92p (ill) – 1 – (ill by hans thoma and hermann daur) – mf#8750 – us Wisconsin U Libr [800]

Johann peter uz : zum hundertsten todestage des dichters / Petzet, Erich – Ansbach: E Bruegel, 1896 [mf ed 1993] – vi/88p/1pl – 1 – mf#7772 – us Wisconsin U Libr [430]

Johann reuchlin : sein leben und seine werke / Geiger, Ludwig – Leipzig: Duncker & Humblot, 1871 – 1mf – 9 – 0-7905-4528-4 – (incl bibl ref) – mf#1988-0528 – us ATLA [920]

Johann reuchlins briefwechsel / ed by Geiger, Ludwig – Stuttgart: Litterarischer Verein, 1875 (Tuebingen: L F Fues) [mf ed 1993] – (= ser Blvs 126) – 372p – 1 – (letters in latin, with exception of a few german and 2 hebrew letters. int and notes in german) – mf#8470 reel 26 – us Wisconsin U Libr [860]

Johann reuchlins briefwechsel / ed by Geiger, Ludwig – Stuttgart: Litterarischer Verein, 1875 (Tuebingen: L F Fues) – us Wisconsin U Libr [860]

Johann rist als weltlicher lyriker / Kern, Oskar – Marburg a.L: N G Elwert 1919 [mf ed 1992] – (= ser Beitraege zur deutschen literaturwissenschaft 15) – 1r – 1 – (incl bibl ref. filmed with: beitraege zur wuerdigung von karl gutzow als lustspieldichter / peter mueller & several other titles) – mf#3098p – us Wisconsin U Libr [430]

Johann rist und das niederdeutsche drama des 17. jahrhunderts : ein beitrag zur deutschen literaturgeschichte / Heins, Otto – Marburg a.L.: Elwert, 1930 – 1r – 1 – (incl bibl ref (3rd-4th prelim. leaves)) – us Wisconsin U Libr [430]

Johann salomo semler : seine bedeutung fuer die theologie, sein streit mit gotthold ephraim lessing / Huber, Fritz – Berlin: R Trenkel, 1906 – 1mf – 9 – 0-7905-7769-0 – (incl bibl ref) – mf#1989-0994 – us ATLA [240]

Johann samuel traugott gehlers physikalisches woerterbuch : neu bearbeitet von brandes, gmelin, horner, muncke, pfaff – Leipzig 1825-45 [mf ed 1993] – 215mf – 9 – €790.00 – 3-89131-164-8 – gw Fischer [055]

Johann sebastian bach : his harmonic equipment and contributions / Redding, Edwyl – U of Rochester 1931 [mf ed 19–] – 3mf – 9 – (with bibl) – mf#fiche 79 – us Sibley [780]

Johann sebastian bach (1685-1750) : complete works / ed by Rust, Wilhelm et al – Leipzig. 47v. 1851-99; 1926 (suppl vol) – 11 – $495.00 set – us Univ Music [780]

Johann tetzel der ablassprediger / Paulus, Nikolaus – Mainz 1899 (mf ed 1995) – 1mf – 9 – €24.00 – 3-8267-3154-9 – mf#DHS-AR 3154 – gw Frankfurter [240]

Johann tobias beck : ein schriftgelehrter zum himmelreich gelehrt / Riggenbach, Bernhard – Basel: C Detloff, 1888 – 1mf – 9 – 0-7905-8566-9 – (incl bibl ref and ind) – mf#1989-1791 – us ATLA [240]

Johann ulrich von koenig : ein beitrag zur litteraturgeschichte des 18. jahrhunderts / Rosenmueller, Max Clemens – [S.l.: s.n.] 1896 (Leipzig-Reudnitz: Druck von A Hoffmann) – 1r – 1 – (incl bibl ref) – us Wisconsin U Libr [430]

Johann von schwarzenberg, das buechlein vom zutrinken / ed by Scheel, Willy – Halle: M Niemeyer, 1900 – 11r – 1 – (incl bibl ref) – us Wisconsin U Libr [430]

Johann von schwarzenberg, trostspruch und abgestorbene freunde / ed by Scheel, Willy – Halle: M Niemeyer, 1907 – 11r – 1 – (incl bibl ref) – us Wisconsin U Libr [430]

Johann von staupitz und die anfaenge der reformation / Keller, Ludwig – Leipzig: S. Hirzel, 1888 – 1mf – 9 – 0-7905-6000-3 – (incl bibl ref) – mf#1988-2000 – us ATLA [242]

Johann wessel : ein bild aus der kirchengeschichte des 15. jahrhunderts / Friedrich, Johann – Regensburg: G J Manz, 1862 – 1mf – 9 – 0-7905-4643-4 – (incl bibl ref) – mf#1988-0643 – us ATLA [920]

Johann wiclif und seine zeit : zum fuenfhundertjaehrigen wiclifjubilaeum, 31. dezember 1884 / Buddensieg, Rudolf – Halle: Verein fuer Reformationsgeschichte. 1885. ([Schriften des Vereins fuer Reformationsgeschichte; Bd. 8-9]) – 1mf – us ATLA [240]

Johann wiclif und seine zeit : zum fuenfhundertjaehrigen wiclifjubilaeum, 31. dezember 1884 / Buddensieg, Rudolf – Halle: Verein fuer Reformationsgeschichte, 1885 – (= ser [Schriften des Vereins fuer Reformationsgeschichte]) – 1mf – 9 – 0-7905-4609-4 – (incl bibl ref) – mf#1988-0609 – us ATLA [240]

Johann wilhelm moellers, doktor der arzneikunde, hofrath des verstorbenen koenigs von pohlen, und mitglied der mineralischen gesellschaft in jena und der naturforschenden gesellschaft – Hamburg 1802 [mf ed Hildesheim 1995-98] – 2mf – 9 – €60.00 – 3-487-28957-1 – gw Olms [920]

Johann wilhelm simler : die rezeption des opitz-barock in der deutschen schweiz / Schumacher, Joachim – Heidelberg, 1933 (mf ed 1994) – 1mf – 9 – €24.00 – 3-89349-789-7 – mf#DHS-AR 789 – gw Frankfurter [430]

Johann wolfgang von goethe in selbstzeugnissen und bilddokumenten / Boerner, Peter – Reinbek bei Hamburg: Rowohlt, 1964 [mf ed 1993] – (= ser Rowohlts monographien 100) – 186p/2pl (ill) – 1 – (incl bibl ref and ind) – mf#8640 – us Wisconsin U Libr [430]

The johannean problem : a resume for english readers / Gilmore, George William – Philadelphia: Presbyterian Board of Publication and Sabbath-School Work, 1895 – 1mf – 9 – 0-8370-3290-3 – (incl ind) – mf#1985-1290 – us ATLA [220]

Die johanneische christologie / Luetgert, Wilhelm – Guetersloh: C Bertelsmann, 1899 – (= ser Beitraege zur foerderung christlicher theologie) – 1mf – 9 – 0-524-07123-3 – mf#1992-1039 – us ATLA [221]

Das johanneische evangelium nach seiner eigenthuemlichkeit / Luthardt, Christoph Ernst – 2. erw mehrfach umgearb. aufl. Nuernberg: C. Geiger, 1875-1876. Chicago: Dep of Photodup, U of Chicago Lib, 1975 (1r); Evanston: American Theol Lib Assoc, 1984 (1r) – 1 – 0-8370-1277-5 – mf#1984-B439 – us ATLA [226]

Das johanneische evangelium nach seiner eigenthuemlichkeit *see* St john's gospel

Der johanneische lehrbegriff in seinem verhaeltnisse zur gesammten biblisch-christlichen lehre / Frommann, Karl – Leipzig: Breitkopf und Haertel, 1839 – 2mf – 9 – 0-7905-1044-8 – (in german and greek. incl bibl ref) – mf#1987-1044 – us ATLA [220]

Der johanneische lehrbegriff in seinen grundzuegen / Weiss, Bernhard – Berlin: Wilhelm Hertz, 1862 – 1mf – 9 – 0-8370-6451-1 – (incl bibl ref) – mf#1986-0451 – us ATLA [220]

Der johanneische ursprung des vierten evangeliums / Luthardt, Christoph Ernst – Leipzig: Doerffling und Franke, 1874 – 1mf – 9 – 0-8370-4206-2 – mf#1985-2206 – us ATLA [225]

Johannes Apostolus *see* Nene karighwiyoston

Johannes a lasco / Bartels, Petrus – Elberfeld: R. L. Friderichs, 1860 – (= ser [Leben und ausgewaehlte schriften der vaeter und begruender der reformirten kirche]) – 1mf – 9 – 0-7905-4247-1 – (incl bibl ref) – mf#1988-0247 – us ATLA [242]

Johannes a lasco und der sacramentsstreit : ein beitrag zur geschichte der reformationszeit / Kruske – Leipzig: Dieterich, 1901 – (= ser Studien zur geschichte der theologie und der kirche) – 1mf – 9 – 0-7905-6482-3 – (incl bibl ref) – mf#1988-2482 – us ATLA [242]

Johannes a s thoma *see*
- Cursus philosophicus thomisticus, secundum exactam, veram et genuinam aristotelis...
- Cursus theologici in primam secundam partem d thomae

Johannes' aabenbaring : indledet og fortolket / Madsen, Peder – 2. gjennemsete udg. Koebenhavn: GEC Gad, 1896 – 2mf – 9 – 0-524-05046-5 – (incl bibl ref) – mf#1992-0299 – us ATLA [220]

Johannes, Adolf *see* Commentar zu der weissagung des propheten obadja

Johannes Agricola aus Eisleben *see*
- Confession vnd bekentnis johanns agricole eisslebens vom gesetze gottes
- Hundert vnd dreissig gemeyner fragestuecke fuer die iungen kinder ynn der deudschen meydlin schule zu eyssleben

Johannes blankenfeld : ein lebensbild aus dem anfaengen der reformation / Schnoering, Wilhelm – Halle a.d. S: Verein fuer Reformationsgeschichte 1905 [mf ed 1990] – (= ser Schriften des vereins fuer reformationsgeschichte 23/86) – 1mf – 9 – 0-7905-5133-0 – mf#1988-1133 – us ATLA [241]

Johannes brahms / Appelbaum, Theodore – U of Rochester 1931 [mf ed 19–] – 2mf – 9 – mf#fiche74 – us Sibley [780]

Johannes brahms (1833-1897) : complete works / ed by Gal, Hans & Mandyczewski, Eusebius – Leipzig: Breitkopf & Haertel. 26v. 1926-28 – 11 – $310.00 set – us Univ Music [780]

Johannes brahms and the french horn / Seiffert, Stephen Lyons – U of Rochester 1968 [mf ed 19–] – 5mf – 9 – (with bibl) – mf#fiche 225, 309 – us Sibley [780]

Johannes brenz : leben und ausgewaehlte schriften / Hartmann, Julius – Elberfeld: R L Friderichs, 1862 – (= ser Leben und ausgewaehlte Schriften der Vaeter und Begruender der lutherischen Kirche) – 1mf – 9 – 0-7905-4807-0 – mf#1988-0807 – us ATLA [240]

Johannes brenz und die reformation in herzogtum wirtemberg [sic] / Hegler, Alfred – Freiburg i B: JCB Mohr, 1899 – 1mf – 9 – 0-7905-4812-7 – mf#1988-0812 – us ATLA [943]

Johannes brinckerinck en zijn klooster te diepenveen / Kuehler, W J – Rotterdam, 1908 – €17.00 – ne Slangenburg [240]

Johannes buenderlin von linz und die oberoesterreichischen taeufergemeinden in den jahren 1525-31 / Nicoladoni, Alexander – Berlin: R Gaertner, 1893 – 1mf – 9 – 0-8370-8927-1 – (incl ind) – mf#1986-2927 – us ATLA [920]

Johannes bugenhagen, pomeranus : leben und ausgewaehlte schriften / Vogt, Karl August Traugott – Elberfeld: RL Friederichs, 1867 – (= ser Leben und ausgewaehlte Schriften der Vaeter und Begruender der lutherischen Kirche) – 2mf – 9 – 0-524-01245-8 – (incl bibl ref) – mf#1990-0384 – us ATLA [240]

Johannes bugenhagens braunschweiger kirchenordnung, 1528 = Braunschweiger kirchenordnung, 1528 / Bugenhagen, Johann; ed by Lietzmann, Hans – Bonn: A Marcus und E Weber, 1912 – (= ser Kleine texte fuer vorlesungen und uebungen) – 1mf – 9 – 0-524-06944-1 – mf#1990-5308 – us ATLA [240]

Johannes calvijn : eene lezing ter gelegenheid van den vierhonderdsten gedenkdag zijner geboorte, 10 juli 1509-1909 / Bavinck, Herman – Kampen: Kok, 1909 – 1mf – 9 – 0-7905-7615-5 – mf#1989-0840 – us ATLA [242]

Johannes calvin : akademischer vortrag / Wernle, Paul – Tuebingen: JCB Mohr, 1909 – 1mf – 9 – 0-7905-7673-2 – mf#1989-0898 – us ATLA [242]

Johannes calvin : festrede bei calvins vierhundertjaehriger geburtstagsfeier / Dalton, Hermann – Berlin: Martin Warneck, 1909 – 1mf – 9 – 0-7905-7622-8 – mf#1989-0847 – us ATLA [242]

Johannes calvin : leben und ausgewaehlte schriften / Staehelin, Ernst – Elberfeld: RL Friderichs, 1863 – (= ser Leben und ausgewaehlte Schriften der Vaeter und Begruender der reformirten Kirche) – 3mf – 9 – 0-524-03431-1 – mf#1990-0985 – us ATLA [242]

Johannes calvin : ein lebensbild zu seinem 400. geburtstag am 10. juli 1909 / Lang, August – Leipzig: Verein fuer Reformationsgeschichte 1909 [mf ed 1990] – (= ser Schriften des vereins fuer reformationsgeschichte 26/99) – 1mf – 9 – 0-7905-4704-X – (incl bibl ref) – mf#1988-0704 – us ATLA [242]

Johannes calvin : rede zur feier der 400. wiederkehr des geburtstages calvins. gehalten in der aula der koeniglichen friedrich-wilhelms-universitaet zu berlin... / Holl, Karl – erw und mit Anmerkungen versehene Ausg. Tuebingen: JCB Mohr, 1909 – 1mf – 9 – 0-7905-5844-0 – (incl bibl ref) – mf#1988-1844 – us ATLA [242]

Johannes calvin und seine bedeutung fuer unsere heutige kultur / Brepohl, Friedrich Wilhelm – Seegefeld: "Das Havelland", 1909 – 1mf – 9 – 0-7905-7617-1 – (incl bibl ref) – mf#1989-0842 – us ATLA [242]

Johannes calvins leben und seine stellung innerhalb der gesamtkirche / Auer, Wilhelm – [S.l.]: W Auer, [1909?] (Ansbach: C Bruegel) – 1mf – 9 – 0-7905-7679-1 – (incl bibl ref) – mf#1989-0904 – us ATLA [242]

Johannes czerski / Czerski, Johannes – Liverpool, England. 1846 – 1r – us UF Libraries [240]

Johannes de Oxenedes *see* Chronica (rs13)

Johannes der taeufer / Procksch, Otto – Berlin: Edwin Runge 1907 [mf ed 1989] – (= ser Biblische zeit- und streitfragen 3/5) – 1mf – 9 – 0-7905-0588-6 – mf#1987-0588 – us ATLA [225]

Johannes gerson, professor der theologie und kanzler der universitaet paris : eine monographie / Schwab, Johann Baptist – Wuerzburg: Stahel, 1858 – 1mf – 9 – 0-524-03660-8 – (incl bibl ref) – mf#1990-1088 – us ATLA [920]

Johannes gossner : ein lebensbild aus der kirche des neunzehnten jahrhunderts / Dalton, Hermann – 3. verm. Aufl. Friedenau-Berlin: Buchhandlung des Gossnerschen Mission, 1898 – 2mf – 9 – 0-7905-5646-4 – mf#1988-1646 – us ATLA [240]

The johannes herbst collection (c. 1752-1812) : the complete collection of manuscripts as found in the archives of the moravian music foundation, winston-salem, north carolina – New York, 1976 – 5,11 – $590.00 mf ed and film ed – (in 4 sections. pt a: congregation music herbst n1-493. pt b: 45 mss of large-scale vocal instrumental works. pt c: misc. scores in 4v. addenda: the book of texts (to pt. a), original mss and typescript transcription. coll contains works of 58 composers, mostly moravian and some baroque and classical masters) – us Univ Music [780]

Johannes huber / Zirngiebl, Eberhard – Gotha: FA Perthes, 1881 – 1mf – 9 – 0-524-00238-X – mf#1989-2938 – us ATLA [920]

Johannes hus : ein lebensbild aus der vorgeschichte der reformation / Lechler, Gotthard Victor – Halle: Verein fuer Reformationsgeschichte, 1889 – (= ser [Schriften des Vereins fuer Reformationsgeschichte]) – 1mf – 9 – 0-7905-4657-4 – (incl bibl ref) – mf#1988-0657 – us ATLA [242]

Johannes Isaaci, Hollandus *see* Opus vegetabile

Johannes kesslers sabbata : st. galler reformationschronik 1523-1539 / Kessler, Johannes – Leipzig: Verein fuer Reformationsgeschichte, 1911. ([Schriften des Vereins fuer Reformationsgeschichte; Bd. 103-104]) – 1mf – (die evangelischen kantone und die waldenser in den jahren 1663 und 1664 / von gerold meyer von knonau) – us ATLA [240]

Johannes kesslers sabbata: st. galler reformationschronik 1523-1539 – die evangelischen kantone und die waldenser in den jahren 1663 und 1664 = Sabbata. selections – Leipzig: Verein fuer Reformationsgeschichte, 1911 – (= ser [Schriften des Vereins fuer Reformationsgeschichte]) – 1mf – 9 – 0-7905-4700-7 – (incl bibl ref) – mf#1988-0700 – us ATLA [949]

Johannes knades selbsterkenntnis : historische erzaehlung aus der zeit der reformation / Quandt, Clara – 3. Aufl. Braunschweig: Grueneberg (Wollermann & Neumeyer) 1889 – 1r – 1 – us Wisconsin U Libr [830]

Johannes maccovius / Kuyper, Abraham – Leiden: D Donner, 1899 – 5mf – 9 – 0-524-07896-3 – (incl bibl ref) – mf#1991-3441 – us ATLA [920]

Johannes mathesius : ein lebens- und sitten-bild aus der reformationszeit / Loesche, Georg – Gotha: F A Perthes, 1895 – 3mf – 9 – 0-7905-5176-4 – (incl bibl ref) – mf#1988-1176 – us ATLA [242]

Johannes Parisiensis *see* De utraque potestate papali et regali

Johannes R Becher *see* Leben und werk

Johannes r becher / ed by Kollektiv fuer Literaturgeschichte im Volkseigenen Verlag Volk und Wissen – Berlin: Volk und Wissen Volkseigener Verlag, 1960 – 1r – 1 – (incl bibl ref) – us Wisconsin U Libr [430]

Johannes ruysbroec : een bijdrage tot de kennis van de ontwikkeling der mystiek / Otterloo, A A V – Amsterdam, 1874 – 7mf – 8 – ne Slangenburg [241]

Das johannes schlaf-buch / ed by Baete, Ludwig et al – Rudolstadt [Thueringen]: Greifenverlag, 1922 [mf ed 1995] – 105p/1pl – 1 – (incl bibl ref) – mf#9268 – us Wisconsin U Libr [430]

Johannes scotus erigena : ein beitrag zur geschichte der philosophie und theologie im mittelalter / Huber, Johannes – Muenchen: J J Lentner, 1861 – 2mf – 9 – 0-7905-7004-1 – (incl bibl ref) – mf#1988-3004 – us ATLA [180]

Johannes scotus erigena und dessen gewaehrsmaenner in seinem werke de divisione naturae libri 5 / Draeseke, Johannes – Leipzig: Dieterich, 1902 – (= ser Studien zur Geschichte der Theologie und der Kirche) – 1mf – 9 – 0-7905-3781-8 – (incl bibl ref) – mf#1989-0274 – us ATLA [240]

Johannes scotus erigena und die wissenschaft seiner zeit / Staudenmaier, A, Fr – Frankfurt a M, 1834 – €18.00 – ne Slangenburg [180]

Johannes scotus erigena und die wissenschraft seiner zeit : mit allegemeinen entwicklungen der hauptwahrheiten auf dem gebiete der philosophie und religion, und grundzuegen zur einer geschichte der speculativen theologie / Staudenmaier, Franz Anton – Frankfurt am Main: Andreaei, 1834 – 5mf – 9 – 0-524-00147-2 – mf#1989-2847 – us ATLA [180]

The johannes steel report on world affairs / ed by Steel, John [d.i. Johannes Stahl] – New York, NY (USA) 1947-1949 jul [gaps] [mf ed 2005] – 1 – (feb ? 1948: report on world affairs; special v4 apr 1947 fr moscow & v22 apr 1947 fr belgrad filmed) – gw Mikrofilm [321]

Johannes tauler und die gottesfreunde / Baehring, Bernhard – Hamburg: Agentur des Rauhen Hauses, 1853 – (= ser Lebensbilder aus der geschichte der inneren mission) – 1mf – 9 – 0-7905-6581-1 – (incl bibl ref) – mf#1988-2581 – us ATLA [240]

Johannes tauler von strassburg : beitrag zur geschichte der mystik und des religioesen lebens im 14. jahrhundert / Schmidt, Charles – Hamburg: F Perthes, 1841 [mf ed 1991] – 1mf – 9 – 0-7905-9628-8 – (incl bibl ref) – mf#1989-1353 – us ATLA [931]

Johannes v. hofmann : ein beitrag zur geschichte der theologischen grundprobleme, der kirchlichen und der politischen bewegungen im 19. jahrhundert / Wapler, Paul – Leipzig: A Deichert, 1914 – 1mf – 9 – 0-7905-3111-9 – (incl bibl ref) – mf#1987-3111 – us ATLA [240]

Johannes volkelts erkenntnistheorie : eine darstellung und kritik / Hallesby, Ole – Erlangen: Junge, 1909 – 1mf – 9 – 0-524-08371-1 – mf#1993-3071 – us ATLA [190]

Johannes von Damascus (John of Damascus, Saint) *see* Genaue darlegung des orthodoxen glaubens (bdk44 1.reihe)

Johannes von damaskus : eine patristische monographie / Langen, Joseph – Gotha: F A Perthes, 1879 – 1mf – 9 – 0-7905-5245-0 – (incl bibl ref) – mf#1988-1245 – us ATLA [240]

Johannes von miquel. sein anteil am ausbau des deutschen reiches bis zur jahrhundertwende / Herzfeld, Hans – Band 1-2. Detmold, 1937 – 1 – gw Mikropress [943]

Johannes von mueller und die franzoesische literatur / Herzog, Peter – Frauenfeld: Huber, 1937 – 1r – 1 – (incl bibl ref) – us Wisconsin U Libr [410]

Johannes wtenbogaert en zijn tijd / Rogge, Hendrik Cornelis – Amsterdam: Y. Rogge, 1874-76. 3v – 1 – us Wisconsin U Libr [920]

Die johannes-apokalypse : textkritische untersuchungen und textherstellung / Weiss, Bernhard – Leipzig: JC Hinrichs, 1891 [mf ed 1986] – (= ser Das neue testament 1/3; Texte und untersuchungen zur geschichte der altchristlichen literatur (tugal) 7/1) – 1mf – 9 – 0-8370-9589-1 – mf#1986-3589 – us ATLA [225]

Die johannes-apokalypse / Weiss, Bernhard – Leipzig, 1891 – (= ser Tugal 1.7/1) – 4mf – 9 – €11.00 – ne Slangenburg [240]

Das johannesbuch der mandaeer / Lidzbarski, Mark – Giessen: Alfred Toepelmann, 1915 – 2mf – 9 – 0-524-05766-4 – (incl bibl ref) – mf#1991-0009 – us ATLA [220]

Johannesburg gazette – Johannesburg. South Africa. 1900-10 – 1 – 1 – sa National [079]

Johannesburg Public LibraryPretoria State Library *see*
– Strange library of africana: author-title catalogue
– Strange library of africana: subject catalogue

[Johannesburg-] race relations news – SA. 1979-84 – 2r – 1 – $100.00 – mf#R63580 – us Library Micro [079]

Johannesburg times – Johannesburg. South Africa. 1895-98 – 15r – 1 – sa National [079]

The johannesburg times – Johannesburg: Johannesburg Times, jan 12 1895-oct 27 1898 – 15r – 1 – us CRL [079]

Johannesburg. University of Witwatersrand. Dept of Commerce *see* Native urban employment

Johannesburg's coloured community / Randall, Peter – Johannesburg, South Africa. 1968 – 1r – us UF Libraries [960]

Das johannes-evangelium : nach der paraphrase des nonnus panopolitanus / Janssen, R – Leipzig, 1903 – (= ser Tugal 2-23/4) – 2mf – 9 – €5.00 – ne Slangenburg [225]

Das johannesevangelium : seine echtheit und glaubwuerdigkeit / Dausch, Petrus – 3. aufl. Muenster i W: Aschendorff 1911 [mf ed 1989] – (= ser Biblische zeitfragen 2/2) – 1mf – 9 – 0-7905-0489-8 – (incl bibl ref) – mf#1987-0489 – us ATLA [225]

Das johannesevangelium : studien zur kritik seiner erforschung / Overbeck, Franz – Tuebingen: J C B Mohr, 1911 [mf ed 1989] – 2mf – 9 – 0-7905-1673-X – (incl bibl ref & ind) – mf#1987-1673 – us ATLA [225]

Das johannesevangelium als einheitliches werk / Weiss, Bernhard – Berlin: Trowitzsch, 1912 – 1mf – 9 – 0-7905-0414-6 – mf#1987-0414 – us ATLA [226]

Das johannes-evangelium als quelle der geschichte jesu / Spitta, Friedrich – Goettingen: Vandenhoeck & Ruprecht, 1910 – 2mf – 9 – 0-7905-2138-5 – mf#1987-2138 – us ATLA [226]

Das johannes-evangelium nach der paraphrase des nonnus panopolitanus / ed by Janssen, Ralph – Leipzig: JC Hinrichs, 1903 – (= ser Tugal) – 1mf – 9 – 0-7905-1719-1 – mf#1987-1719 – us ATLA [226]

Johannes-kommentare aus der griechischen kirche / Reuss, J – Berlin, 1964 – (= ser Tugal 5-89) – 9mf – 9 – €18.00 – ne Slangenburg [240]

Johannesschriften des neuen testaments *see* The johannine writings

Johannessen, Carl L *see* Savannas of interior honduras

Johannesson, Alexander *see* Frumnorraen malfraedi

Johannet, Rene *see* Pan-germanism versus christendom

The johannine books / ed by Benham, William – London: J M Dent; Philadelphia: J B Lippincott 1902 [mf ed 1989] – (= ser The temple bible) – 1mf – 9 – 0-7905-1803-1 – mf#1987-1803 – us ATLA [225]

The johannine epistles / Dodd, C H – Harper. 1946 – 9 – $10.00 – us IRC [240]

The johannine literature and the acts of the apostles / Forbes, Henry Prentiss – New York: G P Putnam, 1907 – 1mf – 9 – 0-8370-3161-3 – (incl ind) – mf#1985-1161 – us ATLA [226]

Johannine perspectives on inclusivity and exclusivity of salvation / Rousseau, Pieter Abraham – Uni of South Africa 2000 [mf ed Johannesburg 2000] – 3mf – 9 – (incl bibl ref; text in afrikaans; incl abstract in afrikaans & english) – mf#mfm14793 – sa Unisa [225]

Johannine problems and modern needs / Purchas, Henry Thomas – London; New York: Macmillan, 1901 – 1mf – 9 – 0-8370-4811-7 – mf#1985-2811 – us ATLA [220]

The johannine theology : a study of the doctrinal contents of the gospel and epistles of the apostle john / Stevens, George Barker – New York:Charles Scribner, 1894 – 1mf – 9 – 0-8370-5406-0 – (incl indes) – mf#1985-3406 – us ATLA [225]

Johannine thoughts : meditations in prose and verse suggested by passages in the fourth gospel / Drummond, James – London: Philip Green, 1909 – 1mf – 9 – 0-7905-3333-2 – mf#1987-3333 – us ATLA [220]

Johannine vocabulary : a comparison of the words of the fourth gospel with those of the three / Abbott, Edwin Abbott – London: Adam and Charles Black, 1905 – 1mf – 9 – 0-8370-2013-1 – (includes appendix on the use of prepositions in the gospels) – mf#1985-0013 – us ATLA [221]

The johannine writings = Johannesschriften des neuen testaments / Schmiedel, Paul Wilhelm – London: Adam & Charles Black 1908 [mf ed 1985] – (= ser Religionsgeschichtliche volksbuecher fuer die deutsche christliche gegenwart) – 1mf – 9 – 0-8370-5119-3 – (incl ind; trans fr german by maurice a canney) – mf#1985-3119 – us ATLA [225]

Johannis burchardi, argentinensis, capelle pontificie sacrorum rituum magistri diarium, sive, rerum urbanarum commentarii (1483-1506) : texte latin publie integralement pour la premiere fois = Diarium / Burchardus, Johannes; ed by Thuasne, Louis – Paris: Ernest Leroux, 1883-1885 – 6mf – 9 – 0-8370-9048-2 – (incl bibl ref and ind) – mf#1986-3048 – us ATLA [240]

Johannis codagnelli annales placentini (mgh7:23.bd) – 1901 – (= ser Monumenta germaniae historica 7: scriptores rerum germanicarum in usum scholarum (mgh7)) – €7.00 – ne Slangenburg [240]

Johannis de wiclif tractatus de officio pastorali e codice vindobonensi = De officio pastorali / Wycliffe, John; ed by Lechler, Gotthard Victor – Lipsiae: Typis A Edelmanni, 1863 – 1mf – 9 – 0-7905-7038-6 – mf#1988-3038 – us ATLA [240]

Johannis pechami quaestiones tractantes de anima / Spettmann, H – 1918 – (= ser Bgphma 19/5-6) – €12.00 – ne Slangenburg [100]

Johannis scoti erigenae de divisione naturae : libri quinque... – rev enl ed. Monasterii Guestphalorum [Muenster in Westfalen]: Typis et sumptibus Librariae Aschendorffianae, 1838 [mf ed 1991] – 2mf – 9 – 0-524-00259-2 – (pref by c b shlueter) – mf#1989-2959 – us ATLA [240]

Johannis uytenbogaerts leven, kerckelijcke bedieninghe ende zedighe verantwoordingh / Wtenbogaert, J – Ed 2. n.p, 1646 – 6mf – 9 – mf#PBA-374 – ne IDC [240]

Johannisburger zeitung – Johannisburg (Jansbork PL), 1922 10 oct-30 dec, 1926 1 jul-1927 30 sep [gaps], 1928 – 6r – 1 – gw Misc Inst [077]

Johanns, P *see* Vers le christ par le vedanta

Johanns von wuerzburg wilhelm von oesterreich / ed by Regel, Ernst – Berlin: Weidmann, 1906 [mf ed 1993] – (= ser Deutsche texte des mittelalters 3) – xxii/333p/2pl – 1 – (incl bibl ref and ind) – mf#8623 reel 1 – us Wisconsin U Libr [430]

Johannsen, Christa *see* An einen juengling im felde

Johannson, J-O *see* Social aspects of sport participation of swedish athletes with disabilities

Johanos, Donald *see* Study of the bartok concerto for orchestra

Johansen, Donald Alexander *see* Plant microtechnique

Johansen, Ernst *see* Ruanda

Johansen, Michelle K *see* Gender differences in walking with respect to movement of the pelvis

Johansson, Claes Elis *see* Die heilige schrift und die negative kritik

Johansson, Johannes *see* Profeten hosea

Johansson, Karl Ferdinand et al *see* Fraemmande religionsurkunder

John a lasco : his earlier life and labours : a contribution to the history of the reformation in poland, germany, and england / Dalton, Hermann – London: Hodder and Stoughton, 1886 – 1mf – us ATLA [242]

John a lasco : his earlier life and labours: a contribution to the history of the reformation in poland, germany, and england = Johannes a lasco / Dalton, Hermann – London: Hodder and Stoughton, 1886 – 1mf – 9 – 0-7905-4285-4 – (incl bibl ref. in english) – mf#1988-0285 – us ATLA [242]

The john a. lent collection on asian mass communications / ed by Lent, John A & Svobodny, Dolly – 1820-1984. 1500 titles. 350mf. Printed card indexes included – 9 – us ATBI [380]

John ainsworth, pioneer kenya administrator, 1864-1946 : being the hitherto unpublished memoirs of colonel john d ainsworth / Ainsworth, John Dawson; ed by Goldsmith, F H – London: Macmillan; New York: St Martin's Press, 1955 – us CRL [920]

John alexander dowie and the christian catholic apostolic church in zion / Harlan, Rolvix – Evansville, WI: Robert M Antes, [1906?] – 1mf – 9 – 0-524-07686-3 – mf#1991-3271 – us ATLA [241]

John amos comenius : bishop of the moravians: his life and educational works / Laurie, Simon Somerville – Syracuse, NY: C W Bardeen, 1893, c1892 [mf ed 1986] – (= ser Standard teachers' library 1) – 1mf – 9 – 0-8370-7561-0 – (incl ind) – mf#1986-1561 – us ATLA [242]

John and qumran / ed by Charlesworth, J H – London, 1972 – 5mf – 8 – €12.00 – ne Slangenburg [226]

John and sebastian cabot : the discovery of north america / Beazley, Charles Raymond – London: Fisher Unwin, 1898 – 4mf – 9 – mf#03632 – cn Canadiana [917]

John and thomas m'avity, hardware merchants : importers and dealers in english, american and german hardware, saint john, new brunswick – [Saint John, NB?: s.n.], 1854 [mf ed 1985] – 1mf – 9 – 0-665-45408-2 – mf#45408 – cn Canadiana [680]

John angell james : a review of his history, character, eloquence, and literary labours / Campbell, John – London: John Snow, 1860 – 1mf – 9 – 0-7905-4169-6 – mf#1988-0169 – us ATLA [920]

John angell james : a review of his history, character, eloquence, and literary labours : with dissertations on the pulpit and the press, academic preaching, college reform, etc. / Campbell, John – London: John Snow, 1860 – 1mf – us ATLA [240]

John b. andrews memorial symposium on labor legislation and social security, memorial union, the university of wisconsin, nov. 4 and 5, 1949 – Proceedings. s.l.: s.n., 1949? – 1 – us Wisconsin U Libr [331]

John b gough : the apostle of cold water / Martyn, William Carlos – New York: Funk & Wagnalls, 1893 – (= ser American reformers) – 1mf – 9 – 0-7905-6241-3 – mf#1988-2241 – us ATLA [975]

John b jervis papers, 1795-1885 – [mf ed ProQuest] – 13r – 1 – us UMI ProQuest [625]

John B Strong (Firm) *see* A catalogue of religious, scientific, illustrated, juvenile, and miscellaneous books (including educational works)

John bachman : the pastor of st john's lutheran church, charleston – Charleston, SC: Walker, Evans and Cogswell, 1888 – 1mf – 9 – 0-524-00502-8 – mf#1990-0002 – us ATLA [242]

John baptist franzelin, s.j : cardinal priest of the title ss. boniface and alexius / Walsh, Nicholas – Dublin: MH Gill, 1895 – 1mf – 9 – 0-524-00661-X – mf#1990-0161 – us ATLA [240]

John barrow's esq reise nach cochinchina in den jahren 1792 und 1793 : nebst nachrichten von diesem koenigreiche und den uebrigen auf dieser reise besuchten laendern – Weimar 1808 [mf ed Hildesheim 1995-98] – 1v on 4mf – 9 – €120.00 – 3-487-26549-4 – (tran fr english) – gw Olms [915]

John Bean of Exeter Family Association *see* Macbeanregister

John bellows : letters and memoir / Bellows, John; ed by Bellows, Elizabeth – London: K Paul, Trench, Truebner, 1904 – 1mf – 9 – 0-524-06762-7 – mf#1991-2769 – us ATLA [240]

John bidwell, pioneer / Bidwell, John – Marcus Benjamin. 1907 – 1 – us Library Micro [978]

John Birch Society *see* Bulletin for...

John birch society bulletin – 1975 apr [n193]-1979 dec, 1980-83, 1984-1989 jun – 3r – 1 – (cont: bulletin [john birch society]) – mf#1006795 – us WHS [071]

John breck family papers *see* Family papers, ms 4675

John breck family papers, 1782-1993 / Breck, John – [mf ed 1995] – 2 ser on 2r – 1 – mf#ms4675 – us Western Res [333]

John brinckmans hoch- und niederdeutsche dichtungen / Rust, Wilhelm – [S.l.: s.n.], 1912 (Rostock: Rats- und Universitaets-Buchdruckerei von Adlers Erben) [mf ed 1989] – 168p – 1 – mf#7088 – us Wisconsin U Libr [810]

John brinckmans plattdeutsche werke / ed by Arbeitsgruppe der Plattdeutschen Gilde zu Rostock – Wolgast (Pommern): P Christiansen, 1924-1934 [mf ed 1989] – 7v – 1 – (incl bibl ref) – mf#7087 – us Wisconsin U Libr [802]

John brinckmans saemtliche werke in fuenf teilen / ed by Weltzien, Otto – Leipzig: Hesse & Becker [1903?] [mf ed 1989] – (= ser Deutsche klassiker-bibliothek. hesses klassiker-ausgaben in neuer ausstattung) – 5v in 1 – 1 – (int and ann by ed) – mf#7086 – us Wisconsin U Libr [802]
John Brown Anti-Klan Committee *see* No kkk – no fascist usa!
John brown letters – [mf ed Chadwyck-Healey] – 1r – 1 – (papers describe the abolitionist leader's personal feelings concerning slavery and the civil strife over that issue. correspondence incl the last letters addressed to his family prior to his execution for treason at harper's ferry, west virginia, in 1859) – uk Chadwyck [976]
John bull – London. -w. 17 Dec 1820-Dec 1833. (13 reels) – 1 – uk British Libr Newspaper [072]
John bull and co : the great colonial branches of the firm, canada, australia, new zealand and south africa / O'Rell, Max – New York: C Webster, 1894 – 4mf – 9 – mf#00236 – cn Canadiana [910]
John c calhoun / Holst, Herman E von – Boston & NY: Houghton, Mifflin & Co, 1899 – (= ser The american statesmen series) – mf – 9 – $7.50 – mf#LLMC 96-028 – us LLMC [975]
John calvin : his life, letters, and work / Reyburn, Hugh Young – London, New York: Hodder & Stoughton 1914 [mf ed 1990] – 1mf – 9 – 0-7905-6007-0 – (incl bibl ref) – mf#1988-2007 – us ATLA [242]
John calvin : the man and the doctrine / Thomson, Alexander – London: Pub for the Congregational Union of England and Wales by Jackson, Walford, and Hodder, 1864 – 1mf – 9 – 0-524-08623-0 – mf#1993-1073 – us ATLA [242]
John calvin : the organiser of reformed protestantism, 1509-1564 / Walker, Williston – New York: Putnam, 1906 – (= ser Heroes of the reformation) – 2mf – 9 – 0-7905-6269-3 – (incl bibl ref) – mf#1988-2269 – us ATLA [242]
John calvin and the genevan reformation : a sketch / Johnson, Thomas Cary – Richmond, Va: Presbyterian Committee of Publication, c1900 – 2mf – 9 – 0-524-07433-X – mf#1991-3093 – us ATLA [242]
John calvin and the twentieth century – Chicago: Boneen Co, 1909 – 1mf – 9 – 0-524-02579-7 – (incl bibl ref) – mf#1990-0631 – us ATLA [242]
John calvin, the geneva reformation, and godly warfare : church and state in the calvinian tradition / Larson, Mark James – 2005 [mf ed 2006] – 1r – 1 – 0-524-10551-0 – (incl bibl ref) – mf#d00012 – us ATLA [230]
John calvin, theologian, preacher, educator, statesman : presented to the reformed churches holding the presbyterian system / Vollmer, Philip et al – Philadelphia: Presbyterian Board of Publication, 1909 – 1mf – 9 – 0-524-07049-0 – mf#1991-2902 – us ATLA [242]
John campbell, esq, of carbrook, called to account by the rev dr... / Thomson, Andrew – Edinburgh, Scotland. 1827 – 1r – us UF Libraries [240]
John campbell of kingsland – Edinburgh, Scotland. 1883 – 1r – us UF Libraries [240]
John cassien / Chadwick, O – Cambridge, 1950 – 4mf – 8 – €11.00 – ne Slangenburg [241]
John, Charles *see* Report of the committee of bishops on the revision of the text and...
John chinaman and a few others / Parker, Edward Harper – 2nd ed. London: John Murray, 1902 [mf ed 1995] – (= ser Yale coll) – xx, 380p (ill) – 1 – 0-524-09521-3 – mf#1995-0521 – us ATLA [306]
John chinaman at home : sketches of men, manners and things in china / Hardy, Edward John – London: T Fisher Unwin [1907] [mf ed 1995] – (= ser Yale coll) – 335p (ill) – 1 – 0-524-09260-5 – mf#1995-0260 – us ATLA [306]
John Chrysostom, Saint, d. 407 *see* Leaves from st. john chrysostom
John company at work : a study of european expansion in india in the late eighteenth century / Furber, Holden – Cambridge: Harvard University Press, 1951, c1948 – (= ser Samp: indian books) – us CRL [330]
John cooper : a study of his ayres / James, Walter Stevens – U of Rochester 1942 [mf ed 1993] – 1r – 1 – mf#pres. film 126 – us Sibley [780]
John crome and his works / Wodderspoon, John – [2nd ed]. Norwich 1876 – (= ser 19th c art & architecture) – 1mf – 9 – mf#4.1.397 – uk Chadwyck [750]
John d spahr papers, 1865-1915 / Spahr, John D – [mf ed 1991] – 1r – 1 – (a daily diary of spahr's service in the 50th ohio volunteer infantry regiment in 1865 during the civil war, & financial & pension docs relating to family life) – us Western Res [976]
John day valley ranger – John Day OR: A R Jones, 1931-48 [wkly] – 1 – (1933-35 incl paper pub by john day high school students; 1936-37 by grant union high school students; cont: east oregon ranger (1930-31); absorbed by: grant county blue mountain eagle (1948-72)) – us Oregon Lib [071]
John de wycliffe, d.d : a monograph / Vaughan, Robert – London: Seeleys, 1853 – 2mf – 9 – 0-7905-7088-2 – mf#1988-3088 – us ATLA [240]
John, Deborah *see* An investigation into accuracy of using rpe to monitor intensity during spinning
John deere union reporter / United Farm Equipment and Metal Workers of America – v1 n4 [1947 jan 29] – 1r – 1 – mf#3629255 – us WHS [331]
John donne : sometime dean of st. paul's, a.d. 1621-1631 / Jessopp, Augustus – Boston: Houghton, Mifflin, 1897 – 1mf – 9 – 0-7905-5162-4 – mf#1988-1162 – us ATLA [240]
John donne : sometime dean of st. paul's, a.d. 1621-1631 / Jessopp, Augustus – Boston: Houghton, Mifflin, 1897 – 1mf – us ATLA [240]
John drayton's gouverneurs und oberkommandanten von sued-carolina, beschreibung von sued-carolina : aus dem englischen, und mit anmerkungen und zusaetzen des uebersetzers begleitet / Drayton, John – Weimar 1808 [mf ed Hildesheim 1995-98] – 1v on 3mf – 9 – €90.00 – ISBN-10: 3-487-26552-4 – ISBN-13: 978-3-487-26552-0 – gw Olms [917]
John edward bruce papers : from the holdings of the schomburg center for research in black culture, manuscripts, archives and rare books division: the new york public library, astor, lenox and tilden foundations – 1995 – ca 4r – 1 – ca $340.00 – (guide sold separately for $20.00 which covers other schomburg center collections) – Dist. us Scholarly Res – us L of C Photodup [070]
John edward bruce papers – Schomburg Center for Research in Black Culture, 1995 – 4r – 1 – $340.00 – (printed guide available for $20.00) – mf#D3305P27 – us NY Public [070]
John, Edward Mills *see* John's american notary and commissioner of deeds manual.
John edwards memorial foundation – Los Angeles. 1965-1985 (1) 1965-1985 (5) 1965-1985 (9) – ISSN: 0021-3632 – mf#7691 – us UMI ProQuest [780]
The john ehrlichman alphabetical subject file, 1969-1973 *see* Papers of the nixon white house
John ehrlichman: notes of meetings with the president *see* Papers of the nixon white house
John englishman – New York. 1755-1755 (1) – mf#3527 – us UMI ProQuest [070]
The john ericsson collection of the american swedish historical foundation / The American Swedish Foundation; ed by Meixner, Esther Chilstrom – 1839-1889 – 8r – 1 – $1040.00 – (with printed guide) – mf#S1848 – us Scholarly Res [355]
John f. kennedey university law review – v1-9. 1988-98 – 9 – $84.00 set – mf#115521 – us Hein [340]
The john f kennedy 1960 campaign – 2pt – (= ser Research colls in american politics) – 1 – (pt1: polls, issues, & strategy 10r isbn 0-89093-917-9 $1795. pt2: speeches, press conferences, & debates 12r isbn 0-89093-918-7 $2145. with p/g) – us UPA [325]
John f kennedy assassination in dallas and the subsequent coverage in the world's major newspapers – 8r – 1 – $400.00 – (incl original dallas daily news as well as national and international papers from the new york irish world to the washington post to the zuerich zeitung) – mf#R05003 – us Library Micro [320]
The john f kennedy national security files, 1961-1963 : africa – (= ser National security files) – 12r – 1 – $2330.00 – 1-55655-001-4 – (printed ind only isbn 1-55655-003-0 $405. 1st suppl 18r isbn 1-55655-905-4 $3485. printed ind onlyisbn 0-88692-598-3 inquire for price) – us UPA [327]
The john f kennedy national security files, 1961-1963 : asia – (= ser National security files) – 1 – (asia & the pacific 10r isbn 1-55655-006-5 $1935; printed ind only isbn 1-55655-005-7 $365. 1st suppl 33r isbn 1-55655-879-1 $6390; printed ind only, inquire. vietnam 7r isbn 1-55655-015-4 $1340; printed ind only isbn 1-55655-016-2 $270. 1st suppl 4r isbn 1-55655-880-5 $770; printed ind only isbn 1-55645-949-6 inquire for price) – us UPA [327]
The john f kennedy national security files, 1961-1963 : latin america – (= ser National security files) – 10r – 1 – $1935.00 – 1-55655-009-X – (printed ind only isbn 1-55655-010-3 $365. 1st suppl 21r isbn 1-55655-926-7 $4070; printed ind only isbn 0-88692-591-6 inquire for price. 1st suppl: cuba 21r isbn 1-55655-904-6 $4070; printed ind only isbn 0-88692-592-4 inquire for price) – us UPA [327]
The john f kennedy national security files, 1961-1963 : the middle east – (= ser National security files) – 3r – 1 – $570.00 – 1-55655-013-8 – (printed ind only isbn 1-55655-014-6 $155. 1st suppl 18r isbn 1-55655-925-9 $3485; printed ind only isbn 0-88692-593-2 inquire for price) – us UPA [327]
The john f kennedy national security files, 1961-1963 : ussr and eastern europe – (= ser National security files) – 3r – 1 – $570.00 – 1-55655-002-2 – (printed ind only isbn 1-55655-008-1 $155. 1st suppl 19r isbn 1-55655-876-7 $3675. printed ind only isbn 0-88692-595-9 inquire for price) – us UPA [327]
The john f kennedy national security files, 1961-1963 : western europe – (= ser National security files) – 10r – 1 – $1935.00 – 1-55655-011-1 – (printed ind only isbn 1-55655-012-X $365. 1st suppl 35r isbn 1-55655-881-3 $6775; printed ind only isbn 0-88692-590-8 inquire for price) – us UPA [327]
The john f kennedy presidential oral history collection – 2pt – (= ser Presidential documents series) – 9 – (pt1: the white house & executive depts 250mf isbn 1-55655-053-7 $1865 or 12r isbn 1-55655-077-4 $1865. pt2: the congress, the judiciary, public figures, & private individuals 325mf isbn 1-55655-054-5 $2345 or 15r isbn 1-55655-078-2 $2345. with p/g) – us UPA [977]
John ferrars limerick chronicle and general advertiser *see* Limerick chronicle
The john fitch papers – 3r – 1 – $105.00 – Dist. us Scholarly Res – us L of C Photodup [623]
John fitzgerald kennedy assassination : a microfilm documentary – 1917-63 – 1r – 1 – $50.00 – (incl articles from us and foreign newspapers and magazines, plus kennedy's biography) – mf#B40016 – us Library Micro [977]
John Fitzgerald Kennedy Library *see* Records of the kennedy administration, 1961-1963
John fletcher hurst : a biography / Osborn, Albert – New York: Eaton & Mains, 1905 – 2mf – 9 – 0-7905-6607-9 – (incl bibl ref) – mf#1988-2607 – us ATLA [920]
John fletcher hurst : a biography / Osborn, Albert Sherman – New York: Eaton & Mains, 1905 – 2mf – us ATLA [240]
John flockhart, esq / Cowan, Robert – Perth, Australia. 1878 – 1r – us UF Libraries [240]
The john foster dulles oral history collection : from the collections of the princeton university libraries – 1994 – 13r – 1 – $1,105.00 – (with printed guide) – mf#D3301 – us L of C Photodup [327]
John foster, (the "essayist,") vindicated from the aspersions of mr... / Anglicanus, Clemens – London, England. 1864 – 1r – us UF Libraries [240]
John g paton / Allen, James T – London, England. 18– – 1r – us UF Libraries [240]
John g. paton : later years and farewell: a sequel to john g. paton – an autobiography / Langridge, Albert Kent & Paton, Frank Hyme Lyall – New York: Hodder and Stoughton, [1910?] – 1mf – 9 – 0-8370-6203-9 – (incl ind) – mf#1986-0203 – us ATLA [920]
John g. paton : missionary to the new hebrides: an autobiography / Paton, John Gibson – new ed. New York: Fleming H Revell, c1898 – 3mf – 9 – 0-8370-6406-6 – mf#1986-0406 – us ATLA [920]
John gill pratt papers / Pratt, John Gill – 1834-99. In Kansas State Historical Society. Guide – 1 – us Kansas [920]
John (gospel, letters, revelation) – Scranton, Penna.: Good News Pub Co, 1902 – 1mf – 9 – 0-524-06916-6 – mf#1992-1009 – us ATLA [225]
John, Griffith *see*
- China
- Griffith john
- Sowing and reaping
- Voice from china

John h. clifford, esq., attorney-general, &c / Clifford, John Henry – Boston: Elder, 1854. 141p. LL-219 – 1 – us L of C Photodup [340]
John h newman : the concept of infallible doctrinal authority / Dibble, R A – Washington, DC, 1955 – 9mf – 8 – €18.00 – ne Slangenburg [241]
John halifax, gentleman / Craik, Dinah Maria – Toronto: Langton & Hall, 1901 – 5mf – 9 – 0-665-72025-4 – (1st publ london, glasgow: collins clear-type press, 1856) – mf#72025 – cn Canadiana [810]
John hall, pastor and preacher : a biography / Hall, Thomas Cuming – New York: F.H. Revell, c1901 – 1mf – us ATLA [240]
John hall, pastor and preacher : a biography / Hall, Thomas Cuming – New York: F.H. Revell, c1901 – 1mf – 9 – 0-7905-4802-X – mf#1988-0802 – us ATLA [240]
The john hay papers – 23r – 1 – $805.00 – Dist. us Scholarly Res – us L of C Photodup [320]
John haynes holmes : opponent of war / Smith, Kenneth Jackson – Chicago, 1949. Chicago: Dep of Photodup, U of Chicago Lib, 1971 (1r); Evanston: American Theol Lib Assoc, 1984 (1r) – 1 – 0-8370-0271-0 – mf#1984-B197 – us ATLA [240]
John hayslip papers *see* Hayslip, john, papers, ms 2944
John henry : a folk-lore study / Chappell, Louis Watson – Jena: Frommann 1933 [mf ed Bloomington IN: Indiana Uni Lib, Preservation Dept 1984] – 144p on 1r – 1 – us Indiana Preservation [390]
John henry / Richardson, Martin D – s.l, s.l? 193? – 1r – us UF Libraries [978]
John henry kardinal newman : ein beitrag zur religioesen entwicklungsgeschichte der gegenwart / Blennerhassett, Charlotte, Lady – Berlin: Gebrueder Paetel, 1904 – 1mf – 9 – 0-7905-7205-2 – mf#1988-3205 – us ATLA [240]
John herling's labor letter – 1950 feb 18-1954 dec 25, 1955 apr 30-may 7, 1956 jan 7, feb 11- 25, may 19-26, dec 22, 1957 feb 16, apr 13, aug 3,24, sep 7,28, nov 1955 jan 1-1959 jun 27, 1959 jul 4-1963 dec 28, 1964 dec 12-1969 dec 29, 1975 jun-1980 dec, 1981-86, 1987 jan 3-1988 sep 17 – 8r – 1 – (cont: chester wright's labor letter) – mf#1544363 – us WHS [331]
John hopkins studies in romance literatures and languages – Baltimore. v1-19. 1923-31 – 1 – $144.00 – mf#0305 – us Brook [440]
John housman's reise durch die noerdlichen gegenden von england : nebst einer beschreibung von cumberland, westmoreland, lancashire und einem theile der westlichen kueste von yorkshire / Housman, John – Weimar 1811 [mf ed Hildesheim 1995-98] – 1v on 2mf – 9 – €60.00 – ISBN-10: 3-487-26539-7 – ISBN-13: 978-3-487-26539-1 – gw Olms [914]
John howard and the prison world of europe / Dickson, Richard W – Webster, MA: Frederick Charlton, 1982 – 5mf – 9 – $7.50 – mf#LLMC 92-112 – us LLMC [360]
John howard and the prison-world of europe : from original and authentic documents / Dixon, William Hepworth – New York: R Carter, 1850 [mf ed 1990] – 1mf – 9 – 0-7905-5651-0 – (int essay by richard w dickinson) – mf#1988-1651 – us ATLA [365]
John Howard Society of Ontario *see* Tocsin
John howland : a mayflower pilgrim / ed by Howland, William – 1926 – 1 – 5.00 – us Southern Baptist [242]
John hunter's "directions for preserving animals and parts of animals for examination" *see* The life of john hunter (1728-93)/ john hunter's "directions for preserving animals and parts of animals for examination"
John hus : the commencement of resistance to papal authority on the part of the inferior clergy / Wratislaw, Albert Henry – London: SPCK; New York: E & J B Young 1882 [mf ed 1990] – (= ser The home library) – 1mf – 9 – 0-7905-6158-1 – mf#1988-2158 – us ATLA [241]
John huss : his life, teachings and death, after five hundred years / Schaff, David Schley – New York: Scribner, 1915 – 1mf – 9 – 0-7905-6361-4 – (incl bibl ref) – mf#1988-2361 – us ATLA [240]
John huss : his life, teachings and death, after five hundred years / Schaff, David Schley – New York: Scribner, 1915 – 1mf – us ATLA [240]
John huss / Mussolini, Benito – Trans. by Clifford Parker. New York: A.& C. Boni, 1929.vi,225p – 1 – us Wisconsin U Libr [920]
John huss / Rashdall, Hastings – Oxford: Thos Shrimpton, 1879 – 1mf – 9 – 0-524-02708-0 – mf#1990-0689 – us ATLA [240]
John huss and the presbyterians and reformed / Good, James Isaac – [S.l.: s.n., 1915?] – 1mf – 9 – 0-7905-5833-5 – mf#1988-1833 – us ATLA [242]
John hyde deforest : missionary, statesman, christian ambassador to japan / Gulick, Sidney Lewis – [s.l: s.n, s.n, 191-?] [mf ed 1995] – (= ser Yale coll) – 32p – 1 – 0-524-09931-6 – mf#1995-0931 – us ATLA [920]
John, I G *see* Hand book of methodist missions
John j mcgilvra papers, 1861-1926 / McGilvra, John Jay – 3.62 cubic ft [9 boxes & 1 oversize vol] – (with finding guide) – us UW Libraries [340]
John jasper : the unmatched negro philosopher and preacher / Hatcher, William Eldridge – New York: FH Revell, c1908 [mf ed 1991] – 1mf – 9 – 0-524-00554-0 – mf#1990-0054 – us ATLA [920]
John jay / Pellew, George – Boston, MA. 1890 – 1r – us UF Libraries [025]
John jay / Pellew, George – Boston & NY: Houghton, Mifflin & Co, 1899 – (= ser The american statesmen series) – 4mf – 9 – $6.00 – mf#LLMC 96-031 – us LLMC [975]

John, John Price Durbin *see*
– Signs of god in the world
– The worth of a man

John keats : manuscripts and papers in keats house, hampstead – 1815-89 [mf ed ProQuest] – 4r – 1 – (coll of letters, poems, & mss written by friends & acquaintances of john keats. bks ann by poet & a few of his mss also incl. with p/ind) – us UMI ProQuest [420]

The john keats memorial volume – Issued by the Keats House Committee, Hampstead. Illus. London, New York: John Lane Company, 1921.xx,276p. 4 pl. 6 facs. including portraits. Ed. by Dr. G.C. Williamson. 1 reel. 1296 – 1 – us Wisconsin U Libr [420]

John keble : a biography / Lock, Walter – Boston: Houghton, Mifflin, 1893 – 1mf – us ATLA [240]

John keble : a biography / Lock, Walter – Boston: Houghton, Mifflin, 1893 – 1mf – 9 – 0-7905-5374-0 – (incl bibl ref) – mf#1988-1374 – us ATLA [920]

John kenneth mackenzie : medical missionary to china / Bryson, Mary Isabella – New York: Fleming H Revell, 1891 – 1mf – 9 – 0-8370-6249-7 – mf#1986-0249 – us ATLA [920]

John kerr papers, 1788-1844 / Kerr, John – [mf ed 1980] – 3r – 1 – mf#ms330 – us Western Res [338]

John know and his 'devout imagination' / Candlish, Robert Smith – Edinburgh, Scotland. 1872 – 1r – us UF Libraries [242]

John knox : a biography / Brown, P Hume – London: A & C Black, 1895 – 2mf – 9 – us ATLA [242]

John knox : a biography / Brown, Peter Hume – London: A and C Black, 1895 – 2mf – 9 – 0-7905-4901-8 – (incl bibl ref) – mf#1988-0901 – us ATLA [242]

John knox : a biography / Macmillan, Donald – London: A Melrose, 1905 – 1mf – 9 – 0-7905-6933-7 – mf#1988-2933 – us ATLA [242]

John knox / Harland, Marion – New York: GP Putnam, 1900 – (= ser Literary Hearthstones) – 1mf – 9 – 0-524-04612-3 – mf#1990-1272 – us ATLA [242]

John knox : the hero of the scottish reformation / Cowan, Henry – New York: G P Putnam, 1905 – (= ser Heroes of the reformation) – 2mf – 9 – 0-7905-4217-X – (incl bibl ref) – mf#1988-0217 – us ATLA [242]

John knox : the hero of the scottish reformation / Cowan, Henry – New York: G P Putnam, 1905 – (= ser Heroes of the Reformation) – 2mf – us ATLA [242]

John knox : his ideas and ideals / Stalker, James – New York: A C Armstrong, 1904 – 1mf – 9 – 0-7905-6017-8 – mf#1988-2017 – us ATLA [242]

John knox : his time, and his work / Candlish, Robert Smith – Edinburgh, Scotland. 1846 – 1r – us UF Libraries [242]

John knox / Innes, Alexander Taylor – Quatercentenary ed. Edinburgh: Oliphant, Anderson & Ferrier, 1905 – 1mf – 9 – 0-7905-6928-0 – (incl bibl ref) – mf#1988-2928 – us ATLA [242]

John knox / Taylor, William Mackergo – New York: A C Armstrong, 1885. Chicago: Dep of Photodup, U of Chicago Lib, 1973 (1r); Evanston: American Theol Lib Assoc, 1984 (1r) – 1 – 0-8370-0297-4 – (incl ind) – mf#1984-B362 – us ATLA [242]

John knox, 1505-1572 : ein erinnerungsblatt zur vierten zentenarfeier / Mulot, Rudolf – Halle a.d.S: Verein fuer Reformationsgeschichte 1904 [mf ed 1990] – (= ser Schriften des vereins fuer reformationsgeschichte 22/84) – 1mf – 9 – 0-7905-5311-2 – (incl bibl ref) – mf#1988-1311 – us ATLA [242]

John knox and the church of england : his work in her pulpit and his influence upon her liturgy, articles, and parties... / Lorimer, Peter – London: Henry S King, 1875 – 1mf – 9 – 0-7905-4889-5 – mf#1988-0889 – us ATLA [242]

John knox and the free church of scotland / Philoknoxus – Glasgow, Scotland. 1873 – 1r – us UF Libraries [242]

John knox and the reformation / Lang, Andrew – London; New York: Longmans, Green, 1905 – 1mf – 9 – 0-7905-4997-2 – mf#1988-0997 – us ATLA [242]

John knox and the scottish reformation : addresses delivered in st david's church, st john, n b...may 22nd, 1905... – [St John, NB?: F Doig], 1905 – 1mf – 9 – 0-665-77124-X – mf#77124 – cn Canadiana [941]

John knox, der reformator schottlands / Brandes, Friedrich – Elberfeld: R L Friderichs, 1862 – (= ser Leben und ausgewaehlte Schriften der Vaeter und Begruender der reformirten Kirche) – 2mf – 9 – 0-524-00512-5 – (incl bibl ref) – mf#1990-0012 – us ATLA [242]

John la farge / Waern, Cecilia – London 1896 – (= ser 19th c art & architecture) – 2mf – 9 – mf#4.2.381 – uk Chadwyck [700]

John leech : his life and work / Frith, William Powell – London 1891 – (= ser 19th c art & architecture) – 8mf – 9 – mf#4.2.785 – uk Chadwyck [740]

John leech, artist and humourist : a biographical sketch / Kitton, Frederic George – London 1883 – (= ser 19th c art & architecture) – 1mf – 9 – mf#4.2.1239 – uk Chadwyck [740]

John luczkiw collection – Thomas Fisher Rare Book Library, University of Toronto [mf ed Markham ON: filmed by Xebec Imaging Services for Slavic & East European Microform Project at CRL 2001-] – 1 – (mf coll of estimated 1270 monographs produced & publ by ukrainian refugees in the displaced persons camps in australia & germany during the post-world war 2 period fr 1945-54; crl has: t3/61/a seemp [39r]) – us CRL [934]

John lyman et son oeuvre – [mf ed 1975] – 3r – 1 – mf#SEM35P124 – cn Bibl Nat [760]

John m. berrien papers / Berrien, John M – University of North Carolina Library. Guide – 1 – $54.00 – us CIS [920]

John m henderson papers, 1810-1892 [1817-1848] / Henderson, John M – [mf ed 1996] – 1r – 1 – (account book, correspondence, election tickets, resolutions, bylaws...of this medical doctor, an early settler of lake co & founder of willoughby medical school) – mf#ms533 – us Western Res [610]

John M. Perkins Foundation for Reconciliation and Development *see* Perkins network

John macgregor ("rob roy") / Macaulay, James – London, England. 18– – 1r – us UF Libraries [240]

John machale, archbishop of tuam : his life, times and correspondence / O'Reilly, Bernard – New York: F Pustet 1890 [mf ed 1990] – 2v on 4mf [ill] – 9 – 0-7905-8174-4 – mf#1988-8057 – us ATLA [241]

John mackenzie : south african missionary and statesman / Mackenzie, William Douglas – New York: A C Armstrong, 1902 – 2mf – 9 – 0-8370-6271-3 – (incl ind) – mf#1986-0271 – us ATLA [920]

John mackenzie, south african missionary and statesman / Mackenzie, William Douglas – New York, NY. 1969 – 1r – us UF Libraries [960]

John marshall : complete constitutional decisions / Dillon, John M – Chicago: Callaghan, 1903 – 3mf – 9 – $13.50 – (edited with annotations historical, critical and legal) – mf#LLMC 84-249 – us LLMC [323]

John marshall : life, character and judicial services / Dillon, John Forrest – Centenary ed. Chicago, Callaghan, 1903. 3 v. LL-1255 – 1 – us L of C Photodup [340]

John marshall and the constitution : a chronicle of the supreme court / Corwin, Edward A – New Haven: Yale UP, 1921 – 3mf – 9 – $4.50 – mf#LLMC 92-234 – us LLMC [323]

John marshall day, celebration by the rhode island bar association and brown university, february 4, 1901: address by hon. le baron bradford colt. / Rhode Island Bar Association – Providence: Rhode Island Printing Co., 1901. 53p. LL-199 – 1 – us L of C Photodup [340]

John marshall in india : notes and observations in bengal, 1668-1672 / Marshall, John; ed by Khan, Shafaat Ahmad – London: Oxford University Press, 1927 – (= ser Samp: indian books) – us CRL [915]

John marshall journal of practice and procedure – Chicago. 1967-1979 [1]; 1971-1979 [5]; 1976-1979 [9] – (cont by: john marshall law review) – ISSN: 0021-7212 – mf#6499 – us UMI ProQuest [340]

John marshall journal of practice and procedure *see* John marshall law review

John marshall law review – Chicago. 1979+ (1) 1979+ (5) 1979+ (9) – (cont: john marshall journal of practice and procedure) – ISSN: 0270-854X – mf#6499,01 – us UMI ProQuest [340]

John marshall law review – 1-34. 1967-2001 – 9 – $533.00 set – (title varies: v1-12 1967-79 as: john marshall journal of practice and procedure) – ISSN: 0270-854X – mf#103801 – us Hein [340]

John mason neale, d.d : a memoir / Towle, Eleanor A – London; New York: Longmans, Green, 1906 – 1mf – 9 – 0-7905-6206-5 – mf#1988-2206 – us ATLA [240]

John mason peck and one hundred years of home missions, 1817-1917 / De Blois, Austen Kennedy & Barnes, Lemeul Call – New York: American Baptist Home Mission Society, 1917 – 1mf – 9 – 0-524-08358-4 – mf#1993-3058 – us ATLA [240]

The john maynard keynes papers in king's college, cambridge : the collected papers of one of the most controversial and influential thinkers of the twentieth century – 1883-1946 [mf ed Chadwyck-Healey] – 170r – 1 – (with printed catalogue) – uk Chadwyck [330]

John melish's reisen durch die vereinten-staaten von america : in den jahren 1806, 1807, 1809, 1810 und 1811 – Weimar 1819 [mf ed Hildesheim 1995-98] – 1v on 3mf – 9 – €90.00 – ISBN-10: 3-487-26512-5 – ISBN-13: 978-3-487-26512-4 – gw Olms [917]

John milton's last thoughts on the trinity : extracted from his posthumous work entitled, "a treatise on christian doctrine, compiled from the holy scriptures alone" = De doctrina christiana. selections / Milton, John – Boston: Wm Crosby and HP Nichols, 1847 – 1mf – 9 – 0-524-08547-1 – mf#1993-2072 – us ATLA [220]

John muir papers, 1858-1957 : the complete papers of one of america's most prominent conservationists / ed by Limbaugh, R H & Lewis, K E – [mf ed Chadwyck-Healey] – 51r 53mf – 1,9 – (with p/g & ind) – uk Chadwyck [333]

John newton – London, England. 18– – 1r – us UF Libraries [240]

John nicholson papers – 1772-1819 (mf ed 1967) – 21r – 1 – silver $630 diazo $420 – (general correspondence of the controversial comptroller general of pennsylvania. with guide compiled by donald h kent et al (1967) $5.50 isbn: 0-911124-21-7) – us Penn Hist [978]

John of Damascus, Saint *see*
– Fragmente vornicaenischer kirchenvaeter aus den sacra parallela
– St john damascene on holy images (pros tous diaballontas tas hagias eikonas)
– Select works – exposition of the orthodox faith

John, of Damascus, Saint *see* St john damascene on holy images (pros tous diaballontas tas hagias eikonas)

John of Ephesus, Bishop of Ephesus *see* The third part of the ecclesiastical history of john, bishop of ephesus

John of Kronstadt, Saint *see* My life in christ, or, moments of spiritual serenity and contemplation, of reverent feeling, of earnest self-amendment, and of peace in god

John of Salisbury, Bishop of Chartres *see* Ioannis saresberiensis episcopi carnotensis policratici..

John of the Cross, Saint *see*
– The complete works of saint john of the cross, doctor of the church
– The spirit of st. john of the cross

John of wycliffe, the morning star of the reformation / Adams, Emma Hildreth – Oakland, Cal[if]: Pacific Press, c1890 – (= ser Young People's Library) – 1mf – 9 – 0-524-04827-4 – mf#1990-1319 – us ATLA [242]

John o'farrell...and william venner...both of the city of quebec, and john simpkins...of the city of new-york, in the united-states of america, plaintiffs : vs alexandre-rene chaussegros de lery...of sainte-marie de la beauce and truman coman...of pittsfield... defendants – Quebec?: s.n, 1867? – 1mf – 9 – mf#11933 – cn Canadiana [347]

John o'groat journal – Wick: P Reid 1837- (wkly) [mf ed 6 jan 1995-] – 1 – (not publ: 26 jun 1959, 4 feb 1983; cont: john o'groat journal and northern miscellany; sister publ to: caithness courier, nov 11 1977-; incl as suppl: john o'groat journal christmas number which is numbered as part of the journal; title varies slightly; imprint varies) – ISSN: 1354-9677 – uk Scotland NatLib [072]

The john osborne sargent papers, 1831-1912 – [mf ed 1965] – 4r – 1 – (with p/g) – us MA Hist [070]

John P Altgeld Memorial Association Of Chicago *see* Dedicatory exercises at the unveiling of bronze tablets in memory...

John p green papers, ms 3379 / Green, John P – 1968-1910 – 6r – 1 – (correspondence, speeches, financial and legal records) – us Western Res [920]

John paterson green papers, 1869-1910 / Green, John Paterson – [mf ed 1972] – 6r – 1 – mf#ms3379 – us Western Res [978]

The john pendleton kennedy papers / Kennedy, John Pendleton; ed by Boles, John B – 1973 – 27r – 1 – $3510.00 – (guide sold separately $10) – mf#S1616 – us Scholarly Res [920]

John penry : the so-called martyr of congregationalism as revealed in the original record of his trial and in documents related thereto / Burrage, Champlin – Oxford: University Press; London: H Frowde, 1913 – 1mf – 9 – 0-7905-5862-9 – mf#1988-1862 – us ATLA [242]

John penry, the pilgrim martyr, 1559-1593 / Waddington, John – London: W & FG Cash, 1854 – 1mf – 9 – 0-524-00660-1 – mf#1990-0160 – us ATLA [240]

John pierpont : a biographical sketch / Ford, Abbie A – Boston: [s.n], 1909 (Jamaica Plain, Mass: J Allen Crosby) – 1mf – 9 – 0-524-04295-0 – mf#1992-2015 – us ATLA [920]

John preston davis papers : from the holdings of the schomburg center for research in black culture, manuscripts, archives and rare books division: the new york public library, astor, lenox and tilden foundations – 1995 – ca 5r – 1 – ca $425.00 – (guide sold separately for $20.00 covers all coll under "literature and the arts" d3305.g6) – mf#D3305P24 – Dist. us Scholarly Res – us L of C Photodup [070]

John price – London, England. 18– – 1r – us UF Libraries [240]

John quincy adams campbell diaries, 1861-1864 / Campbell, John Quincy Adams – [mf ed 1992] – 1r – 1 – (john q a campbell was a native of ohio, relocated to iowa, who served with the 5th iowa volunteer infantry, 1861-1864, during the american civil war, in the western theater of operations) – mf#ms3560 – us Western Res [976]

John Reed Clubs of the Middle West *see* Left front

John rice jones : a brief sketch of the life and public career of the first practising lawyer in illinois; rice jones: a brief memoir of the last representative of randolph county in the general assembly of indiana territory... / Jones, W A Burt – Chicago?: s.n, 1889 – (= ser Fergus' historical series) – 1mf – 9 – mf#11226 – cn Canadiana [920]

John richard brinkley vs. kansas state board of medical registration and examination, et al / Brinkley, John Richard – 1 – us Kansas [610]

John richard brinkley vs. the kansas city star / Brinkley, John Richard – 1 – us Kansas [978]

John robinson : pastor of the pilgrim fathers / Davis, Ozora Stearns – Hartford, CT: Hartford Seminary Press, 1897 [mf ed 1990] – 1mf – 9 – 0-7905-8241-4 – (incl bibl ref) – mf#1988-8104 – us ATLA [242]

John robinson : the pilgrim pastor / Davis, Ozora S – Boston: Pilgrim Press, c1903 – 1mf – us ATLA [240]

John robinson : the pilgrim pastor / Davis, Ozora Stearns – Boston: Pilgrim Press, c1903 – 1mf – 9 – 0-7905-4224-2 – mf#1988-0224 – us ATLA [240]

John rogers / Aston, Thomas H – Birmingham, England. 1863 – 1r – us UF Libraries [240]

John rogers : the compiler of the first authorised english bible / Chester, Joseph Lemuel – London: Longman, Green, Longman and Roberts, 1861 – 2mf – 9 – 0-524-05140-2 – mf#1990-1396 – us ATLA [220]

John rogers : the compiler of the first authorised english bible; the pioneer of the english reformation; and its first martyr...a geneaological account of his family / Chester, Joseph Lemuel – London: Longman, Green, Longman & Roberts 1861 [mf ed 1988] – 1r – 1 – (filmed with: the life of george stephenson, railway engineer / samual smiles (1859)) – mf#7961 – us Wisconsin U Libr [920]

John ruskin / Harrison, Frederic – New York: Macmillan 1903 [mf ed 1992] – (= ser English men of letters (new york, ny)) – 1mf – 9 – 0-524-02183-X – mf#1990-0568 – us ATLA [420]

John ruskin : a sketch of his life, his work, and his opinions / Spielmann, Marion Harry – London 1900 – (= ser 19th c art & architecture) – 3mf – 9 – mf#4.2.1369 – uk Chadwyck [420]

John ruskin, the pre-raphaelite brotherhood and arts and crafts movement : from the john rylands university library, manchester and manchester art gallery – 28r – 1 – (coll of more than 3,700 items with john ruskin at the centre. also covers the pre-raphaelite brotherhood and the artistic and literary culture wh surrounded them. includes a printed guide) – mf#C35-17400 – us Primary [420]

John rutledge papers / Rutledge, John – 1782-1872. University of North Carolina Library. Guide – 1 – $36.00 – us CIS [920]

John Rylands Library Bible Tercentenary Exhibition *see* Catalogue of an exhibition of manuscript and printed copies of the scriptures

John Rylands University Library. Manchester *see* Anti-slavery materials

John Rylands University Library of Manchester *see* Bulletin of the john rylands university library of manchester

John s. brown papers / Brown, John S – 1818-1907. In Kansas State Historical Society – 1 – us Kansas [920]

John S Slater Fund *see* Proceedings and reports

John sewell's memoirs and history of miami, florid... / Sewell, John – Miami, FL. 1938 – 1r – us UF Libraries [978]

John smith, the se-baptist and the pilgrim fathers helwys and baptist origins / Burgess, Walter H – 1911 – 1 – us Southern Baptist [242]

John smith the se-baptist, thomas helwys, and the first baptist church in england : with fresh light upon the pilgrim fathers' church / Burgess, Walter Herbert – London: James Clarke, 1911 – 1mf – 9 – 0-524-07975-7 – (incl bibl ref) – mf#1990-5420 – us ATLA [242]

Sir john soane's museum : the illuminated manuscripts / London. Sir John Soane's Museum – [mf ed Chadwyck-Healey] – 3r – 14 – (with catalogue) – uk Chadwyck [090]

Sir john soane's museum : the italian drawings / London. Sir John Soane's Museum – [mf ed Chadwyck-Healey] – 3r – 14 – (with catalogue) – uk Chadwyck [740]

Sir john soane's musuem : architectural and ornamental drawings / London. Sir John Soane's Museum – [mf ed Chadwyck-Healey] – 55r – 14 – (with printed catalogue & ind ed by margaret richardson. 2 pts of complete coll are also available and listed separately) – uk Chadwyck [720]

John steinbeck collection / Steinbeck, John – Salinas, CA: John Steinbeck Library – 1 – $250.00 – (index and annotated bibliography 1r $50.00 b40502) – mf#B40501 – us Library Micro [420]

John stephenson rowntree, his life and work / Doncaster, Phebe – London: Headley, 1908 – 2mf – 9 – 0-524-06293-5 – (incl bibl ref) – mf#1990-5222 – us ATLA [240]

John stewart, missionary to the wyandots / Love, Nathaniel Barrett Coulson – New York: Missionary Society of the Methodist Episcopal Church, [1900?] – 1mf – 9 – 0-524-07255-8 – mf#1991-2996 – us ATLA [240]

John stuart mill : a criticism / Bain, Alexander – London: Longmans, Green, 1882 – 1mf – 9 – 0-7905-3634-X – mf#1989-0127 – us ATLA [140]

John the baptist : the forerunner of our lord: his life and work / Houghton, Ross C – New York: Hunt & Eaton, 1889 – 1mf – 9 – 0-8370-3675-5 – (incl ind) – mf#1985-1675 – us ATLA [920]

John the baptist / Meyer, Frederick Brotherton – New York:Fleming H. Revell, c1900 – 1mf – 9 – 0-8370-4409-X – mf#1985-2409 – us ATLA [920]

John the loyal : studies in the ministry of the baptist / Robertson, A T – New York: Charles Scribner, 1911 – 1mf – 9 – 0-7905-0222-4 – (incl indes) – mf#1987-0222 – us ATLA [920]

The john thomas papers, 1693-1839 – [mf ed 1976] – 3r – 1 – us MA Hist [355]

John turnbull's reise um die welt : oder eigentlich nach australien, in den jahren 1800, 1801, 1802, 1803 und 1804 – Weimar 1806 [mf ed Hildesheim 1995-98] – 1v on 3mf – 9 – €90.00 – ISBN-10: 3-487-26560-5 – ISBN-13: 978-3-487-26560-5 – gw Olms [919]

John tyler papers – 1691-1918 (mf ed 1958) – (= ser Presidential papers microfilm) – 3r – 1 – us L of C Photodup [975]

John w inglesby papers – c1906-63 – 1 linear ft also on microfilm – 1 – us South Carolina Historical [360]

John walker's courtship : a legend of lauderdale / Albyn [i.e. Andrew Shiels] – Halifax, NS?: J Bowes, 1877 – 9 – mf#06129 – cn Canadiana [830]

John walworth, ashbel walworth papers *see* Walworth, john; walworth, ashbel, papers, ms 1901

John wesley : a lecture / Mason, Arthur James – London: SPCK 1898 [mf ed 1992] – (= ser Church historical society (series) 47) – 1mf – 9 – 0-524-05512-2 – mf#1990-1507 – us ATLA [242]

John wesley / Lelievre, Matthieu – London, England. 1900 – 1r – us UF Libraries [025]

John wesley / McConnell, Francis John – New York: Abingdon Press, c1939 – 1mf – 9 – 0-524-08121-2 – mf#1993-9027 – us ATLA [242]

John wesley / Overton, John Henry – London: Methuen, 1891 – (= ser English Leaders Of Religion) – 1mf – 9 – 0-7905-5782-7 – (incl bibl ref) – mf#1988-1782 – us ATLA [240]

John wesley, an evolutionist / Mills, William Harrison – [S.l.: s.n., 1893?] – 1mf – 9 – 0-524-03325-0 – mf#1990-4685 – us ATLA [242]

John wesley and modern wesleyanism / Hockin, Frederick – 3rd, much enl ed. London: JT Hayes, [1876?] – 1mf – 9 – 0-524-06288-9 – mf#1990-5217 – us ATLA [242]

John wesley, evangelist / Green, Richard – London: Religious Tract Society, 1905 [mf ed 1991] – 2mf – 9 – 0-524-01650-X – mf#1990-0471 – us ATLA [242]

John wesley in company with high churchmen / Holden, Harrington William – 6th rev and enl ed. London: John Hodges, 1874 – 1mf – 9 – 0-524-05588-2 – (incl bibl ref) – mf#1991-2312 – us ATLA [242]

John wesley on toleration of romanism / Wesley, John – Birmingham, England. 18– – 1r – us UF Libraries [242]

John wesley, preacher / Doughty, William Lamplough – London: Epworth, 1955 – 1mf – 9 – 0-524-08101-8 – mf#1993-9007 – us ATLA [242]

John white chadwick / Cameron, Angus deMille – Chicago, 1937. Chicago: Dep of Photodup, U of Chicago Lib, 1971 (1r); Evanston: American Theol Lib Assoc, 1984 (1r) – 1 – 0-8370-0324-5 – mf#1984-B170 – us ATLA [920]

John wiclif : his life, times, and teaching / Pennington, Arthur Robert – London: SPCK; New York: E & J B Young, 1884 – 1mf – 9 – 0-7905-5544-1 – (incl bibl ref) – mf#1988-1544 – us ATLA [941]

John wiclif's polemical works in latin / Wycliffe, John – English ed. London: Published for the Wyclif Society by Truebner, 1883 – 3mf – 9 – 0-524-00223-1 – mf#1989-2923 – us ATLA [240]

John wilhelm rowntree : essays and addresses / Rowntree, John Wilhelm; ed by Rowntree, Joshua – 2nd ed London: Headley, 1906 – 2mf – 9 – 0-8370-9109-8 – mf#1986-3109 – us ATLA [240]

John william burgon : late dean of chichester / Goulburn, Edward Meyrick – London: J Murray, 1892 – 2mf – 9 – 0-7905-5698-7 – mf#1988-1698 – us ATLA [240]

John woolman / Greenwell, Dora – London: FB Kitto, 1871 – 1mf – 9 – 0-7905-7633-3 – mf#1989-0858 – us ATLA [920]

John workman, der zeitungsboy : eine erzaehlung aus der amerikanischen grossindustrie / Dominik, Hans – Leipzig: Koehler & Amelang, 1925 – 1r – 1 – us Wisconsin U Libr [830]

John wycliffe and his english precursors = Johann von wiclif und die vorgeschichte der reformation. selections / Lechler, Gotthard Victor – New rev ed. London: Religious Tract Society, [1884?] – 2mf – 9 – 0-7905-5416-X – (incl bibl ref. in english) – mf#1988-1416 – us ATLA [240]

John wycliffe and the first english bible : an oration / Storrs, Richard Salter – New York: Anson D F Randolph, 1880 [mf ed 1985] – 1mf – 9 – 0-8370-5430-3 – (incl bibl ref) – mf#1985-3430 – us ATLA [220]

John wycliffe and the first english bible / Storrs, Richard S – New York, NY. 1880 – 1r – us UF Libraries [240]

John wycliffe, patriot and reformer : the morning star of the reformation / Wilson, John Laird – New York: Funk & Wagnalls, 1884 – (= ser Standard Library) – 1mf – 9 – 0-524-04631-X – (incl bibl ref) – mf#1990-1291 – us ATLA [240]

John-donkey – New York. 1848-1848 (1) – mf#4379 – us UMI ProQuest [870]

Johnnie courteau : and other poems / Drummond, William Henry – Toronto: Musson, 1905 – 2mf – 9 – 0-665-98239-9 – mf#98239 – cn Canadiana [810]

John's american notary and commissioner of deeds manual. / John, Edward Mills – 4th ed. Chicago, Callaghan, 1931. 523 p. LL-566 – 1 – us L of C Photodup [340]

Johns, Bennett George *see* Moses, not darwin

Johns, Claude Hermann Walter *see*
- Ancient assyria
- Ancient babylonia
- An assyrian doomsday book
- Babylonian and assyrian laws, contracts and letters
- The old testament in the light of the ancient east
- The relations between the laws of babylonia and the laws of the hebrew peoples
- The religious significance of semitic proper names

Johns, David *see* Ny dikisionary malagasy

John's gospel : the greatest book in the world. suggestions for the study of the gospel by individuals and in groups / Speer, Robert Elliott – New York: Fleming H Revell, c1915 – 1mf – 9 – 0-524-04812-6 – mf#1992-0232 – us ATLA [226]

John's gospel : apologetical lectures = Johannes-evangelie / Oosterzee, Johannes Jacobus van – Edinburgh: T and T Clark, 1869 – 1mf – 9 – 0-7905-1616-0 – (incl ind. in english) – mf#1987-1616 – us ATLA [226]

Johns hopkins magazine – Baltimore. 1950+ (1) 1974+ (5) 1974+ (9) – ISSN: 0021-7255 – mf#9102 – us UMI ProQuest [073]

Johns hopkins medical journal – Baltimore. 1889-1982 (1) 1972-1982 (5) 1972-1982 (9) – ISSN: 0021-7263 – mf#7633 – us UMI ProQuest [610]

Johns Hopkins University *see* Studies in historical and political science

Johns hopkins university studies in historical and political science / ed by Adams, Herbert B – Seies 18. Nos. 10-12. Thom, William Taylor. The Struggle for religious freedom in Virginia: The Baptists. 1900. Thomas, David. The Virginia Baptist. 180p – 1 – us Southern Baptist [323]

Johns hopkins university studies in historical and political science – Baltimore. 1882-1995 (1) 1980-1995 (5) 1980-1995 (9) – ISSN: 0075-3904 – mf#2905 – us UMI ProQuest [900]

Johns, J H *see* History of the rock presbyterian church in cecil co., md

Johns, J W *see* The anglican cathedral church of saint james

John's vision of jesus in glory / M'cheyne, Robert Murray – Edinburgh, Scotland. 1857 – 1r – us UF Libraries [240]

Johns, W *see* Appeal

Johnsen, Erik Kristian *see* I kirke

Johnsen, Julia E *see* Selected articles on marriage and divorce

Johnsen, Julia E (Julia Emily) *see* National labor relations act

Johnsen, Julia Emily *see*
- Limitations of power of supreme court to declare acts of congress unconstitutional
- Special legislation for women

Johnsen, Julie Emily *see* Atomic bomb

Johnson *see* Fernandina history

The johnson administration and pacification in vietnam : the robert komer-william leonhart files, 1966-68 – (= ser Vietnam war research colls) – 15r – 1 – $2905.00 – 1-55655-474-5 – (with p/g) – us UPA [934]

The johnson administration's response to anti-vietnam war activities : pt 1: white house central files and aides' files – 25r – 1 – $4840.00 – 1-55655-952-6 – (with p/g) – us UPA [977]

Johnson, Alex R *see* Organization, instruction, and results of evening classes in poult

Johnson, Alice *see* Human personality and its survival of bodily death

Johnson, Allen *see* Readings in american constitutional history, 1776-1876

Johnson, Amandus *see* The swedish settlements on the delaware

Johnson, Andrew *see* Papers

Johnson, Arthur Henry *see* The normans in europe

Johnson, Arthur N *see* British foreign missions, 1837-1897

Johnson, Ashley Sidney *see*
- The holy spirit and the human mind
- Johnson's speeches, hemstead-johnson debate
- The resurrection and the future life

Johnson, Ben *see* Poetaster

Johnson, Brenda M *see* Validation of a modified closed circuit, oxygen dilution residual volume method

Johnson, Burges *see* New rhyming dictionary and poets' handbook

Johnson C Smith University *see* Quarterly review of higher education among negroes

Johnson, Charles C *see* A comparison of the submaximal and maximal responses to upright verus semi-recumbent cycling in males

The johnson citizen – Johnson, NE: Clement L Wilson. 2v. 1898-v2 n1. feb 24 1899 (wkly) [mf ed v1 n12. may 13 1898-feb 24 1899 (gaps) filmed 1976] – 1r – 1 – us NE Hist [071]

Johnson, Clifton *see* The parson's devil

Johnson, Colleen *see* A group intervention programme for adolescents of divorce

Johnson, Colonel *see* Parliamentary observance of the sabbath

Johnson county courier – Sterling, NE: Marion F Packwood. 28v. v53 n14. jan 4 1945-v80 n13. nov 25 1971 (wkly) [mf ed filmed [1974]-1976] – 7r – 1 – (cont: cook weekly courier. absorbed: sterling sun (1892). absorbed by: syracuse journal-democrat) – us NE Hist [071]

Johnson county courier – Sterling, NE: Maverick Media, oct 1982-v8 n43. may 18 1989 (wkly) [mf ed filmed 1989] – 3r – 1 – (split from: syracuse journal-democrat. absorbed by: syracuse journal-democrat) – us NE Hist [071]

Johnson county genealogist – KS. 1973 mar [v1 n1]-1982 – 1r – 1 – mf#660843 – us WHS [929]

Johnson county journal – Tecumseh, NE: Chas D Blauvelt. 14v. v37 n51. feb 3 1916-50th yr n28. sep 26 1929 (wkly) [mf ed with gaps] – 5r – 1 – (cont: johnson county journal=tribunal. absorbed by: tecumseh chieftain) – us NE Hist [071]

Johnson county journal – Tecumseh, NE: J W Barnhart, C W Pool. 31v. mar 13 1879-v31 n40. nov 25 1909 (wkly) [mf ed with gaps)] – 3r – 1 – (merged with: johnson county tribunal to form: johnson county journal=tribunal) – us NE Hist [071]

Johnson county journal=tribunal – Tecumseh, NE: Journal-Tribunal Print Co. 7v. v31 n41. dec 2 1909-v37 n50. jan 27 1916 (wkly) [mf ed lacks apr 12 1912] – 3r – 1 – (formed by the union of: johnson county journal (1879) and: johnson county tribunal. cont by: johnson county journal (1916). issues for dec 2 1909-jan 6 1910 called also v12 n2-7) – us NE Hist [071]

Johnson county press – Franklin, IN. 1865-1869 (1) – mf#62783 – us UMI ProQuest [071]

Johnson county tribunal – Tecumseh, NE: Tribunal Publ Co, nov 1898-v12 n1. nov 26 1909 (wkly) [mf ed with gaps filmed 1974] – 4r – 1 – (merged with: johnson county journal (1879) to form: johnson county journal=tribunal) – us NE Hist [071]

Johnson, D LaMont *see* Computers in the schools

Johnson, Daniel *see* Sketches of field sports as followed by the natives of india

Johnson, Dave R *see* Self-efficacy of male and female golfers at differing ability levels

Johnson, Donald McIntosh *see* Indian hemp: a social menace

Johnson, E Pauline *see* Legends of vancouver

Johnson, Elias Finley *see*
- Elements of the law of negotiable contracts
- Illustrative cases upon the law of bills and notes

Johnson, Elias Henry *see*
- Christian agnosticism as related to christian knowledge
- Ezekiel gilman robinson
- The highest life
- The holy spirit then and now
- Outline of systematic theology – and of ecclesiology
- The religious use of imagination

Johnson, Elizabeth French *see* Weckherlin's eclogues of the seasons...

Johnson, Emory Richard *see*
- Measurement of vessels for the panama canal
- Panama canal traffic and toils

Johnson, Francis *see*
- Affaire guibord
- Miscellaneous collection of 20 vocal and instrumental compositions

Johnson, Francis Godschall *see* Proces de joseph n cardinal et autres

Johnson, Francis Howe *see* God in evolution

Johnson, Frank Ernest et al *see* Ancient arabia

Johnson, Franklin *see*
- The christian's relation to evolution
- The new psychic studies in their relation to christian thought
- The quotations of the new testament from the old

Johnson, G H S *see* Science and natural religion

Johnson, George *see* Information given regarding annexation and other matters

Johnson, George Washington *see*
- Maple leaves
- The public school speller and word-book

Johnson, Gifford *see* Marshall islands resource materials

Johnson, Gisle *see* Konkordiebogen

Johnson, Gordon Allen *see* Harmonic analysis of the requiem by giuseppe verdi

Johnson Graduate School of Management (Cornell University) *see* Cornell business

Johnson, Gyneth *see* How the donkeys came to haiti

Johnson, Helen Mar *see* Canadian wild flowers

Johnson, Herrick *see* Christianity's challenge

Johnson, J *see*
- A journey from india to england through persia, georgia, russia, poland and prussia, in the year 1817
- The oriental voyager

Johnson, J M *see*
- Jonathan c gibbs
- Negro education
- Negro history
- Treasure seekers night of terror

Johnson, James *see*
- An account of a voyage to india, china etc in his majesty's ship caroline
- The recess, or autumnal relaxation in the highlands and lowlands
- Yoruba heathenism

Johnson, James B *see* My land

Johnson, James Weldon *see*
- Autonomie d'haiti
- James weldon and stanton high school

Johnson, Jana A *see* Validation of the maximal met prediction equations on the schwinn air-dyne

Johnson, Jane *see*
- Early impressions
- Essays on some of the testimonies of truth as held by the society of friends

Johnson, Jay R *see* A survey of division 2 athletic and physical education fiscal trends

Johnson, Jesse Harlan *see* Fossil algae from guatemala

Johnson, Jill E *see* Nutritional intakes of older adults embarking on a strength-training program

Johnson, John *see*
- Journal of a tour through parts of france, italy, and switzerland
- Papers
- Plans, sections, and perspective elevation

Johnson, John Butler *see* Theory and practice of modern framed structures

Johnson, John de Monins *see* Transactions of the third international congress for the history of religions

Johnson, John Edgar *see* The monks before christ

Johnson, John Wesley *see* The canadian accountant

Johnson journeys – v1 iss 1-v2 iss 2 [1986 nov-1988] – 1r – 1 – mf#1496843 – us WHS [071]
Johnson, Kandice M *see* Relationship between body image and protective sexual health practices of sexually active heterosexual college women
Johnson, Lyndon B *see* Daily diary of president johnson [1963-69]
Johnson, Martin *see* Papers
Johnson, Mary Coffin *see* Rhoda m. coffin
Johnson, Merton Bainer *see* Bibliography for conductors of college and community orchestras
Johnson, Molly T *see* Studying the role of gender in the federal courts
Johnson, Molly T et al *see*
- Electronic media coverage of federal civil proceedings
- Use of expert testimony, specialized decision makers, and case-management innovations

Johnson, N S *see* Er biblen guds ord?
Johnson national drillers journal – St. Paul. 1950-1955 (1) – ISSN: 0021-0721 – mf#786 – us UMI ProQuest [333]
Johnson news – Johnson, NE: Raymond S Scofield, 1897-v51 n30. aug 27 1942 (wkly) [mf ed v5 n20. feb 12 1897-aug 27 1942 (gaps) filmed 1976] – 12r – 1 – (cont: news. absorbed by: nemaha county herald) – us NE Hist [071]
The johnson news – Johnson, NE: Ray Scofield & Co. v1 n1. oct 14 1892- (wkly) [mf ed filmed 1976] – 1r – 1 – (cont by: news) – us NE Hist [071]
Johnson newsletter – v1-v2 n1 [1976 aug-[1977 aug]] – 1r – 1 – mf#382328 – us WHS [071]
Johnson, Osa *see* Papers
Johnson, Ovid Frazer, Jr *see* Law of mechanics' liens in pennsylvania
Johnson, P B *see*
- Christian re-union

Johnson, R W M *see* Labour economy of the reserve
The johnson rag – Johnson, NE: [s.n.] v1 n1. nov 25 1970 (wkly) [mf ed with gaps filmed 1989-] – 1 – us NE Hist [071]
Johnson reporter – v1 n1-v8 n3 [1981 win-1988 sum] – 1r – 1 – mf#1060597 – us WHS [071]
Johnson, Robert *see*
- A complete treatise on the art of retouching photographic negatives
- Essaies or rather imperfect offers

Johnson, Robert A *see*
- History of the reformation in germany

Johnson, Robert Flynn *see* American prints, 1870-1950
Johnson, Robert James *see* Specimens of early french architecture
Johnson, Ross *see* New guinea patrol reports and related papers
Johnson, Roy Hamlin *see* Twentieth century solo pianoforte sonatina
Johnson, Samuel *see*
- Harleian miscellany
- The odore parker
- Oriental religions and their relation to universal religion
- The worship of jesus in its past and present aspects

Johnson, Samuel K *see* A comparison of ground reaction forces during running and form skipping
Johnson, Scott R *see* The effects on extracurricular participation of academic achievement, self-concept, and locus of control among high school students
Johnson, Susan L *see* The effect of three training methods on the teaching preparation of counselor-teachers in a resident environmental education program
Johnson, Susan M *see* The effect of mutual choice placement on the satisfaction of student and cooperating teachers in physical education
Johnson, Thomas Cary *see*
- Introduction to christian missions
- John calvin and the genevan reformation
- The life and letters of benjamin morgan palmer
- The life and letters of robert lewis dabney
- Life and letters of robert lewis dabney
- Virginia presbyterianism and religious liberty in colonial and revolutionary times

Johnson, Thomas Richard *see* A comprehensive system of book-keeping by single and double entry
Johnson, Tom L *see* Tom l. johnson papers, series 1, ms 3651
Johnson, tom l, papers, ms 4021 – 1901-09 – (= ser Tom l johnson papers) – 1r – 1 – (series 2 of the tom l johnson papers, these consist primarily of correspondence and financial records of the mayor's office, and city departments, featuring constituent service and city power and light concerns) – us Western Res [350]
Johnson, W R *see*
- The history of england, in easy verse
- The history of rome

Johnson, W S *see* Some ores and rocks of southern slocan division, west kootenay, british columbia
Johnson, Walter Seely *see* The clause compromissoire
Johnson weekly – 1968 jul 19, 1974 nov 1 – 1r – 1 – mf#1112742 – us WHS [071]
Johnson, William *see*
- City, rice-swamp, and hill
- Johnson's digest of new york cases, 1799-1836

Johnson, William B *see* The gospel developed through the government and order of the churches of jesus christ
Johnson, William Bishop *see* The scourging of a race
Johnson, William Hallock *see* The christian faith under modern searchlights
Johnson, William Percival *see* Nyasa, the great water
Johnson, William (Tui) Grainger *see* 'Fiji-70 years and one month'
Johnson, Willis Fletcher *see* Four centuries of the panama canal
Johnson's chancery appeals reports / New York. (State) – v1-7. 1814-23 (all publ) – 42mf – 9 – $63.00 – (a pre-nrs title) – mf#LLMC 80-001 – us LLMC [340]
Johnson's digest of new york cases, 1799-1836 / Johnson, William – Philadelphia: E F Backhus. 2nd corr ed. 2v. 1837-38 (all publ) – 15mf – 9 – $22.50 – mf#LLMC 79-520 – us LLMC [348]
Johnson's Island, OH *see*
- Civil war material
- Prison conditions

Johnson's ready legal adviser. / Chase, George – New York: Johnson, 1880. 308p. LL-1618 – 1 – us L of C Photodup [340]
Johnson's speeches, hemstead-johnson debate : thorn grove, tenn., september 16, 17, 1891 / Johnson, Ashley Sidney – Knoxville, TN: Ogden Bros, 1895 – 1mf – 9 – 0-524-07626-X – mf#1991-3233 – us ATLA [240]
Johnson's tennessee harmony – 1821 – 1 – 5.00 – us Southern Baptist [242]
Johnston, Alexander J *see* Johnston's reports of cases determined in the court of appeal of new zealand
Johnston, B D *see* A study of the relationship of "withitness" to alt-pe and monitoring in middle school physical education
Johnston baptist church. edgefield county. south carolina : church records – 1875-1914 – 1 – us Southern Baptist [242]
Johnston, Charles *see*
- Karma
- The memory of past births
- The parables of the kingdom
- The song of life

Johnston, Christopher *see* The epistolary literature of the assyrians and babylonians
Johnston, Christopher N *see* St paul and his mission to the roman empire
Johnston city, illinois. williams prairie baptist church – Church Records, 1872-1903. 288p – 1 – $12.96 – us Southern Baptist [242]
Johnston County Genealogical Society *see* Newsletter
Johnston, David *see*
- Plea for a new english version of the scriptures
- A treatise on the authorship of ecclesiastes

Johnston, David L *see* The effects of functional isometric weight training in conjunction with dynamic weight training on two bench press measurement tests
Johnston family papers : miscellaneous papers, autobiographical memos, family certificates and "sogerinumu" magazine of the sogeri high school, maps and photographs – 1934-1990 – 1r – 1 – mf#PMB1054 – at Pacific Mss [920]
Johnston first -graniteville general six principle baptist church. rohde island : church records – Organized Jun 20, 1771-Sept 1908 (Foster First (Scituate)merged with Johnston First in 1837) 162p – 1 – us Southern Baptist [242]
Johnston, Frances Benjamin *see*
- The carnegie survey of the architecture of the south
- Papers

Johnston, Graydon *see* The marriage laws of new york state
Johnston, Harriet L *see*
- Papers

Johnston, Harry Hamilton *see*
- British central africa
- George grenfell and the congo
- A history of the colonization of africa by alien races
- Pioneers in south africa
- Uganda protectorate

Johnston, Harry Hamilton et al *see* A generation of religious progress
Johnston, Harry Hamilton, Sir *see* Livingstone and the exploration of central africa
Johnston, Howard Agnew *see*
- The beatitudes of christ
- Bible criticism and the average man
- Moses and the pentateuch
- Scientific faith
- Studies for personal workers

Johnston, J Wesley *see*
- The baptism of fire
- The creed and the prayer

Johnston, James *see*
- A century of christian progress and its lessons
- China and formosa
- Missionary points and pictures
- Reality versus romance in south central africa
- Report of the centenary conference on the protestant missions of the world

Johnston, James F *see* The suspending power and the writ of habeas corpus
Johnston, James Finlay Weir *see* The chemistry of common life
Johnston, James Finley Weir *see*
- Relations of geology to agriculture in north-eastern america, vol 1
- Relations of geology to agriculture in north-eastern america, vol 2

Johnston, John *see* Gospel of the kingdom to be universally preached
Johnston, John C *see* Treasury of the scottish covenant
Johnston, John Leslie *see* Some alternatives to jesus christ
Johnston, John Octavius *see* Life and letters of henry parry liddon
Johnston, John Wilson *see* A contribution to the dynamics of racial diet in british india
Johnston, Julia Harriette *see*
- Indian and spanish neighbors
- The life of adoniram judson

Johnston, Kelli D *see* The relationship between blood testosterone levels and body composition in physically active, young adult men
Johnston, Lena E *see* Jin ko-niu
Johnston, Mary *see* The long roll
Johnston, Meta L *see* Jin ko-niu
Johnston, R Keigh *see* Hunting and hunted in the belgian congo
Johnston, R S *see* Report of the debate on the independence of the church, which took...
Johnston, Reginald Fleming *see*
- Buddhist china
- Lion and dragon in northern china

Johnston, Robert *see*
- Presbyterian worship
- Redeemer's last command

Johnston, S *see* Early man in zambia
Johnston, Sir Harry Hamilton *see* A survey of the ethnography of africa and the former racial and tribal migrations in that continent
Johnston, William *see*
- A catalogue of old and new books
- Letter to e.t. smyth, benton county, al, 1854
- Ribbonism, and its remedy

Johnston, Wyatt *see* Syllabus of post mortem methods for the use of students in the montreal general hospital
Johnstone advertiser – [Scotland] Renfrewshire, Johnstone: Landles & Co 19 jul 1890-dec 1950 (wkly) [mf ed 2003] – 4684v on 15r – 1 – (missing: 1895, 1918-19; cont by: advertiser (johnstone, scotland)) – uk Newsplan [072]
Johnstone, Catherine Laura *see*
- The british colony in russia
- The young emigrants

Johnstone, Christian *see* Lives and voyages of drake, cavendish, and dampier
Johnstone gleaner and advertiser – [Scotland] Renfrewshire, Johnstone: R Harper 13 jun-dec 1890 (wkly) [mf ed 2003] – 9r – 1 – uk Newsplan [072]
Johnstone, James *see*
- Conditions of life in the sea
- Few days on the continent
- My experiences in manipur and the naga hills

Johnstone, Joan [formerly Josephine Whiteman] *see* Study of chimbu conjugal relationships, 1972
Johnstone observer – [Scotland] Renfrewshire, Johnstone: T Boyd 6 jan 1899-31 dec 1909 (wkly) [mf ed 2003] – 1762v on 3r – 1 – (began: n1 (1875?)) – uk Newsplan [072]
Johnstone, Peirce De Lacy *see* Muhammad and his power
Johnstone, Robert *see*
- The first epistle of peter
- Lectures exegetical and practical on the epistle of james
- Lectures, exegetical and practical, on the epistle of paul to the philippians

Johnston's reports of cases determined in the court of appeal of new zealand / Johnston, Alexander J – v1-3. 1867-77. Wellington: G Didsbury. 1872-77 (all publ) – 18mf – 9 – $27.00 – mf#LLMC 96-014 – us LLMC [347]
Johnstown free press – Johnstown, PA. 1901. 1 roll – 13 – $25.00r – us IMR [071]
Johst, Hanns *see*
- Ave eva
- Ich glaube!
- Kunterbunt
- Mutter
- Mutter ohne tod – die begegnung
- Wechsler und haendler

Johst, Petra *see* Darstellung des einflusses von erfahrung im umgang mit einem hightech-verfahren in der medizin
Joie fait peur / Girardin, Emile De – Paris, France. 1878 – 1r – us UF Libraries [440]
Joie fait peur / Girardin, Emile De – Paris, France. 1879 – 1r – us UF Libraries [440]
La joie fait peur : comedie en un acte et en prose / Girardin, Emile de, Mme – 3e ed. Paris: M Levy, 1854 [mf ed 1984] – 1mf – 9 – 0-665-18879-X – mf#18879 – cn Canadiana [820]
JOIN Community Union *see* Firing line
Joint army-navy intelligence studies (janis), 1944-1945 / U.S. Strategic Bombing Survey – (= ser Records of the united statgic bombing survey) – 20r – 1 – mf#M1169 – us Nat Archives [355]
Joint Board Fur, Leather & Machine Workers Union *see* Flm joint board tempo
Joint Center for Political Studies [US] et al *see* Focus
Joint Committee on Future Status Subcommittees [TTPI (U.S.)] *see* Fourteen questions
Joint Committee on Future Status [TTPI (U.S.)] *see*
- Draft compact of free association
- Report on 4th round of status negotiations
- Report on 7th round of status negotiations
- Report on 8th round of status negotiations
- Report on the 3rd round of status negotiations
- Summary of future political future status talks

Joint Committee on Future Status [TTPI (U.S.)] Eastern Districts Subcommittee *see*
- Hearings in truk, ponape and the marshall islands, jul 1973
- Report to the 5th congress of micronesia

Joint Committee on Future Status [TTPI (U.S.)] Western Districts Subcommittee *see*
- Hearings in yap, palau and the marianas, jul 1973
- Report to the 5th congress of micronesia

Joint Council of Retail Clerks Unions in Seattle and Vicinty *see* Retail outlook
Joint council of teamsters no 2 news / International Brotherhood of Teamsters, Chauffeurs, Warehousemen, and Helpers of America – v1 n1-v3 n10 [1980 apr-1982 dec/1983 jan] – 1r – 1 – mf#706830 – us WHS [331]
Joint Council of Teamsters No 42 *see* Southern california teamster
Joint East African Board *see* Annual report for the year ended...
Joint East and Central African Board *see*
- Report of the proceedings for the year to...of the joint east african board for promoting the agricultural, commercial and industrial development of kenya, nyasaland, tanganyika, uganda and zanzibar
- The...annual report of the executive council of the joint east african board

Joint expedition with the iraq museum at nuzi / Chiera, E – Paris and Philadelphia, 1927-1934. 1-5v – 11mf – 9 – mf#NE-431 – ne IDC [915]
Joint imc/cbms missionary archives – 12 titles on 2517mf – 9 – (with p/g; material consists mostly of correspondence, memoranda, & other printed material relating especially to two key figures: j h oldham & w paton) – ne IDC [240]
Joint issue – [v1 n1]-v5 n5 [1970 nov 12-1974 may 20] – 1r – 1 – (cont: generation east lansing; bogue street bridge; red apple news; cont by: lansing star weekly) – mf#669085 – us WHS [071]
Joint labor-management trust funds and history of the coverage of non-profit hospitals / U.S. Senate. Committee on Labor and Public Welfare – 2v in 1bk – 9 – $30.00 set – 0-89941-417-6 – mf#201571 – us Hein [360]
Joint legislative report of wisconsin state legislative representatives of the brotherhood of locomotive engineers, order of railway conductors, broth / Brotherhood of Locomotive Engineers [US] et al – 66th-67th, 70th-71st [1943-45, 1951-53] – 1r – 1 – (cont: joint report of wisconsin state legislative representatives of the brotherhood of locomotive engineers, order of railway conductors, brotherhood of railway trainmen, brotherhood of locomotive firemen and enginemen) – mf#3127441 – us WHS [380]
Joint letter of the archbishop and bishops of the ecclesiastical province of halifax : announcing the suppression by the holy see of certain holidays – Halifax: Halifax Print Co, 1893 – 1mf – 9 – mf#03213 – cn Canadiana [241]
Joint letter of the spanish bishops to the bishops of the whole world concerning the war in spain – London, 1938. Fiche W974. (Blodgett Collection of Spanish Civil War Pamphlets) – 9 – us Harvard [946]
Joint letter of the spanish bishops to the bishops of the whole world; the war in spain – N.Y., 1937. Fiche W973. (Blodgett Collection of Spanish Civil War Pamphlets) – 9 – us Harvard [946]

JOINT

Joint participating highway planning and research work program number 2 – WI. 1968-73, 1974-78, 1979-1991/1992 – 3r – 1 – (cont: highway planning & research program; cont by: joint participating intermodal surface transportation planning & research work program, wisconsin) – mf#603622 – us WHS [625]
Joint passenger tariff to the canadian north-west, northern minnesota, dakota and transcontinental points via canadian routes / Canadian Pacific Railway Company – Montreal: Canadian Pacific Railway, [1888]- – 9 – mf#P05148 – cn Canadiana [380]
Joint press reading service, moscow – Moscow, mar 1944-31 dec 1956 – 75r – 1 – us L of C Photodup [073]
Joint regulations of the new hebrides : a consolidated edition of the joint regulations in force on the 18 october 1973 – v1-3. 1973 – 1r – 1 – (available for ref) – mf#pmb doc445 – at Pacific Mss [342]
Joint report of wisconsinstate legislative representatives of the brotherhood of locomotive engineers, order of railway conductors, brotherhood of railway trainmen / Brotherhood of Locomotive Engineers [US] et al – 59th-61st [1929-1933] – 1r – 1 – (cont: report of the legislative representatives of the order of railroad conductors, brotherhood of locomotive firemen and enginemen, brotherhood of railway trainmen, and brotherhood of locomotive engineers; cont by: joint legislative report of wisconsin state legislative representatives of the brotherhood of locomotive engineers, order of railway conductors, brotherhood of railroad trainmen, brotherhood of locomotive firemen and enginemen, brotherhood of railway and steamship clerks) – mf#3127416 – us WHS [071]
Joint resolution regarding the status of the ttpi : message from the president of the us, lyndon b johnson, transmitting same / U.S. Congress – 90th Congr, 1st Sess, Hse Doc n159, aug 21, 1967. Washington: GPO, 1967 – (= ser Micronesia: prelude to the constitutional convention) – 1mf – 9 – $1.50 – mf#LLMC 82-100F, Title 33 – us LLMC [323]
Joint resolution to provide a civil government for the trust territory of the pacific islands (ttpi) / U.S. Congress – Hse jnt rept no 391. Washington: GPO, 1948 – 1mf – 9 – $1.50 – mf#LLMC 82-100F Title 96 – us LLMC [324]
Joint select committee on closer union in east africa : report, minutes of evidence and appendices – v1-3. 1931 – 1r – 1 – mf#95695 – uk Microform Academic [327]
Joint statement of the governments of the united s... – s.l, s.l? 1945 – 1r – us UF Libraries [972]
The joint stock act of connecticut, from the revised statutes. / Connecticut. Laws, Statutes, etc – 3d ed. New Haven: Peck, 1876. 93p. LL-603 – 1 – us L of C Photodup [348]
Joint Strategy and Action Committee [New York NY] *see* Jsac grapevine
Joint Task Force-Southwest Asia *see* Southern watch
The joint trial calendars in the western district of missouri / Steinstra, Donna – Washington: FJC, 1985 – 1mf – 9 – $1.50 – mf#LLMC 95-369 – us LLMC [340]
Joinville esportivo : orgam semanal – Joinvile, SC. 31 jul 1926 – (= ser Ps 19) – mf#UFSC/BPESC – bl Biblioteca [790]
Joinville, Jean, sire de *see* Memoirs of the crusades
Jolibois, Joseph *see* Toujours plus haut!
Jolicur, Philippe Jacques *see* Les freres des ecoles chretiennes
Joliet daily news – Joliet IL. 1908 feb 3, v31 n252, 1908 jul 20, v32 n86 – 1r – 1 – mf#1010898 – us WHS [071]
Joliet signal – Joliet IL. 1877 feb 20, v34 n39 – 1r – 1 – (cont: juliet courier) – mf#871933 – us WHS [071]
Joliet veckoblad – Joliet IL. 1909 feb 17, v1 n3 – 1r – 1 – mf#1010901 – us WHS [071]
Joliette illustre : numero souvenir de ses noces d'or, 1843-93 / Gervais, Albert – Quebec: A Gervais, 1893? – 1mf – 9 – mf#03337 – cn Canadiana [971]
Joliffe, Thomas *see*
– Letters from palestine
Jolivet, A et al *see* Esquisses allemandes
[La jolla-] la jolla breakers – CA. 1906 – 1r – 1 – $60.00 – mf#C03607 – us Library Micro [071]
Jolley, J M *see* Scenic gems of daytona, florida
Jolliffe, Jill *see* East timor question, 1975-2002
Jollivet, Adolphe *see*
– Annexion du texas
– Les etats-unis d'amerique et l'angleterre
Jolly, Alexander *see* Friendly address to the episcopalians of scotland
Jolly, John *see* Gold spring diary
Jolly, Julius *see* Maanava dharma-saastra = the code of manu
Jolly leaves – (New York). 1910 – 1 – us AJPC [830]
Jolly researcher – v1 n1-v6 n4 [1983 jan-1988 oct], ns: v1-3 [1983-85] – 1r – 1 – mf#1060612 – us WHS [071]
Jolly roger / Pringle, Patrick – New York, NY. 1953 – 1r – us UF Libraries [972]
Jolobe, James *see* Amavo
Jolowitz, Heimann *see* Bibliotheca aegyptiaca
Joly, Aylthon Brandao *see* Conheca a vegetacao brasileira
Joly de Lotbiniere, Henri Gustave *see* Aux libres et intelligents electeurs de la province de quebec
Joly, Henri *see*
– The psychology of the saints
– Le socialisme chretien
Joly, Leon *see* Le christianisme et l'extreme orient
Joly, Monique *see* Bibliographie analytique de l'oeuvre de marcel clement
Jom – Warrendale. 1989+ (1) 1989+ (5) 1989+ (9) – (cont: journal of metals) – ISSN: 1047-4838 – mf#531,03 – us UMI ProQuest [660]
Jom – New York. 1974-1976 (1) 1974-1976 (5) 1976-1976 (9) – (cont: journal of metals. cont by: journal of metals) – ISSN: 0098-4558 – mf#531,01 – us UMI ProQuest [660]
Jom journal of occupational medicine – Philadelphia. 1968-1994 (1) 1968-1994 (5) 1979-1994 (9) – (cont: journal of occupational medicine. cont by: journal of occupational and environmental medicine) – ISSN: 0096-1736 – mf#12294,01 – us UMI ProQuest [360]
Joma : der mischnatraktat "versoehnungstag" / ed by Strack, Hermann Leberecht – Berlin: H. Reuther, 1888 – (= ser Schriften des Institutum Judaicum in Berlin) – 1mf – 9 – 0-8370-2085-9 – (incl bibl ref and ind of hebrew words) – mf#1985-0085 – us ATLA [270]
Jomini, Antoine H de *see*
– Napoleon in the other world
– Vie politique et militaire de napoleon
Jommelli, Nicolo *see*
– Astianatte
– Il cajo fabrizio
Jonah – Cambridge, England. 1879 – 1r – us UF Libraries [939]
Jonah : his life, character, and mission viewed in connexion with the prophet's own times and future manifestations of god's mind and will in prophecy / Fairbairn, Patrick – Edinburgh: John Johnstone, 1849 – 1mf – 9 – 0-8370-9862-9 – (incl bibl ref) – mf#1986-3862 – us ATLA [221]
Jonah Ben Abraham Gerondi *see* Sefer ha-yir'ah
Jonah in fact and fancy / Banks, Edgar James – New York: Wilbur B. Ketcham, [1899]. Chicago: Dep of Photodup, U of Chicago Lib, 1975 (1r); Evanston: American Theol Lib Assoc, 1984 (1r) – 1 – 0-8370-0534-5 – mf#1984-B478 – us ATLA [221]
The jonah legend : a suggestion of interpretation / Simpson, William – London: Grant Richards, 1899 – 1mf – 9 – 0-8370-5264-5 – (incl bibl ref and index) – mf#1985-3264 – us ATLA [221]
Jonas, Fritz *see*
– Ansichten ueber aesthetik und literatur
– Schillers briefe
Jonas, J *see*
– Annotationes ivsti ionae
– Das sibend capittel danielis von des tuercken gottes lesterung vnd schrecklicher moerderey mit vnterricht justi jone
Jonas, Johannes Benoni Eduard *see* Burg neideck
Jonas, Justus *see*
– Des 20. psalm auslegung
– Two funeral sermons on the death of dr. martin luther
Jonas, Justus et al *see* Authentische berichte ueber luthers letzte lebensstunden
Jonas king, missionary to syria and greece – New York: American Tract Society, c1879 – 1mf – 9 – 0-524-00544-3 – mf#1990-0044 – us ATLA [240]
Jonas, Ludwig *see*
– Die christliche sitte
– Reden und abhandlungen, der koeniglichen akademie der wissenschaften
Jona's nursing scan in administration – Philadelphia. 1987-1989 (1) 1987-1989 (5) 1987-1989 (9) – (cont by: nursing scan in administration) – ISSN: 0888-6288 – mf#16001 – us UMI ProQuest [610]
Jonas, Richard *see* Gruendlicher bericht des deutschen meistergesangs
Jonathan and his continent : rambles through american society / O'Rell, Max & Allyn, Jack – Toronto: W Bryce, 1887? – 3mf – 9 – (trans by madame paul blouet) – mf#03704 – cn Canadiana [917]
Jonathan c gibbs / Johnson, J M – s.l, s.l? 193-? – 1r – us UF Libraries [978]
Jonathan dickinson and the college of new jersey : or, the rise of colleges in america / Cameron, Henry Clay – Princeton, NJ: CS Robinson, 1880 – 1mf – 9 – 0-524-00983-X – mf#1990-0260 – us ATLA [378]
Jonathan edwards / Allen, Alexander Viets Griswold – Boston: Houghton, Mifflin, 1889 – (= ser American religious leaders) – 1mf – 9 – 0-7905-6160-3 – (incl bibl ref) – mf#1988-2160 – us ATLA [240]
Jonathan edwards : a retrospect / Allen, Alexander Viets Griswold et al; ed by Gardiner, Harry Norman – Boston: Houghton, Mifflin, 1901 – 1mf – 9 – 0-7905-5992-7 – mf#1988-1992 – us ATLA [240]
Jonathan edwards idealismus / MacCracken, John Henry – Halle a S: CA Kaemmerer, 1899 – 1mf – 9 – 0-7905-9322-X – (incl bibl ref) – mf#1989-2547 – us ATLA [140]
Jonathan et son continent : la societe americaine / O'Rell, Max & Allyn, Jack – Paris: C Levy, 1889 – 5mf – 9 – mf#04482 – cn Canadiana [917]
Jonathan swift und g ch lichtenberg : zwei satiriker des achtzehnten jahrhunderts / Meyer, Richard M – Berlin: W Hertz, 1886 [mf ed 1993] – viii/84p – 1 – mf#7593 – us Wisconsin U Libr [410]
Jonathan trumbull papers *see* Trumbull, jonathan, papers, ms 2347
Die jonathan'sche pentateuch-uebersetzung in ihrem verhaeltnisse zur halacha : ein beitrag zur geschichte der aeltesten schriftexegese / Gronemann, S – Leipzig: Robert Friese, 1879 – 1mf – 9 – 0-8370-3404-3 – (incl bibl ref) – mf#1985-1404 – us ATLA [221]
Joncas, Louis Zepherin *see* The sportsman's companion
Jones, A H M *see*
– Cities of the eastern roman provinces
– The cities of the eastern roman provinces
Jones, A M *see*
– Yazini ukuhamba
– Zivai kufamba
Jones, Adam Leroy *see* Early american philosophers
Jones, Alfred Gilpin *see*
– Speeches on the address
– Tao yuan hsi i
Jones, Allen Bailey *see* The spiritual side of our plea
Jones, Allen Bailey et al *see* A symposium on the holy spirit
Jones, Arthur Creech *see* African challenge
Jones, Arthur Gray *see* Thornton rogers sampson, d.d., ll.d., 1852-1915
Jones, Bobby *see* Down the fairway
Jones Bros and Co *see* Latest novelties and season goods
Jones, Burr W *see* The law of evidence in civil cases
Jones, C *see* The living dialect of cardington, shropshire
Jones, Charles *see*
– Calypso and carnival of long ago and today
– Latin american independence
Jones, Charles Colcock *see*
– The history of the church of god during the period of revelation
– The religious instruction of the negroes in the united states
Jones, Charles H *see* Livingstone's and stanley's travels in africa
Jones, Charles P *see*
– An appeal to the sons of africa
– Roman catholicism scripturally considered
Jones, Charles William Frederick *see* A catalogue of the books in the bangor cathedral library
Jones, Chester Lloyd *see*
– Caribbean backgrounds and prospects
– Caribbean since 1900
– Guatemala
Jones, Chester Llyod *see* United states and the carribbean
Jones, Clarence Fielden *see* Symposium on the geography of puerto rico
Jones, Daniel *see*
– Sechuana reader
– Sechuana reader in international orthography
Jones, David *see*
– Funeral sermon
– A journal of two visits made to some nations of indians on the west side of the river ohio, in the years 1772 and 1773
– The welsh church and welsh nationality
Jones, David Lewis *see* British and irish biographies, 1840-1945
Jones, Donald Forsha *see* Genetics in plant and animal improvement
Jones, Dwight Arven *see*
– The business corporations law.
– The business corporations law...and other laws concerning business corporations in the state of new york
– The law and practice under the statutes in the state of new york
– A treatise on the construction or interpretation of commercial and trade contracts
– A treatise on the negligence of municipal corporations
Jones, Eli Stanley *see* Mahatma gandhi
Jones, Emily Elizabeth Constance *see* Lectures on the ethics of t.h. green, mr. herbert spencer, and j. martineau
Jones, Ernest *see* Papers on psycho-analysis
Jones, Eustace Hinton *see* Cross of osiris
Jones family['s] grandchildren / Students for a Democratic Society – v1 n1-v2 n6 [1968 apr-1969:late oct] – 1r – 1 – mf#714256 – us WHS [071]
Jones, Frederick Augustus *see* The dates of genesis
Jones, George *see* Sir francis chantrey...
Jones, George E *see* Tumult in india
Jones, George Heber *see*
– English-korean dictionary
– Korea
– The korean revival
Jones, Gilmer Andrew *see* Jones' quizzer
Jones, Grove B *see*
– Soil survey of hernando county, florida
– Soil survey of jefferson county, florida
– Soil survey of pinellas county, florida
– Soil survey of the jacksonville area, florida
– Soil survey of the marianna area, florida
Jones, H M *see* Report on the 1966 swaziland population census
Jones, H W *see* Reaction of zinc sulfate with the soil
Jones, Harold Spencer *see*
– Greenwich observations in astronomy and magnetism made at the royal observatory, greenwich, the royal greenwich observatory, herstmonceux, and the royal greenwich observatory, abinger, in the year...
– Greenwich observations in astronomy, magnetism and meteorology made at the royal observatory, greenwich, the royal greenwich observatory, herstmonceux, and the royal greenwich observatory, abinger, in the year...
Jones, Harold W *see* Anti-achitophel
Jones, Harry W *see* Woman's piety and its beauty
Jones, Henry *see*
– A critical account of the philosophy of lotze
– Idealism as a practical creed
– The immortality of the soul in the poems of tennyson and browning
– Social powers
Jones, Henry D *see* Purifying hope
Jones, Henry Percy *see* Dictionary of foreign phrases and classical quotations
Jones, Henry, Sir *see* The working faith of the social reformer
Jones, Henry, Sir et al *see* The child and religion
Jones, Henry Stuart *see*
– Companion to roman history
– The roman empire, b.c. 29-a.d. 476
Jones, Hugh *see* The evil of consenting to popery
Jones, Idwal *see* The vineyard
Jones, J A *see* The bunhill memorials
Jones, J M *see* La fin du mandat francais en syrie...
Jones, Jack *see* Daytona beach and environs
Jones, James *see* General legal forms and precedents, for ordinary use and with explanatory changes adapted to special cases.
Jones, James Edward *see* The implementation of a perennial program of evangelism
Jones, Jenkin Lloyd *see* Love and loyalty
Jones, Jesse Henry *see* Know the truth
Jones, Joel *see* The voice of jesus and the coming glory
Jones, John *see* A letter to a friend in the country
Jones, John Cynddylan *see* Primeval revelation
Jones, John Daniel *see*
– The glorious company of the apostles
– The model prayer
– Things most surely believed
Jones, John David Rheinallt *see* Bushmen of the southern kalahari
Jones, John G *see* A concise history of the introduction of protestantism into mississippi and the southwest
Jones, John Griffing *see* Complete history of methodism as connected with...
Jones, John Paul *see*
– Papers
– Papers of john paul jones
Jones, John Peter *see*
– Hinduism and christianity
– India
– India's problem
– India's problem, krishna or christ
– The modern missionary challenge
– The year book of missions in india, burma and ceylon
Jones, John T *see* Journeaux des sieges... espagne
Jones, Kenneth W *see* An analysis of backgrounds of professional baseball players
Jones, Kim D *see* A randomized controlled trail of muscle strengthening versus flexibility training in fibromyalgia
Jones, Leonard Augustus *see*
– Forms in conveyancing, comprising precedents for ordinary use, and clauses adapted to special and unusual cases
– Fraudulent mortgages of merchandise
– Legal forms; including forms in conveyancing, together with general legal and business forms.
– The legal nature of the rolling-stock of railroads

– A treatise on the law of landlord and tenant.
– A treatise on the law of liens, common law, statutory, equitable and maritime
– A treatise on the law of mortgages of real property
– A treatise on the law of mortgages on personal property
– A treatise on the law of pledges, including collateral securities
Jones, Louis Thomas *see* The quakers of iowa
Jones, Margaret Josephine *see* The lure of korea
Jones, Mary *see* The story of mary jones and her bible
Jones [Mrs] *see* An account of the loss of the wesleyan missionaries
Jones, Neville *see*
– Early days and native ways in southern rhodesia
– Guide to the zimbabwe ruins
– Rhodesian genesis
Jones of virginia letter / Genealogical & Historical News [Organization] – 1986 sum-1989/90 win – 1r – 1 – mf#1846856 – us WHS [929]
Jones, Owen *see*
– An apology for the colouring of the greek court
– The church of the living god; also, the swiss and belgian confessions and expositions of the faith
– Designs for mosaic and tessellated pavements
– Examples of chinese ornament
– The grammar of ornament
– Lectures on architecture and the decorative arts
– On the true and the false in decorative arts
Jones, Owen et al *see* Description of the egyptian court
Jones, Philip Lovering *see* A restatement of baptist principles
Jones' quizzer : consisting of north carolina supreme court questions and answers, from september term, 1898, to august term, 1920 / Jones, Gilmer Andrew – Atlanta, Foote & Davies Co., 1921 280 p. LL-925 – 1 – us L of C Photodup [347]
Jones, R O *see* St peter
Jones, Richard *see*
– Friendly address to the receivers of the doctrines of the new jerus...
– Letter to the right honourable sir robert peel, bart
– Remarks on the manner in which tithe should be assessed
Jones, Robert Dorsey *see* With the american fleet from the atlantic to the pacific
Jones, Robert William *see* Journalism in the united states
Jones, Rufus M *see* A boy's religion from memory
Jones, Rufus Matthew *see*
– The abundant life
– A boy's religion
– The double search
– A dynamic faith
– Eli and sybil jones
– The inner life
– Later periods of quakerism
– Practical christianity
– Quakerism
– Social law in the spiritual world
– Spiritual reformers in the 16th and 17th centuries
– Studies in mystical religion
Jones, Rufus Matthew et al *see* The quakers in the american colonies
Jones, Sam Porter *see*
– Lightning flashes and thunderbolts
– Sermons and sayings
– Sermons by sam jones and sam small the noted revivalists
Jones, Sam Porter et al *see* Sermons
Jones, Scervant *see* Arguments for and against a baptist theological school at williamsburg, va
Jones, Singleton Thomas *see* Sermons and addresses of the late rev bishop singleton t jones, dd, of the african methodist episcopal zion church
Jones, Spencer *see* England and the holy see
Jones, T B *see* The effect of chronic exercise stress on hippocampal glucocorticoid and serotonin 1a receptors
Jones, T F E *see* The gold coast and the fantis
Jones, Thomas *see*
– Christian minister
– Natural and spiritual growth
Jones, Thomas Lewis *see* From the gold mine to the pulpit
Jones, Tiberius Gracchus *see* The baptists
Jones, W A Burt *see* John rice jones
Jones, W M *see*
– Amidst timiskiming [sic] and kipawa pines
– Sport and pleasure in the virgin wilds of canada on lakes temiskaming, temagaming, kippewa
Jones, William *see*
– Biographical sketch of the rev. edward irving
– Catholic doctrine of a trinity proved by above an hundred short and...
– Essay on the church
– The jubilee memorial of the religious tract society
– Physiological disquisitions
– Quaker campaigns in peace and war
Jones, William Bence *see* The life's work in ireland of a landlord who tried to do his duty
Jones, William Caswell *see* A practical treatise upon the jurisdiction of, and practice in, the county and probate courts of illinois.
Jones, William F R S *see* The practice of interest
Jones, William Tudor *see*
– An interpretation of rudolf eucken's philosophy
– Present-day ethics in their relations to the spiritual life
– The spiritual ascent of man
Jones, Willoughby, Sir *see* Christianity and common sense
Jones, WN *see* The north carolina manual of law and forms for justices of the peace, county officers, lawyers, and business men.
Jonesboro gazette – Jonesboro IL. 1879 nov 15, v30 n36 – 1r – 1 – (cont: jonesboro weekly gazette) – mf#845807 – us WHS [071]
Jonesborough, Maine. Jonesborough and Addison Baptist Church *see* Records
Jong, Albert Johannes de *see* Afgoderye der oost-indische heydenen
Jong, Hein de *see* Vijftien jaren vrijmetselaar
Jong, Karel Hendrik Eduard de *see* Hegel und plotin
Jong, Ymen Peter de *see* Vragen en definities betreffende de gereformeerde geloofsleer voor catechetisch gebruik
Jonker, Gerrit Jan Abraham *see* De bekeeringsgeschiedenis van een japanner
Jonker, W *see* Een nederlander als baanbreker der zending in tibet
Jonkman, H F *see* Mededeelingen over zuid-afrika
Jonkopings dagblad – Jonkoping, 1872-75 – 9 – sw Kungliga [078]
Jonkopingsbladet – Joenkoeping, 1843-72 – 9 – sw Kungliga [078]
Jonkopingsposten – Jonkoping, Sweden. 1865-1978 – 530r – 1 – sw Kungliga [078]
Jonkopingsposten – Jonkoping, Sweden. 1979- – 1 – (varnamotidningen, 1979-81) – sw Kungliga [078]
Jonquet, A *see* Original sketches for art furniture
Jonson, Linnea M *see* Self-perception and motor proficiency of hearing-impaired children
Jonsson, Sigurur *see* Bibliusoegur og agrip af kirkjusoegunni handa boernum
De joodsche wachter – Rotterdam. v. 1-36. Jan. 5, 1905-May 3, 1940. Johore Bahrue. 1957-1966 – 1 – us NY Public [939]
Joos, Martin *see* Readings in linguistics
Jooste, Jacobus Francious *see* Study of the phytosociology and small mammals of the rolfontein nature reserve, cape province
Joramel / Lainez, Jorge B – San Salvador, El Salvador. 1962 – 1r – us UF Libraries [972]
Jordaan, Bee *see* Splintered crucifix
Jordan : internal affairs and foreign affairs, 1955-1959 / U.S. State Dept – (= ser Confidential u s state department central files) – 10r – 1 – $1935.00 – 1-55655-263-7 – (with p/g) – us UPA [327]
Jordan *see*
– Al-jaridah al-rasmiyah
– Das diakonissenhaus fuer die provinz sachsen zu halle a. saale, 1857-1907
Jordan, A C *see*
– Ingqumbo yeminyanya
– Xhosa course
Jordan, Archibald Currie *see* Practical course in xhosa
Jordan, Claude *see* La clef du cabinet des princes de l'europe
Jordan, Daniel *see* Patria
Jordan, David Starr *see*
– Concerning sea power
– Papers
– Standeth god within the shadow
– The story of a good woman, jane lathrop stanford
Jordan Diaz, Alfredo Alberto *see*
– Canto de soledad y doce poemas crepusculares
– Interludio
Jordan, Dwight Allan *see* Sunday talks on nature topics
Jordan, E L *see* Deutsche kulturgeschichte im abriss
Jordan, Edward *see* An interesting trial of edward jordan and margaret his wife
Jordan, Edwin Oakes *see* Text-book of general bacteriology
Jordan gazette – Dawson, MT. 1914-1922 (1) – mf#64350 – us UMI ProQuest [071]
Jordan, H *see* Armenische irenaeusfragmente
Jordan, Herbert William *see* Converting a business into a private company
Jordan, Hermann *see*
– Jesus und die modernen jesusbilder
– Die mission des christentums und die weltpolitik der nationen
Jordan, Horst W *see* Der einfluss des supreme court auf die politik der u.s.a. von 1789 bis zum ende des zweiten weltkriegs.
Jordan, J *see*
– Appeal to the evangelical clergy against their concurrence in the d...
– Second appeal to the right reverend the lord bishop of oxford, on the divinity of the tract...
Jordan, Joel C *see* The relationship between percent peak oxygen consumption and peak heart rate during deep water running in the adult population
Jordan, John A *see* The grosse-isle tragedy and the monument to the irish fever victims 1847 reprinted
Jordan, John W *see*
– Genealogical and personal history of fayette county pennsylvania, vols 1-3
– History of fayette county, pennsylvania
Jordan, K *see* Die bistumsgruendunen heinrichs des loewen (mgh schriften:3.bd)
Jordan, Lewis G *see* Negro baptist history
Jordan, Lewis Garnett *see*
– Baptist standard church directory and busy pastor's guide
– Pebbles from an african beach
Jordan, Louis Henry *see*
– Comparative religion, its adjuncts and allies
– Comparative religion, its genesis and growth
– Comparative religion, its method and scope
– Comparative religion, its origin and outlook
– Comparative religion, its range and limitations
– Modernism in italy
– The study of religion in the italian universities
Jordan, Mary C *see* A comparison of intermittent exercise and relaxation versus steady state exercise on fitness levels and attitudes in seventh grade girls
Jordan, Richard *see* A journal of the life and religious labours of richard jordan
The jordan river controversy / Khouri, Fred J – Notre Dame, IN: University of Notre Dame Press, 1965 – us CRL [956]
Jordan, Ronald *see* Bodennutzung in venezuela
Jordan, Samuel Alexander *see* Rabbi jochanan bar nappacha
Jordan valley express – Jordan OR: C A Hackney [wkly] – 1 – us Oregon Lib [071]
Jordan, W G *see* Commentary on the book of deuteronomy
Jordan waters conflict / Doherty, Kathryn B – New York: Carnegie Endowment for International Peace, 1965 – us CRL [956]
Jordan, Wilhelm *see*
– Andachten
– Letzte lieder
– Liebe was du lieben darfst
– Schaum
– Die sebalds
– Sein zwillingsbruder
– Tausch enttaeuscht
– Zwei wiegen
Jordan, William George *see*
– Biblical criticism and modern thought
– Commentary on the book of deuteronomy
– The song and the soil
Jordanus de Yano *see* Chronica fratris jordani
Jordbrukarnas foreningsblad – Stockholm, Sweden. 1930-70 – 115r – 1 – sw Kungliga [078]
Jorden och folket – Falkoeping, Sweden. 1913-15 – 1r – 1 – sw Kungliga [078]
Jorden, Ryan A *see* Influence of ankle orthoses on joint motion and postural stability 109=before and after exercise
Jorden, W *see* Das cluniazensische totengedaechtniswesen (910-954)
Jordon high school yearbook – Trail Blazer, Long Beach, CA. 1936-92 – 11r – 1 – $550.00 – mf#R60011 – us Library Micro [373]
Jordon, William F *see* Crusading in the west indies
Jordon, William J *see* William j. jordan papers, ms p.p.
Jorge ricardo bejarano narino, su vida sus infortunios, su talla historica / Bayle, Constantino – Madrid: Razon y Fe, 1939 – 1 – sp Bibl Santa Ana [946]
Jorgensen, Arlo G *see* A new hydrostatic method
Jorgensen, Simon Emanuel *see* Folk og kirke paa madagaskar
Jorgenson, Shane M *see* The cognitive, affective, and behavioral characteristics of students enrolled in physical education activity classes at brigham young university
La jornada – Mexico City, Mexico. 1991 – 12r – 1 – us L of C Photodup [079]
Jornada precisa / Morales, Jorge Luis – Barcelona, Spain. 1962 – 1r – us UF Libraries [972]
Jornadas de educacion especial : 1970 / Centro de Educacion Especial – Caceres: Tip. Extremadura, 1971 – 1 – sp Bibl Santa Ana [370]
Jornadas de mayo – Bogota, Colombia. 1957 – 1r – us UF Libraries [972]
Jornadas divertidas, politicas sentencias y hechos memorables de reyes y heroes de la antiguedad / Gomez, Madeleine A – Madrid. v1-8. 1797 – 1 – $60.00 – mf#0245 – us Brook [880]
Jornais criticos e humoristicos de porto alegre no... / Ferreira, Athos Damasceno – Porto Alegre, Brazil. 1944 – 1r – us UF Libraries [972]
Jornal – Rio de Janeiro Brazil, 10 nov 1939; 16 dec 1944-19 aug 1945; 2 sep, 14 oct, 9 dec 1955 – 6 1/2r – 1 – uk British Libr Newspaper [079]
O jornal – Lisbon: Publicacoes Projornal. ano1-5 n14-250. 1975 agosto 1-1980 feb 7 [mf ed 1984-] – (= ser Portuguese newspapers of the 1970's, a coll) – 2r (ill) – 1 – mf#2014 – us Wisconsin U Libr [074]
O jornal – Rio de Janeiro, RJ: [s.n.] 04-18 nov 1896 – (= ser Ps 19) – mf#P18A,02,22 – bl Biblioteca [073]
O jornal – S Tome: O Jornal, jun 17-sep 5 1923 – us CRL [079]
O jornal batista – Brazil. 1901-20, 1944, 1947 – 1 – us Southern Baptist [242]
Jornal da academia medica homeopatica do brasil – Rio de Janeiro, RJ. jan-fev 1848 – (= ser Ps 19) – mf#P03A,03,18 – bl Biblioteca [615]
Jornal da feira – Feira de Santana, BA: Typ do Jornal da Feira, 01, 06, 09 ago 1884 – (= ser Ps 19) – mf#P18B,02,25 – bl Biblioteca [380]
Jornal da noite – Rio de Janeiro, RJ. 23 nov-dez 1881; mar-24 maio 1882 – (= ser Ps 19) – mf#P05,04,47 – bl Biblioteca [073]
Jornal da parayba – Paraiba: Typ Parahybana, 29 set 1863; fev 1875; jan-maio, jul-set, nov 1888; jan, mar-jun, ago, 14 nov 1889 – (= ser Ps 19) – mf#P11B,04,07 – bl Biblioteca [320]
Jornal da sociedade amante da instrucao – Rio de Janeiro, RJ: Typ do Diario do Rio, 28 ago 1839 – (= ser Ps 19) – mf#P12,05,20 – bl Biblioteca [370]
Jornal da tarde : folha politica e noticiosa – Rio de Janeiro, RJ: Typ Americana, 01 mar 1877-16 mar 1878 – (= ser Ps 19) – mf#P18A,2,34 – bl Biblioteca [321]
Jornal da tarde – Rio de Janeiro, RJ: Typ Americana, 20 nov 1869-28 jun 1872 – (= ser Ps 19) – mf#P18A,2,33 – bl Biblioteca [321]
Jornal das damas : periodico de instruccao e recreio – Recife, PE. 06 dez 1862 – (= ser Ps 19) – bl Biblioteca [073]
Jornal das familias – Paris, Franca: Typ de Simon Racon e Comp, jan-mar 1863; jan 1864-dez 1869; jan-fev, maio-jul 1870; jan 1871-dez 1874; jan-fev, abr-dez 1875; jan-dez 1876 – (= ser Ps 19) – mf#P02B,01,01-17 – bl Biblioteca [640]
Jornal das novidades – Belem, PA: 01 jun-14 ago 1888 – (= ser Ps 19) – mf#P11,05,05 – bl Biblioteca [073]
Jornal de annuncios – Rio de Janeiro, RJ: Typ Real, 05 maio-16 jun 1821 – (= ser Ps 19) – mf#P01,03,04 – bl Biblioteca [079]
Jornal de domingo: literatura, historia e viagens *see* Jornal do recife
Jornal de goyaz : orgam imparcial – Goias, 12 mar 1892-18 dez 1893 – (= ser Ps 19) – bl Biblioteca [073]
Jornal de instruccao e recreacao *see* Revista universal brazileira
Jornal de instruccao e recreio : da associacao litteraria maranhense – Maranhao: Typ Maranhense, 15 fev 1845-20 jan 1846 – (= ser Ps 19) – mf#P02A,03,23 – bl Biblioteca [073]
Jornal de letras – Rio de Janeiro. 1955-58; 1973-78 – 1 – 60.00 – us L of C Photodup [410]
Jornal de letras – Rio de Janeiro. Brazil. -m. Jul 1952-1959; mar 1960-1964; jun-aug 1965 3 1/2r – 1 – uk British Libr Newspaper [072]
Jornal de macau – Macau: [s.n., aug 25 1875] – 1r – 1 – us CRL [079]
Jornal de maceio *see* Partido liberal
Jornal de noticias – Maceio, AL: Typ do Jornal de Noticias, 05 jul, set 1892; 25 fev 1893 – (= ser Ps 19) – mf#P18B,01,33 – bl Biblioteca [073]
Jornal de noticias – San Francisco: PLC Silveira, sep 21 1917-jun 22 1932 – 14r – 1 – us CRL [071]
Jornal de queluz : folha imparcial – Queluz, SP: Typ do Jornal de Queluz, 29 jul 1877; jan 1878; 22 ago 1880 – (= ser Ps 19) – mf#P18,01,88 – bl Biblioteca [320]
Jornal de terentillo arsa – Sao Paulo, SP: Typ Commercial de Antonio Elias da Silva, 08 jul 1877 – (= ser Ps 19) – mf#P17,02,211 – bl Biblioteca [320]
Jornal do acu : politica, commercio, letras e religiao – Acu, RN. 07 abr 1877 – (= ser Ps 19) – bl Biblioteca [073]
Jornal do agricultor : principios praticos de economia rural – Rio de Janeiro, RJ: Typ Carioca, 05 jul 1879-04 nov 1893 – (= ser Ps 19) – mf#P10,02,09-25 – bl Biblioteca [630]
Jornal do amazonas *see* O liberal do para
Jornal do brasil – Rio de Janeiro, Brazil. 1983 aug 16-1988 dec – 46r – (gaps) – us UF Libraries [079]
Jornal do brasil – Rio de Janeiro Brazil; 25 jan 1940; 1 may-4 may 1945; 31 mar 1972-3 apr 1974; 1 may 1974-dec 1975 – 94 3/4r – 1 – uk British Libr Newspaper [079]
Jornal do brasil – Rio de Janeiro, Brazil. -d. March 1972-Dec 1975. 99 reels – 1 – uk British Libr Newspaper [072]
Jornal do brasil – Rio de Janeiro: [s.n.], 1938-43 – 64r – 1 – us CRL [079]

Jornal do brazil – Rio de Janeiro, RJ. 20 jan-jun 1867; 05 dez 1871 – (= ser Ps 19) – mf#DIPER – bl Biblioteca [073]
Jornal do comercio – Lisbon. 21 22 feb 1971-3 sept 1976. (Portuguese Revolution of 1974. Newspapers from Portugal publ. from 21 Feb 1971 to 15 Feb 1980, collected and filmed by University of Wisconsin-Madison libraries.) – 1 – (incomplete) – us Wisconsin U Libr [074]
Jornal do comercio – Lisboa: Jornal do comercio, aug 1944-apr 1945; aug 1-15 1945 – us CRL [380]
Jornal do comercio – Lisbon, Portugal. -d. 5 Oct 1942-31 Dec 1946; 5 Sept-31 Dec 1948; 10 Aug 1949-23 Dec 1950. Imperfect. 26 reels – 1 – uk British Libr Newspaper [072]
Jornal do comercio – Rio de Janeiro: Typ. d'Emile-Seignot Plancheb, [oct 1899-1901]; 1930-apr 1941; may 15 1941-dec 15 1942; 1943-aug 1 1944; may-jul 1945; aug 16 1945-55; 1956-apr 1980; jun 1980-81 – us CRL [380]
Jornal do commercio – Juiz de Fora, MG: [s.n.] 8 abr, set 1897; jun-jul 1898; dez 1901; nov 1912; abr 1913; 25 ago 1920 – (= ser Ps 19) – mf#P11B,03,40 – bl Biblioteca [380]
Jornal do commercio – Rio de Janeiro Brazil, 18 dec 1944-feb 1945; 12 may-19 aug 1945; 27 jan-5 sep 1957 – 5r – 1 – uk British Libr Newspaper [079]
Jornal do domingo : publicacao consagrada aos conhecimentos uteis – Rio de Janeiro, RJ: Typ Economica, 22 maio 1864 – (= ser Ps 19) – mf#P17,04,67 – bl Biblioteca [073]
Jornal do fundao – Fundao: Jornal do Fundao. ano29 n1417. 10 de marco 1974 (wkly) (incomplete) [mf ed 1984] – (= ser Portuguese newspapers of the 1970's, a coll) – 2r – 1 – mf#10021 – us Wisconsin U Libr [074]
Jornal do povo : folha politica, litteraria, commercial e agricola – Rio de Janeiro, RJ: Typ de Quirino & Irmao, 07 abr-12 maio 1862 – (= ser Ps 19) – mf#P25,03,08 n11 – bl Biblioteca [321]
Jornal do recife – Pernambuco: Typ Academica, jan-dez 1859; jan 1889-maio 1935; jul 1937-08 jan 1938 – (= ser Ps 19) – mf#P11,07,07 – bl Biblioteca [073]
Jornal do theatro lucinda – Rio de Janeiro, RJ. 21 set-25 out 1881 – (= ser Ps 19) – mf#P05,04,50 – bl Biblioteca [790]
Jornal dos farmaceuticos : mensario cientifico e de interesses tecnico-profissionais da farmacia e do laboratorio – Florianopolis, SC. nov 1931; ago-set 1932 – (= ser Ps 19) – bl Biblioteca [615]
Jornal junior – Rio de Janeiro, RJ. 28 set 1885 – (= ser Ps 19) – mf#P17,01,172 – bl Biblioteca [073]
Jornal novo – Lisbon. no. 1-1341; 17 apr 1975-26 sept 1979 – 1 – (incomplete. portuguese revolution of 1974. newspapers from portugal publ. from 21 feb 1971 to 15 feb 1980, collected and filmed by university of wisconsin-madison libraries) – us Wisconsin U Libr [072]
Jornal official – Manaus, AM: Typ do Amazonas, 07 jan-28 fev 1882 – (= ser Ps 19) – mf#DIPER – bl Biblioteca [073]
Jornal portugues = Portuguese journal – Alameda, CA: PLC Silveira, jul 1932-1934 – 2r – 1 – us CRL [946]
Jornal revolucionario : orgao official do comando geral das forcas revolucionarias em barbacena – Barbacena, MG. 06-29 out 1930 – (= ser Ps 19) – bl Biblioteca [320]
Jornal unico : celebracao do 4. centenario do descobrimento do caminho maritimo para a india por vasco da gama – Macao: Tpypographias de NT Fernandes e Filhos e Noronha & Ca, 1898 – us CRL [910]
Jorond, Antoine Victor *see* Guadeloupe et ses iles
Jos Schlitz Brewing Co *see* Schlitz news
Jose a saco : estudio y bibliografia / Moreno Fraginals, Manuel – Santa Clara, Cuba. 1960 – 1r – us UF Libraries [972]
Jose Aleixo *see* Euclides da cunha e o socialismo
Jose antonio cortina : epoca y caracter, 1853-1884 / Arce, Luis A – Habana, Cuba. 1953 – 1r – us UF Libraries [972]
Jose Antonio de San Alberto, Archbishop of Rio de la Plata *see* Voces del pastor en el retiro
Jose Antonio de San Alberto, archbishop of Rio de la Plata *see*
– Carta circular, o edicto, de el ilustrisimo
– Carta pastoral, que dirige a los parrocos, sacerdotes y demas fieles de su diocesi
– Carta pastoral que el illustrissimo senor d fr joseph antonio de san alberto, arzobispo de la plata
– Carta pastoral que el illustrissimo senor don fray joseph antonio de san alberto, arzobispo de la plata
– Carta segunda pastoral que el illustrissimo senor d fr joseph antonio de san alberto, arzobispo de la plata
Jose antonio dominguez : su vida y sus obras / Pagoaga, Raul Arturo – Tegucigalpa, Mexico. 1947 – 1r – us UF Libraries [972]
Jose antonio en la carcel de madrid [del 14 de marzo al 6 de junio de 1936] : interesante reportaje con raimundo fernandez-cuesta / Antiguedad, Alfredo R – Cegama: Impr E Gimenez [193-] [mf ed 1977] – (= ser Blodgett coll) – 1mf – 9 – mf#w721 – us Harvard [946]
Jose antonio saco documentos para su vida / Saco, Jose Antonio – Habana, Cuba. 1921 – 1r – us UF Libraries [972]
Jose antonio saco y sus ideas cubanas / Ortiz, Fernando – Habana, Cuba. 1929 – 1r – us UF Libraries [972]
Jose asuncion silva / Miramon, Alberto – Bogota, Colombia. 1957 – 1r – us UF Libraries [972]
Jose asuncion silva; ensayo biografico con documentos ineditos / Miramon, Alberto – Bogota, 1937. Bibliografia general, p191-194 – 1 – us Wisconsin U Libr [972]
Jose bonifacio / Sousa, Octavio Tarquinio de – Rio de Janeiro, Sao Paulo, Brazil: J Olympio 1945 – (= ser Colecao documentos brasileiros 51) – 1r – 1 – us UF Libraries [972]
Jose bonifacio o moco / Faria, Julio Cezar De – Sao Paulo, Brazil. 1944 – 1r – us UF Libraries [972]
Jose de alencar : orgao do club litterario jose de alencar – Maceio, AL: Typ de Amintas de Mendonca, maio-out 1883; maio, jul-ago 1884; maio, 25 jul 1885 – (= ser Ps 19) – mf#P18B,01,34 – bl Biblioteca [440]
Jose de la luz y caballero como educador / Luz Y Caballero, Jose De La – Habana, Cuba. 1931 – 1r – us UF Libraries [972]
Jose de los cubanos / Riveron Hernandez, Francisco – Habana, Cuba. 1960 – 1r – us UF Libraries [972]
Jose eusebio caro / Galvis Salazar, Fernando – Bogota, Colombia. 1955 – 1r – us UF Libraries [972]
Jose eusebio caro, guion de una estirpe / Ospina Ortiz, Jaime – Bogota, Colombia. 1958 – 1r – us UF Libraries [972]
Jose fernandez de madrid y su obra en cuba / Fernandez Madrid, Jose – Habana, Cuba. 1962 – 1r – us UF Libraries [972]
Jose figueres en la evolucionde costa rica / Navarro Bolandi, Hugo – Mexico City?, Mexico. 1953 – 1r – us UF Libraries [972]
Jose joaquin palma / Azcuy Alon, Fanny – Habana, Cuba. 1948 – 1r – us UF Libraries [972]
Jose justo milla / Duron Y Gamero, Romulo Ernesto – Tegucigalpa, Mexico. 1940 – 1r – us UF Libraries [972]
Jose lopez de toro y ramon paz remolar / Barrado Manzano, Arcangel – Madrid: Archivo Ibero-Americano, 1959 – 1 – sp Bibl Santa Ana [946]
Jose ma paranhos, visconde do rio branco / Besouchet, Lidia – Rio de Janeiro, Brazil. 1945 – 1r – us UF Libraries [972]
Jose madriz, diplomatico / Madriz, Jose – Managua, Nicaragua. 1965 – 1r – us UF Libraries [972]
Jose maria vargas / Dominguez, Rafael – Caracas, 1930; Madrid: Razon y Fe, 1931 – 1 – sp Bibl Santa Ana [920]
Jose marti : critico literario / Portuondo, Jose Antonio – Washington, DC. 1953 – 1r – us UF Libraries [972]
Jose marti : escritor americano / Marinello, Juan – Mexico City?, Mexico. 1958 – 1r – us UF Libraries [972]
Jose marti / Machado Bonet, Ofelia – Montevideo, Uruguay. 1942 – 1r – us UF Libraries [972]
Jose marti / Pichardo, Hortensia – Habana, Cuba. 1960 – 1r – us UF Libraries [972]
Jose marti, el santo de america / Rodriguez-Embil, Luis – Habana, Cuba. 1941 – 1r – us UF Libraries [972]
Jose marti y la oratoria cubana / Conte Aguero, Luis – Buenos Aires, Argentina. 1959 – 1r – us UF Libraries [972]
Jose marti y la revolucion cubana / Sanguily Y Garritte, Manuel – New York, NY. 1896 – 1r – us UF Libraries [972]
Jose matias delgado y el movimiento insurgente de... / Baron Castro, Rodolfo – San Salvador, El Salvador. 1962 – 1r – us UF Libraries [972]
Jose mazzini. ensayo italia / Diaz Perez, Nicolas – 1876 – 9 – sp Bibl Santa Ana [920]
Jose moreno nieto / Blanco Garcia, Francisco – Madrid: Saenz de Jubera, 1909 – sp Bibl Santa Ana [440]
Jose, Oiliam *see* Historiografia mineira
Jose p h hernandez : vida y obra / Siaca Rivera, Manuel – San Juan, Puerto Rico. 1965 – 1r – us UF Libraries [440]
Jose sanchez arjona / Blanco Garcia, Francisco – Madrid: Saenz de Jubera, 1909 – sp Bibl Santa Ana [440]
Josef filsers ende : ledzder briefwexel und bolidisches desdamend / Kirschner, Max – Muenchen: F Eher, 1942 – 1r – 1 – us Wisconsin U Libr [920]
Josef viktor widmann : ein lebensbild / Widmann, Elisabeth & Widmann, Max – Frauenfeld; Leipzig: Huber & Co. 2v. 1922-24 – 1 – (incl bibl ref & ind) – us Wisconsin U Libr [430]
Josef von goerres : zum 150. geburtstag (25. januar 1926) / Schellberg, Wilhelm – 2. verb. Aufl. Koeln, 1926 (mf ed 1994) – 1mf – 9 – €24.00 – 3-89349-737-4 – mf#DHS-AR 737 – gw Frankfurter [943]
Josef von goerres : zum 150. geburtstage (25. januar 1926) / Schellberg, Wilhelm – 2. verb. Aufl. Koeln: Gilde-Verlag, 1926 – 1r – 1 – us Wisconsin U Libr [943]
Josef weinheber / Koch, Franz – Muenchen: A Langen, G Mueller, 1942 [mf ed 1992] – 78p – 1 – mf#7760 – us Wisconsin U Libr [430]
Josefa de la Providencia, madre *see* Relacion del origen y fundacion del monasterio del senor san joaquin de religiosas nazarenas carmelitas descalzas de esta ciudad de lima
Josefsohn, Leon *see* Getulio, este desconhecido
Josenhans, Joseph *see* Atlas der evangelischen missions-gesellschaft zu basel
Josenhans, Walther *see* Lord byron und die politik
Joseph : beloved, hated, exalted / Meyer, Frederick Brotherton – New York: F H Revell, [19–?] – (= ser Old Testament Heroes) – 1mf – 9 – 0-524-04472-4 – mf#1992-0141 – us ATLA [221]
Joseph – Breslau (WrocLaw PL), 1879 – 1r – 1 – gw Misc Inst [077]
Joseph / C S – London, England. 18– – 1r – us UF Libraries [240]
Joseph : israelitische jugendzeitung – Breslau: S Freuthal. v1. 1879 [complete] – (= ser German-jewish periodicals...1768-1945, pt 2) – 1r – 1 – $115.00 – mf#B157 – us UPA [939]
Joseph alleine : his companions and times / Stanford, Charles – London: Jackson, Walford, and Hodder, [1861?] – 1mf – 9 – 0-7905-6084-4 – (incl bibl ref) – mf#1988-2084 – us ATLA [941]
Joseph amiot et les derniers survivants de la mission francaise a pekin (1750-1795) / Rochemonteix, Camille de – Paris: Alphonse Picard et Fils, 1915 [mf ed 1995] – (= ser Yale coll) – lxiii/563p – 1 – 0-524-09624-4 – (in french) – mf#1995-0624 – us ATLA [241]
Joseph amiot et les derniers survivants de la mission francaise...pekin (1780-1795) / Rochemonteix, C de – Paris, 1915 – 7mf – 9 – mf#HTM-232 – ne IDC [915]
Joseph and asenath : the confession and prayer of asenath, daughter of pentephres the priest / Brooks, Ernest Walter – London: Society for Promoting Christian Knowledge; New York: The Macmillan Co., 1918 – (= ser Translations Of Early Documents) – 1r – 1 – 0-8370-1521-9 – mf#1984-B388 – us ATLA [270]
Joseph and Co, B H *see* [Trade catalogue of modern jewels]
Joseph and moses, the founders of israel : being their lives as read in the light of the oldest prophetic writings of the bible / Blake, Buchanan – Edinburgh: T & T Clark, 1902 – 1mf – 9 – 0-8370-9444-5 – mf#1986-3444 – us ATLA [221]
Joseph and the land of egypt / Sayce, Archibald Henry – London: J M Dent, [19–?] – (= ser The Temple Series of Bible Handbooks) – 1mf – 9 – 0-524-04811-8 – mf#1992-0231 – us ATLA [221]
Joseph buell family papers, 1785-1956 1810-1890 – [mf 1999] – 2r – 1 – mf#ms3664 – us Western Res [929]
Joseph conrad : centennial essays / Krzyzanowski, Ludwik – New York, NY. 1960 – 1r – us UF Libraries [025]
Joseph costisella : bibliographie descriptive / Martel, Louise – 1964 [mf ed 1979] – (= ser Bibliographies du cours...1947-66) – 1mf – 9 – (with ind) – mf#SEM105P4 – cn Bibl Nat [920]
Joseph, Don *see* Shop talk on spain. the trade unions and the war in spain
Joseph en egypte / Vergote, J – Univ. of Louvain, 1959. In French – 9 – $10.00 – us IRC [240]
Joseph, Eugen *see* Das heidenroeslein
Joseph freiherr von eichendorff : sein leben und seine schriften / Eichendorff, Hermann, Freiherr von – 3. neubearb Aufl. Leipzig: C F Amelang, [1923?] – 1 – (incl ind) – us Wisconsin U Libr [430]
Joseph freiherrn v. eichendorffs werke : in vier baenden = Works / Eichendorff, Joseph, Freiherr von – Leipzig: M Hesse. 4v in 2. [186-?] (mf ed 1990) – 1 – (filmed with: der morgen) – us Wisconsin U Libr [802]
Joseph goerres und die pressepolitik der deutschen reaktion : ein beitrag zur goerresforschung / Poelnitz, Goetz, Freiherr von – Koeln [1930] (mf ed 1992) – 1mf – 9 – €24.00 – 3-89349-064-7 – mf#DHS-AR 26 – gw Frankfurter [430]
Joseph goerres und die pressepolitik der deutschen reaktion : ein beitrag zur goerresforschung / Poelnitz, Goetz, Freiherr von – Koeln: J P Bachem, [1936] – 1r – 1 – us Wisconsin U Libr [943]
Joseph h. choate, new englander, new yorker, lawyer, ambassador / Strong, Theron George – New York, Dodd, Mead, 1917. 390 p. LL-212 – 1 – us L of C Photodup [340]
Joseph, Helen *see*
– If this be treason
– Tomorrow's sun
Joseph herald – Joseph OR: B W Henry, [wkly] – 1 – (began in 1945? cont by: chief joseph herald) – us Oregon Lib [071]
Joseph herald – Joseph OR: Henderson & Henderson, 1902-42 [wkly] – 1 – (cont: silver lake herald (joseph, or); absorbed by: enterprise chieftain) – us Oregon Lib [071]
Joseph hillebrand : sein leben und werk / Schreiber, Hans Ulrich – Giessen: [s.n.], 1937 – 1r – 1 – (incl bibl ref) – us Wisconsin U Libr [920]
Joseph hubert reinkens : ein lebensbild / Reinkens, Joseph Martin – Gotha: F A Perthes, 1906 – 1mf – 9 – 0-7905-8247-3 – mf#1988-8110 – us ATLA [920]
Joseph im schnee : eine erzaehlung / Auerbach, Berthold – 6. aufl. Stuttgart: Cotta, 1874 [mf ed 1993] – 238p – 1 – mf#8464 – us Wisconsin U Libr [880]
Joseph in aegypten / Heyes, Hermann Joseph – 1. & 2. aufl. Muenster i W: Aschendorff 1911 [mf ed 1992] – (= ser Biblische zeitfragen 4/9) – 1mf – 9 – 0-524-05581-5 – (incl bibl ref) – mf#1992-0441 – us ATLA [221]
Joseph, Janice *see* Journal of ethnicity in criminal justice
Joseph jortz : el santo incomparable / Barrado Manzano, Arcangel – Madrid: Arch. Ibero Americano, 1964 – 1 – sp Bibl Santa Ana [240]
Joseph l. bristow papers / Bristow, Joseph L – 1894-1925. In Kansas State Historical Society. Guide – 1 – us Kansas [920]
Joseph lieber joseph mein – (= ser Mssa) – 1mf – 9 – €20.00 – mfchl 451 – gw Fischer [780]
Joseph ludwig colmar. bischof von mainz 1802-1818 : ein zeit- und lebensbild / Selbst, Joseph – Mainz, 1902. 55p – 3-89349-176-7 – gw Frankfurter [240]
Joseph, Morris *see*
– Judaism as creed and life
– The message of judaism
Joseph pilsudski / Merezhkovskii, Dimitrii – Tr. from the Russian by Harriet E. Kennedy. London: S. Low, Marston & Co., 1921. 20p – 1 – us Wisconsin U Libr [920]
Joseph priestley / Thorpe, Thomas Edward – London: JM Dent; New York: EP Dutton, 1906 – (= ser English Men of Science) – 1mf – 9 – 0-524-01096-X – mf#1990-4061 – us ATLA [100]
Joseph s assemani : et la celebration du concile libanais maronite de 1736 / Mafoud, P – Roma, 1965 – €7.00 – ne Slangenburg [243]
Joseph s sewell and his work in madagascar, june 1867 – june 1876 – Antananarivo, 1876 – 1mf – 9 – mf#HT-133 – ne IDC [916]
Joseph, Samuel *see*
– Jewish immigration to the united states
Joseph smith the prophet, his family and his friends : a study based on facts and documents / Wyl, Wilhelm – Salt Lake City: Tribune Print and Publ Co, 1886 – 1mf – 9 – 0-524-04856-8 – mf#1990-1348 – us ATLA [240]
Joseph, Stanislaus *see* The growth of african literature
Joseph the ruler / Royer, Galen Brown – Mt Morris, IL: Brethren Pub House, 1898 – 1mf – 9 – 0-524-03860-0 – mf#1990-4907 – us ATLA [221]
Joseph tuckerman on the elevation of the poor : a selection from his reports as minister at large in boston / Tuckerman, Joseph – Boston: Roberts Brothers, 1874 – 1mf – 9 – 0-7905-6844-6 – mf#1988-2844 – us ATLA [360]
Joseph und wilhelm eichendorffs jugendgedichte : vermehrt durch ungedruckte gedichte aus dem handschriftlichen nachlass / Eichendorff, Joseph, Freiherr von; ed by Pissin, R – Berlin: E Frensdorff, [1906?] – 1 – (incl bibl ref and index of first lines) – us Wisconsin U Libr [810]
Joseph von eichendorff : sein leben und sein werk / Brandenburg, Hans – Muenchen: E H Beck 1922 [mf ed 1989] – 1r – 1 – (incl ind; filmed with: gedichte / joseph freiherrn von eichendorff) – mf#7212 – us Wisconsin U Libr [430]
Joseph von eichendorff : sein leben und seine dichtungen: zur hundertjaehrigen geburtsfeier am 10. maerz 1888 / Keiter, Heinrich – Koeln: J P Bachem, 1887 [mf ed 1989] – 112p – 1 – (incl bibl ref) – mf#7212 – us Wisconsin U Libr [430]

JOURNAL

Joseph von goerres als litterarhistoriker / Wibbelt, Augustin – Koeln: J P Bachem, 1899 [mf ed 1990] – 76p – 1 – (incl bibl ref) – mf#7405 – us Wisconsin U Libr [410]
Joseph von goerres gesammelte schriften / ed by Goerres, Marie – Muenchen: In Commission der literarisch-artistischen anstalt, 1854-74 [mf ed 1989] – (= ser Bibliothek der deutschen literatur 474-491) – 9v – 1 – mf#6989 – us Wisconsin U Libr [940]
Joseph von lassberg : mittler und sammler; aufsaetze zu seinem 100. todestag / ed by Bader, Karl Siegfried – Stuttgart: F Vorwerk: 1955 – 1 – (incl bibl ref) – us Wisconsin U Libr [920]
Joseph williams and the pioneer mission to the southeastern bantu / Holt, Basil Fenelon – Lovedale, South Africa. 1954 – 1r – us UF Libraries [960]
Joseph-Andre, frere *see* Monastere de notre-dame de la trappe du saint esprit, dans le township langevin
O josephense : publicacao semanal – Sao Jose, SC: Imprensa Official, 07 fev 1926 – (= ser Ps 19) – mf#UFSC/BPESC – bl Biblioteca [073]
Josephi medi : collegii christi apud cantabrigienses aliquando socii, opuscula latina ad rem apocalypticam fere spectantia... / Mede, J – Cantabrigiae: Per Thomam Buck, 1652 – 1mf – 9 – mf#PW-18 – ne IDC [240]
Josephine : ein spiel in vier akten / Bahr, Hermann – 2. aufl. Berlin: S Fischer, 1913 [mf ed 1989] – 211p – 1 – mf#6973 – us Wisconsin U Libr [830]
Josephine : ein spiel in vier akten / Bahr, Hermann – 2. aufl. Berlin: S Fischer, 1913 [mf ed 1989] – 211p – 1 – mf#6973 – us Wisconsin U Libr [820]
Joseph-Marie, soeur *see* Bibliographie analytique de l'abbe anselme longpre...du diocese de saint-hyacinthe, premiere partie (1927-1947)
Josephs, Ray *see* Latin america
Joseph...Sancta Barbara *see*
– Het geestelyck kaert-spel met herten troef
– Het geestelyck kaert-spel met herten troef...
– Het gheestelijck kaertspel met herten troef...
Josephus / Feuchtwanger, Lion – New York: The Literary Guild, 1932 – 1r – 1 – us Wisconsin U Libr [830]
Josephus and the jews / Foakes-Jackson, Frederick John – New York, NY. 1930 – 1r – us UF Libraries [939]
Josephus, Flavius *see*
– Flavii iosephi antiquitatum iudaicarum epitome
– Flavii iosephi opera
– Jozef flawjusz dzieje wojny zydowskiej przeciwko rzymianom
– Kadmut ha-yehudim neged apyon
– Yeme 'am 'olam
Josephus, Flavius [Joseph Ben Matthias] *see* Bellum judaicum
Josephus und lucas : der schriftstellerische einfluss des juedischen geschichtsschreibers auf den christlichen / Krenkel, Max – Leipzig: H Haessel, 1894 – 1mf – 9 – 0-8370-3999-1 – (incl bibl ref) – mf#1985-1999 – us ATLA [221]
Joshi, G N *see* The wealth of india
Joshi, Pranshankar Someshwar *see* Verdict on south africa
Joshi, V V *see* The problem of history and historiography
Joshua : the hebrew and greek texts / Holmes, Samuel – Cambridge: University Press; New York: G P Putnam [distributor], 1914 – 1mf – 9 – 0-7905-0954-7 – (in english, greek, and hebrew. incl ind) – mf#1987-0954 – us ATLA [221]
Joshua / Simpson, Albert B – New York: Christian Alliance Pub Co, 1894 [mf ed 1991] – (= ser Christ in the bible 3; Christian and missionary alliance coll) – 1mf – 9 – 0-524-01819-7 – mf#1990-4157 – us ATLA [221]
Joshua and the conquest of palestine / Bennett, William Henry – London: J M Dent; Philadelphia: Lippincott [19-?] [mf ed 1985] – (= ser The temple series of bible handbooks) – 1mf – 9 – 0-8370-2269-X – (incl app & ind) – mf#1985-0269 – us ATLA [221]
Joshua and the land of promise / Meyer, Frederick Brotherton – London: Morgan & Scott, [1893?] – 1mf – 9 – 0-8370-4410-3 – mf#1985-2410 – us ATLA [221]
Joshua, his life and times / Deane, William John – New York: Fleming H Revell, [189?] – 1mf – 9 – 0-8370-9933-1 – (incl bibl ref and index) – mf#1986-3933 – us ATLA [221]
Joshua, judges, ruth / Keil, Carl Friedrich & Delitzsch, Franz – Edinburgh: T & T Clark 1868 [mf ed 1984] – (= ser Biblical commentary on the old testament 4) – 6mf – 9 – 0-8370-0988-X – mf#1984-4328 – us ATLA [221]
Joshua rowntree / Robson, S Elizabeth – London: G Allen & Unwin, 1916 – 1mf – 9 – 0-524-06732-5 – mf#1991-2762 – us ATLA [240]
Josi, Visvanatha Balkrishna *see* Alphabetical index of words occurring in the aitareya braahmanam
Josiah allen's wife as a p.a. and p. i. samentha at the centennial. / Holley, Marietta – Hartford: American, 1891. 580p.illus – 1 – us Wisconsin U Libr [920]
Josiah webster pillsbury, elizabeth dinsmoor pillsbury : memorial discourses. delivered at milford, new hampshire / Rich, Adoniram Judson – [S.l.: s.n., 1902?] – 1mf – 9 – 0-524-04301-9 – mf#1992-2021 – us ATLA [240]
Josiah wedgewood... : his personal history / Smiles, Samuel – London 1894 – (= ser 19th c art & architecture) – 4mf – 9 – mf#4.1.429 – uk Chadwyck [730]
Joslin, James Elliott *see* The formulation and use of a staff policy manual within the greene county baptist association
Josling, Santa *see* 'N terapieprogram ter verbetering van selfkonsep by senior sekondere skoolleerlinge
Josquin, des Prez *see* Missarum josquin liber secundus
Joss, Gottlieb *see* Die vereinigung christlicher kirchen
Josselin De Jong, Jan Petrus Benjamin De *see* Archeological material from saba and st eustatius
Josslyn, William R *see* Ecce regnum
Jost, Cranswick *see* Miracles
Jost, Holger Wilfried *see* Geriatrisches assessment im altersheim unter besonderer beruecksichtigung psychotroper medikation
Jost, Isaak Markus *see* Culturgeschichte der israeliten der ersten halfte...
Jost, Theodor *see* Mechanisierung des lebens und moderne lyrik
Jost, Walter *see* Von ludwig tieck zu e.t.a. hoffmann
Jostes, Franz *see*
– Meister eckhart und seine juenger
– Die tepler bibelubersetzung
– Die waldenser und die vorlutherische deutsche bibeluebersetzung
Josua : eine erzaehlung aus biblischer zeit / Ebers Georg – Stuttgart: Deutsche Verlags-Anstalt, [1893-1897?] [mf ed 1993] – (= ser Georg ebers gesammelte werke 19) – xii/426p – 1 – mf#8554 reel 3 – us Wisconsin U Libr [830]
Josua : eine erzaehlung aus biblischer zeit / Ebers, Georg – Stuttgart: Deutsche Verlags-Anstalt, [1893-97?] [mf ed 1993] – (= ser Georg ebers gesammelte werke 19) – xii/426p – 1 – mf#8554 reel 3 – us Wisconsin U Libr [830]
Josyer, G R *see* History of mysore and the yadava dynasty
Josz, Vergile *see* Rembrandt
La jota aragonesa / Hurtado y Nunez de Arce, G – 1866 – 9 – sp Bibl Santa Ana [830]
Jotham meeker papers / Meeker, Jotham – 1825-64. In Kansas State Historical Society. Guide – 1 – us Kansas [240]
Jottings from japan / Ballard, Susan – Westminster: Society for the Propagation of the Gospel in Foreign Parts, 1909 [mf ed 1995] – (= ser Yale coll) – viii/96p (ill) – 1 – 0-524-09633-3 – mf#1995-0633 – us ATLA [220]
Jottings on the west indies and panama / Radford, Alfred – London, England. 1886 – 1r – us UF Libraries [972]
Jou meng t'ieh / Lu, Yin-ch'uan – Shang-hai: Shih ko yueh pao she, 1934 – (= ser P-k&k period) – us CRL [810]
Joubert, Carl *see* Russia as it really is
Joubert, Christina Helena *see* Personality as predictor of success for mba students
Joubert de LaRue, Jean *see* Lettres d'un sauvage depayse
Joubert, Henri *see* L'espagne de franco
Joubert, William Harry *see* History of the seaboard air line railway company
Joueon, P *see* Notes de lexicographie hebraique
Joueon, Paul *see* Le cantique des cantiques
Jouhaud, Auguste *see*
– Guerre au sexe
– Maison de sante
Jouin, Ernest *see* Le peril judeo-maconnique. 2: la judeomaconnerie a l'eglise catholique. 1e partie, les fideles de la contre-eglise, juifs et macons
Jouin, Louis *see* What christ revealed
The jounal *see* Grahamstown journal
Jounet, Albert *see* Le modernisme et l'infaillibilite
Jouons avec les livres : suggestions de livres et d'activites a faire avec les tout-petits / Gamache, Sylvie – [Montreal]: Communication-Jeunesse; [Quebec]: Ministère des affaires culturelles, 1988 [mf ed 2001] – 1mf – 9 – (with ind) – mf#SEM105P3401 – cn Bibl Nat [080]
Le jour – Beirut: s.n. 1956-sep 14 1963 – 26r – 1 – us CRL [079]
Le jour – Paris. janv-juin 1896 – 1 – fr ACRPP [073]
Le jour *see* l'echo de paris
Le jour de l'an : imite de longfellow, hommage aux lectrices de "l'evenement" / Chapman, William – S.l: s.n, 1881? – 1mf – 9 – mf#58387 – cn Canadiana [880]
Jour l'echo de paris – Paris, France. 10 apr 1940-mar 1942 – 3 1/2r – 1 – uk British Libr Newspaper [072]
Jourdain, Charles *see* De l'influence d'aristote et de ses interpretes sur la decouverte du nouveau monde
Jourdain, Margaret *see* Diderot's early philosophical works
Jourdain, Silvester *see*
– A discovery of the barmudas
– Discovery of the bermudas
Jourdan, A J L *see* Dictionaire des sciences medicales (ael3/16)
Jourdan, George Viviliers *see* The movement towards catholic reform in the early 16 century
Jourgniac de Saint-Meard, Francois *see*
– Memoires sur les journees de septembre 1792
Journal – 1971 apr-1972 dec 1 – 1r – 1 – mf#1107113 – us WHS [071]
Journal / African Methodist Episcopal Church. Virginia Conference – [s.l: s.n] 9th-11th (1875-77) [mf ed 2006] – 1r – 1 – (cont by: african methodist episcopal church. virginia conference. proceedings of the virginia conference,...annual session, african m e church; some pgs lacking, and damaged) – mf#2006-s030 – us ATLA [242]
Journal / Alabama State Teachers Association – 1959 may – 1r – 1 – mf#5266256 – us WHS [370]
Journal – Alexandria, VA. 1950-1984 (1) – mf#66663 – us UMI ProQuest [071]
Journal / Allen Co. Spencerville – (6/1891-6/1895), feb 1896-jun 1897 [wkly] – 2r – 1 – mf#B32511-32512 – us Ohio Hist [071]
Journal / Allen Co. Spencerville – jul 1885-apr 1889 [wkly] – 2r – 1 – mf#B32509-32510 – us Ohio Hist [071]
Journal – Altavista, VA. 1988-2000 (1) – mf#66665 – us UMI ProQuest [071]
Journal / Amalgamated Meat Cutters and Butcher Workmen of North America – 1978 apr 7-1979 jun 22 – 1r – 1 – (cont by: journal [united food and commercial workers international union. local 151 [everett, wa]]) – mf#672487 – us WHS [071]
Journal / American Temperance Union – v. 1-29. Jan 1837-Dec 1865. v. 21, no. 1-5; v. 29, no. 10 wanting – 1 – 81.00 – us L of C Photodup [360]
Journal / The Arizona Academy of Science – v1-6. 1959-Oct 1970 – 1 – us AMS Press [500]
Journal / Ashland Co. Hayesville – 1879,7/82-83,11/87-9/88,89-7/1890 [wkly] – 2r – 1 – mf#B29211-29212 – us Ohio Hist [071]
Journal / Association international des Travailleurs. Section de la Suisse Romande – 1865-66 – 1 – us CRL [330]
Journal / Auglize Co. Waynesfield – v1 n1. aug 1977-jul 1985 [wkly] – 5r – 1 – mf#B32683-32687 – us Ohio Hist [071]
Journal – Azusa, CA. 1924-1929 (1) – mf#62086 – us UMI ProQuest [071]
Journal – Ballston Spa, NY. 1856-1913 (1) – mf#64895 – us UMI ProQuest [071]
Journal – Battle Creek, MI. 1852-1879 (1) – mf#63687 – us UMI ProQuest [071]
Journal – Battle Creek, MI. 1872-1914 (1) – mf#63688 – us UMI ProQuest [071]
Journal – Beacon, NY. 1925-1927 (1) – mf#64905 – us UMI ProQuest [071]
Journal – Beaumont, TX. 1936-1983 (1) – mf#66580 – us UMI ProQuest [071]
Journal – Belgrade, MT. 1906-1944 (1) – mf#64240 – us UMI ProQuest [071]
Journal – Blaine, WA. 1900-1950 (1) – mf#66944 – us UMI ProQuest [071]
Journal – Bossburg, WA. 1897-1901 (1) – mf#69181 – us UMI ProQuest [071]
Journal – Centerville, IA. 1883-1893 (1) – mf#63091 – us UMI ProQuest [071]
Journal – Central Falls, RI. 1899-1900 (1) – mf#66179 – us UMI ProQuest [071]
Journal / Chamber of Commerce. Constantinople – Istanbul, Turkey. -w. Jan. 1907-14 Nov. 1914, 5 Jan. 1918-28 Feb. 1921. 9 reels – 1 – uk British Libr Newspaper [380]
Journal – Cheney, WA. 1928-1945 (1) – mf#69215 – us UMI ProQuest [071]
Journal – Chicago, IL. 1977-1983 (1) – mf#62567 – us UMI ProQuest [071]
Journal – Chinook, MT. 1942-1949 (1) – mf#64317 – us UMI ProQuest [071]
Journal / Clermont Co. Bethel – jan 1983-dec 1988 [wkly] – 6r – 1 – mf#B31038-31043 – us Ohio Hist [071]
Journal – Cleveland Heights, OH. 1938-1939 (1) – mf#65438 – us UMI ProQuest [071]
Journal – Coffeyville, KS. 1972-2000 (1) – mf#68171 – us UMI ProQuest [071]
Journal – Colonial Heights, VA. 1950-1955 (1) – mf#66690 – us UMI ProQuest [071]
Journal / Columbiana Co. New Lisbon – apr 1869-mar 1870+scattered iss [wkly] – 1r – 1 – mf#B30149 – us Ohio Hist [071]
Journal / Columbiana Co. Salem – feb 1866-68, jan-aug 1872 [wkly] – 1r – 1 – mf#B4313 – us Ohio Hist [071]
Journal / Columbiana Co. Salem – jan 1869-dec 1871 [wkly] – 1r – 1 – mf#B6622 – us Ohio Hist [071]
Journal – Conrad, IA. 1880-1899 (1) – mf#63131 – us UMI ProQuest [071]
Journal – Chadron, NE: J W Wright. 1v. v13 n17. feb 19 1897-v13 n52. oct 22 1897 (wkly) – 1r – 1 – (cont: dawes county journal. cont by: chadron journal) – us NE Hist [071]
Journal – Howells, NE: Howells Journal. v96 n32. may 20 1981-v99 n27. apr 1 1987 (wkly) [mf ed filmed 1983-87] – 4r – 1 – (cont: howells journal. cont by: howells journal (1987)) – us NE Hist [071]
Journal – North Freedom WI. 1907 aug 7 – 1r – 1 – (cont: north freedom journal [north freedom, wi: 1903]; cont by: north freedom journal [north freedom, wi: 1908]) – mf#957190 – us WHS [071]
Journal – Oak Creek, South Milwaukee WI. [1912 jan 6/aug 24]-[1948 jan/1950 dec 1] – 11r – 1 – (cont: south milwaukee journal; cont by: reminder-journal press [south milwauke, wi]) – mf#1004491 – us WHS [071]
Journal – Corning, NY. 1847-1905 (1) – mf#64938 – us UMI ProQuest [071]
Journal – Corry, PA. 1970-2001 (1) – mf#61775 – us UMI ProQuest [071]
Journal – Crawfordsville, IN. 1863-1919 (1) – mf#62753 – us UMI ProQuest [071]
Journal – Crawfordsville, IN. 1880-1929 (1) – mf#62754 – us UMI ProQuest [071]
Journal – Crawfordsville, IN. 1887-1887 (1) – mf#62755 – us UMI ProQuest [071]
Journal – Crewe, VA. 1966-1966 (1) – mf#66693 – us UMI ProQuest [071]
Journal / Darke Co. Greenville – 1907-08 1910-jun 1918 [wkly] – 5r – 1 – mf#B9088-9092 – us Ohio Hist [071]
Journal / Darke Co. Greenville – (6/1851-5/1860, 2/1866-12/1906) center shadows [wkly] – 15r – 1 – mf#B7555-7569 – us Ohio Hist [071]
Journal – Dayton, OH. 1940-1948 (1) – mf#65463 – us UMI ProQuest [071]
Journal – East Wenatchee, WA. 1939-1947 (1) – mf#69234 – us UMI ProQuest [071]
Journal / Energy and Chemical Workers Union – 1981 jun-1992 apr – 1r – 1 – (cont by: connections [ottawa, on]; cpu journal [1981]; cep journal) – mf#3282937 – us WHS [331]
Journal – Eureka, MT. 1910-1929 (1) – mf#64372 – us UMI ProQuest [071]
Journal – Evansville, IN. 1871-1901 (1) – mf#62776 – us UMI ProQuest [071]
Journal – Fairfield, IA. 1880-1921 (1) – mf#63210 – us UMI ProQuest [071]
Journal – Falls City, NE. 1993-2000 (1) – mf#61582 – us UMI ProQuest [071]
Journal – Fayetteville, WV. 1900-1937 (1) – mf#67283 – us UMI ProQuest [071]
Journal : (final edition) – Knoxville, TN. 1960-1991 (1) – mf#60582 – us UMI ProQuest [071]
Journal : (final edition) – Shreveport, LA. 1989-1991 (1) – mf#61484 – us UMI ProQuest [071]
Journal / Florida Genealogical Society – 1976-77 – 1r – 1 – (cont: florida genealogical journal [tampa, fl: 1967]; cont by: florida genealogical journal [tampa, fl: 1978]) – mf#2593863 – us WHS [929]
Journal / Florida Genealogical Society – v14 n2-v17 n2 [1978-81] – 1r – 1 – (cont: florida genealogical journal [tampa, fl: 1978]) – mf#2593880 – us WHS [929]
Journal / Forsyth, John R – 1849 – 1 – us Kansas [920]
Journal – Forsyth, MT. 1907-1909 (1) – mf#64382 – us UMI ProQuest [071]
Journal / Fort Smith Historical Society – 1977 sep [v1 n1]-1984 sep – 1r – 1 – mf#966052 – us WHS [978]
Journal – Freeport, IL. 1848-1882 (1) – mf#62618 – us UMI ProQuest [071]
Journal – Freeport, IL. 1856-1913 (1) – mf#62619 – us UMI ProQuest [071]
Journal – Friday Harbor, WA. 1906-1948 (1) – mf#67002 – us UMI ProQuest [071]
Journal – Galata, MT. 1911-1920 (1) – mf#64394 – us UMI ProQuest [071]
Journal / Gallia Co. Gallipolis – jan-jun 1899, jan-dec 1900 [daily] – 2r – 1 – mf#B30389-30390 – us Ohio Hist [071]
Journal / Gallia Co. Gallipolis – (1825-94) [wkly] – 16r – 1 – (request info) – mf#B6174-6189 – us Ohio Hist [071]
Journal / Good, Adolphus Clemens – 1892 – 1 – $50.00 – us Presbyterian [240]
Journal – Grand Coulee, WA. 1935-1936 (1) – mf#69238 – us UMI ProQuest [071]
Journal – Greenwood, SC. 1895-1917 (1) – mf#66498 – us UMI ProQuest [071]
Journal – Havana, NY. 1853-1887 (1) – mf#64994 – us UMI ProQuest [071]
Journal – Haverhill, MA. 1957-1965 (1) – mf#63648 – us UMI ProQuest [071]

Journal – Herrin, IL. 1913-1949 (1) – mf#62631 – us UMI ProQuest [071]
Journal – Huntley, MT. 1912-1913 (1) – mf#64486 – us UMI ProQuest [071]
Journal / Institute for the Study of Nonviolence – n1-9 [1972 nov-1974 jun] – 1r – 1 – mf#1059365 – us WHS [320]
Journal / International brotherhood of boiler makers, iron ship builders and helpers of America – jan 1897-dec 1897 – 1r – 1 – mf#2820160 – us WHS [366]
Journal / International Molders and Allied Workers Union – 1973 feb-1977 dec, 1978 jan-1988 mar/apr – 2r – 1 – (cont: international molders' and allied workers' journal; cont by: gmp horizons) – mf#653318 – us WHS [331]
Journal / International Union of Bricklayers and Allied Craftsmen – 1982 aug-1987 dec, 1988 jan-1994 oct – 2r – 1 – (cont: journal [bricklayers, masons and plasterers international union of america]) – mf#3363050 – us WHS [331]
Journal – Ithaca, NY. 1995+ (1) – mf#61633 – us UMI ProQuest [071]
Journal / Jackson Co. Jackson – jul 1882-jul 1888 – 3r – 1 – mf#B9915-9917 – us Ohio Hist [071]
Journal – Jacksonville, FL. 1922-1988 (1) – mf#60435 – us UMI ProQuest [071]
Journal – Jamestown, NY. 1826-1870 (1) – mf#65013 – us UMI ProQuest [071]
Journal – Joliet, MT. 1904-1909 (1) – mf#64494 – us UMI ProQuest [071]
Journal – Judith Gap, MT. 1908-1924 (1) – mf#64499 – us UMI ProQuest [071]
Journal – Kalispell, MT. 1907-1917 (1) – mf#64505 – us UMI ProQuest [071]
Journal – Kalispell, MT. 1910-1914 (1) – mf#64506 – us UMI ProQuest [071]
Journal / Lac Courte Oreilles Tribe – 1987 jul-1988 jul – 1r – 1 – (cont: lac courte oreilles journal; cont by: news from indian country) – mf#1239991 – us WHS [305]
Journal – Lafayette, IN. 1914-1919 (1) – mf#68591 – us UMI ProQuest [071]
Journal – Lancaster, PA. 1796-1836 (1) – mf#65963 – us UMI ProQuest [071]
Journal – Lansing, MI. 1888-1910 (1) – mf#60497 – us UMI ProQuest [071]
Journal – Lansing, MI. 1892-1896 (1) – mf#61041 – us UMI ProQuest [071]
Journal / Lawrence Co. Ironton – v1 n1. sep 1867-dec 1871 [wkly] – 2r – 1 – mf#B33765-33766 – us Ohio Hist [071]
Journal / Logan Co. DeGraff – 1935, 37-38, 46-1947 (gap fillers) [wkly] – 2r – 1 – mf#B6811-6812 – us Ohio Hist [071]
Journal / Logan Co. DeGraff – (nov 1894-oct 1932, jan-dec 1948) [wkly] – 12r – 1 – mf#B12026-12037 – us Ohio Hist [071]
Journal / Lorain Co. Lorain – jun-jul 1924 (damaged material) [daily] – 1r – 1 – mf#B33260 – us Ohio Hist [071]
Journal / Lucas Co. Toledo – jan 1980-dec 1993 [biwkly, wkly] – 14r – 1 – mf#B34182-34194 – us Ohio Hist [071]
Journal / Lunsford, Isaac – (Methodist Preacher of Tenn.) Nov 1791-95 – 1 – 5.32 – us Southern Baptist [242]
Journal / McKay, Donald – 1870 – 1 – us Kansas [978]
Journal / Mahoning Co. Campbell – 6/1953-10/55,1-12/57,1/60-9/1967 [wkly] – 3r – 1 – mf#B11246-11248 – us Ohio Hist [071]
Journal / Mahoning Co. Struthers – 4/1928-7/51,53-57,60-1976 [wkly] – 18r – 1 – mf#B6975-6992 – us Ohio Hist [071]
Journal / Manitoba. Legislative Assembly – 1870-1900 – (= ser Legislative Assembly) – 6r – 1 – cn Library Assoc [971]
Journal – Manitowoc WI. 1876 jan 23 – 1r – 1 – mf#1109249 – us WHS [071]
Journal – Mansfield, OH. 1924-1932 (1) – mf#65567 – us UMI ProQuest [071]
Journal – Marlington, WV. 1915-1974 (1) – mf#67349 – us UMI ProQuest [071]
Journal / Marquette University – 1985 oct-1988 apr – 1r – 1 – (cont: journal [marquette university]; cont by: marquette journal [milwaukee, wi: 1988]) – mf#1336494 – us WHS [378]
Journal / Marquette University – 1980 oct/nov-1985 spr – 1r – 1 – (cont: marquette journal [milwaukee, wi: 1967]; cont by: journal [marquette university: 1985]) – mf#1224732 – us WHS [378]
Journal – Martinsburg, WV. 1990+ (1) – mf#67358 – us UMI ProQuest [071]
Journal – Massac, IL. 1872-1878 (1) – mf#62648 – us UMI ProQuest [071]
Journal – Mathews, VA. 1905-1937 (1) – mf#68513 – us UMI ProQuest [071]
Journal / Meeker, Jotham – 1832-55 – 1 – us Kansas [978]
Journal / Meigs Co. Pomeroy – jan 1880-aug 1881, dec 1881 [wkly] – 1r – 1 – (in german) – mf#B11267 – us Ohio Hist [071]
Journal / Mercer Co. Fort Recovery – sep 1941-jan 1973 [wkly] – 14r – 1 – mf#B13236-13249 – us Ohio Hist [071]
Journal – Mercersburg, PA. 1857-1916 (1) – mf#65994 – us UMI ProQuest [071]
Journal / Metal Polishers, Buffers, Platers, Brass Molders, and Brass and Silver Workers' International Union of North America – 13 [1904] – 1r – 1 – (cont: journal [metal polishers, buffers, platers, brass molders and brass workers' international union of north america]; cont by: our journal [cincinnati, ohio]) – mf#3164315 – us WHS [071]
Journal / Methodist Episcopal Church. Southwestern Conference – 1870-1905 – 1 – us Kansas [978]
Journal : (metro edition) – Knoxville, TN. 1885-1991 (1) – mf#60583 – us UMI ProQuest [071]
Journal – Milwaukee, WI. 1882-1994 (1) – mf#60617 – us UMI ProQuest [071]
Journal – Missoula, MT. 1904-1907 (1) – mf#64570 – us UMI ProQuest [071]
Journal / Montgomery Co. Dayton – jun 1842-may 1843 [twice wkly] – 1r – 1 – mf#B127 – us Ohio Hist [071]
Journal – Mound City, IL. 1872-1872 (1) – mf#62659 – us UMI ProQuest [071]
Journal – Moundsville, WV. 1910-1915 (1) – mf#67394 – us UMI ProQuest [071]
Journal / National Association of Machinists – v1 n4 [1889 may] – 1r – 1 – (cont: journal of united machinists and mechanical engineers of america; monthly journal of the national association of machinists) – mf#910680 – us WHS [680]
Journal – Nevada, IA. 1895-1900 (1) – mf#63332 – us UMI ProQuest [071]
Journal – New Bedford, MA. 1890-1896 (1) – mf#63657 – us UMI ProQuest [071]
Journal – New Ulm, MN. 1990+ (1) – mf#68562 – us UMI ProQuest [071]
Journal – Newburgh, NY. 1841-1843 (1) – mf#65118 – us UMI ProQuest [071]
Journal / Newfoundland. Legislative Assembly – 1866-1900 – (= ser Legislative Assembly) – 18r – 1 – cn Library Assoc [971]
Journal / Newfoundland. Legislative Council – 1866-1900 – (= ser Legislative Council) – 3r – 1 – ISSN: 1197-964X – cn Library Assoc [971]
Journal / Noble Co. Caldwell – aug 1897-jul 1898 [wkly] – 1r – 1 – mf#B8797 – us Ohio Hist [071]
Journal – Norfolk, VA. 1868-1873 (1) – mf#66776 – us UMI ProQuest [071]
Journal – Norwich, NY. 1816-1830 (1) – mf#61031 – us UMI ProQuest [071]
Journal : the official organ of the commercial telegraphers union of america / Commercial Telegrapher's Union of America – v1 n4-7 [1903 apr-jul] – 1r – 1 – (cont: i u c t journal; cont by: commercial telegraphers' journal) – mf#1231933 – us WHS [380]
Journal : official organ of the international brotherhood of pulp, sulphite, and paper mill workers / International Brotherhood of Pulp, Sulphite, and Paper Mill Workers – 1914 jan-1921 jul/aug – 1r – 1 – (cont by: pulp, sulphite and paper mill workers' journal) – mf#3358004 – us WHS [670]
Journal : official publication of the san francisco economic opportunity council, inc / San Francisco Economic Opportunity Council, Inc – 1967 jul – 1r – 1 – mf#5307211 – us WHS [330]
Journal / Ontario. Legislative Assembly – 1867-1902 – (= ser Legislative Assembly) – 7r – 1 – cn Library Assoc [971]
Journal / Ontario. Legislative Assembly – 1903-23 – (= ser Legislative Assembly) – 7r – 1 – cn Library Assoc [971]
Journal / Order of the Indian Wars – v1 n1-v2 n4 [1980 win-1981 fall] – 1r – 1 – mf#1022811 – us WHS [975]
Journal – Paris/Limoges/Lyons, France. Aug 1914-31 aug 1915; 15 oct 1915-12 aug 1919; 26 mar 1929; 8 aug 1936; 2 aug-27 nov 1940; dec 1940-11 jun 1944 – 20 1/2r – 1 – uk British Libr Newspaper [072]
Journal / Paulding Co. Paulding – mar 1873-mar 1874 [wkly] – 1r – 1 – mf#B1266 – us Ohio Hist [071]
Journal / Peck, John Mason – 1854-56. 114p – 1 – 5.00 – us Southern Baptist [242]
Journal – Pensacola, FL. 1955-1959 (1) – mf#62441 – us UMI ProQuest [071]
Journal – Peoria, IL. 1951-1955 (1) – mf#62678 – us UMI ProQuest [071]
Journal / Portage Co. Garrettsville – jul 1867-dec 1970 [wkly] – 36r – 1 – mf#B3471-3506 – us Ohio Hist [071]
Journal – Poughkeepsie, NY. 1960-1987 (1) – mf#61649 – us UMI ProQuest [071]
Journal – Tours. 1809-15 – 1 – (puis politique et litteraire d'indre-et-loire) – fr ACRPP [073]
Journal – Rensselaer, IN. 1898-1905 (1) – mf#62949 – us UMI ProQuest [071]
Journal – Rock Hill, SC. 1901-1903 (1) – mf#66517 – us UMI ProQuest [071]
Journal – Rockport, IN. 1932-1970 (1) – mf#62968 – us UMI ProQuest [071]
Journal / Sandusky Co. Fremont – (1853-1908) scattered [wkly, semiwkly, wkly] – 22r – 1 – mf#B5010-5031 – us Ohio Hist [071]
Journal / Sandusky Co. Fremont – jan 1855-jan 1857, jan-dec 1861 [wkly] – 1r – 1 – mf#B33271 – us Ohio Hist [071]
Journal – Sarasota, FL. 1952-1982 (1) – mf#62447 – us UMI ProQuest [071]
Journal / Schomburg Center for Research in Black Culture – v1 n1-4 [1976 fall-1978 spr] – 1r – 1 – mf#329511 – us WHS [305]
Journal / Shawnee County. Kansas. Board of Commissioners – 1855-62 – 1 – us Kansas [978]
Journal – Shreveport, LA. 1902-1955 (1) – mf#63543 – us UMI ProQuest [071]
Journal – Sidney, OH. 1863-1905 (1) – mf#65661 – us UMI ProQuest [071]
Journal – Sioux City, IA. 1864-2000 (1) – mf#61439 – us UMI ProQuest [071]
Journal / Snowden, Gilbert T – 1783-85 – 1 – $50.00 – us Presbyterian [920]
Journal / Societe des americanistes de Paris – n.s. v.1-15. 1903-23 – 1 – us L of C Photodup [073]
Journal / La Societe des Amis de la Constitution Monarchique – Par Fontanes. no. 1-27. Paris. dec 1790-juin 1791 – 1 – fr ACRPP [944]
Journal – South Bend, WA. 1890-1956 (1) – mf#67132 – us UMI ProQuest [071]
Journal – South Pasadena, CA. 1963-1964 (1) – mf#62286 – us UMI ProQuest [071]
Journal – Southbridge, MA. 1861-1900 (1) – mf#63662 – us UMI ProQuest [071]
Journal – Springville, NY. 1969-1987 (1) – mf#68381 – us UMI ProQuest [071]
Journal – Sturgis, MI. 1991-2000 (1) – mf#67990 – us UMI ProQuest [071]
Journal / Syndicat canadien de la Fonction publique – v13 n1-v14 n6 [1976 win-1977 dec] – 1r – 1 – (cont by: public employee [ottawa, on]) – mf#1100158 – us WHS [350]
Journal – Frankfurt/M DE, 1832 1 jan-30 jun, 1840-1846 sep, 1849-60 – 57r – 1 – (title varies: 1724: journal in frankfurt am main; 3 jan 1873: frankfurter journal. filmed by misc inst: 1846 oct-1848 [7r], 1866 1 apr-30 dec, 1870 1 aug-1871 2 jul) – gw Mikropress; gw Misc Inst [074]
Journal / Trumbull Co. Kinsman – sep 1977-jan 1982 [wkly] – 2r – 1 – mf#B29494-29495 – us Ohio Hist [071]
Journal – Tupelo, MS. 1882-1935 (1) – mf#64129 – us UMI ProQuest [071]
Journal – Tupelo, MS. 1955-1986 (1) – mf#61552 – us UMI ProQuest [071]
Journal – Turlock, CA. 1904-1919 (1) – mf#62299 – us UMI ProQuest [071]
Journal – Tyler, TX. 1925-1938 (1) – mf#69219 – us UMI ProQuest [071]
Journal / United Church of Christ – 1969 jan – 1r – 1 – (cont by: journal of current social issues) – mf#5265597 – us WHS [242]
Journal / United Food and Commercial Workers International Union – 1979 aug 31-1980 apr 11 – 1r – 1 – (cont: journal [amalgamated meat cutters and butcher workmen of north america. local 151 [everett, wa]]) – mf#672491 – us WHS [331]
Journal / U.S. Army. Quartermaster Corps. St. Louis Arsenal – 1844-50 – 1 – us Kansas [324]
Journal / Utah. Legislative Assembly. House of Representatives – [s.n] [n42-43 1977-1979] – 1 – us CRL [323]
Journal – Utica, NY. 1902-1906 (1) – mf#65251 – us UMI ProQuest [071]
Journal / Vinton Co. McArthur – jan 1858-jul 1862, dec 1862-jan 1863 [wkly] – 2r – 1 – mf#B146-147 – us Ohio Hist [071]
Journal / Walker, William – 1866-69 – 1 – us Kansas [978]
Journal – Walton, NY. 1857-1859 (1) – mf#65266 – us UMI ProQuest [071]
Journal – New Lisbon, OH, apr 19 1867-apr 4 1870 – 2r – 1 – (weekly independent newspaper) – us Western Res [071]
Journal – Westerly, RI. 1888-1891 (1) – mf#66431 – us UMI ProQuest [071]
Journal / Whitsitt, William H – v1. 334p – 1 – us Southern Baptist [242]
Journal / Williams, John Chauner – 1868-72 – 1r – 1 – mf#pmb37 – at Pacific Mss [880]
Journal – Willimantic, CT. 1862-1863 (1) – mf#62376 – us UMI ProQuest [071]
Journal – Wilmington, OH. 1868-1913 (1) – mf#65724 – us UMI ProQuest [071]
Journal – Winchester, VA. 1865-1869 (1) – mf#66912 – us UMI ProQuest [071]
Journal – Windsor, CT. 1973-1983 (1) – mf#62377 – us UMI ProQuest [071]
Journal / Writers Guild of America, West – 1988 nov-1989 nov – 1r – 1 – (cont: wga west newsletter; cont by: written by) – mf#1652659 – us WHS [410]
Journal *see*
– Byrne's emigrants / journal
– Niagara peninsula newspapers, pt 2
Journal.. / Methodist Episcopal Church. Missouri Conference – 1849-64 – 1 – us Kansas [978]
Le journal – n1-96. Paris. 28 juil-31 oct 1848 – 1 – (puis messager du matin) – fr ACRPP [944]
Le journal – Paris. 28 sept 1892-juin 1944 – 1 – (quotidien, litteraire, artistique et politique) – fr ACRPP [073]
The journal – Hemingford, NE: Arthur E Clark. -v9 n35. oct 14 1915 (wkly) [mf ed 1911-15 (gaps)] – 2r – 1 – (cont: hemingford journal. absorbed by: alliance herald) – us NE Hist [071]
The journal : technological horizons in education – Tustin. 1974+ (1,5,9) – ISSN: 0192-592X – mf#10627 – us UMI ProQuest [370]
The journal *see* Miscellaneous newspapers of weld county
Journal, 1856-1892 – journal, 1852, 1870-1907 / Warner, Sarah Gildersleeve & Warner, Thomas Pattison – 1852-1907 – 1r – 1 – 0-8370-1539-1 – mf#1984-B272 – us ATLA [920]
Le journal a un sou – Paris. n1-16. 7-22 dec 1879 – 1 – fr ACRPP [073]
Journal – academy of general dentistry / Academy of General Dentistry – Chicago. 1970-1975 (1) 1973-1975 (5) 1973-1975 (9) – (cont by: general dentistry) – ISSN: 0001-4265 – mf#8038 – us UMI ProQuest [617]
Journal, acts and proceedings of the convention : which formed the constitution of the united states: published under the direction of the president – Boston: Thomas B Wait, 1819 – 6mf – 9 – $9.00 – mf#LLMC 90-359 – us LLMC [342]
Journal aller journale : oder geist der vaterlaendischen zeit-schriften, nebst auszuegen aus den periodischen schriften und besten werken der auslaender / ed by Hess, Jonas Ludwig von – Hamburg 1786-88 – (= ser Dz. abt literatur) – 38mf – 9 – €380.00 – mf#k/n4558 – gw Olms [410]
Journal america – Huntingdon, PA. -w 1859-1870 – 13 – $25.00r – us IMR [071]
Journal american – Georgetown, SC. 1973-1974 (1) – mf#69003 – us UMI ProQuest [071]
Journal american – New York, NY. 1901-1966 (1) – mf#65083 – us UMI ProQuest [071]
Journal amusant : journal illustre, journal d'images, journal comique, critique, satirique et hebdomadaire – Paris. 1866-67, 1878-90 – 1 – fr ACRPP [870]
Journal and account book / Carter, John & Simerwell Elizabeth – 1861-1881 – 1 – us Kansas [920]
Journal and advance news – Ogdensburg, NY. 1999-2000 (1) – mf#69012 – us UMI ProQuest [071]
Journal and advertiser / Montgomery Co. Dayton – 10/1832-8/34, 5/42-12/1856 (damaged) [wkly] – 6r – 1 – mf#B33709-33714 – us Ohio Hist [071]
Journal and advertiser – New York, NY. 1897-1901 (1) – mf#65081 – us UMI ProQuest [071]
Journal and advocate / Select Knights of Canada – Toronto: Select Knights of Canada, [1890?-189- or 19–] – 9 – mf#P05975 – cn Canadiana [360]
Journal and argus – Petaluma, CA. 1864-1873 (1) – mf#62222 – us UMI ProQuest [071]
Journal and correspondence / Gray, Elizabeth (nee McEwen) – 1882-1886 – 1r – 1 – mf#PMB1048 – at Pacific Mss [920]
Journal and courier – Lafayette, IN. 1920+ (1) – mf#61390 – us UMI ProQuest [071]
Journal and evening bulletin – Providence, RI. 1934-1938 (1) – mf#66330 – us UMI ProQuest [071]
Journal and inventories / Belle Springs Creamery. Abilene, Kansas – 1892-98 – 1 – us Kansas [025]
Journal and letters of the late samuel curwen, judge of admiralty, etc : an american refugee in england from 1775 to 1784... / Curwen, Samuel – New York: C S Francis; Boston: J H Francis, 1842 – 7mf – 9 – (incl ind) – mf#48433 – cn Canadiana [920]
Journal and letters of the late samuel curwen, judge of admiralty, etc : a loyalist-refugee in england during the american revolution, to which are added illustrative documents and biographical notices of many loyalists and other prominent men of that period by george atkinson ward – London: Wiley and Putnam; New York: Leavitt, Trow, 1844 – 7mf – 9 – mf#48715 – cn Canadiana [920]
Journal and messenger – Macon, GA. 1823-1869 (1) – mf#68897 – us UMI ProQuest [071]
Journal and messenger, central national baptist paper – 1831-1920 – 1 – us Southern Baptist [242]
Journal and news – Santa Clara, CA. 1966-1970 (1) – mf#62280 – us UMI ProQuest [071]
Journal and noble county leader / Noble Co. Caldwell – dec 1933-jun 1977 [wkly] – 34r – 1 – mf#B6481-6514 – us Ohio Hist [071]
Journal and noble county leader / Noble Co. Caldwell – jan 1 1990-dec 29 1997 – 9r – 1 – mf#B37492-37500 – us Ohio Hist [071]

Journal and noble county leader / Noble Co. Caldwell – jul 1977-dec 1989 [wkly] – 15r – 1 – mf#B34859-34873 – us Ohio Hist [071]

Journal and official gazette / Workmen's Club – Working Men's Club and Institute Union. London. -w. 15 May 1875-9 Feb 1878. (1 reel) – 1 – uk British Libr Newspaper [330]

Journal and other papers / Williams, John et al – 1822-40 – 1r – 1 – mf#pmb35 – at Pacific Mss [920]

Journal and papers / Ogden, William Thomas – 1898-1901 – 1r – 1 – (journal & papers relate to ogden's missionary work in samoa) – mf#pmb111 – at Pacific Mss [243]

Journal and pred – Covington, KY. 1841-1876 (1) – mf#63459 – us UMI ProQuest [071]

Journal and record – Galesville WI. 1873 jan 23-1874 aug 7 – 1r – 1 – (cont: trempealeau county record; galesville journal) – mf#929897 – us WHS [071]

Journal and report of james l. cathcart and james hutton, agents appointed by the secty. of the navy to survey timber resources between the mermentau and mobile rivers, 1818-1819 / U.S. Bureau of Land Management – (= ser Records of the bureau of land management) – 1r – 1 – (with printed guide) – mf#M8 – us Nat Archives [333]

Journal and republican – Watertown, NY. 1929+ (1) – mf#69369 – us UMI ProQuest [071]

Journal and review – Aiken, SC. 1885-1935 (1) – mf#66451 – us UMI ProQuest [071]

Journal and straitsville news / Perry Co. Shawnee – (mar-dec 1878) scattered [wkly] – 1r – 1 – mf#B11581 – us Ohio Hist [071]

Journal and tribune – Logansport, IN. 1876-1920 (1) – mf#62882 – us UMI ProQuest [071]

Journal and weekly news – Newport, RI. 1897-1928 (1) – mf#66223 – us UMI ProQuest [071]

Journal anecdotique de mme campan : ou souvenirs recueillis dans ses entretiens... / Campan, Jeanne Louise Henriette Genest – Paris 1824 [mf ed Hildesheim 1995-98] – 1v on 2mf – 9 – €60.00 – ISBN-10: 3-487-25813-7 – ISBN-13: 978-3-487-25813-3 – gw Olms [880]

Journal ar la : (tw edition) – Shreveport, LA. 1955-1982 (1) – mf#61483 – us UMI ProQuest [071]

Journal asiatique : ou recueil de memoires, d'extraits et de notices relatifs a l'histoire, a la philosophie, aux sciences, a la litterature et aux langues des peuples orientaux – Paris 1822-27 [mf ed Hildesheim 1995-98] – 11v on 33mf – 9 – €330.00 – 3-487-27609-7 – gw Olms [950]

Journal asiatique – Paris, 1822-1946/47. v1-235 – 1952mf – 8 – mf#CH-782c – ne IDC [956]

Journal asiatique ou recueil de memoires, d'extraits et de notices – Paris. 1823-24, 1826-27, 1930-38 – 1 – (devenu: nouveau journal asiatique ou recueil) – fr ACRPP [950]

Journal aus urfstaedt / ed by Knigge, Adolf, Freiherr von – Frankfurt a M 1785/86 [mf ed Hildesheim 1992-98] – (= ser Dz) – 3st on 6mf – 9 – €120.00 – mf#k/n5691 – gw Olms [074]

Journal (bicentennial edition) / Noble Co. Caldwell – july 3 1975 – 1r – 1 – mf#B289 – us Ohio Hist [071]

Journal books of scientific meetings, 1660-1800 *see* Collections from the royal society

Journal britannique / ed by Maty, M – The Hague. 24v. 1750-1757 – 130mf – 8 – mf#H-1382 – ne IDC [450]

Journal – california school library association / California School Library Association – Burlingame. 1995+(1,5,9) – (cont: cmlea cmlea journal) – mf#11980,01 – us UMI ProQuest [020]

Journal canadien de radiographie, radiotherapie, nucleographie – Ottawa. 1970-1973 (1) 1970-1972 (5) (9) – ISSN: 0382-6325 – mf#7148 – us UMI ProQuest [616]

Journal capitol – the osage journal news – Pawhuska, OK. 1996-1999 (1) – mf#65806 – us UMI ProQuest [071]

Le journal (chambly, quebec) – Chambly: [s.n] v1 n1 3 mai 1966-v11 n41 28 mars 1978 (wkly) [mf ed 1986] – 11r – 1 – (cont by: le journal de chambly) – mf#SEM35P255 – cn Bibl Nat [073]

Journal charleroi – Charleroi Belgium, 8 jul-27 sep 1943; 3 oct 1944; 10 jul 1945 – 1r – 1 – uk British Libr Newspaper [074]

Journal Co [Milwaukee WI] et al *see* Newsprint

Journal commercial de la point-a-pitre – Point-a-Pitre, Guadeloupe. 1841-1844 (1) – mf#67955 – us UMI ProQuest [079]

Journal commercial de pointe-a-pitre *see* Journal politique et commercial de la pointe-a-pitre

Journal commercial, economique et maritime de la pointe-a-pitre *see* Journal politique et commercial de la pointe-a-pitre

Journal Communications et al *see* Intercom

Journal Company [Milwaukee WI] et al *see* Imprint

Journal Company Print. Lawrence, Kansas *see* Specimens of custom work

Journal constitution zones – Atlanta, GA. 1999-2000 (1) – mf#69446 – us UMI ProQuest [071]

Journal constructo – Quebec: Journal constructo (1967) inc, [ca 1963]- (biwkly) [mf ed 1988-] – 1 – mf#SEM35P329 – cn Bibl Nat [690]

Journal courier – Groton, NY. 1972-1982 (1) – mf#69294 – us UMI ProQuest [071]

Journal courier – Jacksonville, IL. 1989-2000 (1) – mf#61334 – us UMI ProQuest [071]

Journal d' extreme orient – Saigon, Vietnam. 1956 – 1 – us CRL [079]

Journal d'abbeville et de l'arrondissement – Abbeville. 1842-47 – 1 – (feuille politique, agricole, commerciale, litteraire et d'annonces. devenu: le pilote de la somme. journal d'abbeville et de l'arrondissement) – fr ACRPP [073]

Le journal d'agriculture – Montreal: G E Desbarats, [1877-1879] – 9 – mf#P04222 – cn Canadiana [630]

Le journal d'agriculture canadien = The canadian agricultural journal – Montreal: W Evans, [1844?-1845?] – 9 – ISSN: 0834-485X – mf#P04789 – cn Canadiana [630]

Le journal d'agriculture illustre = The illustrated journal of agriculture – Montreal: E Senecal, [1879-1897] – 9 – mf#P04220 – cn Canadiana [630]

Journal d'alsace = Elsaesser journal – Strasbourg. juin-dec 1873 – 1 – fr ACRPP [073]

Journal d'athenes – Greece. -w. 22 Feb 1879-25 Feb 1883; 7 March-25 May 1886. Imperfect. 1 reel – 1 – uk British Libr Newspaper [949]

Journal de ce qui s'est fait a rome dans l'affaire des cinq propositions / Saint-Amour, L G de – Amsterdam, 1662 – €57.00 – ne Slangenburg [241]

Journal de ce qui s'est passe a la tour du temple : pendant la captivite de louis 16, roi de france / Clery, Jean-Baptiste Cant Hanet – A Londres: de l'impr de Baylis, 1798 [mf ed 1988] – 3mf – 9 – mf#SEM105P952 – cn Bibl Nat [944]

Journal de ce qui s'est passe a la tour du temple : pendant la captivite de louis 16, roi de france / Hanet-Clery, Jean – Londres 1798 [mf ed Hildesheim 1995-98] – 1v on 2mf – 9 – €60.00 – ISBN-10: 3-487-26193-6 – ISBN-13: 978-3-487-26193-5 – gw Olms [944]

Journal de ce qui s'est passe au canada : depuis le mois d'octobre 1755 jusqu'au mois de juin 1756 – Nantes [France]: Chez Joseph Vatar...[1756?] [mf ed 1984] – 1mf – 9 – 0-665-45199-7 – mf#45199 – cn Canadiana [971]

Le journal de chambly – Chambly: [s.n.] v11 n42 4 avril 1978-v20 n38 12 mai 1987 (wkly) [mf ed 1986] – 18r – 1 – (cont: le journal (chambly, quebec); cont by: mon journal de chambly) – mf#SEM35P213 – cn Bibl Nat [971]

Journal de chambly (1989) – Chambly: [s.n.] v22 n32 11 avril 1989- (wkly) [mf ed 1990-] – 1 – (cont: mon journal de chambly) – mf#SEM35P336 – cn Bibl Nat [073]

Journal de chirurgie – Paris. 1968-1980 (1) 1971-1980 (5) 1973-1980 (9) – ISSN: 0021-7697 – mf#3401 – us UMI ProQuest [617]

Journal de comines : et des environs. – Comines. mai 1908-juin 1914 – 1 – fr ACRPP [944]

Journal de conchyliologie – Paris 1850-1943/45 – v1-86 on 682mf – 9 – (ind [1850-92] v1-40) – mf#5620c/2 – ne IDC [590]

Journal de constantinople. echo de l'orient *see*
– Echo de l'orient
– Journal de constantinople et des interets orientaux

Journal de constantinople et des interets orientaux – Constantinople. oct 1843-46, fevr 1848-53 – 1 – (devenu par fusion: journal de constantinople. echo de l'orient) – fr ACRPP [950]

Journal de francfort 1794 – Frankfurt/M DE, 1806-08, 1812, 1817-20 – 1 – (title varies: 1811: gazette du grand-duche de francfort; 1814: journal de francfort) – gw Misc Inst [074]

Journal de geneve – 12r per y – 1 – enquire for prices – us UMI ProQuest [073]

Journal de geneve – Geneva: [s.n.] jul 1938-feb 1948; aug 16 1948-apr 1976 – us CRL [949]

Journal de geneve – Geneve: [s.n., jan 3 1845-mar 3 1846; apr-jun 1850; jun-aug 1 1865; sep 1912-apr 15 1913 – 2r – 1 – us CRL [949]

Journal de geneve – National, politique et litteraire. no. 1-118. Geneve. janv-avr 1917. mq no. 19, 93 – 1 – fr ACRPP [949]

Journal de geneve – Switzerland. -d. 8 Nov 1914-9 Aug 1919; 1 Jan-30 June 1938. 36 reels – 1 – uk British Libr Newspaper [949]

Journal de gynecologie, obstetrique et biologie de la reproduction – Paris. 1978-1980 (1,5,9) – ISSN: 0368-2315 – mf#11543 – us UMI ProQuest [618]

Journal de henri 3, roy de france & de pologne : ou memoires pour servir a l'histoire de france / Estoile, Pierre de l' – La Haye [u.a. 1744 [mf ed Hildesheim 1995-98] – 5v on 36mf – 9 – €360.00 – ISBN-10: 3-487-26117-0 – ISBN-13: 978-3-487-26117-1 – gw Olms [944]

Journal de jean heroard sur l'enfance et la jeunesse de louis 13 (1601-1628) – Paris: F Didot, 1868 [mf ed 1980] – 2v on 1mf – 9 – mf#03762 – cn Canadiana [944]

Journal de jurisprudence – Dedie a Son Altesse serenissime electorale palatine. Bouillon. 1763-mai 1764 (I-VI) – 1 – fr ACRPP [340]

Journal de jurisprudence commerciale et maritime. – Marseille. On film: v1-91, 93-114; 1820-1913, 1915-37. LL-0227 – 1 – us L of C Photodup [346]

Journal de la contre-revolution – Neuwied sur le Rhin. n1, 3. 1791-92 – 1 – fr ACRPP [325]

Journal de la cour et de la ville – Paris 1791-92 [mf ed Hildesheim 1995-98] – 8v on 25mf – 9 – €250.00 – 3-487-26291-6 – gw Olms [073]

Journal de la liberte de la presse – no. 1-43. Paris. sept 1794-avr 1796 – 1 – (devenu: le tribun du peuple ou le defenseur des droits de l'homme) – fr ACRPP [322]

Journal de la librairie ou catalogue hebdomadaire contenant par ordre alphabetique les livres, tant nationaux qu'etrangers – Paris. 1764-79 (II-XVII) – 1 – fr ACRPP [010]

Journal de la marine : recueil mensuel de science et d'histoire; analyses, extraits, fragmens inedits de voyages entrepris dans les diverses parties du monde... – Paris 1833-34 [mf ed Hildesheim 1995-98] – 2v on 6mf – 9 – €120.00 – 3-487-29893-7 – gw Olms [910]

Journal de la marine le yacht – Paris, France. 3 jan 1885-25 dec 1886, 1887-25 dec 1909 – 16r – 1 – (aka: yacht) – uk British Libr Newspaper [790]

Journal de la marine le yacht *see* Yacht

Journal de la marine marchande et de l'empire francais – Paris, France. 21 jan 1943-22 jun 1944; 6 jan-22 dec 1949; 6 jul-27 jul 1950; 9 aug-27 dec 1951; 3 jan-21 feb 1952; 2 dec 1954; 12 apr 1956; 6 mar 1958-1959; 31 mar 1960-28 dec 1961 – 14r – 1 – uk British Libr Newspaper [072]

Journal de la montagne – Paris. juin 1793-nov 1794 – 1 – fr ACRPP [073]

Journal de la parfumerie francaise – Paris, france. 15 jan 1897-10 july 1914. -f – 5r – 1 – uk British Libr Newspaper [660]

Journal de la parfumerie francaise etc – Paris, France. 1897-1911; 10 jan 1912-10 jul 1914 – 5r – 1 – uk British Libr Newspaper [072]

Journal de la reforme sociale *see* L'intelligence

Journal de la republique – Le journal de la guerre. juil 1870-fevr 1871 – 1 – fr ACRPP [944]

Journal de la roer – Aachen DE, 1848-49 – 4r – 1 – (title varies: 18 jan 1814: stadt-aachener zeitung; 2 jan 1849: aachener zeitung. filmed by other misc inst: 1811-1847 31 jan, 1848-68, 1876-1888 30 sep; 1850 [1r]) – gw Misc Inst [074]

Journal de la societe hongroise de statistique / Magyar Statisztikai Tarsasag. Budapest – v1-23. 1923-45 – 1 – us L of C Photodup [314]

Le journal de la solidarite francaise – Paris. n41, 46, 61. juin-dec 1935 – 1 – fr ACRPP [073]

Journal de la vienne, des deux-sevres et de la vendee : gazette des provinces de l'Ouest – Poitiers. 1861 – 1 – fr ACRPP [073]

Le journal de l'acetylene *see* Petit photographe econome

Journal de l'acetylene etc – Paris, France. 1899, 1901, 5 jan 1902-04 – 2r – 1 – (aka: le petit photographe econome) – uk British Libr Newspaper [540]

Journal de l'affaire du canada passee le 8 juillet 1758 entre les troupes du roi : commandees par m le marquis de montcalm... – [Rouen: s.n, 1758] [mf ed 1984] – 1mf – 9 – 0-665-45208-X – mf#45208 – cn Canadiana [971]

Journal de l'agriculture, du commerce, des arts et des finances / ed by Pont, Pierre-Samuel du – jul 1765-nov 1766 – (= ser French economists of the 18th century) – 21mf (24:1) – 9 – $140.00 – us UPA [630]

Journal de l'agriculture, du commerce et des finances – Paris. juil 1765-74, 1780-83 – 1 – fr ACRPP [073]

Journal de l'ain – Bourg. 1848 – 1 – fr ACRPP [073]

Journal de l'eclairage au gaz : du service des eaux et de la salubrite publique – Paris, France 3 apr 1866-20 dec 1881; 5 jan 1889-20 dec 1895; 5 jan 1902-20 dec 1904 [wkly] – 1 – (cont by: journal de l'eclairage au gaz et a l'electricite [5 jan 1905-20 dec 1906]) – uk British Libr Newspaper [621]

Journal de l'eclairage au gaz et à l'electricite – Paris, France 5 jan 1905-20 dec 1906 – 1 – (cont: journal de l'eclairage au gaz, du service des eaux et de la salubrite publique [3 apr 1866-20 dec 1881, 5 jan 1889-20 dec 1895, 5 jan 1902-20 dec 1904]) – uk British Libr Newspaper [621]

Journal de lecture : ou recueil pour les oisifs – Paris, Amsterdam. 12v. 1775-1778 – 86mf – 8 – mf#H-1383 – ne IDC [073]

Journal de l'education – Montreal: J.B. Rolland, [1880] – (= ser Journal de l'instruction publique) – 9 – mf#P04146 – cn Canadiana [370]

Journal de l'education *see* Journal de l'instruction publique

Journal de l'electrolyse, l'aluminium, l'acetylene – Paris, France 15 jan-15 dec 1901 – 1 – (cont: aluminium et l'acetylene [1 nov-15 dec 1900) – uk British Libr Newspaper [660]

Journal de l'Empire *see* Journal des debats et des decrets

Journal de l'exploitation des corps gras industriels – Paris. 15 jan 1878-1 aug 1914, 1927-feb 1928 – 15 1/2r – 1 – (aka: les corps gras industriels) – uk British Libr Newspaper [338]

Journal de l'industriel et du capitaliste – Paris. 1836-40 [mnthly] – 1 – (publ. par une societe d'ingenieurs civils) – fr ACRPP [073]

Journal de l'instruction publique – Montreal: J B Rolland, [1880?-98] – 9 – mf#P04234 – cn Canadiana [370]

Journal de londres – London, UK. 10 Sept-29 Oct 1909 – 1 – uk British Libr Newspaper [072]

Journal de l'ouest – Poitiers. 1901 – 1 – fr ACRPP [073]

Journal de louis 16 et de son peuple : ou le defenseur de l'autel, du trone et de la patrie – Paris, nov 1790-aout 1792 (I-X) – 1 – fr ACRPP [944]

Journal de luneville – Republicain. Luneville. 1923-39 – 1 – fr ACRPP [073]

Journal de ma deportation a la guyane francaise / Laffon De Ladebat, Andre Daniel – Paris, France. 1912 – 1r – us UF Libraries [972]

Journal de madagascar : franco-malgache – Tananarive. fevr 1923-mai 1925 – 1 – fr ACRPP [073]

Journal de marseille – Marseille. 1898 – 1 – fr ACRPP [073]

Journal de medecine, chirurgie, pharmacie *see* Recueil periodique d'observations de medecine, chirurgie, pharmacie etc

Journal de monsieur suleau – Paris, Neuwied sur le Rhin. avr 1791-avr 1792 – 1 – fr ACRPP [073]

Journal de moscou : hebdomadaire politique, economique, social et litteraire – Moscou. avr 1934-avr 1939 – 1 – fr ACRPP [073]

Journal de musique par mr de lagarde, maitre de musique : en survivance des enfants de france / Lagard, N de – Paris: chez l'auteur, chez mr s Prault et Duchesne 1758 [mf ed 1989] – 1v on 1r – 1 – mf#pres. film 74 – us Sibley [780]

Journal de navigation du voyage de la coste de guinee. / Des Marchais, Etienne Renaud – Holograph – 1 – us CRL [916]

Journal de neubourg – Rouen. 1932-36 – 1 – fr ACRPP [073]

Journal de paris – Paris. janv-juin 1869, Janv-juin 1875, janv-avr 1876 – 1 – (national, politique, et litteraire.) – fr ACRPP [073]

Journal de paris : ou poste du soir – Paris, 1784-1793, 1805(may)-1811 – 572mf – 8 – mf#H-1384 – ne IDC [440]

Journal de paris – 1 – (puis de Paris et des departemens. 1777-juin 1827. devenu: Nouveau journal de Paris et des departemens. Feuille administrative, commerciale, industrielle et litteraire. no. 1-679. aout 1827-juin 1829. devenu: La France nouvelle. Nouveau journal de Paris. Politique, litteraire et industriel. no. 680-2109. juin 1829-juin 1833. devenu: Journal de Paris. Nouvelliste du matin et du soir. juin 1833-mai 1840. Paris. 1777-97, 21 mars 1798-22 sept 1800, 21 mai 1801-1839) – fr ACRPP [073]

Journal de paris, ou poste du soir – Paris, 1784-1793, 1805(May)-1811 – 572mf – 8 – mf#H-1384 – ne IDC [700]

Le journal de quebec – Quebec, QC: A Cote, 1842-53, 1862-73 – 31r – 1 – ISSN: 0839-1084 – cn Library Assoc [071]

Journal de radiologie – Paris. 1979-1980 (1,5,9) – (cont: journal de radiologie, d'electrologie, et de medecine nucleaire) – ISSN: 0221-0363 – mf#3399,01 – us UMI ProQuest [616]

Journal de radiologie, d'electrologie, et de medecine nucleaire – Paris. 1914-1978 (1) 1971-1978 (5) 1975-1978 (9) – (cont by: journal de radiologie) – ISSN: 0368-3966 – mf#3399 – us UMI ProQuest [616]

Journal de route d'un caporal de tirailleurs de la mission saharienne, 1898-1900 / Guilleux, Charles – [Belfort: Impr J Spitzmuller] – us CRL [960]

Le journal de royan – Marennes. 22 mai 1859-8 mai 1860, 2 9 juil 1876-79 – 1 – fr ACRPP [073]

Journal de saint-petersbourg – n79-232. Saint-Petersbourg. 1840 – 1 – (mq no. 157) – fr ACRPP [073]

Journal de saverne – Saverne, France. 1817-1940 – 1 – fr ACRPP [073]

Journal de smyrne – Smyrne. 1834-38 – 1 – (commercial, politique et litteraire) – fr ACRPP [073]

Le journal de st-bruno (1977) – St-Bruno: [s.n.] v10 n27 14 sep 1977-v12 n30 7 nov 1979 (wkly) [mf ed 1986] – 3r – 1 – mf#SEM35P278 – cn Bibl Nat [071]

Le journal de st-bruno (1982) – St-Bruno: [s.n.] v16 n10 10 mars 1982- (wkly) [mf ed 1986-] – 1 – mf#SEM35P214 – cn Bibl Nat [071]

Journal de tahiti – 1969-71 – 12r – 1 – mf#pmb doc391 – at Pacific Mss [073]

Le journal de tahiti – Tahiti, jan 1969-dec 1971 – 12r – at Pascoe [079]

Journal de zoologie : comprenant les differentes branches de cette science – Paris 1872-77 – v1-6 on 63mf – 9 – mf#z-2056/2 – ne IDC [590]

Journal d'education – Quebec: L Brousseau, 1881-[1882] – 9 – (incl ind) – mf#P04252 – cn Canadiana [370]

Journal democratique et officiel des ateliers nationaux – Paris: Boule, jun 22/24 1848 – us CRL [320]

Journal der distrikte minden, bielefeld und rinteln – Bielefeld, Osnabrueck DE, 1809-1810 29 dec – 1 – (title varies: 4 apr 1810: oeffentliche anzeigen des weserdepartements. fr 12 oct 1809 publ in osnabrueck) – gw Misc Inst [350]

Journal der neuesten weltbegebenheiten – (Hamburg-) Altona DE, 1795 mar-aug, 1795 nov-1796 jan, 1796 nov-1797 jul, 1798, 1802 feb – 1 – gw Misc Inst [073]

Journal der physik – Halle/Leipzig. 18v. 1790-94 [mf ed 1993] – 41mf – 9 – €290.00 – 3-89131-162-1 – (filmed with: neues journal der physik [leipzig 1795-97] 4v) – gw Fischer [530]

Journal der practischen arzneykunde und wundarzneykunst – v1-27. 1795-1808 [mf ed 1994] – 536mf – 9 – €4000.00 – 3-89131-190-7 – (filmed with: journal der practischen heilkunde [v28-83 1809-83]; c w hufelands journal der practischen heilkunde [v84-98 1837-44]) – gw Fischer [615]

Journal der practischen heilkunde *see* Journal der practischen arzneykunde und wundarzneykunst

Journal der religion, wahrheit und litteratur – Augsburg 1797-99 – (= ser Dz. abt theologie) – 3jge on 20mf – 9 – €200.00 – mf#k/n2277 – gw Olms [200]

Journal der tonkunst – Erfurt: Bey G A Keyser 1795 [mf ed 19–] – 5mf / 3mf – 9 – mf#fiche 497, 744, 031 – us Sibley [780]

Journal des amis – Paris. v1-2. n1-18. janv-juin 1793 – 1 – fr ACRPP [073]

Journal des amis de la constitution – no. 1-41. Paris. nov 1790-sept 1791. – 1 – (en janv 1792 elements repris par: journal des debats de la societe des amis de la constitution) – fr ACRPP [073]

Journal des artistes : revue pittoresque consacree aux artistes et aux gens du monde – Paris. 1827-avr 1848 – 1 – (Subtitle varies.) – fr ACRPP [073]

Journal des arts – Paris, France. 31 jan 1879-5 oct 1888; 13 jul 1895; 5 sep 1896; 23 may 1900 – 3 1/4r – 1 – uk British Libr Newspaper [072]

Le journal des arts : chronique hebdomadaire de l'hotel drouot. – Paris. 31 janv 1879-1900, 1910 – 1 – fr ACRPP [073]

Journal des beaux-arts et des sciences – Par M. l'Abbe Aubert. Paris. 1768 (I-IV) – 1 – fr ACRPP [073]

Journal des blagueurs – Paris: Imp d'A Sirou, No dideochicoquancarflambardino, 1849 – us CRL [074]

Journal des campagnes au canada de 1755 a 1760 / Malartic, Anne-Joseph-Hyppolite de Maures, comte de – Paris: E Plon, Nourrit et Cie, 1890 – 5mf – 9 – mf#09619 – cn Canadiana [971]

Journal des campagnes (edition hebdomadaire) – Quebec: Leger Brousseau. 1re annee n1 9 fevr 1882-20e annee n51 28 dec 1901 (wkly) [mf ed 1989] – 18r – 1 – (cont: courrier du canada (edition hebdomadaire)) – mf#SEM35P337 – cn Bibl Nat [071]

Journal des clubs ou societes patriotiques : dedie aux amis de la constitution, membres des differents clubs francais – Paris. n1-47. nov 1790-sept 1791 – 1 – fr ACRPP [366]

Journal des dames : ou, les souvenirs d'un viellard – New York. 1810-1810 (1) – mf#3997 – us UMI ProQuest [440]

Journal des debats de la societe des amis de la constitution *see* Journal des amis de la constitution

Journal des debats de la societe des amis de la constitution seante aux jacobins a paris – Paris. juin 1791-93 – 1 – (en janv 1792 reprend les elements de: Journal des amis de la Constitution et se divise en 2 s. numerotees separement: Debats et Correspondance) – fr ACRPP [073]

Journal des debats et des decrets – 29 aout 1789-15 juil 1805 – 1 – (devenu: journal de l'empire. 16 juil 1805-31 mars 1814. devenu: journal des debats politiques et litteraires. 1er avr 1814-20 mars 1815. devenu: journal de l'empire. 21 mars-7 juil 1815. devenu: journal des debats politiques et litteraires. 8 juil 1815. paris. 5 mai-1er sept 1789, 29 aout 1789-aout 1944) – fr ACRPP [073]

Journal des debats legislatifs et litteraires du canada – Toronto: [J Blackburn, 1858] – 9 – mf#P05052 – cn Canadiana [323]

Journal des debats politiques et litteraires – Paris: Impr de le Normant, 1919-28 – 32r – 1 – us CRL [073]

Journal des debats politiques et litteraires *see* Journal des debats et des decrets

Journal des debats, politiques et litteraires – Paris. -d 1872-99; 1941-44 – 72 1/2r – 1 – uk British Libr Newspaper [074]

Journal des departements meridionaux : et des debats des amis de la liberte et de l'egalite de marseille – Marseille. mars 1792-mai 1793 – 1 – fr ACRPP [073]

Journal des economistes : revue de la science economique et de la statistique. – Paris. 1850, 1866-1904 – 1 – fr ACRPP [073]

Journal des etats generaux / ed by Hodey, Le – Paris. avr 1789-sept 1791 – 1,5 – fr ACRPP [073]

Journal des fabricants de sucre – Paris, France. -w. 15 apr 1869-15 sep 1870; 7 may-28 dec 1871; 18 jan 1872-1885 – 8r – 1 – uk British Libr Newspaper [660]

Journal des faits – Paris. avr-juin 1850, janv-9 fevr 1854 – 1 – (Tous les journaux dans un.) – fr ACRPP [073]

Journal des familles – Montreal: J E Belair, [1887-1888] – 9 – ISSN: 1190-7673 – mf#P04095 – cn Canadiana [440]

Le journal des familles : publication religieuse, scientifique, industrielle et litteraire – S.l: J V De Lorme, 1840? – 1mf – 9 – mf#56453 – cn Canadiana [440]

Le journal des familles : recueil de litterature – Quebec: G A Lavoie, [1881] – 9 – mf#P04142 – cn Canadiana [440]

Journal des fauborgs – Montmartre [Paris]: Pilloy freres, apr 16 1848 – us CRL [074]

Journal des forets – [Paris: Bureau du Journal 1829]- (mthly) [t1-2(1829)] [mf ed 2005] – 1 – mf#film ms 36388 – us Harvard [634]

Journal des impartiaux – n1-19. Paris. fevr-avr 1790 – 1 – (mq n17) – fr ACRPP [073]

Journal des jacobins – Paris: Schneider, may 14 1848 – us CRL [944]

Journal des luxus und der moden / ed by Bertuch, Friedrich Justin et al – Weimar 1786-1827 – (= ser Dz) – 42v on 271mf – 9 – €1626.00 – (v1: journal der moden; v28: journal fuer luxus, mode und gegenstaende der kunst; v29ff: journal fuer literatur, kunst, luxus und mode) – mf#k/n6564 – gw Olms [110]

Journal des meres et des jeunes filles : recueil religieux et litteraire – Paris. 1844-47 (I-III) – 1 – fr ACRPP [073]

Journal des missions evangeliques – 1826 – 1r – 1 – mf#pmb doc101 – at Pacific Mss [240]

Journal des missions evangeliques – 1827-30 – 1r – 1 – mf#pmb doc102 – at Pacific Mss [240]

Journal des missions evangeliques – 1831-34 – 1r – 1 – mf#pmb doc103 – at Pacific Mss [240]

Journal des missions evangeliques – 1835-37 – 1r – 1 – mf#pmb doc104 – at Pacific Mss [240]

Journal des missions evangeliques – 1838-40 – 1r – 1 – mf#pmb doc105 – at Pacific Mss [240]

Journal des missions evangeliques – 1841-43 – 1r – 1 – mf#pmb doc106 – at Pacific Mss [240]

Journal des missions evangeliques – 1844-46 – 1r – 1 – mf#pmb doc107 – at Pacific Mss [240]

Journal des missions evangeliques – 1847-54 – 4r – 1 – mf#pmb doc108-111 – at Pacific Mss [240]

Journal des missions evangeliques – 1855-57 – 1r – 1 – mf#pmb doc112 – at Pacific Mss [240]

Journal des missions evangeliques – 1858-60 – 1r – 1 – mf#pmb doc113 – at Pacific Mss [240]

Journal des missions evangeliques – 1861-63 – 1r – 1 – mf#pmb doc114 – at Pacific Mss [240]

Journal des missions evangeliques – 1864-66 – 1r – 1 – mf#pmb doc115 – at Pacific Mss [240]

Journal des missions evangeliques – 1867-68 – 1r – 1 – mf#pmb doc116 – at Pacific Mss [240]

Journal des missions evangeliques – 1869-71 – 1r – 1 – mf#pmb doc117 – at Pacific Mss [240]

Journal des missions evangeliques – 1872-84 – 1r – 1 – mf#pmb doc118-123 – at Pacific Mss [240]

Journal des missions evangeliques – 1885-87 – 1r – 1 – mf#pmb doc124 – at Pacific Mss [240]

Journal des missions evangeliques – 1888-90 – 1r – 1 – mf#pmb doc125 – at Pacific Mss [240]

Journal des missions evangeliques – 1891-1940 – 42r – 1 – mf#pmb doc126-166 – at Pacific Mss [240]

Journal des missions evangeliques – 1941-45 – 1r – 1 – mf#pmb doc167 – at Pacific Mss [240]

Journal des missions evangeliques – 1946-48 – 1r – 1 – mf#pmb doc168 – at Pacific Mss [240]

Journal des missions evangeliques – 1949-50 – 1r – 1 – mf#pmb doc169 – at Pacific Mss [240]

Journal des missions evangeliques – 1951-55 – 1r – 1 – mf#pmb doc170 – at Pacific Mss [240]

Journal des missions evangeliques – 1956-59 – 1r – 1 – mf#pmb doc171 – at Pacific Mss [240]

Journal des missions evangeliques – 1960-64 – 1r – 1 – mf#pmb doc172 – at Pacific Mss [240]

Journal des missions evangeliques – 1965-69 – 1r – 1 – mf#pmb doc173 – at Pacific Mss [240]

Journal des missions evangeliques – Paris, 1826-1940. v1-115 – 1326mf – 8 – mf#A-215 – ne IDC [240]

Journal des missions evangeliques / Societe des Missions Evangeliques – Paris. 1826-1969 – 1 – fr ACRPP [242]

Journal des nations – Geneva, Switzerland. -d. 3 June 1936-7 Oct 1938; 9 June 1939-9 May 1940. Imperfect. 5 reels – 1 – uk British Libr Newspaper [949]

Journal des observations physiques, mathematiques et botaniques... / Feuillee, L – Paris, 1714-1725. 3v – 19mf – 9 – mf#H-6179 – ne IDC [910]

Journal des operations de l'armee lors de l'invasion du canada en 1775-76 / Badeaux, Jean Baptiste – Montreal: E Senecal, 1871 – 1mf – 9 – mf#00074 – cn Canadiana [971]

Journal des ouvriers : feuille populaire et economique. – n1-24. Paris. sept-dec 1830. – 1 – (lacking: n3, 19) – fr ACRPP [073]

Journal des pyrenees orientales – Perpignan. 24 mars 1848-8 mai 1852 – 1 – fr ACRPP [073]

Journal des sans-culottes – [Paris]: E Bautrucre, may 28/jun 1 1848-mar 1849 – us CRL [944]

Journal des savans d'italie – Amsterdam. 1748-49 (1-3) – 1 – fr ACRPP [073]

Journal des savants – Paris. janv-mars 1665, 1666-nov 1792, janv-juin 1797, sept 1816-1981 – 1 – fr ACRPP [073]

Journal des spectacles – Paris. juil 1793-janv 1794 (I-III) – 1 – (contenant l'analyse des differentes pieces qu'on a representees sur tous les theatres de Paris) – fr ACRPP [790]

Journal des theatres : ou le nouveau spectateur servant de repertoire universel des spectacles – Paris. avr 1776-juin 1778 – 1 – fr ACRPP [790]

Journal des travailleurs – Paris, France. -w. 4 jan-25 jun 1848 – 1/4r – 1 – uk British Libr Newspaper [331]

Journal des travaux publics *see* Le locateur

Journal des tribunaux – Lausanne. On film: v1-50; 1853-1902. Lacking: v23-24; 1875-76. LL-0240 – 1 – us L of C Photodup [340]

Le journal des trois-rivieres – Trois Rivieres, QC. 1865-73 – 6r – 1 – ISSN: 1180-6397 – cn Library Assoc [971]

Journal des usines a gaz – Paris. -f. 5 may 1877-5 jun 1879; 5 jan 1880-5 dec 1882; 5 jan 1883-5 dec 1884; 5 jan 1885-20 dec 1886; 5 jan 1887-20 dec 1888; 5 jan 1889-20 dec 1890; 5 jan-20 dec 1891; 5 jan-20 dec 1903; 5 jan 1904-5 dec 1905; 5 jan 1906-20 dec 1907; 5 jan 1908-20 dec 1909 – 10r – 1 – uk British Libr Newspaper [338]

Le journal des vedettes – Montreal: Publ independante ltee. v1 24 oct 1954-v25 n38 2/8 juil 1978 [mf ed 1974-89] – 1 – (incl suppls) – mf#SEM35P94 – cn Bibl Nat [073]

Journal des villes et des campagnes, des cures, des maires, des familles – Paris. 3,9,11,13 nov 1842, 1845-54, juil 1855-59, 2 oct 1863-65, 1868-28 fev 1892, 28 fev 1893, 1 mars 1894, 1 mars 1895 – 1 – fr ACRPP [073]

Journal des voyages, decouvertes et navigations modernes : ou archives geographiques du 19e siecle – Paris 1818-29 [mf ed Hildesheim 1995-98] – 44v on 132mf – 9 – €792.00 – 3-487-29899-6 – gw Olms [910]

Journal d'etat et du citoyen – Paris. aout 1789-aout 1790 – 1 – (devenu: mercure national ou journal d'etat et du citoyen) – fr ACRPP [321]

Journal d'extreme orient – Saigon, 1956 – us CRL [950]

Journal die letras – Rio de Janeiro Brazil, jun-sep 1966 – 1/4r – 1 – uk British Libr Newspaper [079]

Journal du camp – London, UK. 24 jul-15 aug 1940 – 1 – (nouvelles de france et du monde, 16 aug-2 sep 1940) – uk British Libr Newspaper [072]

Journal du club des cordeliers – Societe des amis des droits de l'homme et du citoyen. Par Sinties et Momoro. n1-10. Paris. juin-aout 1791 – 1 – fr ACRPP [944]

Journal du commerce – Paris. 1er germinal an VII-6e jour complementaire an VII. 21 mars-21 sept 1799 – 1 – fr ACRPP [380]

Journal du commerce de juil *see* Le constitutionnel

Journal du commerce de la ville de lyon et du departement du rhone – Lyon. dec 1826-aout 1828, dec 1835-juil 1844 – 1 – fr ACRPP [944]

Journal du commerce, politique et litteraire – Paris. 20 dec 1819-21 mars 1848 – 1 – fr ACRPP [073]

Journal du corsaire jean doublet de honfleur : lieutenant de fregate sous louis 14 – Paris: Perrin, 1887 – 4mf – 9 – (int and ann by charles breard) – mf#27037 – cn Canadiana [910]

Journal du departement de la marne – Par une Societe d'amis de la Republique. no. 1-512. Chalons-sur-Marne. 19 juin 1796-25 avr 1800. BM. Reims Per. CH. IX. 2 – 1 – fr ACRPP [944]

Journal du departement des bouches du weser – Bremen DE, feb 2 1812-oct 26 1813 – 2r – 1 – gw Misc Inst [074]

Journal du dimanche – Paris. n1-48 avec 2 prosp. sept 1846-aout 1847 – 1 – (litterature, poesie, histoire, voyages, sciences) – fr ACRPP [073]

Journal du dimanche – Paris. Edite par France-soir. 1953-54 – 1 – fr ACRPP [073]

Journal du droit administratif – v. 1-61. 1853-1913 – 1 – us L of C Photodup [944]

Journal du droit commun *see* L'intelligence

Journal du droit international – v1-63 1874-1936 – 1 – $600.00 – us L of C Photodup [341]

Journal du frap / Front d'action Politique – v1 n2-v2 n4 [1972 nov-1974 apr] – 1r – 1 – mf#1060677 – us WHS [320]

Journal du genie civil, des sciences et des arts a l'usage des ingenieurs – Paris. 1829 – 1 – fr ACRPP [073]

Le journal du maquis – Paris, 1944 – 1 – (in french) – us UMI ProQuest [934]

Le journal du parlement – Paris: Les Imp Lamartine, oct 17 1961; feb 28-mar 1 1962 – us CRL [340]

Journal du petrole – Paris. -m. 10 may 1901-1902; 15 jan-15 dec 1903; 1904-20 dec 1905 – 3r – 1 – uk British Libr Newspaper [660]

Journal du peuple – Paris. 4e-9e annee. 1837-30 avr 1842. mq 23 fevr-6 avr 1841 – 1 – (feuille du dimanche) – fr ACRPP [073]

Journal du peuple – Paris. n1-299. 6 fevr-3 dec 1899 – 1 – (mq no. 169, 185, 277) – fr ACRPP [073]

Journal du peuple – n1-193; prosp. de janv 1792. fevr-aout 1792 – 1 – fr ACRPP [073]

Journal du peuple – Paris. 9 fevr 1916-9 juin 1929 – 1 – (politique, litteraire, artistique et social) – fr ACRPP [073]

Le journal du peuple – Paris. 1er juil-6 sept 1870 [wkly] – 1 – fr ACRPP [073]

Journal du regne de henry 4, roi de france et de navarre : avec des remarques historiques & politiques du chevalier c b a et plusieurs pieces historiques du meme tems / Estoile, Pierre de l' – La Haye 1741 [mf ed Hildesheim 1995-98] – 4v on 26mf – 9 – €260 – ISBN-10: 3-487-26115-4 – ISBN-13: 978-3-487-26115-7 – gw Olms [944]

Le journal du soir – Paris, may 5-6 1871 – us CRL [074]

Journal du travail – Quebec [Province] v1 n1-v8 n1 [1979 avril-1986 fev] – 1r – 1 – mf#1096275 – us WHS [071]

Journal d'un spahi au soudan, 1897-1899 / Herissay, Jacques – Paris, Perrin, 1909 – us CRL [960]

Journal d'un spahi au soudan, 1897-1899 / Labour, Gaston – Paris. 1909 – 1 – us CRL [916]

Journal d'un voyage a temboctou et a jenne, dans l'afrique centrale : precede d'observations faites chez les maures braknas, les nalous et d'autres peuples; pendant les annees 1824, 1825, 1826, 1827, 1828 / Caillie, Rene – Paris 1830 [mf ed Hildesheim 1995-98] – 3v on 9mf – 9 – €180.00 – 3-487-27299-7 – gw Olms [916]

Journal d'un voyage autour du monde : pendant les annees 1816, 1817, 1818 et 1819 / Roquefeuil, Camille de – Paris 1823 [mf ed Hildesheim 1995-98] – 2v on 6mf – 9 – €120.00 – ISBN-10: 3-487-26640-7 – ISBN-13: 978-3-487-26640-4 – gw Olms [910]

Journal d'un voyage dans la turquie-d'asie et la perse, fait en 1807 et 1808 / [Gardane, P A M de] – Paris, Marseille, 1809 – 2mf – 9 – mf#AR-2069 – ne IDC [915]

Journal d'un voyage de constantinople en pologne : fait a la suite de son excellence mr jaq porter, ambassadeur d'angleterre...en 1762 / Boscovic, Ruder – Lausanne 1772 [mf ed Hildesheim 1995-98] – 1v on 2mf – 9 – €60.00 – 3-487-27891-X – gw Olms [880]

Journal d'un voyage en allemagne : fait en 1773 / Guibert, Jacques A de – Paris [u a] 1803 [mf ed Hildesheim 1995-98] – 2v on 4mf – 9 – €120.00 – 3-487-29574-1 – gw Olms [914]

Journal d'un voyage en chine en 1843, 1844, 1845, 1846 / Itier, J – Paris: Dauvin et Fontaine, 1848. 2v – 9mf – 9 – mf#HT-797 – ne IDC [915]

Journal d'un voyage en italie et en suisse : pendant l'annee 1828 / Colomb, Romain – Paris 1833 [mf ed Hildesheim 1995-98] – 1v on 3mf – 9 – €90.00 – 3-487-27745-X – gw Olms [880]

Journal d'un voyage fait aux indes orientales : par une escadre de six vaisseaux commandez par mr du quesne, depuis le 24 fevrier 1690, jusqu'au 20 aout 1691, par ordre de la compag... / Challes, Robert – Rouen 1721 [mf ed Hildesheim 1995-98] – 3v on 15mf – 9 – €150.00 – 3-487-27409-4 – gw Olms [915]

Journal d'un voyage fait dans l'interieur de l'amerique septentrionale : ouvrage dans lequel on donne des details precieux sur l'insurrection des anglo-americains, et sur la chute desastreuse de leur papier-monnoie / Anburey, Thomas – Paris 1793 [mf ed Hildesheim 1995-98] – 2v on 9mf – 9 – €180.00 – 3-487-27069-2 – gw Olms [910]

Journal d'une expedition contre les iroquois en 1687 : lettres et pieces relatives au fort saint-louis des illinois / Baugy, Louis Henri – Paris: E Leroux, 1883 – 3mf – 9 – mf#03503 – cn Canadiana [971]

Journal d'une expedition de d'iberville / Beaudoin, Jean – [Evreux, France: l'Eure], 1900 – 1mf – 9 – (int and notes by auguste gosselin) – mf#03501 – cn Canadiana [917]

Le journal d'une saphiste / Montfort, Charles – Paris: Offenstadt, 1902 – (= ser Les femmes [coll]) – 3mf – 9 – mf#12728 – fr Bibl Nationale [306]

Journal d'urologie – Paris. 1980-1980 (1) (5) 1980-1980 (9) – (cont: journal d'urologie et de nephrologie) – ISSN: 0248-0018 – mf#3400,01 – us UMI ProQuest [616]

Journal d'urologie et de nephrologie – Paris. 1968-1979 (1) 1971-1979 (5) 1976-1979 (9) – (cont by: journal d'urologie) – ISSN: 0021-8200 – mf#3400 – us UMI ProQuest [616]

Journal ecclesiastique : ou Bibliotheque raisonnee des sciences ecclesiastiques – Paris. 1789-juil 1792 – 1 – fr ACRPP [242]

Journal encyclopedique – Liege puis Bouillon, Bruxelles. 1756-93 – 1 – (puis encyclopedique ou universel) – fr ACRPP [073]

Journal et affiches du department de haute-garonne. *see* Affiches, annonces, avis divers de toulouse et du haut-languedoc

Journal etranger – puis ou Notice exacte et detaillee des ouvrages de toutes les nations etrangeres en fait d'arts, de sciences, de litterature, etc.. Paris. avr 1754-58, 1760-sept 1762 – 1 – (puis ou notice exacte et detaillee des ouvrages de toutes les nations etrangeres en fait d'arts, de sciences, de litterature, etc.) – fr ACRPP [073]

Journal every evening – Wilmington, DE. 1935-1960 (1) – mf#62383 – us UMI ProQuest [071]

Journal for nurses in staff development: jnsd – Hagerstown. 1998+ (1,5,9) – (cont: journal of nursing staff development: jnsd) – ISSN: 1098-7886 – mf#14443,01 – us UMI ProQuest [610]

Journal for quality and participation – Cincinnati. 1990+ (1,5,9) – ISSN: 1040-9602 – mf#15804,02 – us UMI ProQuest [650]

Journal for research in mathematics education – Reston. 1970+ (1) 1970+ (5) 1975+ (9) – ISSN: 0021-8251 – mf#6039 – us UMI ProQuest [370]

Journal for special educators – Iowa City. 1978-1983 (1) 1978-1983 (5) 1978-1983 (9) – ISSN: 0197-5323 – mf#2543,02 – us UMI ProQuest [370]

Journal for special educators of the mentally retarded – Richmond Hill. 1969-1978 (1) 1970-1978 (5) 1977-1978 (9) – (cont: digest of the mentally retarded) – ISSN: 0012-2807 – mf#2543,01 – us UMI ProQuest [370]

Journal for specialists in group work – Washington. 1978+ – 1,5,9 – (cont: together) – ISSN: 0193-3922 – mf#10943,01 – us UMI ProQuest [370]

Journal for specialists in pediatric nursing – Philadelphia. 2002+ (1,5,9) – ISSN: 1539-0136 – mf#24906,01 – us UMI ProQuest [610]

Journal for the education of the gifted – Reston. 1983+ – 1,5,9 – ISSN: 0162-3532 – mf#13917 – us UMI ProQuest [370]

Journal for the scientific study of religion – Malden. 1961+ [1]; 1971+ [5]; 1976+ [9] – ISSN: 0021-8294 – mf#2013 – us UMI ProQuest [200]

Journal for the study of the new testament – Sheffield. 1985+ (1,5,9) – ISSN: 0142-064X – mf#15379 – us UMI ProQuest [225]

Journal for the study of the old testament – Sheffield. 1985+ (1,5,9) – ISSN: 0309-0892 – mf#15291 – us UMI ProQuest [221]

Journal for the theory of social behaviour – Oxford. 1983+ (1,5,9) – ISSN: 0021-8308 – mf#13527 – us UMI ProQuest [150]

Journal for truancy and dropout prevention – Long Beach. 1992-1995 – 1,5,9 – (cont: journal of the international association of pupil personnel workers) – mf#11295,01 – us UMI ProQuest [370]

Journal for vocational special needs education – Lincoln. 1983-1996 – 1,5,9 – ISSN: 0195-7597 – mf#14123 – us UMI ProQuest [370]

Journal francais – London, UK. 1 Nov 1910-2 Apr 1912 – 1 – uk British Libr Newspaper [074]

Journal francais – London, UK. 24 Sept 1891-13 Jul 1893 – 1 – uk British Libr Newspaper [072]

Journal from the golden mean society / Golden Mean Society – n1-49 [1980 apr-1984 aug] – 1r – 1 – (cont by: double eagle report; journal-report from the golden mean society) – mf#956956 – us WHS [071]

Journal fuer das gesellige vergnuegen – Strassburg (Strasbourg F), 1797 23 feb-20 mar – 1 – fr ACRPP [074]

Journal fuer deutsche frauen 1805-1806 – [mf ed 1994] – (= ser Hq 16) – 47mf – 9 – €310.00 – 3-89131-128-1 – (filmed with: selene: 1 (1807-2 (1808 wh cont: journal fuer deutsche frauen) – gw Fischer [305]

Journal fuer deutsche frauen von deutschen frauen geschrieben / ed by Wieland, Christoph Martin et al – Leipzig 1805-06 – (= ser Dz) – 2jge[zu je 3bdn] on 19mf – 9 – €190.00 – mf#k/n6550 – gw Olms [305]

Journal fuer die liebhaber der entomologie / ed by Scriba, L G – Frankfurt [1790-]1791 – v1 on 6mf – 9 – mf#z-970/2 – ne IDC [590]

Journal fuer die liebhaber des steinreichs und der konchyliologie – Weimar 1774-80. v1-6 – 101mf – 9 – (cont as: fuer die literatur und kenntniss der naturgeschichte, sonderlich der conchylien und der steine [weimar 1782-85] v1-2; cont as: neue literatur und beytrage zur kenntniss der naturgeschichte, vorzueglich der conchylien und fossilien [leipzig 1784-87] v1-4) – mf#z-903/2 – ne IDC [560]

Journal fuer die neuesten land- und seereisen und das interessanteste aus der voelker- und laenderkunde : zur angenehmen unterhaltung fuer gebildete leser in allen staenden – Berlin 1817-21 [mf ed Hildesheim 1995-98] – 3v on 9mf – 9 – €180.00 – 3-487-26463-3 – gw Olms [910]

Journal fuer die reine und angewandte mathematik – Berlin. 1826+ (1) 1970+ (5) 1970+ (9) – ISSN: 0075-4102 – mf#373 – us UMI ProQuest [510]

Journal fuer kaufleute / ed by Seehusen, Lucas Vincent – Hamburg 1780-81 – (= ser Dz) – 2v(zu je 2st) on 4mf – 9 – €120.00 – (v2.2 ed by j a engelbrecht) – mf#k/n2731 – gw Olms [380]

Journal fuer praktische chemie – Leipzig. 1834-1943 (1) – ISSN: 0021-8383 – mf#539 – us UMI ProQuest [540]

Journal fuer prediger / ed by Niemeyer, David G et al – Halle 1770-1842 – (= ser Dz. abt theologie) – 100v+app on 304mf – 9 – €1824.00 – mf#k/n2099 – gw Olms [240]

Journal fuer sachsen / ed by Schwarz, J N – Dresden 1792-94 – (= ser Dz. historisch-geographische abt) – v1-4[=st1-12] on 8mf – 9 – €160.00 – mf#k/n1267 – gw Olms [074]

Journal gazette – Fort Wayne, IN. 1884+ (1) – mf#60468 – us UMI ProQuest [071]

Journal gazette / Shelby Co. Sidney – jan 8-aug 6 1909 [wkly] – 1r – 1 – mf#B11292 – us Ohio Hist [071]

Journal gazette – Sidney, OH. 1905-1909 (1) – mf#65662 – us UMI ProQuest [071]

Journal gazette (launceston edition) – 1991-92; Jan 7-Jun 24 1993; Jul-Dec 1993; Jan 8-Jun 25 1994; Jul 9-Dec 23 1994; 1995-96 – 5 1/2r – 9 – uk British Libr Newspaper [072]

Journal general – Paris. 1er fevr 1791-10 aout 1792 – 1 – fr ACRPP [073]

Journal general de France *see* Annonces, affiches et avis divers

Journal general de france : supplement – Paris. 1787-6 janv 1790 – 1 – (puis partie d'agriculture et d'economie rurale.) – fr ACRPP [073]

Journal general de la cour et de la ville – Gautier. Paris. sept 1789-aout 1792 – 1 – fr ACRPP [073]

Journal general de la litterature de france ou... – Paris, Strasbourgh. 19-2. 1798-1826 – 271mf – 8 – mf#H-680 – ne IDC [440]

Journal general de l'europe : ou mercure national et etranger. – Liege puis Herve puis Paris. juil 1791-aout 1792 – 1 – fr ACRPP [073]

Journal herald – Dayton, OH. 1949-1986 (1) – mf#60558 – us UMI ProQuest [071]

Journal herald / Jackson Co. Jackson – v1 n1. jun 1974-dec 1982 [twice wkly] – 23r – 1 – mf#B12451-12473 – us Ohio Hist [071]

Journal herald – White Haven, PA. 1993+ [1] – mf#69073 – us UMI ProQuest [071]

Journal historique de l'etablissement des francais a la louisiane / Benard de LaHarpe, Jean B – Nouvelle-Orleans (Etats-Unis) 1831 [mf ed Hildesheim 1995-98] – 1v on 3mf – 9 – €90.00 – 3-487-27146-X – gw Olms [978]

Journal historique des evenemens arrives a saint-eustache : pendant la rebellion du comte du lac des deux montagnes depuis les soulevements commences a la fin de novembre... – Montreal: par J Jones, 1838 [mf ed 1974] – 1r – 5 – mf#SEM16P150 – cn Bibl Nat [971]

Journal historique du dernier voyage que feu m de la sale fit dans le golfe de mexique *see* Mr joutel's journal of his voyage to mexico

Journal historique du voyage de m de lesseps : consul de france, employe dans l'expedition de m le comte de laperouse, en qualite d'interprete du roi... / Lesseps, Jean B de – Paris 1790 [mf ed Hildesheim 1995-98] – 2v on 9mf – 9 – €180.00 – ISBN-10: 3-487-26656-3 – ISBN-13: 978-3-487-26656-5 – gw Olms [910]

Journal historique du voyage fait au cap de bonne-esperance : precede d'un discours sur la vie de l'auteur, suivi de remarques & reflexions sur les coutumes de hottentots & des habitans du cap / Caille, Nicolas Louis de la – a Paris: Chez Guillyn, Libr 1763 – (= ser [Travel descriptions from south africa, 1711-1938]) – 5mf – 9 – mf#zah-35 – ne IDC [916]

Journal historique d'un voyage fait aux iles malouines en 1763 & 1764 : pour les reconnoitre, et y former un etablissement; et de deux voyages au detroit de magellan, avec une relation sur les patagons / Pernety, Antoine – Berlin 1769 [mf ed Hildesheim 1995-98] – 2v on 6mf – 9 – €120.00 – 3-487-26951-? – gw Olms [910]

Journal (hornsey) – London, UK. 15 apr 1983-21 dec 1984; 1985-1 sep 1988 – 13r – 1 – (aka: journal hornsey wood green etc; hornsey and muswell hill journal) – uk British Libr Newspaper [072]

Journal hornsey wood green etc *see* Journal (hornsey)

Journal – human problems in british central africa, 1944-1965 / Rhodes-Livingstone Institute – n1-37 – 70mf – 7 – mf#86309 – uk Microform Academic [960]

Journal in franckfurt am main *see* Journal

Journal index / Manitoba. Legislative Assembly – 1870-1900 – (= ser Legislative Assembly) – 1r – 1 – cn Library Assoc [971]

Journal index / Ontario. Legislative Assembly – 1867-1902 – (= ser Legislative Assembly) – 1r – 1 – cn Library Assoc [971]

Journal index / Prince Edward Island. Legislative Assembly – 1901-20 – (= ser Legislative Assembly) – 1r – 1 – cn Library Assoc [971]

Journal inedit de jallet / Jallet, Jacques – Fontenay le Comte, France. 1871 – 1r – us UF Libraries [025]

Journal inedit du second sejour au senegal (3 decembre 1786 – 25 decembre 1787) / Boufflers, Stanislas-Jean, chevalier de – Paris: Editions de la Revue politique et Litteraire (Revue Bleue)...1905 [mf ed 1977] – 1r – 1 – (int by paul bonnefon) – mf#5000 – us Wisconsin U Libr [916]

Journal international d'archeologie numismatique... = Diethnes ephemeris tes nomimatikes archaiologias – Athenes, 1898-1920/1921. v1-20 – 134mf – 8 – mf#H-681 – ne IDC [700]

Journal international de la jeune republique *see* La rive gauche

Journal inutile : ou melanges politiques et litteraires – New York. 1824-1925 (1) – mf#4468 – us UMI ProQuest [920]

Journal, jan 1888-mar 1894 – Ms. from Furman University. 200p – 1 – 7.00 – us Southern Baptist [242]

Journal, letters / Williams, John Chauner – 1855-74 – 1r – 1 – mf#pmb24 – at Pacific Mss [880]

Journal libre par martel *see* L'orateur du peuple

Journal litteraire – Sallengre, Saint-Hyacinthe. La Haye. mai juin 1713-22, 1728-37 (1-24) – 1 – fr ACRPP [410]

Journal / louisiana state medical society – New Orleans. 1983-1985 (1) 1983-1985 (5) 1983-1985 (9) – (cont: journal of the louisiana state medical society. cont by: journal of the louisiana state medical society) – ISSN: 0024-6921 – mf#5486,01 – us UMI ProQuest [610]

Journal messenger – Manassas, VA. 1885-2000 (1) – mf#66760 – us UMI ProQuest [071]

Journal michigan association of school boards / Michigan Association of School Boards – Lansing. 1991-1995 – 1,5,9 – (cont: masb journal) – ISSN: 1052-2824 – mf#10520,02 – us UMI ProQuest [370]

Journal militaire de henri 4, depuis son depart de la navarre : precede d'un discours sur l'art militaire du temps / Valori, Henri Z de – Paris 1821 [mf ed Hildesheim 1995-98] – 1v on 3mf – 9 – €90.00 – ISBN-10: 3-487-26114-6 – ISBN-13: 978-3-487-26114-0 – gw Olms [355]

Journal, ms 3506 / Indiana. Morgan Raid Commission – Proceedings, Apr 4-Oct 22, 1867 – 1 – us Western Res [976]

Journal – new york state bar association / New York State Bar Association – Albany. 2000+ (1) – (cont: new york state bar journal) – ISSN: 1529-3769 – mf#6506,01 – us UMI ProQuest [340]

Journal news – Evansville, IN. 1901-1920 (1) – mf#62778 – us UMI ProQuest [071]

Journal news (an/ap) – Yorktown/Carmel, NY. 1998-2000 (1) – mf#69443 – us UMI ProQuest [071]

Journal news (rk) – Nyack, NY. 1945-2000 (1) – mf#61642 – us UMI ProQuest [071]

Journal news series / Allen Co. Spencerville – (jul 1894 scattered thru sep 1933) [wkly] – 1r – 1 – mf#B29885 – us Ohio Hist [071]

The journal of... 1774 : from london library / Prevost, Augustine – (= ser British records relating to america in microform) – 1r – 1 – (int by nicholas wainwright) – mf#3896 – uk Microform Academic [920]

Journal of, 1773-1832 / Littlejohn, John – 1 – us Southern Baptist [242]

Journal of a canoe voyage along the kauai palis, made in 1845 / Gilman, Gorham Dummer – Honolulu: Paradise of the Pacific Print, 1908 [mf ed 1995] – (= ser Yale coll; Papers of the hawaiian historical society 14) – 44p (ill) – 1 – 0-524-10026-8 – (bound with: the history of the hawaiian mission press, with bibl of the earlier publ by howard m ballou and george r carter. presented to the society, aug 27 1908) – mf#1995-1026 – us ATLA [919]

Journal of a fourteen days' ride through the bush from quebec to lake st john / Davenport, Mrs – Quebec?: s.n, 1872 – 1mf – 9 – mf#02509 – cn Canadiana [917]

Journal of a harpooner on board the whaling ship massachusetts / Brett, James Warden – 1836-1840 – 1r – 1 – (available for info and research only) – mf#PMB1038 – at Pacific Mss [920]

The journal of a mission to the interior of africa, in the year 1805... / Park, M – London, 1815 – 4mf – 9 – mf#HT-104 – ne IDC [916]

Journal of a missionary tour in india : performed by the rev messrs read and ramsey / Ramsey, William – Philadelphia: J Whetham, 1836 [mf ed 1995] – (= ser Yale coll) – 367p (ill) – 1 – 0-524-09047-5 – mf#1995-0047 – us ATLA [240]

Journal of a nobleman : comprising an account of his travels, and a narrative of his residence at vienna, during the congress / LaGarde, Auguste de – London 1831 [mf ed Hildesheim 1995-98] – 2v on 6mf – 9 – €120.00 – 3-487-29410-9 – gw Olms [880]

Journal of a passage from the pacific to the atlantic : crossing the andes in the northern provinces of peru, and descending the river maranon, or amazon / Maw, Henry – London 1829 [mf ed Hildesheim 1995-98] – 1v on 3mf – 9 – €90.00 – ISBN-10: 3-487-26852-3 – ISBN-13: 978-3-487-26852-1 – gw Olms [918]

Journal of a recent visit to the principal vineyards of spain and france : with some remarks on the very limited quantity of the finest wines produced throughout the world, and their consequent intrinsic value... / Busby, James – London 1834 [mf ed Hildesheim 1995-98] – 2mf – 9 – €60.00 – 3-487-29809-0 – gw Olms [640]

Journal of a residence and tour in the republic of mexico in the year 1826 : with some account of the mines of that country / Lyon, George – London 1828 [mf ed Hildesheim 1995-98] – 2v on 4mf – 9 – €120.00 – 3-487-26981-3 – gw Olms [918]

Journal of a residence and travels in colombia : during the years 1823 and 1824 / Cochrane, Charles – London 1825 [mf ed Hildesheim 1995-98] – 2v on 7mf – 9 – €140.00 – ISBN-10: 3-487-26904-X – ISBN-13: 978-3-487-26904-7 – gw Olms [918]

Journal of a residence at bagdad, during the years 1830 and 1831 / Groves, A N – London, 1832 – 4mf – 9 – mf#HT-55 – ne IDC [915]

Journal of a residence at bagdad, during the years 1830 and 1831 / Groves, Anthony – London 1832 [mf ed Hildesheim 1995-98] – 1v on 2mf – 9 – €60.00 – 3-487-27632-1 – gw Olms [880]

Journal of a residence in ashantee : comprising notes and researches relative to the gold coast, and the interior of western africa; chiefly collected from arabic mss. and information communicated by the moslems of guinea... / Dupuis, Joseph – London 1824 [mf ed Hildesheim 1995-98] – 1v on 5mf – 9 – €100.00 – 3-487-26741-1 – gw Olms [916]

Journal of a residence in ashantee / Dupuis, J – London, 1824 – 17mf – 9 – mf#A-147 – ne IDC [916]

Journal of a residence in ashanti : from the royal commonwealth society library / Dupuis, J – 1824 – 12mf – 7 – mf#2982 – uk Microform Academic [920]

Journal of a residence in china and the neighbouring countries from 1830 to 1833 / Abeel, David – London: James Nisbet, 1835 [mf ed 1996] – (= ser Yale coll) – xxxi/366p – 1 – 0-524-10207-4 – (rev and repr fr american ed. int essay by wriothesley noel) – mf#1996-1207 – us ATLA [880]

Journal of a residence in normandy / Saint John, James – Edinburgh 1831 [mf ed Hildesheim 1995-98] – 2mf – 9 – €60.00 – 3-487-29723-X – gw Olms [920]

Journal of a residence in siam and of a voyage along the coast of china to mantchou tartary / Gutzlaff, K – Canton, 1832 – 1mf – 9 – mf#HT-712 – ne IDC [915]

Journal of a residence in the burmhan empire and more particularly at the court of amarapoorah / Cox, Hiram – London 1821 [mf ed Hildesheim 1995-98] – 1v on 5mf – 9 – €100.00 – 3-487-27407-8 – gw Olms [880]

Journal of a residence in the burmhan empire, and more particularly at the court of amarapoorah / Cox, H – London, 1821 – 5mf – 9 – mf#SE-20139 – ne IDC [915]

Journal of a second expedition into the interior of africa : from the bight of benin to soccatoo / Clapperton, H – London, 1829 – 13mf – 9 – mf#A-292 – ne IDC [916]

Journal of a second expedition into the interior of africa : from the bight of benin to soccatoo / Lander, Richard – London 1829 [mf ed Hildesheim 1995-98] – 1v on 5mf – 9 – €100.00 – 3-487-26735-7 – gw Olms [916]

Journal of a second expedition to africa : from the royal commonwealth society library / Clapperton, Hugh – 1829 – 9mf – 7 – mf#2973 – uk Microform Academic [916]

Journal of a second voyage for the discovery of a north-west passage from the atlantic to the pacific : performed in the years 1821, 1822, 1823 in h m's ships fury and hecla / Parry, W E – London: John Murray, 1824 – 25mf – 9 – mf#N-334 – ne IDC [919]

Journal of a ten months' residence in new zealand / Cruise, Richard – London 1823 [mf ed Hildesheim 1995-98] – 1v on 2mf – 9 – €60.00 – ISBN-10: 3-487-26771-3 – ISBN-13: 978-3-487-26771-5 – gw Olms [920]

Journal of a third voyage for the discovery of a north-west passage from the atlantic to the pacific : performed in the years 1824-1825 in h m's ships hecla and fury / Parry, W E – London, 1826 – 13mf – 9 – mf#N-335 – ne IDC [919]

Journal of a three years' residence in abyssinia : in furtherance of the objects of the church missionary society / Gobat, S – London, 1834 – 5mf – 9 – mf#HTM-66 – ne IDC [916]

Journal of a three years' residence in abyssinia : in furtherance of the objects of the church missionary society / Gobat, Samuel – London: Hatchard, 1834 – 1mf – us ATLA [240]

Journal of a three years' residence in abyssinia, in furtherance of the objects of the church missionary society / Lee, Samuel – London 1834 [mf ed Hildesheim 1995-98] – 1v on 3mf – 9 – €90.00 – 3-487-27232-6 – gw Olms [916]

Journal of a three years' residence in abyssinia: in furtherance of the objects of the church missionary society – a brief history of the church of abyssinia / Gobat, Samuel & Lee, Samuel – London: Hatchard, 1834 – 1mf – 9 – 0-7905-6468-8 – mf#1988-2468 – us ATLA [916]

Journal of a tour and residence in great britain : during the years 1810 and 1811 / Simond, Louis – Edinburgh 1817 [mf ed Hildesheim 1995-98] – 2v on 8mf – 9 – €160.00 – 3-487-28889-3 – gw Olms [914]

Journal of a tour from astrachan to karass, north of the mountains of caucasus : containing remarks on the general appearances of the country, manners of the inhabitants, etc; with the substance of many conversations with effendis, mollas, and other mohammedans, on the questions at issue between them and christians / Glen, William – London 1823 [mf ed Hildesheim 1995-98] – 1v on 2mf – 9 – €60.00 – 3-487-27667-4 – gw Olms [915]

Journal of a tour in asia minor : with comparative remarks on the ancient and modern geography of that country / Leake, William – London 1824 [mf ed Hildesheim 1995-98] – 1v on 3mf – 9 – €90.00 – 3-487-27649-6 – gw Olms [914]

Journal of a tour in europe and the east 1844-1846 / Weston, G F – London, 1894. 3v – 13mf – 9 – mf#HT-156 – ne IDC [910]

Journal of a tour in france : in the years 1816 and 1817 / Carey, Frances – London 1823 [mf ed Hildesheim 1995-98] – 3mf – 9 – €90.00 – 3-487-29785-X – gw Olms [914]

Journal of a tour in france, switzerland, and italy, during the years 1819, 20, and 21 / Colston, Marianne – London 1823 [mf ed Hildesheim 1995-98] – 3v on 5mf – 9 – €100.00 – 3-487-27764-6 – gw Olms [914]

Journal of a tour in france, switzerland, and lombardy, crossing the simplon : and returning by mont cenis to paris, during the autumn of 1818 – Brentford 1821 [mf ed Hildesheim 1995-98] – 2v on 6mf – 9 – €120.00 – 3-487-27509-0 – gw Olms [914]

Journal of a tour in iceland in the summer of 1809 / Hooker, W J – Oxford. 1970+ (1) 1949+ (5) 1975+ (9) – 18mf – 9 – mf#5959 – ne IDC [914]

Journal of a tour in iceland, in the summer of 1809 / Hooker, William – London 1813 [mf ed Hildesheim 1995-98] – 2v on 6mf – 9 – €120.00 – 3-487-28930-X – gw Olms [914]

Journal of a tour in ireland etc : performed in august 1804: with remarks on the character, manners, and customs, of the inhabitants – London 1806 [mf ed Hildesheim 1995-98] – 1v on 2mf [ill] – 9 – €60.00 – 3-487-26450-1 – gw Olms [914]

Journal of a tour in italy : with reflections on the present condition and prospects of religion in that country / Wordsworth, Christopher – London: Rivingtons, 1863 – 2mf – 9 – 0-7905-8257-0 – mf#1988-8120 – us ATLA [914]

Journal of a tour in marocco and the great atlas / Hooker, J D – London, 1878 – 6mf – 9 – mf#9352 – ne IDC [916]

Journal of a tour in the levant / Turner, William – London 1820 [mf ed Hildesheim 1995-98] – 3v on 11mf – 9 – €110.00 – 3-487-27690-9 – gw Olms [914]

Journal of a tour in the state of new york, in the year 1830 : with remarks on agriculture in those parts most eligible for settlers: and return to england by the western islands, in consequence of shipwreck in the robert fulton / Fowler, John – London 1831 [mf ed Hildesheim 1995-98] – 1v on 4mf – 9 – €120.00 – 3-487-27109-5 – gw Olms [917]

Journal of a tour in upper india : performed during the years 1838-39... / French, C J – Simla, 1872 – 2mf – 9 – mf#HT-49 – ne IDC [915]

Journal of a tour made by senor juan de vega, the spanish minstrel of 1828-9 : through great britain and ireland / Cochrane, Charles – London 1830 [mf ed Hildesheim 1995-98] – 2v on 6mf – 9 – €120.00 – 3-487-27968-1 – gw Olms [914]

Journal of a tour made in the years 1828-1829 : through styria, carniola, and italy, whilst accompanying the late sir humphry davy / Tobin, John – London 1832 [mf ed Hildesheim 1995-98] – 1v on 2mf – 9 – €60.00 – 3-487-27761-1 – gw Olms [880]

A journal of a tour of discovery across the blue mountains in new south wales / [Blaxland, G] – London, 1823 – 1mf – 9 – mf#HT-10 – ne IDC [919]

Journal of a tour through part of the snowy range of the himala mountains : and to the sources of the rivers jumna and ganges / Fraser, James – London 1820 [mf ed Hildesheim 1995-98] – 1v on 6mf – 9 – €120.00 – ISBN-10: 3-487-27262-8 – ISBN-13: 978-3-487-27262-7 – gw Olms [915]

Journal of a tour through part of the snowy range of the himalaya mountains and to the sources of the rivers jumna and ganges / Fraser, J B – London, 1820 – 11mf – 9 – mf#H-6141 – ne IDC [915]

Journal of a tour through parts of france, italy, and switzerland : in the years 1823-24 / Johnson, John – London 1827 [mf ed Hildesheim 1995-98] – 1v on 2mf – 9 – €60.00 – 3-487-27748-4 – gw Olms [880]

Journal of a tour through the netherlands to paris, in 1821 / Blessington, Marguerite – London 1822 [mf ed Hildesheim 1995-98] – 1v on 2mf – 9 – €60.00 (3-487-27820-2) – gw Olms [914]

Journal of a tour to the western counties of england : performed in the summer of 1807 – London 1809 [mf ed Hildesheim 1995-98] – 1v on 1mf – 9 – €40.00 – 3-487-26427-7 – gw Olms [914]

Journal of a visit to some parts of ethiopia / Waddington, George – London 1822 [mf ed Hildesheim 1995-98] – 1v on 4mf – 9 – €120.00 – 3-487-27266-0 – gw Olms [916]

Journal of a visit to south africa in 1815 and 1816 / Latrobe, Christian Ignatius – Cape Town, South Africa. 1969 – 1r – us UF Libraries [960]

Journal of a visit to south africa, in 1815, and 1816 : with some account of the missionary settlements of the united brethren, near the cape of good hope / Latrobe, Christian – London 1818 [mf ed Hildesheim 1995-98] – 1v on 5mf – 9 – €100.00 – 3-487-26733-0 – gw Olms [916]

Journal of a visitation-tour in 1843-4 : through part of the western portion of his diocese / Spencer, G T – London, 1845 – 4mf – 9 – mf#HTM-181 – ne IDC [915]

Journal of a voyage / Nugent, Maria – London, England. 1939 – 1r – us UF Libraries [972]

Journal of a voyage for the discovery of a north-west passage from the atlantic to the pacific : performed in the years 1819-1820, in h m's ships hecla and griper / Parry, W E – London: John Murray, 1821 – 20mf – 9 – mf#N-333 – ne IDC [919]

The journal of a voyage from calcutta to van diemen's land : comprising a description of that colony during a six months' residence / Prinsep, Augustus – London 1833 [mf ed Hildesheim 1995-98] – 1v on 1mf – 9 – €40.00 – 3-487-26789-6 – gw Olms [910]

Journal of a voyage from france to new caledonia / Balliere, Achille – 1873 – 1r – 1 – mf#pmb148 – at Pacific Mss [910]

Journal of a voyage in 1811 and 1812, to madras and china : returning by the cape of good hope and st helena... / Wathen, J – London: J Nichols, Son, and Bentley, 1814 – 4mf – 9 – mf#HT-722 – ne IDC [915]

Journal of a voyage, in 1811 and 1812, to madras and china : returning by the cape of good hope and st helena; in the h c s the hope, capt james pendergrass / Wathen, James – London 1814 [mf ed Hildesheim 1995-98] – 1v on 3mf – 9 – €90.00 – 3-487-27248-2 – gw Olms [910]

Journal of a voyage in baffin's bay and barrow straits / Sutherland, P C – London, 1852. 2v – 23mf – 9 – mf#N-408 – ne IDC [917]

A journal of a voyage of discovery to the arctic regions : in his majesty's ships hecla and griper, in the years 1819 and 1820 / Fisher, Alexander – London 1821 [mf ed Hildesheim 1995-98] – 1v on 2mf – 9 – €60.00 – 3-487-27093-5 – gw Olms [919]

Journal of a voyage to greenland : in the year 1821 / Manby, G W – London, 1823 – 5mf – 9 – mf#H-486 – ne IDC [917]

Journal of a voyage to peru : a passage across the cordillera of the andes, in the winter of 1827, performed on foot in the snow; and a journey across the pampas / Brand, Charles – London 1828 [mf ed Hildesheim 1995-98] – 1v on 3mf – 9 – €90.00 – ISBN-10: 3-487-26836-1 – ISBN-13: 978-3-487-26836-1 – gw Olms [918]

Journal of a voyage to the northern whale-fishery : including researches and discoveries on the eastern coast of west greenland, made in the summer of 1822, in the ship baffin of liverpool / Scoresby, William – Edinburgh 1823 [mf ed Hildesheim 1995-98] – 1v on 4mf – 9 – €120.00 – 3-487-27070-6 – gw Olms [919]

Journal of a voyage to the northern whale-fishery / Scoresby, W – Edinburgh, 1823 – 11mf – 9 – mf#H-500 – ne IDC [910]

A journal of a voyage to the south seas : in his majesty's ship the endeavour... / Parkinson, S – London. 1876-1879 (1) – 20mf – 9 – mf#5632 – ne IDC [919]

Journal of a voyage up the mediterranean, principally among the islands of the archipelago, and in asia minor : including many interesting particulars relative to the greek revolution, especially a journey through maina to the camp of ibrahim pacha, together with observations on the antiquities, opinions, and usages of greece, as they now exist / Swan, Charles – London 1826 [mf ed Hildesheim 1995-98] – 2v on 6mf – 9 – €120.00 – 3-487-29055-3 – gw Olms [910]

Journal of a west india proprietor : kept during a residence in the island of jamaica / Lewis, Matthew – London 1834 [mf ed Hildesheim 1995-98] – 1v on 3mf – 9 – €90.00 – ISBN-10: 3-487-26957-0 – ISBN-13: 978-3-487-26957-3 – gw Olms [880]

A journal of a young man of massachusetts : late a surgeon on board an american privateer, who was captured at sea by the british, in may, eighteen hundred and thirteen... / Waterhouse, Benjamin – Boston: printed by Rowe & Hooper, 1816 [mf ed 1984] – 3mf – 9 – 0-665-41868-X – mf#41868 – cn Canadiana [880]

Journal of abnormal child psychology – New York. 1973+ (1) 1973+ (5) 1978+ (9) – ISSN: 0091-0627 – mf#10872 – us UMI ProQuest [150]

Journal of abnormal psychology – Washington. 1906+ (1) 1965+ (5) 1970+ (9) – ISSN: 0021-843X – mf#278 – us UMI ProQuest [150]

Journal of academic librarianship – Ann Arbor. 1975+(1,5,9) – ISSN: 0099-1333 – mf#11117 – us UMI ProQuest [020]

Journal of access services : innovations for electronic and digital library and information resources / ed by Driscoll, Lori – ISSN: 1536-7967 – us Haworth [020]

Journal of accident and emergency medicine – London. 1994-1995 (1,5,9) – (cont: archives of emergency medicine) – ISSN: 1351-0622 – mf#15504,01 – us UMI ProQuest [617]

Journal of accountancy – New York. 1905+ (1) 1969+ (5) 1975+ (9) – ISSN: 0021-8448 – mf#828 – us UMI ProQuest [650]

Journal of accounting and economics – Amsterdam. 1979+ (1) 1979+ (5) 1987+ (9) – ISSN: 0165-4101 – mf#42024 – us UMI ProQuest [650]

Journal of accounting and edp – Pennsauken. 1985-1990 (1,5,9) – (cont by: financial and accounting systems) – ISSN: 8756-5714 – mf#14375 – us UMI ProQuest [000]

Journal of accounting and public policy – New York. 1982+ (1) 1982+ (5) 1987+ (9) – ISSN: 0278-4254 – mf#42416 – us UMI ProQuest [650]

Journal of accounting education – Harrisonburg. 1986+ (1,5,9) – ISSN: 0748-5751 – mf#49521 – us UMI ProQuest [650]

Journal of accounting literature – Gainesville. 1990+ (1,5,9) – ISSN: 0737-4607 – mf#18536 – us UMI ProQuest [650]

Journal of accounting research – Chicago. 1963+ (1) 1988+ (5) 1988+ (9) – ISSN: 0021-8456 – mf#9681 – us UMI ProQuest [650]

Journal of accounting research – v1-7. 1963-69 – 1 – us AMS Press [650]

Journal of acetylene – Paris/London. jun 1901-20 jun 1903 – 1r – 1 – (aka: acetylene) – uk British Libr Newspaper [072]

Journal of acquired immune deficiency syndromes – New York. 1993-1994 (1,5,9) – (cont by: journal of acquired immune deficiency syndromes and human retrovirology) – ISSN: 0894-9255 – mf#18709 – us UMI ProQuest [616]

Journal of acquired immune deficiency syndromes and human retrovirology – Hagerstown. 1995-1999 – 1,5,9 – (cont: journal of acquired immune deficiency syndromes. cont by: journal of acquired immune deficiency syndromes) – ISSN: 1077-9450 – mf#18709,01 – us UMI ProQuest [616]

Journal of acquired immune deficiency syndromes: jaids – Hagerstown. 1999+ (1,5,9) – (cont: journal of acquired immune deficiency syndromes and human retrovirology) – mf#18709,02 – us UMI ProQuest [616]

Journal of addictions and offender counseling – Alexandria. 1990+ (1,5,9) – (cont: journal of offender counseling) – ISSN: 1055-3835 – mf#12744,01 – us UMI ProQuest [360]

Journal of addictive diseases / ed by Stimmel, Barry – v1- 1981- – 1, 9 ($325.00 in US $455.00 outside hardcopy subsc) – us Haworth [360]

Journal of administration overseas – London. 1962-1980 (1) 1974-1980 (5) 1976-1980 (9) – ISSN: 0021-8472 – mf#8677 – us UMI ProQuest [320]

Journal of adolescent and adult literacy – Newark. 1995+ (1) 1995+ (5) 1995+ (9) – (cont: journal of reading) – ISSN: 1081-3004 – mf#1562,01 – us UMI ProQuest [374]

Journal of adolescent health – New York. 1991+ (1,5,9) – (cont: journal of adolescent health care: official publication of the society for adolescent medicine) – ISSN: 1054-139X – mf#42417,01 – us UMI ProQuest [610]

Journal of adolescent health care : official publication of the society for adolescent medicine – New York. 1980-1990 (1) 1980-1990 (5) 1987-1990 (9) – (cont by: journal of adolescent health) – ISSN: 0197-0070 – mf#42417 – us UMI ProQuest [610]

Journal of adolescent research – Tucson. 1986+ (1,5,9) – ISSN: 0743-5584 – mf#16355 – us UMI ProQuest [640]

Journal of advanced nursing – Oxford. 1980+ (1,5,9) – ISSN: 0309-2402 – mf#15552 – us UMI ProQuest [610]

Journal of advanced transportation – Durham. 1979+ (1) 1979+ (5) 1979+ (9) – (cont: high speed ground transportation journal) – ISSN: 0197-6729 – mf#9815,01 – us UMI ProQuest [380]

Journal of advancement in medicine – New York. 1988-1996 (1,5,9) – (cont: journal of holistic medicine) – ISSN: 0894-5888 – mf#16291 – us UMI ProQuest [610]

Journal of advertising – Provo. 1972+ (1) 1972+ (5) 1974+ (9) – ISSN: 0091-3367 – mf#8107 – us UMI ProQuest [650]

Journal of advertising research – New York. 1960+ (1) 1969+ (5) 1975+ (9) – ISSN: 0021-8499 – mf#1735 – us UMI ProQuest [650]

Journal of aerosol science – Oxford. 1970+ (1,5) 1977+ (9) – ISSN: 0021-8502 – mf#49107 – us UMI ProQuest [680]

Journal of aerospace engineering – New York. 1988+ (1,5,9) – ISSN: 0893-1321 – mf#16511 – us UMI ProQuest [629]

Journal of aesthetic education – Champaign. 1966+ [1,5,6] – ISSN: 0021-8510 – mf#6100 – us UMI ProQuest [700]

Journal of aesthetics and art criticism – Philadelphia. 1941+ (1) 1941+ (5) 1941+ (9) – ISSN: 0021-8529 – mf#996 – us UMI ProQuest [700]

Journal of affective disorders – Amsterdam. 1979+ (1) 1979+ (5) 1987+ (9) – ISSN: 0165-0327 – mf#42025 – us UMI ProQuest [610]

Journal of african american history – Washington. 2002+ (1,5,9) – mf#1036,01 – us UMI ProQuest [934]

Journal of african business / ed by Okoroafo, Sam C – mf#1522-8916 – us Haworth [338]

Journal of african civilizations – New Brunswick. 1985-1994 (1,5,9) – ISSN: 0270-2495 – mf#15430 – us UMI ProQuest [960]

Journal of african earth sciences – Oxford. 1983-1987 (1,5,9) – (cont by: journal of african earth sciences (and the middle east)) – ISSN: 0731-7247 – mf#49424 – us UMI ProQuest [550]

Journal of african earth sciences – Oxford, 1994-1994 [1,5,9] – (cont: journal of african earth sciences (and the middle east)) – mf#49424,02 – us UMI ProQuest [550]

Journal of african earth sciences (and the middle east) – Oxford. 1988-1993 (1,5,9) – (cont: journal of african earth sciences) – ISSN: 0899-5362 – mf#49424,01 – us UMI ProQuest [550]

Journal of african history – London. 1960+ (1) 1969+ (5) 1975+ (9) – ISSN: 0021-8537 – mf#2848 – us UMI ProQuest [960]

Journal of african law – v1-44. 1957-2000 – 5,6,9 – $727.00 set – (v1-28 1957-84 on reel $319. v29-44 1985-2000 on mf $408) – ISSN: 0021-8553 – mf#103821 – us Hein [340]

Journal of african studies – Washington. 1980-1988 (1) 1980-1988 (5) 1980-1988 (9) – ISSN: 0095-4993 – mf#12620 – us UMI ProQuest [960]

Journal of aggression, maltreatment and trauma / ed by Geffner, Robert – ISSN: 1092-6771 – us Haworth [150]

Journal of aging and health – Thousand Oaks. 1989+ (1,5,9) – ISSN: 0898-2643 – mf#17054 – us UMI ProQuest [618]

Journal of aging and judaism – New York. 1989-1989 (1) – ISSN: 0884-8688 – mf#15457 – us UMI ProQuest [618]

Journal of aging and pharmacotherapy / ed by Parish, Roy C – ISSN: 1540-5303 – us Haworth [615]

Journal of aging and social policy / ed by Bass, Scott A & Morris, Robert – v1- 1989- – 1, 9 ($175.00 in US $245.00 outside hardcopy subsc) – us Haworth [360]

Journal of aging studies – Greenwich. 1993+ (1,5,9) – ISSN: 0890-4065 – mf#19784 – us UMI ProQuest [618]

Journal of agricultural and applied economics – Commerce. 1993+ (1) 1993+ (5) 1993+ (9) – (cont: southern journal of agricultural economics) – ISSN: 1074-0708 – mf#6719,01 – us UMI ProQuest [630]

Journal of agricultural and food chemistry – v1- 1953- – 1,5,6,9 – us ACS [660]

Journal of agricultural and food information / ed by Frank, Robyn – v4 n1. 1997- – 1,9 – $85.00 in US $119.00 outside hardcopy subsc – us Haworth [630]

Journal of agricultural economics research – Washington. 1987-1994 (1,5,9) – (cont: agricultural economics research) – ISSN: 1043-3309 – mf#1846,01 – us UMI ProQuest [630]

Journal of agricultural education – Carbondale. 1989+ (1,5,9) – (cont: journal of the american association of teacher educators in agriculture) – ISSN: 1042-0541 – mf#11422,01 – us UMI ProQuest [630]

Journal of agricultural research – Washington. 1913-1949 [1] – mf#5772 – us UMI ProQuest [630]

Journal of agricultural science – Cambridge. 1979+ (1,5,9) – ISSN: 0021-8596 – mf#12123 – us UMI ProQuest [630]

Journal of agricultural taxation and law – Boston. 1983-1992 (1) 1983-1992 (5) 1983-1992 (9) – ISSN: 0745-9181 – mf#13454,01 – us UMI ProQuest [343]

Journal of agriculture – Halifax, N.S.: A and W MacKinlay, [1865-1866?] – 9 – (cont by: the nova scotia journal of agriculture) – mf#P04714 – cn Canadiana [630]

Journal of agriculture : western australia – South Perth. 1972-1973 (1) – ISSN: 0021-8626 – mf#7791 – us UMI ProQuest [630]

Journal of agriculture of the university of puerto rico – Rio Piedras. 1917-1996 (1) 1971-1988 (5) 1977-1988 (9) – ISSN: 0041-994X – mf#469 – us UMI ProQuest [630]

Journal of agromedicine : interface of human health and agriculture / ed by Schumann, Stanley H – v3 n1. 1996- – 1,9 – $135.00 in US $189.00 outside hardcopy subsc – us Haworth [614]

Journal of air law *see* Journal of air law and commerce

Journal of air law and commerce – Dallas. 1930+ (1) 1971+ (5) 1976+ (9) – ISSN: 0021-8642 – mf#2980 – us UMI ProQuest [346]

Journal of air law and commerce – Southern Methodist University. v1-65. 1930-2000 – 9 – $1120.00 set – (v1-50 1930-85 on reel $616. v51-65 1985-2000 on mf $504. title varies: v1-9, 1930-38 as journal of air law. suspended oct 1942-jan 1947. cum ind v1-35 1930-69) – ISSN: 0021-8642 – mf#103831 – us Hein [346]

Journal of air transport management – Kidlington. 1994-1995 (1,5,9) – ISSN: 0969-6997 – mf#20748 – us UMI ProQuest [380]

Journal of aircraft – Reston. 1964+ (1) 1969+ (5) 1976+ (9) – ISSN: 0021-8669 – mf#5074 – us UMI ProQuest [629]

Journal of alabama archaeology – Moundville. 1974+ (1) 1974+ (5) 1974+ (9) – ISSN: 0449-2153 – mf#9539 – us UMI ProQuest [930]

Journal of alcohol and drug education – Lansing. 1955+ (1) 1955+ (5) 1955+ (9) – ISSN: 0090-1482 – mf#10570 – us UMI ProQuest [360]

Journal of alcoholism – London. 1971-1976 (1,5) 1976-1976 (9) – (cont by: british journal on alcohol and alcoholism) – ISSN: 0021-8685 – mf#6588 – us UMI ProQuest [616]

Journal of algebraic combinatorics – Boston. 1992-1996 (1,5,9) – ISSN: 0925-9899 – mf#18664 – us UMI ProQuest [510]

Journal of allergy and clinical immunology – St Louis. 1929+ [1]; 1965+ [5]; 1970+ [9] – ISSN: 0091-6749 – mf#1886 – us UMI ProQuest [616]

Journal of allied health – Washington. 1972+ (1) 1972+ (5) 1972+ (9) – ISSN: 0090-7421 – mf#10328 – us UMI ProQuest [360]

Journal of alloys and compounds – Lausanne. 1992+ (1,5,9) – (cont: journal of the less-common metals) – ISSN: 0925-8388 – mf#42292,01 – us UMI ProQuest [540]

Journal of alternative investments – New York. 1998+ (1,5,9) – ISSN: 1520-3255 – mf#32268 – us UMI ProQuest [332]

Journal of ambulatory care management – Gaithersburg. 1978+ (1,5,9) – ISSN: 0148-9917 – mf#12733 – us UMI ProQuest [610]

Journal of ambulatory care marketing / ed by Meadow, H Lee – v1- 1987- – 1, 9 ($75.00 in US $105.00 outside hardcopy subsc) – us Haworth [610]

Journal of american academy of business, cambridge / American academy of business, cambridge – Hollywood. 2001+ (1,5,9) – ISSN: 1540-1200 – mf#32088 – us UMI ProQuest [650]

Journal of american and comparative cultures : studies of civilizations – Bowling Green, 2000+ [1,5,9] – (cont: journal of american culture) – mf#17453,01 – us UMI ProQuest [306]

Journal of american college health / American College Health Association – Washington. 1981+ (1) 1981+ (5) 1981+ (9) – (cont: journal of the american college health association) – ISSN: 0744-8481 – mf#2240,01 – us UMI ProQuest [378]

Journal of american culture – Bowling Green. 1989-1999 (1,5,9) – ISSN: 0191-1813 – mf#17453 – us UMI ProQuest [306]

Journal of american ethnic history – New Brunswick. 1981+ (1,5,9) – ISSN: 0278-5927 – mf#12944 – us UMI ProQuest [305]

Journal of american folk-lore / American Folklore Society – v1,9,12 – 3r – 1 – us WHS [390]

Journal of american folk-lore – Washington. 1888+ (1) 1968+ (5) 1975+ (9) – ISSN: 0021-8715 – mf#902 – us UMI ProQuest [390]

Journal of american history – Bloomington. 1914+ (1) 1968+ (5) 1975+ (9) – ISSN: 0021-8723 – mf#1019 – us UMI ProQuest [975]

Journal of american indian education – Tempe. 1961+ (1) 1971+ (5) 1975+ (9) – ISSN: 0021-8731 – mf#7193 – us UMI ProQuest [370]

Journal of american indian family research – table of contents, 1979-87, 1979 nov-1980 dec, 1981-84, 1985-87 – 3r – 1 – mf#669893 – us WHS [929]

Journal of american insurance – Schaumburg. 1924-1990 (1) 1972-1990 (5) 1975-1990 (9) – ISSN: 0021-874X – mf#6342 – us UMI ProQuest [368]

Journal of american pomological society / American Pomological Society – University Park. 2000+ (1) – (cont: fruit varieties journal) – ISSN: 1527-3741 – mf#2530,01 – us UMI ProQuest [634]

Journal of american studies – Cambridge. 1967+ (1) 1976+ (5) 1976+ (9) – ISSN: 0021-8758 – mf#11035 – us UMI ProQuest [975]

Journal of an embassy from the governor-general of india to the court of ava : in the year 1827 / Crawfurd, John – London 1829 [mf ed Hildesheim 1995-98] – 1v on 7mf – 9 – €140.00 – ISBN-10: 3-487-27253-9 – ISBN-13: 978-3-487-27253-5 – gw Olms [959]

Journal of an embassy from the governor-general of india to the court of ava, in the year 1827 / Crawfurd, J – London, 1829 – 8mf – 9 – mf#SE-20151 – ne IDC [915]

Journal of an embassy from the governor-general of india to the courts of siam and cochin china : exhibiting a view of the actual state of those kingdoms / Crawfurd, John – London 1828 [mf ed Hildesheim 1995-98] – 1v on 7mf – 9 – €140.00 – ISBN-10: 3-487-27252-0 – ISBN-13: 978-3-487-27252-8 – gw Olms [959]

A journal of an embassy from their majesties iwan and peter alexiowitz : czars of muscovy etc over land into china in the years 1693, 1694, and 1695... – Hamburg, 1698 – 1mf – 9 – mf#HT-673 – ne IDC [915]

Journal of an expedition 1400 miles up the orinoco and 300 up the arauca : with an account of the country, the manners of the people, military operations, &c / Robinson, James – London 1822 [mf ed Hildesheim 1995-98] – 1v on 3mf – 9 – €90.00 – ISBN-10: 3-487-26889-2 – ISBN-13: 978-3-487-26889-7 – gw Olms [918]

Journal of an expedition into the interior of tropical australia : in search of a route from sydney to the gulf of carpentaria / Mitchell, T L – London, 1848 – 6mf – 9 – mf#H-6182 – ne IDC [919]

Journal of an expedition to explore the course and termination of the niger : with a narrative of a voyage down that river to its termination / Lander, Richard – London 1832 [mf ed Hildesheim 1995-98] – 3v on 12mf – 9 – €120.00 – 3-487-27200-8 – gw Olms [916]

Journal of an expedition to explore the course and termination of the niger... / Lander, R L & Lander, J – London, 1832. 3v – 20mf – 9 – mf#A-332 – ne IDC [916]

Journal of analytic social work / ed by Brandell, Jerrold R – v1- 1993- – 1,9 ($95.00 in US $133.00 outside hardcopy subsc) – (cont: journal of independent social work) – us Haworth [360]

Journal of analytical and applied pyrolysis – Amsterdam. 1979+ (1) 1979+ (5) 1987+ (9) – ISSN: 0165-2370 – mf#42026 – us UMI ProQuest [540]

Journal of analytical atomic spectrometry – London. 1994-1994 (1,5,9) – ISSN: 0267-9477 – mf#17653 – us UMI ProQuest [540]

Journal of analytical chemistry of the ussr – New York. 1952-1977 (1) 1952-1977 (5) – ISSN: 0021-8766 – mf#10834 – us UMI ProQuest [540]

Journal of analytical toxicology / ed by Baselt, Randall – 1977-96 – 1,5 – $100.00v subsc $270.00 nonsubsc 1 ($75.00v subsc $270.00 nonsubsc 5) – ISSN: 0 – us Preston Publ [615]

Journal of anatomy – Cambridge. 1913-1995 [1]; 1989-1995 [5,9] – ISSN: 0021-8782 – mf#1253 – us UMI ProQuest [611]

Journal of anatomy and physiology – London 1867-1946 – v1-80 on 896mf – 9 – mf#6149c/2 – ne IDC [590]

Journal of anatomy and physiology – London: Macmillan. v1-50. 1867-1916 – 1 – $540.00 – mf#0306 – us Brook [611]

Journal of andrology – Philadelphia. 1980-1992 (1) 1980-1992 (5) 1980-1992 (9) – ISSN: 0196-3635 – mf#12258 – us UMI ProQuest [612]

Journal of animal ecology – Oxford. 1980+ (1,5,9) – ISSN: 0021-8790 – mf#15539 – us UMI ProQuest [574]

Journal of animal science – Savoy. 1942+ (1) 1972+ (5) 1975+ (9) – ISSN: 0021-8812 – mf#7017 – us UMI ProQuest [636]

Journal of anthropological research – Albuquerque. 1973+ (1) 1973+ (5) 1976+ (9) – (cont: southwestern journal of anthropology) – ISSN: 0091-7710 – mf#410,01 – us UMI ProQuest [301]

Journal of anthropology – London. 1870-1871 (1) – mf#2282 – us UMI ProQuest [301]

Journal of antibiotics – Tokyo. 1973+ (1,5,9) – ISSN: 0021-8820 – mf#8704 – us UMI ProQuest [615]

Journal of anxiety disorders – New York. 1987+ (1,5,9) – ISSN: 0887-6185 – mf#49503 – us UMI ProQuest [150]

Journal of applied animal welfare science: jaaws – Mahwah. 1998+ (1) – ISSN: 1088-8705 – mf#28519 – us UMI ProQuest [636]

Journal of applied aquaculture / ed by Webster, Carl David – v6 n1. 1996- – 1,9 – $150.00 in US $210.00 outside hardcopy subsc – us Haworth [630]

Journal of applied bacteriology – London. 1980-1996 (1) 1981-1996 (5) 1980-1996 (9) – (cont by: journal of applied microbiology) – ISSN: 0021-8847 – mf#15553,03 – us UMI ProQuest [576]

Journal of applied behavioral science – Arlington. 1965+ (1) 1971+ (5) 1975+ (9) – ISSN: 0021-8863 – mf#2727 – us UMI ProQuest [150]

Journal of applied biomaterials – New York. 1990-1995 (1,5,9) – ISSN: 1045-4861 – mf#18110 – us UMI ProQuest [574]

Journal of applied business research – Laramie. 1989-1998 (1,5,9) – ISSN: 0892-7626 – mf#18135 – us UMI ProQuest [650]

Journal of applied cardiology – New York. 1986-1991 (1,5,9) – ISSN: 0883-2935 – mf#49504 – us UMI ProQuest [616]

Journal of applied chemistry of the ussr – New York. 1951-1992 (1) 1951-1992 (5) 1989-1992 (9) – (cont by: russian journal of applied chemistry) – ISSN: 0021-888X – mf#10832 – us UMI ProQuest [660]

Journal of applied communication research – Annandale. 1973+ (1,5,9) – ISSN: 0090-9882 – mf#11278 – us UMI ProQuest [380]

Journal of applied ecology – Oxford. 1980+ (1,5,9) – ISSN: 0021-8901 – mf#15554 – us UMI ProQuest [574]

Journal of applied econometrics – Chichester. 1986+ (1,5,9) – ISSN: 0883-7252 – mf#16107 – us UMI ProQuest [510]

Journal of applied electrochemistry – London. 1983-1996 (1,5,9) – ISSN: 0021-891X – mf#14402 – us UMI ProQuest [660]

Journal of applied geophysics – Amsterdam. 1992+ (1,5,9) – (cont: geoexploration) – ISSN: 0926-9851 – mf#42079,01 – us UMI ProQuest [622]

Journal of applied gerontology – Thousand Oaks. 1988+ (1,5,9) – ISSN: 0733-4648 – mf#17055 – us UMI ProQuest [618]

Journal of applied management – Walnut Creek. 1979-1980 (1,5,9) – ISSN: 0149-7901 – mf#12207,01 – us UMI ProQuest [650]

Journal of applied mathematics and decision sciences – Mahwah. 1997+ (1,5,9) – ISSN: 1173-9126 – mf#31732 – us UMI ProQuest [510]

Journal of applied mathematics and mechanics – Oxford. 1958+ (1) 1958+ (5) 1976+ (9) – ISSN: 0021-8928 – mf#49108 – us UMI ProQuest [510]

Journal of applied mechanics – New York. 1950+ (1) 1965+ (5) 1976+ (9) – ISSN: 0021-8936 – mf#575 – us UMI ProQuest [621]

Journal of applied mechanics and technical physics – New York. 1965-1994 (1) 1965-1977 (5) – ISSN: 0021-8944 – mf#10907 – us UMI ProQuest [621]

Journal of applied metalworking – Metals Park. 1979-1987 (1,5,9) – (cont by: journal of materials shaping technology) – ISSN: 0162-9700 – mf#12935 – us UMI ProQuest [660]

Journal of applied microbiology – London. 1997+ (1,5,9) – (cont: journal of applied bacteriology) – ISSN: 1364-5072 – mf#15553,04 – us UMI ProQuest [576]

Journal of applied nutrition – Asheville. 1976+ (1,5,9) – ISSN: 0021-8960 – mf#11312 – us UMI ProQuest [613]

Journal of applied philosophy – Abingdon. 1992-1996 (1,5,9) – ISSN: 0264-3758 – mf#19596 – us UMI ProQuest [100]

Journal of applied photographic engineering – Springfield. 1980-1983 (1,5,9) – (cont by: journal of imaging technology) – ISSN: 0098-7298 – mf#12416 – us UMI ProQuest [621]

Journal of applied phycology – Dordrecht. 1989+ (1,5,9) – ISSN: 0921-8971 – mf#16805 – us UMI ProQuest [580]

Journal of applied physics – v1- 1931- – 1,5,6,9 – us AIP [621]

Journal of applied physiology / American Physiological Society – Washington. 1948-1976 (1) 1965-1976 (5) 1970-1976 (9) – (cont by: journal of applied physiology: respiratory, environmental and exercise physiology) – ISSN: 0021-8987 – mf#777 – us UMI ProQuest [612]

Journal of applied physiology – Bethesda. 1985+ (1) 1985+ (5) 1985+ (9) – (cont: journal of applied physiology: respiratory, environmental and exercise physiology) – ISSN: 8750-7587 – mf#777,02 – us UMI ProQuest [612]

Journal of applied physiology : respiratory, environmental and exercise physiology – Bethesda. 1977-1984 (1) 1977-1984 (5) 1977-1984 (9) – (cont: journal of applied physiology. cont by: journal of applied physiology) – ISSN: 0161-7567 – mf#777,01 – us UMI ProQuest [612]

Journal of applied polymer science : applied polymer symposium – New York. 1965-1985 (1,5,9) – ISSN: 0271-9460 – mf#11401 – us UMI ProQuest [660]

Journal of applied polymer science – New York. 1959+ (1,5,9) – ISSN: 0021-8995 – mf#11056 – us UMI ProQuest [660]

Journal of applied psychology – Washington. 1917+ (1) 1965+ (5) 1970+ (9) – ISSN: 0021-9010 – mf#277 – us UMI ProQuest [150]

Journal of applied rehabilitation counseling – Manassas. 1970+ (1) 1970+ (5) 1975+ (9) – ISSN: 0047-2220 – mf#6566 – us UMI ProQuest [360]

Journal of applied school psychology / ed by Maher, Charles A – ISSN: 1537-7903 – us Haworth [150]

Journal of applied spectroscopy – New York. 1965-1976 (1) 1965-1976 (5) – ISSN: 0021-9037 – mf#10908 – us UMI ProQuest [621]

Journal of applied systems analysis – Lancaster. 1976-1991 (1) 1976-1991 (5) 1976-1991 (9) – (cont: journal of systems engineering) – ISSN: 0308-9541 – mf#9973,01 – us UMI ProQuest [650]

Journal of applied toxicology: jat – Philadelphia. 1981+ (1,5,9) – ISSN: 0260-437X – mf#13309 – us UMI ProQuest [615]

Journal of aquatic ecosystem health – Dordrecht. 1992-1996 (1,5,9) – (cont by: journal of aquatic ecosystem stress and recovery) – ISSN: 0925-1014 – mf#19439 – us UMI ProQuest [574]

Journal of aquatic ecosystem stress and recovery – Dordrecht. 1997+ (1) – (cont: journal of aquatic ecosystem health) – ISSN: 1386-1980 – mf#19439,01 – us UMI ProQuest [574]

Journal of aquatic food product technology : ...an international journal devoted to foods from marine and inland waters of the world / ed by Pigott, George M – v5 n1. 1996- – 1,9 – $150.00 in US $210.00 outside hardcopy subsc – us Haworth [660]

Journal of arab affairs – Fresno. 1981-1993 (1) 1981-1993 (5) 1981-1993 (9) – ISSN: 0275-3588 – mf#12836 – us UMI ProQuest [305]

Journal of architectural education (jae) – Washington. 1975+ (1,5,9) – ISSN: 1046-4883 – mf#10332 – us UMI ProQuest [720]

Journal of architectural engineering – New York. 1995+ (1,5,9) – ISSN: 1076-0431 – mf#21269 – us UMI ProQuest [620]

Journal of architectural research – London. 1974-1980 (1,5,9) – mf#11416,01 – us UMI ProQuest [720]

Journal of archival organization / ed by Frusciano, Thomas J – ISSN: 1533-2748 – us Haworth [020]

Journal of arizona history – Tucson. 1960+ (1) 1972+ (5) 1975+ (9) – ISSN: 0021-9053 – mf#6795 – us UMI ProQuest [978]

Journal of arkansas education – Little Rock. 1972-1975 (1) 1972-1975 (5) (9) – ISSN: 0021-9061 – mf#6797 – us UMI ProQuest [370]

Journal of art – New York. 1988-1991 (1,5,9) – mf#17792 – us UMI ProQuest [700]

Journal of art and entertainment *see* Depaul lca journal of art and entertainment law

Journal of arthroplasty – Edinburgh. 1989-1996 (1) – ISSN: 0883-5403 – mf#15289 – us UMI ProQuest [617]

Journal of artificial intelligence in education – Phoenix. 1989-1997 – 1,5,9 – (cont by: journal of interactive learning research) – ISSN: 1043-1020 – mf#17106 – us UMI ProQuest [370]

Journal of arts management and law – Washington. 1982-1991 (1) 1982-1991 (5) 1982-1991 (9) – (cont: performing arts review. cont by: journal of arts management, law, and society) – ISSN: 0733-5113 – mf#7823,01 – us UMI ProQuest [340]

Journal of arts management and law *see* Journal of arts management, law and society

Journal of arts management, law and society – Helen Dwight Reid Educational Foundation. v1-30. 1971-2001 – 9 – $624.00 set – (v1-13 1971-84 on reel $181. v14-30 1984-2001 on mf $443. title varies: v1-11 1971-81 as performing arts review; v12-21 1982-1992 as journal of arts management and law) – ISSN: 0031-5249 – mf#105071 – us Hein [340]

Journal of arts management, law, and society – Washington. 1992+ (1) 1992+ (5) 1992+ (9) – (cont: journal of arts management and law) – ISSN: 1063-2921 – mf#7823,02 – us UMI ProQuest [340]

Journal of asian american studies – Baltimore. 1998+ (1,5,9) – ISSN: 1097-2129 – mf#33160 – us UMI ProQuest [301]

Journal of asian business – Ann Arbor. 1993+ (1,5,9) – (cont: journal of southeast asia business) – ISSN: 1068-0055 – mf#18576,02 – us UMI ProQuest [337]

Journal of asian earth sciences – New York. 1997+ (1) – (cont: journal of southeast asian earth sciences) – ISSN: 1367-9120 – mf#49489,01 – us UMI ProQuest [550]

Journal of asian history – Wiesbaden. 1967-1995 (1) 1975-1995 (5) 1975-1995 (9) – ISSN: 0021-910X – mf#10754 – us UMI ProQuest [950]

Journal of asian studies – Ann Arbor. 1941+ (1) 1968+ (5) 1975+ (9) – ISSN: 0021-9118 – mf#583 – us UMI ProQuest [950]

Journal of asia-pacific business / ed by Quraeshi, Zahir A – v1 n1. 1995- – 1,9 – $95.00 in US $133.00 outside hardcopy subsc – us Haworth [650]

Journal of asset management – London. 2000+ (1,5,9) – ISSN: 1470-8272 – mf#31749 – us UMI ProQuest [332]

Journal of assisted reproduction and genetics – New York. 1992-1996 (1,5,9) – (cont: journal of in vitro fertilization and embryo transfer: ivf) – ISSN: 1058-0468 – mf#17679,01 – us UMI ProQuest [618]

Journal of asthma – Ossining. 1981-1985 (1,5,9) – (cont: journal of asthma research) – ISSN: 0277-0903 – mf#2537,01 – us UMI ProQuest [610]

Journal of asthma research – Baltimore. 1963-1980 (1) 1970-1980 (5) 1976-1980 (9) – (cont by: journal of asthma) – ISSN: 0021-9134 – mf#2537 – us UMI ProQuest [616]

Journal of athletic training – Dallas. 1992+(1,5,9) – (cont: athletic training) – ISSN: 1062-6050 – mf#10829,01 – us UMI ProQuest [790]

Journal of atmospheric and solar-terrestrial physics – London. 1997+ (1,5,9) – (cont: journal of atmospheric and terrestrial physics) – ISSN: 1364-6826 – mf#49109,01 – us UMI ProQuest [530]

Journal of atmospheric and terrestrial physics – Oxford. 1950-1996 (1) 1950-1996 (5) 1977-1996 (9) – (cont by: journal of atmospheric and solar-terrestrial physics) – ISSN: 0021-9169 – mf#49109 – us UMI ProQuest [530]

Journal of atmospheric chemistry – Dordrecht. 1983-1996 (1,5,9) – ISSN: 0167-7764 – mf#14753 – us UMI ProQuest [540]

Journal of audiovisual media in medicine – Abingdon. 1978-1995 (1) 1978-1995 (5) 1978-1995 (9) – ISSN: 0140-511X – mf#11706 – us UMI ProQuest [610]

Journal of auditory research – Groton. 1960-1987 (1) 1960-1987 (5) 1960-1987 (9) – ISSN: 0021-9177 – mf#11951 – us UMI ProQuest [617]

Journal of autism and childhood schizophrenia – New York. 1971-1978 (1) 1971-1978 (5) 1978-1978 (9) – (cont by: journal of autism and developmental disorders) – ISSN: 0021-9185 – mf#10873 – us UMI ProQuest [150]

Journal of autism and developmental disorders – New York. 1979+ (1,5,9) – (cont: journal of autism and childhood schizophrenia) – ISSN: 0162-3257 – mf#10873,01 – us UMI ProQuest [150]

Journal of automated methods and management in chemistry – London. 1999+ (1) – (cont: journal of automatic chemistry) – ISSN: 1463-9246 – mf#17328,01 – us UMI ProQuest [500]

Journal of automated reasoning – Dordrecht. 1985-1996 (1,5,9) – ISSN: 0168-7433 – mf#14754 – us UMI ProQuest [000]

Journal of automatic chemistry – London. 1989-1995 (1,5,9) – (cont by: journal of automated methods and management in chemistry) – ISSN: 0142-0453 – mf#17328 – us UMI ProQuest [500]

Journal of automation and information sciences – Silver Spring. 1991-1993 (1,5,9) – (cont: soviet journal of automation and information sciences) – ISSN: 1064-2315 – mf#14360,02 – us UMI ProQuest [629]

Journal of autonomic pharmacology – North Ferriby. 1987-1992 (1,5,9) – ISSN: 0144-1795 – mf#16734 – us UMI ProQuest [615]

Journal of back and musculoskeletal rehabilitation – Shannon. 1992+ (1,5,9) – ISSN: 1053-8127 – mf#18972 – us UMI ProQuest [617]

Journal of bacteriology – Washington. 1916+ (1) 1965+ (5) 1966+ (9) – ISSN: 0021-9193 – mf#112 – us UMI ProQuest [576]

Journal of baltic studies – Brooklyn. 1970+ (1) 1972+ (5) 1977+ (9) – ISSN: 0162-9778 – mf#7684 – us UMI ProQuest [949]

Journal of band research – Troy. 1964+ (1) 1971+ (5) 1975+ (9) – ISSN: 0021-9207 – mf#6411 – us UMI ProQuest [780]

Journal of bank accounting and auditing – New York. 1990-1990 (1,5,9) – ISSN: 0895-853X – mf#18371 – us UMI ProQuest [650]

Journal of bank research – Park Ridge. 1970-1987 (1) 1972-1987 (5) 1975-1987 (9) – ISSN: 0021-9215 – mf#6412 – us UMI ProQuest [332]

Journal of bank taxation – New York. 1990-1991 (1,5,9) – (cont by: journal of taxation of financial institutions) – ISSN: 0895-4720 – mf#18372 – us UMI ProQuest [336]

Journal of banking and finance – Amsterdam. 1977+ (1) 1977+ (5) 1987+ (9) – ISSN: 0378-4266 – mf#42237 – us UMI ProQuest [332]

Journal of banking law – New York, NY. v1-7. 1882-88 – 9 – mf#LLMC 84-505 – us LLMC [346]

Journal of basic engineering – New York. 1959-1972 [1]; 1964-1972 [5]; 1970-1972 [9] – ISSN: 0021-9223 – mf#1192 – us UMI ProQuest [621]

Journal of basic writing – New York. 1975+(1,5,9) – ISSN: 0147-1635 – mf#12766 – us UMI ProQuest [400]

Journal of behavior therapy and experimental psychiatry – Oxford. 1970+ (1,5,9) – ISSN: 0005-7916 – mf#49110 – us UMI ProQuest [616]

Journal of behavioral decision making – Chichester. 1988+ (1,5,9) – ISSN: 0894-3257 – mf#16166 – us UMI ProQuest [150]

Journal of behavioral health services and research – Gaithersburg. 1998+ (1) – (cont: journal of mental health administration) – ISSN: 1094-3412 – mf#18341,01 – us UMI ProQuest [360]

Journal of behavioral medicine – New York. 1989+ (1,5,9) – ISSN: 0160-7715 – mf#17671 – us UMI ProQuest [616]

Journal of belles lettres – Philadelphia. 1832-1842 (1) – mf#4861 – us UMI ProQuest [800]

Journal of belles-lettres – Lexington. 1819-1820 (1) – mf#3999 – us UMI ProQuest [420]

Journal of bible and religion – v1-16. 1933-48 [complete] – 3r – 1 – (cont by: american academy of religion journal) – ISSN: 0002-7189 – mf#ATLA S0033 – us ATLA [220]

Journal of biblical literature – 1(1881)-82(1963) – 9 – €810.00 – ne Slangenburg [220]

Journal of biblical literature – Atlanta. 1881+ (1) 1907+ (5) 1907+ (9) – ISSN: 0021-9231 – mf#1089 – us UMI ProQuest [220]

Journal of biochemical and biophysical methods – Amsterdam. 1979+ (1) 1979+ (5) 1987+ (9) – ISSN: 0165-022X – mf#42221 – us UMI ProQuest [574]

Journal of biochemical and molecular toxicology – New York. 1998+ (1) – ISSN: 1095-6670 – mf#24741,01 – us UMI ProQuest [615]

Journal of biocommunication – Durham. 1974+ (1,5,9) – ISSN: 0094-2499 – mf#12816 – us UMI ProQuest [610]

Journal of bioenergetics – London. 1970-1976 (1) 1970-1976 (5) – (cont by: journal of bioenergetics and biomembranes) – ISSN: 0449-5705 – mf#10851 – us UMI ProQuest [574]

Journal of bioenergetics and biomembranes – New York. 1976-1991 (1) 1976-1991 (5) 1980-1991 (9) – (cont: journal of bioenergetics) – ISSN: 0145-479X – mf#10851,01 – us UMI ProQuest [574]

Journal of bioethics – New York. 1982-1984 (1,5,9) – (cont: bioethics quarterly. cont by: journal of medical humanities and bioethics) – ISSN: 0278-9523 – mf#12181,02 – us UMI ProQuest [170]

Journal of biogeography – Oxford. 1980+ (1,5,9) – ISSN: 0305-0270 – mf#15555 – us UMI ProQuest [900]

Journal of biological chemistry – Baltimore. 1905+ (1) 1966+ (5) 1970+ (9) – ISSN: 0021-9258 – mf#510 – us UMI ProQuest [540]

Journal of biological photography – Atlanta. 1980-1998 (1) 1980-1998 (5) 1980-1998 (9) – (cont: journal of the biological photographic association) – ISSN: 0274-497X – mf#780,01 – us UMI ProQuest [574]

Journal of biological psychology – Ann Arbor. 1959-1979 (1) 1970-1979 (5) 1975-1979 (9) – ISSN: 0021-9274 – mf#3466 – us UMI ProQuest [150]

Journal of bioluminescence and chemiluminescence – Chichester. 1986-1998 (1) 1986-1998 (5) 1986-1998 (9) – (cont by: luminescence) – ISSN: 0884-3996 – mf#16108 – us UMI ProQuest [574]

Journal of biomechanical engineering – New York. 1978+ (1,5,9) – ISSN: 0148-0731 – mf#11947 – us UMI ProQuest [620]

Journal of biomechanics – New York. 1968+ (1,5,9) – ISSN: 0021-9290 – mf#49111 – us UMI ProQuest [612]

Journal of biomedical engineering – Guildford. 1979-1993 (1,5,9) – (cont by: medical engineering and physics) – ISSN: 0141-5425 – mf#13334 – us UMI ProQuest [610]

Journal of biomedical materials research – New York. 1967+ (1) 1967+ (5) 1967+ (9) – ISSN: 0021-9304 – mf#11057 – us UMI ProQuest [574]

Journal of biophysical and biochemical cytology – New York. 1955-1961 (1) 1955-1961 (5) 1955-1961 (9) – (cont by: journal of cell biology) – ISSN: 0095-9901 – mf#12246 – us UMI ProQuest [574]

Journal of bioscience and bioengineering – Osaka. 1999+ (1) – (cont: journal of fermentation and bioengineering) – ISSN: 1389-1723 – mf#42593,02 – us UMI ProQuest [576]

Journal of biotechnology – Amsterdam. 1984-1989 (1) 1984-1989 (5) 1987-1989 (9) – ISSN: 0168-1656 – mf#42419 – us UMI ProQuest [574]

Journal of bisexuality / ed by Klein, Fritz – ISSN: 1529-9716 – us Haworth [150]

Journal of black higher education – 1986 sep/oct, 1987 sep/oct-1995 mar/apr – 1r – 1 – mf#4712670 – us WHS [378]

Journal of black poetry – 1967 sum – 1r – 1 – (cont by: kitabu cha jua) – mf#759483 – us WHS [810]

Journal of black psychology – Thousand Oaks. 1979+ (1,5,9) – ISSN: 0095-7984 – mf#12350 – us UMI ProQuest [305]

Journal of black sacred music – Durham. 1989-1989 (1,5,9) – (cont by: black sacred music) – ISSN: 0891-9321 – mf#17596 – us UMI ProQuest [780]

Journal of black studies – Thousand Oaks. 1970+ (1) 1971+ (5) 1975+ (9) – ISSN: 0021-9347 – mf#6595 – us UMI ProQuest [305]

Journal of bone and joint surgery : american volume – Boston. 1919+ (1) 1965+ (5) 1970+ (9) – ISSN: 0021-9355 – mf#374 – us UMI ProQuest [617]

Journal of bone and joint surgery : british volume – London. 1948+ (1) 1975+ (5) 1974+ (9) – ISSN: 0301-620X – mf#572 – us UMI ProQuest [610]

Journal of bridge engineering – New York. 1996+ (1,5,9) – ISSN: 1084-0702 – mf#24573 – us UMI ProQuest [624]

Journal of british studies – Chicago. 1961+ (1) 1961+ (5) 1961+ (9) – ISSN: 0021-9371 – mf#12289 – us UMI ProQuest [941]

Journal of broadcasting – Washington. 1956-1984 (1) 1968-1984 (5) 1975-1984 (9) – (cont by: journal of broadcasting and electronic media) – ISSN: 0021-938X – mf#1457 – us UMI ProQuest [380]

Journal of broadcasting *see* Journal of broadcasting and electronic media

Journal of broadcasting and electronic media – Washington. 1985+ (1) 1985+ (5) 1985+ (9) – (cont: journal of broadcasting) – ISSN: 0883-8151 – mf#1457,01 – us UMI ProQuest [380]

Journal of broadcasting and electronic media – v1-44. 1956-2000 – 5,6,9 – $887.00 set – (v1-28 1956-84 on reel $322. v29-44 1985-2000 on mf $565. title varies: v1-28 1956-84 as journal of broadcasting) – ISSN: 0021-938X – mf#105081 – us Hein [380]

Journal of bryology – Oxford. 1980-1989 (1) 1980-1989 (5) 1980-1989 (9) – ISSN: 0373-6687 – mf#15556,01 – us UMI ProQuest [580]

Journal of burn care and rehabilitation – Lake Forest. 1983+ (1,5,9) – ISSN: 0273-8481 – mf#13590 – us UMI ProQuest [610]

Journal of burnett county – Grantsburg WI. [1895 aug 7/1898 oct 14]-[1909 apr 2/1910 jan 28] [gaps] – 7r – 1 – (cont by: burnett county sentinel; journal of burnett county and burnett county sentinel) – mf#1214404 – us WHS [978]

Journal of burnett county – Grantsburg WI. [1929 apr 4/1930 feb 20]-[1949/1951 aug 9] [small gaps] – 13r – 1 – (cont: journal of burnett county and burnett county sentinel; cont by: times [grantsburg, wi]; journal of burnett county and the times) – mf#1214406 – us WHS [978]

Journal of burnett county – Grantsburg WI. 1955 jun 16-1956, 1957-59, 1960 jan 7-1962 jul 26 – 3r – 1 – (cont: journal of burnett county and the times and burnett county enterprise) – mf#946871 – us WHS [978]

Journal of burnett county and burnett county sentinel – Grantsburg WI. [1910 feb 4/1910 sep 16]-[1928 oct 25/1929 mar 28] [gaps] – 14r – 1 – (cont: journal of burnett county [grantsburg, wi: 1895]; burnett county sentinel; cont by: journal of burnett county [grantsburg, wi: 1929]) – mf#946861 – us WHS [071]

Journal of burnett county and the times – Grantsburg WI. 1951 aug 16-sep 6 – 1r – 1 – (cont: journal of burnett county [grantsburg, wi: 1929]; times [grantsburg, wi]; cont by: burnett county enterprise; journal of burnett county and the times and burnett county enterprise) – mf#946868 – us WHS [978]

Journal of burnett county and the times and burnett county enterprise – Grantsburg WI. 1951 sep 13-dec, 1952-54, 1955 jan-jun 9 – 3r – 1 – (cont: journal of burnett county and the times; burnett county enterprise; cont by: journal of burnett county [grantsburg, wi: 1955]) – us WHS [978]

Journal of business – Chicago. 1928+ (1) 1969+ (5) 1977+ (9) – ISSN: 0021-9398 – mf#478 – us UMI ProQuest [338]

Journal of business – South Orange. 1962-1980 (1) 1971-1980 (5) 1974-1980 (9) – (cont by: mid-atlantic journal of business) – ISSN: 0021-9401 – mf#5137 – us UMI ProQuest [338]

Journal of business – Spokane. 1994+ (1) – ISSN: 1075-6124 – mf#18218 – us UMI ProQuest [650]

Journal of business administration – Vancouver. 1972-1995 (1) 1969-1995 (5) 1976-1995 (9) – (cont by: journal of business administration and policy analysis) – ISSN: 0021-941X – mf#6567 – us UMI ProQuest [650]

Journal of business administration and policy analysis – Vancouver. 1996+ (1) 1996+ (5) 1996+ (9) – (cont: journal of business administration) – mf#6567,01 – us UMI ProQuest [650]

Journal of business and economic statistics – Alexandria. 1990+ – 1,5,9 – ISSN: 0735-0015 – mf#16752 – us UMI ProQuest [310]

Journal of business and finance librarianship / ed by Popovich, Charles J – v1- 1989- – 1, 9 ($75.00 in US $105.00 outside hardcopy subsc) – us Haworth [020]

Journal of business and industrial marketing – Santa Barbara. 1986-1995 (1) 1986-1995 (5) 1986-1995 (9) – ISSN: 0885-8624 – mf#16016 – us UMI ProQuest [650]

Journal of business and psychology – New York. 1986+ (1,5,9) – ISSN: 0889-3268 – mf#15458 – us UMI ProQuest [150]

Journal of business and social studies – Lagos. 1977-1979 (1) 1977-1979 (5) 1977-1979 (9) – ISSN: 0021-9428 – mf#7087 – us UMI ProQuest [338]

Journal of business and technical communication (jbtc) – Thousand Oaks, 1998+ [1,5,9] – ISSN: 1050-6519 – mf#21498,01 – us UMI ProQuest [341]

Journal of business communication – Urbana. 1963+ (1) 1971+ (5) 1977+ (9) – ISSN: 0021-9436 – mf#5740 – us UMI ProQuest [650]

Journal of business education – Washington. 1928-1984 (1) 1968-1984 (5) 1970-1984 (9) – (cont by: journal of education for business) – ISSN: 0021-9444 – mf#866 – us UMI ProQuest [338]

Journal of business ethics: jbe – Dordrecht. 1982+ (1,5,9) – ISSN: 0167-4544 – mf#14755 – us UMI ProQuest [170]

Journal of business finance and accounting – Oxford. 1982+ (1,5,9) – ISSN: 0306-686X – mf#13528 – us UMI ProQuest [650]

Journal of business forecasting methods and systems – Flushing. 1986+ (1,5,9) – ISSN: 0278-6087 – mf#15780 – us UMI ProQuest [650]

Journal of business law – London. 1989+ (1) 1989+ (5) 1989+ (9) – ISSN: 0021-9460 – mf#1355 – us UMI ProQuest [346]

Journal of business logistics – Oak Brook. 1986+ (1,5,9) – ISSN: 0735-3766 – mf#16365 – us UMI ProQuest [338]

Journal of business research – New York. 1973+ (1) 1973+ (5) 1987+ (9) – ISSN: 0148-2963 – mf#42222 – us UMI ProQuest [338]

Journal of business strategy – Boston. 1980+ (1,5,9) – ISSN: 0275-6668 – mf#12554 – us UMI ProQuest [650]

Journal of business venturing – New York. 1986+ (1,5,9) – ISSN: 0883-9026 – mf#42599 – us UMI ProQuest [650]

Journal of business-to-business marketing : ...innovations in applied business and industrial marketing research / ed by Lichtenthal, J David – v3 n1. 1996- – 1,9 – $150.00 in US $210.00 outside hardcopy subsc – us Haworth [650]

Journal of buyouts and acquisitions – San Diego. 1982-1986 (1) 1982-1986 (5) 1982-1986 (9) – (cont by: buyouts and acquisitions) – ISSN: 0736-5527 – mf#14431 – us UMI ProQuest [338]

Journal of canadian petroleum technology – Montreal. 1984-1995 (1,5,9) – ISSN: 0021-9487 – mf#15108 – us UMI ProQuest [550]

Journal of canadian studies/revue d'etudes canadiennes – v. 1-4. 1966-69 – 1 – us AMS Press [800]

Journal of cancer pain and symptom palliation / ed by Smith, Howard S – ISSN: 1543-7671 – us Haworth [615]

Journal of cancer research and clinical oncology – Heidelberg. 1983-1983 (1,5,9) – ISSN: 0171-5216 – mf#13119,02 – us UMI ProQuest [616]

Journal of cannabis therapeutics : studies in endogenous, herbal, and synthetic cannabinoids / ed by Russo, Ethan – ISSN: 1529-9775 – us Haworth [615]

Journal of cardiopulmonary rehabilitation – New York. 1987+ (1,5,9) – ISSN: 0883-9212 – mf#16613,01 – us UMI ProQuest [616]

Journal of cardiovascular nursing – Frederick. 1986+ (1,5,9) – ISSN: 0889-4655 – mf#16004 – us UMI ProQuest [610]

Journal of career development – New York. 1984+ – 1,5,9 – (cont: journal of career education) – ISSN: 0894-8453 – mf#11336,01 – us UMI ProQuest [374]

Journal of career education – Columbia. 1972-1983 (1) 1972-1983 (5) 1972-1983 (9) – (cont by: journal of career development) – ISSN: 0164-2502 – mf#11336 – us UMI ProQuest [374]

Journal of career planning and employment – Bethlehem. 1985-1999 (1) 1985-1999 (5) 1985-1999 (9) – (cont: journal of college placement) – ISSN: 0884-5352 – mf#425,01 – us UMI ProQuest [378]

Journal of cash management – Atlanta. 1990-1993 (1,5,9) – (cont by: tma journal) – ISSN: 0731-1281 – mf#15782 – us UMI ProQuest [650]

Journal of cell biology – New York. 1962+ (1,5,9) – (cont: journal of biophysical and biochemical cytology) – ISSN: 0021-9525 – mf#12246,01 – us UMI ProQuest [611]

Journal of cell science – Cambridge. 1983-1996 (1,5,9) – ISSN: 0021-9533 – mf#13597 – us UMI ProQuest [578]

Journal of cellular biochemistry – New York. 1982+ (1,5,9) – ISSN: 0730-2312 – mf#12888,02 – us UMI ProQuest [574]

Journal of cellular plastics – Lancaster. 1965+ [1]; 1971+ [5]; 1976+ [9] – ISSN: 0021-955X – mf#1921 – us UMI ProQuest [660]

Journal of change management – London. 2000+ (1,5,9) – ISSN: 1469-7017 – mf#31750 – us UMI ProQuest [650]

Journal of charles j deblois, captain's clerk : aboard the u s s macedonian, 1818-1819 – (= ser Naval records coll of the office of naval records and library) – 1r – 1 – (with printed guide) – mf#M876 – us Nat Archives [355]

Journal of charles mason kept during the survey of the mason and dixon line, 1763-1768 / U.S. Dept of State – (= ser Records of miscellaneous civilian agencies) – 1r – 1 – mf#M86 – us Nat Archives [975]

Journal of chaussegros de lery / ed by Stevens, Sylvester K & Kent, Donald H – Harrisburg: Dept of Public Instruction, Pennsylvania Hist Comm, 1940 (mf ed 19–) – [3]/118[i.e. 120]p – (incl index. filmed with: western pennsylvanians / charles alexander rook, editor-in-chief; comp under the direction of the james o jones co. pittsburgh, pa western pennsylvania biographical association, 1923) – mf#ZH-IAG pv656 n3 – us NY Public [917]

Journal of chemical and engineering data – v1- 1956- – 1,5,6,9 – us ACS [660]

Journal of chemical dependency treatment / ed by Finnegan, Dana – v1- 1987- – 1, 9 ($175.00 in US $245.00 outside hardcopy subsc) – us Haworth [615]

Journal of chemical ecology – New York. 1975-1996 (1) 1975-1996 (5) 1978-1996 (9) – ISSN: 0098-0331 – mf#10865 – us UMI ProQuest [540]

Journal of chemical education – Easton. 1924-2000 (1) 1966-2000 (5) 1970-2000 (9) – ISSN: 0021-9584 – mf#2262 – us UMI ProQuest [370]

Journal of chemical industry and engineering (china) : english edition – Elmsford. 1987-1991 (1,5,9) – ISSN: 1000-9027 – mf#49522 – us UMI ProQuest [660]

Journal of chemical information and computer sciences – v1- 1961- – 1,5,6,9 – us ACS [540]

Journal of chemical neuroanatomy – Chichester. 1988-1993 (1,5,9) – ISSN: 0891-0618 – mf#16170 – us UMI ProQuest [540]

The journal of chemical physics – v1-. 1933- – 1,5,6,9 – us AIP [530]

Journal of chemical technology and biotechnology – Oxford. 1996+ (1,5,9) – ISSN: 0268-2575 – mf#15589 – us UMI ProQuest [660]

Journal of chemometrics – Chichester. 1987+ (1,5,9) – ISSN: 0886-9383 – mf#16109 – us UMI ProQuest [540]

Journal of child abuse and the law / ed by Frankel, A Steven & Murphy, Wendy J – ISSN: 1543-771X – us Haworth [345]

Journal of child and adolescent psychiatric and mental health nursing – Philadelphia. 1988-1993 (1,5,9) – (cont by: journal of child and adolescent psychiatric nursing) – ISSN: 0897-9685 – mf#16614 – us UMI ProQuest [610]

Journal of child and adolescent psychiatric nursing – Philadelphia. 1994+ (1,5,9) – (cont: journal of child and adolescent psychiatric and mental health nursing) – ISSN: 1073-6077 – mf#16614,01 – us UMI ProQuest [610]

Journal of child and adolescent substance abuse / ed by DePiano, Frank & Hasselt, Vincent B Van – v1- 1989- – (= ser Journal Of Adolescent Chemical Dependency) – 1, 9 ($145.00 in US $203.00 outside hardcopy subsc) – us Haworth [360]

Journal of child and family studies – New York. 1992+ (1,5,9) – ISSN: 1062-1024 – mf#19615 – us UMI ProQuest [150]

Journal of child custody : research, issues, and practices / ed by Drozd, Leslie – ISSN: 1537-9418 – us Haworth [360]

Journal of child language – Cambridge. 1974+ (1,5,9) – ISSN: 0305-0009 – mf#12124 – us UMI ProQuest [640]

Journal of child neurology – Hamilton. 1989+ (1,5,9) – ISSN: 0883-0738 – mf#16052 – us UMI ProQuest [618]

Journal of child psychology and psychiatry and allied disciplines – Cambridge, 1998+ [1,5,9] – ISSN: 0021-9630 – mf#30051 – us UMI ProQuest [150]

Journal of child sexual abuse : research, treatment and program innovations for victims, survivors and offenders / ed by Geffner, Robert A – v5 n1. 1996- – 1,9 – $85.00 in US $119.00 outside hardcopy subsc – us Haworth [360]

Journal of childhood communication disorders : jccd – Reston. 1987-1995 – 1,5,9 – (cont by: journal of children's communication development: jccd) – ISSN: 0735-3170 – mf#17529 – us UMI ProQuest [370]

Journal of children's communication development : jccd – Reston. 1995-1998 – 1,5,9 – (cont: journal of childhood communication disorders: jccd. cont by: communication disorders quarterly) – ISSN: 1093-5703 – mf#17529,01 – us UMI ProQuest [370]

Journal of chinese law *see* Columbia journal of asian law

Journal of chiropractic – Arlington. 1982-1994 (1,5,9) – (cont: aca journal of chiropractic; cont by: journal of the american chiropractic association) – ISSN: 0744-9984 – mf#10579,01 – us UMI ProQuest [615]

Journal of christian education of the african methodist episcopal church – Nashville. 1982-1992 (1) 1982-1988 (5) 1982-1988 (9) – (cont: journal of religious education of the african methodist episcopal church) – mf#6886,02 – us UMI ProQuest [377]

Journal of christian jurisprudence – Regent University. v1-8. 1980-90 (all publ) – 9 – $90.00 set – (cont by: liberty, life and family) – ISSN: 0741-6075 – mf#108751 – us Hein [340]

Journal of christian nursing – Madison. 1984+ (1,5,9) – ISSN: 0743-2550 – mf#14331 – us UMI ProQuest [610]

Journal of christian philosophy – New York. 1881-1884 (1) – ISSN: 0734-1342 – mf#2906 – us UMI ProQuest [240]

Journal of christian reconstruction – Woodland Hills. 1988-1996 (1,5,9) – ISSN: 0360-1420 – mf#15626 – us UMI ProQuest [240]

Journal of chromatographic science / ed by Gordon, Bert & Walker, John Q. – 1962-96 – 1,5 – $100.00v subsc $240.00v nonsubsc 1 ($75.00v subsc $240 nonsubsc 5) – ISSN: 0 – us Preston Publ [540]

Journal of chromatography – Amsterdam. 1958+ (1) 1958+ (5) 1987+ (9) – ISSN: 0021-9673 – mf#42273 – us UMI ProQuest [540]

Journal of chromatography – Amsterdam. 1992-1993 (1,5,9) – ISSN: 0021-9673 – mf#42721 – us UMI ProQuest [540]

Journal of chromatography a – Amsterdam. 1994-1995 (1,5,9) – mf#42764 – us UMI ProQuest [540]

Journal of chromatography b : biomedical applications – Amsterdam. 1994-1995 (1,5,9) – mf#42765 – us UMI ProQuest [540]

Journal of chronic diseases – Elmsford. 1955-1987 (1) 1955-1987 (5) 1955-1987 (9) – (cont by: journal of clinical epidemiology) – ISSN: 0021-9681 – mf#49113 – us UMI ProQuest [616]

Journal of chronic fatigue syndrome : multidisciplinary innovations in research, theory and clinical practice / ed by Klimas, Nancy – v2 n1. 1996- – 1,9 – $125.00 in US $175.00 outside hardcopy subsc – us Haworth [616]

Journal of church and state – Baylor University. v1-43. 1959-2001 – 5,6,9 – $778.00 set – (v1-26 1959-84 on reel $329. v27-43 1985-2001 on mf $449) – ISSN: 0021-969X – mf#103861 – us Hein [240]

Journal of church and state – Waco. 1993+ (1,5,9) – ISSN: 0021-969X – mf#18428 – us UMI ProQuest [230]

Journal of church music – Philadelphia. 1959-1988 (1) 1972-1988 (5) 1975-1988 (9) – ISSN: 0021-9703 – mf#7323 – us UMI ProQuest [780]

Journal of civil defense / American Civil Defense Association et al – 1976 jan-1981 aug, 1981 oct-1985 jun – 2r – 1 – (cont: survive [starke, fl]) – mf#800230 – us WHS [360]

Journal of classical and sacred philology – Cambridge. 1854-1859 (1) – mf#4724 – us UMI ProQuest [450]

Journal of classroom interaction – Houston. 1976+ (1) 1976+ (5) 1976+ (9) – (cont: classroom interaction newsletter) – ISSN: 0749-4025 – mf#10394,01 – us UMI ProQuest [370]

Journal of climate – Boston, 1998+ [1,5,9] – ISSN: 0894-8755 – mf#23691 – us UMI ProQuest [550]

Journal of climatology – Chichester. 1981-1988 (1) 1981-1988 (5) 1981-1988 (9) – (cont by: international journal of climatology) – ISSN: 0196-1748 – mf#12647 – us UMI ProQuest [550]

Journal of clinical and hospital pharmacy – Oxford. 1980-1986 (1) 1980-1986 (5) 1980-1986 (9) – (cont by: journal of clinical pharmacy and therapeutics) – ISSN: 0143-3180 – mf#15530,01 – us UMI ProQuest [615]

Journal of clinical and pastoral work – Decatur. (1) 1947-1949 (5) (9) – mf#8061 – us UMI ProQuest [240]

Journal of clinical anesthesia – New York. 1988+ (1,5,9) – ISSN: 0952-8180 – mf#17134 – us UMI ProQuest [617]

Journal of clinical child and adolescent psychology – Mahwah. 2002+ (1,5,9) – (cont: journal of clinical child psychology) – ISSN: 1537-4416 – mf#17605,01 – us UMI ProQuest [150]

Journal of clinical child psychology – Hillsdale. 1989+ (1,5,9) – ISSN: 0047-228X – mf#17605 – us UMI ProQuest [150]

Journal of clinical endocrinology and metabolism – Philadelphia. 1941-1977 (1) 1966-1977 (5) 1970-1977 (9) – ISSN: 0021-972X – mf#2358 – us UMI ProQuest [616]

Journal of clinical epidemiology – Elmsford. 1988+ (1,5,9) – (cont: journal of chronic diseases) – ISSN: 0895-4356 – mf#49113,01 – us UMI ProQuest [614]

Journal of clinical gastroenterology – New York. 1993+ (1,5,9) – ISSN: 0192-0790 – mf#18713 – us UMI ProQuest [616]

Journal of clinical immunoassay : official publication of the clinical ligand assay society – Wayne. 1989-1989 (1) – ISSN: 0736-4393 – mf#13579,01 – us UMI ProQuest [574]

Journal of clinical investigation – New York. 1951+ (1) 1924+ (5) 1970+ (9) – ISSN: 0021-9738 – mf#443 – us UMI ProQuest [610]

Journal of clinical microbiology – Washington. 1975+ (1) 1975+ (5) 1976+ (9) – ISSN: 0095-1137 – mf#10397 – us UMI ProQuest [576]

Journal of clinical monitoring – Boston. 1989-1995 (1,5,9) – ISSN: 0748-1977 – mf#14391 – us UMI ProQuest [616]

Journal of clinical monitoring and computing – Dordrecht, 1998+ [1,5,9] – ISSN: 1387-1307 – mf#31505 – us UMI ProQuest [617]

Journal of clinical nursing – Oxford. 1992+ (1,5,9) – ISSN: 0962-1067 – mf#18771 – us UMI ProQuest [610]

Journal of clinical oncology – Philadelphia. 1992+ (1,5,9) – ISSN: 0732-183X – mf#21109 – us UMI ProQuest [616]

Journal of clinical pathology – London. 1947+ (1) 1965+ (5) 1968+ (9) – ISSN: 0021-9746 – mf#1329 – us UMI ProQuest [616]

Journal of clinical pediatric dentistry – Birmingham. 1990+ (1,5,9) – (cont: journal of pedodontics) – ISSN: 1053-4628 – mf#11466,01 – us UMI ProQuest [610]

Journal of clinical pharmacology – Stamford. 1988+ (1,5,9) – ISSN: 0091-2700 – mf#13565,03 – us UMI ProQuest [615]

Journal of clinical pharmacy and therapeutics – Oxford. 1987-1996 (1,5,9) – (cont: journal of clinical and hospital pharmacy) – ISSN: 0269-4727 – mf#15530,02 – us UMI ProQuest [615]

Journal of clinical psychiatry – Memphis. 1978+ (1) 1978+ (5) 1978+ (9) – (cont: diseases of the nervous system) – ISSN: 0160-6689 – mf#1697,01 – us UMI ProQuest [616]

Journal of clinical psychology – Brandon. 1945+ (1) 1965+ (5) 1970+ (9) – ISSN: 0021-9762 – mf#38 – us UMI ProQuest [150]

Journal of clinical psychopharmacology – v1-16. 1981-96 – 16r – 1,5,6,9 – $80.00r – us Lippincott [615]

Journal of clinical research and drug development – New York. 1987-1989 (1) 1987-1989 (5) 1988-1989 (9) – (cont by: journal of clinical research and pharmacoepidemiology) – ISSN: 0889-5813 – mf#42452 – us UMI ProQuest [615]

Journal of clinical research and drug development – New York. 1993-1994 (1,5,9) – (cont: journal of clinical research and pharmacoepidemiology) – ISSN: 1066-7865 – mf#42452,02 – us UMI ProQuest [615]

Journal of clinical research and pharmacoepidemiology – New York. 1990-1992 (1,5,9) – (cont: journal of clinical research and drug development. cont by: journal of clinical research and drug development) – ISSN: 1047-0336 – mf#42452,01 – us UMI ProQuest [615]

Journal of clinical ultrasound – New York. 1973+ (1,5,9) – ISSN: 0091-2751 – mf#11125 – us UMI ProQuest [621]

Journal of clinical virology – Amsterdam. 1998+ – 1 – ISSN: 1386-6532 – mf#42734,01 – us UMI ProQuest [616]

Journal of coated fabrics – London. 1973-1996 (1) 1973-1996 (5) 1974-1996 (9) – (cont: journal of coated fibrous materials) – ISSN: 0093-4658 – mf#6095,01 – us UMI ProQuest [670]

Journal of coated fibrous materials – Westport. 1971-1973 (1) 1971-1972 (5) (9) – (cont by: journal of coated fabrics) – ISSN: 0047-2298 – mf#6095 – us UMI ProQuest [600]

Journal of cognition and development – Mahwah. 2000+ (1,5,9) – ISSN: 1524-8372 – mf#31733 – us UMI ProQuest [150]

Journal of cognitive neuroscience – Cambridge. 1989+ (1,5,9) – ISSN: 0898-929X – mf#17787 – us UMI ProQuest [612]

Journal of cold regions engineering – New York. 1987+ (1,5,9) – ISSN: 0887-381X – mf#16197 – us UMI ProQuest [624]

Journal of college admission – Skokie. 1987+ – 1,5,9 – ISSN: 0734-6670 – mf#16085,02 – us UMI ProQuest [378]

Journal of college and university foodservice / ed by Khan, Mahmood A – v3 n1. 1997- – 1,9 – $60.00 in US $84.00 outside hardcopy subsc – us Haworth [640]

Journal of college and university student housing – Columbus. 1979+ – 1,5,9 – ISSN: 0161-827X – mf#12130 – us UMI ProQuest [378]

Journal of college placement – Bethlehem. 1940-1985 (1) 1970-1985 (5) 1976-1985 (9) – (cont by: journal of career planning and employment) – ISSN: 0021-9770 – mf#425 – us UMI ProQuest [378]

Journal of college science teaching – Washington. 1971+ [1,5]; 1975+ [9] – ISSN: 0047-231X – mf#6500 – us UMI ProQuest [378]

Journal of college student development – Washington. 1988+ (1) 1988+ (5) 1988+ (9) – (cont: journal of college student personnel) – ISSN: 0897-5264 – mf#7998,01 – us UMI ProQuest [378]

Journal of college student personnel – Alexandria. 1959-1987 (1) 1959-1987 (5) 1959-1987 (9) – (cont by: journal of college student development) – ISSN: 0021-9789 – mf#7998 – us UMI ProQuest [378]

Journal of college student psychotherapy / ed by Whitaker, Leighton C – v1- 1986- – 1, 9 ($225.00 in US $315.00 outside hardcopy subsc) – us Haworth [615]

Journal of combustion toxicology – Westport. 1976-1982 (1) 1976-1982 (5) 1976-1982 (9) – (cont: combustion toxicology) – ISSN: 0362-1669 – mf#10458,01 – us UMI ProQuest [360]

Journal of commerce : british columbia edition – British Columbia, CN. 1962-73 – 26r – 1 – cn Commonwealth Imaging [380]

Journal of commerce – Chicago, IL. 1920-1951 (1) – mf#62568 – us UMI ProQuest [071]

Journal of commerce : combined edition – Vancouver, British Columbia, CN. 1974-82; 1983-jun 1989 – 50r – 1 – cn Commonwealth Imaging [380]

Journal of commerce – Milwaukee WI. 1880 mar 3-dec 22 – 1r – 1 – (cont: milwaukee journal of commerce) – mf#1166806 – us WHS [380]

Journal of commerce – Montreal, Canada. -w. 5 oct 1877-4 mar 1919; 1920-1921; 24 feb-30 jun 1922 – 3 1/2 r – 1 – uk British Libr Newspaper [072]

Journal of commerce – New York, NY. 1930-2000 (1) – ISSN: 1088-7407 – mf#60535 – us UMI ProQuest [071]

Journal of commerce – Norfolk, VA. 1900-1905 (1) – mf#66778 – us UMI ProQuest [071]

Journal of commerce : (pacific edition) – New York, NY. 1984-1989 (1) – mf#60531 – us UMI ProQuest [071]

Journal of commerce – Portland OR: Berry & Earle, 1853 [wkly] – 1 – us Oregon Lib [071]

Journal of commerce : prairie edition – British Columbia, CN. 1962-73 – 26r – 1 – cn Commonwealth Imaging [380]

Journal of commerce : weekly edition – British Columbia, CN. 1951-61 – 19r – 1 – cn Commonwealth Imaging [380]

Journal of commerce *see*
- Liverpool journal of commerce
- Natal star / journal of commerce / agriculture / a100

Journal of commerce and commercial bulletin – New York. oct 26 1903-10 – (filmed with: weekly journal of commerce and commercial bulletin) – us CRL [380]

Journal of commerce and commercial bulletin – New York, N.Y. Oct 26, 1903-1910 – 1 – us CRL [380]

Journal of commerce and commercial new york – New York. Jan. 1828-June 30, 1948 – 1 – us NY Public [071]

The journal of commerce and commercial, new york – Commercial precedents, selected from the column of replies and decisions of the New York Journal of Commerce. Hartford, Conn: American Publishing Co., 1881. 588p. LL-403 – 1 – us L of C Photodup [346]

Journal of commerce and shipping telegraph – [NW England] Liverpool jun-aug 1939, oct 1940-1974 – 1 – uk MLA; uk Newsplan [380]

Journal of commerce and shipping telegraph *see* Liverpool journal of commerce

Journal of commerce export bulletin – New York, NY. 1981-1994 (1) – mf#60529 – us UMI ProQuest [071]

Journal of commerce import bulletin – New York, NY. 1981-1995 (1) – mf#60530 – us UMI ProQuest [071]

Journal of commerce of victoria – Melbourne, Australia. 31 mar-dec 1855; 2 jan 1859-1872; 12 jan-28 dec 1888; 5, 19 dec 1893; jan-18 dec 1894; 1895-1917; 13 feb 1918-15 dec 1920 – 25 1/2r – 1 – uk British Libr Newspaper [072]

Journal of commercial bank lending – Philadelphia. 1967-1991 (1) 1967-1991 (5) 1967-1991 (9) – (cont by: journal of commercial lending) – ISSN: 0021-986X – mf#12339,01 – us UMI ProQuest [332]

Journal of commercial biotechnology – London, 1998+ [1,5,9] – ISSN: 1462-8732 – mf#31704,01 – us UMI ProQuest [574]

Journal of commercial lending – Philadelphia. 1991-1995 (1,5,9) – (cont: journal of commercial bank lending. cont by: journal of lending and credit risk management) – ISSN: 1062-6271 – mf#12339,02 – us UMI ProQuest [332]

Journal of common market studies – v1-33. 1962-95 – 9 – $396.00 set – (available on reel only) – ISSN: 0021-9886 – mf#103881 – us Hein [380]

Journal of common market studies – Oxford. 1990+ (1,5,9) – ISSN: 0021-9886 – mf#15805 – us UMI ProQuest [338]

Journal of commonwealth literature – St. Giles. 1965+ (1) 1965+ (5) 1965+ (9) – ISSN: 0021-9894 – mf#9855 – us UMI ProQuest [410]

Journal of communication – Philadelphia. 1951+ (1) 1951+ (5) 1951+ (9) – ISSN: 0021-9916 – mf#10207 – us UMI ProQuest [380]

Journal of communication disorders – New York. 1967+ (1) 1967+ (5) 1987+ (9) – ISSN: 0021-9924 – mf#42100 – us UMI ProQuest [150]

Journal of communications technology and electronics – Silver Spring. 1993-1994 (1,5,9) – (cont: soviet journal of communications technology and electronics) – ISSN: 1064-2269 – mf#14359,03 – us UMI ProQuest [380]

Journal of communist studies – London. 1990-1993 (1,5,9) – ISSN: 0268-4535 – mf#18553 – us UMI ProQuest [335]

Journal of community and applied social psychology – Chichester. 1991+ (1,5,9) – (cont: social behaviour) – ISSN: 1052-9284 – mf#18193 – us UMI ProQuest [301]

Journal of community health – New York. 1975+(1,5,9) – ISSN: 0094-5145 – mf#11181 – us UMI ProQuest [360]

Journal of community health nursing – Hillsdale. 1989+ (1,5,9) – ISSN: 0737-0016 – mf#17606 – us UMI ProQuest [360]

Journal of community practice : multidisciplinary innovations in research, theory and clinical practice sponsored by the association for cummunity organization and social administration (acosa) / ed by Weil, D S W Marie – v3 n1. 1996- – 1,9 – $105.00 in US $147.00 outside hardcopy subsc – us Haworth [366]

Journal of community psychology – Brandon. 1973+ (1,5,9) – ISSN: 0090-4392 – mf#10871 – us UMI ProQuest [150]

Journal of comparative administration – Beverly Hills. 1969-1974 (1,5,9) – (cont by: administration and society) – ISSN: 0021-9932 – mf#12641 – us UMI ProQuest [350]

Journal of comparative and physiological psychology – Arlington. 1948-1982 (1) 1965-1982 (5) 1970-1982 (9) – ISSN: 0021-9940 – mf#1153 – us UMI ProQuest [150]

Journal of comparative business and capital market law – Amsterdam. 1978-1986 (1) 1978-1986 (5) (9) – (cont by: university of pennsylvania journal of international business law) – ISSN: 0167-9333 – mf#42349 – us UMI ProQuest [346]

Journal of comparative business and capital market review *see* University of pennsylvania journal of international economic law

Journal of comparative corporate law and securities regulation *see* University of pennsylvania journal of international economic law

Journal of comparative family studies – Calgary. 1970+ (1) 1972+ (5) 1976+ (9) – ISSN: 0047-2328 – mf#8153 – us UMI ProQuest [301]

Journal of comparative legislation and international law – 1896-1951 – 13r – 1 – $500.00 – us Trans-Media [341]

Journal of comparative physiology – Berlin. 1972-1982 (1) 1972-1982 (5) 1980-1982 (9) – (cont: zeitschrift fuer vergleichende physiologie) – ISSN: 0302-9824 – mf#13120,01 – us UMI ProQuest [612]

Journal of comparative physiology a : sensory, neural, and behavioral physiology – Berlin. 1984-1996 (1,5,9) – ISSN: 0340-7594 – mf#14342 – us UMI ProQuest [612]

Journal of comparative physiology b : biochemical, systemic, and environmental physiology – Berlin. 1984-1996 (1,5,9) – ISSN: 0174-1578 – mf#14343 – us UMI ProQuest [612]

Journal of comparative psychology – Washington. 1983+ (1) 1983+ (5) 1983+ (9) – ISSN: 0735-7036 – mf#1153,01 – us UMI ProQuest [150]

Journal of compensation and benefits – Boston. 1985+ (1,5,9) – ISSN: 0893-780X – mf#15807 – us UMI ProQuest [650]

Journal of composite materials – London. 1967+ (1) 1970+ (5) 1975+ (9) – ISSN: 0021-9983 – mf#6063 – us UMI ProQuest [620]

Journal of composites for construction – New York. 1997+ (1,5,9) – ISSN: 1090-0268 – mf#26622 – us UMI ProQuest [624]

Journal of composites technology and research – Conshohocken. 1989-1989 (1) – ISSN: 0884-6804 – mf#16706,01 – us UMI ProQuest [621]

Journal of computational and applied mathematics – Antwerp. 1983+ (1,5,9) – ISSN: 0377-0427 – mf#42496 – us UMI ProQuest [510]

Journal of computational chemistry – New York. 1980+ (1,5,9) – ISSN: 0192-8651 – mf#11049 – us UMI ProQuest [540]

Journal of computed tomography – Baltimore. 1979-1988 (1) 1979-1988 (5) 1987-1988 (9) – (cont by: clinical imaging) – ISSN: 0149-936X – mf#42411 – us UMI ProQuest [616]

Journal of computer and systems sciences international – Silver Spring. 1992-1996 (1,5,9) – (cont: soviet journal of computer and systems sciences) – ISSN: 1064-2307 – mf#14349,02 – us UMI ProQuest [000]

Journal of computer assisted learning – Oxford. 1985+ – 1,5,9 – ISSN: 0266-4909 – mf#15557 – us UMI ProQuest [370]

Journal of computer assisted tomography – New York. 1993+ (1,5,9) – ISSN: 0363-8715 – mf#18716 – us UMI ProQuest [616]

Journal of computer information systems – Stillwater. 1985+ (1) 1985+ (5) 1985+ (9) – (cont: journal of data education) – ISSN: 0887-4417 – mf#6671,01 – us UMI ProQuest [000]

Journal of computer-based instruction – Bellingham. 1974-1993 (1) 1974-1993 (5) 1974-1993 (9) – ISSN: 0098-597X – mf#11028 – us UMI ProQuest [000]

Journal of computers in mathematics and science teaching – Austin. 1983+ (1,5,9) – ISSN: 0731-9258 – mf#14017 – us UMI ProQuest [510]

Journal of computing in childhood education – Phoenix. 1989-1998 – 1,5,9 – ISSN: 1043-1055 – mf#17105 – us UMI ProQuest [370]

Journal of computing in civil engineering – New York. 1987+ (1,5,9) – ISSN: 0887-3801 – mf#16198 – us UMI ProQuest [624]

Journal of conflict and security law – Oxford, 2000+ [1,5,9] – ISSN: 1467-7954 – mf#31449,01 – us UMI ProQuest [341]

Journal of conflict resolution – Beverly Hills. 1957+ (1) 1971+ (5) 1975+ (9) – ISSN: 0022-0027 – mf#3046 – us UMI ProQuest [150]

Journal of constitutional and parliamentary studies – New Delhi. 1967-1984 (1) 1974-1984 (5) 1974-1984 (9) – ISSN: 0022-0043 – mf#7687 – us UMI ProQuest [323]

Journal of construction engineering and management – New York. 1983+ (1,5,9) – (cont: journal of the construction division) – ISSN: 0733-9364 – mf#8146,01 – us UMI ProQuest [624]

Journal of constructional steel research: jcsr – London. 1982-1991 (1,5,9) – ISSN: 0143-974X – mf#42274 – us UMI ProQuest [624]

Journal of constructivist psychology – Washington. 1994+ (1,5,9) – (cont: international journal of personal construct psychology) – ISSN: 1072-0537 – mf#16657,01 – us UMI ProQuest [150]

Journal of consulting and clinical psychology – Arlington. 1937+ [1]; 1965+ [5]; 1970+ [9] – ISSN: 0022-006X – mf#1154 – us UMI ProQuest [150]

Journal of consumer affairs – Columbia. 1972+ (1) 1961+ (5) 1975+ (9) – ISSN: 0022-0078 – mf#6582 – us UMI ProQuest [380]

Journal of consumer behaviour – London. 2001+ (1,5,9) – mf#31751 – us UMI ProQuest [380]

Journal of consumer credit management – St. Louis. 1969-1980 (1) 1975-1980 (5) 1975-1980 (9) – ISSN: 0022-0086 – mf#9662 – us UMI ProQuest [332]

Journal of consumer health on the internet / ed by Wood, Sandra – ISSN: 1539-8285 – us Haworth [360]

Journal of consumer marketing – Santa Barbara. 1987-1995 (1) 1987-1995 (5) 1987-1995 (9) – ISSN: 0736-3761 – mf#15808 – us UMI ProQuest [650]

Journal of consumer policy – Dordrecht. 1983+ (1,5,9) – ISSN: 0168-7034 – mf#14756,01 – us UMI ProQuest [380]

Journal of consumer product flammability – Westport. 1976-1982 (1) 1976-1982 (5) 1976-1982 (9) – ISSN: 0362-1677 – mf#10457,01 – us UMI ProQuest [360]

Journal of consumer research – Gainesville. 1974+ (1,5,9) – ISSN: 0093-5301 – mf#12972 – us UMI ProQuest [650]

Journal of consumer studies and home economics – Oxford. 1980-1996 (1,5,9) – ISSN: 0309-3891 – mf#15558 – us UMI ProQuest [380]

Journal of contaminant hydrology – Amsterdam. 1989+ (1,5,9) – ISSN: 0169-7722 – mf#42481 – us UMI ProQuest [333]

Journal of contemporary asia – Manila. 1970+ (1) 1974+ (5) 1976+ (9) – ISSN: 0047-2336 – mf#10107 – us UMI ProQuest [338]

Journal of contemporary business – Seattle. 1972-1982 (1) 1975-1982 (5) 1975-1982 (9) – ISSN: 0194-0430 – mf#10542 – us UMI ProQuest [338]

Journal of contemporary china – Princeton, 1998+ [1,5,9] – ISSN: 1067-0564 – mf#22079 – us UMI ProQuest [951]

Journal of contemporary criminal justice – Thousand Oaks. 1991+ (1,5,9) – ISSN: 1043-9862 – mf#19459 – us UMI ProQuest [360]

Journal of contemporary ethnography – Thousand Oaks. 1987+ (1,5,9) – (cont: urban life) – ISSN: 0891-2416 – mf#12645,02 – us UMI ProQuest [301]

Journal of contemporary health law and policy – Catholic University of America, D.C. v1-17. 1985-2001 – 9 – $252.00 set – ISSN: 0882-1046 – mf#109801 – us Hein [344]

Journal of contemporary history – London. 1966+ (1) 1982+ (5) 1982+ (9) – ISSN: 0022-0094 – mf#2770 – us UMI ProQuest [934]

Journal of contemporary law – Salt Lake City. 1974-1998 (1) 1974-1998 (5) 1974-1998 (9) – ISSN: 0097-9937 – mf#10541 – us UMI ProQuest [340]

Journal of contemporary legal issues – University of San Diego. v1-11. 1987-2000 – 9 – $188.00 set – ISSN: 0896-5595 – mf#111681 – us Hein [340]

Journal of contemporary psychotherapy – Forest Hills. 1968+ (1) 1972+ (5) 1975+ (9) – ISSN: 0022-0116 – mf#7086 – us UMI ProQuest [150]

Journal of contemporary revolutions – San Francisco. 1968-1975 (1) 1971-1975 (5) (9) – ISSN: 0449-4741 – mf#6694 – us UMI ProQuest [320]

Journal of contemporary studies – San Francisco. 1981-1985 (1) 1981-1985 (5) 1981-1985 (9) – (cont: taxing and spending) – ISSN: 0272-7595 – mf#12228,01 – us UMI ProQuest [336]

Journal of continuing education in family medicine – Northfield. 1977-1978 (1) 1977-1978 (5) 1977-1978 (9) – (cont: medical digest: the journal of significant medical literature) – ISSN: 0149-0273 – mf#10286,01 – us UMI ProQuest [610]

Journal of continuing education in nursing – Thorofare. 1970+ (1) 1970+ (5) 1976+ (9) – ISSN: 0022-0124 – mf#6037 – us UMI ProQuest [374]

Journal of continuing education in orl and allergy – Northfield. 1977-1978 (1) 1977-1978 (5) 1977-1978 (9) – (cont: orl digest) – ISSN: 0148-5180 – mf#10287,01 – us UMI ProQuest [610]

Journal of continuing education in orthopedics – Northfield. 1978-1979 (1) 1978-1979 (5) 1978-1979 (9) – (cont: orthopedics digest) – ISSN: 0160-7707 – mf#10290,01 – us UMI ProQuest [617]

Journal of continuing education in pediatrics – Northfield. 1978-1978 (1) 1978-1978 (5) 1978-1978 (9) – (cont: pediatrics digest) – ISSN: 0160-7766 – mf#10291,01 – us UMI ProQuest [618]

Journal of continuing education in psychiatry – Northfield. 1977-1979 (1) 1977-1979 (5) 1977-1979 (9) – (cont: psychiatry digest. cont by: psychiatry digest) – ISSN: 0149-0265 – mf#2306,01 – us UMI ProQuest [616]

Journal of continuing education in the health professions – Birmingham. 1988+ – 1,5,9 – (cont: mobius) – ISSN: 0894-1912 – mf#12658,01 – us UMI ProQuest [370]
Journal of continuing education in urology – Northfield. 1977-1979 (1) 1977-1979 (5) 1977-1979 (9) – (cont: urology digest. cont by: urology digest) – ISSN: 0148-5172 – mf#10292,01 – us UMI ProQuest [616]
Journal of controlled release : official journal of the controlled release society – Amsterdam. 1984+ (1,5,9) – ISSN: 0168-3659 – mf#42420 – us UMI ProQuest [540]
Journal of convention and event tourism / ed by Abbott, Je'Anna Lanza – ISSN: 1547-0148 – us Haworth [338]
Journal of convention and exhibition management / ed by Abbott, Je'Anna Lanza – v1 n1. 1997- – 1,9 – $95.00 in US $133.00 outside hardcopy subsc – us Haworth [650]
Journal of cooperative education – Columbia. 1979+ – 1,5,9 – ISSN: 0022-0132 – mf#12299 – us UMI ProQuest [370]
Journal of corporate accounting and finance – New York. 1989+ (1,5,9) – ISSN: 1044-8136 – mf#17460 – us UMI ProQuest [650]
Journal of corporate accounting and finance – v1-4. 1989-93 – 9 – $115.00 set – ISSN: 1044-8136 – mf#112261 – us Hein [650]
Journal of corporate finance – Amsterdam. 1994+ (1,5,9) – ISSN: 0929-1199 – mf#42758 – us UMI ProQuest [332]
Journal of corporate real estate – London, 1998+ [1,5,9] – ISSN: 1463-001X – mf#31705 – us UMI ProQuest [333]
Journal of corporate taxation – New York. 1974-2000 (1) 1974-2000 (5) 1974-2000 (9) – ISSN: 0094-0593 – mf#10073 – us UMI ProQuest [336]
Journal of corporation law – University of Iowa. v1-26. 1976-2001 – 9 – $633.00 set – ISSN: 0360-795X – mf#103921 – us Hein [346]
Journal of correctional education – Glen Mills. 1949+ (1) 1971+ (5) 1977+ (9) – ISSN: 0740-2708 – mf#3366 – us UMI ProQuest [360]
Journal of cost management – Boston. 1992+ (1,5,9) – ISSN: 0899-5141 – mf#19061,01 – us UMI ProQuest [650]
Journal of counseling and development – jcd – Alexandria. 1984+ (1) 1984+ (5) 1984+ (9) – (cont: personnel and guidance journal) – ISSN: 0748-9633 – mf#206,01 – us UMI ProQuest [331]
Journal of counseling psychology – Washington. 1954+ (1) 1968+ (5) 1970+ (9) – ISSN: 0022-0167 – mf#2706 – us UMI ProQuest [150]
Journal of couple and relationship therapy / ed by Wetchler, Joseph L – ISSN: 1533-2691 – us Haworth [360]
Journal of couples therapy / ed by Brothers, Barbara Jo – v1- 1989- – 1,9 ($125.00 in US $175.00 outside hardcopy subsc) – us Haworth [306]
Journal of cranio-maxillo-facial surgery – Stuttgart. 1987-1990 (1) 1987-1990 (5) 1987-1990 (9) – (cont: journal of maxillofacial surgery) – ISSN: 1010-5182 – mf#10164,01 – us UMI ProQuest [617]
Journal of creative behavior – Buffalo. 1967+ (1) 1971+ (5) 1975+ (9) – ISSN: 0022-0175 – mf#2227 – us UMI ProQuest [370]
Journal of creativity in mental health / ed by Duffey, Thelma & Garcia, John L – us Haworth [150]
Journal of crime and justice – Cincinnati. 1985+ (1,5,9) – ISSN: 0735-648X – mf#15309 – us UMI ProQuest [360]
Journal of criminal justice – New York. 1973+ (1,5,9) – ISSN: 0047-2352 – mf#49114 – us UMI ProQuest [360]
Journal of criminal justice education – Academy of Criminal Justice Sciences. v1- – 9 – (filming in process) – ISSN: 1051-1253 – mf#119121 – us Hein [345]
Journal of criminal justice education (jcje) – Highland Heights. 1999+ (1,5,9) – ISSN: 1051-1253 – mf#21257 – us UMI ProQuest [345]
Journal of criminal law and criminology – Chicago. 1989+ (1,5,9) – ISSN: 0091-4169 – mf#16094,03 – us UMI ProQuest [360]
Journal of criminal law and criminology – v1-90. 1910-2000 – 1,5,6 – $2534.00 – (v1-86 1910-96 on reel $2375. v87-90 1996-2000 on mf $159. title varies: v1-31 1910-41 as journal of the american institute of criminal law and criminology. v42-63 1951-72 as journal of criminal law, criminology and police science) – ISSN: 0091-4169 – mf#103941 – us Hein [345]
Journal of criminal law, criminology and police science – American Institute of Criminal Law and Criminology. v1-16. 1910-25/26 – 114mf – 9 – $171.00 – (add vols as copyright expires) – mf#LLMC 95-107 – us LLMC [345]
Journal of criminal law, criminology and police science *see* Journal of criminal law and criminology
Journal of criminal law (london) – v1-64. 1937-2000 – 9 – $1545.00 set – ISSN: 0022-0183 – mf#110421 – us Hein [345]
Journal of criminal profiling – ISSN: 1540-3653 – us Haworth [364]
Journal of critical analysis – Port Jefferson. 1969-1992 (1) 1970-1992 (5) 1975-1992 (9) – ISSN: 0022-0213 – mf#5028 – us UMI ProQuest [370]
Journal of crop improvement / ed by Kang, Manjit S – ISSN: 1542-7528 – us Haworth [635]
Journal of crop production : innovations in practice, theory and research / ed by Basra, Amarjit S – v1 n1. 1997- – 1,9 – $95.00 in US $133.00 outside hardcopy subsc – us Haworth [630]
Journal of cross-cultural gerontology – Dordrecht. 1986-1996 (1) 1986-1996 (5) 1986-1996 (9) – ISSN: 0169-3816 – mf#15259 – us UMI ProQuest [618]
Journal of cross-cultural psychology – Thousand Oaks. 1970+ (1,5,9) – ISSN: 0022-0221 – mf#11974 – us UMI ProQuest [303]
Journal of cryptology – New York. 1988-1995 (1) 1988-1989 (5) 1988-1989 (9) – ISSN: 0933-2790 – mf#17003 – us UMI ProQuest [510]
Journal of crystal and molecular structure – New York. 1971-1977 (1) 1974-1977 (5) – ISSN: 0308-4086 – mf#10853 – us UMI ProQuest [540]
Journal of crystal growth – Amsterdam. 1967+ (1) 1967+ (5) 1987+ (9) – ISSN: 0022-0248 – mf#42211 – us UMI ProQuest [530]
Journal of culinary science and technology / ed by Hegarty, Joseph A & Antun, John M – ISSN: 1542-8052 – us Haworth [660]
Journal of cultural diversity – Lisle. 1998+ (1,5,9) – ISSN: 1071-5568 – mf#26785 – us UMI ProQuest [610]
Journal of cuneiform studies – Cambridge. 1947+ (1) 1971+ (5) 1974+ (9) – ISSN: 0022-0256 – mf#3128 – us UMI ProQuest [930]
Journal of current social issues – New York. 1971-1980 (1) 1962-1980 (5) 1975-1980 (9) – ISSN: 0041-7211 – mf#6380 – us UMI ProQuest [300]
Journal of curriculum and supervision – Alexandria. 1985+ – 1,5,9 – ISSN: 0882-1232 – mf#14843 – us UMI ProQuest [370]
Journal of curriculum studies – London. 1989+ – 1,5,9 – ISSN: 0022-0272 – mf#17303 – us UMI ProQuest [370]
Journal of customer service in marketing and management : innovations of service, quality and value / ed by Winston, William J – v2 n1. 1996- – 1,9 – $125.00 in US $175.00 outside hardcopy subsc – us Haworth [650]
Journal of cybernetics – Washington. 1976-1980 (1,5,9) – (cont by: cybernetics and systems) – ISSN: 0022-0280 – mf#11137 – us UMI ProQuest [621]
Journal of dairy research – Cambridge. 1929+ (1) 1971+ (5) 1976+ (9) – ISSN: 0022-0299 – mf#1364 – us UMI ProQuest [630]
Journal of dairy science – Champaign. 1917+ (1) 1972+ (5) 1977+ (9) – ISSN: 0022-0302 – mf#6341 – us UMI ProQuest [630]
Journal of data education – Stillwater. 1961-1985 (1) 1972-1985 (5) 1976-1985 (9) – (cont by: journal of computer information systems) – ISSN: 0022-0310 – mf#6671 – us UMI ProQuest [000]
Journal of database management – Harrisburg. 1993-1997 (1,5,9) – ISSN: 1063-8016 – mf#19628,01 – us UMI ProQuest [000]
Journal of dental education – Washington. 1949+ (1) 1983+ (5) 1983+ (9) – ISSN: 0022-0337 – mf#691 – us UMI ProQuest [617]
Journal of dental hygiene – Chicago. 1988+ (1) 1988+ (5) 1988+ (9) – (cont: dental hygiene) – ISSN: 1043-254X – mf#2274,02 – us UMI ProQuest [617]
Journal of dental research – Houston. 1919+ (1) 1975+ (5) 1975+ (9) – ISSN: 0022-0345 – mf#10791 – us UMI ProQuest [617]
Journal of dentistry – Kidlington. 1972+ (1) 1973+ (5) 1973+ (9) – ISSN: 0300-5712 – mf#7021 – us UMI ProQuest [617]
Journal of dentistry for children – Chicago. 1933+ (1) 1969+ (5) 1975+ (9) – ISSN: 0022-0353 – mf#3495 – us UMI ProQuest [617]
Journal of dermatological science – Amsterdam. 1990-1992 (1,5,9) – ISSN: 0923-1811 – mf#42615 – us UMI ProQuest [616]
Journal of design and manufactures, 1849-1862 / ed by Cole, Henry – 6v on 19mf – 9 – $140.00 – 0-907006-63-9 – (essential resource for study of early industrial design) – uk Mindata [740]
Journal of design history – Oxford. 1988-1996 (1,5,9) – ISSN: 0952-4649 – mf#17503 – us UMI ProQuest [740]
Journal of development economics – Amsterdam. 1974+ (1,5) 1987+ (9) – ISSN: 0304-3878 – mf#42217 – us UMI ProQuest [337]
Journal of development studies – London. 1989+ (1,5,9) – ISSN: 0022-0388 – mf#15810 – us UMI ProQuest [320]
Journal of developmental and behavioral pediatrics – v2-17. 1981-96 – 1,5,6,9 – $80.00r – us Lippincott [618]
Journal of developmental and physical disabilities – New York. 1991+ – 1,5,9 – (cont: journal of the multihandicapped person) – ISSN: 1056-263X – mf#17690,01 – us UMI ProQuest [370]
Journal of developmental and remedial education – Boone. 1978-1984 – 1,5,9 – (cont by: journal of developmental education) – ISSN: 0738-9701 – mf#11985 – us UMI ProQuest [370]
Journal of developmental education – Boone. 1984+ – 1,5,9 – (cont: journal of developmental and remedial education) – ISSN: 0894-3907 – mf#11985,01 – us UMI ProQuest [370]
Journal of dharma – Bangalore. 1986+ (1,5,9) – ISSN: 0253-7222 – mf#15954 – us UMI ProQuest [280]
Journal of diagnostic medical sonography – PHiladelphia. 1985+ (1,5,9) – ISSN: 8756-4793 – mf#14442 – us UMI ProQuest [616]
Journal of digital and electronic acquisitions / ed by Miko, Chris J – ISSN: 1540-7284 – us Haworth [070]
Journal of direct marketing – New York. 1987-1997 (1,5,9) – (cont by: journal of interactive marketing) – ISSN: 0892-0591 – mf#18111 – us UMI ProQuest [650]
Journal of dispersion science and technology – New York. 1993-1996 (1,5,9) – ISSN: 0193-2691 – mf#14542 – us UMI ProQuest [540]
Journal of dispute resolution – University of Missouri-Columbia, 1984-2001 – 9 – $324.00 set – (title varies: 1984-87 as missouri journal of dispute resolution) – ISSN: 1052-2859 – mf#109441 – us Hein [340]
Journal of district of columbia bar association of the district of columbia *see* Journal of the bar association of the district of columbia
Journal of divorce and remarriage / ed by Everett, Craig A – v1- 1977- – 1,9 ($325.00 in US $455.00 outside hardcopy subsc) – us Haworth [306]
Journal of documentary reproduction – v. 1-5. 1938-42 – 1 – us L of C Photodup [770]
Journal of drug issues – Tallahassee. 1971+ (1) 1971+ (5) 1976+ (9) – ISSN: 0022-0426 – mf#7394 – us UMI ProQuest [360]
Journal of dual diagnosis / ed by Buckley, Peter F – us Haworth [616]
Journal of dynamic systems, measurement, and control – New York. 1971+ (1) 1971+ (5) 1976+ (9) – ISSN: 0022-0434 – mf#7239 – us UMI ProQuest [620]
Journal of early adolescence – Tucson. 1981+ (1,5,9) – ISSN: 0272-4316 – mf#13514 – us UMI ProQuest [150]
Journal of east asian linguistics – Dordrecht. 1992-1996 (1,5,9) – ISSN: 0925-8558 – mf#18658 – us UMI ProQuest [480]
Journal of east-west business / ed by Kaynak, Erdener – v2 n1. 1996 – 1,9 – $95.00 in US $133.00 outside hardcopy subsc – us Haworth [650]
Journal of ecclesiastical history – Cambridge. 1950+ (1) 1950+ (5) 1950+ (9) – ISSN: 0022-0469 – mf#2849 – us UMI ProQuest [240]
Journal of ecology / British Ecological Society – Cambridge 1913-1945/46 – v1-33 on 73mf – 9 – mf#5292c/2 – ne IDC [574]
Journal of ecology – Oxford. 1980+ (1,5,9) – ISSN: 0022-0477 – mf#15531 – us UMI ProQuest [574]
Journal of econometrics – Amsterdam. 1973+ (1) 1973+ (5) 1987+ (9) – ISSN: 0304-4076 – mf#42223 – us UMI ProQuest [330]
Journal of economic behavior and organization – Amsterdam. 1980+ (1) 1980+ (5) 1987+ (9) – ISSN: 0167-2681 – mf#42275 – us UMI ProQuest [330]
Journal of economic dynamics and control – Amsterdam. 1979+ (1) 1979+ (5) 1987+ (9) – ISSN: 0165-1889 – mf#42224 – us UMI ProQuest [330]
Journal of economic education – Washington. 1969+ (1) 1976+ (5) 1976+ (9) – ISSN: 0022-0485 – mf#11098 – us UMI ProQuest [330]
Journal of economic entomology – Lanham. 1909+ [1]; 1969+ [5]; 1976+ [9] – ISSN: 0022-0493 – mf#1450 – us UMI ProQuest [630]
Journal of economic history – Atlanta. 1941+ (1) 1969+ (5) 1975+ (9) – ISSN: 0022-0507 – mf#1071 – us UMI ProQuest [330]
Journal of economic issues – Lincoln. 1967+ (1) 1973+ (5) 1975+ (9) – ISSN: 0021-3624 – mf#9628 – us UMI ProQuest [330]
Journal of economic literature – Nashville. 1963+ (1) 1970+ (5) 1975+ (9) – ISSN: 0022-0515 – mf#1610 – us UMI ProQuest [330]
Journal of economic perspectives – Nashville. 1987+ (1,5,9) – ISSN: 0895-3309 – mf#16356 – us UMI ProQuest [338]
Journal of economic psychology – Amsterdam. 1981+ (1) 1981+ (5) 1987+ (9) – ISSN: 0167-4870 – mf#42276 – us UMI ProQuest [150]
Journal of economic studies – Glasgow. 1992-1995 (1,5,9) – ISSN: 0144-3585 – mf#15811,02 – us UMI ProQuest [330]
Journal of economic surveys – Avon. 1991-1995 (1,5,9) – ISSN: 0950-0804 – mf#17394 – us UMI ProQuest [330]
Journal of economics – Wien. 1992-1992 (1) 1992-1992 (5) 1992-1992 (9) – ISSN: 0931-8658 – mf#13285,01 – us UMI ProQuest [330]
Journal of economics and business / Temple University School of Business Administration – New York. 1949+ (1) 1972+ (5) 1975+ (9) – ISSN: 0148-6195 – mf#6363 – us UMI ProQuest [650]
Journal of economics and management strategy – Cambridge. 1992-1996 (1,5,9) – ISSN: 1058-6407 – mf#19112 – us UMI ProQuest [650]
Journal of ecumenical studies – v1-37. 1964-2000 – 5,6,9 – $672.00 set – (v1-21 1961-84 on reel $260. v22-37 1985-2000 on mf $412) – ISSN: 0022-0558 – mf#103961 – us Hein [340]
Journal of education – Boston. 1875+ (1) 1968+ (5) 1975+ (9) – ISSN: 0022-0574 – mf#913 – us UMI ProQuest [370]
Journal of education – Brooklyn. 1875-1876 – 1 – mf#4787 – us UMI ProQuest [370]
Journal of education – Detroit. 1838-1840 – 1 – mf#4797 – us UMI ProQuest [370]
Journal of education : devoted to education, literature, science and the arts – Montreal: Dept of Public Instruction, [1867-1879?] – 9 – (cont: the journal of education for lower canada) – mf#P04300 – cn Canadiana [370]
Journal of education : nova scotia – [Halifax, NS?: s.n. between 1881 and 1887-19–] – 9 – (cont: the journal of education for the province of nova scotia) – mf#P05112 – cn Canadiana [370]
The journal of education and agriculture for the province of nova scotia – Halifax [N.S.]: A & W Mackinlay, [1858-1860] – 9 – mf#P06004 – cn Canadiana [370]
Journal of education finance – Western Kentucky University. v1-26. 1975-2001 – 5,6,9 – $558.00 set – (v1-10 1975-85 on reel $121. v11-26 1986-2001 on mf $437) – ISSN: 0098-9495 – mf#103971 – us Hein [370]
Journal of education for business – Washington. 1985+ (1) 1985+ (5) 1985+ (9) – ISSN: 0883-2323 – mf#866,01 – us UMI ProQuest [338]
Journal of education for librarianship – State College. 1960-1984 (1) 1970-1984 (5) 1975-1984 (9) – (cont by: journal of education for library and information science) – ISSN: 0022-0604 – mf#5907 – us UMI ProQuest [020]
Journal of education for library and information science – State College. 1984+ (1) 1984+ (5) 1984+ (9) – (cont: journal of education for librarianship) – ISSN: 0748-5786 – mf#5907,01 – us UMI ProQuest [020]
Journal of education for lower canada – Montreal: Dept of Education, [1857-1867] – 9 – (cont by: journal of education. incl ind) – mf#P05086 – cn Canadiana [370]
Journal of education for nova scotia – [S.1: s.n, 1851-1853] – 9 – mf#P05110 – cn Canadiana [370]
Journal of education for ontario – Toronto, ON. 1848-77 – 5r – 1 – cn Library Assoc [370]
Journal of education for social work – New York. 1965-1984 (1) 1975-1984 (5) 1975-1984 (9) – (cont by: journal of social work education) – ISSN: 0022-0612 – mf#9611 – us UMI ProQuest [360]
Journal of education for students placed at risk – Mahwah, 1998+ – 1,5,9 – ISSN: 1082-4669 – mf#25222 – us UMI ProQuest [370]
Journal of education for teaching : jet – Abingdon. 1983-1996 – 1,5,9 – ISSN: 0260-7476 – mf#13404,01 – us UMI ProQuest [370]
Journal of education for the province of nova scotia – [Halifax, NS?: s.n, 1866-between 1881 and 1887] – 9 – (cont by: journal of education, nova scotia) – mf#P05111 – cn Canadiana [370]
Journal of education policy – London. 1991-1996 – 1 – ISSN: 0268-0939 – mf#17304 – us UMI ProQuest [370]
Journal of educational administration – Armidale. 1963-1995 (1) 1975-1995 (5) 1976-1995 (9) – ISSN: 0957-8234 – mf#6874 – us UMI ProQuest [370]
Journal of educational administration and history – Leeds. 1976+ – 1,5,9 – ISSN: 0022-0620 – mf#11099 – us UMI ProQuest [370]
Journal of educational and behavioral statistics – Washington. 1994+ – 1,5,9 – (cont: journal of educational statistics) – ISSN: 1076-9986 – mf#11442,01 – us UMI ProQuest [310]

Journal of educational communication – Camp Hill. 1975-1983 – 1,5,9 – (cont by: journal of educational public relations) – ISSN: 0745-4058 – mf#10942 – us UMI ProQuest [370]

Journal of educational data processing – Soquel. 1972-1979 (1) 1964-1979 (5) 1975-1979 (9) – ISSN: 0022-0647 – mf#6376 – us UMI ProQuest [370]

Journal of educational equity and leadership – Thousand Oaks. 1980-1987 (1) 1980-1987 (5) 1980-1987 (9) – ISSN: 0275-4347 – mf#12827 – us UMI ProQuest [370]

Journal of educational psychology – Washington. 1910+ [1]; 1967+ [5]; 1970+ [9] – ISSN: 0022-0663 – mf#1155 – us UMI ProQuest [150]

The journal of educational psychology – v. 1-47. 1910-56 – 1 – us AMS Press [150]

Journal of educational public relations – Camp Hill. 1984-1995 – 1,5,9 – (cont: journal of educational communication. cont by: journal of educational relations) – ISSN: 0741-3653 – mf#10942,01 – us UMI ProQuest [370]

Journal of educational relations – Camp Hill. 1995-2000 – 1,5,9 – (cont: journal of educational public relations) – ISSN: 1084-726X – mf#10942,02 – us UMI ProQuest [370]

Journal of educational research – Bloomington. 1920+ [1]; 1966+ [5]; 1975+ [9] – ISSN: 0022-0671 – mf#385 – us UMI ProQuest [370]

Journal of educational statistics – Washington. 1976-1994 (1) 1976-1994 (5) 1976-1994 (9) – (cont by: journal of educational and behavioral statistics) – ISSN: 0362-9791 – mf#11442 – us UMI ProQuest [310]

Journal of educational techniques and technologies – Athens. 1989-1989 – 1,5,9 – (cont by: iall journal of language learning technologies) – ISSN: 0891-2521 – mf#12664,03 – us UMI ProQuest [370]

Journal of educational thought (jet) – Calgary. 1967+ (1) 1971+ (5) 1976+ (9) – ISSN: 0022-0701 – mf#3367 – us UMI ProQuest [370]

Journal of e-government / ed by Curtin, Gregory G – ISSN: 1542-4049 – us Haworth [380]

Journal of egyptian archaeology – London, 1914-1923. v1-9 – 54mf – 9 – mf#NE-355 – ne IDC [930]

The journal of egyptian archaeology – v1-. 1914-. London: Egypt Exploration Society. semiannual, 1920-; quarterly, 1914-19; annual, 1976- – 1 – (bibliographies included) – us Wisconsin U Libr [930]

Journal of elasticity – Groningen. 1989-1996 (1,5,9) – ISSN: 0374-3535 – mf#16807 – us UMI ProQuest [620]

Journal of elastomers and plastics – Lancaster. 1974+ (1) 1974+ (5) 1976+ (9) – (cont: journal of elastoplastics) – ISSN: 0095-2443 – mf#6096,01 – us UMI ProQuest [660]

Journal of elastoplastics – Westport. 1969-1973 (1) 1969-1973 (5) (9) – (cont by: journal of elastomers and plastics) – ISSN: 0022-071X – mf#6096 – us UMI ProQuest [660]

Journal of elder abuse and neglect / ed by Wolf, Rosalie S & Aderson, Susan McMurray – v1- 1989- – 1, 9 ($225.00 in US $315.00 outside hardcopy subsc) – us Haworth [360]

Journal of electrical / Washington DC – jan 1926-dec 1930 [mthly] – 3r – 1 – mf#B10895-10897 – us Ohio Hist [331]

Journal of electroanalytical chemistry – Lausanne. 1992+ (1,5,9) – (cont: journal of electroanalytical chemistry and interfacial electrochemistry) – mf#42277,01 – us UMI ProQuest [540]

Journal of electroanalytical chemistry and interfacial electrochemistry – Amsterdam. 1959-1991 (1) 1959-1991 (5) 1987-1991 (9) – (cont by: journal of electroanalytical chemistry) – ISSN: 0022-0728 – mf#42277 – us UMI ProQuest [540]

Journal of electrocardiology – New York. 1968+ (1) 1972+ (5) 1975+ (9) – ISSN: 0022-0736 – mf#6372 – us UMI ProQuest [616]

Journal of electromyography and kinesiology – Oxford. 1993+ (1,5,9) – ISSN: 1050-6411 – mf#18717 – us UMI ProQuest [612]

Journal of electron spectroscopy and related phenomena – Amsterdam. 1972+ (1) 1972+ (5) 1987+ (9) – ISSN: 0368-2048 – mf#42278 – us UMI ProQuest [530]

Journal of electronic materials – Warrendale. 1972-1995 (1) 1972-1990 (5) 1972-1990 (9) – ISSN: 0361-5235 – mf#10870 – us UMI ProQuest [620]

Journal of electronic packaging – New York. 1989+ (1,5,9) – ISSN: 1043-7398 – mf#17274 – us UMI ProQuest [680]

Journal of electronic resources in law libraries / ed by Price, Jeanne Frazier – ISSN: 1545-0422 – us Haworth [020]

Journal of electronic resources in medical libraries / ed by Wood, M. Sandra – ISSN: 1542-4065 – us Haworth [020]

Journal of electrophysiological techniques – Elmsford. 1972-1987 (1,5,9) – ISSN: 0361-0209 – mf#49425 – us UMI ProQuest [612]

Journal of electrostatics – Amsterdam. 1975+ (1) 1975+ (5) 1987+ (9) – ISSN: 0304-3886 – mf#42279 – us UMI ProQuest [621]

Journal of embryology and experimental morphology – Cambridge. 1984-1986 (1,5,9) – (cont by: development) – ISSN: 0022-0752 – mf#13598 – us UMI ProQuest [612]

Journal of emergency medicine – New York. 1984+ (1,5,9) – ISSN: 0736-4679 – mf#49459 – us UMI ProQuest [617]

Journal of emergency nursing – St. Louis. 1975+ (1,5,9) – ISSN: 0099-1767 – mf#11804 – us UMI ProQuest [610]

Journal of emotional abuse : interventions, research, and theories of psychological maltreatment, trauma, and nonphysical aggression / ed by Geffner, Robert A & Rossman, B B Robbie – v1 n1. 1997- – 1,9 – $85.00 in US $119.00 outside hardcopy subsc – us Haworth [150]

Journal of emotional and behavioral disorders – Austin. 1993+ (1,5,9) – ISSN: 1063-4266 – mf#19802 – us UMI ProQuest [150]

Journal of employment counseling – Alexandria. 1964+ (1) 1971+ (5) 1975+ (9) – ISSN: 0022-0787 – mf#3275 – us UMI ProQuest [331]

Journal of end user computing – Harrisburg. 1993-1997 (1,5,9) – ISSN: 1063-2239 – mf#19629,01 – us UMI ProQuest [000]

Journal of endodontics – v9-22. 1983-96 – 1,5,6,9 – $80.00r – us Lippincott [617]

Journal of energy – New York. 1981-1983 (1,5,9) – ISSN: 0146-0412 – mf#13072 – us UMI ProQuest [333]

Journal of energy and development – University of Colorado. v1-25. 1975-2000 – 9 – $417.00 set – ISSN: 0361-4476 – mf#104011 – us Hein [333]

Journal of energy and natural resources law – v1-16. 1983-98 – 9 – $808.00 set – ISSN: 0264-6811 – mf#110101 – us Hein [340]

Journal of energy engineering – New York. 1983+ (1) 1983+ (5) 1983+ (9) – (cont: journal of the energy division) – ISSN: 0733-9402 – mf#8149,02 – us UMI ProQuest [624]

Journal of energy resources technology – New York. 1979+ (1,5,9) – ISSN: 0195-0738 – mf#11948 – us UMI ProQuest [621]

Journal of engineering and applied sciences – Oxford. 1981-1986 (1,5,9) – ISSN: 0191-9539 – mf#49314 – us UMI ProQuest [620]

Journal of engineering and technology management: jet-m – Amsterdam. 1989+ (1,5,9) – (cont: engineering management international) – ISSN: 0923-4748 – mf#42586,01 – us UMI ProQuest [650]

Journal of engineering computing and applications – Boston. 1989-1989 (1) – ISSN: 0887-9796 – mf#16342 – us UMI ProQuest [621]

Journal of engineering education – Washington. 1993+ (1) 1993+ (5) 1993+ (9) – ISSN: 1069-4730 – mf#591,01 – us UMI ProQuest [378]

Journal of engineering for gas turbines and power – New York. 1984+ (1) 1984+ (5) 1984+ (9) – (cont: journal of engineering for power) – ISSN: 0742-4795 – mf#1189,01 – us UMI ProQuest [621]

Journal of engineering for industry – New York. 1959-1996 (1) 1964-1996 (5) 1970-1996 (9) – (cont by: journal of manufacturing science and engineering) – ISSN: 0022-0817 – mf#1190 – us UMI ProQuest [621]

Journal of engineering for power – New York. 1959-1983 (1) 1965-1983 (5) 1970-1983 (9) – (cont by: journal of engineering for gas turbines and power) – ISSN: 0022-0825 – mf#1189 – us UMI ProQuest [621]

Journal of engineering materials and technology – New York. 1973+ (1) 1973+ (5) 1976+ (9) – ISSN: 0094-4289 – mf#7240 – us UMI ProQuest [621]

Journal of engineering mathematics – Alphen aan den Rijn. 1989-1995 (1,5,9) – ISSN: 0022-0833 – mf#16808 – us UMI ProQuest [620]

Journal of engineering mechanics – New York. 1983+ (1,5,9) – (cont: journal of the engineering mechanics division) – ISSN: 0733-9399 – mf#8142,01 – us UMI ProQuest [624]

Journal of engineering physics – New York. 1965-1977 (1) 1965-1977 (5) – ISSN: 0022-0841 – mf#10909 – us UMI ProQuest [530]

Journal of engineering psychology – Ventnor. 1962-1966 (1) – ISSN: 0022-085X – mf#1832 – us UMI ProQuest [150]

Journal of engineering sciences – Riyadh. 1984-1988 (1,5,9) – ISSN: 0377-9254 – mf#14811 – us UMI ProQuest [620]

Journal of engineering tribology *see* Proceedings of the institution of mechanical engineers pt j

Journal of english and germanic philology – Bloomington. v1-45. 1897-1946 – 502mf – 8 – mf#137c – ne IDC [410]

Journal of enterostomal therapy – St. Louis. 1982-1991 (1,5,9) – (cont by: journal of et nursing) – ISSN: 0270-1170 – mf#13045,01 – us UMI ProQuest [617]

Journal of enterprise management – Oxford. 1978-1981 (1) 1978-1981 (5) (9) – ISSN: 0146-6372 – mf#49315 – us UMI ProQuest [650]

Journal of entomology : descriptive and geographical – London 1862-66 – v1-2 on 20mf – 9 – mf#z-969/2 – ne IDC [590]

Journal of environmental education – Madison. 1969+ [1]; 1971+ [5]; 1976+ [9] – ISSN: 0095-8964 – mf#5909 – us UMI ProQuest [333]

Journal of environmental engineering – New York. 1983+ (1,5,9) – (cont: journal of the environmental engineering division) – ISSN: 0733-9372 – mf#8151,01 – us UMI ProQuest [628]

Journal of environmental engineering and science – Ottawa. 2002+ (1,5,9) – mf#31628 – us UMI ProQuest [628]

Journal of environmental health – Denver. 1971+ (1) 1938+ (5) 1975+ (9) – ISSN: 0022-0892 – mf#6087 – us UMI ProQuest [333]

Journal of environmental law and litigation – University of Oregon. v1-15. 1986-2000 – 9 – $166.00 set – ISSN: 1049-0280 – mf#111231 – us Hein [344]

Journal of environmental quality – Madison. 1989+ (1,5,9) – ISSN: 0047-2425 – mf#17535 – us UMI ProQuest [333]

Journal of environmental radioactivity – London. 1990-1990 (1,5,9) – ISSN: 0265-931X – mf#42475 – us UMI ProQuest [530]

Journal of environmental regulation – New York. 1991-1994 (1,5,9) – ISSN: 1055-758X – mf#19151 – us UMI ProQuest [333]

Journal of environmental sciences – Los Angeles. 1959-1989 (1) 1971-1989 (5) 1976-1989 (9) – (cont by: journal of the institute of environmental sciences) – ISSN: 0022-0906 – mf#3170 – us UMI ProQuest [628]

Journal of environmental studies and policy – New Delhi. 1998+ (1) – mf#28253 – us UMI ProQuest [333]

Journal of epidemiology and community health – London. 1978+ (1) 1978+ (5) 1978+ (9) – ISSN: 0143-005X – mf#1334,01 – us UMI ProQuest [610]

Journal of epsilon pi tau – Bowling Green. 1978-1992 – 1,5,9 – (cont by: journal of technology studies) – ISSN: 0887-9532 – mf#11408 – us UMI ProQuest [370]

Journal of esthetic and restorative dentistry – Hamilton, 2001+ [1,5,9] – ISSN: 1496-4155 – mf#21651,01 – us UMI ProQuest [617]

Journal of et nursing – St. Louis. 1991-1993 (1,5,9) – (cont: journal of enterostomal therapy. cont by: journal of wound, ostomy, and continence nursing: wocn) – ISSN: 1055-3045 – mf#13045,02 – us UMI ProQuest [617]

Journal of ethnic and cultural diversity in social work : innovations in theory, research and practice / ed by Anda, Diane de – (former title: journal of multicultural social work) – ISSN: 1531-3204 – us Haworth [360]

Journal of ethnic studies – Bellingham. 1973-1991 (1) 1973-1991 (5) 1973-1991 (9) – ISSN: 0091-3219 – mf#12897 – us UMI ProQuest [305]

Journal of ethnicity in criminal justice / ed by Joseph, Janice – ISSN: 1537-7938 – us Haworth [364]

Journal of ethnicity in substance abuse / ed by Myers, Peter L – ISSN: 1533-2640 – us Haworth [360]

Journal of ethnopharmacology – Lausanne. 1979+ (1) 1979+ (5) 1987+ (9) – ISSN: 0378-8741 – mf#42280 – us UMI ProQuest [615]

Journal of eugenie de guerin = Journal / Guerin, Eugenie de; ed by Trebutien, Guillaume Stanislas – London: Simpkin, Marshall, 1865 – 2mf – 9 – 0-7905-4739-2 – (in english) – mf#1988-0739 – us ATLA [920]

Journal of euromarketing / ed by Kaynak, Erdener – v6 n1. 1996- – 1,9 – $175.00 in US $245.00 outside hardcopy subsc – us Haworth [650]

Journal of european business – New York. 1989-1993 (1,5,9) – ISSN: 1044-002X – mf#18374 – us UMI ProQuest [650]

Journal of european industrial training – Bradford. 1992-1995 (1,5,9) – ISSN: 0309-0590 – mf#15813 – us UMI ProQuest [650]

Journal of european studies – Chalfont St. Giles. 1976+ (1,5,9) – ISSN: 0047-2441 – mf#11327 – us UMI ProQuest [940]

Journal of everett and snohomish county history / Everett Public Library [WA] – n1-6 [1980/81 win-1983 sum] – 1r – 1 – mf#1050456 – us WHS [978]

Journal of evidence-based social work : advances in practice, programming, research, and policy / ed by Feit, Marvin D et al – ISSN: 1543-3714 – us Haworth [360]

Journal of evolutionary biochemistry and physiology – New York. 1969-1976 (1) 1969-1976 (5) – ISSN: 0022-0930 – mf#10881 – us UMI ProQuest [612]

Journal of evolutionary biology – Basel. 1988-1992 (1) – ISSN: 1010-061X – mf#17779 – us UMI ProQuest [580]

Journal of exceptional children – Washington. 1935-1951 (1) 1935-1951 (5) 1935-1951 (9) – (cont: international council for exceptional children council review. cont by: exceptional children) – mf#12546,01 – us UMI ProQuest [640]

Journal of existentialism – San Diego. 1960-1968 [1,5,9] – ISSN: 0449-2498 – mf#1764 – us UMI ProQuest [140]

Journal of experiential education – Boulder. 1978+ – 1,5,9 – ISSN: 1053-8259 – mf#12555 – us UMI ProQuest [370]

Journal of experimental biology – Cambridge. 1983-1996 (1,5,9) – ISSN: 0022-0949 – mf#13599,01 – us UMI ProQuest [574]

Journal of experimental botany – Oxford. 1958+ (1) 1971+ (5) 1976+ (9) – ISSN: 0022-0957 – mf#1245 – us UMI ProQuest [580]

Journal of experimental education – Washington. 1932+ (1) 1966+ (5) 1975+ (9) – ISSN: 0022-0973 – mf#764 – us UMI ProQuest [370]

Journal of experimental marine biology and ecology – Amsterdam. 1967+ (1) 1967+ (5) 1987+ (9) – ISSN: 0022-0981 – mf#42281 – us UMI ProQuest [574]

Journal of experimental medicine – New York. 1896+ (1) 1896+ (5) 1896+ (9) – ISSN: 0022-1007 – mf#12247 – us UMI ProQuest [619]

Journal of experimental pathology – Oxford. 1990 (1,5,9) – (cont: british journal of experimental pathology. cont by: international journal of experimental pathology) – ISSN: 0958-4625 – mf#2517,01 – us UMI ProQuest [619]

Journal of experimental psychology : animal behavior processes – Washington. 1975+(1,5,9) – ISSN: 0097-7403 – mf#10647 – us UMI ProQuest [150]

Journal of experimental psychology : applied – Washington. 1995+ (1,5,9) – ISSN: 1076-898X – mf#21274 – us UMI ProQuest [150]

Journal of experimental psychology : general – Washington. 1975+ (1,5,9) – ISSN: 0096-3445 – mf#276,01 – us UMI ProQuest [150]

Journal of experimental psychology : human learning and memory – Washington. 1975-1981 (1,5,9) – (cont by: journal of experimental psychology: learning, memory, and cognition) – ISSN: 0096-1515 – mf#10645 – us UMI ProQuest [150]

Journal of experimental psychology : human perception and performance – Washington. 1975+(1,5,9) – ISSN: 0096-1523 – mf#10646 – us UMI ProQuest [150]

Journal of experimental psychology : learning, memory, and cognition – Washington. 1982+ (1,5,9) – (cont: journal of experimental psychology: human learning and memory) – ISSN: 0278-7393 – mf#10645,01 – us UMI ProQuest [150]

Journal of experimental psychology – Washington. 1916-1974 (1) 1965-1974 (5) 1970-1974 (9) – ISSN: 0022-1015 – mf#276 – us UMI ProQuest [150]

Journal of experimental zoology – Philadelphia 1904-33 – v1-65 on 662mf – 9 – mf#525/2 – ne IDC [590]

Journal of extension – Madison. 1963-1993 (1) 1963-1993 (5) 1963-1993 (9) – ISSN: 0022-0140 – mf#9485 – us UMI ProQuest [374]

Journal of facilities management – London. 2002+ (1,5,9) – ISSN: 1472-5967 – mf#32086 – us UMI ProQuest [650]

Journal of family and consumer sciences – Alexandria. 1994+ (1) 1994+ (5) 1994+ (9) – (cont: journal of home economics) – ISSN: 1082-1651 – mf#769,01 – us UMI ProQuest [640]

Journal of family and economic issues – New York. 1992+(1,5,9) – (cont: lifestyles) – ISSN: 1058-0476 – mf#14128,02 – us UMI ProQuest [640]

Journal of family communication – Mahwah. 2001+ (1,5,9) – ISSN: 1526-7431 – mf#31734 – us UMI ProQuest [302]

Journal of family counseling – New Hyde Park. 1973-1976 (1) 1973-1976 (5) 1973-1976 (9) – (cont by: international journal of family counseling) – ISSN: 0093-3171 – mf#8240 – us UMI ProQuest [150]

Journal of family history – Thousand Oaks. 1976+ (1,5,9) – ISSN: 0363-1990 – mf#11302 – us UMI ProQuest [929]

Journal of family issues – Thousand Oaks. 1983+ (1,5,9) – ISSN: 0192-513X – mf#14008 – us UMI ProQuest [640]

Journal of family issues – v1-21. 1980-2000 – 5,6,9 – $675.00 set – (v1-5 1980-84 on reel $77. v6-21 1985-2000 on mf $598) – ISSN: 0192-513X – mf#400570 – us Hein [073]

Journal of family law – Louisville. 1961-1992 (1) 1970-1992 (5) 1973-1992 (9) – (cont by: university of louisville journal of family law) – ISSN: 0022-1066 – mf#2205 – us UMI ProQuest [346]

Journal of family nursing – Thousand Oaks. 1995+ (1,5,9) – ISSN: 1074-8407 – mf#21510 – us UMI ProQuest [610]

Journal of family practice – Stanford. 1974+ (1,5,9) – ISSN: 0094-3509 – mf#13094 – us UMI ProQuest [610]

Journal of family psychology: jfp – Washington. 1987+ (1,5,9) – ISSN: 0893-3200 – mf#17056 – us UMI ProQuest [150]

Journal of family psychotherapy : the quarterly journal of case studies, treatment reports, and strategies in clinical practice / ed by Trepper, Terry S – v1- 1990- – 1,9 ($175.00 in US $245.00 outside hardcopy subsc) – (cont: journal of psychotherapy and the family) – us Haworth [616]

Journal of family social work / ed by Bardill, Donald R – v1- 1994- – 1,9 ($60.00 in US $84.00 outside hardcopy subsc) – (cont: journal of social work and human sexuality) – us Haworth [306]

Journal of family therapy – London. 1990+ (1,5,9) – ISSN: 0163-4445 – mf#18102 – us UMI ProQuest [615]

Journal of family violence – New York. 1986+ (1,5,9) – ISSN: 0885-7482 – mf#17676 – us UMI ProQuest [360]

Journal of female liberation : no more fun and games – Medford. 1973-1973 (1) – ISSN: 0029-0815 – mf#7882 – us UMI ProQuest [320]

Journal of feminist family therapy / ed by MacKune-Karrer, Betty – v1- 1989- – 1,9 ($175.00 in US $245.00 outside hardcopy subsc) – us Haworth [305]

Journal of feminist studies in religion – Chico. 1989+ (1,5,9) – ISSN: 8755-4178 – mf#17577 – us UMI ProQuest [230]

Journal of fermentation and bioengineering – Osaka. 1989-1990 (1,5,9) – (cont: journal of fermentation technology. cont by: journal of bioscience and bioengineering) – ISSN: 0922-338X – mf#42593,01 – us UMI ProQuest [576]

Journal of fermentation technology – Osaka. 1986-1988 (1,5,9) – (cont by: journal of fermentation and bioengineering) – ISSN: 0385-6380 – mf#42593 – us UMI ProQuest [576]

Journal of field archaeology – Boston. 1974+ (1) 1977+ (5) 1977+ (9) – ISSN: 0093-4690 – mf#10363 – us UMI ProQuest [930]

Journal of field ornithology – Statesboro. 1980+ (1,5,9) – (cont: bird-banding) – ISSN: 0273-8570 – mf#3255,01 – us UMI ProQuest [590]

Journal of film and video – Los Angeles. 1984+ (1,5,9) – (cont: journal of the university film and video association) – ISSN: 0742-4671 – mf#11567,02 – us UMI ProQuest [790]

Journal of finance – Cambridge. 1946+ (1) 1974+ (5) 1975+ (9) – ISSN: 0022-1082 – mf#10067 – us UMI ProQuest [332]

Journal of financial and quantitative analysis – Seattle. 1966+ (1) 1966+ (5) 1966+ (9) – ISSN: 0022-1090 – mf#6347 – us UMI ProQuest [650]

Journal of financial economics – Amsterdam. 1974+ (1) 1974+ (5) 1986+ (9) – ISSN: 0304-405X – mf#42282 – us UMI ProQuest [332]

Journal of financial markets – Amsterdam. 1998+ (1) – ISSN: 1386-4181 – mf#42808 – us UMI ProQuest [332]

Journal of financial planning / Institute of Certified Financial Planners (US) – Denver. 1988+ (1,5,9) – ISSN: 1040-3981 – mf#16930 – us UMI ProQuest [332]

Journal of financial research – Columbia. 1987+ (1,5,9) – ISSN: 0270-2592 – mf#15814 – us UMI ProQuest [332]

Journal of financial service professionals / Society of Financial Service Professionals – Bryn Mawr. 1998+ (1) 1998+ (5) 1998+ (9) – (cont: journal of the american society of clu and chfc) – mf#7377,03 – us UMI ProQuest [360]

Journal of financial services research – Boston. 1987+ (1,5,9) – ISSN: 0920-8550 – mf#16809 – us UMI ProQuest [332]

Journal of fire and flammability : fire retardant chemistry supplement – Westport. 1974-1975 (1) 1974-1975 (5) 1974-1975 (9) – (cont by: journal of fire retardant chemistry) – ISSN: 0097-0247 – mf#10459 – us UMI ProQuest [360]

Journal of fire and flammability – Lancaster. 1971-1982 (1) 1970-1982 (5) 1975-1982 (9) – ISSN: 0022-1104 – mf#6094 – us UMI ProQuest [360]

Journal of fire retardant chemistry – Lancaster. 1976-1982 (1) 1976-1982 (5) 1976-1982 (9) – (cont: journal of fire and flammability: fire retardant chemistry supplement) – ISSN: 0362-1693 – mf#10459,01 – us UMI ProQuest [360]

Journal of fire sciences – Lancaster. 1983+ (1,5,9) – ISSN: 0734-9041 – mf#13515 – us UMI ProQuest [360]

Journal of fish diseases – Oxford. 1978-1996 (1,5,9) – ISSN: 0140-7775 – mf#15559 – us UMI ProQuest [639]

Journal of fixed income – London. 1991-1999 (1,5,9) – ISSN: 1059-8596 – mf#18544 – us UMI ProQuest [332]

Journal of fluency disorders – New York. 1976+ (1) 1976+ (5) 1987+ (9) – ISSN: 0094-730X – mf#42283 – us UMI ProQuest [150]

Journal of fluid mechanics – Cambridge. 1956+ (1) 1971+ (5) 1976+ (9) – ISSN: 0022-1120 – mf#3030 – us UMI ProQuest [621]

Journal of fluids engineering – New York. 1973+ (1) 1973+ (5) 1976+ (9) – ISSN: 0098-2202 – mf#7241 – us UMI ProQuest [620]

Journal of fluorine chemistry – Lausanne. 1971-1991 (1) 1971-1991 (5) 1986-1991 (9) – ISSN: 0022-1139 – mf#42284 – us UMI ProQuest [540]

Journal of folklore research – Bloomington. 1987+ (1,5,9) – ISSN: 0737-7037 – mf#16283,01 – us UMI ProQuest [390]

Journal of food engineering – London. 1982+ (1,5,9) – ISSN: 0260-8774 – mf#42421 – us UMI ProQuest [660]

Journal of food products marketing : innovations in food advertising, food promotion, food publicity, food sales promotion / ed by Stanton, John L Jr – v3 n1. 1996- – 1,9 – $90.00 in US $126.00 outside hardcopy subsc – us Haworth [650]

Journal of food protection – Des Moines. 1977+ (1) 1977+ (5) 1977+ (9) – (cont: journal of milk and food technology) – ISSN: 0362-028X – mf#2165,01 – us UMI ProQuest [360]

Journal of food science – Chicago. 1936+ (1) 1965+ (5) 1976+ (9) – ISSN: 0022-1147 – mf#41 – us UMI ProQuest [660]

Journal of food technology – Oxford. 1980-1986 (1) 1980-1986 (5) 1980-1986 (9) – (cont by: international journal of food science and technology) – ISSN: 0022-1163 – mf#15560 – us UMI ProQuest [660]

Journal of foodservice business research / ed by Parsa, H G – ISSN: 1537-8020 – us Haworth [338]

Journal of forecasting – Chichester. 1982+ (1,5,9) – ISSN: 0277-6693 – mf#12921 – us UMI ProQuest [650]

Journal of foreign medical science and literature – Philadelphia. 1810-1824 (1) – ISSN: 0001-3404 – mf#4000 – us UMI ProQuest [610]

Journal of forensic identification – Alameda. 1991-1996 (1,5,9) – ISSN: 0895-173X – mf#18592,01 – us UMI ProQuest [360]

Journal of forensic neuropsychology / ed by Hom, Jim – v1- 1999- – 1,9 – $75.00 us $109.00 other – ISSN: 1521-1029 – us Haworth [614]

Journal of forensic psychology practice / ed by Arrigo, Bruce A – v1- 2000- – 1,9 – $95.00 us $138.00 other – ISSN: 1522-8932 – us Haworth [614]

Journal of forensic sciences – Conshohocken. 1956+ (1) 1973+ (5) 1974+ (9) – ISSN: 0022-1198 – mf#10498 – us UMI ProQuest [614]

Journal of forest history – Santa Cruz. 1974-1989 (1) 1974-1989 (5) 1975-1989 (9) – (cont: forest history. cont by: forest and conservation history) – ISSN: 0094-5080 – mf#5895,01 – us UMI ProQuest [634]

Journal of forestry – Bethesda. 1949+ (1) 1970+ (5) 1976+ (9) – ISSN: 0022-1201 – mf#111 – us UMI ProQuest [634]

Journal of forestry and estates management – London: J & W Rider 1878-83 (mthly) [v1-6(may 1877-apr 1883)] [mf ed 2005] – 6v on 3r [ill] – 1 – (cont by: forestry) – mf#film mas 99999 – us Harvard [634]

The journal of frederick horneman's travels : from cairo to mourzouk, the capital of the kingdom of fezzan, in africa in the years 1797-8 / Horneman, F – London, 1802 – 3mf – 9 – mf#H-6132 – ne IDC [916]

Journal of free radicals in biology and medicine – New York. 1985-1986 (1,5,9) – ISSN: 0748-5514 – mf#49488 – us UMI ProQuest [574]

Journal of fuel and heat technology – London. 1968-1972 (1) 1971-1972 (5) – ISSN: 0022-121X – mf#3247 – us UMI ProQuest [690]

Journal of futures markets – New York. 1981+ (1,5,9) – ISSN: 0270-7314 – mf#13059 – us UMI ProQuest [332]

Journal of gambling behavior – New York. 1985-1989 (1) 1985-1989 (5) 1985-1989 (9) – (cont by: journal of gambling studies) – ISSN: 0742-0714 – mf#14481 – us UMI ProQuest [616]

Journal of gambling studies – New York. 1990+ (1,5,9) – (cont: journal of gambling behavior) – ISSN: 1050-5350 – mf#14481,01 – us UMI ProQuest [616]

Journal of garden history – London. 1991-1996 (1) – (cont by: studies in the history of gardens and designed landscapes) – ISSN: 0144-5170 – mf#17324 – us UMI ProQuest [710]

Journal of gastroenterology and hepatology – Melbourne. 1986+ (1,5,9) – ISSN: 0815-9319 – mf#15561 – us UMI ProQuest [616]

Journal of gastrointestinal motility – Cambridge. 1989-1992 (1,5,9) – ISSN: 1043-4518 – mf#18081 – us UMI ProQuest [616]

Journal of gay and lesbian issues in education : an international quarterly devoted to research, policy, and practice / ed by Sears, James T – v1- 2003- – 1,9 – $90.00 us $131.00 other – ISSN: 1541-0889 – us Haworth [305]

Journal of gay and lesbian politics : studies in sexuality, gender, and public policy / ed by Haeberle, Steven H – 1,9 – $75.00 us $109.00 other – ISSN: 1537-9426 – us Haworth [322]

Journal of gay and lesbian psychotherapy / ed by Scasta, David L – v1- 1989- – 1, 9 ($75.00 in US $105.00 outside hardcopy subsc) – us Haworth [150]

Journal of gay and lesbian social services : issues in practice, policy and research / ed by Kelly, James J – v4 n1. 1996- – 1,9 – $60.00 in US $84.00 outside hardcopy subsc – us Haworth [305]

Journal of gemmology – London. 1976-1981 (1) 1976-1981 (5) 1976-1981 (9) – ISSN: 0022-1252 – mf#9728 – us UMI ProQuest [730]

Journal of gender and the law *see* American university journal of gender and the law

Journal of gender, race, and justice – v1-5. 1997-2002 – 9 – $69.00 set – mf#117541 – us Hein [323]

Journal of gender, social policy and the law *see* American university journal of gender, social policy and the law

Journal of general and applied microbiology – Tokyo. 1965+ (1) 1968+ (5) 1970+ (9) – ISSN: 0022-1260 – mf#2986 – us UMI ProQuest [576]

Journal of general chemistry of the ussr – New York. 1949-1991 (1) 1949-1991 (5) 1951-1991 (9) – ISSN: 0022-1279 – mf#10835 – us UMI ProQuest [540]

Journal of general education – University Park. 1946+ (1) 1968+ (5) 1975+ (9) – ISSN: 0021-3667 – mf#994 – us UMI ProQuest [370]

Journal of general internal medicine – Oxford. 1996+ (1,5,9) – ISSN: 0884-8734 – mf#21607 – us UMI ProQuest [610]

Journal of general physiology – New York 1919-1945/46 – v1-29 on 430mf – 9 – mf#2219c/2 – ne IDC [574]

Journal of general psychology – Provincetown. 1987+ (1,5,9) – ISSN: 0022-1309 – mf#16336 – us UMI ProQuest [150]

Journal of genetic psychology – New York. 1983+ (1) 1983+ (5) 1983+ (9) – ISSN: 0022-1325 – mf#2944,02 – us UMI ProQuest [150]

Journal of genetics / ed by Bateson, W & Punnett, R C – Cambridge 1910-42 – v1-43 on 359mf – 9 – (missing: v39-40,42) – mf#359/2 – ne IDC [575]

Journal of geochemical exploration – Amsterdam. 1972+ (1) 1972+ (5) 1977+ (9) – ISSN: 0375-6742 – mf#42285 – us UMI ProQuest [550]

Journal of geodynamics – Amsterdam. 1991-1993 (1,5,9) – ISSN: 0264-3707 – mf#49570 – us UMI ProQuest [550]

Journal of geography – Indiana. 1902+ (1) 1969+ (5) 1975+ (9) – ISSN: 0022-1341 – mf#1047 – us UMI ProQuest [910]

Journal of geological education – Lawrence. 1951-1995 (1) 1972-1995 (5) 1975-1995 (9) – (cont by: journal of geoscience education) – ISSN: 0022-1368 – mf#7663 – us UMI ProQuest [550]

Journal of geology – Chicago. 1893+ (1) 1965+ (5) 1977+ (9) – ISSN: 0022-1376 – mf#483 – us UMI ProQuest [550]

Journal of geophysical research – Printed indexes available – (1896-1900 $25.00 1,5,6,13. 1901-05 $25.00 1,5,6,13. 1906-10 $25.00 1,5,6,13. 1911-15 $25.00 1,5,6,13. 1916-20 $25.00 1,5,6,13. 1921-25 $25.00 1,5,6,13. 1926-30 $25.00 1,5,6,13. 1931-58 $30.00 per 2 yrs 1,5,6,13. 1959-61 v64-66 $50.00y 1,5,6,13. 1962 v67 $75.00 1,5,6,13. 1963 v68 $75.00 1,5,6,13. 1964-67 v69-72 $75.00 1,5,6,13. 1968-69 v73-74 $100.00y 1,5,6,13. 1970-75 v75-80 $100.00y 1,5,6,13. 1971-75 v76-80 $100.00y 9. 1976 v81 $100.00 1,5,6,13 $150.00 9. 1977 v82 $100.00 1,5,6,13 $220.00 9. 1978 v83 $125.00 1,5,6,13 $280.00 9. 1979 v84 $135.00 1,5,6,13 $350.00 9. 1980 v85 $160.00 1,5,6,13 $435.00 9. 1981 v86 $220.00 1,5,6,13 $570.00 9. 1982 v87 $320.00 1,5,6,13 $680.00 9. 1992 v97 $2,555.00. 1993 v98 $2,800.00. 1994v99 $3065.00; 1995 v100 $3510.00) – us AGU [550]

Journal of geophysics = Zeitschrift fuer geophysik – Berlin. 1983-1983 (1,5,9) – ISSN: 0340-062X – mf#13187,01 – us UMI ProQuest [550]

The journal of george fox / Fox, George; ed by Penney, Norman – Cambridge: University Press, 1911 – 3mf – 9 – 0-7905-4640-X – (incl bibl ref) – mf#1988-0640 – us ATLA [920]

Journal of geoscience education – Bellingham. 1996+ (1) 1996+ (5) 1996+ (9) – (cont: journal of geological education) – mf#7663,01 – us UMI ProQuest [550]

Journal of geotechnical and geoenvironmental engineering – New York. 1997+ (1) 1997+ (5) 1997+ (9) – (cont: journal of geotechnical engineering) – ISSN: 1090-0241 – mf#8140,03 – us UMI ProQuest [624]

Journal of geotechnical engineering – New York. 1983-1996 (1,5,9) – (cont: journal of the geotechnical engineering division. cont by: journal of geotechnical and geoenvironmental engineering) – ISSN: 0733-9410 – mf#8140,02 – us UMI ProQuest [624]

Journal of geriatric drug therapy / ed by Cooper, James – v1- 1986- – 1, 9 ($225.00 in US $315.00 outside hardcopy subsc) – us Haworth [615]

Journal of geriatric psychiatry and neurology – Hamilton. 1991-1996 (1,5,9) – ISSN: 0891-9887 – mf#19494 – us UMI ProQuest [618]

Journal of gerontological nursing – Thorofare. 1975+(1,5,9) – ISSN: 0098-9134 – mf#11477 – us UMI ProQuest [610]

Journal of gerontological social work / ed by Dobrof, Rose – v1- 1978- – 1, 9 ($275.00 in US $385.00 outside hardcopy subsc) – us Haworth [360]

Journal of gerontology – Washington. 1946-1994 [1]; 1965-1994 [5]; 1976-1994 [9] – ISSN: 0022-1422 – mf#2012 – us UMI ProQuest [618]

Journal of glass studies – Corning. 1959+ (1) 1972+ (5) 1974+ (9) – ISSN: 0075-4250 – mf#7053 – us UMI ProQuest [740]

Journal of global information management – Harrisburg. 1993-1996 (1,5,9) – ISSN: 1062-7375 – mf#20005 – us UMI ProQuest [380]

Journal of global information technology management – Marietta, 1998+ [1,5,9] – ISSN: 1097-198X – mf#28280 – us UMI ProQuest [000]

Journal of global marketing / ed by Kaynak, Erdener – v1- 1987- – 1, 9 ($200.00 in US $280.00 outside hardcopy subsc) – us Haworth [337]

Journal of government financial management – Arlington, 2001+ [1,5,9] – (cont: government accountants journal) – ISSN: 1533-1385 – mf#10381,02 – us UMI ProQuest [350]

Journal of government information – New York. 1994+ (1,5,9) – (cont: government publications review) – ISSN: 1352-0237 – mf#49078,01 – us UMI ProQuest [350]

Journal of graph theory – New York. 1977+ (1,5,9) – ISSN: 0364-9024 – mf#11774 – us UMI ProQuest [510]

Journal of great lakes research – Ann Arbor. 1977+ (1,5,9) – ISSN: 0380-1330 – mf#11641 – us UMI ProQuest [550]

Journal of group psychotherapy, psychodrama and sociometry – Washington. 1981-1996 (1) 1981-1996 (5) 1981-1996 (9) – (cont: group psychotherapy, psychodrama and sociometry. cont by: international journal of action methods) – ISSN: 0731-1273 – mf#6905,02 – us UMI ProQuest [150]

Journal of guidance and control – New York. 1981-1981 (1,5,9) – (cont by: journal of guidance, control, and dynamics) – ISSN: 0162-3192 – mf#13073 – us UMI ProQuest [629]

Journal of guidance, control, and dynamics – Reston. 1982+ (1,5,9) – (cont: journal of guidance and control) – ISSN: 0731-5090 – mf#13073,01 – us UMI ProQuest [629]

Journal of h m s enterprise 1850-1855 / Collinson, R – London, 1889 – 10mf – 9 – mf#N-168 – ne IDC [910]

Journal of hand surgery : journal of the british society for surgery of the hand – Edinburgh. 1984+ (1,5,9) – (cont: hand) – ISSN: 0266-7681 – mf#13428,01 – us UMI ProQuest [617]

Journal of hand surgery – New York. 1976+ (1,5,9) – ISSN: 0363-5023 – mf#11194 – us UMI ProQuest [617]

Journal of hand therapy – Philadelphia. 1994+ (1,5,9) – ISSN: 0894-1130 – mf#21608 – us UMI ProQuest [615]

Journal of hazardous materials – Amsterdam. 1975+ (1) 1975+ (5) 1987+ (9) – ISSN: 0304-3894 – mf#42286 – us UMI ProQuest [600]

Journal of head trauma rehabilitation – Gaithersburg. 1986+ (1,5,9) – ISSN: 0885-9701 – mf#15593 – us UMI ProQuest [617]

Journal of health : conducted by an association of physicians – Philadelphia. 1829-1833 (1) – mf#4469 – us UMI ProQuest [613]

Journal of health and social behavior – Albany. 1960+ (1) 1971+ (5) 1975+ (9) – ISSN: 0022-1465 – mf#2478 – us UMI ProQuest [360]

Journal of health and social policy / ed by Feit, Marvin D – v1- 1989- – 1, 9 ($200.00 in US $280.00 outside hardcopy subsc) – us Haworth [610]
Journal of health care chaplaincy / ed by Rutz, Kathy – v1- 1987- – 1, 9 ($85.00 in US $119.00 outside hardcopy subsc) – us Haworth [240]
Journal of health care compliance – Gaithersburg. 1999+ (1,5,9) – ISSN: 1520-8303 – mf#32171 – us UMI ProQuest [613]
Journal of health care finance – Gaithersburg. 1994+ (1,5,9) – (cont: topics in health care financing) – ISSN: 1078-6767 – mf#12738,01 – us UMI ProQuest [360]
Journal of health care for the poor and underserved – Nashville. 1990+ (1,5,9) – ISSN: 1049-2089 – mf#18463 – us UMI ProQuest [360]
Journal of health care marketing – Boone. 1985-1996 (1,5,9) – (cont by: marketing health services) – ISSN: 0737-3252 – mf#14916 – us UMI ProQuest [650]
Journal of health economics – Amsterdam. 1982+ (1,5,9) – ISSN: 0167-6296 – mf#42542 – us UMI ProQuest [360]
Journal of health education – Reston. 1991-2000 (1) 1991-2000 (5) 1991-2000 (9) – (cont: health education) – ISSN: 1055-6699 – mf#7254,02 – us UMI ProQuest [360]
Journal of health, physical education, recreation – Washington. 1930-1974 (1) 1969-1974 (5) – (cont by: journal of physical education and recreation) – ISSN: 0022-1473 – mf#772 – us UMI ProQuest [613]
Journal of health politics, policy and law – Durham. 1976+ (1,5,9) – ISSN: 0361-6878 – mf#11145 – us UMI ProQuest [360]
Journal of health politics, policy and law – Duke University. v1-26. 1976-2001 – 5,6,9 – $929.00 set – (v1-9 1976-85 on reel $193. v10-26 1985-2001 on mf $736) – ISSN: 0361-6878 – mf#104021 – us Hein [072]
Journal of healthcare information management: jhim – San Francisco. 1998+ (1) – ISSN: 1099-811X – mf#23949,01 – us UMI ProQuest [650]
Journal of healthcare management – Chicago. 1998+ (1) 1998+ (5) 1998+ (9) – (cont: hospital and health services administration) – ISSN: 1096-9012 – mf#6408,02 – us UMI ProQuest [360]
Journal of heart and lung transplantation – St. Louis. 1991-1996 (1,5,9) – (cont: journal of heart transplantation) – ISSN: 1053-2498 – mf#15985,02 – us UMI ProQuest [617]
Journal of heart transplantation – Newark. 1989-1990 (1) – (cont by: journal of heart and lung transplantation) – ISSN: 0887-2570 – mf#15985,01 – us UMI ProQuest [617]
Journal of heat recovery systems – Oxford. 1981-1986 (1,5,9) – (cont by: heat recovery systems and chp) – ISSN: 0198-7593 – mf#49386 – us UMI ProQuest [530]
Journal of heat transfer – New York. 1959+ [1]; 1965+ [5]; 1970+ [9] – ISSN: 0022-1481 – mf#1191 – us UMI ProQuest [621]
Journal of heat treating – Metals Park. 1979-1991 (1,5,9) – ISSN: 0190-9177 – mf#12936 – us UMI ProQuest [660]
Journal of hellenic studies – London. 1880-1900 (1) – ISSN: 0075-4269 – mf#2907 – us UMI ProQuest [450]
Journal of hendrik jacob wikar (1779) [the] : the journals of jocobus coetse jansz (1760) and willem van reenen [1791] / ed by Mossop, E E – Cape Town: The Van Riebeeck Society 1935 – (= ser [Travel descriptions from south africa, 1711-1938]) – 4mf – 9 – (english trans by a w van der horst, int & footnotes by ed) – mf#zah-70 – ne IDC [916]
Journal of herbal pharmacotherapy : innovations in clinical and applied evidence-based herbal medicinals / ed by Ulbricht, Catherine – v1- 2000- – 1,9 – $75.00 us $109.00 other – ISSN: 1522-8940 – us Haworth [615]
Journal of herbs, spices and medicinal plants / ed by Craker, Lyle E – v4 n1. 1996- – 1,9 – $125.00 in US $175.00 outside hardcopy subsc – us Haworth [580]
Journal of heredity – Washington. 1910+ (1) 1965+ (5) 1970+ (9) – ISSN: 0022-1503 – mf#429 – us UMI ProQuest [575]
Journal of higher education – Columbus. 1930+ (1) 1969+ (5) 1975+ (9) – ISSN: 0022-1546 – mf#849 – us UMI ProQuest [378]
Journal of higher education policy and management – Abingdon. 1997+ – 1 – ISSN: 1360-080X – mf#14719,02 – us UMI ProQuest [378]
Journal of hiroshima university dental society – Hiroshima. 1972-1980 (1) 1975-1980 (5) 1975-1980 (9) – ISSN: 0046-7472 – mf#7809 – us UMI ProQuest [617]
Journal of his voyage round the world in h.m.s. "endeavour", 1768-71 : from the national maritime museum, london, mss. jod/19 and jod/56 / Cook, James – 1r – 1 – (filmed with: narrative account of the voyage of hms "resolution", 1772-73 by richard pickersgill) – mf#97048 – uk Microform Academic [910]
Journal of his voyage round the world in hms "resolution", 1772-1775 : from the national maritime museum, london, ms jod/20.37. ms 1702 / Cook, James – 1r – 1 – mf#97015 – uk Microform Academic [910]
Journal of historical sociology – Oxford. 1988+ (1,5,9) – ISSN: 0952-1909 – mf#17395 – us UMI ProQuest [301]
Journal of historical studies – Princeton. 1967-1969 (1) – mf#3368 – us UMI ProQuest [900]
Journal of hiv/aids and social services : research, practice, and policy, adopted by the national social work aids network (nswan) / ed by Linsk, Nathan L & Gilbert, Dorie J – v1- 2002- – 1,9 – $95.00 us $138.00 other – ISSN: 1538-1501 – us Haworth [360]
Journal of hiv/aids prevention and education for adolescents and children / ed by Morales, Julio & Bok, Marcia – v1 n1. 1997- – 1,9 – $60.00 in US $84.00 outside hardcopy subsc – us Haworth [360]
Journal of holistic medicine – New York. 1980-1986 (1) 1980-1986 (5) 1980-1986 (9) – (cont by: journal of advancement in medicine) – ISSN: 0195-5977 – mf#12187 – us UMI ProQuest [615]
Journal of holistic nursing – Springfield. 1994+ (1,5,9) – ISSN: 0898-0101 – mf#19340 – us UMI ProQuest [610]
Journal of home economics – Washington. 1909-1994 (1) 1969-1994 (5) 1969-1994 (9) – (cont by: journal of family and consumer sciences) – ISSN: 0022-1570 – mf#769 – us UMI ProQuest [640]
Journal of home health care practice – Rockville. 1988-1995 (1,5,9) – (cont by: home health care management and practice) – ISSN: 0897-8018 – mf#16712 – us UMI ProQuest [360]
Journal of homosexuality / ed by Cecco, John P De – v1- 1974- – 1, 9 ($275.00 in US $385.00 outside hardcopy subsc) – us Haworth [305]
Journal of hospital librarianship / ed by Gilbert, Carole M – v1- 2001- – 1,9 – $95.00 us $138.00 other – ISSN: 1532-3269 – us Haworth [020]
Journal of hospital marketing / ed by Winston, William J – v1- 1986- – 1, 9 ($200.00 in US $280.00 outside hardcopy subsc) – us Haworth [650]
Journal of hospital marketing and public relations / ed by Carter, Tony – v14- 2002- – 1,9 – $95.00 us $138.00 other – ISSN: 1539-0942 – us Haworth [650]
Journal of hospitality and leisure marketing : the international forum for research, theory and practice / ed by Knutson, Bonnie – v4 n1. 1996- – 1,9 – $75.00 in US $105.00 outside hardcopy subsc – us Haworth [650]
Journal of housing – Washington. 1944-1994 (1) 1972-1994 (5) 1975-1994 (9) – (cont by: journal of housing and community development) – ISSN: 0272-7374 – mf#6742 – us UMI ProQuest [360]
Journal of housing and community development – Washington. 1995+ (1) 1995+ (5) 1995+ (9) – (cont: journal of housing) – mf#6742,01 – us UMI ProQuest [710]
Journal of housing for the elderly / ed by Pastalan, Leon A – v1- 1983- – 1, 9 ($200.00 in US $280.00 outside hardcopy subsc) – us Haworth [360]
Journal of housing research – Washington. 1995-1995 (1) – ISSN: 1052-7001 – mf#18744 – us UMI ProQuest [360]
Journal of hugh finlay, surveyor of post roads and post offices, 1773-1774 : and accounts of the general post office in philadelphia and of the various deputy postmasters -" the ledger of benjamin franklin"- jan 1775-jan 1780 – (= ser Records of the post office department) – 1r – 1 – mf#T268 – us Nat Archives [380]
Journal of human behavior in the social environment : a professional journal / ed by Feit, Marvin D & Wodarski, John S – v1 n1. 1997- – 1,9 – $45.00 in US $63.00 outside hardcopy subsc – us Haworth [150]
Journal of human nutrition and dietetics – London. 1988-1996 (1,5,9) – ISSN: 0952-3871 – mf#16735 – us UMI ProQuest [613]
Journal of human relations – Wilberforce. 1952-1973 (1) 1971-1973 (5) – ISSN: 0022-1651 – mf#1613 – us UMI ProQuest [301]
Journal of human resources – Madison. 1966+ (1) 1971+ (5) 1975+ (9) – ISSN: 0022-166X – mf#5332 – us UMI ProQuest [331]
Journal of human resources in hospitality and tourism / ed by Adler, Howard – v1- 2000- – 1,9 – $140.00 us $203.00 other – ISSN: 1533-2845 – us Haworth [331]
Journal of human stress – Washington. 1979-1987 (1,5,9) – (cont by: behavioral medicine) – mf#12489 – us UMI ProQuest [150]
Journal of humanistic counseling, education and development – Alexandria. 1998+ – 1,5,9 – (cont: journal of humanistic education and development) – mf#3276,03 – us UMI ProQuest [370]
Journal of humanistic education and development – Falls Church. 1982-1998 (1) 1982-1998 (5) 1982-1998 (9) – (cont: humanist educator. cont by: journal of humanistic counseling, education and development) – ISSN: 0735-6846 – mf#3276,02 – us UMI ProQuest [370]
Journal of humanistic psychology – Beverly Hills. 1961+ (1) 1968+ (5) 1975+ (9) – ISSN: 0022-1678 – mf#3078 – us UMI ProQuest [150]
Journal of hydraulic engineering – New York. 1983+ (1,5,9) – (cont: journal of the hydraulics division) – ISSN: 0733-9429 – mf#8138,01 – us UMI ProQuest [627]
Journal of hydrologic engineering – New York. 1996+ (1,5,9) – ISSN: 1084-0699 – mf#22092 – us UMI ProQuest [550]
Journal of hydrology – Amsterdam. 1963+ (1) 1963+ (5) 1979+ (9) – ISSN: 0022-1694 – mf#42287 – us UMI ProQuest [550]
Journal of hydronautics – New York. 1967-1981 (1) 1971-1981 (5) 1976-1981 (9) – ISSN: 0022-1716 – mf#5076 – us UMI ProQuest [620]
Journal of hygiene – Cambridge. 1979-1986(1,5,9) – (cont by: epidemiology and infection) – ISSN: 0022-1724 – mf#12125 – us UMI ProQuest [613]
Journal of ichthyology – Silver Spring. 1970-1996 (1) 1970-1996 (5) 1970-1996 (9) – ISSN: 0032-9452 – mf#14357,01 – us UMI ProQuest [590]
Journal of imaging science – Springfield. 1985-1991 (1) 1985-1991 (5) 1985-1991 (9) – (cont: photographic science and engineering) – ISSN: 8750-9237 – mf#7955,01 – us UMI ProQuest [770]
Journal of imaging science and technology – Springfield. 1992+ (1,5,9) – ISSN: 1062-3701 – mf#18857 – us UMI ProQuest [621]
Journal of imaging technology – Springfield. 1984-1991 (1,5,9) – (cont: journal of applied photographic engineering) – ISSN: 0747-3583 – mf#12416,01 – us UMI ProQuest [621]
Journal of immigrant and refugee services / ed by Ryan, Angela Shen – v1- 2002- – 1,9 – $85.00 us $123.00 other – ISSN: 1536-2949 – us Haworth [360]
Journal of immunogenetics – Oxford. 1980-1990 (1) 1980-1990 (5) 1980-1990 (9) – (cont by: european journal of immunogenetics) – ISSN: 0305-1811 – mf#15563 – us UMI ProQuest [575]
Journal of immunological methods – Amsterdam. 1971+ (1) 1971+ (5) 1987+ (9) – ISSN: 0022-1759 – mf#42102 – us UMI ProQuest [616]
Journal of immunology – Baltimore. 1982+ (1) 1982+ (5) 1982+ (9) – ISSN: 0022-1767 – mf#79 – us UMI ProQuest [616]
Journal of immunotherapy – New York. 1997+ (1) – mf#18710,03 – us UMI ProQuest [615]
Journal of imperial and commonwealth history – London. 1990-1996 (1,5,9) – ISSN: 0308-6534 – mf#18549 – us UMI ProQuest [941]
Journal of in vitro fertilization and embryo transfer: ivf – New York. 1984-1991 (1,5,9) – (cont by: journal of assisted reproduction and genetics) – ISSN: 0740-7769 – mf#17679 – us UMI ProQuest [618]
Journal of inclusion phenomena and macrocyclic chemistry – Dordrecht. 1999+ (1) – (cont: journal of inclusion phenomena and molecular recognition in chemistry) – mf#14757,02 – us UMI ProQuest [540]
Journal of inclusion phenomena and molecular recognition in chemistry – Dordrecht. 1993-1996 (1,5,9) – (cont by: journal of inclusion phenomena and macrocyclic chemistry) – ISSN: 0923-0750 – mf#14757,01 – us UMI ProQuest [540]
Journal of indian art and industries – London. 1886-1916 (1) – mf#5583 – us UMI ProQuest [700]
The journal of indian art and industry – London, 1884-1916 [mf ed Chadwyck-Healey] – (= ser Art periodicals on microform) – 5r – 1 – uk Chadwyck [700]
Journal of indian philosophy – Dordrecht. 1984+ (1,5,9) – ISSN: 0022-1791 – mf#14758 – us UMI ProQuest [180]
Journal of individual psychology – Austin. 1940-1981 (1) 1973-1981 (5) 1973-1981 (9) – ISSN: 0022-1805 – mf#7717 – us UMI ProQuest [150]
Journal of individual psychology – Austin. 1998+ (1) 1998+ (5) 1998+ (9) – (cont: individual psychology) – mf#7717,02 – us UMI ProQuest [150]
Journal of industrial economics – Oxford. 1952+ [1]; 1982+ [5,9] – ISSN: 0022-1821 – mf#1372 – us UMI ProQuest [338]
Journal of industrial engineering – New York. 1949-1968 (1) – ISSN: 0022-183X – mf#1026 – us UMI ProQuest [620]
Journal of industrial hemp : official journal of the international hemp association / ed by Werf, Hayo M G van der – v7- 2002- – 1,9 – $140.00 us $203.00 other – ISSN: 1537-7881 – us Haworth [660]
Journal of industrial hygiene – Baltimore. 1919-1932 (1) – ISSN: 0095-9022 – mf#44 – us UMI ProQuest [610]
Journal of industrial microbiology – Amsterdam. 1986-1992 (1,5,9) – ISSN: 0169-4146 – mf#42563 – us UMI ProQuest [576]
Journal of industrial progress – Dublin, Ireland. jan, feb, jul, oct-dec 1854; jan, feb 1855 – 1/2r – 1 – uk British Libr Newspaper [072]
Journal of industrial psychology – Ventnor. 1963-1970 [1,5,9] – ISSN: 0022-1848 – mf#1841 – us UMI ProQuest [150]
Journal of industrial teacher education – Blacksburg. 1963+ (1) 1973+ (5) 1973+ (9) – ISSN: 0022-1864 – mf#9770 – us UMI ProQuest [370]
Journal of industrial textiles – London, 1999+ [1,5,9] – (cont: journal of coated fabrics) – ISSN: 1528-0837 – mf#6095,02 – us UMI ProQuest [670]
Journal of infectious disease pharmacotherapy : antimicrobial evaluation, development and clinical application / ed by Bosso, John A – v2 n1. 1996- – 1,9 – $75.00 in US $105.00 outside hardcopy subsc – us Haworth [615]
Journal of infectious diseases – Chicago. 1949+ (1) 1904+ (5) 1977+ (9) – ISSN: 0022-1899 – mf#136 – us UMI ProQuest [616]
Journal of information and image management – Silver Spring. 1983-1986 (1) 1983-1986 (5) 1983-1986 (9) – (cont: journal of micrographics. cont by: inform) – ISSN: 0745-9963 – mf#7820,02 – us UMI ProQuest [020]
Journal of information science – Amsterdam. 1979-1993 (1) 1979-1993 (5) 1987-1993 (9) – ISSN: 1352-7460 – mf#42288 – us UMI ProQuest [020]
Journal of information systems – Oxford. 1991-1993 (1,5,9) – (cont by: information systems journal) – ISSN: 0959-2954 – mf#18084 – us UMI ProQuest [000]
Journal of information systems management – Pennsauken. 1984-1991 (1,5,9) – (cont by: information systems management) – ISSN: 0739-9014 – mf#14372 – us UMI ProQuest [020]
Journal of information technology cases and applications – Marietta. 1999+ (1,5,9) – ISSN: 1522-8053 – mf#31901 – us UMI ProQuest [000]
Journal of infrastructure systems – New York. 1995+ (1,5,9) – ISSN: 1076-0342 – mf#21270 – us UMI ProQuest [624]
Journal of infusion nursing – Hagerstown. 2001+ (1,5,9) – ISSN: 1533-1458 – mf#11473,02 – us UMI ProQuest [610]
Journal of inherited metabolic disease – Lancaster. 1991-1994 (1) 1991-1994 (5) 1991-1994 (9) – ISSN: 0141-8955 – mf#16810 – us UMI ProQuest [618]
Journal of injection molding technology – Brookfield. 1997+ (1) – mf#27000 – us UMI ProQuest [660]
Journal of inorganic and nuclear chemistry – Oxford. 1955-1981 (1) 1955-1981 (5) 1976-1981 (9) – ISSN: 0022-1902 – mf#49115 – us UMI ProQuest [530]
Journal of inorganic biochemistry – New York. 1971+ (1) 1971+ (5) 1987+ (9) – ISSN: 0162-0134 – mf#42289 – us UMI ProQuest [574]
Journal of insect physiology – London. 1957+ (1,5,9) – ISSN: 0022-1910 – mf#49219 – us UMI ProQuest [590]
Journal of instructional development : (jid) / Association for Educational Communications and Technology – Park Forest South. 1977-1988 (1) 1977-1988 (5) 1977-1988 (9) – ISSN: 0162-2641 – mf#12092 – us UMI ProQuest [370]
Journal of instructional psychology – Milwaukee. 1974+ (1,5,9) – ISSN: 0094-1956 – mf#12308 – us UMI ProQuest [150]
Journal of insurance – New York. 1940-1983 (1) 1940-1983 (5) 1940-1983 (9) – (cont by: insurance review) – ISSN: 0022-1929 – mf#991 – us UMI ProQuest [368]
Journal of insurance coverage – New York. 1998+ (1,5,9) – ISSN: 1096-8342 – mf#32172 – us UMI ProQuest [368]
Journal of insurance regulation – Kansas City. 1994-1995 (1,5,9) – ISSN: 0736-248X – mf#19081 – us UMI ProQuest [368]
Journal of integral equations – New York. 1979-1984 (1) 1979-1984 (5) (9) – ISSN: 0163-5549 – mf#42290 – us UMI ProQuest [510]
Journal of intellectual and developmental disability – Abingdon. 1996+ (1,5,9) – (cont: australia and new zealand journal of developmental disabilities) – ISSN: 1326-978X – mf#10741,03 – us UMI ProQuest [150]
Journal of intellectual capital – Bradford. 2001+ (1,5,9) – ISSN: 1469-1930 – mf#31576 – us UMI ProQuest [650]
Journal of intellectual disability research (jidr) – Oxford. 1992+ (1) 1992+ (5) 1992+ (9) – (cont: journal of mental deficiency research) – ISSN: 0964-2633 – mf#8924,01 – us UMI ProQuest [616]

Journal of intelligent and robotic systems – Dordrecht. 1988-1996 (1,5,9) – ISSN: 0921-0296 – mf#16811 – us UMI ProQuest [000]

Journal of intensive care medicine – Boston. 1991+ (1,5,9) – ISSN: 0885-0666 – mf#18090 – us UMI ProQuest [610]

Journal of interactive learning research – Phoenix. 1997+ – 1,5,9 – (cont: journal of artificial intelligence in education) – ISSN: 1093-023X – mf#17106,01 – us UMI ProQuest [370]

Journal of interactive marketing – New York. 1998+ (1,5,9) – (cont: journal of direct marketing) – ISSN: 1094-9968 – mf#18111,01 – us UMI ProQuest [650]

Journal of interamerican studies and world affairs – Beverly Hills. 1959-2000 (1) 1975-2000 (5) 1975-2000 (9) – ISSN: 0022-1937 – mf#10960 – us UMI ProQuest [327]

Journal of interdisciplinary history – Cambridge. 1970+ (1) 1972+ (5) 1973+ (9) – ISSN: 0022-1953 – mf#8425 – us UMI ProQuest [900]

Journal of intergenerational relationships : programs, policy, and research / ed by Newman, Sally – v1- 2003- – 1,9 – $75.00 us $109.00 other – ISSN: 1535-0770 – us Haworth [302]

Journal of intergroup relations – Fort Lauderdale. 1970+ (1) 1970+ (5) 1975+ (9) – ISSN: 0047-2492 – mf#6634 – us UMI ProQuest [302]

Journal of interlibrary loan, document delivery and information supply / ed by Morris, Leslie R – v1- 1990- – 1, 9 ($90.00 in US $126.00 outside hardcopy subsc) – us Haworth [020]

Journal of interlibrary loan, document supply and electronic reserve / ed by Morris, Leslie – v15- fall 2004 – 1,9 – $240.00 us $348.00 other – ISSN: 1072-303X – us Haworth [020]

Journal of intermountain archeology – v1 n1-v5 n1 [1986] – 1r – 1 – mf#1519160 – us WHS [930]

Journal of internal medicine – Oxford. 1989+ (1,5,9) – ISSN: 0954-6820 – mf#17142 – us UMI ProQuest [616]

Journal of international accounting auditing and taxation – Greenwich. 1993+ (1,5,9) – ISSN: 1061-9518 – mf#19457 – us UMI ProQuest [650]

Journal of international accounting research – Sarasota. 2002+ (1,5,9) – ISSN: 1542-6297 – mf#32312 – us UMI ProQuest [650]

Journal of international affairs – New York. 1947+ (1) 1969+ (5) 1975+ (9) – ISSN: 0022-197X – mf#1051 – us UMI ProQuest [327]

Journal of international banking regulation – London. 1999+ (1,5,9) – ISSN: 1465-4830 – mf#31902 – us UMI ProQuest [332]

Journal of international business studies – Washington. 1970+ (1) 1972+ (5) 1975+ (9) – ISSN: 0047-2506 – mf#7898 – us UMI ProQuest [338]

Journal of international consumer marketing / ed by Kaynak, Erdener – v1- 1988- – 1, 9 ($200.00 in US $280.00 outside hardcopy subsc) – us Haworth [380]

Journal of international development – Oxford. 1989-1993 (1,5,9) – ISSN: 0954-1748 – mf#18163 – us UMI ProQuest [300]

Journal of international economics – Amsterdam. 1971+ (1) 1971+ (5) 1987+ (9) – ISSN: 0022-1996 – mf#42291 – us UMI ProQuest [337]

Journal of international financial management and accounting – Oxford. 1989+ (1,5,9) – ISSN: 0954-1314 – mf#17396 – us UMI ProQuest [650]

Journal of international food and agribusiness marketing / ed by Kaynak, Erdener – v1- 1989- – 1, 9 ($150.00 in US $210.00 outside hardcopy subsc) – us Haworth [380]

Journal of international hospitality, leisure and tourism management 1090 a multinational and cross-cultural journal of applied research / ed by Khan, Mahmood A – v1 n1. 1997- – 1,9 – $85.00 in US $119.00 outside hardcopy subsc – us Haworth [650]

Journal of international law and economics – Washington. 1971-1981 (1) 1971-1981 (5) 1977-1981 (9) – (cont by: george washington journal of international law and economics) – ISSN: 0022-2003 – mf#8766 – us UMI ProQuest [341]

Journal of international law and economics *see* George washington journal of international law and economics

Journal of international law and practice *see* Dcl journal of international law

Journal of international marketing – Chicago. 1993+ (1) 1993-1994 (5) 1993-1994 (9) – ISSN: 1069-031X – mf#19540 – us UMI ProQuest [337]

Journal of international money and finance – Kidlington. 1982+ (1,5,9) – ISSN: 0261-5606 – mf#15816 – us UMI ProQuest [332]

Journal of internet cataloging : the international quarterly of digital organization, classification, and access / ed by Carter, Ruth C & Brisson, Roger – v1 n1. 1997- – 1,9 – $65.00 in US $91.00 outside hardcopy subsc – us Haworth [020]

Journal of internet commerce / ed by Berry, Ronald – v1- 2002 – 1,9 – $75.00 us $109.00 other – ISSN: 1533-2861 – us Haworth [000]

Journal of internet law – New York. 1997+ (1,5,9) – ISSN: 1094-2904 – mf#31767 – us UMI ProQuest [346]

Journal of internet law – New York: Aspen Law & Business. v1-3. 1997-2000 – 9 – mf#117701 – us Hein [346]

Journal of interpersonal violence – Beverly Hills. 1989+ (1,5,9) – ISSN: 0886-2605 – mf#16945 – us UMI ProQuest [303]

Journal of intravenous nursing – Hagerstown. 1988+ (1,5,9) – (cont: nita) – ISSN: 0896-5846 – mf#11473,01 – us UMI ProQuest [610]

Journal of investigative dermatology – New York. 1996-1996 (1,5,9) – ISSN: 0022-202X – mf#24727 – us UMI ProQuest [616]

Journal of investment compliance – London. 2000+ (1,5,9) – ISSN: 1528-5812 – mf#32269 – us UMI ProQuest [332]

Journal of irish literature – Newark. 1972-1993 (1) 1972-1993 (5) 1977-1993 (9) – ISSN: 0047-2514 – mf#7278 – us UMI ProQuest [420]

Journal of irreproducible results – Park Forest South. 1980-1996 (1,5,9) – ISSN: 0022-2038 – mf#18091 – us UMI ProQuest [500]

Journal of irrigation and drainage engineering – New York. 1983+ (1,5,9) – (cont: journal of the irrigation and drainage division) – ISSN: 0733-9437 – mf#8147,01 – us UMI ProQuest [627]

Journal of islamic law – 9 – (See: journal of islamic law and culture) – mf#117392 – us Hein [260]

Journal of islamic law and culture – v1-6. 1996-2001 – 9 – $105.00 – ISSN: 1085-7141 – mf#117391 – us Hein [260]

Journal of islamic law and culture *see* Journal of islamic law

Journal of j l : of quebec, merchant – Detroit: Society of Colonial Wars of the State of Michigan, 1911 – 1mf – 9 – 0-665-66531-8 – mf#66531 – cn Canadiana [910]

Journal of james currie, 1776 – Liverpool Public Library – (= ser BRRAM series) – 1r – 1 – £67 / $134 – (int by r s craig) – mf#r95790 – uk Microform Academic [910]

Journal of jazz studies – New Brunswick. 1976-1979 (1,5,9) – ISSN: 0093-3686 – mf#11104 – us UMI ProQuest [780]

Journal of jewish communal service – New York. 1980+ (1,5,9) – ISSN: 0022-2089 – mf#12656,01 – us UMI ProQuest [939]

Journal of jewish education – New York. 1994+ – 1,5,9 – (cont: jewish education) – mf#12037,01 – us UMI ProQuest [370]

Journal of john gabriel stedman / Stedman, John Gabriel – London, England. 1962 – 1r – us UF Libraries [972]

Journal of john landreth on an expedition to the gulf coast, november 15 1818-may 19 1819 – (= ser Naval records coll of the office of naval records and library) – 1r – 1 – mf#T12 – us Nat Archives [917]

The journal of john woolman / Woolman, John – Boston: H Mifflin, c1871 – 1mf – 9 – 0-524-02484-7 – mf#1990-4343 – us ATLA [976]

Journal of joseph pilmore 1769-1774 : typed copied by rev cornelius hudson and mahlon g mazer / Pilmore, Joseph – 1r – 1 – $35.00 – mf#um-7 – us Commission [242]

Journal of joseph tindall / Tindall, Joseph – Cape Town, South Africa. 1959 – 1r – us UF Libraries [960]

Journal of jurisprudence : a new series of the american law journal – Philadelphia. 1821-1821 (1) – mf#4001 – us UMI ProQuest [323]

The journal of jurisprudence – Edinburgh. v1-35. 1857-91 (all publ) – 104mf – 9 – $468.00 – mf#LLMC 82-904 – us LLMC [340]

Journal of juvenile law – University of La Verne. v1-21. 1977-2000 – 9 – $259.00 set – (none publ 1987-88) – ISSN: 0160-2098 – mf#103171 – us Hein [340]

Journal of knee surgery – Thorofare. 2002+ (1,5,9) – (cont: american journal of knee surgery) – mf#17271,01 – us UMI ProQuest [617]

Journal of knowledge management – Kempston. 2001+ (1,5,9) – ISSN: 1367-3270 – mf#27301 – us UMI ProQuest [020]

Journal of korean affairs – Silver Spring. 1971-1974 [1,5] – ISSN: 0047-2522 – mf#6490 – us UMI ProQuest [951]

Journal of korean law – v1. 2001 – (filming in process) – ISSN: 1598-1681 – mf#119141 – us Hein [342]

Journal of labelled compounds – Brussels. 1965-1975 (1) 1965-1975 (5) 1965-1975 (9) – (cont by: journal of labelled compounds and radiopharmaceuticals) – ISSN: 0022-2135 – mf#10800 – us UMI ProQuest [540]

Journal of labelled compounds and radiopharmaceuticals – Chichester. 1976+ (1,5,9) – (cont: journal of labelled compounds) – ISSN: 0362-4803 – mf#10800,01 – us UMI ProQuest [540]

Journal of labor / Atlanta Federation of Trades et al – [v73 n68-v76 n11 [1968 jan 26-1969 dec 5]], 1903 oct 16 [v7 n3]-1907, 1908 jan-1911 dec, 1912-62, 1963-1964 mar 6, 1964 jun 26-1966 sep, 1966 oct-1969, 1970-72, 1973 jan 5-1974 dec 27, 1975 jan 3-1976 dec 24, 1977 jan 7-1978 dec 15, 1979-1985 may 24 – 31r – 1 – mf#382326 – us WHS [331]

Journal of labor – Nashville TN. 1897 mar 5, apr 23 – 1r – 1 – us WHS [331]

Journal of labor – Paducah, Kentucky. v. 2-4. Nov 4 1904-July 20 1907. Incomplete – 1 – us NY Public [331]

Journal of labor – v33 n28 [1934 jan 27]-v35 n3 [1935 sep 14] – 1r – 1 – mf#1277953 – us WHS [331]

Journal of labor and employment law *see* University of pennsylvania journal of labor and employment law

Journal of labor economics – Chicago. 1983+ (1,5,9) – ISSN: 0734-306X – mf#13362 – us UMI ProQuest [331]

Journal of labor research – Fairfax. 1980+ (1,5,9) – ISSN: 0195-3613 – mf#12309 – us UMI ProQuest [331]

Journal of laboratory and clinical medicine – St. Louis. 1915+ (1) 1965+ (5) 1970+ (9) – ISSN: 0022-2143 – mf#1018 – us UMI ProQuest [610]

The journal of lady grace mildmay : ms from northampton central library – 1r – 1 – mf#97377 – uk Microform Academic [920]

Journal of language and social psychology – Thousand Oaks. 1994-1995 (1,5,9) – ISSN: 0261-927X – mf#21666 – us UMI ProQuest [150]

Journal of language, identity, and education – Mahwah. 2002+ (1,5,9) – ISSN: 1534-8458 – mf#31735 – us UMI ProQuest [370]

Journal of laryngology and otology – London. 1961+ (1) 1965+ (5) 1973+ (9) – ISSN: 0022-2151 – mf#1318 – us UMI ProQuest [617]

Journal of latin american studies – Cambridge. 1969+ (1) 1976+ (5) 1976+ (9) – ISSN: 0022-216X – mf#11036 – us UMI ProQuest [972]

Journal of latinos and education – Mahwah. 2002+ (1,5,9) – ISSN: 1534-8431 – mf#31736 – us UMI ProQuest [370]

Journal of law – Philadelphia. v1 1830-31 (all publ) – (= ser Historical legal periodical series) – 1 – $45.00 set – mf#409040 – us Hein [340]

The journal of law – Philadelphia. v1 nos 1-24. 1830-31 (all publ) – 5mf – 9 – $7.50 – mf#LLMC 95-875 – us LLMC [340]

Journal of law and commerce – University of Pittsburgh. v1-19. 1981-2000 – 9 – $400.00 set – ISSN: 0733-2491 – mf#108761 – us Hein [346]

Journal of law and economic development *see* George washington journal of international law and economics

Journal of law and economics – Chicago. 1958+(1,5,9) – ISSN: 0022-2186 – mf#11845 – us UMI ProQuest [340]

Journal of law and economics – University of Chicago. v1-43. 1958-2000 – 9 – $564.00 set – ISSN: 0022-2186 – mf#104051 – us Hein [340]

Journal of law and education – Baltimore. 1972+ – 1,5,9 – ISSN: 0275-6072 – mf#10964 – us UMI ProQuest [370]

Journal of law and environment – University of Southern California. v1-5. 1985-87 (all publ) – 9 – $30.00 set – mf#114281 – us Hein [344]

Journal of law and family studies – Salt Lake City. 1999+ (1,5,9) – (cont: journal of contemporary law) – ISSN: 1529-398X – mf#31606 – us UMI ProQuest [346]

Journal of law and family studies – v1. 1999 – 9 – $15.00 – (supersedes: journal of contemporary law) – mf#118051 – us Hein [346]

Journal of law and policy – Brooklyn Law School. v1-9. 1993-2001 – 9 – $109.00 set – ISSN: 1074-0635 – mf#114981 – us Hein [340]

Journal of law and politics – v1-16. 1983-2000 – 9 – (filming in process) – ISSN: 0749-2227 – mf#109051 – us Hein [320]

Journal of law and religion – Hamline University. v1-15. 1983-2001 – 5,6,9 – $363.00 set – (v1-2 1983-84 on reel $78. v3-15 1985-2001 on mf $285) – ISSN: 0748-0814 – mf#109141 – us Hein [340]

Journal of law and social policy – v1-16. 1985-2001 – 9 – $236.00 set – ISSN: 0829-3929 – mf#115751 – us Hein [341]

Journal of law and society – Oxford. 1988-1996 (1,5,9) – ISSN: 0263-323X – mf#17397,01 – us UMI ProQuest [340]

Journal of law and society – v1-28. 1974-2001 – 9 – $628.00 set – (title varies: v1-8 as british journal of law and society) – ISSN: 0263-323X – mf#112301 – us Hein [340]

Journal of law and technology – Georgetown University. v1-5. 1986-90 (all publ) – 9 – $72.00 set – mf#110331 – us Hein [346]

Journal of law and technology *see* Idea

Journal of law, economics and organization – New Haven. 1990+ (1,5,9) – ISSN: 8756-6222 – mf#17995 – us UMI ProQuest [340]

Journal of law, medicine and ethics – Boston. 1993+ (1,5,9) – (cont: law, medicine and health care) – ISSN: 1073-1105 – mf#12094,02 – us UMI ProQuest [170]

Journal of law, medicine and ethics – v1-29. 1973-2001 – 9 – $459.00 set – (title varies: v1-9 n3, 1973-1981, as medicolegal news; v1-20, 1982-1992, as law, medicine and health care) – ISSN: 1073-1105 – mf#105281 – us Hein [344]

Journal of law reform *see* University of michigan journal of law reform

Journal of leadership and organizational studies – Flint. 2002+ (1,5,9) – mf#32720,01 – us UMI ProQuest [650]

Journal of learning disabilities – Austin. 1968+ (1) 1987+ (5) 1987+ (9) – ISSN: 0022-2194 – mf#16628 – us UMI ProQuest [370]

Journal of legal advocacy and practice – v1-3. 1999-2001 – 9 – $47.00 set – mf#117961 – us Hein [347]

Journal of legal aspects of sport – v1-11. 1991-2001 – 9 – $205.00 – ISSN: 1072-0316 – mf#114511 – us Hein [346]

Journal of legal education – Washington. 1981+ (1,5,9) – ISSN: 0022-2208 – mf#12977 – us UMI ProQuest [340]

Journal of legal medicine – American College of Legal Medicine. v1-22. 1979-2001 – 9 – $588.00 set – ISSN: 0194-7648 – mf#108771 – us Hein [340]

Journal of legal medicine – Bristol. 1979+ (1,5,9) – ISSN: 0194-7648 – mf#12817 – us UMI ProQuest [614]

Journal of legal medicine *see* Legal aspects of medical practice

Journal of legal studies – Chicago. 1972+ (1) 1976+ (5) 1976+ (9) – ISSN: 0047-2530 – mf#11093 – us UMI ProQuest [340]

Journal of legal studies – University of Chicago. v1-29. 1972-2000 – 9 – $624.00 set – ISSN: 0047-2530 – mf#104101 – us Hein [340]

Journal of legal studies education – v1-16. 1983-98 – 9 – $186.00 set – ISSN: 0896-5811 – mf#111991 – us Hein [340]

Journal of legislation – Notre Dame. 1976+ (1) 1976+ (5) 1976+ (9) – (cont: nd journal of legislation) – ISSN: 0146-9584 – mf#10253,01 – us UMI ProQuest [340]

Journal of legislation and public policy *see* New york university journal of legislation and public policy

Journal of leisurability – Islington. 1980-2000 (1) 1980-2000 (5) 1980-2000 (9) – (cont: leisurability) – ISSN: 0711-222X – mf#12714,01 – us UMI ProQuest [790]

Journal of leisure property – London. 2000+ (1,5,9) – ISSN: 1471-549X – mf#31753 – us UMI ProQuest [333]

Journal of leisure research – Arlington. 1969+ (1) 1972+ (5) 1975+ (9) – ISSN: 0022-2216 – mf#7688 – us UMI ProQuest [790]

Journal of lending and credit risk management – Philadelphia. 1995-1999 (1) 1995-1999 (5) 1995-1999 (9) – (cont: journal of commercial lending) – ISSN: 1088-7261 – mf#12339,03 – us UMI ProQuest [332]

Journal of lesbian studies / ed by Rothblum, Esther D – v1 n1. 1997- – 1,9 – $75.00 in US $105.00 outside hardcopy subsc – us Haworth [305]

The journal of liberal religion – v1-9. 1939-49 [complete] – Inquire – 1 – mf#ATLA 1994-S521 – us ATLA [200]

Journal of libertarian studies – New York. 1977-1978 (1,5,9) – ISSN: 0363-2873 – mf#49267 – us UMI ProQuest [320]

Journal of librarianship – London. 1986-1987 (1) 1986-1987 (5) 1986-1987 (9) – ISSN: 0022-2232 – mf#15452 – us UMI ProQuest [020]

Journal of library administration / ed by Lee, Sul H – v1- 1980- – 1, 9 ($115.00 in US $161.00 outside hardcopy subsc) – us Haworth [020]

Journal of library and information services in distance learning / ed by Dew, Stephen H – v1- 2004- – 1,9 – $150.00 us $218.00 other – ISSN: 1533-290X – us Haworth [020]

Journal of library automation – Chicago. 1968-1981 (1) 1970-1981 (5) 1975-1981 (9) – (cont by: information technology and libraries) – ISSN: 0022-2240 – mf#3071 – us UMI ProQuest [020]

Journal of library history (jlh) – Tallahassee. 1966-1987 (1) 1968-1987 (5) 1975-1987 (9) – (cont by: libraries and culture) – ISSN: 0275-3650 – mf#2511 – us UMI ProQuest [020]

Journal of lieutenant charles gauntt : aboard the u s s macedonian july 29 1818-june 18 1821 / U.S. Navy – (= ser Naval records coll of the office of naval records and library) – 1r – 1 – (with printed guide) – mf#M875 – us Nat Archives [355]
The journal of lieutenant commander william b cushing, 1861-1865 – (= ser Naval records coll of the office of naval records and library) – 1r – 1 – (with printed guide) – mf#M1034 – us Nat Archives [355]
Journal of light metals – Oxford, 2001+ [1,5,9] – ISSN: 1471-5317 – mf#42837 – us UMI ProQuest [660]
Journal of linguistics – Cambridge. 1965+ (1) 1976+ (5) 1976+ (9) – ISSN: 0022-2267 – mf#11037 – us UMI ProQuest [400]
Journal of lipid research – Bethesda. 1959+ (1) 1971+ (5) 1976+ (9) – ISSN: 0022-2275 – mf#6193 – us UMI ProQuest [540]
Journal of literacy research : (jlr) – Chicago. 1996+ (1) 1996+ (5) 1996+ (9) – (cont: journal of reading behavior) – ISSN: 1086-296X – mf#8196,01 – us UMI ProQuest [370]
Journal of logic and algebraic programming – New York, 2001+ [1,5,9] – (cont: journal of logic programming) – ISSN: 1567-8326 – mf#42453,01 – us UMI ProQuest [000]
Journal of logic, language and information – Dordrecht. 1992-1994 (1,5,9) – ISSN: 0925-8531 – mf#18659 – us UMI ProQuest [400]
Journal of logic programming – New York. 1984-1999 (1) 1984-1999 (5) 1984-2000 (9) – ISSN: 0743-1066 – mf#42453 – us UMI ProQuest [000]
Journal of long term care administration – Alexandria. 1973-1996 (1) 1973-1996 (5) 1973-1996 (9) – ISSN: 0093-4445 – mf#9138,01 – us UMI ProQuest [360]
Journal of loss prevention in the process industries – Kidlington. 1988-1996 (1,5,9) – ISSN: 0950-4230 – mf#17246 – us UMI ProQuest [660]
Journal of low temperature physics – New York. 1969-1994 (1) 1969-1994 (5) 1994-1994 (9) – ISSN: 0022-2291 – mf#10860 – us UMI ProQuest [530]
Journal of lubrication technology – New York. 1967-1983 (1) 1972-1983 (5) 1976-1983 (9) – (cont by: journal of tribology) – ISSN: 0022-2305 – mf#7543 – us UMI ProQuest [550]
Journal of luminescence – Amsterdam. 1970+ (1) 1970+ (5) 1987+ (9) – ISSN: 0022-2313 – mf#42293 – us UMI ProQuest [580]
Journal of macroeconomics – Baton Rouge. 1979+ (1,5,9) – ISSN: 0164-0704 – mf#12802 – us UMI ProQuest [339]
Journal of macromarketing – Boulder. 1990+ (1,5,9) – ISSN: 0276-1467 – mf#15817 – us UMI ProQuest [650]
Journal of magnetism and magnetic materials – Amsterdam. 1976+ (1) 1976+ (5) 1986+ (9) – ISSN: 0304-8853 – mf#42294 – us UMI ProQuest [530]
Journal of maintenance in the addictions : innovations in research, theory and practice / ed by Payte, J Thomas – v1 n1. 1997- – 1,9 – $60.00 in US $84.00 outside hardcopy subsc – us Haworth [616]
The journal of major geo. washington / Washington, George – 1754 – 9 – 5.00 – us Scholars Facs [975]
Journal of mammalogy – Baltimore. 1919+ (1) 1919+ (5) 1919+ (9) – ISSN: 0022-2372 – mf#718 – us UMI ProQuest [590]
Journal of man – v1 n9, v2 n1,7-8 [1907 nov 15, 1908 jan 1, 1908 apr 1-15] – 1r – 1 – (cont: light of the truth) – mf#701117 – us WHS [071]
Journal of management – Kidlington. 1975+(1,5,9) – ISSN: 0149-2063 – mf#11827 – us UMI ProQuest [650]
Journal of management accounting research – Sarasota. 1991-1996 (1,5,9) – ISSN: 1049-2127 – mf#19062 – us UMI ProQuest [650]
Journal of management consulting – Milwaukee. 1982-1998 (1) 1982-1998 (5) 1982-1998 (9) – (cont by: consulting to management) – ISSN: 0168-7778 – mf#15818 – us UMI ProQuest [650]
Journal of management development – Bradford. 1992-1995 (1,5,9) – ISSN: 0262-1711 – mf#15819 – us UMI ProQuest [650]
Journal of management education – Thousand Oaks. 1994+ (1,5,9) – ISSN: 1052-5629 – mf#21505,03 – us UMI ProQuest [650]
Journal of management in engineering – New York. 1985+ (1,5,9) – ISSN: 0742-597X – mf#14116 – us UMI ProQuest [620]
Journal of management information systems: jmis – Armonk. 1988+ (1,5,9) – ISSN: 0742-1222 – mf#16890 – us UMI ProQuest [650]
Journal of management studies – Oxford. 1982+ (1,5,9) – ISSN: 0022-2380 – mf#13529 – us UMI ProQuest [650]
Journal of managerial issues: jmi – Pittsburg. 1993-1996 (1,5,9) – ISSN: 1045-3695 – mf#19669 – us UMI ProQuest [650]
Journal of managerial psychology – Bradford. 1993-1995 (1,5,9) – ISSN: 0268-3946 – mf#16272 – us UMI ProQuest [150]
Journal of manual medicine – Berlin. 1989-1991 (1) 1989-1991 (5) 1989-1991 (9) – ISSN: 0935-6339 – mf#17006,01 – us UMI ProQuest [610]
Journal of manufacturing and operations management – Amsterdam. 1988-1990 (1,5,9) – ISSN: 0890-2577 – mf#42454 – us UMI ProQuest [650]
Journal of manufacturing processes – Dearborn. 2000+ (1) – ISSN: 1526-6125 – mf#28993 – us UMI ProQuest [670]
Journal of manufacturing science and engineering – New York. 1996+ (1,5,9) – (cont: journal of engineering for industry) – ISSN: 1087-1357 – mf#1190,01 – us UMI ProQuest [621]
Journal of manufacturing systems – Dearborn. 1982+ (1,5,9) – ISSN: 0278-6125 – mf#17545 – us UMI ProQuest [000]
Journal of map and geography libraries : advances in geospatial information, collections and archives / ed by Larsgaard, Mary Lynette & Andrew, Paige G – 1,9 – $200.00 us $290.00 – ISSN: 1542-0353 – us Haworth [900]
Journal of marine research – New Haven. 1980+ (1,5,9) – ISSN: 0022-2402 – mf#3496 – us UMI ProQuest [550]
Journal of marine systems – Amsterdam. 1990+ (1,5,9) – ISSN: 0924-7963 – mf#42622 – us UMI ProQuest [550]
Journal of marital and family therapy – Upland. 1979+ (1,5,9) – (cont: journal of marriage and family counseling) – ISSN: 0194-472X – mf#11303,01 – us UMI ProQuest [360]
Journal of maritime law and commerce – Baltimore. 1977+ (1,5,9) – ISSN: 0022-2410 – mf#11657 – us UMI ProQuest [380]
Journal of marketing – Chicago. 1936+ (1) 1969+ (5) 1975+ (9) – ISSN: 0022-2429 – mf#997 – us UMI ProQuest [650]
Journal of marketing channels : distribution systems, strategy, and management / ed by Rosenbloom, Bert – v5 n1. 1995- – 1,9 – $160.00 in US $224.00 outside hardcopy subsc – us Haworth [650]
Journal of marketing education: jme – Thousand Oaks. 1990+ (1,5,9) – ISSN: 0273-4753 – mf#18136 – us UMI ProQuest [650]
Journal of marketing for higher education / ed by Hayes, Thomas J – v1- 1988- – 1, 9 ($160.00 in US $224.00 outside hardcopy subsc) – us Haworth [380]
Journal of marketing theory and practice – Statesboro. 1992+ (1,5,9) – ISSN: 1069-6679 – mf#20765 – us UMI ProQuest [650]
Journal of marriage and family – Minneapolis. 2001+ (1,5,9) – (cont by: journal of marriage and family) – ISSN: 0022-2445 – mf#1066,01 – us UMI ProQuest [306]
Journal of marriage and family counseling – Claremont. 1975-1978 (1) 1975-1978 (5) 1975-1978 (9) – (cont by: journal of marital and family therapy) – ISSN: 0094-5102 – mf#11303 – us UMI ProQuest [150]
Journal of marriage and the family – Minneapolis. 1939-2000 (1) 1969-2000 (5) 1975-2000 (9) – ISSN: 0022-2445 – mf#1066 – us UMI ProQuest [306]
Journal of mass media ethics: mme – Mahwah. 1993+ (1,5,9) – ISSN: 0890-0523 – mf#19242 – us UMI ProQuest [170]
Journal of mass spectrometry – Chichester. 1995+ (1,5,9) – ISSN: 1076-5174 – mf#21264 – us UMI ProQuest [540]
Journal of materials – Conshohocken. 1966-1972 [1]; 1971-1972 [5,9] – ISSN: 0022-2453 – mf#2513 – us UMI ProQuest [620]
Journal of materials, design and applications *see* Proceedings of the institution of mechanical engineers pt 1
Journal of materials engineering – New York. 1987-1991 (1) 1987-1991 (5) 1987-1991 (9) – (cont: journal of materials for energy systems) – ISSN: 0931-7058 – mf#12937,01 – us UMI ProQuest [660]
Journal of materials engineering and performance – Materials Park. 1992+ (1,5,9) – ISSN: 1059-9495 – mf#19460 – us UMI ProQuest [620]
Journal of materials for energy systems – Metals Park. 1979-1986(1,5,9) – (cont by: journal of materials engineering) – ISSN: 0162-9719 – mf#12937 – us UMI ProQuest [660]
Journal of materials in civil engineering – New York. 1989+ (1,5,9) – ISSN: 0899-1561 – mf#16512 – us UMI ProQuest [624]
Journal of materials processing technology – Amsterdam. 1990+ (1,5,9) – (cont: journal of mechanical working technology) – ISSN: 0924-0136 – mf#42296,01 – us UMI ProQuest [621]
Journal of materials research – Pittsburgh. 1989+ (1,5,9) – ISSN: 0884-2914 – mf#17959 – us UMI ProQuest [620]
Journal of materials science – London. 1983-2000 (1,5,9) – ISSN: 0022-2461 – mf#14403 – us UMI ProQuest [620]
Journal of materials science letters – London. 1982-1999 (1,5,9) – ISSN: 0261-8028 – mf#14404 – us UMI ProQuest [620]
Journal of materials shaping technology – New York. 1987-1991 (1) 1987-1991 (5) 1987-1991 (9) – (cont: journal of applied metalworking) – ISSN: 0931-704X – mf#12935,01 – us UMI ProQuest [660]
Journal of mathematical biology – Wien. 1974-1995 (1) 1974-1995 (5) 1981-1995 (9) – ISSN: 0303-6812 – mf#13188 – us UMI ProQuest [574]
Journal of mathematical economics – Amsterdam. 1974+ (1) 1974+ (5) 1987+ (9) – ISSN: 0304-4068 – mf#42295 – us UMI ProQuest [330]
Journal of mathematical imaging and vision – Boston. 1992-1993 (1,5,9) – ISSN: 0924-9907 – mf#18666 – us UMI ProQuest [510]
Journal of mathematical physics – v1-. 1960- – 1,5,6,9 – us AIP [510]
Journal of maxillofacial surgery – Stuttgart. 1975-1986 (1) 1975-1986 (5) 1975-1986 (9) – (cont by: journal of cranio-maxillo-facial surgery) – ISSN: 0301-0503 – mf#10164 – us UMI ProQuest [617]
Journal of mechanical design – New York. 1990+ (1,5,9) – (cont: journal of mechanisms, transmissions, and automation in design) – ISSN: 1050-0472 – mf#13446,01 – us UMI ProQuest [627]
Journal of mechanical design – New York. 1978-1982 (1,5,9) – ISSN: 0161-8458 – mf#11949 – us UMI ProQuest [621]
Journal of mechanical engineering science – London. 1976-1982 (1,5,9) – ISSN: 0022-2542 – mf#11220 – us UMI ProQuest [621]
Journal of mechanical working technology – Amsterdam. 1977-1990 (1) 1977-1990 (5) 1985-1990 (9) – (cont by: journal of materials processing technology) – ISSN: 0378-3804 – mf#42296 – us UMI ProQuest [621]
Journal of mechanisms, transmissions, and automation in design – New York. 1983-1989 (1) 1983-1989 (5) 1983-1989 (9) – (cont by: journal of mechanical design) – ISSN: 0738-0666 – mf#13446 – us UMI ProQuest [620]
Journal of media and religion – Mahwah. 2002+ (1,5,9) – ISSN: 1534-8423 – mf#33040 – us UMI ProQuest [302]
Journal of medical and veterinary mycology – Abingdon. 1991-1994 (1,5,9) – (cont by: medical mycology) – ISSN: 0268-1218 – mf#18183,01 – us UMI ProQuest [616]
Journal of medical challenge – Evanston. 1973-1979 (1) 1973-1979 (5) 1975-1979 (9) – ISSN: 0190-5333 – mf#6952,02 – us UMI ProQuest [610]
Journal of medical education – Washington. 1926-1988 (1) 1949-1988 (5,9) – (cont by: academic medicine) – ISSN: 0022-2577 – mf#171 – us UMI ProQuest [610]
Journal of medical engineering and technology – London. 1991-1996 (1) – ISSN: 0309-1902 – mf#17329 – us UMI ProQuest [610]
Journal of medical entomology – Lanham. 1975+ (1,5,9) – ISSN: 0022-2585 – mf#10396 – us UMI ProQuest [590]
Journal of medical ethics – London. 1989+ (1,5,9) – ISSN: 0306-6800 – mf#18283 – us UMI ProQuest [170]
Journal of medical genetics – London. 1972+ (1) 1972+ (5) 1972+ (9) – ISSN: 0022-2593 – mf#8651 – us UMI ProQuest [575]
Journal of medical humanities – New York. 1989+ (1,5,9) – (cont: journal of medical humanities and bioethics) – ISSN: 1041-3545 – mf#12181,04 – us UMI ProQuest [170]
Journal of medical humanities and bioethics – New York. 1985-1988 (1,5,9) – (cont: journal of bioethics. cont by: journal of medical humanities) – ISSN: 0882-6498 – mf#12181,03 – us UMI ProQuest [170]
Journal of medical microbiology – London. 1982+ (1,5,9) – ISSN: 0022-2615 – mf#13429 – us UMI ProQuest [576]
Journal of medical practice management – v1-11. 1985-96 – 11r – 1,5,6,9 – $65.00r – us Lippincott [610]
Journal of medical primatology – Basel. 1972-1974 (1) 1972-1972 (5) (9) – ISSN: 0047-2565 – mf#6413 – us UMI ProQuest [610]
Journal of medical systems – New York. 1989-1996 (1,5,9) – ISSN: 0148-5598 – mf#11500 – us UMI ProQuest [610]
Journal of medical technology : official publication of american medical technologists and american society for medical technology – Houston. 1984-1987 (1,5,9) – (cont by: clinical laboratory science) – ISSN: 0741-5397 – mf#14246 – us UMI ProQuest [619]
Journal of medicinal chemistry – v1- 1959- – 1,5,6,9 – us ACS [612]
Journal of medicine – Basel. 1973-1973 (1) – ISSN: 0025-7850 – mf#8400 – us UMI ProQuest [610]
Journal of medicine and law – v1-4. 1997-2000 – 9 – $71.00 – mf#118881 – us Hein [344]
Journal of medicine and philosophy – Dordrecht. 1985+ (1,5,9) – ISSN: 0360-5310 – mf#15260 – us UMI ProQuest [610]
Journal of medieval and early modern studies – Durham. 1996+ (1) 1996+ (5) 1996+ (9) – (cont: journal of medieval and renaissance studies) – ISSN: 1082-9636 – mf#9070,01 – us UMI ProQuest [941]
Journal of medieval and renaissance studies – Durham. 1971-1995 (1) 1975-1995 (5) 1975-1995 (9) – (cont by: journal of medieval and early modern studies) – ISSN: 0047-2573 – mf#9070 – us UMI ProQuest [941]
Journal of medieval history – Amsterdam. 1975+ (1) 1975+ (5) 1987+ (9) – ISSN: 0304-4181 – mf#42297 – us UMI ProQuest [940]
Journal of membrane biology – Heidelberg. 1969-1996 (1) 1969-1996 (5) 1969-1996 (9) – ISSN: 0022-2631 – mf#13189 – us UMI ProQuest [574]
Journal of membrane science – Amsterdam. 1977-1996 (1) 1977-1996 (5) 1987-1996 (9) – ISSN: 0376-7388 – mf#42298 – us UMI ProQuest [540]
Journal of mental deficiency research – Oxford. 1972-1991 (1) 1972-1991 (5) 1972-1991 (9) – (cont by: journal of intellectual disability research: jidr) – ISSN: 0022-264X – mf#8924 – us UMI ProQuest [616]
Journal of mental health administration – Chicago. 1990-1996 (1,5,9) – (cont by: journal of behavioral health services and research) – ISSN: 0092-8623 – mf#18341 – us UMI ProQuest [360]
Journal of mental health counseling – Alexandria. 1987+ (1,5,9) – (cont: amhca journal) – ISSN: 0193-1830 – mf#11726,01 – us UMI ProQuest [150]
Journal of metals – New York. 1950-1974 (1) 1971-1974 (5) – (cont by: jom) – ISSN: 0022-2674 – mf#531 – us UMI ProQuest [660]
Journal of metals – New York. 1977-1988 (1) 1977-1988 (5) 1977-1988 (9) – (cont: jom. cont by: jom) – ISSN: 0148-6608 – mf#531,02 – us UMI ProQuest [660]
Journal of metamorphic geology – Oxford. 1983-1996 (1,5,9) – ISSN: 0263-4929 – mf#15532 – us UMI ProQuest [550]
Journal of mexican american history – Santa Barbara. 1970-1975 (1) 1973-1975 (5) 1975-1975 (9) – ISSN: 0047-2581 – mf#6912 – us UMI ProQuest [972]
Journal of microbiological methods – Amsterdam. 1983+ (1) 1983+ (5) 1984+ (9) – ISSN: 0167-7012 – mf#42422 – us UMI ProQuest [576]
Journal of microencapsulation – London. 1994-1995 (1) – ISSN: 0265-2048 – mf#17330 – us UMI ProQuest [610]
Journal of micrographics – Silver Spring. 1969-1983 (1) 1969-1983 (5) 1969-1983 (9) – (cont: nma journal. cont by: journal of information and image management) – ISSN: 0022-2712 – mf#7820,01 – us UMI ProQuest [020]
Journal of micronutrient analysis – Barking. 1985-1990 (1,5,9) – ISSN: 0266-349X – mf#42530 – us UMI ProQuest [613]
Journal of microscopy – Oxford. 1980-1995 (1,5,9) – ISSN: 0022-2720 – mf#15564,01 – us UMI ProQuest [578]
Journal of midwifery and women's health – New York. 2000+ (1,5,9) – (cont: journal of nurse-midwifery) – ISSN: 1526-9523 – mf#42090,01 – us UMI ProQuest [618]
Journal of military history – Lexington. 1989+ (1,5,9) – ISSN: 0899-3718 – mf#17552,03 – us UMI ProQuest [355]
Journal of milk and food technology – Ames. 1937-1976 (1) 1971-1976 (5) – (cont by: journal of food protection) – ISSN: 0146-3802 – mf#2165 – us UMI ProQuest [660]
Journal of mineral law and policy *see* Journal of natural resources and environmental law
Journal of ministry in addiction and recovery / ed by Albers, Robert H – v3 n1. 1996- – 1,9 – $120.00 in US $168.00 outside hardcopy subsc – us Haworth [616]
Journal of ministry marketing and management / ed by Stevens, Robert E & Loudon, David L – v2 n1. 1996- – 1,9 – $75.00 in US $105.00 outside hardcopy subsc – us Haworth [650]
Journal of minnesota public law *see* Hamline journal of public law and policy
Journal of missions – Boston: American Board of Commissioners for Foreign Missions, 1849- [mf v1-6 1849-55 filmed 2002] – 1r – 1 – (lacks: oct 1949. ceased in 1856. merged with: youth's dayspring to form: journal of missions and youth's dayspring) – mf#2003-s511 – us ATLA [240]
Journal of missions *see* The youth's dayspring
Journal of missions and youth's dayspring *see*
– Journal of missions
– The youth's dayspring
Journal of mississippi history – Jackson. 1979+ (1,5,9) – ISSN: 0022-2771 – mf#12387 – us UMI ProQuest [978]
Journal of modern african studies – Cambridge. 1963+ (1) 1976+ (5) 1976+ (9) – ISSN: 0022-278X – mf#11038 – us UMI ProQuest [960]

Journal of modern history – Chicago. 1929+ (1) 1969+ (5) 1977+ (9) – ISSN: 0022-2801 – mf#486 – us UMI ProQuest [900]

Journal of modern literature – Philadelphia. 1970+ (1) 1973+ (5) 1974+ (9) – ISSN: 0022-281X – mf#7088 – us UMI ProQuest [400]

Journal of molecular catalysis – Amsterdam. 1975-1995 (1) 1975-1995 (5) 1987-1995 (9) – ISSN: 0304-5102 – mf#42081 – us UMI ProQuest [540]

Journal of molecular catalysis a : chemical – Amsterdam. 1996-1996 (1,5,9) – ISSN: 1381-1169 – mf#42771 – us UMI ProQuest [540]

Journal of molecular catalysis b : enzymatic – Amsterdam. 1996-1996 (1,5,9) – ISSN: 1381-1177 – mf#42772 – us UMI ProQuest [540]

Journal of molecular electronics – Chichester. 1985-1991 (1,5,9) – ISSN: 0748-7991 – mf#14812 – us UMI ProQuest [621]

Journal of molecular evolution – Heidelberg. 1981-1996 (1) 1981-1996 (5) 1971-1996 (9) – ISSN: 0022-2844 – mf#13190 – us UMI ProQuest [575]

Journal of molecular graphics – New York. 1987-1994 (1) 1987-1994 (5) 1987-1994 (9) – (cont by: journal of molecular graphics and modelling) – ISSN: 0263-7855 – mf#16648 – us UMI ProQuest [000]

Journal of molecular graphics and modelling – New York. 1997+ (1) – (cont: journal of molecular graphics) – ISSN: 1093-3263 – mf#16648,01 – us UMI ProQuest [000]

Journal of molecular liquids – Amsterdam. 1967+ (1) 1967+ (5) 1987+ (9) – ISSN: 0167-7322 – mf#42104 – us UMI ProQuest [540]

Journal of molecular recognition: jmr – Chichester. 1991+ (1,5,9) – ISSN: 0952-3499 – mf#18160 – us UMI ProQuest [574]

Journal of molecular structure – Amsterdam. 1967-1991 (1) 1967-1991 (5) 1987-1991 (9) – ISSN: 0022-2860 – mf#42082 – us UMI ProQuest [540]

Journal of molecular structure – Amsterdam. 1992-1992 (1,5,9) – ISSN: 0022-2860 – mf#42724 – us UMI ProQuest [540]

Journal of monetary economics – Amsterdam. 1975+ (1) 1975+ (5) 1987+ (9) – ISSN: 0304-3932 – mf#42083 – us UMI ProQuest [332]

Journal of money, credit, and banking – Columbus. 1969+ (1) 1971+ (5) 1975+ (9) – ISSN: 0022-2879 – mf#5753 – us UMI ProQuest [332]

Journal of money laundering control – London, 1997+ [1,5,9] – ISSN: 1368-5201 – mf#31709 – us UMI ProQuest [332]

Journal of moral education – Abingdon. 1971+ (1) 1976+ (5) 1976+ (9) – ISSN: 0305-7240 – mf#10235 – us UMI ProQuest [370]

Journal of morphology – Philadelphia. 1999+ (1,5,9) – ISSN: 0362-2525 – mf#24830,02 – us UMI ProQuest [574]

Journal of motor behavior – Washington. 1969+ (1) 1972+ (5) 1975+ (9) – ISSN: 0022-2895 – mf#6594 – us UMI ProQuest [370]

Journal of multi-body dynamics *see* Proceedings of the institution of mechanical engineers pt k

Journal of multicriteria decision analysis – Chichester. 1992-1992 (1,5,9) – ISSN: 1057-9214 – mf#19121 – us UMI ProQuest [650]

Journal of multicultural counseling and development – Alexandria. 1985+ (1) 1985+ (5) 1985+ (9) – (cont: journal of non-white concerns in personnel and guidance) – ISSN: 0883-8534 – mf#8117,01 – us UMI ProQuest [305]

Journal of multicultural social work / ed by Anda, Diane de – v4 n1. 1996- – 1,9 – $125.00 in US $175.00 outside hardcopy subsc – us Haworth [360]

Journal of multicultural social work *see* Journal of ethnic and cultural diversity in social work

Journal of multistate taxation and incentives – Boston. 1999+ (1) – mf#19013,01 – us UMI ProQuest [336]

Journal of muscle research and cell motility – London. 1989-1989 (1) – ISSN: 0142-4319 – mf#14405 – us UMI ProQuest [612]

Journal of musculoskeletal pain : innovations in research, theory and clinical practice / ed by Russell, I Jon – v4 n1. 1996- – 1,9 – $125.00 in US $175.00 outside hardcopy subsc – us Haworth [616]

Journal of music theory – New Haven. 1957+ (1) 1957+ (5) 1957+ (9) – ISSN: 0022-2909 – mf#12672 – us UMI ProQuest [780]

Journal of music therapy – Silver Spring. 1964+ (1,5,9) – ISSN: 0022-2917 – mf#12860 – us UMI ProQuest [780]

Journal of musick – Baltimore. 1810-1810 (1) – mf#4002 – us UMI ProQuest [780]

Journal of musicology: jm – St. Joseph. 1985+ (1,5,9) – ISSN: 0277-9269 – mf#15680 – us UMI ProQuest [780]

Journal of narrative technique – Ypsilanti. 1971-1998 (1) 1971-1998 (5) 1976-1998 (9) – (cont by: journal of narrative theory: jnt) – ISSN: 0022-2925 – mf#6415 – us UMI ProQuest [400]

Journal of narrative theory (jnt) – Ypsilanti. 1999+ (1) 1999+ (5) 1999+ (9) – (cont: journal of narrative technique) – mf#6415,01 – us UMI ProQuest [400]

Journal of natural fibers – v1- – 1,9 – $160.00 us $232.00 other – ISSN: 1544-0478 – us Haworth [670]

Journal of natural history – London. 1989-1996 (1,5,9) – ISSN: 0022-2933 – mf#17331 – us UMI ProQuest [500]

Journal of natural resources and environmental law – University of Kentucky. v1-15. 1985-2001 – 9 – $230.00 set – (title varies: v1-7 1985-92 as journal of mineral law and policy) – ISSN: 0892-9017 – mf#110911 – us Hein [344]

Journal of near eastern studies – Chicago. 1942+ (1) 1969+ (5) 1978+ (9) – ISSN: 0022-2968 – mf#1053 – us UMI ProQuest [956]

Journal of near-death studies – New York. 1991+ (1) – ISSN: 0891-4494 – mf#16144,01 – us UMI ProQuest [150]

Journal of negro education – Washington. 1932+ (1) 1972+ (5) 1975+ (9) – ISSN: 0022-2984 – mf#6853 – us UMI ProQuest [370]

Journal of negro history – Washington. 1916+ (1) 1965+ (5) 1970+ (9) – ISSN: 0022-2992 – mf#1036 – us UMI ProQuest [305]

Journal of nematology – College Park. 1989-1991 (1) – ISSN: 0022-300X – mf#14060 – us UMI ProQuest [590]

Journal of nervous and mental disease – v120-184. 1954-96 – 1,5,6,9 – $110.00r – us Lippincott [616]

The journal of nervous and mental disease – v. 1-118. 1874-1953 – 1 – us AMS Press [616]

Journal of neural transmission : general section – Wien. 1989-1993 (1,5,9) – ISSN: 0300-9564 – mf#17966 – us UMI ProQuest [612]

Journal of neural transmission – Wien. 1984-1989 (1,5,9) – ISSN: 0300-9564 – mf#13266,02 – us UMI ProQuest [612]

Journal of neuro-aids : a forum devoted to advances in the neurology and neurobiology of human immunodeficiency virus (hiv), aids and related viral infections of the nervous system / ed by Price, Richard W – v1 n1. 1996- – 1,9 – $125.00 in US $175.00 outside hardcopy subsc – us Haworth [616]

Journal of neurobiology – New York. 1969+ – 1,5,9 – ISSN: 0022-3034 – mf#11058 – us UMI ProQuest [612]

Journal of neuroendocrinology – Oxford. 1989-1990 (1,5,9) – ISSN: 0953-8194 – mf#17506 – us UMI ProQuest [616]

Journal of neuroimmunology – Amsterdam. 1981-1996 (1) 1981-1996 (5) 1986-1996 (9) – ISSN: 0165-5728 – mf#42299 – us UMI ProQuest [616]

Journal of neurolinguistics – Tokyo. 1988-1995 (1,5,9) – ISSN: 0911-6044 – mf#49561 – us UMI ProQuest [400]

Journal of neurology = Zeitschrift fuer neurologie – Heidelberg. 1977-1995 (1) 1977-1995 (5) 1974-1995 (9) – (cont: zeitschrift fuer neurologie) – ISSN: 0340-5354 – mf#13121,02 – us UMI ProQuest [616]

Journal of neurology, neurosurgery and psychiatry – London. 1926+ (1) 1971+ (5) 1976+ (9) – ISSN: 0022-3050 – mf#1336 – us UMI ProQuest [616]

Journal of neuro-oncology – Boston. 1991-1994 (1) 1991-1994 (5) 1991-1994 (9) – ISSN: 0167-594X – mf#16812 – us UMI ProQuest [616]

Journal of neuro-ophthalmology – New York. 1994-1996 (1,5,9) – ISSN: 1070-8022 – mf#18714,01 – us UMI ProQuest [617]

Journal of neuropathic pain and symptom palliation / ed by Smith, Howard – 1,9 – $250.00 us $363.00 other – ISSN: 1543-7698 – us Haworth [617]

Journal of neuropathology and experimental neurology – Lawrence. 1949+ (1) 1965+ (5) 1970+ (9) – ISSN: 0022-3069 – mf#103 – us UMI ProQuest [616]

Journal of neurophysiology – Bethesda. 1938+ (1) 1965+ (5) 1970+ (9) – ISSN: 0022-3077 – mf#1619 – us UMI ProQuest [612]

Journal of neuropsychiatry and clinical neurosciences – Washington. 1989+ (1,5,9) – ISSN: 0895-0172 – mf#17519 – us UMI ProQuest [616]

Journal of neuroscience – Baltimore. 1989+ (1,5,9) – ISSN: 0270-6474 – mf#17037 – us UMI ProQuest [612]

Journal of neuroscience methods – Amsterdam. 1979-1992 (1) 1979-1992 (5) 1987-1992 (9) – ISSN: 0165-0270 – mf#42085 – us UMI ProQuest [612]

Journal of neuroscience nursing – Park Ridge. 1986+ (1,5,9) – (cont: journal of neurosurgical nursing) – ISSN: 0888-0395 – mf#12046,01 – us UMI ProQuest [610]

Journal of neurosurgery – Charlotte. 1944+ (1) 1971+ (5) 1976+ (9) – ISSN: 0022-3085 – mf#2129 – us UMI ProQuest [617]

Journal of neurosurgical nursing – Chicago. 1969-1985 (1) 1969-1985 (5) 1969-1985 (9) – (cont by: journal of neuroscience nursing) – ISSN: 0047-2603 – mf#12046 – us UMI ProQuest [610]

Journal of neurosurgical sciences – Torino. 1975-1991 (1) 1975-1991 (5) 1975-1991 (9) – ISSN: 0390-5616 – mf#10040,01 – us UMI ProQuest [617]

Journal of neurotherapy : the official publication of the international society for neuronal regulation / ed by Trudeau, David L – v1- 1995- – 1,9 – $115.00 us $167.00 other – ISSN: 1087-4208 – us Haworth [617]

Journal of new seeds : Innovations in production, biotechnology, quality, and marketing – v1- 1999- – 1,9 – $95.00 us $138.00 other – ISSN: 1522-886X – us Haworth [631]

Journal of non-crystalline solids – Amsterdam. 1968+ (1) 1968+ (5) 1987+ (9) – ISSN: 0022-3093 – mf#42086 – us UMI ProQuest [530]

Journal of nondestructive evaluation – New York. 1994-1995 (1,5,9) – ISSN: 0195-9298 – mf#17680 – us UMI ProQuest [620]

Journal of non-newtonian fluid mechanics – Amsterdam. 1976-1996 (1) 1976-1996 (5) 1987-1996 (9) – ISSN: 0377-0257 – mf#42087 – us UMI ProQuest [530]

Journal of nonprofit and public sector marketing / ed by Self, Donald R – v1- 1992- – 1,9 ($160.00 in US $224.00 outside hardcopy subsc) – (cont: journal of marketing for mental health) – us Haworth [650]

Journal of nonverbal behavior – New York. 1979+ (1,5,9) – (cont: environmental psychology and nonverbal behavior) – ISSN: 0191-5886 – mf#11178,01 – us UMI ProQuest [150]

Journal of non-white concerns in personnel and guidance – Washington. 1972-1985 (1) 1973-1985 (5) 1975-1985 (9) – (cont by: journal of multicultural counseling and development) – ISSN: 0090-5461 – mf#8117 – us UMI ProQuest [305]

Journal of nuclear biology and medicine – Torino. 1976-1976 (1) 1976-1976 (5) 1976-1976 (9) – (cont by: journal of nuclear medicine and allied sciences) – ISSN: 0368-3249 – mf#10023 – us UMI ProQuest [616]

Journal of nuclear biology and medicine – Torino. 1991-1994 (1) 1991-1994 (5) 1991-1994 (9) – (cont: journal of nuclear medicine and allied sciences. cont by: quarterly journal of nuclear medicine) – mf#10023,02 – us UMI ProQuest [616]

Journal of nuclear energy – New York. 1961-1973 (1) 1954-1973 (5) 1973-1973 (9) – (cont by: annals of nuclear science and engineering) – ISSN: 0022-3107 – mf#49010 – us UMI ProQuest [530]

Journal of nuclear materials – Amsterdam. 1959+ (1) 1959+ (5) 1987+ (9) – ISSN: 0022-3115 – mf#42088 – us UMI ProQuest [530]

Journal of nuclear medicine – New York. 1960+ (1) 1971+ (5) 1976+ (9) – ISSN: 0161-5505 – mf#2158 – us UMI ProQuest [616]

Journal of nuclear medicine and allied sciences – Torino. 1977-1990 (1) 1977-1990 (5) 1977-1990 (9) – (cont: journal of nuclear biology and medicine. cont by: journal of nuclear biology and medicine) – ISSN: 0392-0208 – mf#10023,01 – us UMI ProQuest [616]

Journal of nuclear medicine technology – Reston. 1976+ (1,5,9) – ISSN: 0091-4916 – mf#11079 – us UMI ProQuest [616]

Journal of numismatic fine arts – Encino. 1971-1977 (1) 1971-1977 (5) 1976-1977 (9) – ISSN: 0047-2611 – mf#7472 – us UMI ProQuest [930]

Journal of nurse-midwifery – New York. 1975-1999 (1,5,9) – (cont by: journal of midwifery and women's health) – ISSN: 0091-2182 – mf#42090 – us UMI ProQuest [618]

Journal of nursing administration – Philadelphia. 1971+ (1) 1974+ (5) 1974+ (9) – ISSN: 0002-0443 – mf#9994 – us UMI ProQuest [610]

Journal of nursing care – Westport. 1978-1982 (1) 1978-1982 (5) 1978-1982 (9) – (cont: nursing care) – ISSN: 0162-7155 – mf#2564,02 – us UMI ProQuest [610]

Journal of nursing care quality – Gaithersburg. 1991+ (1,5,9) – (cont: journal of nursing quality assurance) – ISSN: 1057-3631 – mf#16005,01 – us UMI ProQuest [610]

Journal of nursing education – Thorofare. 1983+ (1) 1983+ (5) 1983+ (9) – (cont: jne journal of nursing education) – ISSN: 0148-4834 – mf#6469,01 – us UMI ProQuest [610]

Journal of nursing management – Oxford. 1993-1996 (1,5,9) – ISSN: 0966-0429 – mf#19658 – us UMI ProQuest [610]

Journal of nursing quality assurance – Frederick. 1986-1990 (1) 1986-1990 (5) 1986-1990 (9) – (cont by: journal of nursing care quality) – ISSN: 0889-4647 – mf#16005 – us UMI ProQuest [610]

Journal of nursing scholarship – Indianapolis. 2000+ (1,5,9) – (cont: image – the journal of nursing scholarship) – ISSN: 1527-6546 – mf#19463,02 – us UMI ProQuest [610]

Journal of nursing staff development: jnsd – Hagerstown. 1985-1998 (1,5,9) – (cont by: journal for nurses in staff development: jnsd) – ISSN: 0882-0627 – mf#14443 – us UMI ProQuest [610]

Journal of nutraceuticals, functional and medical foods : product development, commercialization, and policy issues / ed by Childs, Nancy M – v1 n1. 1997- – 1,9 – $120.00 in US $168.00 outside hardcopy subsc – us Haworth [660]

Journal of nutrition – Bethesda. 1928+ (1) 1928+ (5) 1975+ (9) – ISSN: 0022-3166 – mf#6568 – us UMI ProQuest [613]

Journal of nutrition education – Hamilton. 1969+ (1) 1972+ (5) 1976+ (9) – ISSN: 0022-3182 – mf#8236 – us UMI ProQuest [613]

Journal of nutrition education and behavior – Hamilton. 2002+ (1,5,9) – ISSN: 1499-4046 – mf#8236,01 – us UMI ProQuest [613]

Journal of nutrition for the elderly / ed by Natow, Annette B – v1- 1980- – 1, 9 ($300.00 in US $420.00 outside hardcopy subsc) – us Haworth [618]

Journal of nutrition in recipe and menu development : innovations in nutritional products, dietary substitutes, and medical issues in food product development / ed by Khan, Mahmood A – v2 n1. 1996- – 1,9 – $125.00 in US $175.00 outside hardcopy subsc – us Haworth [613]

Journal of nutritional biochemistry – New York. 1990+ (1,5,9) – (cont: nutrition reports international) – ISSN: 0955-2863 – mf#17581 – us UMI ProQuest [613]

Journal of nutritional immunology / ed by Spallholz, Julian E – v4 n1. 1996- – 1,9 – $120.00 in US $168.00 outside hardcopy subsc – us Haworth [613]

Journal of obesity and weight regulation – New York. 1982-1989 (1) 1982-1989 (5) 1982-1989 (9) – (cont: obesity and metabolism) – ISSN: 0731-4361 – mf#12190,01 – us UMI ProQuest [616]

Journal of obstetric, gynecologic, and neonatal nursing (jognn) – Philadelphia. 1985+ (1) 1985+ (5) 1985+ (9) – (cont: jogn nursing) – ISSN: 0884-2175 – mf#8778,01 – us UMI ProQuest [610]

Journal of obstetrics and gynaecology – Bristol. 1987-1996 (1,5,9) – ISSN: 0144-3615 – mf#14446 – us UMI ProQuest [618]

Journal of obstetrics and gynaecology of the british commonwealth – Kidlington. 1902-1974 (1) 1954-1974 (5,9) – (cont by: british journal of obstetrics and gynaecology) – ISSN: 0022-3204 – mf#681 – us UMI ProQuest [618]

Journal of occupational accidents – Amsterdam. 1976-1990 (1) 1976-1990 (5) 1987-1990 (9) – (cont by: safety science) – ISSN: 0376-6349 – mf#42089 – us UMI ProQuest [360]

Journal of occupational and environmental medicine – Baltimore. 1995+(1,5,9) – (cont: jom journal of occupational medicine) – ISSN: 1076-2752 – mf#12294,02 – us UMI ProQuest [360]

Journal of occupational and environmental medicine – v28-38. 1986-96 – 11r – 1,5,6,9 – $85.00r – us Lippincott [610]

Journal of occupational and organizational psychology – Leicester. 1992+(1,5,9) – ISSN: 0963-1798 – mf#14708,04 – us UMI ProQuest [150]

Journal of occupational behaviour – Chichester. 1980-1987 (1) 1980-1987 (5) 1980-1987 (9) – (cont by: journal of organizational behavior) – ISSN: 0142-2774 – mf#11998 – us UMI ProQuest [150]

Journal of occupational medicine – New York. 1959-1967 (1) 1959-1967 (5) 1959-1967 (9) – (cont by: jom journal of occupational medicine) – ISSN: 0022-3212 – mf#12294 – us UMI ProQuest [360]

Journal of occurrences – n1-4 [1975 jul-nov/dec] – 1r – 1 – mf#1060798 – us WHS [071]

Journal of offender counseling – Alexandria. 1980-1990 (1) 1980-1990 (5) 1980-1990 (9) – (cont by: journal of addictions and offender counseling) – ISSN: 0275-8598 – mf#12744 – us UMI ProQuest [360]

Journal of offender rehabilitation / ed by Pallone, Nathaniel J – v1- 1976- – 1, 9 ($225.00 in US $315.00 outside hardcopy subsc) – us Haworth [360]

Journal of offshore mechanics and arctic engineering – New York. 1987+ (1,5,9) – ISSN: 0892-7219 – mf#16195 – us UMI ProQuest [627]

Journal of operational psychiatry – Columbia. 1970-1985 (1) 1974-1985 (5) 1976-1985 (9) – ISSN: 0047-2638 – mf#9952 – us UMI ProQuest [616]

Journal of operations management – Columbia. 1993+ (1,5,9) – ISSN: 0272-6963 – mf#42713 – us UMI ProQuest [629]

Journal of optimization theory and applications – New York. 1967-1996 (1) 1967-1996 (5) 1978-1996 (9) – ISSN: 0022-3239 – mf#10861 – us UMI ProQuest [510]

Journal of optometric education – Rockville. 1989-1990 (1) – ISSN: 0098-6917 – mf#12629 – us UMI ProQuest [617]

Journal of oral and maxillofacial surgery : official journal of the american association of oral and maxillofacial surgeons / American Association of Oral and Maxillofacial Surgeons – Philadelphia. 1982+ (1) 1982+ (5) 1982+ (9) – (cont: journal of oral surgery) – ISSN: 0278-2391 – mf#601,01 – us UMI ProQuest [617]

Journal of oral medicine – New York. 1946-1987 (1) 1971-1987 (5) 1976-1987 (9) – ISSN: 0022-3247 – mf#2532 – us UMI ProQuest [617]

Journal of oral rehabilitation – Oxford. 1980-1996 (1,5,9) – ISSN: 0305-182X – mf#15533 – us UMI ProQuest [617]

Journal of oral surgery / American Dental Association – Chicago. 1943-1981 (1) 1975-1981 (5) 1975-1981 (9) – (cont by: journal of oral and maxillofacial surgery : official journal of the american association of oral and maxillofacial surgeons) – ISSN: 0022-3255 – mf#601 – us UMI ProQuest [617]

The journal of organic chemistry – v1- 1936- – 1,5,6,9 – us ACS [540]

Journal of organic chemistry of the ussr – New York. 1965-1991 (1) 1965-1991 (5) 1972-1991 (9) – ISSN: 0022-3271 – mf#10836 – us UMI ProQuest [540]

Journal of organizational behavior – Chichester. 1988+ (1,5,9) – (cont: journal of occupational behaviour) – ISSN: 0894-3796 – mf#11998,01 – us UMI ProQuest [150]

Journal of organizational behavior management / ed by Mawhinney, Thomas C – v1- 1977- – 1, 9 ($250.00 in US $350.00 outside hardcopy subsc) – us Haworth [650]

Journal of organizational excellence – New York. 2000+ (1) – (cont: national productivity review) – ISSN: 1531-1864 – mf#14306,01 – us UMI ProQuest [650]

Journal of organometallic chemistry – Lausanne. 1963+ (1) 1963+ (5) 1987+ (9) – ISSN: 0022-328X – mf#42109 – us UMI ProQuest [540]

Journal of orgonomy – Princeton. 1967+ (1) 1980+ (5) 1980+ (9) – ISSN: 0022-3298 – mf#7681 – us UMI ProQuest [616]

Journal of oriental research – Madras, 1927-1957/1958. v1-27 – 130mf – 9 – mf#I-1167 – ne IDC [240]

Journal of oriental studies – Hong Kong. 1954+ (1) 1972+ (5) 1976+ (9) – ISSN: 0022-331X – mf#7040 – us UMI ProQuest [950]

Journal of orthodontics – Oxford. 2000+ (1,5,9) – ISSN: 1465-3125 – mf#13422,01 – us UMI ProQuest [617]

Journal of orthomolecular medicine – Regina. 1986+ (1,5,9) – (cont: journal of orthomolecular psychiatry) – mf#15489 – us UMI ProQuest [616]

Journal of orthomolecular psychiatry – Regina. 1974-1985 [1]; 1975-1985 [5,9] – (cont by: journal of orthomolecular medicine) – ISSN: 0317-0209 – mf#10538,01 – us UMI ProQuest [616]

Journal of orthomolecular psychiatry – Regina. 1974-1985 (1) 1975-1985 (5) 1975-1985 (9) – (cont: orthomolecular psychiatry) – ISSN: 0317-0209 – mf#10538,01 – us UMI ProQuest [616]

The journal of orthopaedic and sports physical therapy – v1-24. 1979-96 – 1,5,6,9 – $80.00r – us Lippincott [617]

Journal of orthopaedic research – New York. 1993-1996 (1,5,9) – ISSN: 0736-0266 – mf#18721 – us UMI ProQuest [617]

Journal of orthopaedic trauma – New York. 1993+ (1,5,9) – ISSN: 0890-5339 – mf#18722 – us UMI ProQuest [617]

Journal of otolaryngology – Hamilton. 1976+ (1) 1976+ (5) 1976+ (9) – (cont: canadian journal of otolaryngology) – ISSN: 0381-6605 – mf#7640,01 – us UMI ProQuest [617]

Journal of outdoor education – Oregon. 1966-1993 (1) 1966-1993 (5) 1966-1993 (9) – ISSN: 0022-3336 – mf#12767 – us UMI ProQuest [370]

Journal of paediatric dentistry – Oxford. 1985-1990 (1,5,9) – ISSN: 0267-2073 – mf#15565 – us UMI ProQuest [617]

Journal of paediatrics and child health – Melbourne. 1990+ (1,5,9) – (cont: australian paediatric journal) – ISSN: 1034-4810 – mf#8744,01 – us UMI ProQuest [618]

Journal of pain and palliative care pharmacotherapy : advances in acute, chronic, and end-of-life pain and symptom control / ed by Lipman, Arthur G – v16- 2001- – 1,9 – $120.00 us $174.00 other – ISSN: 1536-0288 – us Haworth [615]

Journal of pain and symptom management – Madison. 1989+ (1,5,9) – ISSN: 0885-3924 – mf#42587 – us UMI ProQuest [610]

Journal of paint technology – Blue Bell. 1966-1975 (1) 1971-1975 (5) – (cont by: jct: journal of coatings technology) – ISSN: 0022-3352 – mf#3497 – us UMI ProQuest [660]

Journal of paleolimnology – Dordrecht. 1991-1996 (1,5,9) – ISSN: 0921-2728 – mf#16813 – us UMI ProQuest [560]

Journal of paleontology – Tulsa. 1927+ [1]; 1965+ [5]; 1970+ [9] – ISSN: 0022-3360 – mf#1422 – us UMI ProQuest [560]

Journal of palestine studies – Berkeley. 1971+ (1) 1971+ (5) 1973+ (9) – ISSN: 0377-919X – mf#8772 – us UMI ProQuest [956]

Journal of palliative care – Toronto. 1989+ (1,5,9) – ISSN: 0825-8597 – mf#17049 – us UMI ProQuest [610]

Journal of paralegal education and practice – v1-17. 1983-2001 – 9 – $165.00 – mf#118861 – us Hein [344]

Journal of parapsychology – Durham. 1950+ (1) 1970+ (5) 1976+ (9) – ISSN: 0022-3387 – mf#33 – us UMI ProQuest [150]

Journal of parasitology – Urbana 1914-46 – v1-32 on 232mf – 9 – mf#143c/2 – ne IDC [576]

Journal of partnership taxation – Boston. 1984-1996 (1,5,9) – ISSN: 0749-4513 – mf#14117 – us UMI ProQuest [336]

Journal of pastoral care – Decatur. 1972+ [1]; 1947+ [5]; 1974+ [9] – ISSN: 0022-3409 – mf#6377 – us UMI ProQuest [150]

Journal of pastoral care and counseling (jpcc) – Decatur. 2002+ (1,5,9) – mf#6377,01 – us UMI ProQuest [150]

Journal of pastoral counseling – New Rochelle. 1985+ (1,5,9) – ISSN: 0449-508X – mf#15290 – us UMI ProQuest [240]

Journal of pathology – Chichester. 1976+ (1,5,9) – ISSN: 0022-3417 – mf#14813,01 – us UMI ProQuest [611]

Journal of peace research – Oslo. 1964+ (1,5,9) – ISSN: 0022-3433 – mf#13027 – us UMI ProQuest [327]

Journal of peasant studies – London. 1990+ (1,5,9) – ISSN: 0306-6150 – mf#18551 – us UMI ProQuest [305]

Journal of pediatric health care – St. Louis. 1987+ (1,5,9) – ISSN: 0891-5245 – mf#16185 – us UMI ProQuest [618]

Journal of pediatric nursing – Philadelphia. 1992+ (1,5,9) – ISSN: 0882-5963 – mf#21123 – us UMI ProQuest [618]

Journal of pediatric oncology nursing – Philadelphia. 1992+ (1,5,9) – ISSN: 1043-4542 – mf#21124,01 – us UMI ProQuest [618]

Journal of pediatric ophthalmology – Thorofare. 1971-1977 (1) 1971-1977 (5) 1972-1977 (9) – (cont by: journal of pediatric ophthalmology and strabismus) – ISSN: 0022-345X – mf#6034 – us UMI ProQuest [617]

Journal of pediatric ophthalmology and strabismus – Thorofare. 1978+ (1) 1978+ (5) 1978+ (9) – (cont: journal of pediatric ophthalmology) – ISSN: 0191-3913 – mf#6034,01 – us UMI ProQuest [617]

Journal of pediatric orthopaedics – New York. 1993+ (1,5,9) – ISSN: 0271-6798 – mf#18724 – us UMI ProQuest [617]

Journal of pediatric psychology – New York. 1989+ (1,5,9) – ISSN: 0146-8693 – mf#17681 – us UMI ProQuest [150]

Journal of pediatrics – St. Louis. 1932+ (1) 1965+ (5) 1970+ (9) – ISSN: 0022-3476 – mf#1879 – us UMI ProQuest [618]

Journal of pedodontics – Birmingham. 1976-1990 (1) 1976-1990 (5) 1976-1990 (9) – (cont by: journal of clinical pediatric dentistry) – ISSN: 0145-5508 – mf#11466 – us UMI ProQuest [617]

Journal of pension benefits – v1-7. 1993-2000 – 9 – $239.00 set – ISSN: 1069-4064 – mf#116391 – us Hein [340]

Journal of pension planning and compliance – Greenvale. 1995-1995 (1) 1995-1995 (5) 1995-1995 (9) – ISSN: 0148-2181 – mf#15821,01 – us UMI ProQuest [650]

Journal of pension planning and compliance – v1-26. 1974-2001 – 9 – $664.00 set – (title varies: v1-3 n2 1974-76 as pension and profit sharing tax journal) – ISSN: 0148-2181 – mf#104141 – us Hein [340]

Journal of performance of constructed facilities – New York. 1987+ (1,5,9) – ISSN: 0887-3828 – mf#16199 – us UMI ProQuest [624]

Journal of perinatal and neonatal nursing – Frederick. 1987+ (1,5,9) – ISSN: 0893-2190 – mf#16006 – us UMI ProQuest [610]

Journal of perinatology – Norwalk. 1990+ (1,5,9) – ISSN: 0743-8346 – mf#18338,01 – us UMI ProQuest [618]

Journal of periodontology – Chicago. 1930+ (1) 1970+ (5) 1970+ (9) – ISSN: 0022-3492 – mf#432 – us UMI ProQuest [617]

Journal of personal selling and sales management – New York. 1988+ (1,5,9) – ISSN: 0885-3134 – mf#15822 – us UMI ProQuest [650]

Journal of personality – Cambridge. 1932+ (1) 1965+ (5) 1970+ (9) – ISSN: 0022-3506 – mf#962 – us UMI ProQuest [150]

Journal of personality and social psychology – Washington. 1965+ (1) 1971+ (5) 1975+ (9) – ISSN: 0022-3514 – mf#1681 – us UMI ProQuest [150]

Journal of personality assessment – Mahwah. 1936+ (1) 1971+ (5) 1975+ (9) – ISSN: 0022-3891 – mf#3207 – us UMI ProQuest [150]

Journal of personality disorders – New York. 1990+ (1,5,9) – ISSN: 0885-579X – mf#17421 – us UMI ProQuest [150]

Journal of personnel evaluation in education – Boston. 1989+ (1,5,9) – ISSN: 0920-525X – mf#16814 – us UMI ProQuest [650]

Journal of petroleum science and engineering – Amsterdam. 1987-1994 (1,5,9) – ISSN: 0920-4105 – mf#42539 – us UMI ProQuest [550]

Journal of petrology – Oxford. 1960+ (1) 1975+ (5) 1975+ (9) – ISSN: 0022-3530 – mf#10078 – us UMI ProQuest [550]

Journal of pharmaceutical and biomedical analysis – Oxford. 1983-1995 (1,5,9) – ISSN: 0731-7085 – mf#49426 – us UMI ProQuest [615]

Journal of pharmaceutical care in pain and symptom control : innovations in drug development, evaluation and use / ed by Lipman, Arthur G – v4 n1. 1996- – 1,9 – $125.00 in US $175.00 outside hardcopy subsc – us Haworth [615]

Journal of pharmaceutical finance, economics and policy / ed by Wertheimer, Albert I – v12- 2004- – 1,9 – $120.00 us $174.00 other – ISSN: 1538-5698 – us Haworth [338]

Journal of pharmaceutical marketing and management / ed by Smith, Mickey C – v1- 1986- – 1, 9 ($200.00 in US $280.00 outside hardcopy subsc) – us Haworth [650]

Journal of pharmaceutical medicine – Oxford. 1991-1996 (1,5,9) – ISSN: 0958-0581 – mf#18085 – us UMI ProQuest [615]

Journal of pharmaceutical sciences – Washington. 1912-1995 (1) 1912-1995 (5) 1970-1995 (9) – ISSN: 0022-3549 – mf#9 – us UMI ProQuest [610]

Journal of pharmacoepidemiology / ed by Fincham, Jack E – v1- 1990- – 1, 9 ($125.00 in US $175.00 outside hardcopy subsc) – us Haworth [614]

Journal of pharmacokinetics and biopharmaceutics – New York. 1973-1998 (1) 1973-1998 (5) 1978-1998 (9) – ISSN: 0090-466X – mf#10869 – us UMI ProQuest [615]

Journal of pharmacokinetics and pharmacodynamics – New York, 2001+ [1,5,9] – (cont: journal of pharmacokinetics and biopharmaceutics) – ISSN: 1567-567X – mf#10869,01 – us UMI ProQuest [615]

Journal of pharmacological and toxicological methods – New York. 1992+ (1,5,9) – (cont: journal of pharmacological methods) – ISSN: 1056-8719 – mf#42187,01 – us UMI ProQuest [615]

Journal of pharmacological methods – New York. 1978-1991 (1) 1978-1991 (5) 1987-1991 (9) – (cont by: journal of pharmacological and toxicological methods) – ISSN: 0160-5402 – mf#42187 – us UMI ProQuest [615]

Journal of pharmacology and experimental therapeutics – Baltimore. 1982-1989 (9) – ISSN: 0022-3565 – mf#84 – us UMI ProQuest [615]

Journal of pharmacology and experimental therapeutics – v1-279. 1909-96 – 1,5,6,9 – $125.00r – us Lippincott [615]

Journal of pharmacy and law – Ohio Northern University. v1-6. 1992-96 (all publ) – 9 – $90.00 set – ISSN: 1062-4546 – mf#114231 – us Hein [615]

Journal of pharmacy teaching / ed by Buerki, Robert A – v5 n1. 1996- – 1,9 – $105.00 in US $147.00 outside hardcopy subsc – us Haworth [615]

Journal of pharmacy technology – Cincinnati. 1985+ (1,5,9) – ISSN: 8755-1225 – mf#14851 – us UMI ProQuest [615]

Journal of phase equilibria – Materials Park. 1991-1996 (1,5,9) – ISSN: 1054-9714 – mf#12934,01 – us UMI ProQuest [660]

Journal of phenomenological psychology – Atlantic Highlands. 1970+ (1,5,9) – ISSN: 0047-2662 – mf#12630 – us UMI ProQuest [150]

Journal of philosophical logic – Dordrecht. 1984-1996 (1,5,9) – ISSN: 0022-3611 – mf#14759 – us UMI ProQuest [160]

Journal of photochemistry – Lausanne. 1972-1987 (1) 1972-1987 (5) 1987-1987 (9) – ISSN: 0047-2670 – mf#42188 – us UMI ProQuest [540]

Journal of photochemistry and photobiology a : chemistry – Lausanne. 1987+ (1,5,9) – ISSN: 1010-6030 – mf#42438 – us UMI ProQuest [540]

Journal of photochemistry and photobiology b : biology – Lausanne. 1987+ (1,5,9) – ISSN: 1011-1344 – mf#42439 – us UMI ProQuest [574]

Journal of photochemistry and photobiology, c : photochemistry reviews – Amsterdam, 2000+ [1,5,9] – ISSN: 1389-5567 – mf#42826 – us UMI ProQuest [540]

The journal of photographic science – v1-31. 1953-83 [mf ed Chadwyck-Healey] – 4r + 30mf – 1,9 – (Begun as pt b of: the photographic journal then publ separately fr 1953, the journal of photographic science covers all aspects of medical, scientific and technical photography) – uk Chadwyck [770]

Journal of physical and chemical reference data – v1-. 1972- – 1,5,6 – us AIP [530]

Journal of physical chemistry – Washington. 1896-1906 (1) – ISSN: 0022-3654 – mf#5802 – us UMI ProQuest [530]

The journal of physical chemistry – v1-100. 1896-1996 – 1,5,6,9 – us ACS [540]

Journal of physical chemistry a (molecules) : molecules, spectroscopy, kinetics, environment, and general theory – v101- 1997- – 1,5,6,9 – (cont: the journal of physical chemistry) – us ACS [540]

Journal of physical chemistry b (materials) : condensed matter, materials, surfaces, interfaces, and biophysical chemistry – v101- 1997- – 1,5,6,9 – (cont: the journal of physical chemistry) – us ACS [540]

Journal of physical education and recreation – Reston. 1975-1981 (1) 1975-1981 (5) 1975-1981 (9) – (cont by: journal of physical education, recreation and dance. cont: journal of health, physical education, recreation) – ISSN: 0097-1170 – mf#772,01 – us UMI ProQuest [613]

Journal of physical education new zealand – Wellington. 1993-1996 (1) 1993-1996 (5) 1993-1996 (9) – (cont: new zealand journal of health, physical education and recreation) – mf#7248,01 – us UMI ProQuest [790]

Journal of physical education, recreation and dance – Reston. 1981+ (1) 1981+ (5) 1981+ (9) – (cont: journal of physical education and recreation) – ISSN: 0730-3084 – mf#772,02 – us UMI ProQuest [613]

Journal of physical organic chemistry – Chichester. 1988+ (1,5,9) – ISSN: 0894-3230 – mf#16171 – us UMI ProQuest [540]

Journal of physical therapy education – St Louis, 1999+ [1,5,9] – mf#26716,01 – us UMI ProQuest [617]

Journal of physics and chemistry of solids – Oxford. 1956+ (1) 1956+ (5) 1956+ (9) – ISSN: 0022-3697 – mf#49118 – us UMI ProQuest [540]

Journal of physiology – London. 1960+ (1) 1971+ (5) 1975+ (9) – ISSN: 0022-3751 – mf#1300 – us UMI ProQuest [612]

Journal of pipelines – New York. 1987-1989 (1,5,9) – ISSN: 0166-5324 – mf#42300 – us UMI ProQuest [621]

Journal of plankton research – New York. 1988-1989 (1,5,9) – ISSN: 0142-7873 – mf#16453 – us UMI ProQuest [574]

Journal of plant growth regulation – Heidelberg. 1982+ (1,5,9) – ISSN: 0721-7595 – mf#13191 – us UMI ProQuest [580]

Journal of plasma physics – Cambridge. 1979-1996 (1,5,9) – ISSN: 0022-3778 – mf#12126 – us UMI ProQuest [530]

Journal of podiatric medical education – Philadelphia. 1977-1986 (1) 1977-1986 (5) 1977-1986 (9) – ISSN: 0093-7339 – mf#11666,01 – us UMI ProQuest [617]

Journal of poetry therapy – New York. 1987-1996 – 1,5,9 – ISSN: 0889-3675 – mf#16146 – us UMI ProQuest [410]

Journal of police crisis negotiations / ed by Greenstone, James L – v1- 2001- – 1,9 – $95.00 us $138.00 other – ISSN: 1533-2586 – us Haworth [360]

Journal of police science and administration – Gaithersburg. 1973-1990 (1) 1973-1990 (5) 1973-1990 (9) – ISSN: 0090-9084 – mf#12262 – us UMI ProQuest [360]

Journal of policy analysis and management – New York. 1981+ (1,5,9) – ISSN: 0276-8739 – mf#12887 – us UMI ProQuest [350]

Journal of policy history: jph – University Park. 1989+ (1,5,9) – ISSN: 0898-0306 – mf#16968 – us UMI ProQuest [320]

Journal of policy modeling – New York. 1979+ (1) 1979+ (5) 1987+ (9) – ISSN: 0161-8938 – mf#42189 – us UMI ProQuest [320]

Journal of political and military sociology – DeKalb. 1987+ (1) 1987+ (5) 1987+ (9) – (cont: jpms journal of political and military sociology) – ISSN: 0047-2697 – mf#7475,01 – us UMI ProQuest [320]

Journal of political economy – Chicago. 1892+ (1) 1969+ (5) 1977+ (9) – ISSN: 0022-3808 – mf#489 – us UMI ProQuest [330]

Journal of political marketing : political campaigns in the new millennium / ed by Newman, Bruce I – v1- 2002- – 1,9 – $140.00 us $203.00 other – ISSN: 1537-7857 – us Haworth [650]

Journal of political philosophy – Cambridge. 1993-1995 (1,5,9) – ISSN: 0963-8016 – mf#19835 – us UMI ProQuest [320]
Journal of politics – Malden. 1939+ (1) 1939+ (5) 1975+ (9) – ISSN: 0022-3816 – mf#796 – us UMI ProQuest [320]
Journal of polymer science : [old series] – New York. 1946-1962 (1) 1946-1962 (5) 1946-1962 (9) – ISSN: 0022-3832 – mf#11059 – us UMI ProQuest [540]
Journal of polymer science : polymer chemistry edition – New York. 1966-1986 (1) 1966-1986 (5) 1966-1986 (9) – (cont by: journal of polymer science pt a, polymer chemistry) – ISSN: 0360-6376 – mf#11060 – us UMI ProQuest [540]
Journal of polymer science : polymer letters edition – New York. 1963-1986 (1) 1963-1986 (5) 1963-1986 (9) – (cont by: journal of polymer science pt c, polymer letters) – ISSN: 0360-6384 – mf#11342 – us UMI ProQuest [540]
Journal of polymer science : polymer physics edition – New York. 1966-1986 (1) 1966-1986 (5) 1966-1986 (9) – (cont by: journal of polymer science pt b: polymer physics) – ISSN: 0098-1273 – mf#11343 – us UMI ProQuest [540]
Journal of polymer science : polymer symposia – New York. 1973-1986 (1,5,9) – (cont: journal of polymer science pt c: polymer symposia) – ISSN: 0360-8905 – mf#11344,01 – us UMI ProQuest [540]
Journal of polymer science pt a : general papers – New York. 1963-1965 (1,5,9) – ISSN: 0449-2951 – mf#11357 – us UMI ProQuest [540]
Journal of polymer science pt a : polymer chemistry – New York. 1986+ (1,5,9) – (cont: journal of polymer science: polymer chemistry edition) – ISSN: 0887-624X – mf#11060,01 – us UMI ProQuest [540]
Journal of polymer science pt b : polymer physics – New York. 1986+ (1,5,9) – (cont: journal of polymer science: polymer physics edition) – ISSN: 0887-6266 – mf#11343,01 – us UMI ProQuest [540]
Journal of polymer science pt c : polymer letters – New York. 1986-1990 (1) 1986-1990 (5) 1986-1990 (9) – (cont: journal of polymer science: polymer letters edition) – ISSN: 0887-6258 – mf#11342,01 – us UMI ProQuest [540]
Journal of polymer science pt c : polymer symposia – New York. 1963-1971 (1) 1963-1971 (5) 1962-1971 (9) – (cont by: journal of polymer science: polymer symposia) – ISSN: 0449-2994 – mf#11344 – us UMI ProQuest [540]
Journal of popular culture – Bowling Green. 1967+ (1) 1971+ (5) 1975+ (9) – ISSN: 0022-3840 – mf#2672 – us UMI ProQuest [301]
Journal of popular film – Bowling Green. 1972-1978 (1) 1972-1978 (5) 1977-1978 (9) – (cont by: journal of popular film and television) – ISSN: 0047-2719 – mf#7739 – us UMI ProQuest [790]
Journal of popular film and television – Washington. 1978+ (1) 1978+ (5) 1978+ (9) – (cont: journal of popular film) – ISSN: 0195-6051 – mf#7739,01 – us UMI ProQuest [790]
Journal of population : behavioral, social and environmental issues – New York. 1978-1979 (1,5,9) – (cont by: population and environment) – ISSN: 0146-1052 – mf#11643 – us UMI ProQuest [304]
Journal of population economics – Berlin. 1988-1996 (1,5,9) – ISSN: 0933-1433 – mf#17004 – us UMI ProQuest [304]
Journal of porphyrins and phthalocyanines – Chichester. 1997+ (1) – ISSN: 1088-4246 – mf#25615 – us UMI ProQuest [540]
Journal of portfolio management – London. 1974+ (1) 1978+ (5) 1978+ (9) – ISSN: 0095-4918 – mf#11854 – us UMI ProQuest [332]
Journal of positive behavior interventions – Austin. 1999+ (1) – ISSN: 1098-3007 – mf#28754 – us UMI ProQuest [150]
Journal of post keynesian economics – Armonk. 1989+ (1,5,9) – ISSN: 0160-3477 – mf#16891 – us UMI ProQuest [330]
Journal of post sutler / Fort Harker. Kansas – Nov 1867-Nov 1868 – 1 – us Kansas [025]
Journal of potato production and postharvest handling – v1 n1. 1997- – 1,9 – $75.00 in US $105.00 outside hardcopy subsc – us Haworth [630]
Journal of poverty : innovations on social, political and economic inequalities / ed by Kilty, Keith M et al – v1 n1. 1997- – 1,9 – $45.00 in US $63.00 outside hardcopy subsc – us Haworth [339]
Journal of power sources – Lausanne. 1976+ (1) 1976+ (5) 1987+ (9) – ISSN: 0378-7753 – mf#42190 – us UMI ProQuest [621]
Journal of practical nursing – New York. 1963+ (1,5,9) – ISSN: 0022-3867 – mf#11877,01 – us UMI ProQuest [610]

Journal of pragmatics – Amsterdam. 1977+ (1) 1977+ (5) 1987+ (9) – ISSN: 0378-2166 – mf#42191 – us UMI ProQuest [400]
Journal of presbyterian history – Philadelphia. 1997+ (1,5,9) – (cont: american presbyterians) – ISSN: 0886-5159 – mf#12638,03 – us UMI ProQuest [242]
Journal of presbyterian history – Philadelphia. 1980-1985 (1,5,9) – (cont by: american presbyterians) – ISSN: 0022-3883 – mf#12638,01 – us UMI ProQuest [242]
Journal of presbyterian history – v1-61. 1901-87 – 1 – us Presbyterian [242]
Journal of pressure vessel technology – New York. 1974+ (1) 1974+ (5) 1977+ (9) – ISSN: 0094-9930 – mf#9334 – us UMI ProQuest [621]
Journal of prevention – New York. 1980-1981 (1,5,9) – (cont by: journal of primary prevention) – ISSN: 0163-514X – mf#12188 – us UMI ProQuest [613]
Journal of prevention and intervention in the community / ed by Ferrari, Joseph R – v1- 1981- – (= ser Prevention In Human Services) – 1, 9 ($250.00 in US $350.00 outside hardcopy subsc) – (formerly: prevention in human services) – us Haworth [614]
Journal of preventive dentistry – New York. 1974-1978 (1) 1974-1978 (5) 1974-1978 (9) – ISSN: 0096-2732 – mf#10463 – us UMI ProQuest [617]
Journal of primary prevention – New York. 1981+ (1,5,9) – (cont: journal of prevention) – ISSN: 0278-095X – mf#12188,01 – us UMI ProQuest [613]
Journal of prison and jail health – New York. 1982-1993 (1) 1982-1993 (5) 1982-1993 (9) – (cont: journal of prison health) – ISSN: 0731-8332 – mf#12189,01 – us UMI ProQuest [360]
Journal of prison discipline and philanthropy – Philadelphia. 1845-1855 (1) – (cont by: prison journal) – mf#4003 – us UMI ProQuest [360]
Journal of prison discipline and philanthropy (1845-1920) and the prison journal (1921-1986) – 1972-86 – 6r – 1 – $780.00 – mf#S1867 – us Scholarly Res [365]
Journal of prison health – New York. 1981-1981 (1,5,9) – (cont by: journal of prison and jail health) – ISSN: 0192-7051 – mf#12189 – us UMI ProQuest [360]
Journal of private equity – London. 1997+ (1,5,9) – ISSN: 1096-5572 – mf#32270 – us UMI ProQuest [332]
Journal of proceedings / Kansapolis Association – June 7, 1856 to December 27, 1857 – 1 – us Kansas [978]
Journal of proceedings – Milwaukee WI. 1886-87 – 1r – 1 – us WHS [350]
Journal of proceedings / New York. (City) Board of Estimate and Apportionment – 1879-1941. (Wanting some) – 1 – us L of C Photodup [336]
Journal of proceedings at 39th annual meeting held at toronto, february 4th and 5th, 1897 / list of officers and members / Canadian Press Association – Toronto: Methodist Book and Pub House, 1897 – 1mf – 9 – mf#01442 – cn Canadiana [070]
Journal of proceedings of council / Chippewa and Munsee Indians – 1870-1881 – 1 – us Kansas [305]
Journal of proceedings of the grand lodge, ancient order united workmenof wisconsin, at its...annual session – 2nd-10th [1879-87], 11th-19th [1888-96] – 2r – 1 – (cont: proceedings of the...session of the grand lodge of wisconsin, a o u w, ancient order of united workmen. grand lodge of wisconsin) – mf#2693615 – us WHS [366]
Journal of proceedings of the pattern makers national league of north america...annual session / Pattern Makers National League of North America – 1st [1888], 4th-5th [1891-1892] – 1r – 1 – (cont by: journal of proceedings...regular session, pattern makers league of north america, pattern makers' league of north america) – mf#3372109 – us WHS [366]
Journal of proceedings of the...annual communication of the sovereign grand lodge of the independent order of odd fellows / Independent Order of Odd Fellows – 74th [1898] – 1r – 1 – (cont: proceedings of the...annual communication of the sovereign grand lodge of the independent order of odd fellows, independent order of odd fellows) – mf#1140639 – us WHS [366]
Journal of proceedings of the...annual conference of the african m e church for the district of missouri / African Methodist Episcopal Church. Missouri Conference – Louisville KY: Publ by John M Brown [annual] [mf ed 2004] – 3rd [1857] [complete] on 1r – 1 – (cont by: african methodist episcopal church. missouri conference. session of the missouri annual conference of the african m e church; reel also incl later & unrelated titles: session of the missouri annual conference of the african m e church [2004-s114] and: minutes of the philad'a district annual conference of the african m e zion church in america [2004-s115]) – mf#2004-s113 – us ATLA [242]
Journal of proceedings of the...annual conference of the african methodist episcopal church for the district of indiana / African Methodist Episcopal Church. Indiana Conference – Indianapolis: printed at the Office of the Free Democrat. 14th [1853] [mf ed 2005] – 1r – 1 – (cont by: annual conference of the african methodist episcopal church for the indiana district) – mf#2005-s063 – us ATLA [242]
Journal of proceedings of the...annual convention of the international lodge, a a of i s and t w of north america / Amalgamated Association of Iron, Steel, and Tin Workers of North America – 1024, 1929-36, 1937-41 – 3r – 1 – (cont: journal of the...annual convention of the national lodge, amalgamated association of iron, steel, and tin workers of north america) – mf#3148129 – us WHS [366]
Journal of proceedings of the...annual session / Indiana State Grange – 49th [1920] – 1r – 1 – (cont by: journal of proceedings, indiana state grange) – mf#3424375 – us WHS [071]
Journal of proceedings of the...annual session of illinois state grange, patrons of husbandry / Illinois State Grange – 53rd [1924], 56th-57th [1927-1928] – 1r – 1 – (cont: proceedings of the state grange of illinois at the ... annual session, illinois state grange) – mf#3160872 – us WHS [366]
Journal of proceedings of the...annual session of the california state grange, patrons of husbandry / California State Grange – 18th-22nd [1890-94], 25th-27th [1897-99], 29th-30th [1901-02], 32nd-35th [1904-07] – 1r – 1 – mf#2962706 – us WHS [366]
Journal of proceedings of the...annual session of the colorado state grange of the patrons of husbandry / Colorado State Grange – 14th [1888], 18th [1892], 36th [1910] – 1r – 1 – mf#1113589 – us WHS [366]
Journal of proceedings of the...annual session of the delaware state grange, patrons of husbandry / Patrons of Husbandry – 14th-18th [1888-1892] – 1r – 1 – (cont by: proceedings of the...annual session of the delaware state grange, patrons of husbandry) – mf#3424850 – us WHS [366]
Journal of proceedings of the...annual session of the new york state grange of the patrons of husbandry / New York State Grange – 36th [1909], 39th [1912], 41st [1914], 61st [1934] – 1r – 1 – (cont: proceedings of the...annual session of the new york state grange of the patrons of husbandry, new york state grange) – mf#1436498 – us WHS [366]
Journal of proceedings of the...annual session of the west tennessee conference of the african methodist episcopal church / African Methodist Episcopal Church. West Tennessee Conference – Nashville TN: AMESSU 56th (1932) [mf ed 2004] – 1r – 1 – (cont: minutes of the...annual session of the west tennessee conference of the african methodist episcopal church; cont by: journal of proceedings of the...session of the west tennessee conference of the african methodist episcopal church) – mf#2004-S102 – us ATLA [242]
Journal of proceedings of the...meeting of the synod of the diocese of athabasca / Church of England. Diocese of Athabasca. Synod – Middle Church, Man?: The Synod, 1891 – 9 – mf#A00845 – cn Canadiana [242]
Journal of proceedings of the...session of the baltimore annual conference of the african methodist episcopal church in the united states of america / African Methodist Episcopal Church. Baltimore Conference – Philadelphia: AME Publ House. 89th (1906) [mf ed 2005] – 1r – 1 – (cont: minutes of the...session of the baltimore annual conference of the african methodist episcopal church in the united states of america; cont by: minutes of the...session of the baltimore annual conference of the african methodist episcopal church) – mf#2005-s074 – us ATLA [242]
Journal of proceedings of the...session of the indiana annual conference of the african methodist episcopal church / African Methodist Episcopal Church. Indiana Conference – Nashville TN: AME Sunday School Union. 68th (1906) [mf ed 2005] – 1r – 1 – (cont: annual conference of the african methodist episcopal church for the indiana district; cont by: minutes of the...session of the indiana annual conference of the african methodist episcopal church) – mf#2005-s064 – us ATLA [242]
Journal of proceedings of the...session of the ohio annual conference of the african methodist episcopal church / African Methodist Episcopal Church. Ohio Conference – Nashville TN: AMESS Union [annual] [mf ed 2004] – 81st-88th (1911-18) on 1r – 1 – (lacks: 82nd-87th; damaged: v81st (1911) [front cover]; v88th (1918) p67-68,79; cont: african methodist episcopal church. ohio conference. minutes of the...session of the ohio annual conference of the african m e church; cont by: african methodist episcopal church. ohio conference. journal of the...annual session of the ohio conference of the african methodist episcopal church) – mf#2004-s090 – us ATLA [242]
Journal of proceedings regular session, PatternMakers League of North America – 9th [1900] – 1r – 1 – us WHS [071]
Journal of proceedings...annual encampment / Grand Army of the Republic et al – 35th-38th [1901-05], 39th-40th [1905-06] – 2r – 1 – (cont: proceedings of...annual encampment of the department of wisconsin, grand army of the republic, grand army of the republic; cont by: journal of the...annual encampment of the department of wisconsin, grand army of the republic, grand army of the republic) – mf#2824388 – us WHS [366]
Journal of proceedings...annual session / Patrons of Husbandry – 43rd [1912], 50th [1919], 53rd [1922] – 1r – 1 – (cont: proceedings of the...annual session of the state grange of iowa, patrons of husbandry; cont by: proceedings of the...annual session, patrons of husbandry) – mf#3433236 – us WHS [366]
Journal of proceedings...regular session, pattern makers league of north america / Pattern Makers National League of North America – 9th [1900] – 1r – 1 – (cont: journal of proceedings of the pattern makers national league of north america...annual session, pattern makers national league of north america) – us WHS [366]
Journal of process control – Kidlington. 1991-1994 (1,5,9) – ISSN: 0959-1524 – mf#18242 – us UMI ProQuest [660]
Journal of product and brand management – Santa Barbara. 1993-1994 (1,5,9) – ISSN: 1061-0421 – mf#19823 – us UMI ProQuest [650]
Journal of product innovation management – New York. 1984+ (1,5,9) – ISSN: 0737-6782 – mf#42579 – us UMI ProQuest [338]
Journal of productivity analysis – Norwell. 1989-1996 (1,5,9) – ISSN: 0895-562X – mf#16815 – us UMI ProQuest [338]
Journal of products and toxics liability – New York. 1993-1995 (1,5,9) – (cont: journal of products liability) – ISSN: 0967-2680 – mf#49268,01 – us UMI ProQuest [344]
Journal of products liability – New York. 1977-1992 (1,5,9) – (cont by: journal of products and toxics liability) – ISSN: 0363-0404 – mf#49268 – us UMI ProQuest [344]
Journal of professional issues in engineering – New York. 1983-1990 (1) 1983-1990 (5) 1983-1990 (9) – (cont: issues in engineering. cont by: journal of professional issues in engineering education and practice) – ISSN: 0733-9380 – mf#8148,02 – us UMI ProQuest [620]
Journal of professional issues in engineering education and practice – New York. 1991+ (1) 1991+ (5) 1991+ (9) – (cont: journal of professional issues in engineering) – ISSN: 1052-3928 – mf#8148,03 – us UMI ProQuest [620]
Journal of professional nursing : official journal of the american association of colleges of nursing / American Association of Colleges of Nursing – Philadelphia. 1985+ (1,5,9) – ISSN: 8755-7223 – mf#14733 – us UMI ProQuest [610]
Journal of professional services marketing / ed by Winston, William J – v1- 1985- – 1, 9 ($225.00 in US $315.00 outside hardcopy subsc) – us Haworth [650]
Journal of programmed instruction – New York. 1962-1964 – 1 – mf#1822 – us UMI ProQuest [370]
Journal of progressive human services – v1- 1990- – 1, 9 ($95.00 in US $133.00 outside hardcopy subsc) – us Haworth [360]
Journal of promotion management / ed by Crane, F G – v4 n1. 1996- – 1,9 – $95.00 in US $133.00 outside hardcopy subsc – us Haworth [650]
Journal of property investment and finance – Bradford. 2001+ (1,5,9) – ISSN: 1463-578X – mf#31588,02 – us UMI ProQuest [333]
Journal of property management – Chicago. 1934+ (1) 1975+ (5) 1975+ (9) – ISSN: 0022-3905 – mf#10574 – us UMI ProQuest [333]
Journal of property valuation and taxation – New York, 2001+ [1,5,9) – mf#24964,02 – us UMI ProQuest [341]
Journal of proprietary rights see Intellectual property and technology law journal
Journal of propulsion and power – Reston. 1985+ (1,5,9) – ISSN: 0748-4658 – mf#16127 – us UMI ProQuest [629]
Journal of prosthetic dentistry – St Louis. 1951+ [1]; 1971+ [5]; 1976+ [9] – ISSN: 0022-3913 – mf#1875 – us UMI ProQuest [617]

Journal of prosthetics and orthotics: jpo – Alexandria. 1988+ (1,5,9) – ISSN: 1040-8800 – mf#16759 – us UMI ProQuest [617]
Journal of psychedelic drugs – San Francisco. 1967-1980 (1) 1970-1980 (5) 1976-1980 (9) – (cont by: journal of psychoactive drugs) – ISSN: 0022-393X – mf#3467 – us UMI ProQuest [615]
Journal of psychiatric education – New York. 1977-1988 (1,5,9) – (cont by: academic psychiatry) – ISSN: 0363-1907 – mf#11182 – us UMI ProQuest [616]
Journal of psychiatric nursing and mental health services – Thorofare. 1963-1981 (1) 1971-1981 (5) 1971-1981 (9) – (cont by: journal of psychosocial nursing and mental health services) – ISSN: 0360-5973 – mf#6036 – us UMI ProQuest [610]
Journal of psychiatric research – Oxford. 1963+ (1,5,9) – ISSN: 0022-3956 – mf#49119 – us UMI ProQuest [616]
Journal of psychiatric treatment and evaluation – New York. 1979-1983 (1) 1979-1983 (5) 1979-1983 (9) – ISSN: 0195-8127 – mf#49392 – us UMI ProQuest [616]
Journal of psychiatry and law – New York. 1973+ (1,5,9) – ISSN: 0093-1853 – mf#11141 – us UMI ProQuest [616]
Journal of psychoactive drugs – San Francisco. 1981+ (1) 1981+ (5) 1981+ (9) – (cont: journal of psychedelic drugs) – ISSN: 0279-1072 – mf#3467,01 – us UMI ProQuest [615]
Journal of psychohistory – New York. 1976+ (1) 1976+ (5) 1976+ (9) – (cont: history of childhood quarterly) – ISSN: 0145-3378 – mf#7465,01 – us UMI ProQuest [150]
Journal of psycholinguistic research – New York. 1971+ (1) 1971+ (5) 1989+ (9) – ISSN: 0090-6905 – mf#10864 – us UMI ProQuest [150]
Journal of psychological researches – Madras. 1964-1994 (1) 1972-1994 (5) 1973-1994 (9) – ISSN: 0022-3972 – mf#7273 – us UMI ProQuest [150]
Journal of psychology – Provincetown. 1983+ (1,5,9) – ISSN: 0022-3980 – mf#16337 – us UMI ProQuest [150]
Journal of psychology and christianity – Blue Jay. 1982+ (1,5,9) – (cont: christian association for psychological studies bulletin) – ISSN: 0733-4273 – mf#13580 – us UMI ProQuest [150]
Journal of psychology and financial markets – Mahwah. 2000+ (1,5,9) – ISSN: 1520-8834 – mf#31737 – us UMI ProQuest [150]
Journal of psychology and human sexuality / ed by Coleman, Eli – v1- 1988- – 1, 9 ($225.00 in US $315.00 outside hardcopy subsc) – us Haworth [150]
Journal of psychology and judaism – New York. 1976+ (1,5,9) – ISSN: 0700-9801 – mf#11482 – us UMI ProQuest [150]
Journal of psychology and theology – La Mirada. 1973+ (1,5,9) – ISSN: 0091-6471 – mf#12578 – us UMI ProQuest [150]
Journal of psychopathology and behavioral assessment – New York. 1989-1996 (1,5,9) – ISSN: 0882-2689 – mf#17683,01 – us UMI ProQuest [150]
Journal of psychopharmacology – Margate. 1966-1968 (1) – mf#2200 – us UMI ProQuest [615]
Journal of psychosocial nursing and mental health services – Thorofare. 1981+ (1) 1981+ (5) 1981+ (9) – (cont: journal of psychiatric nursing and mental health services) – ISSN: 0279-3695 – mf#6036,01 – us UMI ProQuest [610]
Journal of psychosocial oncology / ed by Christ, Grace H & Zabora, James R – v1- 1983- – 1, 9 ($275.00 in US $385.00 outside hardcopy subsc) – us Haworth [616]
Journal of psychosomatic research – London. 1956+ (1) 1956+ (5) 1956+ (9) – ISSN: 0022-3999 – mf#49120 – us UMI ProQuest [616]
The journal of psychosophy : a scientific monthly of advanced thought – Toronto: First School of Practical Psychosophy, [1899] – 9 – mf#P04263 – cn Canadiana [130]
Journal of public affairs – London. 2001+ (1,5,9) – mf#31754 – us UMI ProQuest [350]
Journal of public budgeting, accounting and financial management – Boca Raton. 1997+ (1) – ISSN: 1096-3367 – mf#20873,01 – us UMI ProQuest [336]
Journal of public economics – Amsterdam. 1972+ (1) 1972+ (5) 1987+ (9) – ISSN: 0047-2727 – mf#42119 – us UMI ProQuest [338]
Journal of public health dentistry – Raleigh. 1941+ (1) 1966+ (5) 1976+ (9) – ISSN: 0022-4006 – mf#2222 – us UMI ProQuest [617]
Journal of public health management and practice: jphmp – Frederick. 1996+ (1,5,9) – ISSN: 1078-4659 – mf#21634 – us UMI ProQuest [360]
Journal of public law – Atlanta. 1968-1973 (1) 1971-1973 (5) – (cont by: emory law journal) – ISSN: 0022-4014 – mf#3498 – us UMI ProQuest [340]
Journal of public law *see*
– Brigham young university journal of public law
– Emory law journal
Journal of public policy – Cambridge. 1987+ (1,5,9) – ISSN: 0143-814X – mf#16534 – us UMI ProQuest [350]
Journal of public policy and marketing: jpp&m – Chicago. 1990+ (1,5,9) – ISSN: 0743-9156 – mf#14429,01 – us UMI ProQuest [350]
Journal of public procurement – Boca Raton. 2001+ (1,5,9) – ISSN: 1535-0118 – mf#32351 – us UMI ProQuest [350]
Journal of purchasing – New York. 1965-1973 (1) 1971-1973 (5) – (cont by: journal of purchasing and materials management) – ISSN: 0022-4030 – mf#2757 – us UMI ProQuest [650]
Journal of purchasing and materials management – Tempe. 1974-1990 (1) 1974-1990 (5) 1975-1990 (9) – (cont: journal of purchasing. cont by: international journal of purchasing and materials management) – ISSN: 0094-8594 – mf#2757,01 – us UMI ProQuest [650]
Journal of pure and applied algebra – Amsterdam. 1971+ (1,5,9) – ISSN: 0022-4049 – mf#42120 – us UMI ProQuest [510]
Journal of quality assurance in hospitality and tourism : improvements in marketing, management, and development / ed by Pyo, Sungsoo & Hinkin, Timothy R – v1- 2000- – 1,9 – $85.00 us $123.00 other – ISSN: 1528-008X – us Haworth [650]
Journal of quality in clinical practice – Sydney. 1994+ (1,5,9) – (cont: australian clinical review) – ISSN: 1320-5455 – mf#18092,01 – us UMI ProQuest [610]
Journal of quality in maintenance engineering – Bradford. 2001+ (1,5,9) – ISSN: 1355-2511 – mf#31589 – us UMI ProQuest [620]
Journal of quality technology – Milwaukee. 1972+ (1) 1969+ (5) 1977+ (9) – ISSN: 0022-4065 – mf#6223 – us UMI ProQuest [620]
Journal of quantitative anthropology – Dordrecht. 1996-1996 (1) 1995-1996 (5) 1996-1996 (9) – ISSN: 0922-2995 – mf#16816 – us UMI ProQuest [301]
Journal of quantitative criminology – New York. 1985+ (1,5,9) – ISSN: 0748-4518 – mf#17686 – us UMI ProQuest [360]
Journal of quantitative spectroscopy and radiative transfer – Oxford. 1961+ (1) 1961+ (5) 1977+ (9) – ISSN: 0022-4073 – mf#49121 – us UMI ProQuest [530]
Journal of quaternary science: jqs – Chichester. 1989+ (1,5,9) – ISSN: 0267-8179 – mf#17179 – us UMI ProQuest [550]
Journal of radio studies – Broadcast Education Assoc. v1-5. 1992-98 – 9 – mf#117751 – us Hein [380]
Journal of radioanalytical and nuclear chemistry – Lausanne. 1968-1994 (1) 1968-1994 (5) 1986-1994 (9) – ISSN: 0236-5731 – mf#42198 – us UMI ProQuest [540]
Journal of rail and rapid transit *see* Proceedings of the institution of mechanical engineers pt f
Journal of raman spectroscopy – New York. 1973+ (1,5,9) – ISSN: 0377-0486 – mf#13310 – us UMI ProQuest [530]
Journal of range management – Denver. 1972+ [1]; 1948+ [5]; 1975+ [9] – ISSN: 0022-409X – mf#6618 – us UMI ProQuest [630]
Journal of rational emotive therapy : the journal of the institute for rational-emotive therapy – New York. 1983-1987 (1) 1983-1987 (5) 1983-1987 (9) – (cont: rational living. cont by: journal of rational-emotive and cognitive-behavior therapy) – ISSN: 0748-1985 – mf#13998 – us UMI ProQuest [150]
Journal of rational-emotive and cognitive-behavior therapy – New York. 1988+ (1,5,9) – (cont: journal of rational emotive therapy: the journal of the institute for rational-emotive therapy) – ISSN: 0894-9085 – mf#13998,01 – us UMI ProQuest [150]
Journal of reading – Newark. 1957-1994 (1) 1968-1994 (5) 1975-1994 (9) – (cont by: journal of adolescent and adult literacy) – ISSN: 0022-4103 – mf#1562 – us UMI ProQuest [370]
Journal of reading behavior – Chicago. 1969-1995 (1) 1972-1995 (5) 1976-1995 (9) – (cont by: journal of literacy research: jlr) – ISSN: 0022-4111 – mf#8196 – us UMI ProQuest [370]
Journal of reading, writing, and learning disabilities, international – London. 1984-1991 – 1,5,9 – (cont by: reading and writing quarterly) – ISSN: 0748-7630 – mf#16658 – us UMI ProQuest [370]
Journal of real estate finance and economics – Boston. 1988+ (1,5,9) – ISSN: 0895-5638 – mf#16817 – us UMI ProQuest [333]
Journal of real estate literature – Cleveland. 1999+ (1,5,9) – mf#21173 – us UMI ProQuest [333]
Journal of real estate practice and education – Cleveland. 1998+ (1,5,9) – ISSN: 1521-4842 – mf#31926 – us UMI ProQuest [333]
Journal of real estate taxation – New York. 1973+ (1,5,9) – ISSN: 0093-5107 – mf#8700 – us UMI ProQuest [336]
Journal of reform *see* Spiritual universe and journal of reform
Journal of reform judaism – New York. 1978-1991 (1) 1978-1991 (5) 1978-1991 (9) – (cont: ccar journal. cont by: ccar journal) – ISSN: 0149-712X – mf#6963,01 – us UMI ProQuest [270]
Journal of refugee resettlement – Washington. 1980-1981 (1,5,9) – ISSN: 0272-2348 – mf#12803 – us UMI ProQuest [360]
Journal of refugee studies – Oxford. 1988-1995 (1,5,9) – ISSN: 0951-6328 – mf#17507 – us UMI ProQuest [320]
Journal of regional science – Cambridge. 1958+ (1) 1971+ (5) 1975+ (9) – ISSN: 0022-4146 – mf#5333 – us UMI ProQuest [710]
Journal of regulatory economics – Norwell. 1989+ (1,5,9) – ISSN: 0922-680X – mf#16818 – us UMI ProQuest [330]
Journal of rehabilitation – Alexandria. 1989+ (1,5,9) – ISSN: 0022-4154 – mf#17628,01 – us UMI ProQuest [360]
Journal of rehabilitation outcomes measurement – Frederick. 1998+ (1,5,9) – ISSN: 1086-9654 – mf#25926 – us UMI ProQuest [617]
Journal of rehabilitation research and development – Washington. 1983+ (1) 1983+ (5) 1983+ (9) – ISSN: 0748-7711 – mf#9149,01 – us UMI ProQuest [617]
Journal of reinforced plastics and composites – Westport. 1989+ (1) – ISSN: 0731-6844 – mf#14064 – us UMI ProQuest [660]
Journal of relationship marketing : innovations and enhancements for customer service, relations, and satisfaction – v1- 2002- – 1,9 – $140.00 us $$203.00 other – ISSN: 1533-2667 – us Haworth [650]
Journal of religion – Chicago. 1921+ (1) 1969+ (5) 1977+ (9) – ISSN: 0022-4189 – mf#490 – us UMI ProQuest [200]
The journal of religion – 1(1921)-30(1950) – 326mf – 9 – €621.00 – ne Slangenburg [200]
Journal of religion and abuse : advocacy, pastoral care, and prevention / ed by Fortune, Marie M – v1- 1999- – 1,9 – $85.00 us $123.00 other – (simultaneously publ as: remembering conquest: feminist/womanist perspectives on religion, colonization, and sexual violence) – ISSN: 1521-1037 – us Haworth [230]
Journal of religion and health – New York. 1961+ (1) 1972+ (5) 1973+ (9) – ISSN: 0022-4197 – mf#7018 – us UMI ProQuest [230]
Journal of religion and spirituality in social work / ed by Ahearn, Jr, Frederick L – v23- 2004- – 1,9 – $125.00 us $181.00 other – (simultaneously publ as: criminal justice: retribution vs restoration) – ISSN: 1542-6432 – us Haworth [360]
Journal of religion, disability and health : ...bridging clinical practice and spiritual supports / ed by Gaventa William C & Coulter David L – v3- 1999- – 1,9 – $140.00 us $240.00 other – ISSN: 1522-8967 – us Haworth [360]
Journal of religion in disability and rehabilitation : innovations in ministry for independent living / ed by Blair, William A & Davidson, Dana – v3 n1. 1996- – 1,9 – $105.00 in US $147.00 outside hardcopy subsc – us Haworth [360]
Journal of religious and theological information / ed by Miller, William C – v2 n1. 1996- – 1,9 – $48.00 in US $67.20 outside hardcopy subsc – us Haworth [240]
Journal of religious education – Nashville. 1971-1980 (1) 1971-1980 (5) 1976-1980 (9) – (cont by: journal of religious education of the african methodist episcopal church) – ISSN: 0022-4219 – mf#6886 – us UMI ProQuest [377]
Journal of religious education of the african methodist episcopal church / African Methodist Episcopal Church – 1955 jun, 1956 sep, dec – 1r – 1 – (cont by: journal of religious education) – mf#5290313 – us WHS [242]
Journal of religious education of the african methodist episcopal church / African Methodist Episcopal Church. Division of Christian Education – Nashville. 1980-1982 (1) 1980-1982 (5) 1980-1982 (9) – (cont: journal of religious education. cont by: journal of christian education of the african methodist episcopal church) – ISSN: 0276-0770 – mf#6886,01 – us UMI ProQuest [377]
Journal of religious ethics – Malden. 1973+ (1,5,9) – ISSN: 0384-9694 – mf#12867 – us UMI ProQuest [230]
Journal of religious gerontology / ed by Sapp, Stephen – v1- 1990- – 1,9 ($140.00 in US $196.00 outside hardcopy subsc) – (cont: journal of religion and aging and journal of religion and gerontology) – us Haworth [230]
Journal of religious history – Sydney. 1978+ (1,5,9) – ISSN: 0022-4227 – mf#14155 – us UMI ProQuest [200]
Journal of religious psychology – v5-7. 1912-1915 (complete) – 1r – 1 – mf#ATLA 1994-S519 – us ATLA [073]
Journal of religious thought – Washington. 1943+ [1]; 1971+ [5]; 1976+ [9] – ISSN: 0022-4235 – mf#2114 – us UMI ProQuest [200]
Journal of reproductive and infant psychology – Chichester. 1983-1989 (1) 1983-1989 (5) 1983-1989 (9) – ISSN: 0264-6838 – mf#16110 – us UMI ProQuest [150]
Journal of reproductive immunology – Amsterdam. 1979+ (1) 1979+ (5) 1988+ (9) – ISSN: 0165-0378 – mf#42121 – us UMI ProQuest [616]
Journal of reproductive medicine – Chicago. 1968+ (1) 1971+ (5) 1974+ (9) – ISSN: 0024-7758 – mf#3449 – us UMI ProQuest [618]
Journal of research administration – Washington, 2000+ [1,5,9] – (cont: sra journal) – mf#30787 – us UMI ProQuest [650]
Journal of research and development in education – Athens. 1967-2001 (1) 1975-2001 (5) 1975-2001 (9) – ISSN: 0022-426X – mf#11370 – us UMI ProQuest [370]
Journal of research and practice in information technology – Darlinghurst. 2000+ (1) – (cont: australian computer journal) – ISSN: 1443-458X – mf#2758,01 – us UMI ProQuest [000]
Journal of research in childhood education : jrce – Olney. 1991+ – 1,5,9 – ISSN: 0256-8543 – mf#19426 – us UMI ProQuest [370]
Journal of research in crime and delinquency – Beverly Hills. 1964+ (1) 1971+ (5) 1975+ (9) – ISSN: 0022-4278 – mf#1830 – us UMI ProQuest [360]
Journal of research in crime and delinquency – v1-38. 1964-2001 – 5,6,9 – $823.00 set – (v1-21 1964-84 on reel $176. v22-38 1985-2001 on mf $647) – ISSN: 0022-4278 – mf#105091 – us Hein [360]
Journal of research in music education – Reston. 1953+ (1) 1970+ (5) 1976+ (9) – ISSN: 0022-4294 – mf#1592 – us UMI ProQuest [780]
Journal of research in pharmaceutical economics / ed by Smith, Mickey C – v1- 1989- – 1, 9 ($160.00 in US $224.00 outside hardcopy subsc) – us Haworth [330]
Journal of research in reading – Oxford. 1981+ – 1,5,9 – ISSN: 0141-0423 – mf#12908 – us UMI ProQuest [370]
Journal of research in science teaching – New York. 1963+ – 1,5,9 – ISSN: 0022-4308 – mf#11061 – us UMI ProQuest [370]
Journal of research of the national bureau of standards : section a: physics and chemistry / United States National Bureau of Standards – Washington. 1959-1977 (1) 1969-1977 (5) 1970-1977 (9) – ISSN: 0022-4332 – mf#1441 – us UMI ProQuest [540]
Journal of research of the national bureau of standards : section b: mathematical sciences / United States National Bureau of Standards – Washington. 1959-1977 (1) 1971-1977 (5) 1976-1977 (9) – ISSN: 0098-8979 – mf#1442 – us UMI ProQuest [510]
Journal of research of the national bureau of standards : section c: engineering and instrumentation / United States National Bureau of Standards – Washington. 1959-1972 (1) 1970-1972 (5) – ISSN: 0022-4316 – mf#1443 – us UMI ProQuest [621]
Journal of research of the national bureau of standards / United States. National Bureau of Standards – Washington. 1977-1988 (1) 1977-1988 (5) 1977-1988 (9) – (cont by: journal of research of the national institute of standards and technology) – ISSN: 0160-1741 – mf#11504 – us UMI ProQuest [540]
Journal of research of the national bureau of standards / United States National Bureau of Standards – Washington. 1928-1959 (1) – ISSN: 0091-0635 – mf#1169 – us UMI ProQuest [530]
Journal of research of the national institute of standards and technology / United States. National Institute of Standards and Technology – Gaithersburg. 1988+ (1,5,9) – (cont: journal of research of the national bureau of standards) – ISSN: 1044-677X – mf#11504,01 – us UMI ProQuest [540]
Journal of research on adolescence – Mahwah, 1998+ [1,5,9] – ISSN: 1050-8392 – mf#25244 – us UMI ProQuest [150]
Journal of research on computing in education – Eugene. 1986-2000 (1,5,9) – (cont: aeds journal) – ISSN: 0888-6504 – mf#10379,01 – us UMI ProQuest [370]

Journal of research on technology in education (jrte) – Eugene, 2001+ – 1,5,9 – (cont: journal of research on computing in education) – mf#10379,02 – us UMI ProQuest [370]

Journal of researches into the natural history and geology of the countries visited during the voyage of hms beagle round the world : under the command of capt fitz roy / Darwin, Charles Robert – London, 1873 – (= ser 19th c evolution & creation) – 6mf – 9 – mf#1.1.10916 – uk Chadwyck [500]

Journal of residence at the cape of good hope / Bunbury, Charles James Fox – New York, NY. 1969 – 1r – us UF Libraries [960]

Journal of restaurant and foodservice marketing / ed by Bowen, John – v2 n1. 1997- – 1,9 – $60.00 in US $84.00 outside hardcopy subsc – us Haworth [650]

Journal of retail banking – New York. 1979-1995 (1) 1979-1995 (5) 1979-1995 (9) – (cont by: journal of retail banking services: jrbs) – ISSN: 0195-2064 – mf#12991 – us UMI ProQuest [332]

Journal of retail banking services: jrbs – New York. 1995-1999 (1) 1995-1999 (5) 1995-1999 (9) – (cont: journal of retail banking) – mf#12991,01 – us UMI ProQuest [332]

Journal of retailing – New York. 1925+ [1]; 1968+ [5]; 1975+ [9] – ISSN: 0022-4359 – mf#1448 – us UMI ProQuest [650]

Journal of retailing and consumer services – Kidlington. 1997+ (1,5,9) – ISSN: 0969-6989 – mf#20749 – us UMI ProQuest [650]

Journal of rheumatology – Toronto. 1979+ (1,5,9) – ISSN: 0315-162X – mf#12255 – us UMI ProQuest [616]

The journal of richard norwood 1639-1640, surveyor of bermuda / Norwood, Richard – 9 – us Scholars Facs [919]

Journal of risk and insurance – Malvern. 1932+ (1) 1974+ (5) 1976+ (9) – ISSN: 0022-4367 – mf#10318 – us UMI ProQuest [368]

Journal of risk and uncertainty – Boston. 1988-1996 (1,5,9) – ISSN: 0895-5646 – mf#16819 – us UMI ProQuest [330]

Journal of risk finance – London. 1999+ (1,5,9) – ISSN: 1526-5943 – mf#32271 – us UMI ProQuest [332]

Journal of ritual studies – Pittsburgh. 1989+ (1,5,9) – ISSN: 0890-1112 – mf#17537 – us UMI ProQuest [230]

Journal of robotic systems – New York. 1984+ (1,5,9) – ISSN: 0741-2223 – mf#14332 – us UMI ProQuest [629]

Journal of rockingham county history and genealogy – 1976 apr-1980 jun – 1r – 1 – mf#671668 – us WHS [978]

Journal of roman studies *see* The limes tripolitanvs in the light of recent discoveries

Journal of rural studies – Elmsford. 1985+ (1,5,9) – ISSN: 0743-0167 – mf#49476 – us UMI ProQuest [307]

Journal of russian and east european psychology – Armonk. 1992+ (1) – (cont: soviet psychology) – ISSN: 1061-0405 – mf#16899,01 – us UMI ProQuest [150]

Journal of russian studies / Association of Teachers of Russian (Great Britain) – Bristol. 1959-1989 (1) 1959-1989 (5) 1959-1989 (9) – ISSN: 0047-276X – mf#10979 – us UMI ProQuest [460]

Journal of sacred literature – London. 1848-1868 (1) – mf#2782 – us UMI ProQuest [200]

Journal of safety research – Chicago. 1969-1980 (1) 1976-1980 (5) 1976-1980 (9) – (cont by: journal of safety research) – ISSN: 0022-4375 – mf#10451 – us UMI ProQuest [360]

Journal of safety research – Chicago. 1982+ (1,5,9) – (cont: journal of safety research) – ISSN: 0022-4375 – mf#49427 – us UMI ProQuest [360]

Journal of san diego history – San Diego. 1973+ (1) 1974+ (5) 1974+ (9) – ISSN: 0022-4383 – mf#9193 – us UMI ProQuest [978]

Journal of sandwich structures and materials – London, 1999+ [1,5,9] – ISSN: 1099-6362 – mf#32023 – us UMI ProQuest [620]

Journal of scholarly publishing – North York. 1993+ (1,5,9) – (cont: scholarly publishing) – mf#13405,01 – us UMI ProQuest [070]

Journal of school health – Kent. 1930+ (1) 1971+ (5) 1975+ (9) – ISSN: 0022-4391 – mf#3304 – us UMI ProQuest [360]

Journal of school leadership – Lancaster. 1991-1998 – 1,5,9 – ISSN: 1052-6846 – mf#18202 – us UMI ProQuest [370]

Journal of school psychology – New York. 1963-1982 (1) 1974-1982 (5) 1974-1982 (9) – (cont by: journal of school psychology) – ISSN: 0022-4405 – mf#11183 – us UMI ProQuest [150]

Journal of school psychology – New York. 1983+ (1,5,9) – (cont: journal of school psychology) – ISSN: 0022-4405 – mf#49460 – us UMI ProQuest [150]

Journal of school violence : official journal of the international school violence prevention association / ed by Gerler, Jr, Edwin R – 1,9 – $125.00 us $181.00 other – ISSN: 1538-8220 – us Haworth [370]

Journal of science – London. 1864-1885 (1) – mf#2908 – us UMI ProQuest [500]

Journal of science and technology law *see* Boston university journal of science and technology law

Journal of science and the arts – New York. 1817-1818 (1) – mf#4004 – us UMI ProQuest [500]

Journal of science education and technology – New York. 1992+ – 1,5,9 – ISSN: 1059-0145 – mf#19618 – us UMI ProQuest [370]

Journal of scientific and industrial research – New Delhi. 1972-1995 (1) 1972-1995 (5) 1973-1995 (9) – ISSN: 0022-4456 – mf#6913 – us UMI ProQuest [500]

Journal of scientific computing – New York. 1988+ (1,5,9) – ISSN: 0885-7474 – mf#17687 – us UMI ProQuest [500]

Journal of secondary education – Burlingame. 1925-1971 (1) 1969-1971 (5) – ISSN: 0022-4464 – mf#792 – us UMI ProQuest [373]

Journal of sectional proceedings / British Assocation for the Advancement of Science – Montreal: The Association, [1883?-19-?] – 9 – ISSN: 1190-7347 – mf#P04212 – cn Canadiana [500]

Journal of security administration – Miami. 1978+ (1,5,9) – ISSN: 0195-9425 – mf#12444 – us UMI ProQuest [360]

Journal of security education : new directions in education, training, and accreditation / ed by Kostanoski, John I – 1,9 – $95.00 us $138.00 other – ISSN: 1550-7890 – us Haworth [370]

Journal of sedimentary petrology – Tulsa. 1931-1993 (1) 1931-1993 (5) 1931-1993 (9) – ISSN: 0022-4472 – mf#1695 – us UMI ProQuest [550]

Journal of sedimentary research – Tulsa. 1996+ (1,5,9) – mf#25942 – us UMI ProQuest [550]

Journal of sedimentary research section a : sedimentary petrology and processes – Tulsa. 1994-1995 (1,5,9) – ISSN: 1073-130X – mf#20632 – us UMI ProQuest [550]

Journal of sedimentary research section b : stratigraphy and global studies – Tulsa. 1994-1995 (1,5,9) – ISSN: 1073-1318 – mf#20633 – us UMI ProQuest [550]

Journal of segmentation in marketing : innovations in market identification and targeting / ed by Weinstein, Art T – v1 n1. 1997- – 1,9 – $65.00 in US $91.00 outside hardcopy subsc – us Haworth [650]

Journal of semitic studies – 1(1956)-13(1968) – 89mf – 9 – €170.00 – ne Slangenburg [270]

Journal of semitic studies – Oxford. 1956+ (1) 1971+ (5) 1975+ (9) – ISSN: 0022-4480 – mf#1371 – us UMI ProQuest [939]

Journal of service research (jsr) – Thousand Oaks, 1998+ [1,5,9] – ISSN: 1094-6705 – mf#29411 – us UMI ProQuest [650]

Journal of services marketing – Santa Barbara. 1987-1995 (1) 1987-1995 (5) 1987-1995 (9) – ISSN: 0887-6045 – mf#16364 – us UMI ProQuest [650]

Journal of sex and marital therapy – New York. 1974-1997 (1) 1974-1997 (5) 1974-1997 (9) – ISSN: 0092-623X – mf#11184 – us UMI ProQuest [150]

Journal of sex education and therapy – Mount Vernon. 1989-1995 – 1,5,9 – ISSN: 0161-4576 – mf#17422 – us UMI ProQuest [370]

Journal of sex research – New York. 1965+ (1) 1972+ (5) 1974+ (9) – ISSN: 0022-4499 – mf#7135 – us UMI ProQuest [150]

Journal of shoulder and elbow surgery – St Louis, 1998+ [1,5,9] – ISSN: 1058-2746 – mf#19491 – us UMI ProQuest [617]

Journal of sikh studies – Amritsar: Dept of Guru Nanak Studies...v1- 1974- [semiannual] [mf ed 1988] – 1 – (lacks: v1 n1 p79-88) – mf#2003-s0778 – us ATLA [280]

Journal of singing – Jacksonville. 1995+ (1) 1995+ (5) 1995+ (9) – (cont: nats journal) – ISSN: 1086-7732 – mf#1777,02 – us UMI ProQuest [780]

Journal of small and emerging business law – v1-5. 1997-2001 – 9 – $95.00 set – mf#118181 – us Hein [346]

Journal of small business management – Milwaukee. 1964+ (1) 1972+ (5) 1974+ (9) – ISSN: 0047-2778 – mf#6882 – us UMI ProQuest [650]

Journal of small fruit and viticulture / ed by Gough, Robert E – v3 n1. 1995- – 1,9 – $75.00 in US $105.00 outside hardcopy subsc – us Haworth [630]

Journal of social and clinical psychology – New York. 1983+ (1,5,9) – ISSN: 0736-7236 – mf#13835 – us UMI ProQuest [150]

Journal of social history – Pittsburgh. 1967+ (1) 1971+ (5) 1975+ (9) – ISSN: 0022-4529 – mf#2504 – us UMI ProQuest [301]

Journal of social hygiene – v1-40. 1914-54 – 1 – $420.00 – mf#0307 – us Brook [360]

Journal of social issues – New York. 1945+ [1]; 1969+ [5]; 1975+ [9] – ISSN: 0022-4537 – mf#1094 – us UMI ProQuest [150]

Journal of social philosophy – Malden. 1970+ (1) 1972+ (5) 1977+ (9) – ISSN: 0047-2786 – mf#7049 – us UMI ProQuest [100]

Journal of social policy – Cambridge. 1972+ (1,5,9) – ISSN: 0047-2794 – mf#12127 – us UMI ProQuest [350]

Journal of social, political, and economic studies – Washington. 1990+ (1,5,9) – ISSN: 0278-839X – mf#18255,01 – us UMI ProQuest [320]

Journal of social psychology – Washington. 1983+ (1,5,9) – ISSN: 0022-4545 – mf#16464 – us UMI ProQuest [150]

Journal of social science : containing the proceedings of the american association – New York. 1869-1909 (1) – mf#5584 – us UMI ProQuest [300]

Journal of social science – London. 1865-1866 (1) – mf#4850 – us UMI ProQuest [300]

Journal of social service research / ed by Gillespie, David F – v1- 1977- – 1, 9 ($250.00 in US $350.00 outside hardcopy subsc) – us Haworth [360]

Journal of social studies research – Manhattan. 1977+ (1,5,9) – ISSN: 0885-985X – mf#11525 – us UMI ProQuest [300]

Journal of social therapy – Atascadero. 1954-1961 (1) 1954-1961 (5) 1954-1961 (9) – mf#12943 – us UMI ProQuest [615]

Journal of social work education – Washington. 1985+ (1) 1985+ (5) 1985+ (9) – (cont: journal of education for social work) – ISSN: 1043-7797 – mf#9611,01 – us UMI ProQuest [370]

Journal of social work in disability and rehabilitation : new directions in education, training, and accreditation / ed by Pardeck, John T – v1- 2002- – 1,9 – $140.00 us $203.00 other – ISSN: 1536-710X – us Haworth [360]

Journal of social work in long term care / ed by Weiner, Audrey S – v1- 2002- – 1,9 – $85.00 us $123.00 other – ISSN: 1533-2624 – us Haworth [360]

Journal of social work practice in the addictions / ed by Straussner, S Lala Ashenberg – v1- 2001- – 1,9 – $75.00 us $109.00 other – ISSN: 1533-256X – us Haworth [360]

Journal of socio-economics – Greenwich. 1991-1996 (1,5,9) – ISSN: 1053-5357 – mf#18833,01 – us UMI ProQuest [300]

Journal of sociolinguistics – Oxford. 1997+ (1) – ISSN: 1360-6441 – mf#25715 – us UMI ProQuest [400]

Journal of sociology and social welfare – Kalamazoo. 1979+ [1,5,9] – mf#12208 – us UMI ProQuest [301]

Journal of software maintenance – Chichester. 1989-2000 (1,5,9) – ISSN: 1040-550X – mf#17046 – us UMI ProQuest [000]

Journal of software maintenance and evolution – New York, 2001+ [1,5,9] – (cont: journal of software maintenance) – ISSN: 1532-060X – mf#17046,01 – us UMI ProQuest [000]

Journal of soil and water conservation – Ankeny. 1946+ (1) 1975+ (5) 1975+ (9) – ISSN: 0022-4561 – mf#10433 – us UMI ProQuest [630]

Journal of soil science – Oxford. 1949-1993 (1) 1971-1993 (5) 1977-1993 (9) – ISSN: 0022-4588 – mf#1248 – us UMI ProQuest [630]

Journal of solar energy engineering – New York. 1980+ (1,5,9) – (cont: transactions of the american society of mechanical engineers) – ISSN: 0199-6231 – mf#11950 – us UMI ProQuest [333]

Journal of south american earth sciences – Oxford. 1988-1995 (1,5,9) – ISSN: 0895-9811 – mf#49524 – us UMI ProQuest [550]

Journal of south asian literature – East Lansing. 1973-1995 (1) 1977-1995 (5) 1977-1995 (9) – (cont: mahfil) – ISSN: 0091-5637 – mf#7168,01 – us UMI ProQuest [490]

Journal of southeast asia business – Ann Arbor. 1991-1992 (1,5,9) – (cont by: journal of asian business) – ISSN: 1055-2073 – mf#18576,01 – us UMI ProQuest [337]

Journal of southeast asian earth sciences – New York. 1986-1994 (1,5,9) – (cont by: journal of asian earth sciences) – ISSN: 0743-9547 – mf#49489 – us UMI ProQuest [550]

Journal of southeast asian history – Singapore. 1960-1969 (1) – mf#2130 – us UMI ProQuest [959]

Journal of southern african affairs – Publ. of Southern African Research Assoc. and Afro-American Studies, Univ. of Maryland. -q – 1 – us Wisconsin U Libr [960]

Journal of southern african studies – Oxford. 1974+ (1) 1974+ (5) 1977+ (9) – ISSN: 0305-7070 – mf#9856 – us UMI ProQuest [300]

Journal of southern history – Athens. 1935+ (1) 1969+ (5) 1975+ (9) – ISSN: 0022-4642 – mf#1096 – us UMI ProQuest [975]

Journal of southern history / JSTOR [Organization] – v28 [1962] – 1r – 1 – mf#772894 – us WHS [975]

Journal of soviet mathematics – New York. 1973-1977 (1) 1973-1977 (5) – ISSN: 0090-4104 – mf#10886 – us UMI ProQuest [510]

Journal of space law – University of Mississippi. v1-24. 1973-96 – 5,9 – $453.00 set – (v1-12 1973-84 on reel $159. v13-24 1985-96 on mf $294) – ISSN: 0095-7577 – mf#104201 – us Hein [340]

Journal of spacecraft and rockets – Reston. 1964+ (1) 1971+ (5) 1976+ (9) – ISSN: 0022-4650 – mf#5075 – us UMI ProQuest [629]

Journal of spanish studies : twentieth century – Boulder. 1973-1980 (1) 1973-1980 (5) 1976-1980 (9) – ISSN: 0092-1807 – mf#7845 – us UMI ProQuest [440]

Journal of special education – Bensalem. 1987+ – 1,5,9 – ISSN: 0022-4669 – mf#16629 – us UMI ProQuest [370]

Journal of special education technology – Austin. 1978+ – 1,5,9 – ISSN: 0162-6434 – mf#12826 – us UMI ProQuest [370]

Journal of speculative philosophy – University Park. 1867+ (1) 1987+ (5) 1987+ (9) – ISSN: 0891-625X – mf#4134 – us UMI ProQuest [190]

Journal of speech and hearing disorders – Washington. 1948-1990 (1) 1948-1990 (5) 1948-1990 (9) – (cont: journal of speech disorders) – ISSN: 0022-4677 – mf#12777,01 – us UMI ProQuest [617]

Journal of speech and hearing research – Rockville. 1958-1996 (1) 1958-1996 (5) 1958-1996 (9) – (cont by: journal of speech, language and hearing research) – ISSN: 0022-4685 – mf#12778 – us UMI ProQuest [617]

Journal of speech disorders – Danville. 1936-1947 (1) 1936-1947 (5) 1936-1947 (9) – (cont by: journal of speech and hearing disorders) – mf#12777 – us UMI ProQuest [617]

Journal of speech, language and hearing research – Rockville. 1997+ (1,5,9) – (cont: journal of speech and hearing research) – ISSN: 1092-4388 – mf#12778,01 – us UMI ProQuest [617]

Journal of spinal disorders – New York. 1993+ (1,5,9) – ISSN: 0895-0385 – mf#18726 – us UMI ProQuest [617]

Journal of spiritual formation – Pittsburgh. 1994-1994 (1,5,9) – (cont: studies in formative spirituality) – mf#11967,01 – us UMI ProQuest [200]

Journal of sport and social issues – Boston. 1996+ (1,5,9) – ISSN: 0193-7235 – mf#21512 – us UMI ProQuest [790]

Journal of sport behavior – Mobile. 1992+ (1,5,9) – ISSN: 0162-7341 – mf#19208 – us UMI ProQuest [790]

Journal of sports medicine and physical fitness – Torino. 1998+ (1,5,9) – ISSN: 0022-4707 – mf#19246 – us UMI ProQuest [613]

Journal of sports sciences – London. 1983+ (1,5,9) – ISSN: 0264-0414 – mf#14407 – us UMI ProQuest [790]

Journal of staff development – Oxford. 1980+ – 1,5,9 – ISSN: 0276-928X – mf#13064 – us UMI ProQuest [370]

Journal of state government – Lexington. 1986-1992 (1) 1986-1992 (5) 1986-1992 (9) – (cont: state government. cont by: spectrum: the journal of state governments) – ISSN: 1043-2248 – mf#807,01 – us UMI ProQuest [320]

Journal of state government *see* Spectrum

Journal of state taxation – Greenvale. 1993-1996 (1,5,9) – ISSN: 0744-6713 – mf#19029 – us UMI ProQuest [336]

Journal of state taxation – v1-19. 1982-2001 – 9 – $468.00 set – ISSN: 0744-6713 – mf#108791 – us Hein [343]

Journal of statistical physics – New York. 1969+ (1) 1974+ (5) 1991+ (9) – ISSN: 0022-4715 – mf#10863 – us UMI ProQuest [530]

Journal of statistical planning and inference – Amsterdam. 1977+ (1) 1977+ (5) 1988+ (9) – ISSN: 0378-3758 – mf#42122 – us UMI ProQuest [310]

Journal of steel castings research – Des Plaines. 1955-1976 (1) 1972-1976 (5) 1975-1976 (9) – ISSN: 0022-4723 – mf#7659 – us UMI ProQuest [660]

Journal of steroid biochemistry – Oxford. 1969-1990 (1) 1969-1990 (5) 1969-1990 (9) – (cont by: journal of steroid biochemistry and molecular biology) – ISSN: 0022-4731 – mf#49122 – us UMI ProQuest [574]

Journal of steroid biochemistry and molecular biology – Oxford. 1990+ (1,5,9) – (cont: journal of steroid biochemistry) – ISSN: 0960-0760 – mf#49122,01 – us UMI ProQuest [574]

Journal of stored products research – Oxford. 1965+ (1,5,9) – ISSN: 0022-474X – mf#49123 – us UMI ProQuest [630]

Journal of strain analysis for engineering design – London. 1976+ (1,5,9) – ISSN: 0309-3247 – mf#11221,01 – us UMI ProQuest [620]

Journal of strategic change – Chichester. 1992-1993 (1,5,9) – ISSN: 1057-9265 – mf#19122 – us UMI ProQuest [650]

Journal of strategic information systems – Kidlington. 1991+ (1,5,9) – (cont: information age) – ISSN: 0963-8687 – mf#18783 – us UMI ProQuest [000]

Journal of structural chemistry – New York. 1960-1974 (1) 1960-1974 (5) – ISSN: 0022-4766 – mf#10837 – us UMI ProQuest [540]

Journal of structural engineering – New York. 1983+ (1) 1983+ (5) 1983+ (9) – (cont: journal of the structural division) – ISSN: 0733-9445 – mf#8141,01 – us UMI ProQuest [624]

Journal of structural geology – Oxford. 1979+ (1,5,9) – ISSN: 0191-8141 – mf#49316 – us UMI ProQuest [550]

Journal of structured and project finance – New York. 2001+ (1,5,9) – mf#24856,01 – us UMI ProQuest [332]

Journal of student financial aid – Washington. 1976+ (1) 1977+ (5) 1977+ (9) – mf#11007 – us UMI ProQuest [378]

Journal of studies in technical careers – Carbondale. 1982-1995 (1,5,9) – ISSN: 0163-3252 – mf#12909 – us UMI ProQuest [331]

Journal of studies on alcohol – New Brunswick. 1975+ (1) 1975+ (5) 1976+ (9) – (cont: quarterly journal of studies on alcohol) – ISSN: 0096-882X – mf#417,01 – us UMI ProQuest [360]

Journal of studies on alcohol supplement – New Brunswick. 1975-1994 (1) 1975-1994 (5) 1975-1994 (9) – (cont: quarterly journal of studies on alcohol supplement) – ISSN: 0363-468X – mf#8271,01 – us UMI ProQuest [616]

Journal of substance abuse treatment – Elmsford. 1984+ (1,5,9) – ISSN: 0740-5472 – mf#49461 – us UMI ProQuest [360]

Journal of supercomputing – Boston. 1987-1996 (1,5,9) – ISSN: 0920-8542 – mf#16820 – us UMI ProQuest [000]

Journal of superconductivity – New York. 1988-1993 (1,5,9) – ISSN: 0896-1107 – mf#17689 – us UMI ProQuest [530]

Journal of supercritical fluids – Kidlington, 1998+ [1,5,9] – ISSN: 0896-8446 – mf#42794 – us UMI ProQuest [540]

Journal of supply chain management – Tempe. 1999+ (1) 1999+ (5) 1999+ (9) – (cont: international journal of purchasing and materials management) – ISSN: 1523-2409 – mf#2757,03 – us UMI ProQuest [650]

Journal of supramolecular chemistry – Amsterdam. 2001+ (1,5,9) – ISSN: 1472-7862 – mf#42850 – us UMI ProQuest [540]

Journal of surgical oncology – New York. 1981+ (1,5,9) – ISSN: 0022-4790 – mf#12877 – us UMI ProQuest [617]

Journal of surveying engineering – New York. 1983+ (1) 1983+ (5) 1983+ (9) – (cont: journal of the surveying and mapping division) – ISSN: 0733-9453 – mf#8150,01 – us UMI ProQuest [624]

Journal of sustainable agriculture : innovations for long-term and lasting maintenance and enhancement of agricultural resources, production, and environmental quality / ed by Poincelot, Raymond P – v7 n1. 1996- – 1,9 – $120.00 in US $168.00 outside hardcopy subsc – us Haworth [630]

Journal of sustainable forestry / ed by Berlyn, Graeme P – v4 n1. 1997- – 1,9 – $95.00 in US $133.00 outside hardcopy subsc – us Haworth [630]

Journal of symbolic logic – Pasadena. 1985+ (1,5,9) – ISSN: 0022-4812 – mf#15692 – us UMI ProQuest [160]

Journal of systems and software – New York. 1979+ (1) 1979+ (5) 1987+ (9) – ISSN: 0164-1212 – mf#42301 – us UMI ProQuest [000]

Journal of systems architecture – Amsterdam. 1996+ (1,5,9) – (cont: microprocessing and microprogramming) – ISSN: 1383-7621 – mf#42343,01 – us UMI ProQuest [000]

Journal of systems engineering – Lancaster. 1969-1976 (1) 1974-1976 (5) (9) – (cont by: journal of applied systems analysis) – ISSN: 0022-4820 – mf#9973 – us UMI ProQuest [650]

Journal of systems integration – Boston. 1991-1995 (1,5,9) – ISSN: 0925-4676 – mf#18667 – us UMI ProQuest [000]

Journal of systems management – 1950-1996 [1]; 1968-1996 [5]; 1975-1996 [9] – ISSN: 0022-4839 – mf#1797 – us UMI ProQuest [650]

Journal of taxation – New York. 1954+(1,5,9) – ISSN: 0022-4863 – mf#11801 – us UMI ProQuest [336]

Journal of taxation of financial institutions – New York. 2000+ (1) – (cont: journal of bank taxation) – mf#18372,01 – us UMI ProQuest [336]

Journal of taxation of investments – Boston. 1983+ (1,5,9) – ISSN: 0747-9115 – mf#13918 – us UMI ProQuest [332]

Journal of teacher education – Washington. 1950+ (1) 1969+ (5) 1975+ (9) – ISSN: 0022-4871 – mf#1464 – us UMI ProQuest [370]

Journal of teaching in international business / ed by Kaynak, Erdener – v1- 1989- – 1, 9 ($120.00 in US $168.00 outside hardcopy subsc) – us Haworth [370]

Journal of teaching in marriage and family : innovations in family science education / ed by Gentry, Deborah Barnes – v1- 2001- – 1,9 – $75.00 us $109.00 other – ISSN: 1535-0762 – us Haworth [640]

Journal of teaching in social work / ed by Vigilante, Florence & Lewis, Harold – v1- 1987- – 1, 9 ($150.00 in US $210.00 outside hardcopy subsc) – us Haworth [370]

Journal of teaching in the addictions : official journal of the international coalition for addiction studies education (incase) / ed by Taleff, Michael J – v1- 2002- – 1,9 – $65.00 us $94.00 other – ISSN: 1533-2705 – us Haworth [616]

Journal of teaching in travel & tourism : the official journal of the international society of travel and tourism educators (istte) / ed by Hsu, Cathy H C – v1- 2001- – 1,9 – $85.00 us $123.00 other – ISSN: 1531-3220 – us Haworth [338]

Journal of technical topics in civil engineering / ASCE Technical Council on Codes and Standards – New York. 1983-1985 (1,5,9) – (cont: journal of the technical councils of asce) – ISSN: 0733-9461 – mf#11508,01 – us UMI ProQuest [624]

Journal of technology in human services / ed by Schoech, Dick – v16- 1999- – 1,9 – $125.00 us $181.00 other – ISSN: 1522-8835 – us Haworth [331]

Journal of technology studies – Bowling Green. 1993-1996 – 1,5,9 – (cont: journal of epsilon pi tau) – ISSN: 1071-6084 – mf#11408,01 – us UMI ProQuest [370]

Journal of temperance – Prescott [Ont] : "Evangelizer" Office, [1864-1865] – 9 – ISSN: 1190-6782 – mf#P04266 – cn Canadiana [230]

Journal of terramechanics – Oxford. 1964-1995 (1,5,9) – ISSN: 0022-4898 – mf#49124 – us UMI ProQuest [621]

Journal of testing and evaluation – Philadelphia. 1973+ (1) 1973+ (5) 1976+ (9) – ISSN: 0090-3973 – mf#7228 – us UMI ProQuest [620]

Journal of thanatology – New York. 1971-1975 (1,5) 1975-1975 (9) – (cont by: advances in thanatology) – ISSN: 0047-2832 – mf#7826 – us UMI ProQuest [614]

Journal of that faithful servant of christ, charles osborn : containing an account of many of his travels and labors in the work of the ministry, and his trials and exercises in the service of the lord and in defense of the truth, as it is in jesus / Osborn, Charles – Cincinnati: A Pugh, 1854 – 2mf – 9 – 0-524-06475-X – mf#1990-5249 – us ATLA [240]

Journal of the abraham lincoln association – Champaign. 1988-1993 (1) – ISSN: 0898-4212 – mf#16624,01 – us UMI ProQuest [320]

The journal of the academy of religion and psychical research – Bloomfield CT, 1981- [qrterly] [mf v4- 1981- filmed 1987] – 1 – mf0841 – us ATLA [230]

The journal of the academy of religion and psychical research – Bloomfield CT. -1980 [qrterly] [mf v2-3 1979-80 filmed 1987] – 3v on 1r – 1 – (v2 n4 oct 1979 never publ) – mf0841a – us ATLA [230]

The journal of the acoustical society of america – v1-. 1929- – 1,5,6,9 – us AIP [530]

Journal of the aerospace sciences – New York. 1934-1962 (1) – ISSN: 0095-9820 – mf#1481 – us UMI ProQuest [629]

Journal of the air and waste management association – Pittsburgh. 1995+ (1,5,9) – (cont: air and waste) – mf#6210,04 – us UMI ProQuest [628]

Journal of the air and waste management association – Pittsburgh. 1990-1992 (1) 1990-1992 (5) 1990-1992 (9) – (cont: japca. cont by: air and waste) – ISSN: 1047-3289 – mf#6210,02 – us UMI ProQuest [628]

Journal of the alabama academy of science / Alabama Academy of Science – Auburn. 1976+ (1,5,9) – ISSN: 0002-4112 – mf#10793 – us UMI ProQuest [500]

Journal of the alabama dental association / Alabama Dental Association – Birmingham. 1972-1996 (1) 1972-1982 (5) 1975-1982 (9) – ISSN: 0002-4198 – mf#7267 – us UMI ProQuest [617]

Journal of the american academy of audiology / American Academy of Audiology – McLean. 1994+ (1,5,9) – ISSN: 1050-0545 – mf#21650 – us UMI ProQuest [617]

Journal of the american academy of child and adolescent psychiatry – v10-35. 1971-96 – 1,5,6,9 – $110.00r – us Lippincott [616]

Journal of the american academy of matrimonial lawyers – University of Wisconsin. v1-16. 1985-2000 – 9 – $212.00 set – ISSN: 0882-6714 – mf#111751 – us Hein [340]

Journal of the american academy of nurse practitioners / American Academy of Nurse Practitioners – Austin. 1989+ (1,5,9) – ISSN: 1041-2972 – mf#17349 – us UMI ProQuest [610]

Journal of the american academy of physician assistants – Montvale. 1988-1994 (1,5,9) – (cont by: jaapa) – ISSN: 0893-7400 – mf#16286 – us UMI ProQuest [610]

Journal of the american academy of psychiatry and the law – Bloomfield. 1997+ (1,5,9) – (cont: bulletin of the american academy of psychiatry and the law) – ISSN: 1093-6793 – mf#12354,01 – us UMI ProQuest [340]

Journal of the american academy of psychiatry and the law – v1-29. 1973-2001 – 5,6,9 – $707.00 set – (v1-12 1973-84 in reel $138. v13-29 1985-2001 in mf $569. title varies: v1-24 1973-96 as bulletin of the american academy of psychiatry and the law) – ISSN: 0091-634X – mf#100171 – us Hein [344]

Journal of the american association of nurse anesthetists / American Association of Nurse Anesthetists – Chicago. 1933-1973 (1) 1971-1973 (5) – (cont by: aana journal) – ISSN: 0002-7448 – mf#2083 – us UMI ProQuest [610]

Journal of the american board of family practice – Waltham. 1988+ (1,5,9) – ISSN: 0893-8652 – mf#16906 – us UMI ProQuest [610]

Journal of the american breweriana association / American Breweriana Association – n1-10 [1982 nov/dec-1984 may/jun] – 1r – 1 – (cont by: american breweriana journal) – mf#1575518 – us WHS [366]

Journal of the american chemical society – v1- 1879- – 1,5,6,9 – us ACS [540]

Journal of the american chiropractic association / American Chiropractic Association – Arlington. 1995+(1,5,9) – (cont: journal of chiropractic) – ISSN: 1081-7166 – mf#10579,02 – us UMI ProQuest [615]

Journal of the american college health association / American College Health Association – Washington. 1952-1981 (1) 1971-1981 (5) 1976-1981 (9) – (cont by: journal of american college health) – ISSN: 0164-4300 – mf#2240 – us UMI ProQuest [366]

Journal of the american college of cardiology – New York. 1983+ (1) 1983+ (5) 1987+ (9) – ISSN: 0735-1097 – mf#42418 – us UMI ProQuest [616]

Journal of the american college of dentists – Gaithersburg. 1934+ (1) 1971+ (5) 1974+ (9) – ISSN: 0002-7979 – mf#2027 – us UMI ProQuest [617]

Journal of the american college of nutrition / American College of Nutrition – New York. 1982-1991 (1,5,9) – ISSN: 0731-5724 – mf#18109 – us UMI ProQuest [613]

Journal of the american college of surgeons / American College of Surgeons – Chicago. 1994+ (1) 1994+ (5) 1994+ (9) – (cont: surgery, gynecology and obstetrics) – ISSN: 1072-7515 – mf#2284,01 – us UMI ProQuest [617]

Journal of the american concrete institute / American Concrete Institute – Detroit. 1905-1986 (1) 1971-1986 (5,9) – ISSN: 0002-8061 – mf#406 – us UMI ProQuest [690]

Journal of the american dental hygienists' association / American Dental Hygienists' Association – Thorofare. 1927-1972 (1) 1967-1972 (5) 1970-1972 (9) – (cont by: dental hygiene) – ISSN: 0002-8185 – mf#2274 – us UMI ProQuest [617]

Journal of the american geriatrics society – Baltimore. 1953-1991 [1]; 1971-1991 [5]; 1975-1991 [9] – ISSN: 0002-8614 – mf#1689 – us UMI ProQuest [618]

Journal of the american geriatrics society – v39-44. 1991-96 – 6r – 1,5,6,9 – $95.00r – us Lippincott [618]

Journal of the american institute of architects / American Institute of Architects – Harrisburg. 1913-1928 (1) – mf#7811 – us UMI ProQuest [720]

Journal of the american institute of criminal law and criminology *see* Journal of criminal law and criminology

Journal of the american institute of hypnosis / American Institute of Hypnosis – Los Angeles. 1960-1976 [1]; 1972-1976 [5]; 1974-1976 [9] – ISSN: 0002-8975 – mf#6555 – us UMI ProQuest [615]

Journal of the american liszt society – Nashville. 1979+ (1,5,9) – ISSN: 0147-4413 – mf#12105 – us UMI ProQuest [780]

Journal of the american medical informatics association (jamia) / American Medical Informatics Association – Philadelphia. 1997+ (1,5,9) – ISSN: 1067-5027 – mf#21602 – us UMI ProQuest [000]

Journal of the american medical technologists / American Medical Technologists – Park Ridge. 1973-1983 (1) 1974-1983 (5) 1977-1983 (9) – ISSN: 0002-9963 – mf#8705 – us UMI ProQuest [619]

Journal of the american mission of hugh bourne, 1844-1846 – Manchester: Hartley Methodist College – (= ser BRRAM series) – 1r – 1 – £67 / $134 – (int by j t wilkinson) – mf#r96518 – uk Microform Academic [242]

Journal of the american oil chemists' society / American Oil Chemists' Society – Champaign. 1924-1979 (1) 1965-1979 (5) 1970-1979 (9) – (cont by: jaocs, journal of the american oil chemists' society) – ISSN: 0003-021X – mf#131 – us UMI ProQuest [540]

Journal of the american oriental society / American Oriental Society – New Haven. 1843+ (1) 1971+ (5) 1976+ (9) – ISSN: 0003-0279 – mf#2181 – us UMI ProQuest [950]

Journal of the american oriental society – Boston, 1849-1946. v1-66 – 569mf – 8 – mf#I-205c – ne IDC [956]

Journal of the american pharmaceutical association / American Pharmaceutical Association – Washington. 1940-1977 (1) 1971-1977 (5) 1976-1977 (9) – (cont by: american pharmacy) – ISSN: 0003-0465 – mf#18 – us UMI ProQuest [615]

Journal of the american pharmaceutical association : apha / American Pharmaceutical Association – Washington. 1996+ (1,5,9) – (cont: american pharmacy) – ISSN: 1086-5802 – mf#18,02 – us UMI ProQuest [615]

Journal of the american podiatric medical association / American Podiatric Medical Association – Bethesda. 1985+ (1) 1985+ (5) 1985+ (9) – (cont: journal of the american podiatry association) – ISSN: 8750-7315 – mf#6915,01 – us UMI ProQuest [617]

Journal of the american psychiatric nurses association – St. Louis. 1995+ (1,5,9) – ISSN: 1078-3903 – mf#21564 – us UMI ProQuest [610]

Journal of the american real estate and urban economics association / American Real Estate and Urban Economics Association – Bloomington. 1973-1994 (1) 1973-1994 (5) 1973-1994 (9) – (cont by: real estate economics) – ISSN: 1067-8433 – mf#11634 – us UMI ProQuest [333]

Journal of the american society for information science / American Society for Information Science – New York. 1950+ [1]; 1969+ [5]; 1975+ [9] – (cont by: journal of the american society for information science and technology) – ISSN: 0002-8231 – mf#1423 – us UMI ProQuest [020]

Journal of the american society for information science and technology – New York. 2001+ (1) – (cont: journal of the american society for information science) – ISSN: 1532-2882 – mf#1423,01 – us UMI ProQuest [020]

Journal of the american society for mass spectrometry / American Society for Mass Spectrometry – New York. 1990-1993 (1,5,9) – ISSN: 1044-0305 – mf#42447 – us UMI ProQuest [530]

Journal of the american society for psychical research / American Society for Psychical Research – New York. 1907+ (1) 1965+ (5) 1976+ (9) – ISSN: 0003-1070 – mf#10 – us UMI ProQuest [130]

Journal of the american society of clu / American Society of Chartered Life Underwriters – Bryn Mawr. 1984-1986 (1) 1984-1986 (5) 1984-1986 (9) – (cont: clu journal. cont by: journal of the american society of clu and chfc) – ISSN: 0742-9517 – mf#7377,01 – us UMI ProQuest [366]

Journal of the american society of clu and chfc / American Society of CLU & ChFC – Bryn Mawr. 1986-1998 (1) 1986-1998 (5) 1986-1998 (9) – (cont: journal of the american society of clu. cont by: journal of financial service professionals) – ISSN: 1052-2875 – mf#7377,02 – us UMI ProQuest [366]

Journal of the american society of echocardiography / American Society of Echocardiography – St Louis. 1988-1988 (1,5,9) – ISSN: 0894-7317 – mf#16287 – us UMI ProQuest [616]

Journal of the american society of nephrology – v1-7. 1990-96 – 1,5,6,9 – $100.00r – us Lippincott [616]

Journal of the american statistical association / American Statistical Association – Alexandria. 1888+ (1) 1968+ (5) 1975+ (9) – ISSN: 0162-1459 – mf#1009 – us UMI ProQuest [317]

Journal of the american taxation association – Sarasota. 1991+ (1) 1997+ (5) 1998+ (9) – ISSN: 0198-9073 – mf#18573 – us UMI ProQuest [336]

Journal of the american water resources association – Middleburg. 1997+ (1) 1997+ (5) 1997+ (9) – (cont: water resources bulletin) – ISSN: 1093-474X – mf#7388,01 – us UMI ProQuest [333]

Journal of the ancient near eastern society – New York. 1989-1993 (1) – mf#15696,01 – us UMI ProQuest [930]

Journal of the annual convention / Episcopal Church Diocese Of Florida – s.l, s.l? 1839-1842 – 1r – us UF Libraries [978]

Journal of the annual convention / Episcopal Church Diocese Of Florida – s.l, s.l? 1871-1874 – 1r – us UF Libraries [978]

Journal of the annual council / Episcopal Church Diocese Of Florida – s.l, s.l? 1875-1879 – 1r – us UF Libraries [978]

Journal of the annual council / Episcopal Church Diocese Of Florida – s.l, s.l? 1911-1924 – 2r – us UF Libraries [978]

Journal of the annual encampment / Grand Army of the Republic. Dept of the Potomac – 17TH-27TH. 1885-95 – 1 – us L of C Photodup [976]

Journal of the armed forces – Washington. 1863-1963 – 1 – us L of C Photodup [355]

Journal of the asiatic society of bengal – Calcutta 1832-1910 – v1-74 on 1200mf – 9 – (ind [1856] v1-23) – mf#1-101/2 – ne IDC [954]

Journal of the asiatic society of pakistan – [Dacca], 1956-1963. v1-8 – 45mf – 8 – mf#I-215 – ne IDC [954]

Journal of the association for persons with severe handicaps : official publication of the association for persons with severe handicaps – Seattle. 1983+ – 1,5,9 – (cont: journal of the association for the severely handicapped) – ISSN: 0749-1425 – mf#12077,02 – us UMI ProQuest [370]

Journal of the association for the severely handicapped – Baltimore. 1980-1983 – 1,5,9 – (cont by: journal of the association for persons with severe handicaps) – ISSN: 0274-9483 – mf#12077,01 – us UMI ProQuest [370]

Journal of the association of nurses in aids care – Philadelphia. 1992+ (1,5,9) – ISSN: 1055-3290 – mf#19677 – us UMI ProQuest [610]

Journal of the astronautical sciences – v1-33. 1954-86 – 6 – $300.00 set – (in 3 pts: v1-19 1954-june 1972 $100. v20-26 july 1972-78 $100. v27-33 1979-86 $200) – us Univelt [629]

Journal of the australian chiropractors' association – Castlemaine. 1986-1990 (1) 1986-1990 (5) 1986-1990 (9) – (cont by: chiropractic journal of australia) – ISSN: 0045-0359 – mf#15997 – us UMI ProQuest [615]

Journal of the australian institute of agricultural science / Australian Institute of Agricultural Science – Sydney. 1979-1982 (1,5,9) – ISSN: 0045-0545 – mf#49317 – us UMI ProQuest [630]

Journal of the autonomic nervous system – Amsterdam. 1979-1998 (1,5,9) – (cont by: autonomic neuroscience: basic and clinical) – ISSN: 0165-1838 – mf#42212 – us UMI ProQuest [612]

Journal of the bar association of the district of columbia – v1-41. 1934-74 (all publ) – 9 – $655.00 set – (title varies: v1 n1-v2 n3 as bulletin of the bar association of the district of columbia. v2 n4-v7 n6 as journal of district of columbia bar association of the district of columbia. v33 n11-v40 n9 as d.c. bar journal) – mf#101171 – us Hein [340]

Journal of the beverly hills bar association *see* Beverly hills bar association journal

A journal of the bishop's visitation tour through the cape colony, in 1848 : with an account of his visit to the island of st helena, in 1849... / [Gray, R] – London, 1849 – 2mf – 9 – mf#HTM-70 – ne IDC [915]

A journal of the bishop's visitation tour through the cape colony, in 1850 / [Gray, R] – London, 1851 – 3mf – 9 – mf#HTM-71 – ne IDC [916]

Journal of the blantyre mission, 1888-1919 – 2r – 1 – (with int by a c ross) – mf#96606 – uk Microform Academic [240]

Journal of the board of trustees and minutes of committees and inspectors of the freedman's savings and trust company, 1865-1874 – (= ser Records of the office of the comptroller of the currency) – 2r – 1 – (with printed guide) – mf#M874 – us Nat Archives [332]

Journal of the bombay natural history society – Bombay, London, Madras 1886-1967 – v1-64 on 1163mf – 9 – mf#z-279/2 – ne IDC [500]

Journal of the british archaeological association – London, 1845-1894, v1-50; N S, 1895-1920, v1-26 – 716mf – 8 – mf#H-807 – ne IDC [700]

Journal of the british institute of international affairs / British Institute of International Affairs – London. 1922-1926 [1,5,9] – (cont by: journal of the royal institute of international affairs) – mf#1030 – us UMI ProQuest [327]

Journal of the british interplanetary society / British Interplanetary Society – Wallasey. 1934+ (1) 1975+ (5) 1975+ (9) – ISSN: 0007-084X – mf#8658 – us UMI ProQuest [629]

Journal of the british society of dowsers / British Society of Dowsers – London. 1933-1991 (1) 1977-1980 (5) 1975-1980 (9) – ISSN: 0007-179X – mf#8894 – us UMI ProQuest [130]

Journal of the buddhist text society of india – Calcutta: Buddhist Text Society of India, 1893-1906 [mf ed 2001] – (= ser Christianity's encounter with world religions, 1850-1950) – 5r – 1 – (filmed with: journal and of the buddhist text society of india, journal of the buddhist text society of india [1895]; journal of the buddhist text and anthropological society; journal of the buddhist text and research society. in english, some texts in sanskrit, tibetan and pali) – mf#2001-s093-097 – us ATLA [280]

The journal of the burma research society – Rangoon, 1911-1965. v1-V48 – 389mf – 8 – (missing: v32(2); v34(2)) – mf#SE-100 – ne IDC [954]

Journal of the canadian association of radiologists = Journal de l'association canadienne des radiologistes / Canadian Association of Radiologists – Montreal. 1950-1985 (1) 1970-1985 (5) 1974-1985 (9) – (cont by: canadian association of radiologists journal = journal l'association canadienne des radiologistes) – ISSN: 0008-2902 – mf#3032 – us UMI ProQuest [616]

Journal of the canadian chiropractic association / Canadian Chiropractic Association – Toronto. 1986+ (1,5,9) – ISSN: 0008-3194 – mf#16312,01 – us UMI ProQuest [615]

Journal of the canadian linguistic association = Revue de l'association canadienne de linguistique / Canadian Linguistic Association – Montreal. 1954-1961 (1) 1954-1961 (5) 1954-1961 (9) – (cont by: canadian journal of linguistics) – ISSN: 0319-5732 – mf#12024 – us UMI ProQuest [400]

The journal of the canadian mining institute / Canadian Mining Institute – Ottawa: The Institute, 1898?-19– – 9 – mf#A00269 – cn Canadiana [622]

Journal of the chartered institution of building services / Chartered Institution of Building Services – London. 1978-1980 (1,5,9) – ISSN: 0142-3630 – mf#11876 – us UMI ProQuest [690]

Journal of the chemical society / Chemical Society (Great Britain) – London. 1849-1971 (1) 1965-1971 (5) – mf#516 – us UMI ProQuest [540]

Journal of the chemical society : dalton transactions / Chemical Society (Great Britain) – London. 1972-1999 (1) 1972-1999 (5) 1976-1999 (9) – ISSN: 1470-479X – mf#7183 – us UMI ProQuest [540]

Journal of the chemical society : faraday transactions – Cambridge. 1990-1998 (1,5,9) – ISSN: 0956-5000 – mf#17646 – us UMI ProQuest [540]

Journal of the chemical society : faraday transactions 1 / Chemical Society (Great Britain) – London. 1972-1989 (1) 1972-1989 (5) 1976-1989 (9) – ISSN: 0300-9599 – mf#7184 – us UMI ProQuest [540]

Journal of the chemical society : faraday transactions 2 / Chemical Society (Great Britain) – London. 1972-1989 (1) 1972-1989 (5) 1976-1989 (9) – ISSN: 0300-9238 – mf#7185 – us UMI ProQuest [540]

Journal of the chemical society : perkin transactions 1 / Chemical Society (Great Britain) – London. 1972-1999 (1) 1972-1999 (5) 1976-1999 (9) – ISSN: 0300-922X – mf#7186 – us UMI ProQuest [540]

Journal of the chemical society : perkin transactions 2 / Chemical Society (Great Britain) – London. 1972-1999 (1) 1972-1999 (5) 1976-1999 (9) – ISSN: 0300-9580 – mf#7187 – us UMI ProQuest [540]

Journal of the chemical society chemical communications / Chemical Society (Great Britain) – London. 1972-1995 (1) 1972-1995 (5) 1976-1995 (9) – (cont by: chemical communications: chem comm) – ISSN: 0022-4936 – mf#10060 – us UMI ProQuest [540]

Journal of the chemico-agricultural society of ulster – Belfast, Ireland. Oct 1849-Sept 1867. -w. 1 reel – 1 – uk British Libr Newspaper [630]

Journal of the chinese language teachers association / Chinese Language Teachers Association – South Orange. 1966-1996 (1) 1972-1996 (5) 1976-1996 (9) – ISSN: 0009-4595 – mf#6350 – us UMI ProQuest [480]

Journal of the christian medical association of india, pakistan, burma and ceylon / Christian Medical Association of India, Pakistan, Burma and Ceylon – Mysore. 1950-1953 (1) – ISSN: 0009-5443 – mf#642 – us UMI ProQuest [610]

Journal of the cleveland bar association *see* Cleveland bar journal

Journal of the college and university personnel association / College and University Personnel Association – Washington. 1955-1986 (1) 1971-1986 (5) 1976-1986 (9) – (cont by: cupa journal) – ISSN: 0010-0935 – mf#3027 – us UMI ProQuest [378]

Journal of the college of science, king saud university / Jamiat al-Malik Saud. Kulliyat al-Ulum – Riyadh. 1984-1988 (1,5,9) – ISSN: 0735-9799 – mf#14814,03 – us UMI ProQuest [500]

Journal of the community development society / Community Development Society – Columbia. 1984+ (1,5,9) – ISSN: 0010-3829 – mf#14310 – us UMI ProQuest [307]

Journal of the congress of the confederate states of america – 76mf (24:1) – 9 – $350.00 – us UPA [976]

Journal of the construction division / American Society of Civil Engineers. Construction Division – New York. 1973-1982 (1) 1974-1982 (5) 1974-1982 (9) – (cont by: journal of construction engineering and management) – ISSN: 0569-7948 – mf#8146 – us UMI ProQuest [624]

Journal of the copyright society of the u.s.a. / Copyright Society of the USA – v1-48. 1953-2001 – 1,5,6 – $742.00 set – (ind v1-20. v1-41 1953-94 in reel $545. v42-48 1995-2001 in mf $197) – mf#101191 – us Hein [070]

Journal of the derbyshire archaeological and natural history society – London 1879-1920 – v1-42 on 218mf – 9 – (missing: 1886 v8) – mf 860/2 – ne IDC [930]

A journal of the disasters in affghanistan, 1841-1842 / Sale, [F] – London, 1843 – 6mf – 9 – mf#HT-129 – ne IDC [915]

Journal of the early republic – West Lafayette. 1981+ (1,5,9) – ISSN: 0275-1275 – mf#12882 – us UMI ProQuest [975]

Journal of the electrochemical society / Electrochemical Society – New York. 1948+ (1) 1965+ (5) 1970+ (9) – ISSN: 0013-4651 – mf#1156 – us UMI ProQuest [540]

Journal of the energy division / American Society of Civil Engineers. Energy Division – New York. 1979-1982 (1) 1979-1982 (5) 1979-1982 (9) – (cont: journal of the power division. cont by: journal of energy engineering) – ISSN: 0190-8294 – mf#8149,01 – us UMI ProQuest [624]

Journal of the engineering mechanics division / American Society of Civil Engineers. Engineering Mechanics Division – New York. 1973-1982 (1) 1973-1982 (5) 1973-1982 (9) – (cont by: journal of engineering mechanics) – ISSN: 0044-7951 – mf#8142 – us UMI ProQuest [624]

Journal of the entomological society of british columbia / Entomological Society of British Columbia – Victoria. 1911-1993 (1) 1972-1980 (5) 1977-1980 (9) – ISSN: 0071-0733 – mf#6953 – us UMI ProQuest [590]

Journal of the environmental engineering division / American Society of Civil Engineers. Environmental Engineering Division – New York. 1973-1982 (1) 1973-1982 (5) 1973-1982 (9) – (cont by: journal of environmental engineering) – ISSN: 0090-3914 – mf#8151 – us UMI ProQuest [628]

Journal of the evangelical theological society / Evangelical Theological Society – Lynchburg. 1969+ (1,5,9) – (cont: bulletin of the evangelical theological society) – ISSN: 0360-8808 – mf#11887,01 – us UMI ProQuest [242]

Journal of the executive proceedings / U.S. Congress. Senate – v. 1-107. 1789-1965 – 1 – us L of C Photodup [324]

Journal of the executive proceedings of the us senate, 1789-1823 / U.S. Senate – (= ser Records of the united states senate) – 3r – 1 – (with printed guide) – mf#M1252 – us Nat Archives [324]

The journal of the federated canadian mining institute / Federated Canadian Mining Institute – Ottawa, Ont: Secretary-Treasurer, 1896-1898 – 9 – mf#A00187 – cn Canadiana [622]

Journal of the fellowship of st alban and st sergius – 1933-34 – 4mf – 9 – €10.00 – ne Slangenburg [241]

Journal of the florida medical association, inc / Florida Medical Association, Inc – Jacksonville. 1914+ (1) 1971+ (5) 1974+ (9) – ISSN: 0015-4148 – mf#2334 – us UMI ProQuest [610]

Journal of the forum committee on franchising *see* Franchise law journal (aba)

Journal of the franklin institute – Elmsford. 1969+ (1,5,9) – ISSN: 0016-0032 – mf#49074 – us UMI ProQuest [500]

The journal of the french canadian missionary society – Montreal: [s.n, 1872-18– or 19–] – 9 – mf#P04418 – cn Canadiana [242]

Journal of the genealogical society of okaloosa county, florida / Genealogical Society of Okaloosa County, Florida – v1 n1-v2 n1=5 [1977 mar-1978 1st qtr] – 1r – 1 – (cont by: genealogical society of okaloosa county journal) – mf#1018153 – us WHS [978]

Journal of the general assembly : 1st general and 1st special sess, 1965 / Congress of Micronesia – Saipan: Congress of Micronesia. 2v. 1965 – (= ser Micronesia: interim governance) – 34mf – 9 – $51.00 – mf#LLMC 82-100F Title 85 – us LLMC [323]

Journal of the general assembly / Church of the Nazarene – 1st-5th [1907-1919] – 1r – 1 – mf#1054487 – us WHS [071]

The journal of the general mining association of the province of quebec / General Mining Association of Quebec – Ottawa: The Association, 1893?-1895? – 9 – mf#A01850 – cn Canadiana [622]

Journal of the geological society – Oxford. 1980+ (1,5,9) – ISSN: 0016-7649 – mf#15562,01 – us UMI ProQuest [550]

Journal of the geological society of australia / Geological Society of Australia – Sydney. 1980-1983 (1,5,9) – (cont by: australian journal of earth sciences) – ISSN: 0016-7614 – mf#15505 – us UMI ProQuest [550]

Journal of the geotechnical engineering division / American Society of Civil Engineers. Geotechnical Engineering Division – New York. 1974-1982 (1) 1974-1982 (5) 1974-1982 (9) – (cont: journal of the soil mechanics and foundations division. cont by: journal of geotechnical engineering) – ISSN: 0093-6405 – mf#8140,01 – us UMI ProQuest [624]

Journal of the gospel labours of george richardson : a minister in the society of friends: with a biographical sketch of his life and character / Richardson, George – London: A Bennett, 1864 [mf ed 1993] – (= ser Society of friends (quakers) coll) – 1mf – 9 – 0-524-06731-7 – mf#1991-2761 – us ATLA [243]

Journal of the gypsy lore society, 1888-1973 / The Gypsy Lore Society – 1974 – 6r – 1 – $780.00 – mf#S1854 – us Scholarly Res [306]

Journal of the historical society of nigeria – Ibadan: Historical Society of Nigeria. v1-v4 n4. dec 1956-jun 1969; Index 1956-69, 1964-67 – us CRL [960]

Journal of the history of biology – Dordrecht. 1984+ (1,5,9) – ISSN: 0022-5010 – mf#14760 – us UMI ProQuest [574]

Journal of the history of dentistry – Chicago. 1996+ (1) 1996+ (5) 1996+ (9) – (cont: bulletin of the history of dentistry) – ISSN: 0007-5132 – mf#7280,01 – us UMI ProQuest [617]

Journal of the history of economic thought: jhet – Abingdon. 1998+ (1) – ISSN: 1053-8372 – mf#26294,01 – us UMI ProQuest [330]

Journal of the history of ideas – Philadelphia. 1940+ (1) 1940+ (5) 1975+ (9) – ISSN: 0022-5037 – mf#6466 – us UMI ProQuest [000]

Journal of the history of medicine and allied sciences – Eynsham. 1946+ [1]; 1971+ [5]; 1977+ [9] – ISSN: 0022-5045 – mf#1684 – us UMI ProQuest [610]

Journal of the history of philosophy – Berkeley. 1963+ (1) 1971+ (5) 1976+ (9) – ISSN: 0022-5053 – mf#1659 – us UMI ProQuest [100]

Journal of the history of sexuality – Austin. 1990+ (1,5,9) – ISSN: 1043-4070 – mf#19141 – us UMI ProQuest [306]

Journal of the history of the behavioral sciences – Brandon. 1965+ (1) 1972+ (5) 1972+ (9) – ISSN: 0022-5061 – mf#10969 – us UMI ProQuest [150]

Journal of the house of commons – 1982/83- [mf ed Chadwyck-Healey] – (= ser House of commons parliamentary papers, 1901-1974/1975) – 9 – uk Chadwyck [323]

The journal of the house of commons : the official record of the proceedings of the house / Great Britain. House of Commons – 1982/83-1997/98 [mf ed Chadwyck-Healey] – 9 – (available for every session) – uk Chadwyck [323]

Journal of the hydraulics division / American Society of Civil Engineers. Hydraulics Division – New York. 1973-1982 (1) 1973-1982 (5) 1973-1982 (9) – (cont by: journal of hydraulic engineering) – ISSN: 0044-796X – mf#8138 – us UMI ProQuest [627]

Journal of the idaho academy of science – Rexburg, ID: The Academy, v16-21. 1980-1985 – 1r – 1 – us CRL [500]

Journal of the iest / Institute of Environmental Sciences and Technology – Mt. Prospect. 1998+ (1,5,9) – (cont: journal of the institute of environmental sciences) – ISSN: 1098-4321 – mf#3170,02 – us UMI ProQuest [628]

Journal of the illinois state historical society – Springfield. 1998+ (1,5,9) – (cont: illinois historical journal) – ISSN: 1522-1067 – mf#816,02 – us UMI ProQuest [978]

Journal of the illinois state historical society / Illinois State Historical Society – Springfield. 1953-1984 (1) 1971-1984 (5) 1977-1984 (9) – (cont by: illinois historical journal) – ISSN: 0019-2287 – mf#816 – us UMI ProQuest [978]

Journal of the illuminating engineering society / Illuminating Engineering Society – New York. 1971+ (1) 1971+ (5) 1975+ (9) – ISSN: 0099-4480 – mf#6515 – us UMI ProQuest [621]

Journal of the impeachment proceedings before the us senate, 1798-1805 / U.S. Senate – (= ser Records of the united states senate) – 1r – 1 – (with printed guide) – mf#M1253 – us Nat Archives [324]

Journal of the incorporated synod of the church of england in the diocese of toronto – [Toronto?]: The Synod, [1879-189-?] – 9 – (cont: toronto diocesan gazette) – mf#A01555 – cn Canadiana [242]

Journal of the indian chemical society / Indian Chemical Society – Calcutta. 1924+ (1) 1974+ (5) 1974+ (9) – ISSN: 0019-4522 – mf#9507 – us UMI ProQuest [540]

Journal of the indian law institute – v1-38. 1958-96 – 5,6,9 – $883.00 set – (v1-26 1958-84 on reel $579. v27-38 1985-96 on mf $304) – ISSN: 0019-5731 – mf#401261 – us Hein [340]

Journal of the indian medical association / Indian Medical Association – Calcutta. 1931-1990 (1) 1970-1986 (5) 1972-1986 (9) – ISSN: 0019-5847 – mf#673 – us UMI ProQuest [610]

Journal of the indian musicological society / Indian Musicological Society – Baroda. 1970-1980 (1) 1973-1980 (5) 1974-1980 (9) – ISSN: 0251-012X – mf#8805 – us UMI ProQuest [780]

Journal of the indiana state medical association / Indiana State Medical Association – Indianapolis. 1967-1983 (1) 1971-1983 (5) 1977-1983 (9) – (cont by: indiana medicine: the journal of the indiana state medical association) – ISSN: 0019-6770 – mf#2499 – us UMI ProQuest [610]

Journal of the institute for socioeconomic studies – v1-11. 1976-86 – 9 – $145.00 set – (ind v1-6 1976-80) – mf#105101 – us Hein [300]

Journal of the institute for socioeconomic studies *see* Socioeconomic studies

Journal of the institute of actuaries, 1850-1980 – 24r 98mf – 1,9 – mf#8/96503 – uk Microform Academic [310]

Journal of the institute of environmental sciences / Institute of Environmental Sciences – Mt. Prospect. 1990-1997 (1) 1990-1997 (5) 1990-1997 (9) – (cont: journal of environmental sciences. cont by: journal of the iest) – ISSN: 1052-2883 – mf#3170,01 – us UMI ProQuest [628]

Journal of the institute of petroleum / Institute of Petroleum (Great Britain) – London. 1914-1973 (1) 1972-1972 (5) – ISSN: 0020-3068 – mf#625 – us UMI ProQuest [550]

Journal of the institution of nuclear engineers / Institution of Nuclear Engineers – London. 1959-1980 (1) 1972-1980 (5) 1974-1980 (9) – (cont by: nuclear engineer) – ISSN: 0368-2595 – mf#7308 – us UMI ProQuest [621]

Journal of the institution of the rubber industry – London. 1973-1975 (1) 1973-1975 (5) – (cont: journal of the iri) – mf#1277,01 – us UMI ProQuest [670]

Journal of the instiute for the study of legal ethics – Hofstra University Law School. v1-2. 1996-99 – 9 – $31.00 – mf#117821 – us Hein [340]

Journal of the interdenominational theological center / Interdenominational Theological Center – Atlanta. 1993-1995 (1) – ISSN: 0092-6558 – mf#15992 – us UMI ProQuest [230]

Journal of the international association for mathematical geology / International Association for Mathematical Geology – New York. 1969-1977 (1) 1969-1977 (5) – (cont by: mathematical geology) – ISSN: 0020-5958 – mf#10858 – us UMI ProQuest [550]

Journal of the international association of buddhist studies – Berkeley. 1978-1993 (1,5,9) – ISSN: 0193-600X – mf#12078 – us UMI ProQuest [280]

Journal of the international association of pupil personnel workers / International Association of Pupil Personnel Workers – Gaithersburg. 1976-1992 – 1,5,9 – (cont by: journal for truancy and dropout prevention) – ISSN: 0020-6016 – mf#11295 – us UMI ProQuest [370]

Journal of the international phonetic association / International Phonetic Association – Dublin. 1975-1996 (1) 1975-1996 (5) 1975-1996 (9) – ISSN: 0025-1003 – mf#10238 – us UMI ProQuest [400]

Journal of the iowa academy of science (jias) – Cedar Falls. 1988-1996 (1) 1988-1996 (5) 1988-1996 (9) – (cont: proceedings of the iowa academy of science) – ISSN: 0896-8381 – mf#8220,01 – us UMI ProQuest [500]

Journal of the iri / Institution of the Rubber Industry – London. 1967-1973 (1) 1971-1973 (5) – (cont by: journal of the institution of the rubber industry) – mf#1277 – us UMI ProQuest [670]

Journal of the irrigation and drainage division – New York. 1973-1982 (1,5,9) – (cont by: journal of irrigation and drainage engineering) – ISSN: 0044-7978 – mf#8147 – us UMI ProQuest [627]

Journal of the jackson county genealogical society – v1 n1-v4 n4 [1984 win-1987] – 1r – 1 – mf#1528379 – us WHS [071]

Journal of the japan dental association / Japan Dental Association – Tokyo. 1973-1981 (1) 1975-1981 (5) 1975-1981 (9) – ISSN: 0047-1763 – mf#7951 – us UMI ProQuest [617]

Journal of the john bassett moore society of international law *see* Virginia journal of international law

Journal of the kanawha valley genealogical society / Kanawha Valley Genealogical Society – v1 n1-3, v2 n1-8,12, v3 n1-v12 n9 [1977 oct-dec, 1978 jan-aug, dec, 1979 jan-1988 nov/dec] – 1r – 1 – mf#1687093 – us WHS [929]

Journal of the kansas anthropological association – 1979-1985, 1986-1989 may – 2r – 1 – (cont: newsletter [kansas anthropological association]; cont by: kansas anthropolgist) – mf#1009837 – us WHS [301]

Journal of the kansas bar association – v1-70. 1932-2001 – 9 – $1023.00 set – ISSN: 0022-8486 – mf#104251 – us Hein [340]

Journal of the knights of labor – Philadelphia PA, Washington DC. 1889 dec 5-26, 1890-93, 1894-97, 1898-1909, 1910-1917 may – 5r – 1 – (cont: journal of united labor) – mf#1268047 – us WHS [331]

Journal of the korea annual conference, methodist episcopal church, south,...session / Methodist Episcopal Church, South. Korea Conference – Seoul, Korea: YMCA Press, 5th-13th (1922-30) [mf ed 2006] – 1r – 1 – (minutes of the meeting of missionaries for 1922-1926 carries its own numbering (26th-28th), continuing the numbering of the minutes of the korea mission; earlier minutes of the meeting of missionaries are found in: annual mission meeting and meeting of missionaries,... annual session; later minutes found in: minutes of the meeting of missionaries, methodist episcopal church, south,...session; lacks: 8th (1925)) – mf#2006c-s051 – us ATLA [242]

Journal of the law-school and of the mootcourt attached to it : at needham, in virginia – Richmond. 1822-1822 (1) – mf#4005 – us UMI ProQuest [347]

Journal of the legal profession – University of Alabama. v1-25. 1976-2001 – 9 – $297.00 set – ISSN: 0196-7487 – mf#104261 – us Hein [340]

Journal of the legislative proceedings of the us senate, 1789-1817 / U.S. Senate – (= ser Records of the united states senate) – 28r – 1 – (with printed guide) – mf#M1251 – us Nat Archives [324]

Journal of the less-common metals – Amsterdam. 1959-1991 (1) 1959-1991 (5) 1987-1991 (9) – (cont by: journal of alloys and compounds) – ISSN: 0022-5088 – mf#42292 – us UMI ProQuest [540]

Journal of the life and gospel labours of david sands : with extracts from his correspondence / Sands, David – London: C Gilpin, 1848 – 1mf – 9 – 0-524-06779-1 – mf#1991-2786 – us ATLA [240]

Journal of the life and religious labors of sarah hunt, late of west grove, chester county, pennsylvania / Hunt, Sarah – Philadelphia: Friends' Book Assoc, 1892 – 1mf – 9 – 0-524-06906-9 – mf#1991-2819 – us ATLA [240]

Journal of the life and religious labours of elias hicks / Hicks, Elias – New-York: IT Hopper, 1832 – 2mf – 9 – 0-524-02390-5 – mf#1990-4292 – us ATLA [240]

A journal of the life and religious labours of richard jordan : a minister of the gospel in the society of friends, late of newton, in gloucester county, new jersey / Jordan, Richard – Philadelphia: Thomas Kite, 1829 [mf ed 1993] – (= ser Society of friends (quakers) coll) – 1mf – 9 – 0-524-07012-1 – mf#1991-2865 – us ATLA [920]

Journal of the life and religious services of william evans : minister of the gospel in the society of friends / Evans, William – Philadelphia: Friends' Book Store [distributor], 1870 – 2mf – 9 – 0-524-01724-7 – mf#1990-4116 – us ATLA [240]

Journal of the life of john wilbur : a minister of the gospel in the society of friends / Wilbur, John – Providence: GH Whitney, 1859 – 2mf – 9 – 0-7905-8255-4 – mf#1988-8118 – us ATLA [240]

Journal of the life, travels and gospel labors of thomas arnett / Arnett, Thomas – Chicago: Publishing Association of Friends, 1884 – 1mf – 9 – 0-524-06380-X – mf#1991-2502 – us ATLA [240]

A journal of the life, travels, and religious labours of william savery : late of philadelphia, a minister of the gospel of christ, in the society of friends / Savery, William – London: Charles Gilpin, 1844 [mf ed 1992] – (= ser Society of friends (quakers) coll) – 1mf – 9 – 0-524-04064-8 – mf#1990-4972 – us ATLA [920]

Journal of the linnean society – London. v10. 1869 – 12mf – 7 – (incl papers of: m j berkeley, charles darwin, j d hooker, w mitten and others) – mf#5302 – uk Microform Academic [580]

Journal of the lockwood expedition on the north coast of greenland, april 31-june 1, 1882 – (= ser Records Of The Weather Bureau) – 1r – 1 – mf#T298 – us Nat Archives [550]

Journal of the london mathematical society / London Mathematical Society – London. 1991+ (1) – ISSN: 0024-6107 – mf#14631 – us UMI ProQuest [510]

Journal of the louisiana state medical society / Louisiana State Medical Society – New Orleans. 1915-1982 (1) 1971-1982 (5) 1976-1982 (9) – (cont by: journal louisiana state medical society) – ISSN: 0024-6921 – mf#5486 – us UMI ProQuest [610]

Journal of the louisiana state medical society / Louisiana State Medical Society – New Orleans. 1986+ (1) 1986+ (5) 1986+ (9) – (cont: journal louisiana state medical society) – ISSN: 0024-6921 – mf#5486,02 – us UMI ProQuest [610]

Journal of the magoffin county historical society / Magoffin County Historical Society [KY] – v6 n1-v11 n4 [1984 feb-1989 nov] – 1r – 1 – mf#1060841 – us WHS [978]

Journal of the maha-bodhi society / Maha-bodhi society – Ceylon: Maha-bodi Society, 1894-1950 [mf ed 2001] – (= ser Christianity's encounter with world religions, 1850-1950) – 12r – 1 – (filmed with: maha-bodhi and the united buddhist world [1901-23]; maha-bodhi [1924-50]) – mf#2001-s200-202 – us ATLA [280]

Journal of the marine biological association, 1887-1980 – 747mf – 7,9 – mf#153 – uk Microform Academic [580]

Journal of the marine biological association of the united kingdom / Marine Biological Association of the United Kingdom – London 1889-1941/43 – v1-25 on 326mf – 9 – mf#z-316c/2 – ne IDC [574]

Journal of the market research society / Market Research Society – London. 1972-1999 (1) 1972-1999 (5) 1973-1999 (9) – (cont by: faraday discussions of the chemical society) – ISSN: 0025-3618 – mf#7190 – us UMI ProQuest [650]

Journal of the mechanics and physics of solids – London. 1952+ (1,5,9) – ISSN: 0022-5096 – mf#49126 – us UMI ProQuest [530]

Journal of the medical library association / Medical Library Association – Chicago. 2002+ (1,5,9) – ISSN: 1536-5050 – mf#1833,01 – us UMI ProQuest [610]

Journal of the medical society of new jersey / Medical Society of New Jersey – Lawrenceville. 1972-1985 (1) 1972-1985 (5) 1974-1985 (9) – (cont by: new jersey medicine) – ISSN: 0025-7524 – mf#7089 – us UMI ProQuest [610]

Journal of the military service institution of the united states / Military Service Institution of the United States – New York. 1879-1906 (1) – mf#2926 – us UMI ProQuest [355]

Journal of the military service institution of the united states / Military Service Institution of the United States – v5 [1884] – 1r – 1 – mf#1053373 – us WHS [355]

Journal of the mine ventilation society of south africa / Mine Ventilation Society of South Africa – Johannesburg. 1978-1993 (1,5,9) – ISSN: 0026-4504 – mf#11860,01 – us UMI ProQuest [622]

Journal of the minnesota academy of science / Minnesota Academy of Science – Minneapolis. 1972-1996 (1) 1976-1996 (5) 1976-1996 (9) – ISSN: 0026-539X – mf#7207 – us UMI ProQuest [500]

Journal of the mississippi state medical association / Mississippi State Medical Association – Jackson. 1972+ (1) 1972+ (5) 1973+ (9) – ISSN: 0026-6396 – mf#6856 – us UMI ProQuest [366]

Journal of the missouri bar – v1-57. 1945-2001 – 9 – $810.00 set – (cont: missouri bar journal) – ISSN: 0026-6485 – mf#104271 – us Hein [340]

Journal of the missouri dental association – Jefferson City. 1967-1979 (1) 1971-1979 (5) 1975-1979 (9) – (cont by: mda journal) – ISSN: 0026-6523 – mf#2681 – us UMI ProQuest [617]

Journal of the missouri dental association / Missouri Dental Association – Jefferson City. 1980-1981 (1) 1980-1981 (5) 1980-1981 (9) – (cont: mda journal. cont by: missouri dental journal: the journal of the missouri dental association) – ISSN: 0273-3463 – mf#2681,02 – us UMI ProQuest [617]

Journal of the multihandicapped person – New York. 1988-1989 – 1,5,9 – (cont by: journal of developmental and physical disabilities) – ISSN: 0892-7561 – mf#17690 – us UMI ProQuest [370]

Journal of the national association of administrative law judges – v1-20. 1981-2000 – 9 – $243.00 set – ISSN: 0735-0821 – mf#401031 – us Hein [340]

Journal of the national association of colleges and teachers of agriculture – Ruston. 1957-1974 (1) 1972-1972 (5) (9) – (cont by: national association of colleges and teachers of agriculture nacta journal) – ISSN: 0027-8602 – mf#7455 – us UMI ProQuest [630]

Journal of the national association of referees in bankruptcy *see* American bankruptcy law journal

Journal of the national association of women deans and counselors / National Association of Women Deans and Counselors – Washington. 1938-1973 (1) 1971-1973 (5) – (cont by: national association for women deans, administrators and counselors journal) – ISSN: 0027-870X – mf#2383 – us UMI ProQuest [376]

Journal of the national cancer institute / National Cancer Institute (US) – Bethesda. 1940+ (1) 1966+ (5) 1977+ (9) – ISSN: 0027-8874 – mf#2122 – us UMI ProQuest [610]

Journal of the national medical association / National Medical Association (US) – Thorofare. 1988+ (1,5,9) – ISSN: 0027-9684 – mf#16496 – us UMI ProQuest [610]

Journal of the neurological sciences – Amsterdam. 1964+ (1) 1964+ (5) 1987+ (9) – ISSN: 0022-510X – mf#42084 – us UMI ProQuest [617]

Journal of the new african literature and the arts – New York. 1966-1972 (1) – ISSN: 0022-5118 – mf#2801 – us UMI ProQuest [470]

Journal of the new brunswick society... : instituted at fredericton, nb, august 30, 1849 / New Brunswick Society for the Encouragement of Agriculture, Home Manufactures, and Commerce throughout the Province – Fredericton, NB?: J Hogg, 1850 – 6mf – 9 – (v1 only. incl ind) – mf#54480 – cn Canadiana [630]

Journal of the new england water works association / New England Water Works Association – Boston. 1883+ (1) 1970+ (5) 1977+ (9) – ISSN: 0028-4939 – mf#863 – us UMI ProQuest [333]

Journal of the new harbinger / North American Student Cooperative Organization – 1971 oct-1973 mar – 1r – 1 – (cont by: new harbinger) – us WHS [071]

Journal of the new jersey postal history society / New Jersey Postal History Society – 1973 jan-1987 nov – 1r – 1 – mf#1270939 – us WHS [380]

Journal of the new york state nurses association / New York State Nurses Association – Latham. 1973+ (1) 1975+ (5) 1976+ (9) – ISSN: 0028-7644 – mf#7589 – us UMI ProQuest [610]

Journal of the niger : from the royal commonwealth society library / Lander, Richard – 3v. 1832 – 17mf – 7 – mf#2984 – uk Microform Academic [916]

Journal of the nigerian historical society, 1956-1980 – 5r – 1 – mf#97482 – uk Microform Academic [960]

Journal of the oklahoma bar association *see* Oklahoma bar journal

Journal of the operational research society – Elmsford. 1950-1989 (1,5,9) – (cont by: journal of the operational research society) – ISSN: 0160-5682 – mf#49152 – us UMI ProQuest [000]

Journal of the operational research society / perational Research Society (Great Britain) – Houndsmill. 1990+ (1,5,9) – (cont: journal of the operational research society) – ISSN: 0160-5682 – mf#18064,01 – us UMI ProQuest [000]

Journal of the oto-laryngological society of australia / Oto-laryngological Society of Australia – St. Leonards. 1961-1991 (1) 1975-1980 (5) 1975-1980 (9) – ISSN: 0030-6614 – mf#8180 – us UMI ProQuest [617]

Journal of the otto rank association – New York. 1966-1983 (1) 1973-1983 (5) 1973-1983 (9) – ISSN: 0030-6711 – mf#7476 – us UMI ProQuest [150]

Journal of the palestine oriental society – Jerusalem, 1920-1948. v1-21 – 84mf – 9 – mf#H-2822 – ne IDC [956]

Journal of the patent and trademark office society – v1-83. 1918-2001 – 9 – $1571.00 set – (title varies: v1-66 1918-84 as journal of the patent office society) – ISSN: 0882-9098 – mf#104291 – us Hein [346]

Journal of the patent office society *see* Journal of the patent and trademark office society

The journal of the photographic society : the official organ of the royal photographic society of great britain / Great Britain. Royal Photographic Society – v1-123. 1853-1983 [mf ed Chadwyck-Healey] – 48r+51mf – 1,9 – (individual yrs available separately. renamed: the photographic journal in 1876/77) – uk Chadwyck [770]

Journal of the power division / American Society of Civil Engineers. Power Division – New York. 1973-1978 [1,5]; 1976-1978 [9] – (cont by: journal of the energy division) – ISSN: 0569-8030 – mf#8149 – us UMI ProQuest [624]

Journal of the presbyterian historical society of england / Presbyterian Historical Society of England – London. 1914-1972 (1) – ISSN: 0079-5011 – mf#9245 – us UMI ProQuest [242]

Journal of the proceedings of the annual convention / Episcopal Church Docese Of Florida – s.l, s.l? 1844 – 1r – us UF Libraries [978]

Journal of the proceedings of the annual council / Episcopal Church Diocese Of Florida – s.l, s.l? 1875-1879 – 1r – us UF Libraries [978]

Journal of the proceedings of the constitutional c... / Florida Constitutional Convention (1885) – Tallahassee, FL. 1885 – 1r – us UF Libraries [978]

Journal of the proceedings of the grand division of the sons of temperance of the state of wisconsin / Sons of Temperance of North America – 1851, 20th [1868], 22nd [1870], 26th-29th [1874-1877], 31st [1879], 34th-40th [1882-88], 43rd-44th [1891-92], 48th-50th [1896-98] – 1r – 1 – mf#3242049 – us WHS [366]

Journal of the proceedings of the grand division of the sons of temperance of the state of wisconsin / Sons of Temperance of North America – 1851, 20th [1868], 22nd [1870], 26th-29th [1874-77], 31st [1879], 34th-40th [1882-88], 43rd-44th [1891-92], 48th-50th [1896-98] – 1r – 1 – us WHS [366]

Journal of the proceedings of the late embassy to china : comprising a correct narrative of the public transactions of the embassy... / Ellis, Henry – London: printed for John Murray, 1817 – (= ser 19th c books on china) – 6mf – 9 – mf#7.1.1 – uk Chadwyck [915]

Journal of the proceedings of the late embassy to china : comprising a correct narrative of the public transactions of the embassy, of the voyage to and from china... / Ellis, H – London: John Murray, 1817 – 10mf – 9 – mf#HT-781 – ne IDC [915]

Journal of the proceedings of the party from the river mackenzie towards cape bathurst / Pullen, W J S – London, 1852 – 2mf – 9 – (missing:p67-end) – mf#N-354 – ne IDC [917]

Journal of the proceedings...of the sons of temperance...at the quarterly session... / Sons of Temperance of North America. Grand Division of Nova Scotia – [Yarmouth, NS?]: The Division, [1848 or 1849]-1876 (Yarmouth [NS]: H C Flint) – 9 – (cont by: sons of temperance of north america, grand division of nova scotia. journal of proceedings of the grand division, s of t of nova scotia) – mf#P04610 – cn Canadiana [242]

Journal of the rev daniel shute : chaplain in the expedition to canada in 1758 – [s.l: s.n, 18–?] [mf ed 1984] – 1mf – 9 – 0-665-45129-6 – mf#45129 – cn Canadiana [971]

Journal of the rev george champion / Champion, George – Cape Town, South Africa. 1967 – 1r – us UF Libraries [960]

The journal of the rev. john wesley, a.m : sometime fellow of lincoln college, oxford / Wesley, John; ed by Curnock, Nehemiah – standard ed. London: R Culley, [1909?]-1916 – 10mf – 9 – 0-524-05183-6 – mf#1990-5102 – us ATLA [242]

Journal of the rev joseph wolff... : in a series of letters to sir thomas baring... / Wolff, J – London, 1839 – 5mf – 9 – mf#HTM-219 – ne IDC [910]

Journal of the reverend peter jacobs, indian wesleyan missionary : from rice lake to the hudson's bay territory, and returning: commencing may, 1852... – New York: publ for the aut, 1857 [mf ed 1984] – 2mf – 9 – 0-665-45548-8 – mf#45548 – cn Canadiana [242]

Journal of the royal agricultural improvement society of ireland and irish agriculturist – Dublin, Ireland. jun 1852-dec 1855 – 1r – 1 – uk British Libr Newspaper [072]

Journal of the royal anthropological institute / Royal Anthropological Institute of Great Britain and Ireland – Oxford. 1995+ (1,5,9) – (cont: man) – mf#21595 – us UMI ProQuest [301]

Journal of the royal anthropological institute of great britain and ireland / Royal Anthropological Institute of Great Britain and Ireland – London. 1871-1910 [1,5,9] – ISSN: 0307-3114 – mf#1868 – us UMI ProQuest [301]

Journal of the royal army veterinary corps / Royal Army Veterinary Corps – Aldershot. 1950-1953 (1) – ISSN: 0035-8681 – mf#519 – us UMI ProQuest [636]

Journal of the royal asiatic society, 1834-1985 – os: v1-20 [1834-63] on 10r; ns: v1-22 [1864-1890] on 16r & subsequent yrs [1891-1985] on 64r & 24mf – 24mf – 1,9 – £168 / $336 – (90r [r02072]) – mf#f02072 – uk Microform Academic [950]

Journal of the royal asiatic society of great britain and ireland – North China Branch. Shanghai 1858-60 – v-3; ns: 1864-1948 v1-73 on 397mf – 9 – mf#1-564/2 – ne IDC [950]

The journal of the royal asiatic society of great britain and ireland – London, 1834-1863, v1-20; n.s. 1865-1947, v1-79 – 1406mf – 8 – mf#I-101Ac – ne IDC [956]

Journal of the royal astronomical society of canada / Royal Astronomical Society of Canada – Toronto. 1907+ (1) 1972+ (5) 1972+ (9) – ISSN: 0035-827X – mf#7705 – us UMI ProQuest [520]

Journal of the royal australian historical society – 2pt – 1 – (pt1: v1-45 1901-60 [16r] £1500; pt2: v46-91 1960-2005 [15r] £1400; with d/g) – uk Matthew [980]

Journal of the royal college of surgeons of edinburgh – Edinburgh. 1989+ (1,5,9) – ISSN: 0035-8835 – mf#17247 – us UMI ProQuest [617]

Journal of the royal geographical society / Royal Geographical Society – London 1831-33 [mf ed Hildesheim 1995-98] – 3v on 6mf – 9 – €120.00 – 3-487-29945-3 – gw Olms [910]

Journal of the royal horticultural society / Royal Horticultural Society (Great Britain) – London. 1950-1975 (1) 1971-1975 (5) – (cont by: garden) – ISSN: 0035-8924 – mf#500 – us UMI ProQuest [630]

Journal of the royal institute of british architects : [overseas edition] / Royal Institute of British Architects – London. 1987-1993 (1) 1987-1993 (5) 1987-1993 (9) – (cont by: riba journal. cont: architect [overseas ed]) – mf#1383,02 – us UMI ProQuest [720]

Journal of the royal institute of international affairs / Royal Institute of International Affairs – London. 1926-1930 (1) 1926-1930 (5) 1926-1930 (9) – (cont: journal of the british institute of international affairs. cont by: international affairs: journal of the royal institute of international affairs) – mf#1030,01 – us UMI ProQuest [327]

Journal of the royal musical association – Oxford. 1990+ (1,5,9) – ISSN: 0269-0403 – mf#18513,02 – us UMI ProQuest [780]

Journal of the royal naval medical service / Royal Naval Medical Service – Alverstoke. 1973-1996 (1) 1976-1996 (5) 1976-1996 (9) – ISSN: 0035-9033 – mf#9850 – us UMI ProQuest [610]

Journal of the royal society for the promotion of health / Royal Society for the Promotion of Health (Great Britain) – London. 1998+ (1,5,9) – (cont: journal of the royal society of health) – ISSN: 1466-4240 – mf#11352,02 – us UMI ProQuest [366]

Journal of the royal society of arts / Royal Society of Arts (Great Britain) – London. 1952-1987 (1) 1958-1987 (5) 1958-1987 (9) – (cont by: rsa journal) – ISSN: 0035-9114 – mf#862 – us UMI ProQuest [700]

Journal of the royal society of health / Royal Society of Health (Great Britain) – London. 1983-1998 (1,5,9) – (cont: royal society of health journal. cont by: journal of the royal society for the promotion of health) – ISSN: 0264-0325 – mf#11352,01 – us UMI ProQuest [366]

Journal of the royal society of medicine / Royal Society of Medicine (Great Britain) – London. 1978+ (1) 1978+ (5) 1978+ (9) – (cont: proceedings of the royal society of medicine) – ISSN: 0141-0768 – mf#1287,01 – us UMI ProQuest [610]

Journal of the royal statistical society : series b 109 (methodological) / Royal Statistical Society (Great Britain) – London. 1988+ – 1,5,9 – ISSN: 0035-9246 – mf#17410,01 – us UMI ProQuest [310]

Journal of the royal statistical society : series c (applied statistics) / Royal Statistical Society (Great Britain) – London. 1989+ – 1,5,9 – ISSN: 0035-9254 – mf#17411 – us UMI ProQuest [310]

Journal of the royal statistical society series a / Royal Statistical Society (Great Britain) – London. 1992-1996 – 1 – ISSN: 0964-1998 – mf#2835 – us UMI ProQuest [310]

Journal of the science of food and agriculture – London, 1998+ [1,5,9] – ISSN: 0022-5142 – mf#15566 – us UMI ProQuest [630]

Journal of the secretary of the senate, 1789-1845 / U.S. Senate – (= ser Records of the united states senate) – 1r – 1 – (with printed guide) – mf#M1254 – us Nat Archives [324]

Journal of the siege of charleston by the corps of british engineers under the command of capt moncrief : commencing feb 29th and ending may 12th 1780 – [mf ed Spartanburg SC: Reprint Co, 1981] – 1mf – 9 – mf#51-184 – us South Carolina Historical [978]

Journal of the siege of quebec, 1759 : from northumberland county record office / Coates, E – 1r – 1 – mf#96816 – uk Microform Academic [920]

The journal of the sind historical society / Sind Historical Society – Karachi. 8v. 1934-48 – 1 – $23.00r – us L of C Photodup [950]

Journal of the society for the bibliography of natural history – London 1936-43 – v1-2 on 25mf – 9 – mf#7649c/1 – ne IDC [015]

Journal of the society for the propagation of the gospel in foreign parts, 1707-1850 – v1-50 – 17r – 1 – (with app and ind) – mf#95868 – uk Microform Academic [220]

Journal of the society of architectural historians / Society of Architectural Historians – Philadelphia. 1941+ (1) 1941+ (5) 1941+ (9) – ISSN: 0037-9808 – mf#878 – us UMI ProQuest [720]

Journal of the society of dyers and colourists / Society of Dyers and Colourists – Bradford. 1950+ (1) 1971+ (5) 1977+ (9) – ISSN: 0037-9859 – mf#703 – us UMI ProQuest [660]

Journal of the society of environmental engineers / Society of Environmental Engineers (Great Britain) – London. 1968-1988 (1) 1971-1988 (5) 1976-1988 (9) – (cont by: environmental engineering) – ISSN: 0374-356X – mf#3068 – us UMI ProQuest [628]

Journal of the society of occupational medicine – Surrey. 1989-1991 (1) 1989-1991 (5) 1989-1991 (9) – (cont by: occupational medicine) – ISSN: 0301-0023 – mf#13958,02 – us UMI ProQuest [366]

Journal of the society of pediatric nurses – Philadelphia. 1996+ (1,5,9) – (cont: maternal-child nursing journal) – ISSN: 1088-145X – mf#24906 – us UMI ProQuest [610]

Journal of the society of research administrators / Society of Research Administrators – Washington. 1977-1991 (1) 1977-1991 (5) 1977-1991 (9) – (cont by: sra journal) – ISSN: 0038-0024 – mf#11483 – us UMI ProQuest [650]

Journal of the soil mechanics and foundations division / American Society of Civil Engineers. Soil Mechanics and Foundations Division – New York. 1973-1973 (1) 1973-1973 (5) 1973-1973 (9) – (cont by: journal of the geotechnical engineering division) – ISSN: 0044-7994 – mf#8140 – us UMI ProQuest [624]

Journal of the south african chemical institute / South African Chemical Institute – Johannesburg. 1948-1976 (1) 1970-1976 (5) – (cont by: south african journal of chemistry) – ISSN: 0038-2078 – mf#3474 – us UMI ProQuest [540]

Journal of the south african institute of mining and metallurgy / South African Institute of Mining and Metallurgy – Johannesburg. 1969-1996 (1) 1970-1996 (5) 1974-1996 (9) – ISSN: 0038-223X – mf#5336 – us UMI ProQuest [622]

Journal of the south african ornithologist's union – Pretoria 1905/06-1908 – v1-4 on 16mf – 9 – mf#z-2024/2 – ne IDC [590]

Journal of the south african speech and hearing association = Tydskrif van die suid-afrikaanse vereniging vir spraak- en gehoorheelkunde – [Johannesburg]: South African Speech and Hearing Association. v18-23. dec 1971-1976 – 1r – 1 – us CRL [366]

Journal of the south carolina court of general sessions, 1769-1776 – South Carolina: Department of Archives and History, 1995 – 1r – 1 – $130.00 – us Scholarly Res [347]

Journal of the south carolina medical association / South Carolina Medical Association – Columbia. 1905+ (1) 1971+ (5) 1973+ (9) – ISSN: 0038-3139 – mf#2321 – us UMI ProQuest [610]

Journal of the southwest – Tucson. 1987+ (1,5,9) – (cont: arizona and the west) – ISSN: 0894-8410 – mf#2004,01 – us UMI ProQuest [978]

Journal of the straits branch of the royal asiatic society / Royal Asiatic Society of Great Britain and Ireland. Malaysian Branch, Singapore – Singapore 1878-1922 – v1-86 on 304mf – 9 – mf#1-523/2 – ne IDC [950]

Journal of the structural division / American Society of Civil Engineers – New York. 1973-1982 (1) 1973-1982 (5) 1973-1982 (9) – (cont by: journal of structural engineering) – ISSN: 0044-8001 – mf#8141 – us UMI ProQuest [624]

Journal of the suffolk academy of law – v1-14. 1980-2000 – 9 – $125.00 set – ISSN: 0888-2142 – mf#112771 – us Hein [340]

Journal of the surveying and mapping division / American Society of Civil Engineers. Surveying and Mapping Division – New York. 1973-1982 (1) 1973-1982 (5) 1973-1982 (9) – (cont by: journal of surveying engineering) – ISSN: 0569-8073 – mf#8150 – us UMI ProQuest [624]

Journal of the switchmen's union / Switchmen's Union of North America – 1898 nov-1903 oct – 1r – 1 – mf#783112 – us WHS [331]

Journal of the synod of the church of england and ireland in the diocese of toronto – [Toronto?]: The Synod, 1870-1874 – 9 – (title varies slightly; cont by: toronto diocesan gazette) – mf#A01557 – cn Canadiana [242]

Journal of the technical councils of asce – New York. 1977-1982 (1,5,9) – (cont by: journal of technical topics in civil engineering) – ISSN: 0148-9909 – mf#11508 – us UMI ProQuest [624]

Journal of the times – Baltimore. 1818-1819 (1) – mf#3093 – us UMI ProQuest [978]

The journal of the times – Halifax, NS: Macallaster & Paine, [1858-186-?] – 9 – mf#P04750 – cn Canadiana [617]

Journal of the torrey botanical society – Bronx. 1997+ (1) 1997+ (5) 1997+ (9) – (cont: bulletin of the torrey botanical club) – ISSN: 1095-5674 – mf#1472,01 – us UMI ProQuest [580]

A journal of the travels of william colbert, methodist preacher, thro' parts of maryland, pennsylvania, new york, delaware and virginia in 1790, 1, 2, 3, 4, 5, 6, 7, 8 / Colbert, William – 1r – 1 – mf#1993-M000 – us ATLA [242]

Journal of the united hatters of north america – 1898 aug-1905 may – 1r – 1 – mf#1114305 – us WHS [366]

Journal of the united states cavalry association / United States Cavalry Association – v14, v15 – 2r – 1 – (cont by: cavalry journal [washington, dc]) – mf#425466 – us WHS [355]

Journal of the u s continental congress – 179mf (24:1) – 9 – $865.00 – us UPA [323]

Journal of the united states exploring expedition / Hudson, William L – 11 aug 1840-19 feb 1842 – 1r – 1 – mf#pmb416 – at Pacific Mss [910]

Journal of the united states exploring expedition / Hudson, William L – 1838-1840 – 1r – 1 – mf#pmb146 – at Pacific Mss [910]

Journal of the university film and video association – Carbondale. 1982-1983 (1) 1982-1983 (5) 1982-1983 (9) – (cont: journal of the university film association. cont by: journal of film and video) – ISSN: 0734-919X – mf#11567,01 – us UMI ProQuest [790]

Journal of the university film association / University Film Association – Philadelphia. 1977-1981 (1,5,9) – (cont by: journal of the university film and video association) – ISSN: 0041-9311 – mf#11567 – us UMI ProQuest [790]

Journal of the urban planning and development division : proceedings of the american society of civil engineers / American Society of Civil Engineers. Urban Planning and Development Division – New York. 1973-1982 (1) 1973-1982 (5) 1973-1982 (9) – (cont by: journal of urban planning and development) – ISSN: 0569-8081 – mf#8139 – us UMI ProQuest [710]

Journal of the visit of her majesty the queen, to tunis, greece, and palestine / Demont, Louise – London 1821 [mf ed Hildesheim 1995-98] – 1v on 1mf – 9 – €40.00 – ISBN-10: 3-487-26708-X – ISBN-13: 978-3-487-26708-1 – gw Olms [910]

Journal of the voyage of the uss nonsuch up the orinoco, 1819 / U.S. Dept of State – 1r – 1 – mf#M83 – us Nat Archives [355]

A journal of the voyages and travels of a corps of discovery : under the command of captain lewis and captain clarke, of the army of the united states from the mouth of the river missouri, through the interior parts of north america, to the pacific ocean; during the years 1804, 1805, and 1806 / Gass, Patrick – London 1808 [mf ed Hildesheim 1995-98] – 1v on 3mf – 9 – €90.00 – 3-487-27055-2 – gw Olms [910]

Journal of the washington academy of sciences / Washington Academy of Sciences – Washington. 1911-1996 (1) 1972-1996 (5) 1977-1996 (9) – ISSN: 0043-0439 – mf#7669 – us UMI ProQuest [500]

Journal of the water pollution control federation / Water Pollution Control Federation – Alexandria. 1928-1989 (1) 1966-1989 (5) 1975-1989 (9) – ISSN: 0043-1303 – mf#245 – us UMI ProQuest [333]

Journal of the water resources planning and management division / American Society of Civil Engineers. Water Resources Planning and Management Division – New York. 1976-1982 (1,5,9) – (cont by: journal of water resources planning and management) – ISSN: 0145-0743 – mf#11448 – us UMI ProQuest [350]

Journal of the waterway, port, coastal and ocean division / American Society of Civil Engineers. Waterway, Port, Coastal, and Ocean Division – New York. 1977-1982 (1,5,9) – (cont: journal of the waterways, harbors and coastal engineering division. cont by: journal of waterway, port, coastal and ocean engineering) – ISSN: 0148-9895 – mf#8145,01 – us UMI ProQuest [627]

Journal of the waterways, harbors and coastal engineering division / American Society of Civil Engineers. Waterways, Harbors, and Coastal Engineering Division – New York. 1973-1976 (1) 1973-1976 (5) 1973-1976 (9) – (cont by: journal of the waterway, port, coastal and ocean division) – ISSN: 0044-8028 – mf#8145 – us UMI ProQuest [627]

Journal of the western pacific orthopaedic association / Western Pacific Orthopaedic Association – Kowloon. 1972-1992 (1) 1980-1980 (5) 1980-1980 (9) – ISSN: 0043-4019 – mf#7321 – us UMI ProQuest [617]

Journal of the wisconsin annual conference of the methodist episcopal church...session / Methodist Episcopal Church – 45th-46th [1891-1892] – 1r – 1 – (cont: minutes of the wisconsin annual conference of the methodist episcopal church [...session], methodist episcopal church. wisconsin conference]; cont by: minutes of the wisconsin annual conference of the methodist episcopal church...session [1893], methodist episcopal church. wisconsin conference) – mf#2902077 – us WHS [242]

Journal of the...annual convention of the national lodge, a a of i s and t w – 1901-02, 1904-09, 1910-13, 1920 – 4r – 1 – (cont: proceedings of the...annual convention of the national lodge, a a of i & s w, amalgamated association of iron, steel, and tin workers of north america; cont by: journal of proceedings of the...annual convention of the international lodge, a a of i & s w of north america, amalgamated association of iron, steel, and tin workers of north america) – mf#3098712 – us WHS [366]

Journal of the...annual council of colored churchmen, diocese of georgia / Episcopal Church. Diocese of Georgia. Council of Colored Churchmen – Brunswick GA: Wrench Print [annual] [mf ed 2004] – 1r – 1 – (began in 1907? 11th-12th, 14th-15 & 17th-18th iss in combined form. mf: 3rd-18th [1909-23] lacks 4th?) – mf#2003-s003 – us ATLA [242]

Journal of the...annual meeting of the korea mission of the methodist episcopal church / Methodist Episcopal Church. Korea Mission – [s.l: s.n] 14th-15th (1898-99) [mf ed 2006] – 2v on 1r – 1 – (cont: methodist episcopal church. korea mission. official minutes of the...annual meeting of the korea mission of the methodist episcopal church; cont by: methodist episcopal church. korea mission. official minutes of the...annual meeting, korea mission, methodist episcopal church) – mf#2006c-s043 – us ATLA [242]

Journal of the...annual meeting of the korea mission of the methodist episcopal church / Methodist Episcopal Church. Korea Mission – [s.l: s.n] 14th-15th (1898-99) [mf ed 2006] – 2v on 1r – 1 – (cont: methodist episcopal church. korea mission. official minutes of the...annual meeting of the korea mission of the methodist episcopal church; cont by: methodist episcopal church. korea mission. official minutes of the...annual meeting, korea mission, methodist episcopal church) – mf#2006c-s043 – us ATLA [242]

Journal of the...annual session of the council of colored churchmen in the diocese of georgia / Episcopal Church. Diocese of Georgia. Council of Colored Churchmen – [s.l.]: Savannah Journal Print, 1924- [annual] [mf ed 2004] – 1r – 1 – (mf: 19th-27th [1924-32]. iss for 1925-26 contain: minutes of the woman's auxiliary to the national council of the church) – mf#2004-s004 – us ATLA [242]

Journal of the...annual session of the michigan conference of the african methodist episcopal church / African Methodist Episcopal Church. Michigan Conference – [s.l: s.n.] 52nd (1938) [mf ed 2005] – 1r – 1 – (cont: minutes of the...annual session of the michigan conference of the african methodist episcopal church) – mf#2005-s038 – us ATLA [242]

Journal of the...annual session of the ohio conference of the african methodist episcopal church / African Methodist Episcopal Church. Ohio Conference – Xenia OH: Aldine Publ House [mf ed 2004] – 90th (1920) [complete] on 1r – 1 – (cont: african methodist episcopal church. ohio conference. journal of proceedings of the...session of the ohio annual conference of the african methodist episcopal church) – mf#2004-s091 – us ATLA [242]

Journal of the...general conference and the... quadrennial session of the colored methodist episcopal church / Colored Methodist Episcopal Church. General Conference – Jackson TN: C M E Church, 21st-22nd (1946-50) [mf ed 2005] – 1r – 1 – (peculiar; quadrennial session for 1946-50 called 20th-21st; cont by: christian methodist episcopal church. general conference. journal of the...general conference; some pgs damaged) – mf#2005-s002 – us ATLA [242]

Journal of theological studies – 1(1900)-50(1949) – 422mf – 9 – €804.00 – ne Slangenburg [200]

Journal of theological studies – London. 1899+ [1]; 1971+ [5]; 1975+ [9] – ISSN: 0022-5185 – mf#1223 – us UMI ProQuest [200]

Journal of theology / Church of the Lutheran Confession – 1977 jun [v17 n2]-1982, 1983 mar-1988 dec – 2r – 1 – mf#175898 – us WHS [242]

Journal of theology for southern africa – Rondebosch. 1986+ (1,5,9) – ISSN: 0047-2867 – mf#15411 – us UMI ProQuest [240]

Journal of theory construction and testing – Lisle. 1997+ (1) – ISSN: 1086-4431 – mf#26779 – us UMI ProQuest [610]

Journal of the...quadrennial session of the general conference of the african m e church / African Methodist Episcopal Church. General Conference – Philadelphia PA: A M E Book Concern. 21st-30th (1900-1936) (quadrennial) [mf ed 2005] – 3r – 1 – (lacks 25th [1916]; cont: african methodist episcopal church. general conference. journal of the...session and the...quadrennial session of the general conference of the african methodist episcopal church in the united states; cont by: african methodist episcopal church. general conference. official minutes of the...session of the general conference of the african methodist episcopal church) – mf#2005-s090 – us ATLA [242]

Journal of therapeutic humor – New Brunswick. 1980-1981 (1,5,9) – ISSN: 0272-5665 – mf#12223 – us UMI ProQuest [150]

Journal of thermal analysis – London. 1982+ (1,5,9) – ISSN: 0368-4466 – mf#13311 – us UMI ProQuest [621]

Journal of thermal biology – Oxford. 1975+ (1,5,9) – ISSN: 0306-4565 – mf#49127 – us UMI ProQuest [612]

Journal of thermal envelope and building science – Lancaster, 1998+ [1,5,9] – ISSN: 1097-1963 – mf#13823,02 – us UMI ProQuest [620]

Journal of thermal insulation – Lancaster. 1989-1989 (1,5,9) – ISSN: 0148-8287 – mf#13823 – us UMI ProQuest [620]

Journal of thermal stresses – New York. 1978+ (1,5,9) – ISSN: 0149-5739 – mf#11992 – us UMI ProQuest [620]

Journal of thermophysics and heat transfer – Reston. 1987+ (1,5,9) – ISSN: 0887-8722 – mf#16128 – us UMI ProQuest [530]

Journal of the...session (after organization) of the new jersey annual conference of the african methodist episcopal church / African Methodist Episcopal Church – Bridgeton NJ: McCowan & Nichols, Pioneer Book & Job Printers, 13th, 15th-17th. 1886-89 [mf ed 2004] – 1r – 1 – (lacks: 14? 1886-1888 printed in new jersey, 1889- in philadelphia) – mf#2004-s044 – us ATLA [242]

Journal of the...session and the...quadrennial session of the general conference of the african methodist episcopal church in the united states / African Methodist Episcopal Church. General Conference – Xenia OH: Torchlight Print Co. 17th (1880) (quadrennial) [mf ed 2005] – 1r – 1 – (vol for 1880 contains report of the 17th session and 16th quadrennial session; cont: african methodist episcopal church. general conference. session, and the...quadrennial session of the general conference of the african methodist episcopal church; cont by: african methodist episcopal church. general conference. journal of the... quadrennial session of the general conference of the african m e church) – mf#2005-s089 – us ATLA [242]

Journal of the...session of the... / Methodist Episcopal Church. Pacific Japanese Mission – [s.l: s.n] 3rd-5th. 1902-04 [annual] [mf ed 2003] – 3v on 1r – 1 – mf#2003-s114 – us ATLA [242]

Journal of the...session of the alabama annual conference of the african methodist episcopal church / African Methodist Episcopal Church – Nashville TN: A M E Sunday School Union. 57th 1924 (annual) [mf ed 2004] – 1r – 1 – (lacks: p65-66; filmed with: annual report of the executive committee of the institute for the training of colored ministers, at tuskaloosa, alabama, to the general assembly of the presbyterian church in the united states [order052]) – mf#2004-s051 – us ATLA [242]

Journal of the...session of the congo mission conference, methodist episcopal church / Methodist Episcopal Church. Congo Mission Conference – [s l]: Rhodesia Mission Press 1st-20th 1917-38 [mf ed 2005] – 20v on 1r – 1 – (1st session also contains minutes of the 2nd meeting of the congo mission; 18th (1936) mis-numbered as 20th; 19th (1937) mis-numbered as 18th; lacks: 4th-6th, 8th-10th, 15-17th]; cont: methodist episcopal church. congo mission. minutes of the...meeting; cont by: methodist episcopal church. congo mission conference. journal of the...session of the congo mission conference of the methodist episcopal church and the...session of the congo mission conference of the methodist church) – mf#2005c-s037 – us ATLA [242]

Journal of the...session of the congo mission conference of the methodist episcopal church and the...session of the congo mission conference of the methodist church / Methodist Episcopal Church. Congo Mission Conference – Elizabethville, Belgian Congo: Congo Mission Conference 1939- [21st/1st (1939)] [mf ed 2005] – 1 – (also iss on microfilm; cont: methodist episcopal church. congo mission conference. journal of the...session of the congo mission conference, methodist episcopal church) – mf#2005c-s038 – us ATLA [242]

Journal of the...session of the hawaii mission of the methodist church / Methodist Church (US). Hawaii Mission – [Hawaii?]: Hawaiian Print Co, 1947- [annual] [mf ed 2003] – 1r – 1 – (mf: 42nd-45th [1947-50]) – mf#2003-s112 – us ATLA [242]

Journal of third world studies – Americus. 1994+ (1,5,9) – ISSN: 8755-3449 – mf#20586 – us UMI ProQuest [327]

Journal of thoracic and cardiovascular surgery – St Louis. 1931+ [1]; 1965+ [5]; 1970+ [9] – ISSN: 0022-5223 – mf#1877 – us UMI ProQuest [617]

Journal of thoracic imaging – Rockville. 1989-1996 (1) – ISSN: 0883-5993 – mf#14924 – us UMI ProQuest [616]

Journal of thought – DeKalb. 1966+ (1) 1976+ (5) 1976+ (9) – ISSN: 0022-5231 – mf#10400 – us UMI ProQuest [000]

Journal of threat assessment / ed by McCann, Joseph T – 1,9 – $95.00 us $138.00 other – ISSN: 1533-2608 – us Haworth [360]

Journal of three voyages along the coast of china, in 1831, 1832 and 1833 : with notices of siam, corea, and the loo-choo islands / Geutzlaff, Karl Friedrich August – London: Frederick Westley & A H Davis, 1834 [mf ed 1995] – (= ser Yale coll) – vi/xciii/450p/[2]pl – 1 – 0-524-09240-0 – (int essay on the policy, religion, etc of china by the rev w ellis. ill by r sears) – mf#1995-0240 – us ATLA [915]

Journal of three voyages along the coast of china, in 1831, 1832, and 1833 : with notices of siam, corea, and the loo-choo islands / Gutzlaff, K – London: Frederick Westley, A H Davis, 1834 – 6mf – 9 – mf#HT-713 – ne IDC [915]

Journal of three voyages along the coast of china, in 1831, 1832, and 1833 : with notices of siam, corea and the loo-choo islands; to which is prefixed, an introductory essay on the policy, religion, etc of china by the rev w ellis / Guetzlaff, Carl – London 1834 [mf ed Hildesheim 1995-98] – 1v on 6mf – 9 – €120.00 – 3-487-27495-7 – gw Olms [915]

Journal of time series analysis – Avon. 1988+ (1,5,9) – ISSN: 0143-9782 – mf#17398 – us UMI ProQuest [600]

Journal of toxicology and environmental health, pt a – New York. 1975+(1,5,9) – ISSN: 0098-4108 – mf#11138 – us UMI ProQuest [615]

Journal of toxicology and environmental health, pt b : critical reviews – New York. 1998+ (1) – ISSN: 1093-7404 – mf#26724 – us UMI ProQuest [615]

Journal of toxicology clinical toxicology – New York. 1982-1996 (1,5,9) – (cont: clinical toxicology) – ISSN: 0731-3810 – mf#12924,01 – us UMI ProQuest [615]

Journal of transactions : society for promoting the study of religions – n1-8. 1931-34 [complete] – Inquire – 1 – mf#ATLA 1994-S512 – us ATLA [230]

Journal of transnational management development : the official publication of the international management development association / ed by Cordell, Victor V – v2 n1. 1995- – 1,9 – $175.00 in US $245.00 outside hardcopy subsc – us Haworth [650]

Journal of transpersonal psychology – Stanford. 1969+ (1) 1975+ (5) 1976+ (9) – ISSN: 0022-524X – mf#8392 – us UMI ProQuest [150]

Journal of transport history – Manchester. 1989-1996 (1) – ISSN: 0022-5266 – mf#15390 – us UMI ProQuest [380]

Journal of transportation engineering – New York. 1983+ (1) 1983+ (5) 1983+ (9) – (cont: transportation engineering journal of asce: proceedings of the american society of civil engineers) – ISSN: 0733-947X – mf#8144,01 – us UMI ProQuest [624]

Journal of transportation law, logistics and policy – v1-67. 1933-2000 – 5,6,9 – $1464.00 set – (v1-52 1933-85 on reel $913. v53-67 1985-2000 on mf $551. title varies: v1-51 1933-84 as icc practitioners' journal. v52-61 1985-94 as transportation practitioners' journal) – ISSN: 1078-5906 – mf#103201 – us Hein [343]

Journal of trauma : injury, infection and critical care – v1-41. 1961-96 – 1,5,6,9 – $110.00r – (also available on cd-rom: 1985-89 $395.00) – us Lippincott [616]

Journal of trauma & dissociation : the official journal of the international society for the study of dissociation (issd) / ed by Bowman, Elizabeth & Chu, James A – v1- 2000- – 1,9 – $95.00 us $138.00 other – ISSN: 1529-9732 – us Haworth [616]

Journal of trauma practice / ed by Gold, Steven N & Faust, Jan – v1- 2002- – 1,9 – $120.00 us $174.00 other – ISSN: 1536-2922 – us Haworth [617]

Journal of traumatic stress – New York. 1988+ (1,5,9) – ISSN: 0894-9867 – mf#17692 – us UMI ProQuest [150]

Journal of travel and tourism marketing / ed by Chon, K S (Kaye) – v5 n1. 1996- – 1,9 – $80.00 in US $112.00 outside hardcopy subsc – us Haworth [650]

Journal of travel research – Thousand Oaks. 1990+ (1,5,9) – ISSN: 0047-2875 – mf#15826,01 – us UMI ProQuest [910]

A journal of travels in barbary, in the year 1801 : with observations on the gum trade / Curtis, James – London 1803 [mf ed Hildesheim 1995-98] – 1v on 1mf – 9 – €40.00 – 3-487-27343-8 – gw Olms [916]

Journal of travels in the seat of war, during the last two campaigns of russia and turkey : intended as an itinerary through the south of russia, the crimea, georgia, and through persia, koordistan, and asia minor, to constantinople / Armstrong, T B – London 1831 [mf ed Hildesheim 1995-98] – 1v on 2mf [ill] – 9 – €60.00 – 3-487-26696-2 – gw Olms [933]

Journal of travels in the united states of north america, and in lower canada, performed in the year 1817 : containing particulars relating to the prices of land and provisions, remarks on the country and people, interesting anecdotes, and an account of the commerce, trade, and present state of washington, new york, philadelphia, boston, baltimore, albany / Palmer, John – London 1818 [mf ed Hildesheim 1995-98] – 1v on 3mf – 9 – €90.00 – 3-487-27040-4 – gw Olms [917]

A journal of travels into the arkansa territory : during the year 1819 with occasional observations on the manners of the aborigines / Nuttal, Thomas – Philadelphia 1821 [mf ed Hildesheim 1995-98] – 1v on 2mf – 9 – €60.00 – 3-487-27160-5 – gw Olms [917]

Journal of tree fruit production / ed by Autio, Wesley R – v1 n1. 1996- – 1,9 – $95.00 in US $133.00 outside hardcopy subsc – us Haworth [630]

Journal of tribology – New York. 1984+ (1) 1984+ (5) 1984+ (9) – (cont: journal of lubrication technology) – ISSN: 0742-4787 – mf#7543,01 – us UMI ProQuest [550]

Journal of tropical ecology – Cambridge. 1989-1989 (1) – ISSN: 0266-4674 – mf#16535 – us UMI ProQuest [574]

Journal of tropical medicine and hygiene – Oxford. 1980-1995 (1,5,9) – ISSN: 0022-5304 – mf#15541,01 – us UMI ProQuest [614]

Journal of tropical pediatrics – London. 1985+ (1,5,9) – ISSN: 0142-6338 – mf#14339,04 – us UMI ProQuest [618]

Journal of truth / Minister John Muhammad's Temple No 1 – 1984 jun – 1r – 1 – mf#5308486 – us WHS [071]

Journal of turbomachinery – New York. 1986+ (1,5,9) – ISSN: 0889-504X – mf#16196 – us UMI ProQuest [621]

Journal of turfgrass management : developments in basic and applied turfgrass research / ed by Torello, William A – v1 n1. 1995- – 1,9 – $90.00 in US $126.00 outside hardcopy subsc – us Haworth [630]

A journal of two successive tours upon the continent : in the years 1816, 1817, and 1818 / Wilson, James – London 1820 [mf ed Hildesheim 1995-98] – 3v on 12mf – 9 – €120.00 – 3-487-27702-6 – gw Olms [910]

A journal of two visits made to some nations of indians on the west side of the river ohio, in the years 1772 and 1773 : with a biographical notice of the author / Jones, David – New York: Reprinted for Joseph Sabin, 1865. Chicago: Dep of Photodup, U of Chicago Lib, 1969 (1r); Evanston: American Theol Lib Assoc, 1984 (1r) – 1 – 0-8370-0487-X – (sabin's reprints) – mf#1984-B108 – us ATLA [917]

Journal of ukrainian studies = Zhurnal ukrainoznavchykh studii – Toronto. 1989-1993 (1) – ISSN: 0228-1635 – mf#13854,01 – us UMI ProQuest [947]

Journal of ultrasound in medicine – Laurel. 1982+ (1,5,9) – ISSN: 0278-4297 – mf#12939 – us UMI ProQuest [610]

Journal of united labor – Marblehead MA, Philadelphia. 1880 may 15-1885, 1886-89 – 2r – 1 – (cont by: journal of the knights of labor) – mf#1060840 – us WHS [331]

Journal of united machinists and mechanical engineers of america / United Machinists and Mechanical Engineers of America – v1 n1-3 [1889 feb-apr] – 1r – 1 – (cont by: journal [national association of machinists]) – mf#910670 – us WHS [621]

Journal of university studies – Detroit. 1974-1975 (1) 1974-1975 (5) 1974-1975 (9) – (cont: new university thought) – mf#5062,01 – us UMI ProQuest [378]

Journal of urban affairs – Malden. 1992+ (1,5,9) – ISSN: 0735-2166 – mf#19789 – us UMI ProQuest [307]

Journal of urban health – Oxford. 1998+ (1,5,9) – (cont: new york academy of medicine bulletin) – ISSN: 1099-3460 – mf#258,01 – us UMI ProQuest [610]

Journal of urban history – Beverly Hills. 1983+ (1,5,9) – ISSN: 0096-1442 – mf#14009 – us UMI ProQuest [307]

Journal of urban planning and development – New York. 1983+ (1) 1983+ (5) 1983+ (9) – (cont: journal of the urban planning and development division: proceedings of the american society of civil engineers) – ISSN: 0733-9488 – mf#8139,01 – us UMI ProQuest [710]

Journal of urology – v1-156. 1917-96 – 1,5,6,9 – $105.00r – us Lippincott [616]

Journal of vacuum science and technology : b: microelectronics processing and phenomena – v1b-. 1983- – 1,5,6 – us AIP [530]

Journal of vacuum science and technology: a and b – v1A+B-. 1983- – 1,5,6 – us AIP [530]

Journal of vacuum science and technology: a: vacuum, surfaces, and films – v1A-. 1983- – 1,5,6 – us AIP [530]

Journal of value inquiry – The Hague. 1989+ (1,5,9) – ISSN: 0022-5363 – mf#16822 – us UMI ProQuest [170]

Journal of vascular medicine and biology – Cambridge. 1991-1994 (1,5,9) – ISSN: 1042-5268 – mf#18082 – us UMI ProQuest [610]

Journal of vascular nursing – Norwood. 1992+ (1,5,9) – ISSN: 1062-0303 – mf#19565,01 – us UMI ProQuest [610]

Journal of vascular research – Basel. 1994-1996 (1,5,9) – (cont: blood vessels) – ISSN: 1018-1172 – mf#2046,02 – us UMI ProQuest [611]

Journal of vascular surgery – St. Louis. 1984+ (1,5,9) – ISSN: 0741-5214 – mf#13830 – us UMI ProQuest [617]

Journal of vegetable crop production / ed by Robbins, M LeRon – v2 n1. 1997- – 1,9 – $90.00 in US $126.00 outside hardcopy subsc – us Haworth [631]

Journal of venereal disease information – Washington. (1) 1920-1951 (5) (9) – mf#5773 – us UMI ProQuest [616]

Journal of vestibular research – Elmsford. 1990-1998 (1,5,9) – ISSN: 0957-4271 – mf#49582 – us UMI ProQuest [617]

Journal of veterinary medical education – Rockville. 1974-1996 (1) 1974-1996 (5) 1974-1996 (9) – ISSN: 0748-321X – mf#10996 – us UMI ProQuest [636]

Journal of veterinary pharmacology and therapeutics – Oxford. 1978-1994 (1,5,9) – ISSN: 0140-7783 – mf#15567 – us UMI ProQuest [615]

Journal of vibration, acoustics, stress, and reliability in design – New York. 1983-1989 (1) 1983-1989 (5) 1983-1989 (9) – (cont by: journal of vibration and acoustics) – ISSN: 0739-3717 – mf#11949,01 – us UMI ProQuest [621]

Journal of vibration and acoustics – New York. 1990+ (1,5,9) – (cont: journal of vibration, acoustics, stress, and reliability in design) – ISSN: 1048-9002 – mf#11949,02 – us UMI ProQuest [621]

Journal of virginia education – Richmond. 1979-1980 – 1,5,9 – (cont: virginia journal of education. cont by: virginia journal of education) – ISSN: 0198-3504 – mf#1899,01 – us UMI ProQuest [370]

Journal of virological methods – Amsterdam. 1980+ (1) 1980+ (5) 1987+ (9) – ISSN: 0166-0934 – mf#42134 – us UMI ProQuest [576]

Journal of virology – Washington. 1967+ (1) 1971+ (5) 1976+ (9) – ISSN: 0022-538X – mf#2259 – us UMI ProQuest [576]

Journal of visual impairment and blindness – New York. 1977+ (1,5,9) – (cont: new outlook for the blind) – ISSN: 0145-482X – mf#11296,01 – us UMI ProQuest [617]

Journal of visualization and computer animation – Chichester. 1990-1996 (1,5,9) – ISSN: 1049-8907 – mf#18149 – us UMI ProQuest [000]

Journal of vlsi signal processing – Boston. 1989-1995 (1,5,9) – ISSN: 0922-5773 – mf#16821 – us UMI ProQuest [000]

Journal of vlsi signal processing systems for signal, image, and video technology – Boston, 1996-1996 [1,5,9] – (cont: journal of vlsi signal processing) – mf#16821,01 – us UMI ProQuest [621]

Journal of vocational education research – Columbus. 1984+ – 1,5,9 – ISSN: 0739-3369 – mf#14341 – us UMI ProQuest [374]

Journal of vocational rehabilitation – Shannon. 1995+ (1,5,9) – ISSN: 1052-2263 – mf#18973 – us UMI ProQuest [331]

Journal of voice – New York. 1993-1996 (1,5,9) – ISSN: 0892-1997 – mf#18727 – us UMI ProQuest [617]

Journal of volcanology and geothermal research – Amsterdam. 1976+ (1) 1976+ (5) 1987+ (9) – ISSN: 0377-0273 – mf#42135 – us UMI ProQuest [550]

Journal of voluntary action research – San Francisco. 1972-1988 (1) 1972-1988 (5) 1972-1988 (9) – (cont by: nonprofit and voluntary sector quarterly) – ISSN: 0094-0607 – mf#8069 – us UMI ProQuest [303]

Journal of voyage from salem towards st kitts of brig adventure – 1807 – 1r – 1 – cn Library Assoc [910]

A journal of voyages and travels : published for the benefit of the author's orphan daughter / Rees, Thomas – London 1822 [mf ed Hildesheim 1995-98] – 1v on 2mf – 9 – €60.00 – 3-487-27615-1 – gw Olms [910]

Journal of voyages and travels by the rev daniel tyerman and george bennet, esq : ...to visit their various stations in the south sea islands, china, india, &c, between the years 1821 and 1829 / Tyerman, Daniel – London 1831 [mf ed Hildesheim 1995-98] – 2v on 8mf – 9 – €160.00 – ISBN-10: 3-487-26765-9 – ISBN-13: 978-3-487-26765-4 – gw Olms [910]

Journal of voyages and travels...to...the south sea islands, china, india h and c : between the years 1821 and 1829 / Tyerman, D & Bennet, G; ed by Montgomery, J – London, 1831. 2v – 14mf – 9 – mf#HTM-196 – ne IDC [915]

Journal of water resources planning and management – New York. 1983+ (1,5,9) – (cont: journal of the water resources planning and management division) – ISSN: 0733-9496 – mf#11448,01 – us UMI ProQuest [350]

Journal of waterway, port, coastal and ocean engineering – New York, 1983+ [1,5,9] – (cont: journal of the waterway, port, coastal and ocean division) – ISSN: 0733-950X – mf#8145,02 – us UMI ProQuest [624]

Journal of wealth management – London. 2000+ (1,5,9) – ISSN: 1534-7524 – mf#32272,01 – us UMI ProQuest [332]

Journal of website promotion : innovations in internet business research, theory, and practice / ed by Nelson, Richard Alan – 1,9 – us Haworth [000]

Journal of whiplash and related disorders / ed by Centeno, Christopher J & Freeman, Michael D – v1- 2002- – 1,9 – $85.00 us $123.00 other – ISSN: 1533-2888 – us Haworth [617]

Journal of wildlife management – Bethesda. 1937+ (1) 1961+ (5) 1961+ (9) – ISSN: 0022-541X – mf#1941 – us UMI ProQuest [639]

The journal of william de rubruquish a frenchman : of the minorite friers, into tartary and china...1253 / [Ruysbroek, W van] – London, 1705. v1 – 1mf – 9 – mf#HT-673 – ne IDC [915]

Journal of wind engineering and industrial aerodynamics – Amsterdam. 1975+ (1) 1975+ (5) 1987+ (9) – ISSN: 0167-6105 – mf#42110 – us UMI ProQuest [621]

Journal of women and aging / ed by Garner, J Dianne – v1- 1989- – 1, 9 ($175.00 in US $36.00 outside hardcopy subsc) – us Haworth [618]

Journal of women and religion – Berkeley. 1992-1992 (1,5,9) – ISSN: 0888-5621 – mf#19252,02 – us UMI ProQuest [305]

Journal of women and the law *see* William and mary journal of women and the law

Journal of women's history – Bloomington. 1992+ (1,5,9) – ISSN: 1042-7961 – mf#19247 – us UMI ProQuest [305]

Journal of women's imaging – Philadelphia. 1999+ (1,5,9) – ISSN: 1084-824X – mf#23347 – us UMI ProQuest [616]

Journal of workplace learning – Bradford. 2001+ (1,5,9) – ISSN: 1366-5626 – mf#18897,01 – us UMI ProQuest [331]

Journal of world business – Greenwich. 1997+ (1) 1997+ (5) 1997+ (9) – (cont: columbia journal of world business) – ISSN: 1090-9516 – mf#6182,01 – us UMI ProQuest [338]

Journal of world education – Huntington. 1970-1978 (1) 1972-1978 (5) 1975-1978 (9) – ISSN: 0092-2382 – mf#8068 – us UMI ProQuest [370]

Journal of world history – Honolulu. 1993+ (1,5,9) – ISSN: 1045-6007 – mf#18075 – us UMI ProQuest [900]

Journal of world history – Paris. 1953-1972 (1) – ISSN: 0022-5436 – mf#1359 – us UMI ProQuest [900]

Journal of wound, ostomy, and continence nursing – St. Louis. 1994+ (1,5,9) – (cont: journal of et nursing) – ISSN: 1071-5754 – mf#13045,03 – us UMI ProQuest [617]

Journal of youth and adolescence – New York. 1972+ (1) 1974+ (5) 1978+ (9) – ISSN: 0047-2891 – mf#10868 – us UMI ProQuest [640]

Journal of youth services in libraries: joys – Chicago. 1987+ (1,5,9) – (cont: top of the news) – ISSN: 0894-2498 – mf#16698 – us UMI ProQuest [020]

Journal of zoology – London. 1987+ (1) 1987+ (5) 1987+ (9) – ISSN: 0952-8369 – mf#1268,02 – us UMI ProQuest [590]

Journal of zoology : proceedings of the zoological society of london – London. 1958-1984 (1) 1972-1984 (5) 1977-1984 (9) – ISSN: 0022-5460 – mf#1268 – us UMI ProQuest [590]

Journal of zoology series a : proceedings of the zoological society of london – London. 1985-1986 (1,5,9) – ISSN: 0269-364X – mf#1268,01 – us UMI ProQuest [590]

Journal of zoology, series b – London. 1985-1987 (1) 1985-1987 (5) 1985-1987 (9) – ISSN: 0268-196X – mf#15672 – us UMI ProQuest [590]

The journal of...as one of the guard on lord macartney's embassy to china and tartary, 1792-1793 / Holmes, S – London: W Bulmer and Co, 1798 – 3mf – 9 – mf#HT-782 – ne IDC [915]

Journal official / United Arab Republic – 1949-1966. Cairo – 1 – us NY Public [956]

Journal official de la republique algerienne / Algeria – 1970-79. 10 reels – 1 – (1980-. approx.23.00y; 1) – us L of C Photodup [324]

Journal officiel / Algeria – (premier partie. lois et decrets. 1927-juin 1936, 1937-39, avr 1943-44, oct 1957-58. 1. deuxieme partie. decrets, arretes, textes divers. 1927-37, juin 1943-44. 1. troisieme partie. texte arabe. 1927-29. 1) – fr ACRPP [960]

Journal officiel / Algeria – puis de la Republique algerienne; de la Republique algerienne democratique et populaire. juil 1962-70 – 1 – fr ACRPP [960]

Journal officiel / Congo. (Brazzaville) – 1960, 1962-70 – 1 – fr ACRPP [960]

Journal officiel / Dahomey – aout 1894-1928, 1949, 1953-55. Annexes 1953-56 – 1 – fr ACRPP [960]

Journal officiel / Dahomey. Porto Novo – 1957-1967- – 1 – us NY Public [960]

Journal officiel / French Equatorial Africa – A.E.F. mars 1910-27, 1929-38, 1940-45, 1947-aout 1959 – 1 – fr ACRPP [960]

Journal officiel / French Guinea – 1901-26, 1930-44, 1947-sept 1958 – 1 – fr ACRPP [073]

Journal officiel / French Sudan – fevr 1921, 1922-nov 1958 – 1 – fr ACRPP [960]

Journal officiel / French Sudan – Koulouba. 1951-Oct. 1958 – 1 – us NY Public [960]

Journal officiel / French West Africa – A.O.F. oct 1895-1900, 1905-aout 1959 – 1 – fr ACRPP [960]

Journal officiel / French West Africa – Goree. 1952-1959 – 1 – us NY Public [960]

Journal officiel / Gabon – juil 1904-oct 1908 – 1 – fr ACRPP [960]

Journal officiel / Gabon-Congo – juin 1887-88 – 1 – fr ACRPP [960]

Journal officiel / Guinea. French – Conakry. 1948-Sept. 1958 – 1 – us NY Public [960]

Journal officiel / Indochina. French – 1889-1950 – 1 – fr ACRPP [073]

Journal officiel / Ivory Coast – 1895-1927, 1930-38, 1940-42, 1946-58 – 1 – fr ACRPP [960]

Journal officiel / Martinique. French West Indies – Port de France. Jan. 6, 1940-Oct. 12, 1940; 1941-1946 – 1 – us NY Public [324]

Journal officiel / Oubangui-Chari-Tchad – 1907 – 1 – fr ACRPP [960]

Journal officiel / Senegal – 1888-95, 1901-27, 1930-39, 1947-58, 1960-66, Juil 1967-71 – 1 – fr ACRPP [073]

Journal officiel / Silesia. Upper – Oppeln. Commission interalliee de gouvernement et de plebiscite de Haute-Silesie. fevr 1920-juin 1922 – 1 – fr ACRPP [940]

Journal officiel / Togo – 1956-60 – 1 – fr ACRPP [073]

Journal officiel / Tunisia – 1970-79 – 17r – 1 – $360.00; outside North America add $1.25r – (1980-. ca $50.00y) – us L of C Photodup [324]

Le journal officiel / France – 9 – enquire for prices – (yrly mf count varies. various eds available: lois et decrets 1869- . documents de l'assemblee nationale. documents du senat. debats de l'assemblee nationale. debats du senat) – us UMI ProQuest [074]

Le journal officiel associations et fondations d'entreprises – €63.20y – (backfile: upwards 1985- €42.69y) – fr Journal Officiel [366]

Journal officiel de haute-silesie – Oppeln (Opole PL), 1920 [gaps] – 1r – 1 – gw Misc Inst [077]

Journal officiel de la haute-volta / Upper Volta – oct 1919-27 – 1 – fr ACRPP [960]

Journal officiel de la polynesie francaise / Polynesia. French – 1965-70 – 1 – fr ACRPP [980]

Journal officiel de la republic du mali / Mali – apr 1959-68 – 1 – (puis de la federation) – fr ACRPP [960]

Journal officiel de la republique algerienne democratique et populaire – Alger, oct 26 1962-70 – 7r – 1 – us CRL [960]

Journal officiel de la republique centrafricaine / Central African Republic – aout 1959-70 – 1 – fr ACRPP [960]

Journal officiel de la republique de haute-volta / Upper Volta – 1960-70 – 1 – fr ACRPP [960]

Journal officiel de la republique de la cote d'ivoire / Ivory Coast – 1960-70 – 1 – fr ACRPP [960]

Journal officiel de la republique du congo / Congo. Republic – Brazzaville. 1958-1967 – 1 – us NY Public [960]

Journal officiel de la republique du dahomey / Dahomey – dec 1963-nov 1969 – 1 – fr ACRPP [960]

Journal officiel de la republique du niger / Niger – 1959-1967. Niamey – 1 – us NY Public [324]

Journal officiel de la republique du niger / Niger – 1960-70 – 1 – fr ACRPP [960]

Journal officiel de la republique francaise / France – (5 sept 1870-80. 1. ed. du soir. 6 sept 1870-19 mars 1871. 1) – fr ACRPP [323]

Journal officiel de la republique francaise / France – 1881-1924, 25 aout 1944 – 1 – (8 sept 1944-86. 1,5) – fr ACRPP [323]

Journal officiel de la republique francaise / France – La Commune. Ed. du matin. 20 mars-24 mai 1871 et ed. du soir 21 mars-19 mai 1871 – 1 – fr ACRPP [323]

Journal officiel de la republique francaise – Paris. 1941-1971 (1) – mf#2570 – us UMI ProQuest [324]

Journal officiel de la republique gabonaise / Gabon – juil 1959-70 – 1 – fr ACRPP [960]

Journal officiel de la republique islamique de mauritanie / Mauritania – 1959-sept 1969 – 1 – fr ACRPP [960]

Journal officiel de la republique khmere – Phnom-Penh: Secretariat general du Conseil des Ministres 1970- [semiwkly] [mf ed Wooster OH: Micro Photo Division, Bell & Howell [1979]] – 7r – 1 – ("il comprend tous les actes interessant le gouvernement de la republique"; iss for oct 10-dec 1970 filmed with: cambodia. journal offic iel du cambodge, jun-oct 7 1970) – us CRL [342]

Journal officiel de la republique khmere – Phnom Penh, Kampuchea. 1965-1973 (1) – mf#68832 – us UMI ProQuest [079]

Journal officiel de la republique soudanaise / Sudan – janv-sept 1960 – 1 – fr ACRPP [960]

Journal officiel de l'algerie – Alger, 1927-jul 25 1958 – 1 – (pt 1 only: vols 19-24 1945-50 (4r). vols 27-32 1953-jul 25 1958 (6r)) – us CRL [960]

Journal officiel de l'empire francais / France – (ed. du matin. 1869-4 sept 1870. 1. ed. du soir. 1869-5 sept 1870. 1) – fr ACRPP [323]

Journal officiel de l'etat algerien – Alger, jul 6-sep 25 1962 – 2r – 1 – us CRL [960]

Journal officiel de l'union europeennes *see* The official journal of the european union

Journal officiel de madagascar / Malagasy Republic – nov 1957-69 – 1 – fr ACRPP [960]

Le journal officiel des communautes europeenes – 1971- – 9 – enquire for prices – (publ also available in english and other languages of europe) – us UMI ProQuest [341]

Journal officiel des possessions du congo francais / Congo. (French) – juil 1904-10 – 1 – fr ACRPP [960]

Journal officiel du cambodge / Royaume du Cambodge – Phnom-Penh: Secretariat general du Conseil des Ministres, Bureaux des publications [sic] officielles [semiwkly, wkly] [mf ed Wooster OH: Micro Photo Division, Bell & Howell [1979]] – 12r – 1 – ("il comprend tous les actes interessant l'administration du cambodge"; iss for jun-oct 7 1970 filmed with: cambodge. journal officiel de la republique khmere, oct 10-dec 1970; also filmed for the university libraries, university of hawaii at manoa by advanced micro-image systems hawaii, inc 1990 [31r]) – us CRL [342]

Journal officiel du haut-senegal-niger – aout 1906-21 – 1 – fr ACRPP [960]

Journal officiel du tchad / Chad – oct 1959-67 – 1 – fr ACRPP [960]

Journal officiel du territoire du niger / Niger – 1938-39 – 1 – fr ACRPP [960]

Journal – oklahoma state medical association / Oklahoma State Medical Association – Oklahoma City. 1909+ (1) 1971+ (5) 1975+ (9) – ISSN: 0030-1876 – mf#2424 – us UMI ProQuest [610]

JOURNAL

Journal pie – n1-15. Paris. janv-mars 1792 – 1 – (suivi de: Journal royaliste) – fr ACRPP [073]
Journal pie *see* Journal royaliste
Journal politique et commercial de la pointe-a-pitre – Pointe-a-Pitre. 1819-31 – 1 – (devenu: Journal commercial, economique et maritime de la Pointe-a-Pitre. devenu: Journal commercial de Pointe-a-Pitre.) – fr ACRPP [073]
Journal pour tous – Ottawa: P N Bureau, [1878-1880] – 9 – mf#P04925 – cn Canadiana [440]
Le journal pour tous – Montreal: R Villecourt, [1906-1907] – mf#P05216 – cn Canadiana [610]
Journal (prairie city, or) – Prairie City OR: Lester A Wolf, 1937-42 [wkly] – 1 – (cont: grant county journal (prairie city, or); absorbed by: blue mountain eagle (canyon city, or: 19-?-1948)) – us Oregon Lib [071]
Journal press – Blaine, WA. 1925-1938 (1) – mf#66945 – us UMI ProQuest [071]
Journal press – South Milwaukee WI. 1951 mar 8-jun 21 – 1r – 1 – (cont: reminder-journal press [south milwaukee, wi]; cont by: voice [south milwaukee, wi]) – mf#1004494 – us WHS [071]
Journal progressive – Bellingham, WA. 1914-1918 (1) – mf#66941 – us UMI ProQuest [071]
Journal, report and letterbook / Gray, William – 1882-1906 – 1r – mf#PMB1047 – at Pacific Mss [920]
Journal republican – Metropolis, IL. 1916-1917 (1) – mf#62651 – us UMI ProQuest [071]
Journal review – Crawfordsville, IN. 1942-1997 (1) – mf#61378 – us UMI ProQuest [071]
Journal revolutionnaire de toulouse : ou le surveillant du midi – Toulouse. n1-104. sept 1793-sept 1794 – 1 – fr ACRPP [325]
Journal royaliste – Paris. mars-aout 1792 (I-II) – 1 – (suite de: journal pie) – fr ACRPP [073]
Journal royaliste *see* Journal pie
Le journal (saint-bruno-de-montarville, quebec, 1968) – St-Bruno: [s.n.] v1 n1 21 mars 1968-v10 n26 7 sep 1977 (bimthly) [mf ed 1986] – 1 – mf#SEM35P275 – cn Bibl Nat [071]
Le journal (saint-bruno-de-montarville, quebec, 1979) – St-Bruno: [s.n.] v12 n31 14 nov 1979-v16 n9 3 mars 1982 (wkly) [mf ed 1986] – 4r – 1 – mf#SEM35P277 – cn Bibl Nat [071]
Journal series / Clinton Co. Wilmington – (dec 1868-dec 1901, 1903-dec 1913) [wkly] – 18r – 1 – mf#B31282-31299 – us Ohio Hist [071]
Journal series / Gallia Co. Gallipolis – 1895-oct 1899, 1910-sep 1918 [semiwkly, wkly] – 7r – 1 – mf#B10625-10631 – us Ohio Hist [071]
Journal series / Hamilton Co. Cincinnati – jan 1831-dec 1838 [wkly] – 2r – 1 – mf#B29871-29872 – us Ohio Hist [240]
Journal series / Lake Co. Painesville – v1 n1. jul 1871-jun 1874 [wkly] – 1r – 1 – mf#B5525 – us Ohio Hist [071]
Journal series / Mercer Co. Fort Recovery – jan 1935-dec 1941 (fire damaged) [wkly] – 3r – 1 – mf#B6859-6861 – us Ohio Hist [071]
Journal series / Noble Co. Caldwell – (may 1884-apr 1933) (damaged) [wkly] – 22r – 1 – mf#B9480-9501 – us Ohio Hist [071]
Journal,... session of the hawaii mission of the methodist church / Methodist Church (US). Hawaii Mission – Honolulu: [s.n.] 1945 [annual] [mf ed 2003] – 1v on 1r – 1 – (mf: 40th [1945]) – mf#2003-s110 – us ATLA [242]
Journal southern california dental hygienists' association – El Segundo. 1975-1981 (1) 1975-1981 (5) 1975-1981 (9) – ISSN: 0038-3899 – mf#7385 – us UMI ProQuest [617]
Journal standard – Falls Church, VA. 1959-1968 (1) – mf#66706 – us UMI ProQuest [071]
Journal star : (evening edition) – Peoria, IL. 1955-1992 (1) – mf#60459 – us UMI ProQuest [071]
Journal star – Lincoln, NE. 1995-1997 (1) – mf#60693 – us UMI ProQuest [071]
Journal star : (morning edition) – Peoria, IL. 1955+ (1) – mf#60460 – us UMI ProQuest [071]
Journal (state edition) – Knoxville, TN. 1957-1958 (1) – mf#66541 – us UMI ProQuest [071]
Journal (sub edition) – Providence, RI. 1830-1833 (1) – mf#66329 – us UMI ProQuest [071]
Journal town and country advertiser – Providence, RI. 1799-1951 (1) – mf#66331 – us UMI ProQuest [071]
Journal tribune – Blackwell, OK. 1993-2000 (1) – mf#65763 – us UMI ProQuest [071]
Journal tribune – Marysville, OH. 1990-2000 (1) – mf#61722 – us UMI ProQuest [071]
Journal und adler series / Clark Co. Springfield – (sep 1895-oct 1913) [twice wkly, wkly] – 17r – 1 – (in german) – mf#B34544-34560 – us Ohio Hist [071]
Journal universel – Gand. no. 1-20. 14 avr-21 juin 1815 – 1 – (aka: moniteur de gand) – fr ACRPP [073]
Journal universel : ou revolution des royaumes – Paris. nov 1789-oct 1791 – 1 – fr ACRPP [944]
Journal universel de la litterature, de la science et des beaux-arts / L'Athenaeum francais – Paris, 1852-1856 – 152mf – 8 – mf#H-1378 – ne IDC [700]
Journal universel et affiches de toulouse et du languedoc *see* Affiches, annonces, avis divers de toulouse et du haut-languedoc
Journal von auswartigen und deutschen theatern – Vienna, aug 1778-mr 1779 – 1r – 1 – us UMI ProQuest [790]
Journal von brasilien : oder vermischte nachrichten aus brasilien / Eschwege, Wilhelm L von – Weimar 1818 [mf ed Hildesheim 1995-98] – 2v on 4mf – 9 – €120.00 – 3-487-26516-8 – gw Olms [918]
Das journal von tiefurt / ed by Hellen, Eduard von der – Weimar: Goethe-Gesellschaft, 1892 [mf ed 1993] – (= ser Schriften der goethe-gesellschaft 7) – xxxvi/398p/[4]pl – 1 – (int by bernhard suphan) – mf#8657 reel 2 – us Wisconsin U Libr [430]
Journal von und fuer deutschland – Fulda DE, 1784-86, 1787 jul-1790, 1791 jul-1792 apr, 1792 jun-dec – 10r – 1 – gw Misc Inst [943]
Journal von und fuer deutschland / ed by Goeckingk, Ludwig Friedrich Guenter von – Ellrich 1784-92 [mf ed Hildesheim 1992-98] – (= ser Dz) – 9jge[=12st] on 71mf – 9 – €710.00 – mf#k/n5653 – gw Olms [943]
Journal von und fuer franken / ed by Bundschuh, Joh Kaspar et al – Nuernberg 1790-93 – (= ser Dz. historisch-geographische abt) – 6v on 30mf – 9 – €300.00 – mf#k/n1236 – gw Olms [943]
La journal vrai – 1932-34 – 1 – fr ACRPP [073]
The journal, with other writings.. / Woolman, John – London & Toronto: J.M. Dent & sons Ltd.; New York: E.P. Dutton, 1922. xix,250p. (Everyman's Library, ed. by Ernest Rhys. Biography. No.402) – 1 – us Wisconsin U Libr [920]
Journal zur kunstgeschichte und zur allgemeinen litteratur / ed by Murr, Christoph Gottlieb von – Nuernberg 1775-89 – (= ser Dz) – 17pt on 49mf – 9 – €490.00 – mf#k/n4082 – gw Olms [700]
Journal-gazette – Monroe WI. [1898 jul 26/1899 jul 11]-[1926 may 4/1927 jul 15] [gaps] – 34r – 1 – (cont: county journal [monroe, wi]; monroe sun-gazette) – mf#1125630 – us WHS [071]
Journal-herald / Jackson Co. Jackson – jan 1983-jun 1986 [twice wkly] – 7r – 1 – mf#B28698-28704 – us Ohio Hist [071]
Journal-herald / Jackson Co. Jackson – jan 3 1992-dec 30 1994 – 6r – 1 – mf#B41420-41425 – us Ohio Hist [071]
Journal-herald / Jackson Co. Jackson – jul 2, 1986-dec 30, 1991 [twice wkly] – 12r – 1 – mf#B31624-31635 – us Ohio Hist [071]
Journal-herald series / Delaware Co. Delaware – apr 1902-dec 1926 [wkly, semiwkly] – 18r – 1 – mf#B11120-11137 – us Ohio Hist [071]
Journalism abstracts – Columbia. 1971-1993 (1) 1963-1993 (5) 1976-1993 (9) – (cont by: journalism and mass communication abstracts) – ISSN: 0075-4412 – mf#6194 – us UMI ProQuest [070]
Journalism and communication monographs – Columbia. 1999+ (1,5,9) – (cont: journalism and mass communication monographs) – ISSN: 1522-6379 – mf#28805 – us UMI ProQuest [070]
Journalism and mass communication abstracts – Columbia. 1994+ (1) 1994+ (5) 1994+ (9) – (cont: journalism abstracts) – mf#6194,01 – us UMI ProQuest [070]
Journalism and mass communication educator – Columbia. 1995+ (1) 1995+ (5) 1995+ (9) – (cont: journalism educator) – ISSN: 1077-6958 – mf#6195,01 – us UMI ProQuest [378]
Journalism and mass communication monographs – Columbia. 1995-1998 (1) 1995-1998 (5) 1995-1998 (9) – (cont: journalism monographs. cont by: journalism and communication monographs) – ISSN: 1077-6966 – mf#6196,01 – us UMI ProQuest [070]
Journalism and mass communication quarterly – Columbia. 1995+ (1) 1995+ (5) 1995+ (9) – (cont: journalism quarterly) – ISSN: 1077-6990 – mf#241,01 – us UMI ProQuest [070]
Journalism and politics, series 1 : the papers of c p scott, 1846-1932 from the john rylands university library of manchester – [mf ed Marlborough 1992] – 1pt – 1 – (pt1: c p scott's general correspondence c1870-1934, & political diaries, 1911-28 [22r] £2050; with d/g) – uk Matthew [070]
Journalism educator – Columbia. 1971-1995 (1) 1971-1995 (5) 1976-1995 (9) – (cont by: journalism and mass communication educator) – ISSN: 0022-5517 – mf#6195 – us UMI ProQuest [378]
Journalism history – Northridge. 1974+ (1,5,9) – ISSN: 0094-7679 – mf#11684 – us UMI ProQuest [070]
Journalism in the united states / Jones, Robert William – New York, NY. 1947 – 1r – us UF Libraries [025]
Journalism monographs – Columbia. 1971-1994 (1) 1966-1994 (5) 1977-1994 (9) – (cont by: journalism and mass communication monographs) – ISSN: 0022-5525 – mf#6196 – us UMI ProQuest [070]
Journalism quarterly – Columbia. 1924-1994 (1) 1969-1994 (5) 1975-1994 (9) – (cont by: journalism and mass communication quarterly) – ISSN: 0196-3031 – mf#241 – us UMI ProQuest [070]
Die journalisten : lustspiel in vier akten / Freytag, Gustav – 7. Aufl. Leipzig: S Hirzel, 1882 – 1r – 1 – us Wisconsin U Libr [820]
Die journalisten : lustspiel in vier akten / Freytag, Gustav – Leipzig: S Hirzel, 1854 – 1r – 1 – us Wisconsin U Libr [430]
Journalistiek op java : serie "over oost en west" / Wormser, C W – Deventer, 1941 – 1mf – 8 – mf#SE-1447 – ne IDC [070]
Journal-news / Allen Co. Spencerville – jan-may 1913, oct 1919-aug 1948 [wkly] – 13r – 1 – mf#B32661-32673 – us Ohio Hist [071]
Journal-news / Allen Co. Spencerville – sep 1948-dec 1992 [wkly] – 23r – 1 – mf#B32624-32646 – us Ohio Hist [071]
Journal-report from the golden mean society / Golden Mean Society – n50-61 [1984 sep-1985 sep] – 1r – 1 – (cont: double eagle report; journal from the golden mean society; cont by: journal of the golden mean society [1985]; report from the golden mean society) – mf#956957 – us WHS [366]
Journal-republican / Clinton Co. Wilmington – jan 1914-aug 1919 [wkly, semiwkly, wkly] – 4r – 1 – mf#B10996-10999 – us Ohio Hist [071]
Journal-republican – Columbus WI. 1939 oct 13 – 1r – 1 – (cont: columbus journal [columbus, wi]; columbus republican; cont by: columbus journal-republican [columbus, wi: 1939]) – mf#1009963 – us WHS [071]
Journal-republican – Columbus WI. 1968 nov 17-21, 1968 nov 28-1969 dec 4, 1969 dec 11-1970 dec 31, 1971 jan 7- feb 18 – 4r – 1 – (cont: columbus journal-republican [columbus, wi: 1939]; cont by: columbus journal-republican [columbus, wi: 1971]) – mf#1009992 – us WHS [071]
Journals / Alberta. Legislative Assembly – 1906-20 – (= ser Legislative Assembly) – 1r – 1 – (incl ind) – cn Library Assoc [971]
Journals : and sessional papers 1872-75 / British Columbia. Legislative Assembly – 1872-1903 – (= ser Legislative Assembly) – 5r – 1 – (incl ind) – cn Library Assoc [971]
Journals / Bain, Andrew Geddes – Cape Town, South Africa. 1949 – 1r – us UF Libraries [960]
Journals / Dickinson County. Kansas. Board of Commissioners – 1861-83 – 1 – us Kansas [978]
Journals / Ford County. Kansas. Board of Commissioners – 1873-1904 – 1 – us Kansas [978]
Journals / Heintzelman, Samuel – 1854-55 – 1 – us L of C Photodup [920]
Journals / Isham, William S – 1886-1895, Diary entries, copies of letters, and other documents relating to Isham's duties as a constable and businessman in Ohio; Winona, Minnesota; and Superior, Nebraska – 1 – us Kansas [920]
Journals / Manitoba. Legislative Council – 1871-76 – (= ser Legislative Council) – 1r – 1 – cn Library Assoc [971]
Journals / Menkel, Peter – 1875-1897 – 1 – $50.00 – us Presbyterian [920]
Journals / Prince Edward Island. House of Assembly – 1867-1900 – (= ser Legislative Assembly) – 11r – 1 – cn Library Assoc [971]
Journals / Prince Edward Island. Legislative Assembly – 1901-20 – (= ser Legislative Assembly) – 9r – 1 – cn Library Assoc [971]
Journals / Prince Edward Island. Legislative Council – 1867-93 – (= ser Legislative Council) – 2r – 1 – cn Library Assoc [971]
Journals / Saskatchewan. Legislative Assembly – 1906-20 – (= ser Legislative Assembly) – 9r – 1 – (incl ind) – cn Library Assoc [971]
Journals / Walker, William – undated, Journals of William Walker, the provisional governor of Nebraska Territory, edited by William E. Connelley – 1 – us Kansas [920]
Journals, 1875-97; estate, 1897-99 / Sewell, J F – 34 fiche – 9 – sa National [960]
Journals and appendices / Canada. Upper Canada. House of Assembly – 13 jan 1825-10 feb 1840. – 1 – (printed ind available) – cn Commonwealth Imaging [071]
Journals and correspondence / Green, James L – 1874-86 – 1r – 1 – mf#pmb38 – at Pacific Mss [880]
Journals and essays / Todd, John P – 1809-1830 – 1 – $50.00 – us Presbyterian [240]
Journals and letters / Rice, Luther – 1815-23.381p – 1 – us Southern Baptist [242]
Journals and letters of the rev henry martyn... / ed by Wilberforce, S – London, 1837 – 11mf – 9 – mf#HTM-117 – ne IDC [910]
Journals and minutes of the pennsylvania assembly, 1776-1790 / Pennsylvania. Assembly – 1975 – 3r – 1 – $390.00 – mf#S1864 – us Scholarly Res [978]
Journals and papers of chauncy maples... / Maples, C; ed by Maples, E – London, New York, Bombay, 1899 – 4mf – 9 – mf#HTM-111 – ne IDC [910]
Journals and sermons, 1748-1784 / Simpson, Archibald – [mf ed Charleston SC, 1981] [Spartanburg SC: Reprint Co [dist]] – 64mf – 9 – mf#51-155 – us South Carolina Historical [242]
Journals, brigantine hope / Ingraham, Joseph – 1790-92 – 1 – $18.00 – us L of C Photodup [920]
Journals in microform – 1,9 – (catalog avail. on request) – us Alper [073]
[Journals kept in china, 1874-1875] / Sinizininax, E – n.p, 1874-75 – 2v on 3mf – 9 – mf#HT-137 – ne IDC [915]
Journals, ledgers, memo book / Rice, Luther – 1810-26. 758p – 1 – us Southern Baptist [242]
Journals, letters, reports, language study / Roman Catholic Mission, New Hebrides – 1879-1949 – 1r – 1 – mf#pmb52 – at Pacific Mss [241]
Journals of brink and rhenius / Brink, Carel Frederik – Cape Town, South Africa. 1947 – 1r – us UF Libraries [960]
Journals of carl mauch / Mauch, Karl – Salisbury, Zimbabwe. 1969 – 1r – us UF Libraries [960]
Journals of general conventions of the protestant episcopal church in the united states, 1785-1835 / ed by Perry, William Stevens – Claremont, NH: Claremont Manufacturing Company, 1874 – 21mf – 9 – 0-524-07352-X – mf#1990-5389 – us ATLA [242]
Journals of gerontology series a : biological sciences and medical sciences – Washington. 1995+ (1,5,9) – ISSN: 1079-5006 – mf#21128 – us UMI ProQuest [574]
Journals of gerontology series b : psychological sciences and social sciences – Washington. 1995+ (1,5,9) – ISSN: 1079-5014 – mf#21129 – us UMI ProQuest [618]
The journals of lieutenant thomas a dorin : u s navy, 1826-1855 – (= ser Naval records coll of the office of naval records and library) – 1r – 1 – (with printed guide) – mf#M981 – us Nat Archives [355]
Journals of major robert rogers : containing an account of the several excursions he made under the generals who commanded upon the continent of north america, during the late war – London: J Millan, 1765 [mf ed 1988] – 3mf – 9 – mf#SEM105P886 – cn Bibl Nat [971]
The journals of popular music : new musical express and blues and soul – 36r coll – 1 – (titles also listed separately) – mf#C14R-11500 – us Primary [780]
Journals of several expeditions made in western australia : during the years 1829, 1830, 1831, and 1832; under the sanction of the governor, sir james stirling / ed by Cross, Joseph – London 1833 [mf ed Hildesheim 1995-98] – 1v on 2mf – 9 – €60.00 – ISBN-10: 3-487-26809-4 – ISBN-13: 978-3-487-26809-5 – gw Olms [919]
Journals of the assembly of jamaica, 1663-1826 / Jamaica. Assembly – 1972 – 7r – 1 – $910.00 – (with printed guide) – mf#S1843 – us Scholarly Res [972]
Journals of the continental congress / U.S. Continental Congress – v1-34. 1774-89 (all publ) – 167mf – 9 – $250.00 – (with ind) – mf#LLMC 78-080 – us LLMC [323]
Journals of the continental congress, 1774-1789 / U.S. Continental Congress – 1904-37. 1774-80 ONLY – 4r – 1 – $140.00 – Dist. us Scholarly Res – us L of C Photodup [324]
Journals of the democratic left – 1993-1999+ – (= ser Women's periodical and manuscripts coll) – 7r – 1 – £350.00 – mf#JDL – uk World [335]
Journals of the executive proceedings of the u.s. senate / U.S. Congress. Senate – Washington: GPO. v1-140. 1828-1998 – 2026mf – 9 – $3039.00 – (updates planned) – mf#LLMC 78-084 – us LLMC [342]
Journals of the house of burgesses of virginia, 1619-1776 / Virginia. House of Burgesses; ed by Kennedy, John Pendleton & McIlwaine, Henry Read – 1985 – 4r – 1 – $520.00 – (incl printed guide) – mf#S1856 – Virginia State Library – us Scholarly Res [978]

Journals of the labour movement in trade and industry : 20th century – 12r – 1 – £590.00 – (titles: the industrialist 1908-10, the socialist 1902-24, siemens shop stewards' committee journal 1933-38, aircraft shop stewards national council journal (the new propeller / metal worker) 1935-63) – mf#JLM – uk World [331]
Journals of the legislature of the state of yap – 1984-93 – r1-7 – 1 – (available for ref) – mf#pmb doc441 – at Pacific Mss [325]
Journals of the proceedings of the commissioners... / Williamsburg Township. South Carolina. Commissioners – 1788-1811 [mf ed 1981] – 1mf – 9 – mf#51-533 – us South Carolina Historical [350]
Journals of the ship "lloyd", 1767-1772 / Pocock, Nicholas – Greenwich: National Maritime Museum – (= ser BRRAM series) – 1r – 1 – £67 / $134 – (int by w e minchinton) – mf#r95962 – uk Microform Academic [910]
Journals of the u s house of representatives, 1789-1817 / U.S. House of Representatives – (= ser Records of the united states house of representatives) – 17r – 1 – (with printed guide) – mf#M1264 – us Nat Archives [324]
Journals of trip to liberia, 1868 / Pinney, John B – Also: letters pertaining to the liberian college. 1879. 352p – 1 – us Southern Baptist [242]
The journals of william hutchinson, 1768-1793 : from the liverpool record office and the athenaeum, liverpool – 1r – 1 – (with guide. int by philip woodworth) – mf#97584 – uk Microform Academic [920]
Journals...detailing their proceedings in the kingdom of shoa, and journeys in other parts of abyssinia, in the years 1839, 1840, 1841, and 1842... / Isenberg, C W & Krapf, J L – London, 1843 – 8mf – 9 – mf#HTM-90 – ne IDC [916]
Journal-standard – Freeport, IL. 1913-2000 (1) – mf#61328 – us UMI ProQuest [071]
Journal-telephone – Milton Junction, Milton WI. 1912 aug 1-1912 nov 28, 1912 dec 5-1914 sep 17, 1914 sep 24-1916 jul 13, 1916 jul 20-1918 apr 11, 1918 apr 18-1920 jan 29, 1920 feb 5-1922 dec 29, 1922 jan 5-1923 oct 25, 1923 nov 1-1925 aug 20, 1925 aug 27-1927 jul 7, 1927 jul 14-1927 aug 25 – 10r – 1 – (cont: milton journal; telephone [milton junction, wi: 1905]; cont by: milton junction telephone) – mf#1096901 – us WHS [071]
Journal-tribune / Union Co. Marysville – mar 1951-dec 1970 [daily] – 46r – 1 – mf#B12805-12862 – us Ohio Hist [071]
Journaux publies par les prisonniers de guerre allemands en france 1946-1948 / Germany – 11 reels – 1 – (der aufklaerer depot 135 – clermont-ferrand. ausblick depot 94 – angouleme. die auslese depot 164 – nimes. der baustein : der rundblick des depot 141 – saint-fons. bergewart devient der lagerspiegel depot 155 – chambery. die bruecke depot 131 – clermont-ferrand. champ de mars rundschau – la rochelle. depot echo depot 153. depot zeitung – valenciennes. drome-ardeche depot 147 – montelimar. echo depot 121 – saint-paul-d'eyjeaux (haute-vienne). das echo des lagers 301 – cherbourg. die gemeinschaft depot 223 – villeneuve-saint-georges. hautviller zeitung- dijon) – fr ACRPP [943]
Journaux publies par les prisonniers de guerre allemands en france 1946-1948, 2 – 11 bobines cont'd – 1 – (heute und morgen depot 172.- vernet d'ariege. in der fremde. depot 133.- brioude. kamerad (der) aus ambroise des depot pga 41. kleine welt. depot 124. – gueret. kubuck (der). depot 103.- hagueneau. l.i.d. lager informations dienst. depot 15.- bizerte. lager bote. depot 85.- besancon. lager (das) echo. depot 101.- mutzig. lagerbote (der). depot 125.- brantome. lagerbote (der). depot 132.- allier-troncais. lager post (die). depot 171.- camp des sables (haute-savoie). lager post (die). depot 201. – epinal. lager zeitung des lagers 91.- poitiers. lager zeitung des lagers 127.- chateauroux. lager zeitung des p.g. des depot 88. – lons-le-saunier. lager zeitung p.g. depot 158.- gap (hautes-alpes). lager zeitung.- neuland. lager zeitung. depot 82. zaunkoenig. – chalon-sur-saone. lager zeitung. depot 142.- bourg-en-bresse. lager zeitung. p.g. depot 102.- colmar. lager zeitung. depot 151. – marseille. lager zeitung. depot 163.- larzac (aveyron). lager zeitung. depot 105.- strasbourg. lager zeitung (die). depot 189.- bayonne. lager zeitung. depot 104.- mulhouse. lager zeitung 128. depot 402.- thoree. mistral. depot 154.- sorgues (vancluse). n.t.n. neueste tremoulier nachrichter.- tulle. nachricht (die).- grenoble. nachrichter (der). – rouen. nachschlag. lager post.- nevers. neue lager zeitung. depot 15. – bizerte. parole (die). depot 134.- cantal. parole (die). lagerzeitung des depot 221.- cormeilles. ring (der).- laon. riviera bote. depot 159. – nice. rundschau (die). depot. 224.- evreux. stacheldraht (der). depot 152.- aubagne. stacheldraht funk devient der wegweiser.- lens. umschau. depot 22.- beauvais. unser nachrichtenblatt.- belfort. unter uns.- metz. welt (die) und wir. depot 92.- vendee. zaungast. depot 114.- lorient. zeitspiegel (der). depot 183.- st-medard-en-jalles (gironde). zitadelle (die). depot 23.- amiens) – fr ACRPP [943]
Journeaux des sieges...espagne / Jones, John T – 1821 – 9 – sp Bibl Santa Ana [946]
La journee de la petite menagere / Valette, mme – Paris: D Delarue, n.d. – (= ser Les femmes [coll]) – 3mf – 9 – mf#10853 – fr Bibl Nationale [640]
La journee industrielle : quotidien de l'industrie, du commerce et de l'agriculture – Paris. 14 mars 1918-38, janv-juin 1940 – 1 – fr ACRPP [073]
La journee religieuse : ou instructions pratiques / Saint-Pierre, Antoine de – Paris. v1-2. 1869 – v1 5mf v2 4mf – 8 – €18.00 – ne Slangenburg [240]
A journey from aleppo to jerusalem, at easter, a d 1696 / Clayton, Robert – Edinburgh 1812 [mf ed Hildesheim 1995-98] – 1v on 2mf – 9 – €60.00 – 3-487-27678-X – gw Olms [915]
Journey from buenos ayres : through the provinces of cordova, tucuman, and salta, to potosi...and subsequently, to santiago de chili and coquimbounder taken on behalf of the chilian and peruvian mining association, in the years 1825-26 / Andrews, Joseph – London 1827 [mf ed Hildesheim 1995-98] – 2v on 4mf – 9 – €120.00 – 3-487-26924-4 – gw Olms [918]
A journey from india to england through persia, georgia, russia, poland and prussia, in the year 1817 / Johnson, J – London: Longman, Hurst, Rees, Orme, and Brown, 1818 – 5mf – 9 – mf#AR-2045 – ne IDC [910]
A journey from london to odessa : with notices of new russia, etc / Moore, John – Paris 1833 [mf ed Hildesheim 1995-98] – (= ser Fbc) – 2mf – 9 – €60.00 – 3-487-29001-4 – gw Olms [910]
A journey from madras through the countries of mysore, canara, and malabar : performed under the orders of the most noble the marquis wellesley, governor general of india, for the express purpose of investigating the state of agriculture, arts... / Buchanan, Francis – London 1807 [mf ed Hildesheim 1995-98] – 3v on 18mf – 9 – €180.00 – 3-487-27245-8 – gw Olms [915]
A journey from merut in india : to london through arabia, persia, armenia, georgia, russia, austria, switzerland, and france, during the years 1819 and 1820 / Lumsden, Thomas – London 1822 [mf ed Hildesheim 1995-98] – 1v on 2mf – 9 – €60.00 – 3-487-27422-1 – gw Olms [910]
A journey from merut in india, to london, through arabia, persia, armenia, georgia, russia, austria, switzerland and france during the years 1819 and 1820 / Lumsden, T – London, 1822 – 4mf – 9 – mf#AR-2028 – ne IDC [910]
Journey from moscow to constantinople, in the years 1817, 1818 / Macmichael, William – London 1819 [mf ed Hildesheim 1995-98] – 1v on 3mf – 9 – €90.00 – 3-487-27400-0 – gw Olms [910]
A journey from prince of wales's fort in hudson's bay, to the northern ocean / Hearne, S – London, 1795 – 18mf – 9 – mf#N-247 – ne IDC [917]
A journey in carniola, italy, and france : in the years 1817, 1818 containing remarks relating to language, geography, history, antiquities, natural history, science, painting... / Cadell, William A – Edinburgh 1820 [mf ed Hildesheim 1995-98] – (= ser Fbc) – 2v on 7mf – 9 – €140.00 – 3-487-27770-0 – gw Olms [914]
Journey in the caucasus, persia, and turkey in asia / Thielmann, Max Franz Guido, Freiherr von – London: J Murray, 1875 [mf ed 1999] – 2v on 1r – 1 – (trans by charles heneage; incl ind) – mf#28660 – us Harvard [915]
A journey into eastern tartary in 1682 / Verbiest, F – London, 1745-1747. v4 – 2mf – 9 – mf#A-271 – ne IDC [915]
A journey into various parts of europe : and a residence in them, during the years 1818, 1819, 1820, and 1821 with notes, historical and classical / Pennington, Thomas – London 1825 [mf ed Hildesheim 1995-98] – 2v on 9mf – 9 – €180.00 – 3-487-27730-1 – (with memoirs of the grand dukes of the house of medici; of the dynasties of the kings of naples; and of the dukes of milan) – gw Olms [914]
The journey of anthony gaubil, jesuit : from kanton to pe-king, in 1722 – London. v3. 1745-1747 – 1mf – 9 – mf#A-271 – ne IDC [915]
The journey of jean de fontaney : from pe-king to kyang-chew, in the province of shan si and thence to nan king, in 1688 – London, 1745-1747. v3 – 1mf – 9 – mf#A-271 – ne IDC [915]
The journey of joachim bouvet, jesuit : from pe-king to kanton when sent by the emperor kang-hi into europe, in 1693 – London, 1745-1747. v3 – 1mf – 9 – mf#A-271 – ne IDC [915]
The journey of william of rubruck to the eastern parts of the world, 1253-55 : as narrated by himself, with two accounts of the earlier journey of john of pian de carpine = Itinerarium / Ruysbroeck, Willem van – London: Printed for the Hakluyt Society, 1900 – (= ser Works Issued by the Hakluyt Society) – 1mf – 9 – 0-524-01532-5 – (incl bibl ref. in english) – mf#1990-0438 – us ATLA [915]
The journey of william of rubruck to the eastern parts of the world, 1253-55... : with two accounts of the earlier journey of john of pian de carpine – London: The Hakluyt Society, 1900 – 5mf – 9 – mf#AR-1966 – ne IDC [910]
A journey over land : from the gulf of honduras to the great south-sea performed by john cockburn, and five other englishmen, viz thomas rounce, richard banister, john holland, thomas robinson, and john ballman / Withington, Nicholas – London 1734 [mf ed Hildesheim 1995-98] – 1v on 4mf – 9 – €120.00 – 3-487-26757-8 – gw Olms [910]
A journey round the coast of kent : containing remarks on the principal objects worthy of notice throughout the whole of that interesting border, and the contiguous district; including penhurst, and tunbridge-wells / Fussell, L – London 1818 [mf ed Hildesheim 1995-98] – 1v on 2mf – 9 – €60.00 – 3-487-27917-7 – gw Olms [914]
Journey through asia minor, armenia, and koordistan, in the years 1813 and 1814 : with remarks on the marches of alexander, and retreat of the ten thousand / Kinneir, John – London 1818 [mf ed 1995-98] – 1v on 4mf – 9 – €120.00 – 3-487-27650-x – gw Olms [910]
A journey through norway, lapland, and part of sweden : with some remarks on the geology of the country, its climate and scenery...meteorological observations... / Everest, Robert – London 1829 [mf ed Hildesheim 1995-98] – (= ser Fbc) – 3mf – 9 – €90.00 – 3-487-28927-X – gw Olms [914]
A journey through spain in the years 1786 and 1787 : with particular attention to the agriculture, manufactures, commerce, population, taxes, and revenue of that country / Townsend, Joseph – London 1791 [mf ed Hildesheim 1995-98] – 3v on 9mf – 9 – €180.00 – 3-487-29831-7 – gw Olms [914]
Journey through the chinese empire / Huc, Evariste Regis – New York: Harper & Bros, 1855 [mf ed 1995] – (= ser Yale coll) – 2v – 1 – 0-524-09297-4 – mf#1995-0297 – us ATLA [915]
A journey to ashango-land : and further penetration into equatorial africa / Chaillu, P B du – London, 1867 – 6mf – 9 – mf#HT-39 – ne IDC [916]
The journey to be a therapist : personal experiences of ethics in training and therapy / Makena, Paul Tshwarelo – Uni of South Africa 2001 [mf ed Johannesburg 2001] – 2mf – 9 – (incl bibl ref) – mf#mfm14676 – sa Unisa [616]
A journey to canada / Gurney, Jane Tritton – S.l: s.n, 1887? – 1mf – 9 – mf#05133 – cn Canadiana [917]
Journey to emmaus – London, England. 18– – 1r – us UF Libraries [240]
Journey to manaos / Hanson, Earl Parker – New York, NY. 1938 – 1r – us UF Libraries [972]
A journey to marocco : in 1826 / Beauclerk, George R – London 1828 [mf ed Hildesheim 1995-98] – 1v on 3mf – 9 – €90.00 – 3-487-27357-8 – gw Olms [916]
A journey to mequinez : the residence of the present emperor of fez and morocco on the occasion of commodore stewart's embassy thither for the redemption of the british captives in the year 1721 / Windus, J – London: J Tonson, 1725 – (filmed with: blyden, e.w. christianity, islam and the negro race) – us CRL [956]
A journey to rome and naples : performed in 1817 giving an account of the present state of society in italy; and containing observations on the fine arts / Sass, Henry – London 1818 [mf ed Hildesheim 1995-98] – (= ser Fbc) – 3mf – 9 – €90.00 – 3-487-29296-3 – gw Olms [914]
A journey to switzerland : and pedestrian tours in that country including a sketch of its history, and of the manners and customs of its inhabitants / Agassiz, Lewis – London 1833 [mf ed Hildesheim 1995-98] – (= ser Fbc) – 2mf – 9 – €60.00 – 3-487-29349-8 – gw Olms [914]
Journey to the north of india, overland from england, through russia, persia, and affghaunistaun / Conolly, Arthur – London 1834 [mf ed Hildesheim 1995-98] – 2v on 6mf – 9 – €120.00 – 3-487-27464-7 – gw Olms [910]
A journey to two of the oases of upper egypt / Edmonstone, Archibald – London 1822 [mf ed Hildesheim 1995-98] – 1v on 2mf – 9 – €60.00 – 3-487-27347-0 – gw Olms [916]
Journey toward the sunlight / Walker, Stanley – New York, NY. 1947 – 1r – us UF Libraries [972]
A journey with an aids nurse : a narrative study / Thomas, Shireen – Pretoria: Vista University 2003 [mf ed 2003] – 2mf [ill] – 9 – (incl bibl ref) – mf#mfm15265 – sa Unisa [610]
Journey without return / Maufrais, Raymond – London, England. 1953 – 1r – us UF Libraries [972]
Journeyman / Building and Construction Trades Council of Alameda County – v15 n12 [1991 dec]-v18 n12 [1994 dec] – 1r – 1 – (cont: alameda county building and construction trades journeyman) – mf#2604696 – us WHS [690]
The journeyman, and artizans' london and provincial chronicle – London. -w. 12 Jun-4 Sep 1825. (14 ft) – 1 – uk British Libr Newspaper [330]
Journeymen : journal of the washington state men's movement / Journey-Men Collective [Seattle WA] – v1 n1-7 [1978 mar 15-1979] – 1r – 1 – mf#624659 – us WHS [334]
Journey-Men Collective [Seattle WA] *see* Journeymen
Journeymen Stone Cutters' Association of North America *see* Stone cutters' journal
Journeys among the gentle japs in the summer of 1895, with a special chapter on the religions of japan / Thomas, Joseph Llewelyn – London: S. Low, Marston, 1897. x,266p. fold. map – 1 – us Wisconsin U Libr [915]
Journeys in north china, manchuria, and eastern mongolia : with some account of corea / Williamson, A – London, 1870. 2v – 11mf – 9 – mf#HT-157 – ne IDC [915]
Journeys in persia and kurdistan / Bishop, [J F] – London: John Murray, 1891. 2v – 10mf – 9 – mf#AR-2114 – ne IDC [915]
Journeys in time / Niles, Blair – New York, NY. 1946 – 1r – us UF Libraries [972]
Journeys toward progress / Hirschman, Albaert O – Garden City, NY. 1965 – 1r – us UF Libraries [972]
Journi, Jean de *see*
– La dime de penitance
Jours hereux / Puget, Claude-Andre – Paris, France. 1948, c1938 – 1r – us UF Libraries [440]
Les "jours noirs" a la bourse de paris : (du 24 juillet au 7 decembre 1914) / Vidal, Emmanuel – Paris: A Picard, 1919 (mf ed 19–) – 47p – mf#Z-BTZE pv642 n21 – us Harvard [332]
Jours troubles, pages d'epopee africaine / Dufays, Felix – Ixelles, Belgium: R Weverbergh, 1928 – us CRL [960]
Jouslin De La Salle, Armand-Francois *see* Freres feroces
Jouslin De La Salle, M *see* Soldat en retraite
Joutel, Henri *see* Mr joutel's journal of his voyage to mexico
Jouvancy, Joseph de *see* Canadicae missionis relatio ab anno 1611 usque ad annum 1613
Jouve, Joseph *see* Forest of hermanstadt
Jouy, Etienne De *see*
– Mariage de m beaufils
– Prisonniere
– Sylla
– Vestale
Jouy, Victor J de *see*
– Les hermites en liberte
– Les hermites en prison
Joven cadete / Gaud Rodriguez, Santos – San Juan, Puerto Rico. 1951 – 1r – us UF Libraries [972]
Joven de la flecha de oro / Villaverde, Cirilo – Habana, Cuba. 1962 – 1r – us UF Libraries [972]
Joven literatura hispanoamericana / Ugarte, Manuel – Paris, France. 1919 – 1r – us UF Libraries [440]
El joven y el zapatero / Mendoza, Antonio – 1846 – 9 – sp Bibl Santa Ana [830]
La joven y el zapatero / Muscat, Eduardo y – 1846 – 9 – sp Bibl Santa Ana [830]
Jovenes oradores sagrados – Bogota, Colombia. 1936 – 1r – us UF Libraries [960]
Jover, Jose Luis *see* Paisajeh
Jovillos (coplas de estudiante) / Diego, Jose De – Barcelona, Spain. 1916 – 1r – us UF Libraries [972]
Jovinian *see* Iovinianus
Jowett, Benjamin *see*
– The epistles of saint paul to the thessalonians, galatians and romans
– The epistles of st paul to the thessalonians, galatians and romans
– The epistles of st paul to the thessalonians, galatians, romans
– The interpretation of scripture and other essays
– Scripture and truth
– Sermons on faith and doctrine
– Theological essays of the late benjamin jowett

Jowett, John Henry *see*
– The epistles of st peter
– From strength to strength
– The high calling
Joy, Charles Rhind *see*
– Africa of albert schweitzer
– Young people of mexico and central america
Joy in death / Leifchild, J – London, England. 18– – 1r – us UF Libraries [240]
The joy of bible study / Lees, Harrington Clare – London, New York: Longmans, Green 1909 [mf ed 1989] – (= ser Anglican church handbooks) – 1mf – 9 – 0-7905-2047-8 – (incl ind) – mf#1987-2047 – us ATLA [220]
Joy of faith / Massy, Dawson – London, England. 1867? – 1r – us UF Libraries [240]
Joy of jesus newsletter – 1989 fall – 1r – 1 – mf#4027505 – us WHS [071]
The joy of service / Miller, James Russell – New York: Thomas Y Crowell, c1898 – 1mf – 9 – 0-8370-7243-3 – mf#1986-1243 – us ATLA [240]
The joy of youth : [novel] / Phillpotts, Eden – Toronto: McClelland and Goodchild, c1913 – 4mf – 9 – 0-659-90447-0 – mf#9-90447 – cn Canadiana [830]
Joy turned into mourning / Collyer, William Bengo – Camberwell, England. 181-? – 1r – us UF Libraries [240]
Joyau, Auguste *see* Belain d'esnambuc
Joyce, Allen E *see* The atlanta black crackers
Joyce, E D *see* How half a million of the surplus revenue should be invested for the benefit of england and her colonies
Joyce, Gilbert Cunningham *see* The inspiration of prophecy
Joyce, James *see* Indispensable james joyce
Joyce, James Barclay *see* The sword and the keys
Joyce, James Wayland *see*
– Acts of the church, 1531-1885
– Handbook of the convocations, or, provincial synods of the church of england
– The sword and the keys
Joyce, Joseph Asbury *see*
– Adoption and legitimation of children
– A treatise on electric law, covering the law governing all electric corporations, uses and appliances.
– Treatise on the law governing nuisances.
Joyce, Lilian Elwyn *see* Central america
Joyce, Lilian Elwyn (Elliott) *see* Brazil today and tomorrow
Joyce, Patrick Weston *see* The story of ancient irish civilisation
Joyeuse entree de l'empereur maximilien i...gand, en 1508 : (description d'un livre perdu) / Volkaersbeke, P K de – Gand, Bruxelles, Leipzig, 1850 – 1mf – 9 – mf#O-1091 – ne IDC [700]
La joyeuse et magnifique entree de monseigneur francoys fils de france : et frere unicque du roy, par la grace de dieu duc de brabant, d'anjou, alencon, berri etc... – Anvers: Christophle Plantin, 1582 – 1mf – 9 – mf#O-1998 – ne IDC [090]
Joyful heatherby / Erskine, Payne – Toronto: McClelland & Goodchild, 1913 [mf ed 1995] – 6mf – 9 – 0-665-74180-4 – mf#74180 – cn Canadiana [830]
Joyner, A Barry *see* Descriptive and predictive discriminant analysis of the golf ability of college males
Joynt, William Lane *see* Suggestions for the amendment of the arterial drainage laws of ireland
Joyride : manifestations of the 1890s bicycle craze in toronto / Cossarin, Mark A & Morrow, Don – 1993 – 2mf – 9 – $8.00 – us Kinesology [790]
Jozef flawjusz dzieje wojny zydowskiej przeciwko rzymianom / Josephus, Flavius – 1906. 573p. ill – 1 – us Wisconsin U Libr [930]
Jozef pilsudski, jakim go znalem / Wasilewski, Leon – Warszawa: "Roj", 1935. 233p. 1 reel. 1247 – 1 – us Wisconsin U Libr [920]
J'parl' pour parler... : poesies / Narrache, Jean – Montreal: editions Bernard Valiquette: editions ACF, 1939 [mf ed 1974] – 1r – 5 – (ill by simone aubry beaulieu) – mf#SEM16P188 – cn Bibl Nat [810]
Jpcc *see* Journal of pastoral care and counseling (jpcc)
Jpen : journal of parenteral and enteral nutrition – Silver Spring. 1989+ (1,5,9) – ISSN: 0148-6071 – mf#18051 – us UMI ProQuest [613]
Jpl news – Montreal, Quebec. Jan/Feb 1982-Mar 1983 – 1 – us AJPC [071]
Jpma : the journal of the pakistan medical association / Pakistan Medical Association – Karachi. 1975-1991 (1) 1975-1988 (5) 1975-1988 (9) – ISSN: 0030-9982 – mf#9941 – us UMI ProQuest [610]
Jpms journal of political and military sociology – DeKalb. 1973-1987 [1]; 1973-1987 [5]; 1976-1987 [9] – (cont by: journal of political and military sociology) – ISSN: 0047-2697 – mf#7475 – us UMI ProQuest [320]
Jps bookmark / Jewish Publication Society of America – Philadelphia. 1954-1972 (1) – ISSN: 0276-881X – mf#7114 – us UMI ProQuest [939]
Jpt : journal of petroleum technology – Dallas. 1949+ (1) 1949+ (5) 1976+ (9) – ISSN: 0149-2136 – mf#532 – us UMI ProQuest [620]
Jsac grapevine / Joint Strategy and Action Committee [New York NY] – v1 n1-v21 n1 [1989 feb-1989 oct] – 1r – 1 – mf#1060156 – us WHS [360]
Jsk-mitteilungsblatt *see* Sozialistische warte
JSTOR [Organization] *see* Journal of southern history
Jta daily news bulletin – New York: Jewish Telegraphic Agency. v8 n170-v40 n248. jul 2 1941-1973 – (issues for jul 2 1941 filmed with news (jewish telegraphic agency (new york), jan-jun 3 1941; news bulletin (jewish telegraphic agency (new york), jun 4-8 1941; and jta news bulletin, jun 9-jul 1 1941) – us CRL [071]
Jta news bulletin – New York: Jewish Telegraphic Agency. v8 n150-169 (jun 9-jul 1 1941) – (issues for jun 9-jul 1 1941 filmed with news (jewish telegraphic agency (new york), jan-jun 3 1941; news bulletin (jewish telegraphic agency (new york), jun 4-8 1941; and jta daily news bulletin, jul 2-dec 1941) – us CRL [071]
J-Town Collective *see* New dawn
Ju chia che hsueh / Liang, Ch'i-ch'ao – Shang-hai: Chung-hua shu chu, Min kuo 30 [1941] – (= ser P-k&k period) – us CRL [180]
Ju ho chien she hsin chun / Liu, Chih – Ch'ung-ch'ing: Tu li ch'u pan she, Min kuo 28 [1939] – (= ser P-k&k period) – us CRL [355]
Ju hsiang ch'u pu / Lu, Yu-hsi – Shang-hai: Pei hsin shu chu, Min kuo 25 [1936] – (= ser P-k&k period) – us CRL [170]
Ju hua yeh-su hui shih lieh chuan / Pfister, Aloys – Ch'ang-sha: Shang wu yin shu kuan, Min kuo 27 [1938] – (= ser P-k&k period) – us CRL [920]
Ju jui chi / Shen, Ts'ung-wen – Shang-hai: Sheng huo shu tien, Min kuo 23 [1934] – (= ser P-k&k period) – us CRL [480]
Ju ta hsueh che hsu chih / Ho, Ch'ing-Ju – [China]: Chung-hua chih yeh chiao yu she, Min kuo 24 [1935] – (= ser P-k&k period) – us CRL [370]
Ju Thani *see*
– Puras bhnak khluan
– Qanuvacananukram khmaer bises
Ju tz'u huang chun / Fang, Chao – Han-k'ou: Ch'un li shu tien, 1938 – (= ser P-k&k period) – us CRL [951]
Juan bautista picornell y la conspiracion de gual / Lopez, Casto Fulgencio – Caracas, Venezuela. 1955 – 1r – us UF Libraries [972]
Juan bobo y la dama occidente / Marques, Rene – Mexico City?, Mexico. 1956 – 1r – us UF Libraries [972]
Juan carrio y font – Minorca, Spain. v620-623. 1754-1770 – 3r – us UF Libraries [324]
Juan, Ching-ch'ing *see* Hsing ko lei hsing hsueh kai kuan
Juan clemente zenea / Ateneo De La Habana Comite 'Pro-Zenea' – Habana, Cuba. 1919 – 1r – us UF Libraries [972]
Juan clemente zenea, poeta y martir / Carbonell, Jose Manuel – Habana, Cuba. 1929 – 1r – us UF Libraries [972]
Juan criollo / Loveira, Carlos – New York, NY. 1964 – 1r – us UF Libraries [972]
Juan criollo : novela / Loveira, Carlos – Habana, Cuba. 1927 – 1r – us UF Libraries [972]
Juan de borbony battemberg / Estrella Estralla, Jose Emilio – Madrid: Graf. Menor, 1966 – sp Bibl Santa Ana [920]
Juan de dios aranzazu / Henao Mejia, Gabriel – Bogota, Colombia. 1953 – 1r – us UF Libraries [972]
Juan de la Anunciacion, fray *see*
– Doctrina christiana muy cumplida
– Sermon. nica[n]motenehua tenonotzalmachtiliztli, ynic recaquicilos ynfancta tzinco ycenca
– Sermonario en lengua mexicana
Juan de la torre (de j.a. lavalle. informe) / Fernandez Duro, Cesareo – Madrid: Fortanet, 1885. B.R.A.H. vii/pp. 223-228 – 1 – sp Bibl Santa Ana [920]
Juan de los Angeles *see*
– Dialogos de la conquista...dios
– Obras misticas del anotadas y precedidas de una introduccion...por el p. fr. jaime sala. parte 1...
– Obras misticas...obra preparada por el p. fr. jaime sala y revisada por el p. fr. gregorio fuentes...
– El reino de dios
– Triumphos del amor de dios
Juan de los angeles... : en la inmaculada en la literatura franciscano-espanola / Uribe, Angel – Archivo Ibero Americano, 1955 – 1 – sp Bibl Santa Ana [240]
Juan de miralles / Portell Vila, Herminio – Habana, Cuba. 1947 – 1r – us UF Libraries [972]
Juan de padilla / Barrantes Moreno, Vicente – 1855-56. 2 tomas – 9 – sp Bibl Santa Ana [920]
Juan de san esteban o de saucedilla. juan toribio arroya en los custodios y provinciales de la provincia de san jose / Perez, Lorenzo – Archivo Ibero Americano, 1924 – 1 – sp Bibl Santa Ana [946]
Juan de valdes' commentary upon the gospel of st matthew : now for the first time translated from the spanish and never before published in english – lives of the twin brothers, juan and alfonso de valdes / Valdes, Juan de & Boehmer, Edward; ed by Betts, John Thomas – London, Truebner, 1882. Chicago: Dep of photodup, U of Chicago Lib (1r); Evanston: American Theol Lib Assoc, 1984 (1r) – 1 – 0-8370-1476-X – mf#1984-B014 – us ATLA [226]
Juan del Santisimo Sacramento, Fray *see* Viaje y peregrinacion de jerusalen
Juan donoso cortes e la nobilta / Maresca, Giovanni – Madrid: Hidalguia, 1965 – 1 – (ano 13 nov-dic 1965 n73 p729-734) – sp Bibl Santa Ana [920]
Juan fortuna novela / Moyano, Rafael – Madrid, Spain. 1950 – 1r – us UF Libraries [972]
Juan isidro jimenez grullon / Hernandez Franco, Tomas Rafael – Ciudad Trujillo, Dominican Republic. 1945 – 1r – us UF Libraries [972]
Juan isidro perez / Rodriguez Demorizi, Emilio – Ciudad Trujillo, Dominican Republic. 1944 – 1r – us UF Libraries [972]
Juan, Jorge *see* Observaciones astronomicas y phisicas
Juan jose de vertiz y salado, gobernador y virrey de buenos aires. ensayo basado en documentos ineditos del archivo general de indias / Torre Revello, Jose – Buenos Aires, 1932; Madrid: Razon y Fe, 1933 – 1 – sp Bibl Santa Ana [320]
Juan justiniano y arribas / Blanco Garcia, Francisco – Madrid: Saenz de Jubera, 1909 – sp Bibl Santa Ana [440]
Juan macias, inclita gloria de extremadura y de la iglesia / Fernandez Sanchez, Teodoro – Badajoz: Dip. Provincial, 1976. Sep. REE – 1 – sp Bibl Santa Ana [240]
Juan martin roco. senor de campofrio / Solar y Taboada, Antonio & Rujula, Jose de – Badajoz: A. Arqueros, 1928 – 1 – sp Bibl Santa Ana [920]
Juan mascaro y villalonga – Minorca, Spain. v340. 1757-1770 – 1r – us UF Libraries [324]
Juan, mientras la ciudad crecia / Perez, Carlos Federico – Ciudad Trujillo, Dominican Republic. 1960 – 1r – us UF Libraries [972]
Juan quinquin en pueblo mocho / Feijoo, Samuel – Santa Clara, Cuba. 1964 – 1r – us UF Libraries [972]
Juan rafael mora / Echeverria Loria, Arturo – San Jose, Costa Rica. 1964 – 1r – us UF Libraries [972]
Juan ramon jimenez en su obra / Diez Canedo, Enrique – Mexico: El Colegio de Mexico, 1944 – 1 – sp Bibl Santa Ana [920]
Juan ramon molina / Carias Reyes, Marcos – Tegucigalpa, Honduras. 1943 – 1r – us UF Libraries [972]
Juan riudavets – Minorca, Spain. v938. 1752-1771 – 1r – us UF Libraries [324]
Juan tremol – Minorca, Spain. v140-145. 1763-1769? – 3r – (gaps) – us UF Libraries [324]
Juan, Tu-ch'eng *see* Tsu chieh chih tu yu shang-hai kung kung tsu chieh
Juan vazquez de la parra, el extremeno al que amo como a un hijo san martin de porres / Munoz de San Pedro, Miguel – Badajoz: Imp. Diput. Prov. Sep. de la Revista de Estudios Extremenos – 1 – sp Bibl Santa Ana [920]
Juan vazquez en la catedral de plasencia / Gomez Guillen, Roman – Badajoz: Dip. Provincial, 1973. Sep. REE – 1 – sp Bibl Santa Ana [240]
Juan, Wu-ming *see*
– Hsien tai ming chia sui pi ts'ung hsuan
– Jih chi wen hsueh ts'ung hsuan
Juan Y Punal, Arturo De *see* Punaladas
Juana Ines De La Cruz, Sister *see* Obras escogidas
Juana, ou, deux devouements / Muret, Theodore Cesar – Paris, France. 1838 – 1r – us UF Libraries [440]
Juana Sardon, Amalio de la *see* El cerdo de tipo iberico en la provincia de badajoz
Juanini, J *see*
– Cartas escritas a los muy nobles doctores...se dize que el sal azidoy alcali...
– Nueva idea physica natural demostrativa, origen de las materias que mueven las cosas
[Juarez-] el fronterizio – MX. 1967-73 – 53r – 1 – $2650.00 – mf#R04214 – us Library Micro [079]
Juarez, Ernesto *see* Higiene y profilaxis en el medio rural
Juarez Fernandez, Bel *see* En las lomas de el purial
Juarez Sanchez-Rubio, Cipriano *see* Evolucion de la energia electrica en la provincia de badajoz
Juarez Toledo, Enrique *see* Cantamos por la herida
Juarez Y Aragon, J Fernando *see* Esta es guatemala
Juarros, Domingo *see* A statistical and commercial history of the kingdom of guatemala, in spanish america
Jubel-album / Swensson, Carl Aaron & Abrahamson, Laurentius Gustav – Chicago: National Pub Co, 1893 – 2mf – 9 – 0-524-02846-X – mf#1990-4467 – us ATLA [240]
Jubelfestpredigt : am 350. gedachtnisstage der augsburgischen confession den 25. juni 1880, in der dreieinigkeitskirche der evang.-luth. gemeinde zu st. louis, mo / Walther, Carl Ferdinand Walther – St Louis: Luth Concordia-Verlags, 1880 – 1mf – 9 – 0-524-06784-8 – mf#1991-2791 – us ATLA [240]
Jubelschrift zum neunzigsten geburtstag des dr l zunz / Steinschneider, Moritz et al – Berlin: Louis Gerschel, 1884 – 5mf – 9 – 0-8370-3809-X – mf#1985-1809 – us ATLA [270]
Jubenville, Colby B *see* Athletes' perceptions of coaching performance among ncaa division 3 and naia head football coaches in the state of mississippi
Jubera, A *see* Dechado y reformacion de todas las medicinas compuestas, renales...
Jubilaeumsschrift (heft 100) – Leipzig: Verein fuer Reformationsgeschichte 1910 [mf ed 1990] – (= ser Schriften des vereins fuer reformationsgeschichte 27/100) – 1mf – 9 – 0-7905-4734-1 – (incl bibl ref) – mf#1988-0734 – us ATLA [242]
Jubilaris emlekmu / Handler, Rudolf – Lugos, Romania. 1904 – 1r – us UF Libraries [939]
Jubile de calvin a geneve : juillet 1909 / Berguer, Henry et al – [Geneve?: Compagnie des pasteurs de Geneve, 1910?] – 1mf – 9 – 0-524-00248-7 – mf#1989-2948 – us ATLA [242]
Jubilee = Te tiupiri – Wanganui, NZ.1898-1900 – 3r – 1 – mf#43.14 – nz Nat Libr [079]
Jubilee – Washington. 1953-1968 (1) – mf#2427 – us UMI ProQuest [240]
Jubilee anniversary of the pastorate of rev. d.t. fiske, d.d : belleville congregational church, newburyport, mass., 1897 / Fiske, Daniel Taggart et al – [Newburyport, Mass: Belleville Congregational Church, 1897?] – 1mf – 9 – 0-524-08681-8 – mf#1993-3206 – us ATLA [242]
A jubilee essay on imperial confederation as affecting manitoba and the northwest / Attwood, Peter Harold – Winnipeg?: s.n, 1887 – 1mf – 9 – mf#30006 – cn Canadiana [320]
Jubilee lectures : a historical series / Fairbairn, Andrew Martin et al – London: Hodder & Stoughton, 1882 [mf ed 1993] – (= ser Congregational coll) – 2v on 2mf – 9 – 0-524-07563-8 – (incl bibl ref) – mf#1991-3183 – us ATLA [242]
The jubilee memorial of horton college, bradford, containing the sermon preached at the jubilee service, 2 aug 1854; also / Godwin, Benjamin – Historical and biographical sketchBY B. Evans – 1 – 5.00 – us Southern Baptist [242]
Jubilee memorial of the american bible society : being a review of its first fifty years' work / Ferris, Isaac – New York: American Bible Society, 1867 – 1mf – 9 – 0-8370-6042-7 – (incl appendixes) – mf#1986-0042 – us ATLA [220]
The jubilee memorial of the religious tract society : containing a record of its origin, proceedings and results, a.d. 1799 to a.d. 1849 / Jones, William – London: Religious Tract Society, 1850 – 2mf – 9 – 0-8370-6268-3 – (includes appendixes and index) – mf#1986-0268 – us ATLA [240]
Jubilee memorials, 1860-1910 / Fernando, J S A – Colombo: Industrial Home Press, 1910 – 9 – us CRL [240]
Jubilee news – Jubilee Theatre [Fort Worth TX] 1993 feb, 1994 feb-may, dec, 1995 jan, mar, jul, nov, 1996 mar, may, jul, 1997 may, jul, dec, 1998 jan/feb, may – 1r – 1 – (cont by: inside jubilee) – mf#3144847 – us WHS [790]
Jubilee newspaper – 1988 may-1991 jul/aug – 1r – 1 – mf#3974588 – us WHS [071]
Jubilee of "christ church", newbury ont : 1863-1913 / Armstrong, J A – [Newbury, Ont?: s.n, 1913?] – 1mf – 9 – 0-665-98237-2 – mf#98237 – cn Canadiana [242]
Jubilee of the bible, october 4, 1835 / East, John – Bath, England. 1835 – 1r – us UF Libraries [240]
The jubilee of the methodist new connexion : being a grateful memorial of the origin, government, and history of the denomination / Allin, Thomas et al – London: J Bakewell, 1848 – 2mf – 9 – 0-524-06941-7 – mf#1990-5305 – us ATLA [242]

Jubilee papers of the central china presbyterian mission, 1844-1894 : comprising historical sketches of the mission stations at ningpo, shanghai, hangchow, soochow and nanking, with a sketch of the presbyterian mission press – Shanghai: American Presbyterian Mission Press, 1895 [mf ed 1995] – (= ser Yale coll) – iii/116p – 1 – 0-524-09483-7 – (int signed by j c garritt) – mf#1995-0483 – us ATLA [242]

Jubilee philatelist : a monthly magazine devoted to the interests of stamp collectors – Smith's Falls, Ont: Jubilee Stamp and Pub Co, [1899-1900] – 9 – (merged with: the mount royal stamp news to become: the jubilee philatelist and mount royal stamp news) – mf#P04574 – cn Canadiana [760]

Jubilee philatelist and mount royal stamp news – Smith's Falls, Ont: Jubilee Stamp and Pub. Co, [1900] – 9 – (cont: the jubilee philatelist) – mf#P04573 – cn Canadiana [760]

The jubilee philatelist and mount royal stamp news *see* The philatelic advocate

The jubilee prize poem : a loyal ode for 1887, the jubilee year of her majesty queen victoria – [Toronto?]: British Assoc for the Diffusion of Common Sense, [1887?] – 1mf – 9 – 0-665-92378-3 – mf#92378 – cn Canadiana [810]

Jubilee programme : including a brief retrospect of her majesty's reign and other interesting information: demonstration...ottawa, june 30th and july 1st, 1887 – Ottawa?: Daily Citizen, 1887? – 1mf – 9 – mf#11583 – cn Canadiana [971]

The jubilee singers : and their campaign for twenty thousand dollars / Pike, Gustavus D – Boston: Lee & Shepard; New York: Lee, Shepard & Dillingham, 1873 [mf ed 1990] – 1mf – 9 – 0-7905-5551-4 – mf#1988-1551 – us ATLA [780]

Jubilee songs of the anglo-saxon race / Woodruff, John – Ottawa?: s.n, 1897? – 1mf – 9 – mf#26116 – cn Canadiana [780]

A jubilee souvenir, 1876-1926 / Lakshapatiya. Ceylon. St Matthias' Church – Deshiwala: Pearl Press, 1926 – us CRL [954]

The jubilee story of the china inland mission / Broomhall, Marshall – London: Morgan & Scott, 1915 – 1mf – 9 – 0-7905-4249-8 – (with portraits, ill and maps) – mf#1988-0249 – us ATLA [240]

Jubilee volume, 1517-1917 : reformation quadri-centennial / Kline, John Jacob et al; ed by Evangelical Lutheran Ministerium of Pennsylvania and Adjacent States. Norristown Conference (Pennsylvania) – [S.l]: Publ by the Conference, c1917 (Lebanon, Pa: Sowers Print Co) – 1mf – 9 – 0-524-03619-5 – mf#1990-4779 – us ATLA [240]

Jubilee volume...june 22-24, 1885 / Royal Statistical Society. Great Britain – London: E. Stanford, 1885. xv,372p. illus., map, charts, diagrs. Some of the papers in French – 1 – us Wisconsin U Libr [510]

The jubilee year : an oration / Alward, Silas – St John, NB?: Daily Telegraph, 1887 – 1mf – 9 – mf#06077 – cn Canadiana [941]

Jubilee...and memorial exercises / Acadia College. Halifax – Halifax, NS?: Holloway Bros, 1889 – 2mf – 9 – (incl ind) – mf#02385 – cn Canadiana [378]

Jubilejni sjezd : ceskobratrske cirkve evangelicke konany na pamet desateho vyoci jejiho sjednoceni a trvani republiky v praze, ve dnech 7.-9. prosince 1928 – Praze: Nakladem Synodniho vyboru Ceskobratrske Cirkve Evangelicke, 1929 – 1mf – 9 – 0-524-08111-5 – mf#1993-9017 – us ATLA [240]

Jubileum-nummer, 1910-1935 / Sin-Po – Batavia, 1935 (ill) – 6mf – 8 – mf#SE-1614 – ne IDC [959]

Juca Filho, Candido *see* Antonio jose

Juchereau-Duchesnay, Marguerite *see* Bio-bibliographie de monsieur jean vallerand

Juco review / National Junior College Athletic Association – Colorado Springs. 1989+ (1) – ISSN: 0047-2956 – mf#15192 – us UMI ProQuest [790]

Jud, L *see*
– Bewiligung vnd confirmation eines burgermeisters...
– Catechismus
– Catechismus, brevissima christianae religionis formula instituendae iuuentuti tigurinae...
– Des lydens jesu cristi gantze...historia
– Eiu [!] chr[i]stenlich widerfechtug
– Die gantze bibel
– Der kuertzer catechismus
– Nachvolgung christi unnd verschuhung aller ytelkeit dieser welt...
– Rechenschafft des glaubens, der dienst vnnd cerimonien der brueder in behmen vnd mehrern
– Vf entdeckung doctor erasmi vo roterdam der dckische arglisten eynes teutschen buechlins antwurt vn entschuldigung...

Jud, L] *see* Opera d h'i z'ii...partim quidem ab ipso latine conscripta...

[Jud, L] *see* Des hochgelerte erasmi von roterdam vn doctor luthers maynung vom nachtmal...

Jud suess : roman / Feuchtwanger, Lion – Muenchen: Drei Masken Verlag, 1925 – 1r – 1 – us Wisconsin U Libr [830]

Juda home news – Juda Wl. 1907 jul 26-1907 nov 14 – 1r – 1 – mf#927066 – us WHS [071]

Judaea and her rulers : from nebuchadnezzar to vespasian / Bramston, Mary – London: SPCK; New York: E & J B Young [1882?] [mf ed 1989] – (= ser The home library) – 2mf [ill] – 9 – 0-7905-0815-X – mf#1987-0815 – us ATLA [939]

Judaea capta / Charlotte Elizabeth – New York: Baker & Scribner, 1848 – 1mf – 9 – 0-524-04569-0 – mf#1992-0157 – us ATLA [930]

Judaea und die nachbarschaft im jahrhundert vor und nach der geburt christi / Quandt, Ludwig; ed by Dieckmann, R – Guetersloh: C Bertelsmann, 1873 – 1mf – 9 – 0-524-05417-7 – (incl bibl ref) – mf#1992-0427 – us ATLA [915]

Judah *see* Buch kusari des jehuda ha-levi

Judah and israel : or, the kingdom of the god of heaven (dan 2:14) as it is now and the kingdom of the son of david (dan 7:13, 14) as it will be / [Chamberlain, H L] – San Francisco: Bancroft, 1888, c1887 [mf ed 1991] – 1mf – 9 – 0-7905-8771-8 – mf#1989-1996 – us ATLA [240]

Judah, Ha-Levi *see*
– Buch al-chazari
– Mivhar shire yehudah ha-levi

Judah, Henry *see*
– Cadastre abrege de la partie de la seigneurie de bourchemin est...
– Cadastre abrege de la seigneurie de beauharnois...
– Cadastre abrege de la seigneurie de la nouvelle longueuil...
– Cadastre abrege de la seigneurie delery...
– Cadastre abrege du fief coteau st louis

Judaic studies : miqra'ot gedolot – 9 – (v1. torah: genesis. $12.00. v2. torah: exodus. $12.00. v3. torah: leviticus. $10.00. v4. torah: numbers. $10.00. v5. torah: deuteronomy. $10.00. v6. megilloth: song of songs, ruth, lamentations, ecclesiastes, esther. $10.00. v7 former prophets: joshua, judges, 1st & 2nd samuel, 1st & 2nd kings and writings, 1st and 2nd chronicles. $24.00. v8. latter prophets: isaiah, jeremiah. $12.00. v9. latter prophets: ezekiel, book of the twelve. $12.00. v10. writings: psalms, proverbs, job, daniel, ezra, nehemiah. $18.00) – us IRC [270]

Judaica : forschungen zur hellenistisch-juedischen geschichte und litteratur / Willrich, Hugo – Goettingen: Vandenhoeck & Ruprecht, 1900 – 1mf – 9 – 0-8370-5860-0 – (incl ind of names) – mf#1985-3860 – us ATLA [939]

Judaica book news – New York. 1969-1981 (1) 1974-1981 (5) 1974-1981 (9) – ISSN: 0022-5754 – mf#10112 – us UMI ProQuest [939]

Judaica librarianship – New York, NY. Fall 1983-Spring 1985 – 1 – us AJPC [071]

Judaism / Abrahams, Israel – London: Archibald Constable, 1907 [mf ed 1985] – (= ser Religions ancient and modern) – 1mf – 9 – 0-8370-2033-6 – (incl bibl) – mf#1985-0033 – us ATLA [270]

Judaism – New York. 1972+ [1]; 1952+ [5]; 1974+ [9] – ISSN: 0022-5762 – mf#6501 – us UMI ProQuest [270]

Judaism and its history : in two parts = Das judenthum und seine geschichte / Geiger, Abraham – New York: Bloch, c1911 [mf ed 1985] – 1mf – 9 – 0-8370-3243-1 – (english trans fr german by charles newburgh. incl app on renan and strauss) – mf#1985-1243 – us ATLA [939]

Judaism and st paul : two essays / Montefiore, C G – London: Max Goschen, 1914 – 1mf – 9 – 0-7905-2185-7 – mf#1987-2185 – us ATLA [270]

Judaism as a civilization / Kaplan, Mordecai Manahem – New York, NY. 1934 – 1r – us UF Libraries [939]

Judaism as creed and life / Joseph, Morris – London: Macmillan, 1903 – 2mf – 9 – 0-524-03528-8 – (incl bibl ref) – mf#1990-3233 – us ATLA [270]

Judaism at rome : b.c. 76 to a.d. 140 / Huidekoper, Frederic – New York: James Miller, 1877. Beltsville, Md: NCR Corp, 1978 (7mf); Evanston: American Theol Lib Assoc, 1984 (7mf) – 9 – 0-8370-0939-1 – (incl bibl ref and ind) – mf#1984-4306 – us ATLA [270]

Judaism at the world's parliament of religions : comprising papers on judaism read at the parliament, at the jewish denominational congress, and at the jewish presentation – Cincinnati:Publ by the Union of American Hebrew Congregations – 1mf – 9 – 0-8370-5599-7 – (incl ind) – mf#1985-3599 – us ATLA [270]

Judaism in transition / Kaplan, Mordecai Menahem – New York, NY. 1936 – 1r – us UF Libraries [025]

Le judaisme comme race et comme religion : conference faite au cercle saint-simon, le 27 janvier 1883 / Renan, Ernest – [2e ed] Paris: C Levy, 1883 – 1mf – 9 – 0-524-05056-2 – mf#1992-0309 – us ATLA [270]

Le judaisme et le christianisme : identite originelle et separation graduelle / Renan, Ernest – Paris: C. Levy, 1883 – 1mf – 9 – 0-7905-3466-5 – mf#1987-3466 – us ATLA [270]

Judaismo visto con ojos argentinos / Soiza Reilly, Juan Jose De – Buenos Aires, Argentina. 1938 – 1r – us UF Libraries [939]

El judaismo y la cristiandad / Poncins, Leon de – Mexico: [s.n], 1965 (mf ed 1994) – 1r – 1 – (incl bibl ref) – mf#ZZ-34266 – us NY Public [230]

Judaistic christianity : a course of lectures / Hort, Fenton John Antony – Cambridge; New York: Macmillan, 1894 – 1mf – 9 – 0-8370-9791-6 – (incl ind of biblical citations) – mf#1986-3791 – us ATLA [240]

Judas : roman / Strauss und Torney, Lulu von – Stuttgart: Deutsche Verlags-Anstalt, 1922 – 1r – 1 – us Wisconsin U Libr [830]

O judas : jornal pagodista – Rio de Janeiro, RJ: Typ de Teixeira e Cia, 07 abr 1849 – (= ser Ps 19) – mf#P14,02,33 n02 – bl Biblioteca [870]

O judas – Rio de Janeiro, RJ: Typ do Diario, 06 abr 1833 – (= ser Ps 19) – mf#P17,01,100 – bl Biblioteca [320]

Judas, A-C *see* Etude demonstrative de la langue phenicienne et de la langue libyque

Judas atrevido : arranca entranhas – Rio de Janeiro, RJ. 31 mar 1877; 27 mar 1880 – (= ser Ps 19) – mf#P05,04,183 – bl Biblioteca [079]

Judas corsario – Rio de Janeiro, RJ. 16 abr 1881 – (= ser Ps 19) – mf#P05,04,51 – bl Biblioteca [870]

Les judas de la republique – Paris: E Houel, 1848 – us CRL [944]

Judas der erz-schelm fuer ehrliche leut : oder, eigentlicher entwurf und lebens-beschreibung des iscariotischen boeswicht... / Abraham a Sancta Clara – Lindau: J T Stettner, 1872 [mf ed 1993] – 7v on 2r – 1 – mf#8449 – us Wisconsin U Libr [830]

O judas fere fogo : espalha brazas – Rio de Janeiro, RJ. 31 mar 1877 – (= ser Ps 19) – mf#P05,04,184 – bl Biblioteca [870]

Judas iscariote tarte, introduction : la carriere de m tarte d'apres differents auteurs – [Montreal?: s.n.], 1903 – 4mf – 9 – 0-665-65648-3 – mf#65648 – cn Canadiana [971]

Judas maccabaeus and the jewish war of independence / Conder, C R – New ed. London: publ for the Committee of the Palestine Exploration Fund by A P Watt, 1894 – 1mf – 9 – 0-7905-1583-0 – (incl ind) – mf#1987-1583 – us ATLA [939]

O judas malcriado : e o seu testamento em prosa e verso e verso e prosa – Rio de Janeiro, RJ. 17 mar 1880; 16 abr 1881 – (= ser Ps 19) – mf#P05,04,185 – bl Biblioteca [079]

Juda's verhaeltniss zu assyrien in jesaja's zeit : nach keilinschriften und jesajanischen prophetieen / Hildebrandt, August – Marburg: Joh Aug Koch, 1874 – 1mf – 9 – 0-7905-0191-0 – (incl bibl ref) – mf#1987-0191 – us ATLA [221]

Der judasbrief : seine echtheit, abfassungszeit und leser / Maier, Friedrich – Freiburg i B, St Louis MO: Herder 1906 [mf ed 1989] – (= ser Biblische studien 11/1-2) – 1mf – 9 – 0-7905-2171-7 – (incl ind) – mf#1987-2171 – us ATLA [227]

Judd, Charles Hubbard *see* Laboratory equipment for psychological experiments

Judd family diaries – 1861-64 – 1 – us Kansas [978]

Judd, Orrin Bishop *see* Baptism in plain english

Judd, Sylvester *see* The birthright church

Der jude – Berlin. v. 1-10 no. 1. Apr 1916-Mar 1928 – 1 – us NY Public [939]

Der jude – Breslau (WrocLaw PL), 1777 n1-27 – 1 – gw Misc Inst [939]

Der jude – (Hamburg-) Altona DE, 1832-33, 1835 – 1r – 1 – gw Misc Inst [939]

Der jude : ein journal fuer gewissens-freiheit – Altona: Gabriel Riesser. v1. 1835 [complete] – (= ser German-jewish periodicals...1768-1945, pt 2) – 1r – 1 – $125.00 – mf#B165 – us UPA [270]

Der jude – Leipzig DE, 1768-72 – 1r – 1 – gw Misc Inst [939]

Der jude – New York, NY. 1895 – 1 – us AJPC [071]

Der jude : periodische blaetter fuer religion und gewissens-freiheit – Altona: Gabriel Riesser. v1-2. 1832, 1833 [complete] – (= ser German-jewish periodicals...1768-1945, pt 2) – 1r – 1 – $125.00 – mf#B164 – us UPA [270]

Der jude : eine wochenschrift – Leipzig: Gottfried Selig. v1-9. 1768-72 [complete] – (= ser German-jewish periodicals...1768-1945, pt 1) – 1r – 1 – $125.00 – mf#B482 – us UPA [270]

Der jude : eine wochenschrift / ed by Seelig, Gottfried – Leipzig 1768-72 [mf ed Hildesheim 1992] – (= ser Bibliothek des deutschen judentums) – 9v on 50mf – 9 – diazo €224.00 silver €258.00 – gw Olms [270]

Der jude : eine wochenschrift / ed by Selig, Gottfried – Leipzig 1768-72 – (= ser Dz. abt theologie) – 9v on 50mf – 9 – €500.00 – mf#k/n2084 – gw Olms [270]

Der jude an der ostgrenze / Seifert, Hermann Erich – Berlin: Zentralverlag der NSDAP, F Eher, 1942 (mf ed 1994) – (= ser Schriftenreihe der nsdap. gruppe 7, osten europas) – 1r – 1 – mf#ZZ-34501 – us NY Public [939]

Jude in den deutschen dichtungen des 15, 16 und 17 jahrhundertes / Frankl, Oskar – Mahrisch-Ostrau, Czechoslovakia. 1905 – 1r – us UF Libraries [939]

Jude spricht fur deutschland / Bar Eljokum, Schelomo – Frankfurt am Main, Germany. 1949 – 1r – us UF Libraries [939]

Judea from cyrus to titus, 537 b.c.-70 a.d / Latimer, Elizabeth Wormeley – 2nd ed. Chicago: McClurg, 1900 – 1mf – 9 – 0-524-05617-X – (incl bibl ref) – mf#1992-0472 – us ATLA [930]

Judelowitz, Morduch *see* Mahoza

Juden berlins / Wyking, A – Leipzig, Germany. 1891 – 1r – us UF Libraries [939]

Juden der turkei / Trietsch, Davis – Leipzig, Germany. 1915 – 1r – us UF Libraries [939]

Die juden in arabien zur zeit mohammeds / Leszynsky, Rudolf – Berlin: Mayer & Mueller, 1910 – 1mf – 9 – 0-7905-2015-X – (incl bibl ref) – mf#1987-2015 – us ATLA [930]

Juden in new york / Radt, Jenny – Berlin, Germany. 1937 – 1r – us UF Libraries [939]

Juden und ghetto in der deutschen literatur : bis zum ausgang des weltkrieges / Stoffers, Wilhelm – Graz: H Stiasnys Soehne, 1939 [mf ed 1993] – 800p – 1 – (incl bibl ref and ind) – mf#8146 – us Wisconsin U Libr [430]

Juden und samaritaner : die grundlegende scheidung von judentum und heidentum: eine kritische studie zum buche haggai und zur juedischen geschichte im ersten nachexilischen jahrhundert / Rothstein, Johann Wilhelm – Leipzig: J C Hinrichs, 1908 – (= ser Beitraege zur wissenschaft vom alten testament) – 1mf – 9 – 0-7905-3280-8 – mf#1987-3280 – us ATLA [221]

Die juden von barnow / Franzos, Karl Emil – 6e. Auf. Berlin. 1899 – 1 – us CRL [800]

Die juden von barnow : geschichten / Franzos, Karl Emil – 5., stark verm Aufl. Berlin: Concordia Deutsche Verlags-Anstalt, 1894 (mf ed 1990) – 1 – (filmed with: for the right) – us Wisconsin U Libr [830]

Die judenbuche / Droste-Huelshoff, Annette von; ed by Huge, Walter – Stuttgart: Reclam, 1979 – 1r – 1 – (incl bibl ref) – us Wisconsin U Libr [430]

Judenburg, Grundacker von *see* Gundackers von judenburg christi hort

Das judenchristentum im ersten und zweiten jahrhundert / Hoennicke, Gustav – Berlin: Trowitzsch, 1908 – 1mf – 9 – 0-524-02643-2 – (incl bibl ref) – mf#1990-0667 – us ATLA [230]

Das judenchristentum in den pseudoklementinen / Strecker, G – Berlin, 1958 – (= ser Tugal 5-70) – 6mf – 9 – €14.00 – ne Slangenburg [240]

Die "judenfrage" : schriften zur begruendung des modernen antisemitismus 1780 bis 1918 = The "jewish question" / ed by Benz, Wolfgang – [mf ed 2002-03] – 369mf (1:24) – 9 – diazo €3890.00 (silver €4700.00 isbn: 978-3-598-35041-2) – ISBN-10: 3-598-35040-6 – ISBN-13: 978-3-598-35040-5 – (incl guide with int by wolfgang benz) – gw Saur [939]

Die judenfrage : als racen-, sitten-, und culturfrage: mit einer weltgeschichtlichen antwort / Duehring, Eugen Karl – 2. verb. Aufl. Karlsruhe: H Reuther, 1881 – 1mf – 9 – 0-8370-2073-5 – mf#1985-0073 – us ATLA [939]

Das judengrab / aus bimbos seelenwanderungen : zwei erzaehlungen / Huch, Ricarda Octavia – Leipzig: Insel-Verlag [1916?] [mf ed 1990] – 1r – 1 – (filmed with: einer baut einen dom / carl maria holzapfel) – mf#2733p – us Wisconsin U Libr [830]

Judenproblem / Breuer, Isaac – Frankfurt am Main, Germany. 1922 – 1r – us UF Libraries [939]

Der judentaufe : das christentum im lichte der tatsachen / Kern, Karl Peter – Stuttgart: Durch-Verlag F Buehler, 1937 (mf ed 1995) – (= ser Durchbruch-Schriftenreihe) – 1r – 1 – (incl bibl ref) – mf#ZZ-34783 – us NY Public [939]

Das judenthum : in seinen grundzuegen und nach seinen geschichtlichen grundlagen / Guedemann, Moritz – 2. Aufl. Wien: R. Loewit, 1902 – 1mf – 9 – 0-8370-3419-1 – mf#1985-1419 – us ATLA [939]

Judenthum and judenchristenthum : eine nachlese zu der "ketzergeschichte des urchristenthums" / Hilgenfeld, Adolf – Leipzig: Fues, 1886 – 1mf – 9 – 0-7905-3260-3 – mf#1987-3260 – us ATLA [270]

Das judenthum in der vorchristlichen griechischen welt : ein beitrag zur entstehungsgeschichte des christenthums / Friedlaender, Moritz – Wien: M Breitenstein, 1897 – 1mf – 9 – 0-7905-3437-1 – (incl bibl ref) – mf#1987-3437 – us ATLA [270]

Das judenthum in gegenwart und zukunft / Hartmann, Eduard von – 2. durchgesehene aufl. Leipzig: Wilhelm Friedrich, 1885 – 1mf – 9 – 0-8370-3507-4 – (incl bibl ref) – mf#1985-1507 – us ATLA [939]

Judenthum und christenthum im zeitalter der apokryphischen und neutestamentlichen literatur / Holtzmann, Heinrich Julius – Leipzig: Wilhelm Engelmann, 1867 – 1mf – 9 – 0-7905-2412-0 – mf#1987-2412 – us ATLA [230]

Das judenthum und die christliche verkuendigung in den evangelien : ein beitrag zur grundlegung der biblischen theologie und geschichte / Schnedermann, Georg – Leipzig: JC Hinrichs, 1884 – 1mf – 9 – 0-524-05697-8 – mf#1992-0547 – us ATLA [220]

Das judenthum und richard wagner / Friedemann, Edmond – Berlin: Adolf 1869 – 1mf – 9 – mf#wa-30 – ne IDC [320]

Das judenthum und seine geschichte *see* Judaism and its history

Das judentum : von jesus bis zur gegenwart / Fiebig, Paul – Tuebingen: J C B Mohr 1916 [mf ed 1992] – (= ser Religionsgeschichtliche volksbuecher fuer die deutsche christliche gegenwart 2/21-22) – 1mf – 9 – 0-524-04334-5 – (incl bibl ref) – mf#1990-3318 – us ATLA [270]

Judentum als landschaftskundlich-ethnologisches problem / Passarge, Siegfried – Muenchen, Germany. 1929 – 1r – us UF Libraries [939]

Judentum, christentum, germanentum : adventspredigten / Faulhaber, Michael Von – Muenchen, Germany. 1934 – 1r – us UF Libraries [025]

Judentum, christentum, germanentum / Faulhaber, Michael Von – Muenchen, Germany. 1934 – 1r – us UF Libraries [939]

Judentum und christentum / Dienemann, Max – Frankfurt am Main, Germany. 1919 – 1r – us UF Libraries [939]

Judentum und christentum / Somerville, Alexander Neil; ed by Delitzsch, Franz – Erlangen: A Deichert, 1882 – 1mf – 9 – 0-524-03362-5 – mf#1990-0943 – us ATLA [230]

Das judentum und das wesen des christentums : vergleichende studien / Eschelbacher, Joseph – Berlin: Poppelauer, 1905 – 1mf – 9 – 0-8370-3072-2 – (incl ind) – mf#1985-1072 – us ATLA [230]

Judentum und entwicklungslehre / Delitzsch, Friedrich – Berlin, Germany. 1903 – 1r – us UF Libraries [939]

Judentum und judenchristentum im justins dialog mit trypho / Harnack, Adolf von – Leipzig, 1913 – (= ser Tugal 3-39/1b) – 1mf – 9 – €3.00 – ne Slangenburg [240]

Judentum und judenchristentum in justins dialog mit trypho *see* Ist die rede des paulus in athen ein urspruenglicher bestandteil der apostelgeschichte? / judentum und judenchristentum in justins dialog mit trypho

Jude's question discussed / Kingsbury, William – London, England. 1809 – 1r – us UF Libraries [240]

Judex materarum...insingulis / Lopez de Tovar, Gregorio – 1611 – 9 – sp Bibl Santa Ana [946]

Judge / Berea – London, England. 18– – 1r – us UF Libraries [240]

Judge – New York. 1881-1939 (1) – mf#4772 – us UMI ProQuest [340]

Judge advocate general of the army opinions / U.S. Army. Judge Advocate General – Washington: GPO. v1-3. 1917-19 (all publ) – 30mf – 9 – $45.00 – mf#LLMC 84-227 – us LLMC [355]

The judge advocate journal – v1-v2 n3. jun 1944-fall/winter 1945; bulletins 1-48 1948-76 (all publ) – (= ser Military law coll) – 9 – $81.00 – (no issues between 1945-bulletin n1 1948) – mf#LLMC 84-239 – us LLMC [343]

Judge and jury : a popular explanation for the leading topics in the law of the land / Abbott, Benjamin Vaughan – New York: Harper & Bros, 1880 – 5mf – 9 – $7.50 – mf#LLMC 92-102 – us LLMC [360]

Judge burnham's daughters – Boston: D. Lothrop Co., 1888. 339p – 1 – us Wisconsin U Libr [830]

Judge, Charles Joseph *see* An american missionary

Judge, Hugh *see* Memoirs and journal of hugh judge

Judge nothing before the time / Macleod, John – Edinburgh, Scotland. 1894 – 1r – us UF Libraries [240]

Judge or judas? / Jog, Narayan Gopal – Bombay: Thacker & Co, 1945 – (= ser Samp: indian books) – us CRL [954]

Judge, William Quan *see*
- Echoes from the orient
- The yoga aphorisms of patanjali

Judgement of very weak sensory stimuli / Brown, Warner – Berkeley, CA. 1914 – 1r – us UF Libraries [025]

Judgements : together with privy council judgements relating to fiji cases / Fiji Court of Appeal – 1949-1986 – 12r – 1 – mf#pmb1137 – at Pacific Mss [347]

Judges journal (aba) – v1-37. 1962-98 – 9 – $337.00 set – (v1-10 n1 1962-71 as trial judges journal) – ISSN: 0047-2972 – mf#116091 – us Hein [340]

Judges of the united states : published under the auspices of the bicentennial committee of the judicial conference of the united states – 1st ed 1978. Washington: GPO, 1980 – 6mf – 9 – $9.00 – mf#LLMC 95-013A – us LLMC [340]

Judges of the united states : published under the auspices of the bicentennial committee of the judicial conference of the united states – 2nd ed 1983. Washington: GPO, 1983 – 8mf – 9 – $12.00 – mf#LLMC 95-013B – us LLMC [340]

The judge's role in the settlement of civil suits / Lacey, Frederick B – Washington, 1977 – 1mf – 9 – $1.50 – mf#LLMC 95-804 – us LLMC [340]

Judgeship creation in the federal courts : options for reform / Baar, Carl – Washington: FJC, Feb 1981 – 1mf – 9 – $1.50 – mf#LLMC 95-310 – us LLMC [347]

Judgment / Gray, Robert – London: Bell & Daldy, 1864 – 1mf – 9 – 0-7905-6470-X – mf#1988-2470 – us ATLA [240]

Judgment de conseil souverain / Canada. Quebec – v1-4. 1663-1707 (all publ) – 49mf – 9 – $73.00 – (cont by: judgment de conseil superieur) – mf#LLMC 81-069 – us LLMC [340]

Judgment de conseil superieur / Quebec. Canada – v5-6. 1705-16 – 26mf – 9 – $39.00 – (cont: judgment de conseil souverain) – mf#LLMC 81-070 – us LLMC [340]

Judgment delivered by the right hon sir robert phillimore / Phillimore, Robert – London, England. 1870 – 1r – us UF Libraries [240]

Judgment delivered by the right hon sir robert phillimore, dcl / Phillimore, Robert – London, England. 1868 – 1r – us UF Libraries [240]

The judgment of a catholicke englishman living in banishment for his religion / Parsons, Robert – 1608 – 9 – us Scholars Facs [241]

The judgment of the right hon. stephen lushington, d.c.l... : delivered in the consistory court of the bishop of london in the cases of westerton against liddell (clerk) and horne and others, and beal against liddell (clerk) and parke and evans... / ed by Bayford, Augustus Frederick – London: Butterworths, [1856?] – 1mf – 9 – 0-524-05874-1 – mf#1990-5168 – us ATLA [240]

The judgment period preparatory to the establishment of the kingdom of heaven : comprising twelve chapters on the apocalypse / Campbell, David; ed by Chase, Zenas B – Bangor, Me: ZB Chase, c1886 – 1mf – 9 – 0-524-06040-1 – mf#1992-0753 – us ATLA [220]

Judgment records of the u.s. circuit court for the southern district of new york, 1794-1840 / U.S. Circuit and District Courts – (= ser Records of district courts of the united states) – 8r – 1 – (with printed guide) – mf#M882 – us Nat Archives [347]

Judgment records of the u.s. district court for the southern district of new york, 1795-1840 / U.S. District Court – (= ser Records of district courts of the united states) – 16r – 1 – (with printed guide) – mf#M934 – us Nat Archives [347]

The judgments of god upon the nations : pius 9th the last of the popes – New York: Edward H Fletcher, 1855, c1854 [mf ed 1986] – 1mf – 9 – 0-8370-8145-9 – (incl bibl ref) – mf#1986-2145 – us ATLA [241]

Judicature – Chicago. 1917+ (1) 1971+ (5) 1977+ (9) – ISSN: 0022-5800 – mf#179 – us UMI ProQuest [340]

Judicial administration in ethiopia : a reform oriented analysis / Tafara, Worku – 1972 – us CRL [340]

Judicial centennial banquet given at the lenox lyceum, new york, february 4th, 1890 / New York State Bar Association – New York: American Bank Note Co., 1890 29p. LL-373 – 1 – us L of C Photodup [340]

The judicial chronicle, being a list of the judges of the courts of common law and chancery in england and america and of the contemporary reports / Gibbs, George – Cambridge Mass.: Munroe, 1834. 55p. LL-2258 – 1 – us L of C Photodup [347]

The judicial code; being the judiciary act of the congress of the united states, approved march 3, a.d. 1911 / U.S. Laws, Statutes, etc – Chicago, Callaghan, 1911. 254 p. LL-1462 – 1 – us L of C Photodup [348]

The judicial conference and its committee on court administration / Hunter, Elmo B – Washington: FJC, 1986 – 1mf – 9 – $1.50 – mf#LLMC 95-397 – us LLMC [347]

The judicial conference of the u.s., reports and proceedings – the administrative office of the u.s. courts, annual reports – 1940-90 – 254mf – 9 – $381.00 – (lacking: 1943-44 for the judicial conference. 1942 report of the asdministrative office. 1984-88 for both) – mf#LLMC 80-411 – us LLMC [347]

The judicial dictionary of words and phrases judicially interpreted / Stroud, Frederick – 1v. 1835-1912. London: Sweet & Maxwell, 1890 – 11mf – 9 – $16.50 – mf#LLMC 95-448 – us LLMC [340]

Judicial discipline and removal in the u.s. / Wheeler, Russell R & Levin, Leo – Washington: GPO, 1987 – 1mf – 9 – $1.50 – mf#LLMC 95-304 – us LLMC [340]

Judicial dramas : or the romance of french criminal law / Spicer, Henry – London: Tinsley Bros, 1872 – 5mf – 9 – $7.50 – mf#LLMC 92-133 – us LLMC [345]

The judicial interpretation of administrative justice : with specific reference to roman v williams 1997(2) sacr 754(c) / Nemakwarani, Lamson Nditsheni – Uni of South Africa 2000 [mf ed Johannesburg 2000] – 1mf – 9 – (incl bibl ref) – mf#mfm14828 – sa Unisa [342]

Judicial laymens association bulletin – v1-2. 1937, 1941 (all publ) – 1mf – 9 – $1.50 – mf#LLMC 84-508 – us LLMC [340]

The judicial power of the united states / Harris, Robert Jennings – University, LA: Louisiana State University Press, 1940. 238p. LL-217 – 1 – us L of C Photodup [340]

The judicial power of the united states in some of its aspects. / Matthews, Stanley – New Haven, Conn.: Hoggson & Robinson, 1888. 39p. LL-190 – 1 – us L of C Photodup [340]

Judicial reforms proposed by the commission for the codification of the statutes. first report. / Quebec. (Province). Commission to Consolidate the General Statutes – Quebec, 1882. 262p. LL-1663 – 1 – us L of C Photodup [348]

Judicial regulation of attorneys' fees : beginning the process at pretrial / Willging, Thomas E – Washington: GPO, 1985 – 1mf – 9 – $1.50 – mf#LLMC 95-318 – us LLMC [348]

Judicial review under the clean air act and federal water pollution control act / Currie, David P – Washington, 19 Oct 1976 (all publ) – 2mf – 9 – $3.00 – mf#LLMC 94-344 – us LLMC [344]

Judicial sabbaticals / Robbins, Ira P – Washington: FJC, 1987 – 1mf – 9 – $1.50 – mf#LLMC 95-335 – us LLMC [340]

Judicial settlement of controversies between states of american union : cases decided in the supreme court of the united states: collected and edited by james b. scott – New York: Oxford UP, American Branch. 2v. 1918 – 20mf – 9 – $30.00 – mf#LLMC 95-065 – us LLMC [323]

Judicial statistics. / Bannatyne, Andrew – Glasgow, Graham, 1863. 30 p. LL-2255 – 1 – us L of C Photodup [340]

The judicial system of the marathas / Gune, Vithal Trimbak – Poona: Deccan College Postgraduate and Research Institute, 1953 – (= ser Samp: indian books) – us CRL [954]

Judicial writing manual – Washington: FJC, 1991 – 1mf – 9 – $1.50 – mf#LLMC 95-371 – us LLMC [340]

Judiciary act of the republic of the marshall islands, pl 1983-18 – Majuro: Nitijela of the Marshall Islands. 3rd regular sess 1982 [1983] – 1mf – 9 – $1.50 – mf#llmc82-100i, title 10 – us LLMC [348]

Judicium pacifici salomonis christi domini nostri / Bottens, Fulg – Gandavi, 1713 – 5mf – 8 – €12.00 – ne Slangenburg [240]

El judio en la epoca colonial : un aspecto de la historia rioplatense / Lewin, Boleslao – Buenos aires: colegio libre de estudias superiores, 1939 – 158p – 1 – us Wisconsin U Libr [972]

Judio sin careta – Buenos Aires, Argentina. 1936 – 1r – us UF Libraries [939]

Judios en america / Figueroa Fernandez, Cotidio – Santiago, Chile. 1948 – 1r – us UF Libraries [939]

Los judios en extremadura / Munoz de la Pena, Arsenio – Badajoz: (Imprenta Diputacion Provincial), 1970. Sepr. Revista Estudios Extremenos – 1 – sp Bibl Santa Ana [939]

Judische auswanderung aus deutschland / Traub, Michael – Berlin, Germany. 1936 – 1r – us UF Libraries [939]

Judische Gemeinde Zu Berlin *see* Richtlinien fur die nicht voll ausgebauten religionsschulen der jud...

Judische maasssystem und seine beziehungen zum griechischen / Zuckermann, Benedict – Breslau, Germany. 1867 – 1r – us UF Libraries [939]

Judische schulwesen in ungarn unter kaiser josef ii / Mandl, Bernat – Posen, Germany. 1903 – 1r – us UF Libraries [939]

Judische volkslieder / Strauss, Ludwig – Berlin, Germany. 1935 – 1r – us UF Libraries [939]

Judisches Altersheim Und Siechenhaus In Oxtpreussen *see* Bericht uber entstechung des vereins und kassenlegung bis...

Judisches familien-buch / Ehrentheil, Moritz – Budapest, Hungary. 1880 – 1r – us UF Libraries [939]

Judisches genossenschaftswesen in russland / Hillmann, Anselm – Berlin, Germany. 1911 – 1r – us UF Libraries [939]

Judith : ein mitteldeutsches gedicht aus dem 13. jahrhundert / ed by Palgen, Rudolf – Halle (Saale): Verlag von M Niemeyer, 1924 [mf ed 1993] – (= ser Altdeutsche textbibliothek 18) – vii/89p/1pl – 1 – (incl bibl ref) – mf#8193 reel 2 – us Wisconsin U Libr [810]

Judith : an old english epic fragment – Boston, London: D C Heath & Co [1904] [mf ed 1990] – (= ser Belles-lettres series. section 1. english literature...) – xxiv/72p – 1 – (with bibl) – mf#7621 – us Wisconsin U Libr [810]

Judith : an operatic spectacle / Bochsa, Robert Nicolas Charles – [between 1840 and 1859?] [mf ed 1999] – 4 items on 1r – 1 – mf#4622 – us Wisconsin U Libr [780]

Judith / Pentin, Herbert – London: Samuel Bagster, 1908 – 1mf – 9 – 0-8370-9726-6 – (incl bibl ref & ind) – mf#1986-3726 – us ATLA [221]

Judith : eine tragoedie in fuenf aufzuegen / Hebbel, Friedrich – Leipzig: P Reclam [18–] [mf ed 1995] – (= ser Reclams universal-bibliothek 3161) – 1 – mf#8764 – us Wisconsin U Libr [820]

Judith basin news – Lewistown, MT. 1902-1904 (1) – mf#64528 – us UMI ProQuest [071]

Judith basin press – Helena, MT. 1972-1974 (1) – mf#64461 – us UMI ProQuest [071]

Judith basin press – Stanford, MT. 1920-1971 (1) – mf#64653 – us UMI ProQuest [071]

Judith basin star – Hobson, MT. 1908-1954 (1) – mf#64481 – us UMI ProQuest [071]

Judith basintimes – Geyser, MT. 1911-1916 (1) – mf#64399 – us UMI ProQuest [071]

Judith in der deutschen literatur / Baltzer, Otto – Berlin: W de Gruyter & Co, 1930 [mf ed 1993] – (= ser Stoff- und motivgeschichte der deutschen literatur 7) – 62p – 1 – (incl bibl ref) – mf#7840 – us Wisconsin U Libr [430]

Judith, lyric drama / Chadwick, George Whitefield – Text by William Chauncy Langdon. 1900. Holograph score, in ink. At end: score finished Dec. 12, 1900. music 3075 – 1 – us L of C Photodup [780]

Judith "Montefiore" College *see*
- Report for the year
- Report for the year from 1st tamuz, 5651-1891, to 30th sivan,5652-1

Judson, A H *see* An account of the american baptist mission to the burman empire...

Judson, Adoniram *see*
- A dictionary of the burman language
- Dictionary of the burman language, with explanations in english
- Letters to gardner colby, 1846-49

Judson, Ann H *see* An account of the american baptist mission to the burman empire

Judson, Ann Hasseltine *see*
- An account of the american baptist mission to the burman empire
- A particular relation of the american baptist mission to the burman empire

Judson baptist church – Greenville Co, SC. 964p. 1914-21, 1927-oct 1962 – 1 – $43.38 – mf#0900-11 – us Southern Baptist [242]

The judson centennial, 1814-1914 : celebrated in boston, ma, june 24-25, in connection with the centenary of the american baptist foreign mission society / ed by Grose, Howard Benjamin & Haggard, Fred Porter – Philadelphia: publ...[by] American Baptist Publ Soc, c1914 [mf ed 1990] – 1mf – 9 – 0-7905-5151-9 – mf#1988-1151 – us ATLA [242]

The judson centennial celebrations in burma, 1813-1913 / Phinney, Frank Denison [comp] – Rangoon: American Baptist Mission Press, 1914 [mf ed 1993] – (= ser Baptist coll) – 1mf – 9 – 0-524-07108-X – mf#1991-2931 – us ATLA [242]

Judson, E *see* Adoniram judson

Judson, Edward *see*
- Adoniram judson
- The institutional church
- The life of adoniram judson

Judson, Emily Chubbuck *see* Memoir of sarah b judson

Judson, Lyman Spicer V *see* Let's go to guatemala

The judson memorial : intended as a memento of christian sympathy for the loved, departed judsons / ed by Dowling, John – New York: Sheldon, Blakeman, 1858 [mf ed 1993] – (= ser Baptist coll) – 1mf – 9 – 0-524-06404-0 – mf#1991-2526 – us ATLA [242]
The judson offering : intended as a token of christian sympathy with the living, and a memento of christian affection for the dead / ed by Dowling, John – [new ed] New York: L Colby, 1847, c1846 [mf ed 1990] – 1mf – 9 – 0-7905-4789-9 – mf#1988-0789 – us ATLA [242]
Judy – New York. 1846-1847 – 1 – mf#4006 – us UMI ProQuest [073]
Judycki, Zdzislaw *see* Zagadnienia strukturalno-organizacyjne wspolczesnego gospodarstwa narodowego polski
Juedisch-arabische poesien aus vormuhammedischer zeit : ein specimen aus fleischers schule als beitrag zur feier seines jubileums / Delitzsch, Franz – Leipzig: Doerffling & Franke, 1874 – 1mf – 9 – 0-8370-2870-1 – mf#1985-0870 – us ATLA [470]
Die juedisch-aramaeischen papyri von assuan / Staerk, Willy – Bonn: A Marcus & E Weber 1907 [mf ed 1986] – (= ser Kleine texte fuer theologische vorlesungen und uebungen 22-23) – 1mf – 9 – 0-8370-7344-8 – (comm in german; text in aramaic) – mf#1986-1344 – us ATLA [090]
Juedisch-babylonische zaubertexte / ed by Stuebe, Rudolph – Halle (Saale): J Krause 1895 [mf ed 1986] – 1mf – 9 – 0-8370-8388-5 – (comm in german, text in aramaic & german; incl bibl ref) – mf#1986-2388 – us ATLA [470]
Ein juedisch-christliches psalmbuch aus dem ersten jahrhundert / Harnack, Adolf von – Leipzig: J C Hinrichs 1910 [mf ed 1989] – (= ser Tugal 35/4) – 1mf – 9 – 0-7905-1710-8 – (in german & greek; "the odes of solomon, now first publ fr the syriac version by j rendel harris"; incl ind) – mf#1987-1710 – us ATLA; ne Slangenburg [220]
Juedischdeutsche volkslieder aus galizien und russland / ed by Dalman, Gustav Herman – Leipzig: Centralbureau der Instituta Judaica (W Faber), 1888 [mf ed 1993] – (= ser Schriften des institutum judaicum in leipzig 20, 21) – 73p – 1 – mf#8359 – us Wisconsin U Libr [780]
Juedische allgemeine zeitung *see* Juedisch-liberale zeitung
Die juedische apokalyptik : ihre religionsgeschichtliche herkunft und ihre bedeutung fuer das neue testament / Bousset, Wilhelm – Berlin: Reuther & Reichard, 1903 [mf ed 1985] – 1mf – 9 – 0-8370-2798-5 – mf#1985-0798 – us ATLA [270]
Die juedische apokalyptik in ihrer geschichtlichen entwickelung : ein beitrag zur vorgeschichte des christenthums: nebst einem anhang ueber das gnostische system des basilides / Hilgenfeld, Adolf – Jena: Friedrich Mauke, 1857 [mf ed 1989] – 1mf – 9 – 0-7905-1100-2 – (incl bibl ref & ind) – mf#1987-1100 – us ATLA [270]
Juedische arbeits- und wanderfuersorge / ed by Adler-Rudel, Salomon & Kreutzberger, Max – Berlin: George Baum et al. v1-3. 1927-29 [complete] – (= ser German-jewish periodicals...1768-1945, pt 3) – 1r – 1 – mf#B172 – us UPA [939]
Juedische arbeits- und wanderfuersorge – Berlin DE, 1927/28-1929/30 [gaps] – 1 – gw Misc Inst [939]
Die juedische bewegung : gesammelte aufsaetze und ansprachen, 1900-1915 / Buber, Martin – Berlin: Juedischer Verlag, 1916 – 1mf – 9 – 0-524-08154-9 – mf#1991-0284 – us ATLA [270]
Die juedische bibelexegese : vom anfange des zehnten bis zum ende des 15. jahrhunderts / Bacher, Wilhelm – Trier: Sigmund Mayer, 1892 [mf ed 1985] – 1mf – 9 – 0-8370-2539-7 – (incl ind) – mf#1985-0539 – us ATLA [221]
Juedische blaetter – s.l. – 1r – 1 – (individual iss 1842-99 fr holdings of: internationalen zeitungsmuseums der stadt aachen) – gw Misc Inst [939]
Juedische eschatologie von daniel bis akiba / Volz, Paul – Tuebingen: J C B Mohr (Paul Siebeck), 1903 – 1mf – 9 – 0-8370-7274-3 – (incl ind) – mf#1986-1274 – us ATLA [270]
Juedische familien-forschung – Berlin DE, 1925-38 – 1r – 1 – gw Misc Inst [939]
Juedische familien-forschung : mitteilungen – Berlin: Arthur Czellitzer. v1-14. 1925-38 [complete] – (= ser German-jewish periodicals...1768-1945, pt 2) – 1r – 1 – $125.00 – mf#B184 – us UPA [939]
Die juedische frau : ueberparteiliche halbmonatsschrift fuer alle lebensinteressen der juedischen frau – Berlin: Regina Issacson, Anna Beate Nadel. v1-3. 1925-27 [complete] – (= ser German-jewish periodicals...1768-1945, pt 2) – 1r – 1 – $125.00 – mf#B186 – us UPA [939]
Die juedische frau als mutter : ein bild im umbruch? zur historischen stellung und funktion der mutter in der juedischen familie / Herwig, Rachel Monika – (mf ed 1993) – 3mf – 9 – €49.00 – 3-89349-756-0 – mf#DHS 756 – gw Frankfurter [939]
Juedische front – Wien (A), 1932-38 [gaps] – 1 – gw Misc Inst [939]
Die juedische gemeinde von elephantine : und ihre beziehungen zum alten testament / Jirku, Anton – Berlin: E Runge 1912 [mf ed 1989] – (= ser Biblische zeit- und streitfragen 7/11) – 1mf – 9 – 0-7905-3269-7 – mf#1987-3269 – us ATLA [939]
Der juedische gottesdienst in seiner geschichtlichen entwicklung / Elbogen, Ismar – Leipzig: G Fock, 1913 – (= ser Schriften (gesellschaft zur foerderung der wissenschaft des judentums (germany))) – 2mf – 9 – 0-524-03440-0 – mf#1990-3212 – us ATLA [240]
Der juedische handwerker : organ des gesamten juedischen mittelstandes / ed by Basch, Carl et al – Berlin. v1-30. 1909-38 – (= ser German-jewish periodicals...1768-1945, pt 1) – 1r – 1 – $125.00 – (lacking: v1-6 1909-14) – mf#B192 – us UPA [939]
Der juedische handwerker *see* Handwerk und gewerbe
Juedische homiletik : nebst einer auswahl von texten und themen / Maybaum, Siegmund – Berlin: F Duemmler, 1890 – 1mf – 9 – 0-7905-3457-6 – (incl bibl ref) – mf#1987-3457 – us ATLA [270]
Der juedische kirchenstaat in persischer, griechischer und roemischer zeit / Lehmann-Haupt, C F – Tuebingen: J C B Mohr 1911 [mf ed 1989] – (= ser Religionsgeschichtliche volksbuecher fuer die deutsche christliche gegenwart 2/13) – 1mf – 9 – 0-7905-2127-X – (incl bibl ref) – mf#1987-2127 – us ATLA [270]
Das juedische literaturblatt *see* Israelitische wochenschrift
Juedische monatshefte / ed by Breuer, Salomon – Frankfurt a.M: Pinchas Kohn. v1-8. 1914-21 – (= ser German-jewish periodicals...1768-1945, pt 2) – 1r – 1 – $125.00 – (lacking: v5 n2 [may not have been publ]) – mf#B215 – us UPA [270]
Juedische monatshefte – Frankfurt/M DE, 1914-21 – 1r – 1 – (filmed misc inst: 1915-18 [gaps]) – gw Misc Inst [939]
Die juedische presse – Berlin DE, 1872-79, 1881-1914, 1915 2 jul-1916 30 jun, 1918, 1920-1923 30 aug – 20r – 1 – gw Misc Inst [939]
Die juedische presse – Berlin: Samuel Enoch. v1-54. 1869-1923 – (= ser German-jewish periodicals...1768-1945, pt 1) – 18r – 1 – $1980.00 – (lacking: misc iss) – mf#B219 – us UPA [939]
Juedische presse – Wien (A), 1921, 1930-34 – 1 – (with gaps) – gw Misc Inst [939]
Juedische presszentrale : und JuedisBches Familienblatt fuer die Schweiz. – Zurich. jahrg. 1 (1)-23 (1061). 15 Dec 1918-10 May 1940 – 1 – us NY Public [939]
Juedische privatbriefe aus dem jahre 1619 : nach den originalen des k u. k haus-, hof- und staatsarchivs im auftrage der historischen kommission der israelitischen kultusgemeinde in wien / ed by Landau, Alfred & Wachstein, Bernhard – Wien: W Braumueller, 1911 (mf ed 1995) – (= ser Quellen und Forschungen zur Geschichte der Juden in Deutsch-Oesterreich) – 1r – 1 – (hebrew and yiddish text. incl bibl ref and ind) – mf#ZZ-34416 – us NY Public [939]
Juedische reinheitslehre und ihre beschreibung in den evangelien / Brandt, Wilhelm – Giessen: Alfred Toepelmann, 1910 – 1mf – 9 – 0-7905-0255-0 – (incl bibl ref) – mf#1987-0255 – us ATLA [221]
Die juedische religion von der zeit esras bis zum zeitalter christi / Bertholet, Alfred – 1. u 2. Aufl. Tuebingen: J C B Mohr (Paul Siebeck), 1911 – 2mf – 9 – 0-7905-0121-X – (incl bibl ref and indexes) – mf#1987-0121 – us ATLA [270]
Juedische revue – Mukatschewo (Mukacevo UA), 1936 jun-1938 nov – 1r – 1 – gw Misc Inst [077]
Juedische rundschau – Berlin. 1905-10, 1919-20, 1922-Nov. 8, 1938 – 1 – us NY Public [939]
Juedische rundschau – Marburg DE, 1946-48 – 1 – gw Misc Inst [939]
Juedische rundschau *see*
– Berliner vereinsbote
– Rigaer juedische rundschau
Juedische schulzeitung – Hamburg. v. 1-14. 1925-1938 – 1 – us NY Public [939]
Juedische schulzeitung – Mannheim DE, 1934 n1-1936 n12 – 1 – gw Misc Inst [270]
Die juedische stammverschiedenheit : ihr einfluss auf die innere und aeussere entwickelung des judentums / Mosler, Heinrich – Leipzig: W Friedrich, 1884 – 1mf – 9 – 0-524-05622-6 – mf#1992-0477 – us ATLA [939]
Der juedische student – Berlin DE, 1907-08, 1914 n10, 1922 n3-5, 1929 n1-10, 1931-33 [gaps] – 1r – 1 – gw Misc Inst [939]
Juedische theologie : auf grund des talmud und verwandter schriften / Weber, Ferdinand Wilhelm; ed by Delitzsch, Franz & Schnedermann, Georg – 2. verb Aufl. Leipzig: Doerffling & Franke, 1897 – 1mf – 9 – 0-8370-5732-9 – (incl ind) – mf#1985-3732 – us ATLA [270]
Juedische turnzeitung – Berlin DE, 1910, 1912-13 n8 – 1 – gw Misc Inst [790]
Das juedische volk – Berlin DE, 1937-38 – 1r – 1 – gw Misc Inst [939]
Der juedische volkssozialismus / Arlosoroff, C – Berlin, 1919 – 2mf – 9 – mf#J-28-2 – ne IDC [956]
Juedische volksstimme : unabhaengiges, unparteiisches wochenblatt – Bruenn. v1-34? 1900-34? – (= ser German-jewish periodicals...1768-1945, pt 1) – 2r – 1 – $220.00 – (lacking: misc iss) – mf#B240 – us UPA [074]
Juedische volksstimme – Bruenn (Brno CZ), 1900 1 feb-1934 sep [gaps] – 2r – 1 – (with suppl: juedische baeder- und kurortzeitung 1930 n16) – gw Misc Inst [939]
Juedische volkszeitung *see* Juedisches volksblatt
Juedische volkszeitung fuer ostdeutschland *see* Juedisches volksblatt
Juedische welt-rundschau – Paris (F), Jerusalem (IL), 1939 10 mar-1940 20 may – 1r – 1 – (cont: juedischen weltrundschau, berlin) – gw Misc Inst [939]
Der juedische wille – Berlin DE, 1918-1919/20, 1933-37 [gaps] – 1r – 1 – gw Misc Inst [939]
Die juedische witwe : buehnenspiel in fuenf akten / Kaiser, Georg – Potsdam: G Kiepenheuer, 1920 – 1r – 1 – us Wisconsin U Libr [820]
Juedische wochenpost – Bielsko PL, 1934-38 – 1r – 1 – us UMI ProQuest [939]
Juedische wochenschau – Buenos Aires (RA), 1940 26 apr-27 dec, 1941 25 apr-24 nov, 1942-50 [gaps] – 1 – gw Misc Inst [939]
Juedische wohlfahrtspflege und sozialpolitik – Berlin DE, 1929-38 [gaps] – 1r – 1 – gw Misc Inst [939]
Juedische wundergeschichten des neutestamentlichen zeitalters : unter besonderer beruecksichtigung ihres verhaeltnisses zum neuen testament bearbeitet / Fiebig, Paul – Tuebingen: JCB Mohr, 1911 – 1mf – 9 – 0-524-05910-1 – mf#1992-0667 – us ATLA [270]
Juedische zeitschrift fuer wissenschaft und leben / ed by Geiger, Abraham – Breslau 1862-75 [mf ed Hildesheim 1989] – (= ser Bibliothek des deutschen judentums) – 11v on 44mf – 9 – diazo €218.00 silver €238.00 – gw Olms [939]
Juedische zeitung – Duesseldorf DE, jul 1927-oct 1928 [gaps] – 1r – 1 – gw Misc Inst [939]
Juedische zeitung *see* Juedisches volksblatt
Juedische-front : organ of the society of jewish soldiers of the austrian front – Vienna, jan 1933-dec 1938 – 1r – 1 – us UMI ProQuest [074]
Die juedischen baptismen : oder, das religioese waschen und baden im judentum mit einschluss des judenchristentums / Brandt, Wilhelm – Giessen: A Toepelmann, 1910 – 1mf – 9 – 0-7905-0816-8 – mf#1987-0816 – us ATLA [270]
Die juedischen exulanten in babylonien / Klamroth, Erich – Leipzig: J C Hinrichs 1912 [mf ed 1989] – (= ser Beitraege zur wissenschaft vom alten testament 10) – 1mf – 9 – 0-7905-1214-9 – (incl bibl ref & ind) – mf#1987-1214 – us ATLA [270]
Der Juedischen Gemeinde zu Berlin *see* Gemeindeblatt
Juedischer almanach – Berlin, Germany. 1904? – 1r – us UF Libraries [939]
Juedischer bote vom rhein : juedisches wochenblatt fuer die rheinischen lande, bonner gemeindeblatt fuer das rheintal, siegtal, ahrtal, usw / ed by Cohn, Emil & Plawin, Isaak – Bonn. v1-5. 1919-23 – (= ser German-jewish periodicals...1768-1945, pt 3) – 1r – 1 – $165.00 – mf#B433 – us UPA [939]
Juedischer frauenbund von deutschland : blaetter des juedischen frauenbundes fuer frauenarbeit und frauenbewegung / ed by Karminski, Hannah & Ollendorff, Martha – Berlin. v1-14. 1924-38 – (= ser German-jewish periodicals...1768-1945, pt 1) – 1r – 1 – $125.00 – (lacking: misc iss) – mf#B187 – us UPA [305]
Juedischer kulturbund hamburg : monatsblaetter – Hamburg. v1-3. 1836-38 [complete] – (= ser German-jewish periodicals...1768-1945, pt 2) – 1r – 1 – $125.00 – mf#B211 – us UPA [939]
Juedischer kulturbund rhein-ruhr – Koeln DE, 1936-38 [gaps] – 1 – gw Misc Inst [939]
Juedischer volks- und haus-kalender – Wroclaw PL, 1890-1900 – 1r – 1 – us UMI ProQuest [939]
Juedischer volksbote : blaetter zur foerderung der geistigen und wirtschaftlichen interessen der juedischen landbevoelkerung – Frankfurt a.M. n1-37? 1908-15? [complete] – (= ser German-jewish periodicals...1768-1945, pt 1) – 1r – 1 – $125.00 – mf#B237 – us UPA [939]
Juedischer volksbote – Frankfurt/M DE, 1908-15 – 1r – 1 – gw Misc Inst [939]
Juedisches archiv : mitteilungen des komitees "juedisches kriegsarchiv" – Berlin DE, Wien (A), 1915-18, 1920 – 1r – 1 – gw Misc Inst [939]
Juedisches archiv : mitteilungen des komitees...1915-18 – Vienna, Berlin. v1-9. 1920 [complete] – (= ser German-jewish periodicals...1768-1945, pt 2) – 1r – 1 – $125.00 – (filmed with b174) – mf#B173 – us UPA [939]
Juedisches archiv : zeitschrift fuer juedisches museal- und buchwesen /.../ – Wien (A), 1927-29 – 1r – 1 – gw Misc Inst [939]
Juedisches archiv : zeitschrift fuer juedisches museal- und buchwesen, geschichte, volkskunde und familienforschung – Vienna: Leopold Moses. v1-2. 1927-29 [complete] – (= ser German-jewish periodicals...1768-1945, pt 2) – 1r – 1 – $125.00 – (filmed with b173) – mf#B174 – us UPA [939]
Juedisches biographisches archiv (jba1) = Jewish biographical archive (jba1) / Schmuck, Hilmar [comp] – [mf ed 1994-96] – 690mf (1:24) – 9 – diazo €10,060.00 (silver €11,080 isbn: 978-3-598-33603-4) – ISBN-10: 3-598-33590-3 – ISBN-13: 978-3-598-33590-7 – (with printed ind) – gw Saur [939]
Juedisches biographisches archiv (jba1). supplement = Jewish biographical archive (jba1). supplement / Schmuck, Hilmar [comp] – [mf ed 1998] – 127mf – 9 – diazo €2030.00 (silver €2460 isbn: 978-3-598-33516-7) – ISBN-10: 3-598-33513-X – ISBN-13: 978-3-598-33513-6 – (with printed ind) – gw Saur [939]
Juedisches biographisches archiv. neue folge (jba2) = Jewish biographical archive. series 2 (jba2)) / Schmuck, Hilmar [comp] – [mf ed 2001-03] – 610mf (1:24) – 9 – diazo €10,060.00 (silver €11,080 isbn: 978-3-598-34871-6) – ISBN-10: 3-598-34870-3 – ISBN-13: 978-3-598-34870-9 – (with printed ind) – gw Saur [939]
Juedisches gemeindeblatt – Bremen DE, 1932-36 [gaps] – 1 – gw Misc Inst [939]
Juedisches gemeindeblatt – Danzig (Gdansk PL), 1937/38-1938/39 – 1 – gw Misc Inst [939]
Juedisches gemeindeblatt – Koeln DE, 1937-38 [gaps] – 1 – gw Misc Inst [939]
Juedisches gemeindeblatt – Mannheim DE, 1937-38 [gaps] – 1 – gw Misc Inst [270]
Juedisches gemeindeblatt : mitteilungen der juedischen reformgemeinde zu berlin – Berlin. v1-21. 1918-38 – (= ser German-jewish periodicals...1768-1945, pt 1) – 1r – 1 – $125.00 – (lacking: v21 n6-8 1938) – mf#B429 – us UPA [270]
Juedisches gemeindeblatt : mitteilungsblatt der israelitischen gemeinde bremen – Kassel, Mannheim. v1-8. 1929-36 – (= ser German-jewish periodicals...1768-1945, pt 3) – 1r – 1 – $165.00 – (v5 n4 not publ) – mf#B434 – us UPA [939]
Juedisches gemeindeblatt – Dessau DE, 1927-29 – 1r – 1 – (with: leipziger juedische wochenschau, leipzig, 1928-29) – us UMI ProQuest [939]
Juedisches gemeindeblatt *see* Mitteilungen der juedischen reformgemeinde zu berlin
Juedisches gemeindeblatt – allgemeine zeitung der juden in deutschland *see* Allgemeine juedische wochenzeitung
Juedisches gemeindeblatt fuer das gebiet der rheinpfalz : organ des verbandes der israelitischen kultusgemeinden der pfalz / ed by Metzger, Kurt – Landau/Pfalz. v1 n1-12, v2 n1-3. 1937-38. – (= ser German-jewish periodicals...1768-1945, pt 3) – 1r – 1 – $165.00 – mf#B461 – us UPA [939]
Juedisches gemeindeblatt fuer den synagogenbezirk duesseldorf *see* Gemeindezeitung fuer den synagogenbezirk duesseldorf
Juedisches gemeindeblatt fuer die britische zone *see* Allgemeine juedische wochenzeitung
Juedisches gemeindeblatt fuer die israelitische gemeinde zu frankfurt am main – Frankfurt/M DE, 1922/23-1938 n2 – 2r – 1 – gw Misc Inst [939]
Juedisches gemeindeblatt fuer die israelitischen gemeinden wuerttembergs *see* Gemeindezeitung fuer die israelitischen gemeinden wuerttembergs
Juedisches gemeindeblatt fuer die nordrhein-provinz und westfalen *see* Allgemeine juedische wochenzeitung
Juedisches gemeindeblatt fuer die synagogengemeinde breslau *see* Breslauer juedisches gemeindeblatt
Juedisches gemeindeblatt fuer die synagogen-gemeinden in preussen und norddeutschland *see* Verwaltungsblatt des preussischen landesverbandes juedischer gemeinden

Juedisches gemeindeblatt und mitteilungsblatt fuer die israelitischen gemeinden duesseldorf und krefeld – Duesseldorf DE, 1930-1933 19 jan [gaps] – 1r – 1 – gw Misc Inst [270]
Juedisches gemeindeblatt und nachrichtenblatt der gemeindeverwaltung der israelitischen religionsgemeinde zu leipzig – Leipzig DE, 1925 25 sep-1936 26 aug – 2r – 1 – (filmed by other misc inst: 1925-38 [6r]) – gw Misc Inst [270]
Juedisches gemeinde-jahrbuch, 1913/14 [5674] – Berlin: Zionistische Vereinigung fuer Deutschland, 1913/14 – (= ser German-jewish periodicals...1768-1945, pt 3) – 1r – 1 – $165.00 – mf#B191 – us UPA [939]
Juedisches jahrbuch – Berlin: Scherbel. v1-5. 1929-33 – (= ser German-jewish periodicals... 1768-1945, pt 3) – 1r – 1 – $165.00 – mf#B195 – us UPA [939]
Juedisches jahrbuch fuer die schweiz = Annuaire israelite pour la suisse – Lucerne. v1-6. 1916-21 – (= ser German-jewish periodicals...1768-1945, pt 3) – 1r – 1 – $165.00 – (lacking: v2) – mf#B200 – us UPA [939]
Juedisches jahrbuch fuer gross-berlin : ein wegweiser durch die juedischen einrichtungen und organisationen berlins – Berlin-Gruenewald: Scherbel, 1926-33 – (= ser German-jewish periodicals...1768-1945, pt 3) – 1r – 1 – $165.00 – mf#B194 – us UPA [939]
Juedisches leben in koenigsberg/pr. im 20. jahrhundert / Jacoby, Yoram K – Wuerzburg: Holzner Verlag, 1983 – 10r – 1 – (incl bibl ref and index) – us Wisconsin U Libr [943]
Juedisches literaturblatt : zur beleuchtung aller judentum und juden betreffenden literarischen erscheinungen auf dem gebiete der philosophie, geschichte – Magdeburg, Berlin: Ludwig A Rosenthal. v1-37? 1872-1916? [complete] – (= ser German-jewish periodicals...1768-1945, pt 1) – 2r – 1 – $220.00 – mf#B212 – us UPA [939]
Juedisches nachrichtenblatt – Berlin DE, 1938 2 dec-1939 29 sep, 1942 6 jan-1943 22 jan – 2r – 1 – gw Misc Inst [939]
Juedisches nachrichtenblatt – Prag (CZ), 1941-42 [gaps], 1944 n1, 1945 n2 – 1 – gw Misc Inst [939]
Juedisches nachrichtenblatt – Wien (A), 1938-43 [gaps], 1942-1943 jan – 1 – gw Misc Inst [939]
Juedisches nachrichtenblatt : organ der juedischen kultusgemeinde in prag und der zionistischen organisationen in prag = Zidovske listy / ed by Singer, Oskar – Prague. v1-7. 1939-45 – (= ser German-jewish periodicals...1768-1945, pt 3) – 1r – 1 – $165.00 – (lacking: misc iss) – mf#B464 – us UPA [939]
Juedisches schicksal in deutschen gedichten : eine abschliessende anthologie / ed by Kaznelson, Siegmund – Berlin: Juedischer Verlag, c1959 – 1r – 1 – (incl bibl ref and index) – us Wisconsin U Libr [810]
Juedisches volksblatt – Leipzig DE, 1853-1909 – 5r – 1 – us UMI ProQuest [939]
Juedisches volksblatt – Leipzig DE, 1853-66 [gaps] – 1r – 1 – gw Misc Inst [270]
Juedisches volksblatt – Spb., 1888. v8 – 1r – 1 – mf#J-92-4 – ne IDC [077]
Juedisches volksblatt – Breslau (WrocLaw PL), 1900 31 aug-1937 30 apr [gaps] – 4r – 1 – (title varies: jul 31 1914: juedische volkszeitung; feb 13 1925: juedische zeitung fuer ostdeutschland; 8 jan 1932: juedische zeitung. filmed by other misc inst: 1924 22 feb-24 dec, 1929 4 jan-1933) – gw Misc Inst [939]
Juedisches volksblatt : unabhaengiges organ fuer die interessen von gemeinde, schule und haus – Breslau. v1-44. 1896-1937 – (= ser German-jewish periodicals...1768-1945, pt 1) – 4r – 1 – $455.00 – (lacking: 1896-99 and misc iss) – mf#B234 – us UPA [939]
Juedisches volksblatt : zur belehrung und unterhaltung auf juedischem gebiet – Leipzig: Ludwig Philippson. v1-13. 1854-66 – (= ser German-jewish periodicals...1768-1945, pt 1) – 1r – 1 – $125.00 – (lacking: misc iss) – mf#B233 – us UPA [939]
Juedisch-liberale zeitung – Berlin DE, 1924 21 mar-1925, 1927-34, 1936 jan-2 sep [gaps] – 2r – 1 – (filmed by misc inst: 1922-1936 n37 [gaps]. title varies: 7 nov 1934: juedische allgemeine zeitung) – gw Misc Inst [074]
Juedisch-liberale zeitung / ed by Woyda, Bruno – Berlin, 1924-36 – (= ser German-jewish periodicals...1768-1945, pt 1) – 2r – 1 – $220.00 – (lacking: misc iss; cont as: juedische allgemeine zeitung; neue folge der juedisch-liberalen zeitung) – mf#B252 – us UPA [074]
Juego y la vagancia en cuba / Saco, Jose Antonio – Habana, Cuba. 1960 – 1r – us UF Libraries [972]
Juegos antiguos en america / ed by Bayle, Constantino – Madrid: Razon y Fe, 1943 y 1944 – 1 – sp Bibl Santa Ana [740]
Juegos florales centroamericanos y de panama / Quezaltenango (Guatemala) – Quezaltenango. 1957 – 1r – us UF Libraries [972]
Juegos infantiles de extremadura / Hernandez de Soto, Sergio – 1884 – 1 – sp Bibl Santa Ana [946]
Juelicher, Adolf *see*
– Einleitung in das neue testament
– Die gleichnisreden jesu im allgemeinen
– Introduction to the new testament
– Lehrbuch der neutestamentlichen theologie
– Neue linien in der kritik der evangelischen ueberlieferung
– Paulus und jesus
Juelicher volkszeitung – Juelich DE, 1957 2 nov-1959 30 jun – 1 – (bezirksausgabe von aachener volkszeitung, aachen) – gw Misc Inst [074]
Juel-Larsen, Petter *see* Study of left-hand techniques in the piano sonatas of mozart and beethoven
Juenger, Friedrich Georg *see*
– Nietzsche
– Die silberdistelklause
– Der taurus
Die juengere judith : aus der vorauer handschrift / ed by Monecke, Hiltgunt – Tuebingen: M Niemeyer, 1964 [mf ed 1993] – (= ser Altdeutsche textbibliothek n61) – xi/57p – 1 – (incl bibl ref) – mf#8193 reel 5 – us Wisconsin U Libr [810]
Der juengere titurel : [poem] / Scharfenberg, Albrecht von; ed by Wolf, Werner – Bern: A Francke c1952 [mf ed 1998] – (= ser Altdeutsche uebungstexte 14) – 1r – 1 – (filmed with: gotische texte / m szadrowsky [ed]) – mf#4492p – us Wisconsin U Libr [810]
Der juengling / Hasenclever, Walter – Leipzig: K Wolff 1913 [mf ed 1990] – 1r – 1 – (filmed with: der frosch / otto erich hartleben) – mf#2699p – us Wisconsin U Libr [830]
Der juengling – Koenigsberg (Kaliningrad RUS), Mitau/Leipzig, 1768 n1-72 – 1r – 1 – gw Misc Inst [077]
Juengst, Hugo C *see* Flammenzeichen
Juengst, Johannes *see*
– Die evangelische kirche und die separatisten und sektierer der gegenwart
– Kirchengeschichtliches lesebuch
– Kultus- und geschichtsreligion (pelagianismus und augustinismus)
– Der methodismus in deutschland
– Pietisten
– Die quellen der apostelgeschichte
Das juengste gericht / Bullinger, Heinrich – Zuerych, Christoffel Froschouer, 1555 – 2mf – 9 – mf#PBU-184 – ne IDC [240]
Das juengste gericht in der bildenden kunst des fruehen mittelalters / Voss, G – Leipzig, 1884 – €5.00 – ne Slangenburg [700]
Die juengste kritik des galaterbriefes : auf ihre berechtigung / Gloel, Johannes – Erlangen: Andr Deichert, 1890 – 1mf – 9 – 0-8370-9698-7 – (incl bibl ref) – mf#1986-3698 – us ATLA [227]
Der juengste tag – Giessen, Lahn DE, 1848 6 mar-30 dec – 1 – gw Mikrofilm [074]
Juergen Alberts et al *see* Stories fuer uns
Juergens, Cheryl A *see* A kinetic and kinematic comparison of the grab and track starts in competitive swimming
Juergensen, Hans *see* Henrik ibsens einfluss auf hermann sudermann
Juergensen, Jorgen *see* Travels through france and germany, in the years 1815, 1816, et 1817
Juernjakob swehn : der amerikafahrer / Gillhoff, Johannes – Berlin: Verlag der Taeglichen Rundschau, 1920 (mf ed 1990) – 1r – 1 – (filmed with: zwischen den kriegen) – us Wisconsin U Libr [430]
Juerrns, J F *see* Grondig onderwys in de fregoriaansche choorzang of choral, nevens eenige aanmerkingen over di zang-konst
Juez, Antonio *see*
– Defensa ciudadana
– Exposicion juez
– Guerras de retaguardia
– Raza espanola
Juez Nieto, Antonio *see*
– Aldabadas
– Hacia las rutas nuevas
– Luis de morales. el divino
– Por nuestros caminos
– Soy un pobre peregrino...
Juez olaverri y juan canastuj / Aguilar, Octavio – Guatemala. 1956 – 1r – us UF Libraries [972]
Juf news – Chicago, Ill., v6, no. 10 (Dec. 1978)-v7, no. 9 (Nov. 1979); some lacking – us AJPC [939]
Jugantar – Calcutta, India. 12 Jul 1952-1954; 1965-Aug 1966 – 12r – 1 – us L of C Photodup [079]
Jugates / American Political Items Collectors [Organization] – n3-8 [1976 jul/aug-1977 may/jun] – 1r – 1 – mf#382324 – us WHS [071]
Le juge a mabane : (etude historique) / Bois, Louis-Edouard – Quebec?: A Cote, 1881 – 2mf – 9 – mf#07856 – cn Canadiana [340]
Le juge a paix, et officier de paroisse : pour la province de quebec / Burn, Richard – Montreal: Fleury Mesplet, 1789 [mf ed 1971] – 1r – 1 – (trans by joseph-francois perrault) – mf#SEM35P59 – cn Bibl Nat [345]
Jugemens sur quelques ouvrages nouveaux – Avignon. 1744-45 (I-XI) – 1 – fr ACRPP [944]
Jugement des lords du comite judiciaire du conseil prive sur l'appel de dame henriette brown vs les cure et marguilliers de l'oeuvre et fabrique de notre-dame de montreal, au canada, prononce le 21 nov 1874 = Judgments of the lords of the judicial committee of the privy council on the appeal of dame henriette brown v les cure et marguilliers de l'oeuvre et fabrique de notre-dame de montreal... / Grande-Bretagne. Privy Council Judicial Committee – [Montreal?]: [s.n.], [1875] (mf ed 1991) – 1mf – 9 – mf#SEM105P1448 – cn Bibl Nat [340]
Jugement impartial sur napoleon : ou considerations philosophiques sur son caractere, son elevation, sa chute, et les resultats de son gouvernement: suivies d'un parallele entre napoleon et cromwell... / Azais, Hyacinthe – Paris 1820 [mf ed Hildesheim 1995-98] – 1v on 2mf – 9 – €60.00 – 3-487-26234-7 – gw Olms [944]
Jugement rendu souverainement et en dernier ressort, dans l'affaire du canada – A Paris: De l'impr d'Antoine Boudet...1763 – 1mf – 9 – mf#52183 – cn Canadiana [345]
Jugement sur les ministres actuels : ou examen de leur conduite politique et parlementaire pendant et depuis la session de 1828... / Flandin, Jean – Paris 1829 [mf ed Hildesheim 1995-98] – 1v on 3mf – 9 – €90.00 – ISBN-10: 3-487-26138-3 – ISBN-13: 978-3-487-26138-6 – gw Olms [323]
Jugend : ein liebesdrama in drei aufzuegen / Halbe, Max – Berlin: G Bondi, 1911 – 1r – 1 – us Wisconsin U Libr [820]
Jugend – identitaet – sexualitaet : zur ambivalenz von individualisierungsprozessen unter erschwerten lern- und lebensbedingungen / Stange, Helmut – Dortmund: projekt vlg. 1993 (mf ed 1996) – 4mf – 9 – €45.00 – 3-8267-9703-5 – mf#DHS 9703 – gw Frankfurter [150]
Jugend ohne goethe / Kommerell, Max – Frankfurt a.M.: V Klostermann, [1931] – 1r – 1 – us Wisconsin U Libr [430]
Jugend und technik – 1966-1974, 1976-1983 – 425mf – 1 – gw Mikropress [373]
Jugend und volk *see* Freiburger tagespost
Jugend und welt / ed by Arnheim, Rudolf et al – Berlin-Grunewald: Williams & Co c1928 [mf ed 1979] – 1r [ill/pl] – 1 – mf#film mas 8917 – us Harvard [360]
Jugend vor 1914 : [a novel] / Dobiasch, Josef – Berlin: W Limpert, 1939 [mf ed 1989] – 280p – 1 – mf#7178 – us Wisconsin U Libr [830]
Jugend-blaetter – Stuttgart DE, 1852, 1853 jul-dec, 1854 jul-1855 – 2r – 1 – gw Misc Inst [305]
Der jugendbund – Duesseldorf DE, 1925 17 dec-1932 jul – 1r – 1 – gw Misc Inst [074]
Der jugendbund – Duesseldorf: Dr Klein. v1-18. 1925-32? – (= ser German-jewish periodicals... 1768-1945, pt 1) – 1r – 1 – $125.00 – (lacking: misc iss) – mf#B254 – us UPA [939]
Die jugenddichtung friedrich hoelderlins / Grosch, Rudolf – Berlin, 1899 (mf ed 1995) – 1mf – 9 – €24.00 – 3-8267-3134-4 – mf#DHS-AR 3134 – gw Frankfurter [430]
Der jugendfreund – Temeschburg (Timisoara RO), 1928-1932 apr, 1933 may-dec – 1 – gw Misc Inst [077]
Jugendfreunde : lustspiel in vier aufzuegen / Fulda, Ludwig – 3. Aufl. Stuttgart: J G Cotta, 1904 (mf ed 1990) – 1r – 1 – (filmed with: aus der werkstatt) – us Wisconsin U Libr [820]
Jugendgedichte / Borchardt, Rudolf – Berlin: E Rowohlt, 1920 [mf ed 1989] – 127p – 1 – mf#7052 – us Wisconsin U Libr [810]
Jugendlehrere der talmudischen zeit / Wiesner, L – Wien, Austria. 1914 – 1r – us UF Libraries [939]
Jugendliche muenchner illustrierte wochenschrift fuer kunst und leben muenchen (iz5) – Leipzig: Hirth 1896-1940 [mf ed 2002] – (= ser Illustrierte zeitschriften (iz) 5) – 546mf – 9 – diazo €2400 silver €3400 – 3-89131-393-4 – (with contents ind on cd-rom) – gw Fischer [050]
Jugend-post – 1884 sep 27-1887 dec 24, 1890 apr 26, 1893 jul 29, nov 11-18, 1895 dec 28 – 1r – 1 – mf#2948822 – us WHS [071]
Die jugendzeitung – Nuernberg DE, 1841-1844 n1 [gaps] – 1 – gw Misc Inst [305]
Jugend-zeitung – Kehl DE, 1783 12 jul-21 dec – 1r – 1 – gw Misc Inst [305]
Jugie, Martin *see*
– Histoire du canon de l'ancien testament dans l'eglise grecque et l'eglise russe
– Nestorius et la controverse nestorienne
– Theologia dogmatica christanorum orientalium
Jugoslav review – New York, NY. v2 n1. oct 1923 [mthly] – 1r – 1 – (in serbian and/or croatian, slovenian) – us IHRC [073]
Jugoslavia – Chicago: J R Palandech. mar 13 1919-mar 11 1922; may 17 1947-nov 1949 – us CRL [949]
Jugoslovenki svijet – New York NY, jul 23 1908-jun 30 1913; jan 2 1914-jun 20 1920 – 16r – 1r – (croatian newspaper) – us IHRC [071]
Jugoslovenska zastava – Chicago, New York, St Louis, Pittsburgh, 1919* – 1r – 1 – (croatian newspaper) – us IHRC [071]
Jugoslovenski gospodar – Chicago IL, 1907* – 1r – 1 – (slovenian newspaper) – us IHRC [071]
Jugoslovenski komercijalni bilten – Yugoslavia. Jugoslovenski komercijalne novine. -sw. 1960-70. 9 reels – 1 – uk British Libr Newspaper [079]
Jugoslovenski obzor – Milwaukee WI, West Allis. 1933 mar 16-1936 jan 30, 1936 feb 6-1938 oct 27, 1938 nov 10-1943 jan 1, 1943 jan 15-1946 dec 15 – 4r – 1 – (cont: vestnik [milwaukee, wi]) – mf#1131662 – us WHS [071]
Jugoslovenski obzor – Milwaukee WI, 1929, 1938* – 1r – 1 – (slovenian newspaper) – us IHRC [071]
El juguete caido : notas bibliograficas de perez comendador / Vega Mateos, Celestino – Plasencia: editorial sanchez rodrigo, 1970 – sp Bibl Santa Ana [920]
El juguete caido / Vega Mateos, Celestino – Villanueva de la Serena: Tip. Lucas Alonso Garcia, 1940 – sp Bibl Santa Ana [946]
Juhle, Werner *see* Iliamna volcano and its basement
Juhn, Kurt *see* Der hexenhammer
Juhnke, Gerald A *see* Addressing school violence
Juhnke, Richard *see* Wohlau
Juhnke, William E *see* President truman's committee on civil rights
Jui, Chia-jui *see*
– Chien yue chih tu lun
– Chien yue kung ch'ang kuan li fa
– Hsiao hsueh hsing cheng chi tsu chih
Juicio critico de donoso cortes / Perez Cortes y Garcia Camacho, Angel – 1894 – 9 – sp Bibl Santa Ana [440]
Juicio critico del plan de campana titulado de las... – s.l, s.l? 1872 – 1r – us UF Libraries [972]
El juicio de paris verdadero desengano del agua / Fernandez, F – Madrid, 1755 – 2mf – 9 – sp Cultura [615]
Juicio ejecutivo en la legislacion salvadorena / Tomasino, Humberto – San Salvador, El Salvador. 1960 – 1r – us UF Libraries [972]
Juicio particular / Santander Arias, Jorge – Manizales, Colombia. 1960 – 1r – us UF Libraries [972]
Juicio practico sobre las virtudes medicinales... / Fernandez Barea, M – Granada, 1761 – 2mf – 9 – sp Cultura [610]
Juicios verbales / Martinez Escobar, Manuel – Habana, Cuba. 1937 – 1r – us UF Libraries [972]
Juif / Desaugiers, Marc-Antoine – Paris, France. 1824 – 1r – us UF Libraries [440]
Le juif dans la franc-maconnerie / Clarin de la Rive, Abel – Paris: A Pierret 1895 – 5mf – 9 – mf#vrl-134 – ne IDC [366]
Les juifs dans l'empire roman : leur condition juridique, economique et sociale / Juster, Jean – Paris: Paul Geuthner, 1914 [mf ed 1989] – 2mf – 9 – 0-7905-2044-3 – (in french, latin, greek and hebrew) – mf#1987-2044 – us ATLA [939]
Juifs d'aujourd'hui / Eberlin, Elie – Paris, France. 1927 – 1r – us UF Libraries [939]
Juifs de l'afrique du nord / Eisenbeth, Maurice – Alger, Algeria. 1936 – 1r – us UF Libraries [939]
Les juifs en roumanie depuis le traite de berlin (1878) jusqu'a ce jour : les lois et leurs consequences / Sincerus, Edmond – Londres, New York: Macmillan, 1901 (mf ed 1995) – 1r – 1 – (incl bibl ref) – mf#ZZ-34373 – us NY Public [939]
Juifs et l'antisemitisme *see* Israel among the nations
Juigne-Broissiniere, D de *see* Dictionaire theologique, historique, poetique, cosmographique et chronologique (ael1/46)
Juilliard bulletin – New York. 1984-1986 (1) 1984-1986 (5) 1984-1986 (9) – (cont: juilliard news bulletin) – mf#1593,01 – us UMI ProQuest [790]
Juilliard journal – New York. 1985+ (1) 1985-1986 (5) 1985-1986 (9) – mf#16294 – us UMI ProQuest [790]
Juilliard news bulletin – New York. 1962-1984 (1) 1971-1984 (5) 1976-1984 (9) – (cont by: juilliard bulletin) – ISSN: 0022-6173 – mf#1593 – us UMI ProQuest [790]
Juilliard review – New York. 1954-1962 (1) – ISSN: 0449-4016 – mf#1462 – us UMI ProQuest [790]
Juilliard review annual – New York. 1962-1967 (1) – ISSN: 0449-4024 – mf#6416 – us UMI ProQuest [780]

Juin – Paris. n1-47. fevr 1946-7 janv 1947 – 1 – (politique, economique, litteraire.) – fr ACRPP [073]

Juin 36 : trente-six. organe de la federation socialiste de la seine, s.f.i.o. – Paris. fevr 1938-39 – 1 – (puis organe du parti socialiste ouvrier et paysan) – fr ACRPP [325]

Juive / Scribe, Eugene – Paris, France. 1933 – 1r – us UF Libraries [440]

Ju-ju / Case Western Reserve University – 1973 spr, 1975 spr – 1r – 1 – mf#5307087 – us WHS [378]

Jukebox collector newsletter – 1981 jan [iss 41]-1984, 1985-1988 mar [iss 127], incl 1985 [directory], 1982 fall/winter, 1984 spr/sum, 1986 sum, 1987 sum catalogs, 1988 apr-1990 sep – 3r – 1 – (cont: jukebox trader; cont by: jukebox collector) – mf#1060906 – us WHS [071]

Jukebox trader – iss 17-40 [1979 jan-1980 dec] – 1r – 1 – mf#1575535 – us WHS [071]

Jukes, Andrew John *see*
- The characteristic differences of the four gospels
- Drying up of the euphrates
- The law of the offerings in leviticus 1-7
- Mystery of the kingdom traced through the four books of kings part...
- The new man and the eternal life
- The second death and the restitution of all things
- The types of genesis briefly considered as revealing the development of human nature
- The types of genesis briefly considered as revealing the development of human nature in the world within, and without, and in the dispensations

Jules bastien-lepage / Ady, Julia Mary (Cartwright) – London 1894 – (= ser 19th c art & architecture) – 2mf – 9 – mf#4.2.374 – uk Chadwyck [750]

Jules bastien-lepage and his art : a memoir / Theuriet, Claude Adhemar Andre – London 1892 – (= ser 19th c art & architecture) – 2mf – 9 – mf#4.2.213 – uk Chadwyck [750]

Jules cesar / Boissy, Gabriel – Paris, France. 1937 – 1r – us UF Libraries [440]

Jules ferry, 1832-1893 / Reclus, Maurice – Paris: Flammarion, [1947] – us CRL [920]

Jules malou et l'oeuvre congolaise de leopold ii (1876-1886) / Roeykens, Auguste – Bruxelles, Belgium. 1962 – 1r – us UF Libraries [960]

Jules michelet / Monod, Gabriel Jacques Jean – Paris, France. 1905 – 1r – us UF Libraries [025]

Julesburg advocate – Julesburg, CO: Mr & Mrs Ronald B Wilkins. v14 n9. mar 1 1972- (wkly) [mf ed filmed 1975-] – 1 – (cont: julesburg grit-advocate) – us NE Hist [071]

Julesburg grit-advocate – Julesburg, CO: Walter R McKinstry, Jr, 1907-v14 n8. feb 23 1972 (wkly) [mf ed v5 n25. jun 19 1963-feb 23 1972 filmed 1975] – 6r – 1 – (cont: grit-advocate. absorbed: big springs enterprise (1963). cont by: julesburg advocate) – us NE Hist [071]

Julia : ein trauerspiel in drei akten: nebst einer vorrede und einer abhandlung / Hebbel, Friedrich – Leipzig: J J Weber, 1851 [mf ed 1990] – xliv/115p – 1 – mf#7449 – us Wisconsin U Libr [820]

Julia gonzaga : ein lebensbild aus der geschichte der reformation in italien / Benrath, Karl – Halle: Verein fuer Reformationsgeschichte 1900 [mf ed 1990] – (= ser Schriften des vereins fuer reformationsgeschichte 16/65) – 1mf – 9 – 0-7905-5261-2 – (incl bibl ref) – mf#1988-1261 – us ATLA [242]

Julia Marin, Ramon *see* Tierra adentro

Julia soler / Aguero Vives, Eduardo – Habana, Cuba. 1950 – 1r – us UF Libraries [972]

Julian : or, scenes in judea / Ware, William – New York: T R Knox, 1885 – 1r – 1 – 0-8370-1538-3 – mf#1984-B458 – us ATLA [240]

Julian *see*
- Julian's reply to professor rawlinson
- Julian's reply to the lord bishop of ely
- Natural reason versus divine revelations

Julian, Antonio *see* Perla de la america

Julian del casal / Casal, Julian Del – Madrid, Spain. 1916 – 1r – us UF Libraries [972]

Julian der abtruennige / Dahn, Felix – Leipzig: Breitkopf & Haertel. 3v in 2. 1898 – 6r – 1 – us Wisconsin U Libr [830]

Julian der abtruennige : geschichtlicher roman / Dahn, Felix – Leipzig: Breitkopf und Haertel. 3v in 2. 1898 – 1 – us Wisconsin U Libr [830]

The julian gazette – Julian, NE: J C Gentry, may 1899 (wkly) [mf ed v1 n3. may 26 1899-nov 3 1899 (gaps) filmed [1979]] – 1r – 1 – (cont: julian weekly gazette) – us NE Hist [071]

Julian, John *see* A dictionary of hymnology

Julian leader – Julian, NE: W C Ogdon, 1895 (wkly) [mf ed v2 n37. aug 21 1897 filmed [1979] – 1r – 1 – us NE Hist [071]

Julian, philosopher and emperor : and the last struggle of paganism against christianity / Gardner, Alice – New York: GP Putnam, 1906, c1895 – (= ser Heroes of the nations) – 1mf – 9 – 0-7905-4647-7 – (incl bibl ref) – mf#1988-0647 – us ATLA [930]

Julian, philosopher and emperor : and the last struggle of paganism against christianity / Gardner, Alice – New York: G.P. Putnam, 1906, c1895. (Heroes of the nations) – 1mf – us ATLA [240]

[Julian-] sentinel – CA. 1887-89; 1890-92 – 2r – 1 – $120.00 – mf#B02321 – us Library Micro [071]

Julian the emperor : containing gregory nazianzen's two invectives and libanius' monody : with julian's extant theosophical works / King, CW – London: G. Bell, 1888. (Bohn's classical library) – 1mf – us ATLA [240]

Julian the emperor : containing gregory nazianzen's two invectives and libanius' monody: with julian's extant theosophical works – London: G. Bell, 1888 – (= ser Bohn's classical library) – 1mf – 9 – 0-7905-5165-9 – mf#1988-1165 – us ATLA [210]

Julian von eclanum / Bruckner, Albert – Leipzig, 1897 – (= ser Tugal 1-15/3a) – 3mf – 9 – €7.00 – ne Slangenburg [240]

Julian von eclanum : sein leben und seine lehre / Bruckner, Albert – Leipzig: J C Hinrichs, 1897 – (= ser Tugal) – 1mf – 9 – 0-7905-1627-6 – (incl ind) – mf#1987-1627 – us ATLA [220]

Julian weekly gazette – Julian, NE: J C Gentry. v1 n1. may 12 1899-99// (wkly) [mf ed may 12 1899 filmed [1979]] – 1r – 1 – (cont by: julian gazette) – us NE Hist [071]

Julian's indiana radical – Richmond IN. 1872 jun 27 – 1r – 1 – (cont: indiana true republican) – mf#856323 – us WHS [071]

Julian's reply to professor rawlinson / Julian – London, England. 1871 – 1r – us UF Libraries [240]

Julian's reply to the lord bishop of ely / Julian – London, England. 1871 – 1r – us UF Libraries [240]

Julianus apostata in der deutschen literatur / Philip, Kaete – Berlin: W de Gruyter & Co, 1929 – 2r – 1 – (incl bibl ref and index) – us Wisconsin U Libr [430]

Julianus der abtruennige : trauerspiel in fuenf aufzuegen / Boruttau, Carl – Berlin: In Commission bei R Schlingmann, [1865] [mf ed 1989] – 117p – 1 – mf#7053 – us Wisconsin U Libr [820]

Julie : ou, la nouvelle heloise: lettres de deux amants = Nouvelle heloise / Rousseau, Jean-Jacques – Paris: Garnier freres, [188-?] [mf ed 2000] – (= ser Classiques garnier) – 1r – 1 – (filmed with: le meunier d'angibault / george sand [1852?] and: l'alouette / par jean anouilh) – mf#10470 – us Wisconsin U Libr [830]

Julie von bondeli und ihr freundeskreis : wieland, rousseau, zimmermann, lavater, leuchsenring, usteri, sophie laroche, frau v sandoz... / Bodemann, Eduard – Hannover: Hahn, 1874 [mf ed 1989] – 1 – mf#7050 – us Wisconsin U Libr [430]

Julien et justine : ou, encore des ingenus / Desnoyer, Charles – Paris, France. 1828 – 1r – us UF Libraries [440]

Julien, Eugene-Louis *see* Bossuet et les protestants

Julien, Stanislas *see* Vindiciae philologicae in linguam sinicam

Juliet courier – Joliet, Juliet IL. 1840 aug 13, v2 n15, 1842 mar 26, v3 n42 – 1r – 1 – (cont by: joliet signal) – mf#1010897 – us WHS [071]

Juliette : ou, la cle des songes / Neveux, Georges – Paris, France. 1930 – 1r – us UF Libraries [440]

Julio cejador y frauca / Fernandez Larrain, Sergio – Santiago, Chile. v1-2. 1965 – 1r – us UF Libraries [972]

Julio cesar : commentariorum belli gallia – Madrid, SP. S.15 – 1,5 – sp Cultura [945]

Julio sanchez / Marin Canas, Jose – San Jose, Costa Rica. 1972 – 1r – us UF Libraries [972]

Julius cahn's official theatrical guide – New York. v. 1-2, 4-6, 8-9. 1896 97-1897, 1899 1900-1901 02, 1903 04-1904 05 – 1 – us NY Public [790]

Julius echter v. mespelbrunn, fuerstbischof von wuerzburg : ein beitrag zur geschichte der evangelischen kirche in unterfranken / Zeitler, G – Halle a S: Verein fuer Reformationsgeschichte, 1896 – (= ser [Schriften fuer das deutsche Volk]) – 1mf – 9 – 0-524-01955-X – mf#1990-0544 – us ATLA [240]

Julius friedrich : die entstehung der reformatio ecclesiarum hassiae von 1526. eine kirchenrechtliche studie – Giessen 1905 (mf ed 1994) – 2mf – 9 – €31.00 – 3-8267-3009-7 – mf#DHS-AR 3009 – gw Frankfurter [240]

Julius grosses erzaehlende dichtungen / Naegel, Andreas – Zeulenroda: B Sporn, 1938 – 1r – 1 – (incl bibl ref) – us Wisconsin U Libr [430]

Julius leopold klein als dramatiker / Glatzel, Max – Stuttgart: Metzler, 1914 [mf ed 1992] – (= ser Breslauer beitraege zur literaturgeschichte. neue folge 42) – 128p – 1 – mf#8014 reel 5 – us Wisconsin U Libr [430]

Julius mosen : ein deutscher dichter und volksmann / Zimmer, Fritz Alfred – Dresden: Verlag Heimatwerk Sachsen, v. Baensch Stiftung, 1938 – 1r – 1 – (incl bibl ref) – us Wisconsin U Libr [430]

Julius von tarent und die dramatischen fragmente / Leisewitz, Johann Anton; ed by Werner, Richard Maria – Heilbronn: Henninger, 1889 [mf ed 1993] – (= ser Deutsche litteraturdenkmale des 18. und 19. jahrhunderts 32) – lxix/143p – 1 – (incl bibl ref) – mf#8676 reel 4 – us Wisconsin U Libr [820]

Julliard, Emile *see* Tannhaeuser de richard wagner

Jullien, Adolphe *see*
- Causerie a propos de lohengrin
- Richard wagner

Jully, Antony *see* Manuel des dialectes malgaches, comprenant sept dialectes

Julvecourt, Paul de *see* Mes souvenirs de bonheur

July annual kootenay mining standard, 1899 : an illustrated journal showing the beauties and resources of the kootenays – Rossland, BC: Printed and publ by the Standard Pub Co, 1899? – 2mf – 9 – mf#17419 – cn Canadiana [622]

A july up the rhine : with one word to mr bulwer's "england and the english" – London 1834 [mf ed Hildesheim 1995-98] – (= ser Fbc) – 2mf – 9 – €60.00 – 3-487-29557-1 – gw Olms [914]

Jumbo comics – iss n11-20. dec 1939-oct 1940 – 15 – mf#001FH-002FH – us MicroColour [740]

[Jumbo-] miner – NV. apr-jul 1908 [wkly] – 1r – 1 – $60.00 – mf#U04591 – us Library Micro [071]

Jumhuri-i islami – Tehran: Hizb-i Jumhuri-i Islami. sal-i 1, shumarah-'i 1-sal-i 10, shumarah-'i 3902. 9 khurdad 1358-28 aban 1371 [30 may 1979-19 nov 1992] – 58r – 1 – $3074.00 – (many iss missing) – us MEDOC [079]

Jumhuriyyat – Kabul, Afghanistan. 4 aug 1973-20 mar 1974 – 3r – 1 – uk British Libr Newspaper [072]

Jumilhac, Pierre Benoit de *see* La science et la pratique du plain-chant

Jump cut – Berkeley. 1974-2000 (1) 1988-2000 (5) 1988-2000 (9) – ISSN: 0146-5546 – mf#10722 – us UMI ProQuest [790]

Jump n e w s / POW WOW Group – 1989 nov/dec-1990 dec/1991 jan – 1r – 1 – mf#5327623 – us WHS [071]

Junbish – Tehran, 1979-80. shumarah-'i 1-27,51-69,75-88. 20 isfand 1356-21 urdibihisht 1359 [11 mar 1978-11 may 1980] – 1r – 1 – $53.00 – us MEDOC [079]

Juncker, Alfred *see*
- Das christusbild des paulus
- Das gebet bei paulus

Junction city bulletin – Junction City OR: J B Lawrence, -1901 [wkly] – 1 – us Oregon Lib [071]

Junction City. Kansas. Church of the Covenant *see* Parish records

Junction City. Kansas. First Presbyterian Church *see* Records

Junction city times – Junction City OR: S L Moorhead, 1891-1984 [wkly] – 1 – (1925-44 incl newspaper pub by junction city high school students) – us Oregon Lib [071]

Jundt, Andre *see* Le developpement de la pensee religieuse de luther jusqu'en 1517

Jundt, Auguste *see*
- Les amis de dieu au quatorzieme siecle
- Les centuries de magdebourg, ou, la renaissance de l'historiographie ecclesiastique au seizieme siecle
- Histoire du pantheisme populaire au moyen age et au seizieme siecle

Juneau alaska empire – junau, AK: junau Alaska Empire, jul 22 1964-jul 7 1968 – us CRL [975]

Juneau city mining record – Juneau AK. 1891 mar 26 – 1r – 1 – (cont by: alaska mining record) – mf#853297 – us WHS [622]

Juneau county argus – New Lisbon WI. [1858 nov 8/1866 jul 26]-[1894 dec 6] – 12r – 1 – (cont by: new lisbon times; new lisbon times and juneau county argus) – mf#933975 – us WHS [071]

Juneau county chronicle – Elroy WI, Mauston. [1892 dec 7/1893 oct 4]-[1979 jan/dec] [gaps] – 47r – 1 – (cont: elroy chronicle; cont by: new lisbon times and juneau county argus; mauston star [mauston, wi: 1891]; juneau county star-times) – mf#1125708 – us WHS [071]

Juneau County Historical Society *see* Quarterly bulletin

Juneau County Historical Society [Wis.] *see* Juneau county history notes

Juneau county history notes / Juneau County Historical Society [Wis.] – 1965 jan-1972 jul – 1r – 1 – (cont by: quarterly bulletin [juneau county historical society [wis.]]) – mf#637624 – us WHS [071]

Juneau county plaintalker – Elroy WI. 1877 jan 3-1878 jan 25 – 1r – 1 – (cont by: plaintalker) – mf#962625 – us WHS [071]

Juneau county rundschau – Mauston WI. 1886 jul 2, 1892 jul 1 – 2r – 1 – mf#916800 – us WHS [071]

Juneau county star-times – Mauston WI. [1980 jan 3/jun 26]-[2004 dec 15/20] [gaps] – 138r – 1 – (cont: mauston star [mauston, wi: 1891]; new lisbon times and juneau county argus; juneau county chronicle) – mf#1108402 – us WHS [071]

Juneau county sun – Mauston WI. 1885 mar 13-1887 dec 8, 1887 dec 15-1890 jun 11, 1890 jun 18-1890 nov 19 – 3r – 1 – (cont: yellow river lumberman; wonewoc enterprise; cont by: mauston star; mauston star and juneau county sun) – mf#1108116 – us WHS [071]

Juneau independent – Juneau AK. 1955 feb 20-mar 6, 20-apr 10, 1956 feb 19, oct 28 – 1r – 1 – mf#867442 – us WHS [071]

Juneau telephone – Juneau WI [1880 nov 12/1881 jan 14], 1881 jan 21-1884 jul 25, 1884 aug 1-1887 nov 25, 1887 dec 2-1891 mar 27, 1891 apr 3-1894 may 18, 1894 may 25-1897 may 14, 1897 may 21-1900 aug 17, 1900 aug 24-1903 dec 25, 1904 jan 1-1907 jun 28, 1907 jul 5-1909 dec 31, 1910 jan 7-1911 jun 30, 1911 jul 7-1912 dec 27, 1913 jan 3-1914 jun 26, 1914 jul 3-1915 dec 31, 1916 jan 7-1917 jul 13, 1917 jul 20-1918 mar 15 – 16r – 1 – (cont: telephone [mayville, wi]) – mf#1012708 – us WHS [071]

Junee democrat – Junee, jan 1899-mar 1904 – 2r – A$130.33 vesicular A$141.33 silver – at Pascoe [079]

Jung, Andreas *see* Photocycloaddition von aplpha-morpholinoacrylonitril an 8-chinolincarbonsaeuremethylester

Jung, C G *see* Seelenprobleme der gegenwart

Jung, Carl Gustav *see*
- Collected papers on analytical psychology
- The integration of the personality

Jung, Hermann *see* Die beteiligung der religionsgesellschaften an staatlichen aufgaben

Jung, Joh Heinrich et al *see* Der graue mann

Jung juda – Prag (CZ), 1911-13 [gaps] – 1 – gw Misc Inst [270]

Jung kuan chi hsing / Feng, Yu-hsiang – Kuei-lin: San hu t'u shu she, Min kuo 33 [1944] – (= ser P-k&k period) – us CRL [915]

Jung, Lu *see* Fan tz'u chi

Jung ma lien / Yao, Hsueh-yin – Ch'ung-ch'ing: Ta tung shu chu, Min kuo 32 [1943] – (= ser P-k&k period) – us CRL [830]

Jung, Rudolf *see*
- Der boersenterminhandel und der dem reichstage am 19. febr. 1904 vorgelegte "entwurf eines gesetzes betr. die aenderung des abschnitts 4 des boersengesetzes"
- Die englische fluechtlings-gemeinde in frankfurt am main 1554-1559
- Goethes briefwechsel mit antonie brentano 1814-1821
- Studien zur sprachauffassung georg christoph lichtenbergs

Jung, Werner *see* Vom nullpunkt zur wende

Jungbauer, Gustav *see* Da mensch muass a freud habn

Jungbluth, Guenther *see* Untersuchungen zu heinrich von veldeke

Jungborn : jugend-organ des verbandes der bergarbeiter deutschlands – Bochum DE, 1920-22 – 1r – 1 – gw Misc Inst [622]

Der jungdeutsche – Berlin. jan 1929-june 1933 – 1 – us NY Public [325]

Der jung-deutsche orden *see* Zeitung des jungdeutschen ordens

Jungdeutschland-post – Berlin DE, 1915 [single iss], 1916 [gaps], 1917-21, 1922 [gaps] – 1 – gw Misc Inst [074]

Das junge deutschland – Berlin DE, 1926 n10, 12, 1927 n2, 1929-31 [gaps], 1933-43 [gaps] – 1 – gw Misc Inst [074]

Das junge deutschland : texte und dokumente / ed by Hermand, Jost – Stuttgart: P Reclam c1966 [mf ed 1993] – (= ser Reclams universal-bibliothek 8703-07) – 1 – (incl bibl ref) – mf#8244 – us Wisconsin U Libr [430]

Der junge donosco cortes (1809-1836) / Schramm, Edmund – Pp. 248-310 – sp Bibl Santa Ana [920]

Der junge eichendorff : ein beitrag zur geschichte der romantik / Krueger, Hermann Anders – 2. ausg. Leipzig: A Haessel 1904 [mf ed 1989] – 1r [ill] – 1 – (incl bibl ref) – mf#7212 – us Wisconsin U Libr [430]

Der junge eichendorff / Krueger, Hermann Anders – Leipzig, 1898 (mf ed 1995) – 1mf – 9 – €24.00 – 3-8267-3137-9 – mf#DHS-AR 3137 – gw Frankfurter [430]

Die junge front – Duesseldorf, Muenchen DE, 1934-36 – 1 – (title varies: 1 jul 1935: michael; cont: die allgemeine sonntagszeitung. filmed by other misc inst: 1932-1936 26 jan [1r]) – gw Misc Inst [074]

Die junge garde : zentralorgan der sozialistischen jugend deutschlands – Berlin DE, 1918 2 nov-1923 15 aug – 1r – 1 – (with suppls: proletarische jugend im bild 1925-28 [gaps]; unterhaltung und belehrung 1925 n1-1928 n19 [gaps]) – mf#3423 – gw Mikropress [335]

Der junge genosse – (Berlin-Schoeneberg) DE, 1921-24 – 1r – 1 – gw Misc Inst [074]

Der junge goethe : leben und dichtung 1765 bis 1775 / Ibel, Rudolf – Frankfurt/M: M Diesterweg, [1958?] [mf ed 1993] – viii/163p – 1 – (incl bibl ref and ind) – mf#8652 – us Wisconsin U Libr [920]

Der junge goethe : seine briefe und dichtungen von 1764-1776 / ed by Hirzel, Salomon – 2. unveraend Abruck. Leipzig: S Hirzel, 1887 [mf ed 1989] – 3v – 1 – mf#7079 – us Wisconsin U Libr [430]

Der junge goethe : selections / Morris, Max [comp] – neue ausg. Leipzig: Insel-Verlag 1909-12 [mf ed 1993] – 6v on 2r [ill] – 1 – (incl bibl ref & ind) – mf#8593 – us Wisconsin U Libr [430]

Der junge goethe / Vietor, Karl – Leipzig: Quelle & Meyer 1930 [mf ed 1990] – (= ser Wissenschaft und bildung 26) – 1r – 1 – (filmed with: faust als tragodie / benno von wiese) – mf#2678p – us Wisconsin U Libr [430]

Der junge goethe / Vietor, Karl – neue ausg. Bern: A Francke, c1950 [mf ed 1993] – (= ser Sammlung dalp 75) – 190p – 1 – mf#8652 – us Wisconsin U Libr [920]

Der junge goethe als sozialerzieher / List, Friedrich – Giessen: W Huch 1922 [mf ed 1990] – 1r – 1 – (incl bibl ref) – mf#7390 – us Wisconsin U Libr [920]

Der junge haller : nach seinem briefwechsel mit johannes gessner aus den jahren 1728-1738 / Vetter, Ferdinand – Bern: A Francke 1909 [mf ed 1990] – 1r – 1 – (incl bibl ref) – mf#2695p – us Wisconsin U Libr [920]

Der junge heinse / Schurig, Arthur – Muenchen: G Mueller 1912 [mf ed 1990] – 1r – 1 – (incl bibl ref. filmed with: die religion der goethezeit / gustav kruger) – mf#2720p – us Wisconsin U Libr [430]

Der junge herder und winckelmann / Berger, Arnold Erich – Halle a.d.S: M Niemeyer, 1903 [mf ed 1991] – 86p – 1 – mf#7472 – us Wisconsin U Libr [430]

Der junge hutten / Eggers, Kurt – Berlin: G Weise c1938 [mf ed 1989] – 1r [ill] – 1 – (filmed with: die geburt des jahrtausends) – mf#7205 – us Wisconsin U Libr [830]

Der junge josef goerres und friedrich hoelderlins hyperion / Bianchi, Lorenzo – Heidelberg: Weiss, 1926 [mf ed 1990] – 50p – 1 – (incl bibl ref) – mf#7405 – us Wisconsin U Libr [430]

Junge mannschaft : eine symphonie juengster dichtung / ed by Rockenbach, Martin – M Gladbach [i e, Moenchen-Gladbach]; Koeln: Orplid-Verlag, 1924 [mf ed 1993] – 615p – 1 – (incl bibl ref) – mf#8184 – us Wisconsin U Libr [800]

Die junge menschheit *see* Die einigkeit

Junge metall-handwerker – Frankfurt. 1979-1980 (1) 1979-1980 (5) 1979-1980 (9) – ISSN: 0022-6335 – mf#9164 – us UMI ProQuest [660]

Der junge nationalsozialist – Muenchen DE, 1932 jan-oct – 1 – gw Misc Inst [943]

Der junge pionier *see* Die trommel

Der junge schiller am rhein : ein buch von not und kampf / Braun, Max – Neustadt (Haardt): Meininger c1929 [mf ed 1989] – 1r – 1 – (filmed with: abschied von mariampol / rolf brandt) – mf#7063 – us Wisconsin U Libr [830]

Der junge schleiermacher / Haeberle, Alfred – Strassburg i Els: Elsass-Lothringische Druckerei & Lithographie-Anstalt, 1916 [mf ed 1991] – 1mf – 9 – 0-524-00367-X – (incl bibl ref) – mf#1989-3067 – us ATLA [920]

Junge stimme – Stuttgart DE, 1959 may-1971 sep – 1 – gw Misc Inst [305]

Der junge sturmtrupp – Berlin DE, 1931 22 jul-1 nov, 1932 1 jan-15 oct – 1 – gw Misc Inst [355]

Der junge tieck und seine maerchenkomoedien / Brodnitz, Kaethe – Muenchen: Walhalla-Verlag 1912 [mf ed 1991] – 1r – 1 – (incl bibl ref. filmed with: ludwig tieck als uebersetzer mittelhochdeutscher dichtung / joseph bruggemann [comp] & other titles) – mf#2913p – us Wisconsin U Libr [430]

Das junge volk – Prag (CZ), 1930-31 – 1 – gw Misc Inst [943]

Die junge welt – Berlin DE, 1961 22 sep-1968 [gaps], 1970 sep – 2r – 1 – (title varies: 1 nov 1966: welt der jugend: westdeutsche ausgabe der fdj) – gw Mikrofilm [074]

Junge welt 1939 – Berlin DE, 1939 iss3-1944 iss4 – 1r – 1 – gw Mikrofilm [074]

Junge welt 1947 : [main edition] – Berlin DE, 1992- – 5r/yr – 1 – (filmed by other misc inst: 1950 3 jan-1990 5 dec [98r]) – gw Misc Inst [074]

Junge welt / b – Berlin DE, 1952 may-aug – 1r – 1 – gw Misc Inst [305]

Der junge wieland : wesensbestimmung seines geistes / Hoppe, Karl – Leipzig: J J Weber 1930 [mf ed 1992] – 1r – 1 – (incl bibl ref. filmed with: wieland und martin und regula kuenzli / ludwig hirzel) – mf#3050p – us Wisconsin U Libr [430]

Das junge zentrum – Berlin DE, 1924 aug-1933 jul – 2r – 1 – gw Mikrofilm [360]

Jungen der fernen grenze : gedichte aus der kampfzeit / Buchbauer, Oskar – Hallein: Im Selbstverlage des Verfassers 1938 [mf ed 1989] – 1r – 1 – (filmed with: hofische spuren im protestantischen schuldram um 1600 / hildegard schaefer) – mf#7093 – us Wisconsin U Libr [810]

Jungens, maenner und motore / Stoll, Friedrich Albert Robert – Berlin: Volksverband der Buecherfreunde, 1940.86p. illus – 1 – us Wisconsin U Libr [355]

Jungfer, Victor *see* Irka: roman

Jungfern im nebel : und andere luegenhafte geschichten / Blunck, Hans Friedrich – Prag: Noebe, 1944 [mf ed 1989] – (= ser Feldpostreihe noebe 22) – 61p – 1 – mf#7037 – us Wisconsin U Libr [830]

Die jungfrau von orleans in der dichtung / Grenzmann, Wilhelm – Berlin: W de Gruyter & Co, 1929 – 2r – 1 – (incl bibl ref) – us Wisconsin U Libr [430]

Die jungfrau von orleans in der dichtung : shakespeare, voltaire, schiller / Kummer, Carl Ferdinand – Wien: A Hoelder, 1874 – 1r – 1 – (incl bibl ref) – us Wisconsin U Libr [410]

Jungfrauengeburt = The virgin birth / Gruetzmacher, Richard H – New York: Eaton & Mains; Cincinnati: Jennings & Graham, c1907 – 1mf – 9 – 0-8370-3416-7 – (in english. incl bibl ref) – mf#1985-1416 – us ATLA [240]

Jungfreudig volk : gedichte / Diederich, Franz – Berlin: Arbeiterjugend-Verlag, 1925 [mf ed 1989] – 45p – 1 – mf#7176 – us Wisconsin U Libr [810]

Junghans, Hermann August *see* Sebastian brants narrenschiff

Junghuhn, F W *see*
- Java
- Topographische und naturwissenschaftliche reisen durch java

Junghuhn, Franz *see* Rueckreise von java nach europa mit der sogenannten englischen ueberlandpost im september und october 1848

The jungle book / Kipling, Rudyard – London: Macmillan, 1922 – (= ser Samp: indian books) – (ill by j l kipling, wh drake, and p frenzeny) – us CRL [830]

Jungle comics – iss n1-15. jan 1940-mar 1941 – 15 – mf#003FH-005FH – us MicroColour [740]

Jungle days / Beebe, Charles William – New York, NY. 1925 – 1r – us UF Libraries [972]

Jungle days : being the experiences of an american woman doctor in india / Munson, Arley Isabel – New York; London: D Appleton, 1913 [mf ed 1995] – (= ser Yale coll) – viii/297p (ill) – 1 – 0-524-09436-5 – mf#1995-0436 – us ATLA [610]

Jungle gods / Hoffman, Carl von; ed by Lohrke, Eugene – New York: H Holt, [c1929] – us CRL [290]

Jungle gods / Von Hoffman, Carl – London, England. 1929 – 1r – us UF Libraries [960]

Jungle jargon – United States. v1 n173-205 [1945 feb 6-mar 16] – 1r – 1 – mf#3462410 – us WHS [071]

Jungle journey / Waldeck, Jo Besse Mcelween – New York, NY. 1946 – 1r – us UF Libraries [972]

Jungle lore / Corbett, Jim – Bombay: Oxford University Press, 1953 – (= ser Samp: indian books) – us CRL [954]

Jungle man / Pretorius, Philip Jacobus – London, England. 1947 – 1r – us UF Libraries [960]

Jungle peace / Beebe, Charles William – New York, NY. 1918 – 1r – us UF Libraries [972]

Jungle peace / Beebe, William – New York, NY. 1920 – 1r – us UF Libraries [972]

Jungle pioneering in gondland / McMillan, A W – London, 1906 – 2mf – 9 – mf#HT-90 – ne IDC [915]

Jungmaedel auf dem koellingshof / Perzl, Irmgard – Reutlingen: Ensslin & Laiblin, 1941 – 1r – 1 – us Wisconsin U Libr [943]

Jungmann, B *see* Institutiones patrologiae quas denuo recensuit auxit

Jungmann, Bernardus *see*
- Brevis analysis tractatus de deo creatore
- Tractatus de deo uno et trino
- Tractatus de gratia
- Tractatus de novissimis
- Tractatus de vera religione
- Tractatus de verbo incarnato

Jungmann, J A *see*
- Gewordene liturgie
- Die stellung christi im liturgischen gebet. l t q 19/20

Jungmann, Max *see*
- Schlemiel

Jungniss, J *see* Die breslauer ritualien

Jungnitz, Joseph *see* Die breslauer ritualien

Jungsozialisten in der spd – Misc. publications. University of Wisconsin Libraries, Madison, 1986 – 1 – us Wisconsin U Libr [335]

Jungsozialistische blaetter – Nuernberg DE, 1922-1931 jul – 2r – 1 – mf#6142 – gw Mikropress [335]

Jung-Stilling, Johann Heinrich *see* Abhandlungen des staatswirthschaftlichen instituts zu marburg

Jungt-weker *see* Yidisher heftlings-kongres in bergn-belzn

Jungvolk – Berlin DE, 1915 – 1r – 1 – gw Misc Inst [305]

Jungvolk – Muenchen DE, 1932-34 – 1 – gw Misc Inst [943]

Jungvolk – Zentralstelle fuer die arbeitende jugend Deutschlands. 1913-21. illus. -ann. (Serial publications of German trade unions in the Memorial Library, University of Wisconsin-Madison.) – 1 – us Wisconsin U Libr [331]

Junia – Port-au-Prince: Imp V Pierre-Noel. 1ere annee n2-3eme annee n29. juin 1924-mai 1928 – us CRL [972]

Juniata bible lectures : a series of twelve lectures... / Brumbaugh, Martin Grove – [S.l: s.n, 1897?] (Philadelphia: Avil Printing Co) – 1mf – 9 – 0-524-03540-7 – mf#1990-4735 – us ATLA [220]

Juniata herald – Mifflintown, PA. -w 1889-1912 – 13 – $25.00r – us IMR [071]

The juniata herald – Juniata, NE: Wm Knickerbocker, 1876-v43 n5. nov 28 1917 (wkly) [mf ed nov 28 1877-nov 28 1917 (gaps)] – 14r – 1 – us NE Hist [071]

Juniata sentinel – Ebensburg, PA., 1971-1975 – 13 – $25.00r – us IMR [071]

Juniata sentinel – Mifflintown, PA. -w 1860 – 13 – $25.00r – us IMR [071]

Juniata star – Mifflintown, PA. -w 1897-1899 – 13 – $25.00r – us IMR [071]

Juniata tribune – Mifflintown, PA. -w 1891-1896; 1918-1932; 1962-1970 – 13 – $25.00r – us IMR [071]

Juniata true democrat – Mifflintown, PA. Circa 1866 – 13 – $25.00r – us IMR [071]

Junii, Francisci Biturgis *see* Opera theologica

Junior bookshelf – Huddersfield. 1936-1996 (1) 1970-1996 (5) 1975-1996 (9) – ISSN: 0022-6505 – mf#1387 – us UMI ProQuest [070]

Junior boy – 1930-31 – 1 – 9.45 – us Southern Baptist [242]

The junior department of the church school / Athearn, Walter Scott – [Des Moines, Iowa: Dept of Religious Education, Drake University], c1913 – (= ser Outline Studies of the Departments of the Church School) – 1mf – 9 – 0-524-07805-X – mf#1991-3352 – us ATLA [240]

Junior gazette – Salem OR: Wilfred C Hagedorn [wkly] – 1 – us Oregon Lib [071]

Junior girl – 1930-31 – 1 – 9.38 – us Southern Baptist [242]

The junior herald – Lincoln, NE. 1919-20 – 1 – us AJPC [071]

Junior historian newsletter / State Historical Society of Wisconsin – [1968 oct-1979 may] – 1r – 1 – mf#642973 – us WHS [978]

Junior language lessons for first, second, and third classes / Henderson, George E et al – Toronto: Educational Pub Co, 1898 – (= ser School helps series) – 2mf – 9 – 0-665-92811-4 – mf#92811 – cn Canadiana [420]

Junior leader – 1924-55. (Formerly: The Junior Leader's B.Y.P.U. Quarterly. 1925-39) – 1 – us Southern Baptist [242]

The junior league hand-book : devoted to junior league methods of work / Bartlett, S T [comp] – Toronto: W Briggs; Montreal: C W Coates, 1897 – 2mf – 9 – mf#03384 – cn Canadiana [240]

Junior northwestern / Wisconsin Evangelical Lutheran Synod – v59 n1-v63 n12 [1977 jan-1981 dec] – 1r – 1 – mf#593817 – us WHS [242]

Junior notes – 1935 may-nov – 1r – 1 – mf#2699154 – us WHS [071]

Junior projects – Leamington Spa. 1989-1991 (1) 1990-1990 (5) 1990-1990 (9) – ISSN: 0269-9532 – mf#14068,01 – us UMI ProQuest [373]

Junior quarterly – Jan 1930-61 – 1 – us Southern Baptist [242]

Junior scholastic – New York. 1937+ (1) 1970+ (5) 1975+ (9) – ISSN: 0022-6688 – mf#393 – us UMI ProQuest [373]

Junior societies of christian endeavor / Clark, Francis Edward – Boston: United Society of Christian Endeavor, 1888 – 1mf – 9 – mf#27445 – cn Canadiana [366]

Junior teacher's and pupil book : years 1-4 – 1923-28 – 1 – us Southern Baptist [242]

Junior training union quarterly – 1922-61 – 1 – us Southern Baptist [242]

Junior weekly – 1931 – 1 – 5.74 – us Southern Baptist [242]

Junior's portfolio / Wisconsin Christian Endeavor Union – v1 n1-2 [1895 may-jul] – 1r – 1 – mf#1060928 – us WHS [071]

Juniperus : geschichte eines kreuzfahrers / Scheffel, Joseph Viktor von – 6. Aufl. Stuttgart: A Bonz, 1908 (mf ed 1990) – 1r – 1 – (filmed with: goethes faust in urspruenglicher gestalt) – us Wisconsin U Libr [830]

Juniu, Susana *see* Music

Junius : schauspiel in vier akten – [S.l: s.n, 18–?] [mf ed 1993] – 158p – 1 – mf#8456 – us Wisconsin U Libr [820]

Junius, F *see* De pictura veterum libri tres...

Junius, H *see*
- De emblemata van hadrianus junius
- Eiusdem aenigmatum libellus
- Emblemata
- Emblemata, ad d arnoldum cobelium
- Emblemata adriani iunii medici

Junk, Victor *see*
- Alexander
- Rudolfs von ems willehalm von orleans

Junker, H *see*
- Das goetterdekret ueber das abaton
- Die onuris-legende

Junker, Hermann *see* Nubische texte im kenzi-dialekt

Junker, Karl Ludwig *see*
- Tonkunst
- Ueber den werth der tonkunst
- Zwanzig componisten

Die junker, roman / Zobeltitz, Fedor Karl Maria Hermann August von – Berlin: Ullstein, 1918. 443p – 1 – us Wisconsin U Libr [830]

Junker von Langegg, Ferdinand Adalbert *see* Krypto-monotheismus in den religionen der alten chinesen und anderer voelker

Junker, W *see* Reisen in afrika 1875-1886

Junkin, David Xavier *see* The kingdom of god, its constitution and progress

Junkin, David Xavier et al *see* Centenary memorial of the planting and growth of presbyterianism in western pennsylvania and parts adjacent

Jun-lipanj 1968 dokumenti = [Zbornik dokumenata o studentskim zbivanjima u jugoslaviji u junskim danima 1968] – Zagreb: Hrvatsko filozofsko drustvo, 1971 – us CRL [949]

Junod, Henri Alexandre *see*
- Les chants et les contes des ba-ronga de la baie de delagoa
- Elementary grammar of the thonga-shangaan language
- Life of a south african tribe

Junquiera freire / Pires, Homero – Rio de Janeiro, Brazil. 1929 – 1r – us UF Libraries [972]

Junta Civico-Militar Cubana *see* Organ oficial de la junta civico-militar cubana

Junta Cubana De Nueva York *see* Facts about cuba

Junta de agricultura...ensenanza agricola / Paredes Guillen, Vicente – 1871 – 9 – sp Bibl Santa Ana [630]

Junta de aranceles. proyecto de dictamen respecto...al arancel de 23 de marzo de 1906 – Madrid, 1906 – 7mf – 9 – sp Cultura [946]

Junta de Cofradias de Penitencia *see* Horario e itinerario de las procesiones de semana santa...1962

Junta de Cofradias de Penitencia y C.I.T.E. *see*
- Semana santa. badajoz, 1968
- Semana santa. badajoz, 1969, 70

Junta de la habana en 1808 / Ponte Dominguez, Francisco J – Habana, Cuba. 1947 – 1r – us UF Libraries [972]

Junta Diocesana de Accion Catolica. (Coria) *see* Jesucristo redentor

Junta Diocesano de Accion Catolica. (Coria) *see* Reglamento de los secretarios de caridad de accion catolica

La junta para ampliacion de estudios. rectificacion y comentarios / Bayle, Constantino – Madrid: Razon y Fe, 1924 – 1 – sp Bibl Santa Ana [946]

Junta Provincial de Educacion Fisica y Deportes *see*
- Anuario deportivo 1969
- Eleccion al mejor deportista provincial de badajoz-1970

Junta Provincial de Fomento Pecuario *see*
- Labor desarrollada desde su creacion hasta fin de diciembre de 1951
- Memoria-indice de los trabajos efectuados durante los anos 1938-1944. inclusives

Junta Provincial de Fomento Pecuario. Badajoz *see*
- Cursillo sobre explotaciones ovinas en su aspecto de produccion de lana
- Ganado lanar
- Primer concurso provincial de ganado lanar

Junta Provincial de Informacion, Turismo y Educacion Popular *see*
- Caceres. mapa guia turistico provincial
- Ruta de la alta extremadura, cuna de los conquistadores

Junta Provincial de Turismo *see*
- Alcantara
- Caceres
- Catalogo del concurso-exposicion de fotografias sobre temas cacerenos
- Coria
- Guadalupe
- Hervas
- Trujillo, cuna de conquistadores. montanchez, el balcon de extremadura

La junta tribune *see* Miscellaneous newspapers of otero county

Junto selections : essays on the history of pennsylvania / Pennsylvania Historical Junto – Washington, DC: Pennsylvania Historical Junto, 1946 (mf ed 19–) – (= ser Pennsylvania history on microfilm) – 63/[1]p – mf#ZH-IAG pv832 n4 – us NY Public [978]

Jupiter / Boissy, Robert – Paris, France. 1942 – 1r – us UF Libraries [440]

Jupiter island : a grant from the spanish governmen... / Withington, Chester Merrill – s.l, s.l? 1935 – 1r – us UF Libraries [978]

Der jura / Quenstedt, Friedrich A – Tuebingen 1858 [mf ed Hildesheim 1995-98] – 2v on 7mf – 9 – €140.00 – 3-487-29396-X – gw Olms [914]

Jura israelitarum in paelestinam terram chananaeam commentatione in genesin... / Witter, H B – Hildesiae: Pauli Pastore, 1711 – 8mf – 8 – mf#1089 – ne IDC [956]

Jurado E, Gerardo A *see* Conservatismo una ideologia cristiana

Jurado, Ramon H *see*
- Desertores
- San cristobal

Jurandir, Dalcidio *see* Marajo

Juranville, Clarisse *see*
- La civilite des petites filles
- Manuel d'education morale et d'instruction civique

Juras Reales, Baron de *see* El espiritu del siglo

Las jurdes etude de geographia humaine / Legendre, Maurice – Bordeaux: Feret fils, edoteurs, 1927 – 1 – sp Bibl Santa Ana [550]

Las jurdes y su leyendas / Barrantes Moreno, Vicente – 1891 – 9 – sp Bibl Santa Ana [390]

Jurema, Abaelardo *see* Sexta-feira, 13

Jurgensen, Elmer V *see* Papers, 1847-1984

Juridic status of the catholic church and religious orders under the constitutional monarchy – n.p., 193? Fiche W977. (Blodgett Collection of Spanish Civil War Pamphlets) – 9 – us Harvard [946]

Juridical analysis and critical evaluation of ilibolo in a changing zulu society / Dlamini, Charles Robinson Mandlenkosi – U of Zululand 1983 [mf ed S.l: s.n. 1983] – 12mf – 9 – sa Misc Inst [306]

Juridical arguments and collections / Hargrave, Francis – London: Robinson, 1797-99. 2v. LL-993 – 1 – us L of C Photodup [340]

Juridical review – Edinburgh. v1-23. 1889-1912 – 128mf – 9 – $192.00 – (add vols as copyright expires) – mf#LLMC 84-509 – us LLMC [340]

Juridical review – Edinburgh. 1889-1900 (1) – ISSN: 0022-6785 – mf#2909 – us UMI ProQuest [340]

Juridical society papers – Papers read before the Juridical Society. v1-4. 1855-74 (all publ) – 25mf – 9 – $37.50 – mf#LLMC 84-510 – us LLMC [340]

Juridical techniques and the judicial process / Epstein, A L – (= ser Institute for social research, university of zambia. papers 23) – 2mf – 7 – mf#4734 – uk Microform Academic [340]

Juridiction exercee par l'archeveque de rouen / Gosselin, Auguste – Evreux, France?: [Impr de l'Eure], 1895 – 1mf – 9 – mf#03475 – cn Canadiana [241]

Jurieu, P *see*
- Apologie pour la reformation, pour les reformateurs, et pour les reformez
- Histoire critique des dogmes et des cultes, bons et mauvais...
- Lettres pastorales adressees aux fideles de france...
- La politique du clerge de france
- Prejugez legitimes contre le papisme
- Le vray systeme de l'eglise...

[Jurieu, P] *see*
- L'accomplissement des propheties
- Apologie pour la morale des reformez...
- Examen du livre de la reunion du christienisme
- Preservatif contre le changement de religion

Jurimetrics – Chicago. 1959+ (1) 1970+ (5) 1975+ (9) – ISSN: 0897-1277 – mf#1641 – us UMI ProQuest [340]

Jurimetrics journal (aba) – v1-39. 1959-99 – 9 – $549.00 set – (title varies: v1-7 1959-65 as: m u l l modern uses of logic in law) – ISSN: 0897-1277 – mf#104361 – us Hein [340]

Juris consultos cubanos / Valverde Y Maruri, Antonio L – Habana, Cuba. 1932 – 1r – us UF Libraries [972]

Juris doctor – New York. 1971-1977 (1) 1971-1977 (5) 1975-1977 (9) – ISSN: 0047-3014 – mf#8017 – us UMI ProQuest [340]

Jurisdiccao diocesana do bispado de s thome de meliapor nas possessoes inglezas e francezas : averiguacao de successos antigos por occasiao de outros modernos na igreja de royapuram de madrasta / Rivara, Joaquim Heliodoro da Cunha – Nova-Goa: Imprensa Nacional, 1867 [mf ed 1995] – (= ser Yale coll) – 458p – 1 – 0-524-10223-6 – (in portuguese) – mf#1996-1223 – us ATLA [241]

The jurisdiction and practice at large in city judges', mayors', and justices' courts of the state of indiana...2nd ed / Spalding, Hugh Mortimer – Cincinnati, Wilstach, Baldwin, 1876. 789 p. LL-1293 – 1 – us L of C Photodup [347]

Jurisdiction of submerged lands in american samoa, guam, and the virgin islands : hearing before the subcommittee on territorial and insular affairs of the house committee on interior and insular affairs / American Samoa. US Congress – 93rd Congress 1st sess 25 Sept 1973. Washington: GPO, 1974 – 1mf – 9 – $1.50 – mf#LLMC 95-041 – us LLMC [327]

Jurisdiction of the federal courts. rev. ed / Thayer, Amos Madden – St. Louis, Brewer 1900. 52 p. LL-1222 – 1 – us L of C Photodup [347]

Jurisdiction, practice, and peculiar jurisprudence of the courts of the united states / Curtis, Benjamin Robbins – Boston, Little, Brown, 1880. 298 p. LL-1301 – 1 – (2nd ed., boston, little, brown, 1896. 341 p. ll-1276) – us L of C Photodup [347]

Jurispridencia del Tribunal Supremo *see* Codigo penal para las islas de cuba y puerto rico

Jurisprudence canadienne, index analytique des decisions juiciaires rapportees de 1864 a 1871 : dans les volumes 8, 9, 10, 11, 12, 13 et 14 du jurist; 14, 15, 16 et 17 des reports; 1, 2, 3 et 4 du law journal; 1 et 2 de la revue legale / Lusignan, Alphonse – Montreal: s.n, 1872 – 4mf – 9 – (in french and english; incl ind) – mf#10614 – cn Canadiana [348]

Jurisprudence, law and ethics : professional ethics / Kinkead, Edgar Benton – New York: the Banks Publ Co, 1905 – 4mf – 9 – $6.00 – mf#LLMC 95-178 – us LLMC [170]

Jurisprudence medicale : examen medico-legal des proces d'anais toussaint, de joseph berube et de cesaree theriault et precis de procedures a suivre dans les cas d'empoisonnements par l'arsenic et le phosphore / Emery-Coderre, Joseph – [Montreal?: s.n.] 1857 [mf ed 1985] – 9 – 0-665-01720-0 – mf#01720 – cn Canadiana [614]

Jurisprudencia – Maceio, AL. 05 ago, dez 1894; fev-07 mar 1895 – (= ser Ps 19) – mf#P18B,01,13 – bl Biblioteca [340]

Jurisprudencia argentina – Buenos Aires. On film: v1-211; 1918-71; indexes: 1918-66. LL-0285 – 1 – us L of C Photodup [340]

Jurisprudencia civil / Spain. Tribunal Supremo – Madrid. On film: v1-224; 1855-1936. LL-0152 – 1 – us L of C Photodup [347]

Jurisprudencia constitucional de la corte suprema – Bogota, Colombia. v1-2. 1963 – 1r – us UF Libraries [972]

Jurisprudencia criminal / Spain. Tribunal Supremo – Madrid. On film: v1-136; 1871-1937. LL-0151 – 1 – us L of C Photodup [345]

Jurisprudencia de la corte suprema de justicia / Gaitan, Luis Alejandro – Bogota, Colombia. 1942 – 1r – us UF Libraries [972]

Jurisprudencia de los tribunales of colombia / Colombia – Bogota, Colombia. v1-2. 1908-1910 – 1r – us UF Libraries [972]

Jurisprudencia del trabajo / Parisca Mendoza, Carlos – Curacas, Venezuela. 1964 – 1r – us UF Libraries [972]

Jurisprudencia en las republica dominicana / Gaton Richiez, Carlos – Santiago, Dominican Republic. 1943 – 1r – us UF Libraries [972]

Jurisprudencia minera : comentada y explicada / Sarria, Eustorgio – s.l, s.l? 1950? – 1r – us UF Libraries [972]

Jurisprudencia penal e casacion / Barreto Rodriquez, Jesus – Caracas, Venezuela. 1963 – 1r – us UF Libraries [972]

'N jurisprudensiele ontleding van die staatlike paradigma en van staatlike identiteit / Malan, Jacobus Johannes – U of South Africa 2000 [mf ed Johannesburg 2000] – 7mf – 9 – (incl bibl ref; text in afrikaans; abstract in afrikaans & english) – mf#mfm15110 – sa Unisa [320]

Jurist / Catholic University of America – v56 n1-2 [1996] – 1r – 1 – mf#166233 – us WHS [340]

Jurist – London. v1-30. 1837-67 (all publ) – (= ser Historical legal periodical series) – 1 – $1155.00 – mf#409050 – us Hein [340]

Jurist – Washington. 1959+ (1) 1971+ (5) 1977+ (9) – ISSN: 0022-6858 – mf#3051 – us UMI ProQuest [340]

The jurist : a journal for law students and the profession – London. v1-5. 1887-91 (all publ) – 21mf – 9 – $31.50 – (lacking: v5 p93-136) – mf#LLMC 84-512 – us LLMC [340]

The jurist : or quarterly journal of jurisprudence and legislation – London. v1-4. 1827-33 (all publ) – 9 – mf#LLMC 84-513 – us LLMC [340]

The jurist – Suffolk University Law School, 1926-28 (all publ) – 9 – mf#LLMC 84-511 – us LLMC [340]

The jurist, new series : containing reports of cases dtermined in the courts of law and equity, and...in the admiralty and ecclesiastical courts, with a general digest of all the reports published / Great Britain – v1-12. 1855-67. London: Sweet/Stevens, 1856-57 – 276mf – 9 – $414.00 – (all vols have a pt2, serving same purpose as with v6-18 of old series. v4 new series, 1858 sees expansion of coverage to include a section on divorce and probate courts) – mf#LLMC 95-268 – us LLMC [347]

The jurist, old series : containing reports of cases determined in the courts of law and equity, and...in the admiralty and ecclesiastical courts, with a general digest of all reports published / Great Britain – v1-18. 1837-54: London: Sweet/Stevens, 1838-55 (all publ) – 340mf – 9 – $510.00 – (v6-18 have a pt2 containing articles on legal subjects, all important statutes, miscellaneous legal information etc. v11 1847 has a suppl book) – mf#LLMC 95-267 – us LLMC [347]

The juristic status of egypt... / O'Rourke, A – Baltimore, 1935 – 2mf – 9 – mf#ILM-1941 – ne IDC [956]

Juristische abhandlung ueber die floehe / Goethe, Johann Wolfgang von – Altona, 1866 (mf ed 1995) – 1mf – 9 – €24.00 – 3-8267-3119-0 – mf#DHS-AR 3119 – gw Frankfurter [340]

Juristische aufbauschemata im zivil- und arbeitsrecht, strafrecht, staats- und verwaltungsrecht mit kurzkommentierungen / Wittmer, Stephan et al – Neuwied, Frankfurt/Main: Metzner, 1989 (mf ed 1996) – 3mf – 9 – €38.00 – 3-8267-9673-X – mf#DHS 9673 – gw Frankfurter [340]

Juristische bibliothek / ed by Haselberg, Gabriel Peter – Goettingen 1788-90 – (= ser Dz) – 2v on 8mf – 9 – €160.00 – mf#k/n2592 – gw Olms [340]

Juristische fremdwoerter, fachausdruecke und abkuerzungen sowie registerzeichen der ordentlichen gerichtsbarkeit einschl der arbeitsgerichte und des bundesverfassungsgerichts / Meyer, Dieter – Neuwied, Krftel, Berlin: Luchterhand, 1993 (mf ed 1996) – 3mf – 9 – €38.00 – 3-8267-9681-0 – mf#DHS 9681 – gw Frankfurter [340]

Juristische wochenschrift.. – Organ des deutschen anwalt-vereins. Berlin. 1881-1939.Suppl., 1900-02. General register, 1879-1900 – 1 – us L of C Photodup [943]

Juristisches magazin / ed by Siebenkees,Johann Christian – Jena 1782-83 – (= ser Dz) – 2v on 8mf – 9 – €160.00 – mf#k/n2570 – gw Olms [340]

Juristisches magazin fuer die deutschen reichsstaedte / ed by Jaeger, Tobias Ludwig Ulrich – Ulm 1790-97 – (= ser Dz) – 6v on 22mf – 9 – €220.00 – mf#k/n2599 – gw Olms [342]

Juristisches wochenblatt / ed by Schott, August Friedrich – Leipzig 1772-75 – (= ser Dz) – 4jge on 19mf – 9 – €190.00 – (jg4 with title: magazin fuer rechtsgelehrte und geschichtsforscher) – mf#k/n2541 – gw Olms [340]

Juristische-zeitung – Vienna. may 1870-jan 1871 – 1r – 1 – us UMI ProQuest [340]

Jurnal ekuin : harian ekonomi umum – Jakarta: Sistim Multi Media 1981 (daily ex sun) [mf ed Honolulu, HI: University of Hawaii at Manoa, University Libraries 1990] – 5r [apr 30 1981-mar 12 1983 [gaps]] – 1 – (numbering irreg; lacks: 1981 may 5-6,8; 1982 apr 9, jun 7, nov 30) – mf#mf-7182 seam – us CRL [339]

Jurnalul de dimineata – Bucharest. Rumania. -d. 14 Nov 1944-10 Jul 1947. (Imperfect). (5 reels) – 1 – uk British Libr Newspaper [949]

The juror: being a guide to citizens summoned to serve as jurors / Reilly, Andrew Jackson – Philadelphia, Campbell, 1873. 108 p. LL-1280 – 1 – us L of C Photodup [340]

Jurrens, Jay D *see* The effects of hang board exercise on grip strength and climbing performance in college age male indoor rock climbers

Jury selection procedures in u.s. district courts / Bermant, Gordon – Washington: FJC, June 1982 – 1mf – 9 – $1.50 – mf#LLMC 95-372 – us LLMC [347]

Jury service in lengthly civil trials / Cecil, Joe S et al – Washington: GPO, 1988 – 1mf – 9 – $1.50 – mf#LLMC 95-336 – us LLMC [340]

The jury system–defects and proposed remedies / Train, Arthur Cheney – Philadelphia, American Academy of Political and Social Science 1910 p. 175-184. LL-1467 – 1 – us L of C Photodup [340]

The juryman's handbook / Brown, Alec – London: The Harvill Press, 1951 – 2mf – 9 – $3.00 – mf#LLMC 94-280 – us LLMC [340]

Jus majestatis circa sacra... / Apollonius, W – Middelburg, 1642 – 5mfmf – 9 – mf#PBA-123 – ne IDC [240]

Jus populi divinum / Currie, John – Edinburgh, Scotland. 1841 – 1r – us UF Libraries [240]

Jus potandi : oder, deutsches zechrecht: commentbuch des mittelalters nach dem original von 1616 mit einleitung / ed by Oberbreyer, Max – 6. Aufl. Heilbronn: Henninger, [1890?] – 1r – 1 – (includes bilbiographical references) – us Wisconsin U Libr [943]

Jus suffragii / International Woman Suffrage Alliance – Rotterdam. sept 1906-16 [mnthly] – 1 – (cont by: international woman suffrage news) – us Wisconsin U Libr [305]

Juscelino kubitschek / Montevideo Instituto De Cultura Uruguayo-Brasilen – Montevideo, Uruguay. 1958 – 1r – us UF Libraries [972]

Jusdorf, J C *see* Air avec 24 variations pour l'etude de la flute, op 1

Jussawalla, J M *see* Living the vegetarian way

Jusselain, Armand *see* Deporte a cayenne

Jusserand, Jean Adrien Antoine Jules *see*
- English wayfaring life in the middle ages (14th century)
- Shakespeare in france under the ancien regime

Jusserand, Jean Jules *see* English wayfaring life in the middle ages (14th century)

The just and the unjust / Kester, Vaughan – Toronto: McLeod & Allen, c1912 [mf ed 1995] – 5mf – 9 – 0-665-74713-6 – mf#74713 – cn Canadiana [830]

Just before the dawn : the life and work of ninomiya sontoku / Armstrong, Robert Cornell – New York: Published by Macmillan for the Young People's Forward Movement for Missions, 1912 – 1mf – 9 – 0-524-00679-2 – mf#1990-2007 – us ATLA [240]

Just economics / Movement for Economic Justice – v1 n1-v8 n4 [1973 oct/nov-1980 dec] – 1r – 1 – (cont by: organizer [washington, dc]) – mf#674313 – us WHS [330]

Just flesh / Karaka, Dosoo Framjee – Bombay: Thacker & Co, 1941 – (= ser Samp: indian books) – us CRL [830]

Just for us magazine – Dallas TX. 1997 may – 1r – 1 – mf#3907756 – us WHS [071]

Just, Friedr Aug *see* Ossnabrueggische unterhaltungen

Just friedrich wilhelm zachariae und sein renommist : ein beitrag zur litteratur- und kulturgeschichte des 18. jahrhunderts / Zimmer, Hans – Leipzig: Rossberg, 1892 – 1 – (incl bibl ref) – us Wisconsin U Libr [430]

Just, Gustav A *see* Life of luther

Just, Klaus Guenther *see*
- Afrikanische trauerspiele
- Roemische trauerspiele
- Tuerkische trauerspiele
- Wissenschaft als dialog

Just one blue bonnet : the life story of ada florence kinton, artist and salvationist / Kinton, Ada Florence; ed by Randleson, Sara A – Toronto: W Briggs, 1907 [mf ed 1997] – 3mf – 9 – 0-665-83535-3 – mf#83535 – cn Canadiana [242]

Just out : "oregon's lesbian and gay newsmagazine" – Portland OR: Just Out [semimthly] – 1 – us Oregon Lib [305]

Just so stories for little children / Kipling, Rudyard – Toronto: G N Morang, 1902 [mf ed 1995] – 3mf – 9 – 0-665-77321-8 – mf#77321 – cn Canadiana [830]

Just, Thomas Cook *see* The official hand-book of tasmania

Justa repulsa...teatro critico...francisco soto y marne / Feyjoo, B Geronimo – 1749 – 9 – sp Bibl Santa Ana [440]

Justas : y debidas honras, que hicieron, y hacen sus proprias obras, a la m r m maria anna agueda de s ignacio, primera priora... / Villa Sanchez, Juan de – en la Puebla: impr Miguel de Ortega, y Bonilla [1758?] – (= ser Books on religion...1543/44-c1800: ordenes, etc: dominicos) – 1mf – 9 – mf#crl-191 – ne IDC [241]

Justas literarias de san juan organizadas por el excelentisimo ayuntamiento de badajoz / Badajoz – Badajoz: Dip. Prov., 1944 – sp Bibl Santa Ana [946]

Justas poeticas sevillanas del siglo 16 (1531-1542) – Valencia: Edit Catalia, 1955 – sp Bibl Santa Ana [810]

Juster, Jean *see* Les juifs dans l'empire roman

Justi, F *see* Der bundehesh

Justi, Ferdinand *see*
- Der bundehesh
- Geschichte des alten persiens

Justi, Johann Heinrich Gottlieb von *see* Goettingische policey-amts-nachrichten

Justi, Karl *see* Diego velazquez and his times

Justica – Rio de Janeiro, RJ: [s.n.] 1887 – (= ser Ps 19) – mf#P17,02,105 – bl Biblioteca [340]
Justica / Soares, Jose Carlos De Macedo – Paris, France. 1925 – 1r – us UF Libraries [972]
Justice – 1906 jul 13, 1907 apr 27-jun 8,22 – 1r – 1 – mf#3043884 – us WHS [340]
Justice – 1908 oct 8 – 1r – 1 – mf#1154642 – us WHS [340]
Justice – Bloomington, NE: Continent Print & Pub Co (wkly) [mf ed v1 n43. jun 26 1886 filmed 1973] – 1r – 1 – us NE Hist [071]
Justice – Chattanooga, TN: Horn, Wilson & Co, 1887-1888?// [mf ed 1947] – (= ser Negro Newspapers on Microfilm) – 1r – 1 – us L of C Photodup [071]
Justice – Fort-de-France, Martinique. 1965-1987 (1) – mf#68630 – us UMI ProQuest [079]
Justice – Fort-de-France, Martinique. 1991-1996 – 9r – (gaps) – us UF Libraries [079]
Justice / International Ladies Garment Workers' Union – Official Organ. New York. v. 1-48. Jan 18 1919-1968 – 1 – us NY Public [330]
Justice – Jersey City. N.J. Gerechtigkeit. 1919-55 – 1 – us AJPC [071]
Justice – Kingston. 1975-1975 (1) – mf#9708 – us UMI ProQuest [331]
Justice : official organ of the international ladies garment workers union – 1983-87, 1988-92 – 2r – 1 – (cont: ladies' garment worker; cont by: justicia [new york, ny]; 17:labor unity [new york, ny: 1982]; unite!) – mf#1388994 – us WHS [331]
Justice : official organ of the international ladies garment workers union – New York: The Union. v1 n1-3,5-50. 1919 – us CRL [331]
Justice : organ of the social democracy – London, 1884-1925 – 32r – 1 – uk British Libr Newspaper [335]
Justice – Providence, RI. 1894-1895 (1) – mf#66332 – us UMI ProQuest [071]
Justice – v1 n52-v15 n27 [1989 feb 3-1902 sep 3] – 1r – 1 – mf#903956 – us WHS [340]
La justice – Central Falls, RI. 1906-1910 (1) – mf#66181 – us UMI ProQuest [071]
La justice – Paris: L Boure, may 10,12-14,17,19, 1871 – (filmed as part of: commune de paris newspapers) – us CRL [074]
La justice – Holyoke, MA: [s.n.], 1936-1939; 1945-jan 13 1964 – 11r – 1 – us CRL [071]
La justice – Paris. 16 janv 1880-8 nov 1910, 18 fevr-31 dec 1911, 6 mars 1912-29 sept 1913, 20 mars-2 juil 1914, 6 avr-23 dec 1916, 1918-16 janv 1931, 17 janv-18 juil 1939, 1 fevr-8 juin 1940 [daily] – 1 – (journal quot. republicain) – fr ACRPP [073]
La justice : organe des revendications du peuple musulman algerien – Alger. oct 1934-avr 1938 – 1 – fr ACRPP [320]
La justice – Port-au-Prince: J B N Desroches. 1ere annee n34-3eme annee n25. 5 sep 1889-23 mai 1891 – us CRL [972]
La justice – Tamatave. n1-5, 7-9, 11, 13, 15-16, 18. dec 1893-avr 1894 – 1 – fr ACRPP [073]
The justice – (Waltham, Mass.) v40, no. 5 (Oct. 6,1987)-v40, no. 8 (Oct. 17,1987) Lacking: v40, no. 6 (Oct. 13, 1987) – us AJPC [340]
Justice, 1884-1914 : from the british museum – 27r – 1 – mf#15020 – uk Microform Academic [072]
Justice a qui de droit – Levis: impr du journal "Le Quotidien", [1895?] (mf ed 1984) – 1mf – 9 – mf#SEM105P393 – cn Bibl Nat [320]
Justice and authority in england, c1540-c1800 : series 1: cheshire – 3pt-coll – 61r – 1 – (the cheshire quarter sessions provide a detailed legal, economic and social picture of the elizabethan and early stuart period. pt 1: quarter sessions books and ledgers, 1557-1818 18r cl999-17501. pt 2: quarter sessions files, sect a: michaelmas 1571-easter 1603 23r cl999-17502. pt 3: quarter sessions files, sect b: trinity 1603-michaelmas 1616 20r cl999-17503) – mf#CL999-17500 – us Primary [941]
Justice and judgment among the tiv / Bohannan, Paul – London, England. 1957 – 1r – us UF Libraries [025]
Justice and sheriff and attorney's assistant. / Morrison, Charles Robert – Manchester, N.H., 1872. 464p. LL-838 – 1 – (rev. ed. concord, n.h.: sanborn, 1888. 557p. ll-828) – us L of C Photodup [340]
Justice assistance news – Washington. 1985-1985 (1,5,9) – ISSN: 0749-8195 – mf#13092 – us UMI ProQuest [340]
La justice de biddeford – Orono, ME. 1896-1950 (1) – mf#68911 – us UMI ProQuest [071]
La justice de sanford – Sanford, ME. 1925-1928 (1) – mf#63573 – us UMI ProQuest [071]
La justice et le role des magistrats / Toure, Ahmed Sekou – Conakry: Imprimerie nationale "Patrice Lumumba," 1974 – us CRL [340]
Justice in micronesia – v1 n1-2. sep 1977-jan 1978 (all publ?) – 3mf – 9 – $4.50 – mf#LLMC 82-100F, Title 66 – us LLMC [340]
Justice indigene (senegal), 1838-1954 : fonds du senegal colonial, sous serie 6m – [Chicago IL: Cooperative Africana Microfilm Project 1999] – 206r – 1 – (coll of colonial senegal court records, correspondence, etc, covering the period 1838 to 1954; reel guide available in print with title: repertoire numerique sous serie 6m : justice indigene (senégal), 1838-1954) – mf#mf-11933 – Archives du senegal – us CRL [347]
Justice ministries / Institute on the Church in Urban-Industrial Society – n1-17 [1978 sum-1982 sum] – 1r – 1 – mf#648770 – us WHS [230]
Justice of procedure in the free assembly : a reply to mr. taylor innes / Moncreiff, Henry Wellwood, Sir – Edinburgh: John Maclaren, [1881] Princeton: Speer Library, and Dep of Photodup, U of Chicago Lib, 1978 (1r); Evanston: American Theol Lib Assoc, 1984 (1r) – (= ser Case of william robertson smith in the free church of scotland) – 1 – 0-8370-0620-1 – mf#1984-6286 – us ATLA [242]
The justice of the land league / Humphrys, David – London, 1880 – (= ser 19th c ireland) – 1mf – 9 – mf#1.1.1897 – uk Chadwyck [941]
Justice of the peace – London. 1950-1993 (1) 1971-1993 (5) 1976-1993 (9) – (cont by: justice of the peace and local government law) – ISSN: 0022-703X – mf#541 – us UMI ProQuest [340]
Justice of the peace – Croydon. 1998+ (1) – (cont: justice of the peace and local government law) – mf#541,02 – us UMI ProQuest [340]
The justice of the peace – Lancaster Co, PA. v1-8. 1899-oct 1907 (all publ) – 14mf – 9 – $21.00 – (issues for apr 1906-may 1907 were never publ) – mf#LLMC 84-514 – us LLMC [347]
Justice of the peace and county borough : poor law union and parish law recorder – London. v1- 1837- – 9 – mf#LLMC 84-515 – us LLMC [340]
Justice of the peace and local government law – Chichester. 1993-1996 (1) 1993-1996 (5) 1993-1996 (9) – (cont: justice of the peace. cont by: justice of the peace) – ISSN: 1351-5756 – mf#541,01 – us UMI ProQuest [340]
Justice of the peace dockets / Sedgwick County. Kansas. Wichita Township – 1870-73 – 1 – us Kansas [340]
La justice ottomane... / Mandelstam, A N – Paris, 1911 – 3mf – 9 – mf#ILM-390 – ne IDC [956]
Justice quarterly – Academy of Criminal Justice Sciences. v1 – 9 – (filming in process) – mf#119131 – us Hein [340]
Justice quarterly: jq – Omaha. 1984+ (1,5,9) – ISSN: 0741-8825 – mf#14220 – us UMI ProQuest [360]
La justice seigneuriale de notre-dame-des-anges / Roy, Joseph-Edmond – [s.l: s.n, 1890?] [mf ed 1984] – 1mf – 9 – 0-665-12823-1 – mf#12823 – cn Canadiana [343]
La justice sociale – Journal des interets democratiques. Dir. Abbe Naudet. Bordeaux puis Paris. juil 1893-juil 1906 – 1 – fr ACRPP [322]
Justice speaks / Black Workers for Justice – [1986 oct/dec]-[2000 apr/may, aug/sep] [gaps] – 2r – 1 – mf#1073638 – us WHS [071]
Justice suspended / Marsh, Richard – London: Chatto & Eindus, 1913 – 5mf – 9 – $7.50 – mf#LLMC 92-225 – us LLMC [830]
Justice system journal – Denver. 1974+ (1,5,9) – ISSN: 0098-261X – mf#10933 – us UMI ProQuest [340]
Justice system journal – v1-22. 1974-2001 – 9 – $350.00 set – ISSN: 0098-261X – mf#104411 – us Hein [340]
Justice times – 1979 jan-1980 dec, 1981-86, 1987-89 – 3r – 1 – (cont: national tax strike news; cont by: justice [american fork, utah]) – mf#390660 – us WHS [071]
Justice to the jew : the story of what he has done for the world / Peters, Madison Clinton – London, New York: F.T. Neely, c1899 – 1mf – us ATLA [939]
Justice to the jew : the story of what he has done for the world / Peters, Madison Clinton – London; New York: F.T. Neely, c1899 – 1mf – 9 – 0-7905-6353-3 – mf#1988-2353 – us ATLA [270]
Justice to the jew... / Peters, Madison Clinton – New York, NY. 1908 – 1r – us UF Libraries [939]
Justice without law : a reconsideration of the "byoard equitable powers" of the federal courts, 31 aug 1988 – n.p, n.d. – (= ser Office of legal policy, reports to the attorney general) – 2mf – 9 – $3.00 – mf#LLMC 94-366 – us LLMC [347]
O justiceiro – Sao Paulo, SP: Typ do Farol Paulistano, 07 nov 1834-05 mar 1835 – (= ser Ps 19) – mf#P18,02,29 – bl Biblioteca [320]
The justices' manual of statute, judicial and elementary law, with appropriate forms. / Bundy, Charles Smith – Washington, Law Reporter Print, 1880. 496 p. LL-60 – 1 – (2nd ed. washington, byrne, 1896. 435 p. ll-563. 1) – us L of C Photodup [348]
The justices' practice under the laws of maryland / Latrobe, John Hazelhurst Boneval – 2nd ed. Baltimore, Lucas 1835 456 p. LL-948 – 1 – (5th ed. baltimore, lucas 1856 622p ll-774. 8th ed. baltimore, lucas 1889 652p ll-559) – us L of C Photodup [340]
Justicia – 1914-26. Scattered issues wanting – 1 – 87.00 – us L of C Photodup [972]
Justicia / International Ladies' Garment Workers' Union – 1977 feb,1978-1983 nov, 1984 jan-1988 dec, 1989 jan-1992 dec – 3r – 1 – (cont by: justice [new york, ny: 1919]; labor unity [new york, ny: 1982]; unite [new york, ny: 1995]) – mf#713183 – us WHS [331]
La justicia bajo la dictadura / Salazar Alonso, Rafael – Madrid: Editorial Reus, 1930 – 1 – sp Bibl Santa Ana [360]
La justicia revolucionaria en espana / Castilla, Juan de – Buenos Aires, 193? Fiche W 781. (Blodgett Collection of Spanish Civil War Pamphlets) – 9 – us Harvard [946]
Justicia, senor gobernador / Lindo, Hugo – San Salvador, El Salvador. 1960 – 1r – us UF Libraries [972]
Justicia social / Bitetti, Roque – Buenos Aires, Argentina. 1946 – 1r – us UF Libraries [025]
Justicia social en puerto rico / Miranda, Luis Antonio – San Juan, Puerto Rico. 1943 – 1r – us UF Libraries [972]
Justicia y caracter de la guerra nacional espanola / Getino, Luis G – Salamanca, 1937. Fiche W911. (Blodgett Collection of Spanish Civil War Pamphlets) – 9 – us Harvard [946]
Le justicier – Port-au-Prince: G Fouche. 1ere annee n1-2eme annee n26. 24 sep 1903-9 juil 1904 – us CRL [972]
Justificaciones historicas / Bayle, Constantino – Madrid: Razon y Fe, 1924 – 1 – sp Bibl Santa Ana [946]
Justification / Pusey, E B – Oxford, England. 1853 – 1r – us UF Libraries [240]
Justification / Smith, James – London, England. 18– – 1r – us UF Libraries [240]
Justification and peace / Miller, J C – London, England. 1874 – 1r – us UF Libraries [240]
Justification by faith : a charge / McIlvaine, Charles Pettit – Columbus [Ohio]: Isaac N Whiting, 1840 – 1mf – 9 – 0-524-00365-3 – mf#1989-3065 – us ATLA [240]
Justification by faith / Scholefield, James – Cambridge, England. 1832 – 1r – us UF Libraries [240]
Justification by faith : a sermon / Stearns, Jonathan French – New-York: John A Gray, 1852 – 1mf – 9 – 0-524-00110-3 – mf#1989-2810 – us ATLA [240]
Justification by faith as held and taught by lutherans, together with the associated doctrines of sanctification and the union of the soul with christ : or, the lutheran doctrine of the inner life / Harkey, Simeon Walcher – Philadelphia: Lutheran Board of Publ, 1875 – 1mf – 9 – 0-524-04768-5 – mf#1991-2154 – us ATLA [242]
Justification by faith in qumran? / Rorem, Paul Edward – Philadelphia: [s.n.], 1975 – 1r – 1 – 0-8370-0706-2 – mf#1984-T103 – us ATLA [930]
Justification by faith only / Brock, Mourant – London, England. 18– – 1r – us UF Libraries [240]
Justification by faith or justification by grace? : a study of faith and social engagement in luther and lutheranism / Phaswana, Ndanganeni Petrus – Uni of South Africa 2000 [mf ed Johannesburg 2000] – 7mf – 9 – (incl bibl ref) – mf#mfm15056 – sa Unisa [242]
The justification of god : lectures for war-time on a christian theodicy / Forsyth, Peter Taylor – London: Duckworth 1916 [mf ed 1991] – (= ser Studies in theology) – 1mf – 9 – 0-7905-7735-6 – (incl bibl ref) – mf#1989-0960 – us ATLA [210]
Justification of the charges brought against the british and foreig... / Close, F – London, England. 1839 – 1r – us UF Libraries [240]
La justification par la foi : essai de psychologie chretienne / Fulliquet, Georges – Geneve: Charles Schuchardt, 1889 – 1mf – 9 – 0-7905-8793-9 – mf#1989-2018 – us ATLA [150]
Justifier – London, England. 18– – 1r – us UF Libraries [240]
Justin, augustin, bernhard und luther : der entwickelungsgang christlicher wahrheitserfassung in der kirche als beweis fuer die lehre der reformation / Dieckhoff, August Wilhelm – Leipzig: Justus Naumann 1882 [mf ed 1989] – 1mf – 9 – 0-7905-4557-8 – mf#1988-0557 – us ATLA [240]
Justin der maertyrer : eine kirchen-und dogmengeschichtliche monographie / Semisch, K G – Breslau, 1840-1842 – €27.00 – ne Slangenburg [240]
Justin der maertyrer und sein neuester beurtheiler / Staehlin, Adolf von – Leipzig: Doerffling und Franke, 1880 – 1mf – 9 – 0-524-04319-1 – (incl bibl ref) – mf#1990-1245 – us ATLA [240]
Justin, Joseph *see*
– Autour de l'isthme de panama
– Baie de samana
– Conference sur haiti
– De l'organisation judiciaire en haiti
– Differend entre la republique d'haiti
– Etude sur les institutions haitiennes
– Memoire au conseil d'etat
– Question du mole saint-nicolas
– Relations exterieures d'haiti
Justin martyr : the dialogue with trypho / Martyr, Justin, Saint – London: SPCK; New York: Macmillan, 1930 [mf ed 2004] – (= ser Translations of christian literature) – 1r – 1 – 0-524-10497-2 – (incl ind & bibl. trans, int & notes by arthur lukyn williams) – mf#b00712 – us ATLA [240]
Justin the Martyr, Saint *see*
– The apologies of justin martyr
– The apologies of justin martyr. to which is appended the epistle to diognetus
Justin williams, sr. papers / Williams, Justin, Sr – (mf ed 2000) – 43mf – 9 – $4100.00 – (with guide) – University of Maryland – us UMI ProQuest [950]
Justinard, Leopold Victor *see* Un grand chef berbere
Justinian 1 *see* Institutiones, digestum (libri 40-50); novellae constitutiones
Justiniano Arribas, Juan *see*
– Alonso perez de guzman
– Cristobal colon
– Hernan cortes
– Hernando cortes
– Poesias selectas
– Roger de flor
Justinien et la civilisation byzantine au 6e siecle / Diehl, C – Paris, 1901 – 8mf – 9 – mf#H-2927 – ne IDC [956]
Justins des maertyrers lehre von jesus christus : dem messias und dem menschgewordenen sohne gottes / Feder, Alfred Leonhard – Freiburg im Breisgau; St. Louis, Mo.: Herder, 1906 – 1mf – 9 – 0-7905-6169-7 – (incl bibl ref) – mf#1988-2169 – us ATLA [240]
Justinus der Maertyrer *see* Dialog mit dem juden tryphon [bdk33 1.reihe]
Justinus kerners saemtliche poetische werke / Kerner, Justinus; ed by Gaismaier, Josef – Leipzig: M Hesse. 4v. [1905?] – 1 – (incl bibl ref and index) – us Wisconsin U Libr [800]
Justiz Y Del Valle, Tomas Juan De *see*
– Ecos de una guerra a muerte
– Manuel sanguily y garritte
Justiz y Del Valle, Tomas Juan De *see* Elogio del sr nestor leonelo carbonell
Justo arosemena / Mendez Pereira, Octavio – Panama, Panama. 1919 – 1r – us UF Libraries [972]
Justo, Juan Bautista *see* Socialismo
Justo tiempo humano / Padilla, Heberto – Habana, Cuba. 1962 – 1r – us UF Libraries [972]
Justro polski – London, UK. Jul, Nov, Dec 1943 – 1 – uk British Libr Newspaper [072]
Justus falckner : mystic and scholar, devout pietist in germany, hermit on the wissahickon, missionary on the hudson / Sachse, Julius Friedrich – Philadelphia: Printed for the author, 1903 (Lancaster: New Era) – 1mf – 9 – 0-7905-6494-7 – mf#1988-2494 – us ATLA [240]
Justus jonas : nach gleichzeitigen quellen / Pressel, Theodor – Elberfeld: RL Friderichs, 1862 – (= ser Leben und ausgewaehlte Schriften der Vaeter und Begruender der lutherischen Kirche) – 1mf – 9 – 0-524-00586-9 – (incl bibl ref) – mf#1990-0086 – us ATLA [240]
Justus Lunzer, Edler von Lindhausen *see* Ortneit und wolfdietrich
Justus menius, der reformator thueringens : nach archivalischen und andern gleichzeitigen quellen / Schmidt, Gustav Lebrecht – Gotha: FA Perthes, 1867 – 2mf – 9 – 0-7905-8248-1 – mf#1988-8111 – us ATLA [242]
Justus moeser / Kreyssig, Friedrich Alexander Theodor – Berlin: Nicolai, 1857 – 1r – 1 – us Wisconsin U Libr [920]
Justus moeser's approach to history / Bossenbrook, William John – Chicago IL: private ed, dist by Uni of Chicago Libraries 1938 [mf ed 1993] – 1r – 1 – (incl bibl ref; filmed with: von einem, der seine ahnen suchen ging / jassy torrund) – mf#7615 – us Wisconsin U Libr [900]
Justyna *see* Pamietnik justyny
Justyne, William *see* Illustrated guide to kensal green cemetery
Juta's dictionary / Potgieter, Dirk Jacobus – Cape Town, South Africa. 1932 – 1r – us UF Libraries [025]

Juta's first zulu manual with vocabulary / Fox, I – Cape Town, South Africa. 1950 – 1r – us UF Libraries [470]

Jute bulletin – Uniited States. 1975 jan 10-1978 sep 14, 1978 sep 21-1981 jul 9, 1981 jul 16-1983 sep 7, 1983 sep 14-1985 sep 25, 1985 oct 2 -1987 mar 26, 1987 apr 2 -1988 nov 10, 1988 nov 17-1990 jun 28, 1990 jul 3 -1991 feb 27 – 8r – 1 – mf#1607433 – us WHS [071]

Jutnicka – Bautzen DE, 1842; 1848-50 – 1r – 1 – gw Misc Inst [074]

Jutro polski – London, UK. 24 Dec 1944- – 1 – uk British Libr Newspaper [072]

Jutrzenka – Cleveland, OH. 1893-1894 (1) – mf#65424 – us UMI ProQuest [071]

Jutrzenka / Cuyahoga Co. Cleveland – jun 1918-jun 1923 (many damaged) [wkly] – 2r – 1 – (in polish) – mf#B5003-5004 – us Ohio Hist [071]

Juvenal *see*
- Juvenal and persius
- Juvenal's satires
- Saturae 14
- Saturarum libri 5; mit erklarenden anmerkungen von l. friedlaender

Juvenal and persius / Juvenal – London, England. 1924 – 1r – us UF Libraries [025]

Juvenal Rosa, Pedro *see* Masas mandan

Juvenal's satires / Juvenal – London, England. 1954 – 1r – us UF Libraries [025]

O juvenil – Bom Sucesso, MG. 26 maio 1892; maio, nov 1893; maio-ago 1894; mar 1900; dez 1912; abr 1940; jan-jun, ago-4 dez 1979 – (= ser Ps 19) – mf#P17,02,92 – bl Biblioteca [079]

O juvenil – Rio de Janeiro, RJ. 27 jun-15 jul 1904 – (= ser Ps 19) – mf#DIPER – bl Biblioteca [079]

Juvenile and family court journal – Reno. 1978+ (1) 1978+ (5) 1978+ (9) – ISSN: 0161-7109 – mf#6617,01 – us UMI ProQuest [640]

Juvenile and family law digest – Reno. 1981+ (1) 1981+ (5) 1981+ (9) – (cont: juvenile law digest) – ISSN: 0279-2257 – mf#7740,02 – us UMI ProQuest [346]

Juvenile court digest – Reno. 1967-1977 (1) 1972-1977 (5) 1975-1977 (9) – (cont by: juvenile law digest) – ISSN: 0085-2430 – mf#7740 – us UMI ProQuest [347]

Juvenile court laws in the united states : summary by states / Hart, Hastings Hornell – New York: Charities, 1910. 150p. LL-1558 – 1 – us L of C Photodup [347]

The juvenile court laws of the state of colorado / Colorado. Laws, Statutes, etc – Denver: Juvenile Improvement Association of Denver 1905. 80p. LL-353 – 1 – us L of C Photodup [348]

Juvenile court laws of the united states : topical summary of their main provisions / Hiller, Francis Hemperley – New York: National Probation Association, 1933. 82p. LL-1233 – 1 – us L of C Photodup [347]

The juvenile court of denver, concerning its judge, concerning its work, what it has done, what it is doing, what it hopes to do – Denver, 1913 23 p. LL-421 – 1 – us L of C Photodup [347]

Juvenile education in relation to employment after the war, departmental committee on... (lewis report), 1916-1917 : command n8512 – 1mf – 9 – mf#86945 – uk Microform Academic [324]

Juvenile entertainer – Pictou, NS: W Milne, [1831-18–] – 9 – mf#P04921 – cn Canadiana [420]

Juvenile forget-me-not – 1829-37 – (= ser English gift books and literary annuals, 1823-1857) – 27mf – 9 – uk Chadwyck [830]

Juvenile gazette – Providence. 1819-1820 (1) – mf#4007 – us UMI ProQuest [305]

Juvenile justice – Reno. 1949-1977 [1]; 1972-1977 [5]; 1975-1977 [9] – ISSN: 0093-7231 – mf#6617 – us UMI ProQuest [640]

Juvenile justice digest – Washington. 1973+ (1,5,9) – ISSN: 0094-2413 – mf#10589 – us UMI ProQuest [348]

Juvenile keepsake – 1829-30 – (= ser English gift books and literary annuals, 1823-1857) – 3mf – 9 – uk Chadwyck [830]

Juvenile law digest – Reno. 1978-1981 (1) 1978-1981 (5) 1978-1981 (9) – (cont: juvenile court digest. cont by: juvenile and family law digest) – ISSN: 0162-5055 – mf#7740,01 – us UMI ProQuest [347]

Juvenile magazine : or, miscellaneous repository of useful information – Philadelphia. 1802-1803 – 1 – mf#3582 – us UMI ProQuest [073]

Juvenile magazine – Philadelphia. 1811-1813 (1) – mf#4008 – us UMI ProQuest [305]

Juvenile mirror : or, educational magazine – New York. 1812-1812 – 1 – mf#4470 – us UMI ProQuest [370]

Juvenile miscellany – Boston. 1826-1834 (1) – mf#4009 – us UMI ProQuest [305]

Juvenile missionary herald – 1845-1908 – 1 – 763.00 – us Southern Baptist [242]

Juvenile missionary magazine of the united secession church – Glasgow: Robertson & Co 1844-46 [mf ed Cape Town: SA Lib c1944] – 3mf [ill] – 9 – mf#mf.893 – sa National [242]

Juvenile port-folio and literary miscellany – Philadelphia. 1812-1816 (1) – mf#4471 – us UMI ProQuest [305]

The juvenile presbyterian – Montreal: Printed by J Lovell, [1856-186-?] – 9 – (incl ind) – mf#P06041 – cn Canadiana [242]

Juvenile repository – Boston. 1811-1811 (1) – mf#4010 – us UMI ProQuest [640]

Juvenilia / Jastram, Gervais – Paris, France. 1928 – 1r – us UF Libraries [972]

Juvenis *see* An address on the necessity of a liberal education

Juventud – Miami, FL. 1970 mar 22-may 13 – 1r – us UF Libraries [071]

La juventud anarquista: factor determinativo de la guerra y de la revolucion / Briones, Mariano – Barcelona, 1937? Fiche W 763. (Blodgett Collection of Spanish Civil War Pamphlets) – 9 – us Harvard [946]

Juventud de aurelio zaldivar / Hernandez Cata, Alfonso – Barcelona, Spain. 1914? – 1r – us UF Libraries [972]

Juventud de juan gualberto gomez / Perez Cabrera, Jose Manuel – Habana, Cuba. 1945 – 1r – us UF Libraries [972]

La juventud espanola continua su lucha / Claudin, Fernando – Mexico, 1940. Fiche W 802. (Blodgett Collection of Spanish Civil War Pamphlets) – 9 – us Harvard [946]

La juventud: factor de la victoria / Carrillo, Santiago – Barcelona, 1937. Fiche W 776. (Blodgett Collection of Spanish Civil War Pamphlets) – 9 – us Harvard [946]

La juventud que defiende madrid / Munoz Arconada, Felipe – [Valencia: s.n. 1937] – (= ser Blodgett coll) – 9 – mf#w1065 – us Harvard [946]

La juventud y los campesinos : [conferencia nacional de juventudes, enero de 1937] / Alvarez, Segis – Valencia: Lleonart [1937] [mf ed 1980] – (= ser Blodgett coll) – 1mf – 9 – mf#w711 – us Harvard [946]

Juventud y...desencanto / Sanchez-Arjona, Vicente – Sevilla: Imp. Carlos Acuna, 1953 – 1 – sp Bibl Santa Ana [810]

A juventude : orgao litterario da sociedade fraternidade juvenil – Natal, RN: Typ do Conservador, 28 jun, ago-set, 13 nov 1882 – (= ser Ps 19) – mf#P22B,04,187 – bl Biblioteca [440]

Juventus mundi : the gods and men of the heroic age / Gladstone, William Ewart – 2nd ed. London: Macmillan, 1870 – 2mf – 9 – 0-524-02204-6 – mf#1990-2878 – us ATLA [450]

Juwentu *see* Emu be tacifi ilan be hafukiyara manju gisun-i buleku bithe

Juynboll, Th J *see* Chronicon samaritanum

Jwb circle – (New York). 1946-67 – 1 – us AJPC [939]

Jyllands posten – Aarhus, Denmark. jan-13 may 1945; 7,14 jan; 31 may; 9, 15 aug 1964 – 2r – 1 – uk British Libr Newspaper [074]

Jym Bau *see* 100 ganu baky samramn

K : revue de la poesie – Paris. n1-3.juin 1948-mai 1949 – 1 – fr ACRPP [810]

K 8 s"ezdu sovetov, 20/xii 20 g / Narodnyi Komissariat Finansov – M, 1920 – 1mf – 9 – mf#REF-37 – ne IDC [332]

K 50-letiiu preobrazovaniia rumiantsevskogo muzeia v gosudarstvennuiu biblioteku sssr imeni v.i. lenina : sbornik nauchnykh trudov / Nauchnyi sovet po istorii mirovoi kultury Akademii nauk SSSR. Gosudarstvennaia biblioteka SSSR imeni V.I. Lenina – Moskva: Gos bib-ka SSSR im. V.I. Lenina, 1976 – us CRL [947]

K agrarnomu voprosu v rossii : mysli i tsifry / Voblyi, K G – Kiev, 1917 – 30p 1mf – 9 – mf#COR-15 – ne IDC [335]

K and T today / Kearney & Trecker Corporation – n1-3 [1975 nov14-1976 feb] – 1r – 1 – (cont: news-letter [kearney & trecker corporation]; cont by: kt today) – mf#667594 – us WHS [071]

K bibliografii tserkovno-slavianskikh pechatnykh izdanii v rossii / Kaluzhniatskii, E – 1886 – 46p 1mf – 8 – mf#R-4676 – ne IDC [243]

K biografii adama mitiskevicha v 1821-1829 godakh / Wierzbowski, Teodor – Sanktpeterburg: Tip Imp akademii nauk, 1898 [mf ed 2002] – 1r – 1 – (filmed with: croissans-crescens i sredneviekovyia legendy o polovoi metamorfozie / a n veselovskago (1881) & other titles. incl bibl ref) – mf#5239 – us Wisconsin U Libr [460]

K c blaetter – Koeln, Berlin DE, 1910 oct-1933 jan – 2r – 1 – gw Misc Inst [074]

K chemu stremitsia narodno-sotsialisticheskaia (trudovaia) partiia – n.d. – 8p 1mf – 9 – mf#RPP-198 – ne IDC [325]

K G Saur Verlag *see* Tarnschriften 1933 bis 1945

K istorii izdaniia "izvestii i uchenykh zapisok vtorogo otdeleniia imperatorskoi akademii nauk" (1852-1863) / Sreznevskii, V I – Spb., 1905 – 3mf – 9 – mf#R-4826 – ne IDC [077]

K istorii vozniknoveniia partii sotsialistov-revoliutsionerov / Sletov, S – 1917 – 112p 2mf – 9 – mf#RPP-247 – ne IDC [325]

K l von knebel's literarischer nachlass und briefwechsel / ed by Ense, Karl August Varnhagen von & Mundt, Theodor – 2. unveraend usg. Leipzig: Gebrueder Reichenbach, 1840 [mf ed 1993] – 3v in 1 – 1 – mf#8642 – us Wisconsin U Libr [802]

K l von knebel's literarischer nachlass und briefwechsel / ed by Varnhagen von Ense, Karl August & Mundt, Theodor – 2. ausg. Leipzig: Gebrueder Reichenbach, 1840 [mf ed 1993] – 3v in 1 – 1 – mf#8642 – us Wisconsin U Libr [430]

K literaturnoi istorii kamnia very m stefana iavorskogo / Ponomarev, A I – 1905 – 2mf – 8 – mf#R-7781 – ne IDC [947]

K metodike sotsialno-gigiencheskogo izucheniia kogort molodykh semei v sviazi s protsessom rozhdaemosti / Serenko, A – Moskva: In-t konkretnykh sotsialnykh issledovanii AN SSSR, 1972 – us CRL [947]

K mezhdunarodnoi postanovkie evreiskago voprosa / Lazerson, M la – Petrograd, Russia. 1917 – 1r – us UF Libraries [939]

K mneniiu men'shinstva chastnogo soveshchaniia zemskikh deiatelei (noiabria 1904 g) / Shipov, D N et al – M, 1905 – 1mf – 8 – mf#R-3525 – ne IDC [314]

K molodomu pokoleniiu / Kropotkin, P – n.p., 1919 – 30p 1mf – 9 – mf#RPP-84 – ne IDC [325]

K nashei polemikie s staroobriadtsami : dopolneniia i popravki k polemikie otnositelno obshchei eia postanovki i otnositelno glavnieishikh chastnykh punktov raznoglasiia mezhdu nami i staroobriadtsami / Golubinskii, Evgenii Evstignieevich – Izd 2 ispr i dop. Moskva: Tip Ob-va rasprostraneniia poleznykh knig, 1905 – 1mf – 9 – 0-7905-7230-3 – (incl bibl ref) – mf#1988-3230 – us ATLA [240]

K novym beregam – Moscow. no. 1-3. Apr-Aug 1923 – 1 – us NY Public [780]

K obosnovaniiu programmy partii sotsialistov-revoliutsionerov : rechi v m chernova, (tuchkina) na 1-om partiinom sezde / Chernov, V – 1918 – 110p 2mf – 9 – mf#RPP-257 – ne IDC [325]

K oruzhiyu [tyumen'] : izvestiia tiumenskogo gub i uezd ispolnitel'nykh komitetov sovetov krest'ianskikh, rabochikh i krasnoarmejskikh deputatov – Tyumen', Russia 1918 [mf ed Norman Ross] – 1r – 1 – mf#nrp-1851 – us UMI ProQuest [077]

K oruzhiyu [vitebsk] – Vitebsk, Belarus 1918 [mf ed Norman Ross] – 1r – 1 – mf#nrp-2064 – us UMI ProQuest [077]

K pervoi mezhdunarodnoi palinologicheskoi konferentsii, takson, ssha : doklady sovetskikh palinologov: rasshirennye tezisy / Akademiia nauk SSSR, Geologicheskii institut – Moskva: Izd-vo Akademii nauk SSSR, 1962 – us CRL [947]

K ph. moritz' goetterlehre : ein dokument des goetheschen klassizismus / Fahrner, Rudolf – Marburg-Lahn: N G Elwert, 1932 – 1r – 1 – (incl bibl ref (p.[31])) – us Wisconsin U Libr [200]

K reforme gosudarstvennogo banka / Gur'ev, A N – Spb, 1893 – 2mf – 9 – mf#REF-248 – ne IDC [332]

K reforme krest'ianskogo banka / Gur'ev, A N – Spb, 1894 – 1mf – 9 – mf#REF-260 – ne IDC [332]

K samopoznaniiu evreia / Bickermann, Joseph – Parizh, France. 1939 – 1r – us UF Libraries [939]

K Sternwarte [Munich, Germany] *see*
- Annalen der koeniglichen sternwarte bei muenchen
- Astronomische beobachtungen angestellt auf der k sternwarte zu bogenhausen
- Neue annalen der k sternwarte in bogenhausen bei muenchen
- Observationes astronomicae in specula regia monachiensi

K svetu : bespartiin, idein ezhened izd prikhod sovetov g barnaula – Barnaul [Alt gub]: [s n] 1918-19 [1918-] – (= ser Asn 1-3) – n21-36/37 [1918] n6-25 [1919] [gaps] item 189, on reel n39 – 1 – mf#asn-1 189 – ne IDC [077]

K teorii kooperatizma / Evdokimov, A A – 1909 – 32p 1mf – 9 – mf#COR-22 – ne IDC [335]

K toporu – Moscow, Russia: Vyp al'manakha Russkoe Slovo. n5[n.d.] – 1 – mf#mf-12248 [reel 5] – us CRL [077]

K voprosu o munitsipalno-kooperativnoi organizatsii narodnogo prodovolstviia v severnoi oblasti / Pekarskii, V F – 1918 – 48p 1mf – 9 – mf#COR-344 – ne IDC [335]

K voprosu o natsionalizatsii bankov / Sokol'nikov, G – M, 1918 – 1mf – 9 – mf#REF-16 – ne IDC [332]

K voprosu o sotsialisticheskom pereustroistve selskogo khoziaistva: : materialy issled. nkrki ssr / ed by lakovlev, I A – 1928 – 450p 6mf – 9 – mf#COR-476 – ne IDC [335]

K voprosu o stroitelstve kooperatsii : tez k 19 gubpartkonferentsii – 1925 – 24p 1mf – 9 – mf#COR-152 – ne IDC [335]

K voprosu o tsieli revoliutsionnoi raboty v voiskakh – London: Izd avtora, 1903 [mf ed 2004] – 1r – 1 – (filmed with: lessons of the wrecking, diversionist and espionage activities of the japanese-german-trotskyite agents / v m molotov [1937]) – mf#5499p – us Wisconsin U Libr [335]

K voprosu o vzaimnykh otnosheniiakh gubernskikh i uezdnykh zemstv / Shipov, D N – M, 1899 – 2mf – 9 – mf#R-3526 – ne IDC [314]

K voprosu ob obedinenii bunda s rossiiskoi sotsial-demokraticheskoi rabochei partiei – 1906 – 2mf – 9 – (izvlechenie iz protokolov 7-oi konf. bunda (zasedaniia 11,12,13,14 i 15)) – mf#RPP-94 – ne IDC [325]

Ka es atklahju modernisumu (wiltigu mahzibu) starp amerikanu baptisteem un kapehz es nodibinaju anglu-amerikanu missiones beedribu = How i discovered moderism among american baptists, and why i founded the russian missionary society / Fetler, William – Riga: Derigu Rakstu Apgahdneeziba, 1924 – 1r – 1 – (part of an 8-item unit) – us Southern Baptist [242]

Ka euanelio a mataio : oia ka moo olelo hemolele no ko kakou haku e ola'i io iesu kristo; i laweia i olelo hawaii / Matthaeus Apostolus – Rochester NY: Paha ma ka Mea Pai Palapala a Lumiki 1828 – (= ser Whsb) – 2mf – 9 – €30.00 – (incl: ka euanelio a ioane, & ka euanelio a marako) – mf#Hu 455 – gw Fischer [410]

Ka hoaloha – Honolulu: [s.n.] (mar-may 1943) [mf ed 2007] – (= ser Religious periodical literature of the hispanic and indigenous peoples of the americas, 1850-1985) – 1r – 1 – mf#2007i-s019 – us ATLA [242]

Ka khyan pum pran mya / Lha, Lu thu U – Mantale: Lu thu u Lha 1975 [mf ed 1993] – on pt of 1r – 1 – mf#11052 r620 n2 – us Cornell [959]

Ka khyan rui ra su ta padesa / Mai Ja – Ran Kun: A Lan Ron ca pe tuik [196-?] – 1r with other items – 1 – (in burmese) – mf#mf-10289 seam reel 140/2 [§] – us CRL [305]

Lo ka kre mum : [short stories] / Le Mruin – Ran Kun: Nha Lum Lha ca pe tuik 1981 [mf ed 1990] – 1r with other items – 1 – (in burmese) – mf#mf-10289 seam reel 156/1 [§] – us CRL [830]

Ka le lu nay ca pe / Aon Chve, Mon, Takkasuil – Ran Kun: Ca pe bi man a phvai 1974 [mf ed 1990]- – 1r with other items – 1 – (in burmese; incl bibl ref) – mf#mf-10289 seam reel 171/5 [§] – us CRL [400]

Ka lira hawaii : he mau mele himeni a me na melo oli halelu: no na ekalesia o hawaii nei – Honolulu: pai hou ia a mahuahua, 1855 [mf ed 1995] – (= ser Yale coll) – 1 score (161/[15]p) – 1 – 0-524-10236-8 – (english hymns in hawaiian trans with music) – mf#1996-1236 – us ATLA [780]

Ka lympung ri-lum = The hills rally – Shillong: Gilfred S Giri (wkly) [mf ed Chicago IL: Dept of Photoduplication, The University of Chicago Library [for the SEAsian MF Project at CRL] – 1r – 1 – (began in 1956; in khasi) – mf#mf-10018 seam – us CRL [079]

Ka lympung ri-lum = The hills rally – Shillong: Gilfred S Giri, aug 10 1957-oct 18 1958 – 1r – 1 – us CRL [950]

Ka mei ri lum (mother highland) – Shillong, India. 1961-68 – 1 – us CRL [079]

Ka mei rilum = Mother highland – Shillong: Norman Singh Syiem [1962- (wkly) [mf ed Chicago IL: Dept of Photoduplication, The University of Chicago Library [for the SEAsian MF Project at CRL] – 1r – 1 – (in khasi; iss for -dec 25 1968 publ with title: ka mei ri-lum; iss for -dec 25 1968 have english title: mother high land) – mf#mf-10015 seam – us CRL [079]

Ka mei rilum = Mother highland – Shillong: Norman Singh Syiem, feb 8 1962-dec 25 1968 – 1r – 1 – us CRL [960]

Ka pyrta u ri lum (voice of the hillman) – Shillong, India. 1957-66 – 1 – us CRL [079]

Ka pyrta u riewlum – Shillong: N S Syiem (wkly) [mf ed Chicago IL: Dept of Photoduplication, The University of Chicago Library [for the SEAsian MF Project at CRL] – oct 29 1957-jan 26 1966 [gaps] on 1r – 1 – (began in oct 1957; lacking several iss; in khasi; iss for -26 jan 1966 also have title in english: voice of the hillman) – mf#mf-10014 seam – us CRL [079]

Ka pyrta u riewlum – Shillong: N S Syiem, oct 29 1957; aug 8 1958-jan 17 1962; feb 13 1963-jan 26 1966 – 1r – 1 – us CRL [073]

Ka ran rui ra suta padesa : [tuin ran sa lu myui e bhava dhale sutesana pru khyak] / Lan Mrat Kyo, Man – Ran Kun: Mon Cui Mran phran khyi ra thana, Man Bha Khyo 1980 [mf ed 1990] – 1r with other items – 1 – (in burmese; with bibl) – mf#mf-10289 seam reel 140/3 [§] – us CRL [307]

Ka ri wen ha wi / Akwesasne Library Cultural Center – v11 n4-v15 n18 [1982 feb 28-1986 sep 30] – 1r – 1 – mf#1107550 – us WHS [071]

Ka titc tebenimiang jezos, ondaje aking : oom masinaigan ki ojitogoban ka ojitogobanen aiamie tipadjimo8in masinaigan aki enaindibanen / Mathevet, Jean-Claude – Moniang [i.e. Montreal]: O ki magabikickoton John Lovell, ate mekateikonaieikamikong, kanactageng, 1861 [mf ed 1984] – 5mf – 9 – 0-665-46361-8 – mf#46361 – cn Canadiana [225]

Kaapsche grensblad – Grahamstown SA, 18 jul 1844 – 28 dec 1861 (wkly) – 4r – 1 – mf#MS00053 – sa National [079]

Kaapsche handelsblad en zee kronijk – Cape Town SA, 6 jan 1843-[24 mar 1843?] – 1 – mf#MS00360 – sa National [079]

Kaapse moppies / Du Plessis, Izak David – Johannesburg: Perskor-uitgewery 1977 [mf ed Cape Town: SA Lib] – 2mf – 9 – mf#mf.962 – sa National [810]

Kaapse plakkaatboek : afgeskryf en persklaar gemaak / Jeffreys, M K – Kaapstad: Cape Times, 1944-49 – us CRL [960]

Kab, M S *see* Kachestvo i standart v promkooperatsii

Kabak, Aaron Abraham *see* Levadah

Kabanov, A K *see* Smuta moskovskago gosudarstva i nizhnii-novgorod

Kabardinets : ezhedn novosti i telegrammy / ed by Khutsiev, M E et al – Nal'chik [Ter-Dagest krai]: baron Tandefel't 1919 [1919-] – (= ser Asn 1-3) – n17 [1919] item 190, on reel n39 – 1 – mf#asn-1 190 – ne IDC [077]

Kabardino-balkarskaia bednota – Nal'chik, Russia 1921-25 [mf ed Norman Ross] – 4r – 1 – (in russian, kabardian & balkar) – mf#nrp-1190 – us UMI ProQuest [077]

Kabardino-Balkarskaia pravda : organ verkhovnogo soveta kabardino-balkarskoi assr i respublikanskogo komiteta kp rsfsr – Nalchik: [s.n.] [mf ed Minneapolis MN: East View Publ [199-]] – 10r – 1 – (organ of: verkhovnyi sovet kabardino-balkarskoi pespubliki, jan 1, 1993-, parliament and pravitelstvo kabardino-balkarskoi pespubliki, jan 4, 1995-; crl has: mf-11786 seemp [10r] 1991-95; mf-12754 seemp [10r] 1996-2000) – us CRL [077]

Kabardino-balkarskaia pravda – Nal'chik, Russia 1974-88 [mf ed Norman Ross] – 1r – 1 – mf#nrp-1191 – us UMI ProQuest [077]

Kabbala denudata: the kabbala unveiled. – The Book of Concealed Mystery; The Greater Holy Assembly; The Lesser Holy Assembly.Trans. into Eng. from the Latin & collated with the original Chaldee and Hebrew text, by S.L. MacGregor Mathers. London: G. Redway, 1887. viii,359p. plates, tables – 1 – us Wisconsin U Libr [270]

Kabbalah / Franck, Adolphe – New York, NY. 1926 – 1r – us UF Libraries [939]

The kabbalah : its doctrines, development, and literature: an essay / Ginsburg, Christian David – London: Longmans, Green, Reader, and Dyer, 1865 – 1mf – 9 – 0-7905-1400-1 – (in english and hebrew) – mf#1987-1400 – us ATLA [270]

La kabbale : ou, la philosophie religieuse des hebreux / Franck, Adolphe – Paris: L Hachette, 1843 [mf ed 1992] – 1mf – 9 – 0-524-00425-7 – mf#1990-3278 – us ATLA [270]

Der kabbalistisch-bibelsche occident – Hamburg: B S Berendsohn. n1. 1845 [complete] – (= ser German-jewish periodicals...1768-1945, pt 3) – 1r – 1 – $165.00 – mf#B263 – us UPA [270]

Kabel, Michael *see* De betekenis van het woord solitudo in de collationes van cassianus

Kabele, George Philip *see* What lutherans believe

Kaberry, Phyllis Mary *see* Women of the grassfields

Kabhuku kenzanga yaanna musande – Gwelo, Zimbabwe. 1970 – 1r – us UF Libraries [960]

Kabhya manjari : [poems] / Mon Kri, Lay ti Pandita U – Ran Kun: Panna A lan pra Ca up tuik 1323 [1961] [mf ed 1990] – 1r with other items – 1 – (in burmese) – mf#mf-10289 seam reel 199/7 [§] – us CRL [810]

Kabinet van nederlandsche en kleefsche oudheden... / Rademaker, A; ed by Reisig, J H – Amsterdam, 1792-1803. 8v – 65mf – 9 – mf#O-409 – ne IDC [700]

Kabinet voor nederlands antilliaanse en arubaanse – Den Haag, Netherlands. 1988-1989 – 1r – (missing: 1989 apr 5) – us UF Libraries [074]

Kabir *see*
- The bijak of kabir
- One hundred poems of kabir

Kabir and his followers / Keay, Frank Ernest – Calcutta: Association Press; London; New York: Humphrey Milford: Oxford University Press, 1931 – (= ser Samp: indian books) – us CRL [280]

Kabir and the bhagti movement / Singh, Mohan – Lahore: Atma Ram & Sons, Educational Publishers & Booksellers, 1934- – (= ser Samp: indian books) – us CRL [280]

Kabir and the kabir panth / Westcott, G H – Cawnpore: Christ Church Mission Press, 1907 – (= ser Samp: indian books) – us CRL [280]

Kabir and the kabir panth / Westcott, G H – Cawnpore [India]: Christ Church Mission Press, 1907 – 1mf – 9 – 0-524-03536-9 – mf#1990-3241 – us ATLA [280]

Kabir, Humayun *see* Men and rivers

Kabisch, Richard *see* Das vierte buch esra

Kablukov, N *see* Posobie k oznakomleniu s usloviiami i priemami sobraniia i razrabotki svedenii pri zemsko-statisticheskikh issledovaniiakh

Kablukov, N A *see*
- Iz chego vyrosli kooperativy
- Kustarnaia promyshlennost i ee sviazs kooperativami
- Melkoe khoziaistvo i kooperatsiia
- Ob usloviiakh razvitiia krestianskogo khoziaistva v rossii

Kabul : monthly political magazine in pashto – Kabul, 1931-84 – 35r – 1 – us UMI ProQuest [079]

Kabul times – Kabul, Afghanistan. 27 feb 1962-11 jun 1964 – 2 1/2r – 1 – uk British Libr Newspaper [072]

Kabul to kandahar / Diver, Maud – London: P. Davies, 1935. 191p. 2 maps – 1 – us Wisconsin U Libr [915]

Kaccayana *see* Sadda kri path nhan ca cap niyam, sam ranna niyam, bhura kri niyam kyam mya

Kachestvo i standart v promkooperatsii : materialy sektora proizvodstvenno-tekhnicheskoi ratsionalizatsii vsekopromsoiuza / Golubkova, S N & Karnaukhov, D E; ed by Kab, M S – 1931 – 1 – 9 – mf#COR-420 – ne IDC [335]

The kachins : religion and customs / Gilhodes, C – Calcutta: printed by A Rome at the Catholic Orphan Press; London: K Paul, Trench, Truebner, 1922. Chicago: Dep of Photodup, U of Chicago Lib, 1961 (1r); Evanston: American Theol Lib Assoc, 1984 (1r) – 1 – 0-8370-1472-7 – mf#1984-B004 – us ATLA [390]

Kacyzne, Alter *see* Geklibene shriftn

Kaczkowski, Joachim *see* Duo

The kadamba kula : a history of ancient and mediaeval karnataka / Moraes, George M – Bombay: BX Furtado & Sons, 1931 – (= ser Samp: indian books) – (pref by h heras) – us CRL [954]

Kadar of cochin / Ehrenfels, Omar Rolf Leopold Werner, Freiherr von – Madras: University of Madras, 1952 – (= ser Samp: indian books) – us CRL [307]

Kaddisch-gebet / Hubscher, Jacob – Berlin, Germany. 1912 – 1r – us UF Libraries [939]

Kade, E von *see* Streifzuege eines modernen junkers

Kadelburg, Gustav *see*
- Die grosstadtluft
- Im weissen roessl

Kaden, Woldemar *see* Die insel ischia

Kadena falcon – v12 n16-v13 n49 [1981 apr 28-1982 dec 22] – 1r – 1 – (cont by: kadena shogun) – mf#635344 – us WHS [071]

Kadena shogun – Kadena Air Base, Okinawa City. 1983 jan 12-1986 may 28, 1988 jan 15-1989 jun 30 – 2r – 1 – (cont: kadena falcon; cont by: shogun) – mf#1173645 – us WHS [355]

Kades, Charles L *see* Charles l kades, papers

Kadesh-barnea : its importance and probable site... / Trumbull, Henry Clay – 3rd rev ed. Philadelphia: John D Wattles, 1895 [mf ed 1986] – 2mf – 9 – 0-8370-9833-5 – (incl ind) – mf#1986-3833 – us ATLA [939]

Kadesh-barnea / Trumbull, Henry Clay – Scribners, 1884 – 9 – $18.00 – us IRC [930]

Kadets / Latviesu virsnieku apvieniba – 1969-83 – 1r – 1 – mf#2571854 – us WHS [071]

Kadety : konstitutsionno-demokraticheskaia partiia narodnoi svobody / Stalnyi, V – Kharkov, 1930 – 56p 1mf – 9 – mf#RPP-130 – ne IDC [325]

Kadety i evrei / Zaslavskii, D & Ivanovich, S – 1916 – 1mf – 9 – mf#RPP-115 – ne IDC [325]

Kadety v 1905-1906 gg : materialy tsk partii "narodnaia svoboda" krasnyi arkhiv / Grave, B – 1931 (3, p38-68); (4, p112-139) – 1mf – 9 – mf#RPP-112 – ne IDC [325]

Kadiege, Keti Kolobe Herman *see* Emotional and behavioural problems adolescent learners experience without proper pedagocial guidance and assistance

Kadin – Selanik: Asir Matbaasi, 1908-09. Sahib-i Imtiyaz: Mustafa Ibrahim, Mueduer: Enis Avni. n1. 13 tesriniewel 1324 [1908]; 4. 4 tesrinisani 1324 [1908] – (= ser O & t journals) – 1mf – 9 – $25.00 – us MEDOC [956]

Kadin kalbi / Nezihi, Saffet – Istanbul: Aksam Matbaasi, 1927 – (= ser Ottoman literature, writers and the arts) – 6mf – 9 – $90.00 – us MEDOC [470]

Kadlubovskii, A *see* Ocherki po istorii drevnerusskoi literatury zhitii sviatykh

Kadmoniyot ha-'arvim / Yahuda, Abraham Shalom – Jerusalem, Israel. 1894 or 1895 – 1r – us UF Libraries [939]

Kadmut ha-yehudim neged apyon / Josephus, Flavius – Berlin, Germany. 1928 – 1r – us UF Libraries [939]

Kadri, Yakup [Karaosmanoglu] *see* Bir serencam

Kadry gosudartstvennogo i kooperativnogo apparata sssr / Bineman, I M & Kheinman, S A – 1930 – 299p 4mf – 9 – mf#COR-144 – ne IDC [335]

Kaefer, Karl *see* Die betriebsrechnung

Kaegi, Adolf *see* The rigveda

Kaehler, Martin *see*
- Berechtigung und zuversichtlichkeit des bittgebets
- Die bibel, das buch der menschheit
- D gustav warneck, 1834-1910
- Das gewissen
- Die heilsgewissheit
- Jesus und das alte testament
- Das kreuz
- Der lebendige gott
- Der menschensohn und seine sendung an die menschheit
- Die sacramente als gnadenmittel
- Der sogenannte historische jesus und der geschichtliche, biblische christus
- Die universitaeten und das oeffentliche leben
- Unser streit um die bibel
- Wie hermann cremer wurde?: erinnerungen eines genossen – jesu demut: ihre missdeutungen, ihr grund
- Wie studiert man theologie im ersten semester?
- Wiedergeboren durch die auferstehung jesu christi
- Die wissenschaft der christlichen lehre

Kaelin, Frederick Thomas *see* New hydroelectric plant of the shawinigan water and power co

Kaemmel, Otto *see*
- Christian weise
- Die erhebung preussens im jahre 1813 und die rekonstruktion des staates

Kaempchen, Paul Ludwig *see* Die numinose ballade

Kaempf, Saul Isaac *see*
- Das hohelied
- Hohelied aus dem hebraischen originaltext in's deutsche ubertragen
- Die inschrift auf dem denkmal mesa's koenigs von moab

Kaempf, Volker *see* Untersuchungen ueber die angstloesende wirkung von musik

Kaempfe und siege des christentums in der germanischen welt / Uhlhorn, Gerhard – Stuttgart: D Gundert, 1898 – 1mf – 9 – 0-7905-9722-5 – mf#1989-1447 – us ATLA [240]

Kaempfen und lachen : erlebnisse: mit einer selbstdarstellung des verfassers "heimat und herkunft" / Pleyer, Wilhelm – Leipzig: P Reclam, 1944, c1941 – 1r – 1 – us Wisconsin U Libr [920]

Der kaempfende antifaschist – Cordoba (E), 1937 20 mar-17 jun [gaps] – 1 – gw Misc Inst [074]

Der kaempfer – Berlin DE, 1959-61, 1968 1 apr-1981 – 2r – 1 – gw Misc Inst [074]

Der kaempfer – Chemnitz DE, 1921-26, 1927-28 [gaps], 1929-33 – 31r – 1 – (filmed by other misc inst: 1918 30 nov-1919 30 mar, 1920 8 mar-1932 [gaps]. with suppls) – gw Misc Inst [074]

Der kaempfer *see* Westfaelischer kaempfer

Kaempfer, August Hermann *see* Ein fuehrer durch goethes faust

Kaempfer, Englebert *see* East meets west

Die kaempferin – Berlin DE, 1927-32 – 1r – 1 – gw Misc Inst [074]

Die kaempferin – Leipzig DE, 1919-21, 1927-32 – 1 – gw Misc Inst [305]

Kaempffer, Adolf *see*
- Ritt gen mitternacht
- Der tod an der grenze

Kaergel, Hans Christoph *see* Ich blas auf gruenen halmen

Kaerle, Joseph *see* Chrestomathia targumico-chaldaica

Die kaeserei in der vehfreude : eine geschichte aus der schweiz / Gotthelf, Jeremias [Albert Bitzius]; ed by Bloesch, Hans – Erlenbach, Zuerich: E Rentsch, 1922 [mf ed 1993] – (= ser Saemtliche werke in 24 baenden 12) – 366p – 1 – (incl bibl ref) – mf#8522 reel 3 – us Wisconsin U Libr [830]

Kaeslin, Hans *see* Albrecht von hallers sprache in ihrer entwicklung dargestellt

Kaestner, Abraham Gotthelf et al *see* Hamburgisches magazin

Kaestner, Erhart *see* Wahn und wirklichkeit im drama der goethezeit

Kaetchen schoenkopf : eine frauengestalt aus goethes jugendzeit / Vogel, Julius – Leipzig: Klinkhardt & Biermann, 1920 – 1r – 1 – us Wisconsin U Libr [430]

Das kaethchen von heilbronn : grosses historisches schauspiel mit gesang in 5 akten / Kleist, Heinrich von – Aarau: H R Sauerlaender 1899 [mf ed 1991] – 1r – 1 – (filmed with: a sketch of herder and his times / henry nevinson) – mf#2843p – us Wisconsin U Libr [820]

Kaethi die grossmutter / Gotthelf, Jeremias [pseud: Albert Bitzius]; ed by Bohnenblust, Gottfried – Erlenbach, Zuerich: E Rentsch, 1916 [mf ed 1993] – (= ser Saemtliche werke in 24 baenden 10) – 550p – 1 – (incl bibl ref) – mf#8522 reel 3 – us Wisconsin U Libr [830]

Kaethi, die grossmutter : oder, der wahre weg durch jede not: eine erzaehlung fuer das volk / Gotthelf, Jeremias [pseud: Albert Bitzius]; ed by Vetter, Ferdinand – Bern: Schmid & Francke, 1900 [mf ed 1993] – (= ser Volksausgabe seiner werke im urtext 10) – 422p – 1 – mf#8532 reel 2 – us Wisconsin U Libr [880]

Kaev Cinta *see* Subhamangal knun gruasar balakar

Kaev Suvatti *see* Ukna guy

Die kafa-sprache in nordost-afrika / Reinisch, L – Wien, 1888 – 3mf – 9 – mf#NE-20260 – ne IDC [470]

Kafengauz, B B *see* I t pososhkov

Kaffee, zucker und bananen / Key, Helmer – Muenchen, Germany. 1929 – 1r – us UF Libraries [972]

De kaffers aan de zuid kust van afrika : natuur en geschiedkundig beschreven / Alberti, Lodewijk – te Amsterdam : Bij E Maaskamp, koninklijk kunsthandelaar 1810 – (= ser [Travel descriptions from south africa, 1711-1938]) – 4mf – 9 – mf#zah-54 – ne IDC [307]

Kaffir express – Lovedale. South Africa. -m. Oct 1870-Feb, May, Jun, Sep. 1871, May 1872. (11 ft) – 1 – uk British Libr Newspaper [072]

Kaffir express *see* South african outlook

The kaffir express – 1870-1875 – 1r – 1 – mf#MS00315 – sa National [079]

Kaffir folk-lore / Theal, George Mccall – Westport, CT. 1970 – 1r – us UF Libraries [390]

The kaffir war : a letter addressed to the right honourable earl grey...containing remarks on the causes of the present war, and the payment of its expences / Freeman, Joseph John – London, 1851 – (= ser 19th c books on british colonization) – 1mf – 9 – mf#1.1.1006 – uk Chadwyck [960]

Kaffraria, and its inhabitants / Flemyng, Francis Patrick – London, England. 1853 – 1r – us UF Libraries [025]

Kaffrarian – King William's Town SA, 14 may 1864-13 may 1865 – 1r – 1 – mf#MS00318 – sa National [079]

Kaffrarian recorder – King William's Town SA, 15 aug 1863-6 feb 1864 – 1r – 1 – mf#MS00348 – sa National [079]

Kafir phrase book / Stewart, James – Lovedale, South Africa. 1916 – 1r – us UF Libraries [470]

The kafir, the hottentot, and the frontier farmer : passages of missionary life... / Merriman, [N J] – London, 1854 – 3mf – 9 – mf#HTM-128 – ne IDC [910]

Kafir-english dictionary / Appleyard, J W – 1845 – 14 fiche – 9 – sa National [490]

Kafirs of natal and the zulu country / Shooter, Joseph – New York, NY. 1969 – 1r – us UF Libraries [960]

The kafirs of the hindu-kush / Robertson, George Scott – London, Lawrence & Bullen ltd, 1896 [mf ed 1995] – (= ser Yale coll) – xx/658p/pl (ill) – 1 – 0-524-09453-5 – (ill by a d mccormick) – mf#1995-0453 – us ATLA [915]

Kafiya li ibn al-hajib / Uthman ibn 'Umar – [Rome, 1592] – 1mf – 9 – mf#H-8441 – ne IDC [956]

Kafka, Franz *see* Der heizer

Kaftan, Julius *see*
- Das christentum und die philosophie
- Dogmatik
- Drei akademische reden
- Jesus und paulus
- Das leben in christo
- The truth of the christian religion
- Das verhaeltnis des evangelischen glaubens zur logoslehre
- Das wesen der christlichen religion
- Zur dogmatik

Kaftan, Theodor *see*
- Ernst troeltsch
- Der mensch jesus christus, der einige mittler zwischen gott und den menschen
- Moderne theologie des alten glaubens

Kafu, Hazel Bukiwe *see* A historical investigation into black parental involvement in the primary and secondary educational situation

Kagaku shiryo shusei : collection of waka (31 syllable japanese poem), renga (linked verse), haikai (17 syllabled verse), kyoka (comic poem) and senryu (satirical poem) in the holdings of the seikado library, tokyo – 3192v on 252r – 1 – Y2,376,000 – (with 140p guide. in japanese) – ja Yushodo [800]
Kagame, Alexis *see*
– Code des institutions politiques du rwanda precolonial
– Histoire des armees-bovines dans l'ancien rwanda
– Iyo wiliwe nta rungu
– Organisations socio-familiales de l'ancien rwanda
Kagame, P Alegisi *see*
– Indyohesha-birayi
– Isoko y'amajyambere
Kagan, B I *see* Evreiskii rabochii
Kagwa, Apolo *see*
– Basakabaka be buganda
– The book of the ganda clans
– The clans of the baganda
– Ekitabo kye kika nsenene.
– Ekitabo ky'ekika kya nsenene
– Select documents and letters from the collected apolo kagwa papers at makerere college library
Kahan, Louis *see* At the terminals
Kahana, Abraham *see*
– Rabbi yisrael ba'al shem tov
– Sifrut ha-historiya ha-yisraelit
Kahane, Hillel *see* Gelilot ha-arets
Kahane, Ivan *see* Touching all the bases
Kahel : carnet de voyage / Sanderval, Aime Olivier – Paris: F Alcan, 1893 – us CRL [916]
Kahhaleh, Subhi *see* The water problem in israel and its repercussions on the arab-israeli conflict
Kahin, George McTurnan *see* Cambodia
El kahira – Cairo, Egypt. 1885-1889 – 5r – 1 – (aka: el kahira el horra) – uk British Libr Newspaper [079]
El kahira el horra *see* El kahira
Kahle, F Hermann *see* Die geschichte des reiches gottes im alten bunde
Kahle, Paul *see*
– Masoreten des ostens
– Textkritische und lexikalische bemerkungen zum samaritanischen pentateuchtargum
Kahle, Wilhelm *see* Goethe und das christentum
Kahlert, August *see* Erinnerungen an italien
Kahn, Ludwig *see* Social ideals in german literature, 1770-1830
Kahn, Otto Hermann *see* The marketing of american railroad securities
Kahn, Zadoc *see*
– Sermons et allocutions
Kahnis, Karl Friedrich August *see*
– Christenthum und lutherthum
– Die deutsche reformation
– Der gang der kirche in lebensbildern
– Der innere gang des deutschen protestantismus
– Die lutherische dogmatik
– Die moderne wissenschaft des dr. strauss und der glaube unserer kirche
– Rede zum gedaechtniss schleiermacher's am 21. november 1868 in der aula der universitaet leipzig
– Ueber das verhaeltniss der alten philosophie zum christenthum
Kahn-Wallerstein, Carmen *see*
– Aus goethes lebenskreis
– Pegasus im joche
Kahtou / Native Communication Society of B C – 1984 jan 18-1987 dec 28 – 1r – 1 – (cont by: kahtou news) – mf#1277668 – us WHS [071]
Kahun, gurob, and hawara / Petrie, W M – London, 1890 – 2mf – 9 – mf#NE-20339 – ne IDC [956]
Kai chin chi ts'eng cheng chih ching yen t'an / Lin, T'ien-ming – [Fu-chien: Min sheng kung ssu], Min kuo 33 [1944] – (= ser P-k&k period) – us CRL [350]
K'ai fa hsi pei shih yeh chi hua / Chang, Jen-chien – Pei-p'ing: Chu che shu tien, Min kuo 23 [1934] – (= ser P-k&k period) – us CRL [338]
K'ai ko : sung chih-ti tu mu chu chi / Sung, Chih-ti – Shang-hai: Tsa chih kung ssu, Min kuo 35 [1946] – (= ser P-k&k period) – us CRL [820]
K'ai ko kuei : [san mu chu] / Ch'en, Pai-ch'en – [China]: Wen hua ch'u pan she, [1944] – (= ser P-k&k period) – us CRL [951]
Kai ko pi chih yao lan / Yen, Hsiang – Shang-hai: Kuo kuang shu tien, Min kuo 24 [1935] – (= ser P-k&k period) – us CRL [332]
Kai ko ti fang cheng chih ti li lun chi shih shih pan fa / Wu, Yu-hou – [China: Che-chiang sheng min cheng t'ing, Min kuo 26 [1937]] – (= ser P-k&k period) – us CRL [350]
Kai liang chien so i chien shu / Fu, Chen-ch'uan – [China: sn, 1936] – (= ser P-k&k period) – us CRL [360]
K'ai t'o che – Ch'eng-tu: Tung fang shu she, 1939 – (= ser P-k&k period) – us CRL [810]
Kai tsao shih chieh hsin lun / Chiang, Nai-yung – Ch'ung-ch'ing: Chung-kuo kuo min wai chiao hsieh hui, Min kuo 30 [1941] – (= ser P-k&k period) – us CRL [330]
Kai tsou na t'iao lu – Shang-hai: Sheng huo shu tien, Min kuo 24 [1935] – (= ser P-k&k period) – us CRL [170]
Kaiapoi record – 1981-82 – 1 – mf#70.23 – nz Nat Libr [079]
Kaiatonsera ionterennaientak8a ne rosen tharonhiakanere kenha oia sonha 8ahoroke tekaronhianeken : formulaire de prieres / Marcoux, Joseph – Montreal?: J Chapleau, 1879 – 4mf – 9 – (rev and augm by nicholas victor burtin. text in mohawk) – mf#04662 – cn Canadiana [241]
Kaibara, Ekiken *see*
– The way of contentment
– Women and wisdom of japan
Kaibel, George *see* De phrynicho sophista
Kaidanover, Aaron Samuel *see* Sefer she'elot u-teshuvot emunat shemu'el
Das kaidoki : ein reisetagebuch aus der kamakura-zeit / Mittenzwei, Peter – Frankfurt a.M., 1978 – 3mf – 9 – 3-89349-777-3 – gw Frankfurter [950]
K'ai-feng *see* K'ang jih min tsu t'ung i chan hsien chiao ch'eng
Kaifiyats, yadis, etx : containing historical account of certain families of deccan and southern maratha country under the mohammedan and maratha governments / Vad, Ganesh Chimnaji [comp]; ed by Mawjee, Purshotam Vishram & Parasnis, D B – [Bombay]: Purshotam Vishram Mawjee, 1908 – (= ser Samp: indian books) – us CRL [959]
Kaiga soshi – n1-354. 1887-1917 – 15r – 1 – Y225,000 – (in japanese; with 120p guide) – ja Yushodo [700]
K'ai-jen *see* Tsen yang chieh chueh tu shu wen t'i
Kaikini, L V *see* The speeches and writings of sir narayen g chandavarkar
Kaikini, Prabhakar Ramrao *see*
– The recruit
– Shanghai
– Snake in the moon
– Songs of a wanderer
– This civilization
Kaikoura star – 1880-1904; 1906-07; 1909-1914; 1916-17; 1919-1949; 1985-91 – 35r – 1 – (1905, 1908; 1915; 1918 missing) – mf#56.2 – nz Nat Libr [079]
Kailas-manasarovar / Pranavananda, Swami – Calcutta: SP League Ltd, 1949 – (= ser Samp: indian books) – (foreword by pandit jawaharlal nehru) – us CRL [915]
Kain, Joan *see* Report on haiti
Kain und abel in der agada der apokryphen, der hellenistischen, christlichen und muhammedanischen literatur / Aptowitzer, V – Wien, Leipzig: R Loewit, 1922 – 3mf – 9 – mf#J-32-33 – ne IDC [230]
Kain und abel in der deutschen dichtung / Brieger, Auguste – Berlin: W de Gruyter & Co 1934 [mf ed 1992] – 1r – 1 – (incl bibl ref; filmed with: stoff- und motivgeschichte der deutschen literatur / paul merker & gerhard luedtke [ed]) – mf#3000p – us Wisconsin U Libr [430]
Kainai news / Blood Indian Reserve [Alta] – v4 n3-v6 n3 [1971 mar 15-1973 apr 25], 1974 aug 19-1977 dec 23, 1978 jan 16-1979 dec, 1980 jan-1982 jun 2, 1982 jul-1984 jun, 1984 jul-1986 jul, 1986 aug-1987 dec, 1988-1989 jun – 8r – 1 – mf#777591 – us WHS [071]
Kainat : kuetuephane-i tarih / Midhat, Ahmet – Istanbul: Kirk Anbar Matbaasi, 1299 [1881] – (= ser Ottoman histories and historical sources) – 47mf – 9 – $735.00 – us MEDOC [956]
Kaine, Esama *see* Ossomari
Kains, Maurice Grenville *see*
– Ginseng
– Plant propagation
Kairos – [Braamfontein SA]: South African Council of Churches, [1969-81] [mf ed 1980-98] – 2r – 1 – (cont: christian council quarterly. some iss missing) – mf#S0406 – us ATLA [240]
Kairos : an independent quarterly of liberal religion – Taunton MA. n1-33. 1975-85 [complete] [mf ed 1982-84] – 2r – 1 – (publ by ballou channing district (uua) of unitarian universalist societies, aut 1975-sum 1977. publ by kairos inc 1977/78-1983. suspended with winter 1983 iss, resumed publ fall 1983) – ISSN: 0361-2384 – mf#S0409 – us ATLA [243]
Kairos : revista publicada por el grupo de reflexion teologica kairos – Guatemala, CA: El Grupo [n1-23 (enero/jun 1986-jul/dic 1998)] (semiannual) – 2r – 1 – us CRL [240]
Kaisenberg, Moritz von *see* Koenig jerome napoleon
Der kaiser / Ebers, Georg – Stuttgart: Deutsche Verlags-Anstalt, [1893-97?] [mf ed 1993] – (= ser Georg ebers gesammelte werke 11-13) – 3v – 1 – mf#8554 reel 2-3 – us Wisconsin U Libr [830]
Der kaiser : roman / Ebers, Georg – Stuttgart: Deutsche Verlags-Anstalt, [1893-1897?] [mf ed 1993] – (= ser Georg ebers gesammelte werke 11-13) – 3v on 2r – 1 – mf#8554 reel 2-3 – us Wisconsin U Libr [830]
Der kaiser als arzt : genrebild in einem aufzuge / Langer, Anton – Wien: [s.n., 18–?] [mf ed 1995] – 1r3679P – 1 – (filmed with: reichsstaedtische erzaehlugen / herman kurz) – mf#3679p – us Wisconsin U Libr [820]
Kaiser, Bruno *see*
– Die akten ferdinand freiligrath und georg herwegh
– Ausgewaehlte werke
– Der freiheit eine gasse
Kaiser constantin und die christliche kirche : fuenf vortraege / Schwartz, Eduard – Leipzig: B G Teubner 1913 [mf ed 1990] – 1mf – 9 – 0-7905-6885-3 – mf#1988-2885 – us ATLA [240]
Kaiser, Dagmar *see* "Entwicklung ist das zauberwort"
Kaiser, Frederik *see* Eerste onderzoekingen met den mikrometer van airy
Kaiser, Friedrich *see* Die brillantenkoenigin
Kaiser friedrich 2. und papst innocenz 4 : ihr kampf in den jahren 1244 und 1245 / Folz, August – Strassburg i.E: Schlesier & Schweikhardt, 1905 [mf ed 1986] – 1mf – 9 – 0-8370-7862-8 – (incl bibl ref) – mf#1986-1862 – us ATLA [931]
Kaiser friedrich barbarossa : eine tragoedie in fuenf akten / Grabbe, Christian Dietrich – Frankfurt am Main: J C Hermann: G F Kettembeil, 1829 [mf ed 1994] – (= ser Hohenstaufen, ein cyclus von tragoedien 1) – 210p – 1 – mf#8749 – us Wisconsin U Libr [820]
Kaiser, Georg *see*
– Gas
– Im busch
– Die juedische witwe
– Kanzlist krehler
Kaiser heinrich 4 : dramatisches gedicht in zwei abtheilungen / Saar, Ferdinand von – 2., verb. Aufl. in einem Band. Heidelberg: G Weiss, 1872 – 1r – 1 – us Wisconsin U Libr [810]
Kaiser, Ilse *see* Die freunde machen den philosophen, der englaender, der waldbruder
Kaiser in exile : the archive of former kaiser wilhelm 2 of germany during his stay in the netherlands, 1918-1941 / Koen, D T [comp] – [mf ed 2001] – 3776mf – 9 – €27,675.00 set – (map coll separately: 218mf €1200; printed inventory in german; int in english) – mf#m444 – ne MMF Publ [943]
Kaiser julians religioese und philosophische ueberzeugung / Vollert, Wilhelm – Guetersloh: C Bertelsmann, 1899 – (= ser Beitraege zur foerderung christlicher theologie) – 1mf – 9 – 0-7905-8957-5 – (incl bibl ref) – mf#1989-2182 – us ATLA [240]
Kaiser julianus / Geffcken, Johannes – Leipzig: Dieterich, 1914 – (= ser Das Erbe der Alten) – 1mf – 9 – 0-7905-5825-4 – (incl bibl ref) – mf#1988-1825 – us ATLA [930]
Kaiser karl 5. und die roemische curie, 1544-46 / Druffel, August von – Muenchen: Verlag der k. Akademie, in Commission bei G Franz. 4v. 1877-90 – 4mf – 9 – 0-8370-8253-6 – (incl bibl ref) – mf#1986-2253 – us ATLA [240]
Kaiser konstantins taufe : religionstragoedie / Bacmeister, Ernst – Berlin: A Langen, G Mueller, [1937] [mf ed 1989] – 98p/xiii – 1 – mf#6971 – us Wisconsin U Libr [820]
Kaiser max und seine jaeger : dichtung / Baumbach, Rudolf – Leipzig: A G Liebeskind, 1888 [mf ed 1989] – 130p – 1 – mf#6991 – us Wisconsin U Libr [810]
Kaiser napoleon 3 : eine biographische studie / Gottschall, Rudolf von – Liegnitz 1859 [mf ed Hildesheim 1995-98] – 1v on 2mf – 9 – €60.00 – 3-487-26025-5 – gw Olms [944]
Kaiser, Oscar *see* Der dualismus ludwig tiecks als dramatiker und dramaturg
Kaiser, Paul *see*
– Der kirchliche besitz im arrondissement aachen gegen ende des 18. jahrhunderts und seine schicksale in der saekularisation durch die franzoesische herrschaft
– Paul gerhardt
Kaiser rotbart : roman / Strobl, Karl Hans – Budweis: Verlagsanstalt Moldavia, c1935 – 1r – 1 – us Wisconsin U Libr [830]
Der kaiser und die protestanten in den jahren 1537-1539 / Rosenberg, Walter – Halle a. S: Verein fuer Reformationsgeschichte 1903 [mf ed 1990] – (= ser Schriften des vereins fuer reformationsgeschichte 20/77) – 1mf – 9 – 0-7905-5130-6 – (incl bibl ref) – mf#1988-1130 – us ATLA [943]
Kaiser und herzog : kampf zweier geschlechter um deutschland / Beumelburg, Werner – Oldenburg i O: G Stalling, 1937, c1936 [mf ed 1989] – 555p – 1 – mf#7017 – us Wisconsin U Libr [830]
Kaiser, W *see* Die goetterwelt der alten deutschen
Das kaiserbuch : ein epos in drei teilen / Ernst, Paul – Muenchen: M Huebner 1923 [mf ed 1989] – 6v on 1r – 1 – mf#7231 – us Wisconsin U Libr [810]
Kaiserchronik eines regensburger geistlichen (mgh8:1.bd.1.abt) – 1895 – (= ser Monumenta germaniae historica 8: scriptores qui vernacula lingua usi sunt (mgh8)) – €29.00 – (1. bd 2. abt: der trierer silvester. das annolied. 1895) – ne Slangenburg [240]
Die kaiserin, der koenig und ihr offizier : das abenteuerliche leben des johann jakob wuensch / Finckh, Ludwig – Feldpostausg. Muenchen: Deutscher Volksverlag c1944 [mf ed 1989] – 1r – 1 – (filmed with: der deutsche finckh) – mf#7241 – us Wisconsin U Libr [355]
KAISERL AKADEMIE DER WISSENSCHAFTEN IN WIEN *see* Denkschriften der kaiserlichen adakemie
Kaiserl Akademie Der Wissenschaften In Wien *see* Denkschriften der kaiserlichen adakemie
Kaiserlich Deutsches Archaeologisches Institut *see*
– Jahrbuch des kaiserlich deutschen archaeologischen instituts
– Jahrbuch des kaiserlich deutschen archaeologischen instituts. ergänzungsheft
Kaiserliche Universitaets-Sternwarte in Strassburg *see* Annalen der kaiserlichen universitaets-sternwarte in strassburg
Kaiserliches marinekabinett (bestand rm 2) bd 28 / ed by Fleischer, Hans-Heinrich et al – 1987 – xvii/217p – €7.00 – 978-3-89192-011-4 – gw Bundesarchiv [355]
Kaiserliches Statistisches Amt (v1-39) *see* Statistisches jahrbuch fuer das deutsche reich seit 1880
Kaiserlich-koeniglich privilegirte prager oberpostamtszeitung – Prag (CZ), 1807 2 feb-26 jun – 1 – gw Misc Inst [380]
Kaiserlich-koenigliche privilegirte wiener zeitung *see* Wiener zeitung
Das kaiserrech in truemmern : roman / Bergengruen, Werner – Leipzig: K F Koehler, c1927 [mf ed 1989] – 407p – 1 – mf#7009 – us Wisconsin U Libr [830]
Das kaiserreich am scheideweg / Eschenburg, Theodor – Berlin, 1929 – 1 – gw Mikropress [943]
Kaiserslautern american – 1981 may-1982, 1983 jan 7-1984 dec 21, 1985 jan 11-1986 mar 21, 1986 apr 4-1987 dec 18, 1988 jan-1989 feb 24, 1989 mar 3-1990 feb 23, 1990 mar-1991 feb 22 – 7r – 1 – mf#643891 – us WHS [071]
Kaiserswerther nachrichten – Duesseldorf DE, 1930-35 [gaps] – 3r – 1 – (title varies: heimat-zeitung 1934) – gw Misc Inst [074]
Kaisertum und herzogsgewalt im zeitalter friedrichs 1 (mgh schriften:9.bd) : studien zur politischen und verfassungsgeschichte des hohen mittelalters / Mayer, Th et al – 1944 – (= ser Monumenta germaniae historica. schriften (MGH schriften)) – €19.00 – ne Slangenburg [931]
K'ai-yuan *see*
– Shui pien
– Ssu nien chi
Kajawen – [Jakarta: s.n. 1926- (semiwkly, wkly) [mf ed Jakarta: National Library of Indonesia] – 16 [jan 7 1926-1936] – 1 – (ceased in 1942; some iss accompanied by suppls: jagading wanita, and: taman bocah; in javanese (kawi script)) – mf#mf-11890 seam – us CRL [079]
Kak dumaet partiia narodnoi svobody reshit zemelnyi vopros / Cheshikhin, V E – n.d. – 31p 1mf – 9 – mf#RPP-133 – ne IDC [325]
Kak kontrolirovat' gosudarstvennyi i mestnyi biudzhet / ed by Vainshtein, A I – [M, 1925] – 1mf – 9 – mf#REF-49 – ne IDC [332]
Kak kooperatsiia sozdaet krestianskoe bogatstvo / Serebriakov, F S – 1926 – 55p 1mf – 9 – mf#COR-511 – ne IDC [335]
Kak krestianinu obzavestis mashinoi / Vasilevskii, N M – 1926 – 1mf – 9 – mf#COR-467 – ne IDC [335]
Kak organizovat i vesti potrebitelskoe obshchestvo : tekst, zakony, ustavy, instruktsii i pr / Merkulov, A V & Kheisin, M L – 1914 – 322p 4mf – 9 – mf#COR-74 – ne IDC [335]
Kak organizovat kreditnuiu rabotu v promyslovoi kooperatsii / Iakhontov, P V – 1929 – 142p 2mf – 9 – mf#COR-458 – ne IDC [335]
Kak organizovat krestianskoe khoziaistvo v nechernozemnoi polose / Chaianov, A V – 1926 – 46p 1mf – 9 – mf#COR-229 – ne IDC [335]
Kak organizovat selskokhoziaistvennoe kreditnoe tovarishchestvo / Beliaev, V N – 1928 – 88p 1mf – 9 – mf#COR-372 – ne IDC [335]
Kak otkryt selskokhoziaistvennoe tovarishchestvo po pererabotke produktov / Kancher, E S – 1913 – 12p 1mf – 9 – mf#COR-42 – ne IDC [335]
Kak predpolagala nadelit krestian zemlei partiia narodnoi svobody vo vtoroi gosudarstvennoi dume / Shingarev, A I – 1917 – 22p 1mf – 9 – mf#RPP-134 – ne IDC [325]
Kak proshli vybory vo 2-iu gosudarstvennuiu dumu – 1907 – 294p 4mf – 9 – mf#RPP-17 – ne IDC [325]

Kak rabotat revizionnoi komissii selsko-khoziaistvennogo kreditnogo tovarishchestva / Barantsov, M S – 1926 – 126p 2mf – 9 – mf#COR-371 – ne IDC [335]

Kak, Ram Chandra *see* Antiquities of bhimbar and rajauri

Kak sostavit otchet selskokhoziaistvennogo kreditnogo tovarishchestva / Kilchevskii, V A – 1927 – 52p 1mf – 9 – mf#COR-384 – ne IDC [335]

Kak sostavit plan deiatelnosti selskogo obshchestva potrebitelei / Dneprovskii, S P – 1927 – 148p 2mf – 9 – mf#COR-324 – ne IDC [335]

Kak uchredit i otkryt kreditnoe tovarishchestvo / Shmidt, G R – 1924 – 46p 1mf – 9 – mf#COR-409 – ne IDC [335]

Kak ustroit i vesti kreditnoe tovarishchestvo : pisma krestianina ivana bodrogo k diade akimu maksimovichu / Serebriakov, F S; ed by Shefler, M E – 1927 – 97p 2mf – 9 – mf#COR-402 – ne IDC [335]

Kak ustroit' melkii kredit v gorodakh / Borodaevskii, S V – Spb, 1907 – 1mf – 9 – mf#REF-394 – ne IDC [332]

Kak vesti dela kreditnogo tovarishchestva / Kulyzhnyi, A E – Ed 3. Pg, 1914 – 1mf – 9 – mf#REF-402 – ne IDC [332]

Kakaia raznitsa mezhdu chastnym bankom i kreditnym tovarishchestvom : (iz besed na koop. kursakh) / Kil'chevskii, VA – Smolensk, 1913 – 1mf – 9 – mf#COR-47 – ne IDC [332]

The kakamora reporter – Honiara. n4, 7, 10-46. jun 1970-jul 1975 – 1r – 1 – mf#pmb doc414 – at Pacific Mss [079]

Kakati, Banikanta *see*
- Aspects of early assamese literature
- Assamese
- Mother goddess kamakhya
- Visnuite myths and legends

Kalachov, N *see*
- Iuridicheskii vestnik
- Razbor sochineniia g

Kalachov, N V *see*
- Arkhiv istoriko-iuridicheskikh svedenii, otnosiashchikhsia do rossii
- Arteli v drevnei i nyneshnei rossii

Kalahari / Bjerre, Jens – New York, NY. 1960 – 1r – us UF Libraries [960]

Kalahari and its lost city / Clement, A John – Cape Town, South Africa. 1967 – 1r – us UF Libraries [960]

Kalai, David *see* 'Aliyah Ha-Sheniyah

Kalaic corkal – Tirunelveli: Cennai Makanat Tamilccankam, 1936 – (= ser Samp: indian books) – us CRL [056]

Kalamazoo railroad velocipede co, kalamazoo, michigan, usa : manufacturers of steel railroad velocipedes, section, telegraph, push and inspection hand cars... – Chicago?: Rand McNally, 1886? – 1mf – 9 – mf#60706 – cn Canadiana [625]

Kalanda, Mabika *see* Baluba et lulua

Kalata zambe nnom ba mfefe – New York, NY. 1948 – 1r – us UF Libraries [960]

Kalayi, Refi'i *see* The divan project

Kalb, Ernst *see* Kirchen und sekten der gegenwart

Kalb, J S *see* Weight training economy as a function of intensity of the squat and overhead press exercise

Kalbeck, Max *see*
- Das buehnenfestspiel zu bayreuth
- Richard wagners parsifal

Kalbin goezu / Hakki, Ismail [Baltacioglu] – [Istanbul]: Evkaf-i Islamiye Matbaasi, 1338 [1922] – (= ser Ottoman literature, writers and the arts) – 1mf – 9 – $25.00 – us MEDOC [470]

Kalckstein, Carl von et al *see* Nationale und humanistische erziehung

Kalda, Andrea L *see* The effect of upper body excerise on secondary lymphedema following breast cancer treatment

Kaldy, Gyula [Julius] *see* History of hungarian music

Ka'le, M R *see* The dasakumaracharita of dandin

Kale, M R *see*
- Bhavabhuti's malatimadhava
- The nagananda of sri harsha deva
- The priyadarsika of sri harsha-deva
- The ratnavali of sri harsha-deva
- The venisamhar of bhatta narayana

Kale, Moreshvar Ramchandra *see* The abhijnanasakuntala of kalidasa

Kale, Moreshwar Ramachandra *see* Kiratarjuniyam cantos 1-3

Kale, Vaman Govind *see* An introduction to the study of indian economics

Kalee's shrine / Allen, Grant – New York: New Amsterdam Book Co, 1879? – 3mf – 9 – mf#58404 – cn Canadiana [830]

Kaleidoscope – Cambridge, NE: Kaleidoscope Power Print Co. 2v. v12 n19. nov 13 1896-v13 n26. dec 31 1897=whole n592-651 (wkly) [mf ed with gaps] – 2r – 1 – (cont: cambridge kaleidoscope. cont by: cambridge kaleidoscope (1898)) – us NE Hist [071]

Kaleidoscope – Granby MA 1965-67 – 5 – ISSN: 0022-7919 – mf#6757 – us UMI ProQuest [790]

Kaleidoscope – Milwaukee WI. v1967 oct 6-1970 aug 17/23, 1970 aug 17/23-1971 nov 11 – 2r – 1 – mf#1828756 – us WHS [071]

Kaleidoscope : a novel / Mitra, Premendra – Calcutta: Purvasa Ltd, 1945 – (= ser Samp: indian books) – us CRL [830]

Kaleidoscope : or, literary and scientific mirror – La Salle IL 1818-31 – 1 – mf#4725 – us UMI ProQuest [500]

Kaleidoscope chicago – Chicago IL. 1968 nov 22-1969 mar 27 – 1r – 1 – (cont by: chicago kaleidoscope) – mf#1110803 – us WHS [071]

The kalela dance : aspects of social relationships among urban africans in northern rhodesia / Clyde Mitchell, J – (= ser Institute for social research, university of zambia. papers 27) – 2mf – 7 – mf#4734 – uk Microform Academic [960]

Kalelkar, Kaka *see*
- Bapu's letters
- To a gandhian capitalist

Kalem – n1-130. 1324-27 [1908-11] [all publ] – (= ser O & t journals) – 32mf – 9 – $535.00 – us MEDOC [956]

Kalemkiar, Gregoris *see* Eine skizze der literarisch-typgraphischen thaetigkeit der mechitharisten-congregation in wien

Kalendar / American Carpatho-Russian Youth – Pittsburgh. 1949-54, 1958-62, 1964-73 – 1 – us L of C Photodup [360]

Kalendar amerikanskaho russkaho sokola sojedinenija – Homestead, PA: Greek Catholic Union of Russian Brotherhoods of USA, 1919; 1921-36 – us CRL [240]

Kalendar organizacii "svobody" : sostavil ot imeni organizacii greko kaft – Karpato-ruskich spomahajuscich bratstv "Svobody" ..Perth Amboy, NJ: Typografia Vostoka – (filmed with: kalendar "svobody" na rok..1927-28, 1933-38) – us CRL [071]

Kalendar prosvity – 1918-19; 1921-32 – us CRL [071]

Kalendar "sobranija" – McKeesport, PA: United Societies of Greek Catholic Religion of USA, 1935-1938 – us CRL [240]

Kalendar "svobody" na rok... – Perth Amboy, NJ: Gr Kafto russka pravpotporujusca organizacija "Svobody", 1925 – (filmed with: kalendar organizacii "svobody") – us CRL [071]

Kalendaria ecclesiae universae. kalendaria ecclesiae slavicae sive graeco-moschae / Assemanus, J S – Romae. v1-6. 1755 – €151.00 – ne Slangenburg [240]

Die kalendarien von st gallen (tab36) : texte / Munding, E – 1948 – (= ser Texte und arbeiten. beuron (tab). beitraege zur ergruendung des aelteren lateinischen und christlichen schrifttums und gottesdienstes) – €5.00 – ne Slangenburg [241]

Die kalendarien von st gallen (tab37) : untersuchungen / Munding, E – 1951 – (= ser Texte und arbeiten. beuron (tab). beitraege zur ergruendung des aelteren lateinischen und christlichen schrifttums und gottesdienstes) – €11.00 – ne Slangenburg [241]

Kalendarium hortense : or, the gard'ners almanac, directing what he is to do monthly through-out the year and what fruits and flowers are in prime / Evelyn, J – Ed 7. London: T Sawbridge, G Wells and R Bently, 1683 – 2mf – 9 – mf#EJ-3 – ne IDC [700]

Kalendarium hortense : or, the gardners almanac, directing what he is to do monthly through-out the year and what fruits and flowers are in prime / Evelyn, John – London: T Sawbridge, G Wells and R Bently, 1683 – 127p 2mf – 9 – mf#EJ-3 – ne IDC [635]

Kalendarium humanae vitae = [The kalender of mans life] / Farley, R – London: William Hope, 1638 – 2mf – 9 – mf#0-249 – ne IDC [700]

Kalendarium manuale utriusque ecclesiae orientalis et occidentalis / Nilles, N – Oeniponte, 1896-1897 – 2pts – (pt1: immobilia totius anni festa €21. pt2: mobilia totius anni festa €31) – ne Slangenburg [390]

Kalendarium und planetenbuecher *see* Apokalypse / ars moriendi / biblia pauperum / antichrist / fabel vom kranken loewen / kalendarium und planetenbuecher / historia david (mxt2)

Kalendars of scottish saints : with personal notices of those of alba, laudonia, & strathclyde / Forbes, Alexander Penrose – Edinburgh: Edmonston and Douglas, 1872 – 2mf – 9 – 0-7905-8029-2 – (incl bibl ref) – mf#1988-6010 – us ATLA [240]

Kalendarz sw piotra klawera – Krakow: Nakl Sodalicji sw Piotra Klawera 1913- [annual] [mf ed 2005] – v1-37 (1914-51) on 2r – 1 – (publ: salzburg 1916-17; poznan 1918; salzburg 1919; rzym 1920-27; krakow 1928-31; krosno 1932-1939; st louis, mo 1942-51; lacks: v19 (1922) p3-8; v26-27 (1940-41); v33-34 (1947-48); damaged: v12 (1926) p3; v19 (1922) p2,9; v29 (1943) [front cover]; aka: kalendarz swietego piotra klawera; kalendarz misyjny dla wszystkich 1924) – mf#2005c-s075 – us ATLA [241]

Kalender und jahrbuch fuer israeliten – Vienna, Austria 1841-46 [mf ed Norman Ross] – 3r – 1 – mf#nrp-1973 – us UMI ProQuest [939]

Kalendergeschichten / Gotthelf, Jeremias [pseud: Albert Bitzius]; ed by Hunziker, Rudolf & Bloesch, Hans – Erlenbach, Zuerich: Eugen Rentsch, 1931-32 [mf ed 1993] – (= ser Saemtliche werke in 24 baenden 23-24) – 2v – 1 – (incl bibl ref) – mf#8522 reel 6 – us Wisconsin U Libr [830]

Kales, Albert Martin *see*
- Cases on persons and domestic relations
- Conditional and future interests and illegal conditions and restraints in illinois
- Contracts and combinations in restraint of trade

The kalevala : the epic poem of finland – New York: J B Alden 1888 [mf ed Bloomington IN: Indiana Uni Lib, Preservation Dept 1984] – 2v on 1r [ill] – 1 – us Indiana Preservation [390]

Kalfelis, Petros *see* Die auswirkungen der fabrikarbeit auf das traditionale rollenverhalten der frau in megara, griechenland

Kalfus, Radim *see* Moravians

Kalgoorlie miner – Australia. 1901-32.-d. 80mqn reels – 1 – uk British Libr Newspaper [072]

Kalgoorlie miner – Western Australia. 3 apr 1897; 1901; 1902; 3 jul 1905-10 jul 1916; 10 feb 1917-jun 1922; 2 dec 1932; 1 dec 1951-25 apr 1953 – 85 1/2r – 1 – uk British Libr Newspaper [072]

Kalgoorlie western argus – Kalgoolie, Australia 30 jul 1896-9 may 1916 – 1 – (cont by: western argus [16 dec 1919-20 jun 1922]) – uk British Libr Newspaper [079]

Kalhana *see* Rajatarangini, the saga of the kings of kasmir

Kalhoff, Hermann *see* Die drohende spaete metabolische azidose der frueh- und neugeborenen

Kali, the mother / Nivedita, Sister – London: Swan Sonnenschein & Co, 1900 – (= ser Samp: indian books) – us CRL [280]

Kali' wisaks / Oneida Tribe of Indians of Wisconsin – 1974 sep 9-1977 dec 23, 1978 jan 6-1979 aug 3 – 2r – 1 – (cont by: kalihwi' saks) – mf#622809 – us WHS [305]

Der kali-bergmann – Stassfurt DE, 1968 may-1979 jan, 1980 jul-1990 6 mar – 3r – 1 – (with gaps) – gw Misc Inst [074]

Kali-bz *see* Das buendnis

Kalidasa / Ghose, Aurobindo – Calcutta: Arya Sahitya Bhawan, 1929 – (= ser Samp: indian books) – us CRL [280]

Kalidasa *see*
- The abhijnanasakuntala of kalidasa
- The cloud-messenger
- Ghatakarparam oder das zerbrochene gefaess
- Kalidasa's raghuwansha
- Kalidas's meghadutam
- Malavikagnimitra of kalidasa
- The malavikagnimitra of kalidasa
- The meghaduta of kalidasa
- The raghuvanca
- Raghuvansa, kalidasae carmen
- Raksasakavyam satikam
- Ritusamhara
- The sakuntala in hindi
- Urvasia fabula calidasi
- Vikramorvasi
- Vikramorvasie
- Vikramorvasiya

Kalidasa and vikramaditya : a historical and literary diversion to relieve the monotony of retirement / De, S C – Calcutta: SC De, 1928 – (= ser Samp: indian books) – us CRL [490]

Kalidasa's raghuwansha : a mahakavya in 19 cantos with the commentary of mallinatha suri / ed by Panshikar, Vasudev Shastri – Bombay: Nirnay-Sagara Press, 1916 – (= ser Samp: indian books) – (critical and explanatory notes on the text and comm, trans of the text, and an essay on the life and writings of the poet by krishnarao mahadeva joglekar) – us CRL [490]

Kalidas's meghadutam : with vallabhadeva's commentary, english translation, grammatical, critical and explanatory notes, introduction dealing with a few important topics and appendices / ed by Deshpande, R R & Tope, T K – Bombay: Oriental Book House, 1947 – (= ser Samp: indian books) – us CRL [490]

Kaliedoscope eye – iss 1-4 [[1968 jan], 1968 may 20-nov 11] – 1r – 1 – mf#714261 – us WHS [071]

Kalifornier idisher shtern = California jewish star – Los Angeles, San Francisco. 1924 – 1 – us AJPC [071]

Kalifornyer idishe shtime = California jewish voice – San Francisco, CA: The Jewish Pub Assoc. v1 n1. oct 11 1912- (wkly) [mf ed 197-?] – (in yiddish and english) – mf#ZZAN-21684 – us NY Public [071]

Kalihwi' saks = She looks for news / Oneida Tribe of Indians of Wisconsin – 1973 sep 9-1977 dec 23, 1979 aug 17-1980 jan 4, 1980 jan 18-1981 feb 27, 1981 mar 13-1982 mar 26 – 4r – 1 – (cont: kali' wisaks; cont by: kalihwisaks) – mf#622854 – us WHS [305]

Kalihwisaks : the official publication of the oneida tribe of indians of wisconsin = She looks for news / Oneida Tribe of Indians of Wisconsin – v9,11,13,15-18,22 [1986 may 23, jun 23, jul 25, aug 29-oct 10, dec 5], v26,31-33 [1987 feb 6, may 21-jun], 1993 oct 2-1996 dec 19, 1995 [v175-253], 1996, 1997 jan 23-dec 25 [v255-279], 1998, 1999 – 6r – 1 – (cont: kalihwi' saks) – mf#1221833 – us WHS [071]

Kalikristall – Zielitz DE, 1987 may-1989 [gaps] – 1r – 1 – (kalibetrieb zielitz) – gw Misc Inst [622]

Kalinga – Cuttack, India. Jul-Sept 1966 – 1r – 1 – (oriya language) – us L of C Photodup [079]

Kalinin, M I *see* Znachenie selskokhoziaistvennykh kommun v sovetskom stroitelstve

Kalinin, Mikhail Ivanovich *see* Stalin

Kalininskaia pravda – Tver', Russia 1974-88 [mf ed Norman Ross] – 1r – 1 – mf#nrp-1846 – us UMI ProQuest [077]

Kalinychev, F I *see* Gosudarstvennaia duma v rossii

Kalisch, Isador *see* Sketch of the talmud

Kalisch, Isidor *see*
- Sefer yetsirah
- A sketch of the talmud

Kalisch, Marcus Moritz *see*
- The book of jonah
- Genesis
- Leviticus
- The prophecies of balaam (numbers 22 to 24)
- Shemot

Kalischer, Alfred Christlieb *see* Heinrich heine's verhaeltnis zur religion

Kalischer, H *see* Drischath zion, oder zions herstellung

Kalischer, Salomon *see* Goethe als naturforscher und herr du bois-reymond als sein kritiker

Kalischer, Z H *see* Drischath zion, oder zions herstellung

Kalischer, Zevi Hirsch *see* Derishat tsiyon

Kalish, Max *see* Scrapbook, ms p.p.

Kaliszskie gubernskie vedomosti – Kalisz, Poland 1867-1914 [mf ed Norman Ross] – 36r – 1 – mf#nrp-604 – us UMI ProQuest [077]

Kaliuzhnyi, I I *see*
- Narod [vladivostok: 1920]
- Narodnoe delo [ufa: 1918]

The "kalivarjyas" : or, prohibitions in the 'kali' age, their origin and evolution and their present legal bearing / Bhattacharya, Batuknath – Calcutta: University of Calcutta, 1943 – (= ser Samp: indian books) – us CRL [280]

Kali-worship in kerala / Achyuta Menon, Chelnat – [Madras]: University of Madras, 1943- – (= ser Samp: indian books) – us CRL [280]

Kalkar, Christian Andreas Hermann *see* Den danske mission i ostindien i de seneste aar

Kalki : or the future of civilization / Radhakrishnan, Sarvepalli – London; New York: Kegan Paul, Trench, Trubner & Co, 1934 – (= ser Samp: indian books) – us CRL [900]

Kalkoff, Paul *see* Die anfaenge der gegenreformation in den niederlanden

Kalkschmidt, Till *see* Der deutsche frontsoldat

Kalk-werker – Ruebeland DE, 1977 15 dec-1990 31 jul [gaps] – 2r – 1 – (veb harzer kalk- und zementwerk) – gw Misc Inst [690]

The kall of the klan in kentucky / Lougher, E H – [Greenfield, IN: W Mitchell Printing Co], c1924 – us CRL [978]

Kallab, Wolfgang *see* Vasaristudien von wolfgang kallab

Kalle – Duesseldorf DE, 1953 21 aug-1954 3 jul – 1r – 1 – gw Misc Inst [074]

Kallen, G *see* De auctoritate presidendi in concilio generali

Kallen, Horace Meyer *see* William james and henri bergson

Kallenberg, Friedrich *see* Auf dem kriegspfad gegen die massai

Kallikreinaktivitaet im parotisspeichel bei patienten mit parotistumoren / Raff, Alexander – (mf ed 1996) – 1mf – 9 – €30.00 – 3-8267-2314-7 – mf#DHS 2314 – gw Frankfurter [617]

Kallisti / Matrix Productions [US] – v1 iss 11-v2 iss 4 [1988 jan-1989 win] – 1r – 1 – (cont: kallisti komiks) – mf#1520843 – us WHS [071]

Kallisti komiks / Left Track Productions [US] – v1 iss 8-10 [1987 jan-sep] – 1r – 1 – (cont by: kallisti) – mf#1520842 – us WHS [071]

Kallistov, S *see* Narodnoe delo [kazan': 1918]

Kalm, P *see* En resa til norra america...

Kalmar – Kalmar, Sweden. 1864-1918 – 98r – 1 – sw Kungliga [078]

Kalmar laenstidning – Kalmar, Sweden. 1915-18 – 8r – 1 – sw Kungliga [078]

Kalmar lans tidning – Kalmar, Sweden. 1948-78, 1979- – 103r – 1 – sw Kungliga [078]

Kalmarbladet – Oskarshamn, Sweden. 1905-06 – 1r – 1 – sw Kungliga [078]

Kalmar-kalmar laens tidning – Kalmar, Sweden. 1918-47 – 142r – 1 – sw Kungliga [078]

Kal'nev, Mikhail A *see* Sbornik" 17-ti glavneishikh" protovosektantskikh"

Kal'nin, A et al *see* Trud [kharbin: 1918]

Kal'nitskii, IA *see* V bagrovom kol'tse

K'a-lo-sheng *see* Lun kan pu

Kalsakau : records, accounts, notes, correspondence – 1908-69 – 1r – 1 – mf#pmb55 – at Pacific Mss [920]

Das kalte haus, oberschlesien : novellen / Koehler, Willibald – Prag: Noebe 1944 [mf ed 1991] – 1r – 1 – (filmed with: heinrich von kleist / hermann graef) – mf#2772p – us Wisconsin U Libr [830]

Kaltenkirchener nachrichten – Kaltenkirchen DE, 1884-1933 – 76r – 1 – gw Misc Inst [074]

Kaltenkirchener zeitung – Kaltenkirchen DE, 1910 10 sep-1936 29 feb – 1 – gw Misc Inst [074]

Kalthoff, Albert *see*
- Das christus-problem
- Die entstehung des christentums
- Friedrich nietzsche und die kulturprobleme unserer zeit
- Das leben jesu
- Religioese weltanschauung
- Die religioesen probleme in goethes faust
- Schleiermachers vermaechtnis an unsere zeit
- Was wissen wir von jesus?

Kalunga : ein kolonialroman / Coerver, Hubert – Braunschweig: G Westermann, c1940 – 1r – 1 – us Wisconsin U Libr [830]

Kaluzhniatskii, E *see* K bibliografii tserkovno-slavianskikh pechatnykh izdanii v rossii

Kaluzhskaia Kontora Gosstrakha *see* Otchet kaluzhskoi kontory gosstrakha za 1922/23 (operatsionnyi) god 109=deviatomu gubernskomu s"ezdu sovetov r k i k d

Kaluzhskii kooperator – Kaluga, 1926-1927(15) – 8mf – 9 – mf#COR-599 – ne IDC [335]

Kaluzhskij gub ispolnitel'nyj komitet sovetov *see* Izvestiia gubernskogo kaluzhskogo ispolnitel'nogo komiteta sovetov rabochikh, krest'ianskikh i krasnoarmejskikh deputatov

Kalweit, Paul *see* Die stellung der religion im geistesleben

Kalypso : ein vorspiel [zu odysseus] / Wesendonk, Mathilde – Dresden: Schulze [1875] – 1mf – 9 – mf#mw-13 – ne IDC [820]

Kam ko mruin ca tan : [short stories] / Jo Gyi et al – Ran Kun: Pugam ca up tuik 1965 [mf ed 1990] – 1r with other items – 1 – (in burmese) – mf#mf-10289 seam reel 185/2 [§] – us CRL [830]

Kam Man, U' *see* Be dan panna nhan loki padesa kyam

Kamadulski, Mary J *see* A methodological comparison

Kamal, Youssouf *see* Monumenta cartographica africae et aegypti

Kamali, Iris *see* Die relative haeufigkeit des harnblasenkarzinoms an unausgewaehltem biopsiegut der jahre 1977-79...

Kamalist turkey / Anandan, P M – Tellicherry: ACS Bros, 1938 – (= ser Samp: indian books) – us CRL [956]

Kamat, Venkatrao Vithal *see* Measuring intelligence of indian children

Kamba customary law; notes taken in the machakos district of kenya colony / Penwill, D J – Kampala: East African Literature Bureau 1972. 122p. LL-12044 – 1 – us L of C Photodup [340]

Kamba grammar / Farnsworth, E M – s.l, s.l? 1957 – 1r – us UF Libraries [470]

Kambairai – Cape Town, South Africa. 1956 – 1r – us UF Libraries [960]

Kambairai – Salisbury, Zimbabwe. 1966 – 1r – us UF Libraries [960]

Kambha lhann sakhan kuiy to nhuin / Mhuin, Sa khan Kuiy to – Ran Kun: Mon E Mon 1964 – (= ser Sam lvan ca cann 6) – [pl/ill] 1r with other items – 1 – (in burmese) – mf#mf-10289 seam reel 191/4 [§] – us CRL [920]

Kambha ta khvan rok khai can ka / Thavan Phe, U – Yan Kun: Tam Khavan Tuin Cape 1975 [mf ed 1990] – [ill] 1r with other items – 1 – (in burmese) – mf#mf-10289 seam reel 146/7 [§] – us CRL [910]

Kambodscha / Drescher, Max – [Berlin: Deutscher Friedensrat 1961] [mf ed 1990] – (= ser Land und leute) – 1r with other items – 1 – mf#mf-10289 seam reel 125/4 [§] – us CRL [915]

Kambodscha, der sieg des volkes ist gewiss – Koeln: Liga gegen den Imperialismus 1973 [mf ed 1989] – (= ser Der antimperialistische kampf der voelker 3) – 1r with other items – 1 – mf#mf-10289 seam reel 018/04 [§] – us CRL [959]

Kambodscha mit angkor wat, birma, laos / ed by Rathenberg, Erhard et al – Buchenhain vor Munchen: Verlag Volk und Heimat [1969?] [mf ed 1989] – (= ser Mai's auslandstaschenbuch 30) – 1r with other items – 1 – (incl bibl) – mf#mf-10289 seam reel 007/07 [§] – us CRL [915]

Die kambodschanische erfahrung – Berlin: Koenigliche Botschaft Kambodscha in der DDR 1975 [mf ed 1989] – 1r with other items – 1 – ("der erste teil besteht aus bekannten auszugen, exposes und erklarungen von khieu samphan und ieng sary...alle weiteren texte ausser einer reportage wurden durch die 'stimme der nationalen einheitsfront kambodschas' veroffentlicht") – mf#mf-10289 seam reel 017/36 [§] – us CRL [959]

Das kambodschanische volk wird phnom penh befreien : [herausgeber, abteilung fur agitation und propaganda beim zentralvorstand der liga gegen den imperialismus] – Koeln: Internationale Solidaritat-Verlagsgesellschaft [1974] [mf ed 1989] – (= ser Liga aktuell 3) – 1r with other items – 1 – mf#mf-10289 seam reel 017/35 [§] – us CRL [959]

Kambodzha / Verin, Vladimir Petrovich et al – Moskva [Russia]: Gos izd-vo Geograficheskofi lit-ry 1960 [mf ed 1989] – 1r with other items – 1 – mf#mf-10289 seam reel 005/08 [§] – us CRL [959]

Kambra, Karl *see* Three sonatinas for the pianoforte...op 14

Kambuja – Phnom-Penh. 6 janv 1943-27 fevr 1945 – 1 – fr ACRPP [073]

Kambuja (1976) : dassanavatti rupbhab – [Bhnam Ben: s.n.] 1976- [mf ed Ithaca NY: Photo Services Cornell University Lib [1980?] – chnam di 1:lekh 1-12 (1976:makara-dhanu); chnam di 3:lekh 1-6 (1978:makara-mithuna) – 1 – mf#12931 – us Cornell [959]

Kambuja suriya – Phnom-Penh. 1927-46, 1948 – 1 – (missing:1929) – fr ACRPP [073]

Kambuja suriya – Bhnam ben: Impr Nouvelle Albert Portail [mthly, semimthly] [mf ed Paris: Assoc pour la conservation et la reproduction photographique de la presse 1975] – 10r – 1 – (publ varies; description based on: chnam gamrap 1, lekh 7 [1927]; latest iss consulted: chnam di 20, lekh 4 [mesa 1948 [apr 1948]]; mf: v1 n7-v20 n4 [1927-48]; lacks v19 [1947]]; iss irregularly 1927-1940; only 1 iss publ in 1945?; some iss publ in combined form; in khmer; table of contents in khmer & french; iss by: bibliotheque royale du cambodge 1927-1942; iss by: institut bouddhique sep 1943-1948) – us CRL [480]

Kambuja-desa : or, an ancient hindu colony in cambodia... / Majumdar, Ramesh Chandra – Madras: University of Madras 1944 [mf ed 1989] – (= Sir william meyer lectures 1942-43) – 1r with other items – 1 – (with bibl footnotes) – mf#mf-10289 seam reel 006/07 [§] – us CRL [305]

Kambul bran nari : pralom lok knun manosancetana phnaek qaparam / Q'uk S'i Tha – Bhnam Ben: Pannagar Bejr Nil [1968?] [mf ed 1990] – 1r with other items – 1 – (in khmer) – mf#mf-10289 seam reel 108/10 [§] – us CRL [480]

Kamchatskaia pravda – Petropavlovsk-Kamchatskij, Russia 1974-88 [mf ed Norman Ross] – 1r – 1 – mf#nrp-1392 – us UMI ProQuest [077]

Kamchatskii vestnik / ed by Pimenov, A A – Petropavlovsk-na-Kamchatke [Kamchat obl]: [s n] 1919 [1918 14 iiulia-1920 [?]] – (= ser Asn 1-3) – n140-363 [1919] [gaps] item 195, on reel n40,41 – 1 – mf#asn-1 195 – ne IDC [077]

Kamchatskij vestnik – Petropavlovsk-Kamchatsky, Russia 1919 [mf ed Norman Ross] – 1r – 1 – mf#nrp-1393 – us UMI ProQuest [077]

Kamei, Nanmei *see* Confucian analects

Kameleon : oder das thier mit allen farben. eine zeitschrift fuer fuerstentugend und volksglueck / ed by Rebmann, Georg Friedr – Koeln 1798-1801 – (= ser Dz. historisch-politische abt) – 6h on 6mf – 9 – €120.00 – mf#k/n1768 – gw Olms [320]

Kamelhar, Isreal *see* Rabenu eleazar

K'a-men : liu mu she hui chu / T'ien, Han – Shang-hai: Hsien tai hsi chu ch'u pan she, 1941 – (= ser P-k&k period) – us CRL [951]

Kamen very stefana iavorskogo... / Morev, I – 1804 – 7mf – 8 – mf#R-7711 – ne IDC [947]

Kamenets-podolinskie gubernskie vedomosti – Kamenets-Podolsk, Ukraine 1838-1917 [mf ed Norman Ross] – 68r – 1 – mf#nrp-609 – us UMI ProQuest [077]

Kamenev, L B *see*
- Denezhnaia reforma
- O khode denezhnoi reformy (doklad na zasedanii plenuma tsk rkp 31 marta 1924 g)
- Kuda i kak vedet sovetskaia vlast krestianstvo?
- Mezhdu dvumia revoliutsiiami

Kamenskaia mysl' : obshchestv -polit koop gaz / ed by Pakin, P S – Kamen' [Alt gub]: Kul't -prosvet otd soiuza kooperativov 1918-19 [1918 [iiul']-1919 [?]] – (= ser Asn 1-3) – n28-142 [1918] n36-156 [1919] [gaps] item 193, on reel n40 – 1 – mf#asn-1 193 – ne IDC [077]

Kamenskii, P V *see* Veroispovednye i tserkovnye voprosy v gosudarstvennoi dume tretego sozyva i otnoshenie k nim "soiuza 17 oktiabria"

Der kamerad – Liberec, Czechoslovakia. Dec 1936-Aug 1938 – 1r – 1 – us L of C Photodup [077]

Kamerad bursche : ein buch von soldatentreue und soldatenliebe / Berkun, Arthur – Berlin-Wilmersdorf: A Gross, 1941 [mf ed 1989] – 228p – 1 – mf#7010 – us Wisconsin U Libr [880]

Kamerad mit dem haarigen gesicht / Schwab, Guenther – Wien: W Frick, 1941 – 1r – 1 – us Wisconsin U Libr [430]

Kamerad und kameradin : bunte bilder, gedanken und worte aus den morgenfeiern im deutschen rundfunk / Kinau, Rudolf – Hamburg: Quickborn-Verlag, 1941 – 1r – 1 – us Wisconsin U Libr [840]

Kameraden : chronik einer jugend [a novel] / Sturm, Stefan – Karlsbad: A Kraft 1943 [mf ed 1991] – 1r – 1 – (filmed with: verliebte oderfahrt & other titles) – mf#2908p – us Wisconsin U Libr [830]

Kameraden der menschheit : dichtungen zur weltrevolution: eine sammlung / ed by Rubiner, Ludwig – Potsdam: G Kiepenheuer, 1919 – 1r – 1 – (incl bibl ref) – us Wisconsin U Libr [430]

Kameraden unterm spaten / Strauss, Eberhard – Oldenburg i.O.: G Stalling, [1935] – 1r – 1 – us Wisconsin U Libr [830]

Kameradschaft : schriften junger deutscher – Bruessel (B), 1937 nov-1940 feb – 1 – gw Misc Inst [074]

Kamerun als kolonie und missionsfeld / Steiner, Paul – Basel, Switzerland. 1909 – 1r – us UF Libraries [960]

Kamerun times – Victoria, Cameroon. Dec 1960-sep 1961 – 1r – 1 – uk British Libr Newspaper [079]

Kames, Henry Jones *see* Elements of criticism

Kameyama, Masao *see* Yen hsiao lou yu lu

Kamf fun demokratie far menshlakhe rekht / Laserson, Max M – New York, NY. 1942 – 1r – us UF Libraries [939]

Kamf kegn "bund" / Agursky, Samuel – Moskve, Russia. 1932 – 1r – us UF Libraries [939]

Kamf oyf der idisher arbayter gas... / Sachs, Abraham Simchah – New York, NY. 1927 – 1r – us UF Libraries [939]

Kamil, A *see* Afak

Kamil, Ahmet *see* Can vermezler tekkesi

Kaminka, A I *see* Vtoraia gosudarstvennaia duma

Kaminskii, G N *see* Selskokhoziaistvennaia kooperatsiia

Kaminsky, David Cyril *see* The gospel sources of christian-jewish prejudice

Kaminstein legislative history project : a compendium and analytical index of materials leading to the copyright act...1976 – 6v (1981-85) – 9 – $475.00 set – mf#408740 – us Hein [323]

Kamke, Hans-Ulrich *see* Strukturgeschichte brandenburgischer doerfer auf dem barnim in vorindustrieller zeit

Kamla / Jogendra Singh – London: Selwyn & Blount Ltd, 1925 – (= ser Samp: indian books) – us CRL [954]

Kamlah, Wilhelm *see* Christentum und geschichtlichkeit

Kamlamn thmi – Bhnam Ben: [s.n. (daily) [mf ed Ithaca NY: [John M Echols Collection] Cornell University 2001 – 1r – 1 – (in khmer) – mf#mf-12775 seam – us CRL [079]

Kamloops and district mining gazette – Kamloops [BC]: W W Clarke and F E Young, [1899?-1900?] – 9 – mf#P04507 – cn Canadiana [622]

Kamloops daily inland sentinel – Kamloops, British Columbia, CN. may 1880-may 1916 – 17r – 1 – cn Commonwealth Imaging [071]

The kamloops phonographer – Kamloops, BC: St Louis Mission, [1892-1893?] – 9 – ISSN: 1190-6715 – mf#P04498 – cn Canadiana [650]

Kamloops standard and standard sentinel – Kamloops, British Columbia, CN. jul 1897-jul 1924 – 13r – 1 – cn Commonwealth Imaging [071]

"Kamloops wawa" / Renault, Raoul – S.l: s.n, 18-? – 1mf – 9 – mf#15993 – cn Canadiana [490]

Kamloops wawa – [Kamloops, BC: J M R LeJeune, 1891-1923] – 9 – (text in chinook jargon, duployan shorthand. some text in english and french) – mf#P04645 – cn Canadiana [241]

Kamloops wawa directory : january, 1895 – S.l: s.n, 1895? – 1mf – 9 – mf#15429 – cn Canadiana [971]

Kamm pan de qun : [a novel] / Naem Sayyana – Bhnam Ben: Pannagar Yu Hak 1965 [mf ed 1990] – 1r with other items – 1 – (in khmer) – mf#mf-10289 seam reel 111/4 [§] – us CRL [830]

Kammel, Richard *see* Er hilft uns frei aus aller not

Kammer, David J *see* A matter of timing

Kammerer, A *see* Essai sur l'histoire antique d'abyssienie

Kammermusik und klaviermusik [die sammlung der sing-akademie zu berlin...pt 4] = Chamber music and piano music / Fischer, Axel & Kornemann, Matthias; ed by Sing-Akademie zu Berlin – [mf ed 2008] – (= ser Musikhandschriften der staatsbibliothek zu berlin – preussischer kulturbesitz 6) – ca 360mf (1:24) in 2 installments – 9 – silver ca €3300.00 – ISBN-13: 978-3-598-34457-2 – gw Saur [780]

Der kammersaenger : drei szenen / Wedekind, Frank – 4. aufl. Berlin: B Cassirer 1909 [mf ed 1991] – 1r – 1 – (filmed with: weckherlin's eclogues of the seasons / elizabeth friench johnson) – mf#2948p – us Wisconsin U Libr [820]

Kamnitzer, Heinz *see* Das testament des letzten buergers

Kamo and hikurangi echo – 1891-92 – 1r – 1 – mf#12.26 – nz Nat Libr [079]

Kamo wakeikazuchi jinja saida-ke monjo : documents of the saida household of the kamo wakeikazuchi shrine. in the holding of kokugakuin university, tokyo – 889 items on 32r – 1 – Y386,000 – (with 288p guide. in japanese) – ja Yushodo [950]

Kamoes bahasa indonesia-nippon dan nippon indonesia / Adinegoro, D [comp] – Medan, 2602 – 3mf – 9 – mf#SE-2002 mf1-3 – ne IDC [410]

Kamoes harian : nippon-indonesia; indonesia-nippon, tjet 1 / Poerwadarminta, W J S – Djakarta: Toko Boekoe Pendidikan, 2602 – 158p 2mf – 9 – mf#SE-2002 mf. 136-137 – ne IDC [410]

Kamoes harian : nippon-indonesia; indonesia-nippon, tjetakan ketiga, ditambah dan diperbaiki / Poerwadarminta, W J S – Djakarta: Pendidikan (2603) – 191p 3mf – 9 – mf#SE-2002 mf138-140 – ne IDC [410]

Kamoes leutik soenda-indonesia / Satjadibrata, R – Djakarta: Gunseikanbu Kokumin Tosyokyoku (Bale Poestaka), 2605 (B P n1585) – 102p 2mf – 9 – mf#SE-2002 mf152-153 – ne IDC [959]

Kamoes soenda-melajoe / Satjadibrata, R – Djakarta: Gunseikanbu Kokumin Tosyokyoku (Balai Poestaka), 2604 (B P n1561) – 379p 5mf – 9 – mf#SE-2002 mf154-158 – ne IDC [959]

Kamoinge : photographer's journal – 1998 1st qtr – 1r – 1 – mf#5307332 – us WHS [770]

Kamp, A Heinrich *see* Schleiermachers gotteslehre

Kamp, Otto *see* Armeleutslieder

Kamp, Steffen *see* Die treuen haende

Kampen. Netherlands. Ordinances, Local Laws, etc *see* Boeck van rechten der stad kampen: dat gulden boeck

Kampen, Nicolaas Godfried *see* Afrika en deszelfs bewoners, volgens de nieuwste ontdekkingen

Der kampf – Duesseldorf DE, 1921-1933 18 feb – 27r – 1 – (1949 3 jan-1956 17 aug [17r]) – gw Mikropress [074]

Der kampf – Berlin. v. 1-2. july 1927-july 1928 – 1 – us NY Public [073]

Der kampf – Prag [CZ], 1934 may-1938 sep – 2r – 1 – (foreign & austrian ed [1934 jul-1938 apr]; merged with: tribuene; cont: der sozialistische kampf, paris) – gw Misc Inst [335]

Der kampf : suedbayerische tageszeitung der unabhaengigen sozialdemokratie – Muenchen DE, 1919 1 jul-1920 – 1r – 1 – mf#6085 – gw Mikropress [943]

Der kampf – Duesseldorf DE, 1917 28 apr-1918 13 nov [gaps] – 1r – 1 – (title varies: 1 dec 1918: freiheit; 3 jan 1949: freies volk / d, beginning in amsterdam, illegal until 1971? filmed by other misc inst: 1946 1 mar-1948 8 oct [1r], 1949-1956 17 aug) – gw Misc Inst [074]

Der kampf – Vienna. v1-26. oct. 1907-1933 – 1 – us NY Public [073]

Der kampf – Wien (A), 1907-1934 jan – 7r – 1 – mf#2351 – gw Mikropress [074]

Der kampf *see* Mitteilungsblatt des sozialdemokratischen vereins duisburg

Kampf dem tode : roman / Adler, Hans & Frank, Paul – Muenchen: Knorr & Hirth, c1929 [mf ed 1995] – 165p – 1 – mf#8918 – us Wisconsin U Libr [830]

Der kampf der gegenwart : ein dramatischer versuch in 5 akten / Katzer, Friedrich Xavier – Milwaukee: Druck von P V Deuster 1873 – 1 – (with: love and marriage / montaigne, m e) – mf#1986 – us Wisconsin U Libr [820]

Kampf der gestirne / Blunck, Hans Friedrich – Jena: E Diederichs, 1926 [mf ed 1989] – 273p – 1 – mf#7037 – us Wisconsin U Libr [830]

Der kampf der giessener theologischen fakultaet gegen zinzendorf und die bruedergemeine, 1740-1750 : ein beitrag zur kirchengeschichte hessens / Bauman, Irwin Wiegner – 1929 – 1mf – 9 – 0-524-08095-X – (incl bibl ref) – mf#1993-9001 – us ATLA [240]

Der kampf der lutherischen kirche um luthers lehre vom abendmahl im reformationszeitalter : im zusammenhang mit der gesammten lehrentwicklung dieser zeit / Schmid, Heinrich – Leipzig: JC Hinrichs, 1868 – 1mf – 9 – 0-7905-6427-0 – (incl bibl ref) – mf#1988-2427 – us ATLA [242]
Der kampf der lutherischen kirche um luthers lehre vom abendmahl im reformationszeitalter : im zusammenhang mit der gesammten lehrentwicklung dieser zeit / Schmid, Heinrich – Leipzig; J.C. Hinrichs, 1868 – 1mf – us ATLA [242]
Der kampf im spessart : erzaehlung / Schuecking, Levin – Berlin, Leipzig: Enck-Verlag [1924?] [mf ed 1995] – 1r – 1 – (filmed with: sueden und norden / hermann schmid) – mf#3738p – us Wisconsin U Libr [830]
Kampf, Leopold *see* On the eve
Der kampf ludwigs des baiern mit der roemischen curie : ein beitrag zur kirchlichen geschichte des 14. jahrhunderts / Mueller, Carl – Tuebingen: H Laupp, 1879-1880 – 2mf – 9 – 0-8370-9809-2 – mf#1986-3809 – us ATLA [940]
Der kampf mit dem daemon : hoelderlin, kleist, nietzsche / Zweig, Stefan – Leipzig: Insel-Verlag, 1933 [mf ed 1993] – (= ser Die baumeister der welt v2) – 321p – 1 – mf#8233 – us Wisconsin U Libr [430]
Der kampf mit dem drachen : zehn kapitel von der gegenwart des deutschen schrifttums und von der krise des deutschen geisteslebens / Forst de Battaglia, Otto – Berlin: Verlag fuer Zeitkritik 1931 [mf ed 1992] – 1r – 1 – (filmed with: kleine literaturfibel / karl-heinz hofer & other titles) – mf#3206p – us Wisconsin U Libr [430]
Der kampf mit dem freund : oder verwandten in der deutschen literatur bis um 1300 / Harms, Wolfgang – Muenchen: Eidos 1963 [mf ed 1992] – 1r – 1 – (incl bibl ref & ind. filmed with: dichtung und welt im mittelalter / hugo kuhn & other titles) – mf#3221p – us Wisconsin U Libr [430]
Der kampf mit dem freund oder verwandten in der deutschen literatur bis um 1300 / Harms, Wolfgang – Muenchen: Eidos, 1963 [mf ed 1993] – (= ser Medium aevum (munich, germany) 1) – 228p – 1 – (incl bibl ref and ind) – mf#8162 – us Wisconsin U Libr [430]
Der kampf; sozialdemokratische monatsschrift – v. 1-27. 1907-1934. (13 scattered issues wanting) – 1 – us L of C Photodup [943]
Der kampf um die cheopspyramide : roman / Eyth, Max – Berlin: Vier Falken Verlag [19–] [mf ed 1989] – 1r – 1 – (filmed with: blut und eisen) – mf#7228 – us Wisconsin U Libr [830]
Der kampf um die religion / Schmidt, Wilhelm – Guetersloh: C Bertelsmann, 1911 – 1mf – 9 – 0-7905-9629-6 – (incl bibl ref and index) – mf#1989-1354 – us ATLA [210]
Der kampf um die schrift in der deutsch-evangelischen kirche des neunzehnten jahrhunderts / Gennrich, Paul – Berlin: Reuther & Reichard, 1898 – 1mf – 9 – 0-8370-8575-6 – mf#1986-2575 – us ATLA [220]
Der kampf um die seele : vortraege ueber die brennenden fragen der modernen psychologie / Gutberlet, Constantin – 2. verb und verm Aufl. Mainz: F Kirchheim, 1903 – 2mf – 9 – 0-7905-7825-5 – (incl bibl ref) – mf#1989-1050 – us ATLA [150]
Der kampf um die tradition : die deutsche dichtung im europaeischen geistesleben, 1830-1880 / Bieber, Hugo – Stuttgart: J B Metzler, 1928 [mf ed 1993] – (= ser Epochen der deutschen literatur. geschichtliche darstellungen v5) – 646p – 1 – (incl ind) – mf#8213 – us Wisconsin U Libr [430]
Der kampf um einen geistigen lebensinhalt : neue grundlegung einer weltanschauung / Eucken, Rudolf – 2. neugestaltete Aufl. Leipzig: Veit, 1907 – 1mf – 9 – 0-7905-7567-1 – mf#1989-0792 – us ATLA [140]
Der kampf um glatz : aus der geschichte der gegenreformation in der grafschaft glatz / Wiese, Hugo von – Halle: Verein fuer Reformationsgeschichte, 1896 (Schriften des Vereins fuer Reformationsgeschichte; 14. Jahrg., Schrift 54) – 1mf – us ATLA [240]
Der kampf um glatz : aus der geschichte der gegenreformation in der grafschaft glatz / Wiese, Hugo von – Halle: Verein fuer Reformationsgeschichte 1896 [mf ed 1990] – (= ser Schriften des vereins fuer reformationsgeschichte 14/54) – 1mf – 9 – 0-7905-5318-X – (incl bibl ref) – mf#1988-1318 – us ATLA [240]
Kampf um irland : erzaehlung / Auerswald, Annmarie von; ed by Plenzat, Karl – Leipzig: H Eichblatt, c1943 [mf ed 1988] – (= ser Eichblatt-buecher) – 216p – 1 – mf#6969 – us Wisconsin U Libr [880]
Kampf um israel / Buber, Martin – Berlin, Germany. 1933 – 1r – us UF Libraries [025]
Kampf um odilienberg : roman / Ebermayer, Erich – Berlin: P Zsolnay, c1929 [mf ed 1989] – 437p – 1 – mf#7193 – us Wisconsin U Libr [830]
Kampf um preussen : schauspiel / Heynicke, Kurt – Leipzig: Schauspiel-Verlag, 1926 – 1r – 1 – us Wisconsin U Libr [820]
Der kampf um roemer kapitel 7 : eine historisch-exegetische studie / Engel, M R – Gotha: Verlagsbureau, 1902 – 1mf – 9 – 0-524-06126-2 – mf#1992-0793 – us ATLA [227]
Ein kampf um rom / Dahn, Felix – Leipzig: Breitkopf und Haertel. 3v. 1898 – 1 – us Wisconsin U Libr [830]
Ein kampf um rom : historischer roman / Dahn, Felix – Leipzig: Breitkopf & Haertel. 3v. 1898 – 6r – 1 – us Wisconsin U Libr [830]
Ein kampf um rom : historischer roman / Dahn, Felix – Leipzig: Breitkopf & Haertel. 3v. 1904 – 1r – 1 – us Wisconsin U Libr [830]
Ein kampf um rom : historischer roman / Dahn, Felix – Leipzig: Breitkopf und Haertel. 3v. 1904 – 1 – us Wisconsin U Libr [830]
Kampf um spanien : die geschichte der legion condor / Beumelburg, Werner – Berlin: G Stalling. 1942 [mf ed 1980] – 310p/[8]pl (ill) – 1 – mf#44 – us Wisconsin U Libr [946]
Der kampf um teneriffa = Antigueedades des las islas afortunadas de la gran canario / Viana, Antonio de; ed by Loeher, Franz von – Stuttgart: Litterarischer Verein, 1883 (Tuebingen: L F Fues) [mf ed 1993] – (= ser Blvs 165) – 424p – 1 – (spanish text. aft in german) – mf#8470 reel 34 – us Wisconsin U Libr [800]
Der kampf um teneriffa / Viana, Antonio de; ed by Loeher, Franz von – Stuttgart: Litterarischer Verein 1883 (Tuebingen: L F Fues) [mf ed 1993] – (= ser Blvs 165) – 58r – 1 – (spanish text with aft in german) – mf#3420p – us Wisconsin U Libr [946]
Kampf um thurant : ein roman aus dem 13. jahrhundert / Allmers, Robert – [Stuttgart: Deutsche Verlags-Anstalt, 1931] [mf ed 1995] – 271p – 1 – mf#8918 – us Wisconsin U Libr [830]
Der kampf um's dasein : roman / Byr, Robert [Bayer, Robert von] – Jena: H Costenoble, 1869 [mf ed 1989] – 5v in 1 – 1 – (each vol has separate t p) – mf#6992 – us Wisconsin U Libr [830]
Ein kampf ums dasein : lustspiel in drei aufzuegen / Wilbrandt, Adolf – Wien: L Rosner, 1873 [mf ed 1996] – (= ser Neues wiener theater 74) – 116p – 1 – mf#9457 – us Wisconsin U Libr [820]
Kampf und bekenntnis : [poems] / Boehme, Herbert – Muenchen: Deutscher Volksverlag, [1937] [mf ed 1989] – 136p – 1 – mf#7044 – us Wisconsin U Libr [810]
Kampfbereit – Paris (F), Oslo (N), 1937 feb-sep – 1r – 1 – gw Misc Inst [074]
Kampfe im busch : erlebnisse im deutsch-sudwest, 1915-1919 / Raif, Karl – Berlin: Im Deutschen Verlag, 1935 – us CRL [943]
Die kampfe und leiden der evangelischen auf dem eichsfelde waehrend dreier jahrhunderte / Wintzingeroda-Knorr, Levin, Freiherr von – Halle: Verein fuer Reformationsgeschichte, 1892-1893 – (= ser Schriften des Vereins fuer Reformationsgeschichte) – 1mf – 9 – 0-7905-4778-3 – (incl bibl ref) – mf#1988-0778 – us ATLA [943]
Kampfer – Liberec, Czechoslovakia. Aug 1933-Jun 1934 (5 scattered issues) – 1r – 1 – us L of C Photodup [077]
Kampffmeyer, Paul *see* Ocherki iz istorii niemetskoi kul'tury
Kampfjugend : gedichte / Schenk, Walter – 3., erw. Aufl. Berlin: Arbeiterjugend-Verlag, 1927 – 1r – 1 – us Wisconsin U Libr [810]
Der kampfruf – Vienna, Austria jan 1930-dec 1935 [mf ed Norman Ross] – 2r – 1 – mf#nrp-1944 – us UMI ProQuest [074]
Das kampfsignal *see* Die fackel 1931
Kampfsignal – New York NY (USA), 1932-1934 15 nov – 1r – 1 – gw Misc Inst [071]
Kampfsignal – v1-v3 n16 [1932 dec 3-1934 nov 15] – 1r – 1 – mf#1061054 – us WHS [071]
Kamphausen, Adolf *see*
– Die berichtigte lutherbibel
– Die chronologie der hebraeischen koenige
– Einleitung in das alte testament
– Das gebet des herrn
– Das lied moses
– Das verhaeltnis des menschenopfers zur israelitischen religion
Kampioenen des christendoms : tafreelen uit de vervolging der zendelingen in china / Hana, H J – [Rotterdam: J M Bredee, 1902] [mf ed 1995] – (= ser Yale coll; Lichtstralen op den akker der wereld [8. jaarg 1902] 5-6) – 37p (ill) – 1 – 0-524-09663-5 – (in dutch) – mf#1995-0663 – us ATLA [240]
Kampo / Japan. Laws, Statutes, etc – Official gazette. Tokyo. On film: 1887-1930. LL-02048 – 1 – us L of C Photodup [348]
Der kampruf – Wien: Buchdruckerei "Albrecht Durer", sep 16 1930-1932 – 1r – 1 – us CRL [073]
Der kamps um badajoz un fruhjahr / Brodruch, Karl – 1861 – 9 – sp Bibl Santa Ana [946]
Kampschulte, Franz Wilhelm *see*
– Johann calvin
– Die universitaet erfurt in ihrem verhaeltnisse zu dem humanismus und der reformation
Kampschulte, FW *see* Johann calvin
Kampung improvement program : toward a national policy / Indonesia. Direktorat Tata Kota dan Tata Daerah – [Jakarta]: Dept of Public Works & Electric Power, Directorate General of Housing, Building, Planning & Urban Development, Directorate of City & Regional Planning in cooperation with UNDP [1976] [mf ed Ithaca NY: Photo Services Cornell University [1986] – on pt of 1r – 1 – mf#9601 – us Cornell [350]
Kamran gatthapad qaksar silp : thnak di 5 – Bhnam Ben: Krasuan Qaparam 1984 [mf ed 1990] – 1r with other items – 1 – (in khmer) – mf#mf-10289 seam reel 102/02 [§] – us CRL [480]
Kamran gatthapad siksa nin srav jrav gati prajapriy – Bhnam Ben: Samagam Qnak Nibandh Khmaer 1972 [mf ed 1990] – 1r with other items – 1 – (in khmer) – mf#mf-10289 seam reel 120/5 [§] – us CRL [930]
Kamsat ja kun qnak kra : ja ryan lkhon man 5 jhut / Thu Dhan – Kraceh: Pannagar Ghun Thai San 2507 [1964] [mf ed 1990] – 1r with other items – 1 – (in khmer; at head of title: ryan) – mf#mf-10289 seam reel 107/4 [§] – us CRL [959]
Kamsko-volzhskaia gazeta – Kazan, Russia 1872-73 [mf ed Norman Ross] – 1 – mf#nrp-673 – us UMI ProQuest [077]
Kamsko-volzhskaia rech' / ed by Samsonov, V K – Kazan': T-vo P Dubrovin i Ko 1918 [1907 1 dek-1918 [?]] – (= ser Asn 1-3) – n6-14 [1918] item 194, on reel n40 – 1 – (lacks: n7) – mf#asn-1 194 – ne IDC [077]
Kamutata – Mukinge Hill, South Africa. 19– – 1r – us UF Libraries [960]
Kan ch'ing ti yeh ma / Tsang, K'o-chia – Ch'ung-ch'ing: Tang chin ch'u pan she, 1943 (1944 printing) – (= ser P-k&k period) – us CRL [810]
K'an jen chi / Lu-fen – Shang-hai: K'ai ming shu tien, Min kuo 28 [1939] – (= ser P-k&k period) – us CRL [840]
Kan k'ai kuo chin-ling / Fan, Ch'ang-chiang – [Shang-hai]: Ta wen ch'u pan she, Min kuo 27 [1938] – (= ser P-k&k period) – us CRL [951]
Kan Man *see* Jamnuay sgal ti nyn prajajan loe libhab lok
Kan mei ti hui wei / Feng, Tzu-k'ai – Shang-hai: K'ai hua shu chu, Min kuo 29 [1940] – (= ser P-k&k period) – us CRL [840]
Kan pu cheng ts'e / Huang, Hsu-ch'u – Kuei-lin: Wen hua kung ying she, Min kuo 29 [1940] – (= ser P-k&k period) – us CRL [951]
The kan ying pien : the chinese text with introduction, translation and notes = Book of rewards and punishments / Webster, James – Shanghai: Presbyterian Mission Press, 1918 – 1mf – 9 – 0-524-08049-6 – mf#1991-0265 – us ATLA [170]
K'an yueh lou shu hsin : p'u chi pen / Wu, Shu-t'ien – Shang-hai: K'ai ming shu tien, 1931 – (= ser P-k&k period) – us CRL [920]
K'an yun chi / Chou, Tso-jen – Shang-hai: K'ai ming shu tien, Min kuo 21 [1932] – (= ser P-k&k period) – us CRL [840]
K'an yun jen shou chi : ti erh p'i p'ing lun wen chi / Hu, Feng – Ch'ung-ch'ing: Tzu li shu tien, 1944 – (= ser P-k&k period) – us CRL [480]
Kan, Yu-yuan *see*
– Hsiang ts'un min chung chiao yu
– Min chung hsueh hsiao
Kana jawa shimbun – n1-86 jan 1 2604-aug 18 2605 – 10mf – 9 – mf#SE-2002 mf320-329 – ne IDC [079]
Kanaanaeer und hebraeer : untersuchungen zur vorgeschichte des volkstums und der religion israels auf dem boden kanaans / Boehl, Franz Marius Theodor – Greifswald: Julius Abel, 1911 – 1mf – 9 – 0-7905-1743-4 – (incl bibl ref) – mf#1987-1743 – us ATLA [270]
Kanaanaeische inschriften : (moabitisch, althebraeisch, phoenizisch, punisch) / ed by Lidzbarski, Mark – Giessen: Alfred Toepelmann, 1907, c1905 – (= ser Altsemitische Texte) – 1mf – 9 – 0-8370-7403-7 – (incl bibl ref) – mf#1986-1403 – us ATLA [470]
Kanad – Istanbul: Muesterek uel-Menfaa-i Osmani Sirketi Matbaasi, 1910. Mesesisi: Ismail Hami, Sahib-i Imtiyaz: Hueseyin Huesnue, Sermuharriri: Tahsin Nahid. n4. 28 tesrinievvel 1326 [1910] – (= ser O & t journals) – 1mf – 9 – $25.00 – us MEDOC [956]
Kanada *see* The vaiseshika aphorisms of kanada
Kanada kurier – Toronto, Ontario (CDN), 1984 5 jan- – 1 – gw Misc Inst [071]
Kanada shinbun [the canada daily news] and nikkan minshu [the daily people] for 1941 on microfilm : a preservation microfilming project at the university of british columbia library = The canada daily news / the daily people – Vancouver BC: UBCL] 1995 – (original title: kanada nichinichi shinbun) – cn UBC Preservation [071]
Kanadai magyarsag – Canada, 1951-77 – 1 – (in hungarian) – cn Commonwealth Imaging [071]
Kanadai magyarsag – Toronto, Canada. 24 jan 1953-9 nov 1957; 9 may, 7 nov 1959; 4 mar 1961-6 nov 1965 – 2r – 1 – uk British Libr Newspaper [071]
Kanada-kurier – Winnipeg, Manitoba (CDN), 1980 18 sep- – 1 – gw Misc Inst [071]
Kanadeetis / Lettish Friendly Association – n1-22 [1913 jan 30-dec 15] – 1r – 1 – mf#639565 – us WHS [366]
Kanadiisky farmer – Winnipeg, Canada. 1941-30 jun 1948; 8 dec 1948-1970; 9 jan 1971-25 dec 1972; 8 jan-8 oct 1973 – 32r – 1 – uk British Libr Newspaper [630]
Kanadsky gudok *see* Vestnik
Die kanaele von suez und panama / Rheinstrom, H – Borna, Leipzig, 1906 – 1mf – 9 – mf#ILM-2201 – ne IDC [956]
Kanaeva, I *see* Perspektivy razvitiia semi i zhilishcha
Kana'im ha-tse'irim / Churgin, Yaakov – Tel-Aviv, Israel. v1-2. 1935 – 1r – us UF Libraries [939]
Kanakazi kayaya / Phiri, Desmond Dudwa – Cape Town, South Africa. 1959 – 1r – us UF Libraries [960]
Kanakhin, I F *see* Spravochnik po statistike sel'skogo khoziaistva, promyshlennosti i truda
Kanamatsu, Kenryo *see* Naturalness
Kanauri vocabulary / Bailey, T G – 1911 – 1r – 1 – mf#648 – uk Microform Academic [490]
Kanawha Valley Genealogical Society *see* Journal of the kanawha valley genealogical society
Kancher, E S *see*
– Kak otkryt selskokhoziaistvennoe tovarishchestvo po pererabotke produktov
– Rukovodstvo po selskoi kooperatsii
Kand va kav – London. dawrah-'i 1, shumarah-'i 1-8. azar 1353-pa'iz 1357 [nov/dec 1974-aut 1978]-dawrah-'i 2, shumarah-'i 1-3. bahar-zimistan 1359 [spr 1980-win 1980/81] + suppls 1-6 – 1r – 1 – $85.00 – us MEDOC [079]
Kandidov, Boris Pavlovich *see* Sektantstvo i mirovaya voina
Kandler, Gunther *see* Zweitsinn. vorstudien zu einer theorie der sprachlichen
Kane daily republican – Kane, PA., 1894-1899 – 13 – $25.00r – us IMR [071]
Kane, der nordpolfahrer : arktische fahrten und entdeckungen der zweiten grinnell-expedition zur aufsuchung sir john franklins in den jahren 1853, 1854 und 1855 unter dr elisa kent kane / Kane, Elisha – Leipzig 1858 [mf ed Hildesheim 1995-98] – 1v on 2mf – 9 – €60.00 – 3-487-27084-6 – gw Olms [919]
Kane, E K *see* Arctic explorations
Kane, Elisha *see* Kane, der nordpolfahrer
Kane, Elisha Kent *see* Access to an open polar sea
Kane family news notes – [1979 dec]-[1995 jan/sep] [gaps] – 1r – 1 – mf#1114472 – us WHS [929]
Kane leader – Kane, PA., 1885-1888 – 13 – $25.00r – us IMR [071]
Kane republican – Kane, PA., 1894-1979 – 13 – $25.00r – us IMR [071]
Kane, Robert *see* Socialism
Kane, Robert John *see* The industrial resources of ireland
Kane weekly blade – Kane, PA., 1879-1882 – 13 – $25.00r – us IMR [071]
Kaneohe Marine Corps Air Station [Hawaii] *see* Hawaii marine
K'ang chan che hsueh / Feng, Yu-hsiang – Kuei-lin: San hu t'u shu she, Min kuo 31 [1942] – (= ser P-k&k period) – us CRL [951]
K'ang chan cheng chih kung tso kang ling / Chou, En-lai – Shang-hai: Ming ming shu chu, Min kuo 27 [1938] – (= ser P-k&k period) – us CRL [951]
K'ang chan ch'i chien ti wen hsueh / A-ying – Kuang-chou: Chan shih ch'u pan she, 1938 – (= ser P-k&k period) – us CRL [480]
K'ang chan ch'i chung chih fu-chien hsieh ho ta hsueh – [China: sn] – (= ser P-k&k period) – us CRL [951]
K'ang chan ch'i chung ssu-ch'uan liang shih tseng chia ch'an liang chih i chien / Liu, Ta-pei – [Ssu-ch'uan]: Ssu-ch'uan sheng li chiao yu hsueh yuan nung shih shih yen ch'ang, Min kuo 27 [1938] – (= ser P-k&k period) – us CRL [630]
K'ang chan chien kuo kang ling hsuan ch'uan chih tao ta kang – [China: Chung-kuo kuo min tang chung yang chih hsing wei yuan hui hsuan ch'uan pu], Min kuo 27 [1938] – (= ser P-k&k period) – us CRL [951]
K'ang chan chien kuo kang ling wen ta / Shih, Mei & Liu, Shih – Ch'ung-ch'ing: Sheng kuo shu tien, Min kuo 27 [1938] – (= ser P-k&k period) – us CRL [951]
K'ang chan chien kuo lun / Hou, Wai-lu – Han-k'ou: Sheng huo shu tien, Min kuo 27 [1938] – (= ser P-k&k period) – us CRL [951]

K'ang chan chung ching chi chien she chih t'u ching / Mo, Hsuan-yuan – Ch'ung-ch'ing: Chung-kuo wen hua fu wu she, Min kuo 29 [1940] – (= ser P-k&k period) – us CRL [951]

K'ang chan chung ti hai chun wen t'i / Weng, Jen-yuan – Shang-hai: Li ming shu chu, Min kuo 27 [1938] – (= ser P-k&k period) – us CRL [951]

K'ang chan chung ti meng-ku / Hsu, Yung-p'ing – Ch'ung-ch'ing: Tu li ch'u pan she, Min kuo 29 [1940] – (= ser P-k&k period) – us CRL [951]

K'ang chan chung ti nu chan shih / Shen, Tzu-chiu et al – [Kuang-chou]: Chan shih ch'u pan she, [193-?] – (= ser P-k&k period) – us CRL [920]

K'ang chan hou fang ti hsin kuang-hsi / Yu, Po-shun – Han-k'ou: Chien kuo shu tien, Min kuo 27 [1938] – (= ser P-k&k period) – us CRL [951]

Kang chan i lai chih pien chiang / Huang, Fen-sheng – [Sl]: Shih hsueh shu chu, [1944] – (= ser P-k&k period) – us CRL [951]

K'ang chan i nien / Chiang, Kai-shek – [China: sn, 1938] – (= ser P-k&k period) – us CRL [951]

K'ang chan liu nien lai wo kuo kung yeh chi shu chih chin pu – [China: Ching chi pu, 1943] – (= ser P-k&k period) – us CRL [338]

K'ang chan pa nien lai ti pa lu chun yu hsin ssu chun / China Lu chun Ti 18 chi t'uan chun – [China]: Hua-pei hsin hua shu tien, 1945 – (= ser P-k&k period) – us CRL [951]

K'ang chan shih hua / Kao, Yueh-fu – Ch'ung-ch'ing: Tu li ch'u pan she, Min kuo 30 [1941] – (= ser P-k&k period) – us CRL [951]

Kang chan shih ko chi, vol 5 (ccm126) / Feng, Yu-hsiang – Kuei-lin, 1945 [mf ed 198?] – (= ser Ccm 126) – 1 – mf#1984-b500 – us ATLA [810]

K'ang chan ti ch'ien t'u / Li, Ta-kang – [China]: Cheng ch'i she, Min kuo 27 [1938] – (= ser P-k&k period) – us CRL [951]

K'ang chan ti i jih : shang-hai hsueh sheng chi t'i ch'uang tso – [China: sn], 1938 – (= ser P-k&k period) – us CRL [951]

K'ang chan tu mu hsi chu hsuan – Ch'ung-ch'ing: Cheng chung shu chu, Min kuo 31 [1942] – (= ser P-k&k period) – us CRL [951]

K'ang chan wen hsien – Ch'ung-ch'ing: Tu li ch'u pan she, [Min kuo 27 ie 1938] – (= ser P-k&k period) – us CRL [951]

K'ang chan wen hsuan / Pao, Ch'ing-ts'en – [China]: Pa t'i shu tien, Min kuo 27 [1938] – (= ser P-k&k period) – us CRL [480]

K'ang chan wen pien ti 1 chi – Che-chiang Yu-yao: Chan shih tu wu ch'u pan she, 1937 – (= ser P-k&k period) – us CRL [480]

K'ang chan yen lun chi – Shang-hai: Han-k'ou hsien tai ch'u pan she: Shang-hai chin hua shu chu tsung ching shou, Min kuo 26 [1937] – (= ser P-k&k period) – us CRL [951]

K'ang chan ying hsiung chuan chi / Chung-kuo kuo min tang Hsuan ch'uan pu – Ch'ung-ch'ing: Kuo min t'u shu ch'u pan she, Min kuo 32 [1943] – (= ser P-k&k period) – us CRL [951]

K'ang chan yu chien tieh / Huang, ching-chai – Ch'ang-sha: Shang wu yin shu kuan, Min kuo 26 [1937] – (= ser P-k&k period) – us CRL [951]

K'ang chan yu ching chi / Ma, Yin-ch'u – Han-k'ou: Tu li ch'u pan she, Min kuo 27 [1938] – (= ser P-k&k period) – us CRL [951]

K'ang chan yu ching chi t'ung chih = War of resistance and economic control / Chang, Su-min – Ch'ang-sha, Shang wu ying shu kuan, Min kuo 27 [1938] – (= ser P-k&k period) – us CRL [951]

K'ang chan yu chiu chi shih yeh / Chang, Ping-hui – [Ch'ang-sha]: Shang wu yin shu kuan, [1937] – (= ser P-k&k period) – us CRL [951]

K'ang chan yu hou yuan kung tso / T'ao, Pai-ch'uan – Ch'ang-sha: Shang wu yin shu kuan, Min kuo 27 [1938] – (= ser P-k&k period) – us CRL [951]

K'ang chan yu hsi chu – Ch'ung-ch'ing: Tu li ch'u pan she, 1939 – (= ser P-k&k period) – us CRL [820]

K'ang chan yu hsi chu / T'ien, Han – Ch'ang-sha: Shang wu yin shu kuan, Min kuo 27 [1938] – (= ser P-k&k period) – us CRL [820]

K'ang chan yu hsiao fei t'ung chih / Tung, Shih-chin – Ch'ung-ch'ing: Tu li ch'u pan she, [1938] – (= ser P-k&k period) – us CRL [951]

K'ang chan yu hsin wen shih yeh / Wang, Hsin-ch'ang – [Ch'ang-sha]: Shang wu yin shu kuan, [1938] – (= ser P-k&k period) – us CRL [951]

K'ang chan yu hsuan ch'uan – Ch'ung-ch'ing: Tu li ch'u pan she, Min kuo 27 [1938] – (= ser P-k&k period) – us CRL [951]

K'ang chan yu kuo chi hsing shih / Fan, Chung-yun – Shang-hai: Shang wu yin shu kuan, Min kuo 26 [1937] – (= ser P-k&k period) – us CRL [951]

K'ang chan yu min tsu kung yeh / Yang, Chih – Ch'ang-sha: Shang wu yin shu kuan, Min kuo 26 [1937] – (= ser P-k&k period) – us CRL [951]

K'ang chan yu nung ts'un ching chi / Hsu, Hsing-ch'u – Ch'ang-sha: Shang wu yin shu kuan, Min kuo 27 [1938] – (= ser P-k&k period) – us CRL [630]

K'ang chan yu pao chia yun tung / Ch'en, Kao-yung – [Ch'ang-sha]: Shang wu yin shu kuan, [1937] – (= ser P-k&k period) – us CRL [951]

K'ang chan yu she hui wen t'i / Ch'en, Tuan-chih – Ch'ang-sha: Shang wu yin shu kuan, Min kuo 26 [1937] – (= ser P-k&k period) – us CRL [951]

K'ang chan yu sheng ch'an / Ma, Yin-ch'u – Han-k'ou: Tu li ch'u pan she, Min kuo 27 [1938] – (= ser P-k&k period) – us CRL [951]

K'ang chan yu tien ying / Yao, Su-feng – [Ch'ang-sha]: Shang wu yin shu kuan, [1937] – (= ser P-k&k period) – us CRL [951]

K'ang chan yu ts'ai cheng chin jung – Ch'ung-ch'ing: Tu li ch'u pan she, Min kuo 27 [1938] – (= ser P-k&k period) – us CRL [951]

Kang cheng chi yao – [Tsingtao: Ch'ing-tao shih kang wu chu], Min kuo 22 [1933] – (= ser P-k&k period) – us CRL [380]

Kang, H -Y *see* The effect of various lifting intensities in release of human growth hormone

Kang hsiao chieh (ccm327) = The adventures of miss kang – Hong Kong, 1953 [mf ed 198?] – (= ser Ccm 327) – 1 – mf#1984-b500 – us ATLA [830]

Kang hu hua hsueh kung yeh k'ao ch'a chi / Chung-shan ta hsueh (Canton, China) Hua hsueh kung yeh yen chiu so – [Kuang-chou: Kuo li Chung-shan ta hsueh], Min kuo 21 [1932] – (= ser P-k&k period) – us CRL [338]

K'ang jih chiu kuo cheng ts'e / Ch'en, Shao-yu – [Shan-hsi?]: Shan-hsi jen min ch'u pan she, Min kuo 26 [1937] – (= ser P-k&k period) – us CRL [951]

K'ang jih hsien lieh chi / Liu, Hsiang – Han-k'ou: Tu li ch'u pan she, Min kuo 27 [1938] – (= ser P-k&k period) – us CRL [951]

K'ang jih min tsu t'ung i chan hsien chiao ch'eng / K'ai-feng – [China]: Chung-kuo wen hua she, Min kuo 28 [1939] – (= ser P-k&k period) – us CRL [951]

K'ang jih min tsu t'ung i chan hsien chung ti chi ko wen t'i – [China]: K'ang chan pien i she, 1939 – (= ser P-k&k period) – us CRL [335]

K'ang jih ti meng-ku / Te-heng-shan – [Sl]: Chung-kuo pien chiang wen hua ts'u chin hui, 1940 – (= ser P-k&k period) – us CRL [951]

K'ang jih ti mo fan chun jen / Feng, Yu-hsiang – Han-k'ou: San hu t'u shu she, Min kuo 27 [1938] – (= ser P-k&k period) – us CRL [951]

K'ang jih ti ti pa lu chun / Chang, Kuo-p'ing – Shang-hai: K'ang chan ch'u pan she, Min kuo 27 [1938] – (= ser P-k&k period) – us CRL [951]

K'ang jih ti ti pa lu chun / Chao, I-lin – [Shang-ai]: Tzu li ch'u pan she, Min kuo 26 [1937] – (= ser P-k&k period) – us CRL [951]

K'ang jih yu chi chan cheng / Chu, Te – Han-k'ou: Hsin hua jih pao kuan, Min kuo 27 [1938] – (= ser P-k&k period) – us CRL [951]

Kang, Manjit S *see* Journal of crop improvement

K'ang, Min *see*
– Fu lu: tu mu chu chi
– Yung chiu ti p'eng yu

K'ang yuan yin shua chih kuan ch'ang shih chou chi nien k'an – [China]: K'ang yuan yin shua chih kuan ch'ang, Min kuo 23 [1934] – (= ser P-k&k period) – us CRL [951]

Kangaroo out of his element – Australia. 19 Oct-30 Nov 1914. 6 ft – 1 – uk British Libr Newspaper [079]

Kangaroo valley times – Kangaroo Valley, jan 1898-dec 1904 – 2r – A$99.40 vesicular A$110.40 silver – at Pascoe [079]

K'ang-chan ch'i chien ti hsin wen hsuan ch'uan / Jen, Pai-t'ao – [China]: Hsin wen yen chiu she, 1938 – (= ser P-k&k period) – us CRL [070]

The kangchenjunga adventure / Smythe, Francis Sydney – London: Hodder and Stoughton Ltd, 1946 – (= ser Samp: indian books) – us CRL [790]

Kangemi : the impact of rapid culture – Nairobi, Kenya. may 1973 – 1r – us UF Libraries [025]

Kang-hsi, Emperor of China *see*
– The sacred edict
– Sheng yu hsiang chieh

Kangitsheli ngitsho loba ngubani – Salisbury, Zimbabwe. 19–? – 1r – us UF Libraries [960]

K'ang-jih chan-cheng shih-ch'i shen-kan-ning pien-ch'u ts'ai-cheng.. – Collection of historical abstracts on the finance and economy of the Shen-Kan-Ning Border Regions during the war of resistance against Japan – 1 – 83.00 – us Chinese Res [951]

Kangle, R P *see* Priyadarsika of sri harsha

Kangra painting – [London]: Faber and Faber Ltd, [1952] – (= ser Samp: indian books) – (int and notes by w g archer) – us CRL [750]

Kanhistique kansas history and antiques – 1975 may-1980 dec, 1981-86, 1987-92 – 3r – 1 – mf#525462 – us WHS [978]

Kani, Bennett Zolile *see* A case study of the school development functions of a school governing body in a historically disadvantage secondary school

Kanitz, F P *see* Donau-bulgarien und der balkan. historisch-geographisch-etnographische reisestudien aus den jahren 1860-1879

Kanjron cas nin cacak kmen r vivattan gamnit prajadhipateyy khmaer / Duy Ryan – Bhnam Ben: Ron Bumb Maha Labh 1974 [mf ed 1990] – 1r with other items – 1 – (in khmer; with bibl) – mf#mf-10289 seam reel 117/11 [§] – us CRL [959]

Kankakee labor record – 1949 jan 14-1952 mar, 1952 apr 19-1955 aug, 1955 sep 20-1958 dec – 3r – 1 – mf#1061064 – us WHS [331]

Kann ein glaeubiger christ freimaurer sein? : antwort an den herrn dr rudolph sendel, privatdocenten der philosophie zu leipzig / Ketteler, Wilhelm Emmanuel, Freiherr von – 5. aufl. Mainz: Franz Kirchheim 1865 – 1mf – 9 – mf#vrl-203 – ne IDC [366]

Kann, J H *see* Opmerkingen betreffende het beleid van de mandaats-regeering van palestina

Kann po = (Berita pemerintah) – Djakarta: Gunseikanbu, Bahagian Kikakuka, 2602-2605. n1-74 – 42mf – 9 – mf#SE-2002 mf277-319 – ne IDC [079]

Kann, Yu-yuan *see* Hsiang ts'un chiao yu

Kanna kamsat : pralom lok knun manosancetana / Gim Saet – Pat Tam Pan: Pannagar Qyn Hun 2502 [1959] [mf ed 1990] – 1r with other items – 1 – (in khmer) – mf#mf-10289 seam reel 098/04 [§] – us CRL [959]

Kanne, Friedrich August *see* [Wiener] allgemeine musikalische zeitung

Kan-nu *see* Ch'an chuan

Kano, Marshall *see* The relationship between aerobic fitness and diet composition in adult females

Der kanon des alten testaments : ein abriss / Budde, Karl – Giessen: J Ricker (Alfred Toepelmann), 1900 – 1mf – 9 – 0-8370-2512-5 – mf#1985-0512 – us ATLA [221]

Der kanon des alten testaments : nach den ueberlieferungen in talmud und midrasch / Fuerst, Julius – Leipzig: Doerffling & Franke, 1868 – 1r – 1 – 0-8370-0352-0 – mf#1984-B412 – us ATLA [221]

Der kanon des alten testaments : nach den ueberlieferungen in talmud und midrasch / Fuerst, Julius – Leipzig: Doerffling und Franke, 1868 – 1mf – 9 – 0-8370-3220-2 – mf#1985-1220 – us ATLA [221]

Der kanon des alten testaments zur zeit des ben sira / Eberharter, Andreas – Muenster i W: Aschendorff, 1911 [mf ed 1989] – (= ser Alttestamentliche studien 3/3) – 1mf – 9 – 0-7905-0761-7 – (incl bibl ref) – mf#1987-0761 – us ATLA [221]

Der kanon des neuen testamentes / Dausch, Petrus – Muenster i W: Aschendorff 1908 [mf ed 1992] – (= ser Biblische zeitfragen 1/5) – 1mf – 9 – 0-524-05400-2 – (incl bibl ref) – mf#1992-0410 – us ATLA [225]

Der kanon des neuen testaments : biblische zeit- und streitfragen / Ewald, Paul – Berlin: Edwin Runge, 1906 – 1mf – 9 – 0-7905-0491-X – mf#1987-0491 – us ATLA [225]

Der kanon und die kritik des neuen testaments in ihrer gescichtlichen ausbildung und gestaltung / Hilgenfeld, Adolf – Halle: C E M Pfeffer, 1863 – 1mf – 9 – 0-8370-3584-8 – (incl ind) – mf#1985-1584 – us ATLA [225]

Die kanones der wichtigsten altkirchlichen concilien : nebst den apostolischen kanones / ed by Lauchert, Friedrich – Freiburg i B: JCB Mohr, 1896 [mf ed 1992] – (= ser Sammlung ausgewaehlter kirchen- und dogmengeschichtlicher quellenschriften 12) – 1mf – 9 – 0-524-02005-1 – (in greek & latin. int in german. incl bibl ref) – mf#1990-0550 – us ATLA [240]

Kanonikai diataxeis, epistolai, lyseis, thespismata ton hagiotaton patriarchon konstantinoupoleos : apo gregoriou tou theologou mechri dionysiou tou apo adrianoupoleos / Gedeon, Manouel Io – En Konstantinoupolei: ek tou Patriarchikou Typographeiou, 1888-1889 – (= ser Parartema tes "Ekklesiastikes aletheias") – 3mf – 9 – 0-8370-8111-4 – (incl ind) – mf#1986-2111 – us ATLA [240]

Kanonisch und apokryph : ein kapitel aus der geschichte des alttestamentlichen kanons / Hoelscher, Gustav – Naumburg a. S: Lippert (G Paetz), 1905 – 1mf – 9 – 0-8370-3615-1 – mf#1985-1615 – us ATLA [221]

Die kanonissenstifter im deutschen mittelalter (kra43/44) / Schaefer, K H – Stuttgart, 1907 – €14.00 – ne Slangenburg [931]

Kanoradian *see* Miscellaneous kansas state newspapers

Kanorado star *see* Miscellaneous kansas state newspapers

Kansallis-osake-pankki. economic review – Helsinki, Finland 1975-94 – 1,5,9 – ISSN: 0022-8419 – mf#10494 – us UMI ProQuest [330]

Kansan toveri – New York NY. 1897-98 – 1r – 1 – (finnish periodical) – us IHRC [073]

Kansapolis Association *see* Journal of proceedings

Kansas : session laws of american states and territories – 1855-1999 – 9 – $1,622.00 set – mf#402670 – us Hein [348]

Kansas : statutes annotated – Topeka: Office of Revisor of Statutes, 1964-oct 1998 update – 9 – $1,583.00 set – mf#402190 – us Hein [348]

Kansas *see*
– Reports and opinions
– Reports, post-nrs
– Reports, pre-nrs

Kansas 76 / Kansas American Revolution Bicentennial Commission – v1-2 n1 [1973 jun-1976 spr] – 1r – 1 – mf#382332 – us WHS [978]

Kansas American Revolution Bicentennial Commission *see* Kansas 76

Kansas appellate reports / Kansas. Court of Appeals – v1-10. 1895-1900 (all publ) – 33mf – 9 – $148.00 – (no pre-nrs vols. no updates planned) – mf#LLMC 84-136 – us LLMC [340]

Kansas attorney general reports and opinions – 1864-1997 – 6,9 – $222.00 set – (1864-1959, 1961-1979 on reel $105. 1979-97 on mf $117. 1960 not available) – mf#408250 – us Hein [340]

Kansas bar association. journal – Topeka KS 1932+ – 1,5,9 – ISSN: 0022-8486 – mf#2421 – us UMI ProQuest [340]

Kansas bar association reports proceedings : 1st to 49th annual meetings – 1884-1931 (all publ) – 38mf – 9 – $171.00 – (lacking: 1st-2nd meetings) – mf#LLMC 84-516 – us LLMC [340]

Kansas. Board of Statehouse Commissioners *see* Records

Kansas city business journal – Charlotte NC 1987-2000 – 1,5,9 – ISSN: 0734-2748 – mf#16680.01 – us UMI ProQuest [650]

Kansas city daily record – Kansas City, Missouri. On film: v89-149; 1932-62. LL-013 – 1 – us L of C Photodup [340]

Kansas city globe – Kansas City KS. 1992 oct-dec, 1993, 1994 jan-dec, 1994 nov4/11, 1995 jan-dec, 1996 jan-dec, 1997 jan-dec – 10r – 1 – mf#2592833 – us WHS [071]

Kansas city jewish chronicle – Kansas City. Mo. 1920-50 – 1 – us AJPC [071]

Kansas city jewish chronicle – Kansas City. Mo. 1954-67 – 1 – us AJPC [071]

Kansas city journal – Kansas City, MO. v1-1854- (wkly) [mf ed 19–] – 1 – us Eastman [071]

Kansas city labor – Saint Louis MO. 1895 dec 7 – 1r – 1 – us WHS [331]

Kansas city labor beacon – 1968 sep 6-1971 sep 10, 1971 sep 17-1974 jun 28, 1974 jul 5-1977 jul 24, 1977 jul-1979 dec, 1980-86, 1987-1988 apr 8 – 8r – 1 – (cont by: union beacon) – mf#1400138 – us WHS [331]

Kansas city labor bulletin – 1946-49, 1950-53, 1954-1958 aug 15 – 3r – 1 – (cont: labor bulletin [kansas city, ks]) – mf#1466534 – us WHS [331]

Kansas city law reporter – v1 n1-7. 1888 (all publ) – 9 – mf#LLMC 84-517 – us LLMC [340]

Kansas city law review *see* Umkc law review

Kansas city observer – Kansas City, MO: L C Williams, 1896 (wkly) [mf ed 1947] – (= ser Negro Newspapers on Microfilm) – 1r – 1 – us L of C Photodup [071]

Kansas city presse – Kansas City KS (USA), 1931 3 jun-1939 27 sep [gaps] – 5r – 1 – gw Misc Inst [071]

Kansas city review – Ann Arbor MI 1877-85 – 1 – mf#2910 – us UMI ProQuest [330]

Kansas city times – Kansas City MO. 1913 nov 14-1914 jan 26, 1914 aug 3-sep 21, 1914 dec 21-1915 feb 9, 1914 jan 27-mar 15, 1914 jun 15-aug 1, 1914 mar 16-apr 30, 1914 may 1-jun 13, 1914 nov 9-dec 19, 1914 sep 22-nov 7, 1915 apr 1-may 12, 1915 feb 10-mar 31, 1915 may 14-jun 30 – 12r – 1 – mf#841493 – us WHS [071]

Kansas Council of Genealogical Societies *see* Kansas review

Kansas. Court of Appeals *see* Kansas appellate reports

Kansas coyote – Topeka KS. [1981 jun 1, aug 1-dec], [1982 jan 1-dec], [1983 jan-nov], [1984 sep, nov-dec], [1985 feb-mar, may-dec], [1986 mar-apr] – 1r – 1 – (cont by: coyote log) – mf#1221026 – us WHS [071]

Kansas crusader of freedom – Doniphan City KS. 1858 mar 6 – 1r – 1 – us WHS [071]

Kansas daily tribune – Lawrence KS. 1864 aug 24 – 1r – 1 – (cont by: daily kansas tribune) – us WHS [071]

Kansas entomological society. journal – Manhattan KS 1928+ – 1,5,9 – ISSN: 0022-8567 – mf#3031 – us UMI ProQuest [590]

Kansas express – Manhattan City KS. 1859 aug 20 – 1r – 1 – (cont by: manhattan express (manhattan city, ks: 1859)) – us WHS [071]

Kansas farmer / Kansas State Agricultural Society – Topeka KS. v46 n1-v47 n14 [1908 dec 10-1909 apr 3] – 1r – 1 – (cont: farmer's advocate; cont by: farmer's mail and breeze; kansas farmer and mail and breeze) – mf#1051367 – us WHS [071]

Kansas Farmers Union *see* Kansas union farmer

Kansas. Fort Dodge Headquarters *see* Headquarters records of fort dodge, kansas, 1866-1882

Kansas. Fort Scott Headquarters *see* Headquarters records of fort scott, kansas, 1869-1873

Kansas Genealogical Society *see* Tree searcher

Kansas. Governor, 1883-85 (Governor Glick) *see* Governor's correspondence relating to luke short

The kansas headlight – Wichita, KS. -v1 n9. sep 14 1894 [mf ed 1947] – (= ser Negro Newspapers on Microfilm) – 1r – 1 – us L of C Photodup [071]

Kansas historical quarterly – Topeka KS 1931-77 – 1,5,9 – ISSN: 0022-8621 – mf#8411 – us UMI ProQuest [978]

Kansas history – Topeka KS 1978+ – 1,5,9 – ISSN: 0149-9114 – mf#12400 – us UMI ProQuest [978]

Kansas industrial liberator *see* American nonconformist and kansas industrial liberator

Kansas Infantry. 1st Regiment (Colored) *see* Records

Kansas journal of law and public policy – v1-10. 1991-2001 – 9 – $289.00 set – ISSN: 1055-8942 – mf#113841 – us Hein [342]

Kansas journal of sociology – Lawrence KS 1964-75 – 1,5 – ISSN: 0022-8648 – mf#2235 – us UMI ProQuest [301]

Kansas labor weekly / Kansas State Federation of Labor et al – Topeka KS. 1935 jan 17-1937 dec 9, 1946-50, 1951-54, 1955-61, 1962-65, 1966 jan 6-jun 21 – 6r – 1 – (cont: circle) – mf#1037448 – us WHS [331]

Kansas law journal – v1-5. 1885-87 (all publ) – 21mf – 9 – $31.50 – mf#LLMC 84-518 – us LLMC [340]

Kansas library bulletin – Topeka KS 1932-72 – 1,5 – ISSN: 0022-8680 – mf#6548 – us UMI ProQuest [020]

Kansas medical society. journal – Topeka KS 1901-84 – 1,5,9 – (cont by: kansas medicine) – ISSN: 0022-8699 – mf#2555.01 – us UMI ProQuest [610]

Kansas medicine – Topeka KS 1985-97 – 1,5,9 – (cont: journal of the kansas medical society) – ISSN: 8755-0059 – mf#2555,01 – us UMI ProQuest [610]

Kansas nurse – Topeka KS 1970+ – 1,5,9 – ISSN: 0022-8710 – mf#8722 – us UMI ProQuest [610]

Kansas pioneer – Kickapoo City KS. 1854 nov 15-dec 6 – 1r – 1 – us WHS [071]

Kansas review / Kansas Council of Genealogical Societies – 1980 jun-1989 oct – 1r – 1 – (cont: review [garden city, ks]) – mf#1074869 – us WHS [929]

Kansas semi-weekly capital – Topeka KS. 1896 feb 7-dec 29, 1897 jan-sep 3-1898 apr 29, 1898 may 3-1899 feb 28, 1899 mar 3-dec 29, 1900 jan 2-aug 31, 1900 sep 4-1901 may 14, 1901 may 17-1902 jan 21 – 8r – 1 – us WHS [071]

Kansas speech journal – Topeka KS 1949+ – 1,5,9 – mf#5960 – us UMI ProQuest [373]

Kansas State Agricultural Society *see* Kansas farmer

Kansas. State Bar Association *see* Proceedings

Kansas. State Board of Dental Examiners *see* Minutes of meetings

Kansas State Federation of Labor *see*
- Proceedings of the...annual convention of the kansas state federation of labor
- Report of proceedings of the...annual convention of the kansas state federation of labor

Kansas State Federation of Labor et al *see* Kansas labor weekly

Kansas State Historical Society *see*
- Family records: early lyon county settlers
- Historical society mirror

Kansas State Historical Society, Manuscripts Dept *see* Holdings

Kansas state journal – Lawrence KS. 1865 jan 19 – 1r – 1 – (cont: kansas herald of freedom; continued by: daily kansas state journal; lawrence republican; western home journal (ottawa, ks); republican daily journal; western home journal) – us WHS [071]

Kansas. State Penitentiary *see*
- Convicts..
- Records

Kansas State Temperance Union *see* Records

Kansas studies in education – Lawrence KS 1923-72 – 1,5 – ISSN: 0075-496X – mf#2354 – us UMI ProQuest [370]

Kansas. Supreme Court *see*
- Kansas supreme court reports
- Mccahon's territory reports
- Roll of attorneys of the state of kansas

Kansas supreme court reports / Kansas. Supreme Court – v1-144. 1862-1936 – 1424mf – 9 – $2136.00 – (pre-nrs: v1-29 1862-83 236mf $354.00. vols after v144 still in copyright) – mf#LLMC 84-135 – us LLMC [347]

Kansas today / Republican Party of Kansas – v3 n4-v6 n5 [1976 apr-1979 may] – 1r – 1 – mf#667622 – us WHS [325]

Kansas Town and Land Company *see* Letter press volumes from the collection of the kansas town and land co., inc. (july 27, 1887)

Kansas trial brief. / Marshall, John – Kansas City, Mo.: Pipes-Reed Book Co., 1905. 1056p. LL-910 – 1 – us L of C Photodup [340]

Kansas union farmer / Kansas Farmers Union – Salina KS. v26 n4-v29 n26 [1933 sep 14-1936 dec 31] – 1r – 1 – mf#1112800 – us WHS [630]

Kansas university lawyer – v1-2. 1895-96 (all publ) – 9 – mf#LLMC 84-519 – us LLMC [340]

Kansas weekly herald – Leavenworth KS. 1858 feb 27-mar 20 – 1r – 1 – (cont by: weekly leavenworth herald) – us WHS [071]

Kansas weekly herald – Topeka, KS: Rutherford & Eagleson. v3 n3 jan 30 1880-v3 n4 feb 6 1880) (wkly) [mf ed 1947] – (= ser Negro newspapers on microfilm) – 1r – 1 – (cont: colored citizen (fort scott, ks)) – us L of C Photodup [071]

Kansaske rozheledy – Omaha, NE: Jednota Cesko-Americkych Listu. roc1 cis. 1 un 1905-roc10 cis19. 27 kvet 1914 (wkly) [mf ed 1905-11,1913-14 (gaps) filmed 1978] – 1r – 1 – (in czech. absorbed by: osveta americka) – us NE Hist [071]

Kanskii vestnik / ed by Khmelevskii, I A – Kansk [Enis gub]: Enis gub prosvet izdvo 1919 [1919 18 maia-] – (= ser Asn 1-3) – n1-44 [1919] item 196, on reel n41 – 1 – mf#asn-1 196 – ne IDC [077]

Kanskii zemskii golos : vnepartiin organ kans uezd zemstva – Kansk [Enis gub]: Uezd zems uprava 1918-19 [1918 iiul'-1919 [?]] – (= ser Asn 1-3) – n5 [1918]-n131 [1919] [gaps] item 197, on reel n41 – 1 – mf#asn-1 197 – ne IDC [077]

Kan-su chih kung yeh – Lan-chou: Kan-su sheng yin hang tsung hang, Min kuo 33 [1944] – (= ser P-k&k period) – us CRL [338]

Kan-su chih t'e ch'an – Lan-chou: Kan-su sheng yin hang tsung hang, Min kuo 33 [1944] – (= ser P-k&k period) – us CRL [339]

Kan-su hsiang-shih lu – List of successful candidates in the imperial examination in Kansu province: 1873, 1875, 1876, 1888, 1889, 1894, 1900, 1903. 1 reel – 1 – us Chinese Res [951]

Kan-su jih-pao – Lauchow, Kansu. Sep 1, 1949-. Scattered issues missing. 2 reels – 1 – us Chinese Res [079]

Kan-su sheng hsi nan pu pien ch'u k'ao ch'a chi / Wang, Chih-wen – [Lan-chou]: Kan-su sheng yin hang ching chi yen chiu shih, [Min kuo 31 ie 1942] – (= ser P-k&k period) – us CRL [915]

Kan-su sheng ko hsien ching chi kai k'uang – Lan-chou: Kan-su sheng yin hang ching chi yen chiu shih, Min kuo 31 [1942] – (= ser P-k&k period) – us CRL [339]

Kan-su sheng pa hsien shih san nien lai kung wu yuan sheng huo fei chih shu – [China]: Kan-su sheng cheng fu t'ung chi shih, Min kuo 33 [1944] – (= ser P-k&k period) – us CRL [315]

Kan-su sheng yin hang Ching chi yen chiu shih *see* Kan-su sheng yin hang hsiao shih

Kan-su sheng yin hang hsiao shih / Kan-su sheng yin hang Ching chi yen chiu shih – [China]: Kan-su sheng yin hang ching chi yen chiu shih, Min kuo 33 [1944] – (= ser P-k&k period) – us CRL [951]

Kant / Wallace, William – Edinburgh: W Blackwood, 1882 – (= ser Philosophical Classics for English Readers) – 1mf – 9 – 0-7905-8620-7 – mf#1989-1845 – us ATLA [100]

Kant and his english critics : a comparison of critical and empirical philosophy / Watson, John – New York: Macmillan, 1881 – 1mf – 9 – 0-7905-7550-7 – mf#1989-0775 – us ATLA [120]

Kant and spencer : a critical exposition / Bowne, Borden Parker – Boston: Houghton Mifflin, 1912 – 2mf – 9 – 0-7905-3604-8 – mf#1989-0097 – us ATLA [140]

Kant et fichte et le probleme de l'education / Duproix, Paul – Paris: Fischbacher, 1895 [mf ed 1986] – 1mf – 9 – 0-8370-8569-1 – (in french. incl bibl ref) – mf#1986-2569 – us ATLA [190]

Kant, Immanuel *see*
- Dreams of a spirit-seer
- Fundamental principles of the metaphysic of ethics
- Immanuel kant on philosophy in general
- Immanuel kant's critique of pure reason
- Immanuel kant's kleinere schriften zur ethik und religionsphilosophie
- Immanuel kants vorlesungen ueber psychologie
- Kant's cosmogony
- Kant's critique of judgement
- Kritik der reinen vernunft
- Perpetual peace
- The philosophy of kant
- Philosophy of law
- Die religion innerhalb der grenzen der blossen vernunft
- Religion within the boundary of pure reason
- Text-book to kant
- [Ueber den gemeinspruch. 1794]

Kant, lotze, and ritschl : a critical examination / Staehlin, Leonhard – Edinburgh: T & T Clark, 1889 [mf ed 1991] – 1mf – 9 – 0-7905-9678-4 – (in english) – mf#1989-1403 – us ATLA [120]

Kant und die idealismusfrage : eine untersuchung ueber kants widerlegung des idealismus / Jaramillo, Luis E H & Kettner, Matthias – Mainz: Gardez, 1995 (mf ed 1996) – (= ser Philosophie im Gardez) – 3mf – 9 – €38.00 – 3-8267-9653-5 – mf#DHS 9653 – gw Frankfurter [140]

Kanta, Surya *see* Atharva pratisakhya

Kanteerava – Mangalore, India. Apr. 1944-1953; 1957-62 – 9r – 1 – us L of C Photodup [079]

Kanter, Hermann *see* Studien zu den acta apostolorum der chester beatty-papyri

Kanthapura / Raja Rao – London: George Allen & Unwin, 1938 – (= ser Samp: indian books) – us CRL [954]

Kantini / Chafulumira, E W – London, England. 1954 – 1r – us UF Libraries [960]

Kantischer kritizismus und englische philosophie : eine beleuchtung der deutsch-englischen neu-empirismus der gegenwart als beitrag zum centenarium der kritik der reinen vernunft / Pfleiderer, Edmund – Halle: CEM Pfeffer, 1881 – 1mf – 9 – 0-7905-9575-3 – mf#1989-1300 – us ATLA [120]

Kantisch-fries'sche religionsphilosophie und ihre anwendung auf die theologie : zur einleitung in die glaubenslehre fuer studenten der theologie / Otto, Rudolf – Tuebingen: J C B Mohr, 1909 – 1mf – 9 – 0-7905-7994-4 – mf#1989-1279 – us ATLA [240]

Kantor besar badan pembantoe pembelaan / Pradjoerit. Djakarta – 2604-2605 (1-21) – 16mf – 9 – ((with the assistance of booei giyugun siddobu and gun-hoodoobu)) – mf#SE-2002 mf390-407 – ne IDC [959]

Kantor, Cyril *see* Lunch with livingstone

Kantor daerah ditdjen kebudajaan prop sulawesi selatan / Madjalah gelora kebudajaan – Makassar, 1968-1969. v1-2(1-17) – 7mf – 9 – (missing: 1968 v1(1-5)) – mf#SE-1824 – ne IDC [959]

Kantor, Ia *see* Yidishe bafelkerung in ukraine

Kantor, M *see* V pomoshch uchiteliu – stroiteliu derevenskoi kooperatsii

Kantor, M K *see*
- O kooperatsii
- Ocherki teorii i istorii kooperatsii
- Osnovy kooperativnoi politiki rkp(b)

Kantor penerangan agama laporan – Djakarta, 1957 – 6mf – 9 – mf#SE-1407 – ne IDC [950]

Kantor Penghubung Pemerintah Daerah Propinsi Sumatera Barat *see* Tjanang

Kantor penjuluhan perindustrian laporan departemen perindustrian rakjat : indonesia – Djakarta, 1960 – 1mf – 9 – mf#SE-689 – ne IDC [959]

Kantor sensus dan statistik tk i sumatera utara / Sumatera Utara dalam angka – Medan, 1970 – 8mf – 9 – mf#SE-1945 – ne IDC [950]

Kantor urusan pegawai peraturan2 gadji pegawai negeri republik indonesia : indonesia – Djakarta, 1948-1956 – 19mf – 9 – mf#SE-229 – ne IDC [959]

Kantor waligeredja indonesia / Geredja Katolik di Indonesia. buku tahunan – Djakarta, 1962 – 8mf – 9 – mf#SE-1491 – ne IDC [241]

Kantorowicz, Alfred [comp] *see* Du wunderliches kind – bettine und goethe

Kant's cosmogony : as in his essay on the retardation of the rotation of the earth and his natural history and theory of the heavens = Allgemeine naturgeschichte und theorie des himmels / Kant, Immanuel; ed by Hastie, William – Glasgow: J MacLehose, 1900 [mf ed 1991] – 1mf – 9 – 0-7905-7859-X – (english by ed) – mf#1989-1084 – us ATLA [140]

Kant's critical philosophy for english readers / Mahaffy, John Pentland & Bernard, John Henry – new ed. London: Macmillan, 1889 [mf ed 1990] – 2v on 2mf – 9 – 0-7905-7598-1 – (1st ed publ 1872) – mf#1989-0823 – us ATLA [190]

Kant's critique of judgement / Kant, Immanuel – London, England. 1931 – 1r – us UF Libraries [025]

Kant's critique of judgement = Kritik der urtheilskraft / Kant, Immanuel – 2nd rev ed. London: Macmillan, 1914 – 2mf – 9 – 0-7905-7445-4 – (in english) – mf#1989-0670 – us ATLA [160]

Kant's entwickelung vom realismus aus nach dem subjectiven idealismus hin : hauptsaechlich nach der ersten auflage der kritik der reinen vernunft / Grundke, Otto – Breslau: W Koebner, 1889 – 1mf – 9 – 0-7905-7638-4 – (incl bibl ref) – mf#1989-0863 – us ATLA [120]

Kant's ethics : a critical exposition / Porter, Noah – Chicago: S C Griggs 1886 [mf ed 1991] – (= ser German philosophical classics for english readers and students) – 1mf – 9 – 0-7905-9589-3 – mf#1989-1314 – us ATLA [170]

Kant's ethics : a critical exposition / Porter, Noah – Chicago: S.C. Griggs & Co., 1886 – 1 – us Wisconsin U Libr [190]

Kants formbegriff und systementwuerfe / Maier, Dieter – Frankfurt a.M., 1979 (mf ed 1993) – 2mf – 9 – €31.00 – 3-89349-637-8 – mf#DHS-AR 637 – gw Frankfurter [190]

Kant's leben und die grundlagen seiner lehre : drei vortraege / Fischer, Kuno – Mannheim: F Bassermann, 1860 – 1mf – 9 – 0-7905-7579-5 – mf#1989-0804 – us ATLA [140]

Kant's mystische weltanschauung : ein wahn der modernen mystik / Lind, Paul von – Muenchen: M Poessl, [1892?] – 1mf – 9 – 0-7905-8505-7 – (incl bibl ref) – mf#1989-1730 – us ATLA [100]

Kant's philosophy as rectified by schopenhauer / Kelly, Michael – London: S Sonnenschein, 1909 – 1mf – 9 – 0-7905-8670-3 – mf#1989-1895 – us ATLA [100]

Kant's theory of knowledge / Prichard, Harold Arthur – Oxford: Clarendon Press, 1909 – 1mf – 9 – 0-7905-9070-0 – (incl bibl ref) – mf#1989-2295 – us ATLA [120]

Kanu, Andrew J *see* Intention to use condoms for hiv/std prevention

Kanunago, Kalika Ranjana *see*
- Dara shukoh
- History of the jats

Kanun-i esasi – Cairo: Osmanli Ittihad ve Terakki Cemiyeti Misir Suebesi. Muharriri: Seyh Alizade Hoca Muhiddin, 1896-99. Muharriri: Seyh Alizade Hoca Muhiddin. n10. 9 mart 1314 [1898] – (= ser O & t journals) – 1mf – 9 – $25.00 – us MEDOC [956]

Kanunname-yi huemayun ticaret-i bahriye – Istanbul: Tasvir-i Efkar Gazetehanesi, 1280 [1864] – 2mf – 9 – $40.00 – us MEDOC [380]

Kanuri songs : with a translation and introductory note / Patterson, John Robert – Lagos: Govt Painter, 1926 – us CRL [780]

Kanyama, Bester *see* Kutora mifananidzo (photography)

Kanyaro, Ferencz *see* Unitariusok magyarorszagon

Kanzler, Melanie *see* Die einflussnahme der amerikanischen besatzungsmacht auf die berliner kulturpolitik in den nachkriegsjahren, 1945-1947

Der kanzler von tirol : geschichtlicher roman / Schmid, Herman – 2. aufl. Leipzig: Keil [18–?] – (= ser Gesammelte schriften. volks- und familien-ausgabe 10-13) – 4v in 2 – 1 – mf#film mas c438 – us Harvard [830]

Kanzlist krehler : tragikomoedie in drei akten / Kaiser, Georg – Potsdam: G Kiepenheuer, 1922 – 1r – 1 – us Wisconsin U Libr [820]

Kanzo uchimura and the non-church movement / Spink, Harry Neilson – [Philadelphia], 1964. Chicago: Dep of Photodup, U of Chicago Lib, 1965 (1r); Evanston: American Theol Lib Assoc, 1984 (1r) – 1 – 0-8370-0455-1 – mf#1984-B023 – us ATLA [240]

K'ao ch'a ou mei chiao t'ung pao kao / China Chiao t'ung pu K'ao ch'a t'uan – Shang-hai: Shang wu yin shu kuan, Min kuo 24 [1935] – (= ser P-k&k period) – us CRL [380]

Kao, Ch'eng-yuan *see* Kuang-chou wu-han ko min wai chiao wen hsien

Kao chi chung hsueh shih fan k'o chiao k'o shu chiao yu shih – Shang-hai: Shang wu yin shu kuan, Min kuo 23 [1934] – (= ser P-k&k period) – us CRL [370]

Kao, Chih *see* Shu hsia chi

Kao ching nien (ccm201) = To the youth / Hu, I-ku – Shanghai, 1928 [mf ed 198?] – (= ser Ccm 201) – 1 – mf#1984-b500 – us ATLA [230]

Kao fen tzu t'ung hsun = Information bulletin of high polymers – Beijing, China 1959 – 1 – mf#2623 – us UMI ProQuest [660]

Kao, Jim *see* Identifying a collective variable of locomotion

Kao k'ao chih lu / Pin, Yeh-sheng – Ch'ung-ch'ing: Pei tou shu tien, Min kuo 33 [1944] – (= ser P-k&k period) – us CRL [370]
Kao, Ko *see* Wo ti jih chi
Kao, K'o-fu *see* Ti pa lu chun tsai shan-hsi
K'ao ku hsueh pao = Journal of archaeology – Beijing, China 1936-63 – 1 – mf#2624 – us UMI ProQuest [930]
Kao, Nai-t'ung *see* Ts'ai chieh-min hsien sheng chuan lueh
Kao, Shen *see* Chien ch'ai
K'ao shih hsin lun / Shih, Mei-hsuan – Shanghai: Min chih shu chu, Min kuo 22 [1933] – (= ser P-k&k period) – us CRL [350]
K'ao shih yuan kung tso pao kao / China K'ao shih yuan – [China: K'ao shih yuan, Min kuo 24 [1935]] – (= ser P-k&k period) – us CRL [350]
K'ao shih yuan tsung pao kao shu hsu pien – [China]: K'ao shih yuan, Min kuo 24 [1935] – (= ser P-k&k period) – us CRL [350]
Kao, Shu-k'ang *see* Chan shih mao i cheng ts'e
Kao, Tao-yun *see* Sui ch'in lou tsa chu ssu chung
Kao teng k'ao shih tsung pao kao : min kuo erh shih erh nien (1933) – [China]: Kao teng k'ao shih tien shih wei yuan hui, Min kuo 23 [1934] – (= ser P-k&k period) – us CRL [350]
Kao, T'ien *see* Wo men ti sui meng
Kao, Tsung-wu *see* Jih mei kuan hsi kai kuan
Kao, Yueh-fu *see* K'ang chan shih hua
Kao-lo-p'ei *see* Ming mo i seng tung-kao ch'an shih chi k'an
Kao-mien yu liao-kuo = Gaomian yu liaoguo / Lin, Sheng-shih – Tai-pei: Tsung ho chu pan she Min kuo 46 [1957] [mf ed 1989] – 1r with other items – 1 – (at head of title: dong nan ya ju zhu qing zhong zhi) – mf#mf-10289 seam reel 006/05 [§] – us CRL [959]
Kaonde note book / Wright, J L – London, England. 1958 – 1r – us UF Libraries [960]
Kao-pao hu ch'u t'u ti ching chi tiao ch'a pao kao – [China: Tao Huai wei yuan hui, 1933] – (= ser P-k&k period) – us CRL [630]
Das kap und die kaffern : oder mittheilungen ueber meinen fuenfjaehrigen aufenthalt in sued-afrika / Cole, Alfred W – Leipzig 1852 [mf ed Hildesheim 1995-98] – 1v on 2mf – 9 – €60.00 – 3-487-27294-6 – gw Olms [960]
Kapadia, Dinshah D *see* Pahlavi vendidad
Kapadia, Hiralal R *see* The doctrine of karman in jain philosophy
Kapadia, Kanailal Motilal *see* Hindu kinship
Kapadia, Shapurji Aspaniarji *see* The teachings of zoroaster
Kapelrud, Arvid S *see* Baal in the ras shamra texts
Kapff, Sixt Karl *see* Coming of the lord
Kapi mana news – Porirua, NZ. oct 1970-sep 1972; may 1974-1987 – 32r – 1 – mf#47.1 – nz Nat Libr [079]
Der kapitaen : erzaehlungen / Blunck, Barthold – Bayreuth: Gauverlag Bayreuth, 1942 [mf ed 1989] – (= ser Die kleine glockenbuecherei 1) – 111p (ill) – 1 – mf#7036 – us Wisconsin U Libr [880]
Der kapitaine portlock's und dixon's reise um die welt : besonders nach der nordwestlichen kueste von amerika waehrend der jahre 1785 bis 1788 in den schiffen king george und queen charlotte / Beresford, William – Berlin 1790 [mf ed Hildesheim 1995-98] – 1v on 4mf [ill] – 9 – €120.00 – 3-487-26626-1 – (also available in french) – gw Olms [910]
Kapital / Marx, Karl – Chicago, IL. 1965 – 1r – us UF Libraries [025]
Kapital-journal – London, UK. 17 Jun 1912 – 1 – uk British Libr Newspaper [072]
Kapitaly kreditnogo kooperativa : ikh znachenie i organizatsiia / Ivanov, V I – 1927 – 50p 1mf – 9 – mf#COR-380 – ne IDC [335]
Kapitaly krestianskogo khoziaistva i ego kreditovanie pri agrarnoi reforme / Chaianov, A V – 1918 – 32p 1mf – 9 – mf#COR-230 – ne IDC [335]
Kapiti mail – Paraparaumu, NZ. apr 1988-89 – 4r – 1 – mf#46.8 – nz Nat Libr [079]
Kapiti news – Paraparaumu, NZ. 19 feb-20 aug 1981 – 1r – 1 – mf#46.6 – nz Nat Libr [079]
Kapiti observer – Paraparaumu, NZ. sep 1974-1988 – 1 – mf#46.3 – nz Nat Libr [079]
Kapitlen geshikhte fun bund / Rafes, M G – Kiyev, Ukraine. 1929 – 1r – us UF Libraries [939]
Kapituelasyonlar – K D Sancakiyan Matbaasi, 1329 [1913] – 1mf – 9 – $25.00 – us MEDOC [380]
O kapituliatsiiakh v ottomanskoi imperii / Iastrzhembskii, vA – Khar'kov, 1905 – 6mf – 9 – mf#R-9697 – ne IDC [956]
Kaplan, Jacob *see* Papers, 1876-1975 [bulk 1920-1950]
Kaplan, Jacob Hyman *see* Psychology of prophecy
Kaplan, Leah E *see*
– A descriptive analysis of corporate health promotion activity evaluations
– Health beliefs, health values, and preventive health promotion activities of african- and euro-american women
– Variables related to knowledge levels of aging and planning for future aging of texas high school graduates
Kaplan, Linda *see* The effect of wrist weight on the hemodynamic response to exercise in coronary artery disease
Kaplan, Moise N *see* Big game fishermen's paradise
Kaplan, Mordecai Aaron *see* Ruah ha-'et
Kaplan, Mordecai Manahem *see* Judaism as a civilization
Kaplan, Mordecai Menahem *see* Judaism in transition
Kaplan, Pesach *see* Lider-bukh
Kaplan, Susan *see* Occupational therapy in health care
Kaplansky, Solomon *see* Fun onzog tsu fatvirklekhung
Kaplinskii, M A *see* Izvestiia semipalatinskogo soiuza kooperativov
[Kapnist, P] *see* Kratkoe obozrenie napravleniia periodicheskikh izdanii i gazet, i otzyvov ikh po vazhneishim pravitelstvennym i drugim voprosam za 1861 g
Ka-pow : official newsletter of the southeastern wisconsin comics club / Southeastern Wisconsin Comics Club – n11-62 [1978 apr-1983 nov] – 1r – 1 – (cont: swcc news notes; cont by: paper-weight; ken's lightstone journal; ken's paper-weight) – mf#711203 – us WHS [071]
Kapp, Friedrich *see*
– Die deutschen im staate new york waehrend des 18. jahrhunderts
– Geschichte der deutschen im staate new york
Kapp, Julius *see* Wagner
Kappa Alpha Psi Fraternity *see* Confidential bulletin
Kappa delta pi record – Indianapolis IN 1964+ – 1,5,9 – ISSN: 0022-8958 – mf#8067 – us UMI ProQuest [378]
Kappenberg, Hans *see* Der bildliche ausdruck in der prosa eduard moerikes
Kappiya cum thok vatthu mya : [a novel] / Tan Chan – Ran Kun: Siri mangala pum nhip tuik 1958 [mf ed 1990] – 1r with other items – 1 – (in burmese) – mf#mf-10289 seam reel 180/5 [§] – us CRL [830]
Kappler, A *see* Hollaendisch-guiana
Kappler, Charles J *see* Kappler's indian affairs
Kappler, Helmut *see* Der barocke geschichtsbergriff bei andreas gryphius
Kappler's indian affairs : laws and treaties / Kappler, Charles J – Washington: GPO. v1-5. 1903-41 – (= ser Native American coll) – 60mf – 9 – $90.00 – (with suppl vol 1975. also part of llmc's native american collection titles 4130 + 4133) – mf#LLMC 88-003 – us LLMC [324]
Kapt david woodard's geschichte seiner schicksale und seines aufenthalts auf der insel celebes : nebst nachrichten von derselben und ihren bewohnern / Woodard, David – Weimar 1805 [mf ed Hildesheim 1995-98] – 1v on 2mf – 9 – €60.00 – ISBN-10: 3-487-26571-0 – ISBN-13: 978-3-487-26571-2 – gw Olms [915]
Kapt will rob broughton's befehlshabers der koengl grossbritt kriegsschaluppe providence und ihres beischiffs entdeckungsreise in das stille meer : und vorzueglich nach der nordostkueste von asien, gethan in den jahren 1795, 1796, 1797 und 1798 – Weimar 1805 [mf ed Hildesheim 1995-98] – 1v on 3mf – 9 – €90.00 – ISBN-10: 3-487-26581-8 – ISBN-13: 978-3-487-26581-0 – (trans fr english) – gw Olms [910]
Kaptein, A *see* Unie van zuid-afrika
Kapterev, N F *see* Kharakter otnoshenii rossii k pravoslavnomu vostoku v 16 i 17 stoletiiakh
Kapteyn, Jacobus Cornelius *see* Publications of the astronomical laboratory at groningen
De kapucijnen in de nederlanden en het prinsbisdom luik / Hildebrand, P – Antwerpen. v1-10. 1945-1956 – €198.00 – ne Slangenburg [241]
Kapunda herald – Australia. Jan 1885-Jul 1893.-w. 6 reels – 1 – uk British Libr Newspaper [079]
Kar – [Tehran]: Sazman-i Chirik'ha-yi Fada'i-i Khalq, 1979-82. shumarah-'i 1-154. 19 isfand 1357-18 murdad 1361 [10 mar 1979-9 aug 1982] – 2r – 1 – $106.00 – (missing: n146-152) – us MEDOC [079]
Kar *see* Kar (aksariyat)
Kar (aksariyat) – [Tehran]: Sazman-i Fada'iyan-i Khalq. shumarah-'i 62-149. 21 khurdad-28 bahman 1359 [14 jun 1980-17 feb 1981] – 2r – 1 – $106.00 – (missing: n63-64, 66, 90-91, 104-112, 125, 139, 143-144) – us MEDOC [079]
Kar (aksariyat) *see* Kar
Kar international – Arlington, VA: Kar Comm, International Organization of Iranian People's Fedaii Guerillas. n1-3. feb-apr 1981 – 1r – 1 – $53.00 – us MEDOC [079]
Kar international – Los Angeles: Kar International, Kar Comm (Organization of Iranian People's Fedaii Guerrillas). n1-9 special iss. fall 1980, feb 1981-jun 1982, jun 1983 – 1r – 1 – $53.00 – us MEDOC [079]
Kar sandana khmaer-paramn-viat nam = Conversations khmero-franco-viet-namiennes: prononciation romanisee – Bhnam Ben: Sen Nuan H'uat [195-?] [mf ed 1990] – 1r with other items – 1 – mf#mf-10289 seam reel 112/4 [§] – us CRL [480]
Kar siksa vivattan nai qaksar khmaer = Evolution de l'ecriture khmere / Dik Gam – [s.l: s.n.] 2508 [1965] [mf ed 1990] – 1r with other items – 1 – (in khmer) – mf#mf-10289 seam reel 110/6 [§] – us CRL [305]
Kara bela / Kemal, Namik – Istanbul: Mahmud Bey Matbaasi, 1326 [1910] – (= ser Ottoman literature, writers and the arts) – 2mf – 9 – $40.00 – us MEDOC [470]
Kara deniz vaki samsun limani imtiyazine dair huekumet-i osmaniye canibindan imtiyazin muenakasasi muamelat... – Konstantiniye [Istanbul]: Matbaa-yi Ebuezziya – 1mf – 9 – $25.00 – us MEDOC [380]
Kara, I *see* Draysig yor yidishe literatur in rumenye
Kara tehlike / Ileri, Celal Nuri – Dersaadet [Istanbul]: Cemiyet Kitaphanesi, 1334 [1918] – (= ser Ottoman literature, writers and the arts) – 2mf – 9 – $40.00 – us MEDOC [470]
Karabagh : bericht ueber die im sommer 1890 im russischen karabagh...ausgefuehrte reise / Radde, G – Gotha: J Perthes, 1890 – 2mf – 9 – mf#AR-1621 – ne IDC [915]
Karabchevskii, N P *see*
– Obshchedostupnyi ezhenedelenyi zhurnal
– Okolo pravosudiia
Karacelebizade, Abduelaziz *see*
– Ravat uel-ebrar
– Sueleymanname
Karachi commerce – Karachi, Pakistan. -w. Jan 1951-Dec 1956; Jan-Dec 1959. 7 reels – 1 – uk British Libr Newspaper [072]
[Karachi-] outlook – PK. 1972-74 – 3r – 1 – $150.00 – mf#R63576 – us Library Micro [079]
[Karachi-] pakistan economist – PK. 1972-79 – 14r – 1 – $700.00 – mf#R63577 – us Library Micro [330]
[Karachi-] statesman – PK. 1972-79 – 7r – 1 – $350.00 – mf#R04229 – us Library Micro [320]
Karacsay, Fedor *see* Beytraege zur europaeischen laenderkunde, die moldau, wallachey, bessarabien und bukowina
Karadzic, Vuk S *see*
– Mala prostonarodn'a slaveno-serbska pesnarica
– Pismenica serbskoga iezika po govoru prostogo naroda
Karadzic, Vuk Stefanovic *see*
– Dodatak k sanktpeterburgskim sravniteljnim rjecnicima sviju jezika i narjecija
– Vukova prepiska
Karageorgis, Fevronia *see* Perceptions of the importance and achievement of student teaching objectives
Karagoez salnamesi – (= ser Ministry and special interest salnames) – 9 – (1326 [1910] 2mf $100; 1327 [1911] 2mf $100; 1328 [1912] 3mf $100; 1329 [1913] 2mf $100) – us MEDOC [956]
Karaka, Dosabhai Framji *see*
– History of the parsis
– There lay the city
Karaka, Dosoo Framjee *see*
– Betrayal in india
– I've shed my tears
– Just flesh
– Nehru
– New york with its pants down
– Oh! you english
– Out of dust
– This india
– We never die
Karaki aika baibara : aika karakinia aomata ma bai ake a taekinaki n te o tetemanti ake a mana atonaki n te nu tetemanti / Bingham, Minerva Clarissa Brewster – Nu loki, [New York]: E boretiaki iroun te koraki n Amerika...1870 [mf ed 1995] – (= ser Yale coll) – 155p (ill) – 1 – 0-524-10130-2 – (in gilbertese) – mf#1995-1130 – us ATLA [220]
The karamajong cluster / Gulliver, Philip Hugh – London, International African Institute, 1952 – (filmed with his kinship and property among the jie and turkana. london, 1952) – us CRL [960]
Karamania : or a brief description of the south coast of asia-minor and of the remains of antiquity: with plans, views, etc collected...in the years 1811 and 1812 / Beaufort, Francis – London 1817 [mf ed Hildesheim 1995-98] – 1v on 2mf [ill] – 9 – €60.00 – 3-487-27653-4 – gw Olms [915]
Karamanien : oder beschreibung der suedkueste von klein-asien / Beaufort, Francis – Weimar 1821 [mf ed Hildesheim 1995-98] – 1v on 2mf – 9 – €60.00 – 3-487-26499-4 – gw Olms [915]
Karamzin, N *see* Vestnik evropy.
Karamzin, Nikolai M *see* Istoriia gosudarstva rossiiskago
Karamzin, Nikolaj M *see* Briefe eines reisenden russen
Karan nhac sac ku pvai : sui ma hut, ta don ta cvan / Aung Htu, Saw – Pha Pvan: U Co on Thu 1975 [mf ed 1995] – on pt of 1r – 1 – mf#11052 r1951 n7 – us Cornell [959]
Karandikar, S V *see* Hindu exogamy
Karankawa kountry : a publication of calhoun county genealogical society / Calhoun County Genealogical Society – 1982 spr-1988 win – 1r – 1 – (cont: karankawa kountry quarterly) – mf#1609183 – us WHS [929]
Karankawa kountry quarterly / Calhoun County Genealogical Society – 1979 spr-1981 win – 1r – 1 – (cont by: karankawa kountry) – mf#1609173 – us WHS [929]
Karaosmanoglu *see* Bir serencam
Die karas : roman / Vuorio, Anelma Kaarina Kojonen – Wien: W Frick, 1944 – 1r – 1 – us Wisconsin U Libr [830]
Karataev, S I *see* Bibliografiia finansov, promyshlennosti i torgovli
Karatygina, E S *see* Selskokhoziaistvennaia kooperatsiia i gospromyshlennost
Karavaev, V F *see* Bibliograficheskii obzor zemskoi statisticheskoi i otsenochnoi literatury so vremeni uchrezhdeniia zemstv 1864-1903 g
Karavaikin, A *see* Ispolnenie dogovorov
Karawan, Ariel *see* The effects of twelve weeks of walking or exerstriding on upper body muscular strength and endurance
Karawanken-bote – Krainburg (Kranj SLO), 1941 15 nov-1943 31 dec – 2r – 1 – gw Misc Inst [077]
[Karay] *see* Bir avuc sacma
Karbhari, Bhagu Fatehchand *see* The karma philosophy
Die kardinaele und ihre politik um die mitte des 13. jahrhunderts : unter den paepsten innocenze 4., alexander 4., urban 4., clemens 4., (1243-68) / Maubach, Jos – Bonn: Carl Georgi, 1902 [mf ed 1986] – 1mf – 9 – 0-8370-7650-1 – (incl ind) – mf#1986-1650 – us ATLA [241]
Kardinal simon de brion (papst martin 4.) : einleitung und abschnitt 1 und 2 ([teil] 1, 2, 3) / Backes, Nikolaus – Berlin: Hermann Blanke, 1910 – 1mf – 9 – 0-8370-7763-X – (incl bibl ref) – mf#1986-1763 – us ATLA [920]
Kardinal wilhelm sirlets annotationen zum neuen testament : eine verteidigung der vulgata gegen valla und erasmus / Hoepfl, Hildebrand – Freiburg i B, St Louis MO: Herder, 1908 – (= ser Biblische studien) – 1mf – 9 – 0-7905-2415-5 – (incl ind) – mf#1987-2415 – us ATLA [225]
Kardiologiia – Moskva, Meditsina. 1972 n11; 1973 n1,12; 1974 n5; 1978 n6 – us CRL [616]
Kardoo : the hindoo girl / Brittan, Harriette G – 4th ed. New York: William B Bodge, 1869 [mf ed 1995] – (= ser Yale coll) – 183p (ill) – 1 – 0-524-09770-4 – mf#1995-0770 – us ATLA [240]
Kare nhasi mangwana / Patsanza, Peter – Salisbury, Zimbabwe. 1943 – 1r – us UF Libraries [960]
Kare nhasi mangwana / Patsanza, Peter – Salisbury, Zimbabwe. 1948 – 1r – us UF Libraries [960]
Kare nhasi mangwana / Patsanza, Peter – Salisbury, Zimbabwe. 1948 – 1r – us UF Libraries [960]
Kareev, Nikolai Ivanovich *see*
– Gosudarstvo-gorod antichnago mira
– monarkhii drevniago vostoka i greko-rimskago mira
Kare-kare – Cape Town, South Africa. 1956 – 1r – us UF Libraries [960]
Kare-kare – London, England. 1950 – 1r – us UF Libraries [960]
Karelin, A A *see* Obshchinnoe vladenie v rossii
The karen apostle : or, memoir of ko thah-byu, the first karen convert / Mason, Francis – Boston, 1847 – 2mf – 9 – mf#HTM-118 – ne IDC [240]
The karen apostle : or, memoir of ko thah-byu, the first karen convert / Mason, Francis – London: Religious Tract Society, [1880?] – 1mf – 9 – 0-8370-7240-9 – mf#1986-1240 – us ATLA [240]
Karesi – 1305 [1888] – (= ser Vilayet salnames) – 3mf – 9 – $55.00 – us MEDOC [956]
Karet / Institute for Rubber Research and Development – Bogor, [1950]-1964. v1-15(1) – 37mf – 9 – (missing: [1950]-1955, v6(1, 3-12); 1956, v7(1-5, 8-10, 12); 1957, v8(1); 1959, v10(1); 1961, v12(2-6); 1962, v13(1, 5-12); 1963, v14(1-6)) – mf#SE-828 – ne IDC [079]
Karfeld, Kurt Peter *see* South africa in colour
Kargar – [Tehran]: Tribun-i azad-i mardum-i zahmatkash, 1979- . sal-i 1 shumarah-'i 1-10. 1 farvardin-17 murdad 1358 [21 mar-8 aug 1979] – 1r – 1 – $53.00 – us MEDOC [079]

Kargar, bih pish – Tehran: Sazman-i Paykar dar rah-i azadi-i tabaqah-'i kargar. shumarah-'i 1-7 [mar 1979]-30 mihr 1358 [oct 22 1979] – 1r – 1 – $53.00 – (missing: n2) – us MEDOC [079]

Karge, Paul *see*
- Babylonisches im neuen testament
- Geschichte des bundesgedankens im alten testament, 1. haelfte
- Die resultate der neueren ausgrabungen und forschungen in palaestina

Kargopol'skaia kommuna : izdanie kargopol'skogo uezdnogo revoliutsionnogo ispolnitel'nogo komiteta – Kargopol, Russia 1918 [mf ed Norman Ross] – 1r – 1 – mf#nrp-616 – us UMI ProQuest [077]

Karif-english dictionary / Kropf, Albert – Stutterheim, South Africa. 1915 – 1r – us UF Libraries [040]

Karikatuer – Istanbul Matbaa-i Hayriye ve Suerekasi, 1913-14. Mueduer-i Mes'ul: M Hilmi; Mueduer: Turhan. n4,5,8. 20 mart-17 nisan 1330 [1914]) – (= ser O & t journals) – 1mf – 9 – $25.00 – us MEDOC [956]

Karikoga gumiremiseve / Chakaipa, Patrick – Cape Town, South Africa. 1958 – 1r – us UF Libraries [960]

Karin von schweden : novelle / Jensen, Wilhelm – Berlin: Gebrueder Paetel, 1878 [mf ed 1995] – 234p – 1 – mf#8796 – us Wisconsin U Libr [830]

Karis-gerhart collection : from protest to challenge, 1964-1990 – [Chicago, IL: Cooperative Africana Microfilming Project; available from Center for Research Libraries, 1999] – 1 – us CRL [321]

Kariuki, J M *see* Report of the select committee on the disappearance and murder of the late member for nyandarua north

Karkaria, R P *see* The charm of bombay

Karl 5. und die deutsche reformation / Baumgarten, Hermann – Halle: Verein fuer Reformationsgeschichte 1889 [mf ed 1990] – (= ser Schriften des vereins fuer reformationsgeschichte 7/2/27) – 1mf – 9 – 0-7905-4660-4 – mf#1988-0660 – us ATLA [943]

Karl August, Grand Duke of Saxe-Weimar-Eisenach *see* Briefwechsel des grossherzogs carl august von sachsen-weimar-eisenach mit goethe in den jahren von 1775 bis 1828

Karl Baedeker (Firm) *see*
- Austria
- Berlin and its environs
- The dominion of canada with newfoundland and an excursion to alaska
- Great britain
- Nord de la france

Karl beck's literarische entwicklung : ein beitrag zur geschichte der dichtung des vormaerz / Thiel, Ernst – [S.l: s.n.], 1938 (Breslau: Druck K Vater) [mf ed 1989] – 84p – 1 – (incl bibl ref) – mf#7002 – us Wisconsin U Libr [840]

Karl bleibtreu als dramatiker : ein wort an die deutschen buehnenleiter / Merian, Hans – Leipzig: W Friedrich, [1892] [mf ed 1989] – 74p – 1 – mf#7031 – us Wisconsin U Libr [790]

Karl christian planck und die deutsche erneuerungsbewegung nach 1870 / Ruelius, Hermann – Frankfurt a.M., 1938 (mf ed 1993) – 1mf – 9 – €24.00 – 3-89349-318-2 – mf#DHS-AR 174 – gw Frankfurter [430]

Karl Der Grosse [Charles The Great] *see* Das homiliarium karls des grossen auf seine urspruenglichen gestalt hin untersucht

Karl der grosse und die kirche / Ketterer, Johann Adam – Muenchen: R Oldenbourg, 1898 [mf ed 1986] – 1mf – 9 – 0-8370-7879-2 – (incl bibl ref & ind) – mf#1986-1879 – us ATLA [230]

Karl der grosse und die schottischen heiligen / ed by Shaw, Frank – Berlin: Akademie-Verlag, 1981 [mf ed 1993] – (= ser Deutsche texte des mittelalters 71) – xcviii/335p – 1 – (incl bibl ref and ind. poem in middle high german, comm in german) – mf#8623 reel 20 – us Wisconsin U Libr [810]

Karl der grosse von dem stricker / ed by Bartsch, Karl – Quedlinburg, Leipzig: G Basse, 1857 [mf ed 1993] – (= ser Bibliothek der gesammten deutschen national-literatur von der aeltesten bis auf die neuere zeit sect1/35) – viii/xcvi/432p – 1 – (incl bibl ref) – mf#8438 reel 8 – us Wisconsin U Libr [830]

Karl edvard laman's kikongo monograph – [19–] – us CRL [999]

Karl friedrich becker's weltgeschichte / Becker, Karl Friedrich – Berlin, Germany. v1-14. 1936-1938 – 3r – us UF Libraries [025]

Karl friedrich christian fasch / Zelter, Carl Friedrich – Berlin: J F Unger 1801 [mf ed 19–] – 1r – 1 – mf#film 1032, 1304 – us Sibley [780]

Karl goedeke, sein leben und sein werk : ein beitrag zur geschichte der revolution von 1848 im koenigreich hannover / Alpers, Paul – Bremen-Horn: W Dorn, [1948?] [mf ed 1990] – 115p (ill) – 1 – (incl bibl ref) – mf#7311 – us Wisconsin U Libr [943]

Karl gutzkow als dramatiker : mit benuetzung unveroeffentlichter stuecke / Metis, Eduard – Stuttgart: Metzler, 1915 [mf ed 1993] – (= ser Breslauer beitraege zur literaturgeschichte. neue folge 48) – [viii]/189p – 1 – mf#8014 reel 5 – us Wisconsin U Libr [410]

Karl gutzkow as literary critic with special emphasis on the period 1852-1862 / McConkey, Elizabeth – Private ed. Chicago, IL: Distributed by University of Chicago Libraries, 1941 – 1r – 1 – (incl bibl ref) – us Wisconsin U Libr [430]

Karl heinrich : erzaehlung / Meyer-Foerster, Wilhelm – Stuttgart: Deutsche Verlags-Anstalt, 1903 – 1r – 1 – us Wisconsin U Libr [430]

Karl henckell : ein moderner dichter: studie / Blei, Franz – Zuerich: Verlags-Magazin (J Schabelitz), 1895 [mf ed 1990] – 16p – 1 – mf#7471 – us Wisconsin U Libr [430]

Karl llewellyn papers, section f : national conference of commissioners on uniform state laws / Llewellyn, Karl N – 2r – 5 – $295.00 – (price includes twining's "karl llewellyn papers, 1986 (hard copy)) – mf#401980 – us Hein [340]

Karl llewellyn papers, section g : the sacco-vanzetti case / Llewellyn, Karl N – 2r – 5 – $295.00 – (price includes twining's "karl llewellyn papers, 1986 (hard copy)) – mf#401990 – us Hein [340]

Karl llewellyn papers, section i : american indians and primitive law / Llewellyn, Karl N – 9r – 5 – $990.00 – (price includes twining's "karl llewellyn papers", 1986 (hard copy)) – mf#402000 – us Hein [340]

Karl llewellyn papers, section j : the uniform commerical code / Llewellyn, Karl N – 25r – 13 – $1,250.00 – 0-89941-567-9 – (price incl twining's "karl llewellyn papers, 1986 (hard copy)) – mf#400971 – us Hein [348]

Karl llewellyn papers, section r : correspondence / Llewellyn, Karl N – 3r – 5 – $395.00 – (price includes twining's "karl llewellyn papers, 1986 (hard copy)) – mf#402010 – us Hein [340]

Karl marks i agrarnyi vopros : sbornik statei k 50-letiiu so dnia smerti karla marksa / ed by Kuznetsova, Ivan Vasil'evich – Moskva: Mezhdunarodnyi agrarnyi in-t, 1933 [mf ed 2002] – 1r – 1 – (filmed with: morozovskaia stachka 1885 / s predisloviem v i nevskogo (1925). incl bibl ref) – mf#5232 – us Wisconsin U Libr [630]

Karl marx : leben und werk (1818-1883). neurezeption der polis-idee und eschatologie aus juedisch-christlicher geistestradition. beitraege zur politischen theorie / Wittig, Horst E – (mf ed 1993) – 2mf – 9 – €40.00 – 3-89349-794-3 – mf#DHS 794 – gw Frankfurter [320]

Karl marx ueber die polytechnische bildungsidee und der einfluss seiner erziehungsideologie auf die sozialistische paedagogik und vorschulerziehung in der ehem. udssr / Wittig, Horst E – (mf ed 1993) – 2mf – 9 – €40.00 – 3-89349-796-X – mf#DHS 796 – gw Frankfurter [370]

Karl, Mauricio *see* El comunismo en espana. cinco anos en el partido

Karl meinet / ed by Keller, Adelbert von – Stuttgart: Litterarischer Verein, 1858 [mf ed 1993] – (= ser Blvs 45) – 902p – 1 – mf#8470 reel 10 – us Wisconsin U Libr [830]

Karl morgenstern's reise in italien im jahr 1809 – Leipzig 1811-13 [mf ed Hildesheim 1995-98] – 3v on 6mf – 9 – €120.00 – 3-487-29309-9 – gw Olms [914]

Karl nernst's wanderungen durch ruegen – Duesseldorf 1800 [mf ed Hildesheim 1995-98] – 2mf – 9 – €60.00 – 3-487-29538-5 – gw Olms [914]

Karl odebrekt, uhe baptisti elu- ja umberpooramise lugu = Karl odebrekt, the life of a baptist and the story of his turnabout / Ruhl, Gustav – Revel: Tallinna Sinodi kirjastus, 1885 – 1r – 1 – $20.88 – (one part of sa six-part item) – us Southern Baptist [242]

Karl otfried muller's geschichte der griechischen literatur / Muller, Karl Otfried – Stuttgart, Germany. v1-2. 1875 – 1r – us UF Libraries [025]

Karl paultre's franz offiziers bei der leichten artillerie, vormal adjutanten des ober-generals kleber in aegypten kurze geographische nachrichten von syrien : als kommentar zu dessen neuer charte von syrien – Weimar 1804 [mf ed Hildesheim 1995-98] – 1v on 1mf – 9 – €40.00 – ISBN-10: 3-487-26592-3 – ISBN-13: 978-3-487-26592-6 – (trans fr french) – gw Olms [915]

Karl philipp moritz als aesthetiker / Dessoir, Max – Berlin: C Duncker, 1889 – 1r – 1 – (incl bibl ref) – us Wisconsin U Libr [110]

Karl plath : inspektor der gossnerschen mission / Plath, Georg – Schwerin i Meckl: Fr Bahn, 1904 – 1mf – 9 – 0-8370-6601-8 – mf#1986-0601 – us ATLA [240]

Karl rudolf hagenbach : eine friedensgestalt aus der streitenden kirche der gegenwart / Eppler, Christoph Friedrich – Guetersloh: C. Bertelsmann, 1875 – 1mf – us ATLA [240]

Karl rudolf hagenbach : eine friedensgestalt aus der streitenden kirche der gegenwart / Eppler, Christoph Friedrich – Guetersloh: C Bertelsmann, 1875 – 1mf – 9 – 0-7905-4469-5 – mf#1988-0469 – us ATLA [240]

Karl simrocks ausgewaehlte werke in zwoelf baenden / Simrock, Karl Joseph; ed by Klee, Gotthold – Leipzig: Max Hesse, [1907?] – 1 – us Wisconsin U Libr [800]

Karl spindler : ein beitrag zur geschichte des historischen romans und der unterhaltungslektuere in deutschland; nebst einer anzahl bisher ungedruckter briefe spindlers / Koenig, Joseph – Leipzig: Quelle & Meyer 1908 [mf ed 1992] – (= ser Breslauer beitraege zur literaturgeschichte 15) – 1r – 1 – (incl bibl ref. filmed with: das gasel in der deutschen dichtung und das gasel bei platen / hubert tschersig) – mf#3102p – us Wisconsin U Libr [430]

Karl steffensen : gesammelte vortraege und aufsaetze mit einigen erinnerungsblaettern = Selections. 1890 / Steffensen, Karl – Basel: C Detloff, 1890 – 1mf – 9 – 0-7905-8739-4 – mf#1989-1964 – us ATLA [100]

Karl und galie : karlmeinet, teil 1: abdruck der handschrift a (2290) der hessischen landes- und hochschulbibliothek darmstadt und der 8 fragmente / ed by Helm, Dagmar – Berlin: Akademie-Verlag, 1986 [mf ed 1993] – (= ser Deutsche texte des mittelalters 74) – viii/542p – 1 – (incl bibl ref and ind. middle high german text. int in german) – mf#8623 reel 21 – us Wisconsin U Libr [810]

Karl von burgund : ein trauerspiel (nach aeschylus) / Bodmer, Johann Jakob; ed by Seuffert, Bernhard – Heilbronn: Henninger, 1883 [mf ed 1993] – (= ser Deutsche litteraturdenkmale des 18. und 19. jahrhunderts 9) – xii/26p – 1 – mf#8676 reel 1 – us Wisconsin U Libr [820]

Karl von holteis romane : ein beitrag zur geschichte der deutschen unterhaltungs-literatur / Landau, Paul – Leipzig: M Hesse, 1904 [mf ed 1992] – (= ser Breslauer beitraege zur literaturgeschichte 1) – 168p – 1 – mf#8014 reel 1 – us Wisconsin U Libr [430]

Karl von raumer und sein beitrag zur volksbildung im 19. jahrhundert / Dorweiler, Joachim – (mf ed 1994) – 2mf – 9 – €40.00 – 3-8267-2046-6 – mf#DHS 2046 – gw Frankfurter [370]

Karlin, A *see* Divre sefer

Karlmann : roman einer kindheit / Dehnert, Max – Leipzig: H H Kreisel, 1942 [mf ed 1989] – 266p – 1 – mf#7174 – us Wisconsin U Libr [830]

Karl-marx-staedter-blick – Chemnitz DE, 1963 3 jul-1966 30 nov, 1968 10 apr-1970 1 apr – 2r – 1 – (later: blick. v1953-1990 town name karl-marx-stadt) – gw Misc Inst [074]

Karlshamn – Karlshamn, Karlskrona, Sweden. 1871-1928 – 101r – 1 – sw Kungliga [078]

Karlshamns allehanda – Karlshamn, Sweden. 1848-1976 – 1 – (olofstroms nyheter, 1963-64) – sw Kungliga [078]

Karlshamnstidningen – Karlshamn, Sweden. 1928-33 – 19r – 1 – sw Kungliga [078]

Karlshorster anzeiger – Berlin DE, 1925 nov-1926, 1927 apr-dec, 1929 jul-1932, 1933 apr-1934 2 aug – 19r – 1 – (filmed with suppl) – gw Misc Inst [074]

Karlshorster anzeiger *see* Karlshorster lokalanzeiger

Karlshorster lokalanzeiger – Berlin DE, 1937 apr-dec – 1 – (filmed with: karlshorster anzeiger) – gw Misc Inst [074]

Karlskoga bergslagskuriren *see* Orebrokuriren

Karlskoga tidning – Nora, Karlskoga, Sweden. 1883-1985, 1990- – 1 – sw Kungliga [078]

Karlskoga-degerfors allehanda *see* Nerikes allehanda

Karlskogakuriren – Goteborg, 1994- – 9 – sw Kungliga [078]

Karlskogakuriren – Oerebro, 1945-54 – 9 – sw Kungliga [078]

Karlskogakuriren *see*
- Orebro-bergslagskuriren
- Orebrokuriren

Karlskogaposten – Lindesberg, Sweden. 1881-85 – 2r – 1 – sw Kungliga [078]

Karlskrona weckoblad – Karlskrona, Sweden. 1753-1908 – 68r – 1 – (aka: carlscronas wekloblad; nya karlskrona weckoblad) – sw Kungliga [078]

Karlskrona-nisse – Karlskrona, Sweden. 1911 – 1r – 1 – sw Kungliga [078]

Karlskronatidningen – Karlskrona, Sweden. 1908-35 – 80r – 1 – sw Kungliga [078]

Karlsruher anzeiger 1858 – Karlsruhe DE, 1858 5 jan-1876, 1881-82, 1884-85, 1887, 1889-1935 – 1 – (title varies: 2 jun 1863: badischer beobachter) – gw Misc Inst [074]

Karlsruher anzeiger 1967 – Karlsruhe DE, 1967 7 apr-1973 31 aug – 1 – gw Misc Inst [074]

Karlsruher beobachter *see* Karlsruher intelligenz- und wochenblatt

Karlsruher fremdenblatt – Karlsruhe DE, 1916-1933 31 mar [gaps], 1933 1 jul-1935 – 18r – 1 – (title varies: 5 oct 1918: residenz-anzeiger) – gw Misc Inst [074]

Karlsruher intelligenz- und tagblatt *see* Karlsruher intelligenz- und wochenblatt

Karlsruher intelligenz- und wochenblatt – Karlsruhe DE, 1848-49 [single iss] – 1r – 1 – (title varies: 3 jan 1819: karlsruher unterhaltungs- und intelligenzblatt /.../; 1 jan 1833: karlsruher intelligenz- und tageblatt /.../; 1 jan 1843: karlsruher tagblatt. filmed by other misc inst: 1810-1937 30 apr; 1828-37 [4r]. with suppls: karlsruher beobachter (1844-48) 1845 [1r]; karlsruher beobachter (1844-48) 1848 [1r]) – gw Misc Inst [074]

Karlsruher nachrichten *see* Rhein-neckar-zeitung

Karlsruher nachrichten 1870 – Karlsruhe DE, 1870 1 jun-1894 30 jun – 1 – gw Misc Inst [074]

Karlsruher neue zeitung – Karlsruhe DE, 1947 29 jul-1949 – 2r – 1 – (title varies: beginning: sueddeutsche allgemeine) – gw Misc Inst [074]

Karlsruher presse – Karlsruhe DE, 1950 19 aug-7 oct – 1r – 1 – gw Misc Inst [074]

Karlsruher rundschau – Karlsruhe DE, 1949 6 jan-19 may – 1r – 1 – gw Misc Inst [074]

Der karlsruher stadt- und landbote – Karlsruhe DE, 1842 24 dec-1849 24 jun – 3r – 1 – (filmed by other misc inst: 1848-1849 24 jun; title varies: 1 jan 1848: stadt und landbote; filmed with suppls) – gw Misc Inst [074]

Karlsruher tagblatt *see* Karlsruher intelligenz- und wochenblatt

Karlsruher unterhaltungs- und intelligenzblatt *see* Karlsruher intelligenz- und wochenblatt

Karlsruher unterhaltungsblatt *see* Die biene

Karlsruher volksblatt – Karlsruhe DE, 1924 1 feb-1925, 1927-28, 1929 10 apr-1934 18 jan – 1 – (title varies: 2 mar 1925: badische zeitung) – gw Misc Inst [074]

Karlsruher wochenbericht – Karlsruhe DE, 1952 10 oct-1953 6 feb – 1r – 1 – gw Misc Inst [074]

Karlsruher wochenspiegel – Karlsruhe DE, 1953 20 mar-8 aug – 1r – 1 – gw Misc Inst [074]

Karlsson, Elis *see* Cruising off mozambique

Karlstadstidningen – Karlstad, 1917- – 1 – sw Kungliga [078]

Karlstadstidningen – Karlstad, Sweden. 1879-1917 – 34r – 1 – sw Kungliga [078]

Karlstadt, Andreas Rudolff-Bodenstein von *see* Die wittenberger und leisniger kastenordnung, 1522, 1523

Karlstadts schriften aus den jahren 1523-25 / ed by Hertzsch, Erich – Halle: M Niemeyer. 2v. 1956- – (incl bibl ref) – us Wisconsin U Libr [430]

Karlweis, C *see* Das grobe hemd

Karma / Besant, Annie Wood – London; New York: Theosophical Pub Society, 1897 – (= ser Samp: indian books) – us CRL [280]

Karma : works and wisdom / Johnston, Charles – New York: Metaphysical Pub Co, 1900 – 1mf – 9 – 0-524-01835-9 – mf#1990-2670 – us ATLA [280]

Karma and redemption : an essay toward the interpretation of hinduism and the re-statement of christianity / Hogg, Alfred George – 2nd ed. London: Christian Literature Society for India, 1910 – 1mf – 9 – 0-524-01770-0 – (incl bibl ref) – mf#1990-2618 – us ATLA [280]

The karma philosophy / Gandhi, Virchand Raghavji; ed by Karbhari, Bhagu Fatehchand – 1st ed. Bombay: Devchand Lalbhai Pustakoddhar Fund, 1913 – (= ser Devchand Lalbhai Pustakoddhar Fund Series) – 1mf – 9 – 0-524-03252-1 – mf#1990-3182 – us ATLA [180]

Karmakar, R D *see* Mrcchakatika of sudraka

Karmann, Anton *see* Methodische und systematische untersuchungen erkenntnistheoretisch wichtiger sachlicher grundprobleme in der platonischen ideenlehre

Karmarkar, A P *see* The religions of india

Karmarkar, Sumant Vishnu *see* Lessons in the life of christ

Karmin, Otto *see* Michel servet et voltaire

Karminski, Hannah *see* Juedischer frauenbund von deutschland

Karn, Oma *see* Milly and mei kwei, servants of the master

Ein karn voller narren : das ist, etliche blaettel ohn blatt fuers maul... / Abraham...Sancta Clara – Salzburg, 1734 – 1mf – 9 – mf#O-1507 – ne IDC [090]

Karna parva – Calcutta: Bharata Press, 1889 – 1mf – 9 – 0-524-08013-5 – mf#1991-0235 – us ATLA [280]

Karnak : aetude topographique et archeologique avec un appendice comprenant les principaux textes hieroglyphiques / Mariette, A – Leipzig, 1875 – 5mf – 9 – mf#NE-364 – ne IDC [956]

Karnamak i artakhshir papakan / ed by Antia, E E K – Bombay, 1900 – 4mf – 9 – mf#NE-20158 – ne IDC [956]

Karnaukhov, D E *see* Kachestvo i standart v promkooperatsii

Karner, Friedrich Karl *see* Die bedeutung des vergeltungsgedankens fuer die ethik jesu, dargestellt im anschluss an die synoptischen evangelien

Karney, Evelyn Storrs *see* The shining land

Karnovich, E P *see*
- Rodovye prozvaniia i tituly v rossii, i sliianie inozemtsev s russkimi
- Zamechatelnye bogatstva chastnykh lits v rossii

Karo, Jakob *see* Kritische untersuchungen zu levi ben gersons (ralbag) widerlegung d...

Karoff, Julius Marthin *see* Prozess- und ergebnisqualitaet neuer methoden zur flexibilisierung einer kardiologischen rehabilitationsbehandlung

Karoli passaglia sod. e.s.j. commentariorum theologicorum = Commentariorum theologicorum / Passaglia, Carlo – Romae: B Artium, 1850-1851 – 11mf – 9 – 0-7905-8871-4 – mf#1989-2096 – us ATLA [240]

Karoline von guenderode und ihre freunde : und ihre Freunde / Guenderode, Karoline von – Stuttgart: Deutsche Verlags-Anstalt, 1895 [mf ed 2001] – 193p – 1 – mf#10506 – us Wisconsin U Libr [920]

Karori and western suburbs news – Wellington, NZ. 1975-87 – 10r – 1 – mf#41.17 – nz Nat Libr [079]

Der karosseriebauer – Halle S, Aschersleben DE, 1959-1961 nov, 1962-1975 sep, 1976-1989 23 sep – 5r – 1 – (with gaps) – gw Misc Inst [074]

Karpaten-rundschau : wochenschrift fuer gesellschaft, politik, kultur – Kronstadt (Brasov RO), 1976-96 – 21r – 1 – (fr 1996 as suppl to: allgemeinen deutschen zeitung, bukarest; filmed by misc inst: 1969 1 mar-dec 27, 1972-) – gw Mikropress [077]

Karpathenbilder / Hildebrandt, Friedrich – Glogau 1863 [mf ed Hildesheim 1995-98] – 2mf – 9 – €60.00 – 3-487-29080-4 – gw Olms [914]

Karpathen-post – Kaesmark (Kesmarok SK), 1920-21, 1924-41 – 4r – 1 – gw Misc Inst [077]

Karpatorusski kalendar' lemko-soiuza na... – Yonkers, NY: Lemko-Soiuz, 1930-71 – us CRL [520]

Karpatorusskij narodnyj kalendar' – Perth Amboy NJ, 1944-46 – 1r – 1 – (carpatho-rusin periodical) – us IHRC [073]

Karpatorusskije novosti – v1-3, Oct 1943-1945 – 1 – us CRL [073]

Karpato-russkoe slovo = Carpatho-russian word – New York: Carpatho Russian National Committee, jun 1935-38 – us CRL [073]

Karpatorusskoe slovo : ezhened gaz / ed by Labenskii, I A – Omsk [Akmol obl]: TSentr Karpatorus sovet 1918-19 [1918 [20] dek-1919 [?]] – (= ser Asn 1-3) – n1 [1918] n1-25 [1919] item 198, on reel n41 – 1 – mf#asn-1 198 – ne IDC [077]

Karpatska rus = Carpatho-russia – Yonkers, NY: Lemko Association of the United States and Canada, jan 12 1940-69 – us CRL [071]

Karpatska sich – Toronto, Canada: Bratstvo Karpatskykh Sichovykiv, 1950, 1952, 1955-56 – us CRL [071]

Karpats'ka zoria = Carpathian star – New York: Carpathian Star Pub Co, oct 1951-52 – us CRL [073]

Karpeles, Gustav *see*
- Geschichte der juedischen literatur
- Heinrich heine
- Heinrich heine's biographie
- Heinrich heine's life told in his own words
- Jews and judaism

Karr, William Stevens *see*
- Apologetics
- Introduction to christian theology
- System of christian theology

Karrer, O *see* Meister eckehart

Karsch, E *see* Entomologische nachrichten

Die karschin – Muenchen: Muenchener Buchverlag, [1943] – 1r – 1 – us Wisconsin U Libr [800]

Karsen, Fritz *see* Henrik steffens romane

Karsen, Sonja *see* Desenvolvimiento educacional de costa rica con la...

Karskii, E F *see* Listki undolskogo, otryvok kirillovskogo evangelija 11 v fototipicheskoe vosproizvedenie teksta i issledovanie pisma i iazyka

Karslake, William Henry *see* The litany of the english church

Karsthans *see* Die weinsberger ostern

Kart og plan – Norway 1977-81 – 1,5,9 – ISSN: 0047-3278 – mf#8258 – us UMI ProQuest [639]

Kartar Singh *see* Life of guru gobind singh

Kartasheva, K *see* Semia i zhilishche

Kartashov, N S *see* Iz istorii knigi, bibliotechnogo dela i bibliografii v sibiri

Kartell convent deutscher studenten juedischen glaubens : jahrbuch / ed by Weil, Bruno – Strassburg, Leipzig: J Singer, 1906; 1908 – (= ser German-jewish periodicals...1768-1945, pt 3) – 1r – 1 – $165.00 – mf#B262 – us UPA [270]

Die kartelle in der schweizerischen textil-veredlungsindustrie / Schiess, Jakob – Weinfelden: A-G Neuenschwander, 1923 (mf ed 19–) – (= ser Die schweizer industrie und handelsstudien; Harvard social history/business preservation microfilm project) – mf#ZT-TN pv73 n3 – us NY Public [338]

Kartenspiel : aufgabenbuch kleiner prosa / Heyse, Ulrich – Muenchen: K Alber, 1942 – 1r – 1 – us Wisconsin U Libr [830]

Karteria – Athens, Greece. -w. 1 June 1877-9 Nov 1878. 2 reels – 1 – uk British Libr Newspaper [949]

Kartini schools for girls : the archive of the kartini fund, 1912-1960 – [mf ed 2004] – (= ser Women in the netherlands east indies 2) – 803mf – 9 – €5635.00 – (printed inventory in dutch; int in english) – mf#mmp111 – ne Moran [376]

Kartodirdjo, Soejatno *see* Revolution in surakarta 1945-50

Kartosoewirjo, S M *see* Haloean politik islam

Kartschoke, Christopher *see* The difference between participation in intercollegiate athletics and academic performance based on time use

Kartsov, V *see*
- Ekstrennyi biulleten' poslednikh izvestii gazety "trud"; telegrammy
- Trud [minusinsk: 1918-1919]

Kartsov, V S *see* Minusinskii krai

Karumekangu / Chidyausiku, Paul – Salisbury, Zimbabwe. 1970 – 1r – us UF Libraries [960]

Karunakaran, Kotta P *see* India in world affairs, august 1947-january 1950

Karve, D G *see* Historical and economic studies

Karve, Dattatraya Gopal *see*
- Poverty and population in india
- Rnade

Karwas, Marcia R *see* Femininity and masculinity

Karwath, Juliane *see* Die droste

Kas te tahdi irr, baptisti? = Who are they, those baptists? – Jelgawa: J.W. Steffenhagen, 1866. Publ. No. 6298 b. One of five items on a reel – 1 – us Southern Baptist [242]

Kasch, Fritz *see* Leopold f.g. von goeckingk

Kasch-ul-nicab : journal arabe politique hebdomadaire – n1-12. Paris. aout-nov 1894 [wkly] – 1 – fr ACRPP [320]

Kasdoi, Zevi *see* Mamlekhot ararat

Kashi Ram *see* The message of the brahmo samaj

Kashifi, Husayn Vaiz *see* The anvar-i suhaili, or the lights of canopus

Kashin. sovet rk i kd *see* Izvestiia kashinskogo soveta rabochikh, krest'ianskikh i krasnoarmejskikh deputatov

Kashinskii proletari : izdanie uezdnogo komiteta rkp(b) – Kashin, Russia 1919-21 [mf ed Norman Ross] – 3 – 1 – mf#nrp-620 – us UMI ProQuest [077]

Kashiraj *see* An account of the last battle of panipat and of the events leading to it

Kashirin, P *see* Reaktsionnaia sushchnost' religioznoi ideologii

Kashka-dar'inskaia pravda – Karshi, Uzbekistan 1973-88 [mf ed Norman Ross] – 5r – 1 – mf#nrp-617 – us UMI ProQuest [077]

Kashkarov, M *see* Denezhnoe obrashchenie v rossii

Kashkul – Tehran. shumarah-'i 1-40. 15 safar 1325-11 rabi sani 1326 [30 mar 1907-12 may 1908] – 1r – 1 – $53.00 – us MEDOC [079]

Kashmar / Wald, Pine – Buenos Ayres, Argentina. 1929 – 1r – us UF Libraries [939]

Kashmir / Younghusband, Francis Edward – London: A & C Black, 1917 – (= ser Samp: indian books) – (painted by e molyneux) – us CRL [915]

Kashmir *see* The jammu and kashmir government gazette

Kashmir post – Jammu, India. 1962-64 – 3r – 1 – us L of C Photodup [079]

Kashmir times – Jammu, India. 1972-93 – 51r – 1 – us L of C Photodup [079]

Kasi podma choudree – London, England. 18– – 1r – us UF Libraries [240]

Kasovich, Israel Isser *see* Litvisher ingel

Kaspar klee von gerolzhofen : das lebensbild eines elsaessischen evangelischen pfarrers um die wende des 16. zum 17. jahrhundert / Beck, Hermann – Halle: Verein fuer Reformationsgeschichte 1901 [mf ed 1990] – (= ser Schriften des vereins fuer reformationsgeschichte 19/71) – 1mf – 9 – 0-7905-5080-6 – (incl bibl ref) – mf#1988-1080 – us ATLA [242]

Kaspar Ruef et al *see* Der freymuethige

Kasper-ohm un ick / Brinckman, John – 7. aufl. Berlin: W Werther, 1900 [mf ed 1989] – 374p – 1 – (novel in low german) – mf#7086 – us Wisconsin U Libr [830]

Kaspii – Baku, Azerbaijan 1881-1917 [mf ed Norman Ross] – 108r – 1 – mf#nrp-2224 – us UMI ProQuest [077]

Kaspii – Baku: Kaspii [1881- [G1-i n1 (1 ianv 1881 g)- – 44r – 1 – (ceased in 1918?; filmed: g1 n1-g37 n279 (1881-1917) [gaps]) – mf#mr-5 – ne IDC [305]

Kaspii *see* Central asian serials – late 19th- to early 20th-century

Kaspij – Baku, Azerbaijan 1884-87 [mf ed Norman Ross] – 1 – mf#nrp-247 – us UMI ProQuest [077]

Kassel, David *see* Gezang un deklamatyse

Kassel und ahnaberg : studien zur geschichte von stadt und kloster im mittelalter / Buck, Herbert – Frankfurt a.M., 1968 – 3mf – 9 – 3-89349-365-4 – gw Frankfurter [943]

Kasseler journal *see* Hessisches wochenblatt

Kasseler nachrichten – Kassel DE, 1890 25 sep-1892 15 may – 3r – 1 – gw Misc Inst [074]

Kasseler neueste nachrichten – Kassel DE, 1910 4 dec-1943 31 may [gaps] – 82r – 1 – (filmed with suppl) – gw Misc Inst [074]

Kasseler post / stadtausgabe *see* Casseler stadt-anzeiger

Kasseler tageblatt *see* Gewerbliches tageblatt fuer kassel und die umgegend

Kasseler volksblatt – Kassel, 1892-93; 1894-1919; 1921-27; 1928-1933 – 72r – 1 – gw Mikropress [072]

Kasseler volksblatt *see*
- Volksblatt fuer hessen und waldeck

Kasseler zeitung 1851 – Kassel DE, 1851-67 – 17r – 1 – gw Misc Inst [074]

Kasseler zeitung 1881 – Kassel DE, 1881 2 jul-1887 14 aug – 7r – 1 – gw Misc Inst [074]

Kasseler zeitung 1946 – Kassel DE, 1947 24 feb-17 nov, 1948 7 oct-1949 6 oct – 1 – gw Misc Inst [074]

Kasselsche allgemeine zeitung *see* Westphaelischer moniteur

Kasser, Susan L *see* Constraints on functional competence in persons with multiple sclerosis

Kassewitz, Joseph *see* Darlegung der dichterischen technik und litterarhistorischen stellung von goethes elegie "alexis und dora"

Kassimis, Johanna *see* Fremde und fremdes

Kassin, Saul M *see* An empirical study of rule 11 sanctions

Kassovyi otchet ministerstva finansov za 1893 god – Spb, 1894 – 5mf – 9 – mf#REF-192 – ne IDC [332]

Kastalia – Wien (A), 1912 jul-1915 jul – 2r – 1 – gw Mikrofilm [074]

Kastamonu – (= ser Vilayet salnames) – 9 – (1310 [1892] 10mf $165; 1311 [1893] 6mf $90; 1312 [1894] def'a 18 5mf $450; 1314 [1896] def'a 19 7mf $110; 1317 [1899] 6mf $90) – us MEDOC [956]

Kastamonu – Kastamonu: Vilayet Matbaasi. Cikaran: Kastamonu Vilayeti, 1873-? n2174, 2190, 2194, 2197, 2214, 2219, 2222, 2245, 2255, 2267, 2272, 2284, 2778. 31 tesriniewel 1332 [1919]-16 subat 1929 – (= ser O & t journals) – 1mf – 9 – $40.00 – us MEDOC [956]

Het kasteel"de slangenburg" en zijn kunstschatten / Beekman, A W H – Bijdragen en Mededeelingen der Vereeniging "GELRE", deel 48 – €7.00 – ne Slangenburg [240]

Kasthofer, Carl *see*
- Bemerkungen auf einer alpen-reise ueber den bruenig, bragel, kirenzenberg
- Bemerkungen auf einer alpen-reise ueber den susten, gotthard, bernardin

Kastner, Erich *see* Fabian

Kastos, Emiro *see* Mi compadre facundo, y otros cuadros

Kat, A I M *see* De geschiedenis der kerkmuziek in de nederlanden sedert de hervorming

Die katakombengemaelde und ihre alten copien / Wilpert, J – Freiburg im Breisgau, 1891 – 3mf – 9 – mf#H-3055 – ne IDC [700]

Katalaunische schlacht : schauspiel / Bronnen, Arnolt – Berlin: E Rowohlt, 1924 [mf ed 1989] – 120p – 1 – mf#7090 – us Wisconsin U Libr [820]

Katalizatory, soderzhashchie nanesennye kompleksy : materialy simpoziuma / Akademiia nauk SSSR, Sibirskoe otdelenie, Ordena Trudovogo Krasnogo Znameni Institut kataliza – Novosibirsk: Institut, 1980 – us CRL [947]

Katallagete – Nashville TN 1965-90 – 1,5,9 – ISSN: 0022-9288 – mf#5818 – us UMI ProQuest [320]

Katalog, alphabetischer, der dissertationen der universitaetsbibliothek der humboldt-universitaet berlin, bis 1974 – [mf ed Hildesheim 1989] – 458mf – 9 – diazo €1738.00 silver €2100.00 – gw Olms [020]

Katalog, alphabetischer, der kinder- und jugendbuchabteilung der staatsbibliothek zu berlin preussischer kulturbesitz – [mf ed Hildesheim 1994] – 64mf – 9 – diazo €328.00 silver €368.00 – gw Olms [020]

Katalog, alphabetischer, der musiksammlung der staatsbibliothek berlin preussischer kulturbesitz – [mf ed Hildesheim 1990] – (= ser Die europaeische musik) – 537mf (1:42) – 9 – diazo €2840.00 silver €3000.00 – (textregister der musikabteilung 25mf diazo €128 silver €148) – gw Olms [780]

Katalog, alphabetischer, der staatsbibliothek zu berlin preussischer kulturbesitz – [mf ed Hildesheim 1987] – 3199mf – 9 – diazo €13,500.00 silver €15,400.00 – gw Olms [020]

Katalog, alphabetischer, der stadt- und universitaetsbibliothek bern – [mf ed Hildesheim 1991] – 1502mf – 9 – diazo €4600.00 silver €5400.00 – gw Olms [020]

Katalog, alphabetischer, der universitaetsbibliothek graz : druckschriften ab dem erscheinungsjahr 1501 bis zum erwerbungsjahr 1983 – [mf ed Hildesheim 1984] – 711mf – 9 – diazo €3980.00 silver €4800.00 – gw Olms [020]

Katalog, alter, der musikdrucke der oesterreichischen nationalbibliothek – Hildesheim 1985 – (= ser Die europaeische musik) – 249mf – 9 – diazo €2100.00 silver €2380.00 – gw Olms [780]

Katalog arabskikh rukopisei instituta narodov azii an sssr – M, 1960-1965. 3v – 9mf – 9 – mf#R-10974 – ne IDC [956]

Katalog arkheologichesko-artisticheskikh predmetov, tserkovno-slavianskikh rukopisei i staropechatnykh knig kirillovskogo pisma, nakhodiashchikhsia v muzee stavropigiiskogo instituta na den 1(13 marta 1890 / Sharanevych, I – Lviv, 1890 – 40p 1mf – 9 – mf#R-14640 – ne IDC [243]

Katalog arkhivnykh dokumentov po severnoi voine 1700-1721 gg – 1959 – 5mf – 9 – mf#R-10933 – ne IDC [947]

Katalog biblioteki sluzhashchikh v ministerstve finansov – Ed 5. Spb, 1900 – 10mf – 8 – mf#R-5940 – ne IDC [332]

Katalog bon i denznakov rossii, rsfsr, sssr, okrain i obrazovanii, (1769-1927) / ed by Chuchin, F G – Ed 3. M, 1927 – 2mf – 9 – mf#REF-185 – ne IDC [332]

Katalog der alten sammlung der universitaet kopenhagen 1486-1970 – [mf ed Hildesheim 1992] – (= ser Bibliotheca universitatis hafniensis) – 572mf – 9 – diazo €2498.00 silver €3200.00 – gw Olms [020]

Katalog der argelander'schen zonen vom 15. bis 31. grade suedlicher declination in mittleren positionen fuer 1850-0 / Weiss, Edmund – Wien: J N Vernay 1890 [mf ed 2000] – (= ser Annalen der k k universitaets-sternwarte in wien [waehring] 1. supplementband) – 1r – 1 – (new ed of the reduction of argelander's zones made by oeltzen and issued in sitzungsberichte der k akademie der wissenschaften in wien [v26-31]; incl bibl ref) – mf#film mas 29488 – us Harvard [520]

Katalog der argelander'schen zonen vom 45. bis 80. grade noerdlicher deklination in mittleren positionen fuer 1842.0 / Weiss, Edmund – Wien: J N Vernay 1919 [mf ed 2000] – (= ser Annalen der universitaets-sternwarte in wien [waehring] 2. supplementband) – 1r – 1 – (incl bibl ref) – mf#film mas 29488 – us Harvard [520]

Katalog der bestimmungsgroessen fuer 611 bahnen groesser meteore : vorgelegt in der sitzung am 5. maerz 1925 / Niessl von Mayendorf, Gustav; ed by Hoffmeister, Cuno – [s.l: s.n. 1926?] [mf ed 1998] – 1r – 1 – (incl bibl ref; in: denkschriften [akademie der wissenschaften in wien. mathematisch-naturwissenschaftliche klasse] 100. bd) – mf#film mas 28419 – us Harvard [520]

Katalog der bibliothek ponickau : in der universitaets- und landesbibliothek halle / ed by Henning, Marie-Christine & Schnelling, Heiner – [mf ed Hildesheim 2001] – 145mf – 9 – diazo €498.00 – (int by ed) – gw Olms [020]

Katalog der bis 1957 im zentralkatalog baden-wuerttemberg nachgewiesenen werke von martin luther – 5mf – 9 – €42.80 – (pref by manfred mueller) – gw Olms [242]

Katalog der deutschen buecherei leipzig in der deutschen bibliothek : gesamtarchiv des deutschsprachigen schrifttums seit 1913 – [mf ed Hildesheim 1988] – 2532mf – 9 – diazo €10,800.00 silver €12,000.00 – gw Olms [010]

Katalog der hauptbibliothek in den franckeschen stiftungen halle – [mf ed Hildesheim 1993] – 99mf – 9 – diazo €498.00 silver €588.00 – gw Olms [020]

Katalog der koeniglichen bibliothek kopenhagen : auslaendische aeltere sammlung. druckschriften mit den erscheinungsjahren 1454-1949 – [mf ed Hildesheim 1987] – 822mf – 9 – diazo €2498.00 silver €3200.00 – gw Olms [020]

Katalog der libretti / Oesterreichische Nationalbibliothek Wien. Musiksammlung – [mf ed Hildesheim 1985] – (= ser Die europaeische musik) – 52mf – 9 – diazo €388.00 silver €488.00 – gw Olms [780]

Katalog der masonica-sammlung der universitaetsbibliothek poznan/posen – [mf ed Hildesheim 1989] – 46mf – 9 – diazo €168.00 silver €218.00 – gw Olms [020]

Katalog der musikdrucke der nationalbibliothek zu prag-narodni knihorna – [mf ed Hildesheim 1991] – (= ser Die europaeische musik) – 158mf – 9 – diazo €798.00 silver €938.00 – gw Olms [780]

Katalog der musikhandschriften / Oesterreichische Nationalbibliothek Wien. Musiksammlung – Hildesheim 1984 [mf ed Hildesheim 1995-98]] – (= ser Die europaeische musik) – 106mf – 9 – diazo €828.00 silver €938.00 – gw Olms [780]
Katalog der oesterreichischen nationalbibliothek wien : druckschriften 1501-1929 / Oesterreichische Nationalbibliothek Wien – [mf ed Hildesheim 1982] – 808mf – 9 – diazo €3980.00 silver €4800.00 – gw Olms [020]
Katalog der ornamentstich-sammlung des k k oesterreichischen museums fuer kunst und industrie – Wien, 1865 – 1mf – 9 – mf#OA-67 – ne IDC [700]
Katalog der ornamentstich-sammlung des kunstgewerbe-museums, berlin – Leipzig, 1894 – 6mf – 9 – mf#OA-68 – ne IDC [720]
Katalog der seminar-bibliothek / Zuckerman, Benedict – Breslau, Germany. 1870 – 1r – us UF Libraries [939]
Katalog der universitaetsbibliotek wroclaw/ breslau – [mf ed Hildesheim 1990] – alte drucke 720mf neue drucke 847mf – 9 – diazo €3980.00 silver €5000.00 – gw Olms [020]
Katalog der universitaetsbibliothek wien : druckschriften bis zum erscheinungsjahr 1931 – [mf ed Hildesheim 1983] – 693mf – 9 – diazo €2780.00 silver €3200.00 – gw Olms [020]
Katalog der zentralbibliothek der landbauwissenschaft, bonn : alphabetischer gesamtkatalog 1847-april 1986 – schlagwortkatalog 1960-april 1986 = Catalogue of the central library for agricultural science, bonn – [mf ed 1986] – 275mf (1:42) – 9 – silver €3350.00 – ISBN-10: 3-598-30288-6 – ISBN-13: 978-3-598-30288-6 – gw Saur [630]
Katalog dziel tresci przyslowiowej skladajacych bibljoteke ignacego bernsteina / Bernstein, Ignatz – Warszawa: Czcionkami drukarni W Drugulina w Lipsku 1900 [mf ed Bloomington IN: Indiana Uni Lib, Preservation Dept 1984] – 2v on 1r – 1 – (bibliography of proverbs) – us Indiana Preservation [390]
Katalog izdanii imperatorskoi akademii nauk / [Kubasov, I A] – Pg., 1912-1916. v1-3 – 9mf – 9 – mf#R-4828 – ne IDC [077]
Katalog izdanii po kustarnoi promyshlennosti i promyslovoi kooperatsii – 1930 – 32p 1mf – 9 – mf#COR-533 – ne IDC [335]
Katalog prag : musikdrucke katalog der musikdrucke der nationalbibliothek zu parg – narodni knihovna – (mf ed 1991) – (= ser Die europaeische musik) – 168mf (1:42) – 9 – diazo €798.00 silver €938.00 – gw Olms [780]
Katalog rekopisow / Paris. Biblioteka Polska – Krakow, 1939- – 1 – us Wisconsin U Libr [025]
Katalog rossiiskikh rukopisnykh knig, nakhodiashchikhsia v biblioteke novgorodskogo sofiiskogo sobora / Tikhanov, P N – 1881 – (= ser Pamiatniki drevnei pismennosti) – 28p 1mf – 9 – (pamiatniki drevnei pismennosti, v12) – mf#R-11204 – ne IDC [243]
Katalog, systematischer, der staatsbibliothek zu berlin preussischer kulturbesitz, bis 1955 – [mf ed Hildesheim 1988-91] – 6226mf – 9 – diazo €16,400.00 silver €19,800.00 – (subject ind & guide also sold separately €35.80) – gw Olms [020]
Katalog tserkovnoslavianskikh rukopisei i staropechatnykh knig kirillovskogo pisma, nakhodiashchikhsia na arkheologichesko-bibliograficheskoi vystavke v stavropigiiskom zavedenii / Petrushevich, A S – Lvov, 1888 – 45p 1mf – 9 – mf#R-14441 – ne IDC [243]
Katalog uchebnikov, razreshennykh dlia upotrebleniia v tserkovno-prikhodskikh shkolakh – 1899 – 651p 14mf – 8 – mf#R-5951 – ne IDC [243]
Katalog wien : alter katalog der musikdrucke / Oesterreichische Nationalbibliothek Wien. Musiksammlung – (mf ed 1985) – (= ser Die europaeische musik) – 249mf – 9 – diazo €2,100.00 silver €2,380.00 – gw Olms [780]
Katalog wien : katalog der musikhandschriften / Oesterreichische Nationalbibliothek Wien. Musiksammlung – [mf ed Hildesheim 1984] – (= ser Die europaeische musik) – 106mf – 9 – diazo €828.00 silver €938.00 – gw Olms [780]
Katalog zhurnala vestnik evropy za 25 let, 1866-1890, s alfavitnym ukazatelem imen avtorov / ed by Stasiulevich, M M – 1891 – 3mf – 9 – mf#R-237 – ne IDC [077]
Kataloge der bibliothek des zentralinstituts fuer kunstgeschichte in muenchen = Catalogues of the library of the central institute for the history of art in munich – [mf ed 1982-91] – 1025mf (1:42) – 9 – silver €7690.00 – ISBN-10: 3-598-30348-3 – ISBN-13: 978-3-598-30348-7 – (alphabetischer katalog 1982-85 [220mf] isbn: 978-3-598-30349-4 [€1695]; suppl 1989 [94mf] isbn: 978-3-598-30391-3 [€1360]; sachkatalog 1984 [480mf] isbn: 978-3-598-30351-7 [€3050]; aufsatzkatalog (kataloge der unselbstaendigen schriften) 1983-85 [231mf] isbn: 978-3-598-30350-0 [€1585]) – gw Saur [700]
Kataloge der frankfurter und leipziger buchmessen 1594-1860 : (grosse, spaeter weidmann, wigand, avenarius und mendelssohn und avenarius, lamberg, latomus) = Catalogues of the frankfurt and leipzig book fairs / ed by Fabian, Bernhard – [mf ed Hildesheim 1977-85] – 931mf – 9 – diazo €4400.00 silver €5400.00 – (also available separately: michaelismesse 1594-1699, leipzig 1594-1699 (mf ed 1982-85) 211mf diazo €1680 silver €1980. ostermesse 1700-59, leipzig 1700-59 (mf ed 1979) 126mf diazo €698 silver €828. michaelismesse 1759-1800, leipzig 1759-1800 (mf ed 1977) 141mf diazo €938 silver €1148. ostermesse 1801-michaelismesse 1860, leipzig 1801-60 453mf diazo €1248 silver €1498) – gw Olms [070]
Katalogus dari perangko2 republik indonesia – Surabaja, 1961-1967 – 4mf – 9 – (missing: 1962-1963; 1965-1966) – mf#SE-635 – ne IDC [959]
Katanga / Cornet, Rene Jules – Paris, France. 1943 – 1r – us UF Libraries [960]
Katanga – [Elisabethville: s.n., apr 13 1960] – us CRL [079]
Katanga / Elst, Ferdinand Vander – Bruxelles, Belgium. 1913 – 1r – us UF Libraries [960]
Katanga circus / Valahu, Mugur – New York, NY. 1964 – 1r – us UF Libraries [960]
Katanga, pays du cuivre / Lekime, Fernand – Verviers, France. 1966 – 1r – us UF Libraries [960]
Katanga physique / Robert, Maurice – Bruxelles, Belgium. 1927 – 1r – us UF Libraries [960]
Katanga report / Hempstone, Smith – London, England. 1962 – 1r – us UF Libraries [960]
Katanga Secessionist government *see* Moniteur katangais
Katanga, Zaire (Province) *see* Livre blanc du gouvernement katangais sur les activites
Kataoka, S Kanridoo *see* [Djalan jang haroes dilaloei oleh pegawai negeri]
Katastrofale tsaytn un di vaksndige doyres / Schneersohn, Fischel – Berlin, Germany. 1923 – 1r – us UF Libraries [939]
Kate bachman papers *see* Papers
Die katechese in der erzdioezese koeln unter den kurfuersten max heinrich bis max franz. 1650-1801 / Miebach, Peter – Koeln, 1926 (mf ed 1993) – 1mf – 9 – €24.00 – 3-89349-248-8 – mf#DHS-AR 110 – gw Frankfurter [240]
Katechese und predigt : vom anfang des vierten bis zum ende des sechsten jahrhunderts / Probst, Ferdinand – Breslau: F Goerlich, 1884 – 1mf – 9 – 0-524-01466-3 – (incl bibl ref) – mf#1990-0415 – us ATLA [240]
Katechesen (bdk41 1.reihe) / Cyrillus von Jerusalem (Cyril of Jerusalem, Saint) – (= ser Bibliothek der kirchenvaeter. 1. reihe (bdk 1.reihe)) – €15.00 – ne Slangenburg [240]
Katechetisches magazin / ed by Lang, Georg Heinrich – Noerdlingen 1781-84 – (= ser Dz. abt theologie) – 3pt on 6mf – 9 – €120.00 – mf#k/n2154 – gw Olms [240]
Der katechismus als paedagogisches problem / Eberhard, Otto – Berlin: Edwin Runge 1912 [mf ed 1989] – (= ser Biblische zeit- und streitfragen 8/5) – 1mf – 9 – 0-7905-1983-6 – mf#1987-1983 – us ATLA [240]
Katechismus der christkatholischen lehre zum gebrauche der groesseren schueler : (welche den kleinen katechismus gelernt haben) nach anleitung des religion-handbuchs / Overberg, Bernhard – 29. Aufl. Muenster, 1835 (mf ed 1993) – 2mf – 9 – €31.00 – 3-89349-358-1 – mf#DHS-AR 358 – gw Frankfurter [241]
Katechismus der kompositionslehre / Riemann, Hugo – Leipzig: M Hesse 1889 [mf ed 1991] – 2v in 1 on 1r – 1 – (incl ind) – mf#pres. film 102 – us Sibley [780]
Katechismus fuer die katholischen pfarrschulen der vereinigten staaten / Faerber, Wilhelm – St Louis, MO: B Herder, c1895 – 1mf – 9 – 0-8370-7062-7 – mf#1986-1062 – us ATLA [241]
Die kategorien- und bedeutungslehre des duns scotus / Heidegger, M – Tuebingen, 1916 – 5mf – 8 – €12.00 – ne Slangenburg [110]
Die kategorien- und bedeutungslehre des duns scotus / Heidegger, Martin – Tuebingen: JCB Mohr, 1916 – 1mf – 9 – 0-7905-8995-8 – (incl bibl ref) – mf#1989-2220 – us ATLA [100]
Katei bunko : katei collection of edo literature and dramtic works. in the holdings of the tokyo university general library, tokyo – 2032v on 72r – 1 – Y650,000 – (with 464p guide ed by the tokyo university general library) – ja Yushodo [480]
Katekhizis – Nesvizh: Matvei Kavechinskii, Simon Budnyi, Lavrentii Kryshkovskii, 1562 – 10mf – 9 – mf#RHB-21 – ne IDC [460]
Katekisema – Lusaka, Zambia. 1960 – 1r – us UF Libraries [960]
Katekisema – Lusaka, Zambia. 1962 – 1r – us UF Libraries [960]
Katekisima – s.l, s.l? n d – 1r – us UF Libraries [960]
Katekisima thukhu ya pfhundzo ya vakreste / Benedictine Monk Of Termonde (Belgium) – Termonde, Belgium. 1933 – 1r – us UF Libraries [960]
Katekisimo kana chibvunzo chine zwokuziviva na vakristiane / Moro – Mariannhill, South Africa. 1920 – 1r – us UF Libraries [960]
Katekisimo re matrimonio – Chishawasha?, Zimbabwe. 1932 – 1r – us UF Libraries [960]
Katekisimo ye zifundiso ze kirike katolike – Chishawasha, Zimbabwe. 1939 – 1r – us UF Libraries [960]
Katekisimo ye zifundiso ze kirike katolike – Chishawasha, Zimbabwe. 1951 – 1r – us UF Libraries [960]
Katekisimo ye zwifundiso zwe kirike katolike – Chishawasha, Zimbabwe. 1935 – 1r – us UF Libraries [960]
Katekisma kana tsamba ye rudzidziso rwe sangano katolike / Mayr, F – Mariannhill, South Africa. 1910 – 1r – us UF Libraries [960]
Katekisma nedudziro vedzidzo dze dutch reformed church – Morgenster, Zimbabwe. 1937 – 1r – us UF Libraries [960]
Katekisma ya rusangano rukatoliko – Mariannhill, South Africa. 1911 – 1r – us UF Libraries [960]
Katekismo kana chibvunzo chine zwokuziviva – Natal, South Africa. 1900 – 1r – us UF Libraries [960]
Katenlued / Lau, Fritz – 6. Aufl. Hamburg: M Glogau, 1921 – 1r – 1 – us Wisconsin U Libr [830]
Der kater : antifaschistisch-satyrische zeitung – Paris (F), 1933 3 jun – 1r – 1 – (only publ once) – gw Misc Inst [870]
Kateri / Knights of Columbus Club – Kahnawake QC. 1964 win-1984 win – 1r – 1 – mf#1571489 – us WHS [366]
Katerkamp, Theodor *see* Denkwuerdigkeiten aus dem leben der fuerstin amalia von gallitzin
Kath, Lydia *see*
– Der bauernkanzler
– Die schultzen-kathrin
The katha upanisad : an introductory study in the hindu doctrine of god and of human destiny / Rawson, Joseph Nadin – London: Oxford University Press; Calcutta: Association Press, 1934 – (= ser Samp: indian books) – us CRL [280]
The kathakosa : (or treasury of stories) / Tawney, C H – 1895 – (= ser Royal asiatic society translation fund. new series) – 1r – 1 – mf#96731 – uk Microform Academic [830]
Die katharen (mgh schriften:12.bd) / Borst, A – 1953 – (= ser Monumenta germaniae historica. schriften (mgh schriften)) – €17.00 – ne Slangenburg [243]
Katharina die zweite / Brueckner, Alexander – Berlin: G Grote 1883 – (= ser Allgemeine geschichte in einzeldarstellungen 3/10) – [ill/pl/facs] – 1 – mf#film mas c604 – us Harvard [947]
Katharina knie : ein seiltaenzerstueck in vier akten / Zuckmayer, Carl – Berlin: Im Propylaeen-Verlag, c1929 – 1r – 1 – us Wisconsin U Libr [790]
Katharina von bora, martin luthers frau : ein lebens- und charakterbild / Kroker, Ernst – Leipzig: E Haberland [1906?] – (= ser Biographien bedeutender frauen) – 1mf – 9 – 0-524-05954-3 – mf#1991-2354 – us ATLA [920]
The katharist book of perfection / Bessonet, George – Chicago, IL: Katharist Pub Society, c1917 – 1mf – 9 – 0-524-02012-4 – mf#1990-2787 – us ATLA [290]
Katharsis zwischen diesseits und jenseits : eine empirisch-reziproke, psycho-soziale studie dynamischer prozesse zwischen therapie, politik, kultur und religion / Weidkuhn, Wilmar – (mf ed 1993) – 2mf – 9 – €40.00 – 3-89349-782-X – mf#DHS 782 – gw Frankfurter [150]
Hekathemerine – Athena: s.n., 1956-apr 20 1967; sep 15 1974-jun 1987; mar-jun 1989 – us CRL [949]
Kathemerine – Athens. Greece. -d. 1946-55. (28 reels) – 1r – uk British Libr Newspaper [949]
Katherine mansfield in china – 1r – 1 – mf#ZB 06 – nz Nat Libr [920]
Katheterablation von tumornieren mittels ethanol / Roefke, Christian – (mf ed 1997) – 2mf – 9 – €40.00 – 3-8267-2496-8 – mf#DHS 2496 – gw Frankfurter [616]
Katholicismus, protestantismus und unglaube : ein aufruf an alle zur rueckkehr zu christenthum und kirche = Catholicity, protestantism and infidelity / Weninger, Francis Xavier – 4. Aufl. Mainz: Franz Kirchheim, 1864 – 1mf – 9 – 0-8370-8632-9 – (incl bibl ref) – mf#1986-2632 – us ATLA [240]
Katholicismus und protestantismus gegenueber der socialen frage / Uhlhorn, Gerhard – 2. unveraenderte Aufl. Goettingen: Vandenhoeck und Ruprecht, 1887 – 1mf – 9 – 0-7905-6208-1 – mf#1988-2208 – us ATLA [240]
Katholieke Missie in Nederlands Indie *see* Jaarboek
Katholieke Missie in Nederlands Oost-Indië *see* Jaarboek
Katholieke missien – Amsterdam: Society of the Divine Word, 1874-1967 [mf ed 2001] – (= ser Christianity's encounter with world religions, 1850-1950) – 23r – 1 – mf#2001-s504 – us ATLA [241]
Katholikon : tijdschrift voor beschaafde roomsch-katholijken – Breda, 1827-1830 – 38mf – 9 – €73.00 – ne Slangenburg [241]
Katholisch oder jesuitisch? : drei zeitgeschichtliche untersuchungen / Nippold, Friedrich – Leipzig: G Reichardt, 1888 – 1mf – 9 – 0-7905-6419-X – (incl bibl ref) – mf#1988-2419 – us ATLA [241]
Eine katholische antwort auf die paepstliche encyklika vom 5. februar / Michelis, Friedrich – Bonn: P Reusser 1875 [mf ed 1992] – 1mf – 9 – 0-524-03073-1 – mf#1990-4562 – us ATLA [241]
Eine katholische beleuchtung der augsburgischen konfession : polemische studie / Thieme, Karl – Leipzig: Duerr, 1898 – 1mf – 9 – 0-8370-8794-5 – (incl bibl ref) – mf#1986-2794 – us ATLA [241]
Das katholische deutsche kirchenlied in seinem singweisen = German catholic hymn melodies / Baeumker, Wilhelm – Freiburg. 4v. 1883-1911 – 11 – $85.00 set – (covers catholic hymns from earliest times to end of 19th c. based on mss and printed sources. the bibliography section has extensive descriptions of hymnals, and info on poets, composers and editors of hymn books.) – us Univ Music [780]
Das katholische deutsche kirchenlied unter dem einflusse gellerts und klopstocks / Schneiderwirth, Karl – [s.l: s.n.] 1907 [mf ed 1989] – 1r – 1 – (incl bibl ref. filmed with: gellerts lustspiele / wold haynel) – mf#7293 – us Wisconsin U Libr [241]
Katholische dogmatik / Klee, Heinrich – 3., unveraenderte Aufl. Mainz: Kirkheim, Schott und Thielmann, 1844-1845 – 4mf – 9 – 0-524-06022-3 – (incl bibl ref) – mf#1991-2382 – us ATLA [241]
Katholische dogmatik / Kuhn, Johannes von – Tuebingen: H. Laupp, 1846-1857 – 1r – 1 – 0-8370-1563-4 – mf#1984-B479 – us ATLA [241]
Die katholische kirche in armenien / Weber, S – Freiburg Brsg, 1903 – €19.00 – ne Slangenburg [241]
Die katholische kirche in den vereinigten staaten nordamerikas : in vier abschnitten / Hammer, Bonaventure – New York: C Wildermann, [1897?] – 1mf – 9 – 0-524-03614-4 – mf#1990-4774 – us ATLA [241]
Die katholische kirche in den vereinigten staaten von nordamerika / Shea, John Dawson Gilmary et al – Regensburg: GJ Manz, 1864 – 2mf – 9 – 0-524-02696-3 – mf#1990-4403 – us ATLA [241]
Die katholische kirche und ihr recht in den preussischen rheinlanden / Stutz, Ulrich – Bonn, 1915 (mf ed 1992) – 1mf – 9 – €24.00 – 3-89349-074-4 – mf#DHS-AR 48 – gw Frankfurter [241]
Das katholische kirchenrecht / Schulte, Johann Friedrich von – Giessen: Ferber, 1856-1860 – 4mf – 9 – 0-7905-8152-3 – (incl bibl ref) – mf#1988-6099 – us ATLA [241]
Katholische kirchen-zeitung – Duesseldorf, 1924-apr 18 1937 – 25r – 1 – (title varies: katholische kirchenzeitung fuer duesseldorf und umgebung. incl suppl: weltwarte in bild und wort (moenchengladbach) 1925-28, fr 1929: weltwarte (moenchengladbach) 1929-36) – gw Misc Inst [241]
Katholische kirchenzeitung fuer duesseldorf und umgebung *see* Katholische kirchen-zeitung
Der katholische modernismus / Schnitzer, Joseph – Berlin: Protestantischer Schriftenvertrieb, 1912 – (= ser Klassiker der Religion) – 1mf – 9 – 0-7905-6878-0 – (incl bibl ref) – mf#1988-2878 – us ATLA [241]
Katholische religionslehre fuer die vier obersten klassen der gelehrtenschulen und fuer gebildete maenner – 2. verb. Aufl. Regensburg; New York: Friedrich Pustet, 1890-91 – 4mf – 9 – 0-8370-8376-1 – mf#1986-2376 – us ATLA [378]
Katholische rundschau : fur die deutscher katholiken der sudlichen staaten – San Antonio, TX: Rundschau Pub Co. mar 27-aug 7 1918 – 1r – 1 – us CRL [241]
Die katholische sitte der alten kirche in ihrer geschichtlichen entwicklung / Bestmann, Hugo Johannes – Noerdlingen: CH Beck, 1885 – (= ser Geschichte der christlichen sitte) – 2mf – 9 – 0-7905-9238-X – (incl bibl ref) – mf#1989-2463 – us ATLA [170]
Katholische sonntagsschule – Tuebingen DE, 1848 – 1 – gw Misc Inst [241]
Katholische sozialpolitische korrespondenz – Moenchengladbach-Krefeld DE, 1891 15 jul-1896 jun, 1897 [gaps], 1898-1928 18 aug – 3r – 1 – (missing: jul-dec 1887) – mf#2661 – gw Mikropress [241]

Der katholische volksbote – Trier DE, 1848 25 aug-1850 29 jun – 1r – 1 – (title varies: 1 jun 1849: trier'scher volksbote) – gw Misc Inst [241]
Der katholische volkslehrer : eine periodische schrift fuer das unstudierte publikum / ed by Milbiller, J – Nuernberg [Salzburg] 1785 – (= ser Dz) – 1jg[=4st] on 2mf – 9 – €60.00 – mf#k/n5135 – gw Olms [241]
Katholische volkszeitung – Berlin DE, 1892 jan-mar, 1892 1 oct-1893 30 sep, 1894 3 jan-29 jun, 1895, 1896 1 may-1897 sep, 1898-1900 29 jun, 1901-1907 29 jun, 1908-09, 1911-1913 31 aug, 1914-18 – 1 – (bereichsausgabe der germania, berlin; title varies: 1 jan 1908: deutscher volksfreund) – gw Misc Inst [241]
Katholische volkszeitung : ein wochenblatt im interesse der kirche / ed by Kreuzer, Christoph – Baltimore MD: Kreuzer Bros. v1 may 8 1860-v54 oct 1914 [mf ed 1980] – 3r – 1 – (in german, several iss missing) – us Balch [071]
Katholische volkszeitung *see* Mosel-zeitung
Katholische weltanschauung und freie wissenschaft : ein populaerwissenschaftlicher vortrag unter beruecksichtigung des syllabus pius 10 und der enzyklika "pascendi dominici gregis" / Wahrmund, Ludwig – Muenchen: J F Lehmann, 1908 – 1mf – 9 – 0-8370-8719-8 – (incl bibl ref) – mf#1986-2719 – us ATLA [241]
Das katholische zeitungswesen in ostasien und ozeanien / Arens, Bernard – Aachen: Xaverius-Verlag, 1918 [mf ed 1995] – (= ser Yale coll; Abhandlungen aus missions-kunde und missionsgeschichte 5) – 59p – 1 – 0-524-10175-2 – (in german) – mf#1995-1175 – us ATLA [241]
Die katholischen briefe : textkritische untersuchungen und textherstellung – Leipzig: J C Hinrichs, 1892 [mf ed 1986] – (= ser Das neue testament 1/2; Texte und untersuchungen zur geschichte der altchristlichen literatur (tugal) 8/3) – 1mf – 9 – 0-8370-9590-5 – mf#1986-3590 – us ATLA [227]
Die katholischen briefe / Weiss, Bernhard – Leipzig, 1892 – (= ser Tugal 1-8/3) – 4mf – 9 – €11.00 – ne Slangenburg [227]
Die katholischen briefe / Windisch, Hans – Tuebingen: J C B Mohr, 1911 – (= ser Handbuch zum neuen testament) – 1mf – 9 – 0-7905-2038-9 – (includes bibliographies) – mf#1987-2038 – us ATLA [227]
Die katholischen missionen : zeitschrift des internationalen katholischen missionswerkes missio in verbindung mit dem priestermissionsbund – Freiburg, 1947 heft1; 1948 heft1; jahrg67 heft2(1948 dez); jahrg68-69 heft2(1949-50) [bimthly] – 1 – (cont publ with same title iss 1873-1938; cont by: km forum weltkirche; with ind) – ISSN: 0022-9407 – mf#film mas c441 – us Harvard [241]
Die katholischen missionen in indien, china und japan : ihre organisation und das portugiesische patronat vom 15. bis ins 18. jahrhundert / Jann, Adelhelm – Paderborn: F Schoeningh, 1915 – 2mf – 9 – 0-7905-4873-9 – (incl bibl ref) – mf#1988-0873 – us ATLA [241]
Katholischer beobachter / r – Koeln, Koblenz, Recklinghausen, Frankfurt/M DE, 1949 30 aug-1968 – 1 – (title varies: 5 apr 1952: echo der zeit, publ in recklinghausen; 27 sep 1968: publik, publ in frankfurt/m. filmed by misc inst: 1969 3 jan-1971 19 nov [6r]) – gw Mikrofilm; gw Misc Inst [241]
Katholischer bilderbogen – Schmallenberg (-Boedefeld) DE, 1953-54, 1956 7 oct-18 dec, 1958 5 jan-1972 2 apr, 1976, 1978-1992 19 jul, 1992 27 dec – 12r – 1 – (filmed by misc inst: 1957-1958 7 apr. title varies: 1955: neue bildpost; 1992: die neue bildpost) – gw Mikrofilm; gw Misc Inst [241]
Katholischer glaubensbote – Louisville, KY: Wm J Weber, Jr. jul 7 1921-nov 1 1923 – us CRL [241]
Katholischer Krankenhausverband Deutschlands e.V. *see* Krankendienst
Katholischer literaturkalender – v1-5 1891-97, v15 1926 [mf ed 1992] – 45mf – 9 – €290.00 – 3-89131-050-1 – (filmed with: keiters katholischer literaturkalender v6-14 1902-14) – gw Fischer [241]
Katholischer literaturkalender – Regensburg: Selbstverlag des Verfassers 1891- [v1-5] [mf ed 1978] – 1r – 1 – (no iss for 1895, 1896, 1898-1901, 1903, 1904, 1906, 1908, 1915-25; 3. jahrg 1893 publ as: ergaenzungsheft zum 2. jahrg; title varies: 1891-97: katholischer litteraturkalender; 1902-14: keiters katholischer literatur-kalender; 1926-katholischer literaturkalender) – mf#film mas c292 – us Harvard [430]
Katholisches eherecht : mit beruecksichtigung der im deutschen reich, in oesterreich, der schweiz und im gebiete des code civil geltenden staatlichen bestimmungen / Schnitzer, Joseph – Freiburg im Breisgau; St Louis, MO: Herder, 1898 – 2mf – 9 – 0-7905-8151-5 – (incl bibl ref) – mf#1988-6098 – us ATLA [241]
Katholisches wochenblatt – Chicago IL (USA), Omaha NE (USA), 1930 26 jun, 1930 28 aug-1937 7 jan [gaps] – 3r – 1 – gw Misc Inst [241]
Katholisches wochenblatt – Chicago: Rauker, Moninger und Handl. aug 29 1917-aug 1941 – 23r – 1 – us CRL [241]
Die katholisch-theologische fakultaet zu marburg : ein beitrag zur geschichte der katholischen kirche in kurhessen und nassau / Mirbt, Carl – Marburg: N G Elwert, 1905 [mf ed 1990] – (= ser Marburger akademische reden 9) – 1mf – 9 – 0-7905-6245-6 – (incl bibl ref) – mf#1988-2245 – us ATLA [378]
Der katholizismus und das zwanzigste jahrhundert im lichte der kirchlichen entwicklung der neuzeit / Ehrhard, Albert – Neunte bis zwoelfte verm. und verb. Aufl. Stuttgart: Jos. Roth, 1902 – 2mf – 9 – 0-8370-8809-7 – (incl ind) – mf#1986-2809 – us ATLA [241]
Katholizismus und reformation : kritisches referat ueber die wissenschaftlichen leistungen der neueren katholischen theologie auf dem gebiete der reformationsgeschichte / Koehler, Walther – Giessen: A. Toepelmann, 1905 – (= ser Vortraege der theologischen Konferenz zu Giessen) – 1mf – 9 – 0-7905-6198-0 – mf#1988-2198 – us ATLA [240]
Kathpress – Vienna, Austria jan 1951-dec 1978 [mf ed Norman Ross] – 28r – 1 – mf#nrp-1974 – us UMI ProQuest [074]
The kathryn recorder – Kathryn ND: Kathryn Publ Co. sep 1908; -v35 n22 dec 25 1942 (wkly) – 1 – (official paper of barnes co, 1913-1914. official paper village of kathryn, 1920-1942. missing: 1913 jul 10; 1916 dec 28; 1917 apr 19, may 3; 1926 oct 8; 1927 mar 25; 1932 Oct 7) – mf#11050-11051 – us North Dakota [071]
Kathryn weekly star – Kathryn, Barnes Co, ND: J K Dye. v1 n1 jun 13 1902-jul 31 1908?// (wkly) [mf ed with gaps] – (= ser Kathryn star) – 1 – (publisher's block sometimes reads: the kathryn star) – mf#11072 – us North Dakota [071]
Kathwood baptist church – Richland Co, SC – 1 – $41.58 – (deacon's minutes 1964-69; church conference minutes 1967-68, 1970-93) – mf#6818 – us Southern Baptist [242]
Kati kati advertiser – 1977-87 – 8r – 1 – mf#16.18 – nz Nat Libr [079]
[Katib i Rumi] *see* The travels and adventures of the turkish admiral sidi ali reis in india, afghanistan, central asia, and persia, during the years 1553-1556
Katikizima ye sangano re province of south africa – London, England. 1951 – 1r – us UF Libraries [960]
Katilina / Kuernberger, Ferdinand – Hamburg: Hoffmann und Campe, 1855 – 1 – us Wisconsin U Libr [820]
Katilina im drama der weltliteratur : ein beitrag zur vergleichenden stoffgeschichte des roemerdramas / Speck, Hermann Berthold Georg – Leipzig: Hesse, 1906 [mf ed 1992] – (= ser Breslauer beitraege zur literaturgeschichte 4) – 98p – 1 – mf#8014 reel 1 – us Wisconsin U Libr [410]
Katip celebi / Tahir, Mehmed – Dersaadet [Istanbul]: Kanaat Matbaasi, 1331 [1916] – (= ser Ottoman literature, writers and the arts) – 1mf – 9 – $25.00 – us MEDOC [470]
Katipunan national newspaper of the kdp / Union of Democratic Fillipinos [KDP] – v2 n3-v5 n20 [1975 apr-1978 nov], v8 n12, v9 n8-v14 n6 [1981 jul 1/15, 1982 aug-1987 jun] – 2r – 1 – mf#637089 – us WHS [325]
Katkov, M N *see*
– Sobranie peredovykh statei "moskovskikh vedomostei"
– Sobranie statei po poleskomu voprosu, pomeshchavshikhsia v "moskovskikh vedomostiakh, russkom vestnike" i "sovremennoi letopisi." sobranie peredovykh statei "moskovskikh vedomostei", 1863-1864
Kato, Nobutaka *see* Balance control in bipedal animals
Katolicke noviny – Nadas, Slovakia, 1896-1906 – 4r – 1 – us IHRC [077]
Katolicke noviny – Prague/Bratislava, Czechoslovakia. -w. 1953-1970 – 4 1/2r – 1 – uk British Libr Newspaper [072]
Katolicky sokol catholic falcon / Slovak Catholic Sokol [US] – 1977 sep 14-1981, 1982-1984 apr, 1984 may-1986 mar, 1986 apr-1988 jan 13 – 4r – 1 – mf#654980 – us WHS [241]
De katolijke : letterkundig tijdschrift voor godsdienst en wetenschappen – Deventer, 1822-1824 – 37mf – 9 – €71.00 – ne Slangenburg [241]
Katolik – Chicago, IL: Bohemian Benedictine Press, [sep 14 1917-1945]; 1946-dec 19 1975 – us CRL [241]
Katolski posol = Der katholische bote – Bautzen DE, 1863-1939 – 21r – 1 – gw Misc Inst [074]
Katoomba daily – Katoomba. dec 1920-may 1939 – 5r – 9 – A$345.88 vesicular A$373.38 silver – at Pascoe [079]
Katorga i ssylka – Moscow. no. 1-116. 1921-1935. Index for 1921-1930 – 1 – us NY Public [320]
Katre, Sumitra Mangesh *see* Formation of konkani
Katrineholmskuriren – Katrineholm, Sweden. 1916-78 – 297r – 1 – sw Kungliga [078]
Katrineholmskuriren – Katrineholm, Sweden. 1979- – 1 – sw Kungliga [078]
Katrineholmsposten – Katrineholm, Sweden. 1923-28 – 6r – 1 – sw Kungliga [078]
Kats, Vladimir *see* Narodnyi dokhod sssr i ego raspredelenie
Katsenelenbaum, Z S *see*
– Denezhnoe obrashchenie rossii 1914-1924
– Kommercheskie banki i ikh torgovo-komissionnye operatsii
Katsh, A I *see*
– The antonin genizah in the saltykov-shchedrin public library in leningrad
– Hebrew and judeo-arabic mss in the collections of the ussr
Kattenbusch, Ferdinand *see*
– Das apostolische symbol
– Die kirchen und sekten des christentums in der gegenwart
– Lehrbuch der vergleichenden confessionskunde. erster band, prolegomena und erster theil, die orthodoxe anatolische kirche
– Luthers stellung zu den oecumenischen symbolen
Kattengold : verteIln / Fehrs, Johann Hinrich – Garding: H Luehr & Dircks, 1926 – 1r – 1 – us Wisconsin U Libr [880]
Katter, E *see* Entomologische nachrichten
Katterfeld, Anna *see* Leuchtendes leben
Kattowitzer zeitung – Katowice, Poland. Oct 1927-Feb 1938 – 1r – 1 – us L of C Photodup [943]
Kattowitzer zeitung – Kattowitz (Katowice PL), 1923 5 mai-30 sep; 1925 sep-1932; 1935-43 – 53r – 1 – (title varies: oberschlesische zeitung, 1 sep 1942. filmed by other misc inst: 1933 (gaps) & 1935 (gaps) [6r]; 1920 1 jul-31 dec, 1925 1 jan-28 jun, 1933 2 jan-30 jun [4r]) – gw Misc Inst [077]
Katvan u bha gyam nhan su tui a mran / Htwei, Ba – Ran kun Mrui'. Man Ca pe 1975 [mf ed 1994] – on pt of 1r – 1 – mf#11052 r1727 n9 – us Cornell [959]
Katz, Albert *see*
– Biographische charakterbilder aus der judischen geschichte
– Israels feste und gedenktage
– Der wahre talmudjude
Katz, Benzion *see* Le-korot ha-yehudim be-rusya, polin ve-lita
Katz, Bill *see*
– The acquisitions librarian
– The reference librarian
Katz, Esther *see* The margaret sanger papers
Katzen, Leo *see* Gold and the south african economy
Katzenellenbogen, Moses Ben Eleazer Hayim *see* Ohel mosheh
Katzenelson, Gide'on *see* Ha-milhamah ha-sifrutit ben ha-haredim veha-maskilim
Katzer, Ernst *see* Luther und kant
Katzer, Friedrich Xavier *see* Der kampf der gegenwart
Katzew, Henry *see* Solution for south africa
Katzmarzyk, Peter T *see* A familial study of growth and health-related fitness among canadians of aboriginal and european ancestry
Katznelson, B *see* Revolutionary constructivism
Katznelson, Berl *see*
– Ba-mivhan
– Keta'im mi-devarav
Katz/prince collection : from the holdings of the schomburg center for research in black culture, manuscripts, archives and rare books division: the new york public library, astor, lenox and tilden foundations – 1995 – 3r – 1 – $255.00 – (guide which covers all coll under "antebellum america and slavery" sold separately for $20.00 d3305.g2) – Dist. us Scholarly Res – us L of C Photodup [470]
Kauczor, Daniel *see* Bergnubische sprache
Kauder, Hugo *see* Complete works
Kauener zeitung – Kauen (Kaunas, Kowno LT), 1944 3 jan-5 jul – 2r – 1 – (filmed by misc inst: 1942 10 jun-1944 31 mar (gaps)) – uk British Libr Newspaper; gw Misc Inst [077]
Kauf, miethe und verwandte vertraege in dem entwurfe eines buergerlichen gesetzbuches fuer das deutsche reich / Bernhoeft, Franz – Berlin: J Guttentag, 1889 – (= ser Civil law 3 coll; Beitraege zur erlaeuterung und beurtheilung des entwurfes eines buergerlichen gesetzbuches fuer das deutsche reich) – 1mf – 9 – (incl bibl ref) – mf#LLMC 96-595 – us LLMC [346]
Kauffman, Angelica *see* Record of pictures painted in italy, 1781-1798
Kauffman, Daniel *see* Mennonite church history
Kauffmann, Elfriede *see* Wem zeit ist wie ewigkeit
Kauffmann, Friedrich *see*
– Aus der schule des wulfila
– Balder, mythus und sage
– Deutsche mythologie
Kauffmann, Georg Friedrich *see* Harmonische seelen lust, musicalischer goenner und freunde
Kauffmann, Ignatz *see*
– Ignatz kauffmann, 1849-1913
Kaufhold, Anton *see* Spanien wie es gegenwaertig ist
Kaufmaennische hefte / ed by Sinapius, Joh Christian – Altona 1780-81 – (= ser Dz) – 3v on 9mf – 9 – €180.00 – mf#k/n2730 – gw Olms [380]
Kaufman, I I *see*
– Kredit, banki i denezhnoe obrashchenie
– Statistika russkikh bankov
Kaufman, I M *see* Russkie biograficheskie i bibliograficheskie slovari
Kaufman, P J *see* The gospel teacher
Kaufman, Wayne S *see* Validation of the maximal met prediction equations on the schwinn air-dyne
Der kaufmann – Biberach a.d. Riss DE, 1902 1jul-1904 5 nov – 1r – 1 – gw Misc Inst [380]
Kaufmann, Barbara E *see* The impact of "winning weighs" weight control program on perceived body image
Kaufmann, Carl Maria *see* Handbuch der christlichen archaeologie
Kaufmann, David *see*
– Geschichte der attributenlehre
– Die sinne
Kaufmann, Dorothee *see*
– Das aesthetische programm in goethes schriften zur meteorologie
– Einfluesse auf das fruehwerk jakob steinhardts
Kaufmann, Eva *see* Erwartung und angebot
Kaufmann, Frank *see* Arbeitslosigkeit zwischen lohn und effizienz
Kaufmann, Friedrich Wilhelm *see* German dramatists of the 19th century
Kaufmann, Guenter *see* Das kommende deutschland
Kaufmann, Hans *see*
– Krisen und wandlungen der deutschen literatur von wedekind bis feuchtwanger
– Versuch ueber das erbe
Kaufmann, Herman Ezechiel *see* Die anwendung des buches hiob in der rabbinischen agadah
Kaufmann, I I *see* Obzor proektov, vyshedshikh v 1861-78 godakh po voprosu o preobrazovanii kreditnoi denezhnoi sistemy rossii
Kaufmann, M *see* Christ
Kaufmann, Max *see*
– Heines charakter und die moderne seele
– Heines liebesleben
Kaufmann, Moritz *see*
– Christian socialism
– Sermons and lectures on the social duties of the clergy
– Social development under christian influence
– Socialism and communism in their practical application
– Socialism and modern thought
Kaufmann, Richard von *see* Die finanzen frankreichs
Kaufrecht : einschliesslich abzahlungsgeschaefte, agb-gesetz, eigentumsvorbehalt, factoring, finanzierte kaufvertraege, haustuergeschaefte, leasing, pool-vereinbarungen, produzentenhaftung, un-kaufrecht und verbraucherkreditgesetze / Reinicke, Dietrich & Tiedtke, Klaus – Neuwied, Kriftel, Berlin: Luchterhand, 1992 (mf ed 1996) – 6mf – 9 – €59.00 – 3-8267-9676-4 – mf#DHS 9676 – gw Frankfurter [346]
Kaufringer, Heinrich *see*
– Heinrich kaufringers gedichte
Kaukasien, nordkaukasien, aserbeidschan, armenien, georgien, geschichtlicher umriss / Nikuradse, Alexander – Muenchen: Hoheneichen verlag, c1942. 349p. illus. maps – 1 – us Wisconsin U Libr [947]
Die kaukasische militaerstrasse der kuban und die halbinsel taman : erinnerungen aus einer reise von tiflis und der krim / Koch, Karl H – Leipzig 1851 [mf ed Hildesheim 1995-98] – 1v on 2mf – 9 – €60.00 – 3-487-27663-1 – gw Olms [910]
Kaukasische post – Tiflis (Tbilissi GE), 1908-12, 1918-22 – 4r – 1 – (filmed by other misc inst: 1913 6 jan-29 dec) – gw Misc Inst [077]
Kaukasische reisen und studien / Hahn, C von – Leipzig, 1896 – 4mf – 9 – mf#AR-1593 – ne IDC [914]
Die kaukasischen laender und armenien in reiseschilderungen von curzon, k koch, macintosh, spencer und wilbraham / Koch, C – Leipzig, 1855 – 4mf – 9 – mf#AR-1604 – ne IDC [915]
Kaukasus, reisen und forschungen im kaukasischen hochgebirge / Dechy, M von – Berlin, 1905-1907. 3v – 28mf – 9 – mf#AR-1588 – ne IDC [914]
Kaukauna daily times – Kaukauna WI. 1887 jul 27-dec 24 – 1r – 1 – mf#1044403 – us WHS [071]
Kaukauna sun – Kaukauna WI. 1885 jul 16-1887 dec 31, 1888 jan 7-1890 jun 27 – 2r – 1 – (cont by: sun [kaukauna, wi]) – mf#935023 – us WHS [071]

Kaukauna sun – Kaukauna WI. [1895 mar 1/jun 7]-[1917 jun 14/1917 dec 6] [gaps] – 19r – 1 – (cont: sun [kaukauna, wi]; cont by: kaukauna times) – mf#935028 – us WHS [071]
Kaukauna times – Kaukauna WI. [1880 sep 16/1885 sep 4]-[1999 apr/jul 9] [gaps] – 143r – 1 – (cont: kaukauna sun; kaukauna sun; cont by: times [appleton, wi]) – mf#876683 – us WHS [071]
Kaukauna zeitung – Kaukauna WI. 1894 mar 23-1896 sep 11 – 1r – 1 – (cont by: appleton volksfreund) – mf#959839 – us WHS [071]
Kaulen, Franz *see*
- Assyrien und babylonien nach den neuesten entdeckungen
- Der biblische schoepfungsbericht (gen 1, 1 bis 2, 3)
- Einleitung in die heilige schrift, alten und neuen testaments
- Librum jonae prophetae
- Sprachliches handbuch zur biblischen vulgata
- Die sprachverwirrung zu babel

Kaumudi-mahotsava / ed by Sastri, Sakuntala Rao – Bombay: Bharatiya Vidya Bhavan, 1952 – (= ser Samp: indian books) – (trans by ed) – us CRL [954]
Kaunda, Kenneth David *see*
- Take up the challenge
- Zambia shall be free

Kaung Myin, Maung *see* Khet hon khet sac ca krann tuik khai, 1057-1973
Kaups, Richard *see*
- Hea sonum ja eesti baptisti kogudused
- Ilmutustraamatu seletus
- Viiskummend aastat apostlite radadel, 1884-1934

Das kausalprinzip in der philosophie des hl thomas von aquino / Schulemann, G – 1915 – (= ser Bgphma 13/5) – €7.00 – ne Slangenburg [120]
Die kausalsaetze im griechischen bis aristoteles / Nilsson, Martin P – Wuerzburg: A. Stuber, 1907 – (= ser Beitraege zur historischen syntax der griechischen sprache) – 1mf – 9 – 0-8370-1594-4 – (incl bibl ref) – mf#1987-6076 – us ATLA [450]
Kausambi in ancient literature / Law, Bimala Churn – Delhi: Manager of Publications, 1939 – (= ser Samp: indian books) – us CRL [490]
Kausch, Joh Jos *see* Schlesisches bardenopfer
Kausch, Johann Joseph *see* Psychologische abhandlung ueber den einfluss der toene und ins besondere der musik auf die seele
Kausler, E H von *see* Cancioneiro geral
Kausler, Eduard von *see*
- Briefwechsel zwischen christoph, herzog von wuerttemberg, und petrus paulus vergerius

Kautenberger, Peter G *see* Diary
Kautionsfreies kreis-wochenblatt fuer den kreis adenau und umgegend – Adenau DE, 1853 7 jan-1866 – 1 – (title varies: 1855: wochenblatt fuer den kreis adenau und umgegend; 1863: adenauer kreis- und wochenblatt) – gw Misc Inst [074]
Kau-Too, Sau *see* Thesaurus of karen knowledge
Kautschuk : roman aus der industrie / Dominik, Hans – Berlin: Scherl, 1942, c1930 – 1r – 1 – us Wisconsin U Libr [830]
Kautsky, Karl *see*
- Ethik und materialistische geschichtsauffassung
- Ob"edinenie srednei evropy
- T'u ti wen t'i

Kautz, Robert E *see* Comparing tort liability knowledge of future teacher coaches and current practicing teacher coaches
Kautzsch, E *see*
- Biblische theologie des alten testaments
- De veteris testamenti locis a paulo apostolo allegatis

Kautzsch, Emil *see*
- Die apokryphen und pseudepigraphen des alten testaments
- Die aramaismen im alten testament untersucht
- Bibelwissenschaft und religionsunterricht
- Die bleibende bedeutung des alten testaments
- The book of proverbs
- Gesenius' hebrew grammar
- Outline of the history of the literature of the old testament
- Die poesie und die poetischen buecher des alten testaments

Kautzsch, Emil et al *see* Die heilige schrift des alten testaments
Kautzsch, Emil F *see* Die apokryphen und pseudepigraphen des alten testaments
Kavanagh, Julia *see*
- A summer and winter in the two sicilies
- Women of christianity

Kavass, Igor I *see*
- Human rights, european politics, and the helsinki accord
- Human rights, the helsinki accords and the united states

The kaveri, the maukharis and the sangam age / Aravamuthan, T G – Madras: University of Madras, 1925 – (= ser Samp: indian books) – us CRL [954]
Kaverin, V *see* Ispolnenie zhelanii
Kavindabhi Saddhammadharadhaja *see* Bedatthadipani
Kavkaz – Tbilisi, Georgia 1846-1918 [mf ed Norman Ross] – 1 – mf#nrp-1796 – us UMI ProQuest [077]
Kavkazskaia zdravnitsa – Stavropol', Russia 1974-88 [mf ed Norman Ross] – 1r – 1 – mf#nrp-1705 – us UMI ProQuest [077]
Kavkazskii knizhnyi vestnik *see* Ezhemesiachnyi bibliograficheskii zhurnal
Kavkazskii rabochii : organ kavkazskogo kraevogo i tiflisskogo komiteta rsdrp(b) – Tbilisi, Georgia 1917 [mf ed Norman Ross] – 1r – 1 – mf#nrp-1797 – us UMI ProQuest [077]
Kavkazskii rabochii listok – Tbilisi, Georgia 1905 [mf ed Norman Ross] – 1 – mf#nrp-1798 – us UMI ProQuest [077]
Kavkazskii sbornik – Toronto. 1946+ (1) 1970+ (5) 1976+ (9) – 226mf – 9 – mf#1503 – ne IDC [077]
Kavkazskij voin : organ armejskogo komiteta kavkazskoj armii – Erzerum, Turkey 1917 [mf ed Norman Ross] – 1r – 1 – mf#nrp-492 – us UMI ProQuest [079]
Kavussanu, Maria et al *see* The effects of single versus multiple measures of biofeedback on basketball free throw shooting performance
Kawaguchi, Ukichi *see* The bearing of the evolutionary theory on the conception of god
Kawakami, Kiyoshi Karl *see* Japan in world politics
Kawakib al-sinima – Cairo: Najib Fakhr. n1-3. 22 sep-8 oct 1934 – (= ser Arabic journals and popular press) – 1r – 1 – $75.00 – us MEDOC [956]
Kawakibi, 'Abd Al-Rahman *see* Umm al-qura
Das kawallagebiet : oder ost-makedonien unter griechischer herrschaft und bulgarien / Suiuz na bulgarskite ucheni, pisateli i khudozhnitsi; ed by Verband der bulgarischen Gelehrten, Schriftsteller und Kuenstler – [s.l]: Verband der bulgarischen Gelehrten, Schriftsteller und Kuenstler [1941?] [mf ed 2000] – 1r – 1 – mf#29094 – us Harvard [949]
Kawamura, Takayuki *see* Characteristics of current and past participants in the university of wisconsin-la crosse cardiac rehabilitation program with a historical review of cardiac rehabilitation
Kawase, Kazuma [comp] *see*
- Edo bungaku sokan
- Monogatari bungaku sokan

Kawase, Kazuma et al *see* Edo bakufu kankobutsu shusei
[Kaweah-] kaweah commonwealth – CA. – 1r – 1 – $60.00 – mf#B02321 – us Library Micro [071]
Kawerak news / Bering Straits Native Corporation – v1 n1 [1982 aug] – 1r – 1 – (cont by: kawerak nipliksuk) – mf#1048027 – us WHS [071]
Kawerak nipliksuk / Bering Straits Native Corporation – v1 n3-v5 n1 [1982 oct/nov-1986 jan] – 1r – 1 – (cont: kawerak news) – mf#1048055 – us WHS [071]
Kawerau and eastern bay news gazette – 1976-11 mar 1980; 1985-jun 1986; oct 1986-87; 1989 – 20r – 1 – (previously known as: kawerau gazette. aka: eastern bay of plenty picture news and kawerau gazette) – mf#16.11 – nz Nat Libr [079]
Kawerau, G *see* Zwei aelteste katechismen der lutherischen reformation
Kawerau gazette *see* Kawerau and eastern bay news gazette
Kawerau, Gustav *see*
- De digamia episcoporum
- Hieronymus emser
- History of the christian church, a.d. 1517-1648
- Johann agricola von eisleben
- Luther in katholischer beleuchtung
- Luthers fruehentwicklung (bis 1517/9)
- Luthers schriften
- Paul gerhardt
- Die versuche, melanchthon zur katholischen kirche zurueckzufuehren
- Von der winkelmesse und pfaffenweihe

Kawerau, Siegfried *see* Stefan george und rainer maria rilke
Kawerau, Waldemar *see*
- Hans sachs und die reformation
- Hermann sudermann
- Die reformation und die ehe
- Thomas murner und die deutsche reformation
- Thomas murner und die kirche des mittelalters

Kawhia settler – 13 jan 1905-11 apr 1913 – 3r – 1 – mf#15.17 – nz Nat Libr [079]
Kawhia settler and raglan advertiser – 1914-33 – 7r – 1 – mf#15.52 – nz Nat Libr [079]
Kawhia settler and raglan advertiser – New Zealand, 19 Jul 1902-15 Jun 1904 (imperfect) – 1r – 1 – uk British Libr Newspaper [072]
Kawi-djarwa / Poerwadarminta, W J S – Djakarta: Bale Poestaka, 2603 (serie 1446) – 48p 1mf – 9 – mf#SE-2002 mf141 – ne IDC [490]
Kawkab ifriqiya : (kaoukeb ifrika) – Alger. 1908-juil 1914 – 1 – (puis al-kawkab al-gaza-iri) – fr ACRPP [956]
Kawkab-oul-maschrik = l'astre d'orient – Paris. n1-36. juin 1882-mai 1883 – 1 – (mq n33) – fr ACRPP [073]
Kay, George *see*
- Changing patterns of settlement and land use
- Social geography of zambia

Kay, Jason B *see* Predictors of job satisfaction among residential outdoor teachers
Kay, June *see* Wild eden
Kay, Stephen *see* Travels and researches in caffraria
Kay, William *see* Crisis hupfeldiana
Kayak – Santa Cruz CA 1972-84 – 1,5,9 – ISSN: 0022-9555 – mf#6753 – us UMI ProQuest [810]
Kayasare hinda – Bombay, India. 1944-65; 1967-Apr 1981 – 47r – 1 – (english and gujarati languages) – us L of C Photodup [079]
Kayasare hinda1 : the kaiser-i-hind – Bombay: [s.n, [1890-1899] (wkly) – 10r – 1 – us CRL [079]
Kaye, George Rusby *see*
- Astronomical instruments in the delhi museum
- The astronomical observatories of jai singh
- A guide to the old observatories at delhi, jaipur, ujjain, benares
- Hindu astronomy

Kaye, John William *see*
- Christianity in india
- Memorials of indian government

Kaygu – Trabzon, 1918-19. Mueduer-i Mes'ul: Ahmed Sabri. n1-17. 2 kanuniewel 1334-6. eyluel 1335 [1918-19] – (= ser O & t journals) – 3mf – 9 – $57.00 – us MEDOC [956]
Kaygulu *see* The divan project
Kayhan – Tehran. shumarah-'i 10799-11248 13 shahrivar 1358-11 farvardin 1360 [4 sep 1979-31 mar 1981]; shumarah-'i 11537-12117 7 farvardin 1362-28 isfand 1362 [27 mar 1983-18 mar 1984] – 27r – 1 – $1,431.00 – (many iss missing) – us MEDOC [079]
Kayin / Byron, George Gordon, 6th Baron – Warsaw, Poland. 1900 – 1r – us UF Libraries [939]
Kayira, Legson *see* Looming shadow
Kaynak, Erdener *see*
- Journal of east-west business
- Journal of euromarketing
- Journal of global marketing
- Journal of international consumer marketing
- Journal of international food and agribusiness marketing
- Journal of teaching in international business

Kayser, August *see* Die theologie des alten testaments
Kayser, Brigitte *see* Moeglichkeiten und grenzen individueller freiheit
Kayser, C *see* Das buch von der erkenntniss der wahrheit, oder, der ursache aller ursachen
Kayser, Christian Gottlob *see* Vollstaendiges buecher-lexicon 1750-1910
Kayser, G *see* Bibliographie d'ouvrages ayant trait...l'afrique en general dans ses rapports avec l'exploration, et la civilisation de ces contrees
Kayser, Gabriel *see* Bibliographie d'ouvrages ayant trait a l'afrique en general dans ses rapports avec l'exploration and la civilisation de ses contrees...
Kayser, Joh *see* Ueber den sogenannten barnabas-brief
Kayser, W et al *see* Deutsche literatur in unserer zeit
Kayseri – Kayseri. Mesul Mueduerue: Hamdi. n41. 8 mart 1926 – (= ser O & t journals) – 1mf – 9 – $25.00 – us MEDOC [956]
Kayserliche (reichs-) postzeitungen *see* Unvergreiffliche postzeitungen
Kayserlich-privilegirte hamburgische neue zeitung – Hamburg DE, 1793-1797 30 jun, 1797 1 oct-1798 29 dec, 1806-09 – 6r – 1 – (several title changes: 22 aug 1806: hamburgische neue zeitung; 1 jan 1824: hamburgische address-comtoir-nachrichten; 2 jan 1826: hamburgische neue zeitung; 29 nov 1831: neue zeitung; 2 jan 1838: hamburger neue zeitung. filmed by other misc inst: 1834-46 [gaps]; 1767-1811 [55r]) – gw Misc Inst [074]
Kayserling / Neumann, Ede – Budapest, Hungary. 1906 – 1r – us UF Libraries [939]
Kayserling, Meyer *see* Gedenkblatter
Kazach'e ekho : organ zabaik kazachestva / ed by Suslin, G N – Chita [Zabaik obl]: voisk[ovoe] prav[itel'stvo] Zab[aik] kaz[ach'ego] voiska 1920 [1919 26 [13] okt-] – (= ser Asn 1-3) – n38-95 [1920] [gaps] item 191, on reel n40 – 1 – mf#asn-1 191 – ne IDC [077]
Kazach'i dumy – Sofia, Bulgaria. Feb-Jun 1922 – 1r – 1 – us L of C Photodup [949]
Kazach'ia rech' : bespartiin demokrat gaz orenburg kazakov 3-go okr / ed by Starikov, I F – Troitsk [Orenburg gub]: [s n] 1918-19 [1918-] – (= ser Asn 1-3) – n28-43 [1918] n7[1919] [gaps] item 192, on reel n40 – 1 – mf#asn-1 192 – ne IDC [077]
Kazachii spas' : s veroi v boga – volia ili smert'! / Tsentral'noe kazach'e voisko – Moscow, Russia. n1[2][1996]-n5[10][1997], n7[12][1997]-n15[1998] – 1 – mf#mf-12248 [reel 5] – us CRL [077]
Kazak – Miass, Russia 1910-13 [mf ed Norman Ross] – 1 – mf#nrp-1013 – us UMI ProQuest [077]
Den' kazaka: ustraevaemyi sluzhashchimi sluzhby sborov omsk zh d v pol'zu mobilizovannykh kazakov i dobrovol'tsev pri uchastii zhurn "irtysh" : [odnodn gaz] / ed by Bazhenov, A – Omsk [Akmol obl]: [s n] 1919 – (= ser Asn 1-3) – 1v item 117, on reel n27 – 1 – mf#asn-1 117 – ne IDC [077]
[Kazakhstan-] kazakhstanskaya pravda – USSR. 1962-1980 – 20r – 1 – $1000.00 – mf#B63588 – us Library Micro [077]
Kazakov, A *see* Partiia sotsialistov-revoliutsionerov v tambovskom vosstanii 1920-1921 gg
Kazan Universitet. Russia *see*
- Protokoly zasiedanii obshchestva estestvoispytatelei pri imperatorskom kazanskom universitetie
- Uchenye zapiski kazanskogo gosudarstvennogo universiteta

Kazanskie gubernskie vedomosti – Kazan, Russia 1838-1917 [mf ed Norman Ross] – 71r – 1 – mf#nrp-665 – us UMI ProQuest [077]
Kazanskii, P E *see* Narodnost i gosudarstvo
Kazanskii telegraf – Kazan, Russia 1893-1911 [mf ed Norman Ross] – 1 – mf#nrp-675 – us UMI ProQuest [077]
Kazanskii vestnik – Kazan, 1821-1832. v1-36 – 121mf – 9 – (missing: 1830, v28 (jan-aug); 1831, v31 (p 1-174); v32(sep-oct); 1832, v34-35; v36(oct-nov)) – mf#R-18347 – ne IDC [077]
Kazantseva, M G *see* Narodovlastie [tiumen': 1918]
Kazantzakis, Nikos *see*
- Rock garden
- Zorba the greek

Kazbek – Vladikavkaz, Russia 1895-1906 [mf ed Norman Ross] – 25r – 1 – mf#nrp-2226 – us UMI ProQuest [077]
Kazetskii, N N *see* Ekaterinoslavskii vestnik
Kazikli voyvoda / Seyfi, Ali Riza – Istanbul: Resimli Ay Matbaasi, 1928 – (= ser Ottoman literature, writers and the arts) – 3mf – 9 – $55.00 – us MEDOC [470]
Kazim *see* The divan project
Kazimi, Avnullah [Mehmet Selim] *see* Divan-i oerfi ve avnullah kazimi
Al-kazirna – Paris. n1-6 7. dec 1923-mai juin 1924 – 1 – (bilingual) – fr ACRPP [073]
Kazitape / Chafulumira, E W – London, England. 1957 – 1r – us UF Libraries [960]
Kaz'min, A I *see* Finansovoe ozdorovlenie ekonomiki
Kaznelson, B *see*
- The next stage in palestine
- Reaction v progress in palestine

Kaznelson, Siegmund *see* Juedisches schicksal in deutschen gedichten
Kc blaetter : monatsschrift der im kartell-convent der verbindungen deutscher studenten juedischen glaubens vereinigten korporationen / ed by Hochschild, Ernst – Cologne, Berlin. v1-23. 1910-33 [complete] – (= ser German-jewish periodicals...1768-1945, pt 1) – 2r – 1 – $220.00 – mf#B261 – us UPA [270]
Kea, Ray A *see* Ashanti-danish relations
Keach, Benjamin *see*
- Breach repaired in god's worship
- Examination of mr isaac marlow's two papers
- Instructions for children
- Spiritual melody, containing three hundred sacred hymns
- Tropologia: a key to open scripture-metaphors

Keadilan. Madjallah Pemoeda Sosialis Indonesia *see* Markas tertingge pembelaaen "pesindo"
Keane, A H *see* Central and south america
Keane, Augustus Henry *see* Anthropological, philological, geographical, historical
Keane, John *see* Report on irrigation in ceylon
Keane, John J *see* Christian education in america
Kearney and Trecker Corporation *see* Kt today
Kearney and trecker news – 1951 jan-1960 dec, 1961 mar-1970 nov 12 – 2r – 1 – (cont: my buddy [1946]; cont by: news-letter [kearney & trecker corporation]) – mf#911897 – us WHS [071]
Kearney county bee – Minden, NE: W Olds, 1878-v4 n28. apr 21 1882 (wkly) – 1r – 1 – (absorbed: newark herald; cont by: kearney county gazette) – us Bell [071]
Kearney county bee *see* The newark herald
Kearney county democrat – Minden, NE: F C Brobst. -10th yr n21. jun 30 1893 (wkly) – 3r – 1 – (merged with: workman (minden, ne) to form: minden courier) – us Bell [071]
Kearney county gazette – Minden, NE: Williams & Hardman. v4 n29. apr 28 1882-v13 n25. mar 12 1891 (wkly) – 10v on 4r – 1 – (cont: kearney county bee; merged with: minden register to form: minden gazette (1891)) – us Bell [071]

Kearney county news – Minden, NE: Donald Jacobson. 20v. v1 n5. mar 2 1960-v20 n35. oct 1 1980 (wkly) [mf ed mar 2 1960-oct 1 1980 (gaps) filmed 1975-80] – 14r – 1 – (absorbed by: minden courier) – us NE Hist [071]

Kearney county news – Minden, NE: George E Thornton. 3v. v1 n1. nov 1 1901-v3 n18. feb 26 1904 (wkly) – 2r – 1 – (cont: prairie home (heartwell ne); merged with: new gazette (minden ne) to form: kearney county news and the minden new gazette) – us Bell [071]

The kearney county news – Minden, NE: Donald Jacobson. 20v. 1960-v20 n35. oct 1 1980 (wkly) [mf ed v1 n5. mar 2 1960-oct 1 1980 (gaps) filmed 1975-80] – 14r – 1 – us NE Hist [071]

Kearney county news and the minden new gazette – Minden, NE: C F Fordyce. 3v. v3 n19. mar 4 1904-v5 n20. mar 2 1906 (wkly) – 2r – 1 – (formed by the union of: kearney county news and: new gazette (minden, ne); cont by: news-gazette (minden, ne); cont numbering of kearney county news) – us Bell [071]

Kearney daily hub – Kearney, NE: Hub Publ Co, 1888-102nd yr 95 iss. feb 24 1990 (daily ex sun & holidays) [mf ed jun 1 1889-dec 31 1957 (gaps)] – 138r – 1 – (absorbed: kearney morning times, kearney ne, daily news. cont by: kearney hub. weekly ed: kearney weekly hub and central nebraska press, -1892. semi-weekly ed: semi-weekly kearney hub and central nebraska press 1892-1917. weekly ed: kearney weekly hub 1917-) – us NE Hist [071]

Kearney daily news – Kearney, NE: News Co-Operative Co, 1893 (daily) [mf ed 1st yr n5. sep 1-2 1893] – 1r – 1 – (cont by: kearney democrat) – us NE Hist [071]

Kearney daily tribune – Kearney, NE: Vance Beghtol, -v41 n114. may 27 1934 (daily ex mon) [mf ed v41 n1. jan 18-may 27 1934] – 1r – 1 – (cont: kearney weekly tribune (1933). cont by: platte valley sunday tribune) – us NE Hist [071]

Kearney democrat – Kearney, NE: F L Whedon. -v33 n30. jul 29 1926 (wkly) [mf ed v3 n31 aug 13 1896-jul 29 1926 (gaps)] – 9r – 1 – (cont: kearney daily news. cont by: kearney weekly tribune) – us NE Hist [071]

The kearney gait – Kearney, NE: H H Martin. v1 n1. mar 15 1891- (mthly) [mf ed -jun 15 1891 (gaps) filmed [1973]] – 1r – 1 – us NE Hist [071]

Kearney hub – Kearney, NE: Kearney Hub Publishing. 102nd yr 96th iss. feb 26 1990- (daily ex sun & holidays) – 1 – us Bell [071]

Kearney, Kristi D *see* Comparison of bone mineral density in 10- to 13-year-old female gymnasts and swimmers

Kearney morning times – Kearney, NE: Standard Print. 9v. v1 n1. jun 15 1909-v9 n114. oct 31 1917 (daily ex mon) – 10r – 1 – (absorbed by: kearney daily hub) – us NE Hist [071]

Kearney, nebraska, daily news – Kearney, NE: Henry G Kroger, 1939-v48 n191. dec 27 1941 (daily ex sun & mon) [mf ed v46 n20. apr 4 1939-dec 27 1941] – 4r – 1 – (cont: platte valley tribune. absorbed by: kearney daily hub) – us NE Hist [071]

Kearney & Trecker Corporation *see*
- K and T today
- My buddy
- News-letter

Kearney tribune – Kearney, NE: Vance Beghtol. v40 n30. jul 28 1932-v41 n21. jan 27 1933 (wkly) [mf ed with gaps)] – 2r – 1 – (cont: kearney weekly tribune. cont by: kearney weekly tribune (1933). publ as: kearney daily tribune oct-dec 1932) – us NE Hist [071]

Kearney weekly hub – Kearney, NE: [Kearney Hub] 7v. 46th yr 3rd wk. nov 8 1917-52nd yr 35th wk. feb 28 1924 (wkly) [mf ed with gaps filmed 2000] – 2r – 1 – (cont: kearney semi-weekly hub) – us NE Hist [071]

Kearney weekly hub and central nebraska press – Kearney, NE: [s.n.] -v19 n27. mar 31 1892 (wkly) [mf ed nov 7 1889-mar 31 1892 (gaps)] – 1r – 1 – (cont: central nebraska press daily. cont by: semi-weekly kearney hub and central nebraska press) – us NE Hist [071]

Kearney weekly nonpareil – Kearney, NE: Julian Bros (wkly) [mf ed 1879,1881 (gaps)] – 2r – 1 – (cont by: kearney nonpareil) – us NE Hist [071]

Kearney weekly times – Kearney, NE: Standard Print. 4v. v1 n1. jun 17 1909-v4 n14. jan 20 1913 (wkly) [mf ed with gaps] – 2r – 1 – (cont: new era-standard. v1 n1- also called v29 n49-) – us NE Hist [071]

Kearney weekly tribune – Kearney, NE: C N Harris. 8v. v33 n31. aug 5 1926-v40 n29. jul 21 1932 (wkly) [mf ed with gaps)] – 2r – 1 – (cont: kearney democrat. cont by: kearney tribune) – us NE Hist [071]

Kearney weekly tribune – Kearney, NE: Vance Beghtol. v41 n22. feb 3 1933- (wkly) [mf ed -dec 8 1933] – 1r – 1 – (cont: kearney tribune. cont by: kearney daily tribuen) – us NE Hist [071]

Kearsley, C *see* Kearsleys' stranger's guide

Kearsleys' stranger's guide : or companion through london and westminster, and the country round; containing a description of the situation, antiquity, and curiosities of every place, within the circuit of fourteen miles – London [1791] [mf ed Hildesheim 1995-98] – 2mf – 9 – €60.00 – 3-487-27964-9 – gw Olms [914]

Keary, Charles Francis *see*
- The mythology of the eddas
- Outlines of primitive belief among the indo-european races
- The vikings in western christendom, a.d. 789 to 888
- The vikings in western christendom, a.d. 789 to a.d. 888

Keary, W H *see* Report on all the holkham estates, 1851

Keary, Wiliam *see* Continuation of the ampleforth discussion

Keasbey, Edward Quinton *see* The law of electric wires in streets and highways

Keate, George *see*
- An account of the pelew islands
- Relation des iles pelew, situees dans la partie occidentale de l'ocean pacifique

Keate, Henry *see* Guide to marine insurance

Keating, Edward Henry *see* The shubenacadie canal

Keating, John Fitzstephen *see* The agape and the eucharist in the early church

Keating, W H *see* Narrative of an expedition to the source of st peter's river, lake winnepeek, lake of the woods, etc performed in the year 1823

Keating, William *see* Narrative of an expedition to the source of st peter's river, lake winnipeek, lake of the woods, etc

Keatinge, Maurice Walter *see* The great didactic of john amos comenius

Keats / Murry, John Middleton – (Studies in Keats). 4th ed. rev. and enl. London: Cape, (1955). 322p. illus. First ed. publ. in 1930 with title: Studies in Keats – 1 – us Wisconsin U Libr [420]

Keats, leigh hunt & shelley : manuscripts from the additional, ashley, egerton & zweig collections at the british library – 20r – 1 – £1850.00 – (with d/g) – uk Matthew [420]

Keats's great odes : a reconsideration of the five odes which redeem romanticism / Megally, S – Np, 1974 – 2mf – 9 – mf#NE-386 – ne IDC [956]

Keay, Frank Ernest *see* Kabir and his followers

Keay, John Seymour *see* The landlord, the tenant and the taxpayer

Kebajoran baru, kesatuan buruh kerakjatan indonesia : arena berkala – Djakarta, 1957 v1(1) – 1mf – 9 – mf#SE-338 – ne IDC [959]

Kebangoenan : madjallah resmi kaboepaten karo – Kabandjahe, 1946-1947 – 5mf – 9 – (missing: 1946 v1(1-4, 7)) – mf#SE-666 – ne IDC [950]

Kebangsaan poestaka rakjat / Pembaroean. Madjalah boelanan kaoem moeda jang berhaloean merdeka – Djakarta, 1945-1946 – 4mf – 9 – mf#SE-946 – ne IDC [959]

Kebec / Dufresne, Guy – [s.l.]: [s.n.], 1958 [mf ed 1974] – 1r – 1 – mf#SEM35P109 – cn Bibl Nat [890]

Keble, John *see*
- Case of catholic subscription to the thirty-nine articles considered
- Christian year
- Duty of hoping against hope
- Heads of consideration on the case of mr ward
- Letters of spiritual counsel and guidance
- Outlines of instructions or meditations for the church's seasons
- Pastoral letter to the parishioners of hursley
- Postscript to the third edition of the sermon
- Putting on christ
- Sermons, academical and occasional
- Sermons for advent to christmas eve
- Sermons for ascension day to trinity sunday
- Sermons for christmas and epiphany
- Sermons for lent to passiontide
- Sermons for septuagesima to ash-wednesday
- Sermons for the holy week
- Sermons for the saints' days and other festivals
- Sermons for the sundays after trinity, pt 1
- Sermons for the sundays after trinity, pt 2
- Sermons, occasional and parochial
- Sermons preached on various occasions
- Studia sacra
- Village sermons on the baptismal service

Keboedajaan timoer – Djakarta: Keimin Bunka Shidosho, 2603-2605. 3v – 7mf – 9 – (no more publ) – mf#SE-2002 mf330-336 – ne IDC [959]

Keck, Heinrich *see* Ueber das wesen der bildung und den anteil des gymnasiums an demselben

Keck, Karl Heinrich *see* Die gudrunsage

Keck, Mary L B *see* An analysis of the current judging methods used in competitive ballroom dancing as well as comparisons to competitive pairs figure skating and ice dancing

Keck, Stephan *see* Ueber den dual bei den griechischen rednern

Keckermann, B. *see* Opera omnia quae extant.

Keckermann, S B *see* Systema s s theologiae

Keddie, Henrietta [pseud: Sarah Tytler] *see* Modern painters and their paintings

Keding, Karl *see* Feldgeistlicher bei legion condor

Kedney, John Steinfort *see*
- Christian doctrine harmonized and its rationality vindicated
- Hegel's aesthetics
- Mens christi, and other problems in theology and christian ethics
- Problems in ethics

Kedushat levi 'al ha-torah / Levi Isaac Ben Meir – Lublin, Poland. 1927 – 1r – us UF Libraries [939]

Kedutaan besar bulletin switzerland : indonesia – Berne, [1952]-1957. v1-6(4) – 9mf – 9 – (missing: [1952]-1954, v1-3; 1955, v4(1, 4-6); 1956, v5(1-4); 1957, v6(1)) – mf#SE-1669 – ne IDC [959]

Kedutaan besar information bulletin : indonesia – Singapore, 1970-1971 – 2mf – 9 – (missing: 1970(1-7); 1971(11-15)) – mf#SE-1668 – ne IDC [959]

Kedutaan besar republik indonesia / Berita Indonesia – Kuala Lumpur, 1958-1963 – 52mf – 9 – (missing: 1959, v2(5); 1960, v3(1, 22-24); 1961, v4(24); 1962, v5(21-24); 1963, v6(3-24)) – mf#SE-524 – ne IDC [959]

Keeble, Samuel Edward *see* The social teaching of the bible

Keedy, John Lincoln *see* Teachers' book of old testament heroes

Keefer, Samuel *see*
- The cornwall canal
- Report of samuel keefer, esq
- Report on baie verte canal

Keefer, Thomas C *see*
- The canadian pacific railway
- Philosophie des chemins de fer

Keegan, John *see* Six armees en normandie

Keegstra, H *see* Kerkelijk handboek ten dienste der chr ger kerk in noord-amerika

Keel block – 1919 feb-1948 oct – 1r – 1 – mf#1112815 – us WHS [071]

Keele, Alan F *see* Konkordanz zu walter kempowskis "deutscher chronik"

Keele, William Conway *see*
- A brief view of the laws of upper canada up to the present time
- The provincial justice

Keeler, Bronson C *see* A short history of the bible

Keeler, L W *see* Ex summa philippi cancellarii questiones de anima

Keeley, Leslie E *see* Opium, its use, abuse and cure

Keeling, Derek *see* Foundations of photography

Keen, Greenbery *see* War diary while at fort meigs, 1812

Keen, James T *see* A manual for notaries public, justices of the peace, and their employers in massachusetts

Keen, Samuel Ashton *see*
- Faith papers
- Pentecostal papers

Keenagh, Peter *see* Mosquito coast

Keenan, Karen A *see* The effect of heat and cold on ankle stability

Keenan, Stephen *see* Catechism of the christian religion

Keene, Charles Samuel *see* Our people...from the collection of "mr punch"

Keene, Derek *see* Historical gazetteer of london before the great fire

Keene, New Hampshire. Keene Baptist Church *see* Manuscript histories

Keener, John *see*
- Public land laws of the united states

Keener, William Albert *see* A selection of cases on the law of private corporations

Keep strong – 1976 feb-1978 oct, 1979-1980 oct – 2r – 1 – mf#583612 – us WHS [071]

Keep your child healthy = Hou u kind gesond / South Africa. Department of Health [Departement van Gesondheid] [Departement van Gesondheid] – Pretoria: Dept of Health [1980] [mf ed Pretoria, RSA: State Library [199-]] – 59p [ill] on 1r with other items – 5 – mf#op 07357 r25 – us CRL [618]

Keeping a line on 1469 / Retail Clerks International Association – v1 n1-v2 n15 [1960 apr 15-1961 jun] – 1r – 1 – (cont by: keeping score with local 444 [1961]) – mf#563834 – us WHS [338]

Keeping in touch – 1981 mar 9-1984 oct 21, 1982 jan 11, 1985 feb 8-1988 may 27 – 2r – 1 – mf#921814 – us WHS [071]

Keeping posted with ncsy – New York, NY. 1962-83. National Conference of Synagogue Youth – 1 – us AJPC [071]

Keeping score with local 444 / Retail Clerks International Union – v2 n16-51 [1961 jul-1965 jan] – 1r – 1 – (cont: keeping a line on 1469; cont by: keeping score with retail store employees local 444) – mf#563833 – us WHS [331]

Keeping score with local 444 / United Food and Commercial Workers International Union – v3 n21-v14 n69 [1970 feb/mar/apr-1979 win] – 1r – 1 – (cont: keeping score with retail store employees union local 444) – mf#501467 – us WHS [331]

Keeping score with retail store employees union local 444 / Retail Clerks International Union – v6 n52-65 [1965 feb-1966 may], ns: v1 n1- v2 n20 [1966 jun/jul-1969 nov/dec/1970 jan] – 1r – 1 – (cont: keeping score with local 444 [1961]; cont by: keeping score with local 444 [1970]) – mf#562559 – us WHS [331]

Keeping the record straight – Glendale AZ 1959-73 – 1 – ISSN: 0022-9652 – mf#2243 – us UMI ProQuest [320]

Keeping track – Montreal, Canada 1958-74 – 1 – ISSN: 0453-4441 – mf#5558 – us UMI ProQuest [380]

Keepsake – 1828-57 – (= ser English gift books and literary annuals, 1823-1857) – 115mf – 9 – uk Chadwyck [800]

Keer, Dhananjay *see* Savarkar and his times

Keerl, Philipp Friedrich *see*
- Die apokryphen des alten testaments
- Die einheit der biblischen urgeschichte (1 mos 1-3)
- Der gottmensch, das ebenbild des unsichtbaren gottes
- Die schoepfungsgeschichte und die lehre vom paradies
- Das wort gottes und die apokryphen des alten testaments

Kees, Hermann *see* Der opfertanz des aegyptischen koenigs

Keesing's archiv der gegenwart – Essen, Wien (A), Bonn DE, 1931 jul-1993 – 58r – 1 – (title varies: 1946: archiv der gegenwart; fr 1946 publ in bonn-bad godesberg) – mf#3272 – gw Mikropress [934]

Keesing's contemporary archives : records of world events – 379mf – 9 – (pt 1: 1931-75 334mf c39-27961. pt 2: 1976-80 45mf c39-27962. coll covers all important devts in interational and national politics and economics. with detailed indexes) – mf#C39-27960 – us Primary [900]

Keesler news – United States. 1981 may [v43 n17]-1982 jun, 1982 jun 25-1983 apr, 1983 may-1984 jul, 1984 aug-1985 may, 1985 jun-1986 apr, 1986 may 2-dec 19, 1987 jan 9-jul 24, 1987 jul 31-1988 mar 4, 1988 apr-oct, 1988 nov-1989 may – 10r – 1 – mf#646421 – us WHS [071]

Keetch, Anita *see* Effects of adhesive spray and prewrap on taped ankle inversion before and after exercise

Keetmanshooper nachrichten – Keetmanshoop, South West Africa. 1911 – 1r – 1 – sa National [079]

Keetmanshooper zeitung – Keetmanshoop (NAM), 1913 24 apr-1914 23 jul [gaps] – 1 – gw Misc Inst [079]

Keetmanshooper zeitung – Keetmanshoop, South West Africa. 1912-14 – 1r – 1 – sa National [079]

Keezer, Frank H *see* The law of marriage and divorce

Kefar yehoshu'a... – Tel-Aviv, Israel. 1942 – 1r – us UF Libraries [939]

Kegel, Martin *see* Wilhelm vatke und die graf-wellhausensche hypothese

Kegel, Max *see* Press-prozesse

Kehillah – Jacksonville, FL. 1986 dec-1988 jul – 1r – us UF Libraries [071]

Kehler grenzbote – Kehl DE, 1978 6 oct- – ca 9r/yr – 1 – (bezirksausgabe von offenburger tageblatt; filmed by other misc inst: 1942 2 sep-1943. title varies: 1871: kehler wochenblatt; 1898: kehler zeitung; later as regional ed of: offenburger tageblatt) – gw Misc Inst [074]

Kehler grenzbote – Kehl DE, 1978 6 oct- – ca 9r/yr – 1 – (bezirksausgabe von offenburger tageblatts; title varies: 1871: kehler wochenblatt; 1898: kehler zeitung; later as regional ed of: offenburger tageblatt) – gw Misc Inst [074]

Kehler wochenblatt *see*
- Kehler grenzbote

Kehoe, Lawrence *see* Complete works of the most rev john hughes, d d, archbishop of new york

Kehr, C *see* Der christliche religions-unterricht

Kehr, Paul Fridolin *see*
- Italia pontificia

Kehrein, Josef *see* Biographisch-literarisches lexikon der katholischen deutschen dichter, volks-und jungendschriftsteller im 19. jahrhundert

Kehrli, Jakob Otto *see* Die lithographien zu goethes "faust" von eugene delacroix

Kehukee Baptist Association. North Carolina *see* Materials

Kei chan tou che / T'ien, Chien – Kuei-lin: Nan t'ien ch'u pan she, Min kuo 32 [1943] – (= ser P-k&k period) – us CRL [810]

Kei chiu wang t'ung chih ti kung k'ai hsin / Ch'ien, Chun-jui – Han-k'ou: Sheng huo shu tien, Min kuo 27 [1938] – (= ser P-k&k period) – us CRL [951]

KEICHER

Keicher, O *see* Raymundus lullus und seine stellung zur arabischen philosophie
Keidel, Heinrich *see* Die dramatischen versuche des jungen grillparzer
Keighley labour journal *see* Independent labour party newspapers
Keighley labour journal, 1894-1902 : from keighley public library – 2r – 1 – mf#97017 – uk Microform Academic [072]
K-eight – Philadelphia PA 1971-74 – 1,5 – mf#6417 – us UMI ProQuest [370]
Keigwin, Charles Albert *see* Cases in common law actions.
Keil, Carl Friedrich *see*
– The book of the prophet daniel
– The books of ezra, nehemiah, and esther
– Books of ezra, nehemiah, and esther
– The books of the chronicles
– Commentar ueber den brief an die hebraeer
– Commentar ueber die briefe des petrus und judas
– Commentar ueber die buecher der makkaber
– Commentary on the book of joshua
– Commentary on the books of kings
– Joshua, judges, ruth
– Manual of biblical archaeology
– The pentateuch
– The prophecies of jeremiah
– Der tempel salomo's
– The twelve minor prophets
Keil, H *see* Grammatici latini
Keil, J *see* Bericht ueber drei reisen in lydien und der suedlichen aiolis
Keil, Karl August Gottlieb *see* Systematisches verzeichnis derjenigen theo. schrift
Keil, Richard *see* Goethe, weimar und jena im jahre 1806
Keil, Robert *see*
– Aus klassischer zeit
– Goethe, weimar und jena im jahre 1806
Die keilinschriften am eingange der quellgrotte des sebeneh-su / Schrader, Eberhard – Berlin: Koenigl Akademie der Wissenschaften, 1885 – 1mf – 9 – 0-8370-8617-5 – (discussion in german; texts in akkadian and german) – mf#1986-2617 – us ATLA [470]
Die keilinschriften der achaemeniden / ed by Weissbach, Franz Heinrich – Leipzig: J C Hinrichs, 1911 – 1mf – 9 – 0-8370-8319-2 – (incl bibl ref) – mf#1986-2319 – us ATLA [470]
Keilinschriften und das alte testament *see* The cuneiform inscriptions and the old testament
Die keilinschriften und das alte testament von e schrader : mit ausdehnung auf die apokryphen, pseudepigraphen und das neue testament / Winckler, H – Berlin, 1903 – 8mf – 9 – mf#NE-429 – ne IDC [956]
Keilinschriftliche Bibliothek *see* Sammlung von assyrischen und babylonischen texten
Keilinschriftliche bibliothek. band 1-3 : sammlung von assyrischen und babylonischen texten in umschrift und uebersetzung / ed by Schrader, Eberhard – Berlin: H. Reuther, 1889-1892 – 3mf – 9 – 0-7905-3478-9 – (incl bibl ref) – mf#1987-3478 – us ATLA [470]
Der keilinschriftliche sintfluthbericht : eine episode des babylonischen nimrodepos / Haupt, Paul – Leipzig: J C Hinrichs, 1881 – 1mf – 9 – 0-8370-7066-X – mf#1986-1066 – us ATLA [930]
Keilinschriftliches textbuch zum alten testament / Winckler, Hugo – 2. neu bearb aufl. Leipzig: J C Hinrichs, 1903 – 1mf – 9 – 0-8370-7276-X – (akkadian texts and german translations. incl bibl ref) – mf#1986-1276 – us ATLA [221]
Keilmann, Wilhelm *see* Palla toa
Keilschriftliche acten-stuecke aus babylonischen staedten : von steinen und tafeln des berliner museums in autographie, transscription und uebersetzung / Peiser, Felix Ernst – Berlin: Wolf Peiser, 1889 – 1mf – 9 – 0-8370-7320-0 – (discussion in german. texts in german and akkadian) – mf#1986-1320 – us ATLA [470]
Keilschrifttexte : zum gebrauch bei vorlesungen / ed by Abel, Ludwig – Berlin: W Spemann, 1890 – 1mf – 9 – 0-8370-8560-8 – (in akkadian and german) – mf#1986-2560 – us ATLA [470]
Die keilschrifttexte asurbanipals, koenigs von assyrien : (668-626 v. chr.) / ed by Smith, Samuel Alden – Leipzig: Eduard Pfeiffer. 3v. 1887-89 – 3mf – 9 – 0-8370-7680-3 – (texts in german and akkadian; discussion in german) – mf#1986-1680 – us ATLA [470]
Die keilschrifttexte sargons / Sargon 2, King of Assyria; ed by Winckler, Hugo – Leipzig. Eduard Pfeiffer. 2v. 1889 – 3mf – 9 – 0-8370-7825-3 – (texts in akkadian and german; commentary in german) – mf#1986-1825 – us ATLA [470]
Keilschrifturkunden aus boghazkoey / Ehelolf, H – Berlin, 1938-1944 – (= ser Staatliche Museen zu Berlin. Vorderasiatische Abt) – 8mf – 9 – (staatliche museen zu berlin. vorderasiatische abt v29-34) – mf#NE-416 – ne IDC [956]
Keim, Randolph De *see* San domingo
Keim, Theodor *see* Der geschichtliche christus
Keio Gijuku Fukuzawa Memorial Center [comp] *see* Fukuzawa kankei monjo
Keiper, Wilhelm *see* Friedrich leopold stolbergs jugendpoesie
Keiser, C E *see* Selected temple documents of the ur dynasty
Der keiser und der kunige buoch oder die sogenannte kaiserchronik : gedicht des 12. jahrhunderts von 18,578 reimzeilen / ed by Massmann, Hans Ferd – Quedlinburg, Leipzig: G Basse 1849-54 [mf ed 1993] – (= ser Bibliothek der gesammten deutschen national-literatur [von der aeltesten bis auf die neuere zeit]) – 2v on 10r – 1 – mf#3394p – us Wisconsin U Libr [810]
Keita, Abdoulaye *see* Approche ethnolinguistique de la tradition orale wolof
Keiter, Heinrich *see*
– Heinrich heine
– Ida graefin hahn-hahn
– Joseph von eichendorff
– Konfessionelle brunnenvergiftung
Keiters katholischer literaturkalender *see* Katholischer literaturkalender
Keith, A Berridale *see* Speeches and documents on indian policy, 1750-1921
Keith, Alexander *see*
– Demonstration of the truth of the christian religion
– Evidence of the truth of the christian religion
– Evidence of the truth of the christian religion derived from the li...
Keith, Arthur Berriedale *see*
– Buddhist philosophy in india and ceylon
– A history of sanskrit literature
– Imperial unity and the dominions
– Indian logic and atomism
– The religion and philosophy of the veda and upanishads
– The samkhya system
– Vedic index of names and subjects
Keith county news – Ogallala, NE: G F Copper. dec 2 1946 (semiwkly) [mf ed v13 n10. jan 1 1897 (gaps)] – 1 – (cont: keith county news and republican. absorbed: republican argus feb 3 1905, ogallala tribune dec 19 1918, brule citizen apr 10 1941 and: enterprise aug 3 1962. cont by: big springs enterprise (1963)) – us NE Hist [071]
Keith county news and republican – Ogallala, NE: G F Copper, -dec 1896// (wkly) [mf ed v11 n42. aug 16 1895-dec 18 1896 (gaps)] – 1r – 1 – (formed by the union of: keith county news and ogalalla reflector and: keith county republican. cont by: keith county news (1897)) – us NE Hist [071]
Keith, George M *see* A voyage to south america
Keith, George Skene *see* Caution against irreligion and anarchy
Keith, Reuel *see* Lectures on those doctrines in theology usually called calvinistic
Keith, Robert *see* History of the affairs of church and state in scotland from the beginning of the reformation to the year 1568
Keith's collection of instrumental music – containing marches, quicksteps, waltzes, airs, cotillons sic, contra-dances, hornpipes, quadrilles (arranged with figures) Scotch and Irish jigs, reels, strathspeys, arranged for brass, wooden, and strings instruments. No. 1. To be completed in six numbers. Boston: Keith's Music Publishing House, 1844. Music is arranged in four parts. MUSIC 1992 – 1 – us L of C Photodup [780]
Keizertimes – Salem OR: John E Ettinger [wkly] – 1 – (began in 1979) – us Oregon Lib [071]
Kejsardoemets qvinnor : skisser fran det napoleonska paris = Frauen des kaiserreichs / Wachenhusen, Hans – Stockholm: Associations-Boktryckeriet, 1870 – 1r – 1 – (swedish translation from german) – us Wisconsin U Libr [430]
Kekae, D M *see* Prevention and intervention strategies
Kekuatov, K V *see*
– Kooperatsiia i pravo
– Zakon ob ustroitsve selskogo kredita i zemstvo
– Zhurnal zemskoe delo
– Zhurnal "zemskoe delo"
Kelch und schwert : dichtungen / Hartmann, Moritz – 3. sehr verm aufl. Darmstadt: C W Leske, 1851 [mf ed 2001] – viii/279p/1pl (ill) – 1 – mf#10556 – us Wisconsin U Libr [810]
Die kelchbewegung in deutschland und die reform der abendmahlsfeier : mit einer beilage, abbildungen von einzelkelchen / Spitta, Friedrich – Goettingen: Vandenhoeck & Ruprecht, 1904 – 1mf – 9 – 0-524-01954-1 – (incl bibl ref) – mf#1990-0543 – us ATLA [240]
Kelebek – sene 1-2. n1-77. 1339-40 – (= ser O & t journals) – 23mf – 9 – $375.00 – us MEDOC [956]
Keleher, M R *see* Army and navy posts
Kelemina, Jakob *see* Untersuchungen zur tristansage
Kelet – Nyiregyhaza, Hungary. 1962-72 – 16r – 1 – us L of C Photodup [079]
Keleti tanulmanyok goldziher ignacz – Budapest, Hungary. 1910 – 1r – us UF Libraries [939]
Die kelischin-stele und ihre chaldisch-assyrischen keilinschriften / Belck, W – Freienwalde, 1904 – (= ser Anatole, zeitschrift fuer orientforschung 1) – 1mf – 9 – mf#AR-1840 – ne IDC [470]
Kelland, Clarence Budington *see* Quizzer no. 20; being questions and answers on insurance.
Kelland, Jill *see* The level of professionalism among therapeutic recreation practitioners in alberta, canada
Kellar, Harry *see* Keller's variety entertainments
Kelleher-Walsh, Barbara J *see* Development of a child injury data base for use in biomechanics research
Kellems, Jesse Randolph *see* The deity of jesus, and other sermons
Keller, A *see* Meister altswert
Keller, Adelbert von *see*
– Der abenteuerliche simplicissimus und andere schriften
– Amadis
– Augustin tuengers facetiae
– Ayrers dramen
– Decameron
– Das deutsche heldenbuch
– Dyocletianus leben
– Fausts leben
– Die geschichten und taten wilwolts von schaumburg
– Gesta romanorum
– Hans sachs
– Hans Sachs
– Karl meinet
– Martina
– Das nibelungenlied
– Translationen
– Der trojanische krieg
– Von der musica und den meistersaengern
Keller, Adelbert von [comp] *see* Erzaehlungen aus altdeutschen handschriften
Keller, Adolf *see* Der geisteskampf des christentums gegen den islam bis zur zeit der kreuzzuege
Keller, Conrad *see* Reisebilder aus ostafrika und madagaskar
Keller, Ernst *see* Nationalismus und literatur
Keller, Fritz *see* Studien zum phaenomen der angst in der modernen deutschen literatur
Keller, Gabriele *see* Monitoring und geotechnische analyse von massenbewegungen in flysch und falten-molasse des gunzesrieder tals (oberallgaeu)
Keller, Gottfried *see*
– Ausgewaehlte erzaehlungen
– Die drei gerechten kammacher
– Das faehnlein der sieben aufrechten
– Gesammelte gedichte
– Gottfried kellers gesammelte werke
– Der gruene heinrich
– Novellen
– Romeo und julia auf dem dorfe
– Zuericher novellen
Keller, Gottfried [comp] *see* Abraham a sancta clara (1644-1709)
Keller, H *see* Mittelrheinische buchmalereien in handschriften
Keller, Hugo *see* Zur psychologie des volkstuemlichen zahlenbildes
Keller, Ludwig *see*
– Ein apostel der wiedertaeufer
– Geschichte der wiedertaeufer und ihres reichs zu muenster
– Johann von staupitz und die anfaenge der reformation
– Die reformation und die aelteren reformparteien
– Die waldenser und die deutschen bibeluebersetzungen
– Zur geschichte der altevangelischen gemeinden
Keller, Rudi *see* Ueber den begriff der praesupposition
Keller, Tammie L *see* Neuromuscular responses to platform perturbations in endurance versus power trained athletes
Keller, Werner *see* History of the presbyterian church in west cameroon
Keller's variety entertainments / Kellar, Harry – Chicago, IL. 1901 – 1r – us UF Libraries [025]
Kelletat, Alfred *see* Der abenteuerliche simplicissimus
Kellett, Ernest Edward *see* The religion of our northern ancestors
Kellett, F W *see* Pope gregory the great and his relations with gaul
Kelley, B C *see* An examination of a model of burnout in dual-role teacher-coaches
Kelley, Fanny *see* Fanny kelley v. sarah l. larimer, et al
Kelley, Francis Clement *see* Mexico, el pais de los altares ensangrentados. documentos y notas de eber cole byam
Kelley, Henry Smith *see*
– A treatise on criminal law and practice: comprising generally the statutes of missouri.
– A treatise on the law relating to the powers and duties of justices of the peace, constables, etc., etc., in the state of missouri.
Kelley, Kristi S *see* An assessment of the health habits and counseling practices of physicians
Kelleytown baptist church. darlington county. hartsville, south carolina : church records – 1923-72. (Fragments of 1923-65) – 1 – 5.00 – us Southern Baptist [242]
Kellie, E I *see* A history of bethlehem baptist association
Kellison, Barbara *see* The rights of women in the church
Kellner, David *see* Treulicher unterricht im general-bass
Kellner, Karl Adam Heinrich *see*
– Hellenismus und christenthum, oder, die geistige reaktion des antiken heidenthums gegen das christenthum
– Heortology
– Heortology sic
Kellner, Maximilian *see*
– The assyrian monuments illustrating the sermons of isaiah
– The prophecies of isaiah
Kellock, James *see* Mahadev govind ranade
Kellock's list of steamships and sailing vessels for sale – Liverpool, England. Oct 1895-1906 – 1r – 1 – uk British Libr Newspaper [380]
Kellogg, Alfred Hosea *see* Abraham, joseph, and moses in egypt
Kellogg, Frank B *see*
– The frank b kellogg papers, 1923-1937
– Papers
Kellogg, J H *see* Plain facts for old and young
Kellogg, Moses Eastman *see* The vision of the evening and the morning
Kellogg, S H *see* The genesis and growth of religion
Kellogg, Samuel Henry *see*
– Are pre-millennialists right?
– The book of leviticus
– A grammar of the hindi language
– A handbook of comparative religion
– The light of asia and the light of the world
Kelly, Bernard William *see*
– Historical notes on english catholic missions
– Some great catholics of church and state
Kelly, Bruce William *see* Method of forecasting citrus production in the state of florida
Kelly, Edward *see* Diary
Kelly, Erin W *see* The influence of aerobic vs. anaerobic exercise on sex hormone-binding globulin and free testosterone concentration
Kelly, Fred T *see* The strophic structure of habakkuk
Kelly, Helen G *see* A study of individual differences in breathing capacity
Kelly, Herbert *see* The church and religious unity
Kelly, Hugh *see* Letter from the rev hugh kelly
Kelly, irad, papers, ms 485 – 1791-1875 – 2r – 1 – (land deeds and leases, financial accounts and receipts, business papers, clippings, and letters to kelley, a cleveland merchant, postmaster, real estate investor, and politician) – us Western Res [380]
Kelly, James *see*
– The american catalogue of books
– The eternal purpose of god in christ jesus our lord
– Lecture on the general subject of the second advent
Kelly, James J *see* Journal of gay and lesbian social services
Kelly, John *see*
– The divine covenants, their nature and design
– Practical faith
Kelly, John Eogan *see* Pedro de alvarado, conquistador. princeton, 1932
Kelly, K Patrick *see* Effect of chronic cocaine on selected physiological responses during rest and exercise in rats
Kelly, Kathryn D *see* The relationship of pelvic girdle asymmetries and low back pain
Kelly, Luke E *see* Microcomputer assistance for educators in prescribing adapted physical education
Kelly, M T *see* Life of saint francis xavier
Kelly, Mary E H *see* Sand in their craws
Kelly, Michael *see*
– Grand march in the romance of blue beard
– Kant's philosophy as rectified by schopenhauer
Kelly observer – United States. 1981 apr 30 [v16 n17]-1982jun, 1982 jul-1983 oct 13, 1983 nov-1985 mar, 1985 apr-dec 19, 1986 jan-dec 18, 1987 feb 12-dec 17, 1988 jan-aug, 1988 sep-1989 apr, 1989 may-1990 jan, 1990 feb-oct 25 – 10r – 1 – mf#643895 – us WHS [071]
Kelly, W *see* Righteousness of god
Kelly, William *see*
– Elements of prophecy
– An exposition of the gospel of mark
– Lectures introductory to the study of the minor prophets
– Life in victoria
– Notes on the book of daniel
– Notices illustrative of the drama, and other popular amusements
– The revelation
– Sabath and the lord's day
– Three lectures on the book of job

Kelm, Erwin *see* The development of job-related education and training in soweto, 1940-1990
Kelman, John *see* The holy land
Kel'nich, S M *see* Syn otechestva [vech izd]
Keloearga dan roemah tangga nippon : disalin dengan merdeka...oleh tun sri lanang / Akimoto, S – (Medan? Departemen Keboedajaan Soematera Timoer, 2604) – 1mf – 9 – mf#SE-2002 mf4 – ne IDC [640]
Kelowna Board of Trade *see* Kelowna, british columbia
Kelowna, british columbia : the orchard city of the okanagan / Kelowna Board of Trade – [Kelowna, BC?]: Kelowna Board of Trade, 1912 – 1mf – 9 – 0-665-74734-9 – mf#74734 – cn Canadiana [634]
Kelowna, british columbia : with particular reference to lakeside, kelowna's most beautiful residential district – Winnipeg: Grand Pacific Land Co, [1912?] – 1mf – 9 – 0-665-75518-X – mf#75518 – cn Canadiana [917]
Kelsall, Charles *see* Classical excursion from rome to arpino
Kelsall, methodist burial ground – (= ser Cheshire monumental inscriptions) – 1mf – 9 – £2.50 – mf#24 – uk CheshireFHS [929]
Kelsey, D M *see* Life and public services of hon wm e gladstone
Kelsey, Dandridge E *see* Diaries
Kelsey, Elizabeth *see* Trail blazers to radionics
Kelsey [kelso] family bulletin : for the descendants of samuel kelsey [kelso], 1720-1796 and susannah mills – 1984 dec-1987 may 30 – 1r – 1 – mf#1772011 – us WHS [929]
Kelsey, Vera *see*
– Brazil in capitals
– Four keys to guatemala
– Seven keys to brazil
[Kelseyville-] kelseyville sun – CA. 1903-40 (broken series) – 5r – 1 – $300.00 – mf#B02322 – us Library Micro [071]
[Kelseyville-] lake sun – CA. 1976-81; 1987-88 – 8r – 1 – $480.00 – mf#B02323 – us Library Micro [071]
Kelsheimer, E G *see* Ddt treatment for control of mole-crickets in seedbeds
Kelso border mail & gazette for roxburgh, selkirk, berwick & northumberland – Kelso, Scotland 23 may 1945-26 oct 1949 [mf 19 apr 1934-oct 1949] – 1 – (amalg with: kelso chronicle & subsequently publ as: border counties & kelso chronicle & mail; cont: border mail & gazette for roxburgh, selkirk, berwick & northumberland [19 apr 1934-9 may 1945]) – uk British Libr Newspaper [072]
Kelso courier and border counties' advertiser – [Scotland] Kelso: E Murray 12 may 1871-dec 1878 (wkly) [mf ed 2004] – 10v on 3r – 1 – (missing: 1875-76; cont by: kelso courier, melrose and dunse and border counties' advertiser [9 jul 1980-23 feb 1883]) – uk Newsplan [072]
Kelso, James Anderson *see* Hebrew-english vocabulary to the book of genesis
Kelso, John Joseph *see* Some first principles in social welfare work
Kelso, John Russell *see* The bible analyzed in twenty lectures
Kelso mail, & gazette for the counties of roxburgh, selkirk, berwick, & northumberland – Kelso, Scotland 2 jan 1854-12 apr 1934 – 1 – (cont: kelso mail, or, roxburgh, selkirk, berwickshire & northumberland gazette [14 oct 1839, 1 jan-19 dec 1844, 2,6,9 jan, 2 jun-30 nov 1845, jan 1846-29 dec 1853]; cont by: border mail & gazette for roxburgh, selkirk, berwick & northumberland [19 apr 1934-9 may 1945]) – uk British Libr Newspaper [072]
Kelso weekly express & advertiser for roxburghshire, berwickshire & the border counties – [Scotland] Kelso: A W Lyall 2 jan 1863-13 may 1864 (wkly) [mf ed 2004] – 2v on 2r – 1 – uk Newsplan [072]
Y kelt : the london welshman & kelt – London, England 21 jul-29 dec 1906 – 1 – (cont: london welshman: cymro llundain [1 oct 1904-14 jul 1906]; cont by: cymro a'r celt llundain: the london welshman and kelt [5 jan 1905-1 dec 1917]) – uk British Libr Newspaper [072]
Keltic researches; studies in the history and distribution of the ancient goidelic language and peoples / Nicholson, Edward Williams Byron – London, New York: H. Frowde, Oxford University Press, 1904. xviii,211p. Plates,maps – 1 – us Wisconsin U Libr [490]
Keltie, John Scott *see* The partition of africa
Keltie, Sir John Scott *see* History of geography
Keluarga berentjana – Djakarta, 1967-1972(22) – 11mf – 9 – (missing: 1967(1-2)) – mf#SE-1740 – ne IDC [950]
Keluarga kompas : kompas untuk generasi baru – Djakarta, 1951-1954 – 17mf – 9 – (missing: 1951, v1(1, 3-8, 12); 1952, v2; 1953, v3(1-11)) – mf#SE-903 – ne IDC [950]
Keluarga sedjahtera *see* Lembaga keluarga berentjana nasional indonesia
Kelvin (of Largs), William Thomson, Baron *see* Popular lectures and addresses
Kelvin, William Thomson *see* Reprint of papers on electrostatics and magnetism
Kemal, Ahmed *see* Kitab-i subhat uel-ahbar
Kemal, Namik *see*
– Akif bey
– Guelnihal
– Kara bela
– Vatan yahut silistre
Kemalpasazade *see* The divan project
Kemble, Fanny [Frances Ann] *see* Records of later life
Kemble prompt books in the garrick club, london / Garrick Club. London – [mf ed 1986] – 104mf – 9 – $830.00 – 1-900853-70-1 – (complete coll covering period 1710-1821; with library ind cards & p/g) – uk Mindata [790]
Kementerian agama / Penundjuk hadji – Djakarta, 1953 – 2mf – 9 – mf#SE-400 – ne IDC [950]
Kementerian agama, bg penerbitan – Djakarta, 1950-1965 – 57mf – 9 – (missing: 1954, v5(2-12); 1955(2-12); 1965(4-12)) – mf#SE-386 – ne IDC [950]
Kementerian agama penjiaran / Indonesia – Djakarta, 1951 – 7mf – 9 – (missing: 1951(2, 6, 13, 18, 19)) – mf#SE-792 – ne IDC [959]
Kementerian dalam negeri biro pemilihan petundjuk pemilihan daerah / Indonesia – Djakarta, 1956 – 3mf – 9 – mf#SE-230 – ne IDC [959]
Kementerian kesehatan / Indonesia – Jogjakarta, 1950-1956 – 6mf – 9 – mf#SE-1670 – ne IDC [360]
Kementerian kesehatan berita hygiene / Indonesia – Djakarta, 1950-1955 – 8mf – 9 – (missing: 1953(7-12); 1954(1-10, 12)) – mf#SE-853 – ne IDC [360]
Kementerian keuangan / Madjalah keuangan negara – Djakarta, 1957-1960. v1-4(6) – 27mf – 9 – (missing: 1959/1960(9-12)) – mf#SE-301 – ne IDC [959]
Kementerian keuangan nota keuangan negara / Indonesia – Djakarta, 1950-1959 – 21mf – 9 – (missing: 1954-58) – mf#SE-690 – ne IDC [959]
Kementerian keuangan rantjangan anggaran / Indonesia – Djakarta, 1950/1951 – 17mf – 9 – mf#SE-309 – ne IDC [959]
Kementerian luar negeri = Diplomatic and consular list ministry of foreign affairs / Indonesia – Djakarta, 1957-1971 – 31mf – 9 – (missing: 1959; 1962; 1965; 1967) – mf#SE-616 – ne IDC [327]
Kementerian luar negeri direktorat 5 fakta dan dokumen2 untuk menjusun buku "indonesia memasuki gelanggang internasional" / Indonesia – Djakarta – 129mf – 9 – (missing: 1958 v1(1-end); 1958 v3(1, 8, 10, 12); v4(2)) – mf#SE-990 – ne IDC [327]
Kementerian luar negeri direktorat ekonomi antar negara tindakan2 ekonomi : ichtisar berkala disusun berdasarkan peraturan2, indonesia – Djakarta, 1956-1957 – 6mf – 9 – (missing: 1956 v1(6)) – mf#SE-691 – ne IDC [959]
Kementerian penerangan : almanak kempen – Djakarta, 1952 – 11mf – 9 – mf#SE-203 – ne IDC [950]
Kementerian penerangan : arsip dokumentasi – Djakarta, 1951 – 2mf – 9 – mf#SE-770 – ne IDC [950]
Kementerian penerangan / Lukisan Indonesia – Djakarta, 1950-1961 – 17mf – 9 – (several issues missing) – mf#SE-907 – ne IDC [959]
Kementerian penerangan bagian dokumentasi ichtisar peristiwa dalam dan luar negeri / Indonesia – Djakarta, 1949-1960 – 64mf – 9 – (missing: 1950(jul-dec); 1951(jan-jun); 1952(5); 1953(1, 4, 7, 9, 12); 1954(4, 7, 8, 10, 11); 1955(1-7, 9-11); 1956; 1957; 1958; 1959) – mf#SE-375 – ne IDC [959]
Kementerian Penerangan, Biro Dokumentasi, Sedjarah & Research *see* Kronik dokumentasi
Kementerian penerangan dokumenta informasia / Indonesia – Djakarta, 1961-1963 – 15mf – 9 – (missing: [19?]-1960, v1-11; 1961, v12(1-143, 147, 148, 154-157, 164, 165, 167)) – mf#SE-542 – ne IDC [959]
Kementerian penerangan ichtisar indonesia sepekan / Indonesia – Djakarta, 1950/1951 – 24mf – 9 – (missing: 1950 v1(1-8, 10, 14, 16-17, 20-24); 1951 v2(53)) – mf#SE-535 – ne IDC [959]
Kementerian penerangan ichtisar parlemen / Indonesia – Djakarta, 1950-1960. v1-9(85) – 243mf – 9 – (missing: several iss) – mf#SE-231 – ne IDC [959]
Kementerian Penerangan Ichtisar pers *see* Indonesia (republic, 1945-1949)
Kementerian penerangan kepartaian dan parlementaria indonesia / Indonesia – Djakarta, 1950-1954 – 16mf – 9 – (missing: 1952-1953) – mf#SE-530 – ne IDC [959]
Kementerian penerangan penerbitan chusus / Indonesia – Djakarta, 1958-1967(1-478) – 197mf – 9 – (missing: several iss) – mf#SE-559 – ne IDC [959]
Kementerian penerangan repoeblik indonesia : ichtisar isi pers dalam saminggoe teroetama pers di indonesia – Djakarta, 1946 – 6mf – 9 – (missing: 1946 v1(10-12), 16-17) – mf#SE-536 – ne IDC [959]
Kementerian penerangan republik indonesia / Gelombang – Jogjakarta, 1950 – 11mf – 9 – (missing: 1950 v1(2)) – mf#SE-767 – ne IDC [959]
Kementerian penerangan ri : berita knp – Jogjakarta, 1949 – 2mf – 9 – (missing: 1949(1, 13, 14)) – mf#SE-1351 – ne IDC [950]
Kementerian Penerangan Siaran kilat *see* – Indonesia
Kementerian perburuhan laporan / Indonesia – Djakarta. 3pts. 1957 – 3mf – 9 – mf#SE-2784 – ne IDC [959]
Kementerian perburuhan laporan kementerian perburuhan selama 2 tahun kabinet karya, april 1957-april 1959 / Indonesia – Djakarta, 1959 – 2mf – 9 – mf#SE-2785 – ne IDC [959]
Kementerian perburuhan laporan singkat, 1956-1957 / Indonesia – Djakarta, 1957 – 2mf – 9 – mf#SE-2783 – ne IDC [959]
Kementerian perburuhan ri / Masjarakat dan perburuhan – Djakarta, 1950/1951-1952. v1-2(8) – 17mf – 9 – (missing: 1950/1951, v1(2-6); 1952, v2(3-4)) – mf#SE-1796 – ne IDC [950]
Kementerian perburuhan situasi perburuhan dalam dan luar negeri / Indonesia – Djakarta, 1950-1951. v1-2(10/11) – 20mf – 9 – (missing: 1950 v1; 1951 v2(2, 4, 6)) – mf#SE-1672 – ne IDC [959]
Kementerian perhubungan djawatan pelajaran laporan masa 1950 s/d 1952 / Indonesia – [Djakarta, 1953] – 2mf – 9 – mf#SE-5364 – ne IDC [959]
Kementerian pertahanan bagian penhubung masjarakat angkatan perang : kawan tentara – Djakarta, 1949-1951. v1-2(17) – 9mf – 9 – (missing: 1949, v1(2-end); 1950, v2(1-5, 7-9, 14)) – mf#SE-1732 – ne IDC [950]
Kementerian PP & K, Perpustakaan Perguruan *see* Pembimbing pembatja
Kemmer, A *see* Charisma maximum
Kemmer, Ernst *see* Die polare ausdrucksweise in der griechischen literatur
Kemp, A F et al *see* Hand-book of the presbyterian church in canada, 1883
Kemp, Alexander Ferrie *see* The beneficial influence of a well regulated nationality
Kemp, D *see* Nine years at the gold coast
Kemp, D C van der *see* Levensgeschiedenis van den med doctor johannes kemp
Kemp, Dennis *see* Nine years at the gold coast
Kemp, E G *see* Wanderings in chinese turkestan
Kemp, Emily Georgiana *see*
– The face of china
Kemp, Friedhelm *see*
– Ergriffenes dasein
Kemp, H C *see* The x ray self instructor in studies of human nature
Kemp, John *see*
– Character of the apostle paul in some of its features delineated
– "What i remember of the battle of the blue, and incidents connected with it"
Kemp roots – 1985 may/jun-1989 – 1r – 1 – mf#1664567 – us WHS [071]
Kemp, William Webb *see* The support of schools in colonial new york by the society for the propagation of the gospel in foreign parts
Kempener kreisblatt *see* Intelligenzblatt fuer den kreis kempen und dessen umgebung
Kemper, Dirk *see* Missbrauchte aufklaerung?
Kempis commun : ou les quatre livres de l'imitation de jesus-christ... / Poiret, P – Amsterdam, 1683 – 5mf – 9 – mf#PPE-202 – ne IDC [240]
Kemplay, Christopher *see* Comets
Kempner, Alfred *see* Brentanos jugenddichtungen
Kempowski, Walter *see* Konkordanz zu walter kempowskis "deutscher chronik"
Kemp's west london sketcher & theleme – [London & SE] Hammersmith 1 nov 1888-16 aug 1889 [mf ed 2003] – 1r – 1 – uk Newsplan [072]
Kempston – (= ser Bedfordshire parish register series) – 2mf – 9 – £5.00 – uk BedsFHS [929]
Kempt, Robert *see*
– Pencil and palette
– What do you think of the exhibition?
Kemptner zeitung – Kempten DE, 1848-1849 31 mar – 1r – 1 – gw Misc Inst [074]
Kempton rural press – Kempton, PA., 1888-1889 – 13 – $25.00r – us IMR [071]
Kemptville telegram – Ontario, CN. jan 1901-dec 1908 – 3r – 1 – (some iss missing) – cn Commonwealth Imaging [071]
Kemptville weekly advance – Kemptville, Ontario, CN. 1881- – 2r/y – 1 – Can$93.00r – cn Commonwealth Imaging [071]
K'en chih ch'ien shuo / Chiang, Yin-sung – Ch'ung-ch'ing: Cheng chung shu chu, Min kuo 29 [1940] – (= ser P-k&k period) – us CRL [333]
Ken, Thomas *see*
– Directions for prayer
– Manual of prayers for young persons
Ken Vac Sak *see* Citt kramum
Ken Van Sak *see* Guk kam kiles
Ke-na gumi Ra An Chan *see* Angalip
Kenarskii, L V *see* Sotsialist-revolutsioner
Kendall and Wilton keystone – Elroy, Kendall, Wilton WI. 1964 feb 7-1965 jan 7 – 1r – 1 – (cont: kendall keystone; cont by: elroy-kendall-wilton tribune-keystone) – mf#1013664 – us WHS [071]
Kendall, Edward *see* Letters to a friend on the state of ireland, the roman catholic question, and the merits of constitutional religious distinctions
Kendall, Edward Augustus *see* Letters to a friend
Kendall, H E *see* [Modern architecture]
Kendall, Henry Edward Jr *see* Designs for schools and school houses
Kendall hispano – Coral Gables, FL. 1985 dec-1986 apr – 1r – us UF Libraries [071]
Kendall, Holliday Bickerstaffe *see* The origin and history of the primitive methodist church
Kendall, John *see* An elucidation of the principles of english architecture
Kendall keystone – Elroy, Kendall, Wilton WI. 1904-1908: feb 6, 1908 feb 13-1912 apr 18, 1912 apr 25-1915 aug 12, 1915 aug 29-1919 jul 17, 1919 jul 24-1925 aug 27, 1925 sep 3-1932 apr 28, 1932 may 12-1943 oct 1, 1943 oct 8-1946, 1947-61, 1962-1964 jan 24 – 12r – 1 – (cont: kendall & wilton keystone) – mf#1013662 – us WHS [071]
Kendall, Ralph Selwood *see* The luck of the mounted
Kendig, Abby E G *see* Heinrich heine's an essay, read before the monday club, may 21st, 1883
Kendra, Heather A *see* A title 9 source book for today's athletics administrator
Kendrick, A C *see* The life and letters of mrs emily c judson
Kendrick, Asahel Clark *see*
– Commentary on the epistle to the hebrews
– The life and letters of mrs. emily c. judson
– Martin b. anderson, ll. d
– The moral conflict of humanity, and other papers
Kendrick, J R *see* Dueling (a sermon)
Kendrick, Nathaniel *see* A sermon delivered in the chapel of the baptist literary and theological seminary, hamilton, new york, 19 mar 1824
Kendrick, R J *see* A comparison of isometric strength test results between low back injured patients and normals
Kendrick, T D *see* Saint james in spain
Kendrick, Tertius *see* The ionian islands
Kendrick, Zebulon V *see*
– Effects of caloric restriction and resistive exercise on the resting energy expenditure of weight-reduced obese women
– Effects of thymopentin on the responses of hypothalamic-pituitary-adrenal axis to a high intensity dynamic exercise protocol
Kenealy, Dr *see* Dr kenealy's lecture on temperance
Keneder adler – Montreal. (Jewish Daily Eagle). Oct 1908-Dec 1965. Incomplete – 1 – us NY Public [071]
Keneder id = The canadian jew – Montreal, Quebec. 1935; 1938; 1941 – 1 – us AJPC [071]
Kenelm chillingly / Lytton, Edward Bulwer Lytton, Baron – Boston, MA. 189- – 1r – us UF Libraries [025]
Kenesaw citizen – Kenesaw, NE: M DeMotto, 1890-v11 n18. nov 30 1900 (wkly) [mf ed 1891-1900 (gaps)] – 1r – 1 – (cont by: kenesaw weekly citizen) – us NE Hist [071]
Kenesaw citizen – Kenesaw, NE: S H Smith. v12 n29. feb 14 1902- (wkly) [mf ed -1907 (gaps)] – 1r – 1 – (cont: kenesaw weekly citizen. cont by: kenesaw kaleidoscope) – us NE Hist [071]
The kenesaw kaleidoscope – Kenesaw, NE: J A Gardner, -1913// (wkly) [mf ed v22 n5. sep 16 1910-jan 31 1913 (gaps)] – 1r – 1 – (cont: kenesaw citizen (1902). cont by: kenesaw sunbeam) – us NE Hist [071]
Kenesaw progress – Kenesaw, NE: Kenesaw Publ, 1919-v3 n3. mar 6 1919; v30 n28. mar 13 1919-v43 n52. dec 8 1938 (wkly) [mf ed v2 n14. may 23 1918-dec 3 1938 (gaps)] – 12r – 1 – (absorbed: kenesaw sunbeam. cont by: adams county voice) – us NE Hist [071]
Kenesaw sunbeam – Kenesaw, NE: W W Maltman, 1913-mar 6 1919// (wkly) [mf ed v25 n46. jul 9 1914-mar 6 1919 (gaps)] – 2r – 1 – (cont: kenesaw kaleidoscope. absorbed by: kenesaw progress) – us NE Hist [071]
The kenesaw times – Kenesaw, NE: Studley Publ. 2v. v1 n1. mar 1 1974-v2 n23. jul 24 1975 (wkly) [mf ed mar 1 1974-jul 24 1975 (gaps) filmed 1977] – 1r – 1 – (absorbed: minden courier) – us NE Hist [071]

The kenesaw times – Kenesaw, NE: George T Williams. -v5 n24. may 26 1888 (wkly) [mf ed v5 n22. may 12 1888] – 1r – 1 – us NE Hist [071]
Kenesaw weekly citizen – Kenesaw, NE: G D and L J Woods. v11 n19. dec 7 1900-v12 n28.feb 7 1902 (wkly) [mf ed with gaps] – 1r – 1 – (cont: kenesaw citizen. cont by: kenesaw citizen (1902)) – us NE Hist [071]
Keneset – Odessa, 1917 – 6mf – 9 – mf#J-91-45 – ne IDC [077]
Keneset ha-gedolah / ed by Suwalski, I – Warsaw, 1890-1891. v1-4 – 24mf – 9 – mf#J-422-36 – ne IDC [077]
Keneset yisrael – Warsaw, 1886-1889. v1-3 – 30mf – 9 – mf#J-91-46 – ne IDC [077]
Keneset yisra'el be-erets yisra'el, yisudah ve-irgunah / Attias, Moshe – Jerusalem, Israel. 1944 – 1r – us UF Libraries [939]
Kenfig mawdlam, st mary magdalene, monumental inscriptions – 2mf – 9 – £2.50 – uk Glamorgan FHS [929]
Keng k'uan hsing hsueh wen t'i / T'ai, Shuang-ch'iu – Shang-hai: Chiao yu pien i kuan, Min kuo 24 [1935] – (= ser P-k&k period) – us CRL [370]
Keng pa chi; fu, tien ch'ien chi hsing / Lo, Chia-lun – Shang-hai: Shang wu yin shu kuan, Min kuo 35 [1946] – (= ser P-k&k period) – us CRL [480]
Keng tzu chiao hui shou nan chi (ccm240) = The tribulations of the church in china, a d 1900: natives and foreigners / MacGillivray, Donald – Shanghai. 2v. 1901 [mf ed 198?] – (= ser Ccm 240) – 1 – (with ind in english) – mf#1984-b500 – us ATLA [951]
Keng, Wen-t'ien *see* Kuo min ta hui ts'an k'ao tzu liao
Kenig, Refael Tsevi *see* Tsevi kenig (yishai)
Kenilworth observer – Jan 12-Dec 14 1995 – 8r – 9 – uk British Libr Newspaper [072]
Kenilworth weekly news – England. 24 May 1949- – 56+ r – 1 – uk British Libr Newspaper [072]
Kenison, Ervin *see* Mechanical drawing
Kenkmann, P *see* Semia kak agent sotsializatsii v sotsialisticheskom obshchestve
Kennan banner – Kennan WI. 1890 aug-1891 apr – 1r – 1 – mf#927108 – us WHS [071]
Kennan, George *see*
– Campaigning in cuba
– Siberia and the exile system
Kennan, George Frost *see* Soviet foreign policy
Kennard, Barbara A *see* The effects of sports massage upon subsequent quadricep force output, power, and total work
Kennard enterprise – Kennard, NE: E L Tiffainy, nov 1896-v16 n51. jan 17 1913 (wkly) [mf ed 1897-1913 (gaps)] – 3r – 1 – (cont by: enterprise (kennard ne). all issues for nov 1-jan 3 1912 [ie 1913] called v16 n49) – us NE Hist [071]
Kennard, Joseph Spencer *see* Politique et religion chez les juifs au temps de jesus et dans l'eglise primitive
Kennard, Richard *see* Gilfield baptist church, petersburg, virginia, 1803-1903
Kennaway, Adrienne *see* Interviewing japan
Kennaway, C E *see* Offerings of love in the wilderness
Kennedy, Alexander Macpherson *see* Essentials of phonography
Kennedy, Archibald Robert Stirling *see*
– The book of joshua and the book of judges
– Leviticus and numbers
– Samuel
– The second book of moses called exodus
Kennedy, Arnold *see* Story of the west indies
Kennedy, Benjamin Hall *see*
– Christian peaceableness
– Ely lectures on the revised version of the new testament
Kennedy, Charles Malcolm *see* The influence of christianity upon international law
The kennedy clan and tierra redonda / Lynch, Alice Clare – San Francisco: Marnell & Co, jan 1935 – 1r – 1 – $50.00 – mf#B40254 – us Library Micro [978]
Kennedy, David *see* Incidents of pioneer days at guelph and the county of bruce
Kennedy, David Elliott *see* Stylistic differences between haydn and mozart
Kennedy for president washington state campaign records, 1958-1960 – [mf ed 1976] – 5ft – (correspondence, chiefly incoming letters from state campaign workers & copies of outgoing letters from john f kennedy & his staff, memos, and clippings, relating to the 1960 democratic presidential campaign in washington state. correspondents incl lawrence f o'brien, state chairman henry b owen & theodore c sorensen; with finding aid) – us UW Libraries [325]
Kennedy, Gladys *see* Early history of mobile baptists
Kennedy, Harold W *see* Legal support for los angeles county's strict air pollution control program
Kennedy, Harry Angus Alexander *see* Sources of new testament greek
Kennedy, Henry Dawson *see*
– Jacob the wrestler
– Misunderstood
Kennedy, J *see*
– Distinctive principles and present position and duty of the free ch...
– Letter to the members of the free church in the highlands
– Life and work in benares and kumaon 1839-1877
Kennedy, Jack *see* Collected field reports on the phonology of dagaari
Kennedy, James *see*
– Christianity and the religions of india
– Essays ethnological and linguistic
– The great indian mutiny of 1857
– Introduction to biblical hebrew
– Life and work in benares and kumaon, 1839-1877
– Observations on professor w r smith's article "bible" in the encyclopaedia britannica
– Observations on professor w r smith's article "bible" int he encyclopaedia britannica
Kennedy, James Henry *see* The bench and bar of cleveland
Kennedy, James Houghton *see* Natural theology and modern thought
Kennedy, James Shaw *see* Notes on the battle of waterloo
Kennedy, John *see*
– The four gospels
– Hyper-evangelism
– Old testament criticism and the rights of the unlearned
– On the book of jonah
– The pentateuch
– A popular argument for the unity of isaiah
– Reply to dr bonar's defence of hyper-evangelism
– Report on the st lawrence bridge and manufacturing scheme
– The resurrection of jesus christ an historical fact
– The self-revelation of jesus christ
Kennedy, John Curtis *see* Wages and family budgets in the chicago stockyard district
Kennedy, John McFarland *see* The religions and philosophies of the east
Kennedy, John Pendleton *see*
– The john pendleton kennedy papers
– Journals of the house of burgesses of virginia, 1619-1776
Kennedy, John Pitt *see*
– Principles of railway construction analyzed in reference to their financial effects on shareholders and on british and indian interests
– A railway caution!!
Kennedy klues – v1 n1-4 [1975 aug-1976 may] – 1r – 1 – mf#382331 – us WHS [071]
[Kennedy-] nevada new era – NV. 1894 – 1r – 1 – $60.00 – mf#U04592 – us Library Micro [071]
Kennedy, Patrick *see* Fireside stories of ireland
Kennedy, Pringle *see* Arabian society at the time of muhammad
Kennedy, Reginald Frank *see* Africana repository
Kennedy, S O *see* A biomechanical analysis of children's balance behavior
Kennedy, Sir Alexander B W *see* Petra: its history and monuments
Kennedy, Stetson *see*
– Florida keys
– Mister homer
Kennedy, Vans *see*
– Dictionary of the maratha language
– Practical remarks on the proceedings of general courts martial
– Researches into the origin and affinity of the principal languages of asia and europe
– A treatise on the principles and practice of military law
Kennedy, Vans, 1784-1846 *see* A dictionary of the maratha language, in two parts
Kennedy, William Paul McClure *see* Documents of the canadian constitution, 1759-1915
Kennedy, William Sloane *see*
– Henry w. longfellow
– Messianic prophecy and the life of christ
– Oliver wendell holmes
– The plan of union
Kennel, Albert *see* Burleigh, shrewsbury u leicester in schillers maria stuart
Kennel farm poultry yard *see* Irish farm forest and garden
Kennel gazette – Toronto: H B Donovan, 1889-[189–19–] [mf ed v1 n1 feb 1889-v1 n12 [i.e. 11] dec 1889] – 9 – mf#P04176 – cn Canadiana [636]
Kennel gazette *see* Canadian poultry review
Kennen und sammeln : ein verzeichnis der schriften von ursula schlegel / ed by Laschke, Birgit & Welzel, Barbara – (mf ed 1995) – 1mf – 9 – €30.00 – 3-8267-2088-1 – mf#DHS 2088 – gw Frankfurter [700]
Kenner, James B *see* The practice in indiana under the laws for constructing ditches and drains
Kennerly, Clarence Hickman *see* Facts and figures
Kennet, White *see* Christian scholar
Kenneth, H *see* Land tenure: proceedings of the international conference on land tenure and related problems in world agriculture, held at madison, wisconsin, 1951
Kennett 1567-1950 – (= ser Cambridgeshire parish register transcript) – 3mf – 9 – £3.75 – uk CambsFHS [929]
Kennett, R H *see* The composition of the book of isaiah in the light of history and archaeology
Kennett, Robert Hatch *see*
– Early ideals of righteousness
– In our tongues
– The servant of the lord
– A short account of the hebrew tenses
Kenney, John Andrew *see* The negro in medicine
Kennicott, Benjamin *see* Vetus testamentum hebraicum, cum variis lectionibus
Kennington, Anna *see* The relationship of athletic participation and inactivity to fatty food selection in senior citizens
Kennington elector – London, UK. may 1930-oct 1938 – 1/4r – 1 – uk British Libr Newspaper [072]
Kenny letter / Letterkenny Army Depot [PA] – Chambersburg PA. 1981 jul 28 [v32 n20]-1986 jul 22 – 1r – 1 – mf#1099037 – us WHS [071]
Kenny, Michael *see*
– Pedro martinez
– Romance of the floridas
– The romance of the floridas. the finding of the founding. new york, 1934
Die kenose und die moderne protestantische christologie : geschichte und kritik der protestantischen lehre von der selbstentaeusserung christi (phil. 2) und deren anteil an der christologischen frage der gegenwart / Waldhaeuser, Michael – Mainz: Kirchheim, 1912 – 1mf – 9 – 0-8370-9837-8 – (incl indes) – mf#1986-3837 – us ATLA [242]
Kenosha aigredoux and illustrated dime – Kenosha WI. 1857 dec 25, 1861 dec 25 – 1r – 1 – mf#876774 – us WHS [071]
Kenosha Area Chamber of Commerce *see*
– News
– Newsletter
Kenosha County Historical Society *see* Southport newsletter
Kenosha County Historical Society and Museum *see* Bulletin of the kenosha...
Kenosha county republican – Kenosha WI. 1872 jan 24-1873 jul 9 – 1r – 1 – mf#929566 – us WHS [071]
Kenosha county times – Kenosha WI. 1990 jul 31-dec 18, 1991 jan 8-nov 12 – 2r – 1 – (cont: bi-state reporter) – us WHS [071]
Kenosha courier – Kenosha WI. 1880 dec 9-1882 may 25, 1882 jun 1-1884 jan 31 – 2r – 1 – (cont: kenosha democrat [kenosha, wi: 1879]; cont by: kenosha telegraph; telegraph-courier) – mf#927392 – us WHS [071]
Kenosha democrat – Kenosha WI. 1879 oct 18-1880 dec 2 – 1r – 1 – (cont by: kenosha courier) – mf#927388 – us WHS [071]
Kenosha democrat – Kenosha WI. 1850 apr 23-1852 nov 30, 1852 dec 3-1855 may 11, 1955 may 18-1856 dec 2 – 3r – 1 – mf#927248 – us WHS [071]
Kenosha democrat – Kenosha WI. 1859 sep 9-1861 aug 30 – 1r – 1 – mf#927251 – us WHS [071]
Kenosha evening news – Kenosha WI. [1899 aug 17-26]-[1959 may-jun] [gaps] – 358r – 1 – (cont: evening news (kenosha, wi); cont by: kenosha news) – mf#1145414 – us WHS [071]
Kenosha hosiery worker / American Federation of Full Fashioned Hosiery Workers Locked Out by the Allen A Co – Kenosha WI. v1 n1-30 [1928 mar 31-1929 jul 2], v2 n1-2 [1929 jul 23-aug 27], v2 n4 [1930 apr 30] – 1r – 1 – mf#1061302 – us WHS [331]
Kenosha labor / Kenosha Trades and Labor Council – Kenosha WI. [1935 nov 1/1937]-[1991 jan 4/1992 oct 25] [gaps] – 38r – 1 – (cont by: labor paper [kenosha, wi]) – mf#1429042 – us WHS [331]
Kenosha news – Kenosha WI. 1964 feb – 1r – 1 – (cont: kenosha evening news) – mf#1140942 – us WHS [071]
Kenosha telegraph – Kenosha WI. [1860 jun 14/1861 sep 26]-[1945/46] [gaps] – 50r – 1 – (cont: kenosha telegraph and tribune; cont by: kenosha courier; telegraph-courier) – mf#930841 – us WHS [071]
Kenosha telegraph – Kenosha WI. 1850 apr 5-jun 21, 1850 jun 28-1851 jun 20, 1851-1854 dec 28, 1854 feb 14-dec 28 – 4r – 1 – (cont: southport telegraph; cont by: kenosha tribune; kenosha tribune and telegraph [kenosha, wi: 1855]) – mf#936050 – us WHS [071]
Kenosha times – Kenosha WI. 1857 jul 2-1859 dec 15, 1859 dec 22-1863 jan 29 – 2r – 1 – mf#927232 – us WHS [071]
Kenosha times – Kenosha WI. feb-apr, jun, jul, dec 10 1976, and feb 3-apr 4 1977 – 1r – 1 – mf#929560 – us WHS [071]
Kenosha Trades and Labor Council *see* Kenosha labor
Kenosha tribune – Kenosha WI. 1852 jul 8-1854 dec 28 – 1r – 1 – (cont by: kenosha telegraph [kenosha, wi: 1850]; kenosha tribune and telegraph [kenosha, wi: 1855]) – mf#936052 – us WHS [071]
Kenosha tribune and telegraph – Kenosha WI. Vol. 15, n29 [1855 jan 4]-v16, n2 [1855 jun 28] – 1r – 1 – (cont: kenosha tribune; kenosha telegraph [kenosha, wi: 1855]; cont by: tribune and telegraph) – mf#936058 – us WHS [071]
Kenosha tribune and telegraph – Kenosha WI. 1856 oct 2-1858 jun 10 – 1r – 1 – (cont: tribune and telegraph; cont by: kenosha telegraph and tribune) – mf#936062 – us WHS [071]
Kenosha tri-weekly experiment – Kenosha WI. 1850 feb 26 – 1r – 1 – mf#876767 – us WHS [071]
Kenosha union – Kenosha WI. [1866 jun 26/1869 mar 25]-[1907 jun 14/1909 apr 30] – 24r – 1 – mf#930853 – us WHS [071]
Kenosha volksfreund – Kenosha WI. 1893 apr 6-1894, 1895-1896 oct 1 – 2r – 1 – mf#1013501 – us WHS [071]
Kenrick, Francis Patrick *see*
– Diary and visitation record of the rt. rev. francis patrick kenrick
– The historical books of the old testament
– The new testament
– The primacy of the apostolic see vindicated
– Theologia dogmatica
– Theologia moralis
– A treatise on baptism
– A vindication of the catholic church
Kenrick, John *see*
– Biblical essays
– Necessity of revelation to teach the doctrine of a future life
– Obstacles to the diffusion of unitarianism and the prospect of...
– Phoenicia
– The value of the holy scriptures and the right mode of using them
Kenrick, Peter Richard *see*
– The new month of mary
– The validity of anglican ordinations and anglican claims to apostolical succession examined
Kenrick, Samuel *see* Wodrow-kenrick correspondence, c1750-1810
Kensdale, W E N *see* Catalogue of the arabic manuscripts preserved in the university library, ibadan, nigeria
Kensington and hammersmith reporter – [London & SE] Kensington LSC 5 jul 1879-28 dec 1906 – 1 – (incorp: notting hill, acton and shepherds bush news; cont as: west london reporter [2 jan 1892-28 dec 1906]) – uk Newsplan [072]
Kensington churchman and ruridecanal gazette etc – London, UK. apr 1888-20 dec 1890, 1891, mar 1892-5 jan 1895 – 4 1/2r – 1 – (aka: west london church chronicle) – uk British Libr Newspaper [072]
Kensington express notting hill and west london examiner – London, UK. 1892 – 1r – 1 – uk British Libr Newspaper [072]
Kensington gazette and chelsea, fulham and hammersmith chronicle – [London & SE] Hammersmith 9 oct 1868 [mf ed 2003] – 1r – 1 – uk Newsplan [072]
Kensington news and west london times – [London & SE] Kensington LSC 23 jan 1869-13 nov 1869, 23 nov 1872; 1876-14 jul 1972, 21 jul 1972-jun 1974; jul 1974-dec 1977, jan 1978-jun 1982; jul 1982-jun 1992. BLNL jan-nov 1869, 23 nov 1872 – 1 – uk Newsplan [072]
Kensington notting hill shepherds bush bayswater paddington and west london post – London, UK. 3 sep-10 sep 1892 – 1/4r – 1 – uk British Libr Newspaper [072]
Kensington po (kensington news and post) – London. 1918-72.-w. 62mqn reels – 1 – uk British Libr Newspaper [072]
Kensington post and west london star – London, UK. 1918-22 dec 1972; 1976; 1977; jul-dec 1981; 24 jan 1986-1992 – 82r – 1 – uk British Libr Newspaper [072]
Kensington society and london society up to date *see* Kensington weekly advertiser
Kensington weekly advertiser – London, UK. 13 jun-19 dec 1888, 2 jan-24 dec 1889, 1890-11 apr 1896 – 7r – 1 – (aka: kensington weekly advertiser and society journal; kensington society; kensington society and london society up to date) – uk British Libr Newspaper [072]
Kensworth – (= ser Bedfordshire parish register series) – 2mf – 9 – £5.00 – uk BedsFHS [929]
Kensworth, st mary monumental inscriptions the virgin monumental inscriptions – Bedfordshire Family HS 1983 – (= ser Bedfordshire parish register series) – 1mf – 9 – £1.25 – uk BedsFHS [929]
Kent advertiser *see* Chatham newspapers, pt 1

KEPHALIDES

Kent and essex mercury – London, UK. 15 oct 1822-apr 1828, jun 1828; aug 1829-43 – 10 3/4r – 1 – (aka: essex and herts mercury; essex herts and kent mercury; essex herts and suffolk mercury) – uk British Libr Newspaper [072]
Kent and sussex courier – [London & SE] Kent Arts & Lib, Tunbridge Wells Lib 6 jun 1873-1896, 1898-1910, 19 apr-29 nov 1912; East Sussex, Hastings Ref Lib 1980- – 1 – uk Newsplan [072]
Kent archaeological society : records branch – v1-19 + 3 add vols. 1912-66 – (= ser Publications of the english record societies, 1835-1972) – 91mf – 9 – uk Chadwyck [941]
Kent, Charles *see* Vestindiefart
Kent, Charles Foster *see*
– Biblical geography and history
– The founders and rulers of united israel
– The heroes and crises of early hebrew history
– A history of the hebrew people
– A history of the jewish people during the babylonian, persian, and greek periods
– Israel's historical and biographical narratives
– The kings and prophets of israel and judah
– The life and teachings of jesus
– The makers and teachers of judaism
– The messages of israel's lawgivers
– Narratives of the beginnings of hebrew history
– The sermons, epistles and apocalypses of israel's prophets
– The social teachings of the prophets and jesus
– The testing of a nation's ideals
– Twelve studies on the making of a nation
– The wise men of ancient israel and their proverbs
– The work and teachings of the apostles
– The work and teachings of the earlier prophets
Kent, Charles Stanton *see* Tonal expansion in the early romantic period
Kent coast register of hotels, boarding houses and apartments – [London & SE] Kent Arts & Lib, Margate Lib 1893 – 1 – uk Newsplan [640]
Kent coast times and ramsgate and margate observer – [London & SE] Kent Arts & Lib, Ramsgate Lib 17 may 1866-1896 – 1 – (cont as: east kent times & district advertiser [11 mar 1896-12 sep 1932]) – uk Newsplan [072]
Kent County Industrial Union Council, CIO *see* Grand rapids cio news
Kent County Labor Council [MI] *see*
– Grand valley afl-cio news
– Grand Valley labor news
Kent, Donald H *see* Journal of chaussegros de lery
Kent dragonfire – iss 1 [1971] – 1r – 1 – mf#1584079 – us WHS [071]
Kent, Harald Jensen *see* Danske mormoner
Kent herald – Canterbury, England. 1824-70. -w. 18 1/2 reels – 1 – uk British Libr Newspaper [072]
Kent, James *see*
– Commentaries on american law
– Twelve anthems
Kent, Josiah C *see* Northborough history, northborough, mass
Kent, M Chris *see* A satrical interpretation of the history of selected persons, events and organizations in american physical education
Kent messenger and gravesend telegraph & dartford news – England. Kent Messenger – North Kent edition. -w. 1906-69. 103 reels – 1 – uk British Libr Newspaper [072]
Kent messenger and maidstone telegraph – England. -w. 12 Aug 1871-Dec 1891. 20 reels – 1 – uk British Libr Newspaper [072]
Kent, Otis *see* Kent's digest of decisions under the federal safety appliances acts
Kent, Roland Grubb *see* Thirty years of oriental studies
Kent State University *see* Kitabu
Kent state university riots and disorder : newsclippings 1970-1971 – 1r – 1 – mf#B25950 – us Ohio Hist [355]
Kent, Thomas *see* The harp of prophecy
Kent u ze? / Tahitu, A D et al; ed by Stichting Door de Euwen Trouw – Eindhoven, 1966 – 1mf – 9 – mf#SE-1618 – ne IDC [950]
Kent, William *see*
– Drawings and plans for holkham, c1729
– Memoirs and letters of james kent, ll.d., late chancellor of the state of new york
Kent, William Henry *see* Manual of church history
Kentering in de verbondsleer / Hulst, Lammert J – Holland, MI: Holland Print Co, 1917 [mf ed 1993] – (= ser Reformed church coll) – 63p on 1mf – 9 – 0-524-06627-2 – (in dutch) – mf#1991-2682 – us ATLA [242]
Kentish express : all editions – Ashford, England 23 oct 1986- – 1 – uk British Libr Newspaper [072]
Kentish express, ashford and alfred news, hythe gazette – Ashford, England 19 jun-20 nov 1858 – 1 – (cont on kentish express (all ed mf); cont: ashford & alfred news & general advertiser [17 jul 1855-12 jun 1858]; cont by: ashford & alfred news, kentish express, hythe gazette [27 nov 1858]) – uk British Libr Newspaper [072]
Kentish gazette – Canterbury. England. -w. Jan 1772-Dec 1776; Jan 1804-Jul 1806. (4 reels) – 1 – uk British Libr Newspaper [072]
Kentish guardian – Ashford, England. 29 Sept 1868-12 Jan 1869. -w. 1/4 reel – 1 – uk British Libr Newspaper [072]
Kentish independent and kentish mail – London, 1843-1984 [wkly] – 180 3/4 r – 1 – uk British Libr Newspaper [072]
Kentish, John *see* Review of christian doctrine
Kentish mail *see* Kentish independent and kentish mail
Kentish mercury – [London & SE] Lewisham LHAC 1939-64 [wkly] – 1 – (cont: greenwich, woolwich & deptford gazette; aka: south east london mercury; south east london & kentish mercury; deptford & peckham mercury; lewisham mercury; lewisham & catford mercury; lewisham borough mercury) – uk Newsplan [072]
Kentish post, or canterbury news letter – England. -sw. 1 5 Jan 1726-16 20 July 1768. Lacking 1727, 1734, 1735, 1742, 1744. 12 1 2 reels – 1 – uk British Libr Newspaper [072]
Kentish times [bromley & beckenham ed] – Bromley, England 5 oct 2000-20 sep 2002 – 1 – (cont: bromley & beckenham times [7 jun 1990-28 sep 2000]; cont by: bromley times [27 sep 2002-]) – uk British Libr Newspaper [072]
Kentjana *see* Endang
Kenton, Mark A *see* Chronic exercise and the effects on the immune response
Kent's commentaries / Snyder, Emil William – Detroit, Collector 1895 4 v. On film: v. 4. LL-307 – 1 – us L of C Photodup [340]
Kent's digest of decisions under the federal safety appliances acts / Kent, Otis – Washington: GPO, 1910 (all publ) – 1mf – 9 – $4.50 – mf#LLMC 84-391 – us LLMC [344]
Kentucky – (= ser General education board: the early southern program) – 9r – 1 – $1170.00 – us Scholarly Res [370]
Kentucky : attorney general opinions – 1908-98 + 10 yr ind (1980-89) – 9 – $1,209.00 set – (1908-59 on reel $35. 1968-98 + 10 yr ind on mf $1,174) – mf#400870 – us Hein [340]
Kentucky : session laws of american states and territories – 1792-1998 – 9 – $1,911.50 set – mf#402680 – us Hein [348]
Kentucky *see*
– Reports and opinions
– Reports, post-nrs
– Reports, pre-nrs
– Salem on rennix creek
Kentucky Air National Guard *see* Phantom's eye
The kentucky baptist – Franklin, Kentucky. May 17, 1866-June 22, 1867. 192p – 1 – 6.72 – us Southern Baptist [242]
Kentucky Baptist Convention *see* Mountain news and views
Kentucky Baptist Historical Society *see* Publications, no. 3
Kentucky bar association proceedings : 1st to 33rd annual meetings – 1882-84; 1902-34 (all publ) – 38mf – 9 – $171.00 – (cont as: kentucky state bar association proceedings 1st-33rd 1902-34) – mf#LLMC 84-520 – us LLMC [340]
Kentucky bar journal – Frankfort KY 1936-74 – 1,5 – (cont by: kentucky bench & bar) – ISSN: 0362-6113 – mf#3271.02 – us UMI ProQuest [340]
Kentucky bar journal *see* Kentucky bench and bar
Kentucky bench and bar – v1-65. 1936-2001 – 9 – $728.00 set – (title varies: v1-35 n3 1936-71 as kentucky state bar journal. v35 n4-38 1971-74 as kentucky bar journal) – ISSN: 0164-9345 – mf#104471 – us Hein [340]
Kentucky bench & bar – Frankfort KY 1975-94 – 1,5,9 – (cont by: bench & bar) – ISSN: 0164-9345 – mf#3271.02 – us UMI ProQuest [340]
Kentucky bicentennial record / Kentucky Historical Events Celebration Commission – v2 n4-[v3 n2] [[1974 3rd qtr]-1975 sum] – 1r – 1 – mf#382329 – us WHS [071]
Kentucky city / Kentucky Municipal League – 1968 jun-1976 nov – 1r – 1 – (cont: kentucky city bulletin) – mf#4121854 – us WHS [350]
Kentucky conference pulpit : being sermons / Hiner, R – Nashville, Tenn: Pub House of the Methodist Episcopal Church, South, 1874 – 2mf – 9 – 0-524-03845-7 – mf#1990-4892 – us ATLA [240]
Kentucky. Court of Appeals *see*
– Kentucky law reporter
– Kentucky opinions
Kentucky culture : a basic library of kentuckiana – 11,995mf – 9 – (from the holdings of the library of congress, the university of kentucky and the colls of private individuals) – mf#C39-22800 – us Primary [978]
Kentucky foreign language quarterly – Washington DC 1954-66 – 1,5,9 – (cont by: kentucky romance quarterly: krq) – ISSN: 0023-0332 – mf#12702.02 – us UMI ProQuest [400]
Kentucky Genealogical Society *see* Blue grass roots
The kentucky gilpins / Perkins, George Gilpin – Washington, DC: Press of F W Roberts Co, 1927 – 1r – 1 – us Western Res [978]
Kentucky heritage / Kentucky Junior Historical Society – v19 n2-v22 n1 [1978 win-1980 fall] – 1r – 1 – mf#625977 – us WHS [978]
Kentucky Heritage Commission et al *see* Heritage news
Kentucky historical chronicle – 1973 jul 24-1975 dec 22 – 1r – 1 – mf#1061328 – us WHS [978]
Kentucky Historical Events Celebration Commission *see* Kentucky bicentennial record
Kentucky jewish post and opinion – Louisville KY: [s.n.] – 1 – mf#363 – us AJPC [939]
Kentucky Junior Historical Society *see* Kentucky heritage
Kentucky kentucky revised statutes annotated – Charlottesville, Michie Company: 1973 edition thru aug 1999 update – 9 – mf#401780 – us Hein [348]
Kentucky labor news / Kentucky State AFL-CIO [American Federation of Labor and Congress of Industrial Organizations] – 1944 sep 21-1946 apr 11, 1946 apr 18-1950 apr 13, 1950 apr 20-1954 apr 8, 1953 oct 1-1956 mar 21, 1956 mar 28-1958 dec 27, 1959-83, 1984 jan-1989 dec, 1990 jan-1994 dec – 16r – 1 – mf#1061330 – us WHS [331]
Kentucky law journal – v1-14. 1913-1925/26 – 40mf – 9 – $60.00 – (add vols as copyright expires) – mf#LLMC 90-323 – us LLMC [340]
Kentucky law journal – Louisville. v1-2. 1881-82 (all publ) – (= ser Historical legal periodical series) – 1 – $45.00 set – mf#104481 – us Hein [340]
Kentucky law journal – v1-89. 1913-2001 – 5,6,9 – $1463.00 set – (v1-73-1913-85 on reel or mf $973. v74-89 1985-2001 on mf $490) – ISSN: 0023-0264 – mf#104491 – us Hein [340]
Kentucky law reporter : unreported / Kentucky. Court of Appeals – v1-33. 1880-1908 (all publ) – (= ser Kentucky supreme court reports) – 324mf – 9 – $485.00 – (pre-nrs: v1-8 1880-86 67mf $100) – mf#LLMC 84-138 – us LLMC [347]
Kentucky laws made plain. / Morris, Charles Harwood – Sedalia? Mo. Bankers Law, 1906. 100p. LL-251 – 1 – us L of C Photodup [340]
Kentucky. Laws, Statutes, etc *see*
– Acts of the general assembly of the commonwealth of kentucky
– Civil and criminal codes of practice of kentucky; rev. and cor. to july 1, 1908
Kentucky libraries – Frankfort KY 1981+ – 1,5,9 – (cont: kentucky library association bulletin) – ISSN: 0732-5452 – mf#7579.01 – us UMI ProQuest [020]
Kentucky library association bulletin – Frankfort KY 1933+ – 1,5,9 – (cont by: kentucky libraries) – ISSN: 0022-734X – mf#7579.01 – us UMI ProQuest [020]
Kentucky mission monthly – Louisville, Ky. 1902-15. Merged with western recorder, 1919. 2350p – 1 – 82.25 – us Southern Baptist [242]
Kentucky monthly – Frankfort KY 2001+ – 1,5,9 – mf#28165 – us UMI ProQuest [071]
Kentucky Municipal League *see* Kentucky city
Kentucky oil journal – v2 n5-6 [1919 apr 15-may 5] – 1r – 1 – (cont by: oil and mining review) – mf#1097061 – us WHS [622]
Kentucky opinions : unreported / Kentucky. Court of Appeals – v1-13. 1864-86 (all publ) – (= ser Kentucky supreme court reports) – 33mf – 9 – $225.00 – (with 2 digest vols. a pre-nrs title) – mf#LLMC 84-139 – us LLMC [347]
Kentucky pioneer genealogy and records / Society of Kentucky Pioneers – v1 n1-v6 n5 [1979 jan-1983 oct], v6 n1-4 [1985] – 2r – 1 – mf#1009834 – us WHS [929]
The kentucky resolutions of 1798 : an historical study / Warfield, Ethelbert, D – 2nd ed. New York/London: G P Putnam's Sons, 1894 – 3mf – 9 – $4.50 – mf#LLMC 95-089 – us LLMC [323]
The kentucky revival / McNemar, Richard – 1 – 5.53 – us Southern Baptist [242]
Kentucky romance quarterly [krq] – Washington DC 1967-85 – 1,5,9 – (cont: kentucky foreign language quarterly; cont by: romance quarterly) – ISSN: 0364-8664 – mf#12702.02 – us UMI ProQuest [400]
Kentucky school journal – Frankfort KY 1923-72 – 1 – ISSN: 0023-0359 – mf#2042 – us UMI ProQuest [370]
Kentucky sketches – Historical and biographical sketches, Kentucky, 1882-88.With particular coverage of Bourbon, Christian, Fayette, Harrison, Scott, Todd and Trigg Counties. With an index of many more by Bailey F. Davis. Perin ed. 5578p – 1 – us Southern Baptist [920]
Kentucky State AFL-CIO [American Federation of Labor and Congress of Industrial Organizations] *see* Kentucky labor news
Kentucky. State Bar Association *see* Proceedings
Kentucky state bar journal – v1-27. 1936-63 – 29mf – 9 – $130.00 – mf#LLMC 84-521 – us LLMC [340]
Kentucky state bar journal *see* Kentucky bench and bar
Kentucky State Building and Construction Trades Council et al *see* Official proceedings...annual convention of the kentucky state federation of labor
Kentucky state federation of labor *see* Book of laws...
Kentucky state gazetteer and business directory / R L Polk and Co – 1881/82 – 1r – 1 – mf#3537623 – us WHS [978]
Kentucky. Supreme Court *see* Kentucky supreme court reports
Kentucky supreme court reports / Kentucky. Supreme Court – v1-247. 1785-1933 – 2333mf – 9 – $3499.00 – (pre-nrs: v1-84 1785-1886 623mf $934.00. vols after v247 still in copyright) – mf#LLMC 80-807 – us LLMC [347]
Kentucky. Synod (Pres. Church in the USA) *see* Minutes, 1802-1810
Kentzinger, Antoine F de *see* Strasbourg et l'alsace
Kenvyn, Ronald *see* Waterfront wails
Kenwood / Milwaukee Alumni Association. University of Wisconsin – v1 n1-7 [1983 sep-1984 jun], 1984/85 win-1987 sum – 1r – 1 – (cont: alumni news [university of wisconsin. milwaukee]; cont by: uwm today) – mf#1546700 – us WHS [378]
Kenwood United Methodist Church *see* Newsletter
Kenya / Leys, Norman Maclean – London, England. 1925 – 1r – us UF Libraries [960]
Kenya African Affairs Dept *see* Report of the african affairs department
Kenya and the east africa high commission, annual departmental reports relating to... 1903/4-1963 – (= ser Annual departmental reports relating to african countries prior to independence) – 119r – 1 – (with int by h f morris) – mf#97282 – uk Microform Academic [960]
Kenya comment – Nairobi: New Comment, 1957-1958 [wkly] [mf ed [Nairobi]: Kenya National Archives Photographic Service 1970] – n71-v13 n52 [aug 9 1957-dec 26 1958] on 3r – 1 – (incorporating east africa news review [nov 8 1957-dec 26 1958]; iss for nov 8 1957-dec 26 1958 called v12-13 as if vol numbering had been continuous fr 1st iss of east africa news review publ nov 28 1946; ed n71-v13 n4 [1957 9th aug-1958 jan 24] on reel with: new comment [n57-58 [1957 3rd may-10th may], and: new comment for queen and commonwealth [v13 n46-n52 (1958 nov 14-dec 26) on reel with: independent comment, and: independent [v14 n4-n7 (1959 jan 30-feb 20)) – mf#mf-1550 camp r40-42 – us CRL [079]
Kenya Committee on the Organization of Agriculture *see* Report
Kenya (formerly British East Africa). Ministry of Economic Planning and Development. Statistics Division *see* Statistical abstract 1955-1976
Kenya, government publications relating to... 1897-1963 – (= ser Government publications relating to african countries prior to independence) – 134r – 1 – (with int by h f morris) – mf#96995 – uk Microform Academic [960]
Kenya. Land Commission.Carter Commission *see* Minutes of evidence
Kenya. Medical Dept *see* Annual medical report
Kenya Select Committee on the Disappearance and Murder of the Late Member for Nyandarua North *see* Report of the select committee on the disappearance and murder of the late member for nyandarua north
Kenyon, Charles Richard *see* Clive forrester's gold
Kenyon, Frederic G *see* Our bible and the ancient manuscripts
Kenyon, Frederic George *see* The palaeography of greek papyri
Kenyon, Frederic George, Sir *see* Handbook to the textual criticism of the new testament
Kenyon, George Kenyon *see* Observations on the roman catholic question
Kenyon review – Gambier OH 1939-70 – 1,5,9 – (cont by: kenyon review [1979-] #11916) – ISSN: 0163-075X – mf#213, 11916 – us UMI ProQuest [071]
Keola o na mele = The life of music / Musicians' Association of Hawaii – 1984 jul-1985 jul/sep – 1r – 1 – (cont: pahu kani) – mf#1266289 – us WHS [780]
Keough, John *see* Rhymes of a rover
Keough, Walter James *see* The great white banner
Kephalides, August *see* Reise durch italien und sicilien

Kephalometrische untersuchung der skelettalen und dentalen veraenderungen mit der elastoheadgear-apparatur / Emmerich, Kristin – (mf ed 1996) – 1mf – 9 – €30.00 – 3-8267-2306-6 – mf#DHS 2306 – gw Frankfurter [617]

Kephalometrische untersuchung mit einer neuen fernroentgenanalyse nach sergl / Sarabi, Susen Habibi – (mf ed 1996) – 1mf – 9 – €30.00 – 3-8267-2381-3 – mf#DHS 2381 – gw Frankfurter [617]

Kephart, Cyrus Jeffries *see*
– Jesus the nazarene
– The public life of christ

Kephart, Ezekiel Boring *see* A brief treatise on the atonement

Kepler, Johannes *see* Joannis kepleri astronomi opera omnia

Keplinger, Lewis Walter *see* Papers and scrapbook of lewis walter keplinger

Keppel, Frederick, And Company *see* Print-collector's bulletin

Keppel, G *see* Personal narrative of travels in babylonia, assyria, media and scythia in the year 1824

Keppel, George *see* Narrative of a journey across the balcan, by the two passes of selimno and pravadi

Keppler, Paul Wilhelm von *see*
– Die adventsperikopen
– Aus kunst und leben

Ker, James Campbell *see* Political trouble in india, 1907-1917

Ker, Mary Susan *see* Southern women and their families in the 19th century: papers and diaries

Ker, William T *see*
– Church honesty
– Distinctive principles of the free church

Kerala chronicle – Trichur, India. 1962-Apr 1964 – 5r – 1 – us L of C Photodup [079]

Kerala culture : its genesis and early history / Mammen, K – Trivandrum: City Press, 1942 – (= ser Samp: indian books) – (int by a gopala menon) – us CRL [930]

Kerala gazette / Kerala. India – Trivadrum. 1963-1966 – 1 – us NY Public [324]

Kerala. India *see* Kerala gazette

Kerala industry – Trivandrum, India 1980-81 – 1,5,9 – ISSN: 0047-3359 – mf#8121 – us UMI ProQuest [338]

The kerala kaumudi – Trivandrum, India. Jul-Sept 1966 – 1r – 1 – us L of C Photodup [079]

Keramic art of japan / Audsley, George Ashdown – London 1881 – (= ser 19th c art & architecture) – 5mf – 9 – mf#4.2.558 – uk Chadwyck [730]

The keramic gallery : containing several hundred illustrations of rare curious and choice examples of pottery and porcelain from the earliest times to the beginning of the present century / Chaffers, William – London: Chapman & Hall. 4v. 1872 – (= ser 19th c art & architecture) – 7mf – 9 – mf#4.1.205 – uk Chadwyck [730]

Keratry, Emile, comte de *see* Paris exposition 1900 paris-universel english cicerone

Kerbs, Brooke *see* Effects of same-day strength training on shooting skills of female collegiate basketball players

Kerce, Red *see*
– Dummitt orange grove
– Tours 1, 2, and 3

Kerchenskii rabochii – Kerch', Ukraine 1942 [mf ed Norman Ross] – 1 – mf#nrp-681 – us UMI ProQuest [934]

Kerckeliicke historie : ...van het jaer vierhondert of tot in het jaar sestien hondert ende negentien... / Wtenbogaert, J – Rotterdam, 1647 – 14mf – 9 – mf#PBA-373 – ne IDC [240]

Kerckelycke geschiedenissen... / Trigland, J – Leyden, 1650 – 13mf – 9 – mf#PBA-347 – ne IDC [240]

Kerckelycke sermoenen over de feestdaghen... / Bullinger, Heinrich – Amsterdam, Hendrick Laurensz, 1612 – 3mf – 9 – mf#PBU-205 – ne IDC [240]

Kercken ordeninge im lande tho pamern : dorch de dorchluechtigen hochgebarnen foersten unde herren herrn barnim unde herrn philipsen hochloeffliker gedechtnis beide hertogen tho stettin pamern... – Wittenberge: Schwertel 1563 – (= ser Hqab. literatur des 16. jahrh.) – 3mf – 9 – €40.00 – mf#1563b – gw Fischer [780]

Kerckhoffs, August *see* Daniel casper von lohenstein's trauerspiele

Kerdjantara; suara lembaga penelitian daja tenaga dan peralatan perhutanan – Bogor, 1962-1963. nos 1-6 – 4mf – 9 – mf#SE-1743 – ne IDC [950]

Kerem – Varsha, Poland. 1887 – 1r – us UF Libraries [939]

Kerem hemedh – Wien, Austria. 1833-43; 1854-56 – 1 1/2r – 1 – uk British Libr Newspaper [072]

Kerem yisra'el : hu sefer ha-yahas ve-shulshelta di-dahava mi-shene mishpehot rizin ve-tshornobil / Zak, Reuben – Nyu York: S B Rubin, 719 [1958] (mf ed 197-) – 1r – 1 – (repr. originally publ: lublin, 1929 or 1930) – mf#ZZ-16578 – us NY Public [929]

Keren yehoshu'a... / Epshtayn, Yehoshu'a Ben Nahman – Warsaw, Poland. v1-2. 1896 – 1r – us UF Libraries [939]

Keresztje southern cross – Sydney, Australia. 28 feb 1951-15 dec 1956 – 1r – 1 – (aka: fuggetlen magyarorszag free hungary) – uk British Libr Newspaper [079]

Kerfoot, Franklin Howard *see* Abstract of systematic theology

Kerguelen Tremarec, Y J de *see* Relation d'un voyage dans la mer du nord

Keri, hendry county, florida / Huss, Veronica E – s.l, s.l? 193-? – 1r – us UF Libraries [978]

Keri keri chronicle – aug 1978-84 – 6r – 1 – mf#12.13 – nz Nat Libr [079]

Kerista / Abacus, Inc et al – v1 bk 1-v4 bk 2 [1984 sum-1987 aut] – 1r – 1 – (cont by: utopia 2) – mf#1353098 – us WHS [071]

Kerk en eredienst – 1(1945)-13(1958) – 113mf – 9 – €216.00 – ne Slangenburg [240]

Kerk en vrede – v38-46. 1983-91 [complete] – Inquire – 1 – (cont & filmed with: militia christi) – ISSN: 0026-4156 – mf#ATLA S0376A – us ATLA [240]

Kerk en vrede *see* Militia christi

Kerkelijk handboek ten dienste der chr ger kerk in noord-amerika / Dellen, Idzerd van & Keegstra, H – Grand Rapids, MI: Eerdmans-Sevensma, 1915 [mf ed 1993] – (= ser Reformed church coll) – 1mf – 9 – 0-524-06015-0 – (in dutch) – mf#1991-2375 – us ATLA [242]

Kerkelijk leesblad : ten dienste der cleefs- en gelderlandsche catholijken – 1(1800)-2(1801) – 21mf – 9 – €26.00 – ne Slangenburg [241]

Kerkelijke bibliotheek : voornamelijk voor de roomsch-catholyken in nederland – Amsterdam/ Grave, 1(1794)-2(1795) – 8 – €80.00 – (tweede deel. mengelwerk 1795 8mf; uittreksels en beoordelingen 1795 6mf; kerk-nieuws 1795 5mf) – ne Slangenburg [241]

Het kerkelyk en wereltlyk deventer, deel 1 / Dumbar, Gerhard – Arnhem, 1752 – €57.00 – ne Slangenburg [242]

Kerken, Georges Van Der *see* Afrikaanse bevolking van belgisch-kongo en van ruanda-urundi

Kerkenorde der christelijke gereformeerde kerk : zooals herzien en vastgesteld door de synode gehouden to chicago (roseland), ill., den 17 juni, 1914 e.v.d / Heyns, William – Grand Rapids, Mich: Wm B Eerdmans, 1927 – 1mf – 9 – 0-524-06020-7 – mf#1991-2380 – us ATLA [240]

Kerkeraads-protocollen der hollandsche gemeente te londen 1569-1571 (de werken...1/1) / ed by Kuyper, A – Utrecht, 1870 – (= ser De werken der marnix-vereeniging) – €14.00 – ne Slangenburg [242]

Kerkhistorisch archief – 1(1857)-4(1866) – 24mf – 9 – €46.00 – (1870...studien en bijdragen op 't gebied der historische theologie. 1885...archief voor nederlandsche kerkgeschiedenis) – ne Slangenburg [240]

Kerkhistorische studien / Sepp, Christiaan – Leiden: EJ Brill, 1885 – 1mf – 9 – 0-7905-6889-6 – (incl bibl ref) – mf#1988-2889 – us ATLA [240]

Kerkmusiek vir hedendaagse tieners / Tonder, Barend Jacobus van – Uni of South Africa 2001 [mf ed Pretoria: UNISA 2001] – 2mf – 9 – (incl bibl ref; abstract in afrikaans and english; text in afrikaans) – mf#mfm14885 – sa Unisa [240]

Kerl, Georg *see* Robespierres kirchenpolitik

Kerle, Jacobus de *see*
– Egregia cantio...
– Liber psalmorum ad vesperas
– Preces speciales pro salubri generalis concilii
– Sex misse...liber primus

[Kerman-] kerman news – CA. 1971-72; 1981- – 10r – 1 – $600.00 (subs $50y) – mf#B02324 – us Library Micro [071]

La kermesse – Quebec: L Brousseau, [1892-1893] – 9 – (incl ind) – mf#P04485 – cn Canadiana [071]

Kern anmuthiger und zeit-kuerzender [...] wissenschaften [...] – Erfurt DE, 1744-45 – 1r – 1 – gw Misc Inst [074]

Kern, Berthold *see* Gustav freytag, ein publizist

Kern county – 1909-35 – (= ser California telephone directory coll) – 26r – 1 – $1300.00 – mf#P00041 – us Library Micro [917]

[Kern county-] fresno, inyo, kern, merced, san bernardino, stanislaus and tulare counties – CA. 1884-1885 – 2r – 1 – $100.00 – mf#D020 – us Library Micro [978]

Kern County Genealogical Society *see* Kern-gen

[Kern county-] kern county including bakersfield – CA. 1899; 1936-1949 – 14r – 1 – $700.00 – mf#D043 – us Library Micro [978]

[Kern county-] kern, los angeles, san bernardino, san diego, san luis obispo, santa barbara and ventura counties – CA. 1875 – 1r – 1 – $50.00 – mf#D044 – us Library Micro [978]

Kern County Labor Council *see* Union labor journal

Kern county survey / U.S. Library of Congress. Prints and Photographs Division – 415 views from Kern County, CA in the late 1880's. 4 maps. 1 reel. P&P43 – 1 – us L of C Photodup [080]

[Kern county-] taft city directories – CA. 1926; 1949 – 2r – 1 – $100.00 – mf#D045 – us Library Micro [917]

Kern county union labor journal / Central California Non-partisan Alliance et al – [1920 oct 9/1922 feb 10]-[1962/1963 jan 31] – 21r – 1 – (cont: union labor journal; union labor journal; cont by: kern, inyo, and mono counties union labor journal; kern, inyo, and mono counties union labor journal) – mf#629037 – us WHS [331]

Kern, Franz *see*
– Goethes tasso und kuno fischer
– Torquato tasso

Kern, H *see* Geschiedenis van het buddhisme in indie

Kern, Hendrik *see* Manual of indian buddhism

Kern, inyo, and mono counties union labor journal – 1963 feb 7-1965, 1966-1971 feb 25 – 2r – 1 – (cont: kern county union labor journal; cont by: labor journal [bakersfield, ca]) – mf#895346 – us WHS [331]

Kern, inyo, and mono counties union labor journal – v76 n1-v80 n1 [1979 jan-1983 jan/feb] – 1r – 1 – (cont: labor journal [bakersfield, ca]; cont by: union labor journal [1983]) – mf#895369 – us WHS [331]

Kern, Inyo, Mono Counties Building and Construction Trades Council *see* Union labor journal

Kern, Inyo, Mono Counties Central Labor Council *see* Labor journal

Kern, Jack C *see* Tennis racket coefficient of restitution under static and dynamic conditions

Kern, Joh *see* Schwaebisches magazin zur befoerderung der aufklaerung

Kern, Josephus *see* De sacramento extremae unctionis

Kern, Karl Peter *see* Der judentaufe

Kern melodischer wissenschafft : bestehend in den auserlesensten haupt- und grund-lehren der musicalischen setz-kunst oder composition, als ein vorlaeuffer des vollkommenen capellmeisters / Mattheson, Johann – Hamburg: C Herold 1737 [mf ed 19–] – 5mf – 9 – mf#fiche 439 – us Sibley [780]

Kern, O [comp] *see* Orphicorum fragmenta

Kern, Oskar *see* Johann rist als weltlicher lyriker

Kern, Otto *see*
– Beitraege zur geschichte der griechischen philosophie und religion
– Inscriptiones graecae

Kern, R A *see* Rapporten betreffende java afkomstig van r a kern, adviseur inlandse

Kernahan, James *see* A suggestive commentary on st luke

Kernals *see* Jefferson county miscellaneous newspapers

Kernals of political thots n' observations *see* Jefferson county miscellaneous newspapers

Kernel and the husk : letters on spiritual christianity / Abbott, Edwin Abbott – Boston: Roberts Bros, 1887 – 1mf – 9 – 0-8370-2014-X – (incl gloss) – mf#1985-0014 – us ATLA [240]

Kerner, Johann *see* Reise ueber den sund

Kerner, Justinus *see* Justinus kerners saemtliche poetische werke

Kerner, Justinus Andreas Christian *see* The seeress of prevost; being revelations concerning the inner-life of man, and the inter-diffusion of a world of spirits in the one we inhabit

Kern-gen / Kern County Genealogical Society – 1982-89 – 1r – 1 – mf#355988 – us WHS [929]

Kernholt, Otto *see* Vom ghetto zur macht

Kernighan, Robert Kirkland [pseud: The Khan] *see* War poems

Kernisan, Charles Emmanuel *see* Republique d'haiti et la gouvernement democratique

Kernisan, Clovis *see*
– Etrangers et le droit de propriete immobiliere
– Verite ou la mort

Kernot, Henry *see* Bibliotheca diabolica

Keroack, Christopher R *see* The effects of alpha-tocopherol on metabolic determinations in graded exercise

Ker-Porter, R *see* Travels in georgia, persia, armenia, ancient babylonia etc during the years 1817-1820

Kerr, Alastair James *see* Native laws of succession in south africa

Kerr, Alfred *see*
– Clemens brentanos jugenddichtungen
– Godwi

Kerr, Amabel *see* The life of cesare cardinal baronius of the roman oratory

Kerr, Charles H *see* Shcho dumaiut' sotsiialisty?

Kerr, David Shank *see* In the supreme court of new brunswick, (crown side) in the matter of david s kerr, barrister

Kerr, Dianne L *see* An hiv education needs assessment of selected teacher members of the american school health association and the american home economics association

Kerr, James *see*
– Britain's legislation on education
– Christ's testimony to the doctrine of everlasting punishment
– The covenants and the covenanters
– From montreal, in appeal
– Protestant commemoration in 1888 of the defeat of the spanish armad...

Kerr, James Manford *see*
– Kerr's mines and water cases annotated
– Practice on attachment and garnishment of property in the state of ohio
– A selection of adjudicated criminal forms and precedents of indictments and informations.

Kerr, John *see*
– Curling in canada and the united states
– John kerr papers, 1788-1844
– The renascence of worship

Kerr, John Henry *see*
– A harmony of the gospels
– An introduction to the study of the books of the new testament

Kerr, Kathleen A *see* Differentiation of ethnic culture regions using laban movement analysis

Kerr, Malcolm H *see* America's middle east policy

Kerr, R *see* Pioneering in morocco

Kerr, Rev Dr *see* Endowment of romanism in ireland though the "christian brothers'"...

Kerr, Robert *see*
– The consulting architect
– The gentleman's house
– The king of men
– The newleafe discourses on the fine art architecture

Kerr, Robert Pollock *see* Presbyterianism for the people

Kerr, Robert Pollok *see*
– The blue flag
– The people's history of presbyterianism in all ages

Kerr, S R *see* A comparative study of leisure lifestyles

Kerr, Samuel C *see* The jewish church in its relations to the jewish nation and to the "gentiles"

Kerr, Thomas H *see* Critical survey of printed vocal arrangements of afro-american religious folk songs

Kerr, Wilfred Brenton *see* Bermuda and the american revolution

Kerr, William Hastings *see* The fishery question

Kerr, William Henry *see* The king's keys to his kingdom

Kerr-carr collector – v1 n1-v3 n4 [1983 jan-1985 oct] – 1r – 1 – mf#1031506 – us WHS [071]

Kerr's mines and water cases annotated : american, english and canadian / Kerr, James Manford – Chicago: Callaghan. 1v. 1912 (all publ) – 9mf – 9 – $13.50 – mf#LLMC 95-134 – us LLMC [343]

Kerry advocate – Tralee, Ireland. -w. 25 jul 1914-16 may 1916 – 1 1/2r – 1 – uk British Libr Newspaper [072]

Kerry, Esther *see* He is a canadian

Kerry evening post – Tralee, Ireland. -d. 24 may 1813; 22 dec 1824-29 sep 1917 – 89r – 1 – uk British Libr Newspaper [072]

Kerry evening star – Tralee, Ireland. -w. 29 sep 1902-12 mar 1914 – 11 1/2r – 1 – uk British Libr Newspaper [072]

Kerry examiner and munster general observer – Tralee, Ireland. 11 aug 1840-19 oct 1849; 1 mar-20 dec 1850; 1851-8 aug 1854; 2 jan-16 dec 1855; 8 jan-11 mar 1856 – 12r – 1 – uk British Libr Newspaper [072]

Kerry examiner and munster general observer – Tralee, Ireland. -w. 11 Aug 1840-19 Oct 1849, 1 Mar 1850-8 Aug 1854, 5 Jan 1855-11 Mar 1856. (11 reels) – 1 – uk British Libr Newspaper [072]

Kerry independent – Tralee, Ireland. 28 oct 1880-10 jan 1854; 5, 12 jul 1884 (imperfect) – 3r – 1 – uk British Libr Newspaper [072]

Kerry news – Tralee 1939-16th jun 1941 – 1r – 1 – ie National [072]

Kerry news – Tralee, Ireland. 1895; 1896; 1900; 19 feb-15 oct 1930 – 3 1/2r – 1 – uk British Libr Newspaper [072]

Kerry people – Tralee, Ireland. 27 sep 1902-14 mar 1914; 24 nov-dec 1917; 12 jan-9 feb 1918; 22, 29 jun; 28 sep 1918-3 sep 1921; 7 jan-26 aug 1922. -d – 12 3/4r – 1 – uk British Libr Newspaper [072]

Kerry press – Tralee, Ireland. 28 jul 1914-11 may 1916.-w – 1 3/4r – 1 – uk British Libr Newspaper [072]

Kerry reporter *see* Kerry weekly reporter and commercial advertiser

Kerry sentinel – Tralee, Ireland. 26 apr 1878-1896; 1897-1911; 1912-aug 1918 – 39 1/4r – 1 – uk British Libr Newspaper [072]
Kerry star – Tralee, Ireland. 15 may 1861-27 mar 1863.-w – 1 1/4r – 1 – uk British Libr Newspaper [072]
Kerry weekly reporter and commercial advertiser – Tralee, Ireland. 3 feb 1883-14 aug 1920 [wkly] – 33r – 1 – (aka: kerry reporter) – uk British Libr Newspaper [072]
Kerryman – Tralee, Ireland. 1910-12; 1986-93 – 28 1/2r – 1 – uk British Libr Newspaper [072]
Kerrys eye (people power) *see* New kerrys eye
Kersey, Jesse *see* A treatise on fundamental doctrines of the christian religion
Kershaw second baptist church. kershaw, south carolina : church records – 1916-72 – 1 – us Southern Baptist [242]
Kershner, Frederick Doyle *see*
– Christian baptism
– The religion of christ
Kerssebomn, W *see* Essays in political arithmetic
Kerssenbrock, Hermann von *see* Hermanni a kerssenbroch anabaptistici furoris
Kersten *see* Wielands verhaeltnis zu lucian
Kersten, O *see* Baron c c von der decken's reisen in ost-afrika in den jahren 1859 bis 1865
Kersuzan, Francoise Marie *see* Catechisme creole
Kertbeny, Karoly M von *see* Berlin wie es ist
Keruba, the robber – London, England. 18- – 1r – us UF Libraries [240]
Kerval, Leon de *see* Le r. p. hugolin de doullens, ou, la vie d'un frere mineur
Kerwick, Andrew *see* A report of the trials of the caravats and shanavests, at the special commission, for the several counties of tipperary, waterford, and kilkenny.
Das kerygma petri / Dobschuetz, Ernst von – Leipzig: J C Hinrichs, 1893 – (= ser Tugal) – 1mf – 9 – 0-7905-1813-9 – (incl ind) – mf#1987-1813 – us ATLA [240]
Das kerygma petri / Dobschuetz, Ernst von – Leipzig, 1893 – (= ser Tugal 1-11/1) – 3mf – 9 – €7.00 – ne Slangenburg [240]
Kerygma und dogma – 1(1955)-12(1966) – 73mf – 9 – €140.00 – ne Slangenburg [240]
Kerygma und dogma – Goettingen, Germany 1959-63 – 1,5,9 – ISSN: 0023-0707 – mf#1909 – us UMI ProQuest [200]
Kesari – Poona, India. 2 Jul 1943-27 Sept 1949; 17 Jan 1954-Sept 1991 – 128r – 1 – us L of C Photodup [079]
Kesatuan buruh kerakjatan indonesia : berita kbki – Djakarta, 1956-1957 – 5mf – 9 – (missing: 1956(1-17, 19)) – mf#SE-346 – ne IDC [959]
The kesava temple at belur / Narasimhachar, R – Bangalore: Mysore Govt Press, 1919 – (= ser Samp: indian books) – us CRL [720]
Keshab chandra sen and the brahma samaj : being a brief review of indian theism from 1830 to 1884, together with selections from mr. sen's works / Slater, Thomas Ebenezer – Madras: SPCK, 1884 – 1mf – 9 – 0-524-02372-7 – mf#1990-2983 – us ATLA [280]
Keshenev / Korn, Yitshak – Buenos Ayres, Argentina. 1950 – 1r – us UF Libraries [939]
Kesher – Mexico City, Mexico. v1 n1-v9 n119. 1987 jun-1995 – 3r – (gaps) – us UF Libraries [079]
Keshet, Yeshurun *see* Be-doro shel bialik
Keshub chunder sen's england visit / ed by Dobson Collet, Sophia – London: Strahan & Co, Publishers, 1871 – (= ser Samp: indian books) – us CRL [280]
Keshub chunder sen's english visit / Sen, Keshub Chunder; ed by Collet, Sophia Dobson – London: Strahan, 1871 – 2mf – 9 – 0-524-04520-8 – mf#1990-3354 – us ATLA [280]
Keshub chunder sen's essays, theological and ethical. part 2 = Essays. selections / Sen, Keshub Chunder – Calcutta: Brahmo Tract Society, 1886 – 1mf – 9 – 0-524-02545-2 – mf#1990-3040 – us ATLA [280]
Keshub chunder sen's prayers = Prayers. Selections / Sen, Keshub Chunder – Calcutta: Brahmo Tract Society, 1884 – 1mf – 9 – 0-524-02546-0 – mf#1990-3041 – us ATLA [240]
Keskula, DR *see* The reliability of an isokinetic measurement protocol for the posterior rotator cuff musculature
Keskusosuuskunnan tiedonantaja / Co-operative Central Exchange [US] – n2- [1929 dec 19-31] – 1r – 1 – (cont by: tyovaen osuustoimintalehti) – mf#929165 – us WHS [332]
The kesler and ellmore debate : held at jasonville, indiana, september 29 to october 6, 1915, covering the differences between the church of the brethren and the church of christ (disciples) / Kesler, Benjamin Elias – Elgin, Ill: Brethren Pub House, 1916 – 1mf – 9 – 0-524-02829-X – mf#1990-4450 – us ATLA [242]
Kesler, Benjamin Elias *see*
– The kesler and ellmore debate
– The riggle-kesler debate
Kesoema negara : soeara angkatan moeda indonesia – Medan, 1945-1946 – 3mf – 9 – (missing: 1945/1946 v1-2(1-2, 5, 10, 11)) – mf#SE-724 – ne IDC [950]
Kessel, Johann Christian Bertram *see* Unterricht im generalbasse zum gebrauche fuer lehrer und lernende
Kessel, Joseph *see* Marches d'esclaves
Kessle, Gun *see* Ansikte av sten
Keßler, Aloys Wilhelm *see* Die bedeutung der im art. 119 i 2 der reichsverfassung ausgesprochenen gleichberechtigung der geschlechter de lege ferenda fuer die personenrechtliche stellung der ehefrau
Kessler, Edeltraud *see* Schulpaedagogisches repetitorium
Kessler, Franz Josef *see* Sinn und umsetzung eines vertriebswirtschaftlichen marketing-konzeptes im produktionsbetrieb im allgemeinen und speziell am beispiel der kfz-zulieferindustrie
Kessler, Hans *see* Die psalmen
Kessler, Hugo *see* Der fuenffuessige jambus bei christian dietrich grabbe
Kessler, J *see*
– Joachimi vadiani vita
– Sabbata
Kessler, Johannes *see* Johannes kesslers sabbata
Kessler, Lina *see* Vergleichende religionswissenschaft und inspiration der heiligen schrift
Kesson, J *see* The cross and the dragon
Kesten, Hermann *see*
– 24 neue deutsche erzaehler
– Meine freunde die poeten
Kester, Vaughan *see*
– The just and the unjust
– The prodigal judge
Kester, Vaughan, 1869-1911 *see* The fortunes of the landrays
Kestin, Lipe *see* Naye himlen
Kestler Farnes, Maximiliano *see* Introduccion a la teoria constitucional guatemalte
Das kestnerbuch / Kuppers, Paul Erich – Hannover: H Bohme 1919 [mf ed 1981] – 1r [ill] – 1 – mf#116 – us Wisconsin U Libr [760]
Keta'im mi-devarav / Katznelson, Berl – Tel-Aviv, Israel. 1950 – 1r – us UF Libraries [939]
Keta'im mi-mishnato shel b borokhov / Borochov, Ber – Tel-Aviv, Israel. 1957 or 1958 – 1r – us UF Libraries [939]
Ketav bet yisrael / Reicher, Isaiah – Seini, Romania. 1927 – 1r – us UF Libraries [939]
Ketav sofer / Schreiber, Abraham Samuel Benjamin – Budapest, Hungary. 1941 – 1r – us UF Libraries [939]
Ketavim / Nomberg, Hersh David – Warsaw, Poland. 1911 – 1r – us UF Libraries [939]
Ketavim nivharim / Zeitlin, Hillel – Warsaw, Poland. v1-2. 1910 or 11-1919 or 20 – 1r – us UF Libraries [939]
Ketav-yad 'ivri / Bogrov, Grigorii Isaakovich – Piotrkow, Poland. 1900 – 1r – us UF Libraries [939]
Ketcham kables – v1 n1-v6 n4 [1983 sep/nov-1989 apr/jun] – 1r – 1 – mf#1061395 – us WHS [071]
Ketcheson, W G *see* The pilgrim's pilot
Ketchikan alaska chronicle – Ketchikan AK. 1947 mar 29 – 1r – 1 – mf#846214 – us WHS [071]
Ketchikan Indian Corporation *see* Tlin tsim hai
Ketchum, Henry George Clopper *see*
– The chignecto ship railway
– Public opinion on the chignecto ship railway and the baie verte canal
Ketchum, William Quintard *see* Requiescant
Ketelsen, Uwe-K *see* Komoedien des barock
Keter torah / Frank, Zevi Pesah – Yerushalayim, Israel. 1937 – 1r – us UF Libraries [939]
Keterangan saikoo sikikan : tentan hal toeroet mengambil bahagian dalam pemerintahan negeri dan pendjelasan pemerintah / Java. (Japanese Military Administration). Saiko Shikikan – (Djakarta): Djawa-Gunseikanbu (2603) – 14p 1mf – 9 – mf#SE-2002 mf86 – ne IDC [355]
Kethan, J *see* Compendio de la salud humana, zaragoza, 1494
Kethnerum, Leonhardum *see* Die hymni oder geistlichen lobgeseng
Ketkar, Shridhar Venkatesh *see*
– An essay on hinduism
– Evidence of the laws of manu on the social conditions in india during the third century a d
– Hindu law
Ketler, Isaac Conrad *see* The tragedy of paotingfu
Keto, Zodwa Lucy *see* Labour dispute resolution at vista university
Ketteler, Thomas *see* Elektrophysiologische und haemodynamische effekte von magnesium auf spaete reperfusionsarrhythmien bei akutem myokardinfarkt
Ketteler wacht – Koeln DE, 1953-75 – 1 – gw Misc Inst [074]
Ketteler, Wilhelm Emmanuel, Freiherr von *see*
– Das allgemeine concil und seine bedeutung fuer unsere zeit
– Die arbeiterfrage und das christenthum
– Freiheit, autoritaet und kirche
– Die grossen socialen fragen der gegenwart
– Kann ein glaeubiger christ freimaurer sein?
– Die thatsaechliche einfuehrung des bekenntnisslosen protestantismus in die katholische kirche
– Das unfehlbare lehramt des papstes nach der entscheidung des vaticanischen concils
– Die unwahrheiten der roemischen briefe vom concil in der allgemeinen zeitung
Ketten, J M von der *see* Apelles symbolicus exhibens seriem amplissimam symbolorum, poetisque, oratoribus ac verbi dei praedicatoribus conceptus subministrans varios
Ketterer, Johann Adam *see* Karl der grosse und die kirche
Ketterer, Ralf *see* Radio, moebel, volksempfaenger
Kettering evening telegraph *see* Evening telegraph
Kettering leader and observer – [East Midlands] Northamptonshire 20 jan 1890-1895, mar 1898-apr 1973 [mf ed 2003-04] – 82r – 1 – (missing: 1897, 1898, 1911; cont by: kettering leader [jan 1902-jun 1923]; kettering leader & guardian and northamptonshire advertiser [jul 1923-dec 1950]) – uk Newsplan [072]
Kettering-oakwood times / Montgomery Co. Dayton – may 1956-dec 1972 [wkly, semiwkly] – 24r – 1 – mf#B5361-5384 – us Ohio Hist [071]
Kettle moraine index – Delafield, Dousman, Genesee etc WI. 1992 dec, 1993, 1994 jan-jun, 1994 jul-dec, 1995 jan-jun, 1995 jul-dec – 6r – 1 – (cont: index [dousman, wi]) – mf#2634476 – us WHS [071]
Kettlewell, John *see*
– Companion for the penitent
– Office for persons troubled in mind
Kettlewell, Samuel *see*
– The authorship of the de imitatione christi
– Thomas a kempis and the brothers of common life
– Thomas a. kempis and the brothers of the common life
Kettner, Emil *see* Osterreichische nibelungendichtung
Kettner, Gustav *see*
– Goethes nausikaa
– Schillers demetrius
– Ueber den religioesen gehalt von lessings nathan dem weisen
– Ueber lessings emilia galotti
– Ueber lessings minna von barnhelm
Kettner, Matthias *see* Kant und die idealismusfrage
Kettwiger zeitung – Essen DE, 1949 26 nov-1950 1 jun, 1959 19 mar-1960 8 mar, 1960 13 may-31 dec [gaps] – 2r – 1 – (filmed by misc inst: 1950 2 jun-1957 [30r]) – gw Mikrofilm; gw Misc Inst [074]
Die ketzergeschichte des urchristenthums / Hilgenfeld, Adolf – Leipzig: Fues, 1884 – 2mf – 9 – 0-7905-1101-0 – (incl bibl ref and indexes) – mf#1987-1101 – us ATLA [240]
Der ketzer-katalog des bischofs maruta von maipherkat / Harnack, Adolf von – Leipzig, 1899 – (= ser Tugal 2-19/1b) – 1mf – 9 – €3.00 – ne Slangenburg [240]
Der ketzer-katalog des bischofs maruta von maipherkat *see* Die todestage der apostel paulus und petrus
Die ketzertaufangelegenheit in der altchristlichen kirche nach cyprian : mit besonderer beruecksichtigung der konzilien von arles und nicaea / Ernst, Johann – Mainz: F Kirchheim, 1901 – (= ser Forschungen zur christlichen Litteratur- und Dogmengeschichte) – 1mf – 9 – 0-7905-6743-1 – (incl bibl ref) – mf#1988-2743 – us ATLA [240]
Keuangan dan bank / Perbankan Nasional Swasta – Djakarta, 1967/1968. v1-2(1-10) – 7mf – 9 – (missing: 1968 v1(7-8)) – mf#SE-1744 – ne IDC [332]
Keuangan dan bank / Sekretariat Perbana – Djakarta, 1963 – 7mf – 9 – (missing: 1963(1-5, 11)) – mf#SE-297 – ne IDC [332]
Keune, O *see* Maenner, die nahrung schufen
Keur, John Yak *see* Windward children
Keussen, H *see* Die matrikel der universitaet koeln 1389 bis 1559 (pgrg7)
Kevisi, John (?) *see* Solomon islands diary
Kevutsat mikhtavim she-nishlehu le-anshe shem / Wissotzky, Kalonymus Ze'ev... – Warsaw, Poland. 1898 – 1r – us UF Libraries [939]
Kevutsat shirim / Mieses, Fabius – Krako, Poland. 1891 – 1r – us UF Libraries [939]
Kew bulletin : (no longer sold) – 1887-1916 – 1 – $432.00 – ((no longer sold). 1917-74 $1048 [0311]) – mf#0310 – us Brook [580]
Kew bulletin – Norwich, England 1946+ – 1,5,9 – ISSN: 0075-5974 – mf#9888 – us UMI ProQuest [580]
Kew. Royal Botanical Gardens *see* Cumulated index kewensis
Kewanee daily star-courier – Kewanee IL. 1908 oct 7 v14 n181, 1941 dec 7 v35 n294, 1941 dec 8 v35 n295, 1941 dec 9 v35 n296, 1945 may 7 v39 n108, 1945 aug 8 v39 n186, 1945 aug 9 v39 n187, 1945 aug 10 v39 n188, 1945 aug 11 v39 n189, 1945 aug 13 v39 n191, 1945 aug 14 – 1r – 1 – mf#1010908 – us WHS [071]
Kewaskum statesman – Kewaskum WI. [1895 oct 5/1898 jul 30]-[1997] – 60r – 1 – (cont by: statesman [kewaskum, wi]) – mf#1043722 – us WHS [071]
Kewaunee county banner – Kewaunee WI. 1906 feb 1-1909, 1910-13, 1914-17, 1918-21, 1922-1925 aug 20 – 5r – 1 – mf#1012296 – us WHS [071]
Kewaunee county enterprise – Kewaunee WI. 1865 feb 29-dec, 1866 jan 3-1868 jun 10 – 2r – 1 – (cont: kewaunee county enterprize; cont by: kewaunee enterprise) – mf#1045612 – us WHS [071]
Kewaunee county enterprize – Kewaunee WI. 1859 oct 26-1861, 1862-1865 feb 22 – 2r – 1 – (cont: kewaunee enterprize; cont by: kewaunee county enterprise) – mf#1045608 – us WHS [071]
Kewaunee county press – Kewaunee WI. 1918 jun 22, 1919 jul 19, 1921 18-25, nov 12, dec 10, 1922 mar 4, 1923 jul 7-14, nov 24, 1924 feb 16-23, 1925 may 9, 1926 may 28, 1918 nov 2-dec 7,28-1919 jan 18, feb 1, mar 1 [partial], 15, apr 12-19 – 2r – 1 – mf#5491715 – us WHS [071]
Kewaunee county star – Algoma WI. 2002 may 2-dec, 2003 jan-jun, 2003 jul-dec, 2004 jan-jun, 2004 jul-dec – 5r – 1 – (cont: kewaunee star) – us WHS [071]
Kewaunee enterprise – Kewaunee WI. [1869 jan 13/1870]-[2004 oct/dec] [gaps] – 125r – 1 – mf#1032490 – us WHS [071]
Kewaunee enterprize – Kewaunee WI. 1859 jun 22-oct 19 – 1r – 1 – (cont by: kewaunee county enterprize) – mf#1045602 – us WHS [071]
Kewaunee herald – Kewaune WI. 1907 sep 6 – 1r – 1 – us WHS [071]
Kewaunee star – Kewaunee WI. [1948 jul 22/sep 30]-[2002 jan/dec] [gaps] – 54r – 1 – (cont by: kewaunee county star) – mf#935034 – us WHS [071]
Kewaunske listy – Kewaunee WI. 1892-94, 1895-96, 1897-1900, 1901-04, 1905-08, 1909-12, 1913-1917 apr 11 – 7r – 1 – mf#1012311 – us WHS [071]
Key – La Salle IL 1798 – 1 – mf#3528 – us UMI ProQuest [390]
Key / Madison Apartment Association et al – 1980 jul-1982 feb, 1982 mar-jun, sep-1983 apr, 1986 jul-1989 dec – 3r – 1 – (cont: profits and prophets; cont by: wisconsin apartment news) – mf#585260 – us WHS [366]
Key / Opportunities Industrialization Center National Institute et al – 1970 nov, 1971 fall, nov10, dec 10, 1972 jan 10, feb 10 – 1r – 1 – (cont by: key news) – mf#5078433 – us WHS [331]
Key and sounder / Commercial Telegraphers' Union – v6-v9 n3 [1959 jan 23-1962 mar 23] – 1r – 1 – mf#1061398 – us WHS [331]
Key, Ellen *see* Fu nu yun tung
Key family newsletter – v4 n1-v6 n4 [1983 mar 30-1985 dec 31] – 1r – 1 – (cont: key newsletter) – mf#1477157 – us WHS [929]
Key, Fanny *see* Green cove springs, florida, 1816
Key, Francis Scott *see*
– Defence of fort m'henry-star-spangled banner
– The star spangled banner
Key, Helmer *see* Kaffee, zucker und bananen
Key, Joseph *see* Eighteen marches & c
Key, lock and lantern – 1970 sum-1982/83 win – 1r – 1 – mf#1047996 – us WHS [071]
Key look-out – Sarasota, FL. v4 n1-v4 n7. 1959 may 28-1959 aug 20 – 1r – (missing: jun 25-jul 23) – us UF Libraries [071]
Key news / Opportunities Industrialization Centers of America – 1972 apr, 1978 spr, special ed – 1r – 1 – (cont: key [philadelphia, pa: 1970]) – mf#5078434 – us WHS [338]
Key newsletter – v1 n1-v3 n4 [1980 jan 2-1982 dec 23] – 1r – 1 – (cont by: key family newsletter) – mf#1428763 – us WHS [071]
The key note : substitute honest money for fictitious credit... / Griffen, Albert – Philadelphia: S L Griffin, 1896 (mf ed 19–) – 448p – mf#ZT-545 – us Harvard [332]
The key of doctrine and practice / Haweis, Hugh Reginald – "New edition." London: [J. Martin & Son, printers, 1884] Beltsville, Md: NCR Corp, 1978 (3mf); Evanston: American Theol Lib Assoc, 1984 (3mf) – 9 – 0-8370-0910-3 – mf#1984-4262 – us ATLA [240]
Key of the pacific / Colquhoun, Archibald Ross – Westminster, England. 1895 – 1r – us UF Libraries [972]
The key of truth : a manual of the paulician church of armenia: the armenian text / ed by Conybeare, Frederick Cornwallis – 1mf – 9 – 0-8370-8803-8 – mf#1986-2803 – us ATLA [240]

"Key persons" files of the president's commission on the assassination of president Kennedy, 1963-1964 / U.S. President's Commission – (= ser Records of the president's commission on the assassination of president kennedy) – 34r – 1 – mf#M1289 – us Nat Archives [324]

Key place names / Saunders, H J – s.l, s.l? 1937 – 1r – us UF Libraries [978]

Key to baillairge's stereometrical tableau : new system of measuring all bodies...by one and the same rule / Baillairge, Charles P Florent – Quebec: C Darveau, 1874 – 1mf – 9 – mf#02495 – cn Canadiana [510]

Key to health / Gandhi, Mahatma – Ahmedabad: Navajivan Pub House, 1948 – (= ser Samp: indian books) – (trans by sushila nayar) – us CRL [613]

The key to life *see* Jen sheng kai lun (ccm154)

Key to north american birds : containing a concise account of every species of living and fossil bird at present known from the continent north of the mexican and united states boundary, inclusive of greenland / Elliott Coues – 2nd rev ed. Boston: Estes and Lauriat, 1884 – us CRL [590]

Key to north american birds / Coues, Elliott – Boston, MA. v1-2. 1927 – 1r – us UF Libraries [500]

A key to story's equity jurisprudence : containing over eight hundred questions / Guernsey, Rocellus Sheridan – New York, Diossy, 1876. 133 p. LL-270 – 1 – us L of C Photodup [342]

Key to the exercises in the handbook of the venda language / Ziervogel, D – Cape Town, South Africa. 1965 – 1r – us UF Libraries [960]

A key to the exercises of the new method of learning the hebrew language / Herxheimer, S – London: Franz Thimm, 1866 [mf ed 1986] – 1mf – 9 – 0-8370-9247-7 – mf#1986-3247 – us ATLA [470]

Key to the gulf – s.l, s.l? 1908 – 1r – us UF Libraries [978]

Key to the hebrew psalter : a lexicon and concordance combined wherein are all the words and particles contained in the book of psalms / Alcock, George Augustus – London: Elliot Stock, 1903 – 1mf – 9 – 0-8370-9121-7 – mf#1986-3121 – us ATLA [221]

A key to the irish question / Fox, J A – London, 1890 – (= ser 19th c ireland) – 5mf – 9 – mf#1.1.9755 – uk Chadwyck [941]

A key to the knowledge and use of the holy bible / Blunt, John Henry – London: Rivingtons; New York: Pott & Amery, 1868 [mf ed 1989] – 1mf – 9 – 0-7905-0612-2 – (incl ind) – mf#1987-0612 – us ATLA [220]

Key to the massoretic notes, titles, and index generally found in the margin of the hebrew bible / Hahn, August – New York: John Wiley, 1884 – 1mf – 9 – 0-7905-0130-9 – mf#1987-0130 – us ATLA [221]

The key to the missionary problem : thoughts suggested by the report of the ecumenical missionary conference, held in new york, april 1900 / Murray, Andrew – 4th ed. New York: American Tract Society, [pref.1902] Beltsville, Md: NCR Corp, 1978 (3mf); Evanston: American Theol Lib Assoc, 1984 (3mf) – 9 – 0-8370-0996-0 – (incl bibl ref) – mf#1984-4352 – us ATLA [240]

Key to the mystery / Richer, Edouard – Belfast, Northern Ireland. 1853 – 1r – us UF Libraries [240]

A key to the narrative of the acts of the apostles / Norris, John Pilkington – new rev ed. London: Rivingtons, 1877 [mf ed 1985] – 1mf – 9 – 0-8370-4599-1 – (incl bibl ref) – mf#1985-2599 – us ATLA [226]

a key to the pentateuch explanatory of the text and the grammatical forms : pt 1, genesis / Deutsch, Solomon – New York: Holt & Williams 1871 [mf ed 2005] – 1r – 1 – 0-524-10520-0 – (no more publ; a "school ed publ without notes" was publ the same year) – mf#b00732 – us ATLA [221]

Key to the popery of oxford / Maurice, Peter – London, England. 1838 – 1r – us UF Libraries [240]

A key to the psalms : being a tabular arrangement, by which the psalms are exhibited to the eye according to a general rule of composition prevailing in the holy scriptures / Boys, Thomas; ed by Bullinger, Ethelbert William – 2nd ed. London: Eyre & Spottiswoode, 1899 [mf ed 1993] – 1mf – 9 – 0-524-05714-1 – mf#1992-0557 – us ATLA [221]

The key to theosophy : being a clear exposition, in the form of question and answer, of the ethics, science, and philosophy for the study of which the theosophical society has been founded / Blavatsky, Helena Petrovna – London: Theosophical Pub. Society; New York: William Q Judge, c1889 – 1mf – 9 – 0-524-01044-7 – mf#1990-2192 – us ATLA [290]

Key trends and issues impacting local government recreation and park administration in the 1990s / Whyte, Digby N B & Martin, W Donald – 1992 – 4mf – 9 – $16.00 – us Kinesology [790]

Key, V O *see* Southern politics in state and nation

Key west : the gibraltar of america / Harris, Sam – s.l, s.l? 193-? – 1r – us UF Libraries [978]

Key west – s.l, s.l? 193-? – 1r – us UF Libraries [978]

Key west citizen – Key West, FL. 1926 feb 18-1983 dec – 281r – (gaps) – us UF Libraries [071]

Key west enquirer – Key West, FL. 1834 oct 15-1836 sep 17 – 1r – us UF Libraries [071]

Key west fishing – s.l, s.l? 193-? – 1r – us UF Libraries [978]

Key west, florida – a gem of an island / Willis, J A – St Augustine, FL. 1914? – 1r – us UF Libraries [978]

Key west gazette – Key West, FL. 1831 apr 20-1832 sep 15 – 1r – us UF Libraries [071]

Key west papers (various titles) – Key West, FL. n d – 1r – us UF Libraries [071]

Key west wpa strike of december, 1935 – s.l, s.l? 1935 – 1r – us UF Libraries [978]

A key word index to the notes in american negligence cases – Chicago. 1v. 1914 (all publ) – (= ser American negligence cases) – 2mf – 9 – $3.00 – mf#LLMC 84-699E – us LLMC [348]

Keya paha call – Springview, NE: G W Fritz, 1897// (wkly) [mf ed v1 n8. mar 26-sep 3 1897 (gaps) filmed 1974] – 1r – 1 – (merged with: springview herald to form: keya paha call and springview herald-consolidated) – us NE Hist [071]

Keya paha call and springview herald-consolidated – Springview, NE: G W Fritz, 1897 (wkly) [mf ed -1898 (gaps) filmed 1974] – 1r – 1 – (formed by the union of: keya paha call and: springview herald. cont by: springview herald and keya paha call consolidated) – us NE Hist [071]

Keya paha county news – Springview, NE: P Skinner, may 1904 (wkly) [mf ed v1 n2. may 12-nov 24 1904 (gaps) filmed 1998] – 1r – 1 – us NE Hist [071]

Keya paha press – Springview, NE: Barnwell & Oxley, 1885 (wkly) [mf ed v3 n8. feb 18 1887-dec 9 1897 (gaps) filmed 1979] – 1r – 1 – us NE Hist [071]

Keyboard – Manhasset NY 1981+ – 1,5,9 – (cont: contemporary keyboard) – ISSN: 0730-0158 – mf#10650.01 – us UMI ProQuest [780]

Keyboard music, 1600-1820 – 1997 – 326mf – 9 – €2265.00 – mf#M387 – ne MMF Publ [780]

Keyboard works of henry purcell / McRae, James A – U of Rochester 1965 [mf ed 19–] – 2mf – 9 – (with bibl) – mf#fiche 80 – us Sibley [780]

Keyes' appeals reports / New York. (State) – v1-4. 1863-68 (all publ) – 30mf – 9 – $45.00 – (a pre-nrs title) – mf#LLMC 80-011 – us LLMC [340]

The keyes of the kingdom of heaven, and power thereof, according to the word of god / Cotton, John – Boston: Reprinted by Tappan and Dennet, 1843 – 1mf – 9 – 0-524-00965-1 – mf#1990-4023 – us ATLA [240]

Keyes, Wade *see* An essay on the learning of partial, and of future interests in chattels personal

Keyhole for roger williams' key / Ely, William D – 1892 – 1 – $5.00 – us Southern Baptist [242]

Keymer, Nathaniel *see* Notes on genesis

Keyn amerike / Pat, Jacob – Varshe, Poland. 1920 – 1r – us UF Libraries [939]

Keynes, John Maynard *see*
- The john maynard keynes papers in king's college, cambridge
- Treatise on money

Keynes, John Neville *see* Economists' papers

Keynote – New York NY 1972-80 – 1,5,9 – ISSN: 0047-3413 – mf#9652 – us UMI ProQuest [640]

The keys of power : a study of indian ritual and belief / Abbott, John – London: Methuen & Co, 1932 – (= ser Samp: indian books) – us CRL [280]

The keys of saint peter : or, the house of rechab connected with the history of symbolism and idolatry / Bunsen, Ernest de – London: Longmans, Green, 1867 – 2mf – 9 – 0-8370-2531-1 – mf#1985-0531 – us ATLA [270]

The keys of st peter : or, a liberal protestant view of the claims of the papacy / Wendte, Charles William – [s.l]: iss by the First Unitarian Church [ca 1871] [mf ed 1991] – (= ser The oakland unitarian pulpit ser1/3) – 1mf – 9 – 0-524-01142-7 – mf#1990-0356 – us ATLA [241]

The keys of the kingdom : or, the unfailing promise / Moriarty, James Joseph – New York: Catholic Publ Society; London: Burns & Oates, c1885 – 1mf – 9 – 0-8370-7005-8 – mf#1986-1005 – us ATLA [230]

Keys to control : israel's pursuit of arab water resources / Schmida, Leslie – Washington, DC: American Educational Trust, [1983?] – us CRL [327]

Keys to the word : or, help to bible study / Pierson, Arthur Tappan – New York: Anson D F Randolph, c1887 – 1mf – 9 – 0-8370-4752-8 – mf#1985-2752 – us ATLA [220]

Keys, William *see* Capital and labor

Keyser, Cassius Jackson *see* Science and religion

Keyser, H de *see* Architectura moderna ofte bouwinge van onsen tyt

Keyser, Harriette A *see* Bishop potter

Keyser, Leander Sylvester *see*
- Election and conversion
- The rational test
- A system of christian ethics

Keyser, Rudolph *see* The religion of the northmen

Keyserling, Alfred *see* Graf alfred keyserling erzaehlt

Keyserling, E *see* Die arachniden australiens nach der natur beschrieben und abgebildet

Keyserling, H *see* Reisetagebuch eines philosophen

Keyserling, Hermann Alexander, Graf von *see* Amerika, der aufgang einer neuen welt

Keyserling, Hermann, Graf von *see* Hazon-eropah

Keysoe – (= ser Bedfordshire parish register series) – 2mf – 9 – £5.00 – uk BedsFHS [929]

Keysoe (brook end) monumental inscriptions – Bedfordshire Family HS 1977 – (= ser Bedfordshire parish register series) – 1mf – 9 – £1.25 – uk BedsFHS [929]

Keysoe row east baptist monumental inscriptions – Bedfordshire Family HS 2003 – (= ser Bedfordshire parish register series) – 1mf – 9 – £1.25 – uk BedsFHS [929]

Keyssner, Gustav *see* Ausgewaehlte werke

Keystone / Pennsylvania Railroad Technical and Historical Society et al – v9 n3 [1976 sep], v11 n1-v17 n4 [[1978 spr-1984 win] – 1r – 1 – mf#1022466 – us WHS [380]

The keystone : and other essays on freemasonry / Lawrence, John Thomas – London: A Lewis 1913 – 4mf – 9 – mf#vrl-68 – ne IDC [366]

Keystone Alliance *see* Radio activist reporter

The keystone and woman's column – v1-14 n9. 1899-1913 [all publ] – (= ser Periodicals on women and women's rights, series 2) – 2r – 1 – $350.00 – us UPA [305]

Keystone area local news and views / American Postal Workers Union – 1974 dec-1977 nov/dec – 1r – 1 – (cont: harr-penn dispatch) – mf#667614 – us WHS [331]

Keystone baptist – Philadelphia, PA. Pennsylvania Baptist General Convention. 1922-nov 1930 – 1 – us ABHS [242]

Keystone country peddler / Keystone Publishing Co – 1982 oct-1987 aug – 1r – 1 – (cont by: midwest country peddler) – mf#1061410 – us WHS [070]

Keystone folklore – West Chester PA 1973-92 – 1,5,9 – (cont: keystone folklore quarterly) – ISSN: 0149-8444 – mf#6577,01 – us UMI ProQuest [390]

Keystone folklore quarterly – West Chester PA 1956-72 – 1,5 – (cont by: keystone folklore) – ISSN: 0023-0987 – mf#6577.01 – us UMI ProQuest [390]

Keystone kuzzins : bulletin of the erie society for genealogical research / Erie Society for Genealogical Research – v6 n4-v15 n4 [1978 may-1987 may] – 1r – 1 – mf#1611567 – us WHS [929]

Keystone of the sacerdotal system / Stopford, Edward A – Dublin, Ireland. 1870 – 1r – us UF Libraries [240]

Keystone Publishing Co *see* Keystone country peddler

Keystone report : official publication, united food and commercial workers union, local 1036 / United Food and Commercial Workers International Union – 1987 aug-1992 mar/apr – 1r – 1 – (cont: local 1036's newspaper; cont by: solidarity [camarillo, ca]) – mf#1113860 – us WHS [331]

Keystone reporter – Elroy WI. 1997 jan-dec, 1998 jan-dec, 1999 jan-jul – 15r – 1 – (cont: wonewoc reporter; tribune keystone) – mf#3745744 – us WHS [071]

Keystones of faith : or, what and why we believe / Calkins, Wolcott – New York:Baker and Taylor, c1888 – 1mf – 9 – 0-8370-3100-1 – (incl bibl ref and index) – mf#1985-1100 – us ATLA [240]

Keyte, John Charles *see* The passing of the dragon

Keyton, Robert *see* Griffes

Keywords in the teaching of jesus / Robertson, A T – Philadelphia: American Baptist Publ Society, 1906 – 1mf – 9 – 0-7905-0223-2 – (incl indes) – mf#1987-0223 – us ATLA [240]

KFKB Radio, Milford, KS *see* Reception reports and letters received

Kfz-kurier – Magdeburg DE, 1964 16 mar-1990 may – 4r – 1 – (verkehrskombinat) – gw Misc Inst [621]

Kgatle, Selaelo Thias *see* Moloi ga a na mmala

Kha ram nu ron pvan sac : [short stories] – Mantale: Mra Kan sacape 1983 [mf ed 1990] – [ill] 1r with other items – 1 – (in burmese) – mf#mf-10289 seam reel 151/6 [§] – us CRL [830]

Kha, U *see* Yokkya tam khvan kyam

Kha vha tsireledze nwana wavho = Protect your child / South Africa. Department of National Health and Population Development [Departement van Nasionale Gesondheid en Bevolkingsontwikkeling – 2nd ed. Pretoria: Dept of National Health & Population Development 1986 [mf ed Pretoria, RSA: State Library [199-]] – 46p [ill] on 1r with other items – 5 – (in venda; also available in afrikaans, english, northern sotho, sotho, tsonga, tswana, xhosa & zulu) – mf#op 08454 r25 – us CRL [362]

Khabaradara, Aradesara Pharamaji *see*
- The silken tassel
- Zarathushtra

Khabardar, Ardeshir Framji *see* New light on the gathas of holy zarathushtra

Khabar'namah – [Tehran]: Chirik'ha-yi Fada'i-i Khalq-i Iran. shumarah-'i 12-62. 11 bahman 1358-tir 1360 [31 jan 1980-jun 1981] – 1r – 1 – $55.00 – (missing: n1-11, 14, 17-22) – us MEDOC [079]

Khabar'namah-'i jang – [Tehran]: Jabhah-'i Milli-i Iran. shumarah-'i 4,7-9. 29 mihr-10 azar 1359 [21 oct-1 dec 1980] – 1r – 1 – $175.00 – (r also incl: paykar) – us MEDOC [079]

Khabar'namah-i jang *see* Paykar

Khabar'namah-'i jibhah-'i milli-'i iran – Tehran: Jabhah-'i Milli-'i Iran. shumarah-'i 1-36. urdibihisht 1347-farvardin 1353 [apr/may 1968-mar/apr 1974] – 1r – 1 – $53.00 – (missing: n31-35) – us MEDOC [079]

Khach Giang Ho *see*
- Diep vu chong hong quan nhat
- Diep vu duong day hoa luc
- Mac tu khoa dem khong ngu
- Thau cay

Khaketla, B Makalo *see* Moshoeshoe le baruti

Khaki : the canadian army magazine / Ottawa. Minister of National Defence – v1-4 n25. may 5 1943-sep 23 1945// – 2r – 1 – Can$180.00 – cn McLaren [355]

Khalidi, Rashid *see*
- Soviet middle east policy in the wake of camp david
- The soviet union and the middle east in the 1980's

Khalij-i fars – Ahwaz, 1979- . sal-i 1, shumarah-'i 1-16. 19 tir-3 azar 1358 [10 jul-24 nov 1979] – 1r – 1 – $53.00 – us MEDOC [079]

Khaliq – Astrakhan, Russia 1914-15 [mf ed Norman Ross] – 1r – 1 – mf#nrp-210 – us UMI ProQuest [077]

Khalq – sal-i 1, shumarah-'i 15-sal 5, shumarah-'i 23. 10 shahrivar 1304-26 1304 aban [1 sep 1925-17 nov 1925] – 1r – 1 – $53.00 – (cont: aflak; r also incl: aflak, nahid, and sitarah-i subh) – us MEDOC [079]

Khalq *see*
- Nahid
- Sitarah-'i subh

Khalq-i musalman – Tehran: Hizb-i Jumhuri-i Khalq-i Musalman, 1979- . shumarah-'i 1-18. 25 murdad-18 azar 1358 [11 aug-9 dec 1979] [irreg] – 1r – 1 – $53.00 – us MEDOC [079]

Khalsa sewak – Chandigarh, India. May 1967-Mar 1968; Jul 1968-Feb 1970 – 7r – 1 – us L of C Photodup [079]

Khami ruins / Robinson, Keith Sevill Radcliffe – Cambridge, England. 1959 – 1r – us UF Libraries [960]

Khan, Abdul Majid *see* Life and speeches of sardar patel

Khan, Abdul Wajid *see* Financial problems of indian states under federation

Khan Chve U *see* Nrim khyam pa ce

Khan Cin Lhuin, Yuvati *see* Rvhe lham buil

Khan Co, Do *see* Mran ma akkhara phrac po la pum

Khan, Gazanfar Ali *see* With the pilgrims to mecca

Khan Khan *see* Sippam be dan

Khan Khan Cin, Do *see* Pu gam khet yan kye mhu

Khan Khan Le, Da gun *see*
- Ca chui to nan tvan vatthu kri
- Da gun khan khan le e ca chui to vatthu
- Prann tvan chon pa pon khyup

Khan Khan Le, Dagun *see*
- Coc ron khrann
- Phe sann khya

Khan, Mahmood A *see*
- Journal of college and university foodservice
- Journal of international hospitality, leisure and tourism management 1090 a multinational and cross-cultural journal of applied research
- Journal of nutrition in recipe and menu development

Khan Me Nnvan *see* Mran ma' lak cvai

Khan Mon Krann, U *see* Mran ma rhe hon aup khyup re asavan mya

Khan Mon Kri *see* Tuik khan nam pat 2

Khan Mon Kri, U *see* Mran ma re nam

Khan Mon Nnvan *see* Kyon sa lu nya phat ru phav

Khan Nnui, Mon *see* Na lyn dan kham pu gam rvhe prnn

Khan Sak Ve, Do *see* Khyac khyan metta

Khan, Shafa'at Ahmad *see*
- Anglo portuguese negotiations relating to bombay, 1660-1677
- Ideals and realities

Khan, Shafaat Ahmad *see* John marshall in india

Khan, Shahnawaz *see* My memories of ina and its netaji

Khan, Yusuf Husain *see* Nizamu'l-mulk asaf jah i

Khandalavala, Karl *see* The laud ragamala miniatures

Khanna, Radha Krishna *see* India in the new world order

The khan's weekly – [Toronto?: s.n, 1895-189- or 19–] – 9 – mf#P05058 – cn Canadiana [073]

Khanskie iarlyki russkim mitropolitam / Priselkov, M D – 3mf – 8 – mf#U-697 – ne IDC [243]

Khanykov, N *see*
- Note sur le yarligh d'abou-said khan conserve sur les murs de la mosquee d'ani
- Quelques inscriptions musulmanes d'ani et des environs de bakou

Khao co hoc / Hanoi. Institute d'Archeologie. Comite des Sciences Sociales du Viet-Nam – Hanoi. n1-9 10. juin 1969-juin 1971 – 1 – fr ACRPP [930]

Khao phanit – Bangkok, Thailand. 1971; 1973-74; 1983-91 – 40r – 1 – us L of C Photodup [079]

Khaparde, G S *see* Papers, 1914-1919

Kharakter otnoshenii rossii k pravoslavnomu vostoku v 16 i 17 stoletiiakh / Kapterev, N F – 1885 – 7mf – 9 – mf#ILM-2520 – ne IDC [243]

Kharbinskii den' – Kharbin: Iuzhnoe T-vo, dec 17-25, 29-31 1913 (1r); jan-mar 21, apr 2-jun 29, jul 3-sep 13 1914 (3r) – 1 – us CRL [079]

Kharbinskii viestnik – [Kharbin: s.n., jun 10 1903-04; jan 1, jun 15 1905; mar 24 1911-14; 1916-jan 22 1917] – 26r – 1 – us CRL [079]

Kharchova promyslovist : ministerstvo vyshchoi i serednoi spetsialnoi osvity ursr – Kyiv: Tekhnika. v11-17. 1971-73 – us CRL [947]

The kharias / Roy, Sarat Chandra, Rai Bahadur & Roy, Ramesh Chandra – Ranchi: "Man in India" Office, 1937 – (= ser Samp: indian books) – (foreword by r r marett) – us CRL [954]

Khar'kov – Khar'kov: Koop t-vo izd "Khar'kov" 1919 [1919-] – (= ser Asn 1-3) – n2-4 [1919] item 438, on reel n86 – 1 – mf#asn-1 438 – ne IDC [077]

Khar'kovskaia guberniia v sel'sko-khoziaistvennom otnoshenii, 1893 g – Khar'kov, 1893 – 2mf – 8 – mf#RZ-201 – ne IDC [314]

Khar'kovskie gubernskie vedomosti – Khar'kov, Ukraine 1838-1917 [mf ed Norman Ross] – 140r – 1 – mf#nrp-689 – us UMI ProQuest [077]

Khar'kovskii krakh : po povodu protsessa o zloupotrebleniiakh v khar'kovskom zemel'nom i torgovom bankakh / Gertsenshtein, M Ia – Spb, 1903 – 3mf – 9 – mf#REF-507 – ne IDC [332]

Khar'kovskii rabochii – Khar'kov, Ukraine [mf ed Norman Ross] – 1 – mf#nrp-696 – us UMI ProQuest [077]

Khashkes, M Ia *see* Sefer ha-yomi

The Khasis / Gurdon, Philip Richard Thornhagh – London: Macmillan, 1914 – (= ser Samp: indian books) – (int by charles lyall) – us CRL [954]

The khasis / Gurdon, Philip Richard Thornhagh – 2nd ed. London: Macmillan, 1914 [mf ed 1995] – (= ser Yale coll) – xxiv/232p (ill) – 1 – 0-524-09227-3 – (int by charles lyall. incl ind) – mf#1995-0227 – us ATLA [954]

Khaskovska tribuna – Khaskovo, Bulgaria. Sept 1959-1982 – 23r – 1 – us L of C Photodup [949]

Khastagira, Sudhira *see* Dances in lino cut

Khat sac kaby mala / San San Ae, Do – Yan Kun: Sajan Ae Cape 1967 [mf ed 1990] – 1r with other items – 1 – (in burmese) – mf#mf-10289 seam reel 164/4 [§] – us CRL [959]

Khatti, Humayum Sharif *see* The turkish constitution proclaimed under midhat pasha

Khauer, Frank *see* Nationalsozialistische politik in suedosteuropa

Khauke, O A *see*
- Krestianskoe zemelnoe pravo
- Russkoe zemleustroitelnoe zakonodatelstvo

Khaver-pavers mayselakh / Chaver-Paver – NYU York, NY. v1-2. 1925 – 1r – us UF Libraries [939]

Khaybar weekly – Kabul, Afghanistan. 1931; 1932; 28 feb, 14, 21, 28 mar, 11, 25 apr, 2, 9, 16, 26 may, 15, 20 jun 1973 – 1/4r – 1 – uk British Libr Newspaper [072]

Khayelitsha : interpreting a process of social transformation / Conradie, Catharina Maria – U of the Western Cape 1992 [mf ed S.l: s.n. 1992] – 5mf – 9 – (abstract in afrikaans & english; incl bibl) – sa Misc Inst [307]

Khazan, David Moiseevich *see* Light industries of the ussr

Les khazars dans la passion de s abo de tiflis / Peeters, P – 1mf – 8 – (analecta bollandiana, bruxelles, paris 1934 v52) – mf#U-637 – ne IDC [956]

Khdam kamadeb : phnaek jivit sok / Gang Pun Jhyan – Bhnam Ben: Pannagar Qamatah 2513 [1970] [mf ed 1990] – 1r with other items – 1 – (title on spine khtum cam tep; in khmer) – mf#mf-10289 seam reel 103/01 [§] – us CRL [480]

Khedkar, Raghunath Vithal *see*
- Adwaitism and the religions of the east
- A hand book of the vedant philosophy and religion
- Philosophic discussions

Kheinman, S A *see* Kadry gosudartstvennogo i kooperativnogo apparata sssr

Kheiralla, Ibrahim George *see*
- Bab-ed-din
- Beha ullah (the glory of god)
- O christians!
- Facts for behaists

Kheisin, M *see* Rabochii klass i kooperatsiia

Kheisin, M L *see*
- 50 let potrebitelskoi kooperatsii v rossii
- Chto mozhet dat kooperatsiia rabochim
- Istoricheskii ocherk kooperatsii v rossii
- Kak organizovat i vesti potrebitelskoe obshchestvo
- Kooperatsiia v selskom khoziaistve
- Kreditnaia kooperatsiia v rossii
- Organizatsiia upravleniia potrebitelskikh obshchestv i ikh soiuzov
- Rabochii vopros v kooperatsii
- V pomoshch rabotnikam v potrebitelnykh obshchestvakh

Khelemer shtime – Chelm, Poland 1933-39 [mf ed Norman Ross] – 1r – 1 – (in yiddish) – mf#nrp-402 – us UMI ProQuest [939]

Khemradhipti – Bhnam Ben: [s.n.] (daily) [mf ed Ithaca NY: [John M Echols Collection] Cornell University 2001] – 1r – 1 – (in khmer) – mf#mf. – us CRL [480]

Khera, P N *see* British policy towards sindh

Kherakanouthiun hajkakan / Awetik'ean, Gabriel – Wenetik: S Lazar 1815 – (= ser Whsb) – 7mf – 9 – €75.00 – (armenische grammatik) – mf#Hu 227 – gw Fischer [490]

Khet cam pum pran mya : [short stories] / Thippam Mon Va et al – Yan Kun: Pu Gam Ca Aup 1976 [mf ed 1990] – (= ser Pu gam ca aup 250) – 1r with other items – 1 – (in burmese) – mf#mf-10289 seam reel 155/2 [§] – us CRL [830]

Khet hon khet sac ca krann tuik khai, 1057-1973 / Kaung Myin, Maung – Ran Kun: Ca pe Biman 1978 [mf ed 1990] – (= ser Pann su lak cvai ca can) – 1r with other items – 1 – (in burmese; with bibl & ind) – mf#mf-10289 seam reel 165/6 [§] – us CRL [020]

Khet po mran ma kabya = Modern burmese poetry – Yan Kun: Mravati 1978 [mf ed 1990] – (= ser Mravati ca can) – 1r with other items – 1 – (trans by win pe; burmese & english on opp pgs) – mf#mf-10289 seam reel 154/3 [§] – us CRL [810]

Khet sac mran ma ca ka pre lamnnvhan / Lha Samin – [Ran kun: Udan Ca pe 1974] [mf ed 1993] – on pt of 1r – 1 – (in burmese) – mf#11052 r602 n6 – us Cornell [830]

Khet sac mran ma, prann to sa – Ran Kun: Ca pe bi man pum nhip tuik 1954 [mf ed 1990] – [pl/ill] 1r with other items – 1 – (in burmese) – mf#mf-10289 seam reel 148/3 [§] – us CRL [330]

Khet ta kan, lu ta yok, su amran / On Pe, U – Yan Kun: Capay U Cape 1976 [mf ed 1990] – 1r with other items – 1 – (in burmese) – mf#mf-10289 seam reel 164/2 [§] – us CRL [959]

Khetu angalip-mranma abhidhan = Khit u english-burmese dictionary / Lvan U, U – Rankun: Sve pumnhiptuik 1979 [mf ed 1990] – 1r with other items – 1 – mf#mf-10289 seam reel 154/1 [§] – us CRL [040]

Khevenhiller, Chr *see* Annales ferdinandei

Khi chieu giang luoi / Mai Thao – Saigon: Tieng Phuong Dong 1974 [mf ed 1992] – on pt of 1r – 1 – mf#11052 r390 n6 – us Cornell [959]

Khi em hai muoi : truyen dai / Hai Son – [Saigon]: Thien Huong 1974 [mf ed 1993] – on pt of 1r – 1 – mf#11052 r423 n2 – us Cornell [830]

Khiav Jum, Bhikkhu dhammapal *see*
- Buddhasasana prajadhipateyy sadharanaratth
- Nibvan nau ena?

Khieu Samphan *see*
- The economy of cambodia and its problems with industrialization
- Une paix veritable sera retablie au cambodge apres la cessation de toute intervention des imperialistes americains

Khimiia seraorganicheskikh soedinenii, soderzhashchikhsia v neftiakh i nefteproduktakh – Ufa: Bashkirskii filial AN SSSR. v5-8. 1963-68 – us CRL [947]

Khin Myo Chit, Daw *see* Colourful burma

Khiriakov, A *see* Desiat let trudovoi gruppy

Khirimpana / Ratau, J Khathatso – Morija, Zimbabwe. 1955 – 1r – us UF Libraries [960]

Khitrenko, N I *see* Dal'ne-vostochnyi telegraf

Khitrova, M I *see* Lug dukhovnyi

Khitrowo, B *see* Itineraires russes en orient

Khitrowo, B de *see* Itineraires russes en orient

Khivad : daily newspaper in pashto – Kabul, Afghanistan 1956-88 [mf ed Norman Ross] – 15r – 1 – mf#nrp-587 – us UMI ProQuest [079]

Khizhniakov, V V *see*
- Istoriia i organizatsiia soveta vserossiiskikh kooperativnykh sezdov
- Kooperatsiia v shkole
- Reviziia kooperativnykh tovarishchestv
- Sostavlenie khoziaistvennogo plana v promyslovykh tovarishchestvakh i arteliakh
- Spravochnaia knizhka dlia selskokhoziaistvennykh i selskokhoziaistvennykh kreditnykh tovarishchestv
- Uchites revizii!
- Zemstvo i kooperatsiia

Khlac citt nau la pan siak : [a novel] / Lamn Pen Siak – Bhnam Ben: Ron Bumb Camroen Ratth 1971 [mf ed 1990] – 1r with other items – 1 – (in khmer; title on spine: klauch chet naulapansiec) – mf#mf-10289 seam reel 107/3 [§] – us CRL [830]

Khlac qi nyn tha panha jati / Lan Hap Qan – Bhnam Ben: Ron Bumb Sahakaran yoen 2516 [1972] [mf ed 1990] – 1r with other items – 1 – (in khmer) – mf#mf-10289 seam reel 118/2 [§] – us CRL [480]

Khleb i volia – London, Geneva, 1903-1905. nos 1-24 – 4mf – 9 – mf#R-18187 – ne IDC [077]

Khlebnaia selskokhoziaistvennaia kooperatsiia i sotsialisticheskoe pereustroistvo krestianskogo khoziaistva / Bauman, K I – 1929 – 32p 1mf – 9 – mf#COR-461 – ne IDC [335]

Khlopchato-bumazhnye tovary – 1925 – 360p 4mf – 9 – mf#COR-757 – ne IDC [335]

Khmaer 500.000 chnam! / Chatra Prem – Bhnam Ben: Ron Bumb Van Kat [196-?] [mf ed 1990] – 1r with other items – 1 – (in khmer) – mf#mf-10289 seam reel 119/2 [§] – us CRL [959]

Khmaer qangar – [Bhnam Ben: s.n. 1972- (daily) [mf ed Ann Arbor MI]: UMI 1995] – 1r – 1 – mf#mf-12839 seam – us UMI ProQuest [079]

Khmelevskii, I A *see* Kanskii vestnik

Khmel'nitskaia, Evgeniia Semenovna *see* Ocherki perekhodnoi ekonomiki

Le khmer : hebdomadaire independant pour la defense des interets du cambodge – Pnom-Penh: M Leneveu (wkly) [mf ed Paris, France: Association pour la conservation and la reproduction photographique de la presse [19–]] – 1r – 1 – (began in 1935; subtitle varies) – mf#mf-9167 seam – us CRL [959]

Le khmer : hebdomadaire independant pour la defense des interets du cambodge – Phnom-Penh. 28 dec 1935-36 – 1 – fr ACRPP [073]

Khmer armed resistance / Khmer Peace Committee – [n.p.] 1952 [mf ed 1989] – 1r with other items – 1 – mf#mf-10289 seam reel 016/02 [§] – us CRL [335]

[Khmer communist leaders: biographies] – [1975] [mf ed 1989] – 1r with other items – 1 – (contents: khieu samphan, chou chet, sok thuok, son sen, koy thuon, saloth sar, ieng sary) – mf#mf-10289 seam reel 017/30 [§] – us CRL [335]

The khmer crisis : a re-interpretation / Paeng-Meth, A Gaffar – [n.p.] 1970 [mf ed 1989] – 1r with other items – 1 – mf#mf-10289 seam reel 015/37 [§] – us CRL [959]

The khmer king at basan [1371-1373] and the restoration of the cambodian chronology during the 14th and 15th centuries / Wolters, O W – [s.l: s.n.] 1965 [mf ed 1989] – 1r with other items – 1 – (incl bibl ref) – mf#mf-10289 seam reel 022/03 [§] – us CRL [930]

The khmer language = Kkhmerskii iazyk / Gorgoniev, IUrii Aleksandrovich – Moscow [Russia]: Nauka Pub House, Central Dept of Oriental Literature 1966 [mf ed 1989] – (= ser Languages of asia and africa) – 1r with other items – 1 – (trans fr russian by v korotky) – mf#mf-10289 seam reel 029/01 [§] – us CRL [480]

Khmer Peace Committee *see* Khmer armed resistance

Khmer representation at the united nations : a question of law or of politics? / Rasy, Douc – London: Rasy 1974 [mf ed 1989] – 1r with other items – 1 – (french ed...publ in june 1974; stamp ed on t.p: press & information office, embassy of the khmer republic...washington dc) – mf#mf-10289 seam reel 019/04 [§] – us CRL [341]

Khmer republic : speech delivered by long boret to the 27th session of the general assembly of the united nations / Boret, Long – [n.p.] 1972 [mf ed 1989] – 1r with other items – 1 – mf#mf-10289 seam reel 016/29 [§] – us CRL [327]

The khmer republic / Great Britain. Central Office of Information. Reference Division – London [1971] [mf ed 1989] – 1r with other items – 1 – (with bibl) – mf#mf-10289 seam reel 018/05 [§] – us CRL [320]

The khmer republic : a struggle to preserve the nation / Peang-Meth, A Gaffar – [Washington DC: s.n. 1974?] [mf ed 1989] – 1r with other items – 1 – mf#mf-10289 seam reel 017/15 [§] – us CRL [320]

Khmer rouge top secret [5-21] santebal archives / Parti communiste du Kampuchea – Phnom Penh: Yale Uni Library with the Documentation Center of Cambodia [1998]- – 1 – (pt of cooperation with yale's cambodia genocide project and the documentation center of cambodia. seam contributed funds for filming & will receive positive copy. 132 reels expected. reels 114-117 rec'd dec 15 1999. reels 24-29, 31-35, 40-57, 59, 61, 63-90, 92-113, 118-132 rec'd feb 23 2000; with ind) – us CRL [959]

Khmers / Beriault, Raymond – Montreal [Quebec]: Editions Lememac [1957] [mf ed 1989] – 1r with other items – 1 – mf#mf-10289 seam reel 001/08 [§] – us CRL [959]

Les khmers : des origines d'angkor au cambodge d'aujourd'hui / Migot, Andre – [Paris] Le Livre contemporain [1960] [mf ed 1989] – (= ser Coll l'aventure du passe) – 1r with other items – 1 – mf#mf-10289 seam reel 006/01 [§] – us CRL [959]

Les khmers / Thierry, Solange – [Paris] Editions du Seuil [1964?] [mf ed 1989] – (= ser Le temps qui court 33) – 1r with other items – 1 – (with bibl) – mf#mf-10289 seam reel 005/06 [§] – us CRL [959]

Khmoc bray qasurakay : [short stories] / Jjit Khyai – Bhnam Ben: Nagar Dham 1973 [mf ed 1990] – 1r with other items – 1 – (in khmer) – mf#mf-10289 seam reel 105/12 [§] – us CRL [830]

Khnum man dos broh qvi? : [a novel] / Drin Van – Kaban Dham: Drin Sivgim 2505 [1962] [mf ed 1990] – 1r with other items – 1 – (in khmer) – mf#mf-10289 seam reel 098/01 [§] – us CRL [830]

Khnum oy va kham rian : [a play] / Thu Dhan – Bhnam Ben: Pannagar Gim Gi 2507 [1964] [mf ed 1990] – 1r with other items – 1 – (in khmer) – mf#mf-10289 seam reel 111/3 [§] – us CRL [820]

Kho ma pum pran mya / Lha, Lu thu U – Mantale: Lu thu Sa tan ca tuik nhan Kri pva re Pum nhip tuik 1962 [mf ed 1990] – 1r with other items – 1 – (in burmese) – mf#mf-10289 seam reel 202/3 – us CRL [390]

Khoa hoc : revue de vulgarisation agricole artisanale et scientifique en langue annamite – Hanoi. n1-232. 1er juil 1931-juil aout 1940. – 1 – (mq n35-36, n 53., 228-229) – fr ACRPP [073]

Khoa-hoc tap chi = Revue scientifique = ke xue za zhi – Ha Noi: Co-quan truyen-ba cac Khoa-hoc 1931- [n1-232 [jul 1931-aug 1940] – 5r – 1 – (lacks: n85,90,181,209,210) – mf#mf-12545 seam – us CRL [079]

Khobragade, B D *see* Election manifesto of the republican party of india, 1967

O khode denezhnoi reformy (doklad na zasedanii plenuma tsk rkp 31 marta 1924 g) / Kamenev, L B – M, 1924 – 1mf – 9 – mf#REF-54 – ne IDC [332]

Khoetha, Lefau Nathaniel *see* Financing schools through establishment of sound external relations

Khoinatskii, A F *see* Pravoslavie na zapade rossii v svoikh blizhaishikh predstaviteliakh ili paterik volyno-pochaevskii...

Khol divre ha-torah = A vocabulary of the pentateuch / De Solla, Jacob Mendes – Philadelphia: printed for aut by Collins, 1865 [mf ed 1990] – 1mf – 9 – 0-8370-1877-3 – mf#1987-6264 – us ATLA [221]

Khol Gym Chan *see* Sneha pan sanya

Kholmogorskii, F D *see* Obshchimi silami k obshchemu blagu

Kholmskie/sedletskie gubernskie vedomosti – Kholm, Russian Federation 1867-1914; until 1913 Sedletskie [mf ed Norman Ross] – 25r – 1 – mf#nrp-704 – us UMI ProQuest [077]

Kholodkovskogo, N A *see* Faust

Khont-hon-nofer : the lands of ethiopia / Kumm, Hermann Karl Wilhelm – London: Marshall, 1910 – 1mf – 9 – 0-8370-6133-4 – (incl ind and appendix) – mf#1986-0133 – us ATLA [960]

Khorezmskaia pravda – Urgench, Uzbekistan 1986-88 [mf ed Norman Ross] – 1 – mf#nrp-1876 – us UMI ProQuest [077]

Khorovoe i regentskoe delo – St Petersburg, 1909-17 [bimthly] – 45mf – 9 – us UMI ProQuest [780]

Khorsabad, pt 1 : with chapters by henri frankfort and thorkild jacobsen / Loud, Gordon – 1937 – 9 – $8.50f – 0-226-49388-1 – us Oriental [930]

Les khouan : ordres religieux chez les musulmans de l'algerie / Neveu, Edouard de – 3. ed. Alger: A Jourdan, 1913 – 1mf – 9 – 0-524-01858-8 – mf#1990-2693 – us ATLA [260]

Khouri, Fred J *see* The jordan river controversy

Khouri-Saint-Pierre, Anastassia *see* Etude sur les statistiques

Khouw, Eng Tie *see* Etty dan erry

Khozhdenie sv apostola i evangelista ioanna bogoslova : po litsevym rukopisiam 15 i 16 vekov / Likhachev, N P – 1911 – 7mf – 9 – mf#R-10309 – ne IDC [243]

Khoziaistvennoe polozhenie i promysly naseleniia stanits astrakhanskogo kazach'ego voiska statistiko-ekonomicheskoe issledovanie / Makedonov, L V – Spb, 1906 – 7mf – 8 – mf#RZ-109 – ne IDC [314]

Khoziaistvenno-statisticheskie dannye po imeniiam, zalozhennym v obshchestve vzaimnogo pozemel'nogo kredita s 1873 po 1880 god / ed by Skuratov, P Ia – Spb, 1880 – 9mf – 9 – mf#REF-326 – ne IDC [332]

Khoziaistvenno-statisticheskii obzor ufimskoi gubernii za 1903 god – Ufa, 1904 – 24mf – 8 – mf#RZ-99 – ne IDC [314]

Khoziaistvennyi rost i khoziaistvennye zatrudneniia : po materialam nk rki sssr / ed by Miliutin, V & Gol'tsman, A – M, 1926. 104p – 2mf – 9 – mf#RHS-140 – ne IDC [314]

Khozraschet v promkooperatsii / Genkin, D M – 1932 – 160p 2mf – 9 – mf#COR-418 – ne IDC [335]

Khram pravoslavno-khristianskii = The orthodox christian temple / Krasnovskii, N – Moscow, 1888. One of 13 titles on reel – 1 – 86.44 – us Southern Baptist [242]

Khran yuin kat nann ni syann / Man Tan – Ran Kun: Mibhamettha rvhe tam chip pum nhip tuik 1953 – [pl/ill] 1r with other items – 1 – (in burmese) – mf#mf-10289 seam reel 167/6 [§] – us CRL [790]

Khristiani, G G *see* Slavianskaia problema srednei evropy

Khristianin *see* Zhurnal tserkovno-obshchestvennoi zhizni, nauki i literatury

Khristianskaia zhizn' : golos pravoslav prikhoda / ed by Spasskii, N I – Samara: Obed tserkov -prikhod sovet 1918 [1918 7 iiulia [24 iiunia]-] – (= ser Asn 1-3) – n1-10 [1918] item 439, on reel n86 – 1 – mf#asn-1 439 – ne IDC [077]

Khristianski pobornik = Christian advocate – St. Petersburg, 1909-1912 – 1391p – 1 – $55.64 – (Methodist) – us Southern Baptist [242]

Khristianskii vostok – Spb., 1912-1922. v1-6 – 53mf – 9 – mf#AR-1762 – ne IDC [077]

Khristianskoe chtenie, izdavaemoe pri st.-peterburgskoi dukhovnoi akademii – Spb., 1821-1918 – 3234mf – 9 – (missing: 1821(1-4); 1828(29-32); 1853(1); 1867(8); 1869(1); 1916(7-12); 1917-1918(1-12)) – mf#R-2302 – ne IDC [077]

Khristianstvo i aktivnost' cheloveka / Berdiaev, N A – Paris: YMCA Press, [193-] – us CRL [240]

Khronika / Naukove Tovaristvo Imeni Shevchenka – Lemberg. no. 1-59. 1900-1914 – 1 – us NY Public [400]

Khronika evreiskoi zhizni – Spb., 1905-1906 – 75mf – 9 – (missing: 1905(39-40, 45); 1906(2, 4, 35)) – mf#R-1570 – ne IDC [077]

Khronika leningradskogo obshchestva bibliofilov, 5 ianvaria-20 iiunia 1930 g / Leningradskoe obshchestvo bibliofilov – Leningrad: Izd. L.O.B., 1931 – 1 – us Wisconsin U Libr [947]

Khronika Naukovogo tovaristva imeni Shevchenka *see* Levovi

Khronologicheskii ukazatel' postanovlenii soveta gosudarstvennogo kontrolia, 1866-1872 – Spb, 1873 – 6mf – 9 – mf#REF-212 – ne IDC [332]

Khrystyianskyy katekhyzm dlia uzhytku shkilnykh ditei i molodezhy = Christian catechism for the use of schoolchildren and young people / Independent Greek Church (Canada) – Winnipeg: Z drukarn i "Kanadyiskoho farmera" [Canada North West Pub Co] 1904 [mf ed 1996] – 1mf – 9 – 0-665-78316-7 – (in ukrainian & english) – mf#78316 – cn Canadiana [243]

Khuc hat nguoi anh hung / Tran Dang Khoa – Ha-Noi: Phu Nu 1974 [mf ed 1992] – on pt of 1r – 1 – mf#11052 r15 n15 – us Cornell [959]

Khuc sati broh kam tanha : pralom lok phnaek cittasastr / Suddh Pulin – Bhnam Ben: [s.n.] 1965 [mf ed 1990] – 1r with other items – 1 – (text in khmer; pref in french; title in pref: khoch satek pruoh kam tannaha) – mf#mf-10289 seam reel 105/7 [§] – us CRL [480]

Khuda Bukhsh, Salahuddin *see* Essays indian and islamic

Khudiakov, P *see* Deloproizvodstvo i korrespondentsiia selskokhoziaistvennykh kooperativnykh organizatsii

Khudozhestvennaia gazeta – Helsinki. 1949-1973 (1) 1970-1973 (5) 1970-1973 (9) – 24mf – 9 – (missing: 1838-1839; 1841) – mf#1694 – ne IDC [077]

Khudozhestvennii trud – Moscow, 1923-24 – 19mf – 9 – us UMI ProQuest [790]

Khudozhestvenno-bibliograficheskii zhurnal / ed by Burtsev, A – Spb., 1910 – 40mf – 9 – mf#R-4334 – ne IDC [077]

Khudozhestvenno-istoricheskii zhurnal – M., 1914. v1-10 – 12mf – 9 – mf#R-1528 – ne IDC [077]

Khudozhestvenno-literaturnyi ezhemesiachnyi zhurnal – Spb., 1905-1912 – 45mf – 9 – (missing: 1905-1907(1-12); 1908(2, 7-12); 1909(7)) – mf#R-1527 – ne IDC [077]

Khudozhestvenno-literaturnyi ezhemesiachnyi zhurnal – Toronto. 1959-1979 (1) 1971-1979 (5) 1978-1979 (9) – 19mf – 9 – mf#1821 – ne IDC [077]

Khudozhestvenno-literaturnyi zhurnal – Brussels. 1962-1974 (1) 1970-1972 (5) – 19mf – 9 – mf#1734 – ne IDC [077]

Khudozhestvenno-literaturnyi zhurnal – Washington. 1943-1983 (1) 1971-1983 (5) 1975-1983 (9) – 69mf – 9 – mf#2041 – ne IDC [077]

Khudozhestvennye sokrovishcha rossii : a monthly collection of articles on decorative and applied arts = Art treasures of russia / ed by Benois, A & Prakhov, A – St Petersburg: Society for the Encouragement of Artists, 1901-07 [mf ed Norman Ross Publ] – 84 iss on 55mf – 9 – us UMI ProQuest [740]

Khudozhestvennyi zhurnal – M., 1908. v1 – 3mf – 9 – mf#R-3392 – ne IDC [077]

Khudozhestvennyi zhurnal iziashchnykh iskusstv i literatury – Hackensack. 1883-1978 (1) 1971-1978 (5) 1977-1978 (9) – 11mf – 9 – (missing: 1859. nos 1-12) – mf#1919 – ne IDC [077]

Khudozhnik i zritel' – Moscow, 1924 – 11mf – 9 – us UMI ProQuest [790]

Khuen, J C *see* Magnus in ortu

Khuin, Mon, Yu ja na *see*
- Kui kri sa khan
- Metta manduin

Khull, F *see* Zweier deutscher ordensleute

Khull, Ferdinand *see*
- Der kreuziger

Khumalo, Phelios Mtshane *see* Umuzi kawakhiwa kanye

Khumalo, Richard *see* The impact of management styles on staff appraisal system in schools

Khumu kanson phasa thai : kitchakam kanlen prakop kanson / Atchara Chiwaphan – Krung Thep...: Rongrian Sathit chulalongkon Mahawitthayalai 2521 [1978] [mf ed 1993] – on pt of 1r – 1 – mf#11052 r718 n4 – us Cornell [480]

Khun Outtamaprija Chap-Pin et al *see* Ovadapatimokkh

Khun phon kalasing / Chum Na Bangchang – Krung Thep: Phloensan [19–] [mf ed 1993] – on pt of 1r – 1 – mf#11052 r1133 n2 – us Cornell [959]

Khunrath, Heinrich *see* Amphitheatrum sapientiae aeternae

Khurshid – Tashkent, 1906 – 1 – (reel contains short runs of multiple titles. for complete listing of titles on a reel, please inquire) – us UMI ProQuest [077]

Khushwant Singh *see* The mark of vishnu

Khutbat-i garsan da tasi / Tassy, M Garcin de – Deccan: Anjuman-e Taraqqi-i-Urdu Aurangabad, 1935 – us CRL [950]

Khutorskoe khoziaistvo *see* Zhurnal prakticheskogo selskogo khoziaistva i domovodstva

Khutsiev, M E et al *see* Kabardinets

Khvalynskoe Obshchestvo Vzaimnogo Kredita *see* Otchet khvalynskogo obshchestva vzaimnogo kredita

Khve chon cha ra kri e sa ma can ku thum kyam / Cui Tan, U – Ran Kun: Yan kye mhu ca pe 1971 [mf ed 1990] – 1r with other items – 1 – (in burmese) – mf#mf-10289 seam reel 186/7 [§] – us CRL [615]

Khvolson, Daniil Avraamovich *see*
- Beitraege zur entwicklungsgeschichte des judentums con ca 400 v chr bis ca 1000 chr
- Das letzte passamahl christi und der tag seines todes
- The semitic nations

Khwam tok long traiphakhi...khao thai...tripartite agreement between the governments of siam, the united states of america and the united kingdom of great britain and northern ireland. / Thailand. Treaties, etc – Bangkok, 1946 1 v. (various pagings). LL-10019 – 1 – us L of C Photodup [324]

Khyac arum : [a novel] / Ne Van, Takkasuil – Ran Kun: Rup rhan pa de sa Maggajan Tuik 1956 [mf ed 1990] – (= ser Rup rhan pa de sa vatthu can 6) – 1r with other items – 1 – (in burmese; incl: nhon metta / da gun mi mi khan) – mf#mf-10289 seam reel 201/8 [§] – us CRL [830]

Khyac ce nnvhan ro sa la : [a novel] / Rvhe Sacca – Ran Kun: Tan Mon Kri Ca up tuik 1957 [mf ed 1990] – (= ser Kui tan on ma can ca pe thut ve re) – 1r with other items – 1 – (in burmese) – mf#mf-10289 seam reel 203/3 [§] – us CRL [830]

Khyac chan thve la / Mon Mon Kyo, U – Ran Kun: Ca pe sac Ca pa tuik 1971 [mf ed 1990] – 1r with other items – 1 – (in burmese) – mf#mf-10289 seam reel 162/1 [§] – us CRL [959]

Khyac khyan metta : [a novel] / Khan Sak Ve, Do – Ran Kun: s.n.] 1958 [mf ed 1990] – 1r with other items – 1 – (in burmese) – mf#mf-10289 seam reel 200/6 [§] – us CRL [830]

Khyac mi sann van : [a novel] / Tan Tay – Ran Kun: Na Va De Pum nhip tuik [195-?] [mf ed 1990] – (= ser Pvan lan ca can) – 1r with other items – 1 – (in burmese) – mf#mf-10289 seam reel 200/5 [§] – us CRL [830]

Khyac Mon, Sa khan *see*
- A myui sa a ja nann khon chan kri sakhan mra
- Nuin nam re samsara

Khyac nuin nam : [a novel] / Tan Tay – Ran Kun: Mran ma pa de sa Ca pe 1961 [mf ed 1990] – 1r with other items – 1 – (in burmese) – mf#mf-10289 seam reel 190/7 [§] – us CRL [830]

Khyac pan sum pvan / Man Nnvan – Ran Kun: Gun rann Ca up tuik 1315 [1953] [mf ed 1990] – 1r with other items – 1 – (in burmese) – mf#mf-10289 seam reel 192/4 [§] – us CRL [790]

Khyac r : [a novel] / Sin Phe Mran – Ran Kun: Metta rip mrum Ca pe [195-?] [mf ed 1990] – 1r with other items – 1 – (in burmese) – mf#mf-10289 seam reel 190/2 [§] – us CRL [830]

Khyac Sak U *see* Bhava ton tan nhan a khra vatthu tui mya

Khyac tai a kyon mya kui / Mon Mon Thvan – Ran Kun: Do Tan Me sin Ka le Ca pe rip mrum 1973 [mf ed 1990] – 1r with other items – 1 – mf#mf-10289 seam reel 162/3 [§] – us CRL [959]

Khyac to khak tay / Mon Kui, U – Ran Kun: Panna a lan pra ca up tuik 1957 [mf ed 1990] – 1r with other items – 1 – (in burmese) – mf#mf-10289 seam reel 180/4 [§] – us CRL [959]

Khyal kamput tpun : [a novel] / Lamn Pen Siak – Bhnam Ben: Samagam Qnak Nibandh Khmaer 2516 [1972] [mf ed 1990] – 1r with other items – 1 – (in khmer) – mf#mf-10289 seam reel 108/2 [§] – us CRL [830]

Khyal ratuv rin ramn : pralom lok phnaek manosancetana / Dri Sun Mun – Bhnam Ben: Ron Bumb Payan 2518 [1974] [mf ed 1990] – 1r with other items – 1 – mf#mf-10289 seam reel 097/07 [§] – us CRL [480]

Khyan a myui sa myae nay khyai chan kyan re samuin / Do' Cvan, Ma – Ha kha Mrui', Khyan Prannnay, Do Do' Cvan 1976 [mf ed 1995] – on pt of 1r – 1 – mf#11052 r1952 n4 – us Cornell [959]

Khyan rui ra dha le nhan phrat thum – [Ran kun]: Khyan Prann nay Yankye mhu thana Khvai, Yankye mhu Biman u ci thana [1975-] [mf ed 1995] – on pt of 1r – 1 – mf#11052 r1955 n7 – us Cornell [959]

Khyay ri damma vatthu tui mya : [short stories] / Sukha – Ran Kun: Cam Rha Ca pe 1987 [mf ed 1990] – 1r with other items – 1 – (in burmese) – mf#mf-10289 seam reel 165/1 [§] – us CRL [830]

Khyc U Nnui *see* Suvanna bhumi na yai, suvanna bhumi niban

Khytra mekhanika – rev ed. Toronto: Nakladom Ukraonsko-russkoo hrupy Sots, 1917 [mf ed 1999]] – 1mf – 9 – 0-665-97005-6 – (in ukrainian) – mf#97005 – cn Canadiana [890]

Khyui Phru Khan *see* Cin phran pru lup tha so nhalum sa puin rhan mya nhan a khra vatthu tui mya

Kiama examiner – Kiama, jan 1859-dec 1862 – 1r – 9 – A$61.29 vesicular A$66.79 silver – at Pascoe [079]

Kiama independent – Kiama, jul 1863-dec 1968 – 45r – at Pascoe [079]

Kiama reporter – Kiama, jul 1886-dec 1894; apr 1920-dec 1941 – 12r – at Pascoe [079]

Kiamichi Baptist Assembly. Oklahoma *see* Correspondence and records, 1944-55

Kiangsi. China. (Province) *see*
- Chiang-hsi min-cheng kung-pao
- Chiang-hsi sheng-cheng fu kung-pao

Kibardina, A P *see* Piatigorskoe ekho

Kibbutz journal – New York, New York, No. 1 (Jan. 1984)-no. 10 (Mar. 1990) – us AJPC [270]

Kibby pioneer – Kibby, Barnes Co, ND: Pioneer Publ Co. 1882-82?// (mthly) – 1 – mf#11453 – us North Dakota [071]

Kibris – Cyprus. 17 nov-14 dec 1983; 8 feb-18 jul 1984 – 1/4r – 1 – uk British Libr Newspaper [072]

Kibris turk sesi – Cyprus. 17 nov 1961-23 mar 1962 – 1/4r – 1 – uk British Libr Newspaper [072]

Kibulumina bokwe / Lansdown, G N – London, England. 1958 – 1r – us UF Libraries [960]

Kick! – 1996 jun,oct, 1997 apr-jul, 1999 jan-nov – 1r – 1 – (cont by: sbc) – mf#3643817 – us WHS [071]

Kick, Friedrich *see* Die entwicklung der werkzeuge

Kickapoo chief – Wauzeka WI. 1894 sep 20-1896 dec 31, 1897-99, 1900-02, 1903-05, 1906-09 – 5r – 1 – (cont by: wauzeka chief; wauzeka chief) – mf#949035 – us WHS [071]

Kickapoo exchange – 1977 nov-1982 aug – 1r – 1 – mf#611880 – us WHS [334]

Kickapoo papoose – Wauzeka WI. 1968 oct 2-1971 dec 29, 1972 jan 5-1974 dec 25, 1975 jan 1-1977 dec 28, 1978 jan 4-1980 dec 31, 1981 jan 7-1984 jun 6 – 5r – 1 – (cont by: boscobel dial; courier press) – mf#949030 – us WHS [071]

Kickapoo scout – Soldiers Grove WI. [1910 jan 20/1911 jun]-[1979 jan 11/sep 13] [gaps] – 37r – 1 – (cont: kickapoo valley journal) – mf#954666 – us WHS [071]

Kickapoo transcript – Soldiers Grove WI. 1889 aug 23-1893 sep 1 – 1r – 1 – (cont: de soto chronicle; cont by: crawford county advance) – mf#954650 – us WHS [071]

Kickapoo valley journal – Soldiers Grove WI. 1903 jul 24-1904 jul 27, 1904 aug 3-1906 aug 9, 1906 sep 27-1907 oct 10 – 3r – 1 – (cont by: kickapoo scout) – mf#954661 – us WHS [071]

Der kicker – Muenchen, Frankfurt/M, Hamburg, Nuernberg 1951 10 dec-1970 29 jun [gaps], 1972 28 feb-1997, 1998 2 feb-2002 1 jul, 2002 30 dec-2003 [gaps] [mf ed 2005] – 163r – 1 – (with suppl; 10 mai 1954: kicker publ in different locations; various regional ed filmed) – gw Mikrofilm [074]

The kicker – Beatrice, NE: O H Phillips. v1 n1. mar 7 1885- (semimthly) [mf ed mar 7 1885 filmed 1973] – 1r – 1 – us NE Hist [071]

Kid : kriminalroman / Albert, Albrecht – Berlin: Aufwaerts-Verlag M Klieber, c1943 [mf ed 1987] – (= ser Der aufwaerts-kriminal-roman 22) – 239p – 1 – mf#6935 n1 – us Wisconsin U Libr [830]

Kid zoo comics – New York NY 1948 – 1 – mf#6149 – us UMI ProQuest [740]

Kidd, Benjamin *see*
- Principles of western civilisation
- Social evolution

Kidd, Beresford James *see*
- The continental reformation
- Documents illustrative of the continental reformation
- The later mediaeval doctrine of the eucharistic sacrifice
- Selected letters of william bright
- The thirty-nine articles

Kidd, Dudley *see* Savage childhood

Kidd, George Balderston *see* Christophaneia

Kidd, James *see*
- A dissertation on the eternal sonship of christ
- Morality and religion

Kidd, John *see* On the adaptation of external nature to the physical condition of man

Kidder, Daniel *see* The life of rev. richard whatcoat

Kidder, Daniel P *see*
- Brasil e os brasileiros
- Brazil and the brazilians portrayed
- Reminiscencias de viagens e permanencia no brasil

Kidder, Daniel Parish *see*
- The beloved physician
- The greek and eastern churches
- Notices of fuh-chau, and the other open ports of china

Kidder, Ralph W *see*
- Cattle feeding in southern florida
- Fattening steers on winter pasture with ground snapped corn, ground shallu heads, molasses

Kidder, Richard *see* Discourse concerning sins of infirmity and wilful sins

Kidderminster shuttle – [West Midlands] Worcestershire 12 feb 1870-dec 1950 [mf ed 2004] – 89r – 1 – uk Newsplan [072]

Kiddie land souper special – v1 n3-v6 n1 [1979 jan-1984 mar] – 1r – 1 – mf#903568 – us WHS [071]

Kiddle, Henry *see* The dictionary of education and instruction

Kiddusch hachodesch / Maimonides, Moses – Wien, Austria. 1889 – 1r – us UF Libraries [939]
Kideer, Daniel P *see* Brazil and the brazilians
Kidner, F D *see* Sacrifice in the old testament
Kidner, Thomas Bessill *see* Educational handwork
Kidney international – London, England 1972+ – 1,5,9 – ISSN: 0085-2538 – mf#13122 – us UMI ProQuest [616]
Kidson, Frank *see* English country dances
Kidwai, Mushir Hosain *see*
- Islam and socialism
- The miracle of muhammad
- Muhammad, the sign of god
Kieffer, Henry Martyn *see* College chapel sermons
Kieffer, Patricia E *see* The relationship of sport involvement and sex role classification of college females
Kiefhaber, Johann Carl Sigmund *see* Monatliche historisch-litherarisch-artistische anzeigen zur aeltern und neuern geschichte nuernbergs
Kiefiuk, Deborah S *see* Investigation of physical self-perceptions, fitness behavior, and program selection among fitness participants in three fitness club environments
Kiefl, Franz Xaver *see*
- Eid gegen den modernismus
- Der friedensplan des leibniz zur wiedervereinigung der getrennten christlichen kirchen
- Der geschichtliche christus und die moderne philosophie
Kiel, Friedrich *see* 8 short, melodious pieces for pianoforte duet, op 13
Kiel national-zeitung – Kiel WI. 1900 mar 15-1901 mar 7, 1902 mar 13-1903 mar 5, 1903 mar 12-1904 mar 3 – 3r – 1 – (cont by: tri-county record [kiel, wi]) – mf#1270198 – us WHS [071]
Kiel, Rainer-Maria *see* Die amtskalender der fraenkischen fuerstentuemer ansbach und bayreuth [1737-1801]
Kiel tri county record – Kiel WI. [1950 may 16/1952 jun 30]-[2001 jan-may] [gaps] – 64r – 1 – (cont: tri-county record; continued by: chilton spirit; new holstein reporter; tri-county news; tri-county news) – mf#1145385 – us WHS [071]
Kielceskie gubernskie vedomosti – Kel'tsy, 1867-1915 – 23r – 1 – us UMI ProQuest [077]
Kieler anzeigen – Kiel DE, 1946 25 jun-1948 30 aug – 1r – 1 – gw Misc Inst [074]
Kieler correspondenzblatt fuer die herzogthuemer schleswig, holstein und lauenburg – Kiel DE, 1830 11 sep-1861 29 jun – 1 – (title varies: 2 jan 1836: correspondenz-blatt; 9 sep 1848: correspondenz-blatt und kieler wochenblatt; 4 nov 1849: correspondenz-blatt und kieler tageblatt; 1 jul 1850: correspondenz-blatt und kieler wochenblatt) – gw Misc Inst [074]
Kieler demokratisches wochenblatt – Kiel DE, 1848 19 nov-14 dec, 1849 7 jan-1 feb – 1r – 1 – gw Misc Inst [074]
Kieler foerdeblatt – Kiel DE, 1968 27 mar-1970 26 mar – 1r – 1 – gw Misc Inst [074]
Kieler illustrierte zeitung *see* Kieler neueste nachrichten
Kieler kurier – Kiel DE, 1945 25 jul-1946 3 apr – 1r – 1 – gw Misc Inst [074]
Kieler nachrichten – Kiel DE, 1960 1 feb-1963 – ca 9r/yr – 1 – (filmed by other misc inst: 1946 3 apr-1953 30 jun [16r]; 1946 3 apr-1990 [mf nur tw. vorhanden]; 1953 jul-1977 sep; 1968- [ca 9r/yr]) – gw Misc Inst [074]
Kieler nachrichten-blatt der militaerregierung – Kiel DE, 1945 4 jun-14 aug – 1 – gw Misc Inst [355]
Kieler neueste nachrichten – Kiel DE, 1895-1945 2 may – 127r [1913-42] – 1 – (incl suppl: kieler illustrierte zeitung 1908 13 sep-1912 28 dec [2r]) – gw Misc Inst [074]
Die kieler sprotten : die geschichte eines fischereiproduktes am beispiel der standorte kiel-ellerbek und eckernfoerde mit einem schwerpunkt 1871-1914/18 und einem ausblick in die gegenwart / Szadkowski, Karin – (mf ed 1995) – 2mf – 9 – €40.00 – 3-8267-2273-6 – mf#DHS 2273 – gw Frankfurter [639]
Kieler studentenanzeiger – Kiel DE, 1966 feb-1968 feb [gaps] – 1r – 1 – gw Misc Inst [378]
Kieler tageblatt – Kiel DE, 1874 1 oct-1876, 1878, 1882-94 – 1 – gw Misc Inst [074]
Kieler wochenblatt *see* Wochenschrift zum besten der armen in kiel
Kieler zeitung – Kiel DE, 1916 13 jan-1918 19 oct [gaps] – 8r – 1 – (filmed by misc inst: 1895-96 (gaps) [11r]) – uk British Libr Newspaper; gw Misc Inst [074]
Kielisches litteratur-journal – Flensburg DE, 1779 [gaps], 1781 – 1r – 1 – gw Misc Inst [410]
Kielisches magazin vor die geschichte, staatsklugheit und staatenkunde / ed by Heinze, Val Aug – Leipzig 1783-84 – (= ser Dz. historisch-geographische abt) – 2v[zu je 3st] on 6mf – 9 – €120.00 – mf#k/n1143 – gw Olms [323]
Kielkopf, Frieder *see* Individuum und gemeinschaft in den romanen toni morrisons
Kienast, Richard *see* Der sogenannte heinrich von melk
Kiencke, Uwe *see* Reports on industrial information technology
Kiener, Francois-Joseph *see* Le berceau historique des mysteres de la franc-maconnerie
Kienow, Nancy L *see* Death education and death anxiety in student nurse aides
Der kientopp *see* Kinobriefe
Kiepert, H *see* P von tschihatscheff's reisen in kleinasien und armenien 1847-1863
Kier, Martin *see* Soldaten des koenigs
Kier, P O *see* Bedarf es einer besonderen inspirationslehre?
Kieran, John *see* Story of the olympic games, 776 b c-1948 a d
Kierkegaard, Soeren *see*
- Af en endnu levendes papirer
- Afsluttende uvidenskabelig efterskrift til de philosophiske smuler
- Atten opbyggelige taler
- Begrebet angest
- En bladartikel
- Christelige taler
- Dette skal siges, saa vaere det da sagt
- Dymmer selv!
- En literair anmeldelse
- Enten-eller
- Forord
- Gjentagelsen et forsoeg i den experimenterende psychologi
- Guds uforandersighed
- Hvad christus doemmer om officiel christendom
- Indyvelse i christendom
- Kjerlighedens gjerninger
- Lilien paa marken og fuglen under himlen
- Oieblikket
- Om begrebet ironi
- Om min forfatter-virksomhed
- En opbyggelig tale
- Opbyggelige taler i forskjellig aand
- Philosophiske smuler, eller, en smule philosophi
- S kierkegaard's bladartikler
- Stadier paa livets vei
- Sygdommen til doeden
- Synspunket for min forfatter-virksomhed
- Til selvproevelse
- To taler ved altergangen om fredagen
- Tre taler ved taenkte leiligheder
- Tvende ethisk-religieuse smaa-afhandlinger
- Ypperstepraesten, tolderen, synderinden
Kierkegaard, Soren *see* Dommer selv!
Kierkegaard und nietzsche : versuch einer vergleichenden wuerdigung / Sodeur, Gottlieb – Tuebingen: J C B Mohr 1914 [mf ed 1991] – (= ser Religionsgeschichtliche volksbuecher fuer die deutsche christliche gegenwart 5/14) – 1mf – 9 – 0-524-00137-5 – mf#1989-2837 – us ATLA [190]
Kierkegaard's interpretation of luther / Refsell, Lloyd Gerhard – [Chicago], 1964. Chicago: Dep of Photodup, U of Chicago Lib, 1968 (1r); Evanston: American Theol Lib Assoc, 1984 (1r) – 1 – 0-8370-0376-8 – mf#1984-B092 – us ATLA [242]
Kiernan, Ben *see* The samlaut rebellion and its aftermath, 1967-70
Kierunki – Warsaw. v1 n1-v6 n281. May 20 1956-Nov 5 1961. (incomplete) – 1 – us NY Public [073]
Kierzkowski, Leon *see* Trois lettres adressees a r laflamme de la ville de montreal
Kies, William Samuel *see* The liberty loan, a national insurance
Kieser, Dietrich G *see* Das wartburgsfest am 18. october 1817
Kiesewetter, Johann *see* Reise durch einen theil deutschlands, der schweiz, italiens und des suedlichen frankreichs nach paris
Kiesgen, Laurenz *see*
- Heinrich von kleist
- Martin greif
Kiessmann, Rudolf *see* Untersuchungen ueber die motive der robin-hood-balladen
Kietz, Gertraud *see*
- Der ausdrucksgehalt des menschlichen ganges
Kietzell, Philine von *see* Zur proliferativen aktivitaet im fibromuskulaeren stroma der benignen prostatahyperplasie
Kieu / Nguyen, Du – 2eme ed. Hanoi: editions en langues etrangeres 1974 [mf ed 1992] – on pt of 1r – 1 – (trans fr vietnamese into french by nguyen khac vien) – mf#11052 r23 n2 – us Cornell [480]
Kiev. Universytet *see* Zapysky
Kievlianin – Kiev, 1840-1850. v1-3 – 14mf – 9 – mf#R-3199 – ne IDC [077]
Kievlianin – Kiev, Ukraine 1865 [mf ed Norman Ross] – 1 – mf#nrp-738 – us UMI ProQuest [077]
Kievlianin / ed by Shul'gin, V V – Kiev: V V Shul'gin, P V Mogilevskaia 1918-19 [1864 1 iiulia-1919 [dek]] – (= ser Asn 1-3) – n1-7 [1918] n27, 83 [1919] item 199, on reel n41 – 1 – mf#asn-1 199 – ne IDC [077]
Kievskaia mysl' [utr vypusk] / ed by Strzhel'bitskii, A P – Kiev: R K Lubkovskii 1918 [1917 23 maia-1918 [?]] – (= ser Asn 1-3) – n179-184 [1918] item 202, on reel n41 – 1 – mf#asn-1 202 – ne IDC [077]
Kievskaia mysl' [vech vypusk] : ezhedn gaz / ed by Chaikovskii, B K – Kiev: R K Lubkovskii 1918 [1906 30 dek-1918 [?]] – (= ser Asn 1-3) – n1-216 [1918] [gaps] item 201, on reel n41 – 1 – mf#asn-1 201 – ne IDC [077]
Kievskaia zhizn' / ed by Vasilenko, K P et al – Kiev: [s n] 1919 [1919 25 avg [7 sent]-] – (= ser Asn 1-3) – n1, 4, 54 [1919] item 200, on reel n41 – 1 – mf#asn-1 200 – ne IDC [077]
Kievskie gubernskie vedomosti – Kiev, Ukraine 1840-1916 [mf ed Norman Ross] – 136r – 1 – mf#nrp-739 – us UMI ProQuest [077]
Kievskie novosti / Gruppa zhurnalistov – Kiev: [s n] 1920 [1920 9 maia-[10 iiunia]] – (= ser Asn 1-3) – n7,18 [1920] item 203, on reel n41 – 1 – mf#asn-1 203 – ne IDC [077]
Kievskii den' : obshchestv -polit i lit gaz / ed by Vvedenskii, N D – Kiev: Gruppa zhurnalistov 1920 [1920 [mai]-[10 iiunia]] – (= ser Asn 1-3) – n21 [1920] item 204, on reel n41 – 1 – mf#asn-1 204 – ne IDC [077]
Kievskii kinematograf – Kiev, Ukraine 1911 [mf ed Norman Ross] – 1 – mf#nrp-740 – us UMI ProQuest [077]
Kievskii oblastnoi soiuz zemel'nykh sobstvennikov : biulleten' kievskogo obl soiuza zemel'nykh sobstvennikov – Kiev, Ukraine 1917 [mf ed Norman Ross] – 1r – 1 – mf#nrp-742 – us UMI ProQuest [077]
Kievskii telegraf – Kiev, Ukraine 1861 [mf ed Norman Ross] – 1 – mf#nrp-743 – us UMI ProQuest [077]
Kievskij oblastnoj soiuz tekhnikov *see* Izvestiia kievskogo obl soiuza tekhnikov,zemlemerov, chertezhnikov, desiatnikov ok shkolu, topografov i tp
Kiewiet, C W de *see* The anatomy of south african misery
Kiger, John R *see* An examination of the determinants to the overall recreational sports participation among college students
Kigyobetsu shiryo-hen : documents of sclc (the securities coordinating liquidation committees) in the holding of the library of faculty of economics, university of tokyo – (= ser Shoken shori chosei kyogikai shiry b) – 264r – 1 – Y4488,000 – (collected companies' reports. documents collected from the target companies for securities liquidation by sclc, classified by individual companies. among included materials: tables of company research results and business reports; with guide) – ja Yushodo [332]
Kihlman, A O *see* Die expedition nach der halbinsel kola
Kihn, Heinrich *see*
- Die bedeutung der antiochenischen schule auf dem exegetischen gebiete
- Encyklopaedie und methodologie der theologie
- The odor von mopsuestia und junilius africanus als exegeten
- Praktische methode zur erlernung der hebraeischen sprache
- Der ursprung des briefes an diognet
Kiimbila, J K *see* Lila na fila
Kikeriki – Wien (A), 1918 7 jul-1920 – 1r – 1 – gw Mikrofilm [870]
Kiki, Albert Maori *see* Correspondence and papers
Kikuyu / Davidson, Randall Thomas – London: Macmillan, 1915 – 1mf – 9 – 0-7905-5529-8 – mf#1988-1529 – us ATLA [240]
Kilani, Muhammad Sayyid *see* Fi rubu'al-azbakiyah
Kilbarchan magistrate – [Scotland] Glasgow: printed & publ for the Proprietor 21 apr 1881 [mf ed 2004] – 1v on 1r – 1 – uk Newsplan [072]
Kilbon, John Luther *see* A study of the life of jesus the christ
Kilbourn city mirror – Wisconsin Dells WI. 1870 nov 3-dec 22 – 1r – 1 – (cont: wisconsin mirror [wisconsin dells, wi: 1868]; cont by: mirror [wisconsin dells, wi]) – mf#950319 – us WHS [071]
Kilbourn, John Kenyon *see* Faiths of famous men in their own words
Kilbourn weekly events – Kilbourn, Wisconsin Dells WI. 1907 jul 6-1909 nov 18, 1909 nov 25-1913 apr, 1913 may-1916 jul, 1916 aug-1919 oct 16, 1919 oct 23-1922 sep, 1922 oct-1924 dec, 1925 jan-1927 nov, 1927 dec-1928 dec, 1929-30, 1931 jan 1-apr 2 – 10r – 1 – (cont: kilbourn weekly illustrated events; kilbourn weekly illustrated events; mirror gazette; cont by: wisconsin dells events) – mf#1159061 – us WHS [071]
Kilbourn weekly illustrated events – 1905 feb-may – 1r – 1 – (cont: illustrated events; cont by: kilbourn weekly illustrated events [1905 jun 3]) – mf#3354992 – us WHS [071]
Kilbourn weekly illustrated events – Kilbourn, Wisconsin Dells WI. 1905 jun 3-1907 jun 29 – 1r – 1 – (cont: kilbourn weekly illustrated events [1905 may 24]; cont by: kilbourn weekly events) – mf#1159058 – us WHS [071]
Kilbourne, Ernest A *see* The great commission
Kilbourne, John D *see* The thomas penn papers
Kilbourne, John R *see* Building a bridge between athletics and academics
Kilburn times – London UK, 1953; 1986-93 – 21r – 1 – uk British Libr Newspaper [072]
Kilburn & willesden recorder – Brent, England 6 feb-27 mar, 8 may-2 oct, 11 dec 1986-19 sep 1990 – 1 – (cont by: brent recorder [26 sep 1990-5 may 1993]) – uk British Libr Newspaper [072]
Kilchevskii, V A *see*
- Kak sostavit otchet selskokhoziaistvennogo kreditnogo tovarishchestva
- Kooperatsiia i partiia
- Kooperatsiia i revoliutsiia
- Obshchestvenno-kooperativnaia shkola
- Organizatsiia mestnykh kooperativnykh kursov
- Prava i obiazannosti chlena kreditnogo kooperativa
- Robert ouen i osnovy potrebitelskoi kooperatsii
- Selskokhoziaistvennoe kreditnoe tovarishchestvo – organizator raiona:
- Ukazaniia pravleniiam kreditnykh kooperativov dlia samoproverki
Kil'chevskii, VA *see* Kakaia raznitsa mezhdu chastnym bankom i kreditnym tovarishchestvom
Kildare and wicklow chronicle – Athy. Ireland. -w. 17 feb-3 mar 1849 – 1/4r – 1 – uk British Libr Newspaper [072]
Kildare observer – Naas 1879-82, 1913, 1930 – 3r – 1 – ie National [072]
Kildare observer and eastern counties advertiser – Naas, Ireland. 23 oct 1880-1896; 1901; 1902; 4 jan-23 aug 1930 – 18r – 1 – uk British Libr Newspaper [072]
Kilderne til sakses oldhistorie / Olrik, Axel – Kobenhavn: O B Wroblewski 1892-94 [mf ed Bloomington IN: Indiana Uni Lib, Preservation Dept 1984] – 2v in 1r – 1 – us Indiana Preservation [390]
Kileff, Clive *see* Shona customs
Kilgore, Bruce M *see* Wilderness in a changing world anthology
Kilgore, Damon Y *see* The bible in public schools
Kilgore, Jennifer M *see* An investigation of the effects of short-term injuries on psychological readiness for competition
Kilgore's tree – v1-v3 n2 [1975 may-1978 fall/winter] – 1r – 1 – mf#501879 – us WHS [071]
Kilham, Hannah *see*
- Memoir of the late hannah kilham
[Kilian, C] *see* Viridarium moralis philosophiae
Kilian, Eugen *see*
- Adalbert von weislingen
- Goethes faust auf der buehne
Kilian, Werner *see* Herwegh als uebersetzer
Kilianus, C *see* Etymologicum teutonicae linguae sive dictionarium teutonico-latinum
Kilikiya facialari ve urfanin kurtulusu muecadeleleri / Saib, Ali – Ankara, 1340 [1924] – (= ser Ottoman histories and historical sources) – 4mf – 9 – $60.00 – us MEDOC [956]
Kilimandjaro / Geilinger, Walter – Bern, Switzerland. 1930 – 1r – us UF Libraries [960]
Kilimandjaro / Meyer, Hans – Berlin, Germany. 1900 – 1r – us UF Libraries [960]
Kilkeel gazette and south down advertiser – Kilkeel, Ireland. jul 1888-apr 1889 – 1/2r – 1 – uk British Libr Newspaper [072]
Kilkenney people – Ireland. 1896-1920 (missing Aug-Sep 1917).-w. 22 reels – 1 – uk British Libr Newspaper [072]
Kilkenny chronicle – Kilkenny, Ireland. 29 may 1813 – 1/4r – 1 – uk British Libr Newspaper [072]
Kilkenny independent – Kilkenny, oct 1826-jun 1828 – 1r – 1 – ie National [072]
Kilkenny journal : and leinster commercial and literary advertiser – Kilkenny, Ireland. 1832-93, 26 jun 1895-1900 [wkly] – 63r – 1 – (lacking: 23 dec 1893-19 jun 1895) – uk British Libr Newspaper [072]
Kilkenny journal [finn's leinster journal] – Kilkenny 1767-1831, 1830-32 [incomplete], 1847, 1850-51, 1887, 1894, 1901-02, 1915, 1922-35 – 73.25r – 1 – ie National [072]
Kilkenny moderator – Kilkenny, Ireland. -w. 1828-1902 – 75r – 1 – uk British Libr Newspaper [072]
Kilkenny people – Kilkenny, Ireland. 26 oct 1895-24 dec 1897; 1898-14 jul 1917; 13 oct 1917-1922; 1924-18 dec 1926; 1927-29; 1 mar-22 nov 1930; 1986-92 (missing aug-sep 1917) – 52 3/4r – 1 – uk British Libr Newspaper [072]
Killam, Gd *see* Novels of chinua achebe

Killam, Izaak Walton *see* The case against tax-exempt bonds
Killarney echo and south kerry chronicle – Killarny, Ireland. 26 aug 1899-14 aug 1920 – 16 1/2r – 1 – (aka: killarney echo) – uk British Libr Newspaper [072]
Killay & knelston, baptist churches, monumental inscriptions – 1mf – 9 – £1.25 – uk Glamorgan FHS [929]
Killen, James Miller *see* Our friends in heaven
Killen, Thomas Young *see* Ministerial responsibility
Killen, W D *see* The framework of the church
Killen, William Dool *see*
– The ancient church
– The ecclesiastical history of ireland
– The framework of the church
– The ignatian epistles entirely spurious
– The old catholic church
– The unitarian martyr
Killgore, Garry L *see* The effects of surface type on plantar pressure distribution and running kinematics
Killinger, G B *see* Chufas in florida
Killinger, Manfred, Freiherr von *see* Das waren kerle
Killings at kent state / Stone, If – New York, NY. 1971 – 1r – us UF Libraries [025]
Killingsworth, J Alexander *see* Sparks and cinders
Killmer, Karen J *see* The difference in coach role model behaviors for male and female athletes
Killpack, William Bennett *see* The history and antiquities of the collegiate church of southwell
Killy, Walther *see* Grosse deutsche lexika
Kilmarnock advertiser and ayrshire review – Scotland. 22 Aug-12 Dec 1868.-w. 13 ft – 1 – uk British Libr Newspaper [072]
Kilmarnock chronicle, and county advertiser – Scotland. Jan 1854-Aug 1855.-w. 1 reel – 1 – uk British Libr Newspaper [072]
Kilmarnock herald – Scotland. Nov 1882-Apr 1955. 42 reels – 1 – uk British Libr Newspaper [072]
Kilmarnock herald and west county advertiser – Scotland. Jan 1845-May 1848.-w. 1 reel – 1 – uk British Libr Newspaper [072]
Kilmarnock journal and ayrshire advertiser – Scotland. 1844-May 1857. 5 reels – 1 – uk British Libr Newspaper [072]
Kilmarnock pioneer – Scotland. Oct 1870-May 1871.-w. 7 ft – 1 – uk British Libr Newspaper [072]
Kilmarnock standard – Scotland. -w. 1876-91. 6 reels – 1 – uk British Libr Newspaper [072]
Kilmarnock standard, ayrshire weekly news and irvine valley news – Kilmarnock: Scottish & Universal Newspapers Ltd (wkly) [mf ed 6 jan 1995-] – 1 – (not publ: aug 8 1975; formed by union of: kilmarnock standard and ayrshire weekly news and: irvine valley news and galston supplement) – ISSN: 1350-0694 – uk Scotland NatLib [072]
Kilmarnock weekly post – Scotland. Nov 1856-Oct 1865.-w. 8 reels – 1 – uk British Libr Newspaper [072]
Kilobaud – Peterborough NH 1977-78 – 1,5,9 – (cont by: kilobaud: microcomputing) – ISSN: 0192-4583 – mf#11832.02 – us UMI ProQuest [000]
Kilobaud: microcomputing – Peterborough NH 1979-82 – 1,5,9 – (cont: kilobaud; cont by: microcomputing) – ISSN: 0192-4575 – mf#11832.02 – us UMI ProQuest [000]
Kilpin, Ralph Pilkington *see* Parliamentary procedure in south africa
Kilrush herald and kilkee gazette – Kilrush, Ireland. 5 jun 1879-mar 1880, 1889-96, 1900 – 4 3/4r – 1 – (aka: kilrush herald and kilkee gazette kilrush herald) – uk British Libr Newspaper [072]
Kilrush herald and kilkee gazette kilrush herald *see* Kilrush herald and kilkee gazette
Kilsyth chronicle – Kilsyth: J M Duncan 1938-99 [mf ed 1994-99] – 1 – (formed by union of: kilsyth chronicle, denny, bonnybridge and cumbernauld weekly news and: kirkintilloch gazette, lenzie and campsie reporter; cont by: kilsyth chronicle and cumbernauld news; imprint varies; ceased in 1999) – uk Scotland NatLib; uk Newsplan [072]
Kilsyth chronicle and cumbernauld news – Kilsyth: Johnston (Falkirk) Ltd 1999- (wkly) – 1 – (began in 1999; cont: kilsyth chronicle) – uk Scotland NatLib [072]
Kilsyth chronicle, denny bonnybridge and cumbernauld weekly news – [Scotland] North Lanarkshire, Kilsyth: P C Rankin 6 jan 1922-dec 1950 (wkly) [mf ed 2003-04] – 15r – 1 – (merged with: kirkintilloch gazette, lenzie and campsie reporter to form: kilsyth chronicle [jan 1938-dec 1950]; imprint varies) – uk Newsplan [072]
Kilsyth journal and west stirlingshire and east dumbartonshire advertiser – [Scotland] East Dunbartonshire, Kilsyth: Rankin & Mackie 6 jan 1905-31 dec 1936 (wkly) [mf ed 2003] – 29r – 1 – (absorbed by: kirkintilloch herald and lenzie, kilsyth, campsie and cumbernauld press) – uk Newsplan [072]
Kilty, Keith M et al *see* Journal of poverty
Kilvey, glamorgan, parish church of all saints : baptisms 1845-1925 – [Glamorgan]: GFHS [mf ed c2005] – 1mf – 9 – £1.25 – uk Glamorgan FHS [929]
Kilwinning chronicle and north ayrshire record – [Scotland] North Ayrshire, Irvine: C Murchland 15 dec 1893-29 dec 1950 (wkly) [mf ed 2003] – 39r – 1 – (merged with: irvine herald and ayrshire advertiser [3 nov 1881-1 nov 1968] to form: irvine herald, ayrshire advertiser, kilwinning chronicle [8 nov 1968-29 oct 1971]) – uk Newsplan [072]
Kim, Byunghoon *see* Tritheism and divine person as center of consciousness
Kim, Dae Tschong *see* Bertolt brecht und die geisteswelt des fernen ostens
Kim, Dae-Yoong *see* Vision for mission
Kim Dung Tu *see* Tuoi dai
Kim, H D *see* Differential effects of strength training and endurance training on parameters realted to resistance to gravitational forces
Kim, Jong-Il. *see* The effects of glasnost and perestroika on the soviet sport system
Kim, Jong-Kyung *see* Exercise mode comparisons of acute energy expenditure during moderate intensity exercise in obese adults
Kim, Junghoon *see* The influence of force production and eccentric exercise on growth hormone
Kim, Kyoung N *see* The effects of high spatial constraints in determining the nature of the speed-accuracy trade-off in aimed hand movements
Kim, Kyunghee *see* The status of dance in korean higher education
Kim, M P *see* V I lenin i istoriia klassov i politicheskikh partii v rossii
Kim, Sangho *see* An analysis and evaluation of the administrative budget statement between 1984 and 1995 for the south korean ministry of culture and sports
Kim su bang : and other stories of korea / Wagner, Ellasue Canter – Nashville; Dallas: Publ House of the M E Church, South, 1909 [mf ed 1995] – (= ser Yale coll) – 99p (ill) – 1 – 0-524-09513-2 – mf#1995-0513 – us ATLA [306]
Kim, Touch *see*
– La banque nationale du cambodge a quinze ans
– Le regime monetaire et le controle des changes au cambodge
– Le systeme bancaire au royaume du cambodge
Kimball, Charles P *see* Lepidoptera of florida
Kimball, Edward Ancel *see* Christian science and legislation
Kimball family association newsletter – 1981 jan-1986 jul [v17 n4] – 1r – 1 – (cont by: connections [kitterly, me]) – mf#1132657 – us WHS [929]
Kimball, Grayson T *see* Differences in cohesion among starters and non-starters of recreational basketball teams
Kimball, John Calvin *see*
– Mormonism exposed, the other side
– The romance of evolution
Kimball observer – Kimball, NE: C H Randall. 2v. v1 n39-v2 n20. feb 6-sep 24 1886 (wkly) [mf ed with gaps] – 1r – 1 – (cont: nebraska observer (antelopville ne). cont by: western nebraska observer) – us NE Hist [071]
Kimball, Sophie Burt *see* Truths leaf by leaf
Kimber, Edward *see* Relation or journal of a late expedition to the ga...
Kimber, Thomas *see* Historical essays on the worship of god and the ministry of the gospel of our lord and saviour
Kimberley citizen – Kimberley SA, 1897-24 dec 1898 – 1r – 1 – (cont by: kimberley elector; title varies: south african citizen) – mf#MS00291 – sa National [079]
The kimberley colonist – Kimberley SA, 27 mar 1897-20 apr 1897 – 1r – 1 – mf#MS00314 – sa National [079]
The kimberley elector – Cape Town: SA Library, 16 jul 1898-13 aug 1898 – 1r – 1 – (cont: the kimberley citizen; cont by: the citizen) – mf#ms00291 – sa National [079]
Kimberley free press – 28 oct 1905-10 mar 1906 – 1r – 1 – (cont by: kimberley free press and mining journal) – mf#MS00290 – sa National [079]
Kimberley free press and mining journal – 23 jan 1904-21 oct 1905 – 1r – 1 – (cont: kimberley free press; cont by: weekly free press and mining journal) – mf#MS00290 – sa National [079]
[Kimberley-] news – NV. feb-dec 1910 [wkly] – 1r – 1 – $60.00 – mf#U04593 – us Library Micro [071]
Kimberlin kobold – v1 n1-v3 n4 [1981 feb-1983 nov], v4 [1984 nov] – 1r – 1 – mf#1155298 – us WHS [071]
Kimberly and kuruman diocesan magazine – v2-7. 1937-46 [complete] – 1r – 1 – (cont by: highway) – mf#ATLA S0727 – us ATLA [240]
Kimberly free press *see* The weekly free press
Kimbundu grammar / Chatelain, Heli – Ridgewood, NJ. 1964 – 1r – us UF Libraries [470]
Kime, Charles Davidson *see* Study of the cobaltinitrate turbidimetric method for determining ex...
Kime, J *see* Plasma and erythrocyte lactate concentrations in humans after submaximal exercise
Kimhi, David *see* Rabbi david kimchi's commentary upon the prophecies of zechariah
Kimhi, David ben Joseph *see* [Sefer ha-shorashim]
Kimhi, Joseph *see* Sefer zikaron
Kimmel, Ernst Julius *see* Monumenta fidei ecclesiae orientalis
Kimmel, William Breyfogel *see* Polychoral music and the venetian school
Kimmich, Karl *see* Zeichenschule
Kim-Set *see*
– Interpreter and guide
– Toi hoc ch mien
Ki-mtang'ata / Whiteley, Wilfrd Howell – Kampala, Uganda. 1956 – 1r – us UF Libraries [960]
Kimura, Iris F *see*
– Biomechanical comparison of support provided by the airstirrup ankle training brace'm pre- and post-exercise
– The high jump as performed by the 1979 united states outdoor female record holder: a biomechanical analysis
– Influence of the strength shoe and three plyometric drills on the strength, velocity, and jumping ability of high school football players
– Peak torque reliability of biodex b-2000 isokinetic dynamometer during concentric loading of back flexors and extensors
Kimura, Ryukan *see*
– An historical study of the terms hinayana and mahayana and the origin of mahayana buddhism
– The original and developed doctrines of indian buddhism in charts
Kin, caste, and nation among the rhodesian ndebele / Hughes, Arthur John Brodie – Manchester, England. 1956 – 1r – us UF Libraries [960]
Kin in linn / Linn Co. Genealogical Society – 1980 fall-1988 nov – 1r – 1 – mf#1061452 – us WHS [929]
Kina missionaeren – St Paul MN: Foreign Mission Society, v6-25 (1908-25) [mf ed 2006] – 18v on 2r – 1 – (publ by: foreign mission society, 1908-apr 1912; by: china mission society, may 1912-25; cont: luthersk tidskrift feor hednamission och diakoni; cont by: augustana foreign missionary; several pgs damaged) – mf#2006c-s005 – us ATLA [242]
Kinahan, George Henry *see* Irish and canadian rocks, compared
Kinane, Thomas H *see* Mary immaculate, mother of god
Kinart, Chad M *see* Prevalence of migraines in ncaa division 1 men and women basketball players
Kinas religioner : haandbok i den kinesiske religionshistorie / Reichelt, Karl Ludvig – Stavanger: Det Norske Missionsselskaps Boktrykkeri, 1913 [mf ed 1995] – (= ser Yale coll) – 174p – 1 – 0-524-09628-7 – (in norwegian) – mf#1995-0628 – us ATLA [200]
Kinau, Rudolf *see*
– Ein froehlich herz
– Kamerad und kameradin
Kincaid, Charles Augustus *see*
– The anchorite and other stories
– Folk tales of sind and guzarat
– Forty-four years a public servant
– A history of the maratha people
– The indian heroes
– Lakshmibai, rani of jhansi
– The land of "ranji" and "duleep"
– Our hindu friends
– Our parsi friends
– The outlaws of kathiawar
– The tale of the tulsi plants
– Tales from the indian epics
– Tales of old sind
– Teachers of india
Kincaid, Dennis *see* The grand rebel
Kincaid, Randall R *see*
– Cultural practices for root-knot control of annual crops of cigar-wrapper tobacco
– Downy mildew (blue mold) of tobacco
– Effects of certain environmental factors on germination of florida cigar-wrapper tobacco seeds
Kincardine news – Ontario, CN. jan 1978-dec 1980 – 6r – 1 – cn Commonwealth Imaging [071]
Kincardineshire observer – Laurencekirk: A Taylor 1907- (wkly) [mf ed 1 jan 1994-] – 1 – (cont: laurencekirk observer; imprint varies; newsplan 2000 [mf ed 2003] 1r) – uk Scotland NatLib; uk Newsplan [072]
Kincardineshire observer *see* Laurence kirk observer
Kinch family newsletter – iss n1-10 [1984 sep-1987 sep] – 1r – 1 – mf#1477283 – us WHS [929]
Kinchen, Thomas A *see* An investigation of ministerial counseling support
Kinch's henley advertiser – England. Henley Advertiser. -w. 7 May 1870-11 Jan 1908. 28 reels – 1 – uk British Libr Newspaper [072]
Das kind : novelle / Grimm, Hermann Friedrich – New York: E Steiger, [1886?] [mf ed 1990] – (= ser Steiger's deutsche bibliothek 6) – 47p – 1 – mf#7423 – us Wisconsin U Libr [830]
Kind commandment / Vernon, J R – Edinburgh, Scotland. 18– – 1r – us UF Libraries [240]
Das kind der madeleine montcornet / Kraenzlein, Kurt – Hamburg: Hanseatische Verlagsanstalt 1943 [mf ed 1990] – 1r – 1 – (filmed with: kotzebue in england / walter sellier) – mf#2776p – us Wisconsin U Libr [830]
Das kind des torfmachers : eine erzaehlung / Griese, Friedrich – Muenchen: A Langen/G Mueller 1937 [mf ed 1990] – 1r – 1 – (filmed with: mensch aus erde gemacht & other titles) – mf#2688p – us Wisconsin U Libr [830]
A kind of equality : labour and the maori people – 1935-49 – 1r – 1 – mf#ZB 4 – nz Nat Libr [079]
Kind word series – 1866-1929 – 1 – 575.89 – us Southern Baptist [242]
Kindaposten – Vimmerby, Sweden. 1894-1903 – 3r – 1 – sw Kungliga [078]
Kindaposten *see* Vimmerby tidning
Kindelan, Alfredo *see* El derecho de beligerancia
Kinder bibel : der kleine catechismus d martini lutheri, mit schoenen spruechlein beiliger schrifft erkleret gegruendet und bekrefftiget / Opitz, J – [Ursel, 1583 – 4mf – 9 – mf#TH-1 mf 1227-1230 – ne IDC [242]
Kinder bote – v1-50. 1886-1936 – (= ser Mennonite serials coll) – Inquire – 1 – mf#ATLA 1994-S006 – us ATLA [242]
Kinder der eifel : novellen / Viebig, Clara – Berlin: E Fleischel, 1908 – 1r – 1 – us Wisconsin U Libr [830]
Die kinder der excellenz : roman / Wolzogen, Ernst von – Stuttgart: J Engelhorn [1888?] – 1r – 1 – us Wisconsin U Libr [830]
Kinder, Elisabeth *see* Liberale parteien
Kinder ertsihung bay iden / Levin, Moses Elimelech – Montreal, Quebec. 1910 – 1r – us UF Libraries [939]
Kinder ertsihung bay iden, a historishe nakhforshung : a zamlung fun fershidene brilyantene mamorim, talmund bavli, yerushalmi, medresh rabe tan hume... / Levin, Moses A – [Montreal?: s.n.] c1910 [mf ed 1996] – 2mf – 9 – 0-665-81664-2 – mf#81664 – cn Canadiana [270]
Kinder in sonne : ein buch fuer frohe leute / Frank, Ernst – 2. Aufl. Goerlitz: Jungland-Verlag, 1942 (mf ed 1990] – 1r – 1 – (filmed with: trenck) – us Wisconsin U Libr [430]
Kinder khaper fun rusland / Bogrov, Grigorii Isaakovich – New York, NY. 1915 – 1r – us UF Libraries [939]
Kinder, Robin *see* Humanities collections
Kinder- und hausmaerchen : gesammelt durch die brueder grimm – Leipzig: M Hesses Verlag [1907] – 1 – us Wisconsin U Libr [390]
Kinder- und hausmaerchen : gesammelt durch die brueder grimm / ed by Neuburger, Paul – rev ed. Berlin: Deutsches Verlagshaus Bong. 2v. [193-?] – 1 – us Wisconsin U Libr [390]
Kinder- und hausmaerchen / Grimm, Jacob; ed by Riemann, Robert – jubilee ed. Leipzig: Turm-Verlag [1907-09] – 3v (ill) – 1 – (ill by otto abbelohde) – us Wisconsin U Libr [390]
Kinder- und hausmaerchen der brueder grimm – Leipzig: Verlag von hegel & schade, [1927?] – 1 – us Wisconsin U Libr [390]
Kinder zwischen arbeit und schule : kinderarbeit in indien und moegliche bildungspolitische massnahmen zur konfliktloesung / Jessen, Wolfgang – 1997 – 5mf – 9 – 3-8267-2485-2 – mf#DHS 2485 – gw Frankfurter [362]
Der kinderbote – v53-58. 1940-1955 – (= ser Mennonite serials coll) – Inquire – 1 – (lacking: v67 n22) – mf#ATLA 1994-S008 – us ATLA [242]
Kinder-bote and junior-messenger – v51-52. 1937-38 [complete] – (= ser Mennonite serials coll) – Inquire – 1 – mf#ATLA 1994-S007 – us ATLA [242]
De kinderdoop : behoort te huis bij den antichrist. antwoord aan prof hemkes / Smidt, W R – [s.l: s.n, 1913?] [mf ed 1993] – (= ser Reformed church coll) – 18p on 1mf – 9 – 0-524-06666-3 – (in dutch) – mf#1991-2721 – us ATLA [242]
De kinderdoop is niet bijbelsch / Smidt, W R – Grand Rapids, MI: H Verhaar, 1912 [mf ed 1993] – (= ser Reformed church coll) – 16p on 1mf – 9 – 0-524-06104-1 – (in dutch) – mf#1991-2417 – us ATLA [242]
De kinderdoop uit god : apologetisch antwoord op eene onlangs door mr j schoemaker beredeneerde vraag, "is de kinderdoop uit god of uit de menschen?" / Hemkes, Gerrit Klaas – Grand Rapids: Van Dort & Hugenholtz, 1886 [mf ed 1993] – (= ser Reformed church coll) – 115p on 1mf – 9 – 0-524-07008-3 – (in dutch) – mf#1991-2861 – us ATLA [242]

Der kinderfreund – Bremen DE, feb-okt 1856; aug 1857-mar 1958 – 1r – 1 – gw Misc Inst [074]

Der kinderfreund *see* Duesseldorfer arbeiterzeitung

Kindergarten lessons for church sunday schools : a manual for the instruction of beginners – Milwaukee: Young Churchman, 1911 – 1mf – 9 – 0-524-06549-7 – mf#1991-2633 – us ATLA [240]

Kindergarten messenger – Killen TX 1873-77 – 1 – mf#4809 – us UMI ProQuest [370]

Kindergarten messenger and the new education – Killen TX 1877-82 – 1 – mf#4698 – us UMI ProQuest [370]

Kinderland : [a novel] / Voigt-Diederichs, Helene – feldpostausg. Jena: E Diederichs, 1943, c1938 [mf ed 1989] – 83p – 1 – mf#7176 – us Wisconsin U Libr [830]

Kinderlegende / Leitgeb, Josef – Berlin: Wiking Verlag, [1945?] – 1r – 1 – us Wisconsin U Libr [830]

Kinderlieder und geschichten / Seidel, Heinrich – 5. Aufl. Stuttgart: Union Deutsche Verlagsgesellschaft, [190-?] – 1r – 1 – us Wisconsin U Libr [800]

Kindermann, Heinz *see*
- Die deutsche gegenwartsdichtung im kampf um die deutsche lebensform
- Ferdinand raimund
- Hebbel und das wiener theater seiner zeit
- Hoelderlin und das deutsche theater
- Klopstocks entdeckung der nation
- Traum und sendung

Kindermann, Joseph Karl *see* Beitraege zur vaterlandskunde fuer inneroesterreichs einwohner

Die kindermoerderin : ein trauerspiel / Wagner, Heinrich Leopold; ed by Schmidt, Erich – Heilbronn: Henninger, 1883 [mf ed 1993] – (= ser Deutsche litteraturdenkmale des 18. und 19. jahrhunderts 13) – x/116p – 1 – (original ed, leipzig, 1776) – mf#8676 reel 2 – us Wisconsin U Libr [820]

Kindermund : aussprueche und scenen aus dem kinderleben / Schoenthan, Paul von – Leipzig: P. Reclam jun., [1886?] – 1r – 1 – us Wisconsin U Libr [080]

Kinder-post – v1 n1-v5 n52 [1883 jan 14-1887 dec 25], v6 n23-35,41,43-45,48-49,51 [1888 jun 3-aug 26, oct 7, 21-nov 4,25-dec 2,16], v8 n15-16,20,34 [1890 apr 13-20, may 18, aug 24], v9 n51-52 [1891 dec 19-27], v10 n2,17 [1892 jan 10, apr 24], v11 n18,46-47 [1893 apr 30, nov 12-19] – 1r – 1 – mf#2962994 – us WHS [071]

Kinderpost – Essen DE, 1930 apr-1931 mar – 1 – gw Misc Inst [074]

Kinderspel und kinderlust in zuid-nederland / Cock, Alfons de – Gent: A Siffer 1902-08 [mf ed Bloomington IN: Indiana Uni Lib, Preservation Dept 1984] – 8v on 1r – 1 – us Indiana Preservation [390]

Kinderthraenen : zwei erzaehlungen / Wildenbruch, Ernst von – 3. Aufl. Berlin: Freund & Jeckel, 1884 – 1r – 1 – us Wisconsin U Libr [430]

Die kinderwelt – Leipzig DE, 1926-30 – 1 – gw Misc Inst [305]

Kinder-yohren / Feigenberg-Eamri, Rachel – Varsha, Poland. 1909 or 1910 – 1r – us UF Libraries [939]

Kinderzeichnungen : untersuchung von kinderzeichnungen der schuleingangsuntersuchung des hauptgesundheitsamtes bremen... / Leiers, Bastian – (mf ed 1995) – 2mf – 9 – €40.00 – 3-8267-2265-5 – mf#DHS 2265 – gw Frankfurter [150]

Der kindesmord in der literatur der sturm- und drang- periode : ein beitrag zur kultur- und literatur-geschichte des 18. jahrhunderts / Rameckers, Jan Matthias – Rotterdam: Nijgh & Van Ditmar, 1927 [mf ed 1993] – 279p/pl (ill) – 1 – (incl bibl ref and ind) – mf#8210 – us Wisconsin U Libr [410]

Kindex : an index to legal periodical literature concerning children – v1-3. 1965-94 – 9 – $147.00 set – ISSN: 0733-8937 – mf#115941 – us Hein [348]

Kindgerechte bestimmung der transepithelialen potentialdifferenz am respiratorischen epithel der nase / Hofmann, Thomas – (mf ed 1995) – 2mf – 9 – €40.00 – 3-8267-2087-3 – mf#DHS 2087 – gw Frankfurter [611]

Eine kindheit / Carossa, Hans – Leipzig: Insel-Verlag 1929 [mf ed 1989] – 1r – 1 – (filmed with: ungleiche welten) – mf#7145 – us Wisconsin U Libr [430]

Eine kindheit und verwandlungen einer jugend / Carossa, Hans – [Leipzig]: Insel-Verlag 1947 [mf ed 1989] – 1r – 1 – (filmed with: ungleiche welten) – mf#7145 – us Wisconsin U Libr [430]

Kindheitsbegriffe japanischer strafkonzeptionen : zur rezeption westlicher modelle der reformerziehung in der meiji-zeit / Hedenigg, Silvia – (mf ed 1997) – 3mf – 9 – €49.00 – 3-8267-2501-8 – mf#DHS 2501 – gw Frankfurter [370]

Die kindheitsgeschichte unseres herrn jesu christi nach matthaeus und lukas / Nebe, August – Stuttgart: Greiner und Pfeiffer, 1893 – 1mf – 9 – 0-8370-7413-4 – mf#1986-1413 – us ATLA [225]

Kindnytt – Boras, 1992 – 9 – sw Kungliga [078]

Kindom of priests, and, time and eternity / Woodford, James Russell – London, England. 1854 – 1r – us UF Libraries [240]

Kindred and clan in the middle ages and after : a study in the sociology of the teutonic races / Phillpotts, Bertha Surtees – Cambridge: University Press, 1913 [mf ed 1992] – (= ser Cambridge archaeological and ethnological series) – 1mf – 9 – 0-524-03371-4 – mf#1990-3205 – us ATLA [931]

Kindred of the dust / Kyne, Peter B – Toronto: Copp Clark [c1920] [mf ed 1998] – 5mf – 9 – 0-665-99134-7 – mf#99134 – cn Canadiana [830]

Kindred sayings on buddhism / Davids, Caroline Augusta Foley Rhys – Calcutta: University of Calcutta, 1930 – (= ser Samp: indian books) – us CRL [280]

Kindt, B *see*
- Thesaurus asterii amaseni et firmi caesariensis
- Thesaurus basilii caesariensis, 1 et 2
- Thesaurus procopii caesariensis

Kine weekly – 3pt sets. 1907-71 – 1 – £9900.00 set – (pt sets incl: the silent era 1907-28 112r £5000 kws. the golden years 1929-41 43r £2100 kwg. the television era 1942-71 72r £9900 kwt.) – uk World [790]

Kinema – Zuerich (CH), 1913 5 jul-14, 1915 9 jan-14 aug, 1916 29 jul-7 oct, 1918 5 jan-21 dec, 1919 4 jan-25 oct – 4r – 1 – gw Mikrofilm [790]

A kinematic analysis of the developmental sequence of kicking using a direct and angled approach / Bransdorfer, Alfred H – 1998 – 2mf – 9 – $8.00 – mf#PE 3927 – us Kinesology [612]

A kinematic and kinetic analysis of the overgrip giant swing on the uneven parallel bars / Witten, W – 1990 – 2mf – 9 – $8.00 – us Kinesology [790]

Kinematic and kinetic evaluation of high speed backward running / Arata, Alan W – 1999 – 3mf – 9 – $12.00 – mf#PSY 2063 – us Kinesology [612]

The kinematic variables related to the efficiency of throwing : football / Heppe, Robert A & Evans, Gail G – 1992 – 2mf – 9 – $8.00 – us Kinesology [612]

"Der Kinematograph" *see* Der praktikus

Der kinematograph – Duesseldorf, Berlin DE, 1907-1935 30 mar – 23r – 1 – (publ since 1923 in berlin) – gw Mikrofilm [790]

Der kinematographen-operateur *see* Erste internationale film-zeitung 1909

Kinematographische monatshefte – Berlin DE, 1921 apr-1933 jul – 2r – 1 – gw Mikrofilm [790]

Kinematographische rundschau – Berlin DE, Wien (A), 1907 1 feb-1908 15 dec, 1909-1910 28 jul, 1912 7 jul-29 dec – 4r – 1 – gw Mikrofilm [790]

Kinematographische rundschau – Vienna, Austria feb 1907-sep 1921 [mf ed Norman Ross] – 11r – 1 – (missing: 1908-13, 1915) – mf#nrp-1976 – us UMI ProQuest [790]

Kinematographische wochenschau / ed by Gaumont – Berlin DE, 1910 n38-66, 1912 n8-47 [gaps] – 1 – gw Mikrofilm [790]

Kinematoscope-zeitung – Herne DE, 1909 [gaps] – 1 – gw Misc Inst [790]

Kineserne og den kristne mission / Coucheron-Aamot, William – Kristiania: H. Aschehoug, 1894 [mf ed 1995] – (= ser Yale coll) – 79p – 1 – 0-524-09939-1 – (in norwegian) – mf#1995-0939 – us ATLA [951]

Kinesis – Carbondale IL 1968+ – 1,5,9 – ISSN: 0023-1568 – mf#6863 – us UMI ProQuest [100]

Kinesis / Vancouver Status of Women – 1977 may [v6 n6]-1981 jan, 1981 feb-1985 jan – 2r – 1 – (cont: newsletter, vancouver status of women) – mf#524777 – us WHS [305]

Kinesthetic sense and consistency in multijoint movement sequences / Astilla, Michael J – 1998 – 1mf – 9 – $4.00 – mf#PSY 2012 – us Kinesology [612]

A kinetic and kinematic analysis of the Haraihoshi judo technique / Pucsok, Jozsef – 2000 – 76 on 1mf – 9 – $5.00 – mf#PE 4153 – us Kinesology [790]

A kinetic and kinematic comparison of the grab and track starts in competitive swimming / Juergens, Cheryl A – Oregon State University, 1995 – 2mf – 9 – $8.00 – mf#PE 3657 – us Kinesology [612]

A kinetic and kinematic comparison of the grab start and track start in swimming / Allen, David M – 1997 – 1mf – 9 – $4.00 – mf#PE 3815 – us Kinesology [612]

Kinetic and temporal correlates to skillfulness in vertical jumping / Strohmeyer, H Scott – 1995 – 2mf – 9 – $8.00 – mf#PE 3799 – us Kinesology [612]

Kinetics and catalysis – Dordrecht, Netherlands 1960+ – 1,5,9 – ISSN: 0023-1584 – mf#10838 – us UMI ProQuest [540]

Kinetics and kinematics of prepubertal children : participating in osteogenic physical activity / Bauer, Jeremy – 2000 – 97p on 1mf – 9 – $5.00 – mf#PE 41003 – us Kinesology [617]

Kinetics of reactions in solution / Moelwyn-Hughes, Emyr Alun – Oxford, England. 1947 – 1r – us UF Libraries [500]

Kinfolks / Southwest Louisiana Genealogical Society – 1977 apr 28-1989 – 1r – 1 – mf#106146 – us WHS [929]

King, Albert Barnes *see* The purple and scarlet woman and her relatives

King alfred's version of the consolations of boethius = De consolatione philosophiae / Boethius, Anicius Manlius Severinus – Oxford: Clarendon Press, 1900 – 1mf – 9 – 0-7905-3758-3 – (in english) – mf#1989-0251 – us ATLA [180]

King and parliament (a.d. 1603-1714) / Wakeling, George Henry – New York: Scribner 1896 [mf ed 1992] – (= ser Oxford manuals of english history 5) – 1mf – 9 – 0-524-03867-8 – mf#1990-4914 – us ATLA [941]

King, Andrew *see* Plea for union in maintaining the scriptural doctrine of the westmi...

King, Annie Liddon *see* Dr. liddon's tour in egypt and palestine in 1886

King arthur and the table round : tales chiefly after the old french of crestien of troyes, with an account of arthurian romance / Newell, William Wells – Boston, New York: Houghton, Mifflin, 1897. 2v – 1 – us Wisconsin U Libr [830]

King cetywayo zulu dictionary / Samuelson, Robert Charles Azariah – Durban, South Africa. 1923 – 1r – us UF Libraries [470]

King charles 5 : a personal memoir / Gore, John – New York: Charles Scribner's Sons, 1941. xx,464p, front., illus, pl., ports – 1 – us Wisconsin U Libr [941]

King, Charles Daly *see* The oragean version

King, Charles William *see*
- Early christian numismatics
- The gnostics and their remains, ancient and medieval

[King city-] king city herald – CA. 1914-1932 – 6r – 1 – $360.00 – mf#C03257 – us Library Micro [071]

[King city-] the land and its people – CA. Oct 1956-1982; 1986 – 6r – 1 – $360.00 – mf#B02325 – us Library Micro [071]

[King city-] the rustler – CA. May 1901- – 65r – 1 – $3900.00 (subs $90y) – (aka: rustler-herald) – mf#B02326 – us Library Micro [071]

King corporation : a new gospel for strikers according to the acts of the railway presidents – New York: R H Rodda, printer, [1877?] (mf ed 19–) – [4]p – mf#ZT-TN+ pv43 n12 – us Harvard [380]

King country chronicle – Te kuiti, NZ. 26 oct 1906-4 dec 1907; 1908-oct 1917; nov 1918-1939; 18 mar-13 dec 1960; 1976-nov 1980 – 88r – 1 – (incorp in: waitomo news fr dec 1980) – mf#15.15 – nz Nat Libr [079]

King county labor news / AFL-CIO [American Federation of Labor and Congress of Industrial Organizations] – 1965 sep-1968 jun – 1r – 1 – (cont: washington state-labor-news; cont by: scanner and king county labor news; scanner and king county labor news) – mf#806343 – us WHS [331]

King, CW *see* Julian the emperor

King, David *see*
- Baptism
- On civil establishments of christianity
- The principles of geology explained, and viewed in their relations to revealed and natural religion
- State and prospects of jamaica
- Why baptize the little ones?

King, Dougall Macdougall *see* The battle with tuberculosis and how to win it

King, E J *see* The sacred harp

King, Edward *see*
- Duty and conscience
- A letter to the rev charles j elliott...
- The love and wisdom of god
- Sermons and addresses
- Spiritual letters of edward king

King edward 6 : his life and character / Markham, Clements Robert – London: Smith, Elder, 1907 – 1mf – 9 – 0-524-03647-0 – (incl bibl ref) – mf#1990-1075 – us ATLA [941]

King, Edward George *see*
- Early religious poetry of the hebrews
- The yalkut on zechariah

King, Edward, of Blackthorn, Bicester *see* Bliss not riches

King edward the sixth, supreme head : an historical sketch, with an introduction and notes / Lee, Frederick George – London: Burns and Oates; New York: Catholic Publication Society Co, 1886 – 1mf – 9 – 0-7905-6929-9 – (incl bibl ref) – mf#1988-2929 – us ATLA [941]

King, Ernest J *see* Official papers of fleet admiral ernest j king

King, H C *see* Records and documents concerning sergeant's inn, fleet street

King, Helen Jane *see* Critique of three schumann piano works

King, Henry Churchill *see*
- The ethics of jesus
- It's all in the day's work
- The laws of friendship, human and divine
- Letters on the greatness and simplicity of the christian faith
- The moral and religious challenge of our times
- Personal and ideal elements in education
- Rational living
- Reconstruction in theology
- Religion as life
- The seeming unreality of the spiritual life
- Theology and the social consciousness
- The treatment of doubts

King, Henry Melville *see*
- The baptism of roger williams
- The mother church
- Religious liberty
- Why we believe the bible

King, Henry Melville et al *see* Elements in baptist development

King horse / ed by Ryan, John F – Montreal: Canadian National Bureau of Breeding [1911] [mf ed 1994] – 1mf – 9 – 0-665-71309-6 – mf#71309 – cn Canadiana [636]

King, Irving *see*
- The development of religion
- The differentiation of the religious consciousness

King island reporter – King Island, sep 1905-oct 1917 – 1r – A$41.84 vesicular A$47.34 silver – at Pascoe [079]

King, J *see* A voyage to the pacific ocean

King, J R *see* The thirty-nine articles of the church of england

King, James *see*
- Cleopatra's needle
- Recent discoveries on the temple hill at jerusalem

King james 1 of england to king christian 4 of denmark : royal correspondence, 1603-1625 – 11mf – 9 – (coll from the danish national library. with full english translations) – mf#C39-23400 – us Primary [920]

King, James M *see* A statement concerning the first year's active work of the league

King, John *see*
- Character, services and reward of the faithful pastor
- The law of criminal libel
- The law of defamation in canada
- Lectures upon jonah
- Letters from france
- McCaul, Croft, Forneri
- Rest in reversion for the people of god
- Sermons

King, John H *see* The supernatural

King, John Mark *see* Memorial sermon preached in st james' square presbyterian church, toronto

King, Jonas *see* The oriental church and the latin

King, Joseph Hillery *see* Christianity in polynesia

King, Josiah *see* The examination and tryal of old father christmas

King, L W *see*
- Babylonian magic and sorcery...
- The letters and inscriptions of hammurabi, king of babylon...
- The sculptures and inscriptions of darius 1, king of persia, 548-485 bc
- Studies in eastern history

King lazarus / Beti, Mongo – London, England. 1960 – 1r – us UF Libraries [960]

King, Leonard William *see*
- Babylonian magic and sorcery
- Babylonian religion and mythology
- Chronicles concerning early babylonian kings
- First steps in assyrian
- The seven tablets of creation

King leopold's congo / Slade, Ruth M – London, England. 1962 – 1r – us UF Libraries [960]

King leopold's legacy / Anstey, Roger – London, England. 1966 – 1r – us UF Libraries [960]

King, Marie Byrnes *see* Poems

The king of court poets : a study of the work, life and time of lodovico ariosto / Gardner, Edmund Garratt – New York: E P Dutton 1906 – [ill] – 1 – mf#1303 – us Wisconsin U Libr [440]

The king of kings / Zollars, Ely Vaughan – Cincinnati, Ohio: Standard Pub Co, 1911 – 1mf – 9 – 0-524-06509-8 – mf#1991-2609 – us ATLA [220]

The king of men : the blank in his history, its filling up and lessons / Kerr, Robert – Mitchell, IA: R Kerr, 1879 – 1mf – 9 – 0-8370-3891-X – mf#1985-1891 – us ATLA [240]

King of nations and the duty of their rulers to his truth and kingd... / Nixon, William – Edinburgh, Scotland. 1869 – 1r – us UF Libraries [240]

King of tampa bay / Manon, Paul – s.l, s.l? 1936 – 1r – us UF Libraries [978]

The king of the dark chamber / Tagore, Rabindranath – London: Macmillan and Co, 1922 – (= ser Samp: indian books) – (trans into english by aut) – us CRL [830]

King of the hottentots / Cope, John Patrick – Cape Town, South Africa. 1967 – 1r – us UF Libraries [960]
King, Owen C H *see* Character of dr littledale as a controversialist
King, Peter King *see* Life of john locke
King, Peter King, Lord *see* An enquiry into the constitution, discipline, unity, and worship of the primitive church
King, Philip *see* Narrative of a survey of the intertropical and western coasts of australia
King, Richard John *see*
– Handbook to the cathedrals of england
– Handbook to the cathedrals of wales
King, Spencer B *see* Autobiography, 1880-1945
King street trolley – Madison WI. v1 n1-13 [1971 nov17-1972 jun 1 – 1r – 1 – mf#1061468 – us WHS [071]
King, Susan E *see* The enrollment and persistence of african-american doctoral students in physical education and related disciplines
King, T G et al *see* Destitution and suggested remedies
'King taufa" / Collocott, E E V – ca 1930 – 1r – 1 – mf#PMB1029 – at Pacific Mss [980]
King that cometh to save / Muir, William – Edinburgh, Scotland. 1840 – 1r – us UF Libraries [240]
King, the bridegroom, and the tribes of israel – London, England. 18-- – 1r – us UF Libraries [240]
King, Thomas *see*
– The cabinet maker's sketch book
– The modern style of cabinet work
– Shop fronts and exterior doors
– Specimens of furniture
King, Thomas H *see* The study-book of mediaeval architecture and art
King, Thomas Starr *see*
– Spiritual christianity
– Substance and show
King, W W *see* Theological discussion
King, William Harvey *see* History of homeopathy and its institutions in america; their founders, benefactors, faculties, officers, hospitals..
King williams town gazette – King Williams Town SA, 1856-1874 – 18r – 1 – (cont by: king william's town gazette and kaffrarian banner, 1861-1868) – sa National [079]
King williams town gazette *see* The sun
King william's town gazette and border intelligencer : 1856-1860 – 14 aug 1856-28 dec 1860 – 4r – 1 – (cont: king william's town gazette and kaffrarian banner, 1861-1868) – mf#MS00321 – sa National [079]
King william's town gazette and kaffrarian banner : 1861-1868 – 4 jan 1861-30 dec 1868 – 8r – 1 – (cont: king william's town gazette; cont by: king william's town gazette and border intelligencer 1856-1860) – mf#MS00321 – sa National [079]
Kingcrest baptist church (formerly: emmaneul). vancouver, british columbia. canada : church records – 1911-Sep 1969 – 1 – 67.12 – us Southern Baptist [242]
The kingdom and people of siam : with a narrative of the mission to that country in 1855 / Bowring, J – London: John W Parker and Son, 1857 2v – 12mf – 9 – mf#SE-20192 – ne IDC [915]
"Kingdom" and "the church"... – London, England. 18-- – 1r – 1 – us UF Libraries [240]
The kingdom and the church : a reply to a plymouth brethren tract, written by mr r t grant, on the above subject / Carmichael, James – Clinton, Ont?: s.n, 1866 – 1mf – 9 – (incl bibl ref) – mf#64698 – cn Canadiana [240]
The kingdom and the farm / Feeman, Harlan Luther – New York: Fleming H. Revell, c1914 – 1mf – 9 – 0-7905-5390-2 – mf#1988-1390 – us ATLA [240]
The kingdom in history and prophecy / Chafer, Lewis Sperry – New York: Fleming H Revell, c1915 – 1mf – 9 – 0-7905-3818-0 – mf#1989-0311 – us ATLA [220]
The kingdom in india : its progress and its promise / Chamberlain, Jacob – New York: Fleming H Revell, c1908 [mf ed 1986] – 1mf – 9 – 0-8370-6032-X – (incl ind) – mf#1986-0032 – us ATLA [240]
The kingdom in the cradle / Atkins, James – Nashville, Tenn.: Publishing House of the M.E. Church, South, 1905 – 1mf – 9 – 0-7905-5802-5 – mf#1988-1802 – us ATLA [240]
The kingdom in the pacific / Paton, Frank Hume Lyall – London: London Missionary Society, 1913 [mf ed 1995] – (= ser Yale coll) – viii/166p (ill) – 1 – 0-524-09535-3 – mf#1995-0535 – us ATLA [980]
Kingdom now and then – Killarney, Ireland. 20 jun-19 dec 1989; 16 jan-18 dec 1990; 1991; 14 jan-15 dec – 5r – 1 – uk British Libr Newspaper [072]
The kingdom of a heart / Albanesi, Effie Adelaide Maria – London: George Routledge & Sons Ltd, [1899] – (= ser 19th c women writers) – 54mf – 9 – mf#5.1.35 – uk Chadwyck [830]
The kingdom of all-israel : its history, literature, and worship / Sime, James – London: James Nisbet, 1883 – 2mf – 9 – 0-7905-2075-3 – mf#1987-2075 – us ATLA [939]
The kingdom of cambodia / Cambodia. Permanent Mission to the United Nations – New York [1968?] [mf ed 1989] – 1r with other items – 1 – mf#mf-10289 seam reel 015/09 [§] – us CRL [959]
The kingdom of canada : imperial federation, the colonial conferences, the alaska boundary and other essays / Ewart, John Skirving – Toronto: Morang, 1908 – 5mf – 9 – 0-665-73220-1 – (incl bibl ref) – mf#73220 – cn Canadiana [971]
The kingdom of canada : imperial federation, the colonial conferences, the alaska boundary and other essays / Ewart, John Skirving – Toronto: Morang, 1908 [mf ed 1994] – 5mf – 9 – 0-665-73220-1 – (incl bibl ref) – mf#73220 – cn Canadiana [320]
Kingdom of christ / Maurice, Frederick Denison – London, England. v1-2. 1883 – 1r – us UF Libraries [025]
The kingdom of christ on earth : twelve lectures / Harris, Samuel – Andover: WF Draper, 1874 – 1mf – 9 – 0-7905-3897-0 – mf#1989-0390 – us ATLA [240]
The kingdom of god : a course of four lectures / Temple, William – London: Macmillan, 1914 [mf ed 1991] – (= ser Shilling theological library) – 1mf – 9 – 0-7905-8597-9 – (incl bibl ref) – mf#1989-1822 – us ATLA [240]
The kingdom of god : an essay in theology / Schwab, Laurence Henry – New York: E P Dutton, 1897 – 1mf – 9 – 0-8370-5172-X – mf#1985-3172 – us ATLA [240]
The kingdom of god : its origin, nature and duration / Evans, James – Mt Morris, Ill: Brethren at Work Steam Print House, 1882 – 1mf – 9 – 0-524-04210-1 – mf#1990-5001 – us ATLA [240]
The kingdom of god and american life / Brewster, Chauncey Bunce – New York: Thomas Whittaker, 1912 – 1mf – 9 – 0-7905-4100-9 – mf#1988-0100 – us ATLA [240]
The kingdom of god and life therein / Forster, William Rabbeth – Toronto: Musson Book Co, 1906 [mf ed 1995] – 3mf – 9 – 0-665-74192-8 – mf#74192 – cn Canadiana [210]
The kingdom of god and socialism / Wilson, Jackson Stitt – 3rd ed. Berkeley, CA: J Stitt Wilson, 1911 – (= ser Social Crusade Series) – 1mf – 9 – 0-524-03777-9 – mf#1990-1124 – us ATLA [335]
The kingdom of god and the messiah / Scott, Ernest Findlay – Edinburgh: T & T Clark, 1911 – 1mf – 9 – 0-7905-0330-1 – (incl bibl ref and index) – mf#1987-0330 – us ATLA [210]
The kingdom of god in japan : observations and recommendations of a deputation appointed by the american board of commissioners for foreign missions – [Boston], 1918 [mf ed 1996] – (= ser Yale coll) – 86p – 1 – 0-524-10279-1 – mf#1996-1279 – us ATLA [950]
The kingdom of god in the writings of the fathers / Herrick, Henry Martyn – Chicago: University of Chicago Press, 1903 – 1mf – 9 – 0-8370-3570-8 – (incl ind) – mf#1985-1570 – us ATLA [240]
The kingdom of god is within you : christianity not as a mystic religion but as a new theory of life = Tsarstvo bozhie vnutri vas / Tolstoy, Leo, Count – London: William Heinemann, 1894 [mf ed 1986] – 2v on 2mf – 9 – 0-8370-6531-3 – (english trans fr russian by constance garnett) – mf#1986-0531 – us ATLA [240]
The kingdom of god, its constitution and progress : a discourse before the general assembly of the presbyterian church, by appointment at their meeting in rochester, new york... / Junkin, David Xavier – Philadelphia: Board of Domestic Missions, 1860 – 1mf – 9 – 0-524-05545-9 – mf#1990-5149 – us ATLA [242]
Kingdom of god not in word but in power / Hoare, Charles James – London, England. 1829 – 1r – us UF Libraries [240]
The kingdom of god on earth / Belaney, Robert – London: Thomas Baker, 1896 – 1mf – 9 – 0-8370-6962-9 – mf#1986-0962 – us ATLA [230]
The kingdom of god, or, kingdom : a notice of "brief remarks" by r t grant, esq, published january, 1866 on a sermon preached by the rev james carmichael, entitled "the tares and the wheat", published july, 1865 / Carmichael, James – [Clinton, Ont?: s.n.], 1866 – 1mf – 9 – 0-665-89960-2 – mf#89960 – cn Canadiana [220]
The kingdom of god, or, what is the gospel? / Dunn, Henry – London: Simpkin, Marshall, 1868 – 1mf – 9 – 0-8370-8664-7 – (incl indes) – mf#1986-2664 – us ATLA [240]
Kingdom of god the good news, not individual salvation only – Edinburgh, Scotland. 1879 – 1r – us UF Libraries [240]
Kingdom of lesotho second five year development plan, 1975/76-1979/80 – [Maseru?: s.n., 1976?] – (filmed with: lesotho. central planning and development office. lesotho first five-year development plan, 1970/71-1974/75) – us CRL [330]
Kingdom of our lord and of his christ / Maccoll, Dugald – Glasgow, Scotland. 1871 – 1r – us UF Libraries [240]
Kingdom of the world / Carpentier, Alejo – New York, NY. 1957 – 1r – us UF Libraries [972]
The kingdom of the yellow robe : being sketches of the domestic and religious rites and ceremonies of the siamese / Young, Ernest – 3rd ed. London: A Constable, 1907 – 1mf – 9 – 0-524-08167-0 – (incl bibl ref) – mf#1991-0297 – us ATLA [390]
Kingdom preparedness : america's opportunity to serve the world / Kinney, Bruce – New York: Fleming H Revell, c1916 – 1mf – 9 – 0-524-07690-1 – mf#1991-3275 – us ATLA [240]
Kingdom songs : for sunday-school, prayer meeting, christian workers' societies and all seasons of praise – Elgin, IL: Brethren Pub House, 1911 – 3mf – 9 – 0-524-07193-4 – (incl ind) – mf#1990-5351 – us ATLA [780]
Kingdom voice / Yahweh's New Kingdom, Inc – 1988 jul-1991 dec – 1r – 1 – mf#1113751 – us WHS [270]
Kingdom, William *see* America and the british colonies
The kingdoms of africa / Garlake, Peter S – Oxford: Elsevier-Phaidon, 1978. 152p. ill., facsim., maps, plans, ports.Bibliography. Includes index.(The Making of the Past) – 1 – us Wisconsin U Libr [960]
Kingdon, Hollingworth Tully *see*
– God incarnate
– Mis-readings of holy scripture
The king-emperor's english : or, the role of the english language in the free india / Anand, Mulk Raj – Bombay: Hind Kitabs, 1948 – (= ser Samp: indian books) – (aft by maulana abul kalam azad) – us CRL [420]
Kinglake, Alexander *see* The invasion of the crimea
Kinglake, Alexander William *see* Eothen
Kingman Park Civic Association *see* Publications
Kings : introduction, revised version with notes, index and map / ed by Skinner, John – Edinburgh: T C & E C Jack, [1904?] – (= ser The Century Bible) – 2mf – 9 – 0-7905-3073-2 – (incl bibl ref) – mf#1987-3073 – us ATLA [220]
Kings and gods of egypt = Rois et dieux d'egypte / Moret, Alexandre – New York: Putnam, 1912 – 1mf – 9 – 0-524-01203-2 – (english by madame noret) – mf#1990-2279 – us ATLA [290]
Kings and prophets of israel and judah / Simpson, Albert B – New York: Alliance Press Co, c1903 [mf ed 1992] – (= ser Christ in the bible 6; Christian and missionary alliance coll) – 1mf – 9 – 0-524-02155-4 – mf#1990-4221 – us ATLA [221]
The kings and prophets of israel and judah : from the division of the kingdom to the babylonian exile / Kent, Charles Foster – New York: Charles Scribner, 1909 – 1mf – 9 – 0-8370-9959-5 – mf#1986-3959 – us ATLA [221]
King's baptist church – Lexington. 1819-1821 (1) – 1 – $123.75 – mf#4121 – us Southern Baptist [242]
Kings bay periscope / Naval Submarine Base [Kings Bay GA] – Kings Bay GA. 1979 jun 15 [v1 n1]-1984 nov 30, 1985 jan 18-may 24, sep 27, 1986 mar 28-1987 jan 16, may 22-jun 19, dec 8, 1988 jan-1989 jul 28 – 2r – 1 – mf#1520924 – us WHS [071]
King's business / Pierson, Arthur T – London, England. 1892? – 1r – us UF Libraries [240]
The king's business : a study of increased efficiency for women's missionary societies / Raymond, Maud Wotring – West Medford, Mass: Central Committee on the United Study of Foreign Missions, c1913 – 1mf – 9 – 0-524-07710-X – mf#1991-3295 – us ATLA [240]
King's chapel sermons / Peabody, Andrew Preston – Boston: Houghton, Mifflin, 1891 – 1mf – 9 – 0-7905-9566-4 – mf#1989-1291 – us ATLA [240]
Kings college, cambridge, muniments – 3r – 1 – (incl manorial court rolls from the 13th and 14th centuries; village surveys, 1564-1589) – mf#96444/5 – uk Microform Academic [941]
King's college university magazine – [Halifax, NS?: J Bowes, 1871-18– or 19–] – 9 – mf#P04774 – cn Canadiana [378]
King's college, windsor, n s / Milner, William Cochrane – [Halifax, NS?: MacNab, 1909?] – 1mf – 9 – 0-665-75196-6 – mf#75196 – cn Canadiana [378]
King's conflicting civil cases in the texas reports / Texas. Supreme Court – v1-3. 1840-1911 – 17mf – 9 – $25.50 – mf#LLMC 91-030 – us LLMC [346]
King's counsel : journal of the faculty of laws society – King's College, University of London, 1936-85 – 9 – (lacking: 1940-47; 1949-50; 1966; 1968; 1981-82) – mf#LLMC 84-5221 – us LLMC [340]
King's County Agricultural and Industrial Exhibition (1878 : Kentville, NS) *see* General regulations and prize list...
King's county chronicle – Birr, Ireland. 24 Sep 1845-Dec 1921. -w. 27 reels – 1 – uk British Libr Newspaper [072]
Kings county chronicle – Birr, Ireland. 24 sep 1845-1919, 1 jan, 26 feb 1920, 7 oct 1920-29 dec 1921, 4 feb 1922-2 jun 1923, 27 mar-24 dec 1930 – 30 1/2r – 1 – (aka: offaly chronicle) – uk British Libr Newspaper [072]
[Kings county-] hanford city directories – CA. 1906; 1917; 1920; 1926; 1930; 1939; 1943 – 7r – 1 – $350.00 – mf#D047 – us Library Micro [917]
King's county independent *see* Tullamore and king's co independent
[Kings county-] kings county – 1901 – 1r – 1 – $50.00 – mf#D046 – us Library Micro [978]
King's county news – Hampton, NB. 1894-98 – 2r – 1 – cn Library Assoc [071]
The king's english : part 1, its sources and history, part 2, origin and progress of written language, part 3, puzzling peculiarities of english, part 4, spelling reform / Moon, George Washington – London: Hatchards, 1881 – 1mf – 9 – 0-8370-9972-2 – mf#1986-3972 – us ATLA [420]
The king's greatest business / Gilbert, Paul James – New York: Fleming H Revell, 1909 – 1mf – 9 – 0-8370-6120-2 – mf#1986-0120 – us ATLA [240]
Kings grove baptist church. pickens county. south carolina : church records – 1910-59.Minutes, 1910-31. Membership rolls, 1910-1959 – 1 – 8.15 – us Southern Baptist [242]
King's handbook of notable episcopal churches in the united states / Shinn, George Wolfe – Boston MA: Moses King 1889 [mf ed 1990] – 1mf [ill] – 9 – 0-7905-8077-2 – mf#1988-6058 – us ATLA [242]
The king's highway : a study of present conditions on the foreign field / Montgomery, Helen Barrett – West Medford: Central Committee on the United Study of Foreign Missions, [1915] [mf ed 1995] – (= ser Yale coll) – iv/272p (ill); 19 cm – 1 – 0-524-09171-4 – mf#1995-0171 – us ATLA [240]
King's Hill Neighborhood Association *see* King's hill news
King's hill news / King's Hill Neighborhood Association – 1992 oct 3 – 1r – 1 – mf#3912513 – us WHS [366]
The king's indian allies : the rajas and their india / Nihal Singh, Saint – London: Sampson Low, Marston & Co, 1916 – (= ser Samp: indian books) – us CRL [954]
The king's keys to his kingdom : containing a brief line of evidences of the glorious king of heaven and earth... / Kerr, William Henry – Cincinnati: Standard Pub Co, c1914 [mf ed 1993] – (= ser Christian church (disciples of christ) coll) – 1mf – 9 – 0-524-06309-5 – mf#1991-2482 – us ATLA [240]
Kings mountain first baptist church. kings mountain, north carolina : church records – 1944-63. WMU Records. 1917-62 – 1 – 63.95 – us Southern Baptist [242]
Kings mountain second baptist church. kings mountain, north carolina : church records – 1940-63 – 1 – us Southern Baptist [242]
The kings of israel and judah / Rawlinson, George – New York: Anson D F Randolph, [189-?] – 1mf – 9 – 0-8370-9979-X – (incl bibl ref) – mf#1986-3979 – us ATLA [221]
King's proclamation, for the encouragement of piety and virtue / George 3, King – London, England. 1818 – 1r – us UF Libraries [240]
King's secret : being the secret crrespondance of louis 15 / Broglie, Albert, Duc de – London, England. v1-2. 1879 – 1r – us UF Libraries [025]
Kings to esther / Terry, Milton Spenser – New York: Nelson & Phillips; Cincinnati: Hitchcock & Walden 1875 [mf ed 1986] – (= ser Commentary on the old testament 4) – 2mf [ill] – 9 – 0-8370-9510-7 – mf#1986-3510 – us ATLA [221]
[Kingsburg-] the kingsburg recorder – CA. 1904-20; 1925-33; 1937; 1941-43; 1945-46; 1950; 1973- [wkly] – 39r – 1 – $2340.00 (subs $90y) – mf#B02327 – us Library Micro [071]
Kingsbury, Harmon *see*
– The great law book
– Law and government
Kingsbury, William *see*
– Jude's question discussed
– Sermon occasioned by the decease of the rev thomas towle
Kingscote, Adeline Georgina Isabella *see* The english baby in india and how to rear it

Kingscote, Henry *see* Letter to his grace the archbishop of canterbury
Kingsford, Anna Bonus *see*
- The perfect way
- The virgin of the world of hermes mecurius trismegistus

Kingsford, Rupert Etherege *see*
- Evidence and practice at trials in civil cases
- The law relating to executors and administrators in the province of ontario
- Manual of the law of landlord and tenant
- Manual of the law of landlord and tenant for use in the province of ontario

Kingsford, William *see*
- The history of canada
- History, structure, and statistics of plank roads in the united states and canada

The kingship of love / Brooke, Stopford Augustus – London: Isbister, 1903 – 1mf – 9 – 0-7905-9152-9 – mf#1989-2377 – us ATLA [240]
Kingsland, William *see* Esoteric basis of christianity
Kingsley, Angie M *see* The effects of hangboard exercise on climbing performance and grip strength in college age female indoor rock climbers
Kingsley, Calvin *see*
- Resurrection of the dead . . . literal resurrection
- Round the world

Kingsley, Charles *see*
- At last
- Charles kingsley
- The hermits
- Heroes
- Historical lectures and essays
- Poems
- Sermons for the times
- Town and country sermons
- The tutor's story
- Westminster sermons

Kingsley, Florence Morse *see*
- The cross triumphant
- Paul, a herald of the cross
- Stephen

Kingsley, Frances Eliza Grenfell *see* Charles kingsley
Kingsley, M H *see* Travels in west africa, congo francais, corisco and cameroons
Kingsley, st john (+ cemetery) – (= ser Cheshire monumental inscriptions) – 2mf – 9 – £4.00 – mf#25a – uk CheshireFHS [929]
Kingsnorth, G W *see* Africa south of the sahara
Kingston 1599-1950 – (= ser Cambridgeshire parish register transcript) – 3mf – 9 – £3.75 – uk CambsFHS [929]
Kingston 1683-1849 – Oxford, MA (mf ed 1997) – (= ser Massachusetts vital record transcripts to 1850) – 17mf – 9 – 0-87623-273-X – (mf 1t-3t: births & deaths 1683-1859. mf 3t-4t: marriages 1726-68. mf 4t: publishments 1726-69. mf 5t-13t: births & deaths 1744-1856. mf 8t: marriages 1768-1809. mf 8t-9t: publishments 1769-1809. mf 13t-14t: marriages 1809-44. mf 14t-15t: marriage intentions 1809-49. mf 15t-16t: births 1843-49. mf 16t: marriages 1843-49. mf 17t: deaths 1843-49; out-of-town marriages 1727-98) – us Archive [978]
Kingston 1683-1900 – Oxford, MA (mf ed 1992) – (= ser Massachusetts vital records) – 88mf – 9 – 0-87623-154-7 – (mf 1-7: vital records 1683-1882. mf 8-10: vital records 1744-1896. mf 11-13: births & deaths 1758-1894. mf 14-16: marrs & intents 1809-83. mf 17: out-of-town marriages 1727-97. mf 18-22: church records 1720-1880. mf 18-19: vital records 1720-1858. mf 23-31: town records 1717-68. mf 26,31: publishments 1762-69. mf 32-36: town records 1769-95. mf 37-42: town records 1796-1818. mf 43-51: town records 1769-1810. mf 52-58: town records 1818-50. mf 59-66: town records 1838-71. mf 67-68: school records 1838-58. mf 69-70: taxes, voters 1864-84. mf 71: rebellion records 1861-65. mf 72-73: birth index 1843-1921. mf 74-75: marr index 1843-1921. mf 75-76: death index 1843-1921. mf 77-78: vital records 1843-53. mf 79-82: births 1854-1902. mf 83-85: marriages 1854-1902. mf 86-88: deaths 1854-99) – us Archive [978]
Kingston and richmond express and surrey reporter – london, UK. 1889 – 1r – 1 – uk British Libr Newspaper [072]
Kingston and Sherbrooke Gold Mining Co *see* Prospectus, reports and statistics.
Kingston and surbiton guardian and railway guide *see* Kingston and surbiton guardian and surrey record
Kingston and surbiton guardian and surrey record – London, UK. 1/4r – 1 – (aka: kingston and surbiton guardian and railway guide) – uk British Libr Newspaper [072]
Kingston and surbiton news etc – London, UK. 1889; 1897; 1898 – 2 3/4r – 1 – uk British Libr Newspaper [072]
Kingston and the loyalists of the spring fleet of a d 1783 : with reminiscences of early days in connecticut... / Bates, Walter; ed by Raymond, William Odber – St John, NB: Barnes, 1889 – 1mf – 9 – 0-665-03001-0 – mf#03001 – cn Canadiana [917]
Kingston borough guardian – London, UK. 1986-92 – 18r – 1 – (aka: guardian (kingston borough ed); guardian (surrey and kingston ed)) – uk British Libr Newspaper [072]
Kingston borough star *see* Kingston star
Kingston chronicle – Kingston, ON. 1819-32 – 3r – 1 – ISSN: 1181-2443 – cn Library Assoc [071]
Kingston chronicle and gazette – Kingston, ON. 1833-45 – 7r – 1 – cn Library Assoc [071]
Kingston comet extra – London, UK. 5 feb-2 apr 1987 – 1/2r – 1 – uk British Libr Newspaper [072]
The kingston daily news – Kingston, Ontario, CN. oct 1851-dec 1886 – 34r – 1 – cn Commonwealth Imaging [071]
Kingston daily standard *see* Daily standard
Kingston deanery magazine – Sussex, NB: [Deanery of Kingston, 1884?-1889?] [mf ed v3 n1 jan 1886-v3 n6 jun 1886; v3 n8 aug 1886-v3 n11 nov 1886; v6 n1 jan 1889] – 9 – mf#P04523 – cn Canadiana [242]
Kingston gazette – Kingston ON. 1810 sep 25-1816 dec 28, 1817 jan 4-1818 dec 29 – 2r – 1 – (cont by: kingston chronicle [kingston, on]) – mf#3910285 – us WHS [071]
Kingston gazette – Kingston, ON. 1810-18 – 2r – 1 – ISSN: 1101-2435 – cn Library Assoc [071]
Kingston gazette, etc – [Wales] LLGC aug 1869-sep 1907 – 18r – 1 – (missing: 1871, 1873, 1896-97) – uk Newsplan [072]
Kingston, George Templeman *see*
- On the changes of barometric pressure
- On the diurnal and annual variations of temperature at halifax, nova scotia

Kingston informer – London, UK. 1986-23 sep 1988; oct 1988-1992 – 15 1/2r – 1 – uk British Libr Newspaper [072]
Kingston spy – Kingston ON. 1883 may 23-1910 dec, [1911 jan 12-1914 dec 3], 1915 jan-1940 dec 12, 1938 oct 27, nov 10-17, 1943 oct 21-1946 jan 17, 1941, 1942-1943 oct 14 – 6r – 1 – (cont by: kingston tribune) – mf#1137927 – us WHS [071]
Kingston star – London, UK. 9 apr 1987-90 – 12r – 1 – (aka: kingston borough star) – uk British Libr Newspaper [072]
Kingston tribune – Kingston ON. 1846 jan 24-aug 29 – 1r – 1 – (cont: kingston spy; cont by: montello express; montello tribune) – mf#1137936 – us WHS [071]
Kingston whig-standard – Kingston, Ontario. 1 dec 1926-mar 1939.-d – 111r – 1 – uk British Libr Newspaper [071]
Kingston, William Henry Giles *see* Western wanderings
Kingstown and bray gazette – Dun Laoghaire, Ireland. 4 jan-29 mar 1873 – 1r – 1 – uk British Libr Newspaper [072]
Kingstown and bray observer – Dublin, Ireland. 7 may 1870-20 may 1871 – 1r – 1 – (cont as: dublin pictorial advertiser) – uk British Libr Newspaper [072]
Kingstown evening journal and blackrock monkstown dalkey and rathdown advertiser – Dun Laoghaire, Ireland. 5 jun-8 sep 1863 – 1/2r – 1 – (aka: kingstown journal and blackrock monkstown dalkey and rathdown advertiser) – uk British Libr Newspaper [072]
Kingstown evening journal, etc – Wicklow, UK. 5 Jun-8 Sept 1863.-d. 1 reel – 1 – uk British Libr Newspaper [072]
Kingstown gazette and rathdown union advertiser – Dublin, Ireland. 19 dec 1857-9 jan 1858; 2 may 1868-17 jul 1869 – 1/2r – 1 – uk British Libr Newspaper [072]
Kingstown gazette and rathdown union advertiser *see* Kingstown gazette etc
Kingstown gazette etc – Dublin, Ireland. 19 dec 1857-9 jan 1858; 2 may 1868-17 jul 1869 [wkly] – 1/2r – 1 – (incl kingstown gazette and rathdown union advertiser 2 may 1868-17 jul 1869) – uk British Libr Newspaper [072]
Kingstown journal and blackrock monkstown dalkey and rathdown advertiser *see* Kingstown evening journal and blackrock monkstown dalkey and rathdown advertiser
Kingstown monthly – Dun Laoghaire, Ireland. 1894-96 – 1 3/4r – 1 – uk British Libr Newspaper [072]
Kingstown standard bray blackrock and suburban news – Dun Laoghaire, Ireland. 4 jul 1885-2 oct 1886 – 1r – 1 – uk British Libr Newspaper [072]
Kings/tulare counties – 1989- – (= ser California telephone directory coll) – 6r – 1 – $300.00 – mf#P00042 – us Library Micro [917]
Kington gazette, radnorshire chronicle and general advertiser – [Wales] 24 aug 1869-3 sep 1907 [mf ed 2003] – 18r – 1 – (cont as: kington gazette, radnorshire chronicle, general advertiser and intelligencer [jan 1876-dec 1887]; kington gazette and general advertiser [jan 1888-sep 1907]) – uk Newsplan [072]
Kington reporter, north herefordshire and welsh border advertiser – [West Midlands] Herefordshire 5 oct 1907-29 aug 1914 [mf ed 2004] – 6r – 1 – (missing: 1911) – uk Newsplan [072]
Kinh-te tap-chi – n1-19, suppl. Nam-Dinh, Hanoi. 1933-34 – 1 – (lacking: n18) – fr ACRPP [073]
Kinkade, Kristen M *see* The contribution of recreation and sport participation to the quality of life of children with disabilities
Kinkade-Schall, Kristi L *see* Effects of a chair exercise program (sit and be fit tm) for older adults
Kinkead, Edgar Benton *see*
- The complete law quizzer...the ohio supreme court examination questions for admission to the bar
- Jurisprudence, law and ethics
- The law of pleading in civil actions and defenses under the code.
- A treatise on the law of court practice and procedure, civil and criminal.

Kinkead, Thomas L *see* An explanation of the baltimore catechism of christian doctrine
Kinkel, Gottfried *see*
- Otto der schuetz
- Tanagra

Kinkel, Hans *see* Lessings dramen in frankreich
Kinley, Joseph Macy *see* Kinley's american and english precedents.
Kinley's american and english precedents. / Kinley, Joseph Macy – Pt. I, Personal right. vol. I San Francisco, Crocker, 1899. 1030 p. LL-1511 – 1 – us L of C Photodup [340]
Kinloch, Marjory G J *see* Studies in scottish ecclesiastical history in the seventeenth and eighteenth centuries
Kinloch, William Penney *see* The circle of christian doctrine
Kinne, Asa *see*
- Kinne's law compendium
- Questions and answers on law

Kinne, La Vega George *see* Procedure and methods of the courts of final resort of the republic of mexico, the united states of america, and of the several stages and territories of the union
Kinnear, Beverley Oliver *see* Impending judgments on the earth
Kinnear, John Boyd *see* Ireland
Kinnear steel rolling doors / Mussens Ltd – [Montreal?: s.n, between 1907 and 1914] [mf ed 1991] – 1mf – 9 – 0-665-99530-X – mf#99530 – cn Canadiana [670]
Kinnebrew, J H *see* The theology of fatherhood
Kinneir, J M *see* Voyage dans l'asie mineure, l'armenie et le kourdistan, dans les annees 1813 et 1814...
Kinneir, John *see*
- Journey through asia minor, armenia, and koordistan, in the years 1813 and 1814
- Reise durch klein-asien, armenien und kurdistan

Kinne's law compendium / ed by Kinne, Asa – v1-11. 1844-55 (all publ) – 43mf – 9 – $64.50 – (v1-5 under title "questions and answers on law alphabetically arranged". v6-11 cover selected cases from the us and british commonwealth starting 1844. lacking: v7,10,11) – mf#LLMC 95-020 – us LLMC [340]
Kinney, Abbot *see* The conquest of death
Kinney, Bruce *see*
- Frontier missionary problems
- Kingdom preparedness
- Mormonism, the islam of america
- The pith and pathos of frontier missions

Kinnosuke, Adachi *see* Christian missions in japan
Kinns, Samuel *see*
- Graven in the rock
- The harmony of the bible with science
- Moses and geology

Kino *see* Kinobriefe
Kino-adressbuch – Berlin DE, 1917 – 1 – gw Mikrofilm [790]
Der kinobesitzer – Wien (A), 1917 8 sep-1919 31 mar – 1r – 1 – gw Mikrofilm [790]
Kinobriefe – Berlin DE, 1919 5 jan-1921 – 2r – 1 – (title varies: jan 1920: der kientopp; mar 1920: kino) – gw Mikrofilm [790]
Kino-journal *see* Der oesterreichische komet
Kino-kalender – Wien (A), 1918 – 1 – gw Mikrofilm [790]
Kinomatographische rundschau – Wien (A), 1907 feb-dec, 1914 jan-mar, jul-dec, 1916 jan-jun – 4r – 1 – gw Mikrofilm [790]
Kinomusik – Wien (A), 1919 – 1r – 1 – (missing: jul iss) – gw Mikrofilm [780]
Der kino-praktikus *see* Der praktikus
Kinor tsiyon – Varshe, Poland. 1900 – 1r – us UF Libraries [939]
Die kinowoche – Wien (A), 1919 8 6-1922 21 feb – 1r – 1 – gw Mikrofilm [790]
Kinross, John *see* Dogma in religion and creeds in the church
Kinross-shire advertiser – Scotland, UK. 26 Jan 1850-26 Jul 1851. -w. 11 feet – 1 – uk British Libr Newspaper [072]
Kinross-shire advertiser, and general miscellany : ...and the only medium of publicity in the county of kinross – [Scotland] Kinross: G Barnet 2 feb 1847-78, 1889-99, jul 1919-dec 1950 (mthly, wkly) [mf ed 2004] – 38r – 1 – (missing: 1871; cont by: kinross advertiser [jul 1968-dec 1970]; subtitle & imprint vary) – uk Newsplan [072]
Kinross-shire courier and county advertiser – [Scotland] Kinross: D Rintoul 5 jul 1913-27 nov 1915 (wkly) [mf ed 2003] – 125v on 1r – 1 – uk Newsplan [072]
Kinross-shire general record – [Scotland] Kinross: R Annan nov 1838- (mthly) [mf ed 2203] – 1r – 1 – uk Newsplan [072]
Kinross-shire weekly record of local and general news – [Scotland] Perthshire, Kinross: D Brown 3 feb 1872-20 dec 1873 (wkly) [mf ed 2004] – 98v on 1r – 1 – uk Newsplan [072]
Kinsei kindai fuzokushiryo harikomicho – 92v on 14r – 1 – Y210,000 – (coll of modern japanese prints originally collected by zenjiro yasuda (1834-1921) now housed in the tsubouchi memorial theatre museum in waseda university. incl various miscellaneous items such as playbills, posters, broadsheets etc dating from the late edo to early meiji period. with 378p guide by susumu yamamoto in japanese) – ja Yushodo [740]
Kinsei no kaiso shiryo : documents of the marine transportation merchants in the northern parts of japan – 277r – 1 – Y3001,000 – (with 2v guide. in japanese) – ja Yushodo [380]
Kinsey, Clark *see* Clark kinsey collection
Kinsey, Robert Baldwin *see* A serious question
Kinsey, Samuel et al *see* The brethren's reasons for producing and adopting the resolutions of august 24th
Kinship and marriage in early arabia / Smith, William Robertson; ed by Cook, Stanley Arthur – new ed. London: Adam and Charles Black, 1903 – 1mf – 9 – 0-524-01925-8 – mf#1990-2738 – us ATLA [390]
Kinship and property among the jie and turkana / Gulliver, Philip Hugh – London, 1952 – (filmed with his: a premlinary survey of the turkana. [cape town, 1951]- the karamajong cluster. london, 1952) – us CRL [960]
Kinship, genealogical claims and societal integration in ancient khmer society : an interpretation / Kirsch, A Thomas – [Ithaca? NY: Kirsch 1974?] [mf ed 1989] – 1r with other items – 1 – (with bibl) – mf#mf-10289 seam reel 017/39 [§] – us CRL [929]
The kinship of nature / Carman, Bliss – Toronto: Copp, Clark, 1904 – 4mf – 9 – 0-665-73663-0 – mf#73663 – cn Canadiana [840]
Kinship terminology of the south african bantu / Van Warmelo, Nicolaas Jacobus – Pretoria, South Africa. 1931 – 1r – us UF Libraries [960]
Kinship to christ : and other sermons / Tyler, Joseph Zachary – St Louis: J Burns, c1883 – 4mf – 9 – 0-524-07922-6 – mf#1991-3467 – us ATLA [240]
Kinships / South Bay Cities Genealogical Society – v1 n1-v6 n1/2 [1978 [spring]-1983 spr/sum – 1r – 1 – mf#696543 – us WHS [929]
Kinsley, William Wirt *see* Old faiths and new facts
Kinsman / Jesse N. Smith Family Association – 1975 apr-1987 may – 1r – 1 – mf#696546 – us WHS [366]
Kinsman – v1-2 [1897 jun-1898 oct, new series, v1-v2 n23 [1898 nov-1900] – 1r – 1 – mf#1112841 – us WHS [071]
Kinsman, Frederick Joseph *see*
- Catholic and protestant
- The issues before the church
- Principles of anglicanism

Kinsman news : weekly general newspaper – Kinsman, OH. 14 May 1897 – 1r – 1 – us Western Res [071]
Kinsman, Oliver Dorrance *see* Loyal man in florida, 1858-1861
Kinton, Ada Florence *see* Just one blue bonnet
Kinyane kedem / Nissenbaum, Isaac – Warsaw, Poland. v1-2. 1930 – 1r – us UF Libraries [939]
Kinzel, Karl *see*
- Denkmaeler der aelteren deutschen literatur fuer den litteraturgeschichtlichen unterricht an hoeheren lehranstalten
- Geschichte der deutschen literatur
- Walther von der vogelweide und des minnesangs fruehling

Kinzel, Karl [comp] *see*
- Klopstocks messias und oden
- Kunst- und volkslied in der reformationszeit
- Das nibelungenlied im auszuge nach dem urtext

Kinzie, Juliette Augusta *see* Walter ogilby: a novel
Der kinzig-bote – Gengenbach DE, 1942 3 sep-1943 30 mar – 1 – gw Misc Inst [074]

KINZIG-BOTE

Kinzig-bote – Gelnhausen DE, 1889 2 apr-1915 1 may – 26r – 1 – (title varies: 1 jan 1901: gelnhaeuser zeitung. incl suppl) – gw Misc Inst [074]
Kinzigtal-nachrichten – Schluechtern DE, 1983 1 jun- – (= ser Bezirksausgabe von frankfurter neuen presse fuer den hochtaunuskreis) – ca 6r/yr – 1 – gw Misc Inst [074]
Kinzig-wacht – Hanau DE, 1935 19 oct-1943 30 jun – 15r – 1 – (title varies: ed for districts hanau, gelnhausen & schluechtern; publ in frankfurt m) – gw Misc Inst [074]
The kinzua planning newsletter – 1961-65 – (= ser American indian periodicals... 1) – 1mf – 9 – $95.00 – us UPA [305]
Kiokee baptist church. appling, georgia : church records – 1790-1955 – 1 – us Southern Baptist [242]
The kiokio nius *see* Melanesian nius / the kiokio nius
Kiowa Business Committee [OK] *see* Kiowa news
Kiowa indian news = Gkaoy gkoot pigh'gyah – Anadarko, Carnegie OK. 1976 oct [v4 n5]-1981 nov, 1981 dec-1987 sep [v16 n12] – 2r – 1 – (cont: kiowa tribal newsletter) – mf#610625 – us WHS [071]
Kiowa news / Kiowa Business Committee [OK] – v1 n4 [1971 dec] – 1r – 1 – (cont by: kiowa tribal newsletter) – mf#610632 – us WHS [305]
Kiowa tribal newsletter – OK. 1975 mar-1976 sep – 1r – 1 – (cont: kiowa news; cont by: kiowa indian news) – mf#610629 – us WHS [305]
Kip, William Ingraham *see*
– The catacombs of rome as illustrating the church of the first three centuries
– The christmas holydays in rome
– The church of the apostles
– Church of the apostles
– The double witness of the church
– The early conflicts of christianity
– The early days of my episcopate
– Historical scenes from the old jesuit missions
– The unnoticed things of scripture
Kiphuth, R *see* The diagnosis and treatment of postural defects
Kipka, Karl *see* Maria stuart im drama der weltliteratur
Kipling, Rudyard *see*
– Actions and reactions
– The dead king
– The jungle book
– Just so stories for little children
– Letters of marque
– Letters to the family
– Puck of pook's hill
– Rewards and fairies
– Rudyard kipling's verse
– Seven seas
Kiplinger's personal finance – Washington DC 2000+ – 1,5,9 – (cont: kiplinger's personal finance magazine) – ISSN: 1528-9729 – mf#879.02 – us UMI ProQuest [650]
Kiplinger's personal finance magazine – Washington DC 1992-99 – 1,5,9 – (cont: changing times; cont by: kiplinger's personal finance) – ISSN: 1056-697X – mf#879.02 – us UMI ProQuest [650]
Kipp, Catherine G *see* Women on the land; the women's land army
Kippenberg, Anton *see* Goethe und seine welt
Kippenberg, Anton et al *see* Goethes Werke
Kipper, Karl *see* Die entwicklung des arbeitshauses unter besonderer beruecksichtigung der reformatorischen bestrebungen und der vehaeltnisse in westfalen
A kiralyi itelotablak felulvizsgalati tanacsainak elvi jelentosegu hatarozatai / Hungary. Itelotablak – Budapest. On film: v1-19; 1895-1917. LL-0244 – 1 – us L of C Photodup [340]
Kirata-jana-krti = The indo-mongoloids: their contribution to the history and culture of india / Chatterji, Suniti Kumar – Calcutta: Royal Asiatic Society of Bengal, 1951 – (= ser Samp: indian books) – us CRL [305]
Kiratarjuniyam cantos 1-3 : text with mallinath's commentary, prose order of the slokas, notes, translation into english / Kale, Moreshwar Ramachandra – Bombay: Oriental Pub Co, 1916 – (= ser Samp: indian books) – us CRL [305]
Kirbisch, oder, der gendarm, die schande und das glueck : ein episches gedicht / Wildgans, Anton – Leipzig: L Staackmann, 1927 – 1r – 1 – us Wisconsin U Libr [810]
Kirby, Edmund Burgis *see* The ore deposits of rossland, british columbia, canada
Kirby, Elizabeth *see* Chapters on trees
Kirby, James P [comp] *see* Selected articles on criminal justice
Kirby memorial baptist church. cherokee county. gaffney, south carolina : church records – 1954-1973 – 1 – us Southern Baptist [242]
Kirby, Percival Robson *see*
– Andrew smith and natal
– Jacob van reenen and the grosvenor expedition of 1790-1791
– True story of the grosvenor
Kirby reporter – [NW England] Kirby Lib 1964-jul 1979 – 1 – uk MLA; uk Newsplan [072]
Kirby, William *see*
– Le chien d'or
– Counter manifesto to the annexationists of montreal
– On the power wisdom and goodness of god
– The united empire loyalists of canada
Kirby, William Forsell *see*
– Evolution and natural theology
– The hero of esthonia and other studies in the romantic literature of that country
Kirby's reports / Connecticut. Supreme Court – 1v. 1785-1788 (all publ) – (= ser Connecticut Supreme Court Reports) – 6mf – 9 – $9.00 – (a pre-nrs title) – mf#LLMC 94-002 – us LLMC [347]
Kirchbach, Wolfgang *see*
– Deutsche schauspieler und schauspielkunst
– Waiblinger
Die kirche : ihre biblische idee und die formen ihrer geschichtlichen erscheinung in ihrem unterschiede von sekte und haerese / Schmidt, Hermann – Leipzig: Doerffling & Franke, 1884 – 1mf – 9 – 0-7905-9873-6 – mf#1989-1598 – us ATLA [240]
Die kirche der gegenwart : eine monatschrift fuer die reformierte schweiz – 1845-50 [complete] – 2r – 1 – ISSN: 0023-1797 – mf#ATLA B0253 – us ATLA [242]
Die kirche der thomaschristen : ein beitrag zur geschichte der orientalischen kirchen / Germann, Wilhelm – Guetersloh: C Bertelsmann, 1877 – 2mf – 9 – 0-7905-4907-7 – (incl bibl ref) – mf#1988-0907 – us ATLA [240]
Die kirche der wueste 1715 bis 1787 : das wiederaufleben des franzoesischen protestantismus im achtzehnten jahrhundert / Schott, Theodor – Halle: Verein fuer Reformationsgeschichte, 1893 – (= ser [Schriften des Vereins fuer Reformationsgeschichte]) – 1mf – 9 – 0-7905-4709-0 – (incl bibl ref) – mf#1988-0709 – us ATLA [242]
Die kirche deutschlands im neunzehnten jahrhundert : eine einfuehrung in die religioesen, theologischen und kirchlichen fragen der gegenwart / Seeberg, Reinhold – 3. vielfach verb und erw Aufl. Leipzig: A Deichert, 1910 – 1mf – 9 – 0-524-01468-X – mf#1990-0417 – us ATLA [240]
Die kirche gottes und die bischoefe : denkschrift mit ruecksicht auf das angekuendigte allgemeine concilium zur klaerung der religioesen lebensfrage / Liano, Heinrich St A von – Muenchen: J J Lentner, 1869 – 1mf – 9 – 0-8370-8920-4 – (incl bibl ref) – mf#1986-2920 – us ATLA [240]
Die kirche, ihr amt, ihr regiment : grundlegende saetze mit durchgehender bezugnahme auf die symbolischen buecher der lutherischen kirche / Harnack, Theodosius – Nuernberg: U E Sebald, 1862 [mf ed 1990] – 1mf – 9 – 0-7905-4916-6 – mf#1988-0916 – us ATLA [242]
Die kirche im apostolischen zeitalter *see* History of the christian church in the apostolic times
Die kirche im urchristentum : mit durchblicken auf die gegenwart / Scheel, Otto – Tuebingen: J C B Mohr 1912 [mf ed 1989] – (= ser Religionsgeschichtliche volksbuecher fuer die deutsche christliche gegenwart 4/20) – 1mf – 9 – 0-7905-3282-4 – (incl bibl ref) – mf#1987-3282 – us ATLA [240]
Die kirche jerusalems vom jahre 70-130 / Schlatter, Adolf von – Guetersloh: C. Bertelsmann, 1898 – (= ser Beitraege zur foerderung christlicher theologie) – 1mf – 9 – 0-7905-3221-2 – mf#1987-3221 – us ATLA [240]
Kirche, kirchen und sekten : samt deren unterscheidungslehren / Rohnert, Wilhelm – 5., rev und verm Aufl. Leipzig: E Ungleich, 1900 – 1mf – 9 – 0-524-00311-4 – mf#1989-3011 – us ATLA [240]
Kirche und fernsehen – Frankfurt/M DE, 1955-73 – 10r – 1 – gw Mikrofilm [230]
Kirche und gegenwart : vorlesungen / Schaeder, Erich – Guetersloh: Bertelsmann, 1909 – 1mf – 9 – 0-524-00319-X – (incl bibl ref) – mf#1989-3019 – us ATLA [240]
Kirche und kirchen, papstthum und kirchenstaat : historisch-politische betrachtungen / Doellinger, Johann Josef Ignaz von – Muenchen 1861 (mf ed 1992) – 5mf – 9 – €49.00 – 3-89349-045-0 – mf#DHS-AR 73 – gw Frankfurter [240]
Kirche und mann : monatszeitung fuer maennerarbeit der evangelischen kirche in deutschland – v18-29. 1965-76 – 2r – 1 – (ceased: 1976; cont by: evangelisches monatsblatt) – mf#ATLA S0202 – us ATLA [242]
Kirche und papsttum, eine stiftung jesu / Dausch, Petrus – 1. & 2. aufl. Muenster i W: Aschendorff 1911 – (= ser Biblische zeitfragen 4/2) – 1mf – 9 – 0-524-05607-2 – (incl bibl ref) – mf#1992-0462 – us ATLA [220]
Kirche und reich gottes / Dorner, August – Gotha: FA Perthes, 1883 – 1mf – 9 – 0-7905-7290-7 – (incl bibl ref) – mf#1989-0515 – us ATLA [240]
Kirche und rundfunk – Frankfurt/M 1949 21 jan- – 55r [until 2003] – 1 – (1997: epd medien) – gw Mikrofilm [230]
Kirche und welt *see* Germania
Die kirche unterm kreuz : oder, botschafter des heils in christo – 1885 [complete] – (= ser Mennonite serials coll) – 1r – 1 – mf#ATLA 1993-S025 – us ATLA [240]
Kirche + volk : mitteilungsblatt des schweizerischen protestantischen volksbundes – Zuerich: Kantonalsektion, -1991 [bimthly] [mf v16-44 1964-91 filmed 1988-95] – 2r – 1 – (lacks: v43 n5; v44 n1. began with 10.jahrg n1 jan 1958) – mf0856 – us ATLA [242]
Kirche, volk und staat vom standpunkt der evangelischen kirche aus betrachtet : ein erweiterter vortrag / Meyer, Konrad – Leipzig: A Deichert, 1915 – 1mf – 9 – 0-524-02982-2 – mf#1990-0769 – us ATLA [242]
Kirche von kurhessen waldeck : kirchliches amtsblatt – v86-105. 1971-90 – 3r – 1 – (lacks some pp) – mf#atla s0793 – us ATLA [242]
Die kirche von schottland : beitraege zu deren geschichte und beschreibung / Sack, Karl Heinrich – Heidelberg: Karl Winter, 1844-1845 – 1mf – 9 – 0-524-02494-4 – mf#1990-4353 – us ATLA [240]
Kirchen-, address- und intelligenz-zettel der stadt elbing – Elbing (Elblag PL), 1828 – 1r – 1 – gw Misc Inst [077]
Kirche[n] gesaeng : aus dem wittenbergischen, und allen andern den besten gesangbuechern, so biss anhero hin und wider aussgangen, colligirt un[d] gesamlet, in eine feine, richtige und gute ordnung gebracht... – Frankfurt am Mayn: Wolff 1569 – (= ser Hqab. literatur des 16. jahrh.) – 8mf – 9 – €80.00 – mf#1569c – gw Fischer [780]
Kirchen ordnung in meiner gnedigen herrn der marggrauen zu brandenburg : vnd eins erberen rats der stat nuernberg oberkeyt vn[d] gepieten, wie man sich bayde mit der leer vnd ceremonien halten solle – Nuernberg: Gutknecht 1533 – (= ser Hqab. literatur des 16. jahrh.) – 2mf – 9 – €30.00 – mf#1533 – gw Fischer [780]
Die kirchen- und schulvisitation im saechsischen kurkreise vom jahre 1555 / Schmidt, Wilhelm – Halle a d S: Verein fuer Reformationsgeschichte 1906 [mf ed 1990] – (= ser Schriften des vereins fuer reformationsgeschichte 24/90,92) – 1mf – 9 – 0-7905-5131-4 – mf#1988-1131 – us ATLA [240]
Kirchen und sekten der gegenwart / ed by Kalb, Ernst – 2. erw und verb Aufl. Stuttgart: Verlag der Buchh der Evang Gesellschaft, 1907 – 2mf – 9 – 0-524-03402-8 – (incl bibl ref) – mf#1990-0956 – us ATLA [240]
Die kirchen und sekten des christentums in der gegenwart / Kattenbusch, Ferdinand – Tuebingen: J C B Mohr 1909 [mf ed 1990] – (= ser Religionsgeschichtliche volksbuecher fuer die deutsche christliche gegenwart 4/11-12) – 1mf – 9 – 0-7905-4935-2 – (incl bibl ref) – mf#1988-0935 – us ATLA [240]
Kirchenblatt fuer die reformierte schweiz – 1845-54; 1939-86* – 12r – 1 – (filmed with: die zukunft die kirche) – mf#ATLA S0124 – us ATLA [242]
Kirchenbote – Osnabrueck DE, 1958 6 jul-1966 – 1 – gw Misc Inst [240]
Kirchenbuchs and minute book / Trinity Lutheran Church, Otis, KS – 1877-1952 – 1 – us Kansas [240]
Die kirchengemeinde- und synodalordnung fuer die provinzen preussen, brandenburg, pommern, posen, schlesien und sachsen / ed by Uckeley, Alfred – Bonn: A Marcus und E Weber, 1912 – (= ser Kleine texte fuer vorlesungen und uebungen) – 1mf – 9 – 0-524-04713-8 – mf#1990-5065 – us ATLA [240]
Der kirchengesang nach den liturgikern des mittelalters / Schmid, A – Kempten, 1900 – €3.00 – ne Slangenburg [780]
Kirchengesanng teutsch und lateinisch davon in unser angestelter kirchenordnung meldung geschicht... – Nuernberg: Vom Berg [u.a.] 1557 – (= ser Hqab. literatur des 16. jahrh.) – 3mf – 9 – €40.00 – mf#1557b – gw Fischer [780]
Kirchengeschichte : lehrbuch zunaechst fuer akademische vorlesungen / Hase, Karl von – 10. verb Aufl. Leipzig: Breitkopf und Haertel, 1877 – 8mf – 9 – 0-524-03344-7 – (incl bibl ref) – mf#1990-0925 – us ATLA [240]
Kirchengeschichte *see* History of the christian church
Kirchengeschichte, 2. bd (bdk51 1.reihe) / Theodoret von Cyrus (Theodoret of Cyrrhus) – (= ser Bibliothek der kirchenvaeter. 1. reihe (bdk 1.reihe)) – €15.00 – ne Slangenburg [240]
Kirchengeschichte der neueren zeit : von der reformation bis zum ende des achtzehnten jahrhunderts / Baur, Ferdinand Christian; ed by Baur, Ferdinand Friedrich – Tuebingen: L Fr Fues, 1863 – 2mf – 9 – 0-524-03412-5 – (incl bibl ref) – mf#1990-0966 – us ATLA [240]
Die kirchengeschichte der nicephorus callistus xanthopupos und ihre quellen / Gentz, G – Berlin, 1966 – (= ser Tugal 5-98) – 5mf – 9 – €12.00 – ne Slangenburg [240]
Kirchengeschichte der reformierten schweiz / Hadorn, Wilhelm – Zuerich: Schulthess, 1907 – 1mf – 9 – 0-524-03643-8 – mf#1990-1071 – us ATLA [242]
Kirchengeschichte des 18. und 19. jahrhunderts *see* German rationalism
Die kirchengeschichte des eusebius = Ecclesiastical history / Eusebius – Leipzig: J C Hinrichs, 1901 – (= ser Tugal) – 1mf – 9 – 0-7905-1757-4 – (in german) – mf#1987-1757 – us ATLA [240]
Die kirchengeschichte des eusebius / Nestle, E – Leipzig, 1901 – (= ser Tugal 2-21/2) – 5mf – 9 – €12.00 – ne Slangenburg [240]
Kirchengeschichte (gcsej17) / Sozomenus; ed by Bidez, J – 1960 – (= ser Griechische christlichen schriftsteller der ersten jahr-hunderte (gcsej)) – €25.00 – ne Slangenburg [240]
Kirchengeschichte (gcsej18) / Theodoret; ed by Bidez, J – 1911 – (= ser Griechische christlichen schriftsteller der ersten jahr-hunderte (gcsej)) – €21.00 – ne Slangenburg [240]
Kirchengeschichte (gcsej19) / Theodores Anagnostes; ed by Hansen, G C – 1971 – (= ser Griechische christlichen schriftsteller der ersten jahr- hunderte (gcsej)) – €15.00 – ne Slangenburg [240]
Kirchengeschichte heute : geschichtswissenschaft oder theologie / ed by Kottje, R – Trier, 1970 – €7.00 – ne Slangenburg [240]
Kirchengeschichte im grundriss *see* History of christianity
Die kirchengeschichte von spanien / Gams, Pius Bonifatius – Regensburg: GJ Manz, 1862-1879 – 6mf – 9 – 0-7905-4792-9 – (incl bibl ref) – mf#1988-0792 – us ATLA [240]
Kirchengeschichtliche abhandlungen und untersuchungen / Funk, Franz Xaver von – Paderborn: F Schoeningh, 1897-1907 – 4mf – 9 – 0-7905-7048-3 – (incl bibl ref) – mf#1988-3048 – us ATLA [240]
Der kirchengeschichtliche ertrag der exegetischen arbeiten des origenes : 1. teil: hexateuch und richterbuch / Harnack, Adolf von – Leipzig, 1918 – (= ser Tugal 3-42/3a) – 3mf – 9 – €7.00 – ne Slangenburg [221]
Der kirchengeschichtliche ertrag der exegetischen arbeiten des origenes : 2. teil: die beiden testamente mit ausschluss des hexateuchs und des richterbuchs / Harnack, Adolf von – Leipzig, 1919 – (= ser Tugal 3-42/4) – 3mf – 9 – €7.00 – ne Slangenburg [220]
Kirchengeschichtliches lesebuch / ed by Rinn, Heinrich & Juengst, Johannes – grosse Ausg, verm und verb Aufl. Tuebingen: JCB Mohr, 1906 – 1mf – 9 – 0-524-03770-1 – (incl bibl ref) – mf#1990-1117 – us ATLA [240]
Die kirchengeschichtsschreibung : grundzuege ihrer historischen entwicklung / Nigg, W – Muenchen, 1934 – €12.00 – ne Slangenburg [240]
Kirchengeseng : darinnen die heubtartickel des christlichen glaubens kurtz gefasset und ausgelegt sind – s.l. 1566 – (= ser Hqab. literatur des 16. jahrh.) – 2v on 7mf – 9 – €75.00 – mf#1566c – gw Fischer [780]
Kirchengueterfrage und schmalkaldischer bund : ein beitrag zur deutschen reformationsgeschichte / Koerber, Kurt – Leipzig: Verein fuer Reformationsgeschichte 1913 [mf ed 1990] – (= ser Schriften des vereins fuer reformationsgeschichte 30/3,4/111-12) – 1mf – 9 – 0-7905-4767-8 – mf#1988-0767 – us ATLA [242]
Kirchenhistorische anecdota : nebst neuen ausgaben patristischer und kirchlich-mittelalterlicher schriften / ed by Caspari, Carl Paul – Christiania [Oslo]: Malling, 1883 [mf ed 1992] – 1mf – 9 – 0-524-04486-4 – (text in latin, int & notes in german by ed. incl bibl ref. no more publ) – mf#1990-1248 – us ATLA [240]
Das kirchenjahr des christlichen morgen- und abendlandes : mit seinen festen, fasten und bibellectionen historisch dargestellt / Alt, Heinrich – 2. verm erw aufl. Berlin: G W F Mueller, 1860 [mf ed 2003] – (= ser Der christliche cultus 2) – 1r – 1 – (with ind) – mf#b00671 – us ATLA [240]
Der kirchenkampf : the gutteridge-micklem collection held in the bodleian library = Church struggle – [mf ed 1988] – 515mf (1:24) – 9 – silver €4900.00 – ISBN-10: 3-598-32599-1 – ISBN-13: 978-3-598-32599-1 – (accompanying vol with int & ind) – gw Saur [019]

Kirchenlamitzer anzeiger – Kirchenlamitz, Marktleuthen DE, 1935 1 jul-31 dec – 1r – 1 – (in addition: marktleuthener nachrichten) – gw Misc Inst [074]

Kirchenmusikalisches jahrbuch. – Regensburg, New York. 1876-1932, 1950. 3 reels – 1 – 54.00 – us L of C Photodup [780]

Kirchenordnung : wie die vnter dem christlichen koenig auss engelland edward den 6. in der stat londen in der niderlendischen gemeine christi... / Lasco, J – Heidelberg, 1565 – 4mf – 9 – mf#PBA-217 – ne IDC [240]

Kirchenordnung : wie es mit der christlichen leer, raichunge der heiligen sacramenten, ordination der diener des euangelij vnd ordenlicher ceremonien, erhaltung christlicher schulen vnd studien... – [Nuernberg]: [VomBerg & Neuber] 1560 – (= ser Hqab. literatur des 16. jahrh.) – 7mf – 9 – €75.00 – (incl: kirchengesang teutsch vnd lateinisch, dauon in vnser angestelter kirchenordnung meldung geschicht) – mf#1560b – gw Fischer [780]

Kirchenordnung : wie es mit der christlichen lehre, heiligen sacramenten, vnd allerley andern ceremonien in meines gnedigen herrn, herrn otthainrichen, pfaltzgrauen bey rhein, hertzogen im niedern vnd obern bairen...fuerstenthumb gehalten wirt – Nuernberg: Petreius 1543 – (= ser Hqab. literatur des 16. jahrh.) – 5mf – 9 – €60.00 – mf#1543a – gw Fischer [240]

Kirchenordnung : wie es mit der christlichen lehre, heiligen sacramenten, vn[d] cerimonien, in meines gnedigen herrn, herrn otthainrichs, pfaltzgrauen bey rhein, hertzogen in nidern vnd obern bayrn [et]c. fuerstenthumb, gehalten wirdt – Nuernberg: VomBerg & Neuber 1554 – (= ser Hqab. literatur des 16. jahrh.) – 4mf – 9 – €50.00 – mf#1554a – gw Fischer [240]

Kirchenordnung d i form vnd weise : nach welcher die reyne christliche lere, sacramenten, vnd allerley noetige ceremonien, in etlichen fuernemen der augspurg. confession verwandten kirchen...im brauch gewesen vnd noch sind... – Frankfurt: Lechler 1565 – (= ser Hqab. literatur des 16. jahrh.) – 3mf – 9 – €40.00 – mf#1565 – gw Fischer [780]

Kirchenordnung in stedten und wo man schulen hat – Dresden: Stoeckel 1540 – (= ser Hqab. literatur des 16. jahrh.) – 1mf – 9 – €20.00 – (fr: agenda, das ist kyrchenordnung wie sich die pfarrherrn und seelsorger in yhren ampten und diensten halten sollen [dresden 1540]) – mf#1540c – gw Fischer [240]

Kirchenordnung vnnser von gottes genaden julii hertzogen zu braunschweig vnd lueneburg... : wie es mit lehr vnd ceremonien vnsers fuerstenthumbs braunschweig, wulffenbuetlischen theils, auch der selben kirchen anhangenden sachen vnd verrichtungen, hinfurt...gehalten werden sol – Wulffenbuettel: Horn 1569 – (= ser Hqab. literatur des 16. jahrh.) – 7mf – 9 – €75.00 – mf#1569d – gw Fischer [240]

Kirchen-ordnung wie es in weyland...wolffgang's, pfaltzgraffen bey rhein...landen : mit lehr vnd anderen der kirchen nothwendigen stuecken gehalten worden... – Frankfurt: Spiess 1600 – (= ser Hqab. literatur des 16. jahrh.) – 8mf – 9 – €80.00 – (some pgs missing) – mf#1600b – gw Fischer [240]

Kirchenordnung wie es inn des durchleuchtigen, hochgebornen fuersten vnd herrn, herren wolffgangs, pfaltzgrauen bey rhein, hertzogen in baiern...fuerstenthumben vnnd landen : biss anhero mit der christlichen lehr, raichung der heiligen sacramenten... – Nuermberg: Gerlatz 1570 – (= ser Hqab. literatur des 16. jahrh.) – 8mf – 9 – €80.00 – (incl: kirchengesanng teutsch vnd lateinisch, dauon in newburgischer vnd zweybruckischer gleichfoermiger kirchenordnung meldung geschicht) – mf#1570e – gw Fischer [240]

Kirchenordnung wie es mit christlicher lere, reichung der sacrament, ordination der diener des euangelij...im hertzogthumb zu meckelnburg etc gehalten wird – Witteberg: Lufft 1552 – (= ser Hqab. literatur des 16. jahrh.) – 3mf – 9 – €40.00 – mf#1552a – gw Fischer [240]

Kirchenordnung, wie es mit der reinen lehre goettliches worts und ausstheilung der hochwirdigen sacrament... : in den graffschafften lippe, spiegelberg und pyrmont sol eindrechtiglich gehalten werden – Lemgo: Schlodt & Schmidt 1571 – (= ser Hqab. literatur des 16. jahrh.) – 4mf – 9 – €50.00 – mf#1571c – gw Fischer [240]

Kirchenordnunge zum anfang fur die pfarherrn hertzog heinrichs zu sachsen v g h fuerstenthumb – Wittemberg: Lufft 1539 – (= ser Hqab. literatur des 16. jahrh.) – 1mf – 9 – €20.00 – mf#1539 – gw Fischer [780]

Die kirchenpolitik des kanzlers michel de l'hospital / Geuer, F – Duisburg, 1877 (mf ed 1993) – 1mf – 9 – €24.00 – 3-89349-334-4 – mf#DHS-AR 187 – gw Frankfurter [240]

Die kirchenpolitik koenig sigmunds waehrend seines romzuges (1431-1433) / Koch, Max – Leipzig, 1906 (mf ed 1993) – 1mf – 9 – €24.00 – 3-89349-269-0 – mf#DHS-AR 126 – gw Frankfurter [240]

Das kirchenrecht der morgenlaendischen kirche : nach den allgemeinen kirchenrechtsquellen und nach den in den autokephalen kirchen geltenden spezial-gesetzen / Milas, Nikodim – 2., verb und verm Aufl. Mostar: Pacher and Kisic, 1905 – 2mf – 9 – 0-524-01762-X – (incl bibl ref) – mf#1990-0496 – us ATLA [240]

Kirchenrecht. erster band. die geschichtlichen grundlagen / Sohm, Rudolf – Leipzig: Duncker & Humblot, 1892 – (= ser Systematisches Handbuch der deutschen Rechtswissenschaft) – 2mf – 9 – 0-7905-8230-9 – (incl bibl ref) – mf#1988-6130 – us ATLA [340]

Die kirchenrechtsquellen des patriarchats alexandrien / Riedel, Wilhelm – Leipzig: A. Deichert, 1900 – 1mf – 9 – 0-7905-6667-2 – (incl bibl ref) – mf#1988-2667 – us ATLA [240]

Die kirchenvaeter und das neue testament : beitraege zur geschichte der erklaerung der wichtigsten neutestamentlichen stellen / Langen, Joseph – Bonn: Eduard Weber (R Weber & M Hochgueertel), 1874 – 1mf – 9 – 0-8370-4050-7 – (incl bibl ref and indexes) – mf#1985-2050 – us ATLA [225]

Der kirchenvorstand nach dem rechte der evangelisch-lutherischen landeskirche des koenigreichs sachsen / Troll, Alfred – Leipzig, 1913 (mf ed 1993) – 2mf – 9 – €31.00 – 3-89349-352-2 – mf#DHS-AR 352 – gw Frankfurter [242]

Kirchenzeitung fuer das erzbistum koeln – Koeln DE, 1958 17 aug-21 dec, 1959 4 jan-29 mar, 1961 5 nov-1966 – 1 – gw Misc Inst [241]

Kircher, A *see* Lingua aegyptiaca restituta

Kircher, Athanasius *see*
- Athanasii kircheri prodromus coptus sive aegyptiacus
- [Athanasii kircheri]...mvsvrgia vniversalis
- Musurgia universalis
- Phonurgia nova...

Kircher, Konrad *see* Concordantiae veteris testamenti graecae

Der kirchgang des grosswendbauern : novellen / Boehme, Herbert – Muenchen: F Eher, 1936 [mf ed 1989] – (= ser Junges volk 1) – 131p – 1 – mf#7044 – us Wisconsin U Libr [830]

Kirchgeorg, Otto Hermann *see* Die dichterische entwicklung j.f.w. zacharias

Kirchhainer zeitung – Kirchhain DE, 1889 3 apr-1894 30 jun, 1895-1897 31 mar, 1897 6 oct-1933, 1935-44 – 42r – 1 – (title varies: 1 dec 1926: hessische rundschau. filmed with suppls) – gw Misc Inst [074]

Kirchhof, Hans Wilhelm *see*
- Wendunmuth

Kirchhofer, M *see*
- Bertold haller oder die reformation von bern
- Wernher steiner, buerger von zug und zuerich

Kirchhofer, Melchior *see* Sebastian wagner, genannt hofmeister

Kirchhoff, Gustav *see* Untersuchungen ueber das sonnenspectrum und die spectren der chemischen elemente

Kirchliche autoritaet und macht der wissenschaft / Buchmann, Jacob – Breslau [Wroclaw]: Fiedler & Hentschel, 1874 – 1mf – 9 – 0-8370-9127-6 – mf#1986-3127 – us ATLA [210]

Die kirchliche baukunst des abendlandes, historisch und systematisch dargestellt / Dehio, G G & Bezold, G von – Stuttgart, 1887-1901. 2v text, 5v ill – 44mf – 9 – mf#O-222 – ne IDC [720]

Kirchliche benediktionen und ihre verwaltung / Probst, Ferdinand – Tuebingen: Laupp, 1857 – 1mf – 9 – 0-524-01402-7 – (incl bibl ref) – mf#1990-0401 – us ATLA [240]

Der kirchliche besitz im arrondissement aachen gegen ende des 18. jahrhunderts und seine schicksale in der saekularisation durch die franzoesische herrschaft : ein beitrag zur kirchen- und wirtschaftsgeschichte der rheinlande / Kaiser, Paul – Aachen, 1906 (mf ed 1993) – 2mf – 9 – €24.00 – 3-89349-119-8 – mf#DHS-AR 88 – gw Frankfurter [240]

Kirchliche disciplin : in den drei ersten christlichen jahrhunderten / Probst, Ferdinand – Tuebingen: H. Laupp, 1873 – 1mf – 9 – 0-7905-5795-9 – (incl bibl ref) – mf#1988-1795 – us ATLA [240]

Die kirchliche frage und ihre protestantische loesung : im zusammenhange mit den nationalen bestrebungen und mit besonderer beziehung auf die neuesten schriften j.j.j. von doellinger's und bischof von ketteler's / Schenkel, Daniel – Elberfeld: RL Friderichs, 1862 – 1mf – 9 – 0-7905-6679-6 – (incl bibl ref) – mf#1988-2679 – us ATLA [242]

Die kirchliche gesetzgebung des kaisers justinian 1 / Alivizatos, Amilkas S – Berlin: Trowitzsch, 1913 – (= ser Neue Studien zur Geschichte der Theologie und der Kirche) – 2mf – 9 – 0-524-02578-9 – (incl bibl ref) – mf#1990-0630 – us ATLA [240]

Die kirchliche gewalt und ihre traeger / Schneemann, G – Freiburg im Breisgau: Herder, 1867 – (= ser Die Encyclica Papst Pius' 9. Vom 8. Dezember 1864) – 1mf – 9 – 0-8370-8306-0 – mf#1986-2306 – us ATLA [240]

Das kirchliche leben der evangelisch-lutherischen landeskirche des koenigreichs sachsen / Drews, Paul – Tuebingen: JCB Mohr, 1902 – (= ser Evangelische Kirchenkunde) – 1mf – 9 – 0-7905-4461-X – (incl bibl ref) – mf#1988-0461 – us ATLA [242]

Die kirchliche ordnung der taufe / Beckman, J – Stuttgart, 1950 – 1mf – 8 – €3.00 – ne Slangenburg [240]

Kirchliche reformentwuerfe beginnend mit der revision des bibelkanons : ehrerbietige vorlage an das vatikanische concil / Sepp, Prof Dr – Muenchen: J J Lentner, 1870 – 1mf – 9 – 0-8370-9030-X – (incl bibl ref) – mf#1986-3030 – us ATLA [220]

Kirchliche statistik : oder, darstellung der gesammten christlichen kirche. nach ihrem gegenwaertigen aeusseren und inneren zustande / Wiggers, Julius – Hamburg: F und A Perthes, 1842 – 2mf – 9 – 0-524-03435-4 – (incl bibl ref) – mf#1990-0989 – us ATLA [240]

Kirchliche statistik der reformirten schweiz / Finsler, Georg – Zuerich: Meyer und Zeller, 1856 – 8mf – 9 – 0-524-07353-8 – mf#1990-5390 – us ATLA [242]

Kirchliche statistik oder darstellung der gesammten christlichen kirche nach ihrem gegenwaertigen aeusseren und inneren zustande / Wiggers, Julius – Hamburg, Gotha 1842-43. 2pts (mf ed 1992) – 10mf – 9 – €150.00 – 3-89349-024-8 – mf#DHS-AR 64 – gw Frankfurter [240]

Kirchliche und soziale zustaende in bern unmittelbar nach der einfuehrung der reformation (1528-1536) / Quervain, Theodor de – Bern: Gustav Grunau, 1906 – 1mf – 9 – 0-524-05444-4 – (incl bibl ref) – mf#1990-1476 – us ATLA [949]

Kirchliche zeitschrift – Schwerin-Rostock, 1(1854)-6(1859) – 76mf – 9 – €145.00 – ne Slangenburg [240]

Die kirchlichen benediktionen im mittelalter, vol 1-2 / Franz, Albert – Freiburg i.Br. v1-2. 1909 – 2v on 22mf – 8 – €42.00 – ne Slangenburg [241]

Die kirchlichen quatertember : ihre entstehung, entwicklung und bedeutung / Fischer, L – Muenchen, 1914 – 4mf – 8 – €11.00 – ne Slangenburg [240]

Die kirchlichen reunionsbestrebungen waehrend der regierung karls 5 / Pastor, Ludwig, Freiherr von – Freiburg i B; St Louis, MO: Herder, 1879 – 2mf – 9 – 0-7905-6941-8 – (incl bibl ref) – mf#1988-2941 – us ATLA [943]

Die kirchlichen simultanverhaeltnisse der rheinprovinz unter besonderer beruecksichtigung des ryswicker friedens / Beck, Dietrich – Weimar, 1934 (mf ed 1992) – 1mf – 9 – €24.00 – 3-89349-211-9 – mf#DHS-AR 1 – gw Frankfurter [240]

Die kirchlichen zustaende in oesterreich : und das allgemeine konzil in rom / Schoepf, Ignaz – Innsbruck: Wagner, 1869 [mf ed 1986] – 1mf – 9 – 0-8370-8380-X – (incl bibl ref) – mf#1986-2380 – us ATLA [241]

Kirchlicher zentralkatalog beim evangelischen zentralarchiv in berlin / ed by Czubatynski, Uwe – [mf ed 1997] – 216mf (1:42) – 9 – diazo €2500.00 (silver €3000 isbn: 978-3-598-32014-9) – ISBN-10: 3-598-32013-2 – ISBN-13: 978-3-598-32013-2 – (incl suppl vol) – gw Saur [242]

Kirchliches jahrbuch fuer die evangelische kirche in deutschland, 1933-1944 / ed by Beckmann, Joachim – Guetersloh: C Bertelsmann, 1948 [mf ed 2004] – 1r – 1 – 0-524-10500-6 – (incl ind) – mf#b00715 – us ATLA [242]

Die kirchlichkeit der s.g. kirchlichen theologie / Gottschick, Johannes – Freiburg i.B.: J C B Mohr, 1890 – 1mf – 9 – 0-8370-8741-4 – (incl bibl ref) – mf#1986-2741 – us ATLA [240]

Kirchmann, Julius Hermann von *see*
- Erlaeuterungen zu kant's religion innerhalb der grenzen der blossen vernunft
- Friedrich schleiermacher's philosophische sittenlehre
- Immanuel kant's kleinere schriften zur ethik und religionsphilosophie
- Die religion innerhalb der grenzen der blossen vernunft

Kirchner, Aloys *see* Die babylonische kosmogonie und der biblische schoepfungsbericht

Kirchner, Erich *see* Induzierte hypervolaemie und kontrollierte volumenanpassung

Kirchner, Rita *see* Untersuchung zum sprachlichen modeverhalten

Kirchner, T *see*
- Apologia
- Enchiridion d timothei kirchneri jn welchem die fuernembsten hauptstueck der christlichen lehre durch frag vnd antwort auss gottes wort gruendtlich erklaeret
- Histori dess sacramentstreits darinnen klaerlich aussgefuehret wirdt wie diese zwytracht entstanden biss auff vnsere zeit continuiret
- Methodica explicatio praecipvorvm capitvm doctrinae coelestis
- Das die zwey vnd vierzig anhaltische argument wider der vbiquisten trewme noch fest stehen

Kirchner, Werner *see* Der hochverratsprozess gegen sinclai

Kireev, Alexsandr Alekseevich *see* Correspondence on infallability

Kireev, T N *see* Vol'naia kuban'

Kireevskii, Petr Vasil'evich *see* Piesni

Kirgizkoe khoziaistvo v akmolinskoi oblasti – Spb, 1909-1910. v1-5 – 40mf – 8 – mf#RZ-192 – ne IDC [314]

Kirgizskaia stepnaia gazeta – Omsk, Russia 1889-1901 [mf ed Norman Ross] – 1 – mf#nrp-1257 – us UMI ProQuest [077]

Kirgizsko-russkii slovar' – Izd 2. Orenburg : Tipo-litografiia B A Breslina, 1903 [mf ed 2001] – 1r – 1 – (filmed with: Sumnjivo lice / [Branislav Nusi'c] & other titles) – mf#4931 – us Wisconsin U Libr [460]

Kiri, Sarea *see*
- Patrol reports and related papers from the western highlands (enga) and milne bay districts
- Patrol reports and related papers from the western highlands (enga) and milne bay districts, papua new guinea

Kiribati overseas seamens union : archives, 1974-96 – 3r – 1 – (available for reference) – mf#pmb1154 – at Pacific Mss [331]

Kirillov, A *see* Geograficheskо-statisticheskii slovar' amurskoi i primorskoi oblastei

Kirillov, I A *see*
- Lombardy v rossii
- Polgoda raboty vserossiskogo kooperativnogo banka (ianvar'-iiun' 1923 g)

Kirjath sepher – 1(1924)-50(1975) – 709mf – 9 – €1352.00 – ne Slangenburg [930]

Kirk anbar – n1-34. 1290-93 [all publ] – (= ser O & t journals) – 21mf – 9 – $335.00 – us MEDOC [956]

Kirk anderson's valley tan – Salt Lake City UT. 1859 mar 29, apr 19 – 1r – 1 – (cont by: valley tan) – mf#854915 – us WHS [071]

Kirk, Edgar Lee *see* Toward american music

Kirk, Edward Norris *see*
- The church and the college
- The church essential to the republic
- Discourses
- Lectures on revivals

Kirk, Eleanor [comp] *see* Choice recipes

Kirk folk / Anderson, Robert Stuart Guthrie – Toronto: Westminster, 18– – 1mf – 9 – mf#28871 – cn Canadiana [810]

Kirk, Harris Elliott *see* The religion of power

Kirk, Hiram van *see*
- A history of the theology of the disciples of christ
- The rise of the current reformation

Kirk, John *see*
- Annual address of the victoria institute
- Biographies of english catholics in the 18th century
- Gospel letters
- Report by sir john kirk on the disturbances at brass
- Social politics in great britain and ireland

Kirk, S J *see* A contribution towards a bibliography of hosiery and lace, etc

Kirk, W Gordon *see*
- Comparative value of grazing crops for fattening feeder pigs
- Fattening market hogs in dry lot
- Selecting and using beef and veal
- Sugarcane silage
- Weight changes of cattle on a florida range

Kirk, William Alphonso *see* To the honorable house of the commons of england

Kirkby, William West *see* Manual of devotion in the beaver indian dialect

Kirkcaldy mail and dunfermline citizen – [Scotland] Fife, Kirkcaldy: J Bryson 1 jan 1889-dec 1950 (wkly) [mf ed 2004] – 41r – 1 – (missing: 1920; cont by: kirkcaldy mail [jan 1896-dec 1911]; mail [jan 1912-dec 1913]; mail for kirkcaldy, central and west fife [jan 1914-dec 1924]; kirkcaldy mail [jan 1925-dec 1950]) – uk Newsplan [072]

Kirkcaldy observer and county of fife advertiser – [Scotland] Fife, Kirkcaldy: J Crawford 1 nov 1851-17 apr 1852 (wkly) [mf ed 2004] – 13v on 1r – 1 – uk Newsplan [072]

Kirkcaldy times – Kirkcaldy: J Strachan & W G Livingston [1876-1962] (wkly) [mf ed 2002] – 35r – 1 – (started in 1876; ceased in 1962) – uk Scotland NatLib [072]

Kirke, Henry *see*
- The first english conquest of canada
- Twenty-five years in british guiana

Kirkeberg, O L *see* Hvorfor skabte gud mennesket?

Kirke-bladet : organ for det danske, evangelisk-lutherske kirkesamfund i amerika / Danish Evangelical Lutheran Church Association in America – St Paul, NE: Kirke-Blaet. 7de aarg n19. 1ste oct 1884-okt 1896// (3 times a mth) [mf ed with gaps filmed 1975?] – 3r – 1 – (in danish. cont: dansk luthersk kirkeblad (1877). merged with: missions-budet to form: dansk luthersk kirkeblad. publ in blair ne, 15te okt 1885-1 jun 1891; chicago, 15 jun 1891-10 okt 1893; blair ne, 20 dec 1893-96) – us NE Hist [071]

Kirkebygninger og deres udstyr – Stockholm: A Bonnier 1934 [mf ed Bloomington IN: Indiana Uni Lib, Preservation Dept 1984] – 341p on 1r – 1 – us Indiana Preservation [390]

Kirkelig tidende – v1,3-5 [1856, 1859-1861] – 1r – 1 – mf#405105 – us WHS [071]

Kirkens historie fra reformationen til oplysningstidens begyndels / Nielsen, Fredrik – Koebenhavn: Gyldendal, 1908 – 2mf – 9 – 0-524-03769-8 – (incl bibl ref) – mf#1990-1116 – us ATLA [242]

Kirkens historie indtil reformationen / Nielsen, Fredrik – Koebenhavn: Gyldendal, 1902 [mf ed 1992] – 3mf – 9 – 0-524-03401-X – (incl bibl ref) – mf#1990-0955 – us ATLA [240]

Kirke-nyt – v17-38. 1966-86 [complete] – 2r – 1 – mf#ATLA S0853 – us ATLA [242]

Kirk-Greene, Anthony Hamilton Millard *see*
- Modern hausa reader
- This is northern nigeria

Kirkham family newsletter – n1-29 [1985 aug-1990 mar] – 1r – 1 – mf#1703987 – us WHS [929]

Kirkintilloch gazette – Scotland. 15 Oct 1898-12 Aug 1938.-w. 16 reels – 1 – uk British Libr Newspaper [072]

Kirkintilloch herald – Scotland. 18 Jul 1883; 23 Jun 1886-Dec 1974.-w. 74 reels – 1 – uk British Libr Newspaper [072]

Kirkintilloch herald (kirkintilloch, scotland : 1976) : edition of kirkintilloch & bishopbriggs herald – Kirkintilloch: D MacLeod Ltd 1976- (wkly) [mf ed 6 jul 1994-] – 1 – (not publ: 8 jul-12 aug 1981; cont in pt: kirkintilloch herald and bishopbriggs herald; suppl: cumbernauld & kirkintilloch today) – uk Scotland NatLib [072]

Kirkland baptist church. taft, tennessee : church records – 1908-63 – 1 – 8.64 – us Southern Baptist [242]

Kirkland, Caroline Matilda Stansbury *see* The evening book: or, fireside talk on morals and manners, with sketches of western life

Kirkland, J V *see* Apostolic hymns

Kirkland, Jack *see* Route au tabac

Kirkland, R S *see* Apostolic hymns

Kirkman, Thomas Penyngton *see* Clerical intemperance

Kirkor, A *see* Zapiski vilenskoi arkheologicheskoi komissii

Kirkpatrick, Alexander Francis *see*
- The book of psalms
- Critical questions
- The divine library of the old testament
- The doctrine of the prophets
- The first book of samuel
- The second book of samuel

Kirkpatrick, Ellis Lore et al *see* Resettlement and rehabilitation in the central wisconsin nesting area

Kirkpatrick, Ernest Stanley *see* Tales of the st john river

Kirkpatrick, William *see*
- An account of the kingdom of nepaul
- Nachrichten von dem koenigreiche nepaul

Kirkstall abbey / Owen, David E – Leeds, s.a. – 2mf – 8 – €5.00 – ne Slangenburg [241]

The kirkstall abbey chronicals / Taylor, J – Leeds. v62. 1932 – 3mf – 8 – €13.00 – ne Slangenburg [241]

Kirkus reviews [1933] – New York NY 1933-85 – 1,5,9 – (cont by: jim kobak's kirkus reviews) – ISSN: 0042-6598 – mf#13371.02 – us UMI ProQuest [073]

Kirkus reviews [1991] – New York NY 1991+ – 1,5,9 – (cont: jim kobak's kirkus reviews) – mf#13371.02 – us UMI ProQuest [073]

Kirkus, William *see* Religion

Kirkwood, Daniel *see*
- Comets and meteors
- Meteoric astronomy

Kirkwood first baptist church (name later changed to wetzel memorial, still later to kirkwood). kirkwood, missouri : church records – 31 Jul 1870-31 Dec 1900 – 1 – us Southern Baptist [242]

Kirkwood, James *see* Proposals made by rev james kirkwood, (minister of minto) in 1699

Kirkwood, Kenneth *see* Proposed federation of the central african territories

Kirkwood, Robert *see* A plea for the bible

Kirlin, Joseph Louis J *see* Catholicity in philadelphia

Kirn, Otto *see*
- Grundriss der theologischen ethik
- Die leipziger theologische fakultaet in fuenf jahrhunderten
- Schleiermacher und die romantik
- Die sittlichen forderungen jesu
- Vortraege und aufsaetze

Kirnberger, Johann Philipp *see*
- Anleitung zur singekomposition
- Grundsaetze des generalbasses
- Lieder mit melodien
- Die wahren grundsaetze zum gebrauch der harmonie

Kirouac, Jules-Adrien *see* Histoire de la paroisse de saint-malachie

Kirovskaia pravda – Vyatka, Russia 1973-88 [mf ed Norman Ross] – 5r – 1 – mf#nrp-2086 – us UMI ProQuest [077]

Kirpichnikov, A I *see* Sv georgii i egorii khrabryi

Kirriemuir free press and district advertiser – [Scotland] Angus, Kirriemuir: J Gray 21 mar 1884-12 mar 1886 (wkly) [mf ed 2004] – 2r – 1 – (cont by: kirriemuir free press and angus advertiser; circulated in kirriemuir, glamis, ruthven etc) – uk Newsplan [072]

Kirriemuir herald – Forfar: Strachan & Livingston 1977- (wkly) [mf ed 6 jan 1994-] – 1 – (cont: kirriemuir herald, strathmore advertiser, and kirriemuir dispatch; not publ: 13 mar 1980; 9 jul-13 aug 1981) – uk Scotland NatLib [072]

Kirrinnis, Herbert *see* Geschichte der friedrichsschule zu gumbinnen

Kirs, W Gordon *see* Deficiency symptoms in growing pigs fed on a peanut ration

Kirsanoff, Mary T *see* Life stress and social support as predictors of athletic injury

Kirsch, A Thomas *see* Kinship, genealogical claims and societal integration in ancient khmer society

Kirsch, Johann Peter *see*
- The doctrine of the communion of saints in the ancient church
- Die lehre von der gemeinschaft der heiligen im christl. alterthum

Kirschner, Max *see* Josef filsers ende

Kirschweng, Johannes *see* Das tor der freude

Kirsop, Joseph *see* Historic sketches of free methodism

Kirsten, A *see* Skizzen aus den vereinigten staaten von nordamerika

Kirsten, Wulf *see* Die akte detlev von liliencron

Kirtikar, Vasudeva Jagannath *see* Studies in vedanta

Kirtland, Turhand *see* Turhand kirtland papers

Kirtley, James Addison *see* The design of baptism, viewed in its doctrinal relations

Kirtling 1585-1950 – (= ser Cambridgeshire parish register transcript) – 7mf – 9 – £8.75 – uk CambsFHS [929]

Kirunabladet – Lulea, Sweden. 1902 – 1r – 1 – sw Kungliga [078]

Kirwan *see*
- Homely truth for honest men
- Letters to the rt. rev. john hughes, roman catholic bishop of new-york
- Romanism at home

Kirwan, A V *see* Carrington and kirwan's reports

Kirwan unmasked : a review of kirwan, in six letters addressed to the rev. nicholas murray / Hughes, John – 4th ed. New York: Edward Dunigan, 1851 – 1mf – 9 – 0-8370-7954-3 – mf#1986-1954 – us ATLA [240]

Kiryat sefer / Gunzberg, Mordecai Aaron – Warsaw, Poland. 643, 1883 – 1r – us UF Libraries [939]

Kirzhnitz, A *see* Di yidishe prese in der gevezener ruslandisher imperie

Kis ujsag – Budapest. Hungary. -d. 31 Mar 1945-13 Jan 1948, 9 Aug 1949, 10 Aug 1952. (Imperfect). (17 reels) – 1 – uk British Libr Newspaper [072]

Kisa weckoblad – Linkoeping, Sweden. 1871 – 1r – 1 – sw Kungliga [078]

Kisah – Djakarta, 1953-1957 – 81mf – 9 – (cont as: sastra djakarta, 1961-1969. missing: 1954, v2(1-3, 5-6); 1955, v3(1-6); 1961, v1(7, 9-12); 1969, v7(11-12)) – mf#SE-796/1; SE-806 – ne IDC [950]

Kisalfold – Gyor, Hungary. 1962-Jun 1991 – 57r – 1 – (some issues missing) – us L of C Photodup [079]

Kisch, Alexander *see* Neue israelitische zeitung

Kisch, M S *see* Letters and sketches from northern nigeria

Kisch, Wilhelm *see* Elsass-lothringisches landesprivatrecht

Kisebb kiadvanyok *see*
- A m kir konkoly-alapitvanyu astrophysikai observatorium kisebb kiadvanyai
- Kleinere veroeffentlichungen des o-gyallaer astrophysikalischen observatoriums stiftung von konkolys

Kiser, Samuel Ellsworth *see* More sonnets of an office boy

Kishinami, Tsunezo *see* The development of philosophy in japan

Kisliakov, E N *see* Dlia chego nuzhna selskokhoziaistvennaia i promyslovaia kooperatsiia i kak ee organizovat v derevne

Kisling's osnabrueckische anzeigen *see* Woechentliche osnabrueckische anzeigen

Kislov, A A *see* Ideology and politics of the american baptist churches in 1900-1917

Kiss, G R et al [comp] *see* Associative thesaurus of english, an...

Kissimmee gazette – Kissimmee, FL. 1940 may-1987 – 48r – (gaps) – us UF Libraries [071]

Kissin, A A *see* Opyt ratsionalizatsii torgovogo apparata

Kissinger, Kipp R *see* Relationship between reported childhood and adult physical activity

Kissinger saale-zeitung – Bad Kissingen DE, 1976- – ca 9r/yr – 1 – (title varies: 1 jul 1977: saale-zeitung) – gw Misc Inst [074]

Kissling, Johannes Baptist *see* Geschichte des kulturkampfes im deutschen reiche

Kissoon, Freddie *see* Papa, look de priest passing!

Kist, Leopold *see* Amerikanisches

Kistler, Allison Clay *see* History and status of labor in the citrus industry of florida

Kitaab al qadr : materiaux pour servir a l'etude de la doctrine de la predestination dans la theologie musulmane / Vlieger, A de – Leyde: Impr ci-devant EJ Brill, 1903 – 1mf – 9 – 0-524-02673-4 – mf#1990-3103 – us ATLA [260]

Kitab adab al-dunya wa-al-din / Mawardi, Ali ibn Muhammad – Cairo: s.n., 1800? 164p – 1 – us Wisconsin U Libr [260]

Kitab al-'ahd al-'gadid al-mansub ila rabbina 'isa al-masih – Paris: Impr Royale 1819 – (= ser Whsb) – 6mf – 9 – €70.00 – mf#Hu 219 – gw Fischer [460]

Kitab al-ajurrumiya fi'l-nahw ta'lif al-shaik al-imam al-'allamah muhammad ibn daudal-sinhaji al-shahir bi ajurrum / Muhammad ibn Muhammad – Rome, 1592 – 2mf – 9 – mf#H-8440 – ne IDC [956]

Kitab al-fariq al-makhluq wal-khaliq / Bachajizade, 'Abd al-Rahman Bey – Cairo: Mawsu at Bishar Press, 1322 – 1r – 1 – 0-8370-1106-X – mf#1984-6005 – us ATLA [240]

Kitab al-khitat al-maqriziyah / Maqrizi, Ahmad Ibn 'Ali – Misr, Egypt. v1-4. 1907 – 2r – us UF Libraries [025]

Kitab al-qawanin / Israel – 1970-76 – 1 – (1977-. approx.20.00y; 1) – us L of C Photodup [324]

Kita-b al-qudus... : the new testament of our lord and saviour jesus christ, in malay / Burn, Robert & Thomsen, Claudius H – rev ed: Singapore: Brst & For Bible Soc 1831 – (= ser Whsb) – 6mf – 9 – €70.00 – (missing: pt2) – mf#Hu 319 – gw Fischer [490]

Kitab asa al-hamasa maa sarh...abi-zakarija...at-tibrizi wa-arbaa faharis / Abu-Tammam Haabib Ibn-Aus ata-Tai – Bonnae: Baaden 1828 – (= ser Whsb) – 10mf – 9 – €90.00 – (incl ind, latin trans & comm by georg wilhelm freytag) – mf#Hu 213 – gw Fischer [410]

Kitab futuh al-buldan / Baladhuri, Ahmad Ibn Yahya' – Al-Qahirah, Egypt. v1-3. 1956?-1957 – 1r – us UF Libraries [025]

Kitab soerat-menjoerat dalam bahasa indonesia / Zain, S M – Djakarta: Doenia Baroe (2602) – 55p 1mf – 9 – mf#SE-2002 mf188 – ne IDC [490]

Kitab-al-fihrist mit anmerkungen / ed by Fluegel, G – (v1: den text enthaltend von j roediger leipzig, 1871 €21. v2: anmerkungen und indices von a mueller leipzig, 1872 €17) – ne Slangenburg [260]

Kitab-i agah – Tehran: Intisharat-i Agah. jildi 1-4. 1360-62 [1981-83] – 1r – 1 – $65.00 – us MEDOC [079]

Kitab-i cihan-nuema [nesri tarihi] / Nesri; ed by Unat, Faik Resit & Koeymen, Mehmed A – Ankara: Tuerk Tarih Kurumu Basimevi, 1949-57 – (= ser Ottoman histories and historical sources) – 13mf – 9 – $210.00 – us MEDOC [956]

Kitab-i jum'ah – Tehran: Intisharat-i Mazyar, 1979- . sal-i 1, shumarah-'i 1-36. 3 murdad 1358-1 khurdad 1359 [26 jul 1979-22 may 1980] – 5r – 1 – $265.00 – us MEDOC [079]

Kitab-i nuqtatu'l-kaf : being the earliest history of the babis / Jani, Mirza; ed by Browne, Edward Granville – Leyden: EJ Brill, 1910 – (= ser E.J.W. Gibb Memorial Series) – 5mf – 9 – 0-524-07073-3 – mf#1991-0055 – us ATLA [290]

Kitab-i subhat uel-ahbar / Kemal, Ahmed – 1289 [1873] – (= ser Ottoman histories and historical sources) – 1mf – 9 – $25.00 – us MEDOC [956]

Kitabu : institute for african american affairs newsletter / Kent State University – 1990 oct, 1994 mar – 1r – 1 – mf#4851567 – us WHS [378]

Kitabu cha jua – 1975 sum – 1r – 1 – (cont: journal of black poetry) – mf#780306 – us WHS [810]

Kitabuel-ilm uen-nafi fi tahsil-isarf ve nahv-i tuerki : a grammar of the turkish language / Davids, Arthur Lumley – London, 1832 – (= ser 19th c books on linguistics) – 4mf – 9 – mf#2.1.29 – uk Chadwyck [470]

Kitahara, Kinshi *see* Ying yung nung yeh ching chi hsueh

Kitaiskaia biblioteka i uchenye trudy chlenov imp rossiiskoi dukhovnoi i diplomaticheskoi missii v g pekine... / Aleksii – 1889 – 4mf – 8 – mf#R-7051 – ne IDC [243]

Kitamura, S *see* Flora of afghanistan

Kitch, Ehtel May *see* Origin of subjectivity in hindu thought

Kitchen collectibles news – v1 n1-v3 n6 [1984 jan/feb-1986 nov/dec] – 1r – 1 – mf#1154795 – us WHS [640]

Kitchen wisdom : prize essays in the souvenir range literary contest / Cullingham, Aggie – Hamilton [Ont.]: Gurney, Tilden, [1899?] – 1mf – 9 – 0-665-94728-3 – mf#94728 – cn Canadiana [640]

Kitchener daily telegraph *see* Daily telegraph

Kitchener, H H *see* The survey of western palestine

Kitchener journal – Pickering, Ont (CDN), 1972-1980 12 sep – 1 – (cont: kanada-kurier, winnipeg) – gw Misc Inst [071]

Kitchener-waterloo record *see* The record

[Kitchener-waterloo] record – Ontario CN. 1907- – 24r/y – 1 – Can$2130.00 silver Can$1975.00 vesicular – (preceded by: the berlin record. cont by: the record) – cn Commonwealth Imaging [071]

Kitchenrange / Chefs, Cooks, Pastry Cooks and Assistants Union, Local 89 [New York NY] – v26 n5-7 [1974 may-jul/aug] – 1r – 1 – mf#673233 – us WHS [331]

Kitchin, Claude *see* Papers

Kitchin, George William *see* Edward harold browne, d.d

Kitchiner, William *see* The art of invigorating and prolonging life, by food, clothes, air, exercise, wine, sleep, etc

Kite, Thomas *see* Memoirs and letters of thomas kite

Kite, William *see* Memoirs and letters of thomas kite

Kitimat northern sentinel – Kitimat, British Columbia, CN. jan 1954-dec 73 – 20r – 1 – cn Commonwealth Imaging [071]

Kitowski, Karin *see* Moeglichkeiten und grenzen multilateraler handelsvertraege

Kitsap sun – Bremerton WA: Scripps Howard Newspaper c2005- [daily] – v84 n142 [may 22 2005]- – 1 – (cont: sun [bremerton, wa] filmed on same reel) – us UW Libraries [071]

Kitsur divre ha-yamin le-'am yisra'el me-reshit heyoto 'ad ha-yom h... / Goor, Yehudah – Warsaw, Poland. 1900 – 1r – us UF Libraries [939]

Kittanning gazette – Kittanning, PA., 1832-1882 – 13 – $25.00r – us IMR [071]

Kittel, Gerhard *see*
- Jesus und die rabbinen
- Die oden salomos

Kittel, Johann Christian *see*
- 25 chorale, mit achterly general baessen
- Vierstimmige choraele mit vorspielen

Kittel, Rudolf *see*
- Die alttestamentliche wissenschaft in ihren wichtigsten ergebnissen
- Der babel-bibel-streit und die offenbarungsfrage
- Biblia hebraica
- History of the hebrews
- The scientific study of the old testament
- Studien zur hebraeischen archaeologie und religionsgeschichte
- Ueber die notwendigkeit und moeglichkeit einer neuen ausgabe der hebraeischen bibel
- Zur theologie des alten testaments

Kittle, Samuel *see* A concise history of the colony and natives of new south wales

Kittle, Warren Brance *see* Notes on the law of rule days in virginia and west virginia.

Kittlitz, F H von *see* Denkwuerdigkeiten einer reise nach dem russischen amerika, nach mikronesien und durch kamschatka

Kittlitz, Friedrich H von *see* Denkwuerdigkeiten einer reise nach dem russischen amerika, nach mikronesien und durch kamtschatka

Kittlitz, Richard, Freiherr von *see* Schleiermacher's bildungsgang

Kitto, John *see*
- Cyclopaedia of biblical literature
- The history of palestine from the patriarchal age to the present time

Kittoe, Markham *see* Illustrations of indian architecture from the muhammadan conquest downwards

Kitton, Frederic George *see*
- Dickens and his illustrators
- John leech, artist and humourist

Kittredge, Anson Oliver *see* The self-proving accounting system

Kittredge, George Lyman et al *see* Studies in the history of religions

Kitts, Eustace J *see* In the days of the councils

KLEIN

Kitts, Eustace John *see*
– In the days of the councils
– Pope john the twenty-third and master john hus of bohemia
Kituba : basic course / Foreign Service Institute (US) – Washington, DC. 1963 – 1r – us UF Libraries [960]
Kitve ben-ami / Ben-Ami, Mordecai – Odessa, Ukraine. 1913 – 1r – us UF Libraries [939]
Der kitzinger *see* Kitzinger zeitung
Kitzinger zeitung – Kitzingen DE, 1983 1 jun- – ca 7r/yr – 1 – (title varies: 1 aug 1991: der kitzinger) – gw Misc Inst [074]
Kiuner, Nikolai Vasil'evich *see* Lektsii po istorii razvitiia glavnieishikh osnov kitaiskoi material'noi i dukhovnoi kul'tury
Kiurstaker volksstimme – Bruck, Austria. 19 aug 1945; 3, 30 jun 1946; 26 mar 1947 – 2r – 1 – uk British Libr Newspaper [072]
Kivandorlasi ellendoer – Budapest HU, feb-mar 1908 – 1r – 1 – us IHRC [077]
Kivandorlasi ertesitoe – Budapest HU, nov 1903-07 – 2r – 1 – us IHRC [073]
Kivandorlo – Budapest HU, 1904-05 – 1r – 1 – us IHRC [077]
Kivu – Bruxelles, Belgium. 1952 – 1r – us UF Libraries [960]
Ki-vumba / Lambert, H E – Kampala, Uganda. 1957 – 1r – us UF Libraries [960]
Kivu-meer / Watteyne, P J – Antwerpen, Belgium. 1931 – 1r – us UF Libraries [960]
Kiwanis – Indianapolis IN 1917+ – 1,5,9 – ISSN: 0162-5276 – mf#1504 – us UMI ProQuest [360]
Kiwari – Djakarta, 1957/1958. v1(1-7/8) – 6mf – 9 – mf#SE-902 – ne IDC [950]
Kiwi chatter – United States. 1989 jun 23-dec 22, 1990, 1991, 1992 – 4r – 1 – mf#2370338 – us WHS [071]
Kiyevskaya mysl' – Kiev. 1914-1916 [incomplete] – 1 – us NY Public [073]
Kizevetter, A A *see*
– P N miliukov
– Partiia narodnoi svobody i ee ideologiia
– Posadskaia obshchina v rossii 18 st
Kizihrmak – Sivas, 1912-25. Mueduer-i Mes'ul: Haraccizade Ibrahim Hakki, Karaopoladzade Ismet. n30. 25 temmuz 1328 [1912] – (= ser O & t journals) – 1mf – 9 – $25.00 – us MEDOC [956]
Kjemi – Oslo, Norway 1973-81 – 1,5,9 – ISSN: 0023-1983 – mf#8810 – us UMI ProQuest [660]
Kjerlighedens gjerninger : nogle christelige overveielser i talers form / Kierkegaard, Soeren – Kobenhavn: C A Reitzels, 1847 – (= ser Himmelstrup; Arbaugh) – 1mf – 9 – 0-7905-7416-0 – mf#1989-0641 – us ATLA [100]
Kjerstad, Muriel Adele *see* Existing conditions of choral music in the american high school
Klaar, Alfred *see*
– Koenig ottokars glueck und ende
– Die marquise von o–
Klabund *see* Borgia
Kladderadatsch – Berlin DE, 1848 7 may-1886, 1887 3 jul-1940, 1941 18 may-1944 3 sep – 1 – (filmed by misc inst: 1873, 1888, 1927 16 jan-25 dec [3r]) – gw Mikrofilm; gw Misc Inst [870]
Kladderadatsch : humoristisch-satirisches wochenblatt. – may 1848-1906 – 1 – us L of C Photodup [870]
Klaehn, Friedrich Joachim *see*
– Das gastmahl
– Nacht ueber malmaison
– Der sergeant weber
– Timm der tolpatsch
Klaenge aus vergangenen zeiten / Wagner, Philipp – New York: L W Schmidt, 1881 – 1r – 1 – us Wisconsin U Libr [943]
Klaer ende grondich teghen-vertoogh gestelt tegen seker vertoogh der remonstranten... / Trigland, J – Amsterdam, 1617 – 2mf – 9 – mf#H-2500 – ne IDC [240]
Ein klag des frydes... / Erasmus – [Zuerich, Christoph Froschouer, 1521] – 1mf – 9 – mf#PBU-540 – ne IDC [240]
Eine klag vnd trostschrifft : von dem christlichen abschid vnd begrebnusz des philippi melanthonis / Peucer, C – [Nuernberg], 1560 – 1mf – 9 – mf#TH-1 mf 1274 – ne IDC [242]
Die klagelieder *see* The lamentations of jeremiah
Die klagelieder des jeremias nach rabbinischer auslegung : mit vorwort und erlaeuterungen / Schoenfelder, Joseph M – Muenchen: Ernst Stahl, 1887 [mf ed 1985] – 1mf – 9 – 0-8370-5132-0 – (incl bibl ref) – mf#1985-3132 – us ATLA [221]
Die klagelieder des jeremias und der prediger des salomon : im urtext nach neuester kenntniss der sprache behandelt (erstere metrisch) uebersetzt / Raabe, Andreas – Leipzig: L Fernau 1880 [mf ed 1985] – 1mf – 9 – 0-8370-4817-6 – mf#1985-2817 – us ATLA [221]
Die klagelieder jeremiae in der aethiopischen bibeluebersetzung : auf grund handschriftlicher quellen mit textkritischen anmerkungen / ed by Bachmann, Johannes – Halle a S: M Niemeyer, 1893 – 1mf – 9 – 0-8370-1805-6 – mf#1987-6193 – us ATLA [221]
Klages, Ludwig *see* Goethe als seelenforscher
Klah'che'min : squaxin island newsletter / Squaxin Island Tribal Center – 1979 jul-1981 may, nov, 1982 feb, jun – 1r – 1 – (cont by: klah-che-min newsletter) – mf#656869 – us WHS [305]
Klaiber, Theodor *see* Dichtende frauen der gegenwart
Klaic, Vjekoslav *see* Slike iz slavenske povjesti
Klain, Yosef *see* Divre yosef
Klamath basin progress – Klamath Falls OR: Farmers Pub Co Inc, 1933-43 [wkly] – 1 – (cont: klamath basin progress and malin progress (1928-33)) – us Oregon Lib [071]
Klamath basin progress and the malin progress – Klamath Falls OR: Farmers Pub Co Inc, 1928-33 [wkly] – 1 – (cont: malin progress (1926-28); cont by: klamath basin progress (1933-43)) – us Oregon Lib [071]
Klamath chronicle – Klamath Falls OR: Klamath Pub Co [daily ex mon] – 1 – (began in 1910. ceased in 1911. merged with: pioneer express; klamath falls northwestern) – us Oregon Lib [071]
Klamath chronicle *see* Klamath falls northwestern
Klamath county courier – Klamath Falls OR: A L Mallory [wkly] – 1 – us Oregon Lib [071]
Klamath county star – Linkville OR: [Connolly & Haynes] [wkly] – 1 – (cont: linkville weekly star; cont by: klamath star) – us Oregon Lib [071]
Klamath falls express – Klamath Falls OR: David B Worthington, 1892-1911 [wkly] – 1 – (began with apr 28 1892; cont by: morning express (1911-)) – us Oregon Lib [071]
Klamath falls northwestern – Klamath Falls OR: Samuel M Evans [wkly] – 1 – (began in feb 1912. ceased with dec 26 1915? merger of: pioneer press and morning express; klamath chronicle (1910-11)) – us Oregon Lib [071]
Klamath falls northwestern *see* Klamath chronicle
Klamath news – Klamath Falls OR: Klamath News Publ, 1923-41 [daily ex mon] – 1 – (absorbed by: evening herald (klamath falls, or); herald and news (klamath falls, or)) – us Oregon Lib [071]
Klamath news *see*
– Evening herald (klamath falls, or)
– Herald and news
Klamath record – Klamath Falls OR: Klamath Recod Pub Co [daily] – 1 – us Oregon Lib [071]
Klamath republican – Klamath Falls OR: Republican Pub Co [wkly] – 1 – (began apr 26 1896; cont by: semi-weekly herald (klamath falls, or: 1914-)) – us Oregon Lib [071]
Klamath star – Klamath Falls OR: Connolly & Haynes [wkly] – 1 – (cont: klamath county star) – us Oregon Lib [071]
Klamath sun *see* Klamath record
Klamp, Gerhard *see* Ueber die idee einer metaphysik im sinne des kritischen realismus
Klamroth, Erich *see* Die juedischen exulanten in babylonien
The klan unmasked / Simmons, William Joseph – Atlanta, GA: W E Thompson Pub Co, c1923 – us CRL [366]
De klank : en vormleer van het middelnederlandsch dialect der st servatius-legende van heynrijck van veldeken / Leviticus, Felix – Gent: A Siffer, 1892 [mf ed 1993] – 172p – 1 – (incl bibl ref and ind) – mf#8439 – us Wisconsin U Libr [430]
Klansman / Ku Klux Klan – n40,42-131 [1979 apr, jun-1987 dec] – 1r – 1 – mf#1075264 – us WHS [366]
Klanwatch law report / Southern Poverty Law Center – 1985 sum-1987 jan – 1r – 1 – (cont by: law report [montgomery, al]) – mf#1895697 – us WHS [344]
Klapper op de inhoud van de tijdschriften im inheemse talen van het java-instituut – Poesaka Djawi, 1-15, 1922-30; Poesaka Soenda, 1-7, 1922-29; Poesaka Madhoera, 1, 1924. Djogjakarta: Java-Instituut, 1937. Th. Pagaud. 1937 – 1 – us Wisconsin U Libr [959]
Klaproth, Heinrich Julius *see*
– Apercu de l'origine des diverses ecritures de l'ancien monde
– Chrestomathie mandchou
– Comparaison du basque avec les idiomes asiatiques, et principalement avec ceux qu'on apelle semitiques
– East meets west
– Lettre a m champollion le jeune
Klaproth, Heinrich Julius von *see*
– Extrait d'une lettre de m schmidt, de st-petersbourg
– Vocabulaire et grammaire de la langue georgienne
Klaproth, J [H von] *see*
– Asia polyglotta
– Dr j a guldenstadts beschreibung der kaukasischen laender
– Memoires relatifs...l'asie
– Reise in den kaukasus und nach georgien unternommen in den jahren 1807 und 1808...
– Tableau historique, geographique, ethnographique et politique du caucase et des provinces limitrophes entre la russie et la perse
– Voyage au mont caucase et en georgie
Klaproth, J [H] von *see* Beschreibung der russischen provinzen zwischen dem kaspischen und schwarzen meere
Klaproth, Julius *see* Geographisch-historische beschreibung des oestlichen kaukasus
Klaproth, Julius H von *see*
– Memoires relatifs a l'asie
– Voyage au mont caucase et en georgie
Klar, Benjamin *see* Rabi hayim ibn 'atar
Klaralvsbygden – Hagfors, Sweden. 1979-85 – 1 – sw Kungliga [078]
Die klarinette : erzaehlung / Schaffner, Jakob – Stuttgart: Deutsche Volksbuecher, 1942 – 1r – 1 – us Wisconsin U Libr [390]
Die klarinette bei carl maria von weber / Sandner, Wolfgang – Frankfurt a.M., 1971 – 3mf – 9 – 3-89349-721-8 – gw Frankfurter [780]
Klasen, Franz *see* Die innere entwicklung des pelagianismus
Klass, Gert von *see* Die grosse entscheidung
Klass, Sheila Solomon *see* Everyone in this house makes babies
Klassen, N *see* Tovarovedenie
Der klassenkampf – Berlin DE, 1927 oct-1932 1 jul – 2r – 1 – (missing: 16 apr-14 may 1931) – gw Misc Inst [335]
Der klassenkampf : sozialistische politik und wirtschaft – Berlin: [s.n. 1927-32] [jahrg1-6(1927-32)] [mf ed 1980] – 2r – 1 – (no more publ? subtitle varies) – mf#film mas c711 – us Harvard [335]
Der klassenkampf *see* Volksblatt 1920
Klassik und romantik der deutschen / Schultz, Franz – 2nd ed. Stuttgart: J B Metzler, 1959 [mf ed 1993] – (= ser Epochen der deutschen literatur: geschichtliche darstellungen v4/1-2) – 2v – 1 – mf#8237 – us Wisconsin U Libr [430]
Klassik und romantik der deutschen / Schultz, Franz – Stuttgart: J B Metzler. 2v. 1935 – 1r – 1 – (incl bibl ref and ind) – us Wisconsin U Libr [430]
Klassiker heute : die zeit des expressionismus: erste begegnung mit georg heym, georg trakl, gottfried benn, johannes r becher, georg kaiser, alfred doeblin, else laske-schueler, ernst toller / Doerrlamm, Brigitte et al – Frankfurt/Main: Fischer Taschenbuch Verlag, 1982, c1981 [mf ed 1993] – (= ser Fischer taschenbuecher 3026) – 1 – (incl bibl ref) – mf#8272 – us Wisconsin U Libr [430]
Klassiki russkoi literatury : lektsii, chitannye v voskresnom universitete i m g u / Kubikov, Ivan Nikolaevich – Moskva: Izd-vo I-go Moskovskogo gos universiteta, 1930 [mf ed 2004] – 1r – 1 – (filmed with: literatura i kritika: sbornik statei / g v plekhanov (v1 1922). incl bibl ref) – us Wisconsin U Libr [460]
Klassische gestaltung und romantischer einfluss in den dramen heinrichs von kleist / Willige, Wilhelm – Heidelberg: C Winter, 1915 – 1r – 1 – (incl bibl ref) – us Wisconsin U Libr [430]
Klassische schoenheit : johann winckelmann, g ephraim lessing / ed by Gleichen-Russwurm, Alexander von – Jena: E Diederichs, 1906 – 1 – us Wisconsin U Libr [840]
Der klassische wiener franz grillparzer / Ludwig, Alfred Josef – Wien: Amandus-Edition 1946 [mf ed 1990] – 1r [ill] – 1 – (filmed with: grillparzers verhaeltnis zur politischen tendenzliteratur seiner zeit / konrad beste [comp]) – mf#2690p – us Wisconsin U Libr [430]
Klassovyia osnovy izbiratel'nago prava / Chernyshev, Illarion – Petrograd: Kn-vo "Zhizn' i znanie" 1917 [mf ed 2004] – (= ser Deshevaia biblioteka 108-aia) – 1r – 1 – (filmed with: slavianskaia problema srednei evropy / g g khristiani (1919)) – mf#5497p – us Wisconsin U Libr [325]
Klassy i kooperatsiia v derevne sssr : k vopr o razmezhevanii potrebitelskoi i selskokhoziaistvennoi kooperatsii v sovetskoi derevne / Vlasov, M E – 1925 – 103p 2mf – 9 – mf#COR-469 – ne IDC [335]
Klassy i partii v sibiri nakanune i v period velikoi oktiabrskoi sotsialisticheskoi revoliutsii – Tomsk, 1977 – 3mf – 9 – mf#RPP-21 – ne IDC [325]
Klatt, Fritz *see*
– Griechisches erbe; das urbild der antike im widerschein des heutigen lebens
– Hans carossa
Klatzkin, J *see* Krisis und entscheidung im judentum
Klatzkin, Jakob *see* Freie zionistische blaetter
Klatzkin, Naphtali Hirz *see* Ayalah sheluhah
Klauber, E *see* Assyrisches beamtentum nach briefen aus der sargonidenzeit
Klauber, Ernst Georg *see*
– Politisch-religioese texte aus der sargonidenzeit
Klaucke, Paul *see* Erlaeuterungen ausgewaehlter werke goethes
Klauder, Charles Zeller *see* College architecture in america and its part in the development...
Klaus, A *see* Ursprung und verbreitung der dreifaltigkeitsmesse
Klaus groth : zu seinem achtzigsten geburtstage / Bartels, Adolf – Leipzig: E Avenarius, 1899 [mf ed 1990] – 145p/1pl – 1 – mf#7424 – us Wisconsin U Libr [430]
Klaus hinrichs baas : roman / Frenssen, Gustav – Berlin: G Grote 1909 [mf ed 1989] – (= ser Grote'sche sammlung von werken zeitgenoessischer schriftsteller 99) – 1r – 1 – (filmed with: holyland) – mf#7265 – us Wisconsin U Libr [830]
Klauser, T *see* Der ursprung der bischoeflichen insignien und ehrenrechte
Klauser, Th *see* Doctrina duodecim apostolorum. barnabae epistula (fp1)
Klausner, Israel *see*
– Rabi hayim tsevi shne'urson
– Vilna bi-tekufat ha-ga'on
Klausner, Joseph *see*
– Bi-yeme bayit sheni
– Eliiezer ben-yehudah
– Geschichte der neuhebraischen literatur
– Die messianischen vorstellungen des juedischen volkes im zeitalter der tannaiten
– Ob entsiklopedii iudaizma na evreiskom iazykie
– Torat-ha-midot ha-kedumah be-yisrael
– Yahadut ve-enushiyut
– Yeshu ha-notsri
– Yotsrim u-bonim
Klausner, Max Albert *see* Hie babel, hie bibel
Klaveness, Th *see* Bibliusoegur og agrip af kirkjusoegunni handa boernum
Klawon, Dieter *see* Geschichtsphilosophische ansaetze in der fruehromantik
Klaxon : mensario de arte moderna – Sao Paulo, SP: Typ Paulista, 15 maio 1922-jan 1923 – (= ser Ps 19) – mf#P12,03,05 – bl Biblioteca [700]
Kleban, Patricia L *see* The relationship between shared family recreation time and expressed parent-adolescent conflict
Kleber, John Christopher *see* Void judicial and execution sales, and the rights, remedies and liabilities of purchasers thereat.
Klebs, J *see* Die landeskulturgesetzgebung, deren ausfuhrung und erfolge im grossherzogthum posen
Klee, Gotthold *see*
– Karl simrocks ausgewaehlte werke in zwoelf baenden
– Wielands werke
Klee, Heinrich *see* Katholische dogmatik
Kleeblatt / Allen Co. Delphos – may 1891-aug 95, nov 96-jul 1905 [wkly] – 6r – 1 – (in german) – mf#B7165-7170 – us Ohio Hist [071]
Kleeblatt / Vanwert Co. Delphos – may 1891-aug 95, nov 96-jul 1905 [wkly] – 6r – 1 – (in german) – mf#B7165-7170 – us Ohio Hist [071]
Kleef, B A van *see* Geschiedenis van de oud-katholieke kerk van nederland
Kleef, Rald *see* Bestimmung von hla klasse 1 an zur autovaccination bestimmten, epithelmarker-charakterisierten tumorzellen
Kleeman, Karen M *see* A family systems analysis of anxiety, depression, and somatization in graduate nursing students
Kleiber, Joseph *see* Opredelenie orbit meteornykh potokov
Kleiber, Ludwig *see*
– Studien zu goethes egmont
– Zur wirtschaftspolitik oberschwaebischer reichsstaedte im ausgehenden mittelalter
Klein, Albert *see* Be-erot avraham
Klein, Alfred *see* Die akte arno holz
Klein, Anton von *see* Pfalzbyerische beytraege zur gelehrsamkeit
Klein, Augusta *see* The problem of logic
Klein, Ernst Ferdinand *see* Annalen der gesetzgebung und rechtsgelehrsamkeit in den kgl preussischen staaten
Klein, Felix *see*
– America of to-morrow
– In the land of the strenuous life
– Les paraboles de l'evangile
Klein, Frederick Augustus *see* The religion of islam
Klein, Fritz *see* Journal of bisexuality
Klein, Georgette *see* Freiligrath
Klein, Harry *see* Springbok record
Klein, Hermann Joseph *see*
– Das sonnensystem
– Star atlas
Klein, Horst G *see* Das verhalten der telischen verben in den romanischen sprachen eroertert an der interferenz von aspekt und aktionsart
Klein, Johann Joseph *see* Versuch eines lehrbuchs der praktischen musik in systematischer ordnung
Klein, Johannes *see* Walter flex, ein deuter des weltkrieges

Klein, John Frederick *see* Beitraege zu paul heyses novellentechnik
Klein, Judy L *see* Ratings of perceived exertion in college age males and females of high and low fitness levels
Klein, Karl *see* Studien ueber meteoriten
Klein, Karl Kurt *see*
– Die anfaenge der deutschen literatur
– Die lieder oswalds von wolkenstein
Klein, Margarete *see* Stefan george als heldischer dichter unserer zeit
Klein, Samuel *see*
– 'Ever ha-yarden ha-yehudi
– Ma'amarim shonin le-hakirath ha-erez-yisrael
– Toldot ha-yishuv ha-yahudi ha-eraz-yisrael
Klein, Timotheus *see* Das erbe
Klein toggenburger chroniken : mit beilagen und erlaeurterungen / ed by Scherrer, G – St Gallen, Huber, 1874 – 2mf – 9 – mf#PBU-466 – ne IDC [240]
Klein, V K *see* Nadpisi na grobnitsakh v tserkvi nikoly na stolpakh
Kleinasiatische denkmaeler aus pisidien, pamphylien, kappadokien und lykien : studien ueber christliche denkmaeler, pts 5-6 / Rott, H – Leipzig, 1908 – 9mf – 8 – mf#H-649 – ne IDC [720]
Kleinasien und deutschland / Ross, L – Halle, 1850 – 3mf – 9 – mf#AR-1418 – ne IDC [910]
Kleinberg, Alfred *see*
– Die deutsche dichtung in ihren sozialen, zeit- und geistesgeschichtlichen bedingungen
– Ludwig anzengruber
Kleinbort, L M *see* Ocherki narodnoi literatury (1880-1923 g g)
Kleine abentheuer zu wasser und zu lande / Weyland, Philipp C – Hof 1802-11 [mf ed Hildesheim 1995-98] – 12v on 25mf – 9 – €250.00 – 3-487-29931-3 – gw Olms [910]
Kleine altniederdeutsche denkmaeler / ed by Heyne, Moritz – 2. aufl. Paderborn: F Schoeningh, 1877 [mf ed 1993] – (= ser Bibliothek der aeltesten deutschen litteratur-denkmaeler 4; Altniederdeutsche denkmaeler 2) – 1 – mf#8437 reel 1 – us Wisconsin U Libr [430]
Kleine beitrage zur traumlehre / Freud, Sigmund – Wien, Austria. 1925 – 1r – us UF Libraries [025]
Das kleine blatt – Wien: Druck & Verlagsanstalt "Vorwarts" Swoboda & Co, 1928-33 – 24r – 1 – us CRL [074]
Kleine chronik : vier erzaehlungen / Zweig, Stefan – Leipzig: Insel-Verlag, [1929] [mf ed 1992] – (= ser Inselbuecherei n408) – 99p – 1 – mf#7801 – us Wisconsin U Libr [880]
Der kleine coco – Goch DE, 1910-13 – 1 – gw Misc Inst [074]
Kleine deutsche literaturgeschichte / Roos, Carl – Kobenhavn: Gyldendal, 1955 – 1r – 1 – us Wisconsin U Libr [430]
Kleine erzaehlungen / Peters, Friedrich Ernst – Goettingen: Deuerlich, 1941 – 1r – 1 – us Wisconsin U Libr [830]
Die kleine ferne stadt : [short stories] / Blunck, Hans Friedrich – Hamburg: Hanseatische Verlagsanstalt, c1941 [mf ed 1989] – 105p – 1 – mf#7037 – us Wisconsin U Libr [830]
Kleine fussreisen durch die schweiz / Bridel, Jean – Zuerich 1811 [mf ed Hildesheim 1995-98] – 2v on 4mf – 9 – €120.00 – 3-487-29367-6 – gw Olms [914]
Kleine gartenbibliothek / ed by Hirschfeld,Christian Cay Lorenz – Kiel 1790 – (= ser Dz) – 1v on 2mf – 9 – €60.00 – mf#k/n2970 – gw Olms [635]
Kleine geographie des deutschen witzes / Schoeffler, Herbert; ed by Plessner, Helmuth – Goettingen: Vandenboeck & Ruprecht, c1955 – 98p – 1 – (with aft) – us Wisconsin U Libr [430]
Der kleine gerd : humoristisch-militaerische erzaehlung / Baudissin, Wolf Ernst Hugo Emil, Graf von – Berlin: Schreiter, [19–?] [mf ed 1995] – (= ser D r buecher) – 280p – 1 – mf#8972 – us Wisconsin U Libr [880]
Kleine hausapotheke : [verse and prose] / Vegesack, Siegfried von – Hamburg: Hammerich & Lesser 1944 [mf ed 1991] – 1r [ill] – 1 – (filmed with: blumbergshof / siegfried von vegesack) – mf#2944p – us Wisconsin U Libr [800]
Kleine hieroglyphen-grammatik (auszug) nach dem werk h. prof. heinrich brugsch, berling, handschrift / Meyer-Berlin, Richard – Berlin, 1913 (mf ed 1994) – 1mf – 9 – €24.00 – 3-8267-3073-9 – mf#DHS-AR 3073 – gw Frankfurter [470]
Kleine koptische studien 1-58 / Lemm, O von – Leipzig, 1972 – (= ser Subsidia byzantina v1) – 22mf – 8 – €42.00 – ne Slangenburg [243]
Kleine leute : drei novellen / Hopfen, Hans – Berlin: F Schneider 1880 [mf ed 1995] – 1r – 1 – (filmed with: fraenzchens lieder / hoffmann von fallersleben) – mf#3757p – us Wisconsin U Libr [830]
Kleine literarturfibel : eine anleitung zum verstaendnis schoener literatur / Hoefer, Karl-Heinz – Leipzig: Fachbuchverlag, 1967 – 1r – 1 – (incl bibl ref and index) – us Wisconsin U Libr [430]
Der kleine martin / Franzos, Karl Emil – 3e. Auf. Stuttgart and Berlin. n.d – 1 – us CRL [890]
Der kleine missionsfreund – Berlin DE, 1899-1922 – 1 – gw Misc Inst [240]
Kleine mitteilungen aus dem septuaginta-unternehmen / Rahlfs, Alfred – Berlin: Weidmann 1915 [mf ed 1990] – (= ser Mitteilungen des septuaginta-unternehmens 1/7) – 1mf – 9 – 0-8370-1779-3 – (in german & greek) – mf#1987-6167 – us ATLA [221]
Kleine morgenzeitung *see* Breslauer anzeiger
Kleine musik-zeitung – jahrg=v10 n8 [1867 dec 4] – 1r – 1 – mf#1002865 – us WHS [780]
Der kleine nacht-express *see* Nacht-express
Die kleine narrenwelt : [short stories] / Gutzkow, Karl – Frankfurt/Main: Literarische Anstalt, 1856-1857 [mf ed 2001] – 3v – 1 – mf#10522 – us Wisconsin U Libr [830]
Kleine nordische erzaehlungen / Dahn, Felix – Leipzig: Breitkopf & Haertel, 1898 – 4r – 1 – us Wisconsin U Libr [830]
Die kleine passion : roman / Wiechert, Ernst Emil – Berlin: G Grote, 1929 – 1r – 1 – us Wisconsin U Libr [830]
Kleine presse – Frankfurt/M DE, 1885 mai-1888 mai, 1888 jul-1892 nov, 1893-99, 1900 jul-dec, 1901 feb-mai, 1901 jul-1912 jul, 1912 sep-1917 mar – 1 – gw Misc Inst [074]
Kleine prosaische schriften / Schwab, Gustav Benjamin; ed by Kluepfel, K – Freiburg i.Br: J C B Mohr, 1882 – 1r – 1 – (incl bibl ref) – us Wisconsin U Libr [430]
Der kleine rosengarten : volkslieder / Loens, Hermann – Jena: Eugen Diederichs 1919 [mf ed 1990] – 1r – 1 – (filmed with: storbonden og hans sonner) – mf#2829p – us Wisconsin U Libr [780]
Der kleine saemann – v2. 1899 [complete] – (= ser Mennonite serials coll) – 1r – 1 – mf#ATLA 1994-S027 – us ATLA [242]
Kleine schriften / Altaner, Bruno – Berlin, 1967 – (= ser Tugal 5-83) – 11mf – 9 – €21.00 – ne Slangenburg [240]
Kleine schriften / Lotze, Hermann – Leipzig: S Hirzel, 1885-1891 – 5mf – 9 – 0-524-08348-7 – mf#1993-2038 – us ATLA [190]
Kleine schriften / Scherer, Wilhelm; ed by Burdach, Konrad et al – Berlin: Weidmann 1893 [mf ed 1978] – 2v in 1 on 1r – 1 – (v1: kleine schriften zur altdeutschen philologie ed by konrad burdach; v2: kleine schriften zur neueren litteratur, kunst und zeitgeschichte ed by erich schmidt) – mf#film mas 7982 – us Harvard [430]
Kleine schriften / Strauss, David Friedrich – Bonn: Emil Strauss, 1895 – 1 – us Wisconsin U Libr [430]
Kleine schriften 1 : studien zur spaetantiken religions-geschichte / Lietzmann, Hans; ed by Aland, Kurt – Berlin, 1958 – (= ser Tugal 5-67) – 9mf – 9 – €18.00 – ne Slangenburg [240]
Kleine schriften 2 : studien zum neuen testament / Lietzmann, Hans; ed by Aland, Kurt – Berlin, 1958 – (= ser Tugal 5-68) – 6mf – 9 – €14.00 – ne Slangenburg [240]
Kleine schriften 3 : studien zur liturgie- und symbolsgeschichte zur wissenschaftsgeschichte / Lietzmann, Hans; ed by Kommission fuer spaetantike Religionsgeschichte – Berlin, 1962 – (= ser Tugal 5-74) – 7mf – 9 – €15.00 – ne Slangenburg [240]
Kleine schriften religionsgeschichtlichen inhalts / Hausrath, Adolf – Leipzig: S. Hirzel, 1883 – 2mf – 9 – 0-7905-5995-1 – (incl bibl ref) – mf#1988-1995 – us ATLA [240]
Kleine schriften zur kunst / Meyer, Heinrich; ed by Weizsaecker, Paul – Heilbronn: Henninger, 1886 [mf ed 1993] – (= ser Deutsche litteraturdenkmale des 18. und 19. jahrhunderts 25) – clxiii/258p – 1 – (incl bibl ref) – mf#8676 reel 3 – us Wisconsin U Libr [700]
Kleine schriften zur litteratur und kunst / Stahr, Adolf Wilhelm Theodor – Berlin: J Guttentag, 1871-1875 – 1r – 1 – us Wisconsin U Libr [430]
Kleine schriften zur litteratur und kunst / Stahr, Adolf Wilhelm Theodor – Berlin: J Guttentag, Oldenburg, Schulz [1871-75] [mf ed 1979] – 3v on 1r – 1 – (v1: biographische skizzen und nachrufe; v2: biographisches und kritisches; v3: aus dem alten weimar) – mf#film mas 8517 – us Harvard [430]
Kleine schul- und haus-bibel – Berlin, Germany. v1-2. 1928 – 1r – us UF Libraries [270]
Kleine sing- und spielstuecke fuer clavier von verschiedenen meistern : erste [-dritte sammlung] – Berlin: bey Friedrich Wilhelm Birnstiel [1762?-1766] [mf ed 19–] – 3v on 2mf – 9 – mf#fiche 55 – us Sibley [780]
Das kleine volksblatt – Vienna. Austria. -d. 5 Aug 1945-22 Apr 1948, 16 Nov 1948-31 Dec 1953. (Imperfect). (27 reels) – 1 – uk British Libr Newspaper [074]
Kleine volksblatt – Vienna, Austria. 5 aug 1945-20 feb 1948; 14 mar, 4 apr-22 apr, 16 nov-31 dec 1948; 1949-53 – 27r – 1 – uk British Libr Newspaper [072]
Die kleine welt : kurzgeschichten voll humor / Cetto, Gitta von – Berlin-Schildow: E Sicker 1944 [mf ed 1989] – 1r – 1 – (filmed with: die poesie, ihr wesen und ihre formen / moriz carriere) – mf#7146 – us Wisconsin U Libr [830]
Die kleine weltlaterne / Bamm, Peter – Stuttgart: Deutsche Verlags-Anstalt, c1935 [mf ed 1990] – 254p (ill) – 1 – (ill by olaf gulbransson) – mf#7216 – us Wisconsin U Libr [890]
Kleine wiener kriegszeitung – Vienna, Austria sep 1944-apr 1945 [mf ed Norman Ross] – 1r – 1 – mf#nrp-1978 – us UMI ProQuest [074]
Die kleinen propheten / Nowack, Wilhelm – 2. aufl. Goettingen: Vandenhoeck und Ruprecht, 1903 – (= ser Handkommentar zum alten testament) – 1mf – 9 – 0-8370-9494-1 – mf#1986-3494 – us ATLA [221]
Die kleinen prophetischen schriften vor dem exil / Procksch, Otto – Calw: Verlag der Vereinsbuchh., 1910 – (= ser Erlaeuterungen Zum Alten Testament) – 1mf – 9 – 0-7905-1791-4 – mf#1987-1791 – us ATLA [221]
Ein kleiner deutscher : roman / Dittmer, Ernst – Muenchen: Deutscher Volksverlag, c1938 [mf ed 1989] – 245p – 1 – mf#7178 – us Wisconsin U Libr [830]
Kleiner katechismus der christlichen lehre : zum gebrauch fuer katholische schulen / Weninger, Francis Xavier – New York: Benziger, 1866, c1865 – 1mf – 9 – 0-8370-6859-2 – mf#1986-0859 – us ATLA [241]
Kleiner local-anzeiger fuer die kreise dortmund und hoerde – Dortmund DE, 1887 6 apr-1894, 1895 6 jul-1903, 1905 1 jul-30 dec, 1906 2 jul-1911, 1913 1 jul-1915 30 jun, 1917-19, 1920 15 mar-8 jun & 1 jul-30 sep, 1921-1922 29 jul – 1 – (title varies: 7 may 1887: lokal-anzeiger fuer die kreise dortmund und hoerde; 1 jul 1895: dortmunder tageblatt; 3 jan 1921: westfaelische morgenzeitung; n1-9 publ in bochum n10-27 in essen, then dortmund) – gw Misc Inst [074]
Kleiner markt : studien, erzaehlungen, maerchen und gedichte / Anzengruber, Ludwig – Breslau: S Schottlaender, 1883 [mf ed 1993] – 172p – 1 – mf#8459 – us Wisconsin U Libr [800]
Kleiner reiseatlas fuer deutschland : enthaltend 24 karten mit saemmtlichen deutschen eisenbahnen und einigen anderen reiserouten... – Muenchen 1847 [mf ed Hildesheim 1995-98] – (= ser Fbc) – 1mf – 9 – €40.00 – 3-487-29603-9 – gw Olms [914]
Kleiner unitarier-spiegel : kurzer inbegriff der geschichte, der dogmen, der kirchenverfassung und der ceremonien der unitarier-kirche / Ferencz, Jozsef – Wien: Carl Gerold, 1879 – 1mf – 9 – 0-524-03841-4 – mf#1990-4888 – us ATLA [243]
Kleinere deutsche gedichte des 11. und 12. jahrhunderts / ed by Waag, Albert – Halle a. S: M Niemeyer, 1890 [mf ed 1993] – (= ser Altdeutsche textbibliothek 10) – xli/167p – (incl bibl ref) – mf#8193 reel 1 – us Wisconsin U Libr [810]
Kleinere erzaehlungen / Brinckman, John – Berlin: W Werther, [1901?] [mf ed 1989] – 352p – 1 – mf#7088 – us Wisconsin U Libr [830]
Kleinere erzaehlungen / Gotthelf, Jeremias [pseud: Albert Bitzius] – Erlenbach, Zuerich: E Rentsch, 1912-29 [mf ed 1993] – (= ser Saemtliche werke in 24 baenden 16-22) – 7v – 1 – (incl bibl ref) – mf#8522 reels 4-6 – us Wisconsin U Libr [880]
Kleinere erzaehlungen = Short stories / Raabe, Wilhelm Karl – 4. Aufl. Berlin-Grunewald: Hermann Klemm, [189-?] – 1 – us Wisconsin U Libr [830]
Kleinere gedichte / Stricker; ed by Hahn, Karl August – Quedlinburg, Leipzig: G Basse, 1839 [mf ed 1993] – (= ser Bibliothek der gesammten deutschen national-literatur von der aeltesten bis auf die neuere zeit sect1/18) – xx/106p – 1 – mf#8438 reel 5 – us Wisconsin U Libr [810]
Kleinere geistliche gedichte des 12. jahrhunderts / ed by Leitzmann, Albert – Bonn: A Marcus & E Weber 1910 – (= ser Kleine texte fuer theologische und philosophische vorlesungen und uebungen 54) – 1mf – 9 – 0-524-04618-2 – (incl bibl ref; text in middle high german) – mf#1990-1278 – us ATLA [810]
Kleinere mittelhochdeutsche erzaehlungen, fabeln und lehrgedichte – Berlin: Weidmann, 1904-09 [mf ed 1993] – (= ser Deutsche texte des mittelalters 4, 14, 17) – 3v – 1 – (incl bibl ref and ind) – mf#8623 reel 2 – us Wisconsin U Libr [430]
Kleinere mittelhochdeutsche erzaehlungen, fabeln und lehrgedichte – Berlin: Weidmann, 1904-1909 [mf ed 1993] – (= ser Deutsche texte des mittelalters 4, 14, 17) – 3v on 21r – 1 – (incl bibl ref and ind) – mf#8623 reel 2 – us Wisconsin U Libr [430]
Kleinere schriften / Rosenzweig, Franz – Berlin, 1937 – 10mf – 8 – €19.00 – ne Slangenburg [140]
Kleinere schriften zur germanischen heldensage und literatur des mittelalters / Schneider, Hermann – Berlin: De Gruyter, 1962 [mf ed 1993] – (= ser Kleinere schriften zur literatur- und geistesgeschichte) – viii/291p – 1 – (incl bibl ref) – mf#8162 – us Wisconsin U Libr [430]
Kleinere veroeffentlichungen des o-gyallaer astrophysikalischen observatoriums stiftung von konkolys / Konkoly-Alapitvanyu Budapest-Svabhegyi m kir Astrofizikai Obszervatorium – [Ogyalla: s.n.], 1907- (Budapest: Druck von J. Heisler Buch- und Steindruckerei) (irreg) [mf ed 2006] – 1r – 1 – (cont: m kir konkoly-alapitvanyu astrophysikai observatorium kisebb kiadvanyai; ceased with ns 6 (1938)?; in german & hungarian; imprint varies; v12 has title: kleinere veroeffentlichungen des ogyallaer astrophysikalischen observatoriums stiftung v konkoly; v12 has also hungarian title: magyar kir konkoly-alapitvanyu astrophysikai observatorium kisebb kiadvanyai; v14 has title: kleinere veroeffentlichungen des koenig ungarischen astrophysikalischen observatoriums von konkoly's stiftung in o-gyalla; aka: kisebb kiadvanyok) – mf#film mas 37492 – us Harvard [520]
Die kleineren dichtungen heinrichs von muegeln / ed by Stackmann, Karl – Berlin: Akademie-Verlag, 1959- [mf ed 1994] – (= ser Deutsche texte des mittelalters 50-52) – 1 – (incl bibl ref. no more publ?) – mf#8623 reel 13 – us Wisconsin U Libr [810]
Kleinert, Paul *see*
– Abriss der einleitung zum alten testament in tabellenform
– The book of habakkuk
– The book of jonah
– The book of micah
– The book of nahum
– The book of obadiah
– The book of zephaniah
– Das deuteronomium und der deuteronomiker
– Luther im verhaeltniss zur wissenschaft und ihrer lehre
– Die profeten israels in sozialer beziehung
– Zur christlichen kultus- und kulturgeschichte
Ein kleines bild : erzaehlung aus der zeit des deutsch-franzoesischen krieges / Wichert, Ernst – Leipzig: Philipp Reclam jun., [189-?] – 1r – 1 – us Wisconsin U Libr [830]
Kleines deutsches sagenbuch / ed by Peuckert, Will-Erich – Potsdam: Ruetten & Loening, c1939 – us Wisconsin U Libr [390]
Kleines gefluegel *see* Badekuren
Kleines gottsched-denkmal : dem deutschen volke zur mahnung errichtet / Reichel, Eugen – Berlin: Gottsched-Verlag, 1900 – 1r – 1 – (includes selections from gottsched's works) – us Wisconsin U Libr [430]
Kleines klabund-buch : novellen und lieder / Henschke, Alfred (pseud. Klabund) – Leipzig: P Reclam, c1921 – 1r – 1 – us Wisconsin U Libr [800]
Kleines paradies : tiergeschichten / Steguweit, Heinz – 1. aufl. der feldpostausg. Guetersloh: C Bertelsmann 1943 [mf ed 1991] – (= ser Kleine feldpost-reihe) – 1r – 1 – mf#2900p – us Wisconsin U Libr [830]
Das kleingedruckte : allgemeine geschaeftsbedingungen ausgewaehlter verbraucherrechsbereiche / Bultmann. Fritz A – Neuwied, Kriftel, Berlin: Luchterhand 1993 (mf ed 1996) – 3mf – 9 – €38.00 – 3-8267-9680-2 – mf#DHS 9680 – gw Frankfurter [346]
Kleinman, S *see* The perceived influence of participation in intramural sports on purpose, interpersonal relationship, and autonomy
Kleinov, G M *see* Graf s iu vitte
Kleinschmidt, H *see* White liberation
Kleinschmidt, Juergen *see* Untersuchungen zur problematik bei der festsetzung von maximalen immissionskonzentrationen fuer ozon
Kleinschmidt, Karl *see* Friedrich schiller
Kleinschmidt, Lori A *see* Physiological responses to a basketball season
Kleinschnitz, Markus *see* Fazies, diagenese und geochemie des unteren muschelkalks am suedwestrand der querfurter mulde (sachsen-anhalt)
Kleinsorge, John Arnold *see* Beitraege zur geschichte der lehre vom parallelismus der individual- und der gesamtentwicklung
Ein kleinstaatlicher minister des achtzehnten jahrhunderts : leben und wirken friedrich august's, freiherrn von hardenberg / Hardenberg, Friedrich von – Leipzig: Duncker & Humblot 1877 (Pierer'sche Hofbuchdruckerei) [mf ed 1978] – 1r – 1 – (incl selections fr hardenberg's diary & correspondence) – mf#film mas 8161 – us Harvard [920]
Kleinster katechismus zum nothwendigsten unterricht fuer die erste kommunion / Weninger, Francis Xavier – New York: Benziger, 1866, c1865 – 1mf – 9 – 0-8370-6860-6 – mf#1986-0860 – us ATLA [241]

Kleist / Hegeler, Wilhelm – Berlin: Schuster & Loeffler – 1r – 1 – us Wisconsin U Libr [920]
Kleist and hebbel : a comparative study: the novels / Becker, Henrietta K – Chicago: Scott, Foresman, 1904 [mf ed 1991] – 71p – 1 – (incl bibl ref) – mf#7514 – us Wisconsin U Libr [430]
Kleist, Heinrich von *see*
- Amphitryon
- Erzaehlungen
- Die familie ghonorez
- H von kleists werke
- Die hermannsschlacht
- Das kaethchen von heilbronn
- Kleist's hermannsschlacht
- Kleists novellen
- Die marquise von o–
- Penthesilea
- Prinz friedrich von homburg
- Werke
- Der zerbrochne krug

Kleist, Heinrich von et al *see* Phoebus
Kleist, Hugo *see* Bilder aus japan
Kleist's amphitryon / Ruland, Wilhelm – Rostock, 1897 (mf ed 1995) – 1mf – 9 – €24.00 – 3-8267-3130-1 – mf#DHS-AR 3130 – gw Frankfurter [430]
Kleist's hermannsschlacht : ein gedicht auf oesterreich / Kleist, Heinrich von; ed by Mueller-Guttenbrunn, Adam – Wien: Verlag des Kaiserjubilaeums-Stadttheaters, 1898 – 1 – us Wisconsin U Libr [820]
Kleists letzte stunden / Minde-Pouet, Georg – Berlin: Weidmann, 1925 [mf ed 1994] – (= ser Schriften der kleist-gesellschaft 5) – v/63p – 1 – (only v1 publ?) – mf#8707 – us Wisconsin U Libr [920]
Kleists lustspiel "der zerbrochene krug" auf der buehne / Buchtenkirch, Gustav – Heidelberg: C Winter 1914 [mf ed 1991] – (= ser Literatur und theater 2) – 1r – 1 – (incl bibl ref; filmed with: freundesbilder aus goethe's leben / h duntzer) – mf#2867p – us Wisconsin U Libr [790]
Kleists novellen : "michael kohlhaas" und "die heilige caecilie" im wortlaut der ersten fassung / Kleist, Heinrich von – Heidelberg: C Winter, 1926 – 1r – 1 – (incl bibl ref) – us Wisconsin U Libr [830]
Kleists "penthesilea" : oder von der lebendigen form der dichtung / Schulze, Berthold – Leipzig: B G Teubner, 1912 – 1 – (incl bibl ref) – us Wisconsin U Libr [430]
Kleists politisches fragment "zeitgenossen" / Minde-Pouet, Georg – Berlin: Weidmann, 1926 [mf ed 1996] – (= ser Schriften der kleist-gesellschaft 6) – 13p/4pl – 1 – mf#8707 – us Wisconsin U Libr [320]
Kleists und adam muellers freunschaftskrise : zwei ungedruckte briefe adam muellers zur geschichte der zeitschrift "phoebus" / Muehlher, Robert – Wien: Europa-Verlag, c1948 – 1r – 1 – (incl bibl ref) – us Wisconsin U Libr [920]
Klemens brentano : beitraege namentlich zur emmerich-frage / Cardauns, Hermann – Koeln: J P Bachem, 1915 [mf ed 1989] – 130p – 1 – mf#7085 – us Wisconsin U Libr [430]
Klemens von alexandreia und sein hellenisches christentum / Pohlenz, M – Goettingen, 1943 – 2mf – 9 – €5.00 – ne Slangenburg [240]
Klemens von Alexandrien (Clement of Alexandria, Saint) *see*
- Ausgewaehlte schriften, 1. bd (bdk7 2.reihe)
- Ausgewaehlte schriften, 2. bd (bdk8 2.reihe)
- Teppiche, wissenschaftliche darlegungen entsprechend der wahren philosophie (stromateis) (bdk17 2.reihe)
- Teppiche, wissenschaftliche darlegungen entsprechend der wahren philosophie (stromateis) (bdk19 2.reihe)
- Teppiche, wissenschaftliche darlegungen entsprechend der wahren philosophie (stromateis) (bdk20 2.reihe)

Klemens von alexandrien und seine erkenntnisprinzipien / Scherer, Wilhelm – Muenchen: JJ Lentner, 1907 – 1mf – 9 – 0-7905-6780-6 – mf#1988-2780 – us ATLA [120]
Klemens von rom ueber die reise pauli nach spanien : historisch-kritische untersuchung zu klemens von rom, 1 kor 5, 7 / Dubowy, Ernst – Freiburg i B, St Louis MO: Herder, 1914 – (= ser Biblische studien) – 1mf – 9 – 0-7905-1880-5 – (incl bibl ref and index) – mf#1987-1880 – us ATLA [225]
Der klemensroman und seine griechischen quellen / Heintze, W – Leipzig, 1914 – (= ser Tugal 3-40/2) – 3mf – 9 – €7.00 – ne Slangenburg [450]
Die klemensromane : ihre entstehung und ihre tendenzen / Langen, Joseph – Gotha: Friedrich Andreas Perthes, 1890 – 1mf – 9 – 0-7905-6812-8 – (incl bibl ref) – mf#1988-2812 – us ATLA [240]
Die klementinische liturgie aus den constitutiones apostolorum 8 : nebst anhaengen / ed by Lietzmann, Hans – Bonn: A Marcus & E Weber 1910 [mf ed 1992] – (= ser Kleine texte fuer theologische und philologische vorlesungen und uebungen 61) – 1mf – 9 – 0-524-04672-7 – (in greek & latin. int in german) – mf#1990-1299 – us ATLA [240]
Der klementsroman und seine griechischen quellen / Heintze, Werner – Leipzig: J C Hinrichs, 1914 – (= ser Tugal) – 1mf – 9 – 0-7905-3259-X – (incl bibl ref) – mf#1987-3259 – us ATLA [240]
Klemm, Chr Gottlob *see* Das wiener allerley
Klemm, Christian Gottlob *see* Der auf den parnass versetzte gruene hut
Klemm, Frederick Alvin *see* The death problem in the life and works of gerhart hauptmann
Klemm, Guenther *see* Christian morgensterns dichtungen von "ich und du"
Klemm, Ulrich *see* Horst e wittig, personalbibliographie
Klemm, Wilhelm Bernhard *see* Der bertin-altar aus st-omer im kaiser-friedrich-museum zu berlin
Klemmt, Rolf *see* Eine mittelhochdeutsche evangeliensynopse der passion christi
Klemperer, Victor *see* Paul heyse
Klenke, Carsten *see* Ergonomische behandlungskonzepte in der zahnaerztlichen propaedeutik
Klenze, Camillo von *see* The interpretation of italy during the last two centuries
Kleopatra : historischer roman / Ebers, Georg – Stuttgart: Deutsche Verlags-Anstalt, [1893-97?] [mf ed 1993] – (= ser Georg ebers gesammelte werke 26-27) – 2v – 1 – mf#8554 reel 4-5 – us Wisconsin U Libr [830]
Klepikov, S A *see*
- Spisok natsionalizirovannykh predpriiatii rsfsr na 1919 god
- Statisticheskii spravochnik po narodnomu khoziaistvu

Kleppinger, Alison *see* Gender differences in sport orientation and goal orientation
Klett, Ada Martha *see*
- Der streit um "faust 2" seit 1900
- Der streit um "faust 2" seit 1900...1939

Klette, E T *see* Der proces und die acta apollonii
Klette, Emil Theodor *see*
- Die christenkatastrophe unter nero
- Der process und die acta s apollonii

Kletzka, Renate *see* Aesthetik des jahrmarkts
Kleutgen, J *see* Die philosophie der vorzeit
Kleyn, H G *see* Jacobus baradaeus de stichter der syrische monophysietische ker
Kleyn, Hendrik Gerrit *see* Jacobus baradaeus
Kleynhans, Stefan Anton *see* The role of trade usage and the allocation of risk for unauthorized transactions in internet banking
Kliatt – Wellesley MA 1993+ – 1,5,9 – (cont: kliatt young adult paperback book guide) – ISSN: 1065-8602 – mf#9758.02 – us UMI ProQuest [070]
Kliatt paperback book guide – Wellesley MA 1975-77 – 1,5,9 – (cont by: kliatt young adult paperback book guide) – ISSN: 0023-2114 – mf#9758.02 – us UMI ProQuest [070]
Kliatt young adult paperback book guide – Wellesley MA 1978-91 – 1,5,9 – (cont: kliatt paperback book guide; cont by: kliatt) – ISSN: 0199-2376 – mf#9758.02 – us UMI ProQuest [070]
Klibansky, Raymond *see* Continuity of the platonic tradition during the middle ages
Klich – Moscow, Russia. n0[mart'96]-n 1 [mai '96], n 1[3] [1998], n4[6]1998-6[8][apr'98], n7[9] [iiun]-8[10] [iiul'98], spetsvyp 1998, n11[13]-12[14][avg'98], n 15[17] [sen'98], n17[19][noi.1998]-n19[21][1998] – 1 – mf#mf-12248 [reel 5] – us CRL [077]
Klich : organ gorokhovetskogo soveta rabochikh i krest'ianskikh deputatov – Gorokhovets, Russian Federation 1918 [mf ed Norman Ross] – 1r – 1 – mf#nrp-516 – us UMI ProQuest [077]
Klickitat County Historical Society *see* Klickitat heritage
Klickitat heritage / Klickitat County Historical Society – WA. v1 n1-v11 n4=n1-44 [1975 feb-1985 win] – 1r – 1 – mf#930183 – us WHS [978]
Klickmann, Flora *see* How to dress
Kliefoth, Theodor *see*
- Die beichte und absolution
- Christliche eschatologie
- Die confirmation
- Einleitung in die dogmengeschichte
- Liturgische abhandlungen. erster band
- Die urspruengliche gottesdienst-ordnung in den deutschen kirchen lutherischen bekenntnisses

Kliemann, Julian-Matthias *see* Politische und humanistische ideen der medici in der villa poggio a caiano
Klieneberger, H R *see* The christian writers of the inner emigration
Kliente se ervarings van narratiewe terapie met reflekterende groepe / Steyn, Abraham Johannes Christiaan – Uni of Slovo Africa 2000 [mf ed Johannesburg 2000] – 4mf – 9 – (text in afrikaans; abstract in afrikaans and english; incl bibl ref) – mf#mfm14855 – sa Unisa [240]
Klientenzentrierte psychotherapie/beratung und weibliches selbstkonzept / Schumacher, Eva-Maria – (mf ed 1994) – 2mf – 9 – €40.00 – 3-8267-2061-X – mf#DHS 2061 – gw Frankfurter [150]
Klimas, Nancy *see* Journal of chronic fatigue syndrome
Klimasteuerung von karbonatsystemen : fallbeispiele zur biofazies ozeanischer und flachmariner oekosysteme im kaenozoikum / Brachert, Thomas C – (mf ed 1998) – 3mf – 9 – €49.00 – 3-8267-2587-5 – mf#DHS 2587 – gw Frankfurter [550]
Klimatgebiete der sachsengaenger in brandenburg, posen und schlesien / Lezius, Martin – Neudamm 1913 – 1 – gw Mikropress [550]
Klimenko, K I *see* Statisticheskii sbornik cheliabinskoi gubernii za 1920-1923 gg
Klimenko, N K *see* luzhnoe slovo [odessa: 1919]
Klimke, Ansgar *see* Zur bedeutung dopaminerger funktionsstoerungen fuer die psychopathologie und pathophysiologie schizophrener und affektiver psychosen
Klimke, Friedrich *see* Der monismus und seine philosophischen grundlagen
Klimov, Aleksandr Petrovich *see* Sovetskaia potrebitel'skaia kooperatsiia
Kline, John *see* Life and labors of elder john kline, the martyr missionary
Kline, John Jacob et al *see* Jubilee volume, 1517-1917
Kling, Christian Friedrich *see*
- The first epistle of paul to the corinthians
- The second epistle of paul to the corinthians

Klingberg, Fran J *see* "The education of a kansan"
Klingemann, Ernst *see* Kunst und natur
Klingenburg, Georg *see* Das verhaeltnis calvins zu butzer
Klinger, Claudia *see* Tryptophan-biosynthese in aquifex aeolicus, euglena gracilis und saccharomyces cerevisiae
Klinger, Friedrich Maximilian *see*
- Fausts leben, taten und hoellenfahrt
- Faustus, his life, death, and doom
- Otto

Klinger und shakespeare : ein beitrag zur shakespearomanie der sturm- und drangperiode / Jacobowski, Ludwig – Dresden, Leipzig: E Pierson, 1891 – 1 – (incl bibl ref) – us Wisconsin U Libr [410]
Klingman, George Adam *see* Church history for busy people
Klingof, D F *see* Golos truda
Eine klinisch kontrollierte studie zur effektivitaetsbeurteilung des sonic-speed-plaque-remover-instruments bei erwachsenen patienten mit multibandbehandlung / Sliwowska, Beata – (mf ed 1999) – 1mf – 9 – €30.00 – 3-8267-2662-6 – mf#DHS 2662 – gw Frankfurter [617]
Klinisch kontrollierte untersuchungen zur effektivitaet der lokalanaesthetika ultracain#zy 2% suprarenin und ultracain#zy d-s in der kinderzahnheilkunde / Leiers, Christoph – (mf ed 1995) – 1mf – 9 – €30.00 – 3-8267-2216-7 – mf#DHS 2216 – gw Frankfurter [617]
Klinische monatsblaetter fuer augenheilkunde – Stuttgart, Germany 1977-78 – 1,5,9 – mf#10140 – us UMI ProQuest [617]
Klinische untersuchungen zur thermoregulation am beispiel des temperaturmusters der haut und ihre moegliche bedeutung fuer zahnaerztliche diagnostische und therapeutische fragestellungen / Geus, Christoph – (mf ed 1997) – 2mf – 9 – €40.00 – 3-8267-2460-7 – mf#DHS 2460 – gw Frankfurter [617]
Klinische wochenschrift – Dordrecht, Netherlands 1982-83 – 1,5,9 – ISSN: 0023-2173 – mf#13192.02 – us UMI ProQuest [610]
Die klinische-praktische evaluation aerztlicher kompetenz im medizinstudium / Falck-Ytter, Yngve – (mf ed 1997) – 1mf – 9 – €30.00 – 3-8267-2403-8 – mf#DHS 2403 – gw Frankfurter [610]
Klinisch-experimentelle untersuchungen zur wirksamkeit einer systemischen medikation zur prophylaxe der polymorphen lichtdermatose / Sippel, Anke – (mf ed 1996) – 1mf – 9 – €30.00 – 3-8267-2321-X – mf#DHS 2321 – gw Frankfurter [616]
Klinke, Otto *see* E T A hoffmanns leben und werke
Klinkerfues, Wilhelm [Ernst Friedrich Wilhelm] *see* Veroeffentlichungen von der koeniglichen sternwarte zu goettingen
Klinkhamer, G *see*
- Leerzaame zinnebeelden
- Stichtelyke zinnebeelden en bybel-stoffen

Klio : beitraege zur alten geschichte – Leipzig. v1-16. 1901-1920 – 220mf – 8 – mf#H-682 – ne IDC [930]
Klio : monatsschrift fuer franzoesische zeitgeschichte – Leipzig 1795-97 – (= ser Dz. historisch-politische abt) – 3jge on 30mf – 9 – €300.00 – (ab juli 1796 with title: neue klio) – mf#k/n1744 – gw Olms [944]
Klio *see* Beitraege zur alten geschichte
Klippans dagblad *see* Helsingborgs dagblad
Klippans tidning *see* Nordvastra skanes tidningar engelholms tidning
Klippgen, Friedrich *see* Martin luther, saemtliche deutsche geistliche lieder
Klitenik, Sh *see* Verk un shrayber
Klitzman, SH *see* Government in the sunshine act
Kliucharev, N [comp] *see* Statisticheskii ezhegodnik riazanskoi gubernii za 1923-1926 gg
Kliuchevskii, V O *see* Drevnerusskiia zhitiia sviatykh kak istoricheskii istochnik
Kliuchevskii, Vasilii Osipovich *see* Kurs russkoi istorii
Kloeffler, Royce Gerald *see* Telephone communication systems
Kloepper, Albert *see* Der brief an die colosser
Kloepper, Albert et al *see* Theologische studien und skizzen aus ostpreussen
Kloetzer rundschau – Kloetze DE, 1964-66 – 1r – 1 – (publ in magdeburg) – gw Misc Inst [074]
Klondike : mining laws, rules and regulations of the united states and canada applicable to alaska and northwest territory – [Seattle, WA?]: W J Hills & P M Ausherman, c1897 [mf ed 1981] – 2mf – 9 – mf#15227 – cn Canadiana [622]
The klondike : the new gold fields of alaska and the far north-west / Steele, James William – Chicago: Steele Pub Association, 1897 [mf ed 1981] – 1mf – 9 – mf#16206 – cn Canadiana [622]
Klondike and yukon guide : alaska and northwest territory gold fields – Seattle, WA: Seattle-Alaska General Supply Co, [1898] [mf ed 1982] – 1mf – 9 – mf#15370 – cn Canadiana [622]
Klondike gold miners of the alaska-yukon-klondike gold syndicate : capital, $500,000 / Alaska-Yukon-Klondike Gold Syndicate – [Portland, ME?: s.n, 1897?] [mf ed 1981] – 1mf – 9 – mf#15374 – cn Canadiana [622]
Klondike nugget [dawson yt] – Dawson NWT: Metropolitan Print & Binding Co (of Seattle) [semiwkly, wkly] [1898-1900] – 3v filmed with other items – 1 – (cont by: semi-weekly klondike nugget; publ by: allen bros 1899-1900; has occasional suppl & special iss) – us UW Libraries [071]
Klondike pictures / Macdonald, Eustace – [S.l: s.n, 1899?] [mf ed 1981] – 1mf – 9 – 0-665-15544-1 – mf#15544 – cn Canadiana [917]
The klondyke : how the breakman gained his thousands in four months: a complete guide to the gold fields / Clements, James I; ed by James, G Wharton – Los Angeles: B R Baumgardt, 1897 – 2mf – 9 – (with ind) – mf#14697 – cn Canadiana [622]
Klong niras : an analytical and comparative study with other types niras / Sujjapun, Ruenruthai – 1973 – us CRL [950]
Klonierung und charakterisierung der glyzerinaldehyd-3-phosphat-dehydrogenase von onchocera volvulus (leukart, 1893) / Schneider, Erik – (mf ed 2001) – 165p on 2mf – 9 – €40.00 – 3-8267-2762-2 – mf#DHS 2762 – gw Frankfurter [574]
Het klooster te windesheim en zijn invloed / Acquoy, J G R – Utrecht. v1-3. 1875-80 – €40.00 – ne Slangenburg [241]
Kloosterhuis, H *see* Ambon nu!
Klopstock, Friedrich Gottlieb *see*
- Ausgewaehlte dichtungen
- Friedrich gottlieb klopstock
- Klopstocks sammtliche werke in einem bande
- Der messias
- Oden
- Wingolf

Klopstock und schubart : beziehungen im leben und dichten / Bruestle, Wilhelm – Augsburg: Reichel 1917 [mf ed 1990] – 1r – 1 – (incl bibl ref; filmed with: klopstocks leben und werke / karl heinemann) – mf#2770p – us Wisconsin U Libr [430]
Klopstocks entdeckung der nation / Kindermann, Heinz – Danzig: A W Kafemann – Berlin: Junger & Duennhaupt, 1935 – 1r – 1 – (incl bibl ref) – us Wisconsin U Libr [430]
Klopstock's jugendgeschichte und klopstock und der markgraf karl friedrich von baden : bruchstuecke einer klopstockbiographie / Strauss, David Friedrich – Bonn: E Strauss, 1878 – 1r – 1 – (incl bibl ref) – us Wisconsin U Libr [920]
Klopstocks leben und werke / wielands leben und werke / Heinemann, Karl & Boxberger, R – Bielefeld: Velhagen & Klasing, 1899 – 1 – (incl bibl ref) – us Wisconsin U Libr [920]
Klopstocks messias und oden / Kinzel, Karl [comp] – 4. u 5. aufl. Halle/S: Verlag der Buchhandlung des Waisenhauses, 1910 [mf ed 1993] – 144p (ill) – 1 – (incl bibl ref) – mf#8185 – us Wisconsin U Libr [810]

Klopstocks oden : (leipziger periode): ein textkritischer beitrag zur literaturgeschichte seiner zeit / Pawel, Jaro – Wien: C Gerold, 1880 – 1r – 1 – (incl bibl ref) – us Wisconsin U Libr [430]
Klopstocks sammtliche werke in einem bande / Klopstock, Friedrich Gottlieb – Leipzig, Germany. 1840 – 1r – us UF Libraries [430]
Klopstocks sendung / Berger, Arnold Erich – Darmstadt: E Hofmann, 1924 [mf ed 1991] – 39p – 1 – mf#7519 – us Wisconsin U Libr [430]
Kloran : knights of the ku klux klan – [Atlanta, GA.], c1916 – us CRL [366]
Kloran / Ku Klux Klan of Canada – 1st ed. [Toronto, 1925] – 47p on 1 sheet – 1 – Can$75.00 – cn McLaren [366]
Klose, Samuel Benjamin *see*
– Neue litterarische unterhaltungen
– Vermischte beytraege zur philosophie und den schoenen wissenschaften
Klose, W H *see* A study of grillparzer's ahnfrau
Kloss, Cecil Boden *see* In the andamans and nicobars
Kloss, Erich *see* Max kretzer
Kloss, Julius Erich *see* Zwanzig jahre "bayreuth", 1876-1896
Kloss, Richard *see* Saechsisches landesprivatrecht
Kloss, Waldemar *see* Lyra germanica-latina
Das kloster : scheible, 1845-1849 – 188mf – 9 – (a 106-vol coll on central european culture and tradition. included are texts essential for the study of german folk traditions, the reformation, wit and humour and 19th century literature) – mf#C35-14900 – us Primary [430]
Kloster, priestermoench und privatmesse / Nuszbaum, O – Bonn, 1961 – 6mf – 8 – €14.00 – ne Slangenburg [240]
Kloster wendhusen; ursula / Heimburg, W – Stuttgart: Union Deutsche Verlagsgesellschaft, [1890?] – 1r – 1 – us Wisconsin U Libr [430]
Kloster wendhusen; ursula / Heimburg, W – Stuttgart: Union Deutsche Verlagsgesellschaft, [1890?] – 1 – us Wisconsin U Libr [830]
Das klosterland des athos / Schmidtke, Alfred – Leipzig: J C Hinrichs 1903 [mf ed 1990] – 1mf – 9 – 0-7905-6317-7 – mf#1988-2317 – us ATLA [243]
Klostermann, August *see*
– Deuterojesaia
– Ein diplomatischer briefwechsel
– Die hoffnung kuenftiger erloesung aus dem todeszustande bei den frommen des alten testaments
– Korrekturen zur bisherigen erklaerung des roemerbriefes
– Probleme im aposteltexte
– Zur theorie der biblischen weissagung und zur charakteristik des hebraeerbriefs
Klostermann, D et al *see* Die bibelfrage in der gegenwart
Klostermann, Erich *see*
– Agrapha, neue oxyrhynchuslogia
– Analecta zur septuaginta, hexapla und patristik
– Eusebius schrift peri toon topikoon onomatoon
– Evangelien
– Griechische excerpte aus homilien des origenes
– Ignatius von antiochien als christ und theologe – griechische excerpte aus homilien des origenes
– Nachlese zur ueberlieferung der matthaeuserklaerung des origenes
– Origenes, eustathius von antiochien, und gregor von nyssa ueber die hexe von endor
– Origenes homilie 10 ueber den propheten jeremias
– Reste des petrusevangeliums, der petrusapokalypse und des kerygma petri
– Studien zum neuen testament und zur patristik ((tugal5-77
– Ueber des didymus von alexandrien in epistolas canonicas enarratio
– Die ueberlieferung der jeremia-homilien des origenes
– Die ueberlieferung der jeremiahomilien des origenes
– Zur ueberlieferung der matthaeuserklaerung des origenes
Klostermann-Berthold *see* Neue homilien des makarius/symeon
Klostermann's grundstueck : nebst einigen andren begebenheiten, die sich in dessen nachbarschaft zugetragen haben / Rodenberg, Julius – Berlin: Gebrueder Paetel, 1891 – 1 – us Wisconsin U Libr [430]
Klotz, Christian Adolf *see* Deutsche bibliothek der schoenen wissenschaften
Klotz, Christian Adolph et al *see* Neue hallische gelehrte zeitungen
Klotz, Otto *see*
– Earthquakes and the interior of the earth
– Metrology
Klotz, Otto et al *see* Papers on descriptions for deeds read before the association of ontario land surveyors
Klotzsch, J F *see* Die botanischen ergebnisse der reise seiner koenigl hoheit des prinzen waldemar von preussen in den jahren 1845 und 1846
Klotzsch, Johann Friederich et al *see* Sammlung vermischter nachrichten zur saechsischen geschichte
Klub "Sotsializm i narodovlastie" *see* Solntse
Klucke, Walther Gottfried *see*
– Befehl ist befehl
– Liebe mutter
Kluckhohn, Florence Rockwood *see* Variations in value orientations
Kluckhohn, James C *see* Isokinetic evaluation of the knee flexors and extensors of male and female sprinters and distance runners
Kluengelkerl – Dortmund DE, 1976 nov-1987 may [many gaps] – 1mf=2df – 9 – gw Mikrofilm [074]
Kluepfel, Emil Chr *see* Gothaische gelehrte zeitungen
Kluepfel, K *see* Kleine prosaische schriften
Kluepfel, Karl *see*
– N federmanns und h stades reisen in suedamerica
– Nikolaus federmanns und h stades reisen in suedamerica, 1529-1555
– Urkunden zur geschichte des schwaebischen bundes
Klug, Gary A *see*
– [31]p metabolic responses to activity of nonspecifically trained muscle tissue
– The ability of sarcoplasmic reticulum to regulate intracellular calcium following a fatiguing bout of exercise
– A biochemical analysis of the exercise-induced dysfunction of the rat gastrocnemius sarcoplasmic reticulum ca2+-atpase protein
– A comparison of hydrostatic weighing and displacement plethysmography for determining body density of young elite female gymnasts
– Correlation between muscle relaxation and sarcoplasmic reticulum ca2+-atpase during fatigue
Klug, R *see* Jahrbuecher der insektenkunde...
Die kluge bauerntochter : nach dem gleichnamigen maerchen der brueder grimm in fuenf lustigen vorgaengen der jugend uebermittelt / Guembel-Seiling, Max – Leipzig: Breitkopf & Haertel, 1918 – 1r – 1 – us Wisconsin U Libr [820]
Kluge, Hermann *see* Geschichte der deutschen national-literatur
Kluge, Reinhold *see* Lancelot
Kluger, Solomon B *see* Sheelot u-teshuvot tuv ta'am va-da'at mahadura kama
Klugh, A *see* La terre bonne
Kluh, John M *see* The manifestation of ages
Kluizenaars in limburg / Welters, A – Heerlen, 1950 – €11.00 – ne Slangenburg [241]
Klunzinger, C B *see* Die korallthiere des rothen meeres
Klutschak, H W *see* Als eskimo unter den eskimos
Km : the journal of kemetic spiritual study and practice / Maat Djed Nefer Enterprises – 1992 jul-4th qtr – 1r – 1 – mf#5306590 – us WHS [290]
Knaacke, J K F *see* Wider hans worst
Knaake, Andreas *see* Restitution rechter und gesunder christlicher lehre
Knaake, J K F *see* Sendbrief an papst leo 10; von der freiheit eines christenmenschen; warum des papsts und seiner juenger buecher von d. martino luther verbrannt seien
Knaake, Joachim Karl Friedrich *see* Bibliothek knaake
Knabenbauer, Joseph *see*
– Commentarius in danielem prophetam, lamentationes et baruch
– Commentarius in duos libros machabaeorum
– Commentarius in ecclesiasticum
– Commentarius in ezechielem prophetam
– Commentarius in ieremiam prophetam
– Commentarius in librum iob
– Commentarius in prophetas minores
– Commentarius in proverbia
– Commentarius in quatuor s evangelia domini n iesu christi. 2
– Commentarius in quatuor s evangelia domini n iesu christi. 3
– Commentarius in quatuor s evangelia domini n iesu christi. 4
– Commentarius in quatuor s evangelia domini n iesu christi. 1
– Commentarius in s pauli apostoli epistolas. 4
– Commentarius in s pauli apostoli epistolas. 5
– Erklaerung des propheten isaias
– Das zeugniss des menschengeschlechtes fuer die unsterblichkeit der seele
Knackpunkt : jugendliche in der bundesrepublik / ed by Werkstaetten Kassel und Osnabrueck; Werkkreis Literatur der Arbeitswelt – Frankfurt/M: Fischer Taschenbuch Verlag, 1981 – 1r – 1 – us Wisconsin U Libr [430]
Knaepper, Matthias *see* Eg-binnenmarkt und entwicklung von politikfeldern
Knak, G *see* Johann jaenicke
Knanishu, Joseph *see* About persia and its people
Knap, Jan Jacob *see* De heidelbergsche catechismus
Knapp, Bradford *see* Beef cattle improvement in florida
Knapp, Clark D *see* A treatise on the laws of the state of new york relating to the poor, insane, idiots and habitual drunkards.
Knapp, Ferdinand M *see* Ideology, technology and the historical avant garde
Knapp, Jacob *see* Autobiography of elder jacob knapp
Knapp, Martin *see* Albert knapp als dichter und schriftsteller
Knapp, Martin Wells *see*
– Pentecostal aggressiveness, or, why i conducted the meetings of the chesapeake holiness union at bowens, maryland
– Revival kindlings
– Revival tornadoes
Knapp news – Knapp WI. 1902 oct 2-1906 dec 6 – 1r – 1 – mf#927107 – us WHS [071]
Knapp, Samuel Lorenzo *see* American cultural history, 1607-1829
Knappert, Jan *see*
– The religion of israel
– Traditional swahili poetry
Knappert, Laurentius *see*
– De gereformeerde kerk aan den arbeid 1657-1672
– Geschiedenis der nederlandsche hervormde kerk gedurende de 16e en 17e eeuw
– Geschiedenis der nederlandsche hervormde kerk gedurende de 18e en 19e eeuw
– De leidsche vertaling van het oude testament
– De opkomst van het protestantisme in eene noord-nederlandsche stad
Das knappschaftswesen im ruhrkohlenbezirk bis zum allgemeinen preussischen berggesetz vom 24.6.1865 / Buelow, Wilhelm – 1 – gw Mikropress [330]
Knapwell 1598-1950 – (= ser Cambridgeshire parish register transcript) – 3mf – 9 – £3.75 – uk CambsFHS [929]
Knauff, Christopher Wilkinson *see* Doctor tucker, priest-musician
Knaup, Werner *see* Algebraische strukturen in einfachen warteschlangen-netzen
Knauss, Heinz *see* Studien zum stil von grimmelshausens simplicissimus
Knauss, James Owen *see* Farmers' alliance in florida
Knauth, Paul *see*
– Goethes sprache und stil im alter
– Von goethes sprache und stil im alter
Knebel, Johannes *see*
– Die chronik des klosters kaisheim
Knebel, Karl Ludwig von *see* K l von knebel's literarischer nachlass und briefwechsel
Knecht, Friedrich Justus *see*
– Buku duku re masoko anoyera
– Child's bible history
– Testamente
Der knecht gottes andreas nyland : roman / Wiechert, Ernst Emil – Berlin: G Grote 1926 [mf ed 1991] – (= ser Grote'sche sammlung von werken zeitgenoessischer schriftsteller 167) – 1r – 1 – (filmed with: das einfache leben) – mf#3043p – us Wisconsin U Libr [830]
Der knecht gottes bei deuterojesaja / Sellin, Ernst – Leipzig: A Deichert, 1901 – 1mf – 9 – 0-524-06745-7 – mf#1992-0948 – us ATLA [221]
Der knecht gottes in isaias kap 40-55 / Feldmann, Franz – Freiburg i B, St Louis MO: Herder, 1907 – 1mf – 9 – 0-7905-3371-5 – (incl bibl ref) – mf#1987-3371 – us ATLA [221]
Der knecht jahves des deuterojesaia / Giesebrecht, Friedrich – Koenigsburg i. Pr: Thomas & Oppermann, 1902 – 1mf – 9 – 0-8370-3275-X – (incl bibl ref) – mf#1985-1275 – us ATLA [221]
Der knecht jahve's im jesajabuche / Orelli, Conrad von – Berlin: Edwin Runge, 1908 – 1mf – 9 – 0-7905-0510-X – (incl bibl ref) – mf#1987-0510 – us ATLA [221]
Der knecht jehova's im deuterojesaia : eine exegetisch-kritische studie / Oehler, Victor Friedrich – Stuttgart: Chr Belser, 1865 – 1mf – 9 – 0-7905-3086-4 – mf#1987-3086 – us ATLA [221]
Knecht, John *see* The effects of exercise on the strength of the low back
Kneeland, Stillman Foster *see*
– Law, lawyers and lambs
– A treatise on the law of attachments in civil cases.
Kneer, August *see* Die entstehung der konziliaren theorie
Kneile, Karl *see* Die formenlehre bei john lyly
Kneller, George Frederick *see* The educational philosophy of national socialism
[Knesebeck, F J von dem] *see* Dreistaendige sinnbilder zu fruchtbringendem nuetze
Knesebeck, Rosemarie von dem *see* Rudolf pechel und die "deutsche rundschau" 1946-1961
Kneucker, J J *see* Vorlesungen ueber biblische theologie und messianische weissagungen des alten testaments
Kneucker, Johann Jacob *see* Das buch baruch
Knevels, Wilhelm *see* Expressionismus und religion
Knewstub, J *see*
– An aunsweare unto certaine assertions
– A confutation of monstrous and horrible heresies
Kniazev, Vasilii *see* O chem pel kolokol
Knibbs, Henry Herbert *see* Lost farm camp
Der knick im ohr : skizzen / Presber, Rudolf – Berlin: Concordia Deutsche Verlags-Anstalt [190-?] [mf ed 1996] – 1r – 1 – (filmed with: wurzellocker / wilhelm von polenz) – mf#3986p – us Wisconsin U Libr [880]
Knie, Johann *see* Alphabetisch-statistisch-topographische uebersicht aller doerfer, flecken, staedte und andern orte der koenigl preuss provinz schlesien
Kniefall und fall des bischofs wilh. em. freiherrn von ketteler / Reinkens, Joseph Hubert – Bonn: P Reusser, 1877 – 1mf – 9 – 0-524-03022-7 – mf#1990-4544 – us ATLA [240]
Knieschke *see* Das heilige land im lichte der neuesten ausgrabungen und funde
The knife of the higher critic / the judgment of the lord / the burial of an ass / Blake, Samuel Hume – Toronto: L S Haynes Press, [1909?] – 1mf – 9 – 0-665-71449-1 – (incl bibl ref) – mf#71449 – cn Canadiana [220]
Knife world – 1980 jul [v6 n7]-1982 jun, 1982 jul-1984 sep, 1984 oct-1986 dec, 1987-88 – 4r – 1 – mf#639172 – us WHS [071]
Kniga bytiia moego : dnevniki i avtobiograficheskiia zapiski / Uspenskii, P – 1894-1901. v1-7 – 75mf – 8 – mf#R-1812 – ne IDC [243]
Kniga i revolyutziya – St. Petersburg. v. 1-3 no. 1, 3, 4. 1920-23 – 1 – us NY Public [335]
Kniga isusa navina... – Prague, 1518 – 2mf – 9 – mf#RHB-8 – ne IDC [460]
[Kniga o postnichestve] / Vasilii Velikii – Ostrog, 1594 – 22mf – 9 – mf#RHB-32 – ne IDC [460]
Kniga plach : opyt izsledovaniia isagogiko-ekzegeticheskago / Blagoveshchenskii, M – Kiev, 1899 – 8mf – 8 – mf#R-4097 – ne IDC [243]
Kniga proroka daniila v drevneslavjanskom perevode / Evseev, I – Moskva, 1905 – 6mf – 8 – mf#R-172 – ne IDC [243]
Kniga proroka isaii v drevneslavianskom perevode / Evseev, I – 1897 – 6mf – 8 – mf#R-173 – ne IDC [243]
Kniga ruf'... : ezhe byla baba ioseova okta tsaria davydova, z nea ze izydovasha vsi tsari iiudiny, zupolne vylozhena doktorom fransiskom skorinoiu, iz slavnago grada polotska – Prague, 1519 – 1mf – 9 – mf#RHB-10 – ne IDC [460]
Kniga v rossii v pervoi chetverti 18 veka / Luppov, S P – 1973 – 4mf – 9 – mf#R-11172 – ne IDC [947]
Knigge, Adolf, Freiherr von *see* Journal aus urfstaedt
The knight – Masterton, NZ. 1982 – 1r – mf#48.18 – nz Nat Libr [079]
Knight, Adele Ferguson *see* Mademoiselle celeste
Knight, Albion Williamson *see* Lending a hand in cuba
Knight, Alfred Ernest *see*
– Amentet
– India
Knight and Son, T *see* Suggestions for house decoration by t knight and son
Knight, Archibald Patterson *see*
– Introductory physiology and hygiene
– The ontario public school hygiene
Knight, Charles *see* Knight's pictorial gallery of arts
Knight, Charles William Robert *see* Knight in africa
Knight, David et al *see* Nineteenth century books on evolution and creation collection
Knight, E F *see* Madagascar in war time
Knight, Edward Frederick *see*
– Over-sea britain
– Rhodesia of to-day
Knight, Fred Key *see* How to organize and conduct an evening class in citrus culture
Knight, Frederick *see* Knight's vases and ornaments
Knight, George H *see* Patent-office manual, including the law and practice of cases in the united states patent office and the courts holding a revisory relation thereto
Knight, H J C *see* The temptation of our lord considered as related to the ministry and as a revelation of his person
Knight, Helen Cross *see*
– Lady huntington and her friends
– Life of james montgomery
Knight, Henry Gally *see*
– An architectural tour of normandy
– The ecclesiastical architecture of italy
– The normans in sicily
– Saracenic and norman remains to illustrate the normans in sicily
Knight, Henry Joseph Corbett *see* The epistles of paul the apostle to the colossians and to philemon
Knight in africa / Knight, Charles William Robert – London, England. 1937 – 1r – us UF Libraries [960]

Knight, J F *see* Criticism and faith
Knight, John Collyer *see*
– The incredibilities of part 2 of the bishop of natal's work upon the pentateuch
– The pentateuchal narative vindicated
Knight, John George *see* Narrative of the visit of his royal highness the duke of edinburgh to the colony of victoria, australia
Knight, John Thomas Philip *see* Incidentally
Knight letter – 1968 feb-1973 feb – 1r – 1 – mf#1061529 – us WHS [071]
Knight of st george / Catholic Knights of St George [US] – v71 n2-v76 n5 [1977 mar/apr-1981 dec] – 1r – 1 – mf#652306 – us WHS [366]
Knight, Richard *see* History of the general or six principle baptists in europe and america
Knight templar / Knights Templar [Masonic Order] – 1974 mar-1978 aug, 1978 sep-1983 dec, 1984 jan-1987 dec – 3r – 1 – mf#690644 – us WHS [366]
Knight, Thomas Frederick *see* The american war
Knight, W B *see*
– Letter on infant baptism
– Parish priest
Knight watch / Operation Able Sentry et al – 1994 dec 21-1995 may 20 – 1r – 1 – (cont: able sentry 2; cont by: bayonet news) – us WHS [071]
Knight, William *see*
– The arch of titus and the spoils of the temple
– India's plea for men
– Lectures on some of the prophecies concerning the rise and character of the power commonly called antichrist
Knight, William Angus *see*
– Aspects of theism
– The christian ethic
– Colloquia peripatetica
– Essays in philosophy
– Hume
– Inter amicos
– Memorials of coleorton
– Studies in philosophy and literature
– Varia
Knight, William Henry *see* Western australia
Knight-Bruce, George Wyndham Hamilton *see* Gold and the gospel in mashonaland, 1888
Knightley, Thomas Edward *see* Stable architecture
Knighton, Henry *see* Chronicon henrici knighton [rs92]
Knights and Ladies of Honor *see* Proceedings of the supreme lodge, knights and ladies of honor,...regular session
Knights, David B *see* In support of the raison d'etre
[Knights landing-] knights landing news – CA. 1861-1862 – 1r – 1 – $60.00 – mf#C03258 – us Library Micro [071]
Knights of Columbus Club *see* Kateri
Knights of Honor *see* Proceedings of the...annual session of the supreme lodge, knights of honor
Knights of Jericho. Alpha Lodge, No 1 (Brockville, Ont) *see* Constitution, by-laws and rules of order of alpha lodge
Knights of Labor *see*
– Annual convention of the new york protective associations
– Annual report of district assembly no 30, k of l
– Decisions of the general master workman
– Official journal of the new york protective associations affiliated with d a 49
– Proceedings of the general assembly of the knights of labor of america
– Quarterly meeting of d a 30
– Quarterly report of district assembly no 30, k of l
– Record of proceedings of district assembly no 30, of massachusetts
– Record of proceedings of the general assembly of the knights of labor
– Report of the annual session of district assembly no 30, k of l
– Reunion, picnic and games of the new york protective associations under the auspices of district 49 assembly, k of l
– Souvenir journal of the...annual session of the general assembly of the knights of labor
– Union pacific employes' magazine
Knights of labor – v1 n19-v2 n13,21, v4 n4, v6 n198 [1886 aug 14-1887 may 14, oct 8, 1888 aug 4, 1889 oct 19] – 1r – 1 – mf#1061534 – us WHS [071]
Knights of Labor et al *see* Labor enquirer
Knights of Peter Claver *see* Claverite
Knights of Pythias *see*
– Proceedings of the grand lodge of wisconsin
– Pythian international
Knights of Pythias. Grand Lodge of Quebec *see* Constitution and statutes of the grand lodge, knights of pythias of the province of quebec
Knights of Pythias of North America, South America, Europe, Asia, Africa and Australia. Grand Lodge Colored Knights of Pythias of Texas *see* Report of the grand keeper of records & seal at the...annual session of the grand lodge colored knights of pythias
Knights of St Crispin *see* Proceedings of...annual meeting of the international grand lodge of the order of knights of st crispin
The knights of the cross / Sienkiewicz, Henryk – Authorized and unabr. trans. from the Polish by Jeremiah Curtin. Boston: Little, Brown, 1900. 2v – 1 – us Wisconsin U Libr [830]
Knights of the Ku Klux Klan *see* Minutes of the imperial kloncilium
Knights of the labarum : being studies in the lives of judson, duff, mackenzie and mackay / Beach, Harlan Page – Chicago: Student Volunteer Movt for Foreign Missions, 1896 [mf ed 1986] – 1mf – 9 – 0-8370-6323-X – (incl bibl) – mf#1986-0323 – us ATLA [240]
Knights of the Maccabees of the World *see*
– Constitutions and laws of the knights of the maccabees of the world
– Revised laws of the knights of the maccabees of the world
Knight's penny magazine – La Salle IL 1832-46 – 1 – mf#3903 – us UMI ProQuest [073]
Knight's pictorial gallery of arts / Knight, Charles – London [1858-60] – (= ser 19th c art & architecture) – 20mf – 9 – mf#4.1.331 – uk Chadwyck [700]
Knight's quarterly magazine – La Salle IL 1823-24 – 1 – mf#4269 – us UMI ProQuest [420]
Knights Templar [Masonic Order] *see* Knight templar
Knight's vases and ornaments : designed for the use of architects, silversmiths, jewellers, modellers, chasers, die sinkers, founders, carvers, and all ornamental manufacturers / Knight, Frederick – London: F Knight, 1833 – (= ser 19th c art & architecture) – 1mf – 9 – mf#4.1.16 – uk Chadwyck [740]
Knights who fought the dragon / Leslie, Edwin – Toronto: W Briggs, 1906 [mf ed 1997] – 4mf – 9 – 0-665-80981-6 – mf#80981 – cn Canadiana [830]
Den' knigi [chita: 1919] : odnodn gaz, izd zabaik uchenich kooperativom "pervyi shag" im l'va nikolaevicha tolstogo, chit otd-niem petrogr roditel kruzhka, chit org boi i gerl-skautov / ed by Luks, K IA – Chita [Zabaik obl]: [s n] 1919 – (= ser Asn 1-3) – 1v item 119, on reel n27 – 1 – mf#asn-1 119 – ne IDC [077]
Knigi grazhdanskoi pechati 18 v katalog knig, khraniashchikhsia v gosudarstvennoi publichnoi biblioteke ukrainskoi ssr / Petrov, S O – Kiev, 1956 – 8mf – 8 – mf#R-7765 – ne IDC [947]
Den' knigi [krasnoiarsk: 1919] : odnodn gaz, izd garnizon kul't -prosvet k[omi]s i pravl "doma iunoshestva" / ed by Krutovskii, V M – Krasnoiarsk [Enis gub]: [s n] 1919 – (= ser Asn 1-3) – 1v item 118, on reel n27 – 1 – mf#asn-1 118 – ne IDC [077]
Knigi sudei... : ezhe ot evrei nazyvaiutsia shoftim, zupolne vylozheny na ruskii iazyk doktorom fransiskom skorinoiu iz slavnago grada polotska, bogu ko chti liudem pospolitym k nautse... – Prague, 1519 – 2mf – 9 – mf#RHB-9 – ne IDC [460]
Knigi tsarstv... : zupolne vylozheny na ruskii iazyk doktorom fransiskom skorininym synom s polotska – Prague, 1518. v1-4 – 9mf – 9 – mf#RHB-7 – ne IDC [460]
Knigokhranilishche chudova monastyria / Petrov, N P – (= ser Pamiatniki drevnei pismennosti) – 2 – 9 – (pamiatniki drevnei pismennosti, 1879 v5 n4 p141-199) – mf#R-11193 – ne IDC [243]
Knill, Richard *see*
– Happy death-bed
– The missionary's wife
Knipfer, Julius *see* Paul gerhardt
Knippel, Richard *see* Schillers verhaeltnis zur idylle
Knitgoods workers voice / International Ladies' Garment Workers' Union – 1980 jun-1988 april – 1r – 1 – (cont: voice of the knitgoods workers; cont by: knitgoods workers' union voice) – mf#1884846 – us WHS [331]
Der knittelvers des jungen goethe : eine metrische und melodische untersuchung / Feise, Ernst – Leipzig: Roeder & Schunke 1909 [mf ed 1990] – 1r – 1 – (incl bibl ref. filmed with: goethe und schopenhauer / heinrich doll) – mf#7385 – us Wisconsin U Libr [430]
Knitting times – Summit NJ 1960-96 – 1,5,9 – (cont by: american sportswear & knitting times) – ISSN: 0023-2300 – mf#3380.01 – us UMI ProQuest [680]
Knivstabygden – Uppsala, Sweden. 2003- – 1 – sw Kungliga [078]
Knizhitsa v desiati otdelakh – Ostrog, 1598 – 6mf – 9 – mf#RHB-36 – ne IDC [460]
Knizhnaia letopis' : gosudarstvennyi bibiliograficheskii ukazatel' rf – Moscow: Knizhnaia palata 1965- – 9,1 – ISSN: 0869-5962 – mf#m0010 – us East View [460]
Knizhnaia letopise – Oldwick. 1906-1999 (1) 1971-1999 (5) 1976-1999 (9) – 2058mf – 9 – mf#1762 – ne IDC [077]
Knizhnoe obozrenie – Moscow, Russia. n1-52. 1987-1992 – 8r – (gaps) – us UF Libraries [077]
Knizhnoe obozrenie – Moscow: Ministerstvo pechati i informatsii Rossii, 1966- – 9 – us East View [077]
Knizhnoe obozrenie – Moscow, Russia 1966-89 [mf ed Norman Ross] – 17r – 1 – mf#nrp-1074 – us UMI ProQuest [077]
Knizna revue – Bratislava, Slovak Republic 1993-2001 [mf ed Norman Ross] – 1r per y – 1 – (backfile through 1998 $85r) – mf#nrp-357 – us UMI ProQuest [077]
Knjige matice srpske – Collected works, Serbian history – 1 – us Wisconsin U Libr [949]
Knjizevne novine – Beograd. jan 1969-dec 1978 – 1 – 69.00 – us L of C Photodup [073]
Knjizevni jadran – Split. Yugoslavia. -m. Jan 1952-Aug 1953. (1 reel) – 1 – uk British Libr Newspaper [949]
Knjizevnik – Zagreb: [s n] 1928- [mf ed Bloomington IN: Indiana Uni Lib, Preservation Dept 1989] – 3r – 1 – us Indiana Preservation [073]
Knob creek baptist church : church records – Columbia, TN. 1004p. dec 1881-oct 1986 – 1 – $45.18 – (lacking: jan-aug 1958) – us Southern Baptist [242]
Knob creek baptist church : membership rolls and church minutes – Seymour, TN. jan 1911-may 1995 – 1 – $42.57 – mf#6964 – us Southern Baptist [242]
Knobel, Fred H *see* Development of agricultural cooperatives, cambodia
Knobloch, Mary Jo *see* The effects of a farm youth hearing study on parental hearing protection knowledge, attitudes and behavior
Knoblock, Edward *see* Milestones
Knock at a venture / Phillpotts, Eden – Toronto: Morang, 1905 – 4mf – 9 – 0-665-97312-8 – (incl publ's list) – mf#97312 – cn Canadiana [830]
Knockabout club in the tropics / Stephens, Charles Asbury – Boston, MA. 1884 – 1r – us UF Libraries [790]
Knodt, Emil *see* Die bedeutung calvins und des calvinismus fuer die protestantische welt
Knoedler library of art exhibition catalogues – 2ser – 9 – (foll subject groups available: twentieth century north american art [291 catalogues on 306mf]. major national & international expositions & world's fairs [61 catalogues on 171mf]) – uk Chadwyck [700]
Knoefel (Knefel), Johannes *see* Dulcissimae quaedam cantiones numero 32...
Knoes, Anders Erik *see* Kurze darstellung der vornehmsten eigenthuemlichkeiten der schwedischen kirchenverfassung
Knoetel, August *see* System der aegyptischen chronologie
Knoetze, Louis *see*
– Die uitbreiding van kartografiese kennis van die kaapkolonie, 1752-1842
Knoll, Albert *see* Institutiones theologiae dogmaticae generalis seu fundamentalis
Knoop, Kaete *see* Die erzaehlungen eduard von keyserlings
Knopf, R *see* Der erste clemensbrief
Knopf, Rudolf *see*
– Der erste clemensbrief
– Das nachapostolische zeitaltar
– Probleme der paulusforschung
– Der text des neuen testaments
– Die zukunftshoffnungen des urchristentums
Knopf, Rudolph *see* Paulus
Knopf, Sigard Adolphus *see* La tuberculose, maladie du peuple
Knorre, K *see* Der ort des polarsterns
Knortz, Karl *see*
– Deutsches und amerikanisches
– Epigramme
– Goethe und die wertherzeit
– Humoristische gedichte
– Neue gedichte
The knossos notebooks, 1900-29 : from the ashmolean museum, oxford – 3r – 1 – mf#3328 – uk Microform Academic [930]
Knote, Walter *see* Hermann lingg und seine lyrische dichtung
Knots untied : being plain statements on disputed points in religion from the standpoint of an evangelical churchman / Ryle, John Charles – New and improved ed. London: National Protestant Church Union, 1896 – 2mf – 9 – 0-7905-9100-6 – mf#1989-2325 – us ATLA [240]
Knots untied 1090 or ways and by-ways in the hidden life of american detectives / McWatters, George S – Hartford/Chicago: J B Burr & Co, 1874 – 7mf – 9 – $10.50 – mf#LLMC 92-178 – us LLMC [340]
Knott, John Olin *see* Seekers after soul
Knott, R R *see* The new aid to memory
Knott, Raymond Stanley *see* Persoonlikheidsontwikkeling by dames tydens akademiese en militere opleiding
Knotting – (= ser Bedfordshire parish register series) – 1mf – 9 – £3.00 – uk BedsFHS [929]
Know florida – Tallahassee, FL. 1935? – 1r – us UF Libraries [630]
Know news – v5 n1 [1974 jan/feb], lacks title, v7 n2-v13 n1 [1976 jun-1982 spr] – 1r – 1 – mf#29508 – us WHS [071]
Know the truth : a critique on the hamiltonian theory of limitation: including some strictures upon the theories of rev. henry l. mansel and mr. herbert spencer / Jones, Jesse Henry – New York: Hurd and Houghton, 1865 – 1mf – 9 – 0-8370-3797-2 – mf#1985-1797 – us ATLA [210]
Know thyself = Conosci te stesso / Varisco, Bernardino – London: G Allen and Unwin, 1915 – (= ser Library of Philosophy) – 1mf – 9 – 0-7905-9727-6 – (incl bibl ref. in english) – mf#1989-1452 – us ATLA [120]
Know your world – Stamford CT 1967-74 – 1 – (cont by: know your world extra) – ISSN: 0023-2483 – mf#2434.01 – us UMI ProQuest [373]
Know your world extra – Stamford CT 1978-83 – 1,5,9 – (cont: know your world) – ISSN: 0163-4844 – mf#2434.01 – us UMI ProQuest [373]
Knowing the scriptures : rules and methods of bible study / Pierson, Arthur Tappan – New York: Hodder & Stoughton: George H Doran, c1910 – 2mf – 9 – 0-7905-1785-X – mf#1987-1785 – us ATLA [220]
Knowle journal. (knowle journal and solihull advertiser) – England. 8 Apr 1893-9 Feb 1901.-w. 7 reels – 1 – uk British Libr Newspaper [072]
Knowledge [1881] : a monthly record of science – Killen TX 1881-1917 – 1 – mf#2911 – us UMI ProQuest [500]
Knowledge [1983] – Newbury Park CA 1983-95 – 1,5,9 – (cont by: science communication) – ISSN: 0164-0259 – mf#14010.01 – us UMI ProQuest [000]
Knowledge about menopause and attitudes toward menopause among the palestinian women living in the west bank and gaza strip / Haj-Ahmad, Jumana – 1996 – 1mf – 9 – $4.00 – mf#HE 605 – us Kinesology [612]
Knowledge and life = Erkennen und leben / Eucken, Rudolf – London: Williams & Norgate; New York: GP Putman [sic], 1913 – (= ser Crown Theological Library) – 1mf – 9 – 0-7905-7568-X – (in english) – mf#1989-0793 – us ATLA [240]
Knowledge and policy – Piscataway NJ 1992-96 – 1,5,9 – (cont: knowledge in society) – ISSN: 1053-8798 – mf#16504.02 – us UMI ProQuest [301]
Knowledge and reality : a criticism of mr. f.h. bradley's principles of logic / Bosanquet, Bernard – London: K Paul, Trench, 1885 – 1mf – 9 – 0-7905-9141-3 – mf#1989-2366 – us ATLA [160]
Knowledge in society – Piscataway NJ 1988-91 – 1,5,9 – (cont by: knowledge & policy) – ISSN: 0897-1986 – mf#16504.02 – us UMI ProQuest [301]
The knowledge of acute care nurses regarding acute coronary syndromes / Price, Carol G – Uni of South Africa 2000 [mf ed Johannesburg 2000] – 3mf – 9 – (incl bibl ref) – mf#mfm15070 – sa Unisa [610]
The knowledge of critical care nurses : regarding intra-aortic balloonpump counterpulsation therapy / Oosthuizen, Phillippus Johannes – Uni of South Africa 2000 [mf ed Johannesburg 2000] – 3mf – 9 – (incl bibl ref; abstract in english & afrikaans) – mf#mfm15005 – sa Unisa [610]
The knowledge of god : objectively considered / Breckinridge, Robert Jefferson – New York: R Carter; Louisville: A Davidson, 1859, c1857 – 2mf – 9 – 0-7905-3610-2 – mf#1989-0103 – us ATLA [210]
The knowledge of god : subjectively considered / Breckinridge, Robert Jefferson – New York: R Carter, 1859 – 1mf – 9 – 0-7905-8639-8 – mf#1989-1864 – us ATLA [240]
The knowledge of god and its historical development / Gwatkin, Henry Melvill – Edinburgh: T & T Clark, 1906 – (= ser Gifford lectures) – 2mf – 9 – 0-7905-7826-3 – mf#1989-1051 – us ATLA [210]
Knowledge of jesus, the most excellent of the sciences / Carson, Alexander – 2nd ed. New York: Fletcher, 1851 – 1r – 1 – 0-8370-0516-7 – mf#1984-B461 – us ATLA [240]
The knowledge of mary / Concilio, Januarius de – New York: Catholic Publ Society, 1878, c1877 – 1mf – 9 – 0-8370-8333-8 – mf#1986-2333 – us ATLA [240]
Knowledge of the circle : a publication of the boston indian council / Boston Indian Council – v8 n2-8 [1985 jul-1986 jan], 1986 jul – 1r – 1 – (cont: circle [boston, ma]) – mf#1840616 – us WHS [305]
Knowledge quest – Chicago IL 1998+ – 1,5,9 – (cont: school library media quarterly) – ISSN: 1094-9046 – mf#8062.02 – us UMI ProQuest [020]
Knowledge, technology, and policy – Piscataway NJ 1999+ – 1,5,9 – (cont: knowledge & policy) – mf#16504.02 – us UMI ProQuest [301]

Knowledge-based systems – Oxford, England 1987+ – 1,5,9 – ISSN: 0950-7051 – mf#17248 – us UMI ProQuest [000]

Knowles, Archibald Campbell *see* The belief and worship of the anglican church

Knowles, James Davis *see*
- Memoir of ann h judson
- Memoir of roger williams

Knowles, James Purdie *see* Samuel a. purdie

Knowles, James Sheridan *see* Das weib

Knowles, John F R S *see* The life and writings of henry fuseli

Knowles, Joseph *see* Alone in the wilderness

Knowles, Robert Edward *see*
- St cuthbert's
- The singer of the kootenay
- The web of time

Knowling, R J *see* Our lord's virgin birth and the criticism of to-day

Knowling, Richard John *see*
- The epistle of st james: with an introduction and notes
- Literary criticism and the new testament
- Messianic interpretation
- The testimony of st paul to christ viewed in some of its aspects
- The witness of the epistles

Know-nothingism in baltimore, 1854-1860 / Tuska, Benjamin – New York [1925] (mf ed 19–) – 36p – (repr: catholic historical review, jul 1925) – mf#ZH-IAG pv316 n7 – us NY Public [978]

Knox advocate – Knox, ND: Luther H Bratton, nov 11 1899; -v48 n13 dec 13 1946 (wkly) [mf ed with gaps] – 1 – (vol no sequence is irreg; absorbed in pt: pleasant lake news; cont by: leeds herald and the knox advocate) – mf#05620; 03971-03984; 05714 – us North Dakota [071]

Knox, Charles Eugene *see* A year with st paul

Knox church (city hall square, ottawa), anthem book : containing the words of anthems and hymns sung by the choir, for congregational use – [Ottawa?: s.n.], 1887 [mf ed 1993] – 1mf – 9 – 0-665-92809-2 – mf#92809 – cn Canadiana [780]

Knox Co. Centerburg *see* Gazette

Knox Co. Clinton *see* Ohio register

Knox Co. Danville *see* Times series

Knox Co. Fredericktown *see*
- Citizen
- Knox county citizen
- Times series

Knox Co. Gambier *see* Observer

Knox Co. Mount Vernon *see*
- Democratic banner
- Democratic banner series
- Norton's daily true whig
- Ohio register
- Republican
- Times series
- True whig

Knox college : the moderator of the general assembly of the presbyterian church in canada requests the honor of the presence of... – [Toronto?: s.n. 1875?] [mf ed 1983] – 1mf – 9 – 0-665-39749-6 – mf#39749 – cn Canadiana [090]

Knox college monthly – [Toronto?]: Metaphysical and Literary Society, [1883-1887] – 9 – (cont by: the knox college monthly and presbyterian magazine) – mf#P05060 – cn Canadiana [378]

Knox college monthly and presbyterian magazine – Toronto: Alumni Association and the Metaphysical and Literary Society of Knox College, [1887-1896] – 9 – (cont: knox college monthly) – mf#P05059 – cn Canadiana [378]

Knox county atlas, 1896 – 1r – 1 – mf#B27425 – us Ohio Hist [978]

Knox county citizen / Knox Co. Fredericktown – jan 1941-42,(1946-70),85-dec 1993 [wkly] – 15r – 1 – mf#B34625-34639 – us Ohio Hist [071]

Knox county citizen / Knox Co. Fredericktown – jan 1971-dec 1984 [wkly] – 6r – 1 – mf#B9751-9756 – us Ohio Hist [071]

Knox county citizen / Knox Co. Fredericktown – mar-oct 1950 [wkly] – 1r – 1 – mf#B10345 – us Ohio Hist [071]

Knox county, illinois genealogical society quarterly – 1984 mar-1988 dec – 1r – 1 – (cont: knox county genealogical society quarterly; cont by: knox county genealogical society quarterly [galesburg, il: 1993]) – mf#3360360 – us WHS [929]

Knox county news – Niobrara, NE: News Print Co. v1 n1. may 29 1879- (wkly) [mf ed -1880,1885 (gaps) filmed 1958] – 1r – 1 – (cont by: people's news. publ at creighton ne, mar 19 1885- . issues for mar 19 1885- called v1 n8-) – us NE Hist [071]

Knox county recorder – Vverdigre, NE: E H Purcell. v3 n35. dec 1 1892 (wkly) [mf ed 1892,1896 (gaps) filmed [1966]] – 1r – 1 – us NE Hist [071]

Knox, Ellen Mary *see* The girl of the new day

Knox, George William *see*
- Development of religion in japan
- The direct and fundamental proofs of the christian religion
- The gospel of jesus the son of god
- Japanese life in town and country
- Japanese philosopher
- The mystery of life
- The spirit of the orient

Knox, George William et al *see*
- The christian point of view
- The unity of religions

Knox, Henry *see* The henry knox papers, 1719-1825

Knox independent – Knox, Benson Co, N D: M E Delameter. v2 n18 dec 9 1904-v2 n38 apr 28 1905 (wkly) – 1 – (cont: brinsmade blade; cont by: mclean county independent) – mf#05411 – us North Dakota [071]

Knox, John *see*
- An historical journal of the campaigns in north-america
- The history of the reformation of religion within the realm of scotland
- Voyage dans les montagnes de l'ecosse et dans les isles hebrides, fait en 1786

Knox, Kelly M *see* Energy cost of walking with and without arm activity on the cross walk dual motion cross trainer

Knox, Peter *see* Programme of law studies.

Knox, Raymond Collyer *see* Religion in education

Knox republican – Knoxville IL. 1862 jul 23 v6 n43 – 1r – 1 – mf#1147006 – us WHS [071]

Knox, Robert *see*
- Ecclesiastical index
- An historical relation of ceylon
- History of ceylon, from the earliest period to the year 1815

Knox, Ronald Arbuthnott *see* Some loose stones

Knox, Thomas Francis *see*
- The true story of the catholic hierarchy deposed by queen elizabeth
- When does the church speak infallibly?, or, the nature and scope of the church's teaching office

Knox, Thomas Wallace *see* The voyage of the "vivian" to the north pole and beyond

Knox, Vicesimus *see* Narrative of transactions relative to a sermon preached in the pari...

Knox-Heath, Nelly Lloyd *see* Elementary lessons in english for home and school use

Knox-Little, W J *see*
- Christian battle
- Christian watching
- Christian work

Knox-Little, William John *see*
- Christian advance
- Christian joy
- Christian suffering

Knoxville first baptist church. knoxville, tennessee : church records – 1843-1952 – 1 – us Southern Baptist [242]

Knoxville negro world – Knoxville, TN, 1887?- [mf ed 1947] – (= ser Negro newspapers on microfilm) – 1r – 1 – (cont: weekly negro world; cont by: negro world) – us L of C Photodup [071]

Knudson, Albert Cornelius *see* The old testament problem

Knudson-Buresh, Alana D *see* A study of health insurance coverage and health care utilization in north dakota

Knudtzon, J A *see* Assyrische gebete an den sonnengott fuer staat und koenigliches haus

Knudtzon, Joergen Alexander *see* Die el-amarna-tafeln

Der knueppel : satirische arbeiterzeitung – Berlin DE, 1925, 1927 n1-5 – 1 – gw Misc Inst [870]

Knueppelholz, Paul *see* Der monolog in den dramen des andreas gryphius

Knust, Hermann *see*
- Gualteri burlaei liber de vita et moribus philosophorum
- Mittheilungen aus dem eskurial
- Till eulenspiegel

Knutsford division guardian for knutsford, wilmslow and alderley edge etc – [NW England] Knutsford, Cheshire Record Off 1893-95, 1899, feb 1900-1901, 1903-08, 1910, 1914-17, apr 1923-2 oct 1925, 30 oct-dec 1925, 1929, 1931-13 sep 1946 – 1 – (title change: knutsford guardian [20 sep 1946-1949, nov 1950-aug 1953, jun 1954-aug 1964, 1965-77, 1979-83) – uk Newsplan; uk MLA [072]

Knutsford, st cross – [North Cheshire FHS] – (= ser Cheshire monumental inscriptions) – 1mf – 9 – £3.00 – mf#125 – uk CheshireFHS [929]

Knutsford, st john – [North Cheshire FHS] – (= ser Cheshire monumental inscriptions) – 3mf – 9 – £4.00 – mf#126 – uk CheshireFHS [929]

Knutson, Bonnie *see* Journal of hospitality and leisure marketing

Knutson-Kaske, Jill A *see* A description of trends emerging during years three and four

Knuttel, Willem Pieter Cornelis *see*
- Geschiedenis en kritiek der hedendaagsche oud-katholieke beweging in duitschland van juli 1870 tot mei 1877
- Nederlandsche bibliographie van kerkgeschiedenis
- De toestand der nederlandsche katholieken ten tijde der republiek

Ko aotearoa or the maori recorder – 1861-62 – 1r – 1 – mf#11.60 – nz Nat Libr [079]

Ko, Ch'ih-feng *see* Tsang pien ts'ai feng chi

Ko chu chi / Hsiang, Yu – Shang-hai: Ch'en kuang shu tien, 1939 – (= ser P-k&k period) – us CRL [951]

K'o chung hsiao ch'ien lu / Ts'ai, Tung-fan – Shang-hai: Hui wen t'ang hsin chi shu chu, Min kuo 26 [1937] – (= ser P-k&k period) – us CRL [480]

K'o, Chung-p'ing *see* P'ing han lu kung jen p'o huai ta tui ti ch'an sheng

K'o, Hsiang-feng *see* Hsi-k'ang she hui chih niao k'an

Ko, Hsien-ning *see*
- Hai
- Huang ts'un

Ko hsing chiao yu / Fan, Shou-k'ang – Shang-hai: Shang wu yin shu kuan, Min kuo 22 [1933] – (= ser P-k&k period) – us CRL [370]

K'o hsueh che hsueh yu jen sheng / Fang, Tung-mei – Shang-hai: Shang wu yin shu kuan, Min kuo 26 [1937] – (= ser P-k&k period) – us CRL [180]

K'o hsueh kuan li yu hsien tai hsing cheng / Huang, Shou-p'eng – [China]: Chun cheng pu Lu chun ching li tsa chih she, Min kuo 31 [1942] – (= ser P-k&k period) – us CRL [350]

K'o hsueh ta chung = Every day science – Beijing, China 1963-64 – 1 – mf#2625 – us UMI ProQuest [500]

K'o hsueh ti chia t'ing / Lo, Shih-i – Shang-hai: Chung-hua shu chu, Min kuo 25 [1936] – (= ser P-k&k period) – us CRL [306]

K'o hsueh ti i / Ting, Ch'ao-wu – K'un-ming: Chung-hua shu chu, Min-kuo 30 [1941] – (= ser P-k&k period) – us CRL [180]

K'o hsueh ti shih chieh wen hsueh kuan / Hsi-erh-lieh-so – Shang-hai: Chih wen she, Min kuo 29 [1940] – (= ser P-k&k period) – us CRL [480]

K'o hsueh t'ung pao = Science – Ann Arbor MI 1959-64 – 1 – mf#2626 – us UMI ProQuest [500]

Ko, I-hung *see* Tsou: hsien tai tu mu chu hsuan

Ko jen ch'uan tao fan shih (ccm319) = Model personal work / Wang, Yuan-te – Shanghai, 1931 [mf ed 198?] – (= ser Ccm 319) – 1 – mf#1984-b500 – us ATLA [920]

Ko jen pu tao (ccm241) = Individual evangelism / MacNaughtan, W – Shanghai, 1930 [mf ed 198?] – (= ser Ccm 241) – 1 – mf#1984-b500 – us ATLA [240]

Ko kuo ching chi ch'ing k'uang yu ts'e kai shu / Li, Fan – [China]: Kuo min cheng fu chu chi ch'u – (= ser P-k&k period) – us CRL [339]

Ko kuo chung-yang yin hang pi chiao lun / Sun, Tsu-yin – Shang-hai: Shang wu yin shu kuan, Min kuo 23 [1934] – (= ser P-k&k period) – us CRL [332]

Ko kuo hsien fa chi ch'i cheng fu / Sa, Meng-wu – Ch'ung-ch'ing: Nan fang yin shu kuan, 1943 – (= ser P-k&k period) – us CRL [323]

Ko kuo nueh tai hua ch'iao ho li chi yao – [Nan-ching]: Chung yang ch'iao wu wei yuan hui pien yin, min kuo 20 [1931] – (= ser P-k&k period) – us CRL [304]

Ko kuo pi chih / Yang, Yin-p'u – Shang-hai: Shang wu yin shu kuan, Min kuo 23 [1934] – (= ser P-k&k period) – us CRL [332]

Ko kuo ping i hsing cheng kai lun / Ch'en, Ping-yuan – Ch'ung-ch'ing: Chung-kuo wen hua fu wu she, Min kuo 29 [1940] – (= ser P-k&k period) – us CRL [355]

Ko kuo tsung tung yuan kai k'uang / Chang, Kung-hui – [China]: Ta tung shu chu, Min kuo 31 [1942] – (= ser P-k&k period) – us CRL [951]

Ko kuo tui jih chan tung-pei chih p'ing lun – [China]: Kai pu, Min kuo 20 [1931] – (= ser P-k&k period) – us CRL [951]

Ko kuo t'ung huo cheng ts'e yu huo pi chan cheng / Chao, Lan-p'ing – Shang-hai: Hsin Chung-kuo chien she hsueh hui, Min kuo 23 [1934] – (= ser P-k&k period) – us CRL [332]

K'o, Ling *see*
- Shih lou tu ch'ang
- Wang ch'un ts'ao

Ko ming chia shih ch'ao / T'ang, Kuo-ch'uan – Shang-hai: Kuang ming shu chu, Min kuo 23 [1934] – (= ser P-k&k period) – us CRL [810]

Ko ming yu ssu hsiang / Ch'en, Kung-po – Shang-hai: Chung Jih wen hua hsieh hui Shang-hai fen hui, Min kuo 33 [1944] – (= ser P-k&k period) – us CRL [100]

Ko nen-niang / Chu, Sha-lang – Shang-hai: Chung hsueh sheng shu chu, Min kuo 30 [1941] – (= ser P-k&k period) – us CRL [830]

Ko pao tui yu tsui chin yen cheng chih p'ing lun / Yen, cheng tsa – Nan-ching: Chung-shan yin shu kuan, 1934 – (= ser P-k&k period) – us CRL [380]

Ko sheng hsien shih pan li ho tso she teng chi hsu chih – [China]: Shih yeh pu ho tso ssu, Min kuo 25 [1936] – (= ser P-k&k period) – us CRL [334]

Ko sheng shih fan chiao yu she shih chih yen chin / China Chiao yu pu Chung teng chiao yu ssu – [China]: Chiao yu pu, Min kuo 31 [1942] – (= ser P-k&k period) – us CRL [370]

Ko su le dhat pum mhat tam – Ran Kun: Ca pe bi man 1972 [mf ed 1990] – 1r with other items – 1 – (at head of title: mran ma chuirrhay lac lam can pati. cover title: karan prann nay dhat pum mhat tam; in burmese) – mf#mf-10289 seam reel 140/9 [§] – us CRL [915]

Ko, Sui-ch'eng *see* T'ai-p'ing yang wen t'i chih chieh p'ou

K'o, Tun-po *see*
- Sung wen hsueh shih

K'o wai huo tung / Li, Hsiang-hsu et al – Shang-hai: Shang wu yin shu kuan, Min kuo 25 [1936] – (= ser P-k&k period) – us CRL [370]

Ko, Wen-hua *see* Mi; chieh

Koabel-Bagley, Patricia *see* Assessment of the need for certified athletic trainers in new york state high schools

Koain chosa geppo : monthly reports of the board of development of asia: japanese empire – v1-v3 n5. 1940-42 – 10r – 1 – Y90,000 – (in japanese) – ja Yushodo [338]

Kob, Walter *see* Pavan

Kobak, Joseph *see* Jeschurun

Kobayashi, Takiji [comps] *see* Nihon kinsei kayo shiryoshu

Kobbert, Maximilian *see* De verborum 'religio' atque 'religiosus' usu apud romanos

Kobe chronicle – Kobe, Japan. Japan Chronicle. -w. 21 March 1900-22 May 1930; 7 Jan 1937-29 Dec 1938. 60 reels – 1 – uk British Libr Newspaper [072]

Kobel, Erwin *see* Untersuchungen zum gelebten raum in der mittelhochdeutschen dichtung

Kobelt, W *see* Illustriertes conchylienbuch

Kobenhavns universitet. Astronomisk observatorium *see* Publikationer og mindre meddelelser fra kobenhavns observatorium

Kober, Margarete *see* Das deutsche maerchendrama

Koberstein, Astrid Beate *see* Die fundlandschaft guspini, provinz cagliari, sardinien

Kobes, Alois *see* Dictionnaire volof-francais

Kobes, Mgr *see* Dictionnaire volof-francais

Kobijitsu kenkusho : tjandi panataran / Java. (Japanese Military Administration) – Djakarta: Gunseikanbu Kokumin Tosyokyoku, 2605 (B P 1580) – 35p 1mf – 9 – mf#SE-2002 mf69 – ne IDC [959]

Koblenzer volkszeitung *see* Coblenzer volkszeitung

Kobner, Johann *see* Sermons and controversial writings

Koborgher quackbruennla : tausend stueck lauter schlumperliedla, spassreumla und tanzvarschla, zum singa... / Hofmann, Fritz – Hildburghausen: Kesselring, 1857 – 1r – 1 – us Wisconsin U Libr [430]

Kobrin, Leon *see* Yankel boyle

Kocaeli – Izmit, 1920-19? Yayimhyan: Izmit Sancagi. n144. 14 eyluel 1338 [1922] – (= ser O & t journals) – 1mf – 9 – $25.00 – us MEDOC [956]

Koch, Adolf *see* Der semitische infinitiv

Koch, C *see* Die kaukasischen laender und armenien in reiseschilderungen von curzon, k koch, macintosh, spencer und wilbraham

Koch, Carl *see*
- Goethes faust
- Soeren kierkegaard

Koch, Eduard Emil *see* Geschichte des kirchenlieds und kirchengesangs der christlichen

Koch, Ernst *see*
- Griechische schulgrammatik
- Richard wagners buehnenfestspiel der ring des nibelungen in seinem verhaeltniss zur alten sage wie zur modernen nibelungendichtung

Koch, Franz *see*
- Drei goethe-reden
- Geist und leben
- Goethe und die juden
- Goethe und plotin
- Goethes stellung zu tod und unsterblichkeit
- Josef weinheber
- Vergangenheit und gegenwart in eins

Koch, Friedrich C *see*
- Die deutschen colonien in der naehe des saginaw-flusses
- Die mineral-gegenden der vereinigten staaten nord-amerika's am lake superior, michigan, und am obern mississippi, wisconsin, illinois, iowa

Koch, H *see*
- Cyprian und der roemische primat
- Cyprianische untersuchungen
- Pronoia und paideusis
- Pseudo-dionysius areopagita in seine beziehungen zum neuplatonismus und mysterienwesen
- Vincenz von lerin und gennadius
- Virgines christi

Koch, H A *see* Quellenuntersuchungen zu nemesios von emesa
Koch, Hans *see*
- Aufsaetze zur deutschen literaturgeschichte
- Die lyrische gestaltung und die sprachform stefan georges
- Unsere literaturgesellschaft

Koch, Heinrich Christoph *see* Versuch einer anleitung zur composition
Koch, Henri *see* Magie et chasse dans la foret camerounaise
Koch, Herbert *see* Ueber das verhaeltnis von drama und geschichte bei friedrich hebbel
Koch, Hugo *see*
- Cyprian und der roemische primat
- Pseudo-dionysius areopagita in seinen beziehungen zum neuplatonismus und mysterienwesen
- Die unter hippolyts namen ueberlieferte schrift ueber den glauben

Koch, Johannes Guenther *see* Gutzkows theorie des romans in seinem roman "hohenschwangau"
Koch, Julius August *see* Astronomische tafeln zur bestimmung der zeit
Koch, Karl H *see* Die kaukasische militaerstrasse der kuban und die halbinsel taman
Koch, Karl Heinz *see* Paul ernst und das tragische herrscher-ideal
Koch kith 'nkin – v7 copy 2-v11 copy 2 [1978 jun-1981 dec] – 1r – 1 – mf#618146 – us WHS [071]
Koch, L *see* Die arachniden australiens nach der natur beschrieben und abgebildet
Koch, Ludvig *see*
- Fortaellinger af danmarks kirkehistorie fra 1517 til 1848
- Fra grundtvigianismens og den indre missions tid 1848-1898

Koch, M *see* Wie haben wir zaehlen und rechnen gelernt?
Koch, Max *see*
- August graf von platens saemtliche werke in zwoelf baenden
- Chamissos gesammelte werke
- Deutsche vergangenheit in deutscher dichtung
- Franz grillparzer
- Geschichte der deutschen litteratur von den aeltesten zeiten bis zur gegenwart
- Gottsched und die reform der deutschen literatur im achtzehnten jahrhundert
- Helferich peter sturz
- Die kirchenpolitik koenig sigmunds waehrend seines romzuges
- Quellenverhaeltniss von wielands oberon
- Schoenes blumenfeld

Koch, Richard *see* Geld und werthpapiere
Koch, Samuel *see* Papers, 1902-1962
Koch, Stefan *see* Untersuchung und optimierung alternativer zahnfuellungs-composite-materialien auf der basis von polybutadienepoxid
Koch, Wilhelm *see* Die taufe im neuen testament
Koch, Willi *see* Stefan george
Kochanowski, Bodo *see* Die lichten stunden
Kochendoerffer, Karl *see* Tilos von kulm gedicht von siben ingesigeln
Kocher, Pamela L *see* The effects of chromium supplementation and a low carbohydrate diet on high-intensity endurance performance
Kochergin, S M *see* Narodnaia gazeta [omsk: 1918-1919]
Kochheim, Gustav *see* Faust im zeichen des kreuzes
Kochs, Ernst *see* Paul gerhardt
Kochubei, M *see* Ezhemesiachnyi politicheskii organ
Kock, Johannes Hermanus Michiel *see* De roemrijke reis van de zuid-afrikaansche reisbeschrijvers en hunne reizen
Kock, Paul De *see*
- Bouquetiere des champs-elysees
- Laitiere de la foret
- Maitresse dans l'andalousie

Kock, W J de *see* Portuguese ontdekkers om die kaap
Kocka, Juergen et al *see* Acta borussica neue folge
Kocourek, Albert *see* Formative influences of legal development
Koczajowski, Donna L *see* State and trait sport-confidence and physical self-efficacy of professional and amateur female golfers
Kodaikanal Observatory [Kodaikanal, India] *see* Bulletin [kodaikanal observatory [kodaikanal, india]]
Kodak / Milwaukee-Downer College – 1896 nov-1912 dec, 1913 feb-1919 may, 1919 nov-1927 nov, 1927 dec-1933 dec, 1934 mar-1941 nov – 5r – 1 – mf#2973860 – us WHS [378]
Kodama, Yoshio *see* I was defeated
Kodeks zakonov o trude 1922 goda : utverzhden 4 sessiei vserossiiskogo tsentral'nogo ispolnitel'nogo komiteta sovetov 9 sozyva 30 oktiabria 1922 goda / Russian S F S R Vserossiiskii tsentral'nyi ispolnitel'nyi komitet – Moskva: Izd-vo "Voprosy truda", 1926 [mf ed 2004] – 1r – 1 – (filmed with: vzaimnaia pomoshch' sredi zhivotnykh i liudei, kak dvigatel' progressa / p kropotkin (1922)) – us Wisconsin U Libr [344]
Kodiak bear / Kodiak Naval Station – Fort Greely AK. v1 n5 [1942 feb 3] – 1r – 1 – mf#928460 – us WHS [355]
Kodiak cub – Fort Greely AK. 1943 jul 31, 1944 oct 9-16, nov 4-8, extra – 1r – 1 – mf#928454 – us WHS [071]
Kodiak Naval Station *see* Kodiak bear
Kodifikatsiia kreditnogo i valiutnogo zakonodatel'stva / Venediktov, A V – M, 1924 – 1mf – 9 – mf#R-15147 – ne IDC [332]
Kodimo *see* Izdaetsia gruppoi studentov-sionistov
Kodor, I I *see*
- Altaiskii luch
- Svobodnyi luch

Kodumaa – Tallinn, U.S.S.R. -w. Jan 1961-Dec 1970. 8 reels – 1 – uk British Libr Newspaper [947]
Koe makasini a koliji = Tupou college magazine – n10-29. 1875-81 – 1r – 1 – (available for ref) – mf#pmb doc434 – at Pacific Mss [378]
Koe tohi fanogonogo – jun 1929-jul 1982 – 3r – 1 – mf#pmb doc389 – at Pacific Mss [079]
Koebel, William Henry *see*
- Central america
- Great south land

Koeber, Raphael *see* Schopenhauer's erloesungslehre
Koeberle, Justus *see*
- Die alttestamentliche offenbarung
- Das raetsel des leidens
- Suende und gnade im religioesen leben des volkes israel bis auf christum

Koebner, Thomas *see* Tendenzen der deutschen literatur seit 1945
Koeci bey risalesi / Bey, Mustafa Koci – Istanbul: Matba'a-i Ebuezziya, 1303 [1887] – (= ser Ottoman histories and historical sources) – 2mf – 9 – $50.00 – us MEDOC [956]
Koeduktion : oder maedchenbildung? eine vergleichende betrachtung zweier laengsschnittuntersuchungen unter dem aspekt geschlechtsspezifischer unterschiede vor und in der koedukaktiven schule / Amse, Corina – (mf ed 1995) – 3mf – 9 – €49.00 – 3-8267-2155-1 – mf#DHS 2155 – gw Frankfurter [376]
Koegel, Julius *see*
- Christus der herr
- Die gedankeneinheit des ersten briefes petri
- Das gleichnis vom verlorenen sohn
- Zum gleichnis vom ungerechten haushalter
- Der zweck der gleichnisse jesu

Koegel, Rudolf *see* Der brief des jakobus
Koegel, Rudolph *see* Die phantasie als religioeses organ
Koehl, Eduard *see* Die geschichte der festung glatz
Koehler, August *see*
- Die niederlaendische reformirte kirche
- Ueber berechtigung der kritik des alten testaments

Koehler, Georg *see* Entwicklung und erprobung eines computerunterstuetzten curriculums des grundlegenden chemieunterrichtes fuer die schwerpunktthemen "einfuehrung in die chemie" sowie "atombau und chemische bindung"
Koehler, H *see* Von der welt zum himmelreich
Koehler, Joh Georg Wilh *see* Der volksfreund, zur aufklaerung und belehrung des buergers und landmanns
Koehler, Johann *see* Joh david koehlers p p kurtze und gruendliche anleitung zu der alten und mittlern geographie
Koehler, Karen M *see* Development and validation of a questionnaire for assessing habitual physical activity of sixth-grade students
Koehler, Karl *see* Briefe aus amerika fuer deutsche auswanderer
Koehler, Oswald *see* Weltschoepfung und weltuntergang
Koehler, Reinhold *see* Zu heinrich von kleist's werken
Koehler, Robert *see* Correspondence
Koehler, Ruth *see* Die sending der frau in der deutschen geschichte
Koehler, W *see*
- Armenpflege und wohltätigkeit in Zuerich zur zeit Ulrich Zwinglis
- Die geisteswelt ulrich zwinglis
- Huldreich zwinglis
- Huldrych zwinglis bibliothek
- Zwingli und luther

Koehler, Walther *see*
- Beitraege zur geschichte der mystik in der reformationszeit
- Bibliographia brentiana
- Dokumente zum ablassstreit von 1517
- Geist und freiheit
- Die gnosis
- Idee und persoenlichkeit in der kirchengeschichte
- Katholizismus und reformation
- Luther und die kirchengeschichte
- Luther und die luege
- Luthers schrift an den christlichen adel deutscher nation
- Reformation und ketzerprozess
- Wie luther den deutschen das leben jesu erzaehlt hat

Koehler, Willibald *see*
- Hermann stehr
- Das kalte haus, oberschlesien

Koehlerglaube und wissenschaft : eine streitschrift gegen hofrath rudolph wagner in goettingen / Vogt, Karl Christoph – 4., mit einem weiteren Vorwort vermehrte Aufl. Giessen: J Ricker, 1856 – 1mf – 9 – 0-7905-8956-7 – (incl bibl ref) – mf#1989-2181 – us ATLA [210]
Koehring news – v1 n2-v10 n3 [1966 jul-1975 fall 1975, v11 n1 [1977 spr/sum] – 1r – 1 – (cont by: world of koehring) – mf#633529 – us WHS [071]
Koek, Edwin Rowland *see* A table of written law judicially considered by the supreme court of the straights settlements and on appeal therefrom, 1808-1898
Der koeker : mittelniederdeutsches lehrgedicht aus dem anfang der 16. jahrhunderts / Bote, Hermann; ed by Cordes, Gerhard – Tuebingen: M Niemeyer, 1963 [mf ed 1993] – (= ser Altdeutsche textbibliothek n60) – xii/95p – 1 – (text in middle low german. incl bibl ref) – mf#8193 reel 5 – us Wisconsin U Libr [430]
Koelbing, Paul *see* Die geistige einwirkung der person jesu auf paulus
Koelle, Conrad *see* Ernst moritz arndts fragmente ueber menschenbildung in ihrer paedagogischen bedeutung
Koelle, S W *see* Narrative of an expedition into the vy country of west africa
Koelle, Sigismund Wilhelm *see* Mohammed and mohammedanism
Koelling, W *see* Geschichte der arianischen haeresie
Koelling, Wilhelm *see*
- Die lehre von der theopneustie
- Pneumatologie, oder, die lehre von der person des heiligen geistes

Koelmel, Rainer *see* Die geschichte deutsch-juedischer refugees in schottland
Koelner schreinsurkunden des 12. jahrhundert (pgrg1) / ed by Hoeniger, R – Bonn, 1884-1894 – (= ser Publikationen der gesellschaft fuer rheinische geschichtskunde (pgrg)) – 2v – €46.00 – ne Slangenburg [931]
Koelner stadt-anzeiger – Bergisch Gladbach DE, 2 jan-30 jun 1952; 1 oct 1952-2000 – 209r – 1 – (1 apr 1969-1973, 1975-2000 only local ed) – gw Mikrofilm [074]
Koelner stadtanzeiger – Koeln DE, nov 1949-62 – 86r – 1 – (title varies: fr 1960: koelner stadt-anzeiger / k. filmed by misc inst: 1949 29 oct-1967; 1967-1969 14 jun, 1969 15 dec-1975 4 dec, 1976 12 jan-1979 21 dec, 1980 24 jan-31 dec; 1968- [ca 14r/yr]) – mf#12667 – gw Mikropress; gw Misc Inst [074]
Koelner stadt-anzeiger / express *see* Express
Koelner unterhaltungsblatt – Koeln DE, 1859 6 jan-1863 – 1 – gw Misc Inst [074]
Die koelner wirren (1837) : studien zu ihrer geschichte / Schroers, Heinrich – Berlin, Bonn, 1927 (mf ed 1992) – 4mf – 9 – €49.00 – 3-89349-106-6 – mf#DHS-AR 75 – gw Frankfurter [240]
Koelnische blaetter – Koeln DE, 1860 1 apr-1941 31 may – 219r – 1 – (title varies: 1 jan 1869: koelnische volkszeitung; 11 sep 1887: koelnische volkszeitung und handelsblatt. filmed by misc inst: 1920 jan-mar [1r]) – mf#6577 – gw Mikropress; gw Misc Inst [074]
Koelnische rundschau : ausgabe rheinisch bergischer kreis – Bergisch Gladbach DE, 19 mar 1946-25 may 1948; 1 jun 1948-1987 – 280r – 1 – (title varies: 5 oct 1949: bergische rundschau; 21 aug 1952: bergische landeszeitung. filmed by misc inst: 1988- [8r/yr]; 5 nov-31 dec 1957; 21 apr 1958-59 (nur lokalteil); jan 1979 (nur lokalteil). with suppl: zwischen wipper und rhein 1947-57 [1r]) – gw Mikrofilm; gw Misc Inst [074]
Koelnische rundschau – Koeln DE, 1946 19 mar-1980 – 1 – (filmed by other misc inst: 29 nov 1974 [ca 7r/yr]; regional ed also available: bergheim 1988- [8r/yr]; euskirchen 1988- [8r/yr]; siegburg 4 jan 1949- (title varies: siegkreis rundschau [oct 27 1949]; rhein-sieg-rundschau [jan 2 1970])) – gw Misc Inst [074]
Koelnische rundschau – Siegburg DE, 1949 4 jan- – 1 – (title varies: 27 oct 1949: siegkreis rundschau; 2 jan 1970: rhein-sieg-rundschau) – gw Misc Inst [074]
Koelnische rundschau [hauptausgabe [deutschland-ausgabe]] – Cologne, Germany 3 jan 1947-13 aug 1948 (imperfect) – 1 – (cont by: allgemeine koelnische rundschau (deutschland-ausgabe. bundesausgabe) 17 aug 1948-3 jun 1949, 11 oct 1949-30 mar 1950) – uk British Libr Newspaper [074]
Koelnische rundschau [stadt-ausgabe] – Cologne, Germany 19 mar 1946-8 sep 1949, 12 oct 1949-30 mar 1950, 1 nov 1950-30 jun 1951 – 1 – (includes suppl: rundschau am sonntag) – uk British Libr Newspaper [074]
Koelnische volkszeitung – Koeln, 1869-1941 – 203r – 1 – gw Mikropress [074]
Koelnische volkszeitung *see* Koelnische blaetter
Koelnische volkszeitung und handelsblatt *see* Koelnische blaetter
Koelnische zeitung – Koeln DE, 1803-05 [gaps], 1814-15, 1817-1944 sep, 1944 nov-1945 31 jan – 352r – 1 – (also as ed b c, c b, c, west [sun fr 23 sep 1801 & 1846]. filmed by other misc inst: 1920 jan-mar [1r]; 1814-15, 1824, 1829, 1844, 1848-1860 oct, 1861-1865 oct, 1888 mar-dec, 1892 feb-1893 [51r]) – gw Misc Inst [074]
Koelnische zeitung – Koeln: Erben Schauberg [1907-1944] – 1r – 1 – (comprised of iss missing fr mikropress film for 1907-1944) – us CRL [074]
Koelnischer anzeiger : mit koelner fremdenblatt – Koeln DE, 1848-49 – 2r – 1 – (filmed by other misc inst: 1855-65) – gw Misc Inst [074]
Koelnischer kurier / ed by Die Amerikanische Armee – 1945 2 apr-1946 26 feb – 1r – 1 – (fr 23 jun 1945 ed by britische besatzungsbehoerde) – mf#6411 – gw Mikropress [074]
Koen, D T [comp] *see* Kaiser in exile
Koena news – Maseru: Lesotho Dept of Information. v1 n100-v8, n130. aug 6 1967-jul 10 1974 – 5r – 1 – us CRL [960]
Koenekamp, Alfred Heinrich *see* Die preussischen landwirtschaftlichen versuchs- und forschungsanstalten, landsberg/w.
Koeneke, Irene Aniata *see* Letters received
Koenig, Alfred *see* Jesus
Koenig davids wyssagung vom rych messiae im cix. psalmen beschriben... / Wolf, J – Zuerych, Christoffel Froschower, 1560 – 3mf – 9 – mf#PBU-653 – ne IDC [240]
Koenig, Eduard *see*
- Der aeltere prophetismus
- Alttestamentliche kritik und christenglaube
- Babyloniens kultur und die weltgeschichte
- The bible and babylon
- De criticae sacrae argumento e linguae legibus repetito
- Einleitung in das alte testament
- The exiles' book of consolation contained in isaiah 40-66
- Falsche extreme in der neueren kritik des alten testaments
- Fuenf neue arabische landschaftsnamen im alten testament
- Geschichte des reiches gottes bis auf jesus christus
- Der glaubensact des christen
- Die hauptprobleme der altisraelitischen religionsgeschichte
- Hebraeische grammatik fuer den unterricht mit uebungsstuecken und woerterverzeichnissen
- Die moderne pentateuchkritik und ihre neueste bekaempfung
- Neue studien ueber schrift, aussprache und allgemeine formenlehre des aethiopischen
- Neueste prinzipien der alttestamentlichen kritik
- Der offenbarungsbegriff des alten testamentes
- Die originalitaet des neulich entdeckten hebraeischen sirachtextes
- Die poesie des alten testaments
- Prophetenideal judentum und christentum
- The religious history of israel
- Stilistik, rhetorik, poetik in bezug auf die biblische litteratur
- Talmud und neues testament
- Theologie des alten testaments

Koenig eduard und der einsiedler : eine mittelenglische ballade / Kurz, Albert – Erlangen, 1904 (mf ed 1994) – 1mf – 9 – €24.00 – 3-8267-3037-2 – mf#DHS-AR 3037 – gw Frankfurter [420]
Koenig, Fritz *see* Georg buechners "danton"
Koenig geiserich : eine erzaehlung von geiserich und dem zug der wandalen / Blunck, Hans Friedrich – Hamburg: Hanseatische verlagsanstalt, c1936 [mf ed 1989] – 399p – 1 – mf#7037 – us Wisconsin U Libr [880]
Koenig, Gustav *see* The life of luther
Koenig, H *see* Was ist die wahrheit von jesu?
Koenig heinrich 1 und die heilige lanze / Holtzmann, W – Bonn, 1947 – €5.00 – ne Slangenburg [240]
Koenig jerome napoleon : ein zeit- und lebensbild nach briefen der frau von sothen und des reichserzkanzlers von dalberg etc / Kaisenberg, Moritz von – Leipzig, 1899 (mf ed 1993) – 2mf – 9 – €24.00 – 3-89349-118-X – mf#DHS-AR 87 – gw Frankfurter [943]
Koenig, Joerg Udo *see* Der erythroide anionenaustauscher ae1

Koenig, Joh Christoph *see* Der freund der aufklaerung und menschenglueckseligkeit
Koenig, Joseph *see*
- Beitraege zur geschichte der theologischen facultaet in freiburg
- Karl spindler

Koenig, Joseph M *see* A comparison of body density and percent body fat using functional residual capacity and residual volume and development of immersed functional residual capacity and residual volume prediction formulas
Koenig karl : ein trauerspiel in drei aufzuegen mit einem vorspiel "das voelklein auf der heide" / Wolzogen, Ernst von – Darmstadt: A Bergstraesser 1914 [mf ed 1991] – 1r – 1 – (filmed with: volk, ich breche deine kohle! / otto wohlgemuth) – mf#2964p – us Wisconsin U Libr [820]
Koenig, Karl *see* Im kampf um gott und um das eigene ich
Koenig laurins mantel : roman / Dominik, Hans – Berlin: Scherl, 1943, c1928 – 1r – 1 – us Wisconsin U Libr [830]
Koenig laurins mantel : roman / Dominik, Hans – Berlin: Scherl, 1943, c1928 [mf ed 1989] – 321p – 1 – mf#7181 – us Wisconsin U Libr [830]
Koenig lipit-istar's vergoettlichung : ein altsumerisches lied / Zimmern, H – Leipzig, 1916 – (= ser Koenigl Saechs Gesellschaft der Wissenschaften) – 1mf – 9 – (abh koenigl saechs gesellschaft der wissenschaften v68 pt5) – mf#NE-20041 – ne IDC [956]
Der koenig mit dem handgepaeck / Steguweit, Heinz – Hamburg: Hanseatische Verlagsanstalt [1941] [mf ed 1991] – 1r – 1 – (filmed with: frohes leben) – mf#2897p – us Wisconsin U Libr [880]
Koenig nicolo, oder, so ist das leben : schauspiel in drei aufzuegen und neun bildern, mit einem prolog / Wedekind, Frank – Muenchen: G Mueller 1920 [mf ed 1996] – 1r – 1 – (filmed with: die ungleichen schalen / jakob wassermann) – mf#4056p – us Wisconsin U Libr [820]
Koenig ottokars glueck und ende : trauerspiel in fuenf aufzuegen / Grillparzer, Franz; ed by Waniek, Gustav – Wien: F Tempsky, 1903 – 1r – 1 – us Wisconsin U Libr [820]
Koenig ottokars glueck und ende : eine untersuchung ueber die quellen der grillparzer'schen tragoedie / Klaar, Alfred – Leipzig: G Freytag, 1885 – 1r – 1 – (incl bibl ref) – us Wisconsin U Libr [430]
Koenig roderich : ein trauerspiel in fuenf aufzuegen / Dahn, Felix – 2. durchges und veraend ausg. Leipzig: Breitkopf & Haertel, 1876 [mf ed 1989] – 219p – 1 – mf#7169 – us Wisconsin U Libr [820]
Koenig rother – Halle: M Niemeyer, 1884 [mf ed 1993] – (= ser Altdeutsche textbibliothek 6) – iv/162p – 1 – mf#8193 reel 1 – us Wisconsin U Libr [810]
Koenig rother – Halle: M Niemeyer, 1954 – 1r – 1 – (incl bibl ref) – us Wisconsin U Libr [430]
Koenig rother / ed by Rueckert, Heinrich – Leipzig: F A Brockhaus, 1872 – 1r – 1 – (incl bibl ref and ind. middle high german text with introductions in german) – us Wisconsin U Libr [430]
Koenig rother / ed by Vries, Jan de – Heidelberg: C Winter, 1922 – (incl bibl ref and index) – us Wisconsin U Libr [430]
Koenig salomo : ein drama in drei akten / Hardt, Ernst – Leipzig: Insel-Verlag, 1915 – 1r – 1 – us Wisconsin U Libr [820]
Koenig salomon in der tradition : ein historisch-kritischer beitrag zur geschichte der haggada, der tannaiten und amoraeer / Faerber, Rubin – Wien: Jos Schlesinger, 1902 – 1mf – 9 – 0-524-06128-9 – mf#1992-0795 – us ATLA [939]
Koenig tirol, winbeke und winsbekin / ed by Leitzmann, Albert – Halle: M Niemeyer, 1888 [mf ed 1993] – (= ser Altdeutsche textbibliothek 9) – iv/60p – 1 – mf#8193 reel 1 – us Wisconsin U Libr [810]
"Der koenig verdient kein denkmal" : majestaetsbeleidigungen im koenigreich hannover zur zeit koenig georgs 5 (1851-1866) / Reuter, Hans Ulrich – (mf ed 1995) – 2mf – 9 – €40.00 – 3-8267-2191-8 – mf#DHS 2191 – gw Frankfurter [943]
Der koenig von ruecken : geschichten und geschautes / Brehm, Bruno – Karlsbad: A Kraft c1942 [mf ed 1989] – 1r – 1 – (filmed with other titles) – mf#7067 – us Wisconsin U Libr [830]
Der koenig von sidon / Lindau, Paul – Bremen: Verlag der Winking-Buecher [191-?] [mf ed 1995] – 1r – 1 – (filmed with: lichtenberg / paul requadt) – mf#3691p – us Wisconsin U Libr [830]
Der koenig von sion : epische dichtung in zehn gesaengen / Hamerling, Robert – 11. aufl. Hamburg: Verlagsanstalt und Druckerei (vormals J F Richter), 1890 [mf ed 1993] – 336p – 1 – mf#8670 – us Wisconsin U Libr [810]
Die koenigen von saba : oper in vier acten: nach einem text von mosenthal, op 27 / Goldmark, Carl – Bremen: Schwers & Haake [188-?] [mf ed 1992] – 1r – 1 – mf#pres. film 124 – us Sibley [780]
Koeniger, Albert Michael *see*
- Die beicht nach caesarius von heisterbach
- Burchard 1. von worms und die deutsche kirche seiner zeit

Die koenighaeuser : eine erzaehlung aus dem isergebirge / Leutelt, Gustav – 5. Aufl. Reichenberg: F Kraus, 1943 – 1r – 1 – us Wisconsin U Libr [830]
Koenigin sibille (cima26) : farbmikrofiche-edition der handschrift hamburg, staats- und universitaetsbibliothek, cod 12 in scrinio / Scheppel, Huge – (mf ed 1993) – (= ser Codices illuminati medii aevi (cima) 26) – 47p on 3mf – 15 – €260.00 – 3-89219-026-7 – (trans fr the french by elisabeth von nassau-saarbruecken. int & description by jan-dirk mueller) – gw Lengenfelder [090]
Koeniglich Bayerische Akademie der Wissenschaften. Historische Kommission *see* Geschichte der deutschen historiographie
Koeniglich berlinische privilegirte staats- und gelehrte zeitung *see* Berlinische privilegirte zeitung
Koeniglich (genehmigte) west-preussische elbingsche zeitung von staats- und gelehrten sachen – Elbing (Elblag PL), 1825, 1832, 1835, 1837-42, 1844-1852 28 jun – 11r – 1 – (with suppl: elbinger anzeigen jan 8 1825-47 [gaps], 1851-52, 1857-58, 1861-62 (gaps) [21r]) – gw Misc Inst [077]
Koeniglich preussisch pommersche zeitung *see* Koeniglich privilegirte stettinische zeitung 1755
Koeniglich Preussische Akademie der Wissenschaften zu Berlin *see* Sammlung astronomischer tafeln
Koeniglich privilegirte berlinische zeitung von staats- und gelehrten sachen – Berlin, 1812-15 – 3r – 1 – gw Mikropress [074]
Koeniglich privilegirte altonaer adress-comtoir-nachrichten – Hamburg DE, 1822-25, 1827-28, 1850-51 – 3r – 1 – (title varies: 1848 n27: altonaer privilegirte adress-comtoir-nachrichten. filmed by other misc inst: 1775-1854) – gw Misc Inst [943]
Koeniglich privilegirte berlinische zeitung – (Vossiche Zeitung). Berlin. Germany. 1850-1919. (1161 reels) – 1 – uk British Libr Newspaper [072]
Koeniglich privilegirte magdeburgische zeitung *see* Magdeburgische zeitung
Koeniglich privilegirte stettinische zeitung 1755 – Stettin (Szczecin PL), 1759, 1761-63, 1769, 1773, 1780, 1782, 1783 & 1791-92 [many gaps], 1796-97, 1806, 1809, 1812-14, 1816-39, 1832-38, 1839 1 jul-1848 31 mar [gaps] – 21r – 1 – (many with single iss. title varies: 5 nov 1806: stettiner zeitung; 1809: koeniglich preussisch pommersche zeitung; 11 feb 1814: koeniglich preussische stettinische zeitung; 18 jan 1822: koeniglich preussische stettiner zeitung; 1836: koeniglich privilegirte stettiner zeitung; 1 apr 1848: koeniglich privilegirte stettinische zeitung; 2 jul 1852: stettiner zeitung; 13 oct 1856: privilegirte stettiner zeitung; 28 feb 1860: stettiner zeitung) – gw Misc Inst [077]
Koeniglich privilegirter preussischer volksfreund – Berlin DE, 1798-1800 [gaps] – 2r – 1 – gw Misc Inst [943]
Koeniglich privilegirtes intelligenzblatt – Schleswig DE, 1978 1 sep- – ca 7r/yr – 1 – (title varies: 6 jan 1841: koeniglich privilegirtes schleswiger intelligenzblatt; 1 jun 1864: schleswiger nachrichten) – gw Misc Inst [074]
Koenigliche bibliothek kopenhagen : alphabetischer katalog der daenischen und norwegischen abteilung 1474-1959 = Catalog over det store kongelige bibliotheks danske og norske afdeling – [mf ed Hildesheim 1995] – 685mf – 9 – diazo €4980.00 silver €5400.00 – gw Olms [020]
Koenigliche Oeffentliche Bibliothek zu Dresden *see* Der codex boernerianus
Koenigliche Sternwarte zu Berlin *see* Astronomische beobachtungen auf der koeniglichen sternwarte zu berlin
Koenigliche Sternwarte zu Berlin. Rechen-Institut *see* Veroeffentlichungen des rechen-instituts der koeniglichen sternwarte zu berlin
Koenigliche Sternwarte zu Goettingen et al *see* Astronomische mittheilungen von der koeniglichen sternwarte zu goettingen
Koeniglichen preussischen staats- kriegs- und friedens-zeitung *see* Europaeischer mercurius
Der koeniglichen schwedischen akademie der wissenschaftlichen abhandlungen [...] – Hamburg DE, 1739-40, 1779, 1780, 1790 – 1 – (all single iss) – gw Misc Inst [574]
Koenigliches Astronomisches Rechen-Institut zu Berlin *see* Veroeffentlichungen des koeniglichen astronomischen rechen-instituts zu berlin
Koenigliches seelen-panget : das ist: dreyhundert und fuenff-und sechtzig anmuethige monath-gedaechtnussen von dem hochwuerdigisten sacrament desz altars...erster(-zweiter) theil / Theodorus O Cap – Muenchen: In Verlegung Johann Jueckling, 1666. 2v – 9mf – 9 – mf#O-1961 – ne IDC [090]
Koeniglich-preussischer staatsanzeiger *see* Allgemeine preussische staats-zeitung
Koenigsberg. Regierungsbezirk *see* Amtsblatt der preussischen regierung zu koenigsberg
Koenigsberger abend-zeitung – Koenigsberg (Kaliningrad RUS), 1831 [gaps] – 1r – 1 – gw Misc Inst [077]
Koenigsberger allgemeine zeitung *see* Communal-blatt
Koenigsberger allgemeine zeitung 1844 – Koenigsberg (Kaliningrad RUS), 1844 jan-jun, 1845 2 jan-30 sep – 1r – 1 – gw Misc Inst [077]
Koenigsberger apokalypse *see* Apokalypse / koenigsberger apokalypse (cima27)
Koenigsberger, Bernhard *see*
- Aus masorah und talmudkritik
- Monatsblaetter fuer vergangenheit und gegenwart des judentums

Koenigsberger freie presse – Koenigsberg (Kaliningrad RUS), 1877 30 sep-1878 29 sep – 1r – 1 – gw Misc Inst [077]
Koenigsberger hartung'sche zeitung *see*
- Europaeischer mercurius
- Hartungsche kriegszeitung

Koenigsberger intelligenzblatt *see* Koenigsberger intelligenz-zettel
Koenigsberger intelligenz-zettel – Koenigsberg (Kaliningrad RUS), 1816 1 jan-29 mar, 1818 1 apr-1849 [gaps] – 50r – 1 – (title varies: 28 mar 1834: koenigsberger intelligenzblatt) – gw Misc Inst [077]
Koenigsberger kreisblatt – Koenigsberg (Kaliningrad RUS), 1853 2 apr-24 dec – 1 – gw Misc Inst [077]
Der koenigsberger rundfunk – Koenigsberg [Kaliningrad RUS], 1924 2 nov-1935, 1936 28 jun-1939 1 jul – 24r – 1 – (title varies: 1932: koenigsberger und danziger rundfunk; 1 apr 1933: ostfunk) – gw Mikrofilm [380]
Koenigsberger skulpturen und ihre meister 1255-1945 / Muehlpfordt, Herbert Meinhard – Wuerzburg: Holzner Verlag, 1970 – 10r – 1 – (incl bibl ref and ind) – us Wisconsin U Libr [730]
Koenigsberger unterhaltungsblaetter – Koenigsberg (Kaliningrad RUS), 1835 3 oct-1837 30 sep – 1r – 1 – gw Misc Inst [077]
Koenigsberger volkszeitung – Koenigsberg (Kaliningrad RUS), 1932 1 jul-1933 27 feb – 3r – 1 – gw Misc Inst [077]
Koenigsberger wochenblatt – Koenigsberg (Kaliningrad RUS), 1831, 1836, 1841, 1847 – 4r – 1 – gw Misc Inst [077]
Koenigsberger zeitung – Koenigsberg (Chojna PL), 1925 1 jan-31 mar & 1 oct-31 dec, 1926 1 jul-30 sep, 1927 1 sep-1928 30 jun, 1929 1 jan-30 aug, 1930 1 jan-30 mar & 1 oct-31 dec, 1931 1 jul-1932 31 mar, 1933 1 jan-30 sep, 1934 3 jan-30 mar & 3 oct-30 dec, 1935 2 apr-31 dec, 1936 1 apr-30 jun, 1936 1 oct-1937 31 mar, 1937 1 jul-1938 30 sep, 1939 2 jan-31 mar & 1 jul-30 sep, 1940 2 jan-30 mar, 1941 1 oct-1942 30 apr, 1944 3 jan-30 jun – 18r – 1 – gw Misc Inst [077]
Koenigsbergische gelehrte und politische zeitung – Koenigsberg (Kaliningrad RUS), 1764-68, 1771-72 – 1 – (aka: koenigsbergsche gelehrte und politische zeitung) – gw Misc Inst [077]
Koenigsbergsches theaterjournal – Koenigsberg (Kaliningrad RUS), 1782 – 1r – 1 – gw Misc Inst [790]
Die koenigsfanfare / Eich, Hedwig – Prag: Noebe & Co, 1944 (mf ed 1990) – 1r – 1 – (filmed with: der morgen) – us Wisconsin U Libr [430]
Die koenigsfanfare / Eich, Hedwig – Prag: Noebe & Co, 1944 (mf ed 1990) – 1r – 1 – (filmed with: der morgen) – us Wisconsin U Libr [430]
Koenigsgarten, Hugo F *see* Grundvorstellungen der amerikanischen wirtschafts-ethik
Das koenigsideal des alten testaments : rede zur akademischen feier... / Oettli, Samuel – Greifswald: Julius Abel, 1899 – 1mf – 9 – 0-7905-0508-8 – mf#1987-0508 – us ATLA [221]
Der koenigsleutnant : lustspiel in vier aufzuegen / Gutzkow, Karl – 9. aufl. Jena: H Costenoble, [1889] [mf ed 1993] – 111p – 1 – mf#8660 – us Wisconsin U Libr [820]
Koenigsloew, Eva von *see* Das religioese motiv als gestaltende kraft der deutschen volkssagen der gegenwart
Koenigsmark : the legend of the hounds and other poems / Boker, George Henry – Philadelphia: J B Lippincott & Co, 1869 [mf ed 1985] – 244p – 1 – mf#8261 – us Wisconsin U Libr [810]
Die koenigsreihen von juda und israel nach den biblischen berichten und den keilinschriften / Brandes, Heinrich – Leipzig: Alexander Edelmann, [1873?] – 1mf – 9 – 0-7905-0181-3 – mf#1987-0181 – us ATLA [221]
Der koenigstraum : roman / Vogt, Helmut – Stuttgart: Deutsche Verlags-Anstalt c1943 [mf ed 1991] – 1r – 1 – (filmed with: die baltische tragodie / siegfried von vegesack) – mf#2923p – us Wisconsin U Libr [830]
Koenigswusterhausener zeitung – Koenigs Wusterhausen DE, 1940 jan-jun, 1941 jan-jun – 2r – 1 – gw Misc Inst [074]
Koeningsbrief karls d gr an papst hadrian (tab6) : ueber waldo von reichenau-pavia / Munding, E – 1920 – (= ser Texte und arbeiten. beuron (tab). beitraege zur ergruendung des aelteren lateinischen und christlichen schrifttums und gottesdienstes) – €5.00 – ne Slangenburg [860]
Koennecke, Clemens *see* Emendationen zu stellen des neuen testaments
Koennecke, Gustav *see* Schiller, eine biographie in bildern
Koenning, Josef *see* Koerpertherapie mit kindern und jugendlichen
Koepfer, Benno *see* Das stadttor in der fruehen islamisch-arabischen welt
Koepke, Fr Karl *see* Das passional
Koepp, Wilhelm *see* Mystik, gotterserlebnis und protestantismus
Koeppen, Friedrich *see* Schellings lehre oder das ganze der philosophie des absoluten nichts
[Koeppen, K F] *see* Crata repoa
Koepping, Walter *see* Wir tragen ein licht durch die nacht
Koepruelue, Mehmet Fuat *see* Tuerkiye tarihi
Koerber, Kurt *see* Kirchengueterfrage und schmalkaldischer bund
Koerner, Friedrich *see* Sued-afrika
Koerner, Josef *see*
- Geschichte der deutschen sprache und poesie
- Das nibelungenlied
- Die schachtel mit der friedenspuppe

Koerner, R *see* San remo
Koerner, Theodor *see* Theodor koerners briefwechsel mit den seinen
Koeroglu – Kastamonu, 1907-? Mueduer-i Mes'ul: Ismail Sedat, Mueduer: A Nureddin. n14. 19 mart 1909 – (= ser O & t journals) – 1mf – 9 – $25.00 – us MEDOC [956]
Koerpertherapie mit kindern und jugendlichen / Koenning, Josef – (mf ed 1994) – 1mf – 9 – €30.00 – 3-8267-2032-6 – mf#DHS 2032 – gw Frankfurter [150]
Koerte, Franz *see* Moeglinische jahrbuecher der landwirthschaft
Koerting, Heinrich *see* Ueber zwei religioese paraphrasen pierre corneille's
Koester, Albert *see*
- Die briefe der frau rath goethe
- Der briefwechsel zwischen theodor storm und gottfried keller
- Faust
- Die ganze aesthetik in einer nuss

Koester, Friedrich *see* Die christliche glaubenslehre des herrn dr. david friedrich strauss
Koester, H *see* Synoptische ueberlieferung bei den apostolischen vaetern
Koester, H M G *see* Deutsche encyclopaedie
Koester, Heinrich Martin Gottfried *see* Die neuesten religionsbegebenheiten mit unpartheyischen anmerkungen
Koester hemp : lose geschichten von en luetten mann / Bandlow, Heinrich – Leipzig: P Reclam [18-?] [mf ed 1989] – (= ser Reclams universal-bibliothek 4029) – 1 – mf#6972 – us Wisconsin U Libr [830]
Koestering, Johann Friedrich *see* Auswanderung der saechsischen lutheraner im jahre 1838
Die koestliche perle und die innere mission : eine praktisch-theoretische meditation / Wacker, Emil – Guetersloh: C Bertelsmann, 1895 – 1mf – 9 – 0-8370-6785-5 – mf#1986-0785 – us ATLA [240]
Koestlin, Friedrich *see*
- Jesaia und jeremia
- Leitfaden zum unterricht im alten testament fuer hoehere schulen
- Leitfaden zum unterricht im neuen testament fuer hoehere schulen

Koestlin, Heinrich Adolf *see* Geschcihte des christlichen gottesdienstes
Koestlin, Julius *see*
- Die begruendung unserer sittlich-religioesen ueberzeugung
- Friedrich der weise und die schlosskirche zu wittenberg
- Der glaube und seine bedeutung fuer erkenntnis, leben und kirche
- Life of luther
- Luther and j. janssen
- Luthers lehre von der kirche
- Martin luther
- Religion und reich gottes
- Die schottische kirche
- The theology of luther in its historical development and inner harmony
- Das wesen der kirche nach lehre und geschichte des neuen testamentes

Koestlin, Karl *see* Goethes faust
Koestlin, Karl Reinhold *see* Der ursprung und die komposition der synoptischen evangelien
Koestlin, Karl Reinhold von *see* Richard wagner's tondrama
Koethener rundblick – Koethen DE, 1963 22 aug-1964 18 mar – 1r – 1 – gw Misc Inst [074]
Koethke, Ernst *see* Clemens brentanos religioeser werdegang
Koetoekannja boenga srigading / Tan, Boen Soan – Soerabaia: Tan's Drukkery, 1933 [mf ed 1998] – (= ser Penghidoepan 103) – 1r – 1 – (coll as pt of the colloquial malay collection. filmed with: nona olanda sebagi istri tionghoa / [njoo, cheong seng]) – mf#10000 – us Wisconsin U Libr [830]
Koetschau, Karl *see* Goethes schweizer reise 1775
Koetschau, Paul *see*
– Beitraege zur textkritik von origenes' johannescommentar
– Kritische bemerkungen zu meiner ausgabe von origenes' exhortatio, contra celsum, de oratione
– Die textueberlieferung der buecher des origenes gegen celsus
– Die textueberlieferung der buecher des origenes gegen celsus in den handschriften dieses werkes und der philokalia
Koetteritzsch, Georg A *see* Grosse koalition und opposition
Koetzschenbrodaer zeitung – Radebeul DE, 1865 13 dec-1941 29 nov – 89r – 1 – (missing: 1866. later: general-anzeiger des amtsgerichtsbezirks koetzschenbroda. with suppl: die elbaue 1924 mar-1941 feb [2r]) – gw Misc Inst [074]
Koetzschke, R *see*
– Rheinische urbare, bd 2
– Rheinische urbare, bd 3
Koetztinger umschau – Koetzting DE, 1956 may-1973 jul – (= ser Bezirksausgabe von mittelbayerische zeitung, regensburg) – 1 – gw Misc Inst [074]
Koeylue – Izmir, 1908-21. Sahib-i Imtiyaz: Ismail Siddik, Mehmet Refet. n600. 8 agustos 1326 [1910], 601, 1048?. 14 agustos 1334 [1918] – (= ser O & t journals) – 1mf – 9 – $25.00 – us MEDOC [079]
Koeymen, Mehmed A *see* Kitab-i cihan-nuema [nesri tarihi]
Koey-tjoe say ma-tiauw / Bie, L Th – Batavia: Goedang Tjerita, 1949 [mf ed 1998] – (= ser Goedang tjerita 26) – 1r – 1 – (coll as pt of the colloquial malay collection. filmed with: lajangan biroe / oleh im yang tjoe) – mf#10005 – us Wisconsin U Libr [490]
Koff, M *see* The rivonia trial
Kofoed-Hansen, Hans Peter *see*
– Dr s kierkegaard mod dr h martensen
– S kierkegaard mod det bestaaende
– Tegnet fra himmelen
Kofoid, Charles Atwood *see* Termites and termite control
Kogan, Elena I *see* Novyi zhurnal
Kogda razguliaetsia / Pasternak, Boris Leonidovich – Parizh, France. 1959 – 1r – us UF Libraries [025]
Kogutad jutlusi = Selected sermons / Lige, B – Toronto: Oma Press, 1975. Includes sermons by Estonian Baptist pastors; 166p – 1 – 6.64 – us Southern Baptist [242]
Die kohasion innerhalb der thematik durch die motivik : dargestellt an dem werk "histoire du soldat" von igor strawinski / Petri, Hasso Gottfried – [mf ed 2002] – 3mf – 9 – €49.00 – 3-8267-2783-5 – mf#DHS2783 – gw Frankfurter [780]
Kohelet – Leipzig, Germany. 1871 – 1r – us UF Libraries [939]
Kohelet : oder, der salmonischen prediger / Graetz, Heinrich – Leipzig: C F Winter, 1871 – 1mf – 9 – 0-8370-3357-8 – (text in german & hebrew. commentary in german) – mf#1985-1357 – us ATLA [221]
Kohieren van de tiende penning 1543 = Tax rolls for the tenth penny 1543 / Netherlands. General State Archives – 313mf – 9 – ne MMF Publ [949]
Kohl, Christopher C *see* Parental influences in youth sport
Kohl, Horst Ernst Arminius *see* Bismarck-jahrbuch
Kohl, Johann *see*
– Hundert tage auf reisen in den oesterreichischen staaten
– J g kohl's reisen in deutschland
– Nordwestdeutsche skizzen
– Reise nach istrien, dalmatien und montenegro
– Reisen im nordwesten der vereinigten staaten
– Reisen in canada und durch die staaten von new-york und pennsylvanien
– Reisen in den niederlanden.
– Reisen in irland
Kohl, Johann G *see* Die marschen und inseln der herzogthuemer schleswig und holstein
[Kohlard, Johann Joseph] *see* Der unglueckseelige todes-fall caroli 12.
Kohler, C *see* Melanges pour servir...l'histoire de l'orient latin et des croisades
Kohler Co *see* People
Kohler, Erika *see* Liebeskrieg
Kohler, F J et al *see* Hammurabi's gesetz
Kohler, J et al *see* Hammurabi's gesetz
Kohler, Josef *see*
– Lehrbuch des buergerlichen rechts
– Philosophy of law
Kohler, Josef et al *see*
– Archiv fuer buergerliches recht
– Hammurabi's gesetz
Kohler, Kaufmann *see*
– Ausgewaehlte predigten und reden
– Grundriss einer systematischen theologie des judentums auf geschichtlicher grundlage
– Hebrew union college
Kohler, Max James *see* Jewish rights at the congresses of vienna (1814-1815) and aix-la-chapelle (1818)
Kohler, Oswin *see* Study of karibib district (south west africa)
Kohler strike and boycott bulletin / International Union, United Automobile, Aircraft, and Agricultural Implement Workers of America – 1956 apr 30-1959 aug 28 – 1r – 1 – (cont: daily strike bulletin [local 833]) – us WHS [331]
Kohlerian / International Union, United Automobile, Aicraft, and Agricultural Implement Workers of America et al – 1939 aug 3, 1943 sep 16-1946, 1947-50, 1951-52, 1953-54, 1954 mar 25-apr 22, 1955-1956 aug 17 – 7r – 1 – (cont by: uaw local 833 reporter and kohlerian) – mf#3376898 – us WHS [331]
Kohlschmidt, Werner *see*
– Form und innerlichkeit
– Herder-studien
Kohn, Barukh *see* Shevil ha-zahay
Kohn, Hans *see*
– Perakim le-toldot ha-ra'yon ha-tsiyoni
– Zionistische politik
Kohn, Leo *see* Mo'ed va-'atseret
Kohn, S *see* David leib magdeburger
Kohn, Samuel *see*
– De pentateucho samaritano ejusque cum versionibus antiquis nexu
– Die sabbatharier in siebenbuergen
– Samaritanische studien
– Samaritansische studien
– Zur sprache, literatur und dogmatik der samaritaner
Kohn, Tobias *see* 'Aliyat Tuvyah
Kohn-Bistritz, Majer *see* Mannheimer-album
Kohnle, Eduard Hans *see* Studien zu den ordnungsgrundsaetzen mittelhochdeutscher liederhandschriften
Kohn-Zwilling, Isolde *see* Untersuchungen zum dialog bei christopher fry
K'o-hsueh = Science – Shanghai, Jul 1915-Sep 1950 (incomplete) – 25r – 1 – $692.00 – us Chinese Res [500]
Kohut, Adolf *see*
– David friedrich strauss als denker und erzieher
– Heinrich von kleist und die frauen
Kohut, Adolph *see*
– Johann gottfried seume als mensch, dichter, patriot und denker
– The odor koerner
– Ragende gipfel
Kohut, Alexander *see*
– Secular and theological studies
– Ueber die juedische angelologie und daemonologie in ihrer abhaengigkeit vom parsismus
Kohut, George Alexander *see* Semitic studies in memory of rev. dr. alexander kohut
Koik / Ristikivi, Karl – Vadstena, Sweden . 1946 – 1r – us UF Libraries [960]
Koikylides, K M *see* Ai para ton iordanen laurai kalamoonos kai agiou gerasimou
Koinange, Mbiyu *see* The people of kenya speak for themselves
Koinonia kaller / Christian Student Foundation – v1 n1-v2 n1 [1975 may 1-1976 dec] – 1r – 1 – mf#382333 – us WHS [240]
Koinonia Partners *see* Newsletter
Ko-ji-ki : or, records of ancient matters / O, Yasumaro – [s.l: s.n, 1882?] [mf ed 1992] – (= ser [Transactions of the asiatic society of japan] 10) – 2mf – 9 – 0-524-02716-1 – (incl bibl ref) – mf#1990-3119 – us ATLA [390]
Kojisho shusei : collection of japanese classical language dictionaries in the custody of the seikado library, tokyo – 2137v on 178r – 1 – Y1,840,000 – (with 64p guide. in japanese) – ja Yushodo [480]
Kok, K J de *see* Empires of the veld
Kokchetavskii listok : gaz obshchestv i lit / ed by Poimenov, A – Kokchetav [Akmol obl: Upr komendanta g Kokchetava 1918 [1918-] – (= ser Asn 1-3) – n23-47 [1918] [gaps] item 205, on reel n41 – 1 – mf#asn-1 205 – ne IDC [077]
Kokhmanskii, P V *see*
– Ponedel'nik [khar'kov: 1919]
– Rodina [khar'kov: 1919]
Kokhomskii, S V *see* Primechaniya k evangeliyu v oblichenie shtundistov i podobnykh im sektantov
Kokille – Wernigerode DE, 1965 21 mar-1989 23 sep [gaps] – 4r – 1 – (metallgusswerk) – gw Misc Inst [660]
Kokka : an illustrated monthly journal of the fine and applied arts of japan and other eastern countries – n1-1149. 1889-1991 – 48r; 1 col r; 47 col mf – 1,14,15 – Y947,000 – (with 530p guide. in japanese, abstract in english) – ja Yushodo [700]
Kokka – Tokyo, 1889-1977 [mf ed Chadwyck-Healey] – (= ser Art periodicals on microform) – 42r b/w 47col mf – 1,15 – uk Chadwyck [760]
Kokkai kaigiroku : proceedings of the national diet of japan, 1947-1972 – House of Representatives and Councilors, 1st-69th sess – 66r – 1 – Y554,000 – (in japanese) – ja Yushodo [323]
Kokogyo kankei kaisha hokokusho – 50r – 1 – Y850,000 – (reports on industrial, manufacturing, and mining companies for ghq in 1945, in the holding of the library of faculty of economics, university of tokyo / reports produced and submitted in response to ghq's request in the autumn 1945, about japanese industrial and mining companies of which total volume of business in 1944 were over 1000,000 yen; guide carries list of company names with int by haruhito takeda) – ja Yushodo [338]
Kokovin, G A *see* Vsesoiuznaia shkola "primenenie matematicheskikh metodov dlia opisaniia i izucheniia khimicheskikh ravnovesii," g novosibirsk, 9-13 fevralia 1976 g: tezisy dokladov
Kokovtsov, VN *see* Iz moego proshlogo
Kokstad advertiser – Kokstad SA, 1882-1911 – 27r – 1 – mf#MS00297 – sa National [079]
Kokugogaku shiryo shusei : dr k matsui collection of japanese classical language in the custody of the seikado library, tokyo – 769v on 54r – 1 – Y565,000 – (with 40p guide. in japanese) – ja Yushodo [480]
Kol ha t'nuah – New York, NY. Nov 1974-Mar 1985 – 1 – us AJPC [071]
Kol hahinukh ha-tsiyoni / Ben-Yehudah, Barukh – Jerusalem, Israel. 1954 or 1955 – 1r – us UF Libraries [939]
Kol ha-yetsarim / Bar-Yossef, Yehoshua – Jerusalem, Israel. 1937 – 1r – us UF Libraries [939]
Kol, Henri Hubert Van *see* Naar de antillen en venezuela
Kol hillel – Saskatoon, Saskatchewan.Dec 1949; May 1950; Mar 1951; Jan 1958 – 1 – us AJPC [071]
Kol kitve mordekhai tsevi maneh / Mane, Mordecai Zevi – Warsaw, Poland. v1-2. 1897 – 1r – us UF Libraries [939]
Kol kitve s an-ski / An-Ski, S – Vilna, Lithuania. 1930 – 1r – us UF Libraries [939]
Kol kore = Die bibel, der talmud, und das evangelium / Soloweyczyk, Elias – Leipzig: F A Brockhaus, 1877 – 1mf – 9 – 0-8370-5325-0 – (in german) – mf#1985-3325 – us ATLA [220]
Kol mevaser / Hurvits, Shim'on Tsevi – Jerusalem, Israel. 1922 – 1r – us UF Libraries [939]
Kol Shalom Journalism Club [Southern Illinois University at Carbondale] *see* Nonsequitur
Kol simhah / Paltrovitz, Simhah – Jerusalem, Israel. 1906 or 1907 – 1r – us UF Libraries [939]
Kolarovgradska borba – Kolarovgrad, Bulgaria. 1951-63 – 6r – 1 – us L of C Photodup [949]
Kolasker, M B *see* Religious and social reform
Kolb, Georg *see* Wegweiser in die marianische literatur
Kolb, J Chr *see* Das frolockende augspurg
Kolb, John P *see* Selected metabolic functions of chronic low back pain patients with a gender matched control group
Kolb, Peter *see*
– Description du cap de bonne-esperance.
– Description du cap de bonne-esperance
– Naaukeurige en uitvoerige beschryving van de kaap de goede hoop
Kolbasov, Oleg Stepanovich *see* Zakonodatel' stvo o vodopol'zovanii v sssr; problemy sovershenstvovaniia sovetskogo zakonodatel' stva ob ispol'zovanii vodnykh resursov
Kolbe, E *see* Text zu den illustrationen aus reuter's werken
Kolbe, F W *see* Vowels
Kolbe, Henry W *see* Hospital survey of the republic of guatemala
Kolbe, Hermann *see* Electrolysis of organic compounds
Kolbe, Ulrich *see* Analyse zu kardinalsymptomen im langzeitverlauf des morbus meniere
Kolbenheyer, Erwin Guido *see*
– Die begegnung auf dem riesengebirge
– Die bruecke
– Paracelsus
– Weihnachtsgeschichten
Kolberg, A *see* Agenda communis
Kolberg, Joseph *see* Nach ecuador
Kolberger zeitung *see* Gemeinnuetziges colberger wochenblatt
Kolchakovshchina : (khronika) / Dorokhov, Pavel Nikolaevich – Moskva: "Zemlia i fabrika", 1924 [mf ed 2002] – (= ser Biblioteka "zemli i fabriki") – 1r – 1 – (filmed with: rafael' / boris zaitsev (1924)) – mf#5238 – us Wisconsin U Libr [830]
Kolde, Theodor *see*
– Luther und der reichstag zu worms, 1521
– Das religioese leben in erfurt beim ausgange des mittelalters
Koldewey, Friedrich *see*
– Heinz von wolfenbuettel
– Streitgedichte gegen herzog heinrich den juengern von braunschweig
Koldewey, Friedrich E *see* Friedens sieg
Koldewey, Paul *see* Wackenroder und sein einfluss auf tieck
Koldewey, Robert *see* The excavations at babylon
Kolebatelnye sistemy s ogranichennym vozbuzhdeniem / Kononenko, K O – Moskva: Nauka, 1964 – us CRL [947]
Kolek, Iveta *see* Die anwendung der lippenbluetler in der zahnheilkunde von der antike bis heute
Kolenati, F A *see* Reiseerinnerungen
Kolenu – Ithaca, N.Y. – 1 – (v. 1, no. 1 (april 1973); vol. 3, no. 2 (may 1975)-v. 15, no. (fall 1987)) – us AJPC [978]
Kolessa, Filaret, 1871-1947 *see* Studii nad poetichnoiu tvorchistiu t shevchenka
Kolfhaus, W *see* Der verkehr calvins mit bullinger
Kolgespnik ukraini – Kiev, USSR. May 1-Dec 30 1939; Feb 29-Dec 24 1940; Jan 16-Feb 27 1941 – 1r – 1 – us L of C Photodup [077]
Koliupanov, N P *see* Prakticheskoe rukovodstvo k uchrezhdeniiu sel'skikh i remeslennykh bankov po obraztsu nemetskikh ssudnykh tovarishchestv
Kolkhorst, FW *see* The effects of exercise mode on postexercise oxygen consumption, urinary nitrogen, and fat utilization
Kolkhoznaia pravda – Minsk, Belarus 1973 [mf ed Norman Ross] – 8r – 1 – mf#nrp-1030 – us UMI ProQuest [077]
Kolkhoznaia stroika na tereke / Todres, V – [Piatigorsk?]: Gosizdat. Severo-Kavkazskii Otdel, 1930 [mf ed 2004] – 1r – 1 – (filmed with: lessons of the wrecking, diversionist and espionage activities of the japanese-german-trotskyite agents / v m molotov (1937)) – us Wisconsin U Libr [334]
Kolkhoznik za knigoi – Leningrad. n1-10. oct 28 1930-oct 20 1931 – 1 – us NY Public [073]
Koll, Kilian *see*
– Andreas auf der fahrt
– Urlaub auf ehrenwort
Kollarius, A F *see* Analecta monumentorum omnis aevi vindobonensia
Kollbach, Karl *see* Rheinisches wanderbuch
Das kollektiv – Eisleben DE, 1957 19 jan-1960 9 feb [gaps] – 1r – 1 – (kupferbergbau) – gw Misc Inst [334]
Das kollektiv – Weissandt DE, 1966 5 feb-1981 nov [gaps], 1982-1990 8 jan – 4r – 1 – (orbitaplast goelzau) – gw Misc Inst [334]
Kollektiv Eulenspiegel Verlag *see* Geschichten
Kollektiv fuer Literaturgeschichte im Volkseigenen Verlag Volk und Wissen *see* Johannes r becher
Kollektivnoe ispolzovanie selskokhoziaistvennykh mashin i orudii / Debu, K I – 1924 – 66p 1mf – 9 – mf#COR-471 – ne IDC [335]
Kollektivschuld? *see* Nsdap (national socialist german workers party) nazi publications
Koller, Hans Albert *see* Studien zu m von ebner-eschenbach
Koller, W H *see* Faust papers
Kollewijn, Roeland A *see* Ueber den einfluss des hollaendischen dramas auf andreas gryphius
Kollmann, Fritz *see* Novalis "heinrich von ofterdingen" und der "guido" des grafen von loeben
Kolloff, Eduard *see* Schilderungen aus paris
Kollonitz, Paula *see* Eine reise nach mexico im jahre 1864
Kollontai, Aleksandra *see* Hsin fu nu lun
Kollontai, Alexandra *see* La femme nouvelle et la classe ouvriere
Kolmodin, Adolf *see*
– Fran de svartas vaerldsdel
– Fran "soluppgangens land"
Kolmodin, J A *see* Traditions de tsazzegga et hazzegga
Kolnische zeitung – Cologne, Germany. Aug 1906-Apr 1936; Jan-May 1941 – 126r – 1 – us L of C Photodup [074]
Kolns legenden, sagen, und geschichten – Koeln: K A Stauff 1919-21 [mf ed Bloomington IN: Indiana Uni Lib, Preservation Dept 1984] – 2v on 1r – 1 – us Indiana Preservation [390]
Kolo – [Beograd]: Stamparija P Curcica 1901- [mf ed Bloomington IN: Indiana Uni Lib, Preservation Dept 1989] – 2r – 1 – us Indiana Preservation [073]
Kolo – U Beogradu: Stamparija Petra Curcica 1889- [mf ed Bloomington IN: Indiana Uni Lib, Preservation Dept 1989] – 1r – 1 – us Indiana Preservation [073]

Kolo – Zagreb: Matica hrvatska 1963- [mf ed Bloomington IN: Indiana Uni Lib, Preservation Dept 1984] – 1 – us Indiana Preservation [073]

Kolo – Zagreb: Tiskom K P Nar Tiskarne dra hjudevita Gaja 1842-53 [mf ed Bloomington IN: Indiana Uni Lib, Preservation Dept 1989] – 1r – 1 – us Indiana Preservation [073]

Kolokol – London, England 1857-68 [mf ed Norman Ross] – 1 – mf#nrp-869 – us UMI ProQuest [077]

Kolokol / Soiuz russkogo naroda – Volgograd, Russia. n86[17.dek 1994], n88[31.ianv 1995]-n92[28.ianv 1995]; n95[18.fev 1995], n98[11.mar 1995]-n100[25.mar 1995]; n102[8.apr 1995]-n104[22.apr 1995], n107[13.mai.1995]; n23[160][3.avg 1996], n39[176][28.dek 1996]-n123[260][16.okt 1998]; n125[262]-n128[265][20.noia.1998], n130[267][4.dek 1998]-n133[270][25.dek 1998] – 1 – mf#mf-12248 [reel 5-6] – us CRL [077]

Kolokol *see*
- Pribavochnye listy k "poliarnoi zvezde"
- Russkoe pribavlenie

Kolonial missie tijdschrift *see* Onze missien in oost- en west-indien

Koloniales schrifttum in deutschland *see* Nsdap (national socialist german workers party) nazi publications

Kolonialherrschaft und sozialstruktur in deutsch-sudwestafrika / Bley, Helmut – Hamburg, Germany. 1968 – 1r – us UF Libraries [960]

Kolonie – Santa Cruz (BR), 1907 20 apr-1939 1 sep [gaps] – 21r – 1 – gw Misc Inst [079]

Kolonie und heimat 1937 – Berlin DE, 1937 sep-1943 mar – 5r – 1 – gw Misc Inst [943]

Kolonie zeitung : jornal da colonia de dona francisca – Joinville, SC: Typ de C W Boehm, 20 dez 1862-21 maio 1942 – (= ser Ps 19) – mf#UFSC/BPESC – bl Biblioteca [079]

Kolonien des dritten reiches *see* Nsdap (national socialist german workers party) nazi publications

Kolonie-zeitung – Joinville (BR), 1876, 1920 3 feb-1939 17 nov – 10r – 1 – gw Misc Inst [079]

Kolonkarzinomchirurgie im karl-olga-krankenhaus / Stephan, Sabine – (mf ed 1998) – 2mf – 9 – €40.00 – 3-8267-2511-5 – mf#DHS 2511 – gw Frankfurter [617]

Die kolonne (mme2) : zeitung der jungen gruppe dresden – Dresden 1929/30-1932 [mf ed 1998] – (= ser Marbacher mikrofiche-editionen (mme) 2; Kultur – literatur – politik: deutsche zeitschriften des 19./20. jahrhunderts (klp)) – 3mf – 9 – €50.00 – 3-89131-287-3 – gw Fischer [074]

Kolorektale karzinome in der chirurgisch-proktologischen praxis : unter besonderer beruecksichtigung des lynchsyndroms (hnpcc). ergebnisse einer retrospektiven studie / Labonte, Bernd – (mf ed 1998) – 1mf – 9 – €30.00 – 3-8267-2541-7 – mf#DHS 2541 – gw Frankfurter [617]

Kolosov, Evgenii Evgen'evich *see* Ocherki mirovozzrieniia n k mikhailovskago

Kolosov, N A *see* Vazhneishiia russkiia tserkovnyia knigokhranilishcha

Kolozsvari estilap – Cluj, Romania. 1941-42 – 2r – 1 – us L of C Photodup [949]

Kolvoord, John *see* The vision of the evening and the morning

Komando tertinggi abri resimen tjakrabirawa / Tjakrabirawa – Djakarta, 1962/1963 – 8mf – 9 – (missing: 1962/1963(6-end)) – mf#SE-598 – ne IDC [950]

Komarov, A V *see* Krest'ianskii vestnik

Komarovskii, V A *see* Amurskii liman [primor obl]

Kombinationswirkungen in der toxikologie / Thier, Ricarda et al – (mf ed 1998) – 1mf – 9 – €30.00 – 3-8267-2585-9 – mf#DHS 2585 – gw Frankfurter [615]

Kombinierte enzymatische synthese von dtdp-6-dsoxy-4-keto-d-glucose mit saccharose-synthase und dtdp-d-glucose-4, 6-dehydratase / Stein, Andreas – (mf ed 1996) – 2mf – 9 – €40.00 – 3-8267-2311-2 – mf#DHS 2311 – gw Frankfurter [540]

Komdak 7 djaya / Wira dharma – Djakarta, 1969-1970. v1 – 15mf – 9 – mf#SE-2000 – ne IDC [950]

Komertsyal bank / Banco Comercial Israelita – Buenos Ayres, Argentina. 1950 – 1r – us UF Libraries [939]

Der komet : organ zur wahrung der interessen der besitzer von sehenswuerdigkeiten und schaustellungen jeder art – Pirmasens DE, 1894 6 jan-1944 23 sep [gaps] – 74r – 1 – (title varies: 1 may 1943: gemeinschaftsblatt) – gw Mikrofilm [790]

Die kometen und meteore in allgemein fasslicher form dargestellt / Valentiner, Karl Wilhelm Friedrich Johannes – Leipzig: G Freytag 1884 [mf ed 1999] – (= ser Wissen der gegenwart 27) – 1r [ill] – 1 – (incl ind) – mf#film mas 28420 – us Harvard [520]

Kometentanz : astrale pantomime in zwei aufzuegen / Scheerbart, Paul – Leipzig: Insel-Verlag, 1903 – 1r – 1 – us Wisconsin U Libr [820]

Komin bunkai = Huang-ming-wen-hai – 175v in 170 – 24r – 1 – Y240,000 – (in chinese; historical biography comp on the authority of inscription and epitaphs in the ming dynasty, medieval china. in the holdings of the hosokawa family, the head of which formerly was feudal lord of kumamoto) – ja Yushodo [951]

Komin kyoiku : journal of the public education before world war 2 – c2000 items on 120r – 1 – Y330,000 – (in japanese) – ja Yushodo [370]

Komin, V V *see*
- Bankrotstvo burzhuaznykh i melkoburzhuaznykh partii rossii v period podgotovki i pobedy velikoi oktiabrskoi sotsialisticheskoi revoliutsii
- Burzhuaznye i melkoburzhuaznye partii rossii v oktiabrskoi revoliutsii i grazdanskoi voine

Komisarjevsky, Theodore *see* The costume of the theatre

Die komischen elemente der altfranzoesischen chansons de geste / Theodor, Hugo – Halle, 1913 (mf ed 1994) – 2mf – 9 – €31.00 – 3-8267-3018-6 – mf#DHS-AR 3018 – gw Frankfurter [440]

Die komischen mysterien des franzoesischen volkslebens in der provinz : eine sammlung von sittenstudien, komischen und burlesken scenen, volksschwaenken, etc; aus franzoesischen schriftstellern der gegenwart / Baumgarten, Johann – Coburg 1873 [mf ed Hildesheim 1995-98] – 1v on 3mf – 9 – €90.00 – 3-487-25859-5 – gw Olms [390]

Komisi hukum, kongres wanita indonesia : kedudukan wanita indonesia – Djakarta, April, 1959 – 4mf – 9 – mf#SE-779 – ne IDC [959]

Komissiya istorychnoi pisennosty ukrayins'ki narodni dumy / Akademiia Nauk. URSR. Kiev. Istorychna Sektziya – v1. 1927 – 1 – us NY Public [073]

Komite nasional pusat badan pekerdja risalah : indonesia – Djakarta, 1945-1950 – 7mf – 9 – mf#SE-233 – ne IDC [959]

Komite Olahraga Nasional Indonesia *see* Almanak organisasi KONI Pusat

Komite Olympiade Indonesia *see* Bulletin

Komitee der Antifaschistischen Widerstandskaempfer in der DDR *see* Immer bereit fuer die verteidigung der freiheit des volkes

Komitet o Sel'skikh Ssudo-sberegatel'nykh i Promyshlennykh Tovarishchestvakh *see* Trudy 1-go sezda predstavitelei uchrezhdenii melkogo kredita menzelinskogo uezda v g. menzelinske (s 8 po 20 iiunia 1911 goda)

Komitet S"ezda predstavitele' bankov kommercheskogo kredita *see* Stenograficheskii otchet zasedanii obshchikh sobranii ii-go s"ezda predstavitelei obshchestv vzaimnogo kredita v s-peterburge, 10-17 avgusta 1898 goda

Komitet S"ezda predstavitelei aktsionernykh bankov kommercheskogo kredita *see* Statistika kratkosrochnogo kredita

Komitet S"ezda predstavitelei bankov kommercheskogo kredita *see* Russkie banki

Komitet Sezdov predstavitelei uchrezdenii russkogo zemel'nogo kredita *see* Sve deniia ob otsenkakh po ssudam, vydannym aktsionernymi zemel'nymi bankami pod zalog zemel'v 1902-1907 gg

Komitet S"ezdov predstavitelei uchrezhdenii russkogo zemel'nogo kredita *see*
- Stenograficheskii otchet zasedanii komiteta s"ezdov predstavitelei uchrezhdenii russkogo zemel'nogo kredita, 9, 11 i 14 marta 1891 g po voprosu o konversii 6% ssud v 5%
- Stenograficheskii otchet zasedanii xiii-go s"ezda predstavitelei zemel'nykh bankov v oktiabre 1915 g
- Ustav aktsionernykh zemel'nykh bankov s raz"iasneniem voprosov, voznikshikh na praktike pri ego primenenii

Komitet Sezdov predstavitelei uchrezhdenii russkogo zemel'nogo kredita *see*
- Materialy i postanovleniia po voprosam, podlezhashchim obsuzhdeniiu v komitete s'ezdov predstavitelei uchrezhdenii russkogo zemel'nogo kredita, v aprele 1901 g
- Sbornik materialov i postanovlenii po voprosam, podlezuhavshchim obsuzhdeniiu komiteta sezdov predstavitelei zemel'nykh bankov
- Sbornik materialov i postanovlenii po voprosam, podlezuhavshchim obsuzhdeniiu v-go sezda predstavitelei uchrezhdenii russkogo zemel'nogo kredita

Komitet s"ezdov predstavitelei uchrezhdenii russkogo zemel'nogo kredita *see* Statisticheskii sbornik svedenii po zemel'nomu kreditu v rossii

Komitet Vladimirskoi Gubernskoi Kassy Sotsial'nogo Strakhovaniia *see* Otchet komiteta vladimirskoi gubernskoi kassy sotsial'nogo strakhovaniia o rabote strakhovykh organov vladimirskoi gub za 1924-25 operats god

Komm hernieder und hilf uns! : predigt, am missionsfeste, den 6. juli 1886, im dome zu braunschweig, ueber apostelgeschichte 16, 9-10 / Ruperti, Justus – Braunschweig: H Wollermann, 1886 – 1mf – 9 – 0-524-03360-9 – mf#1990-0941 – us ATLA [240]

Komm mit nach madeira : erzaehlung / Zerkaulen, Heinrich – Guetersloh: C Bertelsmann, 1943 – 1r – 1 – us Wisconsin U Libr [430]

Kommando – Mzuzu, Malawi: Friendly Publ [mar 11/24-aug 19/sep 3 1996] (biwkly) – 1r – 1 – us CRL [079]

Das kommende deutschland : die erziehung der jugend im reich adolf hitlers / Kaufmann, Guenter – Berlin: Junker & Duennhaupt 1943 [mf ed 1984] – 1r [ill] – 1 – (with: griechisches erbe / klatt, f) – mf#8556 – us Wisconsin U Libr [943]

Das kommende geschlecht : zeitschrift fuer familienpflege und geschlechtliche volkserziehung auf biologischer und ethischer grundlage – 1921-34 [mf ed 2002] – (= ser Hq 53) – 7v on 16mf – 9 – €110.00 – 3-89131-388-8 – (with v5 known as: zeitschrift fuer eugenetik. ergebnisse der forschung) – gw Fischer [640]

Kommende kirche – Bremen DE, 24 sep 1936-25 dez 1938 – 1r – 1 – gw Misc Inst [074]

Die kommende romantik philipp veit und ernst lieber / Cardauns, Hermann – (mf ed 1999) – 1mf – 9 – €24.00 – 3-8267-3219-7 – mf#DHS-AR 3219 – gw Frankfurter [430]

Die kommenden – Flarchheim DE, 1926-33 – 5r – 1 – gw Misc Inst [074]

Die kommenden : unabhaengige zeitschrift fuer freies geistesleben – Freiburg Br DE, 1946 1 oct-1957 25 dec – 3r – 1 – gw Misc Inst [074]

Kommentar ueber den brief pauli an die epheser / Stoeckhardt, G – St. Louis, MO: Concordia Pub House, 1910 – 1mf – 9 – 0-7905-3292-1 – mf#1987-3292 – us ATLA [227]

Kommentar ueber den ersten brief petri / Stoeckhardt, George – St Louis, MO: Concordia Pub House, 1912 – 1mf – 9 – 0-524-05239-5 – mf#1992-0372 – us ATLA [227]

Kommentar ueber den prediger / Scholz, Anton – Leipzig: L Woerl, 1901 – 1mf – 9 – 0-7905-3477-0 – mf#1987-3477 – us ATLA [220]

Kommentar zu den briefen des hl paulus an die galater und epheser, 8. bd (bdk15 2.reihe) / Chrysostomus (Chrysostom, John, Saint) – (= ser Bibliothek der kirchenvaeter. 2. reihe (bdk 2.reihe)) – €18.00 – ne Slangenburg [227]

Kommentar zu den briefen des hl paulus an die philipper und kolosser, 7. bd (bdk45 1.reihe) / Chrysostomus (Chrysostom, John, Saint) – (= ser Bibliothek der kirchenvaeter. 1. reihe (bdk 1.reihe)) – €17.00 – ne Slangenburg [240]

Ein kommentar zu goethes faust / Boyesen, Hjalmar Hjorth – Leipzig: P Reclam, [1881] (mf ed 1990) – 1r – 1 – (incl bibl ref; filmed with: "old-iniquity": der schluessel zu goethes "faust" / ottomar beta) – mf#7341 – us Wisconsin U Libr [430]

Kommentar zum briefe an die hebrer *see* Commentary on the epistle to the hebrews

Kommentar zum briefe des hl paulus an die roemer, 5. bd 1. teil (bdk39 1.reihe) : homilien 1-15 / Chrysostomus (Chrysostom, John, Saint) – (= ser Bibliothek der kirchenvaeter. 1. reihe (bdk 1.reihe)) – €14.00 – ne Slangenburg [240]

Kommentar zum briefe des hl paulus an die roemer, 6. bd 2. teil (bdk42 1.reihe) : homilien 16-33 / Chrysostomus (Chrysostom, John, Saint) – (= ser Bibliothek der kirchenvaeter. 1. reihe (bdk 1.reihe)) – €14.00 – ne Slangenburg [240]

Kommentar zum evangelium des hl matthaeus, 1. bd (bdk23 1.reihe) : homilien 1-18 / Chrysostomus (Chrysostom, John, Saint) – (= ser Bibliothek der kirchenvaeter. 1. reihe (bdk 1.reihe)) – €15.00 – ne Slangenburg [240]

Kommentar zum evangelium des hl matthaeus, 2. bd (bdk25 1.reihe) : homilien 19-42 / Chrysostomus (Chrysostom, John, Saint) – (= ser Bibliothek der kirchenvaeter. 1. reihe (bdk 1.reihe)) – €15.00 – ne Slangenburg [240]

Kommentar zum evangelium des hl matthaeus, 3. bd (bdk26 1.reihe) : homilien 43-71 / Chrysostomus (Chrysostom, John, Saint) – (= ser Bibliothek der kirchenvaeter. 1. reihe (bdk 1.reihe)) – €17.00 – ne Slangenburg [240]

Kommentar zum evangelium des hl matthaeus, 4. bd (bdk27 1.reihe) : homilien 72-90. ueber das priestertum / Chrysostomus (Chrysostom, John, Saint) – (= ser Bibliothek der kirchenvaeter. 1. reihe (bdk 1.reihe)) – €18.00 – ne Slangenburg [240]

Kommentar zum evangelium johannis = Commentary on the gospel of john / Tholuck, August – Philadelphia: Smith, English, 1859 – 2mf – 9 – 0-7905-0172-4 – (english) – mf#1987-0172 – us ATLA [226]

Kommentar zur eg-verordnung nr. 4064/89 ueber die kontrolle von unternehmenszusammenschluessen / Miersch, Michael – Neuwied, Frankfurt/Main: Luchterhand, 1991 (mf ed 1996) – 3mf – 9 – €38.00 – 3-8267-9674-8 – mf#DHS 9674 – gw Frankfurter [341]

Kommentar zur konkursordnung / Hess, Harald – Neuwied, Kriftel, Berlin: Luchterhand, 1995 (mf ed 1996) – 26mf – 9 – €149.00 – 3-8267-9669-1 – mf#DHS 9669 – gw Frankfurter [338]

Kommercheskaia gazeta – [Saint Petersburg]: Tip Departamenta vnieshnei torgovli, 1835-1839 – 5r – 1 – us CRL [380]

Kommercheskaia gazeta – St Petersburg, Russia 1825-58 [mf ed Norman Ross] – 1 – mf#nrp-1603 – us UMI ProQuest [077]

Kommercheskaia gazeta – [St Petersburg]: Tip Departmenta vnieshnei torgovli, 1827; 1829-34; 1840-60 – 8r – 1 – us CRL [077]

Kommercheskie banki i ikh torgovo-komissionnye operatsii / Katsenelenbaum, Z S – M, 1912 – 2mf – 9 – mf#REF-307 – ne IDC [332]

Kommercheskii bank v kostrome : pravila o tekushchikh schetakh – M, 1871 – 1mf – 9 – mf#REF-283 – ne IDC [332]

Kommerell, Max *see*
- Geist und buchstabe der dichtung
- Jean pauls verhaeltnis zu rousseau
- Jugend ohne goethe

Kommersant (daily) : gazeta izdatel'skogo doma "kommersant" – Moscow: ZAO Kommersant. Izdatel'skii dom 1992- – 1 – us East View [077]

Kommissie van ondersoek na die onluste te paarl op die 21e nov 1962 = Commission of inquiry into the events at paarl, on the 20th to 22nd november 1962 / Snyman, J H – [Minutes of evidence, Johannesburg?, 1962-63] – us CRL [960]

Kommission fuer spaetantike Religionsgeschichte *see* Kleine schriften 3

Kommun aktuellt – 1978-. Veckotidning utgiven av Kommunforbundet. Stockholm: Kommunforbundet, 1978. Forty issues yearly – 1 – us Wisconsin U Libr [948]

Kommunal'nye banki / Derevitskii, V A – M, 1927 – 1mf – 9 – (missing: p33-48) – mf#REF-102 – ne IDC [332]

Kommunar : ezhednevnaia rabochaia gazeta. izdanie tsk rkp(b) – Moscow, Russia 1918-19 [mf ed Norman Ross] – 3r – 1 – mf#nrp-1075 – us UMI ProQuest [077]

Kommunar – Moscow, USSR. Nov 12 1918-Apr 23 1919 – 1r – 1 – us L of C Photodup [077]

Kommunarka ukrainy – Khar'kov, Russia 1921-33 [freq varies] [mf ed Norman Ross] – 152mf – 9 – (in ukranian) – mf#nrp-688 – us UMI ProQuest [305]

Die kommune – Berlin DE, 1927-33 – 1r – 1 – gw Misc Inst [334]

Die kommunikasie van die evangelie aan plattelandse kinders in die verenigende gereformeerde kerk in suider-afrika / Jack, Hendrik Cornelius – Uni of South Africa 2001 [mf ed Johannesburg 2001] – 2mf – 9 – (incl bibl ref; incl abstract in afrikaans & english; text in afrikaans) – mf#mfm14851 – sa Unisa [242]

Kommunikationsgemeinschaft und kontraktualismus : versuch eines grundlegenden beitrags zur staatsphilosophie / Beiner, Marcus – Mainz: Gardez, 1993 (mf ed 1996) – (= ser Philosophie im Gardez) – 1mf – 9 – €24.00 – 3-8267-9667-5 – mf#DHS 9667 – gw Frankfurter [100]

Kommunikationsstoerungen bei kindern mit syndromen / Scheler, Elke Verena Ulrike – (mf ed 1995) – 1mf – 9 – €30.00 – 3-8267-2173-X – mf#DHS 2173 – gw Frankfurter [618]

Kommunikatives management : erwachsenenbildung, weiterbildung, personalentwicklung. eine vermittlungswissenschaftliche perspektive / Merk, Richard – Neuwied, Kriftel, Berlin: Luchterhand 1993 (mf ed 1996) – (= ser Grundlagen der weiterbildung) – 2mf – 9 – €31.00 – 3-8267-9691-8 – mf#DHS 9691 – gw Frankfurter [374]

Kommunismus – Vienna. v. 1-2. no. 32. Feb 1920-Sept 1921. Incomplete – 1 – us NY Public [335]

Kommunismus : zeitschrift der kommunistischen internationale fuer die laender suedosteuropas – Wien. v1 n1/2-v2 n31/32. feb 1920-sep 1921 – (= ser Communist international periodicals from the feltrinelli archives) – 32mf – 9 – $155.00 – (with: zeitschrift der kommunistischen internationale) – us UPA [335]

Der kommunist – Dresden DE, nov 1918-jan 1921 – 1r – 1 – gw Misc Inst [335]

Der kommunist : kp der schweiz – Zuerich (CH), 1919 feb-1922 jul – 1r – 1 – gw Misc Inst [335]

Kommunist – Cherepovets, Russia 1975-85 [mf ed Norman Ross] – 1r – 1 – mf#nrp-408 – us UMI ProQuest [077]

Kommunist – Erivan, U.S.S.R. -d. Jan 1946-May 1948; Jan 1950-Dec 1957; April-Dec 1958; July 1960-Dec 1968. 37 reels – 1 – uk British Libr Newspaper [947]

Kommunist – Moscow. 1924-62 – 1 – 989.00 – us L of C Photodup [335]

Kommunist / Moskovskago Oblastnago Byuro R K P (bol'shevikov) – Moscow. no. 1-2, 4. 1918 – 1 – us NY Public [335]

Kommunist tadzhikistana – Dushanbe, Tajikstan 1974-88 [mf ed Norman Ross] – 1r – 1 – mf#nrp-474 – us UMI ProQuest [077]

Kommunisticheskaia partiia sovetskogo soiuza v rezoliutsiiakh i resheniakh sezdov, konferentsii i plenumov tsk : 1898-1981 – 1970-1984. 14v – 94mf – 9 – mf#R-18,486 – ne IDC [325]

Kommunisticheskaya Akademiya. Moscow *see*
– Plan robat
– Vestnik

Kommunisticheski internatzional – Moscow. May 1919-June 1943. Incomplete – 1 – us NY Public [335]

Kommunisticheskii trud – Moscow, Russia 1920-22 [mf ed Norman Ross] – 12r – 1 – mf#nrp-1076 – us UMI ProQuest [077]

Kommunistische arbeiterzeitung – Berlin DE, 1921 n158-258 – 1r – 1 – gw Misc Inst [331]

Kommunistische arbeiterzeitung – Essen DE, 1922-24 [gaps] – 1r – 1 – gw Misc Inst [335]

Kommunistische gewerkschafter – 1922-1923. Madison: University of Wisconsin, 1977. Includes Supplements – 1 – us Wisconsin U Libr [335]

Die kommunistische internationale – Berlin. v. 1-20, no. 8. Aug 1919-Aug 10 1939. Incomplete – 1 – us NY Public [335]

Die kommunistische internationale – Moskau (RUS), 1919 n1-718 – 1r – 1 – mf#6086 – gw Mikropress [335]

Die kommunistische internationale : zeitschrift des exekutivkomitees der kommunistischen internationale – Basel (CH), Strassburg (Strasbourg F), Paris (F), Stockholm (S), 1933 10 jan-1941 may – 10r – 1 – (publ in berlin before 20 apr 1933) – gw Misc Inst [335]

Kommunistische internationale *see* L'internationale communiste

Kommunistische raete-korrespondenz – Berlin DE, 1919-20 [gaps] – 1r – 1 – gw Misc Inst [335]

Komoediantinnen : roman / Bloem, Walter – Berlin: Ullstein, c1914 [mf ed 1989] – 316p – 1 – mf#7032 – us Wisconsin U Libr [830]

Komoedie der irrungen : ein beitrag zur kulturgeschichte des neunzehnten jahrhunderts / Segesser, H – Zuerich: D Herzog, 1886 – 1r – 1 – us Wisconsin U Libr [430]

Komoedie des verlorenen sohnes / Schmeltzl, Wolfgang; ed by Roessler, Alice – Halle: M Niemeyer, 1955 – us Wisconsin U Libr [430]

Komoedien des barock / ed by Ketelsen, Uwe-K – Reinbek/Hamburg: Rowohlt, 1970 – (incl bibl ref) – us Wisconsin U Libr [430]

Kompaneiskii, B N *see* Obshchii kurs schetovodstva potrebitelskikh obshchestv

Kompanie in polen / Pecher, Erich – Wien: Deutscher Verlag fuer Jugend und Volk, 1942 – 1r – 1 – us Wisconsin U Libr [830]

Kompas – Djakarta, Indonesia. June 28 1965-Dec 1993 – 109r – 1 – us L of C Photodup [079]

Der kompass – Curityba (BR), 1920 5 jan-1939 7 nov – 1 – gw Misc Inst [079]

Kompendium der biblischen theologie des alten und neuen testaments von konstantin schlottmann / Schlottmann, Konstantin; ed by Kuehn, Ernst – Leipzig:Doerffling & Franke, 1889 – 1mf – 9 – 0-8370-5101-0 – (incl ind) – mf#1985-3101 – us ATLA [220]

Kompendium der dogmatik / Luthardt, Christoph Ernst – 8. verbesserte Aufl. Leipzig: Doerffling und Franke, 1889. Beltsville, Md: NCR Corp, 1978 (5mf); Evanston: American Theol Lib Assoc, 1984 (5mf) – 9 – 0-8370-0853-0 – (incl bibl ref and index) – mf#1984-4215 – us ATLA [240]

Kompendium der kirchengeschichte / Heussi, Karl – Tuebingen: J.C.B. Mohr, 1909 – 2mf – 9 – 0-7905-8038-1 – (incl bibl ref) – mf#1988-6019 – us ATLA [240]

Kompendium der palaestinischen altertumskunde / Thomsen, P – Tuebingen, 1913 – 2mf – 9 – mf#H-2870 – ne IDC [930]

Kompendium der palaestinischen altertumskunde / Thomsen, Peter – Tuebingen: J C B Mohr (Paul Siebeck), 1913 – 1mf – 9 – 0-7905-0399-9 – (incl indes) – mf#1987-0399 – us ATLA [930]

Kompert, Leopold *see* Leopold komperts saemtliche werke in zehn baenden

Kompiuternoe modelirovanie dinamicheskikh i strukturnykh svoistv zhidkikh metallov / Polukhin, V A et al – Moskva: In-t vysokikh temperatur AN SSSR, 1979 – us CRL [947]

Die komplementaktivierung in der experimentellen sepsis und waehrend kardiopulmonalen bypass : untersuchungen und besonderer beruecksichtigung des anaphylatoxins c5a / Mohr, Michael – 2000 – 2mf – 9 – 3-8267-2683-9 – mf#DHS 2683 – gw Frankfurter [617]

Die komplementariteit tussen intimiteit en afstand in die terapeutiese verhouding / Eyebers, Cornelia – Uni of South Africa 2000 [mf ed Johannesburg 2000] – 5mf – 9 – (incl bibl ref; text in afrikaans) – mf#mfm14909 – sa Unisa [150]

Komplexbau – Magdeburg DE, 1964 10 feb-1990 oct [gaps] – 6r – 1 – (wohnungsbaukombinat) – gw Misc Inst [690]

Die komplexitaet der bedeutungsexplikation in literarischen dialogen / Dietzel, Uwe – (mf ed 1999) – 3mf – 9 – €49.00 – 3-8267-2663-4 – mf#DHS 2663 – gw Frankfurter [440]

Komplexitaetsanalyse der rr-dynamik im 24-stunden-ekg : entwicklung, visualisierung und anwendung einer klinisch orientierten chaosmetrie / Bettermann, Henrik – (mf ed 1996) – 2mf – 9 – €40.00 – 3-8267-2316-3 – mf#DHS 2316 – gw Frankfurter [510]

Kompong cham, symbole de notre survie!! : deux ans de pourrissement, les sauveurs / Soth Polin – [s.l.]: Nokor Thom 1973 [mf ed 1990] – 1r with other items – 1 – mf#mf-10289 seam reel 017/25 [§] – us CRL [959]

Die komposisies van arthur wegelin tot 1988 / Stanford, Hendrik Josephus – U of the Western Cape 1988 [mf ed S.l: s.n.] 1988 – 7mf – 9 – (summary in afrikaans & english; incl bibl) – sa Misc Inst [780]

Die komposition der genesis / Eerdmans, Bernadus Dirk – Giessen: A Toepelmann, 1908 [mf ed 1989] – (= ser Alttestamentliche studien 1) – 1mf – 9 – 0-7905-0764-1 – (incl bibl ref & ind) – mf#1987-0764 – us ATLA [221]

Die komposition der genesis / Ewald, Heinrich – Braunschweig: Ludwig Lucius, 1823 – 1mf – 9 – 0-7905-3434-7 – (incl bibl ref) – mf#1987-3434 – us ATLA [221]

Die komposition der paulinischen hauptbriefe : 1., der roemer- und galaterbrief / Voelter, Daniel – Tuebingen: J J Heckenhauer, 1890 – 1mf – 9 – 0-8370-9334-1 – (no more publ; superseded by the author's paulus und seine briefe. incl bibl ref) – mf#1986-3334 – us ATLA [227]

Die komposition des aethiopischen henochbuches / Appel, Heinrich – Guetersloh: C Bertelsmann, 1906 – 1mf – 9 – 0-524-05898-9 – mf#1992-0655 – us ATLA [221]

Die komposition des buches jes. c. 28-33 : ein rekonstruktionsversuch / Brueckner, Martin – Halle a S: J Krause, 1897 – 1mf – 9 – 0-8370-2492-7 – (incl bibl ref) – mf#1985-0492 – us ATLA [221]

Komposition und strophenbau / Muller, David Heinrich – Wien, Austria. 1907 – 1r – us UF Libraries [939]

Die kompositionstechnische entropie in der orchesterfassung eines klavierwerkes : dargestellt an "bilder einer ausstellung" von m mussorgski / m ravel / Petri, Hasso Gottfried – (mf ed 2002) – 255p 3mf – 9 – €49.00 – 3-8267-2774-6 – mf#DHS 2774 – gw Frankfurter [780]

Kompositon und entstehungszeit der oracula sibyllina / Geffcken, J – Leipzig, 1902 – (= ser Tugal 2-23/1) – 2mf – 9 – €5.00 – ne Slangenburg [240]

Komsomol i kooperatsiia / Ershov, A – 1925 – 32p 1mf – 9 – mf#COR-151 – ne IDC [335]

Komsomol i kooperatsiia : rezoliutsii i postanovleniia vkp(b), vlksm, pravitelstv i koop organov / Levitas, I et al – 1928 – 192p 3mf – 9 – mf#COR-157 – ne IDC [335]

Komsomol v stroitelstve rabochei kooperatsii : formy i metody koop raboty molodezhi / Milov, V V – 1926 – 76p 1mf – 9 – mf#COR-172 – ne IDC [335]

Komsomolets na kooperativnuiu uchebu / Matveev-Bodryi, N N – 1926 – 35p 1mf – 9 – mf#COR-168 – ne IDC [335]

Komsomol'skaia pravda – Moscow, Russia 1925-89 [mf ed Norman Ross] – 310r 4r/yr [standing order] – 1 – mf#nrp-1076 – us UMI ProQuest [077]

Komsomolskaya pravda – Moscow, U.S.S.R. -d. 1 July 1926; 13 Nov-23 Dec 1927; 18 Sept 1928-30 Dec 1931. 1927, 1928 very imperfect. 7 reels – 1 – uk British Libr Newspaper [947]

Komu na rusi zhitt khorosho... / Nekrasov, Nikolai Alekseevich – Berlin, Germany. 1917 – 1r – us UF Libraries [960]

Komunist – Belgrade, Yugoslavia. May 1957-1960 – 3r – 1 – us L of C Photodup [949]

Komunist – Beograd, 1976-79 – 4r – 1 – gw Mikropress [949]

Komunist – Kiev, USSR. Oct 17 1940-Oct 3 1941; Mar 5-Dec 24 1942 – (= ser Radians'ka Ukraina) – 2r – 1 – (later absorbed by: radians'ka ukraina) – us L of C Photodup [077]

Kondor / Hiller, Kurt – Heidelberg, Germany. 1912 – 1r – us UF Libraries [960]

Kondratenko, V S *see* Altaiskii vestnik

Kondrat'ev, N D *see*
– Kon"iunktura narodnogo khoziaistva sssr i mirovogo khoziaistva
– Narodnoe khoziaistvo rsfsr v 1924-1925 g

Kondratev, N D *see*
– Agrarnyi vopros o zemle i zemelnykh poriadkakh
– Ekonomicheskii biulleten koniunkturnogo instituta pri petrovskoi selsko-khoziaistvennoi akademii tssu sssr
– Mirovoi khlebnyi rynok i perspektivy nashego khlebnogo eksporta
– Perspektivy razvitiia selskogo khoziaistva sssr
– Razvitie khoziaistva kineshemskogo zemstva kostromskoi gubernii
– Rynok khlebov i ego regulirovanie vo vremia voiny i revoliutsii

Koneffke, Gernot *see* Menschenbildung und kinderarbeit bei pestalozzi und owen

Konets azefa / Nikolaevskii, B – 1926 – 79p 1mf – 9 – mf#RPP-240 – ne IDC [325]

Konferentsyah ha-hagit / Histadrut "'Ivriyah" – Krako, Poland. 1909 – 1r – us UF Libraries [939]

Konferenz der Gewerkschaftsvorstaende *see* Protokoll der konferenz...19-23 februar 1906

Der Konferenz der Mennoniten-Bruedergemeinde in Kanada *see* Mennonitische rundschau

Konferenz der Strassenwarter *see* Protokoll der verhandlungen..

Konferenz von Vertretern der gewerkschaftlichen Organisationen und Angestelltenverbande *see* Gesetz ueber den vaterlandischen hilfsdienst

Konfessionelle brunnenvergiftung / Keiter, Heinrich – 2. verb u verm Aufl. Essen (Ruhr): Fredebeul & Koenen, 1908 – 1mf – 9 – 0-8370-6989-0 – mf#1986-0989 – us ATLA [430]

Die konfessionelle schule : ein vortrag / Graeber, H J – Ruhrort: Andreae, 1876 – 1mf – 9 – 0-8370-7633-1 – mf#1986-1633 – us ATLA [377]

Die konflikte des zwinglianismus, luthertums und calvinismus in der bernischen landeskirche von 1532-1558 / Hundeshagen, C B – Bern, Jenni, 1842 – 5mf – 9 – mf#PBU-455 – ne IDC [242]

Konfrontative untersuchungen zur lexikalischen dimension der fachlichkeit von texten / Bruechner, Kathrin – (mf ed 1997) – (= ser Leipziger arbeiten zur fachsprachenforschung) – 1mf – 9 – €30.00 – 3-8267-2454-2 – mf#DHS 2454 – gw Frankfurter [410]

Konfrontative untersuchungen zur semantischen dimension der fachlichkeit von texten / Hrouda, Baerbel – (mf ed 1998) – (= ser Leipziger arbeiten zur fachsprachenforschung) – 2mf – 9 – €40.00 – 3-8267-2591-3 – mf#DHS 2591 – gw Frankfurter [400]

Konfucius / Tschepe, Albert – Yentschoufu [Chefoo]: Druck und Verlag der Katholischen Mission, 1910-1915 – (= ser Studien und Schilderungen aus China) – 2mf – 9 – 0-524-05351-0 – (incl bibl ref) – mf#1990-3472 – us ATLA [180]

Die konfutation des augsburgischen bekenntnisses : ihre erste gestalt und ihre geschichte / Ficker, Johannes – Leipzig: Johann Ambrosius Barth, 1891 – 1mf – 9 – 0-8370-8814-3 – (prolegomonena and notes in german; texts in latin. incl bibl ref) – mf#1986-2814 – us ATLA [240]

Die konfutation des vierstaedtebekenntnisses : ihre entstehung und ihr original – Leipzig: Johann Ambrosius Barth, 1900 – 1mf – 9 – 0-8370-8849-6 – (incl bibl ref) – mf#1986-2849 – us ATLA [240]

Kongelige Bibliotek. Copenhagen *see* Early nineteenth century manuscripts from kumasi, ghana

Kongelige danske videnskabernes selskabs naturvidenskabelige og mathematiske alhandlinger – Kjobenhavn 1828 – v3 on 8mf – 9 – mf#67/2 – ne IDC [500]

Kongelige Norske Frederiks universitet Akademiske collegium *see* Olafs saga helga

Kong-Huot et al *see* Tum-teav

Kongo dia ngunga – Leopoldville: [Ed Nzeza-Lando, nov 15 1959-jan 31 1960 – (issues filmed as pt of: herbert j weiss collection on the belgian congo) – us CRL [960]

Kongo dia ntotila – [Leopoldville: Ed Nzeza-Lando, sep 2 1961 – (issues filmed as pt of: herbert j weiss collection on the belgian congo) – us CRL [960]

Kongoland / Pechuel-Loesche, Eduard – Jena: H Costenoble, 1887 – us CRL [960]

Kongoleesche vertellingen / Struyf, Ivon – Brugge: Excelsior, 1924-26 – us CRL [960]

Kongo-overzee – Antwerp, Belgium 1950-56 – 1 – mf#534 – us UMI ProQuest [301]

Die kongregationalisten oder freien evangelischen gemeinden / Obenhaus, Hermann – Chicago, IL.: German Pilgrim Press, 1913 – 1mf – 9 – 0-524-03017-0 – mf#1990-4539 – us ATLA [240]

Die kongregationalistische kirche : und die erziehung in den vereinigten staaten von nord-amerika / Blome, Rud – Jena: Ant Kaempfe, 1900 [mf ed 1986] – 1mf – 9 – 0-8370-7683-8 – mf#1986-1683 – us ATLA [242]

Kongres buruh seluruh indonesia : berita kbsi – Djakarta, 1953 – 1mf – 9 – mf#SE-347 – ne IDC [959]

Kongres Ilmu Pengetahuan Nasional *see* Laporan madjelis ilmu pengetahuan indonesia

Kongres pendirikan indonesia moeda 28 des 1930-2 jan 1931 (di soerakarta) – Weltevreden, 1931 – 2mf – 8 – mf#SE-1435 – ne IDC [959]

Kongres ra'jat indonesia 1 – Batavia-C, 1940 – 2mf – 8 – mf#SE-1421 – ne IDC [959]

Kongresa grahmata = Congress book / Kronlins, Janis – Speeches and Reports at the annual meeting of the Latvian Baptist Union in Liepja, Sept. 13-16, 1923, etc. Publ. No. 6348b. One of two items on reel. Total Pages 1,228 – 1 – us Southern Baptist [242]

Kongress der Christlichen Gewerkschaften Deutschlands *see* Niederschrift der verhandlungen

Kongress der Gewerkschaften Deutschlands *see*
– Protokoll der verhandlungen..
– Protokoll der verhandlungen des ausserordentlichen kongresses der gewerkschaften deutschlands

Kongress fuer die Freiheit der Kultur *see* Freier geist zwischen oder und elbe

Kongress zeitung – Luzern (CH), 1935 20 aug-8 sep – 1 – gw Misc Inst [323]

Kongress-tribuene – Zuerich (CH), 1937 n1, 3-7 – 1 – gw Misc Inst [074]

Kongress-zeitung – Prag (CZ), 1933 n1-7, 9-13 – 1 – gw Misc Inst [323]

Kongress-zeitung [...] – Duesseldorf DE, 1910 19 jun-23 jun – 1r – 1 – gw Misc Inst [074]

Kongresszeitung – Karlsbad (Karlovy Vary CZ), 1921 n2-13/14 – 1 – gw Misc Inst [077]

Kongresszeitung : organ des 16./15. zionisten-kongresses – Zuerich (CH), 1929 n1-12, 1937 n3-12 – 1 – (1929: organ des 16. zionisten-kongresses; 1937: organ des 20. zionisten-kongresses) – gw Misc Inst [270]

Kongresszeitung – Wien (A), 1925 n1-12 – 1 – gw Misc Inst [074]

Kongreszeitung – Basel (CH), 1927, 1931 – 1 – (wih gaps) – gw Misc Inst [074]

Koni, A F *see* S iu vitte

Konia, la ville des derviches tourneurs : souvenirs d'un voyage en asie mineure / Huart, Clement – Paris: Ernest Leroux, 1897 – 1mf – 9 – 0-524-07784-3 – mf#1991-0161 – us ATLA [915]

Koniaev, A *see* Finansovyi kontrol' v dorevoliutsionnoi rossii

Konig, David Thomas *see* Plymouth court records, 1686-1859

Konigliche Museen Zu Berlin *see* Magnesia am maeander

Koning, M *see* Lexicon hieroglyphicum sacro-profanum

Koning, Robin D *see* The use of physiological and psychophysiological techniques to set exercise intensity in children

Het koninglyk neder-hoog-duitsch en hoog-neder-duitsch dictionnaire (ael2/3) / Kramer, Matthias – Nuernberg 1719 [mf ed 1992] – (= ser Archiv der europaeischen lexikographie: woerterbuecher) – 10mf – 9 – €60.00 – 3-89131-063-3 – (int by laurent bray) – gw Fischer [430]

Koninklijk Instituut van Ingenieurs *see* Voordrachten gehouden voor het koninklijk institut van ingenieurs

Konitzer nachrichten : nachrichtenblatt fuer pommerellen – Konitz (Chojnice PL), 1922 21 mar- 6 dec – 1r – 1 – gw Misc Inst [077]

Konitzer tagblatt – Konitz (Chojnice PL), 1924 23 oct-1936, 1937 10 feb-1939 27 aug – 14r – 1 – gw Misc Inst [077]

Kon"iunktura narodnogo khoziaistva sssr i mirovogo khoziaistva : sbornik obzorov statisticheskikh dannykh po vazhneishim otrasliam kon"iunktury narodnogo i mirovogo khoziaistva v 1925-1926 g / ed by Kondrat'ev, N D – M, 1927. 256p – 3mf – 9 – mf#RHS-29 – ne IDC [314]

Kon"iunkturnaia tovarno-transportnaia statistika – M, 1924-1928. n1-41 – 49mf – 9 – mf#RHS-4 – ne IDC [314]

Kon'iunkturnyi Institut. Moscow *see* Ekonomicheskii biulleten

Konjaev, K V *see* Spectral analysis of random processes and fields

Konjunktur – London, UK. 3 Apr 1909-15 Aug 1914 – 1 – uk British Libr Newspaper [072]

Das konklave pius' 4 : historische abhandlung / Mueller, Theodor – Gotha: Friedrich Andreas Perthes, 1889 – 1mf – 9 – 0-8370-8846-1 – (incl bibl ref) – mf#1986-2846 – us ATLA [920]

Konkoly-Alapitvanyu Budapest-Svabhegyi m kir Astrofizikai Obszervatorium *see* Kleinere veroeffentlichungen des o-gyallaer astrophysikalischen observatoriums stiftung von konkolys

Konkoly-Alapitvanyu Budapest-Svabhegyi m. kir. Astrofizikai Obszervatorium *see* A m kir konkoly-alapitvanyu astrophysikai observatorium kisebb kiadvanyai

Konkordanz zu walter kempowskis "deutscher chronik" / Kempowski, Walter; ed by Keele, Alan F – [mf ed Hildesheim 1994] – (= ser Alpha-omega) – 10v on 34mf – 9 – diazo €158.00 silver €198.00 – gw Olms [430]

Konkordanz zum targum onkelos / Brederek, Emil – Giessen: A Toepelmann, 1906 – 1mf – 9 – 0-7905-1027-8 – (in aramaic and german) – mf#1987-1027 – us ATLA [221]

Konkordiebogen : eller, den evangelisk-lutherske kirkes bekjendelsesskrifter / ed by Caspari, Carl Paul & Johnson, Gisle – Decorah, Iowa: Lutheran Pub House, 1899 – 2mf – 9 – 0-524-07694-4 – (incl bibl ref) – mf#1991-3279 – us ATLA [242]

Konkordieformelens kjerne : paa den evangelisk-lutherske synodalkonferences opfodring forsynet med en historisk indledning og korte oplysende anmaerkyinger og udgivet til det lutherske kristenfolks nytte / Walther, Carl Ferdinand Wilhelm – Decorah, Iowa: Norske Synodes Forlag, 1877 – 2mf – 9 – 0-524-07471-2 – mf#1991-3131 – us ATLA [242]

Konkret – Hamburg DE, 1958-1973 15 nov [gaps], 1974 oct-1984 nov, 1985-87, 1990-92 [gaps] – 24r – 1 – gw Misc Inst [074]

Konkret – Magdeburg DE, 1973 jun-1976 – 1r – 1 – (industriebaukombinat) – gw Misc Inst [620]

Konnersreuther zeitung – Tirschenreuth DE, 1927-28 [gaps] – 1 – gw Misc Inst [074]

Kononenko, K O *see* Kolebatelnye sistemy s ogranichennym vozbuzhdeniem

Konoplev, N *see* Sviatye vologodskago kraia

Konow, Sten *see*
- Central asian fragments of the ashtadasasahasrika prajnaparamita and of an unidentified text
- The religions of india

Konperensi ahli2 perkebunan : seminar karet se-indonesia / 1st, Sei-Karang, 1962 Berita Research Institute of the SPA – Medan, 1962 – 1mf – 9 – mf#SE-1755 – ne IDC [950]

Konrad der Pfaffe *see* Das rolandslied des pfaffen konrad

Konrad, der pfaffe / ed by Bartsch, Karl – Leipzig: F A Brockhaus, 1874 – 1r – 1 – (incl bibl ref and ind. middle high german text with an introduction in german) – us Wisconsin U Libr [430]

Konrad, Martin *see* Zeitbilder oder erinnerungen an meine verewigten wohlthaeter

Konrad, Paul *see* Dr. ambrosius moibanus

Konrad stolles thueringisch-erfurtische chronik / ed by Hesse, Ludwig Friedrich – Stuttgart: Litterarischer Verein, 1854 [mf ed 1993] – (= ser Blvs 32) – 1 – (with biogr sketch of aut) – mf#8470 reel 7 – us Wisconsin U Libr [943]

Konrad von marburg, beichtvater der heiligen elisabeth und inquisitor / Henke, Ernst Ludwig Theodor – Marburg: NG Elwert, 1861 – 1mf – 9 – 0-524-00557-5 – (incl bibl ref) – mf#1990-0057 – us ATLA [240]

Konradin reitet / Gmelin, Otto – Leipzig: P Reclam, 1943, c1938 (mf ed 1990) – 1r – 1 – (filmed with: sommerwind ueber tormohlenhof) – us Wisconsin U Libr [830]

Konrads 'trojanerkrieg' und gottfrieds 'tristan' : vorstudien zum gotischen stil in der dichtung / Green, Dennis Howard – Waldkirch/Br: Waldkircher Verlagsgesellschaft, 1949 [mf ed 1993] – 87p – 1 – (incl bibl ref) – mf#8440 – us Wisconsin U Libr [430]

Konrads von megenberg deutsche sphaera / Sacro Bosco, Joannes de; ed by Matthaei, Otto – Berlin: Weidmann, 1912 [mf ed 1993] – (= ser Deutsche texte des mittelalters 23) – xiv/63p/2pl (ill) – 1 – (incl bibl ref and ind. middle high german trans fr latin) – mf#8623 reel 5 – us Wisconsin U Libr [520]

Konsens und dissens im bildungssystem von baden-wuerttemberg : ein beitrag zum verhaeltnis von partizipation und schulreform nach 1970 / Felgner, Harald – Heidelberg, 1977 (mf ed 1993) – 3mf – 9 – €38.00 – 3-89349-367-0 – mf#DHS-AR 367 – gw Frankfurter [370]

Konservative politik im letzten jahrzehnt des kaiserreiches band i / Westarp, Graf von – 1908-1914 – 1 – gw Mikropress [943]

Konserven-zeitung – Braunschweig (Brunswick DE), 1907-15 – 6r – 1 – uk British Libr Newspaper [660]

Konservnaia i ovoshchesushilnaia promyshlennost – Moskva: Pishchepromizdat, 1981 – us CRL [947]

Konspekt lektsii po selskokhoziaistvennoi kooperatsii / Ostroumov, V I – Omsk, 1929 – 238p 3mf – 9 – mf#COR-269 – ne IDC [335]

De konst der wijsheid / Gracian – Den Haag, 1696 – 5mf – 8 – €12.00 – ne Slangenburg [100]

Konstantin leon'tev / Berdiaev, Nikolai – Paris, France. 1926 – 1r – us UF Libraries [025]

Konstantinopel und st petersburg, der orient und der norden / ed by Murhard, Friedrich Wilhelm August et al – Sankt Petersburg, Penig 1805-06 – (= ser Dz. historisch-geographische abt) – 2jge on 23mf – 9 – €230.00 – mf#k/n1366 – gw Olms [910]

Konstantinopel und st petersburg, der orient und der norden : eine zeitschrift – St Petersburg [u a] 1805-06 [mf ed Hildesheim 1995-98] – 7v on 23mf – 9 – €230.00 – 3-487-26720-9 – gw Olms [910]

Die konstantinopolitanische messliturgie vor dem 9 jahrhundert : uebersichtliche zusammenstellung des wichtigsten quellenmaterials / Baumstark, Anton – Bonn: A Marcus & E Weber 1909 [mf ed 1992] – (= ser Kleine texte fuer theologische und philologische vorlesungen und uebungen 35) – 1mf – 9 – 0-524-04669-7 – (in greek. notes in latin. german int) – mf#1990-1296 – us ATLA [240]

Konstantins des grossen kreuzerscheinung : eine kritische untersuchung / Schroers, Heinrich – Bonn: Hanstein, 1913 – 1mf – us ATLA [240]

Konstantins des grossen kreuzerscheinung : eine kritische untersuchung / Schroers, Heinrich – Bonn: Hanstein, 1913 – 1mf – 9 – 0-7905-6495-5 – (incl bibl ref) – mf#1988-2495 – us ATLA [240]

Konstantins kreuzesvision in ausgewaehlten texten / Aufhauser, Johannes Baptist – Bonn: A Marcus und E Weber, 1912 – (= ser Kleine texte fuer vorlesungen und uebungen) – 1mf – 9 – 0-524-05428-2 – mf#1990-1460 – us ATLA [240]

Konstantios, Patriarch of Constantinople *see* The patriarchate of antioch

Konstanzer altlat propheten- und evang bruchstuecke mit glossen (tab7-9) / Dold, Alban – 1923 – (= ser Texte und arbeiten. beuron (tab). beitraege zur ergruendung des aelteren lateinischen und christlichen schrifttums und gottesdienstes) – €14.00 – ne Slangenburg [242]

Konstanzer politische zeitung *see* Der volksfreund

Konstanzisches intelligenzblatt *see* Der volksfreund

Konstituante RI *see* Res publica

Konstituante risalah perundingan : indonesia – Bandung, 1956-1959 – 134mf – 9 – mf#SE-246 – ne IDC [959]

Die konstitusionele ontwikkeling van botswana / Hough, M – Pretoria, 1968 – us CRL [323]

Das konstitutionelle deutschland – Strassburg [Strasbourg F], 1831 13 may-1832 30 mar – 1 – (title varies: 2 dec 1831: deutschland) – gw Misc Inst [074]

Das konstitutionelle deutschland *see* Beilage zum niederrheinischen kurier fuer das konstitutionelle deutschland

Die konstitutionelle monarchie – Koenigsberg (Kaliningrad, RUS), 1850 jul-1853, 1888 jan-1934 jan-jun & sep-dec – 32r – 1 – (with gaps. filmed by other misc inst: 1869 18 jun, 1877 16 may, 1886 11 jul, 1890 [single iss], 1909 jub-nr, 1922 jul-dec, 1928 31 dec [2r]. title varies: 2 jan 1851: ostpreussische zeitung) – gw Misc Inst [323]

Konstitutsionno-demokraticheskaia partiia : partiia narodnoi svobody postanovleniia 2-go sezda, 5-11 ianv 1906 g i programma – 1906 – 32p 1mf – 9 – mf#RPP-104 – ne IDC [325]

Konstitutsionno-demokraticheskaia partiia : partiia narodnoi svobody: postanovleniia 3-go sezda, 21-25 apr 1906 g i ustav partii – 1906 – 16p 1mf – 9 – mf#RPP-105 – ne IDC [325]

Konstitutsionno-demokraticheskaia partiia : sezd 12-18 okt 1905 g – 1905 – 24p 1mf – 9 – mf#RPP-106 – ne IDC [325]

Konstitutsionno-demokraticheskaia partiia *see* Doklad po evreiskomu voprosu tsentral'nago komiteta partii k-d

Konstitutsionno-demokraticheskaia partiia i zemelnaia reforma / Vikhliaev, P A – 1906 – 32p 1mf – 9 – mf#RPP-111 – ne IDC [325]

Konstitutsionnyi vestnik – Moscow: Konstitutsionnaia kommissiia Verkh. Soveta Rossii 1990-94 – 50mf – 9 – $294.00y – (ceased publ) – us East View [077]

Konstruktion – Heidelberg, Germany 1981-82 – 1,5,9 – ISSN: 0720-5953 – mf#13193.02 – us UMI ProQuest [621]

Konstruktion episomal replizierender vektoren fuer saeugetierzellen und untersuchung ihrer mitosischen stabilitaet / Baiker, Armin – (mf ed 2000) – 1mf – 9 – €30.00 – 3-8267-2714-2 – mf#DHS 2714 – gw Frankfurter [574]

Konstruktion von therapeutische gene tragenden retroviralen vektoren zur kontrolle rasch proliferierender zellen / Mrochen, Stefan H – (mf ed 1998) – 2mf – 9 – €40.00 – 3-8267-2573-5 – mf#DHS 2573 – gw Frankfurter [574]

'N konstruktivistiese beskrywing van veranderende persepsies in 'n welsynsorganisasie / Commerford, Sophia Elizabeth Jacoba – Uni of South Africa 2000 [mf ed Johannesburg 2000] – 3mf – 9 – (text in afrikaans; abstract in afrikaans and english; incl bibl ref) – mf#mfm14842 – sa Unisa [360]

Konsulat djendral bulletin : indonesia – New York, 1967(1-38) – 5mf – 9 – (missing: 1967(19)) – mf#SE-1673 – ne IDC [959]

Konsulat djendral news and views permanent mission to the united nations : indonesia – New York, 1965-1972 – 60mf – 9 – (missing: 1965(48); 1966(203-end); 1967; 1968; 1969(sep-dec); 1970(2-end); 1971(1-2)) – mf#SE-768 – ne IDC [959]

Konsulat djendral rekaman aneka warta indonesia : indonesia – New York, 1966-1967 – 4mf – 9 – (missing 1966(1-5)) – mf#SE-1674 – ne IDC [959]

Konsumentbladet *see* Vi

Kontakt – Magdeburg DE, 1965 10 may-1972 mar, 1972 jun-1984 n21 [gaps] – 3r – 1 – (messgeraetewerk) – gw Misc Inst [621]

Kontinuitaet und bruch : zur genese der malerei hans hofmanns / Hoffmans, Christiane – (mf ed 1999) – 3mf – 9 – €49.00 – 3-8267-2612-X – mf#DHS 2612 – gw Frankfurter [750]

Kontos, Anthony P *see* The effects of perceived risk, risk-taking behaviors, and body size on injury in youth sport

Kontra adventismen / Vig, Peter Sorensen – Blair, NE: Danish Lutheran Pub House, 1905 – 1mf – 9 – 0-524-05777-X – mf#1991-2333 – us ATLA [242]

Kontraktatsiia, sbyt i snabzhenie v selsko-khoziaistvennykh kreditnykh tovarishchestvakh : organizatsiia i finansirovanie / Islankin, F B – 1929 – 195p 3mf – 9 – mf#COR-382 – ne IDC [335]

Kontrastive analysen zur fachsprache der werbung im russischen, ukrainischen und deutschen / Melhorn, Grit – (mf ed 1997) – (= ser Leipziger arbeiten zur fachsprachenforschung) – 1mf – 9 – €40.00 – 3-8267-2465-8 – mf#DHS 2465 – gw Frankfurter [400]

Kontrastive grammatikanalyse und valenztheorie / Schmidt, Wolfgang G A – 2pts(mf ed 1996) – 9 – €56.00v – (pt1: eine einfuehrung in konzeption und methode an hand des deutschen und koreanischen 4mf isbn: 3-8267-2341-4. pt2: anwendungen und beispiele an hand des deutschen und des koreanischen 5mf isbn: 3-8267-2341-5) – mf#DHS 2341 – gw Frankfurter [410]

Kontrolnye tsifry potrebitelskoi kooperatsii sssr na 1927-28 god – 1927 – 184p 2mf – 9 – mf#COR-328 – ne IDC [335]

Kontrol'nyi ezhegodnik za 1913 god *see* Spravochnaia kniga dlia chinov gosudarstvennogo kontrolia

Die kontroverse um die transferproblematik 1924-1929 / Holthaus, Arno – (mf ed 1992) – 2mf – 9 – €49.00 – 3-89349-512-6 – mf#DHS 512 – gw Frankfurter [350]

Konvensie-dagboek van sy edelagbare franciois stephanus malan / Malan, Francois Stephanus – Kaapstad, South Africa. 1951 – 1r – us UF Libraries [960]

Das konverseninstitut des cisterzienserordens / Hoffman, E – Freiburg Schw, 1905 – 2mf – 8 – €5.00 – ne Slangenburg [241]

Konvolut ostdeutscher zeitungen, hauptsaechlich aus koenigsberg – Koenigsberg (Kaliningrad RUS), 1839-1935 [single iss] – 1r – 1 – gw Misc Inst [077]

Konya – (= ser Vilayet salnames) – 9 – (1285 [1868] def'a 1 2mf $190; 1290 [1873] def'a 6 4mf $60; 1291 [1874] def'a 7 2mf $40; 1292 [1875] def'a 8 2mf $40; 1293 [1876] def'a 9 2mf $40; 1294 [1877] def'a 10 2mf $40; 1295 [1878] def'a 11 2mf $40; 1298 [1881] 3mf $55; 1302 [1885] def'a 18 2mf $190; 1303 [1886] def'a 19 3mf $190; 1306 [1889] def'a 22 3mf $55; 1310 [1892] 7mf $110; 1317 [1899] 6mf $90; 1322m [1906] 5mf $75) – us MEDOC [956]

Konya – Konya: Vilayet Matbaasi. Cikaran: Konya Vilayeti, 1871-28. n183(2932). 14 tesrinisani 1928 – (= ser O & t journals) – 1mf – 9 – $25.00 – us MEDOC [956]

Konzentra-blaetter – Berlin DE, 1925 6 jun-1933 [gaps] – 1 – gw Misc Inst [074]

Konzepte einer zentralen europaeischen verkehrswegeplanung : ansatzpunkte, moeglichkeiten und grenzen / Breuer, Joerg – (mf ed 1994) – 1mf – 9 – €30.00 – 3-89349-889-3 – mf#DHS 889 – gw Frankfurter [337]

Der konzeptualismus in der universalienlehre des franziskaner- erzbischofs petrus aureoli (pierre d'auriole) / Dreiling, R – (= ser Bgphma 11/6) – €12.00 – ne Slangenburg [100]

Das konzert : lustspiel in drei akten / Bahr, Hermann – 7. aufl. Berlin: E Reiss, c1909 [mf ed 1998] – (= ser Moderne buehne) – 154p – 1 – mf#9961 – us Wisconsin U Libr [820]

[Konzert-rondo fuer klavier und orchester, k 386] : di wolfgango amadeo mozart mp. vienna 19 octob 1782 / Mozart, Wolfgang Amadeus – [1782] [mf ed 1988] – 1r – 1 – mf#pres. film 7 – us Sibley [780]

Das konzil von nicaea : habilitationsvorlesung / Bernoulli, Carl Albrecht – Freiburg i. B: JCB Mohr, 1896 [mf ed 1990] – 1mf – 9 – 0-7905-6281-2 – (incl bibl ref) – mf#1988-2281 – us ATLA [240]

Das konzil von trient und die universitaeten : festrede zur feier des dreihundertdreiundzwanzigjaehrigen bestehens der koenigl. julius-maximilians-universitaet zu wuerzburg: zugleich zum gedaechtnis von schillers 100. todestage / Merkle, Sebastien – Wuerzburg: H Stuertz, 1905 – 1mf – 9 – 0-7905-3979-9 – mf#1989-0472 – us ATLA [378]

Das konzil zu st. basle : ein beitrag zur lebensgeschichte gerberts von aurillac / Schlockwerder, Karl Theodor – Magdeburg: E. Baensch, 1906 – 1mf – 9 – 0-8370-8063-0 – mf#1986-2063 – us ATLA [240]

Konzilien lexikon, enthaltend : saemmtliche general-, national-, provinzial- und partikular-konzilien vom ersten konzilium zu jerusalem bis auf das konzilium von paris 1811 – Augsburg 1843-44. 2v (mf ed 1995) – 7mf – 9 – 3-8267-3133-6 – mf#DHS-AR 3133 – gw Frankfurter [240]

Konzilienstudien zur geschichte des 13. jahrhunderts : ergaenzungen und berichtigungen zu hefele-knoepfler "conciliengeschichte" band 5 und 6 / Finke, Heinrich – Muenster: Regensberg, 1891 – 1mf – 9 – 0-7905-6171-9 – (incl bibl ref) – mf#1988-2171 – us ATLA [240]

Die konzilspolitik karls 5. in den jahren 1538-1543 / Korte, August – Halle a.d. S: Verein fuer Reformationsgeschichte 1905 [mf ed 1990] – (= ser Schriften des vereins fuer reformationsgeschichte 22/85) – 1mf – 9 – 0-7905-5295-7 – (incl bibl ref) – mf#1988-1295 – us ATLA [230]

Konzilstudien / Schwartz, E – Strassburg, 1914 – €5.00 – ne Slangenburg [240]

Konzument – Maribor YU, 1929* – 1r – 1 – (slovenian periodical) – us IHRC [073]

Koo, Heng Ngo *see* Sesoedanja mati

Koo, V K Wellington *see* The status of aliens in china

Koobaha istatistikada = Statistical abstract / Somalia. Central Statistical Dept – (= ser African official statistical serials, 1867-1982) – 21mf – 9 – uk Chadwyck [316]

Kookogey, William P *see* Patent law in brief

Koondrook / barham bridge – Koondrook/Barham – 17r – 9 – A$1034.75 vesicular A$1128.25 silver – (aka: barham bridge) – at Pascoe [079]

Koones, John Alexander *see* Everybody's law book; legal rights and legal remedies.

Koontz, Harry R *see* The home missions task of the church of the united brethren in christ

Kooperatsiia v rossii / Totomiantis, Vakhan Fomich – Praga: Izd-vo "Nasha Riech'", 1922 [mf ed 2002] – 1r – 1 – (filmed with: ocherki perekhodnoi ekonomiki / a leont'ev i e khmel'nitskaia [1927]) – mf#5231 – us Wisconsin U Libr [337]

Kooperativ v rabote : pravdivaia glava iz istorii vymyshl. potrebitelnogo obshchestva / Orlov, N A – 1918 – 71p 1mf – 9 – mf#COR-85 – ne IDC [335]

Kooperativnaia arenda zemli : o tom, kak soobshcha arenduiut zemliu i uluchaiut khoziaistvo / Maslov, S L – 1914 – 36p 1mf – 9 – mf#COR-69 – ne IDC [335]

Kooperativnaia chainaia : ch 1: organizatsiia chainykh / Armand, L M – 1920 – 1mf – 9 – mf#COR-321 – ne IDC [335]

Kooperativnaia mysl – Kiev, 1916-1917(6) – 7mf – 9 – mf#COR-613 – ne IDC [335]

Kooperativnaia mysl – Vladikavkaz, 1918(1-16) – 2mf – 9 – (missing:1918(1-7,9-11)) – mf#COR-611 – ne IDC [335]

Kooperativnaia mysl – Voronezh, 1921-1923(18) – 17mf – 9 – (missing:1921(1)-1922(3),1922(14),1923(7-10)) – mf#COR-612 – ne IDC [335]

Kooperativnaia mysl – Stavropol, 1923(1-12) – 4mf – 9 – (missing:1923(1)) – mf#COR-614 – ne IDC [335]

Kooperativnaia organizatsiia melkogo kredita / Prokopovich, S N – 1919 – 31p 1mf – 9 – mf#COR-399 – ne IDC [335]

Kooperativnaia sibir – Novonikolaevsk, (Novosibirsk), 1924-1929(3) – 142mf – 9 – (missing:1928(20-21)) – mf#COR-616 – ne IDC [335]

Kooperativnaia viatka – Viatka, 1922(1-8) – 15mf – 9 – (cont as:kooperativnaia zhizn.viatka 1922(1-3[9-11]-1923(1-31).missing:1922(3,5-6)) – mf#COR-602 – ne IDC [335]

Kooperativnaia zhizn – Ekaterinoslav, 1917(1) – 2mf – 9 – mf#COR-603 – ne IDC [335]

Kooperativnaia zhizn – 1912-1920 – 69mf – 9 – (missing:1918(8-end)) – mf#COR-604 – ne IDC [335]

Kooperativnaia zhizn – Novonikolaevsk, 1921-1923(24) – 11mf – 9 – (missing:1921(1)-1922(8,10-18)) – mf#COR-606 – ne IDC [335]

Kooperativnaia zhizn – Iaroslavl, 1921-1924(10) – 18mf – 9 – (missing:1921-1922(1-5)) – mf#COR-609 – ne IDC [335]

Kooperativnaia zhizn – Briansk, 1923(1-12) – 3mf – 9 – (missing:1923(1-7)) – mf#COR-600 – ne IDC [335]

Kooperativnaia zhizn – Samara, 1913-1914(10) – 6mf – 9 – mf#COR-607 – ne IDC [335]

Kooperativnaia zhizn – Tula, 1922(1-6) – 4mf – 9 – mf#COR-608 – ne IDC [335]

Kooperativnaia zhizn – Vladimir, 1923(1-4) – 4mf – 9 – mf#COR-601 – ne IDC [335]

Kooperativnaia zhizn' – Moscow, Russia 1926-31 [mf ed Norman Ross] – 23r – 1 – mf#nrp-110 – us UMI ProQuest [077]

Kooperativnaia zhizn karelii – Petrozavodsk, 1922-1923(16) – 4mf – 9 – (missing:1922-1923(1)) – mf#COR-610 – ne IDC [335]

Kooperativno selo – Sofia, Bulgaria. May 1958-87 – 35r – 1 – us L of C Photodup [949]

Kooperativnoe delo – Orel, 1925-1926(4) – 4mf – 9 – (missing:1925(8)-1926(2)) – mf#COR-617 – ne IDC [335]

Kooperativnoe delo – Tomsk, 1922-1924 – 18mf – 9 – mf#COR-618 – ne IDC [335]

Kooperativnoe dvizhenie v rossii : ego teoriia i praktika / Prokopovich, S N – 1918 – 385p 5mf – 9 – mf#COR-219 – ne IDC [335]

Kooperativnoe dvizhenie v rossii / Prokopovich, S N – 1903 – 243p 3mf – 9 – mf#COR-100 – ne IDC [335]

Kooperativnoe slovo – Kharkov, 1916-1917(39) – 13mf – 9 – (missing:1917(10-12,19-22)) – mf#COR-622 – ne IDC [335]

Kooperativnoe slovo – Chita, 1916-1917(39/40) – 10mf – 9 – (missing:1917(14-22,27-28)) – mf#COR-623 – ne IDC [335]

Kooperativnoe slovo – Smolensk, 1917-1920(21/22) – 14mf – 9 – (missing:1917(4)-1919(1),1919(17-20),1920(11-12)) – mf#COR-621 – ne IDC [335]

Kooperativnoe slovo – Nizhnyi Novgorod, 1918(1-23) – 4mf – 9 – (missing:1918(4-5,8-10,13-14,18-22)) – mf#COR-620 – ne IDC [335]

Kooperativnoe strakhovanie v proshlom i nastoiashchem / Matveev, N G – M, 1923 – 1mf – 9 – mf#REF-132 – ne IDC [332]

Kooperativnoe stroitelstvo – 1926-1929(20) – 95mf – 9 – mf#COR-624 – ne IDC [335]

Kooperativnoe vospitanie detei i vozrozhdenie chelovechestva / Frommett, B R – 1918 – 70p 1mf – 9 – mf#COR-278 – ne IDC [335]

Kooperativnoe zakonodatelstvo / Voitsekhovskii, S F – 1914 – 57p 1mf – 9 – mf#COR-16 – ne IDC [335]

Kooperativnoe zakonodatelstvo s prilozheniem dekretov o vsekh vidakh kooperatsii / Povolotskii, L I – 1926 – 362p 4mf – 9 – mf#COR-300 – ne IDC [335]

Kooperativnoe zhilishche – Kiev, 1925-1926 – 4mf – 9 – mf#COR-619 – ne IDC [335]

O kooperativnom ideale / Tugan-Baranovskii, M I – 1918 – 16p 1mf – 9 – mf#COR-223 – ne IDC [335]

Kooperativnye kruzhki : materialy dlia zaniatii – 1925 – 125p 2mf – 9 – mf#COR-255 – ne IDC [335]

Kooperativnye kruzhki v derevne / Ozerov, N I – 1926 – 96p 2mf – 9 – mf#COR-268 – ne IDC [335]

Kooperativnye obedineniia moskovskoi oblasti – 1919 – 129p 4mf – 9 – mf#COR-717 – ne IDC [335]

Kooperativnye soiuzy v sibiri 1908-1918 gg / Ilimskii, D I – 1919 – 2mf – 9 – mf#COR-248 – ne IDC [335]

Kooperativnye tovarishchestva i ikh klassifikatsiia / Prokopovich, S N – 1919 – 32p 1mf – 9 – mf#COR-220 – ne IDC [335]

Kooperativnyi apparat v sovetskoi rossii v 1920/21 godu – 1921 – 55p 2mf – 9 – mf#COR-718 – ne IDC [335]

Kooperativnyi biulleten – Kaluga, 1921-1923(6) – 10mf – 9 – (missing:1922(1-6,11-end)) – mf#COR-627 – ne IDC [335]

Kooperativnyi biulleten – Vesegonsk, 1924(1) – 1mf – 9 – mf#COR-626 – ne IDC [335]

Kooperativnyi biulleten sumraisoiuza – Sumy, 1923-1924(5/6) – 7mf – 9 – mf#COR-628 – ne IDC [335]

Kooperativnyi byt – Riazan, 1923-1924(13) – 5mf – 9 – mf#COR-625 – ne IDC [335]

Kooperativnyi dvukhnedelnyi zhurnal / Vestnik Nizhegorodskogo soiuza uchrezhdenii melkogo kredita – Nizhnyi Novgorod, 1914-1918 – 15mf – 9 – mf#COR-629 – ne IDC [077]

Kooperativnyi institut i muzei / Antsyferov, A N – Kharkov, 1914 – 1mf – 9 – mf#COR-2 – ne IDC [335]

Kooperativnyi instruktazh i ego metody / Makhov, V N – Novonikolaevsk, 1924 – 309p 3mf – 9 – mf#COR-266 – ne IDC [335]

Kooperativnyi kredit – 1919-1920(1) – 7mf – 9 – mf#COR-630 – ne IDC [335]

Kooperativnyi kredit / Antsyferov, A N – 1918 – 1mf – 9 – mf#COR-369 – ne IDC [335]

Kooperativnyi kredit : polozhenie o kooperativnom kredite 18-go ianvaria 1927 goda, ego osnovy i postateinyi razbor... / Shmidt, G R – 1927 – 131p 2mf – 9 – mf#COR-410 – ne IDC [335]

Kooperativnyi kredit v severo-zapadnom krae – Minsk, 1913-1914(4) – 4mf – 9 – mf#COR-631 – ne IDC [332]

Kooperativnyi listok – Vitebsk, 1924-1926(11) – 5mf – 9 – (missing:1924,1925(1-23)) – mf#COR-632 – ne IDC [335]

Kooperativnyi mir – 1917-1918(6) – 14mf – 9 – mf#COR-633 – ne IDC [335]

Kooperativnyi molot – Tula, 1919(1) – 1mf – 9 – mf#COR-634 – ne IDC [335]

Kooperativnyi put – Odessa, 1927-1928 – 5mf – 9 – mf#COR-635 – ne IDC [335]

Kooperativnyi put – Samara, 1928-1929 – 28mf – 9 – mf#COR-636 – ne IDC [335]

Kooperativnyi sbyt produktov melkogo khoziaistva / Kulyzhnyi, A E – 88p 1mf – 9 – mf#COR-56 – ne IDC [335]

Kooperativnyi sbyt produktov selskogo khoziaistva / Kulyzhnyi, A E – 1918 – 191p 2mf – 9 – mf#COR-480 – ne IDC [335]

Kooperativnyi sbyt produktov selskogo khoziaistva v rossii / Evdokimov, A A – Kharkov, 1911 – 172p 2mf – 9 – mf#COR-23 – ne IDC [335]

Kooperativnyi ural – Ekaterinburg, 1921(1-8) – 2mf – 9 – (missing:1921(1,3-5)) – mf#COR-637 – ne IDC [335]

Kooperativnyi zakon i primernye ustavy obshchestva potrebitelei i soiuza potrebitelskikh obshchestv – 1917 – 44p 1mf – 9 – mf#COR-50 – ne IDC [335]

Kooperativnyi zhurnal – Perm, 1913-17 (2) – 22mf – 9 – (missing: 1914(19-end); 1916(16-21)) – mf#COR-595 – ne IDC [077]

Kooperativy / Bragin, M & Minin, M – 1907 – 79p 1mf – 9 – mf#COR-11 – ne IDC [335]

Kooperativy i sotsializm / Semkovskii, S M – [1917] – 14p 1mf – 9 – mf#COR-106= – ne IDC [335]

Der kooperator : supplement to przeglad spoldzielczy – L'viv, Ukraine 1931-33 [mf ed Norman Ross] – 1r – 1 – (in yiddish) – mf#nrp-888 – us UMI ProQuest [939]

Kooperator – Iaroslavl, 1926-1929(18) – 40mf – 9 – mf#COR-641 – ne IDC [335]

Kooperator – Kursk, 1921-1925(4) – 62mf – 9 – (missing:1921(1-13),1922(2-3,6-10,13-14),1924(19-22)) – mf#COR-638 – ne IDC [335]

Kooperator – Nikolaev, 1921-1922(2) – 2mf – 9 – (missing:1921(1-2,5-6)) – mf#COR-639 – ne IDC [335]

Kooperator – Tambov, 1922-1923 – 6mf – 9 – mf#COR-640 – ne IDC [335]

Kooperatsii : svod trudov mestnykh komitetov po 49 guberniiam evropeiskoi rossii / Borodaevskii, S V – 1904 – 171p 2mf – 9 – mf#COR-51 – ne IDC [335]

O kooperatsii / Kantor, M K – 1927 – 95p 2mf – 9 – mf#COR-198 – ne IDC [335]

O kooperatsii / Tolstoi, L N et al – 1911 – 32p 1mf – 9 – mf#COR-190 – ne IDC [335]

Den' kooperatsii – 21 dekabria 1844 [lxxv] – 21 dekabria 1919 g : odnodn gaz, posviashch semidesiatipiatilet iubileiu mezhdunar koop dvizheniia / ed by Bem, I I – Irkutsk: Irkut sovet koop sezdov 1919 – (= ser Asn 1-3) – 1v item 120, on reel n27 – 1 – mf#asn-1 120 – ne IDC [077]

Kooperatsiia / Korotkov, M A – 1925 – 120p 2mf – 9 – mf#COR-204 – ne IDC [335]

Kooperatsiia : krat ocherki teorii i istorii koop dvizheniia i zadach sov koop / Medvedev, V V – 1930 – 111p 2mf – 9 – mf#COR-211 – ne IDC [335]

Kooperatsiia – 1914-1915(1) – 2mf – 9 – (missing:1914(2-end)) – mf#COR-642 – ne IDC [335]

Kooperatsiia / Nikolaev, A A – 1918 – 79p 1mf – 9 – mf#COR-215 – ne IDC [335]

Kooperatsiia : sbornik statei o potrebitelskoi, selskokhoziaistvennoi i kreditnoi kooperatsii – Tula, 1923 – 95p 2mf – 9 – mf#COR-203 – ne IDC [335]

Kooperatsiia : o tom, kak soobshcha mozhno vygodnee ustraivat svoi khoz dela / Nikolaev, A A – 1908 – 99p 2mf – 9 – mf#COR-77 – ne IDC [335]

Kooperatsiia : ukazatel literatury – Kharkov, 1918 – 137p 2mf – 9 – mf#COR-534 – ne IDC [335]

Kooperatsiia bssr – Minsk, 1927 – 229p 4mf – 9 – mf#COR-719 – ne IDC [335]

Kooperatsiia i agronomiia : sbornik statei – 1919 – 82p 1mf – 9 – mf#COR-479 – ne IDC [335]

Kooperatsiia i borba s dorogoviznoi / Totomiants, V F – 1916 – 12p 1mf – 9 – mf#COR-118 – ne IDC [335]

Kooperatsiia i finansy : sbornik n3-i oktiabr'-dekabr' 1922 g / Bank Potrebitel'skoi Kooperatsii – M, [1922] – 2mf – 9 – mf#REF-76 – ne IDC [332]

Kooperatsiia i gorodskoe samoupravlenie : opyt obosnovaniia munitsipalnoi programmy kooperatorov / Ostrovskii, V V – 1919 – 436p 5mf – 9 – mf#COR-217 – ne IDC [335]

Kooperatsiia i gostorgovlia severo-zapadnoi oblasti i severa sssr na 1925 god – 411p 5mf – 9 – mf#COR-721 – ne IDC [335]

Kooperatsiia i oborona sssr – 1928 – 887p 1mf – 9 – mf#COR-722 – ne IDC [335]

Kooperatsiia i partiia / Kilchevskii, V A – 1919 – 28p 1mf – 9 – mf#COR-155= – ne IDC [335]

Kooperatsiia i pravo : sbornik zakliuchennii iuriskonsulta moskovskogo narodnogo banka / Kekuatov, K V – 1914 – 250p 3mf – 9 – mf#COR-46 – ne IDC [335]

Kooperatsiia i prosveshchenie / Suzdaltsev, K K – 1915 – 47p 1mf – 9 – mf#COR-113 – ne IDC [335]

Kooperatsiia i revoliutsiia / Kilchevskii, V A – 1917 – 16p 1mf – 9 – mf#COR-156 – ne IDC [335]

Kooperatsiia i revoliutsiia : ot staroi k novoi kooperatsii / Shliapnikov, D D – Kazan, 1920 – 47p 1mf – 9 – mf#COR-192 – ne IDC [335]

Kooperatsiia i shkola : sbornik statei – 1925 – 99p 2mf – 9 – mf#COR-258 – ne IDC [335]

Kooperatsiia i sotsializm / Nikolaev, A A – 1918 – 32p 1mf – 9 – mf#COR-175 – ne IDC [335]

Kooperatsiia i sotsializm : sbornik statei / Meshcheriakov, N L – 1920 – 111p 2mf – 9 – mf#COR-171 – ne IDC [335]

Kooperatsiia i vologodskoe artelnoe maslodelie / Stepanovskii, I K – Vologda, 1922 – 67p 1mf – 9 – mf#COR-449 – ne IDC [335]

Kooperatsiia i zhizn – Samara, 1918-1919(3) – 10mf – 9 – (missing:1918(11)) – mf#COR-643 – ne IDC [335]

Kooperatsiia k 15 sezdu vkp(b) – 1927 – 186p 2mf – 9 – mf#COR-724 – ne IDC [335]

Kooperatsiia k 15-mu sezdu vkp(b) – [1927] – 188p 2mf – 9 – mf#COR-160= – ne IDC [335]

Kooperatsiia na 14 konferentsii rossiiskoi kommunisticheskoi partii (bolshevikov) : doklad a i rykova, preniia po dokladu i reshenie konferentsii – 1925 – 107p 2mf – 9 – mf#COR-161 – ne IDC [335]

Kooperatsiia severa – Vologda, 1921-1923(23/24) – 51mf – 9 – (missing:1921(1-2,4)-1922(2),1923(2-3)) – mf#COR-644 – ne IDC [335]

Kooperatsiia sredi slavian / Borodaevskii, S V – 1912 – 82p 1 mf – 9 – mf#COR-10 – ne IDC [335]

Kooperatsiia v 1923-1924 godu i v 1924-1925 godu – 1928 – 230p 3mf – 9 – mf#COR-720 – ne IDC [335]

Kooperatsiia v derevne / Miliutin, V – 1925 – 19p 1mf – 9 – mf#COR-492 – ne IDC [335]

Kooperatsiia v russkoi derevne / Totomiants, V F – 1912 – 448p 5mf – 9 – mf#COR-117 – ne IDC [335]

Kooperatsiia v selskom khoziaistve / Kheisin, M L – 1924 – 74p 1mf – 9 – mf#COR-521 – ne IDC [335]

Kooperatsiia v shkole : khrestomatiinyi sbornik / ed by Iordanskii, N N – 1926 – 128p 2mf – 9 – mf#COR-257 – ne IDC [335]

Kooperatsiia v shkole : prakt ukazaniia po ustroistvu i vedeniiu shkolnogo kooperativa v gorodskikh shkolakh / Khizhniakov, V V – 1924 – 87p 1mf – 9 – mf#COR-280 – ne IDC [335]

Kooperatsiia v sisteme sovetskogo khoziaistva / Sarabianov, V N – 1926 – 92p 1mf – 9 – mf#COR-187 – ne IDC [335]

Kooperatsiia v soiuze ssr / Sevruk, P N – 1926 – 151p 2mf – 9 – mf#COR-274 – ne IDC [335]

Kooperatsiia v sovetskoi derevne / Tsingevatov, I A – 1927 – 156p 2mf – 9 – mf#COR-524 – ne IDC [335]

Kooperatsiia v sovetskoi rossii / Meshcheriakov, N L – 1922 – 72p 1mf – 9 – mf#COR-170 – ne IDC [335]

Kooperatsiia v zapadnoi evrope i rossii : skhema stroitelstva seriia... / Fishgendler, A M – 1923 – 36p 1mf – 9 – mf#COR-224 – ne IDC [335]

O kooperirovanii i kreditovanii bednoty v kustarno-promyslovoi kooperatsii – [1928] – 36p 1mf – 9 – mf#COR-436 – ne IDC [335]

Koopvaardij en visscherij – London, UK. Jul/Aug 1942-Jan/Feb 1945 – 1 – uk British Libr Newspaper [072]

Kootenay mining standard – Rossland, BC: Standard Pub Co, [1898?-189- or 19–] – 9 – mf#P04764 – cn Canadiana [622]

Kootnay lumber worker / Woodworkers' Industrial Union of Canada – v1 n1 [1948 nov17] – 1r – 1 – mf#681703 – us WHS [331]

Kooy, Tijmen van der *see* The distinctive features of the christian school

Kooy-van Zeggelen, M C *see* Ons mooi indie batavia

Kopecky, Lubomir *see* Krystalisace tavenych hornin

Kopenhagener illustrierte – Kopenhagen (DK) 1940 7 jul-29 dec, 1941 6 jul-1942 27 dec – 1 – gw Misc Inst [074]

Koperasi "batari" batik timur asli republik indonesia / Madjallah Batari – Surakarta, 1955-1958(1-8) – 2mf – 9 – (missing: 1955(1-3, 5, 9); 1956(5, 8-9, 11); 1957(3/4)) – mf#SE-1812 – ne IDC [959]

Koperasi penerbitan indonesia – Djakarta, 1964/1965-1965/1966 – 11mf – 9 – (missing: 1964(aug-nov); 1965(7)) – mf#SE-697 – ne IDC [959]

Kopfstein, Marcus *see* Die asaph-psalmen

Kopii s postanovlenii i protokolov po delu o zloupotrebleniiakh v kommercheskom ssudnom banke v moskve, proizvedennomu sudebnym sledovatelem moskovskogo okruzhnogo suda dlia proizvodstva sledstvii po osobo vazhnym delam, globo-mikhalenko – M, 1875-1876. 2v – 17mf – 9 – mf#REF-288 – ne IDC [332]

Kopings tidning – Koping, Sweden. 1855-97 – 18r – 1 – sw Kungliga [078]

Kopitar, Bartholomaeus *see* Grammatik der slavischen sprache in krain, kaernten und steyermark

Kopp, Arthur *see* Die lieder der heidelberger handschrift

Kopp, Frederick Edward *see* Arthur foote

Kopp, Georg Ludwig Karl *see* Ideen zu der organisation der teutschen kirche

Kopp, Josef *see* Zur judenfrage nach den akten prozesses rohling-bloch

Kopp, Thomas Joseph *see* God first – go forward

Koppitz, Alfred *see* Der goettweiger trojanerkrieg

Koppius, F *see* Vitae ac gesta abbatum adwerdensium

Kopplung von schadstoffabbau und nutzstoffproduktion mit halomonas elongata / Fersterra, Holger – (mf ed 2000) – 5mf – 9 – €59.00 – 3-8267-2738-X – mf#DHS 2738 – gw Frankfurter [660]

Kopp-Stache, Juergen *see* Paedagogische dynamik der heimerziehung

Koptische akten zum ephesinischen konzil von jahre 431 / Kraatz, W – Leipzig, 1904 – (= ser Tugal 2-26/2) – 4mf – 9 – €11.00 – ne Slangenburg [243]

Koptische miscellen, 1-148 / Lemm, O von – Leipzig, 1972 – (= ser Subsidia byzantia v11) – 18mf – 8 – €35.00 – ne Slangenburg [243]

Koptische palaeographie : 25 tafeln zur veranschaulichung der schreibstile koptischer schriftdenkmaeler auf papyrus, pergament und papier fuer die zeit des 3.-14. jahrhunderts / Stegemann, Viktor – Heidelberg: Im Selbstverlag von F. Bilabel, 1936. Chicago: Dep of Photodup, U of Chicago Lib, 1973 (1r); Evanston: American Theol Lib Assoc, 1984 (1r) – (= ser Quellen und Studien zur Geschichte und Kultur des Altertums und des Mittelalters) – 1 – 0-8370-1280-5 – (incl bibl ref) – mf#1984-B368 – us ATLA [090]

Der koptische text der kirchenordnung hippolyts / Till, W & Leipoldt, J – Berlin, 1954 – (= ser Tugal 5-58) – 2mf – 9 – €5.00 – ne Slangenburg [240]

Die koptische uebersetzung der vier grossen propheten / Schulte, Adalbert – Muenster i. W: Aschendorff, 1892 – 1mf – 9 – 0-8370-5159-2 – mf#1985-3159 – us ATLA [221]

Koptische untersuchungen / Abel, Carl – Berlin: Ferd Duemler, 1876 – 2mf – 9 – 0-524-03472-9 – mf#1990-3214 – us ATLA [470]

Die koptischen quellen zum konzil von nicaea / Haase, Felix – Paderborn, 1920 – 3mf – 8 – €7.00 – ne Slangenburg [243]

Koptisches handwoerterbuch / Spiegelberg, W – Heidelberg, 1921 – 5mf – 9 – mf#NE-20022 – ne IDC [956]

Koptisch-gnostische schriften (gcsej10) – (= ser Griechische christlichen schriftsteller der ersten jahr- hunderte (gcsej)) – (bd1 ed by i g schmidt 1905 €18) – ne Slangenburg [243]

Koptos / Petrie, W M – London, 1896 – 2mf – 9 – mf#NE-20343 – ne IDC [956]

Kopulasaetze des deutschen und ihre wiedergabe im chinesischen : eine kontrastive analyse aus valenztheoretischer sicht / Schmidt, Wolfgang G A – Frankfurt/Main, Bern, New York: 1986 (mf ed 1996) – 5mf – 9 – €59.00 – 3-8267-2377-5 – mf#DHS 2377 – gw Frankfurter [410]

Die koralle – Berlin DE, 1933-36 – 1 – gw Misc Inst [074]

Korallenkettlin : ein drama in vier akten / Duelberg, Franz – Berlin: E Fleischel 1906 [mf ed 1989] – 1r – 1 – (filmed with: der impressionismus in der lyrik der annette von droste-hulshoff / gerhard fruhbrodt) – mf#7190 – us Wisconsin U Libr [820]

Die korallthiere des rothen meeres / Klunzinger, C B – Berlin, 1877-1879. 3v – 8mf – 9 – mf#Z-2246 – ne IDC [590]

Der koran – Berlin: Brandus Verlagsbuchh, [1916] – 2mf – 9 – 0-524-07728-2 – mf#1991-0150 – us ATLA [260]

Der koran – Crefeld: JH Funcke, 1840 – 6mf – 9 – 0-524-08192-1 – mf#1991-0305 – us ATLA [260]

Der koran : oder, das gesetz der moslemen durch muhammed, den sohn abdallahs – Halle: Gebauer, 1828 – 10mf – 9 – 0-524-08828-4 – mf#1993-4020 – us ATLA [260]
Koran – Philadelphia, PA. 1868 – 1r – us UF Libraries [025]
Le koran : traduction nouvelle faite sur le text arabe / Biberstein-Kasimirski, Albert de – Paris: Charpentier, 1840 – 7mf – 9 – 0-524-07941-2 – mf#1991-0191 – us ATLA [260]
Le koran analyse / Labaume, Jules – Paris: Maisonneuve, 1878 [mf ed 1993] – (= ser Bibliotheque orientale 4) – 2mf – 9 – 0-524-06972-7 – mf#1991-0046 – us ATLA [260]
The koran, commonly called the alcoran of mohammed / Sale, G – 1r – 1 – mf#2174 – uk Microform Academic [260]
Koran English *see* Qur'an
Koran und bibel : ein komparativer versuch / Jaspis, Johannes Sigmund – Leipzig: G Struebig, 1905 – 2mf – 9 – 0-524-02243-7 – mf#1990-2910 – us ATLA [230]
Korana folktales : grammar and texts / Maingard, L F – Johannesburg, South Africa. 1962 – 1r – us UF Libraries [390]
Koranadialekt des hottentottischen / Meinhof, Carl – Berlin, Germany. 1930 – 1r – us UF Libraries [960]
Koranta ea becoana – Mafikeng SA, 1901-1904 [mf ed Cape Town: SA library [198-?]] – 1r – 1 – (text in tswana and english) – mf#MS00298 – sa National [079]
Korbach, Joachim *see* Das ozon/ festbettkatalysator-verfahren bei der stickwasserbehandlung
Kord – Newton MA 1970-72 – 1 – ISSN: 0047-357X – mf#7548 – us UMI ProQuest [780]
Kordofan and the region to the west of the white nile / Sudan. Intelligence Dept – London: H M Stationery Off, 1912 – us CRL [960]
Korea : the land, people, and customs / Jones, George Heber – Cincinnati: Jennings & Graham [1907] [mf ed 1995] – (= ser Yale coll; Little books on missions) – 110p – 1 – 0-524-09606-6 – mf#1995-0606 – us ATLA [950]
Korea / Marshall, Edward Asaph – Chicago: Bible Institute Colportage Association, [1—] [mf ed 1995] – (= ser Yale coll) – 32p – 1 – 0-524-09756-9 – mf#1995-0756 – us ATLA [950]
Korea, 1950-1966 / U.S. State Dept – (= ser Confidential u s state department special files) – 11r – 1 – $1915.00 – 1-55655-198-3 – (1st suppl, 1951-66 12r isbn 1-55655-890-2 $2320. with p/g) – us UPA [327]
Korea commentary / U[/nited/] S[/tates/]-Korea Research and Action Committee – v1 n1-3/4 [1977 may/jun-dec] – 1r – 1 – mf#335947 – us WHS [071]
Korea for christ / Davis, George Thompson Brown – London: Christian Workers' Depot, 1910 [mf ed 1995] – (= ser Yale coll) – 71p (ill) – 1 – 0-524-09910-3 – mf#1995-0910 – us ATLA [950]
Korea for christ / Davis, George Thompson Brown – New York: Fleming H Revell, c1910 – 1mf – 9 – 0-8370-6104-0 – mf#1986-0104 – us ATLA [240]
Korea free press – 1970 nov-1972 jan – 1r – 1 – mf#705464 – us WHS [071]
Korea from its capital : with a chapter on missions / Gilmore, George W – Philadelphia: Presbyterian Board of Publication and Sabbath-school Work, c1892 – 1mf – us ATLA [951]
Korea from its capital : with a chapter on missions / Gilmore, George William – Philadelphia: Presbyterian Board of Publication and Sabbath-school Work, c1892 – 1mf – 9 – 0-7905-6344-4 – mf#1988-2344 – us ATLA [915]
Korea. (Government-General of Chosen, 1910-45) *see* Chosen sotokufu
Korea in transition / Gale, James Scarth – New York: Eaton & Mains; Cincinnati: Jennings & Graham c1909 [mf ed 1986] – (= ser Forward mission study courses) – 1mf – 9 – 0-8370-6186-5 – (incl ind & app) – mf#1986-0186 – us ATLA [951]
The korea pentecost : and other experiences on the mission field / Blair, William Newton – New York: Board of Foreign Missions of the Presbyterian Church in the USA [1—] [mf ed 1995] – (= ser Yale coll) – 51p – 1 – 0-524-10177-9 – mf#1995-1177 – us ATLA [242]
Korea. (Territory under U.S. Occupation, 1945-). Military Governor *see* History of the u.s. army military government in korea, sep 1945-30 jun 1946
Korea. (Territory under U.S. Occupation, 1945-1948). Military Governor *see* Miscellaneous orders, appointments, etc
Korea times : serves the korean communities across canada = Hankukilbo – Toronto. v1 n528- . jun 1 1981- [daily] – 39r – 1 – Can$5265.00 – (cont: canada news [toronto, apr 25 1975-may 30 1981]; canada's only korean language daily; back run [1981-92]) – cn McLaren [071]
The korea times – Seoul, jan 3 1956- – us CRL [079]
Korea week – Arlington VA 1968-77 – 1 – ISSN: 0023-3951 – mf#7952 – us UMI ProQuest [951]
Korean biographical archive (kba) = Koreanisches biographisches archiv (kba) / Frey, Axel [comp] – [mf ed 2000-04] – 345mf (1:24) in 12 installments – 9 – diazo €10.060.00 (silver €11,0800 isbn: 978-3-598-34241-7) – ISBN-10: 3-598-34240-3 – ISBN-13: 978-3-598-34240-0 – (with printed ind) – gw Saur [950]
Korean buddhism : history, condition, art. three lectures / Starr, Frederick – Boston: Marshall Jones 1918 [mf ed 1993] – 1mf [ill] – 9 – 0-524-07735-5 – (incl bibl ref) – mf#1991-0157 – us ATLA [280]
Korean Congress (1st: 1919: Philadelphia) *see* First korean congress
Korean folk songs / Anderson, Sara May – U of Rochester 1940 [mf ed 19–] – 2mf – 9 – mf#fiche328 – us Sibley [780]
Korean mission records : papers of the korean mission, 1889-1986, from birmingham university library – 2pt – 1 – (pt1: minute books, ledgers & correspondence with mission staff, 1908-85 [17r] £1600; pt2: periodicals, pamphlets, press-cuttings & photos 1889-1987 [18r] £1700; with d/g) – uk Matthew [950]
Korean religious tract society. annual report – Seoul, 1894-1940 [mf ed 2001] – (= ser Christianity's encounter with world religions, 1850-1950) – 4r – 1 – (filmed with: korean religious book and tract society. annual report [1917-19]; christian literature society of korea. annual report [1920-40]. in english) – mf#2001-s120-124 – us ATLA [240]
The korean revival : an account of the revival in the korean churches in 1907 / Jones, George Heber & Noble, William Arthur – New York: Board of Foreign Missions of the Methodist Episcopal Church, [1910] [mf ed 1995] – (= ser Yale coll) – 45p (ill) – 1 – 0-524-09981-2 – mf#1995-0981 – us ATLA [242]
The korean situation; the korean situation no. 2 / Federal Council of the Churches of Christ in America – 1919, 1920 – 1 – $50.00 – us Presbyterian [240]
Korean sketches / Gale, James Scarth – Chicago: Fleming H Revell, c1898 – 1mf – 9 – 0-8370-6048-6 – mf#1986-0048 – us ATLA [951]
Korean translation and validation of the eating disorder inventory-2 (edi-2) and the buimia test-revised (bulit-r) / Ryu, Heeseung Roh – 1996 – 2mf – 9 – $8.00 – mf#PSY 1957 – us Kinesology [616]
Korean war studies and after-action reports – (= ser Armed forces oral histories) – 339mf (20:1-29:1) – 9 – $3230.00 – 1-55655-125-8 – (with p/g) – us UPA [355]
Korean-english dictionary / Gale, James Scarth – Yokohama: Printed by Fukuin Printing, 1911 [mf ed 1995] – (= ser Yale coll) – 1 – 0-524-09530-2 – mf#1995-0530 – us ATLA [040]
Korelin, A P *see*
– Banki i razvitie sel'skogo khoziaistva v rossii v kontse 19-nachale 20 vv
– Sel'skokhoziaistvennyi kredit v rossii v kontse 19-nachale 20 v
Koren, B V *see* Sushchnost ucheniia sotsialistov-revoliutsionerov
Koren, Paul *see* Samlede skrifter
Koren, Vilhelm *see* Samlede skrifter
Korenevskaia, N N *see* Biudzhetnye obsledovaniia krest'ianskikh khoziaistv v dorevoliutsionnoi rossii
Koresec, V *see* Hethitische staatsvertraege
Koreshan unity / Shepherd, Rose – s.l, s.l? 1936 – 1r – us UF Libraries [978]
Korespondencja / Mickiewicz, Adam – Paryz: Ksiegarnia Luxemburgska 1870- [mf ed 1986] – 1r – 1 – mf#1920 – us Wisconsin U Libr [920]
Korff, H A *see* Gedichte
Korff, Hermann August *see*
– Die dichtung von sturm und drang im zusammenhange der geistesgeschichte
– Faustischer glaube
– Geist der goethezeit
– Goethes deutsche sendung
Korff, Sergei Aleksandrovich *see* Dvorianstvo i ego soslovnoe upravlenie za stoletie 1762-1855 godov
Korff, Theodor *see* Die auferstehung christi und die radikale theologie
Die koridethi evangelien th038 / ed by Beermann, Gustav & Gregory, Caspar Rene – Leipzig: J C Hinrichs, 1913 [mf ed 1990] – 2mf – 9 – 0-8370-1806-4 – mf#1987-6194 – us ATLA [226]
Korifei, ili kliuche literatury – Lanham. 1909+ (1) 1969+ (5) 1976+ (9) – 29mf – 9 – mf#1450 – ne IDC [077]
Korkunov, N M *see* General theory of law
Korman, Richard I *see* Government organization manuals 1900-1980
Korn, Friedrich *see* Panorama von ofen und pesth
Korn, Heinz *see* Gehetzt uebers meer
Korn, Karl *see* In der stille
Korn, Rachel H *see* Erd
Korn, V E *see* Manuscript collection
Korn, Yitshak *see* Keshenev
Kornatzki, Wolf *see* Das neue lied
Kornemann, Matthias *see*
– Kammermusik und klaviermusik [die sammlung der sing-akademie zu berlin...pt 4]
– Die sammlung der sing-akademie zu berlin
Kornemann, Matthias [comp] *see* Sinfonien, konzerte und ouvertueren [die sammlung der sing-akademie zu berlin...pt 3]
Der kornett des koenigs / Engelkes, Gustav Gerhard – feldpost-ausg. Berlin: Nordland c1940 [mf ed 1989] – 1r – 1 – (filmed with: die kleine weltlaterne / peter bamm) – mf#7216 – us Wisconsin U Libr [880]
Der korngarten code : die erfolgsformel fuer die zukunft / Schuermeyer, Manfred – 1993 – 2mf – 9 – 3-89349-812-5 – gw Frankfurter [330]
Korngold, Ralph *see* Citizen toussaint
Korngreen, Philip *see* Huke ha-mizrah ha-kadmon
Kornweibel, Theodore *see* Federal surveillance of afro-americans (1917-25)
Kornwestheimer zeitung – Kornwestheim DE, 1909 2 jan-1943 31 mar – 18r – 1 – gw Misc Inst [074]
Korobkov, VA *see* Gorodskoe ot ognia strakhovanie so vzaimnoiu mezhdu gorodami garantiei
Koroleff, Alexander *see* Efficacite au debusquage des billots et au maniement des chevaux
Korong vale lance – Korong Vale, Australia. 10 Jan 1903-2 Jun 1906; 10 Jan 1903-24 Dec 1904; 7 Jan 1905-9 Jun 1906.-w. 2 1-2 reels – 1 – uk British Libr Newspaper [072]
Korot ha-torah veha-emunah be-hungariyah / Greenwald, Leopold – Budapest, Hungary. 1921 – 1r – us UF Libraries [939]
Koroth ha-yehudim be-damesek / Rivlin, Eliezer – Tel-Aviv, Israel. 1925 – 1r – us UF Libraries [939]
Korotkov, M A *see*
– Kooperatsiia
– Ocherki po istorii russkoi kooperatsii
Korotkova, N N *see* Rodina [omsk: 1919]
Korotoyakskii uezd : sovet rk i kd. izvestiia korotoiakskogo uezdnogo soveta tabochikh i krest'ianskikh deputatov – Korotoyak, Russia 1918 [mf ed Norman Ross] – 1r – 1 – mf#nrp-773 – us UMI ProQuest [077]
Korrekturen zur bisherigen erklaerung des roemerbriefes / Klostermann, August – Gotha: Friedrich Andreas Perthes, 1881 – 1mf – 9 – 0-8370-9707-X – mf#1986-3707 – us ATLA [227]
Der korrespondent fuer deutschlands buchdrucker und schriftgiesser – Leipzig, Berlin, Stuttgart DE, 1875-78, 1906-07 – 4r – 1 – (later: druck und papier, fr 1933 publ in berlin, later in stuttgart; filmed by misc inst: 1952-54, 1956-57, 1961, 1963 n1 [jub-nr], 1966-1975 8 dec, 1977-88, 1990-96 [25r]) – gw Mikrofilm; gw Misc Inst [070]
Korrespondenz b – Kauen (Kaunas, Kowno LT), 1916 11 oct-1918 20 nov – 4r – 1 – (title varies: fr n111: baltisch-litauische mitteilungen; fr n131: litauische mitteilungen; n17-51 publ in bialystok n131- in wilna) – gw Misc Inst [077]
Korrespondenz der deutschnationalen volkspartei – Berlin DE, 1924 – 1r – 1 – gw Misc Inst [325]
Korrespondenz des reichsverbandes gegen die sozialdemokratie – Berlin DE, 1907 n24-1909 – 1 – gw Misc Inst [320]
Korrespondenzblatt : fuer die mitglieder des gewerkschaftsbundes in bonn-stadt und bonn-land – Bonn, 16 Mar-15 Jun 1946 – 1r – 1 – gw Mikropress [074]
Korrespondenzblatt – Berlin DE, 1905-25, 1927-30, 1933-39 – 1 – (with gaps) – gw Misc Inst [074]
Korrespondenzblatt.. / Allgemeiner deutscher Gewerkschaftsbund – Berlin. v7-21, 23-25, 28, 33-36, 38-40. 1897-1930. (many incomplete) – (= ser Serial publications of german trade unions) – 1 – us Wisconsin U Libr [331]
Korrespondenzblatt des kreises eupen *see* Correspondenzblatt des kreises eupen
Korrespondenz-blatt des verbandes der deutschen juden – Berlin DE, 1907 oct, 1908 may-1914 jun – 1r – 1 – gw Misc Inst [939]
Korrespondenz-blatt des verbandes der deutschen juden – Berlin: Max J Loewenthal. n1-14. 1908-14 [complete] – (= ser German-jewish periodicals...1768-1945, pt 1) – 1r – 1 – $125.00 – mf#B375 – us UPA [939]
Korrespondenzblatt des vereins fuer niederdeutsche sprachforschung / Verein fuer Niederdeutsche Sprachforschung – Hamburg: Der Verein 1876- [v1-68(1876-1961) [mf ed 1979] – 3r – 1 – ISSN: 0342-0752 – mf#film mas c617 – us Harvard [430]
Korrespondenzblatt des vereins zur gruendung und erhaltung einer akademie fuer die wissenschaft des judenthums – Frankfurt/M, Berlin DE, 1920-30 – 1r – 1 – gw Misc Inst [939]
Korrespondenzblatt fuer baecker und konditoren – Duesseldorf DE, 1905 12 jan-1933 1 jul – 3r – 1 – (title varies: 1 oct 1908: solidaritaet) – gw Misc Inst [640]
Korrespondenzblatt fuer die mitglieder des gewerkschaftsbundes in bonn-stadt und bonn-land – Bonn DE, 1946 16 mar-15 jun – 1r – 1 – mf#3867 – gw Mikropress [331]
Korrodi, Eduard *see* Das poetische zuerich
Korsakoff : die geschichte eines heimatlosen / Dwinger, Edwin Erich – Jena: E Diederichs, 1929 [mf ed 1989] – 260p – 1 – mf#7192 – us Wisconsin U Libr [830]
Korsakov, P *see* Trudy uchenykh i literatorov russkikh i inostrannykh
Korshunov, I G *see* Teploprovodnost i temperaturoprovodnost perekhodnykh metallov pri vysokikh temperaturakh
Korsi, Demetrio *see*
– Antologia de panama
– Grillo que canto sobre el canal
Korsinsky, Bernhard *see* Geographisch-statistisch-topographisches lexikon von wuertemberg
Kort begrip van de burgerlijke bouw-konst... / Lauterbach, J B – Ed 2. 's-Gravenhage, 1780 – 2mf – 9 – mf#OA-89 – ne IDC [720]
Kort indlaeg i sagen mellem s kierkegaard og h l martensen : et lejlighedsskrift / Teisen, Niels – Kobenhavn: K Schoenberg, 1884 – 1mf – 9 – 0-524-00394-7 – (incl bibl ref) – mf#1989-3094 – us ATLA [190]
Kort overzicht van de gereformeerde religie in de hollandsche en engelsche taal / Robbert, Jan – Kalamazoo, MI: RE Bartlett, 1904 – 1mf – 9 – 0-524-06661-2 – mf#1991-2716 – us ATLA [240]
Kort udsigt over det lutherske kirkearbeide blandt nordmaendene i amerika / Bothne, Thrond – Chicago: K Taklas 1898 [mf ed 1992] – 1mf – 9 – 0-524-01996-7 – (in english) – mf#1990-4164 – us ATLA [242]
Kort utlaggning af st pauli bref till galaterna / Roos, Magnus Fredrik – Goetheborg: N J Gumperts, 1843 [mf ed 1986] – 1mf – 9 – 0-8370-9652-9 – (in swedish) – mf#1986-3652 – us ATLA [227]
Kort verslag van de handelsvereeniging te soerabaia 1948-1956 – 28mf – 9 – mf#SE-696 – ne IDC [950]
Korte, August *see* Die konzilspolitik karls 5. in den jahren 1538-1543
Korte en getrouwe onderregtinge van de generaal bass *see* Treulicher unterricht im general-bass
Korte en klare instructie... / Melder, G – Amsterdam, 1664 – 4mf – 9 – mf#OA-157 – ne IDC [720]
Korte historie van de chr. geref. kerk te holland, central ave : uitgegeven ter gelegenheid van de herdenking van het 40-jarig bestaan der gemeente... – Holland, MI: H Holkeboer, [1905?] – 1mf – 9 – 0-524-06635-3 – mf#1991-2690 – us ATLA [240]
Korte stellingen : in welke vervat worden de grondstukken van de christelyke leere / Vitringa, C – Delft, 1717 – 7mf – 9 – mf#PBA-384 – ne IDC [240]
Korte verhandeling over het zingen en speelen in de hervormde kerk van nederland : als mede byzondere anmerkingen en eenige regels aangaande het zingen / Beyen, Petrus – Nymegen: H Bronstring 1790 [mf ed 19–] – 1mf – 9 – mf#fiche 5 – us Sibley [780]
Kortenshpiel / Blinken, Meir – NYU York, NY. 19– – 1r – us UF Libraries [939]
Kortfattad redogorelse over arbetet, dess organisation och utveckling samt rapport fran kommittens revisorer / Svenska Hjaelpkommitten for Spanien – Stockholm, 1937 – (= ser Blodgett coll) – 9 – us Harvard [946]
Kortum, Karl Arnold *see* Leben, meynungen und thaten von hieronimus jobs dem kandidaten, und wie er sich weiland viel ruhm erwarb auch endlich als nachtswaechter zu sulzburg starb
Korvin-Piotrovskii, Vladimir L'vovich *see* Beatriche
Korzon, Tadeusz *see* Wewnetrzne dzieje polski za stanislawa augusta, 1764-1794
Kosch, Wilhelm *see*
– Martin greif in seinen werken
– Oesterreich im dichten und denken grillparzers
Koschmieder, Arthur *see* Herders theoretische stellung zum drama
Koscion i teologia – Warszawa: Wydawnictwo Zwiastun, 1965 – 1mf – 9 – 0-524-08150-6 – mf#1993-9056 – us ATLA [240]
Kosegarten, J G L *see* Tribe of hudail
Kosenkov, I N *see* Volia [vladivostok: 1919-1920]
Koseritz, Carlos Von *see*
– Imagens do brasil
Kosgoro – Djakarta, 1963/1964-1964/1965 – 14mf – 9 – mf#SE-857 – ne IDC [950]
Koskolou, Maria D *see* Arterial hypoxemia and performance during intense exercise
Koskull, Harald, Baron von *see* Wielands aufsaetze ueber die franzoesische revolution

Kosmas, Marie *see* Mesures des niveaux d'eau dans le lac des deux montagnes et les exutoires sainte-anne et vaudreuil
Die kosmischen systeme der griechen / Gruppe, Otto Friedrich – Berlin: G Reimer, 1851 [mf ed 1986] – 1mf – 9 – 0-8370-9869-6 – mf#1986-3869 – us ATLA [180]
Die kosmographie in ariosts orlando furioso / Strauch, Alfons – Bonn, 1918 (mf ed 1994) – 1mf – 9 – €24.00 – 3-8267-3058-5 – mf#DHS-AR 3058 – gw Frankfurter [440]
Der kosmopolitische beobachter / ed by Fuchs, A – Mainz 1793 [mf ed Hildesheim 1992-98] – (= ser Dz. historisch-politische abt) – 1jg[=12st] on 2mf – 9 – €60.00 – mf#k/n1728a – gw Olms [320]
Kosmorama – 1954-99+ – 25r – 1 – £820.00 – mf#KOS – uk World [790]
Kosmos – Cobourg, Ont: Science Association of Victoria University, [1885-1886] – 9 – (cont: v p journal) – mf#P04629 – cn Canadiana [500]
Kosmos : entwurf einer physischen weltbeschreibung / Humboldt, Alexander von – Stuttgart: Cotta [1844?] [mf ed 1998] – 4v on 1r [ill] – 1 – (incl bibl ref) – mf#film mas 28291 – us Harvard [520]
Kosova – (= ser Vilayet salnames) – 9 – (1300 [1883] def'a 2 3mf $55; 1311 [1893] 5mf $75; 1314 [1896] 12mf $195; 1318 [1900] 16mf $265) – us MEDOC [956]
Kosova – Skopje, YU. 1877-? n1577. 15 temmuz 1325 [1909] – (= ser O & t journals) – 1mf – 9 – $25.00 – us MEDOC [949]
Kosrae district charter, 1978 : as approved by the seventh congress of micronesia / Federated States of Micronesia – 2nd reg sess. Saipan: the Congress, 14 feb 1978 – 2mf – 9 – $3.00 – mf#LLMC 82-100H Title 3 – us LLMC [324]
Kossenko, M P *see*
– Biulleten' gazety "novoe slovo"
– Novoe slovo
Kossmann, E F *see*
– Fortunati gluecksseckel und wuenschhuetlein
– Das niederlaendische faustspiel des siebzehnten jahrhunderts
Kossovskii, Vladimir V *see*
– Idisher hurbn in rusland
Kossuth, Lajos *see* Memories of my exile
Kost, John *see*
– Field experiments
Kostanoski, John I *see* Journal of security education
Koster, Henry *see*
– Reisen in brasilien
– Viagens ao nordeste do brasil
Koster, S *see* Is het woord "gereformeerd" in het vaandel der hollandsche chr ger kerk in amerika een leugen in hare rechterhand?
Kosters, Willem Hendrick *see* Die wiederherstellung israels in der persischen periode
Kosters, Willem Hendrik *see* De historie-beschouwing van den deuteronomist
Kostiurina, M N *see* Sibirskii listok
Kostnicke jiskry – Prague, Czechoslovakia. - w. 19 may 1949-18 dec 1952; 1953-2 dec 1970 (imperfect) – 6r – 1 – uk British Libr Newspaper [072]
Kostromskiia gubernskiia viedomosti / Russia – 1838-51, 1894-1906 – 1 – $388.00 – us L of C Photodup [947]
Kostromskoe Gubernskogo Zemstvo *see* Issledovanie vkladov v gosudarstvennye sberegatel'nye kassy kostromskoi gubernii za vremia s 1885 po 1911 gg vkliuchitel'no
Kostromskoi kooperator – Kostroma, 1913-1917(3) – 8mf – 9 – (missing:1913(10),1914-1915) – mf#COR-645 – ne IDC [335]
Kostromskoi liberal' / Liberal'no-Demokraticheskaia partiia Rossii – Kostroma, Russia. n1[1][avg 1997], n3[3][okt 1997]-n2[5][1998] – 1 – (suppl to newspaper "LDPR") – mf#mf-12248 [reel 6] – us CRL [077]
Kostrzewa, Frank *see*
– Elemente deskriptiver und inferentieller statistik und ihre vorlaeufer
– Merkmale verstehens- und behaltensfoerdernder kontextueller bedeutungserklaerungen
– Unterricht fuer spaetausgesiedelte kinder und jugendliche
Kostygova, G I *see*
– Obraztsy kalligrafii irana i srednei azii 15-19 vv
– Persidskie i tadzhikskie rukopisi "novoi serii" gpb alfavitnyi katalog
Kotarski, Mark *see* An evaluation of a home-based exercise program involving non-exertional hypoxemic and exertional hypoxemic chronic obstructive pulmonary diseased patients
Kote paunui o aotearo – 24 nov 1894-29 sep 1896 – 1r – 1 – mf#15.49 – nz Nat Libr [079]
Kotei, S I A *see* Politics and government in west africa
Kotei, Samuel Isaac Asharley *see* The social determinants of library development in ghana
Kotel ha-maharavi be-divre yeme yisra'el uve-masorto ve-sifruto / Ben-Zion, S – Tel-Aviv, Israel. 1928 or 1929 – 1r – us UF Libraries [939]
Kotewal, Jehangir Framjee *see* Whither bharat?
Koti, Candlish *see* Incwadi yesixhosa yesiqibi sokuqala
Kotoba no tebiki : nippon-melajoe-djawa-soenda-madoera – (Djakarta) 2603 – 200p 3mf – 9 – mf#SE-2002 mf90-92 – ne IDC [959]
Kotovich, A *see* Dukhovnaia tsenzura v rossii
Kotschy, C G T *see*
– Abbildungen und beschreibungen neuer und seltener thiere und pflanzen in syrien und im westlichen taurus gesammelt
– Die eichen europa's und des orients
– Der libanon und seine alpenflora
– Plantae arabiae in ditionibus hedschas...
– Die sommerflora des antilibanon und hohen hermon
Kotschy, T *see* Plantae binderianae nilotico-aethiopicae
Kottinger, H M *see* Jacob ruffs etter heini uss dem schwizerland
Kottinger, Hermann Marcus *see*
– Jacob ruffs adam und heva
– The youth's liberal guide for their moral culture and religious enlightenment
Kottje, R *see* Kirchengeschichte heute
Kotva – Schuyler, NE: Jos K Sinkule, 1892 (wkly) [mf ed roc1 cis16. 6 pros 1892 filmed 1979] – 1r – 1 – (in czech) – us NE Hist [071]
Kotyska, Vaclav *see* Uplny mistopisny slovnik kralovstfi ceskeho.
Kotze, Tobias Christiaan [comp] *see* Requirements for medical electron accelerators (photon & electron production)
Kotzebue, August Von *see*
– Deux freres
– Etat restitue
– Misanthropie et repentir
Kotzebue, August von *see*
– Une annee memorable de la vie d'auguste de kotzebue
– Die biene oder neue kleine schriften
– Erinnerungen aus paris im jahre 1804 [achtzehnhundertvier]
– Die grille
– Das merkwuerdigste jahr meines lebens
Kotzebue in england : ein beitrag zur geschichte der englischen buehne und der beziehungen der deutschen litteratur zur englischen / Sellier, Walter – [S.l.: s.n.] 1901 (Leipzig: Druck von O. Schmidt) – 1r – 1 – (incl bibl ref) – us Wisconsin U Libr [430]
Kotzebue, M von *see* Reise nach persien mit der russisch kais gesandtschaft im jahre 1817
Kotzebue, Moritz von *see* Moritz v kotzebue's russisch-kaiserlichen hauptmanns im general-stabe, ritters des wladimir- wie auch des persischen sonnen- und loewenordens reise nach persien...
Kotzebue, Otto von *see*
– Entdeckungs-reise in die sued-see und nach der berings-strasse zur erforschung einer nordoestlichen durchfahrt
– Neue reise um die welt
Koumintang party organs and nationalist chinese government gazettes, pre-1949 – Nov 1919-45. 16 titles – 1 – us L of C Photodup [951]
Kouns, Nathan Chapman *see* Arius the libyan
Kova / Lithuanian Socialist Federation of America – 1905 nov 10 [v1 n26], 1910 jan 7 [v6 n1]-may 12, 1911 may 19-dec 29, 1912 jan-1913 may, 1913 jun 6-1914 oct 9, 1914 oct 16-1915 dec 24, 1916 jan-nov – 6r – 1 – mf#612494 – (us WHS [335]
Kovalan and kannaki : (the story of the silappadhikaram) / Panchapakesa Ayyar, Aiylam Subramanier – Madras: C Coomarasawmy Naidu & Sons, [1940] – (= ser Samp: indian books) – us CRL [490]
Kovalenskii, Mikhail N *see* Russkaia revoliutsiia v sudebnykh protsessakh i memuarakh
Kovan'ko, P *see* Glavneishie reformy, provedennye n kh bunge v finansovoi sisteme rossii
Kovarskii, B *see* Ekaterina konstantinovna breshkovskaia
Kovatch-McMillan-Botkin Family *see* Papers
Kovda, V A *see* Problems of soil science
Kovets hokhmat ha-ra'va' / Ibn Ezra, Abraham Ben Meir – Warsaw, Poland. 1922 – 1r – us UF Libraries [939]
Kovets mikhteve eliyahu safir / Saphir, Elijah – Yafo, Israel. 1913 – 1r – us UF Libraries [939]
Kovets sipurav u-ketavav / Feuerberg, Mordecai Zeev – Warsaw, Poland. 192- – 1r – us UF Libraries [939]
Kovodelnik – Prague, Czechoslovakia. Mar 1936-May 1938 (incomplete) – 1r – 1 – us L of C Photodup [077]
Kovsh – Leningrad. v1-4. 1925-1926 – 1 – us NY Public [073]
Kowalski, Ellen M *see* Modeling and motor sequencing strategies of learning disabled boys
Kowan, Theodor Ira *see* The legal stenographer
Kownoer zeitung – Kauen (Kaunas, Kowno LT), 1916 16 aug-1917 30 sep – 2r – 1 – gw Misc Inst [077]
Kozaki, Hiromichi *see*
– Christianity and confucianism
– Essay on the person of christ
– Reasons for faith
Kozaky, Istvan *see* A halaltancok torteneti
Kozel'skii uezd : ispolkom sovetov. izvestiia kozel'skogo ispolnitel'nogo komiteta soveta rabochikh, krest'ianskikh i krasnoarmejskikh deputatov – Kozel'sk, Russia 1918 [mf ed Norman Ross] – 1r – 1 – mf#nrp-778 – us UMI ProQuest [077]
Kozeluch, Leopold *see*
– Trois quatuors concertans pour deux violons, alto et basse
– Trois quatuors, pour deux violons, alto et violoncelle
Kozhanyi, P M *see*
– Rabochaia kooperatsiia i profsoiuzy
– Rabochaia kooperatsiia za 10 let sovetskoi vlasti
Kozheurov, L P *see* Russkii vostok
Kozhevenno obuvnaia promyshlennost' sssr – (Leather and Shoe Industry of the USSR, 1922-40 and Leather and Shoe Industry; a Scientific-Technical and Production Journal, 1959-62) – 1 – us L of C Photodup [947]
Kozierowski, Stanislaw *see* Badania nazw topograficznych starej wielkopolski
Kozigazgatasi dontvenytar. – Budapest. On film: v1-19; 1908-28. LL-0265 – 1 – us L of C Photodup [340]
Kozlony, hivatalos lap – Budapest: [s.n., jun 18-aug 12 1848] – 1r – 1 – us CRL [943]
Kozlov, D E *see* Narodnoe delo [orenburg: 1918]
Kozlovskii, I *see* Silvestr medvedev
Kozlowski, Felix von *see* Gleim und die klassiker goethe, schiller, herder
Kozman, Fr *see* Textes legitatifs touchant le cenobitisme egyptien
Koz'mina, B P *see* Pis'ma egora sozonova k rodnym, 1895-1910 gg / vsesoiuznoe obshchestvo politkatorzhan i ssyl'no-poselentsev
Kozok, Barbara *see* Erklaerungsfunktionalitaet wissensbasierter systeme
Kozubskii, E *see* Ukazatele k 1-20 vypuskam "sbornika materialov dlia opisaniia mestnostei i imen kavkaza" 1881-1894 g
Kpuan dunman yuvajan = Conseils aux jeunes gens / Su Mararet – Bhnam Ben: Pannagar Saravant 2515 [1971] [mf ed 1990] – 1r with other items – 1 – (in khmer) – mf#mf-10289 seam reel 122/8 [§] – us CRL [305]
Kra rann : nan tvan vatthu to kri / Bhui San, Cac kuin U – Ran Kun: Siri mangala Ca pum nhip tuik 1959 [mf ed 1990] – 1r with other items – 1 – (in burmese) – mf#mf-10289 seam reel 201/7 [§] – us CRL [830]
Kra vat rann : [a novel] / Bhui San, Cac kuin U – Ran Kun: Sve Sok kumpani 1959 [mf ed 1990] – (= ser Sve sok thut ca up 57) – 1r with other items – 1 – (in burmese) – mf#mf-10289 seam reel 182/4 [§] – us CRL [830]
Kraatz, G *see* Entomologische monatsblatter
Kraatz, W *see* Koptische akten zum ephesinischen konzil von jahre 431
Krabbe, Otto *see* August neander
Krach im vorderhaus : ein berliner roman / Boettcher, Maximilian – 6. aufl. Berlin: Aufwaerts-Verlag, M Klieber, 1939 [mf ed 1989] – 448p – 1 – mf#7056 – us Wisconsin U Libr [830]
Der krach von wittenberg : blicke auf den religioesen wirrwarr der gegenwart / [Pesch, Tilmann] – Berlin: Verlag der Germania, 1889 [mf ed 1991] – (= ser Christ oder antichrist? 2) – 2mf – 9 – 0-7905-9438-2 – mf#1989-2663 – us ATLA [240]
Kracht, Arnim *see* Die beziehung zwischen elektromyographischen untersuchungen bei sportlicher betaetigung von zerebralparetikern
Krack, Otto *see* Grabbe
Kraehe, Ignaz *see* Alessandro nell'indie
Die kraehen : novellen / Stehr, Hermann – Berlin: S Fischer, 1921 – 1 – us Wisconsin U Libr [830]
Kraelitz-Greifenhorst, F von *see* Studien zum armenisch-tuerkischen
Kraemer, Christoph *see* Das stationaere krankengut des jahres 1988 der universitaetskliniken fuer kinder- und jugendpsychiatrie luebeck und kinderneuropsychiatrie rostock
Kraemer, Gerhard *see* Die roemisch-barocke stilkomponente im werk peter anton von verschaffelts
Kraemer, Henry *see* Scientific and applied pharmacognosy, intended for the use of students in pharmacy,...pharmacists,...food and drug analysts...pharmacologists
Kraemer, Michael *see* Die ueberlieferungsgeschichte der bergpredigt
Kraemer, Wilhelm *see*
– Johann christian guenthers saemtliche werke
– Max dauthendey
– Saemtliche werke
Kraenzlein, Kurt *see* Das kind der madeleine montcornet
Kraetzschmar, Christine *see* Schulpaedagogisches repetitorium
Kraetzschmar, Richard *see*
– Das buch ezechiel
– Die bundesvorstellung im alten testament in ihrer geschichtlichen entwickelung
Kraeuterbuch *see* Das buch der natur [cima33]
Kraeutlein, Jonathan *see* Die sprachlichen verschiedenheiten in den hexateuchquellen
Kraf(f), Michael *see*
– Camoenopaedia sacra concertus vocant 2. 3. 4. 5. 6. 8. vocum
– Liber secundus sacrorum concentuum
– Motectae quibus deo ter optumo maxumo divisque caelitibus
Krafft, C *see* Aufzeichnungen des schweizerischen reformators heinrich bullinger...
Krafft, H *see* Die soziale und praktische weisheit der sprueche salomos
Krafft, Johann Karl *see* Plans, coupes, elevations des plus belles maisons et des hotels construits a paris et dans les environs
Krafft, Wilhelm *see* Die deutsche bibel vor luther
Krafka, Carol et al *see* Stalking the increase in the rate of federal civil appeals
Kraft, Charles H *see* Study of hausa syntax
Kraft der arbeiterklasse *see* Das neue bewusstsein
Kraft, Ruth *see* Das schildbuergerbuch von 1598
Kraft und stoff : oder grundzuege der natuerlichen weltordnung / ed by Roser, Andreas & Buechner, Ludwig – Leipzig: 1880 (mf ed 1997) – (= ser Passauer Texte zur Philosophie) – 5mf – €59.00 – 3-8267-3212-X – mf#DHS 3211 – gw Frankfurter [170]
Der kraft-mayr : ein humoristischer musikanten-roman / Wolzogen, Ernst von – Berlin: J Singer [1933] [mf ed 1993] – 2v in 1 on 1r – 1 – (filmed with: heiteres und weiteres) – mf#7967 – us Wisconsin U Libr [830]
Kraftsman / Thilmany Pulp & Paper Co – 1972 jul, 1976 feb-1985 dec – 1r – 1 – (cont: thilmany kraftsman) – mf#1048053 – us WHS [670]
Der kraftwerker – Vockerode DE, 1967 6 jan-1991 apr [gaps] – 5r – 1 – (energiewerke) – gw Misc Inst [621]
Kragh, Karsten *see* Contributory negligence. a comparative study of the law of torts (u.s.a., england and denmark)
Krahe, Hans *see* Indogermanische sprachwissenschaft
Kraichgau stimme *see* Neue eppinger zeitung
Krainskii, V V *see*
– Agrarnyi vopros i kooperatsiia
– Obshchina i kooperatsiia
Krajan – Bautzen DE, 1868-1938 – 4r – 1 – gw Misc Inst [074]
Krakauer zeitung – Krakau (Krakow PL), 1939 12 nov-1940 10 dec, 1941 2 sep-31 oct, 1944-1945 16 jan – 9r – 1 – (filmed by bnl: 1941 mar, 1942 apr-1945 jan (gaps) [13r]; 1941 jan-sep, 1941 nov-1943 oct, 1944 jan-mar (gaps) [7r]) – gw Mikrofilm; uk British Libr Newspaper; gw Misc Inst [077]
Der krakeeler – Kassel DE, 1866 4 nov-1871 – 1r – 1 – (fr oct 1866-jun 1867 as flyer with variable titles) – gw Misc Inst [074]
Krakivs'ki visti – Krakow, Poland. 1940-43 – 4r – 1 – us L of C Photodup [943]
Krall, William *see* Account books
Kramer, Frederick Ferdinand *see*
– The sources of gnosticism
– The supremacy of the bible
Kramer, Friedrich Oswald *see* Die aethiopische uebersetzung des zacharias, erstes heft
Kramer, George W *see* The what, how, and why of church building
Kramer, Gustav *see*
– August hermann francke
– Beitraege zur geschichte august hermann francke's
– Neue beitraege zur geschichte august hermann francke's
Kramer, Harold Morton *see* The rugged way
The kramer journal – Kramer, Bottineau Co, ND: J H Pittman. v1 n1 mar 29 1917-v1 n29 oct 11 1917 (wkly) – 1 – (cont: russell sentinel. missing: jun 14, jul 5) – mf#11340 – us North Dakota [071]
Kramer, Matthias *see*
– Dictionarium frantzoesicsh-teutsch
– Het koninglyk neder-hoog-duitsch en hoog-neder-duitsch dictionnaire
Kramer, SN *see*
– From the tablets of sumer
– Sumerian mythology
Kramer und friemann : eine lehrzeit / Mueller, Fritz – 3. Aufl. der Feldausg. Guetersloh: C Bertelsmann, 1942 – 1r – 1 – us Wisconsin U Libr [920]
Kramer, W H *see* Elenchus vegetabilium et animalium...
Kramme, Ulrike [comp] *see*
– Arab-islamic biographical archive
– Czech and slovakian biographical archive
– Suedosteuropaeisches biographisches archiv [soba]
– Ungarisches biographisches archiv
Krammer, G *see* Architecture von den funf seulen sambt iren ornamenten und zierden...
Krammer, Mario *see* The odor fontanes engere welt
Kramp, Willy *see* Geist und gesellschaft

Kramrisch, Stella *see*
- Dravida and kerala in the art of travancore
- Indian sculpture
- A survey of painting in the deccan
- The vishnudharmottara+

Kramrisch, Stella et al *see* The arts and crafts of travancore

Kramum khvar ka kamloh slajham : pralom lok knun manosancetena / Hak Chai Huk – Bhnam Ben: Ron Bumb Viriyah 2707 [1965] [mf ed 1990] – 1r with other items – 1 – (in khmer) – mf#mf-10289 seam reel 097/05 [§] – us CRL [480]

Krancke, Adolf *see* Die beiden fassungen von schillers abhandlung "ueber naive und sentimentalische dichtung"

Krane, Vl *see* The relationship between anxiety and athletic performance

Krane, Vikki *see* Relationships among perceived leadership styles, member satisfaction and team cohesion in high school basketball teams

Kraner, Christian Friedrich *see* Ueber schillers unterscheidung von naiver und sentimentalischer dichtung

Kranichfeld, Rudolph *see* Das buch daniel

Krankendienst : zeitschrift fuer katholische krankenpflegegenossenschaften und krankenhaeuser im deutschen sprachgebiet / ed by Katholischer Krankenhausverband Deutschlands e.V. – Freiburg/Br 1920-22, 1941 [mf ed 2006] – (= ser Freie wohlfahrtspflege 12) – 82mf – 9 – €480.00 – 3-89131-483-3 – (subtitles vary; with suppl: Die Anstalt. Wegweiser fuer den Anstaltsbetrieb [1] 1927-10 [1936]) – gw Fischer [610]

Kranken-physiognomik / Baumgaertner, Karl Heinrich – Stuttgart/Leipzig 1839 [mf ed 1991] – 3 b/w + 4 color mf – 9,15 – €130.00 – 3-89131-048-X – (atlas & text in latin/german) – gw Fischer [616]

Die krankheit des apostels paulus / Fischer, Hermann – Berlin: Edwin Runge, 1911 – 1mf – 9 – 0-7905-0492-8 – mf#1987-0492 – us ATLA [225]

Krankheitserfahrungen von krebskranken kindern und ihren familien / Plessen, Christian von – (mf ed 1997) – 2mf – 9 – €40.00 – 3-8267-2394-5 – mf#DHS 2394 – gw Frankfurter [616]

Krann Mra Lvan *see* Mvan bhava dha le' nidan

Krann On *see* La man ta ra kabya mya

Krann U *see* Mran ma rui ra ca ka tha mya

Kranz, Jacob ben Wolf *see* Ohel ya'akov: al hamishah humshe torah.

Kranz, Michael *see* Die verpackungsverordnung

Kranz, P *see* Chronological handbook of the history of china

Krapf, J L *see*
- Journals...detailing their proceedings in the kingdom of shoa, and journeys in other parts of abyssinia, in the years 1839, 1840, 1841, and 1842...
- Reisen in ost-afrika ausgefuehrt in den jahren 1837-1855

Krapf, Johann Ludwig *see* Dictionary of the suahili language

Krapf, Ludwig *see* A nika-english dictionary

Krapotkin, N P *see* Polnyi sbornik protsentnykh bumag

Krappe, Alexander Haggerty *see*
- Etudes de mythologie et de folklore germanique
- La genese des mythes

Krasem *see* Brah raj bidhi dvadasamas

Kraseninnikov, Stepan *see* Histoire de kamtschatka, des isles kurilski, et des contrees voisines

Krasik, A V *see* Krest'ianskii bank i ego deiatel'nost' s 1883 po 1905 gg

Krasinski, Valerian, Count *see* Historical sketch of the rise, progress, and decline of the reformation in poland

Krasnaia armiia : ezhednevnaia voennaia gazeta. izdanie voennogo otdela izdatel'stva vserossijskogo tsentral'nogo ispolnitel'nogo komiteta sovetov rab, krest i kaz deputatov – Moscow, Russia 1918 [mf ed Norman Ross] – 1r – 1 – mf#nrp-1078 – us UMI ProQuest [077]

Krasnaia chuvashiia – Cheboksary, Russia 1973 [mf ed Norman Ross] – 23r – 1 – mf#nrp-399 – us UMI ProQuest [077]

Krasnaia derevnya : organ mikhajlovskogo uezd ispolkoma – Mikhajlov, Russia 1918 [mf ed Norman Ross] – 1r – 1 – mf#nrp-1016 – us UMI ProQuest [077]

Krasnaia kareliia – Petrozavodsk, Russia 1973 [mf ed Norman Ross] – 5r – 1 – mf#nrp-1395 – us UMI ProQuest [077]

Krasnaia moskva, 1917-1920 gg – M, 1920. 744p – 13mf – 9 – mf#RHS-30 – ne IDC [314]

Krasnaia nov' – 1921-31 – 1 – us L of C Photodup [073]

Krasnaia tatariia – Kazan, Russia 1973 [mf ed Norman Ross] – 4r – 1 – mf#nrp-676 – us UMI ProQuest [077]

Krasnaia vaga – Shenkursk, Russia 1919-25 [mf ed Norman Ross] – 9 – 1 – mf#nrp-1549 – us UMI ProQuest [077]

Krasnaia zvezda – Moscow, USSR. Oct 1918-Apr 1919; 1925; 1929 no 160; 1930; 1932-77 – 90r – 1 – (some missing issues) – us L of C Photodup [947]

Krasnaia zvezda : tsentralyi organ narodnogo kommissariata oborony soiuza ssr – Moscow, [1939-45] – 12r – 1 – us CRL [947]

Krasnaia zvezda. literaturno-instruktorskij parokhod "krasnaia zvezda" *see* Izvestiia peredvizhnogo biuro rosta

Krasnaya nov' – Moscow. 1921-38, oct-nov 1939, janv 1940 – 1 – fr ACRPP [073]

Krasnaya nov' – Moscow. nov 1921-nov 1939; mar 1940-may 1941; jan-aug 1942 – 1 – us NY Public [073]

Krasnoarmeets : ezhednevnaia gazeta politotdela voennogo soveta 16-j armii – Mogilev, Belarus 1919-20 [mf ed Norman Ross] – 6r – 1 – mf#nrp-1048 – us UMI ProQuest [077]

Krasnoarmeiskaia gazeta – Moscow, Russia 1942-46 [mf ed Norman Ross] – 1 – mf#nrp-111 – us UMI ProQuest [934]

Krasnoarmeiskoe slovo – (city unknown) Russia 1942-45 [mf ed Norman Ross] – 1 – mf#nrp-112 – us UMI ProQuest [934]

Krasnoe slovo : prof soiuz gazetchikov – Samara, Russia 1918 [mf ed Norman Ross] – 1r – 1 – mf#nrp-1508 – us UMI ProQuest [077]

Krasnoe znamia : organ bogorodskogo uezdnogo komiteta rkp – Bogorodsk, Russia, 1919-20 [mf ed Norman Ross] – 1r – 1 – mf#nrp-349 – us UMI ProQuest [077]

Krasnoff, Joanne B *see* The influence of physical conditioning on the post-menopausal hot flash

Krasnoiarskij komitet obshchestvennoj bezopasnosti. Soedinennoe ispolnitel'noe biuro *see* Biulleten' soedinennogo ispolnitel'nogo biuro komiteta obschestvennoi bezopasnosti i soveta rabochikh, soldatskikh i krest'ianskikh deputatov

Krasnokholmskii uezdnyi isp komitet sovetov : izvestiia krasnokholmskogo uezdnogo ispolnitel'nogo komiteta soveta rabochikh i krest'ianskikh deputatov – Krasny Kholm, Russia 1918 [mf ed Norman Ross] – 1r – 1 – mf#nrp-801 – us UMI ProQuest [077]

Krasnov, Iv *see* Iakutskii oblastnoi vestnik

Krasnovskii, N *see* Khram pravoslavno-khristianskii

Krasnozhen, M *see* Ukazatel literatury tserkovnago prava po 1910 g

Krasnushkin, V A [Viktor Sevskii] *see* Vol'nyi don

Krasnyi arkhiv – Ann Arbor MI 1922-40 – 1 – mf#2665 – us UMI ProQuest [355]

Krasnyi baltiiskii flot – Leningrad, USSR. Aug 20 1919-Sept 20 1921 – 1r – 1 – us L of C Photodup [077]

Krasnyi dneprovets – (city unknown) Russia 1944-45 [mf ed Norman Ross] – 1 – mf#nrp-114 – us UMI ProQuest [934]

Krasnyi flot – Moscow, Russia 1938-52 [mf ed Norman Ross] – 84r – 1 – mf#nrp-1079 – us UMI ProQuest [077]

Krasnyi flot : organ narodnogo kommissariata voenno-morskogo flota sssr – Moskva: Tip "Pravda" imeni Stalina, [jan-aug 29 1941] – us CRL [947]

Krasnyi kavalerist na fronte – (city unknown) Russia 1941-43 [mf ed Norman Ross] – 1 – mf#nrp-115 – us UMI ProQuest [934]

Krasnyi listok : organ ispolnitel'nogo komiteta sezda sovdepov petergofsk u – Petergof, Russia, 1918 [mf ed Norman Ross] – 1r – 1 – mf#nrp-1384 – us UMI ProQuest [077]

Krasnyi nabat : organ vel'skogo komiteta rkp – Vel'sk, Russia 1918 [mf ed Norman Ross] – 1r – 1 – mf#nrp-1920 – us UMI ProQuest [077]

Krasnyi sever : organ ispolnitel'nogo komiteta soveta zh-d deputatov murmanskoj dorogi – Petrozavodsk, Russia 1918 [mf ed Norman Ross] – 1r – 1 – mf#nrp-1396 – us UMI ProQuest [077]

Krasnyi sever – Salekhard, Russia 1974-87 [mf ed Norman Ross] – 5r – 1 – mf#nrp-1506 – us UMI ProQuest [077]

Krasnyi sever [vologda, russia] : obshchestvenno-politicheskaia gazeta – Vologda: Krasnyi Sever [mf ed Minneapolis MN: East View Publ [199-]] – 10r – 1 – (crl has: mf-11784 seemp [10r] 1991-95) – us CRL [077]

Krasnyi ves'egonsk : organ uezd-gor kom ves'egonskoj organizatsii rkp – Ves'egonsk, Russia 1918 [mf ed Norman Ross] – 1r – 1 – mf#nrp-1922 – us UMI ProQuest [077]

Krasnyi voin – Moscow, Russia 1925 [mf ed Norman Ross] – 4r – 1 – mf#nrp-1080 – us UMI ProQuest [077]

Krass, M *see* Zwoelf bilder aus annette von drostes leben und dichtung

Krassem, Maha Bidur *see* Silacarik nagar vatt

Kratander *see* Anti-strauss

Kratikii bolgarsko-russkii slovart – Moskva, Russia. 1959 – 1r – us UF Libraries [025]

Kratkaia istoriia gruszinskoi tserkvi *see* Short history of the georgian church

Kratkie khoziaistvenno-statisticheskie svedeniia po smolenskoi gubernii – Smolensk, 1912 – 6mf – 8 – mf#RZ-87 – ne IDC [314]

Kratkii alfavitnyi katalog / Persidskie i tadzhikskie rukopisi Instituta narodov Azii AN SSSR; ed by Miklukho-Maklai, N D – M, 1964. 2v – 14mf – 9 – mf#R-10984 – ne IDC [956]

Kratkii istoricheskii ocherk deiatel'nosti s-peterburgskogo gorodskogo kreditnogo obshchestva za 40 let, s 1 marta 1862 po 1 marta 1902 goda – Spb, 1902 – 2mf – 9 – mf#REF-375 – ne IDC [332]

Kratkii istoricheskii ocherk i obzor deiatel'nosti nizhegorodskogo gorodskogo lombarda za vremia s 23 fevr 1889 g po 1-e ianv 1914 g dvadtsat' piat' let sushchestvovaniia – N Novgorod, 1914 – 1mf – 9 – mf#REF-465 – ne IDC [332]

Kratkii istoricheskii ocherk vozniknoveniia i razvitiia deiatel'nosti kostromskogo gorodskogo obshchestva vzaimnogo ot ognia strakhovaniia : sostavlen ko dniu xxv-ti-letiia obshchestva, (1884-1909 g) / Krylov, I – Kostroma, 1909 – 1mf – 9 – mf#REF-436 – ne IDC [332]

Kratkii kurs kooperatsii / Chaianov, A V – 1925 – 79p 1mf – 9 – mf#COR-231 – ne IDC [335]

Kratkii kurs kooperatsii : konspekt / Lobov, G Z – Rostov n/D, 1925 – 360p 1mf – 9 – mf#COR-262 – ne IDC [335]

Kratkii kurs po potrebitelskoi kooperatsii / Leonov, V – 1918 – 108p 2mf – 9 – mf#COR-259 – ne IDC [335]

Kratkii kurs tserkovnago prava pravoslavnoi tserkvi / Bernikov, Ilya Stepanovich – A short course of church law of the Orthodox Church. Kazan, 1913 – 1 – 59.28 – us Southern Baptist [242]

Kratkii obzor deiatel'nosti iaroslavskogo gorodskogo obshchestvennogo banka za 50-letie s 1865 goda po 1915 god – Iaroslavl', 1915 – 2mf – 9 – mf#REF-348 – ne IDC [332]

Kratkii obzor istorii i teorii bankov s prilozheniem ucheniia o birzhevykh operatsiiakh / Bishof, A; ed by Levitskii, V – Iaroslavl', 1887 – 2mf – 9 – mf#REF-173 – ne IDC [332]

Kratkii obzor sobraniia rukopisei, prinadlezhavshago preosviashchennomu episkopu porfiriiu, i nyne khraniashchagosia v imp publichnoi biblioteke – 1885 – 185p 4mf – 8 – mf#R-7127 – ne IDC [243]

Kratkii ocherk 50-letiia aktsiznoi sistemy vzimaniia naloga s krepkikh napitkov i 50-letiia deiatel'nosti uchrezhdenii, zavedyiushchikh neokladnymi sborami – Spb, 1913 – 8mf – 9 – mf#REF-206 – ne IDC [332]

Kratkii ocherk deiatel'nosti moskovskogo kupecheskogo obshchestva vzaimnogo kredita za vremia s 11-go noiabria 1869 po 1-oe ianvaria 1894 goda – M, 1894 – 1mf – 9 – mf#REF-369 – ne IDC [332]

Kratkii ocherk deiatelnosti tambovskogo serafimovskogo soiuza russkikh liudei, s 1 okt 1911 goda po 1 okt 1912 g : sedmoi god sushchestvovaniia – Kharkov, 1912 – 13p 1mf – 9 – mf#RPP-165 – ne IDC [325]

Kratkii ocherk dvadtsatipiatiletnego (1874-1898 gg) sushchestvovaniia blagoveshchenskogo gorodskogo obshchestvennogo banka – Blagoveshchensk, 1899 – 1mf – 9 – mf#REF-342 – ne IDC [332]

Kratkii ocherk piatidesiatiletnei deiatel'nosti sushchestvovaniia banka, 1864 – 9 maia 1914 / Nizhegorodskii Nikolaevskii Gorodskoi Obshchestvennyi Bank – N Novgorod, 1914 – 3mf – 9 – mf#REF-347 – ne IDC [332]

Kratkii otchet o deiatel'nosti n k f rsfsr : (za 1923-24 g i pervuiu polovinu 1924-25 g) / Narodnyi Komissariat Finansov – [M, 1925] – 2mf – 9 – mf#REF-39 – ne IDC [332]

Kratkii otchet o rabote za 1923-24 g i za oktiabr'-ianvar' 1924-25 goda i plan kreditovaniia za 1924-25 god / Ural'skii oblastnoi Sel'sko-Khoziaistvennyi Bank "Uralsel'khozbank" – Sverdlovsk, 1925 – 1mf – 9 – mf#REF-113 – ne IDC [332]

Kratkii otchet orabotakh chetvertogo sezda partii sotsialistov-revoliutsionerov : 26 noiabria – 5 dekabria 1917 goda – 1918 – 160p 2mf – 9 – mf#RPP-212 – ne IDC [325]

Kratkii putevoditel po fondam lichnogo proiskhozhdeniia rukopisnogo otdela muzeia istorii religii i ateizma / Gendrikov, V B – [L, 1973] – 1mf – 9 – (ateizm, religiia, sovremennost, 212-223) – mf#R-11116 – ne IDC [243]

Kratkii statisticheskii sbornik uzbekskoi ssr – Tashkent, 1936 – 2mf – 9 – mf#RHS-34 – ne IDC [314]

Kratkii statisticheskii spravochnik – M, 1936. 256p – 3mf – 9 – mf#RHS-32 – ne IDC [314]

Kratkii statisticheskii spravochnik 1935 g – M, 1936. 159p – 2mf – 9 – mf#RHS-31 – ne IDC [314]

Kratkii statisticheskii spravochnik uzbekskoi ssr : osnovnye pokazateli razvitiia khoziaistva i kul'tury uzbekskoi ssr za 1933, 1934, 1935 gg i predvaritel'nye dannye za 1936 g – [n.p.], 1937 – 2mf – 9 – mf#RHS-33 – ne IDC [314]

Kratkiia zhizneopisaniia russkikh sviatykh / Ignatii, Arkhim – 1875. 2v – 5mf – 9 – (missing: 1875 v1) – mf#R-18246 – ne IDC [243]

Kratkoe obozrenie napravleniia periodicheskikh izdanii i gazet, i otzyvov ikh po vazhneishim pravitelstvennym i drugim voprosam za 1861 g / [Kapnist, P] – Spb., 1862 – 1mf – 9 – mf#R-9232 – ne IDC [077]

Kratkoe opisanie po uezdu viatskoi gubernii – Viatka, 1892. 6v – 2mf – 8 – mf#RZ-230 – ne IDC [314]

Kratkoe opisanie rukopisei tserkovnoistoricheskogo drevnekhranilishcha pri bratstve sv bl vel kn aleksandra nevskogo – Vladimir, 1906 – 157p 3mf – 8 – mf#R-7292 – ne IDC [243]

Kratkoe rukovodstvo / Gosudarstvennyi kontrol' i raskhodovanie narodnykh deneg; ed by Izvestiia gosudarstvennogo kontrolia – M, 1918 – 1mf – 9 – mf#REF-45 – ne IDC [332]

Kratkoe uchenie o propovedi = A short study on preaching / Prokhanov, IS – Anapa, Kuban Province, Russia. 1911. 122p – 1 – $5.00 – us Southern Baptist [242]

Kraton ngajogjakarta-hadiningrat / Ngajogjakarta – Jogjakarta, [1952]-1959 v1-7(9) – 8mf – 9 – (missing: [1952]-1959, v(1-5); 1956/1957, v5(7-9); 1957, v6(1-3, 8-9, 11-12)) – mf#SE-1844 – ne IDC [950]

Kratzenberg, Andree *see* Wechselkursrisiko und exportentscheidung

Kratzmann, Ernst *see* Regina sebaldi

Kratzmoeller, Wilhelm *see* Darstellung und kritik der lehre des descartes von der bildung des universums

Kratzsch, Johann *see* Neuestes und gruendlichstes alphabetisches lexicon der saemmtlichen ortschaften der deutschen bundesstaaten

Kratzsch, Konrad [comp] *see* Register der goethe-jahrbuecher 1880-1968

Kraus, Arthur James Israel *see* Sick society

Kraus, Benedikt *see* Die schoepfung

Kraus, Franz Xaver *see*
- Lettere de benedetto 14 al cononico pier francesco peggi bolognese
- Ueber das studium der theologie sonst und jetzt

Kraus, Franz Xaver et al *see* The reformation

Kraus, Hans P *see* Collection(spanish-american documents)

Kraus, J U *see* Heilige augen- und gemueths-lust

Kraus, Joseph Martin *see* Etwas von und ueber musik fuers jahr 1777

Kraus, Konrad *see* Winckelmann und homer

Krause, Bernd Joachim *see* Untersuchungen zur zerebralen repraesentation deklarativer gedaechtnisvorgaenge mittels der positronen-emissions-tomographie und der kernspintomographie

Krause, Carl *see* Der briefwechsel des mutianus rufus

Krause, Christian Gottfried *see* Von der musikalischen poesie...

Krause, J G *see* Neuer buecher-saal der gelehrten welt

Krause, Karl C F *see* Zur geschichte der neueren philosophischen systeme

Krause, Karl Christian Friedrich *see* Zur religionsphilosophie und speculativen theologie

Krausen, Edgar *see* Urkunden des klosters raitenhaslach 1034-1350

Kraushaar, Chr Otto *see* Verfassungsformen der lutherischen kirche amerikas

Kraushar, Alexander *see* Frank ve-adato 1726-1816

Krauskopf, Joseph *see*
- Obituary-address in honor of the late dr samuel hirsch
- Service manual

Krauspenhaar, I *see* Masoko akasarudzwa e wanoyera

Krauss, Alfred Ed *see* Die lehre von der offenbarung

Krauss, Eberhard *see* Funktionelle charakterisierung eines regulativen elementes zwischen transkriptions- und translationsstart im e-kristallipromotor

Krauss, Eugen Adolf Wilhelm *see* Lebensbilder aus der geschichte der christlichen kirche

Krauss, Ferdinand *see* Von der ostsee bis zum nordcap

Krauss, Friedrich S *see* Boehmische korallen aus der goetterwelt

Krauss, Friedrich Salomo *see* Volksglaube und religioeser brauch der suedslaven

Krauss, J U *see* Tapisseries du roy

Krauss, Rudolf *see* Schicksalstage deutscher dichter

Krauss, Samuel *see*
- Griechische und lateinische lehnwoerter im talmud, midrasch und targum
- Das leben jesu nach juedischen quellen
- The mishnah treatise sanhedrin

Krauss, Werner *see* Die franzoesische aufklaerung im spiegel der deutschen literatur des 18. jahrhunderts

Kraussold, Lorenz *see* Die sage vom heiligen gral und parceval

Krausz, Joseph *see* Die goetternamen in den babylonischen siegelcylinder-legenden

Kraut, Wilhelm Theodor *see* Grundriss zu vorlesungen ueber das deutsche privatrecht

Krauth, Charles Porterfield *see*
- Augsburg confession
- Christian liberty in its relation to the usages of the evangelical lutheran church
- Chronicle of the augsburg confession
- The conservative reformation and its theology
- Infant baptism and infant salvation in the calvinistic system

Der krautsteig : die geschichte einer kameradschaft / Jessen, Paul – Darmstadt: L Kichler c1942 – 1 – mf#2743p – us Wisconsin U Libr [830]

Kray tam khvan nhan aukka pyam mya / Su Ta, U – Ran Kun: Ca pe bi man a phvai 1985 [mf ed 1990] – [ill] 1r with other items – 1 – (in khmer; with bibl & ind) – mf#mf-10289 seam reel 205/6 [§] – us CRL [520]

Kre mum = The mirror – Ran' Kun' Mrui: Kre Mum (daily) [mf ed Ithaca NY: [John M Echols Collection] Cornell University 1998] – 5r – 1 – (in burmese) – mf#mf-11857 seam – us CRL [079]

Kreative prozesse im unterricht / Falk, Brigitte – (mf ed 1995) – 1mf – 9 – €30.00 – 3-8267-2196-9 – mf#DHS 2196 – gw Frankfurter [370]

Kreativitaet und historismus : schmuck und entwurf 1848-1870 / Becker, Monika – (mf ed 1997) – 7mf – 9 – €65.00 – 3-8267-2476-3 – mf#DHS 2476 – gw Frankfurter [740]

Kreativitaetsforschung und joy paul guilford (1897-1987) : materialien zu einer erstbegegnung / Pimmer, Hans – (mf ed 1994) – 6mf – 9 – €62.50 – 3-8267-2022-9 – mf#DHS 2022 – gw Frankfurter [150]

Die kreatur – Berlin DE, 1926-1929/30 – 1r – 1 – gw Misc Inst [073]

Die kreatur : eine zeitschrift – Berlin: Martin Buber et al. v1-3. 1926-1929/30 [complete] – (= ser German-jewish periodicals...1768-1945, pt 2) – 1r – 1 – $125.00 – mf#B594 – us UPA [939]

Krebs durch niederfrequente magnetfelder? : eine studie ueber die rolle des melatonins bei blinden und feldexponierten personen / Haubrich, Stefan – (mf ed 1995) – 1mf – 9 – €30.00 – 3-8267-2235-3 – mf#DHS 2235 – gw Frankfurter [616]

Krebs, E *see*
- Meister dietrich, sein leben, seine werke, seine wissenschaft
- Theologie und wissenschaft nach der lehre der hochscholastik

Krebs, Engelbert *see*
- Der logos als heiland im ersten jahrhundert
- Meister dietrich (theodoricus teutonicus de vriberg)
- Das religionsgeschichtliche problem des urchristentums

Krebs, Franz *see*
- Die praepositionen bei polybius
- Die praepositionsadverbien in der spaeteren historischen graecitaet

Krebs, Kurt *see* Saechsische kriegsnot in den jahren 1806 bis 1815

Krebs, Werner *see* Auf der walz vor fuenfzig jahren

Krech, Johannes *see* Die rechte an grundstuecken nach dem entwurfe eines buergerlichen gesetzbuches fuer das deutsche reich

Krechet – Moscow, Russia. n1, 1[2]-4[5] 1996?, n1-n3[1997], n5-6 [1997] – 1 – mf#mf-12248 [reel 6] – us CRL [077]

Krechetov, S A *see* Pered razsvietom

Kredit : (svod trudov mestnykh komitetov po 49 guberniiam evropeiskoi rossii) / Borodaevskii, S V – Spb, 1904 – 8mf – 9 – mf#REF-333 – ne IDC [332]

Kredit, banki i denezhnoe obrashchenie / Kaufman, I I – Spb, 1873 – 11mf – 9 – mf#REF-167 – ne IDC [332]

Kredit i kreditnye operatsii v selskokhoziaistvennykh kooperativakh / Trapeznikov, I F – 1927 – 200p 3mf – 9 – mf#COR-406 – ne IDC [335]

Kredit i potrebitelskaia kooperatsiia / Zelgeim, V N – [1910] – 58p 1mf – 9 – mf#COR-31 – ne IDC [335]

Kredit krest'ianinu : o rabote nizhne-volzhskogo s-kh banka v 1924-25 godu (po saratovskoi gub) / Nizhne-Volzhskoe Oblastnoe Obshchestvo Sel'sko-Khoziaistvennogo Kredita £Nizhvel'bank" – Saratov, 1926 – 1mf – 9 – mf#REF-109 – ne IDC [332]

Kredit und zins / Schmidt, Georg – Leipzig: Duncker & Humblot, 1910 (mf ed 199-) – 52p – (incl bibl) – mf#ZT-230 – us Harvard [332]

Kreditnaia kooperatsiia v rossii : istoricheskii ocherk i sovremennoe polozhenie / Kheisin, M L – 1919 – 199p 3mf – 9 – mf#COR-408 – ne IDC [335]

Kreditnaia kooperatsiia v rossii / Prokopovich, S N – 1923 – 154p 2mf – 9 – mf#COR-400 – ne IDC [335]

Kreditnaia kooperatsiia zauralia : permskogo / Panin, I – Ekaterinburg, 1917 – 159p 2mf – 9 – mf#COR-93 – ne IDC [335]

Kreditnye i komissionno-bankovskie operatsii v selskokhoziaistvennykh kreditnykh tovarishchestvakh / Islankin, F B – 1928 – 224p 3mf – 9 – mf#COR-383 – ne IDC [335]

Kreditnye i ssudo-sberegatelnye tovarishchestva / Kulyzhnyi, A E – 1918 – 31p 1mf – 9 – mf#COR-388 – ne IDC [335]

Kreditnye operatsii selskokhoziaistvennykh tovarishchestv / Merkulov, A V – 1923 – 23p 1mf – 9 – mf#COR-491 – ne IDC [335]

Kreditsicherung : durch schuldbeitritt, buergschaft, patronatserklaerung, garantie, sicherungsuebereignung, sicherungsabtretung, eigentumsvorbehalt, pool-vereinbarungen, pfandrecht an beweglichen sachen und rechten, hypothek und grundschuld / Reinicke, Dietrich & Tiedtke, Klaus – Neuwied, Kriftel, Berlin: Luchterhand, 1994 (mf ed 1996) – 5mf – 9 – €52.00 – 3-8267-9679-9 – mf#DHS 9679 – gw Frankfurter [346]

Krefelder anzeiger – Krefeld DE, 1858 jan-jun, 1859 jan-jun – 1 – gw Misc Inst [074]

Krefelder zeitung – Krefeld DE, 1977- – (= ser Bezirksausgabe von westdeutsche zeitung (wz), duesseldorf) – ca 8r/yr – 1 – gw Misc Inst [074]

Krefting, Achim *see* St michael und st george in ihren geistesgeschichtlichen bezeihung

Krehbiel, Edward B *see* The interdict

Krehbiel, Edward Benjamin *see* Nationalism, war and society

Krehbiel, Henry Edward *see* Afro-american folksongs

Krehbiel, Henry Peter *see* The history of the general conference of the mennonites of north america

Krehbiel, Henry Peter [comp] *see* Mennonite churches of north america

Krehl, Ludolf *see*
- Beitraege zur muhammedanischen dogmatik
- Das leben und die lehre des muhammed
- Ueber die religion der vorislamischen araber

Krehl, Stephan *see* Harmonielehre

Krehm, William *see* Democracias y tiranias en el caribe

Kreibig, Gustav *see* Die versoehnungslehre

Kreider, Eugene G *see* Kreider's index of notes in the annotated cases

Kreider's index of notes in the annotated cases / Kreider, Eugene G – New York, San Francisco: Thompson Co, Bancroft-Whitney. 1v. 1922 (all publ) – 10mf – 9 – $15.00 – (covers all vols of all 3 titles in the annotated cases series) – mf#LLMC 84-695F – us LLMC [348]

Kreidler, Charles Ray *see* Analyzed new york decisions and citations, 1914-1917

Kreihing, J *see* Emblemata ethico-politica carmine explicata

Kreil, Joseph *see* Mnemosyne

Kreiler, Kurt *see* Die schriftstellerrepublik

Der kreis angerburg : ein ostpreussisches heimatbuch / Pfeiffer, Erich [comp] – Angerburg: Selbstverlag der Kreisgemeinschaft 1973 [mf ed 1992] – (= ser Ostdeutsche beitraege aus dem goettinger arbeitskreis 54) – 10r – 1 – (incl bibl ref & ind) – mf#3180p – us Wisconsin U Libr [943]

Der kreis cammin : ein pommersches heimatbuch / Flemming-Benz, Hasso graf von [comp] – Wuerzburg: Holzner Verlag 1970 [mf ed 1992] – (= ser Ostdeutsche beitraege aus dem goettinger arbeitskreis 47) – 10r – 1 – (incl bibl ref) – mf#3180p – us Wisconsin U Libr [914]

Der kreis gerdauen : ein ostpreussisches heimatbuch / Bachor, Oskar-Wilhelm [comp] – Wuerzburg: Holzner Verlag 1968 [mf ed 1992] – (= ser Ostdeutsche beitraege aus dem goettinger arbeitskreis 43) – 10r – 1 – (incl bibl ref & ind) – mf#3180p – us Wisconsin U Libr [914]

Der kreis goldap : ein ostpreussisches heimatbuch / Mignat, Johannes [comp] – Wuerzburg: Holzner Verlag, 1965 [mf ed 1993] – (= ser Ostdeutsche beitraege aus dem goettinger arbeitskreis 36) – 546p (ill) – 1 – (incl bibl ref and ind) – mf#8098 reel 6 – us Wisconsin U Libr [880]

Kreis hammsches wochenblatt – Hamm (Westf) DE, 1824-29, 1831-43, 1845-47, 1850-62, 1864-72, 1874, 1876-78, 1880-98, 1900-21, 1922 1 jul-1935 31 mar, 1935 2 jul-1945 2 apr, 1949 28 oct-1962 25 sep, 1963 8 feb-1964 15 mar, 1964 16 jun-1969 31 mar, 1969 11 sep-1971, 1976 2 jan-29 feb – 1 – (title varies: 2 jan 1825: wochenblatt fuer die stadt und den kreis hamm; 11 dec 1850: westfaelischer anzeiger; 1 nov 1949: westfaelischer anzeiger und kurier; 12 apr 1972: westfaelischer anzeiger. filmed by other misc inst: 1972- [ca 7r/yr]) – gw Misc Inst [074]

Der kreis johannisburg : ein ostpreussisches heimatbuch / ed by Guttzeit, Emil Johannes – Wuerzburg: Holzner 1964 [mf ed 1992] – (= ser Ostdeutsche beitraege aus dem goettinger arbeitskreis 31) – 10r – 1 – (incl bibl ref & ind) – mf#3180p – us Wisconsin U Libr [914]

Der kreis loetzen : ein ostpreussisches heimatbuch / Meyhoefer, Max [comp] – Wuerzburg: Holzner Verlag 1961 [mf ed 1992] – (= ser Ostdeutsche beitraege aus dem goettinger arbeitskreis 20) – 10r – 1 – (incl bibl ref & ind) – mf#3180p – us Wisconsin U Libr [430]

Der kreis (mme4) : zeitschrift fuer kuenstlerische kultur. offizielles organ der hamburger buehne / ed by Benninghoff, Ludwig & Postulart, Wilhelm – Hamburg 1924-33 [mf ed 1998] – (= ser Marbacher mikrofiche-editionen (mme) 4; Kultur – literatur – politik: deutsche zeitschriften des 19./20. jahrhunderts (klp)) – 10v on 48mf – 9 – €400.00 – 3-89131-313-6 – gw Fischer [700]

Der kreis mohrungen : ein ostpreussisches heimatbuch / Wrangel, Wolf, Freiherr von [comp] – Wuerzburg: Holzner Verlag 1967 [mf ed 1992] – (= ser Ostdeutsche beitraege aus dem goettinger arbeitskreis 40) – 10r – 1 – (incl bibl ref & ind) – mf#3180p – us Wisconsin U Libr [914]

Der kreis neustettin : ein pommersches heimatbuch / Stelter, Franz [comp] – Wuerzburg: Holzner Verlag 1972 [mf ed 1992] – (= ser Ostdeutsche beitraege aus dem goettinger arbeitskreis 52) – 10r – 1 – (incl bibl ref & ind) – mf#3180p – us Wisconsin U Libr [914]

Der kreis osterode (ostpr.) : daten zur geschichte seiner ortschaften / Hartmann, Ernst of Osterode – Wuerzburg: Holzner-Verlag 1958 [mf ed 1992] – (= ser Ostdeutsche beitraege aus dem goettinger arbeitskreis 10) – 10r – 1 – (incl ind) – mf#3180p – us Wisconsin U Libr [943]

Der kreis schlossberg : ein ostpreussisches heimatbuch / Mietzner, Franz [comp] – Wuerzburg: Holzner Verlag 1962 [mf ed 1992] – (= ser Ostdeutsche beitraege aus dem goettinger arbeitskreis 24) – 10r – 1 – (incl bibl ref & ind) – mf#3180p – us Wisconsin U Libr [914]

Der kreis sensburg : aus dem nachlass von paul glass / Glass, Paul; ed by Bredenberg, Fritz von – Wuerzburg: Holzner Verlag 1960 [mf ed 1992] – (= ser Ostdeutsche beitraege aus dem goettinger arbeitskreis 15) – 10r – 1 – (incl bibl ref & ind) – mf#3180p – us Wisconsin U Libr [943]

Kreis- und smts- verkuendigungs-blatt fuer triberg und villingen – Villingen-Schwenningen DE, 1869/70 – 1r – 1 – gw Misc Inst [074]

Kreis- und unterhaltungsblatt fuer ahrweiler und dessen umgegend – Bad Neuenahr-Ahrweiler DE, 1834 4 oct-1835 26 sep – 1 – gw Misc Inst [074]

Der kreis wetzlar / Abicht, Friedrich K – Wetzlar 1836/37 [mf ed Hildesheim 1995-98] – 3v on 7mf – 9 – €140.00 – 3-487-29516-4 – gw Olms [914]

Kreis-amtsblatt – Schluechtern DE, 1923 3 apr-30 aug – 1r – 1 – gw Misc Inst [074]

Kreis-anzeiger – Fritzlar DE, 1876-1887 oct [gap 1885], 1888-95, 1899-1914, 1916 16 nov-1944 30 jun – 43r – 1 – (title varies: 2 jul 1885: fritzlarer kreis-anzeiger; 27 mar 1937: kreisblatt fuer den kreis fritzlar-homberg) – gw Misc Inst [074]

Kreisanzeiger fuer wetterau und vogelsberg – Buedingen DE, 1988- – 7r/yr – 1 – gw Misc Inst [074]

Kreis-blatt – Geldern DE, 1868-69, 1873-74, 1879 & 1886, 1891-92, 1917 jul-dec – 7r – 1 – (title varies: 1 jul 1863: amtliches kreisblatt fuer den kreis geldern; 25 mar 1870: geldern'sche zeitung) – gw Misc Inst [074]

Kreis-blatt – Gelnhausen DE, 1876-1910 30 jun, 1910 1 aug-1929, 1930 1 jul-1935 30 nov, 1949 4 aug-1950 2 mar – 66r – 1 – (title varies: 1919 2 jan: gelnhaeuser tageblatt. incl suppl) – gw Misc Inst [074]

Kreisblatt *see*
- Amtliche bekanntmachungen fuer den kreis hofgeismar
- Nachrichten fuer uelzen und die umgegend
- Rotenburger kreisblatt 1857
- Wochenblatt fuer den kreis hoechst

Kreisblatt des hoyerswerdaer kreises – Hoyerswerda DE, 1842 15 oct-1854, 1859-89, 1891-1918, 1920-43 – 36r – 1 – (title varies: 3 apr 1886: hoyerswerdaer kreisblatt; 1 jan 1921: hoyerswerdaer nachrichten) – gw Misc Inst [074]

Kreisblatt des koenigl[ichen] landraths-amt der niederung – Tilsit (Sowjetsk RUS), 1842 8 nov-1846, 1909 5 jan-28 dec, 1915-18 – 3r – 1 – (later titles: amtliches niederunger kreisblatt [tilsit]; niederunger kreisblatt [heinrichswalde]) – gw Misc Inst [077]

Kreisblatt des koeniglichen landraths-amtes des loebauschen kreises zu neumark – Loebau (Lubawa PL), 1835 27 jun-26 dec – 1r – 1 – gw Misc Inst [943]

Kreisblatt des koeniglichen landrathsamts zu graudenz – Graudenz (Grudziadz PL), 1834 1 feb-20 dec, 1836 2 jan-22 dec, 1839 – 1r – 1 – gw Misc Inst [943]

Kreisblatt des koeniglichen preussischen landraths-amtes ortelsburg – Ortelsburg (Szczytno PL), 1858-1863 21 mar – 2r – 1 – (title varies: 2 feb 1861: ortelsburger kreisblatt) – gw Misc Inst [350]

Kreisblatt des kreiees bunzlau – Bunzlau (Boleslawiec PL), 1916 [gaps] – 1r – 1 – gw Misc Inst [350]

Kreisblatt des kreises hirschberg – Hirschberg (Jelenia Gora PL), 1914, 1916-17, 1925, 1929-36 – 6r – 1 – gw Misc Inst [077]

Kreisblatt des saganer kreises – Sagan (Zagan PL), 1849-52, 1903 29 apr-27 may, 1904-05, 1912 & 1915, 1918, 1925-27 – 7r – 1 – (title varies: 29 apr 1903?: saganer kreisblatt) – gw Misc Inst [077]

Kreis-blatt fuer das muensterland – Muenster (Westf) DE, 1852 6-9 may, 1854-56, 1858-62 – 1 – gw Mikrofilm [350]

Kreis-blatt fuer den danziger kreis – Danzig (Gdansk PL), 1861-62 – 1r – 1 – gw Misc Inst [077]

Kreisblatt fuer den kreis achim *see* Neues wochenblatt fuer die amtsbezirke achim und thedinghausen 2 jan 1878

Kreisblatt fuer den kreis frankenberg *see* Kreisblatt fuer den kreis frankenberg-voehl

Kreisblatt fuer den kreis frankenberg-voehl – Frankenberg, Eder DE, 1877-81, 1883 16 jan-1885 22 dec, 1886-1925, 1927-1932 26 nov, 1933-1935 13 dec, 1936-1944 15 dec [gaps] – 53r – 1 – (title varies: 20 apr 1886: kreisblatt fuer den kreis frankenberg; 1912: frankenberger zeitung) – gw Misc Inst [074]

Kreisblatt fuer den kreis fritzlar-homberg *see* Kreis-anzeiger

Kreis-blatt fuer den kreis hoechst a. m *see* Wochenblatt fuer den kreis hoechst

Kreisblatt fuer den kreis hoechst a. m *see* Wochenblatt fuer den kreis hoechst

Kreisblatt fuer den kreis hoechst a. m. und anzeigeblatt fuer die stadt hoechst *see* Wochenblatt fuer den kreis hoechst

Kreisblatt fuer den kreis homberg – Homberg, Bezirk Kassel DE, 1876 1 jan-1 mar, 1878 5 jan-1880 21 apr, 1881-1937 25 mar – 42r – 1 – (title varies: 1878?: homberger kreisblatt) – gw Misc Inst [074]

Kreisblatt fuer den kreis marburg – Marburg DE, 1927 22 dec-1934 – 1r – 1 – gw Misc Inst [074]

Kreisblatt fuer den kreis osterholz *see* Vegesacker wochenblatt

Kreisblatt fuer den kreis rees – Wesel DE, 1848 – 1r – 1 – (title varies: apr 1848: der volksfreund) – gw Misc Inst [074]

Kreis-blatt fuer den kreis und die stadt hoechst a. m *see* Wochenblatt fuer den kreis hoechst

Kreis-blatt fuer den kreis wittgenstein *see* Wittgensteiner kreisblatt und unterhaltungsblatt im sieg-, lahn- und ederthale

Kreis-blatt fuer den kreis ziegenhain – Ziegenhain DE, 1876-1890 27 sep, 1893-1895 29 may – 7r – 1 – (title varies: 2 jan 1889: ziegenhainer kreisblatt) – gw Misc Inst [074]

Kreisblatt fuer die kreise marburg, frankenberg-voehl und kirchhain – Frankenberg, Eder DE, 1872-76 – 1 – gw Misc Inst [943]

Kreisblatt fuer die stadt und den kreis schluechtern – Schluechtern DE, 1876-85, 1890-1915 29 dec – 7r – 1 – gw Misc Inst [350]

Kreisblatt fuer norderdithmarschen *see* Amtliches nachrichtenblatt des kreises norderdithmarschen

Kreisblatt und generalanzeiger : kreis ostprignitz mit pritzwalk und kyritz – Kyritz, Pritzwalk DE, 1933 apr-jun, 1936 apr-jun – 2r – 1 – gw Misc Inst [074]

Kreis-intelligenzblatt fuer euskirchen, rheinbach und ahrweiler *see* Erfa 1840

Kreis-kurrenden-blatt des koeniglichen landrath-amtes in hirschberg – Hirschberg (Jelenia Gora PL), 1840-43, 1847 – 1r – 1 – gw Misc Inst [077]

Kreis-kurrenden-blatt des koeniglichen landrath-amtes zu bunzlau – Bunzlau (Boleslawiec PL), 1847-61 [gaps] – 1 – gw Misc Inst [350]

Kreisler, Emil *see*
- Die dramatischen werke des peter probst
- Hebbels frauengestalten
- Der stoff und die quelle von bauernfelds lustspiel 'das tagebuch'
- Der stoff und die quelle von bauernfelds lustspiel "das tagebuch"

Kreisler, Hugo *see* Schottische reisebilder

Kreisnachrichten – Calw DE, 1983 1 sep- – ca 7r/yr – 1 – gw Misc Inst [074]

Kreisverkuendigungs-blatt fuer den kreis freiburg *see* Breisgauer bote

Kreis-wochenblatt fuer den gesammten freistaedter kreis *see* Freistaedter kreisblatt
Kreis-wochenblatt fuer freistadt und neusalz *see* Freistaedter kreisblatt
Kreis-zeitung – Neutomischl (Nowy Tomysl PL), 1929-39 – 9r – 1 – gw Misc Inst [077]
Kreis-zeitung *see* Nachrichten fuer uelzen und die umgegend
Kreiszeitung : allgemeiner anzeiger fuer die grafschaft hoya – Syke DE, 1976- – ca 11r/yr – 1 – (title varies: 24 aug 1977: kreiszeitung fuer die landkreise diepholz und verden; 22 may 1993: kreiszeitung. syker zeitung) – gw Misc Inst [074]
Kreiszeitung : boeblinger bote – Boeblingen DE, 1987- – 8r/yr – 1 – gw Misc Inst [074]
Kreis-zeitung fuer den kreis rotenburg in hannover *see* Rotenburger anzeiger
Kreiszeitung fuer die landkreise diepholz und verden *see* Kreiszeitung
Kreiszeitung wesermarsch – Nordenham DE, 1978 1 sep- – 6r/yr – 1 – gw Misc Inst [074]
Krell, Leo *see* Deutsche literaturgeschichte
Kremenetzky, Salomon *see* Oyf besere vegn
Krementz, Ph *see* Die offenbarung des hl johannes
Kremer, Alfred, Freiherr von *see*
- Culturgeschichtliche streifzuege auf dem gebiete des islams
- Geschichte der herrschenden ideen des islams

Kremer, Hannes *see*
- "Du, mein volk"
- Der erzbeter
- Der geaechtete
- Moritaten

Kremianskii, I *see* Borba za kachestvo v promkooperatsii
Kremlev, Anatolii et al *see* Piesni rabochikh
Kremmling news *see* Grand county miscellaneous newspapers
Kremmling register *see* Grand county miscellaneous newspapers
Kremnev, G M *see* Razvitie kooperatsii
Kremper marschbote und -zeitung – Krempe DE, 1894 10 mar-1941 31 may, 1946 8 feb-1961 – 1 – (title varies: 8 feb 1946: kremper zeitung) – gw Misc Inst [074]
Kremper zeitung *see* Kremper marschbote und -zeitung
Kremser nachrichten – Krems, Austria. 11 jul 1946-12 feb 1948 – 1r – 1 – uk British Libr Newspaper [072]
Kremser zeitung – Krems, Austria. Aug 1946-12 feb 1948 – 1r – 1 – uk British Libr Newspaper [072]
Krenck *see* Verpondingsquohier zutphen 1640-1650
Krenkel, Ernst Teodorovich *see* Camping at the pole
Krenkel, Max *see*
- Der apostel johannes
- Josephus und lucas
- Paulus, der apostel der heiden
- Vorlesungen ueber die geschichte der messianischen idee

Krepostnoe pravo v votchinakh sheremetevykh, 1708-1885 / Shchepetov, K N – 1947 – 10mf – 8 – mf#R-7836 – ne IDC [947]
Krepostnoi peterburg pushkinskogo vremeni / Iatsevich, Andrei Grigor'evich – Leningrad: Pushkinskoe obshchestvo, 1937 – 1 – us Wisconsin U Libr [947]
Kreppel, Friedrich *see* Zur nationalfeier
Kreshchanovskii, P *see* Vladivostok
Kreshenie armnjan, gruzin, abxazov i alanov svjatym grigoriem : arabskaja versija / Marr, N – 4mf – 8 – (zapiski vostochnogo otdel. imp russ arkh obshchestva. v16 1904-1905 p63-211) – mf#1267 mfB126-B129 – ne IDC [243]
Kress, Johann *see* Das gebet hieremie des propheten
Kressel, Getzel *see*
- 'Ivrit Ba'ma'arav
- Me-hofa'ath "roma we-yerushalayim" 'ad motho shel herzi
- Rebbi yehudah alkalai
- Rishonim

Krestianam i rabochim o partii narodnoi svobody : konstitutsionno-demokraticheskoi – 1906 – 16p 1mf – 9 – mf#RPP-119 – ne IDC [325]
Krest'ianin i rabochii : organ leninskogo (taldomskogo) uezd rk i kd – Taldom, Russia, 1918 [mf ed Norman Ross] – 1r – 1 – mf#nrp-1731 – us UMI ProQuest [077]
Krest'ianin-kooperator – Yaroslavl', Russia 1921-22 [mf ed Norman Ross] – 1r – 1 – mf#nrp-2192 – us UMI ProQuest [077]
Krest'ianskaia gazeta : organ krest'ianskogo soiuza – Chita, Russia 1920 [mf ed Norman Ross] – 1r – 1 – mf#nrp-444 – us UMI ProQuest [077]
Krestianskaia gazeta : izdanie tsentralnogo komiteta partii sotsialistov-revoliutsionerov – Spb., Paris, 1905(1); 1907-1912(1-26) – 11mf – 9 – mf#R-18052 – ne IDC [077]
Krestianskaia gazeta – Moskva: TSK VKP (b). [n1-132/133] jan 4-dec 1930 – us CRL [947]
Krestianskaia kooperatsiia v svobodnoi rossii / Evdokimov, A A – 1917 – 15p 1mf – 9 – mf#COR-24 – ne IDC [335]
Krest'ianskaia mysl' : nar bespartiin gaz / ed by Soldatov, V V – Vladivostok [Primor obl]: [s n] 1919 [1919-] – (= ser Asn 1-3) – n5-20 [1919] [gaps] item 207, on reel n41 – 1 – mf#asn-1 207 – ne IDC [077]
Krestianskaia obshchina : chto ona takoe, k chemu idet, chto daet i chto mozhet dat rossii? / Veniaminov, P – 1908 – 259p 3mf – 9 – mf#RPP-195 – ne IDC [325]
Krest'ianskaia pravda : organ demianskogo uezd komiteta rkp i uezd ispolkoma – Dem'yansk, Russia 1918 [mf ed Norman Ross] – 1r – 1 – mf#nrp-459 – us UMI ProQuest [077]
Krest'ianskaia rossiia – Moscow, Russia 1992- [mf ed Norman Ross] – 2r per y – 1 – $160.00 standing order – (backfile through 1998 $85r) – mf#nrp-1081 – us UMI ProQuest [630]
Krest'ianskaia zhizn' : organ tat uezd kom partii sotsialistov-revoliutsionerov / ed by IAzev, I N – Tatarsk [Akmol obl]: [s n] 1918 [1918 3 avg-okt] – (= ser Asn 1-3) – n1-10 [1918] [gaps] item 206, on reel n41 – 1 – (cont by: narodnoe delo) – mf#asn-1 206 – ne IDC [077]
Krest'ianskie biudzhety voronezhkoi gubernii / Shcherbina, F A – Voronezh, 1900 – 20mf – 8 – mf#RZ-45 – ne IDC [314]
Krest'ianskie i rabochie dumy : organ severo-dvinskogo gubispolkoma – Veliky Ustyug, Russia 1918 [mf ed Norman Ross] – 1r – 1 – mf#nrp-1917 – us UMI ProQuest [077]
Krest'ianskoe obshchestva / Materialy po zemlevladeniiu i zemledeliiu Khar'kovskoi gubernii – Khar'kov, 1886. 1v – 7mf – 8 – mf#RZ-199 – ne IDC [314]
Krest'ianskii bank i budushchnost' russkogo krest'ianstva / Semenov, P N & Saltykov, A A – Spb, 1907 – 1mf – 9 – mf#REF-262 – ne IDC [332]
Krest'ianskii bank i ego deiatel'nost' s 1883 po 1905 gg / Krasik, A V – Iur'ev, 1910 – 3mf – 9 – mf#REF-259 – ne IDC [332]
Krestianskii biudzhet v 1922-23 godu : izd tsentralnogo statisticheskogo upravleniia / Litoshenko, L N – 1923 – 58p 1mf – 9 – mf#COR-207 – ne IDC [335]
Krest'ianskii pozemel'nyi bank : 1883-1910 gg / Zak, A N – M, 1911 – 7mf – 9 – mf#REF-258 – ne IDC [332]
Krest'ianskii pozemel'nyi bank, 1883-1895 gg / Vdovin, VA – M, 1959 – 2mf – 9 – mf#REF-264 – ne IDC [332]
Krest'ianskii pozemel'nyi bank i znachenie ego dlia narodnogo khoziaistva : [doklad chitannyi v zasedanii iii-go otdeleniia 10 ianvaria 1894 g] / Ponomarev, N V – Spb, 1894 – 1mf – 9 – mf#REF-257 – ne IDC [332]
Krest'ianskii vestnik / ed by Komarov, A V – Omsk [Akmol obl]: [s n] 1919 [1919 30 iiulia-] – (= ser Asn 1-3) – n1-35 [1919] [gaps] item 208, on reel n41 – 1 – mf#asn-1 208 – ne IDC [077]
Krestianskii vopros v rossii v 18 i pervoi polovine 19 veka / Semevskii, V I – 1888. 2v – 11mf – 8 – (missing: 1888 v2) – mf#R-129 – ne IDC [947]
Krest'ianskij deputat *see* Izvestiia moskovskogo gub soveta krest'ianskikh deputatov
Krest'ianskoe delo : nar polit i ekon gaz / ed by Shevchenko, E S – Odessa: N K Klimenko 1919 [1919 31 avg-[1920 12 ianv]] – (= ser Asn 1-3) – n2, 22 [1919] item 209, on reel n42 – 1 – mf#asn-1 209 – ne IDC [077]
Krest'ianskoe dvizhenie 1827-1869 / ed by Morokhovets, E A – Moskva: Gos sotsial'no-ekon izd-vo, 1931 [mf ed 2002] – 1r – 1 – (filmed with: ocherki perekhodnoi ekonomiki / a leont'ev i e khmel'nitskaia [1927]. incl bibl ref & ind) – mf#5231 – us Wisconsin U Libr [947]
Krestianskoe khoziaistvo : ocherki ekonomiki melkogo zemledeliia / Maslov, S L – 1918 – 272p 3mf – 9 – mf#COR-484 – ne IDC [335]
Krestianskoe khoziaistvo i ego interesy / Makarov, N P – 1917 – (= ser Liga agrarnykh reform seriia) – 112p 2mf – 9 – (liga agrarnykh reform seriia 5, no 2) – mf#COR-209 – ne IDC [335]
Krestianskoe khoziaistvo i selskokhoziaistvennaia kooperatsiia / Maslov, S L – 1919 – 90p 1mf – 9 – mf#COR-485 – ne IDC [335]
Krestianskoe kooperativnoe dvizhenie v zapadnoi sibiri / Makarov, N P – 1910 – 3mf – 9 – mf#COR-65 – ne IDC [335]
Krestianskoe zemelnoe pravo / Khauke, O A – 1914 – 372p 4mf – 9 – mf#COR-131 – ne IDC [335]
Krest'ianskoe zemlevladenie kazanskoi gubernii – Kazan', 1907-1911. v1-13 – 40mf – 8 – mf#RZ-64 – ne IDC [314]
Krestianstvo i sotsial-demokratiia / Rumiantsev, P – n.d. – 23p 1mf – 9 – (temy zhizni. v8(1)) – mf#RPP-154 – ne IDC [325]
Krestinskii, N *see* Nasha finansovaia politika
Kret, lakiv N *see* Vazhniishi prava kanady
Der kreterkoenig : drama / Baumann, Hans – Jena: E Diederichs, 1944 [mf ed 1989] – 1 – mf#70 – us Wisconsin U Libr [820]
Kretschmer, Elisabeth *see* Gellert als romanschriftsteller
Kretzer, Max *see*
- Assessor lankens verlobung
- Der bassgeiger / das verhexte buch
- Die beiden genossen
- Berliner skizzen
- Die buchhalterin roman
- In frack und arbeitsbluse roman
- Reue roman
- Die verkommenen
- Wenn steine reden

Kreube, Frederic *see* Edmond et caroline
Kreuiter, Allyson *see* The representation of madness in margaret atwood's allas grace
Kreutzberger, Max *see* Juedische arbeits- und wanderfuersorge
Kreutzberger, Max et al *see* Informationsblaetter
Kreutzer, J *see* Zwinglis lehre von der obrigkeit
Kreutzer, Jakob *see* Zwinglis lehre von der obrigkeit
Kreutzer, Rodolphe *see*
- Aristippe
- Deux quatuors pour deux violons, alto et violoncelle
- Pot-pourri
- Trois quatuors pour deux violons, alto et basse
- Trois sonates pour le violon avec accompagnement de basse, lettre b

Das kreuz : grund und mass fuer die christologie / Kaehler, Martin – Guetersloh: C Bertelsmann, 1911 – (= ser Beitraege zur foerderung christlicher theologie) – 1mf – 9 – 0-7905-9291-6 – mf#1989-2516 – us ATLA [240]
Das kreuz im venn : roman / Viebig, Clara – 5. aufl. Berlin: E Fleischel 1908 [mf ed 1989] – 1r – 1 – (filmed with: dilettanten des lebens) – mf#7154 – us Wisconsin U Libr [830]
Das kreuz und die kreuzigung : eine antiquarische untersuchung nebst nachweis der vielen seit lipsius verbreiteten irrthuemer: zugleich vier excurse ueber verwandte gegenstaende / Fulda, Hermann – Breslau: Wilhelm Koebner, 1878 – 1mf – 9 – 0-8370-3216-4 – (includes appendixes on aspects of crucifixion. incl indes) – mf#1985-1216 – us ATLA [240]
Kreuz und halbmond im nillande : nach studienreisen und reisestudien / Boehmer, Julius – Guetersloh: C Bertelsmann, 1910 – 1mf – 9 – 0-8370-6647-6 – mf#1986-0647 – us ATLA [230]
Kreuzberg-abendzeitung *see* Neue kreuzberg-zeitung
Kreuzer, Christoph *see* Katholische volkszeitung
Kreuzer, Wolfgang *see*
- Bau eines rundlaufthermostaten fuer das festkoerperdilatometer und messung des ausdehnungsverhaltens und der volumenrelaxation von polymeren
- Vorhersage der schwingfestigkeit von schweissverbindungen auf der basis des statistischen groesseneinflusses

Der kreuziger / Frankenstein, Johannes von; ed by Khull, Ferdinand – Stuttgart: Litterarischer Verein in Stuttgart, 1882 (Tuebingen: H Laupp) [mf ed 1993] – (= ser Blvs 160) – 428p – 1 – mf#8470 reel 33 – us Wisconsin U Libr [810]
Der kreuziger / Frankenstein, Johannes von; ed by Khull, Ferdinand – Stuttgart: Litterarischer Verein in Stuttgart 1882 (Tuebingen: H Laupp) [mf ed 1993] – (= ser Blvs 160) – 58r – 1 – mf#3420p – us Wisconsin U Libr [830]
Kreuznacher bote und oeffentlicher anzeiger – Bad Kreuznach DE, 1977- – ca 97r/yr – 1 – (filmed by other misc inst: 1848 17 jul-30 dec [gaps]. title varies: 1 oct 1867: oeffentlicher anzeiger fuer bad kreuznach und umgebung; 25 mar 1954: rhein-zeitung / e [main ed in koblenz]; 23 aug 1978: oeffentlicher anzeiger) – gw Misc Inst [074]
Kreuznacher bote und oeffentlicher anzeiger – Bad Kreuznach DE, 1848 17 jul-30 dec [gaps] – 1 – (filmed by other misc inst: 1977- [ca 9r/yr]. title varies: 1 oct 1867: oeffentlicher anzeiger fuer bad kreuznach und umgebung; 25 mar 1954: rhein-zeitung / e; 23 aug 1978: oeffentlicher anzeiger) – gw Misc Inst [074]
Kreuz-zeitung – Strassburg (Strasbourg F), 1884-85 – 1 – (with n4 1884: kreuz-zeitung fuer elsass-lothringen) – gw Misc Inst [074]
Kreuz-zeitung *see* Neue preussische zeitung
Die kreuzzuege und das heilige land : mit 4 kunstbeilagen, 163 abbn und 3 karten / Heyck, E – Bielefeld, Leipzig, 1900 – 3mf – 9 – mf#H-3091 – ne IDC [915]
Krey, August Charles *see* Parallel source problems in medieval history
Krey, Ursula *see* Nachlass friedrich naumann (bestand n 3001) bd 55
Kreyenberg, Gotthold *see* Geschichte der poetischen litteratur der deutschen
Kreyenbuehl, Johannes *see* Die nothwendigkeit und gestalt einer kirchlichen reform
Kreyher, Johannes *see*
- L annaeus seneca und seine beziehungen zum urchristentum
- Die weisheit der brahmanen und das christentum

Kreymborg, Gustav *see* Johann karl wezel
Kreyssig, Friedrich *see* Vorlesungen ueber goethe's faust
Kreyssig, Friedrich Alexander Theodor *see* Justus moeser
Kri Mon, Kre mum *see* Ca nay jan sa muin
Kri Pu, U *see* Ci pva re lam pra aim rhan ma mya lak cvai khyak nann prut nann 500
Kriby, John B *see* New deal agencies and black america
Krichenbauer, Anton *see* Theogonie und astronomie
Krick, Charlotte Virginia *see* Piano concerto
Krickeberg, Walter *see* Antiguas culturas mexicanas
Kricker, Gottfried *see*
- Theodor fontane

Kriebel, Howard Wiegner *see* The schwenkfelders in pennsylvania
Krieck, Ernst *see* Lessing und die erziehung des menschengeschlechts
Der krieg : dichtung / George, Stefan Anton – 2. aufl. Berlin: G Bondi 1917 [mf ed 1989] – 1r – 1 – (filmed with: der siebente ring & other titles) – mf#7294 – us Wisconsin U Libr [810]
Krieg : ein tedeum / Hauptmann, Carl – Leipzig: K Wolff, 1914 – 1r – 1 – us Wisconsin U Libr [810]
Krieg, Cornelius *see* Encyklopaedie der theologischen wissenschaften nebst methodenlehre
Der krieg gibt keinen frieden / Vatsella, Iris – (mf ed 2001) – 280p – 9 – €49.00 – 3-8267-2759-2 – mf#DHS 2759 – gw Frankfurter [410]
Krieg in spanien. barbarei und zivilisation; ein zeitdokument – Barcelona, 1938. Fiche W 979. (Blodgett Collection of Spanish Civil War Pamphlets) – 9 – us Harvard [946]
Der krieg um den wald : eine historie in zwoelf kapiteln / Hartmann, Moritz; ed by Wurzbach, Wolfgang von – Leipzig: M Hesse, [19–] [mf ed 1993] – 180p – 1 – mf#8711 – us Wisconsin U Libr [830]
Der krieg und die infektionskrankheiten : lecture delivered 29 feb 1915 in tuebingen / Mueller, Otfried – Tuebingen: Kloeres 1915 [mf ed 1987] – (= ser Durch kampf zum frieden 9) – 1r – 1 – mf#6840 – us Wisconsin U Libr [933]
Der krieg und die religion : rede am 12. november 1914 / Deissmann, Gustav Adolf – Berlin: C Heymann 1914 [mf ed 1990] – (= ser Deutsche reden in schwerer zeit 9) – 1mf – 9 – 0-7905-5815-7 – mf#1988-1815 – us ATLA [933]
Krieg und frieden : militaerpolitische revue – Paris (F), 1938 oct-1939 sep – 1r – 1 – gw Misc Inst [303]
Krieg und kultur / Smend, Rudolf – Tuebingen: Kloeres, 1915. 20p – 1 – us Wisconsin U Libr [000]
Krieg und kunst / Lange, Konrad – Tuebingen: Kloeres, 1915. 32p – 1 – us Wisconsin U Libr [700]
Kriegbaum heritage – v5 n3-v14 n2/3 [1980 jul-1989 apr/jul] – 1r – 1 – mf#1804971 – us WHS [071]
Krieger, Erhard *see* Ernst theodor amadeus hoffmann
Krieges carte von schlesien 1747-1753 (mcet1) : berlin, staatsbibliothek preussischer kulturbesitz, kartenabteilung gr 2° kart n 15060 / Wrede, Christian Friedrich von – (mf ed 1992) – (= ser Monumenta cartographica et topographica (mcet) 1) – 24 color mf – 15 – €690.00 – 3-89219-500-5 – (int & ind by klaus lindner) – gw Lengenfelder [090]
Kriegs- und reisefahrten / Fischer, Christian A – Leipzig 1821-22 [mf ed Hildesheim 1995-98] – 2v on 6mf – 9 – €120.00 – 3-487-29930-5 – gw Olms [910]
Kriegs- und siegeslieder aus dem 15. jahrhundert / Weber, Veit [pseud for Waechter, Leonhard]; ed by Schreiber, Heinrich – Freiburg: Herder, 1819 [mf ed 1993] – x/108p – 1 – (without music. incl bibl ref) – mf#8361 – us Wisconsin U Libr [780]
Kriegsbilder – Duesseldorf DE, 1915-18 – 1r – 1 – gw Misc Inst [770]
Der kriegsbote – Windhoek: John Meinert, jahrg 1 n1 [jul 1916]-jahrg 4 n55 [jul 1919] – 1r – 1 – (merged with: der welt-krieg to become: allgemeine zeitung) – mf#MS00386 – sa National [079]
Der kriegsbote – Strassburg (Strasbourg F), 1792 7 aug-7 nov [gaps] – 1 – fr ACRPP [074]
Kriegsdichter erzaehlen / ed by Velmede, August Friedrich – Muenchen: A Langen/G Mueller, [1942?] – 1r – 1 – (incl bibl ref) – us Wisconsin U Libr [430]

Kriegsfroemmigkeit : zeugnisse aus dem grossen kriege fuer kirche, schule und haus / Schwencker, Friedrich – Guetersloh: C Bertelsmann, 1915-1916 – 2mf – 9 – 0-524-03772-8 – mf#1990-1119 – us ATLA [240]

Kriegs-kunst... / Quincy, C S & Vauban, [S] – Nuernberg, 1745 – 7mf – 9 – mf#OA-195 – ne IDC [720]

Kriegslieder aus oesterreich / Schaukal, Richard von – Muenchen: G Mueller, 1914- – 1r – 1 – us Wisconsin U Libr [780]

Die kriegsmarine – Berlin DE, 1939-1944 2 mar [gaps] – 1 – gw Misc Inst [355]

Kriegsmesse 1914 : [eine dichtung] / Zerzer, Julius – Jena: E Diederichs, 1914 – 1r – 1 – us Wisconsin U Libr [810]

Kriegssegen / Bahr, Hermann – Muenchen: Delphin-Verlag, 1915 [mf ed 1989] – 71p – 1 – mf#6979 – us Wisconsin U Libr [830]

Kriegssegen / Bahr, Hermann – Muenchen: Delphin-Verlag, 1915 [mf ed 1989] – 71p – 1 – mf#6979 – us Wisconsin U Libr [934]

Kriegstage in suedwest / Willich, C – Pretoria, State Library, 1979. Orig. pub., Oldenburg: Gerhard Stalling, 1916. (Microfiche Reprint Series no.6). 2 fiche. Printed booklet – 9 – sa National [960]

Kriegs-zeitung der 7. armee – Laon (F), 1914 31 oct-1915 16 oct, 1918 2 jun-20 oct – 1r – 1 – gw Misc Inst [355]

Kriegszeitung der vierten armee – Stuttgart DE, 1918 25 apr-3 nov – 1r – 1 – gw Misc Inst [355]

Kriek, Roelof Jacobus *see* Separation of platinum group metals

Kriel, Aletha Catharina *see* Proposed norms and standards for pastoral counsellors/therapists

Kriel, T J *see*
– New english-sesotho dictionary
– Sotho-afrikaanse woordeboek

Krilof and his fables / Krylov, Ivan Andreevich – By W.R.S. Ralston. 3rd ed., enl. London: Strahan, 1871. 268p., ill – 1 – us Wisconsin U Libr [390]

Krilov's mesholim / Krylov, Ivan Andreevich – Byallstok, Poland. 1921 – 1r – us UF Libraries [939]

Krimphove, Frank *see* Globale konvergenz von zufallsstrategien in der optimierung

Krimsky, Joseph *see* Pilgrimage and service

Krinski, Magnus *see* Reshit daat sefat ever

Kripalani, Jiwatram Bhagwandas *see*
– Fateful year
– Gandhi, the statesman
– The gandhian way
– Politics of charkha

Kripalani, Krishna *see* Gandhi, tagore, and nehru

Kripalani, Krishna] *see* Rolland and tagore

Der krippenweg / Schaumann, Ruth – Muenchen: J Koesel & F Pustet c1932 [mf ed 1991] – 1r – 1 – (filmed with: der bluehende stab) – mf#2868p – us Wisconsin U Libr [830]

Krischen, F *see* Die befestigungen von herakleia am latmos

Krischer, Volker *see* Untersuchung der verkernung der jahrringstruktur

Krischna oder christus? : eine religionsgeschichtliche parallele / Dilger, Wilhelm – Basel: Missionsbuchh, 1904 – (= ser Basler missionsstudien) – 1mf – 9 – 0-524-01429-9 – mf#1990-2424 – us ATLA [230]

Die krise der alten welt im 3. jahrhundert n. zw. und ihre ursachen / Altheim, Franz – Berlin-Dahlem: Ahnenerbe-stiftung 1943- [v1,3] [mf ed 1978] – (= ser Deutsches ahnenerbe. reihe b: fachwissenschaftliche untersuchungen. abteilung: arbeiten zur alten geschichte 2-3) – 1r – 1 – (v1: die ausserroemische welt; v2 never publ in this ed; v3: goetter und kaiser; with contr by erika trautmann-nehring, & bibl ref incl in "anmerkungen") – mf#film mas 8190 – us Harvard [900]

Die krise des individuums : ein vergleich zwischen adorno und gehlen / Thies, Christian – (mf ed 1997) – 5mf – 9 – €59.00 – 3-8267-2424-0 – mf#DHS 2424 – gw Frankfurter [140]

Krisen und wandlungen der deutschen literatur von wedekind bis feuchtwanger : fuenfzehn vorlesungen / Kaufmann, Hans – Berlin, Weimar: Aufbau-Verlag, 1966 [mf ed 1993] – 563p – 1 – (incl bibl ref and ind) – mf#8256 – us Wisconsin U Libr [430]

Krishna and the gita : being twelve lectures on the authorship, philosophy and religion of the bhagavadgita / Tattvabhushan, Sitanath – Calcutta: AC Sarkar, [191-?] – (= ser Raja surya rao lectures) – 1mf – 9 – 0-524-02553-3 – mf#1990-3048 – us ATLA [280]

Krishna, Bal *see* Shivaji, the great

Krishna Iyer, E *see* Personalities in present day music

Krishna, Lajwanti Rama *see* Panjabi sufi poets, ad 1460-1900

Krishna Rao, MV *see* The gangas of talkad: a monograph on the history of mysore from the fourth to the close of the eleventh century

Krishna, Roop *see* Art and life

Krishna Sastri, Hosakote *see*
– South-indian images of gods and goddesses
– Two statues of palla kings

Krishna the charioteer : or, the teachings of bhagavad gita / Dhar, Mohini Mohan – London: Theosophical Publ House, 1917 [mf ed 1995] – (= ser Yale coll) – 173p – 1 – 0-524-09250-8 – mf#1995-0250 – us ATLA [280]

Krishna the cowherd : or, a study of the childhood of shri krishna / Dhar, Mohini Mohan – London: Theosophical Publ House, 1917 [mf ed 1995] – (= ser Yale coll) – 110p – 1 – 0-524-09251-6 – mf#1995-0251 – us ATLA [280]

Krishnamurti : the man and his teaching / Fouere, Rene – Bombay: Chetana, c1952 – (= ser Samp: indian books) – us CRL [280]

Krishnamurti and the unity of man / Suares, Carlo – Bombay: Chetana, 1953 – (= ser Samp: indian books) – us CRL [280]

Krishnamurti, Jiddu *see*
– At the feet of the master
– Life in freedom
– Revised report of fourteen talks given by krishnamurti, ommen camp, 1937 and 1938
– The search

Krishnamurti, Y G *see*
– The betrayal of freedom
– Gandhi era in world politics
– Independent india and a new world order
– Jawaharlal nehru
– Reflections on the gandhian revolution
– Sir m visvesvaraya

Krishnarao, Bhavaraju Venkata *see* A history of the early dynasties of andhradesa c 200-625 ad

Krishna's flute and other poems / Thadani, Nanikram Vasanmal – London; New York: Longmans, Green and Co, 1919 – (= ser Samp: indian books) – us CRL [810]

Krishnaswami Aiyangar, Sakkottai *see*
– A history of tirupati
– A little known chapter of vijayanagar history
– Manimekhalai in its historical setting
– Some contributions of south india to indian culture
– South india and her muhammadan invaders

Krishnaswami Aiyangar, Sakkottai et al *see* Sri ramanujacharya

Krishnaswamy Iyengar, Srinivasa *see* Early history of vaishnavism in south india

Krishnayya, Pasupati Gopala *see* Mahatma gandhi and the usa

Die "krisis des christenthums" : protestantismus und katholische kirche / Hettinger, Franz – Freiburg i B; St Louis, MO: Herder, 1881 – 1mf – 9 – 0-7905-7001-7 – (incl bibl ref) – mf#1988-3001 – us ATLA [240]

Die krisis des christenthums in der modernen theologie / Hartmann, Eduard von – Berlin: Carl Duncker, 1880 – 1mf – 9 – 0-8370-8745-7 – (incl bibl ref) – mf#1986-2745 – us ATLA [240]

Krisis und entscheidung im judentum / Klatzkin, J – Berlin, 1921 – 4mf – 9 – mf#J-28-59 – ne IDC [956]

Kristall *see* Nordwestdeutsche hefte

Die kristallkugel : eine altweimarische geschichte / Boehlau, Helene – 4. aufl. Berlin: E Fleischel, 1904 [mf ed 1989] – 135p – 1 – mf#7042 – us Wisconsin U Libr [830]

Kristdemokraten – Stockholm, Sweden. 1988- – 1 – sw Kungliga [078]

Kristeller, Paul *see* Early florentine woodcuts

Kristensen, Evald Tang *see* Vindt molle og dens ejere

Kristianialaget *see* Vikvaeringen

Kristianstads lansdemokraten – Kristianstad, Sweden. 1932-57 – 99r – 1 – sw Kungliga [078]

Kristianstadsbladet – Kristianstad, Sweden. 1979- . Mellersta Skane, 1979-85 – 1 – sw Kungliga [078]

Kristianstadsbladet – Kristianstad, Sweden. 1856-1935, 1979 – 540r – 1 – (mellersta skane: 1971-77) – sw Kungliga [078]

Kristiga balss = The christian voice / Union of Latvian Baptists in America – gads 52 n1/2-gads 56 n11/12 [1981 jan/feb-1985 nov/dec] – 1r – 1 – mf#682005 – us WHS [242]

Kristiga balss-the christian voice – Riga, Latvia. Jul 1920-Aug 1940. Organ of the Latvian Baptist Union. Saweja Balss (The Voice of Isaiah), Jul 1920-Jun 1921 – 1 – us Southern Baptist [242]

Kristigs wehstnesis – Riga, Latvia. 1915, no.10; 1923-29. Ed. by William Fetler – 1 – 83.88 – us Southern Baptist [242]

Kristin : ein deutsches grenzlandmaedel / Lange, Mariluise – Reutlingen: Ensslin & Laiblin, 1943 – 1r – 1 – us Wisconsin U Libr [830]

Den kristna foersamlingen / Waldenstroem, Paul – Stockholm C: Svensk Missionsforbundet, [1914?] – 1mf – 9 – 0-524-05779-6 – mf#1991-2335 – us ATLA [240]

Den kristna tankens tolkning af jesu person : en blick pa den kristologiska utvecklingen intill nutiden / Aulen, Gustaf – Uppsala: Norblad, [1910] [mf ed 1990] – (= ser Hvad aer sanning? 6) – 1mf – 9 – 0-7905-3525-4 – (in swedish) – mf#1989-0018 – us ATLA [240]

Den kristna troslaeran fran metodistisk standpunkt / Sulzberger, Arnold – Stockholm: Metodist-Episkopalkyrkans Foerlag, 1886-1888 – 2mf – 9 – 0-524-08690-7 – mf#1993-3215 – us ATLA [242]

Den kristne buddhist mission – v9-38. 1934-63 – 3r – 1 – (lacking: v17-22 1942-47) – mf#ATLA S0533 – us ATLA [280]

Kriticheskii zhurnal – Raleigh. 1778-1886 (1) – 17mf – 9 – mf#1789 – ne IDC [077]

'N kritiese studie ten opsigte van die status van die plattelandse onderwyser in die ceres-tulbaghgebied / Baartzes, Wesley Barry – U of the Western Cape 1988 [mf ed S.l.: s.n. between 1987 & 1991] – 5mf – 9 – (incl bibl) – sa Misc Inst [370]

Die kritik : theater – kultur – film – Prag (CZ), 1933 aug-1935 jul – 1r – 1 – gw Misc Inst [700]

Kritik der epheser- und kolosserbriefe : auf grund einer analyse ihres verwandtschaftsverhaeltnisses / Holtzmann, Heinrich Julius – Leipzig: Wilhelm Engelmann, 1872 – 1mf – 9 – 0-8370-3633-X – (incl bibl ref) – mf#1985-1633 – us ATLA [227]

Kritik der epheser und kolosserbriefe auf grund einer analyse ihres verwandtschaftsverhaeltnisses / Holtzmann, Heinrich Julius – Leipzig: W Engelmann, 1872 – 1r – 1 – 0-8370-0350-4 – mf#1984-B414 – us ATLA [227]

Kritik der evangelien und geschichte ihres uhrsprungs / Bauer, Bruno – Berlin. v1-2. 1850-1851 – 2v on 11mf – 8 – €21.00 – ne Slangenburg [242]

Kritik der evangelischen geschichte / Bauer, Bruno – Leipzig. v1-3. 1841-1842 – 3v on 21mf – 8 – €38.00 – ne Slangenburg [242]

Kritik der evangelischen geschichte des johannes / Bauer, Bruno – Bremen: Carl Schuenemann, 1840 – 2mf – 9 – 0-8370-9525-5 – (incl bibl ref) – mf#1986-3525 – us ATLA [226]

Kritik der freiheitstheorien : eine abhandlung ueber das problem der willensfreiheit / Mack, Joseph – Leipzig: Johann Ambrosius Barth, 1906 – 1mf – 9 – 0-8370-4260-7 – (incl bibl ref and index) – mf#1985-2260 – us ATLA [210]

Kritik der israelitischen geschichte / Wette, Wilhelm Martin Leberecht de – Halle: Schimmelpfennig, 1807 [mf ed 1992] – (= ser Beitraege zur einleitung in das alte testament 2) – 1mf – 9 – 0-524-02773-0 – (no more publ?) – mf#1987-6467 – us ATLA [221]

Kritik der paulinischen briefe / Bauer, Bruno – 8 – (1.abt: der ursprung der galaterbriefe, berlin 1850 2mf €5. 2.abt: der ursprung des ersten korintherbriefes, berlin 1851 2mf €5. 3.-und letzte abt: der zweiter korintherbrief, berlin 1852 3mf €7) – ne Slangenburg [227]

Kritik der paulinischen briefe / Bauer, Bruno – Berlin: Gustav Hempel. 3v in 1. 1852 – 1mf – 9 – 0-7905-0001-9 – mf#1987-0001 – us ATLA [227]

Kritik der principien der strauss'chen glaubenslehre / Rosenkranz, Karl – Leipzig: G Brauns, 1845 – 1mf – 9 – 0-7905-9469-2 – mf#1989-2694 – us ATLA [240]

Kritik der reinen vernunft / Kant, Immanuel – Berlin: 1919 – xi/609p – 1 – us Wisconsin U Libr [190]

Kritik der schleiermacherschen glaubenslehre / Rosenkranz, Karl – Koenigsberg: AW Unzer, 1836 – 1mf – 9 – 0-7905-3981-0 – mf#1989-0474 – us ATLA [240]

Kritik der socialen reformen frankreichs und ihrer folgen / Stahl, Friedrich Wilhelm – Erlangen 1848 [mf ed Hildesheim 1995-98] – 1v on 1mf – 9 – €40.00 – ISBN-10: 3-487-26022-0 – ISBN-13: 978-3-487-26022-8 – gw Olms [350]

Kritik der theologischen erkenntnis / Lipsius, Friedrich Reinhard – Berlin: CA Schwetschke, 1904 – 1mf – 9 – 0-7905-7523-X – (incl bibl ref) – mf#1989-0748 – us ATLA [120]

Kritik der v. doellinger'schen erklaerung vom 28. maerz d. j / Hergenroether, Joseph – Freiburg im Breisgau: Herder, 1871 – 1mf – 9 – 0-8370-8349-4 – (incl bibl ref) – mf#1986-2349 – us ATLA [240]

Kritik des midrash schir-haschirim / Chowdowski, Salomo – Berlin, Germany. 1877 – 1r – us UF Libraries [939]

Kritik des neuen testaments von einem griechischen philosophen des 3. jahrhunderts / Harnack, Adolf von – Leipzig, 1911 – (= ser Tugal 3-37/4) – 3mf – 9 – €7.00 – ne Slangenburg [225]

Kritik des neuen testaments von einem griechischen philosophen des 3. jahrhunderts : die im apocriticus des macarius magnes enthaltene streitschrift / Macarius Magnes – Leipzig: JC Hinrichs, 1911 [mf ed 1989] – (= ser Tugal 37/4) – 1mf – 9 – 0-7905-1724-8 – (incl bibl ref & ind. in german & greek) – mf#1987-1724 – us ATLA [225]

Kritik und erlaeuterung des goethe'schen faust : nebst einem anhange zur sittlichen beurtheilung goethe's / Weisse, Christian Hermann – Leipzig: Reichenbach, 1837 – 1r – 1 – (incl bibl ref) – us Wisconsin U Libr [430]

Kritik und ueberlieferung auf dem gebiete der erforschung des urchristentums / Krueger, Gustav – 2., um ein Nachwort verm. Abdr. Gieszen: J. Ricker, 1903 – 1mf – 9 – 0-7905-6069-0 – mf#1988-2069 – us ATLA [240]

Kritika – Cracow. v. 8, no. 2-v. 16, no. 6. 1906-1914 – 1 – us NY Public [460]

Kritika operatsii zemel'nykh bankov / Tolvinskii, A I – Vil'na, 1893 – 5mf – 9 – mf#REF-506 – ne IDC [332]

Kritika "teorii" bunda / Pasmanik, D – Odessa, 1906 – 1mf – 9 – mf#RPP-100 – ne IDC [325]

Kritiko-bibliograficheskii ezhemesiachnyi zhurnal – M., 1912-1916 – 47mf – 9 – (missing: 1912(1-5, 7-9); 1913(7); 1914(8); 1915(11-12); 1916(11-12)) – mf#R-4325 – ne IDC [077]

Kritiko-bibliograficheskii ezhenedelenik – Spb., 1907 (1-14) – 19mf – 9 – mf#R-4322 – ne IDC [077]

Kritiko-bibliograficheskii obzor noveishikh trudov po istorii russkoi tserkvi / Titov, F I – Kiev, 1904-1911 – 173p 4mf – 8 – mf#R-7000 – ne IDC [243]

Kritiko-bibliograficheskii zhurnal – London. 1951+ (1) 1969+ (5) 1975+ (9) – 45mf – 9 – mf#1222 – ne IDC [077]

Kritiko-bibliograficheskii zhurnal – M., 1907. nos 1-4 – 2mf – 9 – mf#R-7635 – ne IDC [077]

Kritiko-bibliograficheskii zhurnal – Spb., 1869. v1-3 – 10mf – 9 – mf#R-3500 – ne IDC [077]

Kritiko-bibliograficheskii zhurnal – Spb., 1907 – 3mf – 9 – mf#R-4323 – ne IDC [077]

Kritisch exegetisches handbuch ueber das evangelium des johannes = Critical and exegetical hand-book to the gospel of john / Meyer, Heinrich August Wilhelm – New York: Funk & Wagnalls, 1895, c1884 – 2mf – 9 – 0-7905-0318-2 – (in english. incl ind) – mf#1987-0318 – us ATLA [221]

Kritisch exegetisches handbuch ueber das evangelium des matthaeus / Meyer, Hein. Aug. Wilh – 5. verb. und verm. aufl. Goettingen: Vandenhoeck und Ruprecht, 1864 – (= ser Kritisch Exegetischer Kommentar Ueber Das Neue Testament) – 1mf – 9 – 0-7905-0585-1 – (incl bibl ref) – mf#1987-0585 – us ATLA [226]

Kritisch exegetisches handbuch ueber den hebraeerbrief / Luenemann, Gottlieb – 4. verb verm aufl. Goettingen: Vandenhoeck und Ruprecht, 1878 – (= ser Kritisch exegetischer Kommentar ueber das Neue Testament) – 1mf – 9 – 0-8370-4194-5 – (incl bibl ref) – mf#1985-2194 – us ATLA [227]

Kritisch exegetisches handbuch ueber den roemerbrief = Critical and exegetical hand-book to the epistle to the romans / Meyer, Heinrich August Wilhelm – New York: Funk & Wagnalls, 1889, c1884 – 2mf – 9 – 0-7905-0316-6 – (in english. incl ind) – mf#1987-0316 – us ATLA [227]

Kritisch exegetisches handbuch ueber die apostelgeschichte / Meyer, Heinrich August Wilhelm – 3. verb verm aufl. Goettingen: Vandenhoeck und Ruprecht, 1861 – (= ser Kritisch exegetischer Kommentar ueber das Neue Testament) – 2mf – 9 – 0-7905-0586-X – (incl bibl ref) – mf#1987-0586 – us ATLA [225]

Kritisch exegetisches handbuch ueber die briefe an die philipper, kolosser und an philemon = Critical and exegetical hand-book to the epistles to the philippians and colossians, and to philemon / Meyer, Heinrich August Wilhelm – New York: Funk & Wagnalls, 1889, c1885 – 2mf – 9 – 0-7905-0317-4 – (in english. includes bibliographies and index) – mf#1987-0317 – us ATLA [227]

Kritisch exegetisches handbuch ueber die briefe an die philipper, kolosser und an philemon / Meyer, Heinr. Aug. Wilh – 2. verb verm aufl. Goettingen: Vandenhoeck und Ruprecht, 1859 – (= ser Kritisch Exegetischer Kommentar Ueber Das Neue Testament) – 1mf – 9 – 0-7905-0587-8 – (incl bibl ref) – mf#1987-0587 – us ATLA [227]

Kritisch exegetisches handbuch ueber die briefe an die thessalonicher / Luenemann, Gottlieb – 4. verb verm aufl. Goettingen: Vandenhoeck und Ruprecht, 1878 – (= ser Kritisch exegetischer Kommentar ueber das Neue Testament) – 1mf – 9 – 0-8370-4195-3 – (incl bibl ref) – mf#1985-2195 – us ATLA [227]

Kritisch exegetisches handbuch ueber die briefe an die thessalonicher *see* Critical and exegetical handbook to the epistles of st paul to the thessalonians

Kritisch exegetisches handbuch ueber die briefe an timotheus und titus *see* Critical and exegetical handbook to the epistles of st paul to timothy and titus

Kritisch exegetisches handbuch ueber die evangelien des markus und lukas / Meyer, Heinrich August Wilhelm – 7. aufl. Goettingen: Vandenhoeck und Ruprecht, 1885 – 2mf – 9 – 0-8370-9721-5 – (incl bibl ref) – mf#1986-3721 – us ATLA [225]

Kritische analyse der apostelgeschichte / Wellhausen, Julius – Berlin: Weidmann, 1914 [mf ed 1989] – (= ser Abhandlungen der koeniglichen gesellschaft der wissenschaft zu goettingen. philologisch-historische klasse. neue folge 15/2) – 1mf – 9 – 0-7905-2753-7 – mf#1987-2753 – us ATLA [226]

Kritische beitraege zu den constantin-schriften des eusebius / Heikel, I A – Leipzig, 1911 – (= ser Tugal 3-36/4) – 2mf – 9 – €5.00 – ne Slangenburg [240]

Kritische beitraege zu den constantin-schriften des eusebius (eusebius werke band 1) / Heikel, Ivar August – Leipzig: J C Hinrichs 1911 [mf ed 1989] – (= ser Griechische christliche schriftsteller der ersten drei jahrhunderte (leipzig, germany)) – 1mf – 9 – 0-7905-1989-5 – (in german & greek) – mf#1987-1989 – us ATLA [240]

Kritische beitraege zur geschichte der dichtersprache klopstock's / Petri, Friedrich Karl Wilhelm – Greifswald: J Abel, 1894 – 1r – 1 – (incl bibl ref) – us Wisconsin U Libr [430]

Kritische beleuchtung des c f weitzmann'schen harmoniesystems, und des schriftchens: "die neue harmonielehre im streit mit der alten" / Kunkel, Franz Joseph – Frankfurt am Main: F B Auffarth 1863 [mf ed 19–] – 1r – 1 – mf#pres. film 120 – us Sibley [780]

Kritische bemerkungen zu meiner ausgabe von origenes' exhortatio, contra celsum, de oratione : entgegnung auf die von paul wendland in den goettingischen gelehrten anzeigen 1899 nr. 4. veroeffentlichte kritik / Koetschau, Paul – Leipzig: J.C. Hinrichs, 1899 – 1mf – 9 – 0-7905-6197-2 – mf#1988-2197 – us ATLA [180]

Kritische einleitung in die geschichte und lehrsaetze der alten und neuen musik / Marpurg, Friedrich Wilhelm – Berlin: G A Lange 1759 [mf ed 19–] – 5mf – 9 – (1: vom ursprung der musik bis auf die suendfluth; 2: von der suendfluth bis auf den seezug der argonauten; 3: von dem seezug der argonauten bis auf den anfang der olympiaden; 4: vom anfang der olympischen spiele bis auf die zeiten des pythagoras; 5: eingeschaltetes capitel von der beschaffenheit der alten musik) – mf#fiche 428 – us Sibley [780]

Kritische gaenge / Vischer, Friedrich Theodor – neue folge. Stuttgart: J G Cotta 1861-73 [mf ed 1991] – 6v in 2 on 1r – 1 – mf#2956p – us Wisconsin U Libr [430]

Kritische geschichte der neugriechischen und der russischen kirche : mit besonderer beruecksichtigung ihrer verfassung in der form einer permanenten synode / Schmitt, Herrmann Joseph – Mainz: Kirchheim, Schott und Thielmann, 1840 – 2mf – 9 – 0-7905-8150-7 – (incl bibl ref) – mf#1988-6097 – us ATLA [243]

Kritische geschichte der thalmud-uebersetzungen aller zeiten und zungen / Bischoff, Erich – Frankfurt: J Kauffmann, 1899 – 1mf – 9 – 0-8370-2349-1 – (incl ind of trans of the talmud and mishnah) – mf#1986-0349 – us ATLA [270]

Kritische geschichte der...ersten baslerkonfession... / Hagenbach, K R – Basel, J G Neukirch, 1827 – 4mf – 9 – mf#PBU-461 – ne IDC [242]

Der kritische musicus : ester [-zweiter] theil / ed by Scheibe, Johann Adolph – Hamburg: Thomas von Wierings Erben, 1738 – 2v on 11mf – 9 – mf#fiche 846 – us Sibley [780]

Kritische rundschau – Bremen DE, mar 1 1928-feb 17 1935 – 1r – 1 – gw Misc Inst [074]

Kritische sammlungen zur neuesten geschichte der gelehrsamkeit / ed by Reinhard, Adolf Friedr von – Buetzow, Wismar 1774-84 – (= ser Dz) – 9v on 46mf – 9 – €460.00 – mf#k/n307 – gw Olms [000]

Kritische theorie und metaphysik / Geyer, Carl-Friedrich – Darmstadt, 1980 (mf ed 1992) – 2mf – 9 – €24.00 – 3-89349-025-6 – mf#DHS-AR 65 – gw Frankfurter [110]

Kritische ueberschau der deutschen gesetzgebung und rechtswissenschaft / ed by Arndts, L et al – Muenchen: Verlag der Literarisch-Artistischen Anstalt. v1-6. 1853-59 (all publ) – (= ser Civil law 3 coll) – 31mf – 9 – (merged in 1859 with heidelberger kritische zeitschrift fuer die gesammte rechtswissenschaft to form the following title. this title is indexed in the first index volume of its successor) – mf#LLMC 96-569 – us LLMC [340]

Kritische uebersicht der neuesten schoenen litteratur der deutschen / Heydenreich, Karl Heinrich – Leipzig 1788-89 [mf ed Hildesheim 1995-98] – (= ser Dz) – 2v in 4pt on 5mf – 9 – €100.00 – gw Olms [430]

Kritische uebersicht der neusten schoenen litteratur der deutschen / ed by Heydenreich, Karl Heinrich – Leipzig 1788-89 – (= ser Dz. abt literatur) – 2v[zu je 2st] on 5mf – 9 – €100.00 – mf#k/n4571 – gw Olms [430]

Kritische untersuchungen ueber die evangelien justin's, der clementinischen homilien und marcion's : ein beitrag zur geschichte der aeltesten evangelien-literatur / Hilgenfeld, Adolf – Halle: C A Schwetschke, 1850 – 1mf – 9 – 0-7905-1990-9 – (incl bibl ref) – mf#1987-1990 – us ATLA [220]

Kritische untersuchungen ueber die kanonischen evangelien : ihr verhaeltniss zu einander, ihren charakter und ursprung / Baur, Ferdinand Christian – Tuebingen: L F Fues, 1847 – 6mf – 9 – 0-7905-7557-4 – mf#1989-0782 – us ATLA [220]

Kritische untersuchungen zu levi ben gersons (ralbag) widerlegung d... / Karo, Jakob – Leipzig, Germany. 1935 – 1r – us UF Libraries [939]

Kritische vierteljahrsschrift fuer gesetzgebung und rechtswissenschaft – Muenchen: J B Cotta. Annual, vols 1-44 only, 1859-1903, plus two index vols. covering 1853-77. 1859-1903 – (= ser Civil law 3 coll) – 314mf – 9 – (series continues to vol 68, 1944. original publisher to 1869. later vols. publ: r. oldenbourg, 1870-94; freiburg, mohr, 1895-1903. the first index volume covers all of the preceding title and vols. 1-9 of this title. the second index volume covers vols 10-19 of this title. editors vary. the 44 vols of this title provided by llmc are alternately numbered as: 1st series, vols 1-19, 1859-77; 2nd series, vols 1-17, 1878-94; and 3rd series, vols 1-8, 1895-1903) – mf#LLMC 96-570 – us LLMC [340]

Der kritische wert der altaramischen ahikartexte aus elephantine / Stummer, Friedrich – Muenster i W: Aschendorff, 1914 [mf ed 1989] – (= ser Alttestamentliche abhandlungen 5/5) – 1mf – 9 – 0-7905-2605-0 – mf#1987-2605 – us ATLA [390]

Kritischer commentar zu den psalmen : nebst text und uebersetzung / Graetz, Heinrich – Breslau: S Schottlaender, 1882-83 – 2mf – 9 – 0-8370-1347-X – mf#1987-6052 – us ATLA [220]

Kritischer versuch ueber den ursprung und die geschichtliche entwicklung des pesach- und mazzothfestes : nach den pentateuchischen quellen / Mueller, J – Bonn: Eduard Weber, 1883 – 1mf – 9 – 0-7905-3205-0 – (incl bibl ref) – mf#1987-3205 – us ATLA [270]

Kritischer versuch ueber die glaubwuerdigkeit der buecher der chronik : mit hinsicht auf die geschichte der mosaischen buecher und gesetzgebung / De Wette, Wilhelm Martin Leberecht – Halle: Schimmelpfennig, 1806 [mf ed 1992] – (= ser Beitraege zur einleitung in das alte testament 1) – 1mf – 9 – 0-524-02774-9 – mf#1987-6468 – us ATLA [221]

Kritisches jahrbuch – Hamburg: Actien-Gesellschaft, 1889-1890 – 1 – us Wisconsin U Libr [943]

Kritisches journal der philosophie / ed by Hegel, Georg Wilhelm Friedrich et al – Tuebingen 1802 – (= ser Dz. abt philosophie) – 2v on 6mf – 9 – €120.00 – mf#k/n589 – gw Olms [100]

Kritisch-literarische uebersicht der reisenden in russland bis 1700 : deren berichte bekannt sind / Adelung, F von – Spb, Leipzig, 1846. 2v – 17mf – 9 – mf#U-762 – ne IDC [914]

Kritisch-philosophische untersuchungen. 1. heft, kant's und herbart's metaphysische grundansichten ueber das wesen der seele / Quaebicker, Richard – Berlin: L Heimann, 1870 – 1mf – 9 – 0-7905-9075-1 – mf#1989-2300 – us ATLA [110]

Kritisch-polemische untersuchungen ueber den roemerbrief / Richter, Georg – Guetersloh: C Bertelsmann, 1908 – (= ser Beitraege zur foerderung christlicher theologie) – 1mf – 9 – 0-524-06798-8 – mf#1992-0961 – us ATLA [227]

Kritz, Wilhelm *see* Die evangelische lehre

Kriukov, F D *see* Donskie vedomosti

Krivitsky, Walter G *see* Rusia en espana

Krivopolena, Maria Dmitrievna *see* Babushkiny stariny

Krivosheeva, A *see* Sovetskii fol'klor

Krivosheina, P I *see* Agro-kooperativnyi kruzhok

Kriwalsky, Marcus Stephan *see* Vergleichende volumetrische untersuchungen der kavitaetengestaltung in der primaertherapie mit amalgam-, composite- und goldgussfuellungen

Krizis / Dineson, Jacob – Varsha, Poland. 1905 – 1r – us UF Libraries [939]

Krizis fun der idisher kolonizatsye in argentina / Chasanowitch, Leon – Stanisloy, Ukraine. 1910 – 1r – us UF Libraries [939]

Krochmal, Abraham *see* Agadat ma'amarim

Krockow von Wickerode, Carl *see* Reisen und jagden in nord-ost-afrika 1864-1865

Kroeber, A L *see* Anthropology

Kroeger, Arved *see* Ordensdeutsche

Kroeger, Timm *see*
- Des reiches kommen
- Der einzige und seine liebe
- Hein wieck
- Neun novellen
- Novellen gesamtausgabe
- Der schulmeister von handewitt
- Um den wegzoll

Kroeker, Kate Freiligrath [comp] *see* A century of german lyrics

Kroening, G *see* Das dasein gottes

Die kroenungsopfer : ein mozart-roman / Watzlik, Hans – Karlsbad: A Kraft, 1944 – 1r – 1 – us Wisconsin U Libr [830]

Krogmann, Willy *see*
- Das redentiner osterspiel
- Ulenspegel

Krohn, Barthold Nicolaus *see* Geschichte der fanatischen und enthusiastischen wiedertaeufer vornehmlich in niederdeutschland

Krohn, Julius *see* Suomalaisen kirjallisuuden historia

Kroker, Ernst *see* Katharina von bora, martin luthers frau

Krokodil – 1928-60 – 1 – us L of C Photodup [460]

Krokodil – Moscow, Russia. n1-36. 1987-1988 and 1990-1991 nov – 2r – (missing: 1987 n13-19,21-28 1988 n1,6,10,36 and 1990 n7 1991 n29) – us UF Libraries [077]

Krolewitz, Heinrich von *see* Heinrich's von krolewiz uz missen vater unser

Kroll, Adam *see* Six quartettos for a german flute, violin, viola and violoncello

Kroll, Erwin *see* Ernst theodor amadeus hoffmann

Kroll, J *see* Die lehren des hermes trismegistos

Krolow, Karl *see* Miteinander

Krom mlap qangar : pralom lok phnaek knun pravatti sastr / Hak Chai Huk – Bhnam Ben: Pannagar Trairatan 2509 [1966] [mf ed 1990] – 1r with other items – 1 – mf#mf-10289 seam reel 097/03 [§] – us CRL [959]

Krom, Nicolaas Johannes *see* A short guide to the boro-budur

Krommer, Franz *see* Nouveaux quintetti pour deux violons, deux altos et violoncelle, oeuvre 88 no 1[-2]

Kromrei, Ernst *see* Glaubenslehre und gebraeuche der aelteren abessinischen kirche

Kromsigt, Pieter Johannes *see* De zegen der zending voor de zendende kerk

Kronauer, Fr *see* Die wassernot im emmenthal / fuenf maedchen / dursli der branntweinlaeuter

Kronberg, Nehemias *see* Raschi als exeget

Krone, Rudolf *see* Pfalzgraf wolfgang

Kroneisen, Antonia *see* Antiischaemische und haemodynamische effekte nach akuter und einwoechiger gabe von ramipril im vergleich mit plazebo und isosorbiddinitrat bei patienten mit koronarer herzkrankheit

Kronen zeitung [steirer krone. ausgabe fuer steiermark] : ausgabe fuer steiermark – Vienna, Graz, Austria oct 1972-feb 2 2000 [mf ed Norman Ross] – 252r – 1 – (non-partisan austrian daily) – mf#nrp-2035 – us UMI ProQuest [074]

Kronenberg, Moritz *see* Die all-einheit

Kronik : indonesia kementerian penerangan – Djakarta, 1951-1963 – 122mf – 9 – (missing: several iss) – mf#SE-576 – ne IDC [959]

Kronik der menschheit – Strassburg (Strasbourg F), 1798 24 mar-7 jun [gaps] – 1 – (filmed with: intelligenz nachrichten) – fr ACRPP [900]

Kronik dokumentasi – Djakarta: Kementerian Penerangan, Bag. Dokumentasi, 19(52)-(69).39v. Some vols. also have Dutch or English.Some vols. issued by other subdivisions of the Departemen Penerangan – 1 – us Wisconsin U Libr [959]

Kronik dokumentasi / Kementerian Penerangan, Biro Dokumentasi, Sedjarah & Research – Djakarta, 1966(17/20, 25/26, 30/31-35/36); 1967(1/2-11); 1968(12-20); 1969(1-10); 1970(11-16) – 90mf – 9 – mf#SE-1757 – ne IDC [959]

Kronik pers / Lembaga Pers dan Pendapat Umum – Jogjakarta, 1957-1958 – 3mf – 9 – (missing: 1958) – mf#SE-556 – ne IDC [950]

Kronik tindakan2 ekonomi : peraturan2 pemerintah republik indonesia dilapangan ekonomi / Indonesia. Kementerian perekonomian – Djakarta, 1950-1956 – 13mf – 9 – (missing: 1950/1951(2-4); 1952(6-9)) – mf#SE-292 – ne IDC [330]

Kronika – London, UK. 17 Nov 1962-28 Aug 1971 – 1 – uk British Libr Newspaper [072]

Kronika tygodniowa : organ of the polish democratic association in canada – Toronto. v1-51. feb 22 1941-jan 27 1990// (wkly) – 51r – 1 – Can$2675.00 – (cont: glos pracy, 1932-40; in polish) – cn McLaren [320]

Kronlins, Janis *see*
- Kongresa grahmata
- Muhsu kori un dseesmas: latwijas baptistu draudschu koru wehstures materialu krahjums
- Us augschu!

Kronobergaren – Vaxjo, 1979-93 – 1 – (changes title to: 3 dagar 1993) – sw Kungliga [078]

Kronos : ein archiv der zeit / ed by Rambach, Frierich – Berlin 1801 – (= ser Dz. abt literatur) – v1.2 on 7mf – 9 – €140.00 – mf#k/n4637 – gw Olms [700]

Die kronos-kinder und das reich des zeus / Hartung, Johann Adam – Leipzig: W Engelmann, 1866 – 1mf – 9 – 0-524-04513-5 – (incl bibl ref) – mf#1990-3347 – us ATLA [250]

Kronshtad. sovet rk i kd *see* Izvestiia kronshtadskogo soveta rabochikh, matrosskikh i krasnoarmejskikh deputatov

Kronshtadskii vestnik – Kronshtadt, Russia 1871-95 [mf ed Norman Ross] – 1 – mf#nrp-804 – us UMI ProQuest [077]

Kronstaedter zeitung – Kronstadt (Brasov RO), 1916 10 oct-1917 [gaps], 1921-1941 11 jan – 29r – 1 – (1854-57, 1859, 1861 mar-1864, 1900-1914 17 apr, 1923 jan-jun, 1924-1927 oct, 1928 may-dec) – gw Misc Inst [077]

Kroone der vier hooft-deughden... / Veen, E van – Brussel: Guilliam Scheybels, 1644 – 1mf – 9 – mf#O-3190 – ne IDC [090]

Kroot, Antonius *see* History of the telugu christians

Kropat, Arno *see* Die syntax des autors der chronik verglichen mit der seiner quellen

Kropatscheck, Friedrich *see*
- Der himmel des christen
- Das schriftprinzip der lutherischen kirche
- Die trinitaet

Kropatschek, Friedrich *see* Die furcht vor dem denken – occam und luther

Kropf, Albert *see* Karif-english dictionary

Kropotkin, P *see*
- K molodomu pokoleniiu
- Revoliutsiia v rossii

Kropotkin, P A *see*
- Organ kommunistov-anarkhistov
- Pisma o tekushchikh sobytiiakh
- Vek ozhidaniia

Kropotkin, Petr Alekseevich *see* Modern science and anarchism

Kropotkin, Petr Alekseevich, kniaz' *see* Vzaimnaia pomoshch' sredi zhivotnykh i liudei, kak dvigatel' progressa

Kropper kirchlicher anzeiger – Kropp DE, 1882-85, 1887-90, 1892-1907, 1910-12, 1916-17 – 1 – gw Misc Inst [240]

Kroschewski, Klaus *see* Internationale tendenzen in der tiergesundheitsueberwachung und daraus abgeleitete schlussfolgerungen fuer die anpassung des nationalen tierseuchenberichtssystems

Kross kultures : the magazine about people of color – 1996 nov, 1997 mar-apr/may – 1r – 1 – mf#3862708 – us WHS [305]

Krotoschiner, Hermann Alfred *see* Der schutz des kreditverkehrs vor den gefahren aus zu umfangreichen sicherungsuebereignungen

Krotoschiner zeitung – Krotoschin (Krotoszyn PL), 1922 18 feb-31 aug, 1926 19 jun-1932 28 sep – 4r – 1 – gw Misc Inst [077]

Kroy bel bhlian : kamran ryan bit / Ras Viriya – Paris: Editions Angkor 1979 [mf ed 1990] – 1r with other items – 1 – mf#mf-10289 seam reel 119/7 [§] – us CRL [959]

Krshna : a study in the theory of avataras / Das, Bhagavan – Adyar: Theosophical Pub House, 1929 – (= ser Samp: indian books) – us CRL [280]

Krshnaji Ananta Sabhasada *see* Siva chhatrapati

Krsnadasa Kaviraja Gosvami *see*
- Chaitanya's pilgrimages and teachings
- Sri sri chaitanya-charitamrita of sri sri krishnadasa kaviraja gosvamin

Kruchen, Gottfried *see* Die bibel bernhard overbergs

Kruchenykh, A *see* Apokalipsis v russkoi literature

Kruegel, Rudolf *see* Der begriff des volksgeistes in ernst moritz arndts geschichtsanschauung

Krueger, Anton Robert *see* Science as narrative

Krueger, Auguste *see* Grosses illustriertes frauen-lexikon

Krueger, Bartholomaeus *see* Hans clawerts werckliche historien

Krueger, Deborah L *see* Obstacles adapted physical education specialists encounter when developing transition plans

Krueger, Dieter *see* Dienststellen zur vorbereitung des westdeutschen verteidigungsbeitrages 1950-1955 (bestand bw 9) bd 40

Krueger, Ferdinand *see* Witte liljen

Krueger, Gustav *see*
- Dogma and history
- Das dogma vom neuen testament
- Das dogma von der dreieinigkeit und gottmenschheit in seiner geschichtlichen entwicklung
- Die entstehung des neuen testaments
- History of early christian literature in the first three centuries
- Kritik und ueberlieferung auf dem gebiete der erforschung des urchristentums
- Lucifer, bischof von calaris, und das schisma der luciferianer
- Monophysitische streitigkeiten im zusammenhange mit der reichspolitik
- The papacy
- Das papsttum
- Die religion der goethezeit
- Was heisst und zu welchem ende studiert man dogmengeschichte?

Krueger, Hermann Anders *see*
- Der junge eichendorff

Krueger, Paul *see*
- Hellenismus und judentum im neutestamentlichen zeitalter
- Philo und josephus als apologeten des judentums
- Das syrisch-monophysitische moenchtum im tur-ab(h)din von seinen anfaengen bis zur mitte des 12. jahrhunderts

Krueger, Theodor *see* Richard dehmel als religioes-sittlicher charakter

Krueger, Wilhelm *see*
- Die auferstehung jesu in ihrer bedeutung fuer den christlichen glauben
- Phantasie oder geist?

Krueger-Westend, Herman *see*
- Goethe und der orient
- Goethe und seine eltern

Kruemmer, G *see* Die bergarbeiter-verhaeltnisse in grossbritannien

Kruenitz, Johann Georg *see*
- Oekonomisch-technische enzyklopaedie

Krug, Camille *see* Le certificat d'heritier dans les departements du bas-rhin, du haut-rhin et de la moselle d'apres la loi d'introduction du droit francais du 1er juin 1924

Krug, Hans-Joachim *see* Gotterthrone im urwald

Krug, Marina *see* Die figur als signifikante spur

Krug und tintenfass : gedichte / Baumbach, Rudolf – Leipzig: A G Liebeskind, 1898 [mf ed 1989] – 120p – 1 – mf#6991 – us Wisconsin U Libr [810]

Krug von Nidda, Friedrich *see* Lokal-umrisse kleiner reisen

Krug, Wilhelm Traugott *see*
- Auch eine denkschrift ueber den gegenwaertigen zustand von deutschland
- Schelling und hegel

Kruger, Alet *see* Lexical cohesion and register variation in translation

Kruger, Daniel Wilhelmus *see*
- Making of a nation
- South african parties and policies, 1910-1960

Kruger, J D L *see* Bantustan

Kruger, Jacolien *see* Grondeise van 'n verantwoordbare opvoedkundige sielkundige praktyk

Krüger, Janine *see* Investigating electronic commerce activities of the tourism industry in south africa

Kruger, Martha Elizabeth *see* Die verband tussen depressie en lokus van kontrole jeens skoolwerk by adolessente

Kruger, Matthew J *see* Effects of thick-bar resistance training on strength measures in experienced weightlifters

Kruger, Paul *see* Memoirs of paul kruger

Kruger, Rayne *see* Good-bye dolly gray

Kruger, Stephanus Johannes Paulus *see* The memoirs of paul kruger

Krugger, Tammy Marie *see* An exploratory investigation into the effects of tai chi exercise on balance and gait performance for hip replacement patients

Krugliashova, Vera Petrovna *see* Predaniia reki chusovoi

Kruglov, A V *see* Ezhemesiachnoe obshchedostupnoe izdanie

Kruijer, Gerardus Johannes *see* Suriname en zijn buurlanden

Kruijf, Ernst Frederik *see* Geschiedenis van het nederlandsche zendelinggenootschap en zijne zendingsposten

Het kruis geplant in een onbekend negerland van midden-afrika.. : verhaal van de stichting der sint-antonius-missie in het kroninkrijk oeroendi / Burgt, Joannes Michael M van der – Boxtel-Burgakker: Procure der Witte Paters, 1921 – us CRL [960]

Kruitwagen, E *see* Een missale leodiense gedrukt door joh de westfalia te leuven (overdruck)

Krukenkamp, Christoph *see* Entwicklung einer dna-vakzine gegen das glykoprotein b von varizelle-zoster-virus

Krul, J H *see*
- Eerlycke tytkorting
- Minne-spiegel ende weg.wyser.ter deugden
- Minne-spiegel ter deughden
- Pampiere wereld ofte wereldsche oeffeninge

Krull, Germaine *see* Cidade antiga do brasil, ouro preto

Krumbacher, K *see*
- Geschichte der byzantinischen litteratur
- Der heilige georg in der griechischen ueberlieferung

Krumbacher, Karl *see* Geschichte der byzantinischen litteratur

Krumbholtz, Robert *see* Zwei schriften des muensterschen wiedertaeufers bernhard rothmann

Krummacher, Frederic Adolphus *see* Cornelius the centurion; and, life and character of st. john the evangelist and apostle

Krummacher, Friedrich Wilhelm *see*
- David, the king of israel
- The risen redeemer

Krumme wege zur unfehlbarkeit / Buchmann, Jacob – Breslau [Wroclaw]: Fiedler & Hentschel, 1874 – 1mf – 9 – 0-8370-9128-4 – (incl bibl ref) – mf#1986-3128 – us ATLA [241]

Krummel, Leopold *see*
- Geschichte der boehmischen reformation im fuenfzehnten jahrhundert
- Die religion der alten aegypter
- Utraquisten und taboriten

Krump, Jason G *see* Identification of athletes by athletes

Krunk hayots ashkharhi – Tiflis, 1860-1863 – 24mf – 9 – (missing: 1860 (p161-669); 1860/61 (nov-feb); 1862; 1863 (p18-240)) – mf#AR-356 – ne IDC [077]

Krupnaia burzhuaziia v poreformennoi rossii, 1861-1900 / Laverychev, Vla – M, 1974 – 5mf – 9 – mf#REF-164 – ne IDC [332]

Krupnik, Baruch *see* Rimon

Krupp catalog / Krupp Comic Works – [n1]-n10 [1970 sep-1974 sep] – 1r – 1 – (cont by: krupp dealers' catalog) – mf#659129 – us WHS [740]

Krupp Comic Works *see*
- Krupp catalog
- Krupp dealers' catalog

Krupp dealers' catalog / Krupp Comic Works – n11-41 [1974 sep-1981 fall] – 1r – 1 – (cont: krupp catalog; cont by: catalog [krupp distributing co]) – mf#387130 – us WHS [740]

Krupp mail order catalog – [n1]-n4, n6 [1973 dec-[1974 dec]] – 1r – 1 – mf#660115 – us WHS [010]

Krusch, B *see*
- Fredegarii et aliorum chronica
- Gregorii turonensis opera
- Passiones vitaeque sanctorum aevi merovingici et antiquorum aliquot
- Passiones vitaeque sanctorum aevi merovingici
- Studien zur christlich-mittelalterlichen chronologie

Kruschwitz, William Albin *see* A missions information conference manual for the southern baptist foreign mission board, richmond, virginia

Kruse, Georg Richard *see* Richard wagners tondramen

Kruse, Hans-Joachim *see*
- Die deutsche kriminalerzaehlung von schiller bis zur gegenwart
- Die ursache
- Der verbrecher aus verlorener ehre
- Wer ist schuld?

Kruse, Heinrich *see*
- Brutus
- Das maedchen von byzanz
- Marino faliero

Kruse, Paul *see* Die paedagogischen elemente in der philosophie des jakob balmes

Kruse School of the State of Delawrae (Marshallton DE) *see* Report of the kruse school of the state of delaware

Krushenie (armiia) / Lappo-Danilevskaia, Nadezhda Aleksandrovna (Liutkevich) – (Berlin): Glagol, 1922. 368p. In Cyrillic characters. 1 reel.1246 – 1 – us Wisconsin U Libr [460]

Krusinski, J T *see*
- Prodromus ad historiam revolutionis persicae seu legationis fulgidae portae ad ad persarum regem szach sofi hussein anno d 1720 expeditae relatio...
- Prodromus ad tragicam vertentis belli persici... auctore durri effendi
- Tragica vertentis belli persici historia per repetitas clades, ab anno 1711 ad annum 1728...

Kruske *see* Johannes a lasco und der sacramentsstreit

Kruszka, Waclaw *see* Siedm siedmiolici, czyli pot wieku zycia:...historji polskiej w ameryce

Krut un roeben : rimels / Blum, Max – Berlin: Liebel, [1895?] [mf ed 1989] – 113p – 1 – mf#7035 – us Wisconsin U Libr [430]

Krutovskii, V M *see* Den' knigi [krasnoiarsk: 1919]

De kruys-leer ter zaligheydt : opgerecht door nil virtute prius, [coat of arms with the monogram of e meyster] / [Meyster, E] – t'Amsterdam: Kornelis de Bruyn, 1658 – 2mf – 9 – mf#O-693 – ne IDC [090]

Kruzenstern, Ivan *see* Reise um die welt in den jahren 1803, 1804, 1805 und 1806

Kruzenstern, Ivan F *see* Woerter-sammlungen aus den sprachen einiger voelker des oestlichen asiens und der nordwest-kueste von amerika

Kruzheskiya besedy = Friendly conversations / Pashkov, Vasilii A – St. Petersburg, 1880. 18p. Filmed with: Pamyati A.V. Pashkov, Recollections of A.V. Pashkov, St. Petersburg, 1904. 60p – 1 – $5.00 – us Southern Baptist [242]

Krylov, A *see* Telegrammy telegrafnogo agentstva

Krylov, I *see* Kratkii istoricheskii ocherk vozniknoveniia i razvitiia deiatel'nosti kostromskogo gorodskogo obshchestva vzaimnogo ot ognia strakhovaniia

Krylov, Ivan Andreevich *see*
- Krilof and his fables
- Krilov's mesholim

Krylovskii, A S *see* Sistematicheskii katalog knig biblioteki kievskoi dukhovnoi akademii

Krymov, Vladimir *see* Bog i den'gi

Krymskaia astrofizicheskaia observatoriia *see* Izvestiia krymskoi astrofizicheskoi observatorii

Krymskii kur'er – Yalta, Ukraine 1900 [mf ed Norman Ross] – 1 – mf#nrp-2186 – us UMI ProQuest [077]

Krymskoe khanstvo pod verkhovenstvom otomanskoi porty do nachala 18 veka / Smirnov, D – Spb, 1887 – 15mf – 8 – mf#R-4047 – ne IDC [956]

Krypto-monotheismus in den religionen der alten chinesen und anderer voelker / Junker von Langegg, Ferdinand Adalbert – Leipzig: W Engelmann, 1892 – 1mf – 9 – 0-524-01560-0 – mf#1990-2514 – us ATLA [210]

Krys, Ute *see* Bedeutung von mangelernaehrung im alter unter besonderer beruecksichtigung der therapie mittels perkutaner endoskopischer gastrostomie

Krystalisace tavenych hornin = Crystallization of melted rocks / Kopecky, Lubomir & Voldan, jan – Praha, Nakladatelstvi Ceskoslovenske akademie ved, 1959 – us CRL [550]

Kryzhanovskii, G I *see* Rukopisnye evangeliia v kievskikh knigokhranilishchakh

Krzewinski-Malone, Jeanette A *see* Do american adults know how to exercise for a health benefit?

Krzyk = Outcry – n1-30 [1974 mar-1977 dec] – 1r – 1 – mf#203119 – us WHS [071]

Krzyzanowski, Julian *see* Proza polska wczesnego renesansu

Krzyzanowski, Ludwik *see* Joseph conrad

Ksatriya clans in buddhist india / Law, Bimala Churn – Calcutta: Thacker Spink & Co, 1922 – (= ser Samp: indian books) – (foreword by asutosh mookerjee) – us CRL [280]

Ksemendra *see* The bharatamanjari of kshemendra

Ksenofontov, F *see* Gosudarstvo i pravo (opyt izlozheniia marksistskogo ucheniia o sushchestve gosudarstva i prava). s predisloviem n. v. krylenko

Ksmendra *see* Avadana kalpalata

K-t miller / International Association of Machinists – 1944 oct 2-1945 feb 5 – 1r – 1 – mf#1113331 – us WHS [240]

Kt today / Kearney and Trecker Corporation – n4-21 [1976 apr 9-1979 jun 8] – 1r – 1 – (cont: k & t today) – mf#667596 – us WHS [071]

Ktam srae : pralom lok phnaek jivit sok pradit loen tam manosancetana / Sen Suvatthi et al – Bhnam Ben: Ron Bumb Niyam 2516 [1972] [mf ed 1990] – 1r with other items – 1 – (in khmer) – mf#mf-10289 seam reel 103/10 [§] – us CRL [480]

Kto takie mensheviki : doklad, prochit na sobranii prikreplen. i biuro iacheek – 18 okt 1922 g / Dimanshtein, S – Kharkov, 1922 – 47p 1mf – 9 – mf#RPP-139 – ne IDC [325]

Ku / Lu, Fen – Shang-hai: Wen hua sheng huo ch'u pan she, Min kuo 25 [1936] – (= ser P-k&k period) – us CRL [480]

Ku / Shih, T'o – Shang-hai: Wen hua sheng huo ch'u pan she, Min kuo 26 [1937] – (= ser P-k&k period) – us CRL [480]

K'u ch'a an hsiao hua hsuan / Chou Tso-jen – Shang-hai: Pei hsin shu chu, 1933 – (= ser P-k&k period) – us CRL [870]

K'u ch'a sui pi / Chou, Tso-jen – Shang-hai: Pei hsin shu chu, Min kuo 25 [1936] – (= ser P-k&k period) – us CRL [840]

Ku chi ch'ui tung wu hsueh pao = Vertebrata palasiatica – Beijing, China 1959-64 – 1 – mf#2627 – us UMI ProQuest [611]

Ku chi ch'ui tung wu yu ku jen lei = Paleovertebrate et paleoanthropologie – Ann Arbor MI 1959-60 – 1 – mf#2628 – us UMI ProQuest [560]

Ku ch'ien / Wei, Chu-hsien & Ting, Fu-pao – [Kui-lin?]: Chung yang yin hang ching chi yen chiu ch'u, Min kuo 31 [1942] – (= ser P-k&k period) – us CRL [730]

Ku chin ming chu hsuan / Wu, Mei – Pei-p'ing: Kuo li Pei-ching ta hsueh ch'u pan tsu, 1934 – (= ser P-k&k period) – us CRL [820]

K'u chu tsa chi / Chou, Tso-jen – Shang-hai: Shang-hai liang yu t'u shu, 1936 – (= ser P-k&k period) – us CRL [840]

Ku, Chun-cheng *see* Tsai pei chi ti hsia: k'o hsueh hsiao shuo

Ku, Ch'un-fan *see*
- Lu mei kuan kan
- Shih chieh ching chi shuai ch'en ti chieh p'ou

Ku, Chung-i *see*
- Hsin fu
- San ch'ien chin
- Yeh huo hua

Ku fang chi / Cheng, I-mei – Shang-hai: I hsin shu she, 1932 – (= ser P-k&k period) – us CRL [840]

Ku, Feng-ch'eng et al *see* Hsin wen i tz'u tien

Ku hai yu sheng (ccm292) = Out of the depths / Teng, Shu-k'un – 1st ed. Shanghai, 1931 [mf ed 198?] – (= ser Ccm 292) – 1 – mf#1984-b500 – us ATLA [240]

Ku hsiang / Shu, Hsin-ch'eng – Shang-hai, Chung-hua shu chu, Min kuo 23 [1934] – (= ser P-k&k period) – us CRL [840]

Ku hsiang tsa chi / Mao, Tun – Shang-hai: Chin tai shu tien, 1936 – (= ser P-k&k period) – us CRL [840]

Ku huai meng yu / Yu, P'ing-po – Shang-hai: Shih chieh shu chu, 1936 – (= ser P-k&k period) – us CRL [840]

Ku, Hung Ming *see* The conduct of life

Ku hung ying t'an tz'u : t'an tz'u hsiao shuo: [36 hui] / Li, Tung-yeh – Shang-hai: Hsin min yin shu kuan, 1935 – (= ser P-k&k period) – us CRL [951]

Ku, Hung-ming *see* The story of a chinese oxford movement

Ku, I-ch'iao *see* Yueh fei

Ku Klux Klan *see*
- Klansman
- National kourier
- White patriot
- Wisconsin kourier

The ku klux klan : official, unofficial and anti-klan sources – New York: Andronicus Publ Co, [1977?] – us CRL [366]

Ku Klux Klan [1915-] *see*
- Crusader
- Hawkeye independent

Ku klux klan in prophecy / White, Alma, Bishop – Zarephath, NJ: The Good Citizen, 1925 (mf ed 1973) – 1r – 1 – mf#MF K95 – us Colorado Hist [366]

Ku Klux Klan of Canada *see*
- Kloran
- Manual of the order of citizenship, invisible empire knights and ladies of the ku klux klan of canada

K'u k'ou kan k'ou / Chou, Tso-jen – Shang-hai: T'ai p'ing shu chu, 1944 – (= ser P-k&k period) – us CRL [840]

Ku, Mei *see* Hsien tai chung-kuo chi ch'i chiao yu, i ming, chung-kuo hsin chiao yu pei ching

Ku, Ming-tao *see* T'i chuan lu

Ku msika wa vyawaka / Phiri, Desmond Dudwa – Cape Town, South Africa. 1959 – 1r – us UF Libraries [960]

Ku pei li te tien kuo (ccm152) = Cup of blessing / Ho, Shih-ming – 1st ed. Hong Kong, 1957 [mf ed 198?] – (= ser Ccm 152) – 1 – mf#1984-b500 – us ATLA [240]

Ku Ping-Yuan *see* Chung-kuo lao-tung fa ling hui-pien

Ku, Ping-yuan *see* Hsiu cheng lao tzu cheng i ch'u li fa

Ku sheng wu hsueh pao = Journal of paleontology – Killen TX 1962-64 – 1 – mf#2629 – us UMI ProQuest [560]

Ku sheng yun t'ao lun chi – Pei-p'ing: Hao wang shu tien, Min kuo 22 [1933] – (= ser P-k&k period) – us CRL [480]

Ku shu hsu tzu chi shih : shih chuan / P'ei, Hsueh-hai – Shang-hai: Shang wu yin shu kuan, Min kuo 23 [1934] – (= ser P-k&k period) – us CRL [480]

Ku shu ti hua to, i ming, fan chu-hsien / Tsang, K'o-chia – [China]: Tung fang shu she, Min kuo 37 [1948] – (= ser P-k&k period) – us CRL [810]

Ku sina rum / So Maung, Maung – Ran Kun: Sagara ca pe tuik 1979 [mf ed 1990] – 1r with other items – 1 – (in burmese) – mf#mf-10289 seam reel 138/4 [§] – us CRL [954]

K'u su : kuo nan hsieh chen p'i p'ing yu pao kao / Hsu, Hsiao-t'ien – Shang-hai: Hung yeh shu tien, Min kuo 20 [1931] – (= ser P-k&k period) – us CRL [951]

Ku tai ying hsiung ti shih hsiang : t'ung hua / Yeh, Shao-chun – Shang-hai: K'ai ming shu tien, Min kuo 25 [1936] – (= ser P-k&k period) – us CRL [480]

Ku tao san ch'ung tsou : pu je nao ti hsi chu / Wu, T'ien – Shang-hai: Kuo min shu tien, 1941 – (= ser P-k&k period) – us CRL [820]

[Ku, Wei-chun] *see* Ts'an yu kuo chi lien ho hui tiao ch'a wei yuan hui chung-kuo tai piao ch'u shuo t'ieh

Ku wen chin i chung-kuo ku shih (ccm308) = Chinese classical stories / Wang, Chih-hsin – 1st rev ed. Hong Kong. 2v. 1956 [mf ed 198?] – (= ser Ccm 308) – 1 – mf#1984-b500 – us ATLA [830]

Ku wen tz'u hsueh shih / Wang, Nien-chung – Wu-ch'ang: I shan shu chu, Min kuo 22 [1933] – (= ser P-k&k period) – us CRL [480]

K'u yu chai hsu pa wen / Chou, Tso-jen – Shang-hai: T'ien ma shu tien, Min kuo 23 [1934] – (= ser P-k&k period) – us CRL [840]

Ku, Yu-hsiu et al *see* Shih yeh chi hua tsung ho yen chiu ko lun

Ku yu-tai ko ming shih yen i (ccm103) = Story of the jewish revolt / Chien, Yu-wen – 1st ed. Hong Kong, 1957 [mf ed 198?] – (= ser Ccm 103) – 1 – mf#1984-b500 – us ATLA [270]

Kua fu yuan : ssu mu pei chu / Hsia, Hsia – Shang-hai: Wan hsiang shu wu, Min kuo 31 [1942] – (= ser P-k&k period) – us CRL [820]

Kua tou chi / Chou, Tso-jen – Shang-hai: Yu chou feng she, Min kuo 26 [1937] – (= ser P-k&k period) – us CRL [840]

K'uai le sheng tan ku shih (ccm142) = Happy christmas stories – Hong Kong, 1953 [mf ed 198?] – (= ser Ccm 142) – 1 – mf#1984-b500 – us ATLA [390]

K'uai le te ku shih (ccm143) = Happy stories – Hong Kong. 2v. 1953 [mf ed 198?] – (= ser Ccm 143) – 1 – mf#1984-b500 – us ATLA [801]
K'uai pao = Chinese express – Toronto. an 13 1971-jan 26 1989// (daily) – (= ser Chinese express) – 81r – 1 – Can$9300.00 – (only toronto chinese-language newspaper available complete from first issue to last) – cn McLaren [071]
Kuan, Chi-yu *see* T'ien fu, tu ti ch'en pao, t'u ti shui
Kuan hsia tsai hua wai kuo jen shih shih t'iao li an / China – Nan-ching: [Wai chiao pu], Min kuo 20 [1931] – (= ser P-k&k period) – us CRL [340]
Kuan, Meng-chueh et al *see* Ti-chung-hai wei chi lun
Kuan min lien hsi / Fei-fu-na – [China: sn], Min kuo 30 [1941] – (= ser P-k&k period) – us CRL [350]
Kuan mu chi / Li, Kuang-t'ien – Hsiang-kang: Chien wen shu chu, 1959 – (= ser P-k&k period) – us CRL [480]
Kuan, P'ing *see* Han ch'ieh
Kuan, Tao-chung *see* Erh ch'eng yen chiu
Kuan, wu shu *see* Hai kuan fa kuei hui pien
Kuan yu chu chih-hsin yeh-su shi shen mo tung hsi te tsa p'ing (ccm11) = A critical review of chu chih sin's "what thing is jesus" / Chang, I-ching – Shanghai, 1930 [mf ed 198?] – (= ser Ccm 11) – 1 – mf#1984-b500 – us ATLA [240]
Kuan yu chung-kuo chi-tu-chiao san-tzu ai kuo yun tung te pao kao (ccm334) / Wu, Yao-tsung – Shanghai, 1956? [mf ed 198?] – (= ser Ccm 334) – 1 – mf#1984-b500 – us ATLA [230]
Kuan yu je-ho chih meng yen / Liang, Ching-min – [China]: Meng Tsang wei yuan hui, Min kuo 24 [1935] – (= ser P-k&k period) – us CRL [338]
Kuan yu kung yeh hui chih chung yao wen hsien – [China]: Chung-kuo kung yeh ching chi yen chiu so, Min kuo 33 [1944] – (= ser P-k&k period) – us CRL [331]
Kuan yu lu hsun / Mei, Tzu – Ch'ung-ch'ing: Sheng li ch'u pan she, 1942 – (= ser P-k&k period) – us CRL [920]
Kuan yu nu jen / Ping-hsin – Shang-hai: K'ai ming shu chu, Min kuo 38 [1949] – (= ser P-k&k period) – us CRL [840]
Kuang hua shu chu Pien chi pu *see* Wen i ch'uang tso chiang tso
K'uang huan chih yeh : wu mu feng tz'u hsi chu / Wu, Ch'u-yuan – Yung-an: Ko lin ch'u pan she, Min kuo 31 [1942] – (= ser P-k&k period) – us CRL [820]
Kuang ming ti chuang pei (ccm276) = The armour of light – 1st ed. Hong Kong, 1956 [mf ed 198?] – (= ser Ccm 276) – 1 – (chinese trans fr the english) – mf#1984-b500 – us ATLA [220]
K'uang yeh / Ai, Ch'ing – Ch'eng-tu: Sheng huo shu tien, Min kuo 31 [1942] – (= ser P-k&k period) – us CRL [810]
Kuang yuan lun : san mu chu / Cheng, I-hung – Ch'ung-ch'ing: Tu shu ch'u pan she, 1945 – (= ser P-k&k period) – us CRL [820]
Kuang-chou chih kung yeh – Kuang-chou: Kuang-chou shih li yin hang ching chi tiao ch'a shih, Min kuo 26 [1937] – (= ser P-k&k period) – us CRL [338]
Kuang-chou ch'u ti i che t'ang ying tsao ch'ang kai k'uang – [Kuang-chou: Kuang-chou ch'u ti i che t'ang ying tsao ch'ang], Min kuo 24 [1935] – (= ser P-k&k period) – us CRL [660]
Kuang-chou kung jen chia t'ing chih yen chiu / Yu, Ch'i-chung – [China]: Kuo li Chung-shan ta hsueh fa hsueh yuan ching chi tiao ch'a ch'u, Min kuo 23 [1934] – (= ser P-k&k period) – us CRL [331]
Kuang-chou san yueh erh shih chiu jih ko ming shih / Tsou, Lu – [Ch'ung-ch'ing]: Chung-kuo kuo min tang chung yang chih hsing wei yuan hui hsuan ch'uan pu, Min kuo 33 [1944] – (= ser P-k&k period) – us CRL [951]
Kuang-chou shih cheng fu san nien lai shih cheng pao kao shu / Liu, Chi-wen – Kuang-chou: Kuang-chou shih cheng fu, Min kuo 24 [1935] – (= ser P-k&k period) – us CRL [951]
Kuang-chou shih ts'ai cheng t'ung chi : erh shih liu nien shih i shih erh yueh fen ho k'an – [Kuang-chou]: Kuang-chou shih ts'ai cheng chu, [1937] – (= ser P-k&k period) – us CRL [332]
Kuang-chou wu-han ko min wai chiao wen hsien / Kao, Ch'eng-yuan – [China]: Shen chou kuo kuang she, Min kuo 22 [1933] – (= ser P-k&k period) – us CRL [327]
Kuang-hsi chiao t'ung wen t'i / Ch'en, Hui – Ch'ang-sha: Shang wu yin shu kuan, Min kuo 27 [1938] – (= ser P-k&k period) – us CRL [380]
Kuang-hsi chiao yu kai chin fang an ch'uan kao – [China]: Kuang-hsi chiao yu t'ing chiao yu she chi wei yuan hui, Min kuo 22 [1933] – (= ser P-k&k period) – us CRL [370]
Kuang-hsi hsiang-shih lu – List of successful candidates in the imperial examination in Kwangsi province: 1867, 1893, 1901, 1903. 1 reel – 1 – us Chinese Res [951]
Kuang-hsi jih-pao – Nanning, Kwangsi. mar 1, 1961-sep 25, 1965 – 1r – 1 – (scattered issues missing) – us Chinese Res [079]
Kuang-hsi ko hsien pan li hsiang chen nung ts'un ts'ang k'u hsu chih – Kuang-hsi: Sheng cheng fu min cheng t'ing, 1934 – (= ser P-k&k period) – us CRL [350]
Kuang-hsi liang shih tiao ch'a – [China]: Kuang-hsi sheng cheng fu tsung wu ch'u, Min kuo 27 [1938] – (= ser P-k&k period) – us CRL [315]
Kuang-hsi sheng chiao yu kai k'uang t'ung chi : min kuo erh shih erh nien tu shang hsueh ch'i – Kuang-hsi sheng: Chiao yu t'ing, Min kuo 24 [1935] – (= ser P-k&k period) – us CRL [370]
Kuang-hsi sheng hsiang-hsien tung-nan hsiang hua-lan yao she hui tsu chih / Wang, T'ung-hui – [Shang-hai]: Shang wu yin shu kuan, [1936] – (= ser P-k&k period) – us CRL [305]
Kuang-hsi sheng hsien ti fang k'uan shih hou shen chi pao kao – Kuang-hsi: sheng cheng fu shen chi wei yuan hui, 1936 – (= ser P-k&k period) – us CRL [650]
Kuang-hsi sheng ko hsien p'u t'ung sui ju sui ch'u kai suan shu : chung-hua min kuo erh shih san nien tu – [China: sn, Min kuo 23 [1934]] – (= ser P-k&k period) – us CRL [336]
Kuang-hsi t'ien fu kai yao / Yang, Shih-hsien – Kuang-hsi: Sheng cheng fu ching chi wei yuan hui, 1936 – (= ser P-k&k period) – us CRL [630]
Kuang-tung chih tien tang yeh / Ou, Chi-luan – Kuang-chou, Kuo li Chung-shan ta hsueh fa hsueh yuan ching chi tiao ch'a ch'u, Min kuo 23 [1934] – (= ser P-k&k period) – us CRL [306]
Kuang-tung ching chi nien chien (erh shih chiu nien tu [1940]) hsia ts'e – [Kuang-tung]: Kuang-tung sheng yin hang ching chi yen chiu shih, Min kuo 30 [1941] – (= ser P-k&k period) – us CRL [339]
Kuang-tung ching chi nien chien (erh shih chiu nien tu [1940]) shang ts'e – [Kuang-tung]: Kuang-tung sheng yin hang ching chi yen chiu shih, Min kuo 30 [1941] – (= ser P-k&k period) – us CRL [339]
Kuang-tung hsiang-shih lu – List of sucessful candidates in the imperial examination in Kwangtung province: 1852, 1900. 1 reel – 1 – us Chinese Res [951]
Kuang-tung liang shih wen t'i – [China]: Kuang-tung liang shih t'iao chieh wei yuan hui, 1935 – (= ser P-k&k period) – us CRL [380]
Kuang-tung liang shih wen t'i yen chiu / Huang, P'u-sheng – Kuang-chou: [Chu Yung-chi], Min kuo 26 [1937] – (= ser P-k&k period) – us CRL [630]
Kuang-tung liang shih wen t'i yen chiu – [Kuang-tung: Kuang-tung sheng cheng fu mi shu ch'u pien i shih], 1941 – (= ser P-k&k period) – us CRL [630]
Kuang-tung pi chih yu chin jung / Ch'iu, Pin-ts'un – Shang-hai: Hsin shih tai she, [1941] – (= ser P-k&k period) – us CRL [332]
Kuang-tung sheng cheng fu hsing cheng she chi yu k'ao ho kai k'uang – [Kuang-chou: Kuang-tung sheng cheng fu hsing cheng hsiao lu ts'u chin wei yuan hui, 1941] – (= ser P-k&k period) – us CRL [951]
Kuang-tung sheng cheng fu kung tso pao kao shu – [Kuang-tung: Sheng cheng fu, Min kuo 25 [1936]] – (= ser P-k&k period) – us CRL [951]
Kuang-tung sheng cheng fu san shih nien hsing cheng hui i chi yao / Kwangtung Province Mi shu ch'u – [Kuang-tung: Kuang-tung sheng cheng fu mi shu ch'u, 1941] – (= ser P-k&k period) – us CRL [951]
Kuang-tung sheng ts'ai cheng chi shih / China Kuang-tung ts'ai cheng t'e p'ai yuan kung shu – [China]: Kuang-tung sheng cheng fu ts'ai cheng t'ing, Min kuo 23 [1934] – (= ser P-k&k period) – us CRL [332]
Kuang-tung sheng wu nien lai min cheng kai k'uang : min kuo erh shih nien liu yueh chih erh shih wu nien wu yueh – [China: Kuang-tung min cheng t'ing], 1936 – (= ser P-k&k period) – us CRL [350]
Kuang-tung sheng yin hang erh shih wu nien fen ying yeh pao kao shu / Tu, Mei – [China: Kuang-tung sheng yin hang], 1936 – (= ser P-k&k period) – us CRL [951]
Kuang-tung sheng yin hang min kuo erh shih liu nien fen ying yeh pao kao – [China: Kuang-tung sheng yin hang, 1937] – (= ser P-k&k period) – us CRL [951]
Kuang-tung shih san hang k'ao / Liang, Chia-pin – Shang-hai: Shang wu yin shu kuan, Min kuo 26 [1937] – (= ser P-k&k period) – us CRL [360]
Kuang-tung t'ang yeh yu feng jui / Ch'en, Chao-yu – [Hsiang-kang?: sn, 1937?] – (= ser P-k&k period) – us CRL [951]
Kuang-tz'u i chi / Chiang, Kuang-tz'u – Shang-hai: Hsien tai shu chu, 1932 – (= ser P-k&k period) – us CRL [480]
Kuan-tung lei : chu pen / Ch'en, Yueh – [China: sn, 1932] – (= ser P-k&k period) – us CRL [820]
Kuba field notes / Vansina, Jan – 1929 – us CRL [972]
Kuba field notes / Vansina, Jan – 1979 – 1 – us CRL [960]
Kubanskaia obl. Ispolnitel'nyj komitet *see* vestnik kubanskogo obl ispolnitel'nogo komiteta
Kubanskaia pravda : ezhednevnaia politicheskaia gazeta armavirskogo soveta rab kaz krest'ian ic sold. deputatov – Armavir, Russia 1918 [mf ed Norman Ross] – 1r – 1 – mf#nrp-198 – us UMI ProQuest [077]
Kubanskie vedomosti – Krasnodar, Russia 1887-1904 [mf ed Norman Ross] – 1 – mf#nrp-788 – us UMI ProQuest [077]
Kubanskii kooperator – Krasnodar, 1917-1924(11) – 30mf – 9 – (missing:1917-1922,1923(8)) – mf#COR-646 – ne IDC [335]
Kubanskii put' : ezhedn polit ekon i lit gaz / ed by Fendrikov, F N – Ekaterinodar [Kubano-Chernomor gub]: [s n] 1919 [1919-] – (= ser Asn 1-3) – n32 [1919] item 210, on reel n42 – 1 – mf#asn-1 210 – ne IDC [077]
Kubanskii sbornik : trudy kubanskogo oblastnogo statisticheskogo komiteta – Ekaterinodar, 1883-1916. v1-21 – 113mf – 9 – (missing: 1883 v1 (p.i-vi, 1-79), 1910-1911 v15-16, 1916 v21) – mf#RET-5 – ne IDC [314]
[Kubasov, I A] *see* Katalog izdanii imperatorskoi akademii nauk
Kube, Helga *see* Die industrieansiedlung in ludwigshafen am rhein bis 1892 (chemie und metallverarbeitung)
Kubiak, Rolf *see* Quantitative charakterisierung inhomogener oberflaechen mit mies bei der wechselwirkung von sauerstoff mit ni (100) und cr (100)
Kubika kwakanaka / Preston, Hilary – Gwelo, Zimbabwe. 1969 – 1r – us UF Libraries [960]
Kubikov, Ivan Nikolaevich *see* Klassiki russkoi literatury
Kubo, Hajime *see* Modulation in rat skeletal muscle sarcoplasmic reticulum calcium adenosine triphosphatase transcripts following exercise training
Kubursi, A A *see* Arab economic prospects in the 1980's
Kuchengeta varwere pamusha / Mary Joseph, Sister – Gwelo, Zimbabwe. 19–? – 1r – us UF Libraries [960]
Kuckuck – Vienna, Austria apr 1929-feb 1934 [mf ed Norman Ross] – 5r – 1 – (humor magazine, illustrated) – mf#nrp-1979 – us UMI ProQuest [870]
Der kuckuck und die zwoelf apostel : roman / Beumelburg, Werner – Oldenburg i O: G Stalling, 1931 [mf ed 1989] – 327p – 1 – mf#7017 – us Wisconsin U Libr [830]
Kuczynski Godard, Maximo H *see* La vida en la amazonia peruana. observacion de un medico. lima, 1944
Kuczynski, Juergen *see*
– Bild und begriff
– Die menschenrechte
Kuda i kak vedet sovetskaia vlast krestianstvo? / Kamenev, L B – 1925 – 194p 3mf – 9 – mf#COR-477 – ne IDC [335]
Kuda vremenshchiki vedut soiuz russkogo naroda / Dubrovin, A I – 1910 – 572p 7mf – 9 – mf#RPP-162 – ne IDC [325]
Kudanwa kwedu – Gwelo, Zimbabwe. 1958 – 1r – us UF Libraries [960]
Kuddusi *see* The divan project
Kuder, F C *see* Ramarow
Kudriashov, Konstantin Vasil'evich *see* Poslednii favorit ekateriny 2, platon zubov
Kudriavtsev, Matfii *see* Istoriia pravoslavnago monashestva v severo-vostochnoi rossii, so vremen pred sergiia radonezhskago
Kudriavtsev, V A et al *see* Sibirskii vestnik
Kudrun : schulausgabe mit einem woerterbuche / Bartsch, Karl – 3. aufl. Leipzig: F A Brockhaus, 1874 [mf ed 1993] – (= ser Deutsche classiker des mittelalters 2) – xxviii/357p – 1 – (incl bibl ref and ind) – mf#8189 reel 1 – us Wisconsin U Libr [430]
Kudrun / ed by Symons, Barend – Halle: M Niemeyer, 1883 [mf ed 1993] – (= ser Altdeutsche textbibliothek 5) – vii/306p – 1 – (incl bibl ref) – mf#8193 reel 1 – us Wisconsin U Libr [810]
The kudzu – n1-11. 1968-69 – 1 – us AMS Press [073]
Kuebel, Johannes *see* Geschichte des katholischen modernismus
Kuebel, Robert *see* Ueber den unterschied zwischen der positiven und der liberalen richtung in der modernen theologie
Kuechler, Carl *see* Die faustsage und der goethe'sche faust
Kuechler, Friedrich *see*
– Beitraege zur kenntnis der assyrisch-babylonischen medizin
– Hebraeische volkskunde
Kuechler, Johann *see* Six trios concertant pour deux violons et basse, oeuvre 3
Kuechling, Eduard Hermann *see* Studien zur sprache des jungen grillparzer
Kueck, Eduard *see*
– Die schriften hartmuths von cronberg
– Sickingen und landschad
– Vom alten und neuen gott, glauben und lehre
Kuecuek mecmua – Diyarbakir: Vilayet Matbaasi, 1922-23. Sahib-i Imtiyaz ve Mueduer-i Mes'ul: Ziya Goekalp. n6. 10 temmuz 1338; 7,23. 20 tesrinisani 1338 – (= ser O & t journals) – 1mf – 9 – $25.00 – us MEDOC [956]
Kuecuekcelebizade, Asim Efendi *see* Tarih-i rasit
Kuegelgen, C von *see* Die ethik huldreich zwinglis
Kuegelgen, Constantin von *see*
– Die ethik huldreich zwinglis
– Grundriss der ritschlschen dogmatik
– Immanuel kants auffassung von der bibel und seine auslegung derselben
– Luthers auffassung der gottheit christi
– Die rechtfertigungslehre des johannes brenz
– Schleiermachers reden und kants predigten
Kuegelgen, Wilhelm von *see* Lebenserinnerungen des alten mannes in briefen an seinen bruder gerhard
Kuehl, Ernst *see*
– Der brief des paulus an die roemer
– Die heilsbedeutung des todes christi
– Das selbstbewusstsein jesu
– Das verhaeltniss der massora zur septuaginta im jeremia
Kuehl, Warren F *see* The library of world peace studies
Kuehle, H *see* S alberti magni quaestiones de bono (summa de bono q 1-10) (fp36)
Kuehler, W J *see* Johannes brinckerinck en zijn klooster te diepenveen
Kuehler, Wilhelmus Johannes *see* Het socinianisme in nederland
Kuehlhorn, Walther *see* J A leisewitzens julius von tarent
Kuehme, Tobias *see* Nachsorge bei patienten mit einem kolorektalen karzinom
Kuehn, David *see* Use of the tuba in the operas and musical dramas of richard wagner
Kuehn, Ernst *see* Kompendium der biblischen theologie des alten und neuen testaments von konstantin schlottmann
Kuehn, Julius *see* Die kunst adalbert stifters
Kuehn, Walter *see* Heinrich von kleist und das deutsche theater
Kuehne, August *see* Ueber die faustsage
Kuehne, Benno *see* Unser heiliger vater papst leo 13
Kuehne, F Gustav *see* Wittenberg und rom
Kuehne, Ferdinand *see*
– Mein carneval in berlin 1843 [achtzehnhundertdreiundvierzig]
– Wien in alter und neuer zeit
Kuehne, Johannes *see* Four years in ashantee
Kuehnemann, Eugen *see*
– Auswahl
– Gerhart hauptmann – aus dem leben des deutschen geistes in der gegenwart
– Hermann loens am 20. todestage, 26. september 1934
Kuehner, Raphael *see* Ausfuehrliche grammatik der griechischen sprache
Kuehnhold, Marianne *see* Harmonien
Kuei ch'ao / Ou-yang, Shan – Shang-hai: Liang yu t'u shu kung ssu, 1936 – (= ser P-k&k period) – us CRL [830]
Kuei ch'u lai hsi : [wu mu chu] / Lao, She – Ch'ung-ch'ing: Tso chia shu wu, 1943 – (= ser P-k&k period) – us CRL [820]
Kuei hsiu shih hua / T'iao, Hsi-sheng – Shang-hai: Hsin min shu chu, Min kuo 23 [1934] – (= ser P-k&k period) – us CRL [810]
Kuei i chi-tu tzu shu (ccm346) / Yin, Tzu-heng – Wu-chan, 1931 [mf ed 198?] – (= ser Ccm 346) – 1 – mf#1984-b500 – us ATLA [920]
Kuei kuo yin hsiang / Chang, Cheng-yen – Shang-hai: Sheng huo shu tien, 1933 – (= ser P-k&k period) – us CRL [951]
Kuei lai / Lo, Feng – Shang-hai: Liang yu t'u shu yin shua kung ssu, 1937 – (= ser P-k&k period) – us CRL [830]
Kuei lien / Hsu, Hsu – Shang-hai: Yeh ch'uang shu wu, Min kuo 32 [1943] – (= ser P-k&k period) – us CRL [830]
Kuei ying : tuan p'ien hsiao shuo chi / Lo, Hung – Fu-chien Yung-an: Tien ti ch'u pan she, 1944 – (= ser P-k&k period) – us CRL [480]
Kuei yu jih chi / Pin, Min-kai – [SI: sn, 1938] – (= ser P-k&k period) – us CRL [951]
Kuei-chou ching chi / Chang, Hsiao-mei – Shang-hai: Chung-kuo kuo min ching chi yen chiu so, min kuo 28 [1939] – (= ser P-k&k period) – us CRL [339]
Kuei-chou hsiang-shih lu – List of successful candidates in the imperial examination in Kweichow province: 1870. 1 reel – 1 – us Chinese Res [951]
Kuei-chou jih-pao – Kweiyang, Kweichow. Nov. 28 1949-. Scattered issues missing. 7 reels – 1 – us Chinese Res [079]

Kuei-chou miao i ko yao / Ch'en, Kuo-chun – Kuei-yang: Wen t'ung shu chu, 1942 – (= ser P-k&k period) – us CRL [390]
Kuei-chou sheng pao chia kai k'uang – Kuei-yang: Kuei-chou sheng cheng fu min cheng t'ing, Min kuo 26 [1937] – (= ser P-k&k period) – us CRL [350]
Kuei-chou wei-pi-t'ung ning-chieh-jen huang ti ch'u yu tiao ch'a pao kao / Chang, Chia-ling – [China]: Nung lin pu k'en wu tsung chu, Min kuo 31 [1942] – (= ser P-k&k period) – us CRL [630]
Kuekelhaus, Heinz *see* Thomas der perlenfischer
Kuelfoeldi magyarsag – Budapest HU, nov 1 1920-25 – 4r – 1 – us IHRC [077]
Kuelpe, Oswald *see* The philosophy of the present in germany
Kuen, M *see* Collectio scriptorum
Kuender deutscher einheit : das leben ernst moritz arndts / Breitenkamp, Paul – Berlin: Haude & Spenersche Buchhandlung Max Paschke 1938 [mf ed 1988] – 1r – 1 – (filmed with: ludwig anzengruber / sigismund friedmann) – mf#6954 – us Wisconsin U Libr [943]
Kuenderera mberi kwavatema musouthern rhodesia – Salisbury, Zimbabwe. 1958? – 1r – us UF Libraries [960]
Kuenec artues der guote : das artusbild der hoefischen epik des 12. und 13. jahrhunderts / Guerttler, Karin R – Bonn: Bouvier, 1976 [mf ed 1993] – (= ser Studien zur germanistik, anglistik und komparatistik 52) – 417p – 1 – (incl bibl ref) – mf#8171 – us Wisconsin U Libr [430]
Kuenen, Abraham *see*
- The five books of moses
- Gesammelte abhandlungen zur biblischen wissenschaft
- Histoire critique des livres de l'ancien testament
- Historico-critical inquiry into the origin and composition of the hexateuch (pentateuch and book of joshua)
- National religions and universal religions
- De profetische boeken des ouden verbonds
- The prophets and prophecy in israel
- The religion of israel to the fall of the jewish state

Kuenh uel-ahbar / Ahmed, Mustafa Ali b. (Geliboeluelue) – Istanbul: Takvimhane-i 'Amire. 5v. 1277 [1860] – (= ser Ottoman histories and historical sources) – 21mf – 9 – $400.00 – us MEDOC [956]
Kuenkel, Hans *see*
- Auf den kargen huegeln der neumark
- Schicksal und liebe des niklas von cues

Kuenstle, K *see* Ikonographie der christlichen kunst...
Kuenstle, Karl *see*
- Antipriscilliana
- Eine bibliothek der symbole und theologischer tractate zur bekaempfung des priscillianismus und westgothischen arianismus aus dem 6. jahrhundert

Kuenstler : vier geschichten / Brehm, Bruno – Karlsbad: A Kraft 1944 [mf ed 1989] – 1r – 1 – (filmed with: buch des dankes: bruno brehm zum funfzigsten geburtstag; & other titles) – mf#7082 – us Wisconsin U Libr [830]
Kuenstlerbriefe uebersetzt und erlaeutert von dr ernst guhl / Guhl, E – Ed 2. Berlin, 1880 – 10mf – 9 – mf#O-280 – ne IDC [700]
Kuenstler-inventare : urkunden zur geschichte der hollaendischen kunst des 16ten, 17ten und 18ten jahrhunderts / Bredius, A – Haag, 1915-1922. v5-7, 10-14 – 39mf – 9 – mf#O-518 – ne IDC [700]
Das kuenstlerische element in der metaphysik schleiermachers / Schuetz, Paul – Bremen: Buchdruckerei des Traktathauses, 1914 – 1mf – 9 – 0-524-00341-6 – mf#1989-3041 – us ATLA [110]
Die kuenstlerischen voraussetzungen des genter altars der brueder van eyck / Fuerbringer, Hermann – Weida, 1914 (mf ed 1993) – 1mf – 9 – €24.00 – 3-89349-259-3 – mf#DHS-AR 116 – gw Frankfurter [750]
Kuenstlermonographien des 16. bis 18. jahrhunderts = Biographies of artists from the 16th to 18th centuries / ed by Schuette, Ulrich – [mf ed 2004] – (= ser Nachschlagewerke und quellen zur kunst 6) – 120mf (1:24) in 3 installments – 9 – diazo €1980.00 (silver €1590 isbn: 978-3-598-34590-6) – ISBN-10: 3-598-34591-7 – ISBN-13: 978-3-598-34591-3 – (with guide) – gw Saur [700]
Kuenstlerverein [Bremen, Germany]. Historische Gesellschaft *see* Bremisches jahrbuch
Kuenstliche und wolgerissene figuren, der fuernemhsten evangelien... / Amman, J – Franckfurt am Mayn, 1579 – 2mf – 9 – mf#O-1015 – ne IDC [700]
Kuenstliche...figuren von allerlai jagt und weidwerck, allen liebhabern der maler kunst, auch goltschmieden, bildthawern... / Amman, J – Franckfurt am Mayn, 1592 – 1mf – 9 – mf#O-1519 – ne IDC [700]
Kuentzel, Gerhard *see* Joh. gottfr. herder zwischen riga und bueckeburg
Kuenzel, Carl *see* Ich habe mich rasieren lassen
Kuenzel, Heinrich *see* Drei buecher deutscher prosa in sprach- und stylproben
Kueper, Lic *see* Das priesterthum des alten bundes
Kuepper, Heinz *see* Bibliographie zur tristansage
Kueppers, Karin *see* Die vegetation der chaine de gobanangou
Kuerbs, Harry *see* Studien zur pfahldorfgeschichte aus friedrich theodor vischers roman "auch einer"
Kuerdler / Tarihi ve Ictimai Tedkikat – Istanbul: Kitabhane-i Sudi, 1334 [1915] – (= ser Ottoman histories and historical sources) – 5mf – 9 – $75.00 – us MEDOC [956]
Kuernberger, Ferdinand *see*
- Der amerika-muede
- Ferdinand kuernbergers briefe an eine freundin, 1859-1879
- Fuenfzig feuilletons
- Katilina
- Novellen

Kuerschner, Joseph *see* Pierer's konversations-lexikon (ael1/6.7)
Kuerschners deutscher literaturkalender *see* Allgemeiner deutscher literaturkalender
Kuerschners deutscher literatur-kalender 1922-1988 : 40. jahrgang (1922) – 60. jahrgang (1988), nekrolog 1901-1935 (1936), nekrolog 1936-1970 (1973) – [mf ed 1998] – 240mf (1:24) – 9 – diazo €1260.00 (silver €1690 isbn: 978-3-598-33754-3) – ISBN-10: 3-598-33755-8 – ISBN-13: 978-3-598-33755-0 – (incl suppl; individual years also available – enquire) – gw Saur [430]
Kuerschners deutscher literatur-kalender auf das jahr... – Leipzig: G J Goeschen'sche Verlagshandlung. 22v. [1904]-1934 (annual 1904-17, irreg 1922-34) [mf ed v26-38(1904-16), v40-42(1922-25)] – 1 – (cont: kuerschners deutscher litteratur-kalender auf das jahr...; cont by: kuerschners deutscher literatur-kalender; cont in pt by: kuerschners deutscher gelehrten-kalender; suspended 1918-21) – mf#film mas c437 – us Harvard [430]
Kuersten, Hans *see* Panzer greifen an
Der kuertzer catechismus / Jud, L – Zuerich, C Froschauer, 1538 – 2mf – 9 – mf#PBU-276 – ne IDC [240]
Kuery, Hans *see* Simon grynaeus von basel, 1725-1799
Kuerzinger, J *see* Alfonsus vargas toletanus
Kuestenblick *see* Dz am dienstag
Kuester, K D *see* Des vortrefflichen religionsverbesserer ulrich zwingli...
Kuester, Rudolf *see* Goethes "fischer"
Kuettner, Carl *see*
- Reise durch deutschland, daenemark, schweden, norwegen und einen theil von italien
- Travels through denmark, sweden, austria, and part of italy
- Wanderungen durch die niederlande, deutschland, die schweiz und italien in den jahren 1793 und 1794

Ku-fan *see* Ku-fan ti shih
Ku-fan ti shih / Ku-fan – Ch'ing-tao: Shih ko ch'u pan she, 1936 – (= ser P-k&k period) – us CRL [810]
Kufferath, Maurice *see*
- L'anneau du nibelung
- Les maitres chanteurs de nuremberg de richard wagner
- The parsifal of richard wagner
- Tristan et iseult
- La walkyrie de richard wagner

Kugaevskii, G I *see* Nizov'e amura
Kugath, Steven D *see* The effects of family participation in an outdoor adventure program
Kugelgen, Carlo Von *see* Aus eigener kraft
Kugener, M A *see* Marc le diacre
Kugler, Anna Sarah *see* Dr anna s kugler papers
Kugler, B *see* Geschichte der kreuzzuege
Kugler, Bernhard *see*
- Geschichte der kreuzzuege
- Wallenstein

Kugler, Franz Theodor *see*
- Handbook of painting

Kugler, Franz Xaver *see* Im bannkreis babels
Kugomezgeka / Chirambo, G R – London, England. 1956 – 1r – us UF Libraries [960]
Kuh, Anton *see* Boerne der zeitgenosse
Kuh, Emily *see* Emil kuhs kritische und literarhistorische aufsaetze, 1863-1876
Kuhamba kwomuhambi / Bunyan, John – Harare?, Zimbabwe. 1958 – 1r – us UF Libraries [960]
Kuhlenbeck, L *see* Giordano brunos einfluss auf goethe und schiller
Kuhlenbeck, Ludwig *see* Von den pandekten zum buergerlichen gesetzbuch
Kuhn, Felix *see* Le christianisme de luther
Kuhn, Gottlieb Jacob *see* Die reformatoren berns im16. jahrhundert
Kuhn, Hermann *see* Gepraegte form
Kuhn, Hugo *see* Dichtung und welt im mittelalter
Kuhn, J *see* Musikah be-khitve ha-kodesh, ba-talmud uva-kabalah
Kuhn, Johannes von *see* Katholische dogmatik
Kuhnau, Johann *see*
- Frische clavier fruechte
- Der musicalische quack-salber
- Neuer clavier ubung erster theil
- Neuer clavier uebung andrer theil

Kuhne *see* Four years in ashantee by the missionaries ramseyer and kuehne.
Kuhne, Berthold *see* Neutestamentliches woerterbuch
Kuhne, Kathe *see* Tagebuchblatter beschrieben wahrend der jahre 1891 bis 1895
Kuhn-Foelix, August *see* Heinrich von kleist
Kuhns, Oscar *see* The german and swiss settlements of colonial pennsylvania
Kuhr, Victor *see* Modsigelsens grundsaetning
Kuhring, Otto *see* Das schicksal der westfaelischen domaenenkaeufer in kurhessen
Kui kri sa khan : rup rhan vatthu kri / Khuin, Mon, Yu ja na – Ran Kun: Tan On Ca pe [195-?] [mf ed 1990] – 1r with other items – 1 – (in burmese) – mf#mf-10289 seam reel 200/3 [§] – us CRL [830]
Kui Kui *see* I sac chum
Kui lui ni khte mran ma nuin nam samuin / Ban Mo Tan Aon – Yan Kun: Rhve Kye Cape 1977 [mf ed 1990] – 1r with other items – 1 – (in burmese) – mf#mf-10289 seam reel 139/4 [§] – us CRL [959]
Kui tam na / Mran Chev, U – Ran kun Mrui': Ca khyac su 1975 [mf ed 1993] – on pt of 1r – 1 – mf#11052 r650 n2 – us Cornell [959]
Kui tatk : newsletter of the native american science education association / Native American Science Education Association – v1 n1-v2 n3 [1984 spr-1987 spr] – 1r – 1 – mf#1495120 – us WHS [366]
Kuibyshev, V V *see*
- Lenin i kooperatsiia
- Zadachi vnutrennei torgovli i kooperatsiia

Kuimba, Giles *see* Tambaoga mwanangu
Kuiper, Barend Klaas *see* De janssen kwestie en nog iets
Kuiper, Berens Klaas *see* Ons opmaken en bouwen
Kuiper, Henry J *see* The three points of common grace
Kuiper, Koenrad *see* Onderzoek naar de echtheid van clemens' eersten brief aan de corinthiers
Kuiper, Rienk Bouke *see* Not of the world
Kuiper, Rienk Bouke et al *see* Is jesus god
Kuiy pyok ma : [a novel] / Vajira – Ran Kun: U Bha Ri nhan sami mya Ca up thut ve Thana 1958 [mf ed 1990] – 1r with other items – 1 – (in burmese) – mf#mf-10289 seam reel 195/4 [§] – us CRL [830]
Kujavischer bote – Hohensalza (Inowroclaw PL), 1924 10 jun-1939 27 aug – 1 – gw Misc Inst [077]
Kukarkin, B V *see* Obshchii katalog peremennykh zvezd
Kukolevskii, A G *see* Gonkong v sisteme mirovykh ekonomicheskikh sviazei
Les kuku : possessions alglo-egyptiennes / Vanden Plas, Joseph – Bruxelles: A DeWit [etc], 1910 – us CRL [960]
Kukulka, Carl G *see* Effect of muscle length on motor unit firing behavior in human tibialis anterior muscle
Kukura kwomukristu – Fort Victoria, Zimbabwe. 19– – 1r – us UF Libraries [960]
Kukuwich, Wendee E *see* Selection of exercise intensity using perceptual cues during television distraction
Kulbak, Modhe *see* Yakov frank
Kuleshov, P V *see* Donetskoe slovo
Kulischer, J *see* Periodicheskii organ, posviashchennyi interesam studenchestva iz rossii v germanii
Kulisher, I M *see* Obzor russkogo i inostrannogo zakonodatelstva o kooperativnykh tovarishchestvakh
Kulisher, Iosif Mikhailovich *see* Obzor mirovogo khoziaistva za vremia voiny i posle voiny i sostoianie ego k nachalu 1923 goda
Kulkarni, E D *see* Verbs of movement and their variants in the critical edition of the adiparvan
Kulke, Eduard *see*
- Erinnerungen an friedrich hebbel
- Richard wagner

Kulkielko, Renya *see* Bi-nedudim uva-mahteret
Kull al-haqiqah lil-jamahir – [Lebanon?: Harakat al-Tahrir al-Watani al-Filastini, "Fath", n1-2 may-sep 1970 – 1r – 1 – us CRL [956]
Kullischer, Alexander *see* Lord bikonsfild
Kulm, Tilo von *see* Tilos von kulm gedicht von siben ingesigeln
Kulomzin, A N et al *see* M kh reitern
Kulon ukali = Local news of the reserve – United States. 1977 sep-1985 apr/jun – 1r – 1 – mf#1708623 – us WHS [305]
Kulongisela mukhongelo wa pfuxelelo / Chauke, Joel – Fort Victoria, Zimbabwe. 19–? – 1r – us UF Libraries [960]
Kulsum Nah'nah *see* Customs and manners of the women of persia
Der kult der shang-dynastie im spiegel der orakelinschriften : eine palaeographische studie zur religion im archaischen china / Tsung-tung, Chang – Frankfurt a.M., 1970 – 4mf – 9 – 3-89349-731-5 – gw Frankfurter [290]
Der kultische kalender der babylonier und assyrer / Landsberger, B – Leipzig, 1915 – (= ser Leipziger semitistische Studien) – 2mf – 9 – (leipziger semitistische studien, v6 pt1-2) – mf#NE-20118 – ne IDC [956]
Die kultur – Stuttgart DE, 1952 15 oct-1962 apr – 2r – 1 – gw Misc Inst [074]
Die kultur des managements im kulturmanagement : zu einer theorie des kulturmanagements / Lenders, Britta – (mf ed 1995) – 3mf – 9 – €49.00 – 3-8267-2101-2 – mf#DHS 2101 – gw Frankfurter [650]
Kultur und denken der babylonier und juden / Schneider, Hermann – Leipzig: JC Hinrichs, 1910 [mf ed 1992] – (= ser Entwicklungsgeschichte der menschheit 2) – 2mf – 9 – 0-524-04535-6 – mf#1990-3369 – us ATLA [270]
Kultura – Prague, Czechoslovakia. feb 1957-nov 1958, 1961-15 aug 1968 – 13 1/4r – 1 – (aka: kulturni tvorba) – uk British Libr Newspaper [077]
Kultura i zhittya – Kiev, U.S.S.R. -w. 1967-70. 4 reels – 1 – uk British Libr Newspaper [947]
Kul'tura i zhizn' – Moscow, Russia 1949-51 [mf ed Norman Ross] – 4r – 1 – mf#nrp-1083 – us UMI ProQuest [077]
Kultura i zhizn – Moscow, U.S.S.R. -f. Nov 1948-Feb 1951. 1 reel – 1 – uk British Libr Newspaper [947]
Kul'tura teatra – Moscow, 1921-22 – 10mf – 9 – us UMI ProQuest [790]
Kulturarbeit als bestandteil sozialistischer erwachsenenbildung in der ddr / Pralle, Elka – Frankfurt a.M., 1976 – 3mf – 9 – 3-89349-677-7 – gw Frankfurter [306]
Die kulturaufgaben der reformation : einleitung in eine lutherbiographie / Berger, Arnold Erich – 2., durchgesehene und verm Aufl. Berlin: E Hofmann, 1908 – 2mf – 9 – 0-524-00509-5 – mf#1990-0009 – us ATLA [242]
Die kulturelle entwicklungsfaehigkeit des islam auf geistigem gebiete / Horten, Max – Bonn: F Cohen, 1915 – 1mf – 9 – 0-524-01555-4 – mf#1990-2509 – us ATLA [260]
Kulturelle orientierung und oekologisches dilemma – Dortmund: projekt vlg, 1993 (mf ed 1996) – (= ser Schriftenreihe der uni dortmund; Studium generale) – 2mf – mf#DHS 9717 – gw Frankfurter [306]
Kulturelle ziele im werk gustav frenssens / Braun, Frank Xavier – [s.l: s.n.] 1946 [mf ed 1989] – 1r – 1 – (filmed with: ein glaubensbekenntnis / ferdinand freilingrath) – mf#7269 – us Wisconsin U Libr [430]
Kulturgeschichte der kreuzzuege / Prutz, Hans – Berlin: ES Mittler, 1883 – 2mf – 9 – 0-7905-8224-4 – (incl bibl ref) – mf#1988-6124 – us ATLA [940]
Kulturgeschichte der roemischen kaiserzeit / Grupp, Georg – Muenchen: Allgemeine Verlags-gesellschaft m b h 1903-04 [mf ed 1979] – 2v on 1r [ill] – 1 – (v1: untergang der heidnischen kultur; v2: anfaenge der christlichen kultur) – mf#film mas 9095 – us Harvard [930]
Kulturgeschichtliche bilder aus dem judischen gemeindeleben / Pfeifer, S – Bamberg, Germany. 18– – 1r – us UF Libraries [939]
Der kulturkaempfer – Berlin DE, 1880-83, 1886-88 – 1 – gw Misc Inst [943]
Kulturkampf – Paris (F), 1936-38 – 1 – gw Misc Inst [410]
Die kulturmission unserer dichtkunst : studien zur aesthetik und literatur der gegenwart / Schulze, Paul – Leipzig: F Eckardt, 1908 [mf ed 1993] – 432p – 1 – mf#8241 – us Wisconsin U Libr [430]
Kulturni historie : jeji vznik, rozvoj a posavadni literaturu cizi i ceskou / Zibrt, Cenek – V Praze: Jos R Vilimek, 1892 (mf ed 19–) – (= ser Harvard Slavic humanities preservation microfilm project) – 122p – (incl bibl ref) – mf#ZQ-260 – us NY Public [900]
Kulturni noviny *see* Literarni noviny
Kulturni tvorba *see* Kultura
Kulturni zivot – Bratislava, Czechoslovakia. -w. Jan-Dec 1955; Jan 1957-Dec 1965. 13 reels – 1 – uk British Libr Newspaper [072]
Kulturno-opetestvenite vrski na makedoncite so srbija vo tokot / Kzambazovski, Kliment – Skopje, MACEDONIA . 1960 – 1r – us UF Libraries [025]
Kulturny zivot – Bratislava Czechoslovakia. jan-24 dec 1955; 1957-67, 5 jan-6 sep 1968 – 13r – 1 – uk British Libr Newspaper [072]
Das kulturproblem des minnesangs : studien zur vorgeschichte / Wechssler, Eduard – Halle/S: M Niemeyer, 1909 [mf ed 1993] – xii/502p – 1 – (only 1st vol publ. incl bibl ref and ind) – mf#8173 – us Wisconsin U Libr [930]
Kulturstiftung der Laender *see*
- Deutsche zeitschriften des 18. und 19. jahrhunderts [ergaenzungslieferung]
- Deutsche zeitschriften des 18. und 19. jahrhunderts [hauptlieferung]

Kultur-treger fun der idisher liturgye / Zaludkowski, Elias – Detroit, MI. 1930 – 1r – us UF Libraries [939]
Die kulturverhaeltnisse des deutschen mittelalters : im anschlusz [sic] an die lektuere zur einfuehrung in die deutschen altertuemer im deutschen unterricht / Zehme, Arnold – 2. verb und verm Aufl. Leipzig: G Freytag, 1905 – 1mf – 9 – 0-524-04503-8 – mf#1990-1265 – us ATLA [943]
Kulturwalze / Rothe, Ernst Hermann – Berlin, Germany. 1928 – 1r – us UF Libraries [972]
Die kulturwerte der deutschen literatur des mittelalters / Francke, Kuno – Berlin: Weidmannsche Buchhandlung, 1910 – 1r – 1 – (incl bibl ref) – us Wisconsin U Libr [430]
Die kulturwerte der deutschen literatur in ihrer geschichtlichen entwicklung / Francke, Kuno – Berlin: Weidmannsche Buchhandlung, 1910-1928 – 1 – (incl bibl ref and index) – us Wisconsin U Libr [430]
Die kulturwerte der deutschen literatur von der reformation bis zur aufklaerung / Francke, Kuno – Berlin: Weidmannsche Buchhandlung, 1923 – 1 – (incl bibl ref and index) – us Wisconsin U Libr [430]
Kulturwille – Leipzig DE, 1924-32 – 2r – 1 – gw Mikrofilm [073]
Kultus- und geschichtsreligion (pelagianismus und augustinismus) : ein beitrag zur religioesen psychologie und volkskunde / Juengst, Johannes – Giessen: J Ricker, 1901 – 1mf – 9 – 0-7905-3864-4 – mf#1989-0357 – us ATLA [150]
Kulumba, Ali, Sheikh *see* Ebyafayo by'obusiramu mu uganda
Kulyzhnyi, A E *see*
- Derevenskaia kooperatsiia
- Kak vesti dela kreditnogo tovarishchestva
- Kooperativnyi sbyt produktov melkogo khoziaistva
- Kooperativnyi sbyt produktov selskogo khoziaistva
- Kreditnye i ssudo-sberegatelnye tovarishchestva
- Kurs kreditnoi kooperatsii
- Ocherki po selsko-khoziastvennoi i kreditnoi kooperatsiia, 1900-1918 gg
- Organizatsiia khlebotorgovli i kooperativnyi sbyt khleba v rossii

Kum dhvoe papqun na pan! : pralom lok jivit sok / Hak Chai Huk – Bhnam Ben Ron Viriyah 2508 [1966] [mf ed 1990] – 1r with other items – 1 – (in khmer) – mf#mf-10289 seam reel 097/04 [§] – us CRL [480]
Kumar Maitra, Susil *see* Madhva logic
Kumar nam sar : [a novel] / Lyiv Chin – Bhnam Ben: Pannagar Vayo Thmi 2505 [1962] [mf ed 1990] – 1r with other items – 1 – (in khmer) – mf#mf-10289 seam reel 126/7 [§] – us CRL [830]
Kumarappa, Bharatan *see*
- Capitalism, socialism, or villagism?
- Drink, drugs and gambling
- The hindu conception of the deity as culminating in ramanuja
- The indian struggle for freedom
- My student days in america
- On tour with gandhiji
- Rebuilding our villages
- A righteous struggle
- Towards new education
- Towards non-violent socialism

Kumarappa, J C *see* The nation's voice
Kumarappa, J M *see* Our beggar problem
Kumarappa, Jagadisacandra *see*
- Gandhian economic thought
- Practise and precepts of jesus
- Public finance and our poverty
- A survey of matar taluka
- Swaraj for the masses

Kumarappa, Joseph Cornelius *see*
- The economy of permanence
- Education for life
- The gandhian economy and other essays
- An overall plan for rural development
- Peace and prosperity
- The philosophy of work and other essays
- War

Kumazivandadzoka / Marangwanda, John W – Salisbury, Zimbabwe. 1964 – 1r – us UF Libraries [960]
Kumban yakkha punna ka : [a novel] / Sin Yan, Do – Ran Kun: Aun pan pum nhip tuik 1976 [mf ed 1990] – 1r with other items – 1 – (in burmese) – mf#mf-10289 seam reel 171/2 [§] – us CRL [830]
Kumbler, G *see* Desiat' let raboty moskovskogo gorodskogo banka
Kumm, Hermann Karl Wilhelm *see*
- African missionary heroes and heroines
- Khont-hon-nofer
- The sudan

Kummer, Carl Ferdinand *see* Die jungfrau von orleans in der dichtung
Kummer, Frederick Arnold *see* The brute
Kummer, Friedrich *see*
- Deutsche literaturgeschichte des neunzehnten jahrhunderts

Kummer, Rolf *see* Die frankfurter berichte gustavs v. meyern-hohenberg
Kummerle, Salomon *see* Encyclopedie der evangelischen kirchenmusik
Der kumpel – Halle S DE, 1960-1965 10 sep – 1r – 1 – (bkw "geiseltal mitte") – gw Misc Inst [074]
Der kumpel – (Duisburg-) Walsum DE, 1958-jan 1 1961 – 1r – 1 – gw Misc Inst [331]
Der kumpel ruft – Nachterstedt DE, 1949 15 feb-1951 nov, 1952, 1954-1979 14 may – 5r – 1 – (veb braunkohlenkombinat) – gw Misc Inst [622]
Kumquats / Hume, H Harold – Lake City, FL. 1902 – 1r – us UF Libraries [634]
Kumu hawaii – Honolulu HI. 1834 nov 26 – 1r – 1 – mf#819119 – us WHS [071]
Kun bhon khet cac tam / So, Mon – [Ran kun]: Gun thu Ca pe mha phran khyi sann 1976 [mf ed 1994] – on pt of 1r – 1 – mf#11052 r1716 n9 – us Cornell [959]
Kun bon rha pum to / Nnui Mra, U, Aui ve – Ran Kun: Mran ma Ca pe tuik 1982 [mf ed 1990] – [pl/ill] 1r with other items – 1 – (in burmese) – mf#mf-10289 seam reel 136/4 [§] – us CRL [959]
K'un ch'u hsin tao – Shang-hai: Chung-hua shu chu, Min kuo 20 [1931] – (= ser P-k&k period) – us CRL [820]
K'un ch'ung chih shih = Entomological knowledge – Beijing, China 1959-60 – 1 – mf#2630 – us UMI ProQuest [590]
K'un ch'ung hsueh pao = Journal of entomology – Ann Arbor MI 1959-64 – 1 – mf#2631 – us UMI ProQuest [590]
K'un hsueh chi / Cheng, Chen-to – Ch'ang-sha: Shang wu yin shu kuan, 1941 – (= ser P-k&k period) – us CRL [480]
Kun qakatannu : [a novel] / Qum Samgan – Bhnam Ben: Pannagar Saravant 2505 [1963] [mf ed 1990] – 1r with other items – 1 – (in khmer) – mf#mf-10289 seam reel 108/11 [§] – us CRL [830]
Die kunama-sprache in nordost-afrika / Reinisch, L – Wien, 1881-1889/1890. 3v – 4mf – 9 – (missing: 1881, v1) – mf#NE-20261 – ne IDC [956]
Kunang-kunang / Balai Pustaka – Djakarta, 1949-1951(2) – 2mf – 9 – (missing: 1949(1-2, 4-end); 1950; 1951(1)) – mf#SE-1758 – ne IDC [950]
Kundamala of dinnaga / ed by Shastri, Jai Chandra – Lahore: Punjab Sanskrit Book Depot, 1932 – (= ser Samp: indian books) – (with sanskrit comm by ed; trans into english with int, critical notes, etc by veda vyasa and s d bhanot) – us CRL [490]
Kunene, Mazisi *see*
- An analytical survey of zulu poetry
- Zulu poems

Kunene, Stanley *see* Traits considered important in the selection of a marriage partner among young matriculated blacks
Kunene-sambesi-expedition / Warburg, O – Boston. 1907-1990 (1) 1972-1990 (5) 1973-1990 (9) – 7mf – 9 – mf#8035 – ne IDC [960]
Kung ch'an kuo chi kang ling – [China]: Chung-kuoch'u pan she, Min kuo 28 [1939] – (= ser P-k&k period) – us CRL [335]
Kung ch'ang chien ch'a kai lun / Liu, Chu-huo – Shang-hai: Shang wu yin shu kuan, Min kuo 23 [1934] – (= ser P-k&k period) – us CRL [360]
Kung ch'ang she chi / Ling, Hung-hsun – Shang-hai: Shang wu yin shu kuan, 1934 – (= ser P-k&k period) – us CRL [600]
Kung chen chuan k'an – [China]: Hu-pei shui tsai shan hou wei yuan hui, [1933] – (= ser P-k&k period) – us CRL [360]
Kung chiao lun / Ch'en, Hsiang-po – Shang-hai: Shang wu yin shu kuan, Min kuo 30 [1941] – (= ser P-k&k period) – us CRL [240]
Kung, Chieh *see* Shih ti ch'ou pei tzu chih hui pien
K'ung ch'ueh tan : ssu mu pei chu / Kuo, Mo-jo – Ch'ung-ch'ing: Ch'un i ch'u pan she, Min kuo 34 [1945] – (= ser P-k&k period) – us CRL [951]
K'ung ch'ueh tung nan fei chi ch'i t'a tu mu chu / Yuan, Ch'ang-ying – Ch'ang sha: Shang wu yin shu kuan, Min kuo 29 [1940] – (= ser P-k&k period) – us CRL [820]
K'ung chun k'ang chan chi lueh / Wu, Liang-fu – Nan-ching: Chan cheng ts'ung k'an she, 1937 – (= ser P-k&k period) – us CRL [951]
K'ung, Ch'ung *see* Hsien cheng chien she
Kung ho jih pao = The justice daily news – New York: Kong Wo Yat Bo Pub Co, oct 22 1928-dec 21 1929 – 3r – 1 – us CRL [071]
Kung ho pao = The justice news – New York: Kong Wo Bo Pub Co, jan 4 1930-apr 8 1933 – 2r – 1 – us CRL [071]
K'ung hsiang chu i (ccm105) = Commonism: the 10th chapter of christianity and communism / Chou, Pai-ch'in – 1st ed. Hong Kong, 1953 [mf ed 198?] – (= ser Ccm 105) – 1 – (pref & autobiogr sketch in english) – mf#1984-b500 – us ATLA [360]
K'ung, Hsiang-hsi *see*
- Hsien tsai shih hsing ti so te shui
- Kuo min ching shen tsung tung yuan chih yao i yu ssu wei chih ch'an yang

Kung hsien i tien cheng li pen tang ti i chien / Liu, Chien-ch'un – [China: sn, 1931] – (= ser P-k&k period) – us CRL [951]
Kung, Hsien-ming *see* She hui pao hsien chih li lun yu shih chi
K'ung, Ling-ching *see* Ch'en hsiang chi
Kung min chiao yu / Hsiung, Tzu-jung – Shang-hai: Shang wu yin shu kuan, Min kuo 22 [1933] – (= ser P-k&k period) – us CRL [303]
Kung paano namumuhay at gumagawa ang mga tao : ang "bagong" araling panlipunan / Agno, Lydia Navarro – Quezon City: JMC Press, c1975 [mf ed 1987] – xvi/264p (ill) – 1 – mf#6774 – us Wisconsin U Libr [490]
Kung pao (ccs) = General assembly record – Shanghai. v2-26. 1930-54 [complete] [mf ed 198?] – (= ser Chinese christian serials coll) – 1 – (aka: tsung hui kung pao) – mf0296n – us ATLA [240]
Kung shang pan-yueh k'an – Shanghai, China. 1929-Jan 1936. Semi-monthly economic journal – 1 – 572.00 – us Chinese Res [330]
Kung shang shih liao ti i chi – [China]: Shang-hai chi chih Kuo huo Kung ch'ang lien ho hui fa hsing pu, 1935 – (= ser P-k&k period) – us CRL [338]
Kung shang t'ung chi / Yao, Hsiao-lien – [China: sn, 1943] – (= ser P-k&k period) – us CRL [315]
Kung ssu ts'ai cheng / K'ung, Ti-an – Shang-hai: Shang wu yin shu kuan, Min kuo 24 [1935] – (= ser P-k&k period) – us CRL [332]
K'ung su / Pa, Chin – Shang-hai: Feng huo she, Min kuo 26 [1937] – (= ser P-k&k period) – us CRL [480]
Kung su chieh fa wang ming-tao fan ko ming chi t'uan (ccm294) – Shanghai, 1955 [mf ed 198?] – (= ser Ccm 294) – 1 – mf#1984-b500 – us ATLA [240]
K'ung, Ta-ch'ung *see* Pi chiao ti fang cheng fu t'u piao
K'ung, Ti-an *see* Kung ssu ts'ai cheng
Kung, Tien-min *see*
- Seng lu hsin chu chi
- Tang ch'ao chi-tu chiao chih yen chiu

Kung, Tu *see* Tsui chin jih-pen chih chun pei kai k'uang
Kung tzu li lun chih fa chan / Fan, Hung – Shang-hai: She hui tiao ch'a so: Shang wu yin shu kuan, Min kuo 23 [1934] – (= ser P-k&k period) – us CRL [331]
Kung wu chi yao : min kuo 22 nien fen – [China]: Ch'ing-tao shih kung wu chu, [1933] – (= ser P-k&k period) – us CRL [315]
Kung yeh an ch'uan yu kuan li / Ch'eng, Shou-chung – Shang-hai: Chung hua shu chu tai shou: Sheng huo shu tien tai shou, [1933] – (= ser P-k&k period) – us CRL [650]
Kung yeh chien she yu chin jung cheng ts'e / Yang, Shou-piao – Ch'ung-ch'ing: Shang wu yin shu kuan, Min kuo 34 [1945] – (= ser P-k&k period) – us CRL [330]
Kung yeh yuan liao i chun hsu yuan liao? / Plummer, Alfred – Ch'ang-sha: Shang wu yin shu kuan, Min kuo 28 [1939] – (= ser P-k&k period) – us CRL [338]
Kungaelvsposten – Goeteborg, Moelndal, Sweden. 1958-78, 1979- – 20r – 1 – sw Kungliga [078]
Kungliga Biblioteket-Royal Library. Stockholm *see* Union catalogue of foreign literature in swedish research libraries
Kungsbacka tidning – Kungsbacka, Sweden. 1978- – 1 – sw Kungliga [078]
Kungsbackaposten – Kungsbacka, Sweden. 2004- – 1 – sw Kungliga [078]
Kungsorstidningen karolinen – Koeping, 1907 – 1r – 1 – sw Kungliga [078]
Kung-sun, Chia *see* Feng shou i hou
Kungurskii vestnik – Kungur [Perm gub: Kungur paevoe t-vo 1919 [1919 28 ianv-] – (= ser Asn 1-3) – n2-116 [1919] [gaps] item 211, on reel n42 – 1 – mf#asn-1 211 – ne IDC [077]
Kung-yeh kai-tsao (ccs) = Industrial reform – Shanghai. n8-10 1926; n18 1929 [complete] [mf ed 198?] – (= ser Chinese christian serials coll) – 1 – mf0296o – us ATLA [338]
Kunhardt, George E *see* Lawrence
Kunisch, Richard *see* Bukarest und stambul
Kunitz, J D *see* Surinam und seine bewohner
Kunkel, Franz Joseph *see* Kritische beleuchtung des c f weitzmann'schen harmoniesystems, und des schriftchens: "die neue harmonielehre im streit mit der alten"
Kunkelman, J A *see* The essays, debates, and proceedings
K'un-ming tung ching / Shen, Ts'ung-wen – Kuei-lin: Wen hua sheng huo ch'u pan she, Min kuo 30 [1941] – (= ser P-k&k period) – us CRL [840]
Kunneke, Kathleen Joey *see* The paradigmatic shift of service organisations
Kunos, Ignatz *see* Tuerkce ninniler
Kunow, Hanns von *see* Michel und die schwestern
Kunpei, Gamo *see* Megalithgraeber in yamato aus drei perioden der kofun-zeit
Kunsan Air Base [Korea] *see* Wolf pack warrior
K'un-shan hsien hsien cheng fu pao kao – [China: K'un-shan Hsien cheng fu], 1936 – (= ser P-k&k period) – us CRL [350]
Kunst / ed by Shetelig, Haakon – Stockholm: A Bonnier 1931 [mf ed Bloomington IN: Indiana Uni Lib, Preservation Dept 1984] – 466p on 1r [ill] – 1 – us Indiana Preservation [390]
Kunst – Vienna, Austria jan-dec 1903 [mf ed Norman Ross] – 1r – 1 – mf#nrp-1980 – us UMI ProQuest [700]
Die kunst adalbert stifters / Kuehn, Julius – 2. Aufl. Berlin: Junker und Duennhaupt, 1943 – 1 – (incl bibl ref) – us Wisconsin U Libr [430]
Kunst als medium sozialer konflikte : bilderkampfe von d. spatantike bis z. hussitenrevolution / Bredekamp, Horst – 1. Aufl. Frankfurt am Main: Verlag Suhrkamp, 1975 – us CRL [302]
Die kunst das clavier zu spielen : von verfasser des kritischen musicus an der spree / Marpurg, Friedrich Wilhelm – 3. verb verm aufl, Berlin: Haude & Spener 1760-61 [mf ed 19–] – 2pt on 3mf – 9 – mf#fiche 429 – us Sibley [780]
Die kunst der erzaehlung / Wassermann, Jakob – Berlin: Bard Marquardt, [1904] – 1 – us Wisconsin U Libr [430]
Die kunst der fuge durch herrn johann sebastian bach ehemaligen capellmeister zu leipzig / Bach, Johann Sebastian – [Leipzig 1752] [mf ed 19–] – 1r / 3mf – 1,9 – mf#film 790 / fiche406 – us Sibley [780]
Die kunst der interpretation : studien zur deutschen literaturgeschichte / Staiger, Emil – Zuerich: Atlantis Verlag, c1955 – 1 – (incl bibl ref) – us Wisconsin U Libr [430]
Kunst der nation – Berlin DE, 1933-35 – 1 – gw Misc Inst [700]
Die kunst der reduktionstechnik bei igor strawinsky : dargestellt am werk "le sacre du printemps" / Petri, Hasso Gottfried – [mf ed 2002] – 5mf – 9 – €59.00 – 3-8267-2777-0 – mf#DHS2777 – gw Frankfurter [780]
Die kunst, ihr wesen und ihre gesetze / Holz, Arno – Berlin: Issleib, 1891 – 1r – 1 – us Wisconsin U Libr [100]
Die kunst, ihr wesen und ihre gesetze : neue folge / Holz, Arno – 2. Aufl. Berlin: W Issleib, 1893 – 1r – 1 – us Wisconsin U Libr [100]
Die kunst im dritten reich – Muenchen DE, 1937-39, 1940 [gaps], 1941-42, 1943-44 [gaps] – 1 – gw Misc Inst [934]
Kunst im kino – Berlin DE, 1912 n1, 1913 n2, 4, 6, 8, 9, 13, 31, 1914 n21 – 1 – gw Mikrofilm [790]
Die kunst in den athos-kloestern / Brockhaus, Heinrich – Leipzig: F A Brockhaus 1891 [mf ed 1992] – 4mf [ill] – 9 – 0-524-02973-3 – (incl bibl ref) – mf#1990-0760 – us ATLA [700]
Kunst, mythos, wissenschaft / Rapp, Friedrich – Dortmund: projekt vlg, 1993 (mf ed 1996) – 2mf – 9 – €31.00 – 3-8267-9715-9 – mf#DHS 9715 – gw Frankfurter [700]
Kunst oder tractat der malerei des cennicennini da colle di valdelsa / [Cennini, C] Ilg, A – Wien, 1871. v1 – 3mf – 9 – mf#O-517 – ne IDC [700]
Kunst, religion und kultur : ansprache an die heidelberger studentenschaft gehalten bei der anlaesslich seiner ablehnung des rufs an die berliner universitaet veranstalteten feier / Thode, Henry – Heidelberg, 1901 (mf ed 1993) – 1mf – 9 – €19.00 – 3-89349-295-X – mf#DHS-AR 155 – gw Frankfurter [100]
"Kunst und kuenstler" 1902-1933 : eine zeitschrift in der auseinandersetzung um den impressionismus in deutschland / Paas, Sigrun – Heidelberg, 1976 – 3mf – 9 – 3-89349-390-5 – gw Frankfurter [700]
Kunst und kuenstler am vorabend der reformation : ein bild aus dem erzgebirge – Halle: Verein fuer Reformationsgeschichte, 1890 – (= ser [Schriften des Vereins fuer Reformationsgeschichte]) – 1mf – 9 – 0-7905-4654-X – (incl bibl ref) – mf#1988-0654 – us ATLA [700]
Kunst und leben : vortrage und abhandlungen zur deutschen literatur / Strich, Fritz – Bern: Francke, 1960 [mf ed 1993] – 241p – 1 – mf#8138 – us Wisconsin U Libr [430]
Kunst und literatur – 1966-1983 – 537mf – 1 – gw Mikropress [400]
Kunst und natur : blaetter aus meinem reisetagebuche / Klingemann, Ernst – Braunschweig 1819-28 [mf ed Hildesheim 1995-98] – 3v on 9mf – 9 – €180.00 – 3-487-29604-7 – gw Olms [880]
Kunst- und natur-skizzen aus nord- und sued-europa : ein reise-tagebuch / Woermann, Karl – Duesseldorf 1880 [mf ed Hildesheim 1995-98] – 2v on 6mf – 9 – €120.00 – 3-487-29918-6 – gw Olms [880]
Kunst und reichtum deutscher prosa : von lessing bis nietzsche / Jancke, Oskar [comp] – Muenchen: R Piper, [1942?] [mf ed 1993] – 397p – 1 – mf#8362 – us Wisconsin U Libr [430]

Kunst und religion : untersuchung des problems an schleiermachers reden ueber die religion / Mendelssohn-Bartholdy, Herbert – Erlangen: E Th Jacob, 1912 – 1mf – 9 – 0-7905-9346-7 – (incl bibl ref) – mf#1989-2571 – us ATLA [200]

Kunst- und unterhaltungsblatt fuer stadt und land – Stuttgart DE, 1952-54 – 1r – 1 – gw Misc Inst [074]

Kunst- und volkslied in der reformationszeit / Kinzel, Karl [comp] – Halle/S: Verlag der Buchhandlung des Waisenhauses, 1892 [mf ed 1993] – (= ser Denkmaeler der aelteren deutschen literatur 3/4) – viii/140p – 1 – (incl bibl ref) – mf#8185 – us Wisconsin U Libr [780]

Die kunst zu moduliren und zu praeludiren : ein praktischer beitrag zur harmonielehre / Jadassohn, Salomon – Leipzig: Druck & Verlag von Breitkopf & Haertel 1890 [mf ed 1991] – 1r – 1 – mf#pres. film 102 – us Sibley [780]

Die kunstbestrebungen am bayerischen hofe unter den herzogen albert 5 und wilhelm 5 / Stockbauer, J – Wien. v8. 1874 – 2mf – 9 – mf#O-517 – ne IDC [700]

Kunstbuechlein... / Amman, J – Franckfurt am Mayn, 1599 – 4mf – 9 – mf#O-1017 – ne IDC [700]

Kunsternovellen / Mann, Heinrich – Illustrationen Bert Heller, Auswahl und Nachwort Helga Bemmann, Berlin: Henschelverlag, 1961 – 1 – us Wisconsin U Libr [830]

Die kunstform der althebraeischen poesie / Euringer, Sebastian – 1. & 2. aufl. Muenster i W: Aschendorff 1912 [mf ed 1993] – (= ser Biblische zeitfragen 5/9-10) – 1mf – 9 – 0-524-05720-6 – (incl bibl ref) – mf#1992-0563 – us ATLA [470]

Der kunstfreund – Bozen, 1885-91 – (= ser Architectural periodicals at avery library, columbia university) – 1r – 1 – $155.00 – us UPA [720]

Kunstgeschichtliche forschungen des koeniglich-preussischen historischen instituts in rom – Leipzig, 1910-1912. v1-3 – 20mf – 8 – mf#H-684 – ne IDC [700]

Kunstgeschichtliche grundbegriffe / Woelfflin, H – Muenchen, 1915 – 4mf – 9 – mf#O-466 – ne IDC [700]

Kunstgeschichtliches anschauungsmaterial zu goethes italienischer reise / Ziehen, Julius – 2nd ed. Bielefeld; Leipzig: Velhagen & Klasing, 1923 – 1r – 1 – us Wisconsin U Libr [430]

Kunstgeschichtliches jahrbuch der k k zentral-kommission... – Wien, 1907-1910. v1-4+suppl – 175mf – 9 – (cont as: jahrbuch des kunsthistorischen institutes...[wien 1911-20] v5-14; cont as: jahrbuch fuer kunstgeschichte...[wien 1921/22] v1(15); cont as: wiener jahrbuch fuer kunstgeschichte [wien 1923-37] v2(16)-11) – mf#O-526c – ne IDC [700]

Kunstkamera / Rossiiskaia akademiia nauk. Muzei antropologii i etnografii imeni Petra Velikogo (Kunstkamera), Tentr "Peterburgskoe vostokovedenie" – Sankt-Peterburg: Muzei, n1 1993 – us CRL [301]

Kunstmann, W G *see* Die babylonische gebetsbeschwoerung

Kunstmann, Walter G *see* Die babylonische gebetsbeschwoerung

Ein kunstreych buch von allerley antiquiteten : so zum verstand der fuenff seulen der archltectur gehoerend / [Blum, H] – Zuerich: Froschauer, n.d. – 1mf – 9 – mf#OA-62 – ne IDC [700]

Kunststoffberater rundschau + technik – Ann Arbor MI 1975-78 – 1,5,9 – ISSN: 0340-8442 – mf#7827.01 – us UMI ProQuest [660]

Der kunstwart : Rundschau ueber alle gebiete des schoenen; monatshefte fuer kunst, literatur und leben – Muenchen [1] 1887-[50] 1936/37 [mf ed 2006] – (= ser Kultur – literatur – politik 20) – 646mf – 9 – diazo €3500.00 silver €4200.00 – 3-89131-468-X – (fr 26th 1912/13 variable titles) – gw Fischer [700]

Kunstwart *see* Das froehliche buch

Kunstwerk = Work of art – Stuttgart, Germany 1946-91 – 1,5,9 – ISSN: 0023-561X – mf#2721 – us UMI ProQuest [700]

Das kunstwerk der zukunft und sein meister richard wagner / Schmid, Theodor – Freiburg im Breisgau: Herder 1885 – 3mf – 9 – mf#wa-98 – ne IDC [780]

Das kunstwerk richard wagners / Istel, Edgar – Leipzig: B G Teubner 1910 [mf ed 1991] – (= ser Aus natur und geisteswelt 330) – 1r – 1 – (incl bibl ref. filmed with: wagner und nietzsche / kurt hildebrandt) – mf#3024p – us Wisconsin U Libr [780]

Kunstwerke aus el-amarna / Schaefer, H – Berlin, 1923. 2v – 1mf – 9 – mf#NE-20406 – ne IDC [956]

Kunterbunt / Johst, Hanns – Bielefeld: Velhagen & Klasing, [1940] – 1r – 1 – us Wisconsin U Libr [830]

Kuntres alon bakhut – Szineraaralja, Hungary. 1932 – 1r – us UF Libraries [939]

Kuntres ma'amarim ve-sihot mi-kevod kedosho admormi- lubavitsh / Schneersohn, Joseph Isaac – Warsaw, Poland. 1930 or 1931 – 1r – us UF Libraries [939]

Kuntz, Marthe *see* Ombres et lumieres

Kunz, Dorothea E *see* The effects of a project learning tree workshop on pre-service teachers' attitudes toward teaching environmental education

Kunz, Grete *see* Waechter der heimat

Kunze, August *see* Die pariser boulevards

Kunze, G *see* Die gottesdienstliche schriftlesung 1

Kunze, Johannes *see*
- The apostles' creed and the new testament
- Das apostolische glaubensbekenntnis und das neue testament
- Evangelisches und katholisches schriftprinzip
- Die ewige gottheit jesu christi
- Glaubensregel, heilige schrift und taufbekenntnis
- Die gotteslehre des irenaeus
- Marcus eremita
- Das nicaenisch-konstantinopolitanische symbol
- Die rechtfertigungslehre in der apologie

Kunze, Kurt *see* Die dichtung richard dehmels als ausdruck der zeitseele

Kunzemann, Gertrud *see*
- Lauter kleinigkeiten
- Wiedergeboren

Kuo chi cheng chih yu chung jih wen t'i / Pao, Hua-kuo – Shang-hai: Hua ch'iao t'u shu yin shua kung ssu, Min kuo 23 [1934] – (= ser P-k&k period) – us CRL [327]

Kuo chi chi t'uan ching chi / Chao, I-p'ing – Shang-hai: Sheng huo shu tien, Min kuo 23 [1934] – (= ser P-k&k period) – us CRL [337]

Kuo chi ching chi cheng chih nien pao ti 1 chi / Wang, P'eng-fang – Shang-hai: Shen chou kuo kuang she, 1932 – (= ser P-k&k period) – us CRL [327]

Kuo chi ching chi cheng ts'e, yu ming, chung-kuo tui wai ching chi cheng ts'e chih yen chiu / Ho, Ssu-yuan – Shang-hai: Shang wu yin shu kuan, Min kuo 21 [1932] – (= ser P-k&k period) – us CRL [327]

Kuo chi fa ta kang / Chou, Keng-sheng – Shang-hai: Shang wu yin shu kuan, Min kuo 22 [1933] – (= ser P-k&k period) – us CRL [341]

Kuo chi ho p'ing chi kou ju ho chien li? – Ch'ung-ch'ing: Ta kung pao kuan, Min kuo 34 [1945] – (= ser P-k&k period) – us CRL [951]

Kuo chi hsien shih yu chung-kuo ko ming / Ting, Li-san – Shang-hai: Ta tung shu chu, Min kuo 20 [1931] – (= ser P-k&k period) – us CRL [951]

Kuo chi hsien shih yu k'ang chan wai chiao / Ch'en, Chung-hao – Wu-han: Wu-han jih pao she, Min kuo 27 [1938] – (= ser P-k&k period) – us CRL [951]

Kuo chi hsin wen tz'u tien / Pin, Fu – Kuei-lin: Yueh ch'un shu tien, Min kuo 32 [1943] – (= ser P-k&k period) – us CRL [327]

Kuo chi hsing shih chiang yen chi – Nan-ching: Chun yung t'u shu she, Min kuo 22 [1933] – (= ser P-k&k period) – us CRL [934]

Kuo chi hsing shih yu k'ang chan ch'ien t'u / Kuo, Mo-jo – Han-k'ou: Tzu ch'iang ch'u pan she, Min kuo 27 [1938] – (= ser P-k&k period) – us CRL [951]

Kuo chi kung fa kang yao / Wu, K'un-wu – Shang-hai: Shang wu yin shu kuan, Min kuo 26 [1937] – (= ser P-k&k period) – us CRL [341]

Kuo chi man hsieh / T'ao, Chu-yin – Shang-hai: K'un lun shu tien, Min kuo 28 [1939] – (= ser P-k&k period) – us CRL [934]

Kuo ch'i p'iao tsai ya-ch'ueh-chien / Tsang, K'o-chia – Chung-hua ch'uan kuo wen i chieh k'ang ti hsieh hui Ch'eng-tu fen hui chu pien – (= ser P-k&k period) – us CRL [810]

Kuo chi tien hsin kung yueh : fu shu wu hsien tien kuei tse – Shang-hai: Ya Mei ku fen yu hsien kung ssu, Min kuo 24 [1935] – (= ser P-k&k period) – us CRL [380]

Kuo chi ts'ai chun wen t'i / Chang, Ming-yang – Shang-hai: Chung-hua shu chu, Min kuo 23 [1934] – (= ser P-k&k period) – us CRL [327]

Kuo chi wen t'i chiang hua / Chang, Ch'in-fu – Shang-hai: Sheng huo shu tien, Min kuo 28 [1939] – (= ser P-k&k period) – us CRL [327]

Kuo chi wen t'i tz'u hui / Yang, Li-ch'iao & Chiang, Yin-en – Ch'ang-sha, Shang wu yin shu kuan, Min kuo 30 [1941] – (= ser P-k&k period) – us CRL [327]

Kuo chia chih shang : ssu mu chu / Lao, She – Ch'ung-ch'ing: Nan fang yin shu kuan, 1943 – (= ser P-k&k period) – us CRL [820]

Kuo chia tsung tung yuan ti shih chi wen t'i / Chu, Hsiao-ch'un – Ch'ung-ch'ing: Kuo min t'u shu ch'u pan she, Min kuo 31 [1942] – (= ser P-k&k period) – us CRL [951]

Kuo ch'u ti sheng ming / Chou, Tso-jen – Shang-hai: Pei hsin shu chu, 1933 – (= ser P-k&k period) – us CRL [810]

Kuo, Chung-i *see* Ch'eng jen chih lu [ccm209]

Kuo fang chi pen jen ts'ai chi kuo fang wu tzu kung ch'eng chi shu jen ts'ai hsun lien pan fa ta kang / Tu, Chien-shih – [China]: Kuo fang yen chiu yuan, Min kuo 32 [1943] – (= ser P-k&k period) – us CRL [355]

Kuo fang chih pen i yu tseng ch'iang kuo fang chih t'i ch'ang – [China: sn], Min kuo 23 [1934] – (= ser P-k&k period) – us CRL [355]

Kuo fang hsin lun / Yang, Chieh – [Ch'ung-ch'ing]: Chun shih wei yuan Hui cheng chih pu, Min kuo 31 [1942] – (= ser P-k&k period) – us CRL [355]

Kuo fang k'o hsueh yuan li / Yu, Cheng – Kuei-lin: Kuo fang shu tien, Min kuo 30 [1941] – (= ser P-k&k period) – us CRL [355]

Kuo fang lun / Chiang, Pai-li – [China: sn] – (= ser P-k&k period) – us CRL [355]

Kuo fang ti li / Hu, Huan-yung – [China]: Ch'ing nien shu tien, Min kuo 27 [1938] – (= ser P-k&k period) – us CRL [355]

Kuo fang yu liang shih wen t'i / Yin, I-hsuan – Nan-ching: Cheng chung shu chu, Min kuo 26 [1937] – (= ser P-k&k period) – us CRL [355]

Kuo, Feng-kang *see* Hua hsueh ping ch'i chih yen chiu

Kuo fu ssu hsiang yen chiu / Yang, Ts'an – Chiang-hsi: Shih tai ssu ch'ao she, Min kuo 31 [1942] – (= ser P-k&k period) – us CRL [951]

Kuo, Hou-chueh *see* Min yueh yu ho kuo yu tui chao chi

Kuo, Hsuan *see* Ou-chou chan cheng yu chung-kuo

Kuo hun shih hsuan : shang, chung, hsia ts'e – Shang-hai: Hsin Chung-kuo chien she hsueh hui, Min kuo 23 [1934] – (= ser P-k&k period) – us CRL [810]

Kuo jen chieh yueh-han chien wang ching-wei / Wu, Ching-heng – [Sl: sn] – (= ser P-k&k period) – us CRL [920]

Kuo, Jen-ch'uan *see* Hsiang ts'un hsiao hsueh hsing cheng

Kuo kung chih chien / Tsou, Yang – [China]: Li shih tzu liao kung ying she, Min kuo 24 [1935] – (= ser P-k&k period) – us CRL [325]

Kuo li chi nan ta hsueh chao sheng pao kao tsung pien – [China]: Chi nan ta hsueh, 1933 – (= ser P-k&k period) – us CRL [370]

Kuo li ch'ing hua ta hsueh wen hsueh yuan chung-kuo wen hsueh hsi – [China]: Kuo li ch'ing hua ta hsueh wen hsueh yuan, 1933-1934 – (= ser P-k&k period) – us CRL [370]

Kuo li chung-shan ta hsueh chiao yu yen chiu so kai k'uang – [China]: Kuo li Chung-shan ta hsueh chiao yu yen chiu so, 1934 – (= ser P-k&k period) – us CRL [951]

Kuo li chung-shan ta hsueh chun shih k'o pen chuan 2 – [China: Chung-shan ta hsueh], Min kuo 22 [1933] – (= ser P-k&k period) – us CRL [355]

Kuo li pei-p'ing shih fan ta hsueh i lan – [China]: Kuo li Pei-p'ing shih fan ta hsueh, Min kuo 23 [1934] – (= ser P-k&k period) – us CRL [951]

Kuo li pei-yang kung hsueh yuan chi hsieh kung ch'eng hsueh hsi kai k'uang – [China: Kuo li Pei-yang Kung hsueh yuan, 1935] – (= ser P-k&k period) – us CRL [951]

Kuo li shang-hai i hsueh yuan i lan – [China: Kuo li Shang-hai i hsueh yuan], Min kuo 25 [1936] – (= ser P-k&k period) – us CRL [951]

Kuo lien tiao ch'a t'uan pao kao shu : chung ying wen ho k'an pen: fu, shih chieh ko kuo jen shih chih i chien – Shang-hai: Shang-hai shen she, Min kuo 21 [1932] – (= ser P-k&k period) – us CRL [341]

Kuo min cheng fu cheng li nei wai chai wei yuan hui pao kao shu : min kuo erh shih liu nien erh yueh – [China]: Kuo min cheng fu cheng li nei wai chai wei yuan hui, Min kuo 26 [1937] – (= ser P-k&k period) – us CRL [336]

Kuo min cheng fu ti chan shih t'i chih – [China]: Hsuan ch'uan pu: Chung yang shu pao fa hsing so, Min kuo 32 [1943] – (= ser P-k&k period) – us CRL [951]

Kuo min cheng fu ts'ai cheng kai k'uang lun / Yang, Ju-mei – [Kuang-chou]: Chung-hua shu chu, Min kuo 27 [1938] – (= ser P-k&k period) – us CRL [332]

Kuo min cheng fu t'ung i chieh shih fa ling hsu pien / Yu, Chung-lo & Wu, Hsueh-p'eng – Shang-hai: Shang-hai lu shih kung hui, Min kuo 22 [1933] – (= ser P-k&k period) – us CRL [340]

Kuo min cheng fu t'ung i chieh shih fa ling hsu pien ti erh chi / Yu, Chung-lo & Wu, Hsueh-p'eng – Shang-hai: Shang-hai lu shih kung hui, Min kuo 23 [1934] – (= ser P-k&k period) – us CRL [340]

Kuo min cheng fu t'ung i chieh shih fa ling hui pien / Yu, Chung-lo & Wu, Hsueh-p'eng – Shang-hai: Shang-hai lu shih kung hui, Min kuo 21 [1932] – (= ser P-k&k period) – us CRL [340]

Kuo min ching chi chien she chih t'u ching / Tung, Hsiu-chia – Shang-hai: Sheng huo shu tien, 1936 – (= ser P-k&k period) – us CRL [951]

Kuo min ching chi chien she yun tung / Li, Hsien-yun – Ch'ung-ch'ing: Kuo min t'u shu ch'u pan she, Min kuo 30 [1941] – (= ser P-k&k period) – us CRL [330]

Kuo min ching shen tsung tung yuan chih yao i yu ssu wei chih ch'an yang / K'ung, Hsiang-hsi – Meng Tsang wei yuan hui pien i shih, Min kuo 29 [1940] – (= ser P-k&k period) – us CRL [951]

Kuo min chun hsun / Wang, Hsing-yuan & T'u I-fang – Ch'ung-ch'ing: Tu li ch'u pan she, Min kuo 30 [1941] – (= ser P-k&k period) – us CRL [355]

Kuo min chun shih ch'ang shih / Chiang, K'uei-wu – Shang-hai: Kuo min chun shih ch'ang shih pien i she, Min kuo 26 [1937] – (= ser P-k&k period) – us CRL [350]

Kuo min ta hui tai piao hsuan chu chih nan / China – Shang-hai: Hsin kuang shu chu, 1936 – (= ser P-k&k period) – us CRL [325]

Kuo min ta hui ts'an k'ao tzu liao / Keng, Wen-t'ien – Shang-hai, Chung-hua shu chu, Min kuo 25 [1936] – (= ser P-k&k period) – us CRL [951]

Kuo min tao te lun / Teng, Hsi – Ch'ung-ch'ing: Kuo min t'u shu ch'u pan she, Min kuo 31 [1942] – (= ser P-k&k period) – us CRL [170]

Kuo, Mo-jo *see*
- Chan sheng
- Chin hsi chi
- Fan cheng ch'ien hou
- Hsien ch'in hsueh shuo shu lin
- Hu fu
- K'ung ch'ueh tan
- Kuo chi hsing shih yu k'ang chan ch'ien t'u
- Li shih hsiao pin
- Mo-jo chin chu
- Nan kuan ts'ao, i ming, chin feng chien yu i
- Pei fa
- Po
- P'u chien chi
- Shang-hai k'ang chan chi
- Shih p'i p'an shu
- Shih t'i
- T'ang ti chih hua
- Tsai hung cha chung lai ch'u
- T'ung nien shih tai
- Wen i yu hsuan ch'uan
- Wo ti chieh hun
- Yu shu chi

Kuo, Mo-jo, *see* Min tsu hsing shih shang tui

Kuo mo-jo chuan / Yang, Yin-fu – Shang-hai: Min chung ch'u pan she, Min kuo 27 [1938] – (= ser P-k&k period) – us CRL [920]

Kuo, Mo-jo et al *see* Wen i hsin lun

Kuo mo-jo hsien sheng tsui chin yen lun chi – Kuang-chou: Li sao ch'u pan she, Min kuo 27 [1938] – (= ser P-k&k period) – us CRL [951]

Kuo mo-jo kuei kuo mi chi / Yin, Ch'en – [China]: Yen hsin ch'u pan she, 1945? – (= ser P-k&k period) – us CRL [951]

Kuo mo-jo wen hsuan / Shen, Wen-yao – Shang-hai: Shih tai ch'u pan she, Min kuo 24 [1935] – (= ser P-k&k period) – us CRL [480]

Kuo nei chin shih nien lai chih tsung chiao ssu chao (ccm9) = Religious thought movements in china during the last decade / ed by Chang, Chin-Shih – Peking, 1927 [mf ed 198?] – (= ser Ccm 9) – 1 – mf#1984-b500 – us ATLA [200]

Kuo nei hui tui chi ya hui yeh wu / Chou, Yang-wen – Shang-hai: Shang wu yin shu kuan, Min kuo 24 [1935] – (= ser P-k&k period) – us CRL [332]

Kuo nei t'uan chieh yu kuo wai fan hsiang / Yu, she – Hsiang-kang: Yu she, 1941 – (= ser P-k&k period) – us CRL [951]

Kuo, Shao-yu *see* Yu wen t'ung lun

Kuo shih ta kang / Ch'ien, Mu – [Ch'ung-ch'ing]: Kuo li pien i kuan, min kuo 33 [1944] – (= ser P-k&k period) – us CRL [951]

Kuo, Shou-i *see* Min shih su sung fa

K'uo ta pu tao te hu sheng (ccm255) = The summons to a larger evangelism / Mott, John Raleigh – Shanghai, 1929 [mf ed 198?] – (= ser Ccm) – 1 – (chinese trans fr the english) – mf#1984-b500 – us ATLA [240]

Kuo tsei wang ching-wie : ssu mu k'ang chan chu / Ma, Yen-hsiang – Ch'ung-ch'ing: Ch'ing nien ch'u pan she, Min kuo 30 [1941] – (= ser P-k&k period) – us CRL [480]

Kuo tu shih tai chih ssu hsiang yu chiao yu / Chiang, Meng-lin – Shang-hai: Shang wu yin shu kuan, Min kuo 22 [1933] – (= ser P-k&k period) – us CRL [370]

Kuo, Wei *see*
- Ping i fa shih i
- So te shui chan hsing t'iao li shih i

Kuo wen wen fa : yu wen hui t'ung / T'an, Cheng-pi – Shang-hai: Shih chieh shu chu, Min kuo 33 [1944] – (= ser P-k&k period) – us CRL [480]

Kuo yin hsueh sheng tzu hui / Fang, I & Ma, Ying – Ch'ang-sha: Shang wu yin shu kuan, Min kuo 27 [1938] – (= ser P-k&k period) – us CRL [480]
Kuo ying chao shang chu cheng li pao kao ti i hao – [China: Kuo ying chao shang chu], Min kuo 25 [1936] – (= ser P-k&k period) – us CRL [951]
Kuo ying shih yeh lun / Wu, Pan-nung & Chu, Yun-ying – Ch'ung-ch'ing: Chung-kuo wen hua fu wu she, Min kuo 33 [1944] – (= ser P-k&k period) – us CRL [338]
Kuo ying shih yeh ti fan wei wen t'i / Wu, Pan-nung – Ch'ung-ch'ing: Chung-kuo wen hua fu wu she, Min kuo 30 [1941] – (= ser P-k&k period) – us CRL [338]
Kuo yu chu yin fu hao fa yin fa / Lu, I-yen – Shang-hai: Chung-hua shu chu, 1940 – (= ser P-k&k period) – us CRL [480]
Kuo yu yun tung shih kang / Li, Chin-hsi – [Shang-hai]: Shang wu yin shu kuan, [1935] – (= ser P-k&k period) – us CRL [480]
Kuo, Yuan *see* Chan shih cheng li t'ien fu wen t'i
Kuo, Yuan-hsin et al *see* Shih jen chi
Kuo-chi mao-i ch'ing-pao : (foreign trade bulletin) – Shanghai, China. mar 1936-aug 1937 – 1 – 72.00 – us Chinese Res [380]
Kuo-chi mao-i tao-pao : (foreign trade monthly) – Shanghai, China. apr 1930-ju 1937 – 1 – us Chinese Res [380]
Kupava / Grebenshchikov, Georgii – Southbury, CT. 1936 – 1r – us UF Libraries [960]
Kuper, Hilda *see*
– Bite of hunger
– Indian people in natal
– Swazi
– The swazi
– Witch in my heart
Kuper, Leo *see*
– Durban
– Passive resistance in south africa
Kuperstein, Leib *see* Goral yehude romaniyah
Kuperus, L *see* Die zoekt, die vindt
Kupffer, Julius *see* Goethes faust als erzaehlung
Kupfuya nguruve / Mcnab, A P – Gwelo, Zimbabwe. 1966 – 1r – us UF Libraries [960]
Kupfuyiwa kwehuku – Gwelo, Zimbabwe. 1965 – 1r – us UF Libraries [960]
Kuphika / Chafulumira, E W – London, England. 1952 – 1r – us UF Libraries [960]
K'u-p'ing *see* Tso hsiao
Kuppers, Paul Erich *see* Das kestnerbuch
Die kupplung – Dessau DE, 1950 1 apr-1973 [gaps], 1975 [gaps], 1977-1993 9 dec [gaps] – 8r – 1 – (notes: waggonbau) – gw Misc Inst [621]
Kuppuswami Sastri, S *see*
– Highways and byways of literary criticism in sanskrit
– A primer of indian logic
Kuppuswamy, Bangalore *see* Educational reconstruction
Kuprian, Herman *see* Realidad y espiritualidad
Kupron *see* Stansinos
Kuraishi, Muhammad Hamid *see* List of ancient monuments protected under act 7 of 1904 in the province of bihar and orissa
Kuramoto, Anna K *see* Muscular endurance in women through adulthood
Kuranty = Chimes – Kiev. n1-10. may-oct 1918 – 4mf – 9 – us UMI ProQuest [790]
Kuranzeiger bad sooden-allendorf *see* Badeanzeiger fuer sooden an der werra
Die kurbel *see* Das schwungrad
Kurbykov I M *see* Put' sibiri
Kurdistan / ed by Zadah, Muhammad Sadiq Mufti – Hamadan ['Ikbatan], IR: 'Abd al-Majid Badi al-Zamani. n32-50. dec 1962, jan-apr 1963 (with gaps). ns?: n20,23. apr, nov 1969 – 1r – 1 – $90.00 – (in kurdish) – us MEDOC [079]
Kurdistan and the kurds / Ghassemlou, Abdul Rahman – Prague: Pub. House of the Czechoslovak Academy of Sciences, 1965. 304p. Bibliog – 1 – us Wisconsin U Libr [956]
The kurdistan missionary : [published monthly... in the interest of lutheran missions in persian and turkish kurdistan] – chicago IL: Inter-Synodical Evangelical Lutheran Orient-Mission Society 1910-28 [mthly] [mf ed 2005] – 19v on 2r – 1 – (ceased in 1928? place of publ varies: 1910-jun 1911, chicago; jul 1911-mar/apr 1915, detroit; may 1915-apr 1917, mansfield, ohio; may 1918-sep 1927, minneapolis; oct 1927-nov 1928, columbus, ohio; v3 n10 never publ?; v3 n9 & n11 dated in sequence and paged continuously; v6 n6/7 (mar/apr 1915) iss combined; v19 n10-12 (oct-dec 1927) erroneously numbered v20; several iss lacking; some pgs damaged) – mf#2005c-s062 – us ATLA [242]
Kurdufan – Khartoum, Sudan. jun 30 1990-jan 15 1992 – 1r – 1 – us CRL [960]
Kur'er syzrani : gaz bespartiin / ed by Batrakov, N I – Syzran' [Simb gub]: [s n] 1918 [1918 8 iiunia-] – (= ser Asn 1-3) – n3-21 [1918] [gaps] item 213, on reel n43 – 1 – mf#asn-1 213 – ne IDC [077]
Kureysizade (Mazhar), Mehmed Fevzi Elhac *see* Haber-i sahih
Kurganskaia svobodnaia mysl' : obshchestv -polit bespartiin gaz / ed by Vysotskii, I F – Kurgan [Tobol gub]: T-vo izd i pech dela 1918-19 [1918 [iiun']-1919 [?]] – (= ser Asn 1-3) – n15-164 [1918] n1-168 [1919] [gaps] item 212, on reel n42,43 – 1 – mf#asn-1 212 – ne IDC [077]
Kurgany i kurumy zapadnoi fergany : raskopki, pogrebalnyi obriad v svete etnografii / Litvinskii, B A – Moskva: Nauka, 1972 – us CRL [947]
Das kurgland und die evangelische mission in kurg / ed by Moegling, H & Weitbrecht, Th – Basel: Missionshauses, 1866 [mf ed 1995] – (= ser Yale coll) – viii/334p (ill) – 1 – 0-524-09030-0 – (in german) – mf#1995-0030 – us ATLA [242]
Kurhessen in einer geographisch-statistischhistorischen uebersicht / Appell, Johann – Darmstadt 1851 [mf ed Hildesheim: 1995-98] – (= ser Fbc) – vi/99p on 1mf [ill] – 9 – €40.00 – 3-487-29499-0 – gw Olms [914]
Kurhessische landeszeitung *see* Der sturm 1930
Kurhessische morgenpost – Eschwege DE, 1926-1930 30 jan [gaps], 1930 1 mar-31 dec – 10r – 1 – gw Misc Inst [074]
Der kurier – Berlin, Germany. 1962-66 – 14r – 1 – us L of C Photodup [074]
Der kurier – Berlin DE, 1953 8 oct-31 dec, 1954 10 jun-1955 31 jul, 1956-57 – 1 – (filmed by other misc inst: 1945 12 nov-1946 may, 1946 oct-dec [2r]; 1945 12 nov-1966) – gw Misc Inst [074]
Der kurier – Kiel-Elmschenhagen DE, 1928 6 jan-1940 [gaps], 1950-1963 27 dec [gaps] – 9r – 1 – gw Misc Inst [074]
Der kurier – Halle S DE, 1828 & 1830, 1832-33, 1835-50, 1851 feb-1856, 1870 1 oct-31 dec, 1874-1930 15 apr – 1 – (title varies: 1 jan 1835?: der courier; 1 jan 1851: der hallische courier; 1 jul 1851: hallische zeitung; 1 jul 1893: hallesche zeitung; with suppl) – gw Misc Inst [074]
Kurier : bau- und montagekombinat – Halle S DE, 1966-1991 24 oct [gaps] – 5r – 1 – (title varies: 1975-90 n6: bmk-kurier) – gw Misc Inst [074]
Kurier codzienny – Warsaw, Poland. Jul 1945-May 1953 – 12r – 1 – us L of C Photodup [943]
Kurier fuer niederbayern *see* Tagblatt fuer landshut und umgegend
Kurier polski – London, UK. 14 Dec 1959-16 Jan 1960 – 1 – uk British Libr Newspaper [072]
Kurier polski – Milwaukee WI. 1966 nov5-1968 mar 15 – 1r – 1 – mf#1167201 – us WHS [071]
Kurier polski – Warsaw, Poland. 1958-93 – 74r – 1 – us L of C Photodup [943]
Kurier polski – Warsaw, Poland. -d. Sept 1957-Dec 1967. 31 reels – 1 – uk British Libr Newspaper [943]
Kurier polsko – kanadyjski = The polish canadian courier – dec 1972-nov 1986 [complete] – 4r – 1 – Can$340.00 set – (in polish) – cn Commonwealth Imaging [071]
Kurier sportowy – Bydgoszcz, Poland. Jul-Aug 1945 – 1r – 1 – us L of C Photodup [943]
Kurier, steiermark ausgabe – Vienna, Austria apr 1987-mar 2000 [mf ed Norman Ross] – 85r – 1 – (independent daily for austria) – mf#nrp-1981 – us UMI ProQuest [074]
Kurier szczecinski – Szczecin, Poland. May-Nov 1948; Mar 1950-Apr 1980 – 61r – 1 – us L of C Photodup [943]
Ku-ring-gai courier – jan 1969-feb 1971 – 2r – at Pascoe [079]
Ku-ring-gai courier *see* Courier
Ku-ring-gai observer – 1r – at Pascoe [079]
Ku-ring-gai recorder – feb 1950-dec 1960 – 7r – at Pascoe [079]
Kuriren – Norrkoping, Sweden. 1983-84 – 1 – sw Kungliga [078]
Kuriren – Uddevalla, Sweden. 1921-63 – 153r – 1 – sw Kungliga [078]
Kur'jas – Chicago IL, 1900 – 1r – 1 – (lithuanian newspaper) – us IHRC [071]
Kurjer polski – Warsaw, Poland. Jan-Feb 1919; 1926-Aug 1939 – 28r – 1 – us L of C Photodup [943]
Kurjer poranny – Warsaw, Poland. Jan-Feb 1919 – 1r – 1 – us L of C Photodup [943]
Kurjer warszawski – Warsaw. Jan 1 1909-July 1914; Jan 17 1923-July 29 1939. Not collated – 1 – us NY Public [077]
Kurjer warszawski – Warsaw, Poland. Jan-Feb 1919 – 1r – 1 – us L of C Photodup [943]
Kurkin, P I *see*
– Die semstwo-sanitaetsstatistik des moskauer gouvernements
– Zemskaia sanitarnaia statistika
Kurkjian, Vahan M *see* Armenian kingdom of cilicia
Kurland : reiseeindruecke von land und stadt / Brunier, Ludwig – Leipzig 1862 [mf ed Hildesheim 1995-98] – 2mf – 9 – €60.00 – 3-487-28992-X – gw Olms [914]
Kurland, Roselle *see* Social work with groups
Kurmyshskij uezd : sovet rk i kd. ivestiia kurmyshskogo uezdnogo sovdepa – Kurmysh, Russia 1918 [mf ed Norman Ross] – 1r – 1 – mf#nrp-810 – us UMI ProQuest [077]
Kuroda, Shinto *see* Outlines of the mahayana as taught by buddha
Kuroghlu *see* Specimens of popular poetry of persia
Kurono Maisaichi. (Nippongo) *see* Moedah dan gampang oentoek dapat berbahasa nippon
Kurpfalzbaierische bamberger zeitung *see* Bamberger zeitung 1795
Kurrelmeyer, W *see*
– Die erste deutsche bibel
Kurri kurri times – Kurri Kurri. jan 1905-dec 1907; aug 1926-dec 1937 – 4r – A$266.68 vesicular A$288.68 silver – at Pascoe [079]
Kurs fiziki / ed by Shirokova, I U M – Moskva: Nauka, v2 1980 – us CRL [947]
Kurs kreditnoi kooperatsii / Kulyzhnyi, A E – 1918 – 36p 1mf – 9 – mf#COR-261 – ne IDC [335]
Kurs russkoi istorii / Kliuchevskii, Vasilii Osipovich – Peterburg: Gos izd-vo, 1920-1923 [mf ed 2004] – 5v on 1r – 1 – (incl ind) – us Wisconsin U Libr [947]
Kurs zvezdnoi astronomii : dopushcheno ministerstvom vysshego obrazovaniia sssr v kachestve uchebnika dlia universitetov / Parenago, Pavel Petrovich – 2nd ed. Moskva: Leningrad: Gos izd-vo tekhniko-teoret lit-ry 1946 [mf ed 1998] – 1r [pl/ill] – 1 – (incl bibl ref) – mf#film mas 28408 – us Harvard [520]
Kursachsen und die durchfuehrung des prager friedens 1635 / Duerbeck, Ernst – Leipzig, 1908 (mf ed 1994) – 2mf – 9 – €31.00 – 3-8267-3041-0 – mf#DHS-AR 3041 – gw Frankfurter [943]
Kursbildung am deutschen aktienmarkt unter besonderer beruecksichtigung verhaltensorientierter ueberlegungen / Dette, Guido – (mf ed 1997) – 4mf – 9 – €56.00 – 3-8267-2477-1 – mf#DHS 2477 – gw Frankfurter [332]
Kurskaia bednota : organ kurskogo gub ispolnit kom gorodskogo soveta deputatov, gub i gor kom rkp(b) – Kursk, Russia 1918 [mf ed Norman Ross] – 2r – 1 – mf#nrp-812 – us UMI ProQuest [077]
Kurskaia voennaia gazeta – Kursk, Russia 1917 [mf ed Norman Ross] – 1r – 1 – mf#nrp-812 – us UMI ProQuest [077]
Kurskie gubernskie vedomosti – Kursk, Russia 1838-1918 [mf ed Norman Ross] – 98r – 1 – mf#nrp-811 – us UMI ProQuest [077]
Kurskii kooperator – Kursk, 1928(1-8) – 4mf – 9 – mf#COR-647 – ne IDC [335]
Kurskii paterik : v1: skazanie o zhizni, podvigakh i chudesakh sviatelia i chudotvortsa ioasafa, ep belgorodskago...1911 – Kursk, 1911 – 2mf – 9 – mf#R-18276 – ne IDC [243]
Kurskii, S M *see*
– Altaiskii vestnik
– Zhizn' altaia
Kurskij gor sovet rk i kd *see* Izvestiia soveta rabochikh, koest'ianskikh i krasnoarmejskikh deputatov g kurska i gubernii
Kursun – Manastir: Beynelmilel Ticaret Matbaasi. Mueduer-i Mes'ul: Es'ad, 1911. Mueduer-i Mes'ul: Es'ad. n1-2. 28 mart 1327-4 misan 1327 [1911] – (= ser O & t journals) – 1mf – 9 – $25.00 – us MEDOC [956]
Kursy po kooperatsii / ed by Manuilov, A A – 1912-1913 – 2v 10mf – 9 – mf#COR-59 – ne IDC [335]
Kursy po kooperatsii petrogradskogo obshchestva narodnykh universitetov – [1916] – 4p 1mf – 9 – mf#COR-60 – ne IDC [335]
Kurteev, K K *see* Trudy pervogo sezda inspektorov melkogo kredita i predstavitelei sel'skikh kreditnykh tovarishchestv priamurskogo kraia, sostoiavshegosia v g
Kurth, G *see* Chartes de l'abbaye de saint-hubert en ardenne
Kurtser protokol / World Jewish Congress – Paris, France. 1937 – 1r – us UF Libraries [939]
Kurtz, Benjamin *see* Why are you a lutheran?
Kurtz, Benjamin et al *see* Addresses, inaugurals and charges
Kurtz, Daniel Webster *see*
– Nineteen centuries of the christian church
– An outline of the fundamental doctrines of faith
Kurtz gefasste historische nachrichten – Regensburg DE, 1727-73 – 59r – 1 – gw Misc Inst [900]
Kurtz, Johann Heinrich *see*
– The bible and astronomy
– Biblische geschichte
– Der brief an der hebraeer
– Die einheit der genesis
– History of the old covenant
– Lehrbuch der heiligen geschichte
– Lehrbuch der kirchengeschichte
– Manual of sacred history
– Das mosaische opfer
– Nachweis der einheit von gen 1-4
– Sacrificial worship of the old testament
– Text-book of church history
– Zur theologie der psalmen
Eyn kurtze aber christenliche vslegung fuer die jugend der gebotten gottes... / Grossmann, K – [Basel: Lux Schouber], 1536 – 1mf – 9 – mf#PBU-610 – ne IDC [240]
Kurtze anleytunge vnnd nachrichtunge wie man d egidium hunnen vnd d lucam osiandern sampt jrem anhang examinieren / Huber, S – np, 1597 – 1mf – 9 – mf#TH-1 mf 726 – ne IDC [242]
Kurtze ausslegung der sontags euangelien vnnd catechismi / Mathesius, J – Nuernberg, 1563 – 7mf – 9 – mf#TH-1 mf 991-997 – ne IDC [242]
Kurtze einleitung in einige theile der bergwercks-wissenschaft : anfaengern zum besten / Lehmann, Johann Gottlob – Berlin: bey Christoph Gottlieb Nicolai 1751 [mf ed 1979] – 1r [ill/pl] – 1 – (incl ind) – mf#film mas 8912 – us Harvard [550]
Kurtze erjnnerung von gegenwertigem zweytracht vber die lehre von der gnaden wahl / Huber, S – Muelhausen, 1595 – 1mf – 9 – mf#TH-1 mf 727 – ne IDC [242]
Eyn kurtze klare summ und erklaerung der chirstenen gloubens... / Zwingli, H – Zuerich: Christoffel Froschouer, 1537 – 2mf – 9 – mf#PBU-531 – ne IDC [240]
Ein kurtze replica / Cochlaeus, J – Ingolstadt, 1544 – 1mf – 9 – mf#PBU-696 – ne IDC [240]
Kurtze summarische auslegung der euangelien vnd episteln durch das gantze jar in gebet verfast / Musculus, A – Franckfurd an der Oder, 1574 – 8mf – 9 – mf#TH-1 mf 1203-1210 – ne IDC [242]
Kurtze und warhafftige erzehlung... / Dathenus, P – Heidelberg, 1598 – 3mf – 9 – mf#PBA-163 – ne IDC [240]
Kurtze vorstellung der gantzen civil-bau-kunst : worinnen erstlich die vornehmsten kunstwoerter, so darinnen vorkommen in fuenfferley sprachen angefueret end erklaeret... / Sturm, L C – Augsburg, 1718 – 1mf – 9 – mf#OA-114 – ne IDC [720]
Kurtzer ausszug : der christlichen und catholischen gesaeng, dess ehrwirdigen herrn joannis leisentritij, thumdechants zu budessin... / Leisentrit, Johann – Dilingen: Mayer 1576 – (= ser Hqab. literatur des 16. jahrh.) – 3mf – 9 – €40.00 – mf#1576a – gw Fischer [780]
Kurtzer bericht, gleich aim register oder anweysung in das gantz psalmen buch : was fuer psalmen oder gaistliche gesaeng... in der kirche[n] nach der predig moegen gesungen werden / Mayr, Georg – [Augsburg?]: [Ulhart?] [c1570] – (= ser Hqab. literatur des 16. jahrh.) – 1mf – 9 – €20.00 – (last 6p missing) – mf#1570f – gw Fischer [780]
Kurtzer, einfaeltiger und waarhafter historische bericht... / Utenhoven, J – Herborn, 1608 – 4mf – 9 – mf#PBA-380 – ne IDC [240]
Kurtzer vnd nuetzlicher bericht von dem heilsamen vnd christlichem buch formulae concordiae / Hunnius, A – Magdeburgk, 1593 – 3mf – 9 – mf#TH-1 mf 770-772 – ne IDC [242]
Ein kurtzer vnterricht auff d georgen maiors antwort / Amsdorff, N von – Basel, 1552 – 1mf – 9 – mf#TH-1 mf 15 – ne IDC [242]
Kurtzgefasstes musicalisches lexicon : worinnen eine nuetzliche anleitung und gruendliche begriff von der music enthalten, die termi technici erklaeret... / Walther, Johann Gottfried – Chemnitz: Christoph & Stoessel 1737 [mf ed 19–] – 6mf – 9 – mf#Fiche 587, 748 – us Sibley [780]
Ein kurtzweilig lesen von dil ulenspiegel (mxt4) : in der ausgabe strassburg, johannes grueninger, 1519. farbmikrofiche-edition des exemplars der forschungs- und landesbibliothek gotha, poes 2014/5 rara – (mf ed 1995) – (= ser Monumenta xylographica et typographica (mxt) 4) – 46p on 3 color mf – 15 – €235.00 – 3-89219-404-1 – (int & description by juergen schulz-grobert) – gw Lengenfelder [090]
Kurultay – Eskisehir. Sahib-i Imtiyaz: Idris Vehbi. n21. 22 tesrinievvel 1335 [1919] – (= ser O & t journals) – 1mf – 9 – $25.00 – us MEDOC [956]
Kurultay, Dorduncu *see* Tuerk ocaklari 1927 senesi kurultay zabitleri
Kurun-i cedide ve navolyon'un sukuntuna kadar asr-i hazir mebadisi / Hamid, V Ahmet – Istanbul: Matbaa-i Amire, 1924 – (= ser Ottoman histories and historical sources) – 3mf – 9 – $55.00 – us MEDOC [956]
Kurwa na doto / Farsy, Muhammad Saleh – Dar es Salaam, Tanzania. 1960 – 1r – us UF Libraries [960]
Kuryer bostonski = Polish daily news – Boston, MA: Kuryer Bostonski Pub Co, dec 4 1917-mar 17 1919 – 4r – 1 – us CRL [071]
Kuryer codzienny = Polish daily news – Boston, MA: Kuryer Bostonski Pub Co, mar 18 1819-dec 29 1920; oct 24 1921-oct 30 1922 – 11r – 1 – us CRL [071]
Kuryer glasgowski *see* Voice of poland
Kuryer katolicki = Catholic courier – Toledo, OH: L VSzyperski, mar 14 1918-23; aug-dec 1924 – us CRL [071]

Kuryer polski – Milwaukee WI. [1888 jun 23/1889 may]-[1961 jan 1/1962 sep 23] – 261r – 1 – (cont: krytyka) – mf#852475 – us WHS [071]

Kuryer toledoski – Toledo, OH: Kuryer Toledoski, 1925-jan 14 1926 – 1r – 1 – us CRL [071]

Kuryer zjednoczenia : official organ of the union of poles in america – Cleveland, OH: Kuryer Pub. Co. v17 n33 aug 17 1939-dec 17 1987 – 1 – (formed by the union of: kuryer, and: zjednoczeniec. polish fraternal assoc newspaper in polish and english. weekly 1939-58; semimonthly 1959-82; monthly jul 1982-. other title: polish courier) – us Western Res [071]

Kurz, Albert *see* Koenig eduard und der einsiedler

Kurz, Heinrich *see* Goethes werke

Kurz, Hermann *see*
- Die beiden tubus
- Hermann kurz' saemtliche werke
- Innerhalb etters
- Reichstaedtische erzaehlungen
- Schillers heimatjahre
- Der sonnenwirth

Kurz, Isolde *see*
- Innerhalb etters
- Das leben meines vaters
- Die nacht im teppichsaal

Kurz, Joseph Felix, Freiherr von *see* Prinzessin pumphia

Kurz, L *see* Gregors des grossen lehre von den engeln

Kurz- und langzeiteffekte einer fruehpostnatalen il-1ss- bzw. endotoxin-applikation auf die entwicklung des neuroendokrinen systems sowie auf verhaltensparameter der ratte / Rolletschek, Alexandra – (mf ed 1999) – 2mf – 9 – €40.00 – 3-8267-2601-4 – mf#DHS 2601 – gw Frankfurter [574]

Kurzbach, Herbert *see* Finnische novelle

Kurze anweisung zum klavierspielen : ein auszug aus der groessern klavierschule / Tuerk, Daniel Gottlob – Halle, Leipzig: Auf Kosten des Verfassers: In Kommission bey Schwickert in Leipzig, und bey Hemmerde und Schwetschke in Halle 1792 [mf ed 19–] – 4mf – 9 – mf#fiche 955, 1056 – us Sibley [780]

Kurze bemerkungen auf einer fluechtigen reise am rhein und durch das koenigreich der niederlande im jahre 1828 / Ladenberg, P W von – Koeln am Rhein 1830 [mf ed Hildesheim 1995-98] – 2mf – 9 – €60.00 – 3-487-29555-5 – gw Olms [914]

Kurze beschreibung der gottesdienstlichen gebraeuche : wie solche in...zuerich begangen werden / Herrliberger, D – Zuerich, Eckenstein, 1751 – 1mf – 9 – mf#ZWI-37 – ne IDC [240]

Kurze darstellung der vornehmsten eigenthuemlichkeiten der schwedischen kirchenverfassung : mit hinblicken auf ihre geschichtliche entwicklung / Knoes, Anders Erik – Stuttgart: SG Liesching, 1852 – 3mf – 9 – 0-524-07893-9 – mf#1991-3438 – us ATLA [240]

Kurze darstellung des theologischen studiums zum behuf einleitender vorlesungen entworfen / Schleiermacher, Friedrich [Ernst Daniel] – Berlin: G Reimer 1843 [mf ed 1993] – 2mf – 9 – 0-524-08243-X – (together with: ueber die religion: reden an die gebildeten unter ihren veraechtern; die weihnachtsfeier: ein gespraech) – mf#1993-2018 – us ATLA [240]

Kurze, doch hinlaengliche anweisung zum general-basse : wie man denselben aufs allerleichteste, auch ohne lehrmeister, erlernen kann / ed by Hesse, Johann Heinrich – Hamburg: Gedruckt von M C Bock [1776] [mf ed 19–] – 1mf – 9 – mf#fiche 556, 735 – us Sibley [780]

Kurze erklaerung der briefe des petrus, judas, und jakobus / De Wette, Wilhelm Martin Leberecht – 3. ausg. Leipzig: S Hirzel, 1865 – (= ser Kurzgefasstes exegetisches Handbuch zum Neuen Testament) – 2mf – 9 – 0-8370-6403-1 – mf#1986-0403 – us ATLA [227]

Kurze, G *see*
- Eine rundfahrt durch die londoner suedseemissionen
- Wie die kannibalen von tongoa christen wurden

Kurze, Georg *see* Der engels- und teufelsglaube des apostels paulus

Eine kurze geschichte der baptisten / Ramaker, Albert John – Cleveland, O[hio]: Verlagshaus der deutschen Baptisten, 1906 – 1mf – 9 – 0-524-03942-9 – mf#1990-4936 – us ATLA [242]

Kurze lebens- und regierungsgeschichte ludwig des sechszehnten, koenigs von frankreich : mit umstaendlichen nachrichten von seiner lezten gefangenschaft, verurtheilung und hinrichtung / Scheler, Eugen C von – Stuttgard 1793 [mf ed Hildesheim 1995-98] – 1v on 3mf – 9 – €90.00 – ISBN-10: 3-487-26194-4 – ISBN-13: 978-3-487-26194-2 – gw Olms [944]

Kurze reformations-geschichte : erzaehlt fuer schulen und familien / Redenbacher, Wilhelm – neue verm Aufl. Calw: Vereinsbuchh, 1883 – 1mf – 9 – 0-524-03910-0 – mf#1990-1169 – us ATLA [242]

Kurze und leichte klavierstuecke : mit veraenderten reprisen und beigefuegter fingersetzung fuer anfaenger / Bach, Carl Philipp Emanuel – [1766] [mf ed 19–] – 1r – 1 – mf#pres. film 60, 10 – us Sibley [780]

Kurze und systematische anleitung zum general-bass und der tonkunst ueberhaupt : mit exempeln erlaeutert: zum lehren un lernen entworfen / Bach, Johann Michael – gedruckt zu Cassel: in der Waysenhaus-Buchdruckerey 1780 [mf ed 19–] – 1r / 1mf – 1,9 – mf#film 103 / fiche335 – us Sibley [780]

Kurze, Volker *see* Physikalische charakterisierung ultraflexibler lipid-vesikel

Kurzer arbriss der aegyptischen grammatik / Erman, A – Berlin, 1924 – 2mf – 9 – mf#NE-20380 – ne IDC [470]

Kurzer bericht ueber wissenschaftliche arbeiten und reisen / Baer, K E von – Spb, 1845. v9 – 8mf – 9 – mf#R-1667 – ne IDC [910]

Kurzer bericht von dem unterschied der wahren evangelisch-lutherischen und der reformirten lehre : nebst einem anhang und eroerterung folgender fragen... / Masius, Hector Gottfried – neuer Abdr. St Louis, MO: L Volkening, 1868 – 1mf – 9 – 0-524-05258-1 – mf#1991-2250 – us ATLA [242]

Kurzer bericht von des herren abendmal... / Piscator, J – Herborn, 1589 – 1mf – 9 – mf#PBA-293 – ne IDC [240]

Kurzer und deutlicher unterricht von dem general-bass : in welchem durch deutliche regeln und leichte exempel nach dem neuesten musicalischen stylo gezeiget wird... / Reinhard, Leonhard – Augsburg: J J Lotter 1767 [mf ed 19–] – 2mf – 9 – mf#fiche 929 – us Sibley [780]

Kurzgefasste assyrische grammatik / Meissner, Bruno – Leipzig: J C Hinrichs, 1907 – 1mf – 9 – 0-8370-8457-1 – mf#1986-2457 – us ATLA [470]

Kurzgefasste einleitung in die heiligen schriften : alten und neuen testamentes, zugleich ein hilfsmittel fuer kursorische schriftlektuere / Weber, Ferdinand Wilhelm – in 12. aufl. Muenchen: C H Beck, 1907 – 1mf – 9 – 0-524-06062-2 – mf#1992-0775 – us ATLA [220]

Kurzgefasste geschichte der evangel.-luth. synode von iowa und andern staaten : von der gruendung bis zum jubeljahr 1904 / Deindoerfer, Johannes – Chicago, IL: Wartburg Publishing House, 1904 – 1mf – 9 – 0-524-00966-X – mf#1990-4024 – us ATLA [242]

Kurzgefasste geschichte der lutherischen bibeluebersetzung bis zur gegenwart : mit beruecksichtigung der vorlutherischen deutschen bibel und der in der reformierten schweiz gebrauchten deutschen bibeln: eine denk- und dankschrift zur vierhundertjaehrigen jubelfeier der geburt luthers / Grimm, Wilibald – Jena: Hermann Costenoble, 1884 – 1mf – 9 – 0-8370-3398-5 – (incl bibl ref) – mf#1985-1398 – us ATLA [220]

Kurzgefasste geschichte der lutherischen kirche amerikas / Neve, Juergen Ludwig – 2., verm und ganz umgearb Aufl. Burlington, Iowa: German Literary Board, 1915 – 1mf – 9 – 0-524-01458-2 – (incl bibl ref) – mf#1990-4096 – us ATLA [242]

Kurzgefasste geschichte der lutherischen kirche amerikas *see* A brief history of the lutheran church in america

Kurzgefasste grammatik der biblisch-aramaeischen sprache, literatur, paradigmen, texte und glossar / Marti, K – Berlin, 1925 – (= ser Porta linguarum orientalium. Sammlung von Lehrbuechern fuer das Studium der orientalischen Sprachen) – 3mf – 9 – (porta linguarum orientalium. sammlung von lehrbuechern fuer das studium der orientalischen sprachen v18) – mf#NE-478 – ne IDC [956]

Kurzgefasste methode den generalbass zu erlernen / Albrechtsberger, Johann Georg – Wien: Artaria [1792] [mf ed 19–] – 2mf – 9 – mf#fiche64 – us Sibley [780]

Kurzgefasste syrische grammatik / Noeldeke, Theodor – 2. verb aufl. Leipzig: Chr Herm Tauchnitz 1898 [mf ed 1986] – 1mf – 9 – 0-8370-8603-5 – (in german & syriac) – mf#1986-2603 – us ATLA [470]

Kurzgefasster fuehrer durch goethes faustdichtung : 1. und 2. teil / Straub, Lorenz – Stuttgart: Strecker und Schroeder, 1922 – 1r – 1 – us Wisconsin U Libr [430]

Kurzgefasster ueberblick ueber die babylonischassyrische literatur : nebst einem chronologischen excurs, zwei registern und einem index zu 1700 thontafeln des british museum's / ed by Bezold, Carl – Leipzig: Otto Schulze, 1886 – 1mf – 9 – 0-8370-8404-0 – (in german and akkadian. incl indes) – mf#1986-2404 – us ATLA [470]

Kurzgefasstes exegetisches handbuch zu den apokryphen des alten testamentes / Fritzsche, Otto Fridolin & Grimm, Carl Ludwig Wilibald – Leipzig: Weidmann, 1851-1860 – 5mf – 9 – 0-8370-1741-6 – mf#1987-6137 – us ATLA [221]

Kurzgefasstes lexicon deutscher pseudonymer schriftsteller von der aeltern bis auf die juengste zeit aus allen faechern der wissenschaften : mit einer vorrede ueber die sitte der literarischen verkappung von j w s lindner / Rassmann, F – Leipzig, 1830 (mf ed 1993) – 2mf – 9 – €24.00 – 3-89349-235-6 – mf#DHS-AR 100 – gw Frankfurter [500]

Kurzreiter, Heinrich *see* Ueber die hamburger dramaturgie und corneilles discours

Kurzweil, Benedikt *see* Die bedeutung buergerlicher und kuenstlerischer lebensform fuer goethes leben und werk

Kusamn sin ja taen ti rapas khmaer, truv pragal oy khmaer vin / Samn Mak San – Bhnam Ben: [s.n.] 1973 [mf ed 1990] – 1r with other items – 1 – (in khmer) – mf#mf-10289 seam reel 129/16 [§] – us CRL [480]

Kusasa umngcwabo wakho nami / Mncwango, Leonhard L J – Pietermaritzburg, South Africa. 1953 – 1r – us UF Libraries [960]

The kusasis : a short history / Syme, J K G – [s.l: s.n, 1932] – us CRL [900]

Kuser, John Dryden *see* Haiti

Kush : journal of the sudan antiquities service – Khartoum: The Service: Obtainable from Commissioner for Archaeology, v1-15 1953-1967/68 – 5r – 1 – us CRL [960]

Kushano-sasanian coins / Herzfeld, Ernst – Calcutta: Govt of India, Central Publication Branch, 1930 – (= ser Samp: indian books) – us CRL [730]

Kushch, A D *see* Rech' altaia

Kusheva, Ekaterina Nikolaevna *see* Narody severnogo kavkaza i ikh sviazi s rossiei

Kuskov, P A *see* Nashi idealy

Kuss, E *see* Mitbestimmung und gerechter lohn als elemente einer neuordnung der wirtschaft

Kustanaiskii listok : vnepartiin obshchestv -polit i lit gaz / ed by Berestov, P V – Kustanai [Turg obl]: [s n] 1919 [1919 29 apr-] – (= ser Asn 1-3) – n1-44 [1919] item 214, on reel n43 – 1 – mf#asn-1 214 – ne IDC [077]

Kustanajskij obshchestvennyj komitet *see* izvestiia

Kustar-kooperato – 1925-1927(10) – 26mf – 9 – mf#COR-648 – ne IDC [335]

Kustarnaia promyshlennost i ee sviazs kooperativami / Kablukov, N A – 1915 – 69p 1mf – 9 – mf#COR-38 – ne IDC [335]

Kustarnaia promyshlennost i promyslovaia kooperatsiia : materialy 3 plenuma vsnkh rsfsr, 27 fevr-1 marta 1928 g – 1928 – 136p 2mf – 9 – mf#COR-425 – ne IDC [335]

Kustarnaia promyshlennost i promyslovaia kooperatsiia rsfsr: : sbornik materialov – 1929 – 109p 2mf – 9 – mf#COR-427 – ne IDC [335]

Kustarnaia promyshlennost i promyslovaia kooperatsiia v natsionalnykh respublikakh i oblastiakh sssr / ed by Enbaev, A M – 1928 – 267p 3mf – 9 – mf#COR-426 – ne IDC [335]

Kustarnaia promyshlennost sssr : sbornik statei i materialov / ed by Lereda, S P – 1925 – 167p 2mf – 9 – mf#COR-428 – ne IDC [335]

Kustarno-promyslovaia kooperatsiia na nizhegorodskoi iarmarke 1927 g: : spravochnik vsekopromsoiuza – 1927 – 35p 1mf – 9 – mf#COR-430 – ne IDC [335]

Kustarno-promyslovaia kooperatsiia rsfsr v 1927-28 godu: : statisticheskii ezhegodnik – 1929 – 213p 3mf – 9 – mf#COR-431 – ne IDC [335]

Kustarno-remeslennaia promyshlennost sssr / Feigin, V – 1927 – 128p 2mf – 9 – mf#COR-452 – ne IDC [335]

Kuste von el salvador / Gierloff-Emden, Hans Gunter – Wiesbaden, Germany. 1959 – 1r – us UF Libraries [972]

Kustov, A V *see*
- Biulleteni gazety "golos stepi"
- Golos stepi
- Pavlodarskii telegraf

Kutateladze, S S *see* Teplo- i massoperenos v absorbtsionnykh apparatakh

Kutaura chirungu – Chishawasha, Zimbabwe. 1962 – 1r – us UF Libraries [960]

Kutaura cirungu – Chishawasha, Zimbabwe. 1939 – 1r – us UF Libraries [960]

Kutaura cirungu – Chishawasha, Zimbabwe. 1952 – 1r – us UF Libraries [960]

Kutna, Salomon *see* Gedenkblatter fur oberrabbiner salomon kutna

Kutora mifananidzo (photography) / Kanyama, Bester – Gwelo, Zimbabwe. 1970 – 1r – us UF Libraries [960]

Kutrun : mittelhochdeutsch / ed by Ziemann, Adolf – Quedlinburg: G Basse, 1835 [mf ed 1993] – (= ser Bibliothek der gesammten deutschen national-literatur von den aeltesten bis auf die neuere zeit sect1/1) – x/213p – 1 – mf#8438 reel 1 – us Wisconsin U Libr [430]

Kutscher, Artur *see*
- Gerhart hauptmann
- Loens-brevier
- Das naturgefuehl in goethes lyrik

Kutscher, Austin H *see* Loss, grief and care

Kutter, H *see* Clemens alexandrinus und das neue testament

Kutter, Hermann *see*
- Clemens alexandrinus und das neue testament
- Die revolution des christentums
- Social democracy
- They must
- Wilhelm von st. thierry
- Wir pfarrer

[Kuttner, K G] *see* Reise durch deutschland, daenemark, schweden, norwegen und einen theil von italien, in den jahren 1797, 1798, 1799

Kuttner, Otto *see* Historisch-genetische darstellung von kant's verschiedenen ansichten ueber das wesen der materie

Kutu, Ishmael Gaster *see* Issues related to learners' discipline in schools

Kutzen, Joseph A *see*
- Das deutsche land in seinen charakteristischen zuegen und seinen beziehungen zu geschichte und leben der menschen
- Die grafschaft glatz

Kutzik, Alfred J *see* Social work and jewish values

Kuukausi-julkaisu / Finnish Socialist Organization of the United States – v1 n5 [1911 jul] – 1r – 1 – mf#464946 – us WHS [335]

Kuumba news – Chicago IL. 1972 nov – 1r – 1 – mf#4992540 – us WHS [071]

Kuumba [usa] : [newsletter] – 1994 aut – 1r – 1 – mf#5307254 – us WHS [071]

Kuvalayananda karikas : or, the memorial verses of appaya dikshita's kuvalayananda = Kuvalayananda-karikah / Appayya Diksita; ed by Sarma, P R Subrahmanya – Calcutta: JN Banerjee & Son, 1903 – (= ser Samp: indian books) – (explained with an english tika comm and trans, for the use of english students of sanskrit by ed) – us CRL [490]

Kuwait *see* Al-kuwayt al-yawm

Kuy Yak Hu *see*
- Broh tae prak r tammnak dyk bhnaek gra kroy pamphut
- Citt muay thloem muay
- Gu kamm gu gap

Kuylenstierna, Oswald Fredrik *see* Soeren kierkegaard

Kuyper, A *see*
- Disquisitio historico-theologica exhibens j calvini et de ecclesia sententiarum inter se compositionem
- Kerkeraads-protocollen der hollandsche gemeente te londen 1569-1571 (de werken...1/1)
- Opera tam edita quam inedita

Kuyper, Abraham *see*
- The antithesis between symbolism and revelation
- Calvinism
- Het calvinisme
- Het calvinisme en de kunst
- E voto dordraceno
- Encyclopaedie der heilige godgeleerdheid
- Johannes maccovius
- Die moderne theologie (der modernismus)
- Pantheism's destruction of boundaries
- Pro rege
- Tractaat van den sabbath
- Uit het woord
- Varia americana
- The work of the holy spirit

Kuzadzwa nomweya mutsvene / Jackson, S K – Fort Victoria, Zimbabwe. 19–? – 1r – us UF Libraries [960]

Kuzbass – Kemerovo, Russia 1973-87 [mf ed Norman Ross] – 5r – 1 – mf#nrp-680 – us UMI ProQuest [077]

Kuz'min, V I *see* Obshchee delo

Kuz'min-Karavaev, V D *see* Zemstvo i derevnia 1898-1903

Kuznetskii rabochii – Novokuznetsk, Russia 1974 [mf ed Norman Ross] – 2r – 1 – mf#nrp-1231 – us UMI ProQuest [077]

Kuznetsov, I G *see*
- Omskii vestnik [omsk: 1918: utr vyp]
- Omskii vestnik [omsk: 1918: vech vyp]
- Telegrammy [omskii vestnik]

Kuznetsov, M M *see* Sovremennaia perm'

Kuznetsov, N I *see* Izvestiia gosudarstvennogo instituta opytnoi agronomii

Kuznetsov, V *see* Za narod!

Kuznetsova, Ivan Vasil'evich *see* Karl marks i agrarnyi vopros

Kuznetsov-Gatimurov, A P *see* Vostochnaia okraina

Kuznitza – Moscow. no. 1-3, 7. May-Jul, Dec 1920-Mar 1921 – 1 – us NY Public [460]

Kuznitzky, Hugo *see* Gedenkblatter fur die bruder ehrenvizegrossprasident hugo kuznitzky

Kuzvarwa patsva mupapatisimo / Mavudzi, Emmanuel – Gwelo, Zimbabwe. 1964 – 1r – us UF Libraries [960]

Kuzvipira kwasisita bernadeta woku peramiho – Gwelo, Zimbabwe. 1959 – 1r – us UF Libraries [960]

Kvacala, Jan *see* Thomas campanella

Kvak lap phrann pa nhan a khra vatthu tui mya : [short stories] / Mra Kruin, Mon – Ran Kun: Ca pe Biman 1986 [mf ed 1990] – (= ser Ca pe biman thut prann su lak cvai ca can) – 1r with other items – 1 – (in burmese) – mf#mf-10289 seam reel 161/7 [§] – us CRL [830]

Kvakartidskrift – v1-12. 1974-85 [complete] – 1r – 1 – mf#ATLA S0797 – us ATLA [073]

Kvallsposten – Malmo, Sweden. 1948-78 – 462r – 1 – sw Kungliga [078]

Kvallsposten – Malmo, Sweden. 1979-89 – 1 – sw Kungliga [078]

Kvallsposten – Malmoe, 1995- – 9 – sw Kungliga [078]

Kvallsstunden *see* Vastmanlands nyheter

Kvan, U *see*
- Caphat san kra nann a myui myui
- Mula lan lak tve ca ci ca rum
- Nann sac mran phat ca

Kvartalskrift / Norske Selskab i Amerika – 1905 apr-1921/1922 – 1r – 1 – mf#2892881 – us WHS [071]

Kvelosa loka : romans / Brigadere, Anna – [s.l]: Apgads Sejejs 1949 [mf ed 1984] – 1r – 1 – (filmed with: le probleme de la population en france / rabinowicz, leon) – mf#6559 – us Wisconsin U Libr [460]

Kvety – V Praze: [Vitezslav Halek] 1865- [mf ed Bloomington IN: Indiana Uni Lib, Preservation Dept 1989] – 2r – 1 – us Indiana Preservation [073]

Kvety – V Praze: Vladimir Cech 1879- [mf ed Bloomington IN: Indiana Uni Lib, Preservation Dept 1989] – 34r – 1 – us Indiana Preservation [073]

Kvety americke – Omaha, NE: Jan Rosicky. roc1 cis1. 15 rij 1884 (mthly) [mf ed 1884-87 filmed 1982] – 1r – 1 – (in czech) – us NE Hist [073]

Kvinnan och hemmet – Cedar Rapids, Iowa. 1893-1917 [incomplete] – 1 – sw Kungliga [071]

Kvinnornas tidning – Goteborg, Sweden. 1921-25 – 2r – 1 – sw Kungliga [078]

Kvinnosyn – Karlstad, 1988-94 – 1 – sw Kungliga [078]

Kwa zulu / Cowley, Cecil – Cape Town, South Africa. 1966 – 1r – us UF Libraries [960]

Kwacha angola – London [etc], UNITA office. n5 dec 1966; Special issue nov 1968; n7 jun 1971; Special issue 1972; jan/apr 1973 – us CRL [960]

Kwacha angola / Uniao Nacional para a Indepenencia Total de Angola (UNITA) – [S.l.]: A Uniao, [196]- apr, aug? oct? 1966; jan/feb 1967; mar/jun, Special ed 1972; n12 mar 12 1974 – us CRL [960]

Kwad laerer bibeln om foersoningen? : kort bidrag till swar pa denna fraga / Fjellstedt, Peter – Moline, IL: Swedish Printing Co, 1878 – 1mf – 9 – 0-524-05251-4 – mf#1991-2243 – us ATLA [220]

Kwagandaganda : an archaeozoological case study of the exploitation of animal resources during the early iron age in kwazulu-natal / Beukes, Catharina F – Uni of South Africa 2000 [mf ed Johannesburg 2000] – 7mf – 9 – mf#mfm14896 – sa Unisa [930]

Kwajalein hourglass / U[/nited/] S[/tates/] Army Kwajalein Atoll – 1984 apr 2-1986 mar 28, 1986 apr 21-1987, 1988 – 1r – 1 – mf#1061918 – us WHS [071]

Kwan tong kie hiap = Guan-dong qi xia zhuan / Chang, Ko-nung – [Batavia: Keng Po, 1928?] [mf ed 1998] – 1r – 1 – (coll as pt of the colloquial malay collection. indonesian trans of chinese novel entitled "guandong qixia zhuan" by zhang genong [salmon, claudine. literature in malay by the chinese of indonesia. paris: editions de la maison des sciences de l'homme, c1981]. filmed with: siauw ngo gie / karangannja, weng kuang lou & feng mi tao-jen) – mf#10006 – us Wisconsin U Libr [830]

Kwang tung : or, five years in south china / Turner, John Arthur – 2nd rev ed. London: S W Partridge, 1905 [mf ed 1995] – (= ser Yale coll) – xi/176p (ill) – 1 – 0-524-09145-5 – mf#1995-0145 – us ATLA [951]

Kwangari / Westphal, E O J – London, England. 1958 – 1r – us UF Libraries [960]

Kwang-tung ching-wu chuang-k'uang / China. Kwangtung Provincial Government – Survey of the Police Administration in Kwangtung. Canton, 1928.2v., tables, charts. 1 reel – 1 – us Chinese Res [951]

Kwangtung Province Mi shu ch'u *see* Kuang-tung sheng cheng fu san shih nien hsing cheng hui i chi yao

Kwanyama – Windhoek?, Namibia. 1966 – 1r – us UF Libraries [960]

Kwartalnik historyczny – Krakaw. v.1-47. 1887-1931 – 1 – us L of C Photodup [943]

Kwartir nasional gerakan pramuka / Madjalah pemimpin Pramuka – Djakarta, 1962-1967 – 15mf – 9 – (missing: 1964, v3(4-end); 1965, v4(1, 2, 6-12); 1966, v5(5-end)) – mf#SE-915 – ne IDC [959]

Kwasnicki, Sherri *see* Changes in maternal body composition from month one to month six postpartum in 11 breastfeeding, exercising women

Kwatsha, Linda Loretta *see* Canons of indigenous traditions and western values

Kwawa – Cape Town, South Africa. 1959 – 1r – us UF Libraries [960]

Kweetin, John *see* A hidden jewel

Kweli times : a short history of the apostolic church vanuatu 1946-1965 / Grant, Paul – 2002 – 1r – 1 – mf#pmb doc472 – at Pacific Mss [243]

Kweli, verdad, truth / Arthur A Schomburg I S 201 Educational Complex – 1970 dec – 1r – 1 – mf#4863533 – us WHS [370]

Kwik, Robert Julius *see* An analysis of bultmann's nonform-critical criteria used in evaluating authenticity in the synoptic gospels

Kwo, Lay Yen *see*
- Doea lobang pelor / tjoe bo kim so
- It kie bwee / siauw eng hiong / maoe terbang tida bersajap

Kwp-aktuell : kombinat wassertechnik und projektierung – Halle S DE, 1983-1989 11 dec [gaps] – 1r – 1 – gw Misc Inst [621]

Ky lan – Saigon. n2-69. 18 aout 1928-19 mai 1929 – 1 – (lacking: n50, 64-66, 68) – fr ACRPP [073]

Kya kri raja suik : [a novel] / Mra Cakra, Cha ra kri – Ran Kun: Tan Mon Kri Ca up Tuik [195-?] [mf ed 1990] – 1r with other items – 1 – (in burmese) – mf#mf-10289 seam reel 180/1 [§] – us CRL [830]

Kya pan krui kyu pan kra / San on, a nna mre – Ran kun Mrui: Ca pe Biman 1975 [mf ed 1993] – on pt of 1r – 1 – mf#11052 r630 n4 – us Cornell [959]

Kyan to bava jat kron / Sao Ta Chave – Yan Kun: Gun Thau Cape 1975 [mf ed 1990]- – 1r with other items – 1 – (in burmese) – mf#mf-10289 seam reel 170/3 [§] – us CRL [920]

Kyaw, U, Hsu htu pan *see* Bedan lakkhana

Kybernetes – Bradford, England 2001+ – 1,5,9 – ISSN: 0368-492X – mf#13882.01 – us UMI ProQuest [000]

Kye rva samavayama a san cann kamupade – Ran kun: Pum nhip re nhan ca up thut ve re lup nan ko pui re rhan 1972 cover 1974 [mf ed 1995] – on pt of 1r – 1 – mf#11052 r2002 n2 – us Cornell [959]

Kyemon – Rangoon, Burma. May 1970-73; Mar 1978-Sept 1988 – 53r – 1 – us L of C Photodup [079]

Kyeser, Konrad *see* Bellifortis / feuerwerkbuch (cf-lp3)

Kyi sai Le thap Cha ra to *see* Dhammagan bhat ca nhan puran kyam

Kyin U, U *see* Padeitha thi chin mya

Kyin Win, U *see* Han mien ying ta tju tien [ssu chiao hao ma]

Kyivs'kyi derzhavnyi universytet im. T H Shevchenka. Astronomichna observatoriia *see* Publikatsii kyivskoi astronomichnoi observatorii

Kyk Candatara *see* Snam snehsri kamm

Kyklos : internationale zeitschrift fuer sozialwissenschaften – Oxford, England 1947+ – 1,5,9 – ISSN: 0023-5962 – mf#10696 – us UMI ProQuest [300]

Kyle, Melvin Grove *see* The deciding voice of the monuments in biblical criticism

Kyle, Robert Wood *see*
- Christian equality
- Question, is tractarianism or protestantism true catholicism?

Kyne, Peter B *see* Kindred of the dust

Kynett, Alpha Jefferson *see* Our laity

Kynn Nnvan *see* Samuin van suriya mran ma alan nahn mran ma nuin nam re

Kyo Cin, Dok Ta *see* Cit panna rhu thon a thve thve

Kyo Kyo nay *see* Chu pan lay mha kattipa kyvan

Kyo Man, Tha vay *see*
- Sumthon cam kuiy tve' mhat tam
- Tha vay yan kye mhu a mru te

Kyo Mra San *see* Sayavati thon mha na rai khan mya

Kyo Mran *see* Metta ca le khyak

Kyo Mran Lvan *see* Mran ma lu san jo vit kri

Kyo Mran, U *see* Mran ma nuin nam sa kra lup nan

Kyo Nnin, U *see* Nhac 30 mran ma a sam

Kyo Sa, Mon *see* Ta ce ta con ca re cha ra tui e a kron

Kyo Thvan, Dha nu phru *see* Pi mui nan e vatthu tui a tat panna

Kyo Van, Manussa *see* A khre kham rhe hon su te sa na panna

Kyo Van, Mre lat *see* Sacca metta

Kyogikai shiryo-hen : documents of sclc (the securities coordinating liquidation committees) in the holding of the library of faculty of economics, university of tokyo – jun 1947-jun 1951 – (= ser Shoken shori chosei kyogikai shiry a) – 511v on 100r – 1 – Y1600,000 – (in-house papers of sclc. full record of sclc activities from its establishment in june 1947 through its dissolution in june 1951; with guide) – ja Yushodo [332]

Kyogle examiner – Kyogle, 1912; 1914-15; 1917-78 – at Pascoe [079]

Kyok cim, pattmra, nhan ratana mya / Nnvan E, An va – Ran Kun: Chan sac ca pe pran khyi re 1977 [mf ed 1990] – 1r with other items – 1 – (in burmese; brief study on precious stones, their properties and methods of processing) – mf#mf-10289 seam reel 175/2 [§] – us CRL [550]

Kyok mi kyon suik kri : [a novel] / Mra Cakra, Cha ra kri – Ran Kun: Tan Mon Kri Ca up tuik 1956 [mf ed 1990] [mf ed 1990] – 1r with other items – 1 – (in burmese) – mf#mf-10289 seam reel 179/6 [§] – us CRL [830]

Kyok myak ratana abidan / Nnvan Nuin, Kyi sai – Yan Kun: Amran Sac 1975 [mf ed 1990] – [ill] 1r with other items – 1 – (in burmese; with bibl) – mf#mf-10289 seam reel 176/2 [§] – us CRL [550]

[Kyokuto kokusai gunji saiban kiroku mokuroku oyobi sakuin] / Mori, Kyozo – Tokyo, 1953 – 21,293p – 1 – mf#LL-10030 – us L of C Photodup [340]

Kyon sa lu nya phat ru phav / Khan Mon Nnvan – Ran Kun: Ke Khayn ca pe 1978 [mf ed 1990] – 1r with other items – 1 – (in burmese; with bibl) – mf#mf-10289 seam reel 135/5 [§] – us CRL [900]

Kyoto bijutsu kyokai zasshi : bulletin of kyoto bijutsu kyokai: an organization of traditional art in kyoto – Kyoto bijutsu zasshi: n1-2(oct 1890 & feb 1892); kyoto bijutsu kyokai zasshi: n1-155(jul 1892-jun 1905); kyoto bijutsu: n1-48(sep 1905-dec 1919) – 16r – 1 – Y240,000 – (in japanese; with 80p guide) – ja Yushodo [700]

Kyoto Daigaku. Uchu Butsurigaku Kankyushitsu *see* Contributions from the institute of astrophysics, university of kyoto

Kyowva Genealogical Society *see* Newsletter

Kyowva genealogical society : [newsletter] – v1 n2-v9 n2 [1978 dec-1986 sum] – 1r – 1 – (cont: newsletter [kyowva genealogical society]) – mf#1288304 – us WHS [929]

Kyper, Ralph Edward *see*
- The greatness of human nature
- An inquiry into some of the sources of channing's religious philosophy

Kyphosis in active and sedentary postmenopausal women / Eagan, Marianne S – 1999 – 2mf – 9 – $8.00 – mf#PE 4038 – us Kinesology [611]

Die kyprien : ein hellenisches epos in zwoelf gesaengen neu geschaffen / Scheffer, Thassilo von – 2. Aufl. Wiesbaden: Dieterich, 1947 – 1r – 1 – us Wisconsin U Libr [450]

Kyriakos, A Diomedes *see* Geschichte der orientalischen kirchen von 1453-1898

Kyrieleis, Richard *see* Moritz august von thuemmels roman "reise in die mittaeglichen provinzen von frankreich"

Kyrillos von skythopolis / Schwartz, Eduard – Leipzig, 1939 – (= ser Tugal 4-49/2) – 7mf – 9 – €15.00 – ne Slangenburg [240]

Kyriokos – v1-3. 1976-82 [complete] – 1r – 1 – (cont by: uuwf journal) – mf#ATLA S0342 – us ATLA [242]

Kyrios – 1(1936)-6(1943) – 37mf – 9 – €71.00 – ne Slangenburg [243]

Kyrkans informationcentral : news from the church of sweden – v1-10. 1974-83 [complete] – 1r – 1 – mf#ATLA S0644 – us ATLA [242]

Kyrkans tidning [kt] – Stockholm: Svenska Kyrkans Press AB 2002- [wkly] [mf ed 2004] – n27/28-51/52 (2002) [complete] on reel 1 – 1 – (cont: svenska kyrkans tidning [issn 0280-4603]; with suppls) – ISSN: 0280-4603 – mf1060 – us ATLA [074]

Kyrkans uppgift i fredsarbetet : foeredrag i helga trefaldighet i uppsala midsommarafton 1917 vid allmaenna svenska fredskongressen / Soederblom, Nathan – Stockholm: Svenska andersfoerlaget, 1917 – 1mf – 9 – 0-524-01952-5 – mf#1990-0541 – us ATLA [940]

Kyrklig tidskrift – Stockholm, 1895-1918 [mf ed 2001] – (= ser Christianity's encounter with world religions, 1850-1950) – 6r – 1 – (in swedish) – mf#2001-s160 – Uppsala teologiska fakultet – us ATLA [242]

Kyros der grosse / Prasek, Justin V – Leipzig: JC Hinrichs, 1912 [mf ed 1989] – (= ser Der alte orient 13/3) – 1mf – 9 – 0-7905-2033-8 – mf#1987-2033 – us ATLA [930]

Kytpoe – Larnaca, Cyprus. 21 jun 1888-10 dec 1890 – 1r – 1 – uk British Libr Newspaper [072]

Kyustendilsko delo – Kyustendil, Bulgaria. Jul 1952-Feb 1959 – 2r – 1 – us L of C Photodup [949]

Kyvan to a mran / Samin, Buil mhu – Jeyyavati Mrui: u Tan Mran 1975 [mf ed 1994] – on pt of 1r – 1 – mf#11052 r1717 n1 – us Cornell [959]

Kyvan to e a khyac u / Sin Phe Mran – Ran kun Mrui: Ri Le Ca pe 1974 [mf ed 1993] – on pt of 1r – 1 – mf#11052 r1242 n1 – us Cornell [959]

Kyvan to sa / Lah Thvan Phru – Yan Kun: Capay Cape 1978 [mf ed 1990] – [ill] 1r with other items – 1 – (in burmese) – mf#mf-10289 seam reel 136/3 [§] – us CRL [959]

Kyvan to tui a me : [short stories] / Cin Tan, Takkasuil – Rankun: Nnon ram ca aup tuik 1987 [mf ed 1990] – 1r with other items – 1 – (in burmese) – mf#mf-10289 seam reel 150/6 [§] – us CRL [830]

Kzambazovski, Kliment *see* Kulturno-opetestvenite vrski na makedoncite so srbija vo tokot